Brain's Diseases of the Nervous System

Brain's Diseases of the Nervous System

TWELFTH EDITION

Edited by

Michael Donaghy

Reader in Clinical Neurology, University of Oxford.
Honorary Consultant Neurologist, John Radcliffe Hospital, Oxford
Honorary Civilian Consultant Adviser in Neurology to the Army

OXFORD
UNIVERSITY PRESS

OXFORD
UNIVERSITY PRESS

Great Clarendon Street, Oxford ox2 6DP

Oxford University Press is a department of the University of Oxford.
It furthers the University's objective of excellence in research, scholarship,
and education by publishing worldwide in

Oxford New York

Auckland Cape Town Dar es Salaam Hong Kong Karachi
Kuala Lumpur Madrid Melbourne Mexico City Nairobi
New Delhi Shanghai Taipei Toronto

With offices in

Argentina Austria Brazil Chile Czech Republic France Greece
Guatemala Hungary Italy Japan Poland Portugal Singapore
South Korea Switzerland Thailand Turkey Ukraine Vietnam

Oxford is a registered trade mark of Oxford University Press
in the UK and in certain other countries

Published in the United States
by Oxford University Press Inc., New York

© Oxford University Press 2009

The moral rights of the author have been asserted
Database right Oxford University Press (maker)

First published 1933
Sixth edition published 1962
Seventh edition published 1969
Eighth edition published 1977
Ninth edition published 1985
Tenth edition published 1993
Eleventh edition published 2001
Twelfth edition printed 2009

British Library Cataloguing in Publication Data

Data available

Library of Congress Cataloguing in Publication Data
Data available

Typeset in Cepha Imaging Pvt. Ltd., Bangalore, India
Printed in Italy
on acid-free paper by
LEGO SpA–Lavis, TN

ISBN 978–0–19–856938–1

1 3 5 7 9 10 8 6 4 2

Preface to the twelfth edition

The scope of this 12th edition of Brain's Diseases develops further the wholesale chapter reorganization introduced in the 11th edition. There are new sections on developmental neurology, pain, psychologically determined neurological disorders, and the important matter of assessing the effectiveness of those newly introduced treatments now so common in neurology. Most of the chapters reflect clinical problem areas easily identifiable in everyday neurological practice, and aim to combine the attributes of both textbook and reference book by outlining the clinical approach before coverage of the individual disease entities. Inevitably this leads to some overlap between chapters which the reader is helped to navigate by extensive cross referencing.

To this edition we welcome a number of new authors: David Bates, Peter Clayton, Helen Cross, Janet Eyre, Peter Goadsby, Linda Luxon, Tony McShane, Paul Reading, Peter Rothwell, John Scadding, Neil Scolding, Pamela Shaw, Tom Solomon, Robert Surtees, and Ian Whittle. Tragically, Professor Robert Surtees died in August 2007 before completing his chapter on neurometabolic disorders, a subject on which he was a renowned expert. He is sorely missed both by the authors of Brain's Diseases, and by his many colleagues and friends in paediatric neurology and metabolic disease.

The authors are indebted to the many who have helped with the preparation of this book, particularly Sarah Pendlebury for the sections on the complications of stroke and on its treatment, and Mary Reilly for the classification of inherited polyneuropathies. Many others have advised on the text, illustrations or tables: Rolfe Birch, Nick Beeching, James Byrne, Patrick Chinnery, John Elston, the late Ian Hart, Robin Kennett, Peter Keston, Penny Lewthwaite, Alastair Miller, Roger Mountford, Yesim Parman, Gerardine Quagebeur, Angela Vincent, and David Warrell; their help is gratefully acknowledged. Mrs Joanna Wilkinson has played a huge role in all aspects of the preparation of this book, and I owe her a particular debt of gratitude.

Michael Donaghy
Oxford
December 2008

Preface to the eleventh edition

Dr Russell Brain originally published Brain's Diseases of the Nervous System in 1933 having perceived the need to digest those 20 years of advance which marked the start of neurology as we know it now. Professor John Walton joined Lord Brain in revising the 7th edition, a task which coincided with the founding author's death in 1966. Walton carried Brain's Diseases forward as sole author for its 8th and 9th editions, an increasingly immense and daunting task. In 1989, after a dinner at Green College, Sir John asked if I would consider joining him as the coauthor for the 10th edition of Brain's Diseases. Armed with post-prandial courage, I declined this tempting invitation. It seemed that clinical neurology had become simply too vast in scope, and too sub-specialized, for two authors alone to provide the depth and authority required of a large reference book. So the 10th edition of Brain's Diseases, published in 1993, was the product of a team of 13 authors, revising 24 chapters under the editorship of Lord Walton. Roughly half of the chapters from the previous edition were substantially or completely reorganized and rewritten. Lord Walton decided to step down after the 10th Edition. Neurologists throughout the world owe a debt of gratitude to him for guiding Brain's Diseases into the modern era.

The chapter headings of this, the 11th Edition of Brain's Diseases, have been completely reorganized. Now there are 35 chapters contained within seven sections, and contributed by 14 authors. The text has been completely rewritten with a great increase in the number of illustrations and tables. Most of the chapters represent monographs addressing the clinical problem areas so readily identifiable in neurological practice, such as stroke, epilepsy, polyneuropathy, meningitis, headache, brain tumour, spinal cord disorder, or long-term disability. Such clinical topics will be easily recognized by every doctor possessing a basic knowledge of clinical neurology. Thus readers can consult a single chapter knowing that the entirety of the relevant clinical problem area will be covered by only one author. This avoids the unevenness, lack of perspective, and omissions so often evident when vast and diverse authorships contribute to a medical reference text.

We hope Brain's Diseases will serve two functions. First to continue to be a textbook, thereby providing a programme of advanced instruction in the subject. To this quest, each chapter includes sections outlining a clinical approach to the topic in question. Secondly, to provide a reference book of the depth and completeness required by the busy specialist of today. Some of the chapters address matters of background importance to contemporary clinical practice in neurology. The main neurological investigations are discussed from the perspective of neurologists who use and interpret these tests. Knowledge of the biology of degeneration and regeneration is crucial for understanding the long-term consequences of neurological diseases, and their prospects for repair and recovery. Our introductory chapter describes a practical approach to evaluating clinical problems, the differential diagnosis of ubiquitous symptoms such as muscular weakness, the importance of psychological factors in generating neurological complaints, and the historical development of clinical neurology.

The present authors are cognisant of the disparate forces which have shaped their practice and inspired their understanding of clinical neurology. All of us are deeply grateful for the influence of our mentors, junior colleagues students, research collaborators, and patients for provoking and moulding our thinking. Our secretaries have played an immense role in bringing this book to fruition; in particular Mrs Joanna Wilkinson has been responsible for rendering our piecemeal contributions into a form digestible by Oxford University Press. Warm thanks must go to Dr Robert Surtees of the Institute of Child Health for his extensive work in preparing Tables 4.1 and 4.2 summarizing paediatric neurometabolic disorders. Many of our colleagues have been generous enough to advise upon, or comment on chapters. The following deserve especial thanks for their expert advice on individual topics: Professor Peter Goadsby, Professor Neva Haites, Dr Robin Kennett, Mr Michael Powell, and Dr Kevin Talbot. Also we are immensely grateful to Dr Philip Anslow, Professor Dian Donnai, Dr Darren Forward, Dr Waney Squier, Dr John Stevens, Mr Nick White, and the technical graphics staff of Oxford University Press for their help with the book's many illustrations. Effective teamwork is crucial in producing a book demanding such huge contributions from each individual author. I must add warm and personal thanks to each of my fellow authors for underpinning their skill and scholarship with forbearance and industry, and to the editorial and production staff of Oxford University Press for their unstinting work in bringing this volume to fruition.

Michael Donaghy
Oxford
April 2001

Preface to the first edition

The last twenty years have witnessed a remarkable development in neurology. Investigation of the effects of war injuries of the spinal cord has greatly increased our knowledge of reflex action in man. The appearance of encephalitis lethargica and the multiplication of forms of acute disseminated encephalitis have added a new field to clinical neurology and brought it into relationship with the new branch of bacteriology which studies the filterable viruses. The discovery of important metabolic centres in the hypothalamus has enhanced the importance of neurology to general medicine. Advances in the technique of neurological surgery have aroused fresh interest in the symptoms and in the pathology of intracranial tumours. Other developments, scarcely less important, have occurred.

Much of this new knowledge is physiological, and in one respect I have departed from the traditional arrangement of a textbook of nervous diseases. Neurology is more dependent than many other branches of medicine upon anatomy and physiology. These subjects, the essential basis of neurological diagnosis, are usually dismissed in a few introductory pages, with the result that much clinical neurology is apt to be both unintelligible and uninteresting to the student. In the first part of this book, as an introduction to the subject, I have discussed—at greater length than usual—the application of anatomy and physiology to the interpretation of the physical signs of nervous disease. Elsewhere will be found sections dealing with anatomy and physiology as introductions to clinical sections. In planning the clinical sections I have used what seemed the most practical, if not always the most logical, arrangement, for there is no entirely satisfactory way of arranging subjects, many of which might be placed in more than one group.

Limitations of space restrict the number of references which it is possible to quote. I have, therefore, chosen only those of special interest and those which form the best introduction to a subject, or are themselves useful sources of references. To the many other writers upon whose work I have freely drawn I express my indebtedness. I am indebted also to a number of my colleagues for the loan of illustrations.

Finally, I welcome this opportunity of expressing my gratitude to my colleagues at the London Hospital for their teaching, encouragement, and help, especially to Dr Charles Miller, Professor Arthur Ellis, and Dr George Riddoch, under whom I had the privilege of working on the Medical Unit, and to Mr Hugh Cairns, Dr Dorothy Russell, and Dr S. Phillips Bedson.

W. Russell Brain
London
June 1933

Contents

Contributors

D Bates, FRCP FRCPEdin
Professor of Clinical Neurology in the University of
Newcastle-upon-Tyne, Royal Victoria Infirmary,
Newcastle-upon-Tyne NE1 4LP

DW Chadwick, DM FRCP FMedSci
Professor and Honorary Consultant Neurologist,
Walton Centre for Neurology and Neurosurgery,
Liverpool L9 7LJ

PE Clayton, MD FRCPCH
Professor of Paediatric Metabolic Disease and Hepatology at
University College London, Institute of Child Health,
London WC1N 1EH

DAS Compston, PhD FRCP FMedSci
Professor of Neurology in the University of Cambridge,
Addenbrookes Hospital, Cambridge CB2 0QQ

JH Cross, PhD FRCP FRCPCH
The Prince of Wales's Chair of Childhood Epilepsy at University
College London, Great Ormond Street Hospital for Children,
London WC1N 3JH

MJ Donaghy, DPhil FRCP
Reader in Clinical Neurology in the University of Oxford,
John Radcliffe Hospital, Oxford OX3 9DU

JA Eyre, DPhil FRCP FRCPCH
Professor of Paediatric Neuroscience in the University of
Newcastle-upon-Tyne, Royal Victoria Infirmary,
Newcastle-upon-Tyne NE1 4LP

NA Fletcher, MD FRCP
Consultant Neurologist, Walton Centre for Neurology and
Neurosurgery, Liverpool L9 7LJ

PJ Goadsby, MD PhD DSc FRACP FRCP
Professor of Neurology in the University of California,
San Francisco, CA 94143-0114

R Grant, MD FRCPGlasg FRCPEdin
Consultant Neurologist, Western General Hospital,
Edinburgh EH4 2XU

D Hilton-Jones, MD FRCP FRCPEdin
Consultant Neurologist, John Radcliffe Hospital,
Oxford OX3 9DU

C Kennard, PhD FRCP FMedSci
Professor of Clinical Neurology in the University of Oxford,
John Radcliffe Hospital, Oxford OX3 9DU

LM Luxon, FRCP
Professor of Audiovestibular Medicine at University College
London, The National Hospital for Neurology and Neurosurgery,
London WC1N 3BG

MA McShane, FRCPCH
Consultant Paediatric Neurologist, John Radcliffe Hospital,
Oxford OX3 9DU

PJ Reading, PhD FRCP
Consultant Neurologist, The James Cooke University Hospital,
Middlesborough TS4 3BW

MN Rossor, MD FRCP FMedSci
Professor of Clinical Neurology at University College London,
The National Hospital for Neurology and Neurosurgery,
London WC1N 3BG

PM Rothwell, MD PhD FRCP FMedSci
Professor of Clinical Neurology in the University of Oxford,
John Radcliffe Hospital, Oxford OX2 9DU

JW Scadding, MD FRCP
Consultant Neurologist, The National Hospital for Neurology and
Neurosurgery, London WC1N 3BG

NJ Scolding, PhD FRCP
Burden Professor of Clinical Neurosciences in the University of
Bristol, Frenchay Hospital, Bristol BS16 1LE

PJ Shaw, MD FRCP FMedSci
Professor of Neurology in the University of Sheffield,
Royal Hallamshire Hospital, Sheffield S10 2JF

TS Solomon, PhD FRCP
Professor of Neurological Science in the University of Liverpool,
Walton Centre for Neurology and Neurosurgery,
Liverpool L9 7LJ

†**RAH Surtees**, PhD FRCP FRCPCH
Professor of Paediatric Neurology at University College London,
Great Ormond Street Hospital for Children,
London WC1N 3JH

DT Wade, MD FRCP
Professor of Neurological Rehabilitation in the University of
Oxford, Nuffield Orthopaedic Centre,
Oxford OX3 7LD

IR Whittle, MD PhD FRACS FRCSE(SN) FRCPEdin
Forbes Professor of Surgical Neurology in the University of
Edinburgh, Western General Hospital,
Edinburgh EH4 2XU

Glossary

ACTH	adrenocorticotrophic hormone
AIDS	acquired immunodeficiency syndrome
CSF	cerebrospinal fluid
CT	computed tomography
DNA	deoxyribose nucleic acid
EEG	electroencephalogram
ELISA	enzyme linked immunosorbent assay
ESR	erythrocyte sedimentation rate
HIV	human immunodeficiency virus
HTLV-1	human T-lymphotrophic virus 1
MRI	magnetic resonance imaging
PET	positron emission tomography
RNA	ribose nucleic acid
SPECT	single photon emission computed tomography

SECTION 1

Introduction

CHAPTER 1

A short history of clinical neurology

Alastair Compston

Contents

1.1 The origins of clinical neuroscience

More than any other branch of medicine, the practice of neurology depends on the classical methods of intuitive conversation, structured examination, and selective investigation. We teach the importance of eliciting an accurate neurological history. The key symptoms at onset are identified and their subsequent course defined. For the experienced clinician, this process becomes routine, efficient, and quick. The competent neurologist is the one who instinctively senses relevant components of the history, appreciates the most likely underlying disease mechanisms, reliably elicits the relevant physical signs, knows which investigations are necessary and assesses their relevance in the clinical context, provides a sensible clinical formulation, and communicates the situation accurately and sensitively to the patient and relatives. Rather than slavishly collecting an encyclopaedia of facts, in which the key issues may be lost in a surfeit of redundant information, the critical components are sifted and the subsequent conversation steered down an algorithm that seeks anatomical, physiological, and pathological explanations for what the patient describes.

This system evolved over several centuries during which knowledge accumulated on structure and function, brain and spinal cord localization in health and disease, the reliability of physical signs and ancillary laboratory investigations, and the nosology of neurological disease. That story shows periods of intense activity interspersed with intervals of stagnation: cornerstones of truth supporting the crumbling masonry of discarded hypothesis and unsubstantiated opinion; and the emergence of streams of knowledge that began to organize ideas around defined principles such as anatomical structure, functional systems, clinical syndromes, and clinico-pathological correlations. As for the names of those who created clinical neuroscience, too many candidates qualify and too few can be accommodated in the elite 'first division' team. Several distinguished names do not appear in this brief survey: the references are highly selective; and the account is unashamedly Anglo-centric, focusing, but not exclusively, on pivotal contributions to clinical neuroscience from the United Kingdom. Figure 1.1 identifies a more international crew, from the perspective of US-based clinical neuroscience. Especially, we celebrate the work of Thomas Willis [1621–1675], Robert Whytt [1714–1766], (Sir) Charles Bell [1774–1842], John Hughlings Jackson [1835–1911], (Sir) David Ferrier [1843–1928], (Sir) William Gowers [1845–1915], (Sir) Charles Sherrington [1857–1952], (Sir) Gordon Holmes [1876–1965], and (Russell) Lord Brain [1895–1966]. Many better balanced and more detailed histories of clinical neuroscience are available: *The Human Brain and Spinal Cord* by Edwin Clarke and Kenneth Dewhurst (1968) contains readings and analyses of writings from antiquity to the 20th century with a focus on gross and microscopic anatomical structure in the normal nervous system rather than on function and disorder; *Garrison's History of Neurology* by Lawrence McHenry (1969) is comprehensive but often inaccurate; Haymaker and Schiller's *The Founders of Neurology* (second edition, 1970) contains biographies, and portraits or photographs where available, of all significant neuroscientists arranged by discipline; J.D. Spillane's *The Doctrine of the Nerves* (1981) is detailed and engaging but selective and with significant omissions; *Morton's Medical Bibliography* by Jeremy Norman (fifth edition, 1991) lists all significant publications, books, and articles, in the history of medicine including discoveries relating to the nervous system; *Neurological eponyms* by Peter Koehler, George Bruyn, and John Pearce (2000) appeals to the neurologist who likes to celebrate the individual by identifying accounts of structure and function, physical signs, and descriptions of disease with the name of him or her who first described this or that aspect of clinical neuroscience — although any declaration of dates and details usually risks rival claims for priority.

Individuals who shaped the history of neurology

Comparative Neurology	Neurons	Clinic	Tracts	Physiology	
Flourens	Schwann	Romberg	Foville	Marshall Hall	1847
Baillarger	Purkinje	Duchenne		Weber	
Darwin	Waller	Vulpian	Clarke	Helmholtz	1850
Bischoff	Stilling	Broca	Bouchard	Waller	
Turner	Deiters	Jackson	Burdach	du Bois Raymond	
Zieman	Bethe	Charcot	Henle	Vulpian	
Retzius	Golgi	Wernicke	Meynert	Bernard	
Bethe	His	Freidrich	Deiters	Ludwig	
Cunningham	Betz	Westphal	Kollicker	Fritsch	
Loeb	Kollicker	Gowers	Turk	Hitzig	1875
His	Retzius	Mitchell	Jastrowitz	Ferrier	
Meyer	Walleyer	Erb	Charcot	Fleschig	
Spitzka	Abathy	Ferrier	Fleschig	Goltz	
Elliot Smith	Nissl	Marie	Gowers	Schaefer	
Bolk	Cajal	Bastian		Gaskell	
Brodmann	Campbell	Thomas	Weigert	Lombard	
Edinger	Alzheimer	Oppenheim	Marchi	Mosso	
Donaldson	Bielchowsky	Mills		Bowditch	1900
Herrick	Harrison	Spiller	Bechterew	Langley	
Coghill	Heidenhein	Head	Genuchten	Marey	
Kappers	Brodmann	Liepmann	Cajal	von Frey	
Parker	Economo	Putnam		Lord	
Child	Nicholson	Dana	Monakow	Horsley	
Anthony	Vogt	Dejerine	Mills	Sherrington	
Johnston	Hortega	Collier	Henschen	Pavlov	
Tilney		Economo	Winkler	Magnus	1925
		Foerster		Lucas	
		Babinski		Cannon	
		Wilson		Dale	
		Holmes		Adrian	
				Forbes	
				Berger	1947
				de Baronne	
				Ranson	

From, Stanley Cobb. One hundred Years of progress in neurology, psychiatry and neurosurgery. *Arch. Neurol Psych* 1948: 59; 63

Fig. 1.1 A United States perspective on lineages in the evolution of clinical neuroscience.

Various answers might be given to the question, 'where do the origins of modern neurology lie?' For some, the obvious place to start is the early–mid-19th century when Charles Bell [1774–1842] explored structure and function of the nervous system; John Cooke [1756–1838] wrote the first systematic account of clinical neurology from ancient times to modern; the great clinico-pathological atlases of Robert Hooper [1773–1835], Robert Carswell [1793–1857], Jean Cruveilhier [1791–1874], and James Hope [1801–1841] appeared; the description of clinical syndromes and different diseases that we still acknowledge as entities began to appear; and the clinical system for examination of the nervous system as a means of detecting order and disorder in the nervous system was first formulated. Others would wish to start with the remarkable work of Robert Whytt in Edinburgh who wrote works of lasting value on reflex functions of the nervous system (Whytt 1751), and on clinical neurology (Whytt 1765). But despite these claims, few would probably argue that the step change in understanding had already occurred in that remarkable decade of the brain, the 1660s, during which a series of books condensed the thinking of physician scientists who challenged ideas that had been prevalent for several centuries. Their spokesman, Thomas Willis, coined the term: '*we shall institute the whole neurology or the doctrine of the nerves.*' He took this word from the Greek for sinew, tendon, or bowstring based on observations made whilst dissecting cranial, spinal, and autonomic nerves.

Magic Medicine ignored the sick individual as a source of information. Plato [427–327 BC] believed in health as a state of harmony. Disease was an excess, alteration or re-location of earth, fire, air, and water (the 'body'), and 'soul'. Hippocrates [460–377 BC] internalized medicine, relating disease to the individual and illustrated his ideas with case histories describing the onset, duration and outcome of symptoms. For Galen [130–200 AD], disease had a locus and a pathological process, but function altered first and disordered structure followed. It was a short step to extend the concept of physiology (Francis Glisson [1597–1677]) to the pathological process (Girolamo Fracastoro {Fracastorius} [1478–1553]).

The work of Leonardo da Vinci [1452–1515] moved neurological attention from functions dependent on the contents of the ventricles to a real anatomy. This was much improved by Andreas Vesalius [1514–1564] who considered the human body to be the servant of harmony. But as for how it all worked, Kenelme Digby [1603–1665] was still advancing Platonic ideas on the nervous system in the 17th century; and even William Harvey [1578–1657] left the brain to the attentions of others. Beyond the experimentalists and iatrochemists who fashioned neuroscience in the 1660s, lay the concepts of natural history detected by observation over time (Thomas Sydenham [1624–1689]; and Giorgio Baglivi [1669–1707]), the anatomical seat of disease (Giovanni Battista Morgagni [1682–1771]), and the final transition from disease as alteration in vital properties to the notion of cellular (dis)organization (Rudolph Virchow [1821–1902]). Today, we encapsulate these concepts under the headings of aetiology, mechanisms, and treatment. For all our '-ologies and the -omics', we still respect that tradition. But in 1664, as far as the nervous system was concerned, not much had changed for the group who created our discipline over the millennia since the Greeks first rationalized the concepts of disease formulated by even earlier cultures.

1.2 Structure of the nervous system

It was the need for Renaissance artists better to depict the external form of the human body through an improved understanding of its inner arrangements that first provided an accurate anatomical account of structure in the nervous system. That process started with Leonardo da Vinci (see above) but soon reached full maturity through the work of Andreas Vesalius (1543), Fig. 1.2. It was Vesalius' detailed anatomical depictions of the human body including the brain that finally dislodged the medieval cell doctrine. Vesalius' work epitomizes the role of the printing press in moving science and culture from the medieval to the modern. The seven books that make up *De fabrica* (1543) are famous for their woodcuts portraying accurate anatomical structures, based on the dissections of Vesalius, set in soothing landscapes by their artist, Jean Calcar [1499–1545]. These images were last reprinted from the original pearwood blocks in 1934. That was fortuitous since the woodblocks were soon destroyed by the Allied bombing of Munich. Book VII of *De Fabrica* dealing with the nervous system was first translated into English from the original Renaissance Latin by Charles Singer (1952).

Others quickly borrowed, refined, and extended Vesalius' neuroanatomy. Major schools developed in the Netherlands, Germany, and Italy. In Great Britain, anatomy improved through the work of Thomas Willis, Fig. 1.3, John Browne [1642–1702], William Cowper [1666–1709]—who serially stripped away the muscles leaving his flayed exemplars otherwise represented intact in rural landscapes—and Humphrey Ridley [1653–1708]. Willis improved on primitive depictions of the arterial anastomosis at the base of the brain but the eponymous 'circle of Willis' derives more from his physiological than anatomical observations. In *Cerebri Anatome* (1664: translated 1681), Willis describes a patient dying from carcinoma of the stomach in whom the left carotid artery was occluded; yet the brain had not suffered since the right carotid was increased to three times its normal size. Willis concluded that a connection must exist between the circulations on the two sides. The famous illustration is by Christopher Wren [1632–1723]. Vesalius

Fig. 1.2 Andreas Vesalius [1514–1564].

and Willis were concerned with gross anatomy of the normal brain, nerves, and muscles but an understanding of how the brain works ultimately required knowledge of its cellular structure and organization.

Microscopes were developed in the 1640s and a range of histological stains, suitable for distinguishing cell types within intact tissue, became available during the 19th century. It was Camillo Golgi's [1843–1926], Fig. 1.4, silver stain and its use by Santiago Ramon y Cajal [1852–1934], Fig. 1.5, that culminated in definitive studies of the cellular architecture and organization of the central nervous system. From this work emerged the neurone theory for which Cajal and Golgi were jointly awarded the 1906 Nobel Prize for Physiology or Medicine. Cajal's work is widely quoted but relatively unread in the original. Translations have appeared in recent years and his treatise *Textura del Sistema Nervioso del Hombre y de los Vertebrados*, first published in fasicules between 1899 and 1904, updated and translated into French (1909–1911), has now undergone its second English version (first from the French and now the Spanish text) since 1995. Arguably, Cajal was the most significant neuroscientist of the 20th century. He described accurately the distinguishing features of neurones and glia within discrete regions of the brain and spinal

Fig. 1.3 Thomas Willis [1621–1675].

Fig. 1.4 Camillo Golgi [1844–1926].

cord; he characterized their local organization, supplementing the descriptions with beautiful drawings based on Golgi's silver stains; and he consolidated the neurone doctrine. Preceded in this work by Jan Purkinje [1787–1869], Wilhelm His [1831–1904] and Fridtjof Nansen [1861–1930: neuroscientist, Arctic explorer, and humanist], Cajal perfected Golgi's method and advanced his discoveries through the study of developing nervous systems in order to overcome the limitation of poor silver staining of myelinated fibres. He overturned the reticular theory of neural organization showing histologically the variability of dendritic arborizations and axon terminations. He established that axon cylinders end freely but form contacts. And Cajal conceived that the nerve impulse is conducted between axons, dendrites, and the cell body of neighbouring neurones. Everything we know about structure, function, and physiology in the nervous system at the cellular level, in health and disease, evolves from the concept that organization is through the connectivity of functionally independent neurones and their processes. Cajal's most detailed studies were of the cerebellum but, in time, no part of the brain and spinal cord went unexplored. Paradoxically, Cajal and Golgi disagreed publicly on the neurone doctrine when they gave their 1906 Nobel lectures in Stockholm.

Fig. 1.5 Santiago Ramon y Cajal [1852–1934].

1.3 Linking structure to function in the nervous system

Some of those who first described the structure of the nervous system in the 16th and 17th centuries also formulated primitive concepts of reflex activity—that is, function. Historical guides to knowledge on reflex function are to be found in Franklin Fearing's *Reflex Action: a Study in the History of Physiological Psychology* (1930); *The Historical Development of Experimental Brain and Spinal Cord Physiology before Flourens* by Max Neuburger (1897, translated into English in 1981); and E.G.T. Liddell's *The Discovery of Reflexes* (1960), a tribute to his teacher, Sir Charles Sherrington.

Thomas Willis likened muscle contraction to an explosion of gunpowder with new spirits being supplied to muscle by the blood (1670: translated 1684), extending the ideas of William Croone [1633–1684] who (in *De Ratione Motus Musculorum*, 1664: often wrongly attributed to Thomas Willis) postulated an interaction of spiritous juice from the nerves and blood agitating the space between muscle fibres and transmitting to them a force which expanded their width and shortened their length. Willis later elaborated these ideas (1667: translated 1681): '*the first designation of motion is in the Brain or Cerebel: its transmission. is performed by the spirits within the nerves... implant[ing] a contracture or elastick force*'. Robert Whytt, Fig. 1.6, of Edinburgh consolidated the reflex theory of function in the nervous system, settling a debate polarized by Rene Descartes [1596–1650] and Willis on the dependence of intact reflexes on the integrity of peripheral nerve plexuses. His experiments on the frog (Whytt 1751) established that reflex activity depends on segmental integrity of the spinal cord. He also described the reflex pupillary response to light.

Against this background, the themes which underpin the modern concept of facilitation and inhibition of (spinal) reflexes as the basis for organization of neural systems and behaviour were the recognition that the nervous system consists of nerve cells connected through synapses to their neighbours (see above); evidence for (animal) electricity as the basis for passage of the nerve impulse (work associated especially with Luigi Galvani [1737–1798]); and the accurate designation of afferent and efferent components of the reflex arc. Thus, the contribution of lasting value made by neuroscientists working in the first decades of the 19th century could be summarized as merging the observations of Robert Whytt with the animal studies that revealed the nervous system as an electrical organ, into a concept of reflex function. The emphasis was on functions of the spinal cord, and this work culminated in Sherrington's great synthesis *The Integrative Action of the Nervous System* (1906). But it also spawned an uncontrolled explosion of eponymous descriptions described by Foster Kennedy as 'the open season for the "Hunting of the Reflex"... (generating) impedimenta of variety without variance, shrill claims to be the Prometheus of the pyramidal tracts' (see foreword to *The Examination of Reflexes* by Robert Wartenberg, 1945).

In retelling *The Discovery of Reflexes* (1960), E.G.T. Liddell reduces the story to four critical periods: the nerve cell and the microscope; animal electricity; experimental approaches; and 'Sherrington and his times'. Manuscript lecture notes loosely inserted in Liddell's own copy of *Reflex Activity of the Spinal Cord* (1932) written with R.S. Creed, D. Denny-Brown, J.C. Eccles, and C.S. Sherrington trace how a millennium of post-Galenic darkness was systematically

Fig. 1.6 Robert Whytt [1714–1766].

illuminated across the 15th–19th centuries by Leonardo da Vinci, Robert Whytt, Gilbert Blane [1749–1834], Georg Prochaska [1749–1820], Julien Le Gallois [1770–1840], Pierre Flourens [1794–1867], Marshall Hall [1790–1857], Johannes Müller [1801–1858], Richard Grainger [1801–1865], Eduard Pflüger [1829–1910], Friedrich Goltz [1834–1902], and Alfred Vulpian [1826–1887] before everything came together with the work of Charles Sherrington.

Charles Bell, Fig. 1.7, comes out second best in the Bell–Magendie wrangle. Bell was much excited by his deliberations on the workings of the brain. He had grasped the principles of afferent connections, central processing, and efferent output but attributed both motor and sensory functions to the anterior roots, and oversimplified the central functions of the cerebellum and cerebrum. It is difficult to guess whether Francois Magendie [1783–1855] saw a copy of Bell's *Idea of a New Anatomy of the Brain* (1811) before correctly assigning anterior (motor) and posterior (sensory) nerve root functions (Magendie 1822). Whatever the truth of Magendie's position, Bell was soon putting around modified views which neatly corrected but did not formally retract his earlier error. Marshall Hall (1837) coined the term 'reflex arc' and methodically documented the range and variety of reflex responses in isolated portions of the animal nervous system. Later, he extended these principles to the interpretation of clinical observations including spinal shock—rather

Fig. 1.7 Charles Bell [1774–1842].

Fig. 1.8 John Hughlings Jackson [1835–1911].

unsatisfactorily in the opinion of Spillane (1981). Others share the view that a better synthesis was provided by a less well-known contributor, Richard Grainger (1837), who, following Whytt and based on his own observations and experiments, understood that the reflex functions of the spinal cord occur independently of the 'will', and that the spinal cord is merely the conductor of the volitions of the cerebrum. In work done by himself and later with E.G.T. Liddell, Sherrington warned against allowing '*a tradition to grow up whereby it was scientifically and even clinically acceptable to think of the spinal cord as having an existence quite separate from that of the supraspinal centres*'.

John Hughlings Jackson, Fig. 1.8, sought to understand the organizational principles which determine how the brain works. He was not an experimentalist but used clinical observation to inform his analyses. He learned much from colleagues at the West Riding Lunatic Asylum and the philosopher Herbert Spencer [1820–1903]. Jackson's concept of functional localization was elaborated by Sigmund Freud [1856–1939] and Sir Henry Head [1861–1940] who concluded that function is not normally localized but depends on complex orchestration of the entire nervous system, even though disorder at specific sites yields precise and predictable clusters of clinical deficits. Behind this rather pragmatic view of cerebral localization lay a more sophisticated philosophy which postulated three layers of organization that determine function in the central nervous system. In disease, subservient systems in this hierarchy are released creating syndromes consisting of negative (loss of function) and positive (disinhibited) symptoms. The lowest level is concerned

with vegetative function and operates through reflex activity at the spinal level. The next tier (cortex, striatum, and long tracts) determines movement and sensation. The most elevated system is the pre-frontal region providing higher functions and co-ordination of the more elementary components. Jackson is revered but largely unread. He left no major synthesis of his views and reading his papers (Jackson 1932, edited by James Taylor [1859–1946]) requires determination.

Sir David Ferrier, Fig. 1.9, set out to confirm experimentally, what Jackson had proposed theoretically on destructive and discharging lesions of the nervous system and to deal with the issue of cerebral localization. Using electrical stimulation and local excision of the monkey and canine cortex, together with clinical analyses, Ferrier (1876) concluded that certain areas of the cortex do possess defined functions, and that lesions at different sites will generate predicable syndromes. Inevitably his work was scientifically and socially controversial. Ferrier demonstrated 'Monkey F' to the Physiological section during the 1881 International Medical Congress. Friedrich Goltz [1834–1902] denied that it was possible to study cerebral localization by electrical stimulation, without the need for ablation. He stated that dogs with virtually no cerebrum remaining could

Fig. 1.9 David Ferrier [1843–1928].

Fig. 1.10 Charles Sherrington [1857–1952].

do everything attributed to Ferrier's sensory and motor centres. Ferrier pointed out tartly that a dog is not a monkey. Sir William Gowers was summoned to act as umpire. He concluded that Goltz's dogs had rather more brain *in situ* than he had claimed. After the show, Ferrier was prosecuted for cruelty to animals, but acquitted. Of another of Ferrier's monkeys, the astonished Jean-Martin Charcot [1825–1893] declared *'c'ést un malade'* ('it's a patient'). Ferrier pioneered neurosurgery and his work culminated in the first human operative intracranial procedure carried out (by others, see below) in December 1884.

Only four people attended the last of Sir Charles Sherrington's, Fig. 1.10, celebrated Silliman Lectures in 1904. The published version (Sherrington 1906) offers a summary of functional organization in nervous systems which ranks, in its overall synthesis and stature, with the greatest works in medicine. *The Integrative Action of the Nervous System* is probably the single most significant publication in clinical neuroscience of the 20th century, and the author has been dubbed the 'Harvey of the Nervous System'—although several others have staked this claim, for themselves or on behalf of third parties. Sherrington begins with a restatement of the neurone doctrine: '*nowhere in physiology does the cell theory reveal its presence more frequently in the very framework of the argument than in the study of nervous reactions*'. Against this background, and a certain knowledge of animal electricity and reflex function of the isolated spinal cord, he formulates ideas on propagation of the nerve impulse through synapses, mainly using observations on the scratch reflex in dogs. He defines the efferent and afferent properties of nerve and muscle, and characterizes the physiology of the tendon stretch reflex. Sherrington argues that activities such as walking depend on temporal and spatial regulation of reflex activity. Orchestration of the separate segments, dependent on integrity of the spinal cord, requires interaction of neighbouring segments and reciprocal inhibition and facilitation of agonists and antagonists, as the basis for tonic and phasic contraction of muscles concerned with posture and stepping.

1.4 Pathological anatomy of the nervous system

Giovanni Morgagni [1682–1771] first suggested classifications of pathological anatomy (Morgagni 1761) but not until the early 19th century were anatomical and clinical descriptions of neurological disease systematically correlated and illustrated. Many attended the Parisian clinical demonstrations and dissections in which Philippe Pinel [1745–1826] and Xavier Bichat [1771–1802] consolidated the discipline of pathological anatomy in the first half of the 19th century. This school reached its continental zenith with the 40 livraisons published by Jean Cruveilhier [1791–1874] as *Anatomie pathologique du corps humain; descriptions avec figures lithographiées et coloriées; des diverses alterations morbides dont le corps humain est susceptible* (Cruveilhier 1829–1842). Pathological anatomy also flourished in the United Kingdom during the first few decades of the 19th century because a few clinician scientists with artistic temperaments met in a culture that, culminating with the 1832 Anatomy Act, became liberated with respect to anatomical examination of the sick. The principal activists in this school were Matthew Baillie [1761–1823], Robert Hooper, (Sir) Charles Bell, Richard Bright [1789–1858], Joseph Swan [1791–1874], (Sir) Robert Carswell, and James Hope. Baillie (1793; 1799–1803) established organ-based pathology as a separate science in the United Kingdom. Bell (1802) produced exquisite artistic images of normal structure to illustrate his dissections of the cranial and peripheral nerves. He acknowledged the propensity of the brain to manifest striking disturbances of function on the basis of apparently trivial disorders of structure. Swan (1834) wrote on the normal anatomy of the cranial, spinal, and peripheral nerves. Bright (1827–1831) introduced a classification based on inflammation, pressure, irritation, and inanition but no details were provided of his patients' physical signs in life. In *Pathological Anatomy: Illustrations of the Elementary Forms of Disease*, Robert Carswell (1838) also showed consummate artistry in use of the colour spectrum as organs affected by inflammation depicted by pink, analogous tissues by crimson, atrophy by yellow and ochre (with the first illustrations of multiple sclerosis), hypertrophy by brown, pus by yellow and green, mortification by blue-black, haemorrhage by purple, softening by yellow, melanoma by jet black, carcinoma by orange and green, and tubercle again by red were pictured and described. The first separate work on neuropathology, by Robert Hooper (1826), anticipated Carswell in attempting a classification based on inflammation, tumour, diseased structure and unnatural appearance without tumefaction, and fluid collected around the hemispheres or extravasated, with descriptions of diseases as they affect the meninges, brain, nerves, blood vessels, and sinuses. James Hope (1834) organized his material around the lesions of circulation; nutrition; inflammation of the meninges and cerebral substance; softening; suppuration, abscess, and ulcer; induration, apoplexy, hypertyrophy, atrophy, and anaemia; and tubercle, schirrous, encephaloid, fatty, fibrous, cartilaginous and osseus productions, and hydatids. The 36 figures on four plates are subtly but brightly coloured lithographs '*from nature and revised on stone by Dr Hope*'. Each is supported by a case history: some are borrowed from Cruveilhier; others derive from patients seen at the Marylebone Infirmary, and St George's Hospital, London, and contact with (Sir) James Bardsley [1801–1876] in Manchester.

1.5 **Examination of the nervous system**

Willis described symptoms and made clinical observations; Whytt saw that the illuminated pupil contracted; and James Parkinson [1755–1824] noted the tremor, posture, and slowness of movement of those who shuffled through Hoxton Square. But physical examination did not feature in these early descriptions of neurological disease. Only with the appearance of the first systematic textbooks of the 19th century were the accounts of symptoms complemented by objective evidence for impairments of neurological function; and these methods were fully in place by the time the great treatises, manuals, textbooks, and systems of Moritz Romberg [1795–1873], William Gowers, Carl Wernicke [1848–1905], Hermann Oppenheim [1858–1919: who wrote in a style reminiscent of

S.A. Kinnier Wilson ([1878–1937], see below)—'*I may be permitted to add that my textbook contains all that is essential*'), and Jules Dejerine [1849–1917], duly appeared.

The evolution of the method for physical examination pivots around Joseph Babinski's [1857–1932] presentation to the Biological Society of Paris (Babinski 1896). Other observations based on experimental and clinical work provided the basis for Babinski's description. These included the demonstration of cutaneous and tendon reflexes, and the contributions of Charcot and Alfred Vulpian [1826–1882] in observing different aspects of the upper motor neurone lesion. Charles Edouard Brown-Sequard [1817–1894] knew of a paraplegic American whose valet would relieve spasticity and clonus (Charcot's spinal epilepsy) by forcible flexion of toes; and others undoubtedly stroked the plantar surface in order to induce reflex flexion of the leg. Monographs dedicated to examination of the nervous system first appeared in the 1920s. The Norwegian G.H. Monrad-Krohn [1884–1964] offered no short cuts (Monrad-Krohn 1921): after a history supplemented with leading questions, he advocated a thorough collection of the signs without prior speculation as to the diagnosis, ordered tabulation of the findings, regional diagnosis by applying the rules of functional anatomy, and ideally by pathological interpretation. His book introduced many components of the examination that survive. There is the recall of prime ministers, recitation of serial digits forwards and in reverse, interpretation of proverbs, mental arithmetic on mythical shopping trips, tongue twisters such as *West Register Street* and the *British Constitution*, and sensory examination— preferably spread over two days. Tips on how to test the main muscles and assign their nerve and root supply anticipate the monograph (1943) published by the Medical Research Council (War Memorandum number 7) under the chairmanship of Brigadier (Dr) George Riddoch [1888–1947: Riddoch *et al.* 1943]. The examinee is Dr M.J. (Sean) McArdle [1909–1989] who succeeded Dr William Ritchie Russell [1903–1980]— foundation professor of neurology at the University of Oxford— as neurologist to Scottish Command in 1942. The photographs were taken at Gogarburn with assistance from the department of medical illustration at the University of Edinburgh. Around 20 loose leaf copies were distributed to surgeons in Scotland.

Sir Gordon Holmes [1876–1965], Fig. 1.11, published a system for clinical examination of the nervous system orientated around motor and sensory systems, vision and eye movements, aphasia and related disorders, the mental state, and autonomic function—all based on physiological and anatomical principles of organization— which is still in routine use (Holmes 1946). The three-page practical summary is minimalist but complete. Written at the age of 70, after a professional career in which, like Hughlings Jackson, Holmes had mainly depended for his insights on opportunities made available through clinical neurology, especially gunshot injuries sustained in the First World War, following the example of Silas Weir Mitchell [1829–1914], Holmes' manual synthesizes his many insights into structure and function of the cerebellum, visual system, and spinal cord.

From the earliest times, medical texts were supplemented by illustrations. Whilst sometimes purely decorative, their main purpose has been to provide an additional means of conveying meaning. The principles of medical illustration are nowhere more explicit than in images of the nervous system. Some of the great medical examples are external and internal configurations of the

Fig. 1.11 Gordon Holmes [1876–1965].

brain. In J.D. Spillane's richly illustrated *Atlas of Clinical Neurology* (1968) the subjects tell not only the story of neurological anatomy, clinical phenomena, and pathology but also provide a medico-social window on disorders that were commonly encountered in mid-20th century international and provincial neurological practice. The many photographs by (Professor) Ralph Marshall exploit the power of black and white, use of shadow, creative construction of images such as those depicting scotoma, hemianopia, and trigeminal neuralgia, and the use of stills from filmed sequences to explain the subjective experience of neurological disease organized around head and neck, cranial nerves, acute and chronic polyneuritis, peripheral nerve lesions, muscle and neuromuscular disease, and neurodegeneration. The three great non-photographic artist-illustrators of the brain in health and disease working in the 19th and 20th centuries were A. Kilpatrick Maxwell [1884–1975], Max Brödel [1870–1941], and Frank Netter [1906–1991].

1.6 Investigation of the nervous system

It was not until 1875 that Richard Caton [1842–1926], working in Liverpool, extended knowledge on the electrical basis for nerve

action, discovered by Luigi Galvani [1737–1798] in 1791 and developed by Emil du Bois Raymond [1818–1896] in 1848, to the brain (Caton 1875). Hans Berger [1873–1941] recorded this activity through the intact skull and the technique was perfected by E.D. Adrian [1889–1977] who also developed methods for recording electrical activity from the peripheral nerve. Intracellular recordings later led to elucidation of the conduction of the nerve impulse by Sir Andrew Huxley [born 1917] and Sir Alan Hodgkin [1914–1998]. The exploration of human brain function by evoked potential methods originates from the observations of George Dawson [1912–1983]. Subsequently the group of Martin Halliday [born 1934] and Ian McDonald [1933–2006] moved electrical exploration of conduction in the central nervous system into clinical practice. The most direct method for examining body fluids that reflect brain activity was the introduction of lumbar puncture in life (Domenico Cotugno [1736–1822] removed fluid from cadavers). First used at the Middlesex Hospital in London to treat children with tuberculous meningitis (described by Robert Whytt in 1768) by Walter Essex Wynter [1860–1945], the procedure was routinely applied in neurology by Henirich Iraneaeus Quincke [1842–1922: Quincke 1891]. He measured intracranial pressure and examined the chemical constituents of spinal fluid; qualitative features of the protein content were described by Charles Albert Lange [1883–1959].

Neurologists trained since the introduction of brain scanning in 1973 must find it hard to imagine the confidence needed accurately to localize structural lesions as the sufficient basis for surgical exploration using nothing more than clinical analysis. But such was the belief in anatomical localization that Sir Rickman Godlee [1849–1925], on the advice of Alexander Hughes Bennett [1848–1901], first operated on a cerebral tumour (1885); and, with neurological assistance from Sir William Gowers, Sir Victor Horsley [1857–1916] removed the first spinal cord tumour (1888). Even when neuroradiology was introduced, the procedures offered limited information, showing only the grossest abnormalities, and failing to depict most processes which affect tissue integrity. Definitive textbooks on neuroradiology began to appear within a few years of the demonstration by Wilhelm Roentgen [1845–1923] of the bone X-ray of his wife Bertha's hand. It was an imaginative next step to adapt this technique using substances, including radio-opaque dyes, introduced in and around the brain and spinal cord to define their structure by silhouette. In 1918, Walter Edward Dandy [1886–1946] outlined the outer and inner contours of the brain using air introduced directly into the ventricles or lumbar sac (Dandy 1918). Jean Athanase Sicard [1872–1929] replaced air with iodinized oil and produced images of the spinal canal by myelography (Sicard 1921). The most colourful of these early pioneers was Antoni Caetano de Abreu Egas Moniz [1875–1955] who, after signing the Versailles treaty for Portugal in 1918, returned to neurology and introduced arteriography (Moniz 1927). He was best known for pioneering frontal leucotomy and other forms of psychosurgery, receiving the Nobel prize for Physiology or Medicine in 1949 but nearly losing his life at the hands of a gun-crazed schizophrenic patient in his office.

Although low-resolution radioisotope brain scans had on rare occasions shown large cerebral lesions, radiological techniques capable of routinely identifying disease processes, such as the lesions of multiple sclerosis, with some consistency were not available until the invention in 1971 of computerized axial tomography, or CT, by (Sir) Godfrey Hounsfield [1919–2004]. The idea came to Hounsfield on a country walk. With a background in electronics and radar, it occurred to him that readings from a large quantity of measurements taken randomly of objects in a closed box would reveal their shape when processed. It was a small step to add the principles of using an X-ray beam and sensitive detectors rather than film, constrained to slices that were then reconstructed into a three-dimensional image (Hounsfield 1973). Although Hounsfield was working for Electric and Musical Industries, EMI, where better to apply this technique than in the brain? The first picture of a cerebral cyst was displayed in April 1972—to a standing ovation from 2000 conference attendees (The Times newspaper, 18 August, 2004). The situation changed even more dramatically with the application of nuclear magnetic resonance to biological structures (Lauterbur 1973; Mansfield and Grannell 1973).

No procedure has so revolutionized the everyday practice of medicine and opened up methods for studying normal and abnormal structure as the introduction of brain imaging. Those who witnessed this transition could not have imagined the possibilities (revisiting the glorious age of 18th–19th century phrenology) for demonstrating focal brain activity during a host of behavioural activations, real and imagined, using functional magnetic resonance imaging and positron emission tomography, or the resolution with which brain structure and vascular anatomy can now be depicted in life with three-dimensional reconstruction of computerized tomography and magnetic resonance images. If the Nobel Prize award for Physiology or Medicine is about ingenuity, step-changes in knowledge, promotion of human health, and opening up unimagined opportunities for illuminating medicine—motives that inspired Vesalius and Willis—the recognition of Sir Godfrey Hounsfield (and Allan Cormack) in 1979, and Paul Lauterbur and (Sir) Peter Mansfield (born 1933) who received the Nobel Prize for Physiology or Medicine in 2003, for their unrivalled contributions in the latter half of the 20th century is surely not contested.

1.7 Changing concepts of neurological disease

Thomas Willis is the earliest significant figure in British neurology. In addition to the anatomical contributions, Willis described accurately an astonishing number of general medical and neurological disorders based on experience gathered in his medical practices in Oxford and London. Willis' casebook reveals the origin of the clinical methodology on which modern medicine is based: symptoms as the patients describe them; social aspects of the illness; physical examination; formulation of the problem; a list of pathophysiological alternatives; an approach to treatment; an assessment of prognosis; and communication with patients and their families. That said, the reputation of Willis took some time to become established. His University colleagues in Oxford suggested that much of the work was in reality that of Richard Lower [1631–1691]: '*whatever is anatomical in that Book, the Glory thereof belongs to the said R Lower whose indefatigable industry at Oxon produced that Elaborate Piece. Lower in his travels with Dr Willis made a discovery of the medicinal water at east Throp... the doctor being then, as usually, asleep or in a sleepy condition on horseback*'. Willis' rehabilitation began with the imprimatur of Sir Charles Sherrington who, in *Man on His Nature* (1940), argued that in referring to automatic acts as '*reflex activity*

of animal spirits running up sensory and down motor nerves', Willis put the brain and nervous system on their modern footing.

Willis (1667; translated 1681) proposed that seizures are not an affection of the part that moves but a remote consequence of activity in the brain, albeit in response to a peripheral stimulus, or of the blood entering the brain: '*to wit that the spirits inhabiting it being disposed to explosions, and there being exploded, bring on or cause every Falling Evil*'. He distinguished what would now be classified as complex partial seizures, symptomatic epilepsy, and pseudoseizures, and described several movement disorders. On hysteria, Willis rejected the concept of uterine displacement and the theory of pulmonary congestion and preferred to consider this as a disorder of the brain. This is what he means by 'convulsion' using the term 'epilepsie' for all types of the falling sickness. Willis wrote *On the Soul of Brutes* (1672; translated 1683) as a device for deflecting criticism from the Church on the physical basis for reason and human behaviour—man having both a brutal and rational soul—allowing Willis to get on with his analysis of structure, function and disease in the nervous system, without troubling the Ecclesiastical authorities. In addition to further accounts of epilepsy, he describes headache, apoplexy, neurosyphilis, narcolepsy, mental retardation, paracusis, head and spinal cord trauma, and a range of psychiatric syndromes; '*there is another kind depending on the scarcity of the spirits in which the motion is performed weakly... those being troubled are able to move their arms in the morning... but before noon... they are scarce able to move hand or foot... [and] after long speaking become mute... and [do] not recover the use still after an hour or two*' is taken to be the first description of myasthenia gravis.

Robert Whytt's textbook (1765) is psychiatrically flavoured—the distinction between nervous, hypochondriac, and hysteric disorders being only in the frequency and duration with which his patients experienced somatic manifestations of emotional states. Not until the 19th century did physicians systematically correlate knowledge gathered from pathological anatomy into systems of neurological disease. John Cooke (1820–1823) surveyed clinical neurology from ancient times to modern in three parts dealing with apoplexy, palsy, and epilepsy, respectively, and first drew attention to James Parkinson's description of the shaking palsy. Later, the contributions to an astonishing range of topics in clinical neurology attributable to Jean-Martin Charcot, Fig. 1.12, were faithfully recorded and published by his students (Charcot 1872–1887; translated 1877–1889). The clinical demonstrations, although selectively translated, are still only completely available in the original edition (Charcot 1887, revised and reprinted 1892). Many of his school friends at the Salpêtrière themselves later wrote definitive accounts of clinical and experimental neurology—notably Pierre Marie [1853–1940] and Gilles de la Tourette [1857–1904]. They (and others, including his son Jean-Baptiste Charcot [1867–1936], the Antarctic explorer) are gathered in the painting by Pierre Brouillet of Charcot demonstrating hysteria at La Salpêtrière during one of his Tuesday lectures; Babinski is catching the swooning Blanche Wittmann, one of the many hysterics accommodated long-term at the hospital in return for serial examinations of themselves and, eventually, their tissues.

In the United Kingdom, books that summarized contemporary knowledge on clinical neurology emerged mainly from the National Hospital. Its early history is recorded by Sir Gordon Holmes (1954). Multi-author systems were edited, for example, by John Russell Reynolds [1828–1896] with contributions from many

Fig. 1.12 Jean-Martin Charcot [1825–1893].

Fig. 1.13 William Gowers [1845–1915].

of the foundation staff. Sir William Gowers, Fig. 1.13, wrote the 'bible of nineteenth century neurology'; of him it could reasonably be said that, neurologically speaking, 'what he knew not was not knowledge'. *A Manual of Diseases of the Nervous System* (1886–1888) is decorated with his own line drawings of patients and pathological features of disease. Its production was timely, coinciding with the dawn of descriptive clinical neurology. Few of his accounts can be improved upon and many remain accurate to this day. Gowers' authority was usually his own experience without delving much into the published literature. Organized by anatomical region, the table of contents would do well as a nomenclature of contemporary neurology. The *Manual* summarizes knowledge gathered by arguably the greatest clinical observer in the history of neurology. It appeared at the peak of Gowers' career (according to Foster Kennedy, later his powers declined perhaps through injudicious use of opiates), and after his writings on muscle, ophthalmoscopy (Gowers was not the first systematically to use the ophthalmoscope introduced by Hermann von Helmholz [1821–1894] but he made it popular; it was also much favoured by Hughlings Jackson and Sir Clifford Allbutt [1836–1925] in England, and by Eugène Bouchut [1818–1891] in France), the spinal cord and epilepsy; only a short work on syphilis and the nervous system, some neuro-philsophy, published volumes of lectures, and the distillation of his clinical experience of 'dizzy turns'—the essay on *The Borderland of Epilepsy* (1907)— lay ahead.

Russell Brain [1895–1966], Fig. 1.14, wrote the textbook which has most influenced neurology in the English-speaking world (Brain 1933). Brain wrote six editions between 1933 and 1962; John Walton [born 1922] completed the seventh edition, published in 1969, and the present volume, the second since he relinquished his guiding editorial hand, is the twelfth. Their Lordships Brain and Walton reviewed developments in clinical neurology over a period in which more was learnt about the nervous system in health and disease than at any other time in history. In 1933, the symptoms of Parkinson's disease could best be alleviated by riding in a motor car; in 2007, this requires striatal transplantation of embryonic stem cells. The 900-odd pages of the 1933 edition are remarkable for having first been written by a 38-year-old man who sustained this single-handed effort until his death at the age of 71. Is there any neurologist trained since the 1930s who has not repeatedly used a copy of the contemporary edition? But the book had its critics. John Walton relates how he was advised waspishly by FMR Walshe [1885–1973] to '*put some red cells*' into the 1969 revision. Walshe himself wrote on *Diseases of the Nervous System* for practitioners and students (1940), a book that was popular but not written on the same scale. As a neurologist, Brain's lasting discoveries were the syndromes of median nerve compression in the carpal tunnel, disc prolapse as a cause of cervical myelopathy and paraneoplasia.

Fig. 1.14 Russell Brain [1895–1966].

In some respects, Brain's prestige was short lived for in 1940 appeared the last of the great single-authored textbooks written by Samuel Alexander Kinnier Wilson but seen through the press by Alexander Ninian Bruce [1882–1968] to whose father the book is jointly dedicated (Wilson 1940). Kinnier Wilson was one for the subtleties of symptomatology in early diagnosis—an abdominal reflex that tires, a few kicks of nystagmus—and lightened his text with clinical anecdotes such as the patient who convulsed the ward with a Rabelasian peal of laughter through reference to the condition of his trousers in answer to a question on bladder control; Wilson's penchant for music-hall humour was well known to his colleagues. Unlike Gowers' or Brain's, Wilson's opinions are extensively supported by citations from the (often non-Anglophone) literature of the 1920s and 1930s. Each chapter has a brief but well-researched historical synopsis. Written in an era when infections usually ran their natural course, and substances of abuse were becoming associated with recognizable clinical complications, the emphasis is on toxi-infective disease of the nervous system. Wilson insisted on hepato-lenticular degeneration being known as Kinnier Wilson's disease—and the eponym stuck. In 1940, diseases of uncertain nature included the epilepsies, narcolepsies, headache, and myasthenia gravis. With the exception of Parkinson's disease and the choreas, all movement disorders were considered by Kinnier Wilson to be entirely functional. Out of print since 1954, in an edition revised by Russell Brain and including his own monograph on aphasia, the discerning neurolo-

gist wanting to confirm a clinical fact, not only in the context of diseases which are now less prevalent in everyday neurology, can turn to 'Kinnier Wilson' and invariably come away better informed.

1.8 Postscript

An apocryphal story summarizes the tradition at the National Hospital, Queen Square when clinical neurology was at its zenith: the ideal team to handle a case was Sir Charles Symonds [1890–1978] to take the history, Sir Gordon Holmes to examine the patient, and W.D. Adie [1886–1935] to explain matters to the family. It seems that, then as now, clinical wisdom and a good pastoral approach were not always set on the same pair of magisterial shoulders. But those who practise neurology in the first decade of the new millennium do so courtesy of the systematic discoveries and brilliance of the many who previously struggled to understand the 'doctrine of the nerves'.

References

Babinski J (1896). Sur la réflexe cutané plantaire dans certaines affections organiques du système nerveux central. *Comptes rendus des Seances et Memoires de la Societe de Biologie*, **48**, 207–8.

Baillie M (1793). *The Morbid Anatomy of Some of the Most Important Parts of the Human Body*, pp. 314. Johnson and Nicol, London.

Baillie M (1799–1802). *A series of Engravings Accompanied with Explanations which are Intended to Illustrate the Morbid Anatomy of Some of the Most Important Parts of the Human Body*, pp. 288. Bulmer and Co., London.

Bell C (1802). *The Anatomy of the Brain Explained in a Series of Engravings*, not paginated. Longman and Rees, London.

Bell C (1811). *Idea of a New Anatomy of the Brain*, pp. 36. Strahan and Preston, London.

Bennett AH, Godlee RJ (1885). Case of cerebral tumour. *Medico-Chirurgical Transactions*, **68**, 242–75.

Brain WR (1933). *Diseases of the Nervous System*, pp. 899. Oxford University Press, London.

Bright R (1827–1831). *Reports of Medical Cases, Selected with a View of Illustrating the Symptoms and Cure of Diseases by a Reference to Morbid Anatomy*, pp, 231; 724. Longmans, London.

Cajal SR (1897–1904). *Textura del sistema nervioso del hombre y de los vertebrados*. 3 volumes. Madrid. Translation by Dr L Azoulay (1909–1911) as: *Histologie du systeme nerveux de l'homme et des vertebres*. 2 volumes Paris: and into English by Neely Swanson and Larry W Swanson (1995) *Histology of the nervous system of man and vertebrates*. 2 volumes, pp. 805; 806. Oxford University Press, New York, and by Pedro Pasik and Tauba Pasik (1999–2002) *Texture of the nervous system of man and the vertebrates*. 3 volumes, pp. 631; 666; 661, Springer-Wien.

Carswell R (1838). *Pathological Anatomy; Illustrations of the Elementary Forms of Disease*, not paginated. Longman, Orme, Brown, Green, and Longman, London.

Caton R (1875). The electric currents of the brain. *BMJ*, **2**, 278.

Charcot JM (1872–1887). Lecons sur les Maladies du Systeme Nerveux faites a la Salpêtrière. A Delahaye et E Lecrosnier; Progres Medical, Paris. Translated into English by George Sigerson (1877–1889) as *Lectures on the Diseases of the Nervous System Delivered at the Salpetriere*, pp. 325; 399; 438. The New Sydenham Society, London.

Charcot JM (1887). *Lecons du Mardi a la Salpêtrière*, pp. 638. A Delahaye et Progres Medical. Paris.

Clarke E, Dewhurst K (1968). *An Illustrated History of Brain Function*, pp. 154. Sandford Publications, Oxford.

Cooke J (1820–1823). *A Treatise on Nervous Diseases*, 3 volumes, pp. 469; 215; 235. Longman, Hurst, Rees, Orme and Brown, London.

Creed RS, Denny Brown D, Eccles JC et al. (1932). *Reflex Activity of the Spinal Cord*, pp. 183. Oxford at the Clarendon Press, Oxford.

Croone W (1664). *De ratione motus musculorum*, pp. 34. Hayes, London.

Cruveilhier J (1829–1842). *Anatomie pathologique du corps humain; descriptions avec figures lithographiees et coloriees; des diverses alterations morbides dont le corps humain est susceptible*, 40 livraisons, not paginated. JB Bailliere, Paris.

Dandy WE (1918). Ventriculography following the injection of air into the cerebral ventricles. *Ann Surg*, **68**, 5–11.

Egas Moniz AC de (1927). L'encephalographie arterielle, son importance dans la localisation des tumeurs cerebrales. *Rev Neurol*, **34**, 72–90.

Fearing F (1930). *Reflex Action*, pp. 350. Williams and Wilkins, Baltimore.

Ferrier D (1876). *The Functions of the Brain*, pp. 323. Smith Elder, London.

Gowers WR (1886–1888). *A Manual of Diseases of the Nervous System*, 2 volumes, pp. 463; 975. J and A Churchill, London.

Gowers WR and Horsley V (1888). A case of tumour of the spinal cord. removal; recovery. Medico-Chirurgical Transactions, **71**, 377–431.

Gowers WR (1907). *The Borderland of Epilepsy*, pp. 121. J and A Churchill, London.

Grainger RD (1837). *Observations on the Structure and Functions of the Spinal Cord*, pp. 160. Samuel Highley, London.

Hall M (1837). *Memoirs on the Nervous System*, pp. 113. Sherwood, Gilbert and Piper, London.

Haymaker W, Schiller F (1970). *The Founders of Neurology*, 2nd edition, pp. 616. CC Thomas, Illinois.

Holmes G (1946). *An Introduction to Clinical Neurology*, pp. 183. E and S Livingstone. Edinburgh.

Holmes G (1954). *The National Hospital Queen Square 1860-1948*, pp. 98. E and S Livingstone, Edinburgh.

Hooper R (1826). *The Morbid Anatomy of the Human Brain being Illustrations of the Most Frequent and Important Organic Diseases to which that Viscus is Subject*, pp. 36. Longman, Rees, Orme, Brown, and Green, London.

Hope J (1834). *Principles and Illustrations of Morbid Anatomy*, pp. 299 (95), plates. Whitaker and Co., London.

Hounsfield GN (1973). Computerised transverse axial scanning (tomography). *Br J Radiol*, **46**, 1016–22.

Jackson JH (1932). *Selected Writings*. Ed. James Taylor, pp. 500; 510. Hodder and Stoughton, London.

Koehler PJ, Bruyn GW, Pearce JMS (2000). *Neurological Eponyms*, pp. 386. Oxford University Press, Oxford.

Lauterbur PC (1973). Image formation by induced local interactions: examples of employing nuclear magnetic resonance. *Nature*, **242**,190–1.

Liddell EGT (1960). *The Discovery of Reflexes*, pp. 174. Clarendon Press, Oxford.

Magendie F (1822). Expériences sur les functions des raciness des nerfs rachidiens. *Journal de physiologie expérimentale et pathologique*, **2**, 276–9 and 366–71.

Mansfield P, Grannell PK (1973). NMR 'diffraction' in solids? *Journal of Physics C. Solid State Physics*, **6**, L422–7.

McHenry L (1969). *Garrison's History of Neurology*, pp. 552. CC Thomas, Illinois.

Monrad-Krohn GH (1921). *The Clinical Examination of the Nervous System*, pp. 135. HK Lewis, London.

Morgagni (1761). *De sedibus, et causis morborum per anatomem indigatis libri quinque*, 2 volumes pp. 298, 452. Remondidiana, Venice.

Neuburger M (1981). *The Historical Development of Experimental Brain and Spinal Cord Physiology before Flourens*, pp. 391. Johns Hopkins University Press, Baltimore.

Norman JM (1991). *Morton's Medical Biography*, pp. 1243. Scolar Press, Aldershot.

Quincke HI (1891). Die Lumbarpunction des Hydrocephalus. *Berlin klinisches Weschrift*, **33**, 965–8.

Riddoch G, Rowley Bristow W, Cairns HWB *et al.* (1943). *Aids to Investigation of Peripheral Nerve Injuries*, pp. 54. Medical Research Council, London.

Sherrington CS (1906). *The Integrative Action of the Nervous System*, pp. 411. Constable, London.

Sherrington CS (1940). *Man on His Nature*, pp. 413. Cambridge University Press, Cambridge.

Sicard JA (1921). Methode radiographique d'exploration de la cavite epidurale par la lipodol. *Rev Neurol*, **28**, 1264–6.

Spillane JD (1968). *An Atlas of Clinical Neurology*, pp. 376. Oxford University Press, Oxford.

Spillane JD (1981). *The Doctrine of the Nerves: Chapters in the History of Neurology*, pp. 467. Oxford University Press, Oxford.

Swan J (1834). *A Demonstration of the Nerves of the Human Body*, pp. 98 (82), plates. Longman, Rees, Orme, Brown, Green, and Longman, London.

Vesalius A (1543). *De humani corporis fabrica libri septum*, pp. 663. Oporini, Basel.

Walshe FMR (1940). *Diseases of the Nervous System*, pp. 288. E and S Livingstone, Edinburgh.

Wartenberg R (1945). *The Examination of Reflexes*, pp. 222. Year Book Publishers, Chicago.

Whytt R (1751). *An Essay on the Vital and Involuntary Motions of Animals*, pp. 392. Hamilton, Balfour and Neill, Edinburgh.

Whytt R (1765). *Observations on the Nature Causes and Cure of those Disorders which have been Commonly Called Nervous Hypochondriac or Hysteric etc.*, pp. 520. Becket, Edinburgh.

Willis T (1681). *The Remaining Medical Works etc.*, pp. 178; 192; 106; (30). Dring, Harper, Leigh and Martyn, London.

Willis T (1683). *Two Discourses Concerning the Soul of Brutes*, pp. 234 (8). Dring, Harper and Leigh, London.

Wilson SAK (1940). *Neurology*, 2 volumes, pp. 1838. E Arnold and Co., London.

CHAPTER 2

The clinical approach

Michael Donaghy

Contents

2.1 The frequency of neurological diseases

Some neurological disorders such as stroke are so common and serious that reducing their burden features in the public health goals of many countries. Others are similarly common, and treatable, but often are regarded as having less public health importance for instance epilepsy or migraine. Some common conditions remain unpreventable and incurable such as Alzheimer's disease. The most frequent cause of disability in young adults, multiple sclerosis, is thankfully not all that common. Less common disorders are myriad, of which many are hopelessly incurable, for instance motor neurone disease. Vigilance is required to detect few very rare diseases which are completely treatable, for example Wilson's disease or tetanus. Perhaps the rarest of all, variant Creutzfeldt–Jakob disease, has attracted disproportionate political, economic, and public health concern, at least in the United Kingdom. So, neurologists have to be familiar with a huge range of disorders and balance the challenge of large numbers of patients with common disorders against vigilance for the once-in-a-lifetime patient with a treatable disease who slips in to the end of a busy clinic (MacDonald *et al.* 2000). Some contemporary neurologists need to sub-specialize, so as to provide optimal diagnosis and management of rare disorders such as myasthenia, and also the difficult end of the spectrum of more common disorders such as refractory epilepsy or treatable neuromuscular disease. But it is vital for most neurologists to nurture their general neurological skills, so as to maintain a broad diagnostic perspective, and to teach students and trainees.

2.1.1 Measuring disease frequency

The three traditional measures of disease frequency are mortality, incidence, and prevalence. Choosing which to use depends on the frequency of the disease in question, whether it is likely to be fatal, whether it is an acute one-off event or chronic, and also on logistic and methodological issues to do with recording and coding the disease itself. Because the frequency of most diseases varies by age, and sometimes by sex too, age- and sex-specific rates should be given. Frequencies of disease based on hospital data tend to be hopelessly flawed because one has no idea of the size of the denominator population, or of why some patients were referred to hospital and others not.

Mortality data, the number of deaths from a particular disease per annum in a population of known size, are routinely collected in developed countries from death certificates. However, there are many problems of erratic classification of disease and poor coding practice. Even more problematic is that some diseases like migraine are not fatal; others may be fatal but linger in such a chronic fashion that the patient's death is due to, and is coded as, something quite different; multiple sclerosis patients may die of bronchopneumonia or cancer; and some disease entities have mild forms which are seldom fatal, for instance lacunar stroke. However, mortality data are based on large numbers usually and are likely to be precise as a result. Nonetheless, mortality only crudely approximates to disease frequency and is unhelpful when asking rather specific questions, such as how often do patients with epilepsy die suddenly and to what are their deaths due?

2.1.2 Incidence of neurological disorders

Incidence is the number of new cases of a disease appearing in a defined population of known size per annum. To measure it one must therefore have good census data to define the population denominator, multiple and overlapping case-finding methods to identify all the patients, a clear definition of when the disease actually starts which is easy for stroke but more difficult for gradually progressive disorders such as motor neurone disease, and a large enough number of patients with the disease to calculate precise estimates of frequency over a defined time period. All this is hardly possible outside prospective community-based studies although, if all the community is getting health care in one place and the records system is well organized and funded, such as in Rochester, Minnesota, then retrospective estimates of incidence are reasonably accurate, and probably the only sensible method for rare diseases. Any method relying on scrutiny of patient records would be threatened if society insists that researchers must first obtain the explicit consent of every patient.

Table 2.1 provides some estimates of the incidence of various neurological disorders in the conventional way of number of new cases per 100 000 population per annum, but not stratified by sex. In addition, it emphasizes that even the common neurological disorders such as multiple sclerosis are not all that common in primary care where physicians have to be extraordinarily alert to recognize and diagnose disorders they may never have seen since medical school. In educating students in neurology we should concentrate on the common disorders and on the general principles of recognizing that a patient has a neurological disorder so they can be referred to neurologists for precise diagnosis (Donaghy 2005).

2.1.3 Prevalence of neurological disorders

The prevalence of a disease is the number of patients with that disease at a particular point in time, usually expressed per 100 000 population. Again this requires good census data or some other method of measuring the population denominator, such as a United Kingdom family practice computerized age–sex register, and then finding all the patients with the disease of interest in that population and confirming they actually have the disease. This is surprisingly difficult to do, and tedious, particularly when the disease is rare. Of course, prevalence will tell one nothing about fatalities. Furthermore, for episodic diseases such as transient ischaemic attacks, an episode many years before may well have been forgotten and, even if not, it may be difficult to diagnose in retrospect. Nonetheless, for some chronic and persisting disorders, estimates of prevalence can be illuminating (Table 2.2). This table again shows just how rare many neurological disorders are, even in a family practice of five doctors looking after 10 000 people. If one knows the incidence and the proportion of patients who die, then prevalence can be calculated and does not have to be measured directly.

2.2 Principles of clinical diagnosis

Just under 10 per cent of the population consult their general practitioner about a neurological symptom each year in the United Kingdom. About 10 per cent of these are referred for a specialist opinion, usually to a neurologist. The commonest diseases or clinical problems encountered in a general neurological out-patient clinic are shown in Table 2.3. Together these nine conditions account for roughly 75 per cent of general neurological referrals and are diagnosed initially on purely clinical grounds and frequently managed purely in an out-patient setting (Perkin 1989; Stevens 1989). The remaining 25 per cent of neurological consultations concern the huge range of other neurological disorders, many rare. Such disorders are particularly likely to require highly specialist investigation and treatment, to need in-patient care, and continuing follow-up care. Naturally these broad statistics will vary in different healthcare settings which may not be based upon general practice, and as the demand for, and availability of, neurological services changes.

This section is concerned with a practical, everyday approach to diagnosing neurological disorders. It does not aim for the exhaustive completeness familiar in traditional accounts of the neurological examination. However there are times at which a more detailed approach to clinical assessment is necessary, for instance in elucidating the neuroanatomical site of the lesion responsible for muscle weakness (Section 2.4) or a somatosensory disturbance (Section 2.5). Also it is important to document the different reflexes which can be elicited, giving some indication of their usefulness (Section 2.3). These more traditional clinical approaches to such problems are presented separately, later in this chapter.

2.2.1 History taking

History taking is fundamental to neurological diagnosis. For instance, epilepsy or migraine are diagnosed solely on the basis of the history, with the examination merely ensuring there is no evidence of associated underlying structural disorders of the brain. The history is usually much more informative than examination, which generally is either reassuringly normal or merely confirms features anticipated

Table 2.1 The approximate incidence of various neurological disorders and how often a new case will be seen in primary care by a general practitioner or family doctor with a list size of 2000 people

	Incidence/100 000/annum	Number of years between consecutive new cases seen by a general practitioner with a list size of 2000 people
Stroke	200	0.25
Carpal tunnel syndrome	100	0.5
First epileptic (non-febrile) seizure	50	1.0
Transient ischaemic attack	50	1.0
Bell's palsy	25	2.0
Essential tremor	24	2.1
Parkinson's disease	20	2.5
Primary brain tumour	15	3.3
Secondary brain tumour	14	3.6
Multiple sclerosis (Scotland)	12	4.2
Subarachnoid haemorrhage	10	5.0
Essential tremor	8	6.3
Giant cell arteritis	6	8.3
Migrainous neuralgia	6	8.3
Unexplained motor symptoms	5	10
Trigeminal neuralgia	4	13
Meningococcal meningitis (UK)	3	17
Transient global amnesia	3	17
Guillain–Barré syndrome	2	25
Intracranial vascular malformation	2	25
Motor neurone disease	2	25
Neuralgic amyotrophy	2	25
Progressive supranuclear palsy	1	50
Diabetic amyotrophy	1	50
Benign intracranial hypertension	1	50
Focal dystonia	1	50
Myasthenia gravis	1	50
Polymyositis/dermatomyositis	1	50
Hemifacial spasm	0.8	63
Multiple system atrophy	0.6	83
Gilles de la Tourette syndrome	0.5	100
Pneumococcal meningitis (UK)	0.5	100
Herpes simplex encephalitis	0.2	250
Creutzfeldt–Jakob disease (sporadic)	0.1	500
Tetanus	0.1	500
Subacute sclerosing panencephalitis	0.03	1667
Variant Creutzfeldt–Jakob disease (UK)	0.02	2500

These figures are all very approximate. They have been taken from various more or less sound community-based epidemiological studies in Europe or North America and a large survey of general practice in the United Kingdom (MacDonald *et al.* 2000). When more than one study is available, an approximate average rate has been used. The exact rates will generally depend on the age and sex structure of the population, which varies between communities, the size of the population which will influence the precision of any estimate, and on the precise diagnostic criteria which also vary. However, the rates give a general idea of incidence and how common, or rare, some neurological disorders are.

from the history. However, sometimes examination is crucially helpful. For instance, in localizing the cause of muscle weakness, specific physical signs will reveal whether the lesion affects the upper motor neurone, the lower motor neurone, or the muscle. An unanticipated physical sign such as an extensor plantar response, signifying pyramidal tract damage, or papilloedema, signifying raised intracranial pressure, will alter one's diagnostic view fundamentally if the history

has pointed to diagnoses such as psychologically determined weakness or benign tension headache.

A good history provides a story whose internal direction points intuitively towards a diagnosis. It is much more revealing to treat the history as story telling than to take the utilitarian approach of simply listing symptoms with the expectation that a diagnosis will appear miraculously. Experience reveals that patients often describe

Table 2.2 The approximate prevalence of various neurological disorders and how frequently they are present in an average general practice or family practice of 10 000 people looked after by five doctors

	Prevalence/ 100 000	Number of cases in a general practice with 10 000 people
Migraine	10 000	1000
Chronic tension headache	3000	300
Stroke	800	80
Alzheimer disease	800	80
Active epilepsy	500	50
Essential tremor	300	30
Multiple sclerosis (Scotland)	200	20
Chronic fatigue syndrome	200	20
Parkinson's disease	160	16
Migrainous neuralgia	40	4
Unexplained motor symptoms	38	4
Neurofibromatosis type 1	13	1
Myasthenia gravis	10	1
Hemifacial spasm	10	1
Narcolepsy syndrome	10	1
Huntington's disease	8	<1
Myotonic dystrophy	7	<1
Syringomyelia	7	<1
Progressive supranuclear palsy	5	<1
Motor neurone disease	5	<1
Duchenne muscular dystrophy	4	<1
Fascioscapulohumoral dystrophy	3	<1
Mitochondrial cytopathy	2	<1
Multiple system atrophy	2	<1
Chronic inflammatory demyelinating neuropathy	1	<1
Tuberous sclerosis	1	<1
Wilson's disease	0.4	<1

These figures are all very approximate. They have been taken from various more or less sound community-based epidemiological studies in Europe or North America and a large survey of general practice in the United Kingdom (MacDonald *et al.* 2000). When more than one study is available, an approximate average rate has been used. The exact rates will generally depend on the age and sex structure of the population, which varies between communities, the size of the population which will influence the precision of any estimate, and on the precise diagnostic criteria which also vary. However, the rates give a general idea of prevalence and how common, or rare, some neurological disorders are.

Table 2.3 Commonest conditions seen by neurologists

Headache and face pain
Blackouts and epilepsy
Peripheral nerve and root disorders
Cerebrovascular disease
Multiple sclerosis
Parkinsonism and movement disorders
Dementia
Giddiness and vertigo
Psychologically determined symptoms

the symptoms of certain disorders in a very distinctive way. Intuitive recognition of a characteristic history plays a large part in diagnosis. There is no particular list of questions to ask. It is best to invite the patient to describe their symptoms in the order in which they occurred, with approximate dates. Important detail can be clarified by specific questioning during or after the patient's account. It is helpful to determine whether the patient's symptoms are so 'disabling' as to prevent crucial everyday activities or work, or whether they merely constitute a 'nuisance'. This will guide the decision as to whether a symptom such as headache needs treatment.

Some features of the history provide important clues to the neurological diagnosis. Questions about them should be phrased in open terms which do not influence the patient's response:

Time course. Symptoms of abrupt or instantaneous onset usually indicate epilepsy, with sudden loss of consciousness, or cerebrovascular disease, the instantaneous headache of subarachnoid haemorrhage, or a sudden hemiparesis due to middle cerebral artery embolus. Symptoms that deteriorate subacutely, over hours, days, or even a few weeks are generally caused by inflammatory or demyelinating disorders. Slowly deteriorating symptoms over some weeks, months, or years point to the growth of a tumour, or a neurodegenerative process. Relapsing and remitting symptoms which come and go over weeks are typical of multiple sclerosis whilst recurrent headaches, each lasting 3 h to 3 days, are typical of migraine.

Negative symptoms. These indicate loss of normal neurological functions and are the commonest symptoms of damage to the nervous system. Examples include the hemiparesis due to cerebral hemisphere infarction, memory loss due to Alzheimer's disease, muscle weakness due to motor neurone degeneration, or the loss of micturition control due to a cauda equina tumour.

Positive symptoms. These are novel phenomena which often suggest specific diagnoses. A 'pill-rolling' tremor of the fingers and thumb at rest is characteristic of Parkinson's disease. Flashing lights, photopsia, or zig-zag lines, fortification spectra, preceding a headache are diagnostic of classical migraine. Repetitive twitching of the fingers or the corner of the mouth occurs in focal motor seizures. A hallucination of an odd smell, often like burning rubber, is typical of an epileptic discharge in the temporal lobe. Tingling in the toes and fingers is typical of acquired, rather than inherited, peripheral neuropathy.

Neuroanatomical localization. Sometimes enquiry about other specific symptoms is necessary to anatomically localize the disease process. For example a patient with suspected motor neurone disease should be asked whether there are sensory or sphincter symptoms which might point to the alternative diagnoses of generalized peripheral neuropathy or to spinal cord compression respectively. A patient with sensory symptoms in the legs should be asked whether their hands are also affected; this would be a pointer to a polyneuropathy or cervical myelopathy rather than a focal lesion of the cauda equina or thoracic spinal cord. Determine whether a patient with dysphasia also has impaired spatial abilities, such as a dressing apraxia or getting lost in familiar places; this would point to a generalized dementing process involving both cerebral hemispheres rather than a focal lesion of the left hemisphere causing pure dysphasia. Question a patient with gait unsteadiness about vertigo or double vision, which would imply damage to the brainstem rather than to the cerebellum or somatosensory pathways.

Eye witness descriptions. Patients with blackouts are unaware of what they did whilst unconscious and may not recollect the onset of the blackout. Thus, an eye witness description of a convulsion or automatic behaviour is diagnostic of epilepsy. In a patient with early dementia, it is often the spouse who provides the evidence for loss of intellectual function: forgetting the grandchildren's names, inability to do the usual crossword, or personality change. Patients with motor neurone diseases are often unaware of their limb muscle fasciculations, yet their spouse may have noticed their occurrence whilst in bed.

Previous neurological history. This is vital for establishing the diagnosis of multiple sclerosis, a neurological disorder which is disseminated in space and time. Thus, a history of temporary unilateral visual loss due to optic neuritis a decade previously suggests multiple sclerosis in a 30-year-old woman with unsteady gait and urgency of micturition due to an incomplete spinal cord lesion.

Familial disorders. Many neurological disorders are genetic, although each of these is usually rare. Examination of the relatives of a patient with longstanding muscle wasting and weakness below the knees, and with high foot arches, pes cavus, may reveal autosomal dominant inheritance of a similar disorder so allowing diagnosis of hereditary motor and sensory neuropathy, otherwise known as Charcot–Marie–Tooth disease. First cousin marriage between the parents may be a clue to autosomal recessive disorders in offspring with neurological disease. Sex-linked recessive disorders, transmitted on the X chromosome and occurring in males, will not manifest in the mother, but may be present in the males of earlier or parallel generations.

Contributory general medical disorders. Progressively deteriorating neurological symptoms should prompt questions about possible underlying cancer affecting the nervous system: smoking, weight loss, haemoptysis, bowel symptoms, and recent breast and gynaecological check-ups. In a patient with stroke, a previous history of ischaemic or valvar heart disease, hypertension, diabetes, oral contraceptive usage, migraine, or cocaine abuse may be relevant. Unusual neurological disorders, such as opportunistic infections or lymphoma of the central nervous system, are particularly likely in the increasing numbers of immunosuppressed patients who are HIV infected, or have received organ transplants. Typical neurological side-effects of medicines are headache, giddiness, tremulousness, tinglings, and peripheral neuropathy; a patient's drugs should be checked in a pharmacopeia for side-effects, and the onset of the symptoms related to the introduction of the drug. The travel history may raise the likelihood that a patient's symptoms are due to an underlying infection such as leprosy, schistosomiasis, malaria, diphtheria, or borreliosis. Patients addicted to alcohol or recreational drugs are notorious for underestimating or denying consumption which may be directly relevant to disorders such as ataxia and stroke respectively.

2.2.2 General neurological examination

Present day neurology has acquired a reputation for being both complicated and arcane because of the huge diversity of examination techniques which have been described. This has led in turn to the notion that there is an excessively lengthy entity called 'a full neurological examination' which utilizes this vast panoply of examination manoeuvres. Also, there is a commonly held belief that if one religiously executes all those manoeuvres, a diagnosis will miraculously appear. In reality the diagnostic process is one of intuition which involves devising a selective examination to answer diagnostic hypotheses posed by the symptoms of the particular patient in question. For instance, it is usually pointless to undertake cognitive testing in a patient who has given a cogent history of paraesthesiae in an arm, or to undertake detailed muscle power examination in a patient presenting with cognitive decline.

Many of the described examination manoeuvres are simply alternative methods of detecting the same pathological signature. Therefore to use more than one of them introduces unnecessary redundancy and repetition to the examination. For instance, a cerebellar lesion affecting the arm can be detected by the finger–nose test, by dysdiadochokinesis, by demonstrating 'cerebellar hypotonia', or by showing 'underdamping' when the outstretched arm is displaced with the eyes closed. Experienced neurologists develop sensitive and critical, yet economical, examination skills. They may recognize that a properly conducted finger–nose test is the only test that need be performed to demonstrate a cerebellar disorder of the arm. Valuable physical signs offer proof that an abnormality is present. Less valuable signs merely suggest one. The worth of trying to elicit different signs varies accordingly.

Often tests are of limited usefulness because they do not provide objective evidence of abnormality. Examples include the patient's subjective responses to visual or somatosensory testing, or the influence of psychological factors on exertion of muscle power. Consequently a useful neurological examination will be rich in manoeuvres which can provide unequivocal evidence of pathology. These include inspection for papilloedema, testing pupil-light reflexes and eye movements, examining for cogwheel rigidity, noting muscle wasting and detecting absent tendon reflexes, extensor plantar responses, or sustained ankle clonus. For this reason it is recommended that physicians develop a brief basic neurological examination for routine use which is rich in testing for such unequivocal physical signs (Donaghy 2005). It avoids manoeuvres which are simply repetitive or imprecise ways of detecting the same pathology.

Such basic neurological examinations take only a few minutes. The ensuing example of a quick screening examination includes practical advice such as how to phrase instructions, where to place the hands for best effect, and how to interpret fundamentals such as abnormal reflexes. This basic examination is quite adequate for examining a patient with uncomplicated headache or epilepsy, or as part of a general medical examination for a patient without neurological symptoms. Other tests should be added on to this examination if the patient's symptoms suggest a particular disease, or if abnormalities requiring further assessment are encountered during the basic examination.

It is practical to perform this basic examination in four stages: first during history taking, second whilst the patient is walking, third whilst the patient is sitting facing you, and fourth whilst the patient is lying down (Table 2.4) (Donaghy 2005).

During history taking

Speech and cognition. Abnormalities of speech, thought, or memory raise questions of dysphasia or generalized dementia. Dysarthric speech is slurred. Dysphonic speech is quiet.

Facial expression. An impassive face suggests Parkinson's disease, or occasionally a bilateral facial palsy. A melancholy facial expression occurs in depression. Dementia reduces the use of facial expression and gesture for non-verbal communication.

Involuntary movements. Pill-rolling tremor of the fingers at rest is characteristic of Parkinson's disease. Sudden choreiform movements of the hands, which may look like fidgets, occur in Huntingdon's disease and are often disguised as mannerisms.

Table 2.4 A general neurological examination (after Donaghy 2005)

1.	During history taking note:	Speech and cognition
		Facial expression
		Involuntary movements
2.	With patient standing note:	Gait
		Heel–toe walking
		Romberg Test
3.	With patient sitting note:	
	Cranial nerves:	Fundoscopy (II)
		Visual fields (II)
		Horizontal eye movements (III, VI)
		Pupil-light responses (afferent II; efferent III)
		Facial sensation (V)
		Facial movements (VII)
		Hearing (VIII)
		Palatal movement (X)
		Tongue movement (XII)
	The arms:	Inspection
		Tone
		Power (Shoulder abduction and finger spreading)
		Finger–nose coordination
4.	With the patient lying, note:	
	The arms (cont)	Tendon reflexes (biceps and triceps)
	The legs:	Inspection
		Ankle clonus
		Power (hip flexion and ankle dorsiflexion)
		Tendon reflexes (knee and ankle)
		Plantar responses
5.	Finally examine additional features as required by history or by abnormalities discovered on the basic examination.	

Fig. 2.1 Heel–toe walking.

examiner demonstrates two or three such steps (Fig. 2.1). Patients will stumble to the side if they have ataxia due to cerebellar disease, or loss of leg proprioception due to peripheral neuropathy or dorsal column disease.

Romberg test. This is an excellent test for loss of proprioceptive feedback from the legs in peripheral neuropathy or dorsal column disease. In patients with abnormal heel–toe walking, it differentiates those with ataxia due to loss of proprioceptive feedback from those with ataxia due to cerebellar disease. It is best tested by the examiner instructing 'stand with your feet together like this [whilst demonstrating], get your bearings, and now close your eyes— I won't let you fall' whilst preparing to steady the patient's shoulders with their hands if the patient begins to topple (Fig. 2.2). Romberg test is positive if the patient falls, or is unable to maintain balance without corrective movements of the feet. It is important to realize

Unilateral spasms of eye closure occur in hemifacial spasm. Fixed or spasmodic head rotation to the side occurs in torticollis.

With the patient standing

Walking. In the wide-based gait of ataxia the feet cross more than the usual 0–5 cm apart, and the stride length is irregular. Uniformly small strides occur in the gait apraxia of frontal lobe disease. Difficulty in starting, shuffling, and then progressively lengthening strides occur in Parkinsonism. Arm swing is lost in Parkinson's disease, usually unilaterally early on. Floppy foot drops occur in peripheral nerve or nerve root disease. Stiff foot drops occur in spastic upper motor neurone lesions, or occasionally in dystonia. A waddling gait, with drop of the pelvis on the striding side, occurs in proximal muscle weakness due to myopathy.

Heel-to-toe walking. This is a sensitive screen for cerebellar disease, or sensorimotor abnormalities affecting the limbs. It is best tested by instructing 'Please walk heel-to-toe, like this' whilst the

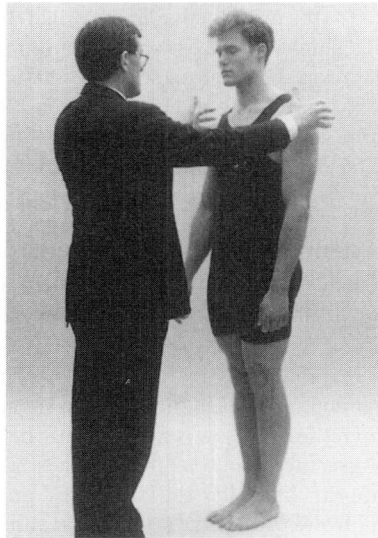

Fig. 2.2 Romberg test.

that a correctly performed Romberg test does not merely test balance with the eyes closed, but is a comparison of stability with and without vision. A truly positive Romberg test takes some moments to develop, with an increasing amplitude of slow swaying until a critical degree of lean occurs, beyond which the patient can no longer remain upright. Not uncommonly one encounters patients who promptly fall in one direction immediately upon closing their eyes; this usually results from lack of confidence or is otherwise psychologically determined, and rarely indicates structural disease of the nervous system.

With the patient sitting

Ophthalmoscopy. Inspect the optic nerve head, also called the optic disc. It is important to understand the location of the optic nerve head, which corresponds to the blind spot, within the visual field and its corresponding position in the eye. This enables the patient's direction of gaze to be aligned so that the examiner can look into their eye confident of looking directly at, or very near to, the optic disc. The blind spot, which represents the optic disc, lies about 20° of visual angle lateral to the point of fixation in each eye. Also it lies just below the horizontal. This determines the 'line of attack' (Fig. 2.3). Therefore the patient should be asked to fixate on a point behind the examiner chosen for height so that he is able to look into their eye comfortably from just below its horizontal meridian. The particular fixation point chosen will depend upon the relative heights of the examiner's and the patient's heads. The ophthalmoscope should be used with one's right eye to examine the patient's right eye, and vice versa for the left, looking into the eye from about 20° lateral to the line of fixation, and from just below the line of sight.

Examine whether the edge of the optic disc is sharply defined as is normal, or has blurred edges suggesting disc swelling due to papilloedema due to raised intracranial pressure. A pale or white disc is due to optic atrophy. Having inspected the optic disc, the vessels and more peripheral parts of the retina can be scrutinized, for instance if diabetic retinopathy is suspected; this is easier if the pupils are dilated. The foveal pit, or macula, can be inspected whilst the patient stares directly at an ophthalmoscope beam adjusted to the small spot.

Visual fields. It is time-consuming and rarely rewarding to carry out detailed examination of the peripheral and central portions of the visual fields of each eye separately unless the patient has symptoms of visual or pituitary disease. The following quick manoeuvre is a simple screen for homonymous hemianopia, an identical visual field deficit in both eyes due to cerebral hemisphere disease, and for sensory inattention due to parietal lobe lesions. The patient is asked to 'keep looking at my nose and point to whichever of my index fingers moves'. The examiner's arms are raised so as to position the index fingers at about 80° peripheral in each visual field. After a moment the tip of one index finger should be moved once whilst keeping the rest of the arm still; the patient should point immediately. If the patient has sensory inattention or a homonymous hemianopia, they will only see movement on one side despite simultaneous movement of the fingers on both sides and further analysis of the deficit can be undertaken. A red pinhead, or perimetry techniques, are often required for accurate detection and delineation of more subtle visual field defects. These include red desaturation monocularly in the temporal field in optic chiasm lesions, or the monocular partial central scotoma so common in optic neuritis.

Sensory inattention may be detected by moving your finger on both sides simultaneously, but the patient will only detect movement on one side. Of course sensory inattention, which reflects a parietal lobe lesion, can only be diagnosed if each visual field is normal when tested separately.

Eye movements. Inspection of the patient's face when gazing straight ahead will show the drooping eyelid of ptosis. Mild to moderate ptosis can be difficult to detect if bilateral. In definite ptosis, the eyelid will overlap the edge of the pupil when looking straight ahead.

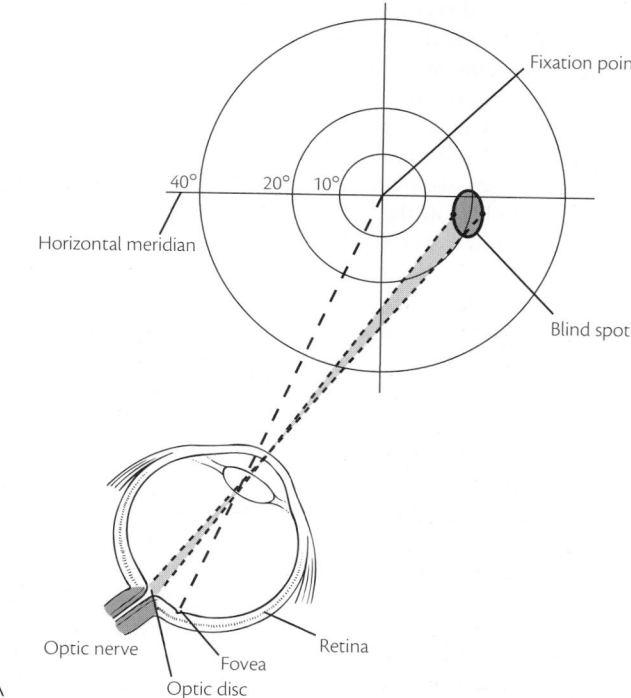

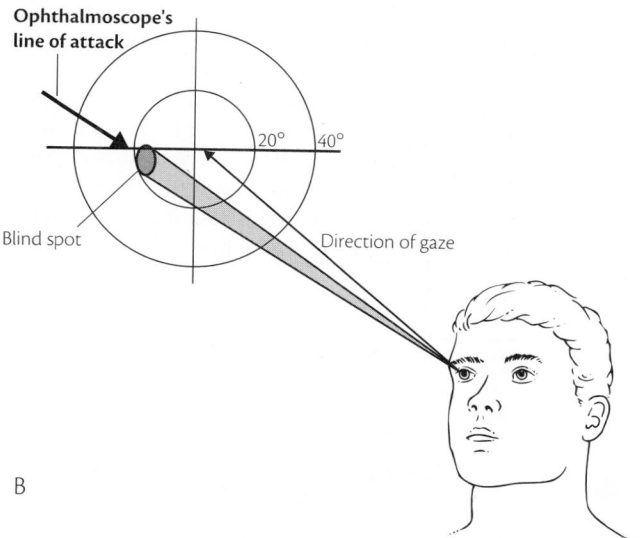

Fig. 2.3 The blind spot of the right eye, its position in the visual field, and its relationship to the optic disc inside the eyeball (A). Determining the test 'line of attack' for ophthalmoscopic examination of the optic disc (B).

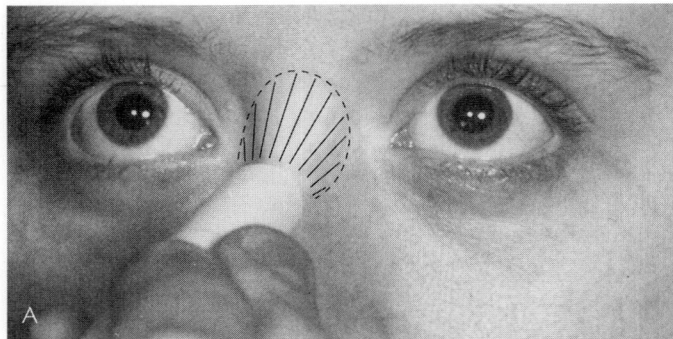

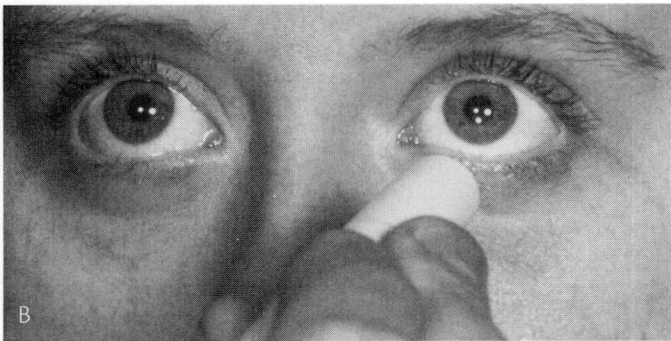

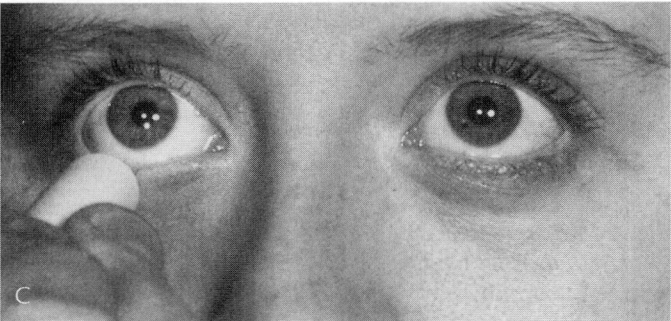

Fig. 2.4 Testing the pupil-light reflex using the 'swinging torch' method.

To test eye movements ask the patient 'to hold your chin with one hand [in order to prevent head movements] and then follow my finger with your eyes'. A finger or a stick is held vertically and moved laterally to about 50 or 60°. After holding it still for a moment, ask the patient whether he sees it as single or double. Simultaneously inspect the eyes carefully to detect nystagmus or any paralysis of ocular movement. Vertical eye movements can be tested similarly by holding a finger horizontally and moving it up and then down by about 45°. Many elderly patients develop clinically insignificant loss of upgaze.

Pupils. To test the pupil-light reflex, the patient should be asked to fixate the examiner's nose whilst he notes the size of the pupils before light stimulation. If it is very difficult to see the pupil because of dim illumination, or a darkly pigmented iris, it helps to shine the torch beam at the bridge of the nose so that light scatter is enough to make the pupil visible, without stimulating the pupil-light response by directly shining the light into the eye (Fig. 2.4). Second, shine the torch directly into the left eye and observe that both pupils constrict equally; this elicits the direct pupil-light response on the left and the indirect, or consensual, response on the right. Third, swing the torch beam quickly across to the right eye and check that there is no further dilatation or constriction of either pupil. This swinging torch test compares the amplitude of the direct and consensual pupil responses of each eye. If there were an optic nerve lesion on the right, both pupils would dilate slightly when the torch was shone in the right eye, compared to their normal constriction following left eye stimulation. This method of comparing the pupil responses has the sensitivity to detect relative, rather than absolute, afferent pupillary defects due to partial optic nerve lesions, as may occur in optic neuritis.

A unilaterally small pupil is most usually due to a cervical sympathetic pathway lesion causing Horner's syndrome, which will be associated with no more than a slight degree of eyelid drooping, ptosis (Fig. 2.5).

A unilaterally, fixedly dilated pupil is typical of an oculomotor nerve lesion (III cranial nerve) (Fig. 2.6), in which also there will be impairment of adduction and vertical eye movement, and the ptosis will be usually much more marked than in Horner's syndrome.

Facial sensation. The fingertips of both the examiner's hands are lightly drawn on both sides simultaneously across the patient's forehead, to the cheek and nose, and then onto the chin. Whilst doing so, he should ask 'do my fingers feel normal and the same on each side?' whilst covering all three territories of the trigeminal nerve, frontal (V_1), maxillary (V_2), and mandibular (V_3) (Fig. 2.7). Any area of altered sensation can be tested in more detail and mapped out using a pin or a wisp of cotton wool.

Testing the corneal reflex with a wisp of cotton wool is not necessary as a routine. The following method is recommended if you do need to test the corneal reflex, for instance, if you suspect potentially harmful loss of corneal sensation or a subtle facial nerve lesion. Make a fine wisp of cotton wool and ask the patient to look upwards whilst warning them that 'I'm going to touch the corner

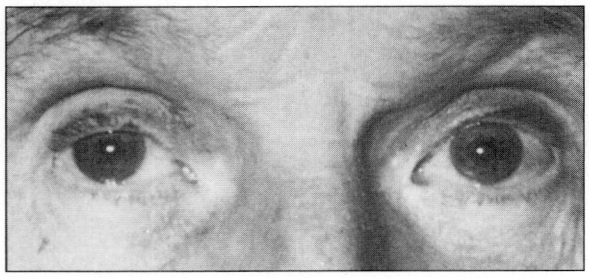

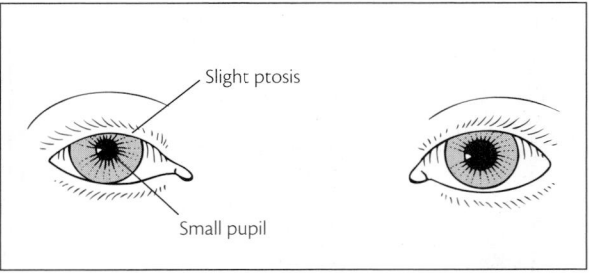

Fig. 2.5 Right-sided Horner's syndrome showing a slight degree of ptosis and a small pupil.

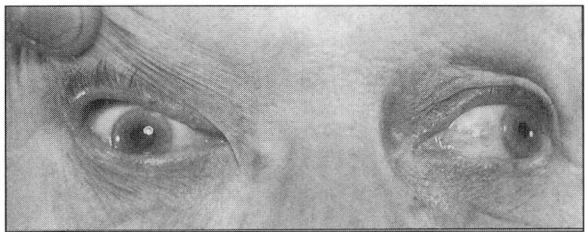 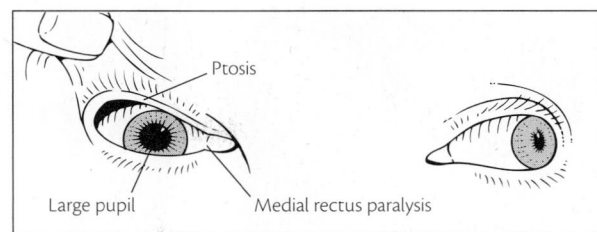

Fig. 2.6 Right-sided third nerve palsy showing marked ptosis, a dilated pupil, and medial rectus weakness.

of your eye with this cotton wool'. Introduce this cotton wisp from below and to the side and brush the junction between the cornea and sclera (Fig. 2.8). Normally a unilateral stimulus provokes bilateral blinking, another example of a 'unilateral afferent-bilateral efferent' reflex. Contact lenses are the commonest cause of seemingly absent corneal reflexes.

Facial movements. Subtle degrees of facial weakness, such as unilateral slowing of movement, are best seen if the examiner steps back a couple of paces so as to see both sides of the patient's mouth simultaneously within their central vision. Voluntary movement of the mouth can be produced by instructing 'show me your teeth like this' or 'give me a smile'. An even better demonstration is provoked by smiling at the patient who usually responds with an involuntary grin which demonstrates facial movements perfectly. Observe whether both sides of the mouth move equally quickly, and produce similar elevation and deepening of the nasolabial skin creases. If the mouth movement is asymmetrical, indicating unilateral weakness, ask the patient to 'raise your eyebrows' to see whether both sides of the frontalis muscle in the forehead contract equally (Fig. 2.9). Lower motor neurone lesions of the seventh nerve will affect movements of both the forehead and the mouth. Unilateral upper motor neurone facial paralysis affects only the mouth and lower face, but not the forehead.

Hearing. To test hearing in the right ear, the patient is instructed 'Could you repeat this number' whilst the examiner lightly rubs the tip of their finger in the other ear to create a masking noise and whispers a number from about 2 ft, and vice versa to test the left ear.

Unilateral or bilateral deafness should be assessed further by Weber's and Rinne's tests (Section 2.2.3) to distinguish between conductive and sensorineural deafness, and by auroscopic examination of the eardrum.

Palatal movements. These are best tested by asking the patient to open their mouth and say 'ah' whilst illuminating the throat with a torch. If the elevation of the palate and uvula is normal and symmetrical, and there is no swallowing difficulty or dysphonia, there is no need for the discomfort of eliciting the gag reflex as a routine. When necessary the gag reflex is elicited by stimulating one side of the soft palate with a stick and watching the resultant rise of both sides of the palate.

Tongue movement. It is best to inspect the tongue for wasting or fasciculations while it is relaxed on the floor of the mouth during examination of the palatal movements. It is misleading to look for fasciculations whilst the tongue is being actively protruded since most normal tongues show ripples and flickers under such circumstances. Tongue movements can be tested by asking the patient to 'stick out your tongue and move it from side to side like this' and demonstrate this movement. A lower motor neurone lesion affecting a hypoglossal nerve causes the tongue to be wasted on, and deviate towards the same side as the lesion. If an upper motor

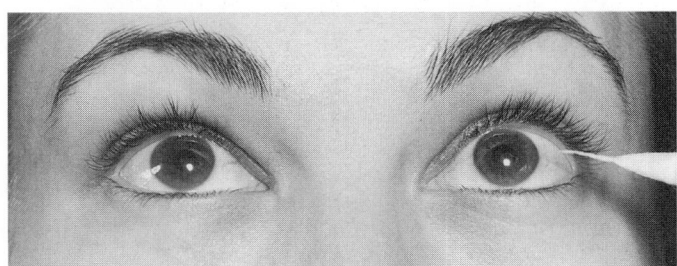

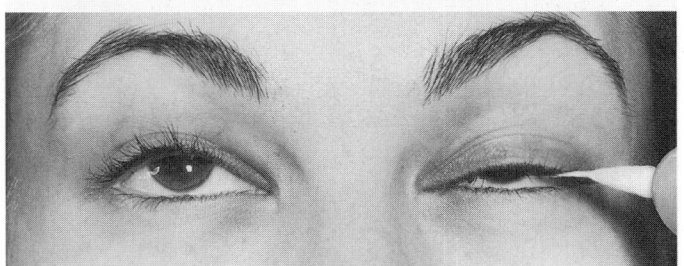

Fig. 2.8 Testing the corneal reflex by touching the edge of the cornea with a wisp of cotton wool on one side (top), and observing blinking of both eyes (bottom).

Fig. 2.7 Testing facial sensation.

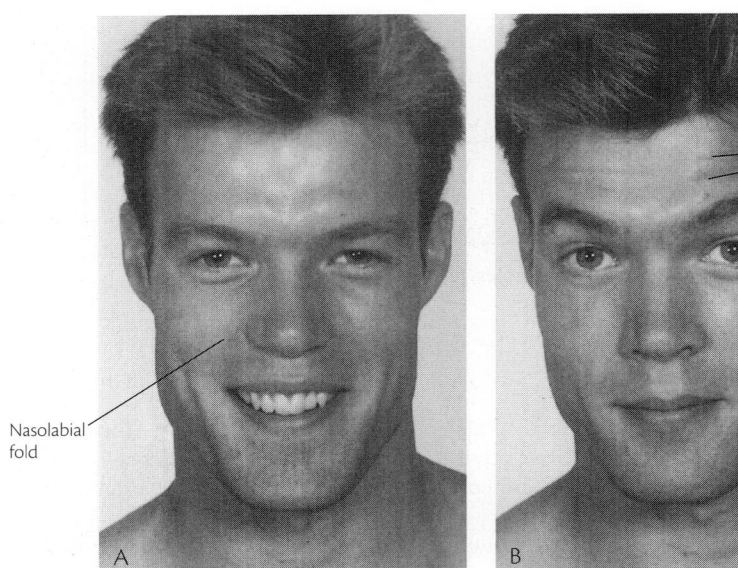

Nasolabial
fold

Frontalis
contraction

Fig. 2.9 Testing facial movements: (A) mouth 'give me a smile'; (B) forehead 'raise your eyebrows'.

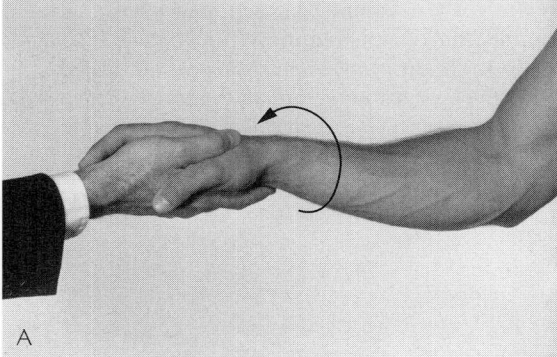

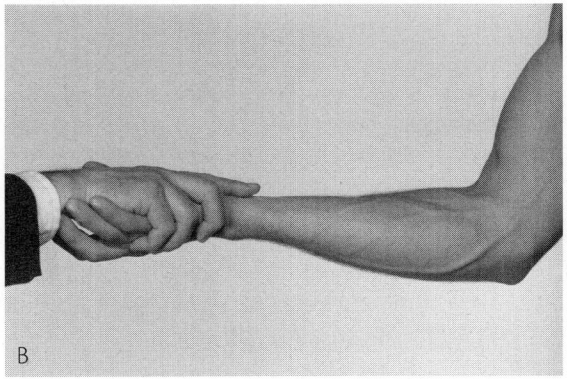

Fig. 2.10 Testing for a pronator catch in the right arm: (A) starting position in pronation; (B) finishing position in supination.

neurone lesion is bilateral, the tongue becomes spastic and square in profile and its movements limited and slow. Tongue power is best tested by asking the patient to push their tongue into the corner of their cheek whilst the examiner palpates. In a cerebellar lesion, alternating tongue movements will be slowed and irregular.

The arms

Inspection. The profile of the upper arms should be inspected for muscle wasting or fasciculations with the patient sitting facing towards the examiner. Then the hands should be inspected for muscle wasting by looking particularly at the first dorsal interosseous muscles on the dorsum of the hand between the thumb and forefinger, innervated by the ulnar nerve, and the abductor pollicis brevis in the lateral part of the thenar eminence, innervated by the median nerve.

Tone. Either the extrapyramidal rigidity of Parkinson's disease, or the spasticity of an upper motor neurone lesion can be detected reliably in the arms. Different techniques are used to test for these tone changes. Which one is chosen should be determined by which of these conditions is suspected. Spasticity should be sought by holding the patient's hand with the elbow flexed, and abruptly supinating the forearm to detect a sudden jerk of spastic resistance known as a 'pronator catch' (Fig. 2.10).

The 'cogwheel rigidity' of Parkinson's disease is best detected by holding the patient's wrist with one hand, and repeatedly flexing and extending the fingers and wrist by gripping the tips of the fingers with the other hand (Fig. 2.11). If mild, sometimes the abnormality can be brought out by the patient simultaneously waving the other arm in the air. The term 'cogwheel rigidity' merely describes leadpipe rigidity with superimposed tremor.

Power. For general screening purposes it is sufficient to test one proximal and one distal muscle in each arm. The best proximal muscle to test is shoulder abduction to 90° by deltoid (C5 root, axillary nerve). A good distal muscle to test is the first dorsal interosseous (TI root, ulnar nerve), which spreads the fingers apart. Its power can be compared with the examiner's own first dorsal interosseous muscle (Fig. 2.12). Additional muscles should be tested if a lesion of a particular peripheral nerve or root is suspected.

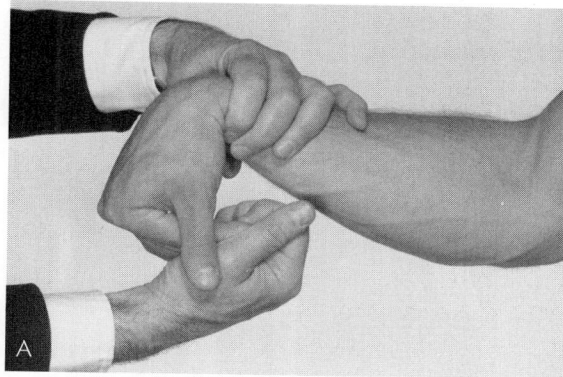

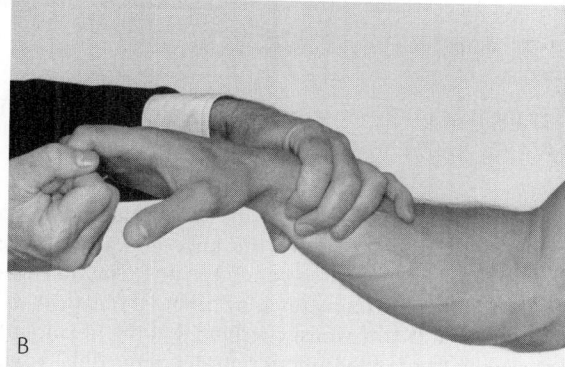

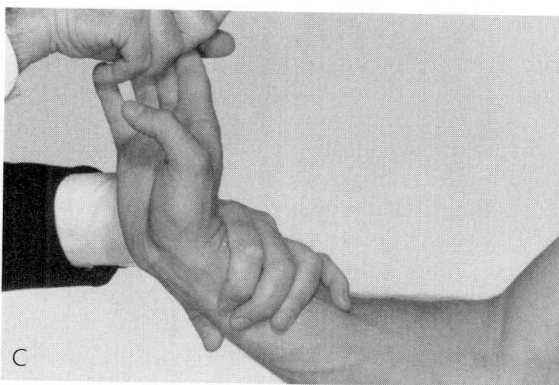

Fig. 2.11 Testing for cog-wheel rigidity in Parkinson's disease.

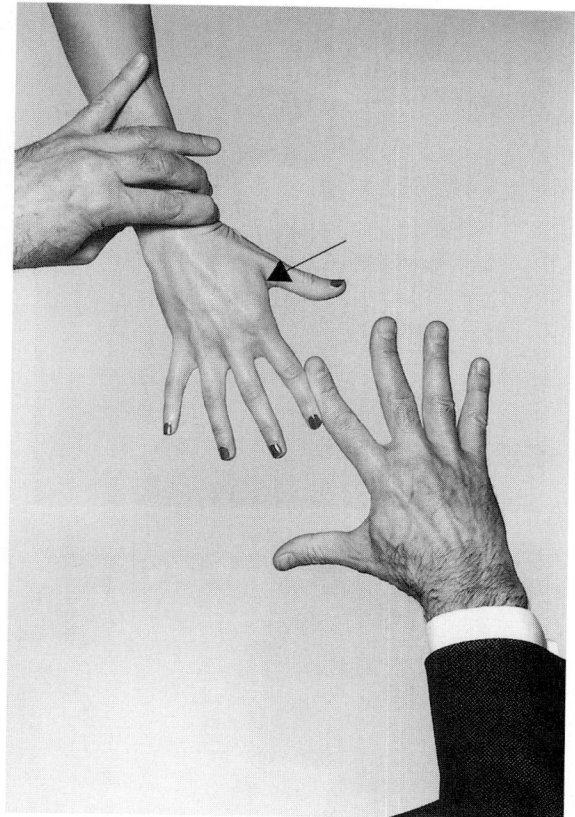

Fig. 2.12 Testing finger abduction power (dorsal interosseous). The examiner compares the power of his abducted index finger with that of the patient. The bulk of the patient's first interosseous muscle is clearly visible (arrow).

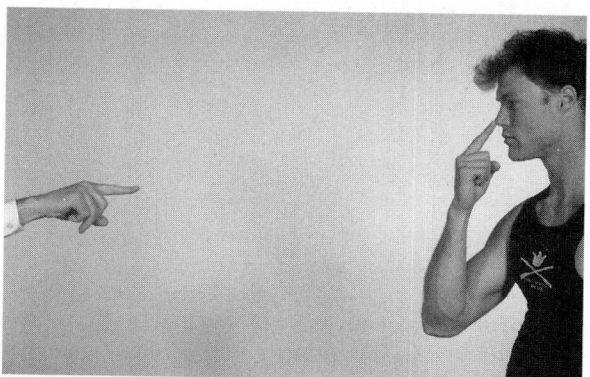

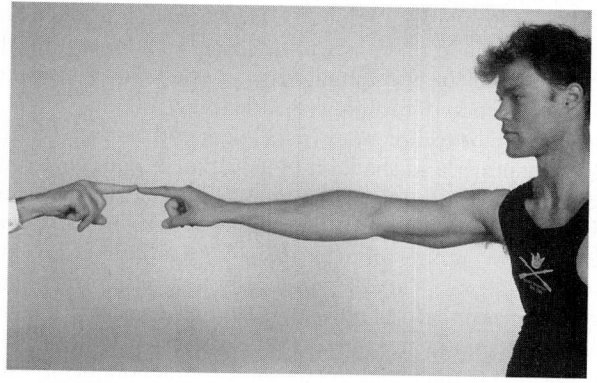

Fig. 2.13 The finger–nose test to detect dysmetria (ataxia).

Coordination. The finger–nose test is the most reliable but is only sensitive if the patient is required to stretch their arm out fully from the shoulder to touch the examiner's target finger (Fig. 2.13). The examiner should stand well behind their own outstretched target finger so as to detect the randomly distributed inaccuracies in the patient's pointing known as ataxia or dysmetria. Dysmetria on this test usually indicates cerebellar disease, cerebellar ataxia, or loss of proprioceptive feedback, sensory ataxia, but can occur in proximal muscle weakness. If ataxia is detected, pseudoathetosis indicative of loss of sensory feedback can be sought by asking the patient to hold out the arms horizontally with the eyes closed, with the fingers extended and spread apart; if pseudoathetosis is present, the fingers wander and fail to remain in position (Fig. 21.47).

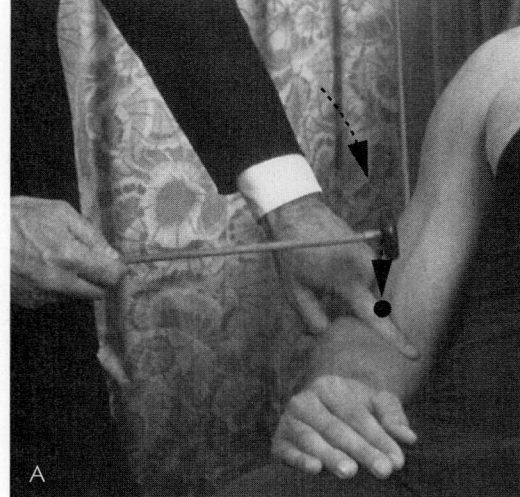

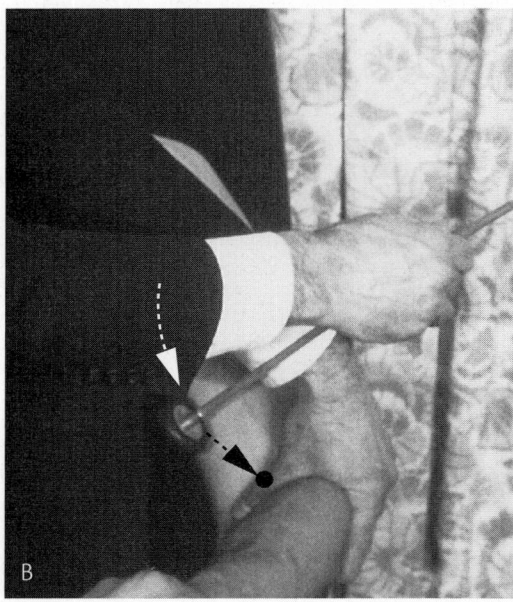

Fig. 2.14 Eliciting (A) the right and then (B) the left biceps tendon reflexes (C5/C6).

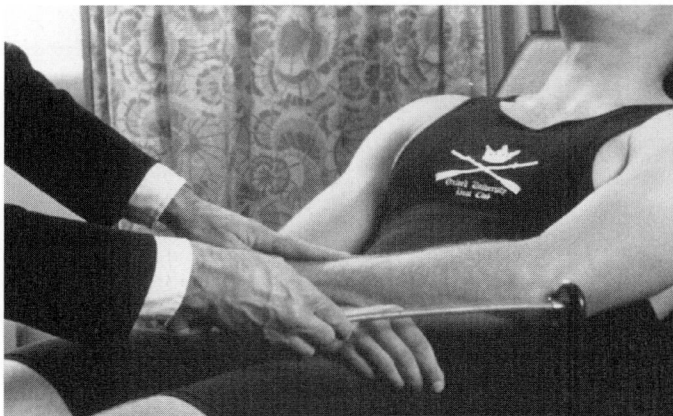

Fig. 2.15 Eliciting the triceps reflex (C7/C8).

With the patient lying down

Arm tendon reflexes. The biceps reflexes (musculocutaneous nerve; 5th and 6th cervical roots) should be tested from the patient's right side. The examiner's thumb should be used to transmit a firm blow from the tendon hammer to the biceps tendon within the cramped space of the antecubital fossa so as to elicit a visible or palpable contraction of the biceps muscle (Fig. 2.14).

When testing the triceps reflex (radial nerve; 7th and 8th cervical roots), the hammerhead should hit the tendon at right angles just above the elbow because the triceps muscle has an extremely short tendon (Fig. 2.15). The brachioradialis reflex (radial nerve; 6th cervical root) is difficult to elicit reliably and rarely adds extra information unless a C6 root lesion is suspected, or the examiner is trying to localize or detect a radial nerve lesion. Tendon reflexes are brisk in upper motor neurone lesions. An absent tendon reflex will be due to a peripheral nerve or nerve root lesion. Before concluding that a reflex is absent reinforcement should be undertaken

by asking the patient to 'bite your teeth together when I say "bite"' whilst you try simultaneously to elicit the reflex.

The legs

Inspection. The bulk of the vastus medialis component of quadriceps just above and medial to the kneecap can be observed by asking the patient to 'tighten your kneecaps'. The bulk of more distal muscles can be demonstrated by asking the patient to 'cock your toes up towards you' whilst checking that the tibialis anterior muscle bulges in front of the anterior border of the tibial bone. The leg muscles should be inspected for fasciculations which are visible flickering contractions within the muscle belly, insufficient to produce movement around the joint, and which signify disease of the lower motor neurone, for instance in motor neurone disease. Sometimes fasciculations are visible in otherwise normal calf muscles of healthy individuals, particular after exercise. Skin ulcers, burns, or disrupted joints, known as Charcot joints, may be trophic changes resulting from loss of protective pain sensation.

Tone. The spasticity of an upper motor neurone lesion, observed as sustained clonus, is the most reliable objective tone change detectable in the legs. Ankle clonus is elicited by externally rotating the foot and holding the knee slightly flexed with one hand, whilst sharply jerking the sole of the foot upwards with the other hand (Fig. 2.16). For a few seconds the foot should be held firmly in sustained dorsiflexion since the rhythmic downward beatings of clonus may take a moment or two to become evident. Sustained

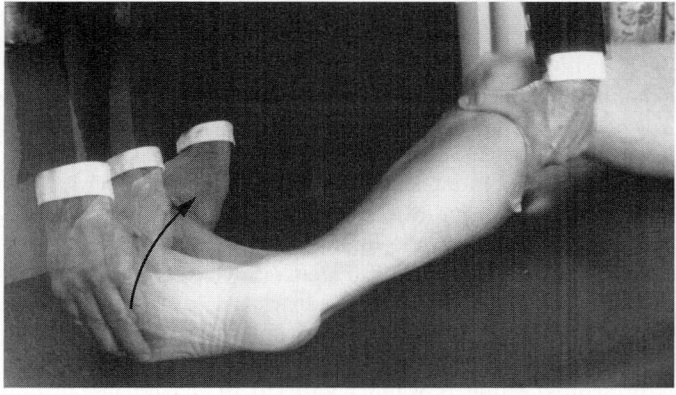

Fig. 2.16 Eliciting ankle clonus.

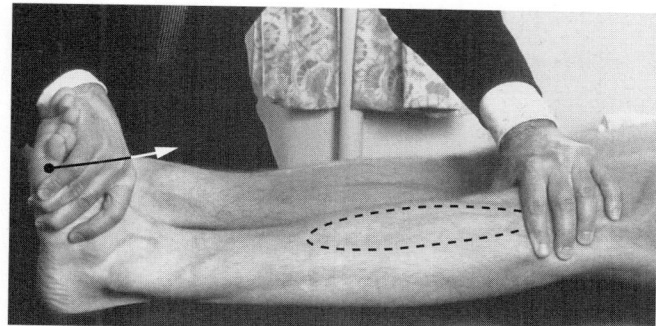

Fig. 2.17 Testing ankle dorsiflexion power (tibialis anterior).

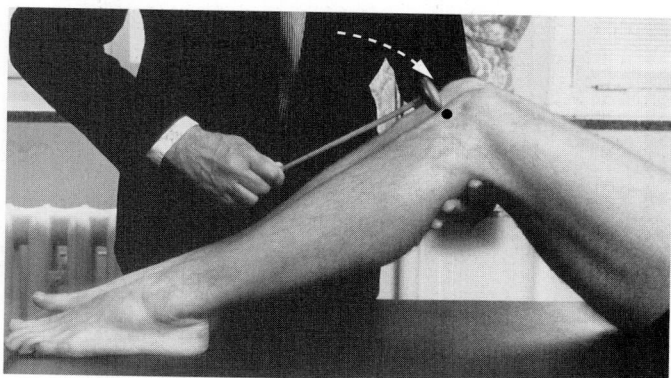

Fig. 2.18 Eliciting the knee jerk (L3/L4).

clonus, or unsustained clonus of more than 6 beats, is generally regarded as definite evidence of an upper motor neurone lesion.

Power. Testing of one proximal and one distal muscle in each leg is sufficient to screen for the weakness of unexpected myopathies (proximal), peripheral neuropathy (distal), or upper motor neurone lesions (both proximal and distal). Proximal leg power is reflected by hip flexion (iliopsoas muscle, 1st and 2nd lumbar roots) best tested by instructing the patient to 'push your leg up to 45 degrees' and then for the examiner to press downwards just above the knee (Fig. 2.28). A distal muscle, tibialis anterior (peroneal nerve, 5th lumbar root) is tested by asking the patient to 'cock your foot up towards you' whilst the examiner tries to overcome this dorsiflexion at the ankle (Fig. 2.17). Tibialis anterior is a particularly valuable muscle to test, since it will be weakened by upper motor neurone lesions, polyneuropathy, common peroneal nerve lesions, and in L5/S1 root lesions due to prolapsed intervertebral disc. Some leg muscles are so naturally powerful that milder degrees of weakness cannot be detected reliably by bedside testing. For instance, mild weakness of knee extension by quadriceps (femoral nerve; 3rd and 4th lumbar roots) may be revealed best by asking a patient to stand up from a chair without using their arms. Ankle plantar flexion by gastrocnaemius (posterior tibial nerve; 1st and 2nd sacral roots) is best tested by asking a patient to stand on tiptoe or even to hop.

Tendon reflexes. The knee jerk or quadriceps tendon reflex (femoral nerve; L3/4) is elicited by lifting and flexing both knees over the

examiner's left arm by 60–90°, and then striking the two patellar tendons in turn to compare the reflex on both sides (Fig. 2.18).

The ankle jerk or gastrocnaemius or Achilles tendon reflex (posterior tibial nerve; S1/S2) is best tested by externally rotating the foot with the knee slightly bent, gently dorsiflexing the knee, and then striking the Achilles tendon firmly with the hammer (Fig. 2.19). Poor technique is often responsible for the ankle jerks appearing to be absent; the examiner may not have struck the Achilles tendon sufficiently firmly, or the patient may be 'helping' by holding the foot rigidly in dorsiflexion. Brisk tendon reflexes point to an upper motor neurone lesion in which case sustained ankle clonus and/or an extensor plantar response would be expected too. Slightly brisk reflexes may occur in anxious, tense patients. The reflexes are absent in peripheral nerve or root lesions. The ankle jerks are absent in many people over the age of 70. As with the arm reflexes, reinforcement should be undertaken before finally declaring a reflex absent.

Plantar responses. An extensor plantar or Babinski response (Fig. 2.20) is a definite sign of an upper motor neurone lesion. It is present from the onset of the upper motor neurone lesion, well before sufficient spasticity has developed to allow clonus or hypereflexia.

Technique is all-important for eliciting the plantar response reliably (Fig. 2.21). The patient should be lying down unable to see

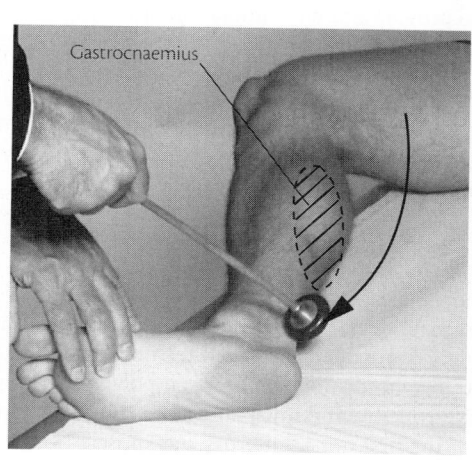

Gastrocnaemius

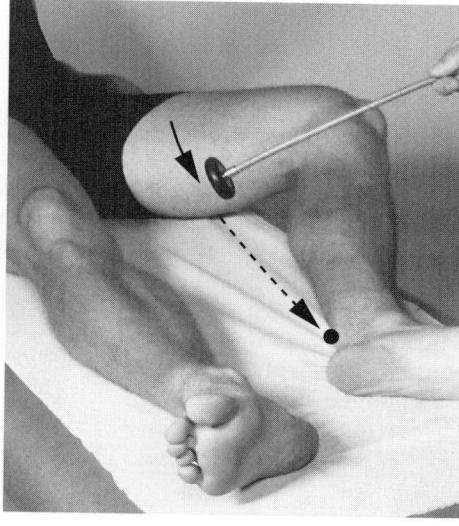

Fig. 2.19 Eliciting the ankle tendon jerks.

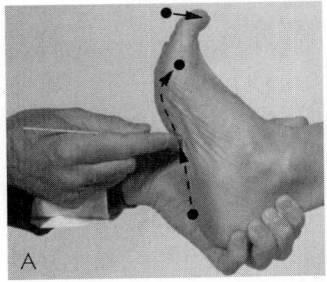
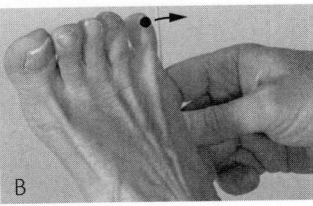

Fig. 2.20 An extensor plantar response, or Babinski sign, showing (A) dorsiflexion of the great toe and (B) fanning of the little toe.

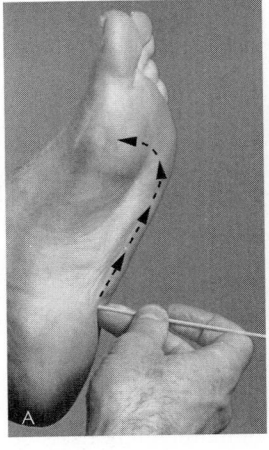
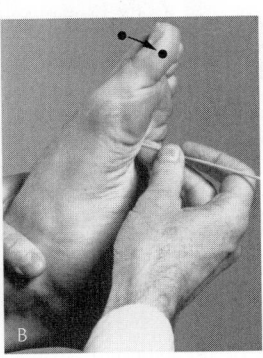

Fig. 2.21 Technique for eliciting the plantar response (A) which produces a normal (flexor) response (B).

their toes. The examiner should passively move the great toe up and down beforehand, both to ensure relaxation, and also to detect hallux rigidis which would mask the toe movement. Then, lightly holding the leg just above the ankle with the left hand, a thin stick, or a key is held in the right hand and slowly but firmly drawn up the outer aspect of the sole and across the ball of the foot. During this the examiner should watch the great toe from the side so as to detect whether its first movement is downwards (flexor and normal) or upwards (extensor and abnormal).

Sensory examination. In a patient without sensory symptoms, such as deadness or tinglings, and whose Romberg test is normal, sensory examination is rarely abnormal. Generally it is not worth performing as a routine if disease of the sensory pathways is not suspected.

2.2.3 Specific clinical circumstances

The following examples show how other examination manoeuvres can be added to the general neurological examination (Section 2.2.2) if the patient's symptoms suggest a specific disorder, or if the basic examination has revealed abnormalities requiring further evaluation. Further details and other examples are given in other chapters.

Individual cranial nerves. The general neurological examination outlined in Section 2.2.2 does not test every cranial nerve in detail. The following main functions of each cranial nerve are easily testable, should the clinical situation require it:

I *Olfactory*: Test the ability to detect a smell in each nostril, whilst the other nostril is blocked, with the eyes closed. It is not necessary that the patient identify a particular smell. Easily available odours such as coffee, soap, or orange peel are quite adequate for testing.

II *Optic*: Fundoscopy, visual acuity using a Snellen chart held at 6 m; pupil-light response (afferent); visual fields can be tested either by confrontation, to detect movement of a finger or by comparing the field within which a hat pin becomes seen as red (usually out to about 40°) when it is moved slowly in from the periphery exactly midway between examiner and patient. Visual fields can be tested with both eyes open when a homonymous field defect due to a lesion of the pathway after the optic chiasm is suspected. The eyes' visual fields must be tested separately if a lesion of the retina, optic nerve, or optic chiasm is suspected.

III *Oculomotor*: Eye movements (horizontal adduction, up, down); eyelid elevation; pupil-light response (efferent).

IV *Trochlear*: Eye movement (down and in) easily tested by asking the patient to 'look towards the tip of your nose'.

V *Trigeminal*: Jaw closure (masseter and temporalis); jaw opening (pterygoids); facial sensation; corneal reflex (afferent) using wisp of cotton wool.

VI *Abducens*: Eye movement (horizontal abduction).

VII *Facial*: Facial muscles (test eyebrow elevation and smiling movements of the corners of the mouth); corneal reflex (efferent).

VIII *Auditory*: Hearing a whisper in each ear; Weber's and Rinne's tests with 512 Hz tuning fork to distinguish between conductive and sensorineural deafness. Weber's test involves putting the vibrating tuning fork on the middle of the forehead and asking 'is it louder on one side or in the middle?'. Normally it is loudest in the middle; in sensorineural deafness it is louder on the opposite side; and in conductive deafness it is louder on the same side. Rinne's test compares the loudness of bone conduction at the mastoid process with air conduction in front of the pinna for each ear. Normally air conduction is louder, whereas bone conduction will be louder in conductive deafness. Auroscopic examination of the eardrum should be carried out if conductive deafness is detected.

IX *Glossopharyngeal*: An orange stick is used to test palatal sensation and to provide afferent stimulation of the gag reflex.

X *Vagus*: Palatal elevation; gag reflex (efferent); vocal cord movement (speaking, sharp cough).

XI *(Spinal) Accessory*: Shoulder shrugging and scapular rotation to abduct the arm beyond 90° (trapezius); head rotation laterally (sternomastoid).

XII *Hypoglossal*: Tongue protrusion.

A weak, areflexic, numb, or painful limb. Commonly patients complain of neurological symptoms affecting a single limb. In such cases, or if abnormalities are found on general neurological examination, a wide range of muscles and sensory territories must be examined using a strategy to distinguish between polyneuropathy, root lesions, mononeuropathy, and myopathy. For example, if a patient has weakness of the first dorsal interosseous muscle (ulnar nerve,

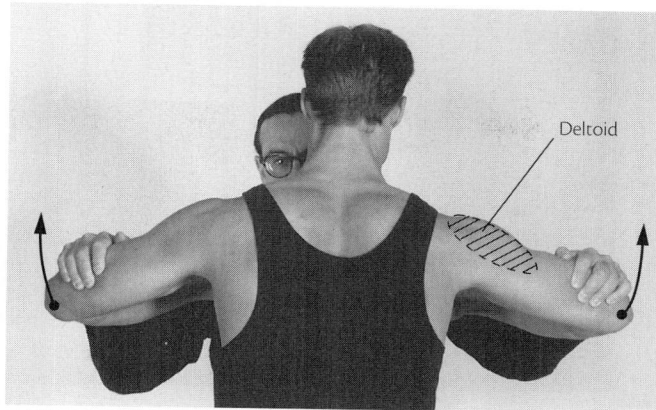

Fig. 2.22 Testing shoulder abduction power (deltoid).

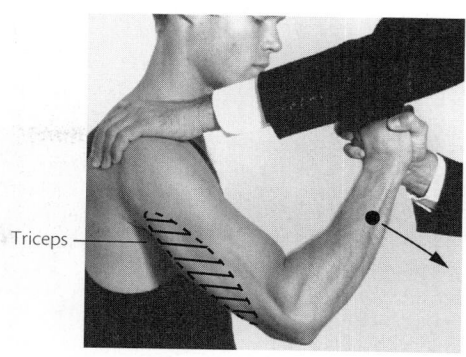

Fig. 2.24 Testing triceps power. Test elbow extension using your own triceps muscle against the patient's triceps muscle. Stabilize the shoulder with your other hand to prevent the shoulder girdle muscles from participating in the pushing action.

T1 root), but abductor pollicis brevis (median nerve, T1) is normally strong, it is clear that there is an ulnar nerve lesion rather than a polyneuropathy or a T1 root lesion. Muscle strength can be assessed particularly sensitively in the arms by testing the strength of individual muscles, such as biceps or finger flexors and extensors, using the identical muscle in the examiner. Full details of muscles innervated by individual peripheral nerves are given in Chapter 22.

Frequently tested muscles in the arm are:

◆ Shoulder abduction (0–15° supraspinatus, suprascapular nerve, C5 root; 15–90° deltoid, axillary nerve, C5 root) (Fig. 2.22).

◆ Biceps: elbow flexion (musculocutaneous nerve, C5/6 root). The patient's power of elbow flexion is compared with that of the examiner's (Fig. 2.23).

◆ Triceps: elbow extension (radial nerve, C7/8). The patient's power of elbow extension is compared with that of the examiner's (Fig. 2.24).

◆ Finger extensors: (radial/posterior interosseous nerve, C7). The examiner can use their own extended fingers to try and overcome the patient's extended fingers (Fig. 2.25).

◆ Flexor digitorum profundus: Terminal interphalangeal joint flexion (anterior interosseous branch of median nerve (index finger) or ulnar nerve (little finger) (C7/8). The power of flexion of the terminal phalanx of the finger can be directly compared with the examiner's (Fig. 2.26).

◆ Dorsal interosseous: Finger abduction (ulnar nerve, T1) (Fig. 2.12).

◆ Abductor pollicis brevis: Abduction of thumb at right angles to the palm (median nerve, T1) (Fig. 2.27).

Useful muscles to test in the leg include:

◆ Iliopsoas: Hip flexion (innervated by lumbar plexus, L1/2 roots) (Fig. 2.28).

◆ Gluteus maximus: Hip extension (inferior gluteal nerve, L5/S1). If hip extension is of normal power, the examiner can lift the patient's buttocks off the bed by lifting upwards at the ankle (Fig. 2.29).

◆ Quadriceps: Knee extension (femoral nerve, L3/4). This very powerful muscle must be tested with the knee starting from a flexed position (Fig. 2.30).

◆ Tibialis anterior: Ankle dorsiflexion (peroneal Nerve, L5) (Fig. 2.17).

◆ Gastrocnaemius: Ankle plantar flexion (tibial nerve, S1/2) (Fig. 2.31).

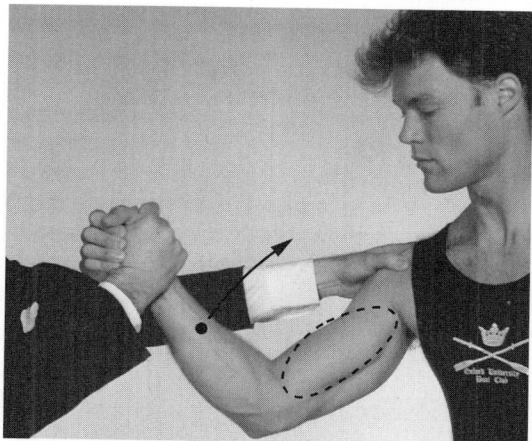

Fig. 2.23 Testing biceps power. Compare elbow flexion using your own biceps muscle to test the patient's biceps muscle. Stabilize the patient's shoulder with your other hand to prevent the trunk muscles from participating in the patient's pulling action.

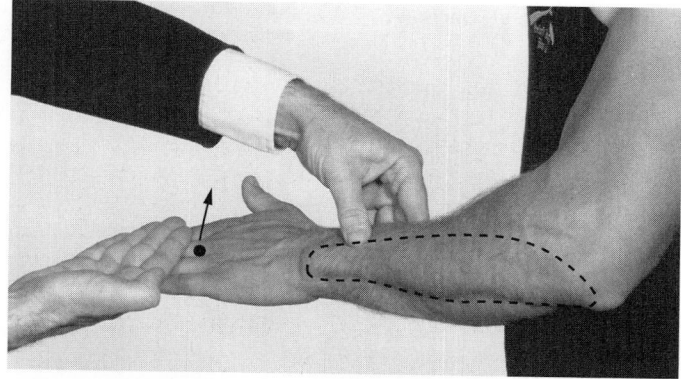

Fig. 2.25 Testing extensor digitorum power. Test finger extension using your own extended fingers for comparison.

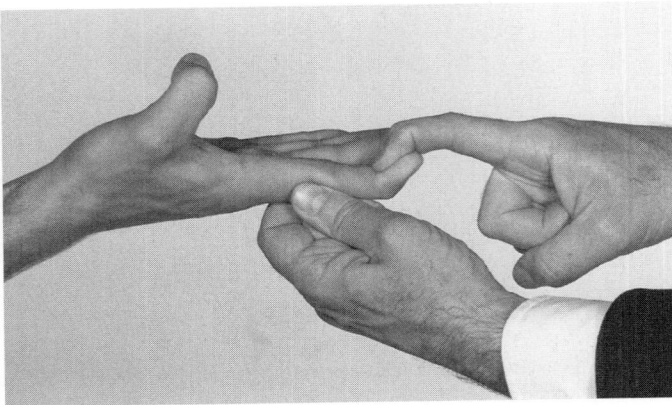

Fig. 2.26 Testing flexor digitorum profundus power. Compare the power of flexion at the distal interphalangeal joint by pulling against the patient's finger.

Speech disorders

◆ *Dysphonia* is an inability to create noise properly from the larynx. The voice is quiet and somewhat featureless because the larynx produces sound inefficiently; patients are unable to shout. Attempts at producing a sharp, explosive cough are 'bovine' because the larynx cannot be tightly closed and then suddenly opened.

◆ *Dysarthria* is an inability to shape that noise accurately into recognizable words. The tongue, pharynx, and lips are uncoordinated on trying to pronounce consonants. This becomes particularly obvious on repeating words rich in consonants, such as 'uNiVeRSiTy' or 'WeST ReGiSTeR STReeT' or 'BRiTiSH CoNSTiTuTioN'. Cerebellar incoordination makes these consonants slurred and slow with 'scanning speech'. A pseudobulbar palsy produces a spastic immobile tongue with 'hot potato speech' or total anarthria, an inability to speak at all.

◆ *Dysphasias* are abnormalities of the understanding of, or the generation of, language. They result from damage to the speech areas of the cerebral hemisphere, usually on the left even in left-handed people. Patients with a receptive, Wernicke, or sensory dysphasia are unable to understand and execute a simple three-stage command such as 'When I clap my hands, please touch your right ear with your left index finger'. Yet their speech is fluent, in that the rate of word production is normal, but meaningless

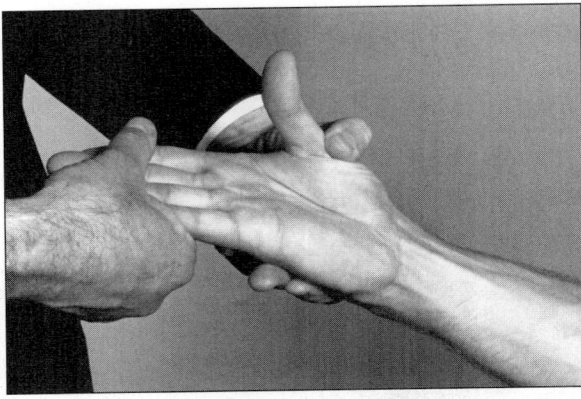

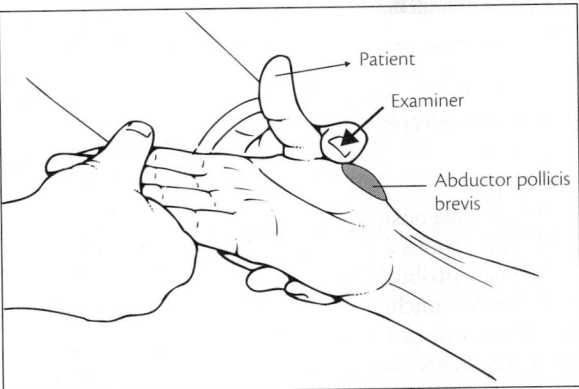

Fig. 2.27 Testing the power of abductor pollicis brevis (median nerve) and observing its bulk. The patient should raise their thumb at right angles from their palm, and the examiner should oppose this movement by pressing down at the base of the proximal phalanx.

because the words are wrong or jumbled up. It should be noted that some patients are unable to execute commands because of dyspraxia, which is common in left cerebral hemisphere lesions. A motor or Broca's dysphasia causes non-fluent speech, with a slowed rate of word production. Also there are obvious difficulties in finding the correct word and gestures are often used to compensate for inability to find the intended word.

Dementia

Dementia is a diffuse loss of cognitive function, particularly involving memory, due to generalized disease of both cerebral

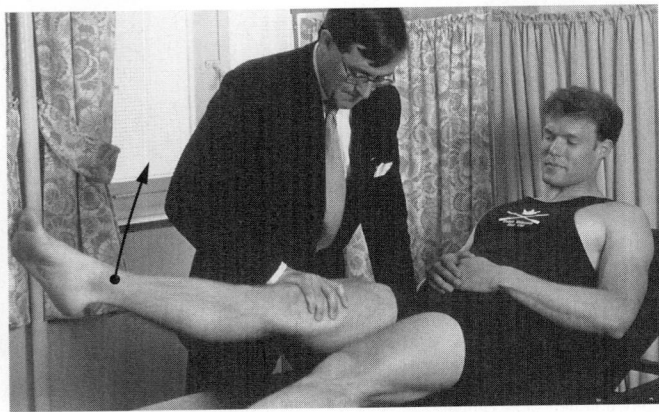

Fig. 2.28 Testing hip flexion power (iliopsoas).

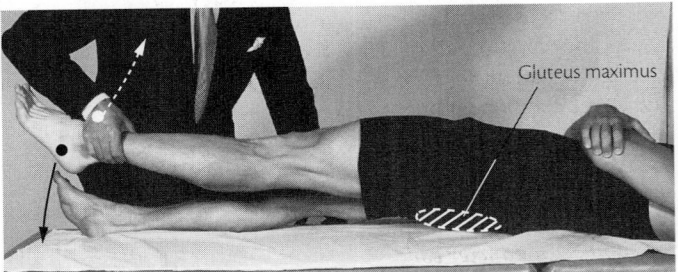

Fig. 2.29 Testing gluteus maximus power. Ask the patient to 'push your leg down into the couch' whilst you pull up on ankle. If the power is normal you can lift the patient's bottom off the couch.

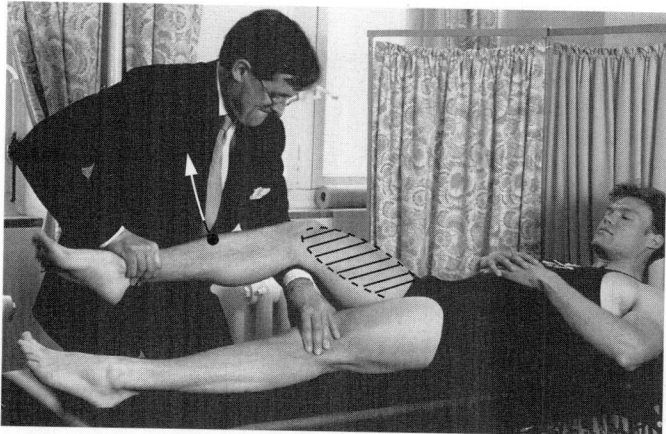

Fig. 2.30 Testing quadriceps power. Ask the patient to 'kick your leg out straight' whilst you push downwards on the ankle. This is a very powerful movement and it helps to put your left arm behind the knee, both to act as a fulcrum and to prevent hip extension from participating in the movement.

Fig. 2.31 Testing gastrocnemius power. A very strong muscle which is best tested by asking the patient to stand on tiptoe, or even to hop, whilst you steady him by holding their forearms.

hemispheres. Diagnosing early dementia can be difficult and the spouse's observations are all-important. Initially minor symptoms may have been attributed to absent-mindedness. Then patients are noted to develop uncharacteristic errors of judgement, inability to perform their customary intellectual tasks such as puzzles or games, loss of interest in hobbies and recreations, and inability to remember the names of friends and family. As the disease becomes more advanced, the personality is lost and patients may become disinhibited about the usual social codes of excretion or sexuality. Ultimately the patient becomes mute and unresponsive, wanders aimlessly, is incontinent, and dependent on feeding.

Demented patients are vague or rambling during history taking although this is sometimes masked by preserved social skills. Simple bedside clues to dementia involve checking the orientation for date and place; orientation for person usually being preserved except in psychiatric disease or simulated dementia. Calculation ability is usefully tested by serial subtraction of 7 from 100. General knowledge of everyday and historical events should be assessed to judge whether it is consistent with the patient's educational and social background. Memory loss may be indicated by impaired immediate and 5-min recall of a simple three-line address or of three objects. Cognitive estimates such as 'Roughly how long is a man's spine?' or 'How many camels are there in Holland?' may be abnormal in frontal lobe disease. If the patient's demeanour is flat or gloomy it suggests depression, which may be a clue to potentially treatable pseudodementia. Self-neglect may be evident, as may be failure to use facial expression and gesture for non-verbal communication. Right parietal lobe spatial functions can be tested by asking the patient to draw or copy a three-dimensional cube. If a dysphasic component prevents understanding of such instructions, impaired spatial functioning may be revealed by dressing apraxia in which the patient is unable to put on a dressing gown or shirt correctly when one of the sleeves has been pulled through the wrong way.

Impaired sphincter control

In a patient with hesitancy or urgency of micturition, or retention or incontinence of urine, the following aspects of examination are crucial. Extensor plantar responses point to a spinal cord lesion affecting the upper motor neurones. Absent ankle jerks point to

cauda equina compression or peripheral neuropathy. Blunted perianal pinprick sensation occurs in cauda equina lesions. The anal reflex can be tested by stroking the anal verge firmly with an orange stick, and looking for a reflex contraction which crinkles the anal skin (Section 2.3.3). This reflex is lost in cauda equina lesions but is difficult to elicit reliably, with the response being particularly uncertain in older patients or those with a patulous anus.

Stroke

Cardiovascular examination is usually more informative than neurological examination to elucidate the underlying cause of stroke or transient ischaemic attacks. Cardiac arrhythmia, particularly atrial fibrillation, or hypertension may be significant. Auscultation may reveal possible sources of cerebral emboli; cardiac murmurs indicative of valvar disease, or the bruit of an internal carotid artery stenosis audible just below the angle of the jaw. When listening for carotid bruits, it is important to ask the patient to hold their breath so that breath sounds are not confused with a bruit. Cranial bruits, for instance due to a vascular malformation, are best heard by applying the stethoscope bell over the closed eyelid, and listening when the patient has opened the other eye and is fixating. This avoids distraction due to eye movement noises. Atrial septal defects, potentially admitting paradoxical emboli from the venous circulation, are suggested by the subtle finding of fixed splitting of the second heart sound during respiration.

Systemic malignancy

Any progressive focal neurological abnormalities of the brain, spinal cord, or nerve roots raises the question of compression or infiltration by tumour. In such cases, the search for a primary systemic tumour should include examination of all lymph node groups, the breasts, testicles, chest including chest X-ray, abdomen, rectum, and prostate or vagina.

Sciatica

Straight leg raising will be limited by pain to less than the normal 80–90° on the side of a prolapsed intervertebral disc affecting the

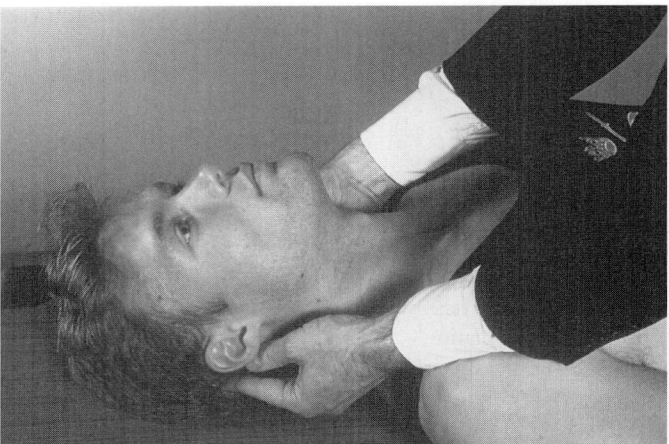

My had a lth. louse the Radcliffe Hfm.y.?

January 1995: Before Treatment

*Mary had a little Lamb
in the Radcliffe infirmary*

July 1997: After L-Dopa

Fig. 2.32 Micrographia in Parkinson's disease.

L5 or S1 nerve roots. Muscles innervated by the different nerve roots under suspicion should be tested, particularly ankle dorsi-flexion (L5). The briskness of the ankle jerks should be compared carefully on the two sides, if necessary from behind with the patient kneeling on a chair. Pinprick sensation in the L5 dermatome on the dorsum of the foot and lateral leg below the knee, and in the S1 dermatome on the sole of the foot and back of the calf, should be compared on the two sides. The lumbar spine should be examined for focal tenderness or deformity which might indicate a tumour deposit or infection in a vertebra.

Parkinsonism

During history taking the patient may exhibit a paucity of facial expression or a characteristic pill-rolling tremor of the finger and thumb at rest. Observe walking for a slow and shuffling start, or loss of arm swing. Unilateral loss of arm swing whilst walking may be the earliest sign of Parkinson's disease. Cogwheel rigidity of the arms is a valuable objective test and is often more pronounced when the patient waves the other arm in the air (Fig. 2.11). Test writing for micrographia (Fig. 2.32) in which the letters get smaller during the writing of a word, but note that some patients have learned to compensate for this by writing long words in segments of a few letters at a time, momentarily stopping or lifting the pen from the paper between these groups of letters.

Coma

A completely different strategy is required to examine uncon-scious patients because of their inability to carry out instructions. The general medical examination may reveal head trauma, cardio-vascular shock, arrhythmia, respiratory failure, pyrexia, alcohol intoxication, or the pinpoint pupils of opiate overdose. Blood sugar testing with Dextrostix will reveal hypo- or hyperglycaemia, and blood should be sent for toxicological analysis and creatinine meas-urement. Neck stiffness due to meningism is usually found in sub-arachnoid haemorrhage or meningitis, careful technique will detect mild neck rigidity (Fig. 2.33). Regular or spontaneous breathing may be disrupted by damage to brainstem respiratory nuclei; Cheyne–Stokes respiration with irregular waxing and waning of respiration is typical of cerebral hemisphere lesions. The depth of unconsciousness is reflected by the extent of any withdrawal response to painful squeezing of the fingernails or toenails. If with-drawal is particularly reduced on one side, it points to a contralat-eral cerebral lesion. The plantar responses are usually bilaterally extensor in unconscious patients and have little specific diagnostic or localizing value. Decerebrate posturing, in which the limbs become stiffly extended, usually indicates a brainstem lesion. Generalized or focal seizures may be evident in status epilepticus or

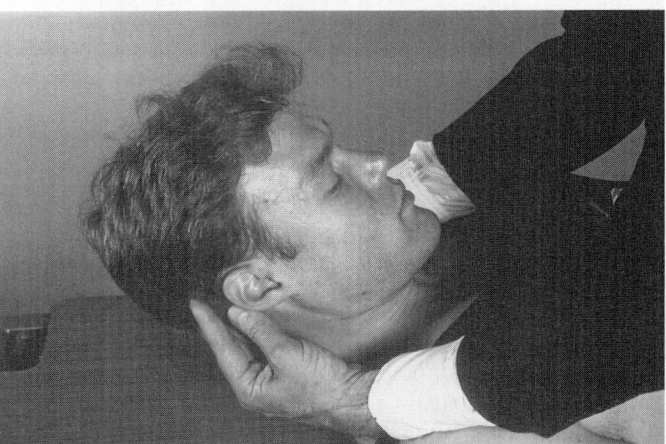

Fig. 2.33 A sensitive method to examine for neck stiffness in suspected meningitis or subarachnoid haemorrhage. The pillow should be removed to extend the neck; the extensor aspect of the examiner's wrists should be rested on the patient's shoulders, while inserting the fingers behind the mastoid processes to assess the degree of resistance while flexing the neck.

encephalitis, and can particularly affect the corner of the mouth, with repetitive twitching. Brainstem function can be tested by elicit-ing the vesibulocular reflex of compensatory eye movement induced by head rotation, or by irrigation of the ears with cold water, the so-called caloric response. Brainstem integrity is also reflected by the corneal reflex of bilateral eye closure when one cornea is stimulated with a wisp of cotton wool, and by the cough and gag responses to laryngeal or pharyngeal stimulation with a sucker or stick.

2.2.4 **False localizing signs**

Although most neurological signs directly reflect pathology affect-ing the corresponding pathway within the nervous system, some reflect secondary pathology at a site remote from the main patho-logical abnormality. These are known as false localizing signs. They mainly result from raised intracranial pressure, or from spinal cord lesions (Larner 2003).

False localizing signs as a result of raised intracranial pressure were first noted by Collier (1904). A sixth nerve palsy is the com-monest false localizing sign of raised intracranial pressure, be it due to intracranial mass, benign intracranial hypertension, or cerebral venous sinus thrombosis. Two pathological anatomical explana-tions for such sixth nerve palsies have been proposed: first stretch-ing of the nerve by downward or backward displacement of the brainstem and second, angulation or compression of the nerve as it passes over the ridge of the petrous temporal bone (Fig. 2.34). Less frequently the trigeminal or facial nerves can be affected by raised intracranial pressure with either positive or negative symptoms. When a supratentorial swelling causes herniation of the medial temporal lobe through the tentorium cerebelli, a third nerve palsy can result. Unilateral pupil dilatation, known as Hutchinson's pupil is usually the first sign, and attributed to the centrifugal location of pupil constrictor axons within the nerve. Usually the dilated pupil is ipsilateral to the cerebral mass lesion but contralateral occurrence also occurs (Larner 2003). Third nerve palsy due to transtentorial herniation of the temporal lobe may be followed by a contralateral hemiparesis attributed to compression of the cerebral peduncle. A less frequent alternative is that of an ipsilateral hemiparesis, attributed to compression of the cerebral peduncle contralateral to the lesion, as a result of lateral displacement of the midbrain against the free edge of the tentorium. This false localizing ipsilateral hemiparesis is known as the Kernohan notch phenomenon and typically results from acute subdural haematoma.

2.2.5 The elderly

A large proportion of those presenting with neurological disorders are the elderly. Stroke, Parkinson's disease, dementia, and cervical spondylotic myelopathy are common disabling neurological conditions in this age group. Furthermore, troublesome neurological symptoms which evade formal diagnosis are common in patients in their seventh decade and beyond. Examples include mild degrees of memory difficulty, dizziness and dysequilibrium, falls (Section 2.6.5), or unwitnessed blackouts. High-level gait disorders (Section 2.6.4), without a demonstrable frontal lobe abnormality on scanning, are a common problem for the very old. These may be difficult to distinguish from the mild gait deterioration which is almost universal by the age of 80. In elderly patients with neurological

disorders particular effort should be made to pursue diagnoses which may lead to improvement or stabilization of the disorder during the patient's natural lifespan. Examples include detection and treatment of subdural haematoma, meningioma, Parkinson's disease, herpes encephalitis, hydrocephalus, idiopathic demyelinating polyneuropathy, lumbar canal stenosis, and myasthenia gravis.

The neurological examination becomes less discriminating in elderly patients. Absent ankle tendon jerks, loss of vibration sense from the feet, mild weakness of ankle dorsiflexion, and general loss of muscle bulk are frequent age-related findings. These only assume clear pathogenic significance if unilateral, or if they emerge at an unexpectedly rapid rate during sequential examinations. Steadily diminishing pupil size, and loss of upgaze or convergence are frequent asymptomatic ocular signs in the elderly. Hearing, smell, and taste all deteriorate with ageing. Gait in the elderly often shows small strides, uncertainty, a widened base, use of a stick, and a tendency to walk carefully around corners. Romberg's test is frequently positive. Heel–toe walking is often impossible for elderly patients yet without any clear relationship to an identifiable and deteriorating disease process.

2.2.6 Identifying syndromes

Traditionally, the clinical approach of neurology has been to detect a particular constellation of physical signs allowing localization of a lesion to a particular site. Examples of such syndromes include:

- cerebellar pontine angle tumour producing ipsilateral hearing loss, facial sensory loss, facial weakness, and ataxia with contralateral hemiparesis and limb sensory disturbance;
- the superior orbital fissure syndrome of ocular motor nerve palsies, frontal trigeminal sensory loss, and retroocular pain;
- the anterior spinal artery syndrome with mid-thoracic level, bilateral upper motor neurone weakness, and spinothalamic sensory loss, but preserved dorsal column sensations;

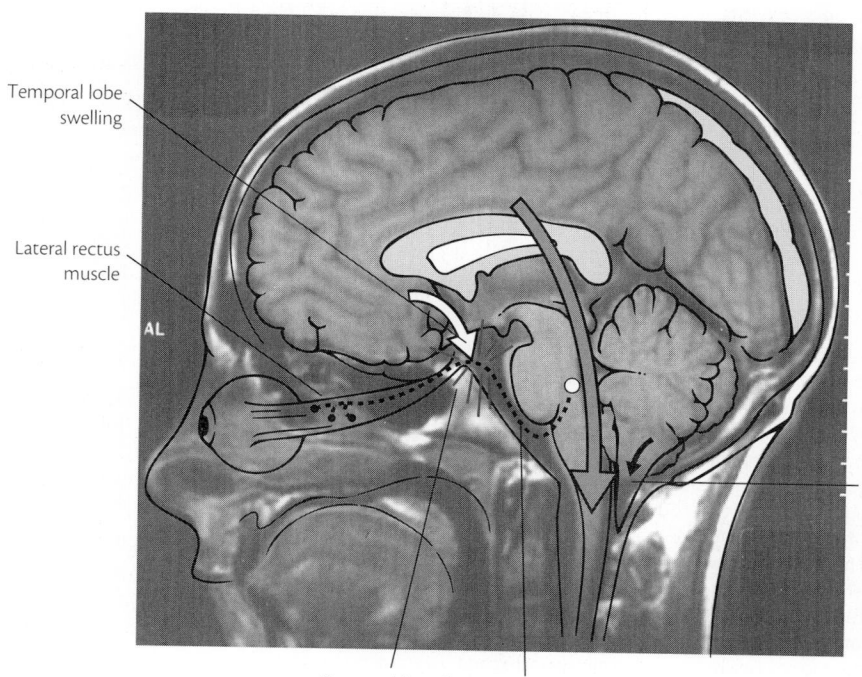

Temporal lobe swelling

Lateral rectus muscle

AL

Petrous ridge of temporal bone

Sixth cranial nerve

Herniation of cerebellor tonsil through foramen magnum

Fig. 2.34 Sixth nerve palsy as false localizing sign in raised intracranial pressure. Diagrammatic illustration of how first the nerve can become stretched by downward displacement of the brainstem during the early stages of a pressure cone, and second how it may become angulated at the petrous ridge of the temporal bone, by downward displacement of the brain stem, or compressed by temporal lobe swelling.

◆ cauda equina compression producing lower motor neurone weakness of both legs, multi-modality sensory loss below the level of the lesion, incontinence, and impotence.

Increasingly nowadays patients seek medical advice before they have progressed to the 'full-house' of symptoms and signs diagnostic of a particular syndrome or disorder, such as those above. The ready availability of high definition imaging allows definitive investigation of such patients at an earlier stage when clinical assessment alone may be unable to provide precise localization. Thus, the clinical approach of modern neurology needs to investigate the possible diseases which could be the cause of such incomplete, early syndromes. For in many instances this will allow treatment of the underlying disorder before irreversible neurological damage has occurred, particularly in the case of spinal cord and cauda equina compression.

2.3 **The reflexes**

2.3.1 **Reflex arcs**

A reflex is the simplest form of involuntary response to a stimulus. The anatomical basis of a reflex arc consists of: (1) a receptor organ; (2) an afferent path running from the periphery to the brainstem or spinal cord; (3) in some reflexes one or more intercalated neurones in the central nervous system link the afferent path to the (4), efferent path which leaves the neuraxis by the lower motor neurone axons to reach (5), the effector organ. Reflexes are elicited by afferent sensory stimuli such as touch, pain, sudden muscle stretch, light, or noise. The efferent response consists of muscular contraction, a modification in muscle tone, or glandular secretion. Important though visceral reflexes are, the neurologist investigating

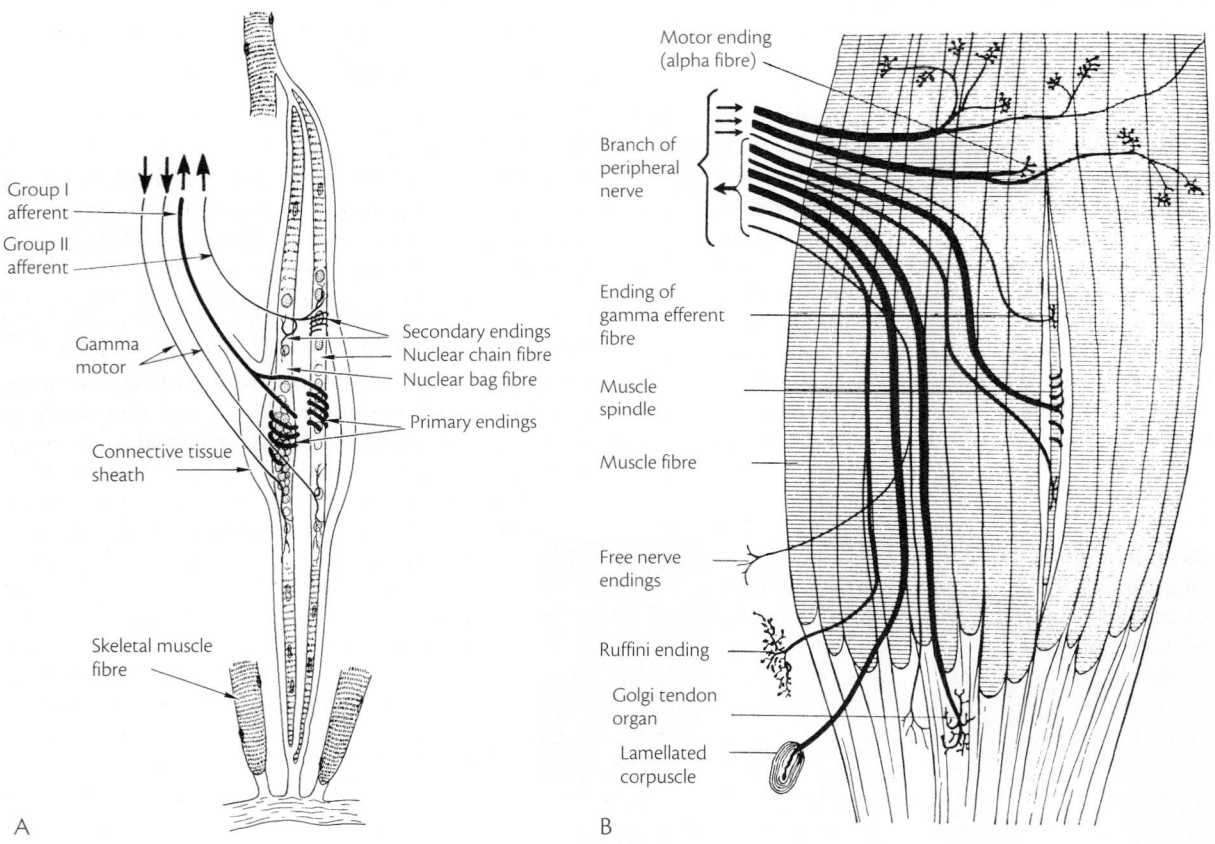

Fig. 2.35 (A) Schematic representation of a muscle spindle. The end parts of three skeletal muscle fibres are as shown (cross-striated, nuclei at edge). Inside the connective-tissue sheath of the spindle are two muscle fibres which are thinner than extrafusal skeletal muscle fibres, with central nuclei, and striations minimal or absent in the region of the sensory endings. Sensory nerve fibres form primary (annulospiral) and secondary (flower-spray) endings, the primary endings arising from the large fibres. Small nerve fibres (gamma efferents) form motor endings at each end of the spindle muscle fibres. Motor discharges over gamma efferents cause the muscle spindle fibres to contract at each end, thus stretching the intervening, non-contractile sensory region, and activating the sensory endings. Arrows indicate direction of impulse conduction. (Courtesy of Gardner (1975).)
(B) Schematic representation of a muscle and its nerve supply. Arrows indicate direction of impulse conduction. Each extrafusal muscle fibre has a motor ending from a large myelinated (alpha) fibre. The intrafusal muscle fibres within the muscle spindle have motor endings from small myelinated (gamma) fibres. Muscle nerves contain many sensory fibres. Some are large myelinated fibres from primary (annulospiral) endings in spindles, from neurotendinous spindles (Golgi tendon organs), and from Pacinian corpuscles in the connective tissue within and external to the muscle. Smaller myelinated and non-myelinated fibres arise from Ruffini endings in the connective tissue in and around muscle, and in joints. Finally there are small myelinated and non-myelinated fibres that form free endings in the connective tissue in and around muscle. (Courtesy of Gardner (1975).)

the state of the nervous system is mainly concerned with reflexes that excite responses in the somatic musculature.

Reflexes play an important role in diagnostic neurology because they reflect the integrity of, or alterations in, the neural structures responsible for their arc. Loss of a reflex may be due to interruption of the afferent path by a lesion involving the first sensory neurone in the peripheral nerves, plexuses, spinal nerves, or dorsal roots, by damage to the central paths of the arc in the brainstem or spinal cord, by lesions of the lower motor neurone at any point between the anterior horn cells and the muscles, of the muscles themselves, or by the neural depression produced by neural shock. In clinical practice, the most useful and oft-elicited reflexes are the tendon reflexes of the limbs, the jaw jerk, the plantar response, the superficial abdominal reflexes, the pupil-light response, and in infants, the Moro reflex. The place of these particular reflexes in the routine neurological examination is outlined in Section 2.2.2. This section describes the elicitation and significance of these reflexes and of a wide variety of others which are used occasionally.

2.3.2 Tendon reflexes

The physiological basis of the tendon reflex is the myotatic reflex, which is the reflex contraction of a muscle or part of a muscle in response to stretch. It is monosynaptic; mediated by a reflex arc consisting of two neurones with one synapse between them (Lloyd 1952). The afferent input of the tendon reflex is transmitted by large myelinated sensory peripheral nerve fibres which innervate the muscle spindles. These in turn are connected in parallel with the main contractile extrafusal muscle fibres. Their nuclear chain fibres signal the actual length of the spindle whilst the nuclear bag fibres detect the velocity of change of length (Fig. 2.35). The overall sensitivity of the muscle spindle is determined by its efferent supply from γ-motor neurones which control contraction of its intrafusal muscle fibres (Fig. 2.35). Tendon reflexes must be distinguished from the tonic stretch reflex which results from slow or prolonged stretch of a muscle and which is a polysynaptic response, probably involving cortical pathways (Marsden *et al.* 1973). The activity of many bulbar and spinal reflexes is profoundly influenced by the state of the muscle spindles and of the gamma efferent system of motor nerve fibres. In conditions causing hypotonia cerebellar lesions the tendon reflexes are depressed. By contrast the hypertonia associated with increased gamma efferent discharge exaggerates reflexes; such enhancement is greater in spasticity than in extrapyramidal rigidity. Anxiety, tensing, and painful conditions may also cause some increase in the deep tendon reflexes. Paradoxically in severe long-standing spinal cord lesions or spastic diplegia, the spasticity is so severe that sometimes the tendon reflexes in the lower limbs may be difficult to elicit. This is due to irreversible muscular contractions resulting from to chronic spasticity, or a dominant flexor withdrawal reflex which inhibits the tendon jerks.

A *tendon reflex* or jerk is a sharp muscular contraction evoked by suddenly stretching the muscle. The sudden stretch may be brought about by tapping the tendon, or by suddenly displacing the segment of a limb into which the muscle is inserted (Fig. 2.36). The response, a muscular contraction, is most evident in the muscle stretched, but may not be confined to this muscle. A tendon reflex is diminished or abolished by a lesion interrupting either the afferent, central, or efferent paths of the reflex arc, or by a disorder which makes the muscle incapable of responding to the nervous impulse.

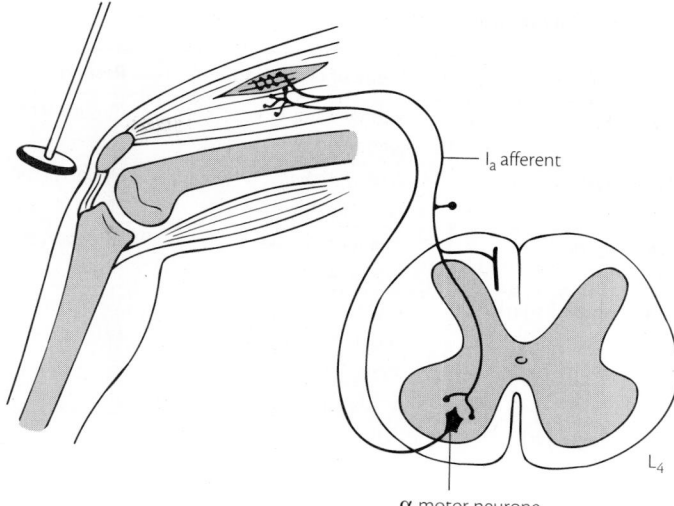

Fig. 2.36 The components of the monosynaptic stretch reflex elicited by percussing the patellar tendon of the quadriceps muscle.

If initially absent, reinforcement of tendon jerks may be achieved by simultaneous voluntary muscle contraction elsewhere in the body such as biting the teeth together, clenching a fist, or by pulling the flexed fingers of the two hands against each other, Jendrassik's manoeuvre. This increases activity in the gamma efferent system. Reflex activity in the legs may be studied electrically by recording the 'H' reflex, a contraction in the calf muscles which can be elicited by stimulating electrically the medial popliteal nerve. The H response, which is a monosynaptic reflex evoked by stimulation of group I afferent fibres in the nerve, follows the so-called M response evoked in the muscle by the direct effect of the nerve stimulus upon alpha efferent fibres. In early polyneuropathy the tendon reflexes may be lost before sensory loss is detectable clinically, although abnormalities of conduction usually are detectable electrically. Rarely, the tendon reflexes are congenitally absent. Some or all of the tendon reflexes may be well-nigh impossible to elicit, despite reinforcement, in well-muscled young men who are completely healthy. Table 2.5 gives the principal tendon reflexes, their mode of elicitation, and their innervation. Additional reflexes are sometimes elicited to localize the level of a spinal cord lesion: deltoid, pectoralis, long finger flexor, thigh adductor, or hamstrings.

Clinical interpretation of tendon reflexes involves two separate considerations; interpretation of an individual reflex relative to the other reflexes, and an absolute judgement as to whether an individual reflex is abnormal. An individual reflex can be regarded as abnormal only if it is absent despite reinforcement, or if its elicitation produces a clonic contraction of the muscle. The varying shades of sluggishness and briskness between these two extremes are not in themselves definitively abnormal. In contrast, comparison of tendon reflexes between different body regions may be helpful in localizing pathology, without any of the reflexes being individually abnormal. Examples include the slightly brisker reflexes unilaterally in a patient with mild hemiparesis, or the brisker reflexes in the legs than in arms in a patient with thoracic spinal cord compression. To document these variations in reflex intensity, various grading systems have been recommended (Dick 2003).

Table 2.5 The tendon reflexes

	Mode of elicitation	Response	Segment	Peripheral nerve
Jaw-jerk	A downwards blow on the chin	Closure of the jaw	Trigeminal	Trigeminal
Biceps-jerk	A blow to the biceps tendon	Flexion of the elbow	Cervical 5-6	Musculocutaneous
Triceps-jerk	A blow upon the triceps tendon	Extension of the elbow	Cervical 6-7	Radial
Brachioradialis-jerk	A blow to the distal end of the radius stretching brachioradialis	Visible contraction of brachioradialis	Cervical 6	Radial
Flexor finger-jerk	A blow upon the palmar surface of the semiflexed fingers	Flexion of the fingers and thumb	Cervical 7-8	Median and ulnar
Knee-jerk	A blow upon the patellar tendon	Extension of the knee	Lumbar 3-4	Femoral
Ankle-jerk	A blow upon the Achilles tendon	Plantar flexion at the ankle	Sacral 1-2	Sciatic/posterior tibial

None of these grading systems is fully satisfactory, and it is recommended that the observations are recorded as: absent despite reinforcement; present only with reinforcement; reduced; normal; brisk without clonus; clonus.

Clonus is a rhythmical series of contractions evoked by maintaining stretch and tension in a muscle. It is associated with increased gamma efferent discharge, and is often elicitable when the tendon reflexes are exaggerated after a corticospinal lesion. Ankle clonus is obtained by sharply dorsiflexing the ankle (Fig. 2.16). Clonus of the quadriceps, patellar clonus, is best elicited by a sudden sharp downward displacement of the patella. Clonus in the finger flexors can sometimes be elicited by suddenly extending the fingers.

Hoffmann's reflex. The patient's hand is pronated and the observer grasps the terminal phalanx of the middle finger between their forefinger and thumb. With a sharp flick the phalanx is passively flexed and suddenly released. A positive response consists of a sharp twitch of adduction and flexion of the thumb and flexion of the fingers. This reflex is physiologically identical with the *flexor finger-jerk*, which is elicited by tapping the palmar surface of the slightly flexed fingers. It is an index of muscular hypertonia rather than proof of a corticospinal lesion as such. It is not always positive in the presence of such a lesion, and may be elicitable in a nervous individual with no organic disease; if present only unilaterally it is likely to be significant.

Reflex spread. In states of muscular hypertonia a reflex response may spread beyond the muscles stretched, as when a tap on the styloid process of the radius elicits a contraction not only of the brachioradialis, but also of the long flexors of the fingers.

Inverted reflexes. In the upper limbs, so-called 'inverted reflexes' may be a useful sign of lesions of the cervical spinal cord. For instance, a lesion at C5-6 may both interrupt the arc for reflexes innervated by that segment, and also compress the corticospinal tracts to give exaggeration of reflexes subserved by lower segments. Thus tapping the biceps tendon may fail to elicit the biceps jerk but gives contraction of triceps, the 'inverted biceps jerk'. Similarly, the brachioradial jerk may be absent but the attempts to elicit it cause finger flexion, the 'inverted radial jerk'. An inverted knee jerk, with contraction of the hamstrings with knee flexion when the quadriceps tendon is tapped, may also be a sign of a spinal-cord lesion at L2-4 (Boyle *et al.* 1979) but is much less common.

2.3.3 Superficial reflexes

The palmomental reflex. To elicit this reflex, an orange stick is firmly scratched across the base of the thenar eminence. A positive response consists of involuntary contraction of the ipsilateral mentalis muscle giving a dimpling of the chin (Owen and Mulley 2002). The reflex may be present bilaterally in normal individuals but exhausts after a few trials. If persistently elicitable, or occurring in response to stimuli outwith the palm, it is likely to indicate cerebral damage.

The superficial abdominal reflexes. These are cutaneous reflexes consisting of a brisk unilateral contraction of the adjacent part of the abdominal wall in response to a corresponding cutaneous stimulus, such as a light stroke with an orange stick (Dick 2003). It is convenient to elicit them at three levels on each side—just below the costal margin (T8), at the level of the umbilicus (T10), and just above the inguinal ligament (T12) using stimuli which run medially along the territory of the dermatome. The abdominal and erector spinae reflexes are polysynaptic, and are reactions of the trunk to potential injury. They are plurisegmental, and lead to a local withdrawal from the stimulus. These reflexes may be absent in normal people, especially older women, in the aged, the obese, those with abdominal scars, and the multiparous.

They are normally dependent upon the integrity of the corticospinal tract for reasons not fully understood. Hence a corticospinal lesion is usually associated with diminution or loss of the superficial abdominal reflexes upon the same side. Loss of the abdominal reflexes is not always proportional to the severity of the lesion.

In multiple sclerosis, for example, they may be lost early, at a stage of the disease when other signs of corticospinal tract dysfunction are slight. In spastic diplegia and motor neurone disease, on the other hand, they are often retained, despite frank spasticity of the legs.

The reflex arcs of the superficial abdominal reflexes are localized in the spinal cord from the seventh to the twelfth dorsal segments. Lesions involving the arcs themselves may produce diminution or loss of the reflexes, for instance a structural lesion of the spinal cord at those segmental levels, or damage to the lower motor neurone by poliomyelitis.

The cremasteric reflex. The cremasteric reflex is a cutaneous reflex closely related to the abdominal reflexes. The stimulus is a light scratch along the inner aspect of the upper part of one thigh. The response is a contraction of the cremaster muscle, with elevation of the testicle on that side. This reflex, mediated by the first lumbar spinal segment, is diminished or abolished by a corticospinal tract lesion or a lesion of the reflex arc. It is usually extremely brisk in children, in whom it may sometimes be elicited by a stimulus applied to any part of the lower limb. It is often diminished or absent on the affected side in a patient with a varicocele.

The gluteal reflex. The gluteal reflex is physiologically akin to the abdominal reflexes. A scratch on the buttock evokes contraction of the glutei. The spinal segments concerned are L4 and 5.

The plantar reflex. The plantar reflex is one of the most important of all reflexes to the neurologist, because, if extensor, it provides unequivocal evidence of an upper motor neurone lesion. The plantar reflex normally remains extensor for the first year of life. After that the normal flexion withdrawal reflexes throughout the leg must be overridden for standing and walking. This plantar reflex is stimulated by a slow, firm, longitudinal scratch up the lateral aspect of the sole of the foot from the heel towards the toes (Fig. 2.21). The normal response is plantar flexion of the toes, sometimes associated with dorsiflexion of the foot at the ankle, contraction of the tensor fasciae latae muscle, and other variable muscular contractions. It is a spinal segmental reflex mediated by the first sacral segment of the cord. As a superficial reflex it is akin to the abdominal cremasteric reflexes (Lance 2002).

The extensor plantar response or Babinski response occurs in the presence of a corticospinal tract lesion. The normal reflex response of the great toe is replaced by an upward, extensor movement (Fig. 2.20). Extension of the great toe is not an isolated phenomenon of the abnormal plantar reflex, but is part of a general reflex flexion of the whole lower limb. This relates to the primitive flexor withdrawal reflex in response to a nociceptive stimulus to the lower limb seen in animals after division of the spinal cord. Both the flexor and extensor plantar responses are nociceptive reflexes, but 'the unique feature of the pathological extensor response is the recruitment of extensor hallucis longus into contraction with tibialis anterior and extensor digitorum longus' (Kugelberg and Hagbarth 1958). The afferent focus, or region from which it is easiest to elicit this reflex, is the outer border of the sole and the transverse arch of the foot but may extend to the leg or thigh in corticospinal tract lesions. The motor focus, or minimal response, is a contraction of the inner hamstring muscles which may be present even when the great toe fails to move. When fully developed, the extensor plantar reflex consists of flexion at all joints of the lower limb with dorsiflexion of the great toe and abduction or fanning of the other toes.

There are several points of practical importance in eliciting the plantar reflex, some of which are discussed in Section 2.2.2.

The stimulus should always be applied first along the outer border of the sole; an extensor response may sometimes be obtained here when the inner border of the sole yields a flexor response. The response is more consistently obtained if the stimulus is then continued medially across the anterior arch of the sole of the foot. Oppenheim's reflex, dorsiflexion of the great toe, evoked by firm moving pressure on the skin over the tibia, is physiologically the same as Babinski's reflex, differing only in the site of the stimulus. The same is true of Chaddock's and Gordon's reflexes. Chaddock's, also called the external malleolar sign, is an extensor plantar response elicited by scratching the skin in the region of the external malleolus. In Gordon's, also called the paradoxical flexor reflex, the stimulus consists of squeezing the calf muscles. The extensor plantar reflex is not an all-or-none reaction: minor degrees of corticospinal tract damage lead to an incomplete flexor response or a failure of the great toe to move either up or down, an 'equivocal' response. In experienced hands such equivocal responses carry reasonably reliable diagnostic implication if clearly unilateral.

Bilateral extensor plantar reflexes are often observed during sleep and deep coma from any cause, and for a short time after an epileptic convulsion. They are usually extensor in the first year of life, when the corticospinal fibres are incompletely developed. An extensor plantar response has been noted sometimes in patients in whom no anatomical lesion of the corticospinal tract was subsequently discovered. Occasionally the plantar response remains clearly flexor despite the presence of such a lesion (Van Gijn 1978). It may occur transiently as a result of physical fatigue. In the presence of a corticospinal tract lesion, the Babinski response may be lost if an associated lower motor neurone lesion paralyses the extensor hallucis muscle. This is a common problem in amyotrophic lateral sclerosis where denervation of the distal leg musculature can obscure the presence of the associated upper motor neurone lesion.

The bulbocavernosus reflex. The bulbocavernosus reflex consists of contraction of the bulbocavernosus muscle, which can be detected by palpation, in response to squeezing the glans penis. The spinal segments concerned are sacral 2, 3, and 4. This reflex is abolished in lesions of the cauda equina.

The anal reflex. The anal reflex consists of contraction of the external sphincter ani in response to a firm scratch on the skin of the anal verge. It tends to be absent in the elderly and those with a patulous anus. The spinal segments concerned are sacral 4 and 5.

2.3.4 Cranial reflexes

Some of the most clinically significant cranial reflexes are *unilateral afferent-bilateral efferent reflexes* in type. In these, a unilateral sensory stimulus evokes a bilateral motor response. Interpretation of the effects of lesions affecting either the afferent or efferent pathways, or the brainstem nucleus is based upon the circuitry (Fig. 2.37 and Table 2.6). *The pupil-light reflex* circuitry is discussed in Section 13.3.1.

The corneal reflex. The stimulus that evokes the corneal reflex is a light touch upon one cornea with a wisp of cotton wool. The normal response is bilateral blinking. The afferent path is through the first division of the fifth cranial nerve. The central path consists of fibres uniting the spinal nucleus of the fifth nerve with both facial nuclei. The efferent path passes through the facial nerves to both orbiculares oculi muscles. A lesion involving the fifth nerve or its spinal nucleus, since it interrupts the afferent path, causes bilateral loss of blinking in response to stimulation of the cornea on the side

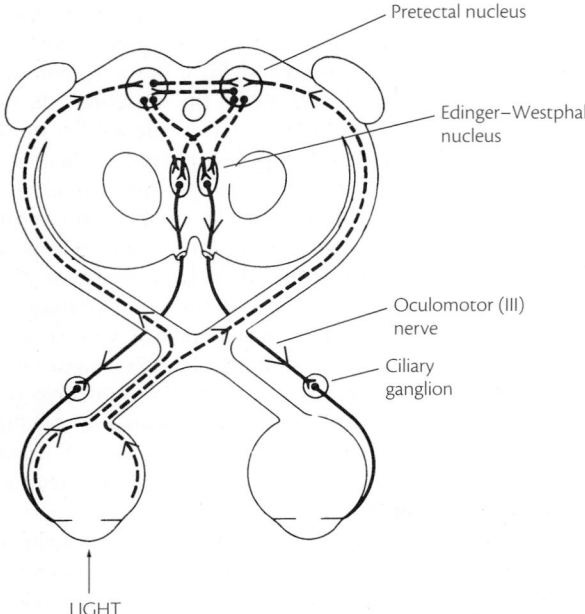

LIGHT

Fig. 2.37 Schematic circuitry for the 'unilateral afferent-bilateral efferent' reflexes illustrated by the pupil-light reflex mediated by the midbrain. The corneal and gag reflexes employ similar circuitry although innervate voluntary muscle.

Table 2.6 Examples of 'unilateral afferent-bilateral efferent' reflexes

	Sensory stimulus (unilateral)	Motor response (bilateral)
Pupil-light reflex	Retinal illumination (Optic Nerve II)	Pupil constriction (Oculomotor Nerve III)
Blink reflex	Corneal touch (Trigeminal Nerve V)	Blinking (Facial Nerve VII)
Gag reflex	Palatal touch (Glossopharyngeal Nerve IX)	Soft palate elevation (Vagus Nerve X)

From Donaghy (2005).

of the lesion. A lesion involving the nucleus or fibres of one seventh nerve interrupts the efferent path and hence causes loss of the reflex on the side of the lesion only, whilst the blink response remains on the other side. Loss of the corneal reflex is often an early sign of a lesion of the fifth nerve and may occur before any cutaneous anaesthesia can be detected. Apart from lesions involving the reflex arc, the corneal reflex is lost in states of deep coma.

The palatal reflex. The palatal or gag reflex consists of bilateral elevation of the soft palate in response to touching it on one side. The afferent path is by the glossopharyngeal nerve; the efferent by the vagus. The prominence of the palatal reflex varies between normal individuals. It is abolished by glossopharyngeal nerve lesions causing anaesthesia of the palate, and by lesions of the vagus nuclei. In lesions of a single vagus nerve, the response is unilateral, irrespective of the side of the stimulus, and the uvula is displaced towards the normal side.

The glabellar tap. A brisk tap on the glabella above the bridge of the nose causes bilateral blinking. In the normal individual, on repeated tapping the blinking ceases after two or three taps, known as the glabellar tap sign, nasopalpebral reflex or blinking reflex (Schott and Rossor 2003). Electrophysiological recordings from the orbicularis oculi have shown that there is an initial low amplitude monosynaptic reflex response, followed by a larger response of longer latency which is clearly polysynaptic. This habituates in normal individuals. In many with Parkinsonism blinks occur continuously in time with the taps, a 'positive glabellar tap sign'. However it is not specific and occurs also in diffuse frontal lobe damage and some normals.

The vestibulo-ocular reflex. This is also known as the oculocephalic reflex. The term 'Doll's head phenomenon' is ambiguous because the eyes of some dolls are fixed, whilst those of others counter-rotate. When the eyelids are held open and the head is rotated sharply from side to side, the eyes initially show conjugate deviation away from the side to which the head is moved. On flexion of the neck they move upwards; after each such movement they return slowly to the mid position even if the head remains rotated or flexed. The reflex persists in blind individuals and after occipital lobectomy (Plum and Posner 1980). It is impaired when there are lesions of the oculomotor nerves. Its absence usually indicates a brainstem lesion.

The caloric or oculovestibular reflex. This has much in common with the vestibulo-ocular reflex. Irrigation of an external auditory meatus with warm or cold water causes nystagmus in normal individuals. As this reflex depends upon the integrity of the vestibular nuclei, its absence may be a valuable sign of pontine damage if there is no reason to suspect a labyrinthine or eighth nerve lesion. Repeated absence of the caloric reflex is an important criterion in diagnosing brainstem death (Section 33.7.2).

The jaw jerk. In response to a tap upon the chin, so as to depress the lower jaw, there is a bilateral contraction of the elevators of the jaw. The jaw jerk is best elicited with the mouth half open, with the examiner resting a finger on the central mandible to cushion the blow of the tendon hammer. Both afferent and efferent paths pass through the trigeminal nerve. This reflex is a muscle stretch reflex, and, like other such reflexes, becomes exaggerated as a result of bilateral corticospinal tract lesions. It is often weakly present in normal individuals, and is only clearly enhanced if notably brisk and of large amplitude, or if it provokes jaw clonus.

Primitive oral reflexes. In the infant the contact of an object with the lips evokes sucking movement of the lips, tongue, and jaw. This *sucking reflex* is lost after infancy but may reappear in states of severe cerebral degeneration, such as senile dementia (Paulson and Gottlieb 1968). It may be unilateral, and associated with a grasp reflex on the same side. The *snout reflex* consists of lip puckering in response to pressure over the base of the nose (Schott and Rossor 2003). The slower *rooting reflex* involves the lips following and seeking out a gentle tactile stimulus on the adjacent cheek, or in response to a visual object, and involves head turning. These sucking, snout, and rooting reflexes are normal in infants, and re-emerge in frontal lobe lesions. It is most accurate to consider the *pout reflex* as a separate entity from the *snout reflex*, although the two terms are often used interchangeably. The pout reflex is evoked by a brisk tap upon a spatula placed on the closed lips, and if positive produces an immediate pouting response. This represents an enhanced myotatic reflex of the orbicularis oris muscle which appears in bilateral corticospinal tract lesions at or above the upper brainstem.

The pharyngeal reflex. This consists of constriction of the pharynx in response to a touch upon the posterior pharyngeal wall. Its afferent path runs in the glossopharyngeal nerve, its efferent path in the vagus. It is abolished by lesions causing pharyngeal anaesthesia and by lesions of the vagus nuclei. In cases of unilateral paralysis of the vagus musculature, the response is confined to the opposite half of the pharynx. Compared to the palatal reflex it is more difficult to elicit and interpret and does not offer additional localizing value.

2.3.5 Postural reflexes

These are reflexes in which the response consists not of a brief muscular contraction but of a sustained modification in the posture of one or more sections of the body.

Tonic neck reflexes. In the decerebrate animal changes in the position of the head relative to the body cause reflex modifications of limb tone and posture. These reflexes, which are excited from the proprioceptors of the cervical spine, are known as tonic neck reflexes and may sometimes be observed in severe cerebral diplegia. Passive turning of the head to one side evokes extension of the arm and leg on the side to which the head is turned; the contralateral limbs flex.

Associated reactions. Associated reactions or associated movements, are automatic modifications of the posture of parts of the body when vigorous voluntary or reflex movement of some other part occurs. They are best observed in the paralysed upper limb in hemiplegia, following a vigorous grasping movement with the sound hand. Other patterns of associated movement occur. Such semi-voluntary activities as yawning, stretching, and coughing often evoke associated movements in the paralysed limbs in hemiplegia, and may arouse false hopes that recovery is occurring.

The Moro reflex. This is normally present at birth and disappears by 20 weeks of age; persistence suggests a diffuse central nervous system disorder. It is elicited by holding the infant supine with the head slightly flexed, and then dropping the head through about 30°. The normal response is of symmetrical abduction, extension, and rotation of the arms. An asymmetrical Moro response occurs in brachial plexus injury or hemiparesis.

The Landau reflex. This reflex should be present by 10 months of age and will be absent in diplegia or tetraplegia. The infant is held prone supported by the examiner's hand. The normal response is extension of the neck, trunk, and legs. If abnormal, the infant tends to collapse into flexion around the examiner's hand.

2.3.6 Grasping reflexes

These are primitive reflexes of the limbs similar to the primitive oral reflexes (Section 2.3.4). Normally present in early development, their later inhibition may be released by frontal lobe damage. Two aspects of the stimulus are required to elicit a grasp reflex (Schott and Rossor 2003). Initially deep pressure over the palm evokes a brief catching movement of the fingers. This only develops into the firm holding phase if the object is then gently pulled away. Grasping reflexes may coexist with utilization behaviour.

Grasp reflex of the hand. Contact of an object such as the examiner's finger with the palmar surface of the fingers, especially in the region between the thumb and the index finger, causes reflex flexion of the fingers and thumb so that the patient's hand involuntarily grasps. The patient is unable to relax this grasp voluntarily, and efforts to pull the object away only cause it to be held more firmly. The patient may even notice that when he is holding an object he is unable to relinquish hold of it. Even an object presented visually may be groped for. Forced grasping and groping, which have been considered a regression to the infantile stage of the function of grasping, usually indicate a lesion involving the upper part of the opposite frontal lobe. A unilateral grasp reflex in a fully conscious patient is of some localizing value. Its localizing value is much less when the reflex is bilateral or the patient semi-comatose.

The grasp reflex of the foot. A similar grasp reflex is sometimes seen in the foot. Light pressure or a stroking movement applied to the distal half of the sole and plantar surface of the toes evokes tonic flexion and adduction of the toes without other associated movements. Like the fingers, the toes may grasp and hold an object. This reflex is present in the normal infant up to the end of the first year, and in 50 per cent of children with Down's syndrome. It may occur either with or without the hand-grasp reflex, and it results from similar lesions.

Utilization behaviour. Patients with frontal lobe lesions may grasp and use everyday implements placed in front of them, without the behaviour being purposeful (Lhermitte 1983). Similar behaviour can also occur when incidental objects are encountered during other activities (Shallice *et al.* 1989). Utilization behaviour typically follows damage to the inferior frontal lobes.

2.4 Diagnosing muscle weakness

Anatomical localization of the lesion responsible for muscle weakness is a common aim of neurological examination.

2.4.1 Symptoms

Surprisingly often patients do not complain of weakness itself, but rather of difficulty in using a limb for certain manoeuvres, or to walk. Particularly for the hand, an integrated motor–sensory organ, the early symptoms of difficulty in manipulating buttons or pens, or dropping things, can be remarkably similar in pure motor and pure sensory disorders. Furthermore, a patient who complains of weakness may be suffering in reality from numbness, disinclination to use a painful limb, incoordination, or even the aesthenic effects of cardiorespiratory disease. Patients may use the term 'weakness' when referring to other motor disorders which do not involve loss of raw muscle power, such as the bradykinesia of Parkinsonism (Section 40.3.1), or apraxia (Section 34.4.3) which is an inability to formulate and execute a complex movement despite intact functioning of the upper and lower motor neurones and muscles. Weakness accompanied by exhaustion may be a complaint of patients with chronic fatigue syndrome. However such patients are capable usually of exerting normal muscle strength, at least momentarily, if adequately encouraged.

Weakness of certain muscles often produces distinctive complaints. Proximal arm muscle weakness usually causes difficulty in doing the hair, hanging out washing, or lifting objects from high shelves. Patients rarely notice isolated weakness of small hand muscles, but sometimes complain of loss of grip, for instance in trying to unscrew bottle tops. Weakness of individual arm muscles can be distinctively symptomatic, such as difficulty in sliding the hand into a pocket with weakness of finger extension due to posterior interosseous nerve lesions, or difficulty in pushing the car gear lever forward with focal triceps weakness. Proximal leg muscle weakness, particularly if it affects quadriceps, leads to difficulty in climbing or descending stairs, standing out of the bath, or arising from sitting without using the arms. Distal leg muscle shows as ankle instability or foot drop, inability to scrunch up the toes into plantar flexion so as to keep loose shoes on, or to grip the edge of a swimming bath so as to dive.

The pattern of the weakness and the presence of other symptoms have considerable implications for localizing the lesion and determining the pathology:

◆ Predominantly proximal muscle weakness points to myopathy or myasthenia.

◆ Weakness developing on usage of muscles, and during the course of the day, suggests myasthenia gravis.

◆ Hemiparesis, affecting both the arm and the leg on one side only, is typical of a cerebral hemisphere lesion.

◆ Weakness of both legs, or paraparesis, points to thoracic spinal cord or cauda equina disease.

◆ Weakness of all four limbs, known as quadriplegia or tetraplegia, suggests cervical spinal cord or brainstem disease, or diffuse neuromuscular disease.

◆ Difficulty in swallowing is typical of motor neurone diseases, myasthenia, inclusion body myositis, and some muscular dystrophies, and acute polyneuropathies such as Guillain–Barré syndrome or diphtheria.

Although patients are usually conscious that their proximal muscles, such as biceps or quadriceps, have wasted, often they are oblivious of advanced atrophy of distal muscles, such as the dorsal interossei.

Concurrent alteration of sphincter control should provoke prompt attempts to diagnose and treat the cause of associated limb weakness. Urgency of micturition with frequent voiding of small quantities, and sometimes incontinence, reflects the small, spastic irritable bladder typical of bilateral upper motor neurone lesions, particularly those affecting the spinal cord. Retention of urine, or sometimes dribbling incontinence, occurs in cauda equina disease. Erectile impotence is an early feature of either spinal cord or cauda equina disease. It is rare for anal sphincter control to be impaired in a manner that is characteristically diagnostic before obvious abnormalities of micturition or potency. Rapidly developing impairment of sphincter control implies compression of the spinal cord or cauda equina and requires emergency investigation.

2.4.2 Differentiating upper and lower motor neurone lesions

If examination reveals muscle weakness, it is necessary to differentiate between lesions of the upper or lower motor neurones, the neuromuscular junction, or primary muscle disease. Also one should recognize that distinctive patterns of fluctuating weakness may be due to psychological factors, loss of kinaesthetic feedback, or pain. The key features to this differential diagnosis are the presence of wasting or fasciculations, the pattern of muscle power loss, changes in tone, tendon reflex abnormalities, plantar responses, and the topography of any associated sensory loss.

Wasting. This is typical of a denervated muscle, or one affected by primary muscle disease. Other causes are much rarer (Section 22.1.4). In polyneuropathy the wasting will be predominantly distal, in myopathy it is predominantly proximal. The wasting follows the distribution dictated by the innervation in individual peripheral nerve lesions (Sections 22.9 and 22.10) or spinal root lesions (Section 29.2). Wasting develops in any muscle which has been significantly denervated for 4–6 weeks. Muscle atrophy is not a feature of myasthenia.

Disuse atrophy. Disuse atrophy of muscles occurs in patients who have been recumbent for general medical reasons. It may affect muscles acting at a diseased joint, such as quadriceps with knee arthritis. Disuse atrophy can be distinguished from the wasting due to lower motor neurone or muscle diseases because the reflexes and tone are normal, and no fasciculations occur. But, most importantly, strength is relatively well-preserved in a disuse-atrophied muscle, whereas a pathologically wasted muscle will be profoundly weakened.

Pseudohypertrophy. It is an unusual physical sign in which a pathologically weakened muscle is hypertrophied. It is a particular sign in the calves in Duchenne muscular dystrophy (Section 24.2.1) and is a rare feature of polyneuropathies such as hereditary motor and sensory neuropathy (Section 21.4) and multifocal motor neuropathy (Section 21.11.3).

Fasciculations. These occur during subacute partial denervation of muscles, and are a particularly common feature of motor neurone disease. A fasciculation is a flickering contraction visible for a moment within the belly of a muscle. It represents simultaneous contraction of all the muscle fibres in the motor unit innervated by a single motor neurone (Fig. 2.38). Fasciculations are most easily

Table 2.7 Physical signs used to differentiate between muscle weakness due to disease of the upper motor neurone, lower motor neurone, or muscle and psychogenic weakness

	Upper motor neuron damage		Lower motor neuron damage	Primary muscle disease	Psychogenic disorder
	Cerebral hemisphere	Spinal cord			
Wasting			Present	Present	
Fasciculations			Present		
Reflexes	Brisk	Brisk	Absent	Normal	Normal
Tone	Spastic	Spastic	Flaccid	Normal	Normal
Plantars	↑↓	↑↑	↓↓	↓↓	↓↓
Sensory loss	Sometimes	Usually	Usually	No	Often
Distribution of weakness	Hemiplegic	Paraplegic or ↑quadriplegic	Individual ↑peripheral nerve or root; distal in polyneuropathy	Proximal	Variable

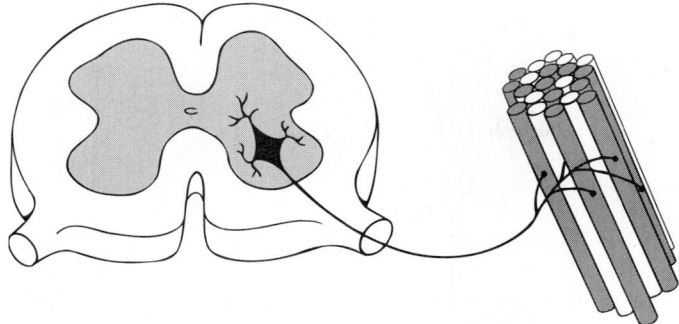

Fig. 2.38 Schematic illustration of a motor unit, consisting of a group of muscle fibres innervated by a single motor neuron. Note that the muscle fibres of a motor unit may be widely scattered throughout the belly of the muscle.

visualized in those muscles with large motor units containing hundreds of muscle fibres, such as powerful proximal limb muscles, rather than in those muscles with small motor units that are used for fine motor control, such as the small hand muscles. However, electromyography will detect fasciculation discharges in such muscles even though they may be invisible to the naked eye. Fasciculations are only definitely pathological if associated with wasting or weakness of the muscle.

Tone. The tone of a muscle is the response it shows to passive stretching. A completely relaxed and resting muscle is not in a state of continuous partial contraction and is electrically silent; it has elasticity, but no tone. Therefore tone can be assessed only when the muscle is stretched or when it is maintaining posture against an applied force such as gravity. Postural tone is the state of partial contraction of certain muscles needed to maintain the posture of body parts.

In neurological practice, tone is usually assessed by moving a limb and observing the reaction which occurs in the muscles that are being stretched. The moment stretch begins, the muscle spindles give out afferent stimuli and reflex partial contraction results. The responses to momentary and to more prolonged stretching are different, the former being responsible for the tendon jerks, the latter eliciting more complex responses, often in the form of tonic contraction. Variations in the sensitivity of these reflexes account for the alterations in tone which occur as a result of nervous disease. Forceful continued contraction of a group of muscles, produced by biting or clenching one fist or pulling firmly with the flexed fingers of both hands, temporarily causes an increased flow of afferent impulses in the sensory fibres from the spindles. In turn this increases the rate of discharge in gamma motor neurones throughout the body, thus causing a generalized slight increase in sensitivity of the spindles to stretch. The tendon reflexes become brisker as the state of contraction of its intrafusal fibres is increased. This phenomenon, also known as reinforcement, or Jendrassik's manoeuvre, is often used to elicit tendon reflexes which at first seem absent. In spasticity and in extrapyramidal rigidity, the 'set' of the spindles is continuously increased.

On stretching, the tone of a muscle may feel to be increased, termed spasticity or rigidity or hypertonia, or reduced, that is flaccidity or hypotonia. These alterations are of great value in neurological diagnosis. Muscle tone is normally regulated by reticulospinal fibres which accompany the pyramidal tract and exert an inhibitory effect upon the stretch reflex. This inhibition balances the background facilitatory impulses conveyed by the pontine reticulospinal and lateral vestibulospinal pathways. In turn these are influenced by multisynaptic reflex arcs traversing the cerebellum, basal ganglia, and brainstem. Dorsal reticulospinal fibres appear specifically to inhibit flexor lower limb reflexes. When lesions of the pyramidal and reticulospinal tracts release stretch reflexes from inhibition, the resultant increase in tone is initially associated with hyperactivity of dynamic fusimotor neurones. If such increased tone persists, termed spasticity, increased alpha-neurone discharge develops so that spasticity may be associated with increases in both gamma, dynamic fusimotor, and alpha-motor neurone activity.

Spasticity. This results from lesions of the pyramidal and often of the reticulospinal pathways. The stretch reflexes become hyperactive because of increased excitability of dynamic fusimotor neurones and alpha neurones which have been released from descending inhibitory influences. If the dorsal reticulospinal system is also damaged, there is disinhibition of afferent flexor reflex pathways. Release of such flexor reflexes may give flexor spasms in the lower limbs in response to stimulation of the legs, bladder, bowels, or skin. The 'extensor' plantar or Babinski reflex (Section 2.3.3) is one component of the primitive flexor withdrawal reflex. Usually in spasticity the affected limb shows increased resistance to passive stretching. This is particularly severe initially, but then 'gives' suddenly as the movement is continued. This sign is seen particularly well in the legs of a patient with a spastic paraplegia due to bilateral pyramidal tract disease and is known as 'clasp-knife' rigidity. Hyperactivity of tendon reflexes is often accompanied by clonus (Section 2.3.2) in which sustained stretch of muscle evokes repetitive contraction and relaxation due to reverberating activity in the hyperexcitable fusimotor system. Spasticity, in the form of sustained ankle clonus, or a pronator catch in the forearm is an incontrovertible sign of pyramidal tract disease.

Extrapyramidal rigidity. This differs from spasticity. It occurs in patients with disease of the basal ganglia such as Parkinsonism. The rigidity is uniform in degree throughout the entire range of passive movement, known as 'plastic' or 'leadpipe' rigidity. If tremor is superimposed it is referred to as 'cogwheel' rigidity. In dystonia there is simultaneous contraction of agonists and antagonists so that the reciprocal inhibition of antagonists is impaired and there is increased alpha-neurone discharge. As a consequence parts of the body become virtually fixed in an abnormal posture.

Decerebrate rigidity. This occurs in animals when a transverse lesion across the midbrain at about the level of the superior colliculus or red nucleus releases the brainstem, cerebellum, and spinal cord from cerebral control. Strong continuous contraction in extensor groups of muscles occurs so that an animal placed upright with support will remain standing, but if pushed over cannot rise. This contraction may occur intermittently, leading to episodes of decerebrate posturing. This predominance of extensor activity is mediated by the reticulospinal and vestibulospinal pathways, and is aroused by a further lesion induced at the level of the vestibular nuclei. In man, a similar state may accompany severe midbrain lesions which usually cause loss of consciousness also. In such a case all four limbs are rigidly extended, the back is arched, and there may be neck retraction. Thus the patient, if lying supine, is virtually supported by the back of the head and the heels, a posture known as opisthotonos. The arching of the back can be increased by any sensory stimulus and there is striking resistance to any attempt at flexing the limbs passively. Tonic neck reflexes can usually be elicited;

turning the head to one side gives extension of the limbs on that side and flexion on the other.

Decorticate rigidity. This usually occurs with lesions of the cerebral white matter, or thalamus and internal capsule. The arm is flexed and adducted whilst the leg is stiffly extended, a posture similar to that of chronic spastic hemiplegia.

Flaccidity. Flaccidity, or hypotonia, is a reduction in tone. It may be due to cerebral or spinal shock resulting from acute and extensive brain or spinal cord lesions which transiently suppress all motor reflex activity. It is a common manifestation of cerebellar disease, associated with diminished gamma efferent activity. It also occurs whenever a lesion of the afferent or efferent pathway interrupts the spinal reflex arc. In severe hypotonia, as in patients with total flaccid paralysis, all resistance to passive stretch is lost and the limbs are limp and flail-like. Lesser degrees of hypotonia in the upper limbs can be elicited by asking the patient to hold out their arms horizontally. The forearms are then tapped briskly. When one limb is hypotonic, the recoil is slowed and the arm oscillates through a wider range as though 'underdamped'. If a hypotonic patient is asked to raise their arms above their head with the palms facing forwards, the palm of a hypotonic limb is seen to be externally rotated. In practice hypotonia has limited diagnostic usefulness, being overshadowed by more prominent features of cerebellar disease, such as dysmetria, or of spinal reflex arc disease, such as areflexia.

Pattern of weakness. Severe upper motor neurone lesions cause complete paralysis of the limb. Less severe upper motor neurone lesions cause distinctive patterns of weakness. In the arm, extensor muscles are most markedly affected: deltoid, triceps, finger, and wrist extensors, and the dorsal interossei. In the leg, hip flexion due to iliopsoas is usually affected earliest, and hamstring and ankle dorsiflexion weakness is often pronounced. As a general rule, weakness is symmetrically distributed distally in polyneuropathy and proximally in myopathies. In mononeuropathy or spinal nerve root lesions, the pattern of weakness follows the innervation pattern.

Tendon reflexes. Tendon reflexes (Section 2.3.2) are crucial to differentiating between upper and lower motor neurone disorders, being brisk in the former, and often absent or hypoactive in the latter. Although areflexia is common in lesions affecting the lower motor neurone reflex arc, this is mainly due to coexisting involvement of the muscle spindle sensory afferent fibres within peripheral nerves or roots. For example, the tendon reflexes are preserved even with quite advanced muscle denervation in motor neurone disease, because the sensory afferent pathways are not affected. Tendon reflexes are preserved in primary muscle disease, except when chronic fibrosis of the muscle or damage to the muscle spindle have occurred. Areflexia occurs in the rare disorder of Eaton–Lambert myasthenic syndrome.

There are only three objective conclusions with clear pathological implications that can be made about any individual tendon reflex viewed alone. Either it is normal, or it is absent despite reinforcement, or it is pathologically brisk in that one or more clonic beats occur during elicitation (Section 2.3.2). A normal reflex may vary from being only obtainable with reinforcement, to being quite brisk in anxious individuals. It is often valuable to compare the briskness of reflexes in different regions of the body. For instance upper motor neurone lesions will produce brisker tendon reflexes on the side of a hemiplegia, or in the legs compared to the arms in thoracic spinal cord lesions. A focally hypoactive reflex occurs in the territory of a diseased spinal nerve root or peripheral nerve. The ankle jerks may be markedly hypoactive compared to the knee and arm reflexes in polyneuropathy.

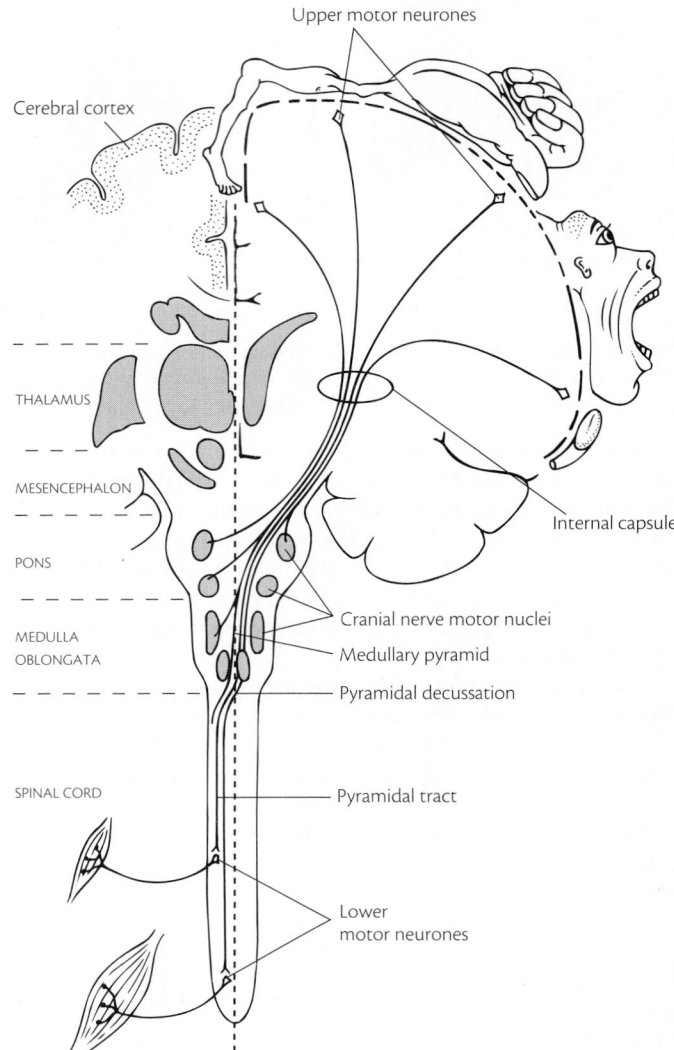

Fig. 2.39 Schematic representation showing the course of the pyramidal tract, the homuncular organization of the motor cortex in the precentral gyrus, the concentration of the motor output within the internal capsule, and the decussation of the pyramidal tract in the medulla oblongata.

An extensor plantar reflex or Babinski response (Section 2.3.3) represents incontrovertible evidence of an upper motor neurone lesion. Circumstantial evidence of an upper motor neurone lesion may be provided by absent superficial abdominal reflexes (Section 2.3.3).

2.4.3 Cerebral hemisphere lesions

Unilateral cerebral hemisphere lesions cause contralateral hemiparesis, often including the lower facial musculature. The forehead, tongue, and bulbar musculature will be spared unless an upper motor neurone lesion is bilateral. Focal lesions affecting only a portion of the motor cortex produce paralysis of the body part represented by that point of the homunculus (Fig. 2.39). An example is the 'cortical hand' in which weakness affects all movement of the hand, including finger extension, flexion, and abduction. Complete hemiplegia commonly results from small lesions in the internal capsule, where

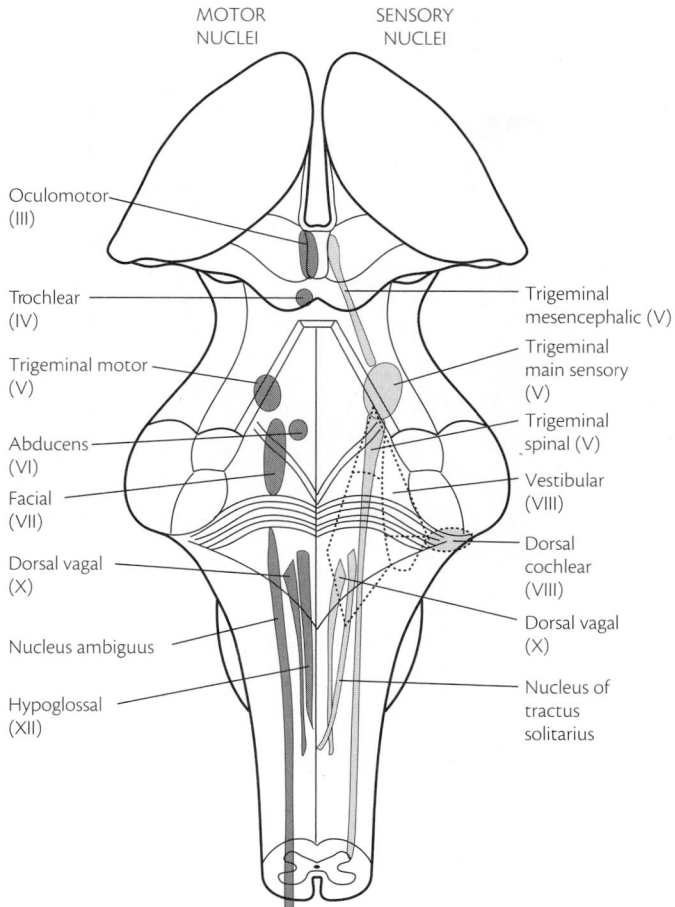

MOTOR NUCLEI

SENSORY NUCLEI

Oculomotor (III)

Trochlear (IV)

Trigeminal motor (V)

Abducens (VI)

Facial (VII)

Dorsal vagal (X)

Nucleus ambiguus

Hypoglossal (XII)

Trigeminal mesencephalic (V)

Trigeminal main sensory (V)

Trigeminal spinal (V)

Vestibular (VIII)

Dorsal cochlear (VIII)

Dorsal vagal (X)

Nucleus of tractus solitarius

Fig. 2.40 Diagram illustrating positions of cranial nerve nuclei as seen from the posterior aspect of the brainstem with the cerebellum removed to reveal the floor of the fourth ventricle.

the corticospinal tract fibres are crowded together. Focal motor, or Jacksonian epileptic attacks often picking out only one side of the mouth, or the finger and thumb, or the great toe, are typical of irritative lesions of the motor cortex. Predominantly proximal or limb girdle patterns of hemiparesis result from high motor cortex lesions and often result from ischaemia in the watershed between middle and anterior cerebral artery territories. Cortical or subcortical lesions often cause an associated cognitive deficit.

By contrast a lesion deep in the white matter, particularly one affecting the internal capsule, usually produces dense hemiplegia without cognitive loss. Milder hemiparesis with prominent dysarthria or ataxia points to lacunar infarction in the posterior limb of the internal capsule. If a hemiparesis is associated with a cranial nerve lesion on the opposite side, this so-called 'alternating hemiplegia' signifies a brainstem location for the lesion.

2.4.4 Brainstem lesions

The clue to a brainstem lesion causing weakness is an associated disorder of a cranial nerve (Fig. 2.40), or of the intrinsic pathways of the brainstem such as the cerebellar or vestibular connections. The majority of lesions of the brainstem are due to ischaemia, with demyelination, haemorrhage, or tumour deposits also occurring. They are often patchy and unstereotyped in their location and effects.

Pyramidal tract damage in the brainstem can be either unilateral or bilateral depending upon the topography of the lesion; asymmetric bilateral lesions are commonest. Because the pyramidal tract decussates in the lower medulla, all brainstem lesions above this level will produce contralateral weakness. Lesions of the cerebellum or its peduncles are often associated with damage to the adjacent brainstem. Cerebellar hemisphere damage causes ipsilateral dysmetria or ataxia, and hypotonicity. Midline damage to the cerebellar vermis causes gait ataxia, truncal ataxia, dysarthria, and slowed and irregular tongue movements. Lesions of the flocculonodular lobe, or vestibulocerebellum produce various eye movement abnormalities, including skew deviation and head tilt, inaccurate saccades, square-wave jerks, loss of smooth pursuit, and opsoclonus.

Midbrain lesions. These cause various syndromes associated with weakness if the cerebral peduncles are involved (Liu *et al.* 1992; Bogousslavsky *et al.* 1994; Silverman *et al.* 1995). Posterior cerebral artery penetrating branch ischaemic vascular lesions are the commonest cause:

- *Weber's syndrome.* A lesion of one cerebral peduncle produces an ipsilateral III nerve lesion, including pupil dilatation, and contralateral hemiparesis, including the face (Fig. 2.41).

- *Benedikt's syndrome.* Lesions affecting the red nucleus region cause ipsilateral III nerve lesions, including pupil dilatation, and contralateral tremor, chorea, or athetosis.

- *Claude's syndrome.* Vascular lesions may cause ipsilateral oculomotor nerve palsy and contralateral cerebellar ataxia.

- *Northnagel's syndrome.* Rarely focal infarction may result in ipsilateral oculomotor palsy, contralateral cerebellar ataxia and trochlear nerve palsies, and nystagmus or facial sensory loss

Lesions of the central and posterior midbrain produce various eye movement abnormalities especially affecting vertical eye movements, pupil reactions, and altered eyelid movements.

- *Parinaud syndrome* is an example involving loss of upgaze, eyelid retraction, dissociation of near-light pupil responses, and convergence–retraction nystagmus.

- A *'top of the basilar' syndrome* usually results from emboli, producing combined infarction of midbrain, thalamus, and occipital and temporal lobes. The clinical features vary in keeping with the varying topography of infarction, but the central features consist of homonymous hemianopia, eye movement disorders, amnesic states, and cerebellar ataxia (Caplan 1980; Mehler 1989).

- A *caudal paramedian midbrain syndrome* can result from a single unilateral lesion affecting the superior cerebellar peduncle decussation, with bilateral cerebellar dysfunction, eye movement disorders, and palatal myoclonus (Mossuto-Agatiello 2006).

Pontine lesions. These are notably variable in their effects. These include dysarthria, contralateral ataxia, trigeminal (V), abducens (VI), facial (VII), or auditory (VIII) nerve lesions (Fig. 2.42). Complex eye movement disorders are common too; ipsilateral gaze palsy, internuclear ophthalmoplegia, one-and-a-half syndrome, and skew deviation (Section 13.2.2). Owing to the higher decussation of the corticofacial fibres, a unilateral corticospinal lesion in the pons does not cause weakness of the opposite side of the face, but only of the opposite limbs. But the lesion may also involve the facial nucleus or the intrapontine fibres of the facial nerve on the same side,

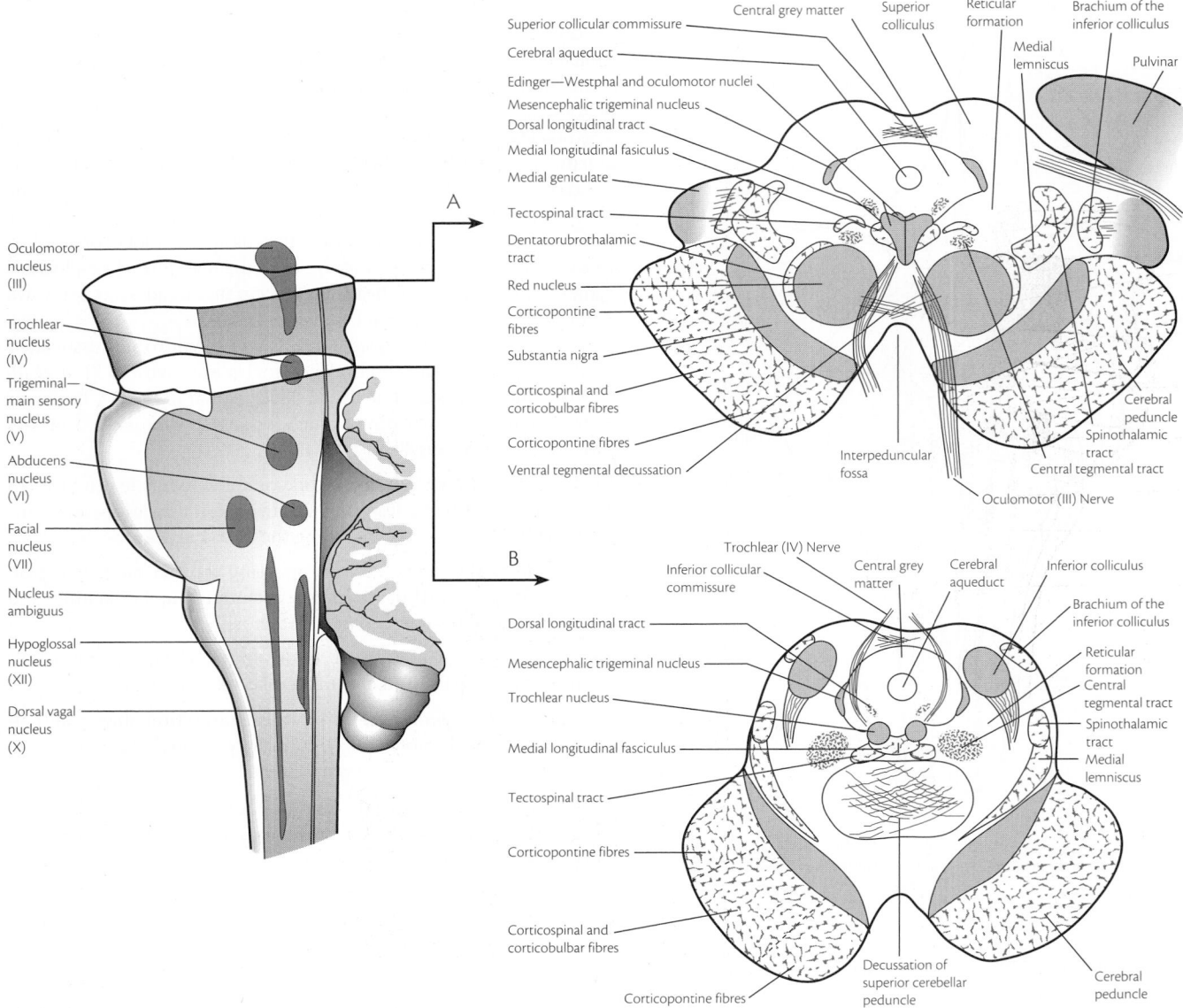

Fig. 2.41 Cross-section of the midbrain at rostral (A), caudal (B) levels.
(A) The rostral midbrain at the level of the superior colliculus and oculomotor nerve (III). Fibres of the oculomotor nerve are leaving to enter the interpeduncular fossa. The medial lemniscus is in the lateral part of the tegmental field. The brachium of the inferior colliculus is entering the medial geniculate, and the brachium of the superior colliculus is entering the superior colliculus. Fibres from the decussation of the superior cerebellar peduncle have formed at the lateral margin of the red nucleus as the dentatorubrothalamic tract; it sends fibres to the red nucleus and to the ventralis lateralis nucleus of the thalamus. Rubrospinal fibres from the red nucleus cross at the ventral tegmental decussation, and at a more caudal level will join the fibres in the central tegmental tract, finally ending in the spinal cord.
(B) The caudal midbrain at the level of the inferior colliculus and the trochlear nerve (IV). Fibres of the trochlear nerve are leaving dorsally. The medial lemniscus is rotating into a dorsoventral position in the lateral tegmental field. The lateral lemniscus is entering the nucleus of the inferior colliculus. Fibres from the cerebellum are crossing though the tegmentum as the decussation of the superior cerebellar peduncle. (Modified from Patton *et al.* (1976).)

thus causing one form of 'crossed hemiplegia'. Different forms of this have been described (Silverman *et al.* 1995):

♦ The *Millard–Gubler syndrome* consists of paralysis of one lateral rectus, due to involvement of the sixth nerve nucleus, with or without lower motor neurone facial paralysis on the same side and supranuclear paralysis of the limbs on the opposite side.

♦ *Foville's syndrome* is similar to the Millard–Gubler syndrome, except that paralysis of the conjugate ocular deviation to the side of the lesion takes the place of lateral rectus paralysis.

Medulla oblongata lesions. These are more likely to cause weakness if medially situated. The lower the medullary lesion the more likely is the involvement of pyramidal decussation. A medial medullary lesion produces an ipsilateral hypoglossal (XII) nerve palsy

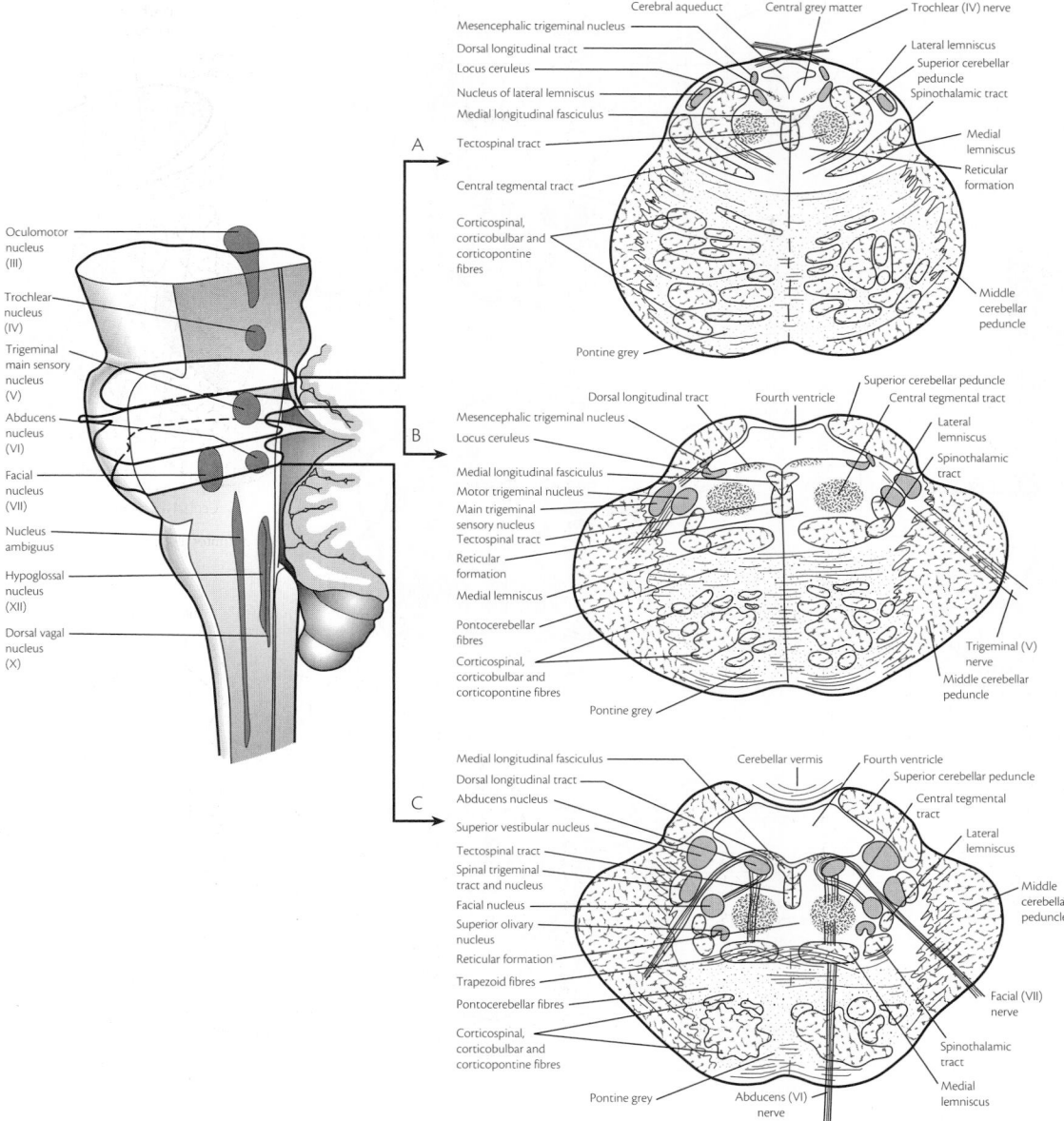

Fig. 2.42 Cross-sections of the pons at rostral (A), mid (B), and caudal (C) levels.

(A) The rostral pons at the isthmus. Fibres of the trochlear nerve (IV) are crossing as they leave dorsally. The medial lemniscus is moving laterally and beginning to rotate to a dorsoventral position. The superior cerebellar peduncle is moving towards the midline. The rostral-most edge of the middle cerebellar peduncle is present. The corticospinal, corticobulbar, and corticopontine fibres, which constitute the cerebellar peduncle, are separating as they plunge into the basilar pontine grey matter.

(B) The midpons at the level of the trigeminal nerve (V). Fibres of the trigeminal nerve separate the main sensory trigeminal and motor trigeminal nuclei. The cell bodies of proprioceptive trigeminal afferents constitute the mesencephalic nucleus. The trigeminal nerve leaves through the middle cerebellar peduncle. The medial lemniscus has begun to move laterally towards the spinothalamic tract. The superior cerebellar peduncle forms the lateral wall of the fourth ventricle as it descends from the cerebellum towards to midbrain tegmentum. Pontocerebellar fibres which receive input from the corticopontine fibres are streaming across the midline to form the middle cerebellar peduncle. The corticospinal, corticobulbar, and corticopontine fibres are scattered throughout the basilar pontine grey matter.

(C) The caudal pons at the level of the abducens (VI) and facial (VII) nerves. The abducens nerve leaves ventrally through the basal pons near the midline; the facial nerve loops medially around the abducens nucleus and then courses laterally to emerge at the caudal edge of the middle cerebellar peduncle. The pontine grey matter is sending pontocerebellar fibres across the midline to form the middle cerebellar peduncle. The superior cerebellar peduncle is projecting towards the midbrain. The medial lemniscus has rotated to a mediolateral position and is obscured by trapezoid fibres of the auditory system that cross the midline; the trapezoid fibres will turn rostrally to ascend in the lateral lemniscus. Primary afferents from the trigeminal nerve have formed the spinal trigeminal tract. (Modified from Patton *et al.* (1976).)

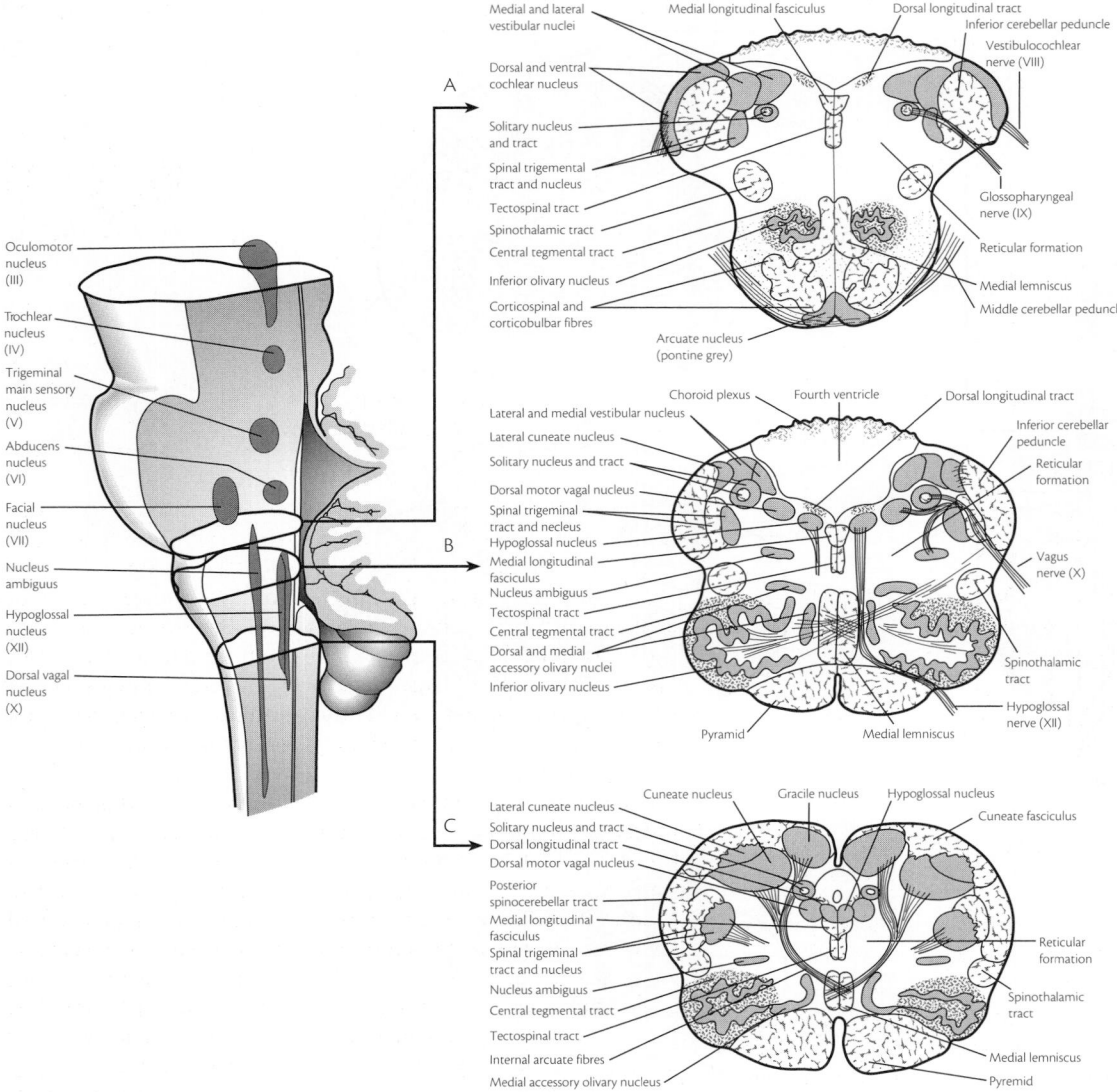

Fig. 2.43 Cross-sections of the medulla oblongata at rostral (A), mid (B), and caudal (C) levels.

(A) The rostral medulla at the level of the vestibulocochlear (VIII) and glossopharyngeal (IX) nerves. The cochlear nuclei cap the lateral surface of the inferior cerebellar peduncle. The medial lemniscus is still situated medially along the midline. Its trigeminolemniscal components (not shown) would be in its most dorsal part; the laterally placed spinothalamic tract would also contain trigeminothalamic components. The rostral pole of the inferior olivary nucleus appears in the course of the descending central tegmental tract, some of whose fibres terminate there; another component of the central tegmental tract will continue its descent to the spinal cord in the rubrospinal tract. The corticospinal and corticobulbar fibres are closely grouped as the pontine grey matter thins out; just cordal to this section they will form the medullary pyramids. The caudalmost edge of the middle cerebellar peduncle is present.

(B) The midmedulla at the level of the vagus (X) and hypoglossal (XII) nerves. The vagus nerve leaves lateral to the inferior olivary nucleus, whereas the hypoglossal nerve does so between it and the pyramid. Motor components of the vagus are shown coming from the dorsal motor vagal nucleus and nucleus ambiguous; visceral afferents are forming the tractus solitarius. The medial lemniscus is orientated dorsoventrally along the midline above the pyramid; the spinothalamic tract is in the lateral part of the tegmental field. Olivocerebellar fibres are crossing and will enter the inferior cerebellar peduncle. The lateral cuneate nucleus is also sending fibres into the inferior cerebellar peduncle. The descending corticospinal and corticobulbar fibres have grouped together to form the pyramids.

(C) The caudal medulla at the level of the sensory decussation. Most of the fibres of the gracile fasciculus have already synapsed in the gracile nucleus. Internal arcuate fibres from the cuneate and gracile nucleus are crossing to form the medial lemniscus. Second-order fibres from the spinal trigeminal nucleus are extending towards the midline. They will cross, some forming a component of the medial lemniscus and others mixing with the spinothalamic fibres. The spinothalamic tract is in the lateral tegmental field. The posterior spinocerebellar tract is lateral to the spinal trigeminal tract and will enter the inferior cerebellar peduncle rostral to this level. (Modified from Patton *et al.* (1976).)

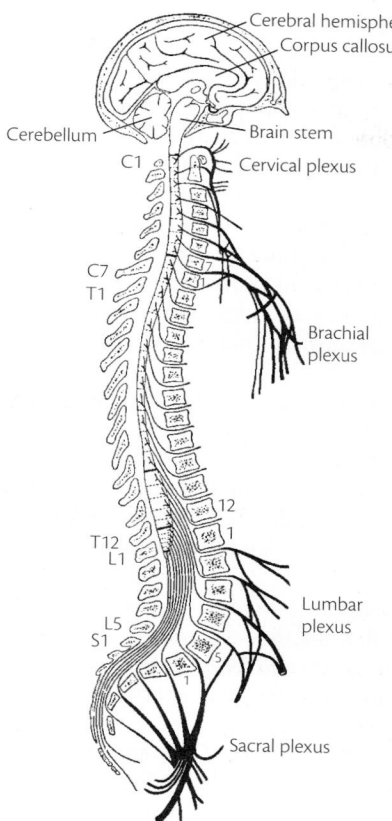

Fig. 2.44 Drawing of the brain and spinal cord *in situ*. The brain is shown sectioned in the median plane. Although not illustrated, the first cervical vertebra articulates with the base of the skull. The letters along the vertebral column indicate cervical, thoracic, lumbar, and sacral. Note that the cord ends at the upper border of the second lumbar vertebra. (Reproduced from Gardner (1975).)

Table 2.8 Mytomes of use in localizing spinal nerve root lesions

Spinal segment	Muscle
C5	Deltoid, spinati
C6	Brachioradialis
C7	Triceps, extensor digitorum
C8	Flexor digitorum profundus
T1	Dorsal interossei
L1, L2	Iliopsoas
L3	Thigh adductors
L3, L4	Quadriceps
L4	Tibialis anterior
L5	Extensor hallicis longus
S1, S2	Gastrocnaemius, soleus

Spinal roots. The myotomal pattern of weakness occurring with spinal nerve root lesion is usually accompanied by dermatomally distributed sensory disturbance and loss of the tendon reflex subserved by that segment. In reality most muscles receive spinal root innervation from at least two segments, and single nerve root lesions usually produce relatively mild degrees of weakness (Table 2.8). Furthermore there is some variation in the exact spinal segment which makes the major contribution to a muscle, depending upon whether the brachial or lumbosacral plexus is pre-fixed or post-fixed (Section 22.5.1).

Peripheral neuropathy. A detailed account of the muscles innervated by each peripheral nerve of the arm (Section 22.9) and leg (Section 22.10) is given elsewhere. Polyneuropathy produces weakness which is symmetrical and predominantly distally located. Additional weakness of proximal muscles may occur in acquired demyelinating polyneuropathies affecting proximal nerve segments and roots. Lesions of the brachial or lumbosacral plexus produce patterns of weakness, and sensory and reflex loss, which cannot be accounted for by lesions either of an individual spinal nerve root or an individual peripheral nerve (Sections 22.5 and 22.6).

Primary muscle disease. This disease characteristically affects the proximal limb muscles symmetrically. Various forms of muscular dystrophy (Section 24.2) pick out specific muscle groups, facio-scapulolumeral, and oculopharyngeal dystrophies being examples. Myasthenia gravis (Section 24.10.1) may present with proximal limb muscle weakness, with pharyngeal and palatal weakness, or with ptosis and eye movement abnormalities; characteristically muscle bulk will be preserved, reflexes are normal, and fatigability is demonstrable. Proximal muscle weakness also occurs in the Lambert–Eaton myasthenic syndrome (Section 24.10.2), but there are associated autonomic features such as dry mouth. Furthermore, the tendon reflexes, although initially absent, show post-tetanic potentiation in which a reflex reappears when retested after sustained maximal contraction of its muscle. Neck extensor muscle weakness is uncommon and occurs in myasthenia, motor neurone diseases, myotonic dystrophy, and some myopathies. Muscle and neuromuscular junction diseases do not produce sensory disturbance.

and contralateral hemiparesis and sensory loss affecting vibration sense and joint position (Fig. 2.43). More lateral medullary lesions cause the *Wallenberg syndrome* of ataxia, vertigo, nystagmus, ipsilateral trigeminal (V) territory sensory loss, Horner's syndrome, nystagmus, dysphagia loss of gag reflex (IX and X cranial nerves), and contralateral spinothalamic tract damage with pain and temperature loss below the face (Sacco *et al.* 1993).

2.4.5 Spinal cord lesions

The distinctive clinical features of a spinal cord lesion are paralysis below the level of the lesion with signs of upper motor neurone damage, impaired sphincter control, and a sensory level. It should be noted that the spinal cord is shorter than the vertebral column and thus the segmental localization of a lesion will correspond to a higher vertebral level (Fig. 2.44). If the dorsal columns are substantially affected, there will be gait ataxia, Rombergism, and altered joint position and vibration sensations in the feet. At the level of the lesion, the tendon reflex arc may be interrupted at the level of the lesion and muscle wasting and weakness may reflect anterior horn cell destruction.

2.4.6 Neuromuscular disease

Lower motor neurone lesions produce patterns of weakness which reflect pathology localized to the anterior horn cells, a spinal nerve root, or an individual peripheral nerve.

2.4.7 Fluctuating weakness

Not uncommonly patients demonstrate momentarily fluctuating, inconsistent, or collapsing patterns of weakness. There are three possible causes of this: loss of sensory feedback, pain or, most usually, psychological factors. Fluctuating weakness due to loss of kinaesthetic feedback usually affects the hand, and is often associated with pseudoathetosis. Normality of underlying muscle power can be demonstrated by asking the patient to look at their hand so as to provide feedback whilst making a simple elementary movement such as abducting the index finger. This will reverse weakness which had been apparent during a more complex movement such as spreading all the fingers apart. Collapsing weakness due to pain in a joint may be accompanied by complaints of discomfort; the raw power of the muscles can be assessed by instructing the patient to 'push as hard as you can just for a moment when I count to three'. Psychologically determined weakness (Section 4.8.1) fluctuates, is inconsistent, may be improved temporarily by firm encouragement, and is discordant with obviously better use of the limb during natural activities such as dressing. If due to malingering, it may be associated with theatrical grunting and sighing in a charade of effort, and the collapsing element may occur more from the trunk than from the limb itself.

2.5 Somatosensory abnormalities

2.5.1 Sensory symptoms

Patients express their symptoms of somatosensory dysfunction in a multitude of ways and only careful enquiry by the neurologist will determine their likely pathophysiological relevance.

Numbness. This is a term that is used confusingly. Most doctors mean by it 'a loss of sensation', but many patients really mean weakness or clumsiness. It is less ambiguous to ask about 'deadness [or loss] of skin sensation'. Polyneuropathy produces numbness in a glove and stocking distribution. When patients describe numbness and/or pins or needles extending on to the trunk, it is most commonly due to myelitis, an inflammation of the spinal cord, which may occur as part of multiple sclerosis. Some patients with numb feet describe a feeling of walking on cotton wool. Those with numb hands may feel as though they are touching things through a plastic bag.

Paraesthesiae. These are spontaneous abnormal sensations, most usually described as 'pins and needles'. They may be physiological, especially during hyperventilation, but in such cases they are generalized, especially periorally, and only intermittent. Continuous paraesthesiae are an important indicator of acquired, rather than congenital disease of the nervous system. They are particularly likely in idiopathic demyelinating polyneuropathy, and less common in lesions of central sensory pathways. Attacks of focal paraesthesiae can occur in focal epilepsy of the sensory cortex. Focal paraesthesiae, occurring intermittently, are common in compressive mononeuropathy, a common example being the finger paraesthesiae at night and after hand usage in carpal tunnel syndrome.

Dysaesthesiae. These are unpleasant distorted sensations resulting from actual sensory stimuli. Usually they occur in focal peripheral nerve damage or polyneuropathies that involve axonal degeneration.

Spontaneous pain. This can occur in association with paraesthesiae in peripheral nerve disorders. It is a particular and early feature in the sensory territory of a nerve affected by vasculitis. Lancinating pain radiating in a dermatomal distribution down a limb, like an electric shock, suggests spinal nerve root compression by prolapsed intervertebral disc. Painful disorders of an internal viscus, joint, or muscle may be referred to an area of skin either overlying or remote to the abnormality. Spontaneous pain in the limbs, trunk, or face

can arise from posterior thalamic lesions and has a particularly unpleasant burning character, often with 'tearing' or 'grinding' qualities. Similar sensations of continuous burning, warmth, or cold may result from a spinothalamic tract lesion, but are clearly localized to within an area of altered skin sensation. Spontaneous pain occurs in causalgia and complex regional pain syndrome (Section 17.5).

Analgesia and thermoanaesthesia. Reduced ability to feel pain and temperature sensations occurs in peripheral neuropathies affecting unmyelinated and small myelinated fibres and in lesions of the spinothalamic tract and posterior thalamus, including syringomyelia. Painless burns, or unfelt wounds may occur in the analgesic area.

Lhermitte's symptom. It consists of an electric shock or strong paraesthesiae radiating down the trunk, and often into the limbs, on sudden flexion of the neck. It is particularly common in myelitis due to multiple sclerosis, and also occurs in cervical spondylitic myelopathy, vitamin B_{12} deficiency, and some sensory neuropathies involving both the central and peripheral axons of the dorsal root ganglia.

Tight bands and size distortions. Tight bands and size distortions, such as feeling that the toes are swollen, are abnormal sensations occurring in the fingers and feet of patients with acquired demyelinating polyneuropathy and lesions of the dorsal columns in the spinal cord.

Clumsiness and Rombergism. These are due to loss of kinaesthetic feedback via large myelinated peripheral nerve sensory fibres or the dorsal column-medial lemniscus system. Hand clumsiness particularly affects the complex motor–sensory integration involved in activities such as doing up buttons or underwear clips, particularly when the eyes cannot monitor the action. 'Dropping things' may be an associated complaint, however this symptom in isolation rarely denotes disease. Loss of joint position sense from the legs causes gait unsteadiness. This is particularly noticeable when the patient tries to walk in the dark or close their eyes in the shower and is known as Rombergism.

Astereognosis. This occurs in patients with parietal lobe lesions. They may complain of being unable to identify coins manually in their pocket or with their eyes closed.

2.5.2 Sensory examination

Many complex and time-consuming methods have been described for semiquantitative assessment of different sensory modalities. In most clinical situations these make little or no extra contribution to diagnosis over and above what can be achieved by simple testing of superficial skin sensation using the fingertips or a pin, by routine testing of vibration, and joint position sense, and by Romberg's test. What is important is to examine each patient with a clear strategy for resolving the diagnostic hypotheses. The neurologist must instruct the patient clearly in how to respond to stimuli, and must present these stimuli unambiguously. In general, superficial sensation is assessed by taking the patients' view as to whether the stimulus 'feels normal'. By contrast, joint position and vibration sensations can be assessed more objectively by using the principles of blinding, in which the patients' eyes are closed, of forced choice in which the patient has to respond 'up' or 'down' or of time-locked response in which the patient has to say 'now' immediately the tuning fork stops vibrating.

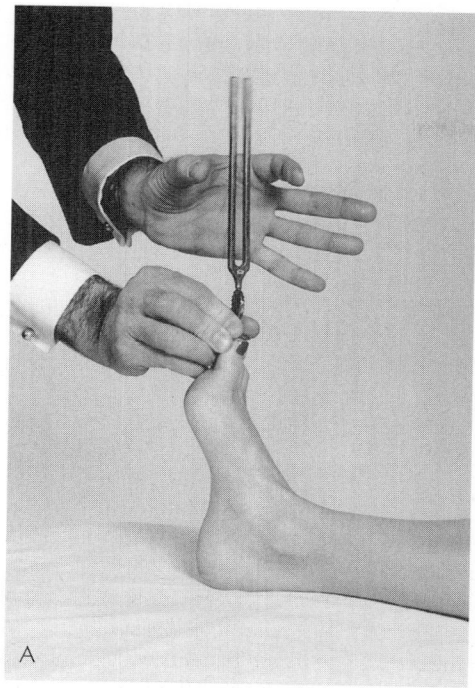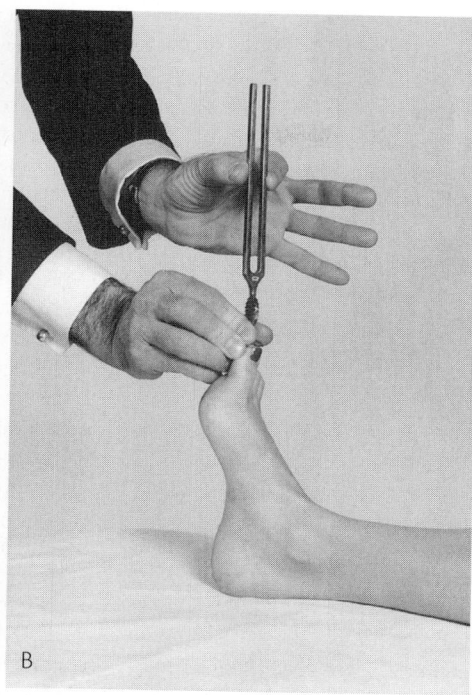

A   B

Fig. 2.45 Testing vibration sense. (A) The fingers of one hand are positioned ready to (B) stop the tuning fork vibrating and the patient is instructed to say 'now' when the tuning fork stops buzzing.

Superficial skin sensation. The boundary of an area of sensory loss is mapped best by starting within the numb area and working outwards until the normal area is reached. Traditionally, an unused pin or a wisp of cotton wool are recommended. However, neither patients nor doctors are familiar with the thresholds for such sensations on different parts of the body. This can make it difficult for the patient to report whether the quality of sensation is altered, unless there is a clear boundary; it is rare for skin sensation to be completely lost. However, everyone is familiar with the feeling of fingertips on every part of their body skin and patients can tell you instantly whether the 'finger feels normal' when you lightly stroke any patch of skin. Furthermore, the examiner can use both his forefingers to present simultaneously comparable stimuli to the two sides of the patient's body. Thus the use of fingertip stroking is recommended for routine testing of superficial sensation and will reveal spinothalamic abnormalities affecting tickle or dorsal column pathway abnormalities impairing pressure sensation. Only rarely it is necessary to test temperature sensation; if so the cold metal of a tuning fork generally provides a sufficient stimulus. A tube of warm water can be used to test warm sensation in those rare occasions when it is necessary to test the unmyelinated thermal fibres in isolation. Occasionally superficial sensory testing produces hyperpathia, in which any background loss of sensation is overshadowed by an abnormal, and sometimes unpleasant, additional quality to the sensory experience. In such situations it can be difficult to determine which is the abnormal side, because of this apparent heightening of sensation.

Vibration sensation. This should be tested using a 128-Hz tuning fork struck in such a way that it does not produce audible high frequency harmonics that can be heard rather than felt by the patient. Place it on the patient's sternum and ask 'can you feel it buzzing?' Then move it to the tip of the great toe, or a finger, and ask the patient to 'close your eyes, and tell me as soon as it stops buzzing'.

The patient should respond promptly when the examiner stops the buzzing prongs with the fingers of their other hand (Fig. 2.45).

If vibration sense is absent from the toes, testing should be repeated more proximally on the ankle malleolus, tibial bone, knee, anterior superior iliac crest, and finally the rib cage. Vibration sense abnormalities usually mean there is a polyneuropathy affecting large myelinated sensory fibres, or a spinal cord lesion. Vibration sense is not affected by a lesion restricted to the somatosensory cerebral cortex. Vibration sense is lost from the lower legs of many elderly people as a natural ageing phenomenon.

Joint position sensation. With the patient's eyes open, move the great toe up and then down, showing to the patient 'this is up and this is down'. Then ask the patient to close their eyes and identify small movements (Fig. 2.46). The distal interphalangeal joints of the fingers can be tested similarly. Usually joint position sensation is lost in similar conditions to vibration sensation. It is particularly likely to be abnormal in patients with sensory ataxia or Rombergism. But, unlike vibration sensation, joint position perception is also lost in lesions of the somatosensory cortex.

Romberg's test and pseudoathetosis. These provide evidence of loss of kinaesthetic sensory feedback from the legs and hands respectively. They are abnormal in disorders of large myelinated sensory peripheral nerve fibres and of the dorsal column —medial lemniscus system. Romberg's test has been described previously (Section 2.2.2). Pseudoathetosis is demonstrated by asking the patient to close their eyes and extend their hands and fingers in front of them. The fingers and wrist 'wander' slowly in random directions, of which the patient is unaware (Fig. 2.47).

Sensory inattention. This occurs with lesions of the parietal lobe insufficient to cause gross cortical sensory loss, but sufficient to cause perceptual rivalry between the two sides of the body. The patient is able to appreciate stimuli when applied simultaneously to both sides of the body. However, when two similar stimuli are

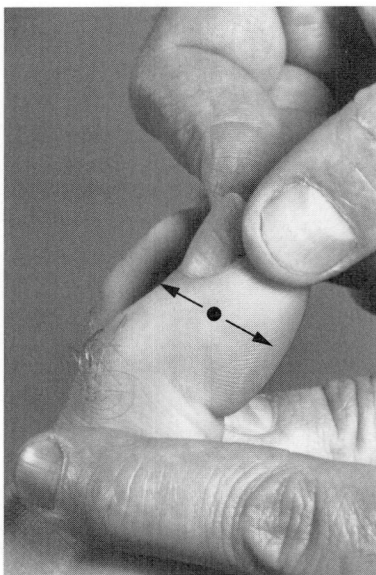

Fig. 2.46 Testing joint position sense at the great toe. The proximal phalanx is steadied with one hand while the toe is moved up and down.

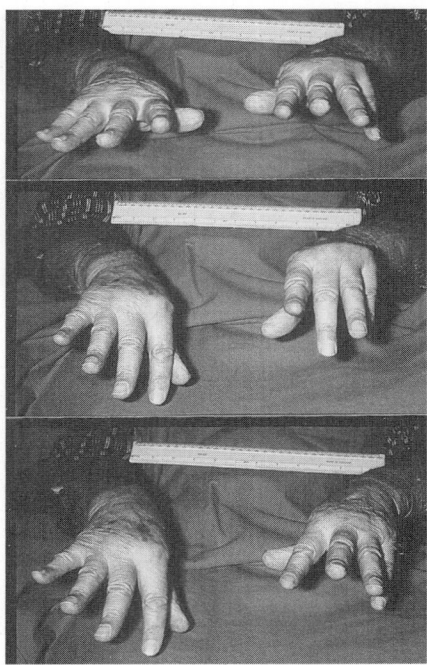

Fig. 2.47 Pseudoathetosis in a patient with sensory ataxic polyneuropathy. Frame intervals at 30 s.

applied simultaneously to the same skin area on each side, one side will be ignored. This finding implies a disturbance in function of the sensory area of the contralateral cerebral cortex. The phenomenon does not occur if the interval between the two contacts is more than 3 s (Critchley 1953).

Astereognosis. This is another sign which may occur in patients with lesions of the arm representation of the opposite sensory cortex. They are unable to appreciate the form and texture of objects placed in the hand with the eyes closed. Correctly this should be called *stereoanaesthesia*, but the term *astereognosis* is more often used. Strictly speaking, the latter term should be reserved for failure to recognize objects, such as coins, when the primary sensory modalities are intact. This is an agnosic defect, due to a disorder of sensory association and akin to the other more complex disorders of parietal lobe function.

Two-point discrimination abnormalities and graphaesthesia. These occur in patients with a lesion of the sensory cortex. The threshold for two-point discrimination is much greater on the abnormal than on the normal side, and there may be inability to recognize figures or letters drawn on the skin. Sensory stimuli are also incorrectly localized on the affected side.

2.5.3 **Patterns of sensory loss**

Polyneuropathy. A distal stocking, and later glove, distribution of diminished skin sensation is typical of polyneuropathy (Fig. 2.48). Usually the border between normal and reduced sensation gradually changes over some centimetres rather than being abrupt.

Focal neuropathy. The pattern of superficial sensory loss corresponds to the territory of innervation of the affected peripheral nerve (Fig. 2.49). The border between normal and reduced sensation is reasonably well-defined although not abrupt. Usually the area of sensory loss for touch is larger than that for pain or temperature.

Spinal root lesions. These cause loss of superficial sensation in the corresponding dermatome (Fig. 2.50). In the earlier stages of a

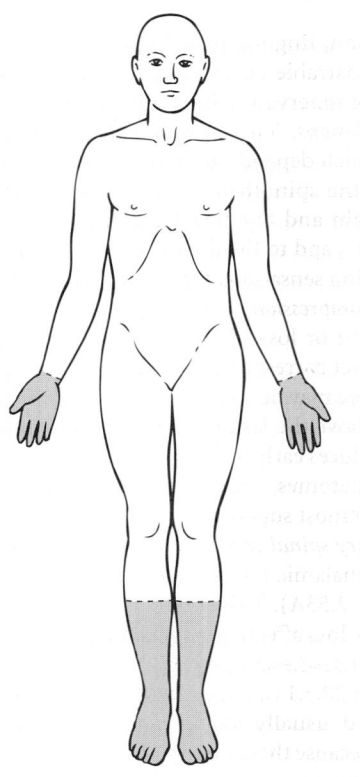

Fig. 2.48 Glove and stocking sensory loss in polyneuropathy.

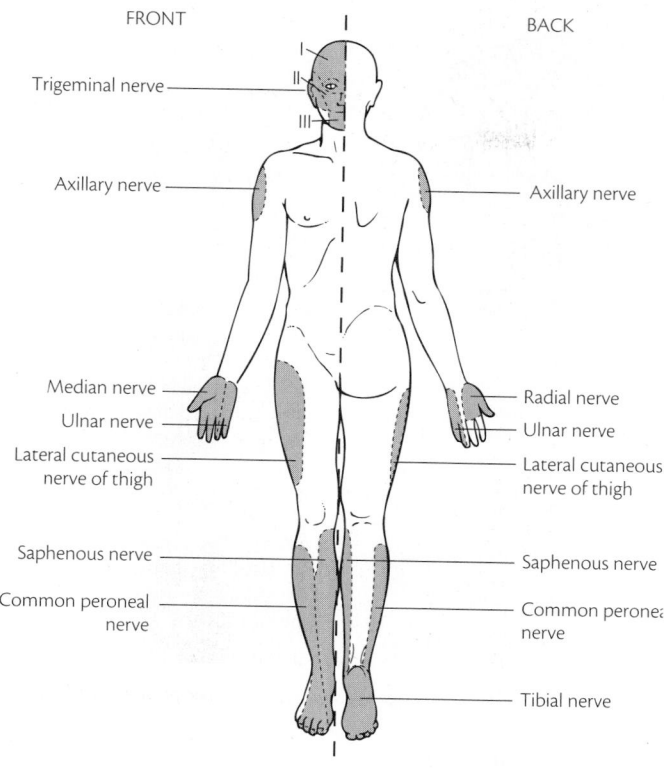

Fig. 2.49 Skin territories of some commonly damaged peripheral nerves. Skin territories of all named cutaneous nerves are shown in Fig. 22.1.

single root lesion, tingling or pain in the dermatome may occur without demonstrable superficial sensory loss, because there is some overlap of innervation from adjacent roots.

Spinal cord lesions. These produce patterns and modalities of sensory loss which depend upon the level of the lesion, the degree of damage to the spinothalamic tracts in the arterolateral cord, which carry pain and temperature sensations from the opposite side (Fig. 2.51A), and to the dorsal columns, which carry vibration and joint position sensations ipsilaterally (Fig. 2.51B).

Spinal cord compression. Spinal cord compression or transection causes reduction or loss of all modalities of sensation below the dermatomal level corresponding to the segment of the transection (Fig. 2.52). There may be a zone of hyperasthesia in the dermatome immediately above the lesion. External compression of the spinal cord often produces early loss of pain and temperature sensation in the sacral dermatomes, because the fibres subserving these lowest segments travel most superficially within the spinothalamic tracts.

Intramedullary spinal cord lesions. These initially affect the decussating spinothalamic tracts within the spinal cord at the level of the lesion (Fig. 2.53A). This produces a 'cape-like' pattern of suspended sensory loss affecting pain and temperature sensations, but not touch or kinaesthesia (Fig. 2.53B); a pattern typical of syringomyelia (Section 28.5.15). Expanding intramedullary lesions within the spinal cord usually spare pain sensation from the sacral dermatomes, because these fibres travel the most superficially in the spinothalamic tracts.

The Brown–Séquard syndrome. This syndrome follows damage to one half of the spinal cord (Fig. 2.54). Below the level of the lesion there is ipsilateral weakness and loss of vibration and joint position sensations and contralateral loss of pain and temperature sensations. Elements of light touch sensation may be preserved bilaterally since it is a composite sensation involving both spinothalamic and dorsal column pathways.

Myelitis. This can produce relatively restricted patterns of sensory loss which characteristically extend onto the trunk, may spare the hand or foot, and may be strictly unilateral (Fig. 2.55). Depending upon the location of the myelitis within the spinal cord, spinothalamic and dorsal column sensations may be differentially affected. Sometimes a Brown–Séquard syndrome is encountered if only one half of the spinal cord is involved. Severe forms of myelitis produce impairment of all modalities of sensation below the level of the lesion.

Foramen magnum. Foramen magnum and high cervical spinal cord compressive lesions can cause loss of vibration sense limited to the arms and upper ribcage if the lesion affects the decussation of the medial lemniscus. External compressive lesions at the foramen magnum may produce symptoms of 'rotating sensory loss' in which sensory symptoms start in one limb, for instance a foot, and later rotate to the other foot and the hand on the same side, before eventually reaching the remaining hand.

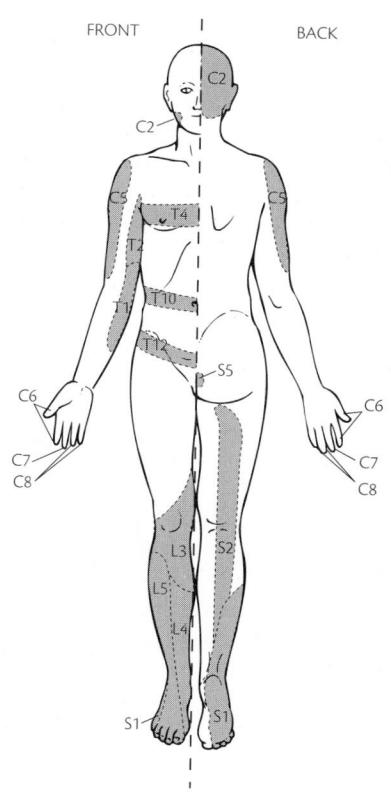

Fig. 2.50 Some memorizable dermatomal landmarks.

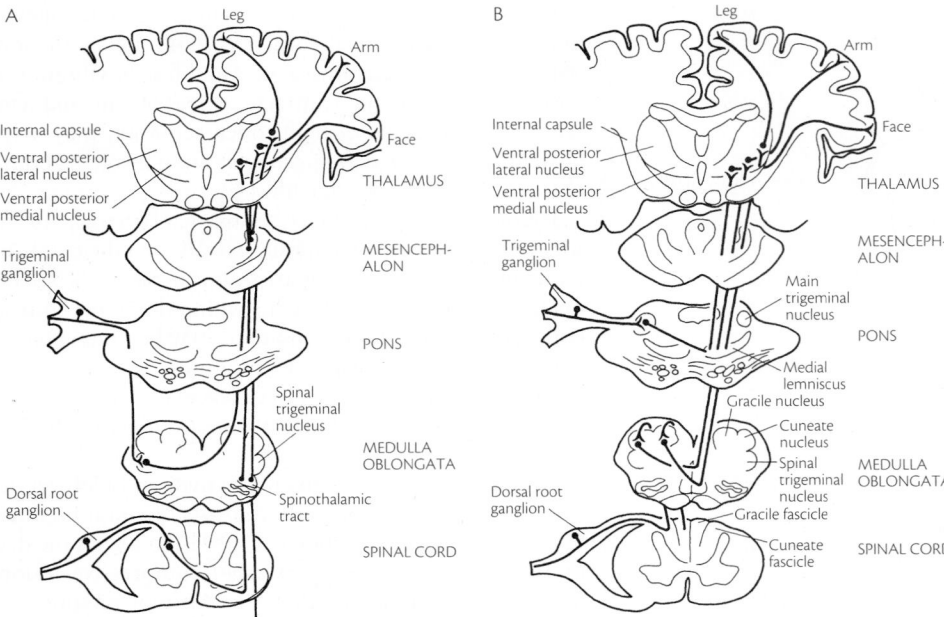

Fig. 2.51 Spinal cord sensory pathways. (A) The spinothalamic tract. This is the main pathway for transmission of signals from nociceptors and thermoreceptors. (B) The dorsal column-medial lemniscus pathway. This is the main pathway for transmission of signals from low-threshold mechanoreceptors. Fibres transmitting impulses from mechanoreceptors in the face join the medial lemniscus in the brainstem.

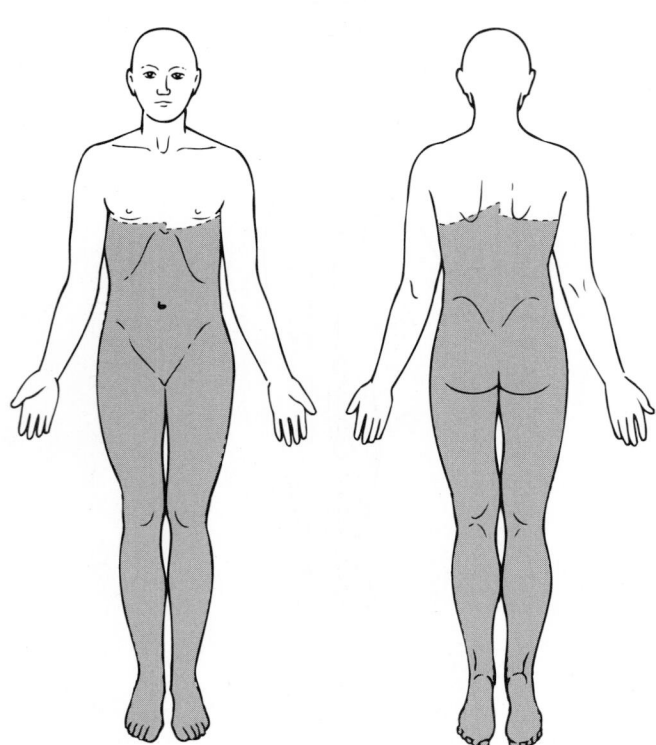

Fig. 2.52 A 'sensory level' in compression of the spinal cord at the T5 level.

Brainstem lesions. These lesions produce sensory abnormalities interpretable on an anatomical basis. Dissociated sensory loss in the face can result from syringobulbia, due to involvement of the descending fibres from the trigeminal nerve. Lesions of the pons and medulla can give facial sensory impairment on one side, due to a lesion of the trigeminal nucleus, with hemianaesthesia and/or hemianalgesia of the trunk and limbs on the opposite side due to involvement of ascending sensory tracts. A lesion of the upper pons or midbrain can give complete contralateral sensory loss. An infarct in the midbrain involving the third nerve nucleus, red nucleus, and medial lemniscus may give a unilateral third nerve palsy with contralateral static tremor, hemianaesthesia known as Benedikt's syndrome (Section 2.4.4), and hemianalgesia. More often such unilateral sensory loss is dissociated, involving only pain and temperature sensation, owing to selective involvement of ascending fibres of the spinothalamic tract, as in the lateral medullary Wallenberg's syndrome due to vertebral or posterior inferior cerebellar artery thrombosis (Section 2.4.4). Occasionally in such cases there is a sensory level on the trunk on the affected side (Matsumoto *et al.* 1988).

Thalamic lesions. These lesions can produce patchy contralateral hemianaesthesia and hemianalgesia. Often there is also spontaneous pain of a peculiar, unpleasant, and disturbing nature on the partially anaesthetic side. Fortunately, this thalamic pain syndrome, usually resulting from infarction, is rare. The discomfort most often affects the face, arm, and foot. Surprisingly, extensive thalamic lesions such as neoplasms usually produce comparatively little sensory loss. Sometimes more anterior thalamic infarcts impair appreciation of posture, passive movement, light touch, and tactile discrimination with little effect upon pain and thermal sensibility. Sharply defined hemisensory loss is an unusual phenomenon, occurring only in lesions of the thalamus or immediately adjacent internal capsule.

Cortical sensory loss. Lesions restricted to the somatosensory cortex characteristically impair joint position sensation and

A

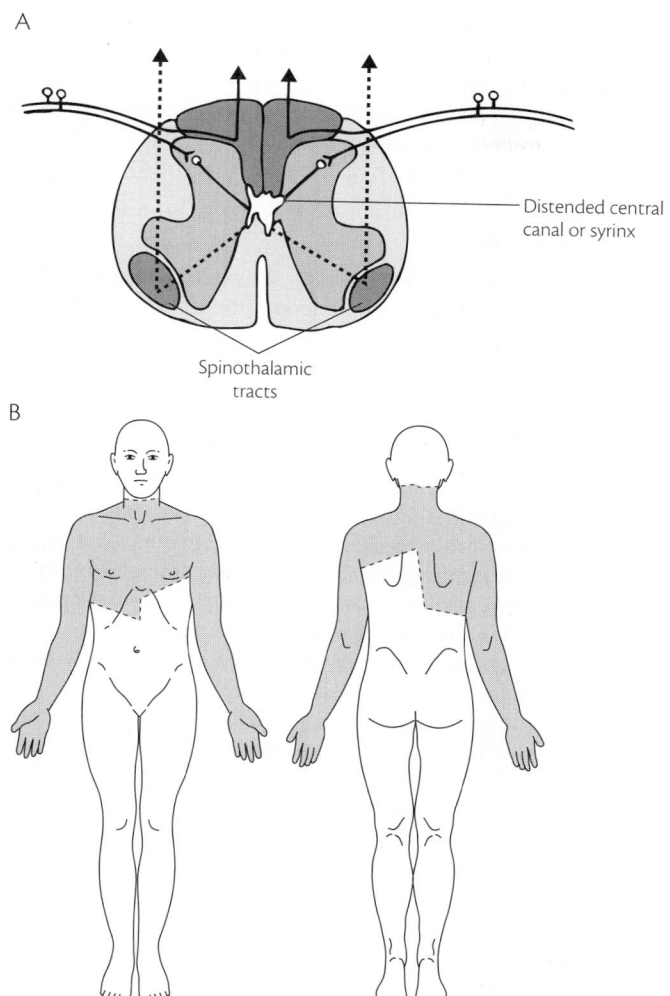

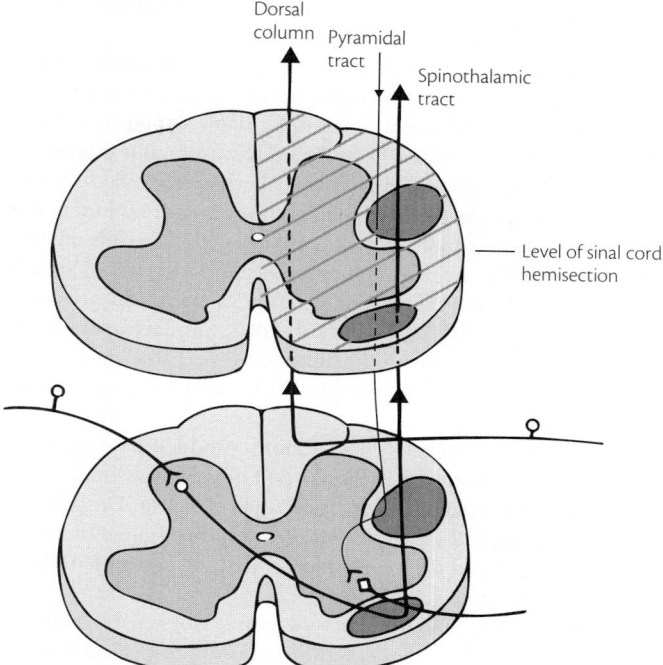

Fig. 2.54 A Brown–Séquard syndrome due to hemisection of the spinal cord. There is ipsilateral weakness due to pyramidal tract damage, ipsilateral loss of joint position and vibration sensations due to dorsal column damage, and contralateral loss of pain and temperature sensation due to spinothalamic tract damage.

Fig. 2.53 Sensory loss in intramedullary spinal cord lesions, such as syringomyelia. (A) Diagram showing how the decussating spinothalamic tract fibres subserving pain and temperature are interrupted by the enlarged central canal or syrinx. The dorsal column fibres transmitting joint position and vibration sense are unaffected. (B) A 'cape-like' pattern of suspended pain and temperature loss in syringomyelia affecting the cervical and upper thoracic segments of the spinal cord.

two-point discrimination, whilst vibration sense is preserved (Gilman 2002). Graphaesthesia is common in cortical lesions and best tested by asking the patient, with closed eyes, to identify numbers inscribed with a stick on the palm of their hand.

Psychologically determined sensory loss. This type of loss (Section 4.8.2) often has implausibly sharply defined boundaries, which may shift in position, and which do not obey anatomical distributions.

2.5.4 Pain

When a patient complains of pain, the neurologist must determine whether it is a neuropathic pain, due to disease directly affecting the nervous system, rather than a musculoskeletal pain. Local tenderness, or discomfort on passively moving a joint, are signs suggestive of disease or strain injuries affecting bones, joints, tendon, muscles, or other organs. Localization of the painful area is diagnostically helpful in spinal root, and cranial or peripheral nerve

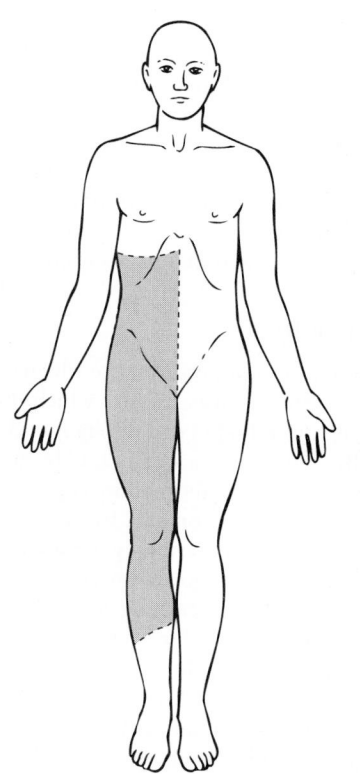

Fig. 2.55 A characteristic pattern of sensory loss in mild myelitis. The exact distribution depends upon the segmental level and pathways of the spinal cord that are affected.

lesions. However, pain may be imprecisely localized in complex regional pain syndrome or cerebral lesions.

Referred pain. Painful lesions of the muscles or viscera sometimes give pain perceived to be in the overlying skin, or in a remote cutaneous area. Such painful sensations seem not to be coming from the viscus involved but from the body surface. But this apparent error in localization, giving what is alternatively called *pseudovisceral pain*, is systematically related to the dermatomes innervated by those dorsal roots that supply the diseased viscus. Thus afferent pain fibres from the myocardium enter the T1-5 dorsal root ganglia and myocardial pain is referred to the anterior chest wall and down the inner aspect of the left or of both arms. Similarly, pain fibres from the diaphragm travel in the phrenic nerve (C3-4) so that diaphragmatic pain is often referred to the C3 and 4 dermatomes in the neck and shoulder.

Cutaneous nerve lesions. These lesions produce pain which is prickly, associated with paraesthesiae and often dysaesthetic to the extent that patients avoid contact or tight clothing. Despite this apparent increase in sensitivity, background thresholds for sensation will be impaired if, for instance, a wisp of cotton is used to compare the two sides.

Spinal nerve root. Spinal nerve root pain is worsened by stretching movements such as the straight leg-raising test in sciatica, by other movements, or by sneezing, all of which increase the degree of compression. The pain radiates in a dermatomal distribution and may be out of all proportion to the degree of demonstrable sensory or motor loss.

Spinal cord and brainstem. These lesions affecting the spinothalamic tracts (Section 17.2) occasionally produce a burning and poorly localized pain associated with demonstrable impairment of pain and temperature sensations.

Thalamic pain. Also known as the Dejerine–Roussy syndrome, this results from lesions within, or just behind, the posterior thalamamus. Appreciation of sensory stimuli is impaired on the contralateral body, but stimuli such as pain, cold, and touch induce marked pain (Section 17.2).

Cortical lesions. These produce pain only rarely. A pseudothalamic burning pain syndrome can result from damage to the deep part of the Sylvian cortex and may be associated with hemiplegia.

2.6 Gait disorders

Walking is a complex motor performance which can be affected by a wide variety of different diseases, and which undergoes natural change with aging. Everybody's gait differs in a characteristic way; one can normally recognize an acquaintance by her walk even when too far away to see her face. Yet mammals generate the rhythmic muscle activities necessary for gait by specialized spinal cord circuits, known as central pattern generators for locomotion (Duysens *et al.* 2000). These locomotor-generating circuits receive important feedback first concerning loading and unloading of limbs to control the onset of limb swinging, and second of hip position to initiate limb swing. Even when these circuits are disconnected from forebrain control, as in much spinal cord injury, robotic devices coupled with a treadmill can be used to retrain a degree of locomotor ability (Dietz *et al.* 2002).

Non-neurological disorders, such as hip joint disease are important in the differential diagnoses of gait abnormalities. Particularly in the elderly, gait disorders may be multifactorial, for instance involving both a neurological disease such as Parkinsonism, the mild gait apraxia of natural senescence, and degenerative arthritis of the hip joints. Gait disorders are conveniently classified hierarchically into low-, middle-, and high-level gait disorders, reflecting the level of the nervous system lesion (Nutt *et al.* 1993). Psychogenic gait disorders are immensely variable (Section 4.8.4) and usually do not result in injury if falls occur.

2.6.1 Normal gait

Normal walking requires equilibrium, so as to maintain a balanced upright posture, coupled with locomotion (Nutt *et al.* 1993). Equilibrium involves various righting reflexes for assumption of the upright posture, antigravity supporting reactions to maintain that posture, postural reflexes to maintain balance during weight transfers, rescue reactions to avoid falling if postural reflexes prove inadequate, and protective reactions if all else fails and falling starts. Locomotion involves gait ignition mechanisms followed by rhythmic stepping. Unsurprisingly, this complex motor task is controlled by many different brain regions. The brainstem appears to have a particular role in righting reflexes, synergizing proximal and trunk muscles to maintain balance, and in gait ignition. The basal ganglia are not involved in the rhythm of walking, but lesions impair postural responses and gait ignition. Chronic cerebellar lesions do not eliminate equilibrium reactions or alter their pattern, but do alter the scaling of responses, generally rendering them too large. The frontal cortex is important for postural responses, and cortical mechanisms are responsible for executive control of whether, when, where, and how fast to walk.

Bedside diagnosis of disordered gait requires analysis of how various features of the patient's walking differ from normal:

Separation of the feet. Normally the feet cross within a few centimetres of each other during a stride leading to occasional scuffs on the inner heel of shoes. In normal efficient gait, the feet loop slightly round each other so that the footprints of the two feet lie almost in a straight line (Fig. 2.56A). A wide-based gait occurs in cerebellar and sensory ataxias (Fig. 2.56B), probably as a compensation to preserve balance and so the feet do not catch on one another when crossing on dysmetric strides. A slightly wide-based gait can occur too in apraxia (Fig. 2.56D) or Parkinson's disease (Fig. 2.56C), but the stride length is short in those conditions. The feet cross in wide arcs in spastic gaits, because the stiffly extended leg and foot cannot be slightly flexed in the normal manner so as to clear the ground during a stride (Fig. 2.56F).

Stride length. People normally hit their standard stride length with their first step and it varies little thereafter (Fig. 2.56A). In ataxia the stride length varies, elongated strides may lead to overbalancing behind a foot which has landed too far ahead, or sometimes shortening causes a stumble over a foot which has gone down too soon (Fig. 2.56B). Shortened, shuffling strides occur in Parkinson's disease. The gait ignition failure of more advanced Parkinson's disease leads to a hesitant start to a walk, and acceleration associated with increasing stride length (Fig. 2.56C). Gait apraxias, or frontal gait disorders, cause short striding of constant length from the start, a *marche à petits pas* (Fig. 2.56D). Such patients have particular difficulty in turning corners, being unable to spin round at a single step, and sometimes getting their feet hopelessly tangled up, or having to walk around corners. (Fig. 2.56D).

Foot drop. Foot drop interferes with normal gait because it prevents the patient from swinging their striding leg through underneath

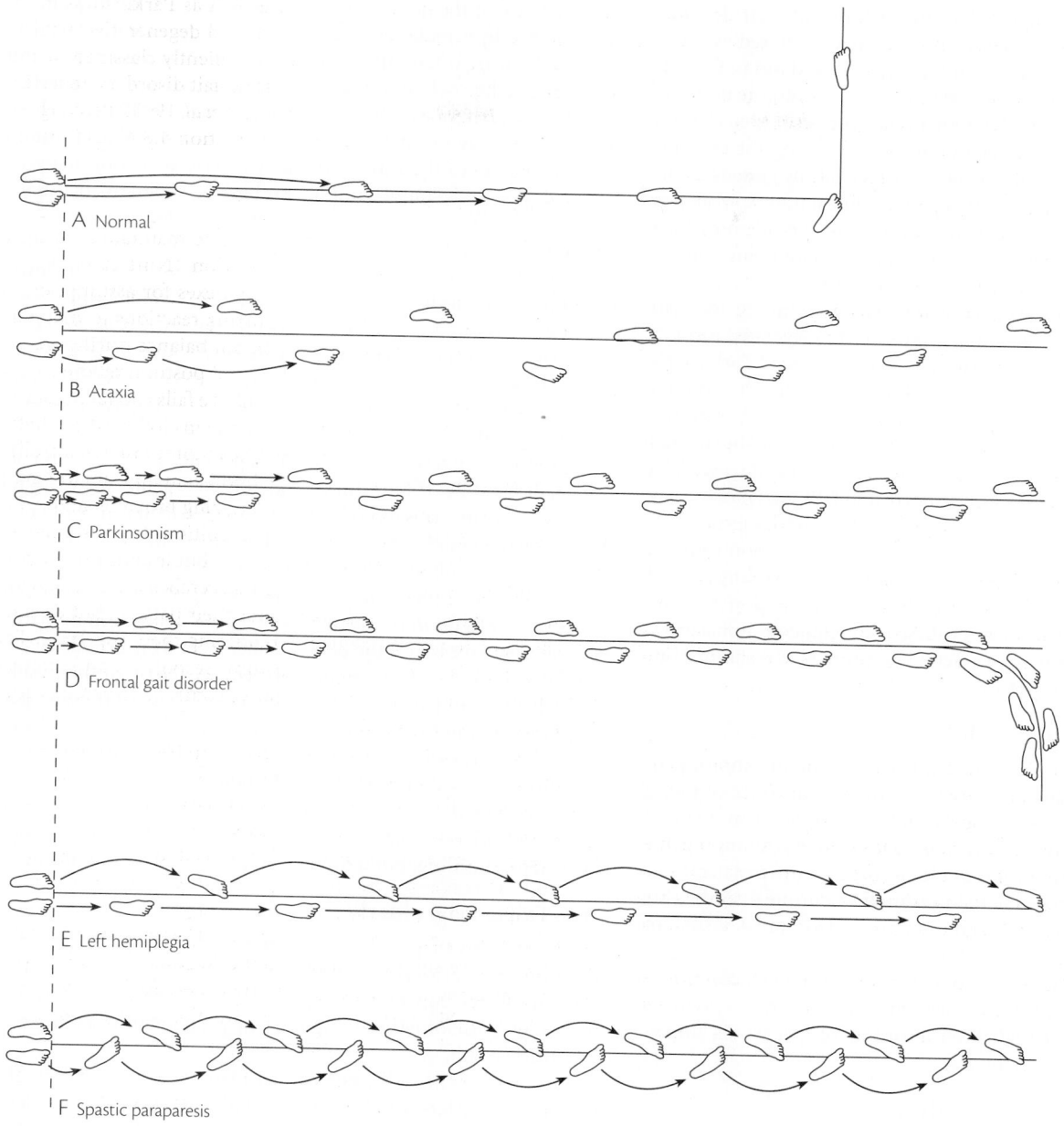

Fig. 2.56 Stride patterns in various gait disorders. (A) The stride pattern in normal gait showing how a fixed stride length is hit immediately, and varies little thereafter, how the feet cross close to one another, and how a corner is taken in a single step by spinning on the foot through a right angle. (B) Ataxia showing wide separation of the feet and an irregular stride length. (C) Parkinsonism showing short strides, with a shuffling or festinant start. (D) Frontal gait disorder, gait apraxia, showing the short steps or *marche à petit pas*, and walking round corners, rather than spinning round them. (E) and (F) Showing a left hemiplegia and spastic paraparesis, respectively, with circumduction of the spastic leg(s).

their body without it catching on the ground. Stiffly held foot drops usually result from spasticity, less often from dystonia. Walking with a spastic leg requires laborious circumduction of the stiffly extended leg and foot in a wide lateral arc so as to avoid catching the toe (Fig. 2.56E), a compensatory manoeuvre required bilaterally in spastic paraparesis (Fig. 2.56F). This scissors gait is exhaustingly inefficient and throws unnatural stresses upon the low back and pelvic girdle which can result in further gait deterioration due to secondary degenerative arthritis. Circumduction of the striding leg is not so marked in dystonic foot drop because usually the patient retains the ability to lift the foot by flexing the leg at the hip and knee and raise the pelvis whilst striding. Floppy foot drops occur with lower motor neurone weakness of tibialis anterior due to motor neurone, spinal root, or peripheral nerve diseases.

In order to swing the dangling foot through during a stride without catching it on the ground, the hip and knee are flexed exaggeratedly, with lifting of the pelvis, and the leg is kicked out in front at the end of the stride before being stamped down onto the ground.

Pelvic tilt. Normally the pelvis remains horizontal, or even elevates slightly, when the weight is taken by only one leg during a stride. This stabilization of pelvic height is achieved principally by gluteus medius contraction on the weight-bearing side. Gluteal muscle weakness will cause the pelvis to flop downwards towards the side on which the leg has been lifted to stride out, giving the gait a waddling appearance. This is the typical gait of proximal myopathy.

Arm swing. Normally balance is assisted by swinging the arm opposite to the striding leg, familiar in its most exaggerated form in military marching. Loss of arm swing is characteristic of Parkinson's disease, occurring unilaterally in early disease. The physician should not make the patient aware that it is their arm swing which is of particular interest. However, later in the examination it should be ensured that the shoulder is not immobile simply due to joint disease. Arm swing of irregular amplitude, sometimes of a slightly wild nature, may occur in ataxia, in part as a balance compensatory mechanism.

Rising and standing. Rising and standing depend upon righting reflexes and supporting responses respectively. Withstanding pushed displacements demonstrates reactive postural responses. The response to larger displacement pushes, or spontaneous imbalances, demonstrates the integrity of rescue and protective reactions, but should be tested with care to avoid injury.

2.6.2 Low-level gait disorders

Central nervous system equilibrium reactions and locomotor generation are well able to compensate for the gait disorder resulting from disorders of the musculoskeletal or peripheral motor and sensory systems. The musculoskeletal gait disorders of limping due to arthritis, prosthetic limbs, waddling due to proximal muscle weakness, and foot drop due to lower motor neurone disorders are easy to recognize. Ataxia due to loss of proprioceptive, visual, or vestibular feedback usually lead to cautious locomotion with efficient compensatory maintenance of equilibrium. Such conditions produce a major threat to independent walking when associated with other middle- or high-level disorders of gait which impair these compensatory mechanisms.

2.6.3 Middle-level gait disorders

The mechanisms for maintaining equilibrium and implementing locomotor control by the cerebral cortex are distorted by disorders of the pyramidal tract, cerebellum, and basal ganglia, and superimposed movement disorders such as chorea or dystonia. For instance cerebellar ataxia causes dysmetria of striding and postural adjustment mechanisms. Early Parkinson's disease and pyramidal tract lesions impair postural responses. It is only when very severe, or conjoined with another gait disorder, that walking ability is completely abolished by middle-level disorders.

2.6.4 High-level gait disorders

These are the least well-understood gait disorders and have attracted varying and confusing descriptive labels. They are defined as disorders of those high level processes, presumably cortical and subcortical in the frontal lobes, responsible for selecting postural responses and locomotor behaviour suitable for the particular circumstance of the patient at that time. By definition they cannot be explained by middle-level gait disorders. In general neurological practice they are often called gait apraxia, *marche à petits pas*, and sometimes atherosclerotic Parkinsonism, but the precise manifestations and disabilities clearly vary widely between patients. These disorders usually occur in the elderly, or in those with acquired bilateral disease of the frontal lobes, due to ischaemia or demyelination. Five useful clinical components have been described, usually some overlap of features is evident (Nutt *et al.* 1993):

Cautious gaits. These are common in the elderly who are often conscious of disequilibrium and a real risk of falling; they walk slowly with small steps, a slightly wide base, and walk carefully round corners. Cautious gait occurs in mild dementia (Jankovic *et al.* 2001).

Subcortical dysequilibrium. This is associated with poor or aberrant postural responses such as neck extension and backwards falling, and are often associated with oculomotor abnormalities, dysarthria, or extrapyramidal signs.

Frontal dysequilibrium. This impairs a patient's ability to stand up, to remain standing or sitting independently, to organize their leg and trunk movements so as to bring their feet under their centre of gravity, or to avoid tangling their feet up when attempting to corner.

Gait ignition failure. This is familiar as part of Parkinsonism. It can be an isolated phenomenon with freezing attacks provoked by diversion of attention or narrow spaces, such as an open doorway. Walking is often easier if the patient concentrates on a rhythmically recurring feature, such as paving stone cracks, the tip of their walking stick, or simply by counting. Primary progressive freezing gait is a manifestation of various underlying neurodegenerative disorders and progresses within a few years to postural instability and eventual loss of walking (Factor *et al.* 2006).

Frontal gait disorders. These result from multiple forebrain lesions and involve varying combinations of a wide-based gait, short steps, shuffling, initiation hesitations, and dysequilibrium.

2.6.5 Falls in the elderly

Thirty per cent of people aged over 65 fall each year, a quarter of whom are seriously injured. One in twenty fracture bones, particularly the hip (Tinetti *et al.* 1988). These injuries are an important cause of either temporary or permanent disability and sometimes death.

Gait disorders are a leading cause of such falls, particularly the high-level disorders which are particularly common in the elderly (Sudarsky 1990; Williams *et al.* 2006). Many such gait disorders are slowly progressive, often presumably due to non-specific neurodegenerative disease which merges with normal age-related changes in gait. Stepwise progression of a gait disorder points to cerebrovascular disease and interventions which may prevent further events should be considered. The gait disorder associated with Parkinson's disease may improve with dopaminergic treatment although the associated impairment of postural adjustment mechanisms often responds poorly. Patients with signs of spinal cord disease should be investigated for potentially operable compressive lesions. Rapidly evolving high-level gait disorders should be investigated for potentially operable structural cerebral disease such as frontal tumours, subdural haematoma, obstructive hydrocephalus, or normal pressure hydrocephalus.

Drop attacks, postural hypotension, inadequate vision, poor illumination, and trips over environmental hazards are the other important cause of falls in the elderly (Sheldon 1960). The common complaint of unsteadiness is often associated with demonstrably impaired vestibular function (Fife and Baloh 1993). Suddenly occurring Tumarkin falls associated with vertiginous episodes can occur with or without the Menière syndrome (Ishiyama *et al.* 2003). *Drop attacks* are momentary losses of postural tone whilst standing, without loss of awareness or consciousness, and usually the patient is aware of the fall before hitting the ground. More than 80 per cent of such patients have become free of drop attacks at a mean follow-up of 6.5 years, whether or not there is an associated medical condition (Meissner *et al.* 1986). The average number of attacks experienced is 11 and they are not a predictor of stroke, suggesting that their pathogenesis is not vascular. Postural hypotension is another important cause of falls in the elderly, often being precipitated by antihypertensive and vasodilator drugs, and occasionally being due to peripheral or central autonomic failure.

2.7 Autonomic disorders

2.7.1 Clinical features

Generally symptoms of autonomic nervous system disease develop insidiously and reflect loss of function, or failure of autonomic regulation. Occasionally paroxysms of *autonomic hyperactivity* occur in diseases such as Guillain–Barré syndrome (Section 21.10.1), causing wide fluctuations in blood pressure and heart rate which predispose to cardiac arrhythmias, episodic skin flushing, sweating disturbances, paralytic ileus, pupil abnormalities, and micturition disturbances. *Autonomic dysreflexia* occurs in high spinal cord lesions. Stimuli originating in the skin, muscles, or internal organs below the level of the lesion lead to hypertension, bradycardia, and sweating. Bladder distension is a noteworthy precipitant of autonomic dysreflexia.

The early symptoms of *autonomic failure* can be relatively unnoticed for years because of compensatory mechanisms. Persistent postural hypotension when standing leads to recurrent faintness, dizziness, episodes of vision draining away and blacking out, feelings of weakness, aching across the shoulders and posterior neck, or attacks of abrupt unconsciousness with falling. Food, alcohol, drugs, hot baths, or exercise can provoke these symptoms. These symptoms are reversed by lying down. Gastrointestinal autonomic dysfunction causes either nocturnal diarrhoeal attacks or pseudo-obstruction. Such obstruction may be generalized, or localized as in achalasia of the cardia or Hirschprung's disease; commonly laparotomy has been performed to exclude mechanical bowel obstruction. Autonomic denervation of the bladder and genitalia impairs voiding, causes erectile impotence, and can allow retrograde ejaculation. Patients are usually unaware that they have lost sweating, but direct questioning can reveal that the palms no longer sweat and the finger skin has lost that moist adhesive quality required for effective gripping of paper. An autonomically denervated foot becomes warm and red due to loss of cutaneous vasoconstriction, and dry due to loss of sweating; the resultant lack of skin lubrication may contribute to cracking, fissuring, and ulcer formation.

Some autonomic disorders are focal, examples being Horner's syndrome (Section 13.3.5), Adie's pupil (Section 13.3.4), gustatory sweating, idiopathic palmar or axillary hyperhidrosis or surgical sympathectomy.

2.7.2 Autonomic function testing

A wide variety of possible investigations of autonomic function have been described (Mathias and Bannister 1999). Most of these are too complex for routine clinical use. They include thermoregulatory and sweat testing, gastrointestinal function testing, sympathetic skin responses, measurement of noradrenaline and renin responses to head-up tilt, urodynamic studies and sphincter electrophysiology, and penile plethysmography.

Simple clinical tests for autonomic failure include:

Sweating. Lack of moistness may be noted on the palms and soles of patients with peripheral neuropathies involving autonomic fibres. Sweating can be provoked by putting a hand in a plastic bag to prevent evaporation and warming it under a light for a few minutes. Indicator dyes, such as Ponzo red, which turn red on becoming wet, can be dusted onto the skin, or placed under pieces of transparent tape stuck to the skin.

Lying and standing blood pressure. Normally on standing the blood pressure is unchanged or rises slightly and the pulse rate increases slightly. In autonomic failure the blood pressure falls and there may be no compensatory tachycardia (Fig. 2.57). Baseline blood pressure tends to fall slightly on repeated testing so it is advisable to measure blood pressure in the sequence lying–standing–lying and to compare the second pair of readings.

Sinus arrhythmia. Normally the heart rate rises during inspiration and falls during expiration. This sinus arrhythmia is mediated

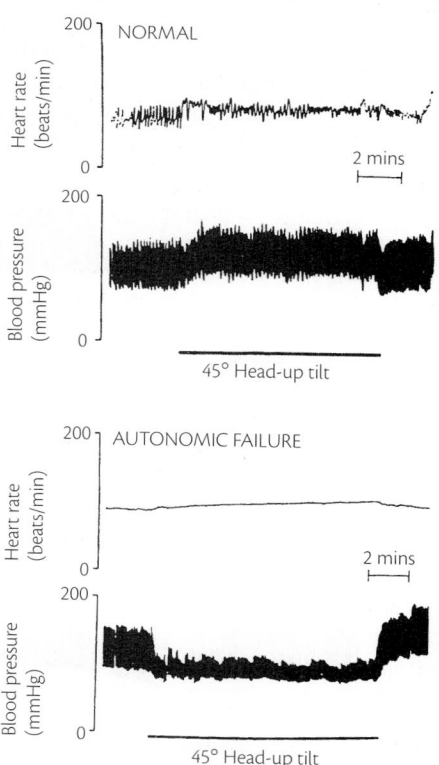

Fig. 2.57 Blood pressure and pulse rate recordings before, during, and after head-up tilt in a normal subject (top) and a patient with pure autonomic failure (bottom). Whereas blood pressure does not fall with upright posture in the normal subject, a marked fall without compensatory tachycardia occurs in autonomic failure.

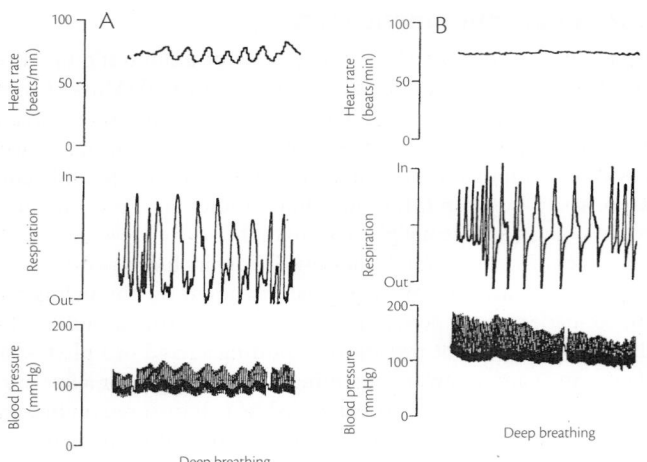

Fig. 2.58 Sinus arrhythmia of the cardiac rate during deep breathing in a normal subject (A). This variation in cardiac rate during respiration is lost in autonomic failure (B).

by the cardiac vagus nerves. Loss of variation in the R–R interval when the electrocardiogram is recorded during deep breathing occurs in autonomic neuropathy (Fig. 2.58).

Valsalva manoeuvre. When a normal subject attempts to exhale forcibly against a voluntarily closed glottis, the blood pressure drops, with loss of venous return to the heart, and the heart rate rises. When this intrathoracic pressure increase is released, the blood pressure overshoots because of continued sympathetic drive, and the heart rate drops below the basal level due to baroreflex activation (Fig. 2.59). Autonomic neuropathies abolish this second blood pressure overshoot and the associated reflex bradycardia (Fig. 2.59). If the baroreflex arc remains intact, as may be the case in high spinal cord lesions and some forms of central autonomic

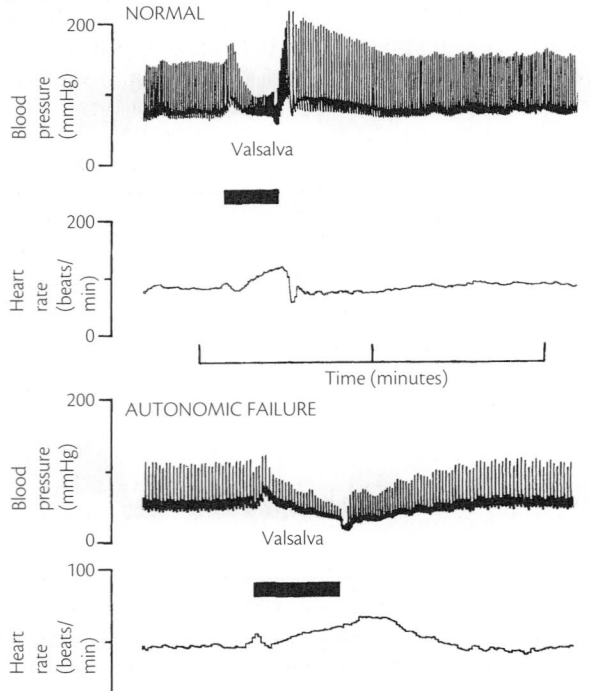

Fig. 2.59 Valsalva manoeuvre. The blood pressure and heart rate responses to exhaling against a resistance of 40 mmHg are shown for a normal subject (top) and a patient with autonomic failure (bottom).

failure, the cardiac rate rises during the initial phase of falling blood pressure. In formal quantitative testing the subject is asked to exhale against a standard resistance of 40 mm Hg.

2.7.3 Causes of autonomic failure

A wide variety of disorders cause autonomic dysfunction, usually associated with other neurological or general medical disorders (Table 2.9). In everyday clinical practice the most common neurological diseases causing autonomic failure are diabetic polyneuropathy and Parkinson's disease. It is usually relatively asymptomatic in both cases.

2.8 Critical illness

2.8.1 Indications for intensive care

Intensive care may be required for patients critically ill either as a result of primary neurological disease, or in those in whom a neurological disorder is a component of, or secondary to, a general medical disorder. Those whose neurological condition is associated with a general medical disorder are generally suffering from hypoxic-ischaemic brain damage, or the complications of, or interactions between, metabolic disease, organ failure, critical illness, and sepsis.

The commonest primary neurological diseases requiring admission to a neurological intensive care unit are myasthenia gravis, Guillain–Barré syndrome, neurological infections, status epilepticus, and stroke (Howard *et al.* 2003). Many such patients will have been fully conscious of the steady decline in their respiratory or bulbar functioning, may remain conscious whilst receiving intensive care, and may remain in an intensive care unit for some weeks before recovery. This can place substantial emotional demands on the patients, relatives, and staff. A regular dialogue should be maintained. In particular, patients with respiratory failure moving inexorably towards assisted ventilation should be forewarned that this is an expected and planned aspect of their treatment. Likewise, avoidance of pain by sympathetic positioning, and its treatment by analgesics is essential.

Indications for admission to neurological intensive care have been defined (Howard *et al.* 2003): impaired consciousness, bulbar muscle failure, severe ventilatory respiratory failure, uncontrolled seizures, severely raised intracranial pressure, some monitoring and interventional treatments, and unforeseen general medical complications. Naturally specific treatments indicated for the particular diagnosis should be instituted along with general intensive care measures.

Particular dilemmas arise in patients with bulbar or ventilatory failure due to progression of an incurable degenerative disorder, such as amyotrophic lateral sclerosis. Patients and their relatives should receive a pre-emptive and frank discussion about the implications of ventilation under such circumstances, stressing the often limited and short-term resultant improvements in the quality of life. Some such patients do receive mechanical ventilation, either as a deliberate decision, or as an ill-considered emergency response when a medical crisis has arisen. Complex medical and legal considerations concern the patient's rights to turn off ventilation by withdrawing consent to an invasive medical procedure, to planned withholding of treatments for pneumonia or other potentially terminal complications, and to the need to prevent the distress of terminal dyspnoea if a decision is taken to discontinue ventilation (Borasio and Voltz 1998; Bradley *et al.* 2002).

Table 2.9 Classification of disorders resulting in autonomic dysfunction (After Mathias and Bannister (1999), with modifications)

Primary (aetiology unknown)

Acute/subacute dysautonomias (Section 21.10.4)
  Pure cholinergic dysautonomia
  Pure pandysautonomia
  Pandysautonomia with neurological features
Chronic autonomic failure syndromes
  Pure autonomic failure (Section 21.11.7)
  Multiple system atrophy (Shy–Drager syndrome) (Section 40.3.8)
  Autonomic failure with Parkinson's disease (Section 40.3.1)

Secondary

Congenital
  Nerve growth factor deficiency
Hereditary
  Familial amyloid neuropathy (Section 21.9.1)
  Porphyria (Section 21.8.6)
  Hereditary sensory and autonomic neuropathies (Section 21.6)
  Familial dysautonomia—Riley–Day syndrome (21.6.3)
  Dopamine β-hydroxylase deficiency
  Aromatic L-amino acid decarboxylase deficiency
  Anderson–Fabry disease (Section 21.8.5)
Metabolic diseases
  Diabetes mellitus (Section 21.17.5)
  Chronic renal failure (Section 21.18.1)
  Chronic liver disease
  Vitamin B_{12} deficiency (Section 21.22.4)
Inflammatory
  Bacterial—Tetanus (Section 42.5.14)
  Viral—HIV infection (Section 43.3.7)
  Parasitic—Trypanosomiasis Cruzi; Chagas' disease (Section 42.11.8)
  Prion—fatal familial insomnia (Section 42.9.8)
Neoplasia
  Brain tumours—especially of third ventricle or posterior fossa
  Paraneoplastic autonomic neuropathy (Section 21.11.7)
  Lambert–Eaton syndrome (Section 24.10.2)
  Primary amyloidosis (Section 21.9.2)
Connective tissue disorders
  Rheumatoid arthritis
  Systemic lupus erythematosus
  Mixed connective tissue disease
Surgery
  Regional sympathectomy—upper limb, splanchnic
  Vagotomy and drainage procedures—'dumping syndrome'
  Organ transplantation—heart, kidney
Trauma
  Spinal cord transection

Neurally mediated syncope

Vasovagal syncope
Carotid sinus hypersensitivity
Micturition syncope
Cough syncope
Swallow syncope
Associated with glossopharyngeal neuralgia

Drugs, chemicals, poisons, and toxins

Alcohol-induced (Section 21.19.1)

2.8.2 Assisted respiration

Ventilatory respiratory failure can result from a wide range of peripheral neuropathies, disorders of neuromuscular transmission, primary muscle diseases, and disordered central respiratory control (Polkey *et al.* 1999). The commonest and most familiar setting is of Guillain–Barré syndrome with progressive reduction of respiratory ventilation over days. Equally important is the risk of aspiration pneumonia or asphyxia by choking due to the development of bulbar muscle failure. Patients at risk of developing neuromuscular respiratory failure should be monitored with regular forced vital capacity measurement, sometimes with measurement of blood oxygenation, particularly if there is pre-existing pulmonary disease. Early warning signs of respiratory failure, such as the patient having to take frequent breaths whilst speaking, or exhibiting the paradoxical abdominal wall movements of diaphragm weakness, should be sought in at-risk patients. Artificial ventilation is generally required when the forced vital capacity falls below 12–15 ml/kg body weight, or there is evidence of significant bulbar muscle failure (Ropper *et al.* 1991). Artificial ventilation may be required for between two and five weeks in patients with Guillain–Barré syndrome. In general an endotracheal tube is regarded as satisfactory for ventilation durations up to two weeks. For more prolonged ventilation a tracheostomy becomes necessary, placed either surgically or percutaneously (Hughes *et al.* 2005). Weaning from ventilation is guided by evidence of recovering strength and respiratory movements, with the patient able to self-ventilate for increasing proportions of the day.

2.8.3 Supportive measures

Deep venous thrombosis is a consequence of leg immobilization due to neuromuscular disease or paraplegia (Weingarden 1992; Henderson *et al.* 2003). This can lead to the potentially fatal consequence of pulmonary embolism. Deep venous thrombosis or pulmonary embolism occur in 16 per cent of patients with spinal cord injury (Weingarden 1992). Prophylaxis against deep venous thrombosis is necessary from the outset by analogy with post-operative surgical prophylaxis. A two-thirds reduction in the occurrence of deep venous thrombosis can be achieved by twelve-hourly subcutaneous heparin 5000 units or elasticated support stockings; both measures being recommended in the paralysed patient (Collins *et al.* 1988; Hughes *et al.* 2005).

Constipation may result from autonomic involvement, diet change, or opiate usage. In severe constipation, gastrointestinal feeding may need to be replaced by parenteral feeding, and nasogastric drainage placed. Adynamic ileus may respond to erythromycin or neostigmine. The immobility of patients in intensive care usually mandates bladder catheterization which should be maintained under strictly sterile conditions. Permanent loss of sphincter control can be anticipated in some conditions, such as traumatic paraplegia, and long-term approaches to sphincter care instituted as early as practicable.

Continuous vigilance is required to detect intercurrent infections such as pneumonia, or metabolic disorders. Decubitus ulcers should be avoided by a regular programme of turning. Contractures of paralysed limbs should be forestalled by regular physiotherapy and intervening limb positioning. A fluid replacement programme should be carefully designed and monitored, and adequately balanced and calorific nutrition provided by nasogastric or parenteral administration (Ropper *et al.* 1991).

2.9 Terminal or chronic disease

Patients are naturally anxious whilst a potentially serious symptom such as muscle weakness is under investigation before a definite diagnosis has been made. It is rarely helpful to discuss an evocative diagnosis such as motor neurone disease while it is still only a possibility. However patients with morbid anxieties may draw their neurologist into detailed discussions of specific diagnoses even while these remain hypothetical. Once the doctor is sure of the diagnosis, most patients seem keen for it to be named so as to resolve the anxiety of uncertainty. In deciding how much to say about the diagnosis of a condition likely to be fatal, such as motor neurone disease, and its progression and complications, one treads a narrow dividing line between brutal honesty and humane economy of truth. Any particular problems likely to occur in an individual patient's disease should be put in perspective before the patient becomes upset by the summary information so readily available from popular sources, such as journalism. Often it is valuable to discuss the diagnosis in stages with the patient, preferably in the presence of a close relative. Well-meaning relatives may try to prevent doctors from telling the patient that they have a fatal disease. But patients ultimately detect this conspiracy of secrecy at a time when death may loom, thereby undermining trust and confidence just when these qualities are of inestimable value.

Many patients become angry with their neurologist soon after hearing of a fatal or life changing disease. This is particularly the case in those too young to have become philosophical about their own mortality. Indeed doctors may bear the brunt of this anger as though they were somehow responsible for the disease's occurrence. But anger should be understood sympathetically as a natural stage in patients' adjustments to incurable or fatal illness. It usually follows an early phase of denial and isolation before being succeeded by bargaining, then depression, and ultimately by acceptance (Kubler-Ross 1969). During the stages of anger and bargaining, the doctor–patient relationship is vulnerable and can only be preserved by patience and understanding, thereby laying the foundations of confidence which enable patients to trust their doctor's advice when miserable problems arise later in the disease.

Patients greatly appreciate the involvement of those who can provide advice and practical help to offset the various disabilities of their disorder. Consultants in neurological disability and rehabilitation should be involved at the first sign of a needy disability, together with care teams of speech therapists, occupational therapists, physiotherapists, and social workers. Charitable organizations, such as the Motor Neurone Disease Association can provide valuable equipment and devices with minimal delay, and often provide psychological support and practical advice to patients.

References

Bogousslavsky J, Maeder P, Regli F et al. (1994). Pure midbrain infarction: clinical syndromes, MRI, and etiologic patterns. *Neurology*, **44**, 2032-40.

Borasio GD, Voltz R (1998). Discontinuation of mechanical ventilation in patients with amyotrophic lateral sclerosis. *J Neurol*, **245**, 717–22.

Boyle RS, Shakir RA, Weir AI et al. (1979). Inverted knee jerk: a neglected localising sign in spinal cord disease. *J Neurol Neurosurg Psychiatry*, **42**, 1005–7.

Bradley MD, Orrell RW, Clarke J et al. (2002). Outcome of ventilatory support for acute respiratory failure in motor neurone disease. *J Neurol Neurosurg Psychiatry*, **72**, 752–6.

Caplan LR (1980). "Top of the basilar" syndrome. *Neurology*, **30**, 72–9.

Collier J (1904). The false localising signs of intracranial tumour. *Brain*, **27**, 490–508.

Collins R, Scrimgeour A, Yusuf S et al. (1988). Reduction in fatal pulmonary embolism and venous thrombosis by perioperative administration of subcutaneous heparin. Overview of results of randomized trials in general, orthopedic, and urologic surgery. *N Engl J Med*, **318**, 1162–73.

Dick JP (2003). The deep tendon and the abdominal reflexes. *J Neurol Neurosurg Psychiatry*, **74**, 150–3.

Dietz V, Muller R, Colombo G (2002). Locomotor activity in spinal man: significance of afferent input from joint and load receptors. *Brain*, **125**, 2626–34.

Donaghy M (2005). *Neurology*. Oxford University Press, Oxford.

Duysens J, Clarac F, Cruse H (2000). Load-regulating mechanisms in gait and posture: comparative aspects. *Physiol Rev*, **80**, 83–133.

Factor SA, Higgins DS, Qian J (2006). Primary progressive freezing gait: a syndrome with many causes. *Neurology*, **66**, 411–4.

Fife TD, Baloh RW (1993). Disequilibrium of unknown cause in older people. *Ann Neurol*, **34**, 694–702.

Gardner, E (1975). *Fundamentals of Neurology*, 6th edn. Saunders, Philadelphia.

Gilman S (2002). Joint position sense and vibration sense: anatomical organisation and assessment. *J Neurol Neurosurg Psychiatry*, **73**, 473–7.

Henderson RD, Lawn ND, Fletcher DD et al. (2003). The morbidity of Guillain-Barre syndrome admitted to the intensive care unit. *Neurology*, **60**, 17–21.

Howard RS, Kullmann DM, Hirsch NP (2003). Admission to neurological intensive care: who, when, and why? *J Neurol Neurosurg Psychiatry*, **74**, iii2–9.

Hughes RA, Wijdicks EF, Benson E et al. (2005). Supportive care for patients with Guillain-Barre syndrome. *Arch Neurol*, **62**, 1194–8.

Ishiyama G, Ishiyama A, Baloh RW (2003). Drop attacks and vertigo secondary to a non-meniere otologic cause. *Arch Neurol*, **60**, 71–5.

Jankovic J, Nutt JG, Sudarsky L (2001). Classification, diagnosis, and etiology of gait disorders. *Adv Neurol*, **87**, 119–33.

Kubler-Ross E (1969). *On Death and Dying*. Tavistock/Routledge, London & New York.

Kugelberg E, Hagbarth KE (1958). Spinal mechanism of the abdominal and erector spinae skin reflexes. *Brain*, **81**, 290–304.

Lance JW (2002). The Babinski sign. *J Neurol Neurosurg Psychiatry*, **73**, 360–2.

Larner AJ (2003). False localising signs. *J Neurol Neurosurg Psychiatry*, **74**, 415–8.

Lhermitte F (1983). 'Utilization behaviour' and its relation to lesions of the frontal lobes. *Brain*, **106**, 237–55.

Liu GT, Crenner CW, Logigian EL et al. (1992). Midbrain syndromes of Benedikt, Claude, and Nothnagel: setting the record straight. *Neurology*, **42**, 1820–2.

Lloyd DP (1952). On reflex actions of muscular origin. *Res Publ Assoc Res Nerv Ment Dis*, **30**, 48–67.

MacDonald BK, Cockerell OC, Sander JW et al. (2000). The incidence and lifetime prevalence of neurological disorders in a prospective community-based study in the UK. *Brain*, **123**, 665–76.

Marsden CD, Merton PA, Morton HB (1973). Is the human stretch reflex cortical rather than spinal? *Lancet*, **1**, 759–61.

Mathias C, Bannister R (1999). *Autonomic Failure*. Oxford University Press, Oxford.

Matsumoto S, Okuda B, Imai T et al. (1988). A sensory level on the trunk in lower lateral brainstem lesions. *Neurology*, **38**, 1515–9.

Mehler MF (1989). The rostral basilar artery syndrome: diagnosis, etiology, prognosis. *Neurology*, **39**, 9–16.

Meissner I, Wiebers DO, Swanson JW et al. (1986). The natural history of drop attacks. *Neurology*, **36**, 1029–34.

Mossuto-Agatiello L (2006). Caudal paramedian midbrain syndrome. *Neurology*, **66**, 1668–71.

Nutt JG, Marsden CD, Thompson PD (1993). Human walking and higher-level gait disorders, particularly in the elderly. *Neurology*, **43**, 268–79.

Owen G, Mulley GP (2002). The palmomental reflex: a useful clinical sign? *J Neurol Neurosurg Psychiatry*, **73**, 113–5.

Patton HD, Sundsten JW, Crill WE *et al.* (1976), *Introduction to Basic Neurology*. Saunders, Philadelphia.

Paulson G, Gottlieb G (1968). Developmental reflexes: the reappearance of foetal and neonatal reflexes in aged patients. *Brain*, **91**, 37–52.

Perkin GD (1989). An analysis of 7836 successive new outpatient referrals. *J Neurol Neurosurg Psychiatry*, **52**, 447–8.

Plum F, Posner J (1980). *The Diagnosis of Stupor and Coma*. Blackwell, Oxford and Philadelphia.

Polkey MI, Lyall RA, Moxham J *et al.* (1999). Respiratory aspects of neurological disease. *J Neurol Neurosurg Psychiatry*, **66**, 5–15.

Ropper A, Wijdicks E, Truax B (1991). *Guillain-Barre Syndrome*. FA Davis Co, Philadelphia.

Sacco RL, Freddo L, Bello JA *et al.* (1993). Wallenberg's lateral medullary syndrome. Clinical-magnetic resonance imaging correlations. *Arch Neurol*, **50**, 609–14.

Schott JM, Rossor MN (2003). The grasp and other primitive reflexes. *J Neurol Neurosurg Psychiatry*, **74**, 558–60.

Shallice T, Burgess PW, Schon F *et al.* (1989). The origins of utilization behaviour. *Brain*, **112**, 1587–98.

Sheldon J (1960). On the natural history of falls in old age. *Br Med J*, **2**, 1685–90.

Silverman IE, Liu GT, Volpe NJ *et al.* (1995). The crossed paralyses. The original brain-stem syndromes of Millard-Gubler, Foville, Weber, and Raymond-Cestan. *Arch Neurol*, **52**, 635–8.

Stevens DL (1989). Neurology in Gloucestershire: the clinical workload of an English neurologist. *J Neurol Neurosurg Psychiatry*, **52**, 439–46.

Sudarsky L (1990). Geriatrics: gait disorders in the elderly. *N Engl J Med*, **322**, 1441–6.

Tinetti ME, Speechley M, Ginter SF (1988). Risk factors for falls among elderly persons living in the community. *N Engl J Med*, **319**, 1701–7.

Van Gijn J (1978). The Babinski sign and the pyramidal syndrome. *J Neurol Neurosurg Psychiatry*, **41**, 865–73.

Weingarden SI (1992). Deep venous thrombosis in spinal cord injury. Overview of the problem. *Chest*, **102**, 636S–9S.

Williams DR, Watt HC, Lees AJ (2006). Predictors of falls and fractures in bradykinetic rigid syndromes: a retrospective study. *J Neurol Neurosurg Psychiatry*, **77**, 468–73.

CHAPTER 3

Investigations

David Chadwick, Alastair Compston,
Michael Donaghy, Nicholas Fletcher,
Robert Grant, David Hilton-Jones,
Martin Rossor, Peter Rothwell, and Neil Scolding

Contents

3.1 Imaging of the nervous system

3.1.1 Computed tomography imaging

Computed tomography, or CT radiographic scanning has been clinically available since the 1970s and was a major advance for anatomical imaging of the brain. New generations of CT scanners have become increasingly rapid, have very high resolution, and image slice thicknesses of as little as 0.4 mm. In addition, it is now possible to create high quality multi-planar images (Fig. 3.1), provide accurate CT angiography (Fig. 3.13) and to interrogate volumes of data and highlight tissues of interest. As a result of these developments, its diagnostic sensitivity and specificity has improved and reduced the requirement for further examinations which are more expensive, such as MRI, or more invasive, for example intra-arterial angiography. In addition, faster speeds of data acquisition have reduced the occurrence of troublesome movement artefact.

The type of scan chosen when trying to exclude an intracranial cause for symptoms will depend on the urgency for scanning, the suspected aetiology, and the availability of scanning modalities. The benefits of CT imaging of the brain over other advanced imaging modalities are that it can be performed quickly to give high

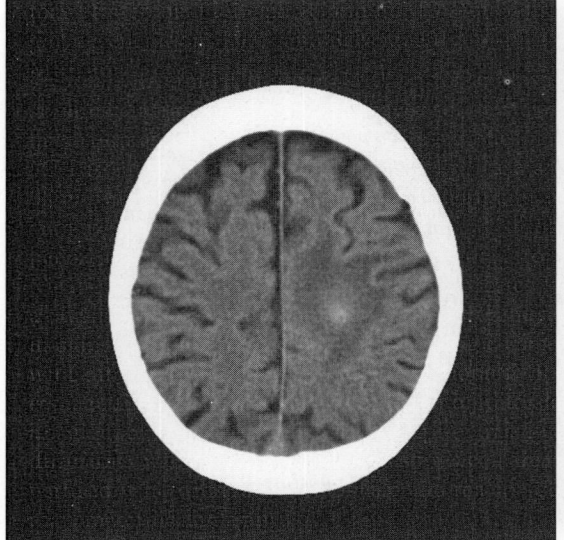
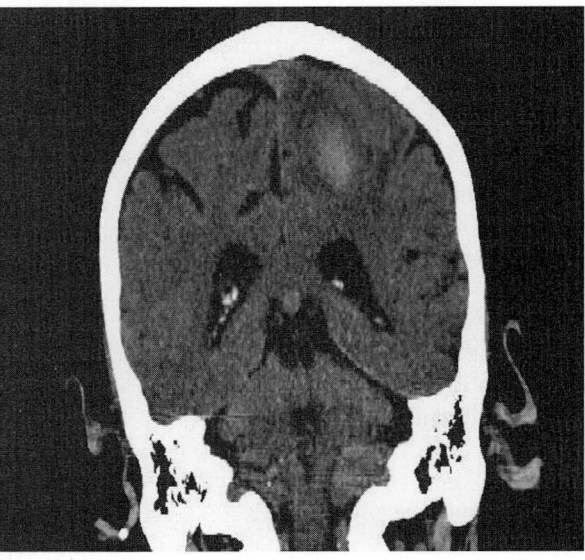

A   B

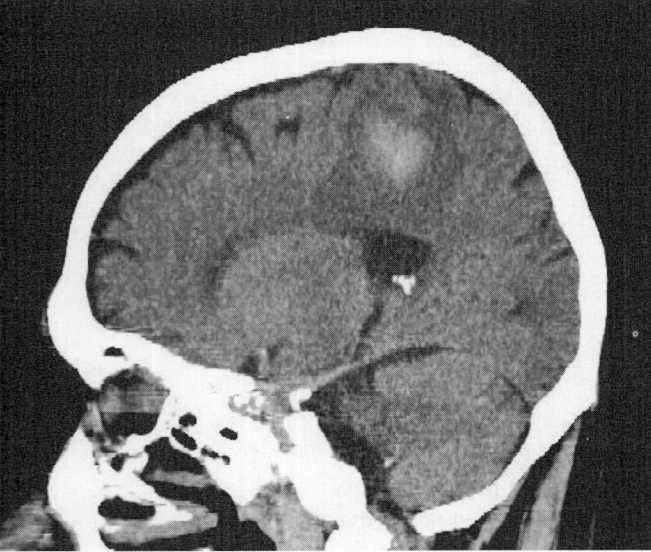

C

Fig. 3.1 Computed tomography, or CT, scan of the brain. These three orthogonal images in the axial (A), coronal (B) and sagittal (C) planes were reconstructed from a series of 1 mm thick axial images. The data from these base images is electronically 'stacked' and reinterpreted to give derived images in any plane desired. This technique can be used for any imaging data set where the base images are thin enough to allow good quality reconstruction.

quality images even when the patient is moving. Image acquisition may take only a few minutes and is more manageable for agitated patients. Urgent cases can usually be scheduled into a busy scanning session without undue disruption. CT scanning is particularly sensitive in identifying acute haemorrhage (Fig. 3.2), bony abnormalities, and calcified lesions. Most intracranial masses above 1cm diameter are visualizable, especially if contrast is given. CT scanning can give important information in patients with craniofacial abnormalities (Gilman 1998), and is valuable when assessing the degree of bone destruction or sclerosis associated with tumours. It is possible to use CT to guide procedures such as percutaneous biopsy of spinal or paraspinal lesions and, in combination with stereotactic surgical equipment, to guide biopsy or excision of intracranial tumours. CT scanning is less expensive and more readily available in most healthcare systems. Intravenous iodine-based contrast agents improve the diagnostic utility of CT scanning. All vascular structures and most malignant tumours enhance with contrast. The region of contrast enhancement corresponds to the active rim of the tumour.

The drawbacks of CT scanning are those first of the administered radiation dose, especially in pregnancy or in young patients who may require frequent repeat CT imaging and second of the frequent requirement to use iodinated contrast agents. The radiation dose from using multi-slice new generation CT scanners is lower than early generation scanners, but nevertheless certain structures should be avoided if possible, particularly the gonads and ocular lens. A single CT study of the head carries a risk of around 1 in 4000 of causing a radiation-induced malignancy. CT imaging should be avoided in pregnant women, if possible, although the radiation dose to the foetus from a cranial CT scan is very low if the abdomen is protected by a lead apron. Nursing mothers should wait 24 h before resuming breast feeding if they have received iodinated contrast agents. Contrast agents can cause flushing or nausea. These should not be considered allergic reactions. Allergic reactions to contrast occur in the form of itching, or

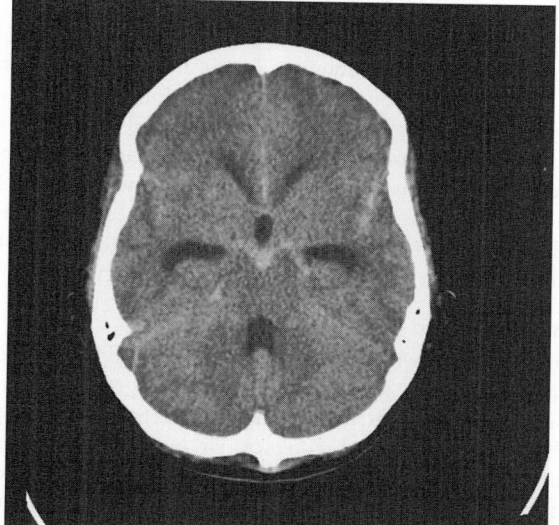

Fig. 3.2 CT is the method of choice for the imaging of intracranial haemorrhage. Coagulated blood has higher attenuation than brain parenchyma and can be seen easily in nearly all cases. This patient has a subarachnoid haemorrhage with secondary hydrocephalus. CT is also the preferred modality to assess ventricular size.

rash, or more severe reactions with dyspnoea, arrhythmia, and anaphylaxis, in less than 1:10 000 cases.

Contraindications to the use of contrast include: previous allergic response to contrast; renal insufficiency, except in dialysis patients prior to their next dialysis session; severe asthma; multiple myeloma; severe cardiac disease or congestive cardiac failure. Extreme care should be taken in patients with food or medication allergies and patients with diabetes who are taking Metformin, particularly if there is any coexisting renal problems. Contrast can interact with Metformin causing a metabolic acidosis. If patients have normal renal function contrast administration is safe, but Metformin should be stopped for 48 h after scanning and serum urea and electrolytes with creatinine should be checked within 48 h of giving contrast.

Most CT scanning couches are designed up to a maximum body weight limit of 150–190 kg, making scanning of heavier patients practically impossible. The aperture of the scanner is such that claustrophobia or patient girth is usually not a problem.

New generation scanners have the capability to acquire contiguous thin sections. Post-processing workstations allow interrogation of these volumes of data as re-formatted cross-sectional images in any plane. This is referred to as multi-planar reformatting. The same workstations can give a view of the whole volume of data and display this as a 3-dimensional model using volume rendering or surface shading. When intravenous contrast agents are used, reformatted images can be adjusted to give a 3D representation of arteries or veins by timing the imaging to coincide with maximal arterial or venous opacification—a CT arteriogram or venogram.

Vascular disease

CT scanning is commonly used as the first diagnostic imaging tool in the management of stroke. It will reliably distinguish infarctive stroke from haemorrhagic stroke. Early on CT scan may be normal in many cases of infarction. CT will identify subarachnoid blood in patients with clinically significant subarachnoid haemorrhage in about 95 per cent of cases (Fig. 3.2). CT angiography is commonly used now for identification of cerebral aneurysms, arteriovenous malformations, or cerebral or neck vessel arterial disease. Similarly, CT venography is useful in the diagnosis of cerebral venous sinus occlusive disease. CT vascular imaging with multi-slice scanners has higher spatial resolution than MRI and is less time consuming. It has fewer risks than catheter angiography. CT angiography is not as sensitive as formal intra-arterial angiography, but its sensitivity to identify aneurysms of diameter exceeding 2 mm is very high.

Demyelination

Routine CT scanning is not sensitive enough to reliably identify plaques of demyelination. In general CT is poorer than MRI at identifying white matter diseases such as inflammation, infections such as encephalitis, or subtle microvascular changes.

Neoplasia

Often suspected brain tumours will be imaged by CT initially, because of its ready availability and speed. In many situations CT scanning will be all that is required. Increasingly however neurosurgeons prefer additional imaging with MRI to exclude multiple lesions before surgery and to obtain gadolinium enhanced or volumetric images to assist with planning of surgical techniques involving stereotactic resection, or operations within the magnet, with a view to getting maximal tumour resection in those cases suitable for such form of surgery.

In clinically suspected cases of brain tumour, contrast should always be given. This increases the chances of detecting a lesion and often gives a clue as to whether it is a malignant or a low grade process. Also it helps target the area most likely to give informative histology on biopsy. The sensitivity and specificity and differential diagnosis is covered in Section 27.4. In the context of tumour recurrence after initial treatment such as surgery or radiotherapy, CT scanning often confirms clinically significant recurrence. However in some cases after radiation therapy, it will remain uncertain how much could be due to radiation necrosis versus tumour recurrence. MRI, single photon emission computed tomography, SPECT or positron emission tomography, PET, scanning may provide more reliable information, although often repeat biopsy is necessary for confirmation and this remains the gold standard where it is feasible and safe.

Raised intracranial pressure

In patients with headache suggestive of raised intracranial pressure or who demonstrate papilloedema, a normal CT scan with contrast and a CT venogram should always be followed by CSF examination to exclude or confirm raised pressure due to idiopathic intracranial hypertension or meningitis. This principle is increasingly important in an era in which primary care physicians have increasingly direct access to CT and MRI imaging facilities for their patients.

Degenerative disease

CT changes can be associated with degenerative conditions such as dementias or focal degenerations. However mesio-temporal sclerosis with atrophy of Ammon's horn is less easily identified or measured with CT than with MRI. The primary role of CT in patients with suspected dementia lies in excluding mass lesions, for instance tumour, subdural haematoma, arteriovenous malformation, chronic abscess, communicating or non-communicating hydrocephalus, and multiple infarcts. If there is a clinical need to get more detailed anatomic information, MRI is more sensitive. For functional information SPECT and PET studies may be required.

Trauma

CT brain scanning is particularly useful in neuro-trauma. In penetrating brain injury, CT can identify any skull fractures, intracerebral haematomas, or subarachnoid blood. CT angiography will identify traumatic intra-cerebral aneurysms, which occur in 0.3 per cent of cases, often accompanied by a haematoma. Following blunt trauma to the head or neck, CT angiography can identify carotid or vertebral arterial dissections or occlusions, detect any accompanying vertebral fractures, and can be particularly helpful in planning surgical intervention. Formal catheter angiography remains the gold standard for assessing penetrating injuries and examining for the cause of intracranial haematoma. However, as a screening tool in the acute situation, CT angiography will often suffice.

Special sites

CT imaging of the para-pituitary area shows the bony structures and any bony erosion well, but MR imaging has the advantage of more clearly defining the soft tissues, nerves, and blood vessels. For the same reasons MRI is generally superior to CT when imaging the spine and nerve plexuses (Fig. 3.3). CT scanning of spine may clarify area of bony overgrowth or bony compression but does not show nerves well, without the addition of intrathecal contrast injection.

3.1.2 Magnetic resonance imaging

Magnetic resonance or MR scanning has been available clinically since the 1980s. As with CT it started as a pure imaging modality,

with the advantage of greater ability to differentiate grey and white matter due to better contrast resolution than CT, and with the ability to image in any plane. These capabilities made it the method of choice for imaging structural abnormalities of the brain parenchyma and spinal cord. Compared to CT it shows improved sensitivity and specificity in the posterior cranial fossa (Fig. 3.4). An MR scanner can be 'programmed' to measure a predetermined variety of physical properties for each voxel. The majority of MR signal is generated by the protons in water molecules and so most MR sequences are dependent on the concentration of water in each voxel, which is the subdivided volume of tissue represented by a pixel on a digital image. It is possible to tune the MR sequences to increase or reduce sensitivity to fat, water, flowing fluid, diffusibility of tissues, flow through tissues, or even to analyse the chemical composition of a volume of tissue. Changes of physical properties of areas of the brain can be used to image changes in levels of local neuronal activity—functional MRI or fMRI.

Structural MR imaging

This generally takes around 4 min for each sequence employed and routine imaging usually requires three or more sequences. The time required to acquire an image renders MRI much more sensitive to patient movement artefact. In practice, it may be difficult to get an urgent MR scan in many countries unless there is a clear clinical need, or the health system has an overprovision of MR scanners, or via a private health care system. However, abnormalities in the posterior fossa are much more clearly demonstrated by MRI as CT has difficulty coping with the dense bone in this region (Teasdale et al. 1989) (Fig. 3.5). Over the years MR imaging has become faster, with more sensitive sequences, and improved resolution, to give more accurate structural imaging. Volumetric scanning has allowed the images from a single acquisition to be reformatted on a workstation in any plane or as an entire volume. This is of great value in measurement of tumour volumes with greater accuracy and in surgical planning. MRI has additional advantages over CT scanning in identifying small masses more easily, especially where they abut bone for instance at the vertex, temporal bone, orbital apex, or internal auditory meati, and the ability to identify small vascular malformations without the need for contrast enhanced imaging. Despite numerous researches on T1 and T2 relaxation times and correlation of different signal intensities with different tumour tissues, MRI cannot predict accurately the tumour type or grade of intracerebral neoplasm. Accordingly, attention has shifted to diffusion and perfusion imaging and MR spectroscopy. Echo-planar imaging sequences on clinical MR scanners, enable rapid acquisition of images with different contrast mechanisms.

Diffusion-weighted MRI

This uses sequences where MR gradients are applied before and after the radiofrequency pulse in gradient echo recalled imaging sequences. Protons that diffuse freely undergo loss of signal on heavily diffusion-weighted sequences so that protons with higher diffusion rates produce less MR signal than those with slower diffusion rates. To reflect the true diffusion rates in tissue a calculated 'map' of diffusibility is derived from images with varied diffusion weighting: the Apparent Diffusion Co-efficient image. This image reflects physical factors such as temperature and viscosity in addition to the restricted motion of the molecules. Differences in Apparent Diffusion Co-efficient map reflect differences in cellularity,

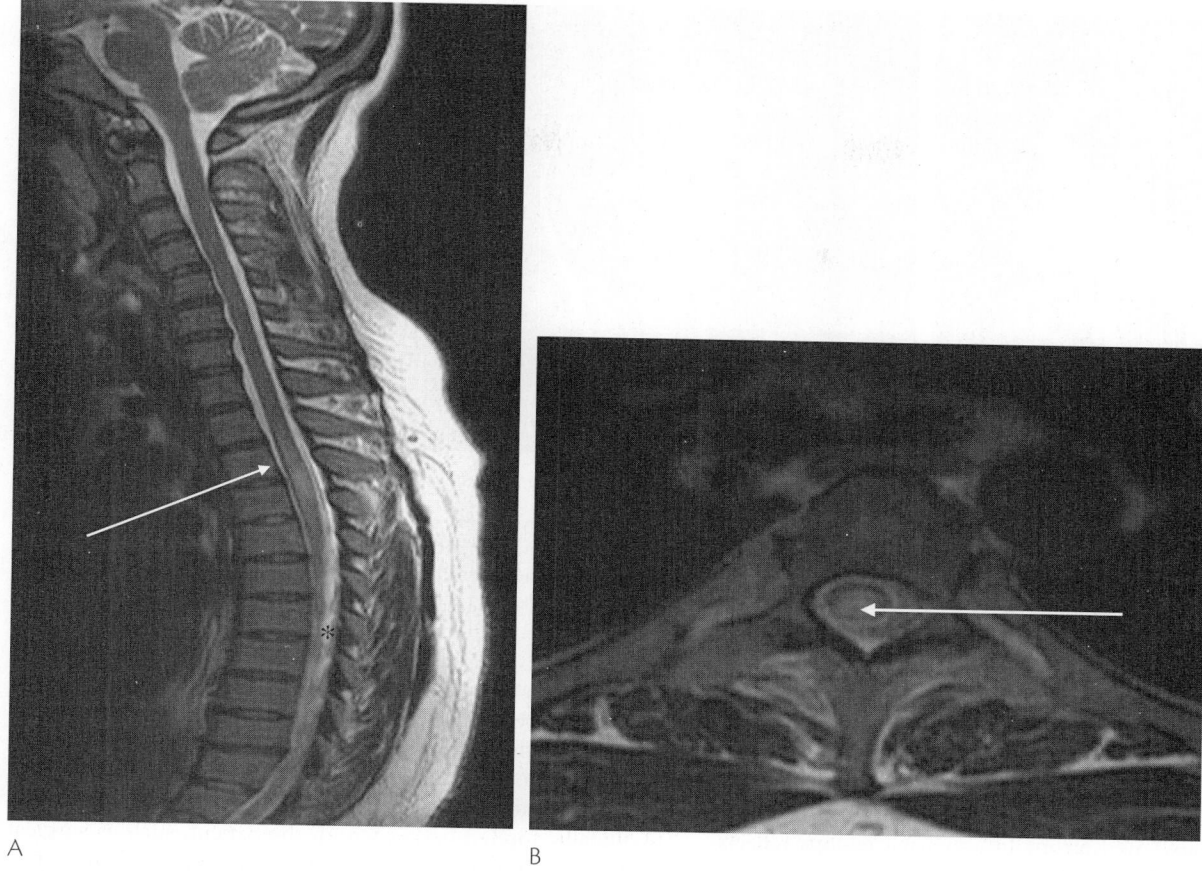

Fig. 3.3 MR imaging of the spine in (A) the sagittal plane and (B) axial plane demonstrates high T2 contrast between CSF and spinal cord / nerve root. This patient has a lesion within the mid thoracic cord (arrow) with minimal mass effect. The characteristics are typical of transverse myelitis. Heterogenous signal in the CSF (*) is due to CSF flow artefact.

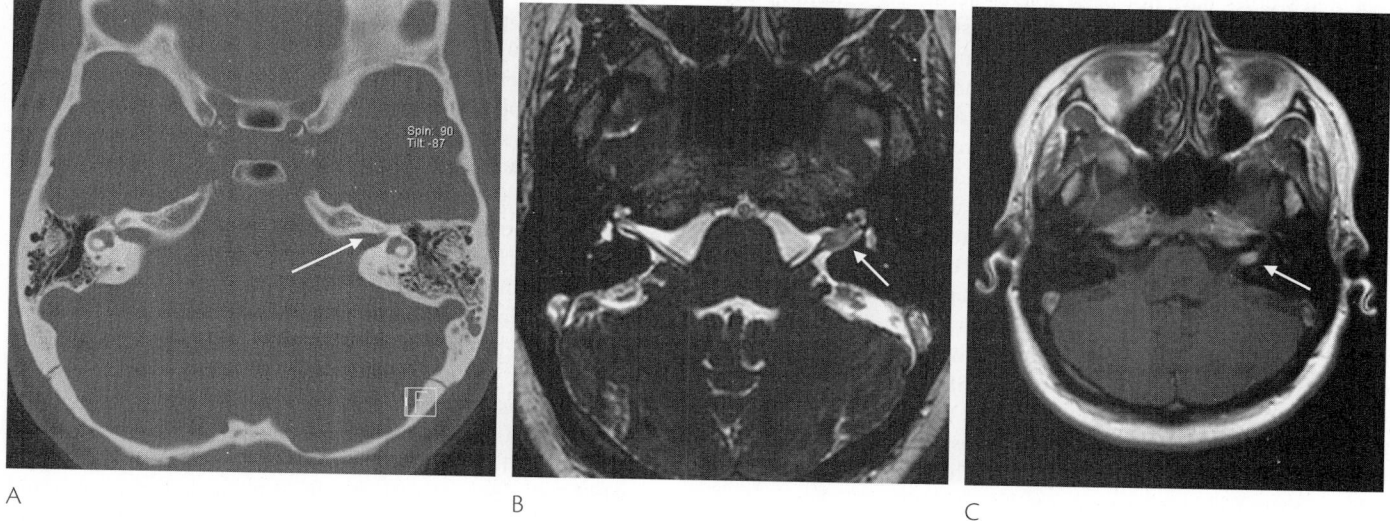

Fig. 3.4 Patient with left-sided sensorineural hearing loss. (A) The initial CT brain scan gives very good bone definition at the skull base but does not reveal the tumour of the left VIII nerve, although some expansion of the internal auditory foramen is evident (arrow). The left VIII cranial nerve schwannoma (arrowed) is clearly seen with (B) high resolution axial and (C) post contrast T1-weighted MR images. In general MR is the preferred modality for imaging cranial nerve lesions.

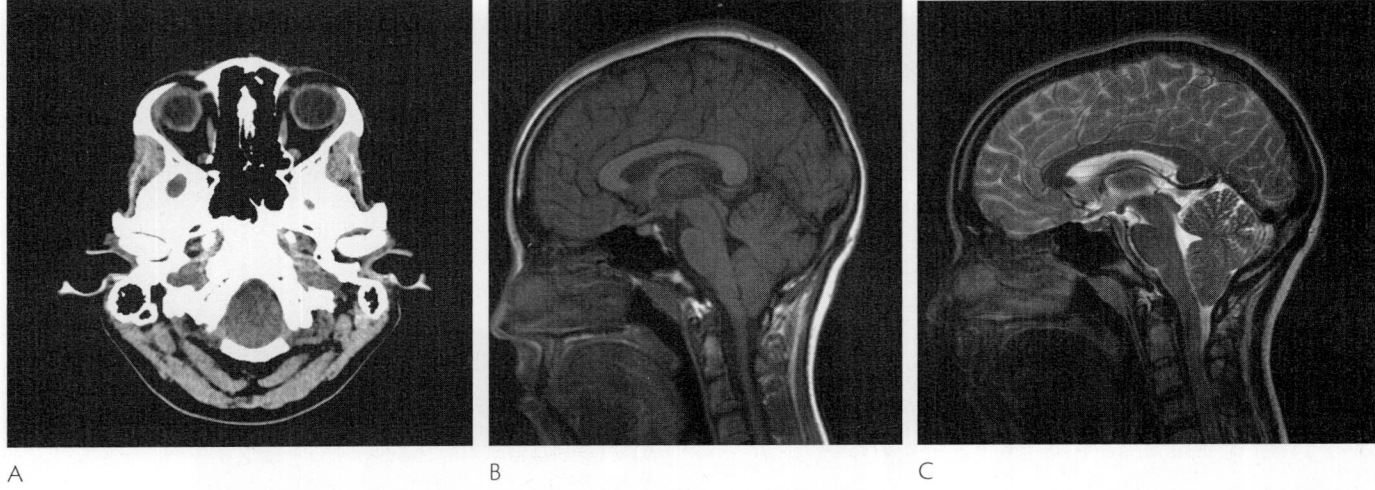

A   B   C

Fig. 3.5 (A) Although modern CT scanners can now resolve much more detail in the posterior fossa, it remains difficult to assess the position of the cerebellar tonsils. (B) Sagittal MR imaging is necessary to reveal the true configuration of the structures at the foramen magnum in a Chiari malformation. (C) T2-weighted MR images give a good impression of the residual CSF space.

cell membrane permeability, intracellular and extracellular diffusion, and tissue structure. Most significantly, diffusion is restricted in acute infarct giving high signal on the diffusion base images and low values, such as dark areas, on the Apparent Diffusion Co-efficient map.

For brain tumours, low isotropy in abnormal regions reflects necrosis, while oedema and cysts return higher values relative to normal white matter. The Apparent Diffusion Co-efficient of regions of tumours has been correlated with malignancy and cellularity. Some authors have found a correlation between lower values in high grade areas and higher values in low grade low cellular areas (Castillo *et al.* 2001). Others have found no difference in Apparent Diffusion Co-efficient values between different types of intracranial neoplasm: glioma, meningioma, or metastases (Kono *et al.* 2001). Diffusion-weighted studies have not been able to accurately determine the limits of tumour by distinguishing neoplastic cell infiltration from peritumoural vasogenic oedema. This technique cannot yet be used to identify the grade or extent of the tumour accurately.

It is possible now to record the average direction of diffusion in organized brain structures such as white matter tracts, and represent this as a 3-dimensional map of diffusion directions. This technique is called diffusion tensor imaging and may provide useful information in diseases disruptive of white matter or in planning surgery so as to avoid injuring important tracts. Techniques, such as MR perfusion and diffusion imaging, have improved the sensitivity and accuracy of diagnosis of early stroke. Dynamic contrast-enhanced susceptibility-weighted perfusion MR imaging utilizes fast echo-planar imaging to give some information about tissue vascularization, reflecting blood volume, transit time, clearance, extraction fraction, and blood flow. MR spectroscopy and functional MR imaging has been of research value, although the clinical value of these techniques remains to be determined.

Contraindications and complications

Since the magnetic field of an MR scanner is strong, there is a risk of attracting metallic foreign objects into the field, which may cause injury to the patient. Large ferromagnetic objects, such as wheelchairs, oxygen cylinders, drip-stands, or patient trolleys, will move towards the bore of the magnet with lethal force. It is essential to check that any metal objects brought into the MRI suite with a patient, or with the supervising clinical staff, are MR compatible. Any metallic foreign bodies within the patient, such as surgical implants or shrapnel, can be moved by the magnetic field. The alternating magnetic gradients and radiofrequency pulses applied can also generate significant electrical eddy currents in even non-magnetic implants. Electronic implants not designed for MRI can malfunction; several such devices actually use magnets for transcutaneous programming.

Absolute contraindications to MR imaging therefore include: cardiac pacemakers; implanted cardiac defibrillators; ferromagnetic aneurysm clips; carotid artery vascular clamps and cochlear implants. Some other implanted devices such as neuro-stimulators or implanted drug infusion pumps are either MRI compatible or will require resetting after imaging but will not be damaged. Orthopaedic implants are generally safe. Most artificial heart valve replacements are safe although it is worth finding the manufacturer and type to check that this is the case.

Pregnancy is a relative contraindication, particularly in the first trimester due to potential teratogenic risk, although there is no evidence for this. MRI should only be performed during pregnancy if there is a significant clinical need, particularly if during the first trimester. Gadolinium enhancement should be avoided.

If gadolinium-based intravenous contrast is to be used, <1 per cent of patients will experience mild nausea or headache after injection. This usually settles quickly without the need for treatment. There is a very low risk of serious allergy to gadolinium, approximately <1 in 100 000.

Around 5 per cent of patients suffer from claustrophobia and cannot tolerate MR scanning. Obese patients may not be able to comfortably fit within the MR scanner as the bore is generally smaller than that of a CT scanner and the weight limit of the scanning table may prohibit scanning.

Structural MRI

While there may be some individual variations between units based on the age and capabilities of the individual machines, there are some generally accepted principles and sequences depending on the site to be imaged and the suspected disease process:

◆ *General structural imaging of the brain.* This includes T1-weighted sagittal sequences along with T2-weighted and FLAIR, or fluid attenuated inversion recovery, axial images. Gadolinium may be given and images repeated if there is concern about infection or tumour or to examine more closely vascular structures within or out-with the brain. Structural problems such as Chiari 1 malformations and platybasia are easily identified by MRI. Gadolinium will enhance blood vessels, areas of breakdown of the blood brain barrier associated with tumour, abscess, active demyelinating plaques or other inflammatory lesions, and the meninges, especially meningeal tumour deposits or inflammation. A particular impact of MRI has been its value in identifying plaques in multiple sclerosis. Early active lesions of multiple sclerosis are hyperintense on T2-weighted imaging with a hypointense ring. Late active lesions are hyperintense on T2, but appear as 'black holes' on T1 scans. Inactive lesions are hypointense on T2 scans and hypointense or normal on T1 scans. The characteristic distribution is multiple bright periventricular and callosal lesions on T2-weighted scans in the white matter with additional lesions in the posterior fossa. The 'open ring' appearance at the grey-white junction of arcuate fibres is also characteristic of multiple sclerosis. Acute plaques commonly enhance with gadolinium.

◆ *Additional sequences in suspected vascular disease.* General sequences may be sufficient to identify previous cerebral ischaemic disease or small vessel disease often associated with hypertension. In the acute presentation where a vascular cause is suspected and active management is contemplated, additional imaging sequences are required to give more information about the site, size, and cause of the stroke:

Cerebral infarction. Diffusion-weighted imaging and a haemorrhage sensitive T2-weighted gradient echo sequence are helpful. Perfusion imaging and MR angiography may be used to further define areas at risk of frank infarction and to demonstrate any large vessel occlusion. If there is a mismatch between these perfusion and diffusion images with the areas on perfusion being much greater than diffusion, then one considers that salvageable brain is at risk and this may influence the speed and type of therapy given. Diffusion-weighted imaging allows distinction between old and new infarcts since the restricted diffusion of recent ischaemia fades after a few weeks.

Cerebral haemorrhage. MRI is not particularly good at imaging intracerebral haemorrhage, as blood undergoes a complex series of signal changes as a clot evolves. MRI is very good at identifying aneurysms by angiographic re-formatting from volumetric acquisitions. Gradient echo sequences may detect the 'footprint' of previous haemorrhage. Acute subarachnoid haemorrhage is often visible on FLAIR sequences.

Arterial dissection. This is probably identified best using axial fat saturated T1-weighted sequences to image thrombus in the lumen of the artery and an MR arteriogram from the aortic arch upwards.

Cerebral venous sinus thrombosis. This can be imaged using a 'time of flight' or phase contrast MR venogram.

◆ *Additional sequences in demyelination.* Although conventional MRI is widely used in diagnosis and monitoring multiple sclerosis, modern quantitative techniques using magnetization transfer and diffusion-weighted MRI allow better quantitative measurement of the extent of structural changes. MR spectroscopy can help monitor inflammatory demyelination and axonal injury (Filippi *et al.* 2002).

◆ *Additional sequences in suspected neoplasm.* MRI with standard T1- and T2-weighted sequences along with fluid attenuated inversion recovery, or FLAIR sequences, will identify intracerebral abnormalities in virtually all patients who have an intracerebral or spinal cord tumour. The difficulty lies in correctly identifying the abnormality as a tumour and not some other underlying process such as inflammation, infection, stroke, arteriovenous malformation, or haematoma. The differential diagnosis, and positive predictive value of MRI in this setting is covered in Section 27.4.3. In cases where the CT scan suggests tumour, volumetric T1-weighted post-contrast MRI is useful for planning image-guided biopsy. Functional MRI paradigms can be helpful in localizing the areas eloquent for language, motor or sensory functions. This information can be used to plan surgical approach to avoid damaging these centres where possible. In certain centres with a special expertise in interpreting functional MRI, MRI within the operating theatre allows the surgeon to update target localization in near real-time to maximize tumour resections; also it can allow early identification of vascular complications related to haemorrhage. This is expensive as all the surgical instruments must be MRI compatible. The physical constraints of operating within an MR scanner also preclude all but straightforward surgical procedures. MR spectroscopy has been used to give more non-invasive information about whether the mass is likely to be a low grade process or a high grade tumour and often can give a better indication of whether it is a primary or secondary tumour. Its diagnostic accuracy is usually not sufficient to obviate the need for biopsy. In situations where a mass recurs after surgery and radiotherapy, MR spectroscopy can often give a non-invasive indication of whether the mass represents tumour recurrence or radiation-induced necrosis.

◆ *Intracranial pressure.* MR imaging cannot directly or indirectly measure intracranial pressure, but can help identify an underlying structural cause such as hydrocephalus, mass lesion, or cerebral venous sinus thrombosis, low intracranial pressure signified by diffuse enhancement and thickening of the dura with sagging of brain and flattening of the ponto-medullary junction and mild prolapse of the cerebellar tonsils downwards through the foramen magnum.

◆ *Neurological infection.* Abscess, cerebritis, and encephalitis are more likely to be identified by MRI than CT scanning. The wall of an abscess can be low signal on T2-weighted images. Pus produces hyperintense signal on diffusion-weighted imaging and reduced Apparent Diffusion Co-efficient values because of high viscosity. This is the opposite of what is usually identified in tumour necrosis where the centre is of low viscosity (Nadal Desbarats *et al.* 2003). However, false positives and some false negatives occur, particularly with brain metastasis and glioblastoma.

Paradoxically, in immunocompromised patients for instance with AIDS, abscesses may demonstrate hypointense signal on diffusion-weighted imaging, whilst lymphomas in this group may display restricted diffusion (Camacho *et al.* 2003).

◆ ***Additional sequences in suspected degenerative diseases.*** For most purposes general MR sequences are all that are required in suspected degenerative disease. They adequately demonstrate distribution of atrophy, for instance frontal versus temporo-parietal; ventricular size relative to cortical atrophy as is relevant to normal pressure hydrocephalus; small vessel white matter change occurring in dementia or multi-infarct/Binswanger disease; extensive confluent white matter change in radiation-induced leucoencephalopathy or associated with HIV; the size of caudate nucleus and putamen in Huntington's disease or of the basal ganglia and brainstem nuclei in Parkinson's Disease, Multi-System Atrophy, Progressive Supranuclear Palsy, Striatonigral Degeneration or Cortico-Basal Degeneration; and atrophy of the cerebellum. MRI can be used to try to differentiate cortico-basal degeneration or progressive supranuclear palsy from multi-system atrophy by calculating regional areas of interest and measuring size allowing positive prediction of over 90 per cent, but these techniques are time consuming and not routine (Yekhlef *et al.* 2003). Genetic tests for Huntington's disease make the imaging changes redundant as a diagnostic tool.

T2-weighted and diffusion-weighted imaging may be helpful in showing abnormal signal of the caudate nucleus and putamen in classic Creutzfeldt–Jakob disease (Collie *et al.* 2001). The 'Pulvinar sign', is seen in >75 per cent of new variant cases, but not in classic Creutzfeldt–Jakob disease (Zeidler *et al.* 2000). Degenerative conditions affecting the basal ganglia, such as striatonigral degeneration, may show putaminal atrophy and hypointensity. The 'cross sign' on T2-weighted and proton-weighted MRI images of the pons is seen in multisystem atrophy. The appearance of dorsolateral hypointensity and slit hyperintensity may be seen on T2-weighted and FLAIR sequences in striatonigral degeneration (Block and Bakshi 2001).

◆ ***Trauma.*** In cranial trauma, MRI or CT may show cerebral swelling, intracerebral haemorrhage, subdural or extradural haematomas, or to be normal despite significant head injury. MRI is superior at identifying areas of diffuse axonal injury, damage to the corpus callosum, brainstem, and white matter damage especially in the frontal and temporal lobes. Delayed neuro-imaging usually identifies only cerebral atrophy and ventricular dilatation. In boxers subjected to repeated head trauma, a cavum septum pellucidum is often seen. MR diffusion-weighted imaging is more sensitive at demonstrating changes associated with diffuse axonal injury than conventional MRI. (Huisman *et al.* 2003). The majority of lesions seen on diffusion-weighted imaging, but not conventional MR, are the result of decreased diffusion. SPECT and PET can also reveal areas of under-perfusion and decreased metabolism and may be more sensitive than anatomical imaging.

◆ ***Epilepsy.*** Whereas CT brain scanning has a low sensitivity at identifying abnormalities in unselected patients with epilepsy, in less than <30 per cent, MRI is much more sensitive and is the imaging investigation of choice, when readily available. The most commonly identified lesion in patients with temporal lobe epilepsy is mesio-temporal sclerosis. MRI shows hippocampal atrophy and increased signal on T2-weighted or FLAIR images. These are seen to greatest effect using angulated coronal imaging at right angles to the long axis of temporal lobe structures. Volumetric scans improve accuracy but the measurements are time consuming and may contribute little to overall management. MRI may identify small vascular malformations or cavernous haemangiomas or diffuse abnormalities such as lissencephaly, band heterotopia, and focal cortical dysplasia (Barkovich *et al.* 2001).

◆ ***Functional neurosurgery.*** Image-guided procedures have allowed major advances in functional neurosurgery over the last 10 years. Stereotactic localization systems mainly rely on pre-operative images which are sensitive to the inevitable brain shifts that accompany craniotomy, whether slight or significant. Shifts associated with stereotactic biopsy, lesional procedures, or placement of stimulators are less of a problem. The main uses are lesions of the thalamus or pallidum for movement disorders or the placement of deep brain stimulation electrodes or implantation of dopaminergic cells for treatment of Parkinson's disease.

◆ ***Meninges, cranial nerves, base of skull, and orbits.*** These regions are best imaged using MR scanning with Gadolinium to increase the sensitivity. Coronal and sagittal imaging are particularly useful when imaging the para-pituitary region and base of skull (Fig. 3.6). Special sequence MRI is rarely necessary, unless in rare circumstances where vascular sequences are required to look for aneurysms or dural arterio-venous malformations.

◆ ***Spinal cord, roots, and plexus.*** MRI is undoubtedly the best imaging modality to use in patients with suspected spinal cord or nerve root problems and is superior to CT scan even with intrathecal contrast. Gadolinium increases the sensitivity and specificity in imaging neoplastic processes or following surgery. Special images are not generally required unless there is considered to be an underlying vascular cause, in which case gradient echo sequences can identify old haemorrhage. It is possible to obtain angiographic images of the spinal cord although the resolution of these images is not yet sufficient to replace catheter angiography in diagnosing spinal arteriovenous malformation.

3.1.3 Functional MRI

Functional MRI, fMRI, depends on the identification of a change in blood oxygenation levels. This is the blood oxygen level dependent 'BOLD' signal. In most situations metabolism and flow are coupled. With neural stimulation these may relatively uncouple with a higher increase in flow compared to oxygen utilization; the delivery of oxygenated blood will exceed the needs of the tissue and the oxygen extraction level falls (Fig. 3.7). Functional MRI signal depends on a change in regional signal from baseline to activated state. It is important that the baseline is stable or the signal may vary and, after averaging, only relative quantification of blood oxygen level dependent signal will be possible. Analysing activation studies may be difficult but statistical parametric mapping analysis using multiple comparison correction and allowances for differences in regional variation may make this easier, depending on the activation paradigm.

In its clinical applications, functional MRI has been able to demonstrate clearly the functional spreading depression of Leao in

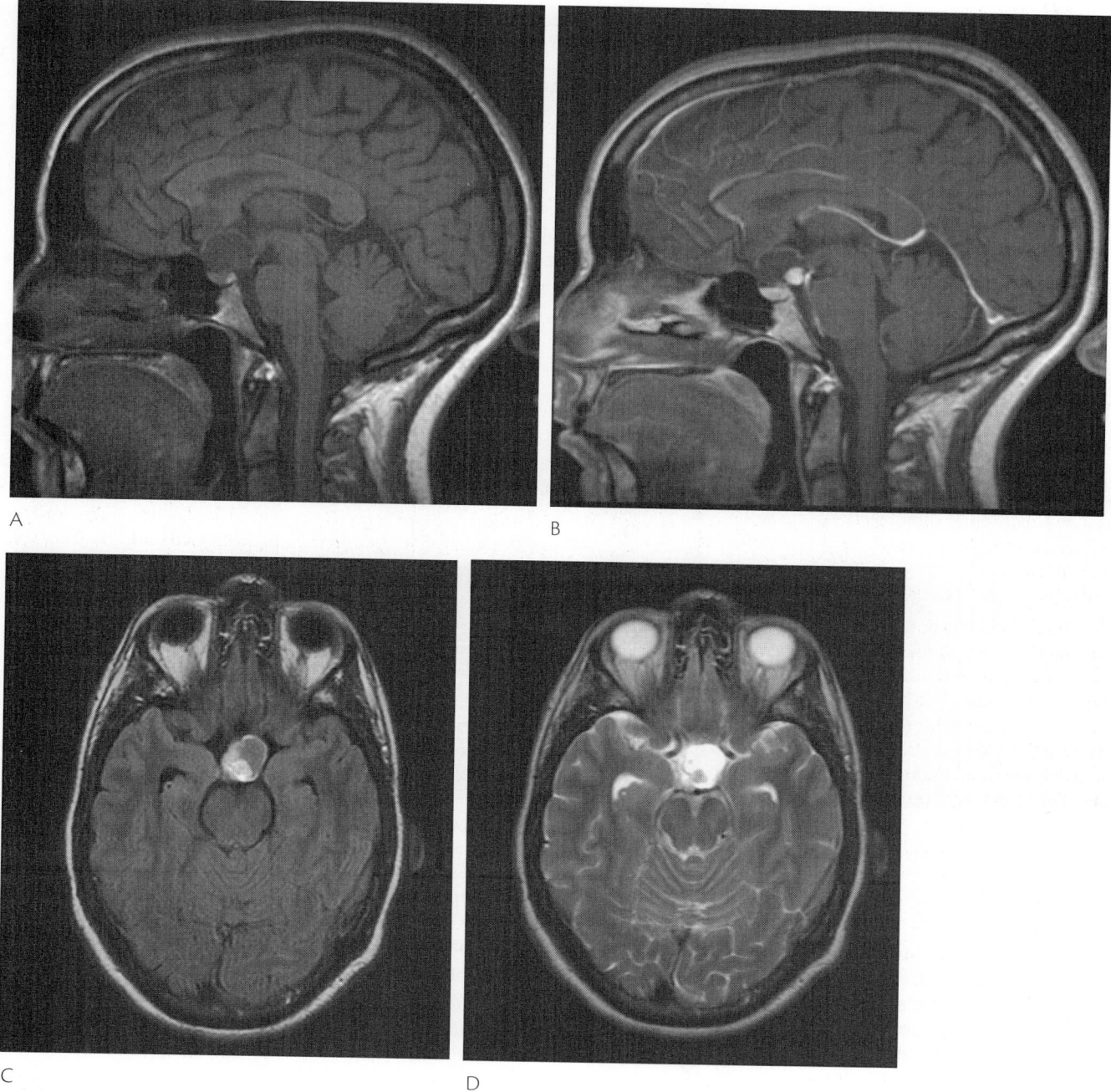

A

B

C

D

Fig. 3.6 The effect of Gadolinium contrast injection is clearly demonstrated on these pre-contrast (A) and post-contrast (B) T1-weighted sagittal MR images of this patient's suprasellar region. There is a well-defined partly cystic lesion with nodular enhancement in the wall. The appearances and position are typical of Craniopharyngioma. The axial T2-weighted image (C) shows the fluid in the cyst has high signal. This does not lose its signal completely on the fluid attenuated, FLAIR, sequence (D). This is due to protein in the tumour cyst fluid.

patients with migraine with aura (Hadjikhani *et al.* 2001). Also, it has been used in the planning of brain surgery for brain tumours. The rationale is that fMRI can identify eloquent areas of the brain to allow a more complete resection of a tumour with a lower likelihood of damaging the functionally important brain area. fMRI has been used alongside intraoperative cortical stimulation for these purposes or with registered fMRI in a neuronavigational system.

Language function is complex and fMRI cannot be used to make critical surgical decisions in the absence of direct brain mapping (Rutten *et al.* 1999). The same is the case for visual areas. fMRI is hindered mainly by the two methodological problems of what the analysis threshold should be and by the choice of paradigm task. Variations in these for language and vision can significantly alter the fMRI activated areas, especially the number of activated areas

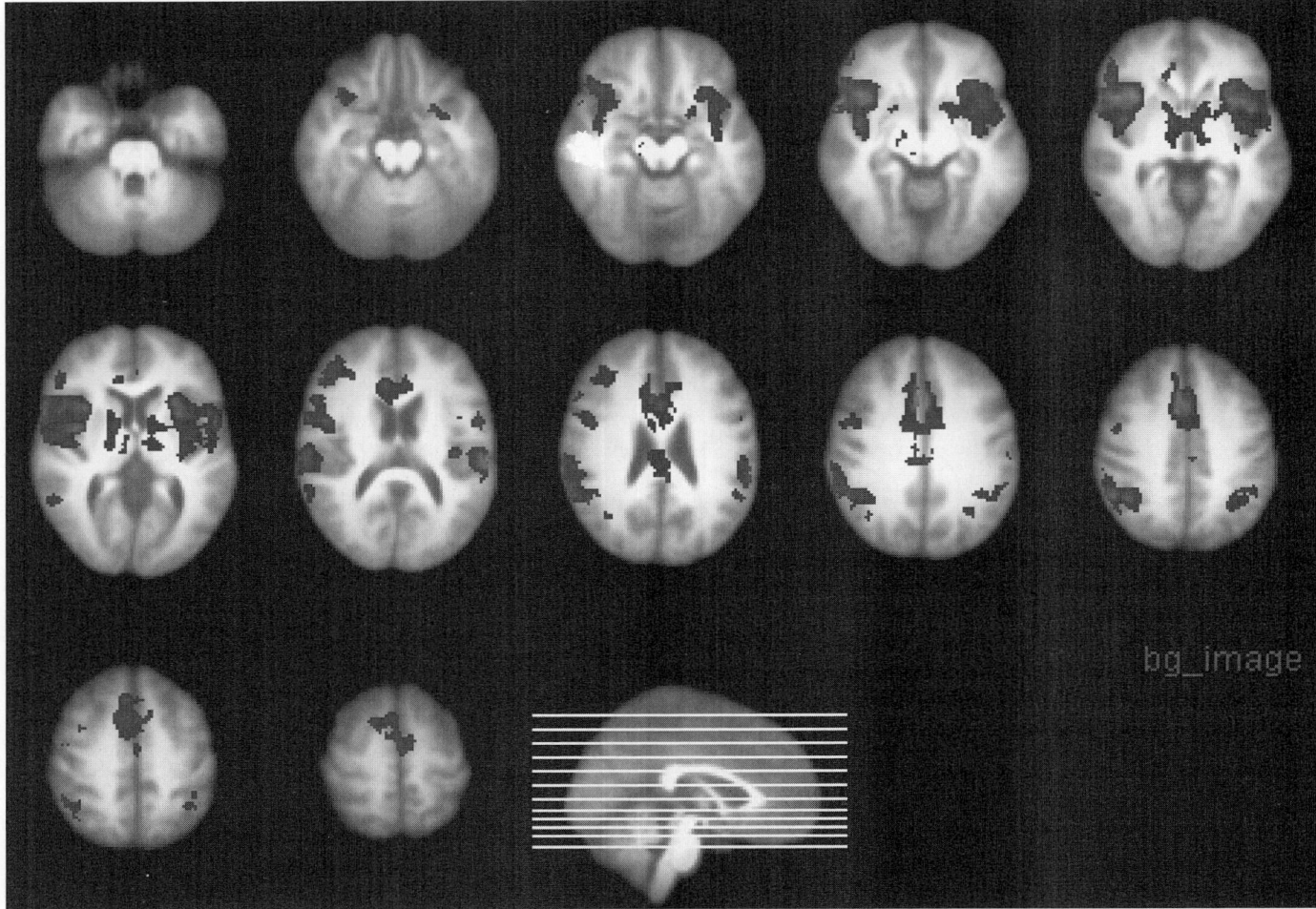

Fig. 3.7 Functional MRI, or fMRI, showing areas of brain activity, as represented by increased blood flow, when a painful heat stimulus is applied to the limb of a patient with rheumatoid arthritis. (Courtesy of Professor I Tracey.) (See Plate 1.)

and in their spatial extent. Multiple different selected fMRI paradigms may improve sensitivity and specificity for detection of essential language areas. Yet at present the clinical value of fMRI remains uncertain as there are no randomized studies of its addition to intra-operative cortical stimulation alone.

3.1.4 Magnetic resonance spectroscopy

MR spectroscopy, MRS, has been used to study regional pathophysiology in various conditions including stroke, tumours, inflammation and infections. It is represented as a number of spectral peaks. The MRS NAA peak is due to *N*-acetyl aspirate and is a neuronal marker. Any process damaging neurons, such as infarct, tumour, epilepsy, or dementia will lead to a reduction in this NAA peak. Glutamate is co-located in the neurone and is an excitatory amino acid. It is released if there is acute damage to the neurone and therefore the glutamate peak is increased in infarction. Creatine is relatively stable irrespective of the state of the brain and acts as a standard with which to measure the other metabolites. It can decrease in hypermetabolic conditions and increase in hypometabolic states. Choline is a precursor for acetylcholine and is also contained in phospholipids. Increased tissue turnover will in turn increase the choline peak. Lactate indicates presence on non-oxidative respiration rather than the usual oxidative process.

◆ *Stroke*: Vascular damage may result in ischaemia and necrosis with resulting high levels of lactate, depressed choline, NAA, and creatine.

◆ *Brain tumours:* MRS is generally abnormal in patients with brain tumours. The characteristics in astrocytoma include a reduction in the NAA, reflecting neuronal loss, a moderate reduction in creatine, reflecting altered metabolism and an elevation in choline, reflecting increased membrane synthesis and cellularity. Elevated choline in the presence of lactate reflects tissue hypoxia and seems to correlate with the grade of the malignancy; lactate is increased in glioblastoma multiforme. The spectroscopic changes with glioblastoma multiforme are similar to those with primary central nervous system lymphoma or metastasis.

◆ *Demyelination:* In the acute stages of demyelination, the NAA is normal, reflecting the lack of permanent axonal damage, but the choline:creatine ratio is increased. In subacute and chronic lesions there is a decrease in the NAA/creatine ratio corresponding with damage to the neurons and in the late stages an increase in free lipids in the plaques relating to destroyed myelin.

◆ *Infection:* The MRS associated with brain abscess is different from that with tumour. There are larger peaks associated with the lactate, pyruvate, and acetate related to microorganisms, enzyme release, or white blood cells.

3.1.5 Single photon emission computed tomography

Single photon emission computed tomography, SPECT, is a radio-isotope based imaging method for examining brain perfusion, reflecting metabolic activity generally 30 min after injection of 740 MBq, 20 mCi of Tc-99m HMPAO using a high resolution dedicated brain SPECT gamma camera. This agent rapidly crosses the blood brain barrier and is taken up by the brain. The concentration in a given volume of brain tissue corresponds closely to the perfusion of that tissue. This information can be used to produce a map of perfusion within the brain. Activity is greatest along the frontal, temporal, parietal, and occipital lobes of the cortical grey matter mantle. Activity is also high in the basal ganglia and thalamus. There is less activity in the regions between the basal ganglia and the convexity corresponding anatomically to cortical white matter and the ventricles. Images can be obtained using multidetector or rotating gamma camera systems. The multidetector system provides a higher sensitivity, better spatial resolution, and is faster. Images from either system can be reconstructed in any plane but the spatial resolution is very poor in comparison to MRI or CT. Perfusion is frequently referred to as regional cerebral blood flow. However there are few real indications for the accurate measurement of regional cerebral blood flow since the methodological limitations mean that the measurements only approximate to regional cerebral blood flow. Generally, imaging differences between brain regions are all that is required to reveal an abnormality.

Other radiopharmaceuticals that can be used for SPECT are: Iodine 123 which is lipophilic and crosses the blood brain barrier well; 99m-Technicium ligands which are lipophilic, rapidly uptaken, but underestimate regional cerebral blood flow; Xenon 133 which is an inert gas which can be used to measure regional cerebral blood flow, and its rapid clearance allows many sequential studies within same day, but it is difficult to image because of poor spatial resolution; and neuroreceptor ligands which have been used to study perfusion or receptor binding by acetylcholine, dopamine, serotonin, or benzodiazepine.

SPECT can be useful for conditions where there has been a disturbance of perfusion, either as the result of structural abnormalities due to tumour, abscess, infarct, or bleed, or from a metabolic functional disturbance such as focal epilepsy, or low perfusion states due to migraine, dementia, Parkinson's disease, or other degenerative disorders. SPECT has been used in suspected Alzheimer's disease, however, in practice its value remain uncertain (Knopman *et al.* 2001). SPECT studies have demonstrated contralateral cerebral hemispheric and ipsilateral hemispheric hypoperfusion in a child with hemiplegic migraine but this was not supported by other studies which were more consistent with prolonged hyperperfusion and oedema (Crawford and Konkol 1997; Barbour *et al.* 2001).

Infection

Initial studies reporting high sensitivity and specificity in differentiating abscess from tumour have been challenged as it seems the size of the lesion primarily determines the accuracy of the differentiating abscess from lymphoma and toxoplasma abscesses in immunocompromised patients (Ruiz *et al.* 1994; Young *et al.* 2005).

Epilepsy

SPECT has been used as part of the evaluation of candidates for temporal lobe resection for epilepsy. Ictal SPECT scans show

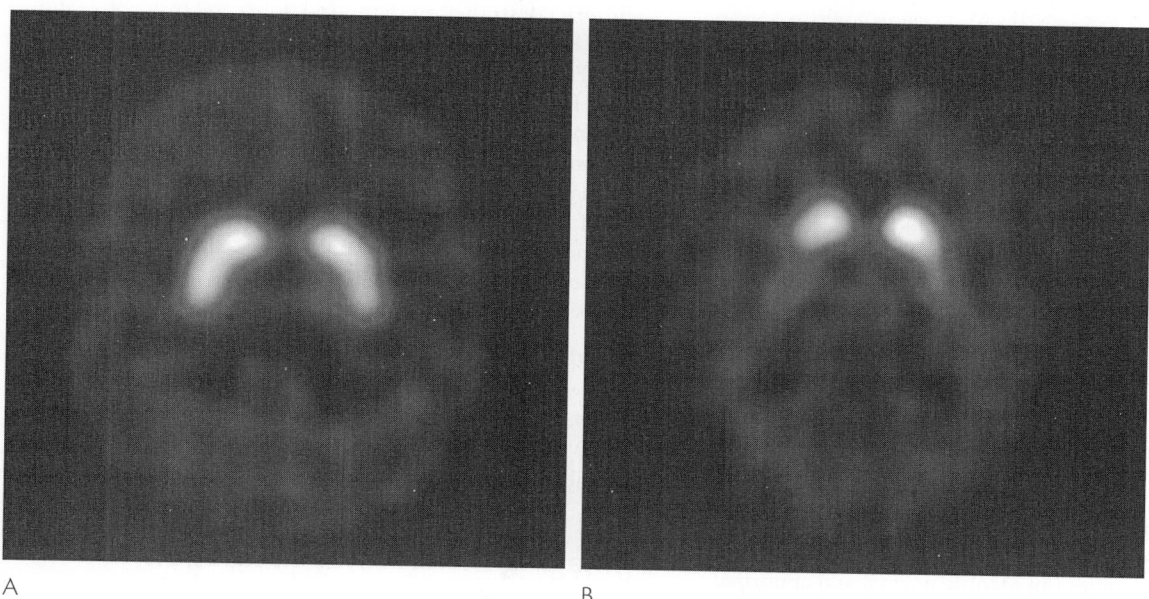

A   B

Fig. 3.8 DAT scan of patient with Parkinson's disease compared with patient with essential tremor. The findings in normal patients or patients with essential tremor show normal signal intensity and isotope uptake in the basal ganglia (A), whereas in Parkinson's disease there is reduced uptake (B), which is often asymmetrically, in unilateral Parkinson's disease. (Courtesy of Dr D Grosset.) (See Plate 2.)

increased uptake in the affected temporal lobe in over 90 per cent of cases of definite temporal lobe epilepsy, and may be very useful too in localizing other brain foci. However this role is often not cost effective, as structural imaging has become more specific.

Movement disorders

SPECT has lower spatial resolution than PET but is still useful in studying large regions of interest such as the basal ganglia. The main neuropathological feature in Parkinson's disease is a severe degeneration of the dopaminergic neurons in the substantia nigra resulting in a loss of the dopamine DA transporters in the striatum (Fig. 3.8). This is translated to decreased striatal I^{123}-N-omega-fluoropropyl-2beta-carbomethoxy-3beta-(4-iodophenyl) nortropane ($[I^{123}]$FP-CIT) binding on SPECT scan. Such I^{123} FP-CIT SPECT, or DAT scan, can be helpful in differentiating idiopathic Parkinson's disease from essential tremor and drug-induced tremors, where the scan shows normal uptake.

It may help also to distinguish young onset Parkinson's Disease from dopa-responsive dyskinesia (Naumann *et al.* 1997). $[I^{123}]$ beta-CIT and [Tc-99m] TRODAT-1 target the synaptic membrane dopamine transporter system, while $[I^{123}]$-IBZM and $[I^{123}]$epidepride target the D2 postsynaptic receptor binding sites. SPECT has been used to differentiate corticobasal degeneration from progressive supranuclear palsy. In the former, there is asymmetrical blood flow to the cortical and basal ganglia regions, whereas in progressive supranuclear palsy the changes are symmetrical (Zhang *et al.* 2001).

Contraindications

The main concern with SPECT imaging is that the tracers are radiopharmaceuticals, therefore they should not be give to pregnant women. Women should stop breast feeding for a few days after the test. Young children should not be scanned unless clearly indicated. The average radiation dose is significantly higher than from a brain CT scan.

3.1.6 Positron emission tomography

Positron Emission Tomography, PET, is another radioisotope imaging modality. The isotopes employed decay by emitting positrons. Positrons immediately annihilate to emit two high energy electrons which travel in opposite directions. These physical properties allow improved spatial resolution and increased receptor sensitivity in comparison to SPECT. PET can be used to measure brain perfusion and brain metabolic activity by using labelled water, using Oxygen-18, and a glucose analogue, [18F]-FDG, respectively (Fig. 3.9). High flow or activity may accompany increased excitation or inhibition since both require energy. Changes in flow may not coincide with changes in local metabolism or neuronal activity under pathological conditions.

PET may be performed in the resting state or under physiological activation. Resting state PET has demonstrated abnormal function in patients with dystonia (Magyar-Lehmann *et al.* 1997; Eidelberg *et al.* 1998) and blepharospasm (Hutchinson *et al.* 2000). Physiological activation studies can identify normal and abnormal brain function not seen under resting state studies. When interpreting physiological activation studies it is assumed that PET counts are linearly related to regional cerebral blood flow and so the estimation of regional responses assumes no global blood flow shift produced by activation. If there are global shifts in cerebral blood flow, regional changes may not give accurate absolute change in local flow or neuronal activity.

PET can measure activity from presynaptic dopaminergic neurons. The dopa analogue [18F] fluorodopa accumulates in the human striatum after administration intravenously. Fluorodopa uptake mainly reflects decarboxylase activity which indirectly indicates residual nigrostriatal neurons. There is reduced uptake of presynaptic dopaminergic nigrostriatal neurons in dystonia. PET can also be used to measure dopaminergic radioligand binding, such as [18F] spiperone. This ligand binds 70 per cent to dopaminergic D-2 like receptors and 30 per cent to serotonergic S2 type receptors in the primate putamen. Putaminal D2 receptor binding is decreased in those people with primary focal dystonia.

[18F] FDG PET is useful in identifying systemic tumours in patients with neurological paraneoplastic syndromes and unknown primary site, although there are some false positive and false negative scans (Rees *et al.* 2001).

PET is still trying to find a cost-effective role in assessment for primary surgery for malignant brain tumour, but has been purported to improve the percentage of tumour safely espectable in image directed surgery when integrated with brain MRI (Pirotte *et al.* 2006). It is also a useful tool in the biological monitoring of response to treatment in glioma using L-[methyl-11C] Methionine PET and in differentiating tumour necrosis from radiation necrosis, although again false positives and false negative scans do occur (Narai *et al.* 2005).

PET imaging has been able to provide supporting evidence for pathogenic mechanisms for certain disorders where there is no apparent structural cause. For instance an area in the dorsal rostral midbrain becomes active at the initiation of a migraine attack in patients without aura and remains active till after the headache phase resolves (Weiller *et al.* 1995). This supports the theory that the brainstem may act as a 'migraine generator' in some patients with migraine.

In interictal studies of temporal lobe epilepsy, PET shows hypometabolic areas in epileptogenetic areas in almost 80 per cent of cases (Engel 2000), but these may be more extensive than those identified in structural imaging or EEG. PET studies have a reducing role in epilepsy, since MR scanning has become so sensitive in detecting mesiotemporal sclerosis or other structural lesions. 11C-flumazanil labelling studies demonstrate reduced binding at the site of an epileptogenic focus and are more specific than FDG PET studies, although the findings may vary depending on how recently seizure activity has taken place, thus limiting its clinical usefulness.

Contraindications

Pregnancy is the only contraindication in view of the radioactive tracers given. Claustrophobia or inability to lie still on the scanner bed for 40 min and insufficient venous access for radioactive tracer are the only other reasons why scanning would not be possible.

3.2 Imaging the cerebral circulation

3.2.1 Introduction

The main clinical indications for imaging the cerebral circulation are:

- transient ischaemic attack so as to identify arterial stenosis;
- acute ischaemic stroke so as to identify vessel occlusion;

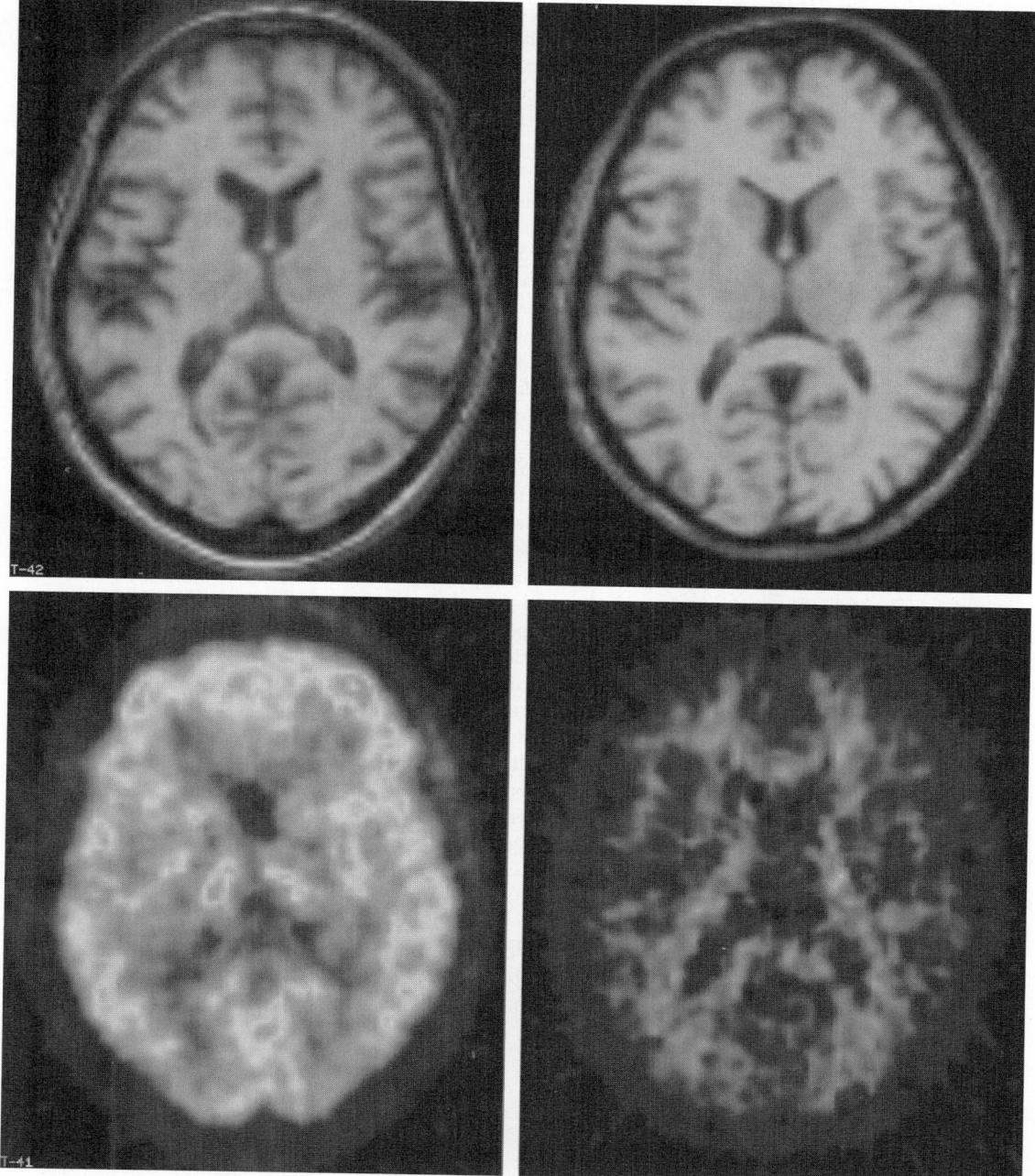

Fig. 3.9 Positron emission tomography, PET, in a 56-year-old patient with Alzheimer's disease (left) and a 55-year-old normal control. Top: MRI brain. Bottom: ¹¹C-PIB PET scan for amyloid. (Courtesy of Professor D Brooks.) (See Plate 3.)

◆ intracerebral haemorrhage in order to identify an underlying vascular malformation; and

◆ to detect possible arterial dissection, fibromuscular dysplasia, or other arteriopathies, cerebral aneurysm, intracranial venous thrombosis, or cerebral vasculitis.

In contrast to pharmaceutical products, diagnostic and imaging technologies are not subject to stringent regulatory control, and no standards are set for validation. Thus the evidence-base on important issues such as diagnostic sensitivity and specificity is often poor. For example, although several hundred studies of carotid imaging have been published over the last few decades, most are undermined by poor design, inadequate sample size, and deficiencies in analysis and presentation of data (Rothwell *et al.* 2000a).

3.2.2 Catheter angiography

Cerebral angiography, introduced by Moniz in Portugal in the 1930s, was the first method to display the cerebral circulation during life. Originally it required the intra-carotid injection of material which was opaque to X-rays. Over the years the technology has improved, with less toxic contrast material, femoral artery catheterization, digital imaging, and catheters that can be controlled and introduced into vessels as small as cortical branches of the middle cerebral artery.

Before the introduction of axial imaging of the brain in the early 1970s, first by CT and then by MRI, catheter angiography was used to identify intracranial mass lesions, hydrocephalus, and other structural abnormalities. Nowadays, with the increased use of CT angiography, MR angiography and ultrasound imaging, catheter angiography is more or less confined to displaying arterial stenosis and occlusions due to vascular disease such as atheroma, vasculitis or dissection; intracranial venous thrombosis; small intracranial aneurysms; and intracranial vascular malformations. The reasons for this diminishing role are that although catheter angiography remains the 'Gold Standard' technique in many situations, it is inconvenient, invasive, uncomfortable, costly, requires hospital admission, and carries a risk. For example, a systematic review of prospective studies of the risks of catheter angiography in patients with cerebrovascular disease reported a 0.1 per cent risk of death and a 1.0 per cent risk of permanent neurological sequelae (Hankey *et al.* 1990), although more recent studies have reported lower risks (Johnston *et al.* 2001). Mechanisms of transient ischaemic attack and stroke complicating angiography include dislodgement of atheromatous plaque by the catheter tip, dissection of the arterial wall, thrombus formation on the catheter tip, and air embolism (Gerraty *et al.* 1996). In addition, there are systemic and allergic adverse effects of the contrast material, particularly during intravenous digital subtraction angiography where large quantities are used. Some patients develop a haematoma, aneurysm, or nerve injury at the site of arterial puncture, which is usually into the femoral artery in the groin. The occasional patient develops *new* or worsened symptoms of peripheral vascular disease in the leg distal to the puncture site, sometimes even leading to amputation.

Compared with cut-film selective intra-arterial catheter angiography recorded directly onto X-ray film, intra-arterial digital subtraction angiography (Fig. 3.10) is quicker, the images are easier to manipulate and store, contrast resolution is better although spatial resolution is less, but there is no evidence that less contrast is used or that it is much safer (Warnock *et al.* 1993). Even for imaging only as far as the carotid bifurcation, neither **i**ntravenous **d**igital **s**ubtraction **a**ngiography, IVDSA, nor arch aortography is a satisfactory alternative to selective intra-arterial angiography (Pelz *et al.* 1985; Rothwell *et al.* 1998; Cuffe and Rothwell 2006).

Even with selective catheter angiography, there can be difficulty in distinguishing occlusion from extreme internal carotid artery stenosis, and then late views are needed to see contrast eventually passing up into the head. Moreover, because of the localized and non-concentric nature of atherosclerotic plaques, biplanar, and preferably triplanar (Jeans *et al.* 1986; Cuffe and Rothwell 2006), views of the carotid bifurcation are required to measure the degree of carotid stenosis accurately. This involves visualising the residual lumen without overlap of other vessels, measurement at the narrowest point, and comparison with a suitable denominator to derive the percentage diameter stenosis.

Catheter angiography can also provide information about ulceration of carotid plaque and complicating luminal thrombosis, albeit with only moderate inter-observer agreement (Streifler *et al.* 1994; Rothwell *et al.* 1998). However, angiographic irregularity and ulceration predict a higher risk of stroke than if the plaque is smooth, given the same degree of stenosis (Eliasziw *et al.* 1994; Rothwell *et al.* 2000b) and there is good correlation between catheter angiographic plaque morphology and histology when the

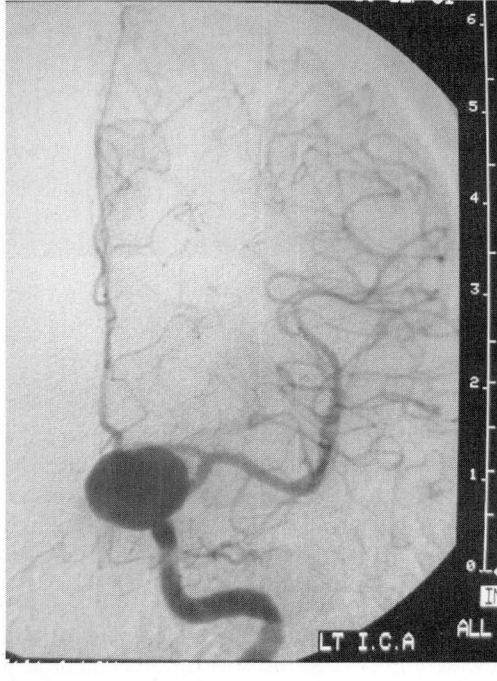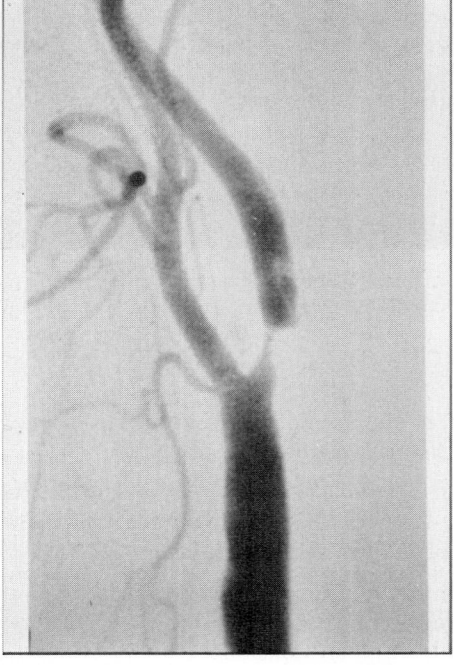

Fig. 3.10 Digitally subtracted arterial angiograms showing a large distal internal carotid artery aneurysm (A) and a severe stenosis of the proximal internal carotid artery (B).

A   B

latter is rigorously evaluated (Lovett *et al.* 2004). Luminal thrombus is relatively unusual, but is thought to be associated with a high risk of stroke (Martin *et al.* 1992).

Catheter angiography is still widely used to reveal a vascular malformation or aneurysm in young patients with intracerebral haemorrhage. In older patients, bleeds due to either hypertension or amyloid angiopathy are predominant. Vascular malformations and aneurysms are most likely if the cerebral haemorrhage is lobar or intraventricular and there is no hypertension (Zhu *et al.* 1997). However, MRI can be more sensitive than catheter angiography in picking up vascular malformations, particularly cavernomas, and MR and CT angiography are increasingly able to pick up aneurysms of a size likely to rebleed.

Until recently, catheter angiography was the standard imaging modality to confirm or exclude carotid or vertebral artery dissection (Fig. 3.11) because ultrasound was neither specific nor sensitive enough. However, there is now a widespread consensus that cross-sectional MRI, to show thrombus within the widened arterial wall, combined with MRA, is the safest and best option.

3.2.3 Catheter angiography versus non-invasive imaging

Although non-invasive methods of imaging continue to improve, catheter angiography remains the 'gold standard' against which other vessel imaging methods must be compared, and is also the underpinning method for the interventional neuroradiological treatment of arterial stenoses, aneurysms, and vascular malformations. However, the clinician is often faced with the question as to whether to base decision-making on non-invasive imaging alone or whether to proceed to catheter angiography. For example, multislice CT angiography is widely used in the screening for cerebral aneurysms because of its speed, tolerability, safety, and potential for 3-dimensional reconstructions. The sensitivity of CT angiography for aneurysms over 3-mm diameter is about 96 per cent but is much less for smaller aneurysms. The sensitivity for detecting ruptured aneurysms with CT angiography with conventional angiography as the gold standard is around 95 per cent (Villablanca *et al.* 2002; Chappell *et al.* 2003; Wintermark *et al.* 2003). MR angiography has similar resolution to CT angiography but is less easy to use in sick patients. Four-vessel catheter angiography may still therefore be required when non-invasive imaging is negative.

A similar trade-off between diagnostic accuracy and risk is necessary when imaging the carotid bifurcation in patients with transient ischaemic attack or ischaemic stroke. Performing intra-arterial catheter angiography in everyone is clearly unacceptable because of the risks and cost. Less than 20 per cent of patients will have an operable carotid stenosis; even if only those with 'cortical' rather than 'lacunar' events are selected (Hankey *et al.* 1991; Hankey and Warlow 1991; Mead *et al.* 1999). Confining angiography to patients with a carotid bifurcation bruit will miss some patients with severe stenosis and still subject too many with mild or moderate stenosis to the risks. Nor will a combination of a cervical bruit with various clinical features do much better (Mead *et al.* 1999).

Thus, non-invasive imaging is required at least as an initial screening tool. In many centres, decisions about endarterectomy are now based solely on non-invasive imaging. However, because benefit from endarterectomy is highly dependent on the degree of symptomatic carotid stenosis as measured on catheter angiography, misclassification of stenosis with non-invasive methods will lead to

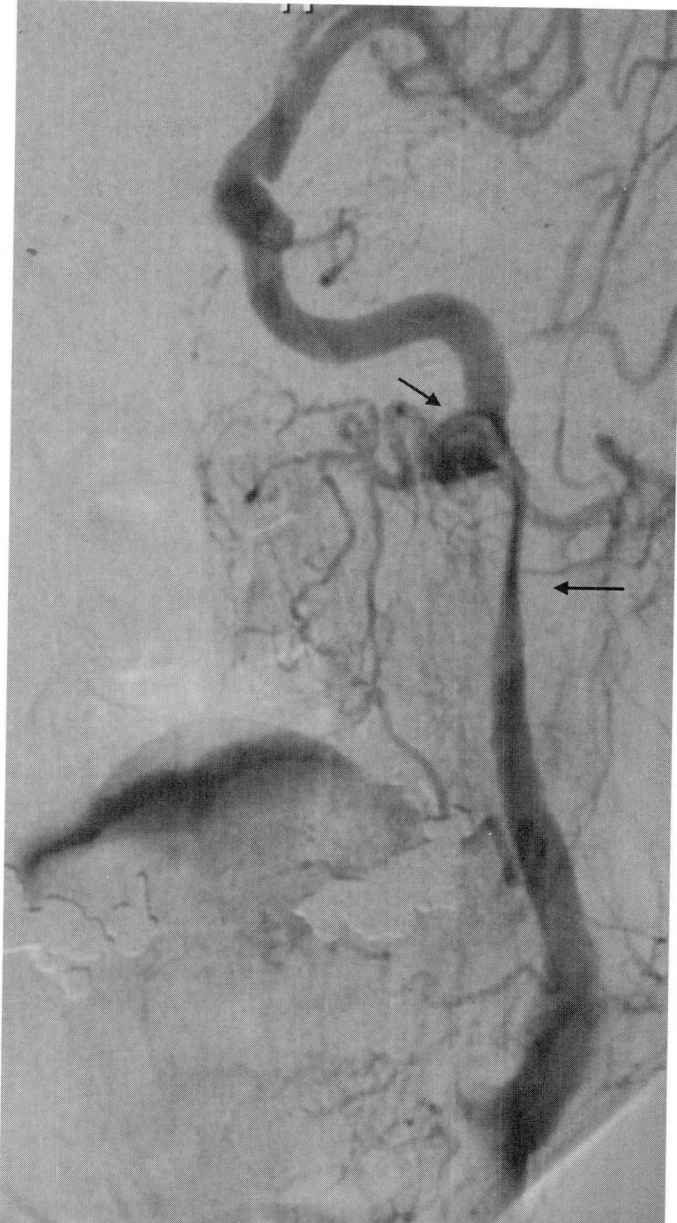

Fig. 3.11 Digitally subtracted arterial angiogram showing a distal internal carotid artery dissection with a tapering lumen (arrow) and a complicating aneurysm (arrow).

some patients being operated unnecessarily, and others being denied appropriate surgery. A meta-analysis of studies of non-invasive carotid imaging published prior to 1995 concluded that non-invasive methods could not substitute for catheter angiography as the sole pre-endarterectomy imaging because of the frequency with which the degree of stenosis was misclassified (Blakeley *et al.* 1995). More recent studies have confirmed this (Johnston and Goldstein 2001; Norris *et al.* 2003; Norris and Halliday 2004; Wardlaw *et al.* 2006). For example, in a comparison of catheter angiography with Doppler ultrasound in 569 consecutive patients in 'accredited' laboratories with experienced radiologists, 28 per cent of decisions about endarterectomy based on Doppler ultrasound alone were inappropriate (Johnston and Goldstein 2001). However, the combination of Doppler ultrasound with another non-invasive method of imaging, such as MR angiography, reduced inappropriate decisions in comparison with catheter angiography to less than 10 per cent in patients in whom the results of Doppler ultrasound and MR angiography are concordant (Johnston and Goldstein 2001). Similar approaches based on two different methods of non-invasive imaging have been shown by other groups to be effective in routine clinical practice (Johnston *et al.* 2002; Barth *et al.* 2006). Catheter angiography is still required in the patients in whom Doppler ultrasound and MR angiography do not produce concordant results.

3.2.4 Duplex sonography

This technique combines real-time ultrasound imaging to display the arterial anatomy with pulsed Doppler flow analysis at any point of interest in the vessel lumen. Its accuracy is enhanced and it is technically easier to carry out if the Doppler signals are colour-coded to show the direction of blood flow and its velocity (Fig. 3.12) Power Doppler and intravenous echocontrast may also help (Droste *et al.* 1999; Wardlaw and Lewis 2005; Gaitini and Soudack 2005). The degree of carotid lumenal stenosis is calculated not only from the real-time ultrasound image, which can be inaccurate when the lesion is echolucent or calcification scatters the ultrasound beam, but also from the blood flow velocities derived from the Doppler signal. If colour Doppler is not available, but only grey-scale duplex, it is usually helpful to first insonate the supraorbital artery with a simple continuous-wave Doppler probe, because inward flow of blood strongly suggests severe internal carotid artery stenosis or occlusion, although not necessarily at the origin.

Although duplex sonography is non-invasive and widely available, there are some difficulties that any ultrasound service must deal with:

- it is very operator dependent and so requires skill, training, and considerable experience to be sure of accurate measurements of stenosis and the avoidance of pitfalls, such as confusing the external with the internal carotid artery;

- it may be difficult to interpret, particularly if there is plaque or periarterial calcification;

- it is not completely reliable in distinguishing very severe stenosis, of which >90 per cent is operable, from occlusion which is inoperable, unless used and interpreted with very great care;

- it is not completely sensitive and specific for severe internal carotid artery stenosis of 70–99 per cent;

- different machines vary in their accuracy in measuring carotid stenosis; and

- it provides little information about the proximal arterial anatomy, although this is seldom affected by disease or relevant to the surgeon, or about distal anatomy.

Nonetheless, with stringent quality control and ideally with confirmation of stenosis by an independent observer, duplex sonography is now the most common way that carotid stenosis severe enough to warrant surgery is diagnosed (Wardlaw *et al.* 2006).

There are no standard and commonly used definitions for the ultrasound appearance of plaques which are soft, hard, or calcified and there is also considerable variation in reporting between and even within the same observers at different times (Arnold *et al.* 1999). Therefore, although unstable and ulcerated plaques are more likely to be symptomatic than stable plaques with fibrous caps, the ultrasound inaccuracy compromises any study of the relationship between plaque characteristics on duplex sonography and the risk of later stroke, and so the selection for carotid surgery (Gronholdt 1999). In asymptomatic stenosis, there is some evidence that a hypoechoic plaque predicts an increased risk of stroke (Polak *et al.* 1998), but this was not confirmed in the Asymptomatic Carotid Surgery Trial (Halliday *et al.* 2004). Until it becomes possible to translate carotid plaque irregularity as seen on catheter angiography, which does add to the risk of stroke over and above the degree of stenosis, into what is seen on duplex sonography, it will remain difficult to use anything other than stenosis to predict stroke risk if only duplex is being used.

Despite these limitations, duplex sonography is a remarkably quick and simple investigation in experienced hands, and it is neither unpleasant nor risky. Very rarely, the pressure of the Doppler probe on the carotid bifurcation can dislodge thrombus, or cause enough carotid sinus stimulation to lead to bradycardia or hypotension (Rosario *et al.* 1987; Friedman 1990). The same conceivably applies to the various arterial compression manoeuvres that may be carried out during transcranial Doppler, and any such compression should be avoided in patients who may have carotid bifurcation disease.

3.2.5 CT angiography and perfusion imaging

CT is now a widely used method for imaging the carotid arteries and cerebral circulation (Brink *et al.* 1997; Bartlett *et al.* 2006). CT is easier to perform in sick patients than MRI and is not contraindicated in those with pacemakers, ferromagnetic implants, mechanical heart valves and those with claustrophobia. However, it does require a large dose of intravenous contrast to outline the arterial lumen, there is X-ray exposure, the images obtained depend on the proficiency of the operator in their selection, and it tends to under-estimate vessel stenosis. Nevertheless, it does provide multiple viewing angles, 3-dimensional reconstruction (Fig. 3.13), and imaging of calcium deposits separately from the vessel lumen

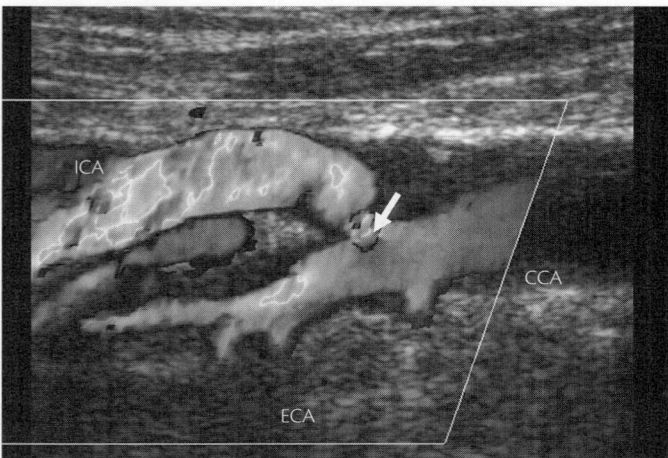

Fig. 3.12 A colour flow Doppler ultrasound of the carotid bifurcation showing a plaque (arrow) at the origin of the internal carotid artery and the resultant stenosis. (See Plate 4.)

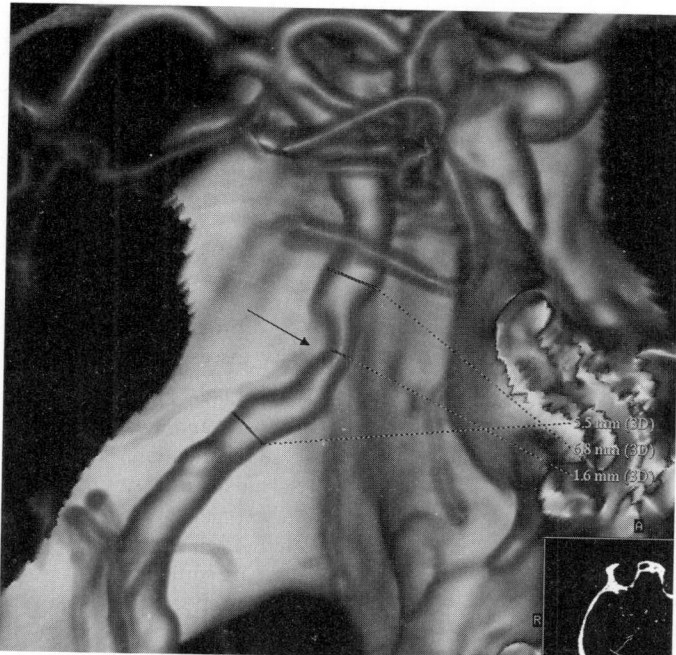

Fig. 3.13 A CT angiogram three-dimensional reconstruction showing a stenosis (arrow) of the distal vertebral artery. (See Plate 5.)

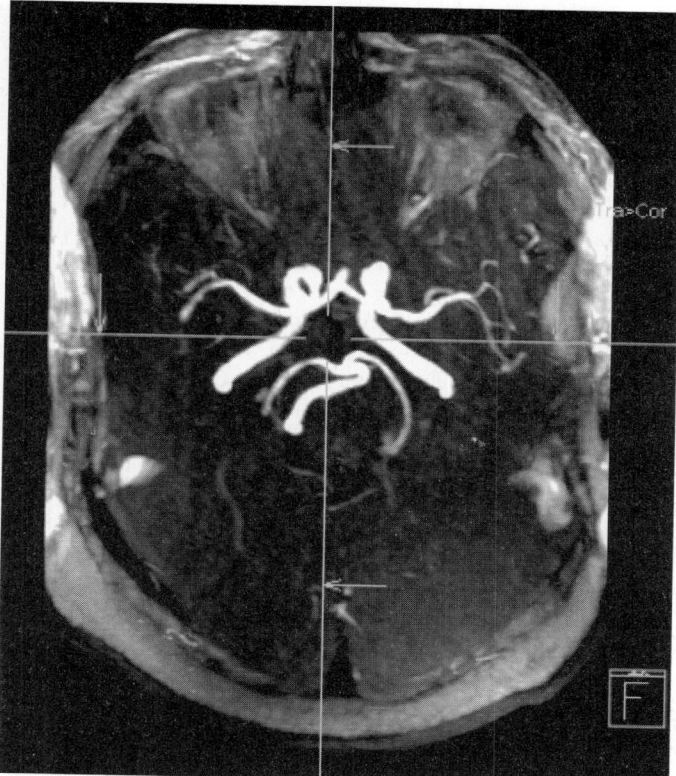

Fig. 3.14 A 'time of flight' magnetic resonance angiogram of the circle of Willis in a patient lacking posterior communicating arteries.

outlined by the contrast (Heiken *et al.* 1993; Leclerc *et al.* 1995; Nandalur *et al.* 2006).

CT perfusion studies can also be used to examine cerebral blood flow (Wintermark and Bogousslavsky 2003) using existing CT technology. CT perfusion using exogenous contrast enables calculation of cerebral blood flow, cerebral blood volume, and mean transit time. However, quantification of cerebral blood flow is problematic, and most studies use ratios comparing values with homologous areas of the contralateral hemisphere which itself may show a reduced cerebral blood flow in acute stroke, for example, due to diaschisis. In addition, current technology allows only limited brain coverage of 2–4 slices which means that large areas of ischaemia are imaged inadequately and small ones may be missed altogether. Nevertheless, there is evidence in acute stroke that perfusion CT values correlate with angiographic findings and can predict infarction and clinical outcome (Nabavi *et al.* 2002; Meuli 2004; Parsons *et al.* 2005; Wintermark *et al.* 2007). It is likely therefore that CT measurements of cerebral blood flow will become increasingly important in selecting patients for acute stroke therapy owing to the speed and relative ease of performing CT as compared to MRI.

3.2.6 Magnetic resonance angiography and perfusion imaging

Magnetic resonance angiography, or MRA, is non-invasive and safe if done without contrast; 'time of flight' imaging (Fig. 3.14). Contrast-enhanced imaging provides improved resolution and reduces problems with flow voids at points of stenosis, and is necessary for detecting cerebral aneurysms or determining the severity of carotid stenosis. However, even with contrast, MRA is unlikely to be accurate enough in estimating carotid stenosis, at least at the present stage of development (Graves 1997; DeMarco *et al.* 2006; Wardlaw *et al.* 2006). The pictures are not always adequate to allow

measurement of the carotid stenosis with movement and swallowing artefacts posing particular problems; the severity of the stenosis tends to be overestimated; there may be a flow gap distal to a stenosis of as little as 60 per cent, making precise stenosis measurement impossible, and even in the posterior part of the carotid bulb, in both cases probably because of loss of laminar flow and increased residence times of the blood; irregularity/ulceration are not well seen; and severe stenosis can be confused with occlusion (Siewert *et al.* 1995; Levi *et al.* 1996; Fox *et al.* 2005). However, image quality and reproducibility of measurement of stenosis are significantly improved with contrast-enhanced MR angiography (Mitra *et al.* 2006; DeMarco *et al.* 2006). So far, there have not been enough methodologically sound comparisons of MRA with catheter angiography (U-king-Im *et al.* 2005; Wardlaw *et al.* 2006). The comparative studies that have been carried out have frequently been overtaken by changes in MR technology.

Perfusion imaging can also be achieved with MRI using an injected contrast agent such as gadolinium or endogenous techniques. The latter have the advantage that they can be used for multiple repeat investigations but at present the level of contrast produced is less that that obtained with exogenous agents. MRI techniques can measure mean transit time, cerebral blood volume, and cerebral blood flow. It has been proposed as a potential tool for use in clinical selection of patients for thrombolysis in acute stroke (Thijs *et al.* 2001; Fiehler *et al.* 2002; Shih 2003), particularly in combination with diffusion-weighted imaging (Kidwell and Hsia 2006; Muir *et al.* 2006), but there is still no consensus on the criteria for patient selection (Hand *et al.* 2006).

3.2.7 Imaging the posterior circulation

Vertebrobasilar transient ischaemic attacks were thought for many years to be associated with a lower risk of stroke than carotid territory transient ischaemic attacks, but recent work has shown that the risk of stroke is at least as high (Flossmann and Rothwell 2003; Flossmann *et al.* 2006). There is therefore an increasing interest in angioplasty and stenting of atherothrombotic stenoses of the vertebral or proximal basilar arteries. MR or CT angiography are the most useful non-invasive methods of imaging the posterior circulation, although catheter angiography is often still necessary to confirm or exclude significant stenosis.

Although asymptomatic subclavian steal is quite common, showing reversed vertebral artery flow detectable by ultrasound or vertebral angiography, symptomatic subclavian steal is rare, presumably because collateral blood flow to the brainstem is enough to compensate for the reversed vertebral artery blood flow distal to ipsilateral subclavian stenosis or occlusion. The clinical syndrome is quite easily recognized by unequal blood pressures between the two arms, a supraclavicular bruit and vertebrobasilar transient ischaemic attacks which may or may not be brought on by exercise of the arm ipsilateral to the subclavian stenosis or occlusion, so increasing blood flow down the vertebral artery from the brainstem to the arm muscles (Bornstein and Norris 1986; Hennerici *et al.* 1988). It is only this sort of symptomatic patient who may require surgery and therefore who has to accept the risk of any preceding angiography. Innominate artery steal is even rarer, with retrograde vertebral artery flow distal to innominate rather than subclavian artery occlusion (Kempczinski and Hermann 1979; Grosveld *et al.* 1988).

3.2.8 Transcranial Doppler sonography

Transcranial Doppler sonography, or TCD, provides information on the velocity of blood flow and its direction in relation to the ultrasound probe, in the major intracranial arteries at the base of the brain, and so whether they are occluded or stenosed (Sloan *et al.* 2004). It is non-invasive, safe, repeatable, not too difficult to perform accurately, can be performed at the bedside, and is not expensive. However, the patient has to keep reasonably still; the examination can take as long as an hour; the skull is impervious to ultrasound in 5–10 per cent of cases, more with increasing age and in females, but less if intravenous echocontrast is used; exact vessel identification may be difficult, but colour-flow real-time imaging makes this easier; spatial resolution is poor; diagnostic criteria vary; and the technique is not always accurate in comparison with cerebral catheter angiography (Baumgartner *et al.* 1997; Baumgartner 1999; Gerriets *et al.* 1999; Markus 1999).

Despite the fact that transcranial Doppler sonography, like positron emission tomongraphy, has increased our knowledge of the cerebral circulation in health and disease, and even though it is inexpensive and quite widely available and repeatable on demand, it still has rather a minor role in routine clinical management. As well as monitoring during carotid endarterectomy, the diagnosis of patent foramen ovale, sickle cell disease, and perhaps in helping define stroke risk (Babikian *et al.* 1997; Molloy and Markus 1999; Dittrich *et al.* 2006; Markus 2006), there are four other possible indications: display of intracranial arterial occlusion and stenosis; emboli detection; assessment of cerebrovascular reactivity; and acceleration of clot lysis following treatment with a thrombolytic agent in acute ischaemic stroke (Alexandrov *et al.* 2004; Kim *et al.* 2005).

Detection of emboli as high-intensity transient signals on the sonogram, so-called microembolic signals, might be of clinical relevance in certain situations. The vast majority of microembolic signals appear to be asymptomatic, their detection may help in distinguishing cardiac and aortic arch from carotid emboli, because with the first two, emboli should be detected in several arterial distributions, whereas with the last in only the one arterial distribution distal to the supposed embolic source (Markus 1994; Sliwka *et al.* 1997; Markus 2006). However, the frequency of microembolic signals can be so frustratingly low and variable that their detection requires prolonged monitoring and automation (Markus *et al.* 1999; Dittrich *et al.* 2006; Markus 2006), and detection is currently therefore used mainly as a research tool, most usefully perhaps as a surrogate outcome in trials of secondary prevention of stroke (Dittrich *et al.* 2006; Markus 2006).

Transcranial Doppler sonography can also be used to assess cerebrovascular reactivity, i.e. the capacity for intracranial vasodilatation in response to acetazolamide, carbon dioxide inhalation or breath-holding, although these three methods do not always produce concordant results (Bishop *et al.* 1986; Markus and Harrison 1992; Dahl *et al.* 1995; Derdeyn *et al.* 2005). However, there is still debate about exactly how to standardize this test and it is not widely used in routine clinical practice. Impaired reactivity may have some prognostic significance for identifying individual patients at particularly high risk of stroke from amongst those with carotid stenosis and internal carotid artery occlusion, although the numbers studied have been small and the situation is not yet clear-cut (Derdeyn *et al.* 1999; Vernieri *et al.* 2001; Derdeyn *et al.* 2005). With time, and presumably increasing collateralization, any impairment of reactivity can return to normal (Kleiser and Widder 1992; Widder *et al.* 1994; Gur *et al.* 1996; Vernieri *et al.* 1999).

3.3 The electroencephalogram

The electroencephalogram, or EEG, averages electrical activity from large numbers of cortical neurones orientated parallel to one another within the cerebral cortex, in both time and space. While it has very good temporal resolution, its spatial resolution is very limited. Large portions of the cortex are too remote for sampling from standard EEG records, such as the medial surfaces of the hemispheres and the basal surfaces of the frontal lobe. Because of the averaging that takes place during recording, the EEG is particularly likely to show changes when the normal, random pattern of cortical activity changes to more synchronized neuronal activity. For this reason, the diagnostic value of the EEG is greatest in disorders such as epilepsy, which is essentially one of hypersynchrony of neuronal hyper-excitability, and in a wide range of other encephalopathies (Daly and Pedley 1990). Further information can be extracted from the EEG by the use of averaging techniques time-locked to sensory stimuli to produce evoked potentials (Section 3.4). The EEG in sleep is covered in Chapter 32.

3.3.1 The normal EEG

Alpha rhythm is the most striking component of the normal EEG. It is of regular and moderate amplitude with a frequency of between 8 and 13 Hz. It occurs during wakefulness with the eyes closed, is

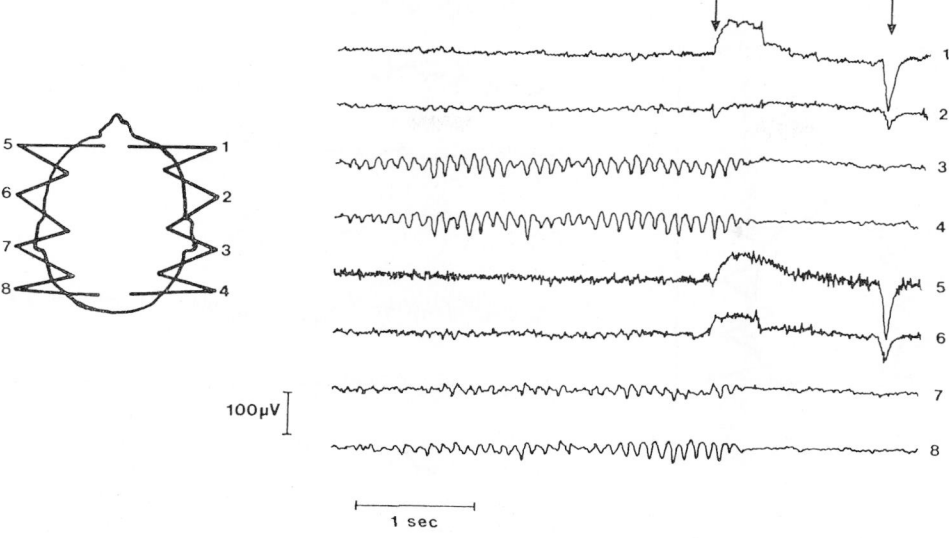

Fig. 3.15 A normal EEG showing alpha rhythm posteriorly and with eye-movement artefacts marked by arrows.

attenuated by eye opening and disappears during sleep, and is most clearly demonstrated in posterior leads (Fig. 3.15).

Beta activity is defined as any rhythmic activity with a frequency of more than 13 Hz. It is usually generalized and may be seen in normal subjects who are tense or anxious about the procedure and also as a drug-induced effect in patients receiving sedative or tranquillizing drugs.

Theta activity has a frequency of 4–7 Hz. It is a normal component of the EEG of children and adolescents, particularly over the temporal regions, but becomes more sparse with maturation. The finding of temporal theta activity represents a non-specific abnormality in adult patients with a variety of neurological disorders.

Artefactual changes

Numerous physiological and technical artefacts may be seen during EEG recording. These include large-amplitude frontal potentials due to eye movements, activity of very short duration due to muscle activity, electrocardiographic and electrode artefacts. The elimination of these problems demands a high level of technical expertise.

3.3.2 Abnormal EEG activity

Delta activity is defined as activity with a frequency of less than 4 Hz. Although delta activity may occur in normal individuals during sleep and in children, its presence in the waking state in adults is abnormal. Generalized delta activity is seen postictally, in patients with metabolic or drug-induced encephalopathy, and in patients with a diffuse encephalitis (Fig. 3.16). When it is localized to one area of the hemisphere it is usually indicative of some form of structural pathology (Fig. 3.17). However, it is impossible to differentiate between tumour and infarction on the basis of localized delta activity.

Spike discharges or sharp waves with a duration of up to 200 ms are the characteristic interictal abnormality found in epileptic patients. When they are generalized, interictal spikes are most commonly associated with following slow waves. When they are localized, although spikes and sharp waves may be associated with slow activity, they may occur independently. Sharp waves may be induced by drugs and alcohol withdrawal.

Paroxysmal activity may be defined as activity of very sudden onset and termination. This obviously includes spike wave activity, but also describes bursts of faster theta activity, seen as part of an epileptic process, other more non-specific activity, and periodic complexes.

3.3.3 EEG responses to activation

Hyperventilation leads to a general slowing of activity often with the development of paroxysmal high voltage delta activity. These changes are most evident in younger patients and diminish with increasing age. Hyperventilation may provoke spike wave discharges in patients with absence and other forms of idiopathic generalized epilepsy. Occasionally hyperventilation may enhance local abnormalities in the EEG.

Photic stimulation is performed with a stroboscopic stimulus usually with the eyes closed and open. This generates some posterior rhythmic activity that is time-locked to the rate of stroboscopic stimulation. In some instances photo-myoclonic responses may be generated with muscle activity most commonly localized around the eyelids, occurring at the rate of stimulation. However, in patients with photosensitive epilepsy, a photo-convulsive response occurs, consisting of bursts of spike-slow wave activity which are usually bilateral and synchronous and which persist briefly after the termination of the stroboscopic stimulus. This abnormality is seen in patients with idiopathic and symptomatic generalized epilepsies.

Historically drugs have been used to 'activate' the EEG. This is now rarely performed, though sodium methohexitone may be used in the specialized investigation of patients being assessed for temporal lobectomy.

Recordings during natural sleep show a significant modification of EEG activity. There tends to be a generalized slowing of activity, with loss of alpha rhythm and quite dramatic activity with K complexes and bursts of sleep spindles may occur; however, a full description of these is beyond the scope of this chapter. The main

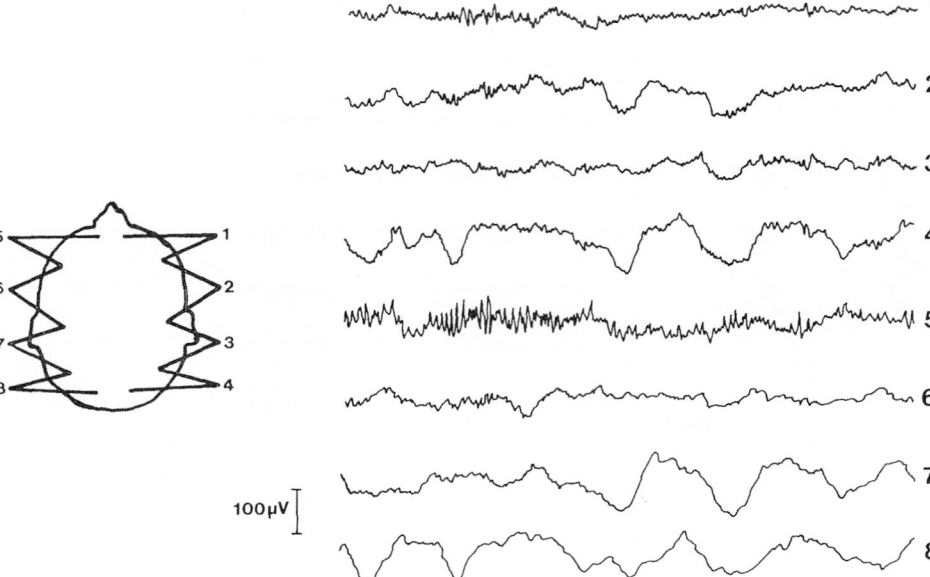

Fig. 3.16 EEG showing generalized slow, or delta, activity of high amplitude, maximal in posterior leads, in a patient in coma due to a viral encephalitis.

100μV

value of sleep recording is in the detection of abnormalities in patients with suspected partial seizures, and in the investigation of sleep apnoea.

3.3.4 The EEG in epilepsy

Synchronous post-synaptic potentials and paroxysmal depolarization shifts will be detected in the scalp EEG when they occur synchronously in large enough populations of neurones. They will produce spikes or sharp waves that are usually electro-negative over the cortex. These are commonly followed by slow waves associated with hyperpolarization of pools of neurones and reduced firing patterns. This is the basic electroencephalographic signature of the epilepsies and may be seen in focal distributions in the partial epilepsies and more generalized of distributions in the generalized epilepsies (Sections 30.4 and 31.4.4).

The patterns of EEG activity occurring during seizures vary considerably with the type of seizure. It is a general rule that diagnostic interictal and ictal EEG abnormalities will be found much more frequently in the generalized than partial epilepsies. For childhood absence epilepsy, an EEG that shows no 3/s spike-wave activity during a period of hyperventilation will for practical purposes exclude the diagnosis, hyperventilation being strongly provocative of spike wave activity in this and other generalized epilepsies. Atypical absences are more commonly associated with slower more irregular spike wave activity associated with a more abnormal EEG background (Fig. 3.18).

In patients with myoclonic seizures, both ictal and interictal discharges, consist of spike and polyspike and wave slow activity (Fig. 3.19) and a high proportion will exhibit a photoconvulsive response with photic stimulation.

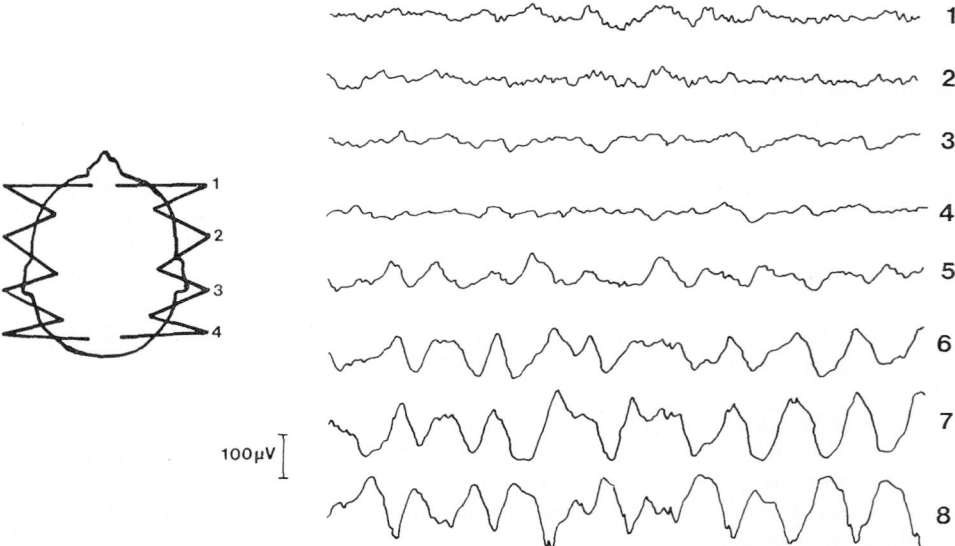

Fig. 3.17 EEG showing slow delta activity of high amplitude arising from the left hemisphere in a patient with a massive cerebral hemisphere infarction.

100μV

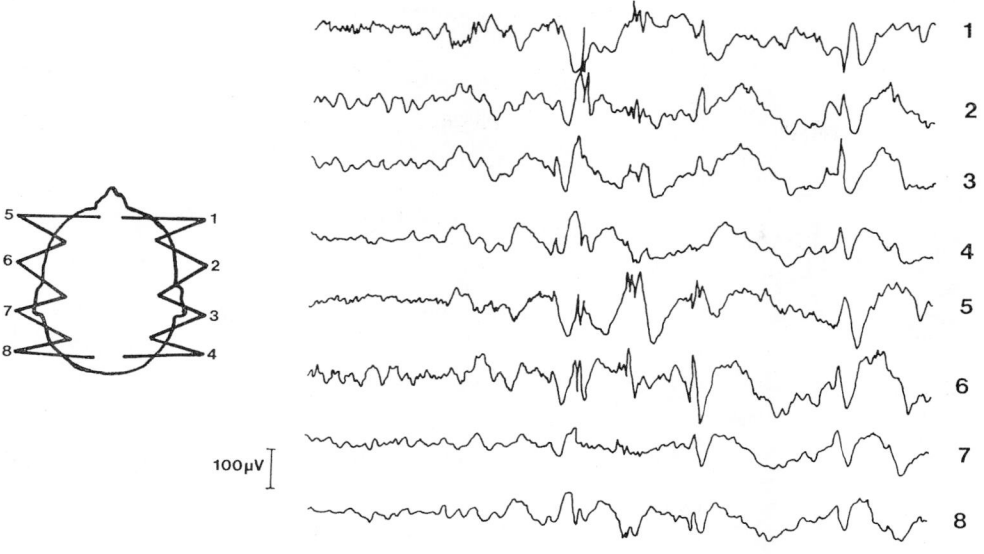

Fig. 3.18 EEG showing irregular, generalized slow spike wave activity in a child with Lennox–Gastaut syndrome.

Tonic-clonic seizures will usually obscure the EEG during the tonic phase, but during the clonic phase rhythmic spike wave activity is seen proceeding to a generally flat postictal EEG. In contrast, tonic and atonic seizures are often associated with low voltage fast activity or electrodecremental events at the onset of seizures.

Well-localized spike wave activity is most commonly seen in the temporal lobe epilepsies (Fig. 3.20). Frontal lobe epilepsies cannot infrequently be associated with relatively normal EEG recordings both ictally and interictally. On other occasions, frontal lobe epilepsies may produce somewhat atypical bifrontal spike wave activity that may be difficult to differentiate from discharges of a generalized epilepsy. Ictal recordings of partial epilepsies usually begin with localized fast activity which builds in amplitude and becomes more rhythmic, eventually transforming a spike wave activity. Seizure activity commonly spreads rapidly to the homotopic area of the contralateral hemisphere, and to other parts of the ipsilateral hemisphere.

Issues concerning the sensitivity and specificity of the EEG for the diagnosis of epilepsy are dealt with elsewhere (Section 31.9.1).

3.3.5 Specialized recording techniques in epilepsy

The usefulness and yield of the EEG in epilepsy can be increased in some patients by the use of specialized techniques, either to prolong recording or to increase the spatial sampling. Ambulatory recording is usually undertaken using a slowly running electromagnetic tape but may alternatively be recorded on other mass storage devices such as flash cards and miniature hard discs. While it is increasingly possible to combine this with synchronized video recording, more commonly ambulatory records are undertaken without such benefit. The main use of ambulatory recording is in differentiating convulsive pseudoseizures from true tonic-clonic seizures and also potentially for studying the links between different states of responsiveness and subtle seizures in individuals with severe epilepsy.

Video-telemetry has in the past allowed more extensive recording with up to 128 channels, with, in addition synchronized video recording to document behavioural correlates of seizure activity. This greatly enhances the diagnostic capabilities for non-convulsive events and allows the use of the EEG for lateralization

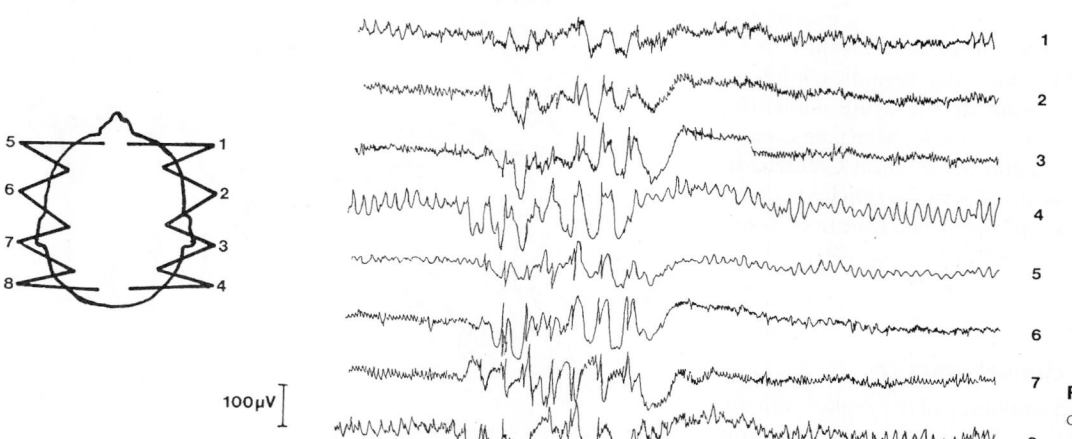

Fig. 3.19 EEG showing a short burst of 4–6 Hz spike and polyspike and wave activity in a patient with juvenile myoclonic epilepsy.

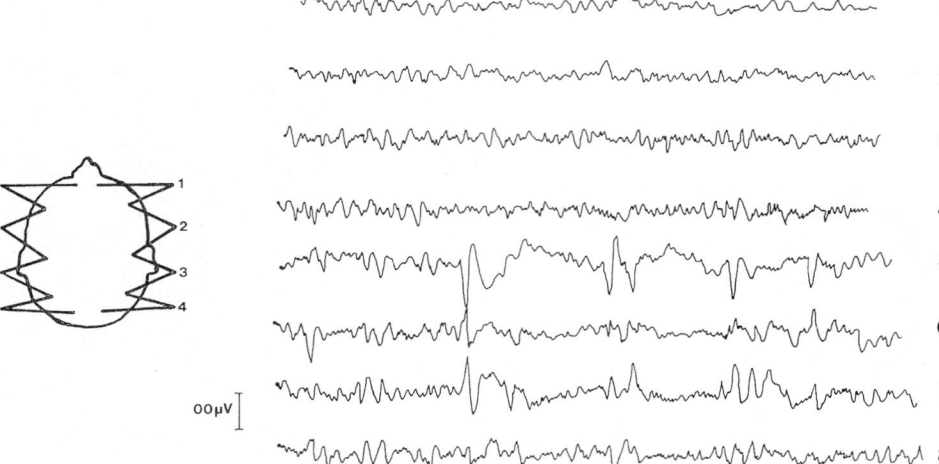

Fig. 3.20 EEG showing focal spike activity showing phase reversal in the left frontotemporal region (between leads 5 and 6) in a patient with complex partial epilepsy.

and localization of seizure onsets in patients being evaluated for surgery.

Intracranial recording may further support the localization of seizure onsets. A number of different techniques are available. Foramen ovale recording is relatively non-invasive and is very sensitive in identifying seizures starting in the medial temporal structures. For extra-temporal epilepsies, subdural mats with or without stereotactically implanted depth electrodes may be used.

Many systems are available which allow the production of iso-potential maps. The production of these maps eliminates the temporal element of EEG recording which is in fact the EEG's greatest strength. It is doubtful that these maps provide any additional information that cannot be derived from a conventional EEG by a reasonably experienced observer.

3.3.6 The EEG and non-epileptic disorders

The EEG will exhibit gross changes of slowing of background activities with ultimately burst suppression types of periodic activity in a number of metabolic, drug-induced, neurodegenerative, infective, and post-hypoxic states. In a few instances, the findings of generalized slow activity or periodic activity may be diagnostically useful within a well-defined clinical setting.

The EEG may be particularly useful in the differentiation of psychogenic coma-like states from true coma. The presence of typical occipital alpha rhythms in an unresponsive subject is for practical purposes diagnostic of pseudocoma. Periodic complexes, when they occur on the background of sub-acute dementing illness, are reasonably specific for sub-acute sclerosing panencephalitis in children and young adults and for classical Creutzfeldt–Jakob disease in older subjects. Localized periodic activity in one or other fronto-temporal regions is a relatively late feature of herpes simplex encephalitis.

3.4 Evoked potentials

3.4.1 Evoked potentials in clinical practice

Abnormalities of the latency and amplitude of the evoked response are seen in a variety of conditions affecting the central nervous system. Evoked potentials were first used to explore the integrity of

myelinated pathways in the central nervous system when it became clear that averaging techniques could be used to distinguish the slowing of conduction in focal demyelination from the near normal physiological properties of surviving fibres in pathways with partial axonal degeneration (Fig. 3.21).

The systematic use of evoked potentials began with the introduction of pattern-reversed stimulation of the visual system

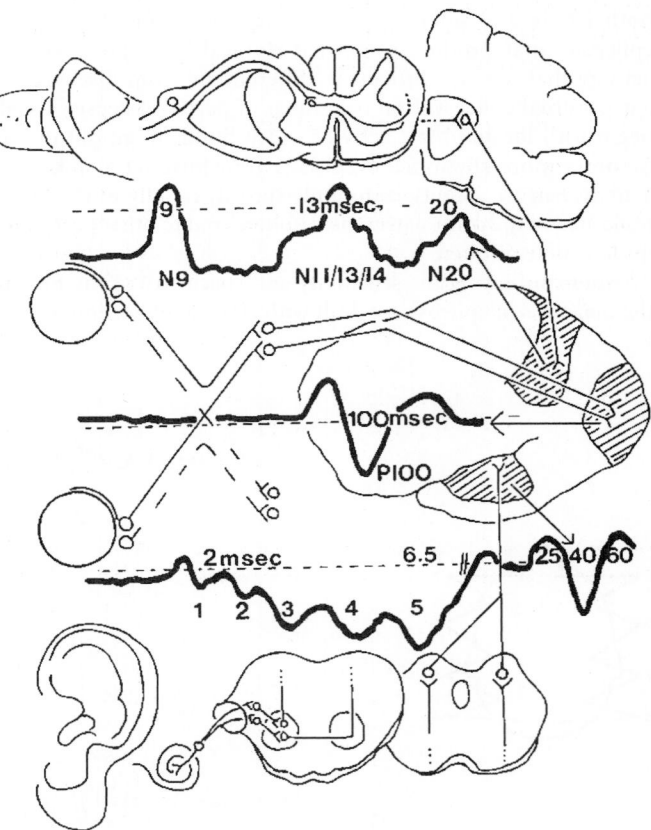

Fig. 3.21 Cartoon to illustrate the principal wave forms obtained by somatosensory (top), visual (middle), and auditory evoked potentials (bottom).

(Halliday *et al.* 1972). This was soon extended to the exploration of other central afferent pathways: sensory (Small *et al.* 1978) and auditory (Robinson and Rudge 1977); descending motor tracts, using electrical (Merton and Morton 1980) and magnetic (Barker *et al.* 1985) stimulation of the cortex; and cognition through event related potentials (see Heinze *et al.* 1999). Because slowing of conduction characterizes both the anatomical site and pathophysiological properties of conduction through myelinated pathways, evoked potentials were soon used routinely to provide laboratory support especially in the context of demyelinating disease and related disorders (Halliday *et al.* 1973; Cowan *et al.* 1984), in compression of the visual pathway, and in neurodegeneration. Now, the ability to depict brain anatomy at high resolution using magnetic resonance imaging; the separation of individual tissue components using magnetic resonance spectroscopy; and the tracing of fibre pathways using diffusion-weighted magnetic resonance imaging have reduced the need for evoked potentials as a means of detecting pathological involvement in parts of the central nervous system that are not implicated by the analysis of symptoms and signs. But evoked potentials are valuable in revealing the likely pathophysiology of tissue injury in terms of demyelination and/or axonal loss; and, in this respect, they add a further dimension to the anatomical cataloguing of brain imaging (Miller *et al.* 2005). That said, the newer magnetic resonance and spectroscopic imaging protocols may also distinguish the separate elements of damage to cellular components of the central nervous system. Evoked potentials are used in the per-operative assessment of spinal surgery which potentially threatens the cord, such as correction of scoliosis. And event related potentials are increasingly used in research studies to register and time cognitive awareness in the context of activity-dependent functional magnetic resonance imaging signals.

3.4.2 **Afferent evoked potentials**

With some variations, the standard technique for visual stimulation is to use a chequer-board pattern of black and white squares that reverses at 2 Hz and where each square subtends approximately 50 min of arc at the retina. The whole stimulus usually occupies 32° of the field. In some situations, half- and 4° central field stimuli are additionally useful in explaining visual symptoms when only a small proportion of central fibres subserving central vision is involved. Apparent delays may be recorded in individuals with a central scotoma, whatever its pathological substrate, if the stimulus predominantly activates the paramacular region. It may not always be possible to identify the separate waveforms and so accurately to document the true latency of responses.

The visual evoked response is transiently decreased in amplitude, or absent if acuity is <6/24, in acute optic neuritis (Fig. 3.22). Typically it stabilizes at a prolonged latency, by about 35 ms, with normal amplitude in the post-acute phase. The latency routinely returns to normal in the majority of affected children (Kriss *et al.* 1988). There is also evidence for progressive shortening of the latency of the evoked response for up to 3 years after an episode of acute optic neuritis in adults, consistent with remyelination or plasticity within the visual pathway, but this is not associated with sustained improvement in vision perhaps through the establishment of irreversible axonal changes in the previously demyelinated nerve (Brusa *et al.* 1999).

Where the absolute latency is not prolonged, demyelination may be marked by an asymmetry in response between the two eyes;

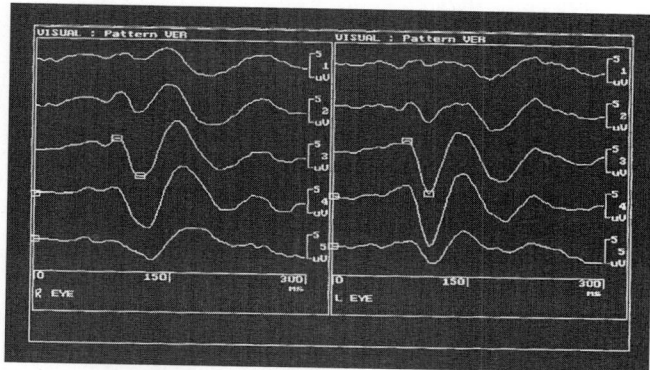

Fig. 3.22 Visual evoked response, VER, to full field pattern reversal stimulation, in a patient with multiple sclerosis, showing a delayed latency of 115 ms and a normal latency of 105 ms from the left and right eyes, respectively, and a larger amplitude from the right of the occiput (lower channels).

differences of >6 ms are usually indicative of damage to one optic nerve. Half-field stimulation can demonstrate demyelinating lesions of the chiasm and posterior visual pathway. Evoked potentials are particularly useful in supporting the diagnosis of non-organic visual loss. Normal amplitude, waveform, and latency with a declared visual acuity of 6/24 or less, not due to refractive error, suggests a functional disturbance of vision. Conversely, delayed latency may guide the interpretation of dubious visual symptoms.

The visual evoked potential is delayed in approximately 70 per cent of patients suspected of having widespread demyelination and in the majority of those with clinically definite multiple sclerosis. Asymptomatic delay in a patient with isolated demyelination affecting the other optic nerve or another part of the central nervous system increases the probability of subsequent clinical conversion to multiple sclerosis. Compression of the optic nerve usually produces a visual evoked potential of irregular form without a convincing increase in latency of the main response (Halliday *et al.* 1976) although compression can produce focal conduction block with slowing in the visual system and spinal cord, thereby spuriously suggesting primary demyelination as the explanation for symptoms.

Visual evoked potentials are the only electrophysiological diagnostic tests retained in the new diagnostic criteria for multiple sclerosis (McDonald *et al.* 2001; Polman *et al.* 2005). Although a delayed evoked potential does not carry the same status as an emerging MRI lesion it is of particular value in the context of myelopathy due to primary progressive multiple sclerosis where brain MRI is often uninformative. However, some genetically determined disorders may be associated with optic pathway and spinal cord involvement, reproducing the clinical and laboratory features of primary progressive multiple sclerosis—a situation in which examination of the cerebrospinal fluid may resolve the diagnostic dilemma.

The standard method for evoking somatosensory potentials is by supra-threshold electrical stimulation of the median or posterior tibial nerves recording over the spinal cord and sensory cortex at 20 and 40 ms, respectively (Fig. 3.23). The somato-sensory evoked potential is reduced or absent in the acute phase of demyelinating myelopathy involving the dorsal part of the cord and this abnormality often persists after clinical recovery. The frequency of

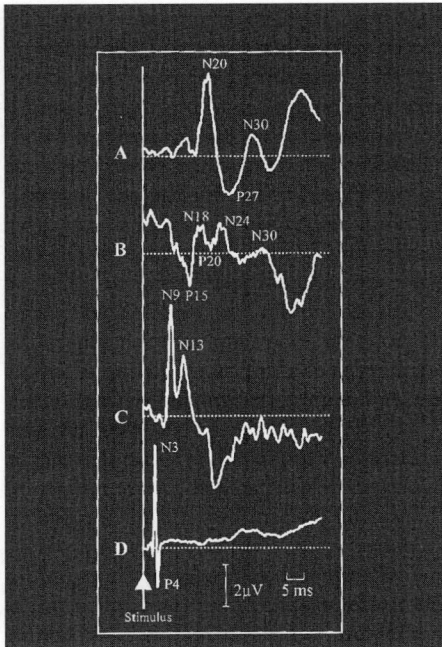

Fig. 3.23 Somatosensory evoked potential, SEP, from stimulation of the median nerve at the wrist, showing normal responses at the elbow (D), Erb's point (C), the neck (B), and scalp (A).

abnormality is around 80 per cent for clinically definite multiple sclerosis but lower in less definite diagnostic categories (Small *et al.* 1978).

Potentials of short-latency, within 10 ms, can be obtained from scalp electrodes after auditory stimulation (Fig. 3.24). Of the five normal waves, I and II originate from the eighth nerve external to the brainstem, III from the cochlear nucleus, and IV and V from the region of the superior olivary complex (McPherson and Starr 1993). Demyelinating disease is characterized by increase in the latency between the first two and later waves (Robinson and

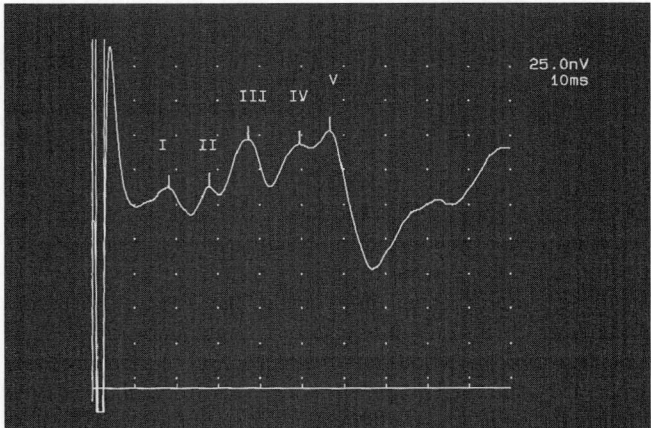

Fig. 3.24 Brainstem auditory evoked potential, BAEP, to stimulation of one ear, showing normal responses from the auditory nerve (I), the cochlear nucleus (II), the superior olive (III), lateral lemniscus (IV), and inferior colliculus (V).

Rudge 1977). The sensitivity in detecting involvement of brainstem pathways is marginally less than for visual or somatosensory systems in clinically definite, about 50–75 per cent, and suspected, about 25 per cent, multiple sclerosis (McPherson and Starr 1993), and end-organ failure is a more common cause of alteration in the auditory evoked potential than in other afferent systems. Long latency auditory evoked responses can be stimulated by complex harmonic tones, and these have a greater sensitivity in multiple sclerosis than short latency responses although this observation may reflect the contribution of disseminated central nervous system lesions rather than those confined to the afferent auditory pathway (Jones *et al.* 2002).

3.4.3 Efferent evoked potentials

Motor evoked responses, following either electrical or magnetic stimulation, are somewhat variable reflecting their passage through multiple synapses between cortex and peripheral limb muscle. Long delays, of about 20 ms, are more likely to result from demyelination than degeneration (Thompson *et al.* 1987) although either may be present in patients with multiple sclerosis. They have rather a low prevalence of asymptomatic abnormality therefore adding little to careful clinical assessment; and when changes in motor evoked potentials are demonstrated in serial studies, these are no more sensitive and discriminating than MR imaging (Kidd *et al.* 1998).

3.4.4 Evoked potentials dependent on reflex functions

Neurophysiological studies can be used to probe the integrity of reflex circuits including their central connections. Mainly these explore the brainstem and depend on recording the early R1 and late R2 blink reflex, the early SP1 and late SP2 masseter inhibitory reflex, and the jaw jerk to chin tapping (Cruccu *et al.* 2005). Although the latter has no independent localizing value in the context of stroke, various combinations can be used to localize lesions to the ventral pons (all three reflexes), the ipsilateral dorsal pons (R1 and SP1), the ipsilateral inferior olive (R2) and the contralateral median pontine tegmentum dorsal-lateral medulla (SP1, SP21 or R2). Vestibular evoked myogenic potentials assess the modulation of tonic electromyographic activity in the sternocleidomastoid muscle in response to a vestibular activation stimulus as another means of detecting brainstem lesions. The frequency of abnormality in multiple sclerosis is claimed to be around 70 per cent (Alpini *et al.* 2004) although not surprisingly this reflex response is less sensitive than either MRI or visual evoked potentials in detecting abnormalities (Bandini *et al.* 2004).

3.4.5 Event related potentials

Event related potentials reflect the summation of cognitive processes including attention, memory, language, and responses (Heinze *et al.* 1999). Being complex, they do not easily define precise cognitive functions and so have a limited role in clinical neurology. The most accessible measurement is the p300 based on auditory stimulation; its latency extends to 700 ms when the visual system is stimulated. The p300 is considered to reflect evaluation more than response time and to depend on activity in several parts of the cerebrum. It is typically delayed in dementia with a specificity of around 70 per cent but poor discrimination across the range of clinical conditions affecting intellectual function.

3.5 Nerve conduction studies and electromyography

Nerve conduction studies localize compressive focal neuropathy and detect polyneuropathy, distinguishing between demyelinating, axonal degeneration, and conduction block neuropathies. Electromyography detects denervation of muscle, helps to distinguish between myopathic and neuropathic weakness, and is diagnostic of myotonias and neuromyotonias (Section 23.7). Single-fibre electromyography and neuromuscular transmission studies are particularly important for diagnosing myasthenia gravis and the Lambert–Eaton myasthenic syndrome (Section 24.10).

3.5.1 Motor conduction studies

Maximal motor conduction velocity is measured by applying a supramaximal stimulus to a motor nerve at two or more different points along its course through bipolar electrodes applied to the skin over the trunk of the nerve (Fig. 3.25).

The evoked compound muscle action potential, referred to as the CMAP, from a muscle supplied by this nerve is recorded. Well-established methods are now available for studying motor conduction in the median, ulnar, and radial nerves in the arm, and in the common peroneal and posterior tibial nerves in the leg, and the normal ranges for motor conduction parameters have been extensively documented (Binnie *et al.* 2004). Temperature has a profound effect upon nerve conduction, so the skin temperature of the limb should be maintained at 25–30°C, if necessary by prior immersion in warm water or by a radiant heat lamp.

Four measurements are derived:

- *Motor conduction velocity*. Measurement of the latency from the stimulus to the initial rise of the compound muscle action potential, and of the distance between the pairs of stimulating electrodes, allows calculation of the conduction velocity in m/s along various segments of the nerve. Motor conduction velocity or MCV normally exceeds 48 m/s in the arm and 40 m/s in the leg nerves.

- *Distal motor latency*. In the case of the median nerve (Fig. 3.25) the interval between the stimulus applied at the wrist and the initial rise of the muscle action potential in the abductor pollicis

brevis is known as the distal motor latency. Distal motor latencies, or DML, vary depending upon the particular nerve, the site of stimulation, and the technique employed. For the median nerve following stimulation at the wrist the normal distal motor latency is less than 4.2 m/s but will be prolonged in carpal tunnel syndrome or generalized demyelinating neuropathy.

- *F-wave latency*. This represents the time taken for the antidromic volley to depolarize the motor neurone cell bodies within the spinal cord, and then for the passage of the resultant action potential to travel orthodromically to the muscle (Fig. 3.26).

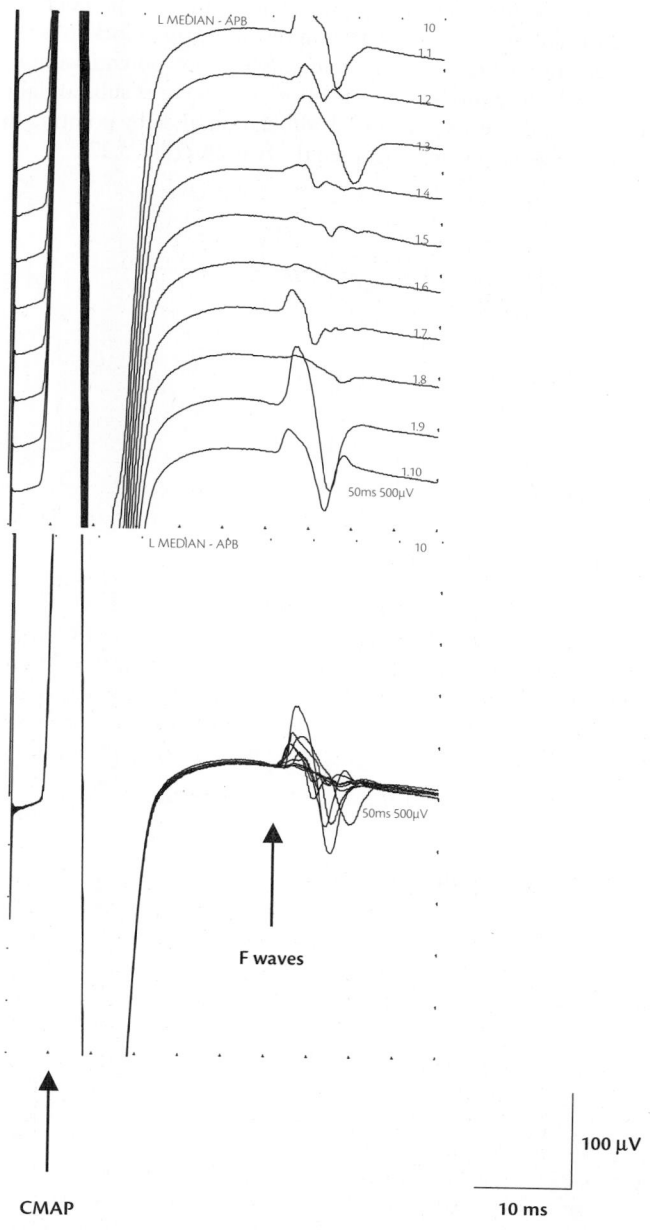

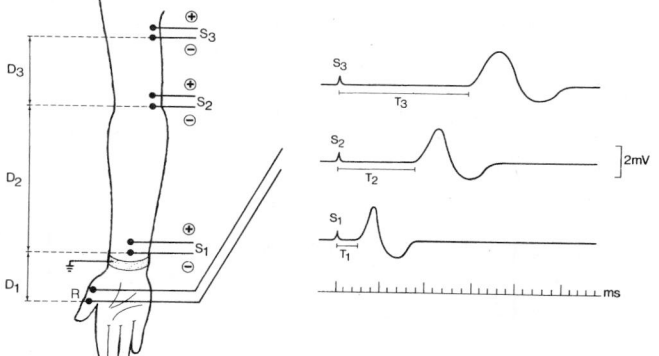

Fig. 3.25 Measuring motor conduction velocity along the median nerve: diagrammatic representation with stimulating electrodes, S1, S2, S3, at various points along the course of the nerve and measurement of the compound muscle action potential using surface electrodes, R, over abductor pollicis brevis (Reproduced from Bradley 1974.)

Fig. 3.26 F waves recorded from median nerve stimulation in a normal individual. The top waves show the varying F wave responses to ten single nerve stimulations. These are superimposed on the bottom trace allowing measurement of a normal F wave latency of 31 ms. Note that the F wave is preceded by a larger electromyographic deflection representing the orthodromic compound muscle action potential, or CMAP. (Courtesy of Dr R Kennett.)

It reflects conduction over proximal segments of the nerve and root, and the normal values are dependent upon height.

♦ *Conduction block.* Blockage of impulse conduction has become recognized as an important cause of weakness due to peripheral nerve disease. This conduction block usually reflects failure of action potential propagation along axons through sites of severe compression or segments which are demyelinated. It is especially seen in acutely demyelinated nerves within the first few weeks before sodium channel redistribution to the denuded internodal segments of the axon allows resumption of conduction, albeit at slowed velocity. Also it is recognized that there are neuropathies in which conduction block appears to be the primary pathophysiological process. Practical detection of conduction block involves demonstrating that the compound muscle action potential amplitude evoked by proximal stimulation of a nerve is substantially lower than that evoked by distal stimulation, thereby pointing to partial conduction block between the two sites (Fig. 3.27).

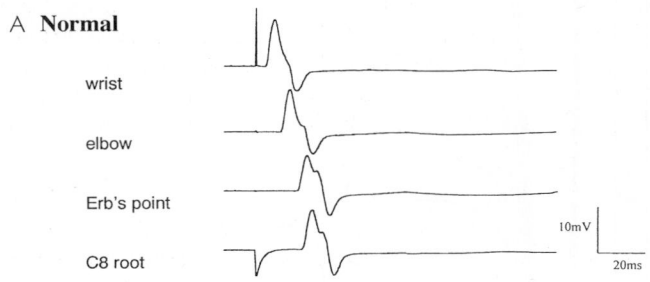

A **Normal**

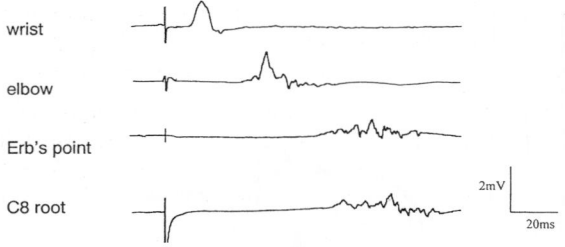

B **Motor nerve slowing**

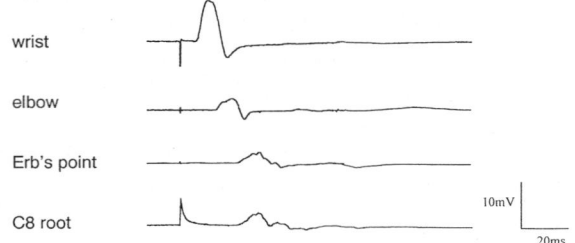

C **Motor nerve conduction block**

Fig. 3.27 Abnormalities in motor conduction along the median nerve demonstrated by surface electrode recording of compound muscle action potentials from abductor pollicis brevis following stimulation at different sites. (A) Normal, showing constant compound muscle action potential amplitude with normal latency. (B) Demyelinating neuropathy, showing slowing and dispersion of compound muscle action potentials. (C) Conduction block in the forearm segment due to multifocal motor neuropathy causing reduced compound muscle action potentials amplitudes following stimulation at the elbow and above, but with preserved conduction velocity. (Courtesy of Dr M Busby.)

3.5.2 Sensory nerve action potentials

Sensory conduction in the median, ulnar, and radial nerves can be measured by applying stimuli through ring electrodes upon a finger (Fig. 3.28) and then recording orthdromic conduction of the sensory nerve action potential, or SNAP, through cutaneous electrodes applied over the trunk of the nerve. For other nerves, such as the sural, the nerve trunk is stimulated and sensory nerve action potentials recorded from surface electrodes, antidromically or orthodromically. The sensory nerve action potential is so small that averaging techniques are required following multiple stimuli and occasionally needle recording electrodes inserted close to the nerve are required (Binnie *et al.* 2004).

Sensory nerve action potential amplitudes generally range from 5 to 50 μV depending upon the particular nerve being studied and drop in amplitude, or absence of the potential, occurs in axonal degenerative, demyelinating, or compressive neuropathies. Thus a diminished sensory nerve action potential cannot in itself distinguish between these types of neuropathy. Sensory nerve action potentials reflect the integrity of the distal axonal branch of the dorsal root ganglion sensory neurones. Sensory nerve conduction latencies and velocities can be measured but generally find little usefulness in diagnostic clinical practice compared to motor nerve velocities. Routine techniques for measuring possible conduction block in sensory nerves have not been established.

Sensory nerve action potentials remain normal in disease such as spondylitic radiculopathy affecting the spinal nerve roots containing the proximal axonal branches because the dorsal root ganglia cell bodies lie outside the intervertebral foramina (Aminoff *et al.* 1985). The only exception to this can occur in the lower lumbar and sacral nerve roots in which the dorsal root ganglia can lie within the intervertebral foramina. Thus loss of foot nerve sensory nerve action potentials may occur with lumbosacral spinal disease as well as with peripheral nerve disease. Radiculopathy may be confirmed by the finding of denervation within limb and paraspinal muscles innervated by the same segment.

3.5.3 Uses of nerve conduction studies

Nerve conduction studies are used principally to diagnose focal mononeuropathies, usually due to nerve compression, and to detect polyneuropathy and determine whether it is due to demyelination

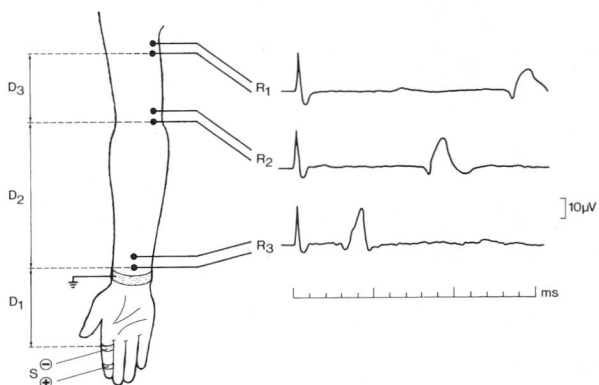

Fig. 3.28 Diagrammatic representation of the technique of measuring the sensory nerve action potential, SNAP, in the median nerve at points R1, R2, R3 after orthodromic stimulation of a finger. (Reproduced from Bradley 1974.)

or axonal degeneration. The additional uses of detecting conduction block neuropathies (Section 21.11.3) and in discriminating nerve root disease from polyneuropathy (Section 22.1.2) are dealt with above. Generally, nerve conduction studies are of more use diagnostically in polyneuropathy than in quantifying the clinical response to immunotherapy.

Focal compressive neuropathy

Electrophysiology can be used to diagnose focal compressive neuropathy affecting most limb nerves and some truncal nerves. It is particularly useful for detecting compression of the median nerve in the carpal tunnel and the ulnar nerve at the elbow. Conduction distal to a neurapraxic lesion of a peripheral nerve may remain normal at a time when its clinical functioning is severely impaired. When a nerve is compressed, as in entrapment neuropathies, motor and sensory conduction across the site of the lesion may be either lost or greatly reduced in speed. This leads to reduction or loss of the sensory action potential in that nerve, and prolongation of the distal motor latency. Localized slowing of motor conduction can be demonstrated at some entrapment sites, such as compression of the ulnar nerve at the elbow and techniques are now available for measuring conduction over short segments of nerve in presumed entrapment neuropathies. Electromyographic sampling of muscles supplied by a trapped nerve will show to what extent axonal degeneration has caused denervation. This generally implies a poorer prospect for recovery of nerve function following surgical release of the compression.

Demyelinating polyneuropathies

Chronic demyelinating polyneuropathy produces marked slowing of conduction along affected nerve trunks. Sometimes difficulty arises because reduced velocities can occur in profound axonal loss when no fast conducting fibres survive to a grossly denervated muscle. If demyelination chiefly affects the proximal segments of motor fibres, it will be associated with normal conduction velocity measurements along distal segments. In this case, prolonged F-wave responses are the clue to proximal conduction slowing. Various criteria have been proposed for defining a neuropathy as demyelinating in nature. Those which have been most widely accepted require three out of four of the following abnormalities affecting two or more nerves (Ad hoc Subcommittee of American Academy of Neurology AIDS Task Force 1991):

- reduction of motor velocity to <80 per cent of the lower limit of normal, e.g. <39 m/s for arm nerves and <34 m/s for leg nerves;

- prolonged distal motor latencies exceeding >125 per cent of the upper limit of normal;

- prolonged F wave latencies >125 per cent of the upper limit of normal;

- partial conduction block of >20 per cent or abnormal temporal dispersion causing >15 per cent change in duration not attributable to an entrapment neuropathy.

Such stringent criteria, whilst useful in established disease, are not met early in the course of mild forms of chronic idiopathic demyelinating polyneuropathy. Furthermore, sometimes these criteria can only be satisfied by exhaustive electrophysiological study of numerous peripheral nerves. Most demyelinating neuropathies also involve sensory fibres with reduced amplitude or absence of sensory nerve action potentials.

Axonal degeneration polyneuropathies

Axonal degeneration polyneuropathies usually involve a dying back process which mainly affects the longest axons. The earliest evidence of axonal polyneuropathy is electromyographic evidence of denervation of foot and hand muscles coupled with reduced amplitude or absence of sensory nerve action potentials in the feet and hands. The muscle denervation is evident on surface electromyography as reduction in the amplitude of compound muscle action potentials, which normally range from 10 to 25 mV in hand muscles. Concentric needle electrodes inserted into denervated muscle will reveal fibrillation potentials and positive sharp waves (Section 3.5.5). Early on the motor conduction velocity in surviving axons will be normal, or only marginally reduced. However, as severe denervation sets in and the large diameter axons are lost, the motor conduction velocity can fall markedly although only rarely to less than 80 per cent of the lower limit of normal, and the distal motor latency can rise, but rarely above 125 per cent of the upper limit of normal (Cornblath *et al.* 1992). This means that primarily demyelinating polyneuropathies can be distinguished electrophysiologically from axonal degeneration. However, in practice many polyneuropathies involve mixed elements of demyelination and axonal degeneration and may evade confident clinical or electrophysiological classification into either category alone.

Conduction block neuropathies

In such neuropathies the block often occurs without sufficient associated conduction slowing to point to underlying demyelination; multifocal neuropathy with conduction block (Section 21.11.3) is an important example. Most frequently, conduction block is partial rather than total. It may be diffusely distributed along a length of nerve, rather than being tightly localized to a particular site. This has led to difficulty and dispute about the quantificative definition of conduction block. Partial conduction block leads to three electrophysiological abnormalities:

- Reduced amplitude and area of the compound muscle action potential evoked by nerve stimulation at proximal sites compared to distal (Fig. 3.27). Different authors quote degrees of reduction ranging from 20 to 60 per cent (Cornblath *et al.* 1991; Ghosh *et al.* 2005; Lewis and Sumner 1982).

- Dispersion of the compound muscle action potential wave-form. However, this abnormal temporal dispersion can lead to phase cancellation of individual motor unit potentials within the compound muscle action potential wave form. This may produce the misleading appearance of amplitude reduction of the compound muscle action potential.

- Absent or sparse F-wave responses if the conduction block affects proximal nerve segments or the nerve roots.

3.5.4 Age and nerve conduction
Infants and children

Infants and children show reduced nerve conduction velocity compared to adults. Motor conduction velocity in the newborn is approximately half of the adult speed and only reaches the adult range at 3–5 years of age (Ouvrier 1990). Compound muscle action potential amplitudes also increase with age. Sensory nerve action potential amplitudes are normally less than half of adult values which are attained by 3 years of age.

The elderly

During their seventh and eighth decades the elderly show variable reduction in nerve conduction velocity and in median and sural sensory nerve action potential amplitudes (Bouche *et al.* 1993). By 80 years of age, all patients will show reduction in these parameters and reduced compound muscle action potential amplitudes.

3.5.5 Electromyography

Routine electromyography, or EMG, records the electrical activity of a muscle at rest, during minimal voluntary contraction, and during full contraction. Two techniques may be used. Surface electrodes may identify which muscle or muscle groups are participating in voluntary movement and allow quantification of the compound muscle action potential during motor conduction studies. But for diagnostic work it is necessary to insert a concentric needle electrode into the muscle so as to detect changes of acute and chronic denervation or myopathy (Mills 2005). The needle electrode records the activity of about 100 muscle fibres in the vicinity, only a few of which represent any single motor unit. The particular use of electromyography to diagnose primary muscle disease is discussed in Section 24.1.5. Modern electromyographic recording techniques allow easy quantification of parameters relating to motor unit wave form and duration. The amplitude and duration of motor unit potentials increase with age.

Normal muscle

Normal muscle at rest shows no electrical activity. On slight voluntary contraction motor-unit potentials of 300–2000 µV in amplitude and 6–10 ms in duration are recorded. These are usually monophasic, biphasic, or triphasic in shape, but 10–25 per cent of potentials recorded from normal muscle may be polyphasic. On vigorous voluntary muscular contraction an interference pattern develops. Since the patient recruits as many motor units as possible, and they fire asynchronously, each one interferes with the waveforms of ones which precede and follow it (Fig. 3.29A).

Denervated muscle

By 3 weeks after complete transection of a nerve, fibrillation potentials can be recorded from the resting muscle fibres (Fig. 3.29B). A fibrillation potential reflects spontaneous contraction of a single denervated muscle fibre, supersensitive to acetylcholine, and is usually 50–100 µV in amplitude, 1–2 ms in duration, and monophasic or biphasic in shape. Total absence of motor unit potentials is evident on attempted voluntary contraction whether the lesion is neurapraxial or due to neurotmesis or axonotmesis (Section 22.2). However, in neurapraxia fibrillation does not occur; hence if no fibrillation potentials appear 2 or 3 weeks after total paralysis of a muscle, the lesion is probably compressive without axonal transection. On the other hand, if fibrillation potentials are recorded it does not necessarily mean that every motor axon has been severed. Another type of spontaneous activity sometimes recorded from resting denervated muscle is that of so-called positive sharp waves or 'saw-tooth' potentials (Fig. 3.29B). The electromyography recorded from partially denervated muscle shows a mixture of positive sharp waves and fibrillation potentials along with an interference pattern of reduced density.

The electromyography of chronic denervation differs from that of acute. Fibrillations and positive sharp waves of acute denervation are absent or sparse if the denervation is very chronic or has become arrested. However, in cases of chronic denervation, fasciculation potentials may be recorded, occurring spontaneously and repetitively when the muscle is at rest (Fig. 3.29C). Fasciculation potentials are morphologically indistinguishable from motor unit action potentials. They often point to a lesion affecting the anterior horn cell. Fasciculation may also be a benign phenomenon (Section 23.1.2). In chronic denervating processes, surviving axons often produce collateral sprouts which re-innervate denervated muscle fibres. As a consequence some surviving motor units become much larger than normal in both amplitude and duration (Fig. 3.29C). These 'giant motor units' are particularly common in muscles previously affected by acute poliomyelitis (Section 23.3.7).

Myopathic muscle

Usually maximum voluntary contraction produces a complete interference pattern despite little force being generated. Spontaneous fibrillation and fasciculations are less commonly seen in primary muscle diseases, although fasciculations can occur in thyrotoxic and hypoparathyroid myopathies. In some cases of myopathy, especially polymyositis, fibrillation potentials do occur because the disease process may affect intramuscular nerve endings, or focal necrosis of part of a muscle fibre may separate the remainder of the fibre from its motor end-plate. In primary muscle disease the duration and amplitude of the motor unit potentials is diminished on slight voluntary contraction. Many motor unit potentials are broken-up or polyphasic because of a reduced content of normal muscle fibres within a motor unit (Fig. 3.29D). On maximal contraction the same number fire and complex polyphasic motor unit action potentials of prolonged duration can fill the oscilloscope interference pattern.

Myotonias, neuromyotonia, and myokymia

Myotonias cause high frequency discharges of single muscle fibres. These are provoked by movement of the exploring needle within the muscle and produce a typical recurring 'dive-bomber' sound in the loud speaker. Myotonic discharges start at a crescendo, and then fade gradually (Fig. 3.30). They occur in dystrophia myotonica, congenital myotonias, proximal myotonic dystrophy, and hypokalaemic periodic paralysis (Mills 2005). They must be distinguished from other bizarre 'pseudomyotonic' high-frequency discharges which begin and end abruptly and give a more constant sound; these occur in disorders as diverse as motor-neurone disease, polyneuropathy, muscular dystrophy, and many metabolic myopathies, and they lack diagnostic specificity unlike true myotonic discharges.

Neuromyotonia causes motor units to fire spontaneously and irregularly as doublets, triplets, or multiplets with high intraburst frequencies of up to 120 Hz (Fig. 23.8). Neuromyotonia occurs in autoantibody mediated anti-voltage gated potassium channelopathy, Morvan's syndrome, and in association with multifocal motor neuropathy with conduction block.

Myokymia consists of flickering within a muscle belly due to spontaneous discharges with groups of motor units. It has a variety of disease associations, including intrinsic tumours of the brainstem and central nervous system demyelination. Myokymia can be distinguished using electromyography from the similar clinical phenomena of fasciculations, neuromyotonia, and myoclonus.

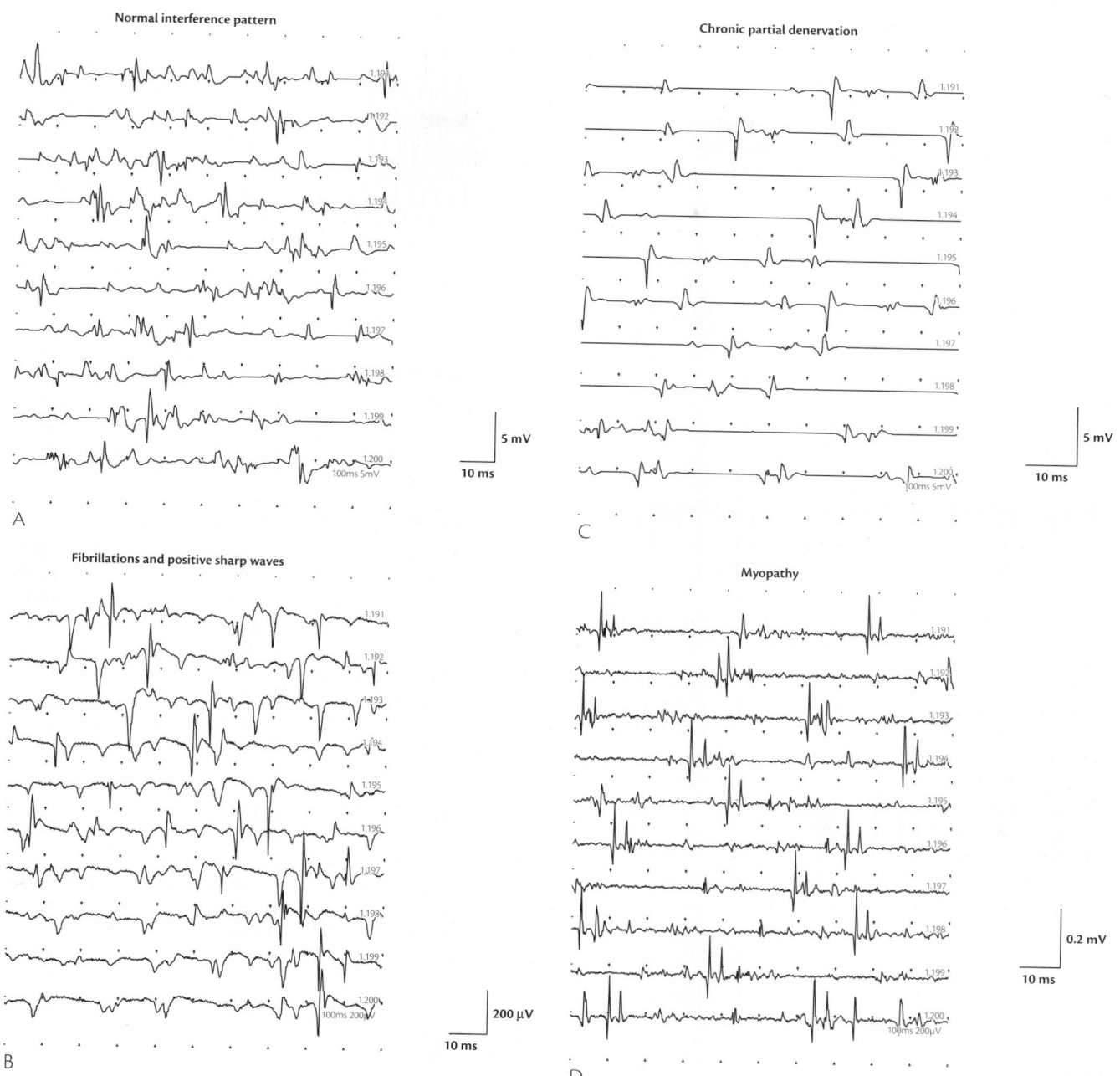

Fig. 3.29 Electromyograms from healthy and diseased muscle: (A) Normal muscle. Interference pattern recorded by concentric needle electrode. The ten traces represent a continuous recording from a single muscle. (B) Fibrillations and positive sharp waves with a sawtooth profile, recorded from muscle with established denervation. Note their tiny amplitude compared to the normal interference pattern or to fasciculations. (C) Chronic partial denervation in motor neurone disease. Interference pattern at rest showing fasciculation potentials, some of which are approximately 5 mV in amplitude. (D) Inflammatory myopathy. Voluntary contraction showing polyphasic motor units of slightly reduced amplitude. (Courtesy of Dr R Kennett.)

3.5.6 Myasthenias

Repetitive stimulation

Normally the amplitude of the compound muscle action potential evoked by nerve stimulation remains relatively constant on repetitive stimulation. In myasthenia gravis the initially normal muscle potential steadily diminishes for the first four or five stimuli delivered to the nerve at 3–5/s (Fig. 3.31A). In Lambert Eaton myasthenic syndrome the muscle potential is initially small, but progressively enlarges after maximal voluntary contraction or at tetanic rates of nerve stimulation between 20 and 50/s (Fig. 3.31C).

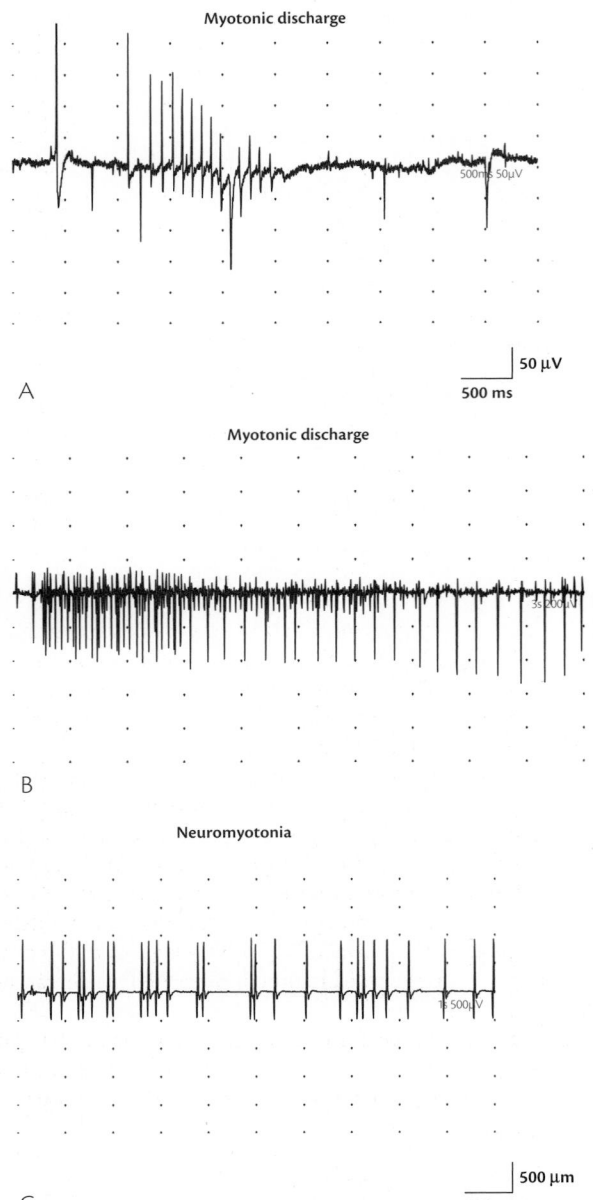

Fig. 3.30 The electromyogram in myotonia. (A) and (B): two traces showing myotonic discharges from a patient with myotonic dystrophy. Bottom (C): neuromyotonia showing groups of repeat motor unit discharges at high frequency within bursts. (Courtesy of Dr R Kennett.)

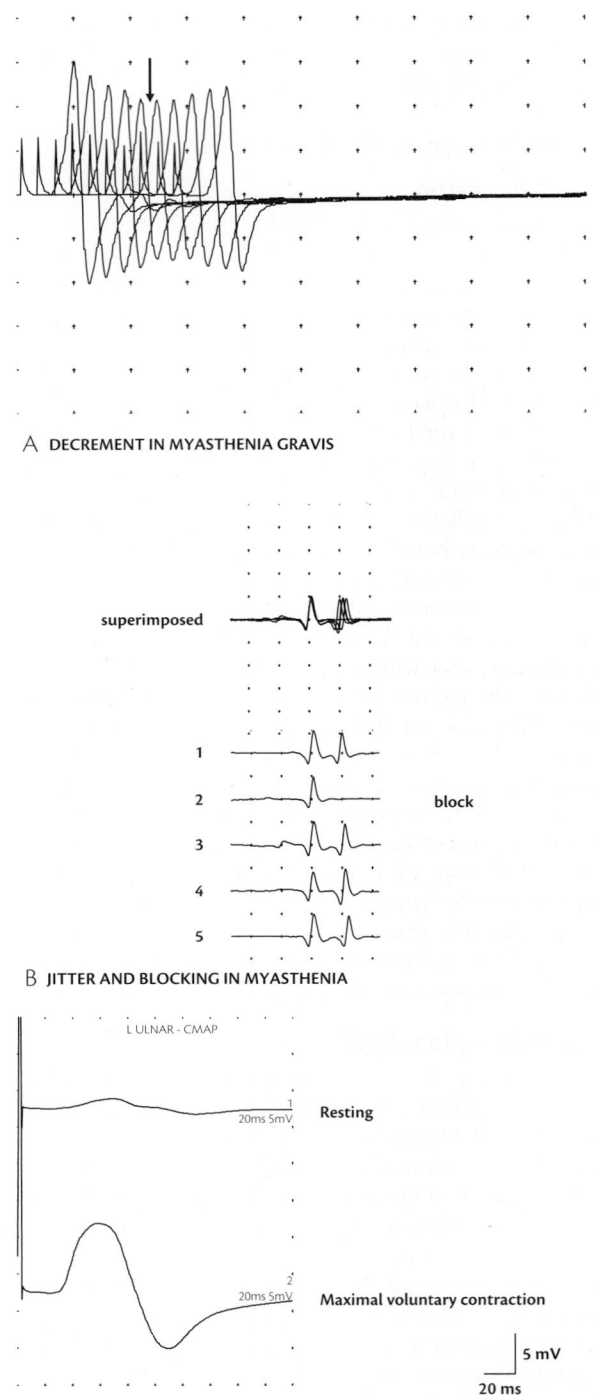

Fig. 3.31 Electromyography in Myasthenias. (A) Myasthenia gravis showing electromyographic decrement in the amplitude of the compound muscle action potential on repetitive nerve stimulation at 3/s. The fifth response shows a 30 per cent decrement compared with the first. (B) Myasthenia gravis. Single fibre electromyograms showing five consecutive discharges of the same motor unit triggered by the first spike. Note from the top superimposed trace that the second spike occurs at a varying interval after the first, a phenomenon known as jitter. Blocking of the second discharge is evident in trace 2. (C) Lambert–Eaton myasthenic syndrome. Compound muscle action potentials recorded before and immediately after maximum voluntary contraction, showing an increment from 2.2 to 17.1 mV. (Courtesy of Dr R Kennett.)

Single-fibre electromyography

The technique of single-fibre electromyography employs a specially constructed electrode for simultaneously recording from a number of individual muscle fibres within the motor unit supplied by a single motor neurone. Normally these adjacent muscle fibres will fire in a close temporal relationship to one another; if this interval varies it is known as jitter (Fig. 3.31B). Blocking is the phenomenon when one of the muscle fibre potentials fails to fire during muscular contraction (Fig. 3.31B). Jitter and blocking generally indicate a failure of neuromuscular transmission at the motor end-plate and occur particularly in myasthenia gravis, but to a lesser extent in neuropathies and motor-neurone disease too.

3.6 Cerebrospinal fluid examination

3.6.1 Introduction

Quinke introduced the technique of diagnostic spinal puncture in 1891 and by the turn of the century it was in widespread clinical usage. For many decades it was one of the few investigative techniques available to neurologists and few patients escaped in-patient admission without encountering its ritual. Later, the development of myelography gave another reason for spinal puncture, and until the advent of MRI many patients, such as those with a paraparesis raising the differential diagnosis of multiple sclerosis or a structural lesion affecting the spinal cord, underwent the process of lumbar puncture with removal of cerebrospinal fluid, CSF, followed by instillation of contrast medium. Arguably, MRI has removed the need for CSF examination in certain instances, including the diagnosis of multiple sclerosis.

CSF is most frequently obtained by lumbar puncture, because that is the easiest and safest site for dural puncture; that is the only technique that will be discussed in any detail. Alternative techniques exist and may be required in a few rare specific circumstances. CSF can be obtained from the cisterna magna by cisternal puncture or the cervical spine by lateral cervical puncture, but it needs little imagination to appreciate the potential hazards, particularly in inexperienced hands. A previous indication was the need to introduce contrast medium for myelography to determine the upper level of a spinal block, but MRI has removed that need. Occasionally CSF examination may be required in a patient in whom the lumbar approach is impossible or contraindicated, for instance due to local sepsis, previous spinal surgery, obesity, arachnoiditis, severe spinal disease.

3.6.2 Cerebrospinal fluid

The CSF is secreted by the choroid plexuses of the ventricles. Flow is caudal, through the fourth ventricle, and out through the foramina of Luschka and Magendie into the subarachnoid space, which covers the entire surface of the brain and spinal cord. The subarachnoid space is contained by the pia mater over the brain and spinal cord and, externally, by the arachnoid membrane. The subarachnoid space is bridged by numerous trabeculae and by blood vessels and nerves, and the surfaces of all of these structures are covered by mesothelial cells. The blood vessels entering the brain and cord carry an extension of the arachnoid and pia mater, forming the Virchow–Robin spaces.

The brain and spinal cord are bathed in CSF. The fluid flows upwards, over the hemispheres, to be absorbed into the venous sinuses through the arachnoid villi. The total CSF volume is about 120 ml. CSF is produced at the rate of about 0.35 ml/min, indicating a turnover of the total volume several times a day.

A major function of the CSF is to provide physical hydrostatic support for the brain. A clinical indication of this function is provided by the headache that develops, due to stretching of pain sensitive structures by the sagging brain, when the CSF volume is reduced, for example following lumbar puncture. The CSF also acts as a sump for cerebral metabolites, although overall their clearance under normal circumstances is probably mainly through the transcapillary route and the venous outflow from the brain. There is also some evidence that CSF may act as an intracerebral transporter, carrying neuroactive substances, such as hormones and releasing factors, from one part of the brain to another.

Although lists are often published of the numerous constituents of the CSF, and their normal concentrations, only very few are assayed in routine clinical practice. These are discussed below.

3.6.3 Indications for spinal puncture

Broadly speaking, spinal puncture, usually lumbar is performed for either diagnostic or therapeutic purposes:

◆ Diagnostic - CSF constituent examination
     - CSF pressure measurement
     - Instillation of radiological contrast media

◆ Therapeutic - Reduction of CSF pressure
     - Instillation of drugs
     - Spinal anaesthesia.

Air encephalography is now redundant, and myelography is largely replaced by MRI. Spinal anaesthesia is not important in the present discussion. Thus, the main indications for spinal puncture are either related to pressure measurement or manipulation, or to the need to obtain CSF for diagnostic purposes.

Although the CSF pressure is raised in numerous pathologies, the only indications to perform spinal puncture specifically in order to measure the pressure are suspected benign intracranial hypertension and some cases of communicating hydrocephalus. Indeed, in most other causes of raised pressure, such as tumour or haemorrhage, spinal puncture is contraindicated. Spinal puncture, with removal of a relatively large quantity of CSF, is an important therapeutic manoeuvre in benign intracranial hypertension.

By far and away the most frequent indication for lumbar puncture is the need to obtain CSF for diagnostic purposes. Broadly speaking, lumbar puncture may be performed in emergency or non-emergency situations:

◆ Emergency - Suspected infection
       • Meningitis: viral, bacterial, fungal
       • Encephalitis
     - Suspected subarachnoid haemorrhage
     - Unexplained confusional state

◆ Non-emergency - Multiple sclerosis
     - Malignancy
       • Carcinomatous meningitis
       • Lymphoma
     - Sarcoidosis
     - Demyelinating polyneuropathy
     - Infection: syphilis, chronic fungal
     - Central nervous system vasculitis
     - Benign intracranial hypertension.

3.6.4 Contraindications to spinal puncture

There are several contraindications to lumbar puncture; some absolute, some relative. By far and away the most dangerous situation is the presence of raised intracranial pressure due to a mass lesion. There is considerable risk of tentorial or cerebellar herniation, which may be fatal. Broadly speaking, the presence of lateralizing abnormal physical signs is a contraindication to lumbar puncture, at least until CT or MRI has excluded a mass lesion causing shift or distortion. The symptoms and signs associated with a spinal canal block may also be exacerbated by the shift caused by removal of CSF from below the block. Local skin sepsis is a contraindication, although in practice it is exceptionally rare for lumbar puncture to give rise to secondary CSF infection, even in the presence of septicaemia. Any disorder of coagulation increases the risk of local haemorrhage with potential cauda equine compression. Contraindications to lumbar puncture are:

- Raised CSF pressure due to an intracranial mass lesion
- Local skin sepsis
- Anticoagulant therapy
- Coagulation defect
- Spinal block.

3.6.5 Lumbar puncture

To learn how to perform a lumbar puncture well, an expert should be watched repeatedly and then the procedure practiced under supervision. There are two types of *'traumatic tap'*. The presence of blood in the CSF is discussed below. The other type summarizes the emotional and physical trauma to the patient, who subsequently tells their friends of the 'lumbar punch' and vows never to let a doctor with a needle near them again. With appropriate clinical skills, this situation should never arise.

Method

The following points note some of the commoner questions, errors, and problems:

- Position – The standard position is the left lateral decubitus; because it is the best position. It is sometimes suggested that if difficulties are encountered, then it is said puncture is easier with the patient sitting up. If that were truly the case, then why are not most lumbar punctures done in this position? If a tyro has failed in the lateral position, that he has at least attempted a few times before, then the chance of success in a position he has never previously attempted is probably even less.

- Posture – The patient lies with their left shoulder, trunk, and hip lying along the near-edge of the bed. The spine, hips, and knees should be flexed as much as is comfortable for the patient; having an assistant forcibly flexing the patient is unnecessary and will only contribute to the patient's discomfort. The spine should be horizontal and not twisted; twisting is avoided by placing a pillow between the patient's knees and making certain that the upper, right, shoulder remains directly above the lower, left, shoulder. Failure to get the patient in the correct position is the major reason for failure of the procedure.

- Level – A line drawn between the anterior superior iliac spines runs through the level of the L3 vertebra. In adults, the lumbar puncture should be performed between either L2 and L3, or

L3 and L4. In children, the spinal cord terminates lower and lumbar puncture should be performed at L3/4 or L4/5.

- Local anaesthesia – It is arguable, and unresearched, whether local anaesthesia is necessary. Only the skin and immediate subcutaneous tissues appear to contribute significant discomfort and it may well be that the discomfort of local anaesthetic infiltration is as great as would be experienced by simply inserting the spinal needle, which is of similar or smaller calibre. If local anaesthesia is used, then only the superficial tissues should be infiltrated, and the volume required is no more than 0.5 ml of lignocaine 1 or 2 per cent. Using larger volumes causes such swelling that it becomes impossible to feel the essential landmarks, the tips of the spinous processes.

- Needle size – As noted below there is a clear correlation between needle size and the incidence of post-lumbar puncture headache. A 26G needle gives rise to a very low incidence of headache, but its very narrow calibre and lack of rigidity gives rise to several problems:

  - insertion, without using a larger diameter short needle to act as a guide, is difficult, particularly for the inexperienced operator;

  - removal of reasonable quantities of CSF may take considerable time; and

  - pressure measurements, particularly 'dynamic' measurements, are difficult because of the slow flow.

In practice, a 26G needle is used by anaesthetists and radiologists for instilling drugs and contrast media, but for diagnostic purposes a 21 or 22G needle is used. The needle should be inserted with the bevel horizontal, as there is evidence that this reduces the incidence of post-lumbar puncture headache, the theory being that this parts rather than severs the longitudinally running fibres of the dura mater.

- The *'careful' LP* – it is sometimes suggested that lumbar puncture may be safer if only a small quantity of CSF is removed, for example in situations of raised intracranial pressure. This is unlikely as in most cases much more CSF probably leaks out from the dural hole after the procedure than is ever removed by the operator.

- Pressure measurement – As noted above, there are only a few specific indications for pressure measurement, which is achieved using a simple manometer. Indeed, in many situations, such as obtaining CSF for protein studies in suspected multiple sclerosis, measuring the pressure is not only unnecessary but the fiddling around with the needle and apparatus increase the chance of getting blood and other contaminants, such as talc from the gloves, into the CSF, invalidating further studies.

- CSF collection – the requisite amount of fluid should be collected in the appropriate tubes. Err on the side of generosity, particularly for cytological studies. If the initial sample is blood stained collect three or more subsequent specimens in separate tubes.

- Bed rest – there is no evidence that bed-rest, whether with the patient lying prone or supine, or with one end of the bed elevated, reduces the incidence of post-lumbar puncture headache (Sudlow and Warlow 2002). Bed-rest for 24 h simply delays the onset of headache by the same period. Therefore, unless there are other indications, there is no reason to confine the patient to bed.

Traumatic tap

This term, in its medical sense, means that blood was introduced into the CSF during the procedure, and its major implication is the confusion that it causes in identifying pre-existing subarachnoid haemorrhage. It also interferes with protein electrophoresis studies. Contrary to popular belief, a traumatic tap is not primarily due to technical difficulties or clumsiness on behalf of the operator, although these might slightly increase its occurrence, but rather to penetration of one of the numerous veins in the region. It occurs in about 1 in 10 punctures and is not necessarily anything to be ashamed about.

The traditional method, and although not foolproof still reasonably reliable, to distinguish traumatic tap is to collect three sequential samples of about 3 ml each and to note by visual inspection, backed up by laboratory red cell counts, that the blood clears. In pre-existing subarachnoid haemorrhage the red cell count in each sample will be the same. Equally important, on obtaining bloodstained CSF, is to centrifuge the sample and observe the supernatent: with a traumatic tap it will be clear, with previous subarachnoid haemorrhage, as long as a reasonable time interval has elapsed, such as 8 h, it will be xanthochromic. Further confirmation, and detection of blood breakdown products of insufficient concentration to render the sample visibly xanthochromic, can be obtained by spectrophotometric analysis, but there are pitfalls limiting sensitivity and specificity (Apperloo *et al.* 2006).

3.6.6 Complications of spinal puncture

The most serious complication, which should generally be avoided by not performing lumbar puncture on unsuitable patients in the first place, is cerebral or cerebellar herniation. Otherwise, serious complications of lumbar puncture are very rare.

Following diagnostic lumbar puncture somewhere between one-quarter and one-third of patients develop headache, but in only a small proportion of these is it severe. There are several risk factors to the development of headache, but needle size is the most important. The cause of headache is CSF hypotension (Section 18.6.7) consequent upon continued CSF leakage through the hole left by the needle. If the headache does not settle with rest and simple analgesics, then almost immediate relief can be obtained by injecting 10 ml of the patient's blood into the epidural space at the level of the previous puncture, a 'blood patch' (Harrington 2004). Complications of this procedure include back and radicular pain resembling sciatica, which occur in up to one-third. Because of these, blood patching can not be recommended as a routine, prophylactic, procedure.

Post-lumbar puncture headache has a number of very specific characteristics that allow it to be distinguished from other non-specific headaches, and these should generally be present before considering blood-patching. The most important feature is its postural dependence. It is exacerbated by sitting-up and standing and typically resolves rapidly on lying. It is often throbbing in character, and may be associated with neck pain and stiffness.

Another complication of persistent low pressure is the formation of subdural fluid collections, typically below the tentorium cerebelli. This may occur following diagnostic lumbar puncture but is also seen in cases of spontaneous CSF hypotension due to dural tears and fistulae.

Rarely, and usually accompanied by significant post-lumbar puncture headache, there may be symptoms and signs of cranial nerve dysfunction including dizziness, tinnitus, deafness, and diplopia due to a VI nerve lesion.

Local intraspinal haemorrhage at the site of puncture is really only seen in patients with disordered clotting. If severe it can lead to paraplegia.

3.6.7 CSF pressure

The opening CSF pressure is measured with a simple manometer. In the conventional position, with the patient in the left lateral decubitus position, the normal CSF pressure is in the range 70–180 mm of CSF. As noted, the pressure needs to be measured in only a few clinical situations. One of those is suspected benign intracranial hypertension. This is most common in obese females. But, importantly, obesity itself is a major cause of a 'false-positive' result. In an otherwise normal obese person, the act of flexing the spine and hips creates pressure on the abdomen, which in turn increases the CSF pressure, up to as much as 300–400 mm CSF. This may be taken to confirm a diagnosis of benign intracranial hypertension in a patient with the not uncommon combination of obesity and tension-type headache. In this situation, instructing the patient to relax by slightly extending their hips and taking the pressure off their abdomen leads to an immediate fall of the CSF pressure.

Measurement of the closing pressure, after CSF sampling and just before removal of the needle, is often performed in patients with benign intracranial hypertension to determine the effect of CSF extraction. It is of dubious validity given that CSF will continue to leak through the dural hole for some time after the procedure.

3.6.8 CSF analysis

The specific findings in individual disorders are discussed throughout the text. Guidelines on CSF analysis have been published (Deisenhammer *et al.* 2006). In routine clinical practice assessment is generally limited to the following areas:

- ◆ Appearance - Naked eye
                 - Spectrophotometric
- ◆ Proteins   - Total protein content
                 - Albumin and globulin
                 - Immunoglobulin assay
                 - Isoelectric focusing
- ◆ Cells   - Lymphocyte and polymorphonuclear cell count
                 - Identification of other cell types, including malignant cells
                 - Cell-typing in certain malignancies
- ◆ Microbiology - Stains: Gram, Ziehl–Neelsen, Indian ink, etc.
                  - Polysaccharide countercurrent electrophoresis, CIE
                  - Polymerase chain reaction, PCR
                  - Fungal antigens
- ◆ Biochemistry - Glucose
                  - Lactate and pyruvate assay
                  - Angiotensin converting enzyme, ACE

Appearance

Normal CSF is often described as being 'gin-clear'; vodka or clean water would be equally descriptive, as would 'crystal-clear'! Purulent CSF in bacterial meningitis is cloudy or turbid, and green-tinged if gross. Following subarachnoid haemorrhage the supernatant following centrifugation of the bloodstained CSF is xanthochromic, or yellow, due to haemoglobin breakdown products, although this may not be evident to the naked eye until upto 12 h after the initial haemorrhage. The value of spectrophotometry is discussed above and elsewhere. After several days the CSF will remain xanthochromic but the red cells will have disappeared. A high protein level also causes xanthochromia. A markedly xanthochromic CSF with a protein content of <1.5 g/l is almost always secondary to previous haemorrhage. The CSF may also appear discoloured in severe jaundice, with dietary hypercarotenaemia, and in the presence of certain drugs such as rifampicin.

Proteins

The normal total protein content of lumbar CSF is in the range 0.1–0.5 g/l. The quoted range varies slightly between laboratories and it is impossible to state an absolute upper limit of normal. Protein levels are lower in ventricular and cisternal CSF.

The total CSF protein is elevated in many disorders and therefore change lacks specificity. The highest levels, sometimes exceeding 10 g/l are seen in purulent meningitis, tuberculous meningitis, tumour particularly if causing spinal block, and arachnoiditis. When the protein level is very high, clots may form.

Quantitative and qualitative assessments of CSF immunoglobulins have achieved importance in providing laboratory support for the diagnosis of multiple sclerosis (Section 37.5.5). Paramount has been the technique of CSF isoelectric focusing (Fig. 3.32).

Cells

Normal CSF contains less than 5 lymphocytes/mm^3, and no other cell type. The cell count is elevated in many conditions. In multiple sclerosis the count rarely exceeds 100, and contains predominantly lymphocytes. Moderately elevated counts, of several hundred/mm^3, are seen in viral meningitis; typically lymphocytes predominate, but polymorphonuclear cells may be the dominant cell type in early stages. Very high counts of thousands/mm^3 are seen in bacterial meningitis, with polymorphonuclear cells predominating, except that in the earliest stages the cell count may be low and lymphocytes predominate; at this stage misdiagnosis suggesting a viral process may be made. In tuberculous meningitis polymorphonuclear cells may predominate in the early stages, but later on lymphocytes gain the ascendancy and typically number up to 500/mm^3. The findings in fungal meningitis are similar to those in tuberculous meningitis.

In the search for malignant cells, CSF sampling may have to be repeated several times before a positive result is obtained. Yield is also increased by taking larger quantities of CSF.

Microbiology

The importance of cell type and count in differentiating different causes of infection is noted above. Specific diagnosis is achieved through one or more of the microbiological methods listed and discussed further below.

Biochemistry

The only frequent biochemical estimation, apart from protein, is glucose. It should be measured in cases of suspected infection or of carcinomatous meningitis. This has to be compared with a simultaneous blood glucose level. The CSF:blood glucose ratio is

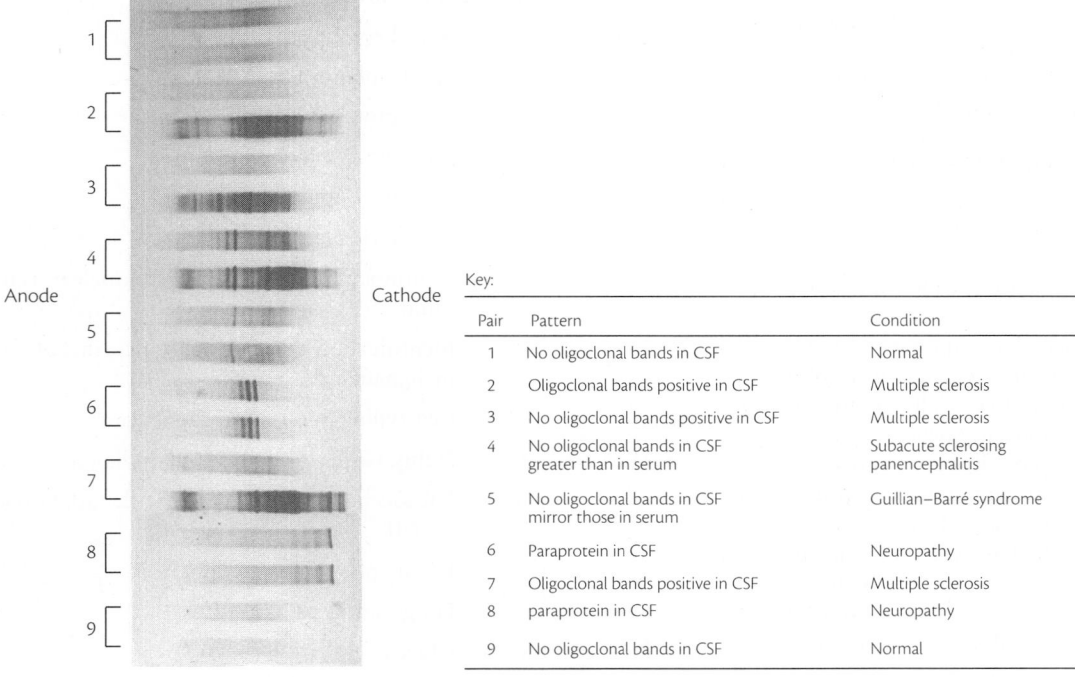

Pair	Pattern	Condition
1	No oligoclonal bands in CSF	Normal
2	Oligoclonal bands positive in CSF	Multiple sclerosis
3	No oligoclonal bands positive in CSF	Multiple sclerosis
4	No oligoclonal bands in CSF greater than in serum	Subacute sclerosing panencephalitis
5	No oligoclonal bands in CSF mirror those in serum	Guillian–Barré syndrome
6	Paraprotein in CSF	Neuropathy
7	Oligoclonal bands positive in CSF	Multiple sclerosis
8	paraprotein in CSF	Neuropathy
9	No oligoclonal bands in CSF	Normal

Fig. 3.32 CSF isoelectric focusing. The figure shows nine pairs of which the upper is serum and the lower is CSF. This illustrates typical banding patterns on isoelectric focusing. (Courtesy of Professor Ed Thompson.)

typically about two-thirds. The ratio falls in the presence of hyperglycaemia, in diabetes mellitus. Ratios between one-third and two-thirds may be normal, values below one-third are invariably abnormal. CSF glucose levels are usually very low in cases of purulent meningitis, and often low in malignant meningitis.

CSF, and indeed serum, lactate estimation is an ancillary investigation in cases of suspected mitochondrial cytopathy. Levels are elevated, as is the lactate:pyruvate ratio, but in general it is other, more specific, investigations that are likely to lead to the correct diagnosis.

3.6.9 Detection of infectious agents

Diagnostic methods include microscopy, culture, serology, polymerase chain reaction detection of genome, and antigen detection. Broadly speaking, bacterial infections are most likely to be identified through microscopy and culture, viral infections through polymerase chain reaction and serology, fungal infections through antigen detection, and parasitic infections through polymerase chain reaction, serology, pathogen detection in stool, or brain biopsy. Details are given elsewhere, when discussing specific infections.

Polymerase chain reaction has become increasingly available and merits specific comment. It allows the detection of low copy numbers of specific nucleic acid sequences and thus has potentially high sensitivity and specificity. As well as being valuable in identifying viruses it is particularly helpful in identifying fastidious bacteria and those that are difficult to culture, such as tuberculosis. False positive and false negative results are well recognized and rarely should diagnosis rely on the polymerase chain reaction result alone. For viral infections, the timing of the sample and the sensitivity of the assay are important factors (Davies *et al.* 2005). There is a surprisingly limited literature concerning polymerase chain reaction and tuberculous meningitis possibly reflecting in part that tuberculosis is still relatively rare in the developed world, although that picture is changing; the laboratory diagnosis of tuberculosis has been reviewed (Garg *et al.* 2003).

3.7 Neurological autoantibodies

3.7.1 The diagnostic use of antibody testing in neurological disease

Testing for antibodies in patients with neurological disease is more frequently done, and rather more useful, than might first be thought. Assaying for immunoglobulin oligoclonal bands in the spinal fluid in a patient suspected of having multiple sclerosis, or as a means of at least contributing to the diagnostic distinction of multiple sclerosis from other forms of brain inflammatory disease, including acute disseminated encephalomyelitis, Behçet's, or sarcoidosis, etc., is a form of antibody testing. Likewise, looking for serum antinuclear antibodies in suspected neurological lupus, for anti-neutrophil cytoplasmic antibodies in suspected vasculitides affecting the nervous system, or for rheumatoid factor in a patient with a peripheral neuropathy, are also forms of neurodiagnostic antibody assay. A number of infectious diseases which can involve the nervous system also are diagnosed by the detection of specific antibodies, from the common Epstein Barr to the rare, in the United Kingdom, West Nile virus. Whilst these disorders, including their diagnosis, are comprehensively dealt with in Chapters 41, 42, and 43, a brief account is worthwhile of one or two more specifically

neurological disorders where antibody testing plays a major role in confirmation or otherwise of the diagnosis (Table 3.1).

Various techniques are exploited for detecting and quantifying specific antibodies in serum or CSF samples, including radioimmunoassays, enzyme immunoassays or 'ELISAs', immunofluorescence and immunohistochemistry, and Western blotting.

Table 3.1 Autoantibodies used in neurological diagnosis

Autoantibody	Associated disorder(s)
Anti-acetylcholine receptor (AChR)	Myasthenia gravis
Anti-muscle specific kinase (MUSK)	Myasthenia gravis seronegative to AChR
Anti-voltage-gated calcium channel (VGCC)	Lambert–Eaton myasthenic syndrome Paraneoplastic cerebellar ataxia
Anti-voltage gated potassium channel (VGKC)	Idiopathic limbic encephalitis Paraneoplastic limbic encephalitis Acquired neuromyotonia, Isaac's syndrome Morvan's syndrome (neuromyotonia + central nervous system symptoms)
Anti-GQIb ganglioside	Miller Fisher syndrome Chronic ataxic neuropathy with ophthalmoplegia M-band, anti-disialyl antibodies, CANOMAD
Anti-GMI ganglioside	Multifocal motor neuropathy
Anti-myelin associated glycoprotein (MAG)	Chronic demyelinating neuropathy
Anti-ganglionic acetylcholine receptor	Autoimmune autonomic neuropathy Paraneoplastic autonomic neuropathy
Anti-glutamic acid decarboxylase (GAD)	Idiopathic stiff person syndrome
Anti-amphiphysin	Polyneuropathy Limbic encephalitis Stiff person syndrome Brainstem encephalitis Encepahlomyelitis
Anti-Ma2	Limbic encephalitis in young adults Myelopathy Cerebellar ataxia Brainstem encephalitis
Anti-CRMP5/CV2	Sensorimotor neuropathy Cerebellar ataxia Optic neuropathy Chorea Dysautonomia
Anti-aquaporin 4	Devic's disease
Paraneoplastic disorders:	
Anti-Hu	Sensory neuronopathy Encephalomyelitis Dysautonomia
Anti-Yo	Paraneoplastic cerebellar degeneration Rarely others
Anti-Ri	Brainstem encephalitis including opsoclonus-myoclonus Polyneuropathies Cerebellar ataxia Others

Enzyme immunoassays are perhaps the commonest approach: target antigen is bound to a solid phase, and exposed to the serum or CSF sample at a particular dilution, allowing any specific antibody present to bind. After washing to remove unbound antibody, commercially obtained enzyme-labelled antibody specifically directed against human immunoglobulin-G is added, which recognizes and sticks to bound patient's antibody. A colour reactant is then added, the substrate of the anti-human IgG antibody's enzyme tag, and colorimetric analysis then allows a relative, quantitative measure to be derived of the concentration of antigen-specific antibody in the patient's serum or CSF sample. For detailed technical descriptions see Meriggioli (2005).

3.7.2 Antibody testing in central nervous system disease

There are perhaps two overlapping areas where antibody testing has assumed primary diagnostic importance: *paraneoplasia*, an area where enormous progress was made in the 1990s, and so-called *autoimmune* or *acquired channelopathies*, very much a topic of the current decade. Whether such antibodies are 'merely' of diagnostic value, or have pathogenetic significance lies beyond the scope of this brief account.

Paraneoplastic neurological disease (Section 38.4) can affect both the peripheral and central nervous system. A range of antigens is described, and with them a number of specific syndromes, though the correlation of phenotype with antibody, and with underlying malignancy, is far less specific than initially considered. New syndromes, antibodies, and associations continue to be reported (Graus *et al.* 2005; Vitaliani *et al.* 2005; Sabater *et al.* 2005; Wieser *et al.* 2005; Iranzo *et al.* 2006). *Stiff person syndrome* (Section 38.3.4) is another, fairly recently described antibody-associated central nervous system disorder, up to 70 per cent patients having anti-glutamic acid decarboxylase antibodies. This syndrome too can be paraneoplastic (Hernandez-Echebarria *et al.* 2006).

Antibodies to ion channels were first found to be associated with disorders primarily affecting the neuromuscular junction, but then recognized also in conjunction with paraneoplastic and non-cancer-associated central nervous system disease. Antibodies to voltage-gated calcium channels are found in paraneoplastic cerebellar ataxia (Graus *et al.* 2002) and indeed may be detectable in both serum and CSF. In both paraneoplastic and idiopathic limbic encephalitis, with or without seizures, antibodies to voltage-gated potassium channels may be found, and such testing may now be recommended as a part of the investigation of patients presenting with acute or sub-acute encephalopathies, with or without seizures (Pozo-Rosich *et al.* 2003; Vincent *et al.* 2004). *Rasmussen's encephalitis* (Section 30.5.6) has also been associated with ion channel antibodies, here directed against the ionotropic glutamate receptor.

Sydenham's chorea (Section 40.5.7) has long been known as a post-streptococcal movement disorder, and an autoimmune basis consequently inferred. More recently, a variety of other movement syndromes, and psychiatric disorders, have been suggested to follow streptococcal infection, including tic disorders and dystonia, and depression and obsessive-compulsive disorder. The acronym *PANDAS*, **p**aediatric **a**utoimmune **n**europsychiatric **d**isorders **a**ssociated with **s**treptococcal **i**nfections, has been offered (Swedo *et al.* 1998; Snider and Swedo 2004), and anti-basal ganglia antibodies have been reported in this group of disorders (Church *et al.* 2002)

(Section 40.10). It is possible that these cross react with streptococcal epitopes (Dale *et al.* 2001).

3.7.3 Antibody testing in peripheral neurological disease

In the peripheral nervous system, it is largely in relation to inflammatory neuropathies and neuromuscular disease that antibody testing is of relevance.

An association of Guillain–Barré syndrome (Section 21.10) with antibodies to gangliosides has been recognized for well over a decade. Certain ganglioside specificities are associated with particular sub-types of neuropathy: for example, anti-GQ1b antibodies with up to 95 per cent specificity and sensitivity in Miller–Fisher syndrome, and anti-GM1 antibody with multifocal motor neuropathy, although in less than 50 per cent of cases). In classical acute inflammatory demyelinating polyradiculoneuropathy, however, a routine diagnostic role for such antibody testing has not emerged (Willison and Yuki 2002). Up to two-thirds of patients with demyelinating neuropathy associated with monoclonal gammopathy of undetermined significance have any antibodies against myelin associated glycoprotein, MAG.

The prototypical and classical antibody-associated (and mediated) disorder is myasthenia gravis (Section 24.10.1), where antibodies to the post-synaptic nicotinic acetylcholine receptor play a key diagnostic role. About 15 per cent of patients are negative when tested for these antibodies; approximately 70 per cent of these have antibodies to muscle-specific kinase, MuSK (Lang and Vincent 2003). Interestingly, an association of autoimmune autonomic neuropathy (Section 21.11.7) with autoantibodies to ganglionic acetylcholine receptors has been recognized (Klein *et al.* 2003; Vernino *et al.* 2000, 2004).

Antibodies to ion channels are of diagnostic value in other neuromuscular junction disorders. In the Lambert–Eaton myasthenic syndrome (Section 24.10.2), anti-voltage-gated calcium channel antibodies are found in over 85 per cent of cases. In acquired neuromyotonia, or Isaac's syndrome (Sections 23.7.1 and 24.10.4), there is antibody specificity towards voltage-gated potassium channels. In Morvan's syndrome, neuromyotonia is associated with central nervous system symptomatology, including hallucinations, amnesia, and insomnia. Such individuals also may harbour anti-voltage-gated potassium channel antibodies (Liguori *et al.* 2001); some indeed have the neuropsychiatric features without neuromyotonia, leading full circle to the spectrum of central nervous system disorders associated with ion channel antibodies.

3.8 Neurogenetics

3.8.1 Patterns of inheritance

The impact of molecular genetics in neurology has been considerable, both in terms of clarifying mechanisms of disease and the development of diagnostic tests. It remains vital that the clinician knows about the main patterns of inheritance at a clinical level. When considering autosomal inheritance, it is the observed phenotypic expression of a gene that is dominant or recessive rather than the gene itself. In other words, pathological phenotypes are dominant or recessive characteristics of mutant genes and produce corresponding patterns of inheritance within families. The recognition of patterns of inheritance can be difficult when environmental factors are important or when the family history is incomplete.

A common problem is the conclusion that a patient's family history is 'negative' when it is in fact unreliable and limited. For example, there may be little or no information about the state of health of relatives who died prematurely or lost contact with the patient years previously. This is common, for example, in families with Huntington's disease.

Terminology in genetics can be confusing. When describing individuals within families, the *proband* or *index case* is the individual by which the family was detected or ascertained. Affected relatives are referred to as *secondary cases*. Nearly all somatic cells contain two copies of autosomal genes and the two versions of a given gene are referred to as *alleles*. Thus an individual may be a *homozygote* with two mutant alleles or a *heterozygote* with one mutant and one normal allele. Individuals with two different mutant alleles are termed *compound heterozygotes*. The *segregation ratio* of a hereditary disorder is the proportion of clinically affected individuals within a generation. This is difficult to observe in single families due to the effects of chance and small numbers of children. Segregation ratios usually need to be calculated from observations of many families.

Autosomal dominant inheritance

In autosomal dominant conditions the effect of the mutant gene is evident in heterozygotes. Homozygotes are only occasionally encountered due to the rarity of the abnormal gene but they may be either more severely affected than heterozygotes or clinically indistinguishable, as in Huntington's disease. The pattern of transmission of an autosomal dominant condition within a family is vertical, from generation to generation, and the risk of the disorder appearing in a child born to an affected individual is 1 in 2. Overall, the segregation ratio among siblings, children, and parents of those affected is 0.5 or 50 per cent. Both sexes are affected and either sex can transmit the condition to children. This classical pattern of autosomal dominant inheritance is not always obvious. *Sporadic cases* may arise due to new mutation within a family. There will be a higher proportion of new mutations in an autosomal dominant disorder if it reduces the ability of affected individuals to reproduce. Therefore, an autosomal dominant disease which is lethal before reproductive age or confers a reproductive disadvantage for any reason is unlikely to be transmitted to children. Accordingly most if not all such cases will arise sporadically because of the constant background rate of mutation of the relevant gene within the population. Conversely, a condition which is associated with normal reproductive ability will usually appear in the children of affected people with only a small proportion of cases being accounted for by new mutation. Other cases may appear to be sporadic if the family history is inaccurate. Only some gene carriers are affected in certain autosomal dominant conditions, with the resultant segregation ratio in children or siblings being lower than 0.5. The proportion of heterozygotes who have the clinical phenotype is the *penetrance* of the gene. In addition, the clinical phenotype itself may be variable in nature or severity; this is referred to as variable *expressivity* of the gene.

Autosomal recessive inheritance

In autosomal recessive inheritance the condition is only seen in those homozygous for the disease related allele, as in Friedreich's ataxia. Sometimes, an autosomal recessive disorder appears in individuals who carry two different disease related alleles, so-called *compound heterozygotes*. In autosomal recessive inheritance, there is horizontal transmission of the disorder with normal parents and one or more affected children; both sexes are affected. Although clinically normal both parents are heterozygous carriers of the mutation. Consequently the chances of a child being homozygous for the mutant allele, and therefore affected, is 1 in 4, of being a heterozygous carrier is 1 in 2, and of being homozygous for the normal allele is 1 in 4. Therefore, with small families, many cases will appear sporadically but among the siblings of affected patients overall, the segregation ratio will be 0.25. Children of affected patients must be heterozygotes since an affected person cannot transmit a normal allele, but are clinically normal. Very rarely, a person with an autosomal recessive condition may transmit the condition to a child if the other parent is a heterozygous carrier, *pseudo-dominance*. The chance of this occurring is small, due to the low frequency of any one mutation in the population. An autosomal recessive disorder is more likely to appear if there is parental consanguinity, due to the increased likelihood of both parents carrying the same mutant allele.

X-linked recessive inheritance

X-linked recessive inheritance is a condition usually seen exclusively in males, in whom the allele on the single X chromosome may be normal or a mutation. It will not usually be seen in heterozygous females because of the corresponding normal allele. An example is Duchenne muscular dystrophy. The affected males have normal parents and children; sons cannot be affected for they must inherit their X chromosomes from their mothers, and all daughters are carriers. The disease is transmitted only by carrier females, with a 50 per cent risk to sons of being affected and 50 per cent to daughters of being carriers. Sometimes daughters of affected males are affected if the mother is by chance a carrier but this is unusual. Females may also be affected because of the normal process of X chromosome inactivation or *Lyonization*. Usually X chromosome inactivation is random so that equal numbers of normal and mutant alleles are inactivated. If, by chance, X inactivation is non-random and mostly normal alleles are inactivated, a female will be affected but not normally as severely as an affected male. Such females are referred to as *manifesting carriers*, as seen in adrenoleucodystrophy.

X-linked dominant inheritance

X-linked dominant inheritance is much less common and resembles autosomal dominant transmission, except that affected males cannot transmit the disorder to sons and always do so to daughters.

Multifactorial inheritance

In multifactorial inheritance, a disease phenotype is the result of the interaction of many different genes and environmental factors. Such disorders are common and although many affected individuals appear to be sporadic cases, the incidence of affected relatives is higher than would be expected by chance. At a clinical level, there may be clues to this type of inheritance. The risks to parents, children and sibs are low but should be similar and approximately $1/\sqrt{q}$, where q = the population prevalence of the disorder. The risks to second and third degree relatives are minimal. The risk to relatives of the less commonly affected sex is higher than that to those related to index cases of the more commonly affected sex. Parental consanguinity will be slightly increased but not as high as that seen with autosomal recessive inheritance. In addition, there will be a greater concordance in monozygotic twins compared with dizygotic pairs; the degree of the concordance increase is a guide to the relative importance of genetic and environmental factors.

Maternal inheritance

In maternal inheritance, a phenotype is transmitted only through the female line; although affected individuals may be male or female, only females may pass on the gene. This is seen with mitochondrial disorders as spermatozoa do not contain mitochondria which are exclusively transmitted in ova.

3.8.2 Chromosomes and genes

The human genome consists of DNA approximately 3×10^9 base pairs in length. *Chromosomes* are huge molecules of genomic DNA in highly compact form, associated with structural proteins of two types, either basic histones or acidic non-histone proteins. The DNA–protein complex is termed chromatin. Human chromosomes are composed of 23 pairs of homologous autosomes and the two sex chromosomes, XX or XY. If metaphase chromosomes are stained with a basic dye, usually Giemsa stain after exposure to trypsin, they are seen to have characteristic bands, 'G-bands', which vary such that the individual chromosomes can be distinguished and identified (Fig. 3.33). The dark stained bands consist of histone-rich heterochromatin and contain mostly structural DNA. The polypeptide encoding genes are mainly located within the unstained bands of non-histone euchromatin. The map positions of genes refer to the number of the chromosome, the part of the chromosome (long arm *q*, short arm *p*, centromere, or telomere), and the band number within which the gene is situated. These positions are approximate as each band contains approximately 5–10 megabases and may contain several genes.

The genes themselves have a complex structure (Fig. 3.34) and are split into coding sequences or *exons* and intervening non-coding *introns*. The transcription process reads the DNA in the 5′ to 3′ direction and a primary messenger RNA, mRNA, transcript is synthesized by RNA polymerase. Subsequently the primary mRNA is processed by the addition of a 7-methyl guanosine cap at the 5′ end, a polyadenylate tail at the 3′ terminal and the removal of the intron sequences by a splicing procedure regulated by small nuclear RNA molecules, snRNAs. The much smaller polypeptide encoding mature mRNA transcript is then transported to the cytoplasm.

The level of transcription of a gene is regulated by sequences upstream of the 5′ end which influence the activity of RNA polymerase; these are referred to as promoters, such as TATA sequences, and enhancer elements, such as CCAAT sequences (Fig. 3.34). Transcription is also affected by chromatin structure and the level of methylation of the cytosine bases within the DNA. Transcription is less likely to occur in histone rich heterochromatin regions and in regions where there is a high degree of DNA methylation.

In addition to protein coding genes, the great majority of the genome contains non-coding DNA. This consists of structural DNA, inactive pseudogenes, and repetitive sequences. A common example of the latter are the numerous Alu sequences, 300 bp in length and containing restriction sites for the enzyme Alu1; the function of these repetitive sequences is unclear.

An important feature of genomic DNA is that it contains frequent variations or *polymorphisms*. These variations take several forms, are stable, and can be inherited in a Mendelian manner. Polymorphisms may be single base changes or variations in the number of consecutive repeats of sequences of DNA, or variable

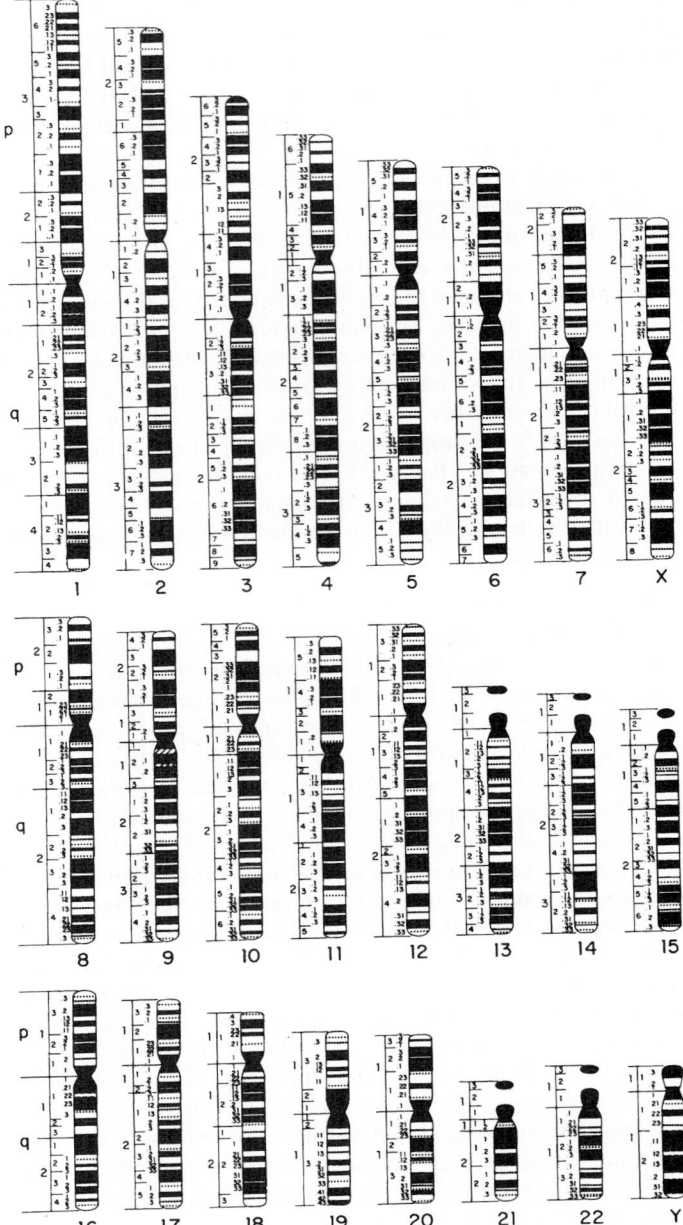

Fig. 3.33 Banding patterns of human chromosomes. Black G bands are those seen at the prometaphase 850 band stage; dashed bands are those seen only in late prophase (1300 band stage) and mid prophase (1700 band stage). (From Vogel and Motulsky (1986).)

number of tandem repeats. Such polymorphisms may be based on repeats of a sequence of about a dozen base pairs so that the different polymorphisms differ by thousands of base pairs, known as minisatellites. Other variable number of tandem repeats are based on repeats of only a few base pairs so that there are polymorphisms differing by only tens of base pairs, microsatellites, or repeats of only a single base.

3.8.3 Mitochondria and genes

Mitochondria are intracellular organelles whose main function is to synthesize ATP utilizing the mitochondrial respiratory chain and

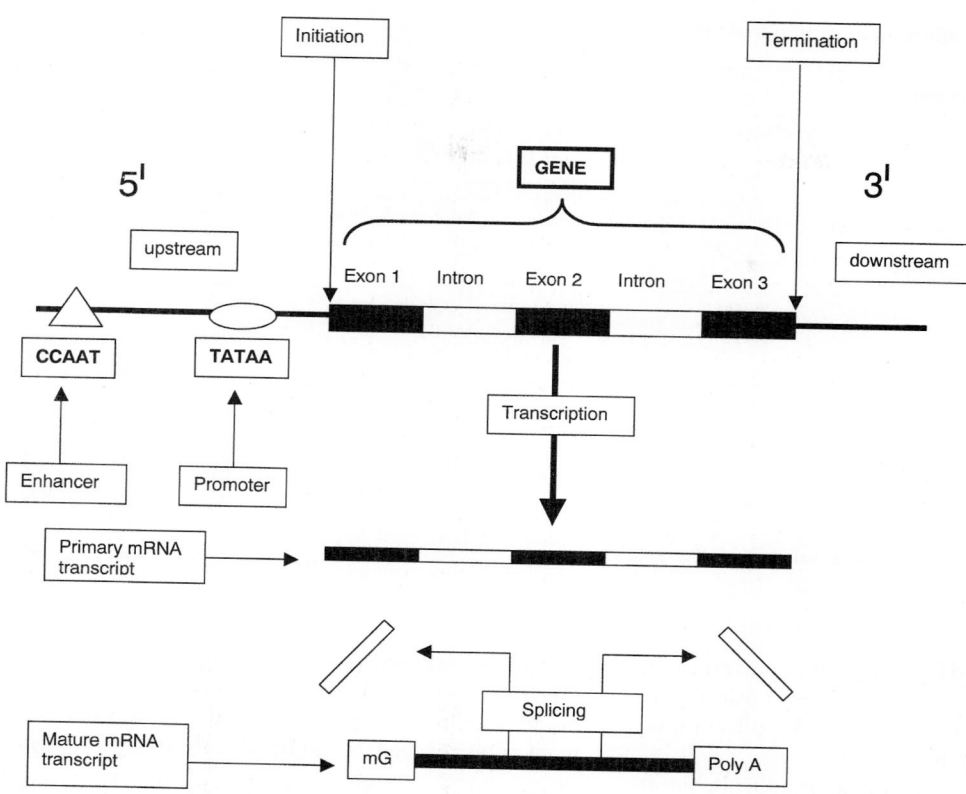

Fig. 3.34 A simplified diagram of a human gene, transcription and RNA processing. mG = methyl guanosine cap, Poly A = polyadenylated tail. Sites of initiation and termination of transcription are indicated.

oxidative phosphorylation system. Each mitochondrion contains 2–10 circular DNA molecules, mtDNA, of about 16 kb. The mtDNA forms about 1 per cent of the total cellular DNA and has a slightly different genetic code, no introns, and very little non-coding DNA. mtDNA replicates with each cell division and both strands contain coding sequences. Mitochondrial genes include:

- 13 protein encoding, *mit*, genes for seven subunits of respiratory complex I;
- three subunits of cytochrome oxidase, COX, complex IV;
- two subunits of ATP synthetase, complex V;
- apocytochrome b, complex III.
- 24 *syn* genes encoding RNAs required for mitochondrial protein synthesis, two ribosomal rRNAs and 22 transfer tRNAs.

All other proteins needed for mitochondrial replication and function are encoded by nuclear genes. There are thousands of copies of mtDNA in most cells which can be variable, heteroplasmy. Mitochondria are not transmitted by spermatozoa, so inheritance of mtDNA is exclusively maternal.

Mutations of mtDNA are the basis of several mitochondrial diseases (Schapira 2006). These tend to affect tissues with low cell turnover, such as the nervous system, muscle, and heart, in which mutant mtDNAs are able to persist, rather than be gradually depleted during repeated mitotic cell division. Mutations of mtDNA are either point mutations or deletions; duplications are unusual. Point mutations are specified by the mtDNA nucleotide position at which they have occurred. A summary of the important mutations and their associated mitochondrial diseases is in Tables 3.2 and 10.9.

Table 3.2 Mutations of mitochondrial DNA and associated disorders (adapted from Schapira 2006)

Inheritance	Disease	RRF in muscle	Mutation	Gene involved
Maternal	LHON	No	11778	Complex I
			3460	Complex I
			14484	Complex I
	NARP	No	8993	ATPase 6
	MELAS	Yes	3243	tRNA
			3271	tRNA
			11084	Complex I
	MERRF	Yes	8344	tRNA
			8356	tRNA
			3243 MELAS	tRNA
	MILS	No	8344	tRNA
			8993	ATPase 6
	CPEO	Varies	Various point	tRNA
	Myopathy		mutations	tRNA
	Cardiomyopathy		including 3243	tRNA
	Deafness		MELAS and	rRNA
	Diabetes		others	
Sporadic	CPEO	Yes	Deletion	Various
	KSS	Yes	Deletion	Various
	Pearson syndrome	Yes	Deletion	Various

CPEO = chronic progressive external ophthalmoplegia; KSS = Kearns Sayre syndrome; LHON = Leber's hereditary optic neuropathy; MELAS = mitochondrial encephalomyopathy, lactic acidosis, and stroke-like episodes; MERRF = myoclonic epilepsy with ragged red fibres; MILS: maternally inherited Leigh's syndrome; NARP = neurogenic atrophy, ataxia, and retinitis pigmentosa; RRF = ragged red fibres.

Table 3.3 Nuclear genetic mutations affecting mitochondrial function

	Nuclear gene	Disorder
mtDNA regulation	C10orf2 'Twinkle'(mtDNA helicase)	Autosomal dominant CPEO; SANDO; myopathy; cardiomyopathy
	POLG	Autosomal dominant / recessive CPEO; SANDO; parkinsonism; Alpers syndrome; infertility; cataract
	Deoxyguanosine / thymidine kinase genes	mtDNA depletion syndrome in infancy with fatal hepatocerebral syndrome; childhood myopathy and encephalopathy.
	Thymidine phosphorylase	MNGIE
Mitochondrial protein function	SCO2; SURF1; COX10; COX15; LRPPRC BSC1L ATP12	Autosomal recessive COX deficiency Leigh syndrome; GRACILE Complex V deficiency (lethal infantile)

POLG = mtDNA polymerase γ

CPEO = chronic progressive external ophthalmoplegia

SANDO = sensory ataxia, neuropathy, dysarthria, ophthalmoplegia

MNGIE = mitochondrial neurogastrointestinal encephalomyopathy

COX = cytochrome oxidase

GRACILE = growth retardation, aminoaciduria, cholestasis, iron overload, lactic acidosis, and early death

There are several disorders caused by mutations of nuclear genes affecting mitochondrial function (Table 3.3). These nuclear gene mutations affect either mtDNA maintenance and replication or respiratory chain p rotein assembly and stability.

Genetic counselling in mitochondrial diseases is difficult, but in general, risks are low to the offspring of women carrying mtDNA deletions which are usually sporadic. For those with point mutations the risks to offspring are low unless the proportion of mutant mtDNA in their cells is very large (Chinnery and Turnbull 1997). Males cannot transmit mtDNA mutations. The situation is different in rare families in which mtDNA or mitochondrial protein abnormalities occur as a result of mutations of nuclear genes; in this situation the mitochondrial disorder is transmitted by either sex as an autosomal dominant or autosomal recessive trait (Table 3.3).

3.8.4 DNA analysis

A comprehensive review of laboratory molecular genetic methods is beyond the scope of this chapter and only a very brief summary of techniques likely to be of interest to clinicians will be attempted here. More detail is available in other texts (Davis and Read 1992; Conneally 1993). DNA is extracted from blood or sometimes other tissues such as muscle and may then be stored at low temperatures for long periods, DNA banking. For a specific region of the DNA to be analysed, many copies of the region of interest are first made using the polymerase chain reaction, or PCR, method.

Polymerase chain reaction is a form of *in vitro* DNA replication to make thousands of copies of a selected DNA region. The DNA to be amplified is designated by the use of primers which are complementary to sequences at its 5′ and 3′ boundaries. The 5′ primer hybridizes at the 5′ end of the sense strand of the designated fragment while the 3′ primer hybridizes to the 5′ end of its complementary antisense strand (Fig. 3.35). The reaction has three steps:

- First, DNA denaturation at high temperature renders the DNA single stranded and therefore able to hybridize with primers;

- Second, hybridization then occurs between the complementary sequences of the primers and the DNA fragment to be amplified;

- Third, DNA synthesis, which can only proceed in a 5′ to 3′ direction, then occurs using a heat stable DNA polymerase. Synthesis has to start from the primer sequences only, therefore yielding many copies of the DNA within the primer hybridization sites.

The resulting amplified fragment must then be analysed by gel electrophoresis; the amplified DNA can be visualized directly in the gel after chemical staining, by mutation scanning or by DNA sequencing, for the detection of polymorphisms and mutations in the amplified sequence.

Standard polymerase chain reaction will amplify DNA fragments of up to 1500 bp. Modifications of the method, *long range polymerase chain reaction*, using alternative polymerases will increase this to approximately 5000 bp.

Fragments larger than this must be analysed using alternative DNA hybridization based methods such as Southern blotting. For this method genomic DNA must be cut into manageable fragments. This is achieved with *restriction endonucleases*. These enzymes cut DNA at certain sequences only, known as *restriction sites*. There are many restriction endonucleases, each recognizing a particular DNA cleavage sequence. The resulting DNA pieces are referred to as restriction fragments within an overall restriction digest. These must then be separated by agarose gel electrophoresis which allows the physical separation of DNA fragments based on their size. Small fragments migrate further through the gel in the direction of the current than larger fragments. Several different DNA samples can be analysed simultaneously in different lanes within the gel. Only fragments up to 20–30 kb can be separated in this way. The DNA can be visualized in the gel by chemical staining but individual fragments cannot be distinguished.

To separate large DNA fragments of up to 30 000 kb in length *pulsed field gel electrophoresis*, which uses an alternating direction electrophoretic current, is required. Such large fragments are produced initially with restriction endonucleases which recognize infrequent widely separated restriction sites. Fragments are then rendered single stranded and transferred from an agarose gel to a membrane by a process known as Southern blotting. The membrane is then hybridized to DNA or RNA *probes* labelled

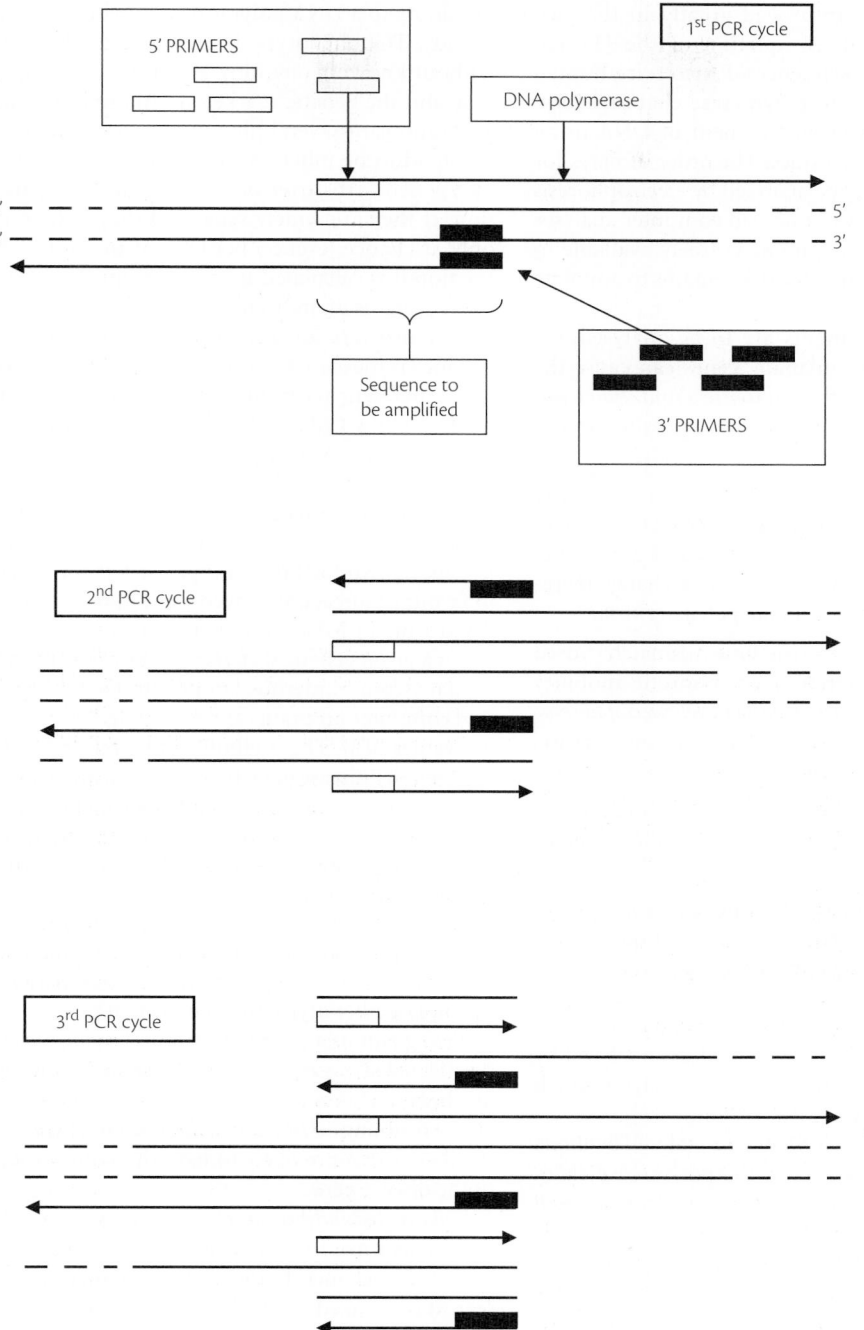

Fig. 3.35 The polymerase chain reaction (PCR). The sense strand of the DNA to be copied runs in a 5′ to 3′ direction, the antisense strand runs 5′ to 3′ in the opposite direction due to the antiparallel arrangement of the DNA double helix. DNA polymerase always proceeds in a 5′ to 3′ direction. The 5′ end of the target sequence on the sense strand is therefore defined by a 5′ primer complementary to the 3′ end of the sequence on the antisense strand (open rectangle). The 3′ end of the target sequence on the sense strand (filled rectangle) is defined by a directly complimentary primer, the 3′ primer. Note that both long and short fragments of DNA are synthesized but that as more cycles are completed, the proportion of short fragments complimentary to the target sequence increases.

chemiluminescently or with a radionucleotide. A probe is a fragment of DNA or RNA with a complementary sequence to that of the DNA fragment of interest and so will hybridize with this fragment and not other fragments with different DNA sequences. In this way the fragment of interest may be detected visually or by autoradiography. Probes may be fragments of DNA with the same sequence as a known gene, complementary or cDNA probes, or RNA transcripts from a given gene, or simply DNA sequences whose chromosomal localization is known. *cDNA* is synthesized from a mature RNA transcript by the enzyme reverse transcriptase; accordingly it contains only the exon DNA sequences of the gene in question.

A variation on this hybridization method is behind the new technology of *DNA microarrays*. Here tens of thousands of single stranded probes are fixed to a small glass slide, a DNA chip, which is then hybridized to labelled fragments of a patient's DNA. The degree of hydrization to each probe is analysed by computer. This is a very powerful technique that allows for the analysis of many genes simultaneously or even a broad analysis of the whole genome.

Equally RNA from specific patient tissues can be used to study gene expression using DNA microarrays.

DNA is analysed for the presence of point mutations (Section 3.8.6) by *DNA sequencing* of the PCR amplified fragments.

DNA sequencing technology has improved greatly in the past 10 years leading to the successful completion of the Human genome project to determine the sequence of the entire human genome. The technology is similar to polymerase chain reaction and uses *in vitro* replication of a given fragment of DNA using modified fluorescently labelled nucleotides. The order of bases for a 500 bp DNA fragment can readily be analysed by electrophoresis through a fine capillary followed by automated computer analysis. The sequence of the whole human genome is freely available to the public (www.ensembl.org) but much work remains to annotate the position of all the genes.

If a large number of DNA fragments are to be analysed, for instance if a gene has many exons or many genes can cause the same disorder, then it is often more efficient to use a *mutation scanning* technique to identify which fragments have possible mutations. Examples include:

- Single stranded DNA shows different electrophoretic mobility with even a single base change, which can be detected by capillary electrophoresis after polymerase chain reaction and denaturing of the DNA to render it single stranded. The base change therefore creates a *single stranded conformational polymorphism*.

- Heteroduplexes in which there is a single base mismatch caused by a base change also show altered electrophoretic mobility through a special type of gel, *denaturing gradient gel electrophoresis*, or will separate into single stands at a different temperature, *temperature gradient gel electrophoresis*.

- Heteroduplexes also show altered properties when analysed by liquid chromatography, *denaturing high pressure liquid chromatography or WAVE analysis*.

Polymerase chain reaction and/or DNA sequencing may also be used to identify *polymorphisms* in DNA. The most common forms of variation that are studied are single nucleotide polymorphisms and microsatellite repeat sequences.

Microsatellite variations produce different sized fragments using the same polymerase chain reaction primers and are analysed by capillary electrophoresis on an automated genetic analyser which can resolve fragments that differ by as little as 2 bp.

Single nucleotide polymorphisms, or SNPs, are analysed by direct DNA sequencing or by using variations of the polymerase chain reaction such as *Amplification resistant mutation system* that will only work if the DNA sequence of the 3′ end of the primer exactly matches the target sequence but not if a single base change is present. Analysis of many thousands of single nucleotide polymorphisms simultaneously is now becoming possible using DNA microarrays.

3.8.5 Gene mapping and linkage analysis

The recent completion of the human genome project has produced a genetic map of all the human chromosomes. Many genes have been localized to this map at particular chromosomal loci using two approaches, either physical mapping or genetic mapping.

Chromosomal abnormalities detectable by cytogenetic methods such as visible deletions, inversions, or translocations, particularly of the X chromosome, can reveal the location of a gene if they are associated with a disease phenotype. This is because the function of a gene is likely to be disrupted if it is located at the site of a deletion or chromosomal break point.

Genetic mapping or *linkage analysis* reveals the chromosomal location of a disease causing gene by establishing close physical proximity to a DNA polymorphism whose map position is known already. The phenotype is used as a marker for the disease gene without knowing anything about its DNA sequence. The phenotype and the genetic marker are tracked together through several generations of several human families. If they are very close they will tend to be inherited together because of the low probability of a genetic crossover between them during meiotic recombination. If they are widely separated they will be inherited together only at chance level 50 per cent of the time. The recombination fraction θ is calculated from the frequency of crossovers between the loci and is an indicator of genetic distance. Thus if recombinations, crossovers, are seen in 10 per cent of meioses, θ = 0.1. Genetic distance is measured in centiMorgans (cM), 1 cM being equivalent to a 1 per cent recombination frequency (θ = 0.01). As a rough guide, 1 cM = 1 Mb (10^6 base pairs) but this is not a reliable calculation because of unequal recombination frequency at different parts of the genome. It should be noted that there is a prior probability of establishing 50 per cent cosegregation of two markers by chance alone. It is therefore necessary to calculate the likelihood of linkage of two loci; this is expressed as the logarithm of the odds in favour of linkage at a given value of θ, referred to as the *lod score*. A lod score of 3.0 is equivalent to a 95 per cent probability of linkage at a given genetic distance. These calculations are usually based on data from numerous families and are complex so that specialized computer programs are used to determine lod scores for different values of θ. The method can be refined by testing for linkage between the disease gene to be mapped and multiple other markers whose chromosomal map positions are known, *multi-locus linkage*, therefore searching for linkage over a wider area of a chromosome. In practice, linkage analysis is limited by the need for multiple highly polymorphic markers, limited human family sizes in which there are only a few informative meioses, and the often limited family histories and available DNA samples from human families.

There are other less accurate methods of linkage analysis using the disease phenotype and genetic markers.

Sibling pair analysis involves searching for allelic markers shared by affected sibs more often than the 25% which would be expected by chance. This is useful in autosomal recessive inheritance where affected siblings share the same mutant allele at a given locus and are therefore more likely to have the same alleles at loci very close to the disease gene.

Linkage disequilibrium refers to the existence of a particular allele at a locus linked to the disease gene more often than would be expected by chance. In the normal population, the particular alleles seen at two linked loci will be determined by the frequency of those alleles in the overall population. If a new pathogenic mutation occurs at one of the loci and causes a disease phenotype which is then transmitted to future generations, it will usually occur in association with the particular alleles that were present at closely linked loci at the time of the mutation. Only after many generations, when there have been frequent recombinations between the linked loci, will this linkage disequilibrium disappear. Therefore, alleles which occur in association with a disease phenotype more often than would be expected from their population frequency may lie close to the disease locus, in linkage disequilibrium with it.

3.8.6 Mechanisms of mutation

- *Point mutations* involve the change of a single DNA base; a transition involves a change of one purine (A or G) or one pyrimidine

(T or C) base to the other while a transversion involves a change of a purine to a pyrimidine. The effects of such subtle mutations can be severe. Pathogenic point mutations alter gene function by damaging the action of 5′ upstream promoter or enhancer elements, transcriptional mutants, by altering a base triplet or codon so that transcription is prematurely terminated, a nonsense mutation, by changing the codon specifying one amino acid to that for another, missense mutation, or by disrupting RNA processing by altering splicing sites.

◆ *A single base deletion* from within an exon alters the whole sequence of base triplets resulting in a protein with a completely different amino acid sequence, a frameshift mutation.

◆ *Deletion of segments of DNA.* Such mutations remove part or even all of one or more genes, for instance as in Duchenne muscular dystrophy.

◆ *Chromosomal reduplication.* In other situations, part of a chromosome is duplicated, disrupting genes involved in the duplicated region, for instance the 17p11.2 duplication in Charcot Marie Tooth disease type IA.

◆ *Repeat expansions.* Some genes contain unstable nucleotide repeats. These are similar to microsatellite variable number tandem repeat polymorphisms but the repeat sequences have expanded into an abnormally large mutation with a repeat number beyond the normal range and consequent disruption of gene function and a deficiency or functional abnormality of the protein product. These mutations are associated with a tendency to undergo greater expansion during meiosis in one sex than in the other; for example the fragile X and myotonic dystrophy mutations tend to undergo greater expansion during maternal transmission, while the expansions causing Huntington's disease or the autosomal dominant cerebellar ataxias are more unstable during paternal meiosis. There

is also a tendency for these expansions to become longer during successive meioses. As greater repeat length tends to be associated with earlier age of disease onset, the phenomenon of *anticipation*, with earlier onset in succeeding generations, is a characteristic clinical feature of trinucleotide repeat disorders. Several diseases are now known to be associated with this mechanism of mutation. In some conditions the expansion is translated into an abnormal protein. For example the CAG repeats of many ataxia mutations are translated into an abnormally large polyglutamine section of protein. In others the expansion is in a non-coding part of the gene, the 3′ or 5′ untranslated regions or an intron, and disrupts gene function (Table 3.4).

3.8.7 Detection of mutations

DNA testing in neurology is carried out for one of two reasons and it is vital that the neurologist is quite clear about which applies when a test is requested. In *diagnostic testing*, a clinically affected patient is investigated to test a diagnostic hypothesis; for example frataxin gene testing in a teenager with ataxia and areflexia. In *presymptomatic* or *predictive testing*, the test is carried out to determine the genetic status of an asymptomatic individual in order to predict whether or not a genetic disease will develop in future; the usual example of this is predictive testing in Huntington's disease. Prenatal testing is a special form of predictive testing usually undertaken by clinical geneticists.

The methods used for detection of mutations in individuals are varied and depend on the type of mutation and whether the gene sequence or only its chromosomal locus is known. If the gene has been sequenced and the type of mutation is known, direct mutation detection is possible. Direct methods include:

◆ Polymerase chain reaction amplification and capillary gel electrophoresis to detect small deletions or expansions, especially

Table 3.4 Repeat expansion disorders

	Inheritance	Gene locus	Gene	Repeat
X-linked bulbospinal neuronopathy (Kennedy syndrome)	XL	Xq13-21	Androgen receptor	CAG
Huntington's disease	AD	4p16.3	*Huntingtin*	CAG
Oculopharyngeal muscular dystrophy	AD / AR	14q11.2-q13	*PABP2*	GCG
Myotonic dystrophy type 1	AD	19q13.2-3	Protein kinase MT-PK (3′)	CTG
Myotonic dystrophy type 2	AD	3q13.3-q24	ZNF9 (intronic)	CCTG
Fragile X syndrome	XL	Xq27.3	*FMR1* (5′)	CGG
Friedreich's ataxia	AR	9q13	*Frataxin* (intronic)	GAA
Progressive myoclonic epilepsy type 1 (EPM1)	AR	21q22.3	*CSTB* (5′)	CCCCGCCCCGCG
Spinocerebellar ataxias:				
SCA1	AD	6p23	*Ataxin 1*	CAG
SCA2	AD	12q24	*Ataxin 2*	CAG
SCA3	AD	14q32	*Ataxin 3*	CAG
SCA6	AD	19p13	α-1A calcium channel	CAG
SCA7	AD	3p12	*Ataxin 7*	CAG
SCA8	AD	13q21	(3′)	CTG
SCA10	AD	22q13	(intronic)	ATTCT
SCA12	AD	5q31-q33	(5′)	CAG
SCA17	AD	6q27		CAG / CAA
DRPLA	AD	12p13	*Atrophin 1*	CAG

AD = autosomal dominant; AR = autosomal recessive; DRPLA = dentatorubropallidoluysian atrophy; SCA = spinocerebellar ataxia; XL = X linked. 3′ indicates expansion located in 3′ flanking region of gene; 5′ indicates expansion located in 5′ flanking region of gene; intronic = expansion in non-coding intron of gene.

trinucleotide repeat mutations, which alter the specified DNA fragment size. Southern blotting and probe hybridization may also be used to reveal abnormally large sized fragments.

- Fluorescence *in situ* hybridization may be used to detect large deletions, for instance in the Angelman or Prader Willi syndromes, or duplications such as of *PMP22* in Charcot Marie Tooth disease or of the *PLP* gene in Pelizaeus Merzbacher disease.

- Homozygous deletions may be detected by failure to amplify the deleted sequence by polymerase chain reaction. This is useful in males with X-linked deletions, for instance the *Dystrophin* gene in Duchenne or Becker muscular dystrophy, or if both copies of the gene are deleted, as for the *SMN1* gene in Spinal Muscular Atrophy.

- Heterozygous deletions or duplications are detected by quantification of the polymerase chain reaction product against a control gene that is not deleted. A novel polymerase chain reaction based technique, *multiplex ligation probe amplification*, allows quantification of 30–40 fragments of DNA simultaneously. Thus all 79 exons of the *Dystrophin* gene can be assessed in two reactions.

- Known specified point mutations may be detected by amplification resistant mutation system.

- Unknown point mutations, which are often variable within the same gene in different families, are detected by DNA sequencing although a mutation scanning method may be used as a preliminary test.

If the sequence of the disease gene and the type of mutation is unknown, as is the case in many disorders in which the gene is mapped but not cloned or sequenced, *indirect* gene tracking is used to detect affected individuals. This uses markers known to be tightly linked to the disease gene, such as single nucleotide polymorphisms or microsatellite polymorphisms. The marker can be analysed in DNA samples from as many family members as possible to determine its transmission through the family. This may reveal whether an individual is likely to have inherited the mutant or normal allele at the disease locus. There are considerable difficulties with this approach: DNA from several individuals in three generations are usually required and may not be available; some individuals in the family may not be heterozygous for the linked marker and are therefore uninformative; an inaccurate family history, non-paternity and reduced gene penetrance may all confound the interpretation of gene marker studies; finally, the possibility of recombination between the disease gene locus and the linked marker introduces some uncertainty into the results. The accuracy of indirect mutation testing can be increased by the use of multiple linked markers known to lie within the gene to be tracked, *intragenic markers*. The indirect approach has been used mainly for the detection of mutations in unaffected individuals who wish to know if they carry a mutation which is segregating within the family; a previous example of this was presymptomatic diagnosis of Huntington's disease. As more genes are discovered and their mutations characterized, direct mutation screening is increasingly used for this purpose, as well as for the detection of specific mutations for diagnosis of affected persons.

3.8.8 Ethical aspects

One of the most important aspects of DNA testing is consent. The patient must understand the nature of the test, the reasons for it, and the potential consequences of a positive result for their family as well as themselves. This enables the patient to make an informed decision about the investigation. In diagnostic testing, there are unlikely to be problems if the patient has given informed consent but there should be information about the implications of the result and about prognosis so that sensible decisions about the future can be made. As with all aspects of patient care, confidentiality is essential. Special care is needed with genetic testing because of interest in the results from relatives, who sometimes request test results via their own doctors, and sometimes employers and insurance companies. Consent and avoidance of adverse psychological and social consequences are much more difficult in predictive testing. Careful counselling is needed before this is done and in some patients, psychiatric evaluation may be required if problems such as suicide are a possibility following a positive test. The first instinct of many relatives of patients with autosomal dominant disorders is to request a DNA test but it is surprising how many reconsider this once the possible consequences of a positive predictive test are thought through in terms of employment, insurance, and personal relationships.

In general, predictive testing should not be undertaken in the neurological outpatient clinic but only in the context of a predictive testing programme conforming to nationally agreed guidelines (World Federation research group on Huntington's disease 1990); such programmes are available in departments of clinical genetics. It is inappropriate to request genetic testing for conditions such as Huntington's disease or autosomal dominant cerebellar ataxia in a patient with vague neurological symptoms unless there is a genuine clinical suspicion of such a disorder. Otherwise a predictive test may inadvertently be carried out with potentially unfortunate consequences. Problems may also arise with predictive testing when a person at 25 per cent risk of an autosomal dominant condition because of an affected grandparent wishes to be tested but the intervening parent, at 50 per cent risk, does not. It may be difficult to resolve this situation but caution is required when considering such tests. Children pose another difficulty; it is not uncommon for patients with hereditary neurological disorders to request that their children are tested for the disorder. However, the child is unable to make an informed decision and in most situations, an adverse result is unlikely to be helpful to the child. For adult-onset conditions for which there is no treatment, predictive testing of minors is not advisable.

3.8.9 Neurological gene information

The enormous expansion of information about neurological genes in recent years makes any tabulation of this information difficult and beyond the scope of this chapter. Comprehensive information is available online at http://www.ncbi.nlm.nih.gov/Omim/. The online Mendelian inheritance in man, OMIM, database has detailed clinical and genetic information about most neurological genes. Reviews of the genetics of many neurological disorders can also be found at www.geneclinics.org.

3.9 Neuropsychology

3.9.1 Neuropsychological assessment

Cognitive assessment is an important component of a neurological examination. It is essential for accurate assessment of disorders of the cerebral cortex (Hodges 1994; Lezak 2004). Classically, identification of the pattern of cognitive impairment has led to topographical diagnosis of which area(s) of the cerebral cortex is involved. Whilst recognition of specific topographical syndromes remains important, the advent of non-invasive imaging has shifted the focus towards diagnosis of the underlying disease process, with an understanding of the cognitive deficits in functional terms and measuring change in cognitive performance.

Neuropsychological tests have become increasingly sophisticated as our understanding of brain function in modular terms has advanced. Any test of clinical value will need to be valid in measuring what it is supposed to do, and will need to be reliable, thereby producing similar results under similar circumstances. The test should be of graded difficulty to avoid ceiling effects, particularly important in subtle disease and, ideally, tests should be of comparative difficulty across different domains. It is important to ensure that a pattern of deficit reflects the underlying disease process and not the pattern of difficulty of the tests used. It follows that interpretation of performance on a neuropsychological test requires knowledge of the normative data concerning how a control population of comparable age, education, and gender to the patient would perform. The neuropsychological tests used in clinical practice vary in the extent to which they meet these requirements.

With a suitable battery of tests, it is possible to identify patterns of deficit which can be diagnostically useful, for example in the diagnosis of frontotemporal degeneration or Alzheimer's disease. It is possible to tell, whether impairment, for example in memory, is outside normal limits and thus indicative of Alzheimer's disease as opposed to normal aging or depression. Identification of an inconsistent pattern of performance may distinguish the patient with abnormal illness behaviour (Section 4.8.5). Measurement of change is also essential for identifying deterioration, which may trigger an intervention, or for measuring the success of such an intervention. Finally, an understanding of the nature of the deficit, for example visual disorientation or dynamic aphasia, may assist in the management of, and an explanation of the deficit for the patient and family (Cipolotti and Warrington 1995).

An adequate neuropsychological assessment should include an assessment of:

- pre-morbid ability and general intelligence;
- memory;
- language and calculation;
- visuoperceptual and visuospatial function;
- problem solving and executive functions; and
- speed and attention.

Whilst a number of tests may be administered by suitably trained neurologists or psychiatrists, comprehensive assessment will be carried out by a qualified clinical psychologist. Routine administration of tests and uncritical acceptance of the resulting numerical data may lead to erroneous conclusions. The choice and interpretation of tests must be made within the overall clinical context and therein lays the contribution of a skilled clinical psychologist. An adequate assessment is unlikely to take less than an hour, and may take much longer depending upon which additional tests are chosen to explore particular deficits, and on the stamina of the patient. Some patients find assessments stressful and, indeed, may have a catastrophic reaction to failure. This is important to recognize and thus avoid over-interpretation of poor performance; reassessment the following day may be necessary.

3.9.2 General intelligence and assessment of pre-morbid function

The Wechsler Adult Intelligence Scale, WAIS, is a cornerstone of the neuropsychological assessment and involves 6 verbal and 5 nonverbal sub-tests exploring various skills. The revised test, WAIS-R, is now the standard test, with normative data available for the elderly (Ryan *et al.* 1990). Other widely used tests of intelligence are the Raven's tests of coloured progressive matrices, the standard progressive matrices, and advanced progressive matrices.

In order to assess deterioration in a patient at first presentation, some impression of pre-morbid intelligence needs to be formed. It is only rarely that pre-morbid test results would be available although some indication may be given by educational attainment and occupation. An additional approach derives from the observation that patients with a variety of disorders, but particularly degenerative diseases such as Alzheimer's disease, have relative preservation of verbal skills and, in particular, reading. Moreover, such skills have an overall association with intelligence in the normal population. Thus, the National Adult Reading Test, NART, assesses ability to read and pronounce correctly 50 irregular words (Nelson and O'Connell 1978). Clearly, such a test is of limited use if lexical skills are selectively affected by a focal degenerative condition.

Memory

Impairment of memory is widely used to refer to failure to remember day-to-day events, or to remember to do something in the future. Whilst this is a salient feature in clinical practice, memory is now seen to be multidimensional with discrete components which may be selectively involved in a disease process. A distinction is made between short-term memory, with a limited short duration capacity of about 30 s, and long-term memory. Both short- and long-term memory may also be divided on the basis of whether visual or verbal information is being retained. Short-term verbal memory is routinely tested with the digit span.

Long-term memory may be divided into episodic memory and semantic memory. Episodic memory deals with day-to-day events and provides the basis of our autobiographical memory. By contrast, semantic memory deals with our lifetime acquisition of knowledge. Semantic memory is usually explored in language tests. Episodic memory is assessed with either recall or recognition test paradigms. Examples of the former would be a story recall or the Rey-Osterreith figure for the verbal and visual domains respectively. An example of recognition memory tests would be the recognition tests for words and faces (Warrington 1984), which provide tests of comparable difficulty, such that patterns of deficit can reflect the left right hemisphere asymmetry in verbal and visual processing of information.

Language and literacy skills

Various tests have been developed to explore the main linguistic components for spoken language, namely phonology, syntax, and semantics. A number of aphasia batteries have been developed and are widely used, such as the Boston Diagnostic Aphasia Examination. However, none of these batteries necessarily explores in depth all the domains of language delineated by recent cognitive psychology.

Written as opposed to spoken language can be explored by tests such as the NART and the Schonell Graded Reading Test. There are similar graded tests for spelling and for calculation.

Visuospatial and visuoperceptual function

Bedside cognitive tests by clinicians will often identify major patterns of memory, language, and frontal dysexecutive failure. Visuoperceptual and visuospatial failure are less easily identified and need to be specifically explored. A variety of tests have been developed which explore visuoperceptual function from early visual processing, such as shape detection, point localization, and colour discrimination through to the ability to form a complex visual percept. Such patients have difficulty with degraded visual stimuli or unusual as opposed to canonical views of common objects.

Frontal executive skills

Frontal lobe deficits have proved difficult to measure reliably. Characteristic frontal behaviour may be only too apparent to the clinician but when the patient is tested formally they may perform surprisingly well on a whole range of tests. Even tests which are sensitive to frontal lobe function may provide variable scores over time, and patients may fail on one particular test but pass well on others. Tests believed to be sensitive to frontal lobe function include cognitive estimates, verbal fluency, Stroop tests, and the Hayling sentence completion test. Quantitative tests of frontal lobe function have proved difficult in the same way that the reliable measurement of praxis has been elusive. Bedside qualitative, as opposed to quantitative assessments of dyspraxia and frontal lobe function remain an important contribution of the clinical neurologist.

In summary, neuropsychological assessment is now a major part of the neurological investigative armamentarium. It has taken the qualitative pattern recognition of classical behavioural neurology to the quantitative assessment of rigorous science. However, like all investigations, the information derived must be interpreted in the light of the clinical picture.

3.10 Assessing treatments

3.10.1 Introduction

It is clearly important that treatments used in neurological disease are properly assessed before being introduced into routine clinical practice. There are many examples throughout medicine of interventions that were considered beneficial on the basis of theory or uncontrolled observational studies, but were subsequently shown to be harmful in randomized controlled trials, and vice versa (Table 3.5).

The justification for randomized trials is not that no worthwhile observations can be made without them, but that important biases can occur in non-randomized comparisons, which are particularly problematic if the benefits of treatment are, in reality, small or absent. For example, a non-randomized comparison of the effect of aspirin dose on the operative risk of carotid endarterectomy (Table 3.6) reported a clinically and statistically significant lower operative risk in patients on high-dose aspirin (1300 mg) versus low-dose aspirin (325 mg or less) (Barnett *et al.* 1998). However a subsequent randomized trial (Taylor *et al.* 1999), performed to confirm this observation, showed that high-dose aspirin was, in fact, harmful (Table 3.6). It is likely that the non-randomized

Table 3.5 A few examples of interventions that were thought on the basis of theory or uncontrolled observational studies to be either beneficial or harmful, but were subsequently shown actually to be harmful or beneficial in randomized controlled trials (Rothwell 2005a)

Considered beneficial, shown to be harmful
◆ High-dose oxygen therapy in neonates
◆ Antiarrhythmic drugs after myocardial infarction
◆ Fluoride treatment for osteoporosis
◆ Bed rest in twin pregnancy
◆ Hormone replacement therapy in vascular prevention
◆ Extracranial to intracranial arterial bypass surgery in stroke prevention
◆ High-dose aspirin for carotid endarterectomy
Considered harmful, shown to be beneficial
◆ Beta-blockers in heart failure
◆ Digoxin after myocardial infarction

comparison had been biased by unmeasured differences between the patients in low-dose and high-dose aspirin groups.

Randomized trials and systematic reviews of trials therefore provide the most reliable data on the effects of treatment. That is not to say, however, that non-randomized studies cannot sometimes provide reliable evidence on the benefits of intervention. Few people would doubt the validity of the observational data on the benefits of antibiotic treatment in bacterial meningitis or the benefits of treatment with levodopa in Parkinson's disease. Similarly, clinical guidelines have been revised worldwide on the basis of the non-randomized evidence of the substantial reduction in the risk of early recurrent stroke (Fig. 3.36) as a result of the urgent initiation of standard secondary prevention (Rothwell *et al.* 2007).

However, such large treatment effects are rare. Most treatments used in medicine have smaller effects that require assessment in randomized controlled trials if they are to be reliably quantified. Specifically, randomization has two main advantages over a non-randomized comparison. First, it ensures that clinicians do not know which treatment the patient will receive, and cannot select certain types of patients for one particular treatment. Second, it tends to result in an equal balance of baseline risk across the treatment groups.

Table 3.6 The relationship between aspirin dose and the risk of stroke and death within 30 days of carotid endarterectomy in a non-randomised comparison within the North American Symptomatic Carotid Endarterectomy Trial NASCET (Barnett *et al.* 1998), and in a subsequent randomized controlled trial (Taylor *et al.* 1999)

	Operative risk of stroke and death			
Aspirin dose:	**< 650 mg**	**>650 mg**	**Relative risk**	**P**
Non-randomized	7.1%	3.9%	1.8	<0.001
Randomized	3.7%	8.2%	0.45	0.002

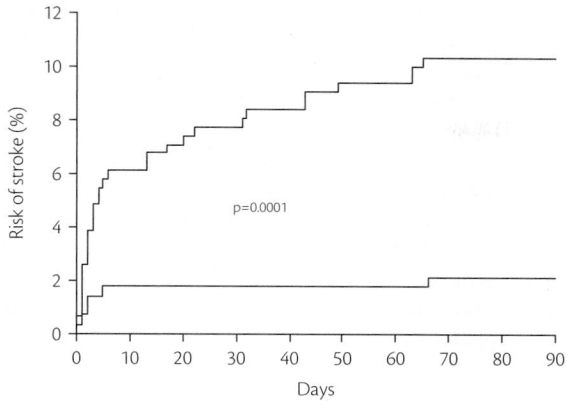

Fig. 3.36 The 90-day risk of recurrent stroke after first seeking medical attention in all patients with transient ischaemic attack or stroke referred to the study clinic in the non-randomised EXPRESS, Early use of Existing Prevention Strategies for Stroke Study (Rothwell *et al.* 2007). The top line represents phase-1 of the study (standard treatment) and the lower line represents phase-2 (urgent treatment).

3.10.2 The internal validity of a randomized controlled trial

Randomized controlled trials have the potential to produce reliable estimates of the effects of treatments, but they will not do so inevitably. There are many potential sources of bias that must be addressed in the design and performance of a trial in order to ensure that results are reliable. The extent to which bias has been avoided is usually termed *internal validity,* the assessment of which is detailed below. The extent to which the results of a trial can be generalized to other settings, usually meaning routine clinical practice, is termed *external validity,* and is considered in Section 3.10.3.

How was randomization performed?

It is important that the method of randomization is actually random. Treatment allocation according to day of the week, date of birth, date of admission, or alternate cases, is not random. The investigator will often know what treatment the patient will get if they enter the trial and so these methods are open to bias. Randomization must be based on tables of random numbers or computer generated random allocation. It is also important that randomization is secure. Central telephone randomization is preferable to other methods, such as sealed envelopes containing the treatment allocation.

Were the treatment groups balanced?

Randomization will not inevitably result in an adequate balance of clinical characteristics and prognostic factors between the treatment groups in a trial, particularly if the sample size is relatively small. Details of the important clinical characteristics of the patients should therefore be reported by treatment group. If a prognostic variable is particularly important, a relatively minor and not necessarily statistically significant imbalance between the treatment groups may have a major effect on the trial result.

Was the trial sufficiently powered?

Sample sizes in randomized controlled trials in neurology may need to be large, either because treatment effects are relatively small, or because the progression of disease is slow. Table 3.7 shows

Table 3.7 Effect of sample size on the reliability of the result of a trial of a hypothetical treatment that is expected to reduce the risk of a poor outcome by 20 per cent i.e. from 10 to 8 per cent

Total Patients	P*	Trial power (%)	Comments on trial size
200	0.99	1	Completely hopeless
400	0.98	2	Still hopeless
800	0.96	4	Completely inadequate
1600	0.90	10	Still inadequate
3200	0.75	25	Not really adequate
6400	0.43	57	Barely adequate
12 800	0.09	91	Probably adequate
20 000	0.01	99	Definitely adequate

* Probability of failing to achieve *P*<0.01 significance if true relative risk reduction is 20 per cent.

the effect of sample size on the reliability of the result of a trial of a hypothetical neurological treatment which is assumed to reduce the risk of a poor outcome by a fifth, from 10 to 8 per cent. The risk of getting the wrong result when a trial has an inadequate sample size is illustrated in Fig. 3.37. In this trial, there was considerable variability in the apparent effect of treatment until several hundred patients had been randomized. If the trial had been small, misleading trends in treatment effect could easily have been reported.

Was the trial stopped early?

A trial may need to be stopped early if a treatment has serious adverse effects, or if there is clear benefit. However, as is seen in Fig. 3.37 the chance fluctuations during the early stages of a trial can easily reach statistical significance at the *P*=0.05 level. If the stopping rule is based on a *P*-value of 0.05, it is quite possible that

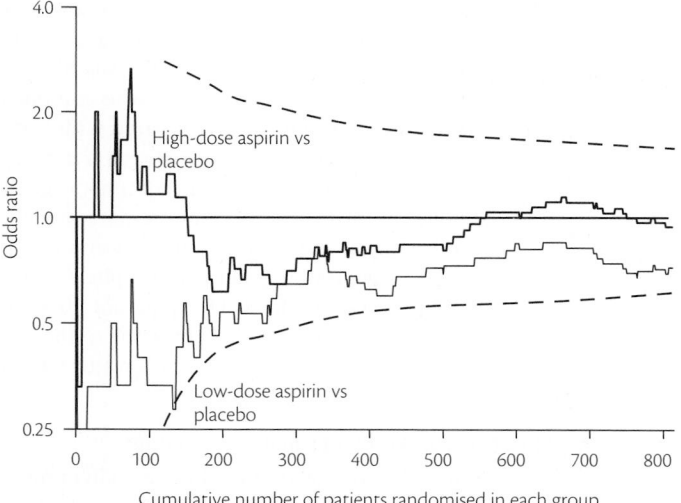

Fig. 3.37 The evolution of the estimates of treatment effect in the UKTIA-Aspirin Trial (Farrell *et al.* 1991) of high-dose aspirin versus low-dose aspirin versus placebo in patients with transient ischaemic attack or minor stroke. The treatment effect (odds ratio) calculated at each point is based on the outcomes at final follow-up for patients randomised to that point (PM Rothwell, unpublished data). The dashed lines represent the level at which the apparent treatment effect approached statistical significance at the *P*=0.05 level.

the trial will be stopped early, and the wrong conclusions drawn. Stopping rules should be based on significance levels of $P<0.01$ or ideally $P<0.001$, and the evolving results should be assessed on only a limited number of pre-specified occasions.

Was outcome assessment blind to treatment allocation?

There are two main reasons for blinding the trial clinicians. First, so that the use of non-trial treatments and interventions is not influenced by a knowledge of whether or not the patients received the trial treatment. Second, so that clinicians are not biased in their assessment of clinical outcomes. The potential for bias depends on the subjectivity of the trial outcome. Biased assessment of neurological impairment and disability was clearly demonstrated in a multiple sclerosis trial in which blind and non-blind outcome assessment produced very different results (Noseworthy *et al.* 1994). Trials with blind assessment should also report whether or not blinding was effective. It is, of course, sometimes impossible to blind clinical assessment, but non-blind trials should report data on non-trial treatments given to patients during follow-up to ensure that these were not biased.

Were serious complications of treatment included in the main outcome?

Some treatments have serious complications which should be included in the primary outcome, rather than relegated to a table of 'side-effects'. Examples include life-threatening gastrointestinal bleeding in trials of antiplatelet agents and anticoagulants.

Was the main analysis an intention-to-treat analysis?

The primary analysis in any randomised trial should be an intention-to-treat analysis. That means patients remain in the treatment group to which they were originally randomized, irrespective of the treatment they eventually received. The alternative, an efficacy analysis, which is confined to patients who complied with the randomized treatment, is prone to bias. This was illustrated by a randomized trial comparing several different lipid-lowering regimens with placebo following myocardial infarction (Coronary Drug Project Research Group 1980). By intention-to-treat analysis, the 5-year mortality in the clofibrate group was 20.0 versus 20.9 per cent in the placebo group. However, when patients who complied with treatment in the clofibrate group were compared with non-compliers the results seemed to suggest that there was a treatment effect: 5-year mortality was 15.0 per cent in the compliers versus 24.6 per cent in the non-compliers. Perhaps clofibrate was beneficial. However, the same analysis in patients in the placebo group showed the same trend: 15.1 per cent mortality in compliers versus 28.2 per cent in non-compliers. The apparent effect of clofibrate in the treatment group was simply a bias due to the fact that patients who do not comply with treatment tend to have a worse prognosis.

Were any patients excluded from the main analysis?

It is not uncommon in reports of trials to find that a certain number of the patients who were randomized are excluded from the final analysis. Common reasons for exclusion are that following randomization it was found that a number of patients did not actually fit the eligibility criteria, so called *protocol violators*, or that some patients never received the randomized treatment because they developed a clear indication for a specific treatment or because they withdrew from the trial for other reasons. However, the interpretation of what is a protocol violation can be rather subjective, and since the decision will often be made towards the end of the trial, and may not be blind to outcome, it is open to abuse. For example, 71 of 1629 patients randomized in a trial of an antiplatelet agent following myocardial infarction were excluded from the final analysis, apparently because they did not meet the eligibility criteria (Anturane Reinfarction Trial Research Group 1980). It subsequently transpired that there was a large excess of deaths in the exclusions from the treatment group compared with the placebo group (Temple and Pledger 1980). Exclusion of these patients led to a bias which had contributed to the statistically significant apparent benefit in the treatment group. A second trial failed to confirm any benefit.

How many patients were lost to follow-up?

Another important potential cause of bias in the analysis of trial results is loss of patients to follow-up. Just as patients who comply with treatment are different from patients who do not, those patients who are lost to follow-up are usually different from those who remain in the trial. For example, it may not be possible to contact patients because they are either incapacitated in some way, or even dead. It is therefore very difficult to interpret the results of a trial with significant loss to follow-up.

3.10.3 The external validity of a randomized controlled trial

Randomized controlled trials must be internally valid so that the design and conduct eliminate the possibility of bias. But to be clinically useful the result must also be relevant to a definable group of patients in a particular clinical setting, in other words they must be externally valid. Lack of external validity is the most frequent criticism by clinicians of trials, systematic reviews, and guidelines, and is one explanation for the widespread under-use in routine practice of many treatments that have been shown to be beneficial in trials and are recommended in guidelines (Rothwell 2005a). Yet medical journals, funding agencies, ethics committees, the pharmaceutical industry, and governmental regulators give external validity a low priority. Admittedly, whereas the determinants of internal validity are intuitive and can generally be worked out from first principles, understanding of the determinants of the external validity requires clinical rather than statistical expertise and often depends on a detailed understanding of the particular clinical condition under study and its management in routine clinical practice. However, reliable judgements about the external validity of randomized trials are essential if treatments are to be used correctly in as many patients as possible in routine clinical practice.

Whether the results of a trial can be applied in routine clinical practice depends to some extent on the type of trial. Generally speaking, explanatory 'phase II' trials measure the effectiveness of treatment, whereas pragmatic 'phase III' trials measure the usefulness of treatment. A treatment may be effective, but may not be useful because it is too poorly tolerated, too expensive, or too complex to administer. Explanatory trials are often small, include a tightly defined group of patients, and frequently have non-clinical and surrogate measures of outcome. Pragmatic trials seek to measure the usefulness of treatments in situations which, as far as possible, mimic normal clinical practice. However, it would be wrong to assume that a pragmatic trial will always have greater external validity than an explanatory trial. For example, although broad eligibility criteria, limited collection of baseline data, and inclusion of centres with a range of expertise and differing patient populations have many advantages, they can also make it very difficult to generalize the effect of treatment to a particular clinical setting. Moreover, no randomized trial or systematic review will ever be relevant to all patients and all settings. However, trials should be designed and reported in a way that allows clinicians to judge to whom the results can reasonably be applied. Table 3.8 lists some of the important potential determinants of external validity, each of which is reviewed briefly below.

What was the setting of the trial?

A detailed understanding of the setting in which a trial is performed, including any peculiarities of the healthcare system in particular counties, can be essential in judging external validity. The potential impact of differences between healthcare systems is illustrated by the analysis of the results of the European Carotid Surgery Trial, a randomized trial of endarterectomy versus medical treatment alone for recently symptomatic carotid stenosis (Fig. 3.38) (Rothwell 2005b). National differences in the speed with which patients were investigated, with a median delay from last symptoms to randomization of greater than 2 months in slow centres in the United Kingdom compared with 3 weeks in fast centres in Belgium and Holland, resulted in very different treatment effects in these different healthcare systems. These differences were simply due to the shortness of the time window for effective prevention of stroke.

Similar differences in performance between healthcare systems will exist for other conditions, and there is, of course, the broader issue of how trials done in the developed world apply in the developing world. Moreover, other differences between countries in the methods of diagnosis and management of disease, which can be substantial, or important racial differences in pathology and natural history of disease, also affect the external validity of trials. A good example is the heterogeneity of results of trials of BCG vaccine in prevention of tuberculosis, with a progressive loss of efficacy ($P<0.0001$) with decreasing latitude (Fine 2005).

How were participating centres selected?

How centres and clinicians were selected to participate in trials is seldom reported, but can also have important implications for external validity. For example, the **A**symptomatic **C**arotid **A**therosclerosis **S**tudy, ACAS, trial of endarterectomy for asymptomatic carotid stenosis only accepted surgeons with an excellent safety record, rejecting 40 per cent of applicants initially, and subsequently barring from further participation those who had adverse operative outcomes in the trial. The benefit from surgery in the trial was due in major part to the consequently low operative risk (Asymptomatic Carotid Atherosclerosis Study Group 1995). A meta-analysis of 46 surgical case series that published operative risks during the 5 years after the trial found operative mortality to be eight times higher and the risk of stroke and death to be about three times higher than in ACAS (Rothwell 2005a). Trials should not include centres that do not have the competence to treat patients safely, but selection should not be so exclusive that the results cannot be generalized to routine clinical practice.

How were patients selected and excluded?

Concern is often expressed about highly selective trial eligibility criteria, but there are often several earlier stages of selection that are rarely recorded or reported but can be more problematic.

Table 3.8 Some of the factors that can affect external validity of randomized controlled trials and should be addressed in reports of the results and be considered by clinicians (Rothwell 2005a)

- ◆ **Setting of the trial**
  - ◇ Healthcare system
  - ◇ Country
  - ◇ Recruitment from primary, secondary or tertiary care
  - ◇ Selection of participating centres
  - ◇ Selection of participating clinicians
- ◆ **Selection of patients**
  - ◇ Methods of pre-randomization diagnosis and investigation
  - ◇ Eligibility criteria
  - ◇ Exclusion criteria
  - ◇ Placebo run-in period
  - ◇ Treatment run-in period
  - ◇ 'Enrichment' strategies
  - ◇ Ratio of randomized patients to eligible non-randomized patients in participating centres
  - ◇ Proportion of patients who declined randomization
- ◆ **Characteristics of randomized patients**
  - ◇ Baseline clinical characteristics
  - ◇ Racial group
  - ◇ Uniformity of underlying pathology
  - ◇ Stage in the natural history of their disease
  - ◇ Severity of disease
  - ◇ Comorbidity
  - ◇ Absolute risks of a poor outcome in the control group
- ◆ **Differences between the trial protocol and routine practice**
  - ◇ Trial intervention
  - ◇ Timing of treatment
  - ◇ Appropriateness/relevance of control intervention
  - ◇ Adequacy of non-trial treatment—both intended and actual
  - ◇ Prohibition of certain non-trial treatments
  - ◇ Therapeutic or diagnostic advances since trial was performed
- ◆ **Outcome measures and follow-up**
  - ◇ Clinical relevance of surrogate outcomes
  - ◇ Clinical relevance, validity, and reproducibility of complex scales
  - ◇ Effect of intervention on most relevant components of composite outcomes
  - ◇ Who measured outcome
  - ◇ Use of patient-centred outcomes
  - ◇ Frequency of follow-up
  - ◇ Adequacy of the length of follow-up
- ◆ **Adverse effects of treatment**
  - ◇ Completeness of reporting of relevant adverse effects
  - ◇ Rates of discontinuation of treatment
  - ◇ Selection of trial centres and/or clinicians on the basis of skill or experience
  - ◇ Exclusion of patients at risk of complications
  - ◇ Exclusion of patients who experienced adverse effects during a run-in period
  - ◇ Intensity of trial safety procedures

For example, consider a trial of a new blood pressure-lowering drug, which like most such trials is performed in a hospital clinic. Fewer than 10 per cent of patients with hypertension are managed in hospital clinics and this group will differ from those managed in primary care. Moreover, only one of the ten physicians who see hypertensive patients in this particular hospital is taking part in the trial, and this physician mainly sees young patients with

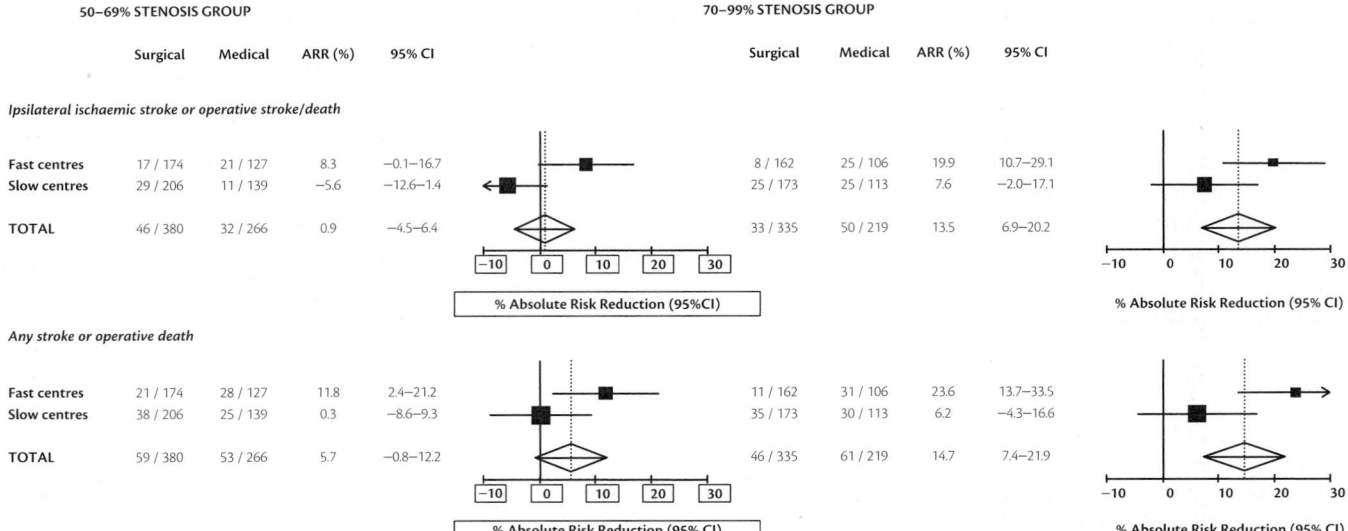

Fig. 3.38 The absolute reductions in the 5-year risks of ipsilateral ischaemic stroke (top) and any stroke or death (bottom) with surgery in European Carotid Surgery Trial centres in which the median delay from last symptomatic event to randomization was ≤ 50 days (Fast centres) compared with centres with a longer delay (Slow centres) (Rothwell 2005a). Data are shown separately for patients with moderate (50–69 per cent) and severe (70–99 per cent) carotid stenosis. ARR: absolute risk reduction, 95% CI; 95% confidence interval.

resistant hypertension. Thus, even before any consideration of eligibility or exclusion criteria, potential recruits are already very unrepresentative of patients in the local community. It is essential, therefore, that where possible trials record and report the pathways to recruitment.

Patients are then further selected according to trial eligibility criteria. Some trials exclude women and many exclude the elderly and/or patients with common comorbidities. One review of 214 drug trials in acute myocardial infarction found that over 60 per cent excluded patients aged over 75 years (Gurwitz *et al.* 1992), despite the fact that over 50 per cent of myocardial infarctions occur in this older age group. A review of 41 US National Institutes of Health randomized trials found an average exclusion rate of 73 per cent (Charlson and Horwitz 1984), but rates can be much higher. One study of an acute stroke treatment trial, found that of the small proportion of patients admitted to hospital sufficiently quickly to be suitable for treatment, 96 per cent were ineligible based on the various other exclusion criteria (Jorgensen *et al.* 1999). One centre in another acute stroke trial had to screen 192 patients over 2 years to find one eligible patient (LaRue *et al.* 1988). Yet, highly selective recruitment is not inevitable. An Italian trial of thrombolysis for acute myocardial infarction, for example, recruited 90 per cent of patients admitted within 12 h of the event with a definite diagnosis and no contra-indications (Gruppo Italiano per lo Studio della Streptochinasi nell'Infarto Miocardico 1986).

Strict eligibility criteria can limit the external validity of trials but physicians should at least be able to select similar patients for treatment in routine practice. Unfortunately, however, reporting of trial eligibility criteria is frequently inadequate. A review of trials leading to clinical alerts by the US National Institutes of Health revealed that of an average of 31 eligibility criteria, only 63 per cent were published in the main trial report and only 19 per cent in the clinical alert (Shapiro *et al.* 2000). Inadequate reporting is also a major problem in secondary publications, such as systematic reviews and clinical guidelines, where the need for a succinct message does not usually allow detailed consideration of the eligibility and exclusion criteria or other determinants of external validity.

Was there a run-in period?

Pre-randomization run-in periods are also often used to select or exclude patients. In a placebo run-in, all eligible patients receive placebo and those who are poorly compliant are excluded. There can be good reasons for doing this, but high rates of exclusion will reduce external validity. Active treatment run-in periods in which patients who have adverse events or show signs that treatment may be ineffective are excluded are more likely to undermine external validity. For example, two trials of carvedilol, a vasodilatory beta-blocker, in chronic heart failure excluded 6 and 9 per cent of eligible patients in treatment run-in periods, mainly because of worsening heart failure and other adverse events, some of which were fatal (Rothwell 2005a). In both trials, the complication rates in the subsequent randomized phase were much lower than in the run-in phase.

Trials also sometimes actively recruit patients who are likely to respond well to treatment; this is often termed 'enrichment'. For example, some trials of antipsychotic drugs have selectively recruited patients who had a good response to antipsychotics previously (Rothwell 2005a). Other trials have excluded non-responders in a run-in phase. One trial of a cholinesterase inhibitor, tacrine, in Alzheimer's disease recruited 632 patients to a 6-week 'enrichment' phase in which they were randomized to different doses of tacrine vs placebo (Davis *et al.* 1992). After a washout-period, only the 215 or 34 per cent of patients who had a measured improvement on tacrine in the 'enrichment' phase were randomized to tacrine, at their best dose, versus placebo in the main phase of the trial. External validity is clearly undermined here.

What were the characteristics of the randomized patients?

Even in large pragmatic trials with very few exclusion criteria recruitment of less than 10 per cent of potentially eligible patients in participating centres is common. Those patients who are recruited generally differ from those who are eligible but not recruited in terms of age, sex, race, severity of disease, educational status, social class, and place of residence (Rothwell 2005a). The outcome in patients included in trials is also usually better than those not in trials, often markedly so, not because of better treatment but because of a better baseline prognosis. Trial reports usually include the baseline clinical characteristics of randomized patients and so it is argued that clinicians can assess external validity by comparison with their own patient(s). However, recorded baseline clinical characteristics often say very little about the real make-up of the trial population, and can be misleading. For example, Table 3.9 shows the baseline clinical characteristics of the patients randomized to warfarin in two trials of secondary prevention of stroke. In one trial, patients were in atrial fibrillation and in the other they were in sinus rhythm, but the characteristics of the two cohorts were otherwise fairly similar. However, the risk of intracranial haemorrhage on warfarin was 19 times higher (P<0.0001) in Stroke Prevention in Reversible Ischaemia Trial than in the European Atrial Fibrillation Trial even after adjustment for differences in baseline clinical characteristics and the intensity of anticoagulation (Gorter 1999). In judging external validity, an understanding of how patients were referred, investigated, and diagnosed, in other words their pathway to recruitment, as well as how they were subsequently selected and excluded is often very much more informative than a list of baseline characteristics.

Table 3.9 The baseline clinical characteristics and haemorrhage outcomes of patients randomized to anticoagulation with warfarin in the European Atrial Fibrillation Trial (EAFT) and the Stroke Prevention in Reversible Ischaemia Trial (SPIRIT) (Gorter 1999)

	SPIRIT (n = 651)	EAFT (n = 225)
Baseline clinical characteristics		
Male sex	66%	55%
Age > 65 years	47%	81%
Hypertension	39%	48%
Angina	9%	11%
Myocardial infarction	9%	7%
Diabetes	11%	12%
Leukoariosis on CT brain scan	7%	14%
Outcomes during trial		
Mean (SD) INR during trial	3.3 (1.1)	2.9 (0.7)
Patient-years of follow-up	735	507
Intracranial haemorrhage	27	0*
Extracranial haemorrhage	26	13
Adjusted hazard ratio (95% CI)*		
Intracranial haemorrhage	19.0 (2.4–250)	
Extracranial haemorrhage	1.9 (0.8–4.7)	

* There were no proven intracranial haemorrhages, but no CT scan was performed in two strokes. For the purpose of calculation of the adjusted hazard ratio for haemorrhage these two strokes were categorized as having been due to intracranial haemorrhage.

Was the intervention, control treatment, and pre-trial or non-trial management appropriate?

External validity can also be affected if trials have protocols that differ from usual clinical practice. For example, prior to randomization in the trials of endarterectomy for symptomatic carotid stenosis patients had to be diagnosed by a neurologist and to have conventional arterial angiography, neither of which are routine in many centres. The trial intervention itself may also differ from that used in current practice, such as in the formulation and bioavailability of a drug, or the type of anaesthetic used for an operation. The same can be true of the treatment in the control group in a trial, which may use a particularly low dose of the comparator drug, or fall short of best current practice in some other way. External validity can also be undermined by too stringent limitations on the use of non-trial treatments. Any prohibition of non-trial treatments should be reported in the main trial publications along with details of relevant non-trial treatments that were used. The timing of many interventions is also critical and should be reported when relevant.

Were the outcome measures appropriate?

The external validity of a trial also depends on whether the outcomes were clinically relevant. Many trials use 'surrogate' outcomes, usually biological or imaging markers that are thought to be indirect measures of the effect of treatment on clinical outcomes. Surrogate outcomes, such as infarct size on CT brain scan in an acute stroke trial or MRI activity in multiple sclerosis, can be useful in explanatory trials because they may be more sensitive to the effects of the treatment than clinical outcomes, and they are readily assessed blind to treatment allocation. However, they do not measure clinical effectiveness, and may sometimes be highly misleading. There are many examples of treatments that had a major beneficial effect on a surrogate outcome, which had been shown to be correlated with a relevant clinical outcome in observational studies, but where the treatments proved ineffective or harmful in subsequent large trials that used these same clinical outcomes (Table 3.10). For example, a trial of three different antiarrhythmic drugs versus placebo after acute myocardial infarction assessed the frequency of ventricular extrasystoles on 24 h ambulatory electrocardiographic monitoring (Cardiac Arrhythmia

Table 3.10 Examples where trials based on surrogate outcomes proved to be misleading predictors of the effect of treatment on clinical outcomes in subsequent pragmatic clinical trials (see Rothwell 2005a for references)

Treatment	Condition	Surrogate outcome	Clinical outcome
Fluoride	Osteoporosis	Increase in bone density	Major increase in fractures
Antiarrhythmic drugs	Post-myocardial infarction	Reduction in ECG abnormalities	Increased mortality
Beta-interferon	Multiple sclerosis	70% reduction in new lesions brain MRI	No convincing effect on disability
Milrinone and Epoprostanol	Heart failure	Improved exercise tolerance	Increased mortality
Ibopamine	Heart failure	Improved ejection fraction and heart rate variability	Increased mortality

Suppression Trial (CAST) Investigators 1989). All three drugs produced a substantial reduction in the frequency of extrasystoles, but the trial was subsequently stopped because of a major excess of deaths in the treatment group (33 versus 9, $P=0.0003$). Similarly, reduced bone density, which is known to be a useful marker for risk of fractures, was used as a surrogate outcome in a trial of sodium flouride in women with osteoporosis (Riggs *et al.* 1990). Sodium fluoride produced a highly statistically significant, and apparently clinically important, increase in bone density. However, further follow-up revealed a 30 per cent increase in vertebral fractures and a three-fold increase in non-vertebral fractures in the sodium fluoride group.

Complex scales, often made up of arbitrary combinations of symptoms and clinical signs, are also problematic. A review of 196 trials in rheumatoid arthritis identified more than 70 different outcome scales (Gøtzsche 1989). A review of 2000 trials in schizophrenia identified 640 outcome scales, many of which were devised for the particular trial and had no supporting data on validity or reliability. These unvalidated scales were more likely to show statistically significant treatment effects than established scales (Marshall *et al.* 2000). Moreover, the clinical meaning of apparent treatment effects is usually impossible to discern, for instance a 2.7 point mean reduction in a 100 point outcome scale made up of various symptoms and signs. Simple clinical outcomes usually have most external validity but even then only if they reflect the priorities of patients. For example, patients with epilepsy are much more interested in the proportion of individuals rendered free of seizures in trials of anticonvulsants than they are in changes in mean seizure frequency. Who actually measured the outcome can also be important. For example, the recorded operative risk of stroke due to carotid endarterectomy is highly dependent on whether patients were assessed by a surgeon or a neurologist (Rothwell and Warlow 1995).

Many trials combine events in their primary outcome measure. This can produce a useful measure of the overall effect of treatment on all the relevant outcomes, and it usually affords greater statistical power, but the outcome that is most important to a particular patient may be affected differently by treatment than the combined outcome. Composite outcomes also sometimes combine events of very different severity and treatment effects can be driven by the least important outcome, which is often the most frequent. Equally problematic is the composite of definite clinical events and episodes of hospitalization. The fact that a patient is in a trial will probably affect the likelihood of hospitalization and it will certainly vary between different healthcare systems.

Was the follow-up sufficient?

Another major problem for the external validity of trials is an inadequate duration of treatment and/or follow-up. For example, although patients with refractory epilepsy or migraine require treatment for many years, most trials of new drugs look at the effect of treatment for only a few weeks. Whether initial response is a good predictor of long-term benefit is unknown. The same problem has been identified in trials in schizophrenia, with fewer than 50 per cent of trials having greater than 6 weeks follow-up and only 20 per cent following patients for longer than six months (Thornley and Adams 1998). The contrast between beneficial effects of treatments in short-term trials and the less encouraging experience of long-term treatment in clinical practice has also

been highlighted by clinicians treating patients with rheumatoid arthritis (Pincus 1998).

How were adverse effects of treatment assessed and reported?

Reporting of adverse effects of treatment in trials and systematic reviews is often poor. In a review of 192 pharmaceutical trials, less then a third had adequate reporting of adverse clinical events or laboratory toxicology (Ioannidis and Contopoulos-Ioannidis 1998). Treatment discontinuation rates provide some guide to tolerability but pharmaceutical trials often use eligibility criteria and run-in periods to exclude patients who might be prone to adverse effects.

Clinicians are usually most concerned about external validity of trials of potentially dangerous treatments. Complications of medical interventions are a leading cause of death in developed countries. Risks can be over-estimated in trials, particularly during the introduction of new treatments when trials are often done in patients with very severe disease. However but stringent selection of patients, confinement to specialist centres, and intensive safety monitoring usually lead to lower risks than in routine clinical practice. Trials of warfarin in non-rheumatic atrial fibrillation are a good example. Prior to 2007 all trials reporting benefit with warfarin had complication rates that were much lower than in routine practice and consequent doubts about external validity were partly to blame for major under-prescribing of warfarin, particularly in the elderly.

3.10.4 **Applying the results of randomized trials to treatment decisions about individual patients**

Many treatments, such as blood pressure lowering in uncontrolled hypertension, are indicated in the vast majority of patients. However, a targeted approach is useful for treatments with modest benefits. These include lipid lowering in primary prevention of vascular disease; costly treatments with moderate overall benefits such as beta-interferon in multiple sclerosis; or if the availability of treatment is limited, as in organ transplantation, or in developing countries with very limited health care budgets; and most importantly, for treatments which although of overall benefit in large trials are associated with a significant risk of harm. The crux of the problem faced by clinicians in these situations is how to use data from large randomized trials and systematic reviews, which provide the most reliable estimates of the overall average effects of treatment, to determine the likely effect of treatment in an individual.

When considering the likely effect of a treatment in an individual patient it is important to consider the overall result of a trial or systematic review as an absolute risk reduction with treatment or the number needed to treat to prevent a poor outcome 'NNT'. An absolute risk reduction tells us what chance an individual has of benefiting from treatment, for instance an absolute risk reduction of 25 per cent indicates that there is a 1-in-4 chance of benefit or an NNT = 4. In contrast, a particular relative risk reduction gives absolutely no information about the likelihood of individual benefit. For example, the relative reductions in the risk of stroke were virtually identical in the Swedish Trial in Old Patients with hypertension, STOP-hypertension, Trial (relative risk = 0.53, 95 per cent CI 0.33–0.86) (Dahlof *et al.* 1991) and the Medical Research Council Trial (0.55, 0.25–0.60) (Medical Research Council Working Party (1985) of blood pressure lowering in primary prevention, but there was a 12-fold difference in absolute risk reduction. All other things being equal, 830 of the young hypertensives in the Medical Research Council trial would have to be treated for 1 year to prevent one stroke compared with 69 of the elderly hypertensives in the STOP-hypertension trial.

Trials should report *subgroup analyses* if there are potentially large differences between groups in the risk of a poor outcome with or without treatment, if there is potential heterogeneity of treatment effect in relation to pathophysiology, if there are practical questions about when to treat related to the stage of disease or the timing of treatment, or if there are doubts about benefit in specific groups, such as the elderly, which are likely to lead to under-treatment (Rothwell 2005c). Analyses must be predefined, carefully justified, and limited to a few clinically important questions, and post-hoc observations should be treated with scepticism irrespective of their statistical significance. Concerns about heterogeneity of treatment effects will often be unfounded, but if they are not addressed they will restrict the use of treatment in routine practice. If important subgroup effects are anticipated, trials should either be powered to detect them reliably or pooled analyses of multiple trials should be undertaken.

Univariate subgroup analysis is of relatively limited value, even when done reliably, in situations where there are multiple determinants of the individual response to treatment. In this situation, targeting treatment using risk models can be useful, particularly in conditions, or for interventions, where benefit is likely to be very dependent on the absolute risk of a poor outcome with or without treatment. Stratification of trial results with independently derived and validated prognostic models can allow clinicians to systematically take into account the characteristics of an individual patient and their interactions, to consider the risks and benefits of interventions separately if required, and to provide patients with personalized estimates of their likelihood of benefit from treatment (Rothwell 2005c).

References

Ad Hoc Subcommittee of the American Academy of Neurology AIDS Task Force (1991). Research criteria for diagnosis of chronic inflammatory demyelinating polyneuropathy (CIDP). *Neurology*, 41, 617–8.

Alexandrov AV, Molina CA, Grotta JC et al. (2004). Ultrasound-enhanced systemic thrombolysis for acute ischemic stroke. *N Engl J Med*, 351, 2170–8.

Alpini D, Pugnetti L, Caputo D et al. (2004). Vestibular evoked myogenic potentials in multiple sclerosis: clinical and imaging correlations. *Mult Scler*, 10, 316–321.

Aminoff MJ, Goodin DS, Parry GJ et al. (1985). Electrophysiologic evaluation of lumbosacral radiculopathies: electromyography, late responses, and somatosensory evoked potentials. *Neurology*, 35, 1514–8.

Anturane Reinfarction Trial Research Group (1980). Sulfinpyrazone in the prevention of sudden death after myocardial infarciton. *N Engl J Med*, 302, 250–56.

Apperloo JJ, van der Graaf F, Dellemijn PL et al. (2006). An improved laboratory protocol to assess subarachnoid haemorrhage in patients with negative cranial CT scan. *Clin Chem Lab Med*, 44, 938–48.

Arnold JA, Modaresi KB, Thomas N et al. (1999). Carotid plaque characterization by duplex scanning: observer error may undermine current clinical trials. *Stroke*, 30, 61–5.

Asymptomatic Carotid Atherosclerosis Study Group (1995). Carotid endarterectomy for patients with asymptomatic internal carotid artery stenosis. *JAMA*, 273, 1421–8.

Babikian VL, Wijman CA, Hyde C et al. (1997). Cerebral microembolism and early recurrent cerebral or retinal ischemic events. *Stroke*, 28, 1314–8.

Bandini F, Beronio A, Ghiglione E et al. (2004). The diagnostic value of vestibular evoked myogenic potentials in multiple sclerosis: a comparative study with MRI and visually evoked potentials. *J Neurol*, 251, 617–21.

Barbour PJ, Castaldo JE, Shoemaker EL (2001). Hemiplegic migraine during pregnancy: unusual magnetic resonance appearance with SPECT scan correlation. *Headache*, 41, 310–6.

Barker AT, Freeston IL, Jalinous R et al. (1985). Magnetic stimulation of the human brain. *J Physiol*, 369, 3P.

Barkovich AJ, Kuzniecky RI, Jackson GD et al. (2001). Classification system for malformations of cortical development: update 2001. *Neurology*, 57, 2168–78.

Barnett HJ, Taylor DW, Eliasziw M et al. (1998). The final results of the NASCET trial. *N Engl J Med*, 339, 1415–25.

Barth A, Arnold M, Mattle HP et al. (2006). Contrast-enhanced 3-D MRA in decision making for carotid endarterectomy: a 6-year experience. *Cerebrovasc Dis*, 21, 393–400.

Bartlett ES, Symons SP, Fox AJ (2006). Correlation of carotid stenosis diameter and cross-sectional areas with CT angiography. *AJNR Am J Neuroradiol*, 27, 638–42.

Baumgartner RW (1999). Transcranial color-coded duplex sonography. *J Neurol*, 246, 637–47.

Baumgartner RW, Baumgartner I, Mattle HP et al. (1997). Transcranial color-coded duplex sonography in the evaluation of collateral flow through the circle of Willis. *AJNR Am J Neuroradiol*, 18, 127–33.

Binnie C, Cooper R, Mauguiere F et al. eds. (2004). *Clinical Neurophysiology*, Vols 1 and 2. Elsevier, Amsterdam.

Bishop CC, Powell S, Insall M et al. (1986). Effect of internal carotid artery occlusion on middle cerebral artery blood flow at rest and in response to hypercapnia. *Lancet*, 1, 710–2.

Blakeley DD, Oddone EZ, Hasselblad V et al. (1995). Noninvasive carotid artery testing. A meta-analytic review. *Ann Intern Med*, 122, 360–7.

Block SA, Bakshi R (2001). FLAIR MRI of striatonigral degeneration. *Neurology*, 56, 1200.

Bornstein NM, Norris JW (1986). Subclavian steal: a harmless haemodynamic phenomenon? *Lancet*, 2, 303–5.

Bouche P, Cattelin F, Saint-Jean O et al. (1993). Clinical and electrophysiological study of the peripheral nervous system in the elderly. *J Neurol*, 240, 263–8.

Bradley WG (1974). *Disorders of Peripheral Nerves*. Blackwell, Oxford.

Brink JA, McFarland EG, Heiken JP (1997). Helical/spiral computed body tomography. *Clin Radiol*, 52, 489–503.

Brusa A, Jones SJ, Kapoor R et al. (1999). Long-term recovery and fellow eye deterioration after optic neuritis, determined by serial visual evoked potentials. *J Neurol*, 246, 776–82.

Camacho DL, Smith JK, Castillo M (2003). Differentiation of toxoplasmosis and lymphoma in AIDS patients by using apparent diffusion coefficients. *Am J Neuroradiol*, 24, 633–7.

Cardiac Arrhythmia Suppression Trial (CAST) Investigators (1989). Preliminary report: effect of encainide and flecainide on mortality in a randomised trial of arrhythmia suppression after myocardial infarction. *N Engl J Med*, 321, 402–12.

Castillo M, Smith JK, Kwock L et al. (2001). Apparent diffusion co-efficients in the evaluation of high grade cerebral gliomas. *Am J Neuroradiol*, 22, 60–4.

Chappell ET, Moure FC, Good MC (2003). Comparison of computed tomographic angiography with digital subtraction angiography in the diagnosis of cerebral aneurysms: a meta-analysis. *Neurosurgery*, 52, 624–31; discussion 630–1.

Charlson ME, Horwitz RI (1984). Applying results of randomised trials to clinical practice: impact of losses before randomisation. *Br Med J (Clin Res Ed)*, 289, 1281–4.

Chinnery PF, Turnbull DM (1997). Clinical features, investigation, and management of patients with defects of mitochondrial DNA. *J Neurol Neurosurg Psychiatry*, 63, 559–63.

Church AJ, Cardoso F, Dale RC *et al.* (2002). Anti-basal ganglia antibodies in acute and persistent Sydenham's chorea. *Neurology*, **59**, 227–31.

Cipolotti L, Warrington EK (1995). Neuropsychological assessment. *J Neurol Neurosurg Psychiatry*, **58**, 655–64.

Collie, DA, Sellar RJ, Zeidler M *et al.* (2001). MRI of Creutzfeld-Jakob disease: imaging features and recommended MRI protocol. *Clin Radiol*, **56**, 726–39.

Conneally M (1993). *Molecular Basis of Neurology.* Blackwell Scientific Publishers, Boston.

Cornblath DR, Kuncl RW, Mellits ED *et al.* (1992). Nerve conduction studies in amyotrophic lateral sclerosis. *Muscle Nerve*, **15**, 1111–5.

Cornblath DR, Sumner AJ, Daube J *et al.* (1991). Conduction block in clinical practice. *Muscle Nerve*, **14**, 869–71; discussion 867–8.

Coronary Drug Project Research Group (1980). Influence of adherence to treatment and response to cholesterol on mortality in the Coronary Drug Project. *N Engl J Med*, **303**, 1038–41.

Cowan JMA, Dick JPR, Day BL *et al.* (1984). Abnormalities in central motor pathway conduction in multiple sclerosis. *Lancet*, **2**, 304–7.

Crawford JS, Konkol RJ (1997). Familial hemiplegic migraine with crossed cerebellar diachisis and unilateral meningeal enhancement. *Headache*, **37**, 590–3.

Cruccu G, Iannetti GD, Marx JJ *et al.* (2005). Brainstem reflex circuits revisitied. *Brain*, **128**, 386–94.

Cuffe RL, Rothwell PM (2006). Effect of nonoptimal imaging on the relationship between the measured degree of symptomatic carotid stenosis and risk of ischemic stroke. *Stroke*, **37**, 1785–91.

Dahl A, Russell D, Rootwelt K *et al.* (1995). Cerebral vasoreactivity assessed with transcranial Doppler and regional cerebral blood flow measurements. Dose, serum concentration, and time course of the response to acetazolamide. *Stroke*, **26**, 2302–6.

Dahlof B, Lindholm LH, Hansson L *et al.* (1991). Morbidity and mortality in the Swedish trial in old patients with hypertension (STOP-hypertension). *Lancet*, **338**, 1281–5.

Dale RC, Church AJ, Cardoso F *et al.* (2001). Poststreptococcal acute disseminated encephalomyelitis with basal ganglia involvement and auto-reactive antibasal ganglia antibodies. *Ann Neurol*, **50**, 588–95.

Daly DD, Pedley TA eds. (1990). *Current Practice of Clinical Electroencephalography*, 2nd edition. Raven Press, New York.

Davies K, Read A (1992). *Molecular Basis of Inherited Disease.* Oxford University Press, Oxford.

Davies NW, Brown LJ, Gonde J *et al.* (2005). Factors influencing PCR detection of viruses in cerebrospinal fluid of patients with suspected CNS infections. *J Neurol, Neurosurg Psychiatry*, **76**, 82–7.

Deisenhammer F, Bartos A, Egg R *et al.* (2006). Guidelines on routine cerebrospinal fluid analysis. Report from an EFNS task force. *Eur J Neurol*, **13**, 913–22.

DeMarco JK, Huston J 3rd, Nash AK (2006). Extracranial carotid MR imaging at 3T. *Magn Reson Imaging Clin N Am*, **14**, 109–21.

Derdeyn CP, Grubb RL Jr, Powers WJ (1999). Cerebral hemodynamic impairment: methods of measurement and association with stroke risk. *Neurology*, **53**, 251–9.

Derdeyn CP, Grubb RL Jr, Powers WJ (2005). Indications for cerebral revascularization for patients with atherosclerotic carotid occlusion. *Skull Base*, **15**, 7–14.

Dittrich R, Ritter MA, Kaps M *et al.* (2006). The use of embolic signal detection in multicenter trials to evaluate antiplatelet efficacy: signal analysis and quality control mechanisms in the CARESS (Clopidogrel and Aspirin for Reduction of Emboli in Symptomatic carotid Stenosis) trial. *Stroke*, **37**, 1065–9.

Droste DW, Jurgens R, Nabavi DG *et al.* (1999). Echocontrast-enhanced ultrasound of extracranial internal carotid artery high-grade stenosis and occlusion. *Stroke*, **30**, 2302–6.

Eidelberg D, Moeller JR, Antonini A *et al.* (1998). Functional brain networks in DYT1 dystonia. *Ann Neurol*, **44**, 303–12.

Eliasziw M, Streifler JY, Fox AJ *et al.* (1994). Significance of plaque ulceration in symptomatic patients with high-grade carotid stenosis. North American Symptomatic Carotid Endarterectomy Trial. *Stroke*, **25**, 304–8.

Engel J (2000). Overview of functional neuroimaging in epilepsy. *Adv Neurol*, **83**, 1–9.

Farrell B, Godwin J, Richards S *et al.* (1991). The United Kingdom transient ischaemic attack (UK-TIA) aspirin trial: final results. *J Neurol Neurosurg Psychiatry*, **54**, 1044–54.

Fiehler J, von Bezold M, Kucinski T *et al.* (2002). Cerebral blood flow predicts lesion growth in acute stroke patients. *Stroke*, **33**, 2421–5.

Filippi M, Tortorella C, Rovaris M. (2002). Magnetic resonance imaging of multiple sclerosis. *J Neuroimaging*, **12**, 289–301.

Fine PEM (1995). Variation in protection by BCG: implications of and for heterologous immunity. *Lancet*, **346**, 1339–45.

Flossmann E, Rothwell PM (2003). Prognosis of vertebrobasilar transient ischaemic attack and minor stroke. *Brain*, **126**, 1940–54.

Flossmann E, Redgrave J, Schulz U *et al.* (2006). Reliability of clinical diagnosis of the symptomatic vascular territory in patients with recent TIA or minor stroke. *Cerebrovasc Dis*, **21**, 18.

Fox AJ, Eliasziw M, Rothwell PM *et al.* (2005). Identification, prognosis, and management of patients with carotid artery near occlusion. *AJNR Am J Neuroradiol*, **26**, 2086–94.

Friedman SG (1990). Transient ischemic attacks resulting from carotid duplex imaging. *Surgery*, **107**, 153–5.

Gaitini D, Soudack M (2005). Diagnosing carotid stenosis by Doppler sonography: state of the art. *J Ultrasound Med*, **24**, 1127–36.

Garg SK, Tiwari RP, Tiwari D *et al.* (2003). Diagnosis of tuberculosis: available technologies, limitations, and possibilities. *J Clin Lab Anal*, **17**, 155–63.

Gerraty RP, Bowser DN, Infeld B *et al.* (1996). Microemboli during carotid angiography. Association with stroke risk factors or subsequent magnetic resonance imaging changes? *Stroke*, **27**, 1543–7.

Gerriets T, Seidel G, Fiss I *et al.* (1999). Contrast-enhanced transcranial color-coded duplex sonography: efficiency and validity. *Neurology*, **52**, 1133–7.

Ghosh A, Busby M, Kennett R *et al.* (2005). A practical definition of conduction block in IvIg responsive multifocal motor neuropathy. *J Neurol Neurosurg Psychiatry*, **76**, 1264–8.

Gilman S (1998). Imaging the brain: first of two parts. *N Engl J Med*, **338**, 812–20.

Gorter JW for the Stroke Prevention in Reversible Ischaemia Trial (SPIRIT) and European Atrial Fibrillation Trial (EAFT) groups (1999). Major bleeding during anticoagulation after cerebral ischaemia: patterns and risk factors. *Neurology*, **53**, 1319–27.

Graus F, Lang B, Pozo-Rosich P *et al.* (2002). P/Q type calcium-channel antibodies in paraneoplastic cerebellar degeneration with lung cancer. *Neurology*, **59**, 764–6.

Graus F, Vincent A, Pozo-Rosich P *et al.* (2005). Anti-glial nuclear antibody: marker of lung cancer-related paraneoplastic neurological syndromes. *J Neuroimmunol*, **165**, 166–71.

Graves MJ (1997). Magnetic resonance angiography. *Br J Radiol*, **70**, 6–28.

Gronholdt ML (1999). Ultrasound and lipoproteins as predictors of lipid-rich, rupture-prone plaques in the carotid artery. *Arterioscler Thromb Vasc Biol*, **19**, 2–13.

Grosveld WJ, Lawson JA, Eikelboom BC *et al.* (1988). Clinical and hemodynamic significance of innominate artery lesions evaluated by ultrasonography and digital angiography. *Stroke*, **19**, 958–62.

Gruppo Italiano per lo Studio della Streptochinasi nell'Infarto Miocardico (GISSI) (1986). Effectiveness of intravenous thrombolytic treatment in acute myocardial infarction. *Lancet*, **1**, 397–402.

Gur AY, Bova I, Bornstein NM (1996). Is impaired cerebral vasomotor reactivity a predictive factor of stroke in asymptomatic patients? *Stroke*, **27**, 2188–90.

Gurwitz JH, Col NF, Avorn J (1992). The exclusion of elderly and women from clinical trials in acute myocardial infarction. *JAMA*, **268**, 1417–22.

Gøtzsche PC (1989). Methodology and overt and hidden bias in reports of 196 double-blind trials of nonsteroidal antiinflammatory drugs in rheumatoid arthritis. *Control Clin Trials*, **10**, 31–56.

Hadjikhani N, Sanchez del Rio M, Wu O *et al.* (2001). Mechanisms of migraine aura revealed by functional MRI in human visual cortex. *PNAS*, **98**, 4687–92.

Halliday AM, Halliday E, Kriss A *et al.* (1976). The pattern-evoked potential in compression of the anterior visual pathways. *Brain*, **99**, 357–374.

Halliday A, Mansfield A, Marro J *et al.* (2004). Prevention of disabling and fatal strokes by successful carotid endarterectomy in patients without recent neurological symptoms: randomised controlled trial. *Lancet*, **363**, 1491–502.

Halliday AM, McDonald WI, Mushin J (1972). Delayed visual evoked response in optic neuritis. *Lancet*, **i**, 982–5.

Halliday AM, McDonald WI, Mushin J (1973). Visual evoked response in diagnosis of multiple sclerosis. *BMJ*, **4**, 661–4.

Hand PJ, Wardlaw JM, Rivers CS *et al.* (2006). MR diffusion-weighted imaging and outcome prediction after ischemic stroke. *Neurology*, **66**, 1159–63.

Hankey GJ, Warlow CP (1990). Symptomatic carotid ischaemic events: safest and most cost effective way of selecting patients for angiography, before carotid endarterectomy. *BMJ*, **300**, 1485–91.

Hankey GJ, Warlow CP (1991). Lacunar transient ischaemic attacks: a clinically useful concept? *Lancet*, **337**, 335–8.

Hankey GJ, Slattery JM, Warlow CP (1991). The prognosis of hospital-referred transient ischaemic attacks. *J Neurol Neurosurg Psychiatry*, **54**, 793–802.

Harrington BE (2004). Postdural puncture headache and the development of the epidural blood patch. *Reg Anesth Pain Med*, **29**, 136–63; discussion 135.

Heiken JP, Brink JA, Vannier MW (1993). Spiral (helical) CT. *Radiology*, **189**, 647–56.

Heinze HJ, Munte TF, Kutas M *et al.* (1999). Cognitive event-related potentials. In Deuschl G, Eisen A, eds. *Recommendations for the Practice of Clinical Neurophysiology. Guidelines of the International Federation of Clinical Neurophysiology* (EEG Suppl. 52), pp. 91–95. Elsevier Science BV, Amsterdam.

Hennerici M, Klemm C, Rautenberg W (1988). The subclavian steal phenomenon: a common vascular disorder with rare neurologic deficits. *Neurology*, **38**, 669–73.

Hernandez-Echebarria L, Saiz A, Ares A *et al.* (2006). Paraneoplastic encephalomyelitis associated with pancreatic tumor and anti-GAD antibodies. *Neurology*, **66**, 450–1.

Hodges JR (1994). *Cognitive Assessment for Clinicians.* Oxford University Press, Oxford.

Huisman TA, Sorensen AG, Hergan K *et al.* (2003). Diffusion weighted imaging for the evaluation of diffuse axonal injury in closed head injury. *J Comput Assist Tomogr*, **27**, 5–11.

Hutchinson M, Nakamura T, Moeller JR *et al.* (2000). The metabolic topography of essential blepharospasm: a focal dystonia with general implications. *Neurology*, **55**, 673–7.

Ioannidis JP, Contopoulos-Ioannidis DG (1998). Reporting of safety data from randomised trials. *Lancet*, **352**, 1752–3.

Iranzo A, Molinuevo JL, Santamaria J *et al.* (2006). Rapid-eye-movement sleep behaviour disorder as an early marker for a neurodegenerative disorder: a descriptive study. *Lancet Neurol*, **5**, 572–7.

Jeans WD, Mackenzie S, Baird RN (1986). Angiography in transient cerebral ischaemia using three views of the carotid bifurcation. *Br J Radiol*, **59**, 135–42.

Johnston DC, Goldstein LB (2001). Clinical carotid endarterectomy decision making: noninvasive vascular imaging versus angiography. *Neurology*, **56**, 1009–15.

Johnston DC, Chapman KM, Goldstein LB (2001). Low rate of complications of cerebral angiography in routine clinical practice. *Neurology*, **57**, 2012–4.

Johnston DC, Eastwood JD, Nguyen T *et al.* (2002). Contrast-enhanced magnetic resonance angiography of carotid arteries: utility in routine clinical practice. *Stroke*, **33**, 2834–8.

Jones SJ, Sprague L, Vaz Pato M (2002). Electrophysiological evidence for a defect in the processing of temporal sound patterns in multiple sclerosis. *J Neurol Neurosurg Psychiatry*, **73**, 561–7.

Jorgensen HS, Nakayama H, Kammersgaard LP *et al.* (1999). Predicted impact of intravenous thrombolysis on prognosis of general population of stroke patients: simulation model. *BMJ*, **319**, 288–9.

Kempczinski R and Hermann G (1979). The innominate steal syndrome. *J Cardiovasc Surg (Torino)*, **20**, 481–6.

Kidd D, Thompson PD, Day BL *et al.* (1998). Central motor conduction time in progressive multiple sclerosis: correlations with MRI and disease activity. *Brain*, **121**, 1109–16.

Kidwell CS, Hsia AW (2006). Imaging of the brain and cerebral vasculature in patients with suspected stroke: advantages and disadvantages of CT and MRI. *Curr Neurol Neurosci Rep*, **6**, 9–16.

Kim YS, Garami Z, Mikulik R *et al.* (2005). Early recanalization rates and clinical outcomes in patients with tandem internal carotid artery/middle cerebral artery occlusion and isolated middle cerebral artery occlusion. *Stroke*, **36**, 869–71.

Klein CM, Vernino S, Lennon VA *et al.* (2003). The spectrum of autoimmune autonomic neuropathies. *Ann Neurol*, **53**, 752–8.

Kleiser B, Widder B (1992). Course of carotid artery occlusions with impaired cerebrovascular reactivity. *Stroke*, **23**, 171–4.

Knopman DS, DeKosky ST, Cummings JL *et al.* (2001). Practice Parameter: diagnosis of dementia (an evidence based review): report of the Quality Standards Subcommittee of the American Academy of Neurology. *Neurology*, **56**, 1143–53.

Kono K, Inoue Y, Nakayama K *et al.* (2001). The role of diffusion weighted imaging in patients with brain tumours. *Am J Neuroradiol*, **22**, 1081–8.

Kriss A, Francis DA, Cuendet F *et al.* (1988) Recovery after optic neuritis in childhood. *J Neurol Neurosurg Psychiatry*, **51**, 1253–8.

Lang B, Vincent A (2003). Autoantibodies to ion channels at the neuromuscular junction. *Autoimmun Rev*, **2**, 94–100.

LaRue LJ. Alter M, Traven ND *et al.* (1988). Acute stroke therapy trials: problems in patient accrual. *Stroke*, **19**, 950–4.

Leao AAP (1944). Spreading depression of activity in the cerebral cortex. *J Neurophysiol*, **7**, 359–70.

Leclerc X, Godefroy O, Pruvo JP *et al.* (1995). Computed tomographic angiography for the evaluation of carotid artery stenosis. *Stroke*, **26**, 1577–81.

Levi C, Mitchell A, Fitt G *et al.* (1996). The accuracy of magnetic resonance angiography in the assessment of extracranial carotid artery occlusive disease. *Cerebrovasc Dis*, **6**, 231–6.

Lewis RA, Sumner AJ (1982). The electrodiagnostic distinctions between chronic familial and acquired demyelinative neuropathies. *Neurology*, **32**, 592–6.

Lezak M (2004). *Neuropsychological Assessment*, 4th edition. Oxford University Press, New York.

Liguori R, Vincent A, Clover L *et al.* (2001). Morvan's syndrome: peripheral and central nervous system and cardiac involvement with antibodies to voltage-gated potassium channels. *Brain*, **124**, 2417–26.

Lovett JK, Gallagher PJ, Hands LJ *et al.* (2004). Histological correlates of carotid plaque surface morphology on lumen contrast imaging. *Circulation*, **110**, 2190–7.

Magyar-Lehmann S, Antonini A, Roelcke U *et al.* (1997). Cerebral glucose metabolism in patients with spasmodic torticolis. *Mov Disord*, **12**, 704–8.

Markus HS (1999). Transcranial Doppler ultrasound. *J Neurol Neurosurg Psychiatry*, **67**, 135–7.

Markus HS (2006). Can microemboli on transcranial Doppler identify patients at increased stroke risk? *Nat Clin Pract Cardiovasc Med*, **3**, 246–7.

Markus HS, Harrison MJ (1992). Estimation of cerebrovascular reactivity using transcranial Doppler, including the use of breath-holding as the vasodilatory stimulus. *Stroke*, **23**, 668–73.

Markus HS, Droste DW, Brown MM (1994). Detection of asymptomatic cerebral embolic signals with Doppler ultrasound. *Lancet*, **343**, 1011–2.

Marshall M, Lockwood A, Bradley C *et al.* (2000). Unpublished rating scales – a major source of bias in randomised controlled trials of treatments for schizophrenia? *Br J Psychiatry*, **176**, 249–52.

Martin R, Bogousslavsky J, Miklossy J *et al.* (1992). Floating thrombus in the innnominate artery as a cause of cerebral infarction in young adults. *Cerebrovasc Dis*, **2**, 177–81.

McDonald WI, Compston A, Edan G *et al.* (2001). Recommended diagnostic criteria for multiple sclerosis: guidelines from the International Panel on the diagnosis of multiple sclerosis. *Ann Neurol*, **50**, 121–7.

McPherson D, Starr A (1993). Auditory evoked potentials in the clinic. In Halliday AM ed. *Evoked Potentials in Clinical Testing*, pp. 359–81. Churchill Livingstone, Edinburgh.

Mead GE, Wardlaw JM, Lewis SC *et al.* (1999). Can simple clinical features be used to identify patients with severe carotid stenosis on Doppler ultrasound? *J Neurol Neurosurg Psychiatry*, **66**, 16–9.

Medical Research Council Working Party (1985). MRC trial of treatment of mild hypertension: principal results. *BMJ*, **291**, 97–104.

Meriggioli MN (2005). Use of immunoassays in neurological diagnosis and research. *Neurol Res*, **27**, 734–40.

Merton PA, Morton HB (1980). Stimulation of the cerebral cortex in the intact human subject. *Nature*, **285**, 227.

Meuli RA (2004). Imaging viable brain tissue with CT scan during acute stroke. *Cerebrovasc Dis*, **17** (Suppl 3), 28–34.

Miller DH, McDonald WI, Smith K (2005). The diagnosis of multiple sclerosis. In Compston DAS ed. *McAlpine's Multiple Sclerosis*, pp. 347–88. 4th edition. Elsevier, London.

Mills KR (2005). The basics of electromyography. *J Neurol Neurosurg Psychiatry*, **76** (Suppl 2), ii32–5.

Mitra D, Connolly D, Jenkins S *et al.* (2006). Comparison of image quality, diagnostic confidence and interobserver variability in contrast enhanced MR angiography and 2D time of flight angiography in evaluation of carotid stenosis. *Br J Radiol*, **79**, 201–7.

Molloy J, Markus HS (1999). Asymptomatic embolization predicts stroke and TIA risk in patients with carotid artery stenosis. *Stroke*, **30**, 1440–3.

Muir KW, Buchan A, von Kummer R *et al.* (2006). Imaging of acute stroke. *Lancet Neurol*, **5**, 755–68.

Nabavi DG, Kloska SP, Nam EM *et al.* (2002). MOSAIC: Multimodal Stroke Assessment Using Computed Tomography: novel diagnostic approach for the prediction of infarction size and clinical outcome. *Stroke*, **33**, 2819–26.

Nadal Desbarats L, Herlidou S, de Marco G *et al.* (2003). Differential MRI diagnosis between brain abscesses and necrotic or cystic brain tumors using the apparent diffusion coefficient and normalised diffusion weighted images. *Magn Reson Images*, **21**, 645–50.

Narai T, Tanaka Y, Wakimoto H *et al.* (2005). Usefulness of L-[methyl-11C] methionine positron emission tomography as a biological monitoring tool in the treatment of glioma. *J Neurosurg*, **103**, 498–507.

Nandalur KR, Baskurt E, Hagspiel KD *et al.* (2006). Carotid artery calcification on CT may independently predict stroke risk. *AJR Am J Roentgenol*, **186**, 547–52.

Naumann M, Pirker W, Reiners K *et al.* (1997). [123I] beta-CIT single-photon emission tomography in DOPA-responsive dystonia. *Mov Disord*, **12**, 448–51.

Nelson HE, and O'Connell A (1978). Dementia: the estimation of premorbid intelligence levels using the New Adult Reading Test. *Cortex*, **14**, 234–44.

Norris JW, Halliday A (2004). Is ultrasound sufficient for vascular imaging prior to carotid endarterectomy? *Stroke*, **35**, 370–1.

Norris JW, Morriello F, Rowed DW *et al.* (2003). Vascular imaging before carotid endarterectomy. *Stroke*, **34**, e16.

Noseworthy JH, Ebers GC, Vandervoort MK *et al.* (1994). The impact of blinding on the results of a randomized, placebo-controlled multiple sclerosis clinical trial. *Neurology*, **44**, 16–20.

Ouvrier R (1990). *Peripheral Neuropathy in Childhood*. Raven, New York.

Parsons MW, Pepper EM, Chan V *et al.* (2005). Perfusion computed tomography: prediction of final infarct extent and stroke outcome. *Ann Neurol*, **58**, 672–9.

Pelz DM, Fox AJ, Vinuela F (1985). Digital subtraction angiography: current clinical applications. *Stroke*, **16**, 528–36.

Pincus T (1998). Rheumatoid arthritis: disappointing long-term outcomes despite successful short-term clinical trials. *J Clin Epidemiol*, **41**, 1037–41.

Pirotte B, Goldman S, Dewitte O *et al.* (2006). Integrated positron emission tomography and magnetic resonance imaging-guided resection of brain tumors: a report of 103 consecutive procedures. *J Neurosurg*, **104**, 238–53.

Polak JF, Shemanski L, O'Leary DH *et al.* (1998). Hypoechoic plaque at US of the carotid artery: an independent risk factor for incident stroke in adults aged 65 years or older. Cardiovascular Health Study. *Radiology*, **208**, 649–54.

Polman CH, Reingold SC, Edan G *et al.* (2005). Diagnostic criteria for multiple sclerosis: 2005 revisions to the 'McDonald Criteria'. *Ann Neurol*, **58**, 840–6.

Pozo-Rosich P, Clover L, Saiz A *et al.* (2003). Voltage-gated potassium channel antibodies in limbic encephalitis. *Ann Neurol*, **54**, 530–3.

Rees J, Hain SF, Johnson MR *et al.* (2001). The role of [18F]fluoro-2-deoxyglucose-PET scanning in the diagnosis of paraneoplastic neurological disorders. *Brain*, **124**, 2223–31.

Riggs BL, Hodgson SF, O'Fallon WM *et al.* (1990). Effect of flouride treatment on fracture rate in postmenopausal women with osteoporosis. *N Engl J Med*, **322**, 802–9.

Robinson K, Rudge P (1977). Abnormalities of the auditory evoked potentials in patients with multiple sclerosis. *Brain*, **100**, 19–40.

Rosario JA, Hachinski VC, Lee DH *et al.* (1987). Adverse reactions to duplex scanning. *Lancet*, **2**, 1023.

Rothwell PM (2005a). External validity of randomised controlled trials: To whom do the results of this trial apply? *Lancet*, **365**, 82–93.

Rothwell PM (2005b). Subgroup analysis in randomised controlled trials: importance, indications and interpretation. *Lancet*, **365**, 176–86.

Rothwell PM (2005c). Subgroup analysis in randomised controlled trials: importance, indications and interpretation. *Lancet*, **365**, 176–86.

Rothwell PM, Warlow CP (1995). Is self-audit reliable? *Lancet*, **346**, 1623.

Rothwell PM, Gibson RJ, Villagra R *et al.* (1998). The effect of angiographic technique and image quality on the reproducibility of measurement of carotid stenosis and assessment of plaque surface morphology. *Clin Radiol*, **53**, 439–43.

Rothwell PM, Gibson R, Warlow CP (2000b). Interrelation between plaque surface morphology and degree of stenosis on carotid angiograms and the risk of ischemic stroke in patients with symptomatic carotid stenosis. On behalf of the European Carotid Surgery Trialists' Collaborative Group. *Stroke*, **31**, 615–21.

Rothwell PM, Giles MF, Chandratheva A *et al.* (2007). Effect of urgent treatment of transient ischemic attack and minor stroke on early recurrent stroke (EXPRESS study): a prospective population based sequential comparison. *Lancet*, **370**: 1432–42.

Rothwell PM, Pendlebury ST, Wardlaw J *et al.* (2000a). Critical appraisal of the design and reporting of studies of imaging and measurement of carotid stenosis. *Stroke*, **31**, 1444–50.

Ruiz A, Ganz WI, Post MJ *et al.* (1994). Use of thallium 201 brain SPECT to differentiate cerebral lymphoma from toxoplasma encephalitis in AIDS patients. *Am J Neuroradiol*, **15**, 1885–94.

Rutten GJ, van Rijen PC, van Vellen CW *et al.* (1999). Language area localization with three dimensional functional magnetic resonance imaging matches intrasulcal electrostimulation in Broca's area. *Ann Neuro*, **46**, 405–8.

Ryan JJ, Paolo AM, Brungardt TM (1990). Standardisation of the Wechsler adult intelligence scale-revised for persons 75 years and older. *Psychological assessments: J Consult Clin Psychol*, **2**, 408–11.

Sabater L, Gomez-Choco M, Saiz A *et al.* (2005). BR serine/threonine kinase 2: a new autoantigen in paraneoplastic limbic encephalitis. *J Neuroimmunol*, **170**, 186–90.

Schapira AH (2006). Mitochondrial disease. *Lancet*, **368**, 70–82.

Shapiro SH, Weijer C, Freedman B (2000). Reporting the study populations of clinical trials. Clear transmission or static on the line? *J Clin Epidemiol*, **53**, 973–9.

Shih LC, Saver JL, Alger JR *et al.* (2003). Perfusion-weighted magnetic resonance imaging thresholds identifying core, irreversibly infarcted tissue. *Stroke*, **34**, 1425–30.

Siewert B, Patel M, Warach S (1995). Magnetic resonance angiography. *Neurologist*, **1**, 167–84.

Sliwka U, Lingnau A, Stohlmann WD *et al.* (1997). Prevalence and time course of microembolic signals in patients with acute stroke. A prospective study. *Stroke*, **28**, 358–63.

Sloan MA, Alexandrov AV, Tegeler CH *et al.* (2004). Assessment: transcranial Doppler ultrasonography: report of the Therapeutics and Technology Assessment Subcommittee of the American Academy of Neurology. *Neurology*, **62**, 1468–81.

Small DG, Matthews WB, Small M (1978). The cervical somatosensory evoked potential in the diagnosis of multiple sclerosis. *J Neurol Sci*, **35**, 211–24.

Snider LA, Swedo SE (2004). PANDAS: current status and directions for research. *Mol Psychiatry*, **9**, 900–7.

Streifler JY, Eliasziw M, Fox AJ *et al.* (1994). Angiographic detection of carotid plaque ulceration. Comparison with surgical observations in a multicenter study. North American Symptomatic Carotid Endarterectomy Trial. *Stroke*, **25**, 1130–2.

Sudlow C, Warlow C (2002). Posture and fluids for preventing post-dural puncture headache. *Cochrane Database Syst Rev*, CD001790.

Swedo SE, Leonard HL, Garvey M *et al.* (1998). Pediatric autoimmune neuropsychiatric disorders associated with streptococcal infections: clinical description of the first 50 cases. *Am J Psychiatry*, **155**, 264–71.

Taylor DW, Barnett HJM, Haynes RB *et al.* (1999). Low dose and high dose acetylsalicylic acid for patients undergoing carotid endarterectomy: a randomised controlled trial. *Lancet*, **353**, 2179–84.

Teasdale GM, Hadley DM, Lawrence A (1989). Comparison of magnetic resonance imaging and computed tomography in suspected lesions in the posterior fossa. *BMJ*, **299**, 349–55.

Temple R, Pledger GW (1980). The FDA's critique of the Anturane Reinfarction Trial. *N Engl J Med*, **303**, 1488–92.

Thijs VN, Adami A, Neumann-Haefelin T *et al.* (2001). Relationship between severity of MR perfusion deficit and DWI lesion evolution. *Neurology*, **57**, 1205–11.

Thompson PD, Day BL, Rothwell JC *et al.* (1987). The interpretation of electromyographic responses to electrical stimulation of the motor cortex in diseases of the upper motor neurone. *J Neurol Sci*, **80**, 91–110.

Thornley B, Adams CE (1998). Content and quality of 2000 controlled trials in schizophrenia over 50 years. *BMJ*, **317**, 1181–4.

U-King-Im J, Hollingworth W, Trivedi RA *et al.* (2005). Cost-effectiveness of diagnostic strategies prior to carotid endarterectomy. *Ann Neurol*, **58**, 506–15.

Vernieri F, Pasqualetti P, Matteis M *et al.* (2001). Effect of collateral blood flow and cerebral vasomotor reactivity on the outcome of carotid artery occlusion. *Stroke*, **32**, 1552–8.

Vernieri F, Pasqualetti P, Passarelli F *et al.* (1999). Outcome of carotid artery occlusion is predicted by cerebrovascular reactivity. *Stroke*, **30**, 593–8.

Vernino S, Ermilov LG, Sha L *et al.* (2004). Passive transfer of autoimmune autonomic neuropathy to mice. *J Neurosci*, **24**, 7037–42.

Vernino S, Low PA, Fealey RD *et al.* (2000). Autoantibodies to ganglionic acetylcholine receptors in autoimmune autonomic neuropathies. *N Engl J Med*, **343**, 847–55.

Villablanca JP, Hooshi P, Martin N *et al.* (2002). Three-dimensional helical computerized tomography angiography in the diagnosis, characterization, and management of middle cerebral artery aneurysms: comparison with conventional angiography and intraoperative findings. *J Neurosurg*, **97**, 1322–32.

Vincent A, Buckley C, Schott JM *et al.* (2004). Potassium channel antibody-associated encephalopathy: a potentially immunotherapy-responsive form of limbic encephalitis. *Brain*, **127**, 701–12.

Vitaliani R, Mason W, Ances B *et al.* (2005). Paraneoplastic encephalitis, psychiatric symptoms, and hypoventilation in ovarian teratoma. *Ann Neurol*, **58**, 594–604.

Vogel F, Motulsky AG, eds. (1986). *Human Genetics*, 2nd edition. Springer Verlag, Berlin.

Wardlaw JM, Lewis S (2005). Carotid stenosis measurement on colour Doppler ultrasound: agreement of ECST, NASCET and CCA methods applied to ultrasound with intra-arterial angiographic stenosis measurement. *Eur J Radiol*, **56**, 205–11.

Wardlaw JM, Chappell FM, Best JJ *et al.* (2006). Non-invasive imaging compared with intra-arterial angiography in the diagnosis of symptomatic carotid stenosis: a meta-analysis. *Lancet*, **367**, 1503–12.

Warnock NG, Gandhi MR, Bergvall U *et al.* (1993). Complications of intraarterial digital subtraction angiography in patients investigated for cerebral vascular disease. *Br J Radiol*, **66**, 855–8.

Warrington EK (1984). *Recognition Memory Test*. NFER-Nelson, Windsor, UK.

Weiller C, May A, Limmrith V *et al.* (1995). Brain Stem activation in human migraine attacks. *Nature Med*, **1**, 658–60.

Went L (1990). Ethical issues policy statement on Huntington's disease molecular genetics predictive test. International Huntington Association. World Federation of Neurology. *J Med Genet*, **27**, 34–8.

Widder B, Kleiser B, Krapf H (1994). Course of cerebrovascular reactivity in patients with carotid artery occlusions. *Stroke*, **25**, 1963–7.

Wieser S, Kelemen A, Barsi P *et al.* (2005). Pilomotor seizures and status in non-paraneoplastic limbic encephalitis. *Epileptic Disord*, **7**, 205–11.

Willison HJ, Yuki N (2002). Peripheral neuropathies and anti-glycolipid antibodies. *Brain*, **125**, 2591–625.

Wintermark M, Bogousslavsky J (2003). Imaging of acute ischemic brain injury: the return of computed tomography. *Curr Opin Neurol*, **16**, 59–63.

Wintermark M, Meuli R, Browaeys P *et al.* (2007). Comparison of CT perfusion and angiography and MRI in selecting stroke patients for acute treatment. *Neurology*, **68**, 694–7.

Wintermark M, Uske A, Chalaron M *et al.* (2003). Multislice computerized tomography angiography in the evaluation of intracranial aneurysms: a comparison with intraarterial digital subtraction angiography. *J Neurosurg*, **98**, 828–36.

World Federation research group on Huntington's disease (1990). Ethical issues policy statement of Huntington's disease molecular genetics predictive test. *J Med Genet*, **27**, 34–8.

Yekhlef F, Ballan G, Macia F *et al.* (2003). Routine MRI for the differential diagnosis of Parkinson's disease MSA, PSP, and CBD. *J Neural Transm*, **110**, 151–69.

Young RJ, Ghesani MV, Kagetsu NJ *et al.* (2005). Lesion size determines accuracy of thallium 201 brain single photon emission tomography in differentiating between intracranial malignancy and infection in AIDS patients. *Am J Neuroradiol*, **26**, 1973–9.

Zeidler M, Sellar RJ, Collie DA *et al..* (2000) The Pulvinar sign on magnetic resonance imaging in Creutzfeldt-Jacob disease. *Lancet*, **355**, 1412–18.

Zhang L, Murata Y, Ishida R *et al.* (2001). Differentiating between progressive supranuclear palsy and corticobasal degeneration by perfusion SPET. *Nucl Med Commun*, **22**, 767–72.

Zhu XL, Chan MS, Poon WS (1997). Spontaneous intracranial hemorrhage: which patients need diagnostic cerebral angiography? A prospective study of 206 cases and review of the literature. *Stroke*, **28**, 1406–9.

CHAPTER 4

Psychologically determined disorders

Michael Donaghy and Martin Rossor

Contents

4.1 Introduction

4.1.1 Terminology

Neurologists frequently see patients with symptoms or inconsistent signs that are not explicable in terms of any recognized neurological disease process. Often it is clear that such symptoms and signs are being manufactured psychologically, either consciously or, more often, by an unconscious process. Such patients are frequently polysymptomatic, and may have a long history of consulting other specialists, particularly abdominal, dental, gynaecological, and otorhinological surgeons. They run the risk of developing secondary abnormalities induced by surgical and other invasive procedures.

Psychologically determined symptoms have been subject to various terminologies, some of which have become codified and pejorative. In reality, frequently one cannot distinguish between hysterical conversion, malingering, somatization, and distorted but otherwise physiological perceptions of the internal bodily state. Furthermore, separate medical and psychiatric classifications have arisen to describe similar conditions. The vast majority of patients with psychologically determined symptoms are seen in general practice and specialist medical settings, such as neurology. By contrast, psychiatry services tend to focus on a minority of such patients, particularly those with disabling and intractable disorders, those with unusual disorders, and those in whom there is coexisting or underlying psychiatric disease. The following terminologies are in general usage (Ron 1994; Sharpe 2002):

- *Medically unexplained symptoms.* This simple term has the advantage of being non-pejorative in that it does not carry any implication that the patient is inventing the disorder. It has two disadvantages. First, it can be interpreted as implying that the doctor is incompetent in having been unable to identify the disease process. Second, it does imply that psychological explanations are not medical in nature.

- *Somatization* is a tendency to express psychological distress in the form of physical symptoms, and to seek help from a physician rather than a psychiatrist. Used accurately, the term should be applied only to those in whom there is an identifiable underlying emotional disorder. That would immediately exclude those many patients in whom there is no such identifiable emotional disorder underlying their psychologically generated symptoms. Also, whilst many patients acknowledge that they have an associated emotional disorder, they usually believe that this results from the distress of their physical symptoms, rather than *vice versa*, citing the relative timings of occurrence as evidence.

- *Somatoform* is a term used by Diagnostic and Statistical Manual of Mental Disorders (DSM 2004), and International Classification of Diseases (ICD 1992). It has obvious parallels with somatization. However, it has a potentially broader utility in that it does not imply that the physical symptoms are an active product of an identifiable underlying emotional disorder.

- *Hysterical conversion disorder* as strictly defined, refers to a loss of function such as weakness, as representing the transformation of a psychologically traumatic experience, often of a sexual nature, into the neurological symptomatology. Yet it is rare in practice to be able to identify the psychological trauma which has acted as a template for the neurological deficit.

- *Functional.* This term has been particularly popular amongst neurologists. Originally it was used to signify that physical symptoms were due to a change in functioning of an otherwise normal nervous system. Its meaning is somewhat confusing because it does not distinguish between bodily symptoms that are a product of psychological processes as against those which are due to variations in bodily physiology. It has become a codified and often pejorative term implying that symptoms are 'all in the mind'.

- *Hypochondriasis.* This is primarily a psychiatric term to describe patients with severe and unjustified worry about underlying disease. Such patients present to neurologists when their hypochondriasis manifests wholly or partly as physical symptomatology. Also, it may present with a morbid anxiety state which misinterprets everyday bodily sensations as being due to a serious underlying disorder, commonly multiple sclerosis.

- *Malingering.* This term applies to that subgroup of patients who deliberately invent physical symptoms or simulate physical signs. The term should be reserved for those proven to be consciously fabricating a psychologically determined disorder. However that proof is diagnostically difficult, particularly in those who may be augmenting or embellishing a genuine physical disorder. When misapplied, the term becomes extremely pejorative.

- *Psychologically determined symptoms.* This term has the advantage of straightforwardly attributing the generation of symptoms and signs to psychological processes, whether they reflect the workings of the normal or the abnormal brain (Donaghy 2005). By accepting the phenomenon it is non-pejorative and does not seek to discriminate between conscious and unconscious processes. It does not carry the connotation that there is some underlying emotional disturbance, painful psychological conflict, or madness. It also provides an introductory substrate when explaining to patients that their symptoms reflect an inaccurate or distorted perception of their internal body state.

4.1.2 The history

The manner in which patients with psychologically generated disorders describe their symptoms may be discordant with the anticipated impact of such symptoms. Relatively trivial symptoms such as tingling may be described in a vivid, florid, and exaggerated manner. A totally paralysed limb may be described in a smiling and unconcerned manner. Or there may be a list of symptoms which cannot possibly be connected by a single pathological process. It is particularly difficult when psychologically determined symptoms occur in someone also suffering from a definite disease. Common examples include an exaggerated gait disorder in multiple sclerosis, pseudoseizures in an epileptic, or exaggerated disability in an injured person seeking compensation. Particularly in these settings, the symptoms often need full investigation before you can come to a definite conclusion that they originate psychologically. This is a difficult diagnostic area which sometimes misleads the most skilful diagnostician. Common psychologically determined symptoms include headache, facial pain, spinal pain, tinglings, patches of sensory loss, numbness, dropping things from the non-dominant hand, giddiness, tremors, blackouts, clumsiness, paralyses, memory blocks, dementia, and gait disorders.

The possibility of underlying psychiatric disease should be considered. Depression underlies many instances of tension headache, facial pain, and apparent dementia. Anxiety states may be responsible for tinglings, tremors, imbalance, blackouts, or memory blocks, and sometimes these symptoms may be mediated by hyperventilation. Typically, anxiety-provoked symptoms fluctuate markedly and may resolve for part of the day. For instance the tingling of a true polyneuropathy does not disappear for half of the day. Malingering takes a number of forms, not all necessarily fully conscious; self-delusion can be prominent. Many patients are seeking compensation for alleged personal injury in road traffic accidents or at the hands of doctors. These usually feign a disabling symptom such as paralysis or pain, which is difficult to disprove. Drug-addicted malingerers may feign intractable pain so as to obtain opiates. Mildly incapacitating symptoms which create a state of dependency can occur in domestic situations, for example in the 'empty-nest syndrome' in middle-aged women whose children are leaving home. True conversion hysteria is uncommon. Such patients seem to accept quite placidly that their paralysis or gait disorder is an incurable condition. Most are female and frequently have a long history of gynaecological and sexual symptoms, and of surgical operations (Ron 1994). Such patients can be permanently wheelchair bound, feature 'la belle indifférence', and frequently no treatment seems to help.

The art of analysing symptoms to decide whether they reflect biomedical disease, or whether they are psychologically manufactured involves attention to the following points (Donaghy 2004). First, a rigorous understanding of neuroanatomy and physiology provides the scientific reference point against which the neurologist can decide whether a patient's story resonates with known pathophysiological principles of biomedical disease. Second, the history should be taken as a story rather than a banal list of presenting symptoms so that one can judge whether the story's internal thread of direction is heading towards a coherent diagnosis. It is very characteristic of histories in those psychologically determined disorders that the story's internal thread is jumbled and wayward, without the clear internal tension that reaches towards a logical pathophysiological conclusion. Third, it is important to gauge the patient's attitudes and reactions to non-controversial matters during a period of small talk and humour before taking the history of the complaint, which may be recounted in a manner entirely at odds with other attitudes the patient has expressed. Fourth, mismatch may be evident between a patient's facial expressions and demeanour compared to the nature and severity of the symptoms and disabilities they are describing. Inconsistencies in the nature, timing, and severity of symptoms, and premature, over eager, or repetitive introduction of secondary gain concepts are noteworthy.

4.1.3 The clinical approach

Patients who bring an illness expressed in the form of symptoms hope the doctor will help resolve the illness either by diagnosing and treating an identifiable pathological disorder, or by providing reassurance that no such disorder is present. For the doctor to start with the attitude that patients with psychologically determined symptoms are wasting his time, and to be dismissive or unkind to such patients, represents a misunderstanding of his role. Such behaviour underlies many instances of a breakdown in the doctor–patient relationship. Over-reliance on a sequence of investigations to prove that there is nothing wrong runs the risk of convincing a patient that you believe they are harbouring disease, even when that is not the case. In such circumstances, kind and authoritative reassurance is much more effective than arranging yet another test.

Individuals vary considerably in their response to illness and in what being ill means to them and their families. Patients wish explanations so as to rationalise their feeling of being ill. Although biomedical scientific description of disease was expected to provide rational explanation of illness and to allow this explanation to connect logically and directly with cures and treatments, all too often science has simply dispersed attempts at all-embracing explanations into a series of subsidiary scientific questions, all equally unresolved. Doctors who try to explain symptoms in purely scientific terms to patients with psychologically determined disorders can be found particularly wanting. Blunt assertions that 'there is nothing wrong with you' usually do not help resolve the patient's concern. Some doctors make the mistake of avoiding discussing the symptoms in terms which are meaningful to the patient, and initiate multiple investigations or refer on to other specialists. Such prevarications can consolidate the notions that the illness has a serious cause, further fuelling anxiety, and engender distrust of doctors whose indecision is interpreted as incompetence. Sometimes the end result is entrenchment of disability.

4.2 Epidemiology

4.2.1 General occurrence

Psychologically determined symptoms are a very common cause of neurological out-patient referral. Retrospective analysis of nearly 8000 out-patient referrals to a single consultant neurologist showed that there was no specific diagnosis in 26.5 per cent, whilst a further 7.5 per cent had tension headache, 3.8 per cent conversion hysteria, 2 per cent hyperventilation, and 1.4 per cent depression (Perkin 1989). A prospective study of 300 new out-patient referrals showed that 11 per cent had symptoms not explicable by organic disease, and in only 43 per cent could the symptomatology be completely explained (Carson et al. 2000b). Of those with psychogenic disorders, pain is the commonest symptom, followed by motor symptoms, gait disturbances, dizziness, blackouts, sensory symptoms, and visual dysfunction (Lempert et al. 1990). Depressive and anxiety disorders are twice as common in those with psychologically determined disorders and those with emotional disorders have a greater number of somatic symptoms but are unenthusiastic about psychiatric treatment (Carson et al. 2000). Over half of patients presenting to neurologists with psychologically determined symptoms have not improved on review 8 months later (Carson et al. 2003). Psychologically determined disorders only lead to severe physical disability in a small proportion, usually those with motor symptoms, chronic fatigue, or blackouts. These patients show a high dependency on physical assistance and welfare benefits, and are relatively resistant to intensive psychiatric and rehabilitative treatments (Allanson et al. 2002). Always there is a danger for the doctor in mistakenly believing that the disorder is psychologically determined, when in reality there is an underlying neurological disease. Gait and movement disorders, and multiple sclerosis provide the particularly common pitfalls.

4.2.2 Outbreaks

Episodes of mass sociogenic illness presenting neurologically are well recognized, with two main types being distinguished (Bartholomew and Wessely 2002). Mass anxiety hysteria is of short duration, typically a day or so, and manifests as sudden extreme anxiety following the perception of a threat which proves false. Mass motor hysteria represents slow accumulation of stress in intolerable social settings with histrionics, dissociative behaviour, and psychomotor abnormalities such as shaking, twitching, and contractures which usually go on for weeks or months. Ever since the 15th century, such disorders have been particularly documented in strict and isolating religious institutions, and were often attributed to demonic possession. More recent examples tended to have occurred under oppressive, isolating, or stressful circumstances in schools and other institutions, at the workplace, or in those whose circumstances made them believe they were vulnerable to biological or chemical weapon exposure.

More contemporary neurological examples have included epidemic neuromyasthenia and an outbreak of a poliomyelitis-like illness in student nurses in a large psychiatric hospital at a time when the polio virus was still prevalent (Shelokov *et al.* 1957). A similar syndrome occurred at an English teaching hospital, with a more protean range of symptomatology including hyperventilation and sensory symptoms; interestingly female workers were 14 times as likely to be affected, and the longer term course was relatively benign (McEvedy and Beard 1970). In 1999 more than 1 in 15 students at a mixed school, exclusively female, developed pseudoseizures during morning prayers with an index case having occurred a few days beforehand (Mkize and Ndabeni 2002). Another experience of mass pseudoseizures in a school setting confirmed the preponderantly female sex of those affected, the considerable institutional disruption which occurred, the subsequent stigmatization of affected students, and the termination of the outbreak by removing the affected individuals (Roach and Langley 2004). A more chronic epidemic of hysterical blackouts in a large school proved difficult to detect because of the diverse medical services which were consulted, and led to some girls receiving antiepileptic treatments (Mohr and Bond 1982). The intentional production of mass hysteria, with physical symptoms based on the template of known disease, is one important aim of bioterrorist and chemoterrorist threats (Section 5.8). Mass hysteria is a well-recognized response to perceived toxin exposure (Section 4.2.7).

The following combination of features are characteristic of mass sociogenic illness: symptoms with no plausible organic basis; transient and benign symptoms; rapid onset and recovery; occurrence in segregated groups; presence of extraordinary anxiety; rapid communication of the symptomatology; spread from older or higher status students to those who are younger; and female preponderance (Bartholomew and Wessely 2002). No particular personality profile has been identified as vulnerable to mass sociogenic illness, apart from the particular occurrence in young females.

4.2.3 Occurrence in established neurological disease

Psychologically determined disorders pose particularly difficult problems of diagnosis and management when they occur in someone already suffering from confirmed disease. The following examples illustrate the almost infinite spectrum of the problem. A patient with definite multiple sclerosis may develop an exaggerated gait disorder and it is only by experience of recognizing truly spastic-ataxic gaits, or by documenting an absence of relevant physical signs, or by observing marked improvement when the patient thinks they are not being observed, that one can diagnose a superadded psychologically determined gait disorder. A flurry of seizures may occur in a previously well-controlled epileptic despite the anticonvulsant blood levels remaining satisfactory; it is only by elucidating a novel variability in the manifestations of the seizures, or by directly visualizing bizarre aspects of these new attacks, and by determining that they tend to occur in evocative situations, that one can determine that they are pseudoseizures. A patient with chronic neck and shoulder pain seeking to increase the quantum of compensation for a whiplash injury may complain of arm weakness. It is only by documenting the lack of signs of nerve root damage, and observing a collapsing pattern of arm muscle weakness often accompanied by theatrical grunting, that one can judge the arm weakness to be determined psychologically. Often it is not difficult to determine that a psychologically determined component is present. But usually it is well-nigh impossible to apportion the relative contributions to the overall level of disability of the disease itself and of the psychological contribution.

In these situations it is all too easy to believe that the superimposition of a psychologically determined disorder represents a conscious manoeuvre with the intent of secondary gain. Undoubtedly such embellishment of disability does occur, particularly in patients attempting to augment the quantum of settlement in personal injury litigation, or in those wishing to increase sympathy and dispensations from family, employers, or benefit agencies. However, one needs to consider also whether some forms of pathological disease affecting the brain might predispose patients to manifest a psychologically determined disorder. Structural disease of the forebrain carries the potential to impair the mechanisms whereby a patient perceives their own internal body state to be normal, or to experience the sense of well-being, or to adapt to emotional stress. Patients with undoubted multiple sclerosis sometimes develop incapacitating hysterical symptoms without any real secondary gain being obvious (Caplan and Nadelson 1980). Conversion hysteria has been observed in more than 30 per cent of patients with severe behaviour disorders after brain injury, and correlated particularly with diffuse brain insults, rather than with severity of injury or family or personal histories of hysterical or other psychiatric disorders (Eames 1992).

4.2.4 Precipitating events

Many seem to express their psychologically determined symptoms by reference to a template. The remarkable stereotypy of presentations in outbreaks of mass sociogenic disease, or in response to a perceived neurotoxic threat are striking examples. Physical or psychiatric illness in other family members may provoke psychologically determined symptoms in some children (Steinhausen *et al.* 1989). Pseudoseizures seem particularly likely in those who have epilepsy, or who have observed epileptic attacks in others. Medically unexplained stroke-like presentations are relatively common, representing 1.6 per cent of all stroke presentations, and tend to be associated with headache, other psychologically determined syndromes, and an absence of vertebrobasilar features (Nazir *et al.* 2005). Prevalence studies of multiple sclerosis in the pre-MRI era found that of the 10 per cent of patients who did not have multiple sclerosis or any other neurological disease, a third were nurses suddenly developing paralysis, generalized sensory disturbance, blindness, aphonia, or amnesia in a stressful situation (Hankey and Stewart-Wynne 1987). Hysterical paraplegia occurs in some patients who experience a spinal injury, either without spinal cord damage, or with incomplete spinal cord damage. Apart from down-going plantar responses, such patients were predominantly paraplegic rather than tetraplegic, had a high incidence of previous psychiatric illness or employment in healthcare professions, and were seeking compensation (Hall *et al.* 1985). Common to all these situations is the probability that a patient has modelled their disease on a template provided by a classmate's attack, another patient with an illness such as multiple sclerosis, or knowledge about the disease from occupation, books, journalism, or the internet, or on their own previous experience of disease. Under stressful circumstances these provide templates for a psychologically determined disorder, be it conscious or unconscious.

4.2.5 Risk factors

Neurological presentations of psychologically determined disorders are unusual before the age of 6 years, and are equally frequent in boys and girls until adolescence, after which females predominate (Ron 1994). Psychologically determined symptoms are commoner in those of low socioeconomic status, probably becoming replaced by depression as communities become more educated and affluent (Escobar and Canino 1989; Nandi *et al.* 1992).

An associated psychiatric disorder seems to be less of a risk factor for psychologically determined disorders in children and adolescents than in adults. Depression is the particular psychiatric association, with a lifetime prevalence of 50 per cent in patients with psychologically determined somatic symptoms; by contrast the lifetime prevalence of panic disorder is 20 per cent (Tomasson *et al.* 1991). A significant depressive illness may be less likely in those with narrowly defined conversion disorder than in the broader group of those with psychologically determined symptoms (Ron 1994). In a prospective cohort study, 47 per cent of new neurology out-patient referrals met formal criteria for anxiety and depressive disorders, with major depression in 26 per cent and strong association of these disorders with higher levels of disability and pain, and greater numbers of somatic symptoms (Carson *et al.* 2000).

4.2.6 Stress and cognitive overload

Information is essential to make decisions but too much information can be disabling. Alvin Toffler introduced the term 'information overload' in his book Future Shock, published in 1970, and this information explosion has accelerated with 24-h television, emails, mobile phones, and blackberries. The demand to keep up, particularly in the office environment, requires multi-tasking, a major cognitive challenge. In general, multi-tasking can only be performed if one of the tasks is automatic such as walking. If a more complex task is demanded then efficiency declines as exemplified by the increased risk of road traffic accidents when driving whilst using a mobile phone. More complex tasks can be handled by rapid task shifting but at an efficiency cost. Modern cognitive demands can commonly lead to anxiety and, with a sense of failure, to the emergence of symptoms of depression. Individuals whose employment is intellectually demanding may often perceive themselves to be failing cognitively when faced with increased informational loads and seek help by requesting investigation. The challenge is that on one hand the accompanying anxiety can blunt performance on cognitive assessment, whilst on the other the 'cognitive stress test' of modern life can unmask an early cognitive disorder. A careful history of work performance, in particular of concerns expressed by colleagues, is important. Often one discovers that the difficulties are only perceived by the patient himself. Nevertheless, it can be difficult to exclude an early degenerative disorder, particularly in the older patient and often follow up will be necessary.

4.2.7 Pseudoneurotoxic disorders

Proven neurotoxic disease, although relatively uncommon, is distinguishable by the distinctive nature of the neurological syndrome, coupled with clear and substantial exposure to a plausible toxin. By contrast, pseudoneurotoxic illness is not associated with a known or plausible neurotoxin, and tends to occur in three settings (Schaumburg and Albers 2005). First, the new onset of a naturally occurring nervous system disease may be mistakenly ascribed to simultaneous exposure to a supposed toxin. Second, worsening of a pre-existing neurological or neuropsychological disorder may be ascribed to simultaneous exposure to a supposed toxin. And

third, an obviously psychogenic illness may follow exposure to a supposed toxin. To varying degrees, the views of physicians non-expert in the field may reinforce patients' beliefs that their symptoms are toxin-induced. Some patients develop long-term disability, underpinned by an entrenched belief that they have been damaged by a toxin, and seek legal redress.

One particularly well-studied form of pseudoneurotoxic illness has followed mass exposures in institutions to substances, usually odorous, which become regarded as toxic following symptoms in an index case. What follows is a form of mass sociogenic illness (see Section 4.2.2). A rapidly enlarging cluster of cases will be followed by demonstration that there has been no significant toxin leak, and that toxicological examination of bodily fluids is negative. The beliefs of others, such as parents, can spread the notion of toxic exposure (Philen *et al.* 1989). Such supposed inhalational exposures tend to produce a characteristic symptom complex including dyspnoea, dizziness, headache, nausea, abdominal pains, visual blurring, and limb weakness (Modan *et al.* 1983; Jones *et al.* 2000). Suspicion of a pseudoneurotoxic illness will be provided by the rapid multiplication of cases following an index occurrence whose illness is attributed to a smelly substance. As with mass sociogenic illness, the majority of patients tend to be young females. Careful clinical observation may show helpful mismatches, such as a patient in marked respiratory distress who is capable of interrupting a dyspnoea to allow normal conversation (Taylor and Werbicki 1993).

The 21st century is already spawning a new form of pseudoneurotoxic mass sociogenic illness in the form of over-exaggerated response to real or perceived bio- or chemical terrorist threats (Bartholomew and Wessely 2002). It is possible that rapid dissemination of such threats by journalism and broadcasting may induce psychologically determined disorders in a broader demographic group than the predominance of young females traditionally affected by mass sociogenic illness.

4.3 Psychophysical basis

4.3.1 The nature of symptoms

Any symptom is simply a perception of some aspect of one's internal body state (Donaghy 2004). It is self-evident there is no such thing as a stable psychophysical threshold for a bodily sensation below which nobody notices anything, and above which everybody experiences identical symptoms. Soldiers injured on a battlefield may feel no pain whatsoever initially; others in civilian life may show histrionic reactions to trivially painful trauma; most people lie somewhere in the midst of this spectrum. Most doctors are familiar with the power of suggestion to produce awareness of previously imperceptible or unnoticed bodily sensations. In students of neurology, a morbid awareness of benign calf muscle fasciculations frequently follows the study of motor neurone disease. It is clear that the perception of one's internal body state occupies a continuum, with the brain having the ability to shift that threshold beyond which some perturbation may be perceived as being abnormal.

The ability of doctors to decipher symptoms is further complicated by the fact that many patients do not have a well-developed vocabulary for describing the state of their body and its functioning. For instance terms such as 'numbness', 'weakness', or 'clumsiness' may be used in an interchangeable sense to describe a wide variety of abnormal feelings in, weakness of, or difficulty in using a limb.

In addition some patients seem to employ collapsing patterns of power production, or wincing and eye deviations during power testing, as non-verbal means of communicating a perceived difficulty with a limb. It can be particularly difficult for doctor and patient to establish common currency in the meaning of terms within the spectrum used to describe muscular function: fatigue, energy, weakness, tiredness, heaviness, lethargy, and listlessness. Yet most people are acutely aware of day-to-day fluctuations in their sense of well-being, energy, and muscular performance, even though it is difficult to communicate these bodily perceptions to others. Most will recognize variations in their mental agility, as the contrast between days when a sluggish mind can never find the right word, and days when the right word springs effortlessly to mind; yet most find it difficult to describe these subjective changes in mental agility.

4.3.2 Hypnotic states

A close relationship between hysteria and hypnosis was realized in the late 19th century by Charcot and Freud (Raz and Shapiro 2002). At the beginning of the 21st century, functional imaging has pointed to common neurological processes in the prefrontal regions underlying both states. In the context of hysteria, it is important to consider hypnosis for three reasons. First, it may illuminate mechanisms whereby a patient entrenches beliefs about the state of their body. Second, experimental paradigms utilizing hypnosis are used in investigation of the neural mechanisms underlying hysterical conversion. Third, hypnosis offers a potential therapy for eliminating hysterical conversion symptoms by its use to normalize a patient's beliefs about the state of their body.

Hypnosis may be defined as an altered state of consciousness involving attentive receptive concentration leading to increased suggestibility and focussed attention (Raz and Shapiro 2002). Hypnosis does not involve a process akin to sleep, and mysterious trance states are not a central component of hypnosis. Hypnotists can alter a subject's voluntary control over behaviour, particularly by stimulating or inhibiting motor acts independently of volition. Sensory reception may be similarly affected, for instance the perception of colour or pain. The ability of hypnosis to alter volitional motor control or conscious perception carries obvious parallels to the state of patients exhibiting conversion hysteria. Taken more broadly, complex processes resembling hypnosis may underlie sporting focus, or the elation of study, prayer, or meditation, all states produced largely by the individual concerned. Similarly hypnosis may be akin to the process of external manipulation of an individual's belief systems by inspirational and charismatic individuals. These considerations provide a paradigm for understanding how a patient's belief about the state of their body could derive similarly from external influences, or from internal psychological processes. They may also explain the phenomenon whereby a patient's perception and control of their body may transmute from consciousness to become embedded in unconsciousness.

Activity in the anterior cingulate cortex of the prefrontal lobe seems to provide a pivotal connection between the brain mechanisms involved in actual perceptions and actions, hypnotic states, and hysterical conversion. By comparison such activity is absent when the same perceptions and movements are being feigned. This cortical region is generally regarded as adjusting the interaction between cognition, sensory perception, and motor control with respect to how these are influenced by altered attentional, motivational, or emotional states (Devinsky *et al.* 1995). Hypnotically induced and nociceptive pain are experienced by activation of a similar network of cortical areas which include the anterior cingulate cortex and insula in addition to the somatosensory cortices; by contrast, imagined pain produces little activation particularly in the anterior cingulate and insula regions (Derbyshire *et al.* 2004). Unperceived stimuli in hysterical anaesthesia activate the rostral anterior cingulate cortex (Mailis-Gagnon *et al.* 2003). Anticipation of pain in normal subjects decreases activation in the anterior ventral cingulate cortex suggesting a top-down modulation of cortical systems for pain perception by the direct influence of cognitive factors (Porro *et al.* 2002). Using hypnosis to reduce pain perception points to a critical role for the anterior cingulate cortex in modifying the cortical networks involved in nociception (Faymonville *et al.* 2003). Study of the motor system by functional imaging shows similar discrepancies between activation patterns associated with hysterical and hypnotic paralysis on the one hand, compared to imagined paralysis on the other (Marshall *et al.* 1997; Halligan *et al.* 2000a; Vuilleumier *et al.* 2001; Ward *et al.* 2003) (Fig. 4.1). In particular, the orbito-frontal cortex seems to play a prominent role in modulating activity in the subcortical motor systems, including the caudate nucleus. This raises the potential for both hysterical and hypnotic alterations of motor output to be gating the activity of subcortical motor control systems in similar manners.

Hypnotism was explored as a treatment for hysterical paralysis by Charcot and Freud in the late 19th century (Koehler 2003). Yet the overall therapeutic value of hypnosis remains unclear with its principal role being the modulation of pain perception. The cognitive modulation of pain by attention and emotion is well established, with the anterior cingulate cortex playing an important modulating role (Villemure and Bushnell 2002). The anti-nociceptive effect of hypnosis involves anterior cingulate cortex activation (Faymonville *et al.* 2003). However, when hypnosis was added to standard intensive management of hysterical motor symptoms, there was only a small additional treatment effect (Moene *et al.* 2002). In general there seem to be two limitations on therapeutic use of hypnosis in treating conversion symptoms. First, the duration of benefit does not generally outlast the period of hypnosis. Second, therapeutic hypnosis is very much directed at the goal of modulating a specific symptom, rather than addressing the more complex psychological processes which are generally regarded as the driver of a hysterical conversion. Indeed, it may be that those doctors and other therapists skilled in the art of healing may have developed an ability to influence patients' belief systems by a more effective holistic, rather than a purely goal-oriented, approach.

4.3.3 Altering the motor output

It is well recognized that the motor output may be substantially modified in altered states of consciousness and by emotions. Whilst the anatomical levels are not known at which the motor output may be gated, the clear existence of this phenomenon is a potential physiological substrate for neural influence by hysterical conversion. Examples of the motor output being inhibited occur in sleep paralysis, cataplexy, and cryptogenic drop attacks. Sleep paralysis (Section 32.4.3) can occur whilst falling asleep or awakening, and involves complete inability to move the limbs despite clear conscious will to do so. This probably represents a spill-over into wakefulness of the inhibition of motor output which prevents one from enacting dreams during rapid eye movement sleep.

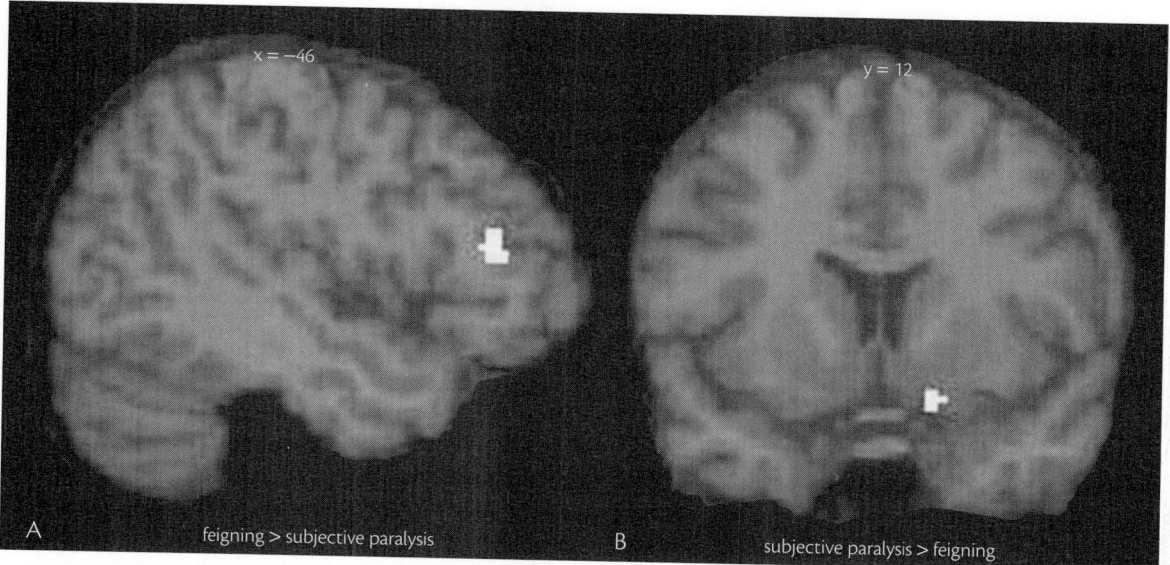

Fig. 4.1 Psychologically determined weakness. Relative differences in brain activity as measured with positron emission tomography whilst attempting to lift the left leg up from supine during two different brain states. During 'feigning', subjects were asked to deceive an examining doctor by pretending that their leg was paralysed. During 'subjective paralysis', subjects underwent hypnotic induction and were told that their leg was temporarily paralysed thus creating the feeling of subjectively real paralysis. A. Feigning (compared to subjective paralysis) was associated with a relative increase in activity in ventrolateral prefrontal cortex, an area involved in conscious volitional inhibition of incipient motor responses. B. Subjective paralysis (compared to feigning) was associated with a relative increase in activity in orbitofrontal cortex and adjacent ventral striatum, areas associated with the interaction between emotion, somatic representations of body state, and volitional decision making. (Courtesy of Dr N. Ward.)

Cataplexy (Section 32.3.1) is an abrupt loss of muscle tone triggered by emotion, shock, or fatigue and can vary in severity from sagging of the facial muscles to falling to the ground. Cryptogenic drop attacks (Section 2.6.5) consist of momentary losses of postural tone whilst standing, without loss of awareness or consciousness, during which the patient is fully aware of falling to the ground. Presumably all such conditions involve aberrant activation of the physiological switch which turns off motor output. By contrast, there are conditions in which the motor output is activated when it would normally be inhibited. Sleepwalking, or somnambulism (Section 32.4.2), is a well-known example which occurs during deep slow wave sleep from which it is difficult to awaken the individual, despite them walking around or exhibiting other purposeful motor behaviours. All these phenomena show the existence of physiological substrates whereby the motor output might be modified during hysterical conversion.

4.3.4 Hyperventilation

The hyperventilation syndrome describes a symptom complex arising from hypocapnea when patients hyperventilate in response to psychological factors, particularly anxiety. Because physical manifestations predominate, the underlying psychological cause may be ignored. Certain emotional states are strongly associated with chronic hyperventilation syndrome, particularly panic and phobic disorders (Halligan *et al.* 2000b). Hyperventilation blows off carbon dioxide, producing a respiratory alkalosis with reduced ionized blood calcium. This increases nerve excitability, resulting in the well-known Chvostek's sign of twitching of the corner of the mouth on percussing the facial nerve over the mandible, or Trousseau's carpo-pedal spasm of the hand. Common neurological symptoms include dizziness, detachment, blurred vision, tinglings of the fingers, perioral tinglings, muscle spasms, and fatigue. Altered consciousness, sometimes misdiagnosed as epilepsy (Section 31.1), is common, but close questioning frequently reveals the duration of altered consciousness to be more prolonged than in discrete epilepsy attacks, that the patient drifts in and out of semi-consciousness during the attack, and that any limb shaking involves a mild quivering rather than a true tonic-clonic convulsion. Paradoxically, patients frequently complain of breathlessness during hyperventilation attacks. Therapeutically, it can be helpful to voluntarily hyperventilate patients in the consulting room, so as to demonstrate that this can generate their symptoms, and then to demonstrate resolution of these symptoms by re-breathing from a paper bag to correct the respiratory alkalosis.

Unilateral somatosensory symptoms, often left-sided, can be the presentation of hyperventilation (O'Sullivan *et al.* 1992). Such patients are predominantly women with a high incidence of preceding stressful life events. This unilateral occurrence of paraesthesiae is not associated with differences in somatosensory evoked potentials between the two sides. It is interpreted as representing a central mechanism reflecting involvement of the right hemisphere in stress and emotional arousal. Hyperventilation is an example whereby emotional or psychiatric disorders may alter behaviour so as to perturb bodily physiology with the resultant production of neurological symptoms.

4.3.5 Autonomic overactivity

The autonomic hyperactivity associated with anxiety states represents another situation in which a psychological condition can precipitate physical symptomatology. Anxiety is the usual underlying factor. The resultant neurological symptoms consist of varying combinations of tinnitus, a sense of blurred vision, non-specific dizziness without true vertigo, and prickling skin sensations. These are often associated with other systemic autonomic symptoms such as palpitations, frequency of micturition, feeling of difficulty in filling the chest with air, and irritable bowel syndrome. Also, patients may notice poor concentration, irritability, noise sensitivity, muscle aching, disturbed sleep, night terrors, and a feeling that sleep is not refreshing. Diagnostic difficulty may arise when patients present with only one, or a limited spectrum of such symptoms rather than a broad variety.

4.4 Clinical syndromes

4.4.1 Briquet's syndrome

Whereas many patients complain of a relatively circumscribed set of psychologically determined symptoms (Section 4.7), with somatization disorder or Briquet's syndrome, there are chronic multiple complaints relating to multiple organ systems. Patients, usually female, acquire voluminous sets of hospital notes involving many different departments. The healthcare costs of this group of patients are substantial (Smith *et al.* 1986). The prognosis is poor but good communication between primary care and hospital departments can help limit over-investigation. The clinician, however, must remain vigilant as patients with Briquet's syndrome are as entitled to develop a neurological disease as any other patient. Cognitive behavioural therapy provided additional benefit to standard care in one randomized trial (Allen *et al.* 2006).

4.4.2 Hypochondriasis

Hypochondriasis, a morbid fear of suffering from a serious disease, is associated with an abnormal interpretation of normal phenomena. This contrasts with Briquet's syndrome where patients seek diagnoses but are usually less fearful, although the two disorders can overlap (Leiborand *et al.* 2000). Some neurological diseases appear frequently in the litany of hypochondriachal concerns such as brain tumours and multiple sclerosis. Explanation of symptoms, and the normal investigations which are sometimes necessary, may provide short-lived reassurance. As with Briquet's syndrome, good communication between primary and secondary care doctors is important in management.

4.4.3 Conversion disorder

A central feature to the diagnosis of conversion disorder or hysteria is that the somatic symptoms and signs develop in the absence of a neurological illness that is sufficient to explain them in their entirety. An associated psychological stressor should be identifiable, but this may depend upon the interpretation of the examining doctor. The negative, as opposed to positive, feature of the diagnosis, for instance the absence of a neurological disease, has always been a major concern as it may only reflect current conceptual thinking and available technology for investigation. An influential study conducted over 40 years ago reviewed patients diagnosed with hysteria and found at follow up that 50 per cent had developed a significant neurological or psychiatric diagnosis (Slater 1965). More recent follow-up studies suggest a much lower emergence of alternative diagnoses (Crimlisk *et al.* 1998) (Section 4.9.5). This change reflects both improved diagnostic techniques and improved understanding of a number of disorders such as spastic dysphonia and writer's cramp which had previously been considered within the conversion hysteria group and, are now firmly established as neurological diseases.

Are there positive as opposed to merely negative features to the diagnosis of a conversion syndrome? The most common manifestations involve the motor system, particularly weakness; the next most common are sensory. Examination will often reveal clear inconsistencies such as co-contraction of antagonist and agonist muscles or sensory deficits which are inconsistent such as loss of vibration sense on either side of the sternal midline. 'Belle indifference' was often said to be a key feature but has not been found to be of predictive value (Sharma and Chaturvedi 1995). A frequent observation has been that conversion symptoms are more common on the left than the right with the suggestion that the underlying pathophysiology may involve disturbed attentional mechanisms. In this regard, there would be some overlap with the anosagnosia that can arise with right parietal lesions. Positron emission tomography in one patient with a hysterical hemiplegia revealed that on attempts to move the paralysed limb, pre-motor areas known to be involved in movement preparation and execution were activated, but not the primary motor cortex. By contrast, significant activation in orbital frontal and anterior cingular cortices suggested an inhibitory effect of these regions (Marshall *et al.* 1997) (Section 4.3.2).

Disorders of cognition present a particularly difficult challenge; amnesic symptoms on either an hysterical or a malingering basis are not uncommon. The *Ganser syndrome* describes patients with apparent gross disturbance of cognition who appear to be unable to answer even the simplest of questions. They often give approximate answers or 'answer past the point', Vorbeireden or Vorbeigehen, examples being 'five' in answer to the question, 'how many legs does a cow have?' Although often considered to be related to malingering, some cases are found following head trauma and many relate to a post-traumatic stress disorder (Dalfen and Feistein 2000). In patients with functional retrograde amnesia, there may be a sharp cut-off chronologically (Schacter *et al.* 1982) and memory testing is very variable (Kopelman 1987).

Psychiatric comorbidity is common with conversion disorder and in particular, depression. It is essential that this is recognized as not only is the treatment of the depression associated with improved prognosis but failure to recognize leads to significant suicide risk. Management of patients requires close collaboration with a psychiatrist experienced in this area.

4.4.4 Body dysmorphic disorder

Body dysmorphic disorder, previously known as dysmorphophobia, is a striking disorder of disturbed body image in which patients become preoccupied by imaginary or minor physical variations. They perceive themselves as ugly, shun society, and will spend at least an hour a day ruminating over their appearance.

Although classified under the somatization disorders (DSM 2004), there are many similarities to obsessive compulsive disorder. Patients will often spend hours in front of the mirror, or avoid mirrors, and will often have elaborate grooming practices or spend time touching the perceived deformity. The face is most commonly affected. Often patients are seen by plastic surgeons or by dermatologists as there is overlap with parasitosis and trichotillomania. However, neurologists do occasionally see patients with this disorder when it can be linked with perceived sensory disturbances, for example, perception that the face is asymmetrical or getting larger on one side and may feel different.

The disorder tends to come on in adolescence and to be chronic with a low remission rate. Patients are often treated with a selective serotonin reuptake inhibitor antidepressant. The suicide rate is high with an annual reported attempt rate of 2.6 per cent in a prospective study, with at least half of individuals reporting suicidal ideation (Phillips and Menard 2006).

4.4.5 Pain disorder

Pain disorder was previously referred to as psychogenic pain or somatoform pain disorder and is classified within the somatoform disorders (DSM 2004). The diagnosis is made when the primary complaint is pain, with major psychological features and the pain cannot be entirely accounted for by a medical condition. The pain dominates the clinical picture, is often associated with helplessness, insomnia, and fatigue and disrupts social activity. Abdominal pain is common in younger patients but the complaints most likely to present to neurologists are those of back pain, headache, face pain, and limb pain. Clearly, however, the diagnosis is difficult since a judgement has to be made about the relative contribution of an underlying medical trigger. Whilst neurologists may become reasonably adept at assessing the contribution to a complaint such as weakness that might be made by a nerve lesion, pain is far more subjective and underlying triggers may remain elusive. Many of the associated emotional concerns may arise from the lack of a clear explanation for the pain and from feelings of being misunderstood or dismissed. As our understanding of the pathophysiology of pain improves, a number of patients who might have been dismissed previously will now have an explanation for their symptoms. For example, paroxysmal extreme pain disorder, a familial condition characterized by bouts of facial, rectal, and limb pain, is now linked to mutations in the gene coding for one of the sodium channels (Fertleman et al. 2006).

4.5 Malingering and illness deception

4.5.1 Occurrence

Not all psychologically determined symptoms are generated unconsciously. Malingering is the process of deliberate deceit, by which symptoms are fabricated. Thus, the entirety of a complex of symptoms and signs may be generated entirely voluntarily. Or as commonly occurs in personal injury claims the overall severity of the complaint may be deliberately augmented. Deception is a common and ingrained human behaviour, often high risk but with the intention of some secondary gain. In the medical context, the feigning of illness is principally evident in claims for alleged personal injury, in attempts to access welfare benefits, as an attempt to obtain earlier retirement benefits, and so as to avoid unwanted military duties. Wessely argues from an historical perspective that malingering

moved into the medical sphere in the early 20th century as a result of permissive social legislation in Bismarkian Germany and Britain (Nimnuan et al. 2000). This allowed claims of dispensations, financial benefits, and avoidance of conscription.

Neurological symptoms are particularly common manifestations of malingering, given that they can fundamentally impair an individual's ability to operate independently in society. Yet their true incidence and prevalence is totally unknown. Indeed, given that they are generated privately by the will of an individual, and given that they may manifest similarly to other unconsciously psychologically determined disorders, it is doubtful that their epidemiology can ever be reliably described. From a practical point of view, the existence of malingering must be considered when symptoms and signs capable of being generated psychologically are encountered in certain situations. Personal injury claims after road traffic and industrial accidents frequently centre around pain states, the existence or severity of which may be impossible to gauge. Many deliberate attempts to access welfare benefits and early retirement seem to involve the augmentation of an existing neurological disorder. Indeed, it is not uncommon for patients with existing neurological disorders, sometimes of trivial severity, to use these as a template for disability. At the softer, non-fiscal end of malingering, the reasons for, and the detection of malingering become much more difficult. Some patients may overemphasize or embellish their symptomatology simply to obtain sympathetic and comprehensive attention from a doctor. In other contexts, symptoms engendered consciously may confer social benefits such as help and sympathy from dependents, friends, and work colleagues. The great difficulty at this softer end of the spectrum lies not so much in determining whether symptoms are being consciously generated, but in recognizing that this may be a legitimate exercise of free will by which a patient wishes to communicate their sense of ill health to doctors and society.

To diagnose malingering has become something of a no go area for doctors. The predominantly scientific conceptual framework of their training renders them unconfident in making this diagnosis. Additionally they fear the opprobrium of disgruntled patients and the undermining of their professional position by complaints and legal procedures. Furthermore, the majority of medical certifications and dispensations are aimed against faceless agencies such as the state, and insurance or pensions companies; little surprise therefore that doctors tend to give the benefit of considerable doubt to the patient sitting in front of them. It is only in the context of the medical evidence underlying personal injury litigation that there are clear demands to define whether malingering is occurring, and if so, its contribution to the disability in question. However, this area is also complicated for two reasons. First, patients who have suffered actual or perceived personal injury may become quite legitimately secondarily depressed or anxious, with those psychiatric states being independent drivers of their psychologically determined symptomatology. Second, it appears that symptoms which were initially generated consciously, may provide the substrate for a slow alteration in a patient's unconscious beliefs about the state of their body, and the expectations they have from it. This process akin to self-delusion or self-hypnosis may lead to symptoms becoming generated unconsciously, despite having been feigned initially.

Denial of severity, or deliberate cleansing of a symptom complex, is a distinctive form of malingering, particularly seen in neurological practice. A very typical example of such reverse malingering occurs in patients with a clear cut history of a seizure disorder. On hearing of a ban from driving, such patients research the clinical features of seizures and request a second opinion. Then they present to the new neurologist with a modified history expunged of those characteristics upon which confident diagnosis of a seizure disorder was initially based. In such circumstances, it is the initial history, openly obtained, upon which the ultimate diagnosis should be based. Preservation of their driving licence is the main stimulus to such reverse malingering, and often involves overoptimistic personal assessment of limb functioning by patients. Patients with cognitive disorders may also become unsafe to drive, but deny this strenuously. The influence of cerebral disease upon their capacity for personal insight and openness to reason are particularly important considerations.

4.5.2 Neural basis

In an experimental setting using functional brain imaging, deception is revealed as an executive task associated with prefrontal activation (Spence et al. 2003). Lying answers to questions are associated with prolonged response times and increased activity bilaterally in the ventrolateral prefrontal cortices. The absolute delay of approximately 200 ms between lying compared to truthful responses has an obvious correlate with the delays in answering questions, or in execution of examination tasks, which are exhibited by malingerers in the clinic. The notion that a frontal area, the ventrolateral prefrontal cortex, has developed an important role in generating lies or withholding the truth is testimony to the importance of lying and deceit within our repertoire of advanced social behaviours. It begs the question as to whether certain brain diseases, particularly those affecting this prefrontal region, might impair ability to lie and deceive. Different brain mechanisms seem to be involved in different paradigms of deception. The anterior cingulate cortex was active, rather than the ventrolateral prefrontal, when subjects had to indicate the opposite of truth; inhibition of the truthful response may be fundamental to intentional deception (Langleben et al. 2002). Yet other prefrontal domains are activated in feigning an autobiographical memory impairment (Reddy et al. 2002).

Comparison of the brain activity in malingering with that in hypnosis has not been explored in detail. Feigned paralysis leads to a relative increase in activation of the ventrolateral prefrontal cortex when compared to subjectively real paralysis, interpreted both as inhibiting the motor act, and in learning an altered association between the stimulus and whether to move (Halligan et al. 2000a; Ward et al. 2003). Because of the difficulty of setting up specific and comparable experimental paradigms, and the remaining uncertainty as to whether a specific executive 'deceiving' frontal area has been identified, it means we are far from using brain imaging to diagnose malingering in everyday clinical practice.

4.5.3 Distinctive clinical features

It is unusual for a single observation during neurological examination to provide clear proof of malingering. More often, the evidence is based on an accumulation of observations during the history taking and examination.

During history taking, malingerers may be inconsistent about the nature, timing, and severity of symptoms. Collateral information from observers may be revealing. Premature, overeager, or repetitive introduction of secondary gain concepts are often noteworthy. Unusual delay in answering can occur, particularly when faced with a question addressing an unrehearsed aspect of the symptoms. Sometimes patients with malingering concentrate on a specific aspect of their disorder, often pain, to preclude discussion of other symptoms which would be expected in association. Other patients with malingering may exhibit the typical general characteristics of psychologically generated disorders, proffering symptom complexes which cannot easily be related to any known pathophysiological disease, or lists of symptoms between which no common pathophysiological thread can be established. Finally, it is common to note mismatch between the patient's facial expressions and demeanour compared to the severity of the symptoms and disabilities being described.

Motor examination often reveals inconsistency in the use of the affected area of the body under different circumstances, particularly when seemingly unobserved. For instance, the patient may be unable to flex the hip against gravity on the couch, but will be observed to stand and flex the hip normally to put on trousers. A patient with weakness of plantar flexion when tested on the couch may be able to stand on tiptoe. Voluntarily generated weakness often involves an enthusiastic collapse of a limb held out strongly until pressure is exerted. Particularly revealing is when this collapse occurs on testing shoulder abduction, with the collapse occurring primarily not from the shoulder abductors, but from a trunk which tilts sideways. The collapsing movement may be associated with copious and theatrical grunting and sighing, and sometimes with eye and head movements towards the collapsing limb, a marker of intentionality.

Voluntarily generated gait disorders tend to show major inconsistencies when the gait is observed at different times. On testing in the clinic the patient may show heavy reliance on a stick, or traverse an open space successfully only to fall and grasp either furniture or the doctor. By contrast, when the patient is leaving the hospital, seemingly unobserved, or at chance encounter in an outside setting, they may be seen to be simply carrying the stick without making use of it, and to be walking either normally, or with a markedly lesser degree of disability.

Skin sensory testing will have the non-anatomical, and implausibly sharply defined boundaries seen in any form of psychologically determined disturbance. The boundaries of the sensory loss may demonstrably shift in location when the stimulus is presented at different speeds, or from different directions, particularly when the patient's eyes are shut. Occasionally one can demonstrate conclusively that the numb limb changes to the other side of the body when the patient is rolled over to compare testing supine and prone. On superficial sensory testing, occasional patients are encountered who are equally adept at answering 'no' when the stimulus is presented in their 'numb' zone as when it is presented in the normal 'yes' areas of skin. Despite adequate training in the task with the eyes open, some patients get joint position sense testing 100 per cent wrong, which cannot occur by chance. Sometimes patients feigning pain wince and withdraw at the merest skin contact, or occasionally at the moment before contact even occurs, which can be interpreted as a marker of intentionality.

Although the intentionally generated physical signs are often bizarre, it should be remembered that the diagnostic conclusion depends upon recording the phenomena calmly. It is generally counterproductive to enter into a battle of wits with the patient, both in terms of trying to help with their perceived complaint, and in terms of being able to achieve a dispassionate analysis of their clinical features.

4.5.4 Factitious disorder

This is relatively uncommonly encountered in neurological practice. The intentional induction of disease by self-inflicted physical injury or self-administered toxins or medication is often not associated with any notion of a secondary external reward. The rare neurological examples include muscular weakness due to administration of heavy metals, such as lead or mercury (Palmer 2003). Also, there is the well-known example of insulin self-administration to cause coma or altered consciousness, often occurring in patients with an occupational background in healthcare.

4.5.5 Munchausen's syndrome

This descriptive term was coined by Asher (1951) for the patient who is admitted to hospital with apparent acute illness supported by a plausible or dramatic history. Usually his story is largely made up of falsehoods; he is found to have attended and deceived an astounding number of hospitals; and he nearly always discharged himself against advice after quarrelling violently with both doctors and nurses. A large number of scars is particularly characteristic of this condition. The most common setting in which neurologists may encounter this malingering disorder is in a patient complaining of severe intractable pain, only relievable by opiate injections.

4.6 Associated psychiatric states

4.6.1 Anxiety

Anxiety is a ubiquitous emotion in response to perceived threat. The combination of fear, apprehension, worry, and accompanying autonomic features (Section 4.3.5) is familiar to all. The perceived threat of a neurological illness means that anxiety in some degree is common when patients first present to the clinic. Anxiety can, however, become maladaptive. Five major types of anxiety disorder are recognized: generalized anxiety disorder; obsessive compulsive disorder; panic disorder; post-traumatic stress disorder; and social anxiety disorder. Patients with generalized anxiety disorder may present to the neurologist if they are focussed on the fear of a neurological disorder. A history of a more pervasive list of anxieties is important as generalized anxiety disorder is commonly often unrecognized but amenable to cognitive behavioural therapy and selective serotonin reuptake inhibitors. Panic disorder is most likely to enter the differential diagnosis of a seizure disorder (Section 31.1.3). Post-traumatic stress disorder is also common and if the history of a traumatic event is not specifically sought can create difficulties, especially since it can present with a delayed onset even years after the event. Anxiety often coexists with depression.

4.6.2 Depression

Depression is very common in patients presenting to neurologists and commonly underpins, or drives psychologically determined neurological symptoms. Some of the symptoms of depression, such as headache, poor sleep, memory difficulties, and even motor slowing, can in themselves result in an initial referral to a neurologist. Depression is also a common reaction to the diagnosis of a disorder which may be chronic, difficult to treat, or life threatening.

In addition, some neurological disorders are particularly associated with depression suggesting a shared biological substrate. Approximately half of patients with stroke fulfil criteria for depression at 3 and 12 months, although slightly less in units with active rehabilitation programmes (Kotila *et al.* 1998). Although there have been reports of a particular association between left hemisphere, and particularly left frontal, strokes and depression this has not been confirmed in a systematic review (Carson *et al.* 2000a). Depression is the most frequent mood disorder in epilepsy with prevalence rates from 20 to 50 per cent. It is most frequently associated with temporal lobe epilepsy with a 50 per cent prevalence compared with only 16 per cent in primary generalized seizures suggesting that it is not merely the reaction to the diagnosis (Perini *et al.* 1996). Depression is also common in multiple sclerosis, perhaps as frequently as in 50 per cent but without any clear relationship to lesion load (Ron and Logsdail 1989).

Degenerative diseases are also associated with high prevalence of depression of about 20–25 per cent in both Parkinson's disease and Alzheimer's disease, depression often preceding the development of the degenerative disorder (Kanner 2005; Rickards 2005).

4.6.3 Schizophrenia

Schizophrenia has had a colourful history with an uncertain nosological status. The cause is multifactorial but with increasing evidence that this is a neurodevelopmental disorder producing disturbance of neurotransmission. It is characterized by impairments in the perception of reality and is often associated with profound social dysfunction. Positive symptoms are characterized by hallucinations, particularly third person auditory hallucinations, disordered thinking and speech, and delusions. Negative symptoms include flattening of affect, anhedonia and impaired memory, attention, and concentration. The disorder usually presents during adolescence or early adulthood and will usually fairly rapidly result in referral to a psychiatrist. Occasionally, however, patients may be first seen by a neurologist. More frequently some of the individual features of schizophrenia such as the hallucinations, delusions, and disordered thinking can occur within the setting of a variety of neurological diseases. Dementia with Lewy Bodies is characterized by hallucinations but most commonly these are visual and are usually silent. When auditory hallucinations do occur they are very rarely in the third person and do not have the accusatory or derogatory flavour of the schizophrenic auditory hallucinations. Episodes of psychosis are seen in a variety of neurological disorders including brain tumours, infectious and inflammatory disorders such as systemic lupus erythematosus and multiple sclerosis, and importantly can be associated with drugs. In particular, amphetamines, cocaine, and LSD can be associated with psychosis and, as more controversially, can cannabis. Anticholinergics and high dosage of anti-histamines can also be associated with psychosis as can alcohol withdrawal in delirium tremens.

4.7 Symptoms

4.7.1 Fatigue

Patients with fatigue and non-specific difficulty in using their limbs are commonly referred to neurological clinics. Most occurrences are sporadic but the phenomenon has been well studied in epidemics, predominantly occurring in female nurses (Shelokov *et al.* 1957; McEvedy and Beard 1970). Unlike the fatigue due to myasthenia gravis, which develops during the course of, and as a result of, muscular exertion, psychogenic fatigue often tends to well up in the hours after exercise has been undertaken. Thereafter it poleaxes the patient who feels quite unable to undertake any sort of exercise for the next few days. Frequently such patients complain of associated myalgia with tenderness. Superficial sensory loss may also occur, often conforming to a non-anatomical territory (McEvedy and Beard 1970). These symptoms may excite diagnostic notions of inflammatory muscle disease or polyneuropathy. Often the fatigue is a central symptom of a wider complex, which may include headache, dizziness, non-specific diplopia, and nausea or vomiting. This spectrum of disease has become known as chronic fatigue syndrome or myalgic encephalomyelitis (Section 24.11.1). Notions of underlying persistent viral infections are rarely borne out. Frank depression may be obvious, whether it be reactive, or the underlying cause. When occurring in specific circumstances, notions of neurotoxin exposure may arise, for instance in the Gulf War syndrome noted by demobilizing active servicemen. It should be noted that fatigue is a common symptom of multiple sclerosis, and after recovery from Guillain–Barré syndrome, but these clinical contexts are usually obvious.

4.7.2 Paralysis

Patients with psychologically determined weakness or paralysis of the limb are rarely specific about which movements are affected, and which tasks have become impossible. Frequently, enquiry about the precise nature of the weakness provokes additional symptoms such as fatigue, pain, or dropping things. A range of mono-, hemi-, and parapareses occur, with a distinctly left-sided emphasis (Binzer *et al.* 1997). Notably, tetraparesis is most uncommon, and the function of at least one arm is usually preserved. A monoparesis globally affecting all the muscles of a limb is the commonest presentation; yet this is a comparatively unusual manifestation of structural disease of the nervous system. Patients with psychologically determined weakness often use a revealing descriptive terminology. The onset may be in combination with panic or some minor physical trauma, or that 'my leg didn't belong to me'. These are interpreted as disassociative phenomena reflecting loss of the normal sense of ownership of one's actions (Stone *et al.* 2002c).

4.7.3 Clumsiness and dropping things

Not infrequently, young or middle-aged adults may complain of clumsiness or fear of dropping things. This predominantly affects the left hand. When asked to cite examples of resultant breakages and spillages, the number is strikingly small in comparison to the concern being devoted to the symptom. The phenomenon particularly seems to be noted by middle-aged women during kitchen and household tasks. Yet, questioning about precise manipulatory tasks, for instance buttons or underwear clips, reveals that there is no elementary loss of function. Sometimes this symptom is conjoined with complaints of weakness, pain, or numbness in the offending limb.

4.7.4 Gait disorders

Apart from the context of hysterical paraplegia, psychologically determined gait disorders are rarely proffered as a symptom.

Yet, they are frequently exhibited as a physical sign, either during formal examination, or when the patient enters the consulting room. Excessive reliance on walking aids is a frequent way of exhibiting this symptom, yet during the course of the consultation, it usually becomes clear that these are not as essential as at first seemed. Apart from weakness, the other psychologically generated symptom affecting walking is dizziness or loss of equilibrium.

4.7.5 Disequilibrium

'Dizziness' and symptoms of disequilibrium are notoriously common. Symptoms of disequilibrium are frequently reported in generalized anxiety disorder, panic attacks, and hyperventilation syndrome. Panic attacks with or without agoraphobia are commonly associated with symptoms of disequilibrium. The clue is the environmental situation that triggers the symptoms, for example social situations, or being in enclosed spaces, on bridges, or in high buildings (Jacob *et al.* 2001).

The true vertigo precipitated by peripheral vestibular disorders can be very distressing and disabling. The suddenness and severity of the symptoms can often lead patients to believe that they are having a stroke or that they are dying. It is not surprising that psychiatric problems are very common in established peripheral vestibular disease, particularly panic disorder and depression (Eagger *et al.* 1992).

4.7.6 Pain

Pain has enormous emotional valence and this will vary depending upon the individual personality and previous experience. The emergence of pain without a clear explanation such as trauma, is also particularly distressing as it may signify an underlying sinister disease process. Many patients are understandably fearful that pain may signify an underlying neoplasm, perhaps the most common presentation for the neurologist being the spectre of a brain tumour. Understanding the psychological contribution to the pain experience is clearly important in neurological practice. Clues of a significant psychological component will be the patient who suffers from an anxiety state or depression. But one has to be cautious as both depression and anxiety can be triggered directly by the pain experience. Important clues are that the pain may increase or decrease with psychological trigger factors and there is often an expression of helplessness and passivity in relation to the pain. Often the pain may be triggered by an illness or an event, particularly an accident, over which the patient feels they have little control. The pain then persists for longer than anticipated. Drug and alcohol abuse are common with pain disorders and these should be specifically sought.

4.7.7 Skin sensory disturbances

Psychologically determined skin sensory disturbance is generally reported as a 'numbness'. Notably, the description rarely involves more diagnostically evocative symptoms, such as the paraesthesiae or dysaesthesiae of acquired polyneuropathies, or the Rombergism or pseudoathetosis occurring with spinal cord lesions or polyneuropathies. Enquiry about the topography of the sensory disturbance usually elicits rather vague and imprecise gestures after a moment's thought. Usually, the key demonstration that the sensory disturbance does not obey anatomical territories is only provable at the later stage of examination. Informal clinical observation, and some retrospective studies, have suggested psychologically determined sensory symptoms are more likely on the left, non-dominant side of the body (Pascuzzi 1994; Rothwell 1994). However meta-analysis suggests this left lateralizing effect to be only slight, and much of this supposed phenomenon is attributed to variable reporting bias (Stone *et al.* 2002a).

4.7.8 Visual loss

Complaints of visual loss or impairment are relatively common and, although usually result in a referral to an ophthalmologist, may present to the neurologist. The normal complaint is of reduced or lost visual acuity with normal fields. Visual field impairments are less commonly reported; a monocular hemianopic disturbance is almost invariably psychologically determined. There are no specific features in the history which direct one towards psychologically determined visual impairment and the key features are often established on examination (see Section 4.8.3). In general, psychologically determined complaints of visual impairment are seen in younger patients with a female gender predominance (Beatty 1999).

4.7.9 Facial pain

A variety of facial pain syndromes present to different specialists, particularly otorhinolaryngologists and dental surgeons (Section 19.6). Patients with poorly localized deep burning or aching that is often bilateral have generally been referred to as atypical facial pain (Section 19.6). Such patients often present or are referred on to neurologists. Typically patients are middle aged and more commonly women. Whilst many such patients do turn out to have a clear explanation, many do not and they often have other body system complaints such as neck and back pain and irritable bowel syndrome (Feinmann 1993). Patients often respond well to anti-depressants although the efficacy may not necessarily be related to the anti-depressant effect (Madland and Feinmann 2001).

Another syndrome which more commonly presents to oral surgeons is the 'burning mouth syndrome' (Section 19.4.5) (Grushka *et al.* 2006). When local oral pathology has been excluded, some such patients may be referred on to neurologists. As with atypical facial pain, a subgroup persist with no clear explanation. The syndrome is seen most commonly in post-menopausal women suggesting that there may be a hormonal trigger.

4.7.10 Headache

Headache is so common as to be the normal experience at some time in one's life. Since it is such a normal phenomenon, the question needs to be asked as to why an individual patient attends clinic often with a long history of variable or even daily headaches (Sections 18.3 and 18.6). The history may reveal a specific trigger which has focussed attention and will often relate to concern about an underlying sinister diagnosis such as a brain tumour. Headache is also common in the setting of patients with pain disorder or with Briquet's syndrome or Somatization disorder. Indeed the 2004 International Classification of headache disorders identifies the presence of pain in other body parts or of non-painful gastrointestinal disturbances as a criterion for diagnosing 'headache attributed to somatization disorder'.

Headache is also very common in depression when it is usually a steady, non-pulsatile generalized headache. As with other depressive symptoms, the headache may be worse in the morning and be accompanied by early morning waking. It is important to exclude hypnic headache in these patients (Section 18.5.5). Depression and anxiety was found to occur in 90 per cent of patients presenting with chronic daily headache (Verri *et al.* 1998).

4.7.11 Memory loss

The more florid psychologically influenced memory impairments are the dissociative amnesias. These can be generalized, often occurring after a particularly stressful event, and present as a fugue state. This usually lasts a few hours or days and, if prolonged, then raises the question of simulation. However, dissociative amnesias may be localized to a particular period in time or selective to a particular location or person (Kopelman 2002).

More commonly, neurologists see patients in whom anxiety and depression are a significant determinant of the complaint of poor memory. Indeed a depressive illness can sufficiently impair cognitive function, and in particular memory, to acquire the label of 'depressive pseudo-dementia' (Section 34.6.1). The patient who is able to attend the clinic unaccompanied is less likely to be suffering from a significant memory impairment. Those patients with significant depression and anxiety will complain bitterly of their deficits and usually more so than their partner. By contrast, in Alzheimer's disease particularly there is often a lack of awareness of a cognitive deficit with far greater concern being shown by the spouse. The patient with early Alzheimer's disease may find it very difficult to give specific examples of their memory problems whereas the anxious and depressed patient will focus on failures and be able to provide a very detailed account of their problems. In one cross-sectional study of patients with memory complaints but demonstrable cognitive impairment, the informant's rating correlated better than the patients' with hippocampal volume and performance on cognitive assessment. Also patients who made use of memory aids were less likely to have significant deficits (Archer *et al.* 2007).

4.7.12 Blackouts and fugue states

Psychologically determined blackouts and fugue states can present a diagnostic challenge. Psychogenic non-epileptic attacks are commonly misdiagnosed as epilepsy and also occur in those with true epileptic seizures (see Section 31.1.3). Convulsive attacks which are resistant to treatment should raise suspicions, as should resistance to eye opening and absence of pupillary dilation during an attack.

Panic attacks may occur in relation to obvious precipitating factors particularly in relation to specific phobias, such as open spaces or heights. They are quite frequent following trauma. Caffeine may precipitate anxiety in normal people and panic attacks in those who are susceptible. They may, however, occur in apparently unprovoked situations. Anxiety and panic attacks usually present in patients in their twenties and are rare as an initial presentation after the age of forty.

Fugue states in which patients wander away from home and on recovery have amnesia for the event may occur in a number of conditions. Wandering may occur as a post-ictal phenomenon, but in this instance it is usually short-lived and the patient will appear confused and will have difficulty with travelling and negotiating roads or public transport. The same comments apply to the rare instances of somnambulism (Section 32.4.2) in which people may leave the home. Transient global amnesia (Section 31.1.5) may involve travelling away from home with amnesia for the event, but in most instances patients travel only familiar routes since they have well preserved topographical memories. They will often appear perplexed during the attacks. By contrast, there are patients who disappear from home to turn up many hours or even days later. They will have successfully negotiated the period of travel and will not have drawn attention to themselves by virtue of confused behaviour. They have amnesia for the event and often a retrograde amnesia which extends back prior to the period of travel, and in some instances appears to be life long. In these situations it is often accompanied by amnesia for personal identity. Such behaviour may be associated with depressive illness and these patients present a suicide risk. In other instances these fugue states are considered as part of the spectrum of conversion disorder.

4.8 Physical signs

4.8.1 Muscle weakness

Before considering muscle weakness to be psychologically determined, it must be shown that muscle bulk, tendon reflexes. and the plantar responses are normal. The manner in which muscle power is exerted can point strongly to weakness being psychologically determined. Variable production of power, often with a marked collapse on first encountering resistance, is characteristic. Exhortation may momentarily improve power production. The collapsing pattern of weakness may be bizarrely diffuse, particularly in malingerers. For instance, apparent weakness of shoulder abduction may be demonstrated as a sideways tilt from the lower trunk, the arm continuing to be strongly abducted from the side at an unchanging angle. Inconsistencies in muscle power may be demonstrated. For instance the patient may be unable to flex the hip against gravity on the couch, but will be observed to stand and flex the hip normally to put on trousers or tights. The patient may have weakness of plantar flexion when tested on the couch, yet be able to stand on tiptoe.

Apparent weakness may be associated with palpable contraction of the antagonist muscle, and this is most usually evident on testing hip or knee movements. Hoover's sign is regarded as the most reliable test for psychologically determined weakness (Koehler and Okun 2004). The test makes use of the crossed extensor reflex which enables the mechanisms of normal walking. It relies on the principle that the contralateral hip is extended during flexion of the other hip. Two ways of eliciting the sign are recommended (Stone *et al.* 2002c). First, a discrepancy can be observed between weakness on voluntary extension of the hip compared to greater power during the involuntary extension of the same hip which occurs when flexion is being tested in the other leg. Second, whilst testing hip flexion in the weak leg, the heel of the normal leg is cupped so as to test hip extension on this side simultaneously; absence of powerful hip extension is taken as a uniform lack of effort. To elicit these signs the examiner must be deft, and take particular care to avoid drawing the patient's attention to the duplicity of testing being carried out. The Hoover sign has been subjected to quantitative myometric study and similar phenomena can be observed on comparison of the two arms (Ziv *et al.* 1998).

4.8.2 Sensory loss

Psychologically determined patches of sensory loss are often implausibly sharply defined with instantaneous transition from complete anaesthesia to normal sensation. It may be possible to demonstrate that these apparently sharp boundaries shift considerably in position when the stimulus is presented at different speeds or from different directions, particularly if the patient's eyes are shut. Sometimes it can be demonstrated that the numb limb changes sides when the patient is rolled over and retested. These sensory findings may change greatly when retested a few days later. Psychologically determined patches of numbness usually do not obey the anatomical territories of peripheral nerves or nerve roots. For instance a sensory boundary may occur at the lower border of the mandible with the facial sensory disturbance extending into the C2 root territory. Apparent unilateral loss to vibration sensation may be observed when tested with a tuning fork on just one side of the sternum. Or patients may be observed to have apparently gross loss of proprioception in the fingers, and yet are able to manipulate an object in their hand successfully when tested for astereognosis. Sharply demarcated hemisensory loss is commonly determined psychologically but it always raises the question of a thalamic lesion (Section 2.5.3). Particularly commonly, loss of sensation from a limb involves a sharply demarcated boundary at the junction with the trunk at shoulder or groin, a distribution rarely observed with structural disease of the nervous system. Although many of these phenomena may tempt an exhaustive exercise in mapping the sensory loss, it is more helpful merely to record the phenomena, quickly and undemonstratively.

4.8.3 Visual disturbances

Although psychologically determined visual impairment tends to present to the ophthalmologist, they may be encountered by the neurologist. A number of bedside assessments can be valuable. The observation of the patient in the consulting room can be useful as they negotiate the environment. The presence of a menace reflex is helpful and if optokinetic nystagmus can be observed then one can infer a visual acuity of greater than 6/60. However, if the complaint is monocular it is important that the good eye is obscured. Joint position testing can also be helpful if one can be confident that proprioception is normal; testing with the eyes open will often result in significant impairment as the patient believes that the task is a visual one. This may be particularly valuable in those cases of more overt simulation of disease. On testing those patients in whom a visual field defect is the key feature, 'tubular' deficits may be observed, in which the width of the field deficit is unchanged with varying distances allowing for variation in target size. One may also be able to demonstrate spiral field defects. Patients may also flinch with increasing illumination, for example on ophthalmoscopy (Beatty 1999). It is always important to refer to an ophthalmologist if that has not already been done for detailed ocular assessment and specialist tests such as fogging and prism shifts.

Both electroretinography and visual evoked potentials can be helpful and there are a number of tests that ophthalmologists can undertake. Although electrophysiological tests can be useful it is important to remember that normal visual evoked potentials can occur in the presence of prominent occipital lobe disease and that patients can decrease the amplitude and prolong the latency of visual evoked potentials by defocussing (Howard and Dorfman 1986).

4.8.4 Gait disorders

Psychologically determined gait disorders are immensely variable. Some reveal themselves by requiring degrees of athleticism, or balleticism, compatible with nothing short of perfect limb functioning. For instance, patients may momentarily balance on one foot in mid-stride, which indicates extremely good motor control. Psychologically determined gaits may improve when the patients think they are not being observed particularly in malingering. Patients with psychologically determined gait disorders are often able to traverse an open space only to fall theatrically once they are able to grasp nearby furniture or an observer. However, bizarre gaits may also be observed in patients with dystonia, paroxysmal kinesigenic choreoathetosis, hereditary spastic paraparesis, or gait apraxia. Similarly, gross truncal ataxia due to midline cerebellar lesions may be misinterpreted because of the normal examination on the couch. It should be noted that of the neurological disorders initially presumed to have a psychological basis, it is particularly for gait disorders that an underlying pathology may emerge (Crimlisk *et al.* 2000).

Various forms of psychologically determined gait disorder have been characterized (Lempert *et al.* 1991). Particularly distinctive is a monoplegic 'dragging' gait in which the whole leg is dragged through like a sack of potatoes, but without the circumduction associated with chronic spasticity. Other commonly observed phenomena include: sudden buckling at the knees, but without falling to the ground; walking cautiously as though on ice and often with the arms held out sideways; and the introduction of seemingly random hesitations whilst walking. Only unusually do psychogenic falls cause injury (Voermans *et al.* 2005).

4.8.5 Cognitive impairment

Psychological factors are frequently important in the presentation of patients with cognitive impairment. In particular, anxiety and depression can both present as cognitive impairment but also they are very frequent accompaniments of organic cognitive syndromes. In the consulting room the depressed patient will often appear flat and pre-occupied by their failures, the interview will frequently be interspersed by 'don't know' answers. However, if one is able to focus on areas of interest and of concern to the patient then a detailed history can often be obtained suggesting a good event memory for matters of importance. The anxious patient may fail cognitive testing in a dramatic way with a catastrophic reaction. However, if one takes the consultation very slowly and gently assesses cognition outside the formal testing format, then good performance can often be obtained. The difficulty is that depression and anxiety can be common early in organic disease. In general, bedside testing of memory is often easier using recognition rather than recall tasks.

Patients may often over-exaggerate their deficits in order to ensure that the doctor takes them seriously. This is far more common than simple simulation. Clues are the variance between test performance and a person's everyday functioning. On recognition memory tasks, the patient may perform below chance. A markedly reduced digit-span is often a clue. This does of course occur with selective impairment of short term or working memory but patients with this syndrome are profoundly impaired. A patient with a digit-span of only two will be unable to follow a simple instruction such as 'place the pen underneath the sheet of paper'. Simulation requires mental effort and when individuals are asked to simulate cognitive impairment they will often take longer to answer and, if hurried and testing maintained over a long period of time, accuracy will improve.

4.8.6 Psychogenic unresponsiveness

Occasionally patients may present to emergency departments in apparent coma, with no verbal responses and no reaction to painful stimuli. However, examination reveals a number of features incompatible with unconsciousness. The oculocephalic responses will resemble those of the conscious patient with the eyes usually moving in the direction of head movement with saccades to refixate. If caloric stimulation is undertaken then nystagmus will be seen; this should be done cautiously as it can induce intense nausea. The roving eye movements of coma are not seen and cannot be mimicked. A useful sign is the deviation of eyes towards the ground when the patient is in a lateral recumbent position, regardless of which side (Henry and Woodruff 1978) (Section 33.4.7).

4.8.7 Tremors

Patients may complain that tremor prevents them using a limb. Two things are striking about such tremors, which are usually exhibited for the doctor's inspection. First, they often involve coarse flexion and extension movements which pick out a single joint of the limb, most commonly the wrist joint. Second, they are immensely variable in severity, and often disappear entirely when the patient's attention is drawn to other matters. This disappearance may be noted either during the examination or questioning, or paradoxically sometimes when a complex task is performed by the affected limb. Although tremor also occurs in this patient population due to anxiety, or as a side effect of anti-depressant drugs, such tremors are characteristic of an underdamped physiological tremor, and predominantly consist of a fine quivering of the fingers, rather than coarser movements around a more proximal joint.

4.9 Management and prognosis

4.9.1 Reassurance and investigation

The vast majority of patients with psychologically determined symptoms can be managed satisfactorily in the neurological clinic, without recourse to specialized treatments or psychiatric referral. Reassurance is critical to resolving patients' symptomatology and their unfounded fear of a medically sinister condition. But doctors vary considerably in their ability to provide the effective reassurance necessary to defuse the vicious circle of concern. The coverage of the history and examination should reassure the patient that their concerns have been completely addressed. Beyond this, it is particularly valuable to enquire whether patients are harbouring morbid anxieties, for instance about underlying multiple sclerosis, and to address the reasons why that is not the diagnosis. Some investigation may be necessary, either to reassure the patient, or the neurologist, or both. However, the expectation that such investigations will be normal should be stressed. Conversely, if the need for investigation is painted in such a way that the patient believes the doctor expects to find pathology, the patient's notion of illness may be entrenched rather than relieved. An excessively materialistic approach with multiple investigations runs the risk of consolidating a patient's belief that they are diseased.

An interesting characteristic of the human mind is its need to provide explanations so as to rationalize events such as illness. Religiosity may be an artefact created by the human to provide definitive, self-contained explanations for events: 'acts of God', or 'fate', or sometimes 'retribution'. Religiosity declined as scientism grew in Western democracies in the 20th century. As a result patients have had to find new ways to understand illness, whether or not due to underlying pathology. The growth of biomedical science was expected to replace religion as a provider of rational explanation, and to allow these explanations to connect logically and directly with cures and treatments. Whilst in some areas of neurology, science has been immensely effective in doing just that, in many others science and multiple investigations have simply dispersed attempts at all-embracing explanations into a series of subsidiary scientific questions, equally unresolved. It is those patients with psychologically generated disorders who particularly risk falling between scientific or spiritual explanations for their symptoms. The contemporary biomedical scientific approach to medicine can be found particularly wanting when trying to explain psychologically determined symptomatology to a patient whose disorder is one of the spirit. This must be one of the reasons for the popularity of so-called 'alternative medicine'. Such practices often provide analyses centred on the symptom, with a correspondingly direct approach to its understanding and treatment, and eschew the need for biomedical explanation. Doctors who have the gift of healing can be highly effective in resolving psychologically determined symptoms, and the concern associated with them. This healing art may depend on exercising the talent of an inspirational influence to alter a patient's belief system about the state of their body, a process akin to hypnosis.

A particular difficulty concerns how to explain to the patient that you believe their symptoms to be generated psychologically. When this notion is implied blamefully by comments such as 'there is nothing wrong with you', or 'you are putting it on', or 'imagining symptoms' it can raise understandable hackles. In a sense this approach is fundamentally wrong since the patient nonetheless feels ill. Surveys suggest that offence is particularly likely to be generated by terms such as 'symptoms are all in the mind' or 'hysterical weakness' or 'psychosomatic weakness' or 'medically unexplained weakness' (Stone *et al.* 2002b). In juxtaposition to an explanation of what is not wrong with them, patients also require an explanation of why they feel ill. Sensitivity and judgement are required, and the precise approach adopted depends upon whether it is believed that understandable psychiatric disorders, such as depression or anxiety lie at the root of the problem. For many patients with mild psychologically determined symptomatology, a simple discussion of the fact that they are misinterpreting their body state, and the ways in which stress, anxiety, or depression may influence that interpretation, often seem effective and are non-controversial. It should be stressed that these are unconscious processes, possibly giving examples such as the differential severity of pain experienced by people injured in different circumstances. The difficult group are malingerers, particularly when there is a clear-cut target of secondary gain through litigation. Such patients can become deeply angered by any hint that you judge their symptoms to be generated psychologically.

4.9.2 Drug treatment

Although a variety of medications are in use, anti-depressants are the only ones which have been shown in randomized controlled trials to be of benefit for psychologically determined disorders. Most frequent are beneficial effects on fatigue and headache (O'Malley *et al.* 1999). Meta-analysis of antidepressants for patients with chronic unexplained pain also showed significant benefit

(Onghena and Van Heudenhove 1992). In both meta-analyses, the effects were seen more rapidly than anticipated from a simple antidepressant effect. In general the trials have been short term and although the long-term benefits have not been demonstrated, it is important for patients to take their medication regularly.

Many patients with psychologically determined neurological disorders are on a large variety of medications, many of which potentially contribute to the problem. Notorious are overuse of analgesics for headache (Section 18.6.4) and anti-convulsants for non-epileptic seizures. A careful review of all medications, including those obtained over the counter, and subsequent rationalization is important. Care of course needs to be taken in patients who may have a mixture of non-epileptic and epileptic seizures when rationalizing anti-convulsant medication.

4.9.3 Psychological therapies

Cognitive behavioural therapy is aimed at achieving specific goals using self-help strategies and avoiding negative behaviours. It has been applied particularly to chronic fatigue and pain syndromes. Meta-analyses of trials comparing cognitive behavioural therapy with standard therapy have demonstrated benefit. However there is less information as to whether this is maintained over the long term (Kroenke and Swindle 2000).

Reattribution techniques involve establishing the link for the patient between their symptoms and a particular behavioural state, for example anxiety and over breathing, and redirecting the patient's focus of attention from a negative significance of symptoms to an explanation in terms of normal physiology. This may be most effective in patients with relatively mild disease and who recognize a significant psychological role in their symptoms (Morriss *et al.* 1998).

4.9.4 Psychiatric referral

Overall few patients with psychologically determined neurological disorders will require referral to the psychiatrist, and certainly not those many patients with mild symptoms and signs who can be managed with sympathetic handling in collaboration with the general practitioner. However, some patients do need specialist involvement from psychiatric services, particularly if they have disabling and persistent symptomatology or there is evidence of a serious contributory psychiatric disease. Depression is common and of particular concern in those patients in whom there is a significant suicide risk. Extended cognitive behavioural therapies are also often best managed within the psychiatric services.

4.9.5 Prognosis and misdiagnosis

The aetiologies for psychologically determined neurological symptoms are varied, and no uniform comment can be made about the overall outcome. Somatic symptoms generated by anxiety or depression, mild misperceptions of the internal body state, severe motor conversion hysteria, and litigation-induced malingering are all likely to have profoundly different prognoses. For an entrenched conversion disorder the prognosis for recovery is regarded as being generally poor and roughly half have no relief from their original symptom at 10-year follow up in a tertiary referral setting (Mace and Trimble 1996). Long-term follow up of motor or sensory conversion disorders shows 80 per cent of patients still have symptoms at a median follow up of 12.5 years, with the majority reporting limited physical functioning, and a third having retired on medical grounds (Stone *et al.* 2003). However this study addressed patients who had required in-patient neurological assessment, and is unlikely to be representative of the as yet unstudied milder end of the spectrum of psychologically determined limb symptomatology. Little is known of the prognosis for patients seen in an out-patient setting who require few, if any, investigations, and who are reassurable. In particular it is unknown whether such patients are prone to represent at a later stage with entirely different symptomatology, their original illness having resolved.

The other aspect of prognosis concerns the frequency with which a diagnosis of psychologically determined symptomatology ultimately turns out to be wrong. In an influential study in the 1950s there was a misdiagnosis rate of about 30 per cent in patients with hysteria (Slater 1965). A subsequent study of patients diagnosed around 1990, once brain imaging had become routine, found a misdiagnosis rate of only 5 per cent (Crimlisk *et al.* 2000). A meta-analysis of patients diagnosed as having motor or sensory conversion symptoms shows a consistent misdiagnosis rate of about 4 per cent since 1970, which predates the routine availability of modern neuroimaging (Stone *et al.* 2005). This is similar to the rate of misdiagnosis in the opposite direction, for instance patients diagnosed with epilepsy who in reality have syncope, and those diagnosed with multiple sclerosis, and later found to have a conversion disorder. Also, many diseases previously thought to be non-neurological, such as spastic dysphonia and writer's cramp, are now considered as neurological diseases.

References

Allanson J, Bass C, Wade DT (2002). Characteristics of patients with persistent severe disability and medically unexplained neurological symptoms: a pilot study. *J Neurol Neurosurg Psychiatry*, **73**, 307–9.

Allen LA, Woolfolk RL, Escobar JI *et al.* (2006). Cognitive-behavioral therapy for somatization disorder: a randomized controlled trial. *Arch Intern Med*, **166**, 1512–8.

Archer HA, McFarlane F, Frost C *et al.* (2007). Symptoms of memory loss as predictors of cognitive impairment? The use and reliability of memory ratings in a clinic population. *Alzheimer Dis Assoc Disord* (in press).

Asher R (1951). Munchausen's syndrome. *Lancet*, **1**, 339–41.

Bartholomew RE, Wessely S (2002). Protean nature of mass sociogenic illness: from possessed nuns to chemical and biological terrorism fears. *Br J Psychiatry*, **180**, 300–6.

Beatty S (1999). Psychogenic medicine: non-organic visual loss. *Postgrad Med J*, **75**, 201–7.

Binzer M, Andersen PM, Kullgren G (1997). Clinical characteristics of patients with motor disability due to conversion disorder: a prospective control group study. *J Neurol Neurosurg Psychiatry*, **63**, 83–8.

Caplan LR, Nadelson T (1980). Multiple sclerosis and hysteria. Lessons learned from their association. *JAMA*, **243**, 2418–21.

Carson AJ, MacHale S, Allen K *et al.* (2000a). Depression after stroke and lesion location: a systematic review. *Lancet*, **356**, 122–6.

Carson AJ, Best S, Postma K *et al.* (2003). The outcome of neurology outpatients with medically unexplained symptoms: a prospective cohort study. *J Neurol Neurosurg Psychiatry*, **74**, 897–900.

Carson AJ, Ringbauer B, Stone J *et al.* (2000b). Do medically unexplained symptoms matter? A prospective cohort study of 300 new referrals to neurology outpatient clinics. *J Neurol Neurosurg Psychiatry*, **68**, 207–10.

Crimlisk HL, Bhatia KP, Cope H *et al.* (2000). Patterns of referral in patients with medically unexplained motor symptoms. *J Psychosom Res*, **49**, 217–9.

Crimlisk HL, Bhatia K, Cope H *et al.* (1998). Slater revisited: 6 year follow up study of patients with medically unexplained motor symptoms. *BMJ*, **316**, 582–6.

Dalfen AK, Feinstein A (2000). Head injury, dissociation and the Ganser syndrome. *Brain Inj*, **14**, 1101–5.

Derbyshire SW, Whalley MG, Stenger VA *et al.* (2004). Cerebral activation during hypnotically induced and imagined pain. *Neuroimage*, **23**, 392–401.

Devinsky O, Morrell MJ, Vogt BA (1995). Contributions of anterior cingulate cortex to behaviour. *Brain*, **118**, 279–306.

DSM (2004). *Diagnostic and Statistical Manual of Mental Disorders*. American Psychiatric Press, Washington, DC.

Donaghy M (2004). Symptoms and the perception of disease. *Clin Med*, **4**, 541–4.

Donaghy M (2005). *Neurology*. Oxford University Press, Oxford.

Eagger S, Luxon LM, Davies RA *et al.* (1992). Psychiatric morbidity in patients with peripheral vestibular disorder: a clinical and neuro-otological study. *J Neurol Neurosurg Psychiatry*, **55**, 383–7.

Eames P (1992). Hysteria following brain injury. *J Neurol Neurosurg Psychiatry*, **55**, 1046–53.

Escobar JI, Canino G (1989). Unexplained physical complaints. Psychopathology and epidemiological correlates. *Br J Psychiatry Suppl*, **4**, 24–7.

Faymonville ME, Roediger L, Del Fiore G *et al.* (2003). Increased cerebral functional connectivity underlying the antinociceptive effects of hypnosis. *Cogn Brain Res*, **17**, 255–62.

Feinmann C (1993). The long-term outcome of facial pain treatment. *J Psychosom Res* **37**, 381–7.

Fertleman CR, Baker, Parker KA *et al.* (2006). SCN9A mutations in paroxysmal extreme pain disorder: allelic variants underlie distinct channel defects and phenotypes. *Neuron*, **52**, 767–74.

Grushka M, Ching V, Epstein J (2006). Burning mouth syndrome. *Adv Otorhinolaryngol*, **63**, 278–87.

Hall S, Bartleson JD, Onofrio BM *et al.* (1985). Lumbar spinal stenosis. Clinical features, diagnostic procedures, and results of surgical treatment in 68 patients. *Ann Intern Med*, **103**, 271–5.

Halligan PW, Athwal BS, Oakley DA *et al.* (2000a). Imaging hypnotic paralysis: implications for conversion hysteria. *Lancet*, **355**, 986–7.

Halligan PW, Bass C, Wade DT (2000b). New approaches to conversion hysteria. *BMJ*, **320**, 1488–9.

Hankey GJ, Stewart-Wynne EG (1987). Pseudo-multiple sclerosis: a clinico-epidemiological study. *Clin Exp Neurol*, **24**, 11–9.

Henry JA, and Woodruff GH (1978). A diagnostic sign in states of apparent unconsciousness. *Lancet*, **2**, 920–1.

Howard JE, Dorfman LJ (1986). Evoked potentials in hysteria and malingering. *J Clin Neuropsysiol*, **3**, 39–49.

ICD (1992). *International Classification of Diseases*, 10th edition. World Health Organization.

Jacob RG, Furman MB (2001). Psychiatric consequences of vestibular dysfunction. *Curr Opin Neurol*, **14**, 41–6.

Jones TF, Craig AS, Hoy D *et al.* (2000). Mass psychogenic illness attributed to toxic exposure at a high school. *N Engl J Med*, **342**, 96–100.

Kanner AM (2005). Depression and the risk of neurological disorders. *Lancet*, **366**, 1147–8.

Koehler PJ (2003). Freud's comparative study of hysterical and organic paralyses: how Charcot's assignment turned out. *Arch Neurol*, **60**, 1646–50.

Koehler PJ, Okun MS (2004). Important observations prior to the description of the Hoover sign. *Neurology*, **63**, 1693–7.

Kopelman MD (1987). Amnesia: organic and psychogenic. *Br J Psychiatry*, **150**, 428–42.

Kopelman MD (2002). Disorders of memory. *Brain*, **125**, 2152–90.

Kotila M, Numminen H, Waltimo O *et al.* (1998). Depression after stroke: results of the FINNSTROKE Study. *Stroke*, **29**, 368–72.

Kroenke K, Swindle R (2000). Cognitive-behavioral therapy for somatization and symptom syndromes: a critical review of controlled clinical trials. *Psychother Psychosom*, **69**, 205–15.

Langleben DD, Schroeder L, Maldjian JA *et al.* (2002). Brain activity during simulated deception: an event-related functional magnetic resonance study. *Neuroimage*, **15**, 727–32.

Lauterbach E. (2000). *Psychiatric Management in Neurological Disease*. American Psychiatric Press, Washington DC.

Leiborand R, Hiller W, Fichter MM (2000). Hypochondriasis and somatisation: two distinct aspects of somatoform disorders? *J Clin Psychol*, **56**, 63–72.

Lempert T, Brandt T, Dieterich M *et al.* (1991). How to identify psychogenic disorders of stance and gait. A video study in 37 patients. *J Neurol*, **238**, 140–6.

Lempert T, Dieterich M, Huppert D *et al.* (1990). Psychogenic disorders in neurology: frequency and clinical spectrum. *Acta Neurol Scand*, **82**, 335–40.

Mace CJ, Trimble MR (1996). Ten-year prognosis of conversion disorder. *Br J Psychiatry*, **169**, 282–8.

Madland G, Feinmann C (2001). Chronic facial pain: a multidisciplinary problem. *J Neurol Neurosurg Psychiatry*, **71**, 716–9.

Mailis-Gagnon A, Giannoylis I, Downar J *et al.* (2003). Altered central somatosensory processing in chronic pain patients with "hysterical" anesthesia. *Neurology*, **60**, 1501–7.

Marshall JC, Halligan PW, Fink GR *et al.* (1997). The functional anatomy of a hysterical paralysis. *Cognition*, **64**, B1–8.

McEvedy CP, Beard AW (1970). Royal Free epidemic of 1955: a reconsideration. *Br Med J*, **1**, 7–11.

Mkize DL, Ndabeni RT (2002). Mass hysteria with pseudoseizures at a South African high school. *S Afr Med J*, **92**, 697–9.

Modan B, Swartz TA, Tirosh M *et al.* (1983). The Arjenyattah epidemic. A mass phenomenon: spread and triggering factors. *Lancet*, **2**, 1472–4.

Moene FC, Spinhoven P, Hoogduin KA *et al.* (2002). A randomised controlled clinical trial on the additional effect of hypnosis in a comprehensive treatment programme for in-patients with conversion disorder of the motor type. *Psychother Psychosom*, **71**, 66–76.

Mohr PD, Bond MJ (1982). A chronic epidemic of hysterical blackouts in a comprehensive school. *Br Med J (Clin Res Ed)*, **284**, 961–2.

Morriss R, Gask L, Ronalds C (1998). Cost-effectiveness of a new treatment for somatised mental disorders taught to GPs. *Fam Pract*, **15**, 119–25.

Nandi DN, Banerjee G, Nandi S *et al.* (1992). Is hysteria on the wane? A community survey in West Bengal, India. *Br J Psychiatry*, **160**, 87–91.

Nazir FS, Lees KR, Bone I (2005). Clinical features associated with medically unexplained stroke-like symptoms presenting to an acute stroke unit. *Eur J Neurol*, **12**, 81–5.

Nimnuan C, Hotopf M, Wessely S (2000). Medically unexplained symptoms: how often and why are they missed? *QJM*, **93**, 21–8.

O'Malley PG, Jackson JL, Santoro J *et al.* (1999). Antidepressant therapy for unexplained symptoms and symptom syndromes. *J Fam Pract*, **48**, 980–90 Review.

O'Sullivan G, Harvey I, Bass C *et al.* (1992). Psychophysiological investigations of patients with unilateral symptoms in the hyperventilation syndrome. *Br J Psychiatry*, **160**, 664–7.

Onghena P, Van Houdenhove B (1992). Antidepressant-induced analgesia in chronic non-malignant pain: a meta-analysis of 39 placebo-controlled studies. *Pain*, **49**, 205–19.

Palmer I (2003). Malingering, shirking, and self-inflicted injuries in the military. In Halligan PW, Bass C, Oakley DA, eds. *Malingering and Illness Deception*, pp. 42–54. Oxford University Press, Oxford.

Pascuzzi RM (1994). Nonphysiological (functional) unilateral motor and sensory syndromes involve the left more often than the right body. *J Nerv Ment Dis*, **182**, 118–20.

Perkin GD (1989). An analysis of 7836 successive new outpatient referrals. *J Neurol Neurosurg Psychiatry*, **52**, 447–8.

Perini GI, Tosin C, Carraro C *et al.* (1996). Interictal mood and personality disorders in temporal lobe epilepsy and juvenile myoclonic epilepsy. *J Neurol Neurosurg Psychiatry*, **61**, 601–5.

Philen RM, Kilbourne EM, McKinley TW *et al.* (1989). Mass sociogenic illness by proxy: parentally reported epidemic in an elementary school. *Lancet*, **2**, 1372–6.

Phillips KA, Menard W (2006). Suicidality in body dysmorphic disorder: a prospective study. *Am J Psychiatry*, **163**, 1280–2.

Porro CA, Baraldi P, Pagnoni G *et al.* (2002). Does anticipation of pain affect cortical nociceptive systems? *J Neurosci*, **22**, 3206–14.

Raz A, Shapiro T (2002). Hypnosis and neuroscience: a cross talk between clinical and cognitive research. *Arch Gen Psychiatry*, **59**, 85–90.

Reddy H, Bendahan D, Lee MA *et al.* (2002). An expanded cortical representation for hand movement after peripheral motor denervation. *J Neurol Neurosurg Psychiatry*, **72**, 203–10.

Rickards H (2005). Depression in neurological disorders: Parkinson's disease, multiple sclerosis, and stroke. *J Neurol Neurosurg Psychiatry*, **76** (Suppl 1), i48–52.

Roach ES, Langley RL (2004). Episodic neurological dysfunction due to mass hysteria. *Arch Neurol*, **61**, 1269–72.

Ron MA (1994). Somatisation in neurological practice. *J Neurol Neurosurg Psychiatry*, **57**, 1161–4.

Ron MA, Logsdail SJ (1989). Psychiatric morbidity in multiple sclerosis: a clinical and MRI study. *Psychol Med*, **19**, 887–95.

Rothwell P (1994). Investigation of unilateral sensory or motor symptoms: frequency of neurological pathology depends on side of symptoms. *J Neurol Neurosurg Psychiatry*, **57**, 1401–2.

Schacter DL, Want PL, Tulving E *et al.* (1982). Functional retrograde amnesia: a quantitative case study. *Neuropsychologia*, **20**, 523–32.

Schaumburg HH, Albers JW (2005). Pseudoneurotoxic disease. *Neurology*, **65**, 22–6.

Sharma P, Chaturvedi S (1995). Conversion disorder revisited. *Acta Psychiatr Scand*, **92**, 301–4.

Sharpe M (2002). Medically unexplained symptoms and syndromes. *Clin Med*, **2**, 501–4.

Shelokov A, Habel K, Verder E *et al.* (1957). Epidemic neuromyasthenia; an outbreak of poliomyelitislike illness in student nurses. *N Engl J Med*, **257**, 345–55.

Slater E (1965). Diagnosis of "Hysteria". *BMJ*, **5447**, 1395–9.

Smith GR Jr, Monson R, Ray D (1986). Patients with multiple unexplained symptoms. Their characteristics, functional health, and health care utilization. *Arch Intern Med*, **146**, 69–72.

Spence S, Farrow T, Leung D *et al.* (2003). Lying as an executive function. In Halligan PW, Bass C, Oakley DA, eds. *Malingering and Illness Deception*, pp. 255–67. Oxford University Press, Oxford.

Steinhausen HC, von Aster M, Pfeiffer E *et al.* (1989). Comparative studies of conversion disorders in childhood and adolescence. *J Child Psychol Psychiatry*, **30**, 615–21.

Stone J, Sharpe M, Carson A *et al.* (2002a). Are functional motor and sensory symptoms really more frequent on the left? A systematic review. *J Neurol Neurosurg Psychiatry*, **73**, 578–81.

Stone J, Sharpe M, Rothwell PM *et al.* (2003). The 12 year prognosis of unilateral functional weakness and sensory disturbance. *J Neurol Neurosurg Psychiatry*, **74**, 591–6.

Stone J, Smyth R, Carson A *et al.* (2005). Systematic review of misdiagnosis of conversion symptoms and "hysteria". *BMJ*, **331**, 989.

Stone J, Wojcik W, Durrance D *et al.* (2002b). What should we say to patients with symptoms unexplained by disease? The "number needed to offend". *BMJ*, **325**, 1449–50.

Stone J, Zeman A, Sharpe M (2002c). Functional weakness and sensory disturbance. *J Neurol Neurosurg Psychiatry*, **73**, 241–5.

Taylor BW, Werbicki JE (1993). Pseudodisaster: a case of mass hysteria involving 19 schoolchildren. *Pediatr Emerg Care*, **9**, 216–7.

Tomasson K, Kent D, Coryell W (1991). Somatization and conversion disorders: comorbidity and demographics at presentation. *Acta Psychiatr Scand*, **84**, 288–93.

Verri AP, Proietti Cecchini A, Galli C *et al.* (1998). Psychiatric comorbidity in chronic daily headache. *Cephalalgia*, **18** (Suppl 21), 45–9.

Villemure C, Bushnell MC (2002). Cognitive modulation of pain: how do attention and emotion influence pain processing? *Pain*, **95**, 195–9.

Voermans NC, Zwarts MJ, van Laar T *et al.* (2005). Fallacious falls. *J Neurol*, **252**, 1271–3.

Vuilleumier P, Chicherio C, Assal F *et al.* (2001). Functional neuroanatomical correlates of hysterical sensorimotor loss. *Brain*, **124**, 1077–90.

Ward NS, Oakley DA, Frackowiak RS *et al.* (2003). Differential brain activations during intentionally simulated and subjectively experienced paralysis. *Cognit Neuropsychiatry*, **8**, 295–312.

Ziv I, Djaldetti R, Zoldan Y *et al.* (1998). Diagnosis of "non-organic" limb paresis by a novel objective motor assessment: the quantitative Hoover's test. *J Neurol*, **245**, 797–802.

CHAPTER 5

Toxic and environmental disorders

Michael Donaghy

Contents

5.1 Introduction

A huge range of toxins can affect the nervous system, most of which are rarely encountered clinically (Spencer and Schaumburg 2000). This chapter addresses those toxins, medical interventions, and environmental insults with common or noteworthy effects on the nervous system. The peripheral nervous system is particularly vulnerable to toxic effects of drugs, metals, and industrial and agricultural poisons; these manifestations are covered in Chapter 21. Overdosage with an enormous range of drugs and chemicals causes acute poisoning syndromes which affect multiple organ systems, including the nervous system. Such systemic poisonings are not covered below and the reader is referred to comprehensive toxicology reference texts (Dart 2004).

5.2 Alcohol toxicity

Acute alcohol intoxication adversely affects judgement, restraint, and co-ordination. In high dosage it starts to have general anaesthetic effects, but at lower dosage affects neurotransmitter systems including GABA-mediated inhibition, increasing opioid effects, and inhibiting glutamate neurotransmission. The behavioural effects contribute to the neurological injury caused by motor vehicle accidents and violent behaviour. Alcohol intoxication predisposes to the acquisition of sexually transmitted infections. In this regard the nervous system may be affected by HIV, herpes simplex virus type II, or syphilis. Pre-existing cerebral conditions such as subdural haematoma or infection dispose the sufferer to apparent intoxication after ingestion of lesser amounts of alcohol than usual. This phenomenon is known as 'pathological drunkenness'.

Habitual heavy drinkers may become physically dependent upon ethanol: alcoholism. There is an inherited predisposition to alcoholism. This leads to a variety of neurological disorders described below. These result from the direct toxic effects of alcohol and its metabolites, or from secondary malnutrition, particularly of thiamine. Many alcoholic patients exhibit combinations of various alcohol-related neurological disorders. Prenatal exposure in alcoholic mothers can lead to a foetal alcohol syndrome with subsequent cognitive and behavioural impairments, and developmental abnormalities of brain structure, most notably microcephaly and dysgenesis of the corpus callosum (Sowell et al. 2001).

5.2.1 Ethanol withdrawal and delirium tremens

Withdrawal symptoms develop in established alcoholics who are starved of alcohol for more than a few hours. Withdrawal symptoms are particularly likely in those deprived of their usual access to alcohol by prostration due to acute infection, accidents, or surgical operations. The 'shakes', a generalized coarse tremor of the face, tongue, and hands, appears earliest and may be the only symptom in mild cases. Frank delirium tremens develops in more serious cases. These patients experience nausea and vomiting, terrifying visual hallucinations often of animals, acute confusion, agitation, tachycardia, sweating, and hyperpyrexia. Generalized tonic-clonic convulsions may occur. These symptoms are maximal about 36 h after alcohol withdrawal. The mortality of delirium tremens is considerable, and is particularly attributable to uncontrolled convulsions and to cardiac arrhythmias caused by autonomic nervous system dysfunction. On recognizing such symptoms, alcoholics usually resume drinking to suppress them. Medical treatment of delirium tremens consists of the administration of sedative drugs, such as 50 mg of Chlordiazepoxide orally every 6 h for 3 days, and benzodiazepines or chlormethiazole to control agitation and reduce the incidence of seizures (Thompson et al. 1975; Saitz and O'Malley 1997), beta-blockers, such as Atenolol to control autonomic manifestations (Kraus et al. 1985), and thiamine parenterally. Neuroleptic drugs are useful adjunctive therapy for troublesome hallucinations and agitation, but may provoke seizures. In addition, any associated infection, dehydration, or hypoglycaemia should be treated. Although delirium tremens is a self-limited disorder, which resolves spontaneously, the majority of patients resume their habit and are vulnerable to further attacks.

5.2.2 Seizures and alcohol

Alcoholics usually develop seizures either as a result of cerebral trauma or due to ethanol withdrawal. Occasionally acute alcohol intoxication provokes seizures within a few hours (Brennan and Lyttle 1987). Seizures due to cerebral trauma sustained during alcoholic binges constitute a diagnosis of epilepsy and should be treated with long-term anticonvulsant therapy. Alcohol withdrawal seizures are generally accompanied by other features of delirium tremens and should be investigated only if the seizures are focal, more than six in number, occur over a period exceeding 6 h, or are associated with protracted postictal confusion, evidence of cranial trauma, or focal neurological signs (Charness et al. 1989). Anticonvulsant drug therapy is not usually recommended for alcohol withdrawal seizures, either over the short- or the long-term (Simon 1988). Other associated features of alcohol withdrawal should be treated as outlined above. Benzodiazepine or chlormathiazole infusions are usually effective for recurrent withdrawal seizures or status epilepticus.

The role of prior alcohol consumption has been studied in an unselected population of patients presenting with their first ever seizure. Alcohol usage is a strong risk factor for a first seizure's occurrence (Leone et al. 1997). Seizures are most frequent within 48 h of last drinking alcohol. However only half of all seizures occur within the conventional period for alcohol withdrawal symptoms, 6–48 h after the cessation of drinking (Ng et al. 1988). Chronic alcoholism alone is not a risk factor for a first symptomatic epileptic seizure (Leone et al. 2002).

5.2.3 Wernicke–Korsakoff syndrome

Wernicke's encephalopathy is a reversible cerebral disorder due to thiamine deficiency. In most patients with Wernicke's encephalopathy, there is an underlying permanent disorder of memory known as Korsakoff's psychosis. Because of the usual concurrence of these two disorders, they are often referred to jointly as the Wernicke–Korsakoff syndrome (Victor et al. 1989).

Wernicke's encephalopathy is a reversible complication of thiamine deficiency in alcoholics, particularly those who are malnourished. Other causes of thiamine deficiency, such as starvation or gastro-intestinal disease, and protracted hyperemesis gravidarum, may also lead to Wernicke's encephalopathy (Reuler et al. 1985). Neurological symptoms develop over hours or days and may be precipitated by a high carbohydrate intake. The typical clinical triad consists of encephalopathy, ataxia, and ophthalmoplegia. The encephalopathy produces somnolence and disorientation and eventually progresses to coma. Ataxia results from the combination of polyneuropathy and cerebellar dysfunction. Abnormal eye movements are crucial to the diagnosis of Wernicke's encephalopathy in life. There may be bilateral lateral rectus palsies, nystagmus, or complex ophthalmoplegias and occasionally the pupils become small and unreactive. Patients may be hypothermic or hypotensive. Atypical presentations are common and post-mortem studies suggest that only 20 per cent of those reaching autopsy had been diagnosed in life (Harper *et al.* 1987). On suspicion of the diagnosis thiamine should be given intravenously and continued regularly thereafter: without treatment mortality approaches 20 per cent. The optimal thiamine dosage has not been established (Day *et al.* 2006); an initial dose of 100 mg is often used. The response to thiamine replacement is dramatically quick. Within 1–6 h, the ocular palsies begin to resolve and conscious level improves.

Underlying Korsakoff's psychosis of variable severity is evident in most patients following thiamine treatment of their associated Wernicke's encephalopathy. Patients exhibit retrograde amnesia, in which they are unable to recall previous information, and anterograde amnesia, in which they cannot register novel information. Confabulation may be present also. Useful recovery from Korsakoff's psychosis occurs in less than a quarter despite adequate treatment of their associated Wernicke's disease with thiamine. Mammillary body atrophy and neuronal loss from the dorsal medial thalamus, periaqueductal grey matter of the midbrain, vagal nuclei and cerebellar vermis are characteristic neuropathological features of Wernicke–Korsakoff disease (Victor *et al.* 1989). Mammillary body atrophy is particularly characteristic and may be demonstrated in life by MRI (Charness and DeLaPaz 1987).

Red cell transketolase enzyme activity is reduced in thiamine deficiency, or there can be an increased thiamine co-enzyme requirement in alcoholics (Heap *et al.* 2002). The slow availability of results of this enzyme assay precludes its use in diagnosing Wernicke–Korsakoff syndrome; treatment with thiamine should be started immediately once the disorder is suspected on clinical grounds. Some patients with Wernicke–Korsakoff disease have an inherited anomaly of the transketolase enzyme, affecting the binding of thiamine pyrophosphate (Blass and Gibson 1977).

5.2.4 Alcoholic cerebellar degeneration

Long-standing alcoholics may develop gait ataxia due to degeneration of cerebellar cortex Purkinje cells (Section 39.11.2). Although generally of gradual onset, alcoholic cerebellar ataxia may evolve relatively acutely, sometimes in the context of Wernicke's encephalopathy. Early on demonstrable ataxia may be limited to the gait alone, but severely affected patients show ataxia if the legs or arms are tested individually. Dysarthria or nystagmus are unusual (Victor *et al.* 1959). In many patients, an alcoholic or thiamine deficiency peripheral neuropathy contributes to the ataxia. The incidence of cerebellar ataxia does not correlate with the extent of lifetime alcohol consumption (Estrin 1987) or with the occurrence of cerebellar atrophy on computed tomography of the brain (Hillbom *et al.* 1986). Quantitative histological studies show Purkinje cell loss from the cerebellum in alcoholics, which is particularly severe in those with additional Wernicke–Korsakoff syndrome (Phillips *et al.* 1987). These observations make it likely that cerebellar degeneration does not only result from the direct toxic effects of alcohol or its metabolites, but may also reflect some other factor, such as thiamine deficiency. Thiamine replacement and prolonged abstinence from alcohol should be recommended in all

patients. Prolonged abstinence decreases the amplitude of body sway associated with alcoholic ataxia suggesting some capacity for the ataxia to improve (Diener *et al.* 1984).

5.2.5 Alcoholic dementia

Cognitive impairment is common in alcoholics. It usually reflects varying combinations of acute intoxication, Wernicke–Korsakoff syndrome, mild delirium tremens, depressive pseudo-dementia, pre-morbid cognitive impairments, previous cerebral trauma, and the diffuse alcoholic brain damage otherwise known as alcoholic dementia. Less frequently the cognitive impairment is due to subdural haematoma, metabolic encephalopathy, nicotinic acid deficiency, or Marchiafava–Bignami disease.

Whether diffuse alcoholic brain damage is an important and frequent cause of dementia in alcoholics (Lishman 1981) is questioned on the grounds that autopsy studies usually show evidence of inactive and chronic Wernicke–Korsakoff Disease (Victor 1994). Moderate regular alcohol consumption, particularly of wine, seems protective against subsequent onset of dementia (Ruitenberg *et al.* 2002; Truelsen *et al.* 2002). Generally it is assumed that alcoholic dementia involves generalized cognitive abnormalities, which distinguishes it from the selective amnesia of the Wernicke–Korsakoff syndrome. Neuropsychological studies show that alcoholic dementia predominantly affects problem solving abilities whereas it is memory which is selectively impaired in the Wernicke–Korsakoff syndrome (Carlen *et al.* 1981). Alcoholic patients with dementia display cortical shrinkage and ventricular dilatation on computed tomography of the brain (Carlen *et al.* 1981; Ron *et al.* 1982). Quantitative neuropathological studies show reduced numbers of neurons within the superior frontal cortex in such patients, despite preserved neuronal populations in the motor cortex (Harper *et al.* 1987). The cerebral volume loss, metabolic, and neuropsychological abnormalities of chronic alcoholism are partly reversible even during the early stages of abstinence, providing motivational feedback for some trying to cease their habit (Bartsch *et al.* 2007).

5.2.6 Central pontine myelinolysis

Alcoholics are particularly prone to central pontine myelinolysis, particularly if they are chronically hyponatraemic (Slager 1986). It can also occur in alcoholics with a normal serum sodium (McKee *et al.* 1988). The clinical picture typically develops an average of 6 days after correction of chronic hyponatraemia with intravenous fluids at rates exceeding 12 mmol/l of sodium per day. Accordingly, if intravenous therapy is deemed necessary, it is recommended that the serum sodium concentration should be increased by less than 8 mmol/l per day (Sterns *et al.* 1986).

Central pontine myelinolysis primarily affects the corticospinal tracts in the central brainstem (Section 37.6). It produces a symmetrical paraparesis or quadriparesis with extensor plantar responses. In some patients the bulbar and facial musculature is also paralysed. Gaze palsies occasionally occur. The full-blown state produces a locked-in syndrome in which the patient is incapable of any voluntary movements except vertical eye movements, yet consciousness is preserved. Many patients die, and the remainder suffer from substantial chronic disability. Worthwhile recovery from severe forms is rare, but does occur from milder forms of pontine myelinolysis. Autopsy studies show a characteristic large area of demyelination within the central pons; axons are spared (Wright *et al.* 1979). Computed tomography is relatively insensitive in detecting the large area of pontine demyelination, but magnetic resonance scanning demonstrates such lesions (Miller *et al.* 1988). MRI brainstem changes compatible with pontine myelinolysis are discovered in about 2 per cent of alcoholics, frequently without symptoms or signs (Uchino *et al.* 2003).

5.2.7 Marchiafava–Bignami disease

In this distinctive disorder, usually associated with underlying cirrhosis, demyelinating lesions develop in the corpus callosum. The disorder was originally noted in malnourished Italian red wine drinkers but is now known to occur also in other groups of alcoholics. Histologically, these lesions are similar to those in the brainstem in central pontine myelinolysis. Patients develop gait apraxia, dementia, spasticity, and dysarthria. Prior to introduction of MR scanning it was thought that most patients died, or survived for many years with severe dementia; recovery was rare. The demyelinating lesion in the corpus callosum and adjacent cerebral white matter is demonstrable by MRI. When based on MR diagnosis *in vivo*, Marchiafava–Bignami disease seems to vary between major and slight initial impairment of consciousness, with a corresponding severity of outcome, depending upon whether the corpus callosum is entirely, or only partially affected (Heinrich *et al.* 2004) and the nature of the signal change (Menegon *et al.* 2005).

5.2.8 Alcohol and stroke

Some have noted heavy alcohol intake to be a risk factor for ischaemic stroke, particularly in young males during or immediately following a bout of acute intoxication (Gill *et al.* 1986). Moderate chronic alcohol consumption is protective against the risk of ischaemic stroke in both men and women, irrespective of racial background (Reynolds *et al.* 2003; Elkind *et al.* 2006). By contrast, the risk of haemorrhagic stroke rises with alcohol consumption (Reynolds *et al.* 2003).

5.2.9 Alcoholic peripheral neuropathy

The peripheral neuropathies due to alcohol or disulfiram (Antabuse) are discussed in Section 21.19 and that due to thiamine deficiency associated with alcoholism, dry beriberi, in Section 21.22.2.

5.2.10 Alcoholic myopathy

Alcoholic myopathy may develop chronically or acutely (Section 24.9.1). Episodes of acute deterioration frequently punctuate an insidious background myopathy. Chronic alcoholic myopathy is a relatively painless affliction predominately affecting proximal muscles. In many alcoholics mild myopathy is an incidental asymptomatic finding on examination. Established myopathy occurs in those chronic alcoholics with a cumulative lifetime consumption exceeding 13 kg ethanol per kg body weight. A standard measure of spirits, wine, or a half-pint of beer contains approximately 10 g of alcohol. Malnutrition or electrolyte imbalance are not thought to be important contributing factors (Urbano-Marquez *et al.* 1989). An associated cardiomyopathy is common. The serum creatine kinase levels are elevated in one-third of the patients. Muscle biopsies show varying degrees of necrosis and atrophy, particularly affecting type II fibres. Electromyography shows non-specific myopathic features; fibrillations occur in the more acute myopathies. Episodes of acute alcoholic muscle weakness due to rhabdomyolysis often follow bouts of massive alcohol ingestion and are associated with dark urine containing myoglobin. Abstinence leads to some improvement. Downhill progression occurs in persistent drinkers (Martin *et al.* 1985).

5.2.11 Methanol poisoning

Consumption of doses of methylated spirits containing more than 30 g of methanol is often fatal. It can be an unnoticed substitute contaminating alcoholic drinks. Methanol causes a toxic confusional state. Misty vision, central scotomata, or blindness are associated with optic disc oedema and optic atrophy eventually develops (Sharpe *et al.* 1982). A Parkinsonian syndrome unresponsive to L-Dopa has been described and involves bilateral infarction of the frontal white matter and putamen (McLean *et al.* 1980). CT or MRI may show putaminal abnormalities. These permanent neurological complications, and death when it occurs, are thought to be due to the accumulation of formic acid. This metabolite of methanol forms within 12 h of ingestion and causes metabolic acidosis. Fomepizole intravenously, 15 mg/kg loading dose followed by 10 mg/kg every 12 h according to whether the blood methanol level remains higher than 0.2 g/l should be administered to patients suspected of methanol poisoning (Mycyk and Leikin 2003). This drug inhibits hepatic alcohol dehydrogenase, thereby reducing formation of toxic metabolites, and is also effective for ethylene glycol poisoning, which can be difficult to differentiate from methanol poisoning. Early treatment with haemodialysis is indicated for mental or visual changes, metabolic acidosis, if the blood methanol level exceeds 0.5 g/l, or following ingestion of more than 30 g of methanol (Lancet 1983).

5.3 Recreational drug abuse

5.3.1 Neurological syndromes

This section addresses the neurological consequences of recreational drug abuse. It does not cover the associated social, epidemiological, or psychiatric aspects. It should be noted that multiple drug abuse is common and may include alcohol; that underlying nutritional deficiency is common; that violent injuries are common in the drugs underworld; that pressure palsies of peripheral nerves may result from periods of stuporous immobility; and that intravenous drug abusers are prone to blood-borne infection, particularly with HIV.

Drug abuse should be considered in young adults developing the following neurological syndromes (Neiman *et al.* 2000):

Stroke. Both haemorrhagic and ischaemic stroke may occur within an hour of drug administration. Cardiovascular effects, embolization from intravascular infection or drug impurities, and rupture of aneurisms or arteriovenous malformations are the likely pathophysiological mechanisms. Most commonly implicated are cocaine, heroin, amphetamines, phenylcyclidine, and LSD.

Headache. Acute severe headache resembling migraine can follow administration of, or withdrawal from cocaine.

Seizures. These generally reflect acute intoxication, usually follow cocaine or amphetamine abuse and can present with status epilepticus. Withdrawal seizures occur from alcohol, barbiturates, or benzodiazepines.

Movement disorders. Various syndromes have been described including choreoathetosis, tremor, and akathisia with cocaine, so-called 'crack dancing', cocaine withdrawal dystonia, a constant choreiform 'jerking sydrome' with amphetamines, and compulsive repetitive behaviours during amphetamine psychosis, known as the 'punding syndrome'.

Encephalopathy. Chronic abuse of cocaine, amphetamine derivatives, or organic solvents can cause diffuse cerebral damage with cognitive deterioration and mild psychiatric symptoms (Ernst *et al.* 2000). A progressive spongiform leukoencephalopathy can follow inhalation of heroin vapour, probably with impurities.

5.3.2 Cocaine

Cocaine is a currently fashionable central nervous system stimulant with vasoconstrictor effects used to induce pleasurable euphoria and hypersexuality. It is usually absorbed through the mucous membranes by sniffing or 'snorting', or by chewing. Highly purified free-base cocaine, 'crack', may be inhaled. Cocaine is occasionally used intravenously, usually in polydrug abusers. Overdosage sometimes follows rupture of cocaine-loaded condoms within body cavities in smugglers. The common neurological consequences of cocaine are seizures and strokes. Cerebrospinal fluid rhinorrhoea has followed protracted cocaine sniffing and poses the risk of meningitis (Sawicka and Trosser 1983). Cocaine may induce choreoathetoid movements, so-called 'crack dancing' (Daras *et al.* 1994).

Seizures may follow acute intoxication with cocaine, generally occurring within 90 min of abuse (Pascual-Leone *et al.* 1990). Most such seizures are generalized but focal attacks do occur. The seizures are generally single and are particularly likely following the use of 'crack'. Persistent neurological features and encephalographic or computed tomographic abnormalities are not subsequently evident. Cocaine provoked seizures are the reason for seeking medical attention in approximately 10 per cent of those with cocaine-induced medical problems. There is an increased frequency of seizures in pre-existing epileptics who abuse cocaine.

Cocaine abuse is a major risk factor for cerebrovascular disease in young adults. It is a potent vasoconstrictor and ischaemic and haemorrhagic strokes occur. Strokes may develop within minutes of 'crack' abuse and are frequently associated with headache. Intracerebral and subarachnoid haemorrhages may derive from pre-existing aneurysms or arteriovenous malformations and be provoked by hypertension during acute cocaine intoxication (Nolte *et al.* 1996). Cocaine-induced cerebral infarction may affect any arterial territory of the brain (Levine *et al.* 1990). Stroke syndromes affecting the thalamomesencephalic regions are noteworthy since they are otherwise uncommon (Rowley *et al.* 1989). The pathogenesis of cocaine-related cerebral infarction is uncertain but probably relates to its powerful vasoconstrictive properties. Although cerebral angiography may reveal narrowed segments and beading of arteries, it is probable that this reflects focal vasospasm rather than a true vasculitis (Aggarwal *et al.* 1996). Habitual cocaine abusers develop computed tomographic evidence of diffuse cerebral atrophy but it is not known whether there is associated dementia (Pascual-Leone *et al.* 1991).

5.3.3 Opiates

Acute overdosage with heroin and other opiates may cause coma associated with pinpoint pupils. Acute ischaemic stroke has been noted either immediately, or within hours of injecting heroin (Caplan *et al.* 1982a). Heroin 'mainlining' may also cause bacterial endocarditis with the attendant risks of haemorrhage due to mycotic cerebral aneurysm and of blood-borne cerebral abscess. Seizures and choreiform movements have been noted soon after heroin administration but resolve spontaneously. Inhalation of poisoned heroin vapours, pyrolysate, has led to a spongiform leucoencephalopathy which initially causes apathy, bradyphrenia, motor restlessness, cerebellar ataxia, and pseudo-bulbar dysarthria; death may follow (Wolters *et al.* 1982). Aspergillosis of the cerebral ventricles (Morrow *et al.* 1983) and cerebral mucormycosis (Masucci *et al.* 1982) have occurred in heroin abusers. Lumbosacral and brachial plexus neuropathies have occurred in those injecting adulterant heroin mixtures. These plexus lesions are thought to represent hypersensitivity reactions and may respond to high-dose steroid therapy (Herdmann *et al.* 1988).

5.3.4 Amphetamines

These central nervous system stimulants are generally consumed orally but can be inhaled or injected. Hypertension and tachycardia follow administration. Amphetamine psychosis with prominent paranoia usually occurs in chronic abusers. Acute neurological side-effects most commonly follow injection. Intracranial haemorrhage is signalled by sudden onset of headache within minutes of amphetamine administration. Both subarachnoid and intracerebral haemorrhage are well-recognized complications. Seizures, ischaemic strokes due to vasospasm, and intracranial infection all occur in intravenous amphetamine abusers (Caplan *et al.* 1982b). Methylphenidate, an amphetamine analogue, has been associated with exacerbation or the onset of Gilles de la Tourette syndrome (Golden 1977).

Ecstasy, MDMA or 3,4-Methylenedioxymethamphetamine, is an orally consumed amphetamine derivative popularly used at 'rave' dance parties to induce euphoria and a sense of familiarity. Sweating, tachycardia, and jaw grinding may accompany its use and hypertensive crises, paranoid psychosis, convulsions, stroke, systemic organ failure, and sudden death may occur (Henry 1992).

5.4 Toxic gases and asphyxia

5.4.1 Carbon monoxide

Carbon monoxide intoxication is a leading cause of death or brain damage due to poisoning. Accidental or suicidal exposure to vehicle exhaust fumes or coal gas leaks, fires, or paint removers may all be responsible. Carbon monoxide replaces the oxygen in haemoglobin with the formation of carboxyhaemoglobin, thus causing hypoxic brain damage. In fatal cases of carbon monoxide poisoning, there is multifocal neuronal loss particularly affecting the cerebral cortex, basal ganglia, and limbic system resembling that in anoxic encephalopathy. Prolonged or permanent neurological sequelae are usually seen only in patients rendered unconscious by the initial exposure. Such patients should be treated immediately with 100 per cent oxygen, and if promptly available, hyperbaric oxygen therapy. Hyperbaric oxygen therapy helps eliminate carboxyhaemoglobin and enhances the oxygen dissolved in plasma, but its practical role in treating carbon monoxide poisoning is uncertain. It should be considered when the carboxyhaemoglobin level exceeds 40 per cent in patients with significant neurological abnormalities within a few hours of the exposure (Dart 2004). Those patients who regain consciousness go through variable periods of restlessness, confusion, disorientation, and amnesia. Multifocal neurological abnormalities may appear and fluctuate considerably: agnosias, dyspraxias, dysphasias, dysgraphias, akinesias, rigidity, a Parkinsonian syndrome, deafness, epilepsy, incontinence, and involuntary movements (Garland and Pearce 1967; Lacey 1981; Klawans *et al.* 1982). Low-density lesions may be evident on brain computed tomography as early as 24 h after exposure, usually bilaterally in the globus pallidus. The presence of such lesions signals a poorer prognosis (Sawada *et al.* 1980). Some neurological recovery occurs in most patients. However, permanent neurological sequelae are common, particularly residual disturbances of gait and memory. An encephalopathy starting some weeks after the initial carbon monoxide exposure has been noted occasionally. This is due to delayed onset of demyelination in the cerebral hemispheres (Plum *et al.* 1962; Sawa *et al.* 1981).

5.4.2 Hypoxic-ischaemic encephalopathy

Diffuse cerebral injury occurs in a variety of anoxic circumstances including temporary cardiorespiratory arrest, anaesthetic accidents, cardiopulmonary bypass operations, near-miss drownings, and attempted strangulation or suffocation. The clinical manifestations closely resemble those of carbon monoxide poisoning (Section 5.4.1). Cessation of oxygenated blood flow to the brain for more than 3–5 min is likely to cause long-term cerebral injury. Diffuse cerebral anoxic injury is unlikely if the patient has not been rendered unconscious by the initial insult. During the first 24 h following anoxia, a poor prognosis for independent daily functioning is signalled by absent pupillary light reflexes, disconjugate and disoriented eye movements, absent or extensor motor responses, and lack of response to commands (Levy *et al.* 1985). Gradual recovery occurs over weeks or months but is of variable extent. Severe anoxic-ischaemic insults may result in a permanent vegetative state in which there is no evidence of cognitive awareness despite recovery of brainstem responses (Dougherty *et al.* 1981). Focal cerebral lesions may occur in patients with pre-existing cerebral vascular disease. Persistent amnesia, Parkinsonism, movement disorders, or action myoclonus can all follow anoxic brain injury. Up to a fifth of children satisfactorily resuscitated from near-miss drownings have minor visuomotor impairments or subtle disparities between verbal and performance intelligence quotients; but hard neurological signs are rare (Pearn 1977). Post-mortem studies of hypoxic-ischaemic brain damage show relatively symmetric multifocal lesions affecting either the cerebral cortex or the cerebral white matter, and may involve the caudate nucleus or cerebellum (Dougherty *et al.* 1981).

Occasionally a secondary neurological deterioration occurs days or a few weeks after the initial cerebral anoxic-ischaemic insult. After a good initial recovery, such patients abruptly become irritable, apathetic, and confused and exhibit a shuffling gait with muscular rigidity (Plum *et al.* 1962). Autopsy studies show demyelination within the cerebral hemispheres.

5.4.3 Nitrous oxide

Chronic repeated recreational inhalation of nitrous oxide can lead to sensorimotor polyneuropathy and a myelopathy, with abnormalities of visual evoked responses and sensory nerve action potentials. Improvement occurs with abstinence (Heyer *et al.* 1986). The neurological abnormalities resemble those seen in vitamin B_{12} deficiency and it is of interest that normally non-toxic doses of nitrous oxide can produce neurological deterioration in patients with pre-existing vitamin B_{12} deficiency (Holloway and Alberico 1990).

5.4.4 Ethylene oxide

Ethylene oxide is used as an industrial chemical precursor and for sterilizing heat sensitive medical equipment. An encephalopathy

manifesting with fatiguability, poor concentration, and impaired co-ordination, or a polyneuropathy, may result from prolonged exposure (Gross *et al.* 1979).

5.4.5 Toluene

Paints and glues containing Toluene have been popular with solvent abusers because of their euphoric effects. Two-thirds of a group of chronic abusers showed cognitive, pyramidal tract, cerebellar, brainstem, or cranial nerve abnormalities (Hormes *et al.* 1986). Accidental massive exposure to Toluene di-isocyanate leads to immediate euphoria, ataxia, and impaired consciousness, with persistent memory, mood, and personality changes (Le Quesne *et al.* 1976).

5.5 Therapeutic and diagnostic agent toxicity

This section addresses noteworthy or permanent neurological side effects of some drugs, and radiographic contrast agents. Many drugs produce mild temporary side effects such as giddiness, headache, or concentration difficulties; these are not covered in this section. Tardive dyskinesia (Section 40.9.2) and acute dystonic reactions (Section 40.4.13) due to neuroleptic drugs, and drug induced peripheral neuropathy (Section 21.19) are covered elsewhere.

5.5.1 Oral contraceptives

Stroke is the commonest serious neurological consequence of oral contraceptive use. A three-fold increased incidence of ischaemic stroke is noted in women using oral contraceptives containing oestrogen (WHO 1996a), with a higher risk for pills containing ≥ 50 μg oestrogen. Hypertension, regular cigarette smoking and age over 35 years are important compounding risk factors for stroke in women using the pill. Haemorrhagic stroke is significantly increased in those pill-taking women aged over 35, with a history of hypertension and who smoke (WHO 1996b). Migraine is also a risk factor for ischaemic, but not haemorrhagic, stroke and this risk is increased in oral contraceptive users, particularly with oestrogen dosages of ≥ 50 μg or higher, and in those who also smoke or have high blood pressure (Chang *et al.* 1999). Cerebral venous sinus thrombosis is also attributable to oral contraceptive use (Atkinson *et al.* 1970).

Chorea may occur in patients taking oral contraceptives (Section 40.5.9). Carpal tunnel syndrome has been reported after oral contraceptive use (Sabour and Fadel 1970).

5.5.2 Neuroleptic malignant syndrome

This life-threatening drug reaction produces fever accompanied by autonomic and extra-pyramidal abnormalities (Section 40.9.1). It is generally under-recognized and can produce permanent neurological abnormalities which may reflect a form of heat-stroke. It usually occurs in patients receiving neuroleptic drugs, acutely or chronically, either for psychiatric disorders, or as anti-emetics, or as premedication (Buckley and Hutchinson 1995). It has also been noted following cessation of dopaminergic therapy for Parkinson's disease. Neuroleptic malignant syndrome usually develops subacutely over 1–3 days, even in those patients who have been taking neuroleptic drugs for a long time. A review of the clinical manifestations of a large number of cases shows that muscular rigidity and hyperthermia, sometimes greater than 41°C, are almost always present (Rosenberg and Green 1989). Other common features include mutism, tachycardia, tachypnoea, sweating, and hypertension. Tremor, mask-like facies, hyporeflexia, and obtundation are less common manifestations. The serum creatine phosphokinase level is elevated in three-quarters, often to extreme levels. Pneumonia and respiratory failure are the commonest life-threatening medical complications of the condition. Prompt recognition of the disorder and initiation of specific therapy greatly diminishes the chance of death, which occurs in up to 30 per cent of untreated patients. The offending causative drug should be stopped, and the patient rehydrated and treated with anti-pyretics. Bromocriptine 5 mg orally or nasogastrically 4 times daily or Dantroline 2–3 mg per kg per day intravenously significantly reduces the recovery time (Rosenberg and Green 1989). Cerebellar degeneration has been described in a patient with a particularly hyperpyrexic form of neuroleptic malignant syndrome and it is proposed that such permanent neurological features may reflect heat-related nervous system injury (Lee *et al.* 1989).

The differential diagnostic considerations in a typical case include heatstroke, idiopathic lethal catatonia, malignant hyperthermia associated with anaesthesia, drug interactions with monoaminoxidase inhibitors, and a central anticholinergic syndrome which can be caused by the anticholinergic effects of several neuroleptic drugs.

5.5.3 Lithium

Lithium carbonate is commonly used to treat bipolar affective disorders. It produces tremor in more than 50 per cent, generally mild in degree. This tremor resolves with reduction or cessation of lithium therapy. Overdosage with lithium can produce peripheral neuropathy (Section 21.19). Seventeen patients have been reported with persisting neurological deficits after lithium therapy, commonly female, often associated with toxic blood levels (Donaldson and Cuningham 1983). These permanent deficits include Parkinsonian syndromes with akinetic hypertonicity or cogwheel rigidity, tremors, drooling, dysarthria, mask-like facies, and a positive glabella tap sign. Less frequent permanent features include choreo-athetosis, corticospinal tract damage, oculogyric crises, opisthotonic attacks, ataxia, impaired ocular conjugation, myoclonus, and grand mal seizures. Downbeat nystagmus in the primary position can persist after cessation of lithium therapy (Williams *et al.* 1988). A subacute dementing syndrome associated with myoclonus has occasionally occurred and is associated with periodic complexes on EEG resembling Creutzfeldt–Jacob disease (Smith and Kocen 1988). Such patients recover after withdrawal of lithium.

5.5.4 Cancer chemotherapy

A number of therapeutic agents used to treat cancer may induce encephalopathies (Verstappen *et al.* 2003). These should be distinguished from cerebral secondary deposits, malignant meningitis, opportunistic infections, metabolic disorders, and paraneoplastic neurological syndromes. Peripheral neuropathy may result from treatment with cisplatinum, misonidazol, taxol, or vincristine (Section 21.19). The spectre of neurological side effects is a frequent limiter of the safer upper dosage of antineoplastic drugs.

Methotrexate, intrathecally or intravenously, may cause three distinct encephalopathies (Glass *et al.* 1986). First, a slowly progressive intellectual loss and personality change, sometimes with seizures and ataxia may occur. This seems particularly common in children, especially if radiotherapy has also been given, and may have a delayed onset. Second, aseptic meningitis, sometimes with acute encephalopathy, may develop within hours of intrathecal methotrexate. Transverse myelopathy is a rare sequel of intrathecal methotrexate. Third, high-dose intravenous methotrexate therapy may cause transient focal neurological abnormalities. These usually develop about 7 days after the second or third administration of methotrexate (Glass *et al.* 1986). Such patients abruptly develop gaze palsies, hemiparesis, focal seizures, sensory deficits, or behavioural abnormalities. These may worsen for up to 3 days before slowly resolving completely; CT scans are normal.

Cytosine arabinocide, used intrathecally to treat leukaemia or lymphoma, can induce aseptic meningitis, myelopathy, or encephalopathy with seizures. Cerebellar dysfunction occurs with cumulative dosage, particularly in the elderly (Hwang *et al.* 1985).

5-Fluorouracil and Levamisole adjuvant therapy for 15–19 weeks for colonic adenocarcinoma has caused encephalopathy which progressively worsens over 2 or 3 weeks associated with MRI and biopsy evidence of central nervous system demyelination (Hook *et al.* 1992). Declining intellect, ataxia, or episodic loss of consciousness have occurred, with subsequent improvement, and the syndrome is most likely to represent 5-Fluorouracil toxicity. If 5-Fluorouracil clearance is delayed because of dihydropyrimidine dehydrogenase deficiency, a comatose encephalopathy may occur.

Ifosfamide can cause an acute encephalopathy with cerebellar and extrapyramidal signs, hallucinations, seizures, and coma. This comes on within a day of infusion, and usually recovers a few days later.

Interferon alpha can induce cognitive dysfunction of mild to moderate severity, often associated with a Parkinsonian syndrome, which is not reversible on stopping the drug (Meyers *et al.* 1991).

5.5.5 Radiological contrast agents

Arteriography. Neurological complications occur in 1–2 per cent of carotid or cerebral angiograms (Willinsky *et al.* 2003). Catheter-induced arterial embolization accounts for most cases of focal cerebral deficit or spinal cord damage and up to a third are left with permanent deficits. The direct toxic effects of angiographic contrast media include seizures, which occur most commonly in patients with an underlying disorder of the blood brain barrier. Spinal myoclonus may occur after selective spinal angiograms (Junck and Marshall 1983). Intravenous administration of contrast agents for computed tomography occasionally causes seizures, most commonly if the blood brain barrier is impaired due to an underlying tumour.

Myelography. Acute or chronic arachnoiditis has been associated with the use of oil-based myelographic contrast media such as iophendylate *Pantopaque* or iophenylundecylate *Myodil* (Keogh 1974; Jorgensen *et al.* 1975; Junck and Marshall 1983). The acute reactions usually involve meningismus associated with CSF pleocytosis and settle in a few days. Chronic reactions produce an adhesive arachnoiditis after an interval of some months or more. Chronic back pain and lumbar or sacral root symptoms occur. Patients may be more vulnerable to chronic arachnoiditis if they received myelograms and operations in close succession, making this a possible cause for the 'failed back surgery syndrome' (Jorgensen *et al.* 1975). MRI of the lumbar spine defines the changes of lumbar arachnoiditis. The most typical changes are clumping of nerve roots into small groups, and adhesion of the nerve roots to the dural tube. The treatment of chronic arachnoiditis is primarily symptomatic. Some recommend attempts to remove any residual contrast medium which is still mobile (Junck and Marshall 1983). *Pantopaque* and *Myodil* were generally replaced as myelographic contrast agents during the early 1980s by the water-based compound Metrizamide. Metrizamide could produce seizures or transient encephalopathy with confusion, hallucinations, asterixis, and myoclonus (Bertoni *et al.* 1981; Junck and Marshall 1983). Metrizamide has since been replaced by less toxic water-based myelographic contrast media such as iohexol. In turn, myelography itself is generally being replaced by non-invasive MR scanning.

5.5.6 Epidural and spinal anaesthesia

Neurological complications follow about 1 in 10 000 epidural, intrathecal and caudal local anaesthetic blocking procedures (Puke *et al.* 1989). Often these neurological problems are not evident until persisting neurological symptoms or signs are noted 12 h or more after the last injection of anaesthetic, by which time the nerve block should have worn off.

Direct needle trauma to a cauda equina roots, or to the conus medullaris of the spinal cord, usually causes immediate neuralgic pain, often in a radicular distribution, often accompanied by sudden involuntary movements of a leg, and is sometimes followed by permanent neurological damage within the distribution of the affected nerve root (Hamandi *et al.* 2002). Spinal epidural haematoma is particularly likely in patients with pre-existing coagulation deficits and usually presents with low back pain associated with progressive leg paralysis over a few hours and loss of sphincter control; urgent scanning is required with a view to early neurosurgical decompression so as to try and prevent permanent neurological damage. Spinal epidural abscess is a rare but potentially catastrophic complication, presenting with back pain, feet, and leg weakness; urgent MRI is required on suspicion of this disorder (Grewal *et al.* 2006).

Accidental puncture of the dura mater occurs during intended epidural anaesthesia in about 2–5 per cent of patients and the subsequent local anaesthetic infusion can lead to total intrathecal blockade with unconsciousness and cardiorespiratory failure; complete recovery is the rule with suitable intensive care. Presumed ischaemic lesions of the spinal cord or cauda equina occur, and may be particularly likely after accidental dural puncture and injection of local anaesthetic mixtures containing adrenaline. However, no pathogenetic mechanism is ever established in many cases of permanent neurological damage following epidural or spinal anaesthesia (Yuen *et al.* 1995). Headache in the upright position due to spinal fluid hypotension, and aseptic meningitis are other recognized transient complications of dural puncture during local anaesthesia.

5.6 Complications of organ transplantation

Neurological disorders make a major contribution to the mortality and morbidity of organ transplantation, often developing many months or years later. A quarter of liver transplant recipients and 15 per cent of haematopoietic progenitor cell transplanted patients develop significant neurological disorders (Lewis and Howdle 2003; Denier *et al.* 2006; Saner *et al.* 2006). The range of neurological disorders is large, but particularly common disorders include

stroke, cerebral lymphoma, intracranial infections, polyneuropathy, and side effects of immunosuppressant drugs. Graft-versus-host disease presents a particular problem after bone marrow transplantation, with malabsorption-induced metabolic encephalopathy and neuromuscular disorders prominent (Sostak *et al.* 2003).

The first diagnostic step requires the patients' neurological syndrome to be categorized (see Table 5.1) (Donaghy 1999).

5.6.1 Diffuse encephalopathy

This may range from mild confusion or ataxia to deep coma, sometimes with headache, seizures, or meningeal irritation depending upon the underlying cause. Cyclosporine or tacrolimus toxicity usually appear within 3 months and may include tremors, seizures, or visual disturbances such as hallucinations or cortical blindness. Listeria meningoencephalitis usually develops more than a month after transplantation and often includes prominent features of brainstem dysfunction, such as abnormal eye movements or dysarthria. Cryptococcal meningitis is usually delayed at least 6 months after transplantation. A syndrome of rejection encephalopathy in young transplant recipients, which includes papilloedema, may reflect cumulative physiological and metabolic insults, including hypertension and electrolyte disorders, rather than representing a

Table 5.1 Neurological syndromes in transplant recipients

Diffuse encephalopathy
Meningitis
Encephalitis
Electrolyte disturbances
Rejection encephalopathy
Hypertensive encephalopathy
Hypoxic, hypotensive encephalopathy
Remote effects of systemic sepsis
Multifocal cerebral lymphoma
Pulmonary, liver, or renal failure
Cyclosporine toxicity
Tacrolimus toxicity
OKT3 antibody meningo-encephalopathy

Focal neurological abnormalities
Cerebral lymphoma
Ischaemic stroke
Intracerebral haematoma
Focal cerebral infection
Central pontine myelinolysis

Seizures
Cyclosporine toxicity
Hypomagnaesaemia
Hyponatraemia
Rejection encephalopathy
Cerebral lymphoma
Meningitis/encephalitis

Neuromuscular disease
Perioperative focal nerve damage
Guillain–Barré Syndrome
Chronic inflammatory demyelinating polyneuropathy
Tacrolimus neuropathy
Critical illness polyneuropathy
Polymyositis
Myasthenia gravis
Rhabdomyolysis

direct consequence of rejection. Cardiac or pulmonary transplant recipients may develop hypoxic-hypotensive encephalopathy perioperatively. Encephalopathy regularly occurs in the weeks following bone marrow or liver transplantation, although the pathogenesis often remains unclear. Patients require brain imaging to detect multiple mass lesions, such as multifocal lymphoma, masquerading as diffuse encephalopathy. If the brain scan is normal, spinal fluid examination will detect infections. As well as being itself a cause of encephalopathy, it should be noted that hyponatraemia may be a secondary feature of other neurological disorders, such as meningitis.

5.6.2 Focal cerebral abnormalities, lymphoma and stroke

Hemiparesis, dysphasia, or homonymous hemianopia are usually due to ischaemic or haemorrhagic stroke, or primary cerebral lymphoma. Focal cerebral infection with Toxoplasma or Aspergillus may become evident as early as 2 weeks post-transplant whereas Nocardia brain abscess tends to present later than 3 months. Central pontine myelinolysis Is particularly likely in hepatic transplant recipients, particularly in the presence of blood sodium disorders (Winnock *et al.* 1993).

The risk of cerebral lymphoma in transplant recipients is estimated at 2 per cent, between 30 and 350 times higher than normal (Patchell 1988). The median interval from transplantation to clinical detection of primary cerebral lymphoma in transplant recipients is 9 months, with a range of 5.5–46 months (Hochberg and Miller 1988). The cerebral lymphoma is multifocal in a third and generally affects the cerebral hemispheres. High dose steroid therapy should be avoided prior to neurosurgical biopsy since dramatic tumour shrinkage can occur within a few days and confuse the histological picture. The treatment of cerebral lymphoma in transplant recipients should follow the usual lines (Section 27.8.3), although the prognosis appears to be poorer than in immunocompetent patients with lymphoma.

Stroke is a major cause of morbidity and mortality both early and late after transplantation. Cerebral ischaemic events occurred in nearly 10 per cent of 10-year survivors in the early days of renal transplantation, but the impression is that these are less frequent; now, high doses of steroid have been replaced by Cyclosporine for the prevention of graft rejection. Perioperative stroke is a particular risk in cardiac transplantation due to air or solid embolism, or cerebral hypoperfusion (Montero and Martinez 1986). Haemorrhagic stroke is a noteworthy problem in bone marrow and liver transplant recipients and can reflect underlying septicaemia, endocarditis, thrombocytopenia, or sickle cell disease (Patchell *et al.* 1985; Wijdicks *et al.* 1995).

5.6.3 Convulsions

A multiplicity of factors is generally responsible for convulsions in transplant recipients. Cyclosporine toxicity is a common cause, sometimes exacerbated by hypomagnaesaemia, particularly early after liver transplants (Kahan *et al.* 1987). Focal cerebral lesions such as lymphoma, infarction, or infection should be sought by scanning if convulsions develop after the immediate post-transplant period. If seizures persist in cyclosporine recipients, despite reducing the dosage if the blood level is high, the choice of an anticonvulsant drug is difficult. Phenytoin, carbamazepine, and phenobarbitone all induce hepatic enzymes which pose difficulties

for achieving adequately immunosuppressive blood levels of Cyclosporine. Sodium valporate is the recommended anticonvulsant in patients simultaneously receiving Cyclosporine (Hillebrand *et al.* 1987).

5.6.4 Neuromuscular disorders

Focal peripheral neuropathies may complicate transplant surgery (Donaghy 1999). Self-retaining retractors in the pelvis can cause femoral nerve palsies in renal transplant recipients. Diabetics undergoing renal transplantation are vulnerable to lumbosacral plexus lesions of presumed ischaemic cause. Phrenic nerve lesions can complicate lung transplantation and prolong ventilator dependence post-operatively. Various mononeuropathies complicate liver transplantation, especially brachial plexus injury due to arm malpositioning.

Acute polyneuropathies of Guillain–Barré type are usually seen in bone marrow or hepatic transplantation and can follow renal transplantation from a cytomegalovirus-infected donor. Chronic inflammatory demyelinating neuropathy can occur in the months following liver transplantation, sometimes after immunosuppression with tacrolimus, and shows the usual good response to steroids, plasma exchange, or intravenous immunoglobulin. Although Cyclosporine often produces tinglings in the fingers and toes, this is a 'hyperexcitability' phenomenon which does not reflect underlying polyneuropathy.

Myopathies occur in bone marrow or liver transplant recipients. Chronic graft-versus-host disease can cause polymyositis or myasthenia gravis. A recoverable quadriplegia can occur in liver transplant recipients; its cause is generally unknown although a few cases are due to rhabdomyolysis.

5.6.5 Cyclosporin and tacrolimus toxicity

These immunosuppressive drugs are used widely because of their effectiveness in preventing rejection of organ transplants. Up to a quarter of patients experience neurological side-effects (Kahan *et al.* 1987; Walker and Brochstein 1988). Tremors are commonest, but seizures, dysaesthesia of the extremities, depression, sleepiness, ataxia, and visual hallucinations have all been reported (Kahan *et al.* 1987; Walker and Brochstein 1988; Steg and Garcia 1991). Occasionally patients with cyclosporin neurotoxicity are hypomagnesaemic (Thompson *et al.* 1984). Some others have toxic blood levels of cyclosporin or its metabolites. Symptoms generally resolve on reducing or stopping the drug but this should only be undertaken by those supervising the organ transplant for fear of precipitating graft rejection. Mild tremor or parasthesiae, the commonest complications, are often tolerated without reduction in drug dosage. If anti-convulsant therapy is needed for seizures, sodium valproate is recommended because of the risk that enzyme-induction by phenytoin, carbamezepine, or phenobarbitone will produce low cyclosporin blood levels (Walker and Brochstein 1988).

An acute encephalopathy with cortical blindness, mutism, or other focal deficits may occur, associated with cerebral white matter hypodensity on computed tomographic scan (Rubin and Kang 1987; Bianco *et al.* 2004). Vasogenic oedema is thought to be the cause. This leukoencephalopathy recovers after drug withdrawal, but sometimes recurs on reintroduction of cyclosporin (Walker and Brochstein 1988).

Tacrolimus provides an alternative immunosuppressant to cyclosporine, and is a useful alternative in case of neurological or other side effects. But Tacrolimus itself produces neurological side effects in up to 30 per cent; speech disturbance, seizures, tremor and ataxia, encephalopathy, nightmares, or agitation have been reported and usually resolve with dosage reduction (Wijdicks *et al.* 1994). A more serious leukoencephalopathy may occur, resembling that caused by cyclosporin, and presents with headache, vomiting, seizures, and visual disturbance (Small *et al.* 1996).

5.7 Metal toxicity

Please see also Section 21.19 for the peripheral neuropathies due to toxicity from gold, and Section 21.20 for neuropathies due to arsenic, lead, mercury, and thallium.

5.7.1 Aluminium

Aluminium has been implicated in the pathogenesis of Alzheimer's disease. X-ray spectrometry shows aluminium accumulation in neuronal fibrillary tangles (Perl and Brody 1980). Increased brain aluminium has been associated with cerebral congophilic angiopathy when dementia developed some 15 years after exposure to high aluminium levels in drinking water (Exley and Esiri 2006). A geographical correlation has been noted between drinking water aluminium concentration and the incidence of dementia, as judged by computed tomography scanning requests (Martyn *et al.* 1989). Despite these findings, a causative relationship between ingested aluminium and Alzheimer's disease is generally regarded as being conjectural.

Aluminium toxicity was responsible for the encephalopathy which used to occur in patients receiving long-term renal dialysis using aluminium-rich dialysis fluids. Such patients developed progressive dementia with noteworthy speech abnormalities, myoclonic jerkings, and epilepsy. They have increased aluminium levels in the cerebral cortex, bone, and blood (Alfrey *et al.* 1976). The incidence of dementia correlates closely both with the incidence of fracturing dialysis osteodystrophy and with the aluminium content of the water used in preparing the dialysis fluids (Parkinson *et al.* 1979). Over the last decade, reduction of the aluminium content in the diasylate has massively reduced the incidence of severe dialysis encephalopathy. However subtle alterations in psychomotor function can still be detected in dialysis patients with only mildly elevated serum aluminium levels (Altmann *et al.* 1989). This has led to the suspicion that dietary sources of aluminium, including gastrointestinal phosphate binders, may also lead to toxic aluminium accumulation in dialysis patients.

5.7.2 Bismuth

Encephalopathy has been noted in patients taking bismuth salts for chronic gastrointestinal disorders, particularly for the control of output from colostomies (Burns *et al.* 1974). Confusion, tremors, myoclonus, and a prominent gait abnormality develop. The blood bismuth level is raised. Recovery occurred when bismuth was withdrawn, sometimes with residual memory deficits.

5.7.3 Lead

Frank toxicity due to inorganic lead usually has followed ingestion of lead-containing paints by children or occupational exposure of adult metal workers. There are public health concerns about the degree to which lead from vehicle exhaust fumes and domestic water supply pipes can cause subtle developmental intellectual abnormalities. Chronic low level environmental exposure, as judged

by bone deposition, is associated with reduced cognitive function (Shih *et al.* 2006). In adults, inorganic lead poisoning leads to a purely or predominantly motor peripheral neuropathy (Section 21.20.2). In children, inorganic lead poisoning causes a subacute encephalopathy with irritability or listlessness, sometimes associated with anaemia. This may be followed by clumsiness, seizures and evidence of elevated intracranial pressure with vomiting, headache and papilloedema (Lidsky and Schneider 2003). Childhood lead poisoning may be fatal and autopsy studies of the brain show exudative oedema and widespread patchy cerebral necrosis (Smith *et al.* 1960). Lead lines may be evident in X-rays of the epiphyseal plates of long bones. Mildly impaired cognitive and psychomotor development in children have been correlated with chronic low-level lead exposure as judged by blood and tooth lead contents (Fulton *et al.* 1987). Indeed intelligence quotient correlates inversely with blood lead level in children even below the 10 µg/dl level which has been defined as an elevated level (Canfield *et al.* 2003). Mild neurobehavioural abnormalities associated with elevated childhood lead levels persist into young adulthood (Needleman *et al.* 1990).

Organic lead intoxication usually follows exposure to tetra-ethyl lead, the anti-knock compound of petroleum. Neurological disease has been reported in industrial workers in the petroleum industry (Cassells and Dodds 1946) and in recreational petroleum inhalers (Kaelan *et al.* 1986). The first symptoms consist of altered sleep patterns, dreams, irritability, and anorexia. Confusion or psychosis subsequently develop. In severe toxicity, myoclonic jerks, ataxia, and hallucinations are evident. Death may occur and autopsies characteristically show loss of neurones from Ammon's horn in the hippocampus and of cerebellar Purkinje and granule cells. Former organolead workers show persisting atrophic brain abnormalities on MRI, which correlate with residual bone lead deposition (Stewart *et al.* 2006).

5.7.4 Manganese

Manganese neurotoxicity has been reported in ore miners, particularly in Chile, and in steel workers and welders (Josephs *et al.* 2005) and in methcathinone abusers (Stepens *et al* 2008). Chronic manganese poisoning generally follows exposure for more than 1 year and produces a clinical picture intermediate between Parkinson's and Wilson's diseases (Cook *et al.* 1974; Huang *et al.* 1989). The initial symptoms consist of psychomotor excitement, somnolence, gait unsteadiness, slurred speech, and manipulatory difficulties. More chronic toxicity produces typical features of a notably low volume speech, oral tremors, dystonias, and neuro-psychiatric abnormalities which may progress even 10 years after ceasing exposure (Huang *et al.* 1998). There is a characteristic gait abnormality in which patients walk on the metatarsophalangeal joints in the talipes equinus position, a so-called 'cock walk' (Cook *et al.* 1974). During exposure, blood and hair manganese levels are elevated (Huang *et al.* 1989). MRI shows characteristically increased T1 signal intensity in the Globus pallidus (Jankovic 2005). Levels of manganese in tissues other than the brain slowly revert to normal after patients are removed from the exposure, with resolution of the MRI signal abnormality, although the neurological syndrome does not improve (Stepens *et al* 2008). Neuronal loss affects the globus pallidus predominantly, and the Lewy bodies characteristic of idiopathic Parkinsonism are not present (Jankovic 2005). Although minor neurological improvements have been noted following therapy after metal chelation therapy with eidetic acid, significantly prolonged benefit does not generally result (Cook *et al.* 1974). Indeed chronic asymptomatic manganese exposure in miners causes subtle movement disorders later in life, such as tremors (Hochberg *et al.* 1996). L-Dopa therapy may produce minor improvements in motor abnormality in some patients (Huang *et al.* 1989). Viewed overall, there is little evidence that either welding or manganese toxicity are relevant to the cause of idiopathic Parkinsonism (Jankovic 2005). The extrapyramidal syndrome known as acquired hepatocerebral degeneration, associated with hepatic cirrhosis, shows similar brain MRI features and elevated manganese levels (Burkhard *et al.* 2003). This MRI T1-signal hyperintensity is associated with a sevenfold increase in pallidal manganese content (Klos *et al.* 2006).

5.7.5 Mercury

Two forms of mercury poisoning occur: exposure to inorganic or elemental forms occurs in the manufacture of mirrors and scientific instruments, whereas inorganic mercurial compounds may be consumed in foods such as fish, which have ingested them, or those such as grain, which have been treated with mercurial fungicide. The peripheral nervous system bears the main brunt of inorganic mercury toxicity (Section 21.20.3). Depression, tremor, emotional outbursts, and insomnia may also occur and chelation therapy may improve symptoms (Hargreaves *et al.* 1988). Methyl mercury poisoning produces paraesthesia in the limbs and mouth, gait ataxia, concentrically restricted visual fields or cortical visual loss, and intellectual loss which may persist until death (Davis *et al.* 1994). Mercury levels were increased in affected cortical areas showing neuronal loss and gliosis. There is no evidence that mercury amalgam in dental fillings, or the thimerosal preservative of vaccines produces either significant elevation of body mercury levels or are associated with neurodegenerative disease (Clarkson *et al.* 2003).

5.7.6 Tin

Triethyl tin and trimethyl tin toxicity have been associated with different syndromes of neurological disease; elemental or inorganic tin compounds are not neurotoxic. Over 200 patients were poisoned when triethyl tin contaminated the anti-bacterial drug stalinon (Alajouanine *et al.* 1958). The main clinical features were raised intracranial pressure, generalized seizures, and muscle weakness; 50 per cent of patients died. Autopsies showed intramyelinic oedema in the brain. Trimethyl tin poisoning has been reported less frequently, recently in six patients exposed to the vapour (Besser *et al.* 1987). Symptoms generally developed 3–5 days after exposure and consisted of deafness, cognitive impairment, behavioural abnormalities, seizures, ataxia, limb sensory disturbances, and hyperphagia. Death may occur and recovery may be incomplete in the more severely affected patients. Urinary organotin levels are elevated for 15–20 days after exposure. Attempts to reduce body tin levels using penicillamine were not thought to be clinically beneficial. Autopsy showed evidence of neuronal damage in the cerebellar Purkinje CCU layers and the amygdala.

5.8 Chemical and biological warfare and pesticides

Chronic peripheral neuropathies due to organophosphates and carbamates insecticides, to the organo-metal rodentocytes arsenic,

thallium, and organic mercury, to the fumigant methylbromide, and the herbicide 2,4-D are covered in Section 21.21.

5.8.1 Organophosphorus compounds

Organophosphorus insecticides are the leading cause of systemic poisoning due to agricultural chemicals. Human disease has been most frequently described following discrete episodes of intense exposure, such as tri-ortho-cresyl phosphate contamination of moonshine whisky in 1930s prohibition America 'Jamaica ginger extract' or of Moroccan cooking oil in the 1950s. Nowadays suicidal consumption is common in the developing world (Agarwal 1993) and accidental agricultural exposures are frequent, especially during crop spraying. The possible neurological effects of repeated low dose exposure have not been defined.

Three distinct phases of neurological illness may follow organophosphorus poisoning. The most common manifestation, which occurs within hours of exposure, consists of an acute cholinergic crisis with weakness, and autonomic and cerebral dysfunction. Occasionally, an intermediate paralytic syndrome develops after 1–4 days. Lastly, after a delay of 1 or 2 weeks, some patients develop a sensorimotor polyneuropathy which progresses over subsequent weeks (Section 21.21.9). It is uncommon for all three phases of organophosphate poisoning to occur in the same patient. There are more than 80 organophosphorus compounds in use with varying degrees of toxicity (Dart 2004).

Acute cholinergic phase. Organophosphates irreversibly phosphorylate acetyl-cholinesterase. This inactivates the enzyme causing build-up of acetylcholine at muscarinic, nicotinic, and central nervous system cholinergic synapses within 12 h of exposure. Muscarinic autonomic symptoms invariably occur: miosis, copious bronchosecretions, salivation and lacrimation, bronchoconstriction, bowel and bladder hyperactivity, bradycardia, and arrhythmias. Roughly half of patients also develop weakness due to depolarization block of neuromuscular transmission. In such patients, fasciculations precede the areflexia and weakness, particularly of proximal muscles. Various central nervous system manifestations may occur: impaired consciousness, agitation, tremors, confusion, ataxia, and convulsions. Respiratory failure is the usual mode of death in untreated patients and results from the combination of broncho-constriction and bronchosecretions, respiratory muscle weakness, and impaired central respiratory drive. Because of the clinical urgency posed by organophosphorus poisoning, decisions concerning specific treatment should be based upon the clinical features and history of possible exposure. A test dose of 1 mg atropine intravenously should confirm the diagnosis within 10 min by producing pupil dilation, tachycardia, confusion, and an ileus. The diagnosis may be confirmed retrospectively by measurement of the red blood cell cholinesterase activity which remains depressed for up to 2 months after intense exposure (Coye *et al.* 1987).

The immediate aim of therapy is to prevent death due to respiratory failure. Endotracheal incubation with suction and assisted ventilation may be necessary. Repeated large doses of atropine should be given parenterally to reduce secretions and bradycardia. Gastric lavage or activated charcoal administration help reduce further organophosphate absorption. Pralidoxime or obidoxime specifically reverse peripheral nervous system cholinesterase inactivation by organophosphorus compounds, by removing the phosphoryl group and should be administered as early as possible. The optimal initial and subsequent dosage of pralidoxime, and its actual clinical effectiveness remain uncertain from the limited human clinical trial evidence (Eddleston *et al.* 2002). Diazepam should be used to treat seizures. Prompt and adequate treatment of the cholinergic phase allows complete recovery in less than 2 weeks.

Carbamate insecticides can produce an acute cholinergic syndrome similar to that caused by organophosphorus compounds. However central nervous system effects are less, inactivation of cholinesterase is less complete and lasts for a shorter duration because the enzyme binding is reversible, and pralidoxime therapy may exacerbate the cholinergic excess.

An intermediate syndrome. This syndrome of muscle paralysis has been described which begins 1–4 days after organophosphorus poisoning, and is separate from the preceding cholinergic crisis (Senanayake and Karalliedde 1987). This delayed syndrome of muscle paralysis typically affects the neck, respiratory, cranial nerve, and proximal limb muscles and may require assisted ventilation. Repetitive nerve stimulation shows a myasthenia-like decrement (He *et al.* 1998). It lasts for less than 20 days. It has been described in patients who have already received pralidoxime for the preceding cholinergic phase of the poisoning. The pathogenesis of this delayed paralytic syndrome is not understood. The delayed respiratory failure of this intermediate syndrome may reflect peripheral neuromuscular disease, compared to the mixed central and peripheral contributions to respiratory failure in the early cholinergic phase (Eddleston *et al.* 2006).

Chronic neurotoxicity. A severe polyneuropathy can appear 1–3 weeks after exposure (Section 21.21.9). There is considerable interest currently in whether permanent central nervous system abnormalities follow organophosphate poisoning. A number of studies show relatively minor long-term impairment on neurobehavioural tests or altered sensory testing thresholds (Steenland *et al.* 1994). Whether chronic subclinical exposure can produce similar impairments has become a contested issue.

5.8.2 Carbon-disulphide-based pesticides

Carbon disulphide and carbon tetrachloride mixtures are extensively used in the grain industry for controlling insects. Abnormal finger tremor at 5–7 Hz has been noted in chronically exposed grain workers and some workers display Parkinsonian syndromes which also include rigidity and gait abnormalities (Chapman *et al.* 1991). Peripheral neuropathy due to carbon disulphide exposure is described in Section 21.21.2.

5.8.3 Strychnine

Strychnine is a plant extract present in some commercial rodenticides. Strychnine blocks inhibitory actions of the neurotransmitter glycine in the central nervous system. Symptoms usually occur within 1 h of poisoning with anxiety, extensor spasms, opisthotonos, and convulsions, usually with preservation of consciousness in the initial stages (O'Callaghan *et al.* 1982). In patients who survive, recovery occurs over a few days. In severely poisoned patients, therapy should include respiratory assistance using endotracheal incubation and neuromuscular blockade, and treatment of seizures with diazepam or barbiturates.

5.8.4 Endrin

Convulsions may occur as an early feature of poisoning by this chlorinated hydrocarbon pesticide of the cyclodiene group.

Small epidemics of poisoning have been reported from Pakistan in which patients, usually children, became ill suddenly within a few hours of consuming food presumed to be contaminated with Endrin (Rowley *et al.* 1987). Vomiting, headache, and muscle fasciculation were noted in some patients in addition to tonic-clonic convulsions. Seizures can be resistant to intravenous therapy with diazepam or phenobarbitone, and death may occur. Blood endrin levels may be elevated.

5.8.5 Chemical warfare agents

Two neurotoxic compounds are considered to have potential utility on the battlefield or by terrorists: cyanide gases and organophophates (Martin and Adams 2003).

Cyanides. Although intense exposure to cyanide gas rapidly causes death, lesser degrees of exposure will bring symptomatic patients to medical attention. Smoke from fires can produce cyanide poisoning. Cyanide blocks aerobic metabolism within the brain and other organs by preventing intracellular ATP synthesis due to blocking the electron transport chain. Accordingly, severe exposures cause unconsciousness, followed by respiratory depression and cardiac arrest within minutes. Less intense exposure causes a slower development of coma, preceded by headache, vertigo, nausea, seizures, and respiratory abnormalities (Martin and Adams 2003). The patient's breath may smell of bitter almonds due to hydrogen cyanide being blown off from the lungs (Martin and Adams 2003). Lactic acidosis, with retention of oxygen in venous blood is characteristic. Apart from essential respiratory and circulatory support, the recommended therapy involves correcting the metabolic acidosis, and administering nitrites to induce the formation of methaemaglobin which acts to decoy the cyanate away from the body's mitochondria. Survivors may develop a delayed Parkinsonian syndrome with MRI abnormalities of the basal ganglia, not dissimilar to carbon monoxide poisoning (Section 5.4.1).

Organophosphate nerve agents. Various organophosphates have been developed for inhalation on the battlefield, either immediately after delivery: Tabun or GA, Sarin or GB, Soman or GD, Cyclosarin or GF, or by delayed evaporation over more than a day, agent VX (Newmark 2004). Within minutes of exposure the characteristic symptoms of the initial cholinergic phase of organophosphate poisoning develop (Section 5.8.1). At-risk military personnel are routinely supplied with atropine and pralidoxime auto-injectors, and also benzodiazepine or midazolam injectors to counter the subsequent risk of seizures (Lee 2003; Newmark 2004).

5.8.6 Biological warfare

This is considered together with chemical warfare because the most likely neurologically active agent, botulinum toxin, would also cause paralysis in the battlefield or terrorist setting. The clinical manifestations of botulinum toxin poisoning are considered in Section 24.10.5. Recognizing the clinical syndrome is the cornerstone of diagnosis: descending flaccid paralysis invariably starting in the bulbar musculature, with other cranial nerve and autonomic symptoms, including blurred vision and pupil dilatation (Donaghy 2006; Martin and Adams 2003). Terrorist attacks could employ oral ingestion or inhalation of one of the family of botulinum neurotoxins A–G which are the most poisonous substances known. Unlike the prompt onset of weakness following organophosphate poisoning, botulism generally produces neurological symptoms after a delay of 12–72 h. Survival depends upon respiratory support. Toxin may be detected retrospectively by bioassay of body fluids. Repetitive nerve stimulation at frequencies of 20–50 Hz produces an incremental electromyographic response similar to the Lambert Eaton syndrome. Antitoxin should be administered within the first 24 h if possible, or if weakness is continuing to progress. Most available antitoxins only cover a proportion of toxin subtypes, and carry the risk of hypersensitivity reactions.

Haemorrhagic meningitis may occur following bioterrorist attack with aerosolized anthrax spores. However, this generally follows the primary clinical presentation of anthrax with systemic and pulmonary illness (Donaghy 2006; Martin and Adams 2003).

A prominent intention of any bioterrorist attack would be the creation of panic amongst civilians, with mass sociogenic illness consequently disabling institutions. Panic and psychologically determined 'me-too' symptomatology would present the largest diagnostic and logistic challenges to neurologists in the case of a bioterrorist attack upon an institution or a random civilian population (Donaghy 2006).

5.9 Radiation damage

The occurrence of radiation-induced damage to the brain and spinal cord depends upon the radiation dosage, the scheduling of fractionation, technical aspects of beam focusing, and different individual susceptibilities (Henson and Urich 1982). Because of these variations, it has proved hard to define 'threshold dosages' for the development of radiation-induced neurological injury. Accepted 'safe' dosage regimens are now in general use with the result that radiotherapy-induced injury to the nervous system is less common nowadays. The pathogenesis of nervous system injury involves prominent endothelial damage, and the resulting radiation-induced vasculopathy seems to be a common feature to the different syndromes discussed below. Initial interest that some such patients improve with anticoagulation has not been borne out by clinical experience (Glantz et al. 1994).

5.9.1 Radiation myelopathy

Various clinical syndromes of radiation-induced spinal cord damage occur. By analogy with animal studies, acute spinal cord damage might be expected within hours or days of inadvertently high dosages of irradiation. An early and benign form of spinal cord damage may develop within 6–18 weeks of treatment to fields that had included the cervical cord (Jones 1964). Paraesthesia may radiate through all four limbs and this syndrome usually resolves within 6 months. Lhermitte's symptom of electric shock sensations in the limbs evoked by neck flexion is common in this benign myelopathy.

Delayed progressive radiation myelopathy may develop at any time from a few months to 6 years after radiotherapy (Godwin-Austen et al. 1975; Henson and Urich 1982). Patients experience progressive deterioration in the sensory, motor, and sphincteric functions of the spinal cord below the irradiated level. Sensory disturbance is the commonest initial symptom. Also, Lhermitte's sign occasionally occurs early on in the progressive form of radiation myelopathy. Early on the signs may be referable to unilateral damage of the spinal cord with monoparesis or a Brown–Séquard syndrome. Ultimately signs of bilateral spinal cord damage develop. Progression over months or years leads to clinically

complete loss of spinal cord function. Stabilization with incomplete spinal cord lesions can occur. The chief differential diagnosis is from spinal cord compression due to recurrence or metastasis of the underlying cancer. MRI of the spinal canal may be required to exclude spinal cord compression. Both these investigations may reveal diffuse spinal cord swelling in post-irradiation myelopathy. Many patients die from the effects of radiation-induced spinal cord disease. Pathological studies reveal necrosis of the spinal cord confined to the irradiated segments. White matter tracts are preferentially affected and the lesions may be patchy. Fibrinoid necrosis or hyaline fibrosis of associated blood vessels suggest a vascular basis for the spinal cord damage in some cases.

Estimating the tolerance of the spinal cord to radiation may provide more general guidance to 'safe' radiation doses to other parts of the nervous system. For palliative therapy the risk of radiation myelopathy increases with size of dose per radiation fraction, although smaller fractionation of the total dosage becomes less convenient in this clinical setting (Macbeth et al. 1996). Estimates of the risk of radiation myelopathy point to < 0.2 per cent if 45 Gy is given in fractions of ≤ 2 Gy. For fractional doses of 3 Gy or less, a 5 per cent incidence of radiation myelopathy occurs with a total dose of 57–61 Gy whilst there is a 50 per cent incidence with approximately 70 Gy (Schultheiss 1990).

5.9.2 Radiation encephalopathy

Radiation-induced brain damage generally follows treatment of cerebral tumours, extracerebral head and neck tumours, or neuraxis irradiation in the treatment of leukaemia. Cerebral and carotid arteries may be damaged by irradiation and occlusive stroke may occur many years later. Angiography in such cases shows arterial narrowings within the previously irradiated field (Murros and Toole 1989). A follow-up of children irradiated for scalp ringworm shows an increased incidence of brain tumours 7–16 years later (Modan et al. 1974). Most of these radiation-induced tumours of the nervous system are not gliomas.

A transient encephalopathy may develop a few weeks after irradiation. This presents with drowsiness, clumsiness, and headache which resolve over subsequent months (Henson and Urich 1982). Bilateral low attenuation areas may be seen in the brain on MRI. Death has resulted from a brainstem form of this early delayed encephalopathy in which demyelination was noted at autopsy. A variant of early transient post-irradiation encephalopathy occurs in leukaemics receiving combined treatment with radiotherapy and chemotherapy. This is known as treatment encephalopathy (Section 5.5.4).

The more usual type of post-irradiation encephalopathy develops months or years after radiotherapy and progressively deteriorates. The onset of symptoms is commonest after an interval of 9 months to 2 years. It is particularly likely if standard brain tumour irradiation doses of 5000–7000 rad are administered in daily fractions exceeding 2 Gy (Martins et al. 1977). The focal neurological deficit in post-irradiation encephalopathy often parallels that of the underlying brain tumour to which radiotherapy had been originally directed. Progressive dysphasia, hemianopia, cognitive dysfunction, or hemiparesis may be features of radiation encephalopathy of the cerebral hemisphere. Focal or secondarily generalized seizures and features of raised intra-cranial pressure may occur (Henson and Urich 1982).

The diagnosis of post-irradiation encephalopathy is easy if such a neurological syndrome develops in a patient without a pre-existing brain tumour, who had received radiotherapy for an extra cerebral tumour of the head or neck. The diagnosis is usually difficult in patients who had received radiotherapy for an underlying brain tumour. Without further brain biopsy, the distinction from tumour regrowth often remains uncertain. MRI may show multifocal lesions with mass effect which contrast-enhance, sometimes in a ring pattern, within the field of previous radiotherapy, and the appearance of these waxes and wanes with time, unlike the steady growth of recurrent tumour (Peterson *et al.* 1995). Neuropathological studies show necrotic areas of brain which particularly involve white matter and associated fibrinoid necrosis, hyaline thickening, or thrombosis of vessels (Martins *et al.* 1977). Without treatment, radiation necrosis of the brain tends to worsen progressively, causing death. Occasionally it stabilizes spontaneously. Dexamethasone therapy may control symptoms (Martins *et al.* 1977). Surgical excision of the swollen area of necrotic brain can lead to permanent improvement (Rottenberg *et al.* 1977). Such surgical treatment is best confined to patients with cerebral necrosis following radiotherapy for extra-cerebral malignancies, and who are not going to be intolerably disabled by extirpation of the affected brain area.

5.9.3 Radiation-induced cranial nerve palsies

Progressive visual failure from optic nerve and chiasm damage has followed external irradiation therapy for pituitary tumours and craniopharyngiomas (Atkinson *et al.* 1979). Such patients may also suffer radiation-induced hypothalamic damage. Any cranial nerve may be compromised as a delayed effect of radiotherapy. The hypoglossal nerve appears to be particularly susceptible following previous radiotherapy for tonsular, pharyngeal or supraglottic pharyngeal tumours and prominent bulbar palsy may develop (Shapiro *et al.* 1996).

5.9.4 Brachial and lumbosacral plexopathy

Irradiation of the brachial or lumbosacral plexuses can lead to delayed onset of a slowly progressive plexopathy which needs to be differentiated from tumour recurrence by MRI. Unlike tumour infiltration, radiation plexopathy is usually painless, and the weakened muscles often show myokymia on electromyography (Section 22.4.3).

5.9.5 Lumbosacral radiculopathy

A predominantly motor disorder can affect the legs following irradiation of the lumbar spinal canal as part of treatment for testicular or other neoplasms. Weakness usually commences between 3 and 25 years after the radiotherapy. The radiotherapy doses exceed 40 Gy, which is above the current treatment recommendation of 35 Gy fractionated over 4 weeks for testicular tumours. The leg muscle involvement is often distal and asymmetrical and associated with areflexia. Despite the purely motor nature of this disorder early on, all patients develop mild sensory symptoms eventually, although sensory nerve action potentials remain normal. Half of them ultimately develop mild sphincter disorders, with lack of appreciation of bladder fullness, dribbling, or occasional incontinence (Bowen *et al.* 1996). In most patients the leg weakness slowly progresses, albeit with periods of stabilization for a year or more,

and severe disability can result. With improved safety of radiotherapy schedules, and the substitution of chemotherapy for radiotherapy in treating testicular cancer, the incidence of this unusual disorder can be expected to lessen.

It was not known whether this disorder reflected irradiation damage to the cauda equina nerve roots, or alternatively to the motor neurones of the conus medullaris. However, a neuropathological autopsy study showed a radiation-induced vasculopathy of the proximal spinal roots within the cauda equina, whilst the spinal cord architecture and motor neuronal cell bodies are preserved (Fig. 5.1) (Bowen *et al.* 1996). MRI shows gadolinium enhancement of the cauda equina in some patients, but this is a relatively non-specific finding.

The differential diagnosis includes neural infiltration by recurrent neoplasm or radiation-induced nerve sheath tumours, for both of which pain, unilaterality, and predominantly proximal weakness are to be expected. Apart from the slow evolution, spinal fluid cytology excludes malignant meningitis. Myokymia may be noted on electromyography but does not serve to distinguish radiation-induced plexopathies from radiculopathies.

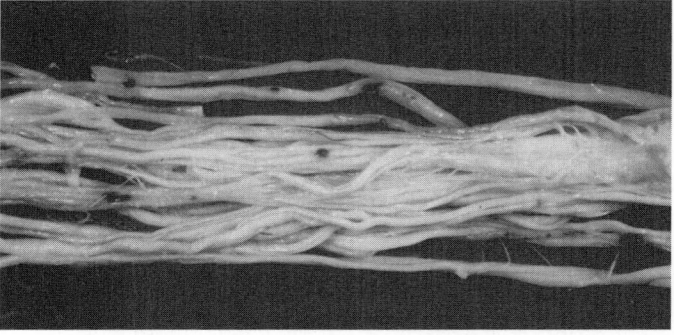

A

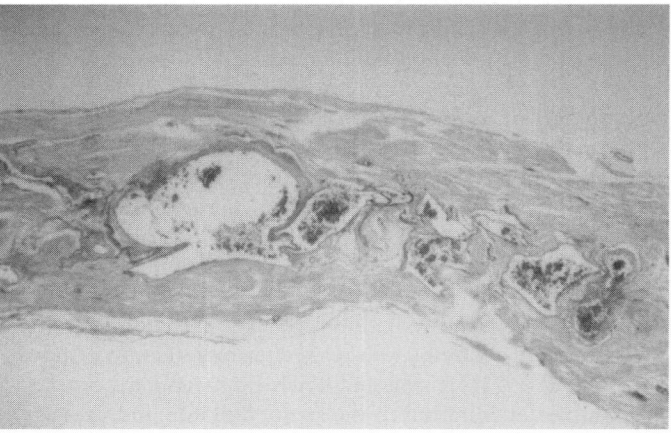

B

Fig. 5.1 Radiation-induced lumbosacral radiculopathy. A. Cauda equina with several thickened roots, some showing focal haemorrhages. B. Longitudinal section of nerve root, showing clusters of dilated vascular channels with thick hyalinized walls. The nerve is fibrosed and distorted by these vessels. Elastic van Giesen Stain.

5.10 Environmental and physical insults

5.10.1 Heatstroke

Heatstroke is defined as a state of acute onset in which the rectal temperature exceeds 40° C, accompanied by hypotension, tachycardia, and hyperventilation, with hot and yet dry skin, and with an associated neurological disturbance (Simon 1993; Yaqub 1987). It occurs in various circumstances including protracted exertions such as marathon running, in unacclimatized visitors to hot climates, in alcoholics, and in those with cardio-vascular disease particularly if elderly and exposed to a heat wave. Various drugs predispose to heatstroke including diuretics, phenothiazines, anti-Parkinsonian drugs, anti-cholinergics, beta-blockers, tricyclic-antidepressants, and amphetamines (Hart et al. 1982). Heat injury may cause some aspects of neurologic dysfunction in the neuroleptic malignant syndrome, particularly in those patients who develop residual neurological abnormalities (Section 5.5.2).

Altered consciousness is the main presenting neurological sign of heatstroke. The pupils are characteristically tightly constricted. Skin temperature may be substantially less than the rectal core temperature which should exceed 40° C. Patients in deep coma may lose brainstem and tendon reflexes and have a poor prognosis even if they receive prompt assisted ventilation and cooling (Yaqub 1987). Convulsions may occur, especially during cooling. The creatine phosphokinase level is elevated and a wide range of metabolic and electrolyte disturbances have been recorded, most notably raised liver enzyme levels or frank hepatic failure (Hart et al. 1982; Yaqub 1987).

Heatstroke constitutes a medical emergency and body cooling should be started immediately. The entire body surface should be exposed and wrapped in a continually moistened sheet in a cool room whilst evaporation is promoted by multiple fans (Yaqub et al. 1986). Such evaporative cooling should be continued until the rectal temperature reaches 38.5°C. A bad prognosis is signalled by an initial rectal temperature exceeding 42°C and failure to achieve cooling within 1 h. Death occurs in approximately 10 per cent of all patients with heatstroke (Yaqub et al. 1986). In the French heatwave of 2003, 60 per cent died of those requiring admission to intensive care. Early predictors of death included high body temperature, prolonged prothrombin times, and requirement for vasoactive drugs (Misset et al. 2006). Outcome was better if detected early and if treated in an intensive care unit with air conditioning. Permanent neurological abnormalities may persist after satisfactory cooling in a few patients. Cerebellar syndromes are commonest (Yaqub 1987) and spinal cord lesions with motor neuron loss also occur (Delgado et al. 1985).

5.10.2 Cold injury

Two forms of tissue injury result from extreme cold. Noteworthy peripheral nerve injury occurs in the trench and immersion foot syndromes which represent non-freezing cold injury resulting from prolonged immersion of the limbs in cold liquid mud in warfare trenches, or in cold waterlogged life rafts (Kennett and Gilliatt 1991). Freezing injury, or frostbite, produces a localized area of generalized tissue necrosis and peripheral nerve injury remains more or less confined to this area.

The clinical features of the trench and immersion foot syndromes are similar, although not identical. An affected limb becomes numb and clumsy. Pain and tingling are uncommon but calf cramps may occur. The skin passes through a hyperaemic red and oedematous phase before becoming 'sickly yellow' or mottled (Ungley et al. 1945). On removing the tight boots encasing oedematous feet and rewarming, the limb goes through a hyperaemic phase lasting up to 10 weeks. During this, signs of a predominantly sensory and autonomic neuropathy are present. Pain and heat sensations are impaired in a glove and stocking distribution and the limb is warm and dry. Over the long term, skin colour and temperature return to normal, sweating returns and may be excessive, and sensory and motor function returns towards normal. Chronic pain is reported by many patients, particularly a burning dysaesthesia in the region of the metatarsal heads exacerbated by walking (Blair et al. 1957). Hyperhidrosis or signs of distal sensorimotor neuropathy may persist. Histopathological studies show Wallerian degeneration in the early stages of cold immersion injury, particularly affecting interdigital nerves.

5.10.3 Altitude sickness

Acute mountain sickness afflicts climbers who rapidly ascend to heights of at least 3000 m without intermediate periods of acclimatization. Symptoms develop a few hours or days after ascent. Headache, ataxia, cognitive impairment, and vomiting due to cerebral oedema may be accompanied by dyspnoea due to pulmonary oedema (Johnson et al. 1984). Headache is the commonest manifestation, in over 80 per cent, and often has characteristics of raised intracranial pressure (Silber et al. 2003). Various other neurological disorders are seen in climbers: transient ischaemic attacks, cerebral venous sinus thrombosis, syncope, and diplopia (Basnyat et al. 2004). Subtle aphasic errors and impaired verbal learning and memory may persist for at least a month after ascents to altitudes above 5000 m (Hornbein et al. 1989). Climbers with a history of repeated conquests of peaks exceeding 8500 m without using supplementary oxygen show long-term impairments of concentration and memory (Regard et al. 1989). Death may result from acute mountain sickness. Acetozolamide pre-treatment reduces the incidence of altitude sickness (Group BMRESMSS 1981). Rapid descent provides the definitive treatment for mountain sickness. However patients may benefit from dexamethasone if descent proves impractical (Levine et al. 1989).

5.10.4 Diving

Decompression sickness, or the 'bends', occurs in deep-sea divers returning to the surface without adequate decompression. It has also occurred in aviators ascending in unpressurized aircraft. At least 24 h should elapse between diving and going to altitude. Compressed air sickness was originally called caisson disease when it was noted following the introduction of high-pressure chambers for underwater work. The neurological illness is generally believed to result from the formation of intravascular gas bubbles, causing arteriolar or venular blockage. The commonest manifestation of acute decompression sickness is 'limb-bends' in which musculoskeletal pains flit from joint to joint. The 'chokes' refers to an acute respiratory decompression sickness which may occur after a latency of several hours. Neurological complications occur in roughly a quarter of patients, particularly if recompression has not been undertaken at the first sign of the 'bends'. Neurological symptoms follow the 'bends' by 1–36 h. Spinal cord damage or 'spinal bends' is the commonest neurological manifestation (Kimbro *et al.* 1997). Minimal limb weakness or paraesthesia may progress to complete paraplegia or tetraplegia in less than 1 h. Minor degrees of spinal cord damage may be discovered at autopsy in medically fit divers dying for unrelated reasons or in divers who have made a full functional recovery from 'spinal bends' as a result of prompt recompression (Palmer *et al.* 1987). Occasionally there is evidence of brain involvement with visual blurring, diplopia, dysarthria, deafness, or cognitive disturbances. Migraine-like symptoms have been described in aviators after descent. Treatment of decompression sickness consists of immediate inhalation of a high concentration of oxygen, and immediate transport to a hyperbaric oxygen chamber for recompression. The nitrogen may be eliminated more quickly if recompression uses a helium–oxygen mixture rather than oxygen alone (Melamed *et al.* 1992).

5.10.5 Electrical and lightning injuries

The site of the neurological injury is mainly determined by the part of the body receiving the electric shock or lightning strike. The immediate consequences of electrical injury include the electrical tinglings familiar to all of us, and for more severe strikes, there may be temporary unconsciousness with retrograde amnesia, temporary tinnitus and deafness, complex visual disturbances, and temporary or permanent cardiorespiratory arrest. A wide variety of longer lasting or permanent neurological sequelae have been recorded following electrical injury: posthypoxic encephalopathy, intracranial haemorrhage, and cerebellar syndromes (Cherington 2003). These may be present from the time of the shock, or develop after delays of days, weeks, or even months. Immediate onset of transient paraplegia with sensory loss has followed lightning strikes, and may recover in less than 24 h. Permanent spastic quadriplegia with small hand muscle wasting has followed electric shock to the arm. The onset of quadriplegia may be delayed for some days after the electric shock, and sometimes eventually recovers partially some months later (Farrell and Starr 1968). Electric shocks or lightning strikes to the head can produce an immediate or delayed onset of hemiparesis, aphasia, or unilateral extra-pyramidal syndromes (Farrell and Starr 1968). Cerebral damage with delayed onset may reflect electrically induced damage to cerebral vessels. Persisting fatigue and concentration difficulties are reported in survivors of lightening strikes sufficient to render them unconscious (van Zomeren *et al.* 1998). Seizures or myoclonic jerks may occur as an immediate sequel of electrical injury. Peripheral nerve damage is usually restricted to the shocked limb. Permanent peripheral nerve damage occurs within the area of generalized tissue burn, most generally affecting the median or ulnar nerves in the hand but more extensive peripheral nerve damage may ensue (Hawkes and Thorpe 1992). Electrical muscle injury may produce substantial subfascial oedema and early fasciotomy may be required to prevent secondary peripheral nerve damage or distal ischaemia (DiVincenti *et al.* 1969).

5.11 Plant and fungus poisoning

A vast range of plants and fungi can produce systemic poisoning syndromes after ingestion which may include autonomic, neurological, or psychiatric features. These are too numerous for comprehensive discussion here and the reader is referred to detailed reference texts for further details (Dart 2004).

5.11.1 Buckthorn

Progressive ascending polyneuropathy resembling Guillain–Barré syndrome has followed consumption of the poisonous Buckthorn shrub, Karwinskia humboldtiana, which grows in Mexico and Texas. Patients develop an areflexic quadriplegia, and may have weakness of respiratory and bulbar muscles. Sensory loss is relatively mild. Patients who survive recover completely over a matter of months. The spinal fluid protein content is typically normal in contrast to Guillain–Barré syndrome. Sural nerve biopsy shows acute segmental demyelination (Calderon-Gonzalez and Rizzi-Hernandez 1967). Supportive treatment should follow that outlined for Guillain–Barré syndrome (Section 21.10.1).

5.11.2 Cicutoxin

Water hemlock, or cicuta, contains the poison cicutoxin. The plant is sometimes accidentally consumed after misidentification as wild parsnip, artichoke, or potato. Different isoforms of cicutoxin have various ion channel blocking effects or block brain GABA receptors (Uwai *et al.* 2000). The symptoms of poisoning reflect cholinergic excess at muscarinic and disinhibition of central nervous system synapses. Abdominal pain, sweating, bronchosecretion, salivation, brachycardia, hypotension, and pupillary abnormalities are common early features. Convulsions are frequent and may lead to status epilepticus. Non-convulsive involuntary movements may occur causing trismus, opisthotomus, and hemiballismus. It is likely that cicutoxin has a direct toxic effect on muscles, causing tenderness and weakness of trunk and proximal limb muscles. Creatine phosphokinase levels are elevated and severe metabolic acidosis may occur. Infusions of thiopentone sodium control the abnormal muscle movements and seizures (Starreveld and Hope 1975). Infusions of atropine, and haemodialytic removal of the circulating toxin are recommended.

5.11.3 Gloriosa

Acute ascending polyneuropathy has followed ingestion of Gloriosa superba, the glory lily, a tuber found in tropical Africa, Asia, and North America (Angunawela and Fernando 1971). Gloriosa contains colchicine which is known to cause a neuromyopathy when given for therapeutic purposes (Section 21.19.5). In massive overdose, colchicine may cause confusion and signs of cerebral oedema leading ultimately to brain death (Heaney *et al.* 1976).

5.11.4 Mushrooms

The wide range of poisoning syndromes that may follow ingestion of different species of mushrooms include hepatorenal failure, gastroenteritis, parasympathomimetic syndromes due to muscarinic effects, and disulfiram-like ethanol sensitivity (Dart 2004). Primarily neurological and psychiatric syndromes follow poisoning with the hallucinogenic mushrooms, of which Psilocybin has been popular for recreational abuse. Patients may develop confusion, visual hallucinations, distorted perceptions, and ataxia, sometimes accompanied by signs of parasympathetic abnormalities. Symptoms usually develop within 90 min of ingestion and resolve within 4–12 h. Seizures or hyperthermia occasionally occur (McCormick *et al.* 1979). Sedation with benzodiazepines may be necessary. Gut decontamination should be considered within the first few hours of large overdoses, particularly in children. Major tranquillizers should be used sparingly if at all to control psychotic features because of their propensity to lower seizure thresholds. Self-injury may occur if the poisoning precipitates aggressive or suicidal behaviour. Disturbing psychiatric 'flash-back' symptoms may persist after the acute poisoning (Benjamin 1979).

5.11.5 Podophyllin

Podophyllin is an antimitotic drug derived from the May Apple, of the genus Podophyllum, which has been used for topical treatment of warts. Human toxicity has followed excessive skin absorption or oral ingestion, including overdosage with herbal laxative tablets containing podophyllin (Dobb and Edis 1984; Filley et al. 1982). Confusion or impaired consciousness, hallucinations, and ataxia may all occur during the first week of toxicity. Evidence of an axonal degeneration sensorimotor peripheral neuropathy commences during the second week, although absent tendon reflexes may have been noted earlier. The neuropathy may worsen for up to 3 months before slowly improving (Filley et al. 1982; Dobb and Edis 1984).

5.11.6 Solanine

Green or sprouting potatoes may contain glycoalkaloids, which causes illness 7–19 h after ingestion. Solanine poisoning from potatoes is uncommon if the green skins are not eaten and if the potato has been boiled thoroughly; baking does not detoxify solanine. Outbreaks of poisoning have occurred in institutions such as schools (McMillan and Thompson 1979). Solanine depresses human pseudocholinesterase activity. Vomiting, diarrhoea, and fever are the commonest symptoms. Some patients develop confusion, delirium, hallucinations, headaches, convulsions, paraesthesia, and muscle spasms. Recovery occurs over a few days but more persistent visual blurring or giddiness have been noted.

5.12 Animal poisons, bites, and stings

Poisons produced by organisms have survival benefit for those organisms either in terms of offence, such as the poison produced by snakes that enables them to immobilize their prey, or defence, such as the toxins produced by sea algae that discourage their consumption. Some biological toxins have highly specific effects upon nerve conduction or synaptic transmission and attract interest as molecular probes of excitable tissues. This section discusses human poisonings due to bungarotoxin and latrotoxin, both of which interfere with cholinergic neurotransmission, and tetrodotoxin, saxitoxin, brevitoxin, and ciguatoxin, all of which interfere with sodium channel function in excitable membranes.

5.12.1 Ciguatera fish poisoning: ciguatoxin

Ciguatoxins occur in certain predatorial fish from tropical reefs in the Atlantic and Pacific: barracuda, red snapper, grouper, and amberjack. The toxin originates in dinoflagellate plankton of the Gambierdiscus genus which are ingested by small fish which, in turn, are themselves eaten by larger predators. Usually within 12 h of a meal, patients develop vomiting, diarrhoea, cramps or myalgias of distal muscles, paraesthesiae, and gait ataxia (Isbister and Kiernan 2005; Pearn 2001). Characteristically, the paraesthesiae start circumorally. The physical signs are of a sensory polyneuropathy, often predominantly affecting small fibre functions. Life-threatening respiratory muscle paralysis occurs in severe cases. Symptoms generally undergo gradual resolution over

approximately 1–2 weeks. There is no specific antidote. Although intravenous mannitol is generally accepted as treatment based on unblinded studies, it was not found effective by a double blind, randomized trial (Schnorf et al. 2002). Fatigue, arthralgias, or polymyositis have been reported as a sequel in some patients.

5.12.2 Puffer fish poisoning: tetrodotoxin

Tetrodotoxin reduces the excitability of nerve and muscle membranes by reducing their permeability to the inflow of sodium ions. Poisoning has usually followed ingestion of internal organs or skin of puffer fish, particularly in Japan and Australia, or occasionally after consumption of porcupine fish. It has also followed envenoming from the bite of the blue ringed octopus (Isbister and Kiernan 2005). Vomiting, dizziness, and a sensation of floating may occur. The initial neurological symptom is paraesthesia, often circumorally. Muscle twitching, generalized flaccid paralysis, repiratory muscle failure in severe cases, ataxia, and hypotension follow (Isbister et al. 2002). Neurophysiology shows axonal excitability to be diminished with conduction slowing, reduced compound muscle action potentials, and sensory action potentials. Detailed neurophysiological analysis showed the axonal membrane changes to be entirely consistent with the known biophysical effects of tetrodotoxin poisoning (Kiernan et al. 2005). There is no specific antidote to tetrodotoxin poisoning and treatment is supportive. It is unclear whether anticholinesterase drugs yield any benefit, despite anecdotal claims.

5.12.3 Shellfish neurotoxicity

Paralytic shellfish poisoning: saxitoxin. Outbreaks of paralytic shellfish poisoning have followed consumption of crabs, mussels, and other shellfish obtained from waters where 'red tides' have been observed, usually in temperate zones. These 'red tides' are due to various toxic dinoflaggelate sea algae, principally of the genera *Alexandrium*, *Pyrodinium*, or *Gymnodium*. Some of these algae contain saxitoxin and are ingested and concentrated by shellfish. Also a blocker of axonal sodium channels, the biophysical and clinical effects of Saxitoxin closely resemble those of tetrodotoxin (Section 5.12.2) except for the hypotension (Isbister and Kiernan 2005; Lehane 2001). Outbreaks of poisoning have occurred in a wide variety of countries bordering on the Atlantic and Pacific oceans. Symptoms usually develop within an hour of the shellfish meal. Paraesthesiae are initially circumoral and later affect the limbs. In more severe cases there is progressive muscular paralysis leading to respiratory muscle failure. There are no specific antidotes to saxitoxin poisoning. Survivors generally recover within a week.

Neurotoxic shellfish poisoning: brevitoxin. The dinoflagellate alga *Gymnodium brevis* produces brevitoxin and is another cause of 'red tides' around the Atlantic coast of the southern USA and New Zealand. The alga is concentrated by shellfish and human disease follows within 3 h of their consumption (Sakamoto et al. 1987). Although the effect on sodium channels is similar to ciguatoxin (Section 5.12.1) there is an initial neuroexcitatory effect (Isbister and Kiernan 2005). Diarrhoea, abdominal pain, rectal burning pain, and circumoral paraesthesiae are the initial symptoms. Tingling later extends to the limbs and trunk. Vertigo, ataxia, and repeated seizures may all occur in severe poisonings. The poisoning is generally milder than paralytic shellfish poisoning due to saxitoxin, and no deaths have been notified.

Amnesic shellfish poisoning: domoic acid. An outbreak of toxic encephalopathy has followed ingestion of mussels contaminated with domoic acid derived from the algal genus *Nitzschia*, which is related to the excitatory transmitter substance glutamate. Patients developed gastrointestinal symptoms within 12 h of consumption. These were followed by various combinations of confusion, altered consciousness, short term memory loss, seizures, myoclonus, unsteadiness, weakness, fasciculations, alternating hemiparesis, and ophthalmoplegia (Perl *et al.* 1990). Many months later survivors displayed anterograde amnesia and evidence of a predominantly motor axonal peripheral neuropathy. Autopsy studies showed hippocampal damage in a pattern resembling that caused by excitotoxins (Teitelbaum *et al.* 1990).

5.12.4 Snake envenoming

Various neurotoxic polypeptides and phospholipases A_2 are present in the venoms of different snakes and are injected into the victim via fangs during a bite. Puncture marks may be visible on the skin. Spitting cobras can spray venom into their victim's eyes without biting. Polypeptide neurotoxins act postsynaptically to block synaptic transmission at the neuromuscular junction. Phospholipases A_2 act presynaptically to deplete the motor nerve terminal of synaptic vesicles. A presynaptic form of blockade is suggested by lack of response to antivenoms or anticholinesterase (Goonetilleke and Harris 2002). Some snake venoms also contain other toxins causing severe bleeding disorders, rhabdomyolysis, renal failure, and hypovolaemic shock and pulmonary oedema due to increased capillary permeability.

Postsynaptic blockade. Krait, genus *Bungarus*, venoms contain bungarotoxins, including α-bungarotoxin which binds powerfully to postsynaptic acetylcholine receptors, thereby blocking neuromuscular transmission (Goonetilleke and Harris 2002) and β-bungarotoxin which acts presynaptically. Kraits are found in Southeast Asia, India, Indonesia, Taiwan, and China, and usually bite their sleeping victims at night (Warrell *et al.* 1983). Not all bites involve envenoming sufficient to result in paralysis. After the bite, the preparalytic phase usually lasts 1–3 h, but delays of up to 12 h have been recorded. Ptosis is usually the earliest sign of impending generalized muscular paralysis. Complete muscular paralysis may occur, with death from respiratory failure unless assisted ventilation is instituted promptly. Muscle fasciculations may be observed after mamba bites. Neither antivenom nor edrophonium show an impressive clinical effect in krait bite victims. With prompt ventilation and adequate supportive care, full recovery occurs within a few days.

Postsynaptic neuromuscular blockade occurs after cobra (*Naja*) envenoming, presenting a similar clinical picture to that of krait envenoming. A myasthenic decrement may be noted neurophysiologically after cobra envenomation, and positive *Tensilon*®, edrophonium, test responses may occur. Unassisted respiratory function may be maintained by infusion of adequate doses of neostigmine (Watt *et al.* 1986).

African mamba bites not only produce severe blockade of neuromuscular transmission by activating presynaptic voltage-gated potassium channels, but also contain toxins blocking muscarinic acetylcholine receptors and inhibiting acetylcholinesterase (Goonetilleke and Harris 2002).

Presynaptic blockade. Russell's viper, common in South Asia, produces a mixed clinical picture after envenoming. In parts of India and Sri Lanka the venom's phospholipase A_2 presynaptic neurotoxins produce external ophthalmoplegia and ptosis, and rarely a descending paralysis resulting in respiratory failure, and limb paralysis. There is no response to *Tensilon*®, edrophonium, suggesting that the neurotoxin inhibits presynaptic release of acetylcholine at the neuromuscular junction, rather than causing postsynaptic blockade. Generalized muscle tenderness and myoglobinuria indicate that the venom also causes rhabdomyolysis (Phillips *et al.* 1988). Generalized rhabdomyolysis is the most serious consequence of bites from a wide variety of other snakes, including sea snakes, the tropical rattlesnake in Brazil and some Australasian snakes: taipan, tiger snake, mulga snake, and small-eyed snake (Phillips *et al.* 1988). It is likely that the phospholipase A group of neurotoxins are responsible for both the presynaptic blockade of neuromuscular transmission and the myopathy. Acute renal failure may complicate the rhabdomyolysis.

The Australasian taipan's venom contains Taipoxin, a phospholipase A_2, which acts on presynaptic nerve endings to abolish transmitter release with a latency of a few hours before the clinical onset of paralysis. Paralysis predominates in the cranial, trunk, and proximal limb muscles, and artificial ventilation may be necessary. Compound muscle action potential amplitudes are low, but repetitive stimulation produces a distinctive brief potentiation in amplitude followed by an enhanced decrement unaffected by edrophonium (Connolly *et al.* 1995).

Stroke. A number of snake venoms possess procoagulant, fibrinolytic anti-platelet and haemorrhagic metalloproteinase activity. This can cause haemorrhagic stroke. Ischaemic cerebral thrombosis is unusual except after envenoming by pit vipers from Martinique and St Lucia. The stroke onset is often delayed to more than 8 hrs following the snakebite, and contributes to the high risk of fatality (Mosquera *et al.* 2003).

Treating snakebites. Measures to delay toxin absorption should be undertaken immediately: immobilization of the affected limb and application of pressure bandages to reduce lymphatic drainage from the site. Analgesic treatment of the severe pain should avoid opiates, which enhance the potential for respiratory depression. The killed snake, or a description may allow species identification. Any signs of systemic envenoming should lead to emergency hospital admission. Antivenoms active against the identified snake, or known local species, should be administered with awareness of potential anaphylaxis. Anticholinesterase and any necessary respiratory support are required if neurotoxic signs emerge (Goonetilleke and Harris 2002).

5.12.5 Spider and scorpion venoms

These venoms are various mixtures of invertebrate nerve ion channel blockers, hyaluronidase, and phospholipases which immobilize invertebrate prey. Human sodium, calcium, and potassium channels are also affected (Goonetilleke and Harris 2002).

Spiders. Envenoming by black- and brown-widow spiders of the genus *Latrodectus* include toxins such as α-latrotoxin, which destroys motor nerve terminals, causing failure of neuromuscular transmission (Okamoto *et al.* 1971). Symptoms in humans also follow bites by funnel web, mouse, and banana spiders. Toxic spider bites are rare in humans but can arise from imported goods. The unlucky victims experience pain at the site of the envenoming, abdominal pain, and leg weakness. A prospective Australian study showed significant pain lasting more than a day in only 6 per cent, and severe neurotoxic effects to be very rare (Isbister and Gray 2002). Recently, it has been concluded that horse serum antivenom does not promote recovery and should be considered only in potentially life-threatening poisonings. The antivenom has the disadvantage of causing allergic reactions (Moss and Binder 1987).

Scorpions. Common in the tropics, the effects of scorpion stings are particularly attributable to their serotonin content, and the various nerve ion channel blockers, which often enhance nerve terminal neurotransmitter release (Goonetilleke and Harris 2002). Severe pain is characteristic at the sting site. Systemic release of neurotransmitters commonly produces autonomic symptoms such as colic and diarrhoea, sweating, priapism, hypertension, and cardiac arrhythmias. Secondary cardiopulmonary damage may result from this autonomic storm.

5.12.6 Tick paralysis

Ascending flaccid paralysis of the limbs, culminating in bulbar and respiratory muscle weakness follows prolonged attachment to the body in the early summer by various species of gravid female tick. This mainly occurs in Northwestern USA, but can occur elsewhere in North America, Australia, South Africa, and Southern Europe. About 40 tick species can cause paralysis. Tick paralysis of animals, particularly sheep, was once of economic importance to farmers. Children are particularly likely to be affected, and tick envenoming enters the differential diagnosis of the acutely weak child. The presence of an attached tick should be sought, usually to be found in the scalp or ear. Ascending paralysis usually occurs after the tick has been attached for 5 days or more. Areflexia and ptosis are frequent. Although paraesthesiae may occur, sensory loss is uncommon. Bulbar paralysis can develop within 2 days of the onset of weakness. Neurophysiological studies show reduced amplitude of compound muscle action potentials but only a moderate slowing of motor nerve conduction and no defect of neuromuscular transmission (Grattan-Smith *et al.* 1997; Vedanarayanan *et al.* 2002). The spinal fluid is usually normal. Clinical and electrophysiological improvement occurs within a few days of removing the tick, and recovery can be complete in less than a week. Guillain–Barré syndrome is the main differential diagnosis but the early clinical and electrophysiological features are not discriminating. Comprehensive search for a tick is critical to diagnosis in children developing ascending paralysis in endemic areas. This form of tick paralysis is due to a toxin and should not be confused with the polyradiculopathy caused by the *Borrelia* infection transmitted by tick bites which is known as Lyme disease or Bannwarth's syndrome (Section 21.14.3).

References

Agarwal SB (1993). A clinical, biochemical, neurobehavioral, and sociopsychological study of 190 patients admitted to hospital as a result of acute organophosphorus poisoning. *Environ Res*, **62**, 63–70.

Aggarwal SK, Williams V, Levine SR *et al.* (1996). Cocaine-associated intracranial hemorrhage: absence of vasculitis in 14 cases. *Neurology*, **46**, 1741–3.

Alajouanine T, Derobert L, Thieffry S (1958). Comprehensive clinical study of 210 cases of poisoning by organic salts of tin. *Rev Neurol (Paris)*, **98**, 85–96.

Alfrey AC, LeGendre GR, Kaehny WD (1976). The dialysis encephalopathy syndrome. Possible aluminum intoxication. *N Engl J Med*, **294**, 184–8.

Altmann P, Dhanesha U, Hamon C *et al.* (1989). Disturbance of cerebral function by aluminium in haemodialysis patients without overt aluminium toxicity. *Lancet*, **2**, 7–12.

Angunawela RM, Fernando HA (1971). Acute ascending polyneuropathy and dermatitis following poisoning by tubers of Gloriosa superba. *Ceylon Med J*, **16**, 233–5.

Atkinson AB, Allen IV, Gordon DS *et al.* (1979). Progressive visual failure in acromegaly following external pituitary irradiation. *Clin Endocrinol (Oxf)*, **10**, 469–79.

Atkinson EA, Fairburn B, Heathfield KW (1970). Intracranial venous thrombosis as complication of oral contraception. *Lancet*, **1**, 914–8.

Bartsch AJ, Homola G, Biller A *et al.* (2007). Manifestations of early brain recovery associated with abstinence from alcoholism. *Brain*, **130**, 36–47.

Basnyat B, Wu T, Gertsch JH (2004). Neurological conditions at altitude that fall outside the usual definition of altitude sickness. *High Alt Med Biol*, **5**, 171–9.

Benjamin C (1979). Persistent psychiatric symptoms after eating psilocybin mushrooms. *BMJ*, **1**, 1319–20.

Bertoni JM, Schwartzman RJ, Van Horn G *et al.* (1981). Asterixis and encephalopathy following metrizamide myelography: investigations into possible mechanisms and review of the literature. *Ann Neurol*, **9**, 366–70.

Besser R, Kramer G, Thumler R *et al.* (1987). Acute trimethyltin limbic-cerebellar syndrome. *Neurology*, **37**, 945–50.

Bianco F, Fattapposta F, Locuratolo N *et al.* (2004). Reversible diffusion MRI abnormalities and transient mutism after liver transplantation. *Neurology*, **62**, 981–3.

Blair JR, Schatzki R, Orr KD (1957). Sequelae to cold injury in one hundred patients; follow-up study four years after occurrence of cold injury. *J Am Med Assoc*, **163**, 1203–8.

Blass JP, Gibson GE (1977). Abnormality of a thiamine-requiring enzyme in patients with Wernicke-Korsakoff syndrome. *N Engl J Med*, **297**, 1367–70.

Bowen J, Gregory R, Squier M *et al.* (1996). The post-irradiation lower motor neuron syndrome neuronopathy or radiculopathy? *Brain*, **119**, 1429–39.

Brennan FN, Lyttle JA (1987). Alcohol and seizures: a review. *J R Soc Med*, **80**, 571–3.

Buckley PF, Hutchinson M (1995). Neuroleptic malignant syndrome. *J Neurol Neurosurg Psychiatry*, **58**, 271–3.

Burkhard PR, Delavelle J, Du Pasquier R *et al.* (2003). Chronic parkinsonism associated with cirrhosis: a distinct subset of acquired hepatocerebral degeneration. *Arch Neurol*, **60**, 521–8.

Burns R, Thomas DW, Barron VJ (1974). Reversible encephalopathy possibly associated with bismuth subgallate ingestion. *BMJ*, **1**, 220–3.

Calderon-Gonzalez R, Rizzi-Hernandez H (1967). Buckthorn polyneuropathy. *N Engl J Med*, **277**, 69–71.

Canfield RL, Henderson CR Jr, Cory-Slechta DA *et al.* (2003). Intellectual impairment in children with blood lead concentrations below 10 microg per deciliter. *N Engl J Med*, **348**, 1517–26.

Caplan LR, Hier DB, Banks G (1982a). Current concepts of cerebrovascular disease—stroke: stroke and drug abuse. *Stroke*, **13**, 869–72.

Caplan LR, Thomas C, Banks G (1982b). Central nervous system complications of addiction to "T's and Blues". *Neurology*, **32**, 623–8.

Carlen PL, Wilkinson DA, Wortzman G *et al.* (1981). Cerebral atrophy and functional deficits in alcoholics without clinically apparent liver disease. *Neurology*, **31**, 377–85.

Cassells D, Dodds E (1946). Tetra-ethyl lead poisoning. *BMJ*, **2**, 681–5.

Chang CL, Donaghy M, Poulter N (1999). Migraine and stroke in young women: case-control study. The World Health Organisation Collaborative Study of Cardiovascular Disease and Steroid Hormone Contraception. *BMJ*, **318**, 13–8.

Chapman LJ, Sauter SL, Henning RA *et al.* (1991). Finger tremor after carbon disulfide-based pesticide exposures. *Arch Neurol*, **48**, 866–70.

Charness ME, DeLaPaz RL (1987). Mamillary body atrophy in Wernicke's encephalopathy: antemortem identification using magnetic resonance imaging. *Ann Neurol*, **22**, 595–600.

Charness ME, Simon RP, Greenberg DA (1989). Ethanol and the nervous system. *N Engl J Med*, **321**, 442–54.

Cherington M (2003). Neurologic manifestations of lightning strikes. *Neurology*, **60**, 182–5.

Clarkson TW, Magos L, Myers GJ (2003). The toxicology of mercury—current exposures and clinical manifestations. *N Engl J Med*, **349**, 1731–7.

Connolly S, Trevett AJ, Nwokolo NC *et al.* (1995). Neuromuscular effects of Papuan Taipan snake venom. *Ann Neurol*, **38**, 916–20.

Cook DG, Fahn S, Brait KA (1974). Chronic manganese intoxication. *Arch Neurol*, **30**, 59–64.

Coye MJ, Barnett PG, Midtling JE *et al.* (1987). Clinical confirmation of organophosphate poisoning by serial cholinesterase analyses. *Arch Intern Med*, **147**, 438–42.

Daras M, Koppel BS, Atos-Radzion E (1994). Cocaine-induced choreoathetoid movements ('crack dancing'). *Neurology*, **44**, 751–2.

Dart R ed. (2004). *Medical Toxicology*. Lippincott Williams & Wilkins, Philadelphia.

Davis LE, Kornfeld M, Mooney HS *et al.* (1994). Methylmercury poisoning: long-term clinical, radiological, toxicological, and pathological studies of an affected family. *Ann Neurol*, **35**, 680–8.

Day E, Bentham P, Callaghan R *et al.* (2006). Thiamine for Wernicke-Korsakoff syndrome in people at risk from alcohol abuse (review). *The Cochrane Library*, 1–12.

Delgado G, Tunon T, Gallego J *et al.* (1985). Spinal cord lesions in heatstroke. *J Neurol Neurosurg Psychiatry*, **48**, 1065–7.

Denier C, Bourhis JH, Lacroix C *et al.* (2006). Spectrum and prognosis of neurologic complications after hematopoietic transplantation. *Neurology*, **67**, 1990–7.

Diener HC, Dichgans J, Bacher M *et al.* (1984). Improvement of ataxia in alcoholic cerebellar atrophy through alcohol abstinence. *J Neurol*, **231**, 258–62.

DiVincenti FC, Moncrief JA, Pruitt BA Jr (1969). Electrical injuries: a review of 65 cases. *J Trauma*, **9**, 497–507.

Dobb GJ, Edis RH (1984). Coma and neuropathy after ingestion of herbal laxative containing podophyllin. *Med J Aust*, **140**, 495–6.

Donaghy M (1999). Neurological considerations. In Ginns B, Cosimi A, Morris P, eds. *Transplantation*. Blackwell Science, Cambridge, Mass.

Donaghy M (2006). Neurologists and the threat of bioterrorism. *J Neurol Sci*, **249**, 55–62.

Donaldson IM, Cuningham J (1983). Persisting neurologic sequelae of lithium carbonate therapy. *Arch Neurol*, **40**, 747–51.

Dougherty JH, Jr., Rawlinson DG, Levy DE *et al.* (1981). Hypoxic-ischemic brain injury and the vegetative state: clinical and neuropathologic correlation. *Neurology*, **31**, 991–7.

Eddleston M, Mohamed F, Davies JO *et al.* (2006). Respiratory failure in acute organophosphorus pesticide self-poisoning. *QJM*, **99**, 513–22.

Eddleston M, Szinicz L, Eyer P *et al.* (2002). Oximes in acute organophosphorus pesticide poisoning: a systematic review of clinical trials. *QJM*, **95**, 275–83.

Elkind MS, Sciacca R, Boden-Albala B *et al.* (2006). Moderate alcohol consumption reduces risk of ischemic stroke: the Northern Manhattan Study. *Stroke*, **37**, 13–9.

Ernst T, Chang L, Leonido-Yee M *et al.* (2000). Evidence for long-term neurotoxicity associated with methamphetamine abuse: A 1H MRS study. *Neurology*, **54**, 1344–9.

Estrin WJ (1987). Alcoholic cerebellar degeneration is not a dose-dependent phenomenon. *Alcohol Clin Exp Res*, **11**, 372–5.

Exley C, Esiri MM (2006). Severe cerebral congophilic angiopathy coincident with increased brain aluminium in a resident of Camelford, Cornwall, UK. *J Neurol Neurosurg Psychiatry*, **77**, 877–9.

Farrell DF, Starr A (1968). Delayed neurological sequelae of electrical injuries. *Neurology*, **18**, 601–6.

Filley CM, Graff-Richard NR, Lacy JR *et al.* (1982). Neurologic manifestations of podophyllin toxicity. *Neurology*, **32**, 308–11.

Fulton M, Raab G, Thomson G *et al.* (1987). Influence of blood lead on the ability and attainment of children in Edinburgh. *Lancet*, **1**, 1221–6.

Garland H, Pearce J (1967). Neurological complications of carbon monoxide poisoning. *QJM*, **36**, 445–55.

Gill JS, Zezulka AV, Shipley MJ *et al.* (1986). Stroke and alcohol consumption. *N Engl J Med*, **315**, 1041–6.

Glantz MJ, Burger PC, Friedman AH *et al.* (1994). Treatment of radiation-induced nervous system injury with heparin and warfarin. *Neurology*, **44**, 2020–7.

Glass JP, Lee YY, Bruner J *et al.* (1986). Treatment-related leukoencephalopathy. A study of three cases and literature review. *Medicine (Baltimore)*, **65**, 154–62.

Godwin-Austen RB, Howell DA, Worthington B (1975). Observations on radiation myelopathy. *Brain*, **98**, 557–68.

Golden GS (1977). The effect of central nervous system stimulants on Tourette syndrome. *Ann Neurol*, **2**, 69–70.

Goonetilleke A, Harris JB (2002). Envenomation and consumption of poisonous seafood. *J Neurol Neurosurg Psychiatry*, **73**, 103–9.

Grattan-Smith PJ, Morris JG, Johnston HM *et al.* (1997). Clinical and neurophysiological features of tick paralysis. *Brain*, **120**, 1975–87.

Grewal S, Hocking G, Wildsmith JA (2006). Epidural abscesses. *Br J Anaesth*, **96**, 292–302.

Gross JA, Haas ML, Swift TR (1979). Ethylene oxide neurotoxicity: report of four cases and review of the literature. *Neurology*, **29**, 978–83.

Group BMRESMSS (1981). Acetozolamide in control of acute mountain sickness. *Lancet*, **1**, 180–3.

Hamandi K, Mottershead J, Lewis T *et al.* (2002). Irreversible damage to the spinal cord following spinal anesthesia. *Neurology*, **59**, 624–6.

Hargreaves RJ, Evans JG, Janota I *et al.* (1988). Persistent mercury in nerve cells 16 years after metallic mercury poisoning. *Neuropathol Appl Neurobiol*, **14**, 443–52.

Harper C, Kril J, Daly J (1987). Are we drinking our neurones away? *BMJ*, **294**, 534–6.

Hart GR, Anderson RJ, Crumpler CP *et al.* (1982). Epidemic classical heatstroke: clinical characteristics and course of 28 patients. *Medicine (Baltimore)*, **61**, 189–97.

Hawkes CH, Thorpe JW (1992). Acute polyneuropathy due to lightning injury. *J Neurol Neurosurg Psychiatry*, **55**, 388–90.

He F, Xu H, Qin F *et al.* (1998). Intermediate myasthenia syndrome following acute organophosphates poisoning——an analysis of 21 cases. *Hum Exp Toxicol*, **17**, 40–5.

Heaney D, Derghazarian CB, Pineo GF *et al.* (1976). Massive colchicine overdose: a report on the toxicity. *Am J Med Sci*, **271**, 233–8.

Heap LC, Pratt OE, Ward RJ *et al.* (2002). Individual susceptibility to Wernicke-Korsakoff syndrome and alcoholism-induced cognitive deficit: impaired thiamine utilization found in alcoholics and alcohol abusers. *Psychiatr Genet*, **12**, 217–24.

Heinrich A, Runge U, Khaw AV (2004). Clinicoradiologic subtypes of Marchiafava-Bignami disease. *J Neurol*, **251**, 1050–9.

Henry JA (1992). Ecstasy and the dance of death. *BMJ*, **305**, 5–6.

Henson R, Urich H (1982). *Cancer and the Nervous System*. Blackwells, Oxford.

Herdmann J, Benecke R, Meyer BU *et al.* (1988). Successful corticoid treatment of lumbosacral plexus neuropathy in heroin abuse. Clinical aspects, electrophysiology, therapy and follow-up. *Nervenarzt*, **59**, 683–6.

Heyer EJ, Simpson DM, Bodis-Wollner I *et al.* (1986). Nitrous oxide: clinical and electrophysiologic investigation of neurologic complications. *Neurology*, **36**, 1618–22.

Hillbom M, Muuronen A, Holm L *et al.* (1986). The clinical versus radiological diagnosis of alcoholic cerebellar degeneration. *J Neurol Sci*, **73**, 45–53.

Hillebrand G, Castro LA, van Scheidt W *et al.* (1987). Valproate for epilepsy in renal transplant recipients receiving cyclosporine. *Transplantation*, **43**, 915–6.

Hochberg F, Miller G, Valenzuela R *et al.* (1996). Late motor deficits of Chilean manganese miners: a blinded control study. *Neurology*, **47**, 788–95.

Hochberg FH, Miller DC (1988). Primary central nervous system lymphoma. *J Neurosurg*, **68**, 835–53.

Holloway KL, Alberico AM (1990). Postoperative myeloneuropathy: a preventable complication in patients with B12 deficiency. *J Neurosurg*, **72**, 732–6.

Hook CC, Kimmel DW, Kvols LK *et al.* (1992). Multifocal inflammatory leukoencephalopathy with 5-fluorouracil and levamisole. *Ann Neurol*, **31**, 262–7.

Hormes JT, Filley CM, Rosenberg NL (1986). Neurologic sequelae of chronic solvent vapor abuse. *Neurology*, **36**, 698–702.

Hornbein TF, Townes BD, Schoene RB *et al.* (1989). The cost to the central nervous system of climbing to extremely high altitude. *N Engl J Med*, **321**, 1714–9.

Huang CC, Chu NS, Lu CS et al. (1998). Long-term progression in chronic manganism: ten years of follow-up. Neurology, 50, 698–700.

Huang CC, Chu NS, Lu CS et al. (1989). Chronic manganese intoxication. Arch Neurol, 46, 1104–6.

Hwang TL, Yung WK, Estey EH et al. (1985). Central nervous system toxicity with high-dose Ara-C. Neurology, 35, 1475–9.

Isbister GK, Gray MR (2002). A prospective study of 750 definite spider bites, with expert spider identification. QJM, 95, 723–31.

Isbister GK, Kiernan MC (2005). Neurotoxic marine poisoning. Lancet Neurol, 4, 219–28.

Isbister GK, Son J, Wang F et al. (2002). Puffer fish poisoning: a potentially life-threatening condition. Med J Aust, 177, 650–3.

Jankovic J (2005). Searching for a relationship between manganese and welding and Parkinson's disease. Neurology, 64, 2021–8.

Johnson TS, Rock PB, Fulco CS et al. (1984). Prevention of acute mountain sickness by dexamethasone. N Engl J Med, 310, 683–6.

Jones A (1964). Transient Radiation Myelopathy (with Reference to Lhermitte's Sign of Electrical Paraesthesia). Br J Radiol, 37, 727–44.

Jorgensen J, Hansen PH, Steenskov V et al. (1975). A clinical and radiological study of chronic lower spinal arachnoiditis. Neuroradiology, 9, 139–44.

Josephs KA, Ahlskog JE, Klos KJ et al. (2005). Neurologic manifestations in welders with pallidal MRI T1 hyperintensity. Neurology, 64, 2033–9.

Junck L, Marshall WH (1983). Neurotoxicity of radiological contrast agents. Ann Neurol, 13, 469–84.

Kaelan C, Harper C, Vieira BI (1986). Acute encephalopathy and death due to petrol sniffing: neuropathological findings. Aust N Z J Med, 16, 804–7.

Kahan BD, Flechner SM, Lorber MI et al. (1987). Complications of cyclosporine-prednisone immunosuppression in 402 renal allograft recipients exclusively followed at a single center for from one to five years. Transplantation, 43, 197–204.

Kennett RP, Gilliatt RW (1991). Nerve conduction studies in experimental non-freezing cold injury: I. Local nerve cooling. Muscle Nerve, 14, 553–62.

Keogh AJ (1974). Meningeal reactions seen with myodil myelography. Clin Radiol, 25, 361–5.

Kiernan MC, Isbister GK, Lin CS et al. (2005). Acute tetrodotoxin-induced neurotoxicity after ingestion of puffer fish. Ann Neurol, 57, 339–48.

Kimbro T, Tom T, Neuman T (1997). A case of spinal cord decompression sickness presenting as partial Brown-Sequard syndrome. Neurology, 48, 1454–6.

Klawans HL, Stein RW, Tanner CM et al. (1982). A pure parkinsonian syndrome following acute carbon monoxide intoxication. Arch Neurol, 39, 302–4.

Klos KJ, Ahlskog JE, Kumar N et al. (2006). Brain metal concentrations in chronic liver failure patients with pallidal T1 MRI hyperintensity. Neurology, 67, 1984–9.

Kraus ML, Gottlieb LD, Horwitz RI et al. (1985). Randomized clinical trial of atenolol in patients with alcohol withdrawal. N Engl J Med, 313, 905–9.

Lacey DJ (1981). Neurologic sequelae of acute carbon monoxide intoxication. Am J Dis Child, 135, 145–7.

Lancet (1983). Methanol poisoning. Lancet, 1, 910–2.

Le Quesne PM, Axford AT, McKerrow CB et al. (1976). Neurological complications after a single severe exposure to toluene di-isocyanate. Br J Ind Med, 33, 72–8.

Lee EC (2003). Clinical manifestations of sarin nerve gas exposure. JAMA, 290, 659–62.

Lee S, Merriam A, Kim TS et al. (1989). Cerebellar degeneration in neuroleptic malignant syndrome: neuropathologic findings and review of the literature concerning heat-related nervous system injury. J Neurol Neurosurg Psychiatry, 52, 387–91.

Lehane L (2001). Paralytic shellfish poisoning: a potential public health problem. Med J Aust, 175, 29–31.

Leone M, Bottacchi E, Beghi E et al. (1997). Alcohol use is a risk factor for a first generalized tonic-clonic seizure. The ALC.E. (Alcohol and Epilepsy) Study Group. Neurology, 48, 614–20.

Leone M, Tonini C, Bogliun G et al. (2002). Chronic alcohol use and first symptomatic epileptic seizures. J Neurol Neurosurg Psychiatry, 73, 495–9.

Levine BD, Yoshimura K, Kobayashi T et al. (1989). Dexamethasone in the treatment of acute mountain sickness. N Engl J Med, 321, 1707–13.

Levine SR, Brust JC, Futrell N et al. (1990). Cerebrovascular complications of the use of the "crack" form of alkaloidal cocaine. N Engl J Med, 323, 699–704.

Levy DE, Caronna JJ, Singer BH et al. (1985). Predicting outcome from hypoxic-ischemic coma. JAMA, 253, 1420–6.

Lewis MB, Howdle PD (2003). Neurologic complications of liver transplantation in adults. Neurology, 61, 1174–8.

Lidsky TI, Schneider JS (2003). Lead neurotoxicity in children: basic mechanisms and clinical correlates. Brain, 126, 5–19.

Lishman WA (1981). Cerebral disorder in alcoholism: syndromes of impairment. Brain, 104, 1–20.

Macbeth FR, Wheldon TE, Girling DJ et al. (1996). Radiation myelopathy: estimates of risk in 1048 patients in three randomized trials of palliative radiotherapy for non-small cell lung cancer. The Medical Research Council Lung Cancer Working Party. Clin Oncol (R Coll Radiol), 8, 176–81.

Martin CO, Adams HP Jr (2003). Neurological aspects of biological and chemical terrorism: a review for neurologists. Arch Neurol, 60, 21–5.

Martin F, Ward K, Slavin G et al. (1985). Alcoholic skeletal myopathy, a clinical and pathological study. QJM, 55, 233–51.

Martins AN, Johnston JS, Henry JM et al. (1977). Delayed radiation necrosis of the brain. J Neurosurg, 47, 336–45.

Martyn CN, Barker DJ, Osmond C et al. (1989). Geographical relation between Alzheimer's disease and aluminum in drinking water. Lancet, 1, 59–62.

Masucci EF, Fabara JA, Saini N et al. (1982). Cerebral mucormycosis (Phycomycosis) in a heroin addict. Arch Neurol, 39, 304–6.

McCormick DJ, Avbel AJ, Gibbons RB (1979). Nonlethal mushroom poisoning. Ann Intern Med, 90, 332–5.

McKee AC, Winkelman MD, Banker BQ (1988). Central pontine myelinolysis in severely burned patients: relationship to serum hyperosmolality. Neurology, 38, 1211–7.

McLean DR, Jacobs H, Mielke BW (1980). Methanol poisoning: a clinical and pathological study. Ann Neurol, 8, 161–7.

McMillan M, Thompson JC (1979). An outbreak of suspected solanine poisoning in schoolboys: Examinations of criteria of solanine poisoning. QJM, 48, 227–43.

Melamed Y, Shupak A, Bitterman H (1992). Medical problems associated with underwater diving. N Engl J Med, 326, 30–5.

Menegon P, Sibon I, Pachai C et al. (2005). Marchiafava-Bignami disease: diffusion-weighted MRI in corpus callosum and cortical lesions. Neurology, 65, 475–7.

Meyers CA, Scheibel RS, Forman AD (1991). Persistent neurotoxicity of systemically administered interferon-alpha. Neurology, 41, 672–6.

Miller GM, Baker HL Jr, Okazaki H et al. (1988). Central pontine myelinolysis and its imitators: MR findings. Radiology, 168, 795–802.

Misset B, De Jonghe B, Bastuji-Garin S et al. (2006). Mortality of patients with heatstroke admitted to intensive care units during the 2003 heat wave in France: a national multiple-center risk-factor study. Crit Care Med, 34, 1087–92.

Modan B, Baidatz D, Mart H et al. (1974). Radiation-induced head and neck tumours. Lancet, 1, 277–9.

Montero CG, Martinez AJ (1986). Neuropathology of heart transplantation: 23 cases. Neurology, 36, 1149–54.

Morrow R, Wong B, Finkelstein WE et al. (1983). Aspergillosis of the cerebral ventricles in a heroin abuser. Case report and review of the literature. Arch Intern Med, 143, 161–4.

Mosquera A, Idrovo LA, Tafur A et al. (2003). Stroke following Bothrops spp. snakebite. Neurology, 60, 1577–80.

Moss HS, Binder LS (1987). A retrospective review of black widow spider envenomation. Ann Emerg Med, 16, 188–92.

Murros KE, Toole JF (1989). The effect of radiation on carotid arteries. A review article. Arch Neurol, 46, 449–55.

Mycyk MB, Leikin JB (2003). Antidote review: fomepizole for methanol poisoning. Am J Ther, 10, 68–70.

Needleman HL, Schell A, Bellinger D et al. (1990). The long-term effects of exposure to low doses of lead in childhood. An 11-year follow-up report. N Engl J Med, 322, 83–8.

Neiman J, Haapaniemi HM, Hillbom M (2000). Neurological complications of drug abuse: pathophysiological mechanisms. *Eur J Neurol*, **7**, 595–606.

Newmark J (2004). Therapy for nerve agent poisoning. *Arch Neurol*, **61**, 649–52.

Ng SK, Hauser WA, Brust JC et al. (1988). Alcohol consumption and withdrawal in new-onset seizures. *N Engl J Med*, **319**, 666–73.

Nolte KB, Brass LM, Fletterick CF (1996). Intracranial hemorrhage associated with cocaine abuse: a prospective autopsy study. *Neurology*, **46**, 1291–6.

O'Callaghan WG, Joyce N, Counihan HE et al. (1982). Unusual strychnine poisoning and its treatment: report of eight cases. *BMJ*, **285**, 478.

Okamoto M, Longenecker HE Jr, Riker WF Jr et al. (1971). Destruction of mammalian motor nerve terminals by black widow spider venom. *Science*, **172**, 733–6.

Palmer AC, Calder IM, Hughes JT (1987). Spinal cord degeneration in divers. *Lancet*, **2**, 1365–6.

Parkinson IS, Ward MK, Feest TG et al. (1979). Fracturing dialysis osteodystrophy and dialysis encephalopathy. An epidemiological survey. *Lancet*, **1**, 406–9.

Pascual-Leone A, Dhuna A, Altafullah I et al. (1990). Cocaine-induced seizures. *Neurology*, **40**, 404–7.

Pascual-Leone A, Dhuna A, Anderson DC (1991). Cerebral atrophy in habitual cocaine abusers: a planimetric CT study. *Neurology*, **41**, 34–8.

Patchell RA (1988). Primary central nervous system lymphoma in the transplant patient. *Neurol Clin*, **6**, 297–303.

Patchell RA, White CL 3rd, Clark AW et al. (1985). Neurologic complications of bone marrow transplantation. *Neurology*, **35**, 300–6.

Pearn J (1977). Neurological and phychometric studies in children surviving freshwater immersion accidents. *Lancet*, **1**, 7–9.

Pearn J (2001). Neurology of ciguatera. *J Neurol Neurosurg Psychiatry*, **70**, 4–8.

Perl DP, Brody AR (1980). Alzheimer's disease: X-ray spectrometric evidence of aluminum accumulation in neurofibrillary tangle-bearing neurons. *Science*, **208**, 297–9.

Perl TM, Bedard L, Kosatsky T et al. (1990). An outbreak of toxic encephalopathy caused by eating mussels contaminated with domoic acid. *N Engl J Med*, **322**, 1775–80.

Peterson K, Clark HB, Hall WA et al. (1995). Multifocal enhancing magnetic resonance imaging lesions following cranial irradiation. *Ann Neurol*, **38**, 237–44.

Phillips RE, Theakston RD, Warrell DA et al. (1988). Paralysis, rhabdomyolysis and haemolysis caused by bites of Russell's viper (Vipera russelli pulchella) in Sri Lanka: failure of Indian (Haffkine) antivenom. *QJM*, **68**, 691–715.

Phillips SC, Harper CG, Kril J (1987). A quantitative histological study of the cerebellar vermis in alcoholic patients. *Brain*, **110**, 301–14.

Plum F, Posner JB, Hain RF (1962). Delayed neurological deterioration after anoxia. *Arch Intern Med*, **110**, 18–25.

Puke M, Arner S, Norlander O (1989). Complications of regional anaesthesia, with special reference to epidural, spinal and caudal anaesthesia. In Nunn J, Utting J, Brown B, eds. General Anaesthesia, Butterworth.

Regard M, Oelz O, Brugger P et al. (1989). Persistent cognitive impairment in climbers after repeated exposure to extreme altitude. *Neurology*, **39**, 210–3.

Reuler JB, Girard DE, Cooney TG (1985). Current concepts. Wernicke's encephalopathy. *N Engl J Med*, **312**, 1035–9.

Reynolds K, Lewis B, Nolen JD et al. (2003). Alcohol consumption and risk of stroke: a meta-analysis. *JAMA*, **289**, 579–88.

Ron MA, Acker W, Shaw GK et al. (1982). Computerized tomography of the brain in chronic alcoholism: a Survey and follow-up study. *Brain*, **105**, 497–514.

Rosenberg MR, Green M (1989). Neuroleptic malignant syndrome. Review of response to therapy. *Arch Intern Med*, **149**, 1927–31.

Rottenberg DA, Chernik NL, Deck MD et al. (1977). Cerebral necrosis following radiotherapy of extracranial neoplasms. *Ann Neurol*, **1**, 339–57.

Rowley DL, Rab MA, Hardjotanojo W et al. (1987). Convulsions caused by endrin poisoning in Pakistan. *Pediatrics*, **79**, 928–34.

Rowley HA, Lowenstein DH, Rowbotham MC et al. (1989). Thalamomesencephalic strokes after cocaine abuse. *Neurology*, **39**, 428–30.

Rubin AM, Kang H (1987). Cerebral blindness and encephalopathy with cyclosporin A toxicity. *Neurology*, **37**, 1072–6.

Ruitenberg A, van Swieten JC, Witteman JC et al. (2002). Alcohol consumption and risk of dementia: the Rotterdam Study. *Lancet*, **359**, 281–6.

Sabour MS, Fadel HE (1970). The carpal tunnel syndrome—a new complication ascribed to the "pill". *Am J Obstet Gynecol*, **107**, 1265–7.

Saitz R, O'Malley SS (1997). Pharmacotherapies for alcohol abuse. Withdrawal and treatment. *Med Clin North Am*, **81**, 881–907.

Sakamoto Y, Lockey RF, Krzanowski JJ Jr (1987). Shellfish and fish poisoning related to the toxic dinoflagellates. *South Med J*, **80**, 866–72.

Saner F, Gu Y, Minouchehr S et al. (2006). Neurological complications after cadaveric and living donor liver transplantation. *J Neurol*, **253**, 612–7.

Sawa GM, Watson CP, Terbrugge K et al. (1981). Delayed encephalopathy following carbon monoxide intoxication. *Can J Neurol Sci*, **8**, 77–9.

Sawada Y, Takahashi M, Ohashi N et al. (1980). Computerised tomography as an indication of long-term outcome after acute carbon monoxide poisoning. *Lancet*, **1**, 783–4.

Sawicka EH, Trosser A (1983). Cerebrospinal fluid rhinorrhoea after cocaine sniffing. *BMJ*, **286**, 1476–7.

Schnorf H, Taurarii M, Cundy T (2002). Ciguatera fish poisoning: a double-blind randomized trial of mannitol therapy. *Neurology*, **58**, 873–80.

Schultheiss TE (1990). Spinal cord radiation "tolerance": doctrine versus data. *Int J Radiat Oncol Biol Phys*, **19**, 219–21.

Senanayake N, Karalliedde L (1987). Neurotoxic effects of organophosphorus insecticides. An intermediate syndrome. *N Engl J Med*, **316**, 761–3.

Shapiro BE, Rordorf G, Schwamm L et al. (1996). Delayed radiation-induced bulbar palsy. *Neurology*, **46**, 1604–6.

Sharpe JA, Hostovsky M, Bilbao JM et al. (1982). Methanol optic neuropathy: a histopathological study. *Neurology*, **32**, 1093–100.

Shih RA, Glass TA, Bandeen-Roche K et al. (2006). Environmental lead exposure and cognitive function in community-dwelling older adults. *Neurology*, **67**, 1556–62.

Silber E, Sonnenberg P, Collier DJ et al. (2003). Clinical features of headache at altitude: a prospective study. *Neurology*, **60**, 1167–71.

Simon HB (1993). Hyperthermia. *N Engl J Med*, **329**, 483–7.

Simon RP (1988). Alcohol and seizures. *N Engl J Med*, **319**, 715–6.

Slager UT (1986). Central pontine myelinolysis and abnormalities in serum sodium. *Clin Neuropathol*, **5**, 252–6.

Small SL, Fukui MB, Bramblett GT et al. (1996). Immunosuppression-induced leukoencephalopathy from tacrolimus (FK506). *Ann Neurol*, **40**, 575–80.

Smith J, McLaurin R, Nichols J et al. (1960). Studies in cerebral oedema and cerebral swelling. 1. The changes in lead encephalopathy in children compared with those in alkyl tin poisoning in animals. *Brain*, **83**, 411–24.

Smith SJ, Kocen RS (1988). A Creutzfeldt-Jakob like syndrome due to lithium toxicity. *J Neurol Neurosurg Psychiatry*, **51**, 120–3.

Sostak P, Padovan CS, Yousry TA et al. (2003). Prospective evaluation of neurological complications after allogeneic bone marrow transplantation. *Neurology*, **60**, 842–8.

Sowell ER, Mattson SN, Thompson PM et al. (2001). Mapping callosal morphology and cognitive correlates: effects of heavy prenatal alcohol exposure. *Neurology*, **57**, 235–44.

Spencer P, Schaumburg H eds. (2000). *Experimental and Clinical Neurotoxicology*. Oxford University Press Inc, New York.

Starreveld E, Hope E (1975). Cicutoxin poisoning (water hemlock). *Neurology*, **25**, 730–4.

Steenland K, Jenkins B, Ames RG et al. (1994). Chronic neurological sequelae to organophosphate pesticide poisoning. *Am J Public Health*, **84**, 731–6.

Steg RE, Garcia EG (1991). Complex visual hallucinations and cyclosporine neurotoxicity. *Neurology*, **41**, 1156.

Stepens A Logina I Ligats V et al (2008). A parkinsonian syndrome in methcathinone users and the role of managanese. *N Engl J Med*, 358, 1009–1017.

Sterns RH, Riggs JE, Schochet SS Jr (1986). Osmotic demyelination syndrome following correction of hyponatremia. *N Engl J Med*, **314**, 1535–42.

Stewart WF, Schwartz BS, Davatzikos C *et al.* (2006). Past adult lead exposure is linked to neurodegeneration measured by brain MRI. *Neurology*, **66**, 1476–84.

Teitelbaum J, Zatorre RJ, Carpenter S *et al.* (1990). Neurological sequelae of domoic acid intoxication. *Can Dis Wkly Rep*, **16** (Suppl 1E), 9–12.

Thompson CB, June CH, Sullivan KM *et al.* (1984). Association between cyclosporin neurotoxicity and hypomagnesaemia. *Lancet*, **2**, 1116–20.

Thompson WL, Johnson AD, Maddrey WL (1975). Diazepam and paraldehyde for treatment of severe delirium tremens. A controlled trial. *Ann Intern Med*, **82**, 175–80.

Truelsen T, Thudium D, Gronbaek M (2002). Amount and type of alcohol and risk of dementia: the Copenhagen City Heart Study. *Neurology*, **59**, 1313–9.

Uchino A, Yuzuriha T, Murakami M *et al.* (2003). Magnetic resonance imaging of sequelae of central pontine myelinolysis in chronic alcohol abusers. *Neuroradiology*, **45**, 877–80.

Ungley C, Channell G, Richards R (1945). The immersion foot syndrome. *Br J Surg*, **33**, 17–31.

Urbano-Marquez A, Estruch R, Navarro-Lopez F *et al.* (1989). The effects of alcoholism on skeletal and cardiac muscle. *N Engl J Med*, **320**, 409–15.

Uwai K, Ohashi K, Takaya Y *et al.* (2000). Exploring the structural basis of neurotoxicity in C(17)-polyacetylenes isolated from water hemlock. *J Med Chem*, **43**, 4508–15.

van Zomeren AH, ten Duis HJ, Minderhoud JM *et al.* (1998). Lightning stroke and neuropsychological impairment: cases and questions. *J Neurol Neurosurg Psychiatry*, **64**, 763–9.

Vedanarayanan VV, Evans OB, Subramony SH (2002). Tick paralysis in children: electrophysiology and possibility of misdiagnosis. *Neurology*, **59**, 1088–90.

Verstappen CC, Heimans JJ, Hoekman K *et al.* (2003). Neurotoxic complications of chemotherapy in patients with cancer: clinical signs and optimal management. *Drugs*, **63**, 1549–63.

Victor M (1994). Alcoholic dementia. *Can J Neurol Sci*, **21**, 88–99.

Victor M, Adams R, Collins G (1989). *The Wernicke-Korsakoff Syndrome and Related Neurologic Disorders due to Alcoholism and Malnutrition.* FA Davis, Philadelphia.

Victor M, Adams R, Manchell E (1959). A restricted form of cerebellar cortical degeneration occurring in alcoholic patients. *Arch Neurol*, **1**, 579–688.

Walker RW, Brochstein JA (1988). Neurologic complications of immunosuppressive agents. *Neurol Clin*, **6**, 261–78.

Warrell DA, Looareesuwan S, White NJ *et al.* (1983). Severe neurotoxic envenoming by the Malayan krait Bungarus candidus (Linnaeus): response to antivenom and anticholinesterase. *BMJ*, **286**, 678–80.

Watt G, Theakston RD, Hayes CG *et al.* (1986). Positive response to edrophonium in patients with neurotoxic envenoming by cobras (Naja naja philippinensis). A placebo-controlled study. *N Engl J Med*, **315**, 1444–8.

WHO (1996a). Ischaemic stroke and combined oral contraceptives: results of an international, multicentre, case-control study. WHO Collaborative Study of Cardiovascular Disease and Steroid Hormone Contraception. *Lancet*, **348**, 498–505.

WHO (1996b). Haemorrhagic stroke, overall stroke risk, and combined oral contraceptives: results of an international, multicentre, case-control study. WHO Collaborative Study of Cardiovascular Disease and Steroid Hormone Contraception. *Lancet*, **348**, 505–10.

Wijdicks EF, de Groen PC, Wiesner RH *et al.* (1995). Intracerebral hemorrhage in liver transplant recipients. *Mayo Clin Proc*, **70**, 443–6.

Wijdicks EF, Wiesner RH, Dahlke LJ *et al.* (1994). FK506-induced neurotoxicity in liver transplantation. *Ann Neurol*, **35**, 498–501.

Williams DP, Troost BT, Rogers J (1988). Lithium-induced downbeat nystagmus. *Arch Neurol*, **45**, 1022–3.

Willinsky RA, Taylor SM, TerBrugge K *et al.* (2003). Neurologic complications of cerebral angiography: prospective analysis of 2,899 procedures and review of the literature. *Radiology*, **227**, 522–8.

Winnock S, Janvier G, Parmentier F *et al.* (1993). Pontine myelinolysis following liver transplantation: a report of two cases. *Transpl Int*, **6**, 26–8.

Wolters EC, van Wijngaarden GK, Stam FC *et al.* (1982). Leucoencephalopathy after inhaling "heroin" pyrolysate. *Lancet*, **2**, 1233–7.

Wright DG, Laureno R, Victor M (1979). Pontine and extrapontine myelinolysis. *Brain*, **102**, 361–85.

Yaqub BA (1987). Neurologic manifestations of heatstroke at the Mecca pilgrimage. *Neurology*, **37**, 1004–6.

Yaqub BA, Al-Harthi SS, Al-Orainey IO *et al.* (1986). heatstroke at the Mekkah pilgrimage: clinical characteristics and course of 30 patients. *QJM*, **59**, 523–30.

Yuen EC, Layzer RB, Weitz SR *et al.* (1995). Neurologic complications of lumbar epidural anesthesia and analgesia. *Neurology*, **45**, 1795–801.

CHAPTER 6

Principles of neurological rehabilitation

Derick Wade

Contents

6.1 Introduction

Neurology has an undeserved reputation for being a speciality where diagnosis requires great intellectual effort, although from which little therapeutic intervention flows. The reader will form their own opinion about the difficulty of making diagnoses, but now neurological rehabilitation can offer all patients great help subsequently. Other chapters discuss the roles of specific medical and surgical treatments in transforming neurological patients' lives; this chapter discusses the role of neurological rehabilitation in focusing primarily on reducing limitations on patient activities rather than by detailing the specific nature of these individual interventions.

Neurological rehabilitation can be defined as a *process* that aims to optimize a person's participation in society and sense of well-being. This definition highlights several important features: rehabilitation is not a particular type of intervention; the focus is on the patient as a person; the goals relate to social functioning, as well as health or well-being; it is not a process restricted to patients who may recover, partially or completely, but applies to all patients left with long-term problems. The contrast to traditional neurology is in the broader scope, extending well away from the underlying pathology but always being fully informed by the paramount importance of the primary diagnosis.

This chapter will start by giving a fuller description of rehabilitation in terms of *structure*, represented by the resources needed, *process*, consisting of what happens, and *outcome*, defined by the goals. Subsequently the general evidence supporting neurological rehabilitation as a process is reviewed. It is not practicable to review the wide range of high class randomized controlled trial evidence investigating different and detailed aspects of the process. Some specific diseases and specific clinical problems are considered in Section 6.4.

Neurological rehabilitation has a sound theoretical and conceptual basis derived from the World Health Organisation's International Classification of Functioning, the WHO ICF (Wade and Halligan 2004) and from a general problem-solving approach (Wade 2005). There is strong evidence supporting its effectiveness as a process, and reasonable evidence in support of some specific treatments. The approach of neurological rehabilitation extends the intellectual challenge of neurology; in most clinical situations the physician and the wider rehabilitation team have to make pragmatic decisions based on incomplete information concerning many important factors.

6.2 A model of illness and rehabilitation

Although the goal of rehabilitation is to optimize social participation, its main objective is to increase the range of activities a person can undertake. Activities are also referred to as functions sometimes; activity is the obverse of disability. This section discusses in detail the analysis of factors that cause disability, a necessary prelude to starting treatment, and then it discusses in detail the nature of rehabilitation. This should provide the necessary basic skills to undertake at least some rehabilitation of any patient.

6.2.1 The WHO ICF model of illness

Rehabilitation focuses on changing behaviour in its broadest sense. In order to change behaviour it is necessary to analyse the factors that determine and potentially limit any person's behaviour, and the way that they interact with their environment. Although many complex models of behaviour exist, most have been derived for other uses and from other backgrounds. Rehabilitation needs a model that is relatively simple and that is relevant in all circumstances appertaining to illness.

The World Health Organisation faced this problem when developing the International Classification of Impairment, Disability and Handicap, WHO ICIDH (World Health Organisation 1980) and used ideas then being developed by sociologists. It was expanded by the World Health Organisation for the International Classification of Functioning (World Health Organisation 2000) and has since been further adapted and modified to be useful in health services generally (Wade and Halligan 2004).

The expanded WHO ICF model will be explained here. In addition to providing a powerful analytic tool, it provides a coherent and consistent terminology which should facilitate communication. An overview is shown in Tables 6.1 and 6.2.

6.2.2 Neurological illness

The starting position is a person who believes that they have a disease and consider themselves ill. The term *illness* will be used here to refer to all aspects of the state of having, or assuming that there is a disease. It encompasses both the symptoms and signs, and any changes in behaviour and role associated with the perception of having a disease. Specifically it encompasses the *sick role* (Parsons 1951) in which a person may absolve themselves legitimately from many expected roles and behaviours; in return society expects that the person will strive to return to full role activity as soon as

Table 6.1 The World Health Organisation's International Classification of Functioning rehabilitation model—the person

	Level of illness	
Term	**Synonym**	**Comment**
Pathology	Disease/diagnosis	Refers to abnormalities or changes in the structure and/or function of an *organ or organ system*
Impairment	Symptoms/signs	Refers to abnormalities or changes in the structure and/or function of the *whole body* set in *personal context*
Activity (was disability)	Function/observed behaviour	Refers to abnormalities, changes, or restrictions in the interaction between a person and his/her environment or *physical context* (i.e. changes in the quality or quantity of behaviour)
Participation (was handicap)	Social positions/ roles	Refers to changes, limitations, or 'abnormalities' in the *position* of the person in their *social context*

Table 6.2 The World Health Organisation's International Classification of Functioning rehabilitation model—the context

	Contextual factors	
Domain	**Examples**	**Comment**
Personal	Previous illness	Primarily refers to *attitudes, beliefs, and expectations* often arising from previous experience of illness in self or others, but also to personal characteristics
Physical	House, local shops, carers	Primarily refers to local physical *structures* but also includes people as *carers* (not as social partners)
Social	Laws, friends, family	Primarily refers to *legal* and local *cultural* setting, including expectations of important others
Temporal	Shortly after a stroke	This covers two separate domains. The first concerns the stage the person is in within their *life trajectory*. The second is the stage within the *illness trajectory*
Choice	Free will	This could be considered either as a sub-category of *personal context*, or as a sub-category of *the person* (i.e. at the same level as impairment). However within an explanatory model it is best conceived of as a separate phenomenon

Note: This model is usually prefaced with the words: 'In the context of illness, . . .'

possible. Ultimately illness can be conceived of as the state of sub-optimal interaction between a person and his or her environment both physical and social.

The challenge is to analyse the illness in a way that allows both an understanding of the situation and identifies the most effective interventions. Two aspects need consideration: the person who is ill, and their context.

6.2.3 The person

Traditionally, using a biomedical model, doctors have collected information about symptoms and looked for signs in order to deduce which organ(s) is malfunctioning and what type of process is affecting it. Recently more specialist investigation has allowed closer examination of organs to facilitate the diagnostic process. However this has been a one-way logic, using symptoms to deduce disease but not considering either the relationships between symptoms and disability or that symptoms may arise in other ways.

An increasing concern by doctors and others with the wider consequences of disease has led to the development of broader models of illness. One well-known model is the *biopsychosocial* model (White 2005) which is widely referred to in the back pain and psychiatric literature with two important features:

- recognition that psychological and social factors are important moderating influences; and

- the concept of *hierarchical systems*.

The WHO ICF expanded model (Wade and Halligan 2004) is a further development of the original biopsychosocial model which:

◆ Specifies four hierarchical levels concerning the ill person;

◆ Introduces the concept of *context* to include four domains; and

◆ Recognizes the importance of the person in terms of *choice* and *quality of life*.

The expanded WHO ICF model has been used successfully to structure national guidelines on the management of conditions such as stroke and multiple sclerosis (ICSWP 2004; NICE 2003), and is now the predominant model within neurological rehabilitation (Table 6.1).

The focus of the WHO ICF model is the ill person, or patient. Changes or abnormalities in the person as part of an illness are referred to as *impairments* and are better known as symptoms and signs.

The person is comprised of a host of organs, and each is itself a complex system that may malfunction. This dysfunction, with or without altered structure, is known as *pathology*, or the disease or diagnosis; *diagnosis* is an ambiguous word as it confuses the process of diagnosis with the outcome arising from that process.

The person will interact with his or her physical environment to achieve goals, and performance within this system of person and physical environment is known as *behaviour* with alterations originally and commonly being referred to as *disability*. Within the WHO ICF the interactions between a person and his or her physical environment are now referred to as *activities* and changes are referred to as limitations in activities.

Finally the person will interact with other people socially, establishing their roles and position within society. Changes in the interaction between a person and his social environment were once referred to as *handicap*, but they are now referred to as restrictions on *social participation*.

It is worth noting that activities are externally verifiable, and require no immediate interpretation. By contrast, participation entirely depends upon the attribution of meaning, by others and by the person, to the activities that are undertaken, especially to vocational activities such as work and leisure.

6.2.4 The person's context

The section above has already introduced the importance of the environment by including the **physical** and the **social** environments as parts of two systems—activities and participation. There are two other important contextual factors: **time**, and the **particular characteristics** of that person (Table 6.2).

The importance of the physical context is obvious, but its scope needs emphasis. It can and should include such factors as clothing, orthoses and prostheses, all equipment, the built environment, geography, and most importantly the presence of other people as those able to provide practical support with activities. In other words the physical context includes other people not in terms of their social interactions, but in terms of their importance in enabling a patient's interaction with the physical environment.

The social environment is also very broad in scope. It starts with the availability of other people for social interactions, whether face to face, or using phone or email. However the social context also encompasses the beliefs and expectations of other people, as part of the local culture, and the legal and social framework of society, which includes any financial assistance available.

The third contextual factor, recognized but not classified within the actual WHO–ICF model is that of the person. This concept encompasses the person's beliefs, expectations, and other personality characteristics. Naturally some aspects of personal context will be determined by direct past experience, whereas other aspects will be determined by genetic and social or cultural factors.

Time is the final contextual factor, which is recognized explicitly in paediatric models of illness but less often in other models. This temporal context refers primarily to two items: the person's stage in their life; and the stage within their illness. Many important factors will be associated with the person's time in their life and illness, such as expectations, personal and social resources available, intrinsic abilities, and likely prognosis. All of these will influence the direction and scope of rehabilitation.

6.2.5 The person—choice and quality of life

Finally any model of illness must ultimately recognize that each person will have their own opinions and make their own judgements on what is important to them, and then will make their own choices concerning, inter alia, health treatments offered. In other words, people exercise their free-will. Their choices naturally will be constrained by many of the other items already discussed, and may be influenced by many of the items already discussed but ultimately a person makes choices between available alternatives.

Quality of life must be included in any complete model of illness. The exact nature and definition of quality of life is difficult to agree, and measurement is fraught with difficulty. However it is probably best conceived of as the person's own judgement concerning their situation; each person will attach different degrees of importance to different aspects of their situation. One person's trivial problem, such as unsightly varicose veins, may dominate another's life.

6.2.6 Rehabilitation—structure, process, and outcome

Rehabilitation may be defined as '*a problem-solving process that requires a multi-professional team which focuses on the person's activities and has the goals of optimising social participation and minimizing distress of both the person, and of others*'. This definition will be enlarged upon, considering the process of rehabilitation, the *goals*, expected outcomes, of rehabilitation and the *structures* or resources needed to succeed in these goals.

Process

People present to rehabilitation with particular problems, usually at the level of activities or participation. Occasionally people may be presented by others because they pose difficulties, although they may not themselves recognize that difficulties exist. The role of a rehabilitation service is to resolve the patient's problems, as far as is possible within any constraints imposed by the underlying disease and the resources available. The problem-solving process employed is no different to that used in any situation involving traditional biomedical care and health service management (see Fig. 6.1).

Two features characterize rehabilitation: the vocabulary, and the focus of attention. Generally in rehabilitation the process of establishing the situation is referred to as *assessment* whereas in neurology it is called *diagnosis*. The words describing interventions in rehabilitation are poorly defined—*therapy, treatment, care, equipment etc.*—in contrast to the specific terms used to describe many medical interventions such as '*sodium valproate 1000 mg daily*', or '*intra-muscular botulinum toxin*'.

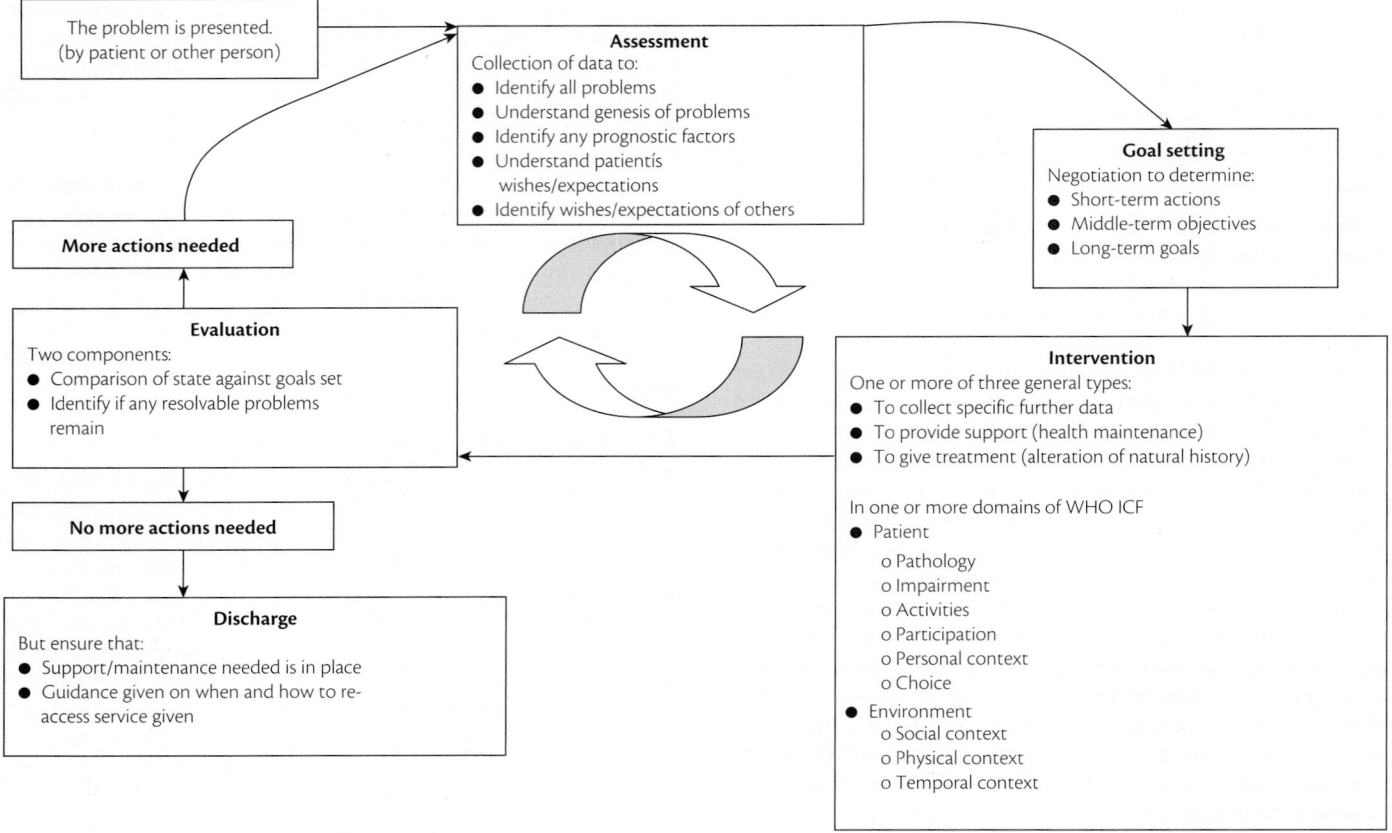

Fig. 6.1 The rehabilitation process.

More importantly the focus of attention, and hence the potential scope of any actions needed also differs. Neurology is primarily concerned with establishing the presence of an underlying pathology and its nature, and then with removing it or minimizing its effects. Rehabilitation has a much broader scope, being closely concerned with a patient's ability to participate in society socially, thus requiring the ability to undertake many activities and the ability to adapt to change. Although the range of potential investigations and treatments in neurology is large, the range of both in rehabilitation is even wider.

Thus rehabilitation starts with collecting data from the patient and others to establish:

◆ The problems, both apparent and unacknowledged;

◆ The causes of, and factors influencing each problem;

◆ Any factors that help establish prognosis of, and the interventions needed for, each problem; and

◆ The wishes and expectations of all interested parties.

The process of assessment is effective (Wade 1998) but only in the context of a treating team. Unsurprisingly, isolated assessment is not in itself effective.

Then goals are set. These need to include both long-term and short-term goals. Long-term goals are usually at the level of participation and the associated extended activities of daily living, such as community skills or work. Whereas short-term goals are

usually at the level of impairment and basic activities. The characteristics of successful goal setting have been established, primarily in fields outside health (Locke and Latham 2002). Goals are more likely to lead to change in patients if:

◆ They are in a domain that the patient agrees is of importance;

◆ Goals are challenging;

◆ Goals are specific, and can be measured in some way to establish successful change;

◆ Feedback on progress is given;

◆ The goal-setting process includes or establishes the links between short- and long-term goals; and

◆ Patients hold a high belief in their own control and ability; their self-efficacy.

The effectiveness of goal setting both in general (Locke and Latham 2002) and in rehabilitation has also been reviewed (Hurn *et al.* 2006; Levack *et al.* 2006), and there is sufficient evidence to encourage its use. Additional benefits might include:

◆ Improved team working, with coordination and ensuring that all problems are covered;

◆ Reduced patient depression and distress with increased patient motivation and self-efficacy;

◆ Incorporation of a personalized outcome measure to report back to funding agencies; goal attainment scaling (Rockwood *et al.* 1997).

The third stage of the rehabilitation process is to intervene or treat. The range of potential interventions is very large, covering all domains of the illness model. The difficulty of classifying treatments has been discussed (Wade 2005). The evidence in favour of many interventions is quite strong, but only a minority of all possible interventions have been subjected to formal trials. While it would be possible to investigate almost any treatment in some way, it should be acknowledged that the benefits of some interventions, such as providing a wheelchair to someone with paraplegia, are so obvious as to not require formal research. Others are for such rare but specific situations that research is practically impossible.

The final phase of rehabilitation is to re-assess the situation, comparing achievement against goals set and checking that all problems have been considered so that further necessary goals can be set. One weakness in many rehabilitation services is that patients are not discharged. Failure to discharge may waste resources, but more importantly it will maintain unrealistic and unachievable expectations in the patient and others. One common reason for failure to discharge is the absence of sufficient long-term support services and social opportunities in the community. Not infrequently attendance for rehabilitation is not infrequently the social highlight of a patient's life, and the only break for the family. However rehabilitation services should identify the needs for support and social activities, and work hard to ensure that they are met; they should not waste their resource and maintain unrealistic hope in others.

Goals and outcome

Rehabilitation is one part of the Health Care System, and the overall goal of most health care services is to improve a person's quality of life. Quality of life is a nebulous concept, difficult to define and measure, and must be judged ultimately by the patient. It seems to incorporate mood and emotional state, the autonomy to make choices, and freedom from pain and distress. In rehabilitation the overarching, global goals are to:

◆ Optimize a person's social participation, considering the person's wishes;

◆ Minimize the person's distress, both emotionally and bodily, for instance pain;

◆ Minimize the somatic and emotional stress as upon, and distress caused to, family and other importantly involved participants. This last goal may often conflict with the first.

The two main subsidiary goals are:

◆ Helping the patient to have the maximum behavioural repertoire that is both achievable in the circumstances, and consistent with his or her goals; and

◆ Giving the patient as much ability to adapt and respond to changes as possible.

Structures

The goals aimed for, and the processes needed to achieve those goals, have major implications for the structures. Specifically the processes cover multiple domains and extend over a long time. Often one process will be contingent upon another, and sometimes only coordinated actions can have any effect. Consequently rehabilitation is crucially dependent upon team work. The main resource needed is a multi-disciplinary team that includes:

◆ All professions to provide the expertise needed to resolve most of the problems posed by the patients likely to be encountered, with the ability to provide most of the interventions required;

◆ Sufficient staff to provide both the care and support needed whilst the patient is being assessed and treated;

◆ Access to all the equipment needed to undertake full diagnostic assessment; and

◆ Access directly or indirectly to all equipment needed by the patient to achieve or maintain their optimal level of activities.

6.2.7 **The utility of this analysis**

The analysis provided by the WHO–ICF model is only important if it has a beneficial effect upon clinical practice, directly or indirectly. There are various lines of evidence concerning its possible utility.

First, it has been used as a basis for several UK National Guidelines agreed by consensus; although the author was frequently a member of such working parties this model was agreed by others as a helpful way to structure recommendations. Second the model is widely and increasingly used in research, especially but not exclusively in rehabilitation research. Third, the model can be used to understand various common problems, so as to see them in a new light that might eventually lead to better management processes. Table 6.3 illustrates some to the lessons that the models teach. The model also allows a classification for all levels of intervention (Table 6.4).

6.3 **Evidence supporting rehabilitation**

'What is the evidence supporting rehabilitation?' is still asked quite frequently. This question can be interpreted in different ways. The first group of questions relate to the *process* of rehabilitation. The next group relate primarily to *specific interventions,* although these are difficult to define and have been likened to Russian dolls, or black boxes. Whatever analogy is preferred, there is agreement that interventions needed to be considered hierarchically. The third group of questions relate to *organization* or management. This section considers the question from three perspectives:

◆ The *process* of rehabilitation;

◆ *Strategies* employed to guide interventions, sometimes referred to as 'approaches' to rehabilitation; and

◆ *Specific interventions* used within the practice of neurological rehabilitation.

The fourth perspective of the best way to organize and deliver rehabilitation services will not be discussed in detail. Suffice it to say that the evidence suggests that organization and expertise are essential for rehabilitation to be effective, but that the site of delivery is less important, be it home, specialist hospital, general hospital, or community hospital, or out-patient department. The location of rehabilitation probably does not affect total resource use, in other words rehabilitation costs no more or less wherever it is delivered, with savings in some costs being countered by increases in other costs.

6.3.1 **Is the rehabilitation process effective?**

One hundred years ago people with complete spinal cord injury had a very short life expectancy, but now it is near normal.

Table 6.3 Lessons from World Health Organisation's International Classification of Functioning model

Observation	Comment
Time	
The time frames are different at each level	Change and management at levels of pathology and impairment are generally quick (hours/days), but change and management at levels of activities and participation are generally slow (weeks/months/years)
	Systems managing different levels should be separated; people with rehabilitation needs are inappropriately placed in an environment focused on disease management
Health services	
Hospitals and health services focus on pathology	Hospitals are environmentally unsupportive of disability; hospital systems are procedurally set in short time frames (hours/days); health service data is usually predicated on a definite diagnosis which is often not available, certainly at presentation
Dependence at the level of disability determines main cost of long-term ill health	Supportive care provided is main resource used in health care, even in acute phase. The resources used are not related reliably to pathological diagnosis.
	Payment for services should not be related to diagnosis; they should relate to dependence and to rehabilitation treatments needed.
Disability and context	
Disability refers not only to 'quantity' (e.g. dependence or otherwise) but also to quality	For some people it matters more how normally they act than whether they can undertake an activity; the social implications of altered behaviour may restrict that behaviour; measures rarely take account of the quality of task performance
	Outcome measures should therefore consider the perspective of the person in addition to that of others
Disability is strongly influenced by the goals of the patient (the personal context)	All behaviour is goal-directed, and so disability cannot be considered 'context free'; many factors including financial considerations may determine the activities undertaken by a patient.
	Patient-centred treatment requires a good understanding of the patient's goals, interests, and concerns
Observed disability also depends upon the physical and social context	How someone behaves is inevitably affected by environmental factors and may be significantly constrained by the environment. The 'environment' includes the capabilities, wishes, and expectations of relevant others
	Interpretation of outcome requires information about context
Relationship between levels	
The nature and extent of the relationships between levels are weak	For example patients may have 'silent' pathology (i.e. disease without symptoms or signs). This gives scope for rehabilitation. It also implies that measures of the extent of pathology are poorly related to the extent of disability in many cases
Causal relationships may extend in any direction, 'up' or 'down' the hierarchy	The relationships are not all one way from pathology through to handicap. Changes in behaviour may 'cause' pathology. For example, electively not moving a shoulder may lead to the pathology of adhesive capsulitis (frozen shoulder)
Not all illness need start from pathology	A systems analysis of the model would predict that illness may start at any level, and interact down the systems as well as up the systems. Abnormal beliefs (part of personal context) may cause as much disability as pathology (abnormal organ structure or function)
	Psychologically determined illness is common in neurological practice, and this model both predicts it and may help understand and manage it
Prognosis depends upon pathology (if present)	The prognostic field for an individual patient is usually determined by the specific disease, but the specific prognosis within that field for a particular patient is usually related to impairments and other factors
Measures should only encompass items from one level	It is invalid to add scores from items or measures covering domains from different levels
Measurement and normality	
'Normal' becomes much less easy to define, and becomes increasingly personal	
Pathology	Structure or function measured against any human, with some allowance for age and gender
Impairment	Structure or function measured against humans matched for age, gender, and other demographic characteristics.
Activities	Behavioural performance and repertoire measured against:
	◆ Socially normative behaviour for some activities
	◆ Previous personal behaviour for some activities
	◆ Desired behaviour for some activities
	◆ Expected (e.g. by family) behaviour for some activities

Table 6.3 (*Cont.*) Lessons from World Health Organisation's International Classification of Functioning model

Observation	Comment
Participation	Social role performance and social position measured against:
	◆ Socially valued and expected roles for whole society
	◆ Culturally valued and expected roles for local, personal society
	◆ Personally valued and expected roles
Miscellaneous	
The terminology used all assumes abnormality	There are currently no good words for the opposite of impairment, disability, or handicap.
	The 'new' terminology of *limitations on* activities and participation overcomes some of this, though there is still no obvious opposite to impairment.
Interventions may occur at many points	While removal of the prime cause of an illness is the ideal, and this prime cause will often be at the level of pathology, interventions at other points are often also effective, especially when there is no pathology or when pathology cannot be altered

This dramatic change arises purely from rehabilitation; it is not due to any specific curative treatment for the underlying damage to the spinal cord. No randomized controlled studies were undertaken to achieve this huge advance, which is universally recognized. More recently the evidence in favour of stroke rehabilitation taking place in stroke units has been shown to be overwhelming (SUTC 2001). Stroke rehabilitation reduces mortality by about 20 per cent, which any pharmaceutical firm would envy, and also reduces morbidity or disability at no extra cost. The process of rehabilitation has been shown to be effective in many other groups of patients such as multiple sclerosis (NICE 2003), moderate head injury (Turner-Stokes 2003),

and motor neurone disease (Van den Berg *et al.* 2005). It should not be surprising that the process of rehabilitation is effective. It is simply a standard problem-solving approach that requires specialist knowledge and skills, as for all other medical specialities and indeed health service managers. There is also some evidence available concerning the different stages of the rehabilitation process:

◆ *Assessment*, which is the equivalent of diagnosis in the medical sphere, is beneficial only if integrated into an overall programme of rehabilitation; isolated identification of problems does not help. However relatively little evidence exists to guide one in the choice of data collection tools or assessment protocols;

Table 6.4 Interventions at different levels within the World Health Organisation's International Classification of Functioning model

Domain	Comment	Example
Patient		
Pathology	Often not reversible or curable, but reduction or control may be important.	Interferon-beta for multiple sclerosis
Impairment	May be reduced directly, or indirectly through activities, or controlled	Botulinum toxin for local spasticity; exercise to increase fitness; analgesia, L-Dopa for Parkinson's Disease
Activities	May be taught (usually new ways of achieving goals, or use of equipment) or practiced	Treadmill gait retraining, using an environmental control system, writing using non-dominant hand
Participation	Cannot give someone new roles, but may suggest possible roles and should facilitate development of new roles (and possibly giving up old roles)	Suggest that a manual worker retrain to be an office clerk, and put in touch with appropriate training course.
Contexts		
Choice or free will	Can be altered through giving information, advice etc.	Explain health advantages of work over being 'off sick'; inform about expectation of health system.
	May be altered through more structured *behaviour modification programme* (structured responses to wanted and unwanted behaviours)	Ignore unwanted behaviour but respond to wanted behaviour
Personal	Changing expectations, beliefs and attitudes, and agreeing goals all may help *motivation* (i.e. willingness to participate in process)	Cognitive behavioural therapy is prime example of therapy aimed at personal context
Physical	Refers to all aspects of physical environment, including adaptation of clothes, altered or new equipment, housing adaptations, and the presence of people as providers of hands-on care or supervision	Provision of orthosis, prostheses, wheelchairs, and adapted cutlery. Also teaching carers how to assist, structuring environment
Social	Altering the social context is usually a slow process as involves changing attitudes, expectations, and beliefs of those interacting with the person	Altering legal framework. Changing population expectations.
Temporal	Providing a predictable structure to day, and ensuring that the person has opportunities to undertake and/or participate in activities throughout the day may be important	Arranging a stable, predictable care routine

◆ The evidence for *goal setting* in general is strong, but there is much less evidence relating to goal setting within the field of neurological rehabilitation (Section 6.2.6); and

◆ There is also some evidence suggesting that *evaluating interventions* in rehabilitation might be achieved using goal attainment scaling (Hurn *et al.* 2006). Apart from this, there is little specific evidence to guide the process of evaluating progress. For example we do not know what proportion of all goals set need to be achieved for significant benefit to accrue.

There is one component of the rehabilitation process where there is no evidence to guide or support clinical practice. That is in the provision of care and support. In other words we do not know the most cost-effective way of providing input to maintain a patient's safety and well-being.

The effectiveness of interventions can be considered in two ways. The first is very specific and focused and asks whether a particular treatment improves outcome, this is discussed later. The second is more general, and possibly more useful because it asks whether using a particular treatment strategy is effective. Evidence concerning the first may only apply to a very specific situation, whereas evidence concerning strategies is more likely to apply to a broader range of situations and patients.

6.3.2 What strategies are most effective?

This section discusses matters that may also be termed treatment 'approaches' or 'techniques'. For example, is the 'Bobath' approach to the management of upper motor neurone weakness better than the 'Motor Relearning Programme' approach, or is treatment focused on optimizing activities better than treatment focused on reducing losses or impairments? The difficulty in answering this question is that there is little agreement on what strategies exist. Furthermore the description of most strategies is very limited, and prone to change over time. At present the evidence suggests that to be effective, rehabilitation requires practice of activities in the most relevant environment possible, rather than undertaking exercises aimed at changing impairments. This is sometimes referred to as task-specific training.

6.3.3 Evidence supporting specific interventions

Neurological rehabilitation faces two problems when considering specific interventions and whether they are effective. The first may seem trivial, but it is important. Many effective interventions are drugs that may be prescribed by any doctor. This leads to debate as to whether the benefits of that treatment can be attributed to 'rehabilitation' rather than 'neurology' or 'general practice'. This does not necessarily matter, but it may be that the drug is only effective if used by someone with adequate expertise in the context of an overall rehabilitation package. Two examples of treating spasticity illustrate this:

Botulinum toxin injections undoubtedly reduce spasticity, but the extent of benefit probably depends upon (a) selection of the appropriate patients, muscles, and doses and (b) concomitant therapy to capitalize upon the benefit obtained by reduction in the spasticity (Francis *et al.* 2004). As an intervention it is probably more effective when used by a rehabilitation specialist in the context of a multi-professional rehabilitation team.

Baclofen also probably controls spasticity, although the evidence is weak (Shakespeare *et al.* 2003). It is common experience that many patients are either not helped with Baclofen, or stop it due to side-effects. Furthermore some patients taking it do not have spasticity, instead having dystonia or choreo-athetoid movements for instance. Unfortunately these practices may arise from limited knowledge and experience. It is best to start baclofen, and indeed other anti-spasticity drugs, at a low dose of 5 mg and to increase by small increments of 5–10 mg at relatively long intervals of 7–14 days, timing the dose if necessary to cover the most troublesome period such as night time spasms. Unfortunately many doctors will commence dosage at 10 mg twice daily, increasing rapidly to 10 mg four times daily, and then stopping when the patient has excessive drowsiness and sees little benefit.

The second problem is also important, but difficult to resolve: how to describe most rehabilitation interventions. Drugs are relatively easy to specify in pharmacological terms, though it is important to consider other factors related to the use and effectiveness of drugs such as their mode of presentation, colour of tablets, or the beliefs and expectations of the recipients (Moerman 2002). Furthermore the interaction between different drugs being taken by the patient, and between drugs and other non-pharmacological interventions must be considered. It is also relatively easy to define pieces of equipment that may be given to, or used by patients, but then it becomes important to consider the characteristics of the patient, what training is offered, and contextual factors such as other pieces of equipment.

It is very difficult to classify and define most specific rehabilitation interventions, not least because each intervention depends upon prior assessment and goal setting and continued monitoring of progress. In other words the intervention is always an integral part of a broader process. The attempts made to describe and classify rehabilitation treatments generally agree that interventions form a hierarchy ranging from the general to the specific. Systematic approaches have been proposed, but at present no system has even been tested or used to any extent.

6.4 Specific diseases

The discussion so far has emphasized the general approach of rehabilitation in considering how to resolve someone's problems related to their activity limitation. Most of these principles, and much of the evidence can apply to any disease. Consequently this section is relatively short, and will only consider a few specific diseases.

In the UK a Government document has been recently published entitled 'the National Service Framework for Long Term Conditions' (DH 2005) which outlines some valuable principles relating to the rehabilitation of people with neurological conditions. This document recognizes that there are probably four different categories of neurological condition:

◆ acute onset disability, with a phase of improvement followed by relative stability (Section 6.4.2);

◆ fluctuating and or unpredictable disability, often with some progression (Section 6.4.3);

◆ progressive disability, at a rate varying between individuals but relatively predictable within an individual (Section 6.4.4);

◆ stable conditions where there may be some change as people age or as circumstances change. This may include people who have had an acute onset disability and stabilized with some residual disability (Section 6.4.5).

Although there may be some diseases that do not fit easily within these four categories, they provide a useful way to categorize rehabilitation and will be adopted here.

6.4.1 The role of the doctor

Before considering each disease, the role of the neurologist as a member of a rehabilitation team will be discussed. The extent to which a neurologist fulfils these roles will depend upon the expertise of any other medical person in the team, and the particular interests of the neurologist.

One central role, which only the doctor can fulfil, is the diagnosis and disease-specific management of any underlying disease. While this may appear simple, particularly later after onset, two points are worth noting. First, most diagnoses retain an element of uncertainty. It is essential to remain alert throughout a patient's involvement with a rehabilitation service to the possibility that the original diagnosis was inaccurate.

Second, a patient may develop new symptoms that require diagnosis. The doctor is needed to evaluate the meaning of any new symptoms, to order investigations if needed, and to initiate and monitor disease-specific treatment if necessary. In practice the main role of the doctor is usually to reassure the patient and others that in fact the new symptoms are of no great significance.

Third, in rehabilitation it is essential to know the prognosis of the underlying disease. Whilst in principle this may be known to other team members, who are able to look it up, in practice it is the neurologist who is likely to have the best knowledge concerning prognosis.

A fourth role for a doctor is that of symptom management using drugs or surgery. There are many symptoms or impairments that can be ameliorated using appropriate medication. Particular skill is needed in managing neurological symptoms in someone with long-term disability.

The main principles to observe are:

◆ Always consider whether medication can be reduced or withdrawn. It is probably more common for patients to be on unnecessary medication than it is for the patient not to be receiving beneficial medication;

◆ To start any new medication at a low dose, and subsequently increase gradually. Patients with long-term neurological disease are often more sensitive to the side-effects of medication;

◆ Avoid changing more than one medication at one time; and

◆ If at all possible, use one medication to control several problems. For example amitriptyline is useful to control pain, to reduce bladder urgency, to help sleep, to reduce depression, to reduce salivation, and in other ways such as reducing anxiety.

Although in many respects the neurologist will be an equal member of the rehabilitation team, in practice it is usually the doctor who has the most experience, and will be most suited to make difficult decisions, and to chair difficult case conferences.

6.4.2 Acute onset disability: stroke and injury to the brain and spinal cord

There are many neurological conditions that start suddenly, reaching their peak within minutes and hours or a few days, and that often leave the patient with a long-term impairment and level of disability. Examples include stroke, spinal cord injury, traumatic brain injury, and most infective central nervous system disorders, Guillain–Barré syndrome and hypoxic brain injury.

There are some general principles that apply to most of these conditions. From the perspective of those involved in rehabilitation, the diagnosis generally is made elsewhere. However it is still important for the rehabilitation team to have access to diagnostic expertise. Occasionally the diagnosis made elsewhere may be wrong. More importantly it is necessary to diagnose any change that may occur in the neurological state: does it reflect recurrence, or is it due to a new condition, or is it an expected part of the natural history? The neurological rehabilitation team will also be particularly interested in the detailed consequences of the specific pathology. They will need to know what impairments are likely, and what impairments are unlikely so that they can undertake an efficient screen of impairments on arrival.

The likely prognosis is the second important feature that will need to be known by the rehabilitation team. Although prognostic factors are not known in detail for all conditions, it is nonetheless likely that a reasonably accurate prognosis can be made by the time someone enters the rehabilitation service. Generally prognosis relates initially to the extent of impairment, but after a few weeks the level of disability in one area is often the best predictor of disability at a later time.

For most of these conditions both the patient and the rehabilitation team can be reassured that any gains achieved will be retained, at least for some years.

Stroke

Stroke is perhaps the most well-researched neurological condition in terms of rehabilitation. There are many systematic reviews, and well-researched and evidence-based guidelines concerning stroke rehabilitation. Only a brief outline will be given here, and readers are strongly recommended to look at a national guideline either for their country, or the one published in the UK (ICSWP 2004).

The main principles underlying stroke rehabilitation are as follows:

◆ The patient should be under the care of a specialist stroke rehabilitation unit whilst in hospital, and a specialist stroke rehabilitation service when back in the community. The evidence supporting these statements is overwhelming.

◆ Therapy should be task oriented. In other words current evidence suggests that practicing an activity is the best way to improve at that activity.

◆ The patient should be set both short- and long-term goals, and those goals should be relatively challenging and set at the level of activities or social participation.

The stroke rehabilitation team will need a doctor with disease specific knowledge and skills. The diagnosis needs to be made as accurately as possible. Furthermore any new events need to be diagnosed in terms of their underlying pathological cause. The diagnosis of stroke is covered elsewhere. Medical treatments need to be initiated and monitored, both for the stroke itself and also importantly for any underlying specific causes. This role will often fall to the neurologist. The third important role for the neurologist is to determine the likely prognosis. Much research has been undertaken into the prognosis after stroke, both in terms of survival and in terms of functional recovery. The best single prognostic indicator for stroke is probably the presence of urinary incontinence in the first few days after the stroke. Patients who are incontinent are more likely to die, are less likely to walk independently, and are more likely to be transferred to a long-term residential setting after leaving hospital. Considering most other measures, the outcome at some future point is usually determined to the greatest extent by the measurement at the earlier point. The best way to predict the future score on a measure is to know the current score on that measure.

Head injury

Traumatic brain injury, which is probably a more accurate term than head injury, is another acute onset disabling neurological condition. The term unfortunately covers an extreme range of clinical severity from the trivial knock on the head which leaves someone dazed for a few minutes and possibly with a headache for a few days through to the patient who is left in a permanent vegetative state for 40 years. This makes it difficult to describe rehabilitation, as the needs vary so greatly. The best single measure of the severity of head injury is the period of post traumatic amnesia; this is the time that elapses from injury to the return of continuous day-to-day memory. Generally this can be determined clinically, by asking the patient and checking against other available contemporaneous information. It is important to be aware that people can have 'islands of memory' within a period of post-traumatic amnesia.

Neurologists should be wary about statements concerning either prognosis or severity. On an individual basis, a prognosis is extremely difficult to provide and it is wise to restrict oneself to outlining a range of potential outcomes. Some people with relatively trivial injuries remain off work indefinitely, whereas other people with apparently severe head injuries, who have been in coma over 24 h and suffered post-traumatic amnesia for over 2 weeks) may return to high-level jobs within 6 months. The word severity has two meanings in the context of head injury. It might refer to the prognosis at the time, for example someone with an extradural haematoma might have a bad prognosis, if not treated. Or it might refer to the actual situation at a particular time. Unfortunately, in a medico-legal context, these two meanings are frequently mixed. For example if someone has returned to work after a head injury involving post-traumatic amnesia of 2–3 weeks, it is misleading to describe the head injury as 'severe' simply because that label has been promulgated or because others with a similar duration of post-traumatic amnesia remain off work.

Neurologists are most likely to see people who have had relatively minor head injuries with a period of post-traumatic amnesia ranging from 0 to 24 h, with the patient complaining of persisting 'post-concussional' symptoms (Section 25.6.1). The evidence suggests that whilst post-concussion symptoms may be related to brain damage for the first 3 months, after that, emotional and other psychological factors assume much greater importance. It is worth noting that formal neuropsychological testing cannot determine the aetiology of such symptoms; the best predictor of neuropsychological performance is the existence of a claim (Binder and Rohling 1996). The best rehabilitation is to offer the patient information, advice, and support as soon as possible after their injury, encouraging a reasonably rapid but not fast return to work. Unfortunately it is not possible to quantify in any detail the speed of return to work.

The other group of patients that neurologists are likely to see are those with much more severe head injury. They may present with refractory epilepsy (Sections 25.5.3 and 31.8.2), or in a vegetative state (Section 33.6). Also the neurologist may be asked to advise on difficult behaviour, or sometimes on specific impairments that are difficult to manage. There is no evidence to guide specific management of aggressive behaviour (Fleminger *et al.* 2006).

Spinal cord injury

In most health care systems acute spinal cord injury (Section 28.4.3) is managed by a specialist service, and in general neurologists will not come into contact with patients who have acute spinal cord injury. The neurological diagnosis is usually obvious, and it is also usually obvious whether or not the spinal cord is completely transacted from the outset. Rehabilitation has several general goals:

- It should aim to minimize the risk of all preventable complications. Although initially this is the responsibility of the health care service, the primary goal of the rehabilitation service is to ensure that the patient is fully aware of how to preserve health and well-being;

- The rehabilitation service needs to teach the patient how to manage in the presence of their impairments and it needs to ensure that all required adaptations and equipment are identified and provided. In other words it needs to ensure a suitable physical context, and to ensure that the patient can use this context;

- It may need to teach others how to provide additional support to the patient if necessary. In general patients with lesions below the cervical level of the spinal cord can live fully independently, whereas patients with cervical spinal-cord lesions will need assistance to a greater or lesser extent.

- In patients with spinal cord injury particular attention needs to be paid to the management of excretion from bowels and bladder, sexual function, and skincare.

- Medical recognition and management of autonomic dysreflexia is important (Section 28.3.7). The primary concern is with the increased blood pressure. The primary management is to identify and treat the precipitating factor, usually bladder stimulation treatable by catheter drainage, and to monitor and reduce blood pressure using, for example, immediate-release nifedipine (Blackmer 2004).

6.4.3 Fluctuant and unpredictable disease: multiple sclerosis

Diseases that fluctuate unpredictably can be the most difficult to manage. Both the patient and the rehabilitation team have to act, at one and the same time, as if there is going to be some stability or improvement whilst also contemplating that matters could get worse in the near future: to 'hope for the best and plan for the worst'. The neurologist has an important role to play. Often there is diagnostic uncertainty, either about the underlying disease or about the specific cause of change. Consequently the neurologist must be prepared to review the diagnosis at unpredictable intervals when matters change suddenly. However the neurologist probably cannot play a definitive role in terms of prognosis as this, by definition, is unknown.

Multiple sclerosis

Multiple sclerosis is the archetypal unpredictable disease (Section 37.5.4). Management is extremely difficult because multiple sclerosis can cause such a wide variety of impairments, of which each patient has their own unique combination. As with stroke, there are national clinical guidelines available within the UK (NICE 2003), and these review evidence in relation to the organization of rehabilitation, and also in relation to each specific common impairment. The neurologist has a vital role in the rehabilitation of individuals who have multiple sclerosis.

A proportion of people presenting to rehabilitation services with a diagnostic label of multiple sclerosis do not in fact have the disease. Most commonly these individuals have non-organic disability, a psychologically determined disorder (Section 4.2.4). More rarely they have an alternative neurological diagnosis. Pursuing an active course of rehabilitation in an individual with a psychologically determined illness may be to their disadvantage because there is no evidence that any particular type of rehabilitation helps, and sharing an environment with other disabled people may reinforce their behaviour and possibly raise unrealistic expectations in family members. The act of incorporating such patients in a service that is focused upon people who do have an underlying disease will reinforce their own belief that they also have a serious underlying disease, whatever one may have told them beforehand. In other patients one may not be alert to likely problems, or one may work on false assumptions about prognosis.

Consequently it is vital that the neurological diagnosis is reconsidered whenever a patient is newly referred to a rehabilitation service, and this requires a doctor with good neurological training.

It is also important, although sometimes difficult, to distinguish between increased dependency resulting from the general effects of an incidental infection, and an increase in dependence related to disease relapse. In patients who worsen with infection, one can expect a reasonably full recovery in a relatively short time after proper treatment of the infection, whereas patients who have relapsed usually recover more slowly and incompletely. However rehabilitation has an important role to play in addition to high-dose steroids (Craig *et al.* 2003) (Section 37.5.9).

Neurologists also have an increasingly important long-term role in treatment. More and more patients are likely to be on disease-modifying drugs (Section 37.5.8), and it is important to have someone with expertise in such drugs as a member of, or advising, the rehabilitation team. An increasing number of drugs can be used to modify one or more of the common impairments, and again it is important for a doctor who is familiar with these impairments, and the drugs used to modify them.

6.4.4 Progressive disorders: neuromuscular disorders

Many neuromuscular disorders are progressive at a faster or slower rate. Motor neurone disease (Section 23.2.1) is one relatively commonly encountered example illustrating some general principles applying to the rehabilitation of most patients with progressive disorders.

First, patients often adapt slowly and progressively to their slow and progressive impairment. Provided they have at least reasonable cognition, then they are able to deduce, either consciously or otherwise, the best way to manage in their situation. Rehabilitation services often have little to offer; sometimes they can advise on how to obtain specific pieces of equipment, or other adaptations. More rarely they can provide practical information about equipment or adaptations that the patient has not considered or found out about themselves.

It is always sensible to inform patients with progressive disorders about special-interest patient support groups. Such groups often

have a higher level of expertise than most general rehabilitation services. Furthermore, with the Internet, patients will often discover information for themselves.

Patients often worry that exercise will in some way exacerbate their condition. They should be specifically reassured that exercise and keeping fit is not known to be harmful and there is evidence in some conditions that it is positively beneficial. Therefore the neurologist or neurological rehabilitation service has an important role to play in informing and reassuring the patient, to alter their beliefs and expectations.

Conversely patients can often become remarkably dependent after a trivial incident that happens render them bedbound, such as an infective episode or a fracture. If they lose skills that they have retained, through a lack of practice, it can become extremely difficult to regain the previous level of independence. Therefore the rehabilitation service may have an important role to play keeping the patient independent through another illness, or helping the patient to reclaim a previous level of independence if this seems possible.

In many progressive disorders it is quite possible to predict how the patient's situation will change, and to take actions and make decisions that will lessen the consequences of these changes. For example somebody with hereditary spastic paraparesis is likely to have increasing difficulty with stairs, and consequently would be well advised to consider moving to a house on one floor if moving house, and to consider buying an automatic car that can be adapted for arm controls.

Motor neurone disease

Motor neurone disease (Section 23.2) straddles many boundaries. In some patients the rate of progress is relatively slow, and the principles outlined above apply. In other patients the rate of progress is too fast for patients to learn how to adapt. Also in motor neurone disease the threat of death is present from the outset, and many patients are aware of this. Consequently it may be important to involve specialist palliative care services, and it is always important to discuss openly how close death might be and to support the patient emotionally and practically. Otherwise the principles are the same and can be applied successfully to improve quality of life (Van den Berg *et al.* 2005).

6.4.5 Stable disease present from childhood: cerebral palsy

Diseases that present at or shortly after birth, or those acquired during childhood pose another set of problems. Their early management will usually be the responsibility of a paediatrician or paediatric neurologist, and for many conditions there will be a prolonged phase of trying to refine the diagnosis. Rehabilitation during childhood has to be undertaken in close liaison with the educational service. Fortunately the close conceptual link between rehabilitation and education usually makes that liaison easy. Rehabilitation in childhood also has to take into account that the central nervous system is maturing, with the consequence that the patient will be gaining extra abilities as part of their normal process of development. Following acute onset conditions, this often means that children have a particularly good prognosis. However in the face of widespread damage before birth, or when the pathology is progressive, significant achievement is often difficult.

Paediatric neurological rehabilitation is outside the scope of this chapter, and is generally well managed by specialist children's services.

However there is one extremely important, yet often poorly managed feature—the handover to adult services. Paediatric rehabilitation services are usually well resourced, and provide a flexible and holistic service whereas, in many countries, the standards of service available to young adults are much lower.

Cerebral palsy

Although cerebral palsy (Section 9.5) is a common condition, and although many children with cerebral palsy survive into adulthood, there is remarkably little research or published evidence concerning its natural history or its rehabilitation.

6.5 Some specific situations

Neurological rehabilitation sits in a hinterland, requiring some knowledge of individual disease and specific treatments while also having an understanding of general aspects of functioning in society, and having the skills to increase activities whatever the cause of their limitation. Neurologists are likely to be involved with or asked about the management of some impairments. The most likely is spasticity.

6.5.1 Spasticity

Spasticity is an extremely common and important impairment, being seen in a very large number of neurological conditions. Before discussing its management, it is worth emphasizing that spasticity refers to a syndrome that includes one or more of the following phenomena:

- increased resistance to passive movement around a joint due to stiffness in the muscle;
- hyperreflexia, consisting of an enhanced reflex response to muscle stretch;
- spasms, consisting of involuntary muscle contractions which cause movement and are often painful;
- pain or discomfort in the muscle;
- adoption or maintenance of a particular posture, especially on effort;
- clumsiness, or reduction of fine motor control; and
- reduced voluntary muscle strength.

Recognition of the complex nature of spasticity is important when considering its management. When assessing someone with spasticity it is important to consider which aspects of the syndrome need to be treated, if any. The evidence available to guide management of patients with spasticity is very weak even for the commonly used drugs. Intra-muscular botulinum toxin injection is probably the best supported intervention. For almost all others the benefit is either not detectable or not present, or is associated with significant side effects. Nonetheless clinical experience suggests that a structured approach such as that recommended in multiple sclerosis (NICE 2003) is beneficial. The approach includes the following options, usually in the following order:

- doing nothing if the spasticity is not troublesome or if its leg stiffness is used to allow walking;
- prevention of complications such as contractures through stretching, evidence for which is lacking, and positioning at rest;

- prevention of spasticity through amelioration of any exacerbating factors such as pain, infection, or anxiety;
- learning self-control, for instance reduction through relaxation;
- simple single drugs, such as baclofen or gabapentin;
- using combinations of drugs;
- using intra-muscular botulinum toxin in conjunction with rehabilitation therapy. This is usually restricted to focal spasticity;
- intra-thecal baclofen delivered by an implanted, programmable pump for more general spasticity especially of the legs; and
- destructive procedures such as tenotomies and phenol injection of nerves.

The main considerations are:

- always to establish the importance of the problem to the patient who has to take the risks associated with any intervention, and to be prepared to do nothing if the patient has no concern;
- to recognize that functional benefit is rarely achieved, and that control of unpleasant or painful problems is usually the main goal;
- to provide a balanced view for each individual patient of the perceived benefits and side-effects; and
- to make changes slowly.

6.5.2 Assessment of activities

This section on the measurement of disability introduces some short simple sets of information that should be useful to any neurological service confronted by a disabled patient. The process of assessment in rehabilitation is not different from the processes used by doctors focused on establishing and managing a specific disease. One needs to identify and quantify phenomena of relevance using data collection tools or techniques.

The important activities to consider are those of personal independence in daily living such as dressing or using the toilet. There are many data collection tools available. Most doctors in the USA and Australia will be familiar with the Functional Independence Measure, a tool that is widely used and often mandated by funding agencies. However the Functional Independence Measure has many disadvantages—it has to be paid for, it requires specific training, and in principle its completion depends upon several team members. It has not been shown to be any more specific or sensitive than other widely available, shorter, and simpler data collection tools (Hobart et al. 2001). It is widely accepted that its non-motor content is unreliable and of little value.

Consequently it is recommended that neurologists should use the Barthel Activities of Daily Living Index (Collin et al. 1988). Several versions exist, but the simplest and most widely used is that shown in Table 6.5; there is no evidence than others are any better. There are some studies comparing the Barthel Activities of Daily Living Index and the Functional Independence Measure, all of which show no significant difference. There seems little benefit in using other scales as the Barthel Index can be used by anyone, can be completed by telephone or post, and only takes a few minutes to complete.

Mobility is the activity that most patients wish to regain if it is affected. There are three simple sets of data that may help. The first is the Rivermead Mobility Index (Table 6.6) (Collen et al. 1991;

Table 6.5 The Barthel Activities of Daily Living Index

Day								
Month								
Year								
Bowels								
0 = Incontinent of faeces (or is given enemas)								
1 = Occasional accident (less than 1x per 24 hours)								
2 = Continent								
Bladder								
0 = Incontinent, or catheterisedcatheterized/convene drain and unable to manage it alone								
1 = Occasional accident (maximum 1x per 24 hours)								
2 = Continent (for last seven days)								
Grooming								
0 = Needs help (supervision, prompts, or practical help)								
1 = Independent in washing face, doing teeth, shaving or putting on make-up, brushing hair								
Toilet use								
0 = Dependent, unable to wipe self								
1 = Needs help, but can wipe self								
2 = Independent in transfers and managing clothes off/on								
Feeding								
0 = Unable; is fed, has gastrostomy, or feeds self minimally								
1 = Needs help cutting food, spreading butter, prompts/supervision etc								
2 = Independent with food provided/selected								
Transfer								
0 = Unable; hoisted and/or unable to sit in wheelchair								
1 = Major help; one or two people, much physical effort								
2 = Minor help; one person, prompts/supervision or minor physical effort								
3 = Independent bed-chair								
Mobility								
0 = Immobile; unable to get from bedroom to dining area								
1 = Wheelchair independent (electric or self-propelled) at least bedroom to dining area								
2 = Walks with help of one person (physical, or prompts/supervision) from bedroom to dining area								
3 = Independent. May use stick, rollator etc if necessary								
Dressing								
0 = Dependent								
1 = Needs help, but does about half (e.g. top or bottom independently, or minor prompts and/or physical help)								
2 = Independent, including shoes, laces, buttons etc.								
Stairs								
0 = Unable								
1 = Needs help, physical or supervision/prompts or carrying equipment								
2 = Independent up and down stairs (any means, including stair lift)								
Bathing								
0 = Dependent								
1 = Independent (bath or shower) including getting in and out, washing, and drying hair								
TOTAL								

Table 6.6 The Rivermead Mobility Index

Topic and Question	0/1	Comment
Turning over in bed		
Do you turn over from your back to your side without help?		
Lying to sitting		
From lying in bed, do you get up to sit on the edge of the bed on your own?		
Sitting balance		
Do you sit on the edge of the bed without holding on for 10 seconds?		
Sitting to standing		
Do you stand up from any chair in less than 15 seconds and stand there for 15 seconds, using hands and/or an aid if necessary?		
Standing unsupported *Ask to stand*		
Observe standing for 10 seconds without any aid		
Transfer		
Do you manage to move from bed to chair and back without any help?		
Walking inside (with an aid if necessary)		
Do you walk 10 metres, with an aid if necessary, but with no standby help?		
Stairs		
Do you manage a flight of stairs without help?		
Walking outside (even ground)		
Do you walk around outside, on pavements, without help?		
Walking inside, with no aid		
Do you walk 10 metres inside, with no caliper, splint, or other aid (including furniture or walls) without help?		
Picking up off floor		
Do you manage to walk five metres, pick something up from the floor, and then walk back without help?		
Walking outside (uneven ground)		
Do you walk over uneven ground (grass, gravel, snow, ice etc) without help?		
Bathing		
Do you get into/out of a bath or shower and to wash yourself unsupervised and without help?		
Up and down four steps		
Do you manage to go up and down four steps with no rail, but using an aid if necessary?		
Running		
Do you run 10 metres without limping in four seconds (fast walk, not limping, is acceptable)?		
TOTAL		

Ask the patient each question. Observe for question 5. Score 1 for 'yes', 0 for 'no'

Rossier and Wade 2001) which covers the whole range of mobility and is short, easily used, and has been shown to detect change and differences.

Other data worth considering include:

◆ number of falls experienced by the person over a defined time. Falls are dangerous and concern the patient, but are rarely recorded in medical notes;

◆ time in seconds taken to walk 10 m using whatever aid is wanted. This is short and simple, and again there is strong evidence concerning its reliability and sensitivity (Rossier and Wade 2001); and

◆ distance walked in 2 min, as a measure of endurance (Rossier and Wade 2001).

Dexterity, the ability to use hands, is of great concern. There are no good questionnaire- based tools that will focus on the affected hand. The set of questions in Table 6.7 cover a range of common problems. Clinically one can ask the patient to quantify a specific activity that they can undertake. Examples include timing how long it takes to move a specific number of small objects such as buttons from one place to another, and counting how many objects can be moved in a defined time. It is best to discuss and identify with the patient a quantifiable activity and to ask them to repeat it as they change treatments.

Table 6.7 Some questions relating to dexterity

Question	Comment
Can you hold and use a knife/fork in your right/left hand?	
Can you do up buttons and zips?	
Can you hold a pen and write?	
Can you use a keyboard?	
Can you drink from a cup?	
Do you spill fluid from a cup?	
Can you clean your teeth (or shave)?	
Can you tie shoelaces?	
Can you pick up a saucepan safely?	

6.6 Conclusion

Neurological rehabilitation has the potential to benefit almost every patient with a long-term neurological condition of any type. This includes dementia (Graff *et al.* 2006), even though it is a disorder which is both progressive and affects cognition, the two features usually thought to render rehabilitation impossible. The keys to success are:

- to use a systematic approach to detecting and analysing the patient's situation;

- to be flexible and thoughtful in the use of any and all potential treatments;

- to remember that the person's overall goals will usually concern social integration, not treatment of symptoms or disease;

- to work as part of a multi-disciplinary team and across all organizational and geographic boundaries.

References

Binder LM, Rohling ML (1996). Money matters: a meta-analytic review of the effects of financial incentives on recovery after closed head injury. *Am J Psychiatry*, **153**, 7–10.

Blackmer J (2004). Rehabilitation medicine: 1. Autonomic dysreflexia. *Can Med Assoc J*, **169**, 931–5.

Collen FM, Wade DT, Robb GF *et al.* (1991). The Rivermead Mobility Index: a further development of the Rivermead Motor Assessment. *Int Disabil Stud*, **13**, 50–4.

Collin C, Wade DT, Davis S *et al.* (1988). The Barthel ADL Index: a reliability study. *Int Disabil Stud*, **10**, 61–3.

Craig J, Young CA, Ennis M *et al.* (2003). A randomised controlled trial comparing rehabilitation against standard therapy in multiple sclerosis patients receiving intravenous steroid treatment. *J Neurol Neurosurg Psychiatry*, **74**, 1225–30.

DH Longterm Conditions NSF Team (2005). *The National Service Framework for Longterm Conditions*. Department of Health, London.

Fleminger S, Greenwood RJ, Oliver DL (2006). Pharmacological management for agitation and aggression in people with acquired brain injury. *Cochrane Database Syst Rev*, Issue 4. Art. No.: CD003299. DOI: 10.1002/14651858.CD003299.pub2.

Francis HP, Wade DT, Turner-Stokes L *et al.* (2004). Does reducing spasticity translate into functional benefit? An exploratory meta-analysis. *J Neurol Neurosurg Psychiatry*, **75**, 1547–51.

Graff MJL, Vernooij-Dassen MJM, Thijssen M *et al.* (2006). Community based occupational therapy for patients with dementia and their care givers: randomised controlled trial. *BMJ*, **333**, 1196–9.

Hobart JC, Lamping DL, Freeman JA (2001). Evidence-based measurement: which disability scale for neurologic rehabilitation? *Neurology*, **57**, 639–44.

Hurn J, Kneebone I, Cropley M (2006). Goal setting as an outcome measure. A systematic review. *Clin Rehabil*, **20**, 756–72.

ICSWP—Intercollegiate Stroke Working Party (2004). National clinical guidelines for stroke (second edition). Clinical Effectiveness and Evaluation Unit, Royal College of Physicians, London.

Levack WMM, Taylor K, Siegert RJ *et al.* (2006). Is goal planning in rehabilitation effective? A systematic review. *Clin Rehabil*, **20**, 739–55.

Locke EA, Latham GP (2002). Building a practically useful theory of goal setting and task motivation. A 35-year Odyssey. *Am Psychol*, **57**, 705–17.

Moerman D (2002). *Meaning, Medicine and the 'Placebo Effect'*. Cambridge University Press, Cambridge.

NICE—National Institute for Clinical Excellence (2003). Clinical Guideline 8. Multiple Sclerosis. National clinical guideline for diagnosis and management in primary and secondary care. National Collaborating Centre for Chronic Conditions. Clinical Effectiveness and Evaluation Unit, Royal College of Physicians, London.

Parsons T (1951). *The Social System*. Free Press, Glencoe, Illinois.

Rockwood KJ, Joyce BM, Stolle P (1997). Use of goal attainment scaling in measuring clinically important change in cognitive rehabilitation patients. *J Clin Epidemiol*, **50**, 581–8.

Rossier P, Wade DT (2001). Validity and reliability comparison of 4 mobility measures in patients presenting with neurologic impairment. *Arch Phys Med Rehabil*, **82**, 9–13.

Shakespeare DT, Boggild M, Young C (2003). Anti-spasticity agents for multiple sclerosis. *Cochrane Database Syst Rev*, Issue 4. Art. No.: CD001332. DOI: 10.1002/14651858.CD001332.

SUTC—Stroke Unit Trialists' Collaboration (2001). Organised inpatient (stroke unit) care for stroke. *Cochrane Database Syst Rev*, Issue 3. Art. No.: CD000197. DOI: 10.1002/14651858.CD000197.

Turner-Stokes L (ed.) (2003). *Rehabilitation Following Acquired Brain Injury*. National Clinical Guidelines. British Society of Rehabilitation Medicine and Royal College of Physicians, London.

Van den Berg JP, Kalmijn S, Lindeman E *et al.* (2005). Multidisciplinary ALS care improves quality of life in patients with ALS. *Neurology*, **65**, 1264–7.

Wade DT (1998). Evidence relating to assessment in rehabilitation. *Clin Rehabil*, **12**, 183–6.

Wade DT (2005). Describing rehabilitation interventions. *Clin Rehabil*, **19**, 811–8.

Wade DT, Halligan PW (2004). Do biomedical models of illness make for good healthcare systems? *BMJ*, **329**, 1398–401.

White P (ed.) (2005). *Biopsychosocial Medicine; an Integrated Approach to Understanding Illness*. Oxford University Press, Oxford.

World Health Organisation (1980). *International Classification of Impairments, Disabilities and Handicaps: A Manual of Classification Relating to the Consequences of Disease*. World Health Organization, Geneva.

World Health Organisation (2000). *International Classification of Functioning, Disability and Health*. http://www3.who.int/icf/icftemplate.cfm

SECTION 2

Developmental neurology

CHAPTER 7

Clinical approach to developmental neurology

Janet Eyre

Contents

7.1 Introduction

The objectives and principles of neurological history and examination in children are the same as those in adults (Section 2.2). This chapter therefore, will not provide an all encompassing description of the neurological assessment of children, but highlights where the approach must differ substantially from that used in adults. Further it aims to provide a practical and useful approach to the examination of children, who may be preverbal and certainly will show less stamina for cooperation than adults. Of course as children get older, the examination can become more conventional and systematized. By adolescence the examination can be the same as the adult examination.

The first and overriding factor for success is to be flexible and to make observations when the opportunity arises rather than to wait for abnormalities to arise during the course of a more systematic approach. Nonetheless a systematic approach to recording these results is essential, so as to bring together related observations made disparately in time. The history is of paramount importance in guiding the examination. Since it is unlikely that you will be able to complete a full examination, it is important to prioritize the observations needed in light of a differential diagnosis before you begin examining. Rather than rushing straight into the examination it is rewarding to gain a young child's confidence by playing briefly with them. Also, instead of insisting on examining the child on a couch, it helps to become adept at examining young children on their parent's or caretaker's knee. Finally, no matter how cooperative a child is, potentially disturbing investigations should be left until last, including tendon reflexes or examination of the tongue, fundi, and ears. Otherwise all subsequent cooperation from the child may be lost after these examinations.

The examination room environment is the key to a successful neurological examination and requires careful thought. There should be sufficient space to accommodate families and for the children to play. The room needs to be friendly and conducive to encouraging play. It needs to be equipped with carefully selected toys, pictures, pencils and paper, and books of interest to children over a wide age range. Observation of the child's play whilst you are taking a history from the parents or caregivers will allow assessment of the child's motor skills and developmental stage. Their use of play material can yield important clues to the nature of a deficit, by revealing ataxia, weakness, involuntary movements, tics, or spasticity. Play also provides an opportunity to assess the child's behaviour, for instance their impulsivity, distractibility, and attention span. Interaction of the child with parents or caregivers can be observed also. If the child participates actively in the history taking, their understanding and contribution to the session allows you to make assessments of their language and intellectual skills.

7.2 The history

In young children the history will need to be obtained from the parents or caregivers. From the age of about 4 or 5 years, information can also be obtained from the child, as long as the questions are simple and specific and more directed than the open questions normally asked of adults. Clearly the older the child the more they can contribute to the history. In addition to the general history taking outlined for adults (Section 2.2.1) information should be obtained also on pregnancy and perinatal history (Section 7.2.1) and on development (Section 7.2.2).

7.2.1 Pregnancy and perinatal history

A careful gestational history is essential. Information should be sought about:

◆ the mother's acute and chronic health problems;

◆ any illicit drug, tobacco, or alcohol consumption;

◆ the duration of gestation;

◆ the dose of medications taken in pregnancy; and

◆ the timing of any infections, acute trauma, or exposure to radiation.

Details should be recorded of the labour and birth including gestation at birth, birth weight, presentation and type of delivery, status at birth, whether there were neonatal seizures and need for intensive or high dependency care.

7.2.2 Development

Knowledge of child development is fundamental to paediatric neurology. It is beyond the scope of this chapter to provide a detailed review of child development; there are many excellent and detailed reviews of normal child development (Egan *et al.* 1969; Sheridan 1973; Illingworth 1980). The development of a child is normally a continuous process and the sequence is the same in all children. Although the rate of development varies between children, there is normally a fairly close parallel within a child between the rates of development in different domains.

The first objective of the developmental history is to ascertain if there has been significant delay in development and whether that delay is restricted to particular areas or whether it is global. Determination of the time course of neurological symptoms may be confounded in a child because contemporaneous normal development, and the consequent increase in skills, may mask progression of the underlying neurological disorder. The second objective is to ascertain whether there has been actual loss of previously acquired developmental skills or a slowing in the rate of development, either of which may indicate a progressive degenerative disorder.

A developmental history is obtained normally by asking the parent or caregiver about the age at which their child achieved major milestones covering the fields of fine and gross motor skills, language, and adaptive-social behaviour. In taking this history it is essential for the questions to be clearly phrased. For example it is not sufficient to ask when a child first sat but to ask when they first sat without support. Table 7.1 is a guide to the milestones which can be used to screen for developmental progress. The questions asked depend on the age of the child. For a child under 2 years most of these milestones are relevant and are likely to be remembered. However for a child aged 6 or 7 years it would be sufficient to ask about sitting without support, walking without support, and the age at first words. For such older children, continue the developmental history by obtaining information on their progress at school and whether they are requiring extra help and support. If parents are uncertain about their child's progress at school it is helpful to write and ask for a report from the school.

Table 7.1 Developmental screen for history taking

Normal age of acquisition	Field	Milestone
4–6 weeks	Social/adaptive	Smiles at mother in response to overtures
2–3 months	Language	Begins to babble and vocalize
5 months	Fine motor	Successfully reaches for an object
6 months	Gross motor	Able to sit without support
9 months	Gross motor	Able to crawl
10 months	Fine motor	Able to use a pinch grip
10 months	Social/adaptive	Plays peek a boo or pat a cake
11–12 months	Language	First words
13 months	Gross motor	Walks without support
15 months	Social/adaptive	Feeds self and uses ordinary cup
18 months	Gross motor	Manages stairs up and down holding onto a rail or hand
2 years	Language	Spontaneously joins 2–3 words together to form a sentence

7.3 Examination of children

Throughout the examination the child's alertness should be evaluated, along with their interest in yourself and their surroundings, their ability to interact with you and to follow instructions, and their ability to learn and cooperate with the examination.

7.3.1 Rapid screening examination

Starting the examination with a rapid 'screening type examination' will delineate the general nature of any deficits at a stage before the child later loses interest or becomes distracted, tired, or oppositional. Whenever possible during the examination each of the actions required should be demonstrated in a manner intended to make it fun for the child. Gait is an important first observation and can often be assessed as the child and their parent or caregiver are escorted from the waiting room to your examining room; look for circumduction of the legs, footdrop, unusual positions of the feet, lack of arm swing, a limp, and waddling. For younger children who arrive in a pushchair, their mode of crawling or shuffling can be observed as they explore the examination room later.

The performance of an older child in a range of gaits such as walking, running, jumping, hopping, heel-to-toe walking, and walking on toes, heels, and sides of feet gives invaluable information. This is usually enjoyed by the child, particularly if the parent or caretaker is involved and it is made into a game. Whenever the history suggests weakness the child should be observed climbing up and down the stairs, if they are sufficiently old and able to do so. If there is a question of proximal weakness, children should also be asked to rise from supine on the floor to see whether they adopt Gower's manoeuvre in which they push their trunk upright by climbing their hands up their legs. The child should then be asked to stand with feet close together, eyes closed, and arms and hands outstretched to assess somatosensory function and involuntary movements. The child can then be asked to perform the finger-nose test; this can be simplified in a very young child by asking them to point from their nose to your nose. All these observations can be made in a relatively short period of time. Then, if the child remains cooperative, a more detailed and

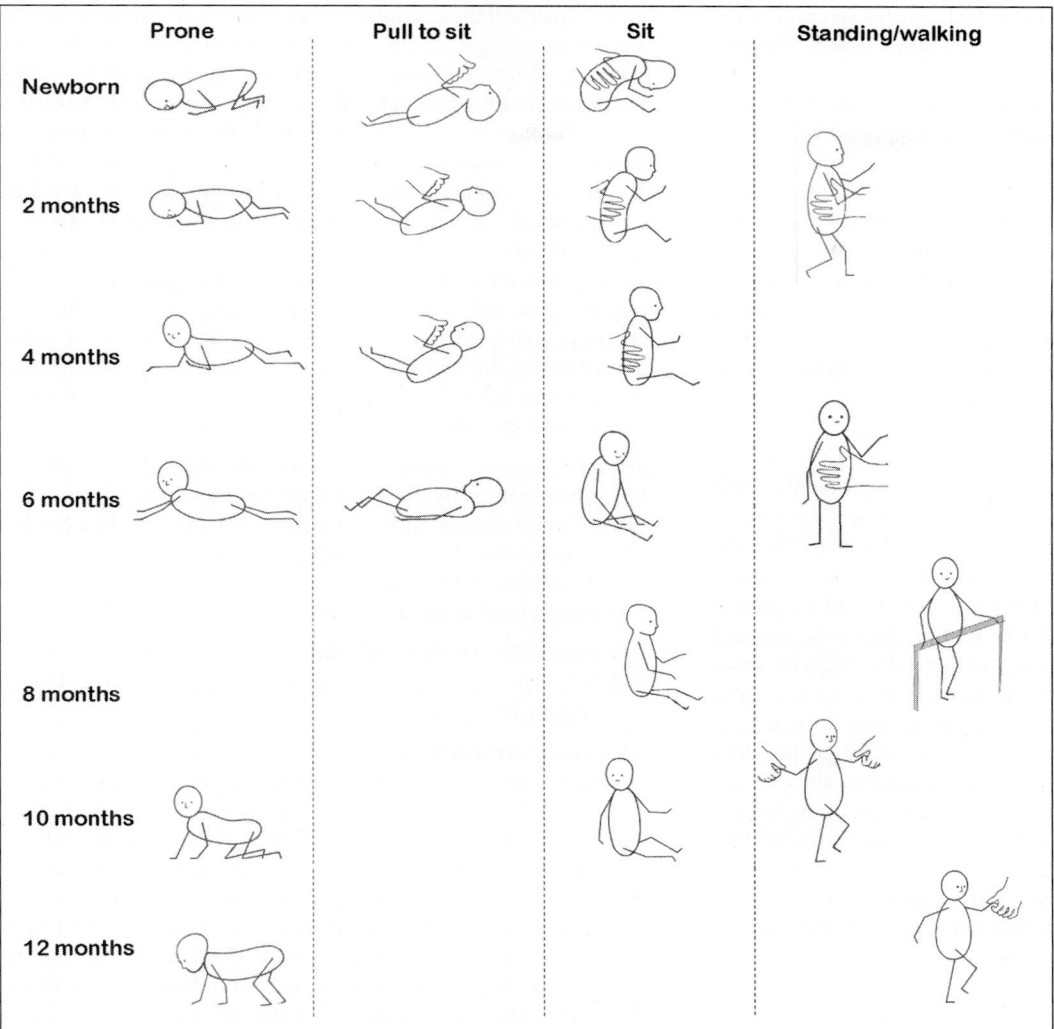

Fig. 7.1 Developmental screen for gross motor milestones.

systematic examination can be undertaken to focus on any suggested abnormalities.

7.3.2 **Developmental screen**

Children aged up to 3 years. A brief examination should be performed to confirm their developmental milestones.

For those under the age of a year, gross motor development is best assessed from:

- the child's posture when prone;
- their head control when pulled to sit;
- their ability to sit; and
- their ability to take weight on their legs when held, to pull themselves to standing, or to walk.

The normal findings in relation to age are summarized in Fig. 7.1. Fine motor control can be assessed by the maturity of the child's grip (Fig. 7.2).

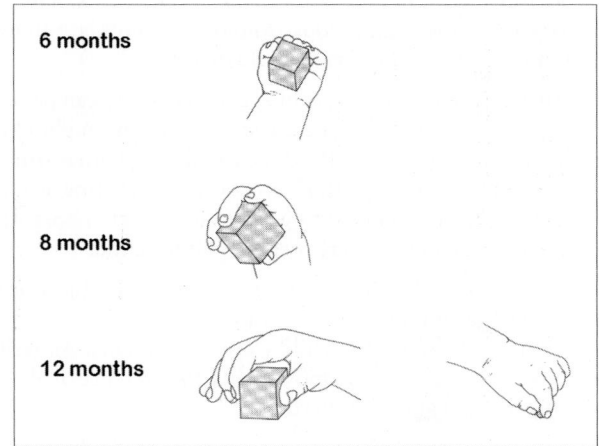

Fig. 7.2 Developmental screen for fine motor milestones.

Language development and social and adaptive development will have already been assessed by observation whilst you are taking the history (Table 7.1).

Children aged from 1–3 years. The milestones summarized in Table 7.1 should be assessed, either during the history taking or whilst examining the child.

If the child fails these development screens, a more detailed assessment will be required, using one of the available standardized developmental assessments: the Griffiths Mental Development Scales—Revised (Griffiths 1996), the Bayley Scales of Infant Development II or BSID-II (Bayley 1993) or the Denver II (Frankenburg *et al.* 1992). Normally this time consuming further assessment requires a special appointment, usually involving referral to a trained assessor.

7.3.3 Occipitofrontal circumference

Among the most important but oft forgotten measurements is that of occipitofrontal circumference. Head circumference is measured over the most prominent part of the occiput and just above the supraorbital ridges using a flexible non-stretchable tape. The measurement should be to the nearest 0.1 cm and should be plotted on a standardized chart in relation to the child's height and weight (Fig. 7.3). If the child has an unusually large or small head for their height, it is useful also to measure their parents' head circumferences, since approximately 50 per cent of normal head size variation is familial (Weaver and Christian 1980). It should be measured in all children at their first examination, since in the absence of hydrocephalus, it provides a measure of brain volume (Winick and Rosso 1969; Bartholomeusz *et al.* 2002). In children under the age of 3 years it is useful to measure the head circumference at every visit since serial measurements provides a measure of the rate of brain growth. Slowing or acceleration in the rate of head growth may be an important diagnostic sign.

7.3.4 Cranial nerves

Examination of the cranial nerves in very young children is done mainly through observation.

◆ *Pupillary responses* can be assessed in much the same manner as in adults (Section 13.3.1).

◆ *Visual fields* can be assessed in young, preverbal children using double simultaneous stimulation by bringing two bright objects or toys into two different areas of the visual fields. This will normally cause the child to look from one to another. Failure to take notice of one may indicate a field defect.

◆ *Visual acuity.* Beyond 3–4 years of age, visual acuity can be tested in the clinic situation using the E test. The child is taught to recognize the E and to discern the direction that the three arms are pointing and to indicate this direction by pointing a finger accordingly. During acuity testing Es of different sizes are rotated in different directions and are presented to the child.

◆ *Eye movements* can be elicited using a small bright object or toy, such as a finger puppet, which is moved slowly and systematically, whilst observing the child's eye movements. As well as determining the range of eye movements, the presence or absence of nystagmus can also be noticed.

◆ *Squints.* Some infants seem to have a squint eye even though their visual axes are straight. This is because the distance between the eyes is small and the epicanthic folds at the inner corner of the eyes are still prominent, so the eyes seem to be too close to the nose (Fig. 7.4). You can determine if there is indeed a squint by noting the position of the light reflection on each cornea when a torch is held in front of the child. Normally the light reflex is symmetrical and slightly nasal to the middle of each pupil (Fig. 7.4). To test for a squint in an older child their attention should be attracted by a small object, such as a finger puppet, placed about 30–50 cm in front of the child at eye level. When their gaze is fixed on the object, each eye is covered in turn to observe the movement of the uncovered eye. Normally the eye not covered does not move when the other eye is covered. In a child with a manifest squint, as the 'good' eye is covered, the 'bad' eye can be seen to move to take up fixation on the object. When the 'bad' eye is covered, there is no movement of the 'good' eye which remains fixed on the object.

◆ *Trigeminal nerve.* Cranial nerve V can be assessed using the jaw jerk reflex, elicited by placing a finger on the subject's chin while the mouth is slightly open and tapping the finger with a tendon hammer (Section 2.3.4). Assessing the corneal reflex is frightening to children and usually only employed in children who have a decreased level of consciousness.

◆ *Facial weakness.* During the history taking and examination the child's face should be observed for facial weakness during laughing, smiling, or crying.

◆ *Hearing.* To test hearing in a child aged under 4 years, a squeaky toy or a bell can be used to generate the sounds. The stimulus should be applied 30–50 cm from the ear and out of sight of the child. Paradoxically this is one test often best performed when a baby or child is crying, since one can observe momentary orientation and quieting in response to the sound. If the baby is quiet, one can stand in front and look for a startle, cry, or blink. From 3–4 months of age onwards one can deliver the sound stimulus from behind the child whilst their parent or caregiver is holding their attention; the child will turn their head towards the sound. In older children, hearing can be tested using a whispered number into each ear and asking the child to repeat what is whispered. Weber's and Rinne's tests using a 512 Hz tuning fork (Section 2.2.3) can usually be undertaken successfully in children older than 5 years.

◆ *Palatal movement.* A gag response can be used to look for palatal movement in young babies and children if this is necessary—however this should be left until last, since it is uncomfortable and you will certainly lose all cooperation afterwards.

◆ *Tongue musculature.* Playing a game involving sticking out of the tongue can be used even in very young children to look for tongue weakness or wasting. It is usually more difficult in young children to look for fasciculation since most find it difficult to open their mouth and let their tongue relax.

7.3.5 Motor assessment

Muscle bulk, reflexes, and muscle tone can be assessed in children in a similar manner to adults (Section 2.2.2). The challenges arise when trying to assess muscle strength in young children aged under 5 years, where problems with cooperation or coordination pose difficulties in assessing the maximum strength of individual muscles. A gross assessment can be obtained by asking a child to climb stairs, to arise from the floor, or to lift up toys of varying weights. From the age of 5 years muscle strength can be assessed as

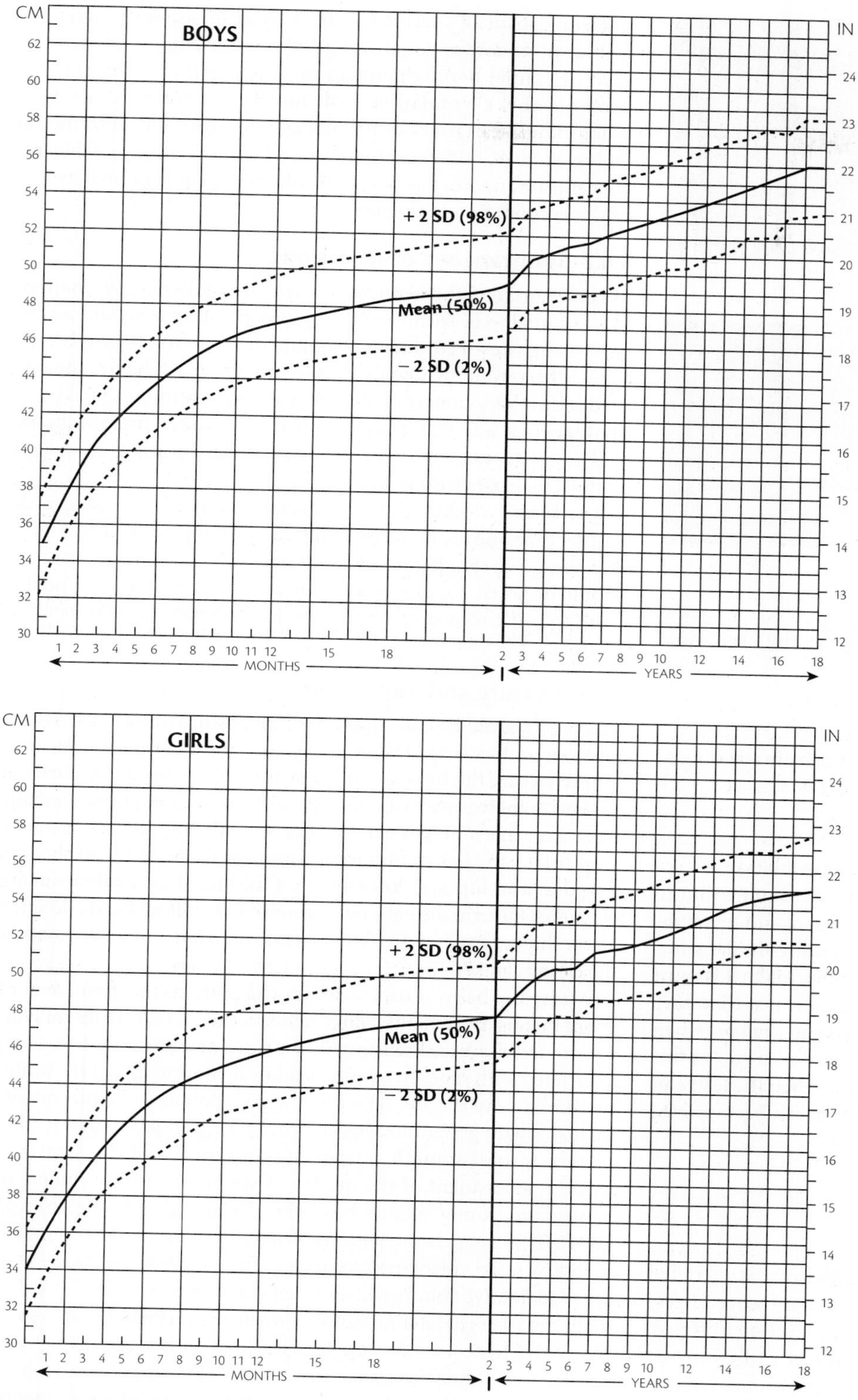

Fig. 7.3 Chart of head circumference vs. age for girls (bottom) and boys (top).

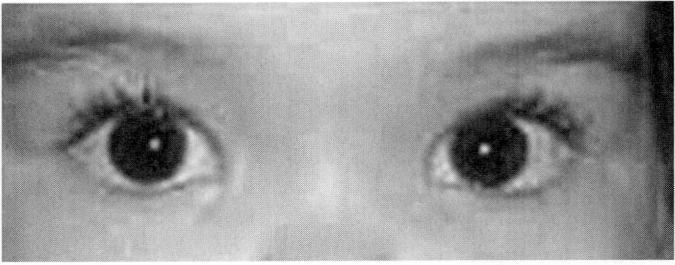

A

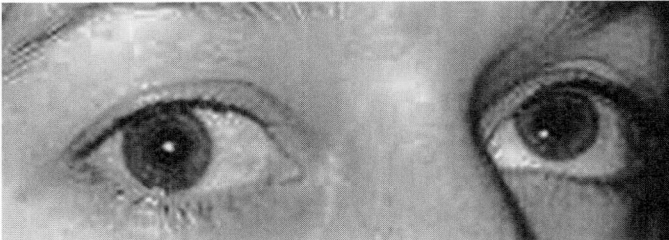

B

Fig. 7.4 A child with apparent squint; note however that the position of the light reflex is symmetrical and slightly nasal to the middle of each pupil in contrast to B where the light reflex is asymmetrically sited in the two eyes of a child with an actual squint.

for an adult by giving simple instructions and by demonstrating to the child what you want them to do (Section 2.2).

7.3.6 Somatosensory testing

In children under the age of 3 years sensory examination is very difficult if not impossible. With patience sensory examination to light finger tip touch, vibration sense, joint position, and temperature can be tested from the age of about 3 years using similar stimuli to those employed in adults (Section 2.5.2). Normally it is most productive to show the children what is expected whilst they have their eyes open—'point to where I have touched you'; 'tell me when the buzzy feeling has stopped'—'point which way your toe or finger has moved'; 'tell me whether it is hot or cold'. Then the tests can be repeated with the child's eyes shut. Cooperation can be improved by trying to make a game out of the assessment.

7.4 Examination of neonates

7.4.1 Introduction

A separate section on examination of the newborn is included because of the particular challenges posed by the influence of gestational age on posture, tone, movements, and level of alertness. Competence in examination of the newborn can really only be obtained by gaining first-hand experience in examining normal babies of varying gestational ages. An examination of the newborn has been devised which can be undertaken by relatively inexperienced examiners (Dubowitz et al. 1999). The revised version of the examination includes 34 items, subdivided into the 6 categories of tone, tone patterns, reflexes, movements, abnormal signs, and behaviour. There is a scoring sheet which provides information on how to assess each of the items. This examination is useful to identify babies at risk for abnormal development, but it is not designed

to be diagnostic of specific lesions or syndromes, since it does not include more conventional aspects of the neurological examination such as cranial nerve examination. It is preferable, therefore, to combine this examination with more conventional aspects of a neurological examination. This chapter does not describe the Dubowitz et al. (1999) examination, which is readily available, but concentrates on how to undertake the more conventional aspects of the examination in the neonate.

7.4.2 Spontaneous observations

When examining the neonate, as with all paediatric examinations, it is essential to be flexible and very observant, so as to take advantage of signs evoked during spontaneous activity. If possible the baby is best examined whilst awake and about 1 h before the next feed is due. Very preterm babies may be fed continuously and their spontaneous waking periods are brief. Even very preterm babies will rouse to external stimulation and the duration of wakeful periods increases with gestation.

The baby's clothes should be removed. Whilst removing the clothes examine the baby for evidence of trauma or malformations and abnormal skin pigmentation. The state of alertness of the baby should be noted, for example whether the baby is apathetic and difficult to rouse or is over-responsive, hyperirritable, and difficult to console.

7.4.3 Posture and movement

The baby should be laid supine and the posture of the head, trunk, and extremities noted (Fig. 7.5). Posture and tone are age dependent, reflecting the increase in flexor tone in the limbs and in axial tone with increasing maturity. Before 32 weeks gestation, the baby lies with arms and legs extended or very slightly flexed. At 32–34 weeks the baby lies in frog leg position while supine with the legs flexed at the hip and knee but not adducted and the arms are extended. By term the baby lies supine with legs flexed and adducted at the hips, flexed at the knees, and the arms are also flexed at the elbow (Fig. 7.5).

Whilst the baby is supine watch the pattern and frequency of spontaneous movements. Movements in the premature infant consist of stretching and twisting of the head, trunk, and limbs, often associated with repetitive wide-amplitude movements of the limbs, resembling myoclonus. The quantity of spontaneous movements increases with gestational age and there is a gradual change of their pattern toward smooth, alternating movements of the arms and legs. An assessment of the quality of spontaneous movements of babies and young infants has been developed (Einspieler et al. 2004). This assesses the pattern of development of general movements by serial videotaped recordings from birth to 16–20 weeks of age. Qualitative abnormalities of general movements are early predictive markers of later neurologic impairment (Prechtl et al. 1997). However, time consuming longitudinal video recording and scoring is not always possible, and intensive training courses lasting 4–5 days are required for correct interpretation of spontaneous behaviour. This assessment, therefore, has limited applicability for routine use in the clinical situation for paediatric neurologists.

7.4.4 Head and spine

The occipitofrontal circumference should be measured (Section 7.3.3) and the head examined for asymmetry, unusual shape, indentations, and protuberances. *Cephalohaematomas* delineated by the

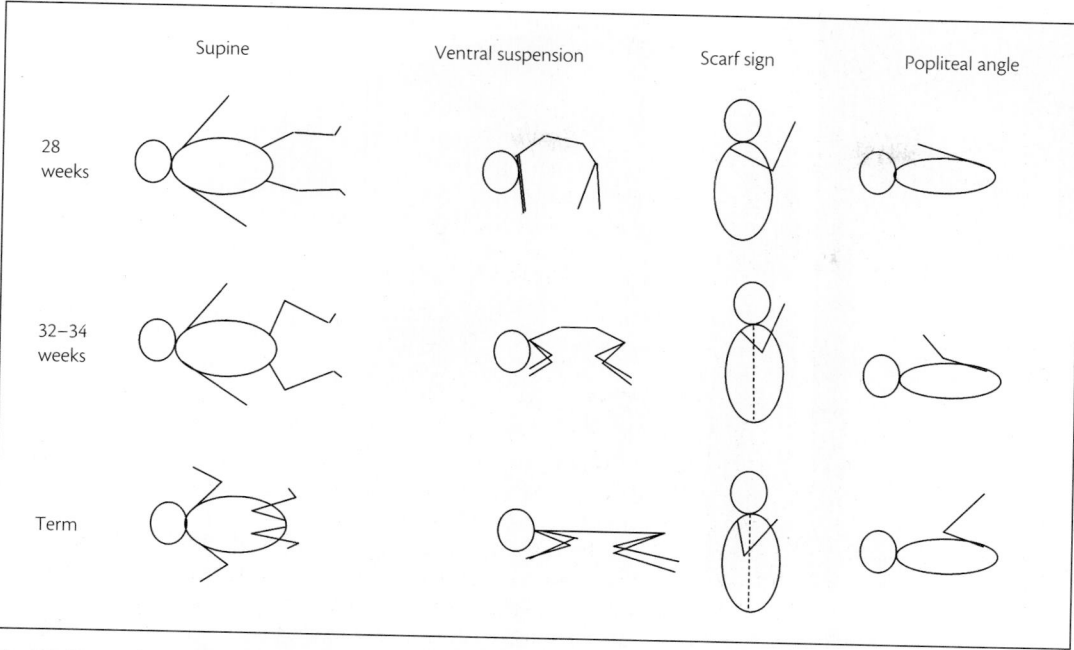

Fig. 7.5 Neonatal examination for trunk and limb tone. For explanation see Section 7.4.6.

periosteum of individual cranial bones, often produce asymmetry (Fig.7.6A). They mostly occur over the parietal bones and therefore may also increase the occipitofrontal circumference. A *caput succedaneum*, which may be present after a ventouse vacuum extraction, can be distinguished from a cephalohaematoma since it extends over two or more cranial bones (Fig. 7.6B) The fontenelle size and pressure should also be assessed. The cranial sutures are readily palpable in the newborn and the examination should include consideration of synostosis or abnormal suture separation. Infants delivered vaginally may manifest overriding of the sutures that resolves over the first week.

Next the cranium, face, palate, and the entire length of the spine including the sacral segment should be carefully examined for midline defects and evidence of spina bifida overt or occult or of dermal sinuses.

7.4.5 Cranial nerves

Pupillary reaction to a bright light can be shown in normal babies from 31 weeks gestation.

Opthalmoscopic examination is important but may be difficult and require dilation of the pupils. The optic discs may show pallor, hypolasia, or abnormalities to suggest septo-optic dysplasia. The retina may show haemorrhages, phakomas, chorioretinitis indicative of intrauterine infection, or pigmentary changes to suggest congenital rubella. An important part is examination of the red reflex. Use an ophthalmoscope about 30 cm from the infant's eyes. Dark spots in the red reflex can be due to cataracts, corneal abnormalities, or opacities in the vitreous. The red reflex may be absent with a dense cataract.

Some preterm infants prior to 32 weeks can focus on a target but they are usually not yet able to track. Therefore in very preterm babies spontaneous eye movements should be observed. Intermittent and slight lapses of conjugate gaze are common but persistently dysconjugate gaze is abnormal. After 32 weeks many babies are able to track horizontally or vertically and by 36 weeks most babies can track even in an arc. The examiner's face is often the best target.

Compensatory vestibulo-ocular movements can be seen when the child's head is turned passively. Movement of the head in the vertical position causes a similar reflex in the vertical plane. The vestibulo-ocular reflex can also be used to assess lateral eye movement. The baby should be held supine on your arm whilst supporting their head with your hand. As you rotate the baby's eyes should be watched for lateral conjugate deviation of the eyes in the direction of rotation.

Ptosis is usually easily recognized and may occur as part of Horner's syndrome, in rare cases of neonatal myasthenia gravis, in other congenital myopathies and as an isolated congenital abnormality.

Facial nerve paralysis is caused by compression of the nerve against the sacral promontory or by trauma resulting from the use of forceps during delivery. Paralysis is usually apparent on the first or second day of life and resolves spontaneously within days. It is usually unilateral and easily detected in a crying baby. Severe bilateral facial weakness is rare and may occur as part of a Mobius' syndrome, myasthenia gravis, myotonic dystrophy, and congenital myopathy.

The response of blinking and startle to a loud noise can be elicited from 27 to 28 weeks postmenstrual age and becomes stronger with increasing gestational age.

Movements of the palate can be observed during crying. Sternomastoid muscle contraction can be assessed when the baby is pulled to sitting to assess head lag. Examination of the tongue may be difficult but it is important since wasting and fasciculation can be seen in Werdnig–Hoffman disease.

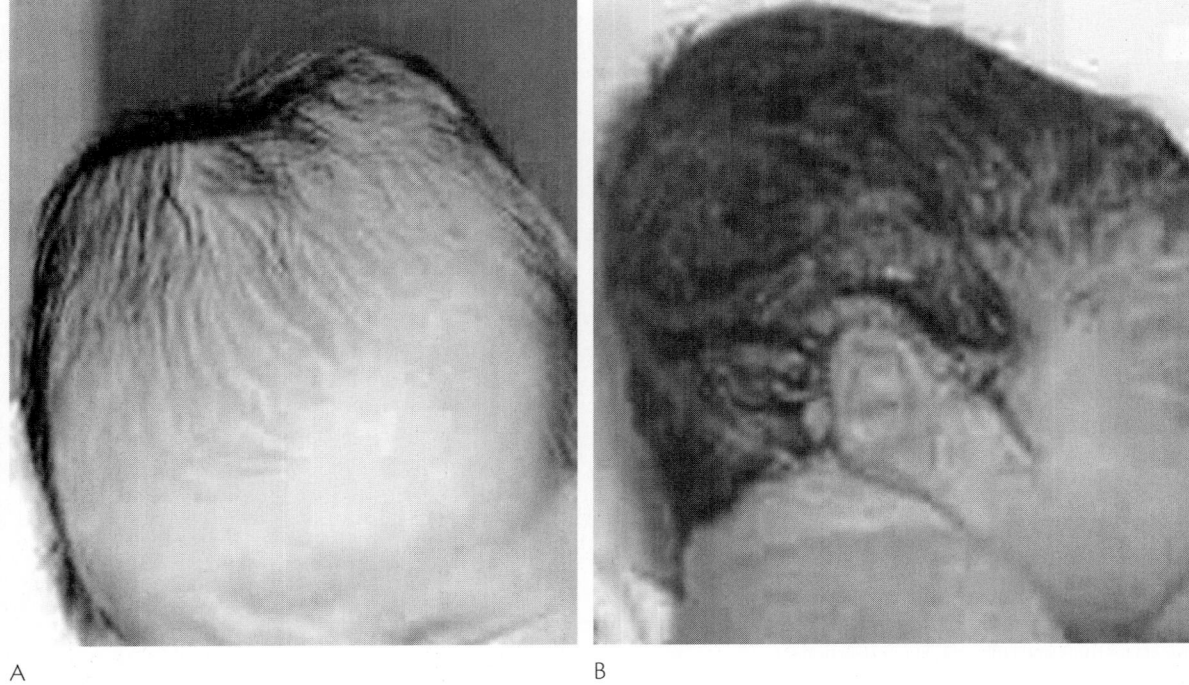

A   B

Fig. 7.6 (A) Cephalohaematoma delineated by the periosteum of an individual cranial bone. (B) Caput succedaneum extending over two or more cranial bones.

7.4.6 Motor assessment

The range of passive movements of the limbs should be examined carefully to look for contractures.

Trunk tone. The preterm baby is very hypotonic. When held prone in ventral suspension by placing your hands around the thorax without providing support for the head or legs, a baby of 28 weeks gestation does not extend the head or spine nor flex the limbs (Fig. 7.5). By term the head is held in line with the thorax and the limbs are flexed. During the traction manoeuvre, when both arms are gently pulled to lift the head off the bed, the head lag is considerable until 30 weeks. By 38 weeks the head follows the trunk.

Limb tone. Limb tone is assessed supine with the head in the midline. If the head is inclined to one side, the asymmetric tonic neck reflex often causes an increase in tone and reflexes on one side. The scarf sign (Fig. 7.5) provides a reproducible measure of upper limb tone. Take the baby's hand and gently try to put it around the neck and as far posteriorly as possible. Note the position of the elbow. Before 32 weeks the elbow approximates the opposite shoulder but by term the elbow does not go beyond the midline. Measurement of the popliteal angle (Fig. 7.5) offers objective evidence for the degree of lower limb tone. It is measured at maximum extension of the leg at the knee with the hip fully flexed and decreases from 180° at 28 weeks to less than 90° at 40 weeks gestation.

Muscle wasting. This can be assessed but muscle strength is impossible to grade. In a floppy infant it is important to look for the presence or absence of antigravity movements. The assessment of tendon reflexes is performed in a similar manner as in the infant and older child. In the newborn there is often variability in the briskness of the reflexes from time to time. Radiation of the reflexes is also observed (O'Sullivan *et al.* 1991) so that adduction of the ipsilateral or contralateral thigh is commonly seen when a knee reflex is elicited, or co-contraction of biceps and triceps when biceps reflex is elicited. These signs are seldom abnormal.

7.4.7 Sensory examination

It is difficult to examine sensation in the newborn, since the baby can only respond to touch or pinprick by crying or moving. Such responses may be unreliable unless painful and thus be distressing to the baby. In spinal cord lesions the trunk incurvation reflex is helpful since it will absent below the level of the lesion.

References

Bartholomeusz HH, Courchesne E, Karns CM (2002). Relationship between head circumference and brain volume in healthy normal toddlers, children, and adults. *Neuropediatrics*, **33**, 239–41.

Bayley N (1993). *Bayley Scales of Infant Development*. The Psychological Corporation: San Antonio.

Dubowitz L, Dubowitz V, Mecuri E (1999). *The Neurological Assessment of the Preterm and Full Term Infant. Clinics in Developmental Medicine*, Vol. 148. McKeith Press: London.

Egan D, Illingworth R, MacKeith R (1969). *Developmental Screening 0-5 Years. Clinics in Developmental Medicine*. William Heinemann Medical Books: London.

Einspieler C, Prechtl H, Bos A *et al.* (2004). *Prechtl's Method on the Qualitative Asssessment of General Movements in Preterm, Term and Young Infants*. MacKeith Press: London.

Frankenburg WK, Dodds J, Archer P *et al.* (1992). The Denver II: a major revision and restandardization of the Denver Developmental Screening Test. *Pediatrics*, **89**, 91–7.

Griffiths R (1996). *The Griffiths Mental Development Scales from Birth to 2 Years Manual*. Henley, association for Research in Infant and Child Development, Test Agency.

Illingworth R (1980). *The Development of the Infant and Young Child, Normal and Abnormal*. Churchill Livingstone: Edinburgh.

O'Sullivan MC, Eyre JA, Miller S (1991). Radiation of phasic stretch reflex in biceps brachii to muscles of the arm in man and its restriction during development. *J Physiol*, **439**, 529–43.

Prechtl HF, Einspieler C, Cioni G *et al.* (1997). An early marker for neurological deficits after perinatal brain lesions. *Lancet*, **349**, 1361–3.

Sheridan M (1973). *Children's Developmental Progress from Birth to Five Years. The Stycar Sequences*. NFER Publishing: London.

Weaver DD, Christian JC (1980). Familial variation of head size and adjustment for parental head circumference. *J Pediatr*, **96**, 990–4.

Winick M, Rosso P (1969). Head circumference and cellular growth of the brain in normal and marasmic children. *J Pediatr*, **74**, 774–8.

CHAPTER 8

Development, degeneration, and regeneration of the central nervous system

Alastair Compston

Contents

8.1 Development in the central nervous system

What does the nervous system do? Primitive organisms respond to threats by reflex withdrawal and explore their environment through goal-directed activities. They sense and respond to their internal environment in order to maintain homeostasis. From these origins emerge more sophisticated forms of discriminative sensation and the acquisition of special senses; precision in the efficiency of movement and coordination between separate elements of motor skills; and cognitive behaviours that anticipate, conceptualize, and enrich physical and social interactions with the environment.

As discussed in Chapter 1, organization in the brain and spinal cord is based on the cell doctrine, the pivotal role of reflexes, integrated facilitatory and inhibitory neuronal activities, and the development of systems and regions having specialized functions. However, it is axiomatic that despite the convergence of sensory and motor traffic in preferred pathways, and the dependence on particular brain regions for activities such as vision, sensation, speech, and movement, the nervous system discharges its responsibilities to the organism in one orchestrated performance.

The structures on which this organization depends are neurons and their axonal appendages, and the network of glia in which they are suspended. Communication between axons and dendrites across synapses provides the basis for connectivity. The 'language' may change resulting in learning and plasticity in the central nervous system. Locally active neurotransmitters, released pre-synaptically in response to the nerve impulse, cross the synapse, bind receptors on the postsynaptic membrane, and activate channels that gate the passage of sodium, potassium, calcium, and chloride ions thereby propagating local current. The main excitatory neurotransmitter is glutamate, whereas glycine and γ-aminobutyric acid, GABA, are inhibitory molecules. Indirect activity of ion channels is mediated through G-proteins and second messenger systems including cyclic adenosine monophosphate, inositol phosphates, arachidonic acid metabolites, and protein kinases. A few synapses rely only on electrical transmission.

Early in development, the embryonic mesoderm induces the overlying dorsal midline ectoderm to form neuroectodermal cells of the neural plate. This invaginates as the neural groove. Its lips then fuse, neuroepithelial cells now comprising the neuroectoderm that makes up the neural tube. Neuroepithelial cells proliferate in the ventricular zone where further division continues. Some then elongate and migrate, spanning the developing nervous system as radial cells. The radial network thus coincides with neurogenesis suggesting that this provides the scaffold for nerve cell migration into the developing cortex.

Neural development is a longitudinal process with the building blocks of the central nervous system assembled precisely in a time specific manner along a spatial grid. Following neural induction, neurons are generated prior to gliogenesis. They adopt regional identities that subserve specialized functions. The hundreds of neuronal sub-types, that together make up neuronal circuits, each possess a 'date and post-code'. Cells acquire their unique 'address' through the interaction of extrinsic and intrinsic mechanisms acting upon neural stem cells. These self-renew through symmetric and asymmetric division, and are capable of generating all three major neural lineages— neurons, astrocytes, and oligodendrocytes (Fig. 8.1). Typically, they express the immature neuroepithelial markers Sox, *nestin* and *musashi* (Gotz and Barde 2005).

Neural stem cells can be isolated from embryonic or adult brain. In the adult mammalian brain, stem cells are restricted to two neurogenic niches—the subventricular and hippocampal subgranular zones. Regulation of neural stem cell proliferation is context dependent with *in vivo* evidence implicating a role for canonical notch, hedgehog and Wnt signalling (Gage 2000). Neural stem cells can also readily be cultured in adherent or substrate free conditions in the presence of basic fibroblast growth factor, FGF-2, and epidermal growth factor, EGF (Ray *et al.* 1993; Vescovi *et al.* 1993). The phenotypic potential of stem cells is influenced by both intrinsic and environmental signals. For example, defined growth factors such as the neurotrophin family and retinoic acid promote neuronal fate and survival (Nieto *et al.* 2001). By contrast, transforming growth factor beta, TGF-β, signalling and activation of the JAK-STAT pathway direct astrocyte generation. Oligodendrocyte specification from neural progenitors is induced by sonic hedgehog while insulin-like growth factor, IGF-1, platelet derived growth factor, PDGF, thyroxine, retinoic acid, and glucocorticoids stimulate oligodendrocyte differentiation and maturation from stem cells or committed precursors *in vitro*.

Issues of topical interest with respect to the application of knowledge on the biology of stem cells to human disease are whether differentiation into neurones, astrocytes, and oligodendrocytes can be manipulated; and whether fate-committed precursors may de-differentiate to a more primitive state reproducing a source of neural stem cells that self-renew and subsequently replenish defined neuronal or glial lineages. Until recently, adult somatic stem cells were regarded as restricted in their phenotypic potential to that of their tissue of origin. Emerging evidence suggests that *ex vivo*

reprogramming of adult somatic cells is feasible, thus raising the prospect of autologous 'neural' stem cells derived from accessible and readily available non-neural sources such as skin (Takahashi and Yamanaka 2006).

8.2 Macroglial lineages

Together with extracellular matrix, astrocytes provide a scaffold for the nervous system. Oligodendrocytes synthesize and maintain the myelin sheath that subserves saltatory conduction along myelinated nerve fibres. Microglia are bone marrow derived cells of the macrophage lineage and provide the nervous system with a degree of immunological independence.

8.2.1 Astrocytes

Recognition that astrocytes express voltage and ligand gated ion channels and neurotransmitter receptors, and behave in the adult brain as neural stem cells has overturned the long held view that astrocytes are essentially passive support cells (Doetsch *et al.* 1999). Through their pleiotropic properties, astrocytes support neurons, define anatomical boundaries, act as a source of growth factors and cytokines, contribute to conduction of the nerve impulse, and participate in the response to injury (Fig. 8.2). The classification of astrocytes has undergone a series of modifications since their original morphology based categorization as protoplasmic, mainly identified in the grey matter, and fibrillary cells, most abundant in white matter. Emerging molecular based characterization suggests regional, developmental, and functional heterogeneity that extends beyond characterization based merely on the expression of glial fibrillary acidic protein, GFAP.

Astrocytes are highly reactive cells. However, the origin of the reactive astrocyte is not resolved. Some have migrated into lesions, whereas others proliferate locally. Most represent resting cells that have undergone an alteration in phenotype. The characteristic feature of astrocytosis is hypertrophy with change from the protoplasmic to fibrillary morphology, and cell proliferation. These changes are accompanied by increased glial fibrillary acidic protein expression

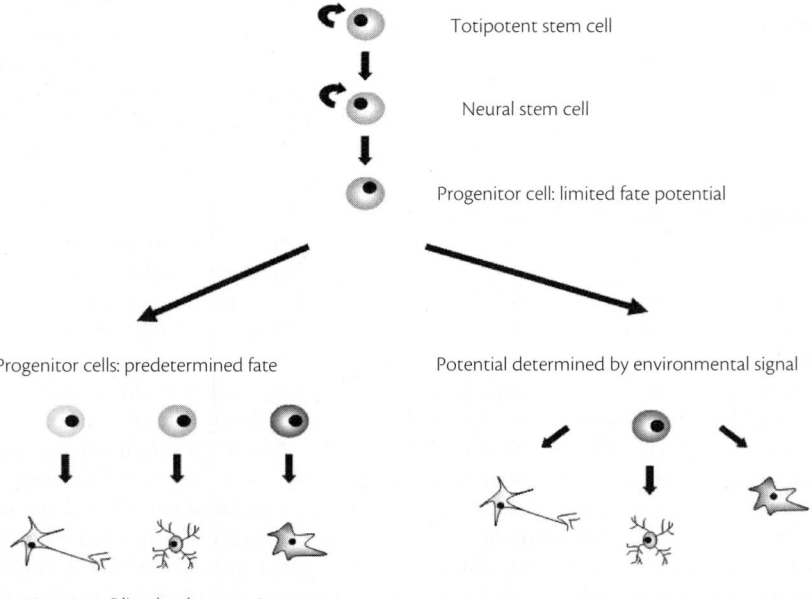

Fig. 8.1 Scheme to show fate determination of stem cells and their progeny in the central nervous system.

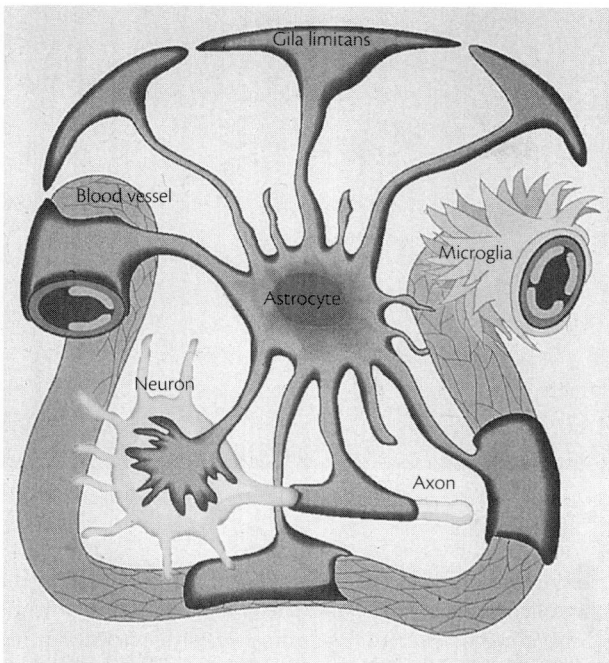

Fig. 8.2 The functions of astrocytes in the central nervous system.

which, in this respect, does serve as a useful marker of the reactive astrocyte. Reactivity involves exposure to extracellular matrix molecules such as chondroitin sulphate proteoglycans. Reactive astrocytes increase their production of trophic factors including insulin-like growth factor I, IGF-1, glial cell derived nerve growth factor, GDNF, and platelet derived growth factor, cytokines, proteases and their inhibitors, and cell surface and extracellular matrix molecules. Several are simultaneously secreted by microglia and it has been suggested that these same cytokines, produced at different stages in the response to injury and by different cell types, mediate both pro- and anti-inflammatory effects.

Astrocyte proliferation and reactivity generally culminate in formation of the astroglial scar. Made up mainly of hypertrophied astrocytes, this also wraps in fibroblasts and meningeal cells which—interacting with surviving astrocytes—form a new glia limitans. Astrogliosis serves an essential role in limiting damage and thus can be considered protective. However, contingent on context, the resulting glial scar may be inhibitory to axonal regeneration and cell migration, therefore providing both a physical and chemical barrier (Rhodes and Fawcett 2004). Following injury, animals that lack reactive astrocytes fail to limit local inflammatory responses, converting small self-limiting lesions to areas of persistent blood–brain barrier disruption, intense inflammation, and tissue destruction with oligodendrocyte and neuronal depletion, and widespread demyelination (Bush et al. 1999). Clearly, the damaged brain needs to steer a clever course between promoting reactive changes in astrocytes, which usefully encapsulate the area of tissue injury and provide mediators for the promotion of recovery, whilst avoiding formation of a dense astrocytic scar that may exclude cells and mediators needed to repair neurons and glia.

8.2.2 Oligodendrocytes

Classification of the oligodendrocyte lineage has also undergone extensive re-evaluation. The oligodendrocyte lineage is the best characterized neural cell population largely due to the availability of discrete stage specific cell surface markers. Initial observations defined the oligodendrocyte-type 2 astrocyte, O-2A, progenitor in the rodent central nervous system. But following the failure to reproduce in vitro staining properties of the oligodendrocyte-type 2 astrocyte cell in vivo, its status has been re-evaluated. Nonetheless, the original studies did provide markers that usefully identify an oligodendrocyte precursor cell, OPC (Raff et al. 1983): the most versatile have proved to be platelet derived growth factor-α chain receptor and the proteoglycan NG2 (Pringle et al. 1992; Nishiyama et al. 1996).

Developmental studies of oligodendrocyte precursor cells have shown that their specification is regionally determined. Serial analysis of the mammalian spinal cord has been particularly informative revealing that oligodendrocytes originate from restricted foci within the ventral neural tube under the influence of opposing gradients of inductive sonic hedgehog and inhibitory transforming growth factor beta morphogenetic signals (Rowitch 2004). The subsequent widespread dispersal of oligodendrocyte precursor cells throughout the central nervous system is largely a consequence of their migratory capacity. Furthermore, recent studies have identified Olig1 and Olig2 basic helix-loop-helix transcription factors as necessary and sufficient for commitment to the oligodendrocyte precursor cell lineage (Lu et al. 2002). Of considerable interest has been the recognition that motor neurones and oligodendrocytes share a common regional precursor domain, pMN. Dependent on association with temporally determined homeodomain factors, motor neurone specification is followed by oligodendrocyte precursor cell commitment (Kessaris et al. 2001; Zhou and Anderson 2002). Recent in vitro and in vivo studies of dorsal cord gliogenesis have also identified an additional hedgehog independent pathway of oligodendrogenesis that implicates a role for fibroblast growth factor-2 and mitogen-activated protein kinase dependent signalling (Miller 2005). Whether regional heterogeneity of oligodendrocyte origins is matched by functional heterogeneity remains unknown.

A first step in differentiation of the oligodendrocyte progenitor is expression of the surface molecule marker recognized by the antibody O4 which stains sulphatide and seminolipid (Fig. 8.3). Proliferation and bipotentiality are maintained at this stage but the cell is no longer motile. A major regulator of oligodendrocyte development is the Notch1 receptor and its ligand Jagged 1 (Wang et al. 1998). Their interaction inhibits the differentiation of oligodendrocyte precursors. The sequential expression of a number of oligodendrocyte- or myelin-specific components follows during subsequent maturation. These include galactocerebroside, GalC, the surface expression of which is generally accepted to mark terminal oligodendrocyte differentiation, signalling the loss of both proliferative and migratory potential. Myelin basic protein, MBP, expression is followed by the intracellular appearance of proteolipid protein, PLP, and the surface expression of myelin oligodendrocyte glycoprotein, MOG, representing the final immunocytochemically identifiable stage of oligodendrocyte maturation. There is an interval between the stage of oligodendrocyte development and the onset of myelination during which the pioneer cells make multiple contacts with axons.

These developmental processes are regulated by growth factors. Platelet derived growth factor is mitogenic for oligodendrocyte progenitors. Although they differentiate after a fixed number of divisions, fibroblast growth factor-2 suspends that maturation and promotes migration. Neuregulins act on oligodendrocyte precursors to promote their proliferation and survival, inhibiting the differentiation of mature oligodendrocytes. Thyroid hormone, erythropoietin,

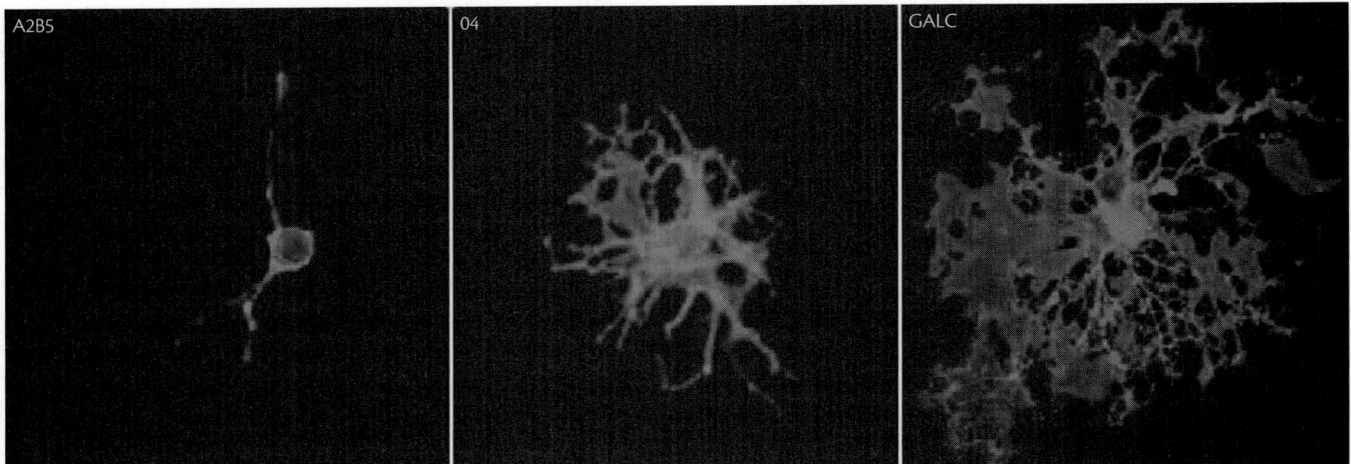

Fig. 8.3 Stages in maturation of the oligodendrocyte lineage from progenitor (A2B5+) to precursor (O4 +) and mature myelinating cell (GalC +). (See Plate 6).

and members of the ciliary neurotrophic factor, CNTF, family are implicated in the late stage of oligodendrocyte development. Survival factors for cells of the oligodendrocyte lineage include insulin growth factors-1 and, -2, leukaemia inhibitory factor, LIF, interleukin 6, IL-6, neurotrophin-3, NT-3, and ciliary neurotrophic factor, CNTF (Barres *et al.* 1992). Oligodendrocyte precursors express a repertoire of extracellular matrix molecules at different stages in development. These exert effects by amplifying the signalling of growth factors such as platelet derived growth factor-α and neuregulin. This glial–neuronal interaction provides the mechanism for target-dependent survival of newly differentiated cells. Only those oligodendrocytes that establish neuronal contact amplify signals from the limiting concentration of available growth factors and so survive, while the remainder will undergo apoptosis (Barres and Raff 1999).

Inevitably, less is known concerning human oligodendrocyte precursors. These are born in many parts of the central nervous system, including the spinal cord and cerebrum differentiating *in vitro* both to pre-oligodendrocytes and fully differentiated oligodendrocytes. Although conservation of developmental principles with rodents is evident, for example, with respect to ventral–dorsal polarity and stage specific markers, there are important inter-species differences. For instance neural precursors derived from the developing human spinal cord do not generate increased numbers of oligodendrocyte precursor cells following treatment with the mitogens fibroblast growth factor-2 and epidermal growth factor (Chandran *et al.* 2004). These, and other observations, suggest caution when extrapolating rodent observations to human studies. Recent studies of human embryonic stem cells offer the prospect of generating unlimited numbers of oligodendrocytes (Nistor *et al.* 2005).

Soon after the oligodendrocyte-type 2 astrocyte precursor was identified *in vitro*, it became clear that cells having the properties of oligodendrocyte precursor cells could also be recovered from the adult rodent nervous system. The same sequence of migration, proliferation, and differentiation operates for adult and neonatal oligodendrocyte precursors, albeit with distinct cell cycle times (Wren *et al.* 1992). Similar cells have also been identified in adult human white matter tissue. Indeed there is some evidence that adult human white matter precursors may also possess comparably

diverse phenotypic potential (Nunes *et al.* 2003). The idea that a putative oligodendrocyte fate-restricted cell has neuronal potential is an observation of potential significance with respect to neural repair. Initial rodent *in vitro* studies have shown that, following treatment with bone morphometric protein, BMP, and fibroblast growth factor-2, oligodendrocyte precursor cells can reactivate the primitive neural precursor gene *Sox2* (Kondo and Raff 2000). These studies have found some indirect support from recent *in vivo* experiments that examine the fate of NG2 precursors (Belachew *et al.* 2003). Although experimentally promising, the question of whether oligodendrocyte precursor cells have *in vivo* neuronal potential remains unresolved.

8.2.3 Schwann cells

Improved understanding of the developmental origin and biology of Schwann cells, the peripheral myelinating cell, has in many ways paralleled that of the oligodendrocyte lineage (Section 8.2.2). Neural crest cells are multipotent stem cells that emerge and migrate from the dorsal neural tube and generate a wider range of cell types ranging from melanocytes and smooth muscle to Schwann cells. Developmental studies have identified three stages in Schwann cell formation beginning with the transition from neural crest stem cell to Schwann cell precursor, subsequently to immature Schwann cell, and finally the differentiation to mature myelinating or non-myelinating Schwann cell (Jessen and Mirsky 1998). These stages are marked by common and unique gene expression along with responses to lineage-specific survival and mitogenic factors. Schwann cells are characterized by expression of intermediate filamentous proteins glial fibrillary acidic protein, S100β, and the glycolipids O4 and GalC. Myelinating Schwann cells suppress glial fibrillary acidic protein and express myelin proteins including P0, PMP22, PLP, and MBP. Unlike oligodendrocytes that myelinate many internodes, Schwann cells have a one-to-one relationship with neurons. Determination of whether a Schwann cell becomes a myelinating or non-myelinating cell depends upon axon derived signals that remain to be characterized. Large diameter axons of greater than 1 μm are myelinated. In contrast to Schwann cell precursors that are dependent on axon derived signals for survival, Schwann cells can survive independent of axons through autocrine signalling, an important asset enabling a relatively improved regenerative response to peripheral nerve injury compared to cells in the central nervous system. Schwann cell derived neurotrophins are important in ensuring support to developing neurons (Riethmacher *et al.* 1997). Unlike myelinating oligodendrocytes, Schwann cells de-differentiate in response to injury adopting the immature and proliferative Schwann cell precursor phenotype. Together with their ability to migrate, express growth promoting factors, and myelinate regenerating axons, these properties of Schwann cells facilitate peripheral nerve regeneration.

8.2.4 Microglia

Microglia represent the primary immunocompetent cell within the central nervous system (Section 38.2.2). In the normal brain, their main function is to take up and digest material that has passed the blood–brain barrier or been liberated within the central nervous system. At least *in vitro*, microglia increase the number of surviving galactocerebroside-positive oligodendrocytes through platelet derived growth factor-α receptor signalling and nuclear factor-κB mediated inhibition of endogenous oligodendrocte precursor apoptosis, and they promote oligodendrocyte differentiation (Nicholas *et al.* 2001). But they also sculpt the developing nervous system by removing material rendered degenerate through the process of apoptosis. In addition to their role in development, microglia contribute to tissue injury and, like astrocytes, they are highly reactive cells. Activated microglia change their morphology, becoming amoeboid and motile, and release pro-inflammatory cytokines, chemokines, and mediators that orchestrate immune functions. Microglia, or bone-marrow-derived macrophages, are the best candidates for a brain-derived antigen-presenting cell, although astrocytes, brain endothelial cells, and pericytes are also all capable of re-stimulating lymphocytes. They have phenotypic features and some functional properties resembling dendritic cells, which also seem capable of migrating into the central nervous system.

8.2.5 The blood–brain and blood–nerve barriers

The central nervous system is separated from the systemic circulation by the blood–brain barrier, composed of endothelial cells abutting a basement membrane formed by layers derived both from endothelial cells and astrocytes and interspersed with pericytes around which a continuous covering is provided by the extended foot processes of astrocytes. Pericytes are adventitial contractile cells forming the smooth muscle cells of larger vessels. Permeable to water and lipid-soluble molecules, the blood–brain barrier prevents the free passage of most solutes from the systemic circulation into the central nervous system. Essential nutrients, including glucose and certain amino acids and ions, are transported by specific carrier mechanisms. The blood–brain barrier allows diffusion gradients to develop between plasma and extracellular space within the central nervous system, which depend upon the size and the charge of the respective molecules. Human foetuses as young as 7 weeks show the presence of tight junctions both between neighbouring choroid plexus epithelial cells and individual cerebral endothelia so that, as soon as these cells differentiate, they have well-formed tight junctions.

The glia limitans is essentially formed by tight layers of astrocyte processes covered by a basal lamina. Ultrastructural analysis of the developing blood–brain barrier reveals a role for aquaporin-4 on the astroglial endfeet, eventually enveloping the endothelium–pericyte layer. Disruption of the blood–brain barrier is associated with reduced expression of aquaporin-4, indicating a close relationship between water transport regulation and blood–brain barrier development. The microglial cell contributes a further component to the perivascular glia limitans, located between the astrocyte foot processes of cerebral microvessels and bone marrow-derived macrophages present in the perivascular space of medium-sized and larger cerebral vessels.

The endothelia of endoneurial capillaries have a similar structure and function to brain microvessels providing a blood–nerve barrier that can be considered similar to the blood–brain barrier. Their tight junctions are also selectively permeable to macromolecules including proteins—perhaps more so than the blood–brain barrier— and may become selectively permeable as a result of disease processes resulting in endothelial fenestrations and discontinuities (Bell and Weddell 1984).

8.3 Axon degeneration and recovery of function

Many disease processes damage axons. Given their central role in structure–function relationships of the central nervous system, these processes lead to clinical deficits. Although the main categories inevitably overlap, immediate alteration of function follows axonal transection: inflammatory mechanisms cause subacute axonal loss; and many neurodegenerative disorders are the result of chronic axonopathies. Whatever the cause, axonal recovery in the central nervous system is limited. Therefore, rather similar strategies need to be deployed in trying to achieve axonal regeneration, whatever its cause. Because they function as one unit, several properties of axons and glia need to be considered together in discussing axonal damage and recovery.

8.3.1 Mechanisms of axonal injury

Clinical observations and related laboratory and experimental studies indicate that acute cytokine release leads to transient conduction block most likely through indirect mechanisms involving nitric oxide. Experimentally, brief exposure to nitric oxide donors produces reversible conduction block in normal or, especially, demyelinated axons. More prolonged exposure to the donors of nitric oxide causes axon degeneration especially if these are electrically active (Redford *et al.* 1997). The mechanism involves nitric oxide inhibition of mitochondrial function leading to ATP depletion, failure of ion exchange mechanisms across the cell membrane, glutamate release, and excitotoxicity consequent upon uncontrolled influx of sodium through voltage-gated channels leading to a lethal increase in intra-axonal calcium levels. Axons are protected by sodium-channel-blocking agents, such as flecainide, phenytoin, and lamotrigine, that prevent the conditions necessary for the reversal of sodium/calcium exchange (Kapoor *et al.* 2003).

Distal or Wallerian degeneration follows axotomy and proximal dying back towards the neuronal cell body in the context of metabolic, toxic, and immunological insults. The same process may be used selectively to sculpt exuberant axonal and neurite outgrowth during development, without compromising neuronal survival. Studies in the Wallerian degeneration slow, Wld, mouse indicate that loss of the distal axon involves an active process in which a nuclear fusion protein affects the synthetic pathway for nicotinamide adenone dinucleotide, NAD, and ubiquitination. Although the inhibitor of apoptosis, Bcl-2, is protective in some experimental models, dying back axonopathy does not necessarily involve apoptosis and caspase-mediated cell death. Thus, axonal and neuronal injury may be dissociated (Raff *et al.* 2002).

Experimental myelin mutant mouse models have established the role of specific structural myelin proteins in determining neuronal and axon survival. In principle, all myelin mutants causing chronic myelin impairment or demyelination are associated with axonal injury. Recent studies of CNP1 mutant mice show dissociation between ultrastructurally normal myelin and progressive axonopathy suggesting that oligodendrocyte–myelin derived signals are necessary for healthy axons independent of myelination (Lappe-Siefke *et al.* 2003). Making the same general point, neurons demonstrate a marked increase in survival when co-cultured directly with soluble factors released from oligodendrocyte precursors and differentiated oligodendrocytes. Attention has focused on insulin-like growth factor-1, since cells of the oligodendrocyte lineage produce insulin-like growth factor-1; and neutralizing antibodies to insulin-like growth factor-1, but not other candidate trophic factors, block the soluble survival effect of oligodendrocytes. Recombinant insulin-like growth factor-1 promotes neuronal survival under identical conditions. Conversely, differentiated oligodendrocytes increase neurofilament phosphorylation and axonal length under the influence of glial cell derived nerve growth factor, GDNF (Wilkins *et al.* 2003). Thus, factors released by oligodendrocyte precursor cells and oligodendrocytes support structural and functional properties of the neuron and its axonal appendage (Fig. 8.4). Rat cortical neurons exposed to a nitric oxide donor, that normally proves lethal, are protected by insulin-like growth factor-1, whereas medium conditioned by oligodendrocyte precursors and glial cell derived nerve growth factor enhance axon survival indicating that different mechanisms underlie neuronal and axonal destructive and protective processes, and suggesting that trophic factors may modulate nitric oxide-mediated neuron/axon destruction via specific pathways (Wilkins and Compston 2005). There are also specific trophic effects of oligodendrocytes and their conditioned medium on the survival of cholinergic neurons attributable to nerve growth factor, NGF, and brain-derived neurotrophic factor, BDNF. These observations highlight the concept of a dynamic dialogue between glia and neurons that is necessary for neuronal homeostasis. It follows, therefore, that loss of glia will result in a compromised and vulnerable neuron.

Mechanisms of axon degeneration in non-inflammatory contexts

The length, polarization, and elaboration of neuronal processes are fundamental to supporting axonal function; and the ability to form complex networks requires the support of an efficient system for intracellular transport. This depends upon the interaction between motor proteins and the cytoskeletal track infrastructure, a high energy dependent process. Disruption to any of these structures may result in axonal degeneration as observed in a wide range of neurological disorders. Several lines of evidence suggest that, despite the diversity of causes implicated in neurodegeneration, there is convergence in mechanism(s) of axonal degeneration centred on disruption to axonal transport, mitochondrial function, and sodium–calcium axonal membrane exchangers (Coleman 2005). Recognition that axonal degeneration, independent of triggering stimulus, occurs by regulated and non-apoptotic processes offers pharmacological targets for axonal protection (Finn *et al.* 2000).

8.3.2 Axonal regeneration

A marked difference exists in the extent to which axon regeneration occurs in the central and peripheral nervous systems. Characterizing and comparing biological properties of the response to axotomy of central and peripheral nerve fibres suggests strategies for manipulating events so as to secure functional recovery in the clinical context. With the exception of some success following reimplantation of avulsed peripheral nerve roots into the spinal cord, most of this work remains experimental but dividends are expected from the application of these advances in neurobiology, especially in the context of spinal cord injury. In one sense, the lack of axonal regeneration that characterizes adult mammalian nervous systems can be considered the price paid for needing an inhibitory environment to guide axons during development, for stabilizing arrangements in the post-mature nervous system, and for inhibiting promiscuous axonal exploration.

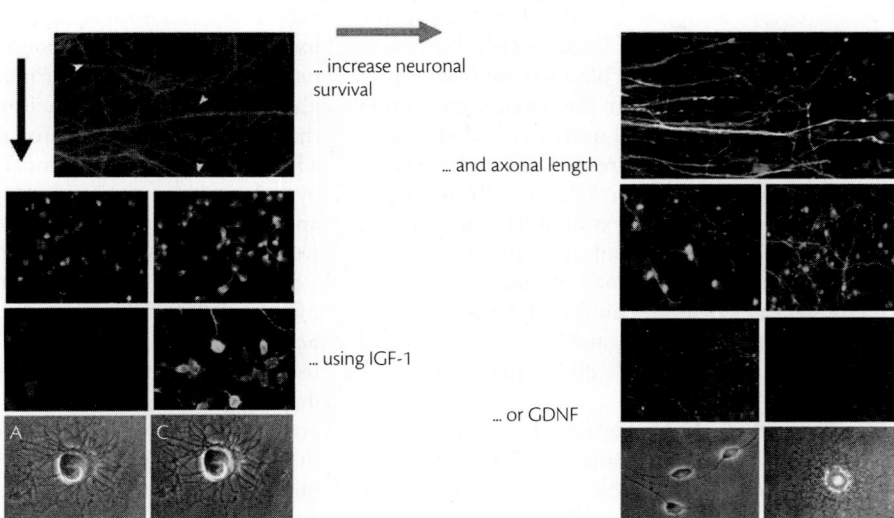

Fig. 8.4 Medium conditioned by cells of the oligodendrocyte lineage supports neuronal survival and axonal growth acting through defined growth factors. (See Plate 7).

The observation that central nervous system axons regenerate in the presence of peripheral glia, Schwann cells, or olfactory ensheathing cells, suggests that the failure of central axons to regenerate following injury to the adult central nervous system reflects a non-permissive environment rather than intrinsic growth-limiting properties. The predominant molecules that are inhibitory to axon regeneration are derived from the astrocytic glial scar and adult oligodendrocytes. Promotion of axon regeneration is thus focussed on overcoming inhibition (Fawcett 2006). Astrocytes block axon regeneration through a mechanism that induces the normal stop signals on growth cones. Extracellular matrix molecules produced by astroglia have growth inhibitory properties that include chondroitin sulphate proteoglycans. The failure of axon regeneration correlates with the amount of local reactive glial extracellular matrix. Much of the inhibitory activity of chondroitin sulphate proteoglycans expressed in the context of injury depends on their glycosaminoglycan chains. These can be removed by chondroitinase ABC. *In vivo* experiments show its efficacy in promoting axon regeneration in the brain and spinal cord. The histological effects correlate with behavioural recovery. Chondroitinase ABC also restores plasticity in the adult cortex and spinal cord.

Axon growth is additionally inhibited by myelin proteins. After fibre tract injury, Nogo spills from the disrupted oligodendrocyte–myelin unit and acts, unhelpfully, to inhibit neurite outgrowth and axon regeneration. Anti-Nogo antibodies placed in the parietal cortex of animals undergoing spinal hemisection partially restore motor function by enhancing recovery of brainstem-spinal or cortico-spinal fibres, and promoting plasticity of surviving serotonergic and adrenergic descending motor neurons. Functional recovery and new connectivity from neighbouring parts of the contralateral cortex are also reported using anti-Nogo antibody in local destructive lesions and experimental stroke (Liu *et al.* 2006).

Myelin-associated glycoprotein, which is involved in the initial adhesion between the processes of myelinating cells and axons, and Nogo bind the same receptor such that anti-Nogo receptor antibody and soluble Nogo-R, each prevent inhibition of neurite outgrowth by myelin-associated glycoprotein. The Nogo receptor mediates the effect of these inhibitory myelin proteins but, in order to enhance axon regeneration, its inhibition must be performed in conjunction with promotion of the neuronal growth programme (Schwab 2004). Oligodendrocyte myelin glycoprotein, Omgp, has recently been characterized as another molecule present on oligodendrocytes that inhibits neurite outgrowth, as does NG2 chondroitin sulphate proteoglycan, expressed on the surface of oligodendrocyte precursors.

Once irreversible axonal injury is established, recovery must presumably partly depend on adaptation and plasticity: at the axonal level through restoration of conduction following increased insertion of sodium channels along the demyelinated axonal surface; through redundancy and rearrangement of connectivity in specific pathways; and through expansion of cortical receptor fields for a degraded axonal signal.

8.4 Myelination

In negotiating stable relationships, several important influences mediated by soluble factors and cell–cell contacts are exchanged between axons and glia. The migration of glial precursors occurs along established axonal tracts that provide a substratum consisting of axons that are aligned in parallel. The movement of migrating cells is influenced by receptor–ligand adhesions with the extracellular matrix and, hence, by discontinuities in adhesiveness, especially involving integrins—one of several classes of adhesion molecules determining cell–cell interactions in the developing and post-mature nervous system. Oligodendrocyte progenitors need to orientate their processes and maximize points of contact with the non-myelinated axon. These alignments coincide with the initial expression of myelin proteins, especially proteolipid protein. Developmental studies have highlighted the importance of neuronal derived signals in matching oligodendrocyte cell number to local axon density during myelination, survival being orchestrated by axonal requirement. Neurons are mitogenic for cells of the oligodendrocyte lineage, an effect dependent upon cell–cell contact and soluble mediators including platelet derived growth factor-α, fibroblast growth factor-2, and neuregulin. Electrical activity in axons also influences oligodendrocyte precursor proliferation.

Myelin accounts for approximately 70 per cent of the dry weight of the mammalian central nervous system, and each interfascicular oligodendrocyte, myelinating up to several tens of internodes along neighbouring axons, produces myelin membranes equivalent to 600–700-fold its own cell body surface membrane. Myelin is composed of lipid, 70–75 per cent dry weight, and protein, 25–30 per cent. Around 25 per cent of the lipid is cholesterol, 40–45 per cent phospholipids, and the remaining 25–30 per cent galactolipid. Galactocerebroside is the major myelin glycolipid, 25 per cent, and both polyclonal and monoclonal anti-galactocerebroside antibodies are specific surface markers of myelin and oligodendrocytes both *in vivo* and *in vitro*. The myelin membrane also expresses many receptors for growth factors and cytokines. Of the minor lipid components of myelin, gangliosides together comprise <1 per cent of total lipid.

Myelin synthesis is triggered when the elongated oligodendrocyte processes make contact with nearby axons and form a cup at the point of contact, extending lengthwise to form a trough whose two lips advance around the circumference of the axon until they meet (Fig. 8.5). One then passes beneath the other to become the inner tongue of the future sheath which rotates many times around the axon to form the multiple membrane layers or lamellae. The developing myelin sheath extends lengthwise in both directions along the axon to form an internodal segment. But at the advancing edge, each layer of the spiral retains a bead of cytoplasm where the two inner leaflets of the surface membrane remain separate. In three dimensions, this bead comprises a ring of cytoplasm around the axon and is termed the lateral loop. Transverse bands, regularly arranged sites of close membrane apposition spaced 10–15 nm apart, later develop between the end of each lateral loop and the underlying axolemma. There are as many lateral loops at the leading edge of the advancing sheath as there are lamellae, and these become stacked in a regular way, those of the outermost lamellae lying outside those of the innermost. The complement of lateral loops at one end of each developing internode almost abuts onto its adjacent counterpart, and together these form the paranodal region next to the node of Ranvier.

During compaction, the cytoplasmic content of all except the inner- and outermost lamellae of the developing spiral sheath is gradually extruded, and the two inner leaflets of the surface membrane lipid bilayer thus become opposed. They then fuse to form the major

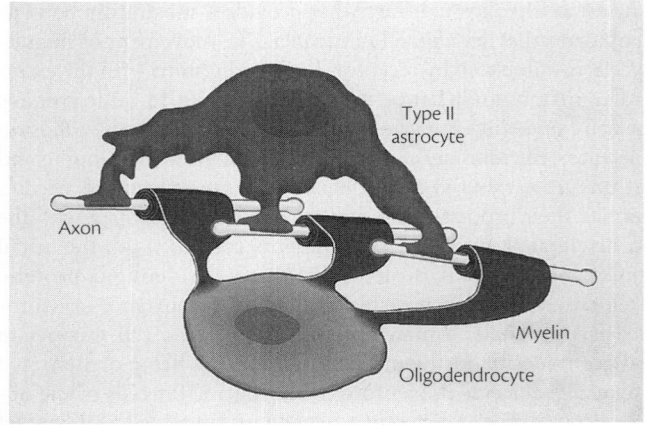

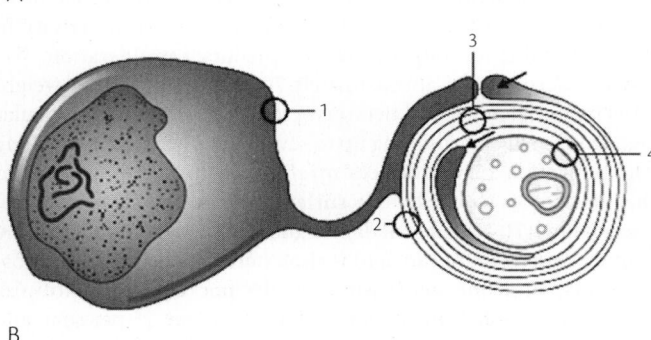

Fig. 8.5 A. The cellular architecture of myelinated axons. B. Schematic drawing of the relationship between an oligodendrocyte and the internodal myelin of a nerve fibre: 1, the oligodendrocyte membrane; 2, the myelin surface; 3, compact myelin; 4, the myelin / axon interface. (Reproduced with permission from McAlpine's Multiple Sclerosis 4th edition 2005). (See Plate 8.)

dense line visible in ultrastructural cross-sections. The two outer leaflets of the adjacent layers of the spiral process are also closely opposed, and although they commonly appear to form only a single, less dense intraperiod line, electron microscopy confirms that this comprises two separate leaflets. Inner and outer tongues of cytoplasm remain where the corresponding central and outermost lamellae have not compacted. Radial components visible in ultrastructural cross-sections of myelin probably correspond to stacks of tight junctions arrayed in lines from outermost to innermost lamellae. These are thought to seal adjacent lamellae together and to anchor the outer cytoplasmic tongue.

The two major proteins are proteolipid protein, approximately 50 per cent of total protein, and myelin basic protein, 30–40 per cent. Myelin basic protein is translated in the myelin sheath whereas proteolipid protein is localized in the cell body. Proteolipid protein and DM20 have roles in formation of the intraperiod line and in maintaining axonal integrity but are also involved in the early stages of axon–oligodendrocyte interaction and in wrapping of the axon. In addition to its role in axon guidance, netrin-1 is widely distributed on oligodendrocytes and may act over short ranges to maintain axon–glial interactions. Both in transgenic mice and people with null mutations of *PLP1*, lack of proteolipid protein does not affect myelination but is associated with length-dependent axonal degeneration. Conversely, the expression of polysialylated

neural cell adhesion molecule, PSA-NCAM, acts as a repellent signal for axon–glial contact and its expression relates inversely to myelination. Connexins may also contribute to the stability of adjacent layers during myelin compaction. Myelin-associated glycoprotein has an important role in stabilizing the initial glial–axon contact as the spiral process begins in anticipation of compaction. Loss of the inhibitory Notch1/Jagged1 signalling in mice also leads to increased myelination and over-abundance of myelin genes during development. The third main fraction of myelin proteins is the group of Wolfgram proteins, which constitute approximately 5 per cent of myelin by dry weight. The higher molecular weight band, W2, is thought to be tubulin, while the doublet of lower molecular weight, W1, corresponds to the myelin-specific enzyme 2′,3′-cyclic nucleotide 3′-phosphohydrolase, CNP-ase.

One function of oligodendrocytes, revealed by disruption in mouse models of defective oligodendrocyte development, is the distribution and clustering of sodium and potassium channels at nodes of Ranvier. Specifically, oligodendroyte-derived soluble factors induce clustering of the Na(v)1.2 channels whereas successful myelination is required for clustering of the Na(v)1.6 channels. With maturation, the diffuse distribution of the Na(v)1.2 channels is rationalized. They are retained along the myelinated segments but replaced by Na(v)1.6 channels at the nodes. Coinciding with the onset of myelination, neurofascin, NF155, clusters at the paranodal portion of the myelin sheath and adheres to the contactin-associated protein, Caspr1, that is present on the axons at the forming node of Ranvier (Charles *et al.* 2002a). Compact myelin thereby consists of a condensed lipid-rich membrane wrapped spirally many times around axons to form a segmented sheath. This is interrupted periodically along the course of the axon at the non-myelinated nodes of Ranvier—areas where electrical resistance can be low because of the high concentration of sodium channels, and where depolarization is thereby facilitated. In myelinated axons, the action potential induced by depolarization generates electrical currents, in turn triggering depolarization at the next node of Ranvier. This saltatory conduction is considerably more rapid than continuous propagation of the nerve impulse.

8.5 Mechanisms of demyelination

The commonest acquired inflammatory demyelinating disease is multiple sclerosis; here, the basis for tissue injury has recently come under revised scrutiny. Ferguson *et al.* (1997) used modern histological techniques of staining for amyloid precursor protein to show that axonal injury occurs as part of the acute demyelinating lesion. Later, the extent and distribution of acute axonal loss were characterized in more detail (Trapp *et al.* 1998). There followed extensive attempts to document axonal pathology at different sites and at different stages in the evolution of tissue damage. The second, and closely related, discovery used magnetic resonance imaging and spectroscopy to infer that chronic axonal loss is probably the substrate for disease progression (Davie *et al.* 1995). Thus, attitudes shifted from the focus on multiple sclerosis as a demyelinating disease to a broader perspective in which the relative contributions of inflammation, demyelination and remyelination, and acute and chronic axonal loss each have to be understood in reaching a coherent account of the pathogenesis. Several formulations can be suggested concerning the relationship between inflammation and axonal degeneration. These range from the idea that these processes are fully independent, to the view that inflammation is an initiating event exposing an intrinsic neurodegenerative vulnerability that results in cumulative axonal injury.

8.5.1 Toxic molecules for oligodendrocytes

Many pathological processes converge on a restricted number of final common pathways. Several involve calcium overload. Neurones and oligodendrocytes are susceptible to anoxia and excitotoxic damage mediated by kainate and glutamate. Both are prevented by AMPA, α-amino-3-hydroxy-5-methylisoxazole-4-propionic acid, receptor antagonists. A number of neuroprotective molecules are provisionally identified but no one has yet translated successfully into clinical practice. Insulin-like growth factor-1, protects oligodendrocyte precursors from excitotoxic injury through inhibition of mitochondrial injury.

Inflammatory mediators are also a major source of oligodendrocyte injury. Oligodendrocytes die from apoptosis after exposure to tumour necrosis factor-α and effects on mitochondrial function, acting through tumour necrosis factor-related apoptosis inducing ligand, TRAIL. Ciliary neurotrophic factor, acts as a neuroprotective growth factor, perhaps expressed in response to injury, that limits the direct effects of pro-inflammatory cytokines on cells of the oligodendrocyte lineage. Leukaemia inhibitory factor, LIF, protects oligodendrocytes in an immune-mediated model of demyelination by a mechanism which is similar to the effect of ciliary neurotrophic factor. The in vitro evidence implicates both these trophic factors as acting through an autocrine loop to block oligodendrocyte toxicity due to tumour necrosis factor-α released by interferon-γ-activated microglia.

Nitric oxide injures oligodendrocytes in addition to its now well-characterized effect on axons. Antagonists of nitric oxide, as well as anti-tumour necrosis factor-α antibodies and transforming growth factor beta-β, each protect rat oligodendrocytes from cell death mediated by nitric oxide—interferon-γ and tumour necrosis factor-α—acting alone or in combination. Death of oligodendrocytes following exposure to interferon-γ and lipopolysaccharide is associated with inducible nitric oxide synthase, iNOS, expression and nitric oxide production within the cultures—effects that are inhibited by interleukins-10 and -4.

8.5.2 Interactions between microglia and oligodendrocytes

Inteferon-γ upregulates the expression of a heterogeneous group of molecules on microglia involved in the phagocytosis of opsonized particles. These include the FcR1 (CD64), FcRII (CD32), and FcRIII (CD16) receptors for immunoglobulin; and complement components C1q, iCR3, and C5a. The activation of microglia by lipopolysaccharide, leading to interactions with oligodendrocytes, involves the toll-like receptor, TLR4. Microglia expressing these receptors are able to attract and engage target cells both in vitro and in vivo.

Activated microglia mainly kill oligodendrocytes by cell–cell contact (Fig. 8.6). Receptor–ligand interactions then allow microglia to deliver a lethal tumour necrosis factor-α signal (Zajicek et al. 1992). Some receptors for C1q, C3b (CR1), and iC3b (CR3) on microglia receptors are constitutively expressed, whereas others require cytokine activation. Complement activation may release membrane-bound and fluid-phase products that determine interactions between oligodendrocytes and microglia (Scolding et al.

1989). Fc receptors are perhaps the more relevant vehicle for cell–cell contacts. Antibody in low concentration, coating the surface of the oligodendrocyte or its myelin sheath, opsonizes the target cell for lytic damage by microglia using their Fc receptors (Scolding and Compston 1991). Demyelinated axons are coated with anti-myelin oligodendrocyte glycoprotein antibody in the lesions of acute multiple sclerosis although these antibodies are directed against linear epitopes that do not induce demyelination in vivo (Genain et al. 1999).

8.6 Remyelination

Spontaneous remyelination in multiple sclerosis has long been recognized, but is limited and ultimately fails, resulting in progressive clinical disability (Fig. 8.7). Together with the pathological observation that remyelinated plaques show no significant axonal injury compared to 'matched' demyelinated plaques this failure of sustained repair provides a compelling argument for developing remyelination strategies that prevent axonal loss (Kornek et al. 2000; Chandran et al. 2007).

8.6.1 Endogenous remyelination

Developmental studies and the use of methods to induce focal or systemic demyelination ranging from immunological and viral mediated insults to gliotoxins, suggest that the adult oligodendrocyte precursor is the remyelinating cell; proliferating cells identified with markers used to identify oligodendrocyte precursor cells remyelinate toxin induced demyelinating lesions (Carroll and Jennings 1994; Gensert and Goldman 1997); oligodendrocyte precursor cells isolated from adult central nervous system remyelinate models of demyelination following transplantation; and serial temporal studies of experimental remyelinating lesions provide compelling but indirect evidence that cells labelled with Olig1/2 and established markers of oligodendrocyte precursor cells are responsible for remyelination (Arnett et al. 2004). The presence of immature oligodendrocyte lineage cells in chronic demyelinated lesions in multiple sclerosis raises the question as to why remyelination ultimately fails (Wolswijk 1998). Several explanations can be considered—failure of sufficient recruitment, failure of differentiation of precursors, and an environment inhibitory to remyelination. Adult oligodendrocyte precursor cells are by definition cells capable of proliferation and whether their proliferative reserve is finite is an important consideration. Experimental studies of focal gliotoxic mediated demyelination suggest that remyelination failure is not due to the inability of oligodendrocyte precursor cells to proliferate and migrate to remyelinate (Chari and Blakemore 2002). However, the finding that, through impairment of both recruitment and differentiation, older animals remyelinate less efficiently has clinical implications (Sim et al. 2002). Recognition that the environment of the adult demyelinated brain and the responsiveness of adult oligodendrocyte precursor cells to such signals may differ compared with their embryonic counterparts suggests that remyelination strategies cannot be considered merely as a simple recapitulation of development. This has been illustrated by the comparative failure of systemically delivered defined factors to promote experimental remyelination (Charles et al. 2002b; Franklin 2002). These and other observations have provided an impetus to develop supplementary interventions aimed at promoting exogenous remyelination.

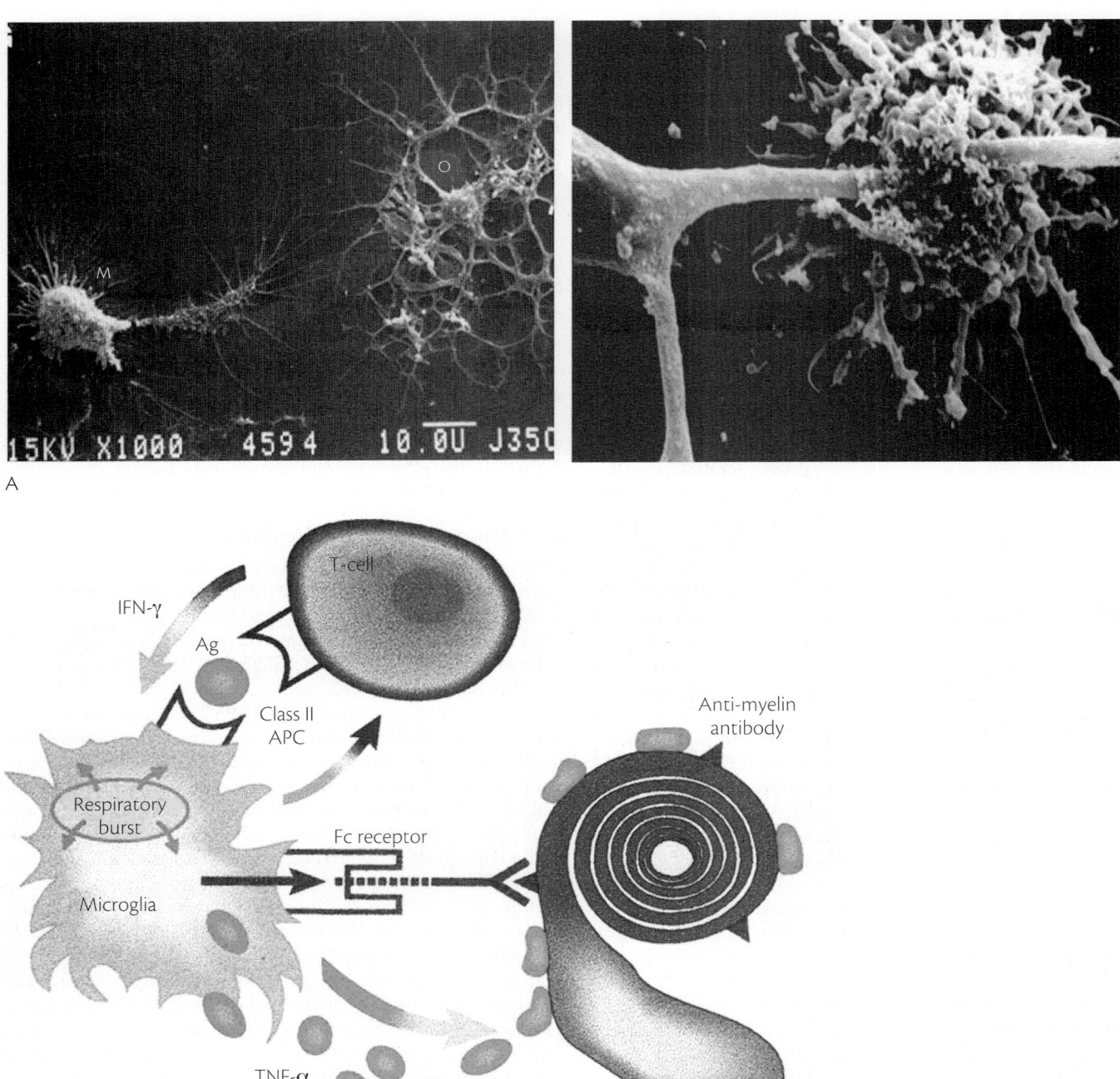

Fig. 8.6 A. Interaction between microglia and oligodendrocytes without (left) and with (right) complement and antibody opsonisation. B. Scheme to represent interactions between activated T cells, microglia, and the oligodendrocyte. (Reproduced with permission from McAlpine's Multiple Sclerosis 4th edition 2005). (See Plate 9.)

8.6.2 **Cell implantation**

The emergence of stem cells potentially as an unlimited source of myelinating cells and the utility of exogenous cells following implantation to remyelinate adult demyelinated axons and restore function raise the stakes for cell implantation as a plausible reparative strategy. Once again experimental models of demyelination have provided important insights into the requirements of the putative myelinating cell. A wide range of cells has demonstrated remyelinating capability including Schwann cells, olfactory ensheathing cells, oligodendrocyte lineage cells, and neural stem or precursor cells. The capacity to remyelinate is likely necessary but not sufficient given the need for cells also to migrate and overcome the pathological astrocytic environment. In this regard the oligodendrocyte precursor cell has proven to be particularly efficient in contrast to more differentiated counterparts of the oligodendrocyte lineage (Blakemore *et al.* 2000).

Studies in the last decade have begun to explore the potential of stem and more recently human derived precursors as remyelinating

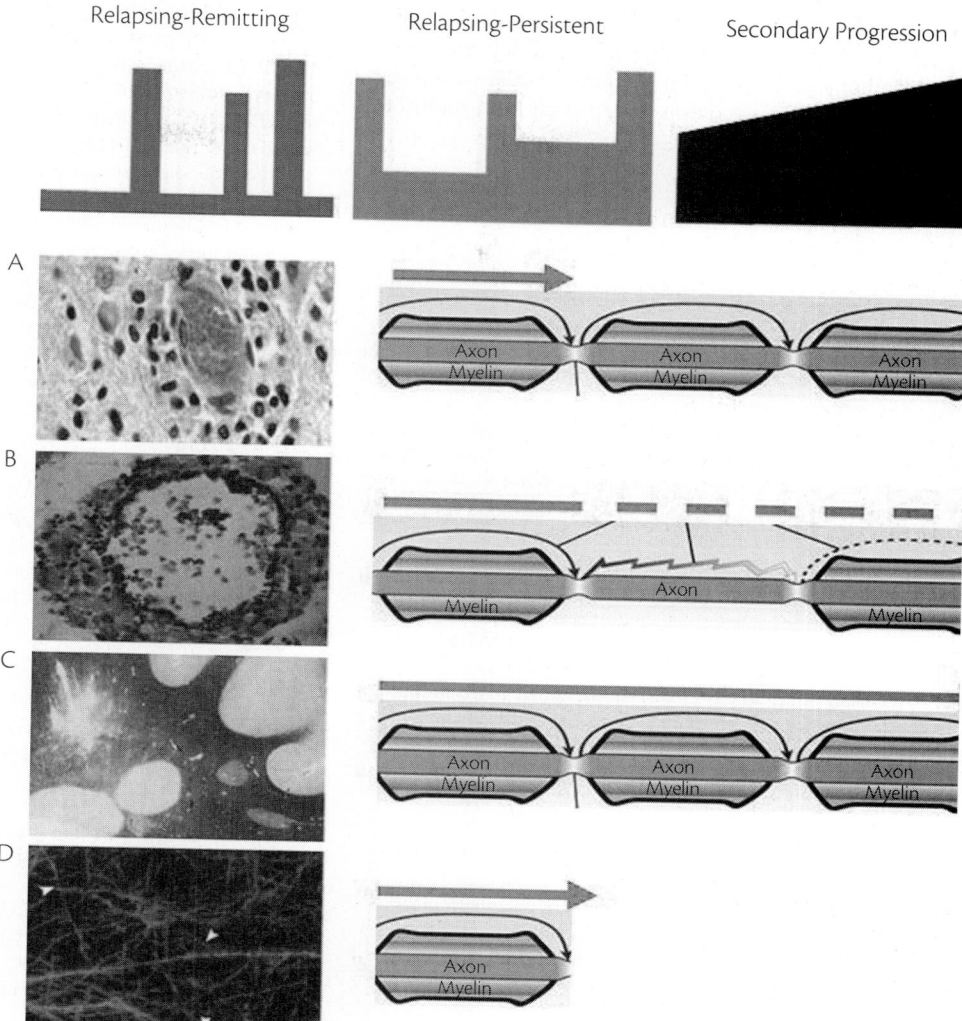

Fig. 8.7 Cartoon to show the stages of injury to myelinated axons and their expression as the clinical course of multiple sclerosis. A. Inflammation causes transient conduction block, with recovery. B. Persistent inflammation leads to acute axonal injury and demyelination. C. Remyelination is associated with recovery of structure and function. D. Persistent demyelination is associated with chronic axonal loss and astrocytosis. (See Plate 10.)

cells. A consensus has emerged from such studies that *ex vivo* directed differentiation to early oligodendrocyte lineage cells is necessary for successful remyelination (Smith and Blakemore 2000). Given the need to work with human derived material, this remains a practical obstacle in view of the current inability to direct foetal and adult human material readily into sufficient numbers of oligodendrocyte precursor cells (Chandran *et al.* 2004). Recent studies on human embryonic stem cells suggest that these difficulties are not insurmountable and offer the prospect of unlimited numbers of myelinating oligodendrocyte precursors (Keirstead *et al.* 2005). An alternative source of potentially readily accessible and autologous human myelinating cells is olfactory ensheathing cells that possess an advantage over Schwann cells in being able to migrate and integrate in a chronic and gliotic environment (Lakatos *et al.* 2000)

In addition to being able to remyelinate, an emerging concept for the utility of cell therapy is centred on the idea that, independent of directed differentiation, cells possess useful promyelinating or neuroprotective abilities. Evidence from several animal models suggests that stem/precursor cells may have a therapeutic role beyond cell replacement and site-specific repair, for example as

cellular sources of trophic factors and/or immune modulators—'therapeutic plasticity' (Martino and Pluchino 2006). The role of inflammation in the aetiology and orchestration of the natural history in multiple sclerosis is complex. The consequences of inflammation are context dependent with evidence implicating a positive effect of inflammation on remyelination (Foote and Blakemore 2005). This raises the idea of 'beneficial inflammation' and the therapeutic role for agents that are immune modulators. Rodent neuronal stem cells can behave as cellular immune modulators capable of crossing the blood–brain barrier and enabling functional recovery independent of exogenous remyelination in experimental immunologically mediated models of multiple sclerosis (Ben Hur *et al.* 2003; Pluchino *et al.* 2003). The precise mechanism of neuroprotection is unknown. Identification of molecular apparatus on stem cells comparable to that used by activated lymphocytes provides a plausible explanation for blood–brain barrier penetration of parenterally delivered stem cells that induce selective apoptotic death of Th1 cells (Pluchino *et al.* 2005). Pathotropism within the injured brain is well recognized and the recent identification of chemokine receptors on neuronal stem cells and oligodendrocyte precursor cells provides a mechanism for migration into areas of neuroinflammation (Tran *et al.* 2004; Krathwohl and Kaiser 2004; Tran *et al.* 2004). Together these findings of blood–brain barrier penetration, pathotropism, and cellular immune modulation suggest that systemic therapy is a rationale approach for a multifocal disease such as multiple sclerosis.

References

Arnett HA, Fancy SP, Alberta JA *et al.* (2004). bHLH transcription factor Olig1 is required to repair demyelinated lesions in the CNS. *Science,* **306**, 2111–5.

Barres BA, Hart IK, Coles HS *et al.* (1992). Cell death and control of cell survival in the oligodendrocyte lineage. *Cell,* **70**, 31–46.

Barres BA, Raff MC (1999). Axonal control of oligodendrocyte development (In Process Citation). *J Cell Biol,* **147**, 1123–8.

Belachew S, Chittajallu R, Aguirre AA *et al.* (2003). Postnatal NG2 proteoglycan-expressing progenitor cells are intrinsically multipotent and generate functional neurons. J Cell Biol, **161**, 169–86.

Bell MA, Weddell AGM (1984). A descriptive study of the blood vessels of the sciatic nerve in the rat, man and other mammals. *Brain,* **107**, 871–98.

Ben Hur T, Einstein O, Mizrachi-Kol R *et al.* (2003). Transplanted multipotential neural precursor cells migrate into the inflamed white matter in response to experimental autoimmune encephalomyelitis. *Glia,* **41**, 73–80.

Blakemore WF, Gilson JM, Crang AJ (2000). Transplanted glial cells migrate over a greater distance and remyelinate demyelinated lesions more rapidly than endogenous remyelinating cells. J Neurosci Res, **61**, 288–94.

Bush TG, Puvanachandra N, Horner CH *et al.* (1999). Leukocyte infiltration, neuronal degeneration, and neurite outgrowth after ablation of scar-forming, reactive astrocytes in adult transgenic mice. *Neuron,* **23**, 297–308.

Carroll WM, Jennings AR (1994). Early recruitment of oligodendrocyte precursors in CNS demyelination. *Brain,* **117**, 563–78.

Chandran S, Compston A, Jauniaux E *et al.* (2004). Differential generation of oligodendrocytes from human and rodent embryonic spinal cord neural precursors. *Glia,* **47**, 314–24.

Chandran S, Hunt D, Joannides A *et al.* (2008). Myelin repair: the role of stem and precursor cells in multiple sclerosis. *Phil Trans R Soc Lond B.* **363**: 171–83.

Chari DM, Blakemore WF (2002). Efficient recolonisation of progenitor-depleted areas of the CNS by adult oligodendrocyte progenitor cells. *Glia,* **37**, 307–13.

Charles P, Reynolds R, Seilhean D *et al.* (2002a). Re-expression of PSA-NCAM by demyelinated axons: an inhibitor of remyelination in multiple sclerosis? *Brain,* **125**, 1972–9.

Charles P, Tait S, Faivre-Sarrailh C *et al.* (2002b). Neurofascin is a glial receptor for the paranodin/Caspr-contactin axonal complex at the axoglial junction. *Curr Biol,* **12**, 217–20.

Coleman M (2005). Axon degeneration mechanisms: commonality amid diversity. *Nat Rev Neurosci,* **6**, 889–98.

Davie CA, Barker GJ, Webb S *et al.* (1995) Persistent functional deficit in multiple sclerosis and autosomal dominant cerebellar ataxia is associated with axon loss. *Brain,* **118**, 1583–92.

Doetsch F, Caille I, Lim DA *et al.* (1999). Subventricular zone astrocytes are neural stem cells in the adult mammalian brain. *Cell,* **97**, 703–716.

Fawcett JW (2006). The glial response to injury and its role in the inhibition of CNS repair. *Adv Exp Med Biol,* **557**, 11–24.

Ferguson B, Matyszak MK, Esiri MM *et al.* (1997). Axonal damage in acute multiple sclerosis lesions. *Brain,* **120**, 393–9.

Finn JT, Weil M, Archer F *et al.* (2000). Evidence that Wallerian degeneration and localized axon degeneration induced by local neurotrophin deprivation do not involve caspases. *J Neurosci,* **20**, 1333–41.

Foote AK, Blakemore WF (2005). Inflammation stimulates remyelination in areas of chronic demyelination. *Brain,* **128**, 528–39.

Franklin RJ (2002). Why does remyelination fail in multiple sclerosis? *Nat Rev Neurosci,* **3**, 705–14.

Gage FH (2000). Mammalian neural stem cells. *Science,* **287**, 1433–8.

Genain CP, Cannella B, Hauser SL *et al.* (1999). Identification of autoantibodies associated with myelin damage in multiple sclerosis. *Nat Med,* **5**, 170–5.

Gensert JM, Goldman JE (1997). Endogenous progenitors remyelinate demyelinated axons in the adult CNS. *Neuron,* **19**, 197–203.

Gotz M, Barde YA (2005). Radial glial cells defined and major intermediates between embryonic stem cells and CNS neurons. *Neuron,* **46**, 369–72.

Jessen KR, Mirsky R (1998). Origin and early development of Schwann cells. *Microsc Res Tech,* **41**, 393–402.

Ji JF, He BP, Dheen ST *et al.* (2004). Expression of chemokine receptors CXCR4, CCR2, CCR5 and CX3CR1 in neural progenitor cells isolated from the subventricular zone of the adult rat brain. *Neurosci Lett,* **355**, 236–40.

Kapoor R, Davies M, Blaker PA *et al.* (2003). Blockers of sodium and calcium entry protect axons from nitric oxide-mediated degeneration. *Ann Neurol,* **53**, 174–80.

Keirstead HS, Nistor G, Bernal G *et al.* (2005). Human embryonic stem cell-derived oligodendrocyte progenitor cell transplants remyelinate and restore locomotion after spinal cord injury. *J Neurosci,* **25**, 4694–705.

Kessaris N, Pringle N, Richardson WD (2001). Ventral neurogenesis and the neuron-glial switch. *Neuron,* **31**, 677–80.

Kondo T, Raff M (2000). Oligodendrocyte precursor cells reprogrammed to become multipotential CNS stem cells. *Science,* **289**, 1754–7.

Kornek B, Storch MK, Weissert R *et al.* (2000). Multiple sclerosis and chronic autoimmune encephalomyelitis: a comparative quantitative study of axonal injury in active, inactive, and remyelinated lesions. *Am J Pathol,* **157**, 267–76.

Krathwohl MD, Kaiser JL (2004). Chemokines promote quiescence and survival of human neural progenitor cells. *Stem Cells,* **22**, 109–18.

Lakatos A, Franklin RJ, Barnett SC (2000). Olfactory ensheathing cells and Schwann cells differ in their in vitro interactions with astrocytes. *Glia,* **32**, 214–25.

Lappe-Siefke C, Goebbels S, Gravel M *et al.* (2003). Disruption of Cnp1 uncouples oligodendroglial functions in axonal support and myelination. *Nat Genet,* **33**, 366–74.

Liu BP, Cafferty WB, Budel SO *et al.* (2006). Extracellular regulators of axonal growth in the adult central nervous system. *Phil Trans R Soc Lond B,* **361**, 1593–610.

Lu QR, Sun T, Zhu Z *et al.* (2002). Common developmental requirement for olig function indicates a motor neuron/oligodendrocyte connection. *Cell,* **109**, 75–86.

Martino G, Pluchino S (2006). The therapeutic potential of neural stem cells. *Nat Rev Neurosci,* **7**, 395–406.

Miller RH (2005). Dorsally derived oligodendrocytes come of age. *Neuron,* **45**, 1–3.

Nicholas RS, Wing MG, Compston A (2001). Nonactivated microglia promote oligodendrocyte precursor survival and maturation through the transcription factor NF-kappa B. *Eur J Neurosci,* **13**, 959–67.

Nieto M, Schuurmans C, Britz O *et al.* (2001). Neural bHLH genes control the neuronal versus glial fate decision in cortical progenitors. *Neuron,* **29**, 401–13.

Nishiyama A, Lin XH, Giese N *et al.* (1996). Co-localization of NG2 proteoglycan and PDGF alpha-receptor on O2A progenitor cells in the developing rat brain. *J Neurosci Res,* **43**, 299–314.

Nistor GI, Totoiu MO, Haque N *et al.* (2005). Human embryonic stem cells differentiate into oligodendrocytes in high purity and myelinate after spinal cord transplantation. *Glia,* **49**, 385–96.

Nunes MC, Roy NS, Keyoung HM *et al.* (2003). Identification and isolation of multipotential neural progenitor cells from the subcortical white matter of the adult human brain. *Nat Med*, **9**, 439–47.

Pluchino S, Quattrini A, Brambilla E *et al.* (2003). Injection of adult neurospheres induces recovery in a chronic model of multiple sclerosis. *Nature*, **422**, 688–94.

Pluchino S, Zanotti L, Rossi B *et al.* (2005). Neurosphere-derived multipotent precursors promote neuroprotection by an immunomodulatory mechanism. *Nature*, **436**, 266–71.

Pringle NP, Mudhar HS, Collarini EJ *et al.* (1992). PDGF receptors in the rat CNS: during late neurogenesis, PDGF alpha- receptor expression appears to be restricted to glial cells of the oligodendrocyte lineage. *Development*, **115**, 535–51.

Raff MC, Miller RH, Noble M (1983). A glial progenitor cell that develops in vitro into an astrocyte or an oligodendrocyte depending on culture medium. *Nature*, **303**, 390–6.

Raff MC, Whitmore AV, Finn JT (2002). Axonal self-destruction and neurodegeneration. *Science*, **296**, 868–71.

Ray J, Peterson DA, Schinstine M *et al.* (1993). Proliferation, differentiation, and long-term culture of primary hippocampal neurons. *Proc Natl Acad Sci USA*, **90**, 3602–6.

Redford EJ, Kapoor R, Smith KJ (1997). Nitric oxide donors reversibly block axonal conduction: demyelinated axons are especially susceptible. *Brain*, **120**, 2149–57.

Rhodes KE, Fawcett JW (2004). Chondroitin sulphate proteoglycans: preventing plasticity or protecting the CNS? *J Anat*, **204**, 33–48.

Riethmacher D, Sonnenberg-Riethmacher E, Brinkmann V *et al.* (1997). Severe neuropathies in mice with targeted mutations in the ErbB3 receptor. *Nature*, **389**, 725–30.

Rowitch DH (2004). Glial specification in the vertebrate neural tube. *Nat Rev Neurosci*, **5**, 409–19.

Schwab ME (2004). Nogo and axon regeneration. *Curr Opin Neurobiol* **14**, 118–24.

Scolding NJ, Compston DA (1991). Oligodendrocyte-macrophage interactions in vitro triggered by specific antibodies. *Immunology*, **72**, 127–32.

Scolding NJ, Morgan BP, Houston WA *et al.* (1989). Vesicular removal by oligodendrocytes of membrane attack complexes formed by activated complement. *Nature*, **339**, 620–2.

Sim FJ, Zhao C, Penderis J *et al.* (2002). The age-related decrease in CNS remyelination efficiency is attributable to an impairment of both oligodendrocyte progenitor recruitment and differentiation. *J Neurosci*, **22**, 2451–9.

Smith PM, Blakemore WF (2000). Porcine neural progenitors require commitment to the oligodendrocyte lineage prior to transplantation in order to achieve significant remyelination of demyelinated lesions in the adult CNS. *Eur J Neurosci*, **12**, 2414–24.

Takahashi K, Yamanaka S (2006). Induction of pluripotent stem cells from mouse embryonic and adult fibroblast cultures by defined factors. *Cell*, **126**, 663–76.

Tran PB, Ren D, Veldhouse TJ *et al.* (2004). Chemokine receptors are expressed widely by embryonic and adult neural progenitor cells. *J Neurosci Res*, **76**, 20–34.

Trapp BD, Peterson J, Ransohof RM *et al.* (1998). Axonal transection in the lesions of multiple sclerosis. *N Engl J Med*, **338**, 278–85.

Vescovi AL, Reynolds BA, Fraser DD *et al.* (1993). bFGF regulates the proliferative fate of unipotent (neuronal) and bipotent (neuronal/astroglial) EGF-generated CNS progenitor cells. *Neuron*, **11**, 951–66.

Wang S, Sdrulla AD, diSibio G *et al.* (1998). Notch receptor activation inhibits oligodendrocyte differentiation. *Neuron*, **21**, 63–75.

Wilkins A, Compston A (2005). Trophic factors attenuate nitric oxide mediated neuronal and axonal injury in vitro: roles and interactions of mitogen-activated protein kinase signalling pathways. *J Neurochem*, **92**, 1487–96.

Wilkins A, Majed H, Layfield R (2003). Oligodendrocytes promote neuronal survival and axonal length by distinct intracellular mechanisms: a novel role for oligodendrocyte-derived glial cell line-derived neurotrophic factor. *J Neurosci*, **23**, 4967–74.

Wolswijk G (1998). Chronic stage multiple sclerosis lesions contain a relatively quiescent population of oligodendrocyte precursor cells. *J Neurosci*, **18**, 601–9.

Wren D, Wolswijk G, Noble M (1992). In vitro analysis of the origin and maintenance of O-2A adult progenitor cells. *J Cell Biol*, **116**, 167–76.

Zajicek JP, Wing M, Scolding NJ (1992). Interactions between oligodendrocytes and microglia. A major role for complement and tumour necrosis factor in oligodendrocyte adherence and killing. *Brain*, **115**, 1611–31.

Zhou Q, Anderson DJ (2002). The bHLH Transcription Factors OLIG2 and OLIG1 Couple Neuronal and Glial Subtype Specification. *Cell*, **109**, 61–73.

CHAPTER 9

Neurodevelopmental disorders

Janet Eyre

Contents

Remarkable advances in the neurosciences, particularly in the fields of genetics, molecular biology, metabolism, and nutrition, have greatly advanced our understanding of how the brain develops and responds to environmental influences. Neurodevelopmental disorders arise from perturbation of these normal developmental processes, by insults from heterogeneous aetiological factors. These factors trigger a sequence of molecular, biochemical, and morphological alterations of the brain, resulting in a morphologically and/or functionally abnormal brain. Rapidly advancing understanding of basic neurodevelopmental processes has direct relevance to understanding human neurodevelopmental disorders, providing insights into pathogenic mechanisms and revealing new pathways that can be exploited in diagnosis and treatment. Conversely the identification of the molecular bases of several neurodevelopmental disorders has also provided invaluable insights into the mechanisms of normal brain development. Technical advances have also improved methods for identifying brain regions involved in developmental disorders, for tracing connections between parts of the brain, for visualizing individual neurons in living brain preparations, for recording the activities of neurons, and for studying the activity of single-ion channels and the receptors for various neurotransmitters. During the past 10 years the genetic basis of an ever increasing number of neurodevelopmental disorders has been discovered and has led to better understanding of the neurobiological basis of even common disorders such as global developmental delay, cerebral palsy, and autism. Current research should reveal their underlying molecular biology and eventually the possibility of targeted chemotherapy and the prevention of many neurodevelopmental disorders.

9.1 Overview of cerebral cortical development

The earliest step in the formation of the nervous system is neural induction, the process by which ectodermal cells adopt a neural identity, called the neuroectoderm. The neuroectoderm is induced to differentiate from the surrounding ectoderm by the presence of the notochord at about 18 days. The nervous system and special sense organs originate from three sources, each derived in turn from specific regions of the neuroecoderm. The first source to delineate is the neural plate, which forms the central nervous system, the somatic motor nerves, and the preganglionic autonomic nerves. The second source is from cells at the perimeter of the neural plate which remove themselves by epithelial/mesenchymal transition from the plate just prior to its fusion into the neural tube; these are the neural crest cells which form nearly all the peripheral nervous system and adrenal and chromaffin cells. The third source is from ectodermal placodes; these are groups of cells which originate at the edge of the neural plate but remain in the surface ectoderm after neural tube formation, undergoing epithelial/mesenchymal transition after neural crest cells have begun their migration. Ectodermal placodes contribute to somatosensory ganglia of the cranial nerves, the hypophysis, and the inner ear.

Concomitantly with neural induction, the induced neural plate becomes patterned along the craniocaudal, dorsoventral, and mediolateral axes. While regional patterning is occurring, the neural plate undergoes morphogenesis: neural folds arise, approach one another in the dorsal midline, and fuse. This process called primary neurulation generates the entire neural tube rostral to the caudal neuropore. It occurs during the third and fourth weeks of development, Carnegie stages CS 8 to 13, and the flat layer of neuroectodermal cells overlying the notochord is transformed into a hollow tube (Fig. 9.1). On day 19, CS 8.5, the border of the neural plate becomes gradually more pronounced and elevated. The neural plate folds longitudinally along the midline of the plate from the head toward the tail to form the neural groove. The folds rise up dorsally, approach each other, and ultimately merge together, forming a tube open at both ends by day 23, CS 10.5. As the neural folds fuse, the cells adjacent to the neural plate also fuse across the midline to become the overlying epidermis. The rostral and caudal openings are called neuropores. The rostral and caudal neuropores close later, on the 26th, CS 12, and 28th, CS 13 days of gestation, respectively (O'Rahilly and Muller 2002).

Prior to the closure of the neural tube the neural folds become expanded considerably in the head region as a first indication of a brain. As a result of unequal growth of different regions three flexures appear in the brain (Fig. 9.2). The flexures delineate the

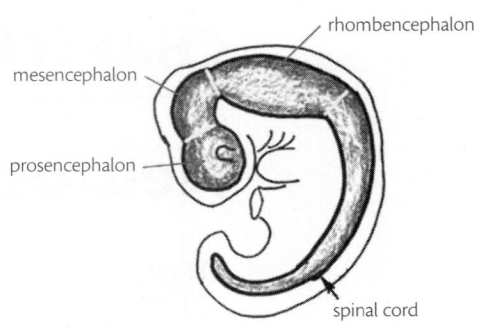

Fig. 9.2 Embryonic development of the brain. (Adapted with permission from Temple University School of Medicine's Department of Anatomy and Biology Neuroanatomy Lab Resource http://isc.temple.edu/neuroanatomy/lab/embryo_new/vse/1/)

hindbrain or rhombencephalon, mesencephalon, and forebrain or prosencephalon. The hindbrain will become the pons, cerebellum, and medulla oblongata. The mesencephalon will become the midbrain and the prosencephalon divides into the telencephalon, which will form the cerebral hemispheres and the diencephalon which will become the thalamus, hypothalamus, and optic vesicles. This regional specification or segmentation of the brain is under the control of a number of genes that encode transcription factors. (Pasini and Wilkinson 2002).

Subsequent to the closure of the rostral neuropore the three primary cerebral vesicles are formed by localized acceleration of growth in the wall of the brain (Fig. 9.3). Following telencephalic cleavage, a layer of proliferative pseudostratified neuroepithelium lines the ventricles of the telencephalic vesicles. These cells will give rise to the neurons and glia of the mature brain.

The development of human cerebral cortex can be divided into three overlapping stages. During the first stage, stem cells proliferate into neuroblasts or glial cells deep in the forebrain, predominantly in the ventricular and subventricular zones lining the cerebral cavity. In the second phase, after their final mitotic division, cortical neurones migrate away from their place of origin in a radial or tangential fashion towards the pial surface, where each successive

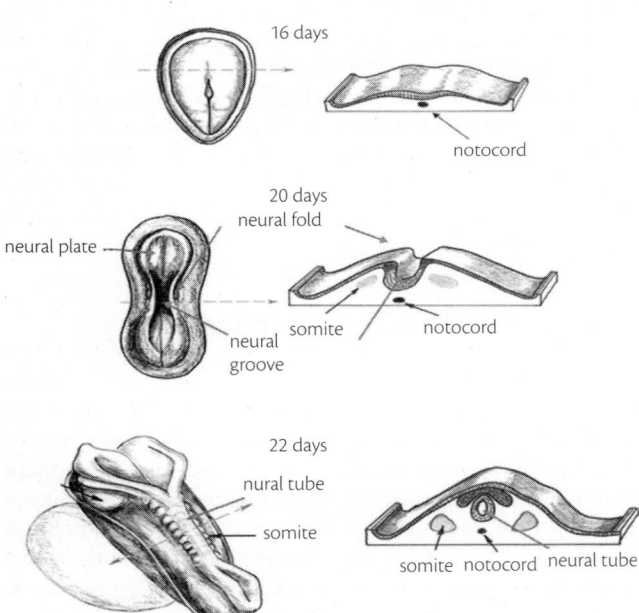

Fig. 9.1 Neural tube development. (Adapted with permission from Temple University School of Medicine's Department of Anatomy and Biology Neuroanatomy Lab Resource http://isc.temple.edu/neuroanatomy/lab/embryo_new/nt/)

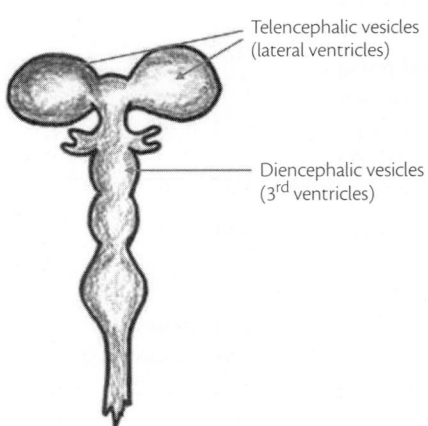

Fig. 9.3 Development of the ventricles. (Adapted with permission from Temple University School of Medicine's Department of Anatomy and Biology Neuroanatomy Lab Resource http://isc.temple.edu/neuroanatomy/lab/embryo_new/vsl/)

generation passes one another and settles in an inside-out pattern within the cortical plate. The third phase represents cortical organization within six layers associated with synaptogenesis and apoptosis. This is a dynamic process and more than one stage may occur simultaneously during several gestational weeks. In humans, the proliferation stage ranges from weeks 5–6 to weeks 16–20, migration from weeks 6–7 to weeks 20–24, and organization from week 16 until well into postnatal life.

At 5 weeks the first neurons are generated and begin to migrate to the cortical surface (Fig. 9.4). Cortical neurons are generated in the ventricular zone of the cortical wall and in the subcortical ganglionic eminence and reach their destination by both radial and tangential migration (Rakic 1972; Anderson *et al.* 1997, 2002). The first postmitotic neurons, Cajal-Retzius cells, accumulate superficial to the neuroepithelium, immediately beneath the pial surface, forming the preplate (Rickmann and Wolff 1981; Stewart and Pearlman 1987). The preplate has also been referred to as the primordial plexiform layer (Marin-Padilla 1972b) or as the pallial anlage (Rickmann *et al.* 1977). Although preplate cells are neurons, they are distinct from those that will populate the definitive cellular layers 2–6 of the mature adult cortex. Layers 2–6 emerge from the neurons of the later generated cortical plate. As these cortical neurons become postmitotic, they migrate superficially and accumulate within the preplate to form the cortical plate. The arrival of the cortical neurons splits the preplate into two zones: the superficial marginal zone, which will later become layer 1, and the deeper subplate, a transient structure present during development. The marginal zone comprises predominantly Cajal-Retzius cells and lies just below the pial surface. The subplate lies below the cortical plate and above the ventricular zone. An intermediate zone develops above the ventricular zone but below the subplate, comprising migrating neurons and efferent and afferent fibres (Marin-Padilla 1971, 1972a; Altman and Bayer 2002). After the arrival of the first cortical plate neurons, formation of the layers within the cortical plate proceeds by progressive adjunction of new immigrant neurons in an inside-out gradient, thus layers 6 and 5 are generated early and layers 2 and 3 are generated last (Angevine and Sidman 1961; Rakic 1974). The marginal zone will eventually form layer 1 of the cortex. The extracellular protein reelin, which is secreted by Cajal-Retzius cells of the marginal zone, is critically involved in this process.

The cortical plate first emerges within the preplate at 8 weeks post-conceptional age (Zecevic *et al.* 1999; Meyer *et al.* 2000; Altman and Bayer 2002). The peak of cell proliferation in the germinal epithelium probably occurs at the end of the first trimester in humans. The neuronal migration period occurs between the 8th and 16th week of post-conceptional age. Cell proliferation in the ventricular zone is evident until 16 weeks. From this time the volume and thickness of the ventricular zone decreases until at the 22nd–24th week when only a few rows of sparse cells can be recognized in the germinal epithelium (McConnell 1995; Zecevic *et al.* 1999). Progressive reduction in the volume of the proliferative zone is accompanied by increasing thickness of the cortical plate and subplate (Simonati *et al.* 1999). The subsequent development of the subplate is a protracted process that extends over several months (Mrzljak *et al.* 1988). The subplate reaches its largest size between 20 and 30 weeks of post-conceptional age, when the subplate is four times the width of the developing cortical plate (Fig. 9.5). During this phase thalamic axons invade the subplate

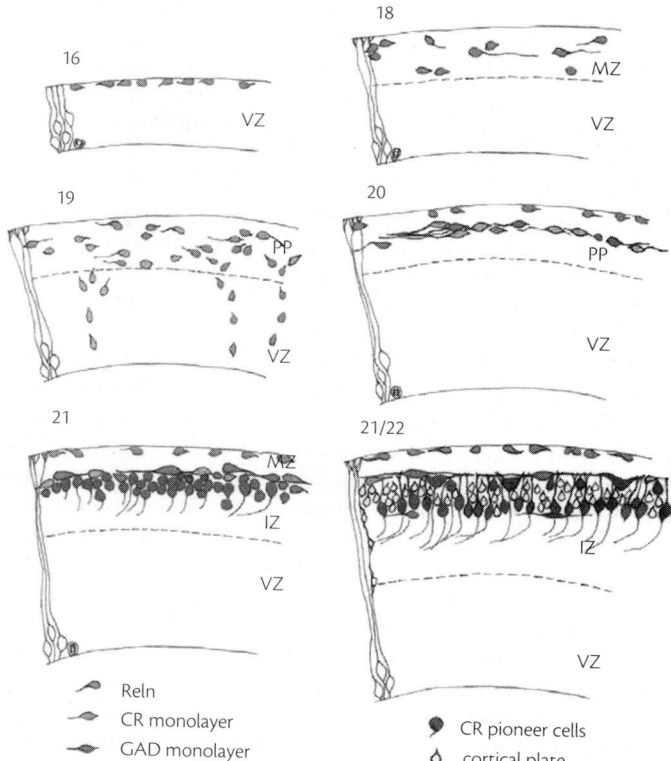

Fig. 9.4 Diagrammatic representation of the developmental events proposed by Meyers and her colleagues for the early development of the human neocortex. All figures were drawn at the same magnitude with the aid of a camera lucida. *Blue*, Reln; *light green*, early calretinin (CR) immunoreactive neurons; *red*, glutamic acid decarboxylase (GAD) immunoreactive neurons; *dark green*, CR immunoreactive 'pioneer cells'; yellow, cortical plate neurons. The first Reln immunoreactive neurons appear at Carnegie stage 16 (5 weeks post-conceptual age) and increase in number from Carnigie stages 17 to 19 (6 to 6.5 weeks post-conceptual age). The first CR immunoreactive neurons appear at Carnigie stage 19 (6.5 weeks post-conceptual age) in what could now be called the preplate. GAD immunoreactive neurons, first appear at stage 20 (7 weeks post-conceptual age). Concurrently, Reln immunoreactive neurons settle in the subpial compartment. At stage 21 (7 weeks post-conceptual age), the pioneer cells send the first corticofugal fibres. The preplate is split apart into a minor superficial component and a large deep component, the subplate, through the first cohorts of the cortical plate, at stages 21 and 22 (8 weeks post-conceptual age). IZ, intermediate zone; MZ, marginal zone; PP, preplate; VZ, ventricular zone. (from Meyer *et al.* 2000). (See Plate 11).

and wait there for several weeks before proceeding into the cortical plate and establishing connections with neurons of layer 4 of the neocortex towards the end of the second trimester (Kostovic and Judas 2002). From 30 weeks post-conceptional age both the germinal matrix and the subzone gradually decrease in size. The subplate is still present however in the full-term newborn, but is not discernible soon after birth (Mrzljak *et al.* 1988; Meyer *et al.* 2000; Kostovic and Judas 2002). Subplate neurons extend the first corticofugal axons from the neocortex into the internal capsule, before many of the neurons of layer 5 and 6 have become postmitotic and begin migration from the ventricular zone (Fig. 9.1) (Luskin and Shatz 1985; De Carlos and O'Leary 1992; Meyer *et al.* 2000). At this time subplate neurons have already assumed a pyramidal shape. They have complex dendritic trees and emit pioneer axons which form a clearly visible fibre tract within the intermediate zone.(Meyer *et al.* 2000) The subplate neurons receive synaptic contacts and are immunoreactive for neuropeptides, neurotransmitters and their receptors, neurotrophin receptors, calcium-binding proteins, and the neuron specific marker, microtubulin associated protein 2, MAP2. The descending axons of the preplate neurons express GAP 43, a protein involved in axonal outgrowth (Kostovic and Rakic 1980, 1990; Allendoerfer and Shatz 1994; Meyer *et al.* 2000). The subplate neurons therefore have a high degree of maturity in comparison to the immature morphology and tight compaction of the cortical plate neurons at this early stage of development. The observation that subplate neurons send the first axons into the internal capsule led to the intriguing suggestion that these axons may play a pioneering role in establishing the corticofugal projections of layer 5 and 6 neurons (McConnell *et al.* 1989; Shatz *et al.* 1991). At the same time that subplate neurons are forming descending projections, their cell bodies are taking part in a functioning network and are playing a role within the neocortex itself. The dendrites of these neurones receive synaptic input from thalamic afferent axons that are 'waiting' in the subplate prior to invading their final targets in layer 4 of the cortical plate (Ghosh and Shatz 1992; Kostovic and Judas 2002). Subplate neurons may relay information into the cortical plate through extensive axon collaterals that project both within the subplate and also upward into the cortical plate and marginal zone (Meyer *et al.* 2000). Ablation studies have revealed that subplate neurons control the ability of thalamic axons to invade the appropriate cortical areas and are involved in refinement of thalamic connections within layer 4, including being critically involved in the formation of ocular dominance columns. When the subplate has been selectively ablated it is remarkable that eye segregation fails even though lateral geniculate nucleus receptive fields are normal and both lateral geniculate nucleus axons and layer 4 neurons are present (Ghosh *et al.* 1990; Ghosh and Shatz 1992; Kanold *et al.* 2003).

The subplate structure is very much larger and more highly developed in phylogenetically more advanced species. The maximum subplate to cortical plate ratio during development in the mouse and rat is only1:2; in the cat,1:1; in the monkey, 3:1, and in humans it reaches a ratio of 4:1 at approximately 25 weeks post-conceptual age. Not only is the subplate larger in extent in human and sub human primates than in cats and rats, but it persists for a much longer period during development as well. In man the subplate is discernible but progressively decreasing in size up until soon after birth (Mrzljak *et al.* 1992; Kostovic and Judas 2002). It has been proposed that subplate neurons are required for the development of a complex cortical organization, since the size of subplate and the extent of its synaptic linkages is more prominent in species with increased radial and tangential cortical connectivity such as cat, monkey, and human (Kostovic and Rakic 1990). The relevance of this proposal to human development lies in the observation that in a neonatal rat model of hypoxic-ischaemic injury that produces the characteristic pattern of subcortical injury associated with human periventricular leucomalaecia, selective subplate neuron death is seen. This may provide an explanation for the high frequency of cognitive, motor, and sensory deficits observed in babies with periventricular leucomalaecia and the fact that with decreasing gestational age, periventricular leucomalacia is associated with more pervasive abnormalities of cortical development (Inder *et al.* 1999).

The adult noecortex is composed of six major layers, which are distinguished by differences in the morphology and density of neurons that constitute them (Brodmann 1909). The developing cortical plate lacks many features that distinguish neocortical areas in the adult, even after all the neurons have been generated and layers begin to differentiate within it. During corticogenesis the laminar destination of cortical neurons appears to be determined early in the life of the neuron, potentially prior to the final mitosis in the ventricular zone (McConnell 1995). Although neurons in different layers have unique dendritic configurations and axonal projections, most, if not all, types of cortical pyramidal neuron initially develop with a common morphology and only later develop the dendritic shape characteristic of their class and layer, by developmental sculpting (Koester and O'Leary 1992).

Different regions of the developing cortex may initially be interchangeable in terms of the axonal connections they develop and maintain, and even in their capacity to form complex and highly organized neuronal assemblies (O'Leary *et al.* 1992). Thus, the neocortical neuroepithelium generates populations of neurons which rely on interactions with intrinsic and extrinsic patterning information, acting both separately and synergistically at different stages of neocortical development, to generate their characteristic area specific features. Intrinsic and extrinsic cortical patterning information include; molecular factors intrinsic to the ventricular zone, to the subplate and to the maturing neocortex; spontaneous activity patterns in the subplate or neocortex; later arriving molecular factors extrinsic to the cortex, such as derived anterogradely from afferent pathways or retrogradely from efferent pathways; and extrinsically driven activity patterns, either sensory inputs or spontaneous activity extrinsic to the cortex (O'Leary and Nakagawa, 2002).

The timing and pattern of development of efferent projections from and afferent projections to the cortex is important in understanding this activity-dependent phase of cortical development. The topographical organization has been shown to be an activity-dependent process for several developing sensory systems, such as

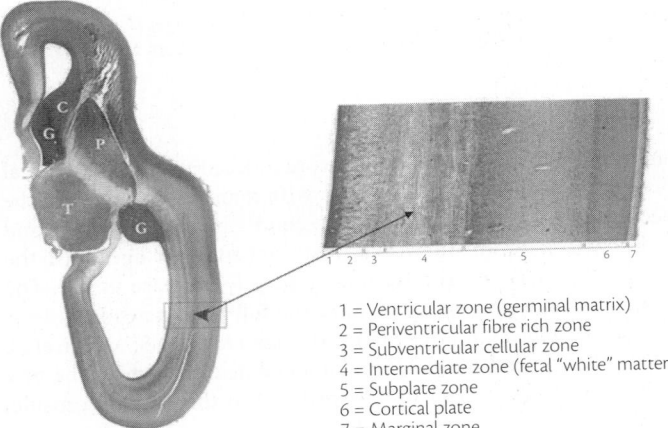

1 = Ventricular zone (germinal matrix)
2 = Periventricular fibre rich zone
3 = Subventricular cellular zone
4 = Intermediate zone (fetal "white" matter)
5 = Subplate zone
6 = Cortical plate
7 = Marginal zone

Fig. 9.5 Horizontal section through the brain in an 18-week post-conceptual age foetus. Low power and high power view stained with cresyl violet. *C,* caudate nucleus; *G,* Ganglionic eminence; *P,* putamen; *T,* thalamus. (Adapted from Kostovic *et al.* 2002.) (See Plate 12.)

visual and somatosensory thalamo-cortical projections (for reviews see (O'Leary *et al.* 1995; Penn and Shatz 1999; Levitt 2003) and for the motor system (for review see Eyre (2005)). If the source of activity is altered during this critical period then the normal patterns of connectivity are disrupted.

9.2 Congenital structural defects

9.2.1 Neural tube defects

Neural tube defects are a group of birth anomalies resulting from defects in fusion of the neural tube. They are the second most common type of birth defect after congenital heart defects but since the early 1980s, estimation of the prevalence of spina bifida in many industrialized countries has been complicated by the availability of prenatal diagnosis and the elective termination of some affected foetuses. The most common neural tube defects are anencephaly and myelomeningocele. Anencephaly, which results from failure of fusion of the cranial neural tube, is characterized by a total or partial absence of the cranial vault and cerebral hemispheres and is uniformly lethal. Myelomeningocele, commonly called spina bifida, results from the failure of fusion in the spinal region of the neural tube and depending on the size and the location of the defect, the patient may suffer no physical impairment through to severe and complex impairments (Fig. 9.6). Failure of closure that involves the entire body axis is known as craniorachischisis, which is an additional, relatively rare, form of dysraphism. Anencephaly and myelomeningocele are referred as 'open' neural tube defects because the affected region is exposed to the body surface. There are also a number of closed or skin-covered conditions that involve the neural tube, including: encephalocele, meningocele, lipomeningocele, also referred to as spina bifida occulta, and sacral agenesis.

The most common presentations, spina bifida and anencephaly, can occur within the same family, raising the question as to whether these phenotypes are related and due to the pleiotropic effect of a common underlying gene or genes (Drainer *et al.* 1991; George *et al.* 1996). For most part these defects do not follow a pattern of simple Mendelian inheritance; the majority of neural tube defects, approximately 70 per cent, occur in isolation and show multifactorial inheritance, with interactions between environmental factors and target genes (Detrait *et al.* 2005). Maternal diabetes has long been recognized to be associated with increased risk of a neural tube defect, but the risk is reduced by tight periconceptional

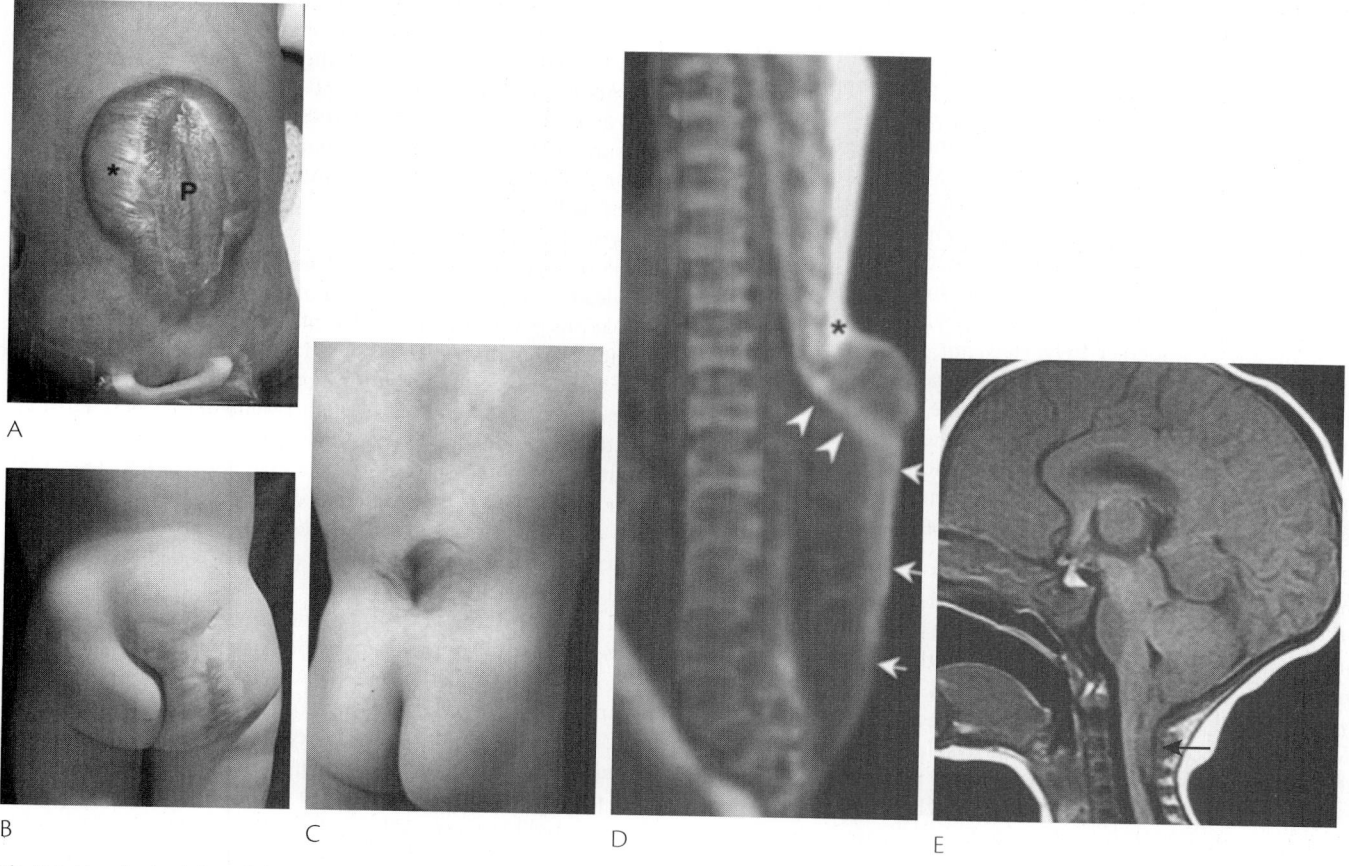

Fig. 9.6 Neural tube defects. (A) Myelomeningocele in a 1-h-old newborn. The wide placode (P) is directly exposed to the environment and is surrounded by partially epithelized skin (*asterisk*). (B) Closed spinal dysraphism with lipomyelocele. (C) 2-year-old patient with diastematomyelia. There is a hairy tuft at high lumbar level. (D) Myelomeningocele, Sagittal T1-weighted MRI of the lumbosacral spine of the baby in A showing the spinal cord (*arrowheads*) crossing the meningeal outpouching and ending with an exposed terminal parietal placode (*arrows*) Dehiscent subcutaneous fat (*asterisk*). (E) Sagittal T1-weighted MRI of the brain of the baby in A showing a Chiari II malformation (*arrow*).
(Adapted from Rossi *et al.* 2004.)

glycemic control (Loeken 2005). Maternal obesity also increases risk by 1.5–3.5-fold (Shaw *et al.* 1996). During the first and second months of pregnancy, exposure to anti-epileptic drugs such as sodium valproate and carbamazepine is associated with a 10–20-fold increased risk (Matalon *et al.* 2002). The teratogenic mechanism is unknown and neural tube defects occur only in a small minority of exposed foetuses, 1–2 per cent; the affected foetuses are therefore likely also to have a genetic predisposition.

Several studies have demonstrated that maternal periconceptional supplementation with folic acid reduces the recurrence risk for neural tube defects by 50–70 per cent, implicating genes involved in the metabolism of folate (Milunsky *et al.* 1991; MRC 1991; Czeizel and Dudas 1992). The risk is not entirely eliminated, suggesting that additional factors are responsible for the development of neural tube defects and these non-folate responsive cases may represent highly genetic cases (Milunsky *et al.* 1991; Chatkupt *et al.* 1994).

Spinal bifida occulta
Spina bifida oculta results from failure of fusion of part of the vertebral arch, usually the lamina, but does not involve the spinal cord or meninges (Fig. 9.6 B, C). Most of these lesions are asymptomatic, identified incidentally on radiological investigations and no treatment is required (Kaplan *et al.* 2005). A minority are sympotomatic, related to myelodysplasia, tethered cord, lipomas of the conus medullaris, or expansion of congenital tissue remnants consisting of dermal or lipomatous tissue. There are cutaneous stigmata in approximately half of patients who present with symptoms. These include lumbosacral cutaneous hemangiomas, dermal sinuses, a midline subcutaneous lipoma, and lumbosacral skin appendages. In addition there may be an open sinus tract which can cause recurrent meningitis (Sattar *et al.* 1996). Occult spina bifida has been found to be associated with various anorectal and urogenital malformations including cloacal exstrophy, imperforate anus, anal atresia, renal dysplasia, and bladder exstrophy (Davidoff *et al.* 1991; Tsakayannis and Shamberger 1995; Heij *et al.* 1996). For those who have progressive symptoms and a radiologically proven diagnosis, surgical intervention is appropriate to release tethers or to remove tissue compressing the cord or nerves. Although natural history data are scant and there are no controlled studies, there is evidence that the intervention prevents further deterioration and may diminish existing symptoms (Guerra *et al.* 2006).

Spina bifida with meningocele
Spina bifida with meningocele is a protrusion of meninges through a deficit in the spine or skull without the inclusion of any underlying neural tissue, usually with a normal epidermis covering the meninges. If these lesions are examined histopathologically however, neural elements are almost always included and the underlying spinal cord or brain tissue is dysplasic. Surgical removal of the meningocoele and repair of the dura mater is indicated to eliminate the risk of injury or infection (Mitchell *et al.* 2004).

Myelomeningocele, the most common form of neural tube defect, is an open lesion in the caudal spine which contains dysplastic spinal cord, often resulting in a lack of neural function below the level of the defect. It may be situated at any spinal level, although lumbosacral involvement is most common (Fig. 9.6 A, D). The location and extent of the defect determines the nature and degree of neurological impairment. Varying degrees of paresis of the legs and sphincter dysfunction are the most common clinical problems.

Congenital dislocation of the hips and congenital deformities of the feet may also occur. Hydrocephalus is a frequently associated defect and is the result of a Chiari Type II malformation, which may itself be symptomatic because distortion of the medulla and midbrain can cause lower cranial nerve palsies and central apnoea (Fig. 9.6E). The Chiari Type II malformation (Section 9.2.6) is most commonly seen in those with thoracicolumbar defects (Rintoul *et al.* 2002).

Maternal serum α-fetoprotein and ultrasound are now routinely used to identify foetuses that have or are likely to have either spina bifida or anencephaly (Drugan *et al.* 2001). Positive findings from either of these two screens would then be followed by detailed ultrasonography. If amniocentesis is also done the foetal karyotype can be examined to rule out chromosomal anomalies. However sonography can identify additional structural malformations that are characteristic of foetuses with chromosomal abnormalities and it has largely superseded amniocentesis (Gray 1999; Sepulveda *et al.* 2004). When a diagnosis of spina bifida is confirmed, ultrasound is used also to assess spontaneous leg and foot motion, leg and spine deformities, the presence of a Chiari II malformation, and other physical defects (Mangels *et al.* 2000). Prenatal MRI, with ultrafast T2-weighted sequences, can also be used to characterize the Chiari Type II and other malformations. Such prenatal imaging studies help to predict neurological deficit and ambulatory potential, but care should be exercised since both ultrasound and MRI have been shown to lead to misdiagnosis of the spinal level by two or more segments in at least 20 per cent of cases (Cochrane *et al.* 1966; Biggio *et al.* 2001; Aaronson *et al.* 2003).

Closure of the spinal lesion is usually undertaken within 48 h of birth to reduce the risk of injury and infection. If there are overt signs of hydrocephalus and imaging studies confirm the presence of ventriculomegaly, a shunt is usually placed at the same time as the lesion is closed. It has been suggested that *in utero* treatment might lead to improved functional outcome. This is based on the observations that in animal models of open neural tube defects significant inflammation in the exposed nerve roots develops *in utero* and is associated with deteriorating neurological function which can be ameliorated by *in utero* repair (Meuli *et al.* 1995). Furthermore there is evidence that a Chiari II malformation could be prevented by *in utero* closure (Osaka *et al.* 1978; Paek *et al.* 2000; Bruner *et al.* 2004) and the outcome data of centres performing in *utero* repairs provide observational evidence of a decrease in the requirement for early shunting of children (Sutton *et al.* 1999; Bruner *et al.* 2004).

The long-term survival of individuals with spina bifida has increased with improvements in medical and surgical management. The most recent population-based data indicate that 1-year survival is about 87 per cent, and that 78 per cent of individuals born with spina bifida survive to the age of 17 years (Wong and Paulozzi 2001; Rintoul *et al.* 2002). Little is known as yet about the additional health problems for adults with spina bifida, but it seems likely that excess morbidity and mortality will continue throughout adult life (McDonnell and McCann 2000; Bowman *et al.* 2001). The coordinated delivery of health services is, therefore, a life-long imperative.

Split cord malformations or diastematomyelia
This refers to splitting of the spinal cord, conus medullaris or terminal filum in the sagittal plane into two parts. Type I malformations are a split cord residing in a common dural tube and

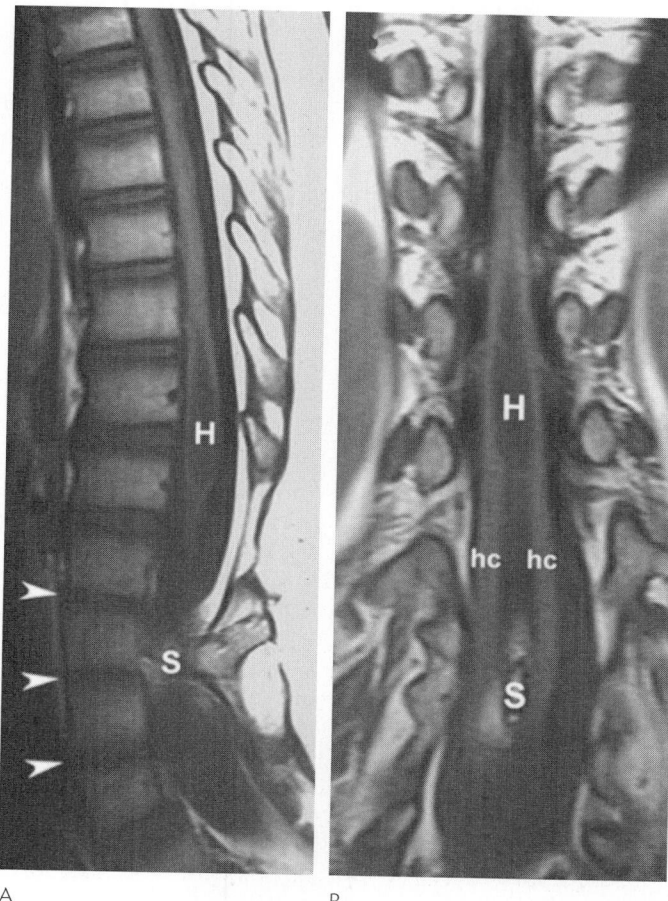

A   B

Fig. 9.7 Diastematomyelia type I with hydromyelia, (H) in a 1-year-old girl. (A) Sagittal and (B) coronal T1-weighted images show bony spur (S) containing high-signal bone marrow and projecting into the spinal canal. The spur is located at the bottom end of the cord splitting. The two hemicords (hc) are visible above the spur. There also is hydromyelia involving the spinal cord above the splitting (H). Multiple vertebral segmentation defects involving the lumbar spine are revealed by rudimentary intervertebral disks (*arrowheads*, A). (Adapted from Rossi *et al.* 2004.)

in Type II malformations the split cord is divided by a bony or cartilaginous spur, with each hemicord residing in a separate dural tube (Fig. 9.7). Most lesions are found in the lumbosacral region but they may occur at other spinal levels. A thick, cutaneous hairy patch usually overlies the region of diastematomyelia (Fig. 9.6C). Bone anomalies are present in 80–90 per cent of cases, scolosis is present in 50 per cent. Progressive neurological signs may develop and this is presumed to be a consequence of associated spinal cord tethering. Surgical intervention is usually indicted based on an expected natural history of continued progression in the absence of treatment (Proctor and Scott 2001; Schijman 2003).

Sacral agenesis

Sacral agenesis is defined as the congenital absence of the whole or part of the sacrum (Fig. 9.8). It is associated with malformations of structures derived from the caudal region of the embryo, that is, the urogenital system, the hindgut, caudal spine and spinal cord, and the lower limbs. Sacral agenesis has a heterogeneous aetiology

but for 15–25 per cent of children their mothers have insulin-dependent diabetes mellitus. The neurological abnormalities range from a minimal detectable deficit of innervation to complete lack of sensory and motor function in the lower extremities, with neurogenic bladder and bowel. There is an autosomal, dominantly inherited sacral agenesis in which a hemisacrum is associated with a presacral mass comprising an anterior meningocele, enteric cyst and/or presacral teratoma, and anorectal stensosis; the first sacral vertebrae are preserved (Fig. 9.8 C, D). This has been named the *Currarino triad* and mutations in the coding sequence of a homeobox gene *HLXB9* have been identified in nearly all cases of familial Currarino syndrome and in approximately 30 per cent of patients with sporadic Currarino syndrome (Hagan *et al.* 2000; Lynch *et al.* 2000; Rodriguez *et al.* 2002).

Encephaloceles

Encephaloceles are defined as a lesion consisting of a failure in closure of the skin, skull, and meninges which contains some neural elements (Fig. 9.9A). The degree of neurological impairment relates to the amount of brain tissue involved and the degree of neural dysplasia of the underlying brain. The most common site for an encephalocele is occipital in 75 per cent followed by frontal in 25 per cent. Encephaloceles may also occur in the midline between the sphenoid and ethmoid bones and present as an intranasal mass. Several disorders have been associated with encephaloceles including chromosomal deletions and trisomes, Meckel–Gruger syndrome, Joubert syndrome, and Walker–Warburg syndrome.

Anencephaly

Anencephaly is characterized by a partial or total absence of cerebral structures and of the cranial vault and abnormal development of the skull base (Fig. 9.9B). Most are still born or die soon after birth. The primary pathology appears to be a defect of neural tube closure with subsequent destruction of the exposed neural tissue (Kashani and Hutchins 2001).

9.2.2 Spinal vertebra anomalies

Congenital kyphosis, kyphoscoliosis, and scoliosis

Congenital kyphosis, kyphoscoliosis, and scoliosis form a spectrum of spinal deformities that develop due to vertebral anomalies that produce a localized imbalance in the longitudinal growth of the spine (Fig. 9.10). The type of deformity that develops depends on whether the impaired spinal growth occurs unilaterally, producing a pure scoliosis, or is anterior or anterolateral to the transverse axis of vertebral rotation in the sagittal plane, producing a kyphosis or a kyphoscoliosis. Congenital kyphosis and kyphoscoliosis are much less common than congenital scoliosis. They are however potentially more serious because they can ultimately result in compression of the spinal cord and paraplegia. Although the scolosis and kyphosis are the result of developmental abnormalities about 50 per cent of scoliotic lesions and nearly all kyphotic lesions progress throughout growth and usually accelerate during the adolescent growth spurt, before stabilizing in adulthood (McMaster and Ohtsuka 1982; McMaster and Singh 1999; Rossi *et al.* 2004). There may be underlying myelodysplasia, syrinx and cord tethering, and associated genitourinary and cardiac abnormalities.

Klippel–Feil syndrome

The designation Klippel–Feil syndrome includes a heterogeneous group of patients unified only by the presence of a congenital

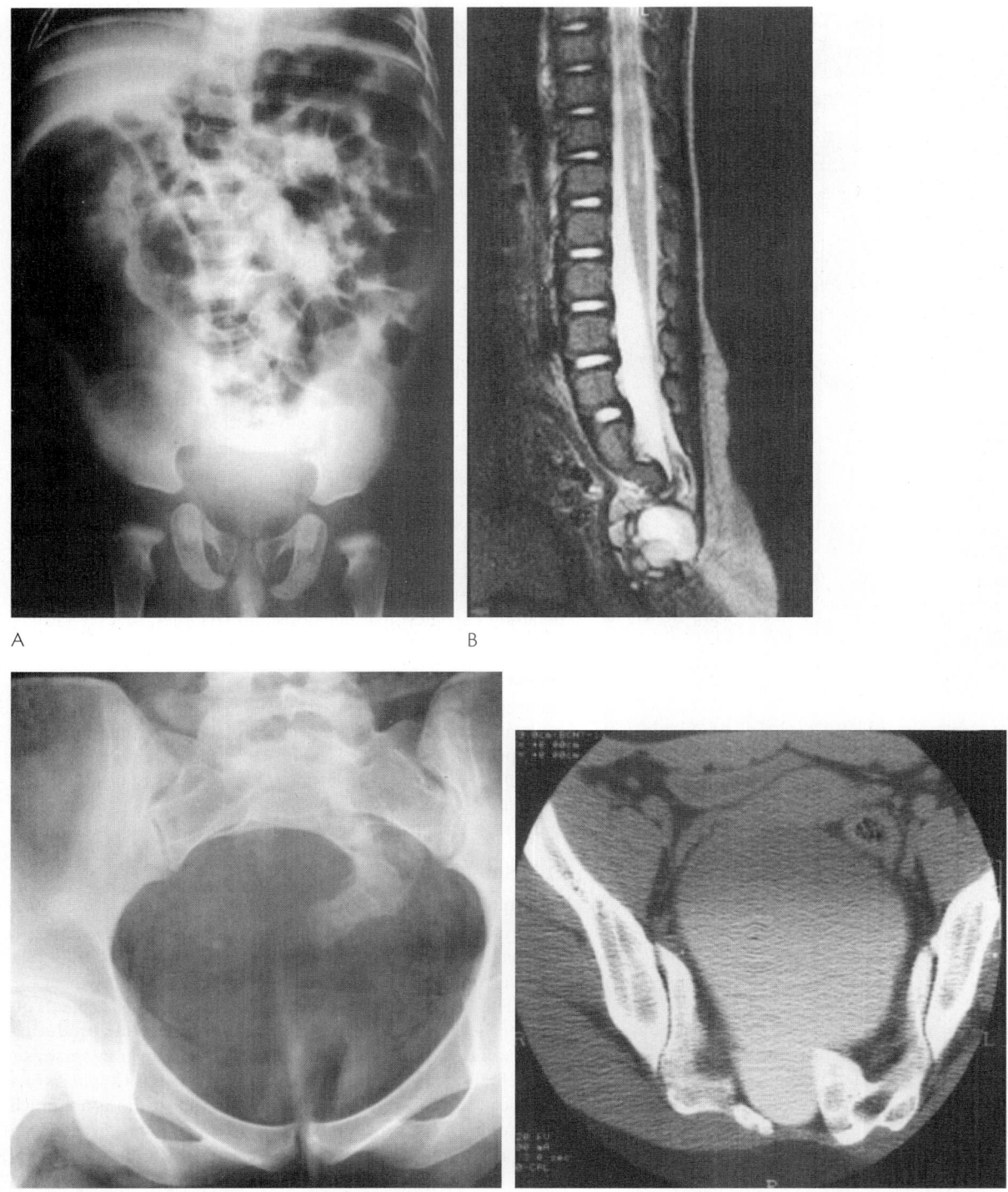

Fig. 9.8 Sacral agenesis or dysgenesis. (A) X-ray of a neonate showing sacral agenesis; (B) MRI of the same baby showing the anchorage medulla with the medullar cone localized at L3, L4 level, and an anterior myelomeningocele. (C) Pelvic X-ray of subject with Currarino Triad showing a right sided defect to the sacrum; (D) CT scan of pelvis showing a large anterior meningocele filling with contrast.
(A and B adapted from Rodriguez *et al.* 2002; C and D adapted from Lynch *et al.* 2000.)

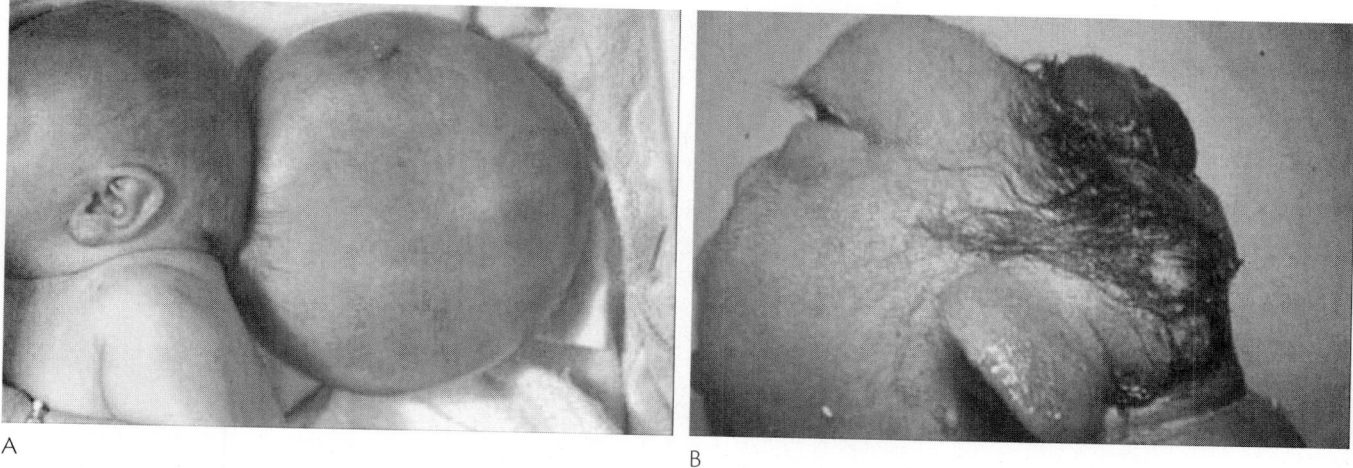

A   B

Fig. 9.9 (A) Neonate with an occipital encephalocele; (B) Neonate with anencephaly. (Reprinted with permission of the Department of Pathology, Virginia Commonwealth University and the VCU Health System.) (See Plate 13.)

synostosis of some or all cervical vertebrae (Fig. 9.11). It occurs in approximately 1:40 000–42 000 births (Thomsen *et al.* 1997; Tracy *et al.* 2004). Cervical fusions may be asymptomatic and identified only incidentally on radiographs obtained for other purposes. Other patients present with a decreased neck range of movement, related to the extent of cervical spinal involvement, neck and/or radicular pain from degenerative changes or hypermobility, cosmetic concerns from a shortened or webbed neck, and problems related to anomalies in other organ systems that are formed embryologically at the same time as the cervical spine, most commonly anomalies

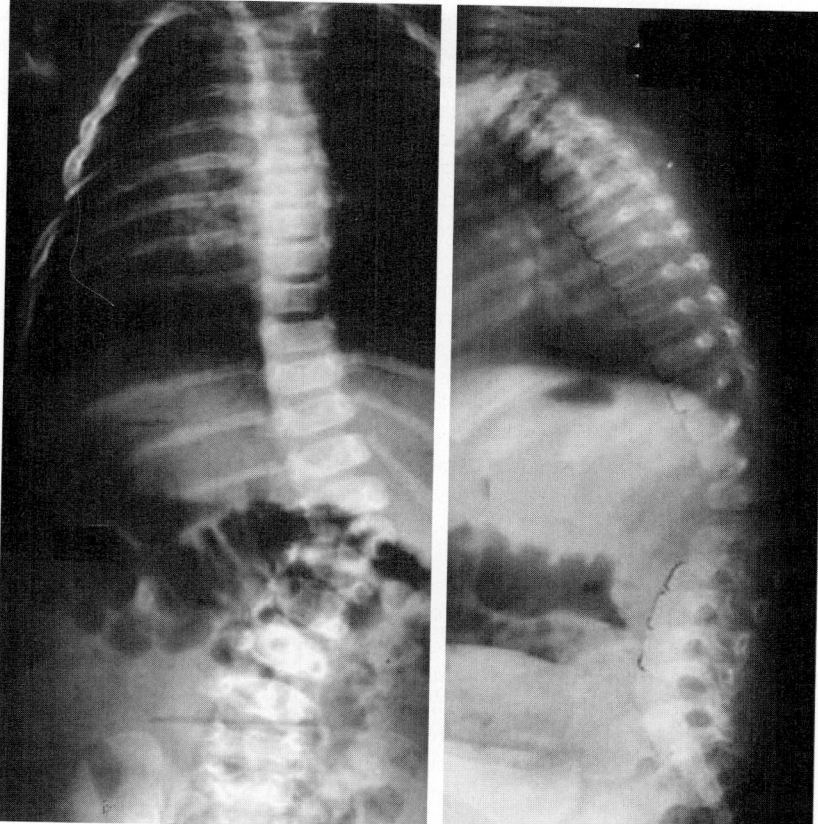

Fig. 9.10 Thoracolumbar kyphoscoliosis due to a posterolateral quadrant vertebra at the first lumbar level. Anteroposterior and lateral X-rays of a 5-month-old infant. (Adapted from McMaster and Singh 1999a.)

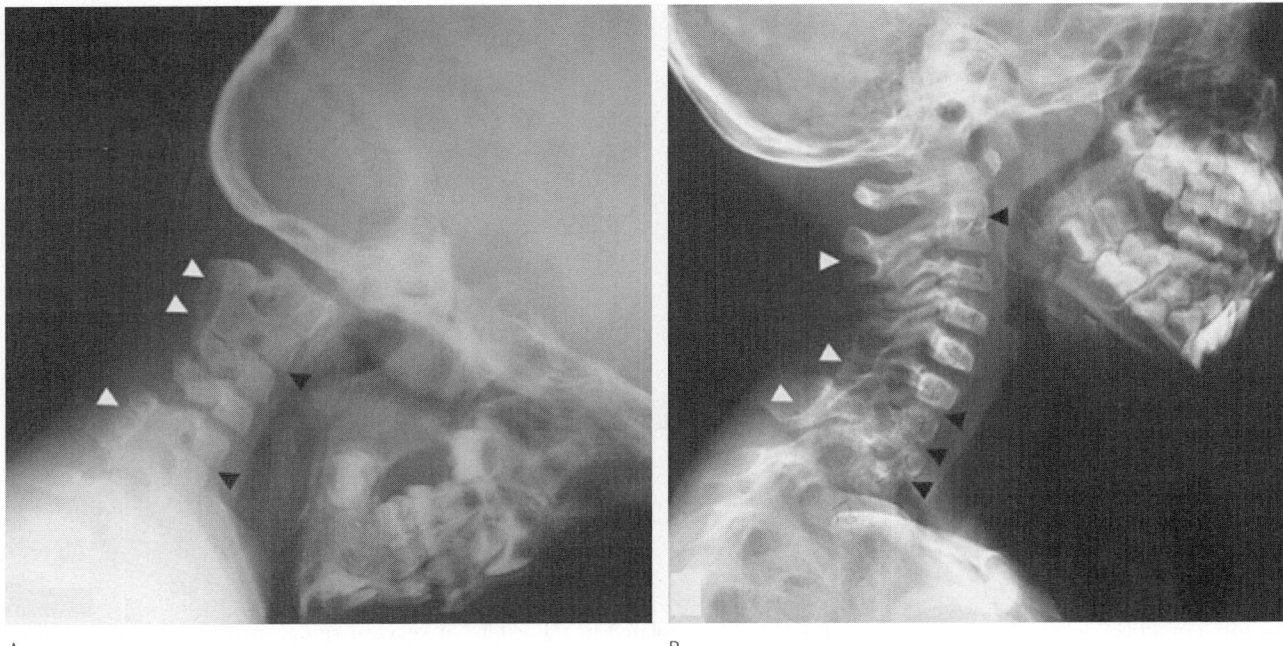

A   B

Fig. 9.11 Klippel–Feil anomaly. (A) and (B) Lateral cervical spine X-rays showing fusion of adjacent vertebral bodies (black arrowheads) and adjacent neural arches (white arrowheads).
(Adapted from Tracy *et al.* 2004.)

of the inner ear, heart, and kidneys. Subjects with Klippel–Feil syndrome also have increased risk of serious neurological injury after relatively minor neck trauma (Smith and Griffin 1992). MRI of the spine is important to evaluate the space in the cervical spine available for the spinal cord, to assess the degree of spinal stenosis caused by the fusion, and to detect other abnormalities such as a syrinx, tethered cord, or diastomyelia (Thomsen *et al.* 1997; Manaligod *et al.* 1999). Patients with progressive symptomatic segmental instability or neurological signs should be considered for surgical stabilization of the abnormal region of the cervical spine. Regular neurological follow-up is necessary to identify progressive cranial nerve abnormalities, cervical radiculopathy, or myelopathy (Herman and Pizzutillo 1999).

9.2.3 Disorders of brain segmentation

Schizencephaly

Schizencephaly is a rare congenital brain abnormality so named to describe clefts in the cerebral mantle extending from the pial surface to the lateral ventricles, which are lined by heterotopic grey matter (Yakovlev and Wadsworth 1946). In Type I or closed-lipped schizencephaly, the edges of the cleft appear to fuse creating a pial-ependymal seam and leaving abnormal cortex on either side. Type II, or open-lipped schizencephaly, is a defect characterized by widely separated grey matter lined clefts in the cerebral mantle communicating with the lateral ventricles, usually accompanied by hydrocephalus (Fig. 9.12).

The causes of schizencephaly are heterogeneous and are rather poorly understood. Established aetiologies include the teratogens, warfarin, alcohol, and cocaine; the *in utero* infections, cytomegalovirus, and herpes virus; maternal trauma of several types and monozygotic twin interactions (Curry *et al.* 2005). All of these agents have a common pathogenic mechanism, hypoxic ischaemic

vascular injury at critical times in neuronal development, which indirectly supports the hypothesis that schizencephaly is a vascular disruptive birth defect. The presence of grey matter lining the schizencephalic clefts suggests that these defects occur early in gestation, prior to the end of neuronal migration. Other rare associations reported in schizencephaly include chromosomal aneuploidy, single gene defects, and a few distinct syndromes. In familial schizencephaly one of the genes implicated is *EMX2* a transcription factor involved in cortical arealization (Brunelli *et al.* 1996; Granata *et al.* 1997). Thus it appears that for some patients the clefts may arise from failure of regional specification of clones of cells in the germinal matrix that are destined to be part of the cortex.

The clefts are typically located in the frontal or parietal regions and may be unilateral or bilateral and symmetrical; they may occur in isolation or be associated with other anomalies of brain development. Schizencephaly can present with a broad range of developmental disabilities and neurological symptoms, the most frequent of which are seizures and hemiparesis. The severity of symptomatology usually correlates with the extent of brain involvement (Liang *et al.* 2002). Unilateral schizencephaly usually presents with hemiparesis and mild delay, whereas bilateral clefts are associated with quadriparesis and severe cognitive impairment (Denis *et al.* 2000; Liang *et al.* 2002).

Holoprosencephaly

Holoprosencephaly is the most common major malformation of human cerebral development, occurring in 1 in 8300 live births. The disorder results from a defect of prosencephalon cleavage, a process normally complete by the fifth week of gestation. It is characterized by a failure of complete separation of the prosencephalon into two cerebral hemispheres; other midline central nervous system structures such as the basal ganglia, hypothalamus,

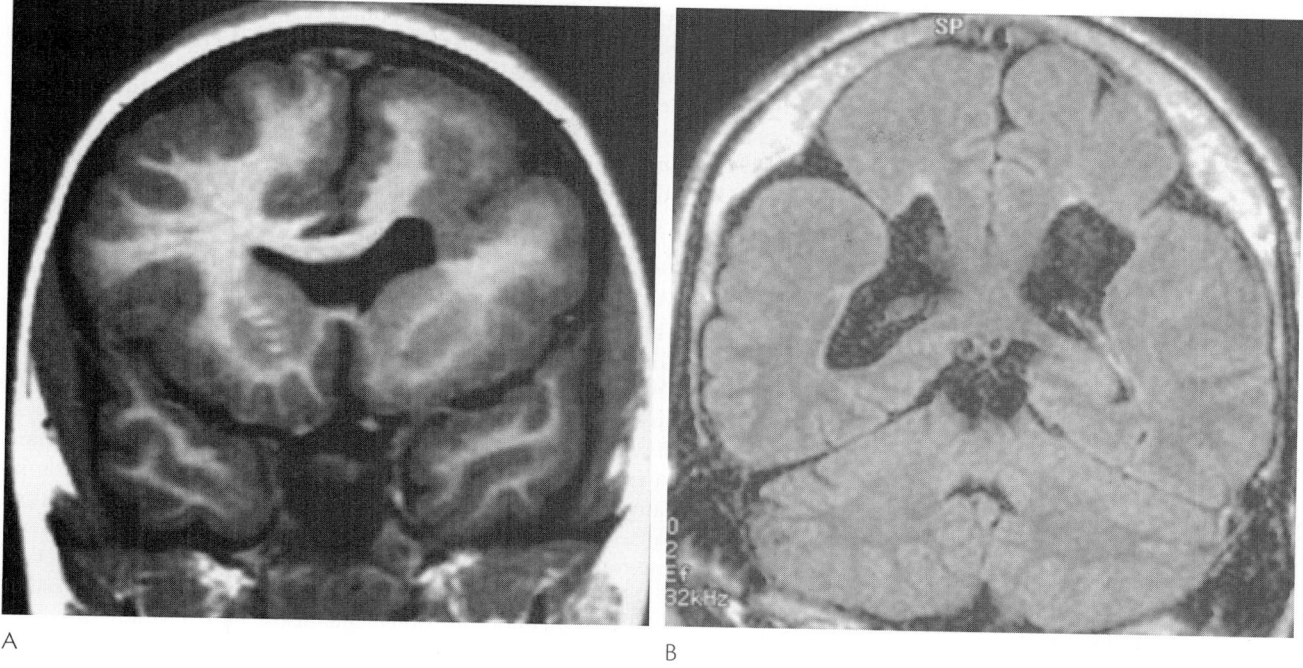

Fig. 9.12 Schizencephaly (A) Type 1 closed-lipped form. MRI shows multiple grey matter lined clefts not communicating with the ventricles. (B) Type 2 Bilateral open-lipped form. MRI shows bilateral clefts communicating with the ventricles. (Adapted from Curry *et al.* 2005.)

pituitary, and thalamus can also be involved. Three forms of this disorder have been described: alobar, semilobar, and lobar. In the alobar form, the telencephalic vesicle completely fails to divide, producing a single horseshoe-shaped ventricle, sometimes with a dorsal cyst, fused thalami, and a malformed cortex (Fig. 9.13). In the semilobar form, the interhemispheric fissure is present posteriorly, but the frontal and, sometimes, parietal lobes continue across the midline. In the lobar form, only minor changes may be seen: the anterior falx and the septum pellucidum usually are absent, the frontal lobes and horns are hypoplastic, and the genu of the corpus callosum may be abnormal. Holoprosencephaly rarely

occurs as an isolated entity. Roughly 80 per cent of cases are associated with varying degrees of facial malformations ranging from mild, the presence of a single upper central incisor, to severe, cyclopia with a proboscis. Pituitary defects may be associated with these malformations and place the child at risk of endocrine dysfunction (Hahn *et al.* 2005).

Holoprosencephaly has been associated with maternal diabetes (Barr *et al.* 1983), retinoic acid exposure, cytomegalovirus, and rubella (Cohen and Lemire 1983). Chromosome abnormalities associated with this disorder include trisomies 13 and 18; duplications of 3p, 13q, and 18q; and deletions in 2p, 7q, 13q,

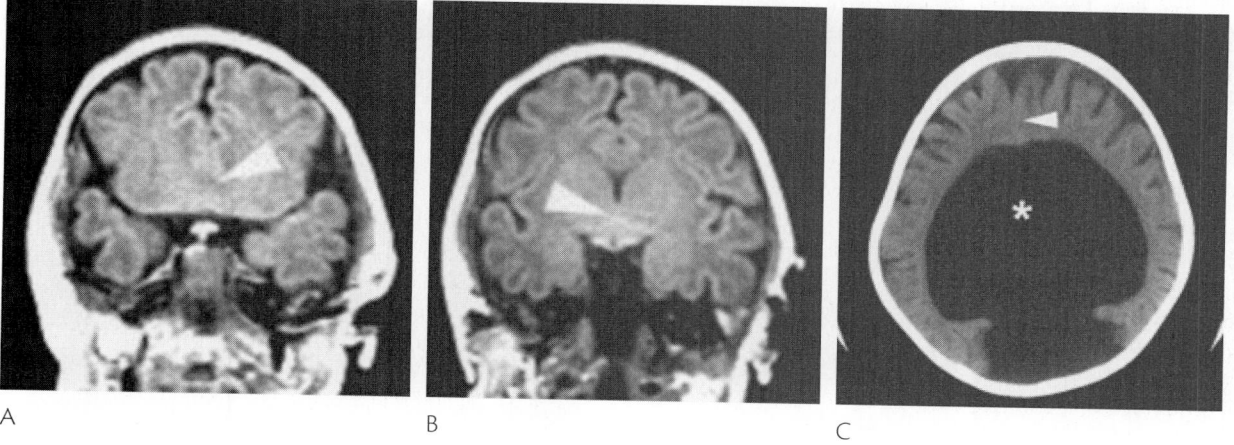

Fig. 9.13 Holoprosencephaly (A) and (B) MRIs from the same patient with semilobar holoprosencephaly showing absent ventral interhemispheric cleavage in A and fused thalami in B (arrows). (C) shows alobar holoprosencephaly. The arrow points to the anterior midline where no interhemispheric fissure is noted. (Adapted from Clark 2004.)

and 18q (Muenke 1989). Several multiple malformation syndromes have been associated with holoprosencephaly including Pallister–Hall, Meckel, and velocardiofacial syndromes. A number of genes have been associated with holoprosencephaly and pedigrees supporting autosomal dominant, autosomal recessive, and possibly X-linked inheritance have been described (Roessler *et al.* 1996; Hahn and Pinter 2002).

The clinical variability within pedigrees can be very great. Only children who have the lobar and semilobar forms are known to survive for more than a few months. An infant affected by the most severe form is microcephalic, hypotonic, and visually inattentive. In infants with the less severe forms of holoprosencephaly, myoclonic seizures frequently develop and, if the infant survives, growth retardation, psychomotor retardation, and atonic or spastic cerebral palsy often are present. Some infants with the lobar form may be affected mildly and, for example, present with a relatively mild spastic diplegia. In the mildest form patients may have only a very subtle abnormality such as single central incisor, hypertelorism, a choroid fissure, or coloboma or anosmia. A careful family history needs therefore to be taken since parents with family trees manifesting these features should be considered to be at high risk for recurrence of holoprosencephaly.

Septo-optic dysplasia

Septo-optic dysplasia or the de Morsier syndrome is a heterogeneous disorder comprising a variable combination of defects of midline brain structures including hypoplasia or absence of the septum pellucidum, optic nerve hypoplasia, and pituitary/hypothalamic dysfunction (Fig. 9.14). It may be associated with agenesis of the corpus callosum. Patients with septo-optic dysplasia commonly also have schizencephaly and/or cortical dysplasia (Aicardi and Goutieres 1981; Polizzi *et al.* 2006). This apparent heterogeneity has resulted in some disagreement as to whether septo-optic dysplasia should be regarded as a single entity or rather as a group of heterogeneous disorders. The aetiology of most cases of septo-optic dysplasia remains unclear. It is most likely that the

syndrome is the end result of multiple different genetic abnormalities and predisposing environmental factors. Patients may present with a wide spectrum of associated disorders including visual defects, eye abnormalities such as microphthalmia and colobomas, seizures, learning difficulties, hemiparesis especially if associated with schizencephaly, quadriparesis or hypothalamic dysfunction, and/or pituitary dysfunction.

9.2.4 Disorders of neuronal and glial proliferation

Microcephaly

This is defined as a small occipitofrontal circumference, below the 2nd centile, and which is disproportionate to the body length (Woods 2004). It is due to abnormal brain growth, but if a small head circumference is observed during early infancy, particularly if the head circumference has been normal at birth, it is important to exclude a normal brain and premature closure of all cranial sutures, or total craniosynostosis (Section 9.2.9). There are multiple causes of microcephaly. During pregnancy maternal and foetal environmental factors such as intrauterine infection, maternal metabolic disorders including phenylketonuria, maternal ingestion of drugs, alcohol and other chemicals, ionizing radiation, and circulatory disturbance are important causes. Postnatally many inborn errors of metabolism are associated with microcephaly. In addition any insult to the brain such as trauma, hypoxia, ischaemia, or hypoglycaemia can result in microcephaly. Many chromosomal disorders including all the common trisomies are associated with microcephaly and it is associated with so many syndromes that it is not a useful search term in dysmorphology data bases.

Megalencephaly and hemimegancephaly

The terms megalencephaly and hemimegalencephaly refer to disorders in which the brain volume is greater than normal and not owing to the abnormal storage of material or to hydrocephalus, leading to an occipital frontal circumference which is greater than the 98th centile and unduly large for body length (Fig. 9.15).

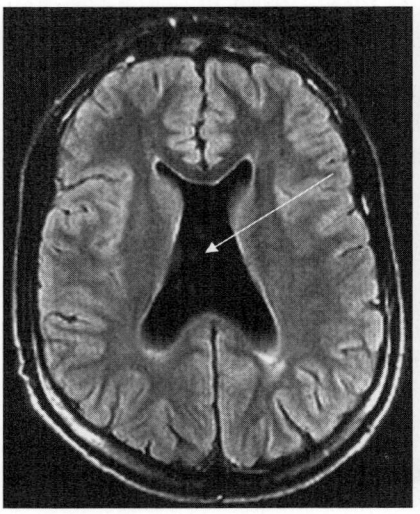 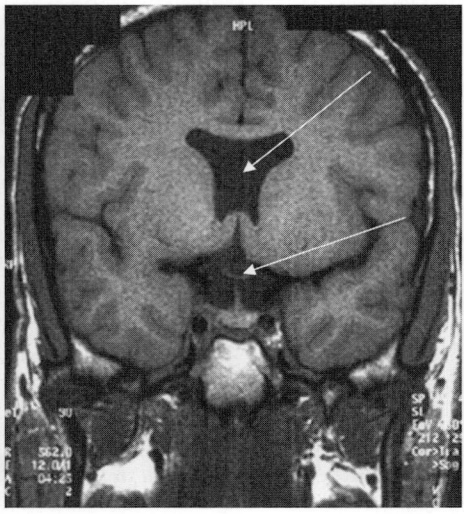 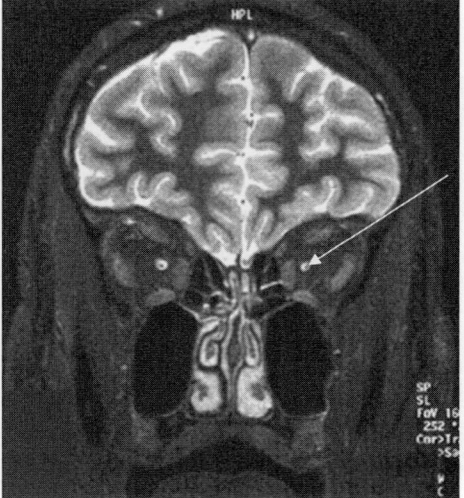

A   B   C

Fig. 9.14 Septo-optic dysplasia. (A and B) MRIs demonstrating absence of the septum pellucidum (arrowed) and hypoplasia of the optic chiasm (arrowed) (B). (C) MRI demonstrating hypoplastic optic nerves (arrowed).
(Reproduced with permission from http://www.uhrad.com/mriarc/mri086.htm)

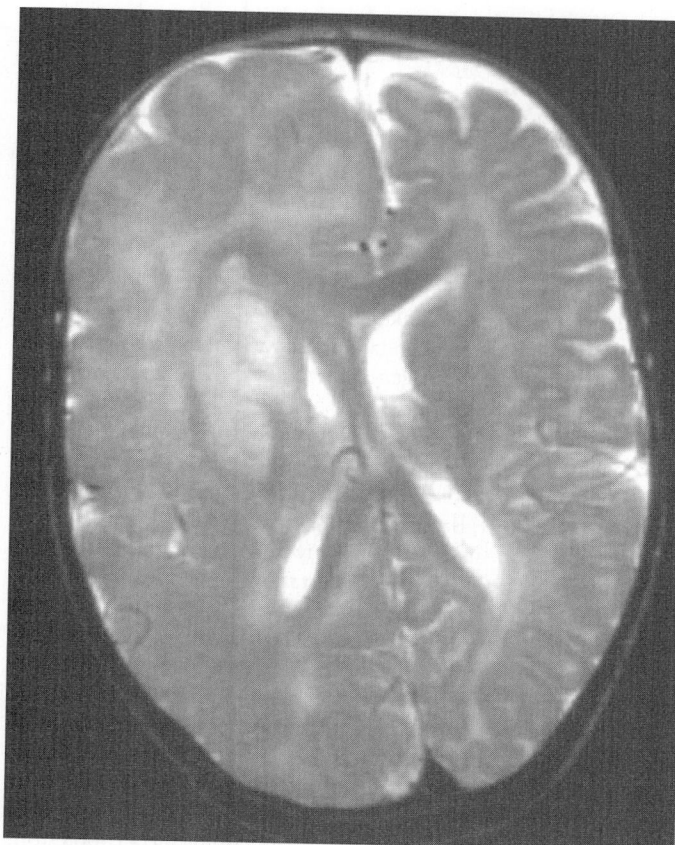

Fig. 9.15 Hemimegencephaly. MRI showing diffuse right hemispheric enlargement and dysplasia. The midline shift, bulging of anterior falx to the left, and compression of the right lateral ventricle suggest a mass effect caused by increased volume of the brain parenchyma. Dysplastic changes are diffuse with thick and disorganized cortex, poor grey–white matter differentiation, and abnormal signal in the white matter. The basal ganglia are also dysplastic with abnormal increased signal.
(Reproduced from Gupta *et al*. 2004.)

Hemimegalencephaly is associated with the dermatological syndromes linear sebaceous nevus syndrome (Hager *et al*. 1991) and hypomelanosis of Ito (Section 11.6.1) or incontinentia pigmentia achromians (Jelinek *et al*. 1973). The neuropathologic and clinical pictures of these associations appear to be identical to the isolated hemimegalencephalies. Megalencephaly is found in overgrowth syndromes such as Sotos syndrome (Sotos 1997; Tatton-Brown *et al*. 2005) or Weaver syndromes (Weaver *et al*. 1974); in haemangiomatosis overgrowth syndromes such as Bannayan–Riley–Ruvalcba syndrome (Riley and Smith 1960), Cowden (Nelen *et al*. 1999), and Proteus syndromes (Section 11.6.4) (Wiedemann *et al*. 1983); in neurofibromatosis (Section 11.2) (Cutting *et al*. 2002) and in several skeletal dysplasias such as achondroplasia (Dennis *et al*. 1961), and thanatophoric dysplasia (Yamaguchi and Honma 2001).

Typically, patients are noted to have large heads at birth and may manifest an accelerated head growth in the first few months of life. The microscopic appearance of the brain is that of cortical dysplasia with an increase in number of cells, both neurons and glia, and in cell size (DeRosa *et al*. 1972). Children with megalencephaly or hemimegalencephaly present with seizures, a developmental disorder, hemihypertrophy, or a hemiparesis. Seizures vary both in onset and in type and usually are the most problematic symptom, often necessitating hemispherectomy or callosotomy in cases of hemimegalencephally (Gupta *et al*. 2004).

9.2.5 Disorders of neural migration
Polymicrogyria
Polymicrogyria is a relatively common cortical malformation characterized by abnormal cortical lamination, excessive cortical infolding resulting in many small gyri and shallow sulci (Fig. 9.16A). Schizencephalic clefts, whether open or closed, are lined by polymicrogyric cortex, implying that polymicrogyria is a necessary component of schizencephaly. Polymicrogyria can be focal or diffuse, unilateral or bilateral. The underlying white matter is thinner than normal. It rarely involves the whole brain and the different forms of polymicrogyria encompass a wide range of clinical, aetiological, and histological findings. The most common clinical problems involve developmental delay and learning difficulties and

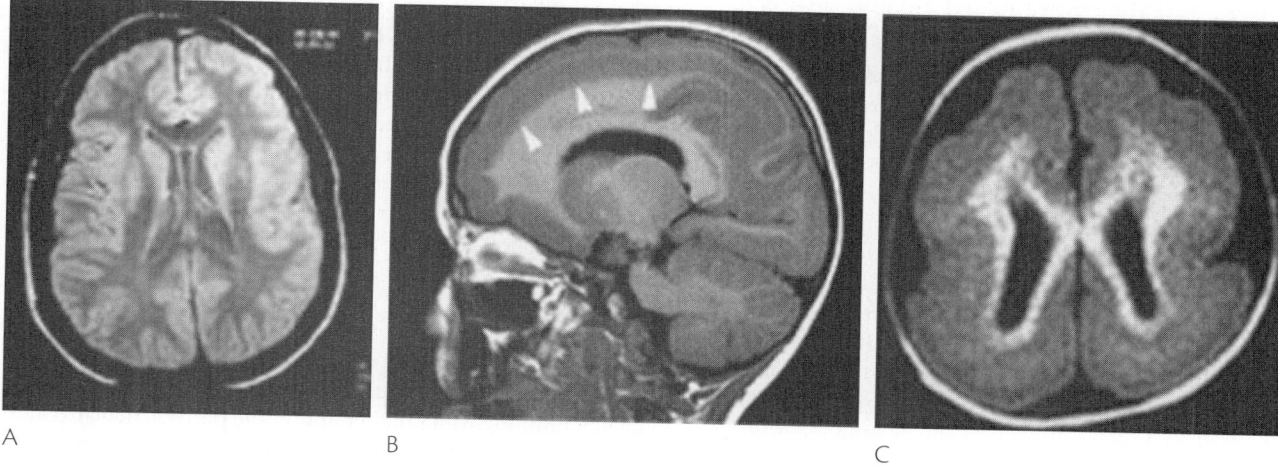

A   B   C

Fig. 9.16 Disorders of neural migration. MR scans showing (A) Polymicrogyria, (B) Subcortical band heterotopia, and (C) Lissencephaly.
(Adapted from Clark 2004.)

associated microcephaly is a poor prognostic indicator. All are at high risk of seizures which are difficult to control.

Advances in imaging have improved the diagnosis and classification of the condition. The molecular basis of polymicrogyria is beginning to be elucidated with the identification of a gene, *GPR56*, for bilateral frontoparietal polymicrogyria. Based on imaging studies, several other region-specific patterns of polymicrogyria have been identified, and there is increasing evidence that these may also have a significant genetic component to their aetiology. In humans, there is some evidence that cytotoxic factors (Barkovich and Lindan 1994) or hypoperfusion (Barkovich *et al.* 1995) in the second trimester can lead to polymicrogyria. Little is known about the factors that contribute to the development of polymicrogyria. There is evidence that extrinsic factors, such as intrauterine cytomegalovirus infection, can be involved in the pathogenesis. However, analysis of the expression pattern of *GPR56*, the first gene identified in polymicrogyria, suggests that the disorder may result from mutations in genes that are involved in the regional patterning of the cerebral cortex at early stages of development, during neuronal proliferation, and migration (Piao *et al.* 2004). The association of polymicrogyria with several genetically determined syndromes such as Zellweger (Section 10.6.2) (Liu *et al.* 1976), Aicardi (Barkovich *et al.* 2001), and Walker–Warburg syndrome (Barkovich 1998), the presence of polymicrogyria in patients with chromosomal abnormalities, and the occurrence of familial cases of polymicrogyria all strongly indicate a genetic component in its development (Jansen and Andermann 2005).

Bilateral perisylvian polymicrogyria is the most common accounting for about two-thirds of patients (Becker *et al.* 1989). It is usually sporadic although there are numerous reports of familial occurrence, suggesting X-linked, autosomal dominant and recessive inheritance. Clinical manifestations include pseudobulbar palsy with diplegia of the facial, pharyngeal, and masticatory muscles, known as facio-pharyngo-glosso-masticatory paresis, pyramidal signs, and seizures. The pseudobulbar involvement results in restricted tongue movements, drooling, feeding problems, and dysarthria. Voluntary and emotional facial movements can be dissociated. Developmental language disorder can be associated and its severity depends on the extent of the cortical damage.

Lissencephaly, agyria-pachygria, and subcortical band heteropia

Lissencephaly, or smooth brain, refers to the external appearance of the cerebral cortex in those disorders in which a neuronal migration aberration leads to a relatively smooth cortical surface with an abnormally thick cortex of four abnormal layers and enlarged dysplasic ventricles (Fig. 9.16C). It encompasses agyria, pachygyria which means a reduced number of broad flat gyri, and subcortical band heterotopia (Fig. 9.16B).

Lissencephaly, pachygyria, and subcortical band heteropia are predominantly genetic in origin (Clark 2004). Severe lissencephaly occurs as part of the *Miller Dieker syndrome* (Section 30.6.2) with dysmorphic facial features and variable other congenital malformations. All have deletions of 17p13.3 which includes the *LIS1* gene. Isolated lissencephaly/pachygyria and subcortical band heterotopia without dysmorphic facial features is predominantly associated with mutations of two genes either *LIS1* or doublecortin, *DCX*, on the X chromosome. *LIS1* mutations cause a more severe malformation in posterior brain regions. Most children have severe developmental delay and infantile spasms, but some may have milder phenotypes, including posterior subcortical band heteropia owing to mosaic mutations of *LIS1*.

DCX mutations usually cause anteriorly predominant lissencephaly in males and subcortical band heteropia in female patients. Mutations of *DCX* have also been found in male patients with anterior subcortical band heteropia and in female relatives with normal brain magnetic resonance imaging. Autosomal recessive lissencephaly with cerebellar hypoplasia, accompanied by severe delay, hypotonia, and seizures, has been associated with mutations of the reelin, *RELN*, gene. X-linked lissencephaly with corpus callosum agenesis and ambiguous genitalia in genotypic males is associated with mutations of the *ARX* gene. Affected boys have severe delay and seizures with suppression-burst EEG. Early death is frequent. Carrier female patients can have isolated corpus callosum agenesis (Guerrini and Marini 2006). Lissencephally can also be a consequence of early congenital infection, particularly cytomegalovirus infection (Hayward *et al.* 1991; Barkovich and Lindan 1994).

Antenatal diagnosis of lissencephaly by ultrasound is not reliable until relatively late, 32–36 weeks gestation, since it cannot be distinguished from normal brain prior to this. However when there is a family history molecular diagnosis is now possible for the majority.

Cobblestone lissencephaly is a severe brain malformation observed in three overlapping syndromes—all genetic with autosomal recessive inheritance (Guerrini and Marini 2006). Fukuyama's congenital muscular dystrophy (Section 24.2.7) due to a mutation of the fukutin gene, muscle-eye-brain disease and *Walker–Warburg syndrome* (Table 10.1 VI.2). The term cobblestone refers to the appearance of the cortical surface upon pathological, rather than MRI, examination. In all these disorders neurons migrate past their stopping point and erupt into the subarachnoid spaces giving a cobblestone appearance to the surface of the cortex at postmortem examination. In addition to lissencephaly they often have cerebellar and brainstem hypoplasia but can be differentiated from lissencephaly with cerebellar hypoplasia by the presence of additional abnormalities. These include muscular dystrophy, ocular anterior chamber abnormalities, retinal dysplasia, obstructive hydrocephalus, and encephaloceles.

The clinical course of all three syndromes consists of profound mental retardation, severe hypotonia, and poor vision. Seizures may occur but are less severe than in isolated lissencephally. *Walker–Warberg syndrome* (Table 10.1 VI.2) has the most severely abnormal clinical findings and few children with Walker–Warberg syndrome live past their third year. Muscle-eye-brain disease and Fukuyama's congenital muscular dystrophy can be distinguished since Fukuyama's congenital muscular dystrophy has severe muscular dystrophy and only minor if any eye abnormality. Both are less severe than Walker–Warberg syndrome and survival into adulthood is possible.

Neuronal heterotopias

Heterotopias are collections of normal appearing neurons in an abnormal location. Heterotopias may be classified by their location: subpial, within the white matter and in the periventricular region (Fig. 9.17). X-linked periventricular neuronal heterotopias can occur and is mainly seen in females because of high rates of embryonic hemizygous male lethality (Fox *et al.* 1998; Moro *et al.* 2002). Mutations of filamin A at Xq28 have been reported in all

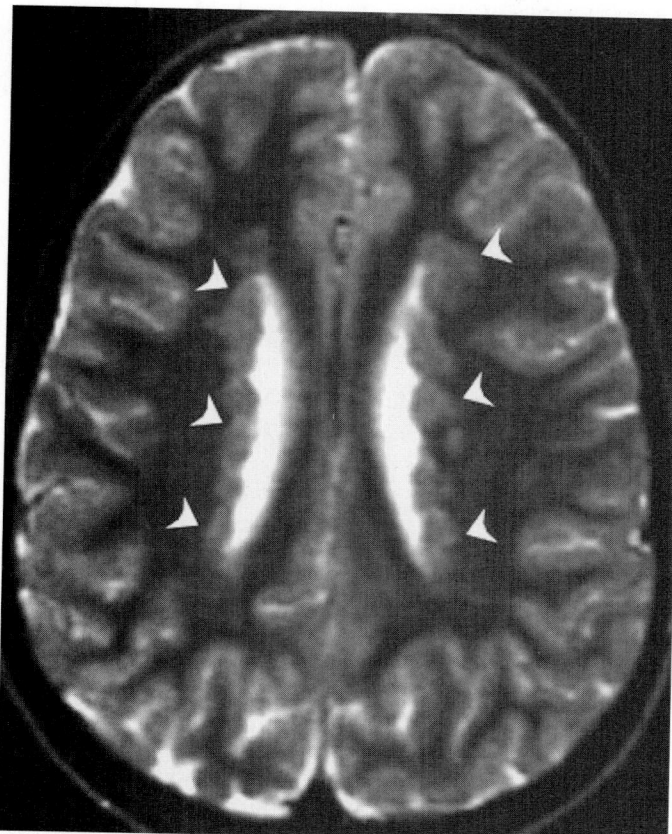

Fig. 9.17 MRI showing bilateral periventricular nodular heterotopia (arrows). (Adapted from Guerrini and Marini 2006.)

familial cases and in about 25 per cent of sporadic patients. A rare recessive form of periventricular neuronal heterotopias due to *ARGEF2* gene mutations has also been reported in children with microcephaly, severe delay, and early seizures (Sheen *et al.* 2004).

Unilateral subependymal heterotopia and sparsely scattered subependymal heterotopia are almost always sporadic. Occasionally heterotopias may be found in a variety of syndromes including neonatal adrenal leucodystrophy, glutaric aciduria type 2, GM1 gangliosidosis, neurocutaneous syndromes, multiple congenital anomaly syndromes chromosomal abnormalities, and foetal toxic exposure.

Heterotopias commonly result in epilepsy (Section 31.8.6), whereas others, discovered as part of family evaluations are asymptomatic. The clinical features of the epilepsy are variable. Both generalized and focal seizures are reported and in many patients multiple seizure types are observed (Dubeau *et al.* 1995; Kothare *et al.* 1998).

9.2.6 Posterior fossa malformations

Malformations of the posterior fossa are less common and less varied than supratentorial anomalies. The three main malformations are Chiari malformations, Dandy–Walker syndrome, and Joubert's syndrome. For more detailed discussion of rarer abnormalities of the posterior fossa see Boltshauser (2004) and Niesen (2002). Cerebellar development is not complete until well into the second year of postnatal life. This prolonged postnatal development period makes the cerebellum susceptible to a variety of insults. A complex hierarchy of genes regulate cerebellar and brainstem development and growth (Boltshauser 2004).

Chiari malformations

The *Chiari I malformation* is defined as an extension of the cerebellar tonsils below the foramen magnum for at least 3–5 mm. The medulla is not displaced and supratentorial anomalies are not present. The association of Chiari I malformation with syringomyelia (Section 28.5.15) means that spinal MRI assessment is recommended. The clinical picture is very variable. Symptoms when present include headache, neck pain, torticollis, lower cranial nerve palsies, nystagmus, and apnea (Cai and Oakes 1997). In recent years since the advent of easy access to MR scanning an increasing number of asymptomatic, doubtfully symptomatic, and minimally symptomatic patients with Chiari malformations have been diagnosed. 22 591 MRI findings have been reviewed in an attempt to estimate the incidence of Chiari malformation (Meadows *et al.* 2000). Herniation of the tonsils greater than 5 mm below the foramen magnum was present in 0.77 per cent of the examinations (Meadows *et al.* 2000). The variability of cerebellar tonsil positioning in relation to age, between 5 months to 89 years of age, has been assessed (Mikulis *et al.*). This concluded that the tonsils rise with age and suggested the following measurements of tonsil herniation below the foramen magnum for diagnosis, based on a distance higher than 2 standard deviation from the normal variation for each decade: First decade, 6 mm below the foramen magnum; Second to 3rd decade, 5 mm; Fourth to 8th decade, 4 mm; Ninth decade, 3 mm below the foramen magnum. Based on these data it is evident that there may be herniation on radiological examination without clinical significance; conversely patients with herniation smaller than 5 mm may have clinical symptoms and signs consistent with Chiari malformation type I, including syringohydromyelia. The radiological criteria for tonsillar herniation must always be related to the clinical context for diagnosis to be made.

The *Chiari II malformation* is clinically more important and is commonly known as the Arnold–Chiari malformation. The Chiari Type II malformation is found only in patients with myelomeningocele and is the leading cause of death in these individuals aged younger than 2 years (Fig. 9.6E). Its cardinal features are a myelomeningocele in the thoraco-lumbar spine, the venting of the intracranial cerebrospinal fluid through the central canal, the hypoplasia of the posterior fossa, herniation of hindbrain into the cervical spinal canal, and the compressive damage to cranial nerves. Early recognition of symptoms of brainstem compression and a subsequent surgical decompression can decrease the high mortality rate among children with Chiari type II malformation (Yumer *et al.* 2006). In these circumstances prompt surgical decompression of the hindbrain prevents serious morbidity and mortality in a patient with myelomeningocele, especially those younger than 2 years old (Yumer *et al.* 2006). Symptomatic Chiari type II malformation in the older child often presents with more subtle findings but rarely in acute crisis (Stevenson 2004).

Dandy–Walker malformation

The Dandy–Walker malformation is characterized by a triad of features: complete or partial agenesis of the cerebellar vermis; cystic dilatation of the fourth ventricle; enlarged posterior fossa (Fig. 9.18). It occurs with an incidence of 1 per 25 to 30 000 live births.

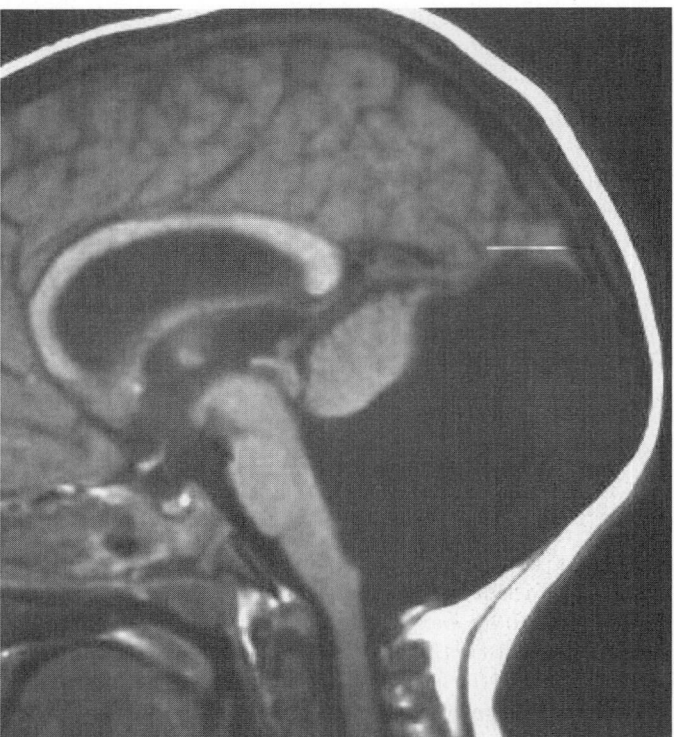

Fig. 9.18 Dandy–Walker Malformation and hydrocephalus, MRI.
(Obtained with permission from the Hirnschadel MRI Library—http://
www.mrx.de/mri-lib/scfe03.html)

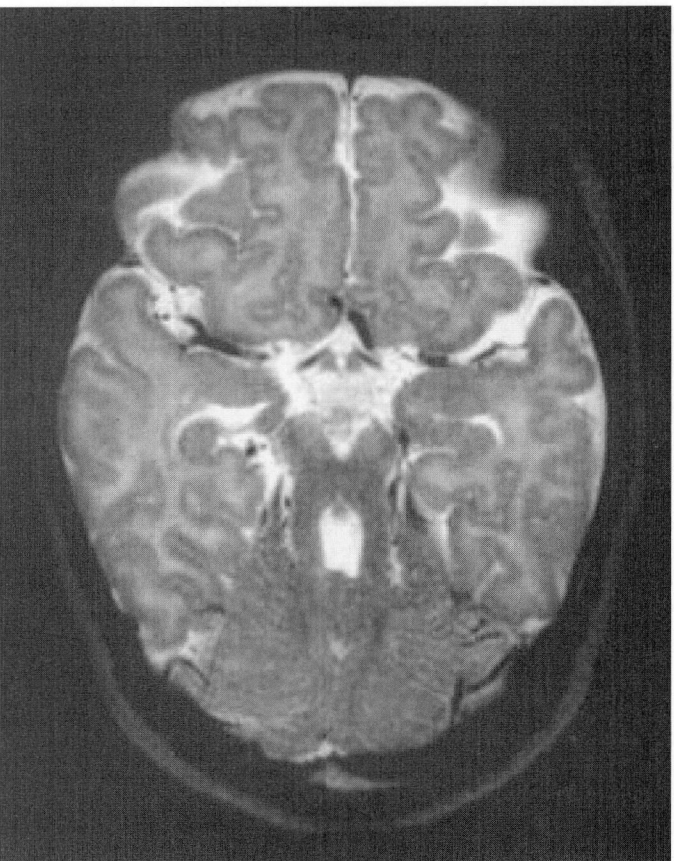

Fig. 9.19 Joubert syndrome in a 4-month-old boy. The MRI illustrates the molar
tooth sign with deep interpedicular fossa and stretched, and thickened superior
cerebellar peduncles.
(Adapted from Boltshauser 2004.)

Dandy–Walker syndrome usually occurs sporadically as an isolated abnormality, where the majority of patients are normal apart from hydrocephalus which occurs in 75 per cent of cases (Maria *et al.* 1987; Notaridis *et al.* 2006). Dandy–Walker syndrome can also occur in Coffin–Siris syndrome, Meckel–Gruber syndrome, trisomy 9, 13, and 18, duplications of chromosome 5p, 8p, and 8q, and congenital infections (Altman *et al.* 1992; Niesen 2002).

Joubert syndrome

This is an autosomal recessively inherited agenesis of the cerebellar vermis. Patients present with neonatal hypotonia, abnormal respiratory pattern, and eye movement disorder comprising nystagmus, dysmetric saccades, or impaired smooth pursuit. The respiratory pattern improves progressively but all patients show developmental delay and a variable degree of cognitive impairment (Joubert *et al.* 1969). The key neuroradiological hallmark of Joubert Syndrome is a complex malformation of the hindbrain–midbrain junction characterized by cerebellar vermis hypoplasia, thick and maloriented superior cerebellar peduncles, and abnormally deep interpeduncular fossa. On axial magnetic resonance imaging sections at the pontomesencephalic level, this malformation causes a peculiar appearance resembling a molar tooth, the so-called molar tooth sign (Fig. 9.19) (Maria *et al.* 1987). Joubert syndrome-related disorders display the neurological features of Joubert syndrome associated with multiorgan involvement, mainly retinal dystrophy and nephronophthisis (Satran *et al.* 1999). Joubert syndrome can also be associated with other central nervous system malformations such as corpus callosum abnormalities,

polymicrogyria, hydrocephalus, and encephalomeningocele (Gleeson *et al.* 2004).

9.2.7 Agenesis of the corpus callosum

Agenesis and dysgenesis of the corpus callosum occurs in 1–3/1000 live births. It is usually sporadic but may be transmitted as an X-linked autosomal dominant or autosomal recessive trait (Davilla-Gutierrez 2002). The spectrum of clinical features that appear with agenesis or dysgenesis of the corpus callosum are very variable and relate mostly to associated cerebral or extracerebral malformations. When it presents as an isolated condition the patient may be normal or have only mild clinical features apart from subtle neuropsychological defects (Francesco *et al.* 2006). Agenesis of the corpus callosum occurs in association with a very large number of other neurodevelopmental anomalies (Hetts *et al.* 2006). At least 46 malformation syndromes and metabolic disorders have been reported in patients with complete agenesis or dysgenesis of the corpus callosum, such as lissencephaly, holoprosencephaly, Chiairi II malformation and neurocutaneous syndromes, suggesting that many commissural anomalies arise as part of an overall cerebral dysgenesis. Furthermore abnormalities in at least 18 different chromosomes have been reported in patients with acallosal defects supporting the conclusion that agenesis and dysgenesis of the corpus callosum lie along a dysgenetic spectrum, as opposed to representing distinct disorders (Kamnasaran 2005).

9.2.8 Hydranencephaly, hemihydranencephaly, and multicystic encephalomalacia

These occur as isolated defects unassociated with malformations elsewhere and virtually all cases are sporadic. Their aetiology has been ascribed to different causes all leading to vascular disruption and varying degrees of neuronal and glial necrosis: foetal infections, irradiation, foetal anoxia, medications, twin–twin transfusion, death of a co-twin *in utero*, cord prolapse (Deshmukh *et al.* 1993; Hahn *et al.* 2003). Hydranencephaly is characterized by complete or almost complete absence of cerebral cortex within a relatively normal-sized cranium and with preservation of meninges, basal ganglia, pons, medulla, cerebellum, and falx. Hemihydranencephaly is extremely rare and occurs when the vascular anomaly is unilateral (Greco *et al.* 2001). Multicystic encephalomalacia is where cortical brain tissue is preserved but there are multiple cysts of varying size (Ferrer and Navarro 1978; Lyen *et al.* 1981; Weiss *et al.* 2004; Karageyim Karsidag *et al.* 2005).

Infants with hydranencephaly are presumed to have a reduced life expectancy, with a survival of several weeks to months. Rarely, patients with prolonged survival have been reported, but these infants may have had other neurologic conditions that mimicked hydranencephaly, such as massive hydrocephalus or holoprosencephaly. However prolonged survival up to 19 years can occur with hydranencephaly, even without rostral brain regions, with isoelectric electroencephalograms, and with absent-evoked potentials (McAbee *et al.* 2000).

9.2.9 Craniosynostosis

Craniosynostosis (or craniostenosis) is the early fusion of skull sutures (Fig. 9.20). Sutures allow growth of the skull bone at right angles to their axis. Skull X-rays show narrow, straight, or obliterated sutures and thickened bone, with increased convolutional markings close the suture (Fig. 9.21). The metopic suture closes atenatally and the rest have fibrous union by 6 months and bony union by 8 years.

Early union of the metopic suture causes a narrow, pointed frontal region with a ridge in the line of the suture and is of little clinical significance. Early closure of the sagittal suture is the most common type of craniosynostosis (Hunter and Rudd 1976). It is more common in boys. It causes a long, narrow head which is recognisable at birth. Although not associated with symptomatic raised intracranial pressure, it is usually treated surgically for cosmetic reasons.

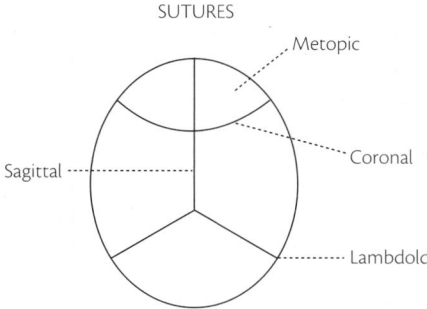

Fig. 9.20 Diagrammatic illustration of the various skull sutures, premature fusion of any of which may lead to craniosynostosis.

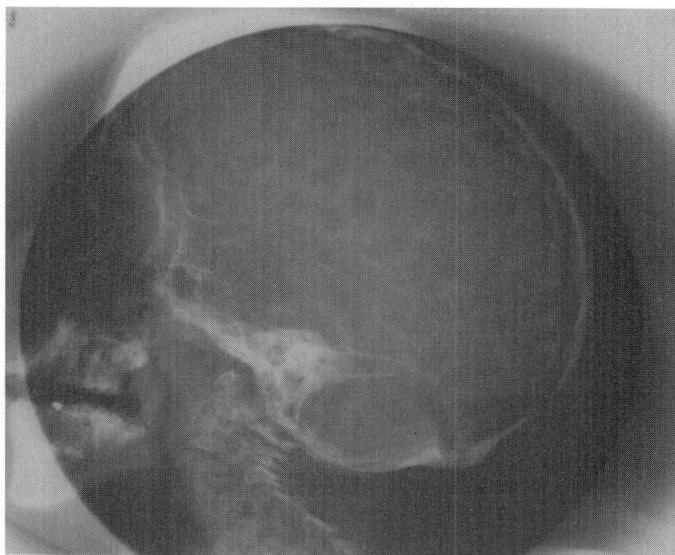

Fig. 9.21 Craniostenosis. Lateral skull X-ray showing copper-beating reflecting the indentation of growing gyri within the skull. A ventriculoperitoneal shunt is *in situ.* (Courtesy of Dr P Anslow.)

Early closure of the coronal suture is more common in girls (Hunter and Rudd 1977). Unilateral closure causes an asymmetric skull deformity, which should be treated surgically for cosmetic reasons but is not usually associated with raised intracranial pressure. Bilateral synostosis may cause raised intracranial pressure and primary or secondary optic atrophy. Although cognitive impairment may occur, this is not purely related to raised pressure. The skull is short and wide. There may be mainly a skull-vault abnormality (Fig. 9.22) or this may be combined with basal suture

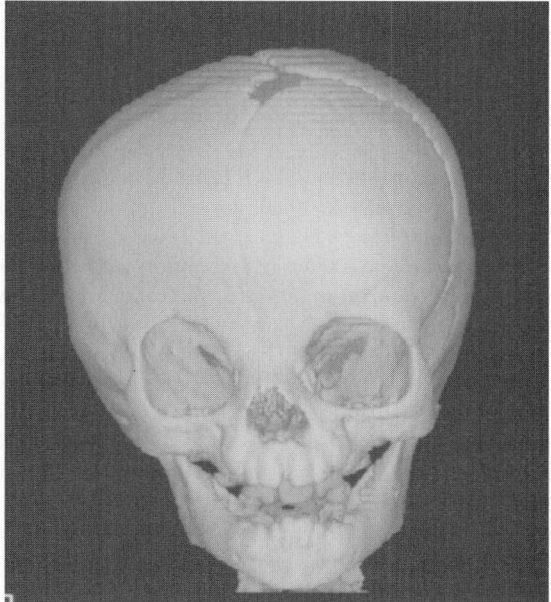

Fig. 9.22 Craniostenosis. CT reconstruction of a skull showing right unicoronal synostosis with a ridge (arrowheads) indicating premature fusion of the right coronal suture, and the early development of a Harlequin eye. (Courtesy of Dr P Anslow.)

synostosis in the coronal plane, in which case a characteristic facial deformity occurs, with widely spaced eyes, shallow orbits with proptosis, and hypoplasia of the maxillae. This is the appearance seen in Crouzon syndrome, one of the many syndromes that include craniosynotosis. Early surgical treatment of the skull vault is used, but for the facial deformities much more extensive reconstruction procedures are now used.

Unilateral lambdoid synotosis is very uncommon and produces a characteristic deformity if not surgically treated. It is important that the very uncommon unilateral coronal and lambdoid synotosis are separated from the common problem of postural plagiocephaly. In this situation the baby has a right or left head turn preference, probably from intrauterine life, and this is associated with flattening of the skull, e.g. left posterior and right anterior flattening for a left preferred head turner, so that the head has developed a predictable postural deformity (Jones *et al.* 1997). In unilateral coronal or lambdoid synotosis, the skull flattening would be on the same side anteriorly and posteriorly. Postural plagiocephaly is more common in relatively inactive babies, in whom a similar deformity of the chest may also occur. Postural plagiocephaly does not require surgical treatment. Total craniosynostotis is rare and causes a small head with very marked convolutional markings on skull X-ray, which distinguish it from primary microcephaly secondary to lack of brain growth in which convolutional markings are diminished or absent. Total craniostenosis requires early surgery. Surgical details of all such procedures are available in specialist texts (Sun and Persing 1999). There is a strong genetic background to craniosynotosis and several discrete syndromes in which multiple additional impairments occur. It is important to recognise that such neurological impairments are not usually caused by raised intracranial pressure but appear to be primary. The gene locations of several of these are known (Aicardi 1998).

9.3 Disorders of CSF flow and the ventricle system

9.3.1 Syringomyelia

Syringomyelia is characterized by the presence of cystic cavities inside the spinal cord, usually in the cervical region but it may involve the whole cord (Section 28.5.15). The underlying cause of all types of syringomyelia is an alteration in physiologic CSF flow dynamics. Although it may occur as an isolated anomaly of the spinal cord, it occurs most often in association with abnormalities of the craniovertebral junction such as Chiari I and II malformations, Klippel–Feil fusion anomaly, and anomalies of the base of the skull. Syringomyelia can also arise from traumatic injury and can be associated with arachnoid cysts and spinal cord tumours (Holly *et al.* 2000; Batzdorf 2005). Conservative treatment is not recommended since the cavity formation is slowly progressive. The destruction of the cord occurs in the area of the anterior white commissure, interrupting crossing fibres of the lateral spinothalamic tracts with resultant progressive loss of temperature and pain sensation. The first step in the surgical treatment is identification of its aetiology to direct the treatment to the underlying cause. Surgical measures that aim to reconstruct the continuity of the subarachnoid space at the site of the block are the first treatment option. Shunting is advocated when reestablishment of the pathways proves impossible or as a second procedure (Di Lorenzo and Cacciola 2005).

9.3.2 Hydrocephalus

Hydrocephalus is a common medical condition that is characterized by abnormalities in the flow or resorption of CSF, resulting in cerebral ventricular dilatation (Section 26.4). CSF is secreted by the choroid plexus and it is also derived from extracellular fluid. CSF flows from the lateral ventricles through the foramen of Munro into the third ventricle. It then traverses the aqueduct of Sylvius into the fourth ventricle, enters the subarachnoid space through the foramina of Luskka and Magendie. CSF is absorbed into the vascular system through arachnoid villi within the aracnoid granulations. Normally rates of secretion and absorption are equal. A blockage to flow of CSF will cause enlargement of the ventricular space rostral to the blockage. Vulnerable sites include the foramen of Munro, the aqueduct of Sylvius, and the foramena of Luschka and Magendie. Common causes are aqueduct stenosis, Arnold–Chiari malformation and tumours. Obstructive hydrocephalus is called non-communicating hydrocephalus. Hydrocephalus also results from impaired absorption of CSF or obstruction within the subarachnoid space. Subarachnoid haemorrhage, meningitis, intrauterine viral infections are common causes. Increased intracranial pressure from sinus venous thrombosis may cause hydrocephalus secondary to decreased CSF absorption. Rarely increased CSF production from the arachnoid plexus papilloma may cause hydrocephalus. All these are classified as communicating hydrocephalus.

Although there is strong evidence for genetic causes (Haverkamp *et al.* 1999), only one hydrocephalus gene, X-linked, has been identified in humans. X-linked hydrocephalus was initially described as different entities—X-linked or recessive congenital hydrocephalus with stenosis of the aqueduct of Sylvius, spastic paraparesis type 1, X-linked agenesis of the corpus callosum, and the MASA syndrome of mental retardation, aphasia, shuffling gait, and adducted thumbs. Molecular genetic studies have revealed that the responsible gene for all these X-linked conditions is at Xq28 encoding for L1CAM (Jouet *et al.* 1993).

Abnormally fast head growth, widely spaced sutures, vomiting, lethargy, and irritability are signs of hydrocephalus in infants. Shunting of CSF fluid to facilitate absorption is usually required for progressive hydrocephalus. These shunts redirect CSF flow into the peritoneal or pleural cavity. Valves, with preset or programmable levels of resistance, reduce over-drainage of CSF and also ensure unidirectional flow from the ventricles. Endoscopic third ventriculoscopy, to create a hole in the floor of the third ventricle, is an alternative to shunts in suitable patients.

9.3.3 Arachnoid cysts

Intracranial arachnoid cysts are benign developmental cysts that occur in the arachnoid membrane and are filled with CSF. The cysts occur most frequently in the Sylvian fissure and can be associated under-development of the underlying anterior superior surface of the temporal lobe. Arachnoid cysts are most commonly asymptomatic, but some may progressively enlarge and cause signs of raised intracranial pressure. Enlarging supracellar cysts can cause endocrine dysfunction, hydrocephalus and optic nerve compression. Enlarging arachnoid cysts clearly require surgical intervention. Relatively minor head trauma may provoke haemorrhage into previously occult arachnoid cysts, with a clinical presentation resembling subdural haematoma (Section 25.2.6). The relationship between arachnoid cysts which are not enlarging and other symptoms including epilepsy remain controversial making decisions difficult about surgical intervention in those with epilepsy but non-enlarging arachnoid cysts (Gandy and Heier 1987; Wang et al. 1998).

Spinal extradural arachnoid cysts are rare. They usually present with progressive signs and symptoms caused by spinal cord compression if they enlarge. Surgical treatment is curative and this rare clinical entity should be considered in the differential diagnosis of spinal extradural lesions (Choi et al. 2006).

9.4 Intracerebral cerebrovascular malformations

9.4.1 Arteriovenous malformation

Arteriovenous malformation is a direct communication between one or more arteries and one or more draining veins without the intervention of a capillary bed. Direct shunting of blood is associated with increased blood flow, distention of the involved arteries with duplication or destruction of the elastica, fibrosis of the media, and focal thinning of the wall. The involved veins also show distention, tortuosity, and secondary changes to their walls. Arteriovenous malformations can vary in size from being very small to giant involving a large part of an entire hemisphere. The vessels may show calcification and spontaneous occlusion. Arteriovenous malformations may occur anywhere in the intracranial spaces, and may appear as an isolated pathology or associated with other diseases (Urgellés et al. 1996).

Apart from the direct effects of the arteriovenous malformation on brain function such as headache, a loud bruit, mass effects, and seizures, the most serious complication of arteriovenous malformation is cerebral haemorrhage, which occurs in about 60 per cent of cases (Arteriovenous Malformation Study Group 1999). Ten per cent of patients with arteriovenous malformations die with the first hemorrhage and another 10 per cent die with the second. Series of long-term follow-up studies report only 20 per cent of patients alive and intact neurologically after 25 years (Troupp et al. 1970). Management of arteriovenous malformation involves surgical resection or embolization of the malformation. Small malformations may be obliterated by radiosurgery.

9.4.2 Venous angiomas

These consist of one or more dilated veins and their tributaries. They can be found in any part of the brain. Very few patients with venous angioma require treatment.

9.4.3 Capillary telangiectases

Capillary telangiectases are composed of dilated capillaries. They are usually small and can occur in any part of the brain or spinal cord. They are of little clinical significance.

9.4.4 Vein of Galen malformations

These are rare congenital malformations that can cause severe morbidity and mortality in neonates and also but less commonly in infants and older children (Johnston et al. 1987). They represent high pressure vascular communications between branches of major cerebral arteries and the vein of Galen. Neonates present in high output cardiac failure, which is frequently fatal (Hoffman et al. 1982; Johnston et al. 1987; Lasjaunias et al. 1989, 1995; Halbach et al. 1998) and obstructive hydrocephalus can occur in later infancy. Neurological symptoms can arise from elevated intracranial pressure and reduced cerebral perfusion which can lead to cerebral ischaemia and infarction. Surgery offers little improvement with fatal outcomes in more than 80 per cent of cases. Endovascular therapy with embolization of the vein of Galen is now the treatment of choice (Gupta et al. 2006; Lasjaunias et al. 2006).

9.5 Cerebral palsy

Cerebral palsy is an umbrella term covering a group of motor impairment syndromes resulting from lesions or anomalies of the brain arising in the early stages of brain development. Degenerative disorders of white and grey matter as well as neoplastic processes and metabolic processes which progressively destroy brain cells are excluded from the definition. Processes affecting only the spinal cord, such as spinal muscular atrophy or myelomeningocele, peripheral nerves, and muscle which result in motor impairment of early onset are also excluded. Although most cases of cerebral palsy are the result of a prenatal or perinatal factor or aetiology, a small number are postnatal in origin. Consensus does not exist regarding the upper age limit for the brain insult for a child to have a diagnosis of cerebral palsy.

It is generally accepted that the incidence of cerebral palsy in developed countries is about 2–2.5/1000 live births. The prevalence of cerebral palsy however depends on how it is defined, the severity threshold, and age of ascertainment. It is well recognized that many abnormal motor signs noted in early infancy resolve over time (Nelson and Ellenberg 1982; Taudorf et al. 1986). Such transient abnormalities do not qualify as cerebral palsy. Many registries therefore have a confirmatory diagnostic age as late as 5 years (SCPE 2000) so that such cases are excluded.

Other neurological disabilities frequently are present in children with cerebral palsy. These include cognitive impairment in 50 per cent, speech and language disorder in 38 per cent, sensory deficits in 60–70 per cent, epilepsy in 20–40 per cent, behaviour problems including attention deficit hyperactivity disorder in 25 per cent, and the severity of these associated disorders varies widely. The frequency of these associated disabilities varies according to the specific type of cerebral palsy and the aetiology of the cerebral palsy (Odding et al. 2006).

Prematurity is the single most important risk factor for cerebral palsy. The birth weight specific prevalence for cerebral palsy ranges from 1.1/1000 neonatal survivors with birth weights greater than 2500 g to 78/1000 neonatal survivors with birth weights less than 1000 g (Colver et al. 2000; Odding et al. 2006). Intrauterine infection, chorioamnionitis, is also an important risk factor. In full-term infants the relative risk following intrauterine infections for cerebral palsy is 4.7 (95 per cent CI 1.3–16.2) and in preterm infants 1.9 (1.4–2.5) (Wu and Colford 2000; Wheather and Rennie 2000; Wu et al. 2003b). Perinatal infection and other risk factors, such as death of a co-twin, placental abruption, and cerebral ischaemia, are thought to trigger a cytokine cascade resulting in damage to the developing brain (O'Shea and Dammann 2000).

There is a clear social class gradient in the prevalence of cerebral palsy as a whole and in the subgroups hemiplegia and diplegia (Dowding and Barry 1990). In the United Kingdom the prevalence of cerebral palsy at birth is 3.33 per 1000 births in the most deprived quintile, and 2.08 in the most affluent quintile. This socio-economic gradient was similar at age five (Dolk *et al.* 2001). There is also an association between birthweight and socio-economic status. Within the normal birthweight category the prevalence of cerebral palsy is 2.42 per 1000 in the most deprived quintile, and 1.29 in the most affluent quintile (Odding *et al.* 2006).

Multiple compared with singleton gestations have a 5–10-fold increased risk of cerebral palsy. The increased risk is partly due to increased incidence of premature birth in multiple pregnancies, however twins born at term also exhibit increased risk (Scher *et al.* 2002; Pharoah 2006). Among twins, the surviving twin of a co-twin that suffered foetal or infant death is at high risk. The increased risk is mainly associated with monochorionic placentation and has been variously ascribed to transfer of thromboplastin or thromboemboli from the dead to the surviving foetus, exsanguination of the surviving foetus into the low pressure reservoir of the dead foetus, or haemodynamic instability with bidirectional shunting of blood between the two foetuses (Pharoah *et al.* 2002). In Europe 5 per cent of infants are the result of *in vitro* fertilization techniques and the latest European data from 2000 showed that 39 per cent of *in vitro* fertilization infants were born as twins (Nyboe Andersen *et al.* 2004; Pinborg *et al.* 2005; Andersen *et al.* 2006). There is an increased risk of cerebral palsy in children after assisted conception, mainly because of the high rate of twins (Strömberg *et al.* 2002; Pinborg *et al.* 2003, 2004). Hence assisted conception is likely to have an impact on the prevalence of cerebral palsy in the developed world.

The classification of cerebral palsy is clinically oriented and based on the observed predominant motor sign: spastic, dyskinetic, ataxic, or mixed. Of the spastic subtypes, there are hemiparetic, quadriparetic, and diplegic variants.

9.5.1 Spastic hemiparesis

Hemiplegic cerebral palsy, with a prevalence of 1 in 1300 live births, is the most frequent form of cerebral palsy (Himmelmann *et al.* 2005). Neuroimaging has led to an appreciation that perinatal stroke is responsible for more than 70 per cent of moderate to severe hemiplegic cerebral palsy (Wu *et al.* 2003a). The left hemisphere is affected in two thirds of subjects for as yet unexplained reasons (Nelson and Lynch 2004). Newborns who suffer stroke are usually otherwise asymptomatic but may present with seizures (Wu *et al.* 2003a). Hemiplegic motor signs evolve slowly over the first 12–18 months, associated sometimes with loss of previously acquired motor skills (Bouza *et al.* 1994). The paucity of signs immediately after the stroke has been attributed to the corticospinal system not assuming a primary role in movement control in the first few months after birth. However there is increasing evidence that this is not so and that the perinatal unilateral lesions initiate changes in the development of the corticospinal system which lead to progressively worsening control of movement of the paretic side (Eyre *et al.* 2007).

Apart from the developing signs of hemiparesis, progressive growth retardation of the paretic limbs also occurs, usually more prominent in the distal arm and hand and distal leg and foot. Growth discrepancy of the paretic leg is important clinically since it can add substantially to the difficulties experienced in walking. Corticosensory impairment of the affected limbs, hemianopia, and hemineglect are common and must be looked for since they too add substantially to the overall impairment. Approximately one-quarter of children with spastic hemiparetic cerebral palsy have associated learning difficulties and about one-third have seizures.

9.5.2 Spastic quadriparesis

Spastic quadriparesis is characterized by four limb involvement and is usually associated with a supranuclear bulbar palsy producing difficulties in swallowing and articulation of speech. The incoordination of oropharyngeal muscles predisposes children to recurrent aspiration pneumonia and upper airways obstruction both when awake and asleep (Wilkinson *et al.* 2006). The children usually have abnormal postural tone, including extensor spasm in response to head movement, which can make the provision of appropriate sitting very challenging at all ages. The motor impairments also lead to other impairments of the musculoskeletal system; for example, children and adolescents with spastic quadriparesis, 75 per cent have hip luxations, 73 per cent contractures, and 72 per cent scoliosis (Odding *et al.* 2006).

Sixty-eight per cent of children with spastic quadriparesis are microcephalic (Edebol-Tysk 1989). The majority have learning impairment (Vargha-Khadem *et al.* 1992). Of those with significant learning difficulties, 47 per cent have severe visual impairments, 94 per cent have epilepsy which is likely to be generalized epilepsy and more than half require two or more anti-epileptic drugs (Edebol-Tysk 1989). Numerous studies have documented failure to thrive in children with spastic quadriparesis (Stallings *et al.* 1993). Non-nutritional factors related to neurological pathophysiology will have an impact on growth (Stevenson *et al.* 1995), however there are persuasive indicators that undernourishment also plays an important role. Signs of undernutrition are common (Stallings *et al.* 1995; Dahl *et al.* 1996), furthermore enteral feeding appears to increase not only the rate of weight gain but also, if started early enough, the rate of linear growth (Sullivan *et al.* 2005). Children with severe spastic quadriparesis have a reduced life span (Strauss *et al.* 1999; Hutton *et al.* 2000). This relates primarily to respiratory difficulties associated with recurrent chest infections and scoliosis leading to progressive lung disease as well as to the complications of epilepsy (Evans and Alberman 1991; Nashef and Brown 1996; Maudsley *et al.* 1999; Reddihough *et al.* 2001).

The ability to sit and to pull themselves to standing by 2 years of age is a strong predictor of whether children with spastic quadriparesis will learn to walk. Among children who were both sitting and pulling themselves to a stand at age 2, 76 per cent walk with or without support by age 6. Among those who achieved full ambulation by age 10, the majority had already done so by age 6 and 96 per cent were at least walking with support by this age (Wu *et al.* 2004).

9.5.3 Spastic diplegia

Spastic diplegia is the most common in children born prematurely (Tang-Wai *et al.* 2006). It is characterized by bilateral lower limb involvement; the upper limbs may be unaffected but most commonly there is some degree of upper limb involvement. The most frequently associated abnormality on imaging is periventricular leukomalacia. Where spastic diplegia occurs in children who are born at term, periventricular leukomalacia is only detected in about half as opposed to 90 per cent of preterm patients

(Koeda *et al.* 1990). In these children it can be associated with various other lesions including schizencephaly and colpocephaly, an abnormal enlargement of the occipital horn of the lateral ventricle (Bozzao *et al.* 2003). There is a lower prevalence of epilepsy, in 20 per cent, and cognitive deficits, 18 per cent have mild learning difficulties, amongst children with spastic diplegia compared to both spastic hemiplegia and spastic quadriplegia.

9.5.4 Extrapyramidal cerebral palsy

Choreoathetotic, dystonic, and ataxic cerebral palsy are usually observed in a mixed picture and rarely occur in isolation (Section 40.4.8). Choreoathetotic and dystonic cerebral palsy are usually seen in association with four limb spasticity. And are most often associated with severe perinatal hypoxic/ischaemic encephalopathy and large lesions in the brain affecting basal ganglia, cortex, and subcortical white matter (Menkes and Curran 1994). Ataxic cerebral palsy is usually associated with spastic diplegia (Section 39.3). It is estimated that approximately 50 per cent of ataxic cerebral palsy is inherited as an autosomal recessive trait (Kvistad *et al.* 1985; McHale *et al.* 2000).

9.6 Neurobehavioural disorders

Neurobiology, genetics, and neuroimaging have led to the appreciation that disorders of behaviour have a significant biological basis. Genetic research in particular has revealed the importance of a genetic predisposition interacting with environmental factors during development for the expression of neurobehaviouraldisorders. Neurophysiological and neuroimaging research indicate that complex behavioural tasks require diffuse networks of neurons and that it is abnormality in the integrated functioning of these networks, rather than isolated lesions which underlies many neurobehavioural disorders (Baird and Santosh 2003).

9.6.1 Learning difficulties and global developmental delay

Learning difficulty is characterized by a significant limitation both in intellectual functioning and in adaptive behaviour, as expressed in conceptual, social, practical, and adaptive skills. As a result of these difficulties many individuals require additional care, education, and medical services. In some cases affected individuals will not achieve independence in adulthood and will need persisting care throughout their lives. Learning difficulties is usually defined as an intelligence quota that is 2 standard deviations below the mean, in other words below 70. The normal intelligence quotient is considered to be above 85; individuals with an intelligence quotient between 71 and 84 are often described as having borderline intellectual functioning. Intelligence quotient between 50 to 70 may represent the lower end of the normal distribution and result in mild to moderate learning difficulties. In contrast severe forms with an intelligence quotient below 50 are usually caused by major illnesses such as perinatal hypoxia or more often by specific genetic factors such as chromosomal aberrations and defects of specific genes. The aetiology is usually easier to establish in individuals with severe learning difficulty, than in those with mild to moderate learning difficulty in whom the underlying cause remains unknown in up to 80 per cent of patients.

It is well recognized that learning difficulties are more common in boys than in girls with ratios varying from 3:1 and 1.9:1. This is mainly explained by the occurrence of specific X-linked conditions in boys. The known causes of learning difficulties are too numerous to be listed here, these include congenital and acquired, social, and environmental factors.

Chromosomal causes

The majority of known causes are due to chromosomal aberrations of which *Down's syndrome*, Trisomy 21 is the most frequent. This occurs on average in 1:650 births. In most cases Down's syndrome occurs as a result of non-dysjunction leading to the presence of a third chromosome 21 in cells. In 8 per cent of cases there is a translocation of a part of chromosome 21. Other trisomy syndromes such as *Edward syndrome* resulting from Trisomy 18, *Patau syndrome* from Trisomy 13, and Trisomy 8 are much less common but still important causes of learning difficulties and are easily recognized by their clinical features. Other chromosomal disorders account for a significant proportion of learning difficulties. These include deletion syndromes such as *Angelman syndrome* from 15q11.2-12, *Prader Willi syndrome* from 15q11-13, and *Cri-du–Chat syndrome* from 5p15.

In recent years, the development and application of various subtelomeric probes for fluorescent *in situ* hybridization and microsatellite markers from these regions, led to the awareness that (sub)telomeric regions are often involved in chromosomal rearrangements not visible by routine cytogenetics. A variety of different rearrangements such as pure truncations, unbalanced translocations, interstitial deletions, and inverted duplications have been detected throughout various screening studies (Rooms *et al.* 2006). The frequency of subtelomeric rearrangements varies considerably from 4 to 35 per cent in reported studies, but the general consensus is that segmental aneusomy of the subtelomeric regions accounts for 6 per cent of patients with learning difficulties or developmental delay and/or non-specific dysmorphic features (Koolen *et al.* 2004).

Following the development of high-resolution new techniques of comparative genomic hybridization arrays and multiplex ligation probe amplification, microdeletions and microduplication can now be detected all over the genome, which might be related with mental retardation (Schoumans *et al.* 2005; Kirchhoff *et al.* 2006; Rauch *et al.* 2006).

Fragile X syndrome

To date more than 60 genes have been identified that cause X-linked mental retardation (Raymond 2006; Ropers 2006). The prevalence of each of these is low except for expansions in *FMR1*, which causes Fragile X syndrome. Systematic screening of all other X-linked genes is currently not feasible in the clinical setting.

Fragile X syndrome is the most common genetic disorder associated with mental retardation. The *FMR1* gene is expressed in high levels in neurons and includes a region near the promoter in which the triplet of bases CGG is repeated multiple times. The exact number of repeats is polymorphic with repeat numbers varying from 5 to 50 in the general population. Subjects with fragile X have more than 200 repeats. Premutations showing no phenotypic effect in fragile X families range in size from 52 to over 200 repeats. All alleles with greater than 52 repeats, including those identified in a normal family, are meiotically unstable and the risk of expansion during oogenesis to the full mutation associated with mental retardation increases with the number of repeats (Verkerk *et al.* 1991). As a result of the pathological expansion of the *FMR1* gene its expression is silenced. Rarely, the syndrome can result from mutations not involving expansion of the CGG repeat region: several patients with the disorder have deletions of *FMR1* or its promoter (Gedeon *et al.* 1992). The absence of its product the protein FMRP leads to cognitive impairment. FMRP protein binds to RNAs within synapses and regulates activity-dependent protein translation. Mouse models of Fragile X show abnormalities in the size and shape of dendritic spines, alterations in synaptic AMPA receptors, and increased metabotropic GluR1-dependent long-term depression (Huber *et al.* 2002; Todd and Malter 2002).

The intelligence quotient scores for affected males with Fragile X syndrome are usually less than 70, The phenotype also includes elongated facies, large prominent ears, macro-orchidism which is present in 90 per cent by age 14 years, and attention deficit or autistic-like behaviour (de Vries *et al.* 1999; van Karnebeek *et al.* 2005). Seizures occur in about 20 per cent of patients who usually have centrotemporal spikes on their EEG, similar to those present in benign focal epilepsy of childhood. Epilepsy resolves during childhood in most people with fragile X syndrome. The centrotemporal spike electroencephalogram pattern seems to be a prognostic factor for remission of epilepsy, as it is in benign focal epilepsy of childhood (Berry-Kravis 2002). Female carriers and males with premutation expansion may present with one or more of the features such as mild cognitive impairment or behavioural problems. Female carriers are at risk for premature ovarian failure, with cessation of periods at age 40 years or younger, or early menopause by age 45 years or younger (Wittenberger *et al.* 2006). There is also a recently recognized neurodegenerative disorder seen in older carriers which consists of intention tremor, gait ataxia, Parkinsonism and autonomic dysfunction, and peripheral neuropathy and cognitive deficits (Jacquemont *et al.* 2007).

9.6.2 Rett syndrome

Rett syndrome, which is a progressive neurodevelopmental disorder that occurs predominantly in females, with an incidence of between 1/10 000 and 1/15 000 live births (Laurvick *et al.* 2006). After apparently normal development until 6–18 months of age, girls with classical Rett syndrome show regression, with deceleration of head growth and loss of speech and acquired motor skills. After a period of pseudo-stabilization and then further deterioration, the condition is mainly characterized by severe learning difficulties, hypotonia, autistic tendency, and abnormalities of fine finger movements and gross movements of the arms. Handwriting, often associated with bruxism and hyperventilation begin and the skill of hand use stabilizes at a 3-month age level. Epilepsy occurs in approximately two-thirds of patients. Atypical forms of Rett syndrome have also been described in females, with heterogeneity for age of onset, severity, and clinical course (Williamson and Christodoulou 2006).

Females with Rett syndrome are usually heterozygous for a *de novo* mutation in *MECP2* (Amir *et al.* 1999) (Table 10.1 XIX.1). Since initially reported, over 2000 mutations have been reported in females with Rett syndrome. Altogether, *MECP2* abnormalities account for more than 95 per cent of sporadic cases of classical Rett syndrome in females (Philippe *et al.* 2006). Attempts to establish genotype–phenotype correlations in females with Rett syndrome initially gave conflicting results, but patterns have recently begun to emerge. Female patients with mutations in *MECP2* that truncate the protein towards its C-terminal end, the late-truncating

mutations, have a phenotype that is less severe, and less typical of classical Rett syndrome, than patients who have missense or N-terminal, or early-truncating, mutations (Charman *et al.* 2005). However, as missense and late-truncating mutations can lead to either classical or atypical Rett syndrome, it has been suggested that genetic background and non-random X-chromosome inactivation in the brain influences the biological consequences of mutations in *MECP2*. Moreover, extreme skewing of X inactivation might account for the existence of rare female carriers with no Rett syndrome symptoms (Couvert *et al.* 2001; Weaving *et al.* 2005).

MECP2 mutations in males were initially thought to be prenatally lethal; however, it has been shown more recently that these mutations occur and cause a broad spectrum of neurodevelopmental disorders which often show a combination of learning difficulties with neurological symptoms (Ravn *et al.* 2003). The proportion of severe learning difficulties in males that is accounted for by mutations in *MECP2* is unclear. In one study, the figure was estimated at 2 per cent (Couvert *et al.* 2001), but another investigation found a frequency of 0.2–0.4 per cent (Yntema *et al.* 2002). Boys with *MECP2* mutations may present with a severe neonatal encephalopathy and die in early childhood because of a central breathing failure. These patients usually carry mutations in *MECP2* that are also found in females with Rett syndrome. Some boys present with symptoms that are similar to classical Rett syndrome in females. These cases result either from somatic mosaicism for mutations in *MECP2* or occur in cases of Klinefelter syndrome in which mutations in *MECP2* are encoded on one X chromosome (Ravn *et al.* 2003). Others have learning difficulties, but the phenotype is heterogeneous: even within the same family, the phenotype ranges from mild to severe (Orrico *et al.* 2000). These patients carry mutations that are inherited from their mothers and have never been found in females with Rett syndrome.

The specific neurodevelopmental effects that are seen in Rett syndrome, together with its onset after a period of apparently normal development, indicate that *MECP2* is not required for the early stages of neurogenesis, neuronal migration, or neuronal maturation. Several studies have analysed the spatio-temporal distribution of *MECP2* in the brains of mice, macaque monkeys (Akbarian 2003), and humans (Shahbazian *et al.* 2002). These data indicate that *MECP2* expression is low or absent in immature neurons and glia then increases during neuronal maturation to reach a high level that is maintained throughout life. Expression studies also indicate that *MECP2* is involved in the differentiation of neuronal cells. Studies of human postmortem tissue have shown less complex dendritic arborizations, smaller neurons and a reduction of dendritic spines of cortical neurons in patients with Rett syndrome which could be due to defects in either the maturation of neurons or the maintenance of their differentiated state (Armstrong *et al.* 1995).

9.6.3 Attention deficit hyperactivity disorder

Attention deficit hyperactivity disorder is a common condition characterized by inattention, impulsivity, and hyperactivity that has no accepted biologic markers. Depending on the definition it affects approximately 6–16 per cent of the population, with the most severe disorder affecting approximately 1 per cent of the population (Barbaresi *et al.* 2004; Dey and Bloom 2005). Diagnosis is clinical and based on observational criteria. Many patients with attention deficit hyperactivity disorder have comorbid conditions including dyspraxia, learning disabilities, obsessive-compulsive problems, depression, anxiety, oppositional defiant, conduct disorders, and tic disorders. The comorbid conditions make diagnosis and treatment more difficult. In addition, comorbid conditions make research more complicated in that studies need to control for the associated conditions. While originally conceived of as a disorder of childhood and adolescence, recent evidence suggest that attention deficit hyperactivity disorder symptoms persist into adult life in up to 40 per cent of childhood cases (Fischer et al. 1993; Biederman et al. 2006). There is now little doubt that attention deficit hyperactivity disorder is a condition in which genetic differences between children make a substantial contribution to the risk of the disorder (Stevenson et al. 2005). Children who have a genetic predisposition will express the disorder when put in a suboptimal environment, typically one characterized by chaotic parenting (Sonuga-Barke et al. 2001; Bor et al. 2002). Children with attention deficit hyperactivity disorder are more likely to have an affected parent. It is possible that such symptoms in parents interfere with effective parenting (Evans et al. 1994; Murphy and RA 1996; Sonuga-Barke et al. 2001).

Epidemiological research indicates that about one-third to one-half of children with epilepsy have significant attention problems or attention deficit hyperactivity disorder, with a lower prevalence of hyperactive-impulsive symptoms than is seen in affected children without epilepsy. Risk factors predicting attention problems include low intelligence quota, other central nervous system dysfunction, subclinical epileptiform activity, and side effects of anticonvulsants (Dunn and Kronenberger 2005).

Structural and functional imaging studies suggest that dysfunction in the fronto-subcortical pathways, as well as imbalances in the dopaminergic and noradrenergic systems, contribute to the pathophysiology of attention deficit hyperactivity disorder, with fMRI studies implicating abnormal functioning of circuits linking prefrontal to the striatum and cerebellum (Frank and Pavlakis 2001). Methylphenidate is well tolerated and has been shown to be effective in the treatment of children and adolescents with attention deficit hyperactivity disorder. Its therapeutic effects are likely to be mediated by increasing endogenous stimulation of α_{2a} adrenoreceptors and dopamine D_1 receptors in the prefrontal cortex (Anderson and Keating 2006; Arnsten 2006). However, an estimated 30–50 per cent either do not respond or do not tolerate treatment with stimulants. The selective noradrenaline reuptake inhibitor atomoxetine may be effective in these children (Wilens 2006).

9.6.4 Autism and autistic spectrum disorders

Autism and autism spectrum disorders, such as Asperger syndrome, are neurodevelopmental conditions diagnosed on the basis of a triad of behavioural impairments: impaired social interaction, impaired communication, and restricted and repetitive interests and activities. Beyond this unifying definition lies extreme clinical heterogeneity, ranging from debilitating impairments to mild personality traits. Hence autism is not a single disease entity, but rather a complex phenotype encompassing either multiple 'autistic disorders' or a continuum of autistic-like traits and behaviours defined as autism spectrum disorder, which includes autistic disorder, childhood disintegrative disorder, pervasive development disorder not otherwise specified 'atypical autism', and Asperger syndrome. The dramatic rise in Autistic spectrum disorder

incidence from 2–5 to 15–60 per 10 000 children during the past two decades can be explained largely by the use of broader diagnostic criteria and increased attention by the medical community (Rutter 2005).

Autism is an entirely behavioural diagnosis, with no population-wide biomarkers. The diagnosis is made by ascertaining whether the child's specific behaviours meet the criteria defined using standardized diagnostic tools. Without biological markers making the diagnosis is challenging since there is increasing evidence that the elements of the autism phenotype might be distributed in the population, with no clear boundary between normal population variation and 'disordered' levels of these behaviours. Overall boys are affected more often than girls with a male:female ratio of 3.8:1 (Williams et al. 2006). The abnormal behaviours commence before 36 months of age and in approximately 25–30 per cent of autistic children developmental regression occurs during the second year of life, involving loss of word use, social withdrawal, loss of eye contact and play interests, and sometimes increased irritability, change of sleep, and eating habits (Kobayashi and Murata 1998; Baron-Cohen et al. 2001; Constantino et al. 2000; Rogers 2004). There is comorbidity with mental retardation in 75 per cent (Rapin 1997), epilepsy in 7–14 per cent (Wong 1993), disruptive behaviours, self-injurious behaviour, hyperactivity, sleep disorders, and tics (Bradley et al. 2004; Fombonne and du Mazaubrun 1992).

Asperger syndrome is a subgroup on the autistic spectrum (Rapin1997; Frith 2004). Children with Asperger syndrome share many of the same features as are seen in autism, but with no history of language delay and have an intellectual ability within the average or above average range. Children with Asperger syndrome have difficulty with social skills and the social use of language.

The aetiology of autism and autistic spectrum disorder is unclear but recent studies provide strong evidence that genes play a major role in conferring susceptibility (Spence 2004; Freitag 2006). Twin studies have reported 60 per cent concordance for autism in monozygotic twins. The occurrence rate in siblings of affected children is approximately 2–8 per cent, and thus much higher than the prevalence rate in the general population (Bailey et al. 1995; Spence 2001; Muhle et al. 2004). To date, genome scans, linkage and association studies, chromosomal rearrangement analyses, and mutation screenings have identified:

◆ genomic regions likely to contain autism susceptibility loci on human chromosomes 1q, 2q, 5q, 6q, 7q, 13q, 15q, 17q, 22q, Xp, and Xq;

◆ genes whose mutations represent a rare cause of 'non-syndromic' autism, *NLGN3* and *NLGN4*, or yield 'syndromic' autism, *FMR1*, *TSC1*, *TSC2*, *NF1*, and *MECP2*;

◆ candidate vulnerability genes, with potential common variants enhancing risk but not causing autism per se (Persico and Bourgeron 2006).

As further supportive evidence for genetically conferred susceptibility, close relatives of children with autism, who themselves do not meet criteria of autism, can have autism-related symptoms such as social and communication deficits and stereotyped behaviours (Dawson et al. 2002). Finally autism can occur as a comorbidity in chromosomal disorders such as in Down's syndrome (Carter et al. 2006).

There is currently no known cure for autism and the mainstay of treatment remains educational teaching/learning approach. There is evidence to suggest that early interventional therapy can improve functioning of autistic children (Bryson *et al.* 2003).

9.6.5 Tourette syndrome and tic disorders

Tics are sudden rapid, purposeless, repetitive stereotyped but non rhythmical movements such as blinking, mouth pouting, mouth opening, or head nodding, sounds such as throat clearing, sniffing or coughing, or vocalizations such as echophenomena which consists of copying what other people say or do, and palilalia in which repetitions of what oneself says are fairly common and very characteristic (Section 40.6.1). Tics can be classified by their anatomical location, number, frequency, intensity, and complexity. The range of motor and phonic tics is huge and virtually any voluntary movement or vocalization can emerge as a tic. Tics can become 'complex' in nature and even appear to be purposeful. In severely affected patients multiple motor tics and vocal or phonic tics can occur simultaneously or in sequence. *Tourette's syndrome* represents the most complex and severe manifestation of the spectrum of tic disorders (Section 40.6.3). Coprolalia was originally described as a pathognomonic symptom by Gilles de la Tourette, but it occurs in only 8–39 per cent of patients with Tourettes, mostly males, and is not required for a diagnosis (Saccomani *et al.* 2005).

Motor and phonic tics occur in bouts and wax and wane in severity over the course of months. The character of tics also changes over time, involving muscles or muscle groups of different parts of the body (Swain and Leckman 2003; Leckman *et al.* 2006a). Anxiety, stress, and fatigue often intensify tics. Tics are usually significantly reduced during sleep or during periods of goal-directed behaviour, especially behaviours that involve both heightened attention and fine motor or vocal control, as occur in musical and athletic performances.

Tics have come to be recognized as a common component of development. Boys are more commonly affected than girls. It has been estimated that as many as one in twenty children has had a tic at some point in the first 10 years of life. Motor tics usually begin between the ages of 3 and 8 years, and if they occur phonic tics begin several years later (Leckman *et al.* 2006b). In uncomplicated cases the long-term outcome is generally favourable; motor and phonic tic severity peaks during early adolescence, with the majority showing a marked reduction in frequency and severity by their twenties.

In addition to tics, patients may also suffer symptoms of attention deficit hyperactivity disorder, oppositional defiant disorders, conduct disorders, and/or obsessive-compulsive disorder. When present these conditions add greatly to the overall morbidity. Furthermore, tics in childhood and early adolescence predict an increased probability of significant obsessive-convulsive symptoms in adulthood (Bloch *et al.* 2006).

Brain imaging, neurophysiological, and postmortem studies support involvement of cortical–striatal–thalamocortical pathways, but the definitive pathophysiological mechanisms or neurotransmitter abnormalities are unknown. Recent evidence, however, suggests a prefrontal dopaminergic abnormality (Singer 2005). Despite evidence that tic disorders may have an inherited susceptibility component, the genes involved have not been identified. Environmental factors are likely also to play important role in the expression of tics and a poststreptococcal autoimmune cause has been proposed, but is unproven.

Tics vary in severity from infrequent and barely noticeable, to nearly continuous and highly disruptive. Treatment of tic disorders depends on the severity of the tics, the distress they cause, and the effects they have on school, work, or daily activities. Many tics do not interfere with school or everyday life and do not require specific treatment, education and reassurance being sufficient. For tics of moderate severity, the alpha$_2$-adrenergic presynaptic agonists clonidine and guanfacine have a reasonable safety profile and are considered the most suitable first-line medications (Qasaymeh and Mink 2006). For more severe tics and in the treatment of Tourette's syndrome potent dopamine D$_2$ antagonists are the most effective in terms of tic reduction but carry the greatest potential for serious side effects. Haloperidol, pimozide, and risperidone are frequently used. However the side effect of tardive dyskinesia, a potentially irreversible neuroleptic-mediated movement disorder characterized by choreoathetoid movements that may be difficult to distinguish from tics, is of particular concern. Most patients with Tourette's syndrome require medication for up to 1–2 years. Finally, for those with coexisting attention deficit hyperactivity disorder effective treatment with stimulant medications does not exacerbate tics in most cases (Kurlan 2003; Qasaymeh and Mink 2006).

References

Aaronson O, Hernanz-Schulman M, Bruner J *et al.* (2003). Myelomeningocele: prenatal evaluation—comparison between transabdominal US and MR imaging. *Radiology*, **227**, 839–43.

Aicardi J, Goutieres F (1981). The syndrome of absence of the septum pellucidum with porencephalies and other developmental defects. *Neuropediatrics*, **12**, 319–29.

Aicardi J (1998). *Diseases of the nervous system in childhood.* MacKeith Press, Oxford.

Akbarian S (2003). The neurobiology of Rett syndrome. *Neuroscientist*, **9**, 57–63.

Allendoerfer K, Shatz C (1994). The subplate, a transient neocortical structure: its role in the development of connections between thalamus and cortex. *Annu Rev Neurosci*, **17**, 185–218.

Altman J, Bayer S (2002). Regional differences in the stratified transitional field and the honeycomb matrix of the developing human cerebral cortex. *J Neurocytol*, **31**, 613–32.

Altman N, MNaidich T, Braffman B (1992). Posterior fossa malformations. *Am J Neuroradiol*, **13**, 691–724.

Amir R, Van den Veyver I, Wan M *et al.* (1999). Rett syndrome is caused by mutations in X-linked MECP2, encoding methyl-CpG-binding protein 2. *Nature Genet*, **23**, 185–8.

Andersen A, Gianaroli L, Felberbaum R *et al.* (2006). The European IVF-monitoring programme (EIM) for the European Society of Human Reproduction and Embryology (ESHRE). *Hum Reprod*, **21**, 1680–97.

Anderson S, Eisenstat D, Shi L *et al.* (1997). Migration of interneuron precursors from the basal forebrain to the neocortex: dependence on Dix-1 and Dix-2. *Science*, **278**, 474–6.

Anderson S, Kaznowski C, Horn C *et al.* (2002). Distinct origins of neocortical projection neurons and interneurons in vivo. *Cereb Cortex*, **12**, 702–9.

Anderson V, Keating G (2006). Methylphenidate controlled-delivery capsules (EquasymXL, Metadate CD): a review of its use in the treatment of children and adolescents with attention-deficit hyperactivity disorder. *Paediatr Drugs*, **8**, 319–33.

Angevine J, Sidman R (1961). Autoradiographic study of cell migration during histogenesis of cerebral cortex in the mouse. *Nature*, **192**, 766–8.

Armstrong D, Dunn JK, Antalffy B *et al.* (1995). Selective dendritic alterations in the cortex of Rett syndrome. *J Neuropathol Exp Neurol*, **54**, 195–201.

Arnsten A (2006). Fundamentals of attention-deficit /hyperactivity disorder: circuits and pathways. *J Clin Psychiatry*, **67**, 7–12.

Arteriovenous Malformation Study Group (1999). Arteriovenous malformation of the brain in adults. *N Engl J Med*, **340**, 1812–18.

Bailey A, Le Couteur A, Gottesman I *et al.* (1995). Autism as a strongly genetic disorder: Evidence from a British twin study. *Psychol Med*, **25**, 63–77.

Baird G, Santosh P (2003). Interface between neurology and psychiatry in childhood. *J Neurol Neurosurg Psychiat*, **74**, i17–22.

Barbaresi W, Katusic S, Colligan R *et al.* (2004). How common is attention-deficit/hyperactivity disorder? Towards resolution of the controversy results from a population-based study. *Acta Paediatr Suppl*, **93**, 55–9.

Barkovich A (1998). Neuroimaging manifestations and classification of congenital muscular dystrophies. *Am J Neuroradiol*, **19**, 1389–96.

Barkovich A, Lindan C (1994). Congenital cytomegalovirus infection of the brain: imaging analysis and embryologic considerations. *AJNR Am J Neuroradiol*, **15**, 703–15.

Barkovich A, Rowley H, Bollen A (1995). Correlation of prenatal events with the development of polymicrogyria. *Am J Neuroradiol*, **16**, 822–7.

Barkovich A, Simon E, Walsh C (2001). Callosal agenesis with cyst: a better understanding and new classification. *Neurology*, **56**, 220–7.

Baron-Cohen S, Wheelwright S, Skinner R *et al.* (2001). The autism-spectrum quotient (AQ): evidence from Asperger syndrome/high-functioning autism, males and females, scientists and mathematicians. *J Autism Dev Disord*, **31**, 5–17.

Barr M, Hansen J, Currey K *et al.* (1983). Holoprosencephaly in infants of diabetic mothers. *J Pediatr*, **102**, 565–8.

Batzdorf U (2005). Primary spinal syringomyelia. Invited submission from the joint section meeting on disorders of the spine and peripheral nerves, March 2005. *J Neurosurg Spine*, **3**, 429–35.

Becker P, Dixon A, Troncoso J (1989). Bilateral opercular polymicrogyria. *Ann Neurol*, **25**, 90–2.

Berry-Kravis E (2002). Epilepsy in fragile X syndrome. *Dev Med Child Neurol*, **44**, 724–8.

Biederman J, Petty C, Fried R *et al.* (2006). Impact of psychometrically defined deficits of executive functioning in adults with attention deficit hyperactivity disorder. *Am J Psychiatry*, **163**, 1730–80.

Biggio J, Owen J, Wenstrom K *et al.* (2001). Can prenatal ultrasound findings predict ambulatory status in fetuses with open spina bifida? *Am J Obstet Gynecol*, **185**, 1016–20.

Bloch M, Peterson B, Scahill L *et al.* (2006). Adulthood outcome of tic and obsessive-compulsive symptom severity in children with Tourette syndrome. *Arch Pediatr Adolesc Med*, **160**, 65–9.

Boltshauser E (2004). Cerebellum-small brain but large confusion: a review of selected cerebellar malformations and disruptions. *Am J Med Genet A*, **126**, 376–85.

Bor W, Sanders M, Markie-Dadds C (2002). The effects of the Triple P-positive parenting program on preschool children with co-occurring disruptive behaviour and attentional/hyperactive difficulties. *J Child Psychol Psychiatry*, **30**, 571–8.

Bouza H, Rutherford M, Acolet D *et al.* (1994). Evolution of early hemiplegic signs in full-term infants with unilateral brain lesions in the neonatal period: a prospective study. *Neuropediatrics*, **25**, 201–7.

Bowman R, McLone D, Grant J *et al.* (2001). Spina bifida outcome: a 25-year prospective. *Pediatr Neurosurg*, **34**, 114–20.

Bozzao A, Di Paolo A, Mazzoleni C *et al.* (2003). Diffusion-weighted MR imaging in the early diagnosis of periventricular leukomalacia. *Eur Rad*, **13**, 1571–6.

Bradley E, Summers J, Wood H *et al.* (2004). Comparing rates of psychiatric and behavior disorders in adolescents and young adults with severe intellectual disability with and without autism. *J Autism Dev Disord*, **34**, 151–61.

Brodmann K (1909). *Vergleichendre Lokalisationslehre der Grosshirnrinde in ihren Prinzipien dargestellt auf Grund des Zellenbaues*. Barth: Leipzig.

Brunelli S, Faiella A, Capra V *et al.* (1996). Germline mutations in the homeobox gene EMX2 in patients with severe schizencephaly. *Nat Genet*, **12**, 94–6.

Bruner J, Tulipan N, Reed G *et al.* (2004). Intrauterine repair of spina bifida: preoperative predictors of shunt-dependent hydrocephalus. *Am J Obstet Gynecol*, **190**, 1305–12.

Bryson S, Rogers S, Fombonne E (2003). Autism spectrum disorders: early detection, intervention, education, and psychopharmacological management. *Can J Psychiatry*, **48**, 506–16.

Cai C, Oakes W (1997). Hindbrain herniation syndromes: the Chiari malformations (I and II). *Semin Pediatr Neurol*, **4**, 209–23.

Carter J, Capone G, Gray R *et al.* (2006). Autistic-spectrum disorders in Down syndrome: Further delineation and distinction from other behavioral abnormalities. *Am J Med Genet B Neuropsychiatr Genet* **[Epub ahead of print]**.

Charman T, Neilson T, Mash V *et al.* (2005). Dimensional phenotypic analysis and functional categorisation of mutations reveal novel genotype-phenotype associations in Rett syndrome. *Eur J Hum Genet*, **13**, 1121–30.

Chatkupt S, Skurnick J, Jaggi M *et al.* (1994). Study of genetics, epidemiology, and vitamin usage in familial spina bifida in the United States in the 1990s. *Neurology*, **44**, 65–9.

Choi J, Kim S, Lee W *et al.* (2006). Spinal extradural arachnoid cyst. *Acta Neurochir (Wien)*, **148**, 579–85.

Clark G (2004). The classification of cortical dysplasia though molecular genetics. *Brain Dev*, **26**, 351–62.

Cochrane D, Wilson R, Steinbok P *et al.* (1966). Prenatal spinal evaluation and functional outcome of patients born with myelomeningocele: information for improved prenatal counselling and outcome prediction. *Fetal Diagn Ther*, **111**, 159–68.

Cohen M, Lemire R (1983). Syndromes with cephaloceles. *Teratology*, **25**, 161–72.

Colver A, Gibson M, Hey E *et al.* (2000). Increasing rates of cerebral palsy across the severity spectrum in north east England 1964-1993. *Arch Dis Child Fetal and Neonat Ed*, **83**, F7–12.

Constantino J, Przybeck T, Friesen D *et al.* (2000) Reciprocal social behavior in children with and without pervasive developmental disorders. *J Dev Behav Pediatr*, **21**, 2–11.

Couvert P, Bienvenu T, Aquaviva C *et al.* (2001). MECP2 is highly mutated in X-linked mental retardation. *Hum Mol Genet*, **10**, 941–6.

Curry C, Lammer E, Nelson V *et al.* (2005). Schizencephaly: heterogeneous etiologies in a population of 4 million California births. *Am J Med Genet A*, **137**, 181.

Cutting L, Cooper K, Koth C *et al.* (2002). Megalencephaly in NF1: predominantly white matter contribution and mitigation by ADHD. *Neurology*, **59**, 1388–94.

Czeizel A, Dudas I (1992). Prevention of the first occurrence of neural-tube defects by periconceptional vitamin supplementation. *N Engl J Med*, **327**, 1832–35.

Dahl M, Thommessen M, Rasmussen M *et al.* (1996). Feeding and nutritional characteristics in children with moderate or severe cerebral palsy. *Acta Paediatr*, **1996**, 697–701.

Davidoff AM, Thompson CV, Grimm JM *et al.* (1991). Occult spinal dysraphism in patients with anal agenesis *J Pediatr Surg*, **26**, 1001–5.

Davila-Gutierrez G (2002). Agenesis and dysgenesis of the corpus callosum. *Sem Ped Neurol*, **9**, 292–301.

Dawson G, Webb S, Schellenberg G *et al.* (2002). Defining the broader phenotype of autism: genetic, brain and behavioral perspectives. *Dev Psychopathol*, **14**, 581–611.

De Carlos J, O'Leary D (1992). Growth and targeting of subplate axons and establishment of major cortical pathways. *J Neurosci*, **12**, 1192–211.

de Vries B, Mohkamsing S, van den Ouweland A *et al.* (1999). Screening for fragile X syndrome among the mentally retarded: a clinical study. The Collaborative Fragile X Study Group. *J Med Genet*, **36**, 467–70.

Denis D, Chateil J, Brun M *et al.* (2000). Schizencephaly: clinical and imaging features in 30 infantile cases. *Brain Dev*, **22**, 475–83.

Dennis JP, Renberg HS, Alvord EC (1961). Megalencephaly, internal hydrocephalus and other neurological aspects of achondroplasia. *Brain*, **84**, 427–45.

DeRosa M, Secor D, Barsom M *et al.* (1972). Neuropathologic findings in surgically treated hemimegalencephaly: immunohistochemical, morphometric and ultrastructural study. *Acta Neuropathol*, **84**, 250–60.

Deshmukh C, Nadkarni U, Nair K *et al.* (1993). Hydranencephaly/ multicystic encephalomalacia: association with congenital rubella infection. *Indian Pediatr*, 1993 Feb;*30(2)*:253-7 **30**, 253–7.

Detrait E, George T, Etchevers H *et al.* (2005). Human neural tube defects: developmental biology, epidemiology, and genetics. *Neurotoxicol Teratol*, **27**, 515–24.

Dey A, Bloom B (2005). Summary health statistics for U.S. children National Health Interview Survey. *Vital Health Stat*, **10**, 1–78.

Di Lorenzo N, Cacciola F (2005). Adult syringomyelia. Classification, pathogenesis and therapeutic approaches. *J Neurosurg Sci*, **49**, 65–72.

Dolk H, Pattenden S, Johnson A (2001). Cerebral palsy, low birthweight and socio-economic deprivation: inequalities in a manor cause of childhood disability. *Paediatr Perinat Epidemiol*, **15**, 359–63.

Dowding V, Barry C (1990). Cerebral palsy: social class differences in prevalence in relation to birthweight and severity of disability. *J Epidemiol Community Health*, **44**, 191–5.

Drainer E, May H, Tolmie J (1991). Do familial neural tube defects breed true? *J Med Genet*, **28**, 605–8.

Drugan A, Weissman A, Evans M (2001). Screening for neural tube defects. *Clin Perinatol*, **28**, 279–87.

Dubeau F, Tampieri D, Lee N *et al.* (1995). Periventricular and subcortical nodular heterotopia A study of 33 patients. *Brain*, **118**, 1273–87.

Dunn D, Kronenberger W (2005). Childhood epilepsy, attention problems, and ADHD: review and practical considerations. *Sem Pediat Neurol*, **12**, 222–8.

Edebol-Tysk K (1989). Epidemiology of spastic tetraplegic cerebral palsy in Sweden. I. Impairments and disabilities. *Neuropediatrics*, **20**, 41–5.

Evans P, Alberman E (1991). Certified cause of death in children and young adults with cerebral palsy. *Arch Dis Child*, **66**, 325–9.

Evans S, Vallano G, Pelham W (1994). Treatment of parenting behaviour with psychostimulant: a case of study of an adult with attention deficit hyperactivity disorder. *J Child Adoles Psychopharm*, **4**, 63–9.

Eyre JA (2005). Developmental aspects of corticospinal projections. In Eisen A, ed. *Motor Neuron Diseases*. Elsevier: Amsterdam.

Eyre J, Smith M, Dabydeen L *et al.* (2007). Is hemiplegic cerebral palsy equivalent to amblyopia of the corticospinal system? *Ann Neurol*, **62**, 493–503.

Ferrer I, Navarro C (1978). Multicystic encephalomalacia of infancy: clinico-pathological report of 7 cases. *J Neurol Sci*, **38**, 179–89.

Fischer M, Barkley R, Flethcher K *et al.* (1993). The adolescent outcome of hyperactive children-predictors of psychiatric, academic, social and emotional adjustment. *J Am Acad Child Adoles Psychiatr*, **32**, 324–32.

Fombonne E, du Mazaubrun C (1992). Prevalence of infantile autism in four French regions. *Soc Psychiatr Psychiatr Epidemiol*, **27**, 203–10.

Fox J, Lamperti E, Eksioglu Y *et al.* (1998). Mutations in filamin 1 prevent migration of cerebral cortical neurons in human periventricular heterotopia. *Neuron*, **21**, 1315–25.

Francesco P, Maria-Edgarda B, Giovanni P *et al.* (2006). Prenatal diagnosis of agenesis of corpus callosum: what is the neurodevelopmental outcome? *Pediatr Int*, **48**, 298–304.

Frank Y, Pavlakis S (2001). Brain imaging in neurobehavioural disorders. *Ped Neurol*, **25**, 278–87.

Freitag C (2006). The genetics of autistic disorders and its clinical relevance: a review of the literature. *Mol Psychiatry* [**Epub ahead of print**].

Frith U (2004). Emanuel Miller lecture: confusions and controversies about Asperger syndrome. *J Child Psychol Psychiatr*, **45**, 672–86.

Gandy S, Heier L (1987). Clinical and magnetic resonance features of primary intracranial arachnoid cysts. *Ann Neurol*, **21**, 342–8.

Gedeon A, Baker E, Robinson H *et al.* (1992). Fragile X syndrome without CCG amplification has an FMR1 deletion. *Nat Genet*, **1**, 341–4.

George T, Wolpert C, Worley G *et al.* (1996). Variable presentation of neural tube defects in three families. *Am J Hum Genet*, **59**, A93.

Ghosh A, Antonini A, McConnell S *et al.* (1990). Requirement for subplate neurons in the formation of thalamocortical connections. *Nature*, **347**, 179–81.

Ghosh A, Shatz C (1992). Involvement of subplate neurons in the formation of ocular dominance columns. *Science*, **255**, 1441–3.

Gleeson J, Keeler L, Parisi M *et al.* (2004). Molar tooth sign of the midbrain-hindbrain junction: occurrence in multiple distinct syndromes. *Am J Med Genet*, **125A**, 125–34.

Granata T, Farina L, Faiella A *et al.* (1997). Familial schizencephaly associated with EMX2 mutation. *Neurology*, **48**, 1403–6.

Gray CLaD (1999). Sensitivity and specificity of ultrasound for the detection of neural tube and ventral wall defects in a high-risk population. *Obstet Gynecol*, **94**, 562–6.

Greco F, Finocchiaro M, Pavone P *et al.* (2001). Hemihydranencephaly: case report and literature review. *J Child Neurol*, **16**, 218–21.

Guerra L, Pike J, Milks J *et al.* (2006). Outcome in patients who underwent tethered cord release for occult spinal dysraphism. *J Urol*, **176**, 1729–32.

Guerrini R, Marini C (2006). Genetic malformations of cortical development. *Exp Brain Res*, **173**, 322–33.

Gupta A, Carreno M, Wyllie E *et al.* (2004). Hemispheric malformations of cortical development. *Neurology*, **62** (**Suppl 3**), S20–6.

Gupta A, Rao V, Varma D *et al.* (2006). Evaluation, management, and long-term follow up of vein of Galen malformations. *J Neurosurg Sci*, **105**, 26–33.

Hagan D, Ross A, Strachan T *et al.* (2000). Mutation analysis and embryonic expression of the HLXB9 Currarino syndrome gene. *Am J Hum Genet*, **66**, 1504–15.

Hager B, Dyme I, Guertin S (1991). Linear nevus sebaceous syndrome: megalencephaly and heterotopic gray matter. *Pediatr Neurol*, **7**, 45–9.

Hahn J, Hahn S, Kammann H *et al.* (2005). Endocrine disorders associated with holoprosencephaly. *J Pediatr Endocrinol Metab*, **18**, 935–41.

Hahn J, Lewis A, Barnes P (2003). Hydranencephaly owing to twin-twin transfusion: serial fetal ultrasonography and magnetic resonance imaging findings. *J Child Neurol*, **18**, 367–70.

Hahn J, Pinter J (2002). Holoprosencephaly: Genetic, neuroradiological and clinical advances. *Semin Ped Neurol*, **9**, 309–19.

Halbach V, Dowd C, Higashida R *et al.* (1998). Endovascular treatment of mural-type vein of Galen malformations. *J Neurosurg Sci*, **89**, 74–80.

Haverkamp F, Wolfle J, Aretz M *et al.* (1999). Congenital hydrocephalus internus and aqueduct stenosis: aetiology and implications for genetic counselling. *Eur J Pediatr*, **158**, 474–8.

Hayward J, Titelbaum D, Clancy R *et al.* (1991). Lissencephaly-pachygyria associated with congenital cytomegalovirus infection. *J Child Neurol*, **6**, 109–14.

Heij H, Nievelstein R, de Zwart I *et al.* (1996). Abnormal anatomy of the lumbosacral region imaged by magnetic resonance in children with anorectal malformations. *Arch Dis Child*, **74**, 441–4.

Herman M, Pizzutillo P (1999). Cervical spine disorders in children. *Orthop Clin North Am*, **30**, 457–66.

Hetts S, Sherr E, Chao S *et al.* (2006). Anomalies of the corpus callosum: an MR analysis of the phenotypic spectrum of associated malformations. *AJR Am J Roentgenol*, **187**, 1343–8.

Himmelmann K, Hagberg G, Beckung E *et al.* (2005). The changing panorama of cerebral palsy in Sweden. IX. Prevalence and origin in the birth-year period 1995-1998. *Acta Paediatr*, **94**, 287–94.

Hoffman H, Chuang S, Hendrick E *et al.* (1982). Aneurysm of the vein of Galen: experience at the hospital for sick children, Toronto. *J Neurosurg Sci*, **57**, 316–22.

Holly L, Johnson J, Masciopinto J et al. (2000). Treatment of posttraumatic syringomyelia with extradural decompressive surgery. Neurosurg Focus, 15, E8.

Huber K, Gallagher S, Warren S et al. (2002). Altered synaptic plasticity in a mouse model of fragile X mental retardation. Proc Natl Acad Sci, 99, 7746–50.

Hunter AGW and Rudd NL (1977). Craniosynotosis. II. Coronal synostosis: its familial characteristics and associated clinical findings in 109 patients lacking bilateral polysyndactyly or synbdactyly. Tyeratology 15, 301–10.

Hutton J, Colver A, Mackie P (2000). Effect of severity of disability on survival in north east England cerebral palsy cohort. Arch Dis Child, 83, 468–74.

Inder T, Huppi S, Zientara G et al. (1999). Early detection of periventricular leukomalacia by diffusion-weighted magnetic resonance imaging techniques. J Pediatr, 134, 631–4.

Jacquemont S, Hagerman R, Hagerman P et al. (2007). Fragile-X syndrome and fragile X-associated tremor/ataxia syndrome: two faces of FMR1. Lancet Neurol, 6, 45–55.

Jansen A, Andermann E (2005). Genetics of the polymicrogyria syndromes J Med Genet, 42, 369–78.

Jelinek J, Bart R, Schiff G (1973). Hypomelanosis of Ito ('incontinentia pigmenti achromians'): report of three cases and review of the literature. Arch Dermatol, 107, 596–601.

Johnston I, Whittle I, Besser M (1987). Vein of Galen malformation: diagnosis and management. Neurosurgery, 20, 747–58.

Jones BM, Hayward R, Evans R and Britto J (1997). Occipital plagiocephaly: an epidemic of craniosynostosis? BMJ, 315, 693–4.

Joubert M, Eisenring J, Robb J et al. (1969). Familial agenesis of the cerebellar vermis. A syndrome of episodic hyperpnea, abnormal eye movements, ataxia, and retardation. Neurology, 19, 813–25.

Jouet M, Rosenthal A, MacFarlane J et al. (1993). A missense mutation confirms the L1 defect in X-linked hydrocephalus (HSAS). Nat Genet, 4, 331–2.

Kamnasaran D (2005). Agenesis of the corpus callosum: lessons from humans and mice. Clin Invest Med, 28, 267–82.

Kanold P, Kara P, Reid R et al. (2003). Role of subplate neurons in functional maturation of visual cortical neurons. Science, 301, 521–5.

Kaplan K, Spivak J, Bendo J (2005). Embryology of the spine and associated congenital abnormalities. Spine J, 5, 564–76.

Karageyim Karsidag A, Kars B, Dansuk R et al. (2005). Brain damage to the survivor within 30 min of co-twin demise in monochorionic twins. Fetal Diagn Ther, 20, 91–5.

Kashani A, Hutchins G (2001). Meningeal-cutaneous relationships in anencephaly: evidence for a primary mesenchymal abnormality. Hum Pathol, 32, 553–8.

Kirchhoff M, Bisgaard A, Bryndorf T et al. (2006). MLPA analysis for a panel of syndromes with mental retardation reveals imbalances in 5.8% of patients with mental retardation and dysmorphic features, including duplications of the Sotos syndrome and Williams-Beuren syndrome regions. Eur J Med Genet [Epub ahead of print].

Kobayashi R, Murata T (1998). Setback phenomenon in autism and long-term prognosis. Acta Psychiatr Scand, 98, 296–303.

Koeda T, Suganuma I, Kohno Y et al. (1990). MR imaging of spastic diplegia. Comparative study between preterm and term infants. Neuroradiology, 32, 187–90.

Koester S, O'Leary D (1992). Functional classes of cortical projection neurons develop dendritic distinctions by class-specific sculpting of an early common pattern. J Neurosci, 12, 1382–92.

Koolen D, Nillesen W, Versteeg M et al. (2004). Screening for subtelomeric rearrangements in 210 patients with unexplained mental retardation using multiplex ligation dependent probe amplification (MLPA). J Med Genet, 41, 892–9.

Kostovic I, Judas M (2002). Correlation between the sequential ingrowth of afferents and transient patterns of cortical lamination in preterm infants. Anat Rec, 267, 1–6.

Kostovic I, Rakic P (1980). Cytology and time of origin of interstitial neurons in the white matter in infant and adult human and monkey telencephalon. J Neurocytol, 9, 219–42.

Kostovic I, Rakic P (1990). Developmental history of the transient subplate zone in the visual and somatosensory cortex of the macaque monkey and human brain. J Comp Neurol, 297, 441–70.

Kothare S, Van Landingham K, Armon C et al. (1998). Seizure onset from periventricular nodular heterotopias: depth-electrode study Neurology, 51, 1723–7.

Kurlan R (2003). Tourette's syndrome: are stimulants safe? Curr Neurol Neurosci Rep, 3, 285–8.

Kvistad P, Dahl A, Skre H (1985). Autosomal recessive non-progressive ataxia with an early childhood debut. Acta Neurol Scand, 71, 295–302.

Lasjaunias P, Chng S, Sachet M et al. (2006). The management of vein of Galen aneurysmal malformations. Neurosurgery, 59, S184–94.

Lasjaunias P, Hui F, Zerah M et al. (1995). Cerebral arteriovenous malformation in children: Management of 179 consecutive cases and review of the literature. Child's Nerv System, 11, 66–79.

Lasjaunias P, Rodesch G, Pruvost P et al. (1989). Treatment of vein of Galen aneurysmal malformation. J Neurosurg Sci, 70, 746–50.

Laurvick C, de Klerk N, Bower C et al. (2006). Rett syndrome in Australia: a review of the epidemiology. J Pediatr, 148, 347–52.

Leckman J, Bloch M, King R et al. (2006a). Phenomenology of tics and natural history of tic disorders. Adv Neurol, 99, 1–16.

Leckman J, Bloch M, Scahill L et al. (2006b). Tourette syndrome: the self under siege. J Child Neurol, 21, 642–9.

Levitt P (2003). Structural and functional maturation of the developing primate brain. J Pediatr, 143 (4 Suppl), S35–45.

Liang J, Lee W, Peng S et al. (2002). Schizencephaly: Correlation between clinical and neuroimaging features. Acta Pediatr, 43, 208–13.

Liu H, Bangaru B, Kidd J et al. (1976). Neuropathological considerations in cerebro-hepato-renal syndrome (Zellweger's syndrome). Acta Neuropathol (Berl), 34, 115–23.

Loeken M (2005). Current perspectives on the causes of neural tube defects resulting from diabetic pregnancy. Am J Med Genet C Semin Med Genet, 135C, 77–87.

Luskin M, Shatz C (1985). Studies of the earliest generated cells of the cat's visual cortex: cogeneration of subplate and marginal zones. J Neurosci, 5, 1062–75.

Lyen K, Lingam S, Butterfill A et al. (1981). Multicystic encephalomalacia due to fetal viral encephalitis. Eur J Pediatr, 137, 11–16.

Lynch S, Wang Y, Strachan T et al. (2000). Autosomal dominant sacral agenesis: Currarino syndrome. J Med Genet, 37, 561–6.

Manaligod J, Bauman N, Menezes A et al. (1999). Cervical vertebral anomalies in patients with anomalies of the head and neck. Ann Otol Rhinol Laryngol, 108, 925–33.

Mangels K, Tulipan N, Tsao L et al. (2000). Fetal MRI in the evaluation of intrauterine myelomeningocele. Pediatr Neurosurg, 32, 124–31.

Maria B, Zinreich S, Carson B et al. (1987). Dandy-Walker syndrome revisited. Pediatr Neurosci, 13, 38–44.

Marin-Padilla M (1971). Early prenatal ontogenesis of the cerebral cortex (neocortex) of the cat (Felis domesticia). A Golgi study. I. The primordial neocortical organisation. Z Anat Entwicklingsgesch, 136, 125–42.

Marin-Padilla M (1972a). Prenatal ontogenesis history of teh principal neurons of the neocortex of the cat (Felis domestica). A Golgi study II. Developmental differences and their significances. Z Anat Entwicklungsgesch, 136, 125–42.

Marin-Padilla M (1972b). Prenatal ontogenesis history of the principal neurons of the neocortex of the cat (Felis domestica). A Golgi study. II. Developmental differences and their significances. Z Anat Entwicklungsgesch, 136, 125–42.

Matalon S, Schechtman S, Goldzweig G et al. (2002). The teratogenic effect of carbamazepine: a meta-analysis of 1255 exposures. Reprod Toxicol, 5, 33–9.

Maudsley G, Hutton J, Pharoah P (1999). Cause of death in cerebral palsy: a descriptive study. *Arch Dis Child*, **81**, 390–4.

McAbee G, Chan A, Erde E (2000). Prolonged survival with hydranencephaly: report of two patients and literature review. *Pediatr Neurol*, **23**, 80–4.

McConnell S (1995). Constructing the cerebral cortex: neurogenesis and fate determination. *Neuron*, **15**, 761–8.

McConnell S, Ghosh A, Shatz C (1989). Subplate neurons pioneer the first axon pathway from the cerebral cortex. *Science*, **245**, 978–82.

McDonnell G, McCann J (2000). Why do adults with spina bifida and hydrocephalus die? A clinic-based study. *Eur J Pediatr Surg*, **10 (Suppl 1)**, 31–2.

McHale D, Jackson A, Campbell *et al.* (2000). A gene for ataxic cerebral palsy maps to chromosome 9p12-q12. *Eur J Hum Genet*, **8**, 267–72.

McMaster M, Ohtsuka K (1982). The natural history of congenital scolosis: a study of two hundred and fifty-one patients. *J Bone Joint Surg AM*, **64**, 1128–47.

McMaster M, Singh H (1999). The natural history of congenitalkyphosis and kyphoscoliosis: a study of one hundred and twelve patients. *J Bone Joint Surg AM*, **81**, 1367–83.

Meadows J, Kraut M, Guarnieri M *et al.* (2000). Asymptomatic Chiari type I malformation identified on magnetic resonance imaging. *J Neurosurg Sci*, **92**, 920–6.

Menkes J, Curran J (1994). Clinical and MR correlates in children with extrapyramidal cerebral palsy. *AJNR Am J Neuroradiol*, **15**, 451–7.

Meuli M, Meuli-Simmen C, Hutchins G *et al.* (1995). In utero surgery rescues neurological function at birth in sheep with spina bifida. *Nat Med*, **1**, 342–7.

Meyer G, Schaaps J, Moreau L *et al.* (2000). Embryonic and early fetal development of the human neocortex. *J Neurosci*, **20**, 1858–68.

Mikulis

Milunsky A, Jick H, Jick S *et al.* (1991). Multivitamin/folic acid supplementation in early pregnancy reduces the prevalence of neural tube defects. *JAMA*, **262**, 2847–52.

Mitchell L, Adzick N, Melchionne J *et al.* (2004). Spina bifida. *Lancet*, **364**, 1885–95.

Moro F, Carrozzo R, Veggiotti P *et al.* (2002). Familial periventricular heteropia: missense and distal truncating mutations of the FLN1 gene. *Neurology*, **58**, 916–21.

MRC (1991). Prevention of neural tube defects: results of the Medical Research Council Vitamin Study. MRC Vitamin Study Research Group. *Lancet*, **338**, 131–7.

Mrzljak L, Uylings H, Kostovic I *et al.* (1988). Prenatal development of neurons in the human prefrontal cortex: I. A qualitative Golgi study. *J Comp Neurol*, **271**, 355–86.

Mrzljak l, Uylings H, Kostovic I *et al.* (1992). Prenatal development of neurons in the human prefrontal cortex. *J Comp Neurol*, **22**, 485–96.

Muenke M (1989). Clinical, cytogenetic, and molecular approaches to the genetic heterogeneity of holoprosencephaly. *Am J Med Genet A*, **34**, 237–45.

Muhle R, Trentacoste S, Rapin I (2004). The genetics of autism. *Pediatrics*, **113**, e472–86.

Murphy KR, RA B (1996). Parents of children with attention-deficit/hyperactivity disorder: psychological and attentional impairment. *Am J Orthopsychiatry*, **66**, 93–102.

Nashef L, Brown S (1996). Epilepsy and sudden death. *Lancet*, **348**, 1324–5.

Nelen M, Kremer H, Konings I *et al.* (1999). Novel PTEN mutations in patients with Cowden disease: absence of clear genotype-phenotype correlations. *Eur J Hum Genet*, **7**, 267–73.

Nelson K, Ellenberg J (1982). Children who "outgrew" cerebral palsy. *Pediatrics*, **69**, 529–36.

Nelson K, Lynch J (2004). Stroke in newborn infants. *Lancet Neurol*, **3**, 150–8.

Niesen C (2002). Malformations of the posterior fossa: current perspectives. *Semin Pediatr Neurol*, **9**, 320–34.

Notaridis G, Ebbing K, Giannakopoulos P *et al.* (2006). Neuropathological analysis of an asymptomatic adult case with Dandy-Walker variant. *Neuropathol Appl Neurobiol*, **32**, 344–50.

Nyboe Andersen A, Gianaroli L, Nygren K (2004). Assisted reproductive technology in Europe, 2000. Results generated from European registers by ESHRE. *Hum Reprod*, **19**, 490–503.

O'Leary D, Nakagawa Y (2002). Patterning centers, regulatory genes and extrinsic mechanisms controlling arealization of the neocortex. *Curr Opin Neurobiol*, **12**, 14–25.

O'Leary D, Borngasser D, Fox K *et al.* (1995). Plasticity in the development of neocortical areas. *Ciba Found Symp*, **193**, 214–30.

O'Leary D, Schlaggar B, Stanfield B (1992). The specification of sensory cortex: lessons from cortical transplantation. *Exp Neurol*, **115**, 121–6.

O'Rahilly R, Muller F (2002). The two sites of fusion of the neural folds and the two neuropores in the human embryo. *Teratology*, **65**, 162–70.

O'Shea T, Dammann O (2000). Antecedents of cerebral palsy in very low-birthweight infants. *Clin Perinatol*, **27**, 285–302.

Odding E, Roebroeck M, Stam H (2006). The epidemiology of cerebral palsy: incidence, impairments and risk factors. *Disabil Rehabil*, **28**, 183–91.

Orrico A, Lam C, Galli L *et al.* (2000). MECP2 mutation in male patients with non-specific X-linked mental retardation. *FEBS Lett*, **481**, 285–8.

Osaka K, Tanimura T, Matsumoto S (1978). Myelomeningocele before birth. *Neurosurgery*, **49**, 711–24.

Paek B, Farmer D, Wilkinson C *et al.* (2000). Hindbrain herniation develops in surgically created myelomeningocele but is absent after repair in fetal lambs. *Am J Public Health*, **183**, 1119–23.

Pasini A, Wilkinson D (2002). Stabilizing the regionalisation of the developing vertebrate central nervous system. *Bioessays*, **24**, 427–38.

Penn AA, Shatz CJ (1999). Brain waves and brain wiring: the role of endogenous and sensory-driven neural activity in developement. *Pediatr Res*, **45**, 447–58.

Persico A, Bourgeron T (2006). Searching for ways out of the autism maze: genetic, epigenetic and environmental clues. *Trends Neurosci*, **29**, 349–58.

Pharoah P (2006). Risk of cerebral palsy in multiple pregnancies. *Clin Perinatol*, **33**, 301–13.

Pharoah P, Price T, Plomin R (2002). Cerebral palsy in twins: a national study. *Arch Dis Child Fetal Neonatal Ed*, **87**, F122–4.

Philippe C, Villard L, De Roux N *et al.* (2006). Spectrum and distribution of MECP2 mutations in 424 Rett syndrome patients: a molecular update. *Eur J Med Genet*, **49**, 9–18.

Piao X, Hill R, Bodell A *et al.* (2004). G protein-coupled receptor-dependent development of human frontal cortex. *Science*, **303**, 2033–6.

Pinborg A, Loft A, Schmidt L *et al.* (2004). Neurological sequelae in twins born after assisted conception: controlled national cohort study. *BMJ*, **329**, 311.

Pinborg A, Loft A, Schmidt L *et al.* (2005). Neurological late sequelae in twins born after in vitro fertilisation—secondary publication. A national cohort study. *Ugeskr Laeger*, **167**, 3051–4.

Pinborg A, Loft A, Schmidt L *et al.* (2003). Morbidity in a Danish national cohort of 472 IVF/ICSI twins, 1132 non-IVF/ICSI twins and 634 IVF/ICSI singletons: health-related and social implications for the children and their families. *Hum Reprod*, **18**, 1234–43.

Polizzi A, Pavone P, Iannetti P *et al.* (2006). Septo-optic dysplasia complex: a heterogeneous malformation syndrome. *Pediatr Neurol*, **34**, 66–71.

Proctor M, Scott R (2001). Long-term outcome for patients with split cord malformation. *Neurosurg Focus*, **10**, e5.

Qasaymeh M, Mink J (2006). New treatments for tic disorders. *Curr Treat Options Neurol*, **8**, 465–73.

Rakic P (1972). Mode of cell migration to the superficial layers of fetal monkey neocortex. *J Comp Neurol*, **145**, 61–84.

Rakic P (1974). Neurons in rhesus monkey visual cortex: systematic relation between time of origin and eventual disposition. *Science*, **183**, 425–7.

Rapin I (1997). Autism. *N Engl J Med*, **337**, 97–104.

Rauch A, Hoyer J, Guth S et al. (2006). Diagnostic yield of various genetic approaches in patients with unexplained developmental delay or mental retardation. *Am J Med Genet A*, **140**, 2063–74.

Ravn K, Nielsen J, Uldall P et al. (2003). No correlation between phenotype and genotype in boys with a truncating MECP2 mutation. *J Med Genet*, **40**, e5.

Raymond F (2006). X Linked mental retardation: a clinical guide. *J Med Genet*, **43**, 193–200.

Reddihough D, Baikie G, Walstab J (2001). Cerebral palsy in Victoria, Australia: mortality and causes of death. *J Paediatr Child Health*, **37**, 183–6.

Rickmann M, Chronwall B, Wolff J (1977). On the development of non-pyramidal neurons and axons outside the cortical plate: the early marginal zone as a pallial anlage. *Anat Embryol*, **151**, 285–307.

Rickmann M, Wolff J (1981). Differentiation of "preplate" neurons in the pallium of the rat. *Bibl Anat*, **19**, 142–6.

Riley H, Smith W (1960). Macrocephaly, pseuudopapilloedema and multiple haemangiomata *Pediatrics*, **26**, 293.

Rintoul N, Sutton L, Hubbard A et al. (2002). A new look at myelomeningceles: functional level, vertebral level, shunting and the implications for fetal intervention. *Pediatrics*, **109**, 409–13.

Rodriguez L, Cuadrado Perez I, Herrera Montes J et al. (2002). Terminal deletion of the chromosome 7(q36-qter) in an infant with sacral agenesis and anterior myelomingocele. *Am J Med Genet*, **110**, 73–7.

Roessler E, Belloni E, Gaudenz K et al. (1996). Mutations in the human Sonic hedgehog gene cause holoprosencephaly. *Nat Genet*, **14**, 357–60.

Rogers S (2004). Developmental regression in autism spectrum disorders. *Ment Retard Dev Disabil Res Rev*, **10**, 139–43.

Rooms L, Reyniers E, Kooy R (2006). Diverse chromosome breakage mechanisms underlie subtelomeric rearrangements, a common cause of mental retardation. *Hum Mutat* [**Epub ahead of print**].

Ropers H (2006). X linked mental retardation: many genes for a complex disorder. *Curr Opin Genet Develop*, **16**, 260–9.

Rossi A, Biancheri R, Cama A et al. (2004). Imaging in spine and spinal cord malformations. *Eur J Radiol*, **50**, 177–200.

Rutter M (2005). Incidence of autism spectrum disorders: changes over time and their meaning. *Acta Paediatr*, **94**, 2–15.

Saccomani L, Fabiana V, Manuela B et al. (2005). Tourette syndrome and chronic tics in a sample of children and adolescents. *Brain Dev*, **27**, 349–52.

Satran D, Pierpont M, Dobyns W (1999). Cerebello-oculo-renal syndromes including Arima, Senior-Loken and COACH syndromes: more than just variants of Joubert syndrome. *Am J Med Genet A*, **86**, 459–69.

Sattar M, Bannister C, Turnbull I (1996). Occult spinal dysraphism—the common combination of lesions and the clinical manifestations in 50 patients. *Eur J Pediatr Surg*, **6**, 10–14.

Scher A, Petterson B, Blair E et al. (2002). The risk of mortality or cerebral palsy in twins: a collaborative population-based study. *Pediatr Res*, **52**, 671–81.

Schijman E (2003). Split spinal cord malformations: report of 22 cases and review of the literature. *Childs Nerv System*, **19**, 96–103.

Schoumans J, Ruivenkamp C, Holmberg E et al. (2005). Detection of chromosomal imbalances in children with idiopathic mental retardation by array based comparative genomic hybridisation (array-CGH). *J Med Genet*, **42**, 699–705.

SCPE SocpiE (2000). Surveillance of cerebral palsy in Europe (SCPE) A collaboration of cerebral palsy surveys and registers. *Dev Med Child Neurol*, **42**, 816–24.

Sepulveda W, Corral E, Ayala C et al. (2004). Chromosomal abnormalities in fetuses with open neural tube defects: prenatal identification with ultrasound. *Ultrasound Obstet Gynecol*, **23**, 352–6.

Shahbazian M, Antalffy B, Armstrong D et al. (2002). Insight into Rett syndrome: MeCP2 levels display tissue- and cell-specific differences and correlate with neu neuronal maturation. *Hum Mol Genet*, **11**, 115–24.

Shatz C, Ghosh A, McConnell S et al. (1991). Pioneer neurons and target selection in cerebral cortical development. *Cold Spring Harb Symp Quant Biol*, **55**, 469–80.

Shaw G, Velie E, Schaffer D (1996). Risk of neural tube defect-affected pregnancies among obese women. *JAMA*, **275**, 1093–6.

Sheen V, Ganesh V, Topcu M et al. (2004). Mutations in ARFGEF2 implicate trafficking in neuronal progenitor proliferation and migration in the human cerebral cortex. *Nat Genet*, **36**, 69–76.

Simonati A, Tosati C, Rosso T et al. (1999). Cell proliferation and death: Morphological evidence during corticogenesis in the developing human brain. *Microscopy Res Tech*, **45**, 341–52.

Singer H (2005). Tourette's syndrome: from behaviour to biology. *Lancet Neurol*, **4**, 149–59.

Smith B, Griffin C (1992). Klippel-Feil syndrome. *Ann Emerg Med*, **21**, 876–9.

Sonuga-Barke E, Daley D, Thompson M et al. (2001). Parent based therapies for preschool attention deficit/hyperactivity disorder: a randomized controlled trial with a community sample. *J Am Acad Child Adoles Psychiatr*, **40**, 402–8.

Sotos J (1997). Overgrowth. Section V. Syndromes and other disorders associated with overgrowth. *Clin Pediatr (Phila)*, **36**, 89–103.

Spence M (2001). The genetics of autism. *Curr Opin Pediatr*, **13**, 561–5.

Spence S (2004). The genetics of autism. *Semin Pediatr Neurol*, **11**, 196–204.

Stallings V, Charney E, Davies J et al. (1993). Nutrition-related growth failure of children with quadriplegic cerebral palsy. *Dev Med Child Neurol*, **35**, 126–38.

Stallings V, Cronk C, Zemel B et al. (1995). Body composition in children with spastic quadriplegic cerebral palsy. *J Pediatr*, **126**, 833–9.

Stevenson J, Asherson P, Hay D et al. (2005). Characterizing the ADHD phenotype for genetic studies. *Dev Sci*, **8**, 115–21.

Stevenson K (2004). Chiari Type II malformation: past, present, and future. *Neurosurg Focus*, **16**, E5.

Stevenson R, Roberts C, Vogtle L (1995). The effects of non-nutritional factors on growth in cerebral palsy. *Dev Med Child Neurol*, **37**, 124–30.

Stewart G, Pearlman A (1987). Fibronectin-like immunoreactivity in the developing cerebral cortex. *J Neurosci*, **7**, 3325–33.

Strauss D, Cable W, Shavelle R (1999). Causes of excess mortality in cerebral palsy. *Dev Med Child Neurol*, **41**, 580–5.

Strömberg B, Dahlquist G, Ericson A et al. (2002). Neurological sequelae in children born after in-vitro fertilisation: a population based study. *Lancet*, **359**, 461–5.

Sullivan P, Juszczak E, Bachlet A et al. (2005). Gastrostomy tube feeding in children with cerebral palsy: a prospective, longitudinal study. *Dev Med Child Neurol*, **47**, 77–85.

Sun PP and Persing JA (1999). Craniosynotosis. In *Principles and practice of pediatric neurosurgery* (ed. AL Albright, IF Pollack and PD Adelson), pp. 219–42. Thième, New York.

Sutton L, Adzick N, Bilaniuk L et al. (1999). Improvement in hindbrain herniation demonstrated by serial fetal magnetic resonance imaging following fetal surgery for myelomeningocele *JAMA*, **282**, 1826–31.

Swain J, Leckman J (2003). Tourette's syndrome in children. *Curr Treat Options Neurol*, **5**, 299–308.

Tang-Wai R, Webster R, Shevell M (2006). A clinical and etiologic profile of spastic diplegia. *Pediatr Neurol*, **34**, 212–18.

Tatton-Brown K, Douglas J, Coleman K et al. (2005). Multiple mechanisms are implicated in the generation of 5q35 microdeletions in Sotos syndrome. *J Med Genet*, **42**, 307–13.

Taudorf K, Hansen F, Melchior J et al. (1986). Spontaneus remission of cerebral palsy. *Neuropediatrics*, **17**, 19–22.

Thomsen M, Schneider U, Weber M et al. (1997). Scoliosis and congenital anomalies associated with Klippel-Feil syndrome types I-III. *Spine*, **22**, 396–401.

Todd P, Malter J (2002). Fragile X mental retardation protein in plasticity and disease. *J Neurosci*, **70**, 623–30.

Tracy M, Dorman J, Kusumi K (2004). Klippel-Feil syndrome: clinical features and current understanding of etiology. *Clin Orthop Relat Res*, **424**, 183–90.

Troupp H, Marttila I, Holonen V (1970). Arteriovenous malformations of the brain: Prognosis without operation. *Acta Neurochir (Wien)*, **22**, 125–8.

Tsakayannis D, Shamberger R (1995). Association of imperforate anus with occult spinal dysraphism. *J Pediatr Surg*, **30**, 1010–12.

Urgellés E, Pascual-Castroviejo I, Roche C *et al.* (1996). Arteriovenous malformation in hypomelanosis of Ito. *Brain Dev*, **18**, 78–80.

van Karnebeek C, Jansweijer M, Leenders A *et al.* (2005). Diagnositic investigations in individuals with mental retardation: a systematic literature review of their usefulness. *Eur J Hum Genet*, **13**, 6–25.

Vargha-Khadem F, Isaacs E, vander Werf S *et al.* (1992). Development of intelligence and memory in children with hemiplegic cerebral palsy. *Brain*, **111**, 315–29.

Verkerk A, Pieretti M, Sutcliffe J *et al.* (1991). Identification of a gene (FMR-1) containing a CGG repeat coincident with a breakpoint cluster region exhibiting length variation in fragile X syndrome. *Cell*, **65**, 905–14.

Wang P, Lin H, Liu H *et al.* (1998). Intracranial arachnoid cysts in children: related signs and associated anomalies. *Pediatr Neurol*, **19**, 100–4.

Weaver D, Graham C, Thomas I *et al.* (1974). A new overgrowth syndrome with accelerated skeletal maturation, unusual facies, and camptodactyly. *J Pediatr*, **84**, 547–52.

Weaving L, Ellaway C, Gecz J *et al.* (2005). Rett syndrome: clinical review and genetic update. *J Med Genet*, **42**, 1–7.

Weiss J, Cleary-Goldman J, Tanji K *et al.* (2004). Multicystic encephalomalacia after first-trimester intrauterine fetal death in monochorionic twins. *Am J Obstet Gynecol*, **190**, 563–5.

Wheather M, Rennie J (2000). Perinatal infection is an important risk factor for cerebral palsy in very-low-birthweight infants. *Dev Med Child Neurol*, **2000**, 364–7.

Wiedemann H, Burgio G, Aldenhoff P *et al.* (1983). The proteus syndrome. Partial gigantism of the hands and/or feet, nevi, hemihypertrophy, subcutaneous tumors, macrocephaly or other skull anomalies and possible accelerated growth and visceral affections. *Eur J Pediatr*, **140**, 5–12.

Wilens T (2006). Mechanism of action of agents used in attention-deficit/hyperactivity disorder. *J Clin Psychiatry*, **67**, 32–8.

Wilkinson D, Baikie G, Berkowitz R *et al.* (2006). Awake upper airway obstruction in children with spastic quadriplegic cerebral palsy. *J Paediatr Child Health*, **42**, 44–8.

Williams J, Higgins J, Brayne C (2006). Systematic review of prevalence studies of autism spectrum disorders. *Arch Dis Child*, **91**, 8–15.

Williamson S, Christodoulou J (2006). Rett syndrome: new clinical and molecular insights. *Eur J Hum Genet*, **14**, 896–903.

Wittenberger M, Hagerman R, Sherman S *et al.* (2006). The FMR1 premutation and reproduction. *Fertil Steril* [**Epub ahead of print**].

Wong L, Paulozzi L (2001). Survival of infants with spina bifida: a population study, 1979–1994. *Paediatr Perinat Epidemiol*, **15**, 374–8.

Wong V (1993). Epilepsy in children with autistic spectrum disorder. *J Child Neurol*, **8**, 316–22.

Woods C (2004). Human microcephaly. *Current Opinion in Neurobiol*, **14**.

Wu Y, Colford JJ (2000). Chorioamnionitis as a risk factor for cerebral palsy. *J Am Med Assoc*, **284**, 1471–24.

Wu Y, Day S, Strauss D *et al.* (2004). Prognosis for ambulation in cerebral palsy: a population-based study. *Pediatrics*, **114**, 1264–71.

Wu Y, Escobar G, Grether J *et al.* (2003a). Chorioamnionitis and cerebral palsy in term and near-term infants. *JAMA*, **290**, 2677–84.

Wu Y, Escobar G, Grether J *et al.* (2003b). Chorioamnionitis and cerebral palsy in tem and near-term infants. *J Am Med Assoc*, **290**, 2677–84.

Yakovlev P, Wadsworth R (1946). Schizencephalies: a study of the congenital clefts in the cerebral mantle. Clefts with hydrocephalus and lips separated. *J Neuropathol Exp Neurol*, **5**, 169–203.

Yamaguchi K, Honma K (2001). Autopsy case of thanatophoric dysplasia: observations on the serial sections of the brain. *Neuropathology*, **21**, 222–8.

Yntema H, Kleefstra T, Oudakker A *et al.* (2002). Low frequency of MECP2 mutations in mentally retarded males. *Eur J Hum Genet*, **10**, 487–90.

Yumer M, Nachev S, Dzhendov T *et al.* (2006). Chiari type II malformation: a case report and review of literature. *Folia Med (Plovdiv)*, **48**, 55–9.

Zecevic N, Milosevic A, Rakic S *et al.* (1999). Early development and composition of the human primordial plexiform layer: an immunohistochemical study. *J Comp Neurol*, **412**, 241–54.

CHAPTER 10

Neurometabolic disorders

Tony McShane, Peter Clayton,
Michael Donaghy, and Robert Surtees

Contents

10.1 Introduction

10.1.1 Introduction to neurometabolic disease

Various disorders result from genetically determined abnormalities of enzymes, the metabolic consequences of which affect the development or functioning of the nervous system. The range of metabolic disturbances is wide, as is the resultant range of clinical syndromes. Although most occur in children, some can present in adult life, and increasing numbers of affected children survive into adult life. In some, specific treatments are possible or are being developed. The last 20 years has seen a considerable expansion in our understanding of the genetic and metabolic basis for many neurological conditions. Particular clinical presentations of neurometabolic disorders include ataxias (Chapter 39), movement disorders (Chapter 40), childhood epilepsies (Chapter 30), or peripheral neuropathy (Chapter 21). Detailed coverage of the entire range of inherited metabolic diseases of the nervous system is available in other texts (Brett 1997; Scriver *et al.* 2001; Menkes *et al.* 2005). The range of neurometabolic disorders is summarized in Tables 10.1 and 10.2.

Treatment is possible for some metabolic diseases. For instance, the devastating neurological effects of phenylketonuria (Table 10.1, I 1a) have been recognized for many years. Neonatal screening for this disorder and dietary modification in the developed world has removed phenylketonuria from the list of important causes of serious neurological disability in children. This success has lead to new challenges in the management of the adult with phenylketonuria and unexpected and devastating effect of the disorder on the unborn child of an untreated Phenylketonuria mother. More recently Biotinidase deficiency (Table 10.1, IX 3a) has been recognized as an important and easily treatable cause of serious neurological disease usually presenting with early onset drug resistant seizures. This and some other neurometabolic diseases can be identified on neonatal blood screening although a full range of screening is not yet routine in the United Kingdom. More disorders are likely to be picked up at an earlier asymptomatic stage as the sophistication of screening tests increases (Wilcken *et al.* 2003; Bodamer *et al.* 2007).

Although individual metabolic disorders are rare, collectively such disorders are relatively common. In reality most clinicians will see an individual condition only rarely in a career. Furthermore, patients with certain rare conditions are often concentrated in specialist referral centres, further reducing the exposure of general and paediatric neurologists to these disorders. A recent study into **p**rogressive **i**ntellectual and **n**eurological **d**eterioration, PIND, gives some information about the relative frequency and distribution of some childhood neurodegenerative diseases in the United Kingdom (Verity *et al.* 2000; Devereux *et al.* 2004). Although primarily designed to identify any childhood cases of variant Creutzfeldt-Jakob disease (Section 42.8.9), the study also provided much information about the distribution of neurometabolic disease in children in the United Kingdom. The commonest five causes of progressive intellectual and neurological deterioration over 5 years were Sanfilippo syndrome, 41 cases, adrenoleukodystrophy, 32 cases, late infantile neuronal ceroid lipofuschinosis, 32 cases, mitochondrial cytopathy, 30 cases, and Rett syndrome, 29 cases. Notably, geographical foci of these disorders were also found and correlate with high rate of consanguinity in some local populations (Fig. 10.1). (Text continues on page 257.)

Fig. 10.1 Total progressive intellectual and neurological deterioration, PIND, in the United Kingdom between May 1997 and September 2002 in children aged up to 16 years (from Devereux *et al.* 2004). (A) The five most frequent diagnoses encountered. Neuronal ceroid lipouscinoses. (B) Map of the United Kingdom showing Public Health Laboratory Service districts. The five highlighted districts are those in which the largest numbers of progressive intellectual and neurological deterioration cases resided and may reflect high rates of consanguinity.

Table 10.1 Neurometabolic disorders without acute episodes of metabolic encephalopathy

I. AMINO ACID DISORDERS

1. Hyperphenylalaninaemia syndromes:

a. Phenylketonuria (phenylalanine hydroxylase deficiency). Identified by neonatal screening, dietary treatment prevents classical phenotype of severe dementia, behaviour disorder and epilepsy. Nevertheless a significantly lowered intelligence quotient, behaviour difficulties and abnormalities of myelin on MRI present in treated young adults. Some adults develop a severe neurological disease if diet is stopped in late childhood. Severe teratogenic effects of untreated maternal phenylketonuria.

Screening tests:	plasma amino acids (raised phenylalanine, reduced tyrosine) phenylalanine and tetrahydrobiopterin loads.
Diagnosis:	liver phenylalanine hydroxylase activity, DNA.

b. Tetrahydrobiopterin (BH$_4$) deficiencies.

BH$_4$ is the cofactor for phenylalanine hydroxylase, tyrosine and tryptophan mono-oxygenases and nitric oxide synthase. Severe forms (usually recessive) present in infancy with hyperphenylalaninaemia and parkinsonism-dystonia (due to dopamine deficiency). Less severe forms (may be dominant) may present later in life with symptoms of dopamine deficiency alone (DOPA-responsive dystonia).

  i. Dihydropteridine reductase and pterin-4a-carbinolamine dehydratase deficiencies prevent recycling of BH4.

 ii. GTP-cyclohydrolase, 6-pyruvoyltetrahydropterin synthase and, in theory, sepiapterin reductase are defects of tetrahydrobiopterin synthesis.

Screening tests:	plasma amino acids, CSF or urine pterin analysis. CSF neurotransmitter amine metabolites.
Diagnosis:	red cell dihydropteridine reductase activity, white cell GTP-cyclohydrolase or 6-pyruvoyltetrahydropterin synthase activity. DNA.

2. Glutathione metabolism defects.

a. Oxoprolinuria (pyroglutamic aciduria). Learning difficulties, spasticity. Haemolysis and severe metabolic acidosis in new-born period.

Screening tests:	urine organic acids (increased oxoproline).
Diagnosis:	red cell glutathione synthetase activity low.

b. Glutamylcysteine synthetase deficiency. Adult spinocerebellar degeneration, haemolysis, peripheral neuropathy, myopathy, generalized aminoaciduria.

Screening tests:	urine amino acids (generalized aminoaciduria).
Diagnosis:	red cell glutamylcysteine synthetase low.

3. Hyperornithinaemia.
Progressive choroidoretinitis (gyrate atrophy) and subcapsular cataract: night blindness in first 10 years, complete blindness in 5th decade. Mild proximal muscular weakness, late white matter degeneration.

Screening tests:	plasma amino acids (raised ornithine, modest decrease in glutamate, glutamine, lysine and creatinine).
Diagnosis:	fibroblast ornithine aminotransferase, DNA.

4. Homocystinuria.

a. Cystathionine ß-synthase deficiency. Learning difficulties, psychiatric illness, osteoporosis, lens dislocation, marfanoid features, vascular disease.

Screening tests:	plasma amino acids (raised methionine; homocystine and mixed disulphides present), plasma total homocysteine, urine amino acids, methionine load.
Diagnosis:	fibroblast cystathionine β-synthase activity.

(continued)

Table 10.1 (Continued) Neurometabolic disorders without acute episodes of metabolic encephalopathy

---

    b. **Inborn errors of folate metabolism**. *Vide infra*. The range of severity of these is great, from severely affected infants to asymptomatic adults. Dementia, motor disorders, seizures and psychiatric.

    c. **Some inborn errors of cobalamin metabolism**. *Vide infra*. The range of severity of these is great, from severely affected infants to adults with a disorder mistaken for multiple sclerosis. Dementia, motor disorders, seizures and psychiatric.

5. <u>**Tyrosinaemia**</u>.

    a. **Type I (fumarylacetoacetase deficiency)**. Liver and renal tubular disease, peripheral neuropathy and porphyric crises.

    b. **Type II (tyrosine aminotransferase deficiency)**. Palmar-plantar keratosis, corneal erosions, mental retardation in 50% (Richner-Hanhart syndrome).

    c. **Type III (4-hydroxyphenylpyruvate dioxygenase deficiency)**. Mild developmental delay, seizures. Delayed maturation can cause benign transient neonatal tyrosinaemia.

    <u>Screening tests</u>:   plasma amino acids (raised tyrosine), urinary succinylacetone (Type 1).

    <u>Diagnosis</u>:   white cell or fibroblast FAA activity, liver TAT and 4HPPD activities, DNA.

6. <u>**Arginase deficiency**</u>. Progressive spastic paraparesis, seizures, learning difficulties, metabolic stroke. May have episodes of encephalopathy.

    <u>Screening tests</u>:   plasma ammonia, amino acids (raised arginine).

    <u>Diagnosis</u>:   red cell arginase activity.

7. <u>**Hyperlysinaemia**</u>. Commonest cause not associated with neurological disease. Rare form with progressive spastic quadriplegia and epilepsy.

8. <u>**Hyperprolinaemia. Type II**</u>. Developmental delay, seizures (pyridoxine-responsive).

    <u>Screening tests</u>:   plasma amino acids (proline 10-15x normal).

    <u>Diagnostic Tests</u>:   fibroblast/leukocyte delta-1-pyrroline 5-carboxylate dehydrogenase activity

                 DNA - *P5CDH* gene.

9. <u>**Serine synthesis disorders**</u>. *Vide infra* under epileptic encephalopathies. Diagnosed from low fasting plasma and CSF serine and glycine. Can result in dysmyelination. One low serine disorder causes peripheral neuropathy.

II. ORGANIC ACID DISORDERS

1. <u>**Canavan's disease**</u>. Macrocephaly, dementia, optic atrophy, demyelination, death in early childhood. Milder variants (juvenile) exist.

    <u>Screening tests</u>:   urine N-acetylaspartate.

    <u>Diagnosis</u>:   fibroblast aspartoacylase activity, DNA.

2. <u>**L2-Hydroxyglutaric aciduria**</u>. Early onset ataxia and mild learning difficulties, later myoclonus, extrapyramidal movement disorder, dementia in teens, characteristic demyelination.

    <u>Screening tests</u>:   urine L2-hydroxyglutaric acid, CSF lysine.

    <u>Diagnosis</u>:   DNA.

2a. <u>**D2-Hydroxyglutaric aciduria**</u>. Early onset intractable epilepsy, severe learning difficulties, cardiomyopathy. Later onset of a variety of neurological problems (ascertainment bias?).

    <u>Screening tests</u>:   urine D2-hydroxyglutaric acid.

    <u>Diagnosis</u>:   DNA.

3. <u>**4-Hydroxybutyric aciduria**</u>. Non-progressive ataxia, hypotonia with depressed reflexes, less commonly hyperkinesis, seizures and supranuclear ophthalmoplegia, psychiatric.

    <u>Screening tests</u>:   urine 4-hydroxybutyric acid.

    <u>Diagnosis</u>:   fibroblast succinate semialdehyde dehydrogenase activity, DNA.

4. <u>**3-Methylglutaconic aciduria**</u>. Some evidence to suspect that, excepting hydratase deficiency, these syndromes may be secondary to mitochondrial oxidative phosphorylation defects.

    a. **Type I**. Delayed speech and macrocephaly.

    <u>Screening tests</u>:   urine 3-methylglutaconic, 3-methylglutaric and 3-hydroxyisovaleric acids.

    <u>Diagnosis</u>:   fibroblast 3-methylglutaconylCoA hydratase activity.

    b. **Type II (Barth's X-linked cardiomyopathy and neutropaenia syndrome)**. Myopathy

    <u>Screening tests</u>:   cholesterol (low), neutrophil count, urine 3-methylglutaconic, 3-methylglutaric acids.

    <u>Diagnosis</u>:   analysis of fibroblast cardiolipins, DNA (*TAZ* gene).

    c. **Type III (Costeff's optic atrophy)**. Early onset optic atrophy, later chorea, paraparesis, ataxia and nystagmus.

    <u>Screening tests</u>:   urine 3-methylglutaconic and 3-methylglutaric acids.

    <u>Diagnosis</u>:   DNA (*OPA3* gene).

    d. **Type IV (unclassified)**.

5. **Glutaric aciduria type 1**. One of the more common metabolic diseases causing neurological disease, often misdiagnosed as cerebral palsy. One-third have gradual onset dystonia in infancy, less commonly seizures and learning difficulties.

Screening tests: urine glutaric, 3-hydroxyglutaric and glutaconic acids, glutarylcarnitine and glutarylglycine. Reduced plasma free carnitine, increased plasma glutarylcarnitine.

Diagnosis: fibroblast or white cell glutaryl-CoA dehydrogenase activity, DNA.

6. **Mevalonic aciduria**. Progressive multisystem disorder. Learning difficulties, progressive ataxia, progressive myopathy and cardiomyopathy, failure to thrive, dysmorphism. Clinical course punctuated by recurrent crises of fever, rash, arthralgia and diarrhoea and vomiting.

Screening tests: urine mevalonic acid, plasma mevalonic acid, creatine kinase (raised) and ubiquinone-10 (reduced).

Diagnosis: fibroblast or white cell mevalonate kinase activity. DNA.

7. **2-Methyl-3-hydroxybutyryl-CoA Dehydrogenase Deficiency**. Delayed psychomotor development, progressive neurodegeneration with hypotonia, choreoathetosis seizures.

Screening tests: urine organic acids—increased 2-methyl-3 hydroxybutyrate, tiglyglycine.

Diagnosis: fibroblast MHBD activity, DNA (*HSD17B10* gene).

III. MITOCHONDRIAL DISORDERS

1. **Pyruvate carboxylase deficiency**. Type A, learning difficulties and lactic acidosis. Type B, neonatal lactic acidosis, seizures, hypotonia, stridor, dystonia, dementia, liver failure and early death (less than three months).

Screening tests: CSF and blood lactate, plasma alanine. Plasma ammonia, citrulline and lysine increased in type B.

Diagnosis: Fibroblast pyruvate carboxylase activity, DNA.

2. **Pyruvate dehydrogenase deficiency**. Variable clinical features. Brain malformation syndromes, through lethal neonatal lactic acidosis, Leigh's disease to carbohydrate-induced episodic ataxia in males.

Screening tests: CSF and blood lactate, plasma alanine.

Diagnosis: Fibroblast pyruvate dehydrogenase activity, DNA.

3. **Citric acid cycle defects**. Variable from neonatal lactic acidosis to Leigh-like presentation.

   a. Succinyl-CoA ligase (ADP-forming subunit) disorders.

   α **subunit** (*SUCLA1/SUCLAG1* gene): Fatal infantile lactic acidosis

   β **subunit** (*SUCLA2* gene): Irritability, inconsolable crying. Severely retarded psychomotor development, muscle hypotonia, impaired hearing. seizures, dystonia. MRI suggestive of Leigh syndrome (with high signal T2 intensity in the putamen bilaterally) and cortical atrophy.

   Screening test: mildly elevated urine methylmalonic acid
   blood C4-dicarboxylic carnitine
   variable lactic acidosis
   mtDNA depletion in muscle biopsy.

   Diagnosis: DNA (*SUCLA2* gene).

   b. Fumarase deficiency. Cutaneous leimyomata in parents. Hypotonia, developmental delay, seizures, microcephaly.

   Screening test: Urine organic acids: fumarate, lactate, pyruvate.

   Diagnosis: Fibroblast fumarase activity, DNA (*FH* gene).

4. **Respiratory chain defects**. A complex group of disorders with both nuclear and mitochondrial inheritance. May have any, and combinations, of: dementia, seizures, spasticity, ataxia, dystonia, retinopathy, optic atrophy, ophthalmoplegia, deafness, peripheral neuropathy, extra-neural symptoms. "Any symptom, any organ, at any age". Recognisable syndromes include Pearson's marrow-pancreas syndrome, Leigh's disease, Leber's hereditary optic neuritis, Kearns-Sayre, MERRF, MELAS, NARP and MNGIE syndromes; considerable overlap occurs in childhood.

   Screening tests: evaluation for multi-organ involvement (e.g. blood count, echocardiogram, liver function tests, tubulopathy markers)
   CSF and blood lactate and alanine, muscle histology and histochemistry
   organic acid analysis.

   Diagnosis: muscle respiratory chain complex assays
   mitochondrial and nuclear DNA.

   a. Pearson's syndrome. Sideroblastic anaemia and exocrine or endocrine pancreatic failure. Variable central nervous system involvement. Survivors may develop Kearns-Sayre syndrome later. Normally large deletion mitochondrial DNA.

   b. Leigh's disease. Originally, a post-mortem diagnosis where necrosis, gliosis and neovascularisation are seen in, predominantly, the basal ganglia and brainstem. In life, the disease may be suspected when there is the stuttering onset of brainstem or extrapyramidal symptoms, raised CSF lactate and neuroimaging findings of symmetrical basal ganglia or periaqueductal lesions. Usually nuclear DNA mutations, particularly *SURF 1* mutations (when cytochrome oxidase activity is deficient in fibroblasts as well as in muscle). Some families have private mitochondrial DNA mutations.

(continued)

Table 10.1   (Continued) Neurometabolic disorders without acute episodes of metabolic encephalopathy

c. **Leber's hereditary optic neuritis**. Subacute onset optic neuritis in early adulthood with permanent visual impairment. Multiple or large scale deletions mitochondrial DNA.

d. **Kearns-Sayre syndrome**. Chronic progressive external ophthalmoplegia, cardiac conduction defects, short stature, myopathy and a variety of central nervous system defects. Most due to a large deletion or duplication mitochondrial DNA.

e. **MERRF**. Myoclonic epilepsy with ragged red fibres on muscle biopsy. Most due to a point mutation in mitochondrial DNA.

f. **MELAS**. Mitochondrial myopathy, encephalopathy, lactic acidosis and stroke-like episodes. Most due to a point mutation in mitochondrial DNA.

g. **MNGIE**. Myoneural, gastrointestinal encephalopathy. Early-onset intestinal pseudo-obstruction followed by myopathy and encephalopathy in adulthood. Some cases are due to thymidine phosphorylase (TP) deficiency and have raised plasma and urine thymidine and reduced white cell TP activity.

h. **Alper's poliodystrophy**. Developmental delay, sudden onset of seizures resistant to medication. progressive deterioration in liver function. Many of these cases have been shown to have mutations in the DNA polymerase -γ responsible for mitochondrial DNA replication.

5.   **Ubiquinone synthesis disorders**.

   Screening   Reduced activity of complexII+II in muscle biopsy, restored by addition of Coenzyme Q white cell ubiquinone.

a. **Mevalonic aciduria**. *Vide supra*.

b. **Prenyl diphosphate synthase (subunit 1) deficiency**. Early-onset deafness, developmental delay, obesity, livedo reticularis, mitral and aortic regurgitation, peripheral neuropathy.

   Diagnosis   DNA (*PDSS1* gene)

c. **Prenyl diphosphate synthase (subunit 2) deficiency**. Leigh syndrome and nephrotic syndrome. Mutations in *PDSS2* gene

d. **Parahydroxybenzoate-polyprenyltransferase**. Nephrotic syndrome, nystagmus, hypotonia, optic atrophy, psychomotor delay followed by regression, renal failure. Improvement in neurological symptoms with CoQ. Mutations in *CoQ2* gene.

e. **Ataxia, early-onset, with oculomotor apraxia and hypo-albuminemia**. Secondary CoQ deficiency but may respond to treatment. Mutations in *Aprataxin (APTX)* gene.

IV. PEROXISOMAL DISORDERS

A group of diseases with either absence of peroxisomes (at least 13 different types) or deficiencies of the oxidative or synthetic enzymes contained within.

1.   **Peroxisomal biogenesis disorders**.

a. **Zellweger's syndrome**. Craniofacial abnormalities (high forehead, upslanting palpebral fissures, hypoplastic supraorbital ridges & epicanthic folds), severe weakness and hypotonia, deafness, neonatal seizures, absent development, retinal pigmentation, optic nerve hypoplasia, cataract, corneal clouding, hepatomegaly, renal cysts. Neuroimaging may show polymicrogyria, pachygyria, neuronal migration defects and abnormal myelination. Reduced eletroretinogram.

   Screening tests: plasma very long chain fatty acids, red cell plasmalogens, plasma and urinary bile acids, X-ray patella/acetabulum (for stippling).

   Diagnosis: studies of peroxisomal function in fibroblasts
                              DNA
                              liver biopsy.

b. **Neonatal adrenoleucodystrophy**. Less severe Zellweger's syndrome with mild adrenal hypofunction.

   Screening tests: As for Zellweger. No calcific stippling/renal cysts.

c. **Infantile Refsum's Disease**. Soft dysmorphic features e.g. high forehead. Motor delay, deafness, pigmentary retinopathy, failure to thrive, liver dysfunction in infancy. May develop leukodystrophy later.

   Screening tests: plasma very long chain fatty acids, phytanate and pristanate, plasma bile acids.

   Diagnosis: studies of peroxisomal function in fibroblasts, DNA.

d. **Childhood/Adult Onset Leukodystrophy**. Deafness, developmental delay in infancy. Ataxia and progressive spastic quadraparesis in childhood.

   Screening tests: as for infantile Refsum's.

   Diagnosis: as for infantile Refsum's.

2.   **Disordered peroxisome enzyme import**.

a. **Rhizomelic chondrodysplasia punctata**. Short proximal limbs, restriction of joint movements, microcephaly, developmental delay, ichthyosis, cataracts. Both phenotypic and genotypic variability.

   Screening tests: X-ray knees (metaphyseal splaying, calcific stippling), plasma phytanate, red cell plasmalogens.

   Diagnosis: platelet dihydroxyacetonephosphate acyltransferase
                              fibroblast studies of peroxisomal function
                              DNA.

3. <u>Defects of single peroxisomal enzymes</u>.

   a. **Deficiencies of acyl-CoA oxidase/D-bifunctional protein (Pseudo-Zellweger syndrome/pseudo neonatal adrenoleukodystrophy)**. Clinical features similar to Zellweger/neonatal adrenoleukodystrophy.

     <u>Screening tests</u>:   plasma very long chain fatty acids, pipecolic acid and bile acid intermediates increased but peroxisomes present on analysis of liver biopsy/fibroblasts. Plasmalogens normal.

     <u>Diagnosis</u>:   fibroblast acyl-CoA oxidase and bifunctional enzyme activities, DNA.

   b. **X-linked Adrenoleucodystrophy**. Severe childhood cerebral form dementia, seizures, motor disorder, adrenal dysfunction; neuroimaging shows a leucodystrophy. Adrenomyeloneuropathy form progressive paraparesis with sphincter involvement, adrenal dysfunction in 60%.

     <u>Screening tests</u>:   plasma very long chain fatty acids, synacthen test.

     <u>Diagnosis</u>:   DNA.

   c. **Refsum's disease**. Tetrad of retinitis pigmentosa, peripheral neuropathy, cerebellar ataxia and increased CSF protein, may have neural deafness, anosmia and ichthyosis.

     <u>Screening tests</u>:   plasma phytanic acid.

     <u>Diagnosis</u>:   fibroblast phytanate α-oxidase, DNA.

   d. **Alpha-methyl-acyl-CoA Racemase Deficiency**. Developmental delay in childhood, sensorimotor neuropathy in adult life.

     <u>Screening tests</u>:   plasma very long chain fatty acids (normal), phytanate (mildly raised) and pristanate (more raised) plasma bile acids (elevated THCA, DHCA, C29-acid).

     <u>Diagnosis</u>:   tests of peroxisomal function in fibroblasts, DNA.

   e. **Dihydroxyacetonephosphate acyltransferase deficiency (Glyceronephosphate O-acyl transferase deficiency)**. Identical presentation to rhizomelic chondrodysplasia punctata in infancy with severe developmental delay or mild developmental delay and chondrodysplasia punctata

   f. **Alkyl-dihydroxyacetone phosphate synthase deficiency**. Identical presentation to rhizomelic chondrodysplasia punctata in infancy

V. LYSOSOMAL ENZYMES

1. <u>Mucopolysaccharidoses</u>.

All develop some features of Hurler syndrome, excrete glycosaminoglycans in urine, and may have white cell inclusions. The clinical features of Hurler's disease include hepatosplenomegaly, characteristic bony features of brachymetacarpal dwarfism, stiff joints, coarse facies, mental deterioration and corneal cloudiness. Behaviour problems are frequent and early.

     <u>Screening tests</u>   skeletal survey, vacuolated lymphocytes, urine glycosaminoglycans.

     <u>Diagnosis</u>:   white cell enzyme (below)
                    DNA.

   a. **MPS I H: Hurler disease (α-L-iduronidase deficiency)**.

   b. **MPS I S: Scheie disease (α-L-iduronidase deficiency)**. Normal central nervous system. Carpal tunnel syndrome.

   c. **MPS I H/S (α-L-iduronidase deficiency)**.

   d. **MPS II XR: Severe Hunter disease (Iduronate sulphate sulphatase deficiency)**. No corneal clouding.

   e. **MPS II XR: Mild Hunter disease (Iduronate sulphate sulphatase deficiency)**. Normal intelligence quotient.

   f. **MPS III A: Sanfilippo disease (Heparan N-sulphatase deficiency)**. Progressive dementia with only mild mucopolysaccharidosis features.

   g. **MPS III B: (N-acetyl-α-D-glucosaminidase deficiency)**.

   h. **MPS III C: (Acetyl-CoA:α-glucosaminide N-acetyl-transferase deficiency)**.

   i. **MPS III D: (Acetyl-CoA:α-glucosaminide 6-sulphatase deficiency)**.

   j. **MPS IV A: Morquio disease (Galactosamine 6-sulphate sulphatase deficiency)**. Skeletal and corneal involvement, normal inkuigence quotient but atlanto-axial dislocation. Some central nervous system, including meningeal, involvement.

   k. **MPS IV B (ß-galactosidase deficiency)**. No central nervous system involvement.

   l. **MPS VI: Severe Maroteaux-Lamy disease (Arylsulphatase B deficiency)**. Mainly skeletal with risk of cervical cord compression.

   m. **MPS VI: Intermediate Maroteaux-Lamy disease (Arylsulphatase B deficiency)**.

   n. **MPS VI: Mild Maroteaux-Lamy disease (Arylsulphatase B deficiency)**.

   o. **MPS VII: Sly disease (ß-glucuronidase deficiency)**. Very variable phenotype.

2. <u>Pompé disease</u>. Infantile progressive proximal muscle weakness, anterior horn cell disease, cardiomyopathy and early death. Later onset progressive proximal muscular weakness.

     <u>Screening tests</u>:   vacuolated white cells, muscle histology.

     <u>Diagnosis</u>:   white cell or muscle acid α-glucosidase, DNA.

(continued)

Table 10.1 (Continued) Neurometabolic disorders without acute episodes of metabolic encephalopathy

---

3. Abnormal lysosomal enzyme phosphorylation. Prevents normal trafficking of enzymes to lysosomes. Hurler-like features with variable time of onset.

Screening tests:	vacuolated white cells, plasma lysosomal enzymes (10–20 fold increase in β-hexosaminidase, arylsulphatase A and iduronate sulphatase), urine glycosaminoglycans (normal).
Diagnosis:	fibroblast UDP-GlcNAc: lysosomal enzyme precursor GlcNAc1-phosphotransferase activity, DNA (*GNPTAB* gene).

   a. I-cell disease (Mucolipidosis II).

   b. Pseudo-Hurler syndrome (Mucolipidosis III).

4. Schindler disease. Severe: onset in second year hypotonia, dementia, upper and lower motor neuropathy, brainstem disease. Mild: mild learning difficulties, angiokeratoma.

Screening tests:	skin nerve electron microscopy (axonal spheroids).
	white cell α-N-acetylgalactosaminidase.
Diagnosis:	white cell α-N-acetylgalactosaminidase, DNA.

5. Abnormal glycoprotein degradation.

   a. α-Mannosidosis. Hurler features, dementia, angiokeratoma.

Screening tests:	vacuolated white cells, urinary oligosaccharides.
Diagnosis:	white cell α-mannosidase A & B.

   b. β-Mannosidosis. Hurler features, deafness, variable neurological phenotype.

Screening tests:	vacuolated white cells, urine oligosaccharides.
Diagnosis:	white cell β-mannosidase.

   c. Fucosidosis. Mild Hurler features, dementia and seizures, retinopathy.

Screening tests:	vacuolated white cells, urine oligosaccharides.
Diagnosis:	white cell α-fucosidase activity.

   d. Sialic Acid Storage Disorders (Salla disease, Sialic acid lysosomal transport deficiency). Dementia, extrapyramidal and cerebellar signs. Infantile and later onset.

Screening tests:	vacuolated white cells, urine sialic acid.
Diagnosis:	white cell sialic acid.

   e. Sialidosis (ML I). Infantile onset and juvenile onset. **i. Normosomic**. Cherry-red spot, action myoclonus, seizures, ataxia, normal facies and bones; burning feet; often no vacuolated lymphocytes. **ii. Dysmorphic (**may be associated with partial ß-galactosidase deficiency**)**. Cherry-red spot, myoclonus, mild MR, coarse Hurler-like facies, bony changes, ataxia.

Screening tests:	vacuolated white cells and foam cells, urine oligosaccharides.
Diagnosis:	white cell neuraminidase activity.

   f. Aspartylglucosaminuria. Mild Hurler features, learning difficulties.

Screening tests:	urine aspartylglucosamine, vacuolated white cells.
Diagnosis:	white cell aspartylglucosaminidase activity.

6. Wolman's disease. Hepatosplenomegaly; severe gastrointestinal illness; neurological deterioration; calcification of adrenals.

Screening tests:	vacuolated lymphocytes and lipid inclusions in bone marrow cells.
	abdominal ultrasound.
Diagnosis:	white cell acid lipase activity.

7. Farber's disease. Infantile onset swollen, very painful joints, joint and tendon nodules; laryngeal involvement; anterior horn cell disease, myopathy, foveal grey spot.

Diagnosis:	white cell ceramidase activity.

8. Niemann-Pick disease A & B. A: hepatosplenomegaly, hypotonia, dementia, foveal grey or cherry red spot. B: hepatosplenomegaly with no central nervous system involvement or mild ataxia alone.

Screening tests:	vacuolated white cells, bone marrow foamy histiocytes.
Diagnosis:	white cell acid sphingomyelinase, DNA.

9. Niemann-Pick disease C. Progressive vertical supranuclear ophthalmoplegia, ataxia, dystonia and dementia. Variable early hepatosplenomegaly and failure to thrive. Neonatal hepatitis in 60%.

Screening tests:	plasma chitotriosidase, foam cells in marrow, white cell sphingomyelinase activity (normal).
Diagnosis:	fibroblast lysosomal cholesterol accumulation, DNA.

10. **Gaucher's disease**.

    a. **Type 1**. Chronic non-neuronopathic form. Can present from infancy to adult life; early splenomegaly; hypersplenism and bone pain; late neurological problems in some.

    b. **Type 2**. Acute neuronopathic form. First year; hepatosplenomegaly; trismus, strabismus, retroflexion of head, progressive spasticity and dementia.

    c. **Type 3**. Subacute neuronopathic form. Half have onset in first 10 years, dementia, spasticity, ataxia, movement disorder, supranuclear horizontal ophthalmoplegia; splenomegaly.

        <u>Screening tests:</u>   raised plasma chitotriosidase, non-tartrate inhibitable acid phosphatase (types 1 and 2), bone marrow for Gaucher cells.

        <u>Diagnosis:</u>   white cell glucocerebrosidase activity.

11. **Krabbe's disease (globoid cell leukodystrophy)**.

    a. **Infantile**. Early irritability, opisthotonus, dementia, peripheral neuropathy, optic atrophy, pyramidal signs, raised CSF protein.

    b. **Juvenile**. Cortical blindness; spasticity or extrapyramidal signs.

        <u>Screening tests:</u>   neuroimaging.

        <u>Diagnosis:</u>   white cell galactocerebrosidase activity.

12. **Metachromatic leukodystrophy (Sulphatide lipidosis)**. Variable time of onset and features; 5 subtypes recognized although only 3 given for simplicity. Occasional rare case of metachromatic leukodystrophy without Arylsulfatase A deficiency. Pseudodeficiency gene is common.

    a. **Infantile**. Ataxia, dementia, pyramidal tract signs and demyelinating peripheral neuropathy.

    b. **Juvenile**. 3–20 years; psychosis, dementia, pyramidal tract signs and dystonia.

    c. **Adult**. Like juvenile, or pure dystonia.

        <u>Screening tests:</u>   neuroimaging, urinary sulphatide excretion, metachromatic inclusions in fresh urinary sediment.

        <u>Diagnosis:</u>   white cell arylsulphatase A activity, pseudodeficiency gene, fibroblast sulphatide load, sphingolipid activating protein.

13. **Fabry's disease**. X-linked recessive. Angiokeratoma corporis diffusum; limb pains; retinopathy, variable neurology, stroke; severe renal disease, ischaemic heart disease.

        <u>Diagnosis:</u>   white cell α-galactosidase A activity.

14. **GM$_1$ gangliosidosis**

    a. **GM$_1$ type I**. Hepatosplenomegaly, Hurler features, dementia, cherry red spot.

        <u>Screening tests:</u>   vacuolated white cells. Bone marrow foam cells.

        <u>Diagnosis:</u>   white cell ß-galactosidase activity.

    b. **GM$_1$ type II—Morquio B**. Later onset; mild bony involvement, ataxia, dementia; spasticity.

        <u>Screening tests:</u>   Vacuolated white cells. Bone marrow foam cells.

        <u>Diagnosis:</u>   white cell ß-galactosidase activity.

15. **GM$_2$ gangliosidosis**.

    a. **GM$_2$ type I—Tay-Sachs disease**.

        i. **Early onset**. Hyperacusis; visual failure, cherry red spot; dementia; hypotonia and pyramidal tract signs.

        ii. **Juvenile**. Dystonic stammer, ataxia, extrapyramidal disease, anterior horn cell disease, dementia.

        <u>Diagnosis:</u>   white cell hexosaminidase A activity.

    b. **Sandhoff disease**. Clinically like Tay-Sachs disease.

        <u>Diagnosis:</u>   white cell hexosaminidase A & B activity.

16. **Batten's disease (ceroid lipofuscinosis)**.

    a. **Infantile (Santavuori)**. Early onset (8 to 12 months); knitting hyperkinesia, extrapyramidal and pyramidal signs, dementia, epilepsy, retinal degeneration, progressive reduction in EEG activity.

        <u>Screening tests:</u>   EEG and electroretinogram/visual evoked potentials.

        <u>Diagnosis:</u>   white cell electron microscopy, neural histopathology on rectal (full thickness) or skin biopsy
                              Fibroblast or white cell palmitoyl-protein thioesterase activity. DNA.

    b. **Late infantile (Jansky-Bielschowsky)**. Myoclonic epilepsy and other seizures; ataxia; dementia; visual failure is a late feature.

        <u>Screening tests:</u>   loss of electroretinogram with large visual evoked
                              potentials to slow rates of flicker.

        <u>Diagnosis:</u>   neural histopathology on rectal (full thickness) biopsy
                              White cell tripeptidyl peptidase I activity (not the variant forms) DNA (not the variant forms).

(continued)

Table 10.1 (Continued) Neurometabolic disorders without acute episodes of metabolic encephalopathy

---

c. **Juvenile (Spielmeyer-Vogt)**. Rapidly progressive retinitis pigmentosa, then disintegrative psychosis and dementia; seizures; late pyramidal and extrapyramidal movement disorder.

    <u>Screening tests:</u>   vacuolated white cells. Electroretinogram/visual evoked potentials.

    <u>Diagnosis:</u>   neural histopathology on rectal (full thickness) biopsy, DNA.

d. **Adult (Kufs)**. May exhibit seizures, psychiatric and extrapyramidal features.

    <u>Diagnosis:</u>   neural histopathology on rectal (full thickness) biopsy.

VI. CONGENITAL DISORDERS OF GLYCOSYLATION

1. <u>Defects of N-Glycosylation of Proteins</u>.

Syndromes defined by the isoelectric focusing pattern of transferrin.

Type 1

    <u>Screening tests:</u>   Transferrrin isoelectric focussing (type I pattern)
                        Plasma albumin, thyroid binding globulin, haptoglobin, specific coagulation factors, proteins C and S, antithrombin III (reduced). Plasma aspartyl-glucosaminidase (increased).

    <u>Differential diagnosis:</u>   Specific enzyme assays; phosphomannomutase activity (type 1a), Fibroblast lipid-linked oligosaccharide (LLO) profiles, DNA.

  i. **Type 1a**. Infantile onset: learning difficulties, squint and supranuclear ophthalmoplegia, hypotonia; abnormal fat distribution, inverted nipples; hepatomegaly, cardiomyopathy, pericardial effusions; cerebellar hypoplasia. Later onset: learning difficulties, ataxia, peripheral neuropathy, strokes, epilepsy; thoracic deformity, hypogonadism; cerebellar hypoplasia.

  ii. **Type 1c**. Axial hypotonia, developmental delay, areflexia, strabismus, seizures, ataxia

  iii. **Type 1d**. Dysmorphic features incl. arthrogryposis. Microcephaly, severe, developmental delay, seizures (with hypsarrhytmia), hypertonia, cerebral atrophy, severe visual impairment with reduced electroretinogram.

  iv. **Type 1e**. Microcephaly, dysmorphia, severe global developmental delay, hypotonia, no visual fixation, seizures, hepatosplenomegaly.

  v. **Type 1f**. Severe encephalopathy with scaly erythematous skin rash.

  vi. **Type 1g**. Facial dysmorphia, hypotonia, developmental delay, progressive microcephaly.

  vii. **Type 1h**. Dysmorphia ± macrocephaly. Diarrhoea, liver and renal dysfunction, oedema, ascites, seizures, progressive hypotonia, head growth deceleration.

  viii. **Type 1i**. Infantile spasms with hypsarrhythmia. dysmyelination on MRI, severe developmental delay, brisk reflexes.

  ix. **Type 1j**. Microcephaly, dysmorphia, hypotonia, developmental delay, infantile spasms, intractable epilepsy.

  x. **Type 1k**. Recurrent refractory seizures, rapidly developing microcephaly, coagulopathy. Early death.

  xi. **Type 1l**. Macrocephaly, central hypotonia, developmental delay seizures. Delayed myelination. Hepatomegaly

  xii. **Type 1m**. Seizures, hypotonia, progressive tetraplegia, acquired microcephaly, hypoglycaemia, ichthyosis, dilated cardiomyopathy.

Type 2. Severe learning difficulties and hypotonia.

    <u>Screening tests:</u>   Transferrin isoelectric focussing (type 2 pattern)
                        Specific coagulation factors.
                        Urine oligosaccharides.

    <u>Differential diag:</u>   Glycan analysis
                        Test for combined N-glycosylation and O-glycosylation defect (apolipoprotein C isoelectric focussing).

    <u>Diagnosis:</u>   DNA.

  i. **Type 2a**. Developmental delay, epilepsy, stereotyped behaviour, dysmorphia, raised AST, normal ALT

  ii. **Type 2b**. Dysmorphia, hypotonia, epilepsy. Abnormal urinary oligosaccharide, transferrin ief normal

  iii. **Type 2c**. Craniofacial dysmorphia, severe psychomotor and growth retardation, recurrent bacterial infections, leucocytosis, leukocyte adhesion defect, transferrin ief normal

  iv. **Type 2d**. Developmental delay, Dandy Walker malformation and myopathy

  v. **Type 2e (COG7 deficiency)***. Perinatal asphyxia, dysmorphia, loose, wrinkled skin, hypotonia, hepatosplenomegaly, jaundice, severe epilepsy.

  vi. **Type 2g (COG1 deficiency)***. Hypotonia, mild developmental delay, progressive microcephaly, mild hepatosplenomegaly

  vii. **Type 2h (COG8 deficiency)***. Hypotonia, developmental delay, axonal neuropathy, acute encephalopathy with loss of skills. Ataxia. Cerebellar and brain stem atrophy.

    * Combined defect of N-glycosylation and O-glycosylation of proteins

2. **Defects of O-Glycosylation of Proteins**: Congenital muscular dystrophies with neuronal migration defects in the brain and ocular abnormalities.

    Screening: Abnormalities of O-glycosylation detected on muscle biopsy (α-dystroglycan staining).

Walker-Warburg syndrome: Severe muscle weakness, absent psychomotor development, ocular abnormalities, death in infancy. Cobblestone lissencephaly, agenesis of corpus callosum, cerebellar hypoplasia, hydrocephalus.

    Diagnosis: DNA (*POMT1, POMT2, fukutin, and fukutin related protein* genes).

Limb Girdle Muscular Dystrophy: Can be caused by milder *POMT1* mutations

Muscle-eye-brain disease: Severe muscle weakness, psychomotor retardation, epilepsy, ocular abnormalities

    Diagnosis: DNA (*POMGnT1* gene).

3. **Defects of Lipid Glycosylation**:

Amish Infantile Onset Epilepsy Syndrome. Irritability, poor feeding, seizures, developmental stagnation, regression. MRI brain showing diffuse atrophy

    Screening: Plasma glycosphingolipid analysis.

    Diagnosis: DNA (*SIAT9 [=GM3 synthase gene]*).

Glycosylphosphatidylinositol Deficiency. Portal and hepatic vein thromboses, recurrent absence seizures.

    Screening: Thrombophilia screen, positive HAMM test.

    Diagnosis: DNA (*PIGM* gene).

VII. PURINE AND PYRIMIDINES

1. **Purine defects**.

    a. **Lesch-Nyhan syndrome**. X-linked. Learning difficulties, hypotonia and chorea in the 1st year, self-mutilation, dystonia and spasticity at some stage in childhood.

      Screening tests: plasma urate, urine urate to creatinine ratio (both increased; urine test more sensitive because of high renal urate clearance).

      Diagnosis: red cell hypoxanthine-guanine phosphoribosyl-transferase activity, DNA.

    b. **Phosphoribosylpyrophosphate synthetase superactivity**. X-linked. Variable neurology. Learning difficulties, deafness.

      Screening tests: plasma urate, urine urate to creatinine ratio (both increased; urine test more sensitive because of high renal urate clearance).

      Diagnosis: red cell phosphoribosylpyrophosphate synthetase activity.

    c. **Adenylosuccinase deficiency**. Severe learning difficulties with autism, later epilepsy. Milder variants occur related to degree enzyme deficiency.

      Screening tests: plasma succinyladenosine and succinylaminoimidazole carboxamide riboside.

      Diagnosis: red cell adenylosuccinase activity, DNA.

    d. **Adenosine deaminase deficiency**. Combined immunodeficiency; spasticity and ataxia.

      Screening tests: white cell count, serum immunoglobulins. Plasma adenosine and deoxyadenosine (both increased), red cell deoxyadenosine triphosphate (increased) and S-adenosylhomocysteine hydrolase activity (decreased).

      Diagnosis: red cell ADA activity
                      DNA.

    e. **Purine nucleoside phosphorylase deficiency**. Combined (mostly cellular) immunodeficiency; spasticity (diplegia), learning difficulties, ataxia and tremor.

      Screening tests: white cell count, serum immunoglobulins. Plasma urate and urine urate to creatinine ratio (both decreased).

      Diagnosis: red cell PNP activity.
                      DNA.

2. **Pyrimidine defects**.

    a. **Dihydropyrimidine dehydrogenase deficiency**. No consistent neurological consequences. Epilepsy, learning difficulties and microcephaly.

      Screening tests: urine thymine, 5-hydroxymethyluracil and uracil (increased).

      Diagnosis: fibroblast or white cell dihydropyrimidine dehydrogenase activity.

    b. **Dihydropyrimidinase deficiency**. No consistent neurological consequences. Epilepsy, learning difficulties and microcephaly.

      Screening tests: urine dihydrothymine, dihydrouracil (increased) and 5-hydroxymethyluracil (not detected).

      Diagnosis: liver dihydropyrimidinase activity.

    c. **Ureidopropionase (β-alanine synthase) deficiency**. Learning difficulties, hypotonia and dystonia.

      Screening tests: urine dihydrothymine, dihydrouracil, ureidopropionate, ureidoisobutyrate (increased) and 5-hydroxymethyluracil (not detected).

      Diagnosis: liver ureidopropionase activity.
                    DNA.

(continued)

Table 10.1 (Continued) Neurometabolic disorders without acute episodes of metabolic encephalopathy

> **d. Hyper-ß-alaninaemia**. Seizures and dementia.
>
> Screening tests: plasma and urine amino acids.
>
> Diagnosis: liver β-alanine-α-ketoglutarate aminotransferase.

VIII. METALS

1.   Copper metabolism.

> **a. Wilson's disease**. Onset over 4 years, usually 2nd/3rd decade. Liver disease; progressive dystonia and other involuntary movements, Kayser Fleischer rings.
>
> Screening tests: plasma copper (total) and caeruloplasmin (reduced), urine copper (increased), penicillamine load.
>
> Diagnosis: liver copper (increased)
> DNA.
>
> **b. Menke's syndrome**. X-linked. Steely or kinky hair; hypothermia; osteoporosis with flared epiphyses, fractures; haemorrhages; seizures, dementia, early death.
>
> Screening tests: plasma copper and caeruloplasmin (reduced).
>
> Diagnosis: fibroblast copper uptake
> DNA.

Milder form: Occipital horn syndrome

2.   Molybdenum metabolism.

Molybdenum cofactor deficiency. Combined sulphite oxidase and xanthine oxidase deficiency, symptoms caused by sulphite oxidase deficiency which may exist as an isolated defect. Neonatal-onset feeding difficulties and seizures, dementia, lens dislocation, pyramidal tract signs.

> Screening tests: plasma urate (reduced), urine sulphite.
>
> Diagnosis: fibroblast sulphite oxidase activity.
> DNA (3 genes).

3.   Magnesium Transport.

Primary hypomagnesaemia (with secondary hypocalcaemia). Defect of intestinal magnesium uptake. Onset before 4 months, irritability, feeding difficulties, tetany, seizures.

> Screening tests: plasma magnesium and calcium (reduced),
> urine magnesium (reduced).
>
> Diagnosis: DNA (*TRPM6* gene).

Renal hypomagnesaemia. Defect affecting renal transport of magnesium. Convulsions and tetany in infancy

> Screening tests: plasma magnesium and calcium (reduced)
> urine magnesium (increased).
>
> Diagnosis: DNA (*paracellin-1 [=CLDN16]* gene)
> dominant form *FXYD2* gene.

Gitelman syndrome. Muscle weakness, cramps, tetany, paraethesiae, convulsions. Polyuria, failure to thrive, alkalosis

> Screening tests: plasma magnesium and potassium (reduced)
> urine magnesium, potassium (increased).
>
> Diagnosis: DNA (*SLC12A3* gene).

IX.   VITAMINS

1.   Tetrahydrofolate (Vitamin B$_9$) metabolism.

> **a. Congenital folate malabsorption**. Infantile onset severe megaloblastic anaemia, learning difficulties, later dementia, pyramidal and extrapyramidal motor disorder, intracranial calcification, demyelination.
>
> Screening tests: plasma total homocysteine. Serum, red cell and CSF 5-methyltetrahydrofolate, oral folate load.
> small bowel folate transporter activity.
>
> Diagnosis: DNA.
>
> **b. 5,10-Methylenetetrahydrofolate reductase deficiency**. Variable severity of symptoms from severely affected infants to asymptomatic adults. Dementia, motor disorders, demyelination, seizures and psychiatric; vascular; no megaloblastic anaemia.
>
> Screening tests: plasma total homocysteine. Serum, red cell and CSF 5-methyltetrahydrofolate.
>
> Diagnosis: fibroblast 5,10-methylenetetrahydrofolate reductase
> activity
> DNA.
>
> **c. Forminoglutamic aciduria**. Variable neurological picture and uncertain whether causes disease. **i. Type 1**: learning difficulties and hypotonia. **ii. Type 2**: mild speech delay.
>
> Screening tests: serum and red cell folate (raised), histidine load.

2. **Cobalamin (vitamin B$_{12}$) metabolism**. Very variable symptoms depending upon site and severity of the metabolic block.

   a. Defects in absorption and transport.

      i. Intrinsic factor deficiency. Onset 2nd to 5th year of life, failure to thrive, irritability, muscular weakness, drowsiness, megaloblastic anaemia.

      ii. Immerslund-Gräsbeck syndrome. Onset after 2nd year, gastrointestinal symptoms and developmental delay, proteinuria.

      iii. Transcobalamin 2 deficiency. Onset 1st year, failure to thrive, weakness, megaloblastic anaemia, developmental stagnation.

      iv. R-binder deficiency. Of uncertain disease-causing status, demyelination in adulthood.

      <u>Screening tests</u>:   blood count, serum cobalamin, plasma total homocysteine, urinary methylmalonate, double isotope Schilling test.

      <u>Diagnosis</u>:   serum transcobalamin 2 activity, fibroblast transcobalamin 2 production.

   b. Defects in intracellular processing.

      i. MethylmalonylCoA mutase defects alone (*cblA* **and** *cblB***)**. Neonatal or late onset forms. Failure to thrive, developmental delay, acute encephalopathy.

      ii. Methionine synthase defects alone (*cblE* **and** *cblG***)**. Onset usually in 1st year, failure to thrive, megaloblastic anaemia, learning difficulties, seizures.

      iii. Combined defects (*cblC, cblD* **and** *cblF* **)**. Infantile and late-onset forms. **Infantile**: microcephaly, failure to thrive, megaloblastic anaemia, learning difficulties, seizures, retinopathy, demyelination, severe systemic vasculitis. **Late onset**: anorexia, irritability, dementia, myelopathy, psychiatric.

      <u>Screening tests</u>:   blood count, serum cobalamin, plasma total homocysteine, urinary methylmalonate.

      <u>Diagnosis</u>:   fibroblast complementation studies.

3. **Biotin (vitamin B$_7$) metabolism**.

   a. Biotinidase deficiency. Seizures, ataxia, hyperventilation or stridor, dementia, deafness, optic atrophy, rash and alopecia.

      <u>Screening tests</u>:   urine lactate, methylcitrate, propionylglycine, 3-hydroxypropionate, 3-methylcrotonylglycine, 3-hydroxyisovalerate may only be apparent late in the course of the disorder. Plasma lactate and alanine. CSF lactate.

      <u>Diagnosis</u>:   plasma biotinidase activity.

   b. Holocarboxylase synthetase deficiency. Neonatal onset, apnoea, hypotonia, seizures, coma.

      <u>Screening tests</u>:   blood acidosis and hyperammonaemia, urine lactate, methylcitrate, propionylglycine, 3-hydroxypropionate, 3-methylcrotonylglycine, 3-hydroxyisovalerate and tiglylglycine.

      <u>Diagnosis</u>:   fibroblast carboxylase activities.

   c. Biotin transporter defect.

4. **Vitamin B$_6$ metabolism**.

   Vide infra under epileptic encephalopathies.

   a. Pyridoxine dependency. Antiquitin deficiency, actually a disorder of valine catabolism. Intractable neonatal seizures, CNS malformations. Untreated can be fatal or associated with severe developmental delay. Milder variants with later onset of treatable seizures may be more common.

   b. Pyridoxal phosphate dependency. Pyridox(am)ine phosphate oxidase deficiency. Progressive neurological dysfunction can occur in untreated patients. Variable neurological problems in treated patients including dystonia and breakthrough seizures.

5. **Nicotinamide (vitamin B$_3$) deficiency**. No enzymatic diagnoses made, but patients present with a photosensitive rash (pellagra-like), learning difficulties and ataxia.

      <u>Screening tests</u>:   urine kynurenine pathway metabolites.

6. **Thiamine (vitamin B$_1$) metabolism**.

   Amish lethal microcephaly (MCPHA). Defect of mitochondrial transport of thiamine pyrophosphate. Severe microcephaly, death before 6 mo.

      <u>Screening tests</u>:   Urine organic acids (2-oxoglutarate).

      <u>Diagnosis</u>:   DNA.

7. **Pantothenate (vitamin B$_5$) Metabolism**.

   Hallervorden-Spatz disease. Pantothenate kinase deficiency. Progressive dystonia, retinopathy and dementia in some.

      <u>Screening tests</u>:   neuroimaging, electroretinogram, acanthocytes, lipoprotein electrophoresis.

      <u>Diagnosis</u>:   DNA (*PANK2* gene).

8. **Riboflavin (vitamin B$_2$ metabolism)**.

   Vide infra—glutaric aciduria type II. Severe form is associated with brain malformation—warty cerebral dysplasia

(*continued*)

Table 10.1 (Continued) Neurometabolic disorders without acute episodes of metabolic encephalopathy

9.	**Isolated Vitamin E Deficiency**.

Neurologic abnormalities similar to those of vitamin E deficiency (retinopathy, neuropathy, ataxia) but no evidence of fat malabsorption.

(contrast abetalipoproteinaemia)

Screening tests:	Low vitamin E.
Diagnosis:	DNA (*TTPA* gene).

X. LIPOPROTEINS

1. Abetalipoproteinaemia. Malabsorption; vitamin E deficiency causing retinopathy; peripheral neuropathy; ataxia.

Screening tests:	acanthocytes on wet blood film, decreased serum cholesterol, triglycerides and Vitamin E.
Diagnosis:	absent apolipoprotein B
	DNA.

2. Tangier disease. Tonsil abnormality; corneal opacification; hepatosplenomegaly; mononeuritis multiplex in adults.

Screening tests:	decreased serum cholesterol, normal or raised triglycerides.
Diagnosis:	absent apolipoprotein A-I.

XI. CHOLESTEROL METABOLISM

1. Cerebrotendinous xanthomatosis (CTX): Neonatal hepatitis. Developmental delay. Late childhood/adult onset dementia/motor dysfunction/psychiatric presentation, tendon xanthomata, atherosclerosis.

Screening tests:	urine bile alcohols,
	plasma bile acid precursors,
	plasma cholestanol.
Diagnosis:	DNA.

2. Peroxisomal disorders: (*vide supra*)

3. Disorders of cholesterol synthesis (pathway from lanosterol to cholesterol): Multiple malformations, dysmorphic features, developmental delay, behaviour problems.

Screening tests:	plasma sterol profile.

a. Smith-Lemli-Opitz syndrome (7-dehydrocholesterol reductase deficiency). Variable severity. Microcephaly, ptosis, anteverted nares, micrognathia, 2-3 syndactyly of toes, hypospadias. Developmental delay, hypotonia in infancy, hypertonia in childhood, autistic features, abnormal sleep pattern, self injury, aggression.

Screening tests::	increased 7-dehydrocholesterol and 8-dehydrocholesterol.
Diagnosis:	DNA (*DHCR7* gene).

b. Conradi-Hunermann syndrome Δ8-Δ7sterol isomerase deficiency. X-linked dominant. Girls: Ichthyosis, alopecia, cataracts, asymmetric limb shortening, chondrodysplasia punctata. May have mild-moderate mental retardation, Dandy-Walker malformation, ventriculomegaly. Boys (mosaics can survive) show severe developmental delay

Screening tests:	increased 8-dehydrocholesterol and 8(9)-cholestenol.
Diagnosis:	DNA (*EBP* gene).

c. Desmosterolosis. Lethal form with multiple formations. Milder form with multiple congenital anomalies, microcephaly, and profound developmental delay.

Screening tests:	increased tissue/plasma desmosterol.
Diagnosis:	DNA (*DHCR24* gene).

d. Lathosterolosis. Sterol C5-desaturase deficiency. Microcephaly, dysmorphia including anteverted nares, poly/syndactyly, intrahepatic cholestasis. Hypotonia, developmental delay.

Screening tests:	increased plasma lathosterol
Diagnosis:	fibroblast 3-beta-hydroxysteroid-delta-5-desaturase
	DNA (*SC5DL* gene)

e. CHILD syndrome. Congenital hemidysplasia with ichthyosiform erythroderma and limb defects. X-linked dominant. NSDHL mutations. Unilateral hypoplasia can include cranial nerves, pons, medulla and spinal cord.

XII. NEUROTRANSMITTER METABOLISM

1. Tyrosine hydroxylase deficiency. Infantile parkinsonism/Segawa syndrome (L-Dopa responsive)

Screening tests:	CSF neurotransmitter amine metabolites
	Low HVA normal 5HIAA and pterins
	Serum prolactin (raised).
Diagnosis:	DNA.

2. <u>**Aromatic amino acid decarboxylase deficiency**</u>. Trunkal hypotonia, hypokinesia, oculogyric crises, ptosis

<u>Screening tests:</u>   CSF neurotransmitter amine metabolites

Low HVA and 5HIAA, raised 3-methoxytyrosine

Serum prolactin (raised).

<u>Diagnosis:</u>   Plasma AADC activity

DNA.

3. <u>**Succinic semialdehyde dehydrogenase deficiency**</u> (4-Hydroxybutyric aciduria) *Vide supra* (organic acid disorders)

XIII.   CREATINE SYNTHESIS AND TRANSPORT

1. <u>**Guanidinoacetate methyltransferase (GAMT) deficiency**</u>: Hypotonia, developmental stagnation, intractable seizures, dystonia. Low plasma creatinine, increased urine guanidinoacetate, absent creatine plus creatine phosphate peak on proton magnetic resonance spectroscopy. White cell guanidinoacetate methyltransferase activity and DNA.

2. <u>**Arginine: glycine amidinotransferase (AGAT) deficiency**</u>: Hypotonia, developmental delay, progressive extrapyramidal movement disorder, ataxia, intractable seizures.

3. <u>**Creatine transporter deficiency**</u>: X-linked. Developmental delay, seizures.

XIV.   TRANSPORT DEFECTS

1. <u>**Lysosomal transport**</u>.

    a.   Sialic acid storage disorders.

       i.   Infantile free sialic acid storage disease. Severe Hurler-like phenotype.

       ii.   Salla disease. Infantile onset learning difficulties and ataxia.

       <u>Screening tests:</u>   vacuolated white cells, urine sialic acid.

       <u>Diagnosis:</u>   white cell sialic acid.

    b.   Cystinosis. Renal failure mid-childhood. Surviving adults may develop myopathy, anterior horn cell disease and central demyelination.

    <u>Screening tests:</u>   corneal crystals on slit-lamp examination.

    <u>Diagnosis:</u>   increased fibroblast or polymorphonucleocyte cystine

    DNA.

2. <u>**Lowe's syndrome**</u>. X-linked. Prenatal cataract and other ocular abnormalities, hypotonia with absent tendon reflexes, learning difficulties.

<u>Screening tests:</u>   renal Fanconi syndrome (bicarbonaturia, renal tubular acidosis, aminoaciduria, phosphaturia, tubular proteinuria, impaired urine concentration).

<u>Diagnosis:</u>   DNA.

3. <u>**Lysinuric protein intolerance**</u>. Can have encephalopathic episodes (*vide infra*) but protein avoidance more usual; hepatosplenomegaly; sparse hair; hypotonia, psychiatric disturbance.

<u>Screening tests:</u>   decreased plasma lysine, ornithine and arginine; increased plasma ammonia, glutamate, alanine, serine, proline, citrulline and glycine. Massive urine lysine excretion.

<u>Diagnosis:</u>   fibroblast cationic amino acid transporter activity

DNA.

4. <u>**Hartnup disease**</u>. Around 10% develop intermittent symptoms of pellagra (photosensitive rash, ataxia).

<u>Screening tests:</u>   neutral aminoaciduria, but not proline, cystine, lysine and ornithine.

XV.   RED CELL GLYCOLYTIC DEFECTS

1. <u>**Triose phosphate isomerase deficiency**</u>. Haemolytic anaemia; cardiomyopathy and early death; jerky dystonia, pyramidal tract, anterior horn cell disease.

<u>Diagnosis:</u>   red cell triose phosphate isomerase activity.

2. <u>**Phosphoglycerate kinase deficiency**</u>. X-linked. Haemolytic anaemia; variable neurology—learning difficulties, dystonia, psychiatric.

<u>Diagnosis:</u>   red cell phosphoglycerate kinase activity.

XVI.   PENTOSE PHOSPHATE PATHWAY DISORDER

1. <u>**Ribose-5-phosphate isomerase deficiency**</u>. Developmental delay in infancy, epilepsy, regression in childhood with cerebellar ataxia, spasticity, optic atrophy, neuropathy.

<u>Screening tests:</u>   urine polyols

pentose phosphate pathway intermediates in blood spots

proton MR spectroscopy of brain.

<u>Diagnosis:</u>   fibroblast ribose-5-phosphate isomerase activity

DNA.

(continued)

Table 10.1 (Continued) Neurometabolic disorders without acute episodes of metabolic encephalopathy

---

XVII. DNA REPAIR DEFECTS

1. **DNA excision repair defects**.

    Screening tests: fibroblast chromosome ultraviolet sensitivity.

    Diagnosis: DNA (in some).

  a. **Cockayne syndrome**. Sun sensitivity, short stature, dysmorphism; microcephaly, demyelination, intracerebral calcification, retinitis pigmentosa, dementia, neuropathy. Milder variant recognised, three complementation groups.

  b. **Xeroderma pigmentosa**. Sun sensitivity and neoplasia; dementia, deafness, ataxia, neuropathy. Seven complementation groups.

  c. **Trichothiodystrophy**. Sulphur-deficient, brittle hair, short stature, learning difficulties, late movement disorder.

2. **Ataxia telangiectasia**. Ataxia, dystonia, supranuclear ophthalmoplegia; immunodeficiency; neoplasia.

    Screening tests: serum α-fetoprotein

        immunoglobulins and IgG subclasses.

    Diagnosis: white cell chromosome radiation sensitivity

        DNA.

XVIII. CLASSICAL WHITE MATTER DISORDERS

    Diagnosis requires expert neuroradiology and magnetic resonance imaging in addition to biochemical and other investigations.

1. **Leukodystrophies**.

  a. **Metachromatic leukodystrophy**. *Vide supra*.

  b. **Krabbe leukodystrophy**. *Vide supra*.

  c. **Adrenoleukodystrophy**. *Vide supra*.

  d. **Canavan's disease**. *Vide supra*.

  e. **Alexander's disease**. Failure to thrive; dementia, leukodystrophy, megalencephaly.

  f. **Aicardi-Goutiere syndrome**. Microcephaly, intracranial calcification, demyelination, CSF pleocytosis, raised CSF α-interferon. DNA.

  g. **Cockayne syndrome**. *Vide supra*.

  h. **Megalencephalic cystic leukoencephalopathy (van der Knaap leukodystrophy)**. Macrocephaly in 1st year, motor disorder by 5th year, seizures and dementia in teens.

  i. **Cerebellar ataxia central hypomyelination/vanishing white matter disease**. Encephalopathy followed by spastic/ataxic motor disorder, late bulbar involvement and optic atrophy.

  j. **Peroxisomal leukodystrophies**. *Vide supra*

2. **Brain dysmyelinating disorders**. Excluding disorders affecting amino acids/B12/folate)

  a. **Pelizaeus-Merzbacher disease**. X-linked. Onset in 1st year nystagmus, spastic paraparesis, movement disorder.

  b. **Merosin-deficient congenital muscular dystrophy**. Weakness and contractures at birth. No central nervous system symptoms.

XIX. OTHERS

1. **Rett syndrome**. Females. From 6 months—2 years; hand stereotypes and loss of hand function, dementia, acquired microcephaly, seizures; later pyramidal tract signs and neuropathy. *MECP2* mutations. Can also occur with *STK9* (*CDKL5*) mutations.

2. **Progressive neuronal degeneration of childhood (Alpers' disease)**. Early onset dementia, myoclonus and seizures; rapid deterioration, late liver involvement. Brain or liver biopsy may show characteristic changes. Some cases due to mitochondrial respiratory chain defects (*vide supra*). Particularly mutations in the mitochondrial DNA polymerase γ gene (*POLG*)

3. **Infantile neuroaxonal dystrophy (Seitelberger's disease)**. Onset at end of 1st year; profound hypotonia, dementia, pyramidal signs, anterior horn cell disease, optic atrophy and squint. Atypical cases include a juvenile variant with progressive myoclonic epilepsy and sometimes retinopathy. Axonal spheroids found on electron microscopy of brain, nerve, conjunctival or skin biopsy. Schindler's disease is a variant (*vide supra*).

4. **Unvericht-Lundborg disease**. Onset 8 to 13 years; progressive myoclonic epilepsy, action myoclonus, dementia.

5. **Lafora-Body disease**. Onset 11 to 18 years; myoclonic seizures, focal occipital seizures, dementia. Lafora bodies in apocrine sweat glands.

6. **Idiopathic torsion dystonias**. Childhood onset progressive dystonia, usually sparing orobulbar musculature. *DYT 1* gene accounts for some dominantly inherited forms, recessive inheritance also occurs.

7. **Hereditary spastic paraplegias**. Childhood onset progressive spastic paraparesis. Usually dominantly inherited with variable penetrance.

8. **Friedreich's ataxia**. Late childhood onset ataxia, axonal neuropathy and cardiomyopathy. DNA.

9. **Sjögren-Larssen syndrome**. Ichthyosis at birth; spastic diplegia and mental retardation developing before 3 years; macular changes. Skin alcohol dehydrogenase activity reduced. DNA.

10. **Chediak-Higashi syndrome**. Partial albinism, hepatosplenomegaly, lymphadenopathy; learning difficulties, cerebellar degeneration, nystagmus, peripheral neuropathy.

Table 10.2 Neurometabolic disorders with episodes of acute metabolic encephalopathy

   I. Carbohydrate disorders (p. 253)
  II. Amino acid disorders (p. 253)
 III. Organic acid disorders (p. 254)
  IV. Mitochondrial fat oxidation defects (p. 254)
   V. Epileptic encephalopathies (p. 255)
  VI. Mitochondrial respiratory chain/PDH/PC disorders (p. 256)
 VII. Peroxisomal disorders (p. 256)
VIII. Congenital disorders of Glycosylation (p. 256)
  IX. Others (p. 256)

I. CARBOHYDRATE DISORDERS

1. **Neuroglycopenia**. Episodes of impaired supply of glucose to the brain cause acute encephalopathy and can cause permanent damage

   a. **Hypoglycaemia**. Comprehensive list of causes not possible here but includes hyperinsulinism, disorders of gluconeogenesis (e.g. fructose-1, 6-bisphosphatase deficiency), glycogen storage diseases, disorders of fatty acid oxidation (*vide infra*), ketone body synthesis and utilisation, ketotic hypoglycaemia etc.

   b. **Glut1 glucose transporter deficiency**. Early onset epilepsy, motor delay with hypotonia/ataxia/dystonia, speech delay. CSF glucose 1.4–2.0 mM when blood sugar normal. Confirmation from DNA analysis.

2. **Galactosaemia**. Neonatal liver failure with hepatomegaly; cerebral oedema; cataracts. Later dyspraxia, variable learning difficulties, later still dystonia.

   Screening tests: urine galactose, red cell galactose-1-phosphate.

   Diagnosis: red cell galactose-1-phosphate-uridyl transferase or UDP-glucose epimerase activity.

3. **Hereditary fructose intolerance**. Drowsiness, apathy; liver disease.

   Liver failure with encephalopathy if large amounts of fructose given.

   Diagnosis: liver fructose-1-phosphate-aldolase deficiency.
   DNA.

II. AMINO ACID DISORDERS

1. **Maple Syrup Urine Disease**. Accumulating keto-acids have an aroma of maple syrup or fenugreek seeds.

   a. **Classical**. Acute new-born presentation with severe ketoacidosis.

   b. **Intermittent**. Ketoacidosis; ataxia, lethargy, slurred speech during attack; neurological handicap varies.

   c. **Intermediate**. Learning difficulties without ketoacidosis.

   d. **Thiamine responsive**. Learning difficulties.

   e. **E$_3$ deficient form**. Progressive encephalopathy in first year.

   Screening tests: raised plasma branched chain amino acids (leucine, isoleucine and valine)
   branched chain 2-oxoacids e.g. 2-oxo-isocaproic acid in urine

   Diagnosis: fibroblast leucine oxidation or branched chain keto-acid decarboxylase activity.

2. **Non-ketotic hyperglycinaemia**. Neonatal encephalopathy with apnoea and myoclonus. Later intractable seizures, pyramidal and extrapyramidal movement disorder; developmental stagnation.

   Screening tests: raised plasma, urine, and CSF glycine and raised CSF/plasma glycine. (>0.09).
   Diagnosis: liver glycine cleavage enzyme activity
   DNA.

3. **Urea Cycle Disorders**. Ataxia, lethargy and coma, especially during decompensation, seizures. May present as Reye-like illness, stroke and intermittent ataxia.

   Screening tests: plasma ammonia and amino acids,
   urine aminoacids, orotic acid.

   Diagnosis: enzyme activity (red cells/fibroblasts/liver)
   DNA.

   a. **N-acetylglutamate synthetase deficiency**. Very severe.

   b. **Carbamoylphosphate synthetase deficiency**. Suspected by normal citrulline, arginosuccinic acid and arginine. No orotic acid in urine.

   c. **Ornithine carbamoyl transferase deficiency**. X-linked, with manifesting carriers; liver disease. Orotic aciduria, low plasma arginine.

   d. **Arginosuccinate synthetase deficiency (citrullinaemia)**. Can be mild.
   Citrulline high, orotic aciduria, no arginosuccinic acid

   e. **Arginosuccinate lyase (argininosuccinic aciduria)**. Metabolic acidosis, hepatomegaly; trichorrhexis nodosa and ataxia. Arginosuccinate in urine. Orotic aciduria.

   f. **Arginase deficiency (argininaemia)**. *Vide supra*.

(continued)

Table 10.2   (Continued) Neurometabolic disorders with episodes of acute metabolic encephalopathy

4.	**Other Amino Acid Disorders with Encephalopathy and Hyperammonaemia**

a. Hyperornithinaemia, Hyperammonaemia, Homocitrullinaemia. Ataxia, growth failure, learning difficulties.

b. Ornithine aminotransferase deficiency. Hyperammonaemia with low plasma ornithine in infancy. Progressive choroidoretinitis (gyrate atrophy) etc. with high plasma ornithine in childhood/adulthood (*Vide supra* Table 10.1 I 3).

c. Citrullinaemia type II. Neonatal cholestatic liver disease. Episodes of encephalopathy starting in 2nd/3rd decade. Symptoms include enuresis, delayed menarche, insomnia, sleep reversal, nocturnal sweats and terrors, recurrent vomiting (especially at night), diarrhea, tremors, episodes of confusion after meals, lethargy, convulsions, delusions, hallucinations, and brief episodes of coma. Avoidance of foods high in carbohydrate.

Screening:	plasma ammonia and citrulline elevated.
Diagnosis:	DNA (*SLC25A13* gene).

III. ORGANIC ACID DISORDERS

1. Propionic acidaemia. Severe acidosis, osteoporosis, neutropaenia, thrombocytopaenia, hyperglycinaemia. Neurological deficits acquired in acute attacks, and as late onset chorea.

Screening tests:	blood count and gases, raised plasma glycine, raised urine methylcitrate, propionylglycine increased propionylcarnitine in blood.
Diagnosis:	white cell propionylCoA carboxylase activity.

2. Methylmalonic aciduria. Severe acidosis, neutropaenia, thrombocytopaenia, hyperglycinaemia. Severe extrapyramidal disorder with low attenuation in globus pallidus following acute decompensation. Cobalamin responsive forms (*cblA* and *cblB*, *vide supra*) do better.

Screening tests:	blood count and gases, raised plasma glycine and methylmalonate, raised urine methylmalonate. increased propionylcarnitine in blood.
Diagnosis:	fibroblast methylmalonylCoA mutase activity.

3. Isovaleric acidaemia. Acquired neurological deficits; sweaty feet smell.

Screening tests:	urine *N*-isovalerylglycine and 3-hydroxyisovaleric acid increased isovalerylcarnitine in blood.
Diagnosis:	white cell or fibroblast isovalerylCoA dehydrogenase.

4. Glutaric aciduria type I. *Vide supra*. Approximately two thirds present with an encephalopathic crisis and on recovery have dystonia and chorea; macrocephaly.

Screening tests:	urine glutaric, 3-hydroxyglutaric and glutaconic acids, glutarylcarnitine and glutarylglycine. Reduced plasma free carnitine, increased glutarylcarnitine.
Diagnosis:	fibroblast or white cell glutarylCoA dehydrogenase activity DNA.

5. Glutaric aciduria type II (multiple acylCoA dehydrogenase deficiency). Dysmorphic features, coma, hypoglycaemia (acidosis, hyperammonaemia); renal cysts; sweaty feet smell. Classically with early death, but milder variants.

Screening tests:	urine lactate, ethylmalonic, glutaric, adipic, 2-hydroxyglutaric, suberic and sebacic acids abnormal carnitine profile with variable increases in isolvaleryl, medium chain and long chain.
Diagnosis:	fibroblast electron transfer flavoprotein or ETF-ubiquinone oxidoreductase activity.

6. 3-Hydroxyisobutyryl-CoA Hydrolase Deficiency. Hypotonia, progressive dystonia with episodes of ketosis and encephalopathy.

Screening tests:	blood hydroxy-C4 carnitine.
Diagnosis:	fibroblast HIBCH activity DNA.

7. 2-Methyl-3-hydroxybutyryl-CoA Dehydrogenase Deficiency. See above. Developmental delay and regression may be associated with episodes of acute metabolic decompensation with hypoglycaemia and/or lactic acidosis.

8. Other organic acidaemias. Acute encephalopathy and neurological damage can also occur in 3-hydroxy-3-methylglutaryl-CoA lyase deficiency, 3-oxoacyl-CoA thiolase deficiency etc.

IV. MITOCHONDRIAL FAT OXIDATION DEFECTS

1. Carnitine transporter defect. Early onset: hypoglycaemia (hypoketotic, hyperammonaemia), myopathy and cardiomyopathy. Late onset: progressive myopathy and cardiomyopathy.

Screening tests:	decreased plasma carnitine increased urine carnitine.
Diagnosis:	white cell or fibroblast carnitine transporter activity.

2. Carnitine palmitoyltransferase I deficiency. Encephalopathy, seizures; hepatomegaly; hypoglycaemia.

Screening tests:	normal or raised plasma carnitine, no abnormal urinary metabolites.
Diagnosis:	white cell or fibroblast CPT I activity.

3. **Carnitine palmitoyltransferase II deficiency**. Early onset: myopathy and cardiomyopathy. Adult: episodic myoglobinuria.

Screening tests: elevation of long chain acyl carnitines (C16, C18:1)
increased C16/C2 and C18:1/C2 ratio.

Diagnosis: muscle, white cell or fibroblast CPT II activity.

4. **Carnitine/acylcarnitine translocase deficiency**. Hypoglycaemia (hyperammonaemia); myopathy and cardiomyopathy.

Screening tests: plasma carnitine low (neonatal presentation)
long chain acyl carnitines elevated.

Diagnosis: fibroblast CAT activity.

5. **Medium chain acylCoA dehydrogenase deficiency**. Fasting induced encephalopathy, Reye's-like illness, no myopathy; hypoglycaemia (hyperammonaemia).

Screening tests: acylcarnitine profile (increased C8, C6, C10:1, low free carnitine,
urine organic acids: hexanoylglycine, C_{6-12} dicarboxylic acids.

Diagnosis: DNA (common mutation in N Europe)
fibroblast MCAD activity.

6. **Very long chain acylCoA dehydrogenase deficiency**. Fasting induced encephalopathy, hypoglycaemia, lethargy, muscle weakness, cardiomyopathy.

Screening tests: blood carnitine species profile - increased C14:1
raised plasma urate,
urine C_{6-10} dicarboxylic acids.

Diagnosis: fibroblast VLCAD activity
DNA.

7. **Short chain acylCoA dehydrogenase deficiency**. Variable failure to thrive, myopathy. May be asymptomatic.

Screening tests: urine ethylmalonate and methylsuccinate
blood butyrylcarnitine.

Diagnosis: fibroblast SCAD activity
DNA (common mutation).

8. **Long chain 3-hydroxyacylCoA dehydrogenase deficiency**. Fasting encephalopathy, myopathy, cardiomyopathy, retinitis pigmentosa, neuropathy.

Screening tests: blood carnitine profile hydroxy-C16 and hydroxy-C18:1 carnitine species
urine C_{6-14} hydroxydicarboxylic acids.

Diagnosis: fibroblast LCHAD activity
DNA.

9. **Short chain 3-hydroxyacylCoA dehydrogenase deficiency**. Most cases have presented with hypoglycaemia due, at least in part, to hyper-insulinism, and responsive to diazoxide. Reye-like presentation also reported.

Screening tests: blood hydroxy-C4 carnitine (elevated)
urine organic acids: 3-hydroxyglutaric acid increased.

Diagnosis: DNA (*SCHAD [=HADH]* gene).

V. EPILEPTIC ENCEPHALOPATHIES

Mostly neonatal onset. Severe developmental delay and may be fatal if untreated or untreatable. Variable learning difficulties in patients whose seizures respond well to treatment.

1. **Pyridoxal phosphate dependency**. (Pyridox(am)ine phosphate oxidase [PNPO] deficiency)

Screening tests: Urine organic acids (vanillactic acid)
CSF amino acids and neurotransmitter amine metabolites
CSF pyridoxal phosphate
Trial of pyridoxal phosphate treatment.

Diagnosis: DNA (*PNPO* gene).

2. **Pyridoxine dependency**. (α-Aminoadipic semialdehyde dehydrogenase [antiquitin] deficiency)

Screening tests: Urine (or plasma or CSF) α-aminoadipic semialdehyde
Plasma or CSF pipecolic acid
Trial of pyridoxine.

Diagnosis: DNA (*antiquitin [ALDH7A1]* gene).

3. **Sulphite oxidase deficiency**. *Vide supra*. Often presents with neonatal onset feeding difficulties and seizures which can be difficult to control.

Screening tests: Urine sulphite (dip stick test)
Urine sulphocysteine.

Diagnosis: Fibroblast sulph]ite oxidase activity
DNA (*SUOX* gene).

(continued)

Table 10.2 (Continued) Neurometabolic disorders with episodes of acute metabolic encephalopathy

4. **Molybdenum cofactor deficiency**. *Vide supra*

5. **Nonketotic hyperglycinaemia**. (*Vide supra*) May present with neonatal epileptic encephalopathy with burst suppression on EEG.

6. **Adenylosuccinate lyase deficiency** .(*Vide supra*) Intractable seizures can occur in early infancy.

7. **Defect of mitochondrial glutamate transporter**. Neonatal hypotonia, early onset myoclonic seizures with EEG showing burst suppression. Mutations in *SLC25A22* gene.

8. **X-Linked EIEE /West syndrome**. Early-onset epileptic encephalopathy with burst suppression on EEG followed by infantile spasms with hypsarrhythmia, and developmental arrest can be caused by mutations in the *ARX* and *STK9 (CDKL5)* genes (see also early onset encephalopathic variant of Rett syndrome below).

9. **Peroxisomal disorders**: Intractable seizures and profound hypotonia in a neonate may be presenting features. Dysmorphic features, ocular abnormalities (especially reduced ERG), epiphysial stippling etc. may be present

10. **Disorders of serine synthesis**.

    Screening tests: fasting plasma and CSF serine and glycine (low).

    Diagnosis: enzymology
         DNA.

    a. Phosphoglycerate dehydrogenase deficiency. Congenital microcephaly, psychomotor retardation, and seizures

    b. Phosphoserine aminotransferase deficiency. Intractable seizures, acquired microcephaly, hypertonia, and psychomotor retardation

11. **Disorders of creatine synthesis and transport**. *Vide supra*. Can cause intractable seizures.

12. **Causes of hypoglycaemia/hypocalcaemia/hypomagnesaemia etc**.

    Vide supra for hypoglycaemia and hypomanganesaemia. Causes of hypocalcaemia include hypoparathyroidism e.g. DiGeorge syndrome, mutations in the calcium-sensing receptor, hypomagnesaemia, vitamin D disorders, bile acid synthesis defects.

13. **Glut1 Deficiency**. *Vide supra*. Can cause anticonvulsant resistant epileptic encephalopathy

14. **Biotinidase deficiency**. *Vide supra*. Can cause neonatal epileptic encephalopathy although later presentation is more common

15. **D2-hydroxyglutaric aciduria**. *Vide supra*. Can present with early onset intractable epilepsy.

16. **Congenital Disorders of Glycosylation**. *Vide supra*. Intractable seizures can be a presenting feature of some congenital disorders of glycosylation.

VI.   MITOCHONDRIAL RESPIRATORY CHAIN/PDH/PC DISORDERS

Acute encephalopathy with lactic acidosis in infancy.

Strokes in MELAS.

VII.   PEROXISOMAL DISORDERS

Eplieptic encephalopathy in infancy—*vide supra*.

Acute encephalopathy in childhood is a rare presentation of peroxisomal disorders (e.g. α-methylacyl-CoA racemase deficiency).

VIII.   CONGENITAL DISORDERS OF GLYCOSYLATION

The following may occur: epileptic encephalopathy (see above), stroke like episodes (CDG 1a), acute encephalopathy with loss of psychomotor abilities (CDG 2h).

IX.   OTHERS

1. **Glycerol kinase deficiency**. X-linked.

    Screening tests: urine glycerol.

    Diagnosis: white cell or fibroblast GK activity.

    a. Juvenile onset. Encephalopathy, vomiting, acidosis.

    b. Adult. Benign (artefactual hypertriglyceridaemia).

    c. Complex. Contiguous gene defect involving Xpter-*adrenal hypoplasia-glycerol kinase-Duchenne muscular dystrophy*-cen.

2. **The porphyrias**. Intermittent neuropathic symptoms in 10%, triggered by drug, hormonal, nutritional or unknown factors. Abdominal pain and vomiting; neuropathic pain and neuropathy (motor, sensory, cranial or autonomic), psychiatric.

    a. δ-aminolevulinic acid dehydratase porphyria. Recessive.

    Screening tests: increased urine δ-aminolevulinic acid, normal porphobilinogen.

    Diagnosis: red cell δ-aminolevulinic acid dehydratase activity.

    b. Acute intermittent porphyria. Dominant.

    Screening tests: increased urine δ-aminolevulinic acid and porphobilinogen.

    Diagnosis: red cell porphobilinogen deaminase activity.

    c. **Hereditary coproporphyria.** Dominant, photosensitivity in 30%.

      <u>Screening tests:</u>   increased urine and faecal coproporphyrinogen III.

      <u>Diagnosis:</u>   hepatic coproporphyrinogen oxidase activity.

    d. **Variegate porphyria.** Dominant, photosensitivity.

      <u>Screening tests:</u>   faecal protoporphyrinogen IX and coproporphyrinogen III.

      <u>Diagnosis:</u>   hepatic protoporphyrinogen oxidase activity.

3.   <u>**Bilirubin encephalopathy**</u> (Kernicterus). In an infant with bilirubin >350μM: Hypotonia, hyporeflexia, athetosis, opisthotonus. May be fatal. Long-term sequelae include hearing loss, cranial nerve palsy, movement disorder, developmental delay.

    Crigler-Najjar syndrome. (Bilirubin UDP glucuronyltransferase deficiency)

    Neonatal haemolytic jaundice. (e.g. glucose-6-phosphate dehydrogenase deficiency)

4.   <u>**Early onset encephalopathic variant of Rett syndrome**</u>. Severe infantile encephalopathy due to *MECP2* and *CDKL5* (*STK9*) mutations. Onset of symptoms in girls with Rett syndrome is usually after the age of 6 months. Boys with an *MECP2* mutation on their X chromosome can have severe neonatal onset severe encephalopathy with profound hypotonia, apnoea/respiratory insufficiency, seizures. Later stereotypic movements, limb rigidity, movement disorder, lack of purposeful movements occur. Similar features can result from *CDKL5* mutations.

10.1.2 When to suspect a neurometabolic disorder

The child with a neurometabolic disease can present in a number of ways. These children typically have a normal birth history and normal early development. Later on development slows, plateaus, and then declines with increasing disability and eventual death. The age at onset and rate of progression depends on the particular disorder. In children this interface between the effects of the disease and the normal developmental process often makes it difficult to recognize the serious underlying nature of the disorder, especially early on. Thus the clinical course of a child with a neurometabolic disorder may, in the early stages, be difficult to distinguish from the child with a relatively 'static' neurodevelopmental disorder, such as Rett syndrome or Autism, or from the child with an epileptic encephalopathy. The term 'static' is used loosely here whilst acknowledging that 'progression' or clinical evolution, either negative or positive, is seen in conditions like Rett syndrome and Autism. This early difficulty in recognizing that a condition is progressive has probably led to underreporting of some disorders, such as juvenile neuronal ceroid lipofuscinosis, in the progressive intellectual and neurological deterioration study.

A neurometabolic disorder should be particularly suspected in an infant or child presenting with neurological symptoms and any of the following:

- A progressively deteriorating neurological or developmental disorder.
- Evidence of diffuse neurological involvement, for instance a combined disorder of the brain and peripheral nerves.
- Parental consanguinity or a known family history.
- Evidence of leukodystrophy or other symmetric abnormalities on initial brain MRI.
- Dysmorphic features (Table 10.3).
- Spinal deformity raises the question of mucopolysaccharidoses (Section 10.4.5) or the newly described congenital disorders of glycosylation (Table 10.1, VI).
- Dysfunction of other organs, for instance the liver in Wilson's disease, or the heart in Pompé disease.

In addition, neurometabolic disorders enter into the differential diagnosis in children with:

- Progressive ataxias of childhood (Table 10.4).
- Dystonias and other movement disorders (Table 10.5).
- Childhood epilepsies, especially if myoclonic (Table 10.2, V) (Table 30.1).
- Presentation with cognitive decline (Table 10.6).

Neurometabolic disorders can be categorized also into those which are associated with episodes of acute encephalopathy (Table 10.2) and those which are not associated with acute encephalopathy (Table 10.1).

10.1.3 Investigation of a possible neurometabolic disorder

A number of variables need to be considered when planning the investigation of a child with a possible neurometabolic disease. Age is of most importance, closely followed by the details of the history and examination. Although some advocate a blanket screen of investigations to consider in any child presenting with specific symptoms there is a need to try and rationalize the approach by undertaking relevant investigations in a logical sequence. Inevitably there is a tension between the desire to reach a timely and accurate diagnosis and to detect important diagnoses with potential genetic and treatment implications, whilst avoiding subjecting the child to a variety of unpleasant, expensive, and potentially unnecessary tests. The approach to investigation will vary between individuals, and is ever changing with the growth of knowledge and the availability of new investigations.

Brain MRI

This is the single investigation that has most helped the practising paediatric neurologist in the past 20 years. It is essential early in the diagnostic process. CT can also be enormously helpful particularly in the infant presenting with an acute encephalopathy, with its more obvious demonstration of calcification and acute haemorrhage. Brain imaging will help to distinguish some of the neurodegenerative diseases and is particularly important in revealing a

Table 10.3 Clinical clues to the diagnosis of metabolic diseases of the nervous system (modified from Menkes *et al.* 2005)

Clue	Diagnosis
Cutaneous abnormalities	
Increased pigmentation	Adrenoleukodystrophy
Telangiectases (conjunctiva, ears, popliteal areas)	Ataxia-telangiectasia
Perioral eruption	Multiple carboxylase deficiency
Abnormal fat distribution	Congenital disorders of glycosylation
Angiokeratoma (red macules or maculopapules) of hips, buttocks, scrotum	Fabry disease, sialidosis, fucosidosis type II
Oculocutaneous albinism	Chédiak–Higashi syndrome
Xanthomas	Cerebrotendinous xanthomatosis
Subcutaneous nodules	Ceramidosis (Farber disease)
Ichthyosis	Sjögren–Larsson syndrome (spasticity, seizures)
	Refsum disease (neuropathy, ataxia, phtylanic acid)
	Dorfman-Chanarin syndrome (lipid storage in muscle, granulocytes)
Inverted nipples	Congenital disorders of glycosylation
Mongolian spots	GM_1 gangliosides
Abnormal urinary or body odour	
Musty	Phenylketonuria
Maple syrup or caramel	Maple syrup urine disease
Sweaty feet or ripe cheese	Isovaleric acidemia
Sweaty feet	Glutaric academia type II
Cat urine	3-methylcrotonyl-CoA carboxylase deficiency
	Multiple carboxylase deficiency
Hair abnormalities	
Alopecia	Multiple carboxylase deficiency
Kinky hair	Menke's kinky hair disease
	Argininosuccinic aciduria
	Multiple carboxylase deficiency
	Giant axonal neuropathy
	Trichothiodystrophy (Pollitt syndrome; mental retardation, seizures)
Unusual facies	
Coarse	Mucopolysaccharidoses (Hunter–Hurler syndrome)
	I-cell disease (mucolipidosis II)
	GM_1 gangliosidosis (infantile)
	Sanfilippo syndrome
Slight coarsening (compared to other family members)	Mucolipidosis III (pseudo-Hurler dystrophy)
	Fucosidosis II
	Mannosidosis
	Sialidosis II
	Aspartylglucosaminuria
High nasal bridge, prominent jaw, large pinnae	Congenital disorders of glycosylation
Ocular abnormalities	
Cataracts	Galactosemia
	Cerebrotendinous xanthomatosis
	Homocystinuria
Corneal clouding	Hurler syndrome
	Hunter syndrome (late in severe cases)
	Morquio syndrome
	Maroteaux–Lamy syndrome
Cherry-red spot	Tay–Sachs, Sandhoff diseases (GM_2 gangliosidosis)
	GM_1 gangliosidosis (infantile)
	Niemann–Pick disease (types A and C)
	Infantile Gaucher disease (type II)
	Sialidosis

Table 10.4 The range of childhood ataxias

Neurometabolic disorders

DNA repair defects (Section 39.7)

   Ataxia-telangiectasia (Sections 11.5 and 39.7.1)

Metabolic ataxias (Section 39.5)

Leukoencephalopathies (Section 10.2)

Mitochondrial disorders (Sections 10.5 and 39.6)

   Coenzyme Q deficiency (Sections 10.5.5 and 39.6)

   Pyruvate dehydrogenase deficiency (Sections 10.5.5 and 39.5.1)

   Leigh's syndrome (Sections 10.5.3 and 39.6)

Hereditary degenerative disorders (Sections 39.3 and 39.4)

Pontocerebellar hypoplasia types I and II (Section 39.3.1)

Congenital early onset ataxias (Section 39.3)

Early onset hereditary degenerative ataxias including Friedreich's ataxia (Section 39.4)

Acquired in childhood (Section 39.8)

Opsoclonus-myoclonus syndrome (Section 38.4.5)

Post-infectious acute cerebellar ataxia (Section 42.3.3)

Toxins (Section 39.11.1)

Gluten sensitivity or coeliac ataxia (39.11.9)

Posterior fossa structural disorders

Table 10.5 Dystonias and other movement disorders in children

Primary torsion dystonias (Section 40.4.2)

Dopa responsive dystonia (Section 40.4.4) and Tetrahydrobiopterin deficiency (Table 10.1, I1b)

Rapid onset dystonia-parkinsonism (Section 40.4.6)

Paroxysmal kinesigenic choreoathetosis (Section 40.4.7)

Paroxysmal dystonic choreoathetosis (Section 40.4.7)

Dyskinetic athetoid cerebral palsy and Kernicterus (Section 40.4.8)

Deafness-dystonia, Mohr Tranebjaerg syndrome (Section 40.4.14)

Ataxia-telangiectasia (Sections 11.5; 39.7.1; 40.4.15)

Huntington's disease, juvenile Westphal variant (Section 40.5.2)

Benign hereditary chorea (Section 40.5.5)

Sydenham's chorea (Section 40.5.7)

Other post-streptococcal movement disorders (Section 40.10)

Gilles de la Tourette's syndrome (Section 40.6.3)

Hereditary hyperekplexia (Section 40.11.11)

Mirror movements (Section 40.11.12)

Wilson's disease (Section 40.8)

Metabolic disorders

Organic acidurias:

   3-methylgutaconic aciduria Type III, Costeff's optic atrophy (Table 10.1, II 4c)

   Glutaric aciduria type I (Table 10.1, II 5)

   Other (Table 10.2, III)

   L2-hydroxyglutaric aciduria (Table 10.1, II 2; Section 10.2.8)

   2-methyl-3-hydroxybutyryl-CoA dehydrogenase deficiency (Table 10.1, II 7)

Mitochondrial disorders (Table 10.1, III 1,3,4)

Lysosomal disorders

   Sialic acid storage disorders (Table 10.1 V5d)

   Niemann–Pick disease type C (Table 10.1 V9)

   Gaucher's disease type 3 (Table 10.1, V 10c)

   GM_2 gangliosidosis (Table 10.1, V 15)

   Neuronal ceroid lipofuscinosis, or Batten's disease (Table 10.1, V 16)

Lesch–Nyan syndrome (Table 10.1, VII 1a; Section 10.3.3)

Ureidopropionase deficiency (Table 10.1, VII 2c)

Caeruloplasmin deficiency (Table 10.1, VIII 1a; Section 40.8)

Congenital folate malabsorption (Table 10.1, IX 1a)

Rett syndrome (Table 10.2, IX 4; Sections 9.6.2; 40.4.15)

Creatine synthesis and transport abnormalities (Table 10.1, XIII)

Red cell glycolytic defects (Table 10.1, XV)

Carbohydrate disorders (Table 10.2, I)

Nonketotic hyperglycinaemia (Table 10.2, II 2)

Pantothenase kinase deficiency (Section 40.4.12)

Neuroferritinopathy (Section 40.5.4)

leukodystrophy (Table 10.7). It is important to remember that the appearance of an MR brain scan changes significantly in the normal child with increasing age. This is mainly due to the process of myelination, of which there is little evidence at birth before an almost normal adult pattern is reached by 5 years. Most of this developmental change in myelination occurs in the first 2 years.

Electroencephalography

The EEG is particularly important in the child with seizures and developmental regression. It can give diagnostic and other specific information concerning Hypsarrhythmia (Section 30.4.2), subacute sclerosing panencephalitis (Section 42.3.7), Battens disease or neuronal lipoid cerdjuscinosis (Section 30.3.4), and Alper's disease (Section 30.3.4). In general, it should be noted that the EEG is of most value in children with the epileptic encephalopathies which occur more commonly than epilepsies due to neurometabolic disease (Section 30.5.2).

CSF

The lumbar puncture remains essential in the differential diagnosis of children with serious neurological disease. The CSF examination should routinely include cell count and culture, with protein, glucose, and lactate. A low CSF glucose might suggest glucose transporter deficiency, a condition with an expanding phenotype (Klepper and Leiendecker 2007). A high lactate suggests a mitochondrial disorder; the CSF lactate is often considered more informative than the blood lactate. CSF amino acid estimations are diagnostically helpful in the infant presenting with an early onset epileptic encephalopathy; with high CSF glycine in non-ketotic hyperglycinemia and low serine in the newly described serine deficiency syndromes (Section 10.7.1). The CSF can be examined for neurotransmitter metabolites, abnormal in Segawa syndrome and other dystonias, and for pterins and folate metabolites (Garcia-Cazorla *et al.* 2007; Pearl *et al.* 2007). In addition to the search for clues to

a neurometabolic disorder the CSF can be examined for evidence of a wide range of infectious agents; for instance subacute sclerosing panencephalitis remains an important cause of degenerative neurological disease worldwide (Section 42.3.7). Inflammatory neurological disease is relatively common in children, such as acute disseminated encephalomyelitis (Section 37.4.1) and occasionally multiple sclerosis (Section 37.5), although the diagnostic utility of CSF oligoclonal bands is limited (Section 3.6.8).

Table 10.6 Some neurometabolic and developmental, conditions which can present with cognitive deterioration

Neurodevelopmental disorders

Hydrocephalus (Section 9.3.2)

Rett syndrome (Sections 9.6.2; 10.7.2)

Autism and childhood disintegrative disorder (Section 9.6.4)

Leukodystrophies

Metachromatic leukodystrophies (Section 10.2.2)

Krabbe disease (Section 10.2.3)

Adrenoleukodystrophies (Section 10.2.4)

Vanishing white matter leukoencephalopathy (Section 10.2.5)

Primary grey matter degenerations

Neuronal ceroid lipofuscinoses (Section 10.3.2)

Mitochondrial disorders

Some (Table 10.1, III.4)

Lysosomal disorders

GM_1 gangliosidoses types 2 and 3 (Table 10.1, V.14; Section 10.4.3)

GM_2 gangliosidosis; Tay Sachs disease (Section 10.4.2)

Niemann Pick disease Type C (Section 10.4.4)

Sanfillipo disease; mucopolysaccharidosis Type III (Section 10.4.5)

Schindler disease (Table 10.1, V.4)

Abnormalities of glycoprotein degradation (Table 10.1, V.5)

Aminoacid disorders

Phenylketonuria (Table 10.1, I.1a)

Homocystinuria (Table 10.1 I, 4b.c)

Other

Huntington's disease, juvenile Westphal variant (Section 40.5.2)

Wilson's disease (Section 40.8)

Cerebrotendinous xanthomatosis (Table 10.1, XI.1; Section 21.8.9)

Xeroderma pigmentosum (Table 10.1, XVII.1b; Section 11.6.7)

Methylenetetrahydrofolate reductase deficiency (Table 10.1 IX, 1b)

Table 10.7 MRI evidence of leukodystrophy in childhood

Metabolic

Metachromatic leukodystrophy (Sections 10.2.2; 21.8.2 ; 37.7.3)

Krabbe disease (Sections 10.2.3; 21.8.3; 37.7.1)

Adrenoleukodystrophy (Sections 10.2.4; 21.8.4; 37.7.2)

Vanishing white matter disease (Sections 10.2.5; 37.7.6)

Cree leukoencephalopathy (Section 10.2.5)

Ovarioleukodystrophy syndrome (Section 10.2.5)

Megaloencephalic leukoencephalopathy with subcortical cyst (Section 10.2.6)

Alexander disease (Section 10.2.7)

L-2-hydroxyglutaric aciduria (Section 10.2.8)

Canavan disease (Section 10.2.9)

Pelizaeus–Merzbacher disease (Section 10.2.10)

Peroxisomal biosynthesis disorders (Section 10.6)

Developmental disorders

Cockayne's syndrome (Section 11.6.8)

Some mitochondrial disorders (Section 10.5)

Other

Acute disseminated encephalomyelitis (Section 37.4.1)

Blood tests

A large range of possible blood tests can be undertaken for suspected neurometabolic disorders. Liver function and coagulation are deranged in many metabolic conditions and the blood glucose and lactate should be available for comparison with CSF values. The glucose, ammonia, pH, and lactate should be measured in any child with encephalopathy. Blood amino acids should be routinely checked in all children with unexplained encephalopathy. Some amino acid disorders can mimic other neurological disease, for instance arginase deficiency can mimic cerebral palsy (Prasad *et al.* 1997) and the polyneuropathy of tyrosinaemia can resemble Guillain Barré syndrome (Noble-Jamieson *et al.* 1994). Biotinidase deficiency should be considered in any child with seizures but can present with other progressive neurology (Rahman *et al.* 1997).

The range of further testing depends on the diagnostic possibilities. An exhaustive listing of the huge number of metabolic diseases that can present in children and adults summarizes the diagnostic tests for each (Tables 10.1 and 10.2). Observation of vacuolated lymphocytes on routine blood film raises the possibility of various inherited metabolic disorders, particularly juvenile forms of neuronal ceroid lipofuscinosis or GM1 gangliosidosis (Fig. 10.2).

Any request for 'leukocyte enzymes' needs to be accompanied by awareness of the diagnostic possibilities to guide the range of conditions for which the laboratory will test. Not all laboratories do the same tests, some offering screens which vary depending on the clinical details which were provided. Often proper targeting of testing is compromised by provision of insufficient clinical information when the tests are requested by an inexperienced clinician. Good communication between the clinician and the laboratory will help ensure appropriate and prompt investigation.

The recent elucidation of the genetics and metabolic causes of pyridoxine-dependent epilepsy will improve identification of this rare but important treatable disorder. All infants with refractory epilepsy should have measurement of pipecolic acid in blood and CSF (Willemsen *et al.* 2005; Bok *et al.* 2007) or α-aminoadipic semialdehyde in urine (Mills *et al.* 2006). Indeed whilst awaiting these results a therapeutic trial with pyridoxine, biotin, and folinic acid should be considered in any infant with an unresponsive epilepsy (Been *et al.* 2005).

Disorders of glycosylation are suggested by the association of dysmorphic features including inverted nipples and subcutaneous fat pads, micro- or macrocephaly, or spinal deformity with a neurological disorder in children aged less than 5 years. These should be investigated by electrophoretic separation of transferrin isoforms, and other blood glycoproteins, and with other investigations (Table 10.1, VI; Section 10.7.2).

Urine

Urine sampling is important in any child presenting with possible neurometabolic disease and should be sent routinely for organic acid estimation. Urine metabolic screening tests are also important particularly in the sick neonate. Urine amino acid screening is not usually necessary in the child without encephalopathy provided blood amino acids and urine organic acids have been checked. The range of possible screening investigations in urine is large. Screening for the mucopolysaccharidoses is essential in all infants with developmental slowing. Other screening tests include urine sulphite in infants with early onset difficult epilepsy. Sulphite oxidase deficiency can mimic hypoxic ischaemic encephalopathy

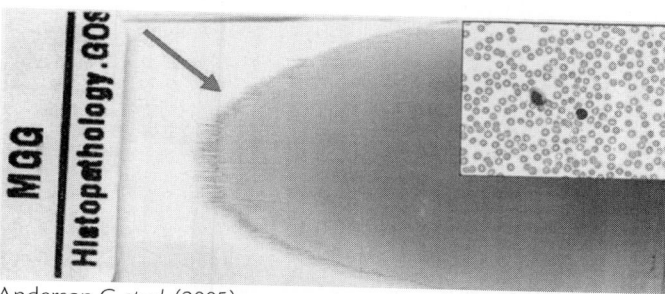

Anderson G *et al.* (2005)

A

Diagnosis	N	% Of total
Juvenile neuronal ceroid lipofuscinosis	49	31.4
GM1 gangliosidosis	14	9
Galactosialidosis	7	4.5
Salla disease	2	1.3
Neuraminidase deficiency	2	1.3
Pompe's disease/adult acid maltase deficiency	24/12	15.4/7.7
Mannosidosis	2	1.3
Fucosidosis	3	1.9
I cell disease	6	3.8
Niemann-Pick A	4	2.6
Mucopolysaccharidosis	7	4.5
No specific confirmation	24	15.4

B

Fig. 10.2 Vacuolated lymphocytes. (A) Low power image of a blood film, illustrating the correct area in which to look for the presence of vacuolated lymphocytes (arrow), and (inset) high power photomicrograph demonstrating a small vacuolated lymphocyte and monocyte (May–Grunwald–Giemsa staining; original magnification, ×400). (See Plate 14.) (B) Table showing diagnoses in 156 patients in whom vacuolated lymphocytes were detected on peripheral blood film examination. (From Anderson *et al.* 2005.)

(Hobson *et al.* 2005). Urine screening is now available for a number of disorders of purine and pyrimidine metabolism and should be considered in the child with severe learning disability with autism, who might have adenylosuccinase deficiency and dystonia, and with possible Lesch–Nyan syndrome. Urine can also be examined for creatine and guanidinoacetate so as to detect recently described disorders of creatine biosynthesis and transport (Table 10.1, XIII). Such disorders present with hypotonia, developmental standstill, and seizures.

10.2 Leukodystrophies and brain dysmyelination

10.2.1 Overview

Leukodystrophies are generally diagnosed on the basis of abnormal white matter on brain MRI (Table 10.7) in children or adults under investigation for developmental delay, cognitive impairment, or demyelinating peripheral neuropathy, often in varying combinations.

Neuropathology can distinguish between:

♦ Demyelination: destruction of normally formed myelin;

♦ Hypomyelination; too little myelin;

♦ Dysmyelination: formation of abnormal myelin; and

♦ Delayed myelination: immature pattern of myelination.

It is more difficult to be sure about these pathological categorizations from MR images. Nevertheless, demyelination is suggested by high signal from the white matter on T2-weighted images with low signal on T1-weighted images; dysmyelination or hypomyelination by high signal from the white matter on T2-weighted images with normal signal on T1-weighted images; and delayed myelination by the pattern of normal myelin signal on T2- and T1-weighted images with respect to age (Fig. 10.3).

10.2.2 Metachromatic leukodystrophy

Metachromatic leukodystrophy is named because of the appearance of accumulation of the sulphatide galactosylceramide-3-O-sulphate in neural tissue. This appears metachromatic using specific histological stains. This disorder is recessively inherited and caused by deficiency of arylsulphatase A.

Traditionally clinical subgroups have been defined based on the age of onset: late infantile, juvenile, and adult (Table 10.1, V 12) (Section 37.7.3). More recent studies have correlated genotype with phenotype, addressing whether mutations result in absence of aryl-sulphatase A activity or whether some residual activity remains. These suggest that the late infantile form of the disease shows little phenotypic variability but that the juvenile and adult forms form a continuum, alternatively described as late onset metachromatic leukodystrophy.

Infantile forms

By far the commonest presentation, occurring in approximately half, is the late infantile form of metachromatic leucodystrophy. Here, first symptoms start towards the end of the first year of life or during the second year. The first manifestation is a gait disturbance due either to the polyneuropathy, as evidenced by hypotonia and absent tendon reflexes (Section 21.8.2), or to cerebral white matter degeneration, manifesting as hypertonia, pathologically brisk reflexes, and extensor plantar responses. These symptoms progress over a few months with the loss of motor skills. At this stage, dementia becomes apparent with loss of language and other cognitive skills. The motor disorder also progresses with the development of ataxia and progressive involvement of the pyramidal system and the child becomes bedridden. The children then become blind and may develop epilepsy. Finally, they become decerebrate (Hagberg 1963). Most children will die within 5 years of disease onset.

Late onset forms

First symptoms of the late onset forms usually occur between 3 to 60 years of age and the disease shows marked clinical heterogeneity in both its presentation and its course. Both juvenile onset, usually between ages of 3 and 10 years, and adult onset, usually after the age of 15 years, are described. There are two common presentations: behavioural and motor and MRI shows leukodystrophy (Fig. 10.4A). The presenting behavioural changes may be a change in personality, loss of social inhibition, dementia, or frank psychosis. These may remain isolated for many years, but gradually an ataxic-pyramidal tract motor disorder and a slowly progressive blindness caused by optic atrophy emerge. Those presenting with a motor disturbance usually have a slowly progressive pyramidal paraparesis or a progressive cerebellar ataxia. Dementia, blindness, and

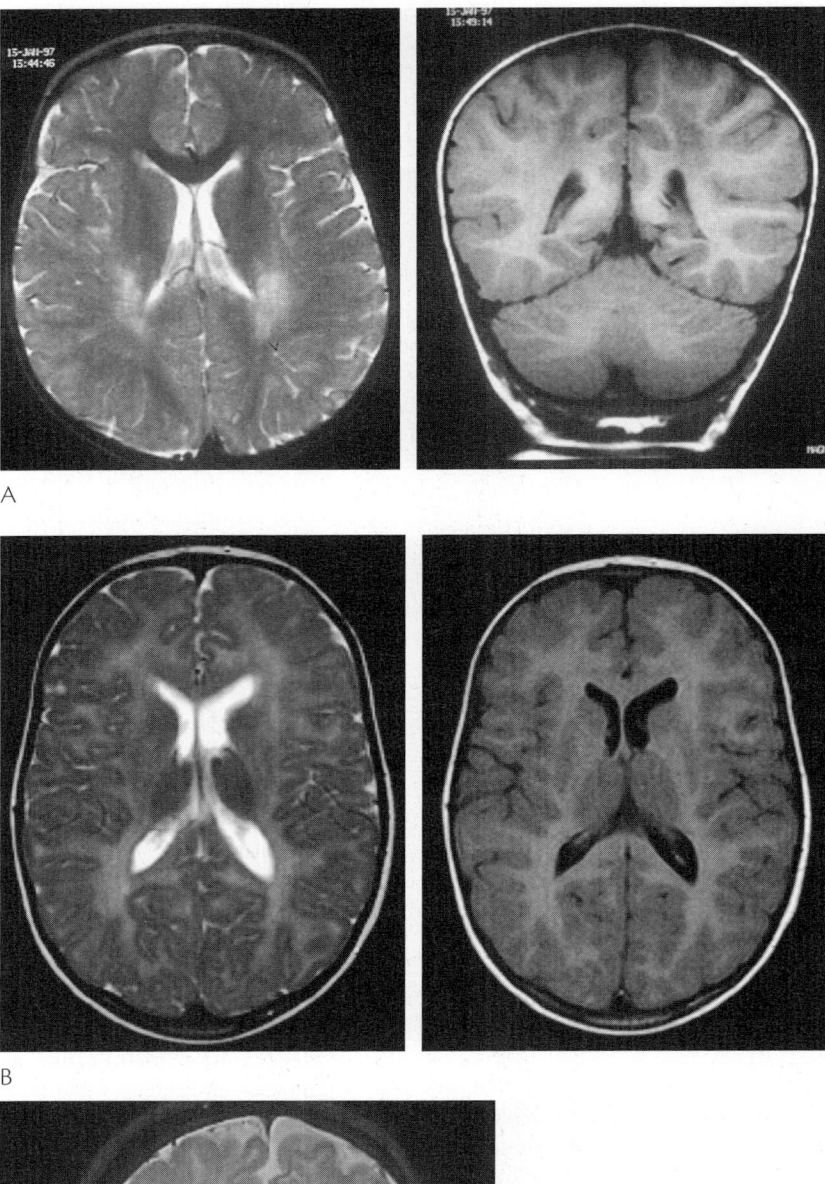

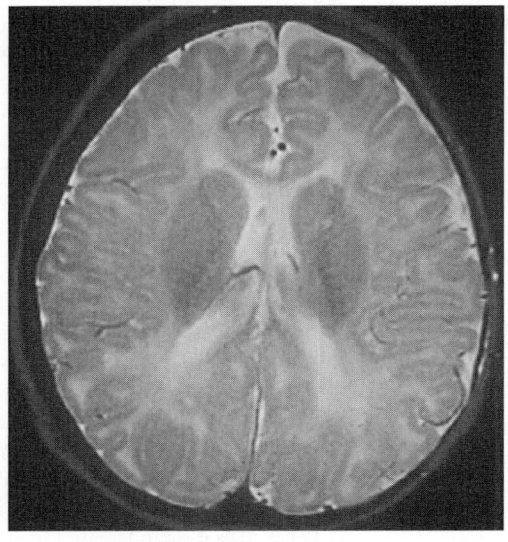

Fig. 10.3 MRI features of white matter pathology. (A) Demyelination. Late infantile Krabbe disease. High signal from posterior white matter on T2-weighted axial image on the left, low signal from parietal white matter on T1-weighted coronal image on the right. (B) Dysmyelination or hypomyelination. Pelizaeus-Merzbacher disease. High signal from hemispheric white matter on T2-weighted axial image on the left, normal signal from hemispheric white matter on T1-weighted axial image on the right. (C) Delayed myelination. Delayed development of unknown aetiology in a 9-month-old infant. The only normal myelin signal is seen in the anterior and posterior limbs of the internal capsule representing an age equivalent to 2–3 months.

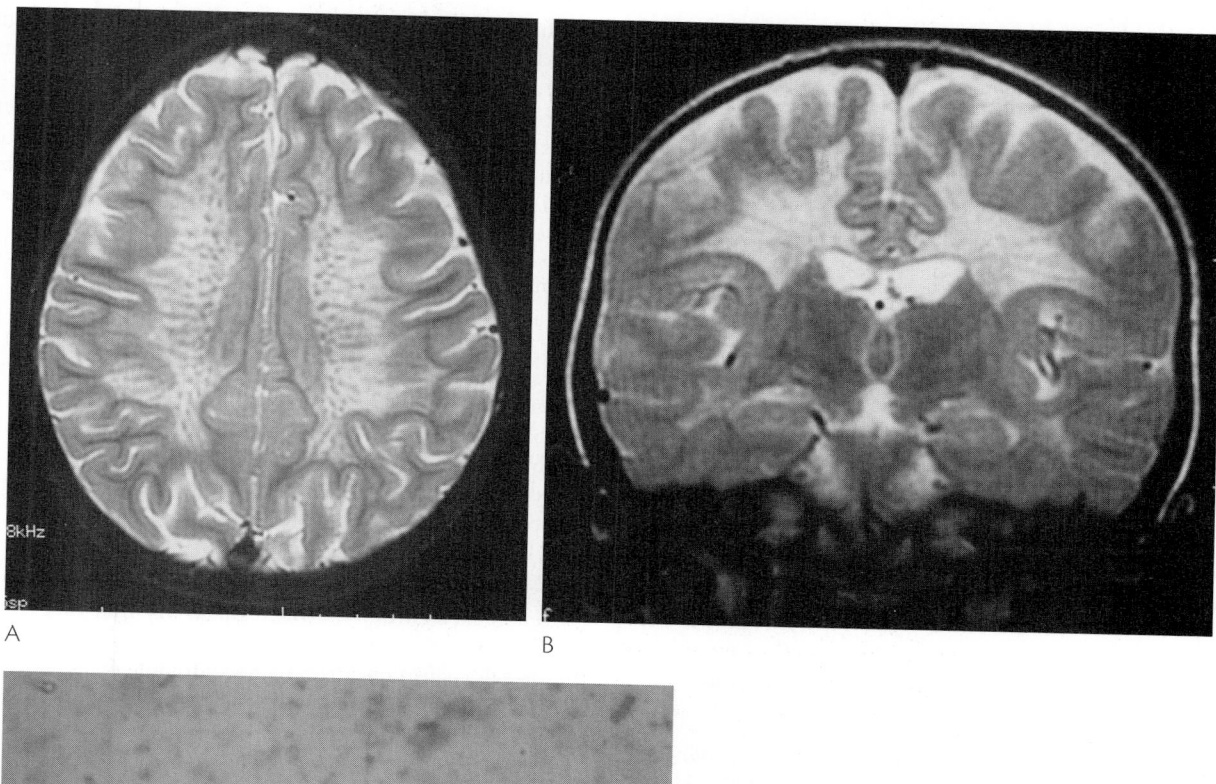

A

B

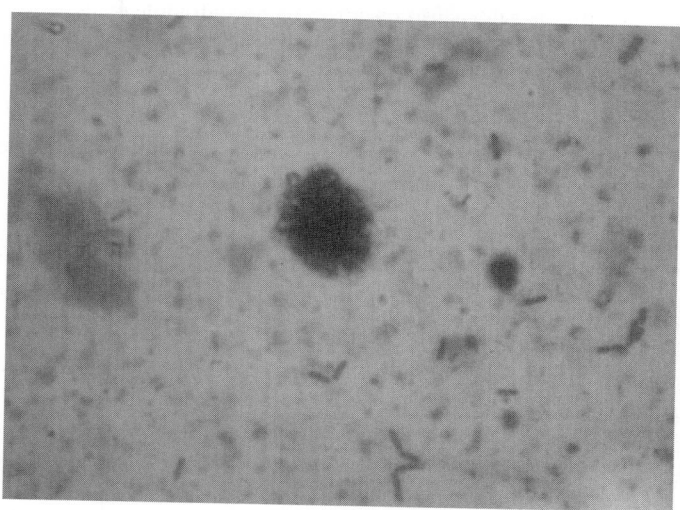

C

Fig. 10.4 Late onset metachromatic leukodystrophy. T2-weighted MRI (A) axial and (B) coronal showing diffusely altered cerebral white matter signal. (Courtesy of Dr P. Anslow). (C) Urinary deposit stained with toluidine blue showing a shed epithelial cell containing brown metachromatic material. (Courtesy Dr W. Squier.) (See Plate 15.)

sometimes epilepsy emerge later. With both presentations a demyelinating peripheral neuropathy is found on nerve conduction studies, but is rarely symptomatic or detectable clinically (Section 21.8.2). The course of the disease is also variable. Some patients have a disorder which evolves rapidly and die within 5 years. In others the course is more protracted course and evolves over a decade or so (Baumann *et al.* 1991; Rauschka *et al.* 2006).

Investigations

MRI of the brain shows features of demyelination, but the pattern of involvement is entirely non-specific. In late infantile onset metachromatic leukodystrophy, the MRI initially shows involvement of the posterior cerebral white matter later spreading to the entire hemispheric white matter, dentate nucleus, and the

brain stem. In late onset metachromatic leukodystrophy it is often the frontal lobe white matter that is initially involved.

In late infantile metachromatic leukodystrophy there is almost invariably a motor sensory demyelinating polyneuropathy with abnormal nerve conduction studies. These nerve conduction abnormalities can precede symptoms.

The diagnosis is made by demonstrating deficient activity of arylsulphatase A in peripheral blood leukocytes or cultured skin fibroblasts. However, this may not be straight-forward because there are two pseudodeficiency alleles present in up to 10 per cent of most populations, which, when combined, can reduce arylsulphatase A activity to approximately 8 per cent of normal. Ancillary methods to help diagnosis are the presence of metachromatic staining cells in a sample of fresh urine (Fig. 10.4C) and urinary sulphatide excretion. Rarely, metachromatic leukodystrophy can be caused by deficiency of saposin B, a sulphatide activating protein.

Treatment

Currently, there is no effective treatment to cure or arrest late-infantile metachromatic leukodystrophy. Haematopoietic stem cell therapy may improve selected patients with late onset metachromatic leukodystrophy (Sevin *et al.* 2007a).

10.2.3 Krabbe disease

Krabbe disease is named after the author of the first clinical and neuropathological description. It is also called globoid cell leukodystrophy because the presence of multinucleated giant macrophages, containing galactocerebroside, in the perivascular spaces of white matter is the histological hallmark of the disorder. It is recessively inherited and caused by deficiency of galactocerebroside β-galactosidase.

Based on age of onset, there are two clinical subgroups: early infantile and juvenile (Table 10.1, V11 (Section 37.7.1)). Early infantile Krabbe disease makes up approximately 90 per cent of cases and symptoms start at 3–4 months of age. Late onset Krabbe disease usually becomes symptomatic before the age of 6 years, but cases presenting in adulthood are increasingly recognized.

Early infantile Krabbe disease

This has a relentlessly progressive course. Affected infants are usually normal for the first 3–4 months of life. Then they become miserable and hypersensitive to tactile and auditory stimuli. Such irritability should alert the clinician to this diagnostic possibility and it may mask or predate the neurological decline. They may develop a fever of central origin and difficulties with feeding causing faltering growth. This period is followed by rapid intellectual and motor deterioration with the development of opisthotonus and marked pyramidal tract signs. Virtually all patients with early infantile Krabbe disease have peripheral nerve involvement and deep tendon reflexes may be absent (Section 21.8.3). The infants become blind with optic atrophy and may develop seizures. The deterioration continues and the infants become decerebrate towards the end of the first year of life and death usually occurs before the age of two (Hagberg *et al.* 1963).

Late onset Krabbe disease

This has a more protracted course with an initially rapid development of a pyramidal hemiplegia or quadriplegia, visual loss, ataxia, and dementia, followed by a more gradual progression of symptoms (Kolodny *et al.* 1991; Lyon *et al.* 1991). Survival is variable, depending upon age of onset, but can be prolonged over decades. Cases presenting in adulthood often have symptoms initially suggestive of a progressive spastic paraparesis or motor neurone disease.

Investigation

MRI of the brain in early infantile Krabbe disease shows signal abnormalities in the cerebellar white matter and dentate nuclei, the basal ganglia, thalamus, and the pyramidal tracts (Fig. 10.3A). There is often prominent atrophy of both grey and white matter structures. By contrast, patients with late onset Krabbe disease have involvement of the pyramidal tracts, parietóoccipital white matter, and corpus callosum (Loes *et al.* 1999). Early infantile Krabbe disease also causes a demyelinating peripheral neuropathy and

virtually all cases have early abnormalities of nerve conduction (Siddiqi *et al.* 2006) (Section 21.8.3). CSF protein concentration is raised in symptomatic early infantile Krabbe disease.

The diagnosis is made by demonstrating deficient activity of galactocerebroside β-galactosidase in peripheral blood leukocytes or cultured skin fibroblasts. Very rarely, Krabbe disease can be caused by deficiency of saposin A, a galactocerebroside β-galactosidase activating protein.

Treatment

Advances in haematopoietic stem cell treatment have led to this becoming the treatment of choice in late onset Krabbe disease in which it can reverse some neurological symptoms. Haematopoietic stem cell treatment is used also in presymptomatic Krabbe disease, most of whom would develop early infantile Krabbe disease, where it greatly ameliorates development of symptoms (Krivit *et al.* 1998; Escolar *et al.* 2005). Unfortunately, symptomatic early infantile Krabbe disease is untreatable.

10.2.4 Adrenoleukodystrophy

Adrenoleukodystrophy is an X-linked disorder that is caused by mutations in the *ABCD1* gene that encodes a peroxisomal membrane ABC half transporter named adrenoleukodystrophy protein (Table 10.1, IV 3b). Mutant adrenoleukodystrophy protein causes many different phenotypes:

- acute cerebral forms of childhood, adolescent and adult onset;
- adrenomyeloneuropathy with or without cerebral involvement in hemizygotes; cerebellar in hemizygotes or heterozygotes;
- adrenal failure alone; and
- asymptomatic in hemizygotes or heterozygotes (Moser *et al.* 2005a).

Acute childhood cerebral adrenoleukodystrophy

This is the commonest manifestation. It presents between three and ten years of age with behavioural changes of attention difficulties and emotional lability. This is followed by increasing difficulties with auditory processing and vision. Affected boys then develop a progressive pyramidal and cerebellar motor disorder, dementia and often epilepsy. The disease rapidly progresses and the children become decerebrate over one to two years. Adolescent and adult onset acute cerebral forms are much rarer but have a similar rapidly progressive course (Section 37.7.2). Approximately 90 per cent of males with acute cerebral adrenoleukodystrophy also have adrenal insufficiency at presentation.

MRI of the brain shows demyelination in the parieto-occipital white matter (Fig. 10.5). There is contrast enhancement at the leading edge of the demyelination. Computed tomography of the brain shows similar white matter lesions and may show calcification within these areas.

Adrenomyeloneuropathy

This is the second most common manifestation. It presents in the second or third decade with a progressive spastic paraparesis. Additionally there may be bladder sphincter involvement, impotence, and signs of a mild sensory axonal peripheral neuropathy (Section 21.8.4). After around 10 years of symptoms, a fifth of males will develop the acute cerebral form. Around 70 per cent of males with adrenomyeloneuropathy have adrenal insufficiency.

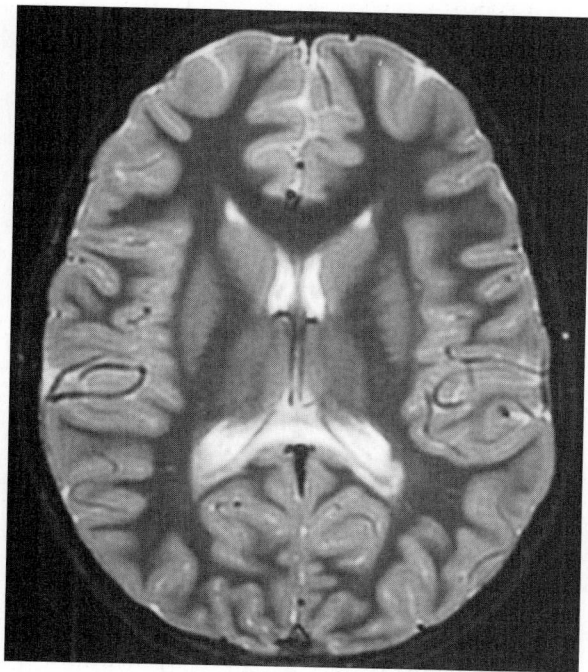

Fig. 10.5 Adrenoleukodystrophy. T2-weighted axial MRI showing predominantly posterior white matter high signal change. (Courtesy of Dr P. Anslow.)

Rarely, adrenomyeloneuropathy can present with a progressive ataxia. It is estimated that approximately half of female heterozygotes will develop adrenomyeloneuropathy; but this is of later onset, does not progress to an acute cerebral form and is not associated with adrenal insufficiency.

MRI of the spinal cord usually shows atrophy only but there may be signal changes in the pyramidal tracts. Nerve conduction studies show a mild axonal polyneuropathy.

Diagnosis

Diagnosis is made in the index patient by demonstrating raised plasma concentrations of the saturated, unbranched very long chain fatty acids, VLCFA. The most usual VLCFA measured is hexacosanoic acid, C26:0; both its absolute concentration and its ratios to docosanoic C22:0 and tetracosanoic C24:0 acids need to be examined. The diagnosis should be confirmed by mutation analysis of the *ABCD1* gene. This is necessary because female heterozygotes may have normal plasma concentrations of VLCFA.

Treatment

There has been consideration of many approaches to the treatment of adrenoleukodystrophy and trials in animal models and humans (Hudspeth and Raymond 2007; Kemp and Wanders 2007). This has led to some limited successes in the management of selected patients with childhood acute cerebral and presymptomatic adrenomyeloneuropathy. The associated adrenal insufficiency responds well to steroid replacement therapy.

There is good evidence that boys with pauci-symptomatic acute cerebral adrenomyeloneuropathy and limited demyelination on magnetic resonance imaging of the brain respond to haematopoietic stem cell therapy (Peters *et al.* 2004), although the mechanism by which this helps is not known. Because of the risks of the procedure and because half of the patients will develop adrenomy-

eloneuropathy rather than the acute cerebral form, prophylactic haematopoietic stem cell therapy is not recommended. There is also good evidence that treatment of presymptomatic boys with normal brain MRI with Lorenzo's oil, a 4:1 mixture of glyceryltrioleate and glyceryltrierucate, reduces the risk of developing MRI abnormalities (Moser *et al.* 2005b). Current practice is to treat presymptomatic boys, identified through family screening or who have isolated adrenal insufficiency, with Lorenzo's oil and to survey them for neurological symptoms, adrenal insufficiency and MRI abnormalities every 6 months. This allows early haematopoietic stem cell therapy at the time of development of symptoms or MRI abnormalities.

10.2.5 Vanishing white matter disease

The eIF2B-related disorders are a group of leukoencephalopathies that are caused by mutations in the genes encoding the five subunits that comprise eukaryocytic initiation factor 2B, eIF2B. Recognized discreet disorders are vanishing white matter leukoencephalopathy, Cree leukoencephalopathy, and ovarioleukodystrophy syndrome. They are recessively inherited.

Vanishing white matter leukoencephalopathy

This has also been called childhood ataxia with central nervous system hypomyelination. Classically it presents between 18 months and 5 years of age; earlier developmental milestones being normal or with minor delays (van der Knapp *et al.* 1997). It is characterized by episodic encephalopathy precipitated by intercurrent illnesses or minor head trauma, or even by fright (Vermeulen *et al.* 2005). This results in a progressive ataxia with progressive pyramidal tract signs. Intellect is said to be intact in the early years of the disorder. Later, optic atrophy and occasionally epilepsy can complicate the disease. Children with vanishing white matter leukoencephalopathy typically become wheel-chair dependent by their teenage years and die in the second and third decades. However, the age of presentation and the evolution of symptoms are very variable. Adolescent and adult onsets are well described with a milder course (van der Knapp *et al.* 1998; Biancheri *et al.* 2003; Ohtake *et al.* 2004) (Section 37.7.6). There are also infantile-onset cases with a rapidly progressive course and death before 2 years of age Francalanci *et al.* 2001), and indeed prenatal onset with multiple organ involvement (van der Knapp *et al.* 2003).

MRI of the brain is very characteristic in vanishing white matter leukoencephalopathy. There is diffuse, symmetrical signal abnormality throughout the cerebral hemisphere white matter and part of the white matter has the signal intensity of cerebrospinal fluid; there is also signal abnormality and atrophy of the dentate nucleus (Fig. 10.6). The diagnosis is made by demonstrating mutations in one of the genes for the five subunits of eIF2B. Treatment of vanishing white matter leukoencephalopathy is symptomatic.

Cree leukoencephalopathy

This is a disease of Cree and Chippewayan North American Indians. Affected infants often have hypotonia and motor delay prior to the onset of an encephalopathy. The encephalopathy occurs around 6 months of age and is precipitated by an incidental infection. The sudden onset of encephalopathy with seizures and pyramidal tract signs is followed by dementia, blindness, increasing pyramidal tract signs, dysautonomia, and stalled brain growth. The infants become decerebrate and die at around 1 year of age (Black *et al.* 1988; Fogli *et al.* 2002).

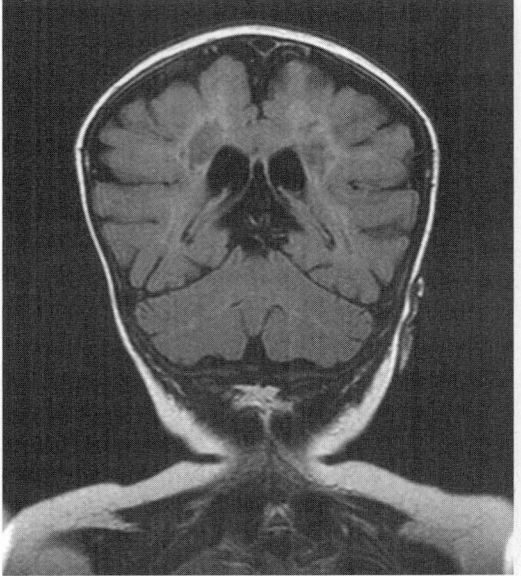 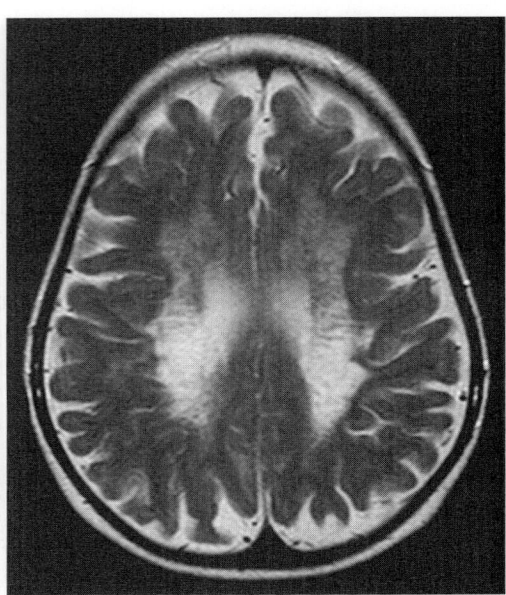

A   B

Fig. 10.6 Vanishing white matter leukoencephalopathy. (A) Coronal flair and (B) Axial T2-weighted MRI showing white matter volume loss with intense high signal indicative of demyelination. (Courtesy of Drs R. Jefferson, S. Jayawant, and P. Anslow.)

MRI of the brain shows a diffuse, symmetrical signal abnormality throughout the cerebral hemisphere white matter, and signal abnormalities in the globi pallida and often the thalamus. The diagnosis is made by demonstrating a mutation in the gene encoding ε-eIF2B (Fogli *et al.* 2002). Treatment of Cree leukoencephalopathy is symptomatic.

Ovarioleukodystrophy syndrome

This is a disorder of young women combining primary ovarian failure with a leukodystrophy. Menarche may be normal, delayed, or absent. There may be a history of learning difficulties in secondary education. Onset of neurological symptoms is in the second or third decade of life, with a slow dementia, pyramidal tract signs, and sometimes optic atrophy (Schiffmann *et al.* 1997; Fogli *et al.* 2003).

MRI of the brain shows a diffuse, symmetrical signal abnormality throughout the cerebral hemisphere white matter which is more prominent frontally and there may be some white matter volume loss posteriorly.

The diagnosis is made by demonstrating mutations in one of the genes for the β-, δ-, or ε-subunits of eIF2B. Treatment of ovarioleukodystrophy syndrome is symptomatic.

10.2.6 Megalencephalic leukoencephalopathy with subcortical cysts

Megalencephalic leukoencephalopathy with subcortical cysts is the preferred terminology for this disorder which previously has been called *megalencephalic cystic leukoencephalopathy* and *van der Knaap disease*. It is inherited recessively, and approximately 80 per cent of cases are caused by mutations in the *MLC1* gene.

In infants affected by megalencephalic leukoencephalopathy with subcortical cysts, macrocephaly is evident at birth or within the first 3 months of life. At this stage, neuroimaging already shows a diffuse leukoencephalopathy affecting cerebral white matter, often with temporal and parietal cysts. There may also be mild

delay in developmental milestones. The first symptoms of the leukoencephalopathy develop usually between 18 months and 10 years of age. These consist of slowly progressive ataxia and pyramidal tract signs. Cognitive involvement is 'discrepantly mild' at this stage. Some children can develop seizures and even coma after mild head injury. Children with megalencephalic leukoencephalopathy with subcortical cysts become wheelchair dependent usually in their teenage years and at this time a slow dementia develops, as can epilepsy. Death usually occurs in the second or third decade. However, the clinical course can be very variable even between affected members of the same family (van der Knapp *et al.* 1995; Singhal *et al.* 1996; Topcu *et al.* 1998).

MRI of the brain shows very distinctive findings (Fig. 10.7). The entire cerebral white matter has a diffuse signal abnormality with subcortical swelling and subcortical cyst formation in the temporal lobes and often the fronto-parietal lobes in addition. In some cases there are signal abnormalities from the cerebellar white matter. With time, the cysts become more extensive and cerebral atrophy develops (van der Knapp *et al.* 1995; Singhal *et al.* 1996; Topcu *et al.* 1998).

The diagnosis is made by detecting mutations in the *MLC1* gene. However 20 to 30 per cent of patients with clinically and radiologically typical megalencephalic leukoencephalopathy with subcortical cysts do not have mutations in *MLC1* and do not link to the *MLC1* region of chromosome 22q. Patients with a similar clinical and neuroimaging picture, but without macrocephaly 'normocephalic leukoencephalopathy with subcortical cysts', also do not link to chromosome 22q.

Treatment of megalencephalic leukoencephalopathy with subcortical cysts is symptomatic.

10.2.7 Alexander disease

Alexander disease is named after the author who first described the clinical and neuropathological findings. It is mostly a sporadic disorder caused by *de novo* dominant mutations in the glial fibrillary

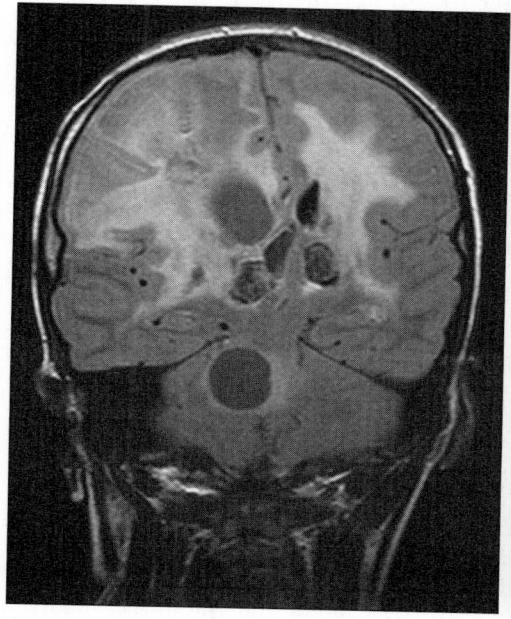 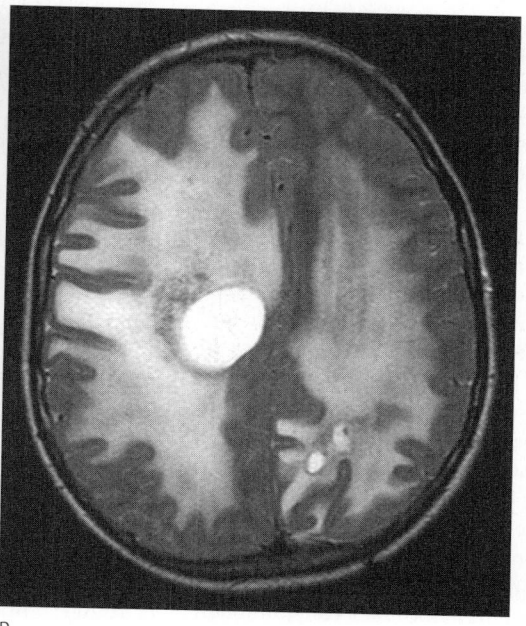

A   B

Fig. 10.7 Megalencephalic leukoencephalopathy with subcortical cysts. Coronal flair (A) and Axial T2-weighted (B) MRI showing multiple cysts, calcification and diffuse white matter loss. (Courtesy of Drs R. Jefferson, S. Jayawant, and P. Anslow.)

acidic protein gene, *GFAP*, although dominant transmission of the adult form is well recognized.

Depending upon the age of onset of symptoms, there are three clinical subgroups:

♦ infantile with onset before 3 years, approximately 65 per cent of cases;

♦ juvenile with onset between 3 and 12 years, in approximately 25 per cent; and

♦ adult with onset after 12 years, in approximately 10 per cent.

Infantile Alexander disease

This presents towards the end of the first year and beginning of the second year of life, usually with seizures. Classically, thereafter the infants develop a rapidly progressive disorder with dementia, macrocephaly, epilepsy, cerebellar ataxia and pyramidal tract signs. Such affected infants survive a few months or years. However, infantile Alexander disease can evolve more slowly with early ataxia and pyramidal tract signs but dementia and macrocephaly occurring in mid-childhood and survival into teenage years (Rodriguez *et al.* 2001).

Juvenile Alexander disease

This commonly presents between 4 and 10 years with a worsening behaviour disorder, bulbar difficulties affecting speech and swallowing or a gait disorder. The disease slowly progresses with increasing bulbar difficulties, breathing problems, ataxia, pyramidal tract signs, and dementia. Survival is variable, with death occurring before 40 years (Gorospe *et al.* 2002; Li *et al.* 2005).

Adult Alexander disease

This is the most variable form clinically. Patients can present with the bulbar symptoms of palatal myoclonus, dysphonia, dysarthria or dysphagia, as well as with pyramidal tract disturbance, cerebellar ataxia, dysautonomia, or sleep apnoea. Symptoms

gradually worsen over the years and decades, but, unlike the infantile and juvenile sub-groups, there appears to be no cognitive involvement, macrocephaly or epilepsy (Namekawa *et al.* 2002; Li *et al.* 2005).

MRI of the brain in infantile Alexander disease is often characteristic. There are diffuse symmetrical signal abnormalities in the hemispheric white matter, more extensive in the frontal regions, signal abnormalities in the basal ganglia, thalamus or brain stem, and contrast enhancement in these areas. MRI in adult Alexander disease often shows no hemispheric white matter involvement, instead showing signal change and later atrophy of the medulla and cervical cord. Diagnosis is confirmed by demonstrating mutations in the *GFAP* gene. The management of Alexander disease is symptomatic.

10.2.8 L-2–Hydroxyglutaric aciduria

L-2-hydroxyglutaric aciduria is a recessively inherited disorder caused by deficiency of L-2-hydroxyglutarate dehydrogenase. This produces a slowly progressive disorder comprising cerebellar ataxia, pyramidal tract signs, dementia, epilepsy, and often a movement disorder and macrocephaly. Early motor and cognitive developmental milestones are often delayed. The children usually present in the first 5 years of life with seizures, mental retardation, or unsteady gait. Thereafter the disease is slowly progressive, but at different rates in different individuals even from the same family (Barth *et al.* 1992; Moroni *et al.* 2000). Survival into late adulthood is not unusual, but a severe neonatal form with death in infancy has also been described (Chen *et al.* 1996; Fujitake *et al.* 1999).

MRI of the brain is quite characteristic. There is subcortical white matter swelling and signal change, signal change in the internal and external capsule, signal change in the putamen and dentate nuclei, and cerebellar atrophy. Peripheral nerve conduction studies are normal.

L-2-hydroxyglutaric aciduria is characterized by persistently high concentrations of L-2-hydroxyglutaric acid in all body fluids, relatively more in CSF than blood and urine. The diagnosis is made by demonstrating deficiency of L-2-hydroxyglutaric dehydrogenase in cultured skin fibroblasts. The disorder is caused by mutations in the *L2HGDH* gene on chromosome 14q22.1.

10.2.9 Canavan disease

Canavan disease is named after an early author of the clinical and neuropathological description; however it was only later that it was recognized as a distinct entity by van Bogaert and Bertrand. It is recessively inherited and caused by deficiency of aspartoacylase.

Symptoms of Canavan disease usually present before 6 months of age. Irritability, hypotonia, and abnormal visual behaviour usually develop after the first month but may be present from birth. Affected infants lose previously acquired developmental skills and may have seizures. As the disease progresses, macrocephaly and pyramidal tract signs appear and the limbs become increasingly rigid, opisthotonus and a pseudobulbar palsy develop. Eventually, the infant becomes decorticate and most die in the first 3 years, although prolonged survival into the third decade is well recognized (Ungar and Goodman 1983; Traeger and Rapin 1998).

MRI of the brain shows diffuse signal abnormality throughout the hemispheric white matter, sometimes sparing the internal and external capsules (Brismar *et al.* 1990). In Canavan disease, the concentration of *N*-acetylaspartate is raised in all body fluids. Increased concentrations of *N*-acetylaspartate are also present in the brain when measured using magnetic resonance spectroscopy. The diagnosis is confirmed by demonstrating deficient activity of aspartoacylase in cultured skin fibroblasts. *N*-acetylaspartate can be detected on standard urine organic acid analysis but it is important to ask the laboratory to search for this particular metabolite if this diagnosis is under consideration (Al-Dirbashi *et al.* 2007).

10.2.10 Pelizaeus–Merzbacher disease

This X-linked syndrome of central nervous system dysmyelination, occurring mainly in boys, characteristically presents in early infancy with involuntary eye movements or nystagmus. The classical form was described by Pelizaeus in 1885 and Merzbacher in 1910 (Koeppen 2005). Pelizaeus–Merzbacher disease is due to mutations affecting the myelin proteolipid protein gene *PLP1* on Xq22 which encodes two proteins expressed in oligodendrocytes, the proteolipid protein which constitutes approximately half of central nervous system myelin protein, and the differently spliced isoform DN20. Duplications of the *PLP1* gene are the commonest cause of Pelizaeus–Merzbacher disease (Inoue 2005). *PLP1* gene deletions or null mutations are associated with more benign disease than that in patients with *PLP1* gene duplications (Garbern 2007). Pelizaeus–Merzbacher disease probably results from accumulation of mutant protein aggregates intracellularly, resulting in oligodendrocyte dysfunction and death. The range of clinical disease associated with *PLP1* mutations is wide, ranging from a relatively pure form of spastic paraplegia type 2 (Section 23.4.2), through classical forms of Pelizaeus–Merzbacher disease to severe connatal or congenital forms involving near complete absence of myelin sheaths and oligodendrocytes from the central nervous system. The early onset and slow progression can lead to an incorrect diagnosis of cerebral palsy. In a personal case we diagnosed a neonate presenting with hypotonia, nystagmus, and stridor by re-examining the neuropathology of a maternal uncle dying in early childhood some 30 years earlier with 'cerebral palsy'.

The classical form

Involuntary eye movements or nystagmus develop in early infancy, associated with poor control of head posture and other motor functions, and tremor or titubation of the head and neck when seated (Garbern 2007). The disorder is slowly progressive, optic atrophy may develop, with paraparesis and cognitive impairment leading to death in the third to seventh decades. Laryngeal stridor, ataxia, athetosis, and spasticity can be prominent. Brain stem auditory evoked potentials are abnormal. This classical form is due to duplications of the *PLP1* gene, which are found in up to 70 per cent of all cases of Pelizaeus–Merzbacher disease.

Connatal forms

A much more severe early onset or congenital form of Pelizaeus–Merzbacher disease occurs with some of the *PLP1* point mutations, or in patients with three or more copies of the *PLP1* gene (Wolf *et al.* 2005). Brain MRI reveals almost complete absence of myelin signal and extreme thinning of the corpus callosum. However brain MRI may be clinically unhelpful in the first few months of life because the normal infant has very little myelin; failure of normal myelination becomes apparent subsequently. Such patients may never achieve stable head control or other motor milestones, have severe mental retardation, suffer paroxysmal episodes which are probably epileptic in nature, and die between infancy and the second decade of life.

Intermediate forms

These occur between the classical and connatal extremes of the disease, and reflect some of more than 100 missense mutations of the *PLP1* gene which account altogether for 15–20 per cent of Pelizaeus–Merzbacher disease pedigrees.

Female carriers

The female carriers of Pelizaeus–Merzbacher disease are either free of clinical disease, or have mild clinical disease compared to boys within the same family (Hurst *et al.* 2006). Clinical signs are rarely present in carrier females from families where the males are affected by mutations causing severe disease. The lowest risk for female disease occurs with *PLP1* gene duplications. Females are most likely to develop significant disease if they have nonsense or null mutations.

Diagnosis

Initially the diagnosis of Pelizaeus–Merzbacher disease is suspected when an infant boy develops slowly progressive eye movement and head control difficulties associated with MRI evidence of abnormal brain myelination. Naturally, the reliability of clinical diagnosis is enhanced if other family members are known to be affected. Molecular genetic testing for *PLP1* gene reduplications or other mutations is diagnostic. Characteristically at autopsy the cerebral white matter shows a 'tigroid' appearance with alternating areas of relatively preserved myelin staining, and others with loss of white matter staining. Peripheral nerve myelination is normal and axons are relatively preserved. A small proportion of patients have Pelizaeus–Merzbacher-like disease, unassociated with *PLP1* mutations; recessive mutations in the *GJA 12*/Cx47 gap junction protein gene have been detected in some such patients (Orthmann-Murphy et al. 2007).

10.2.11 Aicardi–Goutières syndrome

This rare familial leukodystrophy was first described in 1984 and produces a progressive encephalopathy of early onset with characteristic basal ganglia calcification, best seen on CT, and features of leukodystrophy, best seen on MRI. There is an associated chronic CSF lymphocytosis with high CSF levels of interferon-alpha but without any other signs of infection (Goutières 2005). The condition can be mistaken for congenital infection. Mutations of genes encoding the exonuclease, *TREX1*, and the endonuclease complex, *RNASEH2*, have been identified as causes of the disease (Rice et al. 2007).

10.3 Primary grey matter degenerations of infancy and childhood

10.3.1 Overview of grey matter degenerations

These poliodystrophies mainly affect grey matter and include the Neuronal Ceroid Lipofuscinoses, or Batten's disease (Section 10.3.2), the neuroaxonal dystrophies (Section 10.3.5), the neuronal storage diseases such as mucopolysaccarhidoses (Section 10.3.3), Lesch–Nyhan syndrome (Sections 10.3.4 and 40.4.11), and other grey matter disorders. They cause a variety of progressive deficits including cognitive decline, epilepsy, visual loss and motor abnormalities. The reader is also referred to other primary grey matter degenerations of infancy and childhood:

- Menke's syndrome (Section 11.6.6);
- Alper's disease (Section 30.5.7);
- Niemann–Pick disease type C (Sections 10.4.4 and 40.4.11);
- Myoclonic epilepsy with ragged red fibres, MERRF (Section 30.5.7);
- Other primary myoclonic epilepsies (Section 30.5.7);
- Neuroferritinopathy (Section 40.5.4);
- Juvenile Huntington's disease (Section 40.5.2); and
- GM2 gangliosidosis or Tay-Sachs disease (Sections 10.4.2 and 40.4.11).

10.3.2 The neuronal ceroid lipofuscinoses

Frederick Batten (1903) first described two siblings with visual loss and dementia starting in the first decade realizing that these children were clinically distinct from children with Tay Sachs disease. Spelmeyer described a further series in 1905 with further refining of the entity, as being clinically distinct, by Jansky in 1908 and Bielscowsky in 1913. The numerous eponyms used to describe these diseases were confusing (Brett 1997). *Batten's disease* is the term most often used in the United Kingdom. The nature of the material accumulating in the nerve cells was unknown until it was shown that the staining characteristics of the stored material was similar to ceroid or lipofuscin. This led to the term 'neuronal ceroid-lipofuscinoses' which has become the accepted name for this group of conditions (Zeman and Albert 1963). In terms of approach to diagnosis there are 3 main clinical types of neuronal ceroid lipofuscinoses (Table 10.1, V. 16). Recent genetic and biochemical research suggests there are at least 8 distinct disorders within this diagnostic group including the rare adult form also known as Kuf's disease (Table 10.8). The morphological hallmark of this group of diseases is loss of nerve cells and accumulation of neuronal ceroid lipofuscinoses-specific lipopigments. There are different phenotypes and genotypes (Mole et al. 2005).

Late infantile neuronal ceroid lipofuscinosis

This common paediatric neurodegenerative diseases is familiar to most paediatric neurologists (Augestad and Flanders 2006). It is thought to be one of the commonest inherited neurodegenerative diseases with six new cases per year reported in the United Kingdom Progressive Intellectual and Neurological Deterioration study (Fig. 10.1) and a reported incidence in Germany of 0.46 in 100 000 live births (Mole et al. 2005). Development in the first and most of the second year is usually normal before subsequent slowing and then regression of cognition associated with the onset of a severe seizure disorder. Developmental of cognitive slowing usually predates the onset of seizures, perhaps by up to 1 year but its significance may not have been appreciated until the onset of the seizures. Diagnostic confusion may also occur because other relatively more common epilepsies with an impact on cognition can begin at this time including astatic myoclonic epilepsy (Section 30.5.5) and Lennox Gastaut syndrome (Section 30.5.5). Thus tardy diagnosis is relatively common. Multiple seizure types occur including myoclonic seizures with poor response to antiepileptic drug treatment. Unsteadiness and then pyramidal signs follow. Visual loss occurs with pigmentary retinal changes and the development of optic atrophy. A macular cherry red spot is not a feature. The ophthalmological findings will require examination by an experienced ophthalmologist. Steady progression to death is inevitable with a median age of death at 12 years (Augestad and Flanders 2006).

Brain imaging will show progressive cerebral atrophy but is not diagnostically helpful in the early stages. Electrophysiology is most useful in terms of suggesting the diagnosis. All EEGs in children with epilepsy under 5 years should include photic stimulation at slow rates of flash because of the pathognomonic change of grossly enlarged visual evoked potentials in neuronal ceroid lipofuscinosis (Veneselli et al. 2001). The electroretinogram response will be absent.

Most cases of the late infantile form are associated with mutations of the *CLN2* gene. A number of rarer variants are described in association with mutations of the *CLN5*, *CLN6*, and *CLN8* genes. The *CLN8* gene is associated with the Turkish variant and the northern epilepsy variant (Ramirez-Montealegre et al. 2006).

Traditionally the diagnosis required biopsy of nervous tissue; typically examination of the myenteric plexuses in a full thickness rectal biopsy (Table 10.8). Brain biopsy would be diagnostic also but is rarely considered. The neuropathologist needs considerable expertise interpreting the sample and electron microscopy is required to see the characteristic curvilinear bodies. Similar pathological changes can be seen in sweat gland endothelium, smooth muscle, and blood vessel endothelium. Lymphocytes prepared from the blood buffy coat layer will show diagnostic abnormality but electron microscopic examination is required to identify the curvilinear bodies. This blood examination should be undertaken before invasive tissue biopsy. Nowadays the diagnosis should be made on measuring the tripeptidyl peptidase 1 enzyme level in leukocytes and can be confirmed with *CLN2* mutation analysis.

Table 10.8 The neuronal ceroid lipofuscinoses. Electrophysiological and diagnostic tests

	Congenital	Infantile	Late infantile	Juvenile	Adult
Eponym		Santavuori	Jansky–Bielschowsky	Spielmeyer–Vogt	Kuf's
Electrophysiology		EEG shows progressive slowing and reduction in amplitude and absent ERG.	Large VEP at slow rates of flash (may be seen on EEG) and absent ERG	Absent ERG and abnormal EEG (Slow with ill defined spike wave complexes)	Photosensitivity prominent in those with progressive myoclonic epilepsy
Vacuolated lymphocytes	[No info]	–ve	–ve	+ve in 25% (Fig. 10.2)	[No info]
Buffy coat		EM +ve	EM +ve	EM +ve	
Rectal biopsy		PAS +ve sudanophilic and auto-fluorescent neuronal inclusions (Finnish snowballs)	PAS+ve sudanophilic and auto-fluorescent material in neurons. Curvilinear structure on EM	PAS+ve sudanophilic and auto-fluorescent granular material in neurons. Granular appearance on EM (fingerprint bodies)	PAS+ve neuronal inclusions required for diagnosis
Enzyme affected	Cathepsin D	White cell or fibroblast palmitoyl-protein thioesterase	White cell tripeptidyl peptidase1		
Gene identified (chromosome location)	*CTSD* (11p15.5)	*CLN1* (1p32)	*CLN2* (11p15) *CLN8* (8p23) *CLN5* *CLN6*	*CLN3* (16p12)	unknown

EM = electron microscopy

ERG = electroretinogram

LM = light microscopy

VEP = visual evoked potential

Biopsy is no longer required unless this enzyme level is normal. Management is symptomatic with no treatment affecting the course of the disease. Treatment with bone marrow transplantation has not proved successful despite initial promise (Lake *et al.* 1997; Yuza *et al.* 2005).

Juvenile neuronal ceroid lipofuscinosis

A Norwegian study suggested that this was the commonest form of neuronal ceroid lipofuscinosis with 63 affected cases born in Norway between 1957 and 1998 (Augestad and Flanders 2006). This study identified only seven cases of the late infantile form over the same time period. The child with juvenile neuronal lipofuscinosis will present with progressive visual failure in the first decade, usually during the primary school years. A pigmentary retinopathy develops followed by optic atrophy. Initially the diagnosis of a neurodegenerative disease may be missed because the cognitive decline and sometimes associated behavioural problems have been attributed to the visual difficulties. Epilepsy follows usually some years after the onset of visual symptoms, and then dementia occurs with increasing physical impairment. Many children will remain mobile during their early teens but eventual progression will lead to total dependence and then death usually in the third decade.

Diagnosis of this juvenile form is suggested by the clinical presentation with visual loss and finding an absent or reduced electroretinogram with abnormal EEG. The diagnosis can be confirmed by full thickness rectal biopsy where periodic acid Schiff +ve sudanophilic autoflourescence will be seen (Fig. 10.8) and characteristic 'fingerprint like' structures are seen on electron microscopy. In this juvenile form of neuronal ceroid lipofuscinoses 25 per cent of blood lymphocytes may show vacuolation providing

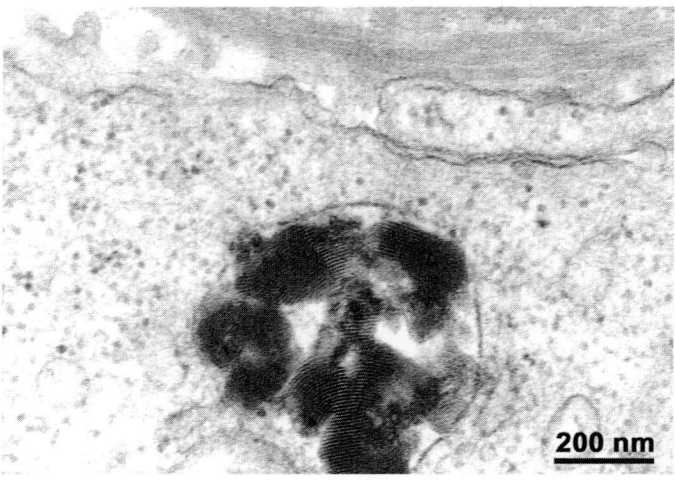

Fig. 10.8 Juvenile neuronal ceroid lipofuscinosis. Electron microscopy of neurone from rectal myenteric plexus showing fingerprint bodies (courtesy of Dr W. Squier).

a useful and relatively easy screening test. The juvenile form is associated with *CLN3* mutations (Table 10.8).

Infantile neuronal ceroid lipofuscinoses

The earlier onset of this rapidly progressive form of neuronal ceroid lipfuscinoses distinguishes it from the late infantile form. It is often referred to by the eponym of Santavuori's disease following a description of a large series in Finland where the condition seems to be more common than elsewhere (Vanhanen *et al.* 1997).

Onset is usually in the first year with progressive mental deterioration, ataxia, and visual failure. Then the child regresses in the second year with loss of social interaction and motor skills and develops hypotonia. The rate of head growth decelerates and microcephaly becomes prominent. Visual loss occurs early and optic atrophy develops. Epilepsy appears relatively late in the course, usually with myoclonic seizures which can be difficult to manage. With progression the child becomes vegetative with increased tone, hyper-reflexia, and extensor plantar responses. The EEG is particularly helpful and will show slowing of rhythmic activity, bursts of sharp and slow waves and spike-wave discharges, and then progressive decrease in amplitude until the record becomes essentially isoelectric when the condition is fully established. The electroretinogram is reduced and then becomes extinguished but the curious abnormality of the visual evoked potentials seen in late infantile neuronal ceroid lipofuscinoses is not seen. The diagnosis is confirmed by finding periodic acid Schiff +ve sudanophilic autofluorescent substance deposited in neurones, smooth muscle cells, and vascular endothelium. These deposits are called 'granular osmophilic deposits' or sometimes 'Finnish snowballs'. Palmitoyl protein thioesterase levels are low in leukocytes and fibroblasts allowing possible diagnosis from assay of blood leukocyte enzymes and thus avoiding tissue biopsy. The infantile form is associated with *CLN1* mutations (Table 10.8). Most cases will present in this classical way but later onset variants of this infantile form associated with *CLN1* mutation has been recognized including some rare cases presenting in adults (Mole *et al.* 2005).

Kuf's disease

Kuf's disease is a rare adult form of neuronal ceroid lipofuscinoses. Two phenotypes have been described, one characterized by epilepsy and the other with dementia (Berkovic *et al.* 1988). Autosomal recessive and dominant forms have been described (Josephson *et al.* 2001). The clinical picture in Kuf's disease is characterized by dementia, seizures, and extrapyramidal features. By contrast with the earlier onset forms, visual loss is not prominent. Deficiency of palmitoyl protein thioesterase suggests the rare adult onset presentation of *CLN1* mutations (Ramadan *et al.* 2007).

Congenital form

These present with severe neurological abnormalities at birth including intractable seizures, spasticity, apnoea, and microcephaly. They are associated with mutations of the *CTSD* gene on chromosome 11 with abnormal cathepsin D enzyme activity (Table 10.8).

10.3.3 Lesch–Nyhan syndrome

This disorder of purine metabolism is due to deficiency of hypoxanthine-guanine phosphoribosyl transferase, HGPRT, activity (Table 10.1, VII 1a). It is an X-linked disorder, first described biochemically in 1964 by Lesch and Nyhan, which produces a characteristic clinical pattern, with progressive slowing of mental and motor development starting at 6 months. Choreoathetoid movements and dystonic spasms usually emerge in the second year (Section 40.4.11). Self injurious behaviour then emerges subsequently and is almost pathognomonic for the disease. The self mutilation seems to be compulsive yet is clearly distressing for both child and carers.

Biochemically there is overproduction of uric acid leading to hyperuricaemia, nephrolithiasis, and gout. Untreated, gouty tophi appear and death from renal failure results. Orange grit in the urine seen on the nappy suggests the diagnosis. Allopurinol reduces the production of uric acid and helps prevent gouty renal failure but does not affect the neurological manifestations of the disease. The condition can be confused with evolving dyskinetic cerebral palsy and survival to adulthood is now usual.

The pathogenesis of the neurological manifestations is not understood. There is evidence that depletion of dopamine in the basal ganglia during foetal brain development may be important (Smith and Jinnah 2007). However postnatal dopaminergic treatment of affected children is not helpful. The clinical phenotype depends on the level of residual HGPRT activity and milder forms are recognized now. The involuntary movements are difficult to treat. The self injurious behaviour poses most challenging and distressing aspect of management. Curiously not all children develop this self injurious problem, although it can emerge late on. The capacity for harm can be reduced by distraction, providing restraints and by dental clearance. There is some evidence that deep brain stimulation may suppress this distressing phenomenon (Cif *et al.* 2007).

10.3.4 Infantile neuroaxonal dystrophy

Also known as Seitelberger's disease, infantile neuroaxonal dystrophy is a neurodegenerative disorder usually of onset between 6 months and 2 years of age. Typically patients develop bilateral upper motor neurone signs with spasticity, sometimes after a hypotonic phase; and the combination of extensor plantar responses with loss of ankle tendon reflexes is typical. Eye movement disturbances, particularly nystagmus, and optic atrophy may be present early on. Impaired hearing can occur. Cognitive deterioration is common, but extrapyramidal disorders and seizures occur less frequently. Electromyography may show signs of denervation with normal motor conduction velocity pointing to anterior horn cell disease and can be helpful diagnostically in distinguishing neuroaxonal dystrophy from other central nervous system degenerations.

MRI typically shows pronounced atrophy of the cerebellum, with diffuse hyperintensity of the cerebellar cortex. Hypointensity of the globus pallidus, subthalamic nuclei, and substantia nigra are variably present, and sometimes the 'eye of the tiger sign' normally associated with the *PANK2* mutation of pantothenate kinase-associated neurodegeneration, or neuroferritinopathy (Section 40.5.4) can be present (Kumar *et al.* 2006). MR spectroscopy of the white matter may show significantly reduced *N*-acetyl aspartate (Khateeb *et al.* 2006).

The disorder is named after the characteristic histological finding of axonal spheroids on biopsy of peripheral or central nervous tissue, and biopsy of nerve and muscle was initially the diagnostic test. However, similar axonal spheroids can be found in a variety of other neurodegenerative disorders, so this finding is not pathognomic.

Mutations of the *PLA2G6* gene on chromosome 22q13.1, encoding phospholipase A2 group VI is associated with infantile neuroaxonal dystrophy (Khateeb *et al.* 2006; Morgan *et al.* 2006). Detection of *PLA2G6* mutations offers the prospect of non-invasive and more specific diagnosis of infantile neuroaxonal dystrophy. The phospholipase A2 enzymes catalyse the release of fatty acids from phospholipids, and seem to promote brain iron deposition by some mechanism as yet not understood.

Infantile neuroaxonal dystrophy carries a wide differential diagnosis, including metachromatic leukodystrophy (Section 10.2.2), neuronal ceroid lipofuscinoses (Section 10.3.2), Leigh's syndrome (Section 10.5.3) and the autosomal recessive disorder, Schindler disease due to deficiency of alpha-*N*-acetylgalactosaminidase (Gordon 2002). In addition neuroferritinopathy due to mutations of the pantothenate kinase-associated neurodegeneration *PANK2* gene, previously known as Hallervorden–Spatz disease, may show similar MRI findings, but generally presents as an extrapyramidal syndrome (Section 40.5.4). Neuroferritinopathy does not show axonal steroids in the peripheral nervous system, and tends to present later in infancy, or during childhood, with patients often surviving into their third decade. By contrast, infantile neuroaxonal dystrophy usually causes death before the age of 10 years.

10.4 Lysosomal disorders affecting the brain

10.4.1 Introduction to the neuronal storage diseases

In these conditions deficiency of a lysosomal enzyme leads to accumulation of various metabolites in neurons with subsequent neurodegeneration (Table 10.1, V 1–16). This accumulation or 'storage' can occur also in other organs and may lead to enlarged liver and spleen or allow diagnosis from examination of blood or bone marrow. Enzyme replacement and gene therapy are being explored as treatments (Beck 2007). These conditions often produce a dysmorphic appearance (Table 10.3) and skeletal deformity, which may lead to secondary spinal cord compression. The gangliosidoses are a group of related disorders in which neuronal storage of gangliosides, which are complex lipids, occurs. The peripheral neuropathy and vascular disorders of Fabry disease, due to α-galactosidase deficiency are considered elsewhere (Table 10.1, V13 and Section 21.8.5). The progressive proximal myopathy of Pompé disease, or α-glucosidase deficiency (Table 10.1, V 2) is considered under muscle disorders (Section 24.6.1).

10.4.2 GM2 gangliosidosis or Tay Sachs disease

First described in London by Warren Tay in 1881 and then in New York by Bernard Sachs in 1887 this condition is caused by deficiency of the enzyme Hexosaminidase allowing massive accumulation of ganglioside in the brain (Table 10.1, V15). This autosomal recessive condition is particularly common among Ashkenazi Jews occurring with a gene frequency in this population in New York of 1:30. The common infantile form of the disease presents in the first year usually before the child has acquired the ability to sit. The rapid development of blindness may initially cause diagnostic confusion. Loss of visual interest and social interaction may lead to initial thought that the child has developed an early and severe autistic-like regression. However, the rapid developmental decline soon suggests the metabolic nature of the underlying problem. Ophthalmological examination will reveal optic atrophy and a cherry red spot in 90 per cent of cases. Once seen this finding is very characteristic and in this clinical setting is diagnostic (Fig. 10.9). Affected infants develop a characteristic exaggerated startle response which usually appears early on in the course of the disease and is suggestive of the diagnosis. Muscle tone is initially reduced but then increases steadily as the disease progresses. The tendon reflexes are brisk and plantar

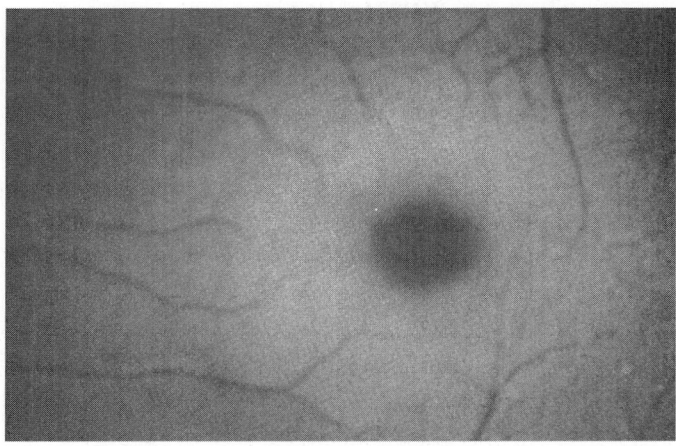

A

- Farber lipogranulomatosis
- Galactosialidosis
- Gaucher disease
- GM$_1$ gangliosidosis
- GM$_2$ gangliosidosis
- Goldberg syndrome
- Metachromatic leukodystrophies
- Niemann-Pick disease
- Dapsone poisoning
- Sialidosis types I and II
- Wolman disease

B

Fig. 10.9 Cherry red spot at the macula on fundoscopy. (A) Gaucher's disease showing a cherry red macular spot with surrounding retinal pallor. (See Plate 16.) (B) The differential diagnosis of retinal cherry red spot. (Adapted from Leavitt and Kotogal 2007.)

responses extensor. This neuronal storage disorder causes a striking increase in head size (Fig. 10.10). Epilepsy is common although the early EEG may show very little abnormality but then deteriorates as the disease progresses.

The electroretinogram is not affected by this disease as compared with the neuronal ceroid lipofuscinoses. The diagnosis is made by measurement of hexosaminadase activity in blood. The enzyme exists as hexosaminidase A and B and 3 patterns of deficiency are recognized:

- Type 1 GM2 gangliosidosis is commonest and involves marked deficiency of hexosaminidase type A and high levels of type B;

- In Type 2, Sandhoff's disease, components A and B are both reduced; and

- In Type 3, the rarest form, the levels of A and B are normal on assay but there is deficiency of an activator protein.

The clinical picture is the same with all 3 types. GM2 ganglioside accumulates in brain and other tissues and the condition can be diagnosed by biopsy although this is not necessary for diagnosis unless the rare activator protein deficiency is suspected. Survival beyond the age of 4 years is rare.

Late onset forms of GM2 gangliosidosis are described but are rare. Recently a series of patients with juvenile or subacute GM2 gangliosidosis has been described (Maegawa *et al.* 2006). A range

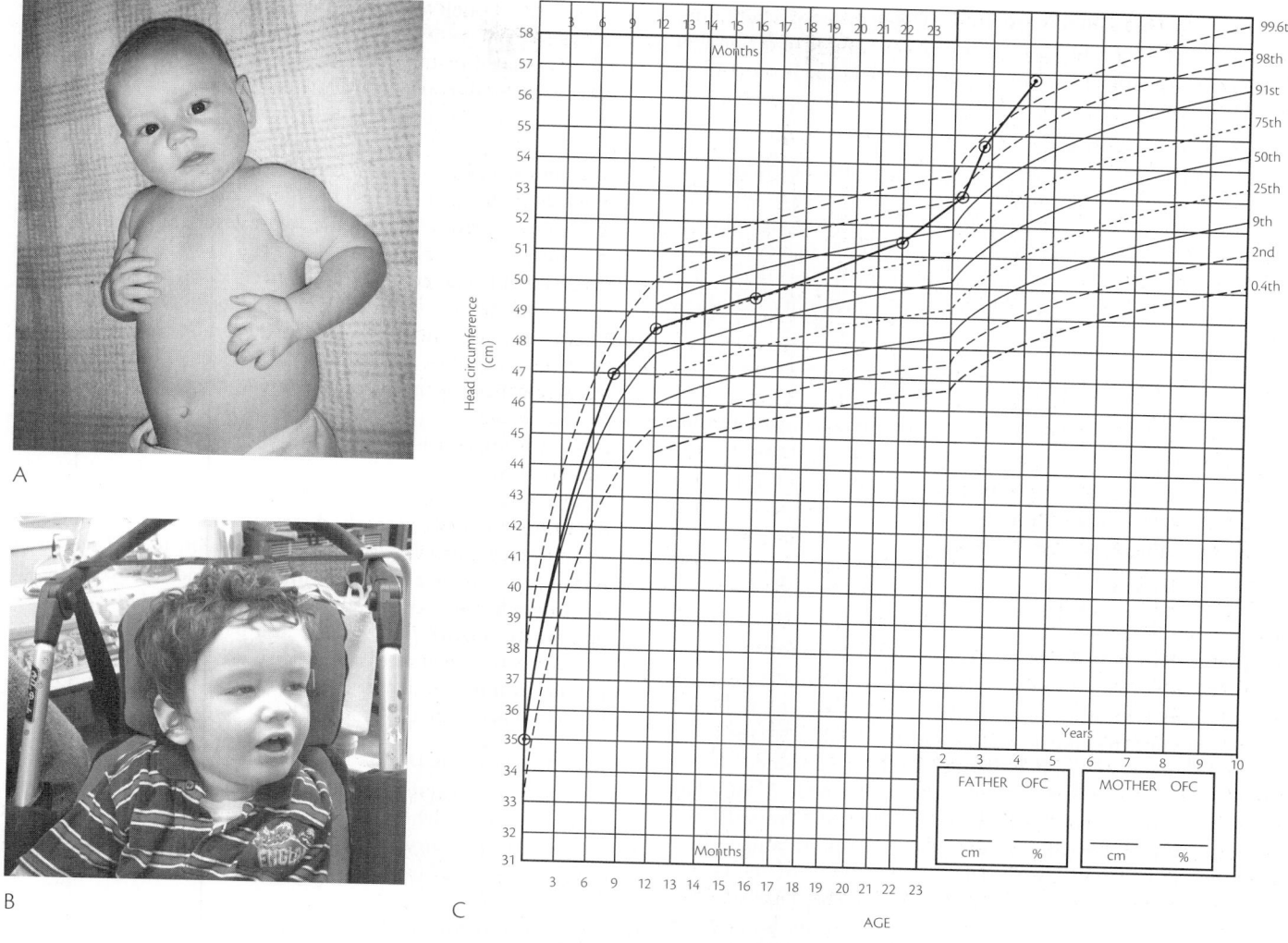

Fig. 10.10 Tay Sachs disease, GM2 gangliosidosis. Family photographs showing (A) normal head size and attentiveness at 3 months age and (B) macrocephaly and diffuse impairment at 4 years of age (C) Head circumference chart showing progressive disproportionate increase in head size. (With kind permission of the family.)

of neurological symptoms and rates of progression are described. In general, the earlier the onset of the disease the more rapid the progression.

10.4.3 GM1 gangliosidosis

This disorder is also known as generalized gangliosidosis, Landings disease, or pseudo-Hurler's disease. The disorder is less common than GM2 gangliosidosis (Section 10.4.2) and usually presents in two forms (Table 10.1, V14 a,b) with an additional rare adult onset variety

Type 1 with early onset

In this, infants present with failure to thrive, abdominal visceromegaly, coarse features, and kyphosis. Affected infants do not feed well, have low tone, and do not make any developmental progress. Extensive Mongolian spots in the newborn have been described in this condition and can suggest the diagnosis before more obvious clinical features have developed (Table 10.3) (Ashrafi *et al.* 2006). The facial dysmorphism increases as the infant grows older and

development is significantly impaired; these infants usually fail to acquire the ability to sit unsupported and show early loss of any acquired skills. Epilepsy occurs and a macular cherry red spot can appear. In the past the term 'Tay-Sachs disease with visceral involvement' was used to describe these children.

The condition is due to deficiency of the enzyme β-galactosidase causing accumulation of the ganglioside GM1 within neurons and other terminal β-galactose residue compounds in the viscera causing significant hepatosplenomegaly. Foamy cells will be seen in the bone marrow and vacuolated lymphocytes are seen in the peripheral blood film. Firm diagnosis depends on enzyme assay in leukocytes or skin fibroblasts. The condition is autosomal recessive and rapidly progressive with most type 1 cases dying before their 3rd birthday, usually with respiratory infection.

Type 2, the late infantile or juvenile form

This presents between 1 and 5 years of age with loss of acquired skills, evolution of pyramidal tract signs, and decerebrate rigidity. The early dysmorphism, hepatosplenomegaly, and severe skeletal deformity are absent in this form of GM1 gangliosidosis but minor skeletal changes are evident on X-ray such as beaking of the lumbar vertebrae as seen in Hurler's syndrome. Death in the first decade is usual (Chen *et al.* 1998).

Type 3, with adult onset

This very rare form shows slower progression with the evolution of both pyramidal and prominent extrapyramidal signs and with dementia. The absence of visceromegaly in the juvenile and adult forms and the non-specific bone marrow findings makes the diagnosis difficult. Diagnosis requires assay of the leukocyte enzymes driven by consideration of this diagnostic possibility (Muthane *et al.* 2004).

10.4.4 Niemann-Pick disease

Niemann–Pick disease refers to a group of lipid storage disorders which are usually classified into two main groups with two subtypes in each (Table 10.1, V8 and 9):

◆ Group 1 includes Niemann–Pick disease types A and B and are due to sphingomyelinase deficiencies

◆ Group 2 includes Niemann–Pick disease types C and D and are due to a defect in cholesterol transport.

Niemann–Pick type A

The infantile form is the classical subtype of the disease and accounts for half of all cases. Onset may be prenatal leading to many infants being born with low birth weight. There is early hepatosplenomegaly and jaundice develops in some. There is failure to thrive and early neurological involvement. Most infants fail to acquire the ability to sit independently. Pyramidal tract signs develop along with 'head retraction', squint, and seizures. Few survive beyond 2 years. There is deficiency of the enzyme sphingomyelinase with toxic accumulation of the fatty substance sphingomyelin in brain and visceral cells. The effects of pulmonary and bony infiltration of the toxin may be seen on X-rays of the chest and long bones. Treatment is symptomatic.

Niemann–Pick type B

This differs from type A in that there is no neuronal storage but massive visceral storage of sphingomyelin. This form of the disease is called visceral, non-neuronopathic Niemann–Pick but is also due to deficiency of spingomyelinase activity. These children present after 2 years with recurrent respiratory symptoms. Pulmonary infiltrates may be seen on chest X-ray. Bone involvement is also seen and haematological problems are common. Survival into adulthood occurs although individuals are prone to infection and may die from liver failure due to cirrhosis.

Intermediate types between A and B are described and the disease is best considered a single entity with a spectrum of phenotypes (Schuchman 2007).

Niemann–Pick type C

This is not due to sphingomyelinase deficiency but due to a defect in cholesterol transport. This disorder is grouped with types A and B because the bone marrow shows similar storage cells. Early jaundice occurs in 60 per cent of cases, often labelled as neonatal hepatitis. This settles and then the child presents with neurological symptoms and splenomegaly. The neurological symptoms can appear at any time during childhood and may be delayed until the teenage years. Epilepsy is common although it is not myoclonic. The most striking clinical finding is a vertical ophthalmoplegia with particular difficulty looking downwards which may lead to problems going downstairs and avoiding objects at ground level. The ocular motor problem is restricted to voluntary eye movements and the vestibulo-ocular reflex is full. Although defects of ocular movement are seen in other conditions, including ataxia telangiectasia (Section 11.5), this vertical supranuclear ophthalmoplegia coupled with a progressive neurodegenerative disorder, is pathognomonic for Niemann–Pick type C. Other neurological features include ataxia, dementia, and in some an akinetic rigid syndrome has been described. Many patients do not have abdominal visceromegaly. The disorder is due to mutations of the *NPC1* gene in 95 per cent of cases and the *NPC2* gene in the remainder. The clinical heterogeneity and wide age range for presentation often causes delay in diagnosis including late into adult life. The biochemical diagnosis of Niemann–Pick disease type C can be determined by assay of skin fibroblasts (Sevin *et al.* 2007b). The diagnosis is suggested by bone marrow examination or rectal biopsy showing cells with characteristic inclusions. The classic Niemann–Pick cell in Types A and B is large with 'foamy' cytoplasm but in type C the vacuoles are not uniform and their contents do not stain with Sudan dyes. Sea blue histiocytes may be seen in the older patient but are rare. Experienced pathological examination is required to confirm this diagnosis histopathologically. The diagnosis can be confirmed on skin fibroblast culture and with mutation analysis (Millat *et al.* 2001).

Niemann–Pick type D

This refers to a specific population of patients with a condition similar to Type C who originated in Nova Scotia. Treatment is symptomatic and survival is dependant on the age of onset and rate of progression. Some do survive into adulthood.

10.4.5 Mucopolysaccharidoses

The mucopolysaccharidoses are a large group of lysosomal storage diseases caused by deficiency of enzymes catalysing the degradation of glycosaminoglycans, or mucopolysaccharides. Children within this group of disorders develop characteristic clinical findings. In all there is developmental slowing with early mental retardation. Corneal clouding may become apparent and there is usually enlargement of the liver and spleen and the development of coarse dysmorphic features. Bony involvement causes skeletal deformity and the hands tend to be broad with short fingers; this can be diagnostically helpful. The tongue often protrudes and growth is restricted. There are a large number of variants within this group (Table 10.1, V1a–o). All the mucopolysaccharidoses are autosomal recessive, apart from Hunter's syndrome which is X-linked recessive, and have varying biochemical bases (Brett 1997; Clarke 2008). Screening for mucopolysaccharidosis is performed by analysis of urinary glycosaminoglycans, GAG. A raised urinary glycosaminoglycan: creatinine ratio is suggestive of mucopolysaccharidosis but this must be followed up by glycosaminoglycans electrophoresis which shows characteristic patterns for the different mucopolysaccharidosis disorders.

Hurler's syndrome or MPS IH

This is the common classical and most severe type. The diagnosis may not be obvious in the first year because development may seem normal or only slightly delayed. Then, there is developmental standstill, before regression with loss of skills and evolution of the dysmorphic features and corneal clouding. Increasing hepatosplenomegaly occurs. Previously their appearance led to these children being described by the derisory term 'gargoyles'. With time, poor growth and skeletal deformity increase, obstructive hydrocephalus can develop and visual impairment and deafness can occur. Excessive respiratory secretions with nasal obstruction and respiratory infection complicate the toddler years. Cardiac and respiratory tract involvement combined with the neurological decline leading to death before the end of the first decade. The deficient enzyme is α-L-iduronidase and there is increased urinary excretion of both heparan and dermatan sulphate.

Scheie's syndrome or MPS IS

This involves deficiency of the same α-L-iduronidase enzyme, but intelligence and growth are not affected with survival well into adult life. Carpal tunnel syndrome may occur.

Hunter's syndrome or MPS II

In this X-linked recessive variant, boys do not develop corneal clouding and progression is slower than that seen in Hurler's syndrome. The range of disease severity is wide. Severe forms involve facial dysmorphism, short stature, hepatosplenomegaly, bone abnormalities, heart valve disease, mental retardation and early death. Mild forms can show no central nervous system involvement and prolonged survival. The severity of the disease is linked to the particular mutation affecting the gene encoding iduronate sulfatase, *IDS,* located at Xq28 with over 300 mutations described (Vafiadaki *et al.* 1998; Muñoz *et al.* 2008).

Sanfilippo disease or MPS III

In this the dysmorphic features are relatively mild and there is no corneal clouding. This diagnosis may be overlooked. However the cognitive decline and often severe behavioural problems should alert the clinician to the possibility (Moog *et al.* 2007).

Morquio's disease or MPS IV

This is a mucopolysaccharidosis without direct neurological involvement unless these are secondary to skeletal complications, a circumstance posing complex management decisions (Giugliani *et al.* 2007). The skeletal involvement is severe but there is no facial dysmorphism. Compression of the spinal cord and medulla from atlanto-axial subluxation pose a challenging neurosurgical decision. Carpal tunnel syndrome can also occur.

Maroteaux–Lamy disease MPS VI

This resembles Hurler's syndrome but there is normal intelligence and urinary excretion of dermatan sulphate is absent. Corneal clouding can occur and skeletal deformity cause spinal cord compression. Several subtypes are described. Management guidelines for this disorder have recently been published (Giugliani *et al.* 2007).

Sly disease MPS VII

This has a variable phenotype ranging from the manifestations of Hunter's to those of Morquio's disease.

Several lysosomas storage disorders are now treatable through a range of therapies. Bone marrow transplantation is effective in MPS I Hurler (Boelens 2006, Orchard *et al.* 2007). If performed early enough in life, < 18 months, sufficient enzyme is produced in the central nervous system to prevent neurological deterioration allowing normal intellectual development. Some aspects of disease such as spinal scoliosis progress despite bone marrow transplantation. The recognition that enzymes are targeted to the lysosome by mannose-6 phosphate has allowed the development of successful enzyme replacement therapy for several lysosomal storage diseases (MPS I Hurler/Scheie and Scheie phenotypes, II, VI, Pompe, Gaucher and Fabry) (Rohrbach *et al.*, 2007). Enzyme replacement therapy, however, does not cross the blood–brain barrier and is not suitable for disorders associated wiith neuroregression. Other new therapies using small molecules have the advantage of blood-brain barrier penetration. These treatments may function a 'chaperones', by 'stabilising a misfolded enzyme' or substrate reduction therapies (Winchester *et al.* 2000, Butters *et al.* 2003). With such approaches to treatment, the long-term outcome of these disorders is likely to be radically altered over the next few years.

10.4.6 Gaucher's disease

First described in 1882 by Gaucher this disease is now recognized to be a group of a disorders which have in common the presence of 'Gaucher cells' in the bone marrow and reticulo-endothelial system. There are both visceral and neuronopathic forms and in all there is a defect in the cleavage of glucose from glucocerebroside due to deficient activity of the enzyme β-glucocerebrosidase (Table 10.1, V10a–c). Four common gene mutations are recognized although other rarer mutations may explain the considerable heterogeneity particularly in Type 1 and Type 3 forms of the disease (Beutler 2006).

Adult Gaucher's disease or Type 1

This is the commonest form and is a slowly progressive disorder with marked hepatosplenomegaly. There is accumulation of glucocerebroside in reticuloendothelial cells. The condition is non-neuronopathic and is rare in children. The disease causes problems because of its effects on the haematological and the skeletal systems. Clinical problems include pain, pathological fractures and hypersplenism.

Acute infantile Gaucher's disease or Type 2

This presents in infancy with poor feeding and failure to thrive. Mental development is slow from the start and then stalls with the development of spasticity, head retraction, and increasing bulbar dysfunction. Occasionally there may be a retinal cherry red spot. Seizures can occur and examination shows enlarged liver and spleen, characteristically with the spleen bigger than the liver. The haematological effects are severe with anaemia and thrombocytopenia. The serum acid phosphatase is raised. Measurement of the enzyme in leukocytes or fibroblasts and finding Gaucher cells in marrow will distinguish this disease from Niemann Pick and other neurovisceral storage diseases. Infants with the severe infantile form will rarely survive beyond 1 year.

Subacute/juvenile Gaucher's disease or Type 3

This is a curious intermediate form of the disease first described in Sweden with a wide range of clinical severity noted even within the same family. Onset can be between birth and adulthood with an average age at diagnosis of 2.5 years. Presentation is often with abdominal distension due to splenomegaly (Brett 1997). Although the splenomegaly can be massive, splenectomy should be avoided for as long as possible because this seems to be associated with a more rapid progression of the disease. The skeletal and haemato-logical effects of the disease are prominent but neurological involvement develops with cognitive slowing. Seizures and oculomotor apraxia are sometimes seen. The prognosis for all the neurological variants is an inevitable decline and eventual death. Survival into adulthood is uncommon.

In recent years enzyme replacement therapy has proved highly effective for patients with Gaucher's disease type 1 (Weinreb *et al.* 2002) and some patients with type 3 disease (Davies *et al.* 2007a). In type 1 disease treatment reduces the size of the liver and spleen and improves the skeletal abnormalities. Enzyme replacement has no effect on neurological progression of type 2 Gaucher disease. Bone marrow transplantation will reverse the non-neurological effects but carries a high mortality and also does not affect the neurological outcome. Enzyme replacement therapy has an uncertain effect on the neurological progression in type 3 disease. Severity scoring tools can be used to monitor the neurological features in type 3 disease and to assess the efficacy of treatment (Davies *et al.* 2007b).

10.5 **Brain mitochondrial diseases**

10.5.1 **Overview of the respiratory chain deficiencies**

Genetic disorders impairing oxidative phosphorylation are responsible for a wide range of neurological, muscular, and systemic disorders presenting in childhood or later. They all impair ATP synthesis by the respiratory chain in mitochondria, and result from mutations of either nuclear or mitochondrial genes encoding mitochondrial respiratory chain proteins (Fig. 10.11). Since mitochondria are passed on through the female germ line, being contained in the ovum, those mitochondrial disorders resulting from mutations of mitochondrial DNA show a maternal pattern of inheritance or are sporadic.

Two important principles underlie clinical suspicion of a mitochondrial disorder:

◆ Usually they are progressive disorders commencing in the first third of life. A notable exception is Leber's hereditary optic neuropathy (Section 12.5.2); and

◆ Because mitochondria are the ubiquitous energy source of the cells in actively metabolizing organs, the diagnosis is particularly suggested on encountering a patient with a constellation of seemingly unconnected clinical abnormalities. Furthermore, patients who present with one discrete symptom, often progress to develop other seemingly unrelated symptoms, thereby providing a clue to a mitochondrial disorder.

Although this section addresses neurological and muscular presentations of mitochondrial disorders, it should be noted that mitochondrial disorders often include, or can present with systemic features of cardiac, renal, nutritional, hepatic, endocrine, haematological, audiological or dermatologic disease, or dysmorphic features. The main neurological and skeletal muscular mitochondrial disorders are summarized in Table 10.9, which directs the reader to the main section in which these individual disorders are covered. Neuromuscular symptoms are the most common reason for referral of patients with mitochondrial disorders (Von Kleist-Retzow *et al.* 1998).

Neurological presentations. Patients may present at birth or in infancy with hypotonia, movement disorders, seizures, reduced alertness, poor sucking, respiratory distress, or lactic acidosis. Other children may develop normally, only to present in childhood with encephalopathy. In older children and young adults, abnormal eye movements, altered respiratory control, ataxia, myoclonic seizures, peripheral neuropathy, or spastic paraplegia are likely to be the main neurological features. MRI evidence of leukodystrophy with psychomotor retardation is common in mitochondrial

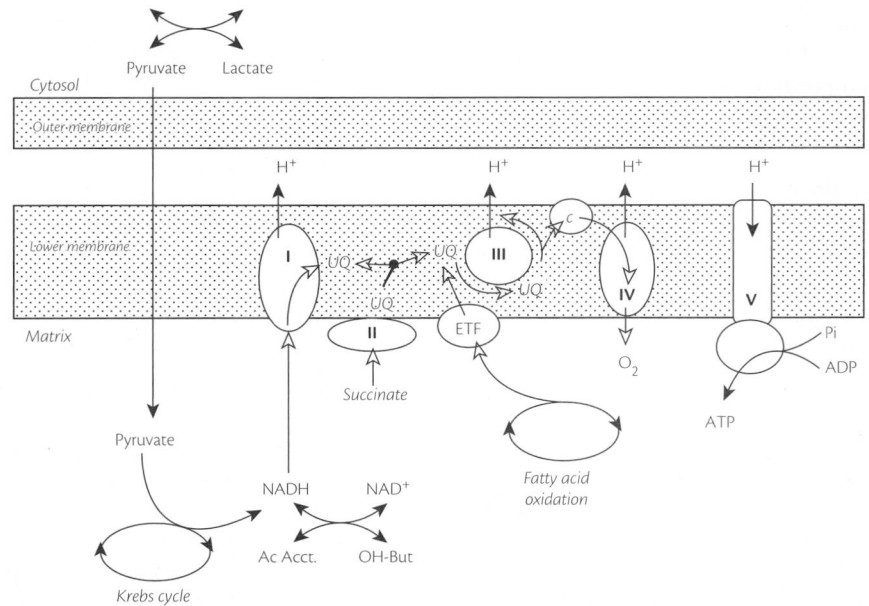

Fig. 10.11 The mitochondrial respiratory chain located in the inner mitochondrial membrane which oxidizes body fuel molecules to create ATP. Of the five complexes, only the peptides of complex 2 are entirely encoded by nuclear DNA. (Adapted from Munnich *et al.* 2000.)
Key:

——▸	electron flux
——◆	proton flux
ETF	electron transfer favoprotein
UQ	quinine
C	cytochrome C

Table 10.9 Inherited mitochondrial disorders causing nervous system or skeletal muscle disease. The characteristic presenting symptom is shown in bolditalics for each syndrome

Syndrome	Neurological features	Systemic features	Inheritance	Locus/gene	Usual age of onset	Section with main coverage
Alpers progressive sclerosing poliodystrophy	**Myoclonic epilepsy** Hypotonia	Hepatic enlargement/insufficiency	AR	POLG (most)	Childhood	30.3.4
Barth syndrome	Myopathy	**Cardiomyopathy**	X-linked	Xq28/tafazzin	Neonatal	
Chronic progressive external ophthalmoplegia	**Progressive external opthalmoplegia**		AD/sporadic/maternal	Various nuclear genes including *POLG1*, *Twinkle*, *ANT1* Also multiple mtDNA mutations	Adults	13.1.5 24.6.3
Hereditary spastic paraplegia	**Leg spasticity**		AR	6q24/*SPG7* – *paraplegin*	Young adults	23.4.2
Kearns Sayre syndrome	**Progressive external ophthalmoplegia** Cerebellar ataxia	Retinitis pigmentosa Heart block Growth retardation	Sporadic/maternal	MtDNA – various mutations	Childhood and adolescence	13.1.5 24.6.3
Leber hereditary optic neuropathy	Subacute visual loss—often sequential Optic atrophy	Cardiac involvement	Maternal	Various MtDNA mutations	Young adults	12.5.2
Leigh's syndrome of subacute necrotising encephalomyopathy	**Progressive psychomotor deterioration** Movement disorders (older children) Neurogenic respiratory failure MRI leukodystrophy and changes in basal ganglia, brain stem		AR Maternal X-linked	Various nuclear mutations including *SURF1* Various mtDNA mutations Pyruvate dehydrogenase E1 gene	Infancy and early childhood	10.5.3
Mitochondrial encephalomyopathy with lactic acidosis and stroke-like episodes 'MELAS'	**Recurrent stroke-like episodes** Migraine-like headaches Proximal weakness Seizures Deafness		Maternal	mtDNA point mutations 3243-tRNALeu (UUR) (in 80%)	Childhood	24.6.3 35.4.8
Mitochondrial, neurogastrointestinal encephalomyopathy 'MNGIE'	**Peripheral neuropathy** MRI leukoencephalopathy Eye movement disorder Ptosis	Thin body Gastrointestinal dysmotility	AR	22q13/Thymidine phosphorylase gene	Adolescents/Young adults	21.7.6
Myoclonic epilepsy with ragged red fibres 'MERRF'	**Myoclonic epilepsy** Ataxia Myopathy		Maternal	mtDNA 8344-tRNALys	Children/Young adults	24.6.3 30.3.4
Neuropathy, Ataxia, Retinitis Pigmentosa 'NARP'	**Polyneuropathy** Cerebellar ataxia	Retinitis pigmentosa	Maternal	mtDNA/nt-8993 ATPase 6	Childhood	10.5.4
Pearson Syndrome	Kearns–Sayre syndrome in survivors of >5 years	**Refractory sideroblastic anaemia** Diarrhoea Diabetes mellitus	Sporadic/Maternal	Similar to Kearns Sayre syndrome	Infancy	10.5.2
Wolfram syndrome	Optic atrophy	Diabetes mellitus Diabetes insipidus Deafness	AR	4p16		

Key:

AD = autosomal dominant

AR = autosomal recessive

POLG = polymerase gamma

neurological disorders (Table 10.7). Some patients initially present, or worsen during intercurrent infections, or in association with a suspected toxin exposure; this may reflect the precarious pre-existing tissue energy metabolism in patients with mitochondrial disorders.

Muscular presentations. Because of its role in translating aerobic metabolism into mechanical energy, muscle is prominently affected in mitochondrial disorders. In older patients, proximal muscle weakness is often a feature of mitochondrial cytopathies. Muscle biopsies, revealing ragged red fibres reflecting aggregations of mitochondria are present in many, although not all, mitochondrial cytopathies (Section 24.1.7). Thus, muscle biopsy can provide a useful clue to the presence of an underlying mitochondrial disorder, not only in those with symptoms and signs of muscle weakness, but also in other patients in whom mitochondrial disease is suspected. Mitochondrial myopathy is often associated with lactic acidosis, or significant elevation in lactic acid levels triggered by exercise. A severe infantile mitochondrial myopathy may be due to cytochrome C oxidase deficiency, sometimes reversibly, but usually leading to death in the first year of life (DiMauro *et al.* 1983). Episodic myalgia or myoglobinuria has been attributed to respiratory chain deficiencies (de Lonlay-Debeney *et al.* 1999).

Investigation of mitochondrial encephalomyopathies. In older children and adults with typical syndromes of mitochondrial disease, specific molecular genetic testing may be ordered (Section 3.8.3). However, with presentations earlier in life, when the full syndrome may not have become evident, targeted molecular genetic testing may not be possible. Mitochondrial genetic testing may be negative on blood, but positive on muscle mitochondria due to the phenomenon of heteroplasmy, in which mutated mitochondrial DNA molecules may be differently distributed in tissues. This reinforces the diagnostic role of muscle biopsy, not only by revealing ragged red fibres, but as a source of genetic material. MRI of the brain commonly shows white matter abnormalities, or diffuse leukodystrophy, in mitochondrial disorders. Alternatively it may show the changes of Leigh's disease in the basal ganglia, thalamus, and brain stem (Section 10.5.3).

Measurement of serum lactate and pyruvate, and CSF lactate' can be especially revealing of a respiratory chain deficiency, especially in children with acute clinical presentations. Blood samples should be taken fasting and 1 hour after feeding, with scrupulous sampling and sample processing techniques to avoid artefactually high levels of lactic acid. Respiratory chain deficiencies are particularly suggested by lactate:pyruvate ratios elevated above 20, in the presence of persistent hyperlactataemia (Munnich *et al.* 1996). Lactate, alanine, and lactate:pyruvate ratios can be measured in the CSF if no elevation of plasma lactate has been observed in a patient suspected of having central nervous system disease due to respiratory chain disorders.

Because mitochondrial disorders often affect other organs as well as the central nervous system, the diagnosis and full assessment of a mitochondrial disorder usually requires careful examination and investigation of muscle, heart, kidney, endocrine system, liver, gut, bone marrow, eyes, and ears. Sometimes the first line biochemical tests, apart from the blood lactate, can suggest a mitochondrial disorder: raised alanine and proline on analysis of plasma amino acids, or raised 3-methylglutaconic acid and 3-methylglutaric acid on urine organic acid analysis.

Treatment. For the vast majority of patients with a mitochondrial disorder, there is no treatment that improves the function of the defective mitochondria and hence leads to clinical improvement in central nervous system disease. Exceptions to this include:

◆ High dose ubiquinone, coenzyme Q_{10}, in some patients in whom reduced activity of complex II + III is secondary to a defect in ubiquinone synthesis.

◆ Riboflavin in some patients with complex I deficiency.

Defective functioning of mitochondria in the central nervous system sometimes leads to secondary problems which can be corrected, such as central nervous system folate deficiency in Kearns Sayre syndrome. In patients with the *MELAS* syndrome of **m**itochondrial **e**ncephalopathy, **l**actic **a**cidosis and **s**troke-like episodes, a beneficial effect of L-arginine on the outcome of stroke-like episodes has been reported. Drugs known to inhibit the respiratory chain, such as valproate and barbiturates should be avoided in patients with mitochondrial disease. Antioxidants such as vitamin E, vitamin C, and ubiquinone are frequently used but firm evidence of their ability to slow the progress of respiratory chain disorders is lacking. The same is true of the use of multivitamin cocktails. Some neuromuscular problems may be treatable including surgery for ptosis, kyphoscoliosis, and feeding difficulties or reflux. Hearing difficulties respond well to aids and cochlear implantation. For problems outside the central nervous system, there are many possibilities for treatment including cardiac pacing, anti-heart failure medications and heart transplantation, and treatment of diabetes, hypoparathyroidism, adrenal insufficiency, correction of electrolyte problems caused by tubulopathy, renal transplantation, pancreatic enzyme replacement, and parenteral nutrition. The ethics surrounding the use of treatments such as transplantation need to be considered carefully given the tendency of mitochondrial disorders to involve progressively more tissues and organs.

10.5.2 Pearson syndrome

The Pearson marrow-pancreas syndrome is of neurological interest because 20 per cent of affected infants have hypotonia, developmental delay, or ataxia, and approximately 10 per cent will develop Kearns Sayre syndrome subsequently (Lee *et al.* 2007). In the first year of life, half of the patients present with a refractory sideroblastic anaemia, often with pancytopaenia, and half present with diarrhoea and failure to thrive, representing exocrine pancreatic dysfunction sometimes associated with subtotal villus atrophy. Bone marrow examination shows vacuolization of erythrocyte prescursors, and on electron microscopy haemosiderin laden mitochondria are present. Insulin dependent diabetes mellitus may develop, lactic acidosis and 3-methyl-glutaconic aciduria are present, and ragged red fibres evident on muscle biopsy. Pearson syndrome is usually due to single large scale deletions or duplications of mitochondrial DNA. Having presented in the first year of life, Pearson syndrome is fatal during infancy or early childhood in more than 60 per cent (Rotig *et al.* 1990, 1995).

Those who survive beyond the age of 5 undergo spontaneous improvement in their haematological disorder, and go on to develop Kearns Sayre syndrome or Leigh's disease. The variable phenotype is illustrated by the cases of a mother and her infant son who harboured identical 5355 mtDNA single deletions and respectively manifested a chronic progressive external ophthalmoplegia accompanied by muscle weakness, and the Pearson syndrome (Shanske *et al.* 2002). In those Pearson patients with neurological manifestations, white matter changes are variably present on brain MRI (Lee *et al.* 2007). Central nervous system folate deficiency should be excluded.

10.5.3 Leigh's syndrome

This encephalopathy usually presents in early infancy, is usually fatal, and is also known as Leigh's encephalopathy, or subacute periventricular necrotizing encephalopathy. The condition is genetically heterogeneous, with a wide variety of mitochondrial and nuclear gene mutations being reported, variously producing enzyme deficiencies in complexes I, II, IV of the mitochondrial respiratory chain. Enzyme defects causing isolated deficiency of complex I seem particularly common (McFarland *et al.* 2004), usually presenting sporadically or with an autosomal recessive pattern of inheritance representing the predominance of nuclear gene mutations. Mutations affecting the mitochondrially encoded complex I subunit genes *ND3, ND4, ND5* and *ND6* are also recognized. Other autosomal recessive cases with complex IV deficiency are associated with mutations of *SURF 1*, a mitochondrial assembly protein gene (Lee *et al.* 1998). Autosomally recessively inherited pyruvate carboxylase deficiency can also cause Leigh's syndrome. Pyruvate dehydrogenase deficiency can be responsible for X-linked or autosomal recessive Leigh's syndrome, and coenzyme Q deficiency can also cause Leigh's syndrome (Section 10.5.5).

Leigh's syndrome commonly presents perinatally or in early infancy, and constitutes the commonest mitochondrial disorder of these age groups. It is characterized by diffuse encephalopathy with poor feeding and aspiration; seizures are relatively uncommon. Neurogenic respiratory failure may result from the brain stem lesions. Infants usually show lost or reduced responsiveness to visual, auditory, and tactile stimuli. They are generally hypotonic, although spastic diplegia can develop. More than 80 per cent of older infants and children develop dystonias, tremors, or cerebellar disorders (Macaya *et al.* 1993). Movement disorders may be the primary presenting feature. Rarely Leigh's syndrome first presents in adulthood, resembling ataxic paraparetic forms of multiple sclerosis (Section 37.5.3) or acute disseminated encephalomyelitis (Section 37.4.1) (Malojcic *et al.* 2004).

The serum lactate is usually elevated in Leigh's syndrome, and if not the CSF lactate. MRI shows perventricular white matter abnormalities, with T_2-weighted hyperintensity in the basal ganglia and thalamus. Although muscle biopsy does not show ragged red fibres, abnormal mitochondria are evident on electron microscopy, and quantitative studies will show abnormalities of mitochondrial respiratory chain enzymes. Neuropathologically there is destruction of the white matter periventricularly with symmetric necrotic lesions in the thalamus, brain stem, and dorsal columns of the spinal cord.

Leigh's syndrome carries a poor prognosis with most infants dying within days or weeks of presentation. Death occurs within months to years in patients of later infantile or childhood onset. A ketogenic diet may be helpful in those with underlying pyruvate dehydrogenase deficiency (Wexler *et al.* 1997).

10.5.4 Neuropathy, ataxia, retinitis pigmentosa syndrome

This syndrome is genetically directly related to Leigh's syndrome through sharing a common point mutation of the *ATPase 6* gene due to a point mutation at the 8993 locus of mtDNA (Holt *et al.* 1990). The resultant inhibition of oxidative phosphorylation leads to increased production of free radical species (Mattiazzi *et al.* 2004).

The onset of sensorimotor polyneuropathy, cerebellar ataxia, and retinitis pigmentosa, NARP, is usually in childhood, with limb weakness and sensory disturbance being followed by loss of night vision. When the heteroplasmy involves mutation of the great majority of mitochondrial DNA, Leigh's syndrome results, whereas when only moderate amounts of mtDNA are affected, NARP results. Correspondingly, mixed syndromes can occur in those with intermediate degrees of mtDNA heteroplasmy.

10.5.5 Deficiencies of pyruvate dehydrogenase and co-enzyme Q10

Although these deficiencies can produce encephalomyopathies resembling Leigh's syndrome, they are considered separately because of the wider range of disorders they can cause, and because both are potentially treatable. Deficiencies of both pyruvate dehydrogenase and co-enzyme Q10 can present with infantile lactic acidosis, and progressive psychomotor retardation and death in infancy. Neurological syndromes of later onset, including cerebellar ataxia and epilepsy, are also described.

Pyruvate dehydrogenase deficiency

Pyruvate dehydrogenase catalyses the irreversible oxidative decarboxylation of pyruvate to acetyl-CoA. Most cases of deficiency are X-linked, with a range of mutations affecting the *PDHA1* gene (Lissens *et al.* 2000). Autosomal recessive inheritance of pyruvate dehydrogenase deficiency can occur as the result of mutations in the *PDHX, DLAT, PDHB, PDHP,* and *DLD* genes. *DLD* mutations, affecting dihydrolipoamide dehydrogenase, produce a combined defect of pyruvate dehydrogenase and branch chain keto acid dehydrogenase, the latter giving rise to an amino acid profile similar to maple syrup urine disease. Autosomal recessive disorders affecting the *PDHAX* gene produce other variants of the pyruvate dehydrogenase complex deficiency (Schiff *et al.* 2006). These various mutations produce a range of clinical disorders involving lactic acidosis, which range from infantile Leigh's syndrome to intermittent ataxias. Roughly equal numbers of affected males and females have been identified associated with the X-linked form, and this is attributed to developmental lethality in some males with severe mutations, and the pattern of X-inactivation in females (Lissens *et al.* 2000). Treatment with a ketogenic diet reportedly improves the lactic acidosis and neurological disorder in Leigh's syndrome resulting from pyruvate dehydrogenase deficiency (Wexler *et al.* 1997).

Co-enzyme Q10 deficiency

A number of neurological syndromes have been associated with reduced levels of co-enzyme Q10 in muscle and may reflect blocks at different stages of the Co-enzyme Q10 synthetic pathway (diMauro *et al.* 2007; Gempel *et al.* 2007):

◆ Myopathy involving proximal limb and trunk muscles, with slow progression (Horvath *et al.* 2006). Muscle biopsies may show ragged red fibres. This may be associated with mutations of the electron-transferring-flavoprotein dehydrogenase gene, *ETFDH* (Gempel *et al.* 2007).

◆ Myopathy with ragged red fibres, recurrent myoglobinuria, and signs of a central nervous system disorder (di Giovanni *et al.* 2001).

◆ Leigh's syndrome presenting from infancy to early adulthood (Rotig *et al.* 2000; Van Maldergem *et al.* 2002). This form is often associated with nephrotic syndrome and subsequent renal failure. Retinitis pigmentosa, optic atrophy, and sensorineural deafness have been noted (Rötig *et al.* 2000).

◆ Cerebellar ataxia occurring on an autosomal recessive basis, and variably associated with other neurological disorders such as seizures or psychomotor delay (Lamperti *et al.* 2003). This ataxic variant is not associated with ragged red fibres on muscle biopsy, and is associated with mutations of the aprataxin gene, *APTX*, which is known to be responsible for ataxia with oculomotor apraxia type 1. This may represent a form of secondary coenzyme Q10 deficiency.

Thus, co-enzyme Q10 deficiencies seem to occur as both primary and secondary disorders. The primary disorders have been attributed to mutations within three of the nine nuclear genes required for synthesis of co-enzyme Q10, *PDSS1*, *PDSS2*, and *COQ2* (DiMauro *et al.* 2007). A primary deficiency of co-enzyme Q10 is particularly suggested by the muscle biopsy findings of ragged red fibres, and severe reduction in co-enzyme Q10 content, to as little as 5 per cent of normal. The importance of diagnosing co-enzyme Q10 deficiency is the often dramatic clinical response to oral ubiquinone supplementation (Rötig *et al.* 2000; Van Maldergem *et al.* 2002).

10.5.6 **Polymerase gamma mutations**

Mutations of the nuclear gene encoding polymerase gamma, *POLG*, are responsible for up to 25 per cent of all mitochondrial diseases (Chinnery and Zeviani 2007). Polymerase gamma is a DNA polymerase which replicates mitochondrial DNA. Mutations can cause both autosomal dominant and recessive disorders. Furthermore the same mutation can cause a wide spectrum of severity of mitochondrial disease, an example being the severe depletion of mitochondrial DNA seen in Alper's syndrome, yet the same genetic abnormality causing mild progressive external ophthalmoplegia of late onset. Autosomal dominant disease associated with *POLG* mutations particularly involve chronic progressive external ophthalmoplegia, and some families show psychiatric disorders, parkinsonism, or primary gonadal failure. By contrast, autosomal recessive disorders associated with *POLG* mutations tend to show cerebellar ataxia and axonal peripheral neuropathy. Alper's syndrome causes seizures, visual failure, and severe hepatic failure and is autosomal recessively inherited also due to *POLG* mutations (Table 10.9 and Section 30.3.4) (Chinnery and Zeviani 2007).

10.6 **Brain peroxisomal biosynthesis disorders**

10.6.1 **The range of peroxisomal disorders**

Peroxisomes are single membrane-lined organelles present in virtually all cells. They contain multiple enzymes involved in a variety of metabolic processes including α- and β-oxidation of certain fatty acids, biosynthesis of bile acids which also involves β-oxidation, and biosynthesis of ether phospholipids,

plasmologens. Thus, tests for peroxisomal disorders include the detection of:

◆ raised concentrations of very long chain fatty acids, pristanic acid and the C27 bile acids, THCA and DHCA which are all substrates for peroxisomal β-oxidation;

◆ a raised concentration of phytanic acid, a substrate of peroxisomal α-oxidation, typical of Refsum disease (Section 21.8.1);

◆ low levels of plasmologens in the red blood cells.

In the past peroxisomal disorders have been classified into those disorders that affect the assembly of peroxisomes, those that affect the import of peroxisomal proteins, and those that affect single peroxisomal enzymes (Table 10.1 IV). The first two groups have been amalgamated in recent years into the disorders of peroxisome biogenesis (Table 10.1 IV, 1–2) (Steinberg *et al.* 2006; Wanders and Waterham 2006). It has become clear that mutations in the same gene can cause disorders of a very wide spectrum of severity, referred to as the Zellweger spectrum. However, the original classification by syndromes such as Zellweger syndrome or infantile Refsum's disease remains important to the clinician who is trying to determine the likely prognosis for a particular child.

Once the first line tests have indicated a peroxisomal disorder, a wide variety of tests can be undertaken in cultured skin fibroblasts to define the disorder at the protein and gene level. This nearly always leads to the possibility of prenatal diagnosis. Peroxisomal disorders show autosomal recessive inheritance with the exception of X-linked adrenoleukodystrophy.

The main clinical syndromes that can occur in the peroxisome biogenesis disorders are:

◆ Zellweger syndrome (Section 10.6.2);

◆ Neonatal adrenoleukodystrophy;

◆ Infantile Refsum's disease (Section 10.6.3);

◆ Childhood onset leukodystrophy due to peroxisome biogenesis defects (Section 10.6.4); and

◆ Rhizomelic chondrodysplasia punctata (Section 10.6.5).

10.6.2 **Zellweger syndrome**

Zellweger syndrome is the most severe disease in this group and is also known as *cerebro-hepato-renal syndrome*. It presents in the newborn period with dysmorphism, including prominent high forehead, widely spaced sutures and large fontanelles, with severe hypotonia, poor sucking, and depressed or absent tendon reflexes. The hypotonia in the neonatal period may be confused with Down's syndrome, Prader–Willi Syndrome, and spinal muscular atrophy. Early seizures occur with poor development, failure to thrive, deafness, and visual impairment. A pigmentary retinopathy may develop with loss of the electroretinogram, and with optic atrophy, sometimes cataracts. Hepatomegaly and deranged liver function are often present. Adrenocortical function is often abnormal. Electromyography and nerve conduction velocities are usually normal. EEG is always abnormal. MR brain scan often shows a neuronal migration abnormality and may also show a leukodystrophy. The kidneys often show small cortical cysts. There is no treatment and most die in the first year. The diagnostic tests are listed in Table 10.1 IV, 1a. Structural brain abnormality combined with involvement of other organs can suggest the diagnosis before birth (Mochel *et al.* 2006) An attempt to treat a child with Zellwegers syndrome with Lorenzo's oil and docosahexaenoic acid has recently been described but despite biochemical improvement the infant died after 144 days (Tanaka *et al.* 2007).

10.6.3 Infantile Refsum disease

Infantile Refsum disease is a milder peroxisome biogenesis disorder. These infants are normal in the neonatal period and then, from 6 months of age, develop nonspecific symptoms including failure to thrive and apparent malabsorption. Hepatomegaly with abnormal liver function, low cholesterol, and facial dysmorphism are apparent on careful evaluation. As toddlers these children show delayed development with hypotonia, ataxia, visual impairment with choroido-retinopathy, and deafness. Severe cognitive impairment and survival to the teenage years can occur. Phytanic acid is elevated but accumulation of this metabolite is not as pathogenetically important as in adult Refsum disease (Section 21.8.1).

10.6.4 Leukodystrophy

In the mildest form of peroxisome biogenesis defect no abnormality, apart from sensorineural deafness, is detected in the first two years of life. However, in the preschool years, ataxia and upper motor neurone signs, particularly in the legs, become apparent and MRI shows white matter changes and often evidence of cerebellar atrophy. Very long chain fatty acids may show minor elevation or be normal; measurements of phytanate, pristanate, and C27 bile acids are important additional pointers to this diagnosis.

10.6.5 Rhizomelic chondrodysplasia punctata

In the classical form of this disorder affected infants have microcephaly and severe developmental delay. Additional features include shortening of the proximal limb bones restricted joint movements and the radiological features of punctuate epiphyseal calcification and metaphysical splaying and cupping. Plasma phytanate is usually elevated if it has been given in feeds (cow's milk-based formula), and red cell plasmalogens are reduced.

10.6.6 Single peroxisomal enzyme or transporter disorders

◆ Acyl-CoA oxidase and D-bifunctional protein deficiencies show phenotypes similar to the Zellweger spectrum (Section 10.6.2);

◆ X-linked adrenoleukodystrophy is discussed in Section 10.2.4;

◆ Refsum's disease is discussed in Section 21.8.1;

◆ Alpha-methyl-acyl-CoA racemase deficiency can present with cholestatic liver disease in infancy, with developmental delay, with episodes of encephalopathy and with an adult onset sensorimotor neuropathy and/or pigmentary retinopathy; and

◆ Dihydroxyacetone phosphate acyl transferase deficiency and alkyl-dihydroxyacetone phosphate synthase deficiency usually present in the neonatal period with features similar to rhizomelic chondrodysplasia punctata. Milder forms may present in early childhood with developmental delay.

10.7 Other inborn errors of metabolism

A wide range of inborn errors of metabolism can cause neurological abnormalities (Table 10.1 I, II), and some are particularly associated with episodes of acute encephalopathy (Table 10.2).

10.7.1 Metabolic causes of epileptic encephalopathy

A number of metabolic diseases can present with epilepsy particularly early in life (Section 30.5.2). These include conditions themselves causing encephalopathy of which some are potentially treatable:

◆ *Pyridoxine dependency* should always be considered as the cause of early onset drug resistant seizures (Table 10.2, V 1 and 2). Neonatal units consider a trial of pyridoxine in an infant with refractory epilepsy and this recommendation is often included in protocols for the management of status epilepticus in infants and young children. Although rare, its treatability makes it an important condition to identify. The biochemical basis for pyridoxine responsive epilepsy is now better understood thereby allowing diagnostic testing by checking pipecolic acid in body fluids, CSF and blood, or urine alpha-aminoadipic semialdehyde levels (Willemsen *et al.* 2005; Bok *et al.* 2007). While awaiting these results a therapeutic trial of pyridoxine can be undertaken, provided that full resuscitation facilities are available. Pyridoxal phosphate is required in infants with a deficiency of pyridoxine phosphate oxidase although the preparation can be difficult to obtain (Mills *et al.* 2005). Hopefully the availability of diagnostic testing will resolve some of the previous uncertainty surrounding the offering of this treatment to infants and children.

◆ *Biotinidase deficiency* is another important treatable cause of refractory seizures. Usually it is associated with skin rash and abnormal hair (Table 10.1, IX 3a). This disorder is now detected by newborn screening in some countries although not in the United Kingdom. Late diagnosis can be associated with visual impairment due to optic atrophy, hearing loss and psychomotor retardation, with pyramidal motor impairment and spasticity (Weber *et al.* 2004). The condition can be detected by checking the biotinidase level in blood, and usually shows a characteristic abnormality on the urine organic acid profile. A trial of biotin should be considered while awaiting the results of diagnostic tests.

◆ Other potentially treatable conditions presenting with early seizures include *folinic acid responsive seizures* and *Glut 1 deficiency* (Frye *et al.* 2003; Klepper and Leiendecker 2007).

Other important metabolic causes of early onset epilepsy are glycine encephalopathy, or non-ketotic hyperglycinaemia, serine deficiency syndromes and sulphite oxidase deficiency:

◆ *Glycine encephalopathy.* The infant with non-ketotic hyperglycinaemia presents with early intractable seizures and severe encephalopathy. These infants often require ventilatory support and many do not survive. Those that do survive will usually have persisting neurodevelopmental problems. Rare later onset variants are recognized that are less severe but are associated with epilepsy. There may be a curious gender difference in terms of survival with boys having a better outcome than girls (Hoover-Fong *et al.* 2004). The diagnosis is suggested by finding a high glycine level in the CSF with a plasma:CSF glycine ratio > 0.08. A transient benign form of neonatal hyperglycinemia is recognized (Lang *et al.* 2008).

◆ *Serine deficiency* (Table 10.2, V 10). Patients with serine deficiency syndromes are diagnosed by finding low levels of serine in blood and CSF. These disorders involve microcephaly, developmental retardation and epilepsy. They are potentially treatable although the limited therapeutic response in reports to date may be due to delay in diagnosis (De Koning and Klomp 2004).

◆ *Sulfite oxidase deficiency* (Table 10.2, V 3) usually presents in early infancy with seizures, irritability, and reduced muscle tone. The condition can resemble birth asphyxia both clinically and radiologically. The diagnosis is suggested by finding elevated urinary levels of sulphite, thiosulphite and S-sulfocysteine. The human sulfite oxidase gene, *SUOX*, is located at 12q13.13 and at least 16 pathogeneic mutations are recognized. The condition is usually fatal with no effective long term treatment. Onset is usually at birth but can be delayed for many months with survival for many years (Tan *et al.* 2005).

10.7.2 Congenital disorders of glycosylation

Congenital disorders of glycosylation are a relatively newly described group of metabolic disorders characterized by defective synthesis of N- and/or O-linked oligosaccharides (Table 10.1, VI). The term 'congenital disorders of glycosylation' is now preferred to the former 'carbohydrate deficient glycoprotein syndrome'. There is considerable phenotypic variation, with more than 20 types being described. The usual screening test for the diagnosis is isoelectric focusing of a glycoprotein, usually serum transferrin (Marklova *et al.* 2007). The clinical presentation can range from early onset severe developmental delay, hypotonia, and multi-organ involvement through to hypoglycaemia and protein losing enteropathy with normal development. The commonest type is 1a and survival to adulthood is recognized. These adults have moderate mental retardation, ataxia, retinitis pigmentosa, peripheral neuropathy, kyphoscoliosis, and endocrinopathies (Krasnewich *et al.* 2007).

Rett syndrome

Rett syndrome is a common neurodevelopmental disorder caused by mutations in the gene encoding methyl-CpG binding protein 2, *MECP2* (Section 9.6.2). The condition was first described in the German literature in 1966 after Andreas Rett saw two girls sitting together in his waiting room. The genetic basis for the disorder was not recognized until 1999. The condition is often listed with neurodegenerative diseases yet there is no biochemical marker and destructive changes in the central nervous system are not seen. The condition is described as a postnatal progressive neurodevelopmental disorder (Chahrour and Zoghbi 2007). Typically the child is initially normal; then development stagnates before regressing with loss of speech, hand skills, and the development of characteristic hand-wringing stereotypies. Affected girls often develop epilepsy, respiratory irregularities, and autonomic dysfunction. Many survive to adulthood and some to old age. Many lose mobility during adolescence and early adult life. Rett syndrome affects 1:10 000 births and is most often sporadic; 95 per cent of cases have a mutation in *MECP2*. Most cases arise in the paternal germ line which is why it is very rare in males. The cyclin-dependent kinase like 5, *CDKL5* gene is another X-linked gene which produces a severe Rett-like phenotype presenting in the first 6 months, often with infantile spasms or severe early onset epilepsy.

References

Al-Dirbashi OY, Rashed MS, Al-Qahtani K *et al.* (2007). Quantification of N-acetylaspartic acid in urine by LC-MS/MS for the diagnosis of Canavan disease. *J Inherit Metab Dis*, 30, 612.

Anderson G, Smith VV, Malone M *et al.* (2005). Blood film examination for vacuolated lymphocytes in the diagnosis of metabolic disorders; retrospective experience of more than 2,500 cases from a single centre. *J Clin Pathol*, 58, 1305–10.

Ashrafi MR, Shabanian R, Mohammadi M *et al.* (2006). Extensive Mongolian spots: a clinical sign merits special attention. *Pediatr Neurol*, 34, 143–5.

Augestad LB, Flanders WD (2006). Occurrence of and mortality from childhood neuronal ceroid lipofuscinoses in Norway. *J Child Neurol*, 21, 917–22.

Barth PG, Hoffmann GF, Jaeken J *et al.* (1992). L-2-hydroxyglutaric acidemia: a novel inherited neurometabolic disease. *Ann Neurol*, 32, 66–71.

Baumann N, Masson M, Carreau V *et al.* (1991). Adult forms of metachromatic leukodystrophy: clinical and biochemical approach. *Dev Neurosci*, 13, 211–5.

Beck M (2007). New therapeutic options for lysosomal storage disorders: enzyme replacement, small molecules and gene therapy. *Hum Genet*, 121, 1–22.

Been JV, Bok LA, Andriessen P *et al.* (2005). Epidemiology of pyridoxine dependent seizures in the Netherlands. *Arch Dis Child*, 90, 1293–6.

Berkovic S, Carpenter S, Andermann F *et al.* (1988). Kufs' disease: a critical reappraisal. *Brain*, 111, 27–62.

Beutler E (2006). Gaucher disease: multiple lessons from a single gene disorder. *Acta Paediatr Suppl*, 95, 103–9.

Biancheri R, Rossi A, Di Rocco M *et al.* (2003). Leukoencephalopathy with vanishing white matter: an adult onset case. *Neurology*, 61, 1818–9.

Black DN, Booth F, Watters GV *et al.* (1988). Leukoencephalopathy among native Indian infants in northern Quebec and Manitoba. *Ann Neurol*, 24, 490–6.

Bodamer OA, Hoffmann GF, Lindner M (2007). Expanded newborn screening in Europe 2007. *J Inherit Metab Dis*, 30, 439–44.

Boelens JJ (2006). Trends in haematopoietic cell transplantation for inborn errors of metabolism. *J Inherit Metab Dis*, 29, 413–20.

Bok LA, Struys E, Willemsen MA *et al.* (2007). Pyridoxine-dependent seizures in Dutch patients: diagnosis by elevated urinary alpha-aminoadipic semialdehyde levels. *Arch Dis Child*, 92, 687–9.

Brett EM ed. (1997). *Paediatric Neurology*, 3rd edition. Churchill Livingstone, New York.

Brismar J, Brismar G, Gascon G *et al.* (1990). Canavan disease: CT and MR imaging of the brain. *AJNR Am J Neuroradiol*, 11, 805–10.

Butters TD, Mellor HR, Narita K, Dwek RA, Platt FM. Small-molecule therapeutics for the treatment of glycolipid lysosomal storage disorders. Philo Trans R Soc Lond B Biol Sci 2003; 358:927–45.

Chabrol B, Mancini J, Chretien D *et al.* (1994). Valproate-induced hepatic failure in a case of cytochrome c oxidase deficiency. *Eur J Pediatr*, 153, 133–5.

Chahrour M, Zoghbi H (2007). The story of Rett syndrome: from clinic to neurobiology. *Neuron*, 56, 422–37.

Chen CY, Zimmerman RA, Lee CC *et al.* (1998). Neuroimaging findings in late infantile GM1 gangliosidosis. *AJNR Am J Neuroradiol*, 19, 1628–30.

Chen E, Nyhan WL, Jakobs C *et al.* (1996). L-2-Hydroxyglutaric aciduria: neuropathological correlations and first report of severe neurodegenerative disease and neonatal death. *J Inherit Metab Dis*, 19, 335–43.

Chinnery PF, Zeviani M (2007). 155th ENMC workshop: Polymerase gamma and disorders of mitochondrial DNA synthesis, 21–23 September 2007, Naarden, The Netherlands. *Neuromuscul Disord*, 18, 259–67.

Cif L, Biolsi B, Gavarini S *et al.* (2007). Antero-ventral internal pallidum stimulation improves behavioral disorders in Lesch-Nyhan disease. *Mov Disord*, 22, 2126–9.

Clarke LA (2008). The mucopolysaccharidoses: a success of molecular medicine. *Expert Rev Mol Med*, 10, e1.

Davies EH, Erikson A, Collin-Histed T *et al.* (2007a). Outcome of type III Gaucher disease on enzyme replacement therapy: review of 55 cases. *J Inherit Metab Dis*, 30, 935–42.

Davies EH, Surtees R, DeVile C *et al.* (2007b). A severity scoring tool to assess the neurological features of neuronopathic Gaucher disease. *J Inherit Metab Dis*, **30**, 768–82.

de Koning TJ, Klomp LW (2004). Serine-deficiency syndromes. *Curr Opin Neurol*, **17**, 197–204.

de Lonlay-Debeney P, Edery P, Cormier-Daire V *et al.* (1999). Respiratory chain deficiency presenting as recurrent myoglobinuria in childhood. *Neuropediatrics*, **30**, 42–4.

Devereux G, Stellitano L, Verity CM *et al.* (2004). Variations in neurodegenerative disease across the UK: findings from the national study of Progressive Intellectual and Neurological Deterioration (PIND). *Arch Dis Child*, **89**, 8–12.

Di Giovanni S, Mirabella M, Spinazzola A *et al.* (2001). Coenzyme Q10 reverses pathological phenotype and reduces apoptosis in familial CoQ10 deficiency. *Neurology*, **57**, 515–8.

DiMauro S, Quinzii CM, Hirano M (2007). Mutations in coenzyme Q10 biosynthetic genes. *J Clin Invest*, **117**, 587–9.

DiMauro S, Nicholson JF, Hays AP *et al.* (1983). Benign infantile mitochondrial myopathy due to reversible cytochrome c oxidase deficiency. *Ann Neurol*, **14**, 226–34.

Escolar ML, Poe MD, Provenzale JM *et al.* (2005). Transplantation of umbilical-cord blood in babies with infantile Krabbe's disease. *N Engl J Med*, **352**, 2069–81.

Fogli A, Rodriguez D, Eymard-Pierre E *et al.* (2003). Ovarian failure related to eukaryotic initiation factor 2B mutations. *Am J Hum Genet*, **72**, 1544–50.

Fogli A, Wong K, Eymard-Pierre E *et al.* (2002). Cree leukoencephalopathy and CACH/VWM disease are allelic at the EIF2B5 locus. *Ann Neurol*, **52**, 506–10.

Francalanci P, Eymard-Pierre E, Dionisi-Vici C *et al.* (2001). Fatal infantile leukodystrophy: a severe variant of CACH/VWM syndrome, allelic to chromosome 3q27. *Neurology*, **57**, 265–70.

Frye RE, Donner E, Golja A *et al.* (2003). Folinic acid-responsive seizures presenting as breakthrough seizures in a 3-month-old boy. *J Child Neurol*, **18**, 562–9.

Fujitake J, Ishikawa Y, Fujii H *et al.* (1999). L-2-hydroxyglutaric aciduria: two Japanese adult cases in one family. *J Neurol*, **246**, 378–82.

Garbern JY (2007). Pelizaeus-Merzbacher disease: genetic and cellular pathogenesis. *Cell Mol Life Sci*, **64**, 50–65.

Garcia-Cazorla A, Serrano M, Perez-Duenas B *et al.* (2007). Secondary abnormalities of neurotransmitters in infants with neurological disorders. *Dev Med Child Neurol*, **49**, 740–4.

Gempel K, Topaloglu H, Talim B *et al.* (2007). The myopathic form of coenzyme Q10 deficiency is caused by mutations in the electron-transferring-flavoprotein dehydrogenase (ETFDH) gene. *Brain*, **130**, 2037–44.

Giugliani R, Harmatz P, Wraith JE (2007). Management guidelines for mucopolysaccharidosis VI. *Pediatrics*, **120**, 405–18.

Gordon N (2002). Infantile neuroaxonal dystrophy (Seitelberger's disease). *Dev Med Child Neurol*, **44**, 849–51.

Gorospe JR, Naidu S, Johnson AB *et al.* (2002). Molecular findings in symptomatic and pre-symptomatic Alexander disease patients. *Neurology*, **58**, 1494–500.

Goutieres F (2005). Aicardi-Goutieres syndrome. *Brain Dev*, **27**, 201–6.

Hagberg B. (1963) Clinical symptoms, signs and tests in metachromatic leukodystrophy. In Folch-Pi J, Bauer H, eds. *Brain Lipids and Lipoproteins and the Leukodystrophies*, pp. 134. Elsevier, Amsterdam.

Hagberg B, Sourander P, Svennerholm L (1963). Diagnosis of Krabbe's infantile leucodystrophy. *J Neurol Neurosurg Psychiatry*, **26**, 195–8.

Hobson EE, Thomas S, Crofton PM *et al.* (2005). Isolated sulphite oxidase deficiency mimics the features of hypoxic ischaemic encephalopathy. *Eur J Pediatr*, **164**, 655–9.

Holt IJ, Harding AE, Petty RK *et al.* (1990). A new mitochondrial disease associated with mitochondrial DNA heteroplasmy. *Am J Hum Genet*, **46**, 428–33.

Hoover-Fong JE, Shah S, Van Hove JL *et al.* (2004). Natural history of nonketotic hyperglycinemia in 65 patients. *Neurology*, **63**, 1847–53.

Horvath R, Schneiderat P, Schoser BG *et al.* (2006). Coenzyme Q10 deficiency and isolated myopathy. *Neurology*, **66**, 253–5.

Hudspeth MP, Raymond GV (2007). Immunopathogenesis of adrenoleukodystrophy: current understanding. *J Neuroimmunol*, **182**, 5–12.

Hurst S, Garbern J, Trepanier A *et al.* (2006). Quantifying the carrier female phenotype in Pelizaeus-Merzbacher disease. *Genet Med*, **8**, 371–8.

Josephson SA, Schmidt RE, Millsap P *et al.* (2001). Autosomal dominant Kufs' disease: a cause of early onset dementia. *J Neurol Sci*, **188**, 51–60.

Inoue K (2005). PLP1-related inherited dysmyelinating disorders: Pelizaeus-Merzbacher disease and spastic paraplegia type 2. *Neurogenetics*, **6**, 1–16.

Kemp S, Wanders RJ (2007). X-linked adrenoleukodystrophy: very long-chain fatty acid metabolism, ABC half-transporters and the complicated route to treatment. *Mol Genet Metab*, **90**, 268–76.

Khateeb S, Flusser H, Ofir R *et al.* (2006). PLA2G6 mutation underlies infantile neuroaxonal dystrophy. *Am J Hum Genet*, **79**, 942–8.

Klepper J, Leiendecker B (2007). GLUT1 deficiency syndrome--2007 update. *Dev Med Child Neurol*, **49**, 707–16.

Koeppen AH (2005). A brief history of Pelizaeus-Merzbacher disease and proteolipid protein. *J Neurol Sci* **228**: 198–200.

Kolodny EH, Raghavan S, Krivit W (1991). Late-onset Krabbe disease (globoid cell leukodystrophy): clinical and biochemical features of 15 cases. *Dev Neurosci*, **13**, 232–9.

Krasnewich D, O'Brien K, Sparks S (2007). Clinical features in adults with congenital disorders of glycosylation type la (CDG-la). *Am J Med Genet C Semin Med Genet* **145C**:302–306.

Krivit W, Shapiro EG, Peters C *et al.* (1998). Hematopoietic stem-cell transplantation in globoid-cell leukodystrophy. *N Engl J Med*, **338**, 1119–26.

Kumar N, Boes CJ, Babovic-Vuksanovic D *et al.* (2006). The "eye-of-the-tiger" sign is not pathognomonic of the PANK2 mutation. *Arch Neurol*, **63**, 292–3.

Lake BD, Steward CG, Oakhill A *et al.* (1997). Bone marrow transplantation in late infantile Batten disease and juvenile Batten disease. *Neuropediatrics*, **28**, 80–1.

Lamperti C, Naini A, Hirano M *et al.* (2003). Cerebellar ataxia and coenzyme Q10 deficiency. *Neurology*, **60**, 1206–8.

Lang TF, Parr JR, Matthews EE *et al.* (2008). Practical difficulties in the diagnosis of transient non-ketotic hyperglycinaemia. *Dev Med Child Neurol*, **50**, 157–9.

Leavitt JA, Kotagal S (2007). The "cherry red" spot. *Pediatr Neurol*, **37**, 74–5.

Lee HF, Lee HJ, Chi CS *et al.* (2007). The neurological evolution of Pearson syndrome: case report and literature review. *Eur J Paediatr Neurol*, **11**, 208–14.

Lee N, Morin C, Mitchell G *et al.* (1998). Saguenay Lac Saint Jean cytochrome oxidase deficiency: sequence analysis of nuclear encoded COX subunits, chromosomal localization and a sequence anomaly in subunit VIc. *Biochim Biophys Acta*, **1406**, 1–4.

Li R, Johnson AB, Salomons G *et al.* (2005). Glial fibrillary acidic protein mutations in infantile, juvenile, and adult forms of Alexander disease. *Ann Neurol*, **57**, 310–26.

Lissens W, De Meirleir L, Seneca S *et al.* (2000). Mutations in the X-linked pyruvate dehydrogenase (E1) alpha subunit gene (PDHA1) in patients with a pyruvate dehydrogenase complex deficiency. *Hum Mutat*, **15**, 209–19.

Loes DJ, Peters C, Krivit W (1999). Globoid cell leukodystrophy: distinguishing early-onset from late-onset disease using a brain MR imaging scoring method. *AJNR Am J Neuroradiol*, **20**, 316–23.

Lyon G, Hagberg B, Evrard P *et al.* (1991). Symptomatology of late onset Krabbe's leukodystrophy: the European experience. *Dev Neurosci*, **13**, 240–4.

Macaya A, Munell F, Burke RE *et al.* (1993). Disorders of movement in Leigh syndrome. *Neuropediatrics*, **24**, 60–7.

Maegawa GH, Stockley T, Tropak M *et al.* (2006). The natural history of juvenile or subacute GM2 gangliosidosis: 21 new cases and literature review of 134 previously reported. *Pediatrics*, **118**, e1550–62.

Malojcic B, Brinar V, Poser C *et al.* (2004). An adult case of Leigh disease. *Clin Neurol Neurosurg*, **106**, 237–40.

Marklova E and Albahri Z (2007). Screening and diagnosis of congenital disorders of glycosylation. *Clin Chim Acta* **385**:6–20.

Mattiazzi M, Vijayvergiya C, Gajewski CD *et al.* (2004). The mtDNA T8993G (NARP) mutation results in an impairment of oxidative phosphorylation that can be improved by antioxidants. *Hum Mol Genet*, **13**, 869–79.

McFarland R, Kirby DM, Fowler KJ *et al.* (2004). De novo mutations in the mitochondrial ND3 gene as a cause of infantile mitochondrial encephalopathy and complex I deficiency. *Ann Neurol*, **55**, 58–64.

Menkes JH, Sarnat HB, Maria BL (2005). *Child Neurology*, 7th edition. Lippincott-Williams and Wilkins, London.

Millat G, Marcais C, Tomasetto C *et al.* (2001). Niemann-Pick C1 disease: correlations between NPC1 mutations, levels of NPC1 protein, and phenotypes emphasize the functional significance of the putative sterol-sensing domain and of the cysteine-rich luminal loop. *Am J Hum Genet*, **68**, 1373–85.

Mills PB, Surtees RA, Champion MP, *et al.* (2005). Neonatal epileptic encephalopathy caused by mutations in the PNPO gene encoding pyridox(am)ine 5'-phosphate oxidase. *Hum Mol Genet* **14**: 1077–1086.

Mills PB, Struys E, Jakobs C *et al.* (2006). Mutations in antiquitin in individuals with pyridoxine-dependent seizures. *Nat Med*, **12**, 307–9.

Mochel F, Grebille AG, Benachi A *et al.* (2006). Contribution of fetal MR imaging in the prenatal diagnosis of Zellweger syndrome. *AJNR Am J Neuroradiol*, **27**, 333–6.

Mole SE, Williams RE, Goebel HH (2005). Correlations between genotype, ultrastructural morphology and clinical phenotype in the neuronal ceroid lipofuscinoses. *Neurogenetics*, **6**, 107–26.

Moog U, van Mierlo I, van Schrojenstein Lantman-de Valk HM *et al.* (2007). Is Sanfilippo type B in your mind when you see adults with mental retardation and behavioral problems? *Am J Med Genet C Semin Med Genet*, **145**, 293–301.

Morgan NV, Westaway SK, Morton JE *et al.* (2006). PLA2G6, encoding a phospholipase A2, is mutated in neurodegenerative disorders with high brain iron. *Nat Genet*, **38**, 752–4.

Moroni I, D'Incerti L, Farina L *et al.* (2000). Clinical, biochemical and neuroradiological findings in L-2-hydroxyglutaric aciduria. *Neurol Sci*, **21**, 103–8.

Moser HW, Raymond GV, Dubey P (2005a). Adrenoleukodystrophy: new approaches to a neurodegenerative disease. *JAMA*, **294**, 3131–4.

Moser HW, Raymond GV, Lu SE *et al.* (2005b). Follow-up of 89 asymptomatic patients with adrenoleukodystrophy treated with Lorenzo's oil. *Arch Neurol*, **62**, 1073–80.

Munnich A, Rotig A, Chretien D *et al.* (1996). Clinical presentation of mitochondrial disorders in childhood. *J Inherit Metab Dis*, **19**, 521–7.

Munnich A, Rötig A, Cormier-Daire V *et al.* (2000). Clinical presentation of respiratory chain deficiency. In Scriver CR, Sly WS, eds. *The Metabolic Bases of Inherited Disease*, pp. 2261–73. McGraw Hill, New York.

Muñoz V, Muenzer J, Martin R *et al.* (2008). Recognition and diagnosis of mucopolysaccharidosis II (Hunter syndrome). *Pediatrics*, **121**, 377–386.

Muthane U, Chickabasaviah Y, Kaneski C *et al.* (2004). Clinical features of adult GM1 gangliosidosis: report of three Indian patients and review of 40 cases. *Mov Disord*, **19**, 1334–41.

Namekawa M, Takiyama Y, Aoki Y *et al.* (2002). Identification of GFAP gene mutation in hereditary adult-onset Alexander's disease. *Ann Neurol*, **52**, 779–85.

Noble-Jamieson G, Jamieson N, Clayton P *et al.* (1994). Neurological crisis in hereditary tyrosinaemia and complete reversal after liver transplantation. *Arch Dis Child*, **70**, 544–5.

Ohtake H, Shimohata T, Terajima K *et al.* (2004). Adult-onset leukoencephalopathy with vanishing white matter with a missense mutation in EIF2B5. *Neurology*, **62**, 1601–3.

Orchard PJ, Blazar BR, Wagner J *et al.* (2007). Hematopoietic cell therapy for metabolic disease. *J Pediatr*, **151**, 340–6.

Orthmann-Murphy JL, Enriquez AD, Abrams CK *et al.* (2007). Loss-of-function GJA12/Connexin47 mutations cause Pelizaeus-Merzbacher-like disease. *Mol Cell Neurosci*, **34**, 629–41.

Pearl PL, Taylor JL, Trzcinski S *et al.* (2007). The pediatric neurotransmitter disorders. *J Child Neurol*, **22**, 606–16.

Peters C, Charnas LR, Tan Y *et al.* (2004). Cerebral X-linked adrenoleukodystrophy: the international hematopoietic cell transplantation experience from 1982 to 1999. *Blood*, **104**, 881–8.

Prasad AN, Breen JC, Ampola MG *et al.* (1997). Argininemia: a treatable genetic cause of progressive spastic diplegia simulating cerebral palsy: case reports and literature review. *J Child Neurol*, **12**, 301–9.

Rahman S, Standing S, Dalton RN *et al.* (1997). Late presentation of biotinidase deficiency with acute visual loss and gait disturbance. *Dev Med Child Neurol*, **39**, 830–1.

Ramadan H, Al-Din AS, Ismail A *et al.* (2007). Adult neuronal ceroid lipofuscinosis caused by deficiency in palmitoyl protein thioesterase 1. *Neurology*, **68**, 387–8.

Ramirez-Montealegre D, Rothberg PG, Pearce DA (2006). Another disorder finds its gene. *Brain*, **129**, 1353–6.

Rauschka H, Colsch B, Baumann N *et al.* (2006). Late-onset metachromatic leukodystrophy: genotype strongly influences phenotype. *Neurology*, **67**, 859–63.

Rice G and Patrick T and Parmar R *et al.* (2007). Clinical and molecular phenotype of Aicardi-Goutieres syndrome. *Am J Hum Genet*, **81**, 713–25.

Rodriguez D, Gauthier F, Bertini E *et al.* (2001). Infantile Alexander disease: spectrum of GFAP mutations and genotype-phenotype correlation. *Am J Hum Genet*, **69**, 1134–40.

Rohrbach M, Clarke JTR (2007). Treatment of lysosomal storage disorders: progress with enzyme replacement therapy. Drugs, vol. 67, no. 18, p. 2697–716, ISSN: 0012–6667.

Rotig A, Cormier V, Blanche S *et al.* (1990). Pearson's marrow-pancreas syndrome. A multisystem mitochondrial disorder in infancy. *J Clin Invest*, **86**, 1601–8.

Rotig A, Bourgeron T, Chretien D *et al.* (1995). Spectrum of mitochondrial DNA rearrangements in the Pearson marrow-pancreas syndrome. *Hum Mol Genet*, **4**, 1327–30.

Rotig A, Appelkvist EL, Geromel V *et al.* (2000). Quinone-responsive multiple respiratory-chain dysfunction due to widespread coenzyme Q10 deficiency. *Lancet*, **356**, 391–5.

Schiffmann R, Tedeschi G, Kinkel RP, *et al.* (1997). Leukodystrophy in patients with ovarian dysgenesis. *Ann Neurol.* **41**:654–661.

Schiff M, Mine M, Brivet M *et al.* (2006). Leigh's disease due to a new mutation in the PDHX gene. *Ann Neurol*, **59**, 709–14.

Schuchman EH (2007). The pathogenesis and treatment of acid sphingomyelinase-deficient Niemann-Pick disease. *J Inherit Metab Dis*, **30**, 654–63.

Scriver C, Beaudet A, Sly W *et al.* (2001). *The Metabolic and Molecular Bases of Inherited Disease*, 8th edition. McGraw-Hill Medical, New York.

Sevin C, Aubourg P, Cartier N (2007a). Enzyme, cell and gene-based therapies for metachromatic leukodystrophy. *J Inherit Metab Dis*, **30**, 175–83.

Sevin M, Lesca G, Baumann N *et al.* (2007b). The adult form of Niemann-Pick disease type C. *Brain*, **130**, 120–33.

Shanske S, Tang Y, Hirano M *et al.* (2002). Identical mitochondrial DNA deletion in a woman with ocular myopathy and in her son with Pearson syndrome. *Am J Hum Genet*, **71**, 679–83.

Siddiqi ZA, Sanders DB, Massey JM (2006). Peripheral neuropathy in Krabbe disease: electrodiagnostic findings. *Neurology*, **67**, 263–7.

Singhal BS, Gursahani RD, Udani VP *et al.* (1996). Megalencephalic leukodystrophy in an Asian Indian ethnic group. *Pediatr Neurol*, **14**, 291–6.

Smith DW, Jinnah HA (2007). Role of neuronal nitric oxide in the dopamine deficit of HPRT-deficient mice. *Metab Brain Dis*, **22**, 39–43.

Steinberg SJ, Dodt G, Raymond GV *et al.* (2006). Peroxisome biogenesis disorders. *Biochim Biophys Acta*, **1763**, 1733–48.

Tan WH, Eichler FS, Hoda S, *et al.* (2005). Isolated sulfite oxidase deficiency: a case report with a novel mutation and review of the literature. *Pediatrics* **116**:757–766.

Tanaka K, Shimizu T, Ohtsuka Y *et al.* (2007). Early dietary treatments with Lorenzo's oil and docosahexaenoic acid for neurological development in a case with Zellweger syndrome. *Brain Dev*, **29**, 586–9.

Topcu M, Saatci I, Topcuoglu MA *et al.* (1998). Megalencephaly and leukodystrophy with mild clinical course: a report on 12 new cases. *Brain Dev*, **20**, 142–53.

Traeger EC, Rapin I (1998). The clinical course of Canavan disease. *Pediatr Neurol*, **18**, 207–12.

Ungar M, Goodman RM (1983). Spongy degeneration of the brain in Israel: a retrospective study. *Clin Genet*, **23**, 23–9.

Vafiadaki E, Cooper A, Heptinstall LE *et al.* (1998). Mutation analysis in 57 unrelated patients with MPS II (Hunter's disease). *Arch Dis Child*, **79**, 237–41.

van der Knaap MS, Barth PG, Gabreels FJ *et al.* (1997). A new leukoencephalopathy with vanishing white matter. *Neurology*, **48**, 845–55.

van der Knaap MS, Barth PG, Stroink H *et al.* (1995). Leukoencephalopathy with swelling and a discrepantly mild clinical course in eight children. *Ann Neurol*, **37**, 324–34.

van der Knaap MS, Kamphorst W, Barth PG *et al.* (1998). Phenotypic variation in leukoencephalopathy with vanishing white matter. *Neurology*, **51**, 540–7.

van der Knaap MS, van Berkel CG, Herms J *et al.* (2003). eIF2B-related disorders: antenatal onset and involvement of multiple organs. *Am J Hum Genet*, **73**, 1199–207.

Vanhanen SL, Sainio K, Lappi M *et al.* (1997). EEG and evoked potentials in infantile neuronal ceroid-lipofuscinosis. *Dev Med Child Neurol*, **39**, 456–63.

Van Maldergem L, Trijbels F, DiMauro S *et al.* (2002). Coenzyme Q-responsive Leigh's encephalopathy in two sisters. *Ann Neurol*, **52**, 750–4.

Verity CM, Nicoll A, Will RG *et al.* (2000). Variant Creutzfeldt-Jakob disease in UK children: a national surveillance study. *Lancet*, **356**, 1224–7.

Veneselli E, Biancheri R, Buoni S *et al.* (2001). Clinical and EEG findings in 18 cases of late infantile neuronal ceroid lipofuscinosis. *Brain Dev*, **23**, 306–11.

Vermeulen G, Seidl R, Mercimek-Mahmutoglu S *et al.* (2005). Fright is a provoking factor in vanishing white matter disease. *Ann Neurol*, **57**, 560–3.

von Kleist-Retzow JC, Cormier-Daire V, de Lonlay P *et al.* (1998). A high rate (20%-30%) of parental consanguinity in cytochrome-oxidase deficiency. *Am J Hum Genet*, **63**, 428–35.

Wanders RJ, Waterham HR (2006). Peroxisomal disorders: the single peroxisomal enzyme deficiencies. *Biochim Biophys Acta*, **1763**, 1707–20.

Wang HS, Kuo MF, Chou ML *et al.* (2005). Pyridoxal phosphate is better than pyridoxine for controlling idiopathic intractable epilepsy. *Arch Dis Child*, **90**, 512–5.

Weber P, Scholl S, Baumgartner ER (2004). Outcome in patients with profound biotinidase deficiency: relevance of newborn screening. *Dev Med Child Neurol*, **46**, 481–4.

Weinreb NJ, Charrow J, Andersson HC *et al.* (2002). Effectiveness of enzyme replacement therapy in 1028 patients with type 1 Gaucher disease after 2 to 5 years of treatment: a report from the Gaucher Registry. *Am J Med*, **113**, 112–9.

Wexler ID, Hemalatha SG, McConnell J *et al.* (1997). Outcome of pyruvate dehydrogenase deficiency treated with ketogenic diets. Studies in patients with identical mutations. *Neurology*, **49**, 1655–61.

Wilcken B, Wiley V, Hammond J *et al.* (2003). Screening newborns for inborn errors of metabolism by tandem mass spectrometry. *N Engl J Med*, **348**, 2304–12.

Willemsen MA, Mavinkurve-Groothuis AM, Wevers RA *et al.* (2005). Pipecolic acid: a diagnostic marker in pyridoxine-dependent epilepsy. *Ann Neurol*, **58**, 653.

Winchester B, Vellodi A, Young E. The molecular basis of lysosomal storage diseases and their treatment. Biochem Sco Trans 2000; 28(2):150–4.

Wolf NI, Sistermans EA, Cundall M *et al.* (2005). Three or more copies of the proteolipid protein gene PLP1 cause severe Pelizaeus-Merzbacher disease. *Brain*, **128**, 743–51.

Yuza Y, Yokoi K, Sakurai K *et al.* (2005). Allogenic bone marrow transplantation for late-infantile neuronal ceroid lipofuscinosis. *Pediatr Int*, **47**, 681–3.

Zeman W and Albert M (1963). On the Nature of the "Stored" Lipid Substances in Juvenile Amaurotic Idiocy (Batten-Spielmeyer-Vogt). *Ann Histochim* **8**:255–257.

CHAPTER 11

Neurocutaneous syndromes

Robert Grant

Contents

11.1 Tuberous sclerosis

11.1.1 Introduction

Tuberous sclerosis, also known as Epiloia or Bournville's Disease, is an autosomal dominant multisystem disease it usually presents in childhood with a characteristic facial rash, adenoma sebaceum, seizures, and sometimes learning difficulties. Central nervous system lesions in tuberous sclerosis are due to a developmental disorder of neurogenesis and neuronal migration. Other organs such as the heart and kidney are less commonly involved. The condition has very variable clinical expression and two-thirds of cases are thought to be new mutations, therefore it is important to examine and screen relatives. Management may involve many specialists and close co-operation between specialists is essential.

11.1.2 Incidence and prevalence

The incidence of tuberous sclerosis is uncertain. The point prevalence is between 1 in 6000 and 1 in 10 000 (Hunt and Lindenbaum 1984).

11.1.3 Genetic factors

The sites of genetic mutation in tuberous sclerosis are now well established. There is locus heterogeneity with one gene on chromosome 9q34, *TSC1*, and a second gene on chromosome 16p13.3, *TSC2*. The *TSC1* gene product, hamartin, and the *TSC2* gene product, tuberin, have been found to function as negative regulators of the mTOR signalling pathway. Tuberin mediates mTOR activity by co-ordinating inputs from growth factors that control cell growth, proliferation, and cell survival. The *TSC1/2* complex may also modulate activity of transforming growth factor beta, TGF beta, and act as a tumour suppressor gene. The loss of function of this tumour suppressor gene may account for the high incidence of tumours in tuberous sclerosis patients (Mak and Yeung 2004). Allelic losses of the tuberous sclerosis genes in the tumours have also been found (Short *et al.* 1995).

11.1.4 Clinical features and imaging diagnosis

Adenoma sebaceum or facial angiofibromas is the most common visible manifestation of this disorder (Fig. 11.1). These facial angiofibromas are most commonly seen over the cheeks and nasolabial folds. The rash can extend to the chin and forehead. The angiofibromas are rather greasy and can be mistaken for acne. Other skin changes include hypopigmented macules, café au lait spots, and 'shagreen patches'. Hypopigmented macules are frequently shaped like an ash leaf, are 1–3 cm in diameter and are most easily identified by shining ultraviolet light over the skin. Subungal fibromas are also found in approximately 50 per cent of cases. Seizures are common in tuberous sclerosis and can be partial, multifocal, or generalized. Seizure type is commonly related to the site of cortical tubers. MRI with gadolinium enhancement is the brain imaging investigation of choice for patients with neurocutaneous syndromes. The cortical dysplastic lesions are hamartomatous and may calcify (Fig. 11.2A). Sometimes clinically and even histologically it can be difficult to differentiate cortical dysplasia from well-defined ganglionic tumours. Subependymal nodules and cortical and white matter tubers characteristic of tuberous sclerosis are readily identified (Fig. 11.2B). The most common tumours associated with tuberous sclerosis are subependymal giant-cell astrocytomas. These account for over 90 per cent of brain tumours. However, immunohistochemical staining may be negative for glial acidic fibrilliary protein, GFAP, and there can be evidence of neuronal differentiation with positive staining with neuronal specific enolase, NSE. Tuberous sclerosis can also be associated with gangliogliomas and pleomorphic xanthoastrocytomas. Subependymal giant cell astrocytomas, gangliogliomas, and pleomorphic xanthoastrocytomas are categorized as WHO Grade 1 astrocytomas and typically alter little in size over several years.

Patients with tuberous sclerosis are at a higher risk of renal disease associated with angiomyolipomas of the kidneys and renal cysts. The gene that accounts for 85 per cent of polycystic kidney disease is situated on chromosome 16p13.3 adjacent to the tuberous sclerosis gene, *TSC2*. Children with large deletions of this region can present with tuberous sclerosis and severe childhood onset polycystic kidney disease (Harris *et al.* 1995). Renal cysts and angiomyolipomas become increasingly common with the passage of time and can progressively enlarge as demonstrated by serial renal imaging (O'Hagan *et al.* 1996). Although bilateral renal angiomyolipomas

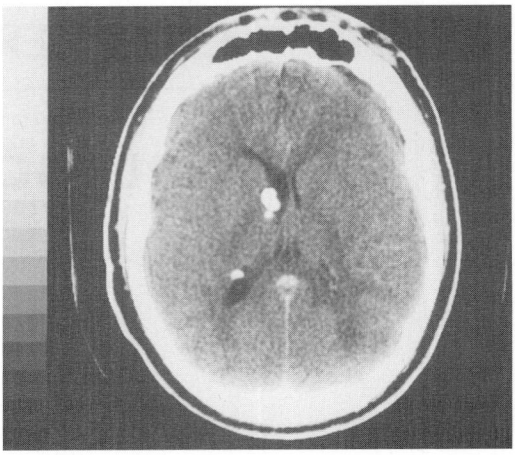

A

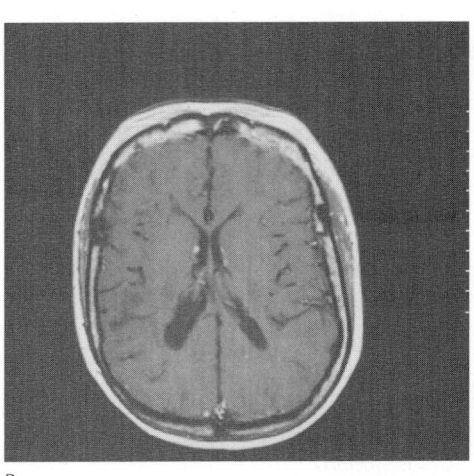

B

Fig. 11.2 A. Axial CT brain scan in a patient with tuberous sclerosis showing calcified hamartomas; B. Axial MRI brain scan in a patient with tuberous sclerosis showing subependymal nodules and cortical and white matter tubers.

are commonly found, chronic renal failure in the absence of cystic disease is uncommon. The imaging appearances of tuberous-sclerosis-related cystic disease of the kidney resemble those of autosomal-dominant polycystic kidney disease; however the histopathological findings are quite different with a hypertrophic, hyperplastic lining to the renal cysts. Hepatic angiomyolipomas are commonly asymptomatic but can occur and present with a abdominal pain followed by malaise and possibly hepatomegaly. The tumour is hyperechoic on ultrasound and of low density by CT scanning, <20 Hounsfield units, with increased vascularity on angiography. There is also an increase in the number and size of retinal hamartomas and cardiac rhabdomyomas with increasing age. The diagnostic criteria are outlined in Table 11.1 (Roach *et al.* 1999).

11.1.5 Management

Adenoma sebaceum can be treated with laser therapy. Epilepsy can be severe and resistant to usual anticonvulsant medications. Carefully selected patients with refractory epilepsy may be suitable for neurosurgical intervention. The success of surgery depends on the clear identification of an epileptogenic focus and identification of a structural abnormality at a corresponding site. Children may be

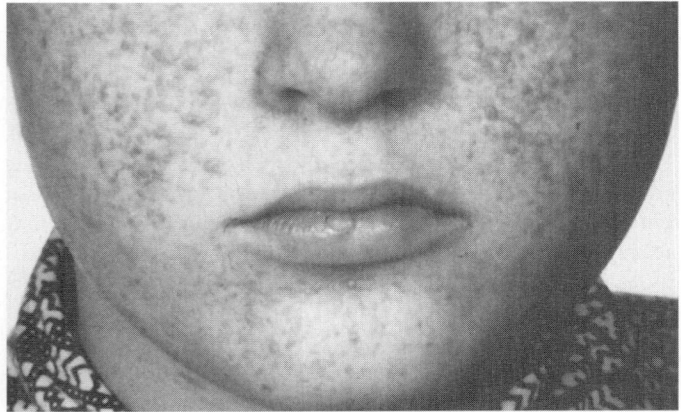

Fig. 11.1 The facial angiofibromas known as adenoma sebaceum in Tuberous sclerosis.

Table 11.1 Diagnostic criteria for Tuberous sclerosis

Major features

1. Renal angiomyolipoma (a)
2. Facial angiofibromas or forehead plaques
3. Non-traumatic ungula or periungual fibroma
4. Hypomelanotic macules (three or more)
5. Shagreen patch (connective tissue nevus)
6. Multiple retinal nodular hamartomas
7. Cortical tuber (b)
8. Subependymal nodule
9. Subependymal giant cell astrocytoma
10. Cardiac rhabdomyoma, single or multiple
11. Lymphangioleimyomatosis (a)

Minor features

1. Multiple renal cysts (c)
2. Non-renal hamartoma (c)
3. Hamartomatous rectal polyps (c)
4. Retinal achromic patch
5. Cerebral white matter radial migration tracts (a,d)
6. Bone cysts (d)
7. Gingival fibromas
8. 'Confetti' skin lesions
9. Multiple, randomly distributed dental enamel pits

DIAGNOSTIC CATEGORIES:

Definite tuberous sclerosis complex

Two major features or one major plus two minor features

Probable tuberous sclerosis complex

One major plus one minor feature

Possible tuberous sclerosis complex

One major feature or two or more minor features

   a) If both renal angiomyolipoma and lymphangioleiomyomatosis are present, other tuberous sclerosis complex features should be present before a definite diagnosis

   b) Coexistent cerebral cortical dysplasia and white matter migration tracts should be counted as one instead of two features

   c) Histological confirmation is suggested

   d) Radiological confirmation is sufficient

found to have learning difficulties and may require special schooling or assistance (Curatolo 1996).

If headaches or focal neurological signs develop, MR scanning of the head is advisable. This may identify cortical tubers, subependymal nodules, or subependymal giant cell astrocytomas, Grade 1 glioma. If a symptomatic cerebral tumour is identified, the best plan of management is to debulk the tumour and follow it up by serial MR scanning annually or if new neurological symptoms develop. The place of radiation therapy is uncertain. Radiation can reduce the size of these tumours, but because of the excellent long-term survival, and the frequency of late radiation-induced side effects such as radiation-induced dementia and leucoencephalopathy, the optimal time for radiation therapy remains uncertain. In general, radiotherapy is withheld until there is symptomatic disease progression and further surgery is not possible. Occasionally, anaplastic variants of astrocytoma can occur and in this situation

early radiation therapy is probably advisable. Close monitoring of renal function in tuberous sclerosis is advisable as chronic renal insufficiency is a cause of morbidity and mortality.

Families should have access to genetic counselling and genetic testing for *TSC1* and *TSC2*. Genetic counselling includes explanation of the risk of a disorder being inherited, the consequences of that risk, the probability of developing or transmitting the disease, and the ways in which transmission can be prevented. Parents and other family members should be clinically examined, including skin examination under Wood's ultraviolet light and ophthalmoscopic examination of the eyes for retinal phakomas. Chest X-ray, MR brain scan, and renal ultrasound should also be performed. CT or MR scans of parents and sibs of apparently sporadic cases will sometimes demonstrate asymptomatic cortical tubers. Even mildly affected parents can have severely affected children. Once any genetic tests or investigative tests have been performed, further counselling about the results is usually necessary.

Cranial imaging every 1–3 years in children has been recommended (Roach *et al.* 1999). Renal ultrasound has been recommended every 1–3 years in all patients. Women should have a chest CT once on reaching adulthood to look for lymphangioleiomatosis. Abdominal ultrasound looking for aneurysms should be performed every 2–3 years.

11.1.6 Prognosis and treatment complications

Severe infantile spasms and other severe forms of epilepsy have a poor prognosis because this is usually a sign of severe brain disease. Treatment of the seizures is frequently ineffective and status epilepticus is common. There is a high mortality rate in infants with infantile spasms, either due to the seizures or as a result of complications occurring during treatment for status. In the absence of severe epilepsy and significant cognitive impairment, prognosis is good, with most patients having a normal life span. If cerebral tumours develop, these are slow growing and have an excellent prognosis, if complete removal can be performed. In most cases however, partial resection is all that is possible and the median survival is approximately 10–20 years (Nagib *et al.* 1984).

11.1.7 Recovery and rehabilitation

Children will require a great deal of medical support when epilepsy is prominent and school/learning support during periods when seizures are quiescent. Advice regarding anticonvulsant medication and unadvisable pastimes should be openly discussed with the patient and family members. Counselling and support helps reintegration into school and society (Curatolo 1996).

11.2 Neurofibromatosis

11.2.1 Introduction and classification

The neurofibromatoses are autosomal-dominant neurocutaneous disorders that can be divided into 'peripheral' and 'central' types, although there is significant overlap (see also Section 14.3.2). The three most common types of neurofibromatosis are (Miller and Sparkes 1977):

- Neurofibromatosis type 1, *NF1*, also called von Recklinghausen's disease or 'peripheral' neurofibromatosis,

- Neurofibromatosis type 2, *NF2*, 'bilateral acoustic neuromas', or 'central' neurofibromatosis,

◆ Localized forms of the disease, 'segmental neurofibromatosis'. Segmental neurofibromatosis is characterized by localized cutaneous neurofibromas and café au lait spots limited to one segment of the body, but which can include underlying intrathoracic or intra-abdominal neurofibromas.

Diagnostic criteria for neurofibromatosis type 1 have been developed (Table 11.2) (National Institutes of Health Consensus Development Conference 1988). A grading system has also been devised for neurofibromatosis type 1 (Table 11.3) (Riccardi and Kleiner 1977). Patients with neurofibromatosis type 2 have a predisposition to develop tumours of the nervous system. Diagnostic criteria for neurofibromatosis type 2 are: bilateral acoustic neuromas; or an affected first degree relative and either a unilateral acoustic neuroma, neurofibroma, glioma, meningioma, schwannoma, or early onset lens opacity. The severity of phenotypes can be defined by age of onset of symptoms, <20 years versus ≥ 20 years, number of associated intracranial tumours, <2 tumours versus ≥ 2 tumours, and whether spinal tumours are present or absent (Evans *et al.* 1992a; Parry *et al.* 1994).

11.2.2 Incidence and prevalence

Neurofibromatosis affects all races and has an estimated frequency of approximately 1 in 3000 of live births and a mutation rate of 1×10^4 per gamete per generation (Crowe *et al.* 1956). The point prevalence of neurofibromatosis type 1 is at least 1 in 4950 ($20.2/10^5$) and neurofibromatosis type 1 accounts for 90 per cent of all cases of neurofibromatosis (Huson *et al.* 1989a). The incidence and prevalence of neurofibromatosis type 2 are uncertain, but it is thought to occur in approximately 1 in 40 000 live births (National Institutes of Health Consensus Statement 1991; Evans *et al.* 1992b).

11.2.3 Genetic factors

Although neurofibromatosis type 1 is an autosomal-dominant condition, about 50 per cent of all cases are new mutations. Its gene, that for von Recklinghausen's disease, was identified in 1991 and is situated on chromosome 17q11.2. It spans over 350 kilobases of genomic DNA and encodes for a protein of 2818 amino acids 'neurofibromin'. One role of neurofibromin is to function as a

Table 11.2 Diagnostic criteria for Neurofibromatosis type 1

Two or more of:
1. Six or more café au lait macules measuring >5mm in greatest diameter in prepubertal individuals and >15mm in greatest diameter in post-pubertal individuals
2. Axillary or inguinal freckling
3. Two or more dermal neurofibromas
4. A plexiform neurofibroma
5. A first degree relative with NF1 (by the NIH consensus statement criteria)
6. Optic nerve glioma
7. Two or more Lisch nodules
8. A distinctive osseous lesion (e.g. sphenoid dysplasia or thinning of the long bone cortex with or without pseudoarthrosis).

(After: National Institutes of Health Consensus Development Conference (1988))

Table 11.3 Grading system for Neurofibromatosis type 1

Grade 1.	Minimal	Café au lait spots only, or with unobtrusive cutaneous neurofibromas
Grade 2.	Mild	Numerous neurocutaneous neurofibromas but without facial disfigurement; small plexiform neurofibromas with no associated problems; asymptomatic osseous lesions; learning difficulties with normal IQ
Grade 3.	Moderate	Numerous neurocutaneous neurofibromas with facial disfigurement; plexiform neurofibromas with modest localized hypertrophy; visceral neurofibromas; mild retardation; scoliosis or pseudoarthrosis requiring surgery; controlled epilepsy
Grade 4.	Severe	Disease complications leading to major health impairment, often requiring surgical intervention, for example, large plexiform neurofibromas with severe secondary problems, CNS tumours, malignancy, aqueduct stenosis, severe mental retardation, phaeochromocytoma, and renal artery stenosis

GTPase-activating protein, or GAP, probably in the same pathway of signal transduction as ras proto-oncogene, therefore being involved in the regulation of cell growth. It is likely that neurofibromin is important in the formation of neurofibrosarcomas (von Deimling *et al.* 1995). It is very likely that tumourigenesis in neurofibromatosis type 1 is a multi-step phenomenon with the 'second hit' in the *NF1* gene initiating tumourigenesis.

Both the *NF1* and *NF2* genes have a high penetrance which is virtually 100 per cent by the age of 5 years. There is no evidence of locus heterogeneity within neurofibromatosis type 1 families, therefore tightly linked polymorphic markers can help determine the risk of a child aged <5 years of developing neurofibromatosis type 1 in the presence of equivocal clinical signs (Goldgar *et al.* 1989). Where the mother has neurofibromatosis type 1, offspring cases are more likely to be severely affected, but there does not appear to be a definite parental age effect or birth order effect (Huson *et al.* 1989b).

The *NF2* tumour suppressor gene is on chromosome 22q12 and encodes for a protein 'merlin' or 'schwannomin' of the 4.1 family of cytoskeletal-associated proteins which may link the cytoskeleton and cell membrane. Most *NF2* alterations result in a truncated, inactivated merlin protein. Specific *NF2* mutations do not always correlate with phenotypic severity, but in general, mutations that lead to premature termination of translation are associated with more aggressive disease. Most of the tumours associated with neurofibromatosis type 2 are benign, such as schwannomas, meningiomas, and ependymomas. These tumours also occur sporadically in the general population. In sporadic cases of acoustic neuroma and some cases of meningioma in neurofibromatosis type 2, there are also aberrations at the *NF2* locus, strongly suggesting that there is a tumour suppressor gene at this site (Kley *et al.* 1995). DNA-based diagnostic testing is now available for neurofibromatosis. Pre-symptomatic diagnosis is possible in multigeneration neurofibromatosis type 2 families, using tightly linked DNA markers and mutational analysis (Bijlsma *et al.* 1995). In 'at risk' individuals who do not carry the *NF2* mutation, DNA testing can exclude the condition and prevent needless clinical investigations (Baser *et al.* 1996).

11.2.4 Clinical features and imaging diagnosis

Neurofibromatosis type 1

The characteristic features of neurofibromatosis type 1 are café au lait spots, neurofibromas (Fig. 11.3), Lisch nodules, osseous lesions, macrocephaly, short stature and mental retardation, axillary freckling, and associations with several different types of tumours. Café au lait spots tend to increase in number and size in the first and second decades. Two spots or more only occur in 0.75 per cent of normal children under the age of 5 years but five spots with a diameter of >0.5 cm is suggestive of the diagnosis of neurofibromatosis type 1. In one very large patient database, 99 per cent of children had six or more café au lait spots greater than 5 mm by the age of one year (DeBella et al. 2000). Children of patients with neurofibromatosis type 1 should be examined annually for cutaneous signs of neurofibromatosis type 1. If by the age of 5 years there are no apparent signs, follow up can be discontinued. In children <5 years of age with equivocal signs where confirmation is sought of whether the child is either unaffected or affected, where there are two or more affected family members available for study, intragenic polymorphic markers can be used to determine the risks of disease.

In adults, where there are six café au lait spots >1.5 cm, neurofibromatosis type 1 is almost always present. Freckling in the axilla, groin, under the breasts, and on the neck are also a helpful associated sign of neurofibromatosis type 1, as are cutaneous or subcutaneous neurofibromas. Cutaneous neurofibromas are soft violet-coloured lesions varying from 0.1 cm to several centimetres in diameter. Subcutaneous neurofibromas commonly appear after the age of six, are present in 48 per cent of 10-year-old children, and in 84 per cent of 20-year-old adults with neurofibromatosis type 1

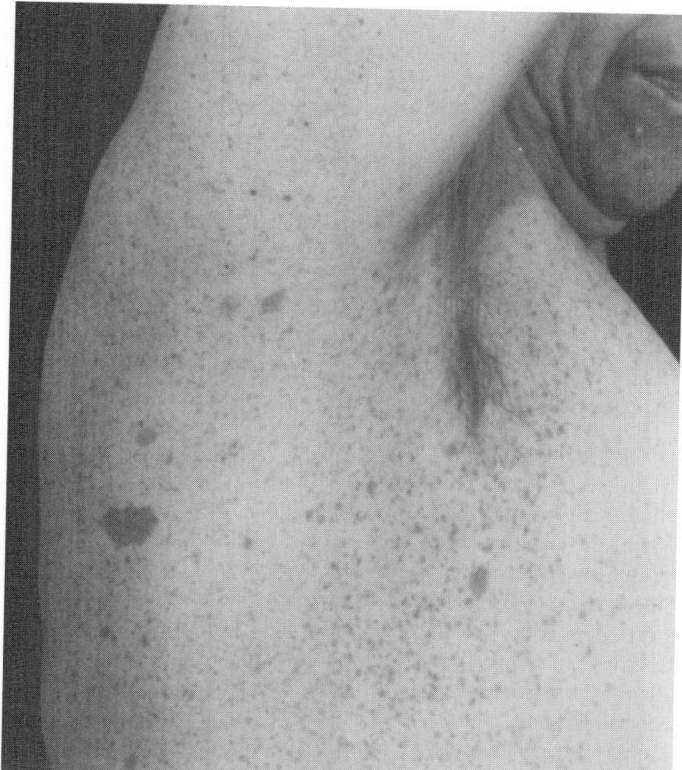

Fig. 11.3 Neurofibromatosis type 1. Café au lait macules, axillary freckling, and neurofibromas.

(DeBella et al. 2000). They are firm spindle cell tumours in the distribution of the trunks of peripheral nerves, arising from the nerve sheath. They increase in number and size, especially during pregnancy or with the use of the oral contraceptive pill. Neurofibromas can be dermal, nodular, or plexiform. Plexiform neurofibromas of the eyelid are frequently associated with glaucoma. Unlike dermal neurofibromas, plexiform neurofibromas can undergo malignant transformation. Peripheral nerve sheath tumours develop in up to 13 per cent of patients with plexiform neurofibromata (Reed and Gutmann 2001; Evans 2002). Sudden enlargement of plexiform neurofibromas or schwannomas or pain should lead to urgent investigation for evidence of malignant peripheral nerve sheath tumour. Complete resection of complex lesions frequently is not possible and treatment by an interdisciplinary team including oncologists, surgeons, and paediatricians is probably desirable (Gutmann et al. 1997). Lisch nodules are melanocytic hamartomas of the iris, and appear as brown nodules. They develop in early childhood and some report their presence in 70 per cent of children with neurofibromatosis type 1 by 10 years of age (DeBella et al. 2000). Others report them to be seen by slit lamp examination in 93–100 per cent of patients by the age of 20 years (Huson et al. 1989a; Lubs et al. 1991).

A four-fold increase in relative risk of cerebral tumours is suggested (Sorensen et al. 1986). Intracerebral tumours occur in 1.5–8 per cent of cases of neurofibromatosis type 1 (Brasfield et al. 1972; Huson et al. 1989b). These are commonly optic nerve or brainstem gliomas or gliosarcomas. Optic nerve glioma associated with neurofibromatosis accounts for almost 10 per cent of all patients with optic nerve gliomas. Optic nerve gliomas are present in 1.5 per cent of patients with neurofibromatosis type 1. These tumours are commonly bilateral or involve the optic chiasm (Font et al. 1972; Listernick et al. 1989). Symptomatic gliomas are present in 1 per cent by 1 year, 4 per cent by 3 years, and one study reports incidental or symptomatic gliomas in 36 per cent of patients by 6 years (DeBella 2000; King et al. 2003). Occasionally, optic nerve gliomas extend into the hypothalamus and cause precocious puberty. Optic nerve gliomas are commonly low grade and may not progress for many years (Listernick et al. 1994). There appears to be an association between plexiform eyelid neurofibromas and optic nerve glioma. Second malignancies occur in some 40 per cent of patients.

The frequency of aqueduct stenosis is increased in neurofibromatosis type 1 (Senveli et al. 1989). Ventriculo-peritoneal shunting or ventriculo-atrial shunting should only be contemplated in symptomatic patients (Spadero 1986). Up to 40 per cent of patients have mild learning difficulties and 6–10 per cent have epilepsy which may be associated with minor abnormalities such as gliosis, neuronal heterotopia, and ependymal overgrowth (Carey et al. 1979; Riccardi 1981). Children should be tested to detect any learning difficulty prior to school entry. Neurocognitive deficits may be subtle (Eldridge et al. 1989). Forty to sixty per cent of patients with neurofibromatosis type 1 have bony anomalies such as scoliosis, bone cysts, bone hypertrophy, or skull and facial deformities. Orthopaedic complications such as scoliosis and pseudoarthrosis of the tibia or fibula occur in about 9 per cent of patients (Akbarnia et al. 1992). Gastrointestinal neurofibromas are usually asymptomatic but can cause abdominal pain. Renal hypertension occurs in 1.5 per cent of affected individuals and can occur as a result of renal artery stenosis. Phaeochromocytoma affects <1 per cent of all cases (Huson 1994). Screening non-hypertensive patients for phaeochromocytoma is not cost-effective (Riccardi et al. 1986).

MRI with gadolinium enhancement is the investigation of choice because it provides better soft tissue contrast. In neurofibromatosis type 1 optic nerve gliomas, astrocytomas, plexiform neurofibromas, and 'unidentified bright objects' may only be identified by MRI. Prenatal risk assessment is only possible in families if there are two or more suitable family members available to check blood to study intragenic polymorphic markers. There is usually no increased risk during pregnancy. If there is significant kypho-scoliosis, labour can be difficult and rarely pelvic neurofibromas can obstruct labour.

Neurofibromatosis type 2

In neurofibromatosis type 2, typical tumours are benign schwannomas of the vestibular portion of the acoustic nerves (Fig. 11.4), although meningiomas frequently coexist. Most commonly patients with neurofibromatosis type 2 have few or no cutaneous manifestations of neurofibromatosis. However, café au lait spots, axillary freckling, and subcutaneous neurofibromas do occur rarely. Ninety five per cent of patients with an acoustic neuroma do not have neurofibromatosis type 2. There may be a family history of acoustic neuroma. More than 95 per cent of people with the *NF2* gene develop bilateral vestibular nerve tumours. Presentation is generally with deafness or tinnitus although headache, vertigo, or unsteadiness related to cerebellar involvement can occur. The characteristic hearing loss pattern in sensorineural hearing loss produces impairment of speech discrimination more so than pure tone loss. There is delayed conduction on brainstem auditory evoked potentials. Bilateral acoustic neuromas of neurofibromatosis type 2 are likely to be identified earlier by MRI than by CT. Two-thirds of tumours are hypointense with brain on T1-weighted images, the remaining third are isointense. All enhance with gadolinium either homogeneously in 66 per cent or patchily in 33 per cent. Multiple cutaneous

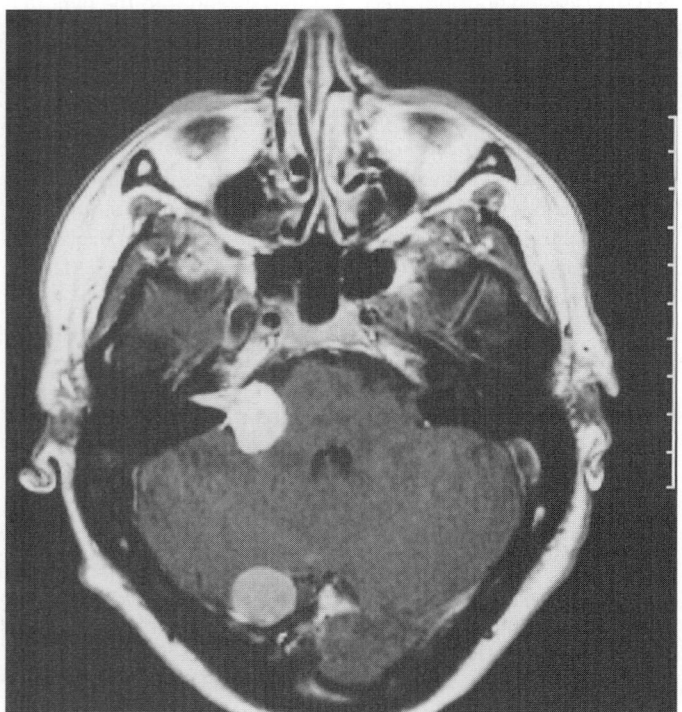

Fig. 11.4 An acoustic neuroma in patient with neurofibromatosis type 2. MRI.

plexiform schwannomas can also occur occasionally. The coexistence of neurofibromatosis and tuberous sclerosis or Von-Hippel–Lindau disease is well recognized.

11.2.5 Management

Neurofibromatosis type 1

Assessment of new patients with neurofibromatosis type 1 should include general physical examination for evidence of spinal deformity or any painful neurofibromas, neuropsychometric assessment in children to look for mild cognitive impairment, and examination of the eye for Lisch nodules by slit lamp examination if necessary. Brain MRI is the most useful diagnostic investigation in patients with cognitive or focal symptoms. High signal areas on T2-weighted MRI scans are seen in roughly 50 per cent of patients with neurofibromatosis type I, but are of no direct relevance in the absence of cognitive or focal symptoms or signs (Duffner *et al.* 1989; Sevick *et al.* 1992). It is advisable that patients with neurofibromatosis type 1 have regular ophthalmological assessment as part of their annual paediatric assessment with particular attention paid to any visual field defects, which are commonly asymptomatic. If there is a history of deterioration of visual acuity or any visual field or visual acuity abnormality is found by examination, a cranial MR scan with optic nerve views is indicated. If asymptomatic lesions are identified, it would seem reasonable to follow up these patients clinically and by MRI and only intervene by biopsy or irradiation if there is clinical or radiological progression. Symptomatic brain lesions should at least be biopsied to confirm diagnosis and determine the grade. Optic nerve gliomas virtually never develop after 30 years of age.

Appropriate neurological imaging of any symptomatic peripheral neurofibromas should be carried out. If hypertension is found, a renal cause or phaeochromocytoma should be actively sought. Measurement of catecholamines or catecholamine derivatives in a 24-h urine collection and renal imaging should be performed. Surgery for phaeochromocytoma offers a good chance of cure of the hypertension and the tumour (Ferner 1994).

Children with abnormal angulation of the long bones should be referred to an interested orthopaedic surgeon for further investigation and management (Morrissy 1982). Patients should be examined for evidence of spinal deformity at follow up visits, particularly during the adolescent growth spurt. If there is evidence of spinal deformity specialist orthopaedic follow up is desirable. Spinal surgery however is not without its complications (Crawford 1989).

Neurofibromatosis type 2

The defining feature of neurofibromatosis type 2 is bilateral vestibular schwannomas or acoustic neuromas. A treatment algorithm has been suggested based on age, hearing status, tumour size, and symptoms (Silverstein *et al.* 1993). There would be a place for considering screening sibs and offspring of patients with acoustic neuroma. This could be done clinically and by audiometry. If there are any signs to suggest neurofibromatosis type 2, then MRI would be the most sensitive investigation. Small asymptomatic acoustic neuromas are being identified increasingly. The best form of management is uncertain and some centres favour a wait and see approach while others advise early stereotactic radiosurgery or even neurosurgery. Treatment options are as for sporadic acoustic neuroma (see Section 27.7.3). There is some evidence to suggest that growth rates are faster in patients presenting young compared with those that present in the elderly (Mautner *et al.* 2002).

11.2.6 Prognosis and complications of treatment

Malignant neoplasms or benign central nervous system tumours occur in 45 per cent of probands. This provides a relative risk of 4.0 (95 per cent confidence intervals 2.8 to 5.6) compared with that expected. The prognosis in patients with tumours affecting the nervous system depends on the age of the patient, the type of tumour, the site of the tumour, and the level of disability at the time of presentation. The complication rate from treatment of these tumours appears to be no higher than in patients with these tumours but without neurofibromatosis. Epilepsy is well controlled with medication in approximately 50–70 per cent of patients.

Hypertension will increase the risk of cerebrovascular disease and there is possibly a higher incidence of cerebral aneurysm and stenosis of intracranial major arteries. Scoliosis and thoracic neurofibromas may lead to thoracic pain, chest infections, and reduced lung volumes. Hydronephrosis due to neurofibromas can lead to renal failure and abdominal pain.

11.2.7 Recovery and rehabilitation

In general, the follow up of patients with neurofibromatosis should be co-ordinated by one specialist. Annual follow up is recommended with regular blood pressure measurement and physical, cognitive, and ophthalmologic evaluation. Neuropsychological assessment should be performed prior to school entry and perhaps every 3–4 years till the age of 12 years to detect subtle learning difficulties and to allow early educational support. Physiotherapy, pain relief, and psychological support are important particularly in the early post-operative stages. Cosmetic advice and occasionally, cosmetic surgery may be necessary.

11.3 Sturge–Weber syndrome

Sturge–Weber syndrome involves a characteristic 'port-wine' facial naevus or angioma associated with an underlying leptomeningeal angioma or other vascular anomaly. It affects approximately 1/20 000 people. There can be seizures, low IQ, and underlying cerebral hemisphere atrophy as a result of chronic state of reduced perfusion and increased oxygen extraction. Patients may present with focal seizures which are generally resistant to anticonvulsant medication and can develop glaucoma. Ninety-eight per cent of people with Sturge–Weber have a cranial port-wine naevus and 52 per cent have extracranial involvement. At least 60 per cent of patients will develop glaucoma, 83 per cent seizures, and 65 per cent have neurological difficulties. Seizures usually start within the first two decades and frequently cause prolonged simple partial motor seizures, often with prolonged hemiparesis lasting days or weeks (Comi 2003). The prolonged 'Todd's paresis' is related to a 'steal phenomenon' and severe local cerebral ischaemia (Namer *et al.* 2005). Perfusion MRI and MR spectroscopy after onset of seizures suggests an impaired venous phase in the region of the angioma (Lin *et al.* 2003). Ictal EEG can be helpful in distinguishing between epileptic deficits and those which are migrainous or ischaemic (Jansen *et al.* 2004). Low-dose aspirin may reduce the incidence of stroke-like episodes (Maria *et al.* 1998). Some cases of familial port-wine stain and arteriovenous malformation have been associated with mutations in the *RASA1* gene (Eerola *et al.* 2003). Other reports suggest a

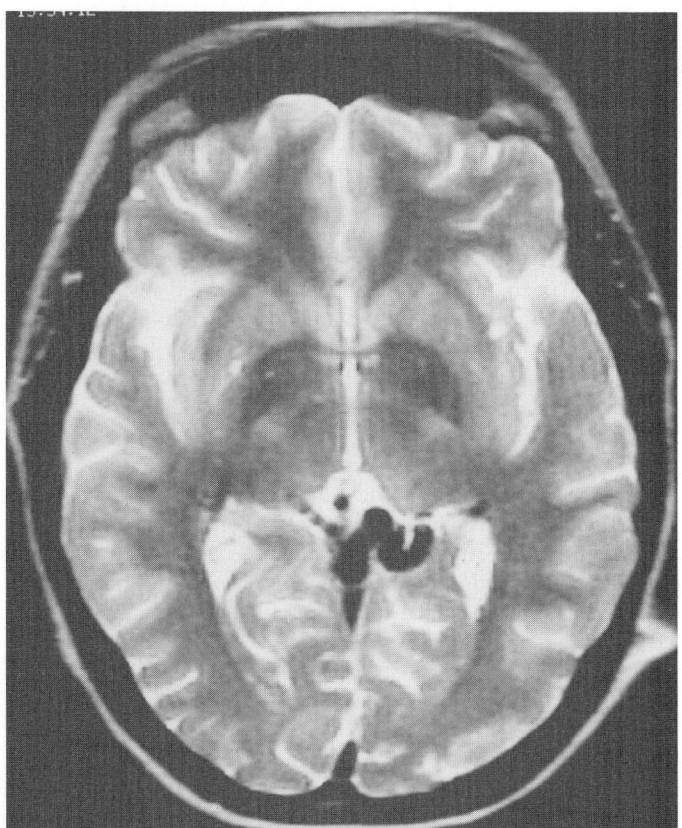

Fig. 11.5 Sturge–Weber syndrome. MRI T2 weighted axial image demonstrating an angiomatous malformation with enlarged draining vein between the cerebral hemispheres posteriorly extending to the occipital horn of the lateral ventricle.

candidate gene on chromosome 17p (Comi *et al.* 2005). Over 40 per cent of these patients will have developmental delay and up to 85 per cent will have emotional and behavioural problems. In cases without epilepsy developmental delay is rare and behavioural problems are less common, in 58 per cent. MRI with gadolinium enhancement is more sensitive than CT and the characteristic features are leptomeningeal angiomatosis, hemiatrophy, cortical calcification and patchy parenchymal gliosis, and demyelination (Adamsbaum *et al.* 1996). Figure 11.5 demonstrates an angiomatous malformation with enlarged draining vein between the cerebral hemispheres posteriorly extending to the occipital horn of the lateral ventricle.

11.4 Von-Hippel–Lindau disease

Von Hippel described angiomas of the retina in 1904 and Lindau described cerebellar and spine angiomas in 1926. Von-Hippel–Lindau disease is one of the most common autosomal-dominant inherited genetic diseases that are associated with familial cancers. Von-Hippel–Lindau disease is characterized by certain types of central nervous system tumours, cerebellar and spinal haemangioblastomas, and retinal angiomas, in conjunction with bilateral renal cysts carcinomas or phaechromocytoma, or pancreatic cysts/islet cell tumours (Neumann and Wiestler 1991). Haemangioblastomas are benign tumours composed of endothelial cells, pericytes, and stromal cells. Retinal angiomas are histologically similar to cerebellar haemangioblastomas except that they do not have cysts. The prevalence is approximately 1:30–40,000. About 30 per cent of haemangioblastomas are due to Von-Hippel–Lindau disease.

11.4.1 Genetics

Von-Hippel–Lindau disease is caused by abnormality of a tumour suppressor gene situated on the short arm of chromosome 3p25-26 proximal to the locus for the RAF-1 oncogene (Seizinger *et al.* 1988). The gene for Von-Hippel–Lindau disease has been cloned (Latif *et al.* 1993). Tumour suppressor genes work on the Knudson 'two hit' hypothesis, that tumours will only develop after both copies of the *VHL* gene are damaged. In families with Von-Hippel–Lindau disease, one damaged gene has been inherited. So far, more than 140 different gene mutations have been identified in *VHL* gene. The second 'hit' to the allele with the normally functioning gene can occur anytime during life. Most of the genes for hereditary tumour syndromes have been shown to be mutated not only in the germlines and tumours but also in their much more common sporadic counterparts. Mutation studies have demonstrated that the tumour suppressor genes are also mutated in the more common sporadic haemangioblastoma and renal carcinoma (Gnarra *et al.* 1994; Kanno *et al.* 1994). Mutations are identified in approximately 80 per cent of Von-Hippel–Lindau disease families. Certain mutations seem to be predictive of the development of phaeochromocytoma especially the 505 point mutation. In Von-Hippel–Lindau disease with phaeochromocytoma type 2, gene alterations consist of missense mutations in almost all patients and in 40 per cent of cases, these are at codon 238. Deactivation of the VHL gene occurs in the majority of cases of cerebellar haemangioblastoma in the context of Von-Hippel–Lindau disease and seems to be associated with increase in vascular endothelial growth factor, VEGF, which in turn stimulates new blood vessel formation or angiogenesis. Increased levels of VEGF are found in the vitreous and in renal cysts in patients with Von-Hippel–Lindau disease. In addition there are high concentrations of vascular endothelial growth factor receptors and VEGF in

haemangioblastomas. Sporadic cerebellar haemangioblastoma may result from a different genetic pathway (Glasker *et al.* 2001).

11.4.2 Associated tumours

Patients with Von-Hippel-Lindau disease are at risk of retinal haemangioblastomas or angiomas, in 57 per cent, central nervous system haemangioblastomas, renal cell carcinoma, in 23 per cent, phaeochromocytomas (19 per cent), or simple cysts of the kidney, pancreas, or liver. Retinal angiomas or haemangioblastomas occur in one-quarter to one-half of patients. They are bilateral in one-third of patients with retinal haemangioblastoma. They can cause progressive unilateral or bilateral blindness from glaucoma, haemorrhage, retinal detachment, or sympathetic ophthalmitis. Symptoms and signs can occur in infancy or in late life. On ophthalmoscopy, haemangioblastomas appear as red masses of any size fed by dilated tortuous arteries (Fig. 11.6).

Central nervous system haemangioblastomas can occur in the cerebellum in 54 per cent, brainstem in 20 per cent, spinal cord in 15 per cent, or rarely in the cerebrum (Fig. 11.7). Haemangioblastomas account for 1–2.5 per cent of all intracranial tumours and 10 per cent of posterior fossa tumours in children. Central nervous system haemangioblastomas may be multiple and imaging of the whole central nervous system is recommended. Although only about 30–40 per cent of patients with haemangioblastoma have Von-Hippel–Lindau disease, the likelihood of haemangioblastoma being associated with Von-Hippel–Lindau approaches 50–60 per cent if diagnosis is made in children or young adults less than 30 years of age. When the haemangioblastoma is of the spinal cord, more than 80 per cent are associated with the Von-Hippel–Lindau disease. These tumours are commonly associated with secondary polycythaemia due to secretion of erythropoietin by the tumour.

Renal lesions are frequently asymptomatic, but may present with haematuria, fever, or pain. Renal carcinoma will develop in 20–25 per cent of patients. Pancreatic lesions can cause abdominal pain and may be associated with diabetes mellitus. Phaeochromocytomas are

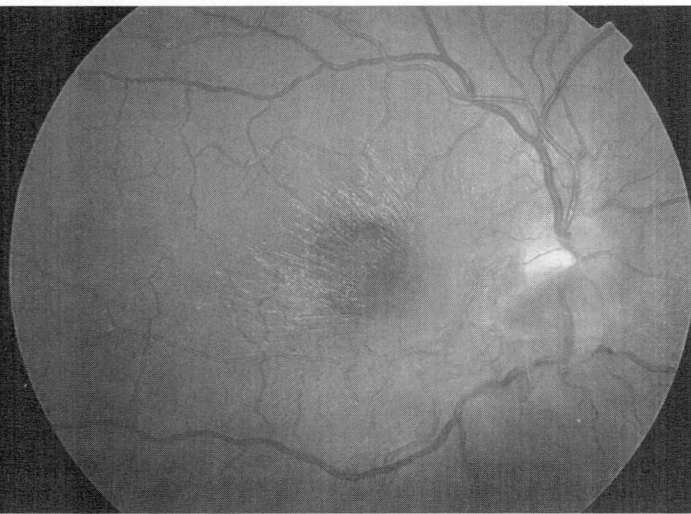
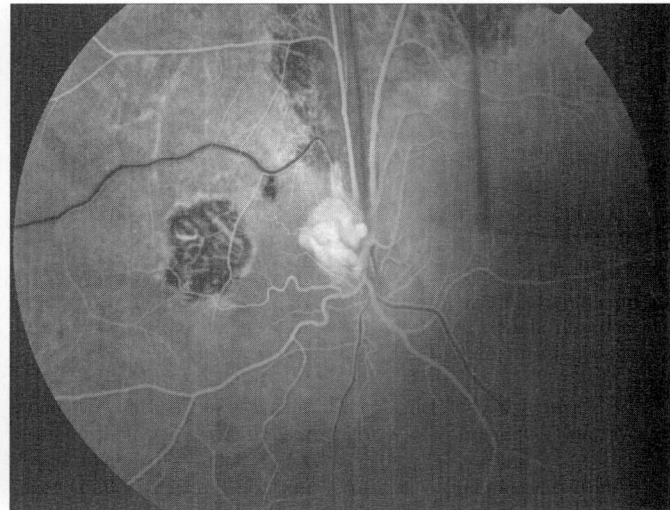

A   B

Fig. 11.6 Von-Hippel–Lindau Disease. A. Haemangioma adjacent to the optic disc with secondary exudates in the macular region. B. Fluorescein angiogram showing two retinal haemangiomas, the one on the left treated. (Courtesy of Dr E. Snodgrass.) (See Plate 17.)

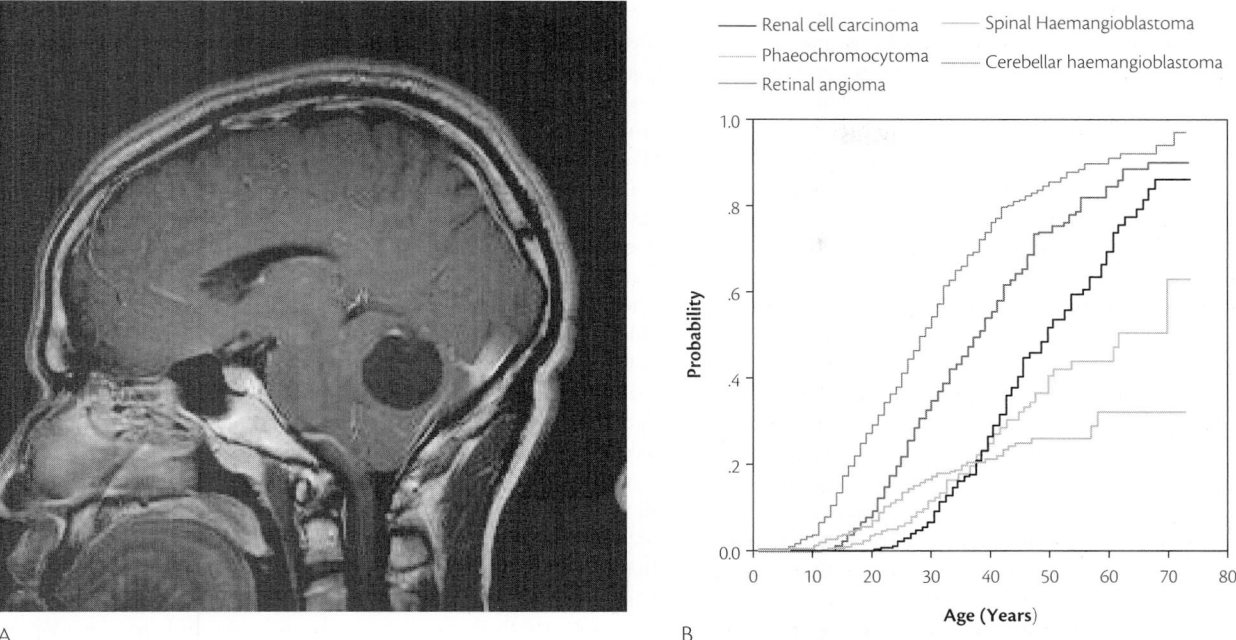

Fig. 11.7 Tumours in Von-Hippel–Lindau disease. A. Sagittal gadolinium-enhanced MRI showing typical cerebellar haemangioblastoma with tumour cyst and enhancing mural nodule. B. Age-related tumour risks in Von-Hippel–Lindau disease.

associated with cerebellar haemangioblastomas and may be unilateral or bilateral. They usually are associated with systemic hypertension.

11.4.3 Diagnosis

Diagnostic criteria for Von-Hippel–Lindau disease are:

- evidence of more than one haemangioblastoma in the central nervous system or retina,

- two types of tumours commonly found in Von-Hippel–Lindau disease in the same patient such as cerebellar haemangioblastoma plus renal carcinoma,

- or a typical tumour and a family history of Von-Hippel–Lindau disease.

Indirect ophthalmoscopy and fluorescein angiography can detect lesions before they are symptomatic. Cerebellar tumours are usually well defined with a cystic or multiloculated component with a mural nodule and the wall enhances with contrast on cranial CT or MRI. MRI is the best technique to examine the posterior fossa and the spine. Spinal and cerebellar tumours are commonly solid and enhance with contrast. Renal, adrenal, and pancreatic lesions are best identified by MRI of the abdomen. Urinary catecholamines and metadrenalins are elevated in phaeochromocytoma. Cerebellar haemangioblastomas may resemble renal clear cell metastasis on histology. The diagnosis is usually clarified by immuno-histochemical studies. Epithelial membrane antigen and cytokeratin stains are positive in renal cell carcinoma and negative in haemangioblastoma.

DNA-based diagnostic testing is now available for Von-Hippel–Lindau disease in cases where there is diagnostic doubt or in cases where early pre-symptomatic diagnosis is desired in Von-Hippel–Lindau families (Kley *et al.* 1995). There is an ethical debate at present about when DNA testing should be carried out. Some favour DNA testing in patients before the age of 5 years, since symptomatic disease, especially retinal angiomas, may start around the age of 5 years. Others favour regular screening of patients from the age of 5 years but to wait till the patient is the age of legal consent before offering genetic testing. Genetic counselling should be performed prior to DNA testing in patients with possible Von-Hippel–Lindau disease or asymptomatic family members. This is best done by a trained counsellor.

11.4.4 Prognosis

Early diagnosis improves prognosis. Following the diagnosis, it is important to screen the patient for other associated tumours and to counsel and screen the family members who are at risk. Mutations are detected in approximately 80 per cent of Von-Hippel–Lindau families. Family members who are not gene carriers on DNA testing, can be discharged from regular follow up. In families where the mutation is not detected, genetic linkage studies can be used to predict carrier status in many cases. Screening of patients and at-risk relatives includes: annual clinical assessment including indirect ophthalomoscopy from the age of 5 years with fluoroscein angiography if there are any suspicious areas, annual urinary metadrenalins from the age of 10 years, biennial cranial imaging ideally by MRI from the age of 15 years, and biennial abdominal scanning by CT or MRI to examine adrenals, kidneys, and pancreas. The identification of an intracranial haemangioblastoma should stimulate the search for spinal haemangioblastomas even in patients without spinal symptoms. Before surgery is contemplated in patients with possible haemangioblastoma, blood and urine tests to exclude a phaeochromocytoma and abdominal CT or MRI scanning should be performed to look for renal or adrenal tumours or cysts.

Prognosis depends on the site and size of the haemangioblastomas or other associated tumours. Patients with retinal haemangioblastoma will usually have peripheral lesions treated at an early stage by photocoagulation or cryotherapy (Annesley *et al.* 1977). Small lesions may be best treated with argon laser. Larger lesions greater than a disc diameter will respond to a combination of cryotherapy and photocoagulation with xenon arc or argon laser. Angiomas in the central part of vision are difficult since the treatment itself can cause some surrounding damage.

Surgical intervention for haemangioblastoma involving the central nervous system is not usually advisable for lesions < 3 cm in diameter. Tumours of the brainstem and spinal cord carry a high post-operative morbidity from haemorrhage. Haemangiomas of the cerebellum are more easily accessible and sometimes pre-operative embolization can reduce the risk of haemorrhage. Operations on cystic cerebellar lesions appear to have fewer complications than operations on solid lesions (Lamiell *et al.* 1989). Complete resection of the haemangioblastoma can be curative. However recurrences or new tumours are common.

Renal tumours in Von-Hippel–Lindau disease grow at a slower rate and are less aggressive than those with sporadic renal small cell carcinoma. Most surgeons practice renal sparing surgery because there is a very high chance that further tumours will develop in the same kidney or the opposite kidney at a later date. It is uncertain when surgery for renal carcinoma should be carried out. Some favour operating only when the tumour reaches approximately 4 cm in diameter, while others support early intervention when the tumour is only 2–3 cm. The difficulty is that metastases are more likely to occur with larger tumours. Most surgeons do not operate for simple renal cysts.

Phaeochromocytomas should be resected when diagnosed. It is believed that there are subtle neuropathological differences in phaeochromocytoma associated with Von-Hippel-Lindau disease and phaeochromocytoma associate with Multiple Endocrine Neoplasia type II (Koch *et al.* 2002).

11.5 Ataxia telangectasia

Ataxia telangiectasia is an autosomal recessive trait in which affected individuals have a progressive cerebellar ataxia starting in early childhood, and telangiectasia of the skin and conjunctiva (see also Section 39.7.1). By the teens, sufferers are significantly ataxic and may have developed choreo-athetosis, facial hypokinesia, and sialorrhoea, and are usually wheelchair bound. Reflexes are usually reduced or lost by the late teenage years and there may be some large fibre neuropathy and spinal muscular atrophy. Telangiectasia may not appear until adolescence and are found in the conjunctiva, nose, ears, neck, and antecubital fossae. Other skin changes such as hypopigmentation or hyperpigmentation and premature greying of hair are commonly found. There is commonly an ocular dyspraxia, with nystagmus and frequent blinking. There is an increased incidence of sinus infections and respiratory infections with bronchiectasis and lung abscesses related to deficiencies in serum immunoglobulins, especially IgA. Hypogonadism, growth failure with normal growth hormone levels, and diabetes mellitus which may be insulin resistant also occur (Woods and Taylor 1992). Sufferers have 100 times the risk of controls for developing

cancer with a particular predisposition to lymphoid malignancies and immunodeficiency (Morrell *et al.* 1990; Shiloh *et al.* 1996). Diagnosis is usually evident on clinical grounds, although the telangiectasias may appear after the ataxia has been present for some time. Occasionally the condition may not become clinically evident till early adulthood.

The ataxia telangiectasia gene is on chromosome 11q22-q23 and has been cloned, the ataxia telangiectasia mutated gene, *ATM*. Its role is to maintain cellular integrity by repairing DNA double strand breaks (Levitt and Hickson 2002). The defective gene resembles that for phosphoinositol-3-kinase and encodes an enzyme responsible for repair of DNA damage and cell control. About 1 per cent of the population are heterozygote carriers for this gene (Morrell *et al.* 1990; Swift *et al.* 1991). The ataxia telangiectasia gene is associated with a sensitivity to ionizing radiation. Homozygotes can get severe tissue necrosis when exposed to conventional therapeutic doses of radiation. This makes the management of cancers that usually respond to radiation therapy particularly difficult. It is also thought that diagnostic or occupational exposure to radiation increases the risk of cancer in these individuals.

Serum alpha-fetoprotein is increased in 90 per cent of cases and some patients also have elevated carcinoembryonic antigen (Woods *et al.* 1992). These can be useful diagnostic tests in young patients who have not yet developed telangiectasia. Serum immunoglobulin levels of IgA, IgE, IgG2 are decreased or absent while that of IgM is normal or elevated (Nowak-Wegrzyn *et al.* 2004). Genetic linkage studies may be helpful if many family members are available. Atrophy in the lateral and superior vermis is seen early on MRI (Tavani *et al.* 2003). Differential diagnosis may include spinocerebellar ataxia and vitamin E deficiency. Diagnosis is usually easy once the telangectasia are recognized.

Management involves treatment of intercurrent infections with physiotherapy and antibiotics. It is important to be aware that if cancers are found, conventional doses of radiation will be highly toxic and are contraindicated. Severe bacterial, viral, or opportunistic infections are uncommon (Nowak-Wegrzyn *et al.* 2004). Vaccination with live virus vaccines should be avoided but pneumococcal-conjugated vaccines may be protective (Pohl *et al.* 1992; Schubert *et al.* 2004).

11.6 Other neurocutaneous syndromes

11.6.1 Hypomelanosis of Ito

Hypomelanosis of Ito, also known as Incontinentia pigmenti achromians, is a rare sporadic multisystem disorder, but it is the third most common neurocutaneous syndrome behind neurofibromatosis type 1 and tuberous sclerosis. The term has been used to describe a condition, in which there are hypopigmented or depigmented streaks or whorls in the skin along the lines of Blaschko (Fig. 11.8). The skin changes are commonly present within the first year of life, in 70 per cent, and the hypopigmented areas are best seen in Caucasians using Wood's light. The extent of skin changes does not correlate with the severity of the systemic disease. Over half of patients have neurological, skeletal, or ocular abnormalities. The neurological problems are due to cerebral and cerebellar developmental abnormalities, arteriovenous malformations, or tumours such as choroid plexus papilloma, and medulloblastoma. The most frequent neurological abnormalities are intellectual and cognitive deficits in 60–70 per cent and seizures in 60–70 per cent

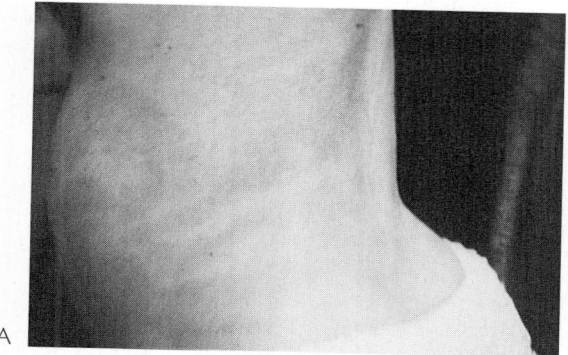

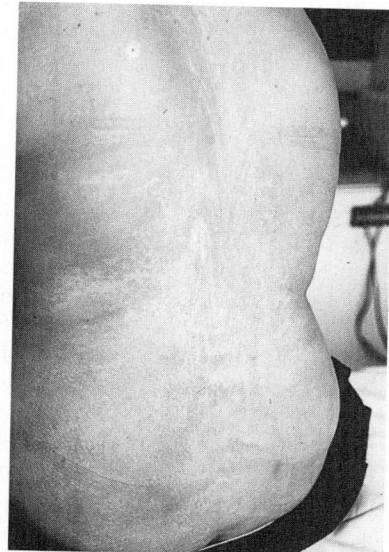

Fig. 11.8 Hypomelanosis of Ito. A. Hypopigmented streaks in the skin of the neck along the lines of Blaschko. B. Depigmented streaks or whorls in the skin of the back along the lines of Blaschko. (Courtesy of Professor D. Donnai.)

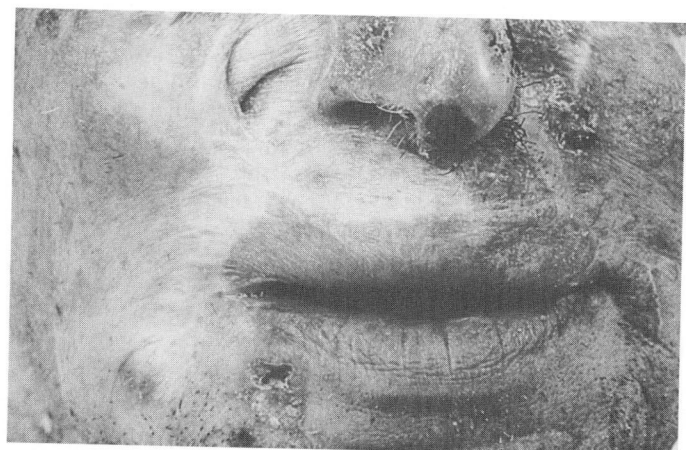

Fig. 11.9 Gorlin syndrome. Multiple facial naevoid basal cell carcinomas. (Courtesy of Professor D. Donnai.)

in early childhood, thought to be related to neuroblast migrational disorders. Oral abnormalities include dental dysplasia and conical teeth. Alopecia and changes in hair colour, to grey-white, may appear prematurely.

Hypomelanosis of Ito is an X-linked genodermatosis (Berlin *et al.* 2002). It is usually lethal in males *in utero*, although some males who survive demonstrate mosaicism at the appropriate region on the X chromosome (Fusco *et al.* 2004). In families where there are sufficient numbers of affected individuals, genetic linkage analysis may be possible using markers from Xq28 (Jouet *et al.* 1997). This region encodes the NF-kappa B essential modulator, NEMO (Jentarra *et al.* 2006).

Diagnosis rests on possibly genetic linkage analysis of the blood or karyotyping of blood and affected skin may detect mosaicism, frequently X: autosomal translocation (Boon *et al.* 1996), and the recognition of skin, nervous system, cardiac, genitor-urinary, musculoskeletal, and other anomalies. EEG and MRI findings are common but non-specific. MRI abnormalities can occur in the absence of neurological signs (Steiner *et al.* 1996).

Management is symptomatic: anticonvulsants for seizures, perhaps steroids for infantile spasms, physiotherapy for motor difficulties, educational support for those with learning difficulties,

and specialist opinion for ocular, dental, and skeletal problems. A first degree relative should be examined and genetic counselling offered to appropriate family members.

11.6.2 Gorlin syndrome

Also known as Naevoid basal cell carcinoma syndrome, Gorlin's syndrome is a rare autosomal-dominant multisystem disorder with variable expression. It is characterized by multiple naevoid basal cell carcinomas (Fig. 11.9), odontogenic keratocysts of the mandible, anomalies of the eye, skeleton, and reproductive system, and medulloblastomas and other neoplasms. Patients usually present with characteristic skin lesions in childhood and seizures. The diagnostic criteria involve two major or one major and one minor criterion.

Major criteria are:

- greater than two basal cell carcinomas in a patient under the age of 30 years,
- odontogenic keratocyst,
- palmar pits,
- falx calcification,
- positive family history.
  Minor criteria are:
- rib or vertebral abnormalities,
- macrocrania,
- fibroma,
- medulloblastoma,
- lymphomesenteric cysts.

The disease is associated with mutations in the *PTCH1* gene (Lindstrom *et al.* 2006). *PTCH1* is thought to act as a tumour suppressor gene. This gene is most likely to be mutated in patients with multiple features of Gorlin's syndrome but may be negative in pedigrees with only multiple or early onset basal cell carcinomas (Klein *et al.* 2005).

11.6.3 Sjogren–Larsson syndrome

Sjogren–Larsson syndrome is an autosomal-recessive disorder mapped to chromosome 17p. The condition is associated with deficiency of fatty aldehyde dehydrogenase (De Laurenzi *et al.* 1996). Fatty aldehyde dehydrogenase is an enzyme involved in long chain fatty alcohol oxidation and the cDNA encoding this enzyme has been cloned and several different mutations have been found (Lacour 1996). The clinical manifestations of this deficiency are congenital icthyosis, mental retardation, speech abnormalities, and spasticity. The skin condition is present in infancy as red scaly skin which becomes thickened and darker as the sufferer grows older. Half of the children have a degeneration of the pigment of the retina. The diagnostic test is a demonstration of mutations in the gene for fatty aldehyde dehydrogenase or measurement of the enzyme activity in cultured fibroblasts or leucocytes. MRI frequently shows deep white matter changes and MR spectroscopy reveals lipid accumulation in the periventricular white matter (Willemsen *et al.* 2004). Some patients are reported to have shown a benefit in spasticity with a medium chain fatty-acid-containing diet. Zileuton, a blocker of synthesis of leukotriene B4 is used in treating pruritis.

11.6.4 **Proteus syndrome**

The Proteus syndrome of multiple hamartomas is a condition where there is partial, usually asymmetrical, enlargement of the hands (Fig. 11.10) or feet, hemiatrophy on one side of the face, body or limbs, pigmented naevi, lipoma, and lymphangioma tumours, skull abnormalities such as cranial exostosis, exostosis of the external auditory meatus, and nasal bridge, macrocephaly or asymmetry of the skull and plantar hyperplasia involving overgrowth of the subcutaneous tissues of the soles of the feet (Fig. 11.11). There may be macrodactyly or syndactyly. Pulmonary and renal abnormalities such as nephrogenic diabetes insipidus have been described. Neurological manifestations may include spinal cord compression from tumour infiltration, cerebral malformations, or spinal stenosis as a result of kyphoscoliosis. There are frequently misdiagnoses of this condition and diagnostic criteria have recently been suggested (Turner *et al.* 2004).

The condition is most likely due to somatic mosaic lethal in the non-mosaic state, although the aetiology is unknown (Cohen 2005). Proteus syndrome probably affects more males than females and does not have a particular racial or geographical pattern (Turner *et al.* 2004). It is distinct from neurofibromatosis but certain features may be similar. Clinical features may be very mild or

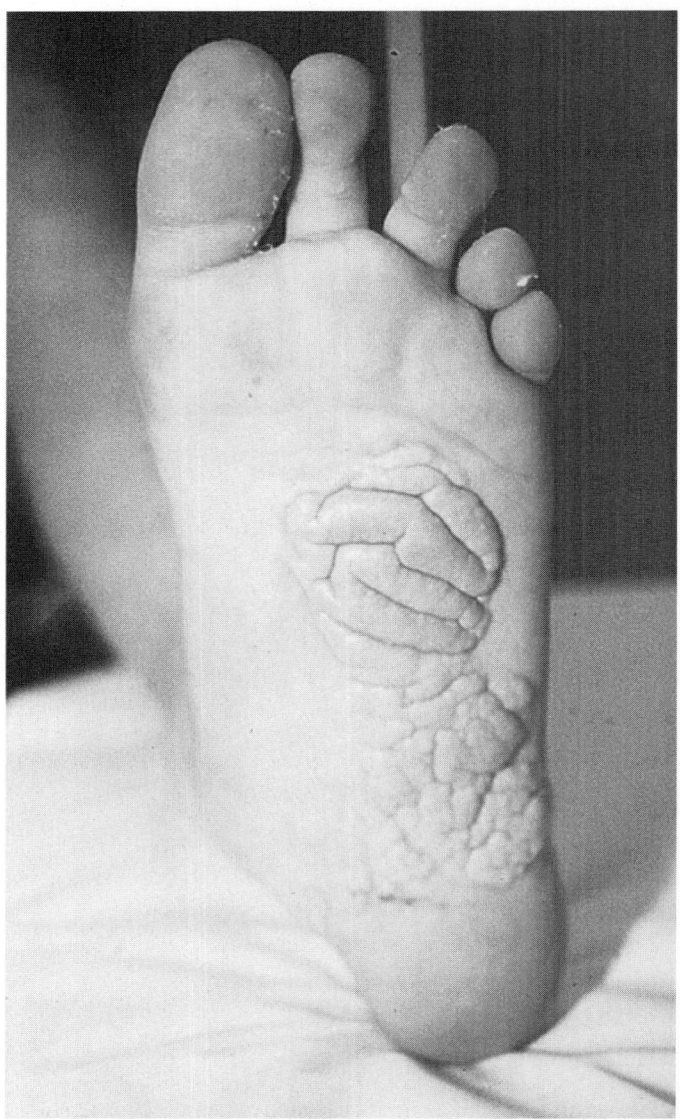

Fig. 11.11 Proteus syndrome. Plantar hyperplasia with overgrowth of the subcutaneous tissues of the soles of the feet. (Courtesy of Professor D. Donnai.)

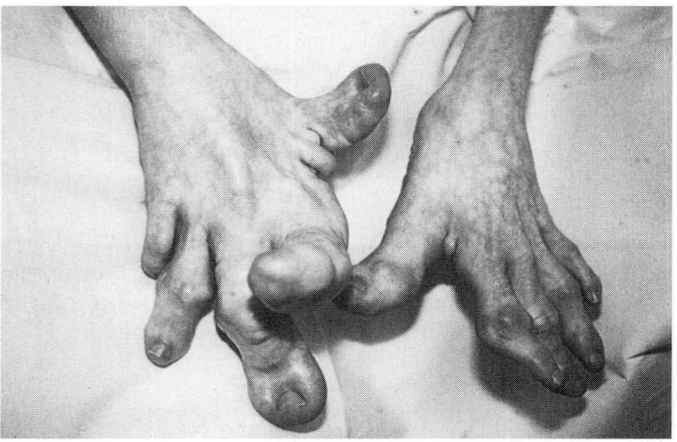

Fig. 11.10 Proteus syndrome. Asymmetrical enlargement of the hands. (Courtesy of Professor D. Donnai.)

severe. It is thought 'the Elephant man', Joseph Merrick had Proteus syndrome.

11.6.5 **Hemiatrophy and hemihypertrophy**

Facial hemiatrophy, also known as the Parry–Romberg syndrome, is a hereditary condition that usually starts in the teens and produces progressive atrophy of the skin and connective tissues of one side of the face or occasionally one side of the body (Malandrini *et al.* 1997). Limb involvement occurs in 19 per cent and epilepsy in 11 per cent (Stone 2003). There may be associated atrophy of the eye and bone and hemicortical atrophy in the ipsilateral cerebral hemisphere. Childhood cases have been described as have cases starting in the elderly. The cause is unknown. Vitiligo, Horner's syndrome, and loss of hair on the same side of the scalp are recognized associations. Seizures, hemianaesthesia, hemianopia, aphasia, migraine, and syringomyelia have also been described. The condition may halt spontaneously and the degree of cosmetic severity varies from case to case. There is no treatment that halts the disease but symptomatic relief of seizures and pain may be helpful. MRI changes may include unilateral focal infarctions in the corpus callosum, cortex thickening, subcortical white matter changes, and ipsilateral leptomeningeal enhancement or calcification (Cory *et al.* 1997).

Hemihypertrophy, also known as the Klippel–Trenaunay–Weber syndrome, is a rare condition which is characterized by a triad of vascular naevi, venous varicosities, and hyperplasia of the soft and hard tissues in the affected area. The syndrome usually affects the extremities causing hypertrophy of the connective tissues and long bones, cutaneous haemangiomas, and varicose veins, but can also affect the craniofacial region. The cause is unknown. Neurological manifestations include seizures. Planned orthopaedic and vascular interventions and plastic surgery are often required (Enjolras *et al.* 2004).

11.6.6 Menke's syndrome

Menke's syndrome, also known as Trichopoliodystrophy and kinky hair disease, is a very rare and fatal X-linked recessive condition. It is caused by mutations in a copper transporting p-type ATPase. It affects 1 in 35 000 live births. The Menke's locus is mapped to Xq13.3 and the defective gene is *ATP7A*; the defective gene in Wilson's disease (Section 40.8) is *ATP7B*. Mutations to the *ATP7A* gene are varied but diagnosis can be made by DNA-based tests (Hsi 2004). In the central nervous system there is often downregulation of myelination, energy, and translational genes (Liu *et al.* 2005).

Patients may be asymptomatic at birth. However, the neonate may present with hypothermia and failure to thrive and the clinical features become apparent. The hair is usually colourless, friable and kinked, curly, and has split shafts (Fig. 11.12). There is focal grey matter damage in the brain and tortuous arteries with damage to the intima. Seizures are common and progressive neurological deterioration then occurs. Levels of serum copper and caeruloplasmin are reduced, the copper content of the liver is low but copper content in fibroblasts is increased. Diagnosis is confirmed by copper studies and genetic testing. The best way to make a molecular diagnosis for Menkes is to first screen DNA samples for all exons using denaturing high performance liquid chromatography and then perform direct sequencing for exons which have an abnormal elution profile in order to detect the mutations. Increased copper content in the fibroblasts

allows intrauterine diagnosis in those with previously affected family members. The urine homovanillic acid/vanillylmandelic acid, HVA/VMA, ratio may be a reasonable screening test in infants, as many infants with Menkes syndrome may initially be asymptomatic (Matsuo *et al.* 2005). Most patients die by the age of 3 years if left untreated. Early parenteral administration of copper may in some cases slow neurological deterioration and lead to a better outcome but most of the patients progress irrespective of treatment.

11.6.7 **Xeroderma pigmentosum**

Xeroderma pigmentosum, Cockayne syndrome, and trichothiodystrophy are all caused by mutations in a set of interacting gene products involved in nucleoside excision repair. Most of the genes have been cloned and many mutations in genes have been identified. Patients with nucleotide excision repair diseases exhibit cancer, neurodegenerative disease, and complex developmental disorders. Xeroderma pigmentosum is associated with cancer; Cockayne syndrome and trichothiodystrophy are associated with neurodegeneration and developmental defects. Xeroderma pigmentosum is a rare autosomal-recessive disorder which affects about 4 people per million of the population. The DNA repair or replication deficiencies in xeroderma pigmentosum involve most of the genome, whereas in Cockayne's syndrome they are confined to actively transcribed genes. At least eight different genetic defects in DNA repair have been identified in xeroderma pigmentosum, A to G and variant V (Copeland *et al.* 1997). The most common types are *XPA*, *XPC*, *XPD*, and *XPV*. *XPA-G* with gene defects on Chromosomes 9, 2, 3, 19, 11, 16, and 13 respectively. Cells from classical xeroderma pigmentosum patients, *XPA-G*, fail to eliminate ultraviolet-light-induced DNA lesions by the nucleotide excision repair mechanism. These defects in DNA repair result in increased sensitivity to ultraviolet light and chemical carcinogens. The cases with xeroderma pigmentosum variant do not have mutations in the nucleotide excision repair mechanism, but instead have defects in post replication repair mechanisms. Xeroderma pigmentosum-V is caused by molecular alterations to the *POLH* gene located on Ch 6p21.1-6p12 (Gratchev *et al.* 2003). Patients with xeroderma pigmentosum present with photosensitivity of the skin severe sunburn in exposed areas and early, frequent, and multiple skin malignancies (Fig. 11.13). Lid freckling and atrophic skin, as

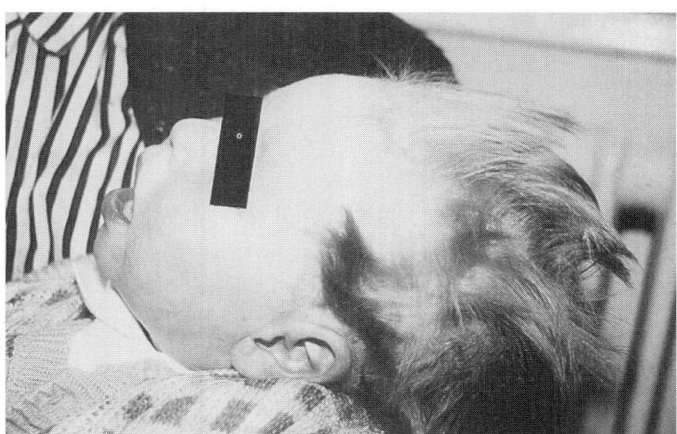

Fig. 11.12 Menke's syndrome. Characteristically colourless, friable and kinked, curly hair with split shafts.

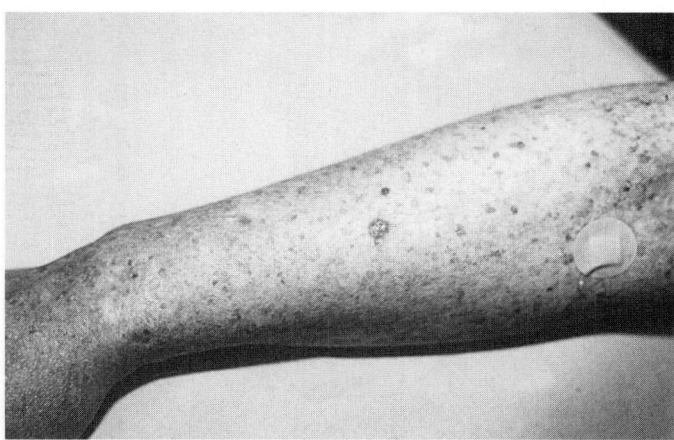

Fig. 11.13 Xeroderma pigmentosum. Photosensitivity and skin cancer on the forearm of a patient with xeroderma pigmentosum. (Courtesy of Professor D. Donnai.)

changes are seen in nearly all cases. There is a 1000-fold increase in the risk of developing basal and squamous cell carcinomas. The average age to develop the first skin cancer is 8 years of age. Melanomas may also appear in early childhood and there is a small increase in frequency of central nervous system malignancies. The eyes are commonly affected with irritation, and redness of the lids with light sensitivity, and early development of cataract. There is a xeroderma pigmentosum/Cockayne syndrome complex with the genetics of xeroderma pigmentosum and a phenotype of Cockayne with hydrocephalus, tigroid-type demyelination, dystrophic calcification, neuronal loss and degeneration, and peripheral neuropathy and myopathy (Lindenbaum *et al*. 2001). About 20 per cent of patients with xeroderma pigmentosum develop neurological degeneration possibly as a result of an accumulation of endogenous un-repaired cellular DNA. The associated neurological syndrome is associated with learning difficulties progressing to a dementia, deafness, cerebellar ataxia, seizures, chorea, dystonia, spasticity, and sensory neuropathy (Lambert *et al*. 1995). Prenatal diagnosis is possible by analysing foetal skin fibroblasts (Cleaver *et al*. 1994). The mainstay of treatment is avoidance of sunlight with clothing or sunscreens. Although *in vitro* studies have led to reconstruction of genetically corrected xeroderma

pigmentosum cells, skin trials of gene therapy in xeroderma pigmentosum patients have yet to show success (Magnaldo 2004).

11.6.8 Cockayne's syndrome

Cockayne's syndrome is a neurocutaneous autosomal recessive condition that is associated with impaired DNA repair, similar to xeroderma pigmentosum, but those affected do not develop cancers. Two defective genes have been identified. CSA gene-excision repair cross complementing group 8, ERCC8, on Chromosome 5 accounts for 25 per cent of cases. CSB gene, excision repair cross complementing group 6, ERCC6, on chromosome 10q11 is associated with 75 per cent of cases. Clinical classification falls into four forms (Neilan 2006):

- Cockayne's syndrome Type I—the 'classic' form,
- Cockayne's syndrome Type II—a severe form with symptoms from birth,
- Cockayne's syndrome Type III—a milder form, and
- Xeroderma pigmentosum–Cockayne's syndrome—an overlap with xeroderma pigmentosum but without the facial characteristics, skeletal problems, or neurological problems of Cockayne's syndrome.

In classical Cockayne's syndrome patients often have a characteristic progeroid appearance with short stature and premature ageing. The skin manifestations consist of photosensitivity in 84 per cent, and the neurological problems include learning difficulties, 'salt and pepper' retinopathy, ataxia, short stature neuropathy, and deafness (Ozdirim *et al*. 1996) (Fig. 11.14). There is frequently slowing on nerve conduction studies and brainstem auditory evoked potentials are abnormal in 70 per cent, reflecting a disorder of peripheral and central dysmyelination. MRI changes of a leucodystrophy are associated with gradual neurological progression. Prenatal diagnosis is possible by analysing foetal skin fibroblasts and showing a slow recovery to radiation therapy (Cleaver *et al*. 1994). Other investigations include ophthalmological exam, nerve conduction studies, MRI of the brain and skeletal X-rays showing thick cranial vault, intracranial calcifications, and marble epiphyses in the terminal phylanges. Cockayne's syndrome type II patients are severely affected from birth and usually die by the age of 7 years, whereas the mean survival of Cockayne's syndrome type I is 12 years although patients may survive to the second or third decade. Type III and Xeroderma pigmentosum–Cockayne's syndrome patients are more mildly affected. There are no effective treatments, but management of neurological symptoms is important.

References

Adamsbaum C, Pinton F, Rolland Y *et al*. (1996). Accelerated myelination in eary Sturge-Weber syndrome: MRI-SPECT correlations. *Pediatr Rad*, **26**, 759–62.

Akbarnia BA, Gabriel KR Beckman E *et al*. (1992). Prevalence of scoliosis in neurofibromatosis. *Spine*, **17**(Suppl 8), S244–8.

Annesley WH, Leonard BC, Shields JA *et al*. (1977). Fifteen year review of treated cases of retinal angiomatosis. *Trans Am Acad Ophthalmol Otolaryngol*, **83**, 446–53.

Baser ME, Mautner VF, Ragge NK *et al*. (1996). Pre-symptomatic diagnosis of neurofibromatosis 2 using linked genetic markers, neuroimaging, and ocular examinations. *Neurology*, **47**, 1269–77.

Berlin AL, Paller AS, Chan LS (2002). Incontinentia pigmenti: a review and update on the molecular basis of pathophysiology. *J Am Acad Dermatol*, **47**, 169–87.

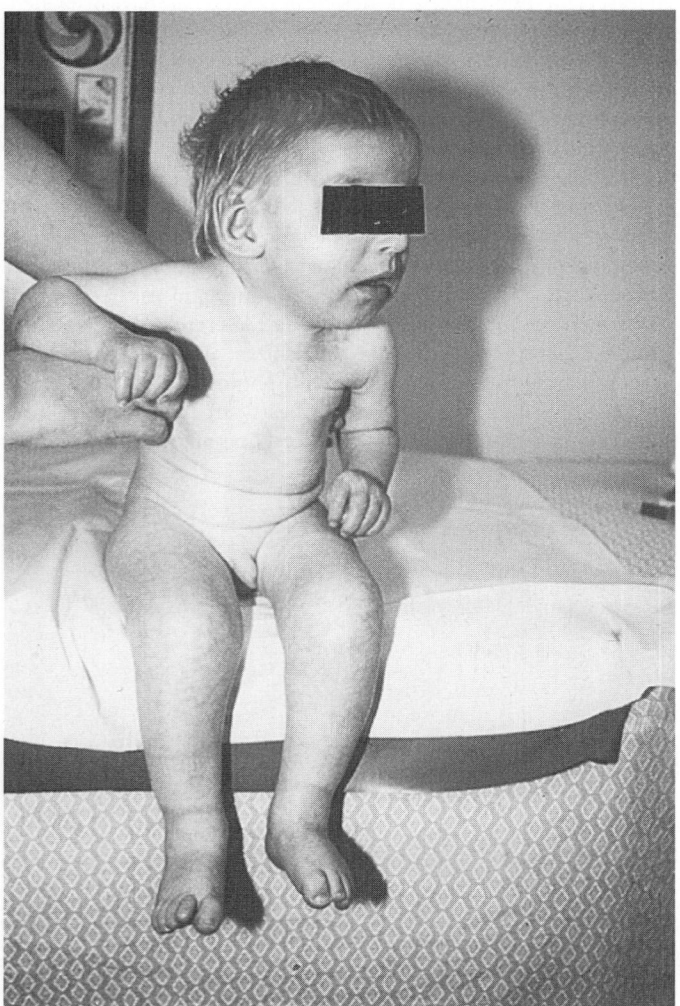

Fig. 11.14 Cockayne's syndrome. Characteristic facial and general appearance.

Bijlsma EK, Merel P, Fleury P *et al.* (1995). Family with neurofibromatosis type 2 and autosomal dominant hearing loss. Identification of carriers of the mutated NF2 gene. *Hum Genet*, **96**, 1–5.

Boon C, Markello T, Jackson-Cook C *et al.* (1996). Partial trisomy 10 mosaicism with cutaneous manifestations: report of a case and review of the literature. *Clin Genet*, **50**, 417–21.

Brasfield RD, Das Gupta TK (1972). Von Recklinghausen's disease: a clinicopathological study. *Ann Surg*, **175**, 86–104

Carey JC, Laub JM, Hall BD (1979). Penetrance and variability in neurofibromatosis: a genetic study of 60 families. *Birth Defects*, **15**, 271–81.

Cleaver JE, Volpe JP, Charles WC *et al.* (1994). Prenatal diagnosis of xeroderma pigmentosum and Cockayne's syndrome. *Prenat Diagn*, **14**, 921–8.

Cohen MM Jr (2005). Proteus Syndrome: an update. *Am J Med Genet C Semin Med Genet*, **137**, 38–52.

Comi AM (2003). Pathophysiology of Sturge-Weber Syndrome. *J Child Neurol*, **18**, 509–16.

Comi AM, Weisz CJ, Highet BH *et al.* (2005). Sturge-Weber Syndrome: altered blood vessel fibronectin expression and morphology. *J Child Neurol*, **20**, 572–7.

Copeland NE, Hanke CW, Michalak JA (1997). The molecular basis of xeroderma pigmentosum. *Dermatol Surg*, **23**, 447–55.

Cory RC, Clayman DA, Faillace WJ *et al.* (1997). Clinical and radiological findings in progressive facial hemiatrophy (Parry-Romberg syndrome). *Am J Neuroradiol*, **18**, 751–7.

Crawford AH (1989). Pitfalls of spinal deformities associated with neurofibromatosis in children. *Clin Ortho Rel Res*, **245**, 29–42.

Crowe FW, Schull WJ, Neel JV (1956). *A Clinical, Pathological and Genetic Study of Multiple Neurofibromatosis*. Charles C Thomas, Springfield, Illinois.

Curatolo P (1996). Neurological manifestations of tuberous sclerosis complex. *Childs Nerv Syst*, **12**, 515–21.

DeBella K, Szudek J, Friedman JM (2000). Use of the National Institutes of Health criteria for diagnosis of neurofibromatosis 1 in children. *Paediatrics*, **105**, 608–14.

De Laurenzi V, Rogers GR, Hamrock DJ *et al.* (1996). Sjogren-Larsson syndrome is caused by mutations in the fatty acid dehydrogenase gene. *Nat Genet*, **12**, 52–7.

Duffner PK, Cohen ME, Seidel FG *et al.* (1989). The significance of MRI abnormalities in children with neurofibromatosis. *Neurology*, **39**, 373–8.

Eerola I, Boon LM, Mulliken JB *et al.* (2003). Capillary malformation-arteriovenous malformation, a new clinical and genetic disorder caused by RASA1 mutations. *Am Hum Genet*, **73**, 1240–9.

Eldridge R, Denckla MB, Bien E *et al.* (1989). Neurofibromatosis Type 1 (Recklinghausen's disease): Neurologic and cognitive assessment with sibling controls. *Am J Dis Child*, **143**, 833–7.

Enjolras O, Chapot R, Merland JJ (2004). Vascular anomalies and the growth of limbs: a review. *J Pediatr Orthop*, **13**, 349–57.

Evans DGR, Huson SM, Donnai D *et al.* (1992a). A clinical study of type 2 neurofibromatosis. *QJM*, **84**, 603–18.

Evans DGR, Huson SM, Donnai D *et al.* (1992b). A genetic study of type 2 neurofibromatosis in the United Kingdom. Prevalence, mutation rate, fitness and confirmation of maternal transmission effect on severity. *J Med Genet*, **29**, 841–6.

Evans DGR, Baser ME, McGaughran J *et al.* (2002). Malignant peripheral nerve sheath tumours in neurofibromatosis type 1. *J Med Genet*, **39**, 311–4.

Ferner RE (1994). Neurofibromatosis I: A pathogenic and clinical overview. In Huson SM, Hughes RAC, eds. *The Neurofibromatoses: A Pathogenic and Clinical Overview*, pp. 316–30. Chapman Hall, London.

Font RL, Ferry AP (1972). Ocular and adenexal tumours. *Int Ophthalmol Clins*, **12**, 1–50.

Fusco F, Bardaro T, Fimiani G *et al.* (2004). Molecular analysis of the genetic defect in a cohort of IP patients and identification of novel NEMO mutations interfering with NF-kappaB activation. *Hum Mol Genet*, **13**, 1763–73.

Glasker S, Bender BU, Apel TW *et al.* (2001). Reconsideration of biallelic inactivation of the VHL tumour suppressor gene in hemangioblastoma of the central nervous system. *J Neurol Neurosurg Psychiatry*, **70**, 644–8.

Gnarra JR, Tory K, Weng Y *et al.* (1994). Mutations of the VHL tumour suppressor gene in renal carcinoma. *Nat Genet*, **7**, 85–90.

Goldgar DE, Green P, Parry DM *et al.* (1989). Multipoint linkage analysis in neurofibromatosis type I: an international collaboration. *Am J Hum Genet*, **44**, 6–12.

Gratchev A, Strein P, Utikal J, Sergij G (2003). Molecular genetics of Xeroderma pigmentosum variant. *Exp Dermatol*, **12**, 529–36.

Gutmann DH, Aylsworth A, Carey JC *et al.* (1997). The diagnostic evaluation and multidisciplinary management of neurofibromatosis 1 and neurofibromatosis 2. *JAMA*, **278**, 51–7.

Harris PC, Ward CJ, Peral B *et al.* (1995). Polycystic kidney disease. 1: Identification and analysis of the primary defect. *J Am Soc Nephrol*, **6**, 1125–33.

Hsi G, Cox DW (2004). A comparison of the mutation spectra of Menkes disease and Wilsons disease. *Hum Genet*, **114**, 165–72.

Hunt A, Lindenbaum RH (1984). Tuberous sclerosis: a new estimate of prevalence within the Oxford region. *J Med Genet*, **21**, 272–7.

Huson SM (1994). Neurofibromatosis 1: A pathogenic and clinical overview. In Huson SM, Hughes RAC, eds. pp. 160–203. Chapman and Hall, London.

Huson SM, Compston DAS, Clark P *et al.* (1989a). A genetic study of von Recklinghausen's neurofibromatosis in south east Wales. 1 Prevalence, fitness, mutation rate, and effect of parental transmission on severity. *J Med Genet*, **26**, 704–11.

Huson SM, Compston DAS, Clark P *et al.* (1989b). A genetic study of von Recklinghausen's neurofibromatosis in south east Wales. II. Guidelines for genetic counselling. *J Med Genet*, **26**, 712–21.

Jansen FE, van der Worp HB, van Huffelen A *et al.* (2004). Sturge-Weber syndrome and paroxysmal hemiparesis: epilepsy or ischemia? *Dev Med Child Neurol*, **46**, 783–6.

Jentarra G, Snyder SL, Narayanan V (2006). Genetic aspects of neurocutaneous disorders. *Semin Pediatr Neurol*, **13**, 43–7.

Jouet M, Stewart H, Landy S, Yates J *et al.* (1997). Linkage analysis in 16 families with Incontinentia Pigmenti. *Eur J Hum Genet*, **5**, 168–70.

Kanno, H, Kondo K, Ito S *et al.* (1994). Somatic mutations of the von Hippel-Lindau tumor suppressor gene in sporadic central nervous system hemangioblastoma. *Cancer Res*, **54**, 4845–7.

King A, Listernick R, Charrow J *et al.* (2003). Optic pathway gliomas in neurofibromatosis type 1: the effect of presenting symptoms on outcome. *Am J Med Genet*, **122A**, 95–9.

Klein RD, Dykas DJ, Bale AE (2005). Clinical testing for nevoid basal cell carcinoma syndrome in a DNA diagnostic laboratory. *Genet Med*, **7**, 611–9.

Kley N, Seizinger BR (1995). The neurofibromatosis 2 (NF2) tumour suppressor gene: implications beyond the hereditary tumour syndrome? *Cancer* Surv, **25**, 207–18.

Kley N, Whaley J, Seizinger BR (1995). Neurofibromatosis type 2 and von Hippel-Lindau disease: from gene cloning to function. *Glia*, **15**, 297–307.

Koch CA, Mauro D, Walther MM *et al.* (2002). Phaeochromocytoma in Von Hippel Lindau disease: distinct histopathological phenotype compared to phaeochromocytoma in multiple endocrine neoplasia type 2. *Endocr Pathol*, **13**, 17–27.

Lacour M (1996). Update on Sjogren-Larsson syndrome. *Dermatology*, **193**, 77–82.

Lambert WC, Kuo HR, Lambert MW (1995). Xeroderma pigmentosum. *Dermatol Clin*, **13**, 169–209.

Lamiell JM, Salazar FG, Hsia E (1989). Von Hippel-Lindau disease affecting 43 members of a single kindred. *Medicine*, **68**, 1–29.

Latif F, Tory K, Gnarra JR *et al.* (1993). Identification of the von Hippel-Lindau disease tumor suppressor gene. *Science*, **260**, 1317–20.

Levitt NC, Hickson ID (2002). Caretaker tumour suppressor genes that defend genomic integrity. *Trends Mol Med*, **8**, 179–86.

Lin DD, Barker PB, Kraut MA *et al.* (2003). Early characteristics of Sturge-Weber syndrome on perfusion MR imaging and proton MR spectroscopic imaging. *Am J Neuroradiol*, **24**, 1912–5.

Lindenbaum Y, Dickson D, Rosenbaum P *et al.* (2001). Xeroderma pigmentosum/Cockayne syndrome complex:first neuropathological study and review of eight other cases. *Eur J Paediatr Neurol*, **5**, 225–42.

Lindstrom E, Shimokawa T, Toftgard R *et al.* (2006). PTCH mutations and analyses. *Hum Mutat*, **27**, 215–9.

Listernick R, Charrow J, Greenwald MJ *et al.* (1989). Optic nerve gliomas in children with neurofibromatosis type 1. *J Pediatr*, **114**, 788–92.

Listernick R, Charrow J, Greenwald M *et al.* (1994). Natural history of optic pathway tumors in children with neurofibromatosis type 1: A longitudinal study. *J Pediatr*, **125**, 63–6.

Liu PC, Chen YW, Centeno JA *et al.* (2005). Downregulation of myelination, energy, and translational genes in Menkes disease brain. *Mol Genet Metab*, **85**, 291–300.

Lubs MLE, Bauer MS, Formas ME *et al.* (1991). Lisch nodules in neurofibromatosis type 1. *N Eng J Med*, **324**, 1264–6.

Magnaldo T (2004). Xeroderma pigmentosum: from genetics to hopes and realities of cutaneous gene therapy. *Expert Opin Biol Ther*, **4**, 169–79.

Mak BC, Yeung RS (2004). The tuberous sclerosis complex genes in tumor development. *Cancer Invest*, 22, 588–603.

Malandrini A, Dotti MT, Federico A (1997). Selective ipsilateral neuromuscular involvement in a case of facial and somatic hemiatrophy. *Muscle Nerv*, **20**, 890–2.

Maria BL, Neufeld JA, Rosainz LC *et al.* (1998). Central nervous system structure and function in Sturge-Weber syndrome: evidence of neurological and radiological progression. *J Child Neurol*, **13**, 606–18.

Matsuo M, Tasaki R, Kodama H, (2005). Screening for Menkes disease using urine HVA/VMA ratio. *J Inherit Metab Dis*, **28**, 89–93.

Mautner VF, Baser ME, Thakker SD *et al.* (2002). Vestibular schwannoma growth in patients with neurofibromatosis type 2; a longitudinal study. *J Neurosurg*, **96**, 223–8.

Miller RM, Sparkes RS (1977). Segmental Neurofibromatosis. *Arch Dermatol*, **113**, 837–8.

Morrell D, Chase CL, Swift M (1990). Cancers in 44 families with ataxia telangiectasia. *Cancer Genet Cytogenet*, **50**, 119–23.

Morrissy RT (1982). Congenital pseudoarthrosis of the tibia. Factors that affect results. *Clin Ortho Rel Res*, **166**, 21–7.

Nagib MG, Haines SJ, Erickson DL *et al.* (1984). Tuberous sclerosis. A review for the neurosurgeon. *Neurosurg*, **14**, 93–8.

Namer IJ, Battaglia F, Hirsch E *et al.* (2005). Subtraction ictal SPECT co-registered to MRI (SISCOM) in Sturge-Weber syndrome. *Clin Nucl Med*, **30**, 39–40.

National Institutes of Health Consensus Development Conference (1988). Neurofibromatosis conference statement. *Arch Neurol*, **45**, 575–8.

National Institutes of Health Consensus Statement (1991). Acoustic Neuroma. Vol. 9.

Neilan G (2006). Cockayne Syndrome. Gene Reviews updated March. (http://www.ncbi.nlm.nih.gov/books/bv.fcgi?call=bv.View. ShowSection&rid=gene.chapter.cockayne&itool=books&referralid=gn d.section.159).

Neumann HPH, Wiestler OD (1991). Clustering of features of Von Hippel-Lindau syndrome: evidence for a complex gene locus. *Lancet*, **337**, 1052–4.

Nowak-Wegrzyn A, Crawford TO, Winkelstein JA *et al.* (2004). Immunodeficiency and infections in Ataxia-Telangectasia. *J Pediatr*, **144**, 505–11.

O'Hagan AR, Ellsworth R, Secic M *et al.* (1996). Renal manifestations of tuberous sclerosis complex. *Clin Pediatr*, **35**, 483–9.

Ozdirim E, Topcu M, Ozon A *et al.* (1996). Cockayne syndrome: review of 25 cases. *Ped Neurol*, **15**, 312–6.

Parry DM, Elridge R, Kaiser-Kupfer MI *et al.* (1994). Neurofibromatosis 2(NF2): clinical characteristics of 63 affected individuals and clinical evidence for heterogeneity. *Am J Med Genet*, **52**, 450–61.

Pohl KR, Farley JD, Jan JE *et al.* (1992). Ataxia telangiectasia in a child with vaccine associated paralytic poliomyelitis. *J Pediatr*, **121**, 405–7.

Reed N, Gutmann DH (2001). Tumorigenesis in neurofibromatosis: new insights and potential therapies. *Trends Mol Med*, **7**, 157–62.

Riccardi VM (1981). von Recklinghausen Neurofibromatosis. *N Engl J Med*, **305**, 161–27.

Riccardi VM, Eichner JE (1986). *Neurofibromatosis: Phenotype, Natural History and Pathogenesis.* Johns Hopkins University Press, Baltimore.

Riccardi VM, Kleiner B (1977). Neurofibromatosis: a neoplastic birth defect with two age peaks of severe problems. *Birth Defects*, **12**, 131–8.

Roach ES, Di Mario FJ, Kandt RS *et al.* (1999). Tuberous Sclerosis Consensus Conference: recommendations for diagnostic evaluation. National Tuberous Sclerosis Association. *J Child Neurol*, **14**, 401–7.

Schubert R, Reichenbach J, Rose M *et al.* (2004). Immunogenicity of the seven valent pneumococcal conjugate vaccine in patients with ataxia-telangiectasia. *Pediatr Infect Dis J*, **23**, 269–70.

Seizinger BR, Roulleau GA, Ozelius LJ *et al.* (1988). Von Hippel-Lindau disease maps to the region of chromosome 3 associated with renal cell carcinoma. *Nature*, **332**, 268–9.

Senveli E, Altinors N, Kars Z *et al.* (1989). Association of von Recklinghausen's neurofibromatosis and aqueductal stenosis. *Neurosurg*, **24**, 99–101.

Sevick RJ, Barkovich AJ, Edwards MS *et al.* (1992). Evaluation of white matter lesions in neurofibromatosis Type I. *Am J Roentgenol*, **159**, 171–5.

Short MP, Richardson EP Jr, Haines JL *et al.* (1995). Clinical, neuropathological and genetic aspectes of the tuberous sclerosis complex. *Brain Pathol*, **5**, 173–9.

Shiloh Y, Rotman G (1996). Ataxia-telangiectasia and the ATM gene: linking neurodegeneration, immunodeficiency and cancer to cell cycle checkpoints. *J Clin Immunology*, **16**, 254–60.

Silverstein H, Rosenberg SI, Flanzer JM *et al.* (1993). An algorithm for the management of acoustic neuromas regarding age, hearing loss, tumor size, and symptoms. *Otolaryngol Head Neck Surg*, **108**, 1–10.

Sorensen SA, Mulvihill JJ, Nielsen A (1986). Long term follow up of von Recklinghausen's neurofibromatosis: Survival and malignant neoplasms. *N Eng J Med Genet*, **314**, 1010–5.

Spadero A (1986). Non-tumoral aqueductal stenosis in children affected by von Recklinghausen's disease. *Surg Neurol*, **26**, 487–95.

Steiner J, Adamsbaum C, Desguerres I *et al.* (1996). Hypomelanosis of Ito and brain abnormalities: MRI findings and literature review. *Pediatr Radiol*, **26**, 763–8.

Stone J (2003). Parry-Romberg Syndrome: a global survey of 205 patients using the internet. *Neurology*, **61**, 674–6.

Swift M, Morrell D, Massey RB *et al.* (1991). Incidence of cancer in families affected by ataxia telangiectasia. *N Engl J Med*, **325**, 1831–6.

Tavani F, Zimmerman RA, Berry GT *et al.* (2003). Ataxia-telangiectasia: the pattern of cerebellar atrophy on MRI. *Neuroradiology*, **45**, 315–9.

Turner JT, Cohen MM Jr, Biesecker LG (2004). Reassessment of the Proteus syndrome literature: application of the diagnostic criteria to published cases. *Am J Med Genet A*, **130**, 111–22.

von Deimling A, Krone W, Menon AG (1995). Neurofibromatosis type 1: pathology, clinical features and molecular genetics. *Brain Pathol*, **5**, 153–62.

Willemsen MA, van der Graaf M, van der Knapp MS *et al.* (2004). MR imaging and proton MR spectroscopic studies in Sjogren-Larsson Syndrome: characterization of the leukoencephalopathy. *Am J Neuroradiol*, **25**, 649–57.

Woods CG, Taylor AM (1992). Ataxia telangiectasia in the British Isles: the clinical and laboratory features of 70 affected individuals. *QJM*, **82**, 169–79.

SECTION 3

Disorders of
special senses

CHAPTER 12

Abnormal vision

Christopher Kennard

Contents

12.1 Introduction

Neuro-ophthalmology is a discipline comprising a wide variety of disorders that overlap the fields of neurology, ophthalmology and general medicine. Diagnosis in this field requires a thorough knowledge of the anatomy and physiology of the visual pathways and ocular motor system, as well as the ability to carry out a thorough neuro-ophthalmological examination. The combination of the history and any abnormalities identified by the examination should enable a detailed differential diagnosis to be reached, leading to appropriate investigations if required, and a final diagnosis.

12.2 Assessment of visual function

12.2.1 Visual acuity

Vision is one of the major sensory inputs in man and defects at any point along the visual pathway, from eye to cortex, often rapidly become obvious to the patient as, impaired visual acuity or localized defects in their field of vision. The first requirement, therefore, when a patient presents with impaired vision is to determine the best corrected visual acuity and by clinical evaluation to attempt to localize the site of the lesion. This is vital if the sophisticated imaging and electrophysiological techniques currently available are to be directed at the appropriate site along the visual pathways.

The visual acuity is determined for each eye using a Snellen chart and is recorded as a fraction where the numerator refers to the distance in feet from which a patient sees the letters, and the denominator being the distance from which a patient with normal vision sees the same letters e.g. 6/12. Normal vision is at least 6/6 and most individuals should be able to read the bottom line 6/5. If there is any impairment then refraction is required. However, a simpler method for bedside use is to retest the acuity with the patient observing the chart through a pinhole. Multiple pin holes with diameters 2.0 to 2.5 mm are best. This improves acuity in uncorrected refractive errors and with abnormalities of the ocular media such as the cornea and its tear film and the lens. It should be remembered that the cornea is the main refractory surface of the eye and damage to it or its tear film by local inflammatory adnexal disease, poor lid coverage or diminished blinking may all result in complaints of impaired vision. Determining visual acuity can offer clues as to the presence of field defects; a patient with a hemianopia may only read one side of the chart, and one with a central scotoma searches the chart to try and see around the scotoma.

If the acuity is still reduced, the eye must be carefully assessed ophthalmoscopically to exclude any opacities in the ocular media or lesions of the fundus. This can only be adequately performed with a properly dilated pupil and the benefits derived from this simple procedure far outweigh the exceedingly small risk of precipitating glaucoma. Although advanced cataract is fairly obvious, subtler early lens changes such as nuclear sclerosis and mildly opacified posterior lens capsules can degrade acuity considerably, with these patients often complaining of seeing a 'ghost' image, which may be misinterpreted as diplopia. These lenticular abnormalities may be identified as irregular retinoscopic reflexes.

If no obvious cause for the visual loss is found in the fundus it is still possible that minimal changes in the retina or retinal pigment epithelium may be the cause. The *brightness test* may be used to differentiate between this and optic nerve disease. A bright light is simply shone into each eye in turn and the patient is asked if the light is of equal brightness. If a difference is noted, this can be roughly quantified by asking the patient to give the normal stimulus a value of 100 per cent, and then give the degree of brightness in the other eye an appropriate value. If a diminution in brightness is found in one eye, this is likely to be due to optic nerve disease rather than macular lesions or pigmentary abnormalities. The *photostress test*, in contrast, is abnormal when retinal lesions are present. In this test the best corrected acuity of the patient is assessed and a bright light is then directed into one eye for 10 seconds whilst the other eye is covered. The visual acuity is reassessed continuously until the patient can read three letters on the Snellen line just above the baseline acuity. The recovery time taken to achieve this is recorded and the test repeated on the other eye. The photostress recovery time is 27 seconds for normal eyes. A comparison of the recovery times for the two eyes is made, and if there is a marked prolongation in one, it is likely to be due to macular receptor disease such as macular oedema, central serous retinopathy and macular degeneration. The test is usually normal in optic nerve disease.

12.2.2 **Pupil light reflex**

A further useful specific indicator of optic nerve disease is the finding of a relative afferent pupillary defect, the Marcus–Gunn pupil, using the *swinging-light test*. The technique for performing this test is described in Section 13.3.2.

12.2.3 **Colour perception**

If an afferent pupillary defect is detected, further confirmation of optic nerve disease is obtained by tests of *colour perception*. This can quickly be performed by asking the patient to compare the red colour, such as of the top of a mydriatic bottle, viewed separately by each eye. Even quite large choroido-retinal lesions fail to produce a gross impairment of colour recognition. Optic nerve disease, however, may result in the red colour appearing 'washed out', faded, grey, orange, or pink. A more formal examination of colour vision can be made using the standard pseudoisochromatic plates of Hardy-Rand-Rittler or Ishihara. The patient has to identify numbers among different coloured dots. Whereas the Ishihara plates were devised to identify congenital dyschromatopsias, the Hardy-Rand-Rittler plates are better for identifying acquired dyschromatopsia due to optic nerve dysfunction. Optic nerve lesions classically produce red–green defects of variable intensity.

12.2.4 **Visual fields**

If these various tests show that there is visual loss due to a neuroretinal lesion the next step is the examination of the *visual fields*. *Simple confrontation tests* can be used as a screening test at the bedside and, if carefully performed, will identify most neurologically produced visual field defects. These allow examination of two important regions; first the central field, especially its fixational area, in the diagnosis of optic nerve disease and second the fields about the vertical meridian in the diagnosis of chiasmal and hemianopic defects. The patient is asked to count the number of fingers presented in each of the quadrants whilst fixating the examiner's nose. Hand or colour comparison is tested, with the examiner placing his two hands, or two equal red objects, in front of the patient on either side of the vertical meridian, first in the upper quadrants and then in the lower ones. The patient then compares them for colour, brightness, and clarity and reports any difference. If the hand or object is abnormally perceived, it is slowly moved towards the vertical meridian and the patient asked to indicate when it appears normal compared with the stationary object. For example, in an early temporal field defect due to a chiasmic lesion, the target will demonstrably 'brighten' or its colour become normal when it passes through the vertical meridian into the nasal field.

The central visual field may be explored by moving a 5 to 10 mm red or white hatpin away from or towards the central point of fixation. The patient is asked if the target changes in colour or brightness, if it disappears, or if it is in a scotoma when it reappears.

Another useful method is the *Amsler grid test*, which consists of a series of lined and patterned grids for testing the central 20° of vision, at a reading distance. Patients report whether the lines appear straight and whether parts of them are blurry or missing. Although mainly used by ophthalmologists for patients with macular disease, these grids can quickly identify small central or paracentral scotomas that occur in optic nerve disease.

More detailed examination of the visual fields may be made using a tangent, Bjerrum or screen for the central 20 to 40° of the visual field. The central and paracentral regions as well as the blindspot can be examined. In addition, both horizontal and vertical 'steps' and the degree of congruity of binocular field defects can be sought. For accurate recording of a field defect and comparison of changes with time, the Goldmann perimeter is best used but its detailed description is beyond the scope of this chapter.

Examination of the visual fields in infants and children may be difficult. Visually elicited movements constitute a useful technique in which a bright or large stimulus, such as a picture of Donald Duck, is presented in the child's peripheral field. This normally initiates a reflex movement, bringing the target on to the fovea, and provides a test of gross function in the peripheral field. This technique is a valuable method for testing infants, but may also be used in partially sighted patients and for bitemporal or homonymous hemianopic field defects. The other method useful for children is the finger counting test, and one may ask them to mimic the number of fingers presented.

The localization of a lesion in the visual pathways producing visual loss can often be achieved by the visual field examination and, although details of specific field defects are described in subsequent sections, a brief summary will be given here (Fig. 12.1).

The first important distinction to be made is between those *field defects* that have a nerve fibre bundle configuration and those that have a clear hemianopic defect. The former, usually involving the central, caecal or arcuate region, indicates a lesion either in the retinal nerve fibre layer or in the anterior part of the optic nerve, whereas the latter suggests a chiasmic or retrochiasmic lesion. If a monocular nerve fibre bundle type field defect is found, the chances that it is due to a compressive lesion are extremely small, approximately 3 per cent. If, however, a unilateral or bilateral hemianopic field defect is detected in association with impaired visual acuity, a compressive, chiasmal lesion is extremely probable.

The ganglion cell axons show a specific pattern which can give rise to specific visual field defects when they are damaged. A central scotoma results from a lesion of the fibres from ganglion cells passing directly from the foveal/macula region to the optic disc, whereas a centrocaecal scotoma involves fibres from both the macula and

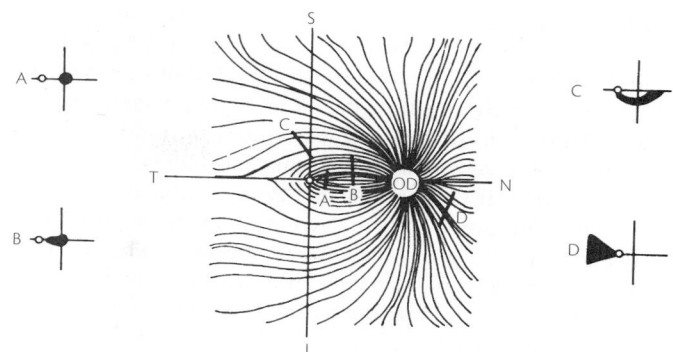

Fig. 12.2 Retinal nerve fibre layer lesions and field defects.

the area between the macula and the optic disc. Nerve fibres from the temporal region course above and below the papillo-macular bundle and a lesion here gives rise to an arcuate scotoma. Finally, if the nasal retina fibres are involved, a wedge shaped scotoma results, with its apex at the blind spot (Fig. 12.2).

The ganglion cell axons from the peripheral retina assume a position deeper within the nerve fibre layer than those arising from ganglion cells closer to the optic nerve. As a result fibres from the posterior pole congregate in the centre of the optic nerve whereas those from the periphery are arrayed around the outer aspects of the nerve.

Hemianopic field defects indicate disease of the chiasmal or retrochiasmal visual pathways. These defects have a border aligned to the vertical meridian, but do not necessarily involve the whole hemifield. The classic bitemporal hemianopia of a lesion of the optic chiasm is rarely symmetrical. In such cases the optic nerve is also often involved resulting in impaired acuity. Various characteristic field defects are found in chiasmic lesions. Lesions involving the retrochiasmic visual pathway mainly produce field defects which are congruent, except when the optic tract or lateral geniculate nucleus is involved. Depending on the extent of involvement of the optic radiation, a homonymous quadrantanopia or hemianopic field defect is produced. Homonymous altitudinal field defects normally result from lesions of the upper or lower calcarine (visual) cortex. The presence of macular sparing of vision indicates that the hemianopia is due to a lesion of the posterior occipital lobe; but it should be noted that lesions of the occipital pole can produce a small central homonymous hemianopic field defect, which may be missed if only the peripheral visual field is examined.

12.3 **Retinal disorders**

12.3.1 **Anatomy**

The retina is a multi-laminated structure which extends from the ora serrata anteriorly to the optic nerve posteriorly. The retinal layers consist of a number of different cell types, which include photoreceptors (rods and cones), pigment epithelial cells, bipolar cells, horizontal cells, amacrine cells, ganglion cells and Müller cells. The retina is strikingly laminated in cross section. The outermost layer consists of the pigment epithelial cells, which send out finger-like processes that interdigitate between the outer segments of the receptor cells. The outer nuclear layer of the retina consists of the inner segments of the receptors which contain many mitochondria

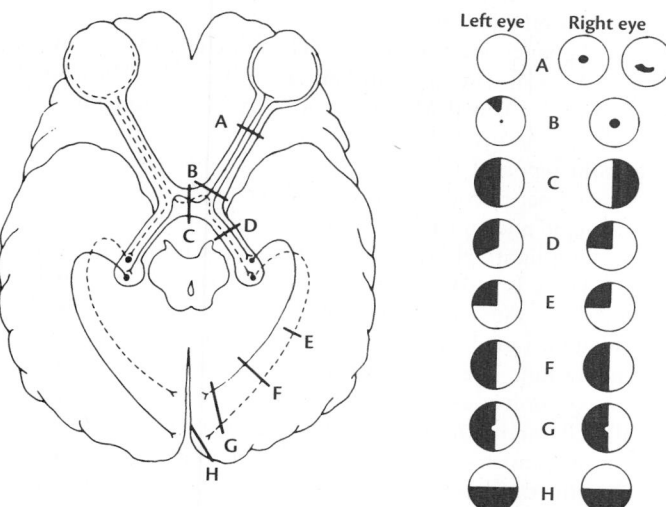

Fig. 12.1 Patterns of visual field loss: (A) Optic nerve lesions result in a central scotoma or arcuate defect. (B) Optic nerve lesions just prior to the chiasm produce junctional scotoma due to ipsilateral optic nerve involvement with the inferior contralateral crossing fibres (dotted line). (C) Chiasm lesions produce a bitemporal hemianopia. (D) Optic tract lesions result in incongruous hemianopic defects. (E) (F) Lesions of the optic radiation result in either homonymous quadrantinopia or hemianopia depending on the extent and location of the lesion (upper quadrant, temporal lobe; lower quadrant, parietal lobe). (G) Lesions of the striate cortex produce a homonymous hemianopia, sometimes with macular sparing particularly with vascular disturbances. (H) Partial lesions of the superior or inferior bank of the striate cortex cause inferior or superior altitudinal field defects respectively.

and nuclei. Moving towards the inner surface of the retina, there is the outer plexiform layer which consists of synaptic contacts between the photoreceptors and the dendrites of bipolar cells, the cell bodies of which constitute the *nuclear layer*. In addition, this layer contains horizontal cells which connect the photoreceptors and the amacrine cells. The latter form connections between the terminals of bipolar cells and the dendrites of ganglion cells. The inner plexiform layer consists of connections between bipolar terminals and dendrites of ganglion cells, the nuclei of which form the *innermost cell layer* of the retina. The inner surface of the retina, the nerve fibre layer, contains the axons of the ganglion cells which course along the inner surface of the retina to enter the optic nerve at the optic disc. The pregenicular visual pathway can simplistically be reviewed as consisting of three cells types, the photo-receptor, bipolar cell and ganglion cell. Injury to the ganglion cells or their axons will result in optic atrophy, whereas damage to the bipolar cells or photo-receptors alone does not lead to disc pallor.

The distribution of the two types of photoreceptors across the retina is non-uniform. Whereas rods are widely dispersed, cones are mainly located in the central macula area. The fovea itself is rod free and consists of some hundred thousand slender cones. The retinal ganglion cells also show a non-uniform distribution, being 4 or 5 cells thick in the parafoveal zone thinning rapidly towards the retinal periphery to a single noncontiguous cell layer.

12.3.2 Retinal degeneration

Degenerative diseases which effect the photoreceptors and/or the retinal pigment epithelium of the outer retina often produce a variety of symptoms for which terms such as retinitis pigmentosa, pigmentary retinopathy, tapeto-retinal degeneration and retinal dystrophy are often used interchangeably (Newsome 1988). A number of other aetiologies such as inflammation, toxic causes, ischaemia and trauma can lead to secondary retinal degeneration. The principal clinical features which are dominated by impaired rod function, are poor night vision or nyctalopia, diminished peripheral fields, narrowed retinal vessels, waxy pallor of the disc and abnormal pigment deposits, called 'bone spicules', observed in the peripheral region of the fundus. This pigmentary retinopathy may be absent in early cases which typically demonstrate fine white spots scattered over the retina known as retinitis punctata albescens, or may be restricted to one sector of the retina, often giving rise to superior temporal visual field defects, sometimes causing confusion with chiasmal compressive lesions. More typically a visual field defect starts as a ring scotoma which gradually enlarges out towards the periphery and in towards fixation.

The aetiology is usually genetic with a variety of patterns of inheritance including autosomal dominant, autosomal recessive, x-linked and mitochondrial.

Although the diagnosis of retinal photoreceptor degeneration can often be made clinically, several investigations may be helpful, including fluorescein angiography and the electroretinogram or ERG, which shows a marked diminution or extinction.

Other inheritable retinal diseases may have a predominant cone disturbance and present with light or glare-induced blindness known as hammarlopia, dyschromatopsia and central vision loss.

Cone dystrophy or cone–rod degenerations

Cone dystrophies are a heterogenous group of inherited conditions that cause dysfunction of the cone photoreceptors. In this condition, a cone dystrophy leads to diminished central acuity and reduced colour vision, and clinical examination reveals a central scotoma, generalized field constriction, ring scotoma, pseudo-altitudinal defects and centro-caecal scotoma, nystagmus and a bullseye macula appearance (Krauss and Heckenlively 1982). Nyctalopia or night blindness may not be present at the outset. The electroretinogram shows a severely depressed photic response.

Leber's congenital amaurosis

In this congenital form of retinitis pigmentosa, infants are born blind or nearly so. Although the fundus may initially appear normal the electroretinogram is extinguished, and subsequently typical bone spicule changes and retinal vessel narrowing appear. This entity accounts for a large proportion of all blindness in children in some countries, but it is not thought to be associated with severe intellectual impairment or poor educability as was once considered.

Retinal degeneration and mitochondrial DNA deletions

A 'salt and pepper' retinopathy is found in Kearns–Sayre syndrome, which in addition is associated with progressive ophthalmoplegia and at least one of either cardiac conduction abnormalities, cerebellar dysfunction or elevated spinal fluid CSF protein. In rare cases of mitochondrial myopathy, encephalopathy, lactic acidosis and stroke-like episodes MELAS, or a pigmentary retinopathy has been observed (Holt *et al.* 1990) (Section 13.1.6).

Paraneoplastic retinal degenerations

A number of distinct entities are encompassed by the term paraneoplastic retinal degeneration each giving rise to impaired vision and retinal function, as a result of a non-metastatic remote effect of a malignancy (Section 38.4.6). These include the commonest Cancer-Associated Retinopathy, CAR, and Diffuse Uveal Melanocytic Proliferation, DUMP, Melanoma-Associated Retinopathy, MAR, and Cancer-Associated Cone Dysfunction, CACD. The principal symptom reported by these patients' impaired vision or a halo of missing peripheral vision. They also complain of bizarre positive entopic phenomena, which include intermittent shimmering or flashing lights photopsias and floaters (Chung and Selhorst 1992). There is usually a progressive decline invision associated with a ring scotoma, and fundoscopy may show retinal arterial attenuation, a mild vitritis, optic disc pallor and peripheral retinal pigmentation. The onset of visual symptoms often precedes the recognition and diagnosis of the malignancy which is usually small cell carcinoma of the lung, breast, uterine and cervical carcinoma and melanoma in MAR. It is likely that the cancer leads to the development of anti photoreceptor antibodies. In the case of CAR an antibody to a 23-kD antigen has been identified, and a different antibody has been found in MAR (Ling and Pavesio 2003).

Specific idiopathic retinal degenerations

In recent years, a number of idiopathic syndromes have been described in which there is rapid visual loss of an area of outer retinal function around the optic disc giving rise to a big blind spot, which are associated with photopsias, minimal fundoscopic abnormalities and usually normal visual acuity (Gass 1993). The abnormalities usually affect one or both eyes and the condition is found predominantly in young women.

In *Acute Idiopathic Big Blind Spot Syndrome*, AIBBSS, there is an acute onset of positive visual phenomena and an enlarged blind spot without marked disc swelling. The focal electroretinogram of the peripapillary region is abnormal. The photopsias usually resolve but the blind spot enlargement persists.

In *Multiple Evanescent White Dot Syndrome*, MEWDS, patients present with acute unilateral visual loss, scotomas and ophthalmoscopy reveals white outer retinal pigment epithelium lesions and vitreous haze which rapidly fade. With time retinal pigment epithelium atrophy develops and the field defects usually persist (Hamed *et al.* 1989).

Acute Zonule Occult Outer Retinopathy, AZOOR, is an umbrella term to include cases of MEWDS, AIBBSS, Acute Macular Neuroretinopathy, AMN, and multifocal choroiditis (Reddy *et al.* 1996).

It is important to be aware of these conditions, which may be incorrectly diagnosed as migraine, optic neuritis, cancer associated retinopathy, papilloedema or acute neuroretinitis.

12.3.3 Vascular occlusive disease

Temporary or partial occlusion of the retinal arteries and veins either separately or jointly may lead to transient monocular blindness known as amaurosis fugax or a permanent, complete or partial monocular visual failure.

Amaurosis fugax

Transient monocular blindness, amaurosis fugax, usually occurs acutely and is described by the patient as a shutter or curtain descending or ascending over their vision in one eye. It is painless, lasts less than 10 minutes, and there are no photopsias or positive phenomena as occurs in migraine. It is usually due to emboli from the origin of the internal carotid artery which pass into the retinal vasculature. Sometimes they may be identified on fundoscopy if the retina is viewed during an acute episode. Otherwise the emboli usually break up and pass distally. The diagnosis and management of patients presenting with amaurosis fugax is discussed in Sections 35.6 and 35.8. Other causes of transient monocular blindness are retinal vasospasm, migraine and antiphospholipid antibody syndrome.

Central retinal artery occlusion

In Central Retinal Artery Occlusion, CRAO, the onset of painless blindness is usually sudden. A relative afferent capillary defect is usually found and fundoscopy reveals, within the first few minutes of the onset, 'cattle truck' segmentation of the venous blood column associated with arterial narrowing. After about an hour the affected retina becomes pale and a cherry red macular spot is observed at the fovea (Figs. 12.3 and 12.4). Such changes fade, usually to be replaced by optic atrophy and a markedly narrowed arterial tree. Several different causes of central retinal artery occlusion have been described which include embolic obstruction from carotid atheroma or calcific cardiac valve disease, local artery atheroma, arteritis, drug abuse and vasospasm associated with migraine.

In some cases consisting of temporary, complete or altitudinal visual loss amaurosis fugax may occur in an eye prior to the onset of CRAO. Retinal emboli are usually composed of cholesterol, platelets or calcium, but occasionally they can be due to fat emboli as a result of long bone fractures or pancreatitis, and foreign bodies such as talc in intravenous drug abusers.

The acute treatment of CRAO is still controversial since irreversible damage to the retina usually occurs within 2 to 4 hours. It is,

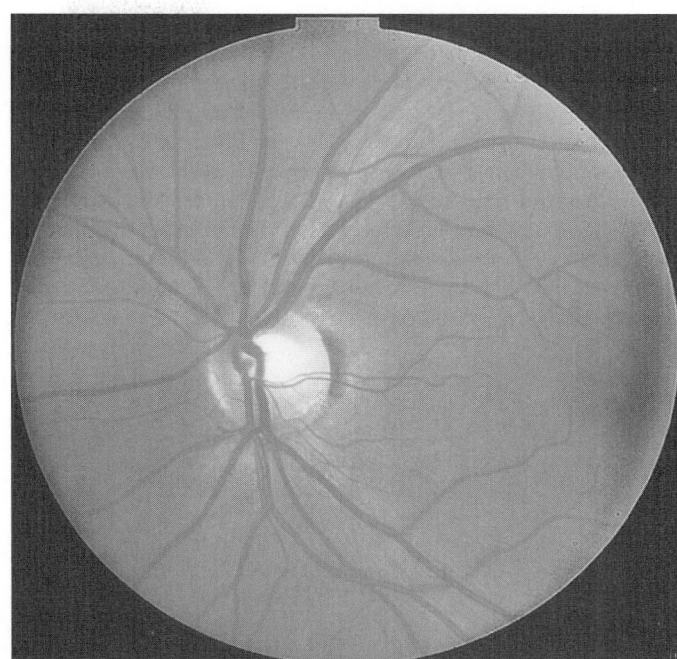

Fig. 12.3 A normal fundus and optic disc. (See Plate 18.)

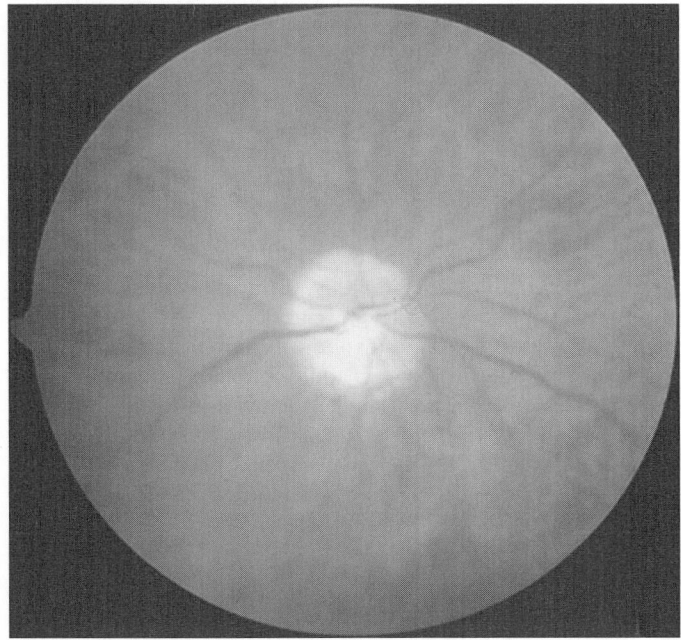

Fig. 12.4 Central retinal artery occlusion. (See Plate 19).

however, worth attempting certain treatments within the first 12 hours. These include ocular massage with the aim of lowering the intraocular pressure in an attempt to dislodge an embolus into the peripheral circulation, intravenous acetazolamide and carbon-dioxide re-breathing. More recently the infusion of thrombolytic agents, such as urokinase, has been used. Unfortunately most reports remain anecdotal due to its low incidence and inadequate power in the reported studies (Rumelt and Brown 2003).

Lesser degrees of visual field loss, usually taking the form of an altitudinal hemianopia or ocular scotoma are usually found with Branch Retinal Artery Occlusions, BRAO. These are usually due to an embolus lodging at a first or second order bifurcation of the central retinal artery. Fundoscopy reveals a pallid retinal oedema in an area which corresponds to the visual field defect. Apart from emboli from cardiac and carotid sources, BRAO can occur in a number of other conditions including infectious retinopathies, e.g catscratch fever and toxoplasmosis, protein S deficiency, temporal arteritis, Susac syndrome, the presence of anticardiolipin and lupus anticoagulant.

Retinal venous occlusion

Central Retinal Vein Occlusion, CRVO, is a common cause of acute visual loss sometimes associated with photopsias, and is associated with optic nerve swelling, cotton-wool spots, tortuosity and distension of the veins, retinal oedema and haemorrhages which pass out to the retinal periphery. Approximately half of these patients will recover vision if at the outset their vision is not severely affected. In those with very poor vision there is a risk that they will develop rubeosis, that is revascularization of the iris and other anterior globe structures, necessitating panretinal photocoagulation, and approximately one-third of them develop chronic open-angle glaucoma. CRVO is associated with hypertension, diabetes mellitus, arteriosclerotic cardiovascular disease, hyperlipidaemia and other thrombotic states. In some instances occlusion of a branch of the retinal vein may occur giving rise to local abnormal retinal signs. The ESR, complete blood count, lipid profile, protean electrophoresis and clotting screen should be undertaken in these patients.

12.3.4 Metabolic storage diseases

Although there is a wide variety of different metabolic storage diseases, the most frequent disorders affecting the retina are the sphingolipidoses and ceroid lipofuscinoses.

In the sphingolipidoses the metabolic by-product accumulates in retinal ganglion cells thereby producing a characteristic cherry red spot in the macula. The cherry red spot may clear as the disease progresses leaving only optic atrophy (Kivlin *et al.* 1985). Many other storage diseases, in addition to the most well known Tay-Sachs disease, produce a similar fundus picture.

The lipofuscinoses characteristically show a retinal pigmentary retinopathy in contrast to the ganglion cell deposition of the sphingolipidoses. Further details of these conditions are to be found in Chapter 10.

12.3.5 Phakomatoses

The phakomatoses are a group of heritable disorders characterized by cutaneous lesions and hamartomatous growths elsewhere in the body. A hamartoma is a tumour composed of tissues normally present in the organ of origin with limited capacity for proliferation. The commonest phakomatoses are tuberose sclerosis or Bourneville's disease, neurofibromatosis types 1 and 2, encephalotrigeminal angiomatosis or Sturge–Weber disease, cerebelloretinal angiomatosis or Von Hippel–Lindau disease and racemose haemangiomas of the retina and midbrain, the Wyburn–Mason syndrome. Most of these conditions have prominent retinal manifestations.

Tuberous sclerosis

This condition, which is inherited as an autosomal dominant trait, is usually associated with the triad of epilepsy, mental retardation and adenoma sebaceum (Section 11.1). However, all tissues are involved especially ocular, renal and the heart. The main retinal abnormality, found in 50 to 90 per cent of patients, are astrocytic hamartomas which are of two types. Those which are typically smooth, dome-shaped semi-translucent greyish and white with a mulberry-like appearance, and flatter, smaller whiter and more transluscent ones. They may be located anywhere in the retina, the first type are preferentially found in the peripapillary region whereas the others are in the posterior pole but more peripheral (Zimmer Galler and Robertson 1995) (Fig. 12.5). In the peripapillary region it is important to differentiate these hamartomas, which obscure underlying retinal vessels, from drusen which do not. They may be calcified and therefore found on CT brain scans and they produce a high signal on T2-weighted MRI.

Neurofibromatosis types 1 and 2

Although orbital and ocular adnexal involvement in neurofibromatosis are common, fundoscopic abnormalities are distinctly uncommon. Neurofibromatosis types 1 and 2 are now considered separate diseases, which map to separate chromosomes—17 for *NF1* and 22 for *NF2*—and developmental hamartomas arise in both types (Ragge 1993). Retinal and choroidal hamartomas are occasionally seen in NF1, but they are more typical features of *NF2*. The ocular features of *NF1* include iris hamartomas or Lisch nodules, congenital glaucoma, anterior sub-capsular cataract, retinal vascular occlusions and optic nerve gliomas (Mustonen *et al.* 1997).

The much rarer Neurofibromatosis type 2 often causes premature visual loss due to cataract, and in addition to the retinal hamartomas

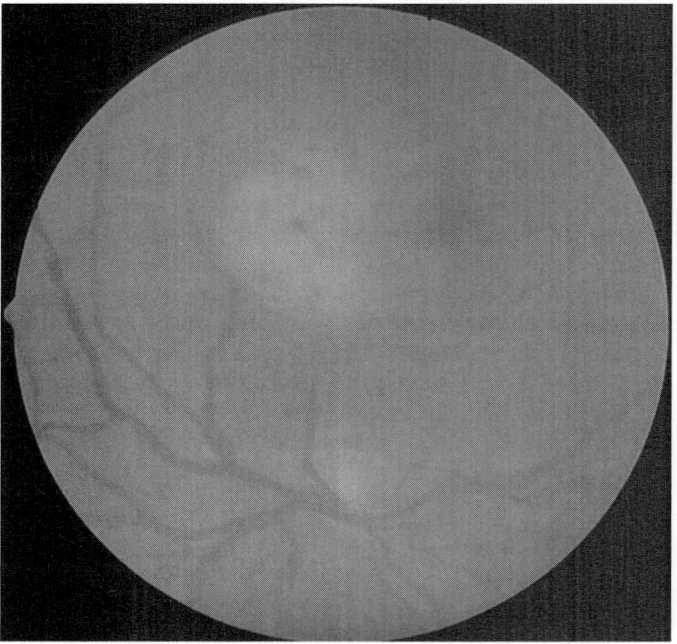

Fig. 12.5 Fundus showing retinal haematoma associated with tuberose sclerosis. (See Plate 20.)

may have epi-retinal membranes, optic disc gliomas, retinal haemangiomas, medullated nerve fibres, posterior subcapsular cataracts, choroidal naevi, uveal melanomas, choroidal hamartomas and bilateral optic nerve gliomas. Further details can be found in Section 11.2.

Sturge–Weber syndrome

This condition is characterized by a facial 'port wine' stain, which is in fact a cavernous haemangioma, usually in the ophthalmic division of the trigeminal nerve (Section 11.3). In addition, leptomeningeal angiomatosis, seizures, cerebral gyriform calcifications, unilateral glaucoma and choroidal haemangioma may occur. Ipsilateral to the facial lesion may be found an intracranial calcifying occipito-parietal lepto-meningeal angiomata.

The ocular manifestations are glaucoma in 60 per cent of patients and choroidal haemangiomas in 55 per cent of patients, which are diffuse, minimally elevated sub-retinal lesions of deep red colour which may easily be overlooked. They may also occur in the conjunctiva and episclera. In some patients, asymptomatic retinal vascular tortuosity, serous retinal detachment, retinal degeneration and iris heterochromia may also be noted (Sullivan et al. 1992).

Von Hippel–Lindau disease

Von Hippel–Lindau disease is an autosomal dominantly inherited disorder with incomplete penetrance due to mutations in VHL1, a tumour suppressor gene located on chromosome 3p. It is characterized by a predisposition to develop haemangioblastomas of the CNS and retina, and in addition, renal cell carcinoma, phaeochromocytoma and renal, pancreatic and ependymal cysts (Section 11.4). As visual symptoms may antedate symptoms due to involvement of other organs, it is imperative that the retinal lesions be properly diagnosed, so that appropriate investigation for systemic involvement can be initiated in a timely manner (Maher et al. 1990).

The most frequent initial manifestation is retinal angiomatosis, which by the age of 60 has developed in more than 70 per cent of patients. These lesions may be incipient or fully developed, the former being frequently located at the equator or beyond, and therefore more effectively observed using indirect ophthalmoscopy (Welch 1970). They are bilateral in about half the patients and around one-third have multiple lesions in one eye. The lesions are about the size of diabetic micro-aneurysms, elevated and globular and consist of a tight network of small vascular channels, often fed by a tortuous arterial feeder and a venous draining vessel. When the lesion is fully developed, these feeding and draining vessels are dilated and tortuous, while the lesion itself is berry-like in configuration. Fluoroscein angiography is required for the detection of nascent preclinical lesions. Exudative retinopathy and retinal detachment usually account for visual morbidity.

In an affected individual, annual ophthalmological examination should be carried out, and when screening at-risk relatives their eyes should be examined from the age of five (with fluorescent angiography from age 10). See Section 11.4.

Racemose haemangiomas

Racemose haemangiomas of the retina may occur, in addition to those involving the thalamus and mid-brain, the Wyburn–Mason syndrome. These uncommon haemangiomas show a massive dilatation and tortuosity of retinal vessels without clear differentiation between arteries and veins. Visual function is often impaired to a point where the eye may be blind. Similar vascular malformations may occur along the visual pathway, including the optic nerve, chiasm, optic tract and mesencephalon.

12.3.6 Acquired immune deficiency syndrome

Patients with acquired immune deficiency syndrome may often show fundoscopic abnormalities, which include cotton wool spots, cytomegalovirus retinitis, conjunctival Kaposi's sarcoma, retinal periphlebitis and acute retinal necrosis (Jabs 1995).

12.3.7 Chronic ocular ischaemia

In patients with occlusion of one carotid artery in association with severe stenosis of the remaining carotid and vertebral arteries, severe chronic ocular ischaemia may occur due to inadequate blood supply via the ophthalmic artery. Initially this presents with venous stasis retinopathy in the posterior pole, but with time leads to anterior segment abnormalities (Riordan-Eva et al. 1994).

In addition to carotid disease, other systemic diseases such as hyperviscosity syndromes, blood dyscrasias or severe anaemia which lead to a reduced oxygen supply to the eye may give rise to a similar condition.

Patients with this condition may complain of blurring of vision, but may also experience a dazzling phenomenon when exposed to bright light, which is thought to be due to hypoxia of the photoreceptors. The initial venous stasis retinopathy has many features which are similar to those found in diabetic retinopathy and central vein occlusion. These include micro-aneurysms, dot-and-blot retinal haemorrhages, venous congestion and irregular calibre retinal veins. In addition, there may be macular oedema and mild disc swelling, but as the condition progresses retinal neovascularization develops, often with iris/angle neovascularization, corneal oedema, uveitis, episcleral vascular congestion and sluggish pupillary light reactions. Light pressure on the eyeball, whilst observing the retinal artery by ophthalmoscopy, will lead to induced pulsation due to the lowered central artery pressure.

Therapy for this condition depends on the ability to restore an adequate blood supply to the ophthalmic artery. It is essential to prevent the ischaemic oculopathy from developing, since visual prognosis by this stage is poor.

Carotid-cavernous fistulas are discussed in Section 13.4.5.

12.4 Abnormalities of the optic disc

12.4.1 Congenital optic disc anomalies

There are a range of congenital optic disc abnormalities most of which have a distinctive appearance. They may be observed coincidentally when examining the fundi of a patient with no visual impairment or may be the cause for a patient's visual complaint. Some of these anomalies may give rise to sudden alterations of vision in adult life. For example, macular detachment may occur in the context of optic drusen.

Optic nerve hypoplasia

Hypoplasia of the optic nerve may be mild or severe, unilateral or bilateral, and may be associated with normal or impaired visual function. It may occur in isolation or be associated with central nervous system anomalies.

When the condition is extreme the disc is small, with a yellow and white de-pigmented halo surrounded by a pigmented rim. In less severe cases, it is important to identify the small amount of nerve tissue in relation to the surrounding, exposed white scleral ring bared of the pigment epithelium—the double-ring sign. There is only a reasonable correlation between the extent of optic disc hypoplasia and the extent of visual acuity impairment. In some instances there may be a peripheral field loss associated with good central vision. In such cases the individual may be unaware of the field defect.

Optic nerve hypoplasia may be isolated, but if bilateral, it is often associated with a variety of congenital ocular syndromes, for example aniridia, ocular colobomas, Duane's syndrome, microphthalmos and hemifacial atrophy. It may also occur in association with congenital suprasellar tumours such as craniopharyngiomas and hypothalamic gliomas. A particular association with intracranial developmental abnormalities is *De Morsier's syndrome of septo-optic dysplasia* (Brodsky 1991). This syndrome is associated with bilateral optic nerve hypoplasia, nystagmus and short stature due to pituitary dwarfism. An important feature is the absence of the septum pellucidum. The condition is frequently associated with growth hormone deficiency and pituitary dysfunction. It is therefore essential for both imaging and endocrine assessment to be undertaken in all cases of bilateral optic nerve hypoplasia. Septo-optic dysplasia is also associated with schizencephaly, a cortical migration anomaly.

Although most cases of optic nerve hypoplasia are idiopathic and sporadic, the condition has been associated with maternal exposure to LSD, crack cocaine use, anticonvulsants, quinine and it may occur in association with the foetal alcohol syndrome.

Optic nerve dysplasia

Optic nerve dysplasia presents with a spectrum of abnormalities, including optic nerve colobomas, optic pits and the morning glory syndrome, all considered to being associated with abnormal closure of the embryonic foetal optic stalk and cup fissure. The abnormalities are usually located inferiorly. They are important to recognize, since they are sometimes associated with basal encephaloceles and other forebrain anomalies (Brodsky and Glaser 1993).

Optic disc colobomas: These are deeply evacuated nerve head anomalies with blood vessels exiting from the margins. Ophthalmoscopically the optic disc appears excavated with sparing of the superior rim and a normally appearing retinal vasculature. These abnormalities are associated with defects in the retinal nerve fibre layer, leading to an appropriate visual field loss, usually superior.

Optic pits: Optic pits are crater-like depressions in the optic disc with a dark grey hue, usually situated in the inferotemporal disc margin with an accompanying nerve fibre layer defect. They may be associated with blind-spot enlargement or arcuate field defects. Optic pits may be associated with a serous detachment of the macula produced by vitreous fluid seeping into the sub-retinal space through the pit.

The morning glory syndrome: In this condition, an enlarged dysplastic disc is associated with an elevated centrally retained mass of glial and embryonic glial and vascular material, which radiates outwards in a sunburst pattern (Pollack 1987). A few cases have been associated with Moyamoya disease.

Tilted discs

An asymmetrically shaped, tilted disc is produced when the optic nerve leaves the globe at an extremely oblique angle. It is often associated with a crescent-shaped zone of exposed sclera along one edge. The vessels may appear to be displaced in the superior division, or appear to originate from the temporal side of the disc, rather than the nasal in situs inversus. The disc may appear hypoplastic and patients with this condition often have moderately high myopia and oblique astigmatism.

However, it is an important condition to recognize, since it results in elevation of the superior disc margin, which may be incorrectly diagnosed as pathological disc swelling. Because the inferonasal margin of the disc is usually deficient, this may result in temporal field defects which do not respect the vertical meridian but may cause confusion with chiasmal compressive processes (Apple *et al.* 1982).

Optic nerve drusen

Drusen of the optic disc can give rise to elevation of the optic nerve head and pseudopapilloedema. They are found in 0.3 to 2 per cent of the population, and in two-thirds they are bilateral. Drusen are intrapapillary, prelaminar refractile concretions which are thought to be due to intracellular axonal debris arising in degenerating nerve fibres damaged as a result of compression as they leave the globe through a congenitally narrow scleral canal. Anomalous discs due to drusen are usually smaller than normal, have an absent central optic disc cup and exhibit an aberrant branching pattern of the central retinal vessels which includes spoke-like appearance secondary to trifurcations of first-order vessels. Initially the drusen are buried with a simple elevation of the disc, but become more apparent in later years when they appear to give rise to a typical lumpy disc with a scalloped margin (Rosenberg *et al.* 1979) (Fig. 12.6). Occasionally there may be a visual deterioration associated with peripapillary splinter haemorrhages and disc oedema.

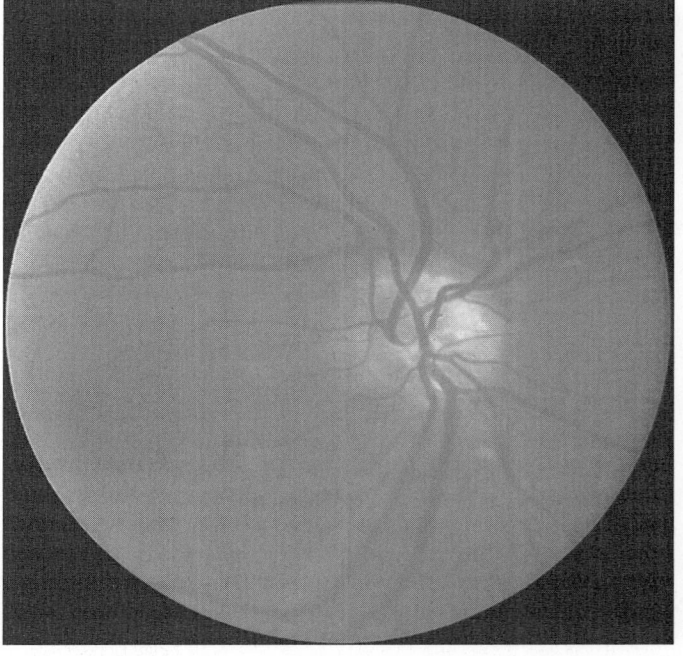

Fig. 12.6 Pseudopapilloedema showing optic nerve head drusen. (See Plate 21.)

There is evidence that disc drusen are inherited as an autosomal dominant trait with incomplete penetrance. It is therefore worthwhile examining the fundi of parents and siblings, when trying to decide whether or not a patient has pseudo-papilloedema due to drusen. Other methods to determine the presence of drusen are to retro-illuminate the disc, which can be achieved by directing the ophthalmoscope beam at the margin rather than the centre of the disc itself, thereby causing the drusen to glow with an opalescent quality. Drusen also exhibit nodular autofluorescence when viewed with a cobalt blue light. Finally, CT scanning through the optic nerve head reveals drusen as calcific densities.

Drusen may sometimes be associated with visual field defects in 70 per cent, usually arcuate nerve fibre bundle defects or an infero-nasal depression, and with haemorrhagic complications (Savino et al. 1979). They may sometimes appear to be progressive but are almost never associated with central visual loss. However, most patients with optic drusen are asymptomatic.

It should be remembered that the presence of drusen does not prevent the patient from harbouring a brain tumour, that chronic papilloedema and optic nerve sheath meningiomas can produce intra-capillary refractile bodies resembling drusen, and finally that occasionally disc drusen can occur in patients with retinitis pigmentosa.

Along with optic disc drusen, other causes of anomalous disc swelling such as tilted discs with asymmetric elevation and the physiologically elevated discs found in hypermetropic eyes may be observed.

12.4.2 Myelinated nerve fibres

In slightly less than 1 per cent of the population, some portions of retinal nerve fibres are myelinated, although normally optic nerve myelination stops at the lamina cribrosa. It appears on fundoscopy as a white area, usually adjacent to the disc, which has a centrifugal feathered edge (Fig. 12.7).

12.4.3 Optic disc swelling

Although optic disc swelling and papilloedema have in the past been used synonymously, it is now usual only to refer to papilloedema as optic disc swelling when it is associated with a raised intracranial pressure (Section 26.5). Other cases of optic disc swelling are either due to local abnormalities in the eye nerve or orbit, or due to congenital anomalies as described above.

Local causes of optic disc swelling are usually associated with impaired visual acuity and colour vision, central arcuate or altitudinal field defects, and an afferent pupillary defect, which contrasts with papilloedema where the acuity remains normal, except in the final stages, and is usually bilateral.

Papilloedema

The evolution of the disc changes in papilloedema due to raised intracranial pressure are usually classified into four stages: early, fully developed, chronic, and atrophic (Neetens and Smets 1989).

In *early* papilloedema, there is disc hyperaemia, mild disc swelling with blurring of the fine peripapillary nerve fibre layer striations, dilatation of retinal veins with loss of spontaneous venous pulsations, and occasionally fine splinter haemorrhages at the disc margin (Fig. 12.8).

In *fully developed* papilloedema, disc elevation is moderate to marked, and there is increased venous distension and tortuosity, increasing number of peripapillary haemorrhages, cotton wool spots, and dilated capillaries on the disc surface. The retinal blood vessels and disc margin become increasingly indistinct, due to the increasing opacification of the retinal nerve fibres. A hemimacular star may be observed in the nasal macula.

In *chronic* papilloedema, there is resolution of the haemorrhages and exudates leaving a dome-shaped 'champagne cork' disc swelling, which often contains hard exudates. White refractile bodies may appear on the disc surface, known as corpora amylacea.

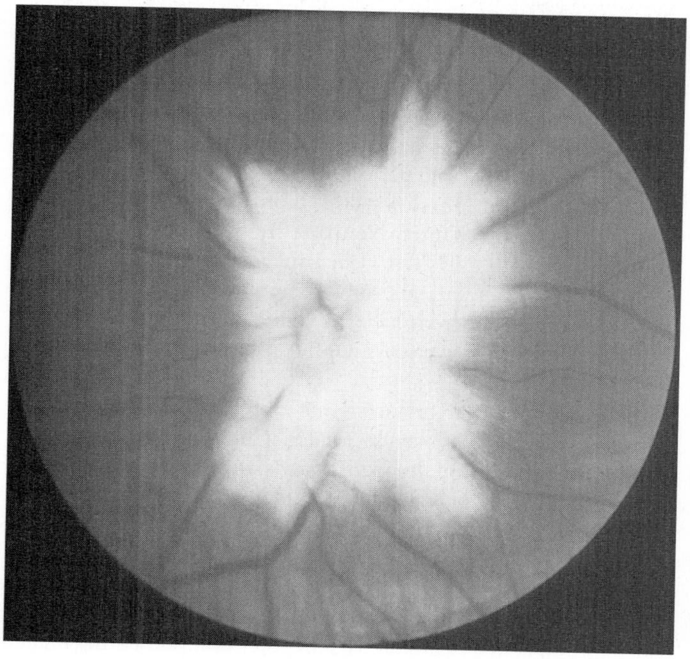

Fig. 12.7 Myelinated retinal nerve fibres. (See Plate 22.)

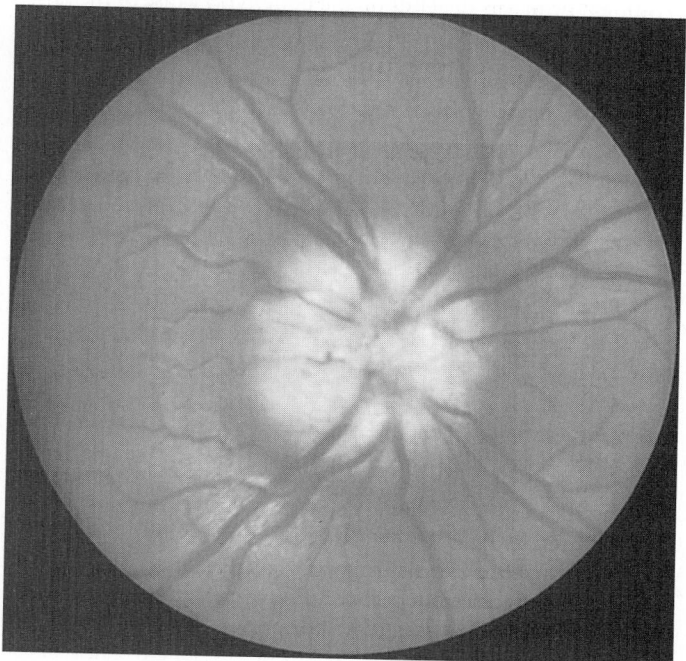

Fig. 12.8 Papilloedema due to raised intracranial pressure. (See Plate 23.)

As time goes on there is increasing nerve fibre attrition, leading to progressive visual field loss.

Finally, the end result is post-papilloedema or consecutive *atrophy*, in which the disc acquires a milky opalescence and the retinal vessels are sheathed.

Clinical features:
Usually papilloedema is bilateral and there is an absence of visual symptoms. However, unilateral or bilateral transient visual obscurations may occur, which last a few seconds and are often associated with postural changes. Obscurations are considered to be due to ischemia of the optic nerve head. Although it has been suggested that such obscurations herald permanent visual loss, there is no evidence to support this view, and they do not correlate with the intracranial pressure, the extent of the visual loss, or the severity of the disc oedema. Because there is compression of the central retinal vein there are no spontaneous venous pulsations, but note 10 per cent of the normal population do not have them either. The longer the papilloedema persists, the more likely there is to be progressive visual field loss, which usually starts as a peripheral field constriction. Occasionally, sudden visual loss occurs in a patient with papilloedema due to ischaemic optic neuropathy (Orcutt *et al.* 1984). In addition, ophthalmoscopy may reveal retinal or choroidal folds, a macular star and haemorrhages.

Pathogenesis:
Papilloedema is due to impairment of axonal transport in the retinal nerve fibres, leading to axonal distension, which is seen as disc swelling at the level of the pre-laminar optic nerve. The raised intracranial pressure compromises elements of retrograde axonal axoplasmic transport at the lamina cribrosa, leading to axonal distension. It is mainly the slow component of axonal transport which is primarily affected (Hayreh 1977).

Aetiology:
There is a vast array of different causes leading to increased intracranial pressure (Table 12.1).

Management:
Treatment primarily depends on the underlying cause of the raised intracranial pressure. If due to a mass lesion which cannot be completely removed, or due to a non-surgically remediable cause then a shunting procedure or medical measures, such as osmotic agents or diuretics such as acetazolamide may be used. Increasingly, optic nerve sheath fenestration is being used for patients with intractable papilloedema who are developing early visual loss. This is particularly so in patients with idiopathic intracranial hypertension (Section 26.5.5). A small window is made in the retro-laminar optic nerve sheath, which may lead to a reduction in the pressure around the optic nerve thereby preventing axoplasmic stasis.

12.4.4 Ischaemic optic neuropathy

Ischaemic optic neuropathy is due to an acute infarction of the optic nerve head, and can either be arteritic as a consequence of giant cell arteritis (Sections 18.7.1; 36.2.8) or non-arteritic anterior ischaemic optic neuropathy, which is the commoner form of the condition.

Non-arteritic anterior ischaemic optic neuropathy

Anterior ischaemic optic neuropathy tends to occur in patients aged between 45 and 80 years (mean 66 years). It is characterized by abrupt, painless and generally non-progressive visual loss, and disc oedema, associated with an arcuate or inferior altitudinal visual field loss and often an afferent pupillary defect (Boghen and

Table 12.1 Causes of optic disc swelling due to raised intracranial pressure

Mass lesions: tumours, aneurysms, granulomas, parasitic cysts
Intracranial haemorrhage: subdural haematoma, epidural haematoma, subarachnoid haemorrhage
Arterio-venous malformations
Intracranial infections: brain abscess, meningitis, encephalitis
Obstructed cranial venous outflow: dural venous sinus thrombosis; dural venous sinus infiltration; jugular vein compression; dural venous sinus arterio-venous malformation
Obstructive hydrocephalus
Brain oedema following trauma
Spinal cord tumours
Benign intracranial hypertension idiopathic secondary to metabolic and endocrine disorders: Addison's disease, diabetic keto-acidosis; thyrotoxicosis; hypoparathyroidism; chronic uraemia secondary to toxic causes: tetracycline; naladixic acid; steroid therapy; lithium, hypervitaminosis A.
Guillain–Barré syndrome
Cranio-stenoses
Mucopolysaccharoidoses
Systemic illness: Behcet's syndrome; status epilepticus; Reye's syndrome; Whipple's disease; systemic lupus erythematosis; systemic hypertension chronic respiratory insufficiency.

Glaser 1975). The majority of patients are elderly and have diabetes and/or hypertension; if they have a small cupless crowded optic nerve head, then they are particularly at risk. In nearly all cases, there is optic disc pallor with oedema which may be diffuse or sectoral, often associated with one or more splinter haemorrhages at the disc margin (Fig. 12.9).

A small proportion of patients with anterior ischaemic optic neuropathy develop progressive stepwise visual loss for as long as 6 weeks. This is rare, although a significant proportion of cases do progress over a 48-hour period. Although previously considered irreversible, as many as 40 per cent of patients may show improvements by three or four lines at 6 months following the acute episode (Ischemic Optic Neuropathy Decompression Trial Research Group IONDTR 1995). Pain is rare but may occur in about 10 per cent of patients. Subsequent involvement of the same eye is unusual, but there is a 40 per cent chance of involvement of the fellow eye within five years. Optic atrophy rapidly ensues after the ischaemic event.

The cause of anterior ischaemic optic neuropathy remains obscure, although it has been associated with haemodynamic shock, carotid artery occlusion, hyperviscosity states and acquired defects of thrombosis and haemostasis regulation (Hayreh *et al.* 1994). Recently attention has been focused on its association with the absence of a cup at the optic nerve head, often associated with hypermetropic eyes (Burde 1993). The visual loss of anterior ischaemic optic neuropathy is often present on awakening in the morning, which has suggested that nocturnal hypotension, sometimes related to medication for hypertension, may be a risk factor.

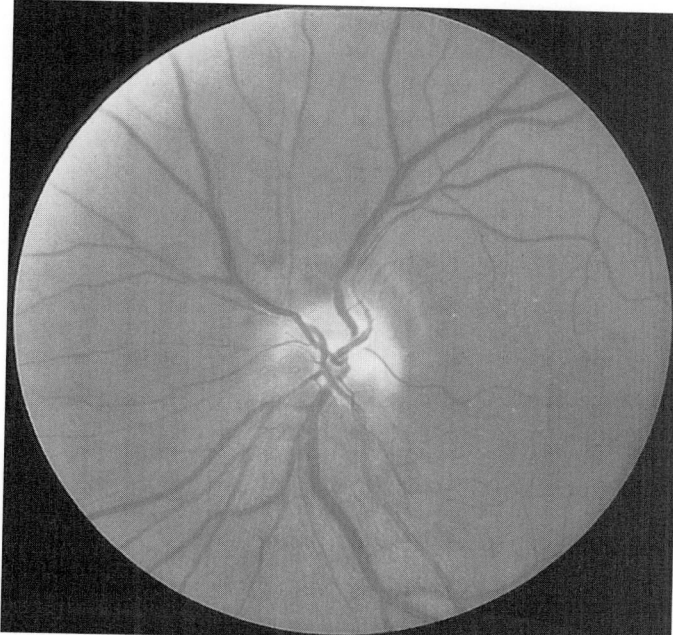

Fig. 12.9 Anterior ischaemic optic neuropathy. (See Plate 24.)

tends to be more prominent. The infarction is believed to be due to occlusion of the short posterior ciliary vessels (Liu *et al.* 1994). In 10 per cent of cases a short period of monocular transient visual loss antedates the ischaemic optic neuropathy.

A high index of suspicion is required for giant cell arteritis, and if suspected an urgent erythrocyte sedimentation rate and temporal artery biopsy should be arranged. At the same time as the blood for the ESR is taken, the patient should be immediately started on systemic steroids: prednisolone 80 mg daily, plus 200 mg iv hydrocortisone immediately or intravenous methylprednisolone 250 mg qds for 3 to 5 days followed by oral prednisolone. This management course is essential to prevent acute visual loss, which occurs in 12 to 50 per cent of biopsy proven cases. If ischaemic optic neuropathy has occurred in one eye and if left untreated, the second eye will also be affected in a third of cases, a third of those within 24 hours and another third within a week.

In most patients the ESR is markedly elevated, as is the C-reactive protein. Occasionally the ESR may be normal. A biopsy of the superficial temporal artery should be obtained as soon as possible after the diagnosis has been considered. The biopsy will not be affected by the use of corticosteroids for up to at least 48 hours. A positive temporary artery biopsy confirms the diagnosis of giant cell arteritis, but in 25 per cent of patients skip areas are found in biopsy specimens, and therefore a negative biopsy can sometimes occur in consequence.

Steroid treatment should not be tapered or withdrawn too early, since a relapse of symptoms is common. The dose of prednisolone can be gradually tapered after 2 to 3 weeks, by which time the ESR should have normalized, to maintain the ESR normal and the patient asymptomatic. Treatment should be continued for at least 6 to 12 months.

Papillo-phlebitis

Papillo-phlebitis, also known as optic disc vasculitis, occurs in healthy young individuals, and is characterized by unilateral disc swelling with venous engorgement and peripapillary haemorrhages, without any significant visual symptoms. The blind spot may be enlarged, with only mild central visual blurring. In most cases this resolves within a few months, although some may progress to profound visual loss.

12.4.5 Optic atrophy

Optic atrophy is the final result of a variety of disturbances that affect the optic nerve or retina. The disc appears pale, and there is an absence of disc vasculature and retinal nerve fibres (Fig. 12.10). An ischaemic process is most likely to have occurred when arteriolar narrowing and sheathing is observed. A cupped atrophic disc usually denotes end-stage glaucoma. Retrograde atrophy from lesions of the optic tract result in band or 'bow tie' atrophy of the contralateral disc with generalized atrophy of the ipsilateral optic disc.

However, as a general maxim the ophthalmoscopic appearance of optic atrophy is usually non-specific and non-diagnostic. If the preceding pathological process is unknown, it is essential to conduct imaging to exclude a compressive lesion.

Optic atrophy results from any disease process which results in death of the retinal ganglion cells, with a dying back of their nerve fibres. This can, therefore, be due to diseases which directly involve the ganglion cells themselves or which damage the axons in the

There is no treatment of proven benefit for patients with anterior ischaemic optic neuropathy. However, if the patient is seen at a very early stage, acetazolomide can be given to raise the optic nerve head perfusion pressure, in association with laying the patient flat. Although in the past few years there has been a vogue for using optic nerve sheath decompression, a recent controlled trial showed it to be of no value (Ischemic Optic Neuropathy Decompression Trial Research Group 1995). In view of their age a number of these patients may be found to have hypertension, but vigorous lowering of the blood pressure can lead to worsening.

The most important aspect of management is to exclude the possibility of the arteritic form, since in such cases the fellow eye is vulnerable to similar involvement. There is no specific treatment available for this condition.

In the 30 to 50 age group, anterior ischaemic optic neuropathy may be difficult to differentiate from optic neuritis since the visual loss, field defects, and disc changes may be found in both conditions. However, anterior ischaemic optic neuropathy is extremely unusual below the age of 30, is usually painless, and is associated with disc swelling, which is unusual in optic neuritis.

Arteritic ischaemic optic neuropathy

The arteritic form of ischaemic optic neuropathy usually occurs in giant cell, cranial or temporal arteritis (Section 36.2.8), but may rarely occur in systemic lupus (Section 36.3.1) and polyarteritis nodosa (Section 36.2.3).

Anyone with arteritic ischaemic optic neuropathy over the age of 50 should be suspected of having giant cell arteritis. This often occurs in the context of headache, malaise, weight loss, anorexia, anaemia, proximal muscle ache or stiffness, temporal artery tenderness, jaw claudication, and fever. These symptoms and signs usually precede the visual loss.

The optic disc infarction is similar to that seen in the non-arteritic form of ischaemic optic neuropathy, except the degree of disc pallor

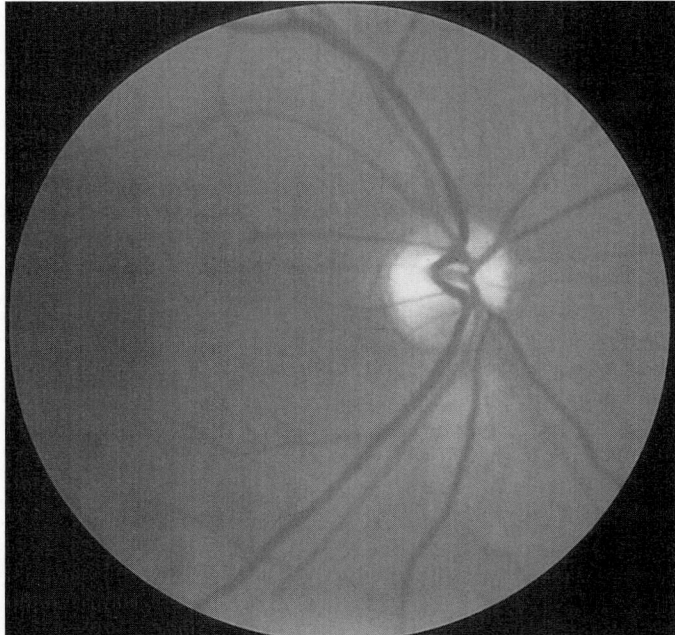

Fig. 12.10 Optic atrophy. (See Plate 25.)

Table 12.2 Causes of optic atrophy

Compression: pituitary tumour, meningioma, other tumours
Deficiency states
Thiamine 'tobacco-alcohol amblyopia'
Vitamin B_{12} pernicious anemia; 'tobacco amblyopia'?
Drugs/toxins
Ethambutol
Chloromycetin
Streptomycin
Isoniazid (INH)
Chlorpropamide
Digitalis
Chloroquine
Placidyl
Antabuse
Heavy metals
Hereditary optic atrophies
Dominant (juvenile)
Leber's hereditary optic neuropathy
Associated heredodegenerative neurological syndromes
Recessive, associated with juvenile diabetes
Demyelination
Graves' disease
Atypical glaucoma
Macular dystrophies

pre-geniculate visual pathway, resulting in retrograde atrophy. The development of optic atrophy is usually slow, dependent on its cause, but takes some 6 weeks to first appear after optic nerve transection due to trauma. In most instances the optic atrophy is bilateral, the disc appearing chalky white in colour with clearly defined margins. Although previously a distinction was made between 'primary' and 'secondary' or 'consecutive' optic atrophy due to subtle ophthalmoscopic differences, it has now been realized that the pathological process is the same in both instances, so the distinction is no longer made.

The differential diagnosis of optic atrophy is considered in Table 12.2.

The ophthalmoscopic appearance of the fundus usually fails to assist in making the diagnosis, although sometimes accompanying blood vessel changes and nerve fibre layer alterations may provide some useful clues. For example, central retinal artery occlusion and anterior ischaemic optic neuropathy produce optic atrophy that is accompanied by narrowing and sheathing of the retinal arterial tree. Retrograde atrophy from lesions of the optic tract may result in band or 'bowtie' atrophy of the contralateral disc and generalized atrophy of the ipsilateral nerve head. Although a cupped optic disc usually accompanies end stage glaucoma it can occasionally result from a compressive lesion.

12.4.6 Infiltrative papillopathy

Rarely the optic nerve head is infiltrated with cells from lymphoma, optic nerve glioma, metastatic carcinoma, leukaemia or sarcoidosis as well as other rare conditions. The clinical presentation is usually of painless monocular loss of vision with a normal or swollen disc. The visual acuity is often reduced in these conditions. Infiltrative papillopathy is usually very steroid-responsive. The presence of

optociliary venous shunt vessels with disc swelling (Fig. 12.11) is typical of optic nerve sheath meningioma.

12.5 **Optic nerve lesions**

12.5.1 **Optic neuritis**

Optic neuritis is a term used to describe an acute or subacute and often painful idiopathic optic neuropathy or one resulting from inflammatory, infectious or a demyelinating aetiology. In the majority of cases the optic disc is normal on ophthalmoscopy and the term retrobulbar neuritis is used. In cases in which the optic nerve head is swollen, the terms papillitis or anterior optic neuritis are used.

Clinical features

It is important to distinguish between the features of typical optic neuritis of idiopathic or demyelinating causation from those of atypical optic neuritis. In typical optic neuritis there is usually acute unilateral loss of visual acuity and of visual field, which may progress over hours or a few days, reaching its maximal effect within one week. Ocular pain is common occurring in 92 per cent, exacerbated by eye movement, and may precede, visual impairment by a few days (Lepore 1991). There may also be some mild tenderness of the globe at onset. The visual loss may range from contrast defects without loss of acuity to no perception of light. The patient is usually aged under 40 years, although optic neuritis may occur at

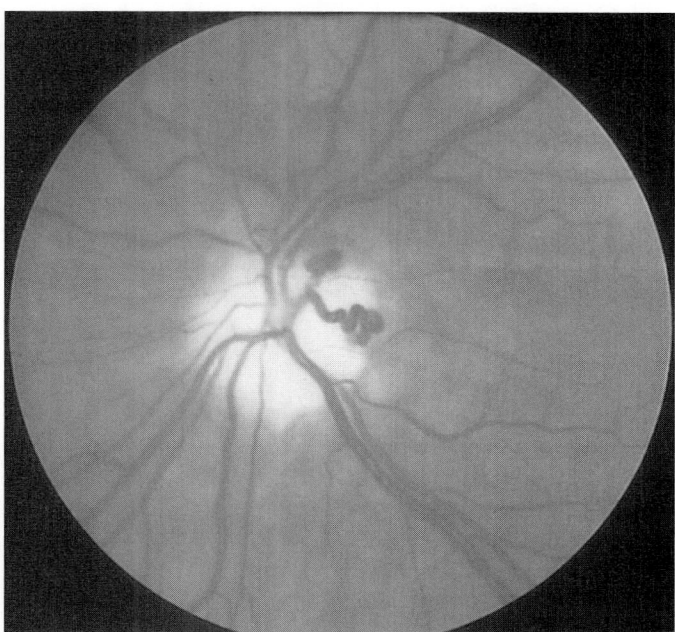

Fig.12.11 Optociliary shunts due to an optic nerve sheath meningioma. (See Plate 26.)

any age. On examination, an afferent pupillary defect is present in patients unless the fellow eye is, or has been, similarly affected. The optic disc often appears normal, disc swelling which is usually mild, occurring in about a third of patients. The commonest visual field defect is a central scotoma, although recent studies have shown that a wide variety of field defects may be found ranging from a central scotoma, to altitudinal and nerve fibre layer defects (Keltner *et al.* 1993). There may be persistent subtle residual defects of colour vision, depth perception and contrast sensitivity, which may continue for several months. Subsequent disc pallor may occur but does not correlate closely with the level of visual recovery (McDonald and Barnes 1992). Improvement commences in 85 per cent of cases within 2 weeks, maximal recovery occurring within 30 days. Normal or near normal, better than 6/9, visual acuity is retained in 90 per cent of patients.

Less common features of the disorder include Uhthoff's phenomenon consisting of transient visual deterioration associated with elevation of body temperature, movement phosphenes light flashes provoked by eye movement and the Pullfrich phenomenon, an illusion of motion.

Atypical optic neuritis may present with monocular or the bilateral simultaneous onset in an adult patient. However, the clinical features differ from those associated with typical optic neuritis. There is often a lack of pain, visual deterioration extending beyond 14 days and there may be other ocular findings suggestive of an inflammatory process, such as an anterior uveitis. Often the patient may be outside the 20 to 50 year age span. They may also have evidence of other systemic conditions, particularly infectious, immune and granulomatous diseases. A number of disorders may be associated with atypical optic neuritis (Table 12.3).

The evaluation of patients with optic neuritis rather depends on whether or not it is a typical or atypical case. Typical optic neuritis probably does not necessitate any additional laboratory investigations,

although an MR brain scan gives some indication of whether subclinical demyelination has taken place elsewhere in the central nervous system. The frequency of this occurring varies between 40–60 per cent in the published series. Of patients presenting with optic neuritis and an abnormal brain MRI from 36 per cent to as high as 82 per cent have been reported to have gone on to develop definite multiple sclerosis within 3 to 5 years. This compares with less than 16 per cent of those with a normal MRI (Morrisey *et al.* 1993).

Those patients with atypical optic neuritis should have a chest X-ray, laboratory tests including a blood count, biochemistry, and tests for collagen and vascular disease and syphilis serology. Examination of the spinal fluid is probably justified in this group of patients.

Management

The clinical management of patients with acute optic neuritis has been significantly enhanced by the multicentre prospective optic neuritis treatment trial in the USA. The trial has also provided the best prospective data regarding the clinical presentation. It was concluded that patients receiving oral prednisolone would not recover any more quickly or achieve a better final acuity than those given oral placebo. Those given iv methylprednisolone did have a more rapid visual recovery, but at the end of 6 months their visual acuity was no better than those receiving placebo (Beck *et al.* 1992). Although there has been some suggestion that patients who receive methylprednisolone had a reduced risk of developing multiple sclerosis, this suggestion has not been accepted by many and awaits a further trial.

As a result of this and other trials, steroid treatment of patients with typical optic neuritis is unnecessary, unless there is severe ocular pain which cannot be managed with analgesics, or if there is already poor vision in the fellow eye due to some other disease process.

Table 12.3 Causes of atypical optic neuritis

Viral infections of childhood (rubella, mumps, chickenpox) with or without encephalitis
Viral encephalitides
Postviral, paraviral infections
Infectious mononucleosis
Herpes zoster
Lyme disease (borellia)
Acquired immunodeficiency syndrome (AIDS)
Contiguous inflammation of meninges, orbit, sinuses
Granulomatous inflammations (syphilis, tuberculosis, cryptococcosis, sarcoidosis)
Intraocular inflammations
Connective tissue disease, e.g SLE, polyarteritis nodosa, Churg–Strauss
Carcinomatous meningitis (lymphoma or metatastic carcinoma) syndrome
Leber's hereditary optic neuropathy
Bee stings
Unknown aetiology

Atypical optic neuritis

Atypical optic neuritis may occur in a variety of different circumstances. It occurs in the context of systemic lupus erythematosis (Section 36.3), when it has been considered that, although elements within the optic nerve may be susceptible to autoantibody-mediated attack (Kupersmith *et al.* 1988), it is likely that the optic neuritis may be due to small vessel occlusive disease. In several other conditions such as Sjögren's syndrome and vascular disease such as polyarteritis nodosa or Churg–Strauss syndrome, inflammation takes place in blood vessel walls leading to ischaemic rather than an inflammatory mechanism. Supposedly immunologically mediated optic neuritis has also been observed following bee stings and the use of interleukin-2 and alpha interferon.

Optic neuritis may also occur following several different viral illnesses including chickenpox, rubella, infectious mononucleosis, and mumps, particularly in children (Selbst *et al.* 1983). Such cases may be bilateral, occur within 7 to 10 days of the onset of the illness and there is usually a spontaneous and full recovery. It may also occur after vaccinations against both bacterial and viral infection.

Outside the context of AIDS, primary infection of the optic nerve is rare (Nichols and Goodwin 1992). However, with AIDS the optic nerve can develop a granulomatous perineuritis due to tuberculosis, aspergillus, syphilis, toxoplasmosis, cryptococcus, borelliosis causing Lyme disease, and *Bartonella henselae* causing cat-scratch fever. Acute retinal necrosis due to herpes viruses is also sometimes encountered. Infectious optic neuritis may occur as a result of spread of bacterial or fungal sinus infection.

Difficulties may sometimes be encountered when atypical optic neuritis presents with visual loss and disc swelling due to infiltration of the intra-orbital nerve sheath or the nerve itself. This may be due to tuberculosis, sarcoidosis, cryptococcus, and idiopathic granulomatous meningeal disease. Acute and chronic presentations may sometimes occur. In such cases of atypical optic neuritis due to granulomatous inflammation, it is necessary to carefully examine the optic media and retina in addition to the optic nerve. Frequently vitreous infiltrates, retinal vasculitis, and choroidal lesions may be observed, but in some cases these are absent. Some of these cases may be steroid dependent, a term used to indicate that following withdrawal of systemic steroid therapy vision usually drops again. This feature may also be observed in patients with infiltrations of the optic nerve head in meningeal involvement by non-Hodgkin's lymphoma or metastatic carcinoma of the breast or lung.

12.5.2 Heredo-fam ilial optic neuropathies

The hereditary optic neuropathies can be divided into those which are autosomal dominant or recessive and those which are due to point mutations in mitochondrial DNA.

Autosomal dominant optic atrophy

This condition is characterized by moderately poor visual acuity of 6/12–6/60, which begins between the ages of 4 and 8 years, and is associated with slow progression and temporal disc pallor (Kjer 1959). There is also central and centro-caecal visual field defects enlargement and blue-yellow dyschromatopsia. In some families there is associated nystagmus. Impairment in this disorder is considerably milder than in either recessive optic atrophy or Leber's optic atrophy. The gene defect has now been isolated to chromosome 3q27–3q28 (Kjer *et al.* 1996; Johnston *et al.* 1997).

Autosomal recessive optic atrophy

A rare condition which may occur in a simple 'isolated' or complex 'associated defects' form, in which marked visual impairment is noted before the age of 4 years, associated with disc pallor and sometimes retinal arterial attenuation. In the complex form optic atrophy is associated with spinocerebellar degenerations, cerebellar ataxia, pyramidal tract dysfunction and mental retardation, known as Behr's syndrome. It also occurs in Wolfram syndrome in association with metabolic disturbances such as diabetes mellitus and diabetes insipidus and hearing loss; Diabetes Insipidus, Diabetes Mellitus, Optic Atrophy and Deafness, **DIDMOAD** (Barrett *et al.* 1995).

Leber's hereditary optic neuropathy

This disease develops primarily in males, only 14 per cent being women, in the second to third decade of life. It is characterized by an abrupt loss of central vision in one eye, followed by a loss of vision in the remaining eye weeks, months, or sometimes years later. Occasionally visual loss may occur simultaneously in the two eyes. However, the visual loss may progress over days or weeks to reach 6/60 or worse. There is no associated pain on eye movement in contrast to acute optic neuritis, and the visual loss is usually permanent with optic atrophy, central or centrocaecal scotomas and dyschromatopsia. The afferent pupillary defect is often subtle. The fundoscopic picture in the acute phase often shows swelling of the peripapillary nerve fibre layer, circumpapillary telangiectatic microangiopathy and tortuosity of the retinal vessels. Fluorescein angiography does not show disc leakage (Riordan *et al.* 1995). Although the visual loss usually occurs in isolation, in some cases there may be associated cardiac dysrhythmias due to pre-excitation syndromes such as Wolff–Parkinson–White.

A maternal pattern of inheritance has been known for some time but there is not a satisfactory explanation as to why there is a preferential involvement of males. A number of point mutations have been identified in mitochondrial DNA, particularly at the 11778 location in more than 50 per cent of cases, and less frequently at 3460, 14484, and 14459 (Mackey 1994). The significance of the point mutation at 14484 is that a much higher percentage, 37 per cent as opposed to 4 per cent, of patients show some visual recovery, which may be delayed for years, when compared with patients who have a defect at 11778. It is therefore, appropriate to carry out genetic testing in those individuals presenting with atypical optic neuritis of the appropriate sex and age, even if a positive family history is not available. There is no effective treatment for this condition, although because of the link to oxidative phosphorylation pathways, the antioxidants vitamins C and E and coenzyme Q have been prescribed. It is of some interest that earlier suggestions of cyanide toxicity in predisposed individuals, which led to the use of hydroxycobalamin, may relate to recent observations that tobacco and alcohol may play a role in phenotypic expression of the mitochondrial mutations.

12.5.3 Nutritional and toxic optic neuropathies

Bilateral, slowly progressive visual loss with central and centrocaecal scotomas, dyschromatopsia, and usually normal or mild temporal atrophic optic discs characterizes optic nerve failure, particularly affecting the maculopapillar bundle, due to either nutritional deficiency or a toxic cause. Once a family history of one of the hereditary familial diseases has been excluded, this disorder should be considered, and is usually due to a combination of alcohol abuse, deficiencies within the B vitamin complex, and a frequently high tobacco consumption. However, there is little convincing evidence for independent toxicity of tobacco. With treatment by early abstinence of the likely toxic agents and vitamin supplementation, recovery of vision usually occurs, unless the condition is so long standing that optic atrophy has intervened.

The Cuban epidemic of bilateral optic neuropathy (Sadun *et al.* 1994) was probably related to multiple dietary deficiencies. A similar epidemic in Nigeria was considered to be due to cyanide toxicity from eating staple cassava.

A wide variety of instances have been cited as causing toxic optic neuropathy including ethambutol, chloramphenicol, methanol, halogenated hyroxyquinolones, lead, isoniazid, and vincristine. The toxic effects may be either dose dependent or idiosyncratic (Grant and Schuman 1993).

12.5.4 Tumours of the optic nerve

Optic nerve sheath meningiomas

Optic nerve sheath meningiomas arise directly from the optic nerve sheath, usually in the orbital regions of the nerve, but meningiomas frequently arise from the tuberculum sellae, sphenoid wing, and olfactory groove leading to secondary invasion or compression of the nerve. Primary optic nerve sheath meningiomas, most frequently found in middle-aged women, are usually unilateral but if bilateral raise the possibility of central neurofibromatosis type 2 (Section 11.2.4). Patients usually complain of progressive visual loss and decreased colour vision over a period of months to years. Although most patients will have mild proptosis of 2–4 mm at the time of their initial consultation, ocular prominence is not their presenting symptom. Usually an afferent pupillary defect is observed, with peripheral field constriction or a centro-caecal scotoma. Transient visual obscurations are common and in some cases amaurosis has been reported. Visual loss progresses over years with optic disc swelling gradually being supplanted by optic atrophy, with the evolution of optociliary venous shunt vessels retino-choroidal anastamoses representing in about 25 per cent (Dutton 1992) (Fig. 12.11).

Meningiomas may invade the orbit by trans-sheath extension, and intracranial spread through the optic canal is always a possible danger, particularly in children, when the tumour sometimes behaves in a more aggressive fashion.

The CT picture in patients with these tumours is most often one of diffuse enlargement of the optic nerve, with bulbous swellings of the nerve in the region of the globe and orbital apex. 'Railroad-track' calcification of the optic nerve sheath in the orbit is a characteristic feature. MRI is now the imaging procedure of choice and has enabled optic nerve sheath meningiomas to be distinguished from optic nerve gliomas, where both may cause a uniform enlargement in the orbital region. The tumour is isointense on T1- and T2-weighted imaging and smoothly enhances with gadolinium. In meningiomas but not gliomas the nerve and optic nerve sheath are readily distinguished (Lindblom *et al.* 1992).

The clinical course for these tumours is of insidious progression leading to complete blindness. Management of patients with optic nerve sheath meningiomas is controversial (Kennerdell *et al.* 1988). While there is general agreement that nerve sheath tumours are most aggressive in children and become progressively more indolent with advancing age, there is no consensus as to the best way to treat these lesions. Clinical resection, particularly when there is intracranial spread, which fortunately is rare, is usually incomplete. Since these patients rarely die from the meningioma, it is probably best to keep them under observation. In some instances radiotherapy has been shown to result in some visual improvement but there is no agreement as to its value. It should probably be reserved for adults in whom there is clear evidence of progression (Moyer *et al.* 2000). Endocrine treatments suitable for optic nerve meningiomas, such as progesterone antagonists, have not been reported as successful.

Optic nerve gliomas

Optic nerve gliomas, which may also involve the chiasm, are of two distinct types. By far the commonest is the benign pilocytic astrocytoma of childhood, and the other the malignant glioblastoma which occurs in adults (Dutton *et al.* 1994). Approximately a quarter of cases occur in the setting of neurofibromatosis NF-1, stigmata of which should be sought.

Benign optic nerve gliomas usually present within the first two decades of life, with a peak incidence from 1 to 6 years of age. The usual presenting manifestations are proptosis and slowly progressive painless visual loss, which may be so mild as to be undetectable, although a profound reduction in acuity is more common. The fundus picture may be that of either papilloedema or optic atrophy, and further diagnostic confusion is sometimes caused by the finding of optociliary shunt vessels, more commonly associated with optic nerve sheath meningiomas. A secondary strabismus is often found accompanying the visual loss.

The clinical course of childhood optic nerve gliomas, which usually present before age 10, is highly variable. In some, tumour enlargement proceeds slowly for a time but then reaches a plateau, while in others the enlargement proceeds unabated (Hoyt and Bagdassarian 1969). Necropsy material shows an absence of mitotic figures, but these are juvenile pilocytic astrocytomas with a potential for significant visual morbidity and a small but significant mortality. Optic nerve gliomas are generally managed conservatively by careful clinical and radiological follow-up. If there are signs of progression in children under the age of 6 years, chemotherapy would be the first-line treatment and in children over that age radiotherapy. When the eye is almost blind, surgical removal of the prechiasmatic portion of the optic nerve is a better option than chemotherapy or radiation (Klug 1982).

Optic nerve gliomas of adulthood on the other hand are malignant gliomas, which usually arise in males aged 40 to 60 years. These patients often present with a rapid onset of visual failure, which on some occasions may mimic acute optic neuritis. The tumour rapidly progresses often leading to complete blindness within weeks. As occurs with glioblastoma multiforme tumours elsewhere in the brain, the patient usually dies within a short period of only months.

Other optic nerve tumour

Metastatic cancer may lead to optic nerve involvement, either as a result of infiltration of the meninges as occurs with cancer of the breast and lung, or by direct tumour infiltration as with lymphoproliferative disorders and certain types of leukaemia and non-Hodgkins lymphoma. Paraneoplastic optic neuropathy has also been described in patients with small cell carcinoma of the lung (Malik *et al.* 1992).

12.5.5 **Diagnosis of acute and chronic visual failure**

When a patient presents with unilateral or bilateral visual failure, as reduced visual acuity, the next step is to try and localize the site of the lesion by taking the history and carrying out a full neuro-ophthalmological examination. The history will give some clues as to the location of the lesion, although in this situation, apart from ascertaining whether the impaired vision is in one or both eyes, it is the examination which is more revealing. There are a number of key points to remember. Lesions in the visual pathway which are located at the optic chiasm or more posteriorly do not cause a reduction in visual acuity unless they are bilateral, and then usually involve the visual cortices. In this situation the pupillary light reflexes are intact, whereas if located more anteriorly in the optic nerve, such lesions would cause a relative afferent pupillary defect, unless there is bilateral optic nerve involvement. It is also necessary to rapidly differentiate optic nerve from macular and retinal disease. Both cause a reduction in the visual acuity and may cause central scotomas on visual field testing. Whereas in optic nerve disease there is impaired colour vision, a relative afferent papillary defect, and a reduced sensitivity to light brightness, in macular disease these signs are not present, and in addition the photostress test may be abnormal. By applying these simple tests it should be possible to localize the lesion.

Ophthalmoscopy may be helpful at this stage. In the acute situation, the optic disc will only be involved if there is involvement of the very proximal optic nerve close to the nerve head. There is, for example, rarely, disc swelling in optic neuritis as the lesion is usually more posterior in the nerve. However, it is important to remember that disc swelling associated with reduced visual acuity indicates optic nerve disease whereas disc swelling with a normal visual acuity usually indicates raised intracranial pressure. In the more slowly progressive visual failure, the optic disc may be observed to be pale and atrophic, indicating some chronic disease affecting the optic nerve. Optic atrophy can also occur some time after an acute lesion of the nerve.

At this stage in the evaluation the history becomes more imperative in determining the likely cause for the visual failure. If it is acute or subacute and unilateral, a compressive lesion or optic neuritis is the most likely cause. A history of periorbital pain and information whether it is made worse by eye movements is valuable in localizing the disorder to the eye itself or the intraorbital optic nerve, and is supportive of optic neuritis, although not invariably so. The most common masses affecting the optic nerve are pituitary tumour, craniopharyngioma, meningioma, aneurysm, and mucocele. For acute visual failure a history of recent infections, insect bites, inoculations, or any previous systemic medical disease or neurological symptoms should be sought. A full list of all current and recent past medications should be obtained along with the dietary intake and smoking and habits.

Chronic bilateral optic neuropathies in which the optic disc appears normal include nutritional optic neuropathy (tobacco–alcohol amblyopia), vitamin B_{12} or folate deficiencies, toxic and drug-related neuropathies. Other considerations include inherited optic neuropathy, bilateral compressive lesions, and rarely bilateral optic neuritis.

12.6 **Disorders of the optic chiasm**

Approximately 25 per cent of all intracranial tumours occur in the chiasmal region and half of these cases initially present with visual loss. The commonest tumours arise from the pituitary gland,

50–55 per cent, followed by craniopharyngiomas, 20–25 per cent, meningiomas, 10 per cent, and gliomas, 7 per cent. Although there are a number of other causes for the chiasmal syndrome, such as trauma and demyelination, these are rare.

12.6.1 **Clinical features**

The neuro-ophthalmological signs of a compressive optic chiasm lesion are primarily a field defect and deterioration of visual acuity, which depend on the relationship of the chiasm to the pituitary and the tumour (Fig. 12.12). An appreciation of the various field abnormalities is important (Hollenhorst and Younge 1973). In 79 per cent of cases the chiasm lies directly over the pituitary, in 17 per cent it is over the tuberculum sellae 'pre-fixed', and in the remaining 4 per cent it is over the dorsum sellae 'post-fixed'. The classical field defect of a chiasmal lesion is a bi-temporal hemianopia. This may be complete or incomplete and may or may not be symmetrical. Sometimes the defect is only paracentral and may, therefore, be missed if the central field is not explored during visual field testing. It is unusual to have a bi-temporal hemianopia without some reduction in central visual acuity in at least one eye, due to the posterior part of the optic nerve being compromised in addition to the chiasm. Whether a bi-temporal hemianopia is more marked in the inferior or superior field is not helpful in deciding whether the compression is from above or below.

The anterior junction syndrome gives rise to another characteristic field defect due to an anterior chiasmic lesion. This is mainly seen with a post-fixed optic chiasm when the optic nerve close to the chiasm is mainly involved. This results in an ipsilateral central scotoma with reduced visual acuity, and a contralateral superior temporal visual field depression. The latter is due to involvement of

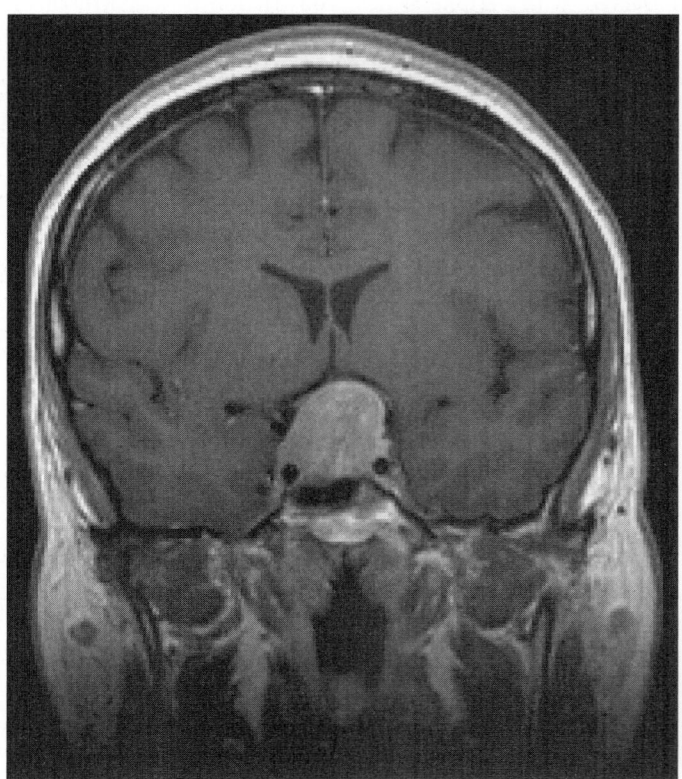

Fig. 12.12 Pituitary adenoma displacing the optic chiasm coronal MRI.

contralateral inferior nasal optic fibres, which decussate and then pass a little way anteriorly into the ipsilateral optic nerve (von Willibrand's knee) before turning into the chiasm and contralateral optic tract. While prechiasmal compression only involves the intracranial portion of the optic nerve, a central scotoma is the only visual field defect. Since there are no differentiating features of such a scotoma resulting either from an inflammatory process (such as optic neuritis) or from compression, apart from temporal compression, the latter possibility must always be considered when this defect is found. However, if the optic tract is mainly involved then an incongruous homonymous hemianopia, with reduced visual acuity in one eye is found, which is particularly suggestive of a prefixed chiasm, and is often found with craniopharyngiomas.

It is important to remember that not all cases of bi-temporal field loss are due to chiasmal disease. Such field defects are also found in centro-caecal scotomas, tilted discs, dysplastic optic discs, refractive scotomas, massive papilloedema due to blind spot enlargement, and overhanging redundant upper eyelids. In large series of patients with pituitary tumours the most common field defect was a bi-temporal hemianopia in 67 per cent, with junctional scotoma, 29 per cent, homonymous hemianopia, 7 per cent, and prechiasmal field loss in 2 per cent (Hollenhorst and Younge 1973). The mechanisms by which field defects occur in chiasmal compression are not certain, but direct pressure on the fibres resulting in impaired axonal transport and interference with the vascular supply producing ischaemia have both been considered. A rapid return of acuity and visual field loss, which sometimes occurs within 48 hours of decompression, supports the former hypothesis. Other signs found in chiasmal lesions include optic disc pallor, which was found in 50 per cent of eyes in one series. Its absence usually denotes a virtual complete return of visual function with successful decompression. Papilloedema is frequently associated with a suprachiasmal tumour but rarely with intrasellar tumours. Involvement of the extraocular nerves may occur in association with parachiasmal lesions such as pituitary tumours, and an unusual phenomenon, seesaw nystagmus, may occur in young patients with a tumour of the chiasm and diencephalic regions.

The main symptom resulting from the chiasmal syndrome is usually deterioration of vision, often with associated dimming of the visual field, particularly temporally. There is usually a progressive deterioration over a period of months or years, except when due to pituitary apoplexy the acuity loss usually precedes the field defect. However, a fairly frequent symptom in patients with the chiasmal syndrome is diplopia. This may be a vertical or horizontal separation of images due to an oculomotor paresis caused by extension of a parachiasmal lesion into the cavernous sinus, producing disturbance of the 3rd, 4th and 6th cranial nerves. However, usually diplopia occurs in the absence of a demonstrable ocular motor paresis. An explanation for this phenomenon is the absence of the temporal field in each eye, which normally acts as a physiological linkage for the two nasal fields and has been called the hemifield slide phenomenon. Minor ocular motor imbalance, which does not normally affect binocular fusion, now results in an inability to maintain the two fields in juxtaposition. Some patients will also complain of a disturbance of depth perception, experiencing problems with such tasks as sewing, threading needles, or using precision tools. This phenomenon, called chiasmic postfixation blindness, is due to the presence of a blind area beyond the fixation point.

The image of objects located in this area falls on the nasal retina which is blind (Kirkham 1972).

12.6.2 Specific causes of the chiasmal syndrome

As has already been mentioned the commonest cause of chiasmal compression is pituitary adenomas (Section 27.7.1), and other tumours include craniopharyngiomas, meningiomas, and gliomas. Other rare non-tumorous causes of chiasmal compression include aneurysms of the circle of Willis, posterior ethmoid or sphenoid sinus mucoceles, pituitary abscess, distension of the third ventricle, and suprasella arachnoid cysts. There are numerous non-compressive causes of bi-temporal hemianopia, some of which will be discussed below.

The empty sella syndrome. In this condition there is an extension of the subarachnoid space into the pituitary fossa, through a deficient diaphragmae sellae, producing enlargement of the sella and compression of the pituitary. This may mimic pituitary disease and may occur either spontaneously, presumably due to a congenital defect, or following surgery or radiation to the sella region and infarction of a pituitary adenoma. It may rarely cause symptoms suggestive of a pituitary tumour, including a moderate hyperprolactinaemia, headaches, and rarely a progressive visual loss due to herniation of the optic nerve, chiasm, and tracts into the sella. (Neelon *et al.* 1973).

Optochiasmal arachnoiditis. This is due to meningitis especially tuberculosis, syphilis, sarcoidosis, or carcinoma, but occurs only rarely. Head trauma may also produce a chiasmal syndrome, the result of a variable combination of contusion, haemorrhage, necrosis, or actual tear. Intrinsic lesions of the optic chiasm are best exemplified by plaques of demyelination found in multiple sclerosis, which despite being an apparent common site of demyelination rarely give rise to appropriate visual field defects. Vascular abnormalities, such as arteriovenous malformations involving the chiasm may occur.

Radionecrosis. Following irradiation to the sella region for pituitary tumours, progressive visual impairment may occur due to radionecrosis of the chiasm, optic tract and intracranial optic nerves (Morris *et al.* 1994). The acute or gradual onset of painless visual loss usually occurs 4 months to 3 years after completion of therapy, with a peak incidence of 12 to 18 months. Both eyes may be involved simultaneously or sequentially. Neuroimaging is required to exclude the possibility of recurrent or secondary tumour. On MRI, radionecrosis appears as an intrinsic enlargement of the chiasm and gadolinium enhancement. Pathological examination of radionecrosis indicates a microvasculopathy and occlusive endarteritis. Patients most at risk are those who have received in excess of 200 cGy per day, a total dose in excess of 4800 cGy, overlapping treatment fields and concurrent chemotherapy. The management of this condition is uncertain, but it has been suggested that anticoagulation with intravenous heparin followed by warfarin treatment given early, may help to arrest further visual loss (Glantz *et al.* 1994). Others have claimed benefits from hyperbaric oxygen therapy.

12.7 Disorders of the optic tract, radiation and occipital lobe

12.7.1 Optic tract lesions

Lesions of the optic tract, although occurring in less than 3 per cent of visual field defects in a series of 100 homonymous hemianopias, often produce specific signs and visual field abnormalities which allow definitive diagnosis (Newman and Miller 1983). The optic tract is the first point in the visual pathways where the ipsilateral temporal and contralateral nasal retinal nerve fibres come together, and so the field defect is usually a partial or complete homonymous hemianopia. When partial there is often gross incongruity between the visual field defects found in each eye, which may also be found with lesions of the lateral geniculate nucleus and more rarely the optic radiations.

Lesions of the optic tract without involvement of the chiasm or optic nerve result in normal visual acuity, but pupillary abnormalities have often been reported. A relative afferent defect may be found in the eye with the temporal field loss (contralateral to the side of the lesion) (Bell and Thompson 1978). In pupillary hemiakinesia described by Wernicke, there is a decreased or absent pupillary reaction when the 'non-seeing' portion of the retina is stimulated compared to the 'seeing' portion. Because of light scatter this phenomena has often been difficult to observe. A second pupillary sign, described by Behr, of the pupil ipsilateral to the homonymous field defect being larger than that contralateral to the field defect has again not been consistently observed.

Ophthalmoscopically optic pallor due to retrograde degeneration may be observed. This takes a characteristic form with band or 'bow tie' atrophy in the eye opposite to the lesion due to loss of nasal retinal fibres. Ipsilateral to the lesion, temporal pallor is found due to loss of arcuate fibres from the temporal hemiretina (Savino *et al*. 1978).

The most frequently encountered lesions causing the optic tract syndrome are aneurysms, craniopharyngiomas, and pituitary tumours. The proximity of the optic tract to the cerebral peduncles and the pituitary, stalk and hypothalamus may lead to associated signs such as contralateral hemiparesis and endocrine disturbances.

12.7.2 Lateral geniculate nucleus

Lesions of the lateral geniculate nucleus have been found to produce incongruous wedge-shaped homonymous field defects, but when the aetiology is ischaemic the defect is usually congruous (Gunderson and Hoyt 1971). The lateral geniculate nucleus receives its blood supply from the distal anterior choroidal and lateral choroidal arteries, and occlusion of one or other artery may result in a specific syndrome. A loss of the upper and lower homonymous sectors in the visual field with corresponding sectoral optic disc pallor characterizes the distal lateral choroidal artery syndrome, whereas a horizontal sectoral defect with appropriate disc pallor is found in the anterior choroidal artery infarction (Frisén *et al.* 1978). Other causes for lateral geniculate nucleus involvement include tumours and demyelination. Since the pupillomotor fibres leave the optic tract to ascend in the superior brachium, a geniculate lesion results in normal pupillary responses.

12.7.3 The optic radiations

As the geniculo-striate fibres leave the lateral geniculate nucleus, the ventral fibres subserving the superior visual field pass anteriorly around the temporal horn of the lateral ventricle to form Meyer's loop. Lesions in this region usually result in a congruous homonymous field defect mainly affecting the superior quadrant 'pie in the sky'. This is often wedge shaped and may extend into the inferior quadrant. The visual acuity and pupillary responses are both normal. The lesions involving the optic radiation are either due to vascular occlusion, tumours (intrinsic or metastatic), or abscesses. Temporal lobectomy for the treatment of epilepsy does not involve fibres of the optic radiation so long as the resection is confined to the anterior 4 to 5 cm only (Tecoma *et al.* 1993). Some degree of incongruity may be found in the resulting field defects.

Although lesions of the dorsal optic radiation in the parietal lobe may result in a homonymous hemianopia, primarily affecting the lower fields, large lesions usually result in a complete homonymous hemianopia with macular splitting. In clinical practise there is considerable variation in the field defects, particularly the degree of congruity. Damage to the parietal or occipitoparietal cortex may result in the phenomenon in the contralateral visual field called unilateral visual inattention or visual extinction (Bender and Furlow 1945). In visual extinction when a test object is presented in this field it is perceived normally but, when an identical object is similarly presented equidistant from the fixation point in the ipsilateral visual field, the stimulus in the field contralateral to the parietal lobe lesion disappears. Visual inattention may be found during the recovery phase of a homonymous hemianopia, but it should be noted that it may also be found rarely in lesions of the frontal lobe, thalamus and mesencephalon (Nachev and Husain 2006).

12.7.4 Occipital lobe

On reaching the occipital lobe there is a high degree of order in the fibres of the optic radiation and lesions, usually due to infarction, trauma, or tumour, produce congruent field defects which are homonymous. The only features of the field defect which help localize the lesion to the occipital lobe, rather than the anterior optic radiation, is the presence of macular sparing, homonymous hemianopic central scotomas and sparing or involvement of the temporal crescent in a homonymous hemianopia.

In macular sparing there is preservation of the visual field within a region of 1 to 2° up to 10° around the fixation point in the hemianopic field. In the more usual situation the hemianopic field is split along the vertical meridian through the fixation point 'macular splitting'. Although it has been argued that macular sparing is a result of poor fixation during visual field testing, this would only account for about 1 to 2° of sparing (Bishoff *et al.* 1995). Despite the continuing controversy concerning the cause of macular sparing, there appears to be two main anatomical factors which may explain this phenomenon. Firstly there is evidence that there is a vertically orientated median strip centred on the fovea in which retinal ganglion cells project either ipsilaterally or contralaterally (Fukuda *et al.* 1989). The macular, therefore, is bilaterally represented but since this strip is at most responsible for 2° of the central field this is insufficient to explain many cases of macular sparing. The second more probable explanation is the rich anastomotic network between terminal branches of the middle cerebral artery and the posterior cerebral artery, which supply the area of the striate cortex containing the macular representation in the occipital pole (Sugishita *et al.* 1993).

Lesions at the pole of the occipital lobe result in small homonymous central scotomas, which may lead the patient to present with reading difficulties and the scotoma may be missed if only the peripheral field is examined to confrontation. The central 10° occupies approximately 60 per cent of the primary visual cortex, the so-called magnification factor (Horton and Hoyt 1991). More anterior lesions of the occipital lobe involving the more anterior area of the calcarine fissure, which contains the representation of the unpaired peripheral nasal retina, results in a monocular defect in the peripheral temporal field, called the 'temporal crescent', between 60 and 90° from the fixation point. However, it should be remembered that the most common cause for such peripheral visual field defects is a retinal lesion rather than an intracranial one. The converse of this defect may be found in which there is sparing of the temporal crescent in a homonymous hemianopia. This usually occurs with a vascular lesion affecting the more posterior striate cortex (Benton *et al.* 1980).

Bilateral lesions of the occipital lobes may result in varying degrees of homonymous hemianopia, ranging from small bilateral central homonymous scotomas to complete blindness. The extent of the abnormality may vary between the two halves, being partial or complete, hemianopic or quadrantic. Sometimes restricted bilateral lesions of the occipital lobes may result in small bilateral homonymous central scotomas (or 'ring' scotomas, if there is some degree of macular sparing in addition). Altitudinal field defects usually occur as a result of trauma (rarely tumours or vascular events) involving both upper and lower occipital poles (Holmes 1918). Inferior altitudinal defects are mainly found, since patients with inferior occipital lobe injury (i.e. a superior altitudinal defect) often die as a result of haemorrhage from lacerated dural sinuses.

Cortical blindness

Cortical blindness usually indicates selective involvement of the occipital visual cortex, but may be difficult to distinguish from bilateral homonymous hemianopia due to bilateral lesions in the optic radiation. Marquis (1934) described the essential features as (1) complete loss of all visual sensation, (2) loss of reflex lid closure to threat, (3) normal pupillary light reactions in contradistinction to blindness due to pregeniculate lesions, (4) normal retina and full extraocular eye movements. These findings often lead to a misdiagnosis of psychogenic blindness. Although the commonest aetiology is hypoxia of the striate cortex, cases have been reported in a number of different conditions (Table 12.4).

Patients with cortical blindness may sometimes be unaware of their visual defect, or anosognosia, and vigorously deny it, whilst at the same time fabricating an imaginary visual environment, known as Anton's syndrome. This may occur with lesions elsewhere causing total blindness. There is no satisfactory explanation for this syndrome and the various hypotheses proposed to explain the syndrome are discussed by Lessell (1975). Various suggestions include an alteration in emotional reactivity, 'psychiatric' denial as an accentuation of a common response to illness, a memory disorder for example in Korsokoff's syndrome, and associated lesions elsewhere in areas of the brain responsible for the recognition and interpretation of visual images.

Transient cortical blindness may occur in a number of different situations, for instance after cardiac or respiratory arrest, head trauma, or meningitis. Greenblatt (1973) has detailed three clinical patterns: (1) in children up to 8 years transient cortical blindness lasting 1–6 hours, and is associated with irritability, somnolence, and vomiting and has a good prognosis; (2) in the age range 8 to 20 years there may be delay before the onset of blindness which recovers over hours with full return of function; (3) in adults transient cortical blindness, with onset often immediately following head trauma, it may last many days and result in some degree of

Table 12.4 Causes of cortical blindness

Vascular
Bilateral posterior cerebral artery infarction
Cerebral angiography
Cardiac surgery
Trauma
Schilders disease
Carbon monoxide poisoning
Meningitis
Air embolism
Neoplasm
Creutzfeld–Jacob disease
Alzheimer's disease (posterior variant)
Tentorial herniation
Cardiac arrest
Progressive multifocal leucoencephalopathy
Systemic lupus erythematosis
Hypertensive encephalopathy
Dialysis disequilibrium

permanent visual loss. Rare cases of transient cortical blindness have also been reported following seizures.

12.8 Disorders of higher visual processing

12.8.1 Residual visual function in hemianopias

In his classic work Holmes (1918) showed that striate cortex damage results in a complete hemianopia. However, incomplete damage to the occipital lobe may result in retention of some aspects of visual perception, the most commonly observed being the ability to perceive small moving objects in the homonymous hemianopia (Riddoch 1917; Zeki and ffytche 1998). Riddoch's phenomenon may be the first evidence of recovery of a homonymous hemianopia. This is then usually followed by perception of static targets and finally colour perception returns. Unfortunately the Riddoch phenomenon is not only found in occipital lesions, but has been reported in patients with lesions in the anterior visual pathways (Safran and Glaser 1980).

The retention of the ability to localize objects in space and limited pattern discrimination in monkeys in whom both striate cortices had been removed, led to interest in the possible visual functions in the hemianopic field of human patients (Weiskrantz 1986). Since these patients are unaware of any residual visual capacity and appear blind by standard clinical perimetric methods this visual capacity has been termed 'blindsight' (Weiskrantz *et al.* 1974; Weiskrantz 1986). Using forced-choice discrimination methods such patients have revealed their ability to locate stimuli both by saccadic eye movements and by pointing (Cowey and Stoerig 1991). The extent of the residual visual capacity is varied amongst the patients so far reported, and as yet there is poor correlation with the precise location of lesions in the occipital lobe (Blythe *et al.* 1986). The most likely explanation for blindsight is the presence of direct projections from the lateral geniculate nucleus to extrastriate visual areas, therefore bypassing the damaged striate cortex.

12.8.2 Functional visual loss in prestriate lesions

There is increasing evidence from electrophysiological studies in primates that once initial processing of visual information has occurred in the striate cortex, segregation of the processing of different properties of the visual stimulus occurs in the extrastriate cortex. This is termed functional cortical specialization (Zeki and Shipp 1988; Zeki 1993). Here there are a number of individual representations of the contralateral hemifield, each containing neurons with a particular response characteristic. For example, some areas contain neurons which are selective for colour, visual area 4 or V4, and another for motion, V5, the middle temporal gyrus, or MT. There appears, therefore, to be parallel processing of different aspects of visual information in these various cortical areas before an organized synthesis of the visual scene can be generated. Specific lesions in one or other of these areas might be expected to give rise to an appropriate specific loss of one such visual modality. In this section such specific losses are described for colour (achromatopsia), movement (akinetopsia) and faces (prosopagnosia). The higher cortical visual areas are divided anatomically and functionally into dorsal and ventral streams. The ventral stream passes forward from the occipital lobe into the temporal lobe and is concerned with object recognition—the 'what' stream. The dorsal stream passes into the parietal lobe and is concerned with spatial orientation—the 'where' stream (Fig. 12.13).

Colour

Acquired disorders of colour vision due to lesions of the central nervous system are of two types. In one the colour sense is normal but the naming and recognition of colour is impaired. This can occur as part of an aphasia such as Wernicke's or anomic, in the

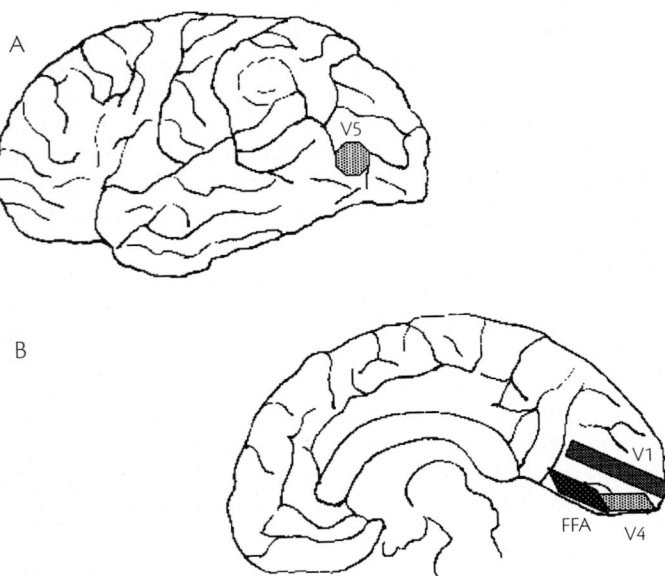

Fig. 12.13 An outline drawing of the lateral (a) and the medial (b) surfaces of the human brain showing some of the specialized visual areas in the extrastriate region around the primary visual area, striate cortex (V1); V4—the colour area; V5—the motion area; FFA—the facial fusiform area.

syndrome of alexia without agraphia, or as one feature of visual agnosia (Section 12.8.3). In the second type there is an inability to see colours, known as dyschromatopsia or achromatopsia (Zeki 1990).

Patients with lesions in the region of the lingual and fusiform gyri, which lies in the anterior inferior region of the occipital lobe and is considered to be the human homologue of the monkey visual area V4, complain that they cannot see colours and that everything looks grey or appears in varying shades of black and white (Meadows 1974a). They are unable to identify the figures on pseudo-isochromatic test plates, although able to correctly name the colours of brightly coloured objects. In addition, they are unable to perform normally on the Farnsell–Munsworth 100-hue test. Patients with cerebral dyschromatopsia may or may not realize that their colour sense is impaired. Other visual functions such as visual acuity, object recognition, and depth perception are all normal, but there is often an associated visual field defect, usually a bilateral superior homonymous quadrantanopia indicative of inferior occipital lobe damage, and it may sometimes be associated with prosopagnosia.

The commonest cause is infarction in the distribution of the posterior temporal or common temporal arteries, branches of the posterior cerebral artery. However, any of the conditions mentioned in Table 12.4 may give rise to prosopagnosia by causing a more focal lesion.

Movement

A case has been reported of a woman who exhibited a selective deficit of movement perception, akinetopsia (Zihl *et al.* 1983). She had no impression of movement in depth and could only discriminate between a stationary and a moving target in the periphery of her otherwise intact visual fields. Akinetopsics describe the deficit as an inability to appreciate the smooth movement of a moving target, with reports that when she poured tea 'the fluid appeared to be frozen like a glacier', and on crossing the road 'when I'm looking at the car first, it seems far away. But then, when I want to cross the road, suddenly the car is very near'. In other words, the moving object appears to jump rather than move smoothly, as experienced by normal individuals when viewing moving objects under stroboscopic illumination, for example at a discotheque. Hemiakinetopsia has also been described, and it is always contralateral to the lesioned side; however, it is not commonly detected because of the masking effect of the concurrent visual field loss. The patient had bilateral lesions involving the lateral occipito-parieto-temporal junction, which positron emission tomography has revealed is specifically activated during motion perception and, therefore, appears to be the human homologue of the monkey visual area V5 (Zeki 1991).

Disorders of stereopsis

The ability to perceive three dimensional images, stereopsis, is one of the most fascinating visual phenomena, but the least well understood. This visual function involves processing disparities between the two dimensional images viewed and perceived from a slightly different angle by the left and right eye. The brain uses these retinal disparities to extrapolate the distance of the objects to guide our vergence eye movements and to provide our perception of stereoscopic depth. This process develops innately but may be impaired by ophthalmological disturbances before the adult level of stereoacuity is achieved. As a result, approximately 5 to 10 percent of the population is unable to appreciate stereopsis, primarily as a result of uncorrected childhood strabismus or amblyopia, presumably due to disruption of the normal development of binocular cortical neurons.

The physiology of stereopsis is rather complex as numerous cortical areas are involved. The current evidence suggests that both the dorsal and ventral streams are involved in stereopsis. The depth information from the dorsal stream is probably used to accurately localize objects in visual space and guide vergence eye movements, and the information from the ventral stream establishes a richer perceptual representation, including detailed 3D representations.

Some patients with astereopsis or stereoblindness complain of difficulty perceiving distance, perspective, depth, or thickness; however, the majority are unaware of any deficits. This could be due to anosognosia, but the ability to use other cues such as motion and linear perspective to compensate for the lack of stereoacuity could explain this observation. Impaired stereopsis in an acquired form has been reported in Alzheimer's disease, tumours, and strokes, following surgical excisions for the treatment of intractable epilepsy and traumatic head injuries.

12.8.3 Visual agnosia

Visual agnosia is characterized by a difficulty to recognize or identify familiar objects using visual information, when this difficulty cannot be explained by other cognitive impairments such as a disorder of intelligence, attention, and language or by impairments of peripheral visual processing affecting acuity, brightness discrimination, depth perception, visual field, and colour vision. Generally a diagnosis of visual object agnosia should be made when a patient misnames an object if they are unable to describe or mime the use of the object presented visually, but are then able to do so correctly when they can feel or hear the object. Teuber (1965) described visual agnosia as a 'percept stripped of its meaning'.

Visual agnosia has been classified in a number of different ways. One classification depends on the specific category of visual material which cannot be recognized. This classification recognizes disturbances in recognition of objects 'object agnosia', faces 'prosopagnosia', and colour 'colour agnosia', which may occur in isolation or in various combinations. Lissauer's (1890) classic dichotomous classification of visual agnosia is, however, still relevant today. When a patient is able to copy and match-to-sample objects that he fails to name or recognize visually, his agnosia is termed associative; but if he fails on all these tasks or demonstrates perceptual abnormalities his agnosia is termed apperceptive (Tranel and Damasio1996).

Apperceptive visual agnosia

Well-documented cases of apperceptive visual agnosia are rare (Warrington and James 1988). They show an inability to copy or match-to-sample drawings which they cannot recognize, and recognition and matching of all other stimuli which demand shape or pattern perception is also affected. Most cases have been associated with diffuse posterior cerebral damage due to cardiac arrest, carbon monoxide or mercury poisoning or bilateral posterior cerebral artery infarction. The typical white matter lesions suggest that it results from some form of disconnection possibly of local intralaminar connections, rather than neuronal loss. Less severe apperceptive disorders are associated with unilateral and generally right cerebral damage.

Associative visual agnosia

Unlike apperceptive agnosia there is no doubting the existence of associative agnosia as a definite neuropsychological syndrome since a number of well-documented cases have been reported (Humphreys and Riddoch 1993).

These cases exhibit the ability to copy and/or match-to-sample items which they fail to identify visually, without any evidence of primary sensory or sensory motor disturbance. The syndrome is commonly associated with colour agnosia, prosopagnosia and alexia in various combinations. This may reflect task and processing similarities between recognition, for instance of faces and objects, resulting from defects of both. Alternatively, lesions giving rise to object agnosia may involve adjacent areas specific for colour or face processing.

Patients with associative visual agnosia show an increasing difficulty in identifying an object when presented as an object itself, or as a picture or a line drawing. Auditory and tactile recognition is usually intact. There is no uniformity about the field defects which are often present. A further commonly found feature is the strong tendency these patients have to perseverate either previously viewed objects or, more commonly, their verbal response to them.

A number of hypotheses to explain visual agnosia have been proposed. Geschwind (1965) suggested that agnosia was not a defect of a unitary process of recognition, but rather a special form of a modality-specific naming defect. Using a disconnection explanation similar to that given for dyslexia without dysgraphia and colour agnosia, he suggested that the confabulatory verbal responses are due to a pathological disconnection of the intact speech area from the intact sensory area. Radcliffe and Newcombe (1982) have argued, however, that since object recognition, as opposed to naming is mediated by the semantic system, disconnection must be a visual-semantic one and not merely visual-verbal. However, patients with surgically sectioned cerebral commisures are able to extract meaning from words and pictures when visual input is restricted to the right hemisphere, making it unlikely that this disconnection of an intact right hemisphere would be sufficient to cause agnosia.

A second hypothesis, proposed by Warrington (1975), suggests that the disorder is due to a disturbance of access to visual semantic information itself, since in her patient 'all links of associations were lost, not just verbal' and hence a visuo-verbal disconnection was not a sufficient explanation. She regarded preservation of the ability to make same/different judgements with respect to photographs of objects taken from different angles as evidence of preserved 'perceptual classification'. However, other authors have suggested a defect of visual categorization in their patients (Albert et al. 1975). It has to be concluded that both the anatomical basis and clinical criteria for associative visual agnosia are still uncertain, but excellent reviews are by Farah (1990) and Tranel and Damasio (1996).

In some patients visual object agnosia may be category specific and cases have been reported in which specific deficits in the identification of fruit and vegetables, or animals have been observed. However, the most common dissociation reported in these patients is an impairment in the recognition of natural living objects relative to man-made non-living objects. Posterior cortical atrophy usually gives rise to various cortical disorders, including visual object agnosia.

The visual association cortex may store the neuronal templates which are required to match a visual stimulus with visual memory, and functional imaging studies, including functional MRI and positron emission tomography, have shown activation in the ventral stream during visual object recognition tasks.

12.8.4 Prosopagnosia

Prosopagnosia is a specific inability to recognize familiar faces despite a normal ability to recognize everyday objects and is therefore, different from visual agnosia (Meadows 1974b). Although facial recognition is a visual pattern discrimination of great complexity, patients with prosopagnosia have no difficulty in discriminating unfamiliar faces and matching faces correctly. Indeed there appears to be no disturbance of visual perception, patients being able to accurately recognize many stimuli which are visually more complex than human faces. It appears that the disorder is not specific to faces but to complex non-verbal visual stimuli that belong to a group where individual members are visually similar and yet individually different. For example, prosopagnosics cannot recognize their own car and do not recognize different makes of car; however, they can distinguish different classes of vehicle, e.g. ambulance or fire engine. Similarly a case has been reported of a farmer suddenly becoming unable to distinguish individual animals within his herd and of a bird watcher developing an inability to recognize different species of birds.

Prosopagnosics appear, therefore, to be unable to match a current visual stimulus within a class such as faces, with the memory traces of other members of this specific class which have been built up from past experience (De Renzi 1997). Pathophysiologically, there is a disorder of visually triggered contextual memory. Under normal circumstances after multiple exposures to a stimulus, a template of the stimulus is stored, perhaps at several levels, but in prosopagnosics there is a defect in activating this template. Recent brain imaging studies have suggested that these processes involve the inferior temporal lobe and also the ventrolateral frontal cortex (Haxby *et al.* 1996). Certainly in monkeys electrophysiological recordings from superior temporal sulcus in monkeys have identified neurones with a response specific for faces, which may well be involved in the facial recognition process (Perrett *et al.* 1984).

Most cases of prosopagnosia are due to infarction, head injury, or hypoxia resulting in bilateral lesions in the ventromedial aspects of the occipitotemporal region (Damasio *et al.* 1982).

12.8.5 Disorder of visuospatial function

Balint's syndrome, a triad of visual defects, was first described in 1890 by Balint, an Austrian-Hungarian neurologist, in a patient with bilateral occipito-parietal lesions due to a stroke. The triad comprises:

- simultanagnosia: an inability to perceive the components of a visual scene as a whole, with variable perception of isolated components.

- ocular apraxia: inability to voluntarily direct gaze towards a new object of interest.

- optic ataxia: impaired target pointing under visual guidance despite normal limb function and joint positional sense.

Most of the subsequent reported cases of Balint's syndrome, either complete or incomplete, have been caused by hypotensive stroke associated with diffuse atherosclerosis or a cardiac bypass operation, multiple emboli or venous infarction, tumour (multiple metastases or a butterfly glioma), trauma, prion disease, HIV infection, corticobasal ganglionic degeneration adrenoleukodystrophy, and Alzheimer's disease (Rafal 1997).

The patients described have had a rather widespread injury to the brain that was almost always bilateral. Recent evidence indicates that it is rather difficult to tie the triad to one single location in the brain. In addition, there have been cases of incomplete triads, for instance, optic ataxia alone without simultanagnosia or ocular apraxia, which certainly supports the idea of different locations responsible for the different components. In general, lesions at the parieto-occipital junction, often bilateral, are frequently found in patients with these visuospatial disorders. Patients with Balint's syndrome are often severely disabled and appear almost blind, requiring assistance to avoid bumping into things. They are not able to direct their gaze towards the new stimuli and do not blink to threat (Rafal 1997).

Although simultanagnosia, contains the term 'agnosia', recent evidence suggests that it has nothing to do with 'agnosia'. In fact, it is a disorder of visual perception where the patient fails to appreciate all the visible items in a complex visual scene and has been termed 'piecemeal vision'. Patients with simultanagnosia appear to have defects in spatial integration and sustained attention, but have relatively intact visual fields on formal testing. They see only with macular vision, which restricts their overall perception of a visual scene and they show unpredictable shifts of focus from region to region. It may occur in Alzheimer's disease and is a common finding in posterior cerebral atrophy.

Ocular apraxia, first described as psychic paralysis of gaze by Balint, is not well described. The most significant feature is the inability to generate voluntary saccades towards a particular object of interest, although reflexive saccades, that is saccades towards a novel target, are intact. Gaze is therefore relatively random and targets are found by chance. Ocular apraxia is considered to be a disorder of visuospatial integration and generation of volitional saccades rather than a 'true' apraxia. Isolated oculomotor apraxia is rare but can be seen in bilateral parietal lobe disorders and Gaucher's disease.

Optic ataxia, also called visuomotor ataxia or defective visual localization, describes a difficulty in reaching for an object by hand under visual guidance. However, patients with optic ataxia can reach to touch their own body parts relatively accurately with their eyes closed, which suggests that this disorder is not a primarily motoric dysfunction of the limbs, nor a feature of cerebellar dysfunction. Accuracies of pointing or grasping in patients with optic ataxia reduce as the object is moved further away from the central vision. Optic ataxia may result from lesions of the dorsal visual association areas (intraparietal sulcus and superior parietal lobe) or from a disconnection of the projections from the visuomotor centres in the parieto-occipital lobes to the frontal lobes, where reaching is programmed before the movement is initiated (Milner *et al.* 1999; Rizzo and Vecera 2002).

12.8.6 Visual illusions

Visual illusions occur when the visually perceived target appears altered in size, shape, colour, position in space, and in number of images (Kölmel 1993). The illusory type of defects may occur in the entire field of vision, or may affect only the object or the background. Illusions of the spatial aspect can be divided into three categories; micropsia in which objects are smaller than reality, macropsia in which objects are perceived larger than reality and metamorphopsia when objects are perceived as a distorted image. Among these, micropsia is probably the most common form of dysmetropsia.

Dysmetropsia usually occurs as a result of retinal disease due to distortion of the relative distance between rods and cones. However, these distortions can also occur as a result of cortical dysfunction, for example in the aura of migraine or epilepsy, chiasmic compression or focal cerebral lesions. Visual allesthesia is a transfer of visual images from one half field to the other (Jacobs 1980). There may also rarely be an inversion of the visual scene or tilting of the environment in patients with the lateral medullary or Wallenberg's syndrome (Hornstein 1974). This relates to a disturbance of the vestibular inputs required for normal visual perception.

12.8.7 Visual hallucinations

Visual hallucinations are visual percepts without real external stimuli. The object is perceived in the absence of an actual object(s), and depending on the degree of alertness and pathology, the observer may or may not be able to appreciate that the seen object does not exist. The quality of the perceived image can range from a simple flash of light or phosphene to a well-formed object, animal, or a person. Movement of the images is often perceived. Patients are often convinced that hallucinations are, in fact, genuine, and under these circumstances a history from relatives or carers refuting this belief is invaluable. Visual hallucinations occur under many circumstances, including confusional states secondary to metabolic derangement such as hypoglycaemia or electrolyte imbalance; adverse drug reaction or drug/alcohol withdrawal, anoxia; migraine; infection; neurodegenerative disorders such as Alzheimer's disease, Parkinson's disease or Huntington's disease; and schizophrenia; in addition to those related to focal neurological disease. Those in the latter category may be unformed, consisting of flashes of light, either coloured or white, lines, simple shapes or they may be complex highly organized hallucinations of people or objects (Kölmel 1993).

Although it is considered that simple visual hallucinations signify involvement of the occipital lobe and complex ones involvement of the temporal lobe this is not always the case, for example, complex hallucinations have been observed in patients with hemianopias due to occipital lobe lesions (Kölmel 1985).

It has long been considered that visual hallucinations could result from irritative foci analogous to epileptic discharges, and certainly electrical stimulation of the occipital and temporal lobes (Penfield and Perot 1963) support this suggestion. Other mechanisms in some cases may be a release phenomenon to be found in the context of sensory deprivation, the Charles Bonnet syndrome. These hallucinations are considered to be due to a release of the visual system, secondary to a reduction of visual information, which allows the emergence into consciousness of endogenous visual activity. Any type of visual loss, including ocular and cerebral pathology, can cause such release hallucinations. Cataracts, senile macular degeneration and diabetic retinopathy are most frequent among the ocular causes and cerebrovascular disease is the commonest cause, secondary to a cerebral insult. They have also been observed to occur after brain resection, ocular enucleation, brimonidine eye drops, and multiple sclerosis. This is the explanation usually given to hallucinations occurring in elderly patients who have impaired vision (Teunisse et al. 1995).

The onset of visual hallucinations can occur immediately after the onset of visual loss, but can also be delayed for as long as 10 years. The duration and frequency of each episode of hallucination can also be extremely variable, ranging from a few episodes lasting 2 to 3 seconds in duration to being continuous for several hours. Although occurring almost exclusively in adults, especially the elderly, they have been reported in young children (Schreir 1999).

The term 'peduncular hallucinations' was first described by Llermitte (1922) to describe visual hallucinations which appear animated, slow moving, cartoon-like and are usually frightening for the patient. This type of hallucination is usually associated with inversion of the sleep–wake cycle, with diurnal somnolescence and nocturnal insomnia, and occurs with lesions in the upper brainstem (McKee et al. 1990).

12.8.8 Palinopsia

Palinopsia is a rare disorder in which there is a persistence or perseveration, or a recurrence of visual images after the exciting stimulus has been removed (Bender et al. 1968). Although in the literature both perseveration and recurrence of visual images have been lumped together under the term palinopsia it has been argued that they may be distinct (Blythe et al. 1986). The images are usually of a real object which has recently been visualized, and is then superimposed on certain parts of the current visual scene. For instance, Meadows and Munro (1977) reported a palinopsic patient who after looking at someone dressed up as Santa Claus saw an image of Santa Claus's face superimposed on the face of other people at the party, thus the location of the persistent image can be contextually specific. Patients with palinopsia may also describe the visual persistence as they turn their gaze away from the object, giving rise to a smeared image. Two types of palinopsia have been reported; an immediate and a delayed type. In the immediate type, the image persists after the disappearance of the actual object or scene and usually persists for several minutes. In the delayed type the perseverated image appears a few seconds after the object has disappeared or the patients gaze has been redirected away from it. Some patients may have both types of palinopsia. The persistent image can occupy any location in the visual field and it usually moves as the eyes move, resembling a retinal afterimage. In some cases, the persistent image multiplies across the entire visual field.

It most commonly occurs during the progressive evolution or resolution of a homonymous hemianopic field defect, usually resulting from a posterior cerebral hemisphere lesion due to neoplasia (Bender et al. 1968), vascular disease or trauma.

Bender et al. (1968) suggested four possible mechanisms for this phenomenon: sensory seizures, psychogenic elaboration or fantasies, visual after-images, or hallucinations. Although some patients with palinopsia have had seizures, most have no evidence of seizure activity on the electroencephalogram and the palinopsia does not respond to treatment with anticonvulsants. Patients with palinopsia show no signs of psychopathology and, therefore, it is unlikely that they are due to psychogenic collaborations. Similarly, there is no evidence that visual after-effects in patients with palinopsia are enhanced and such an explanation would not explain the late recurrence of the image, by some several minutes, which occurs in some patients. However, palinopsia may be a type of release phenomenon as described for visual hallucinations. In favour of this possibility is a fact that formed release hallucinations can occur in patients with palinopsia and that in both conditions there is evidence of an interruption of cortical visual processing.

Specific types of palinoptic phenomena are illusory visual spread and polyopia. In illusory visual spread (Critchley 1951) there is an extension of visual perception over an area greater than that excited by the object presented to the observer. In the time domain visual perseveration of moving objects has also been reported and one patient experienced accelerated movement of a perseverated image.

In instances of usually right-sided occipital lesions patients may experience monocular diplopia or more commonly polyopia, which is the seeing of multiple images, which persist whichever eye is closed. Rare cases of cerebral induced monocular diplopia emphasize the importance of ensuring that this phenomena is not present in patients complaining of diplopia. Other causes for monocular diplopia include ocular causes such as corneal irregularities, iris lesions and retinal detachment.

Certain cases of polyopia may be due to epileptic phenomena (Bender and Sobein 1963) but Bender (1945) in a description of four cases tried to explain the phenomenon as a result of impaired fixation.

References

Albert ML, Reches A, Silverberg R (1975). Associative visual agnosia without alexia. *Neurology*, **25**, 322–6.

Apple DJ, Rabb MF, Walsh PM (1982). Congenital anomalies of the optic disc. *Surv Ophthalmol*, **27**, 3–41.

Barrett TG, Bundley SE, Macleod AF (1995). Neurodegeneration and Diabetes: UK nationwide study of Wolfram (DIDMOAD) syndrome. *Lancet*, **346**, 1458–62.

Beck RW, ONTT Study Group (1992). A randomised, controlled trial of corticosteriods in the treatment of acute optic neuritis. *New Engl J Med*, **326**, 581–8.

Bell RA, Thompson HS (1978). Relative afferent pupillary defect in optic tract hemianopias. *Am J Ophthalmol*, **85**, 538–40.

Bender MB, Feldman M, Sobein AJ (1968). Palinopsia. *Brain*, **91**, 321–38.

Bender MB, Furlow LT (1945). Phenomenon of visual extinction in homonymous fields and psychologic principles involved. *Arch Neurol Psychiatr*, **53**, 29–33.

Bender MB, Sobein AJ (1963). Polyopia and palinopia in homonymous fields of vision. *Trans Am Neurol Assoc*, **88**, 56–7.

Benton S, Levy I, Swash M (1980). Vision in the temporal crescent in occipital infarction. *Brain*, **103**, 83–95.

Bishoff P, Lang J, Huber A (1995). Macular sparing as a perimetric artifact. *Am J Ophthalmol*, **119**, 72–80.

Blythe IM, Bromley, JM, Kennard C et al. (1986). A study of systemic visual perseveration involving central mechanisms. *Brain*, **109**, 661–75.

Boghen DR, Glaser JS (1975). Ischaemic optic neuropathy: the clinical profile and natural history. *Brain*, **98**, 689–708.

Brodsky MC (1991). Septo-optic dysplasia: a reappraisal. *Semin Ophthalmol*, **6**, 227–32.

Brodsky MC, Glaser CM (1993). Optic nerve hypoplasia: clinical significance of associated central nervous system abnormalities on magnetic resonance imaging. *Arch Ophthalmol*, **111**, 66–74.

Burde RM (1993). Optic disc risk factors for nonarteritic anterior ischaemic optic neuropathy. *Am J Ophthalmol*, **115**, 759–63.

Chung SM, Selhorst JB (1992). Cancer associated retinopathy. In Katz B, ed. *Neuro-Ophthlmology in Systemic Disease. Ophthalmology Clinics of North America*, Vol. 5, pp. 587–96. WB Saunders, Philadelphia.

Cowey A, Stoerig P (1991). The neurobiology of blindsight. *Trends Neurosci*, **14**, 140–5.

Critchley M (1951). Types of visual perseveration, palinopsia and illusory visual spread. *Brain*, **74**, 267–99.

Damasio AR, Damasio H, Van Hoesen GW (1982). Prosopagnosia: Anatomic basis and behavioural mechanisms. *Neurology*, **32**, 331–41.

De Renzi E (1997). Prosopagnosia. In Finberg TE, Farah MJ, eds. *Behavioural Neurology and Neuropsychology*, pp. 245–55. McGraw-Hill, New York.

Dutton JJ (1992). Optic nerve sheath meningiomas. *Surv Ophthalmol*, **37**, 167–83.

Dutton JJ (1994). Gliomas of the anterior visual pathway. *Surv Ophthalmol*, **38**, 427–52.

Farah MJ (1990). *Visual Agnosia: Disorders of Optic Recognition and What They tell us about Normal Vision*. MIT Press. Cambridge, Mass.

Frisén L, Holmegaard L, Rosencrantz M (1978). Sectorial optic atrophy and homonymous horizontal sectoranopia: a lateral choroidal artery syndrome? *J Neurol Neurosurg Psychiatry*, **41**, 374–80.

Fukuda Y, Sawai H, Watanbe M et al. (1989). Nasotemporal overlap of crossed and uncrossed retinal ganglion cell projection in the Japanese monkey (*Macaca fuscata*). *J Neurosci*, **9**, 2353–73.

Gass JD (1993). Acute zonal occult outer retinopathy. Donders Lecrure: The Netherlands Ophtalmological Society, Maastricht, Holland June 19 192. *J Clin Neuroophthalmol*, **13**, 79–97.

Geschwind N (1965). Disconnected syndromes in animal and man. *Brain*, **88**, 237–94, 585–644.

Glantz MJ, Burger PC, Friedman AH et al. (1994). Treatment of radiation-induced injury with heparin and warfarin. *Neurology*, **44**, 2020–7.

Grant WM, Schuman JS (1993). *Toxicology of the Eye*, 4th edition. Charles C Thomas, Springfield IL.

Greenblatt SH (1973). Post traumatic transient cerebral blindness: association with migraine and seizure diatheses. *J Am Medical Assoc*, **255**, 1074–6.

Gunderson CH, Hoyt WF (1971). Geniculate hemianopia: Incongruous homonymous field defects in two patients with partial lesions of the lateral geniculate nucleus. *J Neurol Neurosurg Psychiatry*, **34**, 1–6.

Hamed L, Glaser GS, Gass JDM et al. (1989). Protracted enlargement of the blind-spot in multiple evanescent white dot syndrome occurring in the same patients. *Arch Ophthalmol*, **107**, 194–8.

Haxby JV, Ungerleider LG, Horwitz B et al. (1996). Face encoding and recognition in the human brain. *Nat Acad Sci USA*, **93**, 922–7.

Hayreh SS (1977). Optic disc edema in raised intracranial pressure -V. Pathogenesis. *Arch Ophthalmol*, **95**, 1553–65.

Hayreh SS, Joos KM, Podhajsky PA et al. (1994). Systemic diseases associated with nonarteritic anterior ischaemic optic neuropathy. *Am J Ophthalmol*, **118**, 766–80.

Hollenhorst RW, Young BR (1973). Ocular manifestations produced by adenomas of the pituitary gland: Analysis of 1000 cases. In Kohler PO, Ross GT, eds. *Diagnosis and Treatment of Pituitary Tumours*, pp. 53–64. Elsevier, New York.

Holmes G (1918). Disturbances of vision by cerebral lesions. *Br J Ophthalmol*, **2**, 353–84.

Holt IJ, Harding AE, Petty RKH et al. (1990). A new mitochondrial disease associated with mitochondrial DNA heteroplasmy. *Am J Hum Genet*, **46**, 428–33.

Hornstein G (1974). Wallenberg's syndrome. Part I: General symptomatology, with special reference to visual disturbances and imbalance. *Acta Neurol Scand*, **50**, 434–46.

Horton JC, Hoyt WF (1991). The representation of the visual field in human striate cortex: a revision of the classic Holme's map. *Arch Ophthalmol*, **109**, 816–24.

Hoyt WF, Bagdassarian SA (1969). Optic glioma of childhood: natural history and rationale for conservative management. *Br J Ophthalmol*, **53**, 793–8.

Humphreys GW, Riddoch MJ (1993). Object agnosias. In Kennard C, ed. *Visual Perceptual Visual Defects*, pp. 339–59. Baillière Tindell, London.

Ischemic Optic Neuropathy Decompression Trial Research Group (1995). Optic nerve decompression surgery for nonarteritic anterior ischemic optic neuropathy (NAION) is not effective and may be harmful. *JAMA*, **273**, 625–32.

Jabs DA (1995). Ocular manifesations of HIV infection. *Trans Am Ophthalmol Soc*, **93**, 623–83.

Jacobs L (1980). Visual allesthesia. *Neurology*, **30**, 1059–63.

Johnston RL, Burdon MA, Spalton DJ *et al.* (1997). Dominant optic atrophy, Kjer type. Linkage analysis and clinical features in a large British pedigree. *Arch Ophthalmol*, **117**, 100–3.

Keltner JL, Johnson CA, Spurr JO *et al.* (1993). Baseline visual field profile of optic neuritis: the experience of the Optic Neuritis Treatment Trial. *Arch Ophthalmol*, **111**, 231–4.

Kennerdell JS, Maroon JC, Malton M *et al.* (1988). The management of optic nerve sheath meningiomas. *Am J Ophthalmol*, **106**, 450–7.

Kirkham TH (1972). The ocular symtomatology of pituitary tumours. *Proc Roy Soc Med*, **65**, 517–8.

Kivlin JD, Sanborn GE, Myers GG (1985). The cherry-red spot in Tay-Sachs and other storage diseases. *Ann Neurol*, **17**, 356–60.

Kjer B (1959). Infantile optic atrophy with dominant mode of inheritance. A clinical and genetic study of 19 Danish families. *Acta Ophthalmol*, **37**(Suppl 54), 1–46.

Kjer B, Eibergh, Kjer P *et al.* (1996). Dominant optic atrophy mapped to chromosome 3Q region. II Clinical and epidemiological aspects. *Acta Ophthalmol Scand*, **74**, 3–7.

Klug GL (1982). Gliomas of the optic nerve and chiasm in children. *Neuro-Ophthalmol*, **2**, 217–23.

Kölmel HW (1985). Complex visual hallucinations in the hemianopic field. *J Neurol Neurosurg Psychiatry*, **48**, 29–38.

Kölmel HW (1993). Visual illusions and hallucinations. In Kennard C, ed. *Visual Perceptual Defects*, pp. 243–64. Baillière Tindell, London.

Krauss HR, Heckenlively JR (1982). Visual field changes in cone-rod degenerations. *Arch Ophthalmol*, **100**, 1784–90.

Kupersmith MJ, Berenstein A, Choi IS *et al.* (1988). Management of non-traumatic vascular shunts involving the cavernous sinus. *Ophthalmology*, **95**, 121–30.

Lepore FE (1991). The origin of pain in optic neuritis. Determinants of pain in 101 eyes with optic neuritis. *Arch Neurol*, **48**, 748–9.

Lessell S (1975). Higher disorders of visual function: negative phenomena. In Glaser JS, Smith JL, eds. *Neurophthalmology*, pp.1–26. Mosby: St Louis.

Llermitte J (1922). Syndrome de la calotte du pedoncule cerebral: les troubles psychosensoriels dans les lesions du mesocephale. *Revue Neurol (Paris)*, **38**, 359–65.

Lindblom B, Truit CL, Hoyt WF (1992). Optic nerve sheath meningioma: definition of intraorbital, intracanlicular and intracranial components with magnetic resonance imaging. *Ophthalmology*, **99**, 560–6.

Ling CP, Pavesio C (2003). Paraneoplastic syndromes associated with visual loss. *Curr Opin Ophthalmol*, **14**, 426–32.

Lissauer H (1890). Ein fall von Seelenblindheit nebst einem Beitrage zur Theorie derselben. *Arch Psychiatr Nervenkr*, **21**, 22–70.

Liu GT, Glaser JS, Shatz NJ *et al.* (1994). Visual morbidity in giant cell arteritis: Clinical characteristics and prognosis for vision. *Ophthalmology*, **101**, 1779–85.

Mackey DA (1994). Three subgroups of patients from the United Kingdom with Leber's hereditary optic neuropathy. *Eye*, **8**, 431–6.

Maher ER, Yates JR, Harries R *et al.* (1990). Clinical features and natural history of von Hippel-Lindau disease. *QJ Med*, **77**, 1151–63.

Malik A, Furlan AJ, Sweeney PJ *et al.* (1992). Optic neuropathy: a rare paraneoplastic syndrome. *J Clin Neuroophthalmol*, **12**, 137–41.

Marquis DG (1934). Effects of removal of visual cortex in mammals with observations on the retention of light discrimination in dogs. In *Proceedings of the Association for Research in the Nervous and Mental Disease*, pp. 558. William and Wilkins, Baltimore.

McDonald WI, Barnes D (1992). The ocular manifestations of multiple sclerosis. I. Abnormalities of the afferent visual system. *J Neurol Neurosurg Psychiatry*, **55**, 747–52.

McKee AC, Lavine DN, Kowll NW *et al.* (1990). Peduncular hallucinosis associated with isolated infarction of the substantia nigra, pars reticulata. *Ann Neurol*, **27**, 500–4.

Meadows JC (1974a). Disturbed perception of colours associated with localised cerebral lesions. *Brain*, **97**, 615–32.

Meadows JC (1974b). The anatomical basis of prosopagnosia. *J Neurol Neurosurg Psychiatry*, **37**, 489–501.

Meadows JC, Monro SSF (1977). Palinopsia. *J Neurol Neurosurg Psychiatry*, **40**, 5–8.

Milner AD, Paulignan Y, Dijkerman HC *et al.* (1999). A paradoxical improvement of misreaching in optic ataxia: new evidence for two separate neural systems for visual localisation. *Proc R Soc*, **266**, 2225–9.

Morris JGL, Grattam-Smith P, Panegyres PK *et al.* (1994). Delayed cerebral radiation necrosis. *QJ Med*, **87**, 119–29.

Morrisey SP, Miller DH, Kendall BE *et al.* (1993). The significance of brain magnetic resonance imaging abnormalities at presentation with clinically isolated syndromes suggestive of multiple sclerosis. *Brain*, **116**, 135–46.

Moyer PD, Golnik KC, Breneman J (2000). Treatment of optic nerve sheath meningioma with three-dimensional conformal radiation. *Am J Ophthalmol*, **129**, 694–6.

Mustonen E, Poyhonen M, Leisti E-L (1997). Neuro-ophthalmological findings in neurofibro matosis: Clinical and neuroradiological study of 125 patients. *Neuroophthalmology*, **17**, 117–26.

Nachev P, Husain M (2006). Disorders of visual attention and the posterior parietal cortex. *Cortex*, **42**, 766–73.

Neelon FA, Goree JA, Lebovitz HE (1973). A primary empty sella: clinical and radiographic characteristics and endocrine function. *Medicine*, **52**, 73–84.

Neetens A, Smets RM (1989). Papilloedema. *Neuroophthalmology*, **9**, 81–101.

Newman SA, Miller NR (1983). Optic tract syndrome: Neuro-Ophthalmic considerations. *Arch Ophthalmol*, **101**, 1241–50.

Newsome DA (ed.) (1988). *Retinal Dystrophies and Degenerations*. Raven Press, New York.

Nichols JW, Goodwin JA (1992). Neuro-ophthalmic complications of AIDs. *Semin Ophthalmol*, **7**, 24–9.

Orcutt JC, Page NGR Sanders MD (1984). Factors affecting visual loss in benign intracranial hypertension. *Ophthalmology*, **91**, 1303–12.

Penfield W, Perot P (1963). The brain's record of auditory and visual experience. *Brain*, **86**, 596–696.

Perrett DI, Smith PAJ, Potter DD *et al.* (1984). Neurones responsive to faces in the temporal cortex: studies of functional organisation, sensitivity to identity and relation to perception. *Hum Neurobiol*, **3**, 197–208.

Pollack S (1987). The morning glory disc anomaly: contractile movement, classification and embryogenesis. *Doc Ophthalmol*, **65**, 439–60.

Radcliffe G, Newcombe F (1982). Object recognition: Some deductions from the clinical evidence. In Ellis A, ed. *Normality and Pathology of Cognitive Function*, pp.147–71. Academic Press, New York.

Rafal RD (1997). Balint syndrome. In Feinberg TE, Farah MJ, eds. *Behavioral Neurology and Neuropsychology*, pp. 337–56. McGraw-Hill, New York.

Ragge NK (1993). Clinical and genetic patterns of neurofibromatosis 1 and 2. *Br J Ophthalmol*, **77**, 662–72.

Reddy CV, Brown J Jr, Folk JC *et al.* (1996). Enlarged blind-spots in chorioretinal disorders. *Ophthalmology*, **103**, 606–17.

Riddoch G (1917). Dissociation in visual perceptions due to occipital injuries, with special reference to appreciation of movement. *Brain*, **40**, 15–57.

Riordan-Eva P, Restori M Hamilton AMP *et al.* (1994). Orbital ultrasound in the ocular ischaemic syndrome. *Eye*, **8**, 93–6.

Riordan-Eva P, Sanders MD, Govan CG *et al.* (1995). The clinical features of Leber's hereditary optic neuropathy defined by the presence of a pathogenic mitochondrial DNA mutation. *Brain*, **118**, 319–37.

Rizzo M, Vecera SP (2002). Psychoanatomical substrates of Balint's syndrome. *J Neurol Neurosurg Psychiatry*, **72**, 162–78.

Rosenberg MA, Savino PJ, Glaser JS (1979). A clinical analysis of pseudo-papilloedema I. population, laterality, acuity, refractive error, ophthalmoscopic characteristics, and coincident disease. *Arch Ophthalmol*, **97**, 65–70.

Rumelt S, Brown GC (2003). Update on treatment of retinal arterial occlusions. *Curr Opin Ophthalmol*, **14**, 139–41.

Sadun AA, Martone JF, Muci-Mendoza R *et al.* (1994). Epidemic optic neuropathy in Cuba: eye findings. *Arch Ophthalmol*, **112**, 691–9.

Safran AB, Glaser JS (1980). Statokinetic dissociation in lesions of the anterior visual pathways. *Arch Ophthalmol*, **98**, 291–5.

Savino PJ, Glaser JS, Rosenberg MA (1978). A clinical analysis of pseudopapilloedema: II. Visual field defects. *Arch Ophthalmol*, **97**, 71–5.

Schreir HA (1999). Hallucinations in nonpsychotic children: more common than we think. *J Am Acad Child Adolesc Psychiatry*, **38**, 623–5.

Selbst RG, Selhorst JB, Harbison JW *et al.* (1983). Para-infectious optic neuritis. *Arch Neurol*, **40**, 347–50.

Sugishita M, Hemmi I, Sakuma I *et al.* (1993). The problem of macular sparing after unilateral occipital lesions. *J Neurol*, **241**, 1–9.

Sullivan TJ, Clarke MP, Morin JD (1992). The ocular manifestations of Sturge-Weber syndrome. *J Paed Ophthalmol Strabismus*, **29**, 349–56.

Tecoma ES, Laxer KD, Barbaro NM *et al.* (1993). Frequency and characteristics of visual field defects after surgery for mesial temporal sclerosis. *Neurology*, **43**, 1235–8.

Teunisse RJ, Cruysberg JRM, Verbeek A *et al.* (1995). The Charles Bonnet Syndrome. A large prospective study in The Netherlands. *Br J Psychiatry*, **166**, 254–7.

Teuber HL (1965). Somatosensory disorders due to cortical lesions. *Neuropsychologia*, **3**, 287–94.

Tranel D, Damasio AR (1996). Agnosias and apraxias. In Bradley WG, Daroff TCB, Fenichel GH *et al*, eds. *Neurology in Clinical Practice*, pp. 119–29. Butterworth, Heinemann, Boston.

Warrington EK (1975). The selective impairment of semantic memory. *Q J Exp Psychol*, **27**, 635–58.

Warrington EK, James M (1988). Visual aperceptive agnosia: a clinco-anatomical study of three cases. *Cortex*, **24**, 13–32.

Weiskrantz L (1986). *Blindsight: a Case Study and Implications*. Clarendon Press, Oxford.

Weiskrantz L, Warrington EK, Sanders MD *et al.* (1974). Visual capacity in the hemianopic field following a restricted occipital ablation. *Brain*, **97**, 709–28.

Welch RB (1970). Von hippal-Lindau disease: the recognition and treatment of early angiomatosis retinae and the use of cryosurgery as an adjunct to therapy. *Trans Am Ophthalmol Soc*, **68**, 367–424.

Zeki S (1990). A century of cerebral achromatopsia. *Brain*, **113**, 1727–77.

Zeki S (1991). Cerebral akinetopsia (visual motion blindness). A review. *Brain*, **114**, 811–24.

Zeki S (1993). *A Vision of the Brain*. Blackwell, London.

Zeki S, ffytche DH (1998). The Riddoch syndrome: insights into the neurobiology of conscious vision. *Brain*, **121**, 25–45.

Zeki S, Shipp S (1988). The functional logic of cortical connections. *Nature*, **335**, 311–7.

Zihl J, von Cramon D, Mai N (1983). Selective disturbance of movement vision after bilateral brain damage. *Brain*, **106**, 313–40.

Zimmer Galler IE, Robertson DM (1995). Tuberose sclerosis: long term observation of retinal lesions in tuberose sclerosis. *Am J Ophthalmol*, **119**, 318–24.

CHAPTER 13

Ocular motor disorders

Christopher Kennard

Contents

13.1 Disorders of eye movements

A detailed description of the neural control of eye movements and their disorders can be found in the excellent monograph by Leigh and Zee (2006).

13.1.1 Examining eye movements

Actions of the extraocular muscles

Each eye is rotated by six muscles: four recti and two obliques. It should be noted that the actions of the muscles are dependent on the starting position of the eye. For example, the superior rectus, because of the anatomy of its insertion into the sclera, acts as a pure elevator only when the globe is abducted by 23 deg. With increasing adduction of the eye from this position, the superior rectus acts more as an intorter and less as an elevator. Similarly, the superior oblique acts purely as a depressor only when the eye is adducted, and more as an intorter with increasing abduction of the eye. The primary and secondary actions of the different extraocular muscles are shown in Table 13.1.

The other important feature is that the yoke pair of muscles from each eye, for instance right medial rectus and left lateral rectus, or left superior rectus and right inferior oblique, receive equal innervation so that eye movements are conjugate: Hering's law of motor correspondence. It should be noted that the fixating eye determines the innervational input to both eyes. This is of importance in the assessment of the cover test, and in the interpretation of investigations such as the Hess screen test.

The assessment of diplopia

It is first essential to decide whether the patient is complaining of diplopia due to a disparity in retinal stimulation between the two eyes, that is binocular diplopia, or more rarely when it is present in one eye only, monocular diplopia (Shaunak *et al.* 1997). Monocular diplopia occurs, with few exceptions, when there are abnormalities

Table 13.1 Primary and secondary actions of extraocular muscles

	Primary action	Secondary action
Lateral rectus	Abduction	–
Medial rectus	Adduction	–
Superior rectus	Elevation	Intorsion
Inferior rectus	Depression	Extorsion
Superior oblique	Intorsion	Depression
Inferior oblique	Extorsion	Elevation

of the ocular refractive surfaces and media, producing multiple overlapping images on the retina. The commonest cause is myopic astigmatism, but monocular diplopia may occur in early cataracts, especially under conditions of dim illumination. Other causes include abnormalities of the cornea and iris, foreign bodies in the aqueous or vitreous humour, retinal disease, occipital cortex pathology, and psychogenic causes.

If the diplopia is alleviated by covering one eye a systematic approach to evaluation is required. As well as determining the nature of the separation of the two images and the direction of maximal separation, enquiries as to the presence of a family history of strabismus, or a childhood history of orthoptic treatment should be made. If the eyes are misaligned, it should be ascertained at an early stage if one is dealing with a non-comitant or comitant strabismus; the degree of misalignment varies with gaze position in the first, but does not vary with gaze position in the second. Non-comitance suggests a recent paretic or restrictive aetiology. Comitance is characteristic of childhood strabismus, and diplopia in such circumstances is usually due to decompensation of a long-standing *phoria,* a deviation of the visual axes when only one eye is viewing. Normally this is kept in check by fusional mechanisms, a latent deviation. The term *tropia* as used later refers to a manifest deviation of the visual axes when both eyes are viewing, which is not kept in check by fusion.

Patients with diplopia may adopt a compensatory head posture, and the position of the chin, head, and face should therefore be carefully observed. The purpose of the abnormal head posture is to turn the eyes as far as possible from the field of action of the weak muscle. Hence, if one of the muscles that mediates conjugate gaze to the right is underacting, the face will be turned to the right. Underaction of the superior and inferior recti, which act primarily to move the eyes in the vertical plane, is compensated by head flexion and extension respectively. Torsional diplopia usually arises from underaction of the superior and inferior oblique muscles, and patients with this symptom often tilt their head towards the shoulder opposite to that of the weak muscle.

Identification of the paretic muscle

In the *cover/uncover test* the patient, wearing appropriate refractive correction, is asked to fixate a distant target such as a letter on the Snellen chart, with the eyes in the primary position, repeating the test in the nine cardinal positions of gaze and with near fixation (Fig. 13.1). For each position in turn each eye is covered and then uncovered and initially the movements of the uncovered eye are observed. Cover tests rely on the fact that foveation occurs in an eye that is forced to fixate. If the retinal image was not directed on to the fovea before the eye took up fixation, a movement of redress will be noted as the eye fixates, which gives an indication of the degree of misalignment of the visual axes. If the uncovered eye moves to take up fixation, it can be assumed that under binocular viewing conditions the eye was not aligned with fixation, and a manifest deviation was present: a tropia. Inward movement of the uncovered eye indicates an exotropia, and an outward movement an esotropia. A vertical deviation may be either a hypotropia or a hypertropia, depending on whether the eye moves up or down respectively. The examiner should determine whether the tropia is comitant or non-comitant by seeing if the magnitude of the deviation varies with the position of the eye. If no tropia is present, and the uncovered eye is observed to assume

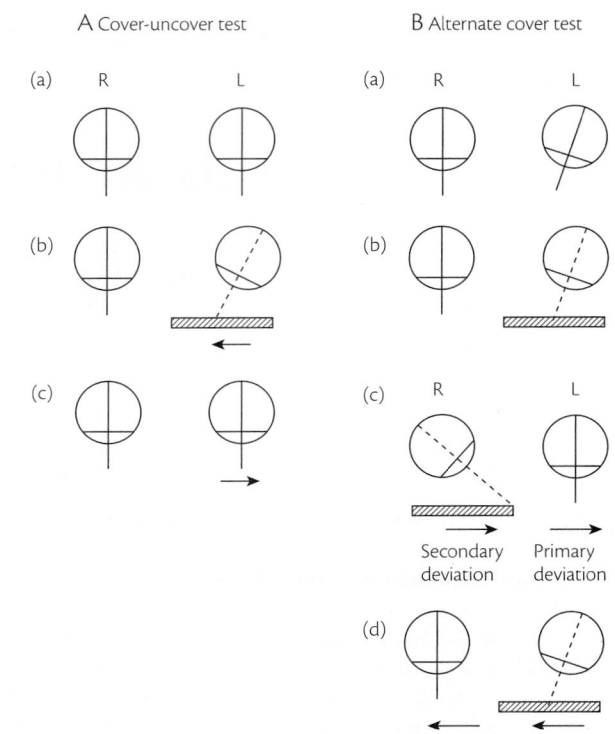

Fig. 13.1 A. Cover/uncover test, showing the presence of an esophoria. Dotted lines indicate the position of the eye when under cover. (a) At rest, the visual axes are aligned correctly. (b) When the cover is placed before the left eye, the eye no longer fixates, and moves inwards. (c) On removal of the cover, the eye moves outwards to take up fixation, indicating an esophoria.
B. The alternate cover test, showing the presence of an esotropia. (a) At rest, with both eyes viewing, there is a manifest inward deviation of the left eye. (b) A cover placed before the non-fixating left eye causes no movement. (c) When the right eye is occluded, the left eye is forced to fixate, and a movement of redress occurs (the primary deviation). The resulting additional innervation to the contralateral yoke muscle leads to deviation of the sound eye under the cover (the secondary deviation). Note that the secondary deviation is greater than the primary deviation. (d) When the cover is transferred to the left eye, both eyes assume their original position.

fixation just after it is uncovered, a latent deviation or heterophoria is present. Depending on the direction of the deviation this may be classified as an exophoria, esophoria, hypophoria, or a hyperphoria. The test is then repeated, and the same observations made while covering the other eye. It should be noted that the convention is that if there is a vertical deviation of the eyes, the higher of the two is referred to as hypertropic/hyperphoric, regardless of which eye is at fault.

The *alternate cover test* is more dissociating than the cover/uncover test, and is used to fully dissociate the eyes and show the maximal deviation. While the patient fixates a target the occluder is quickly switched from eye to eye to prevent binocular viewing, allowing sufficient time for the eyes to settle in their new position after each transfer. The test should be performed in the nine cardinal positions of gaze to determine the direction of gaze that elicits the maximal direction, the eye in which fixation in that field of gaze causes the maximal deviation. Whilst ensuring that the patient is never allowed to regain fixation during transfer of the occluder, the examiner notes the movement of the uncovered eye as the occluder is transferred from one eye to the other. Movement of the uncovered eye may indicate either a heterotropia or a heterophoria, and the alternate cover test will not differentiate between the two. The cover/uncover test should therefore be performed first to determine if a tropia is present.

When there is a vertical deviation of the visual axes the hypertropic 'higher' eye always gives rise to the lower image. This may be due either to a paresis of the depressor muscles of the hypertropic eye or the elevators of the other eye. There are now four possible defective muscles, which can be further reduced to two by asking the patient to look to the left and the right and state in which direction the deviation is maximal. Finally, determining which of these two muscles is paretic is decided by finding whether the deviation is maximal in up or down gaze.

In some instances there may be no differential vertical deviation; this situation occurs with chronic palsies due to an adaptive phenomenon, termed 'spread of comitance'. To work out which vertical muscle is paretic the *Bielschowsky head tilt test* is performed. In this test the vertical deviation is compared with the alternate cover test in right and left head tilt positions. The degree of misalignment will increase when the head is tilted to the side of the paretic muscle if the ipsilateral intorters, superior oblique and superior rectus, are weak, and to the opposite side if the extorting muscles, inferior oblique and inferior rectus, are weak. In practice an increased misalignment on head tilt is usually indicative of an ipsilateral superior oblique palsy. The test is less often positive with palsies of the vertical recti or inferior oblique muscles.

The explanation for the effect lies in the fact that a head tilt to either shoulder induces an ocular counter-rolling, which is mediated by the ipsilateral intorters, superior rectus and superior oblique, and by the contralateral extorters, inferior rectus and inferior oblique. If, for example, the ipsilateral superior oblique is paretic, the superior rectus on the same side receives excessive innervation to intort the eye, and by virtue of its relatively unopposed primary action elevates the eye.

13.1.2 Oculomotor (third) nerve palsies

The oculomotor, or third cranial nerve, in addition to innervating the superior, medial, and inferior rectus, and inferior oblique eye muscles also supplies the levator palpebrae superioris muscle, and carries the parasympathetic nerve fibres to the sphincter muscle of the pupil and the ciliary body. A complete oculomotor palsy is easily recognized by ptosis, a fixed dilated pupil and an eye which is deviated 'down and out' due to the unopposed action of the lateral rectus and superior oblique. Partial palsies are more common (Fig. 13.2). The nerve may be damaged anywhere along its course from nuclear complex to the muscles, and the combination of an oculomotor palsy with other cranial nerve deficits (II, IV, V, VI) and the long tract signs usually enables accurate localization of the site of the lesion.

Common causes

Numerous causes of oculomotor nerve palsies have been described (Table 13.2). In adults, the commonest are either an aneurysm or

Table 13.2 Causes of oculomotor nerve palsy

Nuclear
Congenital hypoplasia
Infarction or haemorrhage
Tumour (metastatic)

Fascicular
Infarction or haemorrhage
Demyelination (rare)
Tumour

Subarachnoid
Aneurysm
Meningitis
Infarction
Tumour
Neurosurgical complication
Post-lumbar puncture

At the tentorial edge
Uncal herniation
Pseudotumour cerebri
Hydrocephalus
Trauma

Cavernous sinus and superior orbital fissure
Aneurysm
Thrombosis
Carotid-cavernous fistula
Tumour (pituitary adenoma, meningioma, nasopharyngeal, and other metastases)
Pituitary apoplexy
Tolosa–Hunt syndrome
Sphenoidal sinusitis and mucoloele
Mucormycosis and other fungal infections
Herpes zoster
Nerve infarction (associated with hypertension and diabetes)

Orbit
Trauma
Inflammatory pseudotumour
Infection (mucormycosis and other fungal infections)

Localization uncertain
Viral infections and infectous mononucleosis
Following immunization
Migraine
Arteritis
Guillain–Barré and Miller Fisher syndromes

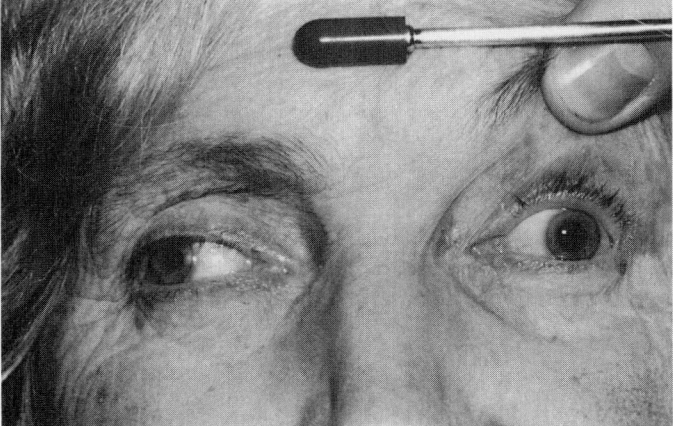

Fig. 13.2 Third cranial nerve palsy. A left oculomotor nerve palsy showing ptosis being held up by the examiner's finger, pupillary dilation, and abduction of the eye due to the unopposed action of the lateral rectus muscle.

presumed peripheral nerve vascular microinfarction, although evidence is growing that fascicular damage in the brainstem may be a more frequent cause than previously thought (Rush and Younge 1981). These infarcts are commonly associated with arteriosclerosis, hypertension, and diabetes mellitus. Infarcts and aneurysms each account for approximately 20 per cent of the total number of oculomotor palsies. The next commonest are tumours or trauma, which each account for 10–15 per cent of cases (Richards *et al.* 1992).

Brainstem lesions The oculomotor nuclear complex is a paired structure lying beneath the aqueductal grey matter of the rostral mid-brain at the level of the superior colliculus. The nuclear complex is divided into distinct motor pools subserving individual extraocular muscles. From the clinical point of view the following features should be noted: the caudal nucleus is a single midline structure supplying the levator palpebrae superioris muscles; the nuclei for each superior rectus muscle lie dorsally, close to the midline, and the axons cross the midline to innervate the contralateral muscle. Two patterns of ocular motility are characteristic of lesions of the oculomotor complex, either an isolated complete bilateral ptosis or a unilateral palsy of the medial rectus, inferior rectus, and inferior oblique together with a contralateral superior rectus palsy. Daroff (1970) has suggested clinical rules to decide whether or not a disorder of the oculomotor nerve is due to a nuclear lesion (Table 13.3).

Infarction of this region of the mid-brain often involves more rostral structures resulting in supranuclear vertical gaze disorders. Isolated nuclear oculomotor nerve palsies of vascular origin usually occur as a result of selective embolic or thrombotic occlusion of small dorsal perforating branches of the mesencephalic portion of the basilar artery, or less often, from occlusion of the distal portion of the basilar artery itself, the *top of the basilar syndrome*. Other aetiologies include haemorrhage, infiltration by tumour, inflammation, and brainstem compression (Bogousslevsky *et al.* 1994).

It has been proposed that both medial rectii subnuclei may be damaged in patients who show bilateral adduction failure with exotropia and loss of convergence. This has been called 'WEBINO': **W**all-**E**yed **B**ilateral **I**nter**N**uclear **O**phthalmoplegia (Daroff and Hoyt 1971).

Table 13.3 Clinical rules to determine whether an oculomotor nerve palsy is due to a nuclear lesion (after Daroff 1970)

1. **Conditions which cannot represent nuclear lesions:**
   unilateral external ophthalmoplegia (with or without pupil involvement) associated with normal contralateral superior rectus function
   unilateral internal ophthalmoplegia
   unilateral ptosis
   isolated unilateral or bilateral medial rectus weakness

2. **Conditions which may be nuclear:**
   bilateral total third nerve palsy
   bilateral ptosis
   bilateral internal ophthalmoplegia
   isolated single muscle involvement (except levator and superior rectus)

3. **Obligatory nuclear lesions:**
   unilateral third nerve palsy with contralateral superior rectus and bilateral partial ptosis
   bilateral third nerve palsy (with or without internal ophthalmoplegia) associated with spared levator function

The fascicles of the oculomotor nerve pass ventrally through the medial longitudinal fasiculus, red nucleus, substantia nigra, and the medial cerebral peduncle. Lesions of the nerve in this location are normally either due to infarction or tumours, and their precise anatomical location determines the associated neurological signs (Section 2.4). Thus, a lesion in the red nucleus results in a contralateral cerebellar tremor *Nothnagel's syndrome*, or is associated with contralateral involuntary movements *Benedikt's syndrome*. If the lesion is in the cerebral peduncle the oculomotor palsy is associated with a contralateral hemiparesis *Weber's syndrome*. When the lesion is more extensive, involving the red nucleus and cerebral peduncle, all these signs are present as a result of *Claude's syndrome*. Palsies of one extraocular muscle may occur due to a mid-brain lesion within the oculomotor nerve fascicles are a result of their topographic organization. It is suggested that the most medial fibres are for the pupil, followed by fibres for the inferior rectus, levator, medial rectus, superior rectus, and most laterally the inferior oblique (Castro *et al.* 1990).

Intradural, extramedullary lesions The fascicles emerge from the mid-brain as several rootlets in the interpeduncular space where basal tumours as well as the rare basilar artery aneurysm may compress the nerve as well as the cerebral peduncle in an extrinsically produced Weber's syndrome. The nerve then passes below the uncus of the temporal lobe lying lateral to the posterior communicating artery. During cerebral herniation due to a unilateral mass lesion the nerve is compressed against the tentorial edge, petroclinoid ligament, or clivus by the uncus. Since the pupillary fibres travel superficially and superonasally in the nerve they are normally affected first during compression leading to mydriasis, to be followed by ptosis and then extraocular muscle weakness. Aneurysms of the posterior communicating artery damage the oculomotor nerve by haemorrhage into the nerve itself or into the aneurysm sac, which causes enlargement and stretching of the nerve to which it is adherent. These aneurysms are often associated with orbital or facial pain, which may precede the oculomotor paresis by up to 2 weeks. It is very unusual for an aneurysm to present with an oculomotor nerve with complete sparing of the pupil i.e. normal pupillary function, and in such cases the pupil usually becomes involved within 7 days (Nadeau and Trobe 1983; Trobe 1988; Cullom *et al.* 1995).

Trauma An important cause for an oculomotor palsy is trauma which usually has to be severe enough to lead to skull fractures and loss of consciousness. The nerve may be damaged at three different locations: rootlet avulsion as it emerges from the mid-brain, in the subarachnoid space where the nerve is fixed as it penetrates the dura, and finally in relation to fractures of the superior orbital fissure. Mild head injury resulting in an oculomotor palsy should always raise the possibility of a tumour in the skull base (Eyster *et al.* 1972). As the nerve lies in the subarachnoid space it is vulnerable to damage from inflammatory processes, particularly infection such as tuberculosis, meningococcus, and syphilis.

Peripheral lesions Shortly after piercing the dura lateral to the posterior clinoid process the oculomotor nerve enters the cavernous sinus where it lies above the trochlear nerve. In this location it is usual for the nerve to be involved with the other ocular motor nerves, and the first and second divisions of the trigeminal nerve. A partial involvement of the nerve, often with sparing of the pupilloconstrictor fibres, is common. Involvement of the nerve in this location may be due to local infection or inflammation, compression by aneurysms of the intracavernous portion of the internal carotid artery, in which case there is often accompanying orbital or facial pain, or by meningiomas or lateral extensions of pituitary tumours.

In the rostral part of the cavernous sinus or in the superior orbital fissure the oculomotor nerve divides into a superior division, which supplies the superior rectus and levator palpebrae superioris muscles, and an inferior division, which supplies the medial and inferior rectii and the inferior oblique muscles, as well as the parasympathetic pupilloconstrictor fibres. Isolated palsy of the superior branch has been described due to intracavernous internal carotid artery aneurysm or mucocoele from the frontal or ethmoid sinuses, and of the inferior branch to trauma and tumour. Presumed post-infectious isolated branch palsies have also been reported. However, since the oculomotor nerve has a divisional topographic arrangement beginning in the brainstem, a divisional oculomotor nerve palsy can also arise from a lesion anywhere along its course, including the brainstem.

Infarction One of the commonest causes for an isolated oculomotor nerve palsy is micro-vascular infarction, which may be associated with diabetes mellitus, hypertension, smoking, hypercholesterolaemia, and collagen vascular disease a 'medical third'. Under these circumstances, there is usually minimal involvement or complete sparing of the pupil (Goldstein and Cogan 1960). Pathological studies have shown focal demyelination without axonal degeneration due to occlusion of intraneural arteries in the intracavernous or subarachnoid segments, with relative sparing of the peripherally located parasympathetic fibres which are supplied by the vasa nervorum (Asbury *et al.* 1970). Occasionally the nerve may be damaged by a mesencephalic infarct. Rarely, the pupil may be involved with vascular oculomotor nerve lesions. Clinically the palsy may be preceded by orbital or facial pain, which may disappear with the onset of the paresis or continue over several days as may the paresis. In diabetes mellitus the palsy may be the presenting symptom, sometimes with involvement of the abducens and trochlear nerve, and ophthalmic division of the trigeminal nerve due to occlusion of branches of the inferolateral trunk arising from the intracavernous carotid artery. However, in an established diabetic, when associated with other ocular motor palsies, a paranasal sinus or orbital infection by mucormycosis must be considered. The normal course is for spontaneous recovery within 8–12 weeks.

Aberrant regeneration Injury of the oculomotor nerve often results in subsequent aberrant regeneration. The commonest clinical findings are lid elevation, when the globe is adducted or depressed, the pseudo-von Graefe phenomenon, with absence of or only partial vertical eye movement. In addition, lid depression on abduction and pupil constriction on adduction or depression despite absent pupillary reflexes may occur (Forster *et al.* 1969). These various combined movements are due to co-contraction of muscles innervated by the oculomotor nerve. Aberrant regeneration may occur after trauma, aneurysm, congenital third nerve palsy, and migraine, but not after micro-infarction due to hypertension or diabetes mellitus, presumably due to preservation of axonal continuity. In addition, it may occur without a history of preceding oculomotor palsy, primary aberrant regeneration. In this case an intracavernous meningioma or carotid aneurysm should be sought (Cox *et al.* 1979). The hypothesis which has traditionally been used to explain this oculomotor synkinesis is that at the site of nerve injury axons become misdirected and eventually innervate muscles for which they were not originally intended. Although there is good support for this hypothesis both experimentally and clinically, other explanations have had to be considered to explain both primary aberrant regeneration, in which there is no acute nerve injury, and the transient nature of the phenomenon in some patients (Lepore and Glaser 1980). These have included ephaptic transmission, in which synkinetic movements may arise from inter-axonal cross activation (ephaptic) at the site of the injury, or as a result of peripheral axonolysis there may be reorganization of neuronal central connections which unmask previously encoded synkinetic movements. Currently there is no single explanation which adequately explains all the clinical features of oculomotor synkinesis.

Childhood lesions Isolated oculomotor palsy in infancy and childhood is rare and most often congenital (Victor 1976; Kodshi and Younge 1992). They probably result from ischaemic or hypoxic insults to the brainstem *in utero*, which may lead to hypoplasia of the nucleus. Some probably result from traction to the subarachnoid portion of the nerve during labour. These palsies are often incomplete and show evidence of aberrant regeneration. The acquired causes include trauma, inflammatory disease, tumour, aneurysms, and ophthalmoplegic migraine (Miller 1977). Slowly progressive isolated palsies should undergo imaging every 2 years with the expectation of eventually detecting a small tumour somewhere along the course of the nerve. An unusual phenomenon is *cyclic oculomotor paresis* which usually occurs in early childhood, but may be noted at birth. In this condition oculomotor paresis alternates every 2 min with the shorter spasms of oculomotor 'overactivity' lasting 10–30 s when the ptotic lid elevates, the globe begins to adduct, the pupil constricts, and accommodation increases. The condition usually persists unchanged throughout life (Loewenfeld and Thompson 1975).

Ophthalmoplegic migraine Ophthalmoplegic migraine has its onset in childhood and occurs in the setting of headache, photophobia, cyclical vomiting, or other migrainous symptoms. A family history of migraine is usually absent. The oculomotor palsy may persist for some time after the headache has resolved, and the palsy may in fact develop as the headache phase abates. The cause of the palsy is considered to be due either to compression of the oculomotor nerve by a swollen dilated carotid or basilar artery, or a delayed ischaemic neuropathy (Walsh and O'Doherty 1960). When such a condition occurs in a young child there is obviously concern about the possibility of an aneurysm. However, presentation of a Berry aneurysm under the age of 10 years is exceptionally rare, and if the palsy recovers along with the other symptoms angiography is not indicated. In a teenager MR angiography is advisable.

Ocular neuromyotonia Patients with ocular neuromyotonia experience paroxysmal diplopia due to involuntary contraction of muscles supplied by the oculomotor nerve, although the abducens and trochlear nerves may also be involved. This condition usually but not always occurs in the context of a patient who has previously received radiotherapy for skull-base or thalamic tumours. The presumed cause is a radiation-induced cranial neuropathy manifesting as a spontaneous discharge from axons with unstable cell membranes. This produces a tonic contraction of one or more muscles, and the paroxysms occur 20–30 times daily. The paroxysms, which last several minutes, may be induced by sudden shifts of gaze into the field of action of the involved muscles. Most patients respond satisfactorily to carbamazepine (Ezra *et al.* 1996).

Investigation and management

The crucial issues regarding the diagnostic procedures, which should be undertaken in adult patients with acute acquired oculomotor nerve palsies, are the patient's age and the presence or absence of pupil involvement. A patient with an acute complete or partial oculomotor nerve palsy with pupil involvement should be urgently investigated for a possible posterior communicating artery aneurysm. On the other hand patients over 50 with an isolated complete pupil sparing oculomotor nerve palsy and vascular risk factors such as diabetes mellitus or hypertension can be observed without resorting immediately to MR scanning. They should be frequently followed up and any worsening in the oculomotor palsy or failure to improve after 8 weeks should lead to MR scanning. In a case of pupil sparing partial oculomotor nerve palsy it has been proposed that scanning is only required if over the next week the pupil becomes involved (Trobe 1988). After this a vasculopathic cause is most likely. However, other causes of pupil sparing oculomotor nerve palsy, such as myasthenia gravis and thyroid eye disease should be considered and the appropriate investigations undertaken.

Isolated oculomotor nerve palsies or in combination should raise the possibility of meningeal processes such as Lyme disease, tuberculous, fungal, carcinomatous, or lymphomatous meningitis, and a CSF examination should be performed along with an MRI with gadolinium enhancement.

The management of patients with oculomotor nerve palsies is complex because of the varied pattern of paresis of the four involved muscles. Initially monocular occlusion or prisms can be used but only after the paresis has been shown to be stable for at least 6 months are ophthalmologists prepared to propose surgery to attempt to produce alignment of the eyes in at least the primary position.

13.1.3 Trochlear (fourth) nerve palsies

The trochlear nerve, or fourth cranial nerve, passes from its nucleus to decussate with the contralateral nerve in the anterior medullary velum, which forms the anterior roof of the fourth ventricle just caudal to the inferior olive. It is unique amongst motor nerves not only for its decussation but for the fact that it emerges from the dorsal surface of the brain, and then passes round the mid-brain tectum, crossing the inferior cerebellar artery to reach the free edge of the tentorium, where it enters the dura to run forward into the cavernous sinus. It finally enters the orbit through the superior orbital fissure to innervate the superior oblique muscle. It is the most slender of all the cranial nerves and has the longest intracranial course.

Clinical features

Clinically the patient with a trochlear nerve paresis usually complains of vertical diplopia with a torsional component, accentuated by looking down. There is usually a compensatory head tilt to the side opposite the affected eye with the chin down. In the majority of cases there is a hypertropia of the affected eye with an increased vertical deviation of the two images, when the head tilted to the side of the paretic muscle, the Bielschowsky manoeuvre. A simple way to identify the side of a suspected trochlear nerve palsy is to ask the patient to view a straight object such as a pen as if is moved horizontally from the primary position downwards. The patient will observe two images, one horizontal and the other tilted, coming together laterally on one side which is the side of the abnormal eye,

Table 13.4 Causes of trochlear nerve palsy

Nuclear and fascicular
Infarction or haemorrhage
Aplasia
Demyelination
Trauma
Subarachnoid
Trauma
Tumour
Neurosurgical procedure
Mastoiditis
Meningitis
Post-lumbar puncture
Cavernous sinus and superior orbital fissure
Tumour
Thrombosis
Aneurysm
Tolosa–Hunt syndrome
Herpes zoster
Orbit
Ethmoidectomy
Ethmoiditis or maxillary sinusitis
Trauma
Localization uncertain
Infarction (associated with hypertension and diabetes)

'the arrow points to the affected fourth nerve'. Torsion greater than 10 deg is usually indicative of a bilateral weakness of the superior obliques, as does a V pattern esotropia.

Causes

A trochlear nerve palsy is the commonest cause for vertical extraocular muscle weakness, but is less common than palsies of the other ocular motor nerves (Richards *et al.* 1992; Tiffin *et al.* 1996). The aetiological causes of trochlear nerve palsies are listed in Table 13.4. However, as with an isolated lateral rectus muscle paresis, other orbital causes for an isolated paresis of the superior oblique muscle must be sought before it can be attributed to a trochlear nerve lesion. These include skew deviation, myasthenia gravis, dysthyroid myopathy, and other restrictive ocular myopathies, childhood strabismus syndromes. In *Brown's superior oblique tendon sheath syndrome* there is restricted elevation of the adducted eye, which may be congenital due to a shortened superior oblique tendon, or acquired when it is caused by a tenosynovitis or neoplastic infiltration of the tendon as it passes through the trochlear pulley (Brown 1973).

Nuclear Lesians

It is impossible to clinically differentiate a lesion of the nucleus from a nerve lesion. The nucleus may be congenitally hypoplastic or damaged by haemorrhage or tumour. The only clue to a lesion of the nucleus or fascicle is the association of a trochlear nerve paresis with a contralateral Horner's syndrome or a contralateral internuclear ophthalmoplegia which localizes the pontomesencephalic junction (Mansour and Reinecke 1986). More commonly the nerve is affected in its long subarachnoid course, especially from head trauma, both severe and minor. The nerve is particularly prone to damage at its posterior decussation, the anterior medullary velum, due to its relationship to the edge of the tentorium. Blunt injury to the forehead or skull leads to a contrecoup contusion of the mid-brain tectum by the free tentorial edge. The resulting trochlear nerve palsy is often bilateral (Lepore 1995). Skull base fractures can also lead to trochlear nerve palsies.

Ischaemic lesions The second commonest cause of an isolated trochlear nerve paresis is microinfarction, particularly in an individual over 50 associated with vasculopathic risk factors such as smoking, hypertension, hypercholesterolaemia, and diabetes mellitus (Keane 1993). This has a better prognosis than a traumatic nerve lesion although unlike oculomotor nerve palsies there has been no clinicopathological correlation. The nerve may be affected in the cavernous sinus and the superior orbital fissure by other diseases such as tumours and aneurysms, but it is usual for the other ocular motor and trigeminal nerves to also be involved.

Muscle underactivity The gradual or occasionally sudden onset of vertical diplopia, without any predisposing cause, should always be considered clinically to be due to an underacting superior oblique muscle, which when occurring from late childhood up to 50 years of age may be due to a decompensated congenital trochlear nerve palsy. The torsional component of the diplopia is less frequent than in the acquired trochlear nerve palsies. Old photographs of the patient may show a head tilt in infancy. These patients tend to have large vertical fusional amplitudes.

In many patients with isolated trochlear nerve pareses no cause can be found even after tensilon testing, MR scanning and a glucose tolerance test. In this situation recovery usually occurs spontaneously. Some patients can be treated by occlusion of the lower half of the spectacle lens over the affected eye with opaque tape. This provides occlusion in the field of action of the impaired superior oblique muscle. Alternatively a base-down prism such as a Fresnel lens may be worn over the affected eye.

Myokymia A rare disorder is superior oblique myokymia in which there are bursts of small amplitude, high frequency torsional oscillations of one eye (Hoyt and Keane 1970, Miller 1996). This results in symptoms of recurring monocular blurring of vision and oscillopsia, vertical or torsional diplopia, and tremulous sensations in the eye. These episodes last less than 10 s and occur many times per day. The attacks may be induced by looking downward, tilting the head toward the side of the affected eye, or by blinking. The oscillations may be difficult to observe on gross examination but can be readily observed with the ophthalmoscope or slit lamp by asking the patient to move the eyes in the direction known to induce the oscillations. Electromyographic studies have suggested that this condition is due to neuronal damage with subsequent regeneration leading to desynchronized contraction of muscle fibres (Komerell and Schaubele 1980). The condition is usually benign, although rare cases have been reported following trochlear nerve palsy, after mild head trauma, associated with multiple sclerosis and after brainstem infarction. Superior oblique myokymia spontaneously resolves in some patients, but may relapse. Whereas in some patients the symptoms are not troublesome, in those in whom their symptoms are distressing a range of medical treatments are available. These include membrane stabilizing drugs such as carbamazepine and phenytoin, baclofen, and systemically administered ß-adrenergic-blocking agents. Gabapentin has also been of benefit in some patients. If drug therapy fails surgical procedures such as superior oblique tenotomy sometimes combined with ipsilateral inferior oblique tenectomy are available.

13.1.4 Abducens (sixth) nerve palsies

An abducens, or sixth cranial nerve palsy, which results in a lateral rectus muscle paresis, is the commonest type of ocular nerve palsy (Fig. 13.3) (Richards *et al.* 1992). Clinically it results in horizontal double vision with the separation of the images increasing with gaze towards the affected side. It is important to differentiate an abducens nerve palsy from disorders of the neuromuscular junction such as myasthenia gravis, or extraocular muscles, such as dysthyroid eye disease and orbital inflammation or pseudotumour. Other conditions which need to be excluded include orbital trauma 'ethmoid blowout', convergence spasm, and the congenital innervation abnormalities of *Duane's* and *Mobius syndromes*. The various causes of abducens nerve paresis at different locations along its course are listed in Table 13. 5.

Causes

The abducens nucleus lies in the floor of the fourth ventricle and contains, in addition to motor neurons that supply the ipsilateral lateral rectus muscle, interneurons whose axons cross the midline to ascend in the medial longitudinal fasciculus to the contralateral oculomotor nucleus or medial rectus subnucleus. Lesions of the nucleus, therefore never produce ipsilateral abduction weakness, but always result in an ipsilateral conjugate gaze palsy often associated with an ipsilateral lower motor neuron palsy of the facial nerve, the fascicles of which course around the abducens nucleus (Henn and Büttner 1982). The abducens nucleus is susceptible to abnormalities of development, such as Duane's syndrome, and acquired lesions mainly due to metastatic tumours and vascular infarction. Patients with Wernicke–Korsakoff syndrome (Section 34.5), often develop horizontal paralysis of conjugate gaze, presumably from a metabolic insult to the abducens nuclei.

Brainstem lesions Involvement of the abducens nerve fascicles in the lateral tegmentum due to infarction in the territory of the anterior inferior cerebellar artery produces abduction paresis, ipsilateral lower motor neuron facial palsy, loss of taste from the anterior two-thirds of the tongue, ipsilateral Horner's syndrome, ipsilateral analgesia of the face, and ipsilateral peripheral deafness: *Foville's syndrome*. As the abducens nerve fascicles pass ventrally through the pontine tegmentum they pass lateral to the pyramidal tract. An infarction at this site, due to thrombotic or embolic occlusion of paramedian penetrating branches of the basilar artery, results in an ipsilateral abducens paresis, ipsilateral facial palsy, and contralateral hemiplegia: *Millard–Gubler syndrome*. Other causes for a fascicular abducens nerve palsy are brainstem glioma or metatasis, and demyelination, which commonly result in bilateral palsies (Silverman *et al.* 1995).

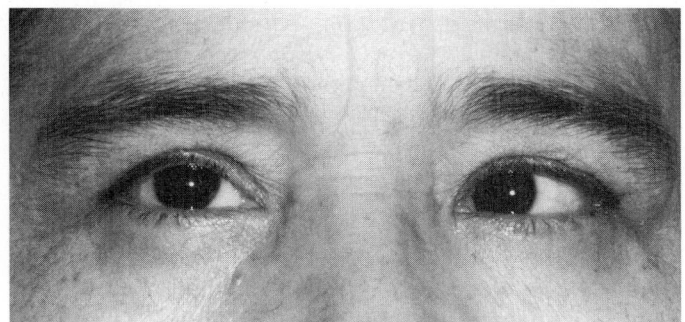

Fig. 13.3 Sixth cranial nerve palsy. A right abducens nerve palsy due to multiple sclerosis, causing failure of abduction.

Table 13.5 Causes of abducens nerve palsy

Nuclear
Congenital, e.g. Möbius syndrome
Duane's syndrome
Tumour
Infarction
Wernicke–Korsakoff syndrome

Fascicular
Demyelination
Infarction
Tumour
Wernicke–Korsakoff syndrome

Subarachnoid
Meningitis
Subarachnoid haemorrhage
Post-infectious
Clivus tumour
Trauma
Compression by aneurysm or ectatic vessels
Sarcoidosis

Petrous
Mastoid or petrous bone tip infection
Fracture of petrous bone
Aneurysm
Thombosis of inferior petrosal sinus

Downward displacement of the brain-stem
Supratentorial mass (raised intracranial pressure)
Following lumbar puncture, epidural anaesthesiae

Trigeminal schwannoma

Cavernous Sinus and Superior Orbital Fissure
Carotid-aneurysm
Tumour (meningioma, nasopharyngeal carcinoma, pituitary adenoma)
Carotid-cavernous fistula
Thombosis
Dural arterio-venous malformation
Tolosa–Hunt syndrome
Herpes zoster
Sinusitis

Orbital
Tumour

Localisation Uncertain
Infarction (often associated with hypertension or diabetes)
Migraine

Intradural, extraaxial lesions The abducens nerve may leave the caudal border of the pons as a single or double trunk, the latter possibility explaining the partial palsies which may occur due to trauma or raised intracranial pressure. The nerve passes almost vertically in front of the clivus where it may be damaged by an enlarged ectatic basilar artery or by tumours such as a chordoma, meningioma, or nasopharyngeal carcinoma (Volpe and Lessell 1993). Minor head trauma resulting in an abducens paresis should raise the suspicion of a clivus or parasellar tumour. In the subarachnoid space the nerve is liable to damage from meningeal inflammation, especially due to secondary carcinoma and any infective organism, particularly tuberculosis. The nerve then enters the dura, medial to the trigeminal nerve, and passes under the petroclinoid ligament in Dorello's canal. It is fixed at this point and hence downward displacement of the brainstem due to a supratentorial lesion may result in unilateral or bilateral abducens palsies a 'false localizing sign'. Partial palsies have been reported after lumbar punctures or myelography, not necessarily associated with raised intracranial pressure (Bell *et al.* 1994). They usually resolve after a few days to weeks after the pressure has been normalized.

Petrons bone lesions Before the nerve passes into the cavernous sinus it lies on the medial tip of the petrous temporal bone where it is susceptible to damage from trauma, particularly as a result of a longitudinal fracture of the temporal bone. The commonest cause for abducens nerve involvement at this site is infection, usually in the middle ear or mastoid, leading to a petrositis with or without thrombosis of the inferior petrosal sinus. The trigeminal ganglion and facial nerve lie nearby, often resulting in associated pain in the face or eye, and a facial paresis may occur in addition to the abducens palsy *Gradinigo's syndrome.*

Cavernous sinus lesions On entering the cavernous sinus the nerve lies lateral to the internal carotid artery and medial to the ophthalmic division of the trigeminal nerve, in close proximity to the oculomotor and trochlear nerves. Despite this, lesions in the sinus often lead to an isolated abducens nerve palsy, probably because it is not tethered to the dural wall. For a short distance the pupillo-sympathetic fibres run with the abducens nerve as they pass from the internal carotid artery to the first division of the trigeminal nerve. Tumour, inflammation, dural arteriovenous malformation, or intracavernous aneurysm of the internal carotid artery may affect the abducens nerve in this location. An isolated palsy may also be the first indication of the contralateral spread of a cavernous sinus thrombosis. When the nerve passes through the superior orbital fissure to innervate the lateral rectus muscle it may be compressed by tumours of the skull base, such as nasopharyngeal carcinoma, often with involvement of the other ocular motor nerves, when facial pain and proptosis may also be a feature. Although isolated chronic abducens nerve palsies require investigation, many have benign causes (Savino *et al.* 1982). Bilateral abducens nerve palsies, in contrast to unilateral palsies, are most commonly due to demyelination, subarachnoid haemorrhage, meningitis, tumours, Wernicke's encephalopathy, and raised intracranial pressure. These palsies need to be carefully differentiated from convergence spasm and divergence paresis.

Management Initial management of a patient with an isolated abducens nerve palsy should be with either monocular occlusion or press-on Fresnel prisms (base-out). Botulinum toxin injection to the ipsilateral medial rectus muscle may also help restore single binocular vision and may prevent the development of medial rectus contracture (Lee 1992). Surgery is usually offered only after 6–12 months with no sign of recovery.

Childhood palsies A transient abducens palsy is occasionally present in the newborn which resolves within approximately 6 weeks (Knox *et al.* 1967). It is important to differentiate this from congenital abnormalities such as Duane's syndrome or congenital esotropia with cross-fixation. Full abduction can be achieved in the latter by the doll's head manoeuvre or by patching one eye for a week. Several diseases in children lead to isolated abducens palsies, which may be the first sign of a posterior fossa tumour (Robertson *et al.* 1970). When it is associated with a gaze palsy a brainstem glioma is suggested, and if associated with cerebellar dysfunction, an astrocyoma, ependymoma, or medulloblastoma. An abducens nerve palsy sometimes follows an upper respiratory tract infection or measles immunization, or develops during a chickenpox infection usually with full recovery. It should, however, always be remembered when a child develops a sudden ocular deviation that this may be due to a unilateral loss of vision resulting from a tumour of the retina or anterior visual pathways, or the battered child syndrome.

13.1.5 Combined ocular motor nerve palsies

It is important to distinguish multiple ocular motor palsies from orbital disease due to dysthyroid eye disease, myasthenia gravis, and progressive myopathy or chronic progressive external ophthalmoplegia. This can usually be achieved by careful consideration of the tempo of progression and associated signs such as pupil involvement, response to edrophonium, and the forced duction test. Unilateral multiple ocular motor nerve palsies are usually associated with lesions involving the cavernous sinus or superior orbital fissure. If bilateral, a wide range of possible diagnoses must be considered (Table 13.6). An isolated or multiple ocular motor nerve palsy associated with pain in or around the eye constitutes the syndrome of painful ophthalmoplegia which again has a wide differential diagnosis (Table 13.7).

Brainstem lesions In the brainstem several conditions can involve the three ocular motor nerves. In particular, Wernicke's

Table 13.6 Causes of multiple ocular motor palsies

Brainstem
Tumour
Infarction or haemorrhage
Motor neuron disease
Leigh disease

Subarachnoid
Meningitis (infective and neoplastic)
Trauma
Clivus tumour
Aneurysm
Wernicke–Korsakoff syndrome

Cavernous Sinus and Superior Orbital Fissure
Aneurysm
Tumour (meningioma, pituitary adenoma with apoplexy; metastases, especially nasopharyngeal carcinoma)
Thrombosis
Tolosa–Hunt syndrome
Herpes zoster
Neurosurgical complication
Infarction
Carotid-cavernous fistula
Mucormycosis and other fungal infections
Sphenoid sinus mucocoele

Orbital
Trauma
Tumour
Inflammatory pseudotumour
Sinusitis

Localization Uncertain
Toxins
Postinflammatory neuropathy (Guillain–Barré and Fisher syndrome)
Arteritis
Behçet disease

encephalopathy in which ophthalmoplegia is usually associated with nystagmus, altered mental status, and ataxia, should always be considered, since administration of thiamine can rapidly reverse the ophthalmoplegia. Other conditions are Bickerstaff's brainstem encephalitis (see below) and rarely in motor neurone disease, usually when the patient's course has been artificially prolonged with long-term ventilatory support.

Meningeal causes Various combinations of involvement of the ocular motor nerves, sometimes with other cranial nerves, in the subarachnoid space may occur with acute and chronic bacterial, fungal, tuberculous, syphilitic or borrelial meningitis, and sarcoid, carcinomatous or lymphomatous meningitis.

Trauma When multiple ocular nerve palsies occur as a result of trauma the head injury is usually severe and associated with fractures of the sphenoid, petrous temporal, or orbital bones (Lepore 1995). They can be confused with blowout fractures of the orbit which lead to restricted eye movements, particularly of upward gaze, due to prolapse of the inferior rectus muscle through the bony defect in the orbital floor.

Skull base lesions Lesions of the skull base can lead to a combination of ocular motor palsies. These include metastatic tumours typically from breast, lung, or prostate primary tumours. Other primary tumours in this region include sphenoid wing or clival meningiomas, chordomas, and chondrosarcomas.

Cavernous sinus lesions Many different disease processes may affect the ocular motor nerves in the cavernous sinus; differentiation between lesions at this site or at the orbital apex is suggested by sensory disturbance in the trigeminal distribution in the former, and by proptosis and visual loss in the latter. The commonest causes for multiple ocular motor palsies in the cavernous sinus are aneurysms and tumours. In a large series of cases of cavernous sinus syndrome Thomas and Yoss (1970) found the differentiation

Table 13.7 Painful ophthalmoplegia syndromes

Subarachnoid
Aneurysm (posterior communicating artery or basilar)
Carcinomatous meningitis

Cavernous Sinus and Superior Orbital Fissure
Aneurysm
Tumour (meningioma, chordoma, pituitary adenoma, nasopharyngeal carcinoma, lymphoma, metastases)
Cavernous sinus thrombosis
Tolosa–Hunt syndrome
Herpes zoster
Carotid-cavernous fistula
Sphenoid sinus carcinoma
Petrositis (Gradenigo's syndrome)

Orbital
Metastatic tumour
Lymphoma
Inflammatory pseudotumour
Contiguous sinusitis
Mucormycosis or other fungus infections

Localization Uncertain
Migrainous ophthalmoplegia
Diabetic ophthalmoplegia
Cranial arteritis

was not possible clinically by analysis of the mode of onset, presence or absence of pain, the pattern of neurologic deficit, or the response to steroids. However, in a series of patients with meningiomas and aneurysms of the cavernous sinus Trobe *et al.* (1978) found that patients with meningiomas tended to be aged over 70 years, systemically healthy and pain free, with a subtle onset of symptoms and insidious progress. Aneurysms presented in patients, who were usually women over 70 years with hypertension or cardiovascular disease, acute severe orbital pain or trigeminal, first and second divisions, dysaesthesiae at onset, early abduction defect due to early involvement of the abducens nerve, and with negligible or an explosive progression. Visual loss secondary to compression of the anterior visual pathway is a late consequence of large aneuryismal expansion. Intracavernous aneurysms, which account for only 2 per cent of intracranial aneurysms, may expand rapidly but rarely rupture, in which case the dural envelope of the cavernous sinus usually contains the haemorrhage and a carotid-cavernous fistula is formed with obvious physical signs. The aneurysm or fistula may be treated electively if symptomatic by endovascular coiling or carotid occlusion.

In the *cavernous sinus syndrome* it is commonly found that when the oculomotor nerve is compressed there appears to be relative pupil sparing. It has been suggested that this is due to coincident sympathetic and parasympathetic paresis. It is often difficult to clinically differentiate between meningioma, intracavernous aneurysm, and nasopharyngeal carcinoma or other metastatic tumours, which are the commonest cause of a cavernous sinus syndrome, occurring in 20 per cent. In the case of multiple ocular motor nerve palsies, with or without pain, in which CT and MR scanning has failed to localize a lesion, a 'blind' nasopharyngeal biopsy may be positive even in the absence of visible nasopharyngeal tumour. Metastases from other sites may infiltrate the cavernous sinus and progressive involvement of the ocular motor and other cranial nerves may be the presenting signs of carcinomatous meningitis.

Pituitary tumours may suddenly expand laterally into the cavernous sinus. This is usually due to infarction of the tumour leading to pituitary apoplexy. Patients with this condition usually present with a sudden onset of severe headache, multiple ocular motor palsies, which are often bilateral, variable degrees of visual loss, and signs of endocrine insufficiency.

Cavernous sinus thrombosis may occur as a complication of infectious and non-infectious processes, and may be life-threatening demanding prompt recognition. Septic thrombosis of the cavernous sinuses is most commonly due to staphylococcal and streptococcal organisms. These organisms gain entry to the cavernous sinus via the valveless veins from the middle third of the face, paranasal and usually sphenoid sinusitis, dental abscess, and less often otitis media. They may also spread from the maxillary and sphenoid sinuses. Fever is a nearly constant feature, but headache may not be prominent. Periorbital oedema, chemosis, proptosis, and limitation of extraocular movements, especially lateral gaze due to abducens nerve involvement, develop in almost all recognized cases. Involvement of the opposite eye frequently appears within 2 days following the onset of unilateral signs when the infection spreads across to the contralateral cavernous sinus. Most patients have elevated peripheral white cell count and positive blood cultures. MRI is the diagnostic procedure of choice. Treatment includes immediate intravenous antibiotics, which should include therapy against penicillinase-resistant staphylococci and anaerobes, and often surgical drainage of the primary site of infection. The use of anticoagulation in this condition is still controversial but early use of heparin may lead to an improved outcome (Levine *et al.* 1988). Less than half the patients recover completely, and the mortality is approximately 30 per cent.

Painful ophthalmoplegia

The development of an acute or subacute painful ophthalmoplegia demands extensive investigation of the patient to exclude an aneurysm, tumour, or one of the rarer causes. In particular, fungal infections should be considered in diabetics and immunocompromised individuals. Mucormycosis and more rarely aspergillosis may rapidly spread from the sinuses to the cavernous sinus and orbit. MRI will delineate the extent of invasion of the fungus. Patients at risk of this infection may require its urgent exclusion by sinus mucosal biopsies, since a favourable outcome with intravenous amphotericin B and surgical debridement, is only possible if treatment is instituted early.

Tolosa–Hunt syndrome Once these other causes have been excluded then the diagnosis of a non-specific granulomatous inflammation in the region of the cavernous sinus resulting in the Tolosa–Hunt syndrome should be considered (Lakke 1962; Kline 1982). The criteria for the diagnosis of the syndrome (Hunt *et al.* 1961) are as follows:

- The pain may precede the ophthalmoplegia, is located behind the eye, which may be proptosed, and has a steady 'boring' or 'gnawing' quality.

- Any combination of ocular motor nerves may be involved with or without the ophthalmic branch of the trigeminal nerve and oculosympathetic nerves, and in about 20–30 per cent of cases the optic nerve or maxillary branch of the trigeminal nerve are involved.

- The symptoms are acute or subacute in onset lasting for days or weeks, and spontaneous remissions may occur with partial or complete regression of deficits.

- The symptoms often rapidly respond to large doses of corticosteroids.

- Attacks may recur at intervals of months or years.

- Exhaustive studies including CT, MRI imaging, and angiography show no evidence of involvement of structures outside the cavernous sinus.

It is clear that the so-called Tolosa–Hunt syndrome may be caused by a spectrum of inflammatory processes, both granulomatous and non-granulomatous inflammation. This condition cannot be clearly distinguished pathologically from the lesions causing the painful superior orbital fissure syndrome and orbital pseudotumour.

MR venography is a useful investigation in these cases often showing obstruction of the cavernous sinus or superior ophthalmic vein. In addition, irregularities of the intracavernous portion of the internal ophthalmic artery may be found. MR imaging often shows an abnormal soft tissue area in the cavernous sinus, with intermediate to high signal on T1-weighted images and enhancement of the abnormal area with gadolinium. These abnormalities reflect the low-grade inflammatory response in the cavernous sinus which has been found pathologically, and which have been shown to disappear with corticosteroids.

It is important to note that a similar systemic steroid responsiveness may be observed with other lesions in the superior orbital fissure and cavernous sinus, which include tumours and aneurysms. It is therefore important that complete neuroradiological investigations are carried out in patients with the syndrome of painful ophthalmoplegia.

The aetiology of the Tolosa–Hunt syndrome is poorly understood. Mathew and Chandy (1970) identified a high prevalence of parasitic infections and tuberculosis in their patients, which suggested that the syndrome may be the result of an unusual immune reaction to endemic infections. Most cases of the condition have no evidence of systemic disease, although some cases have positive serology for systemic lupus erythematosus and a raised ESR. The evidence for a generalized connective tissue disease or endemic infection as being the underlying cause is poorly substantiated. An excellent review of the painful ophthalmoplegia syndrome and Tolosa–Hunt syndrome in particular, has been written by Kline (1982).

Polyneuropathies

Involvement of the third, fourth, and sixth cranial nerves may occur in typical Guillain–Barré syndrome (Section 21.10.1). In the variant of this condition, the *Miller Fisher syndrome*, an external, and often internal, ophthalmoplegia develops in association with ataxia and areflexia (Section 21.10.5). Because the ophthalmoplegia is often incomplete and the resulting paresis symmetrical, suggesting a horizontal or vertical gaze palsy, some authors have suggested that some cases of the syndrome may be due to a central lesion which has been called Bickerstaff's brainstem encephalitis, a monophasic illness. In this condition, typically preceded by an infection or immunization, the patient is stuporosed, and has an ophthalmoparesis associated with ataxia, brisk reflexes, and occasionally a CSF pleocytosis. However, the majority of cases are probably associated with a peripheral demyelinating neuropathy (Berlit and Rakicky 1992). As in Guillain–Barré syndrome, many of these cases have been found to have evidence of *Campylobacter jejuni* infection, and to have autoantibodies against certain gangliosides in their serum, particularly anti-GQ$_1$b IgG antibody (Chiba *et al.* 1992). These antibodies may be found in some patients with chronic ophthalmoplegia of unknown cause (Reddel *et al.* 2000). There is a growing consensus that Miller Fisher syndrome and Bickerstaff's brainstem encephalitis may represent a spectrum of a similar disease process (Al-Din 1987) since both are self-limited and share the anti-GQ$_1$b antibodies. Most patients with the Miller Fisher syndrome improve completely in 8–12 weeks without treatment.

13.1.6 Muscular and neuromuscular junction disorders

Disorders of the neuromuscular junction

A number of different diseases affecting transmission at the neuromuscular junction may produce ocular motor disorders.

Botulism Contaminated food or infected wounds may lead to an elaboration of the toxin of *Clostridium botulinum* (Miller and Moses 1977). This neurotoxin blocks the release of acetycholine from the presynaptic nerve terminals and may lead to varying degrees of internal and external ophthalmoplegia, ptosis, and bilateral facial weakness (Section 24.10.5).

Myasthenia gravis. The commonest disorder affecting the neuromuscular junction is myasthenia gravis, an autoimmune disorder affecting the postsynaptic acetylcholine receptor

(Weinberg *et al.* 1994) (Section 24.10.1). The presenting symptoms are ocular with ptosis or motility disorders, due to weakness of levator palpebrae superioris or extraocular muscles respectively, in about 50 per cent of cases. During its course ocular involvement occurs in 90 per cent (Oosterhuis 1982). Half remain as 'ocular myasthenics' and the other 50 per cent develop generalized features usually within 2 years. The risk of developing generalized involvement after presentation with ocular myasthenia reduces to about 15 per cent after 2 years (Bever *et al.* 1983). Since the disorder is one of muscle fatiguability and spontaneous remissions, it is not surprising that the ocular signs and symptoms fluctuate over hours or weeks. The commonest sign is lid ptosis, which is usually asymmetrical and may be especially pronounced on sustained upgaze. The contralateral lid may be elevated due to the increased innervation required by the ptotic lid, which when covered results in the normal lid returning to normal. Rapid shifts of ptosis from one eye to the other are considered pathognomonic of the disorder (Osserman 1957), as is the lid 'twitch' sign described by Cogan (1965). In this, rapid refixations from downgaze to the primary position result in transient lid retraction followed by a slow droop to the ptotic position or else it twitches several times before settling into a stable position. Forced eyelid closure may lead to fatigue of the orbicularis oculi muscle resulting in the eye 'peeking' at the examiner. Patients with myasthenia gravis have an increased prevalence of thyroid eye disease, which may result in bilateral or unilateral eyelid retraction, the latter without contralateral ptosis.

Myasthenia gravis is the 'great mimicker' of ocular motor disorders and may produce pseudostrabismus, any muscle may be involved to give the appearance of pupil sparing oculomotor, trochlear and/or abducens palsies, and mimic supranuclear conditions which normally are associated with central lesions such as internuclear ophthalmoplegia with abducting nystagmus, one-and-a-half syndrome and conjugate gaze palsies. The medial rectus is the most commonly affected muscle, but muscle fatigue may be seen on sustained upward and lateral gaze. Apart from these ophthalmoplegias, myasthenia can result in a number of saccadic abnormalities including; slow saccades, slowing after repeated refixations, and saccadic dysmetria, as well as increasing nystagmus on sustained lateral gaze. It is still not clear why there is a predilection for the levator and extraocular muscles in myasthenia gravis although several hypotheses have been proposed:

- extraocular muscles show several anatomical and physiological differences from limb muscles;

- these properties of extraocular muscles make them particularly sensitive to a loss of functional acetylcholine receptors;

- the antigenic properties of extraocular muscles may differ from those of skeletal muscle;

- minimal weakness of extraocular muscles are likely to be symptomatic, in contrast to the limb muscles (Kaminsky *et al.* 1990).

It is generally agreed that pupillary reflexes in patients with myasthenia gravis appear clinically normal.

Prolonged ocular involvement in myasthenia gravis may lead to a chronic or 'fixed' ophthalmoplegia which fails to improve with anticholinesterase medication. The resulting symmetrical external ophthalmoplegia, ptosis, and facial weakness may make separation from chronic progressive external ophthalmoplegia difficult, but a slow symmetric progressive course without fluctuations or remissions favours the latter.

When there is a moderate or marked deficit of lid elevation or ocular motility the diagnosis of myasthenia gravis is best confirmed by the edrophonium or Tensilon test. To increase the objective sensitivity of the test the response of the ophthalmoplegia can be assessed by the Hess chart, prisms, or the Lancaster red-green test performed before and 1–2 min after the injection of edrophonium. Ocular deviations may actually get worse if the muscles are differentially responsive to edrophonium, in which case the test is still considered positive. About 50–75 per cent of patients with pure ocular myasthenia were found to have anti-acetylcholine receptor antibody, and abnormal jitter on single fibre electromyographic examination of skeletal muscle was found in 50 per cent of such cases (Kelly *et al.* 1982). Another simple test, the ice pack test (Ertas *et al.* 1994), which has a high degree of sensitivity and specificity and is useful in patients with a cardiac condition in whom the edrophonium test is contraindicated, may be used. Local cooling, using a bag containing ice is placed over the ptotic lid for 2 min and following removal the size of the palpebral fissure is measured and compared with the size before cooling. Patients with myasthenia gravis usually show a difference of greater than 2 mm indicating an improvement in the levator strength. Because a thymic tumour is found in about 10 per cent of patients with myasthenia gravis, part of the evaluation of a patient suspected of having the disease should include a CT scan of the mediastinum.

It is commonly found that the paresis of the extraocular muscles responds poorly to anticholinesterase drugs, although the ptosis may respond more favourably. For this reason steroid therapy has been used in ocular myasthenia and often results in considerable and sometimes complete resolution of ocular symptoms (Oosterhuis 1982). Patients with ocular myasthenia can be commenced on low doses of daily or alternate day steroids (equivalent to 10 mg per day). The dose is gradually increased until the desired effect is achieved. A year later a gradual reduction should be attempted to see if the symptoms reappear (Weinberg *et al.* 1994).

When diplopia becomes troublesome occlusion of one eye is the best initial measure since prisms are unhelpful because of the fluctuations in the angle of the optical axes. The ptosis may be relieved with a ptosis hook attached to spectacles but, if chronic, the patient may be helped by ptosis surgery.

Lambert–Eaton syndrome In contrast to myasthenia gravis, ocular symptoms are rare in Lambert–Eaton syndrome, but mild ptosis and both clinical and subclinical ocular motor involvement does occur in some patients. Autonomic involvement may lead to dry eyes and sluggish pupillary responses (Section 24.10.2).

Ocular myopathies. A progressive limitation of ocular motility, accompanied by ptosis but usually without diplopia or pupillary abnormalities occurs in many diseases (Table 13.8). There are several subgroups of myopathies predominantly affecting the extraocular muscle, *chronic progressive external ophthalmoplegia*, and these are often accompanied by a variety of other findings and have been called 'ophthalmoplegia plus' (Drachman 1968; Petty *et al.* 1986).

Oculopharyngeal dystrophy. This is inherited as an autosomal dominant trait mapped to chromosome 14q11.2-13 in the region of the gene for myosin where a guanine–cytosine–guanine repeat has been demonstrated (Blumen *et al.* 1999) (Section 24.2.6). The marked ptosis with some restriction of ocular motility is associated with wasting of the temporalis muscle and weakness of the bulbar

Table 13.8 Classification of progressive ophthalmoplegia

Site uncertain
Ophthalmoplegia and ptosis, congenital and late forms, sproradic and genetic
Ophthalmoplegia alone
Ptosis alone

Ocular myopathies
Ocular and other cranial muscles
  Oculopharyngeal muscular dystrophy (genetic)
  Oculopharyngeal myopathy (sporadic)
Ocular and proximal limb muscles
Ocular and distal limb muscles
Myotonic dystrophy
Myotubular or centronuclear myopathy
Ophthalmoplegia, glycogen storage, and abnormal mitochondria
Ophthalmopathy of Graves' disease (euthyroid, hypothyroid, hyperthyroid)
Ocular myositis (orbital pseudotumour)
Congenital myopathic ptosis or ophthalmoplegia
  Limb weakness
  Anomalous insertion of ocular muscles
  Some cases of Möbius syndrome

Disorders of neuromuscular junction
Curare-sensitive ocular myopathy
Myasthenia gravis

Neural ophthalmoplegias
Nuclear and supranuclear abnormalities
  Congenital: Möbius syndrome; isolated ophthalmoplegia
  Ophthalmoplegia with central myelopathy or encephalopathy of later onset: mental retardation, hereditary ataxias, hereditary spastic paraplegia, hereditary multisystem disease, dystonia musculorum deformans, abetalipoproteinaemia (Bassen–Kornzweig), progressive supranuclear bulbar palsy (Steele–Richardson–Olszewski)
  Ophthalmoplegia with motor neuron disease: infantile spinal muscular atrophy (Werdnig–Hoffman), juvenile spinal muscular atrophy simulating muscular dystrophy (Wohfart–Kugelbertg–Welander)
  Ophthalmoplegia, retinitis, cardiopathy, and neural disorder (Kearns–Sayre)
Peripheral neuropathies

muscles. The onset is usually in the fifth and sixth decades and mild ptosis usually precedes the dysphagia by years (Murphy and Drachman 1968). Sporadic isolated cases have been reported but may represent poor case ascertainment or reduced penetrance. *Myotonic dystrophy* may give rise to slowed eye movements due to involvement of the extraocular muscles (Ter Bruggen *et al.* 1990) (Section 24.3).

Kearns–Sayre syndrome. Although chronic progressive external ophthalmoplegia may occur in association with a number of other defects those found in the Kearns–Sayre syndrome are the most varied (Kearns and Sayre 1958) (Section 24.6.3). The onset of this condition, in which bilateral ophthalmoparesis is associated with symmetric ptosis, is within the first or second decades, without any family history, is associated with retinal pigmentary degeneration, and at least one of cardiac conduction abnormalities, raised CSF protein to <100 mg/dl, and ataxia. Excessive ragged red fibres are found in peripheral muscle with trichrome staining methods. Other neurological features may be observed (Table 13.9). The sequence

Table 13.9 Features associated with Chronic Progressive External Ophthalmoplegia (CPEO)

Cardinal manifestations

CPEO onset <20 years

Retinal pigmentary degeneration

Heart block

Elevated CSF protein

Negative family history

Myopathy affecting skeletal muscles (ragged-red fibres)

Spongiform encephalopathy

Associated manifestations

Short stature

Hearing loss

Cerebellar ataxia

Corticospinal tract signs

Impaired intellect

Cranial muscle weakness (face, palate, neck)

Peripheral neuropathy

Pendular nystagmus

Corneal clouding

Scrotal tongue

Slowed EEG

Hypogonadism

Endocrine abnormalities (steroid, calcium, glucose metabolism)

Basal ganglia calcification

Elevated creatinine phosphokinase, SGOT, LDH

Abnormal lactate–pyruvate metabolism

of manifestations varies and the cardiomyopathy may be delayed for years. Pathologically there is a spongy degeneration of the brain. The disease is now characterized as a mitochondrial cytopathy in which a deletion from the circular strand of mitochondrial DNA may result in defects of the intracellular respiratory chain. These are large deletions of 1.3–9.1 kb, the commonest of which are from positions 8470 and 13460. Approximately 50 per cent of patients with chronic progressive external ophthalmoplegia and 90 per cent of patients with Kearns–Sayre syndrome have demonstrable mitochondrial deletions (Newman 1992).

Familial varieties of chronic progressive external ophthalmoplegia have been described which are inherited either autosomal dominantly or recessively where multiple mitochondrial deletions have been observed.

Chronic progressive external ophthalmoplegia must be differentiated from a number of conditions. In progressive supranuclear palsy full ocular rotations to oculocephalic manoeuvres are maintained. Chronic ocular myasthenia may be confused with chronic progressive external ophthalmoplegia, especially since there may be a lack of response to edrophonium; but a progressive course lacking fluctuations or remissions favours chronic progressive external ophthalmoplegia. Dysthyroid restrictive myopathy usually has associated lid retraction, proptosis, or congestive conjunctival signs which are absent in chronic progressive external ophthalmoplegia.

13.1.7 Congenital abnormalities of ocular motor innervation

A number of different congenital abnormalities have been described in which there is ocular motor paresis, often associated with synkinesis of movement of other eye and lid muscles. It is important that these congenital conditions should be distinguished from acquired ocular motor disorders so that unnecessary investigations are not undertaken.

Möbius syndrome. In this condition there is a variable degree of facial diplegia associated with a disturbance of horizontal eye movements, most commonly a failure of abduction. In about 25 per cent of cases there is a total external ophthalmoplegia. Other abnormalities include tongue atrophy, cleft palate, and various musculoskeletal dysplasias involving the head and neck, chest, and upper extremities. The diversity of pathological findings in patients with Möbius syndrome suggests that the syndrome is actually a heterogeneous group of congenital disorders which in some cases are due to developmental defects, and in others due to acquired hypoxic or other insults (Towfighti *et al.* 1979).

Duane's retraction syndrome. This syndrome is due to abnormal development of the abducens nucleus, and is so named because of co-contraction of the medial and lateral rectus muscles leading to retraction of the globe with narrowing of the palpebral fissure, which occurs on attempted adduction in association with limited or absent abduction. Duane's syndrome is usually unilateral, the left eye being more frequently affected than the right, and is bilateral in 15–20 per cent of cases. The condition may be familial and sometimes associated with other congenital abnormalities such as Klippel–Fiel anomaly, deafness, urinary tract abnormalities, and cardiac defects.

The condition occurs in three forms (Huber 1974): type I, which is the most common, consists of limited or absent abduction with relatively normal adduction; in type II there is impaired adduction and full abduction, and in type III there is impairment of both adduction and abduction. A number of electromyographic and oculographic studies have indicated abnormal innervation patterns, compatible with the clinicopathological studies which have shown hypoplastic abducens nuclei, and partial or complete innervation of the lateral rectus from branches of the inferior division of the oculomotor nerve (Miller *et al.* 1982). Patients with Duane's syndrome usually have excellent visual adaptation resulting in absence of diplopia, good stereopsis, and fusion in directions of gaze where the visual axes are aligned.

Although Duane's syndrome is usually sporadic it may be familial and one large family showed linkage to the condition at chromosome 2q31 (Appukuttan *et al.* 1999).

Congenital elevator palsies. In the congenital 'double elevator palsy' there is paresis of both the superior rectus and inferior oblique muscles in one eye. Since the eyes are straight in the primary position and the Bell's phenomenon is preserved, it is considered to be a supranuclear paresis of monocular elevation. This condition may develop in later life when it is usually due to a small discrete vascular lesion in the pretectum. Such a lesion would disrupt the efferent fibres from the rostral interstitial nucleus of the medial longitudinal fasciculus to the inferior oblique subnucleus and the contralateral superior sub-nucleus, which innervates the superior rectus muscle contralateral to it.

Marcus Gunn jaw-winking phenomenon. This is an example of anomalous innervation in which a unilateral ptosis of variable extent is noted shortly after birth. When the baby suckles the ptotic lid rhythmically jerks and is intermittently retracted, as it does later with chewing and jaw movements. Two major groups are described: the commonest is external pterygoid-levator synkinesis with lid elevation when the jaw is moved to the opposite side, and internal pterygoid-levator synkinesis with lid elevation on clenching the jaw closed (Sano 1959).

13.1.8 **Assessment of diplopia**

Assessment of a patient complaining of diplopia, which includes taking a history and examining the static eye movements (Section 13.1.1), aims to determine which muscles are involved, whether the cause of the diplopia is due to an ophthalmological, neurological, or old congenital strabismus, and finally the cause. However, it is important to include a neurological examination, including all the cranial nerves and evidence for abnormal long tract signs, to help localize the site of the lesion. Several characteristic ophthalmoparetic or ocular motility patterns may be identified. For example, observing that an eye is in an abducted and slightly depressed position at rest, and that it fails to move in adduction, elevation, and depression, associated with a dilated pupil and ptosis, clearly indicates a complete oculomotor nerve palsy. Similarly the findings of horizontal diplopia at distance, worse on gaze to one side, or of vertical diplopia with a torsional component worse on down gaze and associated with an ipsilateral head tilt are typical of an abducens and trochlear nerve palsy, respectively. Some brainstem lesions leading to diplopia can also give typical pattern of eye movements, for example, an internuclear ophthalmoplegia in which there is slowed or absent ipsilateral adduction with nystagmus in the abducting eye. Disorders of ocular muscle can also lead to diplopia and diplopia worse on up gaze associated with impaired elevation and adduction with conjuctival injection, chemosis, proptosis, and lid retraction is a typical presentation of thyroid eye disease.

If a characteristic ocular paretic picture is not observed it is then necessary to determine which muscles/ocular motor nerves are affected, plus any associated neurological signs, and by applying knowledge of the neuroanatomy of the course of the various nerves, from muscle to brainstem nucleus, it should be possible to determine where the lesion may be located. The history relating to the tempo of onset of the diplopia will also contribute to developing a differential diagnosis. As examples, the combination of oculomotor, trochlear, and abducens nerve palsies locates the lesion to the cavernous sinus which if acute suggests a pituitary apoplexy or cavernous sinus thrombosis or if slowly progressive a mass lesion such as a giant internal carotid aneurysm.

It is important to emphasize the two great mimickers of a wide variety of both central and peripheral oculomotor abnormalities giving rise to diplopia without any accompanying abnormal brainstem signs, myasthenia gravis, and Wernicke's encephalopathy due to thiamine deficiency usually in the context of alcoholism or malnutrition (Section 34.5).

13.2 **Central disorders of eye movements**

Many different disease processes affecting the central nervous system, from the brainstem to the cortex, can give rise to supranuclear disorders of eye movements. Examination of eye movements offers a number of advantages to the neurologist over skeletal movements. These include: eye movements are directly related to the activity of brainstem neurons since the extraocular muscles lack a

stretch reflex; eye movements have limited degrees of freedom so that disordered movements lend themselves to analysis (clinical or quantitative) in three planes, horizontal, vertical, and torsional; finally there are several functional classes of eye movements, each with special physiologic properties that suit a particular purpose and which have a separate and well-segregated neural substrate. This enables the clinician to examine each of these various types of eye movements and identify abnormalities which can then provide information regarding anatomical, physiological, and pharmacological lesions (Leigh and Zee 2006).

13.2.1 **Types of eye movement and their clinical evaluation**

The various types of functional classes of eye movements all subserve the same goal, the acquisition or maintenance of the projection of an image of the object of interest onto the most sensitive part of the retina, the fovea. Rapid conjugate eye movements, saccades, enable the line of gaze to be redirected to bring the image of a new object of interest onto the fovea, and the dysjunctive or vergence eye movements ensure that these images are simultaneously placed on both foveae regardless of their distance from the observer. There is also a need to stabilize the image of the object of interest on the fovea when the object itself moves, performed by the smooth pursuit system, or when the subject's head or body moves as occurs during locomotion when the vestibular and optokinetic ocular motor reflexes are activated. These different functional types of eye movements can each be rapidly tested at the bedside (Shaunak *et al.* 1997)

Saccades. Voluntary saccade initiation should be assessed by instructing the patient to look from side to side and up and down. The patient is then asked to fixate two targets alternately—for example, a pen in one hand and a raised finger of the other—so that for each saccade the location of one or other of the targets has been briefly moved and their distance from each other varied. This generates reflexive saccades towards a novel target, which are tested in the horizontal and vertical planes, and the examiner should observe saccadic variables such as speed of initiation or latency, accuracy, and velocity. Any slowing of saccades can be accentuated by using an optokinetic striped drum or tape, when the repositioning saccades will appear clearly slowed. This is of particular help when showing slowed adducting saccades in a partial internuclear ophthalmoplegia. Another method to accentuate this abnormality is to use oblique targets. Because the velocity is slowed in the horizontal and not the vertical plane, the resulting saccade is L-shaped. Predictive saccades are tested by alternately raising a finger of one hand and then the other in a predictable and regular pattern. The patient is asked to make saccades back and forth to the moving finger. Normally after a few saccades they anticipate the appearance of the stimulus and make a saccade in advance. Finally, the patient should be observed for any head movements or head thrusts or blinks before making a saccade, as occurs in Huntington's disease and ocular motor apraxia.

Smooth pursuit. Smooth pursuit can be tested by asking the patient to track a small target, such as the head of a hat pin, at a distance of about 1 m, whilst keeping their head stationary. Both horizontal and vertical smooth pursuit should be assessed. The target should be moved initially at a slow uniform speed and the pursuit eye movements observed to determine whether they are smooth, or broken up by catch-up saccades. This is a non-specific sign when present in both directions—for example, it may be due to ageing or cerebellar disease—or it may indicate a focal posterior cortical lesion if only present in one direction, in which case the abnormal pursuit is in the direction of the lesion. The speed should be gradually increased, but at high velocities of >50 deg per second of all smooth pursuit eye movements will be broken up by saccades even in normal subjects. The optokinetic nystagmus drum and tape is a useful method to elicit a series of pursuit movements, and does not in fact elicit true optokinetic eye movements.

Optokinetic nystagmus, The optokinetic system cannot be tested as part of the clinical examination, because the optokinetic nystagmus drum and tape commonly used tests smooth pursuit and not the optokinetic system. A full field revolving striped drum is required to elicit true optokinetic nystagmus.

Vestibular system. If the vestibulo-ocular system is functioning normally passive rotation of the patient's head should result in a slow eye movement so that the eyes move in the opposite direction to that of the head movement. This is known as the doll's head or oculocephalic manoeuvre and should be performed both horizontally and vertically. This technique is not only valuable for assessing vestibular function, but also for differentiating infranuclear and nuclear gaze palsies, when the response is absent, from supranuclear gaze palsies in which a normal doll's head response is present. It is also useful in the evaluation of brainstem function in comatose patients. It should be noted that the eye movements elicited in unconscious patients by this procedure largely reflect the integrity of the semicircular canals and their central connections, whereas in conscious patients the effects of visual input on eye movements may influence the response to head rotation.

A rough estimate of any deterioration of vestibular gain, that is head velocity divided by eye velocity, can be obtained by asking the patient to read a Snellen chart while their head is being passively rotated. If there is an abnormality the visual acuity will show a deterioration compared with the acuity obtained when the head is stationary. Another bedside test of the horizontal vestibulo-ocular reflex is for the examiner to observe the patient's optic disc with an ophthalmoscope while the patient tries to fixate a distant object and shake their head from side to side. If the gain of the vestibulo-ocular reflex is normal the optic disc will appear stationary to the examiner, but if abnormal the disc will repeatedly slip from view.

The vestibulo-ocular reflex can be suppressed by activating the smooth pursuit system. This may be tested by asking the patient to fixate their thumbnails with their arms outstretched while rotating their head and trunk in harmony. Impaired cancellation of the vestibulo-ocular reflex and hence abnormal smooth pursuit are shown by observing the eye repeatedly moving off fixation due to the vestibulo-ocular reflex, followed by refixation saccades. This is a particularly useful technique for testing pursuit in patients with gaze-evoked nystagmus.

Further details of tests used to assess the vestibular system are to be found in Section 15.5.

13.2.2 Brainstem and cerebellar disorders

Anatomy and physiology of horizontal and vertical gaze

There are two main features of the brainstem neural control of horizontal and vertical gaze: an anatomic separation so that the neural substrate for horizontal gaze is located in the pons and for vertical gaze in the mid-brain, and the requirement to overcome viscous drag and resist elastic restoring forces in the orbit when making dynamic eye movements. An understanding of the neural mechanisms which generate a horizontal saccade will serve as an illustration of the principles involved. A rapid phasic contraction of the extraocular muscle is required to overcome the orbital viscosity, and a rapid, high frequency burst of nerve impulses, the pulse, is transmitted to the muscle via the ocular motor nerve. The premotor inputs to the motor neurons in the abducens nucleus arise from neurons in a region of the reticular formation which lies ventral and anterior to the nucleus, the paramedian pontine reticular formation. The equivalent premotor region for vertical gaze is the rostral interstitial nucleus of the medial longitudinal fasciculus in the mid-brain, rostral to the oculomotor nucleus at the level of the red nucleus. The pulse, a velocity signal, is generated by cells called *burst neurons*, and must be of an appropriate size to ensure that the fovea of the eye is aligned to the target. Once the saccade has been completed it is necessary to maintain the new position of the eye against orbital viscoelastic restoring forces. The muscle must, therefore, now maintain a sustained tonic contraction to counter these forces and this is achieved by the tonic innervation, the step, which is a position signal the motor neuron receives from so-called *integrator neurons*, which integrate the step in a mathematical sense, lying in the nucleus prepositus hypoglossi and the medial vestibular nucleus. The pulse and step must be perfectly matched to prevent drift of the eye back to the primary position at the end of the saccade. Faulty neural integration leads to an inadequately maintained step, and after a saccade the eye drifts back in an exponential manner due to the unopposed orbital elastic restoring forces, followed by a saccade to refixate the target. This pattern leads to gaze-evoked nystagmus and is observed in cerebellar disease and anticonvulsant or sedative intoxication. An abnormal pulse may either be of reduced duration or of reduced firing frequency. If the step is appropriately matched to the abnormal pulse a reduced duration will result in a reduced amplitude or hypometric saccade, whereas if the firing frequency is reduced a saccade of reduced velocity but of normal amplitude will be generated.

The final neuron in the brainstem involved in saccade generation is the *omnipause neuron*, located in the raphe interpositus nucleus. These neurons are tonically active and pause before saccades in any direction. They are presumed to inhibit the burst neurons from firing except when a saccade is required.

Abnormalities of horizontal eye movements

The abducens nucleus contains two populations of neurons, motor neurons innervating the ipsilateral lateral rectus muscle and interneurons. The abducens nucleus is, therefore, the final common pathway for horizontal gaze. The axons from the interneurons cross the midline and ascend in the medial longitudinal fasciculus to the medial rectus subdivision of the oculomotor nerve nucleus (Fig. 13.4a). The final instructions for horizontal conjugate eye movements, therefore, lie within the abducens nucleus itself, so that its activation results in an ipsilaterally directed horizontal conjugate gaze movement.

Unilateral horizontal gaze palsy. A lesion of the abducens nucleus will result in a horizontal gaze palsy for all types of ipsilateral conjugate eye movements: saccades, pursuit, and vestibular. Vergence movements of the eyes are spared, however, so that adduction is

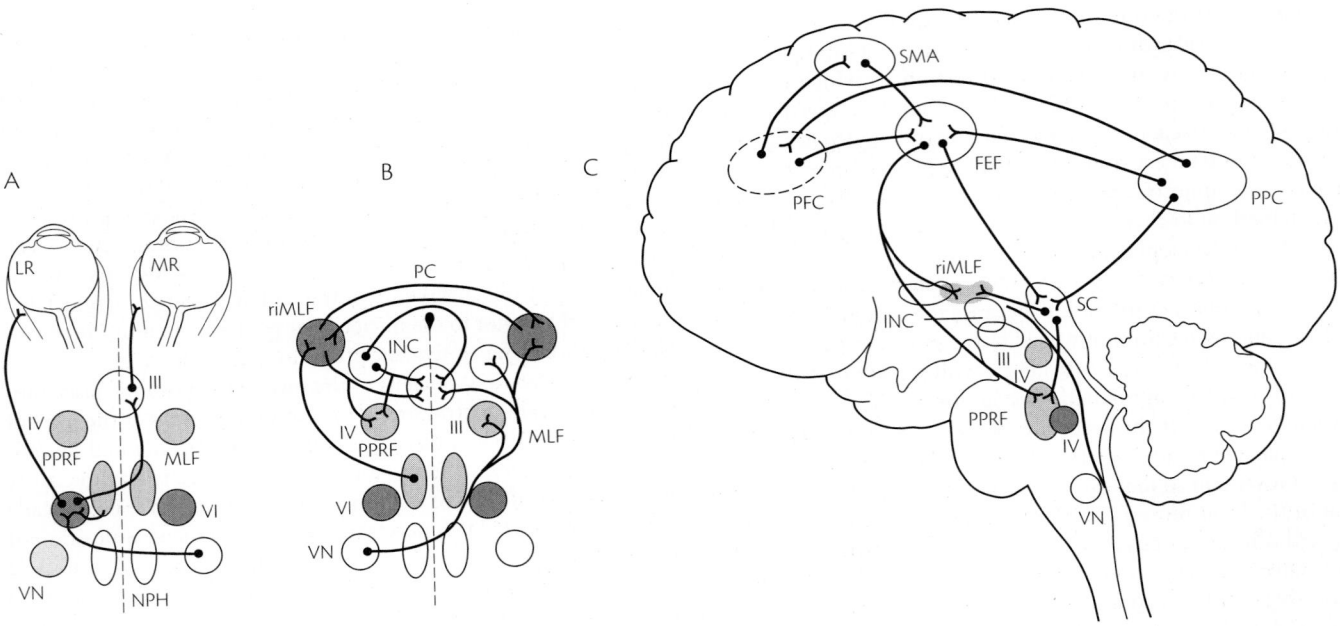

Fig.13.4 Summary of eye movement control. (a) Shows the brainstem pathways for horizontal gaze. Axons from the cell bodies located in the abducens nucleus travel to the ipsilateral lateral rectus muscle (LR), and the axons of abducens internuclear neurons cross the midline and travel in the medial longitudinal fasciculus (MLF) to the portion(s) of the oculomotor nucleus (III) concerned with the medial rectus (MR) function (in the contralateral eye). (b) Shows the brainstem pathways for vertical gaze. Important structures include the rostral interstitial nucleus of the medial longitudinal fasciculus (riMLF), paramedial pontine reticular formation (PPRF), the interstitial nucleus of Cajal (INC), and the posterior commissure (PC). Note that axons from cell bodies located in the vestibular nuclei (VN) travel directly to the abducens nuclei and, mostly via the MLF, to the oculomotor nuclei. IV = trochlear nucleus. (c) Shows the supranuclear connections from the frontal eye fields (FEF) and the posterior parietal cortex (PPC) to the superior colliculus (SC), riMLF, and the PPRF. The FEF and SC are involved in the production of saccades, while the PPC is considered to be important in the production of pursuit as well as saccades.

possible with a near stimulus (Müri *et al.* 1996). The palsy is usually associated with an ipsilateral lower motor neuron facial nerve palsy, due to involvement of the genu of the facial nerve, which passes around the abducens nerve (Fig. 13.5). A selective horizontal gaze palsy involving all saccades, including the quick phases of vestibular and optokinetic nystagmus, occurs when the lesion involves the paramedian pontine reticular formation in isolation, since the vestibular and pursuit inputs pass directly to the abducens nucleus and are therefore spared. The commonest causes for horizontal gaze palsies in adults are either vascular infarction or haemorrhage in the distribution of the pontine paramedian penetrating arteries arising from the basilar artery, demyelination, cavernous angiomas, or trauma. In children medulloblastomas or pontine gliomas are the commonest aetiologies.

Bilateral horizontal gaze palsy. A bilateral pontine lesion involving the paramedian pontine reticular formation can cause a bilateral selective saccadic palsy with preservation of vestibular and optokinetic eye movements (Hanson *et al.* 1986). Such a lesion may impair vertical eye movements since signals for vertical vestibular and smooth pursuit eye movements ascend in the medial longitudinal fasciculus and other pathways through the pons. The commonest causes of a bilateral horizontal gaze palsy, with sparing of vertical gaze, are neurodegenerative diseases such as Huntington's disease or Gaucher's disease. In a patient presenting solely with

a gaze palsy other possible causes including the Miller Fisher variant of Guillian–Barré syndrome, myasthenia gravis, Wernicke's encephalopathy, and thyroid disease.

Internuclear ophthalmoplegia. A lesion of the medial longitudinal fasciculus produces an internuclear ophthalmoplegia, in which there is weakness of adduction ipsilateral to the side of the lesion (Zee 1992) (Fig. 13.6). In a partial internuclear ophthalmoplegia adduction will be slowed, but will be completely absent in a complete lesion. Since the fibres of the medial longitudinal fasciculus carry the horizontal gaze commands subserving all types of conjugate eye movements, this adduction paresis involves not only saccades but pursuit and vestibular eye movements. The presence of intact convergence in the absence of voluntary adduction implies that the medial rectus subdivision of the oculomotor nerve is intact, and that the internuclear ophthalmoplegia is due to a caudal lesion. Cogan (1970) called this a 'posterior' internuclear ophthalmoplegia in contrast to patients with absent convergence which he called 'anterior'. However, such patients do not necessarily have a lesion involving the medial rectus subdivision of the oculomotor nucleus.

The second major feature of an internuclear ophthalmoplegia is the nystagmus on abduction in the contralateral eye. This consists of a centripetal or inward drift, followed by a corrective saccade. Several different mechanisms have been proposed to explain the

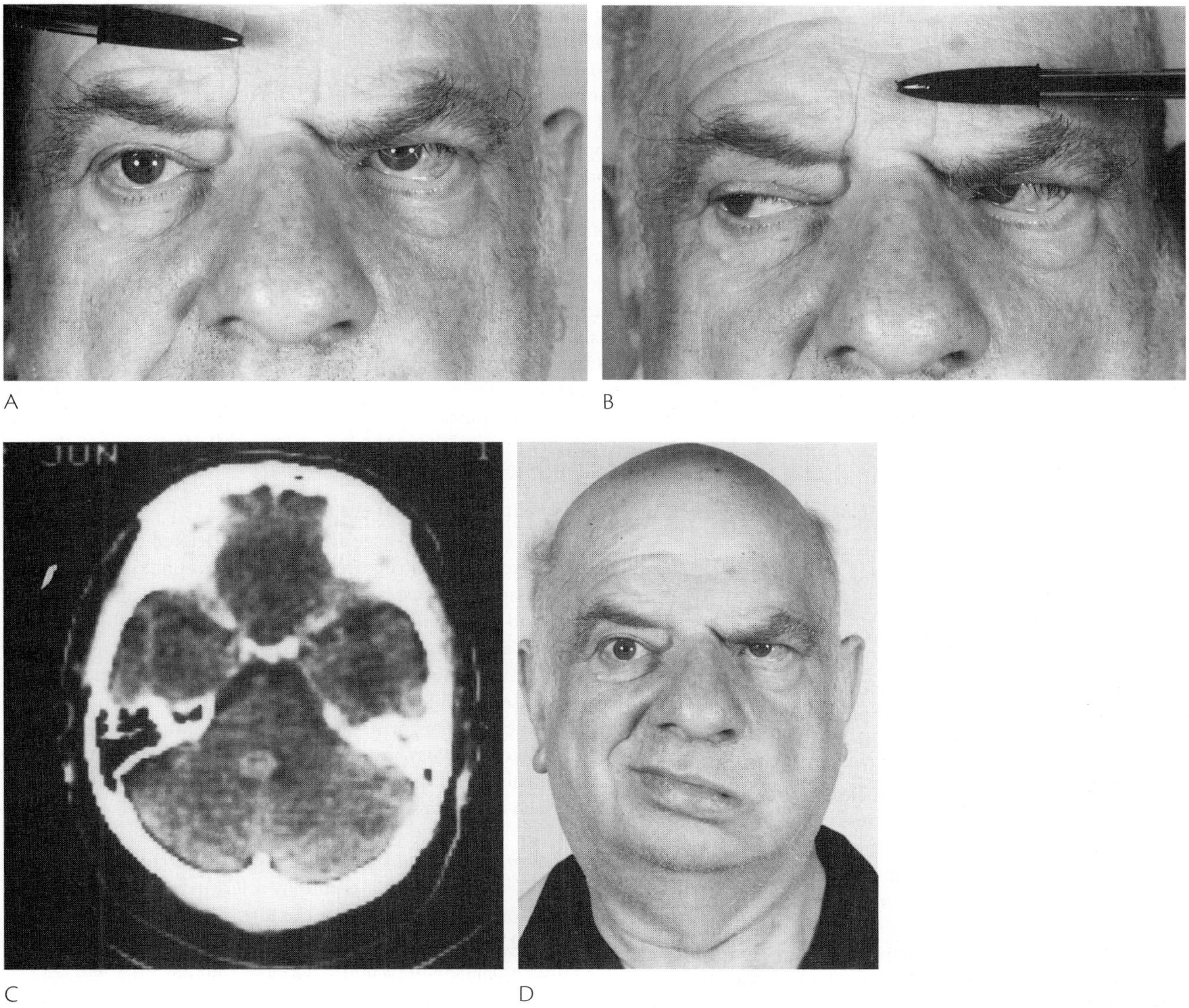

A

B

C

D

Fig. 13.5 A left conjugate gaze palsy (a) due to an arteriovenous malformation involving the left abducens nucleus (shown on CT scan), (c). No eye movement, including vestibulo-ocular, could be made into the left field of gaze. (b) Normal eye movements to the right field or gaze. This was associated with a left lower motor neuron facial nerve palsy (d).

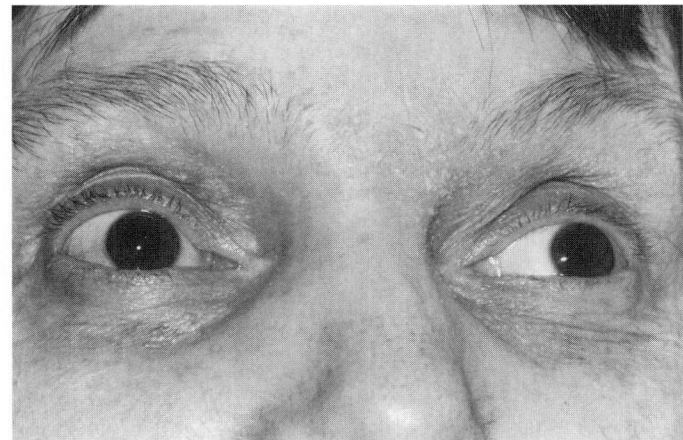

Fig. 13.6 Unilateral right internuclear ophthalmolplegia.

abducting nystagmus (Zee 1992). These include, (a) a gaze-evoked nystagmus, (b) impaired inhibition of the medial rectus contralateral to the lesion, (c) an increase in convergence tone, and (d) in response to the adduction weakness an adaptive increase in innervation to the adducting eye, which because of Hering's law of equal innervation results in a commensurate change in the innervation to the abducting eye, which leads to overshooting and postsaccadic drift giving the appearance of abducting nystagmus. The latter is generally considered the most appropriate explanation.

A skew deviation consists of a vertical misalignment of the visual axes due to a disturbance of prenuclear inputs. It is often observed in patients with a unilateral internuclear ophthalmoplegia, with the higher eye usually on the side of the lesion. Patients with bilateral internuclear ophthalmoplegias have bilateral adduction weakness and abducting nystagmus. In addition, they also have impaired vertical pursuit and vestibular eye movements, and impaired vertical

Table 13.10 Causes of internuclear ophthalmoplegia

Multiple sclerosis

Brainstem infarction or haemorrhage

Brainstem and IV ventricular tumours

Arnold–Chiari malformation and associated hydrocephalus and syringobulbia

Wernicke's encephalopathy

Infection: viral, bacterial and other types of meningioencephalitis

Metabolic disorders: abetalipoproteinaemia, Gabray's disease, hepatic encephalopathy

Drug intoxications: phenothiazines, tricyclic antidepressants, lithium, barbituates

Syphilis

Trauma, subdural haematoma

Carcinoma or paraneoplastic syndrome

Hydrocephalus

Progressive supranuclear palsy

Pseudo internuclear ophthalmolplegia: myasthenia gravis, Miller Fisher syndrome

gaze holding with gaze-evoked nystagmus on looking up or down (Ranalli and Sharpe 1988).

Patients with an internuclear ophthalmoplegia are usually asymptomatic, although if there is a complete adduction failure they may complain of diplopia especially during shifts of horizontal gaze. Occasionally they may complain of oscillopsia. A number of different aetiologies lead to an internuclear ophthalmoplegia (Table 13.10), but if unilateral the commonest is ischemia, and if bilateral, demyelination is associated with multiple sclerosis.

A rarer so-called posterior internuclear ophthalmoplegia of Lutz has been described in which there is an impairment of abduction (not adduction) of saccades and pursuit, but not vestibular eye movements. This is different to the posterior internuclear ophthalmoplegia described by Cogan in which convergence is intact. The pathogenesis of the posterior internuclear ophthalmoplegia of Lutz is unclear (Thömke *et al.* 1992).

One-and-a-half syndrome. A combined lesion of the abducens nucleus or paramedian pontine reticular formation and the adjacent medial longitudinal fasciculus on one side of the brainstem results in an ipsilateral horizontal gaze palsy and internuclear ophthalmoplegia (Wall and Wray 1983). The only preserved horizontal eye movement is abduction of the contralateral eye, and the condition is therefore termed the 'one and a half' syndrome. Although the majority of patients have no deviation or an esotropia in the primary position of gaze, some patients may habitually fixate with the horizontally immobile ipsilesional eye, which results in exotropia of the contralesional eye that has intact abduction. This condition is called paralytic pontine exotropia (Sharpe *et al.* 1974). Convergence is often preserved. Some lesions of the medial longitudinal fasciculus cause an adduction palsy due to internuclear ophthalmoplegia that is bilateral and result in exotropia in the primary position, termed a 'wall-eyed' bilateral internuclear ophthalmoplegia.

The main causes of a one-and-a-half syndrome are brainstem ischemia, haemorrhage, and tumour. The syndrome can be mimicked

by a bilateral internuclear ophthalmoplegia with an ipsilateral abducens nerve palsy.

Lateropulsion. This is a feature of lateral medullary infarction, the Wallenberg syndrome, in which there is a compelling sensation of being pulled toward the side of the lesion, accompanied by appropriate eye movement signs. During voluntary eye closure and sometimes even during blinks, the eyes deviate toward the side of the lesion, and have to make corrective saccades on eye opening to refixate the target. All ipsilaterally directed saccades overshoot the target hypermetrically, and saccades directed away from the side of the lesion undershoot the target hypometrically (Baloh *et al.* 1981). Vertical saccades have a parabolic ipsiversive trajectory. This ipsipulsion is in contrast to the overshooting of contralateral saccades, termed saccadic contrapulsion, observed in patients with infarction in the territory of the superior cerebellar artery. The eye signs of lateropulsion are considered to be due to damage to olivocerebellar projections in the inferior cerebellar peduncle (Solomon *et al.* 1995).

Abnormalities of vertical eye movements

Disturbances of vertical gaze are usually associated with damage to one or more of three structures in the mesencephalon, the posterior commissure, the rostral nucleus of the medial longitudinal fasciculus, and the interstitial nucleus of Cajal (Fig. 13.4b). The only exceptions are an apparent vertical gaze palsy due to mechanical restriction of extraocular muscles in orbital disorders such as thyroid eye disease; large acute pontine lesions involving the paramedial pontine reticular formation bilaterally producing a temporary vertical saccadic palsy, in addition to the permanent horizontal saccadic palsy; and certain degenerative disorders of the nervous system such as progressive supranuclear palsy or adult Niemann–Pick disease.

Dorsal midbrain syndrome also called pretectal syndrome or Parinaud's syndrome. This is due to a lesion which involves the posterior commissure and is associated with a variety of aetiologies (Table 13.11) and clinical features, some of which may not be present in an individual patient (Baloh *et al.* 1985). The essential sign is a loss of upward gaze involving all types of eye movement, although the vestibulo-ocular reflex and Bell's phenomenon may sometimes be spared. When acute, the eyes may be deviated downwards (the setting-sun sign), and may be observed in premature infants following intraventricular haemorrhage, and when a ventricular shunt becomes acutely blocked. Downward saccades and smooth pursuit may be impaired and downbeat nystagmus may be present.

Table 13.11 Causes of disorders of vertical gaze

Tumour: pineal germinoma or teratoma, pineocytoma glioma, metastasis

Hydrocephalus

Vascular: midbrain or thalamic haemorrhage or infarction

Metabolic: e.g. Niemann–Pick variants, Gaucher's disease

Degeneration: progressive supranuclear palsy, Huntington's disease, cortical basal degeneration

Drug-induced: barbituates, carbamazepine, neuroleptics

Miscellaneous: Multiple Sclerosis, Whipple's disease, hypoxia, syphilis

The dorsal mid-brain syndrome may also be associated with disturbances of vergence eye movements including an impairment of convergence, which is usually paralysed but may rarely be excessive and cause convergence spasm, convergence-retraction nystagmus (asynchronous convergent saccades—see Section 8.12.5), eyelid retraction (Collier's sign), and a pupillary light-near dissociation.

Selective vertical gaze palsy due to a rostral nucleus of the medial longitudinal fasciculus lesion. A unilateral or bilateral lesion of the rostral nucleus of the medial longitudinal fasciculus produces a downgaze palsy, mainly affecting saccades, or more rarely a complete vertical gaze palsy (Büttner-Ennever *et al.* 1982). Patients with unilateral midbrain lesions can develop combined upgaze and downgaze palsies, isolated upgaze palsies, an uniocular upward ophthalmoplegia with no primary position hypotropia (monocular double elevator palsy), and a vertical one-and-a-half syndrome which describes the combination of a vertical gaze palsy in one direction and a monocular vertical ophthalmoplegia in the other direction, with no primary position heterotropia (Hommel and Bogousslavsky 1991).

The ocular tilt reaction and lesions of the interstitial nucleus of Cajal. A lesion of the interstitial nucleus of Cajal, which lies immediately caudal to the rostral interstitial nucleus of the medial longitudinal fasciculus and rostral to the oculomotor nucleus, produces two distinct deficits: an ocular tilt reaction, and a deficit in vertical pursuit and vertical gaze holding (Halmagyi *et al.* 1990). The ocular tilt reaction is a head-eye postural synkinesis that consists of a skew deviation with a head tilt towards the side of the hypometric eye, and torsion of the eyes as incyclotropia of the hypermetric eye and excyclotropia of the hypometric eye. Such patients also show a deviation of their subjective vertical. Although the ocular tilt reaction is produced by a lesion of the interstitial nucleus of Cajal it can be found whenever peripheral or central lesions cause an imbalance of otolithic inputs (Brandt and Dieterich 1993).

Abnormalities of horizontal and vertical eye movements due to thalamic lesions

Lesions of the thalamus can give rise to disorders of both horizontal and vertical eye movements (Clark and Albert 1995). Conjugate deviation of the eyes contralateral to the lesion, so-called wrong-way deviation is associated with haemorrhage in the medial thalamus. Thalamic haemorrhage may also lead to forced downward deviation of the eyes, associated with convergence and miosis. Caudal lesions in the thalamus have been associated with esotropia, which although usually associated with a downward gaze deviation may be present as an isolated finding. A paralysis of downgaze is associated with a caudal thalamic infarction, due to occlusion of the proximal portion of the posterior cerebral artery or its perforator branch, the thalamosubthalamic paramedian artery. However, the ocular motor deficit may well be due to damage to the rostral interstitial nucleus of the medial longitudinal fasciculus or its immediate premotor inputs.

The effect of cerebellar lesions upon eye movements

Although it is generally accepted that the cerebellum plays an important role in the control of eye movements in man, pure lesions of the cerebellum without some brainstem involvement are unusual (Lewis and Zee 1993). This creates some difficulty in determining eye movement abnormalities specific for cerebellar dysfunction. It is appropriate to segregate lesions to three main regions of the cerebellum, each of which has a particular ocular motor syndrome: the dorsal vermis and underlying fastigial nucleus, the nodulus and ventral uvula, and the flocculus and paraflocculus. The dorsal vermis and underlying fastigial nucleus are involved in controlling saccadic accuracy and smooth pursuit. Lesions in this region lead to saccadic dysmetria, usually hypermetria, and mild deficits of smooth pursuit. The nodulus and ventral uvula are involved in the control of the low frequency response of the vestibulo-ocular reflex, and disorders in this region give rise to periodic alternating nystagmus, positional nystagmus, and impaired habituation of the vestibulo-ocular reflex, with increased duration of the vestibular responses. The flocculus and paraflocculus are concerned with retinal image stabilization during smooth tracking with the head still, gaze-holding, control of the vestibulo-ocular reflex and its suppression, and pulse-step matching. Lesions of this region, therefore, lead to impaired pursuit and vestibulo-ocular reflex cancellation with gaze-evoked, rebound, centripetal, and downbeat nystagmus; and inappropriate amplitude of the reflex. Other signs which have been associated with cerebellar lesions, although precise localization is not available, include torsional nystagmus during vertical pursuit which occurs with a lesion in the middle cerebellar peduncle, square wave jerks, esotropia with alternating skew deviation, divergent nystagmus, primary position upbeating nystagmus, and centripetal nystagmus.

The cerebellum is also important in generating long-term adaptive responses which enable eye movements to be maintained appropriate to the visual stimulus. For example, when wearing lens corrections there is a magnifying or minifying effect which requires adaptive changes in the gain of the vestibulo-ocular reflex. These changes due to cerebellar adaptation take a few hours to days to occur and explain why some individuals experience difficulties when prescribed new lens.

13.2.3 Disorders of the voluntary control of gaze

Anatomy and physiology of voluntary gaze

The cerebral hemispheres are extremely important for the programming and co-ordination of both saccadic and pursuit conjugate eye movements (Fig. 13.4c). Since different areas are involved in these two types of eye movements they will be dealt with separately, always realizing that for fully effective ocular motor control, co-ordination between these subtypes of eye movement is essential.

Saccadic system

There appear to be four main cortical areas in the cerebral hemispheres involved in the generation of saccades (review Leigh and Kennard 2004). In the frontal lobe in man there is the frontal eye field which lies laterally at the caudal end of the second frontal gyrus in the premotor cortex, Brodmann's area 8, and the supplementary eye field which lies mesially at the anterior region of the supplementary motor area in the first frontal gyrus, Brodmann's area 6. The third area is in the dorsolateral prefrontal cortex, which lies anterior to the frontal eye field in the second frontal gyrus, Brodmann's area 46. Finally, a posterior eye field lies in the parietal lobe, possibly in the superior part of the angular gyrus, Brodmann's area 39, and the adjacent lateral intraparietal sulcus. Studies in monkeys reveal that these areas are all interconnected with each other, and they all appear to send projections to the superior colliculus and the premotor areas in the brainstem-controlling saccades.

It appears that there are two parallel pathways involved in the cortical generation of saccades. An anterior system originating in the frontal eye field projecting both directly, and via the superior colliculus, to the brainstem saccadic generators. This pathway also passes indirectly via the basal ganglia to the colliculus. The second or posterior pathway originates in the posterior eye field passing to the brainstem saccadic generators via the superior colliculus. Only after bilateral lesions to both the frontal eye field and superior colliculus in monkeys is there a failure to trigger saccades.

Although the precise functions of these various cortical areas in saccade generation have not been determined, a number of general statements can be made. The frontal eye field is involved in triggering volitional saccades which, for example, may be predictive in anticipation of the appearance of a target, memory-guided to a previously seen target, or scanning so as to search for a particular target of interest. The posterior eye field could be involved in triggering reflexive saccades to the sudden appearance of novel visual or auditory stimuli, and appears to be involved in visuo-spatial integration and shifting visual attention. The dorsolateral prefrontal cortex may be responsible for maintaining a spatial map of the environment in short-term memory providing spatial information for memory-guided saccades and other volitional saccades as well as playing an important role in antisaccades, when a saccade is made to the mirror image location of a novel visual target, by inhibiting unwanted misdirected reflexive saccades to the target. The supplementary eye field appears to be involved in the generation of sequences of memory-guided saccades and complex ocular motor behaviours.

A subsidiary neural circuit related to saccade generation is from the frontal lobe to the superior colliculus via the basal ganglia. Projections from the frontal cortex pass to the substantia nigra pars reticulata, via a relay in the caudate nucleus. An inhibitory pathway from this nigral nucleus projects directly to the superior colliculus. This appears to be a gating circuit related to volitional saccades, especially of the memory-guided type.

Smooth pursuit system

To maintain foveation of a moving target the smooth pursuit system has developed relatively independently of the saccadic oculomotor system, although there are interconnections between the two. To visually track a target it is first necessary to identify and code its velocity and direction. This is carried out in the extrastriate visual area known as the middle temporal visual area, also called visual area V5, which contains neurons sensitive to visual target motion. In man, this lies immediately posterior to the ascending limb of the inferior temporal sulcus at the occipito-temporal border, Brodmann areas 19/37 junction. The middle temporal area sends this motion signal to the medial superior temporal visual area, which in monkeys is located on the anterior bank of the superior temporal sulcus, but in man is considered to lie superior and a little anterior to middle temporal area within the inferior parietal lobe (Petit and Haxby 1999). Damage to this area results in an impairment of smooth pursuit of targets moving towards the damaged hemisphere. Evidence of a possible contribution of the frontal eye field to the generation of smooth pursuit has recently been obtained in the monkey.

Both areas, medial superior temporal and the frontal eye field, send direct projections to a group of nuclei, which lie in the basis pontis of the pons. In the monkey, the dorsolateral and lateral groups of pontine nuclei receive direct cortical inputs related to smooth pursuit. Lesions of similarly located nuclei in man result in abnormal pursuit. These nuclei transfer the pursuit signal bilaterally to the posterior vermis, contralateral flocculus, and fastigial nuclei of the cerebellum. Finally, the pursuit signal passes from the cerebellum to the brainstem, specifically the medial vestibular nucleus and nucleus prepositus hypoglossi, and thence to the paramedial pontine reticular formation and possibly directly to the ocular motor nuclei. This circuitry, therefore, involves a double decussation, first at the level of the midpons, the pontocerebellar neuron, and second in the lower pons, the vestibulo-abducens neuron.

13.2.4 The diagnosis of specific disorders of eye movements

Disorders of saccadic eye movements

Disorders of saccades can be considered in terms of abnormalities of the saccadic pulse-step innervation pattern. A change in the amplitude of the pulse, either too big or too small, leads to saccadic hypermetria or overshoot, or to hypometric undershoot, respectively. Such a saccadic pulse dysmetria is associated with a lesion of the dorsal vermis in the cerebellum. A decrease in the height of the pulse, which implies disturbed function of the burst neurons in the paramedial pontine reticular formation or medial longitudinal fasciculus, leads to slow saccades. Many causes of slow saccades, several of which involve these areas, have been described (Table 13.12). A mismatch between the size of the pulse and the step (pulse-step mismatch) results in post-saccadic drifts and glissades. They are observed in diseases involving the vestibulo-cerebellum. If the pulse is not followed by a step, called a *saccadic pulse,* the eye drifts back to its previous position in a decreasing velocity exponential smooth eye movement. Both conjugate and monocular saccadic pulses may occur in patients with multiple sclerosis.

Disturbances in the initiation of saccades may lead to a prolonged latency, or the addition of a head movement or blink to initiate the saccade. This may be seen in congenital or acquired oculomotor

Table 13.12 Causes of slow saccades

Spinocerebella ataxias (SCA) especially SCA2Z (Section 39.8.1)
Huntington's disease (Section 40.5.2)
Wilson's disease (Section 40.8)
Parkinson's disease (advanced case) (Section 40.3.1)
Ataxia telangiectasia (Sections 11.5; 39.5.1)
Lipid storage disease
Progressive supranuclear palsy (Section 40.3.9)
Lesions of the paramedial pontine reticular formation
Internuclear ophthalmoplegia
Peripheral nerve palsy or muscle weakness
Drug intoxications, e.g. anticonvulsants
Tetanus (Section 42.5.14)
Paraneoplastic syndromes (Section 38.4)

apraxia, and various degenerative conditions including Parkinson's disease (O'Sullivan and Kennard 1998), Huntington's disease (Lasker and Zee 1997), and Alzheimer's disease (Fletcher and Sharpe 1986).

Saccades may also occur inappropriately, particularly during attempted fixation. *Square wave jerks* are small amplitude saccades of up to 5 deg that take the eyes off fixation, followed some 200 ms later by a corrective saccade. Many normal subjects have these jerks at a low frequency of < 15/min, but elderly subjects often have a higher frequency. They are most prominent in cerebellar disease, progressive supranuclear palsy, multiple system atrophy, and schizophrenia. *Macrosquare wave jerks* (5–40 deg) are encountered in multiple sclerosis and olivopontocerebellar degeneration. Patients with diffuse cerebral cortex damage often exhibit large amplitude saccades away from the object of regard. After an interval of several hundred milliseconds the patient makes a saccade back to the target. These anticipatory saccades are particularly observed in Alzheimer's disease.

Ocular motor apraxia is a term used for failure to generate saccades to commands, and may be of a congenital (Cogan 1952) or acquired type (Pierrot-Deseilligny *et al.* 1988). Congenital types may be recognized shortly after birth when the child does not appear to be fixating upon objects normally. At around 4–6 months the child develops the characteristic thrusting horizontal head movements, sometimes with blinking, when the child wants to change fixation. This manoeuvre serves to use the intact vestibulo-ocular reflex to drive the eyes into an extreme eccentric position in the orbit. As the head moves past the target, the eyes are dragged along in space until they align with the target. The head then rotates back and the vestibulo-ocular reflex ensure that fixation is maintained until the eye is in the primary position (Harris *et al.* 1996). Although the cause of congenital ocular motor apraxia is usually unknown some children are found to have a nonprogressive, noninheritable structural abnormality of the brain either a developmental anomaly or prenatal or perinatal insult, for example cerebellar hypoplasia, Dandy–Walker syndrome, or dysgenesis of the cerebellar vermis or corpus callosum. A variety of genetic disorders with multisystem involvement may present in infancy with congenital ocular motor apraxia including Joubet's syndrome. Patients with congenital ocular motor apraxia usually improve with age. In certain diseases affecting the brainstem an acquired form of ocular motor apraxia similar to congenital types may occur. These include ataxia-telangectasia, cerebral Whipple's disease, Gaucher's disease, Niemann–Pick type C, some of the spinocerebellar ataxias, vitamin E deficiency and many other storage diseases and aminoacidureas.

Disorders of smooth pursuit

A number of different disturbances of smooth pursuit are found (Morrow and Sharpe 1993). The commonest abnormality is a low gain, when gain = eye velocity/target velocity. This appears as deficient pursuit in which pursuit is broken by small catch-up saccades. Low gain pursuit can occur as a result of tiredness and inattention, as a side-effect of medications such as sedatives and anticonvulsants, or due to lesions in the vestibulo-cerebellum. Generally bilateral low gain pursuit has no localizing value. This is not the case with asymmetrical low gain pursuit, which usually occurs as a result of a lesion in the ipsilateral parietal lobe, thalamus, mid-brain tegmentum, dorsolateral nucleus of the pons, and vestibulo-cerebellum (Heide *et al.* 1996). Occasionally a disturbance of pursuit 'tone' or

balance occurs due to cerebral hemisphere lesions, when the eyes drift towards the side of the lesion. Disturbances of direction can occur, for example, in congenital nystagmus in which there is an apparent 'inversion' of pursuit when the eyes move in an opposite direction to the motion of the target.

Disorders of vergence eye movements

The commonest causes of disturbed vergence are congenital abnormalities. Various forms of convergence or divergence excess or insufficiency are usually accompanied by a concomitant strabismus associated with abnormalities of the accommodation-convergence synkinesis. Although this may not give rise to diplopia in childhood it can present as intermittent diplopia later in life. In particular convergence insufficiency is a common disorder in teenagers and university students, the elderly, and after relatively minor head trauma. Such individuals may show impaired phoria adaptation to prisms. It is usually treated by orthoptic exercises or prism therapy. Acquired forms of vergence disorders commonly occur in association with disturbances of vertical gaze as in the dorsal mid-brain syndrome, and in idiopathic Parkinson's disease and particularly in progressive supranuclear palsy. Occasionally acquired cerebral lesions may give rise to impaired stereopsis and poor fusional vergence.

Convergence spasm, or spasm of the near triad, is only rarely due to an organic lesion and is usually a voluntary convergence in patients with a conversion syndrome (Sarkies and Sanders 1985). The organic form occurs most commonly with lesions at the diencephalic-mesencephalic junction, so called thalamic esotropia. This may be due to thalamic haemorrhage, pineal tumours, and mid-brain stroke. It may also rarely occur with lower brainstem and cerebellar disorders. However, the majority of patients presenting with convergence spasm have a psychological disorder. They often complain of discomfort and the convergence, which only lasts for a brief period on each occasion, may be associated with visual blurring, diplopia, and 'eye strain'. It is often misdiagnosed as bilateral sixth nerve palsy, but an important clue to the correct diagnosis is the strong pupillary miosis which accompanies the convergence, and the observation of a full range of eye movements and less pupillary constriction with only one eye viewing. Treatment is best directed toward the underlying psychological factors, although cyclopegic eye drops and refractive measures may be effective.

Disorders of vestibular eye movements

These are covered in Sections 15.4 and 15.5.

13.2.5 The diagnosis of saccadic oscillations and nystagmus

There is an important distinction between saccadic oscillations, which are sustained oscillations that are initiated by fast saccadic eye movements, and nystagmus where the oscillations are initiated by smooth eye movements, thus the fast phase in jerk nystagmus is corrective and not primary.

Saccadic oscillations

Saccadic oscillations are bursts of saccades, which may be intermittent or continuous, causing a disruption of fixation. Two main types can be identified, those with intersaccadic intervals and those composed of back-to-back saccades.

The oscillations with intersaccadic intervals include *square wave oscillations* consisting of sequences of square wave jerks which can occur in Parkinson's disease and progressive supranuclear palsy. *Macrosaccadic oscillations* straddle the intended fixation position and do not occur in the dark. The amplitudes, up to 40 deg, of sequential saccades increase in amplitude and then decrease in a crescendo-decrescendo pattern (Selhorst *et al.* 1976). This type of oscillation is usually observed in acute damage to the fastigial nucleus and its output in the superior cerebellar peduncles as in demyelination, tumour, or haematoma. It can also occur in some forms of spinocerebellar ataxia.

Oscillations without any intersaccadic interval, that is back-to-back, include opsoclonus, ocular flutter, and convergence-retraction saccadic pulses. *Opsoclonus* consists of multidirectional, including oblique and torsional, back-to-back saccades of varying amplitude (Averbuch-Heller and Remler, 1996) (Section 38.4.5). It is often associated with eye blinking, facial twitching, myoclonus, and ataxia. It has been suggested that the disorder arises due to disordered pause cell glycinergic function in the paramedial pontine reticular formation. It can occur in neonates associated with myoclonus producing 'dancing eye and dancing feet'. This appears to be a maturational deficit, which resolves over approximately 6 weeks. In the teens and young adults it is often post-infectious. Other causes of opsoclonus are stroke, trauma, tumours, hyperosmolar nonketotic coma or drug induced by amitriptyline, lithium, phenytoin, cocaine. Opsoclonus is particularly associated with a paraneoplastic (non-metastatic) disorder, which in children is associated with occult neuroblastoma. Fifty per cent of children with opsoclonus have neuroblastoma and thus it is essential to exclude this tumour in all children with the condition. On the other hand only 2 per cent of children with neuroblastoma have opsoclonus. In adults it occurs in association with carcinoma of the lung (small cell), breast, and uterus. A number of anti-neuronal antibodies have been associated with opsoclonus, including anti-Hu, anti-Ri, anti-Yo, anti-Ma1, and anti-amphyphisin antibodies. Treatment may be offered with propranolol, verapamil, clonazepam, verapamil, and thiamine. Intravenous immunoglobulin may benefit those with postinfectious or idiopathic opsoclonus. The condition may disappear following tumour removal in the paraneoplastic variety.

Ocular flutter consists of bursts of back-to-back saccades in the horizontal plane only. It can therefore be observed in patients recovering from opsoclonus. Isolated ocular flutter is most often observed in patients with multiple sclerosis and signs of cerebellar disease. A voluntary form of flutter (voluntary flutter) can be induced by about 8 per cent of the population, usually by convergence. It consists of salvoes of horizontal back-to-back saccades. Lesions of the dorsal mid-brain are often associated with upward gaze palsies and *convergence-retraction nystagmus* (Ochs *et al.* 1979). This is incorrectly termed a nystagmus since it actually consists of adducting saccades and should be redesignated convergence-retraction saccadic pulses. Finally, a further type of saccadic oscillation is *ocular bobbing* (Susac *et al.* 1970). This consists of rhythmic, sudden, downward jerks of the eyes followed by slow return to the midposition, either immediately or after a short delay. The typical type, associated with pontine haemorrhage or infarction, is associated with paralysis of horizontal eye movements. Atypical bobbing is similar except that horizontal eye movements are intact, and occurs in metabolic encephalopathy, obstructive hydrocephalus, or cerebellar haematoma.

When the fast movement is upward followed by a delayed slow return the condition is known as reverse bobbing.

Nystagmus

Nystagmus is an oscillation which is initiated by a slow eye movement. When this slow movement is accompanied by a fast, saccadic, eye movement it is called jerk nystagmus. Although the direction of the nystagmus is conventionally determined by the direction of the quick phases it is important to remember that it is the smooth eye movement imbalance which is responsible for the nystagmus. If both phases are smooth eye movements pendular nystagmus is observed.

Vestibular nystagmus is the commonest form of jerk nystagmus and most frequently results from labyrinthine or vestibular nerve dysfunction. Several different types of central vestibular nystagmus are described, all of which show no change in intensity with the removal of fixation by using Frenzel goggles. This is in contrast to peripheral vestibular nystagmus in which removal of fixation leads to an increased intensity of the nystagmus.

Downbeat nystagmus may or may not be present in the primary position; it beats directly downwards and is often accentuated in downward and lateral gaze (Halmagyi *et al.* 1983). When it is present in the primary position a disturbance of the cerebellar flocculus is found, commonly due to a disturbance at the craniocervical junction such as an Arnold Chiari malformation, type 1 and foramen magnum mass lesions. Other causes include spinocerebellar degenerations, anticonvulsant drugs, lithium toxicity, and intraaxial brainstem lesions. In about one quarter of cases no cause can be found. The pathophysiology of downbeat nystagmus is thought to be an imbalance of the vertical semicircular canal pathways favouring the anterior canal.

Upbeat nystagmus when present in the primary position, is less well localized than downbeat nystagmus but is usually associated with focal brain stem lesions in the tegmental gray matter, either at the pontomesencephalic junction or at the pontomedullary junction, involving the nucleus prepositus hypoglossi or the ventral tegmental pathway of the upward vestibulo-ocular reflex (Fisher *et al.* 1983). It does not usually increase in lateral gaze as does downbeat nystagmus, but follows Alexander's law becoming accentuated on increasing upgaze. Upbeat nystagmus may be influenced by head posture and downbeat nystagmus may convert to upbeat nystagmus in the supine position. Multiple sclerosis, tumour, infarction, Wernicke's encephalopathy, and cerebellar degeneration are the commonest causes.

Torsional nystagmus is a jerk nystagmus around the anteroposterior axis. It is commonly associated with other types of nystagmus. However, when it is pure it indicates a lesion of the lateral medulla involving the vestibular nuclei such as syringobulbia and Wallenberg's syndrome of lateral medullary infarction. Occasionally it may be due to a mid-brain or thalamic lesion, involving the interstitial nucleus of Cajal and medial longitudinal fasciculus.

Periodic alternating nystagmus is a primary position horizontal nystagmus that changes direction in a crescendo-decrescendo manner, characteristically approximately every 90–120 s (Fletcher 1993). Between each directional change there is a null period of 0 to 10 s. There is a congenital form which shows a less regular pattern, and acquired forms are due to Chiari malformations, multiple sclerosis, fourth ventricle tumours, spinocerebellar degenerations, and anticonvulsant intoxication. Ablation of the cerebellar nodulus and uvula which have a velocity-storage role mediated by the neurotransmitter GABA, in monkeys, causes periodic alternating nystagmus. Thus the GABA$_b$ agonist baclofen has been shown to be an effective treatment (Halmagyi *et al.* 1980).

Gaze-evoked nystagmus is a common clinical observation with limited localizing value. It is a jerk nystagmus of amplitude > 4 deg, which is absent in the primary position and is only present and often asymmetric on eccentric gaze. It is due to a disturbance in the gaze-holding neural network, integrator neurons in the paramedian pontine reticular formation or inputs to them. Gaze-evoked nystagmus usually signifies cerebellar parenchymal disease, particularly involving the flocculus or its projections to the brainstem in the region of the medial vestibular nucleus and the nucleus prepositus hypoglossi. Bilateral horizontal, together with vertical, gaze-evoked nystagmus commonly occurs with structural brainstem and cerebellar lesions, diffuse metabolic disorders, and drug intoxication. A variant of gaze-evoked nystagmus is *rebound nystagmus* in which there is a jerk nystagmus that beats away from the previous direction present in eccentric gaze, which appears after the eyes return to the primary position. It usually lasts for 3–25 s, and is associated with parenchymal cerebellar disease.

Pendular nystagmus is either congenital or acquired and is usually due to cerebellar and brainstem disease, most frequently multiple sclerosis and brainstem infarction (Fletcher 1993). Acquired pendular nystagmus may have torsional and both horizontal and vertical components, with the amplitude and phase relationships of the two sine-waves for the horizontal and vertical components determining the final trajectory of the eyes as oblique, circular, or elliptical. This form of nystagmus only has slow phases without any fast phases. It can affect one eye or both, equally or unequally, and is often symptomatic resulting in oscillopsia. It may be associated with oscillations of other structures such as the palate, head, or limbs. When it is present in association with palatal myoclonus at 1–3 Hz, *oculopalatal myoclonus*, the lesion usually occurs several months after an infarction in the region of Mollaret's triangle which consists of the red nucleus, dentate nucleus, and inferior olivary nucleus (Nakada and Kwee 1986). The latter nucleus usually shows pseudohypertrophic degeneration. However, the red nucleus is not known to be involved in eye movements and more recent explanations have proposed an interruption of a pathway from the deep cerebellar nuclei through the superior cerebellar peduncle, which then loops caudally through the central tegmental tract to the inferior olive. A combination of a convergence-induced slow pendular nystagmus 1 Hz and synchronous jaw contractions, called *oculomasticatory myorhythmia*, is characteristic of Whipple's disease (Schwartz *et al.* 1986). However, it may also be observed in brainstem stroke and multiple sclerosis. In *see-saw nystagmus* one eye intorts and rises while the other eye extorts and falls in a rapidly alternating sequence. In this pendular form a bitemporal hemianopia is often present, and the condition is associated with large parasellar masses which have expanded up into the third ventricle and are distorting structures in the mesencephalic-diencephalic region (Daroff 1965).

Congenital nystagmus is almost invariably a horizontal conjugate nystagmus which is unaltered by vertical position. It is generally of jerk type with accelerating slow phases, and has an eccentric null position. Fixation effort enhances congenital nystagmus. Less commonly the nystagmus is of a pendular type. Reversed optokinetic nystagmus, beating in the direction of the target motion, is a

feature of congenital nystagmus. Patients may show a head turn or occasionally a head oscillation (Dell'Osso and Daroff 1975).

Latent nystagmus is a type of congenital nystagmus that is only present on monocular viewing and which then beats toward the viewing eye (Gresty *et al.* 1992). It is absent on binocular viewing. If the patient has amblyopia in one eye latent nystagmus is present with both eyes viewing, then it is called manifest latent nystagmus.

13.3 Disorders of the pupil

The size of the pupil depends on the relative contraction of the iris sphincter and dilator muscles, supplied by the parasympathetic and sympathetic input, respectively. Disruption of the parasympathetic input results in a fixed dilated pupil '*mydriasis*', whereas if the sympathetic input is damaged a small pupil '*miosis*' ensues. *Anisocoria* is a difference in the size of the two pupils and may be physiological or due to under- or over-activity of the parasympathetic or sympathetic inputs.

13.3.1 The pupillary light reflex

The afferent pupillary light reflex pathway is a three-neurone reflex arc originating in the retinal ganglion cells, which project to the pretectal nucleus in the mid-brain (Loewenfeld 1993) (Fig. 13.7). Interneurons from this area project to the Edinger–Westphal subnucleus of the oculomotor nucleus at its rostral end. Passing from

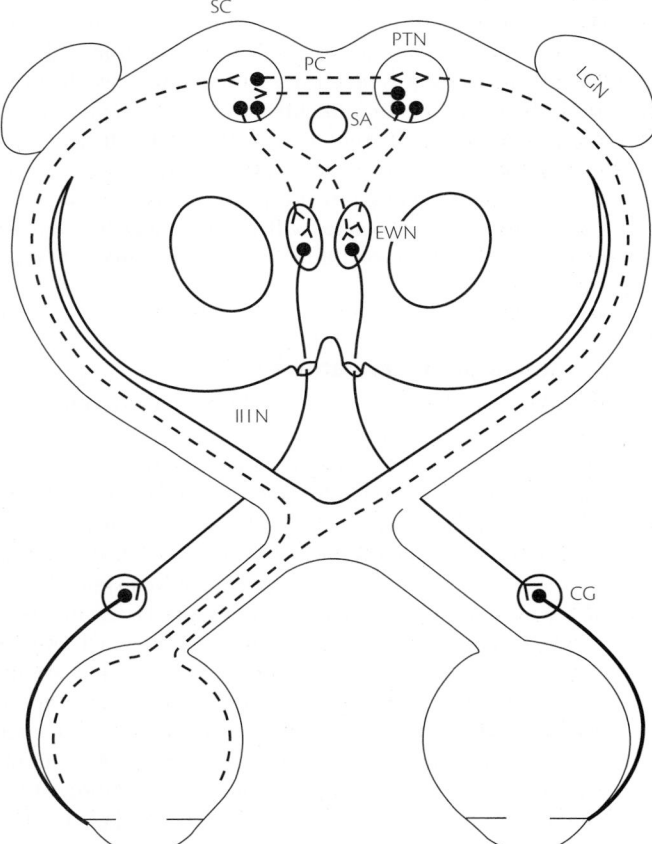

Fig. 13.7 Pupillary light reflex pathway. PC: posterior commissure, PTN: pretectal nucleus, SA: Sylvian aqueduct, LGN: lateral geniculate nucleus, EWN: Edinger-Westphal nucleus, SC: superior colliculus, CG: ciliary ganglion.

the Edinger–Westphal nucleus preganglionic, parasympathetic efferent pupillary fibres lie in the periphery of the oculomotor nerve, where they are particularly susceptible to compression by aneurysms of the posterior communicating artery. They then pass via its inferior division to the ciliary ganglion lying in the floor of the orbit, and reach the iris sphincter muscle via the short ciliary nerves. Interruption of this pathway from the mesencephalon to the sphincter muscle causes pupillary dilatation and decreased speed and amplitude of constriction in response to light.

At the optic chiasm a slightly higher proportion of afferent fibres cross into the contralateral optic tract with a ratio of crossed: uncrossed fibres which has been estimated at 53:47 (Kupfer *et al.* 1967). This may explain the afferent pupillary defect which is sometimes observed in patients with isolated retrochiasmal lesions. In the Edinger–Westphal nucleus there is a functional dissociation with the rostral portion containing mainly efferent neurones relating to accommodation and the caudal neurones involved in pupil constriction.

The pupillary near response results from accommodative effort induced by retinal image blur or conscious near fixation. It is part of the 'near triad' which consists of pupillary constriction, lens accommodation, and convergence of the visual axes. It is possible for a number of neural lesions to give rise to a dissociation of components of the near triad in which there is absent pupillary constriction with the other components remaining intact, but preservation of pupillary constriction with absence of convergence and lens accommodation does not occur.

The size of the pupil is in a constant state of flux adjusting to a variety of external stimuli, such as ambient illumination and fixation distance, as well as psychosensory stimuli. Pupillary diameter tends to be smaller in infants and older adults compared to young adults. A subtle anisocoria is often observed, and a difference in pupillary size of 0.4 mm or more is easily identified clinically in 20 per cent of the normal population. This so-called simple or physiologic anisocoria is associated with a pupil inequality which is the same under all lighting conditions, and the pupillary light reactions are equally brisk.

13.3.2 Afferent pupillary pathway lesions

A unilateral lesion of the afferent pupillary pathway results in an impaired direct light reflex of the affected eye, sparing the consensual response elicited by stimulating the contralateral eye. If there is a complete lesion either in the retina or optic nerve there will be a complete failure of the direct light reflex. However, in most instances there is an impaired response which is best identified clinically by using the *swinging light test*. It is performed by using a hand-held torch with a bright light beam which is moved back and forth from one eye to the other, the light being held on each eye for approximately 1 s. When the light is shining in the normal eye the pupil is constricted and the contralateral consensual response is maximal. When the light is then shifted to the eye with impaired vision the direct light reflex is now reduced in comparison with the former consensual response resulting in further dilatation of the pupil (Thompson 1966). It is best to perform the test in a dimly lit room. The magnitude of the relative afferent pupillary defect may be estimated by placing neutral density filters over the normal eye until the responses from the two eyes are balanced (Thompson *et al.* 1981). A relative afferent pupillary defect does not occur when

the visual loss is a result of refractive errors, opaque optic media, amblyopia, or functional visual loss.

Retina

The degree to which a relative afferent pupillary defect is identified due to retinal disease depends on the degree and location of the lesion. A small retinal detachment involving the macula may not result in an afferent pupillary defect, whereas a complete detachment will certainly do so. A useful clinical guide is that if ophthalmoscopy reveals a normal macula the presence of an afferent pupillary defect is unlikely to be due to retinal disease. Suppression amblyopia does not result in a relative afferent pupillary defect.

Optic nerve

Optic nerve disease is commonly associated with a relative afferent pupillary defect, the magnitude of which correlates closely with the extent of the visual field defect and the visual acuity, particularly when due to optic neuritis (Ellis 1979). The absence of an afferent pupillary defect in a patient with unilateral visual loss and an otherwise normal eye should raise the possibility of bilateral optic nerve disease or non-organic visual loss. Bilateral optic nerve disease is suggested by a dissociation between the direct light reflex and the amplitude of the pupillary near response. Recovery of optic neuritis leads to a reduced degree of relative afferent pupillary defect but not usually to its absence.

Optic tract

A relative afferent pupillary defect may be observed in some patients with optic tract disease in association with a homonymous hemianopia (Bell and Thompson 1978). The defect is in the eye with the temporal field loss, and is thought to be due to the asymmetric decussation of optic nerve fibres at the chiasm as described above.

Pretectal nucleus and brachium of the superior colliculus

The pupillary fibres coming from the ipsilateral optic tract to the pretectal nucleus, via the brachium of the superior colliculus, may be involved by a unilateral lesion such as an arteriovenous malformation, infarction, or tumour (Wilhelm *et al.* 1996). This may produce a contralateral relative afferent pupillary defect without any loss of visual acuity or colour vision and without any visual field defect, although the afferent pupillary defect may occasionally be associated with an ipsilateral or contralateral trochlear nerve paresis.

13.3.3 Central pupillary pathway lesions

Argyll Robertson pupils

The essential features of classic Argyll Robertson pupils, which are usually bilateral and symmetric, are miosis with poor dilation in darkness, absence or marked impairment of the light reflex, and relative preservation of the near response 'light-near dissociation'. In addition the pupil may be irregular due to iris damage, and shows impaired dilatation to mydriatic drugs. For over a century, since the original description by Douglas Argyll Robertson in 1869 of the pupillary abnormalities subsequently shown to be associated with neurosyphilis (Section 42.5.1), there has been controversy regarding the site of the lesion. The most widely held current view (Loewenfeld 1993) is that the Argyll Robertson pupil is the result of neuronal damage in the region of the Sylvian aqueduct in the rostral mid-brain. Diffuse damage around the sylvian aqueduct and the posterior portion of the third ventricle is a prominent finding in patients with Argyll Robertson pupils who have died from tabes or general paralysis. In this location the damage interferes with the light reflex fibres and the supranuclear inhibitory fibres as they approach the visceral oculomotor nuclei.

As the incidence of tertiary syphilis has declined since the introduction of penicillin, the percentage of nonsyphilitic patients with Argyll Robertson pupils has increased. Typically they are observed in patients with diabetes mellitus (Smith and Smith 1983), chronic alcoholism, encephalitis, multiple sclerosis, age-related and degenerative diseases of the central nervous system, some rare mid-brain tumours, and rarely in systemic inflammatory diseases, including sarcoidosis and neuroborreliosis. However, the main differential diagnosis is with bilateral tonic pupils, which after many years become small, unreactive to light, and show light-near dissociation. The major distinguishing feature is the presence of tonicity of the near response in tonic pupils.

Mesencephalic lesions

Pressure on the dorsal mesencephalon may produce Parinaud's syndrome, also known as the dorsal mid-brain syndrome or the Sylvian aqueduct syndrome. This syndrome, due to damage in the region of the posterior commissure, includes a supranuclear vertical gaze palsy, disturbances of pupillary function, accommodation difficulties, and frequently convergence–retraction nystagmus. The pupils are usually midposition to large, fail to constrict to light or do so very poorly, and show relative preservation of the reaction to near vision 'light-near dissociation'. It is considered that it is due to disruption of the ganglion cell axons entering the pretectal region. Dilated pupils due to an impaired light reflex may be the first sign of a pineal or other tumour that compresses or infiltrates the dorsal mid-brain, or from hydrocephalus, particularly if caused by aqueductal stenosis or a blocked shunt.

Mesencephalic lesions in the region of the oculomotor nerve nucleus nearly always damage both the sympathetic and parasympathetic pathways to the eye, resulting in slightly unequal and irregular pupils. Rarely in mid-brain lesions a phenomenon called correctopia occurs in which there is an upward, inward displacement of the pupil (Selhorst *et al.* 1976).

13.3.4 Efferent pupillary pathway lesions

Involvement of the preganglionic parasympathetic fibres located in the periphery of the oculomotor nerve results in a dilated pupil which has an impaired or absent direct, consensual, and near response. Although such a finding suggests significant intracranial pathology, for example, an unruptured posterior communicating artery aneurysm, there is usually in addition some degree of ptosis and ophthalmoplegia. A dilated pupil exposes spherical aberrations of the lens and cornea which may give rise to blurred vision in patients with pupil involving third nerve palsies.

However, a dilated pupil which fails to respond either to light or to the near reflex raises the possibility of accidental or deliberate instillation of a pharmacologically active agent such as scopolamine, atropine, or some plant juices containing belladonna alkaloids. This can be reversed by instilling 1 per cent pilocarpine in contrast to the pupillary dilatation which occurs due to an oculomotor nerve palsy or a tonic pupil. The possibility of acute angle closure glaucoma must be considered in any patient presenting with pupillary inequality with reduced visual acuity or pain.

A rare condition of episodic unilateral transient pupillary dilatation with headache has been described in young women (Edelson and Levy 1974).

The tonic pupil

The commonest cause of abnormal unilateral pupillary light reactions and a dilated pupil is the tonic pupil, due to a lesion involving the preganglionic parasympathetic neuron. The essential feature of a tonic pupil is light-near dissociation, a slow steady near pupil response followed by the pupil holding its contraction for a few seconds and then redilating slowly when the patient is asked to look back into the distance. The tonic pupil is the result of damage to the ciliary ganglion or the short ciliary nerves resulting in denervation and subsequent reinnervation of the iris sphincter and the ciliary muscle. Several different causes have been found which have been classified by Thompson (1979):

- ◆ Holmes–Adie syndrome: associated with tendon areflexia;
- ◆ Local tonic pupils: associated with orbital disease or following orbital surgery; and
- ◆ Neuropathic tonic pupils: associated with peripheral or autonomic neuropathy.

Holmes–Adie syndrome. This is a relatively uncommon syndrome which usually occurs between 20 and 50 years of age. It has a clear predilection for women who constitute 70 per cent of cases. It is unilateral in 80 per cent of cases. The onset is usually acute with the patient often complaining of photophobia, particularly when going outdoors into bright sunlight, blurred near vision when reading, an enlarged pupil and headaches. The essential features of the pupil abnormality are a delayed and reduced amplitude light reaction. When the iris is viewed via a slit lamp in about 90 per cent of cases it can be seen that there is segmental contraction of the sphincter muscle 'vermiform movements' with other segments appearing paralysed 'sectoral paralysis'. This is in contradistinction to a tonic pupil due to pharmacologic anticholinergic blockade in which the entire sphincter is paralysed. The near response which is tonic, however, results in contraction of all the sphincter muscle. The pupillary dilatation is also tonic. As a result of denervation hypersensitivity the tonic pupil constricts with a low concentration of pilocarpine 0.125 per cent, but a normal pupil will not. However, the value of this pharmacologic test has recently been questioned because of the false positive results due to the variable corneal penetration of pilocarpine. Deep-tendon hyporeflexia or areflexia, particularly of the ankle and triceps jerks, can be demonstrated in a substantial number of patients with Holmes–Adie syndrome (Thompson *et al.* 1979b).

The pathology of the Holmes–Adie syndrome is considered to be damage to neurons in the ciliary ganglion or the postganglionic short ciliary nerves, as has been observed in two post-mortem studies. The hyporeflexia or areflexia is probably due to a central lesion within the spinal cord. In one case degeneration was observed in the gracile and cuneate fascicles resulting from a reduction in the neuronal population in the dorsal root ganglia (Selhorst *et al.* 1984).

Two pathophysiological explanations have been proposed for the pupillary abnormalities. In the first Loewenfeld and Thompson (1967) proposed that some of the fibres originally destined for the ciliary muscle resprouted randomly, with some of the fibres reaching the iris sphincter and causing miosis every time the ciliary muscle was innervated. There is also a marked predominance of fibres arising from the ciliary ganglion passing to the ciliary muscle compared to those passing to the iris sphincter in a ratio of 97:3 per cent, making such aberrant reinnervation a likely outcome of damage to the ganglion. This explanation was challenged by Wirtschafter and colleagues (1978), who proposed that the iris sphincter remains permanently dennervated, and that the pupillary near-vision constriction results from acetylcholine released by the accommodative nerve endings in the ciliary muscle. This then diffuses to the pupillary sphincter via the aqueous fluid. On the basis of several clinical observations this hypothesis is not widely supported (Loewenfeld and Thompson 1981).

Follow-up of patients with the Holmes–Adie syndrome have shown the following changes over time (Thompson *et al.* 1979a):

- Recovery of the accommodation paresis;
- Progressive impairment of the pupillary light reaction;
- Increasing hypometria of the deep-tendon reflexes;
- The affected pupil gradually becomes smaller; and
- The other eye may become involved in about 10 per cent of cases.

Local tonic pupils

An acute internal ophthalmoplegia followed by the development of a tonic pupil has been reported following a variety of infections, inflammations, and infiltrative processes which involve the ciliary ganglion. These include infections by herpes zoster, chickenpox, measles, diphtheria, neurosyphilis, rheumatoid arthritis, sarcoidosis, primary and metastatic choroidal and orbital tumours, blunt injury to the orbit and penetrating injuries, as well as following various ocular and orbital surgical procedures (Lowenstein and Loewenfeld, 1965).

Neuropathic tonic pupils

This category consists of tonic pupils which are a result of involvement of the ciliary ganglion or short ciliary nerves as part of a generalized peripheral or autonomic neuropathy. These include those with chronic alcoholism, advanced diabetes mellitus, Guillain Barré syndrome, and the Miller Fisher variant and some hereditary neuropathies such as Charcot–Marie–Tooth disease. Those autonomic neuropathies, which can result in a tonic pupil, include acute pandysautonomia, Shy–Drager and Riley–Day syndromes, and Sjögren's syndrome, in which the pupil abnormality may be the presenting sign.

13.3.5 Sympathetic pathway lesions

A lesion anywhere along the long sympathetic pathway results in a typical Horner's syndrome with miosis and ptosis. The central, first-order, neuron lies in the ipsilateral hypothalamus, and its axon passes to the ciliospinal centre in the intermediolateral gray column via the dorsolateral medulla. Here it synapses with the preganglionic second-order neuron in the upper three dorsal segments of the spinal cord. The axon from the preganglionic neuron exits the spinal cord at this level, passes across the pulmonary apex to ascend to the superior cervical ganglion via the inferior and middle cervical ganglia. The postganglionic, third-order, neuron passes from the superior cervical ganglion up along the internal carotid artery, where it is termed the carotid plexus. It leaves the internal carotid artery in the cavernous sinus, to briefly join the abducens nerve before leaving it to join the ophthalmic division of the trigeminal nerve, entering the orbit with its nasociliary branch.

The ptosis in Horner's syndrome is usually mild, <2 mm, and is due to paralysis of the sympathetically innervated smooth muscle 'Müller's muscle' in the upper eyelid. Similar smooth muscle fibres in the lower eyelid are denervated leading to a slight elevation of the lower lid, producing an 'upside-down' ptosis (Fig. 2.5). Combined these result in a narrowed palpebral fissure and an apparent enophthalmos. The ptosis is suggestive but not diagnostic of Horner's sydrome since there are other causes for a mild ptosis such as senescent levator aponeurosis dehiscence, congenital dystrophy, myasthenia gravis, trauma or long-term contact lens wear. The miosis is due to complete or partial sympathetic denervation of the iris dilator muscle leading to constriction of the iris sphincter producing a small pupil. The weakness of the dilator muscle is greatest in the dark when the anisocoria is most apparent, and may be almost absent in the light. The extent of the anisocoria varies in extent depending on a number of factors which include completeness of the lesion and the extent of reinnervation, the alertness of the patient, the degree of denervation supersensitivity, and the level of circulating adrenergic substances in the blood. The pupil reacts normally to light and to near stimuli.

The paresis of the dilator muscle can be detected by observing a dilation lag of the affected pupil compared to the normal pupil when the lights are turned out. This is best performed observing both pupils by directing a dim torchlight on the eyes from below and turning the room lights out (Loewenfeld 1993). A simultaneous sudden noise accentuates the lag, due to enhanced sympathetic activation of the intact pupil.

Depigmentation of the affected iris is rarely observed in acquired Horner's disease, although hypochromia of the iris is a common finding in the congenital form 'iris heerochromia'.

Horner's syndrome is also associated with characteristic vasomotor and sudomotor changes on the affected side of the face, such as loss of sweating (anhidrosis) and occasionally facial flushing. These changes are most frequently observed following preganglionic lesions, since the fibres for sweating pass onto the external carotid artery from the superior cervical ganglion.

On some occasions the presence or absence of a Horner's syndrome may be in doubt, particularly if ptosis is absent. In this situation pharmacological pupillary testing is advisable. The cocaine test is used to diagnose an oculosympathopareis anywhere along the sympathetic pathway, and the hyroxyamphetamine test is used to determine whether the lesion lies in the central/preganglionic or the postganglionic segments. Cocaine blocks the reuptake of noradrenaline produced tonically by the postganglionic synaptic endings, but only if the entire three-neuron chain is intact. When applied to the eye it leads to a dilatation of the pupil (Thompson 1977). The test is performed by measuring the pupil size in the dark, and then instilling two drops of a 10 per cent solution of cocaine in each eye in turn. After 30 min the pupil size is remeasured in the dark and the affected pupil is found to have dilated less than the normal pupil. This occurs because the sympathetic denervation leads to a reduced release of noradrenaline and a reduced amount accumulates at the receptors of effector cells. A post-cocaine anisocoria of >1 mm or more signifies Horner's syndrome. It is important to observe the diameter of the pupils in a dimmed room since the background ambient illumination may lead to

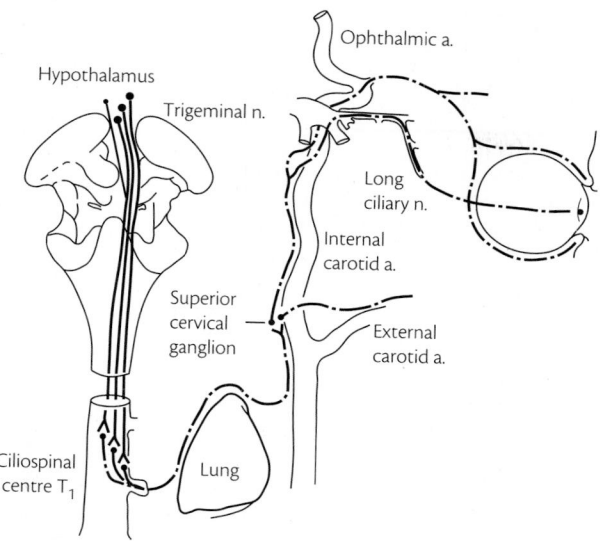

Fig. 13.8 The sympathetic pathway to the pupil. Note the sympathetic fibres to the face pass on to the external carotid artery.

pupillary constriction thereby obscuring the pharmacologically induced anisocoria. If mydriasis has not occurred in either eye an additional drop of cocaine should be instilled and the pupil size reassessed after 30 min.

Localization of the site of damage giving rise to a Horner's syndrome usually depends on the associated clinical findings (Fig. 13.8). Lesions affecting the first-order and second-order neurones may be accompanied by signs of dysfunction of the brainstem and cervicothoracic spinal cord, respectively. Lesions involving the third-order,

Table 13.13 Aetiology of Horner's syndrome

Central (first neuron)
Lateral medullary infarction
Other brain stem infarction
Cerebral infarction
Cerebral haemorrhage
Intracranial tumour
Trauma including surgery
Multiple sclerosis
Syrinx
Transverse myelopathy
Other/unknown

Preganglionic (second neuron)
Thoracic and neck tumour
Trauma—surgical
Trauma—non-surgical
Other/unknown

Postganglionic (third) neuron
Intracranial tumour (cavernous sinus)
Trauma (including surgical)
Carotid artery dissection
Vascular headache
Other unknown

Unknown localization

post-ganglionic neuron are accompanied by signs of lesions affecting structures around the common and internal carotid arteries. The various aetiologies of Horner's syndrome at these different levels are listed in Table 13.13. When trying to determine the location of the sympathetic lesion further pharmacological testing can be of assistance. Hydroxyamphetamine (1 per cent) causes the release of noradrenaline from sympathetic nerve endings and if applied to the normal eye results in pupil dilation (Van der Wiel and van Gijn 1983). It can, therefore, be used to differentiate between a post-ganglionic and a preganglionic or central Horner's syndrome, since in the former the nerve endings are destroyed and there are no noradrenaline stores to release, and there is therefore no mydriatic effect. If the lesion involves the preganglionic or central neuron the pupil will dilate fully since the third-order neuron is intact. The hydroxyamphetamine test is performed in the same way as the cocaine test but at least 24 h should elapse after a cocaine test before it is instilled.

13.3.6 Differentiation of anisocoria

Using some straightforward principles Thompson and Pilley (1976) have described a straightforward approach to determine the cause for anisocoria. If the anisocoria is more apparent in light compared with darkness this suggests that there is a defect in the parasympathetic system or the sphincter muscles, since this implies that both pupils dilate in the dark but one pupil does not respond to light stimulation. If the opposite is observed, that is the anisocoria is more evident in darkness than in light, this suggests that the parasympathetic pathway and the iris sphincter are intact, since both pupils constrict in the light yet one pupil dilates more in darkness than the other.

A flow chart (Fig. 13.9) indicates the various steps required to differentiate between the different causes of anisocoria. The first step is to check the light reaction. A normal light reaction in both eyes suggests that the anisocoria is due either to simple anisocoria or a Horner's syndrome. These two conditions can be differentiated using the cocaine test. A positive cocaine test may call for a hydroxyamphetamine test on another occasion to differentiate a post-ganglionic from a preganglionic and central sympathetic lesion.

If the light reaction in one or both eyes is impaired the patient has a parasympathetic lesion or a damaged iris sphincter. The iris is then viewed with a slit lamp to identify any iris damage. At this point it is necessary to differentiate a pharmacologically blockaded pupil from a neurogenic cause. Here, 0.1 per cent pilocarpine can be used to detect denervation supersensitivity of the parasympathetic system as occurs in a tonic pupil syndrome. If neither pupil constricts then a 1 per cent solution of pilocarpine will identify a dilated pupil due to pharmacological blockade, since the pupil will fail to constrict.

13.4 Orbital disease

13.4.1 Anatomy and examination

The orbit has a pear-like shape with the optic canal as the stem. The orbital walls, which are made up of seven bones (maxillary, frontal, zygomatic, ethmoid, sphenoid, palatine, and lacrimal), are of variable thickness and pierced by several fissures and foramina. The optic canal contains the optic nerve, oculosympathetic nerves, and the ophthalmic artery. The superior orbital fissure, formed by the greater and lesser wings of the sphenoid bone, admits to the orbit

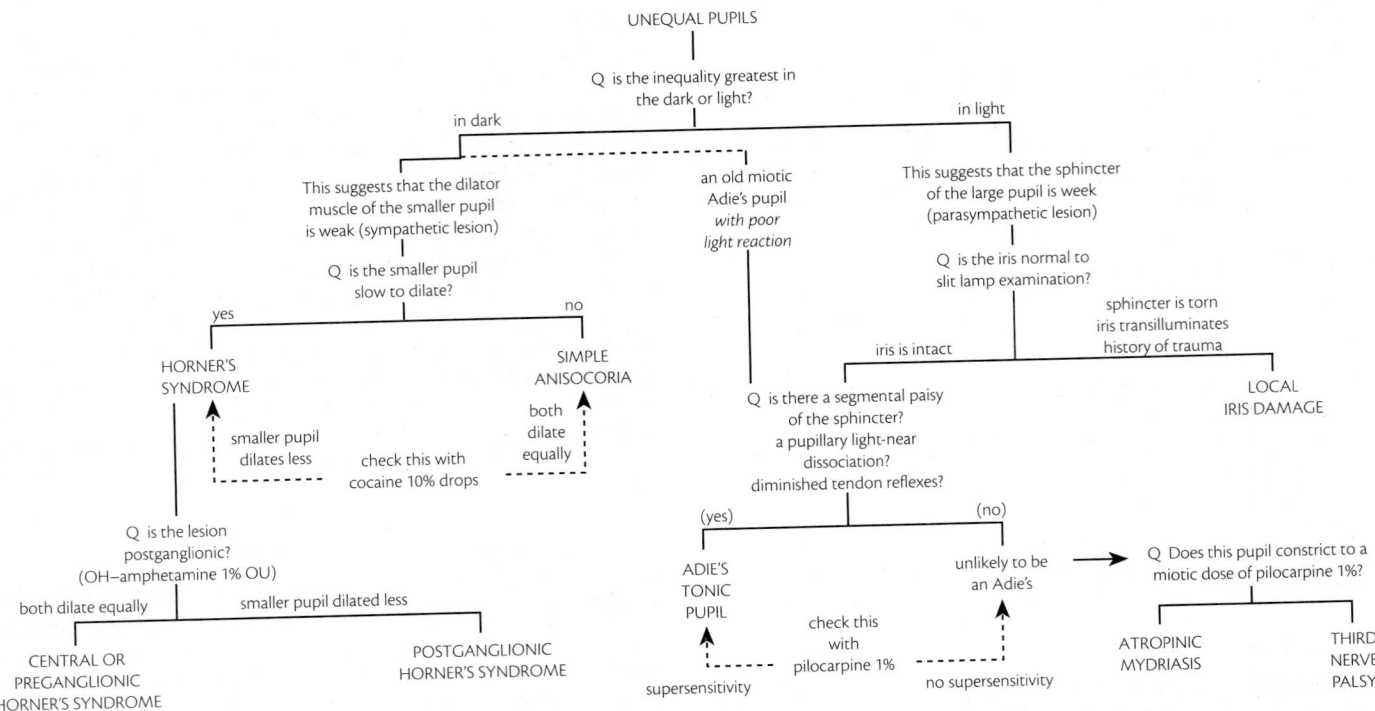

Fig. 13.9 A flow diagram for the clinical assessment of unequal pupils (anisocoria). (From Czarnecki *et al.* 1979.)

the three ocular motor cranial nerves III, IV, and VI, the ophthalmic division of the trigeminal nerve, and some sympathetic fibres. In addition the superior ophthalmic vein, which drains most of the orbit, passes through this fissure. The remainder of the venous drainage passes through the inferior orbital fissure, in the floor of the orbit, to join the pterygoid plexus. This fissure also contains branches of the sphenopalaitine ganglion. Since there are no valves in the orbital venous drainage system a carotico-cavernous fistula will lead to reversed flow in the venous system accounting for the marked venous congestion and orbital oedema.

The apex of the orbit is very crowded with the optic nerve emerging through the canal of Zinn, to which are attached the rectus muscles (intraconal portion of the orbit). This explains why enlargement of the extraocular muscles, as in dysthyroid ophthalmopathy, may lead to optic nerve compression. The ethmoid, sphenoid, maxillary, and frontal sinuses surround the orbit which allows spread of disease, especially infection and tumour, from these spaces to the orbit. The contents within the 25–30 cc of the bony orbit can be divided into the intraconal portion and an extraconal space.

Symptoms

Patients with orbital disease may present with a variety of symptoms including periorbital pain, double vision, blurred vision, swelling, proptosis, and ptosis.

Clinical examination

Although several imaging modalities including ultrasound, CT, and MR scanning are available to aid the localization and diagnosis of orbital disease a systematic clinical examination enables one to use them appropriately. However, it is important to first carry out the standard assessment of visual function including visual acuity, colour vision, and visual fields.

The initial external examination focuses on the assessment of the appearance of the eye looking for any asymmetry, the position of

the globe, swelling, and abnormal eyelid position. The eyelids may show ptosis or retraction, abnormalities of movement, in particular lid lag or lid 'hang-up' on asking the patient to look downwards and fatiguability on sustained upward gaze, and the presence of swelling or a mass. The conjunctivae are next examined for evidence of oedema or chemosis, dilated vessels, or neoplastic infiltration. The globe may show axial or non-axial proptosis. This is best judged by inspection alone and may be difficult to determine because of significant inter-individual and racial variation. Viewing the position of each corneal surface relative to each other from a vantage point above and behind or from below the patient referencing to a boney prominence, the brow or anterior orbital rim, respectively, can be helpful. Up to 2 mm of asymmetry is within normal limits, as measured by the Hertel exophthalmometer. Looking at old photographs can assist in estimating the time of onset of protrusion of the eye. The degree of retrocessability of the globe can be estimated by gentle backward pressure through the closed lid. Retrobulbar mass lesions will reduce the ability to push the globe back into the orbit, whereas blood-filled spaces, such as varices, allow compression and repositioning of the exophthalmic globe. Pain to palpation may accompany orbital infection and inflammation, while being uncommon in dysthyroidism and orbital tumours. Auscultation over the globe with the lids closed may reveal vascular bruits suggestive of arteriovenous malformations. Finally, ophthalmoscopy may reveal disc oedema (only if the first half of the optic nerve is compressed) or optic atrophy with optociliary shunts opening up to allow blood to flow between the retinal and choroidal circulation which occurs if there is obstruction of the central retinal vein and is suggestive of an optic nerve sheath meningioma. Choroidal or choroidoretinal folds and acquired hyperopia usually imply a retrobulbar mass lesion deforming the globe from behind, but can also occur in thyroid-associated ophthalmopathy, inflammatory diseases of the orbits, and mucocoeles.

Table 13.14 Causes of orbital disease

◆ Cellulitis

◆ Sequelae of trauma

◆ Graves' disease

◆ Pseudotumour of orbit

◆ Lymphoma

◆ Cavernous haemangioma

◆ Lacrimal gland tumour

◆ Peripheral nerve tumours

◆ Meningioma

◆ Mucocele

◆ Metastatic and secondary tumours

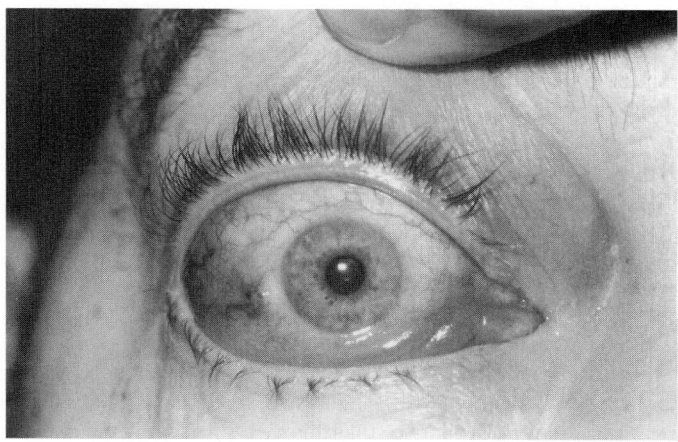

Fig. 13.10 Thyroid eye disease, causing proptosis, chemosis, and lid oedema.

A wide range of disorders can give rise to orbital involvement and are listed in Table 13.14 (Wright 1988).

13.4.2 Dysthyroid eye disease

Dysthyroid eye disease is also known as Grave's orbitopathy, Grave's ophthalmopathy, dysthyroid ophthalmopathy, thyroid orbitopathy, endocrine ophthalmopathy/exophthalmos. It is a self-limiting autoimmune disease usually associated with hyperthyroidism (Char 1997). It is the commonest orbital disorder in adults accounting for 32–47 per cent of cases and is associated with diplopia, ophthalmoparesis, and infiltration of extraocular muscles (Sergott and Glaser, 1981). It is usually accompanied by biochemical and immunological evidence of thyroid dysfunction, although this may not be apparent for months or even years. Pathologic examination of extraocular muscles in patients with dysthyroid eye disease shows infiltration by lymphocytes and plasma cells. The relationship between Grave's disease and dysthroid eye disease is still controversial although it may result from antigenic similarity between the thyroid tissue and orbital tissue. It is thought that Grave's disease results from clonally restricted B lymphocytes in the thyroid gland producing autoantibodies which act on the TSH receptor on thyroid follicles; these autoantibodies may lead to cytokine release resulting in the proliferation of orbital fibroblasts and increase in the synthesis of intracellular matrix proteins, glycosoaminoglycans, in orbital fat and eye muscles. As the disease progresses the infiltration and oedema of the extraocular muscles produce loss of muscle tissue, and the muscles become fibrotic (van der Gaag *et al.* 1996).

The *early symptoms* of this condition may take the form of discomfort, ocular irritation, scratchiness, or 'burning', which is typically worse when the patient first awakes. This may be associated with a feeling of orbital fullness and intermittent vertical diplopia. Photophobia and tearing are usual and blurred vision may be due to corneal exposure due to the proptosis or a central scotoma associated with a compressive optic neuropathy. Significant pain is relatively unusual, the onset is usually gradual with the symptoms developing over weeks, and usually there is evidence of bilateral involvement.

In the acute phase there is usually unilateral or bilateral lid retraction, in which the upper eyelid rests above the superior corneal limbus, accompanied by lid lag on downgaze and lid puffiness

(Fig. 13.10). Conjunctival chemosis and injection overlying the insertions of the horizontal rectus muscles are routine. Periorbital oedema varies in degree and may be extreme. Proptosis is present in approximately two-thirds of patients. These signs usually precede the disturbed ocular motility which is frequently observed in this disease, and is usually due to a restrictive myopathy which has a predeliction for the inferior rectus and the medial rectus muscles leading to impaired ocular elevation and abduction, respectively (Fells *et al.* 1994). In more advanced cases the affected eye becomes hypotropic, and if the medial rectus is similarly affected esotropia may result. If these ocular motility disturbances are found in the context of other typical signs of dysthyroid eye disease the diagnosis is assured. However, in other cases the florid stage is limited and subclinical, and the patient presents with an ophthalmoparesis which has to be differentiated from an ocular motor nerve palsy. To differentiate this from the restrictive myopathy of thyroid eye disease the forced duction test is used. In this test an attempt is made to move the globe with a forceps under topical anaesthesia, into its appropriate field of gaze. CT or MRI of the orbit reveals characteristic enlargement of the rectus muscles, the most frequently affected being the medial and inferior rectii. Although the diagnosis is supported by the demonstration of thyroid-associated autoantibodies and abnormal thyroid function, it is important to remember that these may be absent, and this should not dissuade the clinician from a diagnosis of dysthyroid eye disease if the clinical signs are compatible. Finally it should be remembered that these patients can also develop another auto-immune disease, myasthenia gravis, which can lead to a worsening of the oculomotor abnormalities. There is evidence that cigarette smoking exacerbates dysthyroid eye disease so patients should be advised to quit tobacco (Pfeilschifter and Ziegler 1996).

Management Approximately 2–9 per cent of all patients with dysthyroid eye disease develop visual loss which requires urgent treatment (Neigel *et al.* 1988) However, since it is impossible to predict which patients will develop an orbitopathy, initial management of the condition is based on rendering the patient euthyroid and using symptomatic therapy such as topical drops and ointments for ocular lubrication, and elevation of the head of the bed at night for oedema. The disorder is self-limiting, usually lasting between 18 and 36 months, before spontaneous improvement in two-thirds of patients.

Worsening of the symptoms, with evidence of infiltrative disease causing myopathy and increased orbital congestion can usually be managed by immunosuppression with corticosteroids or radiotherapy. Systemic corticosteroids in doses up to 120 mg daily, or 1 mg to 1.5 mg/kg (Wiersinga 1996), are used. Benefit is usually apparent within 3 weeks. It may be necessary to add steroid sparing drugs such as cyclosporine. There is some evidence that dysthyroid eye disease can be made worse in patients treated with radioactive iodine for hyperthyroidism. If this is contemplated the patient should be treated with steroids.

Lens sparing, low dose orbital radiotherapy at 1500–2000 cGy in divided fractions over 10 days, should be used if there is a failure to adequately respond to steroids. Usually this produces good results in 7 per cent of patients, although the response may take several weeks. There is now evidence that this therapy should be discussed earlier in the course of the disorder than previously was the case (Kazim *et al.* 1991).

Failure to respond to either therapy, particularly if a compressive optic neuropathy or severe proptosis with corneal exposure keratitis are present, requires surgical orbital decompression. The surgical treatment of choice is transantral orbital decompression into the ethmoid and maxillary sinuses, although other approaches have been proposed. It should not be used in the acute management of proptosis unless there is sight-threatening disease. In the fibrotic phase of the disease restrictive myopathy may require extraocular muscle surgery to restore binocular single vision, and retracted lids may require plastic surgery.

13.4.3 Idiopathic orbital inflammation

Idiopathic orbital inflammatory inflammation or orbital pseudotumour are the terms used for an acute syndrome consisting of painful proptosis, orbital congestion, periorbital oedema, conjunctival swelling or chemosis, and injection, diplopia, and sometimes visual loss, for which a specific cause cannot be found after systematic inflammatory conditions such as syphilis, tuberculosis, sarcoidosis, Wegener's granulomatosis, or collagen vascular disease, polyarteritis nodosa, and systemic lupus erythematosis have been excluded (Lakke 1962; Kline 1982). It may be diffuse or selectively affect any orbital structure resulting in one or more specific symptoms or signs. Histopathologically orbital pseudotumour consists of a mixed cellular infiltrate including lymphocytes, neutrophils, and eosinophils. Some accounts of this condition divide it into different entities depending on the structure involved e.g. sclera (posterior scleritis), lacrimal gland (dacryoadenitis), cavernous sinus/superior orbital fissure (Tolosa–Hunt syndrome), extraocular muscle (myositis), and diffuse orbital (idiopathic orbital inflammation), but they are probably all due to the same pathological process. Orbital lymphoma may be particularly difficult to exclude, even with biopsy material.

The condition is usually unilateral, although occasionally it may be bilateral, and it affects both children and adults. Involvement of one or more extraocular muscles may lead to a painful ophthalmoplegia. Many patients experience a general malaise. The condition may run a chronic remitting course with gradual worsening, or spontaneous remissions may occur.

The CT appearance of orbital pseudotumour varies depending on the orbital structures which are preferentially affected, include contrast enhancement, retrobulbar fatty infiltration, proptosis, extraocular muscle enlargement, optic nerve thickening, and uveoscleral thickening. It may also show muscle tendon sheath involvement, which is spared in dysthyroid eye disease. MRI studies show that on T1-weighted images the lesions are hypointense to fat and isointense to muscle. T2-weighted images show lesions which are isointense or only minimally hyperintense to fat, in contradistinction to the markedly hyperintense signal obtained from orbital metastasis, a not infrequent differential diagnosis.

Patients with this constellation of signs, in whom no other cause can be found, should be given a course of systemic corticosteroid treatment consisting of 80–100mg prednisolone. Corticosteroids usually produce a dramatic response within 48 h of commencement and failure to do so requires reevaluation of the diagnosis. When the expected initial response occurs the steroid dose is tapered and may be titrated against the symptoms and signs. If the patient has chronic recurrences after tailing off the corticosteroids, azothiaprine and cyclosporine may be alternatives. In patients who have contraindications to systemic corticosteroids orbital radiotherapy at 1000–2000 cGY can be of value.

13.4.4 Orbital tumours

A variety of tumours may involve structures within the orbit (Shields *et al.* 1984). These include neurogenic tumours (meningiomas, gliomas, and schwannomas), vascular lesions (cavernous haemangioma), cystic lesions (dermoids), lymphoproliferative lesions (lymphomas), and secondary tumours (metastatic and contiguous spread). The location of the mass in the orbit is often a clue to their nature. They may be intraconal (within the cone of the extraocular muscles), extraconal, and periorbital (outside the orbit, but impinging on its structures).

Intraconal tumours cause forward displacement of the globe. These include tumours which may be primary, such as optic nerve sheath meningiomas and optic nerve gliomas (Section 12.5.4), or secondary due to metastases. One of the commonest orbital tumours in adults which has an affinity for the intraconal space, is the *cavernous haemangioma*. They usually present as a painless proptosis and slowly enlarge over a period of many years. Their CT or MRI appearance is of a well-circumscribed homogeneous mass which shows marked enhancement with contrast. Complete surgical excision is the treatment of choice.

Proptosis may be due to venous anomalies or varices. These may exhibit increased proptosis on Valsava manoeuvre or on bending forward. Their CT appearance is characteristic, and nonintervention is appropriate.

Orbital lymphoma is one of the commoner orbital tumours either in its primary form or as a manifestation of a systemic lymphoma. It may be difficult to differentiate from idiopathic orbital inflammation.

The commonest *metastatic tumours* to the orbit come from breast in 42 per cent, lung in 11 per cent, and prostate in 8.3 per cent. Although patients may present with proptosis, scirrhous breast carcinoma may cause enophthalmos. Any proptotic patient who has a history of treatment of cancer must be suspected of having an orbital metastasis. Histologic verification is mandatory before treatment.

Orbital tumours in the extraconal space cause downward or upward displacement of the globe, when located in the superior and inferior orbital spaces respectively. They may arise in the extraconal space, extend into it from surrounding structures, or be metastatic from distant sources. A rapidly developing unilateral proptosis in a child is likely to be an orbital rhabdomyosarcoma.

Tumours frequently located in the superior orbital space are dermoid tumours and mucocoeles, and less frequently lesions of the lacrimal gland and fibrous dysplasia.

13.4.5 Vascular disorders

The main types of neuro-ophthalmic vascular disorders are the high-flow *carotid-cavernous sinus fistulas* and the low-flow spontaneous *dural-cavernous sinus shunts* which produce overlapping clinical syndromes (Keltner *et al.* 1987). The communication between the arterial and venous systems leads to a rise in the venous pressure in the globe resulting in a fall in the arterial perfusion pressure, and a major drop in perfusion pressure.

Fistulas may be classified (Barrow *et al.* 1985), according to the velocity of blood flow through the shunt, into low- and high-flow fistulas; the anatomic origin of the arteries supplying the fistula, and their aetiology which may be spontaneous or traumatic:

- Type A: communication between internal carotid artery and cavernous sinus; high flow; direct tears.

- Type B: communication between meningeal branches of internal carotid artery and cavernous sinus; slow flow; indirect dural arteriovenous malformation.

- Type C: communication between external meningeal branches of external carotid artery and cavernous sinus; slow flow; indirect dural arteriovenous malformation.

- Type D: communication between meningeal branches of internal carotid artery, external carotid artery, and cavernous sinus; slow flow; indirect dural arteriovenous malformation.

Direct carotid cavernous fistula—Type A

Direct fistulas may occur at any location along the length of the intracavernous portion of the internal carotid artery and commonly occur as a result of both penetrating or non-penetrating head trauma and from aneurysmal rupture. There may be a delay of days or even weeks for symptoms to develop following injury. The common signs are proptosis associated with chronic conjunctival injection and oedema, and an audible bruit. The conjunctival veins may be arterialized and appear as corkscrew vessels that go to the limbus. Because of connections between the two cavernous sinus's a unilateral fistula may give rise to bilateral ocular signs. Damage to cranial nerves and sympathetic and parasympathetic fibres in the cavernous sinus may lead to diplopia and pupillary abnormalities. The commonest cause for diplopia is an abducens nerve palsy. The definitive diagnosis of carotid cavernous fistula is made by selective intra-arterial angiography.

There are several potential causes for the visual loss commonly seen in cavernous sinus fistulas, which include: corneal damage due to exposure; retinal artery occlusion; glaucoma due to raised episcleral venous pressure or rarely to iris neovascularization; macula involvement due to ischaemia, haemorrhage or cystoid oedema; anterior segment ischaemia; retinal artery occlusion; optic nerve ischaemia, and corneal decompensartion due to exposure. Ophthalmoscopy reveals features of a slow flow retinopathy, which include blot haemorrhages, microaneurysms, mild disc swelling, and venous congestion and tortuosity. Occasionally a picture similar to a complete central vein occlusion may occur (Brosnaham *et al.* 1992).

Without treatment the ocular abnormalities will progress, leading to blindness. A variety of endovascular procedures have been developed for cavernous sinus fistulas with the aim of closing the fistula and maintaining the patency of the distal internal carotid artery. These include electrometallic thrombosis using coils and detachable balloon occlusion.

Spontaneous dural-cavernous sinus shunts—Types B, C, and D

These slow flow indirect cavernous fistulas are thought to be due to rupture of congenital arteriovenous anomalies, which usually occur in middle-aged women. The clinical manifestations depend on the direction of venous outflow from the fistula. If anterior then the signs are similar, but less dramatic, to those which occur with direct carotid cavernous fistula, including conjunctival venous arterialization, orbital congestion, and proptosis. Posterior drainage may

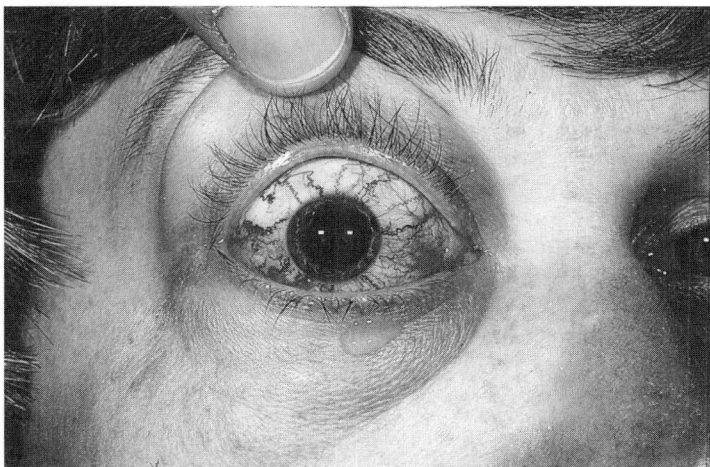
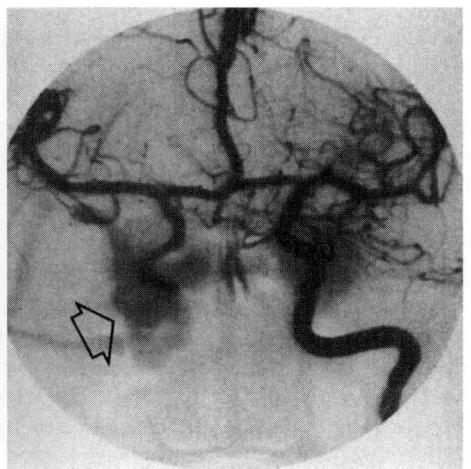

A   B

Fig. 13.11 Carotid-cavernous sinus fistula. (a) Chronic conjunctural injection and oedema with arterialisation of the conjunctival veins. (b) Left carotid arteriogram illustrating a right direct carotid cavernous sinus fistula (arrow).

cause cranial nerve palsies, most commonly of the abducens nerve. A bruit may not be present. Less commonly, choroidal detachment and angle closure glaucoma develop as a result of altered venous outflow.

Treatment

Treatment should be conservative, as between 20 and 50 per cent of spontaneous dural carotid cavernous fistulas close spontaneously, not infrequently after carotid angiography. If vision is threatened then endovascular therapy is indicated (Kupersmith *et al.* 1988).

13.4.6 **Orbital infections**

Orbital infections are potentially sight- and life-threatening, so require urgent evaluation and treatment. The presentation may be acute as occurs with bacteria or more insidious when due to fungal infection. Patients with orbital cellulitis present with proptosis, periorbital swelling, and ophthalmoplegia associated with pyrexia and a peripheral leucocytosis. Most patients will have contiguous sinus disease commonly of the ethmoid sinus, with the frontal and maxillary sinuses being involved less frequently. Organisms can spread into the orbit through valveless veins from the face, teeth, and neck. If the cellulitis occurs in the context of trauma, scanning must be performed to locate any foreign body, which must then be removed.

The causative agent must be sought by culture of both the blood and any purulent wound drainage. The most likely organisms are *Haemophilus influenzae* in children, *Staphylococcus aureus*, and *Streptococcus pneumoniae*. As soon as blood cultures have been obtained the patient should be treated with intravenous broad-spectrum antibiotics to cover the possible causative organisms, until the specific organism has been identified. Surgical drainage may be necessary if despite antibiotics there is worsening proptosis or visual loss suggestive of an orbital abscess.

Patients presenting with an orbital cellulites who either have diabetes mellitus or are immunocompromised, should be considered to have mucormycosis until proven otherwise (Section 42.3.7). Early symptoms include sinusitis, orbital pain, and sudden visual loss (Gass 1961). The organism causes an obliterative arteritis which results in necrotic lesions appearing in the skin, orbit, nasal mucosa, or palate, although this is infrequently present at the onset of symptoms. If mucormycosis infection is suspected, a complete nasal examination should be undertaken to identify the typical black eschars of mucor. Prompt treatment with amphotericin B and surgical debridement ensuring adequate sinus and orbital drainage is essential.

13.4.7 **The diagnosis of proptosis**

The clinical approach to diagnosing a patient presenting with proptosis depends on aspects of the history as well as a careful examination (Section 13.4.1). The patients' past medical history may provide evidence of previous thyroid disease suggesting thyroid eye disease, diabetes mellitus raising the possibility of a mucormycosis infection, or of previous tumour surgery suggesting possible metastatic infiltration. The family history may also provide clues; for example, if members of the family have a history of skin lesions, epilepsy, or brain tumours this raises the possibility of an optic nerve glioma being the cause of the proptosis due to neurofibromatosis.

The tempo of development of the proptosis and the presence or absence of pain are also important features to ascertain. A rapidly developing proptosis in a child would immediately raise the suspicion of a malignant tumour such as a rhabdomyosarcoma or metastatic neuroblastoma. Such a rapid painless onset in adults suggests a metastatic tumour, but if associated with pain the diagnosis includes orbital cellulitis and inflammatory orbital pseudotumour, all of which may be associated with diplopia. If there is a history of intermittent painful proptosis the most likely aetiologies are venous varices or lymphangiomas.

When the tempo of development of the proptosis is unclear evidence for earlier proptosis than is often recognized by the patient can be derived from viewing antecedent photographs. If the history or photographs suggest a more slowly progressive painless course then thyroid eye disease or an orbital tumour should be considered. If the tumour is associated with visual impairment due to associated involvement of the optic nerve, which occurs some time after the onset of the proptosis, the mass is probably within the anterior or mid-third of the orbit. If, however, the visual loss precedes the proptosis the mass is more likely to lie in the posterior third of the orbit. More anterior lesions tend to cause horizontal diplopia, whereas apical or posterior masses cause vertical diplopia due to impaired vertical gaze.

Other symptoms should be sought, such as swishing noises in the head suggestive of a carotid cavernous fistula, a history of chemosis and lid swelling upon rising in the morning associated with photophobia, lacrimation, and burning suggestive of thyroid eye disease, and increasing proptosis with raised intra-abdominal pressure during the Valsalva manoeuvre or change in position suggestive of an orbital varix or mucocele.

Painful proptosis is relatively uncommon and the differential diagnosis includes acute orbital inflammation, metastases, acute thrombosis of orbital varices and of the enlarged veins associated with an AV fistula. It should be remembered that there are other causes for orbital pain including the superior orbital fissure/cavernous sinus syndrome or referral from the dura.

Once a satisfactory history has been obtained an examination of the orbit and its contents, as described above, should be undertaken. This should help to refute or substantiate clues to the diagnosis which have been obtained from the history and so enable the clinician to order the appropriate investigations necessary to confirm the diagnosis.

References

Al-Din AS (1986). Controversy about Fisher's syndrome. *Arch Neurol*, **43**, 543–4.

Appukuttan B, Gillanders E, Juo SH *et al.* (1999). Localization of a gene for Duane retraction syndrome to chromosome 2q21. *Am J Hum Genet*, **65**, 1639–46.

Asbury AK, Aldredge H, Hershberg R *et al.* (1970). Oculomotor palsy in diabetes mellitus, A clinico-pathological study. *Brain*, **93**, 555–66.

Averbuch-Heller L, Remler B (1996). Opsoclonus. *Semin Neurol*, **16**, 21–6.

Baloh RW, Furman JN, Yee RD (1985). Dorsal midbrain syndrome, clinical and oculographic finding. *Neurology*, **35**, 54–60.

Baloh RW, Yee RD, Honrubia V (1981). Eye movements with Wallenberg's syndrome. *Ann NY Acad Sci*, **374**, 600–13.

Barrow DL, Spector RH, Braun IF *et al.* (1985). Classification and treatment of spontaneous carotid-cavernous sinus fistulas. *J Neurosurg*, **62**, 248–56.

Bell HA, McIlllwain GG, O'Neill D (1994). The iatrogenic lateral rectus nerve palsies, a series of post-myelographic cases. *J Neuroophthalmol*, **14**, 205–9.

Bell RA, Thompson HS (1978). Relative afferent pupillary defect in optic tract hemianopias. *Am J Ophthalmol*, **85**, 538–40.

Berlit P, Rakicky J (1992). The Miller Fisher syndrome, review of the literature. *J Clin Neuroophthalmol*, **12**, 57–63.

Bever CT, Aquino AV, Penn AS et al. (1983). Prognosis in ocular myasthenia. *Ann Neurol*, **14**, 516–9.

Blumen SC, Brais B, Korczyn AD et al. (1999). Homozygotes for oculopharyngeal muscular dystrophy have a severe form of the disease. *Ann Neurol*, **46**, 115–8.

Bogousslevsky J, Maeder P, Regli F et al. (1994). Pure mid-brain infarction, clinical syndromes, MRI and aetiologic patterns. *Neurology*, **44**, 2032–40.

Brandt T, Dieterich M (1993). Skew deviation with ocular torsion, a vestibular brainstem sign of topographic diagnostic value. *Ann Neurol*, **33**, 528–34.

Brosnaham D, McFadzean RM, Teesdale E (1992). Neuro-ophthalmic features of carotid cavernous fistulas and their treatment by endoarterial balloon embolisation. *J Neurol Neurosurg Psychiatry*, **55**, 553–6.

Brown HW (1973). True and simulated superior oblique tendon sheath syndromes. *Doc Ophthalmol*, **34**, 123–36.

Büttner-Ennever JA, Büttner U, Cohen B et al. (1982). Vertical gaze paralysis and the rostral interstitial nucleus of the medial longitudinal fasiculus. *Brain*, **105**, 125–49.

Castro O, Johnson LN, Mamourian AC (1990). Isolated inferior oculomotor paresis from brainstem infarction. Prospective oculomotor fascicular organisation in the ventral mid-brain pigmentum. *Arch Neurol*, **47**, 235–7.

Char DH (1997). Normal thyroid gland and mechanisms of hyperthyroidism. In *Throid Eye Disease*, 3rd edition, pp. 5–23. Butterworth-Heinemann, Boston.

Chiba A, Kusunoki S, Shimuzi T et al. (1992). Serum IgG antibody to ganglioside GQ1b is a possible marker of Miller Fisher syndrome. *Ann Neurol*, **31**, 677–9.

Clark JM, Albert GW (1995). Vertical gaze palsies from medial thalamic infarctions without mid-brain involvement. *Stroke*, **26**, 1467–70.

Cogan CG (1952). A type of congenital ocular motor apraxia presenting with jerky head movements. *Trans Am Acad Ophthalmol*, **56**, 853–62.

Cogan DG (1965). Myasthenia gravis, a review of the disease and a description of lid twitch as a characteristic sign. *Arch Ophthalmol*, **74**, 217–21.

Cogan TG (1970). Internuclear ophthalmoplegia, typical and atypical. *Arch Ophthalmol*, **84**, 583–9.

Cox TA, Wurster JB, Godfrey WA (1979). Primary aberrant oculomotor degeneration due to intracranial aneurysm. *Arch Neurol*, **36**, 570–1.

Cullom ME, Savino PJ, Sergott RC et al. (1995). Relative pupillary sparing 3rd nerve palsies; arteriogram or not? *J Clin Neuroophthalmol*, **15**, 136–141 (also see commentary).

Daroff RB (1965). See-saw nystagmus. *Neurology*, **15**, 874–7.

Daroff RB (1970). Ocular motor manifestations of brainstem and cerebellar dysfunction. In Smith JL, ed. *Neuroophthalmology*, Vol. 5, pp. 104–118. Hoffmann, Florida.

Daroff RB, Hoyt WF (1971). Supranuclear disorders of ocular control systems in man. Clinical, anatomical and physiological correlations-1969. In Bach-y-Rita P, Collins CC, Hyde JE, eds. *The Control of Eye Movements*. Academic Press, New York.

Dell'Osso LF, Daroff RB (1975). Congenitital nystagmus waveforms and foveation strategy. *Doc Ophthalmol*, **39**, 155–82.

Drachman DA (1968). Ophthalmolplegia plus, the neurodegenerative disorders associated with progressive external ophthalmolplegia. *Arch Neurol*, **18**, 654–74.

Edelson RN, Levy DE (1974). Transient benign unilateral pupillary dilation in young adults. *Arch Neurol*, **31**, 12–4.

Ellis CJK (1979). The afferent pupillary defect in acute optic neuritis. *J Neurol Neurosurg Psychiaty*, **42**, 1008–17.

Ertas M, Arac N, Kumral K et al. (1994). Ice test as a simple diagnostic test for myasthenia gravis. *Acta Neurol Scand*, **89**, 227–9.

Eyster EF, Hoyt WF, Wilson CB (1972). Oculomotor palsy from minor head trauma. *JAMA*, **220**, 1083–6.

Ezra E, Spalton D, Sanders MD (1996). Ocular neuromyotonia. *Br J Ophthalmol*, **80**, 350–5.

Fells P, Kousoulides L, Pappa A et al. (1994). Extraocular muscle problems in thyroid eye disease. *Eye*, **8**, 497–505.

Fisher A, Gresty M, Chambers B et al. (1983). Primary position up-beating nystagmus. A variety of central positional nystagmus. *Brain*, **106**, 949–64.

Fletcher WA (1993). Nystagmus, an overview. In Sharpe JA, Barber HO, eds. *The Vestibular-Ocular-Reflex and Vertigo*, pp. 195–215. Raven Press, New York.

Fletcher WA, Sharpe JA (1986). Saccadic eye movement dysfunction in Alzheimer's disease. *Ann Neurol*, **20**, 464–71.

Forster RK, Shatz NJ, Smith JL (1969). A subtle eyelid sign in aberrant regeneration of the 3rd nerve. *Am J Ophthalmol*, **67**, 696–8.

Gass JDM (1961). Acute orbital mucormycosis. *Arch Ophthalmol*, **65**, 214–20.

Goldstein JE, Cogan DG (1960). Diabetic ophthalmoplegia with special reference to the pupil. *Arch Ophthalmol*, **64**, 592–600.

Gresty MA, Metcalfe T, Timms C et al. (1992). Neurology of latent nystagmus. *Brain*, **115**, 1303–21.

Halmagyi GM, Brandt T, Dieterich M et al. (1990). Tonic controversive ocular tilt reaction due to unilateral mesodiencephalic lesions. *Neurology*, **40**, 1503–9.

Halmagyi GM, Rudge P, Gresty MA et al. (1980). Treatment of periodic alternating nystagmus. *Ann Neurol*, **8**, 609–11.

Halmagyi GM, Rudge P, Gresty MA et al. (1983). Down-beating nystagmus, a review of 62 cases. *Arch Neurol*, **40**, 777–84.

Hanson MR, Hamid MA, Thomsak RL et al. (1986). Selective saccadic palsy caused by pontine lesions, clinical, physiological and pathological correlations. *Ann Neurol*, **20**, 209–17.

Harris C, Shawkat F, Russell-Eggitt I et al. (1986). Intermittent horizontal saccade failure, 'ocular motor apraxia', in children. *Br J Ophthalmol*, **80**, 151–8.

Heide W, Kurzidin K, Kömpf D (1996). Deficits in smooth pursuit eye movements after frontal and parietal lesions. *Brain*, **119**, 1951–69.

Henn V, Büttner U (1982). Disorders of horizontal gaze, functional basis of ocular motility disorders. In Lennerstrand G, Zee DS, Keller E, eds. pp. 239–45. Pergamen Press, Oxford.

Hommel B, Bogousslavsky J (1991). The spectrum of vertical gaze palsy following unilateral brainstem stroke. *Neurology*, **41**, 1229–34.

Hoyt WF, Keane JR (1970). Superior oblique myokymia. *Arch Ophthalmol*, **84**, 461–7.

Huber A (1974). Electrophysiology of the retraction syndromes. *Br J Ophthalmol*, **58**, 293–300.

Hunt WE, Meagher JN, LeFever HE et al. (1961). Painful ophthalmoplegia. *Neurology*, **11**, 56–62.

Kaminsky HJ, Maas E, Spiegel P et al. (1990). Why are eye muscles frequently involved in myasthenia gravis? *Neurology*, **40**, 1663–9.

Kazim M, Trokel S, Moore S (1991). Treatment of acute Graves' orbitopathy. *Ophthalmology*, **98**, 1143–448.

Keane JR (1993). Forth nerve palsy, historical review and study of 215 inpatients. *Neurology*, **43**, 2439–43.

Kearns TP, Sayre GP (1958). Retinitis pigmentosa, external ophthalmoplegia and complete heart block. *Arch Ophthalmol*, **60**, 280–9.

Kelly JJ, Daube JR, Lennon et al. (1982). The laboratory diagnosis of mild myasthenia gravis. *Ann Neurol*, **12**, 238–42.

Keltner JL, Satterfield D, Dublin AB *et al.* (1987). Dural and carotid cavernous sinus fistulas: diagnosis, management and complications. *Ophthalmology*, **94**, 585–1600.

Kline LB (1982). The Tolosa-Hunt syndrome. *Surv Ophthalmol*, **27**, 79–95.

Knox D, Clark D, Schuster F (1967). Benign VI nerve palsies in children. *Paediatrics*, **40**, 560–4.

Kodski SR, Younge BR (1992). Acquired oculomotor, trochlear and abducens cranial nerve palsies in pediatric patients. *Am J Ophthalmol*, **114**, 568–74.

Komerell G, Schaubele G (1980). Superior oblique myokymia. An electromyographic analysis. *Trans Ophthalmol Soc U K*, **100**, 504–6.

Kupersmith MJ, Burde RM, Warren FA *et al.* (1988). Auto-immune optic neuropathy, evaluation and treatment. *J Neurol Neurosurg Psychiatry*, **51**, 1381–6.

Kupfer, C, Chumbley L, Downer J (1967). Quantitative histology of optic nerve and optic tract nucleus of man. *J Anat*, **101**, 393–402.

Lakke JPWF (1962). Superior orbital fissure syndrome, report of a case caused by local pachymeningitis. *Arch Neurol*, **7**, 289–300.

Lasker AG, Zee DS (1997). Ocular motor abnormalities in Huntington's disease. *Vis Res*, **37**, 3639–45.

Lee J (1992). Modern management of sixth nerve palsy. *Aust NZ J Ophthalmol*, **20**, 41–6.

Leigh RJ, Kennard C (2004). Using Saccades as a research tool in the clinical neurosciences. *Brain*, **127**, 460–77.

Leigh RJ, Zee DS (2006). *The Neurology of Eye Movements*, 4th edition. Oxford University Press, Oxford.

Lepore FE (1995). Disorders of ocularmotility following head trauma. *Arch Neurol*, **52**, 924–6.

Lepore FE, Glaser JS (1980). Misdirection revisited, a critical appraisal of acquired ocularmotor nerve synkinesis. *Arch Ophthalmol*, **98**, 2206–9.

Levine SR, Twyman RE, Gilman S (1988). The role of anticoagulation in cavernous sinus thrombosis. *Neurology*, **38**, 517–22.

Lewis RF, Zee DS (1993). Ocular motor disorders associated with cerebellar lesions, pathophysiology and topical diagnosis. *Revue Neurol*, **149**, 665–77.

Loewenfeld IE (1993). The pupil, anatomy, physiology and clinical applications. In Ames IA, Detroit MR, eds. Iowa State University Press, in collaboration with Wayne State University Press.

Loewenfeld IE, Thompson HS (1967). The tonic pupil, a reevaluation. *Am J Ophthalmol B*, 46–87.

Loewenfeld IE, Thompson HS (1975). Ocular motor paresis with cyclic spasms. A critical review of the literature and a new case. *Surv Ophthalmol*, **20**, 81–124.

Loewenfeld IE, Thompson HS (1981). Mechanism of tonic pupil. *Ann Neurol*, **10**, 275–6.

Lowenstein O, Loewenfeld IE (1965). Pupillotonic pseudo-tabes. *Surv Ophthalmol*, **10**, 130–85.

Mansour AM, Reinecke RD (1986). Central trochlear palsy. *Surv Ophthalmol*, **30**, 279–97.

Mathew NT, Chandy J (1970). Painful ophthalmoplegia. *J Neurol Sci*, **11**, 243–56.

Miller NR (1977). Solitary oculomotor nerve palsy in childhood. *Am J Ophthalmol*, **83**, 106–11.

Miller NR (1996). The clinical manifestations, natural history and results with treatment of superior oblique myokymia. *Am Orthopt J*, **46**, 189–94.

Miller NR, Kiel SM, Green WR *et al.* (1982). Unilateral Duane's retraction syndrome (Type 1). *Arch Ophthalmol*, **100**, 1468–72.

Miller NR, Moses H (1977). Ocular involvement in wound botulism. *Arch Ophthalmol*, **95**, 1788–9.

Morrow MJ, Sharpe JA (1993). Smooth pursuit eye movements. In Sharpe JA, Barber HO, eds. *The Vestibular-Ocular Reflex and Vertigo*, pp. 141–62. Raven Press, New York.

Müri RM, Chermann JF, Kohen L *et al.* (1996). Ocular motor consequences of damage to the abducens nucleus area in humans. *J Neurol Ophthalmol*, **16**, 191–5.

Murphy SF, Drachman DM (1968). The oculopharyngeal syndrome. *J Am Med Assoc*, **203**, 1003–8.

Nadeau SE, Trobe JD (1983). Pupil sparing in oculomotor palsy, a brief review. *Ann Neurol*, **13**, 143–8.

Nakada T, Kwee IL (1986). Oculo-palato-myoclonus. *Brain*, **109**, 431–41.

Neigel JM, Rootman J, Belkin RI *et al.* (1988). Dysthyroid optic neuropathy. The crowded orbital apex syndrome. *Ophthalmology*, **95**, 1515–21.

Newman NJ (1992). Mitochondrial disease and the eye. *Ophthalmol Clin North Am*, **55**, 405–24.

O'Sullivan EP, Kennard C (1998). Neuro-ophthalmology of movement disorders. In Jankovic J, Tolosa E, eds. *Parkinson's Disease and Movement Disorders*, pp. 869–86. Williams Wilkins, Baltimore.

Ochs AL, Stark L, Hoyt WF *et al.* (1979). Opposed adducting saccades in convergence-retraction nystagmus. A patient with Sylvian aqueduct syndrome. *Brain*, **102**, 497–508.

Oosterhuis HJGH (1982). The ocular signs and symptoms of myasthenia gravis. *Doc Ophthalmol*, **52**, 363–78.

Osserman KE (1957). *Myasthenia Gravis*. Grune and Stratton, New York.

Petit L, Haxby JV (1999). Functional anatomy of pursuit eye movements in humans as revealed by fMRI. *J Neurophysiol*, **81**, 463–71.

Petty RK, Harding AE, Morgan-Hughes JA (1986). The clinical features of mitochondrial myopathy. *Brain*, **109**, 915–38.

Pfeilschifter J, Ziegler R (1996). Smoking and endocrine ophthalmopathy: Impact of smoking severity and current vs lifetime cigarette consumption. *Clin Endocrinol*, **45**, 477–81.

Pierrot-Deseilligny C, Gautier JC, Loron P (1988). Acquired ocular motor apraxia due to bilateral fronto-parietal infarcts. *Ann Neurol*, **23**, 199–202.

Ranalli PJ, Sharpe, JA (1988). Vertical vestibulo-ocular reflex smooth pursuit and eye head tracking dysfunction in internuclear ophthalm-oplegia. *Brain*, **111**, 1299–317.

Reddel SW, Barnett MH, Yan WX *et al.* (2000). Chronic ophthalmoplegia with anti-GQ1b antibody. *Neurology*, **54**, 1000–2.

Richards BW, Jones FR, Younge BR (1992). Causes and prognosis in 4,278 cases of the oculomotor, trochlear and abducens cranial nerve palsies. *Am J Ophthalmol*, **113**, 489–96.

Robertson DN, Heinz JD, Rucker CW (1970). Acquired sixth nerve paresis in children. *Arch Ophthalmol*, **83**, 574–9.

Rush JA, Younge BR (1981). Paralysis of cranial nerves III, IV and VI. Cause and prognosis in 1000 cases. *Arch Ophthalmol*, **99**, 76–9.

Sano K (1959). Trigemino-oculomotor synkineses. *Neurologia*, **1**, 29–51.

Sarkies NJC, Sanders MD (1985). Convergence spasm. *Trans Ophthalmol Soc*, **104**, 782–6.

Savino PJ, Hilliker JK, Cassell GH *et al.* (1982). Chronic sixth nerve palsies, are they really harbingers of serious intracranial disease? *Arch Ophthalmol*, **100**, 1442–4.

Schwartz MA, Selhorst JB, Ochs AL *et al.* (1986). Oculomasticatory myorhythmia, a unique movement disorder occurring in Whipple's disease. *Ann Neurol*, **20**, 677–83.

Selhorst JB, Hoyt WF, Feinsord M *et al.* (1976). Midbrain correctopia. *Arch Neurol*, **33**, 193–5.

Selhorst JB, Madge G, Ghatack N (1984). The neuropathology of the Holmes-Adie syndrome. *Ann Neurol*, **16**, 138.

Sergott RC, Glaser JS (1981). Grave's ophthalmopathy. A clinical and immunologic review. *Surv Ophthalmol*, **26**, 1–21.

Sharpe JA, Rosenberg MA, Hoyt WF *et al.* (1974). Paralytic pontine exotropia. A sign of acute unilateral gaze palsy and internuclear ophthalmo plegia. *Neurology*, **24**, 1076–81.

Shaunak S, O'Sullivan E, Kennard C (1997). Eye movements. In Hughes JAC, ed. *Neurological Investigations*, pp. 253–82. *BMJ*, London.

Shields JA, Bakewell B, Augsberger JJ et al. (1984). Classification and incidence of space-occupying lesions of the orbit-a survey of 145 biopsies. *Arch Ophthalmol*, **102**, 1606–11.

Silverman IE, Liu GT, Volpe NJ et al. (1995). The crossed paralysis, the original brainstem syndromes of Millard-Gubler, Foville, Weber and Raymond-Cestan. *Arch Neurol*, **52**, 625–38.

Smith SA, Smith SE (1983). Evidence for a neuropathic aetiology in the small pupil of diabetes mellitus. *Arch J Ophthalmol*, **67**, 89–93.

Solomon D, Galetta SL, Liu GT (1995). Possible mechanisms for horizontal gaze deviation and lateropulsion in the lateral medullary syndrome. *J Neuroophthalmol*, **15**, 26–30.

Susac JO, Hoyt WF, Daroff RB et al. (1970). Clinical spectrum of ocular bobbing. *J Neurol Neurosurg Psychiatry*, **33**, 771–5.

Ter Bruggen JP, Bastiaensen AK, Turssen CC et al. (1990). Disorders of eye movement in myotonic dystrophy. *Brain*, **113**, 463–73.

Thomas JE, Yoss RE (1970). The parasellar syndrome, Problems in determining etiology. *Mayo Clin Proc*, **45**, 617–23.

Thömke F, Hopf HC, Krämer G (1992). Internuclear ophthalmoplegia of abduction, clinical and electrophysiological data on the existence of abduction paresis of prenuclear origin. *J Neurol Neurosurg Psychiatry*, **55**, 105–11.

Thompson HS (1966). Afferent pupillary defects. *Am J Ophthalmol*, **62**, 860–73.

Thompson HS (1977). Diagnosing Horner's syndrome. *Trans Am Ophthalmol Otolaryngol*, **83**, 840–2.

Thompson HS (1979). A classification of tonic 'tonic pupils'. In Thompson HS, Daroff R, Friesen L et al. eds. *Topics in Neuroophthalmology*, pp. 95–6. Williams and Wilkins, Baltimore.

Thompson HS, Bell RA, Bourgon P (1979a). The natural history of Adies syndrome. In Thompson, HS, Daroff R, Frisen L et al. eds. *Topics in Neuroophthalmology*, pp. 96–9. Williams and Wilkins, Baltimore.

Thompson HS, Bourgon P, van Allen MW (1979b). The tendon reflex in Adie's syndrome. In Thompson HS, Daroff R, Frisen L et al. eds. *Topics in Neuroophthalmology*, pp. 104–13.Williams and Wilkins, Baltimore.

Thompson HS, Pilley SFJ (1976). Unequal pupils, a flow chart for sorting out the anisocorias. *Surv Ophthalmol*, **21**, 45–8.

Thompson JS, Corbett JJ, Cox TA (1981). How to measure the relative afferent pupillary defect. *Surv Ophthalmol*, **26**, 39–42.

Tiffin PA, MacEwen CJ, Craig EA et al. (1996). Acquired palsy of the oculomotor, trochlear and abducens nerves. *Eye*, **10**, 377–84.

Towfighti J, Marks K, Palmer E et al. (1979). Möbius syndrome, neuropathologic observations. *Acta Neuropathol*, **48**, 11–7.

Trobe JD (1988). Third nerve palsy and the pupil. *Arch Ophthalmol*, **106**, 601–2.

Trobe JD, Glaser JS, Post JD (1978). Meningiomas and aneurysms of the cavernous sinus. *Arch Ophthalmol*, **96**, 457–67.

van der Gaag R, Schmidt ED, Zonneveld FW et al. (1996). Orbital pathology in thyroid-associated ophthalmopathy. *Orbit*, **15**, 109–17.

Van der Wiel AL, van Gijn J (1983). Localisation of Horner's syndrome. Use and limitations of the hydroxyamphetamine test. *J Neurol Sci*, **59**, 229–35.

Victor DI (1976). The diagnosis of congenital unilateral third nerve palsy. *Brain*, **99**, 711–7.

Volpe NJ, Lessell S (1993). Remitting sixth nerve palsy in skull based tumours. *Arch Ophthalmol*, **111**, 1391–5 (Published erratum appears in *Arch Ophthalmol*, 112, 1118, 1994).

Wall M, Wray SH (1983). The one-and-a-half syndrome-a unilateral disorder of the pontine tegmentum, a study of 20 cases and review of the literature. *Neurology*, **33**, 971–80.

Walsh FB, O'Doherty DS (1960). A possible explanation of the mechanism of ophthalmoplegic migraine. *Neurology*, **10**, 1079–84.

Weinberg DA, Lesser RL, Vollmer TL (1994). Ocular myasthenia, A proteam disorder. *Surv Ophthalmol*, **39**, 169–210.

Wiersinga WM (1996). Advances in medical therapy of thyroid-associated ophthalmopathy. *Orbit*, **15**, 177–86.

Wilhelm H, Wilhelm B, Petersen D et al. (1996). Relative afferent pupillary defects in patients with retrogeniculate lesions. *Neuroophthalmology*, **16**, 219–24.

Wirtschafter JD, Volk C, Sawchuk RJ (1978). Transaqueous diffusion of acetylcholine to denervated iris sphincter muscle, a mechanism for the tonic pupil syndrome. *Ann Neurol*, **4**, 1–5.

Wright JE (1988). Doyne Lecture, Current concepts in orbital disease. *Eye*, **2**, 1–11.

Zee DS (1992). Internuclear ophthalmoplegia, clinical and pathophysiological consideration. In Büttner U, Brandt T, eds. *Ocular Motor Disorders in the Brainstem*, pp. 455–70. WB Saunders, London.

CHAPTER 14

Disorders of hearing

Linda Luxon

Contents

14.1 Introduction

Hearing loss is the commonest sensory disability worldwide, and the World Health Organisation has estimated that 278 million people suffer a moderate to profound hearing loss in both ears, with 80 per cent of deaf and hearing-impaired people living in low- and middle-income countries (WHO 2006). Tinnitus affects approximately 10 per cent of developed populations (Coles 1984) and of these, 5 per cent find the symptom troublesome and seek help (Davis 1995). Tinnitus and hearing loss are primary symptoms of disordered cochlear function, but may also present as a result of central auditory pathology with normal cochlear function. Pathology affecting the central auditory pathways characteristically presents as difficulty hearing in conditions of poor signal-to-noise ratio, for example, in a classroom in the presence of background noise, listening to transmitted sound, for example on the telephone or on a television, and sound localization. As a consequence of multiple relays and bilateral representation above the level of the cochlear nuclei, central auditory dysfunction does not present with hearing loss. Hearing loss and/or tinnitus, with or without associated vestibular abnormalities, will most commonly be the result of otological pathology. However, importantly for the neurologist cochlear, VIII nerve, or central auditory dysfunction may be part of the clinical presentation of a neurological disorder.

14.1.1 Definitions

Hearing loss may be consequent upon pathology in the external, middle, or internal ear (Fig. 14.1), and is judged by the threshold of hearing across a standard frequency range of 250–8000 Hz. For clinical purposes, threshold values better than 25 dBHL are considered to be normal, and averages across the 500–4000 Hz range are used to define significant to profound impairment, including deafness (see Table 14.1).

Tinnitus is defined as the perception of a sound that originates from within the body rather than the external world. *Subjective tinnitus* refers to the situation in which tinnitus is perceived only by the patient, while *objective tinnitus* refers to a sound that may be audible externally and has a physical source such as palatal myoclonus, an arteriovenous fistula, or turbulent blood flow through a stenotic artery. A distinction must be made between tinnitus presence and tinnitus complaint. Occasional tinnitus is an almost universal perception, while tinnitus complaint and intrusiveness has been shown to have no direct correlation with psychoacoustic changes in the inner ear, but a strong correlation with psychological factors (Hinchcliffe and King 1992).

Dysacusis refers to any deviation from normal auditory perception, and includes a variety of phenomena (Hinchcliffe 2002).

Hyperacusis is an intensity-related dysacusis, and may be defined as a reduced tolerance to noise, or an increased sensitivity to sounds in levels that would not cause discomfort in a normal individual, and in this way differs from loudness recruitment, which refers to oversensitivity to loud sounds.

Paracusis refers to auditory dysfunction in which the perception of volume, pitch, timbre, or other quality of sound may be altered. Most commonly, paracuses are associated with abnormalities of the peripheral auditory system, but they may also occur with involvement of the auditory pathways by central nervous system pathology (Ghosh 1990).

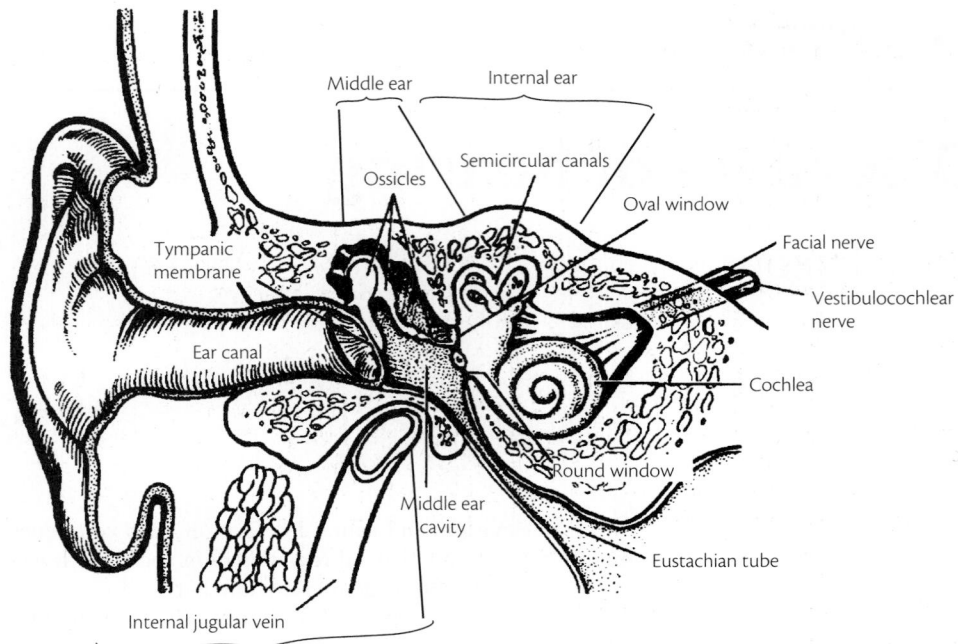

Fig. 14.1 Diagram to illustrate the anatomy of the external, middle, and internal ear.

Table 14.1 The World Health Organisation grading of hearing impairment

Grade	Corresponding audiometric ISO value	Performance	Recommendations
0 – No impairment	25 dB or better (better ear)	No or very slight hearing problems. Able to hear whispers.	
1 – Slight/mild impairment	26–40 dB (better ear)	Able to hear and repeat words spoken in normal voice at 1 m.	Counselling. Hearing aids may be needed.
2 – Moderate impairment	41–60 dB (better ear)	Able to hear and repeat words spoken in raised voice at 1 m.	Hearing aids usually recommended.
3 – Severe impairment	61–80 dB (better ear)	Able to hear some words when shouted into better ear.	Hearing aids needed. If no hearing aids available, lip-reading and signing should be taught.
4 – Profound impairment including deafness	81 dB or greater (better ear)	Unable to hear and understand even a shouted voice.	Hearing aids may help understanding words. Additional rehabilitation needed. Lip-reading and sometimes signing essential.

Grades 2, 3, and 4 are classified as *disabling hearing impairment*.

The audiometric ISO values are averages of values at 500, 1000, 2000, and 4000 Hz.

(http://www.who.int/pbd/deafness/hearing_impairment_grades/en/index.html -viewed 03/06/2006)

Diplacusis is a phenomenon in which the pitch of a single tone is heard doubly, as two tones. This phenomenon may occur as binaural diplacusis in which the same tone is perceived differently in each ear, and this is most commonly seen in Menière's disease, or as monaural diplacusis, which is much rarer, and may be associated with spontaneous cochlear emissions, which interact with external sounds (Long 1998). This phenomenon usually results from alterations in peripheral sound processing, but rarely may be associated with central auditory pathology.

Dysstereoacusis is an impairment of spatial analysis of sounds, and manifests as impairment of the perception of sound localization. It may result from abnormal processing of sound intensity and/or phase differences, either peripherally or centrally.

14.1.2 Epidemiology

In the United Kingdom, 1.1 in a thousand children are born with a permanent bilateral hearing impairment (Hall 2004), and approximately the same number of children up to the age of 16 become hearing-impaired annually. Twenty-five per cent of all children up to 4 years of age suffer from otitis media with effusion and associated hearing loss (Haggard and Hughes 1991), and up to 10 per cent of children may have some degree of auditory processing disorder (Auditory Processing Disorders Special Interest Group 2004). Thirty-five per cent of the UK population over 50 have at least a mild hearing loss (Davis 1993), and one-third of those over 60 years have hearing loss of 25 dB or more (Steel 1998).

Prevalence statistics for tinnitus vary widely in the literature depending upon the criteria for definition, as if all forms of tinnitus are considered, the prevalence of the symptom is probably 100 per cent (Heller and Bergmann 1953). It is a common symptom in the developed world, while it is virtually unreported in the developing world (Gill and Sharma 1967), and, interestingly, an overwhelming preponderance of patients in a tinnitus clinic are from the higher socio-economic groups (Shulman 1988). In the United Kingdom, based on the National Study of Hearing (Coles 1984),

15 per cent of the adult population have prolonged 'spontaneous' tinnitus, i.e. with no obvious trigger and a symptom duration of over 5 min, while in 8 per cent, tinnitus interferes with sleep and causes moderate or severe annoyance, but only 0.5 per cent describe tinnitus as severely intrusive, affecting the sufferer's ability to lead a normal life. Similar estimates of prevalence have been reported in the United States (Cooper 1994) and in a literature review of seven large studies from four different countries (Davis and Rafaie 2000). Tinnitus prevalence is positively correlated with age (Coles 1984), and slightly more females than males are affected (Coles 1984; Cooper 1994). Tinnitus prevalence also increases with increasing hearing loss of all aetiologies (Coles *et al.* 1990; Collet *et al.* 1990).

There are few data on the prevalence of other forms of dysacuses, but hyperacusis is reported in 6–7 per cent of the general population (Andersson *et al.* 2002), and in the majority of cases it is associated with normal hearing. This symptom is also reported in a significant number of individuals with tinnitus, but the figure varies between 40 and 80 per cent of this group (Jastreboff *et al.* 1996; Dauman and Bouscau-Faure 2005).

14.1.3 Social and economic consequences

Disorders of the ear represent 24 per cent of all disabilities in the adult population—the second highest cause of all disabilities (HMSO 1997). Hearing impairment and deafness have profound consequences, which may impose a significant social and economic burden on individuals, families, and society. Without appropriate care, children with severe to profound hearing loss often fail to develop normal speech, language, and cognitive skills (Downs and Yoshinago-Itano 1999), which in turn impacts upon educational progress and employment. Adults with hearing impairment and deafness are less likely to be in skilled employment, and find it difficult to obtain, maintain, and progress in the occupational environment. Both children (Hindley 1997) and adults (Fellinger *et al.* 2005) may suffer from social stigmatization, isolation, and the incidence of psychological disorders in severely hearing-impaired adults and children is higher than in the normal population. Thus throughout the life-span, hearing disorders constitute an important compromise to the health of the population.

14.2 Anatomy and physiology

The ear may be divided into the external, the middle, and the internal ear (Fig. 14.1).

In the *external ear,* the pinna plays a small role in the collection of sound, and loss of this structure results in worsening of the hearing level by about 5 dB across the speech frequencies. The external ear canal is approximately 2.5 cm long, with the lateral third comprised of cartilaginous walls, and the medial two-thirds of bony walls. The ear canal has a more important role in hearing, protecting the tympanic membrane from direct damage by the curvature of the canal, and enhancing sound pressure levels at the ear drum over a range of frequencies. There is no amplification of sound within the external canal, but a redistribution of energy in the form of resonant peaks, which produce an enhancement over the mid-frequency range of hearing, with a gain of up to 15 dB at the tympanic membrane, across the 2–6 kHz region.

The *middle ear* is comprised of the tympanic membrane and the middle ear cleft. Laterally, the structure is bounded by the oval tympanic membrane, while a number of important structures lie

in close proximity to the medial wall: the facial nerve, the lateral semicircular canal, the oval window, the basal turn of the cochlea, and the round window. Inferiorly, the eustachian tube links the middle ear to the nasopharynx.

The *inner ear* is embedded in a thin, but dense bony structure: the bony labyrinth. It is filled with perilymph, and can be divided into three anatomical and functional regions: the semi-circular canals, the vestibule, and the cochlea. Within the bony labyrinth lies the membranous labyrinth, which is filled with endolymph, and contains the sensory cells of both hearing and balance.

The *bony cochlea* resembles a snail shell. This is triangular in cross section and spirals through approximately two-and-a-half turns of the cochlea from base to apex. The cochlear duct is approximately 35 mm in length, with a flat floor known as the spiral lamina, a side wall which is mainly comprised of the stria vascularis and a sloping diagonal 'roof' known as Reissner's membrane (Fig. 14.2). The spiral lamina coils around the central bony core, the modiolus, but at the apex is not attached to the modiolus, forming the helicotrema—a gap between the scala vestibulae and the scala tympani.

The *organ of Corti* is situated on the basilar membrane, and contains the auditory sensory receptor cells. These cells, known as hair cells, have stereocilia projecting from their upper endolymphatic surface. There are two types of cells, the inner hair cells and the outer hair cells. Each inner hair cell has a flask-shaped structure, and is surrounded by supporting cells with approximately

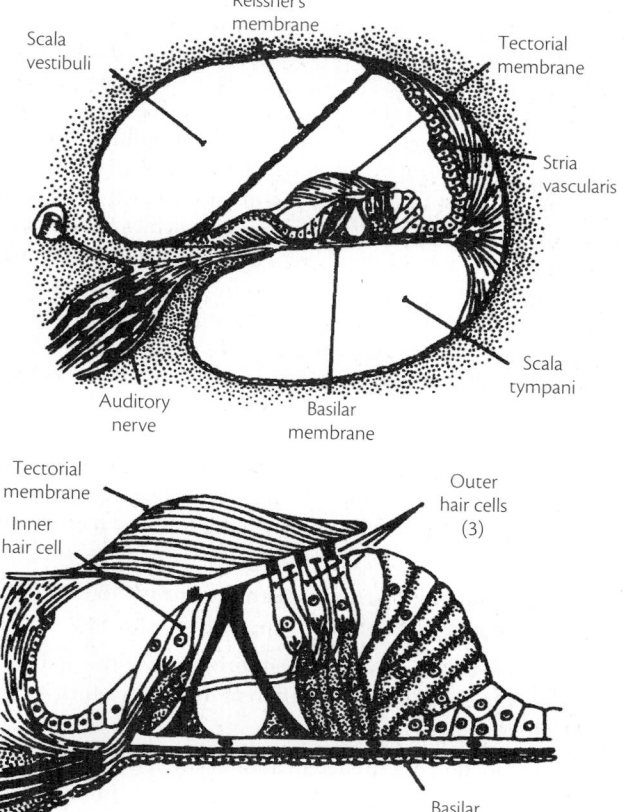

Fig. 14.2 Diagrams to illustrate a cross section through one turn of the cochlea, with magnification of the organ of Corti.

10 separate afferent auditory nerve fibres, making a synaptic connection with the base of the cell. There are approximately 3500 inner hair cells in the healthy human cochlea.

The outer hair cells are cylindrical in shape, and at the lower ends are supported by cup-like processes of Deiter's cells. The phalangeal processes extend to the upper endolymphatic surface of the organ of Corti, where they spread out to form a supporting network surrounding the outer hair cells. The outer hair cells are arranged in three parallel rows, and there are approximately 12 000 in each organ of Corti. These cells receive input from the efferent olivocochlear bundle, but contribute little to the afferent innervation arising from the cochlea.

The stereocilia of the cochlear hair cells are not true cilia, but structures with a core of active molecules packed in a paracrystalline array. The cilia do not bend when displaced, but pivot about their insertions into the thickened upper surface of the hair cell. Each of the stereocilia of one hair cell is linked by fine bands to the adjacent stereocilia, and these links are thought to be responsible for the opening of the ion channels during auditory stimulation.

Between the inner and outer hair cells is a triangular space called the tunnel of Corti. The roof of this is formed by an arch of the processes of the inner and outer pillar cells, and these give rigidity to the overall structure. The lateral wall of the cochlear duct is predominantly made up of the stria vascularis, which is thought to maintain the composition of the endolymph with a high concentration of potassium, approximately 140 mmol, and a high positive endocochlear potential of + 80 mV.

The *auditory afferent pathway* commences with the inner hair cells of the organ of Corti transforming the acoustical information from mechanical energy to electrical activity that is conveyed to the type 1 auditory afferent fibres. The outer hair cells, on the other hand, are thought to act as a modulator and amplifier capable of fine tuning the receptor function of the cochlea (Santos Sacchi 2001). The auditory signal from the organ of Corti then travels through the auditory nerve to the ipsilateral cochlear nucleus, and from there the majority of the afferent auditory fibres project to the contralateral superior olivary complex, the lateral leminiscus, the inferior colliculus, the medial geniculate body, and thence to the auditory cortex (Fig. 14.3A) (Chermak and Musiek 1997).

The *auditory efferent pathway* (Fig. 14.3B) arises in the auditory cortex and descends parallel to the afferent tracts to the level of the cochlear nuclei (Suga *et al.* 2000). The anatomy of the higher efferent auditory system remains ill-defined, but within the brainstem the olivocochlear bundle projects from the superior olivary complex to the cochlea (reviewed by Warr 1992), and has two main pathways:

◆ medial olivocochlear system that projects mainly to the contralateral cochlea and connects to the outer hair cells; and

◆ lateral olivocochlear system that projects to the ipsilateral cochlea and ends on the type 1 afferent dendrites which connect to the inner hair cells.

The exact function of the efferent auditory system remains poorly understood, but it is thought that there is an autoregulatory feedback mechanism that is mainly inhibitory but may also be excitatory at different levels, and thus adjusts and improves the processing of the auditory signal (Suga *et al.* 2000) (Fig. 14.4).

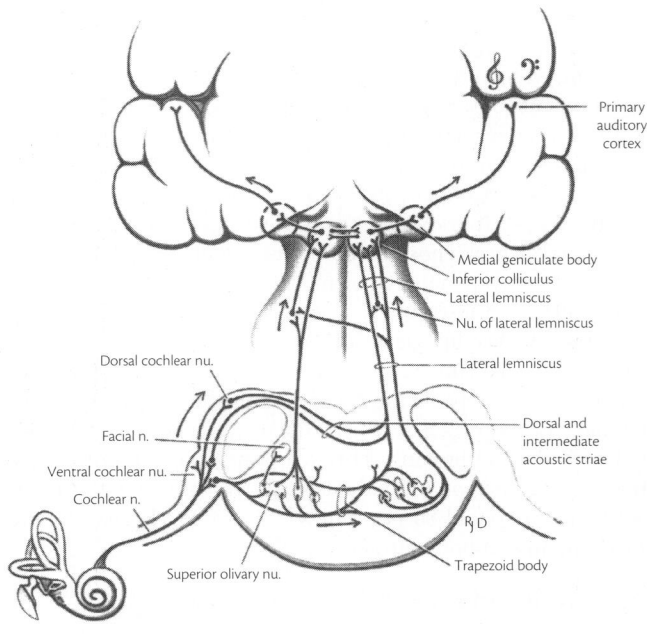

A

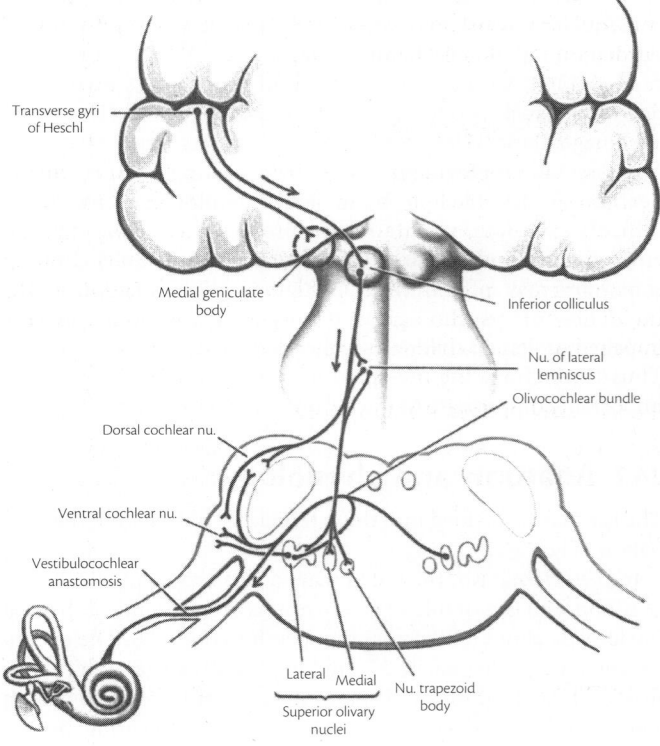

B

Fig. 14.3 Diagrams of auditory pathways: A. ascending afferent auditory pathway; B. descending efferent auditory pathway. (From Noback C.R. and Demarest R.J., 1981 The Human Nervous System 3rd Edition. McGraw Hill, New York.)

ORGAN OF CORTI

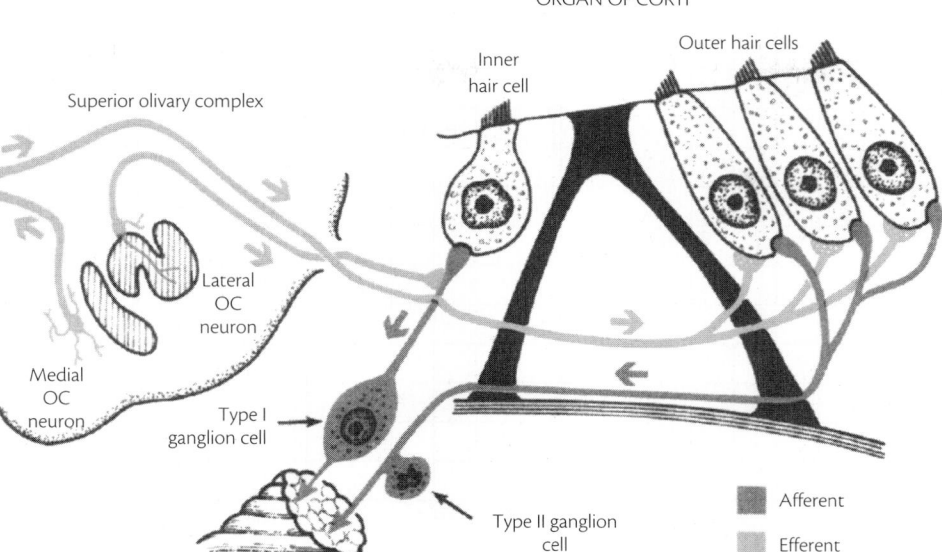

Fig. 14.4 Diagram to illustrate efferent supply to inner and outer hair cells from medial and lateral olivocochlear bundles. (Schuknecht HF (1993). Pathology of the ear. Lea & Febiger, Philadelphia, p. 67.)

The efferent fibres leave the brainstem in the superior division of the vestibular nerve (for a detailed description of cochlear physiology and mechanics, see Pickles 2007).

14.3 Auditory dysfunction

The external ear is important in the localization of sound, and funnels sound to the tympanic membrane, while the middle ear, by means of the tympanic membrane and ossicular chain transfers airborne changes in sound pressure to the fluid-filled compartment of the membranous labyrinth via the oval window. Within the internal ear, the mechanical activity at the oval window is transduced into neural responses within the organ of Corti.

14.3.1 Conductive hearing loss

Pathology affecting the external and middle ear may give rise to abnormalities of the mechanical transmission of sound waves from the environment to the cochlea, known as conductive hearing loss (Fig. 14.5A), and may be associated with:

- impacted wax in the external canal;

- middle ear disease, such as chronic infection or otitis media with effusion); and

- ossicular chain dysfunction, for example secondary to trauma or otosclerosis.

Conductive hearing loss is significantly less common than causes of sensorineural hearing loss in the adult population. In children, 80 per cent of the under-fours develop otitis media with effusion at some stage. There is a bimodal presentation peaking at just under 2 and 4.5 years of age. Although hearing aids may be useful in persistent cases of hearing loss due to middle ear effusion in children, hearing loss is not the only symptom of this condition, with parental concerns including respiratory or ear symptoms, speech, language, and communication problems, difficulties at school, and behavioural problems. About 1 or 2 per cent of children have

serious long-term problems, while 2–5 per cent require ventilation tubes, or grommets, before the age of seven (Haggard 2004). Conductive hearing loss is common in certain syndromes with neurological abnormalities, such as Down's syndrome.

In the adult, conductive hearing loss may be seen in head injury, following haemotympanum, ossicular discontinuity, temporal bone fractures, and middle ear adhesions (Wennmo and Svensson 1989; Davies and Luxon 1995; Ishman and Friedland 2004). Unilateral or asymmetric conductive hearing loss in an adult with normal middle ear pressure and tympanic membrane compliance should raise the suspicion of otosclerosis, which may be genetic in origin, with a typical family history (Menger and Tange 2003), or present as a sporadic case (Holt 2003). Chronic suppurative otitis media is relatively rare in the indigenous British population, but should be sought in immigrants, as this is one of the commonest and preventable causes of mixed conductive and sensorineural hearing loss in the developing world (Acuin 2004). This condition falls into two categories, those with and without cholesteatoma. The latter requires surgical intervention. The diagnosis of tuberculous otitis media should be considered when a chronic perforation and discharge is associated with progressive and profound hearing loss, particularly when there is no response to routine treatment. This condition is frequently complicated by facial paralysis, and while there may be evidence of tuberculosis elsewhere, commonly ear disease may be the only manifestation.

The diagnosis and treatment of suppurative otitis media is important because of the intratemporal and intracranial complications (Table 14.2) (Youngs 1998). Rarely, a vascular malformation such as a glomus tumour or benign or malignant primary or secondary tumours may invade the middle ear and present with a conductive hearing loss. Frequently these conditions will be associated with tinnitus, and a mass may be seen behind the tympanic membrane, which may be white, as in congenital cholesteatoma, or pink, as in middle ear adenoma, or blue, as in glomus tumour or aberrant vascular structure. Detailed CT and MR imaging are required to aid diagnosis.

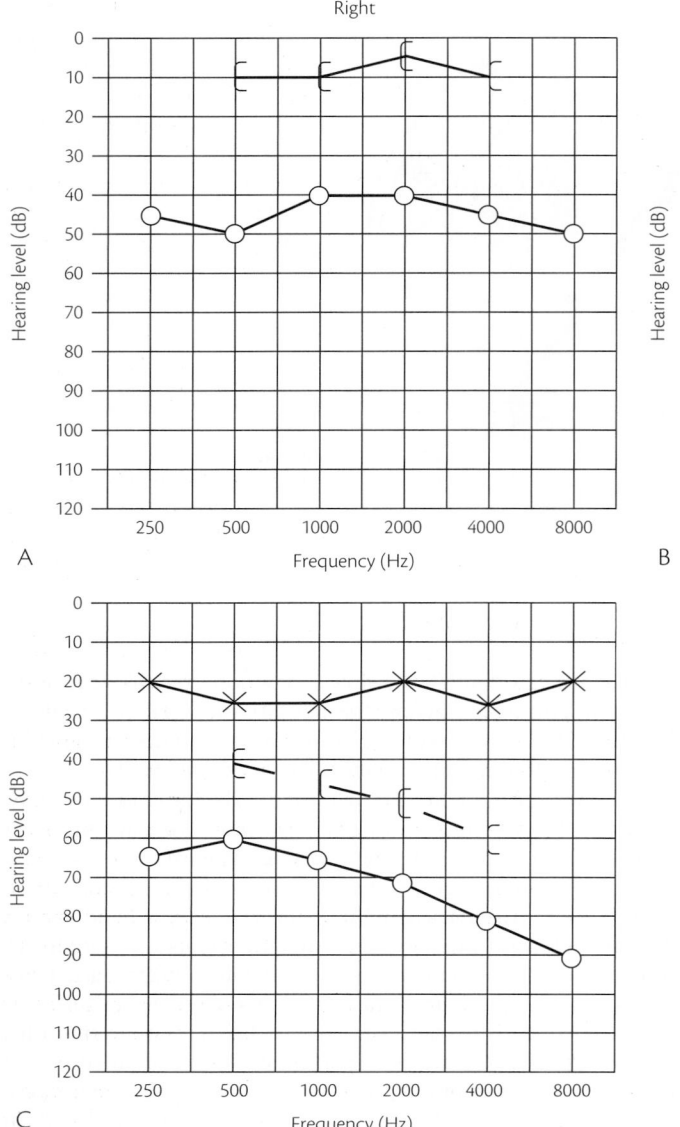

Fig. 14.5 Patterns of hearing loss: A. Right conductive hearing loss; B. Right sensorineural hearing loss; C. Mixed right hearing loss, normal left hearing. (Key: o, x – air conduction threshold; [– masked right bone conduction threshold.)

14.3.2 **Sensorineural hearing loss**

Pathology of the cochlea and VIII cranial nerve characteristically gives rise to a sensorineural hearing loss in which there is an inability to transduce mechanical energy of sound waves into electrical activity within the cochlea or transmit the signals along the VIII nerve (Fig. 14.5B). There is, in addition, an inability to perceive both bone- and air-conducted sounds, and both the intensity of sound and the frequency resolution of complex sounds are impaired. Many different conditions may result in sensorineural hearing loss (Table 14.3), but certain pathologies involving both the middle and internal ears, for example chronic middle ear disease, otosclerosis with involvement of the ossicles and otic capsule and physical trauma may give rise to a mixed hearing loss in which there is both a conductive and sensorineural component (Fig. 14.5C).

Sensorineural hearing loss may be further divided into that of cochlear origin and that of neural origin on the basis of two pathophysiological phenomena:

◆ *Loudness recruitment* is defined as an abnormally rapid increase in loudness with an increase of intensity of stimulus. This is characteristic of disorders affecting the hair cells of the organ of Corti, but is absent in pathology of the VIII nerve and brainstem.

◆ *Abnormal auditory adaptation* is a decline in discharge frequency with time, observed following an initial burst of neural activity in

Table 14.2 Complications of suppurative otitis media

Intracranial complications	Intratemporal bone complications
Extradural abscess	Facial nerve paralysis
Subdural abscess	Suppurative labyrinthitis
Lateral sinus thrombophlebitis	Labyrinthine fistula
Meningitis	Acute mastoiditis
Brain abscess	Subperiosteal abscess
Otitic hydrocephalus	Postauricular fistula
	Petrositis fistula

Table 14.3 Causes of sensorineural hearing loss

Genetic	Nonsyndromal
	Syndromal
Trauma	Physical
	Barotrauma
	Acoustic trauma
Vascular	Malformation
	Cardiovascular ischaemia
	Cerebrovascular ischaemia
Autoimmune	Isolated inner ear disease
	Systemic disorder, e.g. Systemic Lupus
	erythematosus, Polyarteritis nodosa
Infection	Bacterial
	Viral
	Fungal
Degenerative	Cochlear
	Neuropathy
	Neurological
Iatrogenic	Drugs
	Surgical
	Radiotherapy
	Organic chemicals

response to an adequate continuing stimulus applied to the organ of Corti. This phenomenon is characteristic of neural auditory dysfunction arising in both the VIII nerve and brainstem.

Retrocochlear hearing loss is characterized by abnormal adaptation, and is the term used to classify hearing loss caused by lesions more proximal to the brain than the cochlea, namely of the VIII nerve and lower brainstem. Lesions above this level do not give rise to hearing loss as a result of bilateral representation and multiple commissures of the central auditory pathways. A typical example of retrocochlear hearing loss is that associated with vestibular schwannomas which arise on the vestibular division of the VIII nerve and present with auditory dysfunction as a result of compression of the acoustic branch of that nerve. In 1996, Starr and co-workers defined the entity of auditory neuropathy in their seminal publication. Subsequent to this, a plethora of literature has defined this concept further, and there may be a wide range of pure-tone audiometric in this condition. Auditory neuropathy has been renamed *auditory dyssynchrony* (Berlin *et al.* 2001), partly in recognition that this entity includes a heterogeneous group of conditions which may involve three different sites of lesions:

◆ impairment of electrical and mechanical transduction of the inner hair cells;

◆ impairment of axons, cell bodies, and/or myelin sheaths; and

◆ alteration in efferent influences through the olivocochlear efferent pathway (Davies, in press).

Criteria for the diagnosis of auditory neuropathy or dyssynchrony include normal otoacoustic emissions and a normal cochlear microphonic, together with absent or abnormal auditory brainstem-evoked responses, with absent stapedial reflexes and absent contralateral suppression of otoacoustic emissions via the efferent

pathway. Auditory neuropathy is now recognized as a significant cause of hearing impairment in children, and in a large screening study of hearing in children, the incidence of auditory neuropathy was found to be 2 per cent (Rance *et al.* 1999), while the prevalence of auditory neuropathy in a population of children with permanent hearing loss was reported as 12 per cent by Berlin and co-workers (1994). The prevalence of auditory neuropathy in adults has not been defined, but it may be associated with congenital and inherited disorders such as the Arnold–Chiari malformation with elongation of the VIII nerve, Friedreich's ataxia, trauma, infection such as meningitis, the Ramsay–Hunt syndrome, vascular disease, neoplasia, temporal bone disorders compressing the VIII nerve, and toxic disorders such as vincristine ototoxicity, alcohol, and both lead and mercury poisoning, autoimmune disorders, and demyelination.

Aetiology of sensorineural hearing loss. In the developed world interest in the aetiology of sensorineural hearing loss has risen in part due to the earlier detection of infants with profound hearing loss through the newborn hearing screening programmes (Morton and Nance 2006) and in part due to effective prosthetic, pharmacological, and genetic interventions.

Genetic hearing loss. In the last two decades there have been enormous strides in the understanding of genetic and environmental causes of hearing loss (Nance 2003). Currently, 67 loci for autosomal-recessive inheritance of hearing loss have been defined, while 54 for autosomal-dominant inheritance have been reported, in addition to mitochondrial mutations and genetic aberrations giving rise to X-linked hearing loss (Hereditary Hearing Loss Homepage). Many of these forms of genetic hearing impairment may present in a nonsyndromal and syndromal pattern, and the majority result in cochlear abnormalities. Age-related hearing loss, which is characterized by progressive deterioration of auditory sensitivity with age, and is the leading cause of adult auditory impairment, has been previously attributed to a variety of factors including genetic, nutritional, socioeconomic, and environmental variables. However, recent work has suggested that this condition may in fact represent inherited late-onset progressive hearing loss, and that specific genes may predispose to environmental triggers affecting various molecular mechanisms underlying changes in auditory function (Pickles 2004; Seidman *et al.* 2004). Specifically, a mitochondrial mutation which is associated with aminoglycoside-induced hearing loss (Prezant *et al.* 1993) has also been reported.

In general, *autosomal-recessive hearing loss* manifests as profound, congenital/ prelingual, and stable impairment, and accounts for approximately 40 per cent of all cases of childhood hearing loss. There may be marked intrafamilial variation in the severity of the loss, and radiology is generally normal. Three recessive loci, DFNB2, 4, and 12 have been associated with vestibular dysfunction. The genes causing recessive hearing loss have been demonstrated to code for transcription factors, *POU* genes; motor molecules unconventional myosins myo7A, myo15; gap junction proteins ion transporters GJB2, GJB3, GJB6, KCNQ4, PDS, and Prestin and matrix proteins Tecta and Col11A2. *GJB2* gene encodes the gap junction protein connexin 26, which has been shown to cause 50 per cent of autosomal-recessive sensorineural hearing loss in Caucasian and European populations. The commonest mutation is 35delG, and the overall carrier frequency of this mutation has been identified as 1:51 in Europe. The prevalence of *GJB2* mutation in autosomal-recessive sensorineural hearing loss has led to routine clinical screening for this mutation in families with an autosomal-recessive presentation.

In addition to genetic dysfunction of the cochlea, recent studies have defined genetic abnormalities giving rise to auditory neuropathy. As noted above, auditory neuropathy may be associated with syndromes, such as the Charcot–Marie–Tooth disease, and in this latter condition, hearing loss has been shown to be associated with genetic changes, including one at 8q23 in HSMN-Lom, and a mutation of Thr124Met in the myelin protein zero gene (Table 21.7, Section 21.4). A number of children have autosomal-recessive auditory neuropathy without any other abnormality, and both linkage studies and mutation analysis have identified the otoferlin gene as being responsible for this type of hearing loss (Varga *et al.* 2003). A more recent study (Wang *et al.* 2005) has demonstrated that mutations in 12SrRna gene is also associated with auditory neuropathy.

Autosomal-dominant sensorineural hearing loss is uncommon in prelingually profound hearing impairment, but is well recognized in families with various audiometric configurations, differing ages of onset, and differing rates of progression. Commonly there is good correlation between phenotype and genotype, and vestibular involvement has been identified in *DFNA9* and *DFNA11*. A wide variety of different configurations of hearing loss in autosomal-dominant genetic hearing loss have been reported, including unilateral loss, low-frequency sensorineural hearing loss both of stable and progressive type, mid-frequency loss, high-frequency loss, and associated progressive vestibulo-cochlear dysfunction.

More than 100 syndromes have been reported with associated hearing impairment, but the more common syndromes, with their predominant features are shown in Table 14.4. Many of these present in childhood, but hearing loss associated with certain syndromes can progress or indeed become apparent in adult life.

A number of syndromes associated with hearing impairment are of particular neurological interest:

Hereditary neuropathies

As early as 1951, Denny Brown reported a case of hereditary sensorimotor neuropathy with deafness, which at post-mortem examination revealed 'thin auditory nerves'. Hearing loss has subsequently been reported in a range of hereditary motor and sensory neuropathies, but more recent work suggests that it is more common than was previously reported (Raglan *et al.* 1987; Rungby *et al.* 1998).

Charcot–Marie–Tooth disease

It is a form of hereditary motor and sensory neuropathy with autosomal-dominant, autosomal-recessive, and X-linked forms of inheritance. The types of sensorineural hearing impairment in Charcot–Marie–Tooth disease type 1a (Section 21.41) and hereditary neuropathy with liability to pressure palsies (Section 21.5) have been characterized (Verhagen *et al.* 2005). This latter group comprised half with the common *PMP22* deletion, and half with a *PMP22* frameshift mutation. A 'substantial sensorineural hearing impairment' in all three groups was identified. The commonest audiometric configuration was a dome-shaped loss with both a mild to moderate low-frequency and high-frequency loss. In the group with hereditary neuropathy with liability to pressure palsy, the data suggested 'excessive presbyacusis', but, for the group as a whole, it was concluded that *PMP22* gene overexpression in Charcot–Marie–Tooth 1a or underexpression in the hereditary liability to pressure palsies mutation or deletion is linked with peripheral nerve dysfunction on the basis of demyelination. As the gene product of *PMP22* is more abundantly expressed in the cochlea than in the brain, it may be that the neural elements of the end-organ are the primary focus of associated dysfunction. This might explain the abnormalities of brainstem-evoked responses observed in this group of patients (Raglan *et al.* 1987).

Refsum disease

This is characterized by defective peroxisomal alpha-oxidation of phytanic acid, with clinical features that include retinitis pigmentosa, polyneuropathy, anosmia, and hearing loss (Section 21.8.1). Adult Refsum disease is a recessive disorder of phytanoyl CoA hydroxylase, *PAHX*, gene on chromosome 10p30. Although hearing loss in Refsum's disease is common, there are few detailed assessments. A recent report demonstrated that 7 out of 9 adults with this disorder suffered a mild to moderate sensorineural hearing loss, predominantly of high-frequency type (Bamiou *et al.* 2003). Subtle auditory nerve involvement was identified in 6 out of the 7 patients with hearing loss, and in one of the two patients with normal pure-tone audiometry on the basis of auditory-evoked brainstem responses.

Mitochondrial disorders

These are responsible for a variety of neurological syndromes. Children with mitochondrial encephalopathies including classical Kearns–Sayre syndrome and the MELAS syndrome of mitochondrial myopathy, encephalopathy, lactic acidosis, and stroke-like episodes were found to have progressive hearing loss of both cochlear and retrocochlear origin. No correlation between the type and severity of hearing loss and number or severity of other clinical neurological findings was demonstrated (Zwirner and Wilichowski 2001). In a series of 66 patients with mitochondrial cytopathy, 25 per cent were found to suffer from a hearing impairment (Morgan-Hughes *et al.* 1982), while in a more recent study of a 3-year follow-up of adult patients with 3243A < G mutation, 16 underwent detailed audiological follow-up, of whom 81 per cent had a sensorineural hearing loss. The loss was shown to be progressive over the 3-year period, and both this feature and the severity were correlated with the mutation heteroplasmy in muscle at entry and at end of the 3-year follow-up (Majamaa-Voltti *et al.* 2006).

Inherited muscle disorders

These can be associated with hearing loss, including *facioscapulohumeral dystrophy* (Voit *et al.* 1986) and *myotonic dystrophy* (Huygen *et al.* 1994) (Sections 24.2.4; 24.3). In this latter condition, the hearing loss is reported to resemble 'precocious presbyacusis', or an excessive high-frequency hearing loss characteristic of presbyacusis, and genetic anticipation. In addition, a significant increase in the I–V interpeak interval of auditory brainstem-evoked responses was noted, suggesting a retrocochlear component.

Arnold–Chiari malformation

This is a congenital deformity in which the brainstem and the cerebellum are elongated downwards into the cervical canal. A study of 77 patients with Chiari-1 malformation identified that 50 per cent had auditory dysfunction, two-thirds of which was unilateral sensorineural hearing loss, while a third had bilateral involvement (Kumar *et al.* 2002). An earlier study by Rydell and Pulec (1971) suggested that the primary pathology was stretching of the VIII cranial nerve or pressure of the VIII cranial nerve, as it is bent over the edge of the porus acusticus.

Table 14.4 Characteristics of common syndromes associated with hearing loss

Syndrome	Mode of inheritance	Predominant features	Auditory and vestibular features	CNS status
Treacher–Collins	AD	Mandibulofacial dysostasis	Severe dysplasia middle and internal	N
Branchio-oto-renal	AD	Branchial cysts/fistulae Structural +/or functional renal abnormalities	Anomalies of external ear CHL 20% SNHL 30% Mixed 50%	N
CHARGE	Occasional AD Rare AR Most sporadic	Coloboma, Heart defect, Atresia of choanae, Retarded growth and development, Genital hypoplasia, Ear anomalies.	Structural labyrinth (A + V) dysplasia VIII nerve involvement Absent semi- circular canals SNHL +/- CHL	Impaired IQ and development
Usher	AR	Retinitis pigmentosa	Type I – A + V failure Type II – A failure Type III – Variable	N
Alstrom	AR	Pigmentary retinopathy Diabetes mellitus Obesity	SNHL	N
Apert	Mainly sporadic. Some AD	Craniosynostosis + oral manifestations. Brachydactyly	CHL ME anomalies	Frequent low IQ CNS manifestations
Crouzon	AD	Craniosynostosis Shallow orbits Ocular proptosis	CHL ME anomalies	Generally N
Osteogenesis imperfecta	AD	Blue sclera Opalescent teeth Deformities of long bone and spine Joint hyperextensibility	Type 1 Mild CHL > 10 years of age Progressive mixed hearing loss	Generally N
Stickler	AD	Flat midface and cleft palate High myopia and retinal detachment and cataracts Arthopathy Spondyloepiphyseal dysplasia	Progressive SNHL	N
Wildervanck	Sporadic	Fused cervical vertebrae Abducens palsy and retracted globe	SNHL, CHL or mixed in ~ 30%. Vestibular failure common.	Usually N
Alport	Usually X- linked dominant, rarely AR	Progressive glomerulonephritis Bilateral anterior lenticonus and macular flecks or peripheral coalescing flecks	Progressive SNHL in >10 years in approximately 50% of cases	Usually N
Jervell Lange Nielsen	AR	Prolonged Q-T interval Fainting spells/sudden death	Profound SNHL Scheibe anomaly	N
Pendred	AR	Goitre	Mondini defect Dilated vestibular aqueduct Severe-profound SNHL Vestibular failure 30%	N
Mucopolysaccharidoses • Hurlers • Hunter	AR except Hunter which is X-linked	Growth failure. Death <10 years Craniofacial dysmorphism Lysomal Storage Excess excretion of dermatomes and heparin sulphates. Severe form → death at 4–14 years Mild form → less severe than Hurlers	Progressive CHL Central auditory dysfunction Mixed HL in ~ 50%	Impaired IQ and development
Refsum	AR	Retinitis pigmentosa Polyneuropathy (Cardiac enlargement/dysrhythmias Icthyosis)	Progressive SNHL Vestibular Function - Normal	Anosmia Progressive motor weakness legs/arms + sensory neuropathy. Cerebellar signs.
Down	Trisomy 21	Short stature Dysmorphic facial features Hypotonia Single palmar crease Cardiac anomalies Delayed mental development	Abnormal external ear morphology CHL and/or SNHL in majority	Impaired IQ and development

Key: A—Auditory; AD—Autosomal dominant; AR—Autosomal recessive; CHL—Conductive hearing loss; CNS—Central nervous system; IQ—Intelligence quotient; ME—Middle ear; N—Normal; SNHL—Sensorineural hearing loss; V—Vestibular.

Multiple system atrophy

This includes a number of degenerative neurological disorders, including progressive autonomic failure, atypical Parkinsonism and cerebellar ataxia associated with a multiplicity of other features, including optic nerve atrophy, peripheral neuropathy, dementia, epilepsy, myoclonus, hearing loss, and vestibular dysfunction (Duvoisin 1987; Harding 1981a) (Section 40.3.8).

Friedreich's ataxia

This is the most common type of inherited ataxia, and is frequently caused by a large expansion of an intronic GAA repeat, resulting in decreased expression of the target frataxin (Pandolfo 1999) (Section 39.4). The condition is usually inherited in an autosomal-recessive pattern, and as frataxin is a mitochondrial protein, it suggests that dysfunction in Friedreich's ataxia is caused by mitochondrial abnormality and free-radical toxicity. In a large study of 115 patients from 90 families, Harding (1981a) demonstrated hearing impairment as an associated but rarer feature. Ell and co-workers (1984), examining the neuro-otological abnormalities in Friedreich's ataxia in 10 patients, reported that hearing impairment was found in the majority of patients with abnormal brainstem auditory-evoked potentials. A more recent study (Lopez-Diaz-de-Léon et al. 2003) reported abnormal auditory-evoked brainstem responses with normal otoacoustic emissions, suggesting auditory neuropathy, in two cases of Friedreich's ataxia. Hearing loss in association with dominantly-inherited late-onset **cerebellar ataxia** has also been reported, but is rare (Harding 1982), whereas 14 per cent of patients with idiopathic late-onset cerebellar ataxia had a hearing loss (Harding 1981b).

Infections

Bacterial, viral, and myototic infections may give rise to hearing impairment by direct invasion, blood-borne transmission, or by transfer along the nerves from the cerebrospinal compartment.

Viral infections

In adults, sudden sensorineural hearing loss is commonly presumed to be viral in origin. Westmoore et al. (1979) reported the detection of mumps virus in the perilymph after sudden-onset deafness, and there is circumstantial evidence to suggest that sudden hearing loss may be associated with a variety of viruses. The commonest condition is the Ramsay–Hunt syndrome caused by a Herpes Zoster infection and characterized by facial palsy, hearing loss, and the characteristic vesicles of a herpetic infection around the pinna and in the external auditory meatus. Sensorineural hearing loss occurs in between 24 and 69 per cent of cases (Wayman et al. 1990; Devriese 1968) and may be the result of cochlear or retrocochlear involvement (Kuhweide et al. 2002).

HIV AIDS

A variety of auditory abnormalities have been reported in AIDS, ranging from conductive to sensorineural hearing loss, with mild audiometric changes, abnormalities in the brainstem-evoked response and central auditory dysfunction. A recent study of patients with HIV AIDS in an outpatient clinic in South Africa revealed a prevalence rate of hearing impairment of 23 per cent, but the authors concluded that the hearing loss may have been associated with opportunistic infections such as otosyphilis, cyclomegalovirus, or streptococcal meningitis or treatment (Khoza and Ross 2002).

Syphilis

Labyrinthine involvement is more common in late acquired syphilis than congenital syphilis. Various presentations have been observed, including sudden sensorineural hearing loss and a pattern suggestive of Ménière's disease. Importantly, this is one form of treatable progressive sensorineural hearing loss (Darmstadt and Harris 1989). More recent papers have emphasized the need to check syphilitic serology in a patient with unexplained sensorineural hearing loss, and this is particularly the case with an increase in opportunistic infections in patients with HIV AIDS. Treatment of otosyphilis with penicillin and corticosteroids has achieved improvement in hearing, tinnitus and vertigo in 25 per cent of a group of 16 patients (Linstrom and Gleich 1993).

Meningitis

Sensorineural hearing loss in children with bacterial meningitis is well documented, with a prevalence of between five and 30 per cent. In a group of 15 hearing-impaired children there was no evidence that any one of the common bacterial organisms in this condition conferred a greater loss of hearing (Fortnum and Davis (1993). A recent systematic review confirmed that treatment with antibiotics and adjuvant corticosteroids resulted in not only lower fatality and lower long-term neurological sequelae, but also lower rates of severe hearing loss (van de Beek et al. 2003). Although tuberculous meningitis is now rare in Western societies, it should be considered in immigrant, debilitated, alcoholic and immunosuppressed populations (Kotnis and Simo 2001).

Lyme disease

This is an infection with *borrelia burgdorferi*, a spirochaete transmitted by the bite of a tick (Section 41.5.5). Hanner et al. (1989) found elevated antibodies to borrelia antigen in 17 per cent of 98 subjects with a unilateral sudden or fluctuating sensorineural hearing loss. Treatment with intravenous penicillin resulted in improvement of high-frequency hearing loss in five patients. This condition as a treatable cause of sensorineural hearing loss has recently been re-emphasized (Peltomaa et al. 2000).

Vascular disease

There is a wealth of literature on the possible relationship of vascular risk factors and sensorineural hearing loss. To date, no definitive association has been demonstrated between sensorineural hearing loss and essential hypertension or postural hypotension, hyperlipidaemias (Jones and Davis 2001) or diabetes mellitus. However, of note, a recent study of young women with cerebral small vessel disease together with a subclinical sensorineural hearing loss suggested that such findings are diagnostic of the Susac syndrome of retinocochleocerebral vasculopathy (Ringelstein and Knecht 2006).

Evaluation of 1501 participants using a questionnaire assessment of angina, myocardial infarction, and stroke showed no association between these conditions and cochlear impairment in men (Torre et al. 2006). However women, with a self-reported history of myocardial infarction, were twice as likely to have cochlear impairment as women without a history of myocardial infarction. However, further confirmation is required, in the light of the earlier conflicting literature.

Much debate has ensued as to whether vascular loops may compress the VIII cranial nerve, giving rise to vestibular or auditory symptoms. It is well recognized that vascular loops crossing the VIII nerve are a normal variant, but a recent study of 47 patients with unexplained tinnitus has identified that high resolution MRI T2-weighted CISS images showed a significantly higher number of vascular loops in the internal auditory canal in patients with arterial pulsatile tinnitus compared to patients with non-pulsatile tinnitus (Nowe *et al.* 2004). Moreover, a correlation between the clinical presentation of tinnitus and hearing loss was found.

Sudden deafness has been reported in a number of isolated cases of inferior collicular infarction, vertebral artery dissection, infarction of the anterior-inferior cerebellar artery and migrainous infarction. Eight per cent of 364 cases of vertebrobasilar insufficiency demonstrated sensorineural hearing loss, the majority ($N = 27$) being unilateral as opposed to bilateral ($N = 2$) (Lee and Baloh 2005). Vertigo was associated with hearing loss in all cases except one. Approximately 50 per cent of cases showed a cochlear-type of hearing loss, with approximately half of those improving symptomatically, over a follow-up period of 1 year.

Autoimmune inner-ear disease

Immune-mediated ear disease may present as an isolated entity, but also includes all inner-ear pathology associated with autoimmune disorders, including *Cogan's syndrome* in which profound auditory and vestibular failure is associated with ocular pathology (Schuknecht 1994), and *Wegener's granulomatos* is in which 30–50 per cent of patients suffer otological complications, including both middle-ear pathology with chronic attacks of otitis media and conductive hearing loss and sensorineural hearing impairment (Stephens *et al.* 1982; Mcdonald and De Remee 1983). Hearing loss is reported in association with all autoimmune disorders, including rheumatoid arthritis, systemic lupus erythematosus, polyarthritis nodosa, systemic sclerosis, and dermatomyositis (Ruckenstein 2004).

Of particular interest to neurologists, *sarcoidosis* is a common multi-system disease presenting with audiovestibular involvement. In a recent survey (Colvin 2006), hearing loss was found bilaterally in 75 per cent of patients, with three-quarters of these demonstrating asymmetry. Fluctuation was noted, and the hearing loss varied between mild to profound. In approximately half the cases, vestibular testing had been performed, and was abnormal in all but one. Eighty-one per cent of the patients evaluated had additional features of neurosarcoidosis.

A similar review of 62 patients with *Behçet's syndrome* showed hearing loss in approximately a third of patients, predominantly in the high frequencies, with normal auditory brainstem-evoked responses suggesting clear dysfunction. No peripheral vestibular abnormalities were demonstrated, but subtle eye movement abnormalities were defined in 6 per cent of cases involving dysmetric saccades and smooth-pursuit dysfunction (Kulahli *et al.* 2005). However, in distinction to this study, an earlier study reported 54 per cent of patients with Behçet's syndrome suffering from auditory loss of cochlear type, with approximately a third of patients suffering vestibular dysfunction, the preponderance being unilateral peripheral vestibular abnormalities. No correlation was identified between the audiovestibular lesion and the presence of pathology in other lesions, disease duration, age, or sex of the patient (Pollak *et al.* 2001).

In addition to multisystem autoimmune disorders, Lehnhardt (1958) was the first to postulate that bilateral sensorineural hearing may be the result of isolated autoimmune dysfunction within the inner ear. In 1979, McCabe reported several cases of sensorineural hearing loss that were successfully treated with dexamethasone and cyclophosphamide, and introduced the clinical entity of *autoimmune-mediated sensorineural hearing loss*. The importance of this diagnosis is that it represents a treatable and thus reversible form of hearing impairment. However, there are no uniform criteria for diagnosing immune-mediated inner-ear disease, and evidence-based diagnostic criteria and assessment methods remain to be established (Agrup and Luxon 2006). The presentation of autoimmune inner-ear disease is characteristically bilateral sequential or simultaneous auditory and/or vestibular loss. Fluctuation in symptomatology may occur, but overall there is relentless deterioration in both auditory and vestibular function.

Metabolic disease

A plethora of literature exists on *diabetes mellitus* as an aetiological factor of hearing loss, but despite this, controversy remains, and there is no clear evidence indicating whether or not patients with diabetes mellitus suffer auditory and/or vestibular abnormalities as a consequence of neuropathy, angiopathy, or both pathologies (Maia and Campos 2005). Recent genetic studies have defined the relationship of diabetes and hearing loss in mitochondrial mutations (Maassen 2002) and in mutations of the *WFS1* gene in the Wolfman syndrome of non-syndromic hearing impairment, diabetes mellitus, and psychiatric disease (Cryns *et al.* 2003).

Hearing loss is a common finding in patients with *renal failure*, and uraemic toxins. Ototoxicity from both disease and drugs and axonal uraemic neuropathy have all been suggested as possible factors, while both dialysis and renal transplantation have been reported to be associated with recovery of hearing impairment (Anteunis and Mooy 1987). A more recent paper using auditory-evoked brainstem responses has supported a possible underlying neuropathy with abnormalities suggesting retrocochlear involvement (Kustel *et al.* 1993).

Neoplasia

Vestibular schwannomas. These account for 10 per cent of intracranial tumours, and more than 75 per cent of cerebellopontine angle lesions (Gonzalez-Revilla 1948). Although most of these tumours arise on the superior division of the vestibular nerve, the most common presenting features are deafness and tinnitus (Pulec *et al.* 1971). A review of 200 patients with vestibular schwannoma revealed that only 42 per cent demonstrated a classical retrocochlear hearing impairment (Johnson 1968), while brainstem-evoked responses are reported to be abnormal in 95 per cent of surgically proven cases (Josey *et al.* 1980). In all patients with a unilateral sensorineural hearing impairment, asymmetric bilateral sensorineural loss, or unilateral tinnitus, it is mandatory to consider and exclude the presence of a small vestibular schwannoma, the diagnostic modality of choice being magnetic resonance imaging.

Neurofibromatosis types 1 and 2. These are autosomal-dominant neurocutaneous disorders.(Section 11.2). The *NF1* and *NF2* genes encode for neurofibromin and merlin respectively, which act as tumour-suppressor genes. Neurofibromatosis types 1 and 2 are very different disorders, with a few overlapping features. Type 1 is characterized by predominantly cutaneous abnormalities, in addition to nerve sheath tumours, optic gliomas, and other CNS tumours. Type 2 presents primarily with tumours of the central and peripheral nervous system, particularly schwannomas, with few cutaneous or non-nervous system abnormalities. Strict diagnostic abnormalities have been clarified (Yohay 2006). Bilateral VIII nerve tumours may occur with both forms of neurofibromatosis, either the peripheral von Recklinghausen disease of type 1 or central neurofibromatosis of type 2. The latter condition is much rarer, and is most commonly seen in young adults with bilateral

Table 14.5 Tumours of the temporal bone.

Benign neoplasms
Glomus jugulare
Glomus tympanicum
Adenoma
Facial nerve neuroma
Meningioma
Haemangioma
Dermoid
Chondroma
Glioma

Malignant neoplasms
Rhabdymyosarcoma
Adenocarcinoma
Chondrosarcoma
Squamous cell carcinoma
Basal cell carcinoma
Nasopharyngeal carcinoma
Metastasis

hearing loss. Historically, the treatment of unilateral vestibular schwannomas has been surgical resection, but recent studies have defined the slow growth of many tumours and a more conservative approach to management is therefore advocated (Rosenberg 2000; Smouha *et al.* 2005). In addition to surgical intervention, stereotactic radiosurgery has been used in the treatment of schwannomas, with favourable tumour control and preservation of nerve function (Lunsford *et al.* 2005).

Tumours other than schwannomas (Table 14.5) account for 25 per cent of all cerebellopontine angle lesions, and include meningiomas, epydermoid cysts, neuromas of the V and VII cranial nerves, and brainstem gliomas. In addition the VIII cranial nerve may be compressed by schwannomas on the V, VII, IX or XI cranial nerves.

Metastatic tumours. These are relatively common within the temporal bone, including secondary deposits from the breast, kidneys, lungs, stomach, larynx, prostate, and thyroid gland. Moreover, leukaemic infiltrates of the VIII cranial nerve may give rise to auditory symptoms (Papparella *et al.* 1973). In addition, *paraneoplastic syndromes* may manifest with auditory loss (Gulya 1993). *Treatment of malignancy* may also give rise to post-radiation, and drug-induced ototoxicity with hearing loss, as demonstrated by recent reports in patients with nasopharyngeal carcinoma (Low *et al.* 2006a) and parotid gland tumours (van der Putten *et al.* 2006). In the former group, a recent study has identified good post-cochlear implant hearing outcomes for four patients following radiation for nasopharyngeal tumours (Low *et al.* 2006b).

Carcinomatous meningitis. This is a well-documented cause of both auditory and vestibular symptomatology. The VIII nerve is involved in up to 10 per cent of patients (Alberts and Terrence 1978; Zeller *et al.* 2002). Characteristically, this pathology gives rise to the simultaneous occurrence of symptoms and signs in more than one area of the neural axis, and diagnosis is confirmed by the presence of malignant cells in the cerebrospinal fluid.

Trauma

Hearing loss may result from acoustic trauma in the form of noise, physical trauma resulting from head injury and whiplash, or iatrogenic trauma including drugs, radiotherapy (see above), and surgery.

Noise-induced permanent threshold shift. It is one of the commonest and most easily preventable causes of sensorineural hearing loss. It is commonly the consequence of hazardous occupational exposure to noise, but may also be associated with acoustic trauma, for example gun fire and explosions. Characteristically the maximal loss is at 4000 Hz, with a notched configuration to the audiogram (Fig. 14.6). With the passage of time, the adjacent frequencies gradually deteriorate, but it is rare for a hearing loss greater than 70 dB to be the result of occupational noise exposure. The diagnosis of noise-induced hearing loss

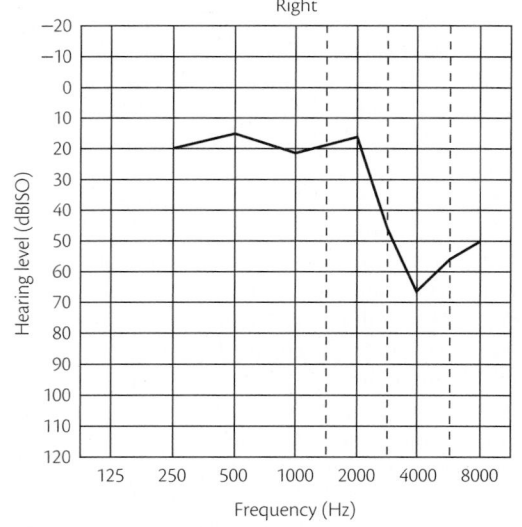
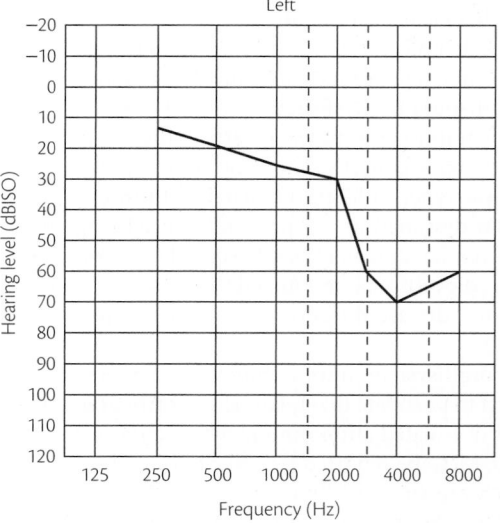

Fig. 14.6 Diagram to illustrate characteristic configuration of noise-induced hearing loss with a 'notch' at 4 kHz. (From: Luxon LM (1988). Clinical diagnosis of noise induced hearing loss. In: Prasher DK and Luxon, LM (eds). *Advances in Noise Research Vol l: Biological Effects*. Whurr Publishers, London. p. 96.)

is by exclusion of other causes, but as an aetiological diagnosis cannot be made in up to two-thirds of patients with sensorineural hearing loss, this presents a diagnostic dilemma. The American College of Occupational Medicine (1989) have therefore devised clear diagnostic criteria for occupational noise-induced hearing loss. A variety of factors should be considered in making this diagnosis, but in all cases of high-frequency sensorineural hearing loss, a history of noise exposure should be sought (Luxon 1998).

Acute barotrauma. This is associated with diving (Newton 2001; Shupak *et al.* 2003), unpressurized air travel (Mirza and Richardson 2005), and explosions (Persaud *et al.* 2003), which may give rise to tympanic membrane haemorrhage into the middle ear, a conductive hearing loss, or perilymph fistula, which is commonly associated with vestibular symptoms.

Head injury. This may lead to middle ear, inner ear, VIII nerve, and central auditory loss. Trauma may give rise to various pathologies: labyrinthine concussion with or without fracture, and fractures of the temporal bone:

- *longitudinal*, with a fracture extending through the middle ear cavity with concomitant conductive hearing loss; and

- *transverse* fractures which may result in section of the VIII nerve accompanied by facial paralysis and haemotympanum. Frequently there is a profound sensorineural loss and acute vertigo.

Labyrinthine concussion may produce inner ear symptoms, without any obvious clinical neurological signs. Head trauma may also give rise to auditory processing disorders and cortical deafness or auditory agnosia, as outlined below. In the author's experience, 38 per cent of patients with mild and moderate head injuries complain of hearing loss and/or tinnitus (Davies and Luxon 1995), whereas the prevalence of auditory abnormalities attributed to trauma on detailed testing was 53 per cent. The commonest configuration of sensorineural hearing loss without a fracture is a bilateral high-tone sensorineural hearing loss (Toglia and Katinsky 1976; Davies and Luxon 1995), although a variety of other configurations, including asymmetric and unilateral loss, may be observed. A notch-shaped hearing loss at 4 Hz has been reported by Schuknecht (1969), and confirmed by subsequent workers (Wennmo and Svensson 1989). A recent study has shown progression of sensorineural hearing loss after closed head injury in 74 per cent of cases (Bergemalm 2003).

Following whiplash injury, auditory symptoms have been reported in association with tinnitus, although the mechanism and aetiology remain obscure. A recent study of 153 patients identified subtle auditory abnormalities in a subset of this group, but no correlation was established between objective auditory deficits and tinnitus (Tjell *et al.* 1999).

Toxic disorders

Drugs, alcohol, and chemicals may all give rise to auditory deficit. Lead and mercury poisoning may both give rise to auditory and vestibular symptoms, although the underlying pathophysiological mechanism remains unclear. Early work in animals demonstrated segmental demyelination and axonal degeneration of the VIIIth cranial in 75 per cent of animals receiving intra-peritoneal injections of 1 per cent lead acetate solution, but end organ/ganglion cell abnormalities were not observed (Gozdzick-Zolnierkiewicz and Moszynski 1969). Human temporal bone studies in lead and mercury poisoning have reported eighth nerve changes.

Many *drugs* produce ototoxicity, with the most common being quinine, loop diuretics, and aminoglycosides (Shine and Coates 2005). Platinum-based chemotherapeutic agents, in addition to the aminoglycosides have been shown to damage the hair cells of the inner ear, while vincristine sulphate has been shown to produce bilateral cochlear nerve damage (Mahajan *et al.* 1981). Salicylates, which are highly concentrated in the perilymph, may interfere with the enzymatic activity of the hair cells or the cochlear neurones, or both (Silverstein *et al.* 1967). Thalidomide has been demonstrated to produce aplasia of the VIII cranial nerve in association with a Michel aplasia of the inner ear (Jorgensen *et al.* 1964). The aminoglycosides and platinum-based chemotherapeutic agents are both ototoxic and postulated to produce cochlear damage as a result of the production of reactive oxygen species in the cochlea. Recent work has focused on the possible prevention of such ototoxicity by the administration of anti-oxidant drugs (Rybak and Whitworth 2005).

Extensive degeneration of both myelinated and unmyelinated nerve fibres in the cochlear and vestibular divisions of VIII cranial nerve, has been reported in a chronic alcoholic patient with a marked peripheral neuropathy (Ylikoski *et al.* 1981). However, cranial nerve involvement is rare in chronic alcoholic neuropathy and it has been suggested that such neuropathy is perhaps more related to malnutrition than to a direct toxic effect of alcohol.

Menière disease

Menière disease is a clinical diagnosis based on the triad of hearing loss, which is frequently fluctuant in the early stages of the disorder, acute episodes of vertigo, and ipsilateral tinnitus. Commonly, the clinical picture is associated with pressure or fullness in the affected ear. This presentation is based on the idiopathic syndrome of endolymphatic hydrops. Despite a vast literature on the topic of Menière disease, the diagnosis, pathology, and aetiology remain poorly understood (Semaan *et al.* 2005; Gates 2006).

The diagnosis of Menière disease should be based on strict criteria defined by the American Academy of Ophthalmology and Otolaryngology (Committee on Hearing and Equilibrium, AAOO 1995). The hearing loss is characteristically a low tone, sensorineural fluctuating loss in the early stages of the disorder, but with the passage of time the loss may become plateau across all frequencies or 'tent-shaped', with preservation of the mid-frequencies, in configuration. Occasionally, sensorineural hearing loss may progress for some months or years before the onset of acute vertiginous episodes.

Treatment options remain empirical rather than evidence based (Bamiou and Luxon 2003), in view of the diagnostic difficulties and the natural history of relapses and remissions over many months or years.

Demyelination

Hearing loss is an unusual presentation of demyelination, although reports of sudden, unilateral, sensorineural loss as both part of the clinical presentation and also as the initial symptom are noted (de Seze *et al.* 2001) (Section 37.5.3). The lesion is proposed to be in the intramedullary auditory nerve or cochlear nucleus (Barratt *et al.* 1988) and the usual clinical course is of resolution of the hearing impairment. Miller and co-workers (1988) have shown enhanced magnetic resonance imaging lesions at both route entry zones in a patient with multiple sclerosis who had acute deafness and MRI is the gold standard for diagnosis (Cadoni *et al.* 2006).

More rarely, midline brainstem plaque involvement may give rise to auditory processing deficits.

14.3.3 **Auditory processing disorders**

Auditory processing disorder is defined as difficulty in processing non-speech sounds, and is attributed to abnormal function within the central nervous system. A variety of aetiologies, including trauma, neoplasia, degenerative disease, metabolic dysfunction, neurotoxicity, infection, and/or iatrogenic lesions, may impair the neurological processing of auditory information in adults, while in children such deficits may also result from benign central nervous system dysfunction, including neuromaturational delays in the auditory nervous system. The prevalence of auditory processing disorders in children is considered to be between 2 and 39 per cent (Chermak and Musiek 1997b), while in older populations it is estimated to be considerably higher, depending upon screening for peripheral auditory and/or cognitive deficits (Baran 1997).

Brainstem dysfunction. Brainstem hearing loss is rarely encountered in clinical practice because of the multiplicity of auditory pathways and decussations above the cochlear nuclei, and the symmetrical tonotopic organization subserved by the auditory nuclei at all levels.

The audiometric configuration associated with focal brainstem lesions is the subject of some debate. Bilateral symmetrical high-frequency sloping configuration has been reported (Dix and Hood 1973), but animal studies have suggested a low-frequency loss. This latter finding was documented in a study of well-defined midline brainstem lesions in man (Cohen *et al.* 1996). The most effective tests in identifying brainstem auditory pathology include auditory brainstem-evoked responses, acoustic reflex thresholds, and masking level difference. These techniques are highly sensitive in the absence of a peripheral auditory deficit, but their application may be limited if there is significant cochlear or VIII nerve pathology.

Efferent auditory function may also be impaired in brainstem pathology. Dysfunction of the olivocochlear bundle in the floor of the IV ventricle may release the outer hair cells of the organ of Corti from normal inhibitory influences. Recent work has demonstrated a lack of efferent suppression of transient-evoked otoacoustic emissions with contralateral masking in brainstem pathology (Prasher *et al.* 1994 Coelho et al., 2007).

Cortical hearing impairment. Cortical hearing loss is rare, but is most commonly associated with vascular disease (Kaga *et al.* 2000) or trauma (Wirkowski *et al.* 2006) affecting both temporal lobes (Musiek and Lee 1998). Additional and more dramatic neurological sequelae, including hemiparesis and dysphasia are the rule. The primary auditory cortex lies in the anterior/posterior transverse temporal gyrus of Heschl. Each ear has bilateral representation in the auditory cortex, and thus it is possible to remove the non-dominant hemisphere in man without significant effect on either the pure-tone audiogram or the discrimination of distorted speech.

Various auditory deficits result from cerebral lesions, and have previously been divided into *auditory agnosias* (Pan *et al.* 2004) in which pure-tone audiometry is normal or only minimally affected, and *cortical deafness* in which there is a severe if not total hearing loss (Leussink *et al.* 2005). However, more recent work in patients with insular strokes has demonstrated defects of central processing with subtle auditory symptoms (Bamiou *et al.* 2006).

Auditory agnosia was defined originally as a selective disorder of sound recognition: 'I can hear you talking, but I cannot translate it'. This group can be further subdivided into two different clinical presentations, patients who are unable to recognize different types of sound, speech, music, and environmental noises, and patients who are unable to recognize particular types of sound, for example verbal or non-verbal. Most of the cases reported in the literature correspond to the wider definition with impairment of all three modalities of auditory function (Lechevalier *et al.* 1984). Nonetheless there are also reports of verbal auditory agnosia or 'word deafness' in which speech perception is severely impaired in distinction to other linguistic skills, while the recognition of non-verbal material such as musical tunes in environmental noises remains intact (Peretz *et al.* 1995).

In some cases, the primary auditory deficit predominates, and these cases are described as true cortical deafness. In this situation, a patient may present with 'no subjective experience of hearing', and demonstrate profound hearing loss on pure-tone audiometry. This may be misdiagnosed as a peripheral loss if electroacoustic and electrophysiological testing of the auditory system are not conducted, for example otoacoustic emissions and brainstem auditory-evoked responses will demonstrate normal peripheral auditory function (Musiek and Lee 1995). However, abnormal central auditory function will be identified by the later auditory-evoked potentials, specifically the middle latency response N1 and P2 waves.

Not infrequently, the deafness may improve, leaving a residual agnosia for speech and other sounds, with only a minor audiometric deficit (Tanaka *et al.* 1991; Leussink *et al.* 2005). Thus the relationship between auditory agnosia and cortical hearing loss may represent a continuum, dependent upon the precise site of pathology in the primary auditory cortex, the association areas, the auditory radiation, and the medial geniculate body.

Interhemispheric lesions. Patients with surgical section of the posterior portion of the corpus callosum demonstrate a typical pattern of auditory processing test results termed the Auditory Disconnection Profile (Musiek *et al.* 1984). Characteristically the patients demonstrate:

◆ a normal performance on monaural low-redundancy speech tests;

◆ left ear deficits on dichotic speech tests; and

◆ bilateral deficits on temporal pattern testing.

These abnormalities may be explained on the basis of Kimura's model (Fig. 14.7), which highlights the dominance of the contralateral pathway from the right ear to the left hemisphere and the left ear to the right hemisphere in dichotic tasks. In this situation, in which sounds are presented to both ears, the ipsilateral pathway from left ear to left hemisphere and right ear to right hemisphere tends to be suppressed. Section of the corpus callosum interferes with labelling of speech signals delivered to the right hemisphere from the left ear, which require transfer, through the corpus callosum to the left hemisphere for verbal reporting. Thus, monaural tasks are normal, as in the diotic paradigm the ipsilateral pathways are effective, but there is a left ear deficit on dichotic speech tests, and bilateral deficits on temporal pattern testing, which requires interaction of the two hemispheres, again by way of the corpus callosum. This pattern of abnormalities may be seen in a variety of patients with pathology involving the corpus callosum (Musiek *et al.* 1994).

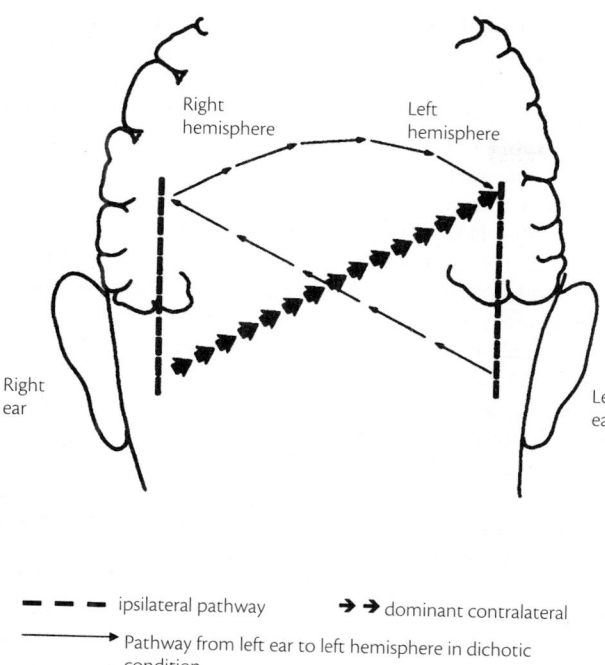

Fig. 14.7 Diagram to illustrate Kimura's model of preponderance of contralateral auditory pathway in dichotic listening tasks.

- - - ipsilateral pathway → → dominant contralateral

→ Pathway from left ear to left hemisphere in dichotic condition

14.3.4 Auditory hallucinations

Auditory hallucinations have been defined as the perception of an auditory experience in the absence of an acoustic stimulus over which the individual has no voluntary control. Tinnitus differs markedly from auditory hallucinations by the absence of the organization of content. Hallucinations may include elementary sounds and complex sounds such as speech, music, and voices, and have been reported with temporal lobe pathology, including tumours, vascular lesions, and epilepsy (Penfield and Rasmussen 1950), but have also been reported in pontine lesions (Adams and Victor 1981).

Most commonly, hallucinations are reported by the elderly, psychiatric patients, or by patients suffering from drug or alcohol intoxication. The experiences may be relatively constant or intermittent. The origin of auditory hallucinations remains ill-defined, but recent functional imaging data has demonstrated activation in the left inferior frontal and right middle temporal gyri occurring 6–9 s before the onset of the experience (Shergill et al. 2004). After the onset of the hallucination, there is activation bilaterally of the middle and superior temporal gyri, the left insula, and the left inferior frontal area. In addition, other studies have reported activation of Heschl's gyrus and the speech areas of the brain during auditory hallucinations (Dierks et al. 1999).

There is also evidence to suggest that auditory deprivation is associated with hallucinations, in that patients with deafness or severe and profound hearing loss are reported to demonstrate a higher prevalence of auditory hallucinations (Wengel et al. 1989; Evers and Ellger 2004). Nonetheless, the pathophysiology of auditory hallucinations remains poorly understood.

Sporadic and familial cases of autosomal-dominant partial epilepsy with auditory symptoms, which may be simple or complex, have been reported. The seizures often begin in childhood and LGI1 mutations are a common cause of inherited partial epilepsy with auditory features, such as humming, which always precede a seizure. In sporadic cases (Bisulli et al. 2004) auditory auras occur either in isolation in about half the cases or in association with visual, psychic, or aphasic symptoms. The onset of symptoms occurs from childhood to the end of the fourth decade and there is low seizure frequency initially and good drug responsiveness. Characteristically there is no cerebral lesion on imaging.

Auditory aura, which lead to 'ear plugging' in children, are also recognized in epilepsy with a superior temporal gyrus focus (Clarke et al. 2003), while musical auditory hallucinations in adults (Roberts et al. 2001) may herald underlying pathology such as intracranial aneurysms but are well recognized in association with temporal lobe abnormalities. Conversely musicogenic epilepsy has been described in right temporal lobe lesions especially those involving the transverse temporal gyri (Shibata et al. 2006).

Transient cortical deafness in partial seizures (Ghosh et al. 2001) with a focus centred in the temporal lobe has also been described and it is hypothesized that an epileptic focus around the primary auditory cortex, dampens its receptive ability and may manifest as cortical deafness. Recent work has also identified auditory processing deficits in patients with temporal lobe epilepsy (Meneguello et al. 2006).

14.3.5 Tinnitus

Tinnitus may result from many different pathologies, and in a review of 411 consecutive patients, Axelsson (1992) identified that approximately two-thirds of his patient group in an otorhinolaryngological clinic suffered ear disease, while almost 6 per cent suffered neurological disease.

A variety of different mechanisms have been hypothesized to explain the pathophysiology of tinnitus, all of which are assumed to lead to an alteration in the spontaneous activity arising from changes in the balance between excitation and inhibition within the auditory system. This alteration may result from:

◆ Abnormal afferent excitation at the cochlea level:

  (a) tinnitus based on spontaneous cochlear oscillations;

  (b) glutamate neuro-excitotoxicity;

  (c) modulation (enhanced sensitivity) of NMDA and non-NMDA receptors; and

  (d) abnormal ion channel conductance/calcium channel dysfunction.

◆ Efferent dysfunction/reduction of GABA effect;

◆ Stress/psychological disorders (Ceranic and Luxon in press);

◆ Decoupling of the stereocilia of the hair cells (Holgers and Barrenas 2003);

◆ Misinterpretation of auditory neural activity in higher auditory centres (Jastrehoff 1994); and

◆ Self-sustaining oscillation of the basilar membrane (Ceranic and Luxon, in press).

Based on the discovery of otoacoustic emissions from within the cochlea, it was considered that spontaneous otoacoustic emissions might provide an explanation for many cases of tinnitus

(Kemp 1978). However, despite the demonstration of such a link, the association has proved to be uncommon (Penner 1990).

Other proposed theories of tinnitus generation rely upon abnormalities of neural function. An abnormality of spontaneous resting activity of primary auditory nerve fibres, either secondary to hypo- or hyper-excitability of damaged hair cells, or as a direct consequence of derangement of the primary neurons themselves, may give rise to the symptom. Møller (1984), suggested ephaptic transmission, or cross talk, between adjacent nerve fibres, due to damage to the myelin sheath between auditory nerve fibres. An alternative proposal has been derangement of efferent activity within the vestibular cochlear nerve, producing aberrant auditory behaviour.

It must be emphasized that the majority of patients who suffer from tinnitus do not find the symptom intrusive. Psychological factors are highly significant in patients with tinnitus complaint, as noted above, and the onset of tinnitus has been related to negative life events such as retirement, redundancy, bereavement, and divorce (Holgers and Barrenas 2003).

14.4 Clinical assessment

14.4.1 Hearing loss

Childhood permanent hearing impairment falls into two broad categories: prelingual and postlingual:

Prelingual hearing loss may be prenatal, perinatal, or postnatal in origin, and children with particular risk factors (Table 14.6) are more likely to present as a result of parental observation, and more rarely as a result of screening programmes for children with risk factors. The introduction of the neonatal hearing screening programme using otoacoustic emissions is now well established (Bamford *et al.* 2005). Moreover, this approach to infant screening of hearing is becoming universally available even in developing countries (Olusanya *et al.* 2005). Despite the efficacy of neonatal hearing screening, it must be recalled that an equal number of children to those identified in infancy with significant hearing impairment develop hearing loss by the age of 16 (Fortnum *et al.* 2001), and thus ongoing evaluation of hearing is required, and behavioural and attentional problems or problems at school should signal appropriate audiological investigation to identify underlying hearing impairment. *Postlingual hearing loss* may result from a range of causes (Table 14.3) and without appropriate intervention will lead to a degradation of speech and educational difficulties.

In adults, the clinical presentation of hearing loss may provide diagnostic clues. Hearing loss may be of sudden or progressive

Table 14.6 Risk factors for sensorineural hearing loss in children

- ◆ Neonatal Intensive Care Unit for >48 h (10-fold risk of hearing problems)
  - Includes use of aminoglycoside antibiotics
  - Includes severe jaundics/metabolic hearing impairment
- ◆ Family history of permanent childhood hearing impairment (1.2–1.9 relative risk of hearing problems)
- ◆ Syndrome associated with hearing impairment
- ◆ Craniofacial abnormality such as cleft palate
- ◆ Congenital infection
- ◆ Bacterial me ningitis

type. The former is most commonly sensorineural in type, and associated with trauma, infection, vascular pathology, or autoimmune disease, while the latter condition is often associated with fluctuation, lack of concordant vestibular symptoms, and variation between ears. There is no consistent definition of sudden sensorineural hearing loss in the literature, and this is compounded by the fact that homogeneous groups of patients are not studied, making comparison of aetiology and configurations of auditory loss difficult (Hughes *et al.* 1996). Nonetheless, the majority of cases of sudden sensorineural hearing loss defy precise aetiological diagnosis, and approximately 60 per cent recover spontaneously, with maximal improvement in the first few weeks after onset (Mattox and Simmons 1977).

Slowly progressive hearing loss is most commonly associated with so-called 'presbyacusis'. However, recent work has led to the view that presbyacusis merely represents a genetic predisposition to environmental triggers (Garringer *et al.* 2006), together with the cumulative effects of multiple otological insults over a lifetime. Specific risk factors for cochlear damage include hazardous noise exposure, barotrauma, head injury, ototoxic drugs, and infections such as syphilis. More rapidly progressive hearing loss is frightening for the patient, and often frustrating for the doctor. A significant proportion of progressive hearing loss is now recognized to be of genetic origin (Martini and Prosser 2003), both as an isolated finding or as a part of a syndrome associated with neurological disease, as outlined above.

14.4.2 Clinical examination

The clinical examination of the auditory system requires careful assessment of the head and anatomy of the external ear to define visible signs of congenital ear disease such as pits, tags, nodules, or malformations and evidence of craniofacial features which may suggest the presence of a syndrome associated with hearing impairment. Wax or debris obstructing the external auditory meatus should be removed by, or under the supervision of, a clinician with experience in this field. Syringing should not be undertaken in the presence of an infection, or if it is unknown whether the tympanic membrane may be perforated. Equally, care must be taken when there is a healed atelectatic area of the drum. A detailed examination of the tympanic membrane is required, to note the presence of a light reflex from the lower tympanic membrane, scarring, the ossicular chain, and any obvious abnormalities from within the middle ear, for example tumours or fluid levels.

Tuning fork tests remain the most valuable clinical test of auditory function, and frequently enable a clinician to distinguish a conductive loss from a sensorineural hearing loss (Fig. 14.8) and, on rare occasions, may enable the identification of a psychologically determined hearing loss. Tuning fork tests rely on two physiological phenomena: first, the inner ear is normally more sensitive to sound conducted by air, as a consequence of the redistribution of frequencies in the external canal, and the amplification sound through the middle ear mechanism; and second, in the presence of a purely conductive hearing loss, the affected ear is subject to less environmental noise, making it more sensitive to bone-conducted sound. It should be emphasized that in a profoundly deaf patient, it is imperative to obtain the services of an interpreter, for example using British sign language, to ensure that an appropriate history and full examination may be undertaken.

A general medical examination is required to define evidence of systemic disease.

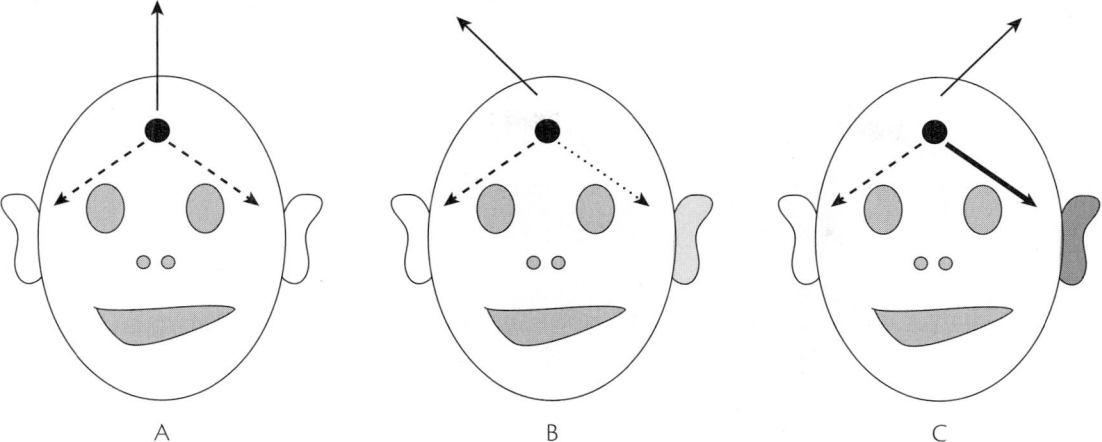

Fig. 14.8 Diagram illustrating the results observed during the Weber test. In: A. a normal subject, the perceived sound is heard in the middle of the head; B. a patient with a left sensorineural hearing loss, the perceived sound is heard towards the better i.e. right ear; and C. a patient with a left conductive hearing loss, the perceived sound is heard towards the left, i.e. the affected side. ┄➤ Reduced perception by cochlea secondary to sensorineural loss ┄➤ Normal perception of tuning fork by cochlea ──➤ Direction of perceived sound ━➤ Enhanced perception of bone-conducted sound as a consequence of reduction of air-conducted environmental sound due to conductive hearing loss ▫ Conductive hearing loss ▪ Sensorineural hearing loss.

14.5 Investigations

The aim of auditory investigations, which may be psychoacoustic or behavioural in type, is to define the presence of pathology in the auditory system, and site the level of the lesion. A battery of *audiological tests* is required to:

- quantify the audiometric threshold at each frequency;
- differentiate a conductive from a sensorineural hearing loss;
- differentiate a cochlear from a retrocochlear abnormality;
- identify central auditory dysfunction in the brainstem, midbrain, or auditory cortex; and
- identify a non-organic hearing impairment.

Tests may be either *subjective* and depend upon patient cooperation, or *objective* in that they do not rely on patient cooperation. In the differentiation of a sensorineural hearing loss of cochlear origin from that of VIII nerve dysfunction or neurological impairment, loudness recruitment and abnormal auditory adaptation are important.

Pure-tone audiometry. This is the most widely available subjective, quantitative test of auditory threshold (British Society of Audiology 1991). The technique, performed in a sound-proofed room according to standardized protocols allows the severity, symmetry, and configuration of hearing loss to be defined across frequencies between 125 and 8000 Hz in each ear. Electrically generated pure-tones are delivered by headphones, and the subject is required to respond to the quietest tone. Sound may be delivered by air conduction or, if the tones are delivered via a bone vibrator on the mastoid process, by bone conduction. In this latter condition, because the intra-aural attenuation for a bone-conducted sound is negligible, the ear which is not being tested must be masked with narrow-band noise centred on the test frequency. Bone conduction thresholds, which are significantly better than air conduction thresholds, indicate a disorder affecting the transmission

of sound waves through the middle ear into the inner ear, i.e. conductive hearing loss, whereas similar bone conduction and air conduction thresholds imply a sensorineural hearing loss (see Fig. 14.5).

Acoustic impedance measurements (Fig. 14.9A). These provide information about the middle and internal ears, in addition to the VIII nerve and brainstem function. Passive measurements are made of the change in acoustic impedance or immitance of the tympanic membrane as a function of the pressure in the sealed external acoustic meatus (British Society of Audiology 1992). Dynamic changes resulting from the contraction of the stapedius muscle, acoustic reflex thresholds, (Fig. 14.9B) in response to stimuli of 500, 1000, 2000, and 4000 Hz, at intensities of 70–100 dB sound pressure level are also measured. These values provide objective evidence of recruitment and abnormal auditory adaptation, and allow an assessment of middle ear, cochlear, VIII nerve, and brainstem auditory function (Fig. 14.9C).

Otoacoustic emissions. These represent weak signals generated by the contractile properties of the outer hair cells in the cochlea in response to acoustic stimuli (Fig. 14.10). These responses are measured in the external auditory canal, and provide direct objective information about the integrity of the cochlea. Efferent auditory function can be assessed by otoacoustic emission suppression brought about by the application of noise to the contralateral ear.

Speech audiometry (Fig. 14.11). This is concerned with the assessment of auditory discrimination as opposed to the assessment of auditory acuity. The test is subjective and requires the subject to repeat standard lists of words delivered at varying intensities through headphones. The responses are scored, and provide an assessment of auditory discrimination which, together with other tests, may be of value in distinguishing conductive, sensory, and neural hearing impairment, but this test is particularly of value in assessing the efficacy of hearing aid provision.

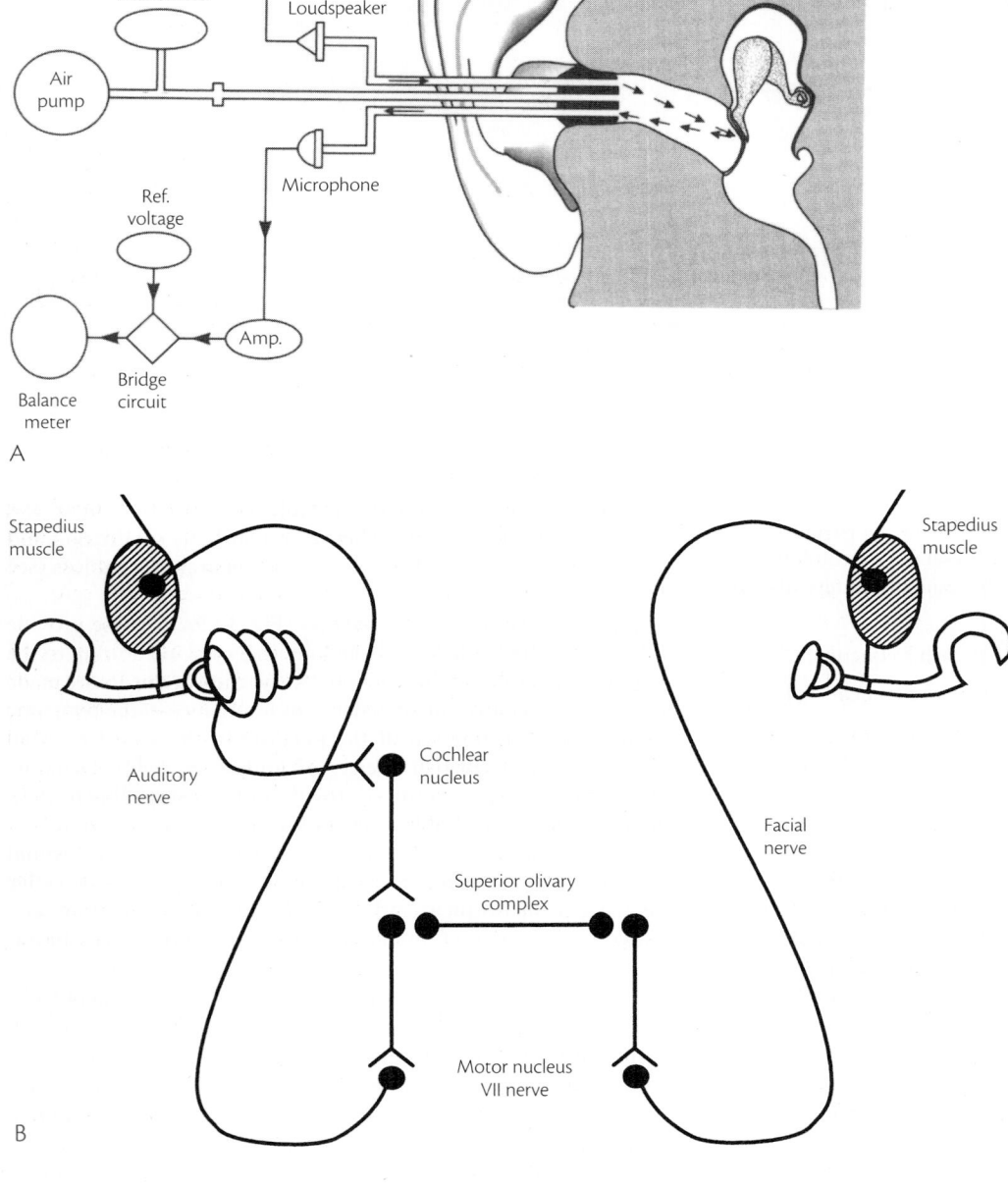

Fig. 14.9 Diagrams to illustrate acoustic impedance measurements: A. Diagram of components of electroacoustic impedance bridge; (Ludman H (1998). Basic acoustics and hearing tests. In: Ludman H and Wright T (eds). *Diseases of the Ear* 6th Edn. Arnold, London, pp. 58–86.) B. Diagram to illustrate anatomical pathway of the bilateral acoustically-induced stapedius reflex.

Electrophysiological tests

These provide the most objective technique of assessing auditory function and siting pathology in the auditory system. *Electrocochleography* is the measurement of the electrical output of the cochlea and VIII cranial nerve in response to auditory stimulation. It is most commonly used in the diagnosis of Menière disease, when the summating potential:action potential ratio is greater than 30 per cent, whereas in the normal population it is significantly smaller.

Brainstem-evoked responses (Fig. 14.12A). These are a series of neurogenic potentials which are recorded using surface electrodes in response to click stimuli, in the 10 s immediately after the stimulus. As a diagnostic tool, brainstem auditory evoked responses are of particular value in discriminating between cochlea and VIII nerve/brainstem dysfunction. Analysis of the waveform (Fig. 14.12B) must be undertaken in the knowledge of pure-tone audiometric thresholds, if appropriate and valid conclusions are to be drawn.

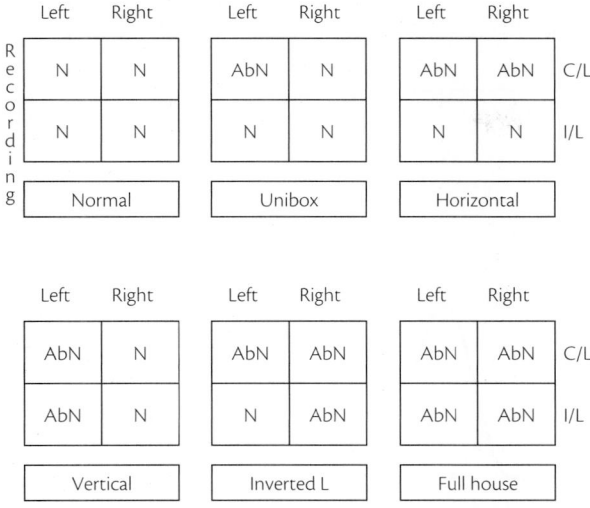

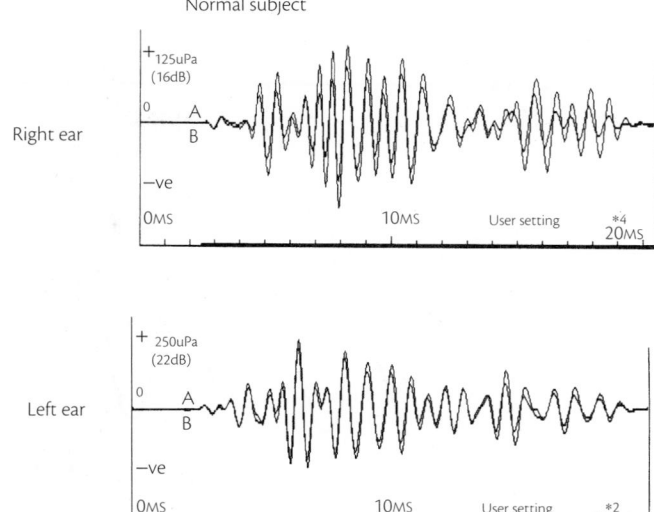

Fig. 14.9 (*Cont.*) C. Tabulation of patterns of stapedial reflex responses in different sites of auditory pathology. (Key: N—normal; AbN—abnormal; C/L—contralateral; I/L—Ipsilateral; Unibox—small unilateral brainstem lesion medial to cochlear nucleus; Horizontal—midline brainstem lesion; Vertical—left VIIIth nerve lesion; Inverted L—intra-axial brainstem lesion plus extension to the cochlear nucleus or VIIIth nerve on the affected side (NB, a conductive lesion may also present in this way); Full-house—a midline brainstem lesion with extension to involve the cochlear nuclei and/or VIIIth nerves (a bilateral conductive disorder requires exclusion)).

Fig. 14.10 Otoacoustic emissions recorded from each ear in a normal subject.

Prolongation of the I–III interval can be seen in auditory nerve, and cochlear nucleus pathology. Prolongation of the III–V is usually indicated when pathology is sited above the level of the cochlear nucleus, while absent IV and/or V waves are found in cases with involvement of the mid-upper pons. In severe brainstem pathology, waves III–V may be absent. Inter-aural latency comparisons of wave V are of value in diagnosis of acoustic neurinoma (Fig. 14.12C), but may not be useful in detecting brainstem involvement (Weinstein 1994). In general, while the abnormality in brainstem lesions may be ipsilateral, with respect to the acoustic stimulus, or bilateral, contralateral findings are rare (Musiek and Lee 1995). The sensitivity and specificity of the auditory brainstem response in identifying brainstem lesions depends on the site of lesion, i.e. more caudal intra-axial structural brainstem lesions are identified but the auditory brainstem response is only moderately sensitive to degenerative disorders or rostral lesion involving the brainstem. Overall, the sensitivity/specificity of the auditory brainstem response for a variety of brainstem lesions is around 80 per cent, which is less than that noted for acoustic tumours (Musiek and Lee 1995).

Middle latency response. These generator sites are thought to be in the thalamocortical pathway in the auditory cortex (Kraus *et al.* 1994). There is much intersubject variability in both latency and amplitude measurements of the middle latency response, but in general the most effective measurements are intra-subject comparisons of the electrode effect and the ear effect (Musiek *et al.* 1994). Maturation of the middle latency response is relatively long, and very variable, such that this test cannot be applied reliably to children under the age of 10 years. The sensitivity and specificity of the middle latency response for central auditory pathology is reasonably good, and it is therefore a valid objective test in the assessment

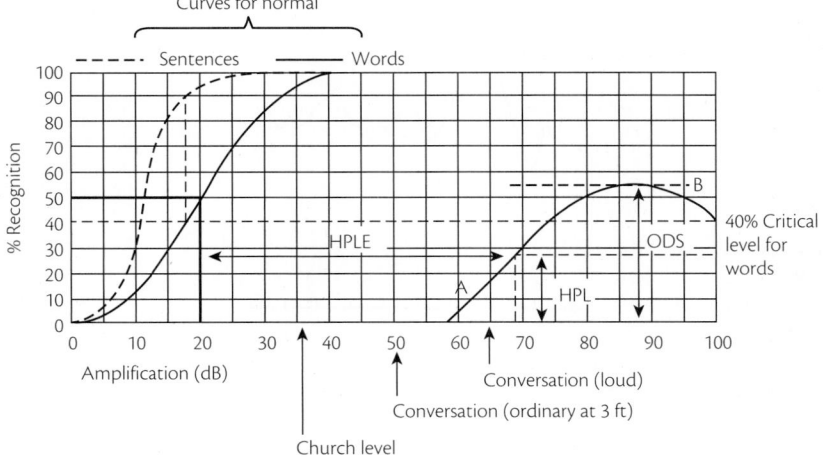

Fig. 14.11 Speech audiogram: AB—speech audiogram of a sensorineural hearing loss; HPL—half-peak level; HPLE—half-peak level elevation; ODS—optimal discrimination score. (Luxon LM (1988). Methods of examination: audiological and vestibular. In Ludman H, Mawson S, eds. *Diseases of the Ear*, 5th edition. Oxford University Press, Oxford.)

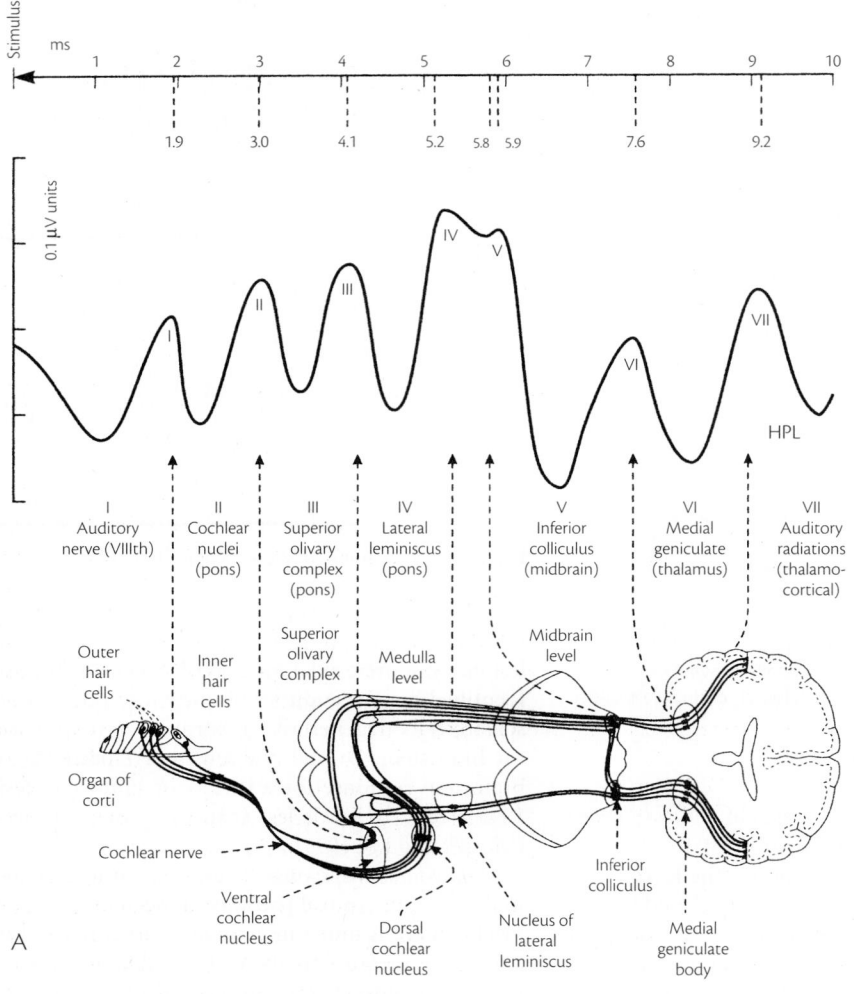

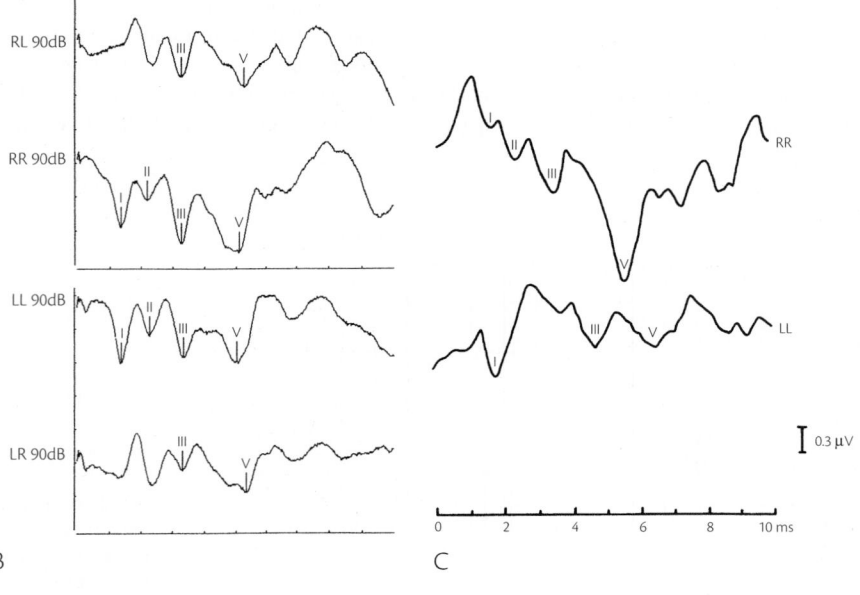

Fig. 14.12 Auditory brainstem-evoked responses: A. Anatomical correlates of the waves I–VII observed in auditory brainstem-evoked responses. (Reproduced with kind permission from Duane 1977, Central Auditory Dysfunction. Ed: Keith R.W. Grune and Stratton New York.); B. Diagram to illustrate auditory brainstem-evoked responses in a normal subject; and C. Diagram to illustrate delay of wave III and wave V in a small left acoustic neuroma compared with normal response from the right ear. (Key: LL—ipsilateral left recording; RR—ipsilateral right recording; LR—Left stimulation, right recording RL—Right stimulation, left recording.)

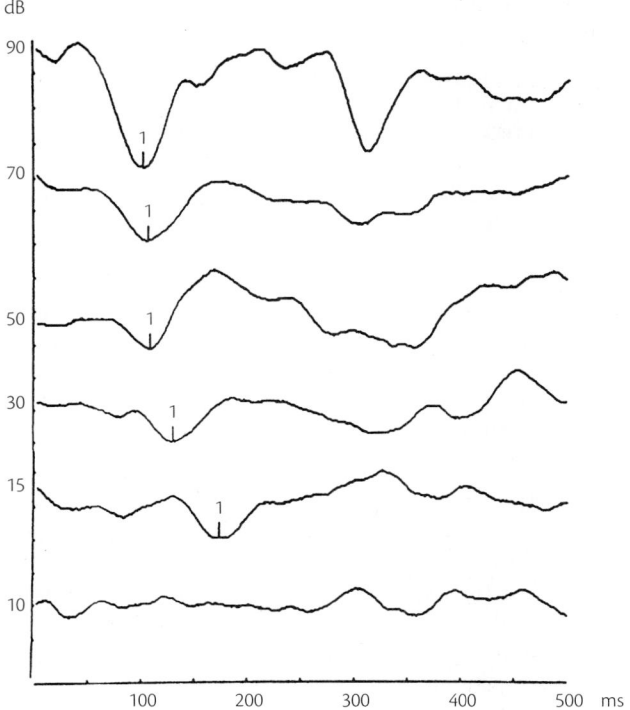

Figure 14.13 Tracings of cortical-evoked responses recorded to threshold in a normal subject. (From Luxon LM and Cohen M (1997). Central Auditory Dysfunction In: *Scott Brown's Diseases of the Ear Nose and Throat*, 6th edition. Kerr AG Vol 9: Adult Audiology Ed: Stephens SDG. Butterworths, London.)

of central auditory dysfunction (Musiek *et al.* 1999; Kileny *et al.* 1987), but sleep and/or sedation may affect the response.

Cortical- or late-evoked auditory responses. These are the most effective method of defining auditory threshold at each frequency (Fig. 14.13) in a patient who is unable or unwilling to cooperate,

and are essential in legal cases in which non-organic loss should always be excluded.

Behavioural tests. These may be grouped together into tests which share a common parameter such as mode of presentation, for example monaural or binaural signals, or type of stimulus employed, for example tones or speech. The brainstem auditory nuclei are important in terms of auditory processing for two reasons:

◆ the extraction of signals from a background of noise, which has led to the development of auditory separation tasks; and

◆ binaural integration of auditory information which has led to the development of binaural interaction tasks.

Behavioural tests which are of particular value in the assessment of brainstem disorders therefore include masked speech, the synthetic sentence identification with ipsilateral computing message test, the masking level difference test (Fig. 14.14), and the binaural fusion test.

Behavioural tests which assess cortical function employ speech or speech-like stimuli. In the dichotic paradigm, different stimuli are presented simultaneously to the two ears, and the listener is required to respond to one or both stimuli, for example the dichotic digit test and the threshold of interference test (Fig. 14.15). In the dichotic digit test, two digit pairs, or four digits in total, are presented to each ear simultaneously at 50 dB sensation level. The digits are carefully aligned in terms of their stimulus onset, and include the numbers from 1 to 10 with the exception of 'seven'. The patient is asked to repeat all digits heard, but they may do this in any order, and they are encouraged to guess if unsure of a response. The percentage of correct score is derived for each ear, and compared to age-appropriate normal data.

As noted above, in the dichotic situation, the weaker ipsilateral auditory path tends to be suppressed, and neural impulses travel up the contralateral pathway to reach auditory reception areas in the auditory cortex (Fig. 14.7). Contralateral ear effects are observed in cases of auditory cortical involvement, while left ear deficits are

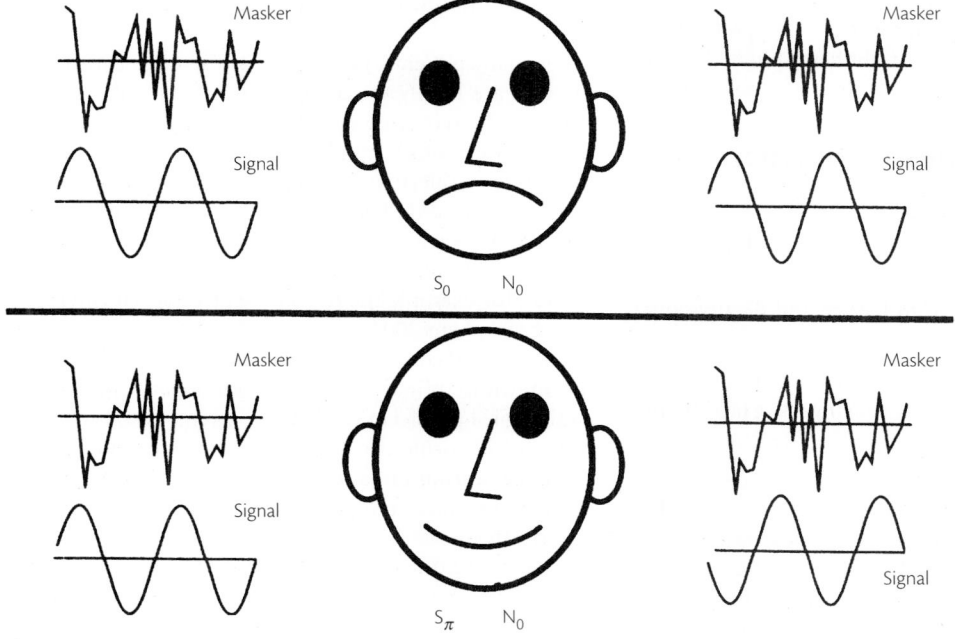

Fig. 14.14 Diagram to illustrate the masking level difference test. (From From Luxon LM and Cohen M (1997). Central Auditory Dysfunction In: *Scott Brown's Diseases of the Ear Nose and Throat*, 6th edition. Kerr AG Vol 9: Adult Audiology Ed: Stephens SDG. Butterworths, London.)

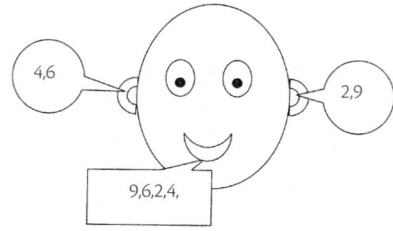

Fig. 14.15 The dichotic digit test.

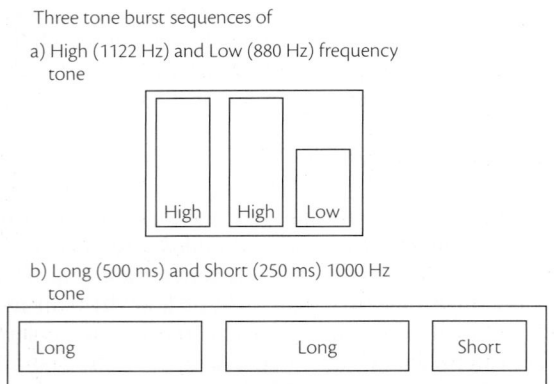

Three tone burst sequences of

a) High (1122 Hz) and Low (880 Hz) frequency tone

b) Long (500 ms) and Short (250 ms) 1000 Hz tone

Fig. 14.16 Diagrammatic illustration of frequency and duration discrimination tasks.

observed in lesions involving the interhemispheric pathways, for example the corpus callosum. In brainstem pathology, ipsilateral ear abnormalities are common, with extra-axial lesions, whereas bilateral, contralateral, or ipsilateral ear effects may be observed in patients with intra-axial lesions.

Pattern recognition tests. Temporal ordering tasks use non-verbal stimuli in order to assess the auditory processes of feature detection, frequency, or duration discrimination (Fig. 14.16). All of these tasks are largely dependent on an intact right hemisphere auditory function.

14.6 **Management of auditory disorders**

14.6.1 **General principles**

Management of hearing impairment may be divided into three aspects:

- *prevention* to ensure protection from leisure and occupational noise hazards; and ototoxic drugs;

- *medical management*, to ensure treatment of systemic medical conditions which may be impacting, causing, or exacerbating auditory dysfunction; and

- *auditory rehabilitation* which is a problem-solving exercise centred on each individual patient, assessing both the auditory disability and the relevance of this to the patient's lifestyle and other important people in his or her life. Both the auditory impairment and communication skills, including lip-reading ability, the use of visual cues, and the level of speech and language,

together with psychological and sociological factors must all be considered.

In a highly motivated individual with uncomplicated hearing loss, the remedial process may be relatively straightforward. Frequently however, the process becomes significantly more complicated by factors such as arthritis, making manipulation of the hearing aid difficult, or failure to appreciate the impact hearing impairment is having on family members. The particular problem in each individual must be addressed if optimal use of a hearing aid is to be achieved. In patients who have a strongly negative view of hearing aids, prior to introducing a hearing aid, environmental aids and instruction in communication skills, may facilitate long-term rehabilitation. In general the provision of a hearing aid is only effective when the patient himself wishes to pursue this line of management rather than it being the aim of well-meaning family members.

Hearing aids

Although hearing aids play a pivotal role in audiological rehabilitation, a detailed description of their prescription is outside the scope of this chapter (reviewed by Gatehouse 2003). Conventional aids may be divided into those that are body-worn and head-worn by mounting in spectacles, in addition to being mounted in or around the ear. The major advantage of body-worn aids is the very high gain and maximum output which can be achieved, whereas the disadvantage is the obvious and unsightly nature of the device, and the poor microphone placement.

The general principles of hearing aid provision include the fitting of a comfortable ear mould, which provides a secure mounting for the aid, and a good acoustic connection between the aid and the ear canal. Hearing aid selection requires matching the amplification requirement from the aid at specific frequencies with that of the user. The majority of hearing-impaired people complain, in particular, of difficulty hearing speech in the presence of background noise, and although programmable digital processing devices are of some help in this situation, conventional analogue aids provide selective amplification across speech frequencies with minimal amplification at the peak frequencies of background noise.

For many patients, post-aural, in-the-ear, or in-the-canal hearing aids are highly effective, but some patients will also require additional environmental aids. These include assisted listening devices, for example amplification systems attached to televisions or telephones, alerting warning devices, for example flashing lights connected to a door bell or an alarm clock. Other sensory substitution systems such as the visual signals generated by auditory cues may be particularly helpful to a profoundly hearing-impaired person, for example the telephone or doorbell ringing, or a baby crying (Stephens 2003).

The value of counselling for the hearing-impaired subject within the multidisciplinary team of audiologists, psychologists, and audiological physician cannot be overemphasized. Simple hearing tactics can enormously improve communication, such as encouraging the hearing-impaired person to ensure that light is always on the speaker's face, and placement so that the better ear is always towards the speaker, sitting close to the sound source and minimizing background noise. For those with profound hearing loss, the need for psychological support should always be considered, and moreover, social and occupational support should be offered.

14.6.2 Conductive hearing loss

In the management of conductive loss, care must be taken to ensure that there is no obstruction to the transmission of sound through the external ear by a foreign body, wax, a polyp, tumour, or infection. Acute otitis externa requires suction clearance under microscopy, culture and sensitivity of the organism, and appropriate medication, with or without steroids (Rosenfeld *et al.* 2006).

The treatment of acute otitis media in both children and adults depends upon the relief of pain, re-establishment of Eustachian tube function using nasal drops, inhalations or decongestants, mucolytics, and the prescription of systemic antibiotics. The drug of choice is amoxycillin, or in cases of sensitivity to penicillin, erythromycin. Myringotomy is indicated if the drum is bulging. In recurrent acute otitis media, particularly in children, an underlying immunodeficiency should be excluded, and a focus of infection within the upper respiratory tract should be sought. To counter antibiotic-resistant bacteria, microbial vaccines have been introduced against *S.pneumoniae*, *N.catarrhalis*, respiratory syncytial virus, adenovirus, influenza A, and parainfluenza viruses (Giebink 1999). Otitis media with effusion remains a common, and in a percentage of cases, difficult problem to treat. About 2–5 per cent of children were previously subjected to grommet insertion, but current opinion is moving towards 'watchful waiting', while considering that children who suffer with long periods of glue ear with conductive loss and delay in speech development should be offered amplification (Lous *et al.* 2005). The aim of treatment of chronic suppurative otitis media is to eliminate infection using antibiotics, and when the ear is healthy, repair aural damage, such as a perforated ear drum or damage to the ossicles, which may prevent reinfection and improve sound transmission (Robinson 1998; Raglan 2003).

Conductive hearing loss caused by otosclerosis or the hereditary osseous dysplasias may be managed conservatively using hearing aids, or surgically by stapedectomy. The procedure carries a small risk of complication of late sudden sensorineural hearing loss, and for this reason stapedectomy is not generally undertaken on both ears. Congenital malformations of the auditory canal and the middle ear may be treated conservatively with bone conduction or bone-anchored hearing aids (Snik *et al.* 2005), or may be surgically remediable.

14.6.3 Sensorineural hearing loss

The management of sensorineural hearing loss is determined by the time course of the loss, be it sudden, progressive, or chronic.

Sudden hearing loss is a medical emergency requiring hospital admission, bed rest, and investigation of possible causes. In the case of bilateral hearing loss, psychological and aggressive auditory rehabilitation are required. There is no universally accepted form of treatment, and a spontaneous recovery rate of approximately 65 per cent is reported (Haberkamp and Tanyeri 1999).

Frequently, an aetiological diagnosis is not determined in this condition, and a variety of different interventions have been proposed, including inhalation of carbigen, hyperbaric oxygen, antiviral treatment, immunosuppression, calcium channel blockers, steroids, blood volume expanders, and various combinations of these different treatment strategies (Haberkamp and Tanyeri 1999; Zadeh *et al.* 2003; Sismanis 2005). There is no consensus on effective treatment of this condition. Although randomized control studies have demonstrated the efficacy of systemic steroids (Haberkamp and Tanyeri 1999), follow-up studies have questioned the benefit of this approach. In particular, steroid therapy is contraindicated in the presence of bacterial infection, recent surgery, peptic ulceration, a history of tuberculosis, poorly controlled hypertension, or diabetes.

Progressive hearing impairment may be the result of various pathologies which may be treatable either medically or surgically. Subsequent rehabilitation reduces disability. Specific conditions which should be treated include syphilitic labyrinthitis with steroids and penicillin, and large acoustic neurinoma by surgical intervention or laser therapy. However, there is now a consensus that in the majority of cases of small tumours a 'watch, wait, and monitor' policy is effective, and avoids intervention for many extremely slowly progressive benign tumours (BAOL-HNS 2002).

Immune-mediated sensorineural hearing loss requires urgent management with steroids and/or immunosuppressives following diagnosis (Agrup and Luxon 2006).

There is a plethora of treatments for Menière disease, but no definitive therapy has been supported by a double-blind control trial. Trials of therapy in this condition are particularly difficult, because of the lack of diagnostic certainty, the relapsing and remitting nature of the disorder, and the variability of frequency and severity of the different symptoms within the diagnostic triad between and within patients. Therapies can be divided into medical and surgical regimens. Medical therapy includes dietary modifications, pharmacological interventions including diuretics, vestibular sedative drugs, drugs aimed at improving the circulation of the inner ear, and immunosuppressives, psychological support, physiotherapy, and auditory rehabilitation (Cohen-Kerem *et al.* 2004; Doyle *et al.* 2004; James and Thorp 2004; Minor *et al.* 2004). A recent low-pressure pulse generator (Meniett device) has been advocated as a non-invasive effective treatment for Menière disease, but there is no definite evidence available (Gates 2005). Conventional surgical treatment is considered when medical management has failed to control vertigo, and is not an intervention advocated for control of progression of hearing impairment. Chemical labyrinthectomy using intratympanic gentamicin has been popularized as a method of controlling severe vestibular symptoms, although no clear treatment protocol has been established and cochleotoxicity is a significant risk (Cohen-Kerem *et al.* 2004) Surgical procedures may be divided into 'therapeutic' procedures such as endolymphatic sac decompression or destructive procedures which may seek to control vertigo by section of the vestibular nerve, or in cases of severe/profound hearing loss, labyrinthectomy may be undertaken (Van de Heyning *et al.* 2005). Chronic symptoms of dizziness due to vestibular dysfunction and hearing impairment may be treated with auditory and vestibular rehabilitation.

Chronic sensorineural hearing impairment may be managed along two complementary approaches: specific aetiological treatment to rectify or halt the progression of the causative condition, and audiological rehabilitation of residual hearing. The aim is to minimize disability, prevent handicap, and facilitate optimal physical, mental, and social potential. Goldstein and Stephens (1981) described the components of an auditory rehabilitation programme: evaluation including the degree of hearing impairment and presence of auditory discomfort, disability, handicap, communication status, mobility, degree of manual dexterity, and related aural pathology, and remediation, which emcompasses personal instrumentation such as a hearing aid, tactile aid, or cochlear implant, as well as general instrumentation using assistive listening devices such as telephone or TV attachment and alerting warning systems or alarms.

The selection and fitting of hearing aids is considered briefly above, but is a key element of rehabilitation for the majority of patients with hearing impairment (Luxon and Raglan 2006). In addition to the aids outlined above, bone conduction aids, which are in general used for conductive hearing loss in cases in which an ear mould cannot be placed in the ear canal, for example because of active ear discharge, implantable hearing aids (see below), and assistive listening devices may all be of value. Hearing aid benefit may be assessed by a number of questionnaires of which the Glasgow Hearing Aid Benefit Profile (Gatehouse 2000) is of particular value and well validated.

Cochlear implants, either in one or both ears, have revolutionized the rehabilitation of children with profound hearing impairments, and adults who become profoundly hearing-impaired from conditions such as meningitis, superficial siderosis, mitochondrial disease, and head trauma (Cohen *et al.* 2005; Das and Buchman 2005). Current reviews of cochlear implantation reveal that hearing, language, and social development outcomes in children have been positive, and both audiological and quality of life measures have been 'excellent' in older patients (Buchman *et al.* 1999; Balkany *et al.* 2002). As with all hearing aid provision, the patient requires long-term auditory training. Auditory *brainstem implants* have been used in profoundly deaf patients in whom there is complete dysfunction of the VIII cranial nerve, for example in neurofibromatosis type 2 (Tatagiba and Gharabaghi 2005). *Middle ear implants* (vibrators on one of the ossicles or on the tympanic membrane) may be used with all types of hearing loss, although their value is not yet clearly defined (Magnan 2005).

References

Acuin J (2004). Chronic suppurative otitis media. *Clin Evid*, **12**, 710–29.

Adams RD, Victor M (1981). Time to understand. A case study of word deafness with reference to the role of time in auditory comprehension. *Brain*, **97**, 373–84.

Agrup C, Luxon LM (2006). Immune-mediated inner ear disorders in Neuro-otology. *Curr Opin Neurol*, **19**, 26–32.

Alberts MC, Terrence CF (1978). Hearing loss in carcinomatous meningitis. *J Laryngol Otol*, **92**, 233–41.

American College of Occupational Medicine Noise and Hearing Conservation Committee (1989). Occupational noise induced hearing loss. *J Occup Med*, **31**, 996.

Andersson G, Lindvall N, Hursti T *et al.* (2002). Hypersensitivity to sound (hyperacusis): a prevalence study conducted via the internet and post. *Int J Audiol*, **41**, 545–54.

Anteunis LJ, Mooy JM (1987). Hearing loss in a uraemic patient: indications of involvement of the VIIIth nerve. *J Laryngol Otol*, **101**, 492–6.

APD (Auditory Processing Disorders) BSA Special Interest Group Definitions http://www.thebsa.org.uk/apd/Home.htm#working%20def -viewed 12.6.06.

Axelsson A (1992). Causes of tinnitus. In Aran JM, Dauman R, eds. *Tinnitus 91. Proceedings of the IV International Tinnitus Seminar*, pp. 275–7. Bordeaux, Kugler Publications, Amsterdam/New York.

Balkany TJ, Hodges AV, Eshraghi AA *et al.* (2002). Cochlear implants in children—a review. *Acta Otolaryngol*, **122**, 356–62.

Bamford J, Uus K, Davis A (2005). Screening for hearing loss in childhood: issues, evidence and current approaches in the UK. *J Med Screen*, **12**, 119–24.

Bamiou D, Luxon LM (2003). Medical management of vestibular disorders and vestibular rehabilitation. In Luxon LM, Stephens SDG, Martini A, Furman J, eds. *Textbook of Audiological Medicine*, pp 889–916. Martin Dunitz, London.

Bamiou DE, Spraggs PR, Gibberd FB *et al.* (2003). Hearing loss in adult Refsum's disease. *Clin Otolaryngol*, **28**, 227–30.

Bamiou DE, Musiek FE, Stow I *et al.* (2006). Auditory temporal processing deficits in patients with insular stroke. *Neurology*, **67**, 614–9.

Baran JA (1997). Speech perception test materials for central auditory processing assessment. In Mendel LL, Danhauer JL, eds. *Audiologic Evaluation and Management and Speech Perception Assessment*, pp. 147–68. Singular Publishing Group, San Diego.

Barratt HJ, Miller D, Rudge P (1988). The site of the lesion causing deafness in multiple sclerosis. *Scand Audiol*, **17**, 67–71.

Bergemalm PO (2003). Progressive hearing loss after closed head injury: a predictable outcome? *Acta Otolaryngol*, **123**, 836–45.

Berlin C, Hood L, Rose K (2001). On renaming auditory neuropathy as auditory dys-synchrony. Audiology Today, **13**, 15–7.

Berlin CI, Hood LJ, Hurley A *et al.* (1994). Contralateral suppression of otoacoustic emissions: An index of the function of the olivocochlear system. *Otolaryngol Head Neck Surg*, **110**, 3–21.

British Association of Otolaryngologists - Head and Neck Surgeons (2002). *Clinical Effectiveness Guidelines: Acoustic neuroma (Vestibular Schwannoma) BAOL-HNS Document 5.* British Association of Otolaryngologists - Head and Neck Surgeons, London.

British Society of Audiology (1981). Recommended procedure for pure tone audiometry using a manually operated instrument. *Br J Audiol*, **15**, 213–6.

British Society of Audiology (1992). Recommended procedure for tympanometry. *Br J Audiol*, **26**, 255–7.

Buchman CA, Fucci MJ, Luxford WM (1999). Cochlear implants in the geriatric population: benefits outweigh risks. *Ear Nose Throat J*, **78**, 489–94.

Bisulli F, Tinuper P, Avoni P *et al.* (2004). Idiopathic partial epilepsy with auditory features (IPEAF): a clinical and genetic study of 53 sporadic cases. *Brain*, **127**, 1343–52.

Cadoni G, Cianfoni A, Agostino S *et al.* (2006). Magnetic resonance imaging findings in sudden sensorineural hearing loss. *J Otolaryngol*, **35**, 310–6.

Campuzano V, Montermini L, Moltó M (1996). Friedreich ataxia: autosomal recessive disease caused by an intronic GAA triplet repeat expansion. *Science*, **271**, 1423–7.

Ceranic BJ, Prasher DK, Raglan E *et al.* (1998). Tinnitus after head injury: evidence from otoacoustic emissions. *J Neurol Neurosurg Psychiatry*, **65**, 523–9.

Ceranic B, Luxon LM (2002). Disorders of the auditory system. In Asbury A, McKhann G, McDonald WI, *et al.* eds. *Diseases of the Nervous System*, 3rd edition, pp. 658–77. Cambridge University Press, Cambridge.

Ceranic B, Luxon LM (In press) Tinnitus and other dysacuses In Gleeson M, ed. Browning G, Luxon LM Section eds. *Scott Brown's Otolaryngology*, 7th edition. Hodder, London.

Chermak GD, Musiek FE (1997). Neurobiology of the central auditory nervous system relevant to central auditory processing. In Chermak GD, Musiek FE, eds. *Central Auditory Processing Disorders New Perspectives*, pp. 27–70. Singular Publishing Group, San Diego.

Clarke DF, Otsubo H, Weiss SK *et al.* (2001). The significance of ear plugging in localization-related epilepsy. *Epilepsia*, **44**, 1562–7.

Cohen N, Ramos A, Ramsden R *et al.* (2005). International consensus on meningitis and cochlear implants. *Acta Otolaryngol*, 125, 916–7.

Cohen M, LuxonLM, Rudge P (1996). Auditory deficits and hearing loss associated with focal brainstem haemorrhage. *Scand Audiol*, **25**, 133–41.

Cohen-Kerem R, Kisilevsky V, Einarson TR *et al.* (2004). Intratympanic gentamicin for Meniere's disease: a meta-analysis. *Laryngoscope*, **114**, 2085–91.

Coles RRA (1984). Epidemiology of tinnitus: (1) Prevalence and (2) Demographic and clinical features. *J Laryngol Otol Suppl*, **9**(1) 7–15, (2) 195–202.

Coles RR, Davis A, Smith O (1990). Tinnitus: Its epidemiology and management. In Jensen JH, ed. *Proceedings XIV Danavox Symposium, Copenhagen*, pp. 377–402. Danavox Jubilee Foundation.

Collet L, Kemp DT, Veuillet E *et al.* (1990). Effect of contralateral auditory stimuli on active cochlear micromechanical properties in human subjects. *Hear Res*, **43**, 251–62.

Colvin IB (2006). Audiovestibular manifestations of sarcoidosis: a review of the literature. *Laryngoscope*, **116**, 75–82.

Committee on Hearing and Equilibrium, the American Academy of Ophthalmology and Otolaryngology (1995). Committee on hearing and equilibrium guidelines for diagnosis and evaluation of therapy in Menière's disease. *Otolaryngol Head Neck Surg*, **113**, 181–5.

Cooper JC Jr (1994). Tinnitus, subjective hearing loss and well-being. Health and Nutrition Examination Survey of 1971–75: Part II. *J Am Acad Audiol*, **5**, 37–43.

Cryns K, Sivakumaran TA, Van den Ouweland JM *et al.* (2003). Mutational spectrum of the WFS1 gene in Wolfram syndrome, nonsyndromic hearing impairment, diabetes mellitus, and psychiatric disease. *Hum Mutat*, **22**, 275–87.

Darmstadt GL, Harris JP (1989). Luetic hearing loss: clinical presentation, diagnosis, and treatment. *Am J Otolaryngol*, **10**, 410–21.

Das S, Buchman CA (2005). Bilateral cochlear implantation: current concepts. *Curr Opin Otolaryngol Head Neck Surg*, **13**, 290–3.

Dauman R, Bouscau-Faure F (2005). Assessment and amelioration of hyperacusis in tinnitus patients. *Acta Otolaryngol*, **125**, 503–9.

Davis A, Rafaie EA (2000). Epidemiology of tinnitus. In Tyler R, ed. *Tinnitus Handbook*, pp. 1–23. Singular, San Diego.

Davis A (1993). The prevalence of deafness. In Ballantyne J, Martin A, Martin M, eds. *Deafness*. Whurr, London.

Davies RA, Luxon LM (1995). Dizziness following head injury: a neuro-otological study. *J Neurol*, **242**, 222–30.

Davies RA (In press). Retrocochlear hearing loss. In Gleason M, ed. *Scott Brown's Otolaryngology*, 7th edition. Hodder, London.

Denny-Brown D (1951). Hereditary sensory radiculopathy. *J Neurol Neurosurg Psychiatry*, **14**, 237–52.

de Seze J, Assouad R, Stojkovic T *et al.* (2001). Hearing loss in multiple sclerosis: clinical, electrophysiologic and radiological study. *Rev Neurol (Paris)*, **157**, 1403–9.

Devriese PP (1968). Facial paralysis in cephalic herpes zoster. *Ann Otol Rhinol Laryngol*, **77**, 1101–19.

Dierks T, Linden D, Jandl M *et al.* (1999). Activation of Heschl's gyrus during auditory hallucinations. *Neuron*, **22**, 615–21.

Dix MR, Hood JD (1973). Symmetrical hearing loss in brainstem lesions. *Acta Otolaryngol*, **75**, 1677.

Downs MP, Yoshinago-Itano C (1999). The efficacy of early identification and intervention for children with hearing impairment. *Paediatric Clinics of North America*, **46**, 79–87.

Doyle KJ, Bauch C, Battista R *et al.* (2004). Intratympanic steroid treatment: a review. *Otol Neurotol*, **25**, 1034–9.

Duane BB (1977). *Central Auditory Dysfunction*, Keith RW, ed. Grune and Stratton, New York.

Duvoisin RC (1987). The olivopontocerebellar atrophies. In Marsden CD, Fahn S, eds. *Movement Disorders II*, pp. 249–71. Butterworth Publishers, London.

Ell J, Prasher D, Rudge P (1984). Neuro-otological abnormalities in Friedreich's ataxia. *J Neurol Neurosurg Psychiatry*, **47**, 26–32.

Evers S, Ellger T (2004). The clinical spectrum of musical hallucinations. *J Neurol Sci*, **227**, 55–65.

Fellinger J, Holzinger D, Dobner U *et al.* (2005). Mental distress and quality of life in a deaf population. *Soc Psychiatry Psychiatr Epidemiol*, **40**, 737–42.

Fortnum H, Davis A (1993). Hearing impairment in children after bacterial meningitis: Incidence and resource implications. *Br J Audiol*, **27**, 43–52.

Fortnum HM, Summerfield AQ, Marshall DH *et al.* (2001). Prevalence of permanent hearing impairment in the United Kingdom and implications for universal neonatal hearing screening: questionnaire based ascertainment study. *Br Med J*, **323**, 536–40.

Garringer HJ, Pankratz ND, Nichols WC *et al.* (2006). Hearing impairment susceptibility in elderly men and the DFNA18 locus. *Arch Otolaryngol Head Neck Surg*, **132**, 506–10.

Gatehouse S (2000). The Glasgow hearing aid benefit profile and what it measures and how to use it. *The Hearing Journal*, **53**, 10–8.

Gatehouse S (2003). Auditory amplification in adults. In Luxon LM, Martini A, Furman J, Stephens SDG, eds. *A Textbook of Audiological Medicine*, pp. 533–53. Martin Dunitz, London.

Gates GA (2005). Treatment of Meniere's disease with the low-pressure pulse generator (Meniett device). *Expert Review of Medical Devices*, **2**, 533–7.

Gates GA (2006). Meniere's disease review 2005. *J Am Acad Audiol*, **17**, 16–26.

Ghosh D, Mohanty G, Prabhakar S (2001). Ictal deafness—a report of three cases. *Seizure*, **10**, 130–3.

Ghosh P (1990). Central diplacusis. *Eur Arch Otorhinolaryngol*, **247**, 48–50.

Giebink S (1999). Prevention. In Rosenfeld R, Bluestone C, eds. *Evidence-based Otitis Media*, pp. 223–34. BC Decker, Hamilton.

Gill BS, Sharma DN (1967). Level of hearing above 50 years of age. *Indian J Otolaryngol*, **19**, 112–8.

Goldstein D, Stephens SDG (1981). Audiological rehabilitation: management model. *Audiology*, **20**, 432–52.

Gonzalez-Ravilla A (1948). Differential diagnosis of tumours at the cerebellar recess. *Bull Johns Hopkins Hosp*, **83**, 187–212.

Gozdzick-Zolnierkiewicz T, Moszynski B (1969). VIII nerve in experimental lead poisoning. *Acta Otolaryngol*, **68**, 85–9.

Gulya AJ (1993). Neurologic paraneoplastic syndromes with neurotologic manifestations. *Laryngoscope*, **103**, 754–61.

Haberkamp TJ, Tanyeri M (1999). Management of idiopathic sensorineural hearing loss. *Am J Otol*, **20**, 587–93.

Haggard M (2004). *Presentation to Children's Audiology Services meeting*, RCPCH.

Haggard M, Hughes E (1991). *Screening Children's Hearing*, HMSO, London.

Hall D (2004). Children's Audiology Services. www.rcpch.ac.uk/publications/recent_publications/Audiology.pdf (viewed 15.5.06).

Hanner P, Rosenhall U, Edstrom S *et al.* (1989). Hearing impairment in patients with antibody production against Borrelia burgdorferi antigen. *Lancet*, **7**, 13–5.

Harding AE (1981a). Friedreich's ataxia: a clinical and genetic study of 90 families with an analysis of early diagnostic criteria and intrafamilial clustering of clinical features. *Brain*, **104**, 589–620.

Harding AE (1981b). 'Idiopathic' late onset cerebellar ataxia. A clinical and genetic study of 36 cases. *J Neurol Sci*, **51**, 259–71.

Harding AE (1982). The clinical features and classification of the late onset autosomal dominant cerebellar ataxias. A study of 11 families, including descendants of the 'the Drew family of Walworth'. *Brain*, **105**, 1–28.

Heller MF, Bergmann M (1953). Tinnitus in normally hearing persons. *Ann Otol*, **62**, 73–83.

Hereditary Hearing Loss Homepage http://webhost.ua.ac.be/hhh/ viewed 2.6.06.

Hinchcliffe R (2002). Aspects of paracuses. In Luxon L, ed. *Textbook of Audiological Medicine*, pp. 271–87. Martin Dunitz, London.

Hinchcliffe R, King P (1992). Medicolegal aspects of tinnitus. I: Medicolegal position and current state of knowledge. *Journal of Audiological Medicine*, **1**, 38–58.

Hindley P (1997). Psychiatric aspects of hearing impairments. *J Child Psychol Psychiatry*, **38**, 101–17.

HMSO (HM Stationery Office) Health Survey for England (1997). Department of Health, London.

Holgers K-M, Barrenäs M-L (2003). The pathophysiology and assessment of tinnitus. In Luxon LM, Furman JM, Martini A, Stephens SDG, eds. *Textbook of Audiological Medicine*, pp. 555–70. Martin Dunitz, London.

Holt JJ (2003). Cholesteatoma and otosclerosis: two slowly progressive causes of hearing loss treatable through corrective surgery. *Clin Med Res*, **1**, 151–4.

Hughes GB, Freedman MA, Haberkamp TJ *et al.* (1996). Sudden sensorineural hearing loss. *Otolaryngol Clin North Am*, **29**, 393–405.

Huygen PL, Verhagen WI, Noten JF (1994). Auditory abnormalities, including 'precocious presbyacusis', in myotonic dystrophy. *Audiology*, **33**, 73–84.

Ishman SL, Friedland DR (2004). Temporal bone fractures: traditional classification and clinical relevance. *Laryngoscope*, **114**, 1734–41.

James A, Thorp M (2004). Meniere's disease. *Clin Evid*, **12**, 742–50.

Jastrehoff PJ (1994). An animal model of tinnitus: A decade of development. *Am J Otol*, **15**, 19–27.

Jastreboff PJ, Gray WC, Gold SL (1996). Neurophysiological approach to tinnitus patients. *Am J Otol*, **17**, 236–40.

Johnson EW (1968). Auditory findings in 200 cases of acoustic neuromas. *Arch Otolaryngol*, **88**, 598–604.

Jones NS, Davis A (2001). A prospective case-control study of 50 consecutive patients presenting with hyperlipidaemia. *Clin Otolaryngol*, **26**, 189–96.

Jorgensen MB, Kristensen HK, Buch NH (1964). Thalidomide-induced aplasia of the inner ear. *J Laryngol Otol*, **78**, 1095–101.

Josey AF Jackson CG, Glasscock ME 3rd (1980). Brainstem evoked response audiometry in confirmed eighth nerve tumors. *Am J Otolaryngol*, **1**, 285–90.

Kaga K, Shindo M, Tanaka Y *et al.* (2000). Neuropathology of auditory agnosia following bilateral temporal lobe lesions: a case study. *Acta Otolaryngol*, **120**, 259–62.

Kemp DT (1978). Stimulated acoustic emissions from within the human auditory system. *J Acoust Soc Am*, **64**, 1386–91.

Khoza K, Ross E (2002). Auditory function in a group of adults infected with HIV/AIDS in Gauteng, South Africa. *S Afr J Commun Disord*, **49**, 17–27.

Kileny P, Paccioretti D, Wilson AF (1987). Effects of cortical lesions on middle latency auditory evoked responses (MLR). *Electroencephalogr Clin Neurophysiol*, **66**, 108–20.

Kotnis R, Simo R (2001). Tuberculous meningitis presenting as sensorineural hearing loss. *J Laryngol Otol*, **115**, 491–2.

Kraus N, Kileny P, McGee T (1994). Middle latency auditory evoked potentials. In Katz J, ed. *Handbook of Clinical Audiology*, 4th edition, pp. 487–505. Williams & Wilkins, Baltimore.

Kuhweide R, Van de Steene V, Vlaminck S *et al.* (2002). Ramsay Hunt syndrome: pathophysiology of cochleovestibular symptoms. *J Laryngol Otol*, **116**, 844–8.

Kulahli I, Balci K, Koseoglu E *et al.* (2005). Audio-vestibular disturbances in Behcet's patients: report of 62 cases. *Hear Res*, **203**, 28–31.

Kumar A, Patni AH, Charbel F (2002). The Chiari I malformation and the neurotologist. *Otol Neurotol*, **23**, 727–35.

Kustel M, Buki B, Gyimesi J *et al.* (1993). Auditory brainstem potentials in uraemia. *ORL Journal of Otorhinolaryngology and Related Specialties*, **55**, 89–92.

Lechevalier B, Rossa Y, Eustache F *et al.* (1984). Un cas de surdité corticale épargnant en partie la musique. *Revue de Neurologie (Paris)*, **140**, 190–201.

Lee H, Baloh RW (2005). Sudden deafness in vertebrobasilar ischemia: clinical features, vascular topographical patterns and long-term outcome. *J Neurol Sci*, **228**, 99–104.

Lehnhardt E (1958). Sudden hearing disorders occurring simultaneously or successively on both sides. *Z Laryngol Rhinol Otol*, **37**, 1–16.

Leussink V, Andermann P, Reiners K *et al.* (2005). Sudden deafness from stroke. *Neurology*, **64**, 1817–8.

Linstrom CJ, Gleich LL (1993). Otosyphilis: diagnostic and therapeutic update. *J Otolaryngol*, **22**, 401–8.

Long G (1998). Perceptual consequences of the interaction between spontaneous otoacoustic emissions and external sounds. *Hear Res*, **119**, 49–60.

López-Díaz-de-León E, Silva-Rojas A, Ysunza A *et al.* (2003). Auditory neuropathy in Friedreich ataxia. A report of two cases. *Int J Pediatr Otorhinolaryngol*, **67**, 641–8.

Lous J, Burton MJ, Felding JU *et al.* (2005). Grommets (ventilation tubes) for hearing loss associated with otitis media with effusion in children. *Cochrane Database Syst Rev*, **25**, CD001801.

Low WK, Gopal K, Goh LK *et al.* (2006a) Cochlear implantation in postirradiated ears: outcomes and challenges. *Laryngoscope*, **116**, 1258–62.

Low WK, Toh ST, Wee J *et al.* (2006b). Sensorineural hearing loss after radiotherapy and chemoradiotherapy: a single, blinded, randomized study. *J Clin Oncol*, **24**, 1904–9.

Ludman H (1998). Basic acoustics and hearing tests. In Ludman H, Wright T, eds. *Diseases of the Ear*, 6th edition, pp 58–86. Arnold, London.

Lunsford LD, Niranjan A, Flickinger JC *et al.* (2005). Radiosurgery of vestibular schwannomas: summary of experience in 829 cases. *J Neurosurg*, **102** (Suppl), 195–9.

Luxon LM (1988). Methods of examination: audiological and vestibular. In Ludman H, Mawson S, eds. *Diseases of the Ear*, 5th edition. Oxford University Press, Oxford.

Luxon LM (1998). Clinical diagnosis of noise induced hearing loss. In Prasher DK, Luxon LM, eds. *Advances in Noise Research Vol l: Biological Effects*. Whurr Publishers, London.

Luxon LM, Cohen M (1997). Central auditory dysfunction. In *Scott Brown's Diseases of the Ear Nose and Throat*, Vol. 9, 6th edition. Eds: Kerr A G: Adult Audiology Ed: Stephens S D G Butterworths, London.

Luxon LM, Raglan E (2006). Deafness and tinnitus. In Noseworthy JH, ed. *Neurological Therapeutics: Principles and Practice*, 2nd edition, pp. 2157–81. Informa Healthcare Abingdon, UK.

Maassen JA (2002). Mitochondrial diabetes: pathophysiology, clinical presentation, and genetic analysis. *Am J Med Genet*, **115**, 66–70.

Magnan J, Manrique M, Dillier N *et al.* (2005). International consensus on middle ear implants. *Acta Otolaryngol*, **125**, 920–1.

Mahajan SL, Ikeda Y, Myers TJ *et al.* (1981). Acute acoustic nerve palsy associated with vincristine therapy. *Cancer*, **47**, 2404–6.

Maia CA, Campos CA (2005). Diabetes mellitus as etiological factor of hearing loss. *Review Brazilian Otorrinolaringologica*, **71**, 208–14.

Majamaa-Voltti KA, Winqvist S, Remes AM *et al.* (2006). A 3-year clinical follow-up of adult patients with 3243A>G in mitochondrial DNA. *Neurology*, **66**, 1470–5.

Martini A, Prosser S (2003). Disorders of the inner ear in adults In Luxon LM, Martini A, Furman J, Stephens SDG, eds. *A Textbook of Audiological Medicine*, pp. 451–75. Martin Dunitz, London.

Mattox DE, Simmons FB (1977). Natural history of sudden sensorineural hearing loss. *Ann Otol Rhinol Laryngol*, **86**, 463–80.

McCabe BF (1979). Autoimmune sensorineural hearing loss. *Ann Otol Rhinol Laryngol*, **88**, 585–90.

McDonald TJ, DeRemee RA (1983). Wegener's Granulomatosis. *Laryngoscope*, **93**, 220–31.

Meneguello J, Leonhardt FD, Pereira LD (2006). Auditory processing in patients with temporal lobe epilepsy. *Rev Bras Otorinolaringol* (English ed.), **72**, 496–504.

Menger DJ, Tange RA (2003). The aetiology of otosclerosis: a review of the literature. *Clin Otolaryngol*, **28**, 112–20.

Miller DH, Rudge P, Johnson G *et al.* (1988). Serial gadolinium enhanced magnetic resonance imaging in multiple sclerosis. *Brain*, **111**, 927–39.

Minor LB, Schessel DA, Carey JP (2004). Meniere's disease. *Curr Opin Neurol*, **17**, 9–16.

Mirza S, Richardson H (2005). Otic barotrauma from air travel. *J Laryngol Otol*, **119**, 366–70.

Møller AR (1984). Pathophysiology of tinnitus. *Ann Otol Rhinol Laryngol*, **93**, 39–44.

Morgan-Hughes JA, Hayes DJ, Clark JB (1982). Mitochondrial encephalomyopathies: biochemical studies in two cases revealing defects in the respiratory chain. *Brain*, **105**, 553–82.

Morton CC, Nance WE (2006). Newborn hearing screening - a silent revolution. *New Engl J Med*, **354**, 2151–64.

Musiek FE, Baran JA, Pinheiro ML (1994). *Neuroaudiology: Case Studies*. Singular Publishing Group, San Diego.

Musiek FE, Charette L, Kelly T *et al.* (1999). Hit and false-positive rates for the middle latency response in patients with central nervous system involvement. *J Am Acad Audiol*, **10**, 124–32.

Musiek FE, Kibbe K, Baran JA (1984). Neuroaudiological results from split-brain patients. *Seminars in Hearing*, **5**, 219–29.

Musiek FE, Lee WW (1995). The auditory brainstem response in patients with brainstem and cochlear pathology. *Ear Hear*, **16**, 631–6.

Musiek FE, Lee WW (1998). Neuroanatomical correlates to central deafness. *Scand Audiol*, **27**, 18–25.

Nance WE (2003). The genetics of deafness. *Ment Retard Dev Disabil Res Rev*, **9**, 109–19.

Newton HB (2001). Neurologic complications of scuba diving. *Am Fam Physician*, **63**, 2211–8.

Noback CR, Demarest RJ (1981). *The Human Nervous System*, 3rd edition. McGraw Hill, New York.

Nowe V, De Ridder D, Van de Heyning PH *et al.* (2004). Does the location of a vascular loop in the cerebellopontine angle explain pulsatile and non-pulsatile tinnitus? *Eur Radiol*, **14**, 2282–9.

Olusanya,B, Luxon LM, Wirz S (2005). Screening for early childhood hearing loss in Nigeria. *J Med Screen*, **12**, 115–8.

Pan CL, Kuo MF, Hsieh ST (2004). Auditory agnosia caused by a tectal germinoma. *Neurology*, **63**, 2387–9.

Pandolfo M (1999). Molecular pathogenesis of Friedreich ataxia. *Arch Neurol*, **56**, 1201–8.

Paparella MM, Berlinger NT, Oda M *et al.* (1973). Otological manifestations of leukemia. *Laryngoscope*, **83**, 1510–26.

Peltomaa M, Pyykko I, Sappala I *et al.* (2000). Lyme borreliosis, an etiological factor in sensorineural hearing loss? *Eur Arch Otorhinolaryngol*, **257**, 317–22.

Penfield W, Rasmussen T (1950). *The Cerebral Cortex of Man*. The MacMillan Co, New York.

Penner MJ (1990). An estimate of the prevalence of tinnitus caused by spontaneous otoacoustic emissions. Arch Otolaryngol Head Neck Surg, **116**, 418–23.

Peretz I, Babai M, Lussier I *et al.* (1995). Corpus d'extraits musicaux:indices relatifs à la familiarité, à l'age acquisition et aux evocations verbales. Canadian Journal of experimental psychology. *Can J Exp Psychol*, **49**, 211–39.

Persaud R, Hajioff D, Wareing M *et al.* (2003). Otological trauma resulting from the Soho Nail Bomb in London, April 1999. *Clin Otolaryngol*, **28**, 203–6.

Pickles JO (2004). Mutation in mitochondrial DNA as a cause of presbyacusis. *Audiol Neurootol*, **9**, 23–33.

Pickles JO (2007). Physiology of the auditory system. In Gleason M, ed. *Scott Brown's Otolaryngology*, 7th edition. Hodder, London.

Pollak l, Haskard D, Luxon LM (2001). Labyrinthine involvement in Behcet Syndrome. *J Laryngol Otol*, **115**, 522–9.

Prasher D, Ryan S, Luxon LM (1994). Contralateral suppression of transient evoked otoacoustic emissions in neuro-otology. *Br J Audiol*, **28**, 247–54.

Prezant TR, Agapian JV, Bohlman MC *et al.* (1993). Mitochondrial ribosomal RNA mutation associated with both antibiotic-induced and non-syndromic deafness. *Nat Genet*, **4**, 289–94.

Pulec JL, House WF, Britten BH Jr (1971). A system of management of acoustic neuroma based on 364 cases. *Trans Am Acad Ophthalmol Otolaryngol*, **75**, 48–55.

Raglan E (2003). Otitis media with effusion in children. In Luxon LM, Martini A, Furman J, Stephens SDG, eds. *A Textbook of Audiological Medicine*, pp. 381–92. Martin Dunitz, London.

Raglan E, Prasher DK, Trinder E *et al.* (1987). Auditory function in hereditary motor and sensory neuropathy (Charcot-Marie-Tooth disease). *Acta Otolaryngol*, **103**, 50–5.

Rance G, Beer DE, Cone-Wesson B, Shepherd, RK *et al.* (1999). Clinical findings for a group of infants and young children with auditory neuropathy. *Ear Hear*, **20**, 238–52.

Ringelstein EB, Knecht S (2006). Cerebral small vessel diseases: manifestations in young women. *Curr Opin Neurol*, **19**, 55–62.

Roberts DL, Tatini U, Zimmerman RS *et al.* (2001). Musical hallucinations associated with seizures originating from an intracranial aneurysm. *Mayo Clin Proc*, **76**, 423–6.

Robinson J (1998). Reconstruction of the middle ear. In Ludman H, Wright A, eds. *Diseases of the Ear*, 6th edition, pp. 429–38. Arnold, London.

Rosenberg SI (2000). Natural history of acoustic neuromas. *Laryngoscope*, **110**, 497–508.

Rosenfeld RM, Singer M, Wasserman JM *et al.* (2006). Systematic review of topical antimicrobial therapy for acute otitis externa. *Otolaryngol Head Neck Surg*, **134**, S24–48.

Ruckenstein M (2004). Autoimmune inner ear disease. *Curr Opin Otoalryngol Head Neck Surg* **12**, 426–30.

Rungby JA, Skibsted R, Johnsen T *et al.* (1999). Hearing loss in hereditary motor and sensory neuropathy: a review. *J Audiol Med*, **8**, 131–41.

Rybak LP, Whitworth CA (2005). Ototoxity: therapeutic opportunities. *Drug Discov Today*, **10**, 1313–21.

Rydell RE, Pulec JL (1971). Arnold-Chiari malformation. Neuro-otologic symptoms. *Arch Otolaryngol*, **94**, 8–12.

Santos-Sacchi J (2001). Cochlear physiology. In Jahn AF, Santos-Sacchi J, eds. *Physiology of the Ear*, pp. 357–91. Singular, Australia.

Schuknecht HF (1969). Mechanisms of inner ear injury from blows to the head. *Ann Otol*, **78**, 253–62.

Schuknecht HF (1994). Temporal bone pathology in a case of Cogan's Syndrome. *Laryngoscope*, **104**, 1135–42.

Seidman MD, Ahmad N, Joshi D *et al.* (2004). Age-related hearing loss and its association with reactive oxygen species and mitochondrial DNA damage. *Acta Otolaryngol Suppl*, **552**, 16–24.

Semaan MT, Alagramam KN, Megerian CA (2005). The basic science of Meniere's disease and endolymphatic hydrops. *Curr Opin Otolaryngol Head Neck Surg*, **13**, 301–7.

Shergill S, Brammer M, Amaro E *et al.* (2004). Temporal course of auditory hallucinations. *Br J Psychiatry*, **185**, 516–7.

Shibata N, Kubota F, Kikuchi S (2006). The origin of the focal spike in musicogenic epilepsy. *Epileptic Disord*,**8**, 131–5

Shine NP, Coates H (2005). Systemic ototoxicity: a review. *East Afr Med J,*, **82**, 536–9.

Shulman A (1988). Introduction: Definition and classification of tinnitus. In Kitahara M, ed. *Tinnitus: Pathophysiology and Management*, pp. 1–6. Igaku-Shoin, Tokyo.

Shupak A, Gil A, Nachum Z *et al.* (2003). Inner ear decompression sickness and inner ear barotrauma in recreational divers: a long-term follow-up. *Laryngoscope*, **113**, 2141–7.

Silverstein H, Bernstein JM, Davies DG (1967). Salicylate ototoxicity. A biochemical and electrophysiological study. *Ann Otol Rhinol Laryngol*, **76**, 118–28.

Sismanis A (2005). Diagnostic and management dilemma of sudden hearing loss. *Arch Otolaryngol Head Neck Surg*, **131**, 733–4.

Smouha EE, Yoo M, Mohr K *et al.* (2005). Conservative management of acoustic neuroma: a meta-analysis and proposed treatment algorithm. *Laryngoscope*, **115**, 450–4.

Snik AF, Mylanus EA, Proops DW *et al.* (2005). Consensus statements on the BAHA system: where do we stand at present? *Ann Otol Rhinol Laryngol Suppl*, **195**, 2–12.

Starr A, Picton TW, Sininger Y *et al.* (1996). Auditory Neuropathy. *Brain*, **119**, 741–53.

Steel KP (1998). New interventions in hearing impairment. *Science*, **279**, 1870–1.

Stephens SDG, Luxon LM, Hinchcliffe R (1982). Immunological disorders and auditory lesions. *Audiology*, **21**, 128–48.

Stephens SDG (2003). Audiological rehabilitation. In Luxon LM, Martini A, Furman J, Stephens SDG, eds. *A Textbook of Audiological Medicine*, pp. 451–75. Martin Dunitz, London.

Suga N, Gao E *et al.* (2000). The corticofugal system for hearing: recent progress. *Proc Natl Acad Sci USA*, **97**, 11807–14.

Tanaka Y, Kamo T, Yoshida M *et al.* (1991). So called cortical deafness: clinical neurophysiological and radiological observations. *Brain*, **114**, 2385–2340.

Tatagiba M, Gharabaghi A (2005). Electrically evoked hearing perception by functional neurostimulation of the central auditory system. *Acta Neurochir Suppl*, **93**, 93–5.

Tjell C, Tenenbaum A, Rosenhall U (1999). Auditory function in whiplash-associated disorders. *Scand Audiol*, **28**, 203–9.

Toglia JU, Katinsky S (1976). Neuro-otological aspects of closed head injury. In *Handbook of Clinical Neurology*, pp. 119–40. Elsevier, New York.

Torre P 3rd, Cruickshanks KJ, Klein BE *et al.* (2006). The association between cardiovascular disease and cochlear function in older adults. *J Speech Lang Hear Res*, **48**, 473–81.

van de Beek D, de Gans J, McIntyre P *et al.* (2003). Corticosteroids in acute bacterial meningitis. *Cochrane Database Syst Rev*, 3, CD004405.

Van de Heyning PH, Wuyts F, Boudewyns A (2005). Surgical treatment of Meniere's disease. *Curr Opin Neurol*, **18**, 23–8.

van der Putten L, de Bree R, Plukker JT *et al.* (2006). Permanent unilateral hearing loss after radiotherapy for parotid gland tumors. *Head Neck*, **28**, 902–8.

Varga R, Kelley PM, Keats BJ *et al.* (2003). Non-syndromic recessive auditory neuropathy is the result of mutations in the otoferlin (OTOF) gene. *Am J Med Genet*, **40**, 45–50.

Verhagen WIM, ter Bruggen JP, Huygen PLM (1992). Oculomotor, auditory, and vestibular responses in myotonic dystrophy. *Arch Neurol*, **49**, 954–60.

Verhagen WIM, Huygen PLM, Gabreëls-Festen AA *et al.* (2005). Sensorineural hearing impairment in patients with PMP22 duplication, deletion, and frameshift mutations. *Otol Neurotol*, **26**, 405–14.

Voit T, Lamprecht A, Lenard HG *et al.* (1986). Hearing loss in facioscapulohumeral dystrophy. *Eur J Pediatr*, **145**, 280–5.

Wang Q, Li R, Zhao H *et al.* (2005). Clinical and molecular characterization of a Chinese patient with auditory neuropathy associated with mitochondrial 12S rRNA T1095C mutation. *Am J Med Genet*, **133**, 27–30.

Warr WB (1992). Organization of the olivocochlear efferent systems in mammals. In Webster DB, Popper AN, Fay RR, eds. *Mamalian Auditory Pathway: Neuroanatomy*, pp. 410–48. Springer-Verlang, New York.

Wayman DM, Pham HN, Byl FM *et al.* (1990). Audiological manifestations of Ramsay Hunt syndrome. *J Laryngol Otol*, **104**, 104–8.

Weinstein BE (1994). Presbycusis. In Katz J, ed. *Handbook of Clinical Audiology*, 4th edition, pp. 553–67. Williams & Wilkins, Baltimore.

Wengel SP, Burke WJ, Holeman D (1989). Musical hallucinations. The sources of silence? *J Am Geriatr Soc*, **37**, 163–6.

Wennmo C, Svensson C (1989). Temporal bone fractures. *Acta Otolaryngol Suppl*, **468**, 379–83.

Westmore CA, Pickard BH, Stern H (1979). Isolation of mumps virus from the inner ear after sudden deafness. *Br Med J*, **655**, 14–5.

WHO (2006). http://www.who.int/mediacentre/factsheets/fs300/en/index.html viewed 1.6.06.

Wirkowski E, Echausse N, Overby C *et al.* (2006). I can hear you yet cannot comprehend: a case of pure word deafness. *J Emerg Med*, **30**, 53–5.

Ylikoski JS, House JW, Hernandez I (1981). Eighth nerve alcoholic neuropathy: a case report with light and electron microscopic findings. *J Laryngol Otol*, **95**, 631–42.

Yohay KH (2006). The genetic and molecular pathogenesis of NF1 and NF2. *Semin Pediatr Neurol*, **13**, 21–6.

Youngs R (1998). Complications of suppurative otitis media In Ludman H, Wright T, eds. *Diseases of the Ear*, 6th edition, pp. 394–416. Arnold, London.

Zadeh MH, Storper IS, Spitzer JB (2003). Diagnosis and treatment of sudden-onset sensorineural hearing loss: a study of 51 patients. *Otolaryngol Head Neck Surg*, **128**, 92–8.

Zeller JA, Zunker P, Witt K *et al.* (2002). Unusual presentation of carcinomatous meningitis: case report and review of typical CSF findings. *Neurol Res*, **24**, 652–4.

Zwirner P, Wilichowski E (2001). Progressive sensorineural hearing loss in children with mitochondrial encephalomyopathies. *Laryngoscope*, **111**, 515–21.

CHAPTER 15

Vertigo and imbalance

Linda Luxon

Contents

15.1 Introduction

The mechanism for maintaining balance in man is complex. Vision, proprioception, and vestibular inputs are integrated in the central nervous system, and modulated by activity from the cerebellum, the extrapyramidal system, the reticular formation, and the cortex (Fig. 15.1). This integrated, modulated information provides one mechanism for control of oculomotor activity, controls posture, gait, and motor skills and allows perception of the head and body in space. Recent evidence also supports an effect upon autonomic function, cognition, and emotion. The complexity of the system is such that pathology in a variety of different bodily systems, including the endocrine system, the cardiovascular system, and the haemopoietic system, can impact upon vestibular activity, in addition to primary otological and neurological pathology.

Patients with dysfunction in the vestibular end-organs or vestibular pathways commonly complain of symptoms of dizziness, vertigo, unsteadiness, light-headedness, imbalance, and a plethora of synonyms associated with a sense of instability. Not infrequently, in an attempt to define their 'unphysiological' experience, patients use rather vague and imprecise semantics. The clinical distinction between dizziness, a symptom of non-specific pathological significance, and vertigo, a hallucination or illusion of movement, is rarely made, although the latter is a cardinal manifestation of a disorder of the vestibular system (Dix 1973). Ten to 20 per cent of all 'dizzy' patients are reportedly seen in neurology clinics (Dieterish 2004), therefore it behoves the neurologist to have a clear diagnostic strategy, including knowledge of detailed neuro-otological examination, to enable appropriate diagnosis and management of the patient with vestibular symptoms.

15.1.1 Prevalence and incidence

The common symptoms of vertigo, dizziness, and disequilibrium represent 5 to 10 percent of all patients seen in general practice and 10 to 20 per cent of all patients seen by neurologists and otolaryngologists (Dieterish 2004). Yardley and co-workers (1998) in a community study of a random sample registered in four north London general practices, identified that more than one in five of the population had experienced dizziness in the past month, and 30 per cent had suffered symptoms of dizziness for more than 5 years. The German Health Questionnaire Study (Neuhauser *et al.* 2005)

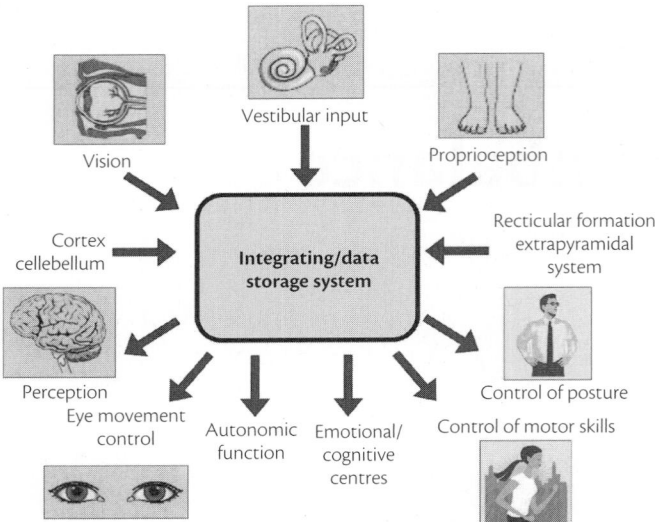

Fig. 15.1 Mechanism of balance in man.

reported vestibular vertigo as common in the general population affecting more than 5 per cent of adults in a year, and estimated a life-time prevalence of vestibular vertigo at 7.8 per cent. They concluded that the frequency and healthcare impact of vestibular symptoms at the population level have been underestimated. One-third of the population has suffered symptoms of balance disorder by the age of 65 years (Roydhouse 1974), while by the age of 88 to 90 years, 45 to 50 per cent of the population suffer symptoms of balance dysfunction (Jonsson *et al.* 2004). In children, balance symptoms are generally considered to be less frequent, but a recent study has reported that 8 per cent of 1- to 15-year-olds in the general population have suffered dizziness and vertigo, and of these, 27 per cent reported symptoms which were sufficiently severe to interrupt activity (Niemensivu *et al.* 2006).

15.1.2 **Social and economic consequences**

Balance disorders in children are frequently ignored, as they are dismissed as problems of behaviour or lack of coordination (Tusa *et al.* 1994). However the failure to diagnose vestibular pathology and the paucity of vestibular information related to children may be explained by:

◆ Variable presentations of vestibular dysfunction, for example cyclical vomiting

◆ Inability of infants to describe their symptoms

◆ Lack of centres with paediatric vestibular expertise

◆ Difficulty in interpreting vestibular results, consequent upon variable maturation

◆ The wide range of normality and

◆ The misattribution of symptoms to behavioural problems.

Notwithstanding this, undiagnosed vestibular disorders in children are associated with psychological dysfunction, loss of time from school, and considerable family anxiety.

In adults, peripheral vestibular disorder of sufficient severity for the patient to present to a tertiary neuro-otological service led to

85 per cent of the population taking an average of nine months off work, despite the mean age of the patients being only 38 years (Eagger *et al.* 1992). The same study reported that two-thirds of patients in this group had suffered psychiatric symptoms of depression, anxiety, panic attacks, and avoidance behaviour in a three-to four-year review period.

Moreover, vestibular symptoms are the commonest cause of failure to return to work after head or whiplash injury (Luxon 1996), and Neuhauser and colleagues (2004) reported that 80 per cent of individuals with vertigo suffered interruption of daily activities or sick leave, and required a medical consultation. Thus both the occupational and healthcare economic impact of this symptom complex are significant.

In terms of healthcare expense, the low level of training in vestibular medicine leads to multiple medical attendances prior to diagnosis. Of patients seen in a London teaching hospital neuro-otology clinic, 63 per cent had had two or more previous specialist consultations prior to a neuro-otological diagnosis. The National Institutes of Health have reported that a patient with peripheral vestibular pathology visits a mean of 4.5 physicians prior to receiving a correct diagnosis. The poor diagnostic pathways for vertigo lead to unwarranted use of expensive investigations such as MRI (Halmagyi *et al.* in press; MacDonald and Melhem 1997).

The cost to the Health Service of disorders of balance in the elderly is still greater. In a typical primary care trust catchment of 150,000 population, 30,000 will be over the age of 65, and one-third of this group will fall each year, with half of them falling repeatedly, i.e. 5,000 high-risk fallers (American Geriatric Society, British Geriatric Society and American Academy of Orthopaedic Surgeons 2001). Eighty per cent of older people presenting to an emergency department with an unexplained fall had symptoms of vestibular impairment (Pothula *et al.* 2004); it is well recognized that falls are the commonest cause of accidental death in over 75-year-olds (Downton 1993). Thus in the elderly, the social and economic impact of vestibular dysfunction is high.

An interesting study by one US airline between 1995 and 2000 illustrated that neurological symptoms comprised the single largest category of in-flight incidents requiring medical consultation (Sirven *et al.* 2002). Of these 2042 incidents, just over 15 per cent resulted in unscheduled emergency landings or diversions with 'dizziness and vertigo' being the commonest precipitants in the

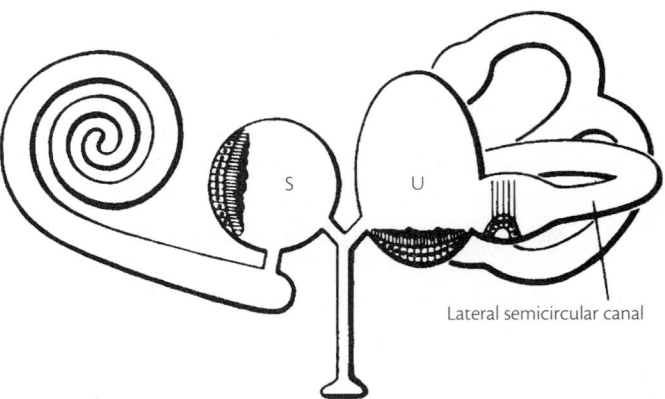

Fig. 15.2 Schematic drawing of the inner ear. (From Frenzel H 1955.)

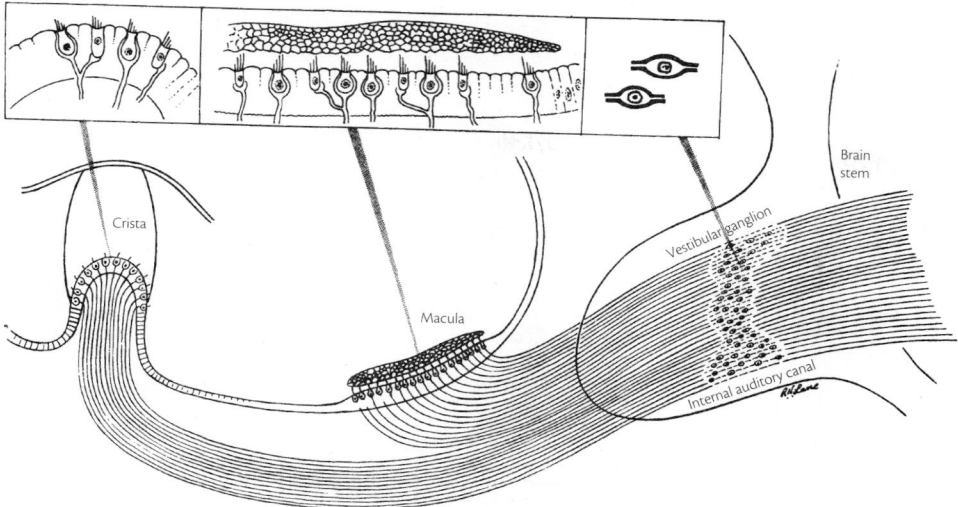

Fig. 15.3 Diagram of the vestibular receptor organ. (From Ballantyne 1979.)

neurological category, with an estimated total cost of approximately three million US dollars.

15.2 Anatomy and physiology

15.2.1 Vestibular system

The internal ear comprises the membranous labyrinth, which lies within the bony labyrinth, buried within the petrous temporal bone. The inner ear includes both the cochlea—the auditory end-organ (Section 14.2)—and five vestibular receptors, one in each of the three semi-circular canals, and two otolith organs, the utricle and saccule, within the vestibule (Fig. 15.2). The vestibular sensory epithelium is comprised of type 1 and type 2 hair cells, together with supporting cells, which are covered with a gelatinous membrane, the cupula, in the semi-circular canals, and the otolithic membrane, which in addition contains calcium carbonate crystals termed otoconia, in the saccule and utricle (Fig. 15.3).

The vestibular receptors, or cristae, of the semi-circular canals respond to angular acceleration in the three planes of space, but are insensitive to gravity or head position. The otolith organs, the saccule and utricle, lie approximately vertically and horizontally within the vestibule, and respond to linear head accelerations. A force parallel to the surface of the sensory epithelium of each of the vestibular receptors provides the maximal stimulus.

Physiologically, the vestibular system on either side of the head works in parallel in a push–pull mechanism Thus, when the head is turned to the right, the right horizontal semi-circular canal increases its firing rate, whereas the left decreases. This excitation/inhibition in vestibular neural activity is transmitted to the vestibular nuclei, modulated specifically by the cerebellum which controls the amplitude and timing of movements, and passes to the thalamus, and from thence to the parieto-insular vestibular cortex.

A data bank of vestibular, visual, and proprioceptive signals, associated with every movement, is established in the reticular formation of the brain stem (Wuyts and Boudewyns in press). The sensory inputs generated by movements are compared against this database. Under normal circumstances there is no discrepancy, and

reflex oculomotor and motor responses result in appropriate eye and body movement at a subconscious level (Fig. 15.4). Mismatch of data because of an abnormal sensory input may lead to the generation of a perception of imbalance, an abnormal vestibulo-ocular reflex response manifesting as nystagmus, or abnormal vestibulo-spinal activity with falling to one side or veering when walking.

In addition, there is extensive convergence of vestibular and autonomic afferent information in the brain stem and cerebellum, which allows for coordination of the motor response with autonomic responses during movement or changes in posture (Pagarkar and Luxon in press). Moreover, connections at various levels of the central vestibular system with the locus coeruleus, the limbic system and other brain regions which control affective responses, mood, and arousal may underlie the observed overlap between psychiatric and vestibular disorders (Balaban 2002; Furman *et al.* 2005).

15.2.2 Balance

Balance is maintained as a consequence of the postural body schema, which is an internal representation of body posture, including body geometry, movements, and orientation with respect to gravity. The body is orientated with respect to the upright posture by the gravity receptors in the labyrinth, vision, and possibly body gravity receptors. Postural tone depends not only on tonic labyrinthine input, which regulates the tonic activity of the postural extensor muscles, controlling the joints of the limbs, and orientating the head in space, but is also modulated through the myotatic reflex loop, neck reflexes, lumbar reflexes, and positive supporting reactions. Vision and vestibular inputs enable orientation of the head in space, while somatosensory receptors provide information with respect to head and body orientation. The multiple sensory inputs used for balance are integrated within the central nervous system to interpret the body's orientation and allow preservation of balance with dynamic equilibrium. As with the efferent copy of sensory inputs highlighted above, the information for balance is compared to the internal model of the body such that motor commands are generated to maintain and regain body equilibrium.

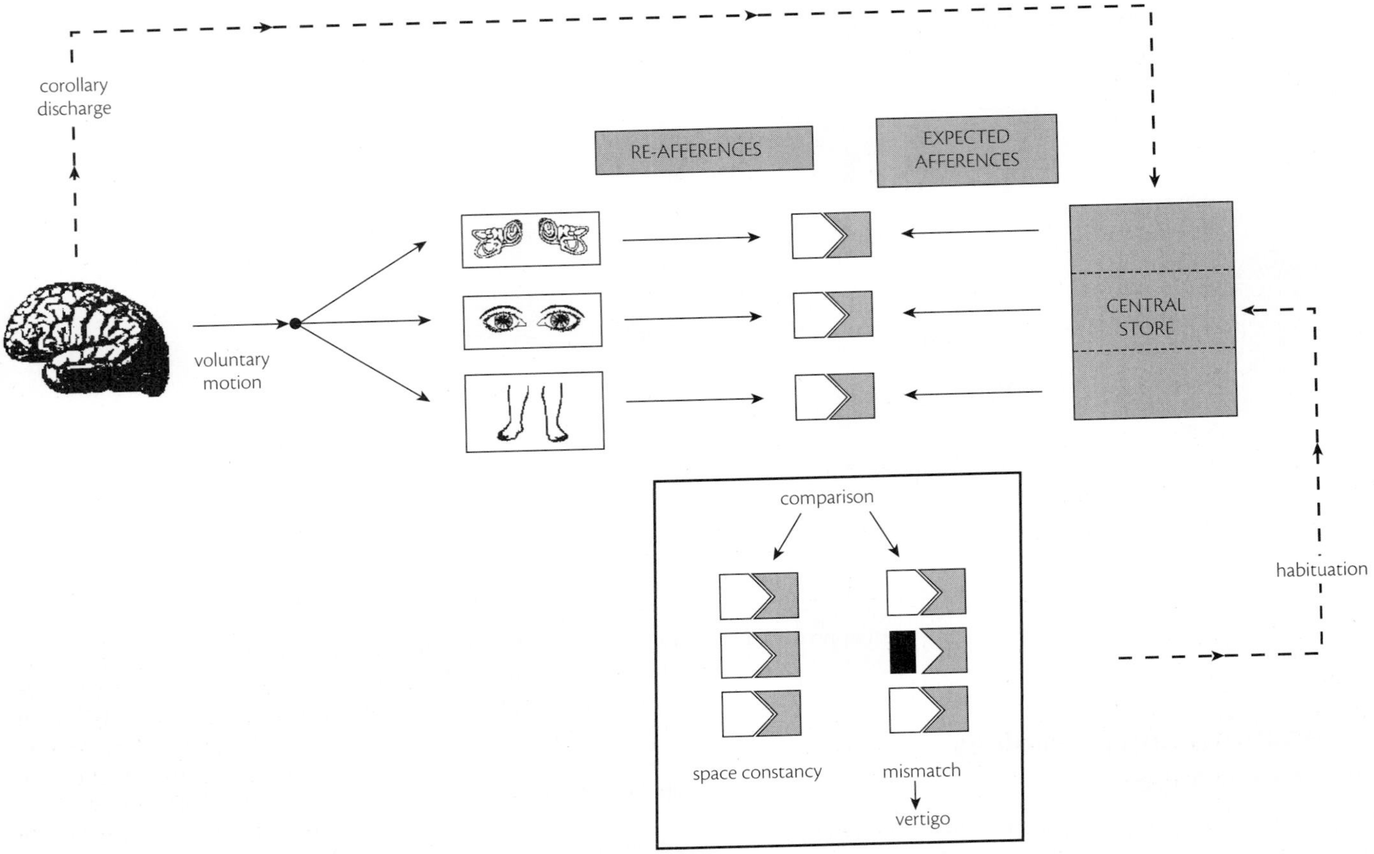

Fig. 15.4 Schematic diagram of the sensory conflict or neural mismatch concept of vertigo. (From Brandt, 1999.)

To remain upright in the presence of gravity, the central nervous system regulates the relationship between the centre of mass of the body and the base of support, the feet (Horstmann and Dietz 1990). Stability demands that two conditions are met:

- The centre of mass must be positioned over the feet, and the outline of the area of the feet represents the static stability limits. The maintenance of balance can be modelled about the body as an inverted pendulum, with rotation at the ankle joint (Fig. 15.5).

- The control of the momentum associated with movement of the centre of mass is also effective.

A fall will result when there is a disturbance of postural equilibrium, for example a perturbation that precipitates the loss of balance and failure of the balance control system to compensate adequately for that perturbation (Maki and McIlroy 2003).

Postural control is not organized as a single unit, but independent control of the position and orientation of the head, the trunk and forearm segments have been shown to exist. Movements which destabilize posture have been shown to be preceded by activation of postural muscles known as anticipatory postural adjustments. So in addition to feedback processes, the postural system is also supported by anticipatory actions which bring about displacements of the centre of body mass to meet environmental conditions (Massion and Woollacott 1996). Balance control can be improved by training and learning, and varies widely between normal healthy subjects.

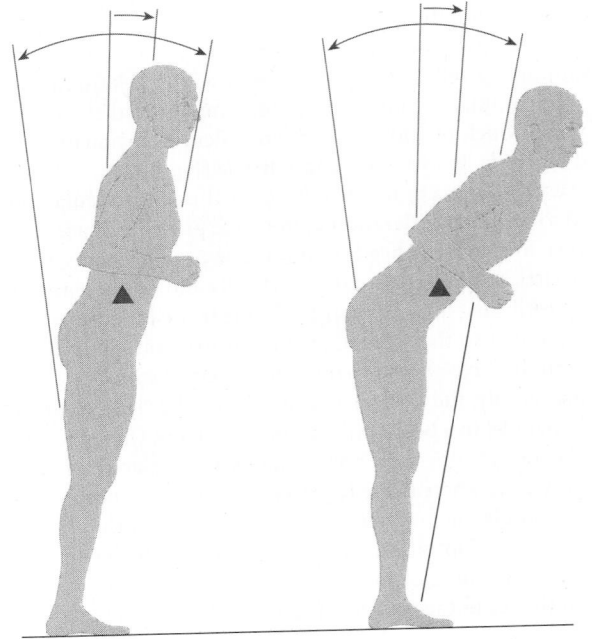

Fig. 15.5 Centre of gravity sway angle in relation to 'limits of stability' cone. The figure on the left is moving about the ankles, whereas the figure on the right is moving about the hips. The triangles represent the body centre of gravity positions. (From Nashner 1996.)

15.2.3 **Mechanisms of vestibular symptoms**

Physiological. Dizziness is a normal response to stimulation of the vestibular apparatus or as a consequence of unfamiliar visual, somatosensory, and vestibular interactions that underpin normal balance. For example a child on a roundabout or a swing may feel disorientated, an elderly person altering their visual input with new glasses may feel disorientated and a passenger in a train watching the train next to them move, may feel that they are moving rather than the adjacent train.

Moreover, motion sickness, a sensation of nausea together with vomiting precipitated by unfamiliar and unusual motion with atypical sensory inputs, for example, reading during a car journey or riding on an elephant, or travelling in a boat are all well recognized physiological syndromes of vestibular stimulation, with 'mismatch' of the sensory signals for balance (Brandt 1999a). Space sickness, in which the normal gravitational force on the otoliths is absent results in a 'mismatch' of semicircular canal and otolith inputs giving rise to a similar motion sickness phenomenon (Lackner and Graybiel 1986).

Pathological mechanisms giving rise to the symptom complexes characteristic of vestibular dysfunction can be broadly divided into 'vestibular', 'medical', and 'neurological' causations.

Vestibular vertigo

Following an acute unilateral vestibular deafferentation, as may occur with many pathologies, the patient presents with acute vertigo, nausea, vomiting, and ataxia, but is gradually rendered asymptomatic over a period of approximately 6 weeks to 6 months, by mechanisms collectively known as cerebral compensation. The structures underpinning compensation include the brain stem, cerebellum, and cortical structures (Gonshor and Melvill Jones 1976; Ito 1984; Smith and Curthoys 1989; Curthoys and Halmagyi 1995). In addition, for optimal recovery, all the sensory inputs, including vision, somatosensory afferents, and remaining labyrinthine function, are required (Lacour and Xerri 1981; Luxon 1997; Halmagyi *et al.* 2003a). Furthermore, integrity of both the

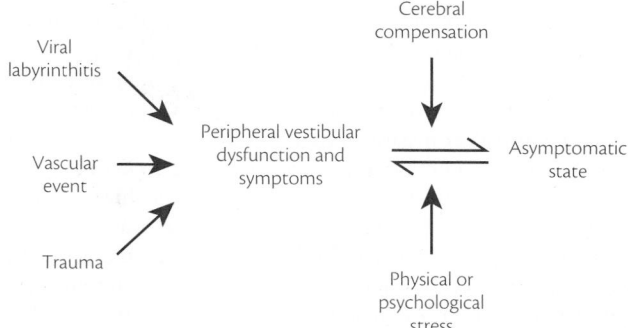

NATURAL HISTORY OF PERIPHERAL VESTIBULAR PATHOLOGY

Fig. 15.7 The balance between compensation and decompensation resulting from a peripheral vestibular disorder (From Luxon LM 1997).

vestibular nerve (Cass and Goshgarian 1991) and of the central vestibular connections is necessary (Petrone *et al.* 1991).

Physiologically, compensation depends upon physical activity (Lacour *et al.* 1976; Igarashi *et al.* 1981) and vision (Courjon *et al.* 1977). The effect of abolition of movement using a plaster cast immediately after vestibular neurectomy in baboons delays the recovery of balance (Fig. 15.6). In addition, it has been demonstrated that the occipital lobe is necessary for compensation (Fetter *et al.* 1988), and loss of proprioception by transection of the cervical spinal cord also delays vestibular recovery (Schaeffer and Meyer 1973).

The majority of cases of unilateral peripheral vestibular dysfunction recover effectively and spontaneously by cerebral compensation (Fig. 15.7). However, a percentage of patients fail to improve spontaneously, and require vestibular rehabilitation with physiotherapy. This intervention relies upon promoting visual, proprioceptive, and vestibular stimulation by means of systemic or customized exercises. A number of factors have been identified, which predispose to failure of compensation (Fig. 15.8) or decompensation

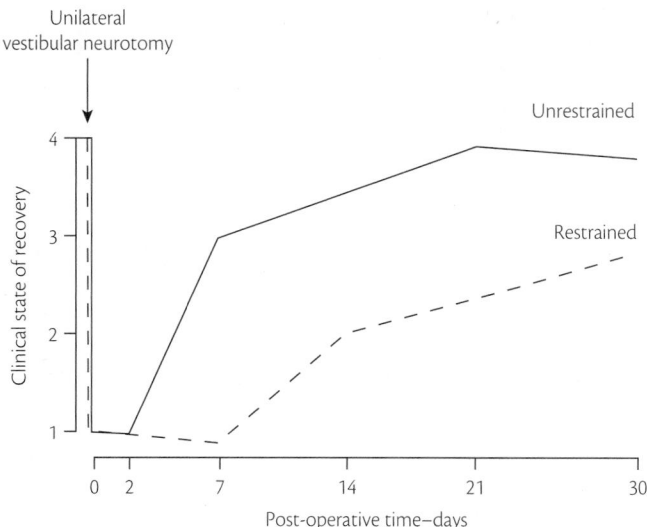

Fig. 15.6 Functional balance recovery in restrained and unrestrained baboons following unilateral vestibular neurotomy. (After Lacour *et al.* 1976.)

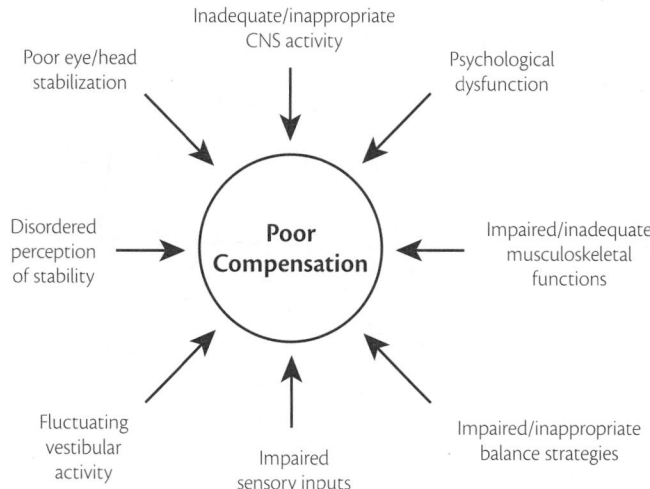

Fig. 15.8 Factors predisposing to decompensation after a peripheral vestibular disorder. (After Shumway-Cook and Horak 1998.)

from a previously recovered state. In animals, there is some evidence of a critical period following vestibular loss, during which, if appropriate stimuli are not received, recovery is delayed, and the adaptive and substitution mechanisms of compensation do not become established (Lacour 1984). Conversely, Pavlou *et al.* 2004 evaluating clinical populations have shown that neither duration of symptoms nor age have been identified as negative prognostic factors in outcome for vestibular rehabilitation programmes (Shepard *et al.* 1993). However, financial compensation, head injury, and severe postural control abnormalities were reported to be poor prognostic indicators.

Failure of compensation may follow a single acute episode of unilateral labyrinthine loss, and results in the patient presenting as a vestibular invalid with constant disorientation and instability (Fig. 15.9A). Alternatively, there may be some compensation with marked setbacks and fluctuations and little overall recovery (Fig. 15.9B). An alternative pattern of events is the patient who recovers normally but then suffers repeated episodes of decompensation (Fig. 15.9C). All three patterns are commonly associated with patients who develop psychological symptoms in association with peripheral vestibular pathology (Jacob *et al.* 2003; Eagger *et al.* 1992).

If both labyrinths are destroyed sequentially, and compensation has occurred for the first unilateral loss, a second acute peripheral vestibular syndrome ensues. However, if both labyrinths are lost in rapid succession, or prior to compensation for the first loss, there is no acute vertiginous syndrome, as there is no left–right asymmetry in vestibular nucleus activity. However the long-term effects of

bilateral vestibular failure are the same, irrespective of whether the two labyrinths became impaired simultaneously or sequentially. The patient will experience chronic vestibular dysfunction known as Dandy's syndrome (Syms and House 1997). This syndrome is characterized by three key factors, resulting from reduced afferent input to the vestibulo-ocular, vestibulospinal, and vestibulocortical pathways:

◆ The patient cannot walk confidently in the dark, particularly if the ground is uneven, because of reduced input to vestibulospinal pathways.

◆ The patient develops 'bobbing oscillopsia', and cannot see clearly when moving the head, because there is reduced input to the vestibulo-ocular pathways.

◆ The patient feels disorientated when visual and proprioceptive inputs are ambiguous.

Nevertheless, patients with bilateral vestibular hypofunction or failure can compensate remarkably well, and benefit from intensive vestibular rehabilitation programmes (Brown *et al.* 2001)

Medical causes may be associated with vascular pathologies giving rise to presyncope, the drop in blood pressure. This in turn may result in ischaemia of vestibular nuclei or the labyrinth with a sense of light-headedness or vertigo. This situation may appertain in cases of orthostatic hypotension, simple vasovagal attacks, cardiac conditions with low cardiac output, and dysrrhythmias.

Certain *haematological disorders* such as anaemia and hyperviscosity syndromes may alter blood flow and reduce oxygen supply

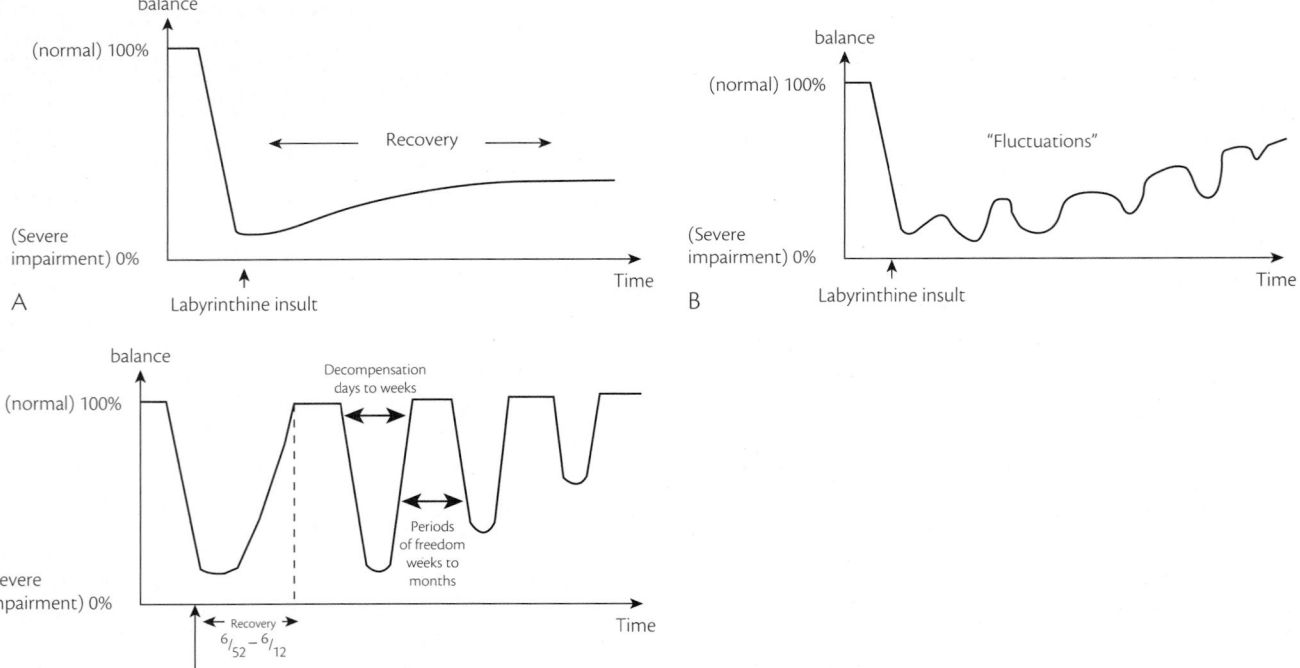

Fig. 15.9 Patterns of decompensation: (A) Failure of vestibular compensation characterized by progressive but incomplete recovery; (B) Failure of vestibular compensation characterized by fluctuating symptoms with slow but overall improvement. (C) Vestibular compensation followed by recurrent episodes of decompensation characterized by no interval symptoms. (From Luxon LM, 1997.)

to the vestibular nuclei presenting with light-headedness, faintness, or vertigo.

Hypoglycaemia is most commonly associated with hypoglycaemic drug treatment in diabetics, but may occur with insulinomas or in association with fasting or post-prandial functional hypoglycaemia.

Drug induced dizziness may result from a variety of mechanisms and many drugs list 'dizziness' as a side effect. However, specific care should be taken with drugs used in the treatment of cardiovascular disease, for example, beta blockers, anti-anginal agents, hypotensive agents, anti-dysrhythmics, all of which may alter blood flow and produce presyncope. Drugs used in the treatment of diabetes, hormonal preparations, both steroids and non-steroidal anti-inflammatory agents, psychotropic drugs including anticonvulsants, sedatives, tranquillizers and antidepressants, and certain analgesics also are commonly associated with disequilibrium. Recreational drugs including alcohol, narcotics and 'soft' drugs may all present with disordered balance. The ototoxic effect of certain chemotherapeutic and aminoglycoside antibiotics, particularly gentamicin are well documented.

Neurological disorders

Neurological causes of imbalance may be associated with primary neurological disease affecting the vestibular pathways infra-tentorially within the brainstem and cerebellum, or supra-tentorially and the majority of neurological disorders may be accompanied by dizziness or vertigo, or imbalance. Many psychological disorders may also include symptoms of disorientation and dissociation or depersonalization including acute anxiety, panic attacks, depression, phobias, and avoidance behaviour. It should be emphasized that these psychological conditions may also present following the onset of a vestibular disorder (Jacob *et al.* 2003).

Many of these conditions are noted in this chapter, as are the causes of 'vestibular' mechanisms which include labyrinthine and eighth nerve pathology, leading to an asymmetry in vestibular afferents, and thus, the generation of a perception of instability/vertigo, and imbalance.

15.2.4 Ageing

Ageing is a physiological process which occurs throughout life, and care should be taken not to attribute vestibular symptoms in the elderly to this process alone, although undoubtedly balance disorders are more common in the older age group, and complaints of dizziness and imbalance increase with age (Jonsson *et al.* 2004).

A plethora of morphological changes have been described in the vestibular organs and their neural pathways, including degeneration of the cristae ampullaris, degeneration of the maculae, hair cell alterations, otoconial changes, vestibular nerve degeneration, and degeneration of the central vestibular system (Bergström 1973; Engström *et al.* 1974; Nakayama *et al.* 1994). Physiological studies have demonstrated declining vestibular responses with increasing age, although there is poor correlation with anatomical changes (Peterka *et al.* 1990).

Neural degeneration within the central nervous system affects central integration of postural information, which becomes less efficient with age (Perrin *et al.* 1997). In addition, the musculoskeletal system decreases in strength, and this together with neural delay results in balance impairments (Konrad *et al.* 1999). Many studies have demonstrated an increase in body sway in old age

using posturography, and this is more marked in those who complain of balance disorders (Baloh *et al.* 1995). Normal ageing changes may impair overall balance but rarely give rise to acute dizziness, which depends upon an asymmetry of vestibular input. However, elderly people may decompensate from previously acquired vestibular disorders, for which they may compensate in their youth, but in old age these compensatory mechanisms are less efficient. Moreover, the elderly are prone to co-morbidity, the multi-sensory dizziness syndrome (Drachman and Hart 1972) and polypharmacy, any of which may compound disorders of balance (Luxon 1984).

15.3 Vestibular dysfunction

The majority of vestibular disorders are due to peripheral labyrinthine or VIII nerve pathology (Table 15.1)

The commonest disorders in the adult population are benign paroxysmal positional vertigo, vestibular neuritis, and migrainous vertigo. A majority of neuro-otological units report 5 to 10 per cent of patients with vestibular symptoms suffering from central neurological involvement, while in studies of 'dizzy patients', approximately 75 per cent of patients suffer vestibular dysfunction, predominantly due to peripheral pathology, while 25 per cent suffer vestibular symptoms arising usually as a consequence of systemic pathology.

15.3.1 Common vestibular disorders

Benign paroxysmal positional vertigo

Benign paroxysmal positional vertigo was characterized by Dix and Hallpike in their seminal work in 1952 on patients with vertigo. The condition represents the single most common cause of vertigo in adults, and is characterized by brief but severe attacks of vertigo associated with nystagmus induced by changes in head position (Furman and Cass 1999; Parnes *et al.* 2003). The importance of diagnosis of this condition lies in its common occurrence, the ease of treatment with complete resolution of symptoms, and the differentiation of central positional nystagmus (Table 15.2).

Two pathophysiological mechanisms have been proposed to explain this condition: cupulolithiasis and canalithiasis. Both mechanisms depend on otoconia, calcium carbonate crystals embedded in the otolithic membranes of the utricle and saccule, which become free-floating within the vestibular system. They may become attached to the cupula of the semicircular canal, giving rise to a heavy 'cupula' or cupulolithiasis (Schucknecht 1969) or may float within the lumen of the canal, canalithiasis (Hall *et al.* 1979). Changes in head position cause movement of the heavy cupula, or allow the free-floating débris to fall under gravity through the narrow lumen of the canal and act as a plunger drawing the cupula behind it (Fig. 15.10). Either mechanism results in movement of the cupula, with stimulation of the hair cells of the cristae and resultant vertigo, together with nystagmus in the plane of the stimulated semicircular canal. In the vertical canals, the nystagmus is therefore vertical/torsional, while in the horizontal canal, the nystagmus is in the horizontal plane (Table 15.3).

The condition is most commonly idiopathic, and is more frequently observed in older patients (Baloh *et al.* 1987). It may also be associated with various inner ear pathologies, but is particularly common after head trauma, giving rise to labyrinthine concussion (Brandt 1999b). The condition may affect any of the three

Table 15.1 Causes of dizziness and vertigo

General medical	Neurological	Otological
Haematological	**Supratentorial**	Benign paroxysmal positional vertigo
Anaemia	Trauma	**Infection**
Polycythaemia	Neoplasia	Vestibular neuritis
Hyperviscosity syndromes	Epilepsy	Labyrinthitis
Cardiovascular	Cerebrovascular disease	Syphilitic
Postural hypotension	Syncope	HIV-AIDS
Carotid sinus syndrome	Psychogenic	Lyme disease
Dysrhythmias	**Infratentorial**	**Congenital anomalies**
Mechanical dysfunction	Vascular disease	Isolated inner ear
Shock	Vertebrobasilar insufficiency	Superior semicircular canal dehiscence
Metabolic/Endocrine	Subclavian steal syndrome	Syndromes
Hypo- and hyperglycaemia	Wallenberg's syndrome	**Genetic conditions**
Thyroid disease	Anterior inferior cerebellar artery syndrome	Inner ear dysplasias
Chronic renal failure	Degenerative disorders including	Bilateral vestibular failures
Alcohol	Spinocerebellar degenerations	**Menière disease**
	Neuropathy	**Post-traumatic syndromes**
	Tumour, including those of the vestibulo-cochlear nerves	**Otosclerosis and Paget's disease**
	Infective disorders	**Vascular events**
	Meningitis	Migranous vertigo
	Ramsay–Hunt	Ischaemic labyrinthitis
	Neurosyphilis	**Auto-immune disorders**
	Tuberculosis	**Tumours**
	HIV-AIDS	**Drug intoxication**
	Trauma	
	Foramen magnum abnormalities	
	Basal ganglion disease	
	Multiple sclerosis	

semicircular canals, but is most commonly associated with the posterior semicircular canal (Baloh *et al.* 1987; Baloh *et al.* 1993).

Characteristically, the patient complains of brief, severe episodes of vertigo induced by turning over in bed, tipping the head backwards to look at the sky, or reaching for something on a high shelf, i.e 'top-shelf vertigo'. The diagnosis is made by the Dix Hallpike positioning test, with observation of characteristic features of paroxysmal symptoms (Fig. 15.11), which are currently considered to be explained most satisfactorily by the pathophysiological mechanism of canalithiasis outlined above. The observation of positional nystagmus may be the only abnormal clinical sign in a dizzy patient and thus the Dix Hallpike manoeuvre should therefore be performed routinely as part of the neuro-otological examination.

Treatment using a particle repositioning procedure (Section 15.6.3) is highly effective.

Viral vestibular neuritis or labyrinthitis

Acute vertigo associated with nausea and vomiting, but with no cochlear or neurological symptoms, is a common presentation in all age groups. The attacks are often unexpected and unprecipitated, but are commonly ascribed to a viral infection, and termed vestibular neuritis, vestibular neuronitis, or acute vestibulopathy. The term labyrinthitis is often preferred for a similar presentation with associated hearing loss. The evidence of underlying viral infection with total or subtotal loss of afferent vestibular input is

Table 15.2 Distinguishing features of peripheral and central positional nystagmus

Characteristic feature	Peripheral positional nystagmus SCC-BPPV	Central positional nystagmus
Latent period	2–20 seconds	None
Nystagmus	Rotational and geotropic	Any
Adaptation	15–60 seconds	None
Fatiguability	Yes	No
Symptoms	Strongly present	Variable

Table 15.3 Distinguishing features of semicircular canal responsible for Benign paroxycural positional vertigo

Characteristic feature	Posterior canal	Anterior canal	Horizontal canal
Test manoeuvre	Hallpike test	Hallpike test	Flat supine- head to lateral position (R or L)
Direction of nystagmus	Geotropic rotational	Ageotropic rotational	Horizontal towards side of head position
Latency (seconds)	5–20	5–20	Almost immediate
Duration (seconds)	< 30	< 30	> 30
Fatiguability	Yes	Yes	No

Key: R = right; L = left

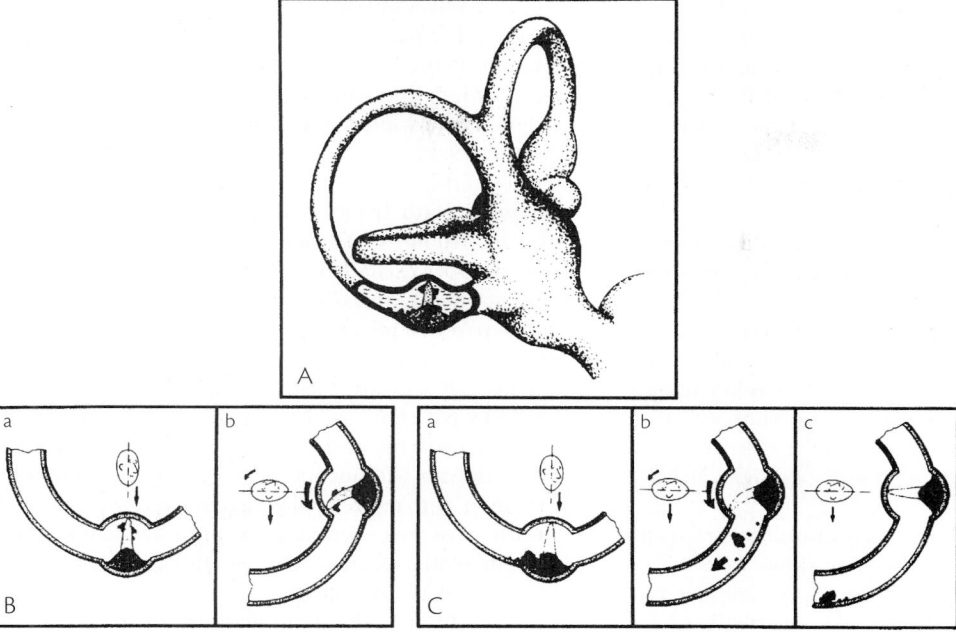

Fig. 15.10 Anatomical position of the posterior semicircular canal in the upright posture (A). The proposed pathophysiological mechanism of cupulolithiasis (B) and canalolithiasis (C). (From Brandt T and Steddin S 1993.)

usually circumstantial. A post-mortem study has identified selective neuronal loss in the vestibular ganglia and atrophy of the associated vestibular sensory epithelia, which would be consistent with a viral infection of the vestibular nerve (Baloh *et al.* 1996). In addition, viral DNA of herpes simplex type 1 has been detected in the superior and inferior vestibular ganglia, suggesting latent infection (Arbusow *et al.* 1999).

The majority of cases of vestibular neuritis affect the superior vestibular nerve (Fetter and Dichgans 1996; Aw *et al.* 2001), and it has been postulated that the superior division of the vestibular nerve travels through a bony canal, which is longer and narrower than that traversed by the inferior division, and thus renders it more susceptible to compression when inflamed. This hypothesis may explain the differential involvement of the superior rather than the inferior division in this condition (Goebel *et al.* 2001). Thus, it is frequently possible to obtain a normal saccular response,

as judged by vestibular evoked myogenic potentials, which depend upon normal inferior vestibular nerve function. In approximately 25 per cent of patients with vestibular neuritis, benign paroxsymal positional vertigo of the posterior canal variant may develop, and this would be compatible with preservation of the superior division of the vestibular nerve in those cases of inferior nerve vestibular neuritis.

The diagnosis of vestibular neuritis is a clinical diagnosis, based on the acute vestibular symptoms of vertigo, nausea, vomiting, and instability, which are exacerbated by head movement, but gradually abate over a period of one or two days. Within a period of two to three weeks, the majority of patients are rendered asymptomatic (Baloh 2003).

The acute symptoms of unilateral vestibular failure are accompanied by:

◆ A partial or complete ocular tilt reaction to the affected side.

◆ Spontaneous horizontal nystagmus, with the fast phases directed to the opposite side, and enhanced by the removal of optic fixation, for example with Frenzel glasses, video nystagmography.

◆ Rotation towards the affected side when marching on the spot with eyes closed, or drift to the affected side when walking with eyes closed.

◆ A positive horizontal head impulse test to the affected side, which frequently remains when all other symptoms and signs have improved.

The vestibular system has been shown to be extremely adaptable, (Gonshor and Melvill Jones 1976), and symptomatic recovery occurs as a result of cerebral compensation. This process may be facilitated and expedited by physiotherapy intervention (Section 15.6.3).

Fig. 15.11 The clinical Dix–Hallpike manoeuvre for eliciting positional nystagmus.

Caloric testing in general reveals a canal paresis, which usually persists, despite recovery of symptoms, although recent work has demonstrated recovery of the caloric abnormality in up to 50 per cent of patients (Bergenius and Perols 1999; Schmid-Priscoveanu *et al.* 2001; Okinaka *et al.* 1993). Electronystagmography demonstrates unidirectional horizontal spontaneous nystagmus in all directions of gaze initially, but with recovery, first-degree nystagmus only, directed towards the unaffected ear will be observed in the absence of optic fixation and ultimately a directional preponderance of induced nystagmus or no abnormality will be found. Importantly, central vestibular abnormalities, such as abnormal vestibulo-ocular reflex suppression, deranged optokinetic responses or smooth pursuit, will not be observed although in the initial phase of the illness, a directional preponderance of optokinetic and induced nystagmus may be seen, because of the superimposition of spontaneous nystagmus upon the responses. In the absence of any central visuo-vestibular abnormality, or neurological signs, an MRI scan is not required.

In an acute unilateral peripheral vestibular disorder, assessment of the subjective visual vertical and horizontal shows a deviation from the true vertical or horizontal by 20° or more (Böhmer and Rickenmann 1995), while a normal subject can consistently adjust the light bar to within 2 to 3° of the true gravitational horizontal or vertical.

The differential diagnosis of vestibular neuritis involves the differentiation of hypothesized viral effects from a vascular event in the older patient and vascular risk factors should be evaluated. Labyrinthine infarction caused by occlusion of the internal auditory artery, may result in an acute vestibular syndrome, accompanied by hearing loss. Nonetheless, patients with acute vestibular lesions without hearing loss as a consequence of occlusion of branches of the internal auditory artery may mimic a vestibular neuritis (Kim *et al.* 1999).

From a neurological perspective, in the absence of a negative head thrust test, in which there are no catch-up saccades, and inability to stand with acute vertigo, cerebellar infarction should be considered. An MR scan should be performed in this clinical situation, as approximately a third of patients with this disorder will develop oedema with potentially fatal compression of the posterior fossa structures (Huang and Yu 1985; Iwase *et al.* 2001).

Migrainous vertigo

Migraine is the commonest neurological condition in the United Kingdom and is the commonest cause of dizziness or vertigo in children, and a common presentation, both with and without headache, of dizziness in adults. Symptoms of disequilibrium are reported in 50 to 70 per cent of migraineurs, and conversely, there is a high prevalence of migraine in patients presenting with vertigo (Savundra *et al.* 1997). The presentation of vertigo in association with migraine is very variable with episodes lasting seconds to hours, to prolonged instability lasting days (Dieterich and Brandt 1999).

Migraine, although a disorder characterized primarily by episodic headache, may be associated with a variety of different symptoms (Harker 1996; Silberstein 2004) (Section 18.2.1). While various types of presentation of migraine are described by the International Headache Society (2004), migrainous vertigo is not included in the criteria, but Neuhauser and Lempert (2004) have reviewed the topic extensively, and proposed criteria, which have been expanded by Brantberg and co-workers (2005).

The mechanism by which migraine gives rise to vertigo remains speculative (Furman *et al.* 2003), but hypotheses include complex interactions between the vestibular nuclei, trigeminal system, and thalamocortical processing centres, and a manifestation of a brainstem aura due to a spreading wave of neural depression. (Dieterich and Brandt 1999).

In cases of migrinous vertigo, there is frequently a family history of migraine, with troublesome motion sickness in childhood (Cutrer and Baloh 1992). In addition to vertigo, classical symptoms of sensory hyper-excitability may be present, including photophobia, phonophobia, and osmophobia. Some patients demonstrate vertigo as part of a migrainous aura and go on to develop a typical hemicranial headache, while others have vertigo that develops simultaneously with the headache, or appears after the headache phase. Most patients have attacks of vertigo unaccompanied by associated headache, and the picture is very similar to that of Menière's disease, without auditory symptoms (Baloh 1997). There is acute onset, with severe vertigo, and frequently vomiting. Tinnitus and hearing loss are uncommon except with basilar migraine, making the differential diagnosis between this condition and Menière's syndrome particularly difficult.

In young children, the manifestations of migraine are diverse and, commonly, headache is absent (Al-Twaijri and Shevell 2002). Presentations include cyclical vomiting, attacks of abdominal pain, ophthalmoplegia, and instability. Basser (1964) described an episodic disorder in young children under the age of 4 years, termed *benign paroxysmal vertigo*, which is characterized by the sudden onset of anxiety, crying, clinging to the parent, staggering, pallor, and vomiting. Typically the attack is brief, lasting only a few minutes, and the symptoms are exacerbated by head movement. Nystagmus and/or torticollis may be present. The child rapidly returns to normal, and although the attacks may occur several times a month under the age of 4 years, they gradually reduce in frequency and severity, disappearing by the age of 7 or 8 years. Frequently, these children develop classical migraine in adult life (Lanzi *et al.* 1994).

Review of the literature would suggest that benign recurrent vertigo described by Slater in 1979 represents a migrainous equivalent. Moreover, evaluation of patients with 'vestibular Menière's syndrome' led to the conclusion that a significant percentage (approximately half) suffered vestibular migraine rather than Menière's disease (Rassekh and Harker 1992). Diagnosis of migrainous vertigo remains difficult, as the diagnostic framework is based on a combination of the International Headache Society Criteria for Migraine, the presence of specific other symptoms, and the exclusion of other pathology (Neuhauser *et al.* 2001; Furman *et al.* 2003). There are no diagnostic clinical signs, nor vestibular test abnormalities.

15.3.2 Peripheral labyrinthine disorders

Genetic and developmental disorders

About 1:1000 children are born with a significant hearing loss and, while this group have been extensively investigated in terms of auditory function, little attention has been paid to the associated vestibular dysfunction. Vestibular loss is now recognized as a feature of both genetic hearing loss and a number of syndromes associated with hearing impairment (Table 14.4). Children with congenital vestibular failure compensate extremely well, and the sole pointer to pathology may be delayed motor milestones.

In general *recessively inherited hearing loss* presents as congenital profound loss with, rarely, involvement of vestibular abnormalities, although vestibular dysfunction has been reported in DFNB2, 4 and 12 (Nance 2003). *Dominantly inherited hearing loss* is less severe, often progressive in adult life and may be associated with vestibular dysfunction in DFNA11 and DFNA9. The latter condition presents with a late onset, progressive hearing loss, together with vestibular failure and may mimic the clinical presentation of Menière's syndrome (Fransen *et al.* 1999). *X-linked hearing loss* may also be associated with total vestibular failure and this finding corresponds with novel radiologic abnormalities. (Fig. 15.12) (Phelps *et al.* 1991).

A wide range of inherited *syndromes* are associated with auditory and vestibular dysfunction. *Usher syndrome* is characterized by sensorineural hearing loss and retinitis pigmentosa. Clinically three forms are recognized:

♦ type I with congenital profound loss and vestibular failure

♦ type II with progressive sensorineural hearing loss and normal vestibular function

♦ type III with variable presentation of hearing loss and vestibular function.

However, recent genotypic, phenotypic studies have shown that the differentiation of Usher types is not as clear-cut as proposed clinically (Kremer *et al.* 2006).

The *Jervell and Lange-Nielsen syndrome*, which is characterized by a long QT interval on electrocardiogram is recessively inherited and demonstrates both congenital, auditory, and vestibular failure, together with cardiac defects. Diagnosis is essential in the early years of life to prevent sudden cardiac death from dysrhythmia.

Pendred syndrome is an autosomal recessive disorder characterized by congenital deafness and goitre. Dilated vestibular aqeducts with or without a Mondini-like deformity of the cochlea (Fig. 15.13) have been described, and the condition is reported to occur with normal and abnormal vestibular function (Luxon *et al.* 2003).

CHARGE syndrome is characterized by Coloboma, Heart Anomalies, coanal atrisia, Retardation of growth and development, and

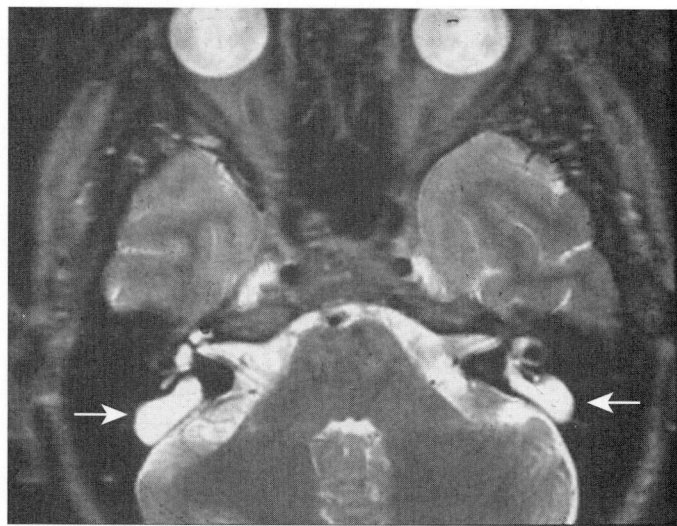

Fig. 15.13 Axial MRI scan showing dilated bilateral vestibular aqueducts (arrowed) in a case of Pendred syndrome. (From Luxon LM *et al.* 2003.)

Genital and Ear anomalies. The disorder is an autosomal dominant condition with genotypic heterogenarity. The underlying genetic abnormality is a mutational deletion of the chromodomain helicase DNA-binding protein-7 *CHD7* gene in the majority of cases. CT scan of the temporal bone demonstrates partial or complete semicircular canal hypoplasia, and characteristically, children have late motor milestones (Morimoto *et al.* 2006).

Mitochrondrial diseases are a clinically heterogenous group of disorders that give rise to dysfunction of the mitochondrial respiratory chain. Mitochrondrial mutations have been associated with syndromic and non-syndromic hearing loss (Xing *et al.* 2007), with limited reporting of vestibular function. However, vestibular loss has been reported in Kearns Sayre Syndrome (Weidauer and Lenarz 1984;), large deletions of the mitochondrial genome (Zeviani *et al.* 1990) and children with mitochrondrial disease (Lenard *et al.* 1992).

Menière's disease

Menière's disease is a clinical diagnosis based on the classic triad of fluctuating hearing loss, tinnitus, and vertigo (Menière 1861). It may occur at any age, but most commonly the initial presentation occurs between the ages of 30 and 60 years. It is rare both in children and in those over the age of 60 years. There is a family history in approximately 10 per cent of patients (Paparella and Djalilian 2002). The literature abounds with controversy on all aspects of this disorder, and the diagnosis should be based on the strict American Academy of Otolaryngology Head and Neck Surgery Committee on Hearing and Equilibrium Guidelines (1995).

The underlying pathophysiological mechanism is postulated to be overproduction or malabsorption of endolymph, resulting in endolymphatic hydrops or hypertension (Fig. 15.14). The underlying trigger has not been defined, but genetic, autoimmune, and environmental factors have all been postulated (Minor 2004). Menière's attacks are considered to be the result of intermittent ruptures of the membranous labyrinth, with leakage of potassium-rich endolymph into the perilymph. The initial irritative phase of the attack is attributed to excitation of the hair cells

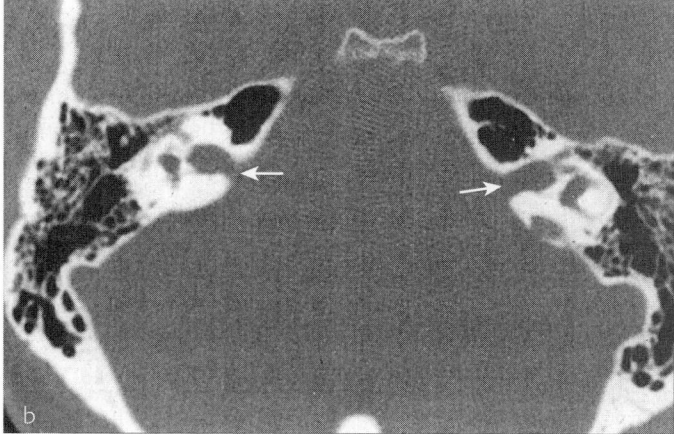

Fig. 15.12 Axial thin section high-resolution CT scan of the petrous temporal bones, to illustrate the characteristic internal bulbous internal auditory meatus seen in X-linked hearing loss. (From Phelps *et al.* 1991.)

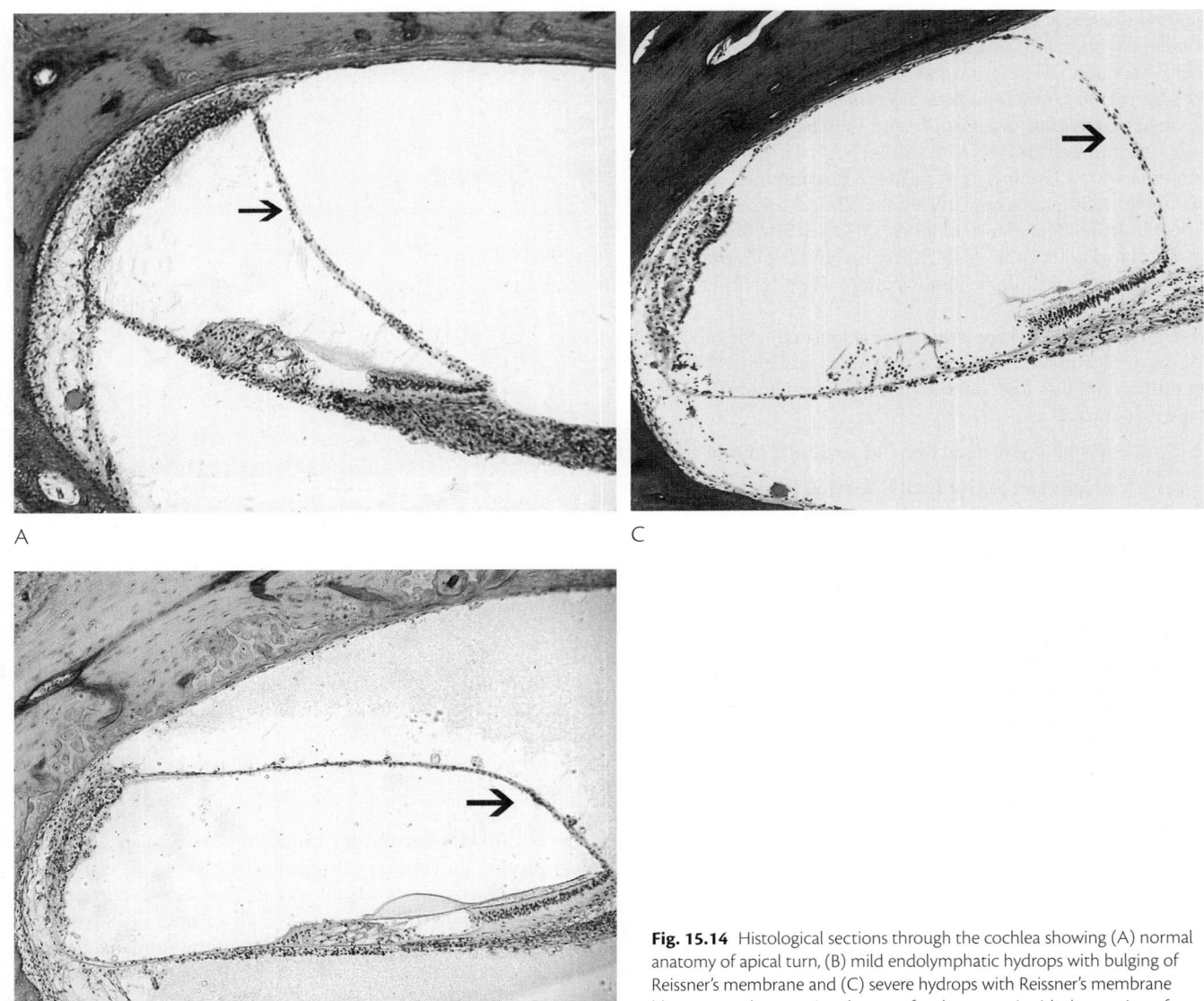

A

B

C

Fig. 15.14 Histological sections through the cochlea showing (A) normal anatomy of apical turn, (B) mild endolymphatic hydrops with bulging of Reissner's membrane and (C) severe hydrops with Reissner's membrane blown up and contacting the top of scala tympani, with destruction of the organ of Corti in a case of Menière disease. Arrows points to Reissner's membrane. (Courtesy of Professor Leslie Michaels.) (See Plate 27.)

by the increased potassium concentration around their basal surfaces, while the subsequent paretic phase is thought to result from a blockade of neurotransmitter release. Closure of the rupture is thought to allow recovery, with return of the normal chemical composition of the endolymph and perilymph (Tonndorf 1983).

Endolymphatic hydrops has also been documented in a variety of internal ear conditions, including syphilis, mumps, Cogan's syndrome, trauma, chronic suppurative otitis media, and congenital ear disease. In this latter condition, longstanding hearing impairment is accompanied by the onset of acute vestibular episodes characteristic of Menière's disease. This condition is sometimes called Menière syndrome, secondary Menière disease, or delayed endolymphatic hydrops (Schuknecht *et al.* 1990; Harcourt and Brooks 1995).

The clinical presentation of Menière's disease is characteristically one of vertiginous attacks lasting for 1 to 8 hours, with tinnitus, hearing loss, and fullness in one ear, and this latter symptom may precede and outlast the acute vertigo. There is profound nausea and vomiting. Both vestibular and cochlear symptoms develop in approximately 60 per cent of those affected within six months of the onset of disease, although clinically it is not unusual to see a patient with fluctuating hearing loss or a single acute episode of dizziness unaccompanied by cochlear symptoms prior to the full-blown clinical presentation.

The episodes of vertigo tend to occur in clusters with intervals of freedom, and in the initial phase of the disorder both vestibular and cochlear symptoms recover, such that vestibular investigations and audiometry may be normal between attacks. However, as the disease progresses, a low-frequency sensorineural hearing loss is observed, and in an older patient with presbyacusis, the audiogram may assume a tent 'shape' (Fig. 15.15). With continuing progression, a plateau hearing loss emerges. Moreover, in the vestibular system, a canal paresis or caloric testing will be documented, with or without an accompanying spontaneous nystagmus or directional preponderance. With progressive attacks, interval disorientation and imbalance are common, as a result of fluctuating vestibular function with poor compensation. Some patients, especially in the later stages of the disease may develop drop attacks known as Tumarkin or otolithic crises (Ishiyama *et al.* 2001). In these attacks the patient drops to the ground without any warning, and frequently suffers injury. There may be a sense of being pushed to the ground by an external force, but frequently there is no associated vertigo, and no loss of consciousness.

The differential diagnosis of Menière's disease includes perilymph fistula, vestibular neuritis with repeated decompensation, and vestibular migraine, which is a particularly difficult diagnosis to exclude, given the increased incidence of Menière's disease in migrainous subjects and the increased incidence of migraine in patients with Menière's disease (Kayan and Hood 1983; Neuhauser *et al.* 2001). Autoimmune inner ear disease may, in the early stages, mimic Menière's disease, but progresses much more rapidly, leading to bilateral severe hearing loss and vestibular dysfunction within a relatively short time-span (Agrup and Luxon 2006).

Autoimmune inner ear disease

Autoimmune inner ear disease may present as an isolated phenomenon or as part of a systemic autoimmune disorder (Broughton *et al.* 2004; Agrup and Luxon 2006). It is a rare, but important diagnosis, as rapid progressive, stepwise bilateral loss of auditory and vestibular function requires urgent medical treatment in an attempt to prevent or limit progression. Systemic autoimmune disorders with reported cochleo-vestibular involvement include polyarteritis

nodosa, Wegener's granulomatosis, systemic lupus erythematosus, Behçet's disease, and Cogan's syndrome. The latter condition is particularly catastrophic with interstitial keratitis and acute sudden audiovestibular failure. As noted above, initial symptoms may mimic Menière's disease, with fluctuation of hearing and ear pressure, but unlike this latter condition, the symptoms progress rapidly over weeks or months to involve both ears. Treatment requires high-dose steroids, with or without the addition of immunosuppressive drugs. A variety of immunosuppressive regimes have been evaluated, but whilst steroids may result in improvement of hearing loss, recent work has shown no significant benefit of methotrexate, and cyclophosphamide use is restricted by significant side effects (Ruckenstein 2004).

Labyrinthine trauma

Damage to the vestibular apparatus may occur as a consequence of barotrauma, acoustic trauma, and physical trauma.

Over-exertion, scuba diving, and flying in unpressurized aircraft are common causes of otitic *barotraumas* (Luxon 1996), which, if sufficiently severe, may result in a perilymph fistula, with disruption of the oval or round window (Kohut *et al.* 1988). This leads to acute loss of hearing and vertigo, due to leakage of perilymph. Clinically, there may be a positive fistula sign, in which eye movement is induced in association with change in pressure in the external ear canal using a Segal's speculum, with vertigo and nystagmus induced by pressure change in the external canal, or CSF as a result of coughing or straining.

Audiometry may show a sensorineural hearing loss, and electronystagmography may reveal peripheral vestibular nystagmus, with a canal paresis on caloric testing. However, the diagnosis of this condition is notoriously difficult and even observation at surgery may not be definitive. Management is conservative, with bed-rest, head elevation, and symptomatic treatment. Surgical exploration is

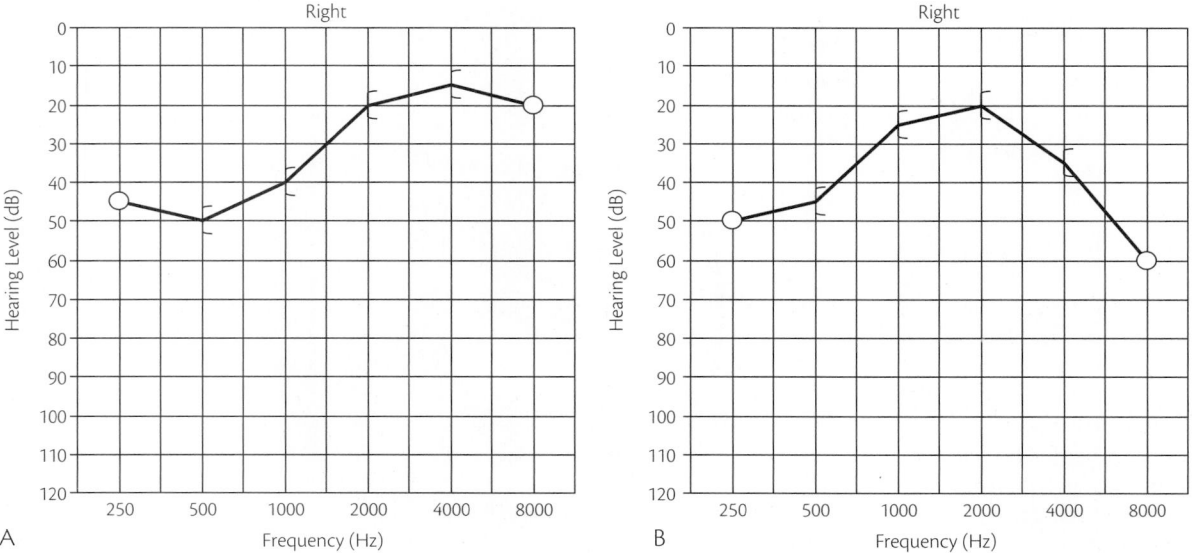

Fig. 15.15 Pure-tone audiograms to illustrate characteristic configurations of auditory loss in Menière's disease. (A) low-frequency right hearing loss which frequently fluctuates in the early stage of the disease; (B) tent-shaped audiogram illustrating a fixed low-frequency right sensorineural hearing loss and a high-frequency loss of presbyacusis in an older patient, giving rise to the characteristic 'tent-shaped' audiogram.

indicated if symptoms persist, or there is a clear relationship of the onset of symptoms to trauma (Ludman 2003).

Acoustic trauma of sufficient intensity, for example an explosion, may rupture the tympanic membrane and give rise to a perilymph fistula. However, acoustic trauma in the form of hazardous occupational noise, gun-fire, or amplified music does not give rise to vertigo (Hinchcliffe *et al.* 1992) except in superior canal dehiscence.

Physical trauma to the head, with or without skull fracture, may give rise to three different clinical vestibular presentations:

- benign paroxysmal positional vertigo (Section 15.3.1);
- labyrinthine concussion;
- vestibular failure.

Labyrinthine concussion with or without unilateral hearing loss is common after head injury of any severity (Davies and Luxon 1995). Acute vertigo with evidence of a canal paresis on caloric testing, and spontaneous nystagmus directed towards the unaffected ear, are common (Luxon 1996). The natural history of the disorder is characteristic of any acute unilateral vestibular loss, with gradual improvement in symptoms, but commonly, persistence of vestibular impairment on caloric testing. Benign paroxysmal vertigo may occur as a late consequence of labyrinthine trauma, some weeks or months after the head injury—a finding of particular importance in the medicolegal context. Associated central vestibular pathology or cerebellar dysfunction may preclude full recovery by vestibular compensation, with persistence of symptoms.

Dizziness or vertigo is reported in over three-quarters of patients with temporal bone fractures (Fig. 15.16) (Wennmo and Svensson 1989), although frequently other more serious neurological symptoms take priority. Thus with significant head injuries, vestibular involvement may be overlooked. Longitudinal fractures, with involvement of the middle ear, tend to result in less severe vestibular symptoms associated with labyrinthine concussion. A conductive hearing loss is commonly present, and this may or may not be associated with a sensorineural loss (Luxon 1996). Positional vertigo is common. Transverse skull fractures frequently involve the labyrinth or internal auditory meatus, with profound permanent sensorineural hearing loss, and severe vertigo, nausea, and vomiting. The vestibular symptoms gradually improve as a result of cerebral compensation, providing there is no neurological dysfunction, while the hearing loss does not recover. Commonly this type of fracture is also associated with involvement of the facial nerve.

Labyrinthine infarction

The labyrinth is supplied by the internal auditory artery, a branch of the anterior/inferior cerebellar artery (Fig. 15.17). Infarction of the internal auditory artery, therefore, commonly gives rise to profound auditory loss in addition to acute vestibular failure with vertigo and imbalance. The associated hearing loss suggests differentiation from acute vestibular neuritis, although the vestibular presentation may be very similar. Frequently there is associated infarction in the brainstem and/or cerebellum with preceding episodes of transient ischaemia within the vertebrobasilar circulation (Oas and Baloh 1992). Occasionally acute unilateral vestibular failure may occur in isolation and this has been attributed to occlusion of a branch of the internal auditory artery supplying the vestibular apparatus while sparing the cochlea. The diagnosis should be considered in older patients with vascular risk factors. Intra-labyrinthine haemorrhage may occur in patients with leading disorders, such as leukaemia (Schuknecht 1993).

Superior semicircular canal dehiscence

Superior semicircular canal dehiscence is a disorder in which the bone between the apex of the superior semicircular canal and the

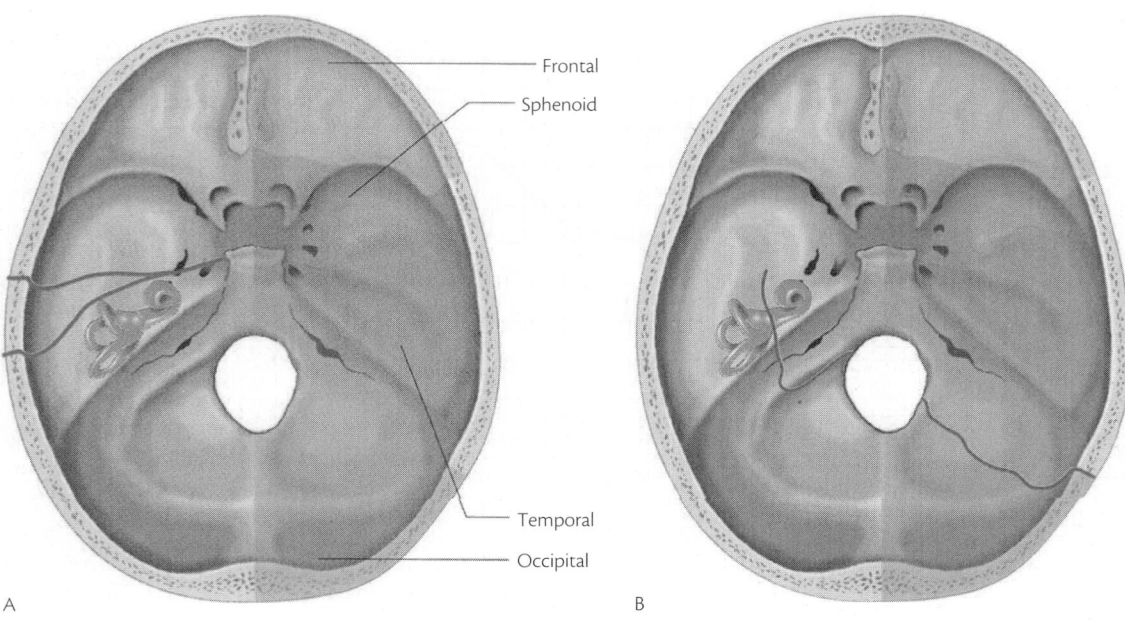

Fig. 15.16 Diagram to illustrate (A) left-sided longitudinal temporal bone fractures sparing the bony labyrinth; and (B) transverse temporal bone fracture across the labyrinth. (From Backous, 2003.)

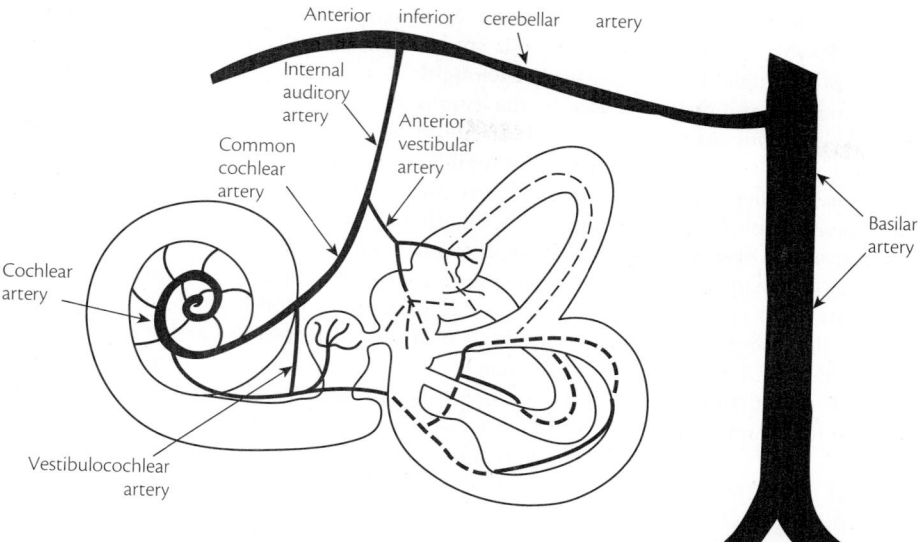

Fig. 15.17 Blood supply to the labyrinth.

middle cranial fossa is absent or attenuated (Fig. 15.18). The characteristic symptom associated with this anomaly is the Tullio phenomenon, in which vertigo and imbalance are precipitated by loud sound. However, Calvert's sign, consisting of dizziness provoked by change in pressure in the external auditory canal, or Valsalva-induced vertigo, may also be associated with this condition (Minor 2005). Semicircular canal dehiscence is considered to be a congenital or developmental anomaly, as in many cases the abnormality is bilateral (Hirvonen *et al.* 2003), although the clinical presentation is often unilateral. The trigger for the unilateral presentation is

unknown, but it has been postulated that abnormally thin bone may be disrupted as the result of a traumatic injury.

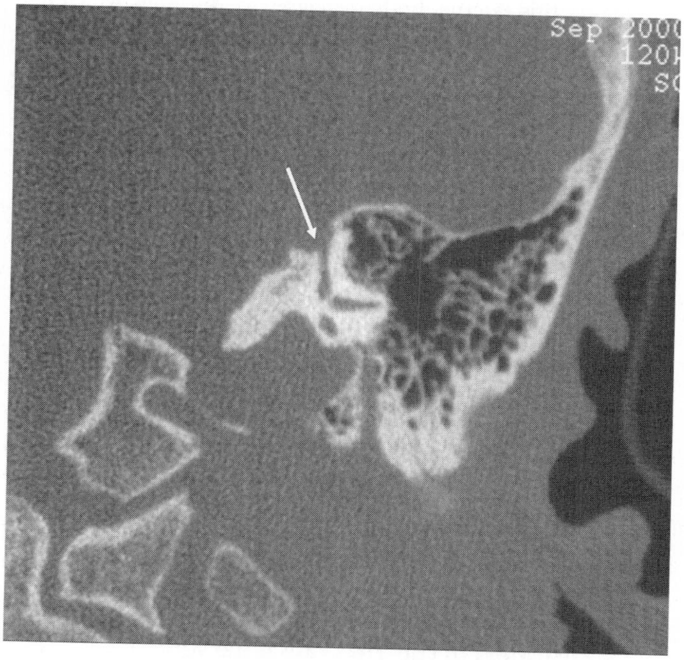

Frequently, patients complain of dizziness and oscillopsia in response to sound at particular frequency and intensity, but the symptom may also be generated by coughing, sneezing, or straining. Commonly there is hyperacusis and an apparent air-bone gap, despite normal air conduction thresholds (Minor *et al.* 2003). Patients often complain of hearing themselves walk and swallow, and report a perception of rumbling or rustling in the ear.

The diagnosis is made by characteristic nystagmus, aligned to the plane of the dehiscent semicircular canal, in response to sound. The direction of the nystagmus may change depending on whether pressure is raised within the cerebrospinal fluid, for example by the Valsalva manoeuvre, or raised externally through the external auditory meatus. Diagnostically, vestibular-evoked myogenic potentials in response to clicks have also been shown to demonstrate a lower threshold in patients with superior canal dehiscence (Halmagyi *et al.* 2003b). Confirmation of the structural abnormality is made with high-resolution computerized tomography, with images reconstructed in a plane of the semicircular canal (Belden *et al.* 2003).

Bilateral vestibular failure

Bilateral vestibular failure may result from genetic abnormalities, meningitis, trauma, autoimmune disease, or most commonly ototoxicity (Bronstein *et al.* 1994). Most cases of acquired vestibular loss in childhood are associated with hearing loss, the commonest cause is pre, peri or postnatal infections with toxoplasmosis, herpes, cytomegalovirus, or rubella.

In adult-acquired vestibular failure, there is profound bilateral loss of afferent vestibular input, with postural imbalance, and 'bobbing' oscillopsia during head movements (Rinne *et al.* 1998). Aminoglycoside antibiotics are well-recognized to give rise to vestibular, or more rarely to auditory toxicity, but the effect is frequently not appreciated until the seriously ill patient begins to mobilize, and complains of dizziness, ataxia, and oscillopsia. The various antibiotics vary in their predilection for the auditory or vestibular system, but gentamicin is the most common culprit, with respect to the vestibular apparatus. Aminoglycoside antibiotics are

Fig. 15.18 Coronal section of CT scan showing dehiscence of superior semicircular canal, with deficient bony plate between apex of the canal (arrow) and the middle cranial fossa.

not metabolized, but are excreted by glomerular filtration, and thus patients with renal impairment are at particular risk, with accumulation of the aminoglycoside in the blood and inner ear fluids. The ototoxic damage occurs in the hair cells of the inner ear, and may continue because of high levels of aminoglycoside in the inner ear fluids, even after the drug has been discontinued.

The diagnosis of bilateral vestibular failure can be clarified clinically by an inability to walk across a foam mattress with eyes closed, degradation of visual acuity with head movement, and by observation of catch-up saccades on the head impulse test (Halmagyi and Curthoys 1988). Confirmation of the abnormality, especially for medicolegal purposes, may be achieved by absent caloric responses, even with irrigation of the external auditory canals with ice-cold water, or water at 20°C for 1 minute, and by marked reduction or absence of vestibulo-ocular reflexes on impulsive rotation testing (Brandt *et al.* 1996a).

The differential diagnosis of bilateral vestibular failure requires consideration of a range of disorders. Postural imbalance is characteristically observed with both otological and neurological conditions, including severe unilateral vestibular failure, peripheral neuropathy, cerebellar disease, hydrocephalus, and extrapyramidal disorders such as progressive supranuclear palsy or Parkinson's disease, while oscillopsia is common with central nystagmus syndromes such as vertical nystagmus, see-saw nystagmus, and periodic alternating nystagmus. However, these latter conditions are not particularly exacerbated by movement, unlike bilateral vestibular failure.

Neoplasia

Tumours invading the petrous temporal bone and labyrinth are rare, but labyrinthine erosion with vestibular symptoms tends to be a poor prognostic sign. Malignant tumours include squamous cell carcinomas, arising in the external auditory meatus, middle ear, and mastoid in the elderly and aggressive primary temporal bone osteogenic sarcoma and chondrosarcoma in older children and young adults. Rhabdomyosarcomas are common in infants under 5 years, while metastatic temporal bone deposits are relatively common in adults.

Spontaneous labyrinthine fistula is almost always the result of bone erosion by cholesteatoma (Smith and Danner 2006), but rarely may be the result of syphilitic osteitis, tuberculous otitis media, chronic perilabyrinthine osteomyelitis or glomus vagale, and jugulare tumours. This is a life threatening situation, with acute vertigo, profound hearing loss. and the possible introduction of infection. Intralabyrinthine schwannoma are rare, but present with both hearing loss and vertigo. Detailed MR imaging allows diagnosis (Kennedy *et al.* 2004)

Infections

Viruses, bacteria, syphilis, and fungi may all give rise to inner ear pathology.

Labyrinthitis is an infection or inflammation of the labyrinth. Its symptoms are similar to vestibular neuritis, but additionally include sensorineural hearing loss. Three types are described: serous labyrinthitis, otogenic suppurative labyrinthithis, and meningogenic suppurative labyrinthitis. Serous labyrinthitis represents an irritation of the inner ear without bacterial or viral invasion. Toxins may spread into the inner ear as a result of otitis media or as a result of labyrinth surgery, from the middle ear through the round or oval window. Otogenic suppurative labyrinthitis involves bacterial invasion of the inner ear from contiguous areas within the temporal bone. Meningogenic suppurative labyrinthitis occurs when bacteria spread from the subarachnoid space into the inner ear during meningitis. A specific type of viral labyrinthitis is that caused by the mumps virus. The hearing loss resolves completely or partially in about 50 per cent of cases, while compensation brings about resolution of vestibular symptoms.

Viral infections may reach the inner ear via the blood stream, the meninges, the eighth nerve, or the inner ear. The evidence for viral infection in the inner ear is circumstantial. Westmore (1979) detected the mumps virus in the perilymph after sudden onset of hearing loss and changes in the otolith organs have been observed histologically in relationship to measles and mumps. In animals vestibular changes have been documented in infections of the inner ear with mumps, rubeola, measles, influenza A and B, cytomegalovirus and herpes simplex (Davis 1993). Thus, by extrapolation, it is considered that sudden vertigo may be caused by viral infections.

Acquired immunodeficiency syndrome is associated with both peripheral and central cochleovestibular symptoms, although hearing loss is more common than vertigo and balance. Vestibular changes have been demonstrated histologically (Pappas *et al.* 1995); antiviral agents have been reported to cause vertigo (Fantry and Staeker 2002); and opportunistic infections such as otosyphilis can give vestibular symptoms (Song *et al.* 2005).

Bacterial infection of the middle ear cleft is most commonly associated with chronic middle ear disease, which if erosive gives rise to a labyrinthine fistula with vertigo and sudden sensorineural hearing loss. This is a medical emergency with the risk of cerebral infection and abscess formation (Penido *et al.* 2005). Good public health and antibiotic therapy have significantly reduced the prevalence of these complications in developed countries, but complications of otitis media remain a serious problem in the Developing World.

Two acute bacterial infections are worthy of note: petrositis and malignant otitis externa. Both are aggressive disorders which may involve the inner ear and/or the eighth nerve. Petrositis is a perilabyrinthine infection which spreads to involve the apex of the petrous temporal bone. It may present as Gradenigo's syndrome (1893) with otitis media with involvement of the trigeminal ganglion giving pain behind the ipsilateral eye, paralysis of the lateral rectus muscle, acute vertigo, and hearing loss.

Malignant otitis externa occurs in debilitated or immunosuppressed patients with *Pseudomonas aeruginosa* infection, which invades the surrounding tissues and gives rise to vertigo and hearing loss. Prolonged antibiotic treatment has improved the previously poor prognosis.

Otological syphilis is increasing in prevalence and 50 per cent of cases complain of vertigo and imbalance (Yimtae *et al.* 2007). The clinical course of the early acquired and late congenital forms of this condition are similar: sudden or rapidly progressive bilateral sensorineural hearing loss with mild vestibular symptoms (Garcia-Berrocal *et al.* 2006). Vestibular investigations in otosyphilis have demonstrated peripheral as opposed to central pathology and vestibular abnormalities were more marked in the congenital form of the infection (Wilson and Zoller 1981).

15.3.3 Disorders of the VIII cranial nerve

Congenital disorders

Aplasia or hypoplasia of the VIII nerve is a rare condition in children, but has been reported in 4 per cent of profoundly deaf children (Bamiou *et al.* 2001). If the vestibular nerve is involved, either in isolation or in association with hearing impairment, there may be delayed motor milestones, and frequently a syndromic diagnosis will be made. The VIII nerve can be reliably visualized on high-resolution MRI (Casselman *et al.* 1997).

Inherited VIII nerve disorders, with hearing loss and absent vestibular responses, have been reported in an autosomal recessive hereditary motor and sensory neuropathy (Butinar *et al.* 1999). *Friedreich's ataxia*, is characterized by severe loss of both cochlear and vestibular neurons, but preservation of the sensory epithelia (Spoendlin 1974). Both auditory and vestibular neuropathy may be associated with peripheral neuropathy in *olivopontocerebellar* and *spinocerebellar degenerations* (Starr 2001) *Cerebro-oculo-facial-skeletal syndrome* is a rare autosomal recessive disorder in which dysmorphic features, hypotonia, osteoporosis, and neural degeneration are associated with accelerated cochlear and vestibular nerve degeneration (Fish *et al.* 2001). The vestibular nerve has been cited as abnormal in a range of inherited neurological disorders (Huygen and Verhagen 1994; Verhagen and Huygen 1994).

Compression of the VIII nerve in the internal auditory canal may be a consequence of osteopetrosis (Hanson and Parnes 1995) or craniodiaphyseal dysplasia (Himi *et al.* 1993). There is some controversy as to whether the VIII nerve is compressed in Paget's disease, despite the common occurrence of vestibular symptoms (Khetarpal and Schuknecht 1990).

Infections

Vestibular neuritis (Section 15.3.1) A number of aetiologies have been postulated, but viral eighth nerve involvement is the preferred hypothesis (Strupp and Arbusow 2001).

The Ramsay–Hunt syndrome. The Ramsay–Hunt syndrome refers to a clinical presentation of herpes zoster oticus with facial palsy, auricular vesicular rash, hearing loss, and acute vertigo (Sweeney and Gilden 2001), which is now known to be caused by reactivation of the varicella zoster virus (Sections 20.2.2; 42.3.3). Abramovich and Prasher (1986) reported vertigo in 85 per cent of their series. The primary site of pathology remains unclear with studies having defined both labyrinthine and VIII nerve pathology (OMahoney and Luxon 1997). Recent work supports the use of steroids and acyclovir for prompt treatment and optimal outcome (Morrow 2000), although the presence of vertigo, diabetes mellitus, essential hypertension, and age were all negative prognostic factors (Yeo *et al.* 2006).

Bell's Palsy. Vestibular dysfunction has been reported in Bell's palsy or idiopathic facial palsy (Yagi *et al.* 1988; Watanabe and Suzuki 2006) (Section 20.2.3). A variety of mechanisms of vestibular dysfunction in Bell's palsy have been postulated, including compression of the VIII nerve by an oedematous VII nerve, and involvement of both VII and VIII cranial nerves in the same viral process (OMahoney and Luxon 1997).

Lyme disease. Lyme disease is a tick-borne spirochaete infection with protean multisystem manifestations (Wormser 2006) (Sections 21.14.3; 42.5.2). Both auditory and vestibular symptoms may occur, and 12 per cent of patients have been reported to suffer dizziness (Lesser *et al.* 1990). Conversely, Rosenhall and colleagues (1988) studied 73 patients with vertigo and found serological evidence of Lyme disease in 14 per cent. Vestibular studies have identified both central and peripheral vestibular abnormalities, which have been interpreted as due to neurological and VIII nerve pathology (Ishizaki *et al.* 1993). *Borrelia burgdorferi*, the causative agent, is a particularly complex bacterium, which is frequently resistant to standard antibiotic regimes, and concern has been expressed about the need for better diagnosis and management of this disease (Stricker *et al.* 2006).

Meningitis. Vestibular loss occurs more commonly than deafness following bacterial meningitis, although the site of lesion, i.e labyrinth or VIII nerve, is unclear. Moreover, it may be difficult to distinguish infective from drug-induced vestibular damage in some cases. Rasmussen and colleagues (1991) have reported that of 94 people who survived pneumococcal meningitis, 9 suffered vertigo and 14 per cent demonstrated bilateral vestibular failure on formal vestibular assessment, 4 to 16 years after the illness, while Naess and colleagues (1994) found audiovestibular abnormalities in a similar number (14 per cent) who had suffered meningococcal meningitis one year previously. Post-meningitic vestibular loss in young children manifests as regression in motor milestones, although long-term follow-up into adulthood indicates that, although vestibular symptoms compensate very well, residual symptoms of imbalance may remain whilst walking in the dark (Hugosson *et al.* 1997). In adults and older children, poor balance and oscillopsia are the common complaints of vestibular failure. MR scan may indicate labyrinthitis ossificans of both the semicircular canals and the cochlea (Fig. 15.19). Other causes of basilar meningitis which may involve the VIII nerve include tuberculosis, cryptococcosis, and coccidioidomycosis (Baloh and Honrubia 2001).

Vascular disorders

Infarction. The vertebrobasilar circulation supplies the labyrinth, VIII nerve, vestibular nuclei, and vestibulo-cerebellar connections. Amarenco and Hauw (1990) have reported auditory and/or vestibular symptoms and signs in over half of patients with anterior inferior cerebellar artery territory infarction. Histologically, half of the patients had involvement of the VIII nerve, and in one this was the only abnormality. As this internal auditory artery arises from the anterior inferior cerebellar artery, this finding may be expected.

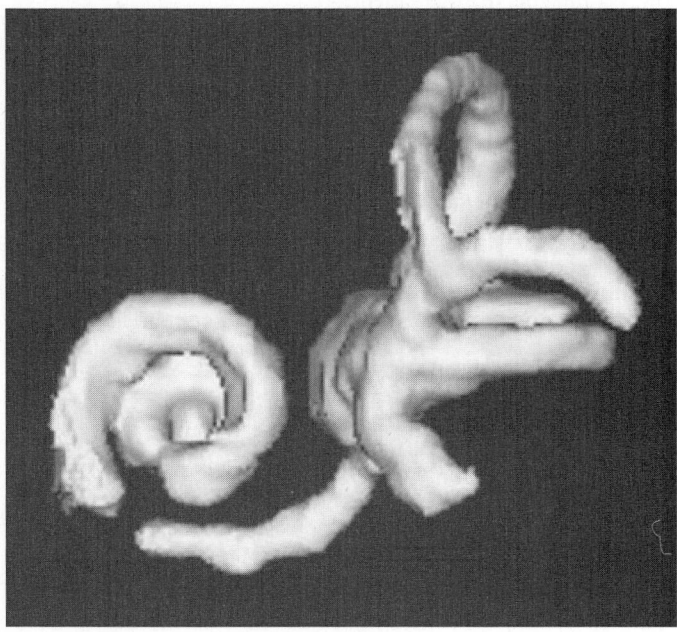

Fig. 15.19 Simultaneous segmentation and registration MRI illustrating involvement of the cochlea and semicircular canals by labyrinthitis ossificans after meningitis. (Courtesy of Dr Catherine Ludman.)

Moreover, aneurysms on this vessel may also involve the VIII nerve (Porter and Eyster 1973).

Vascular loops. Neurovascular compression of the VIII nerve remains a controversial entity. The close relationship of a blood vessel to the VIII nerve on brain MRI does not prove the existence of a causative relationship, even in the presence of vertigo and auditory symptoms. This MRI finding has been reported in 12.5 per cent of otherwise normal scans (Parnes *et al.* 1990). Nonetheless, reports of improvement in 'disabling' vestibular symptoms and signs following VIII nerve microvascular decompression have been made (Jannetta *et al.* 1984). The term 'vestibular paroxysmia' has been coined to describe brief spells of vertigo, often attributed to vascular loops frequently provoked by head position and commonly controlled by carbamazepine (Brandt and Dieterich 1994).

Vascular compression syndromes. Basilar artery ectasia may give rise to VIII nerve symptoms as a consequence of compression or ischaemic events (Passer and Nuti 1996; Passero and Filosomi 1998). Moreover, bilateral vestibular failure has been reported in this condition (Nuti *et al.* 1996).

Neoplasia

Cerebellopontine angle lesions and, in particular, acoustic neurinomas are a rare cause of vestibular symptoms. Despite the misnomer, acoustic neurinomas arise mainly on the vestibular division of the VIII cranial nerve. *Vestibular schwannoma* represents 6 per cent of intracranial tumours, with 13 newly diagnosed cases per million population per year (Sections 11.2 and 27.7.3). The most common presentation is of progressive unilateral hearing loss in 85 per cent, accompanied by tinnitus in 70 per cent and vertigo in 20 per cent (British Association of Otolaryngologists, Head and Neck Surgeons 2002). However, about half will complain of 'imbalance' (Selesnick *et al.* 1993). The clinical presentation is described in five stages (Ramsden 1997) (Table 15.4). As the tumour expands into the cerebellopontine angle, there is involvement of the V and VII cranial nerves, together with ipsilateral cerebellar signs and, ultimately, lower cranial nerve involvement.

While auditory, particularly auditory brainstem-evoked responses, and vestibular investigations are suggestive of the diagnosis, gadolinium-enhanced MRI is the gold standard investigation

Table 15.4 Five stages of clinical presentation of cerebellopontine angle tumours (after Ramsden 1997)

Stage	Features + pathology
1. **Otological stage**	Audiovestibular ± facial nerve involvement usually unilateral hearing loss + tinnitus ± mild instability —includes intrameatal and extrameatal tumours < 2 cm diameter
2. **Trigeminal nerve involvement**	Additional or isolated facial numbness and loss of corneal reflex—suggestive of tumours >2 cm. diameter
3. **Brainstem and cerebellar compression**	Neurological presentation with ataxia, central nystagmus, long tract signs, and lower cranial nerve involvement
4. **Rising intracranial pressure**	Neurological presentation with above symptoms and headache, vomiting, and visual disturbance
5. **Terminal presentation**	Tonsillar herniation and death

(Zealley *et al.* 2000). Recent work has advocated conservative management (Smouha *et al.* 2005) with repeated follow-up scanning for small tumours, but gamma knife intervention and microsurgery may also be appropriate (Yamakami *et al.* 2003). Any rapid increase in size with signs of brainstem compression requires urgent surgical intervention.

Neurofibromatosis 1, von Recklinghausen's disease or *NF1*, and neurofibromatosis 2 or *NF2* are clinically and genetically distinct disorders (Gutmann *et al.* 1997; Mrugala *et al.* 2005; Yohay 2006) (Section 11.2). Autosomal dominant *NF2* is due to loss of function of a tumour-suppressor gene on 22q12, whilst the *NF1* gene is located on chromosome 17q11.2. Management options include expectant policy, with interval scanning, surgical removal, and stereotactic radiosurgery/radiotherapy. The decision regarding management depends on age, tumour size, health status, patient preference, and surgical considerations, but the benefits of each treatment option remain ill-defined (British Association of Otolaryngologists, Head and Neck Surgeons 2002). Postoperatively, vestibular rehabilitation physiotherapy frequently improves postural stability and the perception of disequilibrium (Herdman *et al.* 1995), but vestibular symptoms persist in about one-third of cases postoperatively (Driscoll *et al.* 1998). The differential diagnosis of cerebellopontine angle lesions includes meningiomas, lipomas, haemangiomas, granulomas, and hamartomas (Ramsden 1997).

Metastatic carcinoma in the temporal bone is uncommon, but the internal auditory canal is the second most common site for temporal bone metastases, from primary lesions of the breast, lung, kidney, stomach, larynx, prostate, and thyroid gland (Streitmann and Sismanis 1996). Moreover, cochlear and vestibular symptoms have been reported in 10 per cent of patients with carcinomatous meningitis (Alberts and Terrence 1978).

Paraneoplastic syndrome. The non-metastatic complications of carcinomatous encephalomyelitis may involve the vestibular nerve (Gulya 1993). Almost all tumours have been associated with non-metastatic complications, but small cell carcinoma of the lung is most common. The presence of antineuronal antibodies provides a diagnosis and directs the search for the primary tumour.

Immunological disorders

Audiovestibular problems are commonly associated with neural involvement in sarcoidosis. Five per cent of patients with sarcoidosis develop a granulomatous meningitis, which directly infiltrates the cranial nerves (Section 36.4). The VIII nerve is the fourth most commonly affected and vestibular symptoms may be the first manifestation of neurosarcoidosis (Jahrsdoerfer *et al.* 1981) Steroid treatment may improve audiovestibular symptom, provided it is commenced before permanent damage ensues (Brihaye and Halame 1993).

Involvement of the VIII nerve has also been reported in Hashimoto's disease and rheumatoid arthritis (Stephens 1970), but the site of vestibular symptoms remains poorly clarified in the many systemic autoimmune disorders in which vestibular symptoms have been reported.

15.3.4 Central vestibular disorders

Central vestibular disorders, which include developmental, ischaemic, degenerative, neoplastic, traumatic, or infective and inflammatory pathologies, present with distinct clinical neurological, and oculo-motor signs, which provide accurate indicators for site of lesion diagnosis.

Developmental and genetic disorders

Disorders of the craniocervical junction. Pathology at the cranio-cervical junction can give rise to brainstem and lower cranial nerve symptoms, including vertigo, hearing loss, tinnitus, swallowing difficulties, hoarseness, and rarely airway obstruction. Such symptoms may result from brainstem compression, stretching of the lower cranial nerves or vertebrobasilar ischaemia.

The *Arnold Chiari malformation* is a developmental abnormality in which the brainstem and cerebellum are elongated downwards into the cervical canal (Fig. 15.20) (Section 9.3). The less severe Chiari type presents in adult life and is characterized by oscillopsia, associated with downbeat nystagmus, and gait unsteadiness (Plaza Mayor *et al.* 2006). In addition, cough headache is a common symptom and lower cranial nerve palsies, including hearing loss, and both gait and limb ataxia are typically noted on examination. Diagnosis is by means of MR scanning. Sub-occipital decompression of the foramen magnum may alleviate progression of the neuro-otological abnormalities, but rarely brings about symptomatic improvement (Cristante *et al.* 1994). Recent work has suggested a genetic basis for this condition (Boyles *et al.* 2006). The more severe Chiari type II malformation commonly presents in the first few months of life and is associated with hydrocephalus, spina bifida, and other nervous system malformations (Stevenson 2004).

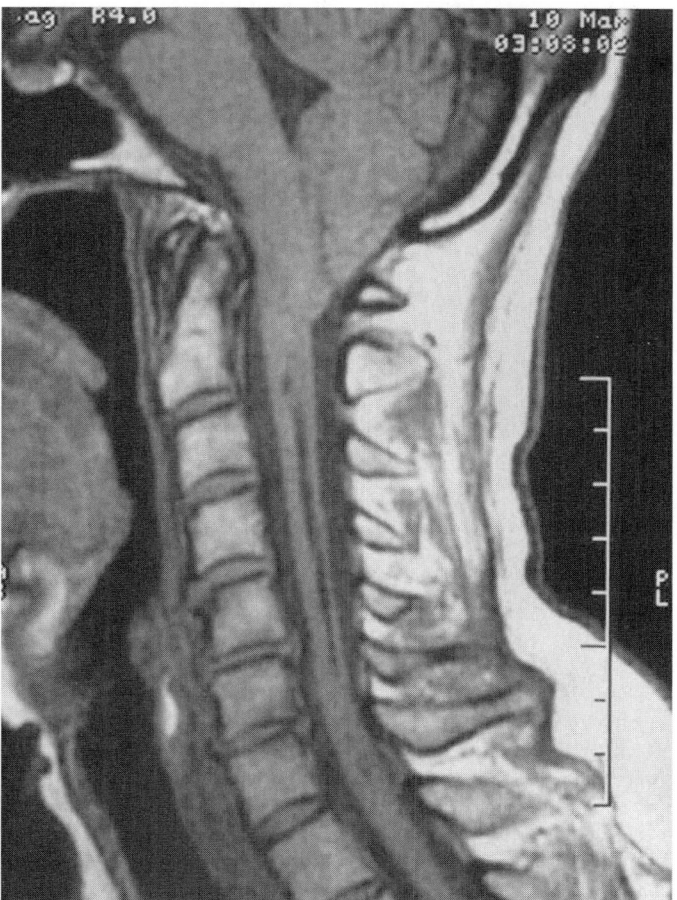

Fig. 15.20 Sagittal MRI showing Arnold Chiari malformation with cerebellar tonsils herniated through the foramen magnum into the cervical spinal canal. (Courtesy of Dr J Stevens.)

Syringomyelia and syringobulbia. Syrinx formation in the brainstem is commonly associated with Arnold Chiari type I malformations and syringobulbia describes the extension of the syrinx into the brainstem. The precise symptoms and signs depend on the course of the syrinx, but commonly patients present with involvement of the lower cranial nerve nuclei, particularly the XII nerve and descending tract and nucleus of V (Aryan *et al.* 2004). Symptoms include atrophy and fasciculations of the tongue, together with loss of pain and temperature sensation in one or both sides of the face. Dysphonia and dysphagia are common with involvement of the IX and X nerve nuclei. Central positional nystagmus is a characteristic finding (Thrush and Foster 1973) and may be the only abnormal neurological sign.

Basilar impression is an upward indentation or invagination of the rigid cervical spine into the convex skull base (Goel *et al.* 1998), with intracranial projection of the odontoid compressing the medulla and the cerebellum being compressed posteriorly by the upper cervical vertebrae. A similar acquired presentation may be seen in the elderly in Paget's disease, and more rarely, it may also occur with rheumtaoid arthritis, osteomalacia, osteogenesis imperfecta, cretinism, and rickets (Menezes *et al.* 1980).

Abnormalities of the cranio-cervical junction are diagnosed by imaging. Basilar impression is confirmed on lateral radiography of the skull with the tip of the odontoid peg extending above Chamberlain's line, a line drawn from the posterior edge of the hard palate to the posterior lip of the foramen magnum. MRI defines the presence of syringobulbia.

Hereditary ataxias

Spinocerebellar ataxias. The hereditary cerebellar ataxias (Chapter 39) are a heterogenous group of inherited degenerative disorders in which the cerebellum is primarily involved, together with its afferent and efferent connections (Schöls *et al.* 2004). With advances in genetics, the spinocerebellar ataxias, SCA, which are included in this category are numbered according to the order in which the associated genetic abnormalities were identified. About one-third of the spinocerebellar ataxias are inherited as autosomal dominant disorders and present with gait ataxia, dysarthria, dysphagia, dysmetria, and intention tremor (Schöls *et al.* 1997). The brainstem and spinal cord, together with the basal ganglia, peripheral nerves, optic nerve, retina, and cerebrum may also be involved in specific syndromes.

Neuro-otologically, auditory and vestibular symptoms are common in many of the hereditary spinocerebellar ataxia syndromes, though cerebellar findings frequently overshadow the loss of vestibular function. Vertigo is frequently not a feature of the progressive cerebellar ataxias as vestibular function is lost slowly, progressively, and symmetrically, but is a common feature of the *episodic ataxia syndromes.* Oculomotor abnormalities are common (Burk *et al.* 1999) and a detailed examination of eye movements may allow identification of the phenotype prior to genetic testing (Buttner *et al.* 1998). Cerebellar eye movement abnormalities, including failure of smooth pursuit function and suppression of the vestibular oculo-reflex with optic fixation, frequently give rise to head movement induced oscillopsia. This latter symptom may also be consequent upon bilateral vestibular failure (Rinne *et al.* 1998). A variety of forms of central pathological nystagmus are observed with these conditions, including bidirectional gaze evoked nystagmus, vertical nystagmus, both upbeat and downbeat, rebound nystagmus, and central positonal nystagmus (Table 15.2).

Specifically, patients with SCA6 may present with spontaneous vertigo that pre-dates the onset of ataxia and may be responsive to treatment with acetazolamide (Jen *et al.* 1998). This condition may also present with downbeat nystagmus (Takahashi *et al.* 2004) and periodic alternating nystagmus (Hashimoto *et al.* 2003). In SCA1 and SCA2, saccades are often dysmetric, but may also be abnormally slow (Buttner *et al.* 1998). In SCA3 there may be a progressive supranuclear ophthalmoplegia. Bilateral vestibular failure has been documented in SCA1 (Buttner *et al.* 1998) and SCA3 (Gordon *et al.* 2003).

Friedreich's ataxia is the most common hereditary ataxia and is autosomal recessive (Section 39.4.1). It is characterized by progressive ataxia, sensory loss, and muscle weakness often with scoliosis, pes cavus, and heart disease (Friedreich 1863). The phenotypic expression of Friedreich's ataxia shows marked intra- and interfamilial variability (Montermini *et al.* 1997). Auditory (Section 14.3.2) and vestibular loss are common features (Ell *et al.* 1984), particularly in the later stages of the disease. Oculomotor findings characteristically include saccadic dysmetria, ocular flutter, and square wave jerks. A variety of other eye movement disorders are reported (Bhidayasiri *et al.* 2005; Spieker *et al.* 1995). The *episodic ataxias* together with SCA6 are channelopathies associated with mutations in ion channel genes (Cannon 2006). They are characterized by attacks of ataxia, beginning in early childhood or adulthood, with essentially normal neurological function in between the acute episodes (Baloh and Jen 2000). Episodic ataxia type II is commonly associated with acute episodes of vertigo, nausea, and vomiting, associated with marked unsteadiness, and with the progression of time, interval nystagmus and cerebellar eye movement abnormalities are observed (Baloh *et al.* 1997). The disorder is responsive to acetazolamide and valproic acid (Scoggan *et al.* 2006).

Cerebrovascular disease

The peripheral vestibular system and central vestibular connections in the brainstem and cerebellum are supplied by the vertebrobasilar system (Fig. 15.21). Ischaemia in this territory frequently results in vestibular symptoms as the vestibular nuclei, which occupy a large area in the lateral zone of the brainstem (Gillilan 1964), are particularly susceptible to a reduction in the blood flow of the main basilar artery. Thus, vertigo or dizziness has been reported as the first and most frequent symptom of vertebrobasilar insufficiency and may occur, not only due to primary vertebrobasilar disease, but also as a result of hypotension, cardiac arrhythmia, or cerebral vasospasm.

Transient vertebrobasilar ischaemia. In 1955, Millikan and Siekert proposed a definition of vertebrobasilar insufficiency: 'a state of transient decrease in the cerebral blood flow, without actual infarction resulting in transient inability to meet the metabolic requirements of the brain'. Episodic vertigo and oculomotor abnormalities are the commonest early symptoms of reduced blood flow in the vertebrobasilar territory (Williams and Wilson 1962). Transient ischaemia gives rise to brief episodes of vertigo, usually of a few minutes' duration and is associated with one or more of the constellation of brainstem symptoms and signs characteristic of ischaemia in the posterior circulation (Grad and Baloh 1989). By definition, symptoms and signs associated with transient

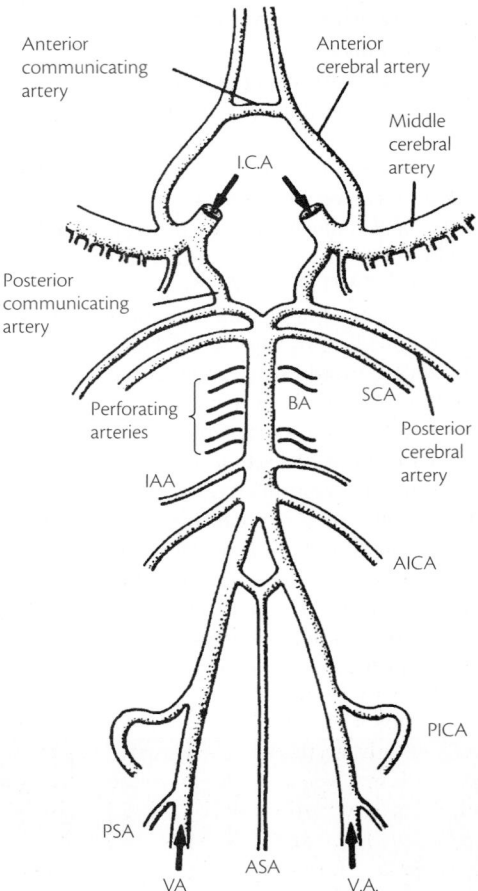

Fig. 15.21 Diagram of vertebrobasilar circulation with internal auditory artery (IAA) supplying the labyrinth and perforating vessels supplying the vestibular nuclei in the brainstem. ICA = internal carotid artery; AICA = anterior inferior cerebellar artery; BA = basilar artery; PICA = posterior inferior cerebellar artery; SCA = superior cerebellar artery; VA = vertebral artery; ASA = anterior spinal artery; PSA = posterior spinal artery.

ischaemic attacks are of less than 24 hours duration. Recurrent vertigo unassociated with additional neurological symptoms should not be diagnosed as ischaemic events (Fisher 1967; Luxon 1990).

The commonest cause of basilar insufficiency is atherosclerosis of the subclavian and vertebral or basilar arteries. Dissection, arteritis, emboli, polycythaemia, and hyperviscosity syndromes may also present in this way. It must be emphasized that whilst cervical spondylosis is common in older patients, mechanical compression of the extracranial vertebral artery is extremely rare in this condition (Baloh and Honrubia 2001). In transient ischaemic attacks MRI is frequently normal, but may show evidence of old infarcts, or diffuse cerebrovascular disease.

Occlusion of the vertebral artery associated with voluntary turning of the head is known as *Bow Hunter's stroke* and is an extremely rare cause of vertebrobasilar ischaemia. A recent review has highlighted the success of vertebral artery decompression (Netuka *et al.* 2005). Unlike benign paroxymal positional vertigo, in this condition, vertigo is not positional, but is precipitated in the upright position by turning the head to right or left.

Total occlusion or stenosis of the subclavian or innominate artery may give rise to the disorder known as the *subclavian steal syndrome*, which occurs in 3 per cent of patients with vertebrobasilar syndromes. Occlusion of the proximal subclavian artery results in reversal of blood flow in the vertebral artery, which then acts as a collateral to the upper limb, and blood is siphoned from the vertebrobasilar system into the distal subclavian artery to maintain adequate blood flow during exercise (Taylor *et al.* 2002). This diagnosis should be considered when claudication or fatigue of the upper limb is accompanied by vertigo and other vertebrobasilar symptoms, although vertigo may be the sole symptom (Wheeler and Vincent 1980). A systolic bruit in the supraclavicular fossa and a disparity of blood pressure between the two arms are the characteristic physical signs of this disorder and surgical intervention is highly effective (Smith *et al.* 1994).

Strokes. A stroke, by definition, involves a neurological deficit present for more than 24 hours or leads to death (Warlow *et al.* 2003). It may be consequent upon infarction or haemorrhage, and vertigo with or without hearing loss may be the main clinical presentation (Murakami 2006).

Brainstem/cerebellar infarction Brainstem and cerebellar infarction may be the consequence of primary atherosclerotic disease in the vertebrobasilar territory (Vilela and Goulao 2005). It is seen also with vertebral artery dissection, and is often traumatic in origin, for instance associated with chiropractic neck manipulation (Saeed *et al.* 2000). Infarction may also occur as a result of small vessel disease with involvement of the deep perforating arteries of the vertebrobasilar circulation. Embolism may also give rise to brainstem or cerebellar infarction as a consequence of cardiogenic embolism, often associated with dysrhythmias such as atrial fibrillation (Hart *et al.* 2000), more proximal artery occlusion (Caplan *et al.* 1992) or paradoxical embolism from a patent foramen ovale (Mas 2003). Brainstem infarction may occur with or without cerebellar involvement as a consequence of the excellent collateral circulation of the major cerebellar arteries (Amarenco 1991). Brainstem and cerebellar haemorrhage are commonly associated with hypertension, but may be seen in association with arteriovenous malformations, coagulopathies, cavernous angiomas, aneurysms, or tumours (Sutherland and Aurer 2006).

Vertigo is a common presenting feature of both brainstem and cerebellar infarction and may be associated with profound nausea, vomiting, and gross postural imbalance with inability to stand or sit. The accompanying symptoms and signs allow differentiation between posterior/inferior cerebellar artery syndrome, anterior inferior cerebellar syndrome, or Wallenberg's syndrome (Section 2.4.4), and superior cerebellar artery infarction.

Lateral medullary infarction. The Wallenberg or lateral medullary syndrome has been ascribed to occlusion of the posterior inferior cerebellar artery, although Fisher (1967) has reported that it is more commonly associated with primary occlusion of the ipsilateral vertebral artery (Fig. 15.22). In young adults, traumatic aortic dissection should be considered (Frumkin and Baloh 1990). The patient presents with sudden onset vertigo, nausea, vomiting, severe imbalance, ipsilateral facial numbness and weakness, diplopia, dysphagia, and dysphonia. There is a motor disturbance which causes both a tendency to deviate towards the site of the lesion, as if pulled by a strong external force, and excessively large voluntary and involuntary saccades directed towards the site of the lesion with hypermetric saccades directed away from the lesion (Kommerell

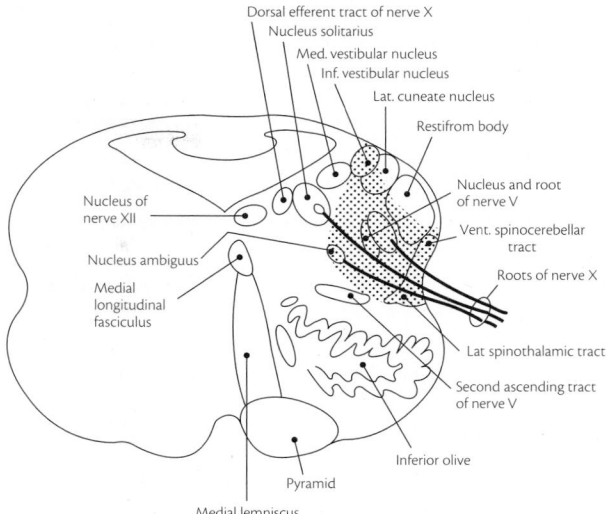

Fig. 15.22 Cross-section of the medulla illustrating the zone of infarction with Wallenburg's syndrome (stippled area). (From Baloh RW and Honrubia V 2001.)

and Hoyt 1973). In addition, there is an ocular tilt reaction with ipsilateral head tilt, the ipsilateral eye being lower than the contralateral eye and ocular torsion with the upper pole of the eye rotated towards the site of the lesion. On neurological examination, there is an ipsilateral dissociated sensory loss in the distribution of the facial nerve, with contralateral truncal loss, and ipsilateral cerebellar ataxia, bulbar palsy, and Horner's syndrome.

Lateral ponto-medullary infarction. The anterior inferior cerebellar artery syndrome, or lateral ponto-medullary infarction (Oas and Baloh 1992) is also characterized by vertigo. However, as the labyrinthine artery arises from the anterior inferior cerebellar artery, infarction of the membranous labyrinth is commonly associated with this condition. Thus, distinguishing between the lateral medullary syndrome and the lateral ponto-medullary syndrome characteristically depends upon the profound unilateral hearing loss in the latter as a result of infarction of the labyrinth, VIII nerve or VIII nerve root entry zone (Table 15.5). Neurological signs include unilateral facial paralysis, cerebellar ataxia and ipsilateral loss of pain and temperature sensation on the face due to involvement of the trigeminal nucleus and tract, together with a contralateral decrease in pain and temperature sensation on the body, due to involvement of the crossed spinothalamic tract.

Cerebellar infarction. Cerebellar infarction may be confused with acute peripheral labyrinthine dysfunction, as the presenting symptoms may be primarily vertigo, vomiting and ataxia, without brainstem signs (Rubenstein *et al.* 1980; Huang and Yu 1985). The diagnosis should be considered in all patients under 50 years presenting with acute dizziness and headache (Savitz *et al.* 2007). The distinction between these two conditions lies in the severe truncal ataxia and direction changing gaze evoked nystagmus associated with cerebellar infarction. The presence of downbeat nystagmus localizes pathology to the caudal midline cerebellum.

The three main cerebellar arteries, the posterior inferior, anterior inferior, and superior, supply branches to the brainstem near their

Table 15.5 Similarities and differences in distinguishing presentations and site of pathology in anterior inferior cerebellar artery (AICA) and posterior inferior cerebellar artery (PICA) syndromes

Clinical features	AICA syndrome	PICA syndrome
Vertigo + nystagmus	Labyrinth, vestibular nerve, flocculus	Vestibular nuclei, posterior inferior cerebellum
Hearing Loss, tinnitus	Cochlea, VIIIn, coclear nucleus	*No involvement*
Gait + limb ataxia	Middle cerebellar peduncle, anterior inferior cerebellum	Ventral spinocerebellar tract, posterior inferior cerebellum
Dysphagia, decreased gag	*No involvement*	Vagal nuclei and nerve
Facial hamianaesthesia	Fifth nerve and nucleus	Fifth nerve and nucleus
Facial paralysis	Seventh nerve	Seventh nerve
Crossed hemisensory loss	Spinothalamic tract	Spinothalamic tract
Horner syndrome	Descending sympathetic fibres	Descending sympathetic fibres

(After Baloh and Honrubia 2001)

origin and form a rich anastomotic network across the surface of the cerebellum. Thus, occlusion of any of these arteries near their origin may result in only brainstem infarction with sparing of the cerebellum. However, more distal pathology gives rise to a cerebellar infarction, which may give rise to progressive symptomatology as a consequence of swelling and compression of the brainstem or may produce hydrocephalus (Sypert and Alvord 1975) with fatal consequences. Urgent medical or surgical intervention is required (Jensen and St Louis 2005).

Brainstem and cerebellar haemorrhage Acute sudden onset vertigo with severe headache may be a prodrome for brainstem (Barinagarrementeria and Cantu 1994) or cerebellar haemorrhage (Elkind and Mohr 1997) associated with hypertension in approximately two-thirds of patients. Although vertigo may be the presenting symptom, there is frequently a rapid deterioration characterized by coma, flaccid quadriplegia, loss of horizontal eye movements, pin point reactive pupils, and ocular bobbing in pontine lesions with rapid cardio-respiratory failure and death in medullary lesions.

Cerebellar haemorrhage deserves particular mention as accurate diagnosis, together with neurosurgical intervention may be life-saving (Jensen and St Louis 2005). Like cerebellar infarction the diagnosis may be confused with acute peripheral vestibulopathy, but the distinguishing features include nucal rigidity and prominent cerebellar signs, together with ipsilateral, facial, and gaze paralysis. Midline cerebellar haemorrhage may be particularly difficult to diagnose, but a characteristic feature is profound inability to stand, which is never associated with peripheral vestibulopathy. Without surgical intervention the prognosis is poor with 50 per cent of patients losing consciousness within 24 hours and 75 per cent becoming comatose within one week.

Imaging is essential to define pathology and should be carried out in any patient with a presentation atypical of a peripheral vestibulopathy CT scanning is superior to MRI for identifying intraparenchymal blood. Any patient therefore who presents with evidence of

cerebellar involvement and vertigo should undergo a CT scan and if this is negative, an MR scan (Amarenco *et al.* 1993). Imaging may also allow delineation of the intra- and extra-cranial circulation in order to identify focal vascular lesions (Vilela and Goulao 2005). Vertebral artery dissection may be identified by the presence of haemosiderin the wall of the vertebral artery, on fat saturated T1 weighted MR images of the neck, while Doppler studies may be useful for demonstrating reversal of blood flow in the vertebral artery in patients with subclavian steel syndrome. MR angiography is of value in assessing the vertebrobasilar circulation, but formal angiography is indicated with severe transient ischaemic attacks without obvious risk factors or in posterior circulation ischaemic events following trauma or neck manipulation. Notwithstanding this, the main risk of contrast angiography is infarction within the distribution of the injected vessel (Sections 3.2.2 and 5.5.5).

Degenerative disorders

Multiple sclerosis Multiple sclerosis is a degenerative disease, in which plaques of demyelination are scattered in time and space throughout the central nervous system (Section 37.5). The demyelination is confined to central nervous system myelin, and therefore plaques involving the vestibular (Gass *et al.* 1998) and auditory root entry zones may present with sudden hearing loss or vertigo (Commins and Chen 1997). Vestibular symptoms are common and may be the presenting feature (Grenman 1985; Rae-Grant *et al.* 1999).

Acute vertigo may present in up to 5 per cent of cases of multiple sclerosis, but may occur at some point in the disease in up to 50 per cent (Grenman 1985). Hearing loss (Section 14.3.2) occurs in about 10 per cent of patients and commonly presents as an acute unilateral loss with gradual recovery, while midline brainstem plaques may give rise to bilateral loss and more proximal plaques may give rise to auditory processing dysfunction (Section 14.3.3). Multiple sclerosis may mimic acute vestibular neuritis or labyrinthitis and should be included in the differential diagnosis of these conditions (Sasaki *et al.* 1994; Thomke and Hopf 1999).

The clinical presentation varies depending on the distribution of cerebral lesions, but brainstem and cerebellar symptoms and signs are common, in addition to optic atrophy secondary to optic neuritis. Oculomotor findings are almost the rule (Alpini *et al.* 2001) and disordered pursuit, hypermetric saccades, and a range of central nystagmus, including bidirectional gaze evoked, rebound, periodic alternating, vertical, torsional, see-saw ataxic nystagmus, internuclear ophthalmoplegia, and pendular nystagmus have all been reported in multiple sclerosis, with the latter two conditions being particularly suggestive of this diagnosis. Positional vertigo and central positional nystagmus are common in multiple sclerosis particularly with plaques in the floor of the fourth ventricle (Buttner *et al.* 1999).

The gold standard for diagnosis is T2-weighted MRI showing characteristic white matter plaques in approximately 95 per cent of patients (Pyhtinen *et al.* 2006). Clinical examination may reveal a diversity of signs including involvement of the pyramidal tracts, cerebellum, and sensory tracts, but characteristically patients will demonstrate addition oculomotor abnormalities, for example, internuclear ophthalmoplegia, bidirectional gaze evoked nystagmus, central positional nystagmus, together with disordered pursuit, optokinetic responses, and saccades, with the exact abnormalities depending upon the site of lesions within the brainstem and cerebellum.

Neoplasia

Dizziness and/or vertigo are early or initial symptoms in 25 per cent of brainstem tumours. In later life, metastases are the most common neoplasms involving the brainstem and/or cerebellum, which give rise to vestibular dysfunction. Brain-stem lesions typically present with progressive cranial nerve palsies together with long tract signs, while midline cerebellar lesions give rise to truncal ataxia and oculomotor abnormalities, including impaired smooth pursuit, saccadic dysmetria, and rebound nystagmus. Hemispheric cerebellar lesions cause ataxia of the ipsilateral limbs, with truncal ataxia. Infratentorial tumours account for 60 to 70 per cent of intrinsic brain tumours in children and present with gait abnormalities, ataxia, dizziness, cranial nerve palsies, symptoms of increased intracranial tension, irritability, and decline in school performance. Temporal lobe tumours give rise to 'disequilibrium' more frequently than in any other cortical site. This is not surprising given that the temporal lobes exert a modifying influence upon the vestibular nuclei.

Infection

Middle ear infections can give rise to life-threatening intracranial complications including extradural abscess, subdural abscess, sigmoid sinus thrombosis, meningitis, brain abscess and otitic hydrocephalus and in about one third of patients there are multiple complications within this group (Ludman 1997). Meningitis remains the most common complication, but in all complications imbalance or vertigo are overshadowed by fever, headache, vomiting and frequently drowsiness.

Cerebral trauma

Post-traumatic vertigo is most commonly the result of a head injury, whiplash injury, or more rarely barotrauma. With mild head injuries, dizziness is reported to occur within one week in 53 per cent (Levin *et al.* 1987), but persists for at least two years in 18 per cent (Cartlidge 1978). In some patients dizziness and imbalance may remain intrusive for more than 4 years after injury (Edna and Cappelen 1987). The relevance of these figures lies in the very high percentage of the population in civilized societies suffering head injury and reports of even mild or moderate head injury (Davis and Luxon 1995) giving rise to significant disability, with a significant number of injured people failing to return to their pre-accident or equivalent level of work within 5 years (Eide and Tysnes 1992; Berman and Fredrickson 1978).

Blunt head injury is the commonest cause of post-traumatic vertigo and recent radiographic classification of otic capsule violating fractures versus otic capsule sparing fractures has enabled the prediction of temporal bone complications, including facial nerve injury, labyrinthine involvement, and cerebrospinal fluid otorrhoea (Little and Kesser 2006). In general, pathology disrupting the otic capsule gives rise to profound sensorineural hearing loss and vestibular failure, while fractures that do not encroach upon the otic capsule give rise to vestibular symptoms as a consequence of labyrinthine concussion, perilymph fistula, and canalolithiasis (Section 15.3.2). In major head injury, post-traumatic vertigo is frequently overshadowed by other more life-threatening neurological abnormalities. Evidence of the skull fracture should be sought, progressive ecchymoses of the mastoid region, the Battle sign, may be associated with longitudinal fractures extending into the mastoid region. In this group of patients, dizziness or vertigo has been reported in up to 93 per cent (Wennmo and Svenson 1989). With transverse fractures there is commonly involvement of the VIII cranial nerve with severe vertigo, nausea, vomiting, and profound hearing loss. Vestibular symptoms tend to be less dramatic with longitudinal fractures in which laceration of the tympanic membrane with cerebrospinal fluid or bloody otorrhoea or haematympanum give rise to a conductive hearing loss. Neurologically, cerebral concussion is commonly associated with dizziness and vertigo, and a detailed vestibular investigation is required to differentiate the underlying pathology (Ernst *et al.* 2005; Luxon 1996). Six weeks after cerebral concussion approximately 50 per cent of patients have totally recovered but the remainder complain of symptoms and 14 per cent report dizziness (Rutherford 1977). Psychosocial factors are important in the persistence and severity of symptoms (Fenton 1996).

Whiplash injury describes the mechanism of hyper-extension, followed by flexion of the neck. Imbalance, including vertigo, is second only to cervical pain and headache following whiplash injury and is reported in up to 85 per cent of patients following rear-end road traffic accidents (Oosterveld *et al.* 1991). The symptoms of whiplash injuries develop rapidly and within 24 hours of injury, 93 per cent of patients present with neck pain. Hence, dizziness presenting weeks or months after injury is unlikely to be the consequence of the injury—an important medicolegal consideration. The mechanism of vertigo following whiplash injury is unclear, but a variety of mechanisms have been suggested (Luxon 1996) to explain the presence of both peripheral and central vestibular abnormalities on formal neuro-otological testing, including injuries to cervical muscles and discs (Endo *et al.* 2006), involvement of cervical sympathetic nerve supply and cervical nerve roots, damage to the otolith organs, brainstem damage, spinal cord contusion/haemorrhages, and alterations in blood flow.

Epilepsy

Vertigo may present as an aura of epilepsy (Gowers 1907), or as part of a temporal lobe seizure, but has also been identified with other forms of epilepsy (Lennox 1960; Schneider *et al.* 1968). Individual cases are reported throughout the literature, but relatively little advance has been made in the inter-relationship of the phenomenon for a number of reasons: the absence of pathophysiological data, the vague description of the patient's symptoms and the lack of objective electrophysiological recordings contemporaneous with the vertigo. (Brandt 1999a). Smith and Docherty (1982) reported a case of temporal lobe epilepsy with oscillopsia and nystagmus, Furman and coworkers (1990) described a woman with episodic ataxia and nystagmus in association with EEG changes in the temporo-parietal region and more recently Kluge and colleagues (2000) reported a 5-year-old boy with left frontocentral onset of epileptic discharges accompanied by complaints of vertigo. Thus epileptic foci in a range of sites may be accompanied by vertigo. Epileptic vertigo must be distinguished from 'vestibulogenic epilepsy' a very controversial entity in which labyrinthine stimulation has been reported to precipitate an epileptic attack. There is little evidence to support this theory (Brandt 1999a). Notwithstanding the rare occurrence of epileptic vertigo in adults, the diagnosis of epilepsy in 'funny turns' and vertigo in children is more common and requires a high index of suspicion (Murphy and Dehkharghani 1994).

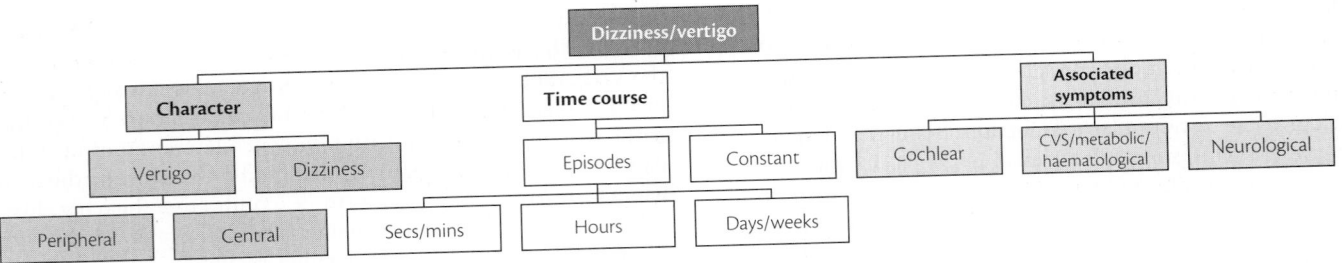

Fig. 15.23 Diagram to illustrate key points in the vestibular clinical history.

Cervical vertigo

Cervical vertigo is defined as vertigo induced by changes of position of the neck in relation to the body, but there is much controversy as to the diagnosis and underlying pathophysiology of this entity (Brandt 1996b). Sympathetic irritation resulting in vertebro-basilar ischaemia, intermittent vertebral artery compression by osteophytes caused by cervical spondylosis, and deranged sensory input from the cervical kinesthetic receptors have all been postulated; there are no clear-cut clinical signs.

15.4 Clinical assessment

Good management relies upon accurate diagnosis. Neuro-otological assessment may enable the identification of a single pathology in many cases, but certain generalizations deserve mention:

- Some pathologies give rise to symptoms as a result of involvement of more than one site, for example, cardiovascular disease with ischaemic labyrinthitis and brain stem or cerebellar involvement.

- Patients who develop an acute peripheral vestibular disorder, for example vestibular neuritis, may make an excellent initial recovery, but after either physical or psychological stress, for example influenza, bereavement, redundancy, they may 'decompensate' and suffer a recurrence of symptoms (Fig. 15.9).

- In the elderly population, disequilibrium may be the result of an interaction among multiple pathologies. For example, visual impairment associated with cataracts, proprioceptive loss due to arthritis, and vestibular impairment due to vascular disease. In isolation each such impairment may be trivial, but together they give rise to a significant balance problem which is commonly termed the *multisensory dizziness syndrome* (Drachman and Hart 1972).

15.4.1 Clinical history

A full general medical history, with specific reference to the cardiovascular and neurological systems should be obtained. In addition, the following neuro-otological history should be sought (Fig. 15.23).

Character of symptoms

Vertigo is a cardinal manifestation of a disordered vestibular system, whereas dizziness, light-headedness, faintness, or giddiness, are more common in general medical pathologies. Importantly, patients may report bizarre complaints, which tend to be dismissed by the non-expert, for example 'my brain is sloshing around inside my skull': 'I feel like a goldfish in a bowl', and 'I feel that my brain is lagging behind my head'. Symptoms of depersonalization, difficulty concentrating, and impaired memory are all common with vestibular pathology. In addition, visual vertigo with nausea precipitated by strong visual stimuli, such as busy stations, supermarkets, escalators, patterned carpet, fast-moving television scenes, or scrolling computer screens, is common (Bronstein 2005) (Fig. 15.23).

Acute vertigo. Classically, vertigo of acute peripheral labyrinthine origin is unprecipitated, lasts less than a day, with rotational movement of the world, known as objective vertigo, or movement of the patient, subjective vertigo. Commonly there is accompanying nausea and vomiting, and more rarely diarrhoea. The patient becomes pale, sweats profusely, and is often acutely anxious.

Falls. In the case of falls, the history requires the evaluation of risk factors such as drug intake, neurological conditions, cognitive function, or general medical disorders. Environmental factors should be taken into account including an uneven or slippery surface, poor lighting, or an unfamiliar environment. Potentially treatable predisposing factors should be identified and corrected. A patient who suffers a single uncomplicated fall with no sequelae may be reassured and discharged, but recurrent falls with or without injury require careful assessment (Fig. 15.24) and a detailed evaluation of balance. For any particular fall there is a *liability* to fall, and an *opportunity* to fall factor. Both should be identified.

Instability A complaint of instability, unsteadiness or 'weakness' of the legs requires neurological investigation specifically to exclude cerebellar and brainstem disease associated with ataxia, proprioceptive loss, and neuropathy.

Time course of disequilibrium

The time course should be defined for both the entire illness and the duration of individual episodes Firstly, it should be clarified whether the symptoms are constant or episodic, and if the latter, whether the episodes occur erratically or in clusters, with relapses and remissions.

'*Short episodes*' Attacks of acute rotational vertigo of less than one minute's duration, which occur in clusters, are most commonly associated with benign paroxysmal positional vertigo (Section 15.3.1), which is particularly common after head injury and in the older age group. The periods of freedom from episodes may last many weeks or months. Particularly in the elderly, such brief episodes of dizziness may be misattributed to vertebrobasilar ischaemia, but in the absence of any concomitant neurological symptoms or signs, this diagnosis is unlikely.

'*Medium episodes*' Episodes of vertigo lasting several hours are common with migraine and Ménière's disease. The latter may usually be distinguished by fluctuating auditory symptoms and

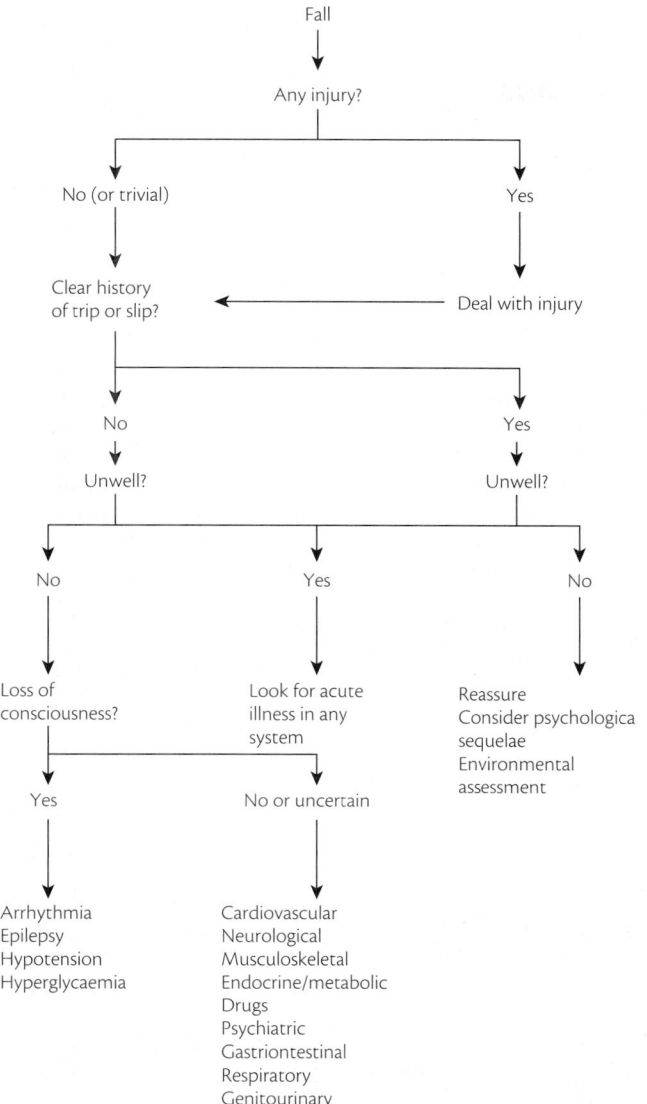

Fig. 15.24 Flow chart to analyse causation of falls. (From Downton 1969.)

movement abnormalities, are the rule. Within the brainstem, vestibular symptoms are commonly associated with eye movement abnormalities, and other cranial nerve and long-tract symptoms and signs. It should be noted that bilateral vestibular failure may also present with chronic disequilibrium, associated with bobbing oscillopsia and reduction of visual acuity on movement.

Associated Symptoms

Cochlear, general medical, and neurological symptoms occurring concurrently with the vestibular symptoms should also be sought. Pathology within the VIII nerve and labyrinth commonly involves both vestibular and auditory systems. However, in the elderly, who consider hearing loss as part of the ageing process, cochlear symptoms, including tinnitus, hearing loss and dysacusis remain unreported in association with vertigo unless specifically sought. A plethora of general medical disorders, including haematological, cardiovascular, and endocrine conditions may be associated with dizziness, and relevant systematic questioning should be undertaken.

Precipitating factors

Triggers for a patient's symptom of disequilibrium should be identified. These vary depending upon underlying pathology. For example dizziness on standing from lying may suggest postural hypotension, while dizziness on tipping the head backwards may suggest benign positional vertigo. Oscillopsia and unsteadiness, particularly on uneven surfaces or in the dark, will suggest vestibular failure or a sensory ataxia due to loss of proprioceptive feedback.

Drug history

The importance of a detailed drug history cannot be overemphasized, as many drugs give rise to dizziness as a side effect (Table 15.6). The ototoxic effects of aminoglycoside antibiotics, and loop diuretics, should be borne in mind.

15.4.2 Clinical examination

A full general medical examination with particular reference to the fundi, visual fields, visual acuity, and neurological and cardiovascular systems is essential. The specific clinical examination for vestibular and balance disorders includes:

- examination of the external ear and tympanic membrane, together with clinical tests of auditory acuity and tuning fork tests;
- assessment of eye movements, including smooth pursuit, saccades, and optokinetic nystagmus;
- assessment of vestibulo-ocular function by inspecting for spontaneous and positional nystagmus;
- assessment of vestibulospinal function: Romberg test, stance, and gait.

A basic understanding of vestibular physiology and pathology is of key importance for a suitable neuro-otological examination.

Otological examination. In all patients with episodic vertigo, active chronic middle-ear disease with labyrinthine erosion must be excluded, and obvious otological abnormalities evaluated in the context of the clinical history and other signs. An abnormal appearance of the tympanic membrane requires an expert otological opinion. The presence of a perforation precludes caloric testing. A fluid level, cholesteatoma, or the bluish bulge of a glomus jugulare

accompanying tinnitus, while in migraine a clear family history or past history of characteristic headache may facilitate diagnosis.

'Long episodes' Acute vertigo lasting more than 24 hours raises the possibility of an acute vestibular failure, due to a range of pathologies including vestibular neuritis, labyrinthitis, a vascular event, or trauma.

'Chronic vertigo' As noted above, following an acute vestibular event, symptomatic recovery occurs over a period of several weeks. Visual and somatosensory inputs and the cerebellum play a crucial role in recovery. The commonest cause of chronic vertigo is failure of compensation or intermittent decompensation after an acute labyrinthitis. The factors giving rise to failure of compensation (Fig. 15.8) should be specifically sought and excluded in patients presenting with chronic vertigo. More rarely, chronic persistent vertigo may result from central neurological disease such as multiple sclerosis, spinocerebellar degeneration, or vascular disease. Aassociated neurological symptoms or signs, particularly eye

Table 15.6 Common drugs with the side-effect of dizziness

Psychotropic drugs	
Antidepressants	Tricyclics, monoamine oxidase inhibitors, selective serotonin reuptake inhibitors
Tranquillizers	Benzodiazepines, phenothiazines
Anticonvulsants	Phenytoin, carbamazepine, gabapentine, lamotrigine
Analgesics	Paracetamol, acetylsalicylate, nonsteroidal anti-inflammatory drugs, opioids
Cardiovascular drugs	
Antihypertensives	Diuretics (thiazides and loop), β-blockers, calcium-channel blockers, angiotensin converting enzyme inhibitors, methyldopa, hydralazine
Anti-arrhythmic	β-blockers, verapamil, mexiletine, flecainide, miodarone, disopyramide
Anti-angina	Nitrates, calcium-channel blockers, β-blockers, potassium-channel activators
Anti-microbials	Aminoglycosides, tetracyclines, macrolides, chloroquine, isoniazid
Anti-allergic drugs	Non-sedating and sedating antihistamines
Hormone replacement/ substitute	Hypoglycaemics, corticosteroids, hormone replacement therapy
Chemotherapeutic agents	Cisplatin, busulfan, cyclophosphamide, vinblastine, methotrexate

tumour should be noted. The presence of an auditory deficit as judged by tuning fork tests or whispered-voice test may suggest labyrinthine or VIII nerve pathology, but audiometry is the definitive test to exclude such disorders.

15.4.3 Eye movements

For the purposes of a vestibular examination, it is important to evaluate saccades, smooth pursuit, and optokinetic nystagmus (Section 13.2.1).

Saccades are rapid eye movements which may be assessed clinically by asking the patient to alternating fixate between targets at angle of 30° to right and left, and 30° up and down from the mid-position of gaze, to assess horizontal and vertical saccades respectively. Delayed initiation, inaccuracy of fixation, or slow velocity of the saccades may all indicate neurological disease. These will impair the vestibular nystagmic response, as the fast phases of nystagmus are involuntary saccades.

Smooth pursuit is a slow eye movement which is intimately related to the mechanism of suppression of vestibular responses by optic fixation. It is assessed clinically by asking the patient to follow a target moving in the horizontal and then the vertical plane, and assessing 'smoothness' of the eye movement. 'Broken' pursuit implies catch-up saccadic intrusions to enable the subject to maintain gaze on the moving target, if the smooth pursuit mechanism is deranged. Asymmetrically deranged pursuit almost always indicates neurological pathology, but symmetrically disordered pursuit may be pathological, for instance due to cerebellar or brainstem dysfunction. It may be associated with fatigue, old age, or psychotropic drugs including anticonvulsants, transquillisers, and antidepressants.

Optokinetic nystagmus is a reflex oscillation of the eyes in response to movement of the visual surround. For example, it may be seen in the eyes of train passengers who are watching objects passing outside the carriage window. The pathways subserving the optokinetic response include a cortical pathway, which is intimately related to smooth pursuit, and a subcortical pathway which passes through the vestibular nuclei (Fig. 15.25). Clinically, the response may be elicited by using a hand-held striped drum, or a piece of striped material such as school scarf or tie which is moved to the right and left, or up and down in front of the patient's eyes, to examine horizontal and vertical optokinetic responses respectively. Asymmetry of horizontal or vertical nystagmus is sought. The importance of this test lies in the differentiation of peripheral from central vestibular pathology. In peripheral vestibular disease, the cortical pathway overrides, and no abnormality of optokinetic nystagmus is observed, except in the acute phase of a peripheral vestibular lesion, when spontaneous nystagmus may confound the optokinetic response. However, in neurological disease giving rise to disequilibrium, optokinetic abnormalities are seen in 60 to 75 per cent of patients, depending on the precise site of lesion (Table 15.7).

15.4.4 Spontaneous nystagmus

Detailed evaluation of spontaneous nystagmus is the key to defining the presence and location of vestibular pathology. Clinically, it

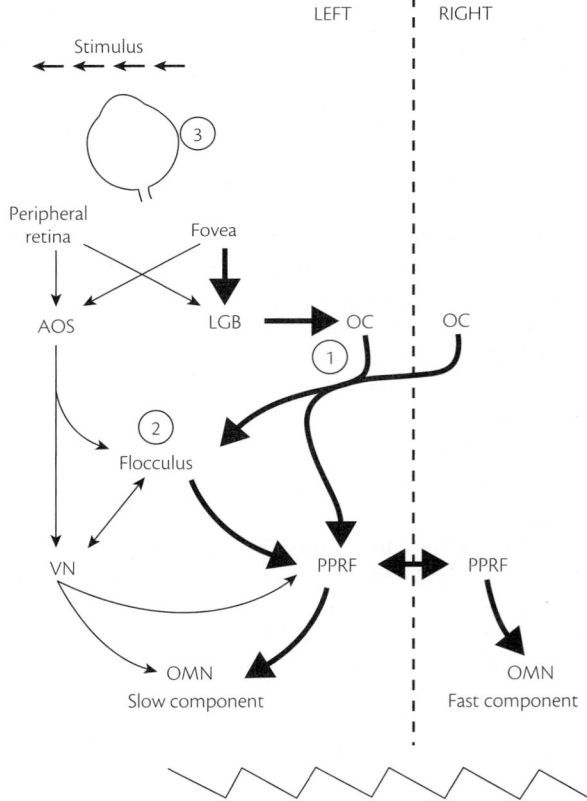

Fig. 15.25 Diagram to illustrate the cortical and subcortical pathway of optokinetic response. AOS = accessory optic tract; LGB = lateral geniculate body; OC = occipital cortex; VN = vestibular nuclei; PPRF = parapontine reticular formation; OMN = oculomotor nuclei; 1 = smooth pursuit pathway from visual cortex to flocculus, 2 = cerebellum; 3 = globe of eye (Yee *et al.* 1982).

Table 15.7 Optokinetic and smooth pursuit abnormalities in patients with pathologies associated with vestibular symptoms

Site of lesion	Optokinetic nystagmus		Smooth pursuit		No. of patients
	Abnormal (%)	Normal (%)	Abnormal (%)	Normal (%)	
Brainstem	59	41	81	19	126
Cerebellum	72	28	94	6	99
Basal ganglia	58	42	75	25	36
Labyrinth/VIIIn	1	99	3	97	384
Parietal lobe	25	77	75	25	8
Other lobes	0	100	0	100	5
Cranial nerve palsy	0	100	25	75	4
Ocular myopathy	0	92	8	92	24

From Honrubia and Luxon (1984).

is important to ensure that the eyes are conjugate and identify latent nystagmus, by performing a cover test, which may confound the interpretation of vestibular nystagmus. The presence of involuntary eye movements such as square-waves or saccadic flutter should be sought, and spontaneous nystagmus should be evaluated in the primary position of gaze, with eyes deviated to no more than 30°, to exclude the possibility of physiological endpoint nystagmus. Both horizontal and vertical nystagmus should be sought and observed, both with and without optic fixation using either Frenzel's glasses or observing the eyes in a blackened room with an infra-red viewer.

Unilateral loss of afferent vestibular input gives rise to spontaneous nystagmus, which is directed, as defined by the direction of the fast phases, towards the normal ear. Irritative lesions may cause nystagmus with the quick component towards the disordered ear. Horizontal nystagmus of peripheral, labyrinthine, or VIII nerve origin obeys *Alexander's Law*, which states that the nystagmus is always in one direction, irrespective of the direction of gaze, and that the intensity of the nystagmus is greatest when the eyes are deviated in the direction of the fast phase (Fig. 15.26). Nystagmus is described as:

◆ first-degree when the fast phases are in the same direction as the direction of gaze;

◆ second-degree when it beats in one direction in the primary position of gaze;

◆ third-degree when the fast phases are directed in the opposite direction to the direction of gaze.

In an acute vestibular lesion, one would expect to see third-degree nystagmus immediately after the onset of loss of vestibular function, but with the passage of time the nystagmus will gradually abate, such that it becomes second-degree and then first-degree, and finally is only observed in the absence of optic fixation. The following types of nystagmus arise from central neurological disorders, and imply the need for further neurological investigation (Table 15.8):

Table 15.8 Types of pathological spontaneous nystagmus and site of lesion

	Site of lesion
Unidirectional horizontal (obeying Alexander's law)	Labyrinthine, VIII nerve, or vestibular nuclei
Gaze evoked (uni-or bidirectional)	Brainstem or cerebellar flocculus
Rebound	Cerebellar flocculus
Brun's (ipsilateral gaze evoked + contralateral vestibular)	Ipsilateral CP angle lesion involving VIII nerve + cerebellum
Vertical	
Upbeat	Perihypoglossal nucleus, dorsal cerebellar vermis, pontomedullary/ pontomesencephalic tegmentum
Downbeat	Cerebellar uvula and flocculonodular lobes
Torsional	Medulla, pontomedullary junction, midbrain
Ataxic dysconjugate	Medial longitudinal fasciculus
Dissociated (Torsional, horizontal, vertical varying in each eye)	Lesions of Posterior fossa
Seesaw	Lesions near optic chiasma ?compression of midbrain tegmentum
Periodic alternating	Caudal brainstem
Convergence retraction	Diencephalic midbrain junction
Pendular	Dentate nucleus, superior cerebellar peduncle, inferior olive

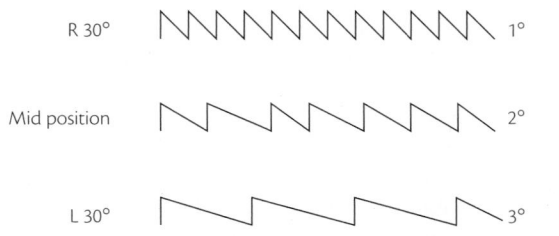

Fig. 15.26 Diagram to illustrate Alexander's law, with the direction of the nystagmus fast phase always being in the same direction irrespective of the angle of gaze.

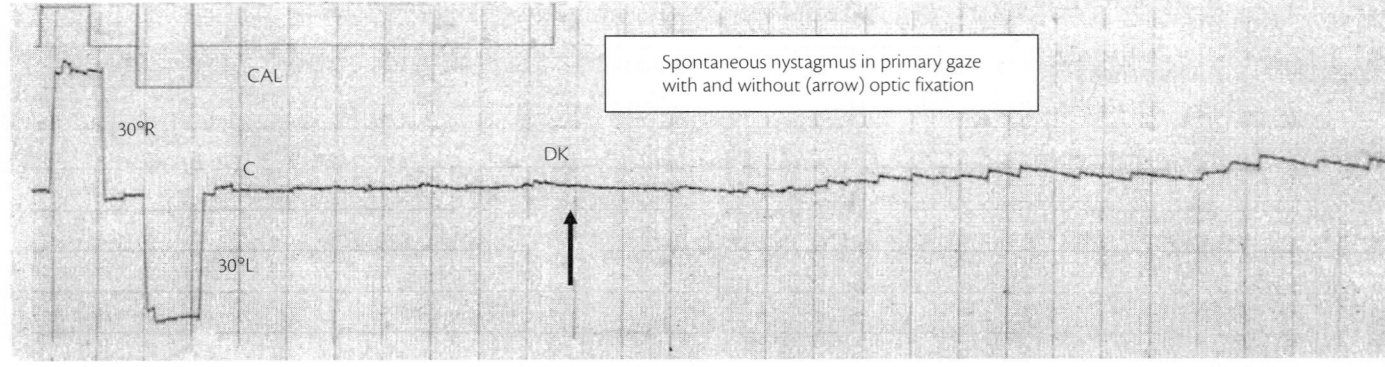

A

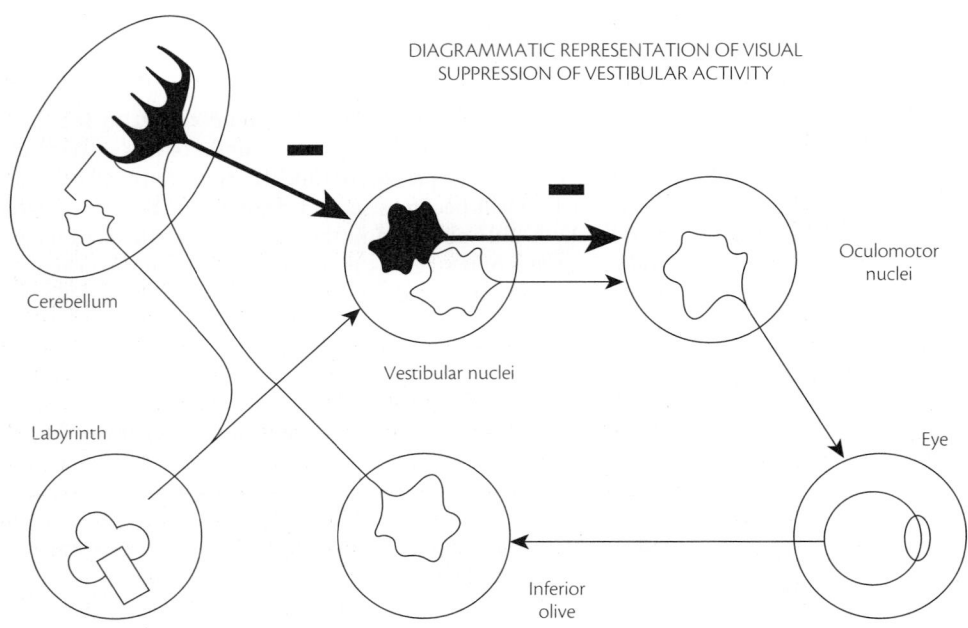

DIAGRAMMATIC REPRESENTATION OF VISUAL
SUPPRESSION OF VESTIBULAR ACTIVITY

B

Fig. 15.27 (A) Electronystagmographic trace showing suppression of vestibular nystagmus by optic fixation (B) Diagram to illustrate pathways subserving visual suppression of vestibular activity. → = afferent pathway of vestibulo-ocular reflex; ⟶ = inhibitory pathway from cerebellum via vestibular nuclei. (After Baloh RW and Honrubia V, 2001.)

◆ bidirectional, or direction-changing nystagmus, for example, first-degree nystagmus to the right on looking to the right, and first-degree nystagmus to the left on looking to the left;

◆ vertical nystagmus, as upbeat or downbeat nystagmus or both;

◆ disconjugate nystagmus, a differing nystagmic response in each eye.

Under normal circumstances, the nystagmus generated by a peripheral labyrinthine stimulus, be it pathological or physiological, may be suppressed by visual fixation upon a target (Fig. 15.27). In attempting to differentiate peripheral from central unidirectional nystagmus, the effect of optic fixation is invaluable. In the absence of optic fixation, spontaneous nystagmus of peripheral type will be enhanced, whereas in the absence of optic fixation, spontaneous nystagmus of central origin will be attenuated, demonstrating a lower frequency and velocity (Fig. 15.28).

15.4.5 Positional nystagmus

Positional nystagmus is an important sign, as it may be the only abnormality observed on clinical examination. Thus in every dizzy patient, a briskly performed *Hallpike manoeuvre* should be carried out (Fig. 15.11). The principle of the test should be explained to the patient, who should be told to observe the bridge of the examiner's nose throughout the test. The head is turned 45° to the right or left, the side being determined by any suggestion of the patient as to which side is more likely to precipitate symptoms. The subject is then taken rapidly backwards with the head tipped over the edge of the bed at an angle of approximately 45°. The patient's eyes are observed for the development of any nystagmic movement. If positional nystagmus develops, it is observed until it disappears, or for 2 or 3 minutes, after which it may be assumed that the nystagmus is persistent. The patient is then returned to the upright position, and the eyes are carefully observed for any reversal of a positional response. The procedure is then repeated with the head turned in the opposite direction.

In general terms, positional nystagmus may be divided into two main types, as outlined in Table 15.3, although their differentiation may prove difficult (Buttner *et al.* 1999). Nystagmus of peripheral type demonstrates a latent period of up to 30 seconds, followed by

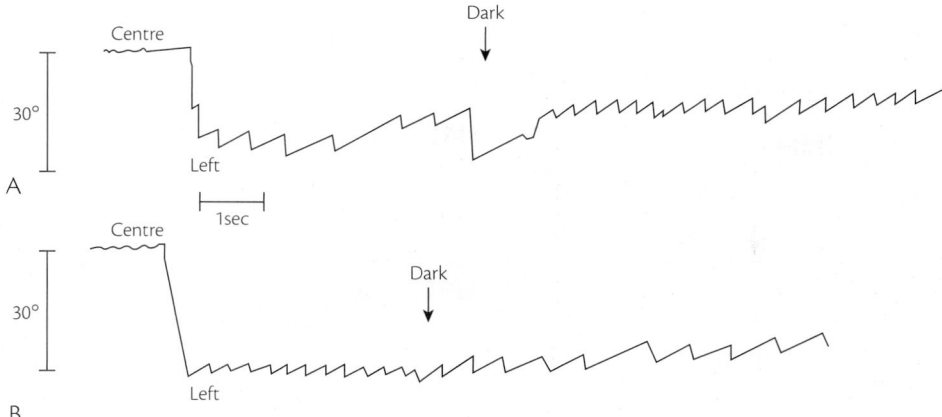

Fig. 15.28 Electronystagmographic trace to illustrate (A) First-degree spontaneous vestibular nystagmus to the left, with enhancement following the removal of optic fixation (dark) (i.e. 'peripheral'); and (B) Attenuation of gaze-evoked (ie 'central') left beating spontaneous nystagmus with removal of optic fixation (dark).

severe vertigo with rotatory nystagmus most commonly directed towards the undermost ear. During this period the patient may be distressed and feel acutely dizzy and nauseated. The symptoms and signs then gradually adapt, and if the patient is sat up, a reversal of the nystagmic response may be observed. If the procedure is then repeated in the same direction, the nystagmus frequently is less marked or absent, known as fatiguability. Care must be taken to carry out the procedure correctly on the first attempt, as it is possible that if the patient shuts their eyes and the examiner cannot observe the response, it will not be present on a second attempt.

The importance of the differentiation of positional nystagmus lies in the fact that transient positional nystagmus of the type described above is almost always of benign peripheral type, and easily treatable. By contrast positional nystagmus without a latent period, and which is vertical (Bertholon *et al.* 2002), or direction-changing, or which cannot be attributed to one of the forms of benign paroxysmal positional vertigo (Section 15.3.1) affecting each of the semicircular canals, is found to be of neurological origin in approximately 40 per cent of patients.

15.4.6 Gait and stance

Vestibulospinal function cannot be assessed directly, and clinical examinations of balance are non-specific and insensitive in comparison with assessment of vestibulo-ocular function. Nonetheless, stance and gait assessments provide invaluable information with respect to a patient's disability.

The *Romberg test* (1846) is performed by asking the patient to stand in the upright position with feet together, arms by the side, head straight forward and eyes closed. A tendency to sway to one side usually suggests peripheral vestibular pathology. An initial inability to stand with the feet together with the eyes open is more characteristic of cerebellar dysfunction. However, it must be emphasized that the Romberg test was initially developed to assess posterior column loss in tabes dorsalis, only becoming positive when the eyes are closed and depends upon a variety of sensory and motor systems. Thus, it is non-specific for vestibular disease. Patients with a psychological overlay to their vestibular disorders, often tend to fall promptly backwards like a wooden soldier on performing the Romberg test (Section 2.2.2).

The Unterberger test (Fig. 15.29) involves the patient marching up and down on the spot, with their arms closed and their hands clasped at an arm's length in front of them. Normal subjects show little tendency to deviate to right or left, whereas patients with peripheral or central vestibular disorders tend to move linearly forwards or backwards, in addition to rotating to right and left. However, the value of the test in correlating findings with side and location of lesion is very limited.

Gait testing is assessed by asking a patient to walk towards a fixed point with eyes first open and then closed. In peripheral vestibular lesions there is a tendency to veer towards the side of the lesion, whereas in central neurological disorders and bilateral vestibular loss there is a tendency to walk on a wide-based gait. A variety of other gait disorders associated with balance dysfunction may be observed and provide valuable diagnostic information, for example, the slow shuffling parkinsonian gait, the high-stepping foot-dropping gait of posterior column loss or peripheral neuropathy and bizarre non-organic gait patterns that are frequently observed in patients with vestibular dysfunction, in whom a diagnosis has not been made and psychological overlay becomes prominent (Section 2.6).

15.4.7 Psychological correlates

The psychological correlates of vestibular disease deserve special consideration. In patients with persistent dizziness, psychiatric disorders have been demonstrated to be the second most common cause of symptoms, occurring in 10 to 25 per cent of patients (Jacob *et al.* 2003). Conversely, complaints of dizziness and feelings of loss of balance are extremely common in psychiatric patients, especially those with panic and other anxiety disorders such as agrophobia. Recent studies have demonstrated that the commonest psychiatric disorders associated with symptoms of disequilibrium are panic disorder, generalized anxiety disorder, phobic anxiety disorder, and depression (Furman and Jacob 2001). More specific syndromes such as space phobia (Marks 1981) and the motorist disorientation syndrome (Page and Gresty 1985) have been described in patients with both peripheral and central vestibular disorders. In a small percentage of patients, hyperventilation may compound the presentation of dizziness.

For the clinician evaluating a patient with disequilibrium, the common co-morbidity of psychological symptoms with vestibular disorders is important. In the past, patients have frequently been dismissed as simply being 'stressed' when their psychological symptoms have been precipitated by an underlying vestibular

A

B

Fig. 15.29 The Unterberger test (for explanation see Section 15.4.6).

disorder (Eagger *et al*. 1992; Jacob *et al*. 2003). It is well recognized that vestibular compensation is rarely effective in the presence of psychological symptoms, and it therefore behoves the clinician to be aware of the relationship between psychological illness and vestibular disease (Fig. 15.30).

Self-rating questionnaires, such as the Hospital Anxiety and Depression Scale, the Beck Depression Inventory, the Fear Questionnaire, which detects phobic anxiety and avoidance behaviour, and the General Health Questionnaire, are of value in identifying psychiatric dysfunction as part of the overall clinical assessment. It is now well established that an effective treatment package for chronic disequilibrium addresses both the physical and psychiatric components of the disorder (Yardley and Luxon 1994).

15.5 **Investigations**

Vestibular investigations define the presence and side of vestibular pathology and localize it to the peripheral, labyrinthine or VIII nerve, or central nervous system pathways. Classically, vestibular investigations have quantified the response of the horizontal semicircular canal to caloric, or thermal, stimulation and rotation testing, but the recent development of vestibular-evoked myogenic potentials also allows assessment of saccular function, while posturography provides information about the patient's overall ability to balance and the strategy used for balance. None of the tests provides aetiological information regarding pathology.

15.5.1 **Caloric test**

The caloric test is widely regarded as the cornerstone of vestibular testing, and was first described by Barany in 1906. Fitzgerald and

Hallpike (1942) popularized the technique using water 7°C below and above body temperature to irrigate the external canal for 40 seconds with the head raised 30° to the horizontal, to bring the horizontal semicircular canal into the vertical plane. The thermal gradient induced across the two limbs of the horizontal canal (Fig. 15.31) results in convection currents of the fluid within the canal, giving rise to displacement of the cupula, and stimulation of the crista. The nystagmic response generated is quantified, either by direct observation of its duration, or by electro-oculographic or videonystagmographic recording of the maximal slow-phase velocity.

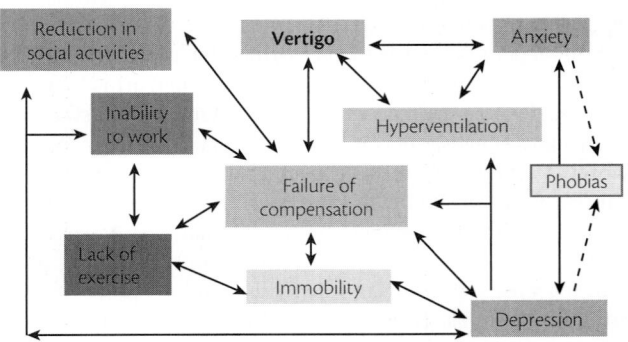

Fig. 15.30 Diagram to show the interrelationship of psychological, vestibular, and physical factors in the generation and prolongation of vertigo.

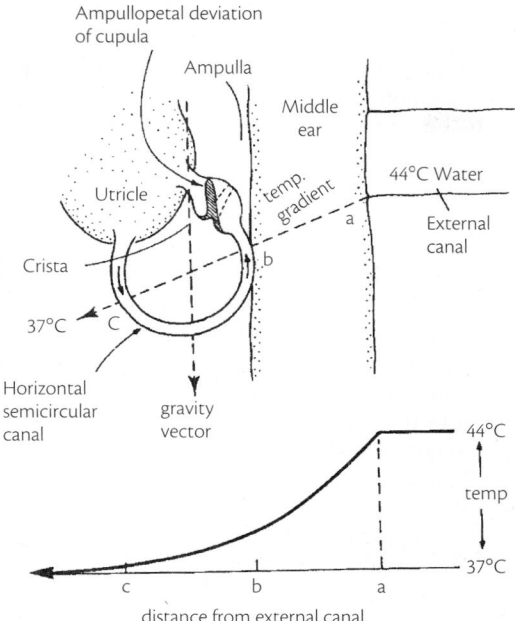

Fig. 15.31 Mechanism of caloric stimulation of the horizontal semicircular canal. (From Baloh and Honrubia 2001.)

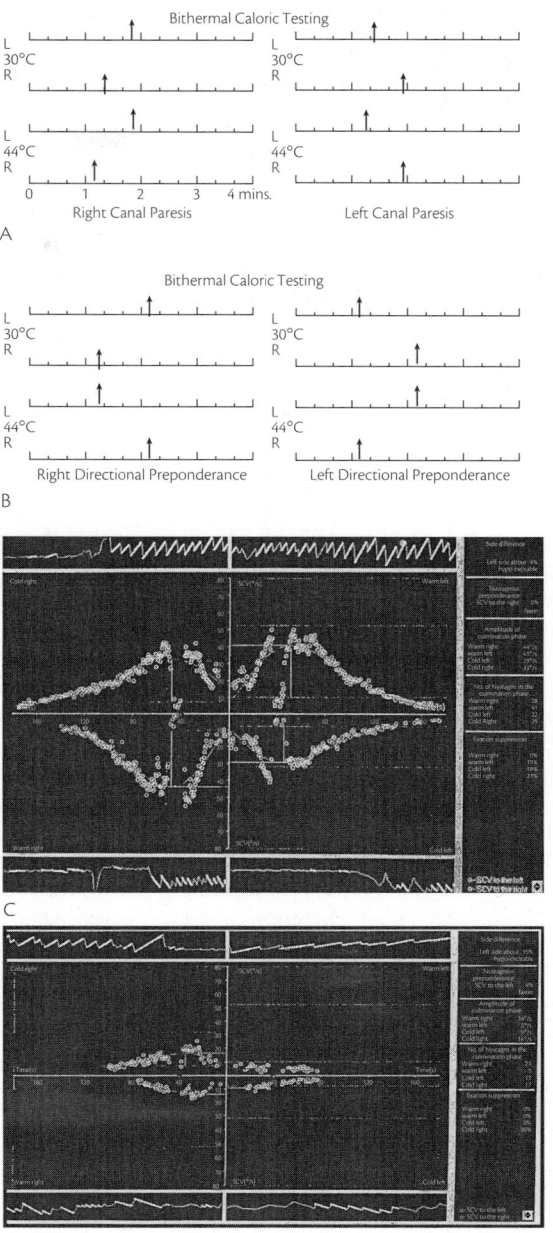

Fig. 15.32 A— Canal paresis on caloric testing as judged by duration parameter; B—Directional preponderance on caloric testing as judged by duration parameter; C—Normal caloric test as judged by slow-component velocity measurements; D—Left canal paresis on caloric testing as judged by slow-component velocity measurements.

Using this *Fitzgerald Hallpike technique*, the duration of the nystagmic response is observed with optic fixation. When the nystagmic response ceases, the room is totally darkened, and the eyes observed with an infra-red viewer, or Frenzel's glasses are applied to the patient's eyes to remove optic fixation, and the duration of the enhanced response in the absence of optic fixation is timed. Using the electro-oculographic or videonystagmographic response, the initial part of the caloric response is carried out in the absence of optic fixation, and at the peak of the response, the light is turned on to observe the effect of optic fixation.

Using either method, two main abnormalities are sought:

- a pattern abnormality, consisting of a canal paresis or a directional preponderance (Fig. 15.32);

- the effect of optic fixation on the nystagmic response (Fig. 15.33).

Two types of patterns are commonly observed. A *canal paresis* is one in which the responses generated by stimulation of the left ear are greater than the right ear, or vice versa. A *directional preponderance* is one in which the generation of nystagmus beating to the right is greater than the generation of nystagmus beating to the left, or vice versa. It should be recalled that irrigation of the ear with cool water generates nystagmus with beating of fast phases in the opposite direction, while irrigation of the ear with warm water generates nystagmus in the direction towards the irrigated ear. A mnemonic for recalling this pattern of results is COWS, Cold Opposite Warm the Same.

The degree of canal paresis and directional preponderance is calculated as a percentage by using the Jongkee's formulae (1962):

Canal paresis percentage equals:

$$\frac{(L30+L40)-(R30+R44)}{L30+L44+R30+R44}$$

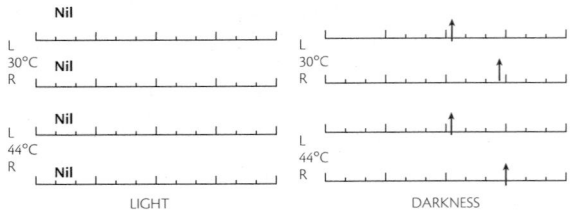

Fig. 15.33 Illustration of effect of optic fixation (light) on caloric-induced nystagmus in a patient with a left peripheral vestibular disorder, which becomes apparent in the absence of optic fixation (darkness).

Directional preponderance (percentage):

$$\frac{(L30 + R44) - (R30 + L44)}{L30 + R44 + R30 + L44}$$

where L = left, R = right; 30 and 44 represent water temperature and each integer, e.g. L30 and R44 represents the duration of nystagmus in seconds or the slow component velocity in degrees per second.

However, combined patterns can also be observed. For instance, in the case of a left labyrinthine lesion, a left canal paresis is associated with a right spontaneous nystagmus, and, therefore, a right directional preponderance. Hence the caloric result will give a left canal paresis, together with a right directional preponderance. This combined pattern will demonstrate a marked discrepancy between the hot responses, but little if any discrepancy between the cold responses. This pattern can easily be understood by summating algebraically the results that one would expect to find with a left canal paresis and a right directional preponderance (Fig. 15.34). However, central pathology, for example pathology in the left zone of the VIII nerve, may affect both the ipsilateral cerebellar vestibular connections and the left VIII nerve, giving rise to a left

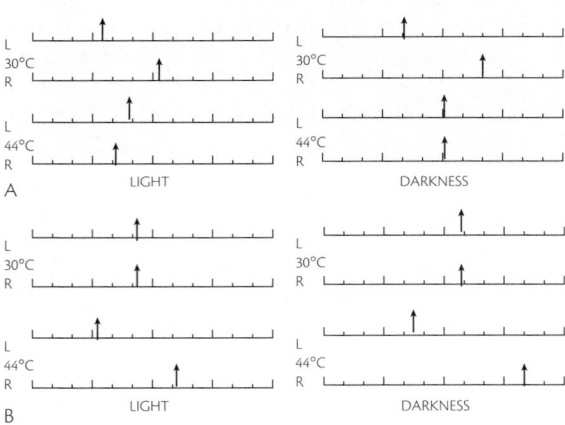

Fig. 15.34 A—Diagram to illustrate a combined caloric response of a left canal paresis together with a right directional proponderance, showing widely disparate cold responses with very similar warm responses, both with (light) and without (darkness) optic fixation; B—Caloric responses demonstrating a left canal paresis together with a right directional preponderance, illustrating the markedly different warm responses and very similar cold responses, both with (light) and without (darkness) optic fixation.

canal paresis and a left directional preponderance. In this scenario, the cold responses will be widely disparate, whereas the hot responses will be very similar. These examples illustrate that for correct interpretation of the caloric test it is necessary to obtain all four irrigations, and that the widespread use of cold or hot irrigation as a means of obtaining rapid results will lead to an erroneous interpretation.

Bilaterally symmetrical increased or decreased caloric responses are difficult to detect, because of the wide range of normal values. A response of less than 90 seconds' duration without optic fixation is generally considered to represent hypofunction. If no nystagmic response is observed with standard irrigation, irrigation at 20°C for 1 minute is usually undertaken to determine whether or not there is any significant residual function.

In addition to the pattern abnormality, the ability to suppress vestibular nystagmus with optic fixation provides information as to whether there is impaired integration of visual and vestibular responses at the level of the vestibular nuclei or cerebellum.

$$\text{Fixation index} = \frac{\text{Maximum slow-phase velocity without fixation}}{\text{Maimimum slow-phase velocity with fixation}}$$

An index greater than 1 indicates normality or a peripheral vestibular disorder, with normal optic fixation suppression of the vestibular response. Values equal to or less than 1 indicate central nervous system pathology.

The patterns of abnormal results on caloric testing may, in general terms, be interpreted as follows:

◆ *A canal paresis* almost always indicates peripheral vestibular, that is labyrinthine or VIII nerve pathology. Rarely, an isolated canal paresis may be found in a vestibular nuclei lesion (Francis *et al.* 1992), but this distinction can usually be made by the presence or absence of visual suppression of the vestibular response.

◆ *A directional preponderance* indicates an asymmetry within the vestibular system. The optic fixation index aids discrimination between peripheral or central vestibular pathology.

◆ *A combined canal paresis and directional preponderance in opposite directions*, for example, a right canal paresis and a left directional preponderance, is most likely to represent a right peripheral vestibular disorder. However rarely, with poor visual suppression of the vestibular response this may indicate a vestibular nuclei lesion. A canal paresis and directional preponderance in the same direction, for example, a left canal paresis and a left directional preponderance, may represent an 'overcompensated' left peripheral vestibular lesion or a left vestibular nuclei lesion with involvement of the left cerebellar connections. Again, it is usually possible to make this distinction on the basis of visuovestibular interactions, particularly the effect of optic fixation on vestibular responses.

◆ *Bilaterally reduced or absent vestibular responses in the presence of optic fixation* may represent habituation in a normal subject, who repeatedly stimulates the vestibular apparatus, for example a ballet dancer or acrobat. In the absence of optic fixation, normal responses will be observed. In a patient, absent responses in the presence of optic fixation may be observed because of marked habituation secondary to repeated acute attacks of vertigo, and in this situation, an abnormal pattern is likely to be seen in the absence of optic fixation.

◆ *Bilateral vestibular loss* is diagnosed by the absence of nystagmic response with *and* without fixation despite irrigation for one minute using water at 20°C.

The caloric test is simple, widely available, and provides information about the severity and location of vestibular dysfunction, based on horizontal semicircular canal function. Moreover, in a very simple format, it allows each ear to be assessed independently. The disadvantages of the caloric test are that it is not well tolerated by some patients and it is difficult to interpret in the presence of middle ear disease, for example it cannot be undertaken in the presence of a perforation and with a middle ear effusion, the transmission of the thermal gradient is uncertain. Perhaps most importantly, the caloric test with unilateral stimulation of the labyrinth is unphysiological, particularly in comparison with rotational testing.

15.5.2 Rotational tests

Rotational tests (Fig. 15.35) enable stimulation of both peripheral labyrinths by the application of multiple graded accelerations. This contrasts with the caloric test which stimulates each labyrinth in turn. Moreover, rotational testing is a physiological stimulus, enabling correlation of the stimulus to the semicircular canal response.

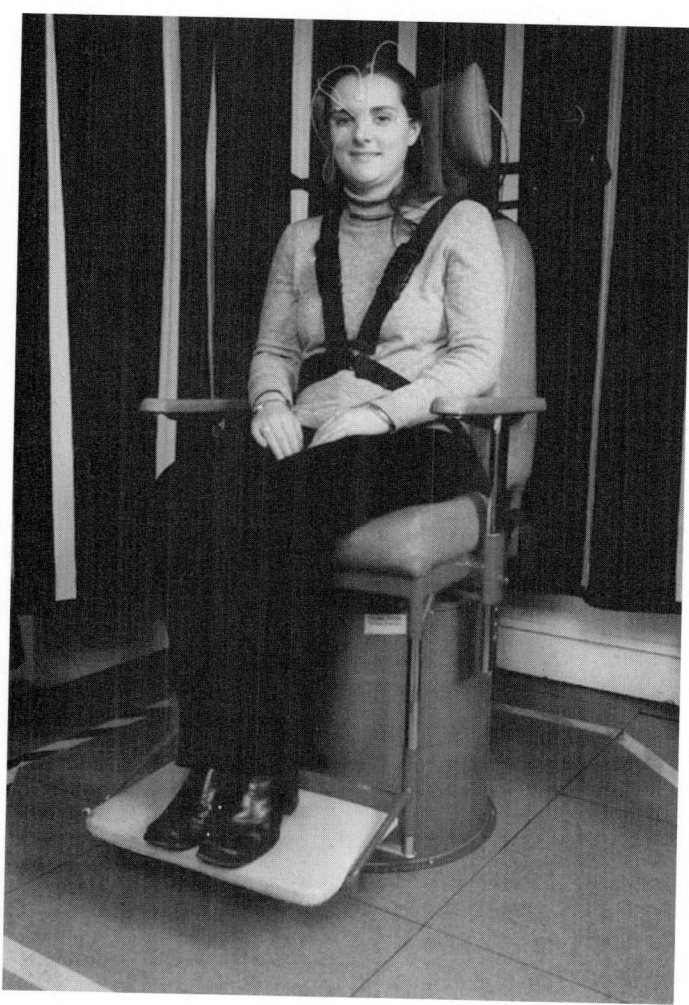

Fig. 15.35 Rotation testing.

For all practical purposes, current rotation tests assess the function of the horizontal semicircular canals, although by altering the axis of rotation it is possible to stimulate the horizontal or the vertical canals, with or without otolith stimulation in addition (Table 15.9). The relationship between the stimulus and response can be described by three calculated parameters:

◆ gain: eye velocity/stimulus velocity;

◆ phase: time lag between response and stimulus; and symmetry between clockwise and counter-clockwise rotations;

◆ the vestibulo-ocular reflex which functions optimally between 0.1 and 5 Hz rotations (Wilson and Melville 1979), and across this range of frequencies the gain and phase of the vestibulo-ocular reflex are approximately 1 and 0°, respectively.

Thus, within this range of stimulation, vestibular-induced eye movements are compensatory, that is they are virtually instantaneous and precisely match head movement. This range of 0.5 to 5 Hz matches the predominant frequencies of rotational head movements during walking and running, (Grossman *et al.* 1988). Thus ideally, vestibulo-ocular reflex measurements would include testing across a similar stimulus range, but the construction of a rotational chair (Fig. 15.35) is such that the usual frequency range of stimulation varies between 0.01 to 1 Hz, with a maximum velocity of 50 to 60° per second. The majority of patients, both children and adults, tolerate the test well, and valuable data can be obtained.

The most common test paradigms are sinusoidal harmonic acceleration and a velocity step. Sinusoidal harmonic acceleration comprises a succession of rotations at different frequencies, with the disadvantages being long test duration and resultant difficulty in maintaining patient alertness. The velocity step paradigm consists of a rapid acceleration from a velocity of 0 to 60° in less than 1 second with recording of the nystagmus response until cessation, followed by an abrupt stop. This provides a more physiologically relevant high-frequency stimulus and is a less time-consuming test. However, during rotational tests, alertness is vital, and the patient should be asked to perform mental arithmetic. Moreover, if the gain of the response is low because of vestibular hypofunction, measures of symmetry and phase are likely to be inaccurate.

Active head-only rotational testing has been introduced as an inexpensive and transportable piece of equipment (Fig. 15.36) (O'Leary and Davis 1990). Eye movements are recorded by standard electro-oculography (Section 15.5.3), and a head accelerometer measures head movement. An auditory cue from a metronome allows the patient to keep time to generate head movements of variable frequencies and velocities from 0.1 to 6 Hz. Both horizontal and vertical movements can be recorded, although electro-oculography is

Table 15.9 Stimulation of vestibular receptors in different rotation paradigms

Test	Horizontal canal	Vertical canals	Otoliths
Conventional	+		
Upright pitch		+	+
Onside pitch		+	
Eccentric	+		+
Off-vertical	+		+

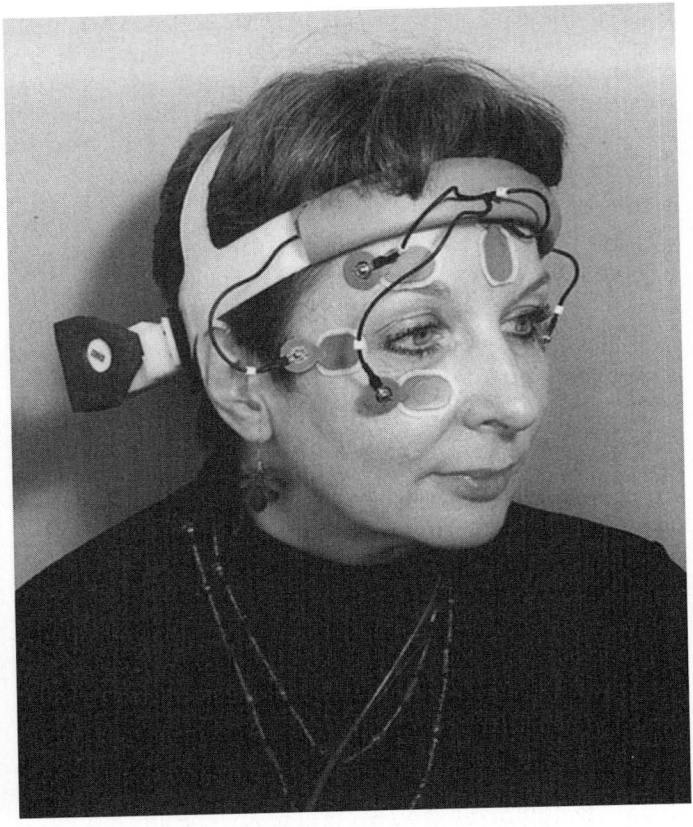

Fig. 15.36 Active head only rotation testing (vestibular auto-rotation test).

an unsatisfactory technique for the measurement of eye movement recordings. A computer program analyses the head and eye movement responses to generate gain and phase measurements. A major advantage is that auto-rotation testing provides a measurement of gain of the vestibulo-ocular reflex above 1 Hz, which lies in the normal physiological range, but its disadvantage is that patients often find it quite difficult to undertake the task. Also, at rapid head movements, it is difficult to stabilize the accelerometer on the headband, with resultant poor test/retest reliability of gain measurements.

15.5.3 Eye movement recording techniques

Eye movement recording allows documentation and analysis of oculomotor responses to visual and vestibular stimuli. A number of techniques are available with advantages and disadvantages depending on the requirements of the clinician (Table 15.10)

Electronystagmography also called electro-oculography is the mainstay of clinical eye movement recording. A potential difference exists between the retina and the cornea, with the retina being negative with respect to the cornea. In the normal situation of conjugate eye movements, electrodes are placed at either outer canthus, and as the eyes rotate towards the right, the right electrode becomes positive with respect to the left, and vice versa. Deviation of the eyes in the horizontal plane causes a potential difference between the two active electrodes, and a ground electrode is placed on the forehead. The magnitude of the potential difference is proportional to the magnitude of the eye movement. This voltage difference activates a pen recording system and provides a trace of the eye movement. Conventionally, an upward movement of the pen

Table 15.10 Comparison between electro-oculography (EOG), video-oculography (VNG) and scleral coil (SCR) eye movement recordings

Characteristic	EOG	VNG	SCR
Recording device	Ag/AgCl electrodes	Video camera	Copper coil embedded in silicon ring
Principle	Corneo-retinal dipole potential	Image processing	Voltage changes induced in coil of wire moving in oscillating magnetic field
Recording of horizontal eye movement	Good	Very good	Excellent
Recording of vertical eye movement	Unacceptable	Very good	Excellent
Recording of torsional eye movement	Not possible	2D—not possible 3D—Good	Excellent
Approximate accuracy	1-2°	1°	0.02°
Head movement artefact	High	Low	Nil
Sampling rate	>150Hz	Most clinical systems: 50 Hz Research systems: 200 Hz	1000 Hz
Amplifier drift	Yes	No	No
Calibration	Repeatedly	Once	Once
Errors due to: Blinks Changes in room lighting Myogenic activity	Yes Yes Yes	Yes No No	No No No
Patient tolerance	> 1 hour	30–45 minutes	>30 minutes
Value	Simple, inexpensive, readily available, clinical tool	More expensive, 2D recording relatively simple, good clinical tool except for saccades	Excellent research technique, too invasive for routine clinical use

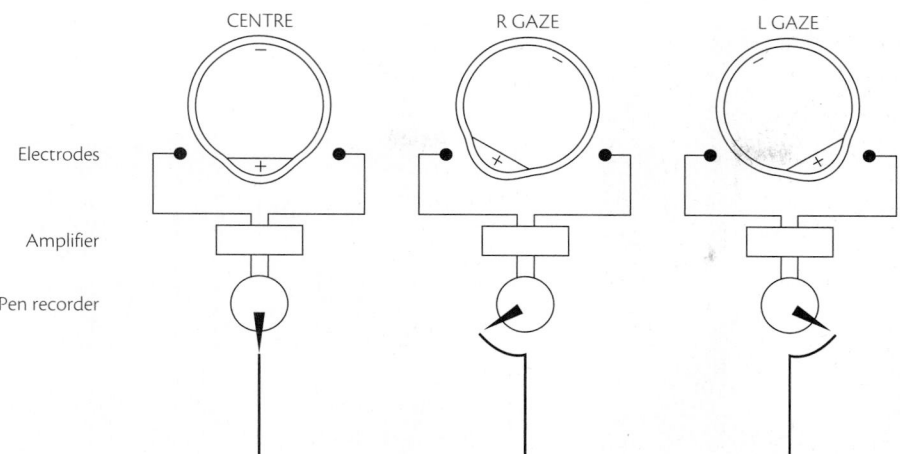

CENTRE   R GAZE   L GAZE

Electrodes

Amplifier

Pen recorder

Fig. 15.37 Principle of electro-oculography recording. (From OMahoney and Davies (1998).

represents deviation to the right, and a downward movement deviation to the left (Fig. 15.37).

Electro-oculography is a good technique for horizontal eye movement recording, but lid artefact impairs adequate vertical recording. The technique is relatively inexpensive, widely available, and provides an accurate quantified measurement of eye movement recordings, both spontaneous and in response to visual and vestibular stimuli. The major disadvantages of the technique are the need for scrupulous skin cleansing and careful application of the electrodes, with low electrode skin impedance; the confounding effect of eyelid artefact and muscular activity near the electrodes, produced for example by blinks, and the variability of the corneo-retinal potential with respect to ambient light. Thus repeated calibrations throughout eye movement recordings are essential.

Video-oculography is a video recording of eye movements using a small video camera that is placed in front of the eyes on a specific device. It is rapidly becoming a widely available clinical tool, and is reliable for both horizontal and vertical eye movement recording. The movement of the eye is determined by image processing, using either two-dimensional or three-dimensional systems. Most systems function in total darkness using infra-red light, and the majority of commercial systems use a standard frame-rate of 50 to 60 Hz, limiting the use of video-oculography for recording saccades. The technique avoids the confounding factors of changes in room lighting, myogenic activity, and repeated calibration, but is in general less well tolerated by the patient than electronystagmography.

Scleral search coil is the most accurate form of eye movement recording and is now considered 'the gold standard'. However, it is an invasive technique that requires a wired contact lens to be placed on the patient's eyes. It causes discomfort and is not routinely employed for clinical purposes.

15.5.4 Galvanic testing

Galvanic stimulation refers to the delivery of a small electric current through the vestibular labyrinth using surface electrodes and recording either eye movements or postural movements (Fig. 15.38). The technique has been recognized for more than 100 years, but has not gained clinical acceptance, initially because of the painful electric currents required to produce recordable results, and more recently because it is unclear exactly which vestibular end-organ is

responsible for the eye movement or postural response. The advantages of the technique are that galvanic stimulation tests each labyrinth separately, and is thought to excite the synapse between the hair cell and the VIII nerve afferent. It was therefore considered a possible tool to evaluate 'neural' versus 'sensory' function and could, in theory, be used to evaluate unilateral VIII nerve disorders. More recent work has suggested that this may prove to become a valuable clinical tool.

15.5.5 Vestibular-evoked myogenic potentials

Strong acoustic stimulation may lead to short-latency muscle contractions, known as vestibular-evoked myogenic potentials, which may be recorded from the tensed sternocleidomastoid muscle using a surface electrode (Fig. 15.39). The pathway for this response is considered to arise in the saccule, and passes along the inferior vestibular nerve to the lateral vestibular nucleus and from there via the vestibulospinal reflex to the sternocleidomastoid muscle (Murofushi *et al.* 1995; Zhou and Cox 2004).

The value of this test is that it provides a simple technique for evaluating otolith as opposed to semicircular canal function, and moreover can be undertaken in patients in whom the vestibulo-ocular reflex cannot be studied accurately, for example, uncooperative

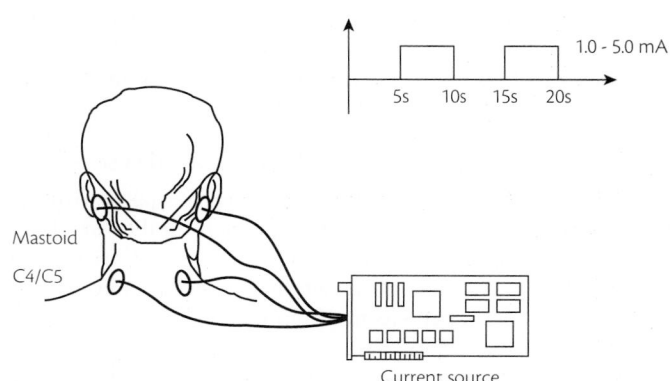

1.0 - 5.0 mA

5s 10s 15s 20s

Mastoid

C4/C5

Current source

Fig. 15.38 Galvanic stimulation of the labyrinth and VIII nerve at the mastoid, with stimulation at C4/C5 level for comparison of non-vestibular effects. Pulsed DC current of 2 to 4 mA is used. (From Brandt 1999.)

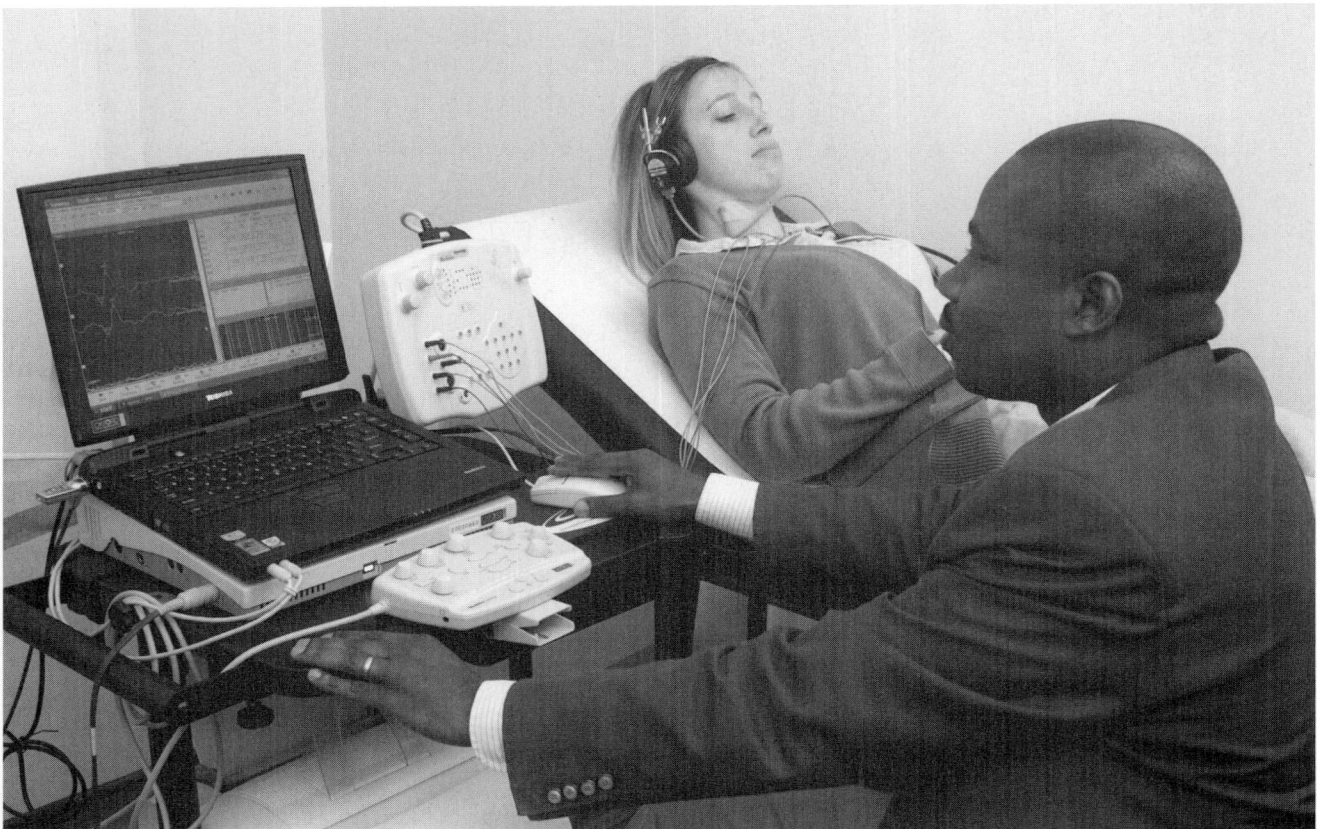

Fig. 15.39 Recording of vestibular-evoked myogenic potentials.

children, subjects who are blind or who have congenital nystagmus, or have other eye movement disorders. Moreover, it allows evaluation of the inferior as opposed to the more commonly studied superior vestibular nerve. There are two uncertainties of the technique:

- whether the vestibular-evoked myogenic potential signifies a pure vestibular response (McCue and Guinan 1994);
- the technical challenge of standardizing the pre-activation of the muscle generating the response, to limit variability of the amplitude of the response, which is dependent upon background muscle activity (Kingma 2006).

To date, no single recording technique has been agreed, although many studies have reported abnormalities of vestibular-evoked myogenic potentials in unilateral VIII nerve and labyrinthine disorders (Welgampola and Colebatch 2005; Osei-Lah *et al.* in press).

15.5.6 Subjective Visual Vertical–Horizontal test

The Subjective Visual Vertical and Subjective Visual Horizontal tests evaluate otolith function, with influence from the semicircular canals (Pavlou *et al.* 2003) Such perceptual tasks of orientation with respect to gravity provide an alternative means of evaluating vestibular function in patients in whom it is not possible to assess the vestibulo-ocular reflex. Using simple instruments, a normal subject can align a laser beam in relation to the earth horizontal and vertical in a totally darkened room, without visual cues, in a very reproducible manner (Friedmann 1970, 1971). Following an acute peripheral vestibular disorder, deviation towards the affected side is commonly observed. However, with compensation, the subjective visual vertical and horizontal tests frequently return to within the normal range. Central vestibular disorders produce abnormal subjective visual vertical and horizontal test results consistently with caudal brainstem lesions causing ipsiversive tilts of the subjective visual vertical, whereas upper brainstem lesions cause contraversive tilts. Recent work has suggested that standardization of the test technique may result in better diagnostic accuracy in peripheral vestibular disorders (Pagarkar *et al.* in press).

15.5.7 Posturography

Assessment of vestibulospinal function in isolation is difficult because of the interaction of sensory and motor systems in maintaining balance. *Static-force platforms* measure the position of the centre of gravity, and allow measurement of body sway. However, the major limiting factors of such devices relate to the combination of systems used in maintaining stability and upright stance, and the absence of any stimulus response measures, with respect to vestibulospinal function. Thus, static-force platforms merely record spontaneous body movements, and while this may be of value in clarifying a dysfunction in an individual, it is rarely of value in defining vestibular pathology.

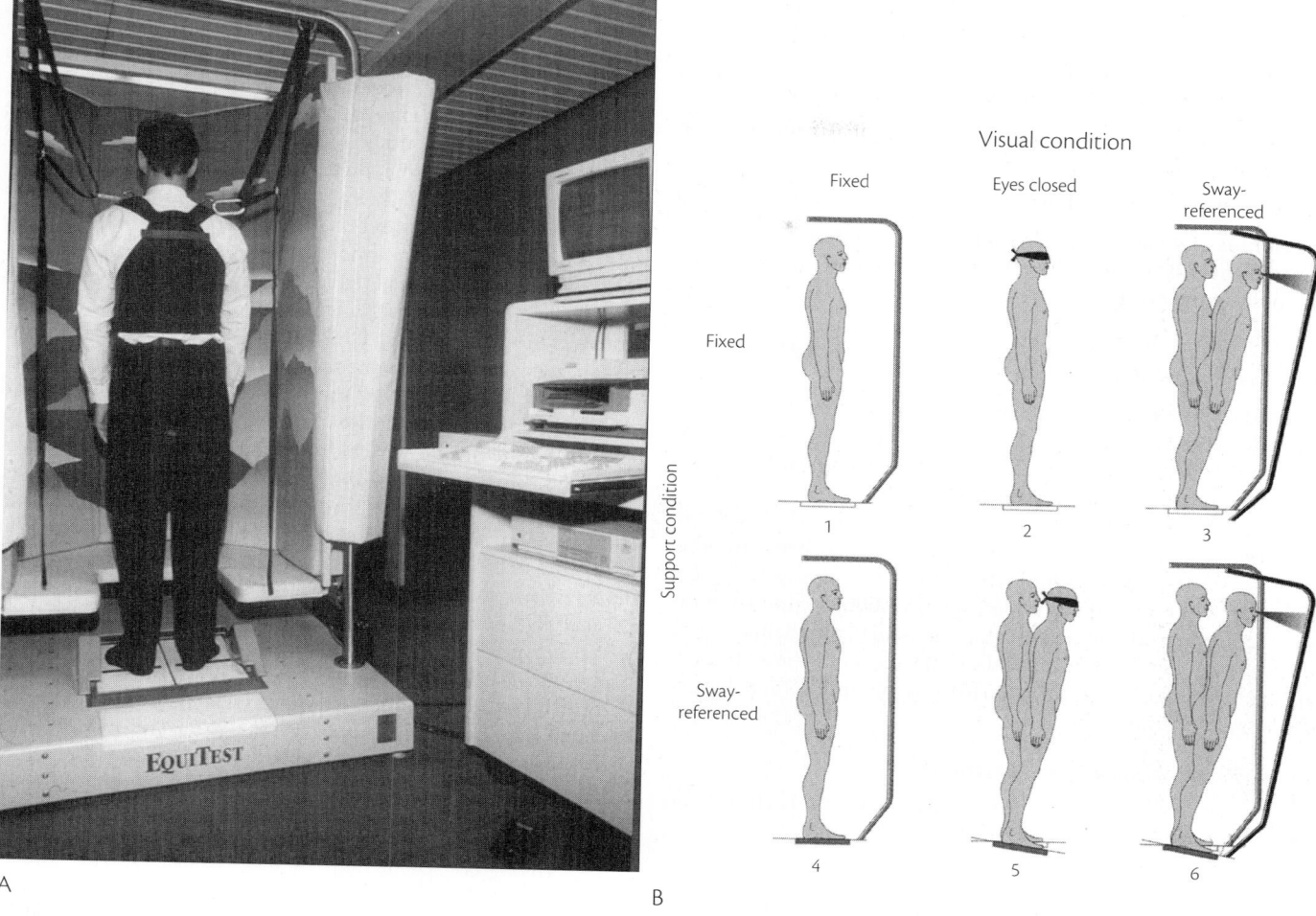

Fig. 15.40 (A) Dynamic posturography; (B) The six test paradigms of the sensory organization test using dynamic posturography. (From Nashner 1996.)

Moving platform, or dynamic, posturography (Fig. 15.40) was devised to overcome these limitations, by:

- controlling the relative contributions of visual, somatosensory, and vestibular inputs required to maintain upright posture;

- incorporating stimulus response measurements.

Dynamic posturography relies upon coupling of the platform to the sway of the subject, and by maintaining the angle between the foot and the lower leg at a constant value, the input of the somatosensory system to postural control is reduced. The system also allows visual information to be 'sway-referenced' to the subject, so that visual input can be reduced by movement of the visual surround in the same direction as any body sway, thus providing inappropriate visual information to the subject. These posturographic techniques have been valuable in defining the contribution of the different sensorimotor components to postural control. The Report of the Therapeutics and Technology Assessment Committee of the American Academy of Neurology (1993), highlights that posturography is not a diagnostic test, but a method to quantify balance dysfunction under different sensory conditions. Thus, this technique has advanced the knowledge of how postural mechanisms

work, and how posture is impaired in patient groups, but adds little to the diagnosis of an individual patient complaining of a balance disorder (Bronstein 2003).

15.6 Management of vestibular disorders

15.6.1 General principles

Successful management of patients with vestibular dysfunction depends upon accurate diagnosis, an understanding of vestibular physiology, the physician's awareness of the overlap between the vestibular system and the autonomic and limbic systems, and the psychological manifestations that frequently accompany vestibular disorders. Evaluation of progress and efficacy of intervention require careful monitoring, not least as vestibular signs and test results do not correlate with symptoms. Consequently, it is of value to use validated questionnaires documenting vestibular symptomatology, disability, and handicap, for example the *Dizziness Handicap Inventory* (Jacobson and Newman 1990) or the *Vertigo Symptom Scale* (Yardley and Putnam 1992), in addition to validated psychological questionnaires such as the *Beck Anxiety and Depression Scale* and the *Short Form 36*. Objectively, posturography results may provide clear guidance as to suitable physical therapy intervention in terms of balancing strategies and may provide objective quantification of functional improvement.

Treatment may be classified under five main categories:

- general medical evaluation, with correction or amelioration of associated disorders;
- specific pharmacological therapies for vestibular disorders;
- vestibular rehabilitation physiotherapy, including both structured and customized exercise regimes, together with particle-repositioning procedures when necessary;
- psychological support;
- surgical interventions.

Based on the diagnosis, a rehabilitation plan should be devised for each patient, and specific care should be taken to ensure understanding of the vestibular condition to ensure compliance with the rehabilitation programme (Table 15.11).

A general medical examination will identify comorbid systemic conditions such as hypertension, vascular disease, diabetes, autoimmune disorders, ophthalmological disease, and psychological pathology, all of which may impact upon vestibular compensation and require appropriate treatment. In particular, ophthalmological and rheumatological or orthopaedic problems should be addressed, to ensure optimal visual and somatosensory input to enable vestibular rehabilitation.

15.6.2 Pharmacological treatment

Pharmacological intervention for vestibular disease is chosen for one of three reasons:

- Treatment of acute vestibular symptoms;
- specific treatment of a condition that causes vestibular symptoms, for example, Menière's disease, migraine or epilepsy;
- non-specific empirical treatment of central vestibular dysfunction.

Recent research has led to a better understanding of the neurochemistry of the vestibular system and thus a more rational basis for the treatment of vestibular pathology. Nonetheless, there has been no development of new drugs and the treatment of vestibular disorders remains empirical in the absence of well designed drug trials (Bamiou and Luxon in press).

Table 15.11 Vestibular rehabilitation programme

Investigation and diagnosis
Explanation of symptoms: diagnosis and rationale for management
Rehabilitation plan: • Correction of remediable medical problems   • Review of medication   • General fitness programme   • Physical exercise regime   • Psychological assessment and intervention   • Realistic family/social/occupational goals
Monitoring/feedback/follow-up
Discharge

Symptomatic treatment of acute vestibular episode

Symptoms of vertigo, nausea, vomiting, sweating, pallor, and diarrhoea are extremely alarming for the patient, who commonly fears a brain tumour or stroke, and simple reassurance is effective for the majority of patients. The nature of the symptoms should be explained and anti-emetics should be administered, for example hyoscine, prochlorperazine, promethazine, cyclizine, dimenhydrinate or metaclopramide. Buccal administration of prochlorperazine is frequently effective, but intramuscular treatment can be used, if oral preparations cannot be tolerated because of vomiting. Hyoscine may be administered transdermally. Anti-emetic drugs block the afferent pathways from the chemoreceptor zone in the area postrema, the gastrointestinal tract, and the labyrinth (Timmerman 1994; Takeda et al. 1993) to the medullary vomiting centre.

Vestibular sedative drugs should then be administered and include the anticholinergics hyoscine and scopolamine, the antihistamines promethazine, prochlorperazine, cyclizine, metoclopramide, and dimenhydrinate, and the calcium channel antagonists, cinnarazine and flunarizine. The latter two drugs may give rise to extrapyramidal side effects, and should be used only for a very limited period in the elderly (Daniel and Mauro 1995).
Diazepam has no specific action on the vestibular system, but acts by reducing neural activity and causing inhibition throughout the central nervous system, including the vestibular nerve and nuclei (Smith and Darlington 1998). While the role of this drug in the treatment of vertigo is controversial, it is widely used for its anxiolytic activity in acute vestibular crises (Foster and Baloh 1996). There is some recent evidence to hypothesize efficacy for steroid treatment in promoting recovery of labyrinthine function in the initial phase after an acute vestibular episode (Straube 2005).

Specific treatment of vestibular disorders

Menière's disease. The treatment of Menière's disease remains controversial and empirical, not least because the underlying pathophysiological mechanism of this disorder defies definition. A very high placebo response has been reported (Claes and van de Heyning 1997), and there are a few double-blind randomized trials assessing treatment efficacy. The aim of pharmacological intervention is the control of endolymphatic hydrops or treatment of the postulated immunological mechanisms responsible for endolymphatic hydrops. Failure of medical intervention may lead to surgery, including chemical labyrinthectomy with transtympanic installation of gentamicin or surgical destruction of the labyrinth or vestibular neurectomy.

Treatments aimed at influencing endolymphatic hydrops include diet, diuretics (Santos et al. 1993), and the use of betahistine, a histamine analogue reported to improve circulation in the stria vascularis (Lacour and Stercker 2001). No double-blind trials have reported the efficacy of a low-salt diet, 1 to 2 mg/day, but the author's experience is that this is a highly effective strategy in patients who can be persuaded to follow the regime strictly. Patient compliance depends upon education regarding the high levels of salt in many prepared foods, such as cornflakes.

There is evidence that diuretics are effective in the long-term control of vertigo, but not of hearing impairment, although there is no good randomized double blind trial (Thirlwall and Kundu 2006). More potent loop diuretics such as frusemide should be avoided, due to potential ototoxicity. Bendroflumethazide, diazide (Van Deelen and Huizing 1986), a potassium-conserving diuretic, and chlorthalidone (Klockhoff *et al.* 1974) are all reportedly effective. Despite a number of trials using betahistine, none have demonstrated convincing efficacy (James and Burton 2001).

Treatments aimed at immunological suppression (Hamman and Arnold 1999) are based on the assumed autoimmune pathogenesis of Menière's disease (Silverstein *et al.* 1998). Steroids have been administered topically and systemically (Parnes *et al.* 1999), but no double-blind studies have demonstrated clinical efficacy. Various alternative treatment regimes have been suggested in this condition including alternobaric oxygen therapy (Fattori *et al.* 2002) and intermittent micropressure pulses to the inner ear through a tympanostomy tube with the Meniett device (Densert and Sass 2001). A prospective randomised placebo controlled, multicentre trial of this device reported good results (Odkvist *et al.* 2000).

Not infrequently, in non-specialist units, Menière's disease is treated by straightforward symptomatic management with vestibular sedative drugs and anti-emetics. Failure of treatment leads to the consideration of destructive procedures, and intratympanic gentamicin treatment is currently the preferred option (Silverstein *et al.* 2003). It has been reported that there is a high efficacy in controlling vertigo with this management strategy (Bottril *et al.* 2003). There is no agreed optimal protocol regarding the dose, technique, and administration or end-point of therapy, and profound sensorineural hearing loss may develop in up to 30 per cent of treated cases (Blakley 2000). Moreover, recurrence of vertigo has been reported in up to 30 per cent of treated cases within 24 months of treatment (Wu and Minor 2003).

Surgical interventions (Ludman 2003) may also include theoretical prophylactic measures such as endolymphatic sac decompression, but there is no firm evidence of the efficacy of this procedure (Thomsen *et al.* 1996). Destructive surgical procedures may be divided into those that aim to preserve auditory function, such as vestibular neurectomy, and those aimed at removing the labyrinth if the disease process has caused profound hearing loss, in addition to the intractable vertigo. In the latter situation, labyrinthectomy is commonly advocated, if medical treatment has failed. However, destructive procedures should be undertaken with extreme caution in view of the possible bilateral loss of auditory and vestibular function in Menière's disease, and the possible failure of compensation for a total unilateral vestibular loss.

Migraine. Migraine affects approximately one-fifth of the population. Although the association between migraine and vestibular signs and symptoms is well described, it is poorly understood (Eggers 2006). The treatment of migrainous vertigo parallels the treatment of migrainous headache (Bikhazi *et al.* 1997). This includes dietary measures, life-style adaptation, and stress reduction techniques, together with psychological treatment and vestibular exercises in the 25 to 30 per cent of patients with migrainous vertigo who demonstrate peripheral vestibular dysfunction.

In the absence of clear diagnostic criteria for migrainous vertigo and limited randomized controlled trials, clinical experience suggests that pharmacological treatment of migrainous vertigo may result in resolution or improvement of acute vertiginous episodes in over 90 per cent of those treated cases (Johnson 1998). Pharmacological treatment includes both symptomatic and prophylactic measures. Symptomatic treatment may include both antivertiginous and anti-emetic drugs, together with specific treatment for headache; for example aspirin, paracetamol, ibubrufen, or non-steroidal anti-inflammatory analgesics (Baloh 1997). Triptans

(Neuhauser *et al.* 2003), ergot drugs, and acetazolamide may all also be of value as acute treatments (Bamiou and Luxon in press).

Prophylactic treatment can be considered for frequent acute episodes of vertigo or episodes of sufficient severity that symptomatic treatment is inadequate (Baloh 1997). Betablockers such as propranolol; calcium channel blockers such as cinnarizine, and serotonin reuptake inhibitors such as pizotifen, in addition to tricyclic antidepressants, such as amitriptyline have all been effective in some cases (Bamiou and Luxon in press).

Episodic ataxia type 2 may present with acute vertigo and ataxia, with or without interval symptoms in both adults and children. Although rare by comparison with migraine, this diagnosis should be considered in intractable recurrent vertiginous episodes. Acetazolamide or 4-aminopyridine may be effective in the management of episodic ataxia type 2 (Strupp and Brandt 2006), while a case with a novel mutation in the CACNA1A gene, also recorded a good therapeutic response to acetazolamide coupled with valproic acid.

Treatment of central vestibular dysfunction

Central vertigo giving rise to chronic vestibular imbalance and dizziness is the most difficult condition to treat. It is often associated with disordered eye movements including central forms of nystagmus: vertical, period alternating, see-saw, and pendulum nystagmus. No single treatment is of benefit to all patients, but current understanding of the neurochemistry of the central vestibular system has enabled a rational approach to specific treatments for some of these disorders (Rucker 2005). Frequently, treatment of the eye movement disorder may lessen the sense of disorientation and associated nausea. Baclofen has been effective in treating periodic alternating nystagmus (Halmagyi *et al.* 1980) while acetazolamide, a drug which stabilizes the transient dysfunction of abnormal calcium channels is an effective treatment for episodic familial ataxia type II (Baloh 2002). Downbeat nystagmus may respond to clonazepam (Young and Huang 2001) whilst 3,4-D-diaminopyridine, a potassium channel blocker has also been suggested in the treatment of this condition (Strupp *et al.* 2003). Gabapentin (Averbuch-Heller *et al.* 1997) and memantine, a glutamate antagonist with NMDA blocking action (Starck *et al.* 1997), have both been reported to be effective in the treatment of acquired pendular nystagmus due to multiple sclerosis. There are no specific treatment regimes and frequently these drugs are used on a therapeutic trial basis, titrating the dose against symptoms and side effects. For patients with chronic instability due to central vestibular disorders there is now some evidence to suggest that intensive vestibular rehabilitation physiotherapy and gait retraining strategies may be effective in providing greater confidence and reducing disability (Shepard *et al.* 1993).

15.6.3 **Vestibular rehabilitation physiotherapy**

The majority of patients with chronic vertigo suffer from uncompensated peripheral vestibular pathology. Progressive or persistent vestibular pathology, for example due to acoustic neurinoma, migraine, or autoimmune inner ear disease, may present with chronic vertiginous symptoms. Absence of vestibular function and central vestibular disorders may present similarly. Uncompensated unilateral peripheral vestibular dysfunction and bilaterally reduced or absent vestibular function with chronic symptoms should be managed in a similar manner. Patients with fluctuating pathology, such as Menière's disease, benign paroxysmal positional vertigo, or migraine require management of acute episodes prior to standard vestibular rehabilitation (Black *et al.* 2000). Anti-emetics or vestibular sedative drugs should not be used in the management of this group, as they may impair vestibular compensation (Zee 1988). Unfortunately such drugs are commonly prescribed for chronic vertigo, both in primary and tertiary care.

Plasticity of the central nervous system underpins vestibular rehabilitation and symptomatic vestibular compensation. During the Second World War, Sir Terence Cawthorne, an otolaryngologist, and Dr Harold Cooksey, a rheumatologist, noticed that servicemen with head injuries and dizziness recovered more quickly, if they were active. The *Cawthorne–Cooksey exercises* were subsequently developed on an empirical basis. The programme comprised a systematic graduated set of exercises aimed at stimulating vision, vestibular, and proprioceptive inputs to enhance recovery. In the 1970s and 1980s, animal models of vestibular compensation demonstrated the value of vision and motor activity (Igarashi *et al.* 1975, 1981; Lacour 1976; Lacour *et al.* 1979; Lacour and Xerri 1980) in symptomatic recovery from unilateral vestibular disorders. This contributed to the evidence-base underpinning the current approach to vestibular rehabilitation. Vestibular compensation relies on a range of physiological mechanisms involving adaptation and habituation, resulting in recalibration of the vestibular reflexes, and sensory and motor substitutions, including for example greater reliance on visual inputs for balance, as well as 'new' predictive oculomotor responses. A range of vestibular rehabilitation programmes have been devised (Norre and De Weerdt 1980; Shepard and Telian 1995; Herdman and Whitney 2000) to promote compensation. All contain key elements:

- detailed explanation of the rationale of the exercises and aim of regime to ensure patient motivation and compliance;

- graded approach increasing sensory input and speed of task to enable patients to progress with the regime which initially increases their symptoms;

- emphasis upon exercises which are functionally relevant, for instance provocation of dizziness in the individual patient;

- repeated short but frequent repetitions of individual exercises to promote compensation. For example, any given exercise should be repeated five to ten times during a two to three minute session, three times a day;

- outcome measures to quantify improvement for clinician and to provide 'objective' evidence of progress to the patient.

Nonetheless, the disparity between vestibular symptoms, signs, and test results (Hallam *et al.* 1988; Stephens *et al.* 1991) requires a range of measures to be undertaken before and after therapy to evaluate efficacy. Standardized self-report questionnaires are useful including the dizziness handicap inventory (Jacobson and Newman 1990) and the vertigo symptoms scale (Yardley and Putnam 1992), in addition to assessor scored balance assessment tests such as the dynamic gait index (Hall *et al.* 2004) and dynamic posturography results (Badke *et al.* 2005). Lack of motivation, attention, and effort have all been recognized as reasons for poor outcome (Hecker *et al.* 1974) as has poor performance of exercises (Norre and De Weerdt 1980). Moreover, both physical factors such as intercurrent illness, fatigue, and vestibular suppressant use and psychological factors including anxiety, depression, and avoidance behaviour may impact upon a good outcome (Luxon 1997).

Recent research has shown the efficacy of 'customized' exercises in which a programme is devised for each patient based on their individual vestibular limitations, and there is good evidence that this is more effective than systematic exercise regimes (Szturm *et al.* 1994; Mruzek *et al.* 1995; Shepard and Telian 1995). Moreover, recent studies of 'mechanical' exercise programmes, utilizing optokinetic stimulation, virtual reality, and rotational stimuli, may all enhance optimal vestibular compensation (Pavlou *et al.* 2004).

Traditionally, vestibular rehabilitation has been used in the management of stable unilateral peripheral vestibular disorders which have failed to compensate spontaneously. However, there is also evidence of the value of this technique in the management of imbalance of migraine (Whitney *et al.* 2000), bilateral vestibular failure (Bronstein and Hood 1987; Herdman and Clendaniel 2000), and brainstem pathology giving rise to vestibular disorders (Shepard *et al.* 1993), although patients with cerebellar lesions respond less well (Shepard and Asher 2000). A particularly important group of patients that are difficult to rehabilitate are those with visually induced symptoms, i.e. visual vertigo in isolation or in addition to movement induced symptoms. Nonetheless recent work has suggested that visual motion desensitization in these patients with optokinetic training may be of value (Guerraz *et al.* 2001). In this group, virtual reality stimulation may be of value (Viirre *et al.* 2002). A further group of patients frequently encountered in 'dizzy clinics' is those with vestibular pathology and associated psychological symptoms. This group has also been shown to benefit from vestibular rehabilitation physiotherapy (Jacob *et al.* 2001).

As compliance and active collaboration with vestibular rehabilitation programmes are required, an individual approach may be more effective than a generic regime, not least because a detailed explanation of the mechanisms of balance and vestibular compensation enables each patient to understand why physiotherapy, rather than medication or surgery, is helpful. In addition, there is some information in the literature that early intervention is beneficial. Of note, recent work has suggested that older age *per se* is not a negative prognostic factor.

15.6.4 Particle repositioning procedures

Benign paroxysmal positional vertigo requires treatment with specific particle repositioning procedures aimed at treating the underlying pathophysiological mechanism of cupulo or canalithiasis (see Section 15.3.1). The importance of diagnosing this condition correctly is two fold:

- the highly successful outcomes of simple particle repositioning procedures;

- the importance of differentiating benign paroxysmal positional vertigo arising from any one of the three semicircular canals, as distinct from central positional nystagmus. This differentiation may be facilitated by observation of nystagmus direction, latency, duration, and time course, and fatigueability (Table 15.2) (Brandt 1999b; Buttner *et al.* 1999).

The symptoms associated with benign paroxysmal positional vertigo commonly abate spontaneously within a few weeks, but in as many as 30 per cent of untreated cases, the symptoms may persist for months (Casani *et al.* 2002) with significant disability and patient distress (Furman and Cass 1999). A number of specific regimes have been described for the treatment of benign paroxysmal positional vertigo but the three most commonly used are the Brandt Daroff

Fig. 15.41 Brandt-Daroff exercises. (From Brandt and Daroff 1980.)

exercises (1980), the Semont liberatory manoeuvre (Semont *et al.* 1988), and the Epley particle repositioning procedure (1992).

The Brandt Daroff exercises (1980) were devised on the basis of cupulolithiasis and consist of rapid movements of the body and head from one lateral position to the other (Fig. 15.41). This management option is particularly effective for patients who are not able to seek appropriate professional help or are unresponsive to one of the particle repositioning procedures. Brandt and Daroff (1980) reported that 98 per cent of patients obtained relief from benign paroxysmal positional vertigo within 14 days using this technique, although a controlled trial showed resolution in only 23 per cent of patients after one week (Radtke *et al.* 1999).

Semont's manoeuvre (1988) requires the patient to lie on the affected side with the face turned 45° upwards towards the ceiling (Fig. 15.42B). The patient is then quickly swung through the sitting position to the opposite side with the face turned downwards by 45° (Fig. 1542C). This latter position should be maintained for five minutes prior to the patient being brought slowly up to the sitting position (Fig. 15.42D). The authors reported that 92 per cent of their patients treated at least once with this manoeuvre were rendered asymptomatic.

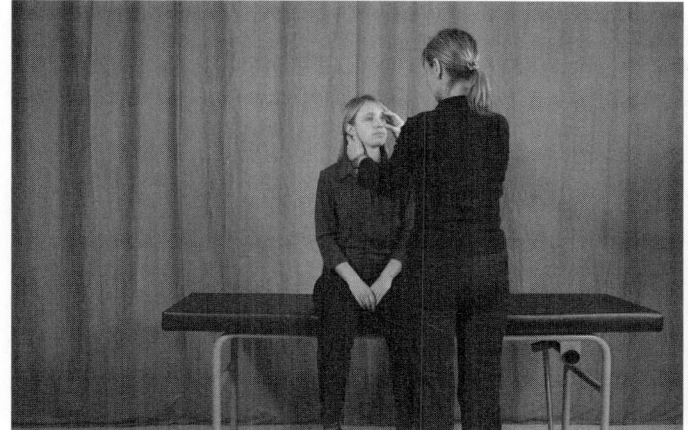

A

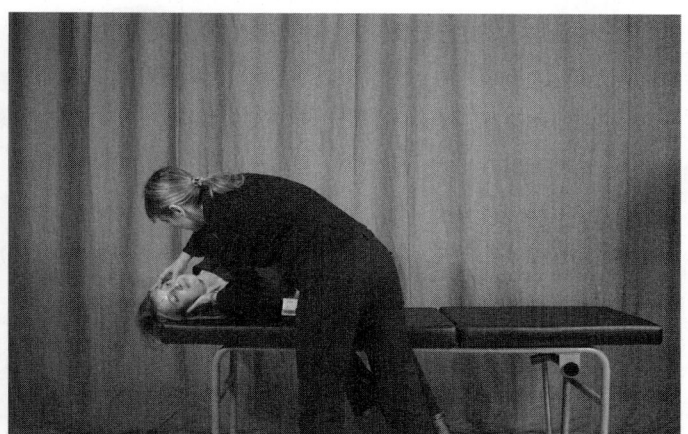

B

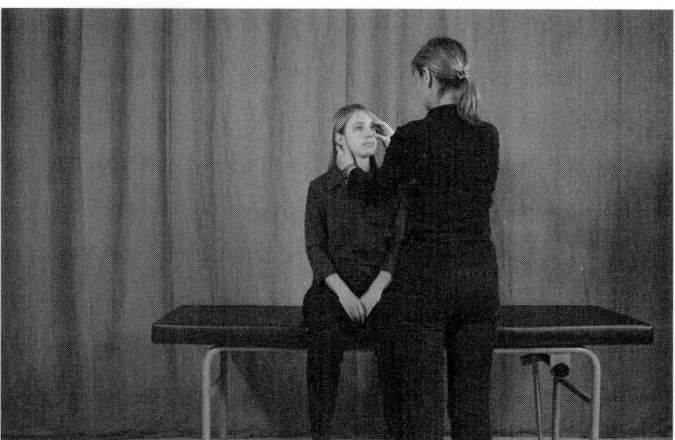

C

D

Fig. 15.42 Semont's manoeuvre stages A to D (for explanation see Section 15.6.4). (From Bamiou and Luxon, in press.)

The Epley particle repositioning procedure (Fig. 15.43) is based on the theory of canalithiasis. It has been adapted to treat both the anterior and lateral canal conditions as opposed to the much more common posterior canal benign paroxysmal positional vertigo. The manoeuvre aims to move the patient through a series of positions starting with the Hallpike manoeuvre to determine the side and characteristic features of benign paroxysmal positional vertigo. The manoeuvre allows the labyrinth to be positioned in different directions allowing the otoliths to fall along the lumen of the semicircular canal under gravity and out into the vestibule where the critical head position does not permit them to fall onto the cupula of the posterior canal.

Both the Epley and Semont manoeuvres have been reported to be highly effective in trials evaluating subjective outcome, in addition to objective presence of positional nystagmus, with up to 90 per cent of patients becoming symptom-free after repeated manoeuvres (Herdman *et al.* 1993; Froehling *et al.* 2000).

There is no clear evidence to support the application of mastoid vibration in improving the efficacy of particle repositioning procedures (Hain *et al.* 2000), nor the adoption of specific sleeping position, such as keeping the head up and wearing a collar immediately after the procedure (Massoud *et al.* 1996). There is some evidence to suggest that traumatic benign paroxysmal positional vertigo has a lower success rate with treatment than idiopathic forms (Harvey *et al.* 1994). Conversely patients in whom the condition is thought to be related to vestibular neuritis appeared to have a better prognosis than patients with other pathologies (Herdman 1996). Moreover, there is a recurrence rate of approximately 50 per cent at 5 years (Hain *et al.* 2000). It is also important to note that particle repositioning procedures may, on occasions, convert a posterior canal benign paroxysmal positional vertigo into a horizontal or an anterior canal benign paroxysmal positional vertigo. Therefore, careful observation of the positional nystagmic in 'failed' particle repositioning procedures is required to ensure

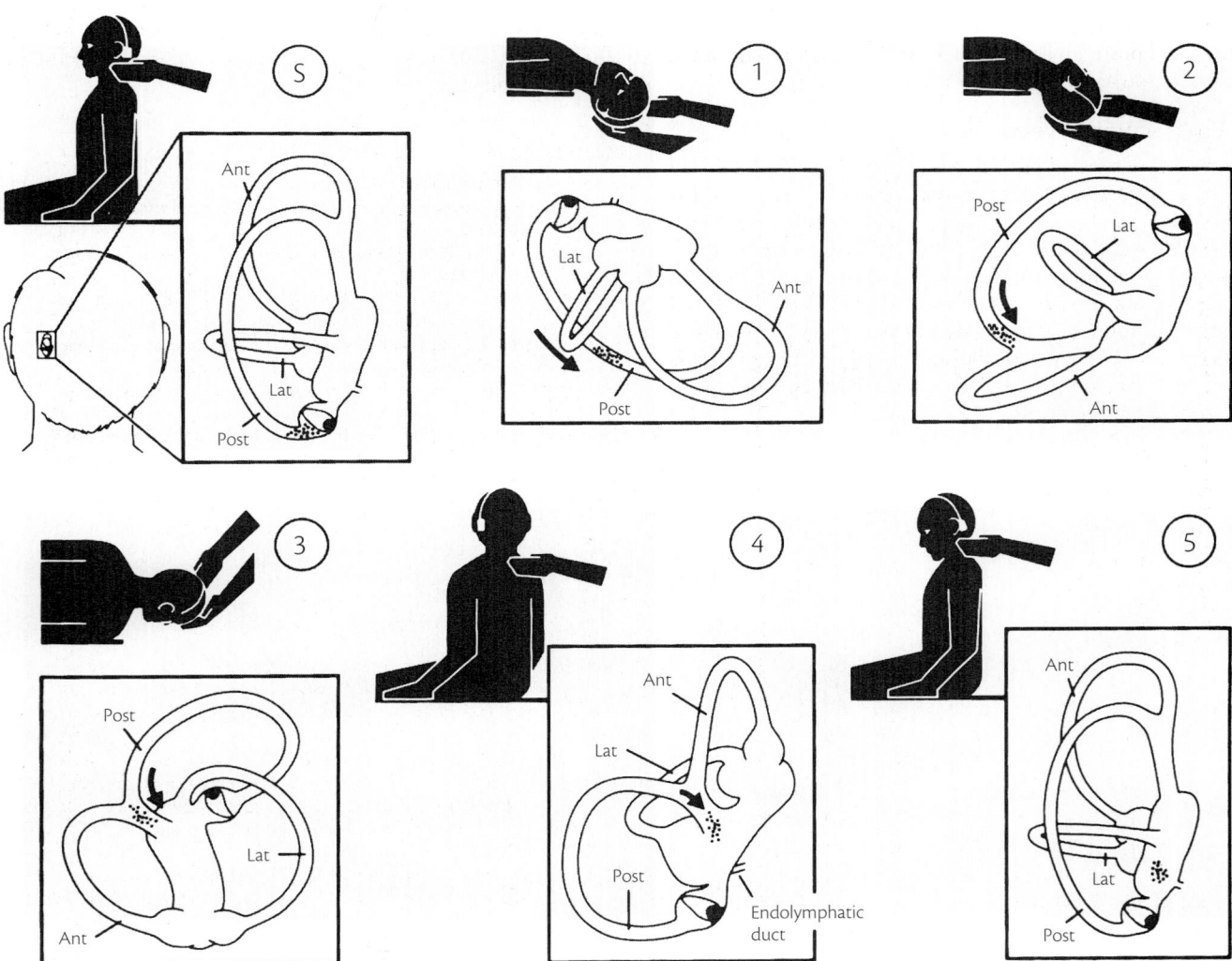

Fig. 15.43 Epley manoeuvre (for explanation see Section 15.6.4). (From Epley 1992.)

that the appropriate 'second' procedure is carried out (Herdman and Tusa 1996).

Specific treatments for horizontal and anterior canal benign paroxysmal positional vertigo have been described. Forced prolonged positioning on the healthy side for 12 hours has been advocated (Vannuchi et al. 1997); an adapted Epley manoeuvre (Lempert and Tiel-Wilck 1996) and 360° rotation (Fife et al. 1998) have all been reported to render patients asymptomatic from horizontal benign paroxysmal positional vertigo. Anterior canal benign paroxysmal positional vertigo has been reported to be treated with a reverse canal repositioning procedure, in other words a right anterior benign paroxysmal positional vertigo may be treated with a left canal repositioning procedure and visa versa) (Kim et al. 2005), while Brandt (1999b) has reported that the Brandt Daroff exercises were effective.

In conclusion, the medical treatment of benign paroxysmal positional vertigo is highly effective and as one of the commonest most disabling vestibular presentations it behoves the clinician to become familiar with the appropriate treatment strategies. Less than 1 per cent of patients with benign paroxysmal positional vertigo may be considered for surgical treatment due to failure of a particle repositioning procedure. Surgical correction of posterior benign paroxysmal positional vertigo may be achieved with singular neurectomy, occlusion of the posterior semicircular canal, or partitioning of the labyrinth (Bamiou and Luxon in press).

15.6.5 Psychological treatment

The interaction of psychological factors cannot be underestimated in both the exacerbation of symptoms of peripheral vestibular disorder and failure of compensation from acute vestibular pathology (Eagger et al. 1992; Furman and Jacob 2001). Many studies have highlighted that patients with vestibular symptom may have associated agoraphobia, anxiety states, panic attacks, depression, and avoidance behaviour, together with situational phobias such as space and motion phobia (Jacob et al. 2003). The clinical overlap between vestibular and psychiatric disorders can be consequent upon a variety of mechanisms:

◆ a chance concurrence of two separate pathologies;

◆ a vestibular disorder with psychiatric overlay;

◆ a causative relationship between the two disorders;

◆ somato-psychic effects of balance dysfunction;

◆ a neurological link between the two disorders

◆ psychogenic dizziness (Section 4.7.5).

Recent work has suggested a neurological basis for the relationship between anxiety and vestibular symptoms (Balaban 2002; Furman et al. 2005). Effective management of the balanced disordered patient frequently requires a multi-disciplinary approach with specific attention to both psychological and vestibular symptoms and disability. An appropriate complementary approach using cognitive behavioural therapy to diminish the fear of dizziness in parallel with vestibular rehabilitation physiotherapy to promote vestibular compensation and reduce dizziness is frequently highly effective (Jacob et al. 2001). Most importantly, the clinician needs to understand the interaction between vestibular and psychological influences and to explain how a patient's symptoms are compounded by the interaction of these two aspects of the illness.

In young and otherwise healthy patients who fail to compensate from an apparently straightforward peripheral vestibular disorder the index of suspicion of an underlying psychological problem should be high.

15.6.6 Surgical interventions

Surgical intervention for the treatment of vertigo is relatively rare (Ludman 2003). Conditions requiring surgery include life threatening complications of chronic middle ear disease, neoplasia such as acoustic schwannomas or other cerebello-poatine angle tumours, and perilymph fistula. Destructive procedures may be required in cases of failure of medical management of recurrent incapacitating vertigo.

Therapeutic procedures such as saccus decompression have been advocated for the management of Menière's disease, but there is little evidence to support the efficacy of such procedures aimed at presumed cause of hydrops (Thomsen et al. 1996). Plugging of the relevant semicircular canal with bone wax or bone paté has been advocated for intractable benign paroxysmal positional vertigo (Parnes et al. 1996). Destructive procedures such as vestibular nerve section, surgical labyrinthectomy, and chemical labyrinthectomy with gentamicin may be indicated rarely. However, the possibility of developing bilateral disease coupled with the need to postulate better vestibular compensation from a total destruction of the labyrinth as opposed to a partial pathological disorder must be weighed up carefully.

References

Abramovich S, Prasher DK (1986). Electrocochleography and brain-stem potentials in Ramsay Hunt syndrome. *Arch Otolaryngol Head Neck Surg*, **112**, 925–8.

Agrup C, Luxon LM (2006). Autoimmune inner ear disease. *Curr Opin Neurol*, **19**, 26–32.

Alberts MC, Terrence CF (1978). Hearing loss in cases of carcinomatous meningitis. *J Laryngol Otol*, **92**, 233–41.

Alpini D, Caputo D, Pugnetti L et al. (2001). Vertigo and multiple sclerosis: aspects of differential diagnosis. *Neurol Sci*, **22**, S84–7.

Amarenco P (1991). The spectrum of cerebellar infarctions. *Neurology*, **41**, 973–9.

Amarenco P, Hauw JJ (1990). Cerebellar infarction in the territory of the anterior inferior cerebellar artery. A clinicopathological study of 20 cases. *Brain*, **113**, 139–55.

Amarenco P, Kase CS, Rosengart A et al. (1993). Very small (border zone) cerebellar infarcts: distribution, causes, mechanisms and clinical features. *Brain*, **116**, 161.

American Academy of Otolaryngology and Head and Neck Surgery Committee of Hearing and Equilibrium (1995). Guidelines for the diagnosis and evaluation of therapy in Meniere's disease. *Otolaryngol Head Neck Surg*, **113**, 181–5.

American Academy of Neurology (1993). Assessment posturography. Report of the therapeutics and technology assessment subcommittee of the American Academy of Neurology. *Neurology*, **43**, 1261–4.

American Geriatric Society, British Geriatric Society and American Academy of Orthopaedic Surgeons Panel on Falls Prevention (2001). Guidelines for the Prevention of Falls in Older Persons. *J Am Geriatr Soc*, **49**, 664–72.

Arbusow V, Schulz P, Strupp M et al. (1999). Distribution of herpes simplex virus type in human geniculate and vestibular ganglia: implications for vestibular neuritis. *Ann Neurol*, **46**, 416–9.

Al-Twaijri WA, Shevell MI (2002). Pediatric migraine equivalents: occurrence and clinical features in practice. *Pediatr Neurol*, **26**, 365–8.

Aryan HE, Yanni DS, Nakaji P et al. (2004). Syringocephaly. *J Clin Neurosci*, **11**, 421–3.

Averbuch-Heller L, Tusa RJ, Fuhry L, et al. (1997). A double-blind controlled study of gabapentin and baclofen as treatment for acquired nystagmus. *Ann Neurol*, **41**, 818–25.

Aw ST, Fetter M, Cremer PD *et al.* (2001). Individual semicircular canal function in superior and inferior vestibular neuritis. *Neurology*, **57**, 768–74.

Backous D (2003). External auditory canal and temporal bone trauma. In Lustig LR, Niparko JK, eds. *Clinical Neuro-otology: Diagnosing and Managing Disorders of Hearing, Balance and the Facil Nerve*. Martin Dunitz, London.

Badke MB, Miedaner JA, Shea TA *et al.* (2005). Effects of vestibular and balance rehabilitation on sensory organization and dizziness handicap. Ann Otol Rhinol Laryngol, **114**, 48–54.

Balaban CD (2002). Neural substrates linking balance control and anxiety. *Physiol Behav*, **77**, 469–75.

Ballantyne J (1979). Anatomy of the ear. In Ballantyne J, Groves J, eds. *Scott-Brown's Diseases of the Ear, Nose and Throat*, Vol. 1, Basic Sciences. Butterworth's, London.

Baloh RW (1997). *Neuro-otology of Migraine Headache*, **37**, 615–21.

Baloh RW (2002). Episodic vertigo: central nervous system causes. *Curr Opin Neurol*, **15**, 17–21.

Baloh RW (2003). Vestibular neuritis. *New Engl J Med*, **348**, 1027–32.

Baloh RW, Honrubia V (2001). Vascular disosrders. In Baloh RW, Honrubia V, eds. *Clinical Neurophysiology of the Vestibular System*, 3rd edition. Oxford University Press, New York.

Baloh RW, Honrubia V, Jacobson K (1987). Benign positional vertigo: clinical and oculographic features in 240 cases. *Neurology*, **37**, 371–8.

Baloh RW, Ishyama A, Wackym PA *et al.* (1996). Vestibular neuritis: clinical-pathologic correlation. *Otolaryngol Head Neck Surg*, **114**, 586–92.

Baloh RW, Jacobson K, Honrubia V (1993). Horizontal semicircular canal variant of benign positional vertigo. *Neurology*, **43**, 2542–9.

Baloh RW, Jen JC (2000). Episodic ataxia type 2/spinocerebellar ataxia type 6. In Klockgether T, ed. *Neurological Ataxia*. Marcel Dekker, New York.

Baloh RW, Spain S, Sochotch TM *et al.* (1995). Posturography and balance problems in older people. *J Am Geriatr Soc*, **43**, 638–44.

Baloh RW, Yue Q, Furma JM, Nelson SF (1997). Familial episodic ataxia: clinical heterogeneity in four families linked to chromosome 19p. *Ann Neurol*, **41**, 8.

Bamiou D, Luxon LM (In press). Vertigo—Clinical management and rehabilitation. In Gleeson M, ed. *Scott Brown's Otolaryngology*, 7th edition. Hodder, London.

Bamiou DE, Worth S, Phelps P *et al.* (2001). Eighth nerve aplasia and hypoplasia in cochlear implant candidates: the clinical perspective. *Otol Neurotol*, **22**, 492–6.

Bárány R (1906). Untersuchungen über den vom Vestibularapparat des Ohres reflektorisch ausgelösten rhythmischen Nystagmus und seine Begleiterscheinungen *Mschr Ohrenheilk*, **40**, 193–297.

Barinagarrementeria F, Cantu C (1994). Primary medullary hemorrhage. Report of four cases and review of the literature. *Stroke*, **25**, 1684–7.

Basser LS (1964). Benign paroxysmal vertigo of childhood (a variety of vestibular neuronitis). *Brain*, **87**, 141–52.

Belden CJ, Weg N, Minor LB *et al.* (2003). CT evaluation of bone dehiscence of the superior semicircular canal as a cause of sound and/or pressure-induced vertigo. *Radiology*, **226**, 337–43.

Bergenius J, Perols O (1999). Vestibular neuritis: a follow-up study. *Acta Otolaryngol*, **119**, 895–9.

Bergström B (1973). Morphology of the vestibular nerve. *Acta Otolaryngol*, **76**, 173–9 and 331–8.

Berman JM, Fredrickson JM (1978). Vertigo after head injury—a five year follow up. *J Otolaryngol*, **7**, 237–45.

Bertholon P, Bronstein AM, Davies RA *et al.* (2002). Positional downbeating nystagmus in 50 patients: cerebellar disorders and possible anterior semicircular canalithiasis. *J Neurol Neuropsychiatry and Psychiatry*, **72**, 366–72.

Bhidayasiri R, Perlman SL, Pulst SM *et al.* (2005). Late-onset Friedreich ataxia: phenotypic analysis, magnetic resonance imaging findings, and review of the literature. *Arch Neurol*, **62**, 1865–9.

Bikhazi P, Jackson C, Ruckenstein MJ (1997). Efficacy of antimigrainous therapy in the treatment of migraine associated dizziness. *Am J Otol*, **18**, 350–4.

Black FO, Angel CR, Pesznecker SC *et al.* (2000). Outcome analysis of individualized vestibular rehabilitation protocols. *Am J Otol*, **21**, 543–51.

Blakley BW (2000). Update on intratympanic gentamicin for Menière's disease. *Laryngoscope*, **110**, 236–40.

Bohmer A, Rickenmann J (1995). The subjective visual vertical as a clinical parameter of vestibular function in peripheral vestibular diseases. *J Vestib Res*, **5**, 35–45.

Bottril I, Wills A, Mitchell AL (2003). Intratympanic gentamicin for unilateral Meniere's disease: results of therapy. *Clin Otolaryngol*, **28**, 133–41.

Boyles AL, Enterline DS, Hammock PH *et al.* (2006). Phenotypic definition of Chiari type I malformation coupled with high-density SNP genome screen shows significant evidence for linkage to regions on chromosomes 9 and 15. *Am J Med Genet*, **140**, 2776–85.

Brandt T (1999). *Vertigo:Its Multisensory Syndromes*, 2nd edition, pp. 5. Springer-Verlag, London.

Brandt T (1996a). Bilateral vestibulopathy revisited. *Eur J Med Res*, **1**, 361–8.

Brandt T (1996b). Cervical vertigo—reality or fiction? *Audiol Neurootol*, **1**, 187–96.

Brandt T (1999a). Introduction. In *Vertigo: Its Multisensory Syndromes*, 2nd edition, pp. 251–84. Springer, London.

Brandt T (1999b). Benign paroxysmal positional vertigo. In *Vertigo Its Multisensory Syndromes*, 2nd edition, pp. 251–84. Springer, London.

Brandt T, Daroff RB (1980). Physical therapy for paroxysmal positional vertigo. *Arch Otolaryngol*, **106**, 484–5.

Brandt T, Dieterich M, Danek A (1994). Vestibular paroxysmia. *Baillieres Clin Neurol*, **3**, 565.

Brant T, Steddin S (1993). Current view of the mechanism of benign paroxysmal positional vertigo: cupulolithiasis or canalothiasis? *J Vestib Res*, **3**, 373–82.

Brantberg K, Trees N, Baloh RW (2005). Migraine-associated vertigo. *Acta Otolaryngol*, **125**, 276–9.

Brihaye P, Halame AR (1993). Fluctuating hearing loss in sarcoidosis. *Acta Otorhinolaryngologica Belgica*, 47, 23–6.

British Association of Otolaryngologists, Head and Neck Surgeons (2002). Clinical Effectiveness Guidelines: Acoustic Neuroma (Vestibular Schwannoma) *BAO- NHS, London (http://www.entuk.org/publications/)*.

Bronstein A (2003). Posturography. In Luxon LM, Furman J, Martini A, Stephens SDG, eds. *A Textbook of Audiological Medicine. Clinical Aspects of Hearing and Balance*. Martin Dunitz, London.

Bronstein A (2005). Visual symptoms and vertigo. *Neurol Clin*, **23**, 705–13.

Bronstein AM, Hood JD (1987). Oscillopsia of peripheral vestibular origin. Central and cervical compensatory mechanisms. *Acta Otolaryngol*, 104, 207–14.

Bronstein A, Rinne T, Gresty M *et al.* (1994). Bilateral loss of vestibular function. *Acta Otolaryngol Suppl*, 520, 247–50.

Broughton SS, Meyerhoff WE, Cohen SB (2004). Immune-mediated inner ear disease: 10-year experience. *Semin Arthritis Rheum*, **34**, 544–8.

Brown KE, Whitney SL, Wrisley DM *et al.* (2001). Physical therapy outcomes for persons with bilateral vestibular loss. *Laryngoscope*, **111**, 1812–7.

Bürk K, Fetter M, Abele M *et al.* (1999). Autosomal dominant cerebellar ataxia type 1: oculomotor abnormalities in families with SCA1, SCA2, and SCA3. *J Neurol*, **246**, 789–97.

Butinar D, Zidar J, Leonardis L *et al.* (1999). Hereditary auditory, vestibular, motor, and sensory neuropathy in a Slovenian Roma (Gypsy) kindred. *Ann Neurol*, **46**, 36–44.

Buttner N, Geschwind D, Jen JC *et al.* (1998). Oculomotor phenotypes in autosomal dominant ataxias. *Arch Neurol*, **55**, 1353–7.

Buttner U, Helmchen C, Brandt T (1999). Diagnostic criteria for central versus peripheral positioning nystagmus and vertigo: a review. *Acta Otolaryngol*, **119**, 1–5.

Cannon SC (2006). Pathomechanisms in channelopathies of skeletal muscle and brain. *Annu Rev Neurosci*, Mar 21 [Epub ahead of print].

Caplan LR, Amarenco P, Rosengart A *et al.* (1992). Embolism from vertebral artery origin occlusive disease. *Neurology*, **42**, 1505–12.

Cartlidge NEF (1978). Postconcussional syndrome. *Scott Med*, **23**, 103.

Casani AP, Vannucci G, Fattori B *et al.* (2002). The treatment of horizontal canal positional vertigo: our experience of 66 cases. *Laryngoscope*, **112**, 172–8.

Cass SP, Goshgarian HG (1991). Vestibular compensation after labyrinthectomy and vestibular neurectomy in cats. *Otolaryngol Head Neck Surg*, **104**, 14–9.

Casselman JW, Offeciers FE, Govaerts PJ et al. (1997). Aplasia and hypoplasia of the vestibulocochlear nerve: diagnosis with MR imaging. *Radiology*, **202**, 773–81.

Claes J Van de Heyning PH (1997). Medical treatment of Menière's disease: a review of literature. *Acta Otolaryngol Suppl*, **526**, 37–42.

Commins DJ, Chen JM (1997). Multiple sclerosis: a consideration in acute cranial nerve palsies. *Am J Otol*, **18**, 590–5.

Courjon JH, Jeannerod M, Ossuzio I et al. (1977). The role of vision in compensation of vestibule-ocular reflex after hemilabyrinthectomy in the cat. *Exp Brain Res*, **28**, 235–48.

Cristante L, Westphal M, Herrmann HD (1994). Cranio-cervical decompression for Chiari I malformation. A retrospective evaluation of functional outcome with particular attention to the motor deficits. *Acta Neurochir*, **130**, 94–10.

Curthoys IS, Halmagyi GM (1995). Vestibular compensation: a review of the oculomotor, neural and clinical consequences of unilateral vestibular loss. *J Vestib Res*, **5**, 67–107.

Cutrer FM, Baloh RW (1992). Migraine associated dizziness. *Headache*, **32**, 300–4.

Daniel JR, Mauro VF (1995). Extrapyramidal symptoms associated with calcium-channel blockers. *Ann Pharmacother*, **29**, 73–5.

Davies RA, Luxon LM (1995). Dizziness following head injury: a neuro-otological study. *J Neurol*, **242**, 222–30.

Davis LE (1993). Viruses and vestibular neuritis—review of human and animal studies. *Acta Otolaryngol Suppl*, **503**, 70–3.

Densert B, Sass K (2001). Control of symptoms in patients with Meniere's disease using middle ear pressure applications: two years follow-up. *Acta Otolaryngol*, 121, 616–21.

Dieterish M (2004). Easy, inexpensive and effective: vestibular exercises for balance control. *Ann Intern Med*, **141**, 641–3.

Dieterich M, Brandt T (1999). Episodic vertigo related to migraine (90 cases): vestibular migraine? *J Neurol*, **246**, 883–92.

Dix MR (1973). Vertigo. *Practitioner*, **211**, 295–303.

Dix MR, Hallpike CS (1952). Pathology, symptomatology and diagnosis of certain common disorders of the vestibular system. *Proc R Soc Med*, **45**, 341–54.

Downton JH (1993). *Falls in the Elderly*. Edward Arnold, London.

Drachman DA, Hart C (1972). A new approach to the dizzy patient. *Neurology*, **22**, 323–34.

Driscoll CL, Lynn SG, Harner SG et al. (1998). Preoperative identification of patients at risk of developing persistent dysequilibrium after acoustic neuroma removal. *Am J Otol*, **19**, 491–5.

Eagger S, Luxon LM, Davies RA et al. (1992). Psychiatric morbidity in patients with peripheral vestibular disorder: a clinical and neuro-otological study. *J Neurol Neurosurg Psychiatry*, **55**, 383–7.

Edna T-H, Cappelen J (1987). Late postconcussional syndromes in traumatic head injury: an analysis of frequency and risk factors. *Acta Neurochir*, **86**, 12–7.

Eggers SD (2006). Migraine-related vertigo: diagnosis and treatment. *Curr Neurol Neurosci Rep*, **6**, 106–15.

Eide PK, Tysnes O-B (1992). Early and late outcome in head injury patients with radiological evidence of brain damage. *Acta Neurol Scand*, **86**, 194–8.

Elkind MS, Mohr JP (1997). Cerebellar hemorrhage. *New Horiz*, **5**, 352–8.

Ell J, Prasher D, Rudge P (1984). Neuro-otological abnormalities in Friedreich's ataxia. *J Neurol, Neurosurg Psychiatry*, **47**, 26–32.

Endo K, Ichimaru K, Komagata M et al. (2006). Cervical vertigo and dizziness after whiplash injury. *Eur Spine J*, **15**, 886–90.

Engström H, Bergström B, Rosenhall U (1974). The vestibular sensory epithelia. *Arch Otolaryngol*, **100**, 411–8.

Epley JM (1992). The canalith repositioning procedure: for treatment of benign paroxysmal positional vertigo. *Otolaryngol Head Neck Surg*, **107**, 399–404.

Ernst A, Basta D, Seidl RO, et al. (2005). Management of posttraumatic vertigo. *Otolaryngol Head Neck Surg*, **132**, 554–8.

Fantry LE, Staeker HV (2002). Vertigo and abacavir. *AIDS Patient Care STDS*, **1**, 5–7.

Fattori B, De Iaco G, Nacci A et al. (2002). Alternobaric oxygen therapy in long-term treatment of Meniere's disease. *Undersea Hyperb Med*, **29**, 260–70.

Fenton GW (1996). The postconcussional syndrome reappraised. *Clin Electroencephal*, **27**, 174–82.

Fetter M, Zee DS, Proctor LR (1988). Effect of lack of vision and of occipital lobectomy upon recovery from unilateral labyrinthectomy in rhesus monkey. *J Neurophysiol*, **59**, 394–407.

Fetter M, Dichgans J (1996). Vestibular neuritis spares the inferior division of the vestibular nerve. *Brain*, **119**, 755–63.

Fife TD (1998). Recognition and management of horizontal canal benign positional vertigo. *Am J Otol*, **19**, 345–51.

Fish JH 3rd, Scholtz AW, Hussl B et al. (2001). Cerebro-oculo-facio-skeletal syndrome as a human example for accelerated cochlear nerve degeneration. *Otol Neurotol*, **22**, 170–7.

Fisher CM (1967). Vertigo and cerebrovascular disease. *Arch Otolaryngol*, **85**, 529–34.

Fitzgerald G, Hallpike CS (1942). Studies in human vestibular function: 1. Observations on the directional preponderance of caloric nystagmus resulting from cerebral lesions. *Brain*, **65**, 115–37.

Foster C, Baloh RW (1996). Drug therapy for vertigo. In Baloh RW, Halmagyi GM, eds. *Disorders of the Vestibular System*. Oxford University Press, New York.

Francis DA, Bronstein AM, Rudge P et al. (1992). The site of brainstem lesions causing semicircular canal paresis: an MRI study. *J Neurol Neurosur Psychiatry*, **55**, 446–9.

Fransen E, Verstreken M, Verhagen WI et al. (1999). High prevalence of symptoms of Menière's disease in three families with a mutation in the COCH gene. *Hum Mol Genet*, **8**, 1425.

Frenzel H (1955). *Spontan-und- Provokations-Nystagmus als Krankheitssymptom*. Springer, Berlin/Gottingen/ Heidelberg.

Friedmann G (1970). The judgement of the visual vertical and horizontal with peripheral and central vestibular lesions. *Brain*, **93**, 313–28.

Friedmann G (1971). The influence of unilateral labyrinthectomy on orientation in space. *Acta Otolaryngol*, **71**, 289–98.

Friedreich N (1863). Über degenerative Atrophie der spinalen Hinterstränge. *Virchows Arch Pathol Anat*, **27**, 1.

Froehling DA, Bowen JM, Mohr DN et al. (2000). The canalith repositioning procedure for the treatment of benign paroxysmal positional vertigo: a randomized controlled trial. *Mayo Clin Proc*, **75**, 695–700.

Frumkin LR, Baloh RW (1990) Wallenberg's syndrome following neck manipulation. *Neurology*, **40**, 611.

Furman JM, Balaban CD, Jacob RG et al. (2005). Migraine-anxiety related dizziness (MARD): a new disorder? *J Neurol Neurosurg Psychiatry*, **76**, 1–8.

Furman JM, Crumrine PK, Reinmuth OM (1990). Epileptic nystagmus. *Ann Neurol*, **27**, 686–8.

Furman, JM, Marcus DA, Balaban CD (2003). Migrainous vertigo: development of a pathogenetic model and structured diagnostic interview. *Curr opin Neurol*, **16**, 5–13.

Furman JM, Cass SP (1999). Benign paroxysmal positional vertigo. *New Engl J Med*, **341**, 1590–6.

Furman JM, Jacob RG (2001). A clinical taxonomy of dizziness and anxiety in the otoneurological setting. *J Anxiety Disord*, **15**, 9–26.

Garcia-Berrocal JR, Gorriz C, Ramirez-Camacho R et al. (2006). Otosyphilis mimics immune disorders of the inner ear. *Acta Otolaryngol*, **126**, 679–84.

Gass A, Steinke W, Schwartz A et al. (1998). High resolution magnetic resonance imaging in peripheral vestibular dysfunction in multiple sclerosis. *J Neurol Neurosurg Psychiatry*, **65**, 945.

Gillilan LL (1964). The correlation of the blood supply to the human brainstem with clinical brainstem lesions. *J Neuropathol Exp Neurol*, **23**, 78–108.

Goebel JA, O'Mara W, Gianoli G (2001). Anatomic considerations in vestibular neuritis. *Otol Neurotol*, **22**, 512–8.

Gonshor A, Melvill Jones G (1976). Extreme vestibule-ocular adaptation induced by prolonged optical reversion of vision. *J Physiol*, **256**, 381–414.

Goel A, Bhatjiwale M, Desai K (1998). Basilar invagination: a study based on 190 surgically treated patients. *J Neurosurg*, **88**, 962.

Gordon CR, Joffe V, Vainstein G et al. (2003). Vestibulo-ocular arreflexia in families with spinocerebellar ataxia type 3 (Machado-Joseph disease). *J Neurol Neurosurg Psychiatry*, **74**, 1403–6.

Gowers WR (1907). *Epilepsy and Other Chronic Convulsive Diseases: their Causes Symptoms and Treatment.* Dover Publications Incorporation, New York, 1964.

Grad A, Baloh RW (1989). Vertigo of vascular origin: clinical and electronystagmographic features in 84 cases. *Arch Neurol*, **46**, 281–4.

Gradenigo G (1893). On the clinical signs of affectations of the auditory nerve. *Arch Otolaryngol*, **22**, 213–30.

Grenman R (1985). Involvement of the audiovestibular system in multiple sclerosis: an otoneurologic and audiologic study. *Acta Otolaryngol Suppl (Stockh)*, 420–9.

Grossman GE, Leigh RJ, Abel LA et al. (1988). Frequency and velocity of rotational head perturbations during locomotion. *Exp Brain Res*, **70**, 470–6.

Guerraz M, Yardley L, Bertholon P et al. (2001). Visual vertigo: symptom assessment, spatial orientation and postural control. *Brain*, **124**, 1646–56.

Gulya AJ (1993). Neurologic paraneoplastic syndromes with neurotologic manifestations. *Laryngoscope*, **103**, 754–61.

Gutmann DH, Aylsworth A, Carey JC et al. (1997). The diagnostic evaluation and multidisciplinary management of neurofibromatosis 1 and neurofibromatosis 2. *JAMA*, **278**, 51–7.

Hain TC, Helminski JO, Reis IL et al. (2000). Vibration does not improve results of the canalith repositioning procedure. *Arch Otolaryngol Head Neck Surg*, **126**, 617–22.

Hall CD, Schubert MC, Herdman SJ (2004). Prediction of fall risk reduction as measured by dynamic gait index in individuals with unilateral vestibular hypofunction. *Otol Neurotol*, **25**, 746–51.

Hall SF, Ruby RR, MClure JA (1979). The mechanics of benign paroxysmal vertigo. *J Otolaryngol*, **8**, 151–8.

Hallam RS, Beyts J, Jakes SC (1988). Symptom reporting and objective test results. *Adv Audiol*, **5**, 129–36.

Halmagyi GM, Thurtell MT, Curthoys IS (in press). Vertigo clinical syndromes. In Gleeson M, ed. *Scott Brown's Otolaryngology*, 7th edition, Hodder Arnold, London.

Halmagyi GM, Cremer PD, Curthoys IS (2003a). Peripheral vestibular disorders and diseases in adults. In Luxon LM, Furman JM, Martini A, Stephens D, eds. *Textbook of Audiological Medicine*, pp. 797–818. Martin–Dunitz, Taylor & Francis Group, London.

Halmagyi GM, Curthoys IS (1988). A clinical sign of canal paresis. *Arch Neurol*, **45**, 737–9.

Halmagyi GM, McGarvie LA, Aw ST et al. (2003b). The click-evoked vestibulo-ocular reflex in superior semicircular canal dehiscence. *Neurology*, **60**, 1172–5.

Halmagyi GM, Rudge P, Gresty MA et al. (1980). Treatment of periodic alternating nystagmus. *Ann Neurol*, **8**, 609–11.

Hamman KF, Arnold WW (1999). Menière's disease. In Buttner U, ed. *Vestibular Dysfunction and its Therapy. Advances in Otorhinolaryngology*, pp. 55, 195–227. Basel, Karger.

Hanson W, Parnes LS (1995). Vestibular nerve compression in Camurati-Engelmann disease. *Ann Otol, Rhinol Laryngol*, **104**, 823–5.

Harcourt JP, Brookes GB (1995). Delayed endolymphatic hydrops: clinical manifestations and treatment outcome. *Clin Otolaryngol*, **20**, 318–22.

Harker LA (1996). Migraine associated vertigo. In Baloh RW, Halmagyi GM, eds. *Disorders of the Vestibular System*, pp. 407–17. Oxford University Press, Oxford.

Hart RG, Pearce LA, Rothbart RM et al. (2000). Stroke with intermittent atrial fibrillation: incidence and predictors during aspirin therapy. *J Am Coll Cardiol*, **35**, 183–7.

Harvey SA, Hain TC, Adamiec LC (1994). Modified liberatory maneuver: effective treatment for benign paroxysma positional vertigo. *Laryngoscope*, **104**, 1206–12.

Hashimoto T, Sasaki O, Yoshida K et al. (2003). Periodic alternating nystagmus and rebound nystagmus in spinocerebellar ataxia type 6. *Mov Disord*, **18**, 1201–4.

Hecker HC, Haug CO, Herndon JW (1974). Treatment of the vertiginous patient using Cawthorne's vestibular exercises. *Laryngoscope*, **84**, 2065–72.

Herdman SJ, Clendaniel RA (2000). Assessment and treatment of complete vestibular loss. In Herdman SJ, ed. *Vestibular Rehabilitation*. FA Davis, Philadelphia.

Herdman SJ, Clendaniel RA, Mattox DE et al. (1995). Vestibular adaptation exercises and recovery: acute stage after acoustic neuroma resection. *Otolaryngol Head Neck Surg*, **113**, 77–87.

Herdman SJ, Tusa R, Zee DS et al. (1993). Single treatment approaches to benign paroxysmal positional vertigo. *Arch Otolaryngol Head Neck Surg*, **119**, 450–4.

Herdman SJ, Tusa R (1996). Complications of the canalith repositioning procedure. *Arch Otolaryngol Head Neck Surg*, **122**, 281–6.

Herdman SJ, Whitney SL (2000). Treatment of vestibular hypofunction 2000 In Herdman SJ, ed. *Vestibular Rehabilitation*, FA Davis, Philadelphia.

Himi T, Igarashi M, Kataura A et al. (1993). Temporal bone findings in craniodiaphyseal dysplasia. *Auris Nasus Larynx*, **20**, 255–61.

Hinchcliffe R, Coles RR, King P (1992). Occupational noise induced vestibular malfunction? *Br J Ind Med*, **49**, 63–5.

Hirvonen TP, Weg N, Zinreich SJ et al. (2003). High-resolution CT findings suggest a developmental abnormality underlying superior canal dehiscence syndrome. *Acta Otolaryngol*, **123**, 477–81.

Honrubia V and Luxon LM (1984). Optokinetic nystagmus with reference to smooth pursuit function. In: Oosterveld WJ ed. *Otoneurology*. John Wiley and Sons Ltd, Chichester.

Horstmann GA, Dietz V (1990). A basic postural control mechanism: the stabilization of the centre of gravity. *Electroencephalogr Clin Neurophysiol*, **76**, 165–76.

Huang CY, Yu YL (1985). Small cerebellar strokes may mimic labyrinthine lesions. *J Neurol Neurosurg Psychiatry*, **48**, 263–5.

Hugosson S, Carlsson E, Borg E et al. (1997). Audiovestibular and neuropsychological outcome of adults who had recovered from childhood bacterial meningitis. *Int J Pediatr Otorhinolaryngol*, **42**, 149–67.

Huygen PLM, Verhagen WIM (1994). Peripheral vestibular and vestibule-cochlear dysfunction in hereditary disorders. *J Vestib Res*, **4**, 81–104.

Igarashi M, Levy JK, O-Uchi T et al. (1981). Further study of physical exercise and locomotor balance compensation after unilateral vestibular neurotomy. *Acta Otolaryngol*, **92**, 101–5.

Igarashi M, Alford BR, Kato Y et al. (1975). Effect of physical exercise upon nystagmus and locomotor disequilibrium and labyrinthectomy in experimental primates. *Acta OtoLaryngol*, **79**, 214–20.

International Headache Society (2004). *The International Classification of Headache Disorders*, 2nd edition, pp. 1–160; *Cephalalgia* **24** (Suppl 1).

Ishiyama G, Ishiyama A, Jacobson K et al. (2001). Drop attacks in older patients secondary to an otologic cause. *Neurology*, **57**, 1103–6.

Ishizaki H, Pyykko I, Nozue N (1993). Neuroborreliosis in the aetiology of vestibular neuritis. *Acta Otolaryngol Suppl*, **503**, 67–9.

Ito M (1984). *The Cerebellum and Neural Control*. Raven Press, New York.

Iwase H, Kobayashi M, Kurata A et al. (2001). Clinically unidentified dissection of vertebral artery as a cause of cerebellar infarction. *Stroke*, **32**, 1422–4.

Jacob RG, Furman JM, Cass SP (2003). Psychiatric consequences of vestibular dysfunction. In Luxon LM, Furman J, Martini A, Stephens SDG, eds. *A Textbook of Audiological Medicine. Clinical Aspects of Hearing and Balance*. Martin Dunitz, London.

Jacob RG, Whitney SL, Detweiler-Shostak G et al. (2001). Vestibular rehabilitation for patients with agoraphobia and vestibular dysfunction: a pilot study. *J Anxiety Disord*, 15, 131–46.

Jacobson GP, Newman CW (1990). The development of the dizziness handicap inventory. *Arch Otolaryngol Head Neck Surg*, 116, 424–7.

Jahrsdoerfer RA, Thompson EG, Johns MM et al. (1981). Sarcoidosis and fluctuating hearing loss. *Ann Otol Rhinol Laryngol*, 90, 161–3.

James AL, Burton MJ (2001). Betahistine for Menière disease or syndrome. *Cochrane Database Syst Rev*, 1, CD001873.

Jannetta PJ, Moller MB, Moller ARC (1984). Disabling positional vertigo. *New Engl J Med*, 310, 1700.

Jen JC, Yue Q, Karrim J et al. (1998). Spinocerebellar ataxia type 6 with positional vertigo and acetazolamide responsive episodic ataxia. *J Neurol Neurosurg Psychiatry*, **65**, 565–8.

Jensen MB, St Louis EK (2005). Management of acute cerebellar stroke. *Arch Neurol*, **62**, 537–44.

Johnson G (1998). Medical management of migraine-related dizziness and vertigo. *Laryngoscope*, **108**, 1–28.

Jonsson R, Sixt E, Landahl S et al. (2004). Prevalence of dizziness and vertigo in an urban elderly population. *J Vestib Res*, **14**, 47–52.

Kayan A, Hood JD (1983). Neuro-otological manifestations of migraine. *Brain*, **107**, 1123–42.

Kennedy RJ, Shelton C, Salzman KL *et al.* (2004). Intralabyrinthine schwannomas: diagnosis, management, and a new classification system. *Otol Neurotol*, **25**, 160–7.

Khetarpal U, Schuknecht H (1990). In search of pathologic correlates for hearing loss and vertigo in Paget's disease. A clinical and histopathologic study of 26 temporal bones. *Ann Otol Rhinol Laryngol Suppl*, **145**, 1–16.

Kim JS, Lopez I, DiPatre PL *et al.* (1999). Internal auditory artery infarction: clinicopathologic correlation. *Neurology*, **52**, 40–4.

Kim YK, Shin JE, Chung JW (2005). The effect of canalith repositioning for anterior semicircular canal canalithiasis. *ORL J Otorhinolaryngol Relat Spec*, **67**, 56–60.

Kingma H (2006). Function tests of the otolith or statolith system. *Curr Opin Neurol*, **19**, 21–5.

Klockhoff I, Lindblom U, Stahle J (1974). Diuretic treatment of Menière's disease. Long term results with chlorthalidone. *Arch Otolaryngol*, **100**, 262–5.

Kluge M, Beyenburg S, Fernandez G *et al.* (2000). Epileptic vertigo: evidence for vestibular representation in human frontal cortex. *Neurology*, **55**, 1906–8.

Kohut RI, Hinojosa R, Ryu JH (1988). Perilymphatic fistulae: a single-blind clinical histopathological study. *Adv Otorhinolaryngol*, **42**, 148–52.

Kommerell G, Hoyt WF (1973). Lateropulsion of saccadic eye movements; electro-oculographic studies in a patient with Wallenberg's syndrome. *Arch Neurol*, **28**, 313.

Konrad HR, Girardi M, Helfert R (1999). Balance and ageing. *Laryngoscope*, **109**, 1454–60.

Kremer H, van Wijk E, Marker T *et al.* (2006). Usher syndrome: molecular links of pathogenesis, proteins and pathways. *Hum Mol Genet*, **15** Spec No 2, R262–70.

Lackner JR, Graybiel A (1986). Head movements in non-terrestrial force environments elicit motion sickness: implications for the aetiology of space motion sickness. *Aviat Space Environ Med*, **57**, 443.

Lacour M (1984). Relearning and critical postoperative period in the restoration of nerve function. Example of vestibular compensation and clinical implications. Annals Otolaryngology Chir Cervicofac, **101**, 177–87 [French].

Lacour M, Roll JP, Appix M (1976). Modifications and development of spinal reflexes in the alert baboon (Papio papio) following an unilateral vestibular neurotomy. *Brain Res*, **113**, 255–69.

Lacour M, Sterckers O (2001). Histamine and betahistine in the treatment of vertigo: elucidation of mechanisms of action. *CNS Drugs*, **15**, 853–70.

Lacour M, Xerri C (1981). Lesion induced neuronal plasticity in sensorineural systems. In Flohr H, Precht W, eds. *Vestibular Compensation: New Perspectives*, pp. 240–53. Springer, Berlin, Heidelberg, New York.

Lacour M, Xerri C (1980). Vestibular compensation: new perspectives. In Flohr H, Precht W, eds. *Lesion Induced Neuronal Plasticity in Sensorimotor Systems*, pp. 240–253. Springer, Berlin, Heidelberg, New York.

Lacour M, Xerri C, Hugon M (1979). Compensation of postural reactions to fall in the vestibular neurectomized monkey. Role of the reamining labyrinthine afferences. *Exp Brain Res*, **37**, 563–80.

Lanzi G, Balottin U, Fazzi E *et al.* (1994). Benign paroxysmal vertigo of childhood: a long-term follow-up. *Cephalalgia*, **14**, 458–60.

Lempert T, Tiel-Wilk K (1996). A positional maneuver for treatment of horizontal canal benign positional vertigo. *Laryngoscope*, 476–8.

Lenard HG, Voit T, Lamprecht A *et al.* (1992). Sudden loss of hearing and vestibular function, muscular weakness, and multiple white matter lesions in preschool children. *Neuropediatrics*, **23**, 221–4.

Lennox LG (1960). *Epilepsy and Related Disorders*. Little Brown and Company, Boston.

Lesser THJ, Dort JC, Simmen DPB (1990). Ear nose and throat manifestations of Lyme Disease. *J Laryngol Otol*, **104**, 301–4.

Levin HS, Mattis S, Ruff RM (1987). Neurobehavioural outcome following minor head injury: a three-center study. *J Neurol*, **66**, 234–43.

Little SC, Kesser BW (2006). Radiographic classification of temporal bone fractures: clinical predictability using a new system. *Arch Otolaryngol Head Neck Surg*, **132**, 1300–4.

Ludman H (1997). Complications of suppurative otitis media In Kerr A, Booth J, eds. *Scott Brown's Otolaryngology Volume 3: Otology*, pp3/12/1–29. Butterworth, London.

Ludman H (2003). Role of surgery in the management of the dizzy patient. *Textbook of Audiological Medicine*, **55**, 917–27.

Luxon LM (1984). Vertigo in old age. In Dix MR, Hood JD, eds. *Vertigo*, Vol. 14, pp. 291–319. Wiley & Sons, Chichester.

Luxon LM (1990). Signs and symptoms of vertebrobasilar insufficiency. In Hofferberth B, Brune GG, Sitzer G, Weger H-D, eds. *Vascular Brain Stem Diseases*, pp. 93–111. Karger, Basel.

Luxon LM (1996). Post-traumatic vertigo. In Baloh RW, Halmagyi M, eds. *Disorders of the Vestibular System*, pp. 381–95. Oxford University Press, New York.

Luxon LM (1997). Vestibular compensation. In Luxon LM, Davies RA, eds. *Handbook of Vestibular Rehabilitation*, pp. 17–29. Whurr Publishers, London.

Luxon LM Cohen M, Coffey RA *et al.* (2003). Neuro-otological findings in Pendred syndrome. *Int J Audiol*, **42**, 82–8.

MacDonald CB, Melhem ER (1997). An approach to imaging the dizzy patient. *J Neuroimaging*, 7, 180–6.

Maki BE, McIlroy WE (2003). Effects of aging on control of stability. In Luxon LM, Furman J, Martini A, Stephens SDG, eds. *A Textbook of Audiological Medicine. Clinical Aspects of Hearing and Balance*. Martin Dunitz, London.

Marks I (1981). Space 'phobia': a pseudo-agoraphobic syndrome. *J Neurol Neurosurg Psychiatry*, 44, 387–91.

Mas JL (2003). Patent foramen ovale and stroke. *Pract Neurol*, **3**, 4–11.

Massion J, Woollacott MH (1996). In Bronstein AM, Brandt T, Woollacott MH, eds. *Clinical Disorders of Balance Posture and Gait*. Arnold, London.

Massoud EAS, Ireland DJ (1996). Post-treatment instructions in the non-surgical management of benign paroxysmal positional vertigo. *J Otolaryngol*, **25**, 121–5.

McCue MP, Guinan JJ (1994). Acoustically responsive fibres in the vestibular nerve of the cat. *J Neurosci*, 14, 6058–70.

Menezes AH, Van Gilder JC, Graf CJ *et al.* (1980). Cranocervical abnormalities: a comprehensive surgical approach. *J Neurosurg*, **53**, 444–55.

Menière P (1861). Mémoire sur les lésions de l'oreille interne donnant lieu a des symptomes de congestion cérébrale apoplectiforme. *Gazette Medicale de Paris*, **16**, 597–601.

Millikan CH, Siekert R (1955). Studies in cerebrovascular disease 1. The syndrome of intermittent insufficiency of the basilar arterial system. *Proceedings of the Staff Meeting at the Mayo Clinic*, **4**, 61–5.

Minor LB, Schessel DA, Carey JP (2004). Meniere's disease. *Curr Opin Neurol*, **17**, 9–16.

Minor LB (2005). Clinical manifestations of superior semicircular canal dehiscence. *Laryngoscope*, **115**, 1717–27.

Minor LB, Carey JP, Cremer PD *et al.* (2003). Dehiscence of bone overlying the superior canal as a cause of apparent conductive hearing loss. *Otol Neurotol*, **24**, 270–8.

Montermini L, Richter A, Morgan K *et al.* (1997). Phenotypic variability in Friedreich ataxia: role of the associated GAA triplet repeat expansion. *Ann Neurol*, **41**, 675.

Morimoto AK, Wiggins RH 3rd, Hudgins PA *et al.* (2006). Absent semicircular canals in CHARGE syndrome: radiologic spectrum of findings. *Am J Neuroradiol*, **27**, 1663–71.

Morrow MJ (2000). Bell's palsy and herpes zoster oticus. *Curr Treat Options Neurol*, **2**, 407–16.

Mrugala MM, Batchelor TT, Plotkin SR (2005). Peripheral and cranial nerve sheath tumours. *Curr Opin Neurol*, **18**, 604–10.

Mruzek M, Barin K, Nicholas DS *et al.* (1995). Effects of vestibular rehabilitation and social reinforcement on recovery following ablative vestibular surgery. *Laryngoscope*, **105**, 686–92.

Murakami T, Nakayasu H, Doi M *et al.* (2006). Anterior and posterior inferior cerebellar artery infarction with sudden deafness and vertigo. *J Clin Neurosci*, **13**, 1051–4.

Murofushi T, Curthoys IS, Topple AN *et al.* (1995). Responses of guinea pig primary vestibular neurons to clicks. *Exp Brain Res*, **103**, 174–8.

Murphy JV, Dehkharghani F (1994). Diagnosis of childhood seizure disorders. *Epilepsia*, **35**, S7–17.

Naess A, Halstensen A, Nyland H *et al.* (1994). Sequelae one year after meningococcal disease. *Acta Neurol Scand*, **89**, 139–42.

Nakayama M, Helfert RH, Konrad HR *et al.* (1994). Scanning electron microscopic evaluation of age-related changes in the rat vestibular epithelium. *Otolaryngol Head Neck Surg*, **111**, 799–806.

Nance WE (2003). The genetics of deafness. *Ment Retard Dev Disabil Res Rev*, 9, 109–19.

Nashner LM (1996). Practical biomechanics and physiology of balance. In Jacobsen GP, Newman CW, Kartush JM eds. *Handbook of Balance Function Testing*. Mosby, St Louis.

Netuka D, Benes V, Mikulik R *et al.* (2005). Symptomatic rotational occlusion of the vertebral artery: case report and review of the literature. *Zentralbl Neurochir*, 4, 217–22.

Neuhauser H, Radtke A, von Brevern M, Lempert T (2003). Zolmitriptan for treatment of migrainous vertigo: a pilot randomized placebo-controlled trial. *Neurology*, 60, 882–3.

Neuhauser NK, von Brevern M, Radtke A *et al.* (2005). Epidemiology of vestibular vertigo. A neurotologic survey of the general population. *Neurology*, 65, 898–904.

Neuhauser H, Leopold M, von Brevern M *et al.* (2001). The interrelations of migraine, vertigo and migrainous vertigo. *Neurology*, 56, 436–41.

Neuhauser H, Lempert T (2004). Vertigo and dizziness related to migraine: a diagnostic challenge. *Cephalalgia*, 24, 83–91.

Niemensivu R, Pyykko I, Wiener-Vacher SR *et al.* (2006). Vertigo and balance problems in children—an epidemiological study in Finland. *Int J Pediatr Otorhinolaryngol*, 70, 259–65.

Norre ME, De Weerdt W (1980). Treatment of vertigo based on habituation. *J Laryngol Otol*, 94, 971–7.

Nuti D, Passero S, Di Girolaamo S (1996). Bilateral vestibular loss in vertebrobasilar dolichoectasia. *J Vestib Res*, 6, 85.

Oas JG, Baloh RW (1992). Vertigo and the anterior inferior cerebellar artery syndrome. *Neurology*, 42, 2274.

Odkvist LM, Arlinger S, Billermark E *et al.* (2000). Effects of middle ear pressure changes on clinical symptoms in patients with Meniere's disease—a clinical multicentre placebo-controlled study. *Acta OtoLaryngol*, 543, 99–101.

Okinaka Y, Sekitani T, Okazaki H *et al.* (1993). Progress of caloric response of vestibular neuronitis. *Acta Otolaryngol Suppl*, 503, 18–22.

O'Leary DP, Davis LL (1990). High frequency autorotational testing of the vestibulo-ocular reflex. *Neurological Clin*, 8, 297.

OMahoney C, Davies RA (1998). Vestibular investigations. In Ludman H, Wright T, eds. *Diseases of the Ear*, 6th edition. Arnold Publishers, London.

OMahoney CF, Luxon LM (1997). Causes of balance disorders. *Scott-Brown's Otolaryngology*, 20, 1–58.

Oosterveld WJ, Kortschot HW, Kingma GG *et al.* (1991). Electonystagmographic findings following cervical whiplash injuries. *Acta Otolaryngol*, 111, 201–5.

Osei-Lah V, Ceranic B, Luxon LM (in press). Tone burst VEMPs in acute and stable Meniere's disease at threshold. *Audiol Neurootol*.

Pagarkar W, Bamiou D-E, Ridout D *et al.* (in press). Subjective visual vertical and horizontal—correlation with preset angle. *Arch Otolaryngol Head Neck Surg*.

Pagarkar W, Luxon LM (in press). Autonomic vestibular dysfunction. In Mathias C Bannister R, eds. *Autonomic Failure*, 5th edition. Oxford University Press, Oxford.

Page NG, Gresty MA (1985). Motorist's vestibular disorientation syndrome. *J Neurol Neurosurg Psychiatry*, 48, 729–35.

Paparella MM, Djalilian HR (2002). Etiology, pathophysiology of symptoms and pathogenesis of Meniere's disease. *Otolaryngol Clin North Am*, 35, 529–45.

Pappas DG Jr, Roland JT Jr, Lim J *et al.* (1995). Ultrastructural findings in the vestibular end-organs of AIDS cases. *Am J Otol*, 16, 140–5.

Parnes LS (1996). Update on posterior semicircular canal occlusion for benign paroxysmal positional vertigo. *Otolaryngol Clin North Am*, 29, 333–42.

Parnes LS, Agrawal SK, Atlas J (2003). Diagnosis and management of benign paroxysmal positional vertigo. *Can Med Assoc J*, 169, 681–93.

Parnes LS, Shimotakahara SG, Pelz D *et al.* (1990). Vascular relationships of the vestibulocochlear nerve on magnetic resonance imaging. *Am J Otol*, 11, 278–81.

Parnes LS, Sun Ah, Freeman DJ (1999). Corticosteroid pharmacokinetics in the inner ear fluids: an animal study followed by clinical application. *Laryngoscope*, 109, 1–17.

Passero S, Filosomi G (1998). Posterior circulation infarcts in patients with vertebrobasilar dolichoectasia. *Stroke*, 29, 65.

Passer S, Nuti D (1996). Auditory and vestibular findings in patients with vertebrobasilar dolichoectasia. *Acta Neurol Scand*, 93, 50.

Pavlou M, Wijnberg N, Faldon M *et al.* (2003). Effect of semicircular canal stimulation on the perception of the visual vertical. *J Neurophysiol*, 90, 622–30.

Pavlou M, Lingeswaran A, Davies RA *et al.* (2004). Simulator based rehabilitation in refractory dizziness. *J Neurol*, 251, 983–95.

Penido Nde O, Borin A, Iha LC *et al.* (2005). Intracranial complications of otitis media: 15 years of experience in 33 patients. *Otolaryngol Head Neck Surg*, 132, 37–42.

Perrin PP, Jeandel C, Perrin CA *et al.* (1997). Influence of visual control, conduction and central integration on static and dynamic balance in healthy older adults. *Gerontology*, 3, 233–31.

Peterka RJ, Black FO, Schoenhoff MB (1990). Age related changes in human vestibulo-ocular reflexes: sinusoidal rotation and caloric tests. *J Vestib Res*, 1, 49–59.

Petrone D, De Beneditis G, De Candia N (1991). Experimental research on vestibular compensation using posturography. *Bollettino – Societa Italiana Biologia Sperimentale*, 67, 731–7.

Phelps PD, Reardon W, Pembery M *et al.* (199l). X-Linked deafness, stapes gushers and a distinctive defect of the inner ear. *Neuroradiology*, 33, 326–30.

Plaza Mayor G, Baron Rubio M, (2006). Neuro-otological manifestations as presentacion of type I Chiari malformation. *An Otorrinolaringol Ibero Am*, 33, 613–22.

Porter RJ, Eyster E (1973). Aneurysm in the anterior inferior cerebellar artery at the internal acoustic meatus: report of a case. *Surg Neurol*, 1, 27–8.

Pothula VB, Chew F, Lesser TH *et al.* (2004). Falls and vestibular impairment. *Clin Otolaryngol*, 29, 179–82.

Pyhtinen J, Karttunen A, Tikkakoski T (2006). Increasing benefit of magnetic resonance imaging in multiple sclerosis. *Acta Radiol*, 47, 960–71.

Radtke A, Neuhauser H, von Brevern M *et al.* (1999). A modified Epley's procedure for self-treatment of benign paroxysmal positional vertigo. *Neurology*, 53, 1358–60.

Rae-Grant AD, Eckert NJ, Bartz S *et al.* (1999). Sensory symptoms of multiple sclerosis: a hidden reservoir of morbidity. *Mult Scler*, 5, 179.

Ramsden RT (1997). Vestibular schwannoma. In Kerr A, Booth J, eds. *Scott Brown's Otolaryngology Volume 3:Otology*, pp. 3/21/1–38. Butterworth, London.

Rassekh CH, Harker LA (1992). The prevalence of migraine in Menière's disease. *Laryngoscope*, 102, 135–8.

Rasmussen N, Johnsen NJ, Bohr VA. (1991). Otologic sequelae after pneumococcal meningitis: a survey of 164 consecutive cases with a follow-up of 94 survivors. *Laryngoscope*, 101, 876–82.

Rinne T, Bronstein AM, Rudge P *et al.* (1998). Bilateral Loss of Vestibular Function: Clinical Findings in 53 Patients. *J Neurol*, 245, 314–21.

Romberg MH (1846). *Lerbuch der NervenKrankheiten des Menschen*. A Duncker, Berlin.

Rosenhall U, Hanner P, Kaijser B (1988). Borrelia infection and vertigo. *Acta Otolaryngol*, 106, 111–6.

Roydhouse N (1974). Vertigo and its treatment. *Drugs*, 7, 297–309.

Rubenstein RL, Normal DM, Schindler RA *et al.* (1980). Cerebellar infarction – a presentation of vertigo. *Laryngoscope*, 90, 505.

Ruckenstein MJ (2004). Autoimmune inner ear disease. *Curr Opin Otolaryngol Head Neck Surg*, 12, 426–30.

Rucker JC (2005). Current treatment of nystagmus. *Curr Treat Options Neurol*, 7, 69–77.

Rutherford WH (1977). Sequelae of concussion caused by minor head injuries. *Lancet*, 1;1, 1–4.

Saeed AB, Shuaib A, Al-Sulaiti G *et al.* (2000). Vertebral artery dissection: warning symptoms, clinical features and prognosis in 26 patients. *Can J Neurol Sci*, 27, 292–6.

Santos PM, Hall RA, Snyder JM *et al.* (1993). Diuretic and diet effect on Menière disease evaluated by the 1985 Committee on Hearing and equilibrium guidelines. *Otolaryngol Head Neck Surg*, 109, 680–9.

Sasaki O, Ootsuka K, Taguchi K *et al.* (1994). Multiple sclerosis presented acute hearing loss and vertigo. *J Otorhinolaryngol Relat Spec*, 56, 55–9.

Savitz SI, Caplan LR, Edlow JA (2007). Pitfalls in the diagnosis of cerebellar infarction. *Acad Emerg Med*, 14, 63–8.

Savundra PA, Carroll JD, Davies RA *et al.* (1997). Migraine-associated vertigo. *Cephalagia*, 17, 2–7.

Schaeffer KP, Meyer DL (1973). Compensatory mechanisms following labyrinthine lesions in the guinea pig. A simple model of learning. In Zippel HP, ed. *Memory and Transfer of Information*, pp. 203–232. Plenum, New York.

Schmid-Priscoveanu A, Bohmer A, Obzina H *et al.* (2001). Caloric and search-coil head-impulse testing in patients after vestibular neuritis. *J Assoc Res Otolaryngol*, 2, 72–8.

Schneider RC, Calhoun HD, Crosby EC (1968). Vertigo and rotational movement in cortical and subcortical lesions. *J Neurol Sci*, 6, 493–516.

Schöls L, Amoiridis G, Buttner T *et al.* (1997). Autosomal dominant cerebellar ataxia: phenotypic differences in genetically defined subtypes? *Ann Neurol*, 42, 924.

Schöls L, Bauer P, Schmidt T *et al.* (2004). Autosomal dominant cerebellar ataxias: clinical features, genetics, and pathogenesis. *Lancet Neurol*, 3, 291–304.

Schuknecht HF (1969). Cupulolithiasis. *Arch Otolaryngol*, 90, 765–78.

Schuknecht HF, Suzuka Y, Zimmermann C (1990). Delayed endolymphatic hydrops and its relationship to Meniere's disease. *Ann Otol Rhinol Laryngol*, 99, 843–53.

Schuknecht H (1993). *Pathology of the Ear*, 2nd edition. Lea and Febiger, Philadelphia.

Scoggan KA, Friedman JH, Bulman DE (2006). CACNA1A mutation in a EA-2 patient responsive to acetazolamide and valproic acid. *Can J Neurol Sci*, 33, 68–72.

Selesnick SH, Jackler RK, Pitts LW (1993). The changing clinical presentation of acoustic tumours in the MRI era. *Laryngoscope*, 103, 431–6.

Semont A, Freyss G, Vitte E (1988). Curing the BPPV with a liberatory manoeuvre. *Adv Otolaryngol*, 42, 290–3.

Shepard NT, Asher A (2000). Treatment of patients with non-vestibular dizziness and dysequilirbium. In Herdman SJ, ed. *Vestibular Rehabilitation*, 2nd edition. FA Davis, Philadelphia.

Shepard NT, Telian SA (1995). Programmatic vestibular rehabilitation. *Otolaryngol Head Neck Surg*, 112, 173–82.

Shepard NT, Telian SA, Smith-Wheelock M *et al.* (1993). Vestibular and balance rehabilitation therapy. *Ann Otol Rhinol Laryngol*, 102, 198–205.

Shumway-Cook A, Horak FB (1998). Rehabilitation strategies for patients with vestibular deficits. *Neurol Clin North Am*, 8, 441–57.

Silberstein SD (2004). Migraine. *Lancet*, 363, 381–91.

Silverstein H, Lewis WB, Jackson LE *et al.* (2003). Changing trends in the surgical treatment of Meniere's disease: results of a 10-year survey. *Ear Nose Throat J*, 82, 185–7, 191–4.

Silverstein H, Isaakson JE, Olds MJ *et al.* (1998). Dexamethasone inner ear perfusion for the treatment of Meniere's disease: a prospective, randomized, double-blind, crossover trial. *Am J Otol*, 19, 196–201.

Sirven JI, Caypool DW, Sahs KL *et al.* (2002). Is there a neurologist on this flight? *Neurology*, 58, 1739–44.

Slater R (1979). Benign recurrent vertigo. *JNNP*, 42, 363–7.

Smith CL, Darlington PF (1998). Drug treatment for vertigo and dizziness. *N Z Med J*, 111, 332–4.

Smith JA, Danner CJ (2006). Complications of chronic otitis media and cholesteatoma. *Otolaryngol Clin North Am*, 39, 1237–55.

Smith JM, Koury HI, Hafner CD *et al.* (1994). Subclavian steal syndrome. A review of 59 consecutive cases. *J Cardiovasc Surg*, 35, 11–4.

Smith NJ, Docherty TB (1982). Case report: nystagmus: an unusual manifestation of temporal lobe epilepsy. *Journal of Electrophysiological Technology*, 8, 7–13.

Smith PF, Curthoys IS (1989). Mechanisms of recovery following unilateral labyrinthectomy: A review. *Brain Res Rev*, 14, 155–80.

Smouha EE, Yoo M, Mohr K *et al.* (2005). Conservative management of acoustic neuroma: a meta-analysis and proposed treatment algorithm. *Laryngoscope*, 115, 450–4.

Song JJ, Lee HM, Chae SW *et al.* (2005). Bilateral otosyphilis in a patient with HIV infection. *Eur Arch Otorhinolaryngol*, 262, 972–4.

Spieker S, Schulz JB, Petersen D *et al.* (1995). Fixation instability and oculomotor abnormalities in Friedreich's ataxia. *J Neurol*, 242, 517–21.

Spoendlin H (1974). Optic cochleovestibular degenerations in hereditary ataxias. II. Temporal bone pathology in two cases of Friedreich's ataxia with vestibulo-cochlear disorders. *Brain*, 97, 41–8.

Starck M, Albrecht H, Pollmann W *et al.* (1997). Drug therapy for acquired pendular nystagmus in multiple sclerosis. *J Neurol*, 244, 9–16.

Starr A, Sininger Y, Nguyen T *et al.* (2001). Cochlear receptor (microphonic and summating potentials, otoacoustic emissions) and auditory pathway (auditory brain stem potentials) activity in auditory neuropathy. *Ear Hear*, 22, 91–9.

Stephens SDG (1970). Temporary threshold drift in myxoedema. *J Laryngol Otol*, 84, 317–21.

Stephens SDG, Hogan S, Meredith R (1991). The desynchrony between complaints and signs of vestibular disorders. *Acta Otolaryngol*, 111, 188–92.

Stevenson KL (2004). Chiari type II malformation: past, present, and future. *Neurosurgery Focus*, 16, E5.

Straube A (2005). Pharmacology of vertigo/nystagmus/oscillopsia. *Curr Opin Neurol*, 18, 11–4.

Streitmann MJ, Sismanis A (1996). Metastatic carcinoma of the temporal bone. *Am J Otol*, 17, 780–3.

Stricker RB, Lautin A, Burrascano J (2006). Clinical aspects of neuroborreliosis and post-Lyme disease syndrome in adult patients. *Int J Med Microbiol*, 296, 11–6.

Strupp M, Arbusow V (2001). Acute vestibulopathy. *Curr Opin Neurol*, 14, 11–20.

Strupp M, Brandt T (2006). Pharmacological advances in the treatment of neuro-otological and eye movement disorders. *Curr Opin Neurol*, 19, 33–40.

Strupp M, Schuler O, Krafczyk S *et al.* (2003). Treatment of downbeat nystagmus with 3,4-diaminopyridine: a placebo-controlled study. *Neurology*, 61, 165–70.

Sutherland GR and Auer RN (2006). Primary intracerebral hemorrhage. *J Clin Neurosci*, 13, 511–7.

Sweeny CJ, Gilden DH (2001). Ramsay hunt syndrome. *J Neurol Neurosurg Psychiatry*, 71, 149–54.

Syms CA III, House JW (1997). Idiopathic dandy's syndrome. *Otolaryngol Head Neck Surgery*, 116, 75–8.

Sypert GW, Alvord EC Jr (1975). Cerebellar infarction: a clinicopathological study. *Arch Neurol*, 32, 357.

Szturm T, Ireland DJ, Lessing-Turner M (1994). Comparison of different exercise programs in the rehabilitation of patients with chronic peripheral vestibular dysfunction (Clinical trial. Journal Article. Randomized Controlled Trial). *J Vestib Res*, 4, 461–79.

Takahashi H, Ishikawa K, Tsutsumi T *et al.* (2004). A clinical and genetic study in a large cohort of patients with spinocerebellar ataxia type 6. *J Hum Genet*, 49, 256–64.

Takeda N, Morita M, Hasegawa S *et al.* (1993). Neuropharnacology of motion sickness and emesis. *Acta Otolaryngol Suppl*, 501, 10–15.

Taylor CL, Selman WR, Ratcheson RA (2002). Steal affecting the central nervous system. *Neurosurgery*, 50, 679–88.

Thirlwall AS, Kundu S (2006). Diuretics for Menière disease or syndrome. *Cochrane Database Syst Rev*, 19;3: CD003599.

Thomke F, Hopf (1999). Pontine lesions mimicking acute peripheral vestibulopathy. *J Neurol Neurosurg Psychiatry*, 66, 340.

Thomsen J, Kerr A, Bretlau P *et al.* (1996). Endolymphatic sac surgery: why we do not do it. The non-specific effect of sac surgery. *Clin Otolaryngol*, 21, 208–11.

Thrush DC, Foster JB (1973). An analysis of nystagmus in 100 consecutive patients with communicating syringomyelia. *J Neurol Sci*, 20, 381–6.

Timmerman H (1994). Pharmacotherapy of vertigo: any news to be expected? *Acta Otolaryngol Suppl*, 513, 28–32.

Tonndorf J (1983). Vestibular signs and symptoms in Meniere's disorder: mechanical considerations. *Acta Otolaryngol*, 95, 421–30.

Tusa RJ, Saada AA Jr, Niparko JK (1994). Dizziness in childhood. *J Child Neurol*, 9, 261–74.

Van Deelen GW, Huizing EH (1986). Use of a diuretic (dyazide) in the treatment of Menière's disease. A double-blind cross-over placebo-controlled study. *J Otolaryngol Relat Spec*, 48, 287–92.

Vannuchi P, Giannoni B, Pagnini P (1997). Treatment of horizontal semicircular canal benign paroxysmal positional vertigo. *J Vestib Res*, 7:

Verhagen WIM, Huygen PLM (1994). Central vestibular, vestibulo-acoustic and oculomotor dysfunction in hereditary disorders. *J Vestib Res*, 4, 105–35.

Vilela P, Goulão A (2005). Ischemic stroke: carotid and vertebral artery disease. *Eur Radiol*, 15, 427–33.

Viirre E, Sitarz R (2002). Vestibular rehabilitation using visual displays: preliminary study (Clinical trial, Journal Article. Randomized Controlled Trial). *Laryngoscope*, **112**, 500–3.

Warlow C, Sudlow C, Dennis M *et al.* (2003). Stroke. *Lancet*, **362**, 1211–24.

Watanabe T, Suzuki M (2006). Equilibrium test findings in patients with Bell's palsy. *Auris Nasus Larynx*, **33**, 143–7.

Weidauer H, Lenarz T (1984). Kearns-Sayre syndrome from the otorhinolaryngologic viewpoint. *Laryngol Rhinol Otol (Stuttgart)*, **63**, 141–6.

Welgampola M, Colebatch J (2005). Characteristics and clinical applications of vestibular-evoked myogenic potentials. *Neurology*, **64**, 1682–8.

Wennmo C, Svensson C (1989). Temporal bone fractures. *Acta Otolaryngol Suppl*, **468**, 379–83.

Westmore GA, Pickard CH, Stern H (1979). Isolation of mumps virus from the inner ear after sudden deafness. *Br Med J*, **1**, 14–5.

Wheeler SD, Vincent FM (1980). Vertigo as the only symptom of subclavian steal syndrome. *J Postgrad Med*, **67**, 180–1, 184.

Whitney SL, Wrisley DM, Brown KE *et al.* (2000). Physical therapy for migraine-related vestibulopathy and vestibular dysfunction with history of migraine. *Laryngoscope*, **110**, 1528–34.

Williams D, Wilson TG (1962). The diagnosis of the major and minor syndromes of basilar insufficiency. *Brain*, **85**, 741–74.

Wilson BJ, Melville Jones G (1979). *Mammalian Vestibular Physiology*. Plenum Press, New York.

Wilson WR, Zoller M (1981). Electronystagmography in congenital and acquired syphilitic otitis. *Ann Otol*, **90**, 21–4.

Wormser GP (2006). Clinical practice. Early Lyme disease. *New Engl J Med*, **354**, 2794–801.

Wu LC, Minor LB (2003). Long term hearing outcome in patients receiving intratympanic gentamicin for Ménière's disease. *Laryngoscope*, **113**, 815–20.

Wuyts F, Boudewyns A (In press). Physiology of equilibrium. In Gleason M, ed. *Scott Brown's Otolaryngology*, 7th edition. Hodder, London.

Xing G, Chen Z, Cao X (2007). Mitochondrial rRNA and tRNA and hearing function *Cell Res*. Jan 2; [Epub ahead of print]

Yagi T, Yamaguchi J, Nonaka M (1988). Neurotological findings in Bell's palsy and Hunt's syndrome. *Acta Otolaryngol Suppl*, **446**, 97–100.

Yamakami I, Uchino Y, Kobayashi E *et al.* (2003). Conservative management, gamma-knife radiosurgery, and microsurgery for acoustic neurinomas: a systematic review of outcome and risk of three therapeutic options. *Neurol Res*, **25**, 682–90.

Yardley L, Luxon LM (1994). Treating dizziness with vestibular rehabilitation. *Br Med J*, **308**, 1252.

Yardley L, Owen O, Nazareth I *et al.* (1998). Prevalence and presentation of dizziness in a general practice community sample of working age people. *Br J Gen Pract*, **48**, 1131–5.

Yardley L, Putnam J (1992). Quantitative analysis of factors contributing to handicap and distress in vertiginous patients: a questionnaire study. *Clin Otolaryngol*, **17**, 231–6.

Yee RD, Baloh RW, Honrubia V *et al.* (1982). Pathophsiology of optokinetic nystagmus. In *Nystagmus and Vertigo. Clinical Approaches to the Patient with Dizziness*. Academic Press, New York.

Yeo SW, Lee DH, Jun BC *et al.* (2006). Analysis of prognostic factors in Bell's palsy and Ramsay Hunt syndrome. *Auris Nasus Larynx*, Oct, 18; [Epub ahead of print]

Yimtae K, Srirompotong S, Lertsukprasert K. (2007). Otosyphilis: a review of 85 cases. *Otolaryngol Head Neck Surg*, **136**, 67–71.

Yohay K (2006). Neurofibromatosis types 1 and 2. *Neurologist*, **12**, 86–93.

Young YH, Huang TW (2001). Role of clonazepam in the treatment of idiopathic downbeat nystagmus. *Laryngoscope*, **111**, 1490–3.

Zealley IA, Cooper RC, Clifford KM *et al.* (2000). MRI screening for acoustic neuroma: a comparison of fast spin echo and contrast enhanced imaging in 1233 patients. *Br J Radiol*, **73**, 1129–30.

Zee D (1988). The management of patients with vestibular disorders. In Barber HO, Sharpe JA, eds. *Vestibular Disorders*, pp. 254–74. Year Book Medical Publishers, Chicago.

Zeviani M, Bresolin N, Gellera C *et al.* (1990). Nucleus-driven multiple large-scale deletions of the human mitochondrial genome: a new autosomal dominant disease. *Am J Hum Genet*, **47**, 904–14.

Zhou G, Cox LC (2004). Vestibular evoked myogenic potentials: history and overview. *Am J Audiol*, **13**, 135–43.

CHAPTER 16

Abnormalities of smell and taste

Christopher Kennard

Contents

Since both the sensation of smell, olfaction, and taste, gustation, rely on chemical stimuli to excite their receptors, they are known as the chemosensory system (Smith and Shepherd 1999). Both of these senses are interdependent together providing the sensation of flavour of food and drink, but dysfunction of one may be misinterpreted as an abnormality of the other. Although loss of either sensation is rarely a major handicap, they are essential to detect noxious odours, such as smoke or gas, and to avoid spoiled food or potential poisons. Their loss could, therefore, have serious consequences. In addition, loss of smell or taste may indicate serious intracranial or systemic disease.

16.1 Olfaction

16.1.1 Anatomy and physiology of olfaction

Odours, which must be volatile and soluble in water, are detected by specialized olfactory receptor cells in the olfactory epithelium, located in the mucous membrane of the upper and posterior parts of the nasal cavity, the superior turbinates and nasal septum, which measures 2–5 cm^2, and by the free nerve endings of the trigeminal nerve. The olfactory epithelium contains three cell types, the olfactory sensory neurons, approximately 6–10 million in each nasal cavity, the sustentacular or supporting cells which maintain the electrolyte concentration in the extracellular milieu, especially K$^+$, and basal cells which are the source of new receptor and sustentacular cells, since the former have a life span of only 4–8 weeks (Fig. 16.1).

The olfactory receptor cell is a bipolar sensory neuron with a thin, single dendritic knob which extends into the mucous layer of the nasal cavity. The mucous layer contains immunoglobulins A and M, lactoferrin, lysoenzyme, and odorant-binding proteins. These molecules are thought to prevent the passage of noxious pathogens into the intracranial cavity via the olfactory nerve. From the knob protrude 10–30 non-motile cilia which bear the specific membrane receptor proteins and where signal transduction is initiated. When an odour binds to a receptor, there is activation of a membrane-bound GTP-dependent adenyl cyclase, a G protein, which then activates a second messenger leading to conformational changes in the transmembrane receptor and a series of intracellular events leading to the generation of axon potentials (Shepherd 1994; Smith and Shepherd 1999). Each olfactory sensory neuron expresses a single functional odorant receptor gene and those neurons with the same odour receptor are randomly dispersed within one olfactory epithelial area. Consistent with their ability to detect and discriminate diverse odorants, mammals have as many as 1000 different odour receptors that vary in protein sequence and are used combinatorially to detect different odours and encode their unique identities (Mombaerts 2004). Most mammals possess a second olfactory system—the accessory olfactory system or vomeronasal system. The mammalian vomeronasal organ is generally considered to specialize in pheromone detection. Very thin unmyelinated nerve axons leave the receptor cells and converge into small fascicles, enwrapped by Schwann cells, which pass through the cribriform plate of the ethmoid bone to the olfactory bulb. These axons collectively constitute the olfactory or first cranial nerve, and terminate within the olfactory glomeruli of the olfactory bulb. Here they form synaptic contacts with interneurons that have processes restricted to the bulb and with output neurons, the mitral and internal tufted cells, that contribute axons to the lateral olfactory tract. From the olfactory tract axons project to terminate in primitive cortical areas, known as the primary olfactory cortex. In humans, this probably includes small portions of the uncus, hippocampal gyrus, amygdaloid complex, and entorhinal cortex (Fig. 16.1).

16.1.2 Classification of olfactory disorders

Disturbances of olfaction can be grouped into four main subtypes:

1 Quantitative abnormalities: total or general anosmia and partial abilities to detect olfactory sensations. There may also be a

complete general, hyposmia, or incomplete, partial hyposmia, insensitivity to odorants, or heightened sensitivity, partial or total hyperosmia.

2 Qualitative abnormalities: distortions or illusions of smell known as dysosmia or parosmia.

3 Olfactory delusions or hallucinations associated with disorders of the temporal lobe and psychiatric disease.

4 Olfactory agnosia in which there is an inability to recognize an odour sensation despite intact olfactory sensory processing, language, and general intellectual function.

16.1.3 Evaluation of olfactory function

In the clinical examination of olfactory function it is necessary to discriminate between deficits due to nasal obstruction, which prevent the access of volatile substances from reaching the olfactory epithelium, transport olfactory loss, and neurogenic loss which may be due to abnormalities of the receptors or their axons, sensory olfactory loss, or to pathological processes affecting the central pathways. Transport olfactory loss can result from a variety of causes, which include rhinitis, upper respiratory infection, polyps, sinusitis, and neoplasms. The symptoms of impaired olfactory detection, discrimination, or distortion of normal smells are no

different to those accompanying sensory olfactory loss, which may be due to impaired receptor cell turnover resulting from radiation or chemotherapeutic drugs, or damage to the olfactory axons due to closed head injury, toxic substances, and viral infection.

It is therefore essential that in addition to taking a full history and testing smell and taste, careful examination of the nose, mouth, and nasopharynx is undertaken. Examination of smell is usually carried out using a variety of familiar odiferous substances such as coffee, oil of peppermint, tobacco, oil of cloves, and vanilla. A bottle of each is held under each nostril with the other being occluded by a finger. The patient is asked whether or not he can detect an odour and if so whether or not it can be identified. If the odour can be detected even if it cannot be described, it may be assumed that the olfactory nerves are relatively intact. Malingering can be detected by using ammonia, which stimulates the trigeminal nerve. If the patient denies noticing the stimulus, the anosmia is likely to be bogus.

A more refined assessment of olfaction can be performed using the 40-item 'scratch 'n sniff' test developed and standardized by Doty and colleagues (1984), the University of Pennsylvania Smell Identification Test, or UPSIT. This test is highly reliable and allows the classification of patients into discrete categories of dysfunction.

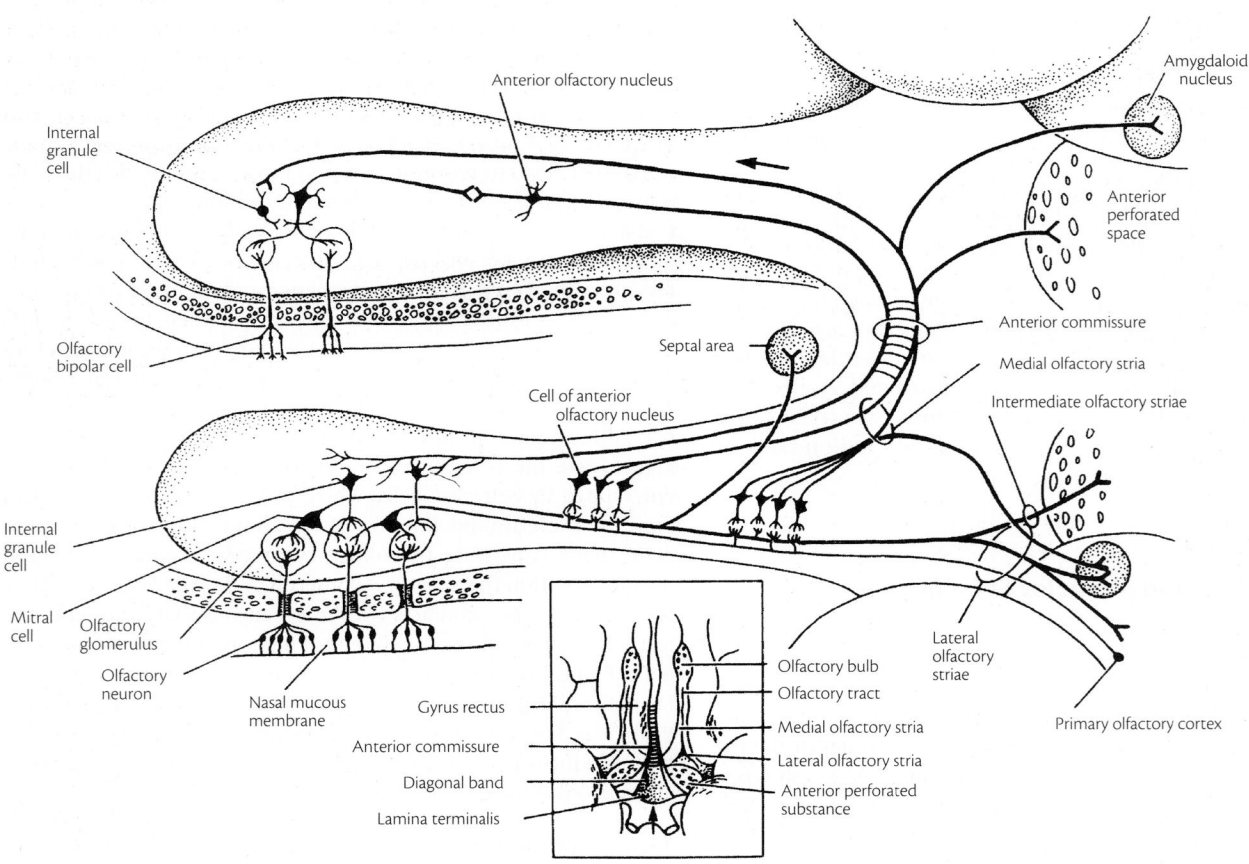

Fig. 16.1 Diagram illustrating the relationships between the olfactory receptors in the nasal mucosa and neurons in the olfactory bulb and tract. Cells of the anterior olfactory nucleus are found in scattered groups caudal to the olfactory bulb. Cells of the anterior olfactory nucleus make immediate connections with the olfactory structures via the anterior commissure. Inset: diagram of the olfactory structures on the inferior surface of the brain. (From Adams, Victor, Ropper 1997, with permission.)

16.1.4 Disorders of olfaction and their management

These are reviewed by Doty (2003). The commonest cause of permanent olfactory loss appears to be a severe upper respiratory infection, usually viral in origin, in which the neuroepithelium is damaged (Deems *et al.* 1991). The second most common cause of olfactory dysfunction is head injury. The third commonest cause is hypertrophy and hyperaemia of the nasal passages, from whatever cause, which leads to hyposmia or anosmia, due to the inability of odours to reach the olfactory epithelium. Chronic rhinitis and sinusitis of allergic, infective, or vasomotor origin are frequent causes. Nutritional and endocrinological disorders, such as thiamine deficiency, adrenal insufficiency, vitamin A deficiency, cirrhosis, renal failure, hypothyroidism and Cushing's syndrome may also have similar effects, due to sensorineural dysfunction (Doty *et al.* 1992). A frequent cause of hyposmia is heavy smoking. Infections due to influenza, herpes simplex, and hepatitis viruses can lead to hyposmia or anosmia due to destruction of the receptor cells, and recovery may not occur if the basal cells are also destroyed. There are several congenital diseases in which the receptor cells are absent or hypoplastic, including Kallman's syndrome of anosmia and hypogonadotrophic hypogonadism, Turner's syndrome, and albinism.

Loss of smell in head trauma is usually due to the severing of the delicate axons of the receptor cells. The incidence of smell dysfunction following head trauma is 7–15 per cent and is proportional to the severity of the injury. Anosmia or microsmia may be unilateral or bilateral. The loss is due to shearing of the olfactory filaments as they pass through the cribriform plate, although contusions to the frontal and temporal poles can also be present (Doty *et al.* 1997). Recovery of smell occurs in about a third of cases but is unlikely to occur if the loss of smell has been present for more than one year after injury (Sumner 1967).

The olfactory epithelium can be damaged by a variety of toxic agents including organic solvents such as benzene, and drugs such as antimicrobial agents (ampicillin, griseofulvin, streptomycin, tetracyclines), anti-inflammatory agents (allopurinol colchicine, gold, D-penicillamine, phenylbutazone), antiproliferative agents (methotrexate, vincristine, doxorubicin), and other drugs including phenindione, amphetamines, cocaine, and corticosteroids.

Olfaction shows a gradual deterioration with age with greater and earlier loss occurring in men than in women. Approximately half of the population between the ages of 8 to 85-years have meaningful olfactory loss (Hoffman *et al.* 1998). Impaired odour detection and/or discrimination may be one of the first signs of a number of neurological disorders including Alzheimer's disease, idiopathic Parkinson's disease, and schizophrenia (Mesholam *et al.* 1998; Doty 2003). In Parkinson's disease the prevalence of olfactory impairment is higher than some of the cardinal neurological signs. Several investigators have shown that scores on a simple three-item microencapsulated odour test differentiate better between Alzheimer's disease and depression than scores on the Mini-Mental State Examination (McCaffrey *et al.* 2000). Alcoholics with Korsakoff's psychosis have a defect of odour discrimination, as have some patients with temporal lobe epilepsy. A similar deficit is found in patients in whom anterior temporal lobe or orbitofrontal cortical excision has been performed. Anosmia may be the first symptom of an olfactory groove meningioma which may involve the olfactory bulb and tract, and extend posteriorly to involve the optic nerve leading to atrophy, the Foster Kennedy syndrome.

There is no specific treatment for patients with hyposmia or anosmia, unless there is a local or systemic remediable cause. In some patients a brief course of systemic steroid therapy—definitely not longer-term—can help to distinguish between conductive and sensorineural olfactory loss, with the former responding to treatment. It is important to realize that many patients with taste and smell disorders experience considerable depression which requires treatment. However, these patients are at potential risk from inhaling noxious fumes and failing to detect burning, so it is important to advise them of the necessary precautions. These should include the use of domestic smoke and gas detectors, and the provision of adequate ventilation in enclosed areas in which toxic solvents are being used.

The commonest cause of *parosmia*, or *dysosmia*, the distortion of normal smell, is a local nasopharyngeal condition such as sinusitis. Other causes include temporal lobe seizure, partial injuries of the olfactory bulb, and depression. The majority of patients have associated hyposmia or anosmia.

Olfactory hallucinations are always of central origin, and are most often due to temporal lobe seizures known as uncinate seizures. Other causes include Alzheimer's disease, endogenous depression, schizophrenia, and alcohol withdrawal (Pryse-Phillips 1975).

16.2 Gustation

The tongue can identify a wide array of tastes that can be classified as sweet, bitter, salty, or sour, with umami being increasingly recognized as a fifth taste modality. Disturbances of taste are far less frequent than disorders of smell, and frequently patients with lack or loss of taste turn out to have impaired olfaction with normal taste sensation. Several disorders of taste are recognized: ageusia or loss of taste, hypogeusia or diminished sensitivity, dysgeusia or parageusia which are distortions of normal taste, and finally gustatory hallucinations.

16.2.1 Anatomy and physiology of taste

The peripheral receptors for taste—the taste buds—are mainly located on the surface of the tongue and in smaller numbers over the soft palate, the pharynx, larynx, and oesophagus (Fig. 16.2). Each taste bud consists of about 200 vertically orientated receptor cells,

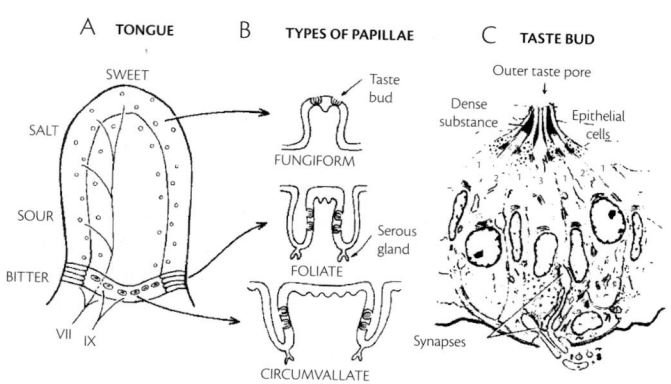

Fig. 16.2 A. Distribution of taste buds, innervation pattern, and lowest threshold regions for different tastes in the human tongue. B. Main types of taste papillae, containing taste buds. C. Fine structure of a taste bud. (From Murray 1973).

such that the superficial portion of the bud is marked by an excavation, the taste pit or pore, into which the microvilli of the receptor cells project. Receptor cells have a limited lifespan of about 10 days and undergo constant replacement from adjacent basal epithelial cells. Fine unmyelinated sensory fibres pass up through the base of the bud to innervate the receptor cells, which have no axons. Taste afferent fibres from the anterior two-thirds of the tongue course through the lingual nerve, a branch of the trigeminal nerve, which they leave via the corda tympani to join the facial nerve, the nervus intermedia portion, with their cell bodies in the geniculate ganglion. The posterior third of the tongue and pharynx, are supplied by the glossopharyngeal nerve, with cell bodies in the nodose ganglion. Afferents from taste buds on the palate travel with the superficial petrosal nerve, and from the larynx and oesophagus via the vagal afferents. Gustatory fibres from these nerves project to the ipsilateral solitary tract, from which they project via the gustatory lemniscus to the thalamus and then to the postrolandic sensory cortex.

The tongue taste receptors respond to chemical substances in solution, the four primary taste sensations being salty, bitter, sweet, and sour. More complex taste sensations are derived from combinations of these four basic tastes and from olfaction (Smith and Shepherd 1999). For a full description of the molecular basis for taste see reviews (Breslin and Huang 2006; Small 2006).

16.2.2 Clinical assessment

Patients presenting with a disturbance of taste should be asked about any associated disorder of smell, and any associated medical conditions or medications (Doty *et al.* 1992). In addition to testing smell and taste it is essential to carefully examine the oral cavity, noting any evidence of infection, masses, and atrophy and dryness of the tongue, gums, and dentition.

Taste can then be assessed using standard solutions of sugar, sodium chloride, acetic acid, and quinine. Electrical stimulation of the tongue, electrogustometry, can also be used as a sour stimulus by simply applying a low-voltage direct current (Stillman *et al.* 2003). If the taste loss is bilateral the solutions can be swished around the mouth and the patient asked to identify the taste. The solution is then spat out and the mouth rinsed with water before the next solution is tried. If the taste loss is unilateral or focal the tongue is protruded and gently held with a piece of gauze. Crystals of salt or sugar are then placed on the tongue and the patient asked to identify the taste.

16.2.3 Disorders of gustation and management

Disturbances of taste are due either to local causes involving the tongue and/or taste buds or to damage to the peripheral or central neural pathways. The commonest associated cause of hypogeusia is an upper respiratory tract infection and smoking (Henkin *et al.* 1975). Deficiency of saliva as in Sjogren's syndrome, or hyperviscosity as in cystic fibrosis, pandysautonomia or post irradiation of the head and neck results in dryness of the mouth, xerostomia, this leads to disturbed taste, because taste stimuli are only effective in a fluid medium. There may also be an accompanying reduction in the number of papillae and taste buds, which may possibly be due either to the loss of the lubricating effect of saliva or of trophic factors contained within it. Unfortunately, artificial saliva or regular water mouthwashes do not appear to restore normal taste

in patients with xerostomia. Other causes of ageusia or hypogeusia include scleroderma, hypothyroidism, adrenocortical insufficiency, Cushing's syndrome, diabetes mellitus, chronic renal failure, liver cirrhosis, niacin or vitamin B_6 deficiency, zinc deficiency, and neoplasia of the oral cavity and base of skull. Reduced or distorted function of taste not infrequently occurs following influenza-like infections. A unilateral loss of taste is often found in cases of Bell's palsy.

Post-traumatic ageusia is far less common than post-traumatic anosmia, occurring in less than 1 per cent of serious head injuries (Sumner 1967). However, it always occurs in association with anosmia, and often the ageusia resolves within a few weeks. The cause for such ageusia is unclear. Although bilateral lesions near the frontal operculum and paralimbic areas would result in both ageusia and anosmia, this would not explain the frequent recovery of ageusia in advance of the anosmia. It is likely that many cases of ageusia are in fact mislabelled cases of anosmia.

Gustatory hallucinations occur much less frequently than olfactory ones in association with epileptic seizures. Such an aura, which may represent a primary taste such as sweet, bitter, or a peculiar and rotten one, usually occurs in a seizure originating from the frontoparietal or suprasylvian cortex or the uncal region (Hausser-Hauw and Bancaud 1987).

Management of taste disorders includes the use of various salivary substitutes for xerostomia, the correction of any nutritional deficiency, and in some cases flavour enhancers.

References

Adams D, Victor M, Roper A (1997). *Principles of Neurology* 6th edn. McGraw-Hill, New York.

Breslin PA, Huang L (2006). Human taste: peripheral anatomy, taste transduction and coding. *Adv Otorhinolaryngol*, **63**, 152–90.

Deems DA, Doty RL, Settle RG, *et al.* (1991). Smell and taste disorders, a study of 750 patients from the University of Pennsylvania Smell and Taste Centre. *Arch Otolaryngol Head Neck Surg*, **117**, 519–28.

Doty RL (2003). Odor perception in neurodegenerative diseases. In Doty RL, ed. *Handbook of Olfaction and Gustation.* pp. 479–502. New York, Marcel Dekker.

Doty RL, Shaman P, Dann M (1984). Development of the University of Pennsylvania Smell Identification Test: a standardized microencapsulated test of olfactory function. *Physiol Behav*, **32(3)**, 489–502.

Doty RL, Kimmelman CP, Lesser R (1992). Smell and taste and their disorders. In Asbury AK, McKhann GM and MacDonald WI. Saunders, ed. *Diseases of the Nervous System*, pp. 390–403. Philadelphia.

Doty RL, Yousem DM, Pham LT *et al.* (1997). Olfactory dysfunction in patients with head trauma. *Arch Neurol*, **54**, 1131–1140.

Hausser-Hauw C, Bancoud J (1987). Gustatory hallucinations in epileptic seizures. *Brain*, **110**, 339–60.

Henkin RI, Larson AL, Powell RD (1975). Hypogeusia, dysgeusia, hyposmia and dysosmia following influenza like infection. *Ann Otol Rhin Laryngol*, **84**, 672–9.

Hoffman HJ, Ishii EJ, Macturk RH (1998). Age-related changes in the prevalence of smell/taste problems among the United States adult population. Results of the 1994 disability supplement to the National Health Interview Survey (NHIS). *Ann NY Acad Sci*, **30 (855)**, 716–22.

McCaffrey RJ, Duff K, Solomon GS (2000). Olfactory dysfunction discriminates probable Alzheimer's dementia from major depression: a cross-validation and extension. *J Neuropsychiatry Clin Neurosci*, **12**, 29–33.

Mesholam RI, Moberg PHJ, Mahr *et al.* (1998). Olfaction in neurodegenerative disease: a meta-analysis of olfactory functioning in Alzheimer's and Parkinson's diseases. *Arch Neurol*, **55**, 84–90.

Mombaerts P (2004). Genes and ligands for odorant, vomeronasal and taste receptors. *Nat Rev Neurosci*, **5**, 263–78.

Murray RG (1973). The ultrastructure of taste buds. In *The Ultrastructure of sensory organs*. pp.1–81. Friedmann I. Elsevier, New York.

Pryse-Phillips W (1975). Disturbances in the sense of smell in psychiatric patients. *Proc R Soc Med*, **68**, 26–32.

Shepherd GM (1994). Discrimination of molecular signals by the olfactory receptor neurone. *Neurone*, **13**, 771–90.

Small DM (2006). Central gustatory processing in humans. *Adv Otorhinolaryngol*, **63**, 191–220.

Smith DV, Shepherd GM (1999). Chemical senses: taste and olfaction. In MJ Zigmond, FE Bloom, SC Landis *et al*. *Fundamental Neuroscience*, pp. 719–59. Academic Press, Santiago.

Stillman JA, Morton RP, Hay KD *et al*. (2003). Electrogustometry: strengths, weaknesses, and clinical evidence of stimulus boundaries. *Clin Otolaryngol Allied Sci*, **5**, 406–10.

Sumner D (1967). Post-traumatic ageusia. *Brain*, **90**, 187–97.

SECTION 4

Pain

CHAPTER 17

Neuropathic pain

John Scadding

Contents

17.1 Introduction

Pain signalled by a normal sensory system, nociceptive pain, serves a vital protective function. The peripheral and central nervous somatosensory systems permit rapid localization and identification of the nature of painful stimuli, prior to appropriate action to minimize or avoid potentially tissue damaging events. A reduction or absence of pain resulting from neurological disease emphasizes the importance of this normal protective function of pain. For example, tissue destruction occurs frequently in peripheral nerve diseases which cause severe sensory loss such as leprosy, and in central disorders such as syringomyelia. Neuropathic pain results from damage to somatosensory pathways and serves no protective function. This chapter provides an overview of neuropathic pain, considering its context, clinical features, pathophysiology, and treatment.

In the peripheral nervous system, neuropathic pain is caused by conditions affecting small nerve fibres, and in the central nervous system by lesions of the spinothalamic tract and thalamus, and rarely by subcortical and cortical lesions. The clinical feature common to virtually all conditions leading to the development of neuropathic pain is the perception of pain in an area of sensory impairment, an apparently paradoxical situation. The exception is trigeminal neuralgia (Sections 19.2.1; 20.1.4).

Neuropathic pain is heterogeneous clinically, aetiologically, and pathophysiologically. Within a given diagnostic category, whether defined clinically or aetiologically, there are wide variations in reports of pain by patients. This heterogeneity poses one of the greatest challenges in understanding the mechanisms of neuropathic pain. Knowledge of the pathophysiology is an obvious pre-requisite to the development of effective treatments. The goal of a pathophysiologically based understanding of the symptoms and signs of neuropathic pain is, of course, just such a rational and specific approach to treatment. While this is not yet achievable, clinical-pathophysiological correlations have led to some recent advances in treatment.

17.1.1 Epidemiology of neuropathic pain

Pain is a common symptom in patients with neurological disease; neuropathic and nociceptive pains frequently coexist. Reliable epidemiological data on neuropathic pain are difficult to obtain, due to problems of case ascertainment, retrospective rather than prospective studies, inclusion variations and biases in different case series, small sample size, differing defining thresholds for considering pain a leading symptom, and historical disagreements about the definition of neuropathic pain. In many patients with neurological disease, pain is but one of several symptoms and not necessarily the leading symptom. Some conditions are characteristically painful, while others are not, and it is in the latter category that the incidence and prevalence of neuropathic pain have tended to be underestimated.

For example, while the prevalence of peripheral neuropathies is 2.4 per cent in the general population, rising to 8 per cent with age (Martyn and Hughes 1998), the prevalence of neuropathic pain within this group is uncertain. One study has calculated a point prevalence for peripheral neuropathic pain in the general population as high as 5 per cent (Daousi et al. 2004). Studies of the incidence of postherpetic neuralgia demonstrate the problem of an agreed definition of the condition, including the interval following the acute attack of shingles, and underline the need for prospective studies.

In relation to central nervous system disease, several studies have indicated a prevalence rate of neuropathic pain of around 71 per cent in spinal cord injury (Bonica 1991). Epidural cord compression by metastatic carcinoma is a common example of a pathology that may cause both neuropathic pain, be it myelopathic or radicular, and nociceptive somatic pain due to involvement of skeletal structures and soft tissues. In one study, pain was a first symptom in 96 per cent of such patients (Gilbert et al. 1978). Estimates of neuropathic pain as a symptom in patients with multiple sclerosis vary widely, from 14.5 to 82.1 per cent (Clifford and Trotter 1984; Vermote et al. 1986; Kassirer and Osterberg 1987; Moulin et al. 1988), emphasizing variations in symptom characterization, definition of pain type, and case selection bias. Finally, the incidence of central post-stroke pain, formerly known as thalamic pain, was found in a prospective study to be as high as 8.4 per cent in a group of 191 patients. In the sub-group with sensory deficits, 42 per cent of the stroke group, 18 per cent of patients had such pain (Andersen et al. 1995).

17.1.2 Terminology and definitions

There is a continuing debate about the definition of neuropathic pain. Matters are not helped by the use of a multiplicity of terms, some with identical or overlapping meanings. This situation exists partly for historical reasons, partly because a relatively recent change in definition broadened the definition of neuropathic pain (IASP 1994) and partly because neuropathic pain is defined on a clinical basis rather than pathophysiologically. A mechanism-based understanding of the symptoms and signs of neuropathic pain will clarify the nosology of neuropathic pain and assist in the development of specific treatments (Woolf et al. 1998; Woolf and Mannion 1999; Scadding and Koltzenburg 2005).

Neurogenic pain. This refers to all neurological causes of pain, both peripheral and central. It is defined as 'pain initiated or caused by a primary lesion, dysfunction, or transitory perturbation in the peripheral or central nervous system' (IASP 1994). Although an acceptable term, its use has declined, possibly because of the inclusion of states of 'transitory perturbation' and 'dysfunction' of the nervous system, for neither of which is there a definition or agreed description. This broad definition of neurogenic pain, with ill-defined limits, has questionable clinical usefulness. For example, it has led some to consider conditions such as fibromyalgia, as neuropathic pains.

Neuropathic pain. This was the term originally restricted to refer to pain due to peripheral neuropathies and plexopathies. As part of an initiative to produce a comprehensive taxonomy of terms used to describe all painful conditions, the definition was broadened to include 'pain initiated or caused by a primary lesion or dysfunction of the nervous system' (IASP 1994). Under this definition, neurogenic and neuropathic are now virtually synonymous, but as with neurogenic pain, the 1994 definition includes a category of 'dysfunction of the nervous system'. This would include complex regional pain syndrome type 1, known previously as reflex sympathetic dystrophy (Section 17.5), which shares many clinical features with pain that undoubtedly has a primary nervous system cause. These features include severe pain, allodynia, hyperalgesia, sometimes accompanied by vasomotor and sudomotor disturbances. Within the current broad definition, neuropathic pain is subdivided into peripheral and central, to denote the location of the causative lesion.

Neuralgia. This term describes 'pain arising in the distribution of a nerve or nerves' (IASP 1994). It has become restricted to describe neuropathic pain due to lesions of specific nerves, for example intercostal, sciatic, femoral, or trigeminal, and of roots in the case of postherpetic neuralgia. Neuralgia is thus a sub-category of neuropathic pain.

Central pain. Behan (1914) was the first to use the term central pain. Riddoch's (1938) later refined description of pain arising from central nervous system lesions is applicable to this day. Central pain is subsumed within the broader definition of neuropathic pain.

Working definition of neuropathic pain. In the light of this confusing terminology and in order to avoid ambiguity, the working definition of neuropathic pain used in this chapter is 'pain arising as a direct consequence of a lesion affecting the somatosensory system'. This reflects a growing consensus among pain clinicians and scientists. Importantly, this working definition excludes any reference to vague 'dysfunction' of the nervous system, an aspect of the current definition that has proved so contentious.

17.1.3 Classification of neuropathic pain

In the absence of a comprehensive mechanism-based nosology of neuropathic pain, classifications remain anatomical and aetiological. For peripheral neuropathic pains, various classification schemes have been proposed. The anatomical distribution pattern of the affected nerves provides valuable differential diagnostic clues as to possible underlying causes. Neuropathic pain comprises stimulus-independent, or ongoing, and stimulus-dependent pains. Mechanisms underlying these are now partly established, so that an anatomical and aetiological classification can be supplemented, but not yet replaced, by pathophysiological data (Section 17.4). These data are more complete and reliable in relation to peripheral than for central neuropathic pain (Woolf et al. 1988; Koltzenburg 1996; Boivie 2005; Scadding and Koltzenburg 2005; Siddall 2005).

17.2 Causes of neuropathic pain

Tables 17.1 and 17.2 list the many causes of neuropathic pain, classified anatomically. Descriptions of many of the conditions listed in the tables are to be found elsewhere in this book. Postherpetic neuralgia is discussed in Section 19.2.3.

17.3 Clinical features of neuropathic pain

17.3.1 A complex of symptoms

Neuropathic pain is multi-dimensional, comprising various types of ongoing stimulus-independent pains, and the evoked stimulus-dependent pains of allodynia, hyperalgesia, and hyperpathia. The latter are frequently major components of the overall pain complaint. The presence and clinical differentiation of stimulus-independent and stimulus-dependent pains, and their relative contributions to a patient's complaint of pain is often indicated by the history. Examination includes careful evaluation of the evoked sensory phenomena that frequently accompany neuropathic pain, in addition to the usual assessment of modality specific sensory impairment (Bennett 2001; Rasmussen *et al.* 2004). Neuropathic pain is frequently associated with comorbidities which also require careful evaluation. The combination of neuropathic pain and its comorbidities, together with associated neurological deficits in many patients, often results in major impairment of quality of life.

17.3.2 Stimulus-independent, ongoing pain

Patients with neuropathic pain often find it difficult to characterize the qualities of their painful symptoms, because these fall outside their previous lifelong experience of nociceptive pain. The most commonly described ongoing symptoms are deep aching in the extremities and a superficial burning, stinging, or pricking pain. Verbal descriptors from the McGill Pain Questionnaire that are used significantly more frequently by patients with neuropathic pain than those with nociceptive pain include electric shocks, burning, tingling, itching, or pricking. Descriptors such as dull, heavy, and tiring are more commonly reported by patients with nociceptive pains (Boureau *et al.* 1990). Accompanying paroxysmal, shock-like, or lancinating pains, sometimes radiating through a whole limb are fairly common. Stimulus-independent pains may be continuous, intermittent, or paroxysmal.

Paroxysmal pains are characteristic of certain types of neuropathic pain, for example trigeminal neuralgia, in which pains are

Table 17.1 Painful peripheral neuropathies

Traumatic mononeuropathies	Disulfiram
Causalgia	Ethambutol
Amputation stump pain (nerve transaction, partial or complete)	Isoniazid
Post-thoracotomy neuralgia	Nitrofurantoin
Entrapment neuropathies	Thalidomide
Morton's neuralgia (plantar digital nerve entrapment)	Thiouracil
Mastectomy	Vincristine
Painful scars	*Toxins*
	Acrylamide
Other mononeuropathies and multiple mononeuropathies	Arsenic
Postherpetic neuralgia	Clioquinol
Diabetic mononeuropathy	Dinitrophenol
Proximal diabetic neuropathy	Ethylene oxide
Malignant plexus invasion	Pentachlorophenol
Radiation plexopathy	Thallium
Neuralgic amyotrophy	*Hereditary*
Plexus neuritis (idiopathic, hereditary)	Amyloid neuropathy
Trigeminal and Glossopharyngeal neuralgia	Fabry's disease
Borreliosis	Charcot Marie Tooth disease type V, type 2B
Connective tissue disease (vasculitis)	Hereditary sensory and autonomic neuropathy, type I, type IB
Herpes simplex	*Malignant*
Polyneuropathies	Paraneoplastic
Metabolic/nutritional	Myeloma
Alcoholic	*Infective /post-infective/immune*
Diabetic	Guillain–Barré syndrome
Amyloid	Borreliosis
Beriberi	HIV
Burning feet syndrome	*Other polyneuropathies*
Cuban neuropathy	Erythermalgia (synonym: Erythromelalgia)
Pellagra	Idiopathic small fibre neuropathy
Strachan's syndrome	Trench foot (cold injury)
Tanzanian neuropathy	
Drugs	
Antiretrovirals	
Cisplatin	

Table 17.2 Causes of central neuropathic pain by anatomical location

Spinal root/dorsal root ganglion	
Prolapsed disc	Root avulsion
Arachnoiditis	Tumour
Trigeminal neuralgia	Postherpetic neuralgia
Surgical rhizotomy	
Spinal cord	
Trauma including compression	HIV
Syringomyelia	Multiple sclerosis
Dysraphism	Vitamin B$_{12}$ deficiency
Vascular: infarction, haemorrhage,	
arteriovenous malformation	by anatamical location
Anterolateral cordotomy	Syphilis
Brainstem	
Lateral medullary syndrome	
Syringobulbia	
Multiple sclerosis	
Tumours	
Tuberculoma	
Thalamus	
Infarction	
Haemorrhage	
Tumours	
Surgical thalamotomy	
Sub-cortical and cortical	
Infarction	Trauma
Arteriovenous malformation	Tumour

both spontaneous and evoked, the lightning pains of tabes dorsalis and the painful crises of the neuropathy in Fabry's disease. They are also common in other neuropathic pains, for example stump and phantom limb pains in amputees and may occur as part of almost any type of neuropathic pain, of peripheral or central cause. Painful paraesthesiae, ongoing or evoked, often accompany neuropathic pain.

17.3.3 Stimulus-dependent pain

Stimulus-dependent pains include allodynia, hyperalgesia, and hyperpathia, evoked by mechanical, thermal, or chemical stimulation. Stimulus-dependent pain is a hallmark of both inflammatory states and of neuropathic pain (Kilo *et al.* 1994; Koltzenburg *et al.* 1994; Woolf and Mannion 1999). Several terms are used to describe stimulus-dependent pains:

Allodynia is pain resulting from a stimulus that does not normally provoke pain.

Hyperalgesia is an increased response to a stimulus that is normally painful; noxious stimuli are often associated with a lowering of the pain threshold, together with an exaggerated perception of pain. In clinical practice, the term hyperalgesia tends to be loosely used to describe abnormally painful responses to stimuli that are normally not painful, so that these really fall into the category of allodynia rather than hyperalgesia. Several subdivisions of hyperalgesia are recognized. In *static hyperalgesia*, gentle pressure on the skin causes pain. In punctate hyperalgesia, stimuli such as pinprick evoke pain. In *dynamic hyperalgesia*, light brushing of the skin evokes pain; strictly, this is a form of allodynia rather than hyperalgesia. In *heat and cold hyperalgesia*, warm and cool stimuli respectively evoke pain. In the physical examination to elicit these signs,

hot and cold stimuli are used that are not normally painful, at 40 and 20°C respectively. Painful reactions to these stimuli are again further examples of allodynia.

The basis for all these types of hyperalgesia is sensitization of nociceptors. Brush-evoked allodynia is mediated by an A beta fibre input, but depends on a state of central sensitization, albeit initially established by a nociceptor input (Table 17.3).

In *hyperpathia* there is a raised sensory threshold, delay in perception of a stimulus, an abnormally painful reaction with summation causing increasing pain to a repetitive stimulus, and a painful after-sensation, sometimes longlasting. Hyperpathia is often severe and frequently has an explosive character, due to rapid summation. The raised sensory threshold of hyperpathia results from partial loss of afferent input, while the summation and after-sensation are due predominantly to central sensitization. So-called wind-up pain can be induced by repetitive C fibre stimulation in normal human skin, or by normally innocuous stimulation in hyperalgesia due to inflammation, as well as in neuropathic pain (Mendell and Wall 1965; Gottrup *et al.* 1998). Wind-up and painful after-sensations are thought to be the result of abnormal activity in wide dynamic range neurons in the dorsal horn of the spinal cord, this effect being mediated by *N*-methyl D-aspartate, NMDA receptors (Dickenson and Sullivan 1987).

17.3.4 Pain rating scales

In clinical practice, pain scales are not routinely used, though visual analogue scales can provide information about spontaneous fluctuations of pain severity and treatment-related changes. The McGill Pain Questionnaire is widely used as a research tool (Melzack 1975); the short form McGill Pain Questionnaire is robust and well validated (Melzack and Katz 1999). A pain scale specific for neuropathic pain has been developed (Galer and Jensen 1997).

17.3.5 Radiation of neuropathic pain

Neuropathic pain is often confined to the anatomical area corresponding to the causative neurological lesion, either peripheral or central, and this is helpful diagnostically. However, neuropathic pain may radiate beyond the causative neural territory, either spontaneously, or more commonly as an evoked sensation. There is a direct relationship between the severity of pain and the extent of radiation (Laursen *et al.* 1997). Recruitment of wide dynamic range neurons over several segments of the spinal cord may be the basis for this clinical phenomenon (Lamotte *et al.* 1991; Jensen and Gottrup 2003).

17.3.6 Non-neurological pain in patients with neuropathic pain

Neuropathic and nociceptive pains frequently coexist. It is necessary to characterize the different components comprising a patient's pain, as investigation and treatment are guided by accurate diagnostic assessment. A common example is cervical and lumbar spine disease, in which pain is often of both musculoskeletal, nociceptive, and neuropathic types. In this situation, the description of the pain and its distribution may not discriminate between the two types, particularly when the pain is unilateral. The presence of a neurological deficit, together with the results of electrophysiological testing and imaging may be needed to elucidate such presentations. Even following these investigations, there can be continuing

Table 17.3 Mechanisms of peripheral neuropathic pain

Abnormal property	Increased by	Decreased by	Resulting painful symptom /sign
Ectopic impulse generation (EIG) in primary sensory afferent fibres (due to type III sodium channel expression)	Mechanical stimulation Catecholamines ATP Nitric oxide Cytokines Prostaglandins Bradykinin Ischaemia / hypoxia	LA Alpha receptor blockade GDNF Colchicine Vincristine Corticosteroid Carbamazepine Phenytoin Glycerol	Ongoing pain Tinel sign with mechanical stimulation Sympathetically maintained pain
Sympathetic–sensory fibre coupling	Sympathetic stimulation	LA Alpha block— weak effect	Ongoing pain, exacerbated by anxiety and emotion
Primary sensory fibre sensitization (see Table 17.4)			
After-discharge in damaged sensory fibres			AHH
Crossed after-discharge in damaged sensory fibres		LA	AHH
Ephapses (presence in man uncertain)		LA	AHH
EIG in DRG neurons with damaged peripheral axons		LA	Paroxysmal pains
Nerve trunk pain (nociceptive in type)			Ongoing pain
DH cell wind up by C fibre input (central sensitization)			Nerve lesion local tenderness
DH cell-activity-induced calcium release, PG and NO synthesis		NMDA blockade	Pain summation / hyperpathia
DRG cell death secondary to peripheral axonal damage		NMDA blockade	Pain summation / hyperpathia
Neurotransmitter excitotoxicity			Central disinhibition of DH neurons —deafferentation pain
GAP expression leading to inappropriate regeneration with altered connectivity in DH; low threshold afferents cause pain			Central DH cell death —deafferentation pain Ongoing and evoked pains
Reduced inhibitions in spinal cord: surround, segmental, descending			Ongoing and evoked pains

Abbreviations; AHH: allodynia, hyperalgesia and hyperpathia; ATP: adenosine triphosphate; DH: dorsal horn; DRG: dorsal root garglion; EIG: ectopic impulse generation; GAP: growth-associated protein; GDNF: glial-cell-derived neurotrophic factor; LA: local anaesthetic; NMDA: *N* methyl D aspartate; NO: nitrous oxide; PG: prostoglandin

doubt about the relative contributions of different types of pain, and these situations are further complicated when previous cervical or lumbar surgery has been performed.

Other painful consequences of neurological disease include arthropathies, skeletal deformities, spasticity, contractures, and dystonia. For example, in patients with painful diabetic peripheral neuropathy, distal pain in the lower legs is neuropathic, but pain may also arise from vascular insufficiency, foot and ankle arthropathies, or diabetic skin ulceration. This emphasizes the need for clinical awareness and careful diagnostic assessment, supported by investigations, in order to identify the various pathologies contributing to a patient's overall complaint of pain.

17.3.7 Sensory loss

Sensory loss is usually found in a distribution that corresponds anatomically to the causative lesion, but as described above, neuropathic pain and sensory signs may radiate well beyond the expected anatomical area. Furthermore, sensory impairment can be mild and difficult to detect, particularly when overshadowed by allodynia, hyperalgesia, or hyperpathia. The degree of sensory loss and the severity of ongoing and evoked neuropathic pain are not closely related. Trigeminal neuralgia (Section 19.2.2) is the only

neuropathic pain in which cutaneous sensation is characteristically normal on routine physical examination.

Impairment of pinprick and temperature, small fibre functions sometimes highly selectively, is found in peripheral neuropathies causing neuropathic pain (Scadding and Koltzenburg 2005), though involvement of small fibres may not be selective. In the case of central lesions in the spinal cord or brain, sensory loss is of spinothalamic type, again sometimes selectively (Jensen and Lenz 1995; Vestergaard *et al.* 1995).

17.3.8 Sympathetic activity and neuropathic pain

The role of sympathetic efferent activity as a causal factor in the initiation and maintenance of persistent pain in man remains controversial. There is substantial evidence for a sympathetic influence in nerve injury in animal experiments (Devor 2005). Disturbances of vasomotor and sudomotor activity are frequently seen in causalgia, complex regional pain syndrome type 2, and in complex regional pain syndrome type 1, formerly known as reflex sympathetic dystrophy (Baron 2005). Signs of altered sympathetic activity in these conditions, and occasionally in neuropathic pain states other than causalgia, include swelling, smooth glossy skin, excessive sweating, and vasomotor instability ranging from warm extremities with vasodilatation and redness of the skin, to cool vasoconstricted extremities (Fig. 17.1). Sympathetic symptoms and signs, and the role of the sympathetic nervous system in the generation and maintenance of neuropathic pain and in the pathogenesis of complex regional pain syndrome type 1 are considered in Section 17.5.8.

17.3.9 Sensory examination

The sensory examination in patients with neuropathic pain requires care, but can be completed using simple standard equipment, without the need for quantitative sensory testing in the great majority of patients. Thresholds and suprathreshold responses to light touch and pin prick punctuate stimulation are assessed, before dynamic and repetitive stimulation to assess the presence of allodynia and hyperpathia.

Cold and heat are best tested either with metal rollers kept at 20 and 40°C respectively, or with water-filled tubes. These also test touch punctuate and dynamic mechanical stimulation respectively. To overcome this, radiant heat sources or lasers can be used for selective thermal testing in selected patients. Thermal threshold testing using a thermocouple with a Peltier element is now performed routinely in many neurophysiological laboratories, supplementing clinical examination.

Examination of vibration and joint position helps with anatomical localization, and together with cutaneous sensory modalities, provides information about the density and selectivity of sensory loss. It is important to identify the degree of deafferentation and the balance of peripheral and central mechanisms responsible for neuropathic pain, as this influences the approach to treatment.

17.3.10 Comorbidities of neuropathic pain

The comorbidities of chronic pain contribute to loss of function and impair quality of life in many patients with neuropathic pain. Depression is very common, present in up to 100 per cent of patients in some series (Romano and Turner 1985). The coexistence of pain and depression in neurological disease is well established (Fishbain et al. 1997; Williams et al. 2004). For example, in a study of patients with peripheral neuropathies, depression, anxiety, altered sleep patterns, social isolation, and reduced employment status were important comorbidities (Meyer-Rosberg et al. 2001).

Pain can be a symptom of primarily psychiatric disease, for example atypical facial pain and somatoform disorders, and in patients with chronic pain there is an association with somatoform disorders (Fishbain 1995) (see Section 4.7.6). Anxiety is a frequent feature, related to pain severity, delayed diagnosis, poor response to treatment, fear of progression, and social and financial consequences of the illness. In patients presenting with various types of pain, anxiety is a significantly associated symptom (Gureje et al. 1998). Conversely, in panic disorder, pain is a presenting complaint in up to 81 per cent of patients (Katon 1984). Substance abuse is an important cause of comorbidity in patients with chronic pain (Fishbain et al. 1992). Adverse effects of prescribed medications also frequently produce comorbidity: sedation, fatigue, dysphoria, and depression are the common complaints.

A wide variety of scales measuring the comorbidities of chronic pain are available, used mainly as research tools. Comorbidity measurement is reviewed by Williams (1999). The Beck Depression Inventory (Beck et al. 1961) and the Hospital Anxiety and Depression Scale (Zigmond and Snaith 1983) provide quantitative information about mood and affect. Of the wider multidimensional scales, the most commonly used are the Short Form 36 of Medical Outcomes Study (Ware et al. 1993), the Sickness Impact Profile (Bergner et al. 1981) and the Multidimensional Pain Inventory (Kerns et al. 1985). The Oswestry Low Back Questionnaire (Fairbank et al. 1980) has been validated in patients with back pain.

17.4 Pathophysiology of neuropathic pain

Detailed discussion of the mechanisms contributing to the development of neuropathic pain lies beyond the scope of this chapter. As mentioned earlier, symptoms and signs in patients with neuropathic pain cannot yet be tightly linked with underlying cellular and neuropharmacological pathophysiologies. Animal experiments have yielded a great number of candidate mechanisms that may underlie neuropathic pain, and increasingly, investigative techniques such as microneurography and skin biopsy in peripheral neuropathies are contributing data of direct relevance to human disease.

Table 17.3 outlines the major pathophysiological changes likely to contribute to peripheral neuropathic pain. These include: abnormal impulse generation in damaged primary afferent axons and their cell bodies in the dorsal root ganglia, with the development of mechanical and noradrenergic sensitivities; fibre interactions in damaged nerves; and secondary central effects in the spinal cord. Table 17.4 summarizes the relationship between symptoms, primary afferent type, and pathophysiological properties. Full accounts of these changes can be found in McMahon and Koltzenburg (2005).

In relation to mechanisms of central neuropathic pain, knowledge is much less complete. Table 17.5 summarizes the main likely contributing properties. Data concerning the factors governing the development of neuropathic pain with cerebral lesions have reached the level of brain region excitation/inhibition anatomicophysiological correlations. And while positron emission tomography and functional MRI have opened up the area of pain perception, in both cognitive and emotional aspects and now permit psychophysical investigation in painful states in man, neuropathic pain has as yet been relatively little studied. The major changes responsible for the development of central neuropathic pain include: loss of normal inhibitory controls at cord and thalamic levels; inflammatory changes in spinal cord neurons leading to increased excitability; increased excitatory amino acid activation and abnormal expression of sodium and calcium channels; and imbalance of activity in central pathways. However, it should be appreciated that the links between pathophysiological changes and symptoms in central neuropathic pain are much less certain than for those established for peripheral neuropathic pain. McMahon and Koltzenburg (2005) provide a detailed account of the current understanding of the mechanisms of central neuropathic pain.

17.5 Complex regional pain syndrome

Complex regional pain syndrome, often referred to as CRPS, is the term introduced recently (Boas 1996) to refer to a group of conditions previously known as reflex sympathetic dystrophy, causalgia, algodystrophy, Sudeck's atrophy, and a number of other diagnostic terms (Table 17.6). Clinical features common to all these conditions, in very variable proportion, include pain associated with allodynia and hyperalgesia, autonomic disturbances, trophic changes,

Table 17.4 Mechanisms of peripheral neuropathic pain related to sensory fibre type

	Afferent fibres involved	Mechanism
Stimulus-independent pain		
Ongoing pain	Nociceptors, A delta and C fibres	EIG in damaged and regenerating peripheral axons and DRG cells
Sympathetically maintained pain	A beta, A delta and C fibres	Sensitization due to expression of alpha adrenergic receptors in regenerating axons
Deafferentaion pain—partial, e.g. PHN complete, e.g. BPA	None	Loss of primary afferent fibres leading to disinhibition of DH cells in spinal cord
Pain radiation	A beta, A delta, or C fibres	Recruitment of dorsal horn WDR neurons over several spinal cord segments
Stimulus-dependent pain		
Light touch	C fibres	Sensitized C fibres
Light brush /stroking-dynamic hyperalgesia, allodynia	A beta fibres	Sensitization of dorsal horn WDR neurons, by nociceptor input (NMDA mediated) and maintained by low threshold fibre input
Pinprick-punctuate hyperalgesia	A delta fibres	Central sensitization initiated but not maintained by nociceptor input
Cold hyperalgesia	Cold sensitive C fibres	Central disinhibition and probably also peripheral sensitization
Heat hyperalgesia	C fibres	Sensitization of peripheral nociceptors
Hyperpathia	Nociceptors or A beta fibres	Recruitment of dorsal horn WDR neurons over several segments of the spinal cord

Abbreviations: BPA: brachial plexus avulsion; DH: dorsal horn; EIG: ectopic impulse generation; NMDA: *N* methyl D aspartate; PHN: postherpetic neuralgia; WDR: wide dynamic range neurons

Table 17.5 Mechanisms of central neuropathic pain

	Pathological change	Contribution to central pain
Spinal cord	Deafferentation disinhibition	Ongoing pain
	Glial activation, GAP expression, inappropriate connectivity	Innocuous peripheral stimuli cause pain in partial lesions. Border zone allodynia and hyperalgesia; below-lesion pain
	Loss of GABA-containing inhibitory neurons in DH of spinal cord	Reduced effectiveness of local and descending inhibitions in spinal cord
	Inflammatory changes in DH WDR: increased intracellular calcium, synthesis of NO and PG	Sensitization of dorsal horn neurons
	Increased excitatory amino acid receptor activation, particularly glutamate	Central sensitization (spinal cord generation of pain)
	Abnormal expression of sodium and calcium channels	Central sensitization
	Imbalance in central pathways: DC and STT, and STT and spinoreticulothalamic	Ongoing and evoked pain
	Denervation of central neuromatrix	Burning or shooting pain
	Thalamic reorganization secondary to caudal changes: altered receptive fields bursting activity in thalamic neurons	Ongoing and paroxysmal pains
Brain	Thalamic lesions: ventroposterior nuclei reticular nucleus medial/intralaminar nuclei	Lesions in these nuclei most likely to cause central pain (CPSP)
	Thermosensory disinhibition: loss of cold activated STT projections that normally inhibit burning sensations	Ongoing pain— following thalamic and spinal lesions
	Thalamocortical lesions	Occasional cause of central pain
	Reactivation of 'memory' of deafferented region and long-term potentiation; possibly due to NMDA receptor and calcium channel activation	Delayed onset of central pain, e.g. CPSP

Abbreviations: CPSP: central post-stroke pain; DC: dorsal columns; DH: dorsal horn; GABA: gamma amino butyric acid; GAP: growth-associated protein; NMDA: *N* methyl D aspartate; NO: nitric oxide; PG: prostaglandins; STT: spinothalamic tract; WDR: wide dynamic range neurons

oedema, and loss of function of the affected part, usually a limb. The term *causalgia* literally means burning pain and was used by Weir Mitchell to describe the severe burning pain, hyperaesthesia, glossy skin, and colour changes in the limbs of soldiers following injury to major nerves, sustained in the American Civil War (Mitchell *et al.* 1864; Richards 1967). Later, it became clear that limb injuries not involving nerves could produce a very similar clinical picture, becoming known as *reflex sympathetic dystrophy* (Evans 1946). The definition and nosology of these conditions remains almost as problematic as it was in the 1940s (Livingston 1943), underlining the

Table 17.6 Previously described conditions now included in complex regional pain syndrome

Reflex sympathetic dystrophy
Post-traumatic sympathetic dystrophy
Algodystrophy
Causalgia major
Causalgia minor
Sudeck's atrophy
Transient osteoporosis
Migratory osteolysis
Post-traumatic painful osteoporosis
Acute bone atrophy
Shoulder-hand syndrome
Post-traumatic vasomotor syndrome

fact that, as with neuropathic pain, it is not possible to tightly link symptoms and signs with specific pathophysiological properties (Woolf and Mannion, 1999). A diagnosis of complex regional pain syndrome describes a clinical state, without making unjustifiable pathophysiological assumptions, as was the case with the older term reflex sympathetic dystrophy.

17.5.1 **Definition**

Complex regional pain syndrome describes a variety of painful conditions that usually follow injury, occur regionally, have a distal predominance of abnormal findings, exceed both in magnitude and duration the expected clinical course of the inciting event, often result in significant impairment of motor function, and show variable progression over time (Boas 1996). Complex regional pain syndrome is divided into types 1 and 2:

In *complex regional pain syndrome type 1*:

- There is an initiating noxious event.

- Ongoing pain and /or allodynia and hyperalgesia occur beyond the territory of a single peripheral nerve, and are disproportionate to the inciting event.

- There is or has been evidence of oedema, skin blood flow abnormality, or abnormal sudomotor activity, in the region of the pain since the inciting event.

- The diagnosis is excluded by the existence of conditions that would otherwise account for the degree of pain and dysfunction.

Complex regional pain syndrome type 2 follows nerve injury and is synonymous with causalgia. It is similar in other respects to type 1, with the following features:

- It is a more regionally confined presentation about a joint or area, provoked by a nerve injury.

- Ongoing pain and /or allodynia and hyperalgesia are usually limited to the nerve area involved, but may spread distally or proximally, outside the territory of the affected peripheral nerve.

- Intermittent and variable oedema, skin blood flow change, abnormal sudomotor activity, and motor dysfunction, disproportionate to the inciting event, are present about the area involved.

These descriptive definitions, based on symptoms and signs, avoid unwarranted inclusion of pathophysiology, but lead to difficulty in

recognizing the clinical limits of these conditions, particularly complex regional pain syndrome type 1. Confusion has also resulted from making complex regional pain syndrome type 2 and causalgia synonymous, and emphasizes the problems caused by applying a term originally intended to refer to a single symptom of burning pain, to a clinical syndrome. For the moment, however, this terminology prevails.

17.5.2 **Causes**

The causes of complex regional pain syndrome are listed in Table 17.7. The great majority of cases are secondary to peripheral tissue injury such as fractures and soft tissue injury. Complex regional pain syndrome type 2 represents a small minority of cases. Rarely, complex regional pain syndrome occurs with central nervous system lesions.

17.5.3 **Clinical features**

The large prospective study of Veldman *et al.* (1993) provides an indication of the relative frequency of the many symptoms and signs of complex regional pain syndrome, at two time intervals following the inciting event (Table 17.8), divided into four categories. Such prospective studies are few, but suggest an under-recognition of the syndrome, while also raising important questions about the limits of the diagnosis (Bickerstaff and Kanis 1994). Spontaneous and evoked pains are common to all patients at some time. Pain quality is often burning, aching, or throbbing. In complex regional pain syndrome type 2, additional paroxysmal pains are common. Allodynia, hyperalgesia, and hyperpathia are often very severe, leading to immobilization and avoidance of any skin contact or pressure on the affected limb. Immobilization itself may in turn result in

Table 17.7 Causes of complex regional pain syndrome

Peripheral Tissues
Fractures and dislocations
Soft tissue injury
Fasciitis
Tendonitis
Bursitis
Ligamentous strain
Arthritis
Deep vein thrombosis
Immobilization
Post-mastectomy
Peripheral nerve and dorsal root
Peripheral nerve trauma[#]
Brachial plexus lesions
Postherpetic neuralgia
Spinal root lesions
Central nervous system
Spinal cord lesions, particularly trauma
Head injury
Cerebral infarction
Cerebral tumour
Viscera
Abdominal disease
Myocardial infarction
Idiopathic

[#] This is complex regional pain syndrome type 2. All other conditions lead to complex regional pain syndrome type 1 (see text)

secondary muscle wasting and joint stiffness, over and above the dystrophic changes that occur as part of the condition.

Autonomic signs are variable in complex regional pain syndrome, but occur at some time in the course of the illness. They may be subtle in some patients and gross in others. Abnormalities of colour, temperature, and sweating are frequently described. Oedema is common but not universal (Fig. 17.1) (Table 17.8).

Objective motor signs are difficult to elicit, because both passive and active movements provoke severe pain. Tendon reflex examination is often impossible. Wasting and weakness are common. Tremor, incoordination, muscle spasms, and dystonia affect some patients.

Dystrophic changes include skin thinning, sometimes with a shiny appearance, or thickened, flaky skin. Hair may either be lost or become coarse, and nails may become thickened. Osteoporosis is common, recognized in its extreme form as *Sudeck's atrophy*.

Less common features of complex regional pain syndrome include a migratory or relapsing pattern, recurrent skin infections associated with chronic oedema, increased skin pigmentation, nodular fasciitis of the palmar or plantar skin, and nail clubbing (Bentley and Hameroff 1980; Veldman *et al.* 1993).

17.5.4 Psychological factors

The disproportionate pain and loss of function of complex regional pain syndrome, together with lack of clarity about the pathogenesis of the condition have led to consideration of a psychological contribution to its development and perpetuation. Patients with conversion disorder or factitious illnesses may present with symptoms that can closely resemble complex regional pain syndrome. Not all

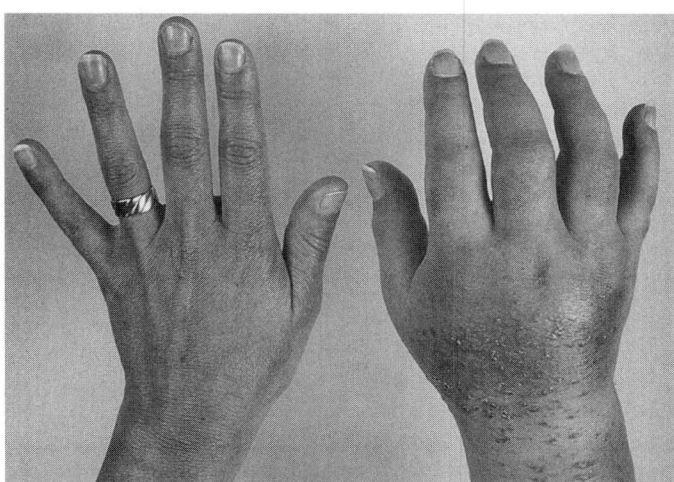

Fig. 17.1 Complex regional pain syndrome type 1 affecting the right hand, showing swelling, discolouration, and skin changes. (Courtesy of Dr. C Glynn.)

the clinical features listed in Table 17.8 need to be present in order to establish the diagnosis, and it is perhaps not surprising that some patients with primary psychiatric morbidity are erroneously diagnosed as suffering from complex regional pain syndrome.

The severe pain and loss of function in complex regional pain syndrome, together with frequent delay in diagnosis causes anxiety, fear, and depression in many patients. Issues of secondary gain may be raised. However, there is no evidence that the condition is primarily psychologically determined (Covington 1996). Some of the patients reported by Ochoa and Verdugo (1995) as having pseudoneuropathy, presenting with clinical features mimicking complex regional pain syndrome, were likely to have suffered from primarily psychiatric illnesses. This emphasizes the need for repeated careful evaluation of patients in whom there is continuing diagnostic doubt.

17.5.5 Staging of clinical features

It has been suggested that three clinical stages of complex regional pain syndrome can be recognized (Blumberg and Janig 1994):

- an acute warm phase, in which oedema is a major feature;
- a dystrophic phase characterized by muscle wasting and vasomotor instability;
- a later atrophic phase, characterized by bone and skin changes.

However, not all patients follow this course, the duration of these phases is very variable and not all patients progress through all three stages (Veldman *et al.* 1993).

17.5.6 Diagnostic tests

As already stated, the diagnosis of complex regional pain syndrome is based on clinical features. Claims have been made for the usefulness of three-phase bone scans as a diagnostic test, but while this investigation is frequently abnormal (Goldsmith *et al.* 1989), a normal scan does not exclude the diagnosis.

17.5.7 Incidence

The clinical limits of complex regional pain syndrome are difficult to define and so data concerning the incidence of the condition are unreliable. Careful prospective studies, for example the series of 829 patients reported by Veldman *et al.* (1993), and the study of

Table 17.8 Clinical features of complex regional pain syndrome (CRPS) (data adapted from Veldman *et al.* 1993)

Clinical feature category	Duration of CRPS 2–6 months (%)	Duration of CRPS >12 months (%)
Inflammatory		
Pain	88	97
Colour difference	96	84
Temperature difference	91	91
Limited movement	90	83
Exacerbation with exercise	95	97
Oedema	80	55
Neurological		
Allodynia/hyperalgesia	75	85
Hyperpathia	79	81
Incoordination	47	61
Tremor	44	50
Involuntary spasms	24	47
Muscle spasms	13	42
Paresis	93	97
Pseudoparesis	7	26
Dystrophic		
Skin	37	44
Nails	23	36
Muscle	50	67
Bone	41	52
Sympathetic		
Hyperhidrosis	56	40
Abnormal hair growth	71	35
Abnormal nail growth	60	52

274 patients with a Colles fracture of Bickerstaff and Kanis (1994), indicate the prevalence of symptoms at various intervals following the inciting injury. At what point the symptoms and signs are judged to be disproportionate in severity and duration to the inciting event is a matter for debate in the different conditions that may lead to complex regional pain syndrome, and this once again emphasizes the imprecision of the current diagnostic criteria.

17.5.8 Pathophysiology

Involvement of the sympathetic nervous system: **sympathetically maintained pain**. Leriche (1916) described the relief of causalgia in a patient with a brachial plexus injury and thrombosis of the brachial artery, by periarterial surgical sympathectomy. Pain relief was accompanied by an improvement in discolouration and sweating changes following sympathectomy in a further series of patients (Leriche 1939), and pre-ganglionic sympathectomy became established as standard treatment for painful nerve injuries.

In relation to neuropathic pain, including complex regional pain syndrome type 2, three sites of sympathetic-sensory interaction after nerve injury have been identified in experimental animal studies: the region of nerve damage itself, undamaged fibres distal to the nerve lesion, and the dorsal root ganglion (Devor 2005). The findings are consistent with expression of alpha adrenoreceptors on regenerating and partly damaged nerve fibres. Denervation supersensitivity may contribute to the magnitude of the agonist effect of circulating and locally released catecholamines.

There is also evidence of a sympathetic influence on neuropathic pain in man. For example, intraoperative stimulation of the sympathetic chain exacerbates causalgia (Walker and Nulsen 1948). In patients with successfully treated causalgia, intracutaneous injection of noradrenaline rekindled their original pain (Wallin *et al.* 1976). Intracutaneous injection of adrenaline or noradrenaline in dermatomes affected by postherpetic neuralgia increases both ongoing pain and allodynia (Choi and Rowbotham 1997). Temporary relief of neuropathic pain by sympathetic blockade has often been reported, usually of peripheral origin but also some central neuropathic pains (Bonica 1990; Arner 1991). However, the existence of this influence has been questioned, on the basis of poor design of some studies, including the lack of proper controls and an underestimation of psychological factors (Ochoa *et al.* 1994).

By definition, complex regional pain syndrome type 1 develops in the absence of an initiating nerve injury. As described below, there is substantial evidence of inflammation in the pathogenesis of this condition. Some components of cutaneous inflammation and hyperalgesia are enhanced by alpha adrenoreceptor stimulation (Drummond 1995; Kinman *et al.* 1997). In addition, noradrenaline causes release of prostaglandins, which are major mediators of inflammatory responses. This effect may be dependent on the pre-existing state of nociceptor activity (Baron *et al.* 1999). Finally, there is an increased alpha adrenoreceptor density in skin biopsies form patients with complex regional pain syndrome type 1 (Drummond *et al.* 1996).

In studies on patients with complex regional pain syndrome type 1 whose pain is relieved by temporary sympathetic block the severity of ongoing pain and allodynia is increased by sympathetic stimulation produced by whole body cooling, indicating a peripheral noradrenergic interaction with primary sensory afferents (Baron *et al.* 2002). This and other evidence indicate a peripheral sympathetic influence in complex regional pain syndrome type 1 (Baron 2005). The term sympathetically maintained pain is now widely used to describe the sympathetic agonist effect in chronic pain states.

Central autonomic dysregulation. Several observations are consistent with a central disturbance of autonomic control in patients with complex regional pain syndrome type 1. Hyperhidrosis is found in many patients, both in resting states and in response to physiological stimulation (Birklein *et al.* 1997). Studies of centrally mediated sympathetic reflexes show that complex regional pain syndrome type 1 is associated with an abnormal unilateral inhibition of cutaneous sympathetic vasoconstrictor neurons, leading to the development of a warm limb, present in some patients. In patients with cold limbs, temperature and vascular perfusion are both reduced in the affected limb during the full range of sympathetic stimulation procedures (Baron 2005).

Inflammation and immune stimulation. Clinical signs of inflammation are obvious in many patients with complex regional pain syndrome type 1 (Table 17.8). Response of type 1 to systemic corticosteroids, in the early stages, at less than 13 weeks duration, has been reported (Christensen *et al.* 1982), though complex regional pain syndrome is refractory to steroids at later stages.

Neurogenic inflammation is an important component of complex regional pain syndrome, leading to oedema, vasodilatation, and increased sweating. For example, plasma extravasation of radiolabelled immunoglobulins has been shown (Oyen *et al.* 1993); there is evidence of neurogenic inflammation in joints (Weber *et al.* 2001); and in the fluid of artificially produced blisters, higher levels of interleukin-6 and TNF alpha have been reported in the affected limb, compared with the unaffected contralateral limb (Huygen *et al.* 2002).

The possibility that the inflammatory features of complex regional pain syndrome type 1 might represent a chronic post-infectious state is suggested by the presence of higher titres of serum antibodies to intestinal pathogens than in healthy controls (Goebel 2001). In addition, Borrelia infection has been reported in association with complex regional pain syndrome with marked bone dystrophy (Sudeck's atrophy) (Bruckbauer *et al.* 1997). A preliminary report indicates response of the pain to treatment with intravenous human immunoglobulin (Goebel *et al.* 2002), though an analgesic effect was also found in patients with a variety of other types of pain, both nociceptive and neuropathic, indicating a rather non-specific effect of the non-blinded treatment in this study.

Antibodies to *Campylobacter jejuni* and increased tissue specific antibodies in sera from patients with complex regional pain syndrome, with disease duration of less than 1.5 years, are further evidence of immune activation resulting from antecedent infection (Goebel *et al.* 2005). Finally, it has been shown that serum from a patient with complex regional pain syndrome type 1, whose pain responded on three separate occasions to IVIG, produced pain behavioural changes when injected into mice (Goebel *et al.* 2005).

While these must be regarded as preliminary observations, they indicate the probable importance of inflammatory processes and possibly of immune-mediated changes in the pathogenesis of complex regional pain syndrome.

Central sensitization. As in peripheral neuropathic pain, prolonged noxious inputs from peripheral tissues in complex regional pain syndrome type 1 will lead to central sensitization (Tables 17.3 and 17.4). In addition, functional MRI studies have demonstrated adaptive changes in the thalamus (Fukomoto *et al.* 1999) and cortex (Maihofner *et al.* 2003), the latter resolving with successful treatment of the pain (Maihofner *et al.* 2004).

Motor abnormalities. The prospective investigation of Veldman *et al.* (1993) emphasized the frequency of motor abnormalities in complex regional pain syndrome: tremor, paresis, and dystonia. Kinematic analysis of certain motor functions indicates that the problem lies in central motor processing (Baron 2005).

17.5.9 Treatment of complex regional pain syndrome

The treatment of causalgia, complex regional pain syndrome type 2, is as for neuropathic pain (Section 17.6). The following comments relate to the treatment of complex regional pain syndrome type 1. It seems intuitive that early treatment improves outcomes. However, there are difficulties in identifying the onset of the condition, and there is no controlled study addressing this important issue. Clearly, the management of conditions known to have the potential of developing into complex regional pain syndrome should be optimized in the early stages, but there is no evidence that it is exclusively poorly managed patients who develop the syndrome. Immobilization and disuse probably contribute to the development of complex regional pain syndrome type 1, so it makes sense to minimize these factors with physiotherapy and aid restoration of function at as early stage after injury as possible.

Sympatholytic procedures. Temporary sympathetic block can be achieved either by injection of local anaesthetic around sympathetic ganglia or by intravenous regional block, using guanethidine or other noradrenaline-depleting drugs (Hannington-Kiff 1974). These techniques were used for many years, for neuropathic pains of various types and for complex regional pain syndrome type 1, before being subjected to controlled study. Most of the studies cited here have included patients with neuropathic pain, as well as patients with complex regional pain syndrome type 1.

There are seven randomized controlled trials of intravenous regional block (Jadad *et al.* 1995). In most of these, guanethidine was used, but some studies included reserpine, bretylium, droperidol, or ketanserin. In four of these, no analgesic effect was found. Recruiting patients who had responded to guanethidine in open trial, to a controlled trial of the same treatment did not reveal a therapeutic effect (Jadad *et al.* 1995). Sixty patients randomized to treatment with intravenous regional block using guanethidine and lignocaine, versus saline and lignocaine, showed no difference in pain relief between the two groups (Ramamurthy and Hoffman 1995). Interestingly, both groups showed improvements in oedema, sudomotor, vasomotor, and trophic changes. A tourniquet inflated to suprasystolic pressure alone relieves hyperalgesia, but temperature sensation is not altered, indicating that hyperalgesia is mediated by A beta mechanoreceptor fibres, these being most susceptible to pressure block by the cuff. In patients with complex regional pain syndrome type 1, four intravenous regional block with guanethidine and pilocarpine versus placebo were ineffective (Livingstone and Atkins 2002). A controlled trial of sympathetic ganglion local anaesthetic block in complex regional pain syndrome type 1 failed to demonstrate any immediate effect, though at 24 h, the lignocaine-treated patients were better than controls (Price *et al.* 1998).

Systematic reviews of sympatholysis in complex regional pain syndrome type 1 and neuropathic pain have concluded that the treatment is ineffective (Kingery 1997; Perez *et al.* 2001).

Surgical sympathectomy can no longer be recommended; no study has demonstrated lasting analgesia (Baron 2005). A return of pain, sometimes worse than the original pain for which the procedure was performed, at an interval following sympathectomy has been frequently observed. The initial warm limb produced by sympathectomy is replaced by a cold limb, despite evidence of no sympathetic reinnervation (Baron and Maier 1996). This change, and the recurrence of pain are probably due to the development of denervation supersensitivity to circulating catecholamines.

Systemic drugs. The evidence base for systemic drug therapy of complex regional pain syndrome type 1 is very limited. Drawing on the limited number of published trials available, not all of which are randomized controlled trials (Baron 2005), only tentative conclusions can be reached at present.

Mild standard analgesic drugs such as paracetamol have only marginal effect. Non-steroidal anti-inflammatory drugs are often used and found to be useful by some patients. Neither class of drug has been investigated in controlled trials. The same applies to opioids, but in routine clinical practice these drugs are found to be helpful by many patients and may be the only treatment that has an effect on the severe pain. A trial of tramadol, followed if ineffective by either slow release morphine or fentanyl patches is justified.

If the development of complex regional pain syndrome type 1 can be confidently identified at less than 3 months, a trial of prednisolone, initially at high dose, can be recommended (Christensen *et al.* 1982). Treatment at later times is ineffective. There is some evidence that calcium-regulating drugs have an effect. These include bisphosphonates ((Varenna *et al.* 2000; Manicourt *et al.* 2004), and intranasal calcitonin (Gobelet *et al.* 1992). Two trials of gabapentin have shown an analgesic effect in complex regional pain syndrome type 1 (Mellick and Mellick 1995; van de Vusse *et al.* 2004). There are reports that free radical scavengers are effective, either oral *N*-acetyl cysteine or topical dimethylsulfoxide (Perez *et al.* 2003). Uncontrolled studies have suggested therapeutic effects for phenoxybenzamine, tricyclic antidepressants, phenytoin, and nifedipine (Scadding 1999).

There are many clinical similarities in the nature of the ongoing and evoked pains of complex regional pain syndrome type 1 and of neuropathic pain. Thus despite the absence of firm evidence of effectiveness, it is justifiable to recommend that the drugs shown to be effective for neuropathic pain should also be tried in patients with complex regional pain syndrome, with a clear understanding on the part of both clinician and patient that the drugs in question are being used for an unlicensed indication. This recommendation reflects both the severity of the pain in complex regional pain syndrome and the current very restricted armamentarium of effective treatments.

Epidural and intrathecal treatment. Intrathecal morphine is effective in the treatment of severe pain due to complex regional pain syndrome, refractory to all other measures (Becker *et al.* 1995). Epidural clonidine, an alpha 2 receptor agonist, has been reported to be effective for pain in complex regional pain syndrome affecting either upper or lower limbs (Rauck *et al.* 1993). Intrathecal baclofen may help the dystonia sometimes associated with complex regional pain syndrome (van Hilten *et al.* 2000).

Electrical stimulation. Uncontrolled observations in clinical practice indicate that transcutaneous electronical nerve stimulation has a useful effect in some patients. The effect of spinal cord stimulation is unpredictable, but this can be effective (Kemler *et al.* 2000).

Physiotherapy. Efforts to achieve pain relief should always be accompanied by attempts to mobilize the affected part and restore function; the clinical features of complex regional pain syndrome type 1 are probably to an extent perpetuated by disuse and immobility. Physiotherapy and occupational therapy have both been shown to be helpful (Oerlemans *et al.* 2000).

Attention and mirror visual feedback. Distraction from pain by various means has been advocated and exploited as a therapeutic intervention for many years. The mechanisms and potential for treatment have recently come under scrutiny (McCabe *et al.* 2005). Mirror visual feedback, a method in which patients can utilize the movements of their normal limb to promote active movement of a limb affected by complex regional pain syndrome type 1, has shown promise in short-term treatment (Ramachandran 2005). Further studies are awaited.

Psychological measures. The comments concerning cognitive behavioural therapy (Section 17.6.7) in relation to the treatment of neuropathic pain apply equally to complex regional pain syndrome type 1. Combined physical therapy and coggitive behavioural therapy has been shown to be effective (Lee *et al.* 2002).

Amputation. Patients with very severe intractable pain in a limb rendered useless by complex regional pain syndrome type 1 sometimes raise the question of amputation. While amputation will rid them of a painful hyperaesthetic limb, they should be warned that there is a high risk of developing stump and phantom limb pain. Although it has not been proven, there is a strong clinical impression that those with painful limbs prior to amputation are more likely to develop these new problems, which for some patients may be just as incapacitating as their original pain.

17.5.10 Prognosis

Information concerning prognosis of complex regional pain syndrome is limited and it is difficult to compare directly results of different studies, due to case inclusion bias and initial severity. Many patients remain incapacitated by their symptoms for years and in a proportion, problems persist for decades or even lifelong. In one study, at follow up after 5.5 years, 62 per cent of patients remained markedly troubled by pain and impaired function (Geertzen *et al.* 1994). Factors adversely affecting outcome include severity at the time of diagnosis, female gender, and affection of the lower limb. It is in these patients that the most serious complications are likely to occur, including skin ulceration, infection, chronic oedema, and dystonia (van der Laan *et al.* 1998).

17.5.11 Occurance in children

Complex regional pain syndrome type 1 is an uncommon condition in children and thus prone to delayed diagnosis. The lower limb is much more frequently affected than the upper limb in a ratio of about 5:1, and girls are more commonly affected than boys, in a ratio of approximately 4:1. Those affected are typically pubertal adolescent girls. Many children with complex regional pain syndrome have participated in competitive sports and other physical activities, placing them at increased risk of musculoskeletal injury. There has been interest in the possible psychological gain to a child of a persistent injury enabling escape from the stress of competition and parental expectations, as a factor influencing the development of the condition. The prognosis for complete recovery with physiotherapy, transcutaneous nerve stimulation and cognitive behavioural therapy is very much better in children than in adults (Wilder 1996; Sherry and Weisman 1988).

17.6 Treatment of neuropathic pain

17.6.1 General considerations

Neuropathic pain is much more difficult to treat than most nociceptive pains and the range of specific therapies is still limited. Once established, neuropathic pain is often a lifelong complaint. Regardless of the presence or absence of additional neurological deficits, neuropathic pain and its comorbidities impose a substantial burden for many patients.

When neuropathic pain arises from compressive lesions of peripheral nerves, spinal roots and sometimes of the spinal cord, surgical decompression can partly or completely relieve pain. However, this is certainly not always the case, particularly if compression has been prolonged. There is a poor correlation between the severity of the neurological deficit and the degree of pain, both pre- and post-surgery. The place of other surgical interventions for neuropathic pain is very limited. Ablative operations that produce new lesions in the somatosensory system, designed to relieve neuropathic pain, may do so temporarily, but are themselves potent causes of neuropathic pain. Such pain develops at variable intervals following surgery, ranging from days to years.

When neuropathic pain is anatomically limited in extent and associated with painful evoked symptoms—allodynia, hyperalgesia and hyperpathia—every effort should be made to employ local measures. Not only may these be partially effective, but they will also reduce or obviate the need to consider systemic drug treatment. All classes of drugs given for neuropathic pain may cause treatment-limiting adverse effects and while by no means confined to older patients, these represent a particular problem in this age group.

Many patients with chronic neuropathic pain require a multimodality approach to their treatment. This reflects both the multidimensional nature of neuropathic pain and the partial effectiveness of individual treatments. Support and follow up over long periods are needed. Many patients are most appropriately managed in an integrated multidisciplinary setting that includes input from neurologists, anaesthetists, psychologists, physiotherapists, and occupational therapists.

17.6.2 Local treatments

Local anaesthetic injections. Local anaesthetic blocks of peripheral nerves, plexuses, or spinal roots may be helpful diagnostically, though do not always permit accurate localization of causative lesions. For example, an effective sensory root block may mask a more peripheral lesion. The effect of a local anaesthetic block is usually shortlived and thus of limited therapeutic value.

Topical local anaesthetic. In patients with limited areas of allodynia, local-anaesthetic-impregnated patches, or EMLA cream are indicated. A less elaborate but sometimes surprisingly effective alternative is simple 5 per cent lignocaine ointment. Topical local anaesthetic has been shown to be beneficial in painful polyneuropathy, postherpetic neuralgia, and a variety of other neuropathic pains (Galer *et al.* 1999; 2002).

Topical capsaicin. This pungent extract of chilli peppers, binds to vanilloid receptors and causes depolarization of afferent C fibres, with release of substance P. This agonist action leads to burning pain on first application, and this frequently limits its clinical use. However, following repeated topical application, there is prolonged depletion of substance P and this probably accounts for desensitization of afferent C fibres. Topical capsaicin 0.075 per cent can be helpful in patients with neuropathic pain associated with limited areas of allodynia. It has been shown to be effective in painful diabetic neuropathy (Capsaicin Study Group 1991), postherpetic neuralgia (Watson *et al.* 1993), and post-surgical pain (Ellison *et al.* 1997), but not in HIV neuropathy (Paice *et al.* 2000).

17.6.3 Systemic drug treatment

Until relatively recently, the literature abounded with therapeutic claims for a wide variety of systemic drugs in neuropathic pain, based on anecdotal, open label reports or poorly controlled studies using limited pain outcome measures. An appreciation of the need to consider the multidimensional aspects of neuropathic pain, and improved trial methodology has led to clinical trial results of greater relevance to patients with neuropathic pain (discussed in several recent systematic reviews: Kingery 1997; McQuay and Moore 1998; Sindrup and Jensen 1999, 2000; Finnerup *et al.* 2005). Nonetheless, comparison of data from different trials remains problematic.

A useful calculated measure of effectiveness of treatments, now widely used, is the number needed to treat, NNT, defined as the number of patients who have to be treated to produce pain relief in one patient (McQuay and Moore 1998). The degree of analgesia is usually defined as 50 or 30 per cent , the latter figure equating to a value patients describe as at least 'moderate' pain relief. Similarly, the number needed to harm, NNH, provides a useful measure of safety and acceptability of a drug. This is defined as the number of patients who need to be treated for one patient to drop out due to adverse effects. NNT and NNH can only be calculated from trials with dichotomous data.

Calculation of NNT permits data from different trials to be pooled, but does not necessarily give a consistent measure of relative effectiveness of different drugs. This is due to methodological factors including trial design, whether parallel or cross-over, different drug dosages, variation in the scales and instruments used for assessment of pain severity, baseline pain scores at the time of entry to the trial, variable inclusion of comorbidity data, heterogeneous patient inclusion, incomplete reporting of results including adverse effects, and variable magnitude of placebo effects in different trials. NNT data thus need to be interpreted with caution. Additional confounding factors include an absence of the most relevant head-to-head trials in certain instances, and under-reporting of trials produces negative results.

Multiple mechanisms contribute to the development of neuropathic pain (Tables 17.3–17.5), so it should not be surprising that a single drug rarely produces more than partial analgesia. In clinical practice this frequently leads to treatment with drug combinations, associated with a substantial incidence of adverse effects.

Table 17.9 summarizes NNT values for systemic drug treatment of peripheral, central, and mixed neuropathic pain, together with NNH values where these have been calculated.

Antidepressant drugs. The mechanism by which tricyclic antidepressant drugs exert an analgesic effect in neuropathic pain remains uncertain. A serotoninergic action, enhancing the inhibitory effect of the descending pathway from brainstem to the dorsal horn of the spinal cord (Fields and Basbaum 1978) is often invoked. However this is unlikely to be the sole mechanism, as the selective serotonin reuptake inhibitors, which should theoretically be more potent analgesics due to this action, have a weaker analgesic effect

Table 17.9 Drug treatment of neuropathic pain

Drug	Peripheral neuropathic pain NNT	Central neuropathic pain NNT	Mixed neuropathic pain NNT	NNH
Antidepressants:				
TCAD	2.1–2.8	4.0	-	14.7
SSRI	6.8	-	-	-
SNRI	5.5	-	-	-
Combined	3.3	4.0	1.6	16.7
Anticonvulsant drugs:				
Phenytoin	2.1	-	-	-
Carbamazepin	2.3	-	-	-
GBP and PGB	3.9–4.6	-	-	-
Lamotrigine	4.0–5.4	-	-	-
Valproate	2.1–2.4	-	-	-
Topiramate	7.4	-	-	6.3
Combined	4.2	-	10.0	10.6
Opioids:				
Strong opioids	2.3–3.0	-	2.1	17.1
Tramadol	3.5–4.8	-	-	9.0
NMDA antagonists:				
Dextromethorphan	2.5–3.4	-	-	8.8
Antiarrhythmics:				
Mexiletine	2.2–7.8	-	-	-
Topical lignocaine	4.4	-	-	-
Cannabinoids	-	3.4	9.5	-
Topical Capsaicin	3.2–11.0	-	-	11.5

Pooled data, adapted from Finnerup *et al.* (2005)

Abbreviations: GBP: gabapentin; NNT: number needed to treat (see text for explanation); NNH: number needed to harm (see text for explanation); PGB: pregabalin; SNRI: selective noradrenaline and serotonin re-uptake inhibitors; SSRI: selective serotonin re-uptake inhibitors; TCAD: tricyclic antidepressant drugs

than tricyclics in neuropathic pain. Tricyclics also block the uptake of noradrenaline, and may potentiate noradrenergic inhibitory mechanisms at the spinal level (Rang *et al.* 2001).

Chronic pain and depression frequently coexist, with a complex relationship between the two conditions (Monks and Merskey 1999). That the mechanisms of action of tricyclics in chronic pain and depression are different and distinct is indicated by an earlier onset of an analgesic than an antidepressant action (Langohr *et al.* 1982), an analgesic effect without relief of depression in some patients (Lascelles 1966), and an analgesic effect in patients who are not depressed (Lance and Curran 1964).

Amitriptyline, nortriptyline, imipramine, desipramine, clomipramine, and maprotiline have all been demonstrated to reduce neuropathic pain, with a combined NNT of 3.1 and a combined NNH of 14.7 (Finnerup *et al.* 2005). Adverse effects, particularly the anticholinergic actions, often limit the dose that is tolerated. There is little to choose between the different tricyclics for the treatment of neuropathic pain, based on current evidence.

Of the selective serotonin reuptake inhibitors, paroxetine and citalopram have been shown to be weakly effective in peripheral neuropathic pain, with a combined NNT of 6.8 (Finnerup *et al.* 2005), but in one trial fluoxetine was no more effective than placebo in patients with painful diabetic neuropathy (Max *et al.* 1992). The more recently available serotonin and noradrenaline reuptake inhibitors, venlafaxine and duloxetine, appear to be of similar efficacy to selective serotonin reuptake inhibitors in painful peripheral neuropathy, with a combined NNT of 5.5 (Finnerup *et al.* 2005).

Neuroleptic drugs. These were reported to have analgesic effects in both peripheral and central neuropathic pain in the 1950s (Sadove *et al.* 1955; Margolis and Gianascol 1956; Sigwald *et al.* 1959). There is limited evidence of efficacy of neuroleptics when used in combination with antidepressants, for both nociceptive and neuropathic pains (Monks and Merskey 1999), but overall, the evidence for a substantial independent analgesic effect of neuroleptics in neuropathic pain is slender. The serious long-term adverse effects of neuroleptics, together with the relative lack of evidence of efficacy, has led to a reluctance to use neuroleptics for the treatment of chronic neuropathic pain.

Anticonvulsant drugs. Carbamazepine and *phenytoin* have membrane-stabilizing actions mediated by non-specific sodium channel blockade, and theoretically, might have multiple sites of action in relation to the mechanisms underlying neuropathic pain.

The remarkable effect of carbamazepine in trigeminal neuralgia (Blom 1963) led to the hope that this and other antiepileptic drugs might be as effective generally in neuropathic pain. Sadly, this has not proved to be the case, either for peripheral or central neuropathic pain (McQuay *et al.* 1995; Finnerup *et al.* 2005). Although early trials suggested efficacy, these were methodologically flawed. Open trial in numerous patients over the last three decades has demonstrated the very limited usefulness of both drugs as analgesics in both peripheral and central neuropathic pains, with the notable exception of trigeminal neuralgia.

Gabapentin and *pregabalin* bind to the alpha 2 delta subunit of voltage-dependent calcium channels. Their mode of action in neuropathic pain is uncertain, but they may modulate neurotransmitter release from primary afferent terminals, via an action on interneurones in the dorsal horn of the spinal cord. Both drugs have been extensively studied in painful diabetic neuropathy and postherpetic neuralgia in large, well-controlled trials that have included multidimensional pain measures. A combined NNT for gabapentin is 4.7, with an NNH of 17.8 (Finnerup *et al.* 2005). With pregabalin, the NNT for postherpetic neuralgia is 5.5, and for painful diabetic neuropathy 7.7 (Dworkin *et al.* 2003; Rosenstock *et al.* 2004; Sabatowski *et al.* 2004; Richter *et al.* 2005).

Lamotrigine exerts an inhibitory effect on voltage-sensitive sodium channels, and also inhibits release of the excitatory amino acids glutamate and aspartate. By these actions it stabilizes neuronal membranes. In painful diabetic neuropathy, Eisenberg *et al.* (2001) reported an analgesic effect, with an NNT of 4.0. In HIV neuropathy, lamotrigine was shown to have a weak analgesic effect in only one of two trials (Simpson *et al.* 2003). In central post-stroke pain, lamotrigine has a weak effect (Vestergaard *et al.* 2001), but no effect in spinal cord injury pain (Finnerup *et al.* 2002).

Sodium valproate has several actions which indicate that it might exert an analgesic effect in neuropathic pain, including increased synthesis and release of GABA, sodium channel blockade, and reduced neuronal excitability to glutamate (Locher 1999). Although an analgesic effect was reported in three trials in diabetic neuropathy and postherpetic neuralgia, this was not found in another trial in painful poyneuropathy including patients with diabetic neuropathy, and valproate was ineffective in a trial in patients with spinal cord injury (see Finnerup *et al.* 2005).

Topiramate stabilizes neuronal membranes through various mechanisms, including anti-glutamate effects, sodium channel blockade, and enhancement of GABA-mediated inhibitory actions (Shank *et al.* 2000). Three out of four trials of topiramate in painful diabetic neuropathy showed no effect, and while the other trial did indicate an analgesic effect, NNT 7.4, all four trials were associated with high drop out rates due to adverse effects, producing a combined NNH of 6.3 (Raskin *et al.* 2004; Thienel *et al.* 2004).

Opioids. For many years, it was received wisdom that opioids were ineffective in neuropathic pain. This, combined with a reluctance to use opioids in patients with chronic non-malignant pain, resulted in a reappraisal of opioids in neuropathic pain only in recent years (Portenoy 1990). Finnerup *et al.* (2005) summarize the effectiveness of opioids in different neuropathic pain types. Morphine, oxycodone, or the weaker opioid tramadol have been shown to produce analgesia in peripheral neuropathic pain, including diabetic neuropathy, postherpetic neuralgia, and phantom limb pain, with combined NNT of 2.5–3.9, and NNH of 7.9–11.3.

NMDA antagonists. The action of N methyl Daspartate antagonists in blocking the afferent C fibre wind-up of dorsal horn neurons (Table 17.3) led to hopes that drugs of this class might be effective for neuropathic pain. Intravenous infusions of NMDA antagonists, including ketamine, do produce pain relief (Sang 2000), but often with a limited duration of minutes or hours. Moreover, the treatment is frequently associated with marked sedation and other unpleasant adverse effects including hallucinations. Two trials of dextromethorphan have shown an effect in painful polyneuropathy, including diabetic (Nelson *et al.* 1997; Sang *et al.* 2002). However, 11 other trials of dextromethorphan, memantine, or riluzole in painful neuropathy, postherpetic neuralgia, phantom limb pain, or mixed types of neuropathic pain have been negative (Finnerup *et al.* 2005).

Antarrhythmics. Lignocaine, a non-specific sodium channel blocker, given intravenously, produces short-lived relief of neuropathic pain, for example painful diabetic neuropathy (Kastrup *et al.* 1987). The oral analogue of lignocaine, mexiletine has been extensively trialed but found to be effective in only two studies, one in painful diabetic neuropathy (Dejgard *et al.* 1988) and the other in painful peripheral nerve injury (Chabal *et al.* 1992). Seven other studies have shown a lack of effect in mixed painful peripheral neuropathies, HIV neuropathy, and spinal cord injury (Finnerup *et al.* 2005).

Cannabinoids. The cannabinoid receptor CB_1 is widely distributed in the central nervous system. Of particular relevance to pain processing and the analgesic effects of cannabinoids is the expression of CB_1 receptors in the thalamus, periaqueductal grey, and rostroventromedial medulla. In the spinal cord, receptors are expressed in the superficial dorsal horn and dorsolateral funiculus. Several endogenous ligands are now known, the endocannabinoids. The history of development of cannabis as an analgesic and the analgesic mechanisms of cannabinoids are reviewed by Rice (2005).

In a study of pain related to brachial plexus avulsion, Berman *et al.* (2004), demonstrated a modest analgesic effect with two cannabis extracts. Using the synthetic cannabinoid CT-3, Karst *et al.* (2003) also showed a mild analgesic effect in a mixed group of patients with neuropathic pain. Pain was not a primary end-point of a large study of cannabinoids in patients with multiple sclerosis (Zajicek *et al.* 2003), but 30–50 per cent of patients reported some improvement in their pain. A preliminary study examined the effect of cannabinoids on a wide range of intractable neurogenic symptoms, finding a reduction of pain scores in the order of 30–35 per cent , though with a 21 per cent placebo response rate (Wade *et al.* 2003). Adverse effects led to 17 per cent of patients withdrawing from this study.

Conclusion and drug treatment recommendations

Comparison of systemic drug trial data is difficult for the reasons already outlined and it is thus hard to make firm recommendations. Trials have been far greater in number for peripheral than for central neuropathic pain. Finnerup *et al.* (2005) have drawn attention to the criteria relevant to setting an order of preference for prescribing drugs to patients. These include consistent outcome in high quality trials, low NNT and high NNH values, prolonged effectiveness, effect on quality of life measures, and low cost.

Not all the data necessary to satisfy these criteria are available for all classes of medication. Bearing this in mind, the following sequence for treating *peripheral neuropathic pain* is suggested:

1 Tricyclic antidepressant, or if contraindicated, gabapentin or pregabalin.

2 Gabapentin or pregabalin. There is currently no evidence to indicate superiority of one of these drugs above the other.

3 Serotonin and noradrenaline reuptake inhibitors.

4 Tramadol.

5 Oxycodone or other strong opioid.

For *central neuropathic pain*, recommendations are less securely evidence based, but the following order is suggested:

1 Gabapentin or pregabalin.

2 Tricyclic antidepressants, if not contraindicated.

3 Lamotrigine, tramadol, cannabinoids, and strong opioids. There is insufficient evidence at the moment to rank these drug classes.

17.6.4 **Intrathecal and epidural drug treatment**

The place of epidural analgesia in acute pain management for childbirth, thoracic, and abdominal surgery is well established (Breivik 2005). For chronic pain, the indications are limited. Some patients with severe intractable cancer pain, mostly of non-neuropathic type, are effectively treated with chronic epidural infusions. The main neuropathic pain indication for epidural analgesia is severe spinal root pain that is resistant to all other measures. Injections of local anaesthetic and corticosteroid may relieve chronic root pain refractory to other treatments, sometimes for days or even weeks, and this may be a useful form of treatment in a few patients.

There is a synergistic analgesic action of local anaesthetic and opioids such as fentanyl given via the epidural route. The duration of action of this combination can be substantially prolonged by the addition of adrenaline, which reduces the absorption of the other two drugs. This occurs without causing any reduction in spinal cord blood flow. Epidural infusions can be very effective for the management of severe root pain, but over time, epidural fibrosis and adhesions develop, thus limiting the usefulness of the technique for long-term treatment (Breivik 2005).

Epidural treatment of chronic pain requires an experienced team with good facilities for monitoring the treatment. Accurate placement of the epidural catheter is critical to achieving good analgesia; loss of an initial good effect can be the result of catheter tip migration. Adverse effects include haemodynamic disturbances, sensory loss, and motor paralysis, loss of bladder function, respiratory problems due to rostral spread of the infused opioid, infection, granuloma formation, and intraspinal bleeding with haematoma formation leading to cord compression. For these reasons, epidural and intrathecal local anaesthetic and fentanyl are only rarely employed in the treatment of chronic severe non-malignant pain.

17.6.5 **Sympatholysis and sympathectomy**

This topic has been considered earlier in relation to the pathophysiology and treatment of complex regional pain syndrome (Section 17.5.9).

17.6.6 **Neural stimulation**

Acupuncture, transcutaneous electrical nerve stimulation or TENS, and other forms of peripheral counter-stimulation, together with a variety of central nervous system stimulation techniques, are relatively poorly validated treatments for pain. The problems of adequate blinding of these modalities in clinical trials are obvious. However, neurophysiological mechanisms of analgesia have been established for all these treatments, and there is a limited evidence base for efficacy in patients with both nociceptive and neuropathic pains. The potential value of electrical stimulation and acupuncture has been recognized since ancient times (Kane and Taub 1975; Lu and Needham 1980). The gate control theory (Melzack and Wall 1965), which proposed a neuromodulating effect of peripheral large fibre activity on the forward transmission of noxious inputs, in the dorsal horn of the spinal cord, reawakened interest in the scientific investigation of counter-stimulation as a means of pain control.

Transcutaneous electrical nerve stimulation, TENS. Physiological evidence indicates that the non-painful stimulation used therapeutically in TENS inhibits nociceptive transmission in the spinal cord by both pre- and post-synaptic mechanisms (Sluka and Walsh 2003), and possibly via long-range thalamic inhibition (Olausson *et al.* 2002). Recruitment of these inhibitions is dependent on stimulus frequency and intensity. High intensity, high frequency TENS activates A delta and C fibres, and is itself painful. It appears to be more effective in activating spinal meachanisms. Analgesia induced by this form of stimulation is probably mediated by endogenous opioids, and there is some evidence that low frequency TENS also exerts its effect partly through a similar mechanism (Han 2003). It is interesting to note that low intensity TENS, which selectively stimulates A beta fibres, can relieve pain associated with brush-evoked pain, allodynia, that is itself mediated by A beta fibres. However, in such patients, TENS usually cannot be tolerated within the area of allodynia but can be effective when applied to adjacent areas.

There is a body of evidence indicating efficacy of TENS in a variety of nociceptive pains, notwithstanding the methodological difficulties of clinical trials (Barlas and Lundeberg 2005). In patients with peripheral neuropathic pain, there are several investigations that strongly indicate a therapeutic action (Hansson and Lundeberg 1999). In central neuropathic pain, the evidence is weaker (Davis and Lentini 1975; Leijon and Boivie 1989).

From a practical point of view, a trial of TENS is worthwhile in many patients with neuropathic pain, particularly when the pain is relatively localized, which is more likely when the cause is related to peripheral nerve, plexus, or root pathology. The most common adverse effect is skin reaction to the electrodes, usually mild, but occasionally treatment-limiting. TENS should not be used in patients with cardiac pacemakers, and caution is advised in pregnancy, particularly during the first trimester.

An important consideration in the use of TENS is that it provides patients with active personal involvement and control in the treatment of their pain in a way that most other treatments do not. This helps to counter feelings of helplessness that are so frequently a feature of chronic neuropathic pain.

Acupuncture. This activates A delta and C fibres and is usually perceived as painful by patients. There is strong evidence that acupuncture analgesia is mediated via endogenous opioid mechanisms. Acupuncture is associated with an increase in CSF endorphin concentrations and can be blocked or reversed with naloxone (Pomeranz and Chiu 1976; Han and Terenius 1982). There have been few well-controlled clinical trials of acupuncture. Sham acupuncture is difficult but not impossible. Acupuncture may have a place in the treatment of tension type headache (Melchart *et al.* 2003). A Cochrane systematic review of acupuncture for the treatment of non-specific low back pain concluded that there was insufficient evidence of benefit (van Tulder *et al.* 1999). Acupuncture has been shown to have some effect in patients with painful diabetic neuropathy (Abuaisha *et al.* 1998), but there is otherwise, as yet, no evidence base to support its use in neuropathic pain.

Vibration. This activates superficial and deep mechanoreceptors and the primary endings of muscle spindles connected to large fibre afferents (Eklund and Hagbarth 1965; Vallbo and Hagbarth 1968). The best evidence of therapeutic efficacy of vibration derives from studies in nociceptive pain, particularly musculoskeletal (Hansson and Lundeberg 1999). For neuropathic pain, in a single randomized placebo-controlled trial, vibration was reported to relieve stump and phantom limb pain (Lundeberg 1985). Other counterstimulation methods, including hot and cold packs and massage are sometimes found to be helpful by patients with neuropathic pain, particularly cold packs in postherpetic neuralgia. However, none of these modalities has been subjected to rigorous assessment in clinical trials.

Peripheral nerve stimulation. This has been advocated for the treatment of neuropathic pain due to lesions of single peripheral nerves. It is associated with technical difficulties and is little used in clinical practice. Peripheral nerve stimulation can be effective in neuropathic pain resulting from mononeuropathies of limb nerves (Waisbrod *et al.* 1985) and for occipital neuralgia (Weiner and Reed 1999) (Section 19.2.6).

Spinal cord stimulation. This occurs on the dorsal aspect of the cord and produces both anterograde ascending impulse activity and retrograde descending activity in the dorsal columns. Both may be important in inducing analgesia (Simpson *et al.* 2005). In experimental studies, transection of the spinal cord rostral to the site of stimulation abolishes its effect (Roberts and Rees 1994). Descending impulses suppress activity in dorsal horn wide dynamic range neurons below the site of stimulation (Linderoth and Foreman 1999). The effectiveness of spinal cord stimulation is dependent upon GABA and possibly other neuromodulating substances, including 5-hydroxytryptamine, substance P, adenosine, and glycine (Linderoth and Foreman 1999).

There is evidence indicating an effect of spinal cord stimulation in symptom relief and restoration of the microcirculation in chronic limb ischaemia (Ubbink *et al.* 1999). It is now generally agreed that it has a role in the treatment of neuropathic pain, particularly the failed back surgery syndrome, in which there is usually a mixture of nociceptive and neuropathic pain components (North *et al.* 1991; Gybels *et al.* 1998). A recent study indicates that spinal cord stimulation is preferable to re-operation in failed back surgery syndrome (North *et al.* 2005). This has become the major indication for spinal cord stimulation, but robust evidence of efficacy is limited. A systematic review reached the conclusion that unequivocal evidence of efficacy was lacking, and was critical of the methods of many studies (Turner *et al.* 1995).

In clinical practice, spinal cord stimulation should only be considered for patients with failed back surgery syndrome and other intractable radicular pains, including postherpetic neuralgia, and stump and phantom limb pain, when all other measures have failed. There is a high incidence of technical problems with spinal cord stimulation, and loss of an initial analgesic effect after weeks or months is common, either for technical reasons or from presumed physiological adaptation.

Brain stimulation. Despite reports of pain relief from brain stimulation since the 1960s (Mazars *et al.* 1979), the development of therapeutic brain stimulation has been hampered by a paucity of well-controlled studies and by observer bias resulting from nonindependent patient evaluations in many trials (see Simpson *et al.* 2005). Three target regions have attracted greatest interest: the sensory thalamus, the periaqueductal/periventricular grey, and most recently, the motor cortex.

Sensory thalamic stimulation. The mechanism of analgesia produced by stimulation of the main thalamic sensory nuclei remains uncertain, though experimental observations point to a type of supraspinal gating. For example, it has been shown that stimulation in the ventrobasal complex in monkeys reduces responses of dorsal horn neurones to noxious peripheral stimuli (Gerhart *et al.* 1983); and in a rat model of sciatic nerve injury, thalamic stimulation reduces hypersensitivity of the affected hind-paw (Kupers and Gybels 1993).

Thalamic sensory nucleus stimulation has been tried for many intractable pains, both nociceptive and neuropathic. A metaanalysis, based on reports of large series, indicates that neuropathic pain responds much better than nociceptive pain (Bendok and Levy 1998; and see also Hosobuchi 1986). However, central post-stroke pain usually does not respond. Suggested indications include failed back surgery syndrome when there is a major neuropathic component, stump and phantom pain, and facial anaesthesia dolorosa resulting most often from surgical deafferentation procedures performed for trigeminal neuralgia (Section 19.2.1). However, evidence of efficacy is weak. Although there are case reports of long-term pain relief, loss of effect in many patients after weeks or months is widely recognized.

Periaqueductal/perivantricular grey stimulation. Following the demonstration that periaqueductal grey stimulation and periventricular grey stimulation in the upper midbrain and medial thalamus exerts a powerful analgesic effect in experimental animals (Reynolds 1969), it was later found to be an endogenous opioid-mediated effect, associated with raised CSF concentrations of beta-endorphin, and reversible with naloxone (Richardson 1995). Such stimulation is more effective in nociceptive than in neuropathic pain. Periventricular stimulation is preferred to peri-aqueductal grey stimulation because of unwanted effects of the latter including dysphoria and diplopia (Hosobuchi 1986; Kumar et al. 1997).

Motor cortex stimulation. This has been reported to be effective for central post-stroke pain and trigeminal anaesthesia dolorosa (Nguyen et al. 2003). The mechanism is uncertain, but an inhibitory action in several areas of importance in pain perception seems likely. This is based on the finding in positron emission tomography studies that motor cortex stimulation leads to an increase in blood flow in the ipsilateral ventral lateral thalamus, cingulate gyrus, insula, and brainstem. Stimulation is effective at a threshold below that of motor activation, and tolerance does not develop (Brown 2004). The limited evidence available at the moment does not permit firm conclusions to be drawn about the place of motor cortex stimulation in the treatment of intractable pain.

17.6.7 Cognitive behavioural therapy

The major comorbidities of chronic pain have already been discussed. Those distressed and disabled by their chronic pain, particularly when it fails to respond to standard treatments, are those most likely to benefit from psychological interventions. However, it is important to emphasize that one should not wait until all other treatment options have been exhausted before considering psychological therapy. The latter can be invaluable either in combination with other treatment modalities or on its own.

A frequent sequence of events is that pain with limited responsiveness to treatment is accompanied by reduced physical activity, associated with fatigue, poor sleep, social and family isolation, depression, anger, frustration, and a fear of making the pain worse, particularly through physical exertion 'catastrophizing', and an increasing dependence on medical services. It is important to recognize this symptom constellation at an early stage, before it becomes entrenched, and consider psychological intervention. The aims of treatment must be tailored to the individual patient, but are likely to include improved physical activity and fitness, reduction in fear and catastrophizing, improved adaptive and coping behaviour, relief of depressive symptoms, and return to work.

Many psychological techniques have been employed, but the treatment of choice is cognitive behavioural therapy. A detailed description of the rationale and methods is beyond the scope of this chapter (see Eccleston et al. 2003). Sifting the published reports of cognitive behavioural therapy to assess evidence of efficacy presents a host of methodological difficulties (Eccleston et al. 2003). However, a meta-analysis of its use in chronic pain, without distinguishing the nociceptive, neuropathic, or mixed nature of the pain, yields some evidence of efficacy (Morley et al. 1999). Further studies are needed.

17.6.8 Surgical ablative treatment

The value of decompressive surgery of peripheral nerves, plexuses, spinal roots, and sensory cranial nerves for the treatment of neuropathic pain and associated deficits is clearly established. The indications for ablative neurosurgery, on the other hand, are very limited. Neuropathic pain, both peripheral and central, is caused by lesions of the somatosensory system; thus surgical procedures designed to interrupt some part of this system are themselves at risk of leading to the development of neuropathic pain.

Periphal neurectomy. Resection of painful peripheral nerve neuromas is inevitably followed by regrowth of the neuroma. In nerve injury, nerve repair and grafting can alleviate pain, but regenerating nerves partially innervating their original territory are often associated with allodynia and hyperalgesia, so that while such surgery may improve motor and sensory function, it may also lead to the development of these additional painful symptoms, transiently or permanently, depending on the eventual success of peripheral tissue reinnervation.

Painful neuromas are often sited in positions where they are subject to repeated minor physical trauma due to tethering and traction with limb movement, or because they are subject to pressure. In such circumstances, surgery to relocate neuromas can be helpful, though it is by no means always successful, as it is difficult to fashion appropriate environments for exquisitely mechanically sensitive neuromas in many anatomical situations. Nonetheless, in highly selected patients the results can be excellent (Patil and Campbell 2005). Resection of the affected plantar digital nerve is routinely performed for the treatment of Morton's neuralgia (Section 22.10.2). The nerve lesion has neuromatous elements, but is aetiologically an entrapment neuropathy, often histologically severe (Scadding and Klenerman 1987). Resection leads to neuroma formation, so that even with resection far enough proximally, away from the area of most intense pressure on the nerve end, 33 per cent of patients continue to experience pain in the long term (Johnson et al. 1988).

Dorsal rhizotomy and ganglionectomy. There is large overlap of the territories of sensory spinal segmental innervation, and sensory afferents entering through sensory roots at one level may terminate over several spinal segments (White and Kjellberg 1973). Thus in order to denervate a painful area, dorsal rhizotomy at several levels may be needed. Rhizotomy interrupts large as well as small nociceptive afferents, leading to unwanted sensory loss, including proprioceptive, and in the sacral segments also to impaired bladder and bowel function. In addition, the deafferentation produced surgically may cause the later development of central neuropathic pain. There have been numerous reports of rhizotomy and ganglionectomy for a wide range of pains, both nociceptive and neuropathic, with mixed results (Patil and Campbell 2005). Given the unpredictable pain relief and complications, these procedures have been largely abandoned, though there may be a limited place in the treatment of cancer pain in patients with a short prognosis.

Dorsal root entry zone lesioning. Of all the ablative operations for the treatment of chronic pain, dorsal root entry zone lesioning, the Nashold procedure (Nashold and Ostdahl 1979), has the best evidence base for efficacy. The operation ablates the dorsal root entry zone, including the dorsal horn containing the cells on which the majority of small afferent fibres terminate, and the more deeply situated wide dynamic range neurons, abnormal activity of which is an important component of central sensitization, and likely to contribute to the development of neuropathic pain. In deafferentation states, typified by brachial plexus avulsion, bursting activity of dorsal horn neurons develops in the deafferented spinal segments (Loeser and Ward 1967). Ablation of these cells, either by surgical section or by radio-frequency heat lesioning relieves pain in many patients with brachial plexus avulsion. The extent of lesioning is not easy to control, and other spinal cord structures are sometimes affected, including long tracts. This may lead to pyramidal deficits and impaired bladder function (Friedman and Bullitt 1988).

Indications for dorsal root entry zone lesioning include neuropathic pain, unresponsive to all medical measures, due to severe brachial plexus lesions, particularly avulsion and malignant pain, for example due to Pancoast tumours. In spinal cord injury, lesioning may relieve segmental pain at the upper extent of the lesion, but not myelopathic pain below the level of the lesion. Dorsal root entry zone lesioning has been used with variable success for amputation pain and for postherpetic neuralgia, for which it appears to be most effective when paroxysmal pain, allodynia, and hyperalgesia are prominent features. The procedure has also been used for the treatment of pain associated with disabling hyperspastic states (Patil and Campbell 2005).

Anterolateral cordotomy. Interruption of the spinothalamic tract in the cervical cord leads to contralateral loss of pain and temperature sensation. The only indication for the procedure is intractable cancer pain, though with improved methods of pain control it is now performed infrequently. It is contraindicated for the treatment of chronic pain of non-malignant origin for two reasons. First, because of the limited duration of analgesia of not more than 2–3 years (Nathan 1963), indicating a remarkable plasticity of the central nervous system. And second, because of the development of central neuropathic pain due to the surgical lesion itself, after months or years.

Anterolateral cordotomy performed at open operation was superseded by percutaneous radio-frequency heat lesioning at C1 and 2 (Rosomoff *et al.* 1965), a more acceptable procedure in terminally ill patients. Complications include Horner's syndrome, mild pyramidal deficit, ataxia, and paraesthesiae. Bladder, sexual, and respiratory problems are more likely with bilateral cordotomies. Lipton (1989) reviewed a series of 300 cordotomies; 75 per cent of patients had complete pain relief and a further 8 per cent had partial relief. Transient weakness was common, but persistent at 1 month post-procedure in only 2 per cent. Lahuerta *et al.* (1985) reported complete pain relief in 64 per cent of patients and partial relief in 23 per cent . Mortality was 6 per cent in this series, due to respiratory complications.

Midline myelotomy. The rationale of midline myelotomy in the treatment of chronic pain is interruption of decussating spinothalamic fibres as they cross the midline in the anterior white commissure of the cord to form the anterolateral spinothalamic tract. The operation is performed at about three segmental levels above the level of the pain. Lesions produce bilateral hypoalgesia just below the level of the myelotomy (Sourek 1977). Intended to be highly selective, the adverse effects of this procedure can include sensory loss other than of spinothalamic type, paraesthesiae, ataxia, weakness, and sphincter disturbance. The operation has been recommended for bilateral, centrally situated pain of abdominal or pelvic visceral origin, usually due to malignant disease. It is not now commonly performed, partly because of improved medical methods of pain control in palliative cancer care, and partly because of the associated morbidity. However, recent case reports and a review indicate that the procedure probably still has a limited place (Hwang *et al.* 2004).

Mesencephalotomy. The rationale for making surgical lesions of ascending pathways in the midbrain is that quintothalamic fibres

from the face and projections from the lower brainstem reticular formation to the thalamus can be interrupted, together with spinothalamic tract fibres. Mesencephalotomy has been performed for unilateral or bilateral pain caused by cancer of the head and neck, though there are reports of its use for a wide variety of intractable pains, including brachial plexus avulsion and other neuropathic pains (Nashold *et al.* 1977). The treatment target is extremely small: the spinothalamic tract occupies an area of about 0.65 mm^2 and the medially placed quintothalamic tract is smaller. Stimulation prior to lesioning is thus advised (Nashold *et al.* 1977). Despite this, mesencephalotomy is frequently associated with problems. Lasting analgesia is achieved in only 30–50 per cent of patients, mortality ranges from 3 to 10 per cent and morbidity is as high as 37 per cent (Nashold *et al.* 1977). Complications include ocular palsies, nystagmus, disabling contralateral dysaesthesiae, and occasionally contralateral hemiparesis. The operation is more effective for nociceptive than neuropathic pains (Tasker 1990). It is now rarely performed.

Thalamotomy and other supratentorial targets. Numerous supratentorial targets for ablative procedures have been proposed over many years, but evidence of efficacy is weak and these operations are now rarely performed. As brain lesioning has declined, so neuroaugmentation by brain stimulation has increased.

The thalamic region in which ablative surgery is most likely to produce selective analgesia is the intralaminar group of nuclei. Lesions at this site minimize the risk of accompanying loss of tactile and proprioceptive sensation. In a large systematic review of medial thalamotomy, Tasker (1990) found an overall pain relief rate of nociceptive pains of up to 57 per cent , but with a 50 per cent recurrence rate of pain. For neuropathic pains, up to 67 per cent of patients obtained some degree of pain relief. Complications occurred in up to 20 per cent of patients and included confusion, dysphasia, other cognitive deficits, ocular palsies, and dysaesthesiae. Stimulation and physiological recordings undertaken in awake patients during these procedures have provided important psychophysical insights into the perception of chronic pain (Gybels and Tasker 1999, 2005).

Other brain areas lesioned for intractable pain include the dorsomedian nucleus, which projects to the cingulum, frontal lobes, and limbic system, and the medial and lateral pulvinar nuclei. Both produce only transient analgesia. Lesions of the frontothalamic connections and the frontal lobes themselves produces analgesia, but at the expense of a change in personality, albeit mild (Hitchcock 1977).

References

Abuaisha BB, Costanzi JB, Boulton AJ (1998). Acupuncture for the treatment of chronic painful peripheral diabetic neuropathy: a long-term study. *Diabetes Res Clin Pract*, **39**, 115–21.

Andersen G, Vestergaard K, Ingeman-Nielsen M *et al.* (1995). Incidence of central post-stroke pain. *Pain*, **61**, 187–93.

Arner S (1991). Intravenous phentolamine test: diagnostic and prognostic use in reflex sympathetic dystrophy. *Pain*, **46**, 17–22.

Barlas P, Lundeberg T (2005). Transcutaneous electrical nerve stimulation and acupuncture. In McMahon SB, Koltzenburg M, eds. *Wall and Melzack's Textbook of Pain*, Chapter 38, 5th edition, pp. 583–90. Elsevier, Amsterdam.

Baron R (2005). Complex regional pain syndromes. In McMahon SB, Koltzenburg M, eds. *Wall and Melzack's Textbook of Pain*, Chapter 64, 5th edition, pp. 1011–27. Elsevier, Amsterdam.

Baron R, Maier C (1996). Reflex sympathetic dystrophy: skin blood flow, vasoconstrictor reflexes and pain before and after surgical sympathectomy. *Pain*, **67**, 317–26.

Baron R, Levine JD, Fields HL (1999). Causalgia and reflex sympathetic dystrophy: does the sympathetic nervous system contribute to the generation of pain? *Muscle Nerve*, **22**, 678–95.

Baron R, Schattschneider J, Binder A *et al.* (2002). Relation between sympathetic vasoconstrictor activity and pain and hyperalgesia in complex regional pain syndromes: a case-control study. *Lancet*, **359**, 1655–60.

Beck AT, Ward CH, Mendelson M *et al.* (1961). An inventory for measuring depression. *Arch Gen Psychiatry*, **4**, 561–71.

Becker WJ, Ablett DP, Harris CJ *et al.* (1995). Long term treatment of intractable reflex sympathetic dystrophy with intrathecal morphine. *Can J Neurol* Sci, **22**, 153–9.

Behan RJ (1914). *Pain: its Origin, Conduction, Perception and Diagnostic Significance*, pp. 198–203. V Appleton, New York.

Bendok B, Levy RM (1998). Brain stimulation for persistent pain management. In GildenbergPL, TaskerRR, eds. *Textbook of Stereotactic and Functional Neurosurgery*, p. 1539. McGraw-Hill, New York.

Bennett M (2001). The LANSS Pain Scale: the Leeds assessment of neuropathic symptoms and signs. *Pain*, **92**, 147–57.

Bentley JB, Hammeroff SR (1980). Diffuse reflex sympathetic dystrophy. *Anaesthesiology*, **53**, 256–7.

Bergner M, Bobbitt RA, Carter WB *et al.* (1981). The sickness impact profile: development and final revision of a health status measure. *Med Care*, **19**, 787–805.

Berman JS, Symonds C, Birch R (2004). Efficacy of two cannabis based medicinal extracts for relief of central neuropathic pain from brachial plexus avulsion: results of a randomised controlled trial. *Pain* **112**, 299–306.

Blom S. (1963). Tic douloureux treated with a new anticonvulsant: experiences with G 32883. *Arch Neurol*, **30**, 285–90.

Blumberg H, Janig W (1994). Clinical manifestation of reflex sympathetic dystrophy and sympathetically maintained pain. In Wall PD, Melzack R, ed. *Textbook of Pain*, 3rd edition, pp. 685–98. Churchill Livingstone, Edinburgh.

Bickerstaff DR, Kanis JA (1994). Algodystrophy: an under-recognised complication of minor trauma. *Brit J Rheumatol*, **33**, 240–8.

Birklein F, Sittle R, Spitzer A *et al.* (1997). Sudomotor function in sympathetic reflex dystrophy. *Pain*, **69**, 49–54.

Boas RA (1996). Complex regional pain syndromes: symptoms, signs and differential diagnosis. In Stanton-Hicks M, Janig W, eds. *Reflex Sympathetic Dystrophy: a Reappraisal. Progress in Pain Research and Management*, Vol. 6, pp. 79–92. IASP Press, Seattle.

Boivie J (2005) Central pain. In McMahon SB, Koltzenburg M, eds. *Wall and Melzack's Textbook of Pain*, Chapter 67, 5th edition, pp. 1057–74. Elsevier, Amsterdam.

Bonica JJ (1990). Causalgia and other reflex sympathetic dystrophies. In Bonica JJ, ed. *The Management of Pain*, 2nd edition, pp. 230–43. Lea and Febiger, Philadelphia.

Bonica JJ (1991). Introduction: semantic, epidemiologic, and educational issues. In *Pain and Central Nervous System Disease. The Central Pain Syndromes*, pp. 13–30. Raven Press, New York.

Boureau F, Doubrere JF, Luu M (1990). Study of verbal description in neuropathic pain. *Pain*, **42**, 145–52.

Breivik H (2005). Local anaesthetic blocks and epidurals. In McMahon SB, Koltzenburg M, eds. *Wall and Melzack's Textbook of Pain*, Chapter 33, 5th edition, pp. 507–20. Elsevier, Philadelphia.

Brown JA (2004). Motor cortex stimulation. *Seminars in Neurosurgery*, **15**, 177–82.

Bruckbauer HR, Preac Mursic V, Herzer P *et al.* (1997). Sudeck's atrophy in Lyme borreliosis. *Infection*, **25**, 372–6.

Capsaicin Study Group (1991). Treatment of painful diabetic neuropathy with topical capsaicin. A multicenter, double-blind, vehicle-controlled study. The Capsaicin Study Group. *Arch Intern Med*, **151**, 2225–9.

Chabal C, Jacobson L, Mariano A *et al.* (1992). The use of oral mexiletine for the treatment of pain after nerve injury. *Anesthesiology*, **76**, 513–7.

Choi B, Rowbotham MC (1997). Effects of adrenergic receptor activation on post-herpetic neuralgia pain and sensory disturbances. *Pain*, **69**, 55–63.

Christensen K, Jensen EM, Noer I (1982). The reflex sympathetic dystrophy response to treatment with systemic corticosteroids. *Acta Chir Scand*, **148**, 653–5.

Clifford DB, Trotter JL (1984). Pain in multiple sclerosis. *Arch Neurol*, **41**, 1270–2.

Covington EC (1996). Psychological issues in reflex sympathetic dystrophy. In Stanton-Hicks M, Janig W, eds. Reflex sympathetic dystrophy: a reappraisal. *Prog. Pain Res. and Manage.* Vol. 6, pp. 191–216. IASP Press, Seattle.

Daousi C, MacFarlane IA, Woodward A *et al.* (2004). Chronic painful peripheral neuropathy in an urban community: a controlled comparison of people with and without diabetes. *Diabet Med*, **21**, 976–82.

Davis R, Lentini R (1975). Transcutaneous nerve stimulation for treatment of pain in patients with spinal cord injury. *Surg Neurol*, **1**, 100–1.

Dejgard A, Petersen P, Kastrup L (1988). Mexiletine for treatment of chronic painful diabetic neuropathy. *Lancet*, **1**, 9–11.

Devor M (2005). Response of nerves to injury in relation to pain. In McMahon SB, Koltzenburg M, ed. *Wall and Melzack's Textbook of Pain*, Chapter 58, 5th edition, pp. 905–28. Elsevier, Amsterdam.

Dickenson AH, Sullivan AF (1987). Evidence for a role of the NMDA receptor in the frequency dependent potentiation of rat dorsal horn nociceptive neurones following C fibre stimulation. *Neuropharmacology*, **26**, 1235–8.

Drummond PD (1995). Noradrenaline increases hyperalgesia to heat in skin sensitised by capsaicin. *Pain*, **60**, 311–5.

Drummond PD, Skipworth S, Finch PM (1996). Alpha 1-adrenoceptors in normal and hyperalgesic human skin. *Clin Sci*, **91**, 73–7.

Dworkin RH, Corbin AE, Young JP *et al.* (2003). Pregabalin for the treatment of postherpetic neuralgia: a randomized placebo-controlled trial. *Neurology*, **60**, 1274–83.

Eccleston C, Williams A, Morley S (2003). Cognitive-behaviour therapy for chronic pain in adults. In Jensen TS, Wilson PR, Rice ASC, eds. *Clinical Pain Management: Chronic Pain*, Chapter 25, pp. 325–34. Arnold, London.

Eklund G, Hagbarth KE (1965). Motor effects of vibratory muscle stimuli in man. *Electroencephalogr Clin Neurophysiol*, **19**, 619.

Eisenberg E, Lurie Y, Braker C *et al.* (2001). Lamotrigine reduces painful diabetic neuropathy: a randomised controlled trial. *Neurology*, **57**, 505–9.

Ellison N, Loprinzi CL, Kugler J *et al.* (1997). Phase III placebo controlled trial of capsaicin cream in the management of surgical neuropathic pain in cancer patients. *J Clin Oncol*, **15**, 2974–80.

Evans JA (1946). Reflex sympathetic dystrophy. *J Surg Gynaecol Obstet*, **82**, 36–43.

Fairbank JCT, Couper J, Davies JB *et al.* (1980). The Oswestry low back pain disability questionnaire. *Physiotherapy*, **66**, 271–3.

Fields HL, Basbaum AI (1978). Brain stem control of spinal pain transmission neurons. *Annu Rev Physiol*, **40**, 193–221.

Finnerup NB, Sindrup SH, Bach FW *et al.* (2002). Lamotrigine in spinal cord injury pain: a randomized controlled trial. *Pain*, **96**, 375–83.

Finnerup NB, Otto M, McQuay HJ *et al.* (2005). Algorithm for neuropathic pain treatment: an evidence based proposal. *Pain*, **118**, 289–305.

Fishbain DA (1995). DSM-IV: implications and issues for the pain clinician. American Pain Society Bulletin, 6–18.

Fishbain DA, Cutler R, Rosomoff HL *et al.* (1997). Chronic pain-associated depression: antecedent or consequence of chronic pain: a review. *Clin J Pain*, **13**, 116–37.

Fishbain DA, Steele-Rosomoff R, Rosomoff HL (1992). Drug abuse, dependence, and addiction in chronic pain patients. *Clin J Pain*, **8**, 77–85.

Friedman AH, Bullitt E (1988). Dorsal root entry zone lesions in the treatment of pain following brachial plexus avulsion, spinal cord injury and herpes zoster. *Appl Neurophysiol*, **51**, 164–9.

Fukomoto M, Ushida T, Zinchuk VS *et al.* (1999). Contralateral thalamic perfusion in patients with reflex sympathetic dystrophy syndrome. *Lancet*, **354**, 1790–1.

Galer BS, Jensen MP (1997). Development and preliminary validation of a pain measure specific to neuropathic pain: the neuropathic pain scale. *Neurology*, **48**, 332–8.

Galer BS, Rowbotham MC, Perander J *et al.* (1999). Topical lidocaine patch relieves postherpetic neuralgia more effectively than a vehicle topical patch: results of an enriched enrolment study. *Pain*, **80**, 533–8.

Galer BS, Jensen MP, Ma T *et al.* (2002). The lidocaine patch 5 per cent effectively treats all neuropathic pain qualities: results of a randomized, double-blind, vehicle-controlled, 3-week efficacy study with use of the neuropathic pain scale. *Clin J Pain*, **18**, 297–301.

Geertzen JH, de Bruijn H, de Bruijn-Kofman AT *et al.* (1994). Reflex sympathetic dystrophy: early treatment and psychological aspects. *Arch Phys Med Rehabil*, **75**, 442–6.

Gerhart KD, Yeziersky RP, Fang ZR *et al.* (1983). Inhibition of primate spinothalamic tract neurons by stimulation in ventral posterior lateral (VPL) thalamic nucleus: possible mechanisms. *J Neurophysiol*, **49**, 406–23.

Gilbert RW, Kim JH, Posner JB (1978). Epidural spinal cord compression from metastatic tumor: diagnosis and treatment. *Ann Neurol*, **3**, 40.

Gobelet C, Waldburger M, Meier JL (1992). The effect of adding calcitonin to physical treatment on reflex sympathetic dystrophy. *Pain*, **48**, 171–5.

Goebel A (2001). Screening of patients with complex regional pain syndrome for antecedent infections. *Clin J Pain*, **17**, 378–9.

Goebel A, Netal S, Schedel R *et al.* (2002). Human pooled immunoglobulin in the treatment of chronic pain syndromes. *Pain Med*, **3**, 119–27.

Goebel A, Vogel H, Caneris O *et al.* (2005). Immune responses to Campylobacter and serum antibodies in patients with complex regional pain syndrome. *J Neuroimmunol*, **162**, 184–9.

Goldsmith DP, Vivino FB, Athreya BH *et al.* (1989). Nuclear imaging and clinical features of childhood reflex neurovascular dystrophy: comparison with adults. *Arthritis Rheum*, **32**, 480–5.

Gottrup H, Nielsen J, Arendt-Nielsen L *et al.* (1998). The relationship between sensory thresholds and mechanical hyperalgesia in nerve injury. *Pain*, **75**, 321–9.

Gureje O, VonKorff M, Sim GE *et al.* (1998). Persistent pain and well-being: a World Health Organisation study in primary care. *J Am Med Assoc*, **280**, 147–51.

Gybels J, Erdine S, Maeyaert J *et al.* (1998). Neuromodulation of pain: a consensus statement prepared in Brussels 16–18 January 1998 by the task force of the European Federation of IASP Chapters (EFIC). *Eur J Pain*, **2**, 203–9.

Gybels JM, Tasker RR (1999). Central neurosurgery. In Wall PD, Melzack R, eds. *Textbook of Pain*, Chapter 57, 4th edition, pp. 1307–40. Churchill Livingstone, Edinburgh.

Gybels JM, Tasker RR (2005). Supratentorial neurosurgery for the treatment of pain. In McMahon SB, Koltzenburg M, eds. *Wall and Melzack's Textbook of Pain*, Chapter 36, 5th edition, pp. 553–561. Elsevier, Amsterdam.

Han JS (2003). Acupuncture: neuropeptide release produced by electrical stimulation of different frequencies. *Trends Neurosci*, **26**, 17–22.

Han JS, Terenius L (1982). Neurochemical basis of acupuncture analgesia. *Annu Rev Pharmacol Toxicol*, **22**, 193–220.

Hannington-Kiff JG (1974). Intravenous regional sympathetic block with guanethidine. *Lancet*, **1**, 1019–20.

Hansson P, Lundeberg T (1999). Transcutaneous electrical nerve stimulation, vibration and acupuncture as pain-relieving measures. In Wall PD, Melzack R, eds. *Textbook of Pain*, Chapter 58, 4th edition, pp. 1341–52. Churchill Livingstone, Edinburgh.

Hitchcock ER (1977). Small frontal lesions for intractable pain. In Krayenbul H, Maspes PE, Sweet WH, eds. *Progress in Neurological Surgery*, Vol. 8, pp. 114–131. Karger, Basel.

Hosobuchi Y (1986). Subcortical electrical stimulation for control of intractable pain in humans. Report of 122 cases (1970-1984). *J Neurosurg*, **64**, 543–53.

Huygen FJ, de Bruijn AG, Bruijn MT *et al.* (2002). Evidence for local inflammation in complex regional pain syndrome type 1. *Mediators Inflamm*, **11**, 47–51.

Hwang SL, Lin CL, Lieu A *et al.* (2004). Punctate midline myelotomy for intractable visceral pain caused by hepatobiliary or pancreatic cancer. *J Pain Symptom Manage*, **27**, 79–84.

IASP (1994). Classification of chronic pain. In Merskey H, Bogduk N, eds. 2nd edition. IASP Press, Seattle.

Jadad AR, Carroll D, Glynn CL *et al.* (1995). Intravenous regional sympathetic blockade for pain relief in reflex sympathetic dystrophy:a systematic review and a randomised, double-blind crossover study. *J Pain Symptom Manage*, **10**, 13–20.

Jensen TS, Gottrup H (2003). Assessment of neuropathic pain. In Jensen TS, Wilson PR, Rice ASC, eds. *Clinical Pain Management: Chronic Pain*, pp. 113–24. Arnold, London.

Jensen TS, Lenz FA (1995). Central post-stroke pain: a challenge for the scientist and clinician. *Pain*, **61**, 161–4.

Johnson JE, Johnson KA, Unni KK *et al.* (1988). Persistent pain after excision of an interdigital neuroma. Results of reoperation. *J Bone Joint Surg*, **70**, 651–7.

Kane K, Taub A (1975). A history of local electrical analgesia. *Pain*, **1**, 125–38.

Karst M, Salim K, Burstein S *et al.* (2003). Analgesic effect of the s ynthetic cannabinoid CT-3 on chronic neuropathic pain. *JAMA*, **290**, 1757–62.

Kassirer MR, Osterberg DH (1987). Pain in chronic multiple sclerosis. *J Pain Symptom Manage*, **2**, 95–7.

Kastrup J, Petersen P, Dejgerd A *et al.* (1987). Intravenous lignocaine infusion –n a new treatment of chronic painful diabetic neuropathy? *Pain*, **28**, 69–75.

Katon W (1984). Panic disorder and somatization: review of 55 cases. *Am J Med*, **77**, 101–106.

Kemler MA, Barendse GA, van Kleef M *et al.* (2000). Spinal cord stimulation in patients with chronic reflex sympathetic dystrophy. *N Eng J Med*, **343**, 618–24.

Kerns RD, Turk DC, Rudy TE (1985). The West Haven-Yale Multidimensional Pain Inventory (WHYMPI). *Pain*, **23**, 345–56.

Kilo S, Schmelz M, Koltzenburg M *et al.* (1994). Different patterns of hyperalgesia induced by experimental inflammation in human skin. *Brain*, **117**, 385–96.

Kingery WS (1997). A critical review of controlled clinical trials for peripheral neuropathic pain and complex regional pain syndromes. *Pain*, **73**, 123–39.

Kinman E, Nygards EB, Hausson P (1997). Peripheral alpha-adrenoreceptors are involved in the development of capsaicin induced ongoing and stimulus evoked pain in humans. *Pain*, **69**, 79–85.

Koltzenburg M, Torebjörk HE, Wahren LK (1994). Nocicetor modulated central sensitization causes mechanical hyperalgesia in acute chemogenic and neuropathic pain. *Brain*, **117**, 579–91.

Koltzenburg M (1996). *Afferent mechanisms mediating pain and hyperalgesia in neuralgia.* In Janig W, Stanton-Hicks M eds. Reflex sympathetic dystrophy: a reappraisal pp. 123–150. IASP Press, Seattle.

Kumar K, Toth C, Nath RK (1997). Deep brain stimulation for intractable pain: a 15-year experience. *Neurosurgery*, **40**, 736–47.

Kupers R, Gybels J (1993). Electrical stimulation of the ventroposterolateral thalamic nucleus (VPL) reduces mechanical allodynia in a rat model of neuropathic pain. *Neurosci Lett*, **150**, 95–8.

Lahuerta J, Lipton S, Wells JCD (1985). Percutaneous cervical cordotomy: results and complications in a recent series of 100 patients. *Ann R Coll Surg Eng*, **67**, 41–4.

Lamotte RH, Shain CN, Simone DA *et al.* (1991). Neurogenic hyperalgesia: psychophysical studies of underlying mechanisms. *J Neurophysiol*, **66**, 190–211.

Lance JW, Curran DA (1964). Treatment of chronic tension headache. *Lancet*, **42**, 1236–9.

Langohr HD, Stohr M, Petruch F (1982). An open and double-blind crossover study of clomipramine (Anafranil) in patients with painful mono- and polyneuropathies. *Eur Neurol*, **21**, 309–17.

Lascelles RG (1966). Atypical facial pain and depression. *Br J Psychiatry*, **112**, 651–9.

Laursen RJ, Graven-Nielsen T, Jensen TS *et al.* (1997). Referred pain is dependent on sensory input from the periphery: a psycho-physical study. *Eur J Pain*, **1**, 261–9.

Lee BH, Scharff L, Sethna NF *et al.* (2002). Physical therapy and cognitive-behavioural treatment for complex regional pain syndromes. *J Pediatr*, **141**, 135–40.

Leijon G, Boivie J (1989). Central post-stroke pain – the effect of high and low frequency TENS. *Pain,* **38**, 187–191.

Leriche R (1916). De la causalgie envisagee comme au nevrite de sympathique et de son traitement par la denudation et l'excision des plexus nerveaux peri-arteriels. *Presse Med*, **24**, 177–80.

Leriche R (1939). The *Surgery of Pain* (translated by A Young). Bailliere, Tindall and Cox, London.

Linderoth B, Foreman RD (1999). Physiology of spinal cord stimulation: review and update. *Neuromodulation*, **2**, 150–64.

Lipton S (1989). Percutaneous cordotomy. In Wall PD, Melzack R, eds. *Textbook of Pain*, 2nd edition, pp. 832–9. Churchill Livingstone, Edinburgh.

Livingston WK (1943). *Pain Mechanisms*. MacMillan Co, New York.

Livingstone JA, Atkins RM (2002). Intravenous regional guanethidine blockade in the treatment of post-traumatic complex regional pain syndrome type 1 (algodystrophy) of the hand. *J Bone Joint Surg*, **84**, 380–6.

Locher W (1999). Valproate: a reappraisal of its pharmacodynamic properties and mechanisms of action. *Prog Neurobiol*, **58**, 31–59.

Loeser JD, Ward AA (1967). Some effects of deafferentation on neurons of the cat spinal cord. *Arch Neurol*, **17**, 629–36.

Lu GD, Needham J (1980). *Celestial Lancets: a History and Rationale of Acupuncture and Moxa*, Cambridge University Press, Cambridge.

Lundeberg T (1985). Relief of pain from a phantom limb by peripheral stimulation. *J Neurol*, **232**, 79–82.

Maihofner C, Handwerker HO, Neundorfer B *et al.* (2003). Patterns of cortical reorganization in complex regional pain syndrome. *Neurology*, **61**, 1707–15.

Maihofner C, Handwerker HO, Neundorfer B *et al.* (2004). Cortical reorganization during recovery from complex regional pain syndrome. *Neurology*, **24**, 693–701.

Manicourt DH, Brasseur JP, Boutsen Y *et al.* (2004). Role of alendronate in therapy for posttraumatic complex regional pain syndrome type 1 of the lower extremity. *Arthritis Rheum*, **50**, 3690–7.

Martyn C, Hughes RAC (1998). Peripheral neuropathies. In Martyn CN, Hughes RAC, eds. *The Epidemiology of Neurological Disorders*, pp. 96–117. BMJ Books, London.

Margolis LH, Gianascol AJ (1956). Chlorpromazine in thalamic pain syndrome. *Neurology*, **6**, 302–4.

Max MB, Lynch SA, Muir J *et al.* (1992). Effects of desipramine, amitriptyline, and fluoxetine on pain in diabetic neuropathy. *N Engl J Med*, **326**, 1250–6.

Mazars GJ, Merriene L, Cioloca C (1979). Comparative study of electrical stimulation of posterior thalamic nuclei, periaqueductal grey and other midline mesencephalic structures in man. In Bonica JJ, Liebeskind J, Albe-Fessard D, eds. *Adv Pain Res Ther*, Vol. 3, pp. 541–6. Raven Press, New York.

McMahon SB, Koltzenburg M. eds. (2005). *Wall and Melzack's Textbook of Pain*, 5th edition, Elsevier, Amsterdam.

McQuay H, Carroll D, Jadad AR, Wiffen P, Moore A (1995). Anticonvulsant drugs for management of pain: a systematic review. *Brit Med J*, **311**, 1047–52.

McQuay HJ, Moore RA (1998). *An Evidence-based Resource for Pain Relief*, Oxford University Press, Oxford.

Melchart D, Linde K, Fischer P *et al.* (2003). Acupuncture for idiopathic headache (Cochrane review). In the Cochrane Library, issue 3,. Update Software, Oxford.

Mellick GA, Mellick LB (1995). Gabapentin in the management of reflex sympathetic dystrophy. *J Pain Symptom Manage*, **10**, 265–6.

Melzack R (1975). The McGill pain questionnaire: major properties and scoring methods. *Pain*, **1**, 277–99.

Melzack R, Katz J (1999). Pain measurement in persons in pain. In Wall PD, Melzack R, eds. *Textbook of Pain*, Chapter 17, 4th edition, pp. 409–26. Churchill Livingstone, Edinburgh.

Melzack R, Wall PD (1965). Pain mechanisms: a new theory. *Science*, **150**, 971–9.

Mendell LM, Wall PD (1965). Responses of single dorsal horn cells to peripheral cutaneous unmyelinated fibres. *Nature*, **206**, 97–9.

Meyer-Rosberg K, Kvarnstrom A, Kinnman E *et al.* (2001). Peripheral neuropathic pain - a multidimensional burden for patients. *Eur J Pain*, **5**, 379–89.

Mitchell SW, Morehouse GR, Keen WW (1864). *Gunshot Wounds and Other Injuries of Nerves*, Lippincott, Philadelphia.

Monks R, Merskey H (1999). Psychotropic drugs. In Wall PD, Melzack R eds. *Textbook of Pain*, Chapter 50, 4th edition, pp. 1155–86. Churchill Livingstone, Edinburgh.

Morley S, Eccleston C, Williams AcdeC (1999). Systematic review and meta-analysis of randomized controlled trials of cognitive behaviour therapy for chronic pain in adults, excluding headache. *Pain*, **80**, 1–13.

Moulin DE, Foley KM, Ebers GC (1988). Pain syndromes in multiple sclerosis. *Neurology*, **38**, 1830–1834.

Nashold BS, Ostdahl RH (1979). Dorsal root entry zone lesions for pain relief. *J Neurosurg*, **51**, 59–69.

Nashold BS, Slaughter DG, Wilson WP *et al.* (1977). Stereotactic mesencephalotomy. In Krayenbuhl H, Maspses PE, Sweet WH, eds. *Prog Neurol Surg*, Vol. 5, pp. 35–49. Karger, Basel.

Nathan PW (1963). Results of anterolateral cordotomy for pain in cancer. *J Neurol Neurosurg Psychiatry*, **26**, 353–62.

Nelson KA, Park KM, Robinovitz C *et al.* (1997). High-dose dextromethorphan versus placebo in painful diabetic neuropathy and postherpetic neuralgia. *Neurology*, **48**, 1212–8.

Nguyen J-P, Lefaucheur JP, Decq P *et al.* (2003). Motor cortex stimulation. In Simpson BA, ed. *Electrical Stimulation and the Relief of Pain*, Chapter 13. Elsevier, Amsterdam.

North RB, Ewend MG, Lawton MT *et al.* (1991). Spinal cord stimulation for chronic intractable pain:superiority of "multi-channel" devices. *Pain*, **44**, 119–30.

North RB, Kidd DH, Piantadosi SA (2005). Spinal cord stimulation versus repeat lumbosacral spine surgery for chronic pain: a randomized controlled trial. *Neurosurgery*, **56**, 98–107.

Ochoa JL, Verdugo RJ, Campero M (1994). Pathophysiological spectrum of organic and psychogenic disorders in neuropathic pain patients fitting the description of causalgia or reflex sympathetic dystrophy. *Progress in Pain Research and Management*, **2**, 483–94.

Ochoa JL, Verdugo RJ (1995). Reflex sympathetic dystrophy. A common clinical avenue for somatoform expression. *Neurol Clin*, **13**, 351–63.

Oerlmans HM, Oostendorp RA, de Boo T *et al.* (2000) Adjuvant physical therapy versus occupational therapy in patients with reflex sympathetic dystrophy/complex regional pain syndrome type 1. *Arch Phys Med Rehabil*, **81**, 49–56.

Olausson H, Lamarre Y, Backlund H *et al.* (2002). Unmyelinated tactile afferents signal touch and project to insular cortex. *Nat Neurosci*, **5**, 900–4.

Oyen WJ, Arntz IE, Claessens RM *et al.* (1993). Reflex sympathetic dystrophy of the hand: an excessive inflammatory response? *Pain*, **55**, 151–7.

Paice JA, Ferrans CE, Lashley FR *et al.* (2000). Topical capsaicin in the management of HIV-associated peripheral neuropathy. *J Pain Symptom Manage*, **19**, 45–52.

Patil PG, Campbell JN (2005). Peripheral and central nervous system surgery for pain. In McMahon SB, Koltzenburg M, eds. *Wall and Melzack's Textbook of Pain*, Chapter 39, 5th edition, pp. 591–602. Elsevier, Amsterdam.

Perez RS, Kwakkel G, Zuurmond WW *et al.* (2001). Treatment of reflex sympathetic dystrophy (CRPS type 1): a research synthesis of 21 randomised clinical trials. *J Pain Symptom Manage*, **21**, 511–26.

Perez RS, Zuurmond WW, Bezemer PD *et al.* (2003). The treatment of complex regional pain syndrome type 1with free radical scavengers: a randomized controlled study. *Pain*, **102**, 297–307.

Pomeranz B, Chiu D (1976). Naloxone blockage of acupuncture analgesia: endorphin implicated. *Life Sci*, **19**, 1757–62.

Portenoy RK (1990). Chronic opioid therapy in non-malignant pain. *J Pain Symptom Manage*, **5** (Suppl), S46–62.

Price DD, Long S, Wilsey B *et al.* (1998). Analysis of peak magnitude and duration of analgesia produced by local anesthetics injected into sympathetic ganglia of complex regional pain syndrome patients. *Clin J Pain*, **14**, 216–26.

Ramamurthy S, Hoffman J (1995). Intravenous regional guanethidine in the treatment of reflex sympathetic dystrophy/causalgia: a randomised, double –blind study. *Anesth Analg*, **81**, 718–23.

Rang HP, Dale MM, Ritter JM (2001). Analgesic drugs. In *Pharmacology*, 4th edition, pp. 579–603. Churchill Livingstone, Edinburgh.

Raskin P, Donofrio PD, Rosenthal NR *et al.* CAPSS-141 Study Group (2004). Topiramate versus placebo in painful diabetic neuropathy: analgesic and metabolic effects. *Neurology*, **63**, 865–873.

Rasmussen PV, Sindrup SH, Jensen TS *et al.* (2004). Symptoms and signs in patients with suspected neuropathic pain. *Pain*, **110**, 461–9.

Rauck RL, Eisenach JC, Jackson K *et al.* (1993). Epidural clonidine treatment for refractory reflex sympathetic dystrophy. *Anesthesiology*, **79**, 1163–9.

Reynolds DV (1969). Surgery in the rat during electrical analgesia induced by focal brain stimulation. *Science*, **164**, 444–5.

Richards R (1967). Causalgia. A centennial review. *Arch Neurol*, **16**, 339–50.

Rice ASC (2005). Cannabinoids. In McMahon SB, Koltzenburg M, eds. *Wall and Melzack's Textbook of Pain*, Chapter 34, 5th edition, pp. 521–40. Elsevier, Amsterdam.

Richardson DE (1995). Deep brain stimulation for the relief of chronic pain. *Neurosurg Clin N Am*, **6**, 135–44.

Richter RW, Portenoy R, Sharma U, *et al.* (2005). Relief of painful diabetic neuropathy with pregabalin: a randomized, placebo-controlled trial. *J Pain*, **4**, 253–260.

Riddoch G (1938). The clinical features of central pain. *Lancet*, **234**, 1093–98; 1150–6; 1205–9.

Roberts MHT, Rees H (1994). Physiological basis of spinal cord stimulation. *Pain Rev*, **1**, 184–98.

Romano JM, Turner JA (1985). Chronic pain and depression: does the evidence support a relationship? *Psychol Bull*, **97**, 18–34.

Rosenstock J, Tuchmann M, LaMoreaux L *et al.* (2004). Pregabalin for the treatment of painful diabetic neuropathy: a double-blind placebo-controlled trial. *Pain*, **110**, 628–38.

Rosomoff HL, Caroll F, Brown J *et al.* (1965). Percutaneous radiofrequency cervical cordotomy technique. *J Neurosurg*, **23**, 639–44.

Sabatowski R, Galvez R, Cherry DA *et al.* (2004). Pregabalin reduces pain and improves sleep and mood disturbances in patients with post-herpetic neuralgia: results of a randomised, placebo-controlled clinical trial. *Pain*, **109**, 26–35.

Sadove MS, Rose RF, Balagot RC, Reyes R (1955). Chlorpromazine in the management of pain. *Mod Med*, **23**, 117–20.

Sang CN (2000). NMDA-receptor antagonists in neuropathic pain: experimental methods to clinical trials. *J Pain Symptom Manage*, **19**, S21–5.

Sang CN, Booher S, Gilron I *et al.* (2002). Dextromethorphan and memantine in painful diabetic neuropathy and postherpetic neuralgia. Efficacy and dose-response trials. *Anesthesiology*, **96**, 1053–61.

Scadding JW (1999). Complex regional pain syndrome. In Wall PD, Melzack R, eds. *Textbook of Pain*, Chapter 36, 4th edition, pp. 835–50. Churchill Livingstone, Edinburgh.

Scadding JW, Koltzenburg M (2005). Painful peripheral neuropathies. In McMahon SB, Koltzenburg M, eds. *Wall and Melzack's Textbook of Pain*, Chapter 62, 5th edition, pp. 973–1000. Elsevier, Amsterdam.

Scadding JW, Klenerman LE (1987). Light and electron microscopic observations on Morton's neuralgia. *Pain*, **5**(Suppl 4), 246.

Shank RP, Gardocki J, Streeter A *et al.* (2000). An overview of the preclinical aspects of topiramate: pharmacology, pharmacokinetics, and mechanisms of action. *Epilepsia*, **41** (Suppl 1), S3–9.

Sherry DD, Weisman R (1988). Psychologic aspects of childhood reflex neurovascular dystrophy. *Pediatrics*, **81**, 572–8.

Siddall PJ (2005). Pain following spinal cord injury. In McMahon SB, Koltzenburg M, eds. *Wall and Melzack's Textbook of Pain*, Chapter 66, 5th edition, pp. 1043–56. Elsevier, Amsterdam.

Sigwald J, Bouttier D, Caille F (1959). [The treatment of zona and of its associated pains. Study of the results obtained with levomepromazine]. *Therapie*, **14**, 818–24. French.

Simpson BA, Meyerson BA, Linderoth B (2005). Spinal cord and brain stimulation. In McMahon SB, Koltzenburg M, eds. *Wall and Melzack's Textbook of Pain*, Chapter 37, 5th edition, pp. 563–82. Elsevier, Amsterdam.

Simpson DM, McArthur JC, Olney R *et al.* (2003). Lamotrigine for HIV-associated painful sensory neuropathies: a placebo-controlled trial. *Neurology*, **60**, 1508–14.

Sindrup SH, Jensen TS (1999). Efficacy of pharmacological treatments of neuropathic pain: an update and effect related to mechanism of drug action. *Pain*, **83**, 85–90.

Sindrup SH, Jensen TS (2000). Pharmacological treatment of pain in polyneuropathy. *Neurology*, **55**, 915–20.

Sluka KA, Walsh DM (2003). Transcutaneous electrical nerve stimulation: basic science mechanisms and clinical effectiveness. *J Pain*, **4**, 109–21.

Sourek K (1977). Mediolongitudinal myelotomy. In Krayenbuhl M, Maspes PE, Sweet WH. eds. *Progressin Neurological Surgery*, Vol. 8, pp. 201–59. Karger, Basel.

Tasker RR (1990). Management of nociceptive, deafferentation and central pain by surgical intervention. In Fields HL, ed. *Pain Syndromes in Neurology*, pp. 143–200. Butterworths, London.

Thienel U, Neto W, Schwabe SK *et al.* (2004). The topiramate Diabetic Neuropathic Pain Study Group. Topiramate in painful diabetic polyneuropathy: findings from three double-blind placebo-controlled trials. *Acta Neurologica Scandinavica*, **56**, 221–31.

Turner JA, Loeser JD, Deyo RA *et al.* (1995). Spinal cord stimulation for patients with failed back surgery syndrome or complex regional pain syndrome: a systematic review of effectiveness and complications. *Pain*, **108**, 137–47.

Ubbink DT, Spincemaille GH, Prins MH *et al.* (1999). Microcirculatory investigations to determine the effect of spinal cord stimulation for critical leg ischaemia: the Dutch multicentre randomized controlled trial. *J Vasc Surg*, **30**, 236–44.

van der Laan L, Veldman PH, Goris RJ (1998). Severe complications of reflex sympathetic dystrophy: infection, ulcers, chronic edema, dystonia, and myoclonus. *Arch Phys Med Rehabil*, **79**, 424–9.

van de Vusse AC, Stomp-van den Berg SG, Kessels AH, Weber WE (2004). Randomised controlled trial of gabapentin in complex regional pain syndrome type 1. *BMC Neurology*, **4**, 13.

van Hilten JJ, van de Beek WJ, Hoff JI *et al.* (2000). Intrathecal baclofen for the treatment of dystonia in patients with reflex sympathetic dystrophy. *N Engl J Med*, **343**, 625–30.

Van Tulder MW, Cherkin DC, Berman B *et al.* (1999). The effectiveness of acupuncture in the management of acute and chronic low back pain: a sysytematic review within the framework of the Cochrane Collaboration Back Review Group. *Spine*, **24**, 1113–23.

Vallbo AB, Hagbarth KE (1968). Activity from skin mechanoreceptors recorded percutaneously in awake human subjects. *Exp Neurol*, **21**, 270–89.

Varenna M, Zucchi F, Ghiringhelli D *et al.* (2000). Intravenous clodronate in the treatment of reflex sympathetic dystrophy syndrome. A randomized, double blind, placebo controlled syudy. *J Rheumatol*, **27**, 1477–83.

Veldman PH, Reynen HM, Arntz IE *et al.* (1993). Signs and symptoms of reflex sympathetic dystrophy: prospective study of 829 patients. *Lancet*, **342**, 1012–1016.

Vermote R, Ketelaer P, Carton H (1986). Pain in multiple sclerosis patients. *Clin Neurol and Neurosurg*, **88**, 87–93.

Vestergaard K, Nielsen J, Andersen G *et al.* (1995). Sensory abnormalities in consecutive, unselected patients with central post-stroke pain. *Pain*, **61**, 177–86.

Vestergaard K, Andersen G, Gottrup H *et al.* (2001). Lamotrigine for central poststroke pain: a randomized controlled trial. *Neurology*, **56**, 184–90.

Wade DT, Robson P, House H *et al.* (2003). A preliminary controlled study to determine whether whole-plant cannabis extracts can improve intractable neurogenic symptoms. *Clin Rehabil*, **17**, 18–26.

Waisbrod H, Panhans C, Hansen D *et al.* (1985). Direct nerve stimulation for painful peripheral neuropathies. *J Bone Joint Surg*, **67**, 470–2.

Walker AE, Nulsen F (1948). Electrical stimulation of the upper thoracic portion of the sympathetic chain in man. *Arch Neurol Psychiatry*, **59**, 559–60.

Wallin BG, Torebjork HE, Hallin RG (1976). Preliminary observations on the pathophysiology of hyperalgesia in the causalgic pain syndrome. In Zotterman Y, ed. *Sensory Function of the Skin in Primates*, pp. 489–99. Pergamon, Oxford.

Ware JE, Snow KK, Kosinski M *et al.* (1993). SF-36 Health Survey: manual and interpretation guide. Health Institute, New England Medical Center, Boston.

Watson CP, Tyler KL, Bickers DR *et al.* (1993). A randomized vehicle-controlled trial of topical capsaicin in the treatment of postherpetic neuralgia. *Clin Ther*, **15**, 510–26.

Weber M, Birklein F, Neundorfer B *et al.* (2001). Facilitated neurogenic inflammation in complex regional pain syndrome. *Pain*, **91**, 251–7.

Weiner RL, Reed KL (1999). Peripheral neurostimulation for control of intractable occipital neuralgia. *Neuromodulation*, **2**, 217–21.

White JC, Kjellberg RN (1973). Posterior spinal rhizotomy; a substitute for cordotomy in the relief of localised pain in patients with normal life expectancy. *Neurochirurgia*, **16**, 141–70.

Wilder RT (1996). Reflex sympathetic dystrophy in children and adolescents: differences from adults. In Janig W, Stanton-Hicks M, eds. *Reflex Sympathetic Dystrophy: a Reappraisal. Progress in Pain research and Management*, Vol. 6, pp. 67–77. IASP Press, Seattle.

Williams LS, Jones WJ, Shen J *et al.* (2004). Outcomes of newly referred neurology outpatients with depression and pain. *Neurology*, **63**, 674–7.

Williams AcdeC (1999). Measures of function and psychology. In Wall PD, Melzack R, eds. *Textbook of Pain*, Chapter 18, 4th edition, pp. 427–44. Churchill Livingstone, Edinburgh.

Woolf CJ, Bennett GJ, Doherty M *et al.* (1988). Towards a mechanism-based classification of pain? *Pain,* **77**, 227–9.

Woolf CJ, Mannion RJ (1999). Neuropathic pain: aetiology, symptoms, mechanisms and management. *Lancet*, **353**, 1959–64.

Zigmond AS, Snaith RP (1983). The Hospital Anxiety and Depression Scale. *Acta Psychiatrica Scandinavica*, **67**, 361–70.

Zajicek J, Fox P, Sanders H *et al.* (2003). Cannabinoids for treatment of spasticity and other symproms related to multiple sclerosis (CAMS study): multicentre randomised placebo-controlled trial. *Lancet*, **362**, 1517–26.

CHAPTER 18

Headache

Peter Goadsby

Contents

Migraine is an episodic brain disorder that affects about 15 per cent of the population (Lipton *et al.* 2001; Steiner *et al.* 2003), can be highly disabling (Menken *et al.* 2000), and has been estimated to be the most costly neurological disorder in the European Community at more than €27 billion per year (Andlin-Sobocki *et al.* 2005). It is the most common reason for neurological referral in the United Kingdom, estimated by the Association of British Neurologists to drive 20 per cent of referrals in outpatients; epilepsy is next at 12 per cent. Unfortunately, there is a tacit assumption that doctors in general just understand headache, and that neurologists in particular have special knowledge and training in the field. Sadly this is most often not the case and they *learn on the job* often perpetuating mistakes of their supervisors. To manage headache can be a source of extreme frustration or undiluted pleasure; the difference simply reflects how much one knows about the subject. Readers encouraged either by this text or by their clinical experience can look more deeply into headache with detailed texts (Goadsby and Silberstein 1997; Silberstein *et al.* 2002; Lance and Goadsby 2005; Olesen *et al.* 2005).

18.1 General principles

A formal nosology for headache disorders exists in the second edition of the International Classification of Headache Disorders (Headache Classification Committee of The International Headache Society 2004). This system will largely be employed here. There are many types of headache, and diagnosis is the key to proper management. The International Headache Society system is explicit, in the sense that it uses characteristic features of the headache to make the diagnosis, summing these features to make the diagnosis more certain. The general concept is that there are primary and secondary forms of headache, following the generic medical principle that clinical syndromes may be caused by something exogenous, secondary, or may manifest *de novo* as the primary disease process. Such a system is outlined in Table 18.1.

Table 18.1 Common causes of headache†

Primary headache		Secondary headache	
Type	Prevalence (%)	Type	Prevalence (%)
Migraine	16	Systemic infection	63
Tension-type	69	Head injury	4
Cluster headache	0.1	Sub-arachnoid haemorrhage	<1
Idiopathic stabbing	2	Vascular disorders	1
Exertional	1	Brain tumour	0.1

†after Olesen *et al.* (2005).

Broadly, primary headaches are those in which headache and its associated features are the disease in themselves, and secondary headaches are those caused exogenously, such as the headache commonly associated with fever. Mild secondary headache, such as that seen in association with upper respiratory tract infections is common but only rarely worrisome. The clinical dilemma remains that while life-threatening headache is relatively uncommon in the Western society, it occurs and its detection requires suitable vigilance by the doctors. Primary headache in contrast often confers considerable disability over time and while not life-threatening certainly robs patients of quality of life. Primary headache is a staple of the clinical neurologist's diet.

18.1.1 Primary headache syndromes

The primary headaches are a group of fascinating disorders in which headache and associated features are seen in the absence of any exogenous cause. The common syndromes (Table 18.1) are tension-type headache, migraine, and cluster headache. The collection of headaches known as primary chronic daily headache form the greatest part of the neurologist's burden. Some other less well-known, indeed rarer syndromes will be mentioned because they are easily treated when diagnosed.

18.1.2 Anatomy and physiology

The disabling primary headaches, migraine, and cluster headache, have been studied extensively in recent times and they are now relatively well-understood insofar as neurological disorders that involve the brain are concerned. In experimental animals the detailed anatomy of the connections of the pain-producing intracranial extracerebral vessels and the dura mater has been built on the classical human observations of Wolff and others. It is these structures, and not the brain itself, that are involved in head pain, although it is not at all clear to what extent there is nociceptive activation or the perception of that activation.

The key structures involved are:

◆ the large intracranial vessels and dura mater;

◆ the peripheral terminals of the trigeminal nerve that innervate these structures;

◆ the central terminals and second order neurons of the caudal trigeminal nucleus and dorsal horns of C_1 and C_2, trigeminocervical complex;

◆ higher centre processing in the thalamus, ventroposteromedial and posterior thalamus, and cortex; and

◆ modulatory centres in the diencephalon and brainstem, such as periaqueductal grey matter, locus coeruleus, and parts of the hypothalamus.

The innervation of the large intracranial vessels and dura mater by the trigeminal nerve is known as the trigeminovascular system. The cranial parasympathetic autonomic innervation provides the basis for symptoms, such as lacrimation and nasal stuffiness, which are prominent in cluster headache and paroxysmal hemicrania, although they may also be seen in migraine. It is clear from human functional imaging studies that vascular changes in migraine and cluster headache are driven by these neural vasodilator systems so that these headaches should be regarded as *neurovascular*. The concept of a primary *vascular* headache should be abandoned since it neither explains the pathogenesis of what are complex central nervous system disorders, nor does it necessarily predict treatment outcomes. The term vascular headache has no place in modern neurological practice when referring to primary headache.

Migraine is an episodic syndrome of headache with sensory sensitivity, such as to light, sound, and head movement, probably due to dysfunction of aminergic brainstem/diencephalic sensory control systems (Fig. 18.1). The first of the migraine genes has been identified for familial hemiplegic migraine, and includes mutations in the *CACNA1A* gene for the $Ca_V2.1$ (α_{1A}) subunit of the neuronal P/Q voltage-gated calcium channel, the Na/K ATP pump α_2 subunit gene *ATP1A2*, and the voltage-gated sodium channel *SCN1A*. These findings and the clinical features of migraine suggest it might be part of the spectrum of diseases known as channelopathies, or now ionopathies, disorders involving dysfunction of ion channel fluxes. Functional neuroimaging has suggested that brainstem regions in migraine (Fig. 18.2), and the posterior hypothalamic grey matter site of the human circadian pacemaker cells of the suprachiasmatic nucleus, in cluster headache (Fig. 18.3), are good candidates for specific involvement in primary headache (Cohen and Goadsby 2006a).

18.1.3 Secondary headache

It is imperative to establish in the patient presenting with any form of head pain whether there is an important secondary headache that is declaring itself. The headaches of subarachnoid haemorrhage (Section 35.16.2), meningitis (Section 41.2.1), giant cell arteritis (Sections 18.7.1; 36.2.8), and raised intracranial pressure (Section 26.5), are important examples of medically sinister headaches. Perhaps the most crucial clinical feature to elicit is the length of the history. Patients with a short history require prompt attention and may require quick investigation and management. Patients with a longer history generally require time and patience rather than alacrity. There are some important general features, including associated fever or sudden onset of pain (Table 18.2); these demand attention. Patients with a history of recent onset headache or neurological signs need a positive diagnosis of a benign disorder or require brain imaging with CT or MRI. Patients with a history of recurrent headaches over a period of 1 year or more, fulfilling International Headache Society criteria for migraine (Table 18.3) and with a normal physical examination, have positive brain imaging in only about 1/1000 images. In general it should be noted that brain tumour is a rare cause of headache, and rarely a cause of isolated long-term histories of headache. A notable exception to the general rules about secondary headache is

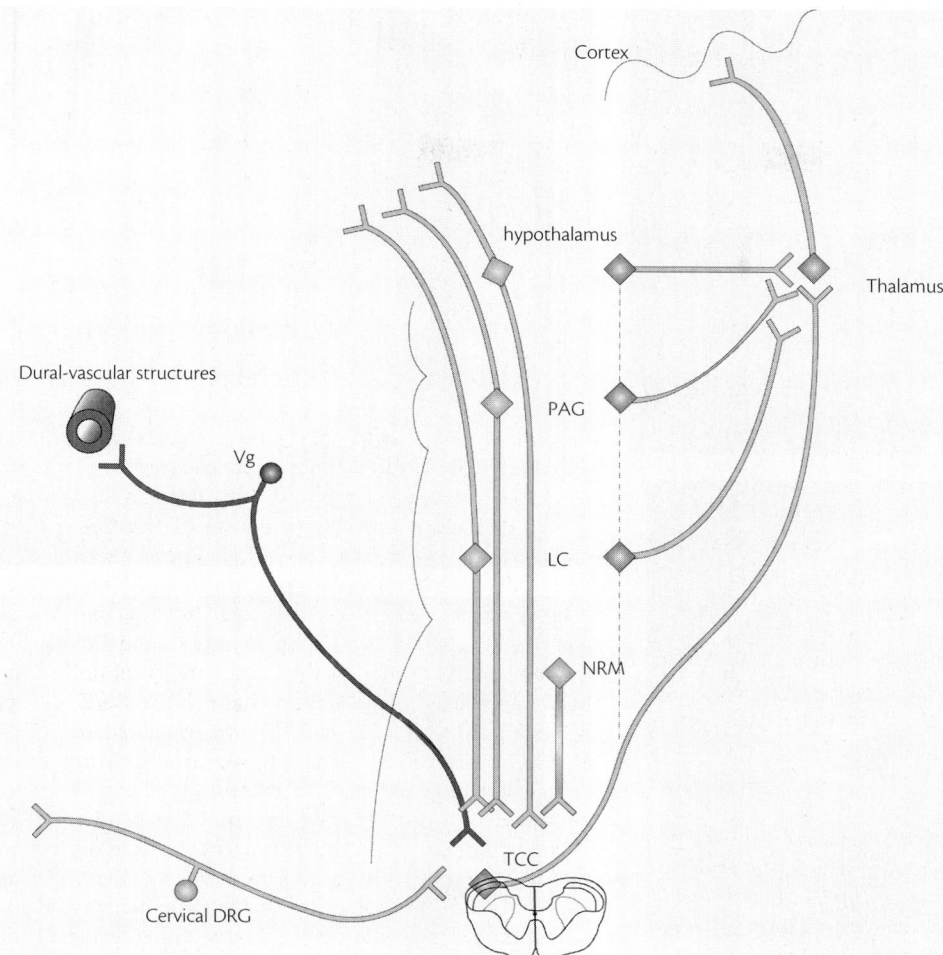

Fig. 18.1 Pathophysiology of migraine. Diagram of some structures involved in the transmission of trigeminovascular nociceptive input and the modulation of that input that form the basis of a model of the pathophysiology of migraine (Goadsby 2005). Afferents from dural-vascular structures innervated predominantly by branches of the first (ophthalmic division) of the trigeminal nerve whose cell bodies are found in the trigeminal ganglion (Vg) project to second order neurons in the trigeminocervical complex (TCC). The TCC extends from trigeminal nucleus caudalis to the caudal portion of the dorsal horn of the C_2 spinal cord. Input from cervical structures, such as joints or muscle, project through cell bodes in the upper cervical dorsal root ganglia (DRGs) to the TCC. TCC neurons project to ventrobasal thalamus (thalamus) and thence to cortex. Sensory modulation can occur by descending influences onto the TCC that largely respect the midline (dashed line), such as those from hypothalamus, midbrain periaqueductal grey (PAG), pontine locus coeruleus (LC), and nucleus raphe magnus (NRM). These influences are cartooned as being direct but both direct and indirect projections are recognized. In addition sensory modulation can occur from at least LC, PAG, and hypothalamic projects to thalamus nuclei as ascending systems again that largely respect the midline.

Table 18.2 Warning signs in head pain

- ◆ Sudden onset of pain
- ◆ Fever
- ◆ Marked change in pain character or timing
- ◆ Neck stiffness
- ◆ Pain associated with higher centre complaints
- ◆ Pain associated with neurological disturbance, such as clumsiness or weakness
- ◆ Pain associated with local tenderness, such as of the temporal artery

pituitary tumour, which can trigger underlying primary headache biologies, and should always be considered, especially in the differential diagnosis of trigeminal autonomic cephalalgias (Levy *et al.* 2005).

The management of secondary headache is generally self-evident: treatment of the underlying condition, such as an infection or mass lesion. An exception is the condition of chronic post-traumatic headache in which pain persists for long periods after head injury (Section 25.6.1). This is an interesting generic problem that may be seen after central nervous system infection, trauma, both blunt and surgical, intracranial bleeds, and other precipitants. While the syndrome is generally self-limiting up to 3–5 years after the event,

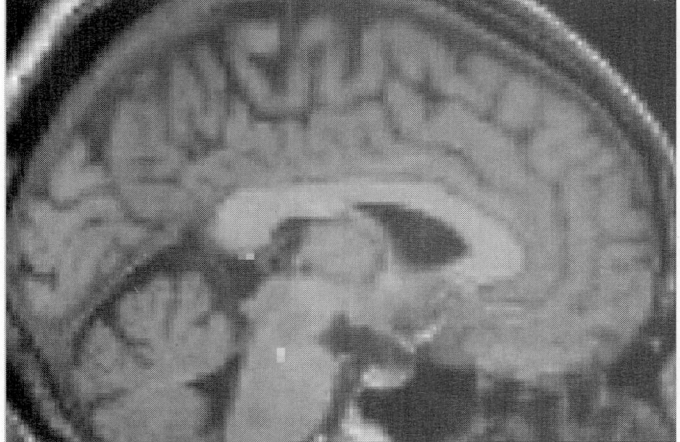

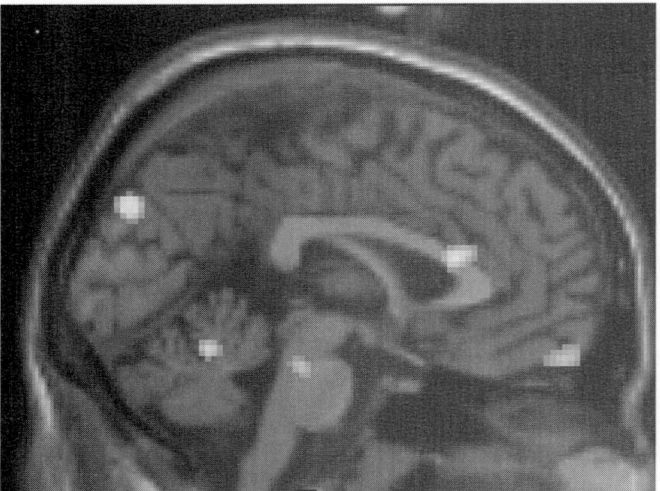

Fig. 18.2 Activation on positron emission tomography in a patient with cluster headache and migraine (top) who experienced a migraine without aura during the scan and demonstrated activation in the rostral ventral pons (Bahra *et al.* 2001). Similar activations are shown in patients with chronic migraine who were scanned during an attack (below) (Matharu *et al.* 2004a).

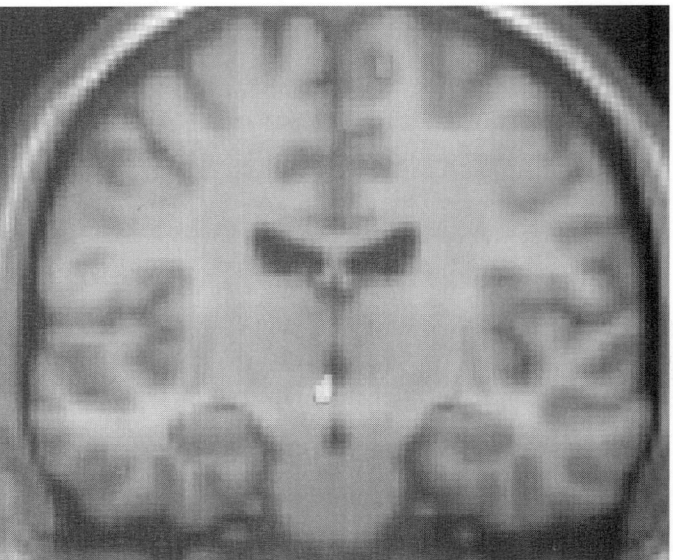

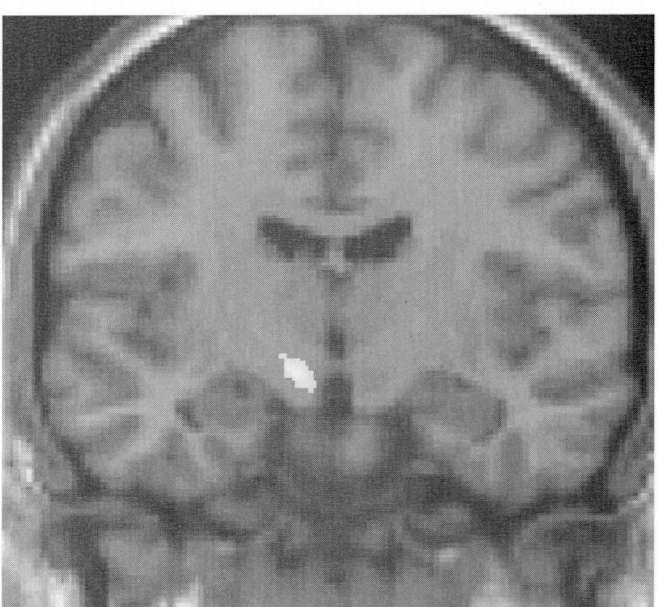

Fig. 18.3 Activation on positron emission tomography in the posterior hypothalamic grey matter in patients with acute cluster headache (top). The activation demonstrated is lateralized to the side of the pain (May *et al.* 1998). When comparing the brains of patients with cluster headache with a control population using an automatic anatomical technique known as voxel-based morphometry that employs high-resolution T1-weighted MRI a similar region is demonstrated (bottom) and has increased grey matter (May *et al.* 1999a).

treatment of the headache may be required if it is disabling (Section 18.3).

18.2 Migraine

18.2.1 Clinical features

Migraine is generally an episodic headache with certain associated features, such as sensitivity to light, sound, or movement, and often with nausea or vomiting accompanying the headache (Table 18.3). None of the features is compulsory, and given that the migraine aura, visual disturbances with flashing lights or zig-zag lines moving across the fields, or other neurological symptoms, is reported in only about 20 per cent of patients, a high index of suspicion is required to diagnose migraine. In a blinded, controlled study of patients presenting to General Practitioners with a main complaint of headache it was migraine on more than 90 per cent of occasions; thus it seems clear that most headache patients seen in neurology clinics probably have migraine as the underlying problem.

A headache diary can often be helpful in making the diagnosis although usually the diary helps more in assessing disability or recording how often patients use acute attack treatments. Phenotyping headache remains a clinical art of mixing experience with an understanding of the problems likely to present. In differentiating the two main primary headache syndromes seen in clinical practice:

◆ migraine at its most simple level is headache with associated features, and tension-type headache is headache that is featureless,

Table 18.3 Simplified diagnostic criteria for migraine adapted from the Headache Classification Committee of The International Headache Society (2004)

Repeated attacks of headache lasting 4–72 h that have these features, normal physical examination and no other reasonable cause for the headache:

At least 2 of	At least 1 of
◆ Unilateral pain	◆ Nausea/vomiting
◆ Throbbing pain	◆ Photophobia and phonophobia
◆ Aggravation by movement	
◆ Moderate or severe intensity	

by features is meant throbbing pain; or sensitivity to sensory stimuli: visual, auditory, olfactory; or to head movement itself.

◆ most disabling headache is probably migrainous in biology.

18.2.2 Frequent migraine

If headache with associated features describes migraine attacks, then *headachy* describes the migraine sufferer over their lifetime. It is important to realize that the word migraine can both describe the attacks using standard criteria (Table 18.3), and describe the disorder itself, which is more than just the attacks themselves. The migraine sufferer inherits a tendency to have headache that is amplified at various times by their interaction with their environment, the much-discussed triggers. The brain of the migraineur seems more sensitive to sensory stimuli and to change; and this tendency is even more notably amplified in females during their menstrual cycle. The migraine sufferer does not habituate to sensory stimuli easily and so can be unfairly and often stimulated in

the world in which they live and work. Migraine sufferers may have headache when they oversleep, when tired, when they skip meals, when stressed, or when relaxed. They are less tolerant to change and part of successful management is to advise them to maintain regularity in their lives in the knowledge of this fluctuating biology. It is this biology that marks migraine and which in clinical practice must override the phenotype of individual headaches.

It has been said that migraine can never occur daily, but few biological issues respect absolute rules. Chronic migraine very definitely occurs and in neurology or headache practice, is the very largest part of the group of headaches known collectively as chronic daily headache (Section 18.6). It is simply the most severe end of a complex biology and unsurprisingly often requires neurological input. Only development of disease markers will resolve diagnostic issues around daily headache. After making a diagnosis the second step in the clinical process is to be sure that the disease burden has been captured: how much headache does the patient have and more important, what cannot the patient do; what is their degree of disability? One can ask the patient directly to get a flavour for this, keep a diary or get a quick but accurate estimate using the Migraine Disability Assessment Scale, which is well-validated and very easy to use in practice (Fig. 18.4).

18.2.3 Principles of management

After diagnosis the management of migraine begins with an explanation of some aspects of the disorder to the patient. It is useful to explain that:

◆ migraine is an inherited tendency to headache; this is caused by the patient's genes, therefore it cannot be cured *but*;

◆ migraine can be modified and controlled by lifestyle adjustment and the use of medicines; and

INSTRUCTIONS: Please answer the following questions about ALL your headaches you have had over the last 3 months. Write your answer in the box next to each question. Write zero if you did not do the activity in the last 3 months (Please refer to the calendar below, if necessary)

1. On how many days in the last 3 months did you miss work or school because of your headaches? .... |__|__| days

2. How many days in the last 3 months was your productivity at work or school reduced by half or more because of your headaches (*Do not include days you counted in question 1 where you missed work or school*)? .... |__|__| days

3. On how many days in the last 3 months did you **not** do household work because of your headaches? .... |__|__| days

4. How many days in the last 3 months was your productivity in household work reduced by half or more because of your headaches (*Do not include days you counted in question 3 where you did not do household work*)? .... |__|__| days

5. On how many days in the last 3 months did you miss family, social, or leisure activities because of your headaches? .... |__|__| days

A. On how many days in the last 3 months did you have a headache? (If a headache lasted more than one day, count each day) .... |__|__| days

B. On a scale of 0 - 10, on average how painful were these headaches? (*where 0 = no pain at all, and 10 = pain as bad as it can be*) .... |__|__|

Version 3.0   © Innovative Medical Research 1997

Fig. 18.4 Migraine Disability Assessment Score. (MIDAS) Questionnaire.

◆ migraine is not life threatening nor associated with serious illness with the exception of females who smoke and use oestrogenic oral contraceptives, but migraine can make life a misery; and

◆ migraine management takes time and co-operation when, for example a headache diary has to be collected, or inquiry made concerning the disability.

18.2.4 Non-pharmacological management

This approach aims to help the migrainous patient identify things that make the problem worse and encourage them to modify these. Patient associations can be very helpful with pamphlets for this form of education, and those of the Migraine Trust and Migraine Action Association in the United Kingdom, and the American Council for Headache Education 'ACHE' are recommended. Many patients will not find this approach rewarding and should not be pilloried for this. Patients need to know that the brain sensitivity in migraine varies, so that the effect of triggers will vary. This knowledge will remove considerable frustration on the patient's part, will ring true to most as they have had the experience, and is biologically plausible, since it is exactly what one would predict from the channelopathic/ionopathic theory of migraine pathogenesis. The crucial lifestyle advice is to explain to the patient that migraine is a state of brain sensitivity to change. This implies that the migraine sufferer needs to regulate their lives: healthy diet, regular exercise, regular sleep patterns, avoiding excess caffeine, and alcohol and, as far as practical, modifying or minimizing changes in stress. The balanced life with less highs and lows will benefit most migraine sufferers.

18.2.5 Preventive treatments

The decision to start a patient on preventive treatment requires crucial input from the migraineur. The patient needs to have come to terms with the fact that they have an inherited, non-curable but manageable problem, and that they have sufficient disability to wish to take a medicine to reduce the affects of the disease on their life. Only then can the doctor explain the choices available and their relative merits. The basis of considering preventive treatment from a medical viewpoint is a combination of acute attack frequency and attack tractability. Attacks that are unresponsive to abortive medications are easily considered for prevention, while simply treated attacks may be less obviously candidates for prevention. The other part of the equation relates to what is happening with time. If a patient diary shows a clear trend of an increasing frequency of attacks it is better to introduce prevention early than to wait for the problem to become chronic.

A simple rule for frequency might be that for 1–2 headaches a month there is usually no need to start a preventive drug, for 3–4 it may be needed but not necessarily, and for 5 or more a month prevention should definitely be on the agenda for discussion. Options available for treatment are covered in detail in Table 18.4 and vary somewhat by country. The problem with preventives is not that there are none, but that they have fallen into usage for migraine from other indications. Often the doses required to reduce headache frequency produce marked and intolerable side effects. It is not clear how preventives work although it seems likely that they modify the brain sensitivity that underlies migraine. Another key clinical point is that each drug should be started at a low dose and

gradually increased to a reasonable maximum if there is going to be a clinical effect.

Little has been done in terms of systematic study of patients with more intractable forms of migraine. Neuromodulation approaches are promising, largely by stimulation of the occipital nerve, and show that central processing of pain signals in migraine in the thalamus may be modified (Matharu *et al.* 2004a). This is an exciting and developing area.

18.2.6 Acute attack therapies

Acute attack treatments for migraine can be usefully divided into disease non-specific treatments, analgesics, and non-steroidal anti-inflammatory drugs, and disease-specific treatments, ergot-related compounds, and triptans (Table 18.5). It must be said at the outset that most acute attack medications seem to have a propensity to aggravate headache frequency and can induce a state of refractory daily or near-daily headache, known as medication overuse headache. As evidence is gathered this seems to occur in patients with migraine: either a previous clear history or a family or personal history of *headachiness*. Codeine-containing compound analgesics are a particularly pernicious problem when available in over-the-counter preparations. One should advise patients with migraine who have two headache days a week or more to avoid their regular use. A proportion of patients who stop taking regular analgesics will have substantial improvement in their headache with a reduction in frequency, however, for some it will not make any difference. It is crucial to emphasize to the patient that standard preventive medications often simply do not work in the presence of regular analgesic use.

Treatment strategies. Given the array of options to control an acute attack of migraine, how does one start? The simplest approach to treatment has been described as *Stepped care*. In this model all patients are treated, assuming no contraindications, with the simplest treatment, such as aspirin 900 mg or paracetamol 1000 mg with an anti-emetic. Aspirin is an effective strategy, has been proven so in double-blind controlled clinical trials, and is best used in its most soluble formulations. The alternative would be a strategy known as *Stratified care*, by which the physician determines, or stratifies, treatment at the start based on likelihood of response to levels of care. An intermediate option may be described as stratified care by attack. The latter is what many headache authorities suggest and what patients often do when they have the options. Patients use simpler options for their less severe attacks relying on more potent options when their attacks or circumstances demand them (Table 18.5).

Non-specific acute migraine attack treatments. Since simple drugs, such as aspirin and paracetamol, are cheap and can be very effective, they can be employed in many patients. Dosages should be adequate and the addition of domperidone 10 mg orally or metoclopramide 10 mg orally can be very helpful. Non-steroidal anti-inflammatory drugs can very useful when tolerated. Their success is often limited by inappropriate dosing, and adequate doses of naproxen 500–1000 mg orally or rectally with an anti-emetic, ibuprofen 400–800 mg orally or tolfenamic acid 200 mg orally can be extremely effective.

Specific acute migraine attack treatments. When simple analgesic measures fail or more aggressive treatment is required, the specific anti-migrainous treatments are required (Table 18.6). While ergotamine remains a useful anti-migraine compound it can no longer be considered the treatment of choice in acute migraine

Table 18.4 Preventive drug treatments in migraine†

	Dose	Selected side effects
Pizotifen	0.5–2 mg daily	Weight gain Drowsiness
β-Blocker Propranolol	40–120 mg bd	Reduced energy Tiredness Postural symptoms *Contraindicated in asthma*
Tricyclics ◆ Amitriptyline ◆ Dosulepin ◆ Nortriptyline	25–75 mg nocte	Drowsiness *Note*: some patients are very sensitive and may only need a total dose of 10 mg, although generally 1–1.5 mg/kg body weight is required
Anticonvulsants ◆ Valproate	400–600 mg bd	Drowsiness Weight gain Tremor Hair loss Foetal abnormalities Haematological or liver abnormalities
◆ Topiramate	50–200 mg/day	Parasthesiae Cognitive dysfunction Weight loss Care with a family history of glaucoma Nephrolithiasis
◆ Gabapentin	900–3600 mg daily	Dizziness Sedation
Methysergide	1–6 mg daily	Drowsiness Leg cramps Hair loss Retroperitoneal fibrosis (one month drug holiday is required every 6 months)
Flunarizine	5–15 mg daily	Drowsiness Weight gain Depression Parkinsonism
Single studies‡ ◆ Lisinopril ◆ Candasartan	20 mg daily 16 mg daily	Cough Dizziness
Neutriceuticals‡‡ ◆ Riboflavin ◆ Coenzyme Q10 ◆ Butterburr ◆ Feverfew	400 mg daily 100 mg tid 75 mg bd 6.25 mg tid	GI upset
No Convincing Controlled Evidence ◆ Verapamil		
Controlled trials to demonstrate no effect ◆ Nimodipine ◆ Clonidine ◆ SSRIs: fluoxetine		

†Commonly used preventives are listed with reasonable doses and common side effects. The local national formulary should be consulted for detailed information.

‡Compounds not widely considered as mainstream but with a positive randomized control trial against placebo.

‡‡Non-pharmaceuticals with at least one positive randomized controlled trial against placebo.

Table 18.5 Oral acute migraine treatments

Non-specific treatments	Specific treatments
(often used with anti-emetic/ prokinetics, such as domperidone (10 mg) or metoclopramide (10 mg))	
Aspirin (900 mg)	*Ergot derivatives*
	◆ Ergotamine (1–2 mg)
Paracetamol (1000 mg)	
	Triptans
NSAIDs	◆ Sumatriptan (50 or 100 mg)
◆ Naproxen (500–1000 mg)	◆ Naratriptan (2.5 mg)
◆ Ibuprofen (400–800 mg)	◆ Rizatriptan (10 mg)
◆ Tolfenamic acid (200 mg)	◆ Zolmitriptan (2.5 or 5 mg)
	◆ Eletriptan (40 or 80 mg)
	◆ Almotriptan (12.5 mg)
	◆ Frovatriptan (2.5 mg)

NSAIDS: non-steroidal antiinflammatory drugs

Table 18.6 Clinical stratification of acute specific migraine treatments, given orally unless otherwise indicated

Clinical situation	Treatment options
Failed analgesics/non-steroidal antiinflammatory drugs	*First tier*
	Sumatriptan 50 mg or 100 mg
	Almotriptan 12.5 mg
	Rizatriptan 10 mg
	Eletriptan 40 mg
	Zolmitriptan 2.5 mg
	Slower effect/better tolerability
	Naratriptan 2.5 mg
	Frovatriptan 2.5 mg
	Infrequent headache
	Ergotamine 1–2 mg
	Dihydroergotamine nasal spray 2 mg
Early nausea or difficulties taking tablets	Zolmitriptan 5 mg nasal spray
	Sumatriptan 20 mg nasal spray
	Rizatriptan 10 mg MLT wafer
Headache recurrence	Ergotamine 2 mg (most effective rectally /usually with caffeine)
	Naratriptan 2.5 mg
	Almotriptan 12.5 mg
	Eletriptan 40 mg
Tolerating acute treatments poorly	Naratriptan 2.5 mg
	Almotriptan 12.5 mg
Early vomiting	Zolmitriptan 5 mg nasal spray
	Sumatriptan 25 mg rectally
	Sumatriptan 6 mg subcutaneously
Menstrually related headache	*Prevention*
	Ergotamine nocte
	Oestrogen patches
	Treatment
	Triptans
	Dihydroergotamine nasal spray
Very rapidly developing symptoms	Zolmitriptan 5 mg nasal spray
	Sumatriptan 6 mg subcutaneously
	Dihydroergotamine 1 mg intramuscularly

(Tfelt-Hansen *et al.* 2000). There are particular situations in which ergotamine is very useful, but its use must be strictly controlled as ergotamine overuse produces dreadful headache in addition to various vascular problems. The triptans have revolutionized the life of many patients with migraine and are clearly the most powerful option available to stop a migraine attack. They can be rationally applied by considering their pharmacological, physicochemical, and pharmacokinetic features (Goadsby 2000), as well as the formulations that are available (Goadsby *et al.* 2002).

18.3 Tension-type headache

18.3.1 Clinical features

As its name suggests, tension-type headache tension-type headache is a term that describes the headache form most in need of pathophysiological understanding. Consider for a moment how hard it is to study something that is commonly considered to be well-understood, and then ask what is the essence of tension-type headache? Tension-type headache is diagnosed commonly, and while the phenotype is common much of the disabling headache that goes under the name tension-type headache is likely to be chronic migraine in terms of its biology. Tension-type headache has two forms:

◆ episodic tension-type headache, where attacks occur on less than 15 days a month

◆ chronic tension-type headache where attacks, on average over time, are seen on 15 days or more a month.

This is part of the broader clinical syndrome of chronic daily headache, but these terms are not equal.

Tension-type headache has been defined by the International Headache Society both for its episodic and chronic forms, although the admixture of symptoms allowed has consistency problems. A useful clinical approach is to diagnose tension-type headache when the headache is completely featureless: no nausea, no vomiting, no photophobia, no phonophobia, no osmophobia, no throbbing, and no aggravation with movement. Such an approach neatly divides migraine, which has one of more of these features and is the main differential diagnosis, from tension-type headache.

18.3.2 Pathophysiology

The pathophysiology of tension-type headache is very poorly understood. This results from the fact that the name implies to most that it is a product of *nervous tension*, for which there is no clear evidence, and the definitions employed have undoubtedly admitted patients with migraine to the studies. Moreover, the concept that tension-type headache in some way involves muscle contraction is spurious since the evidence is that muscle contraction is no more likely that it is in migraine. It seems likely that tension-type headache will be due to a primary disorder of central nervous system pain modulation alone, to contrast it with migraine, which is a much more generalized disturbance of sensory modulation. There are data suggesting a genetic contribution to tension-type headache but one must question these since they have been gathered with probably faulty diagnostic criteria.

18.3.3 Management

Adopting the clinical approach to tension-type headache outlined above results in diagnosing a headache form that is usually less disabling, and more in the category of irritating. Its episodic form is generally amenable to simple analgesics, paracetamol, acetaminophen, aspirin, or non-steroidal antiinflammatory drugs, which can be purchased over the counter. There are clear clinical studies to demonstrate that triptans in tension-type headache alone are not helpful, although germane to the above discussion, triptans are effective in tension-type headache where the patient also has migraine. Amitriptyline is the only treatment for chronic tension-type headache with a clear evidence base; the other tricylics, selective serotonin reuptake inhibitors or the benzodiazepines have not been shown in controlled trials to be effective. Similarly, there is no controlled evidence for the use of electromyographic biofeedback, relaxation therapy, or acupuncture. Botulinum toxin has been shown reasonably clearly to be ineffective. Stress management has been shown to be an effective approach in a controlled trial.

18.4 Trigeminal-autonomic cephalalgias

18.4.1 Cluster headache

Cluster headache is a rare form of primary headache with a population frequency of 0.1 per cent. It is covered in a specialized book (Olesen and Goadsby 1999). It is about as common as Multiple Sclerosis in the United Kingdom, and must be regarded as a disorder best managed by neurologists. It is perhaps the most painful condition in humans; in the cohort of more than 800 patients seen at the National Hospital not a single one has had a more painful experience, including childbirth, multiple fractures of the limbs, or renal stones. It is one of a group of conditions known now as Trigeminal-autonomic cephalalgias, and needs to be differentiated from other trigeminal-autonomic cephalalgias (Goadsby and Lipton 1997) and the short-lasting headaches without cranial autonomic symptoms, such as lacrimation or conjunctival injection (Table 18.7).

The core feature of cluster headache is periodicity, be it circadian or in terms of active and inactive bouts over weeks and months (Table 18.8). The typical cluster headache patient is male, with a 3:1 predominance, who has bouts of 1–2 attacks of relatively short duration unilateral pain every day for bouts of 8–10 weeks a year.

Table 18.7 Primary headache—cluster headache, other trigeminal autonomic cephalalgias, and short-lasting headaches*

Trigeminal autonomic cephalalgias	Other short-lasting headaches
◆ Cluster headache	◆ Primary stabbing headache
◆ Paroxysmal hemicrania	◆ Trigeminal neuralgia
◆ SUNCT† syndrome	◆ Primary cough headache
	◆ Primary exertional headache
	◆ Primary sex headache
	◆ Hypnic headache

*Beware of pituitary tumour-related headache in the differential diagnosis of these trigeminal autonomic cephalalgias.

†SUNCTs Short-lasting unilateral neuralgiform headache attacks with conjunctival injection and tearing..

They are generally perfectly well between times. Patients with cluster headache tend to move about during attacks, pacing, rocking, or even rubbing their head for relief. The pain is usually retro-orbital boring and very severe. It is associated with ipsilateral symptoms of cranial parasympathetic autonomic activation: a red or watering eye, the nose running or blocking, or cranial sympathetic dysfunction: eyelid droop. Cluster headache is likely to be a disorder involving central pace-maker regions of the posterior hypothalamus (Fig. 18.2) (May et al. 1998; 1999a)

The trigeminal-autonomic cephalalgias: cluster headache, paroxysmal hemicrania, and the syndrome of shortlasting unilateral neuralgiform headache attacks with conjunctival injection or tearing known as SUNCT, present a distinct group to be differentiated from short-lasting headaches that do not have prominent cranial autonomic syndromes, notably trigeminal neuralgia, idiopathic or primary stabbing headache, and hypnic headache (Goadsby 2002a). By determining the cycling pattern, length of attack, frequency of attack, and timing of the attacks, most patients can be usefully classified. The importance of clinical classification of this group is threefold. First, the clinical phenotype determines the likely secondary causes that must be considered and appropriate investigations ordered. Secondly, the appropriate classification gives clarity to the patient with a clear diagnosis and allows the physician to

Table 18.8 Diagnostic criteria for cluster headache (Headache Classification Committee of The International Headache Society 2004)

3.1 **Diagnostic criteria**:

A. At least 5 attacks fulfilling B–D;

B. Severe or very severe unilateral orbital, supraorbital, and/or temporal pain lasting 15–180 min if untreated;

C. Headache is accompanied by at least one of the following:

1. ipsilateral conjunctival injection and/or lacrimation;
2. ipsilateral nasal congestion and/or rhinorrhoea;
3. forehead and facial sweating;
4. ipsilateral eyelid oedema;
5. ipsilateral forehead and facial sweating;
5. ipsilateral miosis and/or ptosis; and
6. a sense of restlessness or agitation.

D. Attacks have a frequency from 1 every other day to 8 per day; and

E. Not attributed to another disorder.

3.1.1 **Episodic cluster headache**

Description: Occurs in periods lasting 7 days to 1 year separated by pain-free periods lasting 1 month or more

Diagnostic criteria:

A. All fulfilling criteria A–E of 3.1.

B. At least 2 cluster periods lasting from 7 to 365 days and separated by pain-free remissions of ≥ 1 month.

3.1.2 **Chronic cluster headache**

Description: Attacks occur for more than 1 year without remission or with remissions lasting less than 1 month.

Diagnostic criteria:

A. All alphabetical headings of 3.1; and

B. Attacks recur over > 1 year without remission periods or with remission periods < 1 month.

draw on available literature to comment on natural history. Thirdly, the correct diagnosis determines therapy that can be very different in these conditions, being very good if the diagnosis is correct but probably ineffective if it is not (Table 18.9).

18.4.2 Managing cluster headache

Cluster headache is managed using acute attack treatments and preventive agents. Acute attack treatments are usually required by all cluster headache patients at some time, while preventives can seem almost life-saving for the patients with chronic cluster headache and are often needed to shorten the active periods in patients with the episodic form of the disorder.

Preventive treatments. The options for preventive treatment in cluster headache depend on the bout length (Table 18.10). Patients with short bouts require medicines that act quickly but will not necessarily be taken for long periods, whereas those with long bouts or indeed those with chronic cluster headache require safe, effective medicines that can be taken often for long periods. Most experts now favour verapamil as the first-line preventive treatment when the bout is prolonged, or in chronic cluster headache. By contrast, limited courses of oral corticosteroids or methysergide can be very useful strategies when the bout is relatively short.

Verapamil has been suggested as a useful option for the last decade and compares favourably with lithium. What has clearly emerged from clinical practice is the need to use higher doses than had initially been considered and certainly higher than those used in cardiological indications. Although most patients will start on doses as low as 40–80 mg twice daily, doses up to 960 mg daily and beyond are now employed (Olesen and Goadsby 1999). Side effects, such as gingival hyperplasia, constipation, and leg swelling, can be a problem, but more difficult is the issue of cardiovascular safety. Verapamil can cause heart block by slowing conduction in the atrioventricular node as demonstrated by prolongation of the A–H interval. Given that the PR interval on the electrocardiogram is made up of atrial conduction, A–H and His bundle conduction, it

Table 18.10 Preventive management of cluster headache

Short-term prevention	Long-term prevention
Episodic cluster headache	*Episodic cluster headache and prolonged chronic cluster headache*
◆ Prednisolone	◆ Verapamil
◆ Methysergide	◆ Lithium
◆ Verapamil	◆ Methysergide
◆ Greater occipital nerve injection	◆ Melatonin
◆ (Daily nocturnal ergotamine)	◆ ?Topiramate
	◆ ?Gabapentin

? = unproven but promising.

may be difficult to monitor subtle early effects as verapamil dose is increased. Given that the effects on the atrioventricular node take up to 10 days to manifest, 2-week intervals are recommended between dose changes on the first exposure, with electrocardiograms prior the next escalation, and 6-monthly electrocardiograms after the dose is established.

Acute attack treatment. Cluster headache attacks often peak rapidly and thus require a treatment with quick onset. Many patients with acute cluster headache respond very well to treatment with oxygen inhalation. This should be given as 100 per cent oxygen at 10–12 l/min for 15–20 min. It is important to have a high flow and high oxygen content. Injectable sumatriptan 6 mg has been a boon for many patients with cluster headache. It is effective, rapid in onset, and with no evidence of tachyphylaxis. Sumatriptan 20 mg and Zolmitriptan 5 mg nasal sprays are effective in acute cluster headache in controlled trials, and offer a useful option for patients who may not wish to self-inject daily. Sumatriptan is not effective when given pre-emptively as 100 mg orally three times daily, and there is no evidence that it is useful when used orally in the acute treatment of cluster headache; indeed it can be associated with medication overuse headache problems.

Table 18.9 Differential diagnosis of short-lasting headaches

Feature	Cluster headache	Paroxysmal hemicrania	SUNCT*	Primary stabbing headache	Trigeminal neuralgia*	Hypnic headache
Gender	M>F 3:1	F~M	M~F	F>M	F>M	M=F
Pain						
–type	Boring/stabbing	Boring/throbbing	Stabbing	Stabbing	Stabbing	Throbbing
–severity	Very severe	Very severe	Severe	Severe	Very severe	Moderate
–location	Orbital	Orbital	Orbital	Any	V2/V3>V1	Generalized
Duration	15–180 min	1–45 min	15–600 s	Seconds–3 min	<5 s	15–30 min
Frequency	1–8 per day	1–40 per day	1/day–30/h	Any	Any	1–3/night
Autonomic	+	+	+	-	-	-
Alcohol	+	One-third	-	-	-	-
Cutaneous trigger to attacks	-	-	+	-	+	-
Indomethacin	-	+	-	+	-	-

SUNCT, Short-lasting neuralgiform headache attacks with conjunctival injection and tearing.

*SUNCT generally has no refractory period to trigger additional attacks, while this is a very common feature of trigeminal neuralgia.

Surgical treatment. The surgical treatment of cluster headache has been completely revolutionized with the introduction of neuromodulation techniques. Surgical treatment of cluster headache is reserved for the most refractory patients, typically with chronic cluster headache. Destructive procedures such as pterygopalatinectomy or radiofrequency lesions of the trigeminal ganglion have been used, the former without clear effects, and the latter being helpful but often at significant cost, including ocular complications or anaesthesia dolorosa. Trigeminal rhizotomy has also been employed, with all the complications of radiofrequency lesions and the occasional death. Set against this is the functional imaging describing activations in the posterior hypothalamic region (May *et al.* 1998) directly leading to deep brain stimulation approaches in the same region that seem highly effective (Leone *et al.* 2004). A further approach is that of occipital nerve stimulation, which is very promising as a non-invasive approach to the management of intractable chronic cluster headache (Burns *et al.* 2007).

18.4.3 Paroxysmal hemicrania

Sjaastad and Dale (1976) first reported eight cases of a frequent unilateral severe but short-lasting headache without remission coining the term chronic paroxysmal hemicrania. The mean daily frequency of attacks varied from 7 to 22 with the pain persisting from 5 to 45 min on each occasion. The site and associated autonomic phenomena were similar to cluster headache, but the attacks were suppressed completely by indomethacin. A subsequent review of 84 cases showed a history of remission in 35 cases whereas 49 were chronic. By analogy with cluster headache the patients with remission have been referred to as episodic paroxysmal hemicrania (Kudrow *et al.* 1987) and those with the non-remitting form chronic paroxysmal hemicrania; the overall syndrome can be simply called paroxysmal hemicrania.

The essential features of paroxysmal hemicrania are (Table 18.11):

- unilateral, usually fronto-temporal, very severe pain;

- short-lasting attacks of 2–45 min;

- very frequent attacks, usually more than 5 a day;

- marked autonomic features ipsilateral to the pain; and

- robust, quick and excellent response to indomethacin in less than 72 h.

The pathophysiology of paroxysmal hemicrania is marked by activations on positron emission tomography in the contralateral posterior hypothalamus and contralateral ventral midbrain (Matharu *et al.* 2006). The posterior hypothalamic activity is shared with cluster headache, SUNCT, and hemicrania continua, while the ventral midbrain activity is only seen in hemicrania continua (Section 18.5.7), which remarkably is also an indomethacin-sensitive primary headache.

The therapy of paroxysmal hemicrania is complicated by gastrointestinal side effects seen with indomethacin, although thus far there is no reliable alternative option. Piroxicam has been suggested to be helpful, although not as effective as indomethacin. By analogy with cluster headache verapamil has been used in paroxysmal hemicrania, although the response is not spectacular; higher doses require exploration. Paroxysmal hemicrania can co-exist with trigeminal neuralgia, the paroxysmal hemicrania-tic syndrome, just as in cluster-tic syndrome, and each component requires separate treatment.

Table 18.11 Paroxysmal hemicrania (Headache Classification Committee of The International Headache Society 2004)

3.2 Diagnostic criteria:

A. At least 20 attacks fulfilling B–D.

B. Severe unilateral orbital, supraorbital, or temporal pain lasting 2–30 min.

C. Headache is accompanied by at least one of the following:
   1. ipsilateral conjunctival injection and/or lacrimation
   2. ipsilateral nasal congestion and/or rhinorrhoea
   3. forehead and facial sweating
   4. ipsilateral eyelid oedema
   5. ipsilateral forehead and facial sweating
   5. ipsilateral miosis and/or ptosis

D. Attacks have a frequency above 5 per day for more than half the time, although periods with lower frequency may occur

E. Attacks are prevented completely by therapeutic doses of indomethacin

F. Not attributed to another disorder

3.2.1 Episodic paroxysmal headache

Description: Occurs in periods lasting 7 days to one year separated by pain free periods lasting one month or more

3.2.2 Chronic paroxysmal headache

Description: Attacks occur for more than one year without remission or with remissions lasting less than one month.

Secondary paroxysmal hemicrania has been reported with lesions in the region of the sella turcica, an arteriovenous malformation, cavernous sinus meningioma, and a parotid epidermoid. Secondary paroxysmal hemicrania is more likely if the patient requires high doses' >200 mg/day, of indomethacin. Raised CSF pressure should be suspected in apparent bilateral paroxysmal hemicrania. It is worth noting that indomethacin reduces CSF pressure by an unknown mechanism. It is appropriate to image patients, with MRI when practical, when a diagnosis of paroxysmal hemicrania is being considered.

18.4.4 Short-lasting unilateral neuralgiform headache attacks with conjunctival injection and tearing or cranial autonomic activation: SUNCT/SUNA

Sjaastad (1989) reported three male patients whose brief attacks of pain in and around one eye were associated with sudden conjunctival injection and other autonomic features of cluster headache. The attacks lasted only 15–60 s and recurred 5–30 times per hour, and could be precipitated by chewing or eating certain foods, such as citrus fruits. They were not abolished by indomethacin. Brain imaging has suggested that they share with cluster headache and paroxysmal hemicrania the feature on activation studies of involvement of the posterior hypothalamic region (May *et al.* 1999b). Of the patients recognized with this problem males dominate slightly and the paroxysms of pain may last between 5 and 300 s, although longer duller interictal pains are recognized, as are longer attacks with a saw-tooth pattern (Cohen and Goadsby 2006b). The conjunctival injection seen with SUNCT is often the most prominent autonomic feature and tearing may be very obvious. If one of either conjunctival injection or tearing are absent, or neither are present but another cranial autonomic symptom is seen, the term SUNA is used (Table 18.12). The two clinical features of SUNCT/SUNA are that the attacks are triggerable and when triggerable there is no refractory period to triggering further attacks. The latter serves as a very useful distinction between SUNCT/SUNA and trigeminal neuralgia. SUNCT/SUNA can be treated very often with lamotrigine, and if that is unhelpful topiramate or gabapentin. Carbamazepine often has a useful but incomplete effect.

Secondary SUNCT and associations. The literature reports a number of patients with secondary SUNCT syndromes that invariably have lesions involving the posterior fossa. Two reported patients had homolateral cerebellopontine angle arteriovenous malformations diagnosed on MRI, while another had a cavernous hemangioma of the brainstem seen only on MRI. Structural deformity involving the posterior fossa, including osteogenesis imperfecta and craneosynostosis, have presented as SUNCT-like syndromes, as have pituitary tumours. A posterior fossa lesion causing otherwise typical SUNCT has also been noted in HIV/AIDS.

Table 18.12 Proposed diagnostic criteria for short-lasting unilateral neuralgiform headache attacks with conjunctival injection and tearing (SUNCT) or cranial autonomic features (SUNA) (Headache Classification Committee of The International Headache Society 2004)

3.3R **Diagnostic criteria:**

  A. At least 20 attacks fulfilling criteria B–E

  B. Attacks of short-lasting (1–600 s) unilateral head pain

    a. orbital, supraorbital, temporal or other trigeminal distribution of moderate or severe pain

    b. occurring as

      i. single stabs

      ii. groups of stabs

      iii. in a saw-tooth pattern

    c. Triggerable without a refractory period

  C. Pain is accompanied ipsilaterally by either:

    a. Conjunctival injection and Tearing (SUNCT), or,

    b. One or more of the following cranial Autonomic symptoms (SUNA)

      i. conjunctival injection, or tearing, but not both

      ii. nasal congestion and/or rhinorrhoea

      iii. eyelid oedema

      iv. ipsilateral sense of aural fullness or peri-aural swelling

      v. ipsilateral forehead and facial sweating

      vi. ipsilateral miosis and/or ptosis

  D. Attacks occur with a frequency of ≥1 per day for more than half the time when the disorder is active

  E. Not attributed to another disorder

A 3.3.1 **Episodic SUNCT/SUNA**

Description: SUNA attacks occurring for 7 days to 1 year with pain free intervals longer than 1 month

---

A 3.3.2 **Chronic SUNCT/SUNA**

Description: At least 2 attack periods last 7 days to 1 year separated by remission periods of less than one month (untreated).

These cases highlight the need for cranial MRI in the diagnostic evaluation of SUNCT. Cases with both SUNCT and trigeminal neuralgia have been reported. Given that the attacks are short this can be a challenging clinical problem. The differential diagnosis turns around the degree of cranial autonomic activation, which may be seen to some degree in trigeminal neuralgia but is very prominent in SUNCT, and the lack of a refractory period to triggering of attacks in SUNCT.

18.5 Other primary headaches

18.5.1 Stabbing headache

Short-lived jabs of pain, defined as primary stabbing headache (Headache Classification Committee of The International Headache Society 2004) are well documented in association with most types of primary headache.

The essential clinical features are:

- pain confined to the head, although rarely is it facial;

- stabbing pain lasting from 1 to many seconds and occurring as a single stab or a series of stabs; and

- recurring at irregular intervals of hours to days.

Raskin and Schwartz (1980) described sharp, jabbing pains about the head resembling a stab from an ice-pick, nail, or needle. They compared the prevalence of such pains in 100 migrainous patients and 100 headache-free controls. Only three of the control subjects had experienced ice-pick pains compared with 42 of the migraine patients, of whom 60 per cent had more than one attack per month. The pains affected the temple or orbit more often than the parietal and occipital areas and often occurred before or during migraine headaches. The sites of these pains generally coincide with the site of the patients' habitual headache. Retroauricular and occipital region pains are also well described and these respond promptly to indomethacin. Stabbing headaches have been described in conjunction with cluster headaches, and generally are experienced in the same area as the cluster pain. Sjaastad described 'jabs and jolts' lasting less than a minute in patients with chronic paroxysmal hemicrania (Antonaci and Sjaastad 1989). These longer attacks are probably part of the spectrum of stabbing headache. It is of interest that stabbing pains generally are not accompanied by cranial autonomic symptoms. The response of idiopathic stabbing headache to indomethacin, 25–50 mg twice to three times daily, is generally excellent. As a general rule the symptoms wax and wane and after a period of control on indomethacin it is appropriate to withdraw treatment and observe the outcome. Most patients will not want treatment when the nature of the problem is unexplained and they are reassured that the attacks are not sinister in any way.

18.5.2 Cough headache

Sharp pain in the head on coughing, sneezing, straining, laughing, or stooping has long been regarded as a symptom of organic intracranial disease, commonly associated with obstruction of the CSF pathways. The presence of an Arnold–Chiari malformation or any lesion causing obstruction of CSF pathways or displacing cerebral structures must be excluded before cough headache is assumed to be benign. Cerebral aneurysm, carotid stenosis, and vertebrobasilar disease may also present with cough or exertional headache as the initial symptom. The term 'Benign Valsalva's manoeuvre-related headache' covers the headaches provoked by coughing, straining, or stooping but *cough headache* is more succinct and so widely used it is unlikely to be displaced.

The essential clinical features of benign and primary cough headache are:

◆ bilateral headache of sudden onset, lasting minutes, precipitated by coughing;

◆ may be prevented by avoiding coughing; and

◆ diagnosed only after structural lesions, such as posterior fossa tumour, have been excluded by neuroimaging.

Comparing benign cough with benign exertional headache Pascual and colleagues (1996) reported that the average age of their patients with benign cough headache was 43 years, and thus older than their patients with exertional headache. Indomethacin is the medical treatment of choice in cough headache. Raskin (1995) followed up an observation of Sir Charles Symonds reporting that some patients with cough headache are relieved by lumbar puncture. This is a simple option when compared to prolonged use of indomethacin. The mechanism of this response remains unclear.

18.5.3 Exertional headache

The relationship of this form of headache to cough headache is unclear and certainly much is shared. Indeed the relationship to migraine also requires delineation. Credit must be given to Hippocrates for first recognizing this syndrome when he wrote: 'one should be able to recognise those who have headache from gymnastic exercises, or walking, or running, or any other unseasonable labour, or from immoderate venery'".

The clinical features are:

◆ pain specifically brought on by physical exercise;

◆ bilateral and throbbing in nature at onset and may develop migrainous features in those patients susceptible to migraine;

◆ lasts from 5 min to 24 h; and

◆ prevented by avoiding excessive exertion, particularly in hot weather or at high altitude.

The acute onset of headache with straining and breath holding, as in weightlifter's headache, may be explained by acute venous distension. The development of headache after sustained exertion, particularly on a hot day, is more difficult to understand. Anginal pain may be referred to the head, probably by central connections of vagal afferents and may present as exertional headache, so called cardiac cephalgia. The link to exercise is the important clinical clue. Pheochromocytoma may occasionally be responsible for exertional headache. Intracranial lesions or stenosis of the carotid arteries may have to be excluded as for benign cough headache. Headache may be precipitated by any form of exercise and often has the pulsatile quality of migraine. The most obvious form of treatment is to take exercise gradually and progressively whenever possible. Indomethacin at daily doses varying from 25 to 150 mg is generally very effective in benign exertional headache. Indomethacin 50 mg, ergotamine tartrate 1–2 mg orally, dihydroergotamine by nasal spray, or methysergide 1–2 mg orally given 30–45 min before exercise are useful prophylactic measures.

18.5.4 Sex headache

Sex headache may be precipitated by masturbation or coitus and usually starts as a dull bilateral ache while sexual excitement increases, suddenly becoming intense at orgasm. The term orgasmic cephalgia is not accurate since not all sex headaches require orgasm. Three types of sex headache have been discussed, a dull ache in the head and neck that intensifies as sexual excitement increases, a sudden severe 'explosive' headache occurring at orgasm, and a postural headache. This last is now recognized as being due to low CSF volume due to a CSF leak developing after coitus and is usefully considered with 'new daily persistent headache' as a 'secondary chronic daily headache' (Section 18.6.7).

The essential clinical features of sex headache are:

◆ precipitation by sexual excitement;

◆ bilateral at onset; and

◆ prevented or eased by ceasing sexual activity before orgasm.

Headaches developing at the time of orgasm are not always benign. Subarachnoid haemorrhage (Section 35.16) was precipitated by sexual intercourse in 5 per cent of 66 cases reported by Fisher (1968) and 12 per cent of 50 cases studied by Lundberg and Osterman (1974). One young man reported developed a brainstem thrombosis and another a left hemisphere infarction. Sex headache is reported by men more often than women and may occur at any time during the years of sexual activity. It may develop on several occasions in succession and then not trouble the patient again, despite no obvious change in sexual technique. In patients who stop sexual activity when headache is first noticed, it may subside within a period of 5 min to 2 h, and it is recognized that more frequent orgasms can aggravate established sex headache. About one-third of the patients with sex headache have a history of exertional headaches, but there is no excess of cough headache in patients with sex headache. In about 50 per cent of patients sex headache will settle in 6 months. Migraine is reported in about 25 per cent of patients with sex headache.

Primary sex headaches are usually irregular and infrequent in recurrence, so management can often be limited to reassurance and advice about ceasing sexual activity if a milder, warning headache develops. When the condition recurs regularly or frequently, it can be prevented by the administration of propranolol, but the dosage required varies from 40 to 200 mg daily. An alternative is the calcium channel-blocking agent diltiazem 60 mg three times daily. Ergotamine 1–2 mg or indomethacin 25–50 mg taken about 30–45 min prior to sexual activity can also be helpful.

18.5.5 Hypnic headache

This syndrome was first described in patients aged from 67 to 84 who had headache of a moderately severe nature that typically came on a few hours after going to sleep (Raskin 1988). These headaches last from 15 to 30 min, are typically generalized, although may be unilateral, and can be throbbing. Patients may report falling back to sleep only to be awoken by a further attack a few hours later with up to three repetitions of this pattern over the night. In a large series of 19 patients, 84 per cent were female and the mean age at onset was 61±9 years (Dodick *et al.* 1998). Headaches were bilateral in two-thirds and in 80 per cent of cases mild or moderate. Three patients reported similar headaches when falling asleep during the day. None had photophobia or phonophobia and nausea is unusual.

Patients with this form of headache generally respond to a bedtime dose of lithium carbonate 200–600 mg and in those that do not tolerate this, verapamil or methysergide at bedtime may be alternative strategies. Two patients who responded to flunarizine 5 mg at night have now been reported. One to two cups of coffee or caffeine 60 mg orally at bedtime was reported to be (Dodick *et al.* 1998) helpful. This is a simple approach that is effective in about one-third of patients. A patient poorly tolerant of lithium has been controlled using verapamil 160 mg at night.

18.5.6 Thunderclap headache

Sudden onset of severe headache may occur in the absence of sexual activity and the differential diagnosis includes the sentinel bleed of an intracranial aneurysm, cervicocephalic arterial dissection, and cerebral venous thrombosis. Headaches of explosive onset may also be caused by the ingestion of sympathomimetic drugs or tyramine-containing foods in a patient who is taking monoamine oxidase inhibitors, and can also be a symptom of pheochromocytoma. Whether thunderclap headache can be the presentation of an unruptured cerebral aneurysm is unclear. Day and Raskin (1986) reported a woman with three episodes of sudden onset of very severe headache who was found to have an unruptured aneurysm of the internal carotid artery, with adjacent areas of segmental vasospasm. In the absence of CT scan or CSF evidence of subarachnoid haemorrhage, studies indicate that such patients do very well, and there indeed seems to be a form of benign or primary thunderclap headache.

Wijdicks *et al.* (1988) followed up 71 patients whose CT scans and CSF findings were negative for an average of 3.3 years. Twelve patients had further such headache, and 31, 44 per cent, later had regular episodes of migraine or tension-type headache. Factors identified as precipitating the headache were sexual intercourse in three cases, coughing in four, and exertion in 12, while the remainder had no obvious cause. A history of hypertension was found in 11 and of previous headache in 22. Markus (1991) compared the presentation of 37 patients with subarachnoid haemorrhage and 189 with a similar thunderclap headache but normal CSF examination and could not discern any characteristic to distinguish the two conditions on clinical grounds.

Investigation of any sudden onset of severe headache, be it in the context of sexual excitement or isolated thunderclap headache, should be driven by the clinical context. The first presentation should be vigorously investigated with CT and CSF examination, and if possible MR imaging and angiography. Formal cerebral angiography should be performed if no primary diagnosis is forthcoming, and the clinical situation is particularly suggestive of intracranial aneurysm. Bearing in mind the entity of diffuse multifocal reversible cerebral vasospasm, which may be seen in apparent primary thunderclap headache without there being an intracranial aneurysm, caution in interpretation of angiographic findings is crucial.

18.5.7 Hemicrania continua

Two patients were initially reported with this syndrome, a woman aged 63 years and a man of 53. They developed unilateral headache without obvious cause. One of these patients noticed redness, lacrimation, and sensitivity to light in the eye on the affected side. Both patients were relieved completely by indomethacin while other non-steroidal antiinflammatory drugs were of little or no benefit. 24 previously reported cases, with the addition of 10 new cases, including some with pronounced autonomic features resembling cluster headache, have been reviewed (Newman *et al.* 1992). Their case histories were divided into remitting and unremitting forms. Of the 34 patients reviewed, 22 were women and 12 men with the age of onset ranging from 11 to 58 years. The symptoms were controlled by indomethacin 75–150 mg daily. The essential features of hemicrania continua are:

- unilateral pain;
- pain is moderate and continuous but with fluctuations;
- complete resolution of pain with indomethacin; and
- exacerbations may be associated with autonomic features.

Apart from analgesic overuse as an aggravating factor, and a report in an HIV-infected patient, the status of secondary hemicrania continua is unclear. The 'indotest' has been proposed by which the intramuscular injection of indomethacin 50 mg could be used as a diagnostic tool (Antonaci *et al.* 1998). In hemicrania continua, pain was relieved in 73 ± 66 min and the pain-free period was 13 ± 8 h. A placebo-controlled modification of this test is preferred to the open-label version. Using the latter method in conjunction with positron emission tomography, it has been shown that there is activation of the contralateral posterior hypothalamus and ipsilateral dorsal rostral pons in association with the headache of hemicrania continua, as well as activation of the ipsilateral ventrolateral midbrain (Matharu *et al.* 2004b). The alternative is a trial of oral indomethacin, initially 25 mg three times daily, then 50 mg three times daily, and then 75 mg three times daily. One should allow up to 2 weeks for any dose to have a useful effect. Acute treatment with sumatriptan has been employed and reported to be of no benefit in hemicrania continua. Cyclooxygenase II antagonists seem effective in hemicrania continua, and topiramate is helpful in some patients, as is greater occipital nerve block.

18.6 Chronic daily headache

18.6.1 Range of conditions

Each of the above primary headache forms can occur very frequently. When a patient experiences headache on 15 days or more a month one can apply the broad diagnosis of chronic daily headache. Chronic daily headache is not one thing but a collection of very different problems with different management strategies. Crucially not all daily headache is simply tension-type headache (Table 18.13). This is the commonest clinical misconception in headache of confusing the clinical phenotype with the headache *biotype*. Population-based estimates of daily headache are remarkable, demonstrating that 4.5–4.8 per cent of the Western populations have daily or near daily headache. Daily headache may be primary or secondary, and it seems clinically useful to consider the possibilities in this way when making management decisions (Table 18.13). Population-based studies bear out clinical practice in that a large group of refractory daily headache patients overuse various over-the-counter preparations.

Table 18.13 Classification of chronic daily headache

Primary		Secondary
> 4 h daily	< 4 h daily	
Chronic migraine†	Chronic cluster headache‡	Post-traumatic
		◆ Head injury
		◆ Iatrogenic
		◆ Post-infectious
Chronic tension-type headache†	Chronic paroxysmal hemicrania	Inflammatory, such as
		◆ Giant cell arteritis
		◆ Sarcoidosis
		◆ Behcets syndrome
Hemicrania continua†	SUNCT	Chronic central nervous system infection
New daily persistent headache†	Hypnic headache	Substance abuse headache

†May be complicated by analgesic overuse. In the case of substance abuse headache, the headache is completely resolved after the substance abuse is controlled (Headache Classification Committee of The International Headache Society 2004). Clinical experience suggests that many patients continue to have headache even after cessation of analgesic use. The residual headache probably represents the underlying headache biology.

‡Chronic cluster headache patients may have more than 4 h per day of headache. The inclusion of the syndrome here is to emphasize that, by and large, the attacks themselves are less than 4-h duration.

18.6.2 Chronic migraine

While it is widely accepted that some of the primary headaches, tension-type headache, cluster headache, and paroxysmal hemicrania, have chronic varieties, this question seems to have become unnecessarily troublesome for migraine. Few headache authorities would argue that migraine can *never ever* be chronic in terms of frequency, but the issue of whether patients with frequent headache, some of which fulfils standard criteria for migraine and some for tension-type headache, have a single migrainous biology is a very vexed one. Given that tension-type headache describes a phenomenology that is indistinct at best it seems unlikely that all its phenotype will have a single biological generator.

The concept behind chronic migraine is that some patients who inherit a migrainous biology end up with chronic daily headache. The typical patient will have daily headache of a dull, non-specific type, punctuated by more severe attacks that would often, in isolation, fulfil standard criteria for migraine. In headache speciality clinics this group is dominant, with about 90 per cent of patients having chronic migraine usually with analgesic overuse. It could be suggested that they have a biologically more difficult problem and this is the basis for their over-representation in referral centres.

If one applies the concepts outlined for tension-type headache (Section 18.3) then the diagnosis of chronic tension-type headache chronic tension-type headache is made when the patient has 15 days or more a month of entirely featureless generalized dull or pressure-like pain. When *any* of the attacks on *some* days have migrainous features: nausea, photophobia, phonophobia, throbbing, or aggravation with movement, then chronic migraine is diagnosed. The problem is not that both chronic migraine and chronic tension-type headache do not exist, but that some patients must simply have chronic tension-type headache and episodic migraine, two conditions; it is, however, simply impossible on clinical or other grounds to determine who they are. The approach outlined over-diagnoses chronic migraine, taking that to be a biological entity, and under-diagnoses the co-existence of chronic tension-type headache and episodic migraine. The converse would be true, if one diagnoses all as chronic tension-type headache and episodic migraine, then chronic migraine is missed. In clinical practice the concept of chronic migraine is particularly helpful. Given that the life-style advice is identical for both tension-type headache and migraine, and that the range of therapeutic options for preventive treatment in migraine is so much greater, the clinician loses nothing by diagnosing chronic migraine, and the patient has much to gain.

18.6.3 Management

The management of chronic daily headache can be very rewarding. Most patients overusing analgesics respond very sensibly when the problem is explained.

The keys to managing daily headache are:

◆ exclude treatable causes (Table 18.13);

◆ obtain a clear analgesic history; and

◆ make a diagnosis of the primary headache type involved.

18.6.4 Medication overuse

For outpatients it is essential that analgesic use be reduced and eliminated. Patients can reduce their use either by, as an example, 10 per cent every week or two, depending on their circumstances, or if they wish, and there is no contraindication, by immediate cessation of use. Either approach can be facilitated by first keeping a careful diary over a month or two to be sure of the size of the problem. A small dose of an non-steroidal antiinflammatory drug, such as naproxen 500 mg bd if tolerated, will take the edge off the pain as the analgesic use is reduced. It is useful to note that non-steroidal antiinflammatory drug overuse does not seem to be a common cause of daily headache with once or twice daily dosage, whereas with more frequent dosing problems may develop. When the patient has reduced their analgesic use substantially a preventive should be introduced. It must be emphasized that preventive therapies often do not work in the presence of analgesic overuse. Thus the patient must reduce the analgesics or the entire attempt to use the preventive is largely wasted, although this helpful rule must have some limitations worthy of study. The most common cause of intractability to treatment is the use of a preventive when analgesics continue to be used regularly. For some patients this poses a difficult problem in management and often one must be blunt that some degree of pain is inevitable in the first instance if the problem is to be controlled ultimately.

Some patients with medication overuse will require admission for detoxification. Broadly they consist of two groups, those who fail outpatient withdrawal, or those who have a significant complicating medical indication, such as brittle diabetes mellitus, or complicating medicines, such as opioids, where withdrawal may be problematic as an outpatient. When such patients are admitted acute medications are withdrawn completely on the first day, unless there is some contraindication. Anti-emetics, such as domperidone oral or suppositories, and fluids are administered as required, as well as clonidine for opioid withdrawal symptoms. For acute intolerable pain during the waking hours intravenous aspirin 1g is useful and at night chlorpromazine by injection, ensuring adequate hydration. If the patient does not settle over 3–5 days a course of intravenous dihydroergotamine can be employed as Raskin described (Raskin 1986). As time goes by one feels that dihydroergotamine is indispensable in this setting; administered 8-hourly for 3 days, it can induce a significant remission that allows a preventive treatment to be established. Often 5-HT$_3$ antagonists, such as ondansetron or granisetron, will be required with dihydroergotamine as it is essential to ensure that the patient does not have significant nausea.

18.6.5 Preventive treatments

The tricyclics, amitriptyline or dothiepin, at doses up to 1 mg/kg are very useful in patients with chronic daily headache. Tricyclics are started in low dosage of 10–25mg daily and best given 12 h prior to when the patient wishes to wake up so as to avoid excessive morning sleepiness. The other (excessive) useful medications for these patients are the anticonvulsants, such as valproate, topiramate, and gabapentin. Valproate doses up to 1500 mg daily are used, starting at 200 mg bd and increasing to 400 mg or 600 mg bd as tolerated over 2–4-week intervals. The blood count and liver enzymes should be checked at baseline and the various side effects explained to patients, especially the foetal abnormalities to females. For topiramate one can start at 25 mg nightly and increase by 25 mg every 10–14 days to aim for 50 mg twice daily (Table 18.4). For gabapentin the dose is 1800–3600 mg daily; it is very well tolerated, although probably less effective from a population viewpoint. For some patients flunarizine can be very effective, as can methysergide or phenelzine.

18.6.6 New daily persistent headache

New daily persistent headache is a clinically distinct syndrome with a range of important possible causes (Table 18.14). From a nosological point of view all entities mentioned here could be placed at various categories by the Headache Classification Committee of The International Headache Society (2004). However, the term serves both patients and clinicians by highlighting a group of conditions some of which are curable. New daily persistent headache can have both primary and secondary forms (Table 18.14) and neurologists will be called on to diagnose and treat these patients.

The patient with new daily persistent headache presents with a history of headache on most if not all days that lasts from one day to the next. The onset of headache is abrupt, often from one moment to the next, although the current definition accepts a period of onset over three days. The typical history is for the patient

Table 18.14 Differential diagnosis of new daily persistent headache

Primary	Secondary
◆ Migrainous-type	◆ Sub-arachnoid hemorrhage
◆ Featureless (tension-type)	◆ Low CSF volume headache
	◆ Raised CSF pressure headache
	◆ Post-traumatic headache*
	◆ Chronic meningitis

*Includes post-infective forms.

to recall the exact day and circumstances, so from one moment to the next a headache develops that never leaves them. This presentation provokes certain key questions about the onset and behaviour of the pain. These need to be woven with the more generic questions that one asks a patient with persistent headache, to form a provisional diagnosis. The pressing issues arise from considering the differential diagnosis, particularly of the secondary headache forms. Although subarachnoid haemorrhage is listed for some logical consistency, as the headache may certainly come on from one moment to the next, it is not likely to produce diagnostic confusion in this group of patients. The issues surrounding late imaging and management of unruptured aneurysms are covered elsewhere (Section 35.16.6). Suffice to say that subarachnoid haemorrhage is so important that it must always be considered if only to be excluded, either by history or appropriate investigation.

Primary new daily persistent headache. Initial descriptions of primary new daily persistent headache recognized it to occur in both males and females. Migrainous features were common, with unilateral headache in about one-third and throbbing pain in about one-third. Nausea was reported in about half the patients, as was photophobia and phonophobia observed again in about half. A number of these patients have a previous history of migraine but not more than one might expect given the population prevalence of migraine. It is remarkable that the initial report noted that 86 per cent of patients were headache free at 24 months. It is general experience amongst headache specialist that primary new daily persistent headache is perhaps the most intractable and least therapeutically rewarding form of headache. In general one can classify the dominant phenotype, migraine or tension-type headache, and treat with preventives according to that sub-classification, as for patients with chronic daily headache. Primary new daily persistent headache with a tension-type headache phenotype is very unresponsive to treatment.

Secondary new daily persistent headache. The secondary causes of the syndrome of new daily persistent headache are worthy of consideration, as they have distinctive clinical pictures that can guide investigation (Table 18.14).

18.6.7 Low CSF volume headache

The syndrome of persistent low CSF volume headache is an important diagnosis not to miss. The more immediately obvious form of this problem is encountered commonly in neurology after lumbar puncture. In that situation the headache usually settles rapidly with bedrest. In the chronic situation the patient typically presents with a history of headache from one day to the next. The pain is generally not present on waking, worsens during the day, and is relieved by lying down. Recumbency usually improves the headache in minutes, and it takes only minutes to an hour for the pain to return when the patient is upright again. The patient may give a history of an index event: lumbar puncture or epidural injection, or a vigorous Valsalva, such as with lifting, straining, coughing, clearing the eustachian tubes in an aeroplane, or multiple orgasms. Patients may volunteer, or a history may be obtained, that soft drinks with caffeine provide temporary respite. Spontaneous leaks are recognized, and the clinician should not be put off the diagnosis if the headache history is typical but there is no obvious index event. As time passes from the index event the postural nature may be less obvious; certainly cases whose index event was several years prior to the eventual diagnosis are recognized.

The term low volume rather than low pressure is used, since there is no clear evidence at which point the pressure can be called low. While low pressures, such as 0–5 cm CSF are usually identified, a pressure of 16 cm CSF has been recorded with a documented leak. One should be aware of the possibility of the development of subdural collections in patients with low CSF volume headaches, which makes imaging before any invasive studies all the more important.

The investigation of choice is MRI with gadolinium (Fig. 18.5), which produces a striking pattern of diffuse pachymeningeal enhancement, although in about 10 per cent of cases a leak can be documented without enhancement. This finding of diffuse meningeal enhancement is so typical that in clinical context immediate treatment is indicated. It is also common to see Chiari malformations on MRI with some degree of descent of the cerebellar tonsils. This is important from the neurologist's viewpoint since surgery in such settings simply makes the headache problem worse. Any patient being considered for such surgery for a headache indication should be reviewed by a neurologist first. Alternatively the CSF pressure may be determined, or a leak sought with ^{111}In-DPTA CSF studies that can demonstrate the leak and any early emptying of tracer into the bladder, indicative of a leak.

Treatment is bedrest in the first instance. False positive transient improvement in persistent low CSF volume headache with chiropractic and other similar therapies is recognized where the treatment necessitated the patient lying down for a prolonged period. Intravenous caffeine, 500 mg in 500 ml saline administered over 2 h, is the standard and often very efficacious treatment. The electrocardiogram should be checked for any arrthymia prior to administration. A reasonable practice is to carry out at least two infusions separated by 4 weeks after obtaining the suggestive clinical history and MRI with enhancement. Since intravenous caffeine is safe, and can be curative, by an unknown mechanism, it spares many patients the need for further tests. If that is unsuccessful, an abdominal binder may be helpful. If a leak can be identified, either by the radioisotope study, or by CT myelogram, or spinal T2-weighted MRI, an autologous blood patch is usually curative. In more intractable situations theophylline is a useful alternative that allows out-patient management.

18.6.8 Raised CSF pressure headache

As is the case for low CSF pressure states, raised CSF pressure as a cause of headache is well recognized by neurologists. Brain imaging can often reveal the cause, such as raised pressure due to a space-occupying lesion. The particular setting in which patients enter the spectrum of new daily persistent headache are those with idiopathic intracranial hypertension (Section 26.5.6) who present with headache without visual problems, particularly with normal fundi. It is recognized that intractable chronic migraine can be triggered by persistently raised intracranial pressure. These patients typically give a history of generalized headache that is present on waking, and gets better as the day goes on. It is generally worse with recumbency. Visual obscurations are frequently reported. Fundal changes on raised intracranial pressure would make the diagnosis relatively straightforward but it is in those without such changes that the history must drive investigation. Patients often report a curious whooshing sensation in the occipital region.

Brain imaging is mandatory if raised pressure is suspected, and it is most simple in the long run to obtain an MRI, and include MR venography. The CSF pressure should be measured by lumbar puncture taking care to do so when the patient is symptomatic, so that both the pressure and response to removal of 20 ml of CSF can be determined. A raised pressure and improvement in headache with removal of CSF is diagnostic of the problem. The visual fields

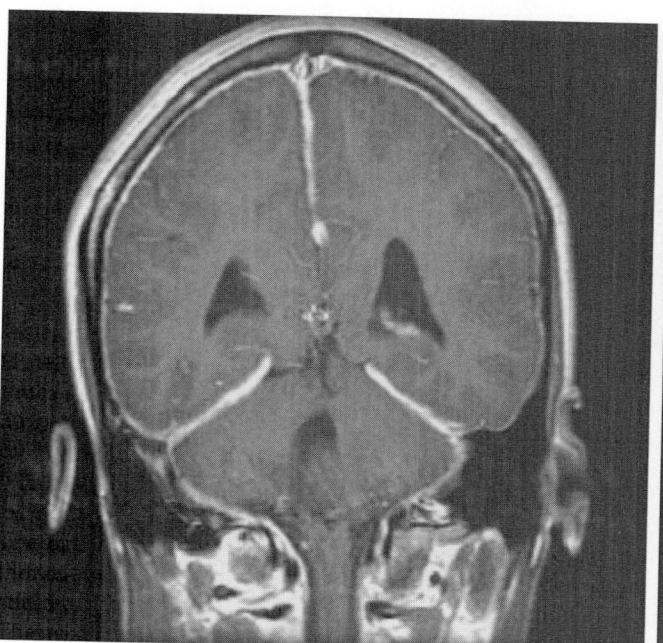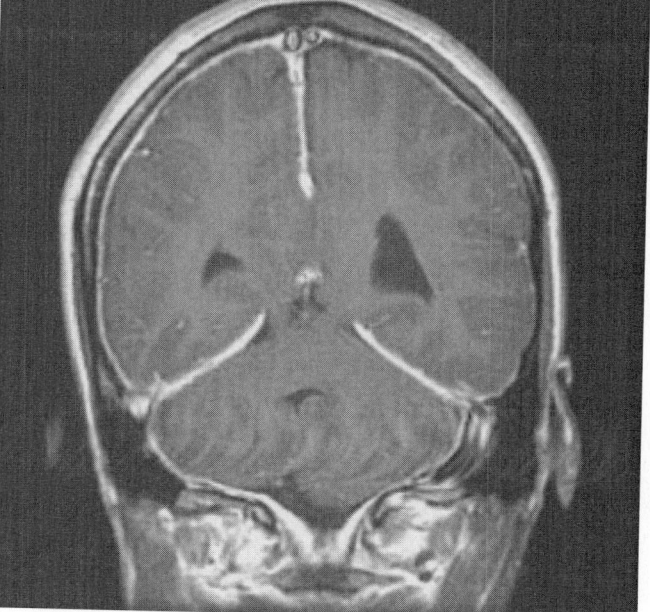

Fig. 18.5 MRI showing diffuse meningeal enhancement after gadolinium administration in a patient with low CSF volume headache.

should be formally documented even in the absence of overt ophthalmic involvement. Initial treatment can be with acetazolamide 250–500 mg twice daily. The patient may respond in weeks with improvement in headache. If this is not effective topiramate has many actions that may be useful in this setting: carbonic anhydrase inhibition, weight loss, and neuronal membrane stabilization probably through actions on phosphorylation pathways. A small number of severely disabled patients who do not respond to medical treatment will come to intracranial pressure monitoring and even shunting. This is exceptional and not to be undertaken without careful work-up.

18.6.9 Post-traumatic headache

The issue of post-traumatic headache is vexed. The existence of such a syndrome is accepted (Headache Classification Committee of The International Headache Society 2004). Much of the scientific discussion becomes marred by the often-quoted medico-legal morass concerning delayed effects of head injury (Section 25.6.1). The term is used here to indicate trauma in a very broad way. New daily persistent headache may be seen after a blow to the head but more commonly after an infective episode, typically viral, or in one case malarial meningitis. A recent series identified one-third of all patients with new daily persistent headache reported the headache starting after a flu-like illness. The patient may note a period in which they had a significant infection: fever, neck stiffness, photophobia, and marked malaise. The headache starts during that period and never stops. Investigation reveals no on going cause for the headache. It has been suggested that some patients with this syndrome have a persistent Epstein–Barr infection, but such a syndrome is anything but clearly delineated. A complicating factor will often be that the patient had a lumbar puncture during that illness, so a persistent low CSF volume headache needs to be considered initially. Post-traumatic headache may be seen after carotid artery dissection, sub-arachnoid hemorrhage, and following intracranial surgery for a benign mass. The underlying theme seems to be that a traumatic event involving the dura mater can trigger a headache process that lasts for many years after that event.

The treatment of this form of NDPH is substantially empirical. Tricyclics, notably amitriptyline, and anticonvulsants, valproate, topiramate, and gabapentin, have been used with good effects. The MAOI phenelzine may also be useful in carefully selected patients. On the positive side the headache seems to run a limited course of 3–5 years, so will eventually settle. It can certainly be very disabling in that period.

18.7 Other forms of secondary headache

These are summarized in Table 18.15.

18.7.1 Giant cell arteritis

This is an important cause of headache because delay in steroid treatment may result in blindness due to retinal artery ischaemia (Section 12.4.4). It is also known as temporal arteritis or cranial arteritis. Patients are usually elderly with focal tenderness of the

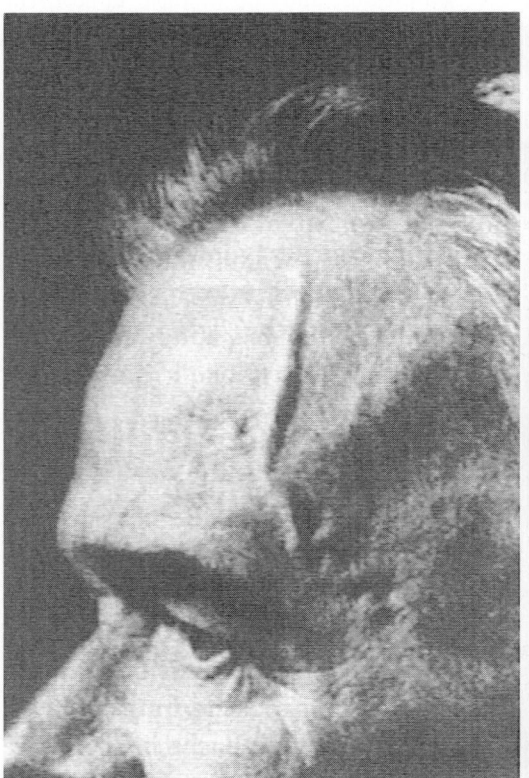
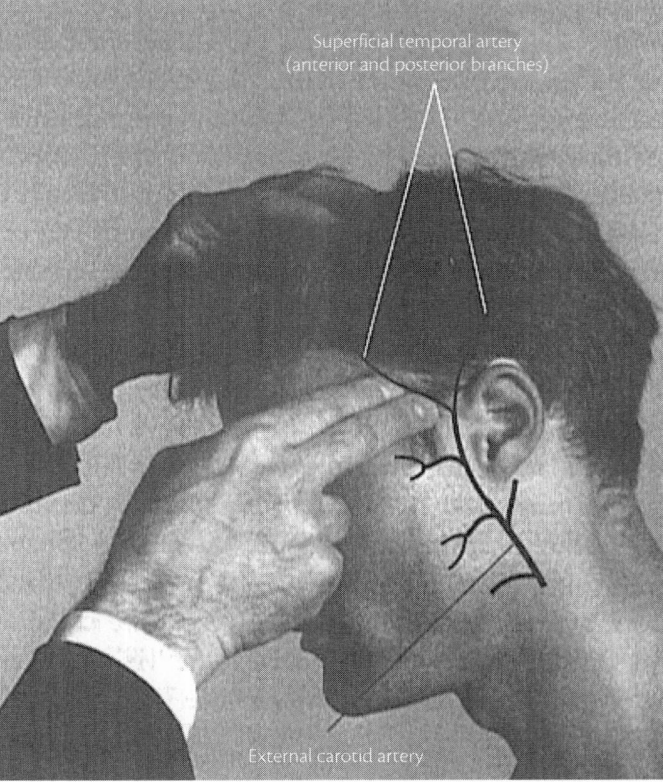

Fig. 18.6 (A) Indurated, tender temporal artery in cranial arteritis; (B) How to palpate the temporal artery.

Table 18.15 Other secondary headaches

- Giant cell arteritis (Sections 18.7.1 and 36.2.8)
- Cervicogenic headache (Section 18.7.2)
- Reader's paratrigeminal neuralgia (Section 18.7.3)
- Tolosa–Hunt syndrome (Sections 13.1.5 and 19.2.8)
- Headache as a presentation of cervical dystonia
- Headache in temporomandibular dysfunction (Section 19.3.1)
- Cardiac cephalalgia (Lance and Lambros 1998)
- Headache with endocrine disturbance, particularly pituitary tumour (Levy *et al.* 2005)
- Neck–tongue syndrome (Bogduk 1981)
- Red-ear syndrome (Lance 1996)

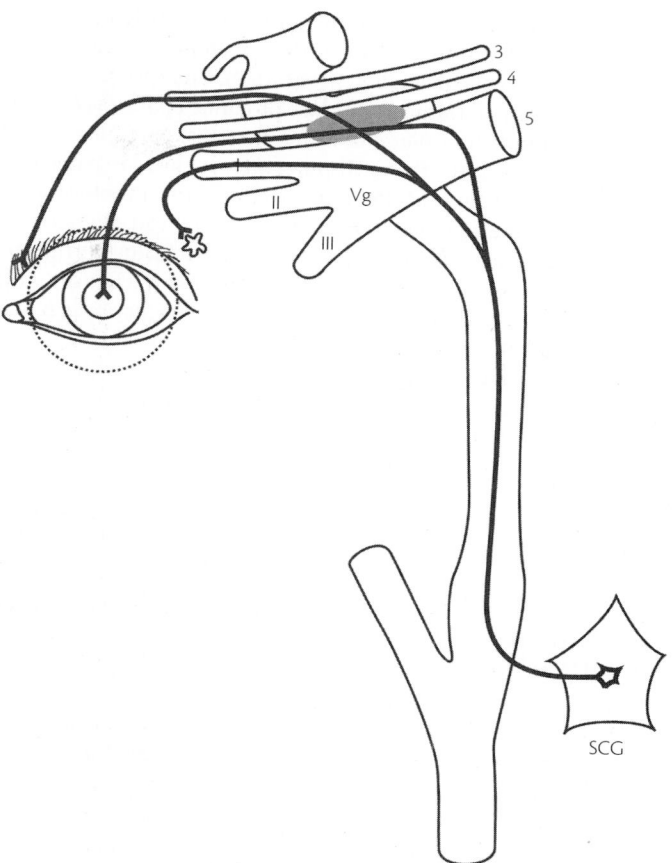

Fig. 18.7 Drawing of the anatomy of the paratrigeminal oculo-sympathetic syndrome (Goadsby 2002b). Sympathetic fibres from the superior cervical ganglion (SCG) travel with the internal carotid artery in the neck branching to innervate the sweat glands, levator palpebrae superioris (specifically Müller's muscle), and the pupilo-dilator fibres. The latter fibres travel near the ophthalmic portion of the trigeminal nerve, so that in this paratrigeminal region (grey shade) lesions produce the classic syndrome. *Abbreviations*: 3, oculomotor n.; 4, trochlear n.; 5, trigeminal nerve; Vg, trigeminal ganglion, with first (I), second (II), and third (III) branches.

scalp which may be provoked markedly by resting the head on the pillow. Jaw claudication provoked by chewing is a characteristic but relatively uncommon feature. Constitutional symptoms are common, particularly weight loss, malaise, or polymyalgia rheumatica. An elevated ESR is a strong pointer to the diagnosis. The temporal artery may be tenderly inflamed, swollen, or pulseless (Fig. 18.6). On suspicion of this diagnosis steroid treatment should be started pending the result of temporal artery biopsy. The clinical features and pathogenesis of this vasculitic condition are considered in Section 36.2.8.

18.7.2 Cervicogenic headache

It is a time-honoured concept that the neck is responsible for much of what is seen in headache referral practice. Unfortunately as with much of history the good story is often ruined by the facts. Whilst there is little doubt that there is a rich overlap between the innervation of intracranial pain-producing structures by the ophthalmic division of the trigeminal nerve, and the posterior fossa and high cervical innervation by branches especially of the C_2 dorsal root (Bartsch and Goadsby 2005), causality is complex. The Headache Classification Committee recognizes that head pain can arise from the neck and labels this cervicogenic headache (Headache Classification Committee of The International Headache Society 2004). The term has been used by others to define a syndrome (Antonaci *et al.* 2001) that is so confusing as to be useless in practice (Goadsby 2004). Most patients with neck discomfort and headache referred to specialty practice will have migraine. They will have neck stiffness or discomfort as a premonitory symptom that can clearly persist in all stages of the attack (Giffin *et al.* 2003). They may respond to local therapies, such as greater occipital nerve injection (Afridi *et al.* 2006), however, this implies no more than triggering, and is to be expected. The pursuit of neck pathology, and the treatment of patients who have migraine by manipulative or physical means has no support in the controlled literature, and is rarely of long-lasting value.

18.7.3 Raeder's Syndrome

J.G. Raeder (1924) wrote a classical clinical-anatomical colocalization paper describing five patients with two key features: involvement of the trigeminal nerve and the oculo-sympathetic nerves. He sought to differentiate the restricted oculo-sympathetic findings from the classical Horner's syndrome: cervical sympathetic dysfunction characterized by ptosis, miosis, anhidrosis, and enophthalmos. Since that time various terms have been employed, meanings defined, and classifications developed (Mokri 1982; Vijayan and Watson 1986). Raeder made an interesting clinical observation that pointed out the likely localization of a lesion adjacent the trigeminal nerve in the middle cranial fossa (Fig. 18.7). Solomon and Lustig (2000) recently set out the clinical cases that in many respects have illustrated the trigemino-sympathetic anatomy of the carotid artery, concluding that the use of the term Raeder's paratrigeminal neuralgia had become corrupted to the point of being useless by careless attribution of cases that did not respect the anatomy. Conditions such as carotid disease, particularly dissection may give rise to pain and Horner's syndrome, and cluster headache may lead to oculo-sympathetic loss and impaired sympathetic facial sweating (Drummond 1988a, b). In both situations forehead sweating may be impaired.

The key anatomical feature to be understood for Raeder's syndrome is the relationship between the trigeminal nerve and the oculo-pupillary sympathetic fibres (Fig. 18.7). The trigeminal nerve lies in middle cranial fossa and in close proximity, *paratrigeminally*, there are a number of other cranial nerves. Most particularly for a short course the fibres that will innervate the levator palpebrae superioris, specifically Müller's muscle, and the pupilo-dilator fibres without the sudomotor fibres for the forehead. A restricted lesion in middle cranial fossa might cause the syndrome of trigeminal nerve involvement, neuralgic pain or sensory change, with ptosis or miosis, or both, but no anhidrosis. Such a paratrigeminal oculo-sympathetic syndrome usefully reminds clinicians to pursue vigorously possible lesions of the middle cranial fossa with careful, and possibly repeated, imaging studies. Whether Raeder's name should remain attached is open to debate (Goadsby 2002b).

References

Afridi SK, Shields KG, Bhola R *et al.* (2006). Greater occipital nerve injection in primary headache syndromes-prolonged effects from a single injection. *Pain*, **122**, 126–9.

Andlin-Sobocki P, Jonsson B, Wittchen HU *et al.* (2005). Cost of disorders of the brain in Europe. *Eur J Neurol*, **12**, 1–27.

Antonaci F, Fredriksen T, Sjaastad O (2001). Cervicogenic headache: clinical presentation, diagnostic criteria, and differential diagnosis. *Curr Pain Headache Rep*, **5**, 387–92.

Antonaci F, Pareja JA, Caminero AB *et al.* (1998). Chronic paroxysmal hemicrania and hemicrania continua. Parenteral indomethacin: the 'Indotest'. *Headache*, **38**, 122–8.

Antonaci F, Sjaastad O (1989). Chronic paroxysmal hemicrania (CPH): a review of the clinical manifestations. *Headache*, **29**, 648–56.

Bahra A, Matharu MS, Buchel C *et al.* (2001). Brainstem activation specific to migraine headache. *The Lancet*, **357**, 1016–7.

Bartsch T, Goadsby PJ (2005). Anatomy and physiology of pain referral in primary and cervicogenic headache disorders. *Headache Currents*, **2**, 42–8.

Bogduk N (1981). An anatomical basis for the neck-tongue syndrome. *J Neurol Neurosurg Psychiatry*, **44**, 202–8.

Burns B, Watkins L, Goadsby PJ (2007). Treatment of medically intractable cluster headache by occipital nerve stimulation: long-term follow-up of eight patients. *Lancet*, **369**, 1099–106.

Cohen AS, Goadsby PJ (2006a). Functional neuroimaging of primary headache disorders. *Expert Rev Neurother*, **6**, 1159–72.

Cohen AS, Goadsby PJ (2006b). Short-lasting Unilateral Neuralgiform Headache Attacks with Conjunctival injection and Tearing (SUNCT) or cranial Autonomic features (SUNA). A prospective clinical study of SUNCT and SUNA. *Brain*, **129**, 2746–60.

Day JW, Raskin NH (1986). Thunderclap headache: symptom of unruptured cerebral aneurysm. *Lancet*, **2**, 1247–8.

Dodick DW, Mosek AC, Campbell JK (1998). The hypnic ('alarm clock') headache syndrome. *Cephalalgia*, **18**, 152–6.

Drummond PD (1988a). Autonomic disturbance in cluster headache. *Brain*, **111**, 1199–209.

Drummond PD (1988b). Dysfunction of the sympathetic nervous system in cluster headache. *Cephalalgia*, **8**, 181–6.

Fisher CM (1968). Headache in cerebrovascular disease. In Vinken PJ, Bruyn GW, eds. *Handbook of Clinical Neurology*, Vol. 5, pp. 124–6. Elsevier, Amsterdam.

Giffin NJ, Ruggiero L, Lipton RB *et al.* (2003). Premonitory symptoms in migraine: an electronic diary study. *Neurology*, **60**, 935–40.

Goadsby PJ (2000). The pharmacology of headache. *Prog Neurobiol*, **62**, 509–25.

Goadsby PJ (2002a). Pathophysiology of cluster headache: a trigeminal autonomic cephalgia. *Lancet Neurol*, **1**, 37–43.

Goadsby PJ (2002b). Raeders Syndrome: 'Paratrigeminal' paralysis of oculo-pupillary sympathetic. *J Neurol Neurosurg Psychiatry*, **72**, 297–9.

Goadsby PJ (2004). A critical view of cervicogenic headache. In Sjaastad O, Fredriksen TA, Bono G, Nappi G, eds. *Cervicogenic Headache*, pp. 131–6. Smith-Gordon, London.

Goadsby PJ (2005). Can we develop neurally-acting drugs for the treatment of migraine? *Nat Rev Drug Discov*, **4**, 741–50.

Goadsby PJ, Lipton RB (1997). A review of paroxysmal hemicranias, SUNCT syndrome and other short-lasting headaches with autonomic features, including new cases. *Brain*, **120**, 193–209.

Goadsby PJ, Lipton RB, Ferrari MD (2002). Migraine- current understanding and treatment. *N Engl J Med*, **346**, 257–70.

Goadsby PJ, Silberstein SD, eds. (1997). *Headache*, Vol. 17, Butterworth-Heinemann, New York.

Headache Classification Committee of The International Headache Society (2004). The International classification of headache disorders (second edition). *Cephalalgia*, **24**, 1–160.

Kudrow L, Esperanca P, Vijayan N (1987). Episodic paroxysmal hemicrania? *Cephalalgia*, **7**, 197–201.

Lance JW (1996). The red ear syndrome. *Neurology*, **47**, 617–20.

Lance JW, Goadsby PJ (2005). *Mechanism and Management of Headache*, New York, Elsevier.

Lance JW, Lambros J (1998). Headache associated with cardiac ischemia. *Headache*, **38**, 315–6.

Leone M, Franzini A, Broggi G *et al.* (2004). Long-term follow-up of bilateral hypothalamic stimulation for intractable cluster headache. *Brain*, **127**, 2259–64.

Levy M, Matharu MS, Meeran K *et al.* (2005). The clinical characteristics of headache in patients with pituitary tumours. *Brain*, **128**, 1921–30.

Lipton RB, Stewart WF, Diamond S *et al.* (2001). Prevalence and burden of migraine in the United States: data from the American Migraine Study II. *Headache*, **41**, 646–57.

Lundberg PO, Osterman PO (1974). The benign and malignant forms of orgasmic cephalgia. *Headache*, **14**, 164–5.

Markus HS (1991). A prospective follow-up of thunderclap headache mimicking subarachnoid haemorrhage. *J Neurol, Neurosurg Psychiatry*, **54**, 1117–25.

Matharu MS, Bartsch T, Ward N *et al.* (2004a). Central neuromodulation in chronic migraine patients with suboccipital stimulators: a PET study. *Brain*, **127**, 220–30.

Matharu MS, Cohen AS, Frackowiak RSJ *et al.* (2006). Posterior hypothalamic activation in paroxysmal hemicrania. *Ann Neurol*, **59**, 535–45.

Matharu MS, Cohen AS, McGonigle DJ *et al.* (2004b). Posterior hypothalamic and brainstem activation in hemicrania continua. *Headache*, **44**, 747–61.

May A, Ashburner J, Buchel C *et al.* (1999a). Correlation between structural and functional changes in brain in an idiopathic headache syndrome. *Nat Med*, **5**, 836–8.

May A, Bahra A, Buchel C *et al.* (1998). Hypothalamic activation in cluster headache attacks. *Lancet*, **352**, 275–8.

May A, Bahra A, Buchel C *et al.* (1999b). Functional MRI in spontaneous attacks of SUNCT: short-lasting neuralgiform headache with conjunctival injection and tearing. *Ann Neurol*, **46**, 791–3.

Menken M, Munsat TL, Toole JF (2000). The global burden of disease study - implications for neurology. *Arch Neurol*, **57**, 418–20.

Mokri B (1982). Raeder's paratrigeminal neuralgia. Original concept and subsequent deviations. *Arch Neurol*, **39**, 395–9.

Newman LC, Gordon ML, Lipton RB *et al.* (1992). Episodic paroxysmal hemicrania: two new cases and a literature review. *Neurology*, **42**, 964–6.

Olesen J, Goadsby PJ (1999). *Cluster Headache and Related Conditions*, Vol. 9. Oxford University Press, Oxford.

Olesen J, Tfelt-Hansen P, Ramadan N *et al.* (2005). *The Headaches*. Lippincott, Williams & Wilkins, Philadelphia.

Pascual P, Iglesias F, Oterino A *et al.* (1996). Cough, exertional, and sexual headache. *Neurology*, **46**, 1520–4.

Raeder JG (1924). "Paratrigeminal" paralysis of the oculo-pupillary sympathetic. *Brain*, **47**:149–58.

Raskin NH (1986). Repetitive intravenous dihydroergotamine as therapy for intractable migraine. *Neurology*, **36**, 995–7.

Raskin NH (1988). The hypnic headache syndrome. *Headache*, **28**, 534–6.

Raskin NH (1995). The cough headache syndrome: treatment. *Neurology*, **45**, 1784.

Raskin NH, Schwartz RK (1980). Icepick-like pain. *Neurology*, **30**, 203–5.

Silberstein SD, Lipton RB, Goadsby PJ (2002). *Headache in Clinical Practice*. Martin Dunitz, London.

Sjaastad O, Dale I (1976). A new (?) clinical headache entity 'chronic paroxysmal hemicrania'. *Acta Neurol Scand*, **54**, 140–59.

Sjaastad O, Saunte C, Salvesen R *et al.* (1989). Shortlasting unilateral neuralgiform headache attacks with conjunctival injection, tearing, sweating, and rhinorrhea. *Cephalalgia*, **9**, 147–56.

Solomon S, Lustig P (2001). Benign Raeder's syndrome is probably a manifestation of carotid artery disease. *Cephalalgia*, **21**, 1–11.

Steiner TJ, Scher AI, Stewart WF *et al.* (2003). The prevalence and disability burden of adult migraine in England and their relationships to age, gender and ethnicity. *Cephalalgia*, **23**, 519–27.

Tfelt-Hansen P, Saxena PR, Dahlof C *et al.* (2000). Ergotamine in the acute treatment of migraine- a review and European consensus. *Brain*, **123**, 9–18.

Vijayan N, Watson C (1986). Raeder's syndrome, pericarotid syndrome and carotidynia. In Rose FC, ed. *Headache*, Vol. 48, pp. 329–41. Elsevier Science Publishers, BV, Amsterdam.

Wijdicks EFM, Kerkhoff H, van Gijn J (1988). Long-term follow up of 71 patients with thunderclap headache mimicking subarachnoid haemorrhage. *Lancet*, **2**, 68–70.

CHAPTER 19

Craniofacial pain

John Scadding

Contents

19.1 Clinical approach and differential diagnosis

Craniofacial pain, excluding the headache disorders, comprises a heterogeneous group of conditions. Some fall within the sphere of the neurologist, but many call for other specialist skills in diagnosis and management. The site of pain is not always a good guide to either the nature of the pain or the tissue of origin. Patients with craniofacial pain are frequently referred to the neurologist on an assumption that the pain is neuralgic, and for this reason neurologists need to be aware of the many potential causes and be prepared to collaborate closely with other specialists, including the ear, nose, and throat surgeon, ophthalmologist, dental surgeon, oral medicine physician, and psychiatrist. An accurate and detailed history is essential, because in many patients with facial pain, there will be no abnormal physical signs and the diagnosis rests entirely on the history. The causes of craniofacial pain are listed in Table 19.1, in a classification partly anatomical and partly pathological.

The somatosensory innervation of craniofacial structures is through the sensory components of the trigeminal, facial, glossopharyngeal, and vagus nerves, and the upper cervical roots, C2 and C3. The trigeminal innervation is extensive, involving deep tissues as well as skin (Table 19.2). Pathological processes affecting non-cutaneous tissues innervated by the trigeminal nerve can cause pain that is localized to the affected area, or diffuse pain radiating onto the face, leading to diagnostic difficulty. However, processes affecting trigeminal innervated structures cause facial pain that is usually experienced within the confines of trigeminal territory. Furthermore, with a few exceptions, facial pain with a serious underlying cause, calling for neurological or other investigations, is nearly always strictly unilateral. Pain that extends beyond trigeminal territory or is bilateral is much less likely to have an underlying neurological basis.

Accurate diagnosis of craniofacial pain can be either very straightforward or extremely challenging; liaison with specialists in the relevant fields listed above is frequently advisable. The neurologist must be aware of the broad range of causes of craniofacial pain, should take a comprehensive and detailed history, and perform a careful physical examination. The examination should include not only a full neurological assessment, but additionally, depending on the context and distribution of symptoms, the following may also require assessment:

◆ temporal arteries and scalp tenderness;

◆ temporomandibular joint movement and tenderness;

Table 19.1 Causes of craniofacial pain excluding headache disorders

Craniofacial neuralgias	Idiopathic trigeminal neuralgia
	Symptomatic trigeminal neuralgia
	Glossopharyngeal neuralgia
	Herpes zoster neuralgia
	C2 and C3 neuralgia
	Occipital neuralgia
	Posttraumatic neuralgia
	Painful ophthalmoplegia (Table 19.4)
	Nervus intermedius or 'geniculate' neuralgia
Cavernous sinus	Carotid aneurysm
	Meningioma
	Pituitary tumour
Meningeal	Tumour: lymphoma, cancer
	Tuberculosis
	Syphilis
Trigeminal Root	Compression—tumour: acoustic neuroma, meningioma
	—vascular: arteriovenous malformation
Central nervous system	Facial anaesthesia dolorosa
	Thalamic infarcts; central poststroke pain
	Brainstem tumours
	Brainstem demyelination
	Posterior inferior cerebellar artery occlusion
	Syrinx
Musculoskeletal	Temporomandibular joint disorders
	Eagle's syndrome of elongated stylohyoid process
	Facial dyskinesia
	Temporal bone lesions: glomus jugulare tumours
	metastases
	Paget's disease
Otological	Otitis externa and media
	Cholesteatoma
Sinus disease	Infection
	Sinus outlet obstruction
Dental and oral	Odontalgia
	Atypical odontalgia
	Burning mouth syndrome
Nasopharynx	Tumour: nasopharyngeal carcinoma
Salivary glands	Infection
	Inflammation and granulomatous disease
	Duct obstruction
	Tumour
Vascular	Giant cell arteritis
	Carotid and vertebral artery dissection
Referred	Ophthalmic disease
	Cervical spine disease
	Thoracic outlet syndrome
	Myocardial ischaemia
Psychogenic	Atypical facial pain

- teeth and gums for evidence of overt disease, including testing for local tenderness using a wooden spatula;
- the oro-pharynx;
- the sinuses for evidence of tenderness over frontal and maxillary sinuses, and nasal obstruction;
- the external auditory meatus, tympanic membrane, and pinna;
- the salivary glands.

In patients presenting with localized cranial pain posterior to the trigeminal territory, scalp tenderness, movements of the cervical spine, and posterior nuchal muscle tenderness should be assessed. The many causes of craniofacial pain (Table 19.1) emphasize the importance of such a broad and comprehensive approach to examination.

19.2 Neuralgias of the face and head

19.2.1 Trigeminal neuralgia

Trigeminal neuralgia, or tic douloureux, has unique and distinctive characteristics. It is pain of abrupt onset, occurring unilaterally in severe brief paroxysms, in the distribution of one or more branches of the trigeminal nerve (Section 20.2.4). In the great majority of patients there are no physical signs and the diagnosis rests entirely on an accurate history. There is a tendency for trigeminal neuralgia to be over-diagnosed, as a result of failure to recognize and apply the strict clinical criteria for the diagnosis. Erroneous diagnosis has major implications for treatment, leading potentially to unsuitable medical and worse still, surgical treatment.

Trigeminal neuralgia is sometimes referred to as 'idiopathic trigeminal neuralgia', but as described below, there is now good evidence from imaging, surgical, and pathological studies that subtle vascular compression can be demonstrated or inferred in the majority of patients.

'*Symptomatic trigeminal neuralgia*'. This refers to clinically indistinguishable pain which is shown to be due to structural lesions of the trigeminal root other than subtle vascular compression, such as tumours and arteriovenous malformations. In fact, it is rare for lesions of this type to cause neuralgia which truly has the characteristics of 'idiopathic' trigeminal neuralgia, and there are frequently other clues to the presence of a gross lesion in patients with 'symptomatic' trigeminal neuralgia, in particular facial sensory loss. Trigeminal neuralgia also occurs in multiple sclerosis. In this setting, in the early stages, pain is often indistinguishable from 'idiopathic' neuralgia, but tends to occur in younger patients, to change over time, and responds progressively less well to treatment.

'*Atypical trigeminal neuralgia*' is a term sometimes used to describe pain that has some but not all the characteristic features of the condition. However, although there may be unusual or additional features in some patients, one should guard against making a diagnosis of trigeminal neuralgia on inadequate grounds, for the reasons stated above. The term 'atypical trigeminal neuralgia' can be unclear in its meaning and thus misleading, so that care is needed when assessing reports of patients included under this diagnosis.

Epidemiology

Estimates of the incidence of trigeminal neuralgia vary. It increases with age, being commonest at age 75 years and above, at 11 per 100 000, with a female to male ratio of 1.17:1.0 in one study (Rothman and Monson 1973). Another investigation identified an annual incidence in women of 5.7 per 100 000 and in men of 2.5 per 100 000 (Katusic *et al.* 1990). In a population study in the United Kingdom,

Table 19.2 Non-cutaneous trigeminal sensory innervation

Ophthalmic	Cornea
	Mucosa of frontal sinus and upper nose
	Dura in anterior part of head
	Cerebral arteries and venous sinuses in anterior part of head; note the posterior dura and vessels of head are supplied by upper cervical dorsal roots
Maxillary	Lateral wall and floor of nasal cavity
	Upper jaw and teeth
	Roof of mouth
	Mucosa of maxillary sinus
Mandibular	Anterior wall of external auditory meatus
	Tympanic membrane
	Lower jaw and teeth
	Floor of mouth
	Anterior two-thirds of tongue

a lifetime prevalence of 0.07 per 1000 was calculated (MacDonald *et al.* 2000). The condition occurs rarely in young adults, and although the appearance of trigeminal neuralgia in those aged in the third to fifth decades should increase the suspicion of symptomatic trigeminal neuralgia, as defined above, many younger patients fall into the 'idiopathic' category. The overall relative risk of trigeminal neuralgia in multiple sclerosis, not age related, has been calculated as 20 (95 per cent confidence interval 4.1–58.6) (Katusic *et al.* 1990).

Clinical features

The International Headache Society (2004) has suggested that the five key characteristics of trigeminal neuralgia are:

A. Paroxysmal attacks of pain lasting from a fraction of a second to 2 min, affecting one or more divisions of the trigeminal nerve and fulfilling criteria B and C.

B. Pain has at least one of the following characteristics:

   intense, sharp, superficial, or stabbing

   precipitated from trigger areas or by trigger factors

C. Attacks are stereotyped in the individual patient.

D. There is no clinically evident neurological deficit.

E. The pain is not attributable to another disorder.

These clinical features are elaborated in further detail in Table 19.3. Trigeminal neuralgia is unilateral in all but 4 per cent of patients, and in the latter, bilateral simultaneous neuralgia is rare (Loeser 1989). The pain is nearly always felt within the second and third divisions of the trigeminal, the two most common sites being the upper lip and nares radiating over the medial cheek towards the eye, and the angle of the mouth radiating along the mandible towards the ear. In some patients, the pain is perceived to originate in the teeth of the upper or lower jaw; these patients are likely to present initially to a dentist. Pain may also be felt along the side of the tongue and in the buccal mucosa and gingiva. Trigeminal neuralgia affecting the ophthalmic division is very uncommon; pain in this area, unless characteristic of trigeminal neuralgia should always raise suspicion of an alternative cause.

Paroxysms of pain in trigeminal neuralgia occur either unprovoked or triggered by a variety of innocuous cutaneous and oral stimuli, including touching and washing the face, shaving, wind blowing on the face, facial movement, chewing, and hot or cold liquids in the mouth (Rasmussen 1991). Paroxysms usually last 15–60 s, but sometimes as long as 2 min. A history of triggering of pain varies between patients, but it is always present.

If asked whether the pain is continuous, some patients will reply that it is, but in trigeminal neuralgia this refers to the fact that there may be very short pain-free intervals between numerous, but distinct paroxysms. Careful elucidation of this feature of the history is needed. The quality of the pain is most often described as shooting, shock-like, like lightning or like electricity, and it is very often severe. During a bout of neuralgia, patients are often reluctant to eat or drink, for fear of provoking the pain, they lose weight and may sometimes become severely dehydrated.

Particularly in the early stages, neuralgia occurs in bouts lasting days to weeks, then may remit for months or even years (Kurland 1958). Over time, there is a tendency for bouts to last longer and for periods of remission to become shorter, so that eventually for some patients, the symptoms become chronic and persistent. Severe pain at night is unusual (Rasmussen 1990, 1991).

Facial flushing occasionally accompanies trigeminal neuralgia, and appears to be particularly a feature of ophthalmic division neuralgia (Nurmikko *et al.* 2000).

Another symptom commonly reported by patients with long-standing neuralgia is a milder continuous pain between paroxysms, occurring particularly following a series of rapidly repeated paroxysms with very brief pain-free intervals. This continuous pain may have a burning or dull aching quality, and is felt in the same distribution as the paroxysmal pain. Such pain is not a diagnostic feature of trigeminal neuralgia and indeed the presence of

Table 19.3 Clinical features of trigeminal neuralgia

Site	Maxillary—upper lip, nares, radiating over medial cheek to eye
	upper jaw teeth
	Mandibular—corner of mouth, radiating over lower jaw and cheek to ear
	Ophthalmic—eye and forehead (rare)
Nature	Paroxysmal: shooting, shock-like
Frequency	Up to several times per minute
	Often brief pain-free intervals
	Rare during sleep
Duration	Seconds to 2 min
Severity	Mild to very severe
Triggers	Any innocuous trigeminal cutaneous or oral stimuli
	Movement of face or jaw
Periodicity	Bouts lasting for days to months
	Tendency for periods of remission to become shorter and lost over time
Other features	Anorexia
	Weight loss
	Dehydration
	Depression

continuous pain at an early stage of the condition should always cause the clinician to question the diagnosis. However, in patients who have undoubtedly suffered from trigeminal neuralgia for several years, a background milder continuous pain is a fairly common feature (Zakrzewska *et al*. 1999). It does not negate the diagnosis in patients with otherwise typical symptoms and response to treatment.

Depression is a common comorbidity of chronic trigeminal neuralgia (Zakrzewska and Thomas 1993).

Examination. In idiopathic trigeminal neuralgia, there should be no sensory loss on standard clinical testing, though paroxysms of pain may be provoked during facial and oral sensory examination. The obvious severity of the pain and its duration usually leave the examiner in little doubt about the diagnosis. Patients in an active phase of the condition may be reluctant to allow an examination.

Quantitative sensory testing reveals subtle sensory impairment in more than 50 per cent of patients, both in the affected division and in the adjacent division (Nurmikko 1991). Patients should be re-examined periodically; although rare, symptomatic trigeminal neuralgia may be associated with an initially normal sensory examination and sensory loss may not become evident for up to several years (Cheng *et al*. 1993).

Investigation

Ideally, all patients presenting with trigeminal neuralgia should be investigated with MRI. This is sensitive in detecting blood vessels in contact with the nerve root, but about one-third of patients will be found to have a similar change on the asymptomatic side (Patel *et al*. 2003); this change is frequently reported in the scans of patients without trigeminal neuralgia. This emphasizes the fact that the diagnosis of trigeminal neuralgia is made on clinical grounds. MRI is obviously of importance when a diagnosis of multiple sclerosis is being considered. In patients with a short-lived initial bout of neuralgia, it is reasonable to not investigate and to await recurrence, which may not occur for several years.

The place of laser-evoked potentials and other quantitative methods in the assessment of patients with trigeminal neuralgia remains uncertain (Cruccu *et al*. 2002).

Cause and pathophysiology of pain

Dandy (1934) identified compression of the trigeminal root entry zone in more than 40 per cent of patients with trigeminal neuralgia coming to operation. The abnormalities causing compression included arterial loops and less commonly, tumours, cysts, aneurysms, or arteriovenous malformations. Later, Gardner and Miklos (1959) drew attention to decompression of the trigeminal root as a treatment for trigeminal neuralgia. However, it was not until Jannetta (1976) reported subtle vascular abnormalities in the root entry zone in 88 out of 100 consecutive patients that the importance of vascular compression in the aetiology of trigeminal neuralgia became widely appreciated. Among the remaining 12 patients in Jannetta's series, 4 had tumours and 2 had arteriovenous malformations compressing the trigeminal root. Of the 6 patients with multiple sclerosis, 4 had identifiable plaques and 2 had atrophic areas in the trigeminal root entry zone.

Vascular compression is usually caused by the superior cerebellar or anterior inferior cerebellar arteries; the ability of veins to produce compression sufficient to cause trigeminal neuralgia remains controversial. Not all reported series have found such a high incidence of identifiable structural lesions, with estimates of vascular compression ranging from about 65 to 85 per cent (Bederson and Wilson 1989; Loeser 1989).

The importance of vascular compression in the aetiology of trigeminal neuralgia is strongly supported by the success of microvascular decompression, also known as the Jannetta procedure, described below.

Control observations for the intra-operative findings reported in trigeminal neuralgia are hard to obtain. Direct comparison with post-mortem appearances is difficult because the blood vessels collapse, and there is distortion of the contents of the posterior fossa. Histological studies have revealed small areas of demyelination in the trigeminal root and ultrastructural changes in the Gasserian ganglion (Beaver 1967; Kerr 1967). More recent studies have confirmed and expanded these findings. Using electron microscopy, focal demyelination, dysmyelination, and close apposition of demyelinated axons has been observed in biopsies from the region of root compression in patients with trigeminal neuralgia (Love and Coakham 2001). Similar changes are present in patients with multiple sclerosis and trigeminal neuralgia (Love *et al*. 2001).

Hypotheses of the pathophysiology of trigeminal neuralgia rely heavily on experimental models of nerve injury, though bursting activity in trigeminal axons has been recorded at the time of operation in man (Baumann and Burchiel 1997). Ectopic impulse generation in damaged sensory fibres and dorsal root ganglion cell bodies with damaged peripheral axons is a well-established property (Devor 2005; and see Section 17.4). This may occur as an ongoing phenomenon or triggered by stimulation, either physical or chemical. Such triggered activity may lead to prolonged bursts of impulse activity, termed afterdischarges. These discharges may in turn recruit similar activity in neighbouring damaged afferent fibres, either by tight electrical transmission between adjacent axons, known as ephapses, or by a non-synaptic, non-ephaptic coupling that occurs at sites of nerve injury and in sensory ganglia, so-called crossed afterdischarge (Lisney and Devor 1987). Crossed afterdischarge probably depends on release of neurotransmitter substances or potassium ions into the narrow interstitial spaces between adjacent demyelinated axons (Amir and Devor 1996). Crossed afterdischarge occurs both at sites of nerve damage and in dorsal root ganglion cells with damaged peripheral axons (Devor and Wall 1990). Ephaptic transmission occurs between relatively few axons in areas of nerve injury (Seltzer and Devor 1979). It has often been inferred as a mechanism of neuropathic pain in man, though never directly demonstrated.

Crossed afterdischarge, causing mass recruitment of activity in sufficient numbers of trigeminal axons, is a more likely mechanism to explain the paroxysmal pain of trigeminal neuralgia. This property forms the basis of the ignition hypothesis (Devor *et al*. 2002).

Both ephaptic crosstalk and crossed afterdischarge lead to excitation of small nociceptive axons by low threshold A beta afferents (Amir and Devor 2000). This would explain the characteristic triggering of painful paroxysms of pain by innocuous peripheral stimuli such as brushing the skin. Interestingly, painful stimuli in trigger areas are less likely to induce painful paroxysms in trigeminal neuralgia (Kugelberg and Lindblom 1959). This clinical finding is matched by the experimental observation that crossed afterdischarge in sensory ganglia is more easily produced by A beta fibre than by C fibre stimulation (Devor and Dubner 1988; Devor and Wall, 1990).

Paroxysms of trigeminal neuralgia are brief, usually lasting 20–30 s, though occasionally as long as 2 min. During a burst of high frequency firing, calcium ions enter the axon, activating calcium-activated potassium channels. Potassiums ions flow out through these channels leading to a hyperpolarized state in which the bursting activity ceases and the axon is refractory for a period (Amir and Devor 1997). This mechanism is likely to explain the short-lived nature of each paroxysm of neuralgia.

In a minority of patients coming to posterior fossa surgery, no vascular or other trigeminal root compression is found. The ignition hypothesis outlined above depends on pathophysiological changes in both the trigeminal root and ganglion. It is possible that in patients in whom no macroscopic root pathology is observed at operation, the major pathology is situated in the ganglion rather than the root. This is supported by observations from the era before treatment with microvascular decompression, when trigeminal ganglionectomy was a standard surgical procedure for the treatment of trigeminal neuralgia; pathological changes in the ganglion were reported in virtually all cases (Beaver 1967; Kerr 1967).

The pathophysiological basis for the remissions that occur in the majority of patients in the earlier course of the condition is more difficult to explain. It is possible that partial remyelination of root axons could raise the thresholds sufficiently to prevent ectopic impulse generation and afterdischarges, but this is speculative. Furthermore, with vascular compression that is presumably fixed, it is difficult to envisage the conditions under which remyelination might occur.

Medical treatment

Numerous drugs have been advocated for the treatment of trigeminal neuralgia. With the exception of carbamazepine, for which a systematic review is possible, only single trials of other drugs are available, many of these including relatively few patients. Comparison of the results of these trials is difficult, for the reasons outlined in relation to the drug treatment of neuropathic pain (Section 17.6.3). Medical treatment of trigeminal neuralgia is reviewed by McQuay et al. (1995), Wiffen et al. (2000), and Zakrzewska and Lopez (2004).

By far the most effective drug is carbamazepine (Blom 1963), particularly in the early course of the condition. Overall, the drug relieves trigeminal neuralgia partially or completely in about 70 per cent of patients. Treatment should start with low doses in the region of 300 mg daily, because in many patients the pain responds to low doses and some of the adverse effects are minimized. Drug blood levels are often helpful in determining therapeutic dosage and adverse effect sensitivity. Slow release preparations provide sustained therapeutic blood levels and reduce adverse effects; this is particularly helpful in older patients, who are more prone to develop adverse effects. An allergic rash develops in up to 10 per cent of patients; other serious adverse effects are less common. The latter include blood dyscrasias, a low white cell count being the commonest, abnormal liver function, fluid retention, and hyponatraemia. A systematic review, including reports of 315 patients, calculated a number needed to treat, NNT, of 2.6, with a number needed to harm, NNH, of only 3.7, emphasizing the potential for the drug to produce unacceptable adverse effects, particularly in the elderly (Wiffen et al. 2000).

A common pattern in trigeminal neuralgia is for periods of remission to become shorter and in many patients to disappear. In addition, progressively larger doses of carbamazepine are frequently required to control pain.

Lamotrigine, baclofen, and tizanidine have been assessed in single controlled trials, including only small numbers of patients. Their efficacy remains unproven. Although effective in trigeminal neuralgia, tocainide was withdrawn due to serious adverse effects (Lindstrom and Lindblom 1987). Other drugs advocated for use in trigeminal neuralgia, but never subjected to adequate clinical trials include phenytoin, oxcarbazepine, clonazepam, sodium valproate, gabapentin, and topiramate (Zakrzewska and Lopez 2005). Pimozide is effective, but causes unacceptable adverse effects (Lechin et al. 1989).

Improvements in the surgical treatment of trigeminal neuralgia mean that patients should be considered for one of the procedures described below when carbamazepine begins to lose its effect, when the adverse effects of the drug impair quality of life, and in some patients, as primary treatment. Exhaustive trials of the drugs discussed above, prior to consideration of surgical treatment are no longer appropriate.

Surgical treatment

Surgical approaches to the treatment of trigeminal neuralgia include several neuroablative procedures and one operation aimed at treating the cause.

◆ *Alcohol injections* of either the peripheral branches within the affected division or the Gasserian ganglion, produce pain relief. However, these usually lasts less than 1 year and the accompanying anaesthesia, dysaesthesiae, and neuroparalytic keratitis can be troublesome. For these reasons, alcohol injections are no longer performed.

◆ *Cryosurgery,* in which freezing lesions of the peripheral branches are produced following surgical exposure of the nerves, leads to pain relief in 46 per cent of patients for more than 6 months, though in only 16 per cent of patients for longer than 1 year (Barnard et al. 1981). Nerve regeneration following freezing lesions is surprisingly good, with restoration of normal sensation in about 6 weeks. Thus the analgesia produced by the procedure far outlasts the sensory loss.

Three procedures performed at the level of the Gasserian ganglion, via an approach through a needle introduced into the foramen ovale, have been shown to relieve trigeminal neuralgia:

◆ *In controlled radiofrequency thermocoagulation,* cycles of radiofrequency-induced heat are applied to the appropriate part of the trigeminal division during brief periods of general anaesthesia, until partial sensory loss can be detected. The lesion thus produced is associated with pain relief. In a review of published reports in over 5000 patients, Loeser (1989) found that 80 per cent of patients obtained pain relief for at least 1 year, and over 50 per cent had pain relief at 5 years. The degree of analgesia is related to the severity of sensory loss induced by the procedure (Lopez et al. 2004). However, the sensory loss can be troublesome. Although recurrent pain can be treated by repeating the procedure, this inevitably increases the density of facial anaesthesia and the risk of producing anaesthesia dolorosa.

◆ *In glycerol gangliolysis,* a small volume of absolute glycerol is injected into the arachnoid cistern of the ganglion (Hakanson 1981). Glycerol is mildly neurotoxic, and analgesia, usually associated with mild sensory loss, develops over several days.

◆ *In balloon microcompression of the Gasserian ganglion,* a balloon is introduced via a needle and inflated briefly, causing partial damage to the ganglion, again leading to analgesia, together with a variable degree of sensory loss (Burchiel 1996).

- The ganglion may also be partially lesioned non-invasively by *gamma knife radiosurgery*, following which analgesia and accompanying sensory loss develop slowly over 2–3 months (Maesawa *et al.* 2001).

All these procedures have the advantage of being minimally invasive, but none offers the prospect of permanent pain relief and all are associated with sensory loss, of variable severity, together with a variable degree of weakness of the masseter muscle. Repeat procedures can be performed, but increase the risk of sensory loss, with all the attendant problems including the development of anaesthesia dolorosa. A systematic review found that overall, 40–50 per cent of patients treated by these methods have recurrent neuralgia at 36 months (Lopez *et al.* 2004). There appears to be little to choose between the procedures, though neuroparalytic keratitis is more common after radiofrequency lesioning, and experience of gamma knife gangliolysis is limited.

- The older operations of *complete trigeminal rhizotomy* and *descending trigeminal tractotomy* relieve trigeminal neuralgia, but are associated with dense sensory loss and a high incidence of anaesthesia dolorosa. The problems produced by *medullary tractotomy* are described (Section 17.6.8).

- *Microvascular decompression,* also now known as the *Jannetta procedure,* is directed to relieving the cause of trigeminal neuralgia, as already discussed. The operation consists of dissecting the offending compressing blood vessel from the nerve root, and interposing a pad of non-absorbable material between the two. Perioperative mortality associated with the posterior craniectomy required is up to 0.6 per cent (Kalkanis *et al.* 2003). Complications include CSF leak, eighth cranial nerve and cerebellar damage (McLaughlin *et al.* 1999). Trigeminal sensation is only rarely impaired after microvascular decompression (Barker *et al.* 1997). Pain relief is immediate and more long-lasting than any of the neuroablative procedures described above. At 1–2 years post-operatively, 80 per cent of patients are pain free, and at 8–10 years, approximately 60 per cent remain pain free and up to 12 per cent have recurrent neuralgia that is mild (Barker *et al.* 1996).

- In patients undergoing microvascular decompression in whom no vascular root compression is found, many surgeons perform a *partial rhizotomy,* including the caudal lateral part of the root. This usually produces pain relief with surprisingly little sensory loss (Zakrzewska and Lopez 2005).

Surgical treatment for trigeminal neuralgia is usually considered only when medical treatment fails to control the pain adequately, or is associated with sufficiently unpleasant adverse effects. However, in younger otherwise fit patients, surgery may be considered at an early stage. The choice of procedure depends on the patient's age, fitness for general anaesthesia, presence of medical comorbidities, and patient choice. Microvascular decompression offers the best prospect of long-term pain relief, with a low incidence of complications, and is usually now recommended for younger patients and those fit to undergo a posterior fossa craniectomy. Of the neuroablative procedures, radiofrequency lesioning is most often performed. It is particularly suitable for the very elderly and those with associated severe medical problems, which can include multiple sclerosis.

19.2.2 **Glossopharyngeal neuralgia**

Glossopharyngeal neuralgia is a paroxysmal pain occurring in the distribution of the glossopharyngeal nerve, with many similarities to trigeminal neuralgia (Section 20.3.3). It is a rare condition, with an incidence of 0.8 per 100 000, and an average age of onset of 50 years. It is bilateral in about 5 per cent of patients (Rushton *et al.* 1981; Katusic *et al.* 1991).

Clinical features

The pain is felt in the posterior part of the tongue, tonsillar fossa, pharynx, or beneath the angle of the jaw, or in the ear. It is paroxysmal, lasting from seconds to 2 min, and is triggered by swallowing, chewing, talking, coughing, and yawning. Attacks are stereotyped in individual patients. Some report a sensation of a foreign body in the throat. Examination reveals no neurological deficit. Glossopharyngeal neuralgia is episodic, in the same way as trigeminal neuralgia, though spontaneous remissions are common (Katusic *et al.* 1991). It is sometimes associated with sick sinus syndrome, syncope, severe bradycardia, and occasionally asystole (Ferrante *et al.* 1995).

The distribution of the pain and the associated autonomic features are explained by the complex anatomy of this region. The glossopharyngeal nerve contains motor, somatosensory, visceral sensory, and parasympathetic components, and communicates with the facial and vagus nerves and the sympathetic trunk. The somatosensory innervation provided by the nerve has two parts: the auricular/tympanic branch, supplying the external auditory meatus and tympanic membrane, part of the pinna and the mastoid; and the pharyngeal branch, innervating the pharynx. There is variable communication between pharyngeal and vagal afferents, leading to varying territories of innervation, but together, these nerves supply the soft palate, tonsil, and posterior part of the tongue. The primary sensory afferents from this distribution terminate in the spinal nucleus of the trigeminal, and there are connections between this nucleus and autonomic centres in the medulla. The glossopharyngeal nerve emerges from the anterior part of the jugular foramen, medial to the styloid process, and then curves around the posterior border of the process at the level of the origin of the stylohyoid muscle.

Cause, diagnosis, and investigation

There are similarities to trigeminal neuralgia in relation to the cause of glossopharyngeal neuralgia. High-resolution MRI often reveals vascular compression (Patel *et al.* 2002; Fiscbach *et al.* 2003), but other compressive causes such as tumour or an elongated styloid process are sometimes found, and glossopharyngeal neuralgia can also occur in multiple sclerosis (Bruyn 1983).

The characteristic paroxysmal nature of the pain in glossopharyngeal neuralgia, including triggering, in the absence of neurological deficit may lead to a confident diagnosis. However, when the history leaves doubt, a careful otorhinolaryngological assessment is mandatory, to exclude the many other important potential causes of pain in this region. MRI is the best form of imaging, though of unknown sensitivity and specificity (Zakrzewska and Lopez 2005).

Treatment

Because of the rarity of the condition, there have been no randomized controlled trials of any modality of treatment. However, it is generally accepted that carbamazepine is often effective; in addition, phenytoin, baclofen, gabapentin, and lamotrigine have all been advocated (Nurmikko and Jensen 2005). Surgical treatments reported to be effective include microvascular decompression

(Patel *et al.* 2002) and rhizotomy of glossopharyngeal and sometimes also vagal roots (Kondo 1998), though with perioperative mortality rates of up to 5 per cent and complications including dysphagia and dysphonia.

19.2.3 Postherpetic neuralgia

The ophthalmic division of the trigeminal nerve and the mid-thoracic sensory roots are by far the most common dermatomes affected by acute herpes zoster, known as shingles (Section 21.14.5). The maxillary and mandibular trigeminal divisions may be involved together with the ophthalmic division, but uncommonly in isolation. In other dermatomes on the head and neck, C2 and C3 shingles is unusual. The following sections describe the clinical features, pathology and treatment of postherpetic neuralgia in general. Geniculate herpes zoster, the Ramsay–Hunt syndrome, is described in Section 19.2.4.

Incidence and natural history

The incidence of herpes zoster in immune competent people is about 0.2 per cent in those younger than 50 years and about 1 per cent in those older than 80 years (Kost and Straus 1996). Women are affected more often than men in a ratio of approximately 3 to 2 (Hope-Simpson 1965; Watson *et al.* 1988a). Postherpetic neuralgia is defined as pain that persists after healing of the acute rash, but it decreases and resolves gradually in many patients over time. At 3 months following the acute eruption, about 10 per cent of patients will experience postherpetic neuralgia (Jackson *et al.* 1997). Stratification by age in a large study of postherpetic neuralgia revealed that in more than 400 patients, 3 months after the onset, postherpetic neuralgia was present in fewer than 2 per cent of patients younger than 60 years, and in 10 per cent of patients over 60 years. At 1 year following the acute eruption, none of the patients under 60 years had postherpetic neuralgia, and in those older than 60 years, only 3 per cent had postherpetic neuralgia.

At even longer intervals, there is a continuing gradual resolution of postherpetic neuralgia. In a large follow-up study in which more than half of the patients recruited had experienced postherpetic neuralgia for more than 1 year, at a median follow up of 3 years with a range of 3 months to 12 years, 56 per cent of patients had either no pain or pain which had decreased to an intensity of being no longer troublesome (Watson *et al.* 1988a).

These studies demonstrate the tendency for postherpetic neuralgia to improve in many patients even at long intervals after the acute eruption. This under-appreciated natural history has confounded the design of trials of treatment for postherpetic neuralgia and emphasizes the need for recruitment of large numbers of patients to trials in order to reach valid conclusions about treatment efficacy. In contrast to the general trend towards gradual improvement in postherpetic neuralgia over time, some patients' pain worsens with time, despite all efforts to relieve the pain (Watson *et al.* 1991a).

Clinical features

Pre-eruptive pain is a frequent symptom of acute herpes zoster, usually of 1–2 days' duration, but on occasions lasting as long as 7 days, and rarely longer than this, leading to delay in diagnosis. The appearance of the characteristic vesicular rash in an appropriate anatomical distribution establishes the diagnosis. The rash is variable in severity and there is a poor correlation between rash severity and either acute neuralgia or postherpetic neuralgia. Making a diagnosis of acute zoster neuralgia in a patient without a rash at any stage should be resisted; it is extremely rare and difficult to prove. Alternative diagnoses should always be considered in such patients.

The pain of acute shingles is often severe and associated with systemic upset, including marked general malaise, anorexia, and sometimes fever. An associated myelitis is an uncommon but well-recognized feature. The painful symptoms of acute shingles and postherpetic neuralgia are qualitatively indistinguishable, the one merging into the other.

Although the primary focus of zoster reactivation is the dorsal root ganglion, and the consequences are predominantly sensory, anterior horn cells are often involved at the same spinal segmental level. This is not clinically obvious when zoster affects mid-thoracic dermatomes, but on the relatively uncommon occasions that upper or lower limb dermatomes are affected, it is not unusual for weakness, wasting and, at relevant levels, reflex loss to occur. The recovery of such motor features tends to take place over weeks or a few months following an episode of acute shingles. When acute shingles affects an abdominal dermatome, motor involvement may be manifest by bulging of the abdominal wall, best seen when the patient is standing.

The ongoing, stimulus-independent pain of postherpetic neuralgia is most often described as burning, raw, gnawing, or tearing. Superimposed stabbing, shock-like paroxysmal pains are common. In addition, evoked stimulus-dependent pains are frequently worse than the ongoing pain. The lightest brushing of clothes against the skin can produce pain of such intensity that patients become immobilized and avoid skin contact in the affected area.

Careful examination reveals that there is hypoaesthesia in scarred, often hypopigmented areas of skin, and severe allodynia in normal appearing skin. The allodynia comprises three types of mechanical hyperalgesia: touch-evoked pain is usually the most evident, but pin prick hyperalgesia and pressure hyperalgesia may also be present (Pappagallo *et al.* 2000). Mechanical hyperalgesia is common in neuropathic pain from many different causes, but it is particularly severe in postherpetic neuralgia (Scadding and Koltzenburg 2005; Section 17.4).

Comorbidities, particularly depression and sleep disturbance, are major accompaniments of postherpetic neuralgia.

Pathology

Early investigators documented the pathological changes in the dorsal root ganglion, sensory roots, and peripheral nerves, with demyelination, axonal loss, and lymphocytic infiltration (Head and Campbell 1900; Lhermitte and Nicholas 1924; Denny-Brown *et al.* 1944; Zachs *et al.* 1964). Noordenbos (1959) reported a relative loss of large myelinated fibres in intercostal nerves in affected dermatomes in patients with postherpetic neuralgia. This finding supported the idea that an imbalance between activity in small and large fibre sensory input might lead to pain, further elaborated in the gate control theory of Melzack and Wall (1965).

These pathological observations have been extended in two more recent studies. Watson *et al.* (1988b) reported the autopsy findings in a 67-year-old man who had experienced postherpetic neuralgia in a right T7-8 distribution for 5 years before death. There was right-sided dorsal horn atrophy of the cord from T4 to T8 and fibrosis of the T8 dorsal root ganglion and dorsal root. Unmyelinated fibres appeared normal and this was supported by biochemical marker measures, transmitter levels, and receptor densities. In a subsequent autopsy study, Watson *et al.* (1991b) examined five patients affected *in vivo* by acute shingles. Of these, three had experienced postherpetic neuralgia and two had had no pain. Dorsal horn atrophy was present in the three subjects who had had postherpetic neuralgia, while in all five subjects, the peripheral nerve in the affected segment showed severe loss of myelinated axons, particularly larger axons. Staining for substance P and calcium gene-related peptide, CGRP, was absent in the dorsal root ganglia of two subjects with postherpetic neuralgia, but normal in the dorsal horn; however, quantification of unmyelinated fibres was not performed. In one subject who had experienced postherpetic neuralgia for 22 months before death, inflammatory changes and lymphocytic infiltration were found bilaterally in the dorsal root ganglia of four adjacent spinal segments and in their peripheral nerves. This raises the interesting possibility that a chronic inflammatory process can develop following zoster reactivation in some patients, and could provide an explanation for the gradual worsening of pain in a small minority of patients, as reported by Watson et al. (1988a).

Pathogenesis

In summary, no single pathological change has been identified that is unique to patients with postherpetic neuralgia. The nature and extent of the damage due to zoster reactivation indicate both peripheral and central pathology, and it is thus likely that there are both peripheral and central generators contributing to the neuropathic pain of postherpetic neuralgia (Section 17.4). In a minority of patients with severe sensory loss, indicating marked deafferentation and loss of peripheral sensory nerve fibres, dorsal root ganglion cells, and dorsal root fibres, postherpetic neuralgia is predominantly of central origin. However, in the great majority, severe skin sensitivity is a prominent feature, indicating partial, albeit abnormal peripheral input.

Psychophysical studies in patients with postherpetic neuralgia and severe touch-evoked pain, using measures of afferent C fibre function and density of innervation determined by skin biopsy, have demonstrated chronic abnormal sensitization of unmyelinated nerve fibre terminals in some patients. However, in other patients, pain is associated with marked loss of small fibre functions and partially preserved large fibre functions (Fields et al. 1998). These observations indicate that mechanisms of pain in postherpetic neuralgia are heterogeneous, both peripherally and centrally.

Preventive treatment

Demonstration of the efficacy of treatments in preventing the development of postherpetic neuralgia following acute shingles must take into account the natural history of the condition. Many studies undertaken prior to the full elucidation of this natural history are flawed for this reason. For example, it was suggested that topical idoxuridine in DMSO applied in the acute stage prevented the development of postherpetic neuralgia (Juel-Jensen et al. 1970). Two trials suggested effectiveness of corticosteroids (Eaglestein et al. 1970; Keczkes and Basheer 1980), though there is a risk of viral dissemination (Merselis et al. 1964). Hopes that acyclovir would prevent the development of postherpetic neuralgia have not been realized, though this treatment may shorten the period of acute zoster neuralgia and promote more rapid healing of the acute rash (Bean et al. 1982; Esman et al. 1982; Balfour et al. 1983), so the drug is indicated in acute shingles. A report that postherpetic neuralgia could be prevented by sympathetic blockade performed during acute shingles (Colding 1969) has not been substantiated.

In a recent study, more than 38 000 immune-competent people over the age of 60 years were immunized with a live attenuated zoster vaccine (Oxman et al. 2005). The results indicate a reduction both of the incidence of acute shingles and subsequent postherpetic neuralgia. Analysis of the data from this large study indicates that approximately 60 individuals need to be treated to prevent one episode of acute shingles, and more than 350 individuals need to be immunized to prevent one person developing postherpetic neuralgia.

Treatment of established postherpetic neuralgia

Treatment is along the lines set out for the treatment of neuropathic pain in general (Section 17.6). The mechanical hyperalgesia component presents a major therapeutic challenge. Application of cold packs is often helpful, though the partial analgesia produced is shortlived. Topical lignocaine and capsaicin offer partial relief to some, though patients should be warned that initial applications of capsaicin may temporarily exacerbate their pain. Capsaicin is contraindicated near the eye and is thus inappropriate for ophthalmic postherpetic neuralgia. Acupuncture is ineffective (Lewith et al. 1983), and transcutaneous electrical nerve stimulation, TENS, helps some patients but is unpredictable in its effect (Portenoy et al. 1986; Watson et al. 1988a).

Systemic drug therapy should be tried (Section 17.6.3). Surgical treatment has a poor record: dorsal rhizotomy will relieve the peripheral mechanical hyperalgesia, but carries the considerable risk of increasing the component of central neuropathic pain, due to exacerbation of deafferentation, with an overall increase in the intensity of pain.

The effect of intrathecal methyl prednisolone, given once weekly for 4 weeks, to patients with postherpetic neuralgia of at least 1 year's duration, was reported in a recent controlled trial (Kotani et al. 2000). Pain relief in the treated group of 89 patients was reported as being 'good' or 'excellent' at follow up 2 years after this treatment. This strikingly good therapeutic effect requires confirmation in other studies.

For ophthalmic postherpetic neuralgia, several surgical procedures have been advocated, including trigeminal rhizotomy, trigeminal tractotomy, cryocoagulation, or alcohol injection of the supraorbital nerve, but none consistently relieves the pain (Loeser 1986; Portenoy et al. 1986). There is also a substantial risk of producing permanent anaesthesia dolorosa with trigeminal rhizotomy and tractotomy, as discussed in relation to the treatment of trigeminal neuralgia. Other treatments which have been helpful in some patients, but not subjected to controlled clinical trials, include thalamic stimulation (Loeser 1986) and nucleus caudalis dorsal root entry lesioning (Bernard et al. 1987) (Section 17.6.8).

19.2.4 Geniculate herpes zoster: Ramsay–Hunt syndrome

Acute herpes zoster affecting the geniculate ganglion causes severe pain deep in the ear, often with retro-auricular radiation. There is continuing debate as to the exact site of the infection, and this may vary between patients. Clinically, the neurological deficit is frequently more extensive than can be explained by involvement solely of the geniculate ganglion, and it has been suggested that the focus of infection may be in the brainstem in some patients.

The sensory root of the geniculate ganglion, the nervus intermedius, supplies the middle and inner ear, the posterior wall of the external auditory canal, part of the pinna, the eustachean tube, and the mastoid air cells. Geniculate zoster is predominantly a condition of late middle and old age. As with zoster elsewhere, pre-eruptive pain may precede the appearance of a rash by several days, often leading to diagnostic difficulty in this situation. The rash affects the external auditory canal and part of the pinna, but is often discreet and is easily missed. Other parts of the facial nerve are affected: facial palsy occurs in almost all patients, with loss of taste on the anterior two-thirds of the tongue due to involvement of the chorda tympani. Involvement of the eighth cranial nerve causes deafness and vertigo. Marked general malaise and a low grade fever are common.

A high index of suspicion is needed to make the diagnosis at an early stage. Acyclovir should be given, but because the syndrome is rare, it is unknown whether or not early treatment leads to a better outcome. Recovery of the facial palsy is often incomplete.

19.2.5 Nervus intermedius neuralgia

The distribution of sensation supplied by the nervus intermedius is described in Section 19.2.4. Nervus intermedius neuralgia, also known as geniculate neuralgia, is extremely rare. It is characterized by paroxysms of pain lasting for a few seconds or minutes felt deep in the ear, and sometimes also in the posterior pharynx. There is often a trigger area on the posterior wall of the external auditory canal (Furlow 1942). The underlying cause is unknown, and the diagnosis is established by the clinical features, the absence of signs other than a trigger zone in some patients, and exclusion of other causes of otalgia. The condition is sufficiently rare that there are no controlled trials of treatment, which should be along the lines recommended for peripheral neuropathic pain (Section 17.6). In view of the paroxysmal nature and triggering of the pain in some patients, a trial of carbamazepine is reasonable. Various surgical approaches have been reported, with variable success; these include procedures to the nervus intermedius and branches of the glossopharyngeal and vagus nerves (Lovely and Jannetta 1997).

19.2.6 C2 and occipital neuralgias

The C2 spinal nerve root runs adjacent to the lateral atlanto-axial joint. It may become involved in inflammatory conditions affecting the joint, by tumours, usually neurofibromas or meningiomas, angiomas, arterial loops, and in subluxation of the atlanto-axial joint, most often seen in rheumatoid disease. Pain due to C2 root compression is felt within the distribution of the root, over the back of the head. It may be intermittent, occurring as hemicranial attacks, or persistent (Jansen *et al.* 1989a, b), and there is associated sensory impairment.

C2 neuralgia. This describes a condition of unknown cause, which presents with a characteristic clinical picture. There is intermittent, lancinating unilateral occipital pain, often associated with ipsilateral lacrymation and redness of the eye. The pain may occur several times per day and remissions lasting for months are common. Imaging and other investigations are normal. The diagnostic test is a C2 root block using local anaesthetic, which temporarily abolishes the pain. Longer term relief of the pain with root thermo-coagulation has been reported (Jansen *et al.* 1989a, b), but there are no controlled trials.

Occipital neuralgia. This is assumed to be due to damage to or entrapment of the greater or lesser occipital nerves. It can follow whiplash-type neck injuries and may also result from chronic contraction of the posterior nuchal and scalp muscles, though the basis for the condition in many patients is uncertain (Behrman 1983). Pain may be intermittent or persistent, described as a shooting pain starting in the occipital region and radiating towards the vertex, or as a dull, deep, aching pain. It is sometimes provoked by neck movement. There may be local tenderness over the occipital nerves together with nuchal muscle tenderness and mild occipital sensory impairment. Local anaesthetic injections temporarily relieve the pain and in some patients prolonged relief may be obtained with injected corticosteroid.

Neck–tongue syndrome. In the neck–tongue syndrome, sudden turning of the head can produce backwards subluxation of the lateral atlanto-axial joint, stretching the C2 nerve root. This causes episodes of pain in the occipital region lasting seconds or minutes, associated with numbness or paraesthesiae on the ipsilateral side of the tongue, the latter symptoms being due to compression of proprioceptive afferent fibres from the tongue, passing from the ansa hypoglossi to the C2 ventral ramus (Lance and Anthony 1980; Bogduk 1981). The condition usually occurs in normal subjects, but also in patients with rheumatoid disease or congenital joint laxity (Lance and Anthony 1980; Bertoft and Westerberg 1985).

19.2.7 Posttraumatic facial neuralgias

The supraorbital and infraorbital nerves may be damaged by direct trauma to the face, with or without frontal and maxillary fractures. The inferior alveolar nerve is occasionally damaged during wisdom tooth extractions, and the lingual nerve may also be damaged as a result of dental procedures. Pain in all these situations is not paroxysmal, as in trigeminal neuralgia, but is usually continuous, fluctuating in severity, with associated numbness and tingling paraesthesiae within the nerve territory. There is often tenderness and a Tinel sign at the site of damage to the affected nerve. Treatment is as for other peripheral neuropathic pain (Section 17.6).

19.2.8 Painful ophthalmoplegia

Painful ophthalmoplegia is the term preferred to include all conditions in which there is some form of ophthalmoplegia associated with pain. In all these conditions pain is felt in the orbital region, but may radiate widely onto the face and the head. The cavernous sinus is densely populated by a number of important structures including the carotid artery, first division of the trigeminal nerve, third, fourth, and sixth cranial nerves, and the sympathetic and parasympathetic supply to the eye. Painful ophthalmoplegia can be produced by a wide range of pathological causes affecting one or more of these structures, including vascular, neoplastic, granulomatous, and infective processes (Table 19.4).

Various terms have been employed to describe these conditions, including superior orbital fissure syndrome, orbital apex syndrome, cavernous sinus syndrome, parasellar syndrome, and Tolosa–Hunt syndrome. While the individual anatomical descriptors remian appropriate, the eponymous title is no longer justifiable. The all-embracing term painful ophthalmoplegia recognizes the frequent anatomical overlap that occurs in individual patients and the lack of pathological specificity previously implied for some of the

Table 19.4 Causes of painful ophthalmoplegia

Neurological	Ophthalmoplegic migraine
	Diabetic third nerve palsy
Vascular	Giant cell arteritis
	Carotid, middle meningeal, and posterior communicating artery aneurysms
	Carotico-cavernous fistula
	Cavernous sinus thrombosis
Tumour	Pituitary tumours
	Retrobulbar tumours
	Skull base tumours
	Lymphoma
	Nasopharyngeal carcinoma
Infectious	Tuberculosis
	Aspergillosis
	Actinomycosis
	Syphilis
Inflammatory	Sarcoidosis
	Orbital pseudotumour
	Systemic lupus erythematosus

conditions. It also encourages a logical and systematic approach to diagnosis.

The Tolosa–Hunt syndrome was thought to be exclusively the result of granulomatous infiltration of the cavernous sinus (Hunt *et al.* 1961; Tolosa 1965) (Section 13.4.3). It is still listed in the International Headache Society taxonomy (Headache Classification, IHS 2004), with clinical diagnostic criteria as follows:

♦ One or more episodes of unilateral orbital pain persisting for weeks if untreated.

♦ Paresis of one or more of the third, fourth, and/or sixth cranial nerves, and/or demonstration of granuloma by magnetic resonance imaging or biopsy.

♦ Paresis coincides with the onset of pain or follows it within 2 weeks.

♦ Pain and paresis resolve within 72 h when treated adequately with corticosteroids.

♦ Other causes have been excluded by appropriate investigation.

Although previously regarded as a clinically and pathologically distinct condition caused by presumed granulomatous infiltration of the cavernous sinus, the criteria for clinical diagnosis include failure to demonstrate a structural cause, a relapsing course, and a response to corticosteroids. However, although these clinical limits of the syndrome seem to be clearly defined, several variants have been described which cast doubt on the nosological separation of Tolosa–Hunt syndrome as a distinct entity. These include *Raeder's syndrome* (Raeder 1924) (Section 18.7.3), the combination of Horner's syndrome, pain and parasellar cranial nerve involvement, and *Gradenigo's syndrome*, a sixth cranial nerve palsy with pain due to lesions at the apex of the petrous temporal bone. In Tolosa–Hunt syndrome itself, optic nerve involvement has been described, indicating anterior extension of the lesion responsible from the cavernous sinus, and the involvement of the maxillary division of the trigeminal nerve in some cases indicates posterior extension of the causative lesion (Smith and Taxdal 1966). Seventh cranial nerve involvement has also been described (Swerdlow 1980), and occasionally the eighth, ninth, tenth, or twelfth cranial nerves (Bogduk 2005).

Furthermore, there are uncertainties about the granulomatous pathological specificity of Tolosa–Hunt syndrome and about the specificity of CT, angiographic, and phlebographic abnormalities said to be characteristic of the condition (Bogduk 2005). Finally, a rapid response to corticosteroid treatment is not unique to Tolosa–Hunt syndrome; it has been reported in patients with painful ophthalmoplegia caused by aneurysms and tumours, including lymphoma, nasopharyngeal carcinoma, pituitary tumours, and metastases, and by fungal infection in the cavernous sinus (Bogduk 2005). For all these reasons it seems appropriate to include all the previously described syndromes under the heading of painful ophthalmoplegia.

19.2.9 Facial anaesthesia dolorosa

Anaesthesia dolorosa refers to pain felt in an area of decreased sensation, due to denervation from either peripheral or central lesions. Thus strictly speaking, central poststroke pain is an example of anaesthesia dolorosa, affecting the face in about 35 per cent of patients (Boivie 2005). However, anaesthesia dolorosa is a term usually used to denote pain due to loss of peripheral sensory input, that is, deafferentation pain. Loss of sensation in the face is particularly likely to lead to anaesthesia dolorosa, compared to other body areas. The pain is described as deep, often diffuse, burning, gnawing, or raw in nature, and when present is often severe. Trigeminal anaesthesia dolorosa now occurs relatively infrequently, as it was seen largely as a result of ablative procedures performed for the treatment of trigeminal neuralgia, including trigeminal rhizotomy and medullary tractotomy, operations now superseded by microvascular decompression. However, as discussed above, it can occur following radiofrequency lesioning (Lopez *et al.* 2004), and is related to the degree of sensory impairment produced by the lesion.

19.3 Musculoskeletal craniofacial pains

19.3.1 Temporomandibular joint disorders

Pain in the region of the temporomandibular joint is common and is usually transient. In two population surveys conducted by questionnaire, the prevalence of temporomandibular joint pain has been reported as 9.1 and 12 per cent, and associated symptoms of limitation of jaw opening and clicking or popping noises are common (Agerberg and Carlsson 1972; Locker and Grushka 1987).

Terminology. Costen (1934) provided an early description of temporomandibular joint pain and the diagnostic term *Costen's syndrome* is still used today. However, three other non-eponymous diagnostic terms are now more commonly employed to describe the syndromes of teamporomandibular joint pain; these are *temporomandibular pain and dysfunction syndrome* (International Association for the Study of Pain 1994), *oromandibular dysfunction* (International Headache Society 1988), both of which are associated with dysfunction of the teamporomandibular joint, as their names suggest; and *facial arthromyalgia*, in which dysfunction is a variable feature (Harris 1974).

Clinical features. In the context of neurological practice, the clinical features of the three temporomandibular joint pain syndromes can be considered together. There is aching in the muscles of mastication, exacerbated by chewing, associated with restriction of jaw movement and clicking or popping sounds. Other associated features include jaw clenching and gnashing of the teeth, jaw locking on opening, and other oral 'parafunction', including biting of the tongue, lips, or cheek (International Association for the Study of Pain 1994; International Headache Society 1988). In facial arthromyalgia, temporomandibular joint pain is not necessarily associated with dysfunction or oral parafunction (Harris 1974). Among those seeking medical attention, women outnumber men, and the conditions may present at any age during adult life.

The pain in all these conditions is similar: it is dull and aching and centred on the temporomandibular joint and masticatory muscles, but can radiate widely over the face, to the ear, and occasionally onto the neck. Pain may have been present for weeks to years, it is unilateral or bilateral, it is usually described as continuous and is often, but not always, exacerbated by jaw movement. Mental stress may provoke the pain in some patients. Signs include tenderness of the temporomandibular joint and masticatory muscles, trismus, clicking of the joint on movement, subluxation, and evidence of bruxism.

Pathology. Three main pathophysiological mechanisms have been proposed: psychogenic, meniscal displacement, and malocclusion. It is possible that these are not mutually exclusive.

The psychogenic theory proposes that psychological factors including adverse life events, sleep disturbance, anxiety, and stress, lead to masticatory muscle overactivity and pain (Schwartz 1959; Laskin 1980), and it has been suggested that facial arthromyalgia is more common in those with vulnerable personality types (Feinmann *et al.* 1984).

The lateral pterygoid muscle alters the position of the meniscus within the temporomandibular joint. It has been proposed that psychological stressors provoke hyperactivity of this muscle, causing the meniscus to be displaced anteromedially in the joint, with loss of attachment to the lateral pole of the condyle, leading to instability in the joint and the development of pain (Juniper 1984). However, anterior displacement of the meniscus, as demonstrated by MRI, is present in 32 per cent of asymptomatic subjects (Kircos *et al.* 1987), casting doubt on this hypothesis.

Costen (1934) proposed that malocclusion causes pain in the temporomandibular joint. It was later reported that occlusal equilibration relieved pain (Ramijford 1961; Magnusson and Carlsson 1983), though this was based on uncontrolled observations. Subsequent controlled studies have not demonstrated any clear therapeutic effect (Goodman *et al.* 1976; Dao *et al.* 1994), and there is no difference in the incidence of malocclusion in patients with facial arthromyalgia and control subjects (Thomson 1971).

Treatment. Many treatments including physical, pharmacological, and psychological measures have been advocated for the treatment of the temporomandibular joint pain disorders (Zakrzewska and Harrison 2003). Of these, there is evidence from randomized controlled trials of a therapeutic effect with antidepressants (Feinmann *et al.* 1984), diazepam (Jagger 1973), and cognitive behavioural therapy (Harrison *et al.* 1997).

19.3.2 Facial dyskinesias

Hemifacial spasm (Section 20.2.4) is sometimes painful, and patients with longstanding facial palsy sometimes complain of pain on the affected side of the face. This is particularly likely when contracture develops following lower motor neurone palsy, or when hemifacial spasm develops. Oro-facial dyskinesia affecting facial, jaw, and occasionally lingual muscles may also be painful.

19.4 Lesions of the ear, sinuses, and mouth

19.4.1 Otalgia

Otalgia due to otitis externa and otitis media will usually not present to neurologists, though chronic otitis media can cause widely radiating pain. Malignant otitis media, a condition virtually always occurring in patients with diabetes mellitus, is a chronic condition caused by pseudomonas aeruginosa. It presents with otalgia, followed at intervals of up to 3 months by facial palsy and sometimes other cranial nerve palsies.

Cholesteatomas lead to destructive changes in the middle and inner ear, with secondary infection and bone erosion. They are sometimes painful.

Otalgia as the presenting symptom of geniculate herpes zoster has already been discussed (Section 19.2.4).

Lesions of the temporal bone may present with poorly localized pain in the region of the ear, together with conductive deafness if the middle ear is involved, as for example with glomus jugulare tumours (Section 27.6.3). Occasionally, metastatic deposits in this region of the skull lead to otalgia; this occurs particularly with prostatic cancer, though any of the malignancies commonly metastasizing to bone can cause painful skull deposits. Paget's disease of bone involving the skull bones may cause localized pain, with a characteristic external appearance and radiological features (Section 27.6.6).

19.4.2 Sinus disease

Sinusitis is by far the commonest cause of sinus pain. Bacterial maxillary sinusitis, unilateral or bilateral, usually follows viral upper respiratory infections, associated with nasal obstruction and discharge. Pain is localized over the maxillary antrum and is sometimes also felt in the upper jaw teeth, with tenderness on chewing. Maxillary sinusitis is occasionally caused by a periapical dental abscess. Frontal sinusitis causes pain and tenderness in the supraorbital region. Sphenoid and ethmoid sinusitis produce pain between or behind the eyes. Pain is characteristically exacerbated by bending, and relieved by spontaneous or surgical drainage of the infected sinus. Plain skull radiographs, or with greater sensitivity, CT scans show opacification of the affected sinuses and sometimes a fluid level.

Fungal infection of the sinuses is most frequently, but not exclusively, seen in immunocompromised patients. In the rhinocerebral syndrome there is fungal infection progressively involving the sinuses, orbits, and brain, occurring most often in patients with diabetes mellitus (DeShazo *et al.* 1997) (Section 43.2.7).

Tumours within the sinuses lead to sinus pain, and nasal polyps and severe allergic rhinitis can lead to obstruction of the sinuses causing pain in the absence of infection. A markedly deviated nasal septum can obstruct sinus drainage and also cause pain due to pressure on one of the bony turbinates.

19.4.3 Odontalgia

Toothache is the province of the dental surgeon. However, although dental pain is usually well localized, it can be diffuse and lead to diagnostic difficulty. Neurologists should be aware of the leading causes of odontalgia (International Association for the Study of Pain 1994; Zakrzewska and Harrison 2003).

Dentinoenamel defects, due to caries or trauma, cause shortlasting and sometimes diffuse orofacial pain. Local dental stimuli evoke the pain, which does not usually occur in the absence of stimulation.

Pulpitis is infection of the tooth pulp due to deep caries. Pain may again be diffuse and occurs without local stimulation. A history of exacerbation by chewing and hot and cold liquids may lead to an erroneous diagnosis of trigeminal neuralgia and the pain in the upper jaw may also mimic maxillary sinusitis. Without treatment, pulpitis can progress to periapical periodontitis and abscess.

Periapical periodontitis and abscess cause severe pain in the affected tooth and adjacent gingiva, and sometimes widely radiating pain. Cursory examination of the teeth may not reveal obvious disease, but the gingiva are often inflamed and a gum boil may discharge into the mouth.

Gingival pain due to local trauma or infection can usually be easily identified but when associated with pericoronitis, a bacterial infection affecting supporting tissues surrounding an impacted or erupting tooth, may lead to diffuse pain.

Cracked tooth syndrome, due to a crack in a tooth, causes local pain provoked by chewing; part of the tooth cusp may fracture.

Dry socket refers to a painful condition following tooth extraction, usually from the lower jaw. It is due to localized osteitis and there is often associated submandibular lymphadenitis. After tooth extraction, clotted blood normally fills the socket, but this is inhibited if excessive adrenaline is used with the local anaesthetic. If the clot is washed out or broken down by infection, this may also result in a dry socket. Food impaction in the socket causes severe pain and halitosis. Treatment consists of washing out and packing the socket.

19.4.4 Atypical odontalgia

Atypical odontalgia occurs predominantly in women and is defined as severe throbbing pain in a tooth or teeth in the absence of major pathology. In the condition, teeth become both spontaneously painful and hypersensitive to stimuli, particularly heat and cold. Pain may radiate widely, as far as the temporomandibular joints, and associated oral dysaesthetic symptoms are common. The condition is considered to be of psychological origin, akin to atypical facial pain (Section 19.6), with which there is overlap clinically, associated either with depression or representing a monosymptomatic hypochondriacal disorder (Rees and Harris 1979).

19.4.5 Burning mouth syndrome

Burning mouth syndrome, also known as glossodynia, burning tongue, or oral dysaesthesia, is characterized by burning pain, present predominantly on the tip and lateral borders of the tongue, but also sometimes on the palate, alveolar mucosa, and lips. It nearly always affects post-menopausal women over the age of 50 years. Associated symptoms include xerostomia, dysgeusia, and thirst. Anxiety and depression are common comorbidities. Symptoms are exacerbated by emotion, fatigue, and hot drinks. Temporary relief is obtained by sleeping, eating, consuming cold drinks, and alcohol.

It is important to exclude treatable conditions that can cause diffuse pain in the mouth, including bacterial and fungal infection, allergies, oesophageal reflux, xerostomia as part of Sjogren's syndrome, iron, vitamin B12, and folate deficiencies, and diabetes mellitus.

Until recently, the pathogenesis of the condition has been obscure and trials of treatment on the basis of presumed vitamin deficiency or infection have been disappointing; these treatments include iron, vitamins, zinc, and antifungal drugs. However, tongue biopsies in patients with burning mouth syndrome have shown lower densities of unmyelinated epithelial nerve fibres compared with controls, with evidence of axonal degeneration (Lauria *et al.* 2005). This evidence suggests that burning mouth syndrome may be a trigeminal small fibre neuropathy.

Tricyclic antidepressant drugs are often given, but a recent Cochrane review did not find convincing evidence of a therapeutic effect for these or any other class of drug in burning mouth syndrome (Zakrzewska *et al.* 2005).

19.4.6 Salivary gland disease

Submandibular gland disease includes duct obstruction, inflammation, infection, and rarely tumour. These all lead to local pain, swelling, and tenderness of the gland, causing few diagnostic problems. Lesions of the parotid can be more difficult: pain is often diffuse, affecting most of the side of the face. In addition, mild to moderate gland swelling may not be clinically obvious. The development of a facial palsy suggests a mixed parotid tumour. Painful symmetrical swelling of the salivary glands occurs in mumps infection, and enlargement of one or both parotids is a feature of sarcoidosis. When facial pain is considered to be due to a parotid lesion, imaging with CT or MR is helpful and a surgical opinion should be obtained.

19.5 Suboccipital and cervical disease

19.5.1 Carotid and vertebral artery dissection

Pain is often the first symptom of both internal carotid and vertebral artery dissection and may precede the onset of symptoms and signs of cerebral ischaemia by hours or days (Section 35.4.4). In carotid dissection, pain is usually ipsilateral to the dissection and is felt in the face, head or neck. In a minority, pain is bilateral. In vertebral artery dissection, unilateral or bilateral neck pain and headache occur, sometimes with facial pain, though the last may be a specific localizing symptom of brainstem ischaemia produced by the dissection. Typically, focal symptoms and signs develop at an interval following the onset of neck pain and headache.

19.5.2 The Styloid process or Eagle's syndrome

Eagle (1937, 1958) described a constellation of symptoms resulting from elongation of the styloid process or calcification or ossification of the stylohyoid ligament, now known as Eagle's syndrome. Although rare, neurologists should be aware of this condition, particularly as recognition leads to surgical treatment that is often successful. The styloid process originates from the temporal bone medial and anterior to the stylomastoid foramen. The process points anteromedially and is bordered on medial and lateral sides by the internal and external carotid arteries respectively. Three muscles are attached to the process: stylopharyngeus inervated by the glossopharyngeal nerve, stylohyoid, the facial nerve, and styloglossus, the hypoglossal nerve. The stylohyoid and stylomandibular ligaments also originate from the process. The internal jugular vein and the glossopharyngeal, vagus, and hypoglossal nerves lie medial to the process.

Patients with Eagle's syndrome are typically aged 30–50 years, with a slight female preponderance. Symptoms may be intermittent or continuous and include pain in the throat, sensation of a foreign body in the pharynx, dysphagia, otalgia, mandibular and facial pain, vertigo, and syncope. Head turning towards the side of the pain, with the neck flexed, may provoke the pain, and is a symptom well worth enquiring about in patients with pharyngeal pain of obscure origin. Less commonly, symptoms arise due to mechanical irritation of the external and internal carotid arteries, carotidynia, with a wide distribution of pain. With external carotid artery irritation, the pain radiation includes the eye, ear, mandible, face, soft palate, and nose; when the internal carotid artery is involved, pain may radiate to the whole head (Correll *et al.* 1979).

The symptoms can often be provoked by palpation of the styloid process, which may be obviously elongated. Head turning towards the side of the pain, with neck flexion may elicit the pain. Plain X-rays or CT scans demonstrate an elongated styloid process, allow measurement of length, and may also show mineralization of the stylohyoid complex.

The cause of elongation of the styloid process and abnormal bone mineralization in Eagle's syndrome is unknown in the majority of patients. It is considered by some otorhinolaryngologists that trauma from tonsillectomy may induce bone formation, leading to an elongated styloid process or ossified stylohyoid ligament. Recurrent trauma to the stylohyoid ligament due to neck movement is thought to be an aetiological factor in some patients (Salamone *et al.* 2004).

The wide distribution of symptoms and their non-specific nature can make diagnosis of this uncommon condition difficult. The differential diagnosis is extensive and can include chronic pharyngotonsillitis, otitis media, mastoiditis, dental pain, pharyngeal foreign body, submandibular salivary gland disease, and tumours of the pharynx or base of the tongue. It is common for patients to be seen by several doctors, including psychiatrists, before the diagnosis is established (Beder *et al.* 2005).

Treatment is excision of the elongated part of the styloid process, and this is curative in the majority of patients (Beder *et al.* 2005; Mendelsohn *et al.* 2006).

19.6 **Atypical facial pain**

The term atypical facial pain is used in the United Kingdom to refer to pain of psychological origin (Section 4.7.9). However this usage is not universally adopted, and the term is sometimes used in the clinical literature to denote atypical forms of organically determined conditions. The International Headache Society defines atypical facial pain as persistent facial pain that does not have the characteristics of the cranial neuralgias and is not associated with physical signs or a demonstrable organic cause. Pain may be initiated by an operative procedure or injury to the face, teeth, or oral tissues (International Headache Society 1988). While this definition indicates an absence of organic disease, it does not go as far as stating positively that the condition has a psychologically determined basis.

There is a history of previous dental treatment or injury prior to the onset of symptoms in about 50 per cent of patients (Mock *et al.* 1985). Some studies have drawn attention to the frequency of depression and anxiety in atypical facial pain (Lesse 1956; Moore and Nally 1975).

Atypical facial pain is commonest in women, usually presenting in middle age. However, robust epidemiological data are lacking. Patients present with poorly localized pain affecting non-muscular parts of the face. There is a tendency for the pain to spread over time, and symptoms may persist for years. Pain usually starts on one side of the face, but often becomes bilateral and frequently extends beyond trigeminal territory on the head and upper neck. It may be provoked by fatigue and psychological stressors, and physical provoking factors are described by some patients, though without the characteristic features of triggering as found in trigeminal neuralgia. The pain is often described as deep, aching, and throbbing. It is frequently severe and may markedly restrict normal activities. Symptoms of anxiety and depression are commonly present, though patients often deny feeling depressed. There may be a history of repeated dental treatments, undertaken in attempts to relieve the symptoms. The overlap with atypical odontalgia has already been mentioned.

Examination reveals no physical signs except deep tenderness of the face in some patients. Patients may appear agitated and distressed and there may be obvious depressive features. All investigations are normal.

Treatment with antidepressant medication is often effective (Lascelles 1966; Harrison *et al.* 1997), and psychiatric assessment is advisable for the majority of patients. Diagnostic review is essential for those patients not responding to medication.

19.7 **Referred pain**

Ophthalmic disease causes pain that is usually centred on the eye, but which can radiate widely onto the face and head. *Cervical spine degenerative disease* frequently leads to pain both in the spine and in the head, particularly posteriorly, often associated with nuchal muscle tenderness. *Myocardial ischaemia* may present with pain in the lower jaw, anterior neck, and throat, in the absence of the usual distribution of central chest pain, with radiation to the left arm. In *thoracic outlet syndrome* (Section 22.5.3) pain in the root of the neck, radiating down the arm is the characteristic distribution, referred pain may be experienced on the ipsilateral side of the face and head by a minority of patients.

References

Agerberg G, Carlsson GE (1972). Functional disorders of the masticatory system. *Acta Odontologica Scandinavica*, **32**, 597–613.
Amir R, Devor M (1996). Chemically-mediated cross-excitation in rat dorsal root ganglia. *J Neurosci*, **16**, 4733–41.
Amir R, Devor M (1997). Spike-evoked suppression and burst patterning in dorsal root ganglion neurons. *J Physiol (London)*, **501**, 183–96.

Amir R, Devor M (2000). Functional cross-excitation between afferent A- and C-neurons in dorsal root ganglia. *Neuroscience*, **95**, 189–95.

Balfour HH, Bean B, Laskin OL *et al.* (1983). Acyclovir halts progression of herpes zoster in immunocompromised patients. *N Engl J Med*, **308**, 1453.

Barker FG, Jannetta PJ *et al.* (1996). The long-term outcome of microvascular decompression for trigeminal neuralgia. *N Engl J Med*, **334**, 1077–83.

Barker FG, Jannetta PJ *et al.* (1997). Trigeminal numbness and tic relief after microvascular decompression for typical trigeminal neuralgia. *Neurosurgery*, **40**, 39–45.

Barnard JDW, Lloyd JW, Evans J (1981). Cryoanalgesia in the management of chronic facial pain. *J Maxillofac Surg*, **9**, 101–2.

Baumann TK, Burchiel KJ (1997). Intraoperative microneurographic recordings in patients with trigeminal neuralgia. In Jensen TS, Turner JA, Wiesenfeld-Hallin Z, eds. *Proceedings of the 8th World Congress on Pain. Progress in Pain Research and Therapy 8*, pp. 459–67. IASP Press, Seattle.

Bean B, Braun C, Balfour HH (1982). Acyclovir therapy for acute herpes zoster. *Lancet*, **ii**, 118–21.

Beaver DL (1967). Electron microscopy of the Gasserian ganglion in trigeminal neuralgia. *J Neurosurg*, **26**, 138–50.

Beder E, Ozgursoy OB, Ozgursoy SK (2005). Current diagnosis and transoral surgical treatment of Eagle's syndrome. *J Oral Maxillofac Surg*, **63**, 1742–5.

Bederson JB, Wilson CB (1989). Evaluation of microvascular decompression and partial sensory rhizotomy in 252 cases of trigeminal neuralgia. *J Neurosurg*, **71**, 359–67.

Berhman S (1983). Traumatic neuropathy of second cervical spinal nerves. *Br Med J*, **286**, 1312–3.

Bernard EJ, Nashold BS, Caputi F (1987). Clinical review of nucleus caudalis dorsal root entry zone lesions for facial pain. *Appl Neurophysiol*, **51**, 218–24.

Bertoft ES, Westerberg CE (1985). Further observations on the neck-tongue syndrome. *Cephalalgia*, **5** (Suppl 3), 312–3.

Blom S (1963). Tic douloureux treated with a new anticonvulsant: experiences with G 32883. *Arch Neurol*, **30**, 285–90.

Bogduk N (1981). An anatomical basis for neck tongue syndrome. *J Neurol Neurosurg Psychiatry*, **44**, 202–8.

Bogduk N (2005). Pain of cranial nerve origin other than primary neuralgias. In Olesen J, Goadsby PJ, Ramadan NM, Tfelt-Hansen P, eds. *The Headaches*, Chapter 126, 3rd edition, pp. 1043–51. Lippincott Williams and Wilkins, Philadelphia.

Boivie J (2005). Central pain. In McMahon SB, Koltzenburg M, eds. *Wall and Melzack's Textbook of Pain*, Chapter 67, 5th edition pp. 1057–74. Elsevier, Amsterdam.

Bruyn GW (1983). Glossopharyngeal neuralgia. *Cephalalgia*, **3**, 143–57.

Burchiel KJ (1996). Pain in neurology and neurosurgery: tic douloureux (trigeminal neuralgia). In Campbell JN, ed. *Pain 1996—an updated review*, pp. 41–60. IASP Press, Seattle.

Cheng TM, Cascino TL, Onofrio BM (1993). Comprehensive study of diagnosis and treatment of trigeminal neuralgia secondary to tumors. *Neurology*, **43**, 2298–302.

Colding A (1969). The effect of sympathetic blocks on herpes zoster. *Acta Anaesthesiol Scand*, **13**, 113–41.

Correll RW, Jenson JL, Taylor JB (1979). Mineralization of the styloid-stylomandibular ligament complex. *Oral Surg Oral Med Oral Pathol*, **48**, 286–91.

Costen JB (1934). A syndrome of ear and sinus symptoms dependent upon disturbed function of the temporomandibular joint. *Ann Rhinol Laryngol*, **43**, 1–15.

Cruccu G, Galeotti F, Iannetti GD *et al.* (2002). Trigeminal neuralgia: update on reflex and evoked potential studies. *Mov Disord*, **17**, S37–40.

Dandy WE (1934). Trigeminal neuralgia. *Am J Surg*, **24**, 447–55.

Dao TT, Lavigne GJ, Charbonneau A *et al.* (1994). The efficacy of oral splints in the treatment of myofascial pain of the jaw muscles: a controlled clinical trial. *Pain*, **56**, 85–94.

Denny-Brown D, Adams RD, Fitzgerald PJ (1944). Pathologic features of herpes zoster: a note on 'geniculate herpes'. *Arc Neurol Psychiatry*, **77**, 337–49.

DeShazo R, Chapin K, Swain R (1997). Fungal sinusitis. *New Engl J Med*, **337**, 254–9.

Devor M (2005). Response of nerves to injury in relation to pain. In McMahon SB, Koltzenburg M, eds. *Wall and Melzack's Textbook of Pain*, Chapter 58, 5th edition, pp. 905–28. Elsevier, Amsterdam.

Devor M, Dubner R (1988). Centrifugal activity in afferent C-fibers influences the spontaneous afferent barrage generated in nerve end neuromas. *Brain Res*, **446**, 396–400.

Devor M, Amir R, Rappaport ZH (2002). Pathophysiology of trigeminal neuralgia: the ignition hypothesis. *Clin J Pain*, **18**, 4–13.

Devor M, Wall PD (1990). Cross excitation among dorsal root ganglion neurons in nerve injured and intact rats. *J Neurophysiol*, **64**, 1733–46.

Eagle WW (1937). Elongated styloid process: report of 2 cases. *Arch Otolaryngol*, **25**, 584–7.

Eagle WW (1958). Elongated styloid process: symptoms and treatment. *Arch Otolaryngol*, **67**, 172–6.

Eaglstein WH, Katz R, Brown JA (1970). The effects of early corticosteroid therapy on the skin eruption and pain of herpes zoster. *JAMA*, **211**, 1681–3.

Esman V, Ipsen J, Peterslund NA *et al.* (1982). Therapy of acute herpes zoster with acyclovir in the non-immunocompromised host. *Am J Med*, **73**, 320–5.

Feinmann C, Harris M, Cawley R (1984). Psychogenic facial pain: presentation and treatment. *Br Med J*, **288**, 436–8.

Ferrante L, Artico M, Nardacci P *et al.* (1995). Glossopharyngeal neuralgia with cardiac syncope. *Neurosurgery*, **36**, 58–63.

Fields HL, Rowbotham M, Baron R (1998). Postherpetic neuralgia: irritable nociceptors and deafferentation. *Neurobiol Dis*, **5**, 209–27.

Fiscbach F, Lehmann TN, Ricke J *et al.* (2003). Vascular compression in glossopharyngeal neuralgia: demonstration by high-resolution MRI at 3 tesla. *Neuroradiology*, **45**, 810–1.

Furlow LP (1942). Tic douloureux of the nervus intermedius. *JAMA*, **119**, 255.

Gardner WJ, Miklos MV (1959). Response of trigeminal neuralgia to 'decompression' of the sensory root. *JAMA*, **170**, 1773–6.

Goodman P, Greene CS, Laskin DM (1976). Response of patients with myofascial pain-dysfunction syndrome to mock equilibration. *J Am Dent Assoc*, **92**, 755–8.

Hakanson S (1981). Trigeminal neuralgia treated by the injection of glycerol into the trigeminal cistern. *Neurosurgery*, **9**, 638–46.

Harris M (1974). Psychogenic aspects of facial pain. *Br Dent J*, **136**, 199–202.

Harrison SD, Glover L, Feinmann C *et al.* (1997). A comparison of antidepressant medication alone and in conjunction with cognitive behavioural therapy for chronic idiopathic facial pain. In Jensen TS, Turner JA, Weisenfeld-Hallin Z, eds. *Proceedings of the 8th World Congress on Pain. Progress in Pain research and Management*, **8**, 663–72. IASP Press, Seattle.

Head H, Campbell AW (1900). The pathology of herpes zoster and its bearing on sensory localisation. *Brain*, **23**, 353–523.

Hope-Simpson RE (1965). The nature of herpes zoster: a long term study and a new hypothesis. *Proc R Soc Med*, **58**, 9–20.

Hunt WE, Meagher JN, LeFever HE *et al.* (1961). Painful ophthalmoplegia: its relation to indolent inflammation of the cavernous sinus. *Neurology*, **11**, 56–62.

International Headache Society (1988). Classification and diagnostic criteria for headache disorders, cranial neuralgias and facial pain. *Cephalalgia*, **8** (Suppl 7), 1–96.

International Association for the Study of Pain (1994). Classification of chronic pain.. In Merskey H, Bogduk N, eds. *Descriptors of Chronic Pain Syndromes and Definitions of Pain Terms*, 2nd edition. IASP Press, Seattle.

International Headache Society (2004). The international classification of headache disorders. 2nd edition. *Cephalalgia*, **24** (Suppl 1), 1–160.

Jackson JL, Gibbons R, Meyer G, Inouye L (1997). The effect of treating herpes zoster with oral acyclovir in preventing postherpetic neuralgia. A meta-analysis. *Arch Intern Med*, **157**, 909–12.

Jagger RG (1973). Diazepam in the treatment of temporomandibular joint dysfunction syndrome – a double blind trial. *J Dent*, **2**, 37–40.

Jannetta PJ (1976). Microsurgical approach to the trigeminal nerve for tic douloureux. In Krayenbuhl H, Maspes PE, Sweet WH, eds. *Prog Neurol Psychiatry*, **7**, 180–200. Karger, Basel.

Jansen J, Bardosi A, Hildebrandt J *et al.* (1989a). Cervicogenic, hemicranial attacks associated with vascular irritation or compression of the cervical nerve root C2. Clinical manifestations and morphological findings. *Pain*, **39**, 203–12.

Jansen J, Markakis E, Rama B *et al.* (1989b). Hemicranial attacks or permanent hemicrania – a sequel of upper cervical root compression. *Cephalalgia*, **9**, 123–30.

Juel-Jensen BE, MacCallum FO, MacKenzie AMR, Pike MC (1970). Treatment of zoster with idoxuridine in dimethyl sulphoxide. Results of two double blind controlled trials. *Br Med J*, **iv**, 776–80.

Juniper RP (1984). Temporomandibular joint dysfunction: theory based upon electromyographic studies of the lateral pterygoid muscle. *Br J Oral Maxillofac Surg*, **22**, 1–8.

Kalkanis SN, Eskandar EN, Carter BS, Barker FG (2003). Microvascular decompression surgery in the United States, 1996-2000: mortality rates, morbidity rates, and the effects of hospital and surgeon volumes. *Neurosurgery*, **52**, 1251–1262.

Katusic S, Beard CM, Bergstralh E, Kurland LT (1990). Incidence and clinical features of trigeminal neuralgia, Rochester, Minnesota, 1945-1984. *Ann Neurol*, **27**, 89–95.

Katusic S, Williams DB, Beard CM *et al.* (1991). Incidence and clinical features of glossopharyngeal neuralgia, Rochester, Minnesota, 1945-1984. *Neuroepidemiology*, **10**, 266–75.

Keczkes K, Basheer AM (1980). Do corticosteroids prevent post-herpetic neuralgia? *Br J Dermatol*, **102**, 551–5.

Kerr FLW (1967). Pathology of trigeminal neuralgia: light and electron microscopic observations. *J Neurosurg*, **26**, 151–6.

Kircos LT, Ortendahl DA, Mark AS, Arakawa M (1987). Magnetic resonance imaging of the TMJ disc in asymptomatic volunteers. *J Oral Maxillofac Surg*, **45**, 852–854.

Kondo A (1998). Follow-up results using microvasular decompression for treatment of glossopharyngeal neuralgia. *J Neurosurg*, **88**, 221–5.

Kost RG, Straus SE (1996). Postherpetic neuralgia - pathogenesis, treatment, and prevention. *New Engl J Med*, **335**, 32–42.

Kotani N, Kushikata T, Hashimoto H *et al.* (2000). Intrathecal methylprednisolone for intractable postherpetic neuralgia. *New Engl J Med*, **343**, 1514–9.

Kugelberg E, Lindblom U (1959). The mechanism of pain in trigeminal neuralgia. *J Neurol Neurosurg Psychiatry*, **22**, 36–43.

Kurland LT (1958). Descriptive epidemiology of selected neurological and myopathic disorders with particular reference to a survey in Rochester, Minnesota. *J Chronic Dis*, **8**, 378–418.

Lance JW, Anthony M (1980). Neck tongue syndrome on sudden turning of the head. *J Neurol Neurosurg Psychiatry*, **43**, 97–101.

Lascelles RG (1966). Atypical facial pain and depression. *Br J Psychiatry*, **112**, 651–9.

Laskin DM (1980). Myofascial pain syndrome: etiology. In Sarnat BG, Laskin DM, eds. *The Temporomandibular Joint: a Biological basis for Clinical Practice*, 3rd edition. Charles C Thomas, Springfield, Illinois.

Lauria G, Majorana A, Borgna M *et al.* (2005). Trigeminal small-fiber sensory neuropathy causes burning mouth syndrome. *Pain*, **115**, 332–7.

Lechin F, van der Dijs, Lechin ME *et al.* (1989). Pimozide therapy for trigeminal neuralgia. *Acta Neurol Scand*, **69**, 960–3.

Lesse S (1956). Atypical facial pain syndrome of psychogenic origin. *J Nerv Ment Dis*, **124**, 341–63.

Lewith GT, Field J, Machin D (1983). Acupuncture compared with placebo in post-herpetic pain. *Pain*, **17**, 361–8.

Lhermitte J, Nicholas M (1924). Les lesions spinales du zona. La myelite zosterienne. *Revue Neurol*, **1**, 361–4.

Lindstrom P, Lindblom U (1987). The analgesic effect of tocainide in trigeminal neuralgia. *Pain*, **28**, 45–50.

Lisney SJW, Devor M (1987). Afterdischarge and interactions among fibers in damaged peripheral nerve in the rat. *Brain Res*, **415**, 122–36.

Locker D, Grushka M (1987). Prevalence of oral and facial pain and discomfort: preliminary results of a mail survey. *Community Dentistry and Oral Epidemiol*, **15**, 169–72.

Loeser JD (1986). Herpes zoster and postherpetic neuralgia. *Pain*, **25**, 149–64.

Loeser JD (1989). Trigeminal neuralgia and atypical facial pain. In Wall PD, Melzack R, eds. *Textbook of Pain*, 2nd edition, pp. 535–43. Churchill Livingstone, Edinburgh.

Lopez BC, Hamlyn PJ, Zakrzewska JM (2004). Systematic review of ablative neurosurgical techniques for the treatment of trigeminal neuralgia. *Neurosurgery*, **54**, 973–82.

Love S, Coakham HB (2001). Trigeminal neuralgia. Pathology and pathogenesis. *Brain*, **124**, 2347–60.

Love S, Gradidge T, Coakham HB (2001). Trigeminal neuralgia due to multiple sclerosis: ultrastructural findings in trigeminal rhizotomy specimens. *Neuropathol Appl Neurobiol*, **27**, 1–8.

Lovely TJ, Jannetta PJ (1997). Surgical management of geniculate neuralgia. *Am J Otol*, **18**, 512–7.

MacDonald BK, Cockerell OC, Sander JWAS *et al.* (2000). The incidence and lifetime prevalence of neurological disorders in a prospective community-based study in the UK. *Brain*, **123**, 665–76.

Maesawa S, Salame C, Flickinger JC *et al.* (2001). Clinical outcomes after stereotactic radiosurgery for idiopathic trigeminal neuralgia. *J Neurosurg*, **94**, 14–20.

Magnusson T, Carlsson GE (1983). Occlusal adjustment in patients with residual or recurrent signs of mandibular dysfunction. *J Prosthet Dent*, **49**, 706–10.

McLaughlin MR, Jannetta PJ *et al.* (1999). Microvascular decompression of cranial nerves: lessons learned after 4400 operations. *J Neurosurg*, **90**, 1–8.

McQuay HJ, Carroll D, Jadad AR *et al.* (1995). Anticonvulsant drugs for management of pain: a systematic review. *Br Med J*, **311**, 1047–52.

Melzack R, Wall PD (1965). Pain mechanisms: a new theory. *Science*, **150**, 971–9.

Mendelssohn AH, Berke GS, Chhetri DK (2006). Heterogeneity in the clinical presentation of Eagle's syndrome. *Otolaryngol Head Neck Surg*, **134**, 389–93.

Merselis JG, Kaye D, Hook EW (1964). Disseminated herpes zoster. *Arch Intern Med*, **113**, 679–86.

Mock D, Frydman W, Gordon AS (1985). Atypical facial pain: a retrospective study. *Oral Surg Oral Med Oral Pathol*, **59**, 472–4.

Moore DS, Nally FF (1975). Atypical facial pain: an analysis of 100 patients with discussion. *J Can Dent Assoc*, **41**, 396–401.

Noordenbos W (1959). *Pain*. Elsevier, Amsterdam.

Nurmikko TJ (1991). Altered cutaneous sensation in trigeminal neuralgia. *Arch Neurol*, **48**, 523–7.

Nurmikko TJ, Haggett CE, Miles J (2000). Neurogenic vasodilatation in trigeminal neuralgia. In Devor M, Rowbotham MC, Wiesenfeld-Hallin Z, eds. *Proceedings of the 9th World Congress of Pain*. IASP Press, Seattle. 747–55.

Nurmikko TJ, Jensen TS (2005). Trigeminal neuralgia and other facial neuralgias. In Olesen J, Goadsby PJ, Ramadan NM, Tfelt-Hansen P, eds. *The Headaches*, 3rd edition, Chapter 127, pp. 1053–62. Lippincott Williams and Wilkins, Philadelphia.

Oxman MN, Levin MJ, Johnson GR *et al.* (2005). A vaccine to prevent herpes zoster and postherpetic neuralgia in older adults. New Engl J Med, **352**, 2271–84.

Pappagallo M, Oaklander AL, Quatrano-Piacentini AL, Clark MR, Raja SN (2000). Heterogeneous patterns of sensory dysfunction in postherpetic neuralgia suggest multiple pathophysiologic mechanisms. *Anesthesiology*, **92**, 691–8.

Patel A, Kassam A, Horowitz M *et al.* (2002). Microvascular decompression in the management of glossopharyngeal neuralgia: analysis of 217 cases. *Neurosurgery*, **50**, 705–11.

Patel NK, Aquilina K, Clarke Y *et al.* (2003). How accurate is magnetis resonance angiography in predicting neurovascular compression in patients with trigeminal neuralgia? A prospective single-blinded comparative study. *Br J Neurosurg*, **17**, 60–4.

Portenoy RK, Duma C, Foley KM (1986). Acute herpetic and postherpetic neuralgia: clinical review and current management. *Ann Neurol*, **20**, 651–64.

Raeder JG (1924). Paratrigeminal paralysis of oculo-pupillary sympathetic. *Brain*, **47**, 149–58.

Ramijford SP (1961). Dysfunctional temporomandibular joint and muscle pain. *J Prosthet Dent*, **11**, 354–74.

Rasmussen P (1990). Facial pain. II. A prospective survey of 1052 patients with a view of: character of the attacks, onset, course, and character of pain. *Acta Neurochir (Wien)*, **107**, 121–8.

Rasmussen P (1991). Facial pain. IV. A prospective study of 1052 patients with a view of precipitating factors, associated symptoms, objective psychiatric and neurological symptoms. *Acta Neurochir (Wien)*, **108**, 100–9.

Rees RT, Harris M (1979). Atypical odontalgia. *Br J Oral Surg*, **16**, 212–8.

Rothman KJ, Monson RR (1973). Survival in trigeminal neuralgia. *J Chronic Dis*, **26**, 303–9.

Rushton JG, Stevens JC, Miller RH (1981). Glossopharyngeal (vagoglossopharyngeal) neuralgia: a study of 217 cases. *Arch Neurol*, **38**, 201–15.

Salamone FN, Falciglia M, Steward DL (2004). Eagle's syndrome reconsidered as a manifestation of heterotopic ossification: woman presenting with a neck mass. *Otolaryngol Head Neck Surg*, **130**, 501–3.

Scadding JW, Koltzenburg M (2005). Painful peripheral neuropathies. In McMahon SB, Koltzenburg M, eds. *Wall and Melzack's Textbook of Pain*, 5th edition, Chapter 62, pp. 973–1000. Elsevier, Amsterdam.

Schwartz L (1959). *Disorders of the Temporomandibular Joint and Muscle Pain*. WB Saunders, Philadelphia, PA.

Seltzer Z, Devor M (1979). Ephaptic transmission in chronically damaged peripheral nerves. *Neurology*, **29**, 1061–4.

Smith JL, Taxdal DSR (1966). Painful ophthalmoplegia: the Tolosa-Hunt syndrome. *Am J Ophthalmol*, **61**, 1466–72.

Swerdlow B (1980). Tolosa-Hunt syndrome: a case with associated facial nerve palsy. *Ann Neurol*, **8**, 542–3.

Thomson H (1971). Mandibular dysfunction syndrome. *Br Dent J*, **130**, 187–93.

Tolosa E (1965). Periarteritic lesions of the carotid siphon with the clinical features of a carotid infraclinoidal aneurysm. *J Neurol Neurosurg Psychiatry*, **17**, 300–2.

Watson CPN, Evans RJ, Watt VR, Birkett N (1988a). Post-herpetic neuralgia: 208 cases. *Pain*, **35**, 289–97.

Watson CPN, Morshead C, Van der Koog D, Deck JH, Evans RJ (1988b). Post-herpetic neuralgia: post-mortem analysis of a case. *Pain*, **34**, 129–38.

Watson CPN, Watt VR, Chipman M, Birkett N, Evans RJ (1991a). The prognosis with post-herpetic neuralgia. *Pain*, **46**, 195–9.

Watson CPN, Deck JH, Morshead C, Van der Koog D, Evans RJ (1991b). Post-herpetic neuralgia: further post-mortem studies of cases with and without pain. *Pain*, **44**, 105–17.

Wiffen P, McQuay HJ, Carroll D *et al.* (2000). Anticonvulsant drugs for acute and chronic pain. *Cochrane Database Syst Rev*, CD001133.

Zachs SI, Langfit TW, Elliot FA (1964). Herpetic neuritis: a light and electron microscopic study. *Neurology*, **14**, 644–750.

Zakrzewska JM, Thomas DGT (1993). Patients' assessment of outcome after three surgical procedures for the management of trigeminal neuralgia. *Acta Neurochir (Wien)*, **122**, 225–30.

Zakrzewska JM, Sawsan J, Bulman JS (1999). A prospective, longitudinal study on patients who underwent radiofrequency thermocoagulation of the Gasserian ganglion. *Pain*, **79**, 51–8.

Zakrzewska JM, Harrison SD (2003). Facial pain. In Jensen TS, Wilson PR, Rice ASC, eds. *Clinical Pain Management: Chronic Pain*, Chapter 37, pp. 481–504. Arnold, London.

Zakrzewska JM, Lopez BC (2004). Trigeminal neuralgia. *Clinical Evidence*, **11**, 1755–65.

Zakrzewska JM, Lopez BC (2005). Trigeminal and glossopharyngeal neuralgia. In McMahon SB and Kotzenburg M, eds. *Wall and Melzack's Textbook of Pain*, Chapter 63, 5th edition, pp 1001–10. Elsevier, Amsterdam.

Zakrzewska JM, Forssell H, Glenny A (2005). Interventions for the treatment of burning mouth syndrome. *Cochrane Database Syst Rev*, CD002779.

SECTION 5

Nerve and muscle disease

CHAPTER 20

The lower cranial nerves and dysphagia

Pamela Shaw and David Hilton-Jones

Contents

20.1 Cranial nerve V: trigeminal

20.1.1 Clinical neuroanatomy

The trigeminal nerve contains both sensory afferent and motor efferent fibres. Sensory information is conveyed from the skin of the face and forehead (Fig. 20.1), from the mucous membranes of the nasal sinuses and oral cavities, from the teeth, and from the dura of the anterior and middle cranial fossae. Efferent fibres innervate the muscles of mastication consisting of the masseter, temporalis, and pterygoid muscles, as well as the tensor

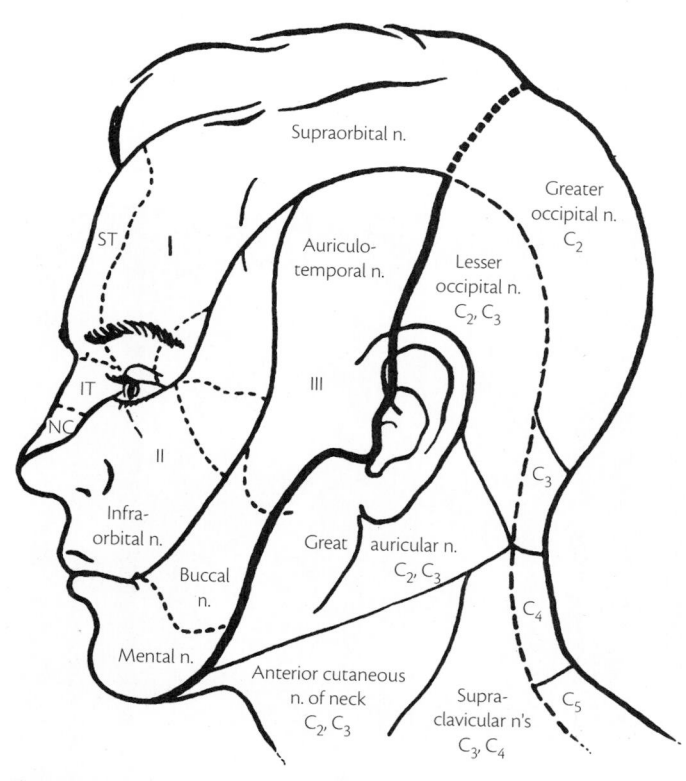

Fig. 20.1 The cutaneous distribution of the trigeminal nerve and its branches. ST, supratrochlear nerve; IT, infratrochlear nerve; NC, nasociliary nerve. (Reproduced from Brodal (1981). *Neurological Anatomy*. Oxford University Press.)

veli palatine and tensor tympani muscles. Whereas the organization of the motor pathways is relatively straightforward, the sensory system is complex with three major peripheral branches, each with several smaller branches, and two central pathways each subserving different sensory modalities. The peripheral distribution will be discussed first followed by a description of the central connections.

Peripheral sensory pathways

The cell bodies of the afferent fibres lie in the trigeminal ganglion, also known as the gasserian or semilunar ganglion. This is located in a dural cavity, Meckel's cave, near the tip of the petrous apex bone (Fig. 20.2). The peripheral processes of these unipolar ganglion cells give rise to the three main divisions of the trigeminal nerve; the ophthalmic, maxillary, and mandibular branches, often conveniently referred to as V_1, V_2, and V_3. The central processes form the sensory root which enters the pons on its lateral aspect.

The **ophthalmic nerve, V_1**, passes forwards in the lateral wall of the cavernous sinus, where it lies close to the ocular motor nerves, cranial nerves III, IV, and VI, and enters the orbit through the superior orbital fissure. Branches of the nerve supply sensation to the scalp, forehead, and nose, to the mucous membrane of the frontal sinus and upper part of the nasal cavity, and to the conjunctiva and cornea of the eye, the latter providing the afferent limb of the corneal reflex. The area of cutaneous distribution is clearly defined (Fig. 20.1) but often misunderstood. Posteriorly, the nerve supplies the scalp as far back as the lambdoidal suture. Psychologically determined sensory loss over the face and forehead often extends only to the hair-line, but care must be taken in interpretation because such apparently non-anatomical distribution of sensory loss can be seen in organic disease. The lateral extent of the cutaneous distribution and the boundaries with the areas of supply of the maxillary and mandibular divisions are shown in Fig. 20.1.

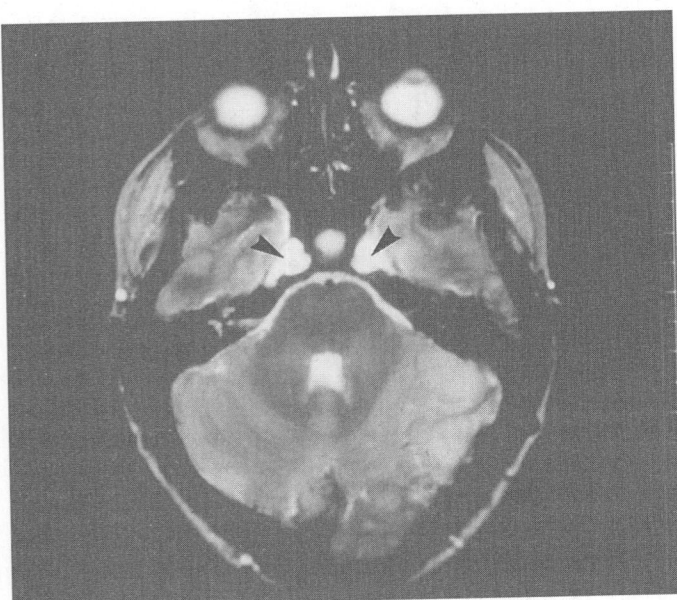

Fig. 20.2 Axial T2-weighted MRI showing the position of Meckel's cave (arrows) where the trigeminal ganglion is located. (Courtesy of Dr. P. Anslow.)

The **maxillary nerve, V_2**, passes through the inferior part of the cavernous sinus and leaves the skull through the foramen rotundum. A major branch, the infra-orbital nerve, enters the orbit through the inferior orbital fissure and exits through the infra-orbital foramen. The area of cutaneous supply is shown in Fig. 20.1. The nerve also supplies the mucous membrane of the upper lip, hard palate, and anterior part of the soft palate, the mucous membrane of the maxillary sinus and of the lower nasal cavity, and the teeth of the upper jaw.

The **mandibular nerve, V_3**, fuses with the motor root and leaves the skull through the foramen ovale. The cutaneous distribution is shown in Fig. 20.1. The angle of the jaw is not supplied by the mandibular nerve, but this area is often affected in psychologically determined facial sensory loss. The sensory territory includes part of the pinna, external auditory meatus, and tympanic membrane. The nerve also supplies the mucous membrane of the cheek, floor of the mouth, and anterior two-thirds of the tongue, and the teeth of the lower jaw. The lingual branch carries taste fibres from the anterior two-thirds of the tongue, which then joins the facial nerve via the chorda tympani.

Peripheral motor pathways

The motor fibres emerging from the pons pass under the trigeminal ganglion and fuse with sensory fibres to form the mandibular nerve, which leaves the skull through the foramen ovale. They innervate the temporalis, masseter, and medial and lateral pterygoid muscles. Other muscles supplied by the mandibular nerve include the tensor veli palatini and tensor tympani which cannot be tested at the bedside and their involvement in isolation is not associated with specific clinical features.

Central connections

The sensory root, formed by the central processes of the trigeminal ganglion cells, enters the lateral pons and divides into short ascending and long descending branches. The short ascending fibres terminate in the principal sensory nucleus of the trigeminal nerve which lies in the substantia gelatinosa of the lateral tegmentum of the upper pons. These fibres subserve tactile and pressure sensation; in clinical practice, lesions impair light-touch sensation and the corneal reflex. The long descending fibres form the spinal trigeminal tract which descends down through the brainstem, extending as far as the C2 level of the upper cervical spinal cord. As the tract descends, the fibres gradually terminate in the medially placed substantia gelatinosa forming the spinal trigeminal nucleus which therefore can also be seen to extend from the pons down to the upper cervical cord. This pathway subserves pain and thermal sensation; in clinical practice, lesions impair pin-prick and temperature sensation. There is a specific but complex topographical arrangement of fibres within the trigeminal tract; fibres from the ophthalmic division lie most ventrally, and those from the mandibular division most dorsally.

The secondary trigeminal pathways arise from the primary sensory nucleus and from the spinal trigeminal nucleus. Crossed and uncrossed fibres are important in various reflex pathways, discussed below, but the major sensory pathways involve crossed fibres. The secondary fibres arising from the principal sensory nucleus cross the midline in the pons and ascend as the quintothalamic tract, in close association with the medial lemniscus. Secondary fibres arising from the long spinal trigeminal nucleus cross the midline raphe and ascend in association with the medial lemniscus. Secondary trigeminal pathways, from both the principal sensory nucleus and the spinal trigeminal nucleus, terminate in the thalamus.

Trigeminal reflexes

Secondary crossed and uncrossed fibres from the two trigeminal nuclei are involved in several reflex pathways including tearing, sneezing, and vomiting. In clinical practice the most important are the corneal reflex and the jaw jerk (Section 2.3.4).

Stimulation of the cornea sends afferent impulses via the ophthalmic division to the trigeminal nucleus. Secondary fibres project bilaterally to the facial nuclei which, with the facial nerves, complete the reflex arc. Unilateral corneal sensation evokes bilateral blinking. In the presence of a facial nerve palsy, ipsilateral corneal stimulation will cause contralateral blinking. Corneal stimulation on the side of an ophthalmic nerve lesion will produce no response, but contralateral corneal stimulation will produce bilateral blinking.

The jaw jerk is a monosynaptic muscle stretch reflex. Jaw tapping invokes bilateral contraction of the masseter and temporalis muscles. The reflex is not noticeably affected by a unilateral upper or lower motor neurone lesion of the trigeminal nerve, but is exaggerated in the presence of bilateral upper motor neurone lesions, often to the point of clonus.

20.1.2 Lesions of the trigeminal nerve

Trigeminal nerve function may be affected by supranuclear, nuclear, or peripheral lesions. Because of the wide anatomical distribution of the components of the trigeminal nerve, complete interruption of both the motor and sensory parts is rarely observed in practice. However, partial involvement of the trigeminal nerve, particularly the sensory component, is relatively common, the main symptoms being numbness and pain.

Supranuclear lesions

The motor nuclei receive bilateral upper motor neurone or corticobulbar innervation. A unilateral upper motor neurone lesion, for instance a hemispheric stroke, causes no clinically discernible weakness of the trigeminal nerve-innervated muscles. Bilateral upper motor neurone lesions result in a pseudobulbar palsy with dysarthria, dysphagia, and a brisk jaw jerk. Common causes include bilateral hemispheric strokes, occurring simultaneously or consecutively, or an upper brainstem lesion affecting both corticobulbar pathways.

Nuclear lesions

Lesions involving the motor nucleus in the pons will cause ipsilateral weakness and wasting of the muscles of mastication. On jaw opening, the jaw will deviate towards the affected side because of weakness of the pterygoid muscles (Fig. 20.3A). On jaw closure the ipsilateral masseter and temporalis may be visibly and palpably wasted (Fig. 20.3B).

Isolated involvement of the pontine primary sensory nucleus might be expected to produce ipsilateral loss of facial light-touch sensation, with preservation of pin-prick sensation. However in practice such lesions invariably also involve descending fibres and both sensory modalities are usually impaired. Lesions in this area affecting trigeminal motor and sensory function also frequently cause contralateral hemiplegia and spinothalamic sensory loss. Common pathologies include vascular disease, demyelination, and tumour. Rarer causes include vascular malformations and syringobulbia.

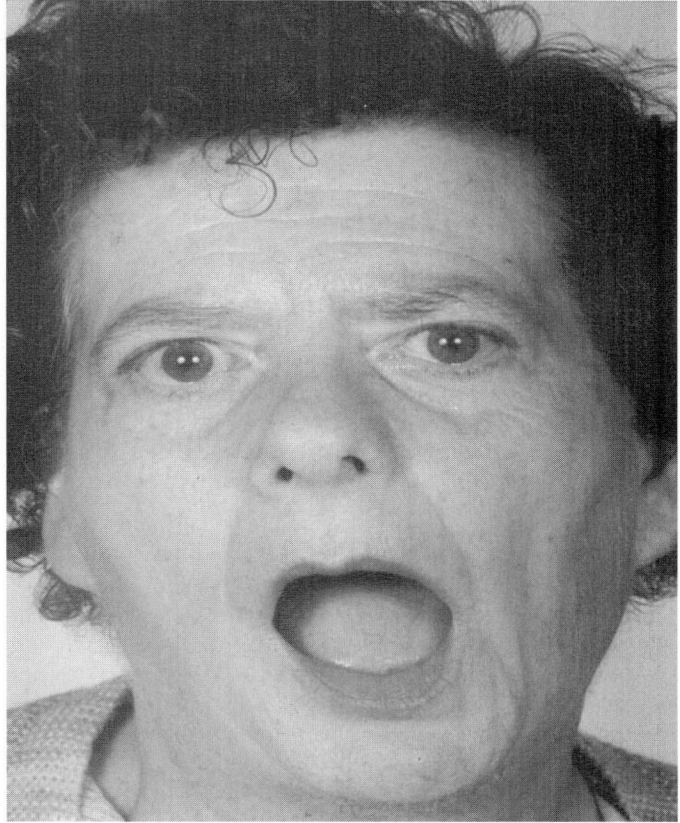

A

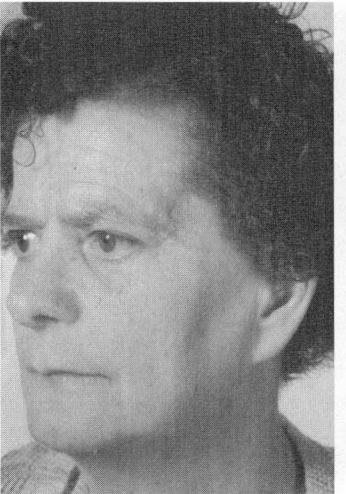

B

Fig. 20.3 A patient with left-sided trigeminal motor neuropathy. A. Opening the mouth causes the jaw to deviate towards the side of the lesion. B. Wasting of the masseter and temporalis muscles on the left side of the face.

Lesions in the medulla and upper cervical cord may affect the spinal trigeminal tract and nucleus, causing ipsilateral loss of facial pin-prick and temperature sensation, with preservation of light-touch and the corneal reflex. The commonest cause is infarction of the lateral medulla secondary to occlusion of the vertebral artery or the posterior inferior cerebellar artery, *the lateral medullary syndrome of Wallenberg* (Section 2.4.4). Additional symptoms include hiccoughs, dizziness, dysarthria, and dysphagia. Ipsilateral signs, in addition to the facial sensory loss, include Horner's syndrome, palatal and vocal cord paresis, and cerebellar ataxia. Contralateral signs include limb and trunk spinothalamic sensory loss affecting pain and temperature.

In syringomyelia, the cavity within the central part of the cervical spinal cord may gradually extend upwards into the medulla, when it is referred to as syringobulbia, and possibly as far as the pons (Section 28.5.15). Spinal cord tumours may progress in a similar way. In the presence of this type of pathology, there is bilateral involvement of the trigeminal tract and nuclei. Due to the topographical organization of trigeminal sensory pathways within the brainstem, a particular pattern of facial sensory loss evolves, affecting pain and temperature, which has been likened to an onion-skin or the wearing of a balaclava helmet. Thus, the sensory loss gradually progresses in a series of layers forwards and medially towards the nose.

Peripheral lesions

Multiple pathologies can affect the intracranial parts of the trigeminal nerve complex: the motor and sensory roots, trigeminal ganglion, and the three major nerve divisions. These include: tumours, metastases, carcinomatous meningitis, acoustic neuromas, trigeminal neuromas, meningiomas, or nasopharyngeal carcinoma; infections, viral, acute, and chronic meningitis, abscesses, osteitis; Paget's disease; trauma; aneurysms, and granulomatous processes. Depending upon the anatomical site of the lesion other cranial nerves may be involved and particular syndromes can be identified (Table 20.1).

A lesion simultaneously affecting both the sympathetic nerve fibres around the internal carotid artery and the trigeminal ganglion may produce a Horner's syndrome, but without anhydrosis as sudomotor fibres travel along the external carotid artery, and trigeminal nerve involvement, either with pain alone or with a demonstrable sensori-motor neuropathy. This combination is referred to as *Raeder's paratrigeminal syndrome* (Section 18.7.3) and causes include carotid aneurysm, infection, tumours, and

trauma. Such lesions may also involve the optic nerve and cranial nerves III, IV, and VI in the parasellar region.

More peripheral branches of these three main nerves may be damaged by blunt, penetrating or surgical trauma resulting in areas of sensory disturbance, sometimes accompanied by continuous or neuralgic pain. These problems do not often come to the attention of neurologists. The most superficial branches of the trigeminal nerve—the supratrochlear, supraorbital, and infraorbital branches—are most commonly affected. The rather distinctive numb cheek and numb chin syndromes are discussed below.

20.1.3 Trigeminal herpes zoster

Shingles is due to reactivation of herpes zoster virus lying dormant in a sensory root ganglion. In youth it most frequently affects the trunk, less often than limbs (Section 21.14.5), but with increasing age facial involvement becomes more common. Reactivation of virus in the trigeminal ganglion usually affects only the ophthalmic division, producing herpes zoster ophthalmicus. Early ophthalmic complications and later neurological sequelae are common. Most cases occur apparently spontaneously but sometimes the trigger appears to be an intercurrent infection, or a state of drug or disease-induced immunosuppression. Unilateral pain in the distribution of the ophthalmic nerve is followed within hours or days by a vesicular skin eruption (Fig. 20.4). The rash resolves spontaneously. Treatment is based on systemic acyclovir with the addition of topical steroids if there is evidence of anterior segment involvement, such as iritis.

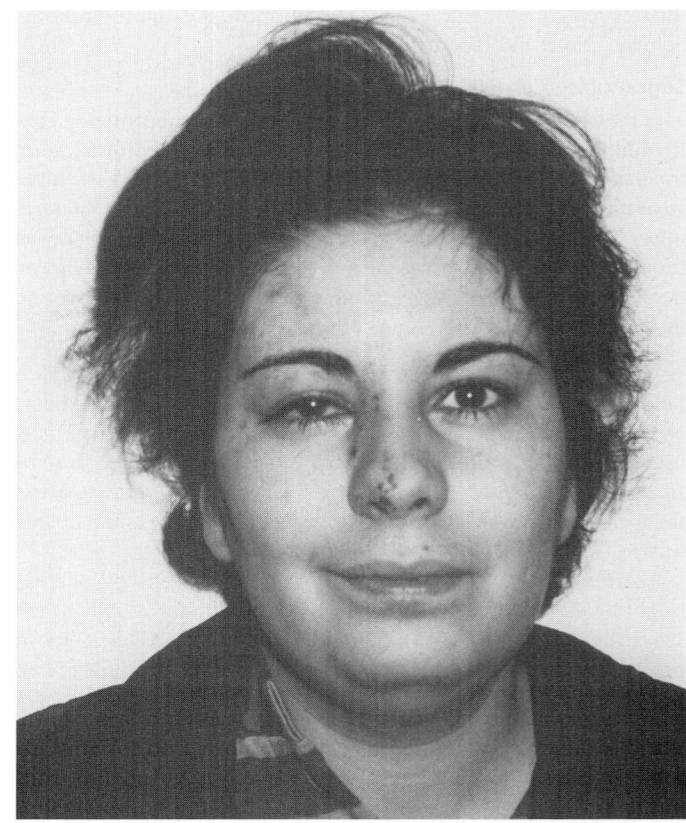

Fig. 20.4 Herpes zoster ophthalmicus on the right side.

Table 20.1 Cranial nerve syndromes involving the trigeminal nerve

Cranial nerves	Site of lesion	Common causes
III, IV, V$_1$, VI	Superior orbital fissure	Tumours Carotid aneurysm Granulomata
III, IV, V$_1$, VI Occasionally V$_2$	Cavernous sinus	Carotid aneurysm Cavernous sinus thrombosis Tumours Granulomata
V, VI	Apex of petrous bone	Tumour Osteitis (Gradenigo's syndrome)
V, VII, VIII, rarely IX	Cerebellopontine angle	Acoustic neuroma Meningioma Embryonic tumours

Important sequelae of herpes zoster ophthalmicus include ocular complications, post-herpetic neuralgia, cranial nerve lesions, and cerebral involvement.

Ocular complications

These occur in up to 50 per cent of patients and include corneal perforation, uveitis, keratitis, scleritis, entropion, and glaucoma (Womack and Liesegang 1983; Severson *et al.* 2003). Orbital myositis has also been described preceding the vesicular eruption (Kawasaki and Borruat 2003).

Post-herpetic neuralgia

This is the commonest neurological complication of herpes zoster ophthalmicus, occurring in over 10 per cent of cases, and much more frequently in the elderly (Section 19.2.3). The initial acute pain and the rash resolve leaving the patient complaining of a continuous severe pain. The pain of post-herpetic neuralgia is usually described as a constant burning pain, with superimposed waves of lancinating pain and the skin in the area of the preceding eruption is often extremely sensitive to tactile stimuli. The pathophysiology has not been clearly defined, though peripheral and central demyelination as well as neuronal destruction are considered important (Christo *et al.* 2007). Functional MRI studies have shown that overall brain activity for the spontaneous pain of post-herpetic neuralgia involved affective and sensory-discriminative areas including the thalamus, primary and secondary somatosensory, insula and anterior cingulate cortices, as well as areas involved in emotion, hedonics, reward and punishment including the ventral striatum, amygdala, orbital frontal cortex, and the ventral tegmental area (Geha *et al.* 2007).

The pain is notoriously resistant to treatment (Watson 1995). Early administration of anti-viral therapy in the acute phase may reduce the risk of the development of post-herpetic neuralgia (Volpi *et al.* 2005). Current evidence indicates that various therapies are effective in reducing the pain associated with post-herpetic neuralgia including tricyclic anti-depressants, anti-epileptics such as carbamazepine, sodium valproate, and gabapentin, pregabalin, opioids, as well as topical lignocaine and capsaicin (Attal *et al.* 2006; Young 2006; Christo *et al.* 2007).

Cranial neuropathy

Cranial nerve lesions may develop several weeks after the acute episode. The ocular motor nerves, III, IV, and VI, are most frequently affected, either alone or in combination, sometimes resulting in an orbital apex or superior orbital fissure syndrome (Womack and Liesegang 1983; Gupta and Vishwakaram 1987; Harding *et al.* 1987). The pathogenesis is unclear and the extent of recovery, very variable.

Cerebral involvement

Rarely, at any time up to 4 weeks after the acute episode, patients develop contralateral hemiparesis, dysphasia, or hemianopia (Womack and Liesegang 1983). Angiography shows segmental arteritis of the carotid artery and granulomatous arteritis with viral particles in the smooth muscle cells of the arterial wall (Linnemann and Alvira 1980).

20.1.4 Trigeminal neuralgia

Trigeminal neuralgia, or Tic Douloureux, is the most frequently encountered disorder of the trigeminal nerve (Section 19.2.1). It may be symptomatic of an underlying structural disorder affecting the nerve, but in the majority of patients it is an idiopathic disorder. It is commoner in the second half of life, cases in younger people more often being symptomatic, is slightly more frequent in women and has an overall prevalence of the order of 3–5 per hundred thousand population.

Clinical features

These are highly characteristic, but despite this the diagnostic label is frequently applied erroneously to many other causes of facial pain, particularly atypical facial pain and dental disease. Trigeminal neuralgia consists of paroxysms of intense, stabbing pain in the distribution of the mandibular or maxillary divisions of the fifth cranial nerve. The pain seldom lasts more than a few seconds, but may be so severe that the patient winces involuntarily. Affected individuals often provide graphic descriptions which indicate the severity and quality of the pain—like 'a dagger', or 'red hot needle'. The paroxysms recur frequently, both day and night, for several weeks at a time. In the early stages, spontaneous remission for months or years may occur. Unfortunately permanent remission is rare and with time the bouts of pain become more frequent. When attacks are frequent, secondary depression is common. Many patients identify one or more triggers for their attacks. These include touching a specific part of the face, a cold draught, talking, swallowing, chewing, and brushing the teeth. Tactile triggers may prevent the patient washing their face, shaving, or eating. The pain is strictly within the trigeminal distribution, most commonly in the maxillary and mandibular divisions. The ophthalmic division is involved in less than 10 per cent of cases. In later stages both the mandibular and maxillary areas may become involved, but spread to the ophthalmic area is unusual.

Between the paroxysms, particularly if they are frequent, there may be a dull background ache that is not severe. Trigeminal neuralgia never causes continuous discomfort without the characteristic paroxysms. Occasional patients develop typical symptoms bilaterally but do not experience bilateral pain at the same time. Physical examination is normal in idiopathic trigeminal neuralgia. Abnormal physical signs suggest symptomatic trigeminal neuralgia.

Aetiology

It is recognized that trigeminal neuralgia may be symptomatic of underlying disease. Thus, about 4 per cent of patients with multiple sclerosis experience it, although it is very rare as a presenting symptom. Primary tumours of the trigeminal nerve and compression of the nerve by tumour or aneurysm rarely produce symptoms identical to those of trigeminal neuralgia but more commonly produce complaints of continuous pain or numbness, and are accompanied by abnormal physical signs on examination.

Excluding these rare causes of symptomatic trigeminal neuralgia one is left with a majority of patients in whom no physical cause is readily apparent, and thus the disorder might be considered to be one of altered function rather than structure. However, it has been suggested that in over 90 per cent of these patients the cause is a misdirected or ectatic blood vessel in the posterior fossa compressing the trigeminal sensory roots, and that symptomatic improvement can be gained by surgically separating the root from the aberrant blood vessels (Jannetta 1977; Haines *et al.* 1980). In another series such vascular compression was found in only 11 per cent of patients (Adams *et al.* 1982). Modern high definition MRI techniques may shed further light on this issue as scans may show vessels apparently

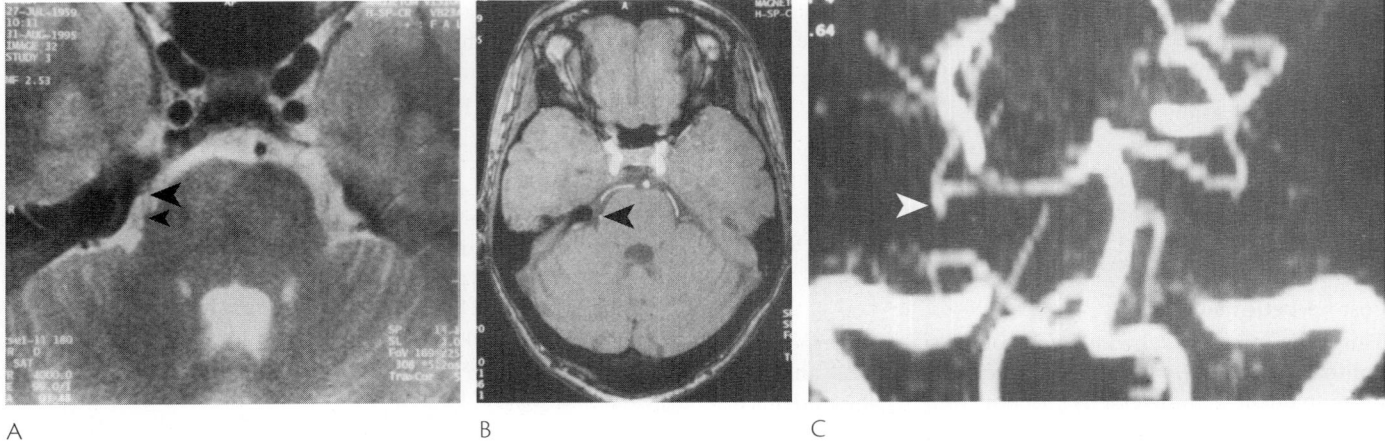

A   B   C

Fig. 20.5 Trigeminal neuralgia. MRI showing blood vessel 'impinging' on the trigeminal nerve. A. High-resolution axial T$_2$ sequence, showing the motor and sensory roots of the trigeminal nerve (large arrowhead). Immediately next to these is a blood vessel (small arrowhead), seen as a flow-void. That it is a blood vessel is confirmed on magnetic resonance angiography: B. axial image from MR angiography sequence. The aberrant blood vessel appears white (arrowhead). C. MR angiography (MIP: projection). The looping blood vessel can be seen (arrowhead). These findings were confirmed at surgery. Separation of the vessel and nerve proved therapeutic. (courtesy of Dr. P. Anslow).

impinging on the trigeminal nerve (Fig. 20.5) but a cause–effect relationship has not been definitively proven.

Differential diagnosis

The rare symptomatic causes of trigeminal neuralgia discussed above are often accompanied by abnormal physical signs. A substantial number of patients initially diagnosed as having trigeminal neuralgia prove to have other conditions. By far the commonest source of diagnostic error centres around the teeth. Dental disease, such as apical abscess, may cause paroxysmal as well as continuous pain, but the specific clinical features and triggers for the pain should readily distinguish this from trigeminal neuralgia. It is not uncommon for patients with trigeminal neuralgia to have healthy teeth removed.

Referred facial pain may also be caused by sinus disease and eye disease such as glaucoma (Section 19.4). Other causes of trigeminal nerve-related pain, which can be distinguished from trigeminal neuralgia on the basis of the history and physical signs, include brainstem lesions, post-herpetic neuralgia, and tabes dorsalis. Local irritative lesions and trauma in the regions of exit from the skull of the supraorbital and infraorbital nerves can cause localized neuralgic pain. Facial pain may also be caused by temporomandibular joint dysfunction (Section 19.3.1), Costen's syndrome, and maladjustment of dental occlusion. Atypical facial pain is a characteristic disorder seen mainly in young and middle-aged women (Section 19.6). They complain of a dull constant ache in the upper jaw/cheek region which may extend to the whole of the side of the head and down into the neck. Often, but not always, there is clear evidence of an anxiety or depressive disorder. This disorder may improve with antidepressant medication.

Cluster headache, or migrainous neuralgia (Section 18.4.1), is a highly characteristic condition that really should not be confused with trigeminal neuralgia, but sometimes is. The duration, distribution, and characteristics of the pain, the different triggering factors, the accompanying symptoms, and the pattern of attacks distinguish the condition from trigeminal neuralgia. Glossopharyngeal neuralgia (Sections 19.2.2, 20.3.3) causes attacks of identical character but in a different distribution.

Treatment

Various drugs have been tried for preventing trigeminal neuralgia (Section 19.2.1). Carbamazepine gives good or excellent symptomatic relief in up to 70 per cent of patients. A reasonable starting dose is 200–300 mg daily increasing, as is often required, over a 1–2-week period to either the lowest effective dose or the maximal tolerable dose. Common dose-related side-effects include nausea, unsteadiness, and visual disturbance. Up to 10 per cent of patients develop an idiosyncratic drug rash which necessitates stopping the drug.

If carbamazepine does not work or cannot be tolerated other drugs that can be tried include sodium valproate, phenytoin, lamotrigine, clonazepam, and baclofen, but success rates are lower than with carbamazepine. Amitriptyline may be beneficial particularly in combination with an anti-convulsant agent. Intranasal lignocaine spray may provide some transient pain relief in patients with second-division trigeminal neuralgia (Kanai *et al.* 2006). It is apparent from systematic literature reviews that there is a relative paucity of high quality randomized controlled trials of therapies for trigeminal neuralgia (He *et al.* 2006; Chole *et al.* 2007).

Spontaneous remission may occur, especially in the early stages. Therefore, if drug treatment leads to resolution of symptoms it is appropriate to attempt discontinuing treatment when the patient has been pain free for several weeks.

In some patients even very determined attempts with drug treatment prove unsuccessful, whilst in others there may be partial or complete relief but only at the cost of unacceptable side-effects. In such circumstances some form of surgical intervention should be considered.

Patients with intractable pain may require surgical management or some form of nerve root destructive procedure. A detailed discussion of the procedures which have been used is in Section 19.2.1. The commonly used procedures include:

◆ *Stereotactically controlled thermocoagulation* of the trigeminal ganglion or roots using a radiofrequency generator (Sweet and Wepsic 1974; Taha and Tew 1996) or similarly applied focussed gamma irradiation.

• *Posterior fossa exploration and microvascular decompression* (Breeze and Ignelzi 1982). Barker and coworkers reported that 70 per cent of 1185 patients were relieved of pain by repositioning a vessel compressing the trigeminal nerve at the root entry zone (Barker *et al.* 1996). Benefit was sustained in most patients, with only 1 per cent developing recurrent pain over a 10-year period. The advantage of microvascular decompression is that facial sensation is spared. Procedures which alleviate pain, but which result in facial sensory loss may be complicated by distressing dysaesthesiae in the anaesthetic area of the face, anaesthesia dolorosa, which is very resistant to treatment. A further complication is that of exposure keratitis, if the surgical treatment produces anaesthesia in the ophthalmic nerve territory.

• *Gamma knife irradiation*, usually to the retrogasserian cisternal portion of the trigeminal nerve, is emerging as a less invasive alternative to microvascular decompression with an acceptable side effect profile (Balamucki *et al.* 2006; Gorgulho and De Salles 2006; Regis *et al.* 2006), but the full therapeutic effects may take several months.

20.1.5 Trigeminal sensory neuropathy

This term describes a syndrome with rather variable clinical features and several pathological causes. The only constant feature is sensory disturbance confined to the distribution of the trigeminal nerve. This is usually numbness, with or without pain, and may affect one or more divisions of the nerve, unilaterally or bilaterally, symmetrically or asymmetrically. The onset may be acute or insidious. Ipsilateral taste sensation may be affected. Motor involvement is rare. The corneal reflex may be impaired. Trophic ulceration or self-mutilation may be seen in the anaesthetic area (Lecky *et al.* 1987; Hagen *et al.* 1990). There is a strong association with connective tissue disorders (Hagen *et al.* 1990; Forster *et al.* 1996), particularly Sjogren's syndrome (Mori *et al.* 2005) systemic sclerosis, and mixed connective tissue disease, and with autoantibodies (Lecky *et al.* 1987) (Section 36.3).

Occasionally patients with dorsolateral pontine segemental infarction may present with isolated trigeminal sensory disturbance (Ishii *et al.* 1998). Burning mouth syndrome, a disorder most commonly occurring in women in the 5th to 7th decade is caused by a trigeminal small-fibre sensory neuropathy which can be detected by quantifying nerve fibre density in a superficial biopsy of the tongue (Section 19.4.5) (Lauria *et al.* 2005).

20.1.6 Numb cheek and chin syndromes

The development of numbness or paraesthesia affecting the chin, lips, or cheek, and usually unilaterally, should immediately raise concern because of the strong association with tumours, particularly breast, prostate, and lung metastases, and haematological malignancies (Horton *et al.* 1973; Massey *et al.* 1981; Lossos and Siegal 1992). Several mechanisms may produce these symptoms. Meningeal involvement in the region of the trigeminal roots may not be associated initially with other features of carcinomatous meningitis. Basal skull deposits may affect the trigeminal ganglion and the major branches of the trigeminal nerve in their exit foramina. The mandibular division is that most frequently involved. Metastases in the mandible may damage the mental nerve causing localized chin numbness (Horton *et al.* 1973).

Careful investigation is required but an underlying cause is not always found. In the elderly, simple bone atrophy leading to mental

foramen stenosis may lead to sensory alteration over the chin area (Furukawa 1990).

20.2 Cranial nerve VII: facial

20.2.1 Clinical neuroanatomy

The facial nerve has two roots (Fig. 20.6). The larger contains the motor nerve fibres which supply the ipsilateral facial muscles. The smaller root, the intermediate nerve, contains fibres conveying taste sensation from the anterior two-thirds of the tongue, cutaneous sensory fibres from the posterior part of the ear, and preganglionic parasympathetic fibres that innervate the lacrimal and submandibular and sublingual salivary glands.

Motor pathways

The facial nerve motor nucleus lies in the ventrolateral tegmentum of the pons. The efferent fibres arising from the nucleus sweep dorsomedially, to the floor of the fourth ventricle, loop sharply around the sixth nerve nucleus, and then pass ventrolaterally to emerge from the lateral border of the caudal pons, at the cerebellopontine angle.

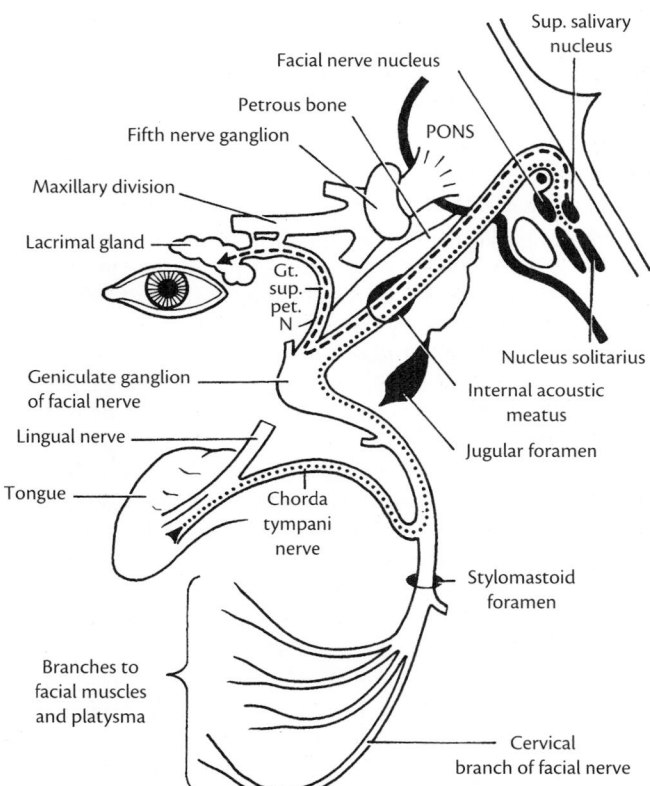

Fig. 20.6 The facial nerve. Lesions of the facial nerve nucleus or fibres within the pons are also likely to involve the VI cranial nerve which lies in close proximity. Lesions involving the facial nerve trunk above the geniculate ganglion will involve loss of lacrimation (Greater superficial petrosal nerve) and loss of taste over the anterior two-thirds of the tongue (chorda tympani), as well as paralysis of both upper and lower facial muscles. Lesions between the geniculate ganglion and the point where the chorda tympani leaves the facial nerve will cause facial paralysis and loss of taste sensation over the anterior two-thirds of the tongue, but lacrimation will be spared. Lesions below the point where the chorda tympani leaves the facial nerve will cause paralysis of the facial muscles, but both taste and lacrimation will be preserved. (Redrawn from an original drawing by Mr. Charles Keogh.)

In view of their proximity in the floor of the upper IV ventricle, the 6th and 7th cranial nerves are often affected together by a vascular or infltrative pontine lesion. After exiting from the pons, the facial nerve is medial to the eighth nerve and between the two lies the intermediate nerve. These three nerves pass through the internal auditory meatus and then the facial nerve and intermediate nerve enter the facial canal. In the facial canal, on the medial side of the middle ear, the facial nerve turns sharply as the 'genu' of the facial nerve, moving posteriorly and inferiorly, gives off a branch to the stapedius muscle, and exits from the skull through the stylomastoid foramen.

After leaving the stylomastoid foramen the facial nerve sends branches to the stylohyoid muscle and to the posterior belly of the digastric, and then passes through the parotid gland, dividing into several branches which supply platysma and all of the muscles of facial expression, excluding levator palpebrae superioris which is supplied by the oculomotor nerve. Thus, facial nerve lesions do not cause ptosis.

Corticobulbar pathways

The facial nucleus is composed of a number of distinct cell groups, each innervating specific facial muscles. Those supplying the upper facial muscles receive bilateral supranuclear innervation, whereas those supplying the lower facial muscles receive mainly crossed fibres from the contralateral hemisphere. In addition to direct corticobulbar fibres there are several indirect pathways between the cortex and facial nuclei, involving the thalamus and reticular formation.

Sensory pathways

In the facial canal there is an expansion of the facial nerve as it makes its sharp backwards turn, at the genu. This expansion is the geniculate ganglion and is formed by the cell bodies of the nerves that give rise to the two sensory components of the facial nerve.

Special visceral afferent fibres convey taste sensation from the anterior two-thirds of the tongue. From the tongue these fibres travel first in the lingual nerve, and then in the chorda tympani which enters the skull, crosses the tympanic cavity, and joins the facial nerve in the facial canal. From the geniculate ganglion, the central connections pass via the intermediate nerve and terminate in the nucleus of the tractus solitarius in the medulla.

Somatic afferent nerve fibres arise from a small area of skin which includes the posterior part of the external auditory meatus and the skin behind the ear and in front of the mastoid. They enter the facial canal just proximal to the stylomastoid foramen. From the cell bodies in the geniculate ganglion the fibres pass centrally in the intermediate nerve and terminate in the spinal trigeminal tract. Sensory loss is not a clinically detectable feature of facial nerve lesions, presumably because of overlap from adjacent cutaneous nerve territories, but the presence of this sensory pathway probably explains the symptom of pain in the mastoid region which is so common in patients with Bell's palsy (Section 20.2.3).

Autonomic pathways

Preganglionic parasympathetic fibres arise from the superior salivary nucleus, in the dorsolateral reticular formation. They travel in the intermediate nerve and at the genu of the facial nerve divide into two groups. One group passes with the greater superficial petrosal nerve to the pterygopalatine ganglion. Post-ganglionic fibres innervate the lacrimal gland and mucous membrane of the nose and mouth. The other group of fibres travels in the chorda tympani and terminates in the submandibular ganglion. Post-ganglionic fibres innervate the submandibular and sublingual salivary glands.

Facial nerve reflexes

In clinical practice the corneal reflex (Sections 2.3.4 and 20.1.1) and, to a lesser extent, the glabellar tap reflex (Section 2.3.4), are of value. The glabellar tap reflex is polysynaptic and elicited by tapping the forehead over the bridge of the nose and observing contraction of orbicularis oculi, as blinking, bilaterally. After several taps there is habituation and blinking stops. In early childhood and in Parkinsonian syndromes there is failure of habituation and blinking continues in time with the tapping.

Other reflexes include the naso-lacrimal reflex of tearing in response to stimulation of the nasal mucosa, the naso-mental reflex which, on tapping the side of the nose, causes elevation of the upper lip and the stapedius reflex in which stapedius contracts in response to a loud noise. These and other similar reflexes are of little importance at the bedside but some, such as the stapedius reflex, can help with localization of facial nerve lesions but require laboratory study.

20.2.2 Lesions of the facial nerve

Lesions of the facial nerve, its nucleus, or supranuclear pathways may produce facial muscle weakness. Only peripheral lesions, affecting the facial nerve itself, affect taste sensation and autonomic function. As noted above, numbness is not an expected finding in facial nerve lesions although symptomatic complaints of sensory disturbance are common in Bell's palsy.

Supranuclear lesions

The upper facial muscles have almost equal bilateral cortical representation, whereas the lower facial muscles receive mainly crossed fibres from the contralateral hemisphere. Thus, a unilateral upper motor neurone lesion causes contralateral facial weakness with the lower part of the face being relatively more affected than the upper (Fig. 20.7). As noted above there is more than one pathway of supranuclear innervation and depending upon the site of the lesion spontaneous emotional movements may be more affected than voluntary movements, and vice versa. Muscles innervated by the facial nerve may be affected by lesions of the supranuclear pathways which cause dysfunction of brainstem reflex activity. In the condition referred to as apraxia of eyelid closure, the patient cannot sustain eyelid closure voluntarily, but can close the eyes in reflex response to stimulation of the supraorbital branch of the V nerve.

Nuclear and peripheral lesions

A lesion of the nucleus or facial nerve generally causes equal weakness of all ipsilateral facial muscles and the clinical features are exemplified by Bell's palsy, described below. Occasionally, partial lesions of the nucleus or nerve may selectively affect the lower facial muscles thus mimicking the appearance seen with an upper motor neurone lesion.

Considering the origins and sites of union with the facial nerve of the greater superficial petrosal nerve, the nerve to stapedius and the chorda tympani (Fig. 20.6), the presence of impaired lacrimation, hyperacusis or an impaired stapedius reflex, or altered taste sensation can help in localizing the site of a facial nerve lesion. Such features are difficult to elicit reliably and their absence is of limited localizing value.

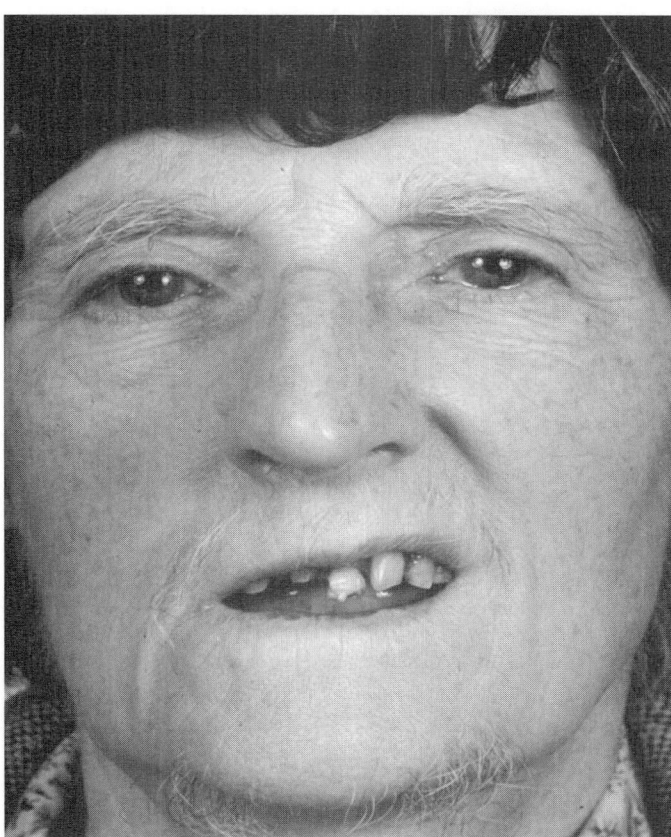

Fig. 20.7 Right-sided upper motor neurone weakness following a cerebrovascular accident. Note that the lower part of the face is more affected than the upper facial muscles, which have bilateral cortical representation.

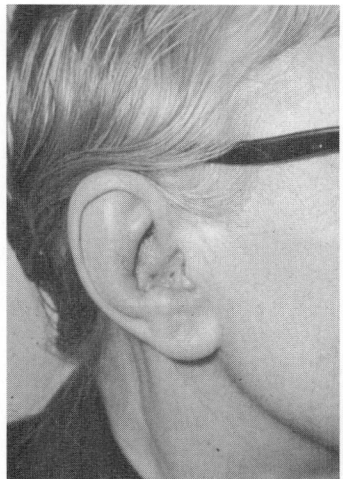

Fig. 20.8 Ramsay Hunt syndrome due to herpes zoster infection of the geniculate ganglion of the facial nerve. The patient has right-sided facial palsy and a vesicular rash over the external auditory meatus.

The facial nucleus may be affected by pontine lesions, and the nerve by lesions in the cerebellopontine angle, within the petrous temporal bone and outside the skull.

Pontine lesions. These rarely affect the facial nucleus or nerve fibres in isolation and associated features include ipsilateral lateral rectus or conjugate gaze palsy, trigeminal motor and sensory involvement, and contralateral hemiparesis and hemisensory loss. Common pathologies include vascular lesions, multiple sclerosis, and tumours, less common disorders being brainstem encephalitis, syringobulbia, and poliomyelitis. Bilateral facial paralysis due to agenesis of the facial nuclei, Möbius' syndrome, is a rare disorder that may be associated with other cranial nerve lesions and dysmorphic features (Section 13.1.7).

Cerebellopontine angle lesions. The commonest lesions at this site, which affect the facial nerve, intermediate nerve, and eighth nerve, are acoustic neuromas and meningiomas (Section 27.7). Less common lesions include secondary tumours, nasopharyngeal carcinoma, developmental tumours, cholesteatomata, and any basal meningitic process such as sarcoidosis or carcinomatous menigitis.

Petrous temporal bone lesions. In the facial canal the nerve may be affected by infection spreading from the middle ear or mastoid, or by surgical procedures in that area. Inflammation and swelling of the facial nerve in the facial canal and at the stylomastoid foramen is presumed to be present in Bell's palsy and in Ramsay Hunt syndrome.

Lesions outside the skull. Benign and malignant lesions of the parotid gland may involve some or all of the branches of the facial nerve.

Infections involving the facial nerve

In the Ramsay Hunt Syndrome swelling of the geniculate ganglion due to reactivation of latent herpes zoster infection may compress the motor fibres, or there may be direct infection of the motor nerve (Section 19.2.4). The resultant facial palsy is accompanied by a rash, typically seen in the external auditory meatus (Fig. 20.8). The rash is often more extensive than this involving the trigeminal distribution, as seen in the anterior pillar of the fauces, or cervical dermatomes. There is usually pain around the ear. Often the VIII cranial nerve is also involved leading to deafness, vertigo, and nausea. It has been shown that virus can be detected even before the emergence of typical vesicles by performing polymerase chain reaction on exudate from the skin of the pinna (Murakami *et al.* 1998). Treatment with prednisolone, 60 mg per day for 5 days, and acyclovir, 800 mg 5× daily for 7 days, has been recommended. (Sweeney Gilden, 2001).

Lyme disease commonly causes facial nerve involvement and should be considered, particularly if there is a history of tick bite, erythema migrans, and arthritis (Section 21.14.3). HIV is another common infectious cause of facial palsy. In both neuroborreliosis and HIV, the facial palsy is likely to be accompanied by CSF pleocytosis. Tuberculous infection of the middle ear, mastoid, or petrous bone may cause facial palsy. Seventh cranial nerve involvement may also be seen in leprosy, infectious mononucleosis, poliomyelitis, and following chicken pox.

Bilateral facial palsy

Bilateral, as well as unilateral, lower motor neurone facial weakness may be seen in Guillain–Barré syndrome (Section 21.10.1), in sarcoidosis due to basal meningeal or parotid involvement (Section 36.4.2), in Lyme disease and, often accompanied by facial rash and induration, in HIV infection at the time of seroconversion. In Melkersson's syndrome recurrent episodes of unilateral or bilateral facial swelling and facial palsy are associated with a deeply furrowed tongue. Acute onset of facial weakness in sarcoidosis is referred to as Heerfordt syndrome or uveoparotid fever.

20.2.3 Bell's palsy

Bell's palsy is the most common disorder of the facial nerve. It is generally accepted that there is inflammation and oedema of the nerve in the facial canal but, not surprisingly, there have been few pathological studies. A viral aetiology is suspected. There are conflicting reports of clustering of cases, suggesting an infective aetiology and recurring reports implicating herpes viruses (Morgan *et al.* 1995; Bauer and Coker 1996). The genome of herpes simplex virus has been identified in the geniculate ganglion of an elderly man who died 6 weeks after the onset of Bell's palsy (Burgess *et al.* 1994). Viral genomic sequences of type 1 herpes simplex virus were identified in the endoneurial fluid of 11/14 cases of Bell's palsy obtained during surgical decompression of the nerve, while control patients with fracture or infection of the temporal bone did not yield viral sequences (Murakami *et al.* 1996).

The incidence of Bell's palsy is about 23/100 000/year (Hauser *et al.* 1971). It affects both sexes equally and is less frequent in children than adults. It shows relatively weak associations with hypertension and diabetes, particularly in older patients (Hauser *et al.* 1971). Recurrence, on the same or opposite side, is relatively common, occurring in approximately 8 per cent of cases, whereas simultaneous bilateral idiopathic, or Bell's, palsy is very rare. Pedigrees with familial Bell's palsy have been described, sometimes associated with eye movement abnormalities (Yanagihara *et al.* 1988; Hageman *et al.* 1990; Hemminki and Sundquist 2007).

Clinical features

The entire course of Bell's palsy may be painless but frequently patients complain of pain behind the ipsilateral ear in the mastoid region, for a day or two before the onset of weakness and this may continue for a week or more. Paralysis develops rapidly and may reach maximum severity within a few hours. Continuing progression for 24–48 h is not uncommon and rarely may be over as long as five days.

All of the muscles on the affected side of the face are involved but the degree of weakness may range from mild to being complete in about 70 per cent of patients. The appearance (Fig. 20.9), of even an incomplete palsy, is striking and it is not surprising that it causes the patient considerable alarm. In elderly patients, presumably due to greater laxity of supporting tissues, the resultant facial deformity is more evident than in younger patients. The eyebrow droops and cannot be elevated, and the brow loses its furrows and becomes smooth. The lower eyelid everts into an ectropion impairing drainage of the tears which overflow onto the cheek. The eye cannot close voluntarily or on blinking and the palpebral fissure is widened. On attempted eye closure there will be some lowering of the upper lid due to reflex inhibition of levator palpebrae superioris and both eyes will roll upwards in Bell's phenomenon. The nasolabial fold becomes flattened, the angle of the mouth droops and cannot be retracted, the cheek billows on respiration and food tends to accumulate between the cheek and teeth. There is mild dysarthria particularly affecting labial consonants. If the nerve is involved proximal to the point where it is joined by the chorda tympani, or even more proximally affecting the nerve to stapedius, then the patient may complain of impaired taste sensation or hyperacusis, an unpleasant quality to loud sounds.

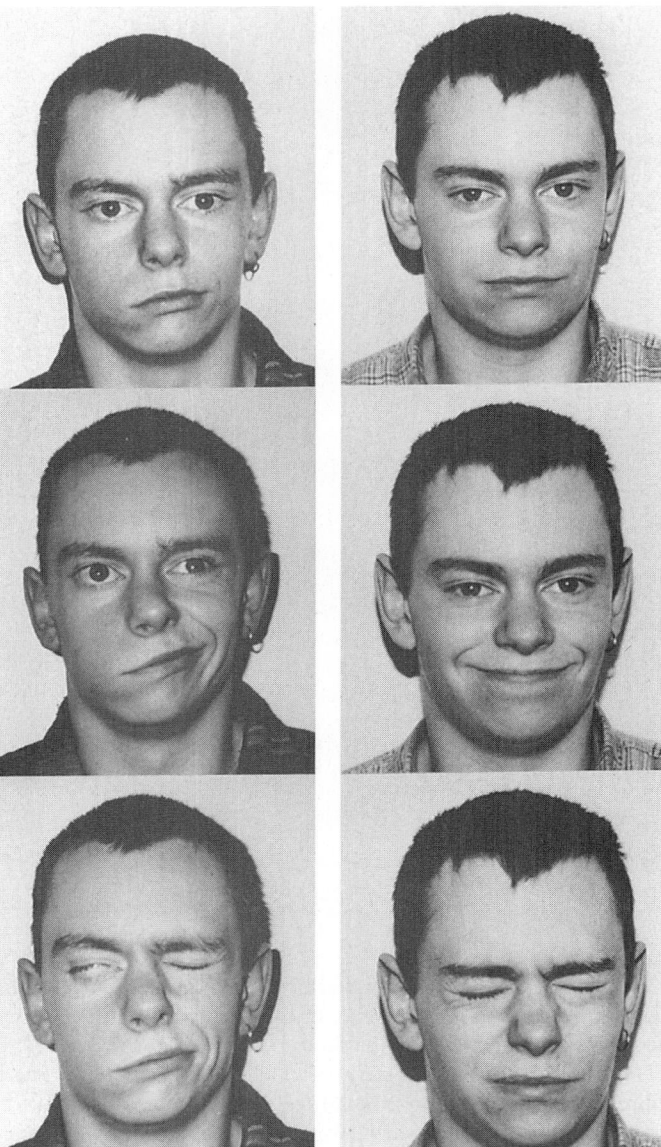

Fig. 20.9 Bell's palsy. Left panel: 5 days after onset; right panel: 3 months later. From above down: face at rest, smiling, forceful eye closure.

Many patients complain of numbness over the affected side of the face and, sometimes, tongue. This may be objective, in the sense that the patient will say that light-tough and pin-prick sensation are less on the affected side. The corneal reflex is always preserved. There is no obvious anatomical explanation for such sensory symptoms and they are usually attributed to distorted perception caused by the drooping musculature, skin, and associated tissues.

The facial nerve in Bell's palsy often displays enhancement on gadolinium-enhanced MRI. There may be mild CSF pleocytosis in some cases.

Prognosis

In about 80 per cent of patients improvement starts early and there is full recovery within a few weeks from the onset. Pathologically it is presumed that the weakness in these cases is due entirely to conduction block, from segmental demyelination, from which recovery is rapid. In the remaining 20 per cent of patients, in addition to conduction block, there is Wallerian degeneration of some or all of the axons and full recovery will not occur, although most patients have a satisfactory cosmetic outcome eventually. Nerve regeneration starts from the point of interruption but reinnervation, and thus functional recovery, does not develop until at least 3 months after the onset of the palsy and is never complete, leaving some residual weakness. Some of the regenerating axons become mis-directed and innervate muscles that they did not originally supply. Thus, movement in one area may be accompanied by associated movement elsewhere, known as synkinesis. This and other complications are discussed below.

The most favourable prognostic signs are an incomplete palsy and early recovery of motor function within the first 5–7 days. Adverse features may include advanced age, diabetes, hypertension, severe pain, loss of taste, and hyperacusis but none is a reliable indicator of prognosis. Neurophysiological studies, particularly if performed more than 1 week after onset, may offer prognostic information. Electromyography may be of value in distinguishing temporary conduction block from axonal degeneration. If there is evidence of denervation 10 days after onset, then recovery can be expected to be prolonged. Whether the findings indicate a good or poor prognosis does not alter management and so neurophysiological studies are rarely performed in clinical practice.

Complications. Associated synkinetic movements are the result of aberrant reinnervation by regenerating axons (Fig. 20.10). Common patterns include eye closure on lip movement, elevation of the angle of the mouth on blinking or when the eyebrow is raised, or the jaw winking phenomenon, which is also called Wartenberg's or the inverse Marcus–Gunn sign, in which jaw movement, especially laterally, causes ipsilateral eye closure. Occasionally the synkinetic movements may be very extensive.

Also there may be aberrant parasympathetic nerve reinnervation giving rise to the phenomenon of crocodile tears—profuse watering of the affected eye when eating. The simplest explanation is that regenerating fibres destined to innervate the submandibular and sublingual salivary glands become misdirected and reach the lacrimal gland. An alternative explanation is that glossopharyngeal nerve fibres, destined for the parotid gland, in the lesser superficial petrosal nerve send branches to the greater superficial petrosal nerve, where they lie close together, and then innervate the lacrimal gland.

When recovery is incomplete there is often some residual contracture of the affected muscles. This may be evident as narrowing of the palpebral fissure or deepening of the nasolabial fold.

Treatment

It is likely that the eventual outcome of Bell's palsy is determined within days of the onset. If the inflammation and oedema of the facial nerve cause conduction block alone, then full recovery will occur independently of treatment. If the inflammatory process is more severe and results in axonal degeneration, then recovery will be incomplete. Thus, the aim of treatment must be to reduce the oedema and self-compression of the nerve within its bony canal before axonal degeneration occurs. It is impossible, on available evidence, to commend surgical decompression. To be effective it would have to be performed very rapidly after the onset of the palsy. There are no clinical or neurophysiological pointers in those first few days to indicate which patients are going to have a poor prognosis and it would clearly be unjustified to operate on all patients given that the vast majority will have a satisfactory outcome in any case.

Whether corticosteroids are helpful is not certain (Matthews 1993). A systematic review in 2004 concluded that there was no definite evidence of benefit from administration of corticosteroids in Bell's palsy, but the available evidence from randonized controlled trials was relatively weak (Salinas *et al.* 2004). In the absence of contraindications, it is very common practice for patients seen within 1 week of onset of the palsy to be given a short course of oral steroids. A typical regimen might start with prednisolone 1mg/kg body weight/day, with gradually diminishing doses over the next 10–14 days. The finding of viral material in the vicinity of the facial nerve raises the question of whether anti-viral agents may be beneficial. Several small trials have suggested that acyclovir, when used alone, is no more effective than corticosteroids (De Diego *et al.* 1998). A systematic review on the use of acyclovir or valaciclovir for Bell's palsy concluded that more data are needed from large multicentre randomized controlled and blinded studies with at least 12 months follow up before definitive informed recommendations could be made (Allen and Dunn 2004). However, evidence is now emerging indicating benefit from the combination of antiviral and steroid therapy. A retrospective study provided evidence for a better outcome in 94 patients who were treated with prednisolone and acyclovir compared to 386 patients treated with prednisolone alone (Hato *et al.* 2003). A multicentre, randomized, placebo-controlled study has been reported of 221 patients who were treated within 7 days of onset with valacyclovir, at 1000 mg/day for 5 days, plus prednisolone versus placebo plus prednisolone, with follow up until complete recovery or for more than 6 months in those with a poor prognosis. Cases of zoster sine herpete were excluded. The combined antiviral and steroid group had a better rate of recovery, overall of 96.5 per cent versus 89.7 per cent, overall and also better recovery in the group with complete or severe palsy, of 95.7 per cent versus 86.6 per cent ($p<0.05$) (Hato *et al.* 2007).

The normal ocular tear film is disturbed in Bell's palsy. Despite anatomical considerations a dry eye or under-production of tears due to denervation of the lacrimal gland is very uncommon.

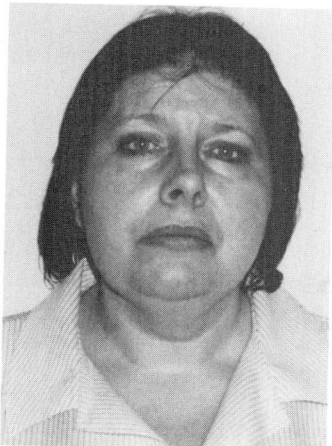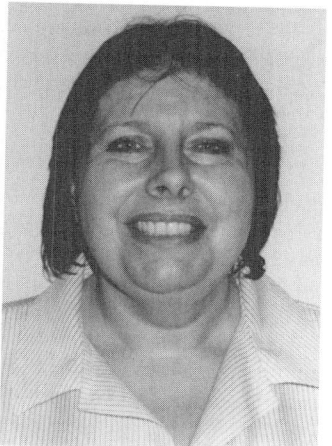

Fig. 20.10 Synkinetic movements due to aberrant re-innervation following previous left Bell's palsy. Smiling causes narrowing of the left palpebral fissure.

Rather, tear drainage is affected due to the ectropion and this, together with reduced blinking, often causes mistiness of vision and associated patient anxiety. Corneal sensation is normal and corneal damage is rare in Bell's palsy. Tarsorrhaphy is rarely required, but the patient may find it more comfortable to tape the eye closed during sleep, and to use glasses to protect the eye from dust and wind. If tear production is impaired methylcellulose eye drops should be used.

Until recently there was no effective treatment for synkinetic movements. Some benefit has recently been reported for low-dose subcutaneous injection of botulinum type A toxin for facial synkinesis and hyperlacrimation following Bell's palsy (Ito *et al.* 2007). In those few patients with severe residual weakness various plastic surgery procedures can improve the cosmetic appearance. Because of the slow rate of nerve regeneration no surgical intervention, except occasionally tarsorrhaphy, should be considered until at least 6, and probably 12, months after the onset of the palsy. Crocodile tears may be treated by section of the tympanic nerve which carries the glossopharyngeal salivary fibres and by botulinum toxin A administration.

A major element in the management of Bell's palsy is reassurance of the patient with detailed explanation of what has happened and its generally favourable prognosis. There is no good evidence that physiotherapy or electrical stimulation are of specific value. However, some patients wish to take an active part in the management of their problem, for example by regularly massaging and attempting to exercise their facial muscles.

20.2.4 Hemifacial spasm

This painless but distressing condition develops in middle age or later and is commoner in women. It is characterized by unilateral repetitive involuntary contractions of the facial muscles (Fig. 20.11) (Section 40.10.6).

Aetiology

The condition may be symptomatic of an irritative lesion of the facial nerve. Rarely, this may be a mass lesion such as a tumour or aneurysm. Very rarely there is an antecedent history of Bell's palsy. Although initially controversial, it is now accepted that the irritative lesion is often an aberrantly placed artery compressing the nerve (Kaye and Adams 1981; Barker *et al.* 1995). It frequently proves to be a tortuous branch of the basilar artery that lies under the ventral surface of the pons and forms a loop under the proximal part of the facial nerve. Multiple studies have confirmed the efficacy of microsurgical decompression, with interposition of a pledget between the aberrant blood vessel and the facial nerve (Barker *et al.* 1995; Illingworth *et al.* 1996). The pathophysiology is believed to involve focal demyelination at the site of compression of the facial nerve root. The demyelinated axons may undergo spontaneous ectopic excitation or activation of adjacent nerve fibres by ephaptic transmission.

Patients should be investigated with MR imaging. A recent report has indicated that fusion of 3-D MR cisternograms and 3-DE MR angiograms is useful in the pre- and post-operative assessment of microvascular decompression in patients with hemifacial spasm (Satoh *et al.* 2007). Electromyography shows rhythmical high frequency discharges during an attack and evidence of synkinesis between episodes. The likely pathophysiological basis is ephaptic excitation of neighbouring nerve fibres.

Fig. 20.11 An episode of left-sided hemi-facial spasm.

Clinical features

The disorder evolves slowly over many years. In the early stages the involuntary contractions are confined to one area such as the corner of the mouth or, most often, the lower followed by the upper eyelid causing spasmodic eye closure. The spasms are worsened by fatigue and emotional stress. The extent of the contraction slowly spreads until the whole of the side of the face and platysma are involved. Individual contractions, although sometimes very frequent, are usually brief, lasting a few seconds, although occasionally may persist for several minutes. As the condition advances it is clear, between attacks, that there is weakness of the facial muscles on the affected side and synkinetic movements are apparent. Spontaneous remission rarely occurs.

Treatment

Drug treatment is usually disappointing and should not be persisted with over-long before trying more effective methods. Carbamazepine, sodium valproate, baclofen, and anticholinergic drugs have been reported occasionally to be of some benefit, but adequate and sustained improvement is rarely achieved, or if so only at doses causing unacceptable side-effects.

Currently, the main treatment choice is between injection of botulinum toxin into the affected muscles, or posterior fossa exploration. Botulinum toxin injections are effective, with a benefit rate of 76–100 per cent , and have a relatively low incidence of side-effects, particularly ptosis and diplopia, but have to be repeated indefinitely, every 3–6 months (Elston 1986; Costa *et al.* 2005). The rationale behind posterior fossa exploratory surgery is to decompress the nerve from aberrant blood vessels (Barker *et al.* 1995). Such compressive lesions are not always found but it appears that simply moving and wrapping the nerve may be an effective treatment (Kaye and Adams 1981). Surgery is clearly not without some degree of risk including post-operative facial weakness, deafness, and CSF leak and, in some patients, hemifacial spasm recurs.

20.2.5 Other involuntary facial movements

Bilateral involuntary facial movements are seen in a number of dyskinetic syndromes, particularly orofacial dyskinesia and oromandibular dystonia (Section 40.4.2), and facial chorea may be striking in Huntington's disease (Section 40.5.2). Focal epilepsy can give rise to unilateral facial contractions. Gross fasciculation of the facial muscles, particularly in the mentalis muscle, is seen in the X-linked bulbospinal neuronopathy, Kennedy's disease (Section 23.5.1).

In facial myokymia there is an irregular writhing movement of the facial muscles typically of the cheek, sometimes rather graphically described as 'creeping flesh'. It is usually unilateral and most commonly due to multiple sclerosis, when it is transient, or to a brainstem glioma, when it progresses. Bilateral myokymia may rarely be seen in Guillain–Barré syndrome.

20.3 Cranial nerve IX: glossopharyngeal

Although the glossopharyngeal nerve contains sensory, motor, and parasympathetic fibres, only its sensory function is testable at the bedside and for all practical purposes its other functions can be ignored. Functionally and anatomically it is closely related to the vagus nerve and together with the accessory nerve all three nerves may be affected by lesions at the jugular foramen or extracranially by dissections of the internal carotid artery. Isolated glossopharyngeal nerve lesions are rare, but this nerve alone is affected in the rare syndrome of glossopharyngeal neuralgia.

20.3.1 Clinical neuroanatomy

Three components of the glossopharyngeal nerve are of little interest in everyday clinical neurology:

- Special visceral afferent fibres subserve taste sensation from the posterior one-third of the tongue. However, symptomatic loss of taste sensation from lesions of the nerve is not seen and there are no practical tests of this function at the bedside.

- General visceral efferent fibres give rise to parasympathetic fibres which stimulate secretion from the parotid gland.

- Special visceral efferent fibres innervate stylopharyngeus which contributes to elevation of the pharynx, but this muscle cannot be assessed clinically.

Peripheral course

The glossopharyngeal nerve is formed by a series of rootlets which enter and leave the medulla, in the posterior lateral sulcus, rostral to the vagus nerve. The nerve crosses the posterior fossa and leaves the skull through the jugular foramen, together with the vagus and accessory nerves. On leaving the skull it crosses in front of the internal carotid artery to reach the lateral wall of the pharynx. The nerve contains two peripheral ganglia, the superior ganglion which lies in the jugular foramen and the inferior petrosal ganglion which is extracranial. The ganglia contain the cell bodies of primary sensory neurones that subserve general somatic and general visceral sensation.

Sensory pathways

General visceral afferent fibres convey tactile, thermal, and pain impulses from the posterior one-third of the tongue, tonsil, part of the soft palate, posterior wall of the upper pharynx, and eustachian tube, via the inferior ganglion, to the solitary fasciculus and its nucleus. Clinically, this particular sensory pathway is the most important function of the nerve and the only function readily assessed at the bedside. General somatic afferent fibres carry sensation from the posterior part of the ear, via the superior ganglion, and terminate in the spinal trigeminal tract and nucleus. Within the glossopharyngeal nerve are afferent fibres from baroreceptors in the wall of the carotid sinus involved in the regulation of blood pressure and from chemoreceptors in the carotid body responsible for the ventilatory response to hypoxia.

20.3.2 Lesions of the glossopharyngeal nerve

As noted above, clinically the most important function of the glossopharyngeal nerve is its provision of sensory input from the upper pharynx. It thus provides the afferent limb of the gag reflex, the efferent limb of which is provided by the vagus. This reflex does not provide a very sensitive assessment of glossopharyngeal nerve function. If a lesion is suspected, the sensation on each side of the posterior pharyngeal wall should be tested using an instrument such as an orange stick. Most patients will gag, and the palate will be seen to move, but what is more important is for the patient to state whether the sensation is normal and the same on both sides.

Supranuclear and nuclear lesions

Supranuclear lesions have no clinically discernible effect on glossopharyngeal nerve function, although involvement of stylopharyngeus may contribute to pseudobulbar palsy. Nuclear lesions in isolation are rarely, if ever, seen and other cranial nerve nuclei, particularly the vagus, are usually also involved. The commonest cause is a vascular lesion, with other causes including primary and secondary neoplasia and syringobulbia.

Peripheral lesions

Between the medulla and jugular foramen the nerve may be affected by meningeal disease, such as inflammatory and neoplastic processes, and metastases. In the jugular foramen the commonest lesion, which of course may also affect the vagus and accessory nerves, is a glomus tumour (Section 27.6.3). Neuromas of any of these three nerves may arise at or near the jugular foramen and each nerve may be affected by basal skull fracture and basilar invagination. Metastatic disease may affect the nerve anywhere along its course, intracranially or extracranially. Outside the skull, the IX nerve lies adjacent to the internal carotid artery, where it may be damaged by a dissection of that vessel.

Damage to the lingual branch of the glossophayngeal nerve may cause the complication of post-operative distortion of taste following tonsillectomy (Goins and Pitovski 2004).

20.3.3 Glossopharyngeal neuralgia

Although much rarer, glossopharyngeal neuralgia (Section 19.2.2) shares many similarities with trigeminal neuralgia with respect to aetiology, treatment, and the characteristic features of the paroxysms of pain (Pearce 2006). Most cases are idiopathic, but neuralgia may be symptomatic of a lesion affecting the glossopharyngeal nerve, particularly tumours, and there is evidence that some cases are caused by compression of the nerve by an aberrantly placed artery (Laha and Jannetta 1977). Modern imaging techniques may be useful in demonstrating vascular compression (Fischbach *et al.* 2003; Karibe *et al.* 2004).

Clinical features

The paroxysms of pain may occur in clusters, with long periods of remission, or may be chronic. The pain is experienced in the back of the throat, below the angle of the jaw, and within the ear. The stabbing or lancinating quality is similar to that occurring in trigeminal neuralgia. Precipitants include eating, swallowing, talking, head turning, coughing, sneezing, and touching the outer ear. Temporary blocking of the pain by anaesthetizing the tonsillar fauces and posterior pharynx with lignocaine spray can be of help in confirming the diagnosis. Syncope may occur in association with pain and is due to sinus bradycardia or asystole (Jacobson and Ross-Russell 1979; Ozenci *et al.* 2003), reflecting the intimate association between the glossopharyngeal and vagus nerves.

Treatment

The treatment of choice is carbamazepine but if this does not work, or is not well tolerated by the patient, then the same drug alternatives as used in trigeminal neuralgia may be tried: phenytoin, sodium valproate, baclofen, lamotrigine, amitriptyline, and pregabalin. Thereafter it is necessary to resort to surgical techniques including micro-vascular decompression, nerve section, and gamma knife surgery (Stieber *et al.* 2005). Several series have reported that microvascular decompression is a safe, effective treatment for glossopharyngeal neuralgia, with sustained benefit (Resnick *et al.* 1995; Sampson *et al.* 2004). The main post-operative complications, reported in approximately 11 per cent of patients, include hoarseness, dysphagia, and facial paresis which are generally mild (Sampson *et al.* 2004).

20.4 Cranial nerve X: vagus

20.4.1 Clinical neuroanatomy

The vagus nerve has the most extensive course of any of the cranial nerves and is anatomically complex, with different courses for the main nerve trunks and their branches on each side of the body. The nerve carries motor, sensory, and autonomic fibres but with respect to structural lesions only the motor pathways are of major clinical importance. Disturbances of autonomic function are discussed in Section 2.7.

Sensory and autonomic pathways

General somatic afferent fibres subserve sensation from the skin over the back of the ear and the posterior wall of the external auditory meatus. The cell bodies are situated in the *superior ganglion*, which sits in or just below the jugular foramen, and centrally the fibres enter the spinal trigeminal tract in the medulla. General visceral afferent fibres, from the pharynx, larynx, trachea, oesophagus, and thoracic and abdominal viscera have their cell bodies in the *inferior ganglion*, and centrally the fibres enter the nucleus and tractus solitarius.

Preganglionic parasympathetic fibres, general visceral efferents, arise from the *dorsal motor nucleus* of the vagus nerve, situated in the floor of the fourth ventricle and are destined to innervate the heart, thoracic, and abdominal viscera.

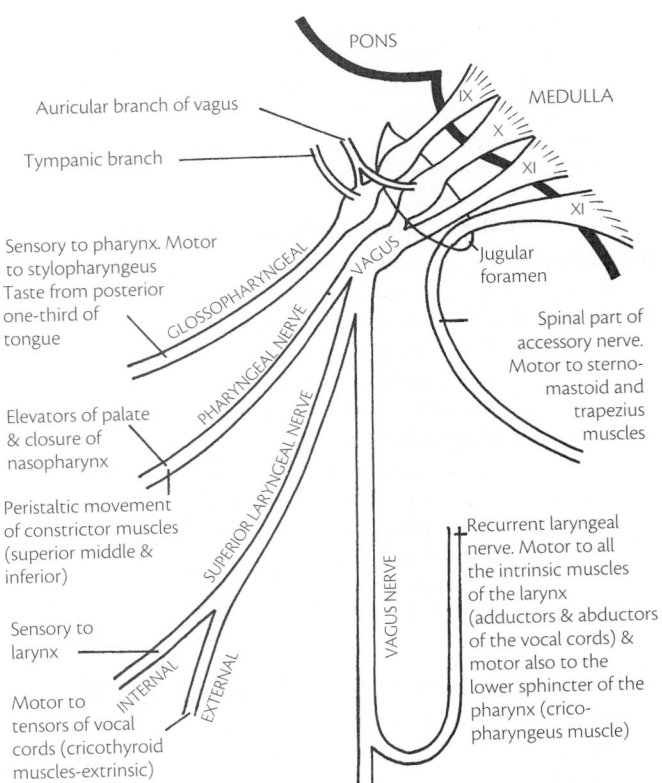

Fig. 20.12 The motor and sensory nerves supplying the pharynx and larynx, explaining the various patterns of paralysis commonly met with. (Redrawn from an original drawing by Mr Charles Keogh.)

Motor pathways

The somatic motor fibres of the vagus nerve originate from the *nucleus ambiguus* which lies in the medullary reticular formation between the inferior olive and the spinal trigeminal nucleus and which supplies the striated muscles of the palate, pharynx, and larynx.

Peripheral course

The trunk of the vagus nerve is formed by a series of rootlets which emerge from the medulla, anterior to the inferior cerebellar peduncle, in line with the rootlets of the glossopharyngeal and accessory nerves (Fig. 20.12). The nerve leaves the skull through the jugular foramen, intimately associated with the accessory nerve and separated from the glossopharyngeal nerve only by a fibrous septum. In the neck it lies within the carotid sheath, initially between the internal carotid artery and internal jugular vein, and then between the common carotid artery and internal jugular vein. Below the root of the neck the course of the nerve is different on the two sides of the body.

On the right, the nerve crosses the subclavian artery and descends through the superior mediastinum posterior to the brachiocephalic vein and to the right of the trachea, to reach the posterior aspect of the lung root. On the left, the nerve passes between the common carotid and subclavian arteries to enter the thorax. It descends through the superior mediastinum, behind the phrenic nerve and brachiocephalic vein and crosses the left side of the aortic arch, reaching the posterior surface of the lung root.

The nerves branch behind the lung roots and these branches unite with fibres from thoracic sympathetic ganglia to form the right and left posterior pulmonary plexuses. Fibres from these form the posterior and anterior oesophageal plexuses respectively. Trunks, containing fibres from both vagus nerves, are re-formed from these plexuses and pass into the abdomen, through the oesophageal opening, where they undergo complex further branching before supplying the abdominal viscera.

On each side, the vagus nerve has important branches arising in the jugular foramen, neck, and thorax:

- A *meningeal branch* arises from the superior ganglion which innervates the dura in the posterior fossa,

- An *auricular branch* which subserves sensation from the posterior auricle and external auditory meatus,

- The *pharyngeal branch* arises from the inferior ganglion and is the main motor nerve to the pharynx and soft palate,

- The *superior laryngeal nerve* also arises from the inferior ganglion. It has two branches, the internal which is the main sensory nerve of the larynx, and the external which is motor to the inferior pharyngeal constrictor and cricothyroid muscles.

- The *recurrent laryngeal* nerve has a different origin and course on each side of the body. On the right it arises from the vagus at the root of the neck, in front of the subclavian artery. It winds below and behind that vessel and ascends beside the trachea and behind the common carotid artery. At the level of the thyroid gland the nerve is closely related to the inferior thyroid artery. On the left the nerve arises from the vagus at the level of the aortic arch. It winds under the arch and then ascends along the side of the trachea. On both sides the recurrent laryngeal nerves ascend in a groove between the oesophagus and trachea, pass in close proximity to the medial surface of the thyroid gland, and enter the larynx to supply all of the laryngeal muscles except cricothyroid.

20.4.2 Lesions of the vagus nerve and its branches

Supranuclear, nuclear, nerve trunk, and branch lesions may affect swallowing and phonation, the exact pattern of symptoms depending upon the site and chronicity of the lesion.

Supranuclear lesions

Because the nuclei receive both crossed and uncrossed corticobulbar fibres, unilateral supranuclear lesions do not usually cause persisting problems with phonation and swallowing although dysphagia may be prominent following an acute hemispheric stroke. Bilateral lesions are associated with the syndrome of pseudobulbar palsy in which dysphagia and dysarthria are due to disordered movement of the pharyngeal, laryngeal and tongue muscles rather than frank paralysis. Common causes include upper brainstem or bilateral hemispheric strokes, motor neurone disease, and demyelination.

Nuclear lesions

A unilateral nuclear lesion will cause ipsilateral palatal, pharyngeal, and laryngeal paralysis. On phonation the soft palate does not rise on the affected side and the uvula is drawn to the normal side (Fig. 20.13). Dysphagia is variable but usually mild. Phonation is affected not only because of laryngeal muscle weakness but also by accumulation of secretions near the opening of the oesophagus and which overflow into the larynx as a result of impaired pharyngeal empty-

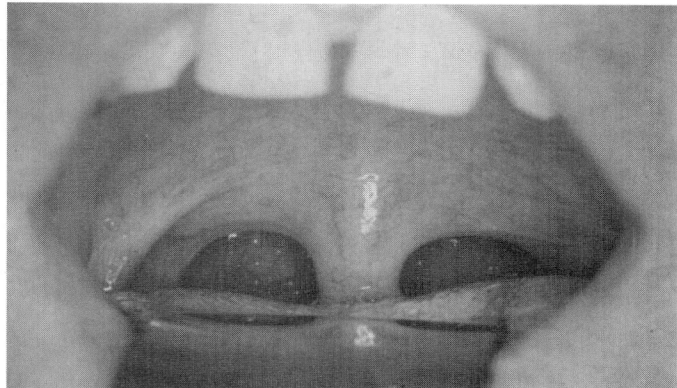
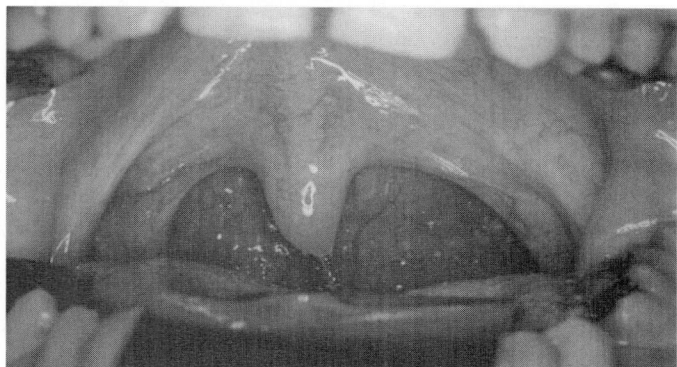

Fig. 20.13 Patient with a left vagus nerve lesion. At rest the palate is symmetrical (top), but on phonation the uvula is pulled towards the normal side and away from the side of the lesion (bottom).

ing. The patient is hoarse, the voice and cough are weak, and there is difficulty clearing the throat of secretions. Bilateral lesions, in the syndrome of bulbar palsy, produce much more severe symptoms. The speech is nasal and on attempting to swallow fluids there is nasal regurgitation due to palatal weakness. There may be snoring and inspiratory stridor. Coughing is paralysed leading to a high risk of bronchial aspiration. Acute bulbar palsy is life-threatening and tracheostomy is required.

Unilateral nuclear lesions rarely occur in isolation and are usually accompanied by involvement of other cranial nerve nuclei and long tracts. Causes include vascular lesions such as the lateral medullary syndrome of Wallenburg (Section 2.4.4) and tumour. Bilateral nuclear lesions, sometimes asymmetric, may be due to vascular lesions, motor neurone disease, tumour, syringobulbia, encephalitis, poliomyelitis, and rabies.

Nerve trunk lesions

The clinical features of a lesion affecting the vagus nerve trunk between its origin and the jugular foramen are as described above for a unilateral nuclear lesion. Thus, there is unilateral paralysis of the soft palate, pharynx, and larynx. Causes include glomus jugulare, meningioma and secondary tumour, meningitic processes, and basal skull fracture. There is frequently involvement of the neighbouring cranial nerves IX, XI, and XII.

The inferior ganglion lies just below the jugular foramen and from it arise the pharyngeal and superior laryngeal nerves, which supply the muscles of the soft palate and pharynx, and the cricothyroid tensors of the vocal cords. A lesion of the vagus nerve trunk below the inferior ganglion thus spares these muscles and the pattern of laryngeal paralysis is the same as is seen with an isolated lesion of the recurrent laryngeal nerve.

The vagus nerve may be involved in severe cases of diabetic or alcoholic neuropathy and in Guillain–Barré syndrome.

Vagal nerve stimulation is an accepted therapy for patients with refractory epilepsy (Section 31.11.5). Serious side effects are rare, but it is relatively common for patients to experience cough, hoarseness, and voice alteration during the 'on' phase of stimulation, though these effects tend to decrease over time (Ben-Menachem 2001).

Recurrent laryngeal nerve lesions

In practice, isolated lesions of the vagus nerve trunk are rare, but recurrent laryngeal nerve palsies are common. A unilateral lesion causes paralysis of the ipsilateral larynx and lower sphincter of the pharynx. The vocal cord is immobile and lies near the midline. Dysphagia is not a major feature, because the pharyngeal nerve is unaffected, although following an acute lesion there may be transient difficulties in swallowing fluids. In the acute phase there may also be dysphonia but despite the paralysed vocal cord compensatory mechanisms are so efficient that the voice may remain or soon return to normal.

The left recurrent laryngeal nerve is more commonly involved than the right, as a result of mediastinal lesions, particularly neoplasia. Less frequently it may be compressed by an aortic aneurysm or enlargement of the left atrium. In the neck the nerve may be affected unilaterally or bilaterally by trauma, surgery, cervical lymph gland enlargement, thyroid enlargement, and oesophageal carcinoma. Up to one-third of cases of recurrent laryngeal nerve palsy are idiopathic. In these idiopathic cases the deficit may be persistent or show partial or complete recovery (Blau and Kapadia 1972).

20.5 **Cranial nerve XI: accessory**

This is a purely motor nerve. It is formed from cranial and spinal roots which, as the accessory nerve, run together for only a very short distance. The cranial component is essentially part of the vagus nerve and is distributed mainly to the pharyngeal and recurrent laryngeal branches of that nerve, whereas the spinal component innervates the sternomastoid and trapezius muscles.

20.5.1 **Clinical neuroanatomy**

Spinal and cranial origins

The spinal nucleus of the accessory nerve lies in the lateral part of the anterior horn grey matter and extends from the pyramidal decussation to the fifth cervical segment. The fibres arising from it emerge from the lateral aspect of the cord, between the dorsal and ventral roots. They unite to form a trunk which ascends posterior to the denticulate ligament and ascends into the skull through the foramen magnum, dorsal to the vertebral artery. The cranial root is formed by nerve fibres arising from the lower part of the nucleus ambiguus. Rootlets emerge from the lateral medulla, below the origin of the vagus.

Peripheral course

The cranial and spinal components unite for a short distance and leave the skull through the jugular foramen, in close relationship to the vagus nerve to which the cranial root fibres are distributed. The spinal part runs backwards and laterally between the internal jugular vein and internal carotid artery, and crosses the transverse process of the atlas. It passes deep to the sternomastoid muscles, which it supplies, and emerges from its posterior border from where it crosses the posterior triangle, lying on levator scapulae. In this part of its course it is quite superficial, thus subject to trauma, and it is also related to cervical lymph nodes. The nerve then passes under the anterior border of trapezius and unites with branches of the third and fourth cervical nerves, C3 and C4, to form a plexus which innervates the muscle. The pattern of innervation of the different parts of trapezius is probably quite variable but in general the accessory nerve appears to supply the upper part of the muscle and fibres derived from C3 and C4 motor roots supply the lower part.

20.5.2 **Lesions of the accessory nerve**

Unilateral sternomastoid weakness (Fig. 20.14) is asymptomatic because it normally acts in concert with other cervical muscles which can compensate. Bilateral, but otherwise isolated, nerve lesions must be vanishingly rare. Bilateral sternomastoid weakness, with symptomatic weakness of neck flexion, is seen in myotonic dystrophy, inflammatory myopathies, various muscular dystrophies, myasthenia gravis, and motor neurone disease, but in all of these cases other cervical muscles are also involved.

Trapezius weakness is symptomatic (Fig. 20.15). The shoulder droops slightly and there is mild scapular winging at rest with the scapular rotated outwards and downwards. The winging is exacerbated by abduction of the arm, whereas winging due to serratus anterior weakness is most evident on forward flexion of the arm. The patient notices difficulty shrugging the shoulder, abducting the arm above 90° (Fig. 20.15) and carrying the extended arm backwards.

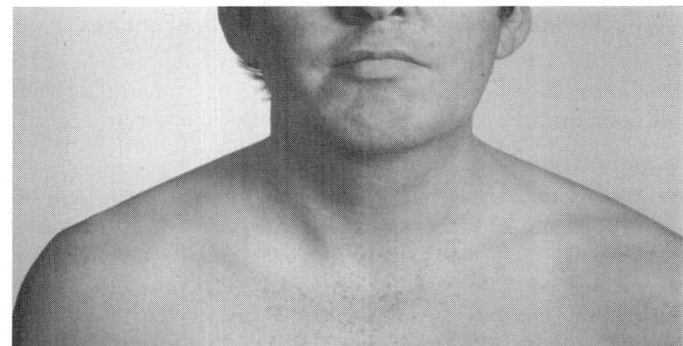
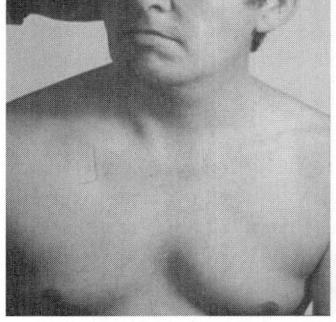
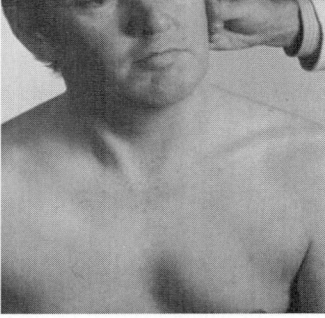

Fig. 20.14 Weakness and wasting of the left sternomastoid muscle in a patient with an accessory nerve lesion caused by a tumour at the jugular foramen.

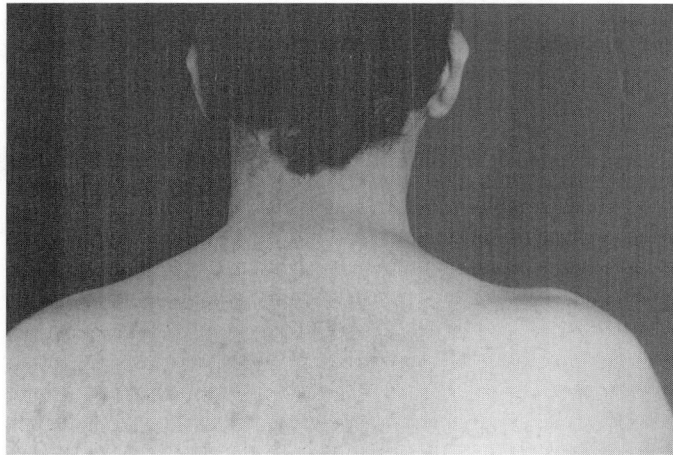

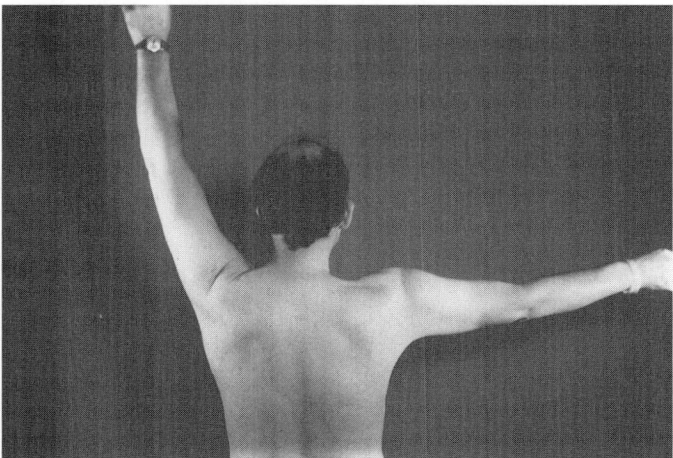

Fig. 20.15 Trapezius wasting and weakness on the right side from a skull base tumour affecting the accessory nerve.

Use of the outstretched arm may provoke pain around the shoulder. Bilateral trapezius weakness causes the head to fall forwards. This is rarely the result of bilateral accessory nerve trunk involvement, but is seen in myasthenia gravis, motor neurone disease, and various myopathies.

Supranuclear lesions

The cortical representation is mainly ipsilateral for sternomastoid and contralateral for trapezius. Thus, following a major hemisphere stroke trapezius is weak on the paralysed side, but sternomastoid is weak on the side of the hemispheric event.

Nuclear and nerve trunk lesions

The spinal cord nucleus is rarely involved in isolation. The anterior horn cells may be affected by motor neurone disease and poliomyelitis, and the nuclei may be compressed by upper cervical cord tumours and syringomyelia.

In the posterior fossa lesions affecting the accessory nerve often also involve cranial nerves IX, X, and XII. Common pathologies include primary and secondary tumours, meningitic processes, and basal skull fracture through the jugular foramen.

Outside the skull the commonest site of damage is in the posterior triangle, giving rise to trapezius but not sternomastoid

weakness. The nerve may be damaged by trauma or during surgical procedures such as removal of cervical lymph glands and carotid endarterectomy (Sweeney and Wilbourn 1992; Prim *et al.* 2006).

Up to one-third of accessory nerve palsies are idiopathic, probably often arising as a forme fruste of acute brachial neuritis (Section 22.5.4). The symptoms begin with pain in the lower lateral part of the neck which subsides within a few days and is then followed by weakness and atrophy of the sternomastoid and trapezius muscles (Eisen and Bertrand 1972). A recurrent form of spontaneous accessory neuropathy has been described (Chalk and Isaacs 1990).

20.6 Cranial nerve XII: hypoglossal

This motor nerve innervates all of the muscles of the tongue through general somatic efferent fibres. Although it contains some afferent fibres their function is unclear and they are of no importance in clinical practice.

20.6.1 Clinical neuroanatomy

Nucleus

The nucleus, which is nearly 2 cm long, lies in the central grey matter of the medial eminence and extends from the stria medullaris to the most caudal part of the medulla. The axons arising from it pass ventro-laterally and emerge as a series of rootlets on the ventral aspect of the medulla between the inferior olivary complex and the pyramid.

Peripheral course

The rootlets pass behind the vertebral artery and unite as they exit the skull through the hypoglossal canal, which lies about 1 cm anterior, inferior, and medial to the jugular foramen. Immediately outside the skull the nerve is in close proximity to cranial nerves IX, X, and XI, the internal jugular vein, and the internal carotid artery. At the level of the angle of the mandible it sweeps antero-laterally looping below the occipital artery and crossing the external carotid artery and then the loop of the lingual artery just above the hyoid bone. It then passes deep to the digastric muscle and terminates through multiple branches in the intrinsic and extrinsic tongue muscles, supplying genioglossus which acts to protrude the tongue; styloglossus which retracts and elevates the root of the tongue, and hypoglossus which causes the upper surface of the tongue to adopt a convex shape.

20.6.2 Lesions of the hypoglossal nerve

A unilateral lesion of the nucleus or nerve trunk causes ipsilateral wasting and weakness of the tongue (Figs 20.16 and 20.17). Fasciculation may be prominent, especially in infantile spinal muscular atrophy and classical motor neurone disease (Chapter 23) (Fig. 20.18). In the acute stage articulation and swallowing may be slightly impaired, but chronic lesions are typically asymptomatic. At rest in the floor of the mouth the tongue curves slightly to the healthy side, but on protrusion it deviates towards the affected side because of the unopposed contraction of the genioglossus muscle on the contralateral side. The degree of weakness can be judged by pressure against the tongue pushed against the inside of the cheek. Bilateral lower motor neurone lesions cause weakness of both sides of the tongue with inability to protrude the tongue, marked dysarthria, and mild swallowing difficulties. Such bilateral lesions are rare in isolation and are usually part of the syndrome of bulbar

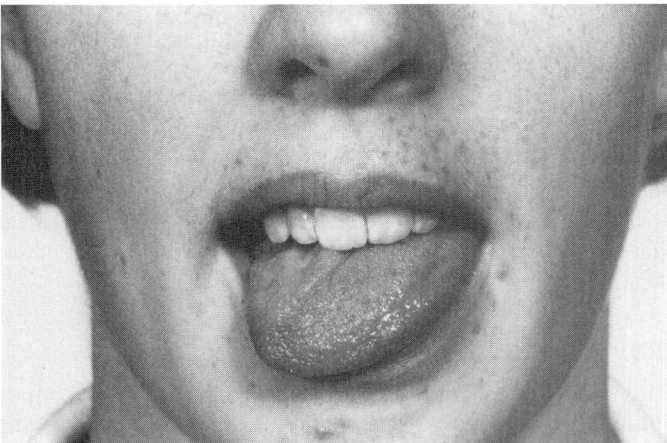

Fig. 20.16 Idiopathic right hypoglossal nerve palsy. Deviation of the protruded tongue to the affected side.

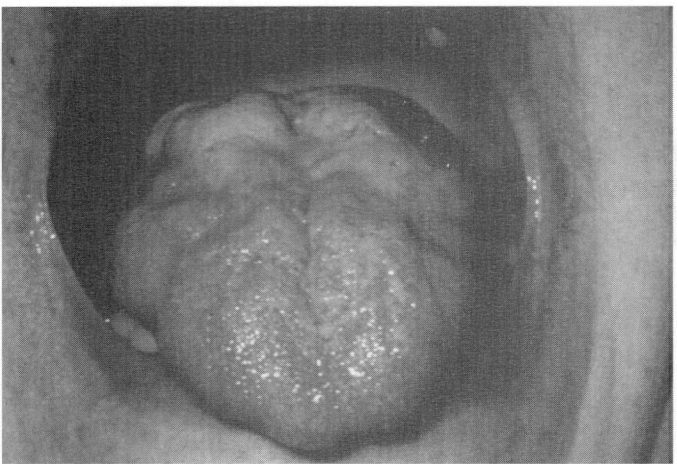

Fig. 20.18 Bilateral wasting and fasciculation of the tongue in a patient with motor neurone disease.

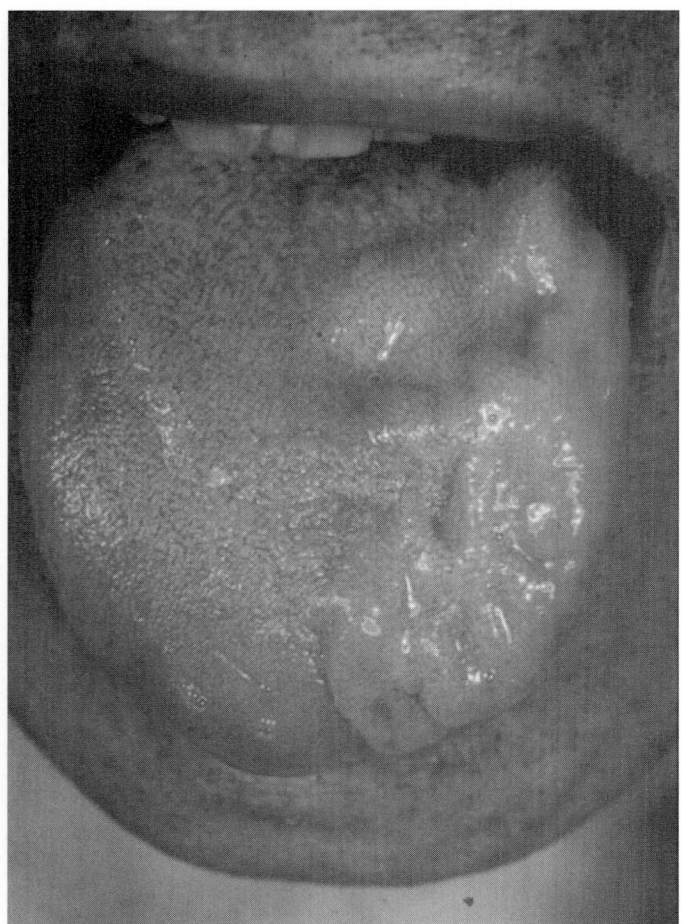

Fig. 20.17 Unilateral tongue wasting in a patient with a left hypoglossal nerve palsy.

palsy in which other bulbar muscles are affected and in which there is significant dysphagia. In one large series of cases of hypoglossal nerve palsy, 49 of 100 cases were due to tumour (Keane 1996b).

Supranuclear lesions

The nuclei have bilateral cortical representation so that a unilateral upper motor neurone lesion may have no observable effect, though occasionally the tongue may deviate to the contra-lateral side. Bilateral upper motor neurone involvement is seen as part of the syndrome of pseudobulbar palsy. The tongue is weak, clumsy, and contracted secondary to spasticity, but not wasted. Causes include bilateral hemispheric vascular disease, upper brainstem stroke, and tumours, multiple sclerosis, and motor neurone disease.

Nuclear lesions

Unilateral nuclear damage may be caused by tumours or vascular lesions, in both of these cases other structures are usually involved. Thus, a vascular event in the lower medulla might cause a unilateral hypoglossal nerve palsy and contralateral hemiplegia due to corticospinal tract involvement. Bilateral, but sometimes asymmetric, hypoglossal nuclear lesions may result from vascular lesions, tumours, syringobulbia, spinal muscular atrophy, motor neurone disease, and poliomyelitis.

Nerve trunk lesions

In the posterior fossa the hypoglossal nerve rootlets may be affected, often together with cranial nerves IX, X, and XI, by glomus jugulare, meningioma, and secondary neoplasms and basal meningitic processes. Unilateral or bilateral palsies may arise as a result of congenital or acquired bony abnormalities around the foramen magnum such as basilar impression or Paget's disease.

In the neck the nerve may be damaged by external trauma, during surgery including carotid endarterectomy, by tumours, and as a late consequence of regional radiotherapy. Vascular causes include aberrantly located arteries, carotid artery dissection (Baumgartner and Bogousslavsky 2005), and as a complication of central venous catheterization. It has been emphasized that the hypoglossal nerve may be subject to entrapment by crossing points with at least 3 blood vessels during its course between the brainstem and the tongue, the so-called 'triple cross' (Bademci *et al.* 2006). Cases of idiopathic hypoglossal nerve palsy have been described. One report described that in the absence of surgery, radiotherapy, and trauma, up to 50 per cent of cases of isolated hypoglossal nerve palsy are idiopathic; 20 per cent are vascular, and 20 per cent are caused by malignancy (Khoo *et al.* 2007). MR imaging to assess the entire pathway of the hypoglossal nerve is the investigation of choice.

The *neck–tongue syndrome* is the term coined for the simultaneous development of pain in the neck and occipital region, with numbness of the ipsilateral tongue caused by a sudden sharp turning movement of the head. It is thought to be caused by compression in the atlanto-axial space of the C2 root, which carries some of the sensory fibres from the tongue via the hypoglossal nerve to the C2 segment of the cervical cord.

20.7 Multiple cranial nerve palsies

Multiple cranial nerves may be affected by a single disease process located within the brainstem or extra-axially. Extra-axial pathologies involving multiple cranial nerves are outlined in Table 20.2. Various eponyms have been given to the groups of cranial nerves involved at specific anatomical locations. Lesions within the brainstem affecting multiple cranial nerves often involve the ascending sensory and descending motor pathways, with crossed motor and/or sensory symptoms and signs. There are multiple distinctive brainstem syndromes to which eponymous labels have also been given. Those involving the lower cranial nerves are outlined in Table 20.3.

Exramedullary involvement of multiple cranial nerves may result from meningeal processes such as carcinomatous or lymphomatous meningitis, granulomatous processes such as sarcoidosis or Wegener's granulomatosis; or chronic infections such as tuberculosis, syphilis, fungal, or borrelia infections. Involvement of the VIII cranial nerve and the sequential painless involvement of multiple cranial nerves, with a subacute time course, are particularly characteristic features of malignant meningitis. Infection of the cranial nerves is usually suspected with the involvement of V, VII, or VIII and there may be an associated CSF pleocytosis. Infective agents which have been implicated, either directly or on a parainfectious basis, include herpes zoster, HIV, cytomegalovirus in the context of HIV/AIDS, varicella, measles, rubella, mumps, mycoplasma, and listeria. Bilateral facial nerve involvement is common as part of Guillain–Barré syndrome and has been described in infectious mononucleosis.

Solid tumours involving multiple cranial nerves in the base of the skull include solid tumour metastases and local invasion by chordoma involving the clivus, nasopharyngeal tumour, sarcoma, or cholesteatoma. Compression by neurofibromas, meningiomas, and schwannomas including acoustic neuroma, can involve multiple cranial nerves. Other structural bony problems of the skull base including basilar invagination, Arnold–Chiari malformations, and Paget's disease can also cause damage of multiple lower cranial nerves. Vascular pathology such as carotid artery dissection or jugular vein thrombosis may also cause multiple lower cranial nerve lesions.

20.7.1 Cavernous sinus syndrome

This consists of various combinations of oculomotor nerve palsies and upper facial sensory loss, usually accompanied by periorbital

Table 20.2 Syndromes of multiple cranial nerve palsies involving the lower cranial nerves due to extra-axial pathologies at different anatomical locations

	Eponym	Cranial nerves involved	Usual pathology
Sphenoidal fissure	Foix	III, IV, V_1, VI	◆ Aneurysms ◆ Invasive tumours of the sphenoid bone
Cavernous sinus (lateral wall)	Tolosa–Hunt	III, IV, V_1, (V_{II}), VI	◆ Aneurysms within or thrombosis of the cavernous sinus ◆ Tumour invasion e.g. from sinuses or sella turcica ◆ Granulomatous processes which may be relapsing and steroid responsive
Retrosphenoidal space	Jaccoud	II, III, IV, V, VI	◆ Middle cranial fossa tumours
Petrous apex	Gradenigo	V, VI	◆ Tumours of the petrous bone ◆ Apical petrositis
Internal auditory meatus	–	VII, VIII	◆ Acoustic neuroma ◆ Tumours of the petrous bone
Cerebellopontine angle	–	V, VII, VIII, (IX)	◆ Acoustic neuroma ◆ Meningioma
Jugular foramen	Vernet	IX, X, XI	◆ Tumours ◆ Aneurysms, including carotid dissection
Posterior lateral condylar space	Collet–Sicard	IX, X, XI, XII	◆ Carotid dissection ◆ Carotid body, parotid, or lymph node tumours
Posterior retroparotid space	Villaret	IX, X, XI, XII + Horner's syndrome	◆ Tumours of the parotid, carotid body, lymph nodes ◆ Carotid dissection ◆ Sarcoidosis and other granulomatous processes
Posterior retroparotid space	Tapia	X, XII (+/– XI)	◆ Tumours within the upper part of the neck or parotid ◆ Penetrating trauma

Table 20.3 Brainstem syndromes involving the lower cranial nerves

	Eponym	Cranial nerves and other structures involved	Clinical signs	Common pathology
Base of pons	Millard–Gubler syndrome Raymond–Foville syndrome	VII (+/− VI) Corticospinal tract	◆ Facial and abducens palsy ◆ Contralateral hemiparesis ◆ May be a gaze palsy on the side of the brainstem lesion	Infarction or tumour
Medullary tegmentum	Avellis syndrome	X Spinothalamic tract (+/− descending sympathetic fibres)	◆ Paralysis of the soft palate and vocal cord ◆ Contralateral hemisensory disturbance ◆ +/− Horner's syndrome	Infarction or tumour
Medullary tegmentum	Jackson syndrome	X, XII Spinothalamic tract	◆ As for Avellis syndrome, but with ispilateral tongue paralysis	Infarction or tumour
Lateral medulla	Wallenberg syndrome	Spinal V, IX, X, XI Lateral spinothalamic tract Descending sympathetic fibres Spinocerebellar and olivocerebellar tracts	◆ Ipsilateral V, IX, X, XI palsies ◆ Horner's syndrome ◆ Ataxia ◆ Contralateral loss of pain and temperature sensation	Occlusion of the vertebral or posterior inferior cerebellar artery

pain and chemosis. The main underlying causes are septic, often in the context of diabetes mellitus, or aseptic thrombosis of the sinus due to clotting abnormalities, adjacent infection, trauma, malignant infiltration, or carotid aneurysm. In a series of 151 patients with cavernous sinus syndrome the most commonly associated pathologies were trauma and surgical procedures, and tumours of the nasopharynx, pituitary, metastases, or lymphomas (Keane 1996a).

20.7.2 Tolosa–Hunt syndrome

This is an idiopathic, steroid-responsive granulomatous condition involving the anterior portion of the cavernous sinus and/or the adjacent superior orbital fissure (Section 13.1.5). The patient develops unilateral painful opthalmoplegia which may be accompanied by trigeminal sensory loss, particularly over the ophthalmic division. The inflammatory changes may be visible with gadolinium-enhanced MRI, especially on coronal views. However, sarcoidosis, lymphomas, or a small meningioma may produce a similar appearance.

20.7.3 Jugular foramen syndrome

Tumours, aneurysms of the vertebral or carotid arteries, or carotid dissection may involve the IX, X, and XI cranial nerves either before or after they exit through the skull base via the jugular foramen. The characteristic clinical picture consists of a patient with hoarseness due to vocal cord paralysis, a degree of dysphagia, abnormal movement of the soft palate on phonation, with deviation towards the normal side, loss of sensation over the posterior wall of the pharynx, and weakness of the sternomastoid and upper trapezius muscles. Glomus tumours are a particular cause (Section 27.6.3).

20.7.4 Polyneuritis cranialis

Polyneuritis cranialis is a term which tends to be used for multiple cranial neruopathies of undetermined cause. Some reported cases are considered to be due to disseminated herpes zoster, sometimes in the absence of the characteristic vesicular rash, zoster sine

herpete (Osaki *et al.* 1995; Mehta *et al.* 2002). Other cases have been considered to be part of the spectrum of Guillain–Barré or Miller Fisher syndromes (McFarland 1976; Polo *et al.* 1992; Chowdhury *et al.* 2006) (Section 21.10).

In some cases the underlying cause remains uncertain. Fourteen patients with multiple cranial neuropathy of uncertain cause typically presented with headache and facial pain, followed within a few days by evolving cranial nerve palsies most commonly unilateral involving VI, III, V, and VII, with less frequent involvement of VII, IX, and X (Juncos and Beal 1987). Some patients had raised CSF protein and pleocytosis. Prompt relief of pain was achieved with steroid administration and recovery usually occurred over a period of months. The authors concluded that this condition overlapped with, and was of similar pathogenesis to, the Tolosa–Hunt syndrome. This condition appears to be similar to the description by other authors of idiopathic hypertrophic cranial pachymeningitis, where meningeal biopsy has shown only non-specific inflammatory features (Phanthumchinda *et al.* 1997). Imaging may show contrast enhancement of several cranial nerves (Morosini *et al.* 2003).

In some cases there may be a more chronic course, with sequential involvement of cranial nerves over a period of years. Some of these cases may eventually receive a confirmed diagnosis of sarcoidosis.

Before a diagnosis of idiopathic polyneuritis cranialis can be made, other conditions should be carefully excluded by screening for infections and granulomatous conditions, imaging, and CSF analysis, including cytology. Meningeal biopsy may need to be considered and a trial of steroids is likely to be warranted in cases where no definite underlying cause has been established.

20.8 Dysphagia

Difficulty in swallowing, or dysphagia, is a common neurological problem and the most important consequences include aspiration and malnutrition (Wiles 1991). Often it is these complications, particularly aspiration, that first draw the clinician's attention to the possibility of a swallowing disorder. Dysphagia may be due to mechanical factors, upper and lower motor neurone disorders, myasthenic syndromes, and myopathy.

20.8.1 Swallowing mechanisms

The process of swallowing is a complex neuromuscular activity, which allows the safe transport of material from the mouth to the stomach for digestion, without compromising the airway. It involves the synergistic action of at least 32 pairs of muscles and depends on the integrity of sensory and motor pathways of several cranial nerves; V, VII, IX, X, and XII. There are three phases involved in swallowing: oral, pharyngeal, and oesophageal. In the oral stage food is contained within the mouth, with mastication, mixing with saliva, and the formation of a bolus. The bolus is cupped by the anterior portion of the tongue, compressed against the hard palate, and pushed into the oropharynx. In the pharyngeal phase, the swallowing reflex is initiated as the bolus passes between the pillars of the fauces. The individual elements of this reflex include elevation of the soft palate, thus preventing nasal regurgitation, elevation of the larynx, closure of the entry to the trachea by the epiglottis, and peristaltic propulsion of the bolus through the cricopharyngeal sphincter into the oesophagus. During the pharyngeal phase, respiration is momentarily halted during expiration. In the oesophageal phase, co-ordinated peristalsis carries the bolus through the lower oesophageal sphincter, into the stomach. The whole process of swallowing lasts approximately 12 s.

The central neural control of swallowing consists of three basic elements:

- The afferent system consists of input from sensory components of the V, IX, and X cranial nerves, with the superior laryngeal branch of X being particularly important. These afferent fibres terminate in the tractus solitarius and nucleus of the spinal trigeminal system in the brainstem.

- The efferent system comprises the motor nuclei and nerve fibres of cranial nerves V, VII, X, and XII. Influenced by inputs from suprabulbar regions and the brainstem control pattern generator, they provide a patterned sequential discharge coordinating the orderly sequence of the oral, pharyngeal, and oesophageal swallowing phases.

- There is good evidence for the existence of a brainstem swallowing centre located in the medullary reticular formation on either side of the midline and just dorsal to the inferior olives. This centre integrates descending and peripheral information before activating a programmed sequence of responses which determine the pattern of swallowing.

Various cortical regions and the basal ganglia are important in the control of swallowing. Swallow-related cortical activity is multidimensional, recruiting brain areas implicated in the processing of motor, sensory, attentional, and affective aspects of the task. Functional imaging studies have provided evidence that there is significant interhemisheric asymmetry in the motor control of swallowing (Hamdy et al. 1999).

20.8.2 Causes of dysphagia

In neurological practice dysphagia is most often seen in association with other, obvious, neurological problems. Apart from in

Table 20.4 Causes of dysphagia

	Typical Causes
Mechanical	Oropharyngeal or oesophageal tumour
	Goitre
	Anterior cervical osteophytes
	Scleroderma
	Strictures
	Post-surgical
	Hiatus hernia
	Diverticulum
Pseudobulbar palsy	Cerebrovascular disease
	Upper brainstem tumours
	Motor neurone disease
	Demyelination
Bulbar palsy	Motor neurone disease
	Lower brainstem tumour
	Bilateral medullary infarction
	Syringobulbia
	Polyneuropathy
	Poliomyelitis
Neuromuscular disorders	Myasthenia gravis
	Oculopharyngeal muscular dystrophy
	Inflammatory myopathies
	Inclusion body myositis
	Myotonic dystrophy
Extrapyramidal, cerebellar, and autonomic disorders	Spinocerebellar degenerations
	Parkinson's disease
	Other Parkinsonian disorders e.g. progressive supranuclear palsy
	Autonomic neuropathies

oculopharyngeal muscular dystrophy (Section 24.2.6), it is relatively rare as a sole presenting symptom although occasionally this is seen in motor neurone disease, myasthenia gravis, and inclusion body myositis. Conversely, in general medical practice, there are many mechanical or structural disorders which may have dysphagia as the presenting feature. In some of the disorders listed in Table 20.4, notably motor neurone disease, both upper and lower motor neurone dysfunction may contribute to the dysphagia.

20.8.3 Clinical features of neurological dysphagia

The patient's description of the swallowing difficulty may give a clear indication of the level of the problem. A defect in the initiation of swallowing is usually caused by weakness of the tongue and may be caused by motor neurone disease, myasthenia gravis, or hypoglossal nerve lesions. Oropharyngeal disorders cause symptoms on swallowing or immediately after; there may be nasal regurgitation, coughing, and choking due to aspiration, a sensation of blockage in the neck, and sometimes pain. With oesophageal problems symptoms develop a little later and patients locate the site of blockage and discomfort to the lower throat or retrosternal region. As a general rule obstructive causes of dysphagia, such as oesophageal carcinoma, initially give rise to greater problems with solids than fluids, whereas in neuromuscular disorders dysphagia for fluids may be a relatively early feature.

When considering individual cranial nerves, unilateral upper or lower motor lesions generally do not cause major problems with dysphagia, particularly if of gradual onset. An important exception is the dysphagia, sometimes severe, seen transiently following hemispheric stroke. This is a major contributing factor to aspiration pneumonia.

It cannot be overemphasized that aspiration may be silent, without symptoms of coughing and choking, and without abnormal physical signs on bedside examination. If a patient with a disorder that might cause swallowing problems develops a chest infection then, even in the absence of specific features pointing towards such a problem, further investigation, such as videofluoroscopy, should be performed. Aspiration may be caused by any one of the following problems:

◆ Weakness of the pharyngeal muscles caused by vagal neuropathy, myopathy such as oculopharyngeal dystrophy or polymyositis, or a neuromuscular disorder such as myasthenia gravis or motor neurone disease.

◆ A brainstem lesion affecting cranial motor nuclei or descending motor pathways. This could be caused by vascular or inflammatory lesions, motor neurone disease, corticopsinal syringobulbia or tumour.

◆ Incoordination or slowness of the swallowing mechanism arising from corticospinal or basal ganglia disease. This allows the airway to remain open as the food bolus is in transit through the pharynx.

Frequently patients with neurological dysphagia will have accompanying symptoms including dysarthria, drooling, difficulty in coughing and clearing secretions from the throat, emotional lability, and weight loss. These and the presence of other symptoms and signs outside the bulbar territory such as ocular or limb neuromuscular symptoms may be helpful in establishing the underlying diagnosis.

20.8.4 Investigation of dysphagia

Often the history and findings on examination, particularly those related to the lower cranial nerves, will identify the neurological disorder causing the dysphagia, or will at least point to the likely nature of the problem, examples being myopathy, neurogenic problem, or upper motor neurone disorder. This directs further investigation. Physical examination may be normal in patients with mechanical oesophageal problems.

Endoscopy is of limited value in neurological practice but invaluable for the investigation of structural disorders of the oesophagus. Videofluoroscopy is a combination of a barium swallow with video recording, allowing frame-by-frame playback and detailed analysis of the swallowing process. It is useful in identifying the presence of aspiration during swallowing and in differentiating several types of dysphagia. Neurophysiological examination provides evidence for the neuromuscular disorders outlined in Table 20.4. Oesophageal manometry remains largely a research tool.

20.8.5 Management of dysphagia

Once dysphagia has been identified as a real or potential problem, the patient should undergo expert evaluation by a clinician and a speech therapist, prior to any attempt at feeding. Videofluoroscopy may be required. If there is any doubt it is best to achieve adequate nutrition through the use of a fine-bore nasogastric tube and to periodically reassess swallowing.

If the degree of dysphagia is slight, patients may be able to achieve an adequate nutritional intake, safely, by eating food of suitable consistency, perhaps with the addition of high-energy food supplements. They will require close supervision by a speech therapist and dietician. With a slightly greater degree of dysphagia some patients may cope with a combination of nasogastric feeding and limited oral intake.

More active therapeutic intervention for the dysphagia should be considered when the following problems are apparent:

◆ continuing weight loss of more than 10–20 per cent of the normal body weight despite the above measures,

◆ dehydration,

◆ aspiration with resultant respiratory infection,

◆ meal times have become very prolonged and tiring or intolerable due to frequent choking spells.

Percutaneous endoscopic gastrostomy is the procedure of choice for long-term enteral feeding. The placement of a percutaneous endoscopic gastrostomy is a relatively straight-forward procedure which can be performed under local anaesthetic. After placement many patients report great relief and increased well-being, though as yet no large-scale quality of life studies have been conducted. Percutaneous endoscopic gastrostomy feeding in motor neurone disease has been the subject of a recent systematic review (Langmore *et al.* 2006), and represents one of the major advances in symptomatic care for patients, leading to weight stabilization and adequate nutrional and fluid intake, although a survival benefit has not yet been convincingly shown. The need for percutaneous endoscopic gastrostomy feeding should be anticipated as the risks of the procedure are higher once the patient's forced vital capacity falls below 50 per cent (Chio *et al.* 2004).

In some patients, technical difficulties may be experienced in the insertion of a percutaneous endoscopic gastrostomy tube and in this situation, a radiologically guided method may be used (Thornton *et al.* 2002). In patients with a low vital capacity, the use of non-invasive ventilation during percutaneous endoscopic gastrostomy insertion has been shown to improve tolerance and safety of the procedure (Gregory *et al.* 2002). A full strength feeding regime providing 1500–2000 calories per day can be introduced over 48 h following percutaneous endoscopic gastrostomy tube insertion. Feeding can be managed by syringe boluses at normal meal-times, or by continuous infusion by pump which can be given overnight.

In the case of patients who have been malnourished for a prolonged period prior to percutaneous endoscopic gastrostomy tube insertion, vigilance is required for the possibility of the re-feeding syndrome as a post-operative complication (Fotheringham *et al.* 2005).

Anticholinergic drugs may be helpful to reduce problems with excess saliva and drooling that occur in patients with neurological dysphagia. Hyoscine can be given as a transdermal patch and amitriptyline or atropine are alternative agents which can be given either orally, via a percutaneous endoscopic gastrostomy or sublingually in the case of atropine. Staged low dose irradiation can be given to the parotid glands, but this carries the risk of making the mouth too dry. Injections of botulinum toxin into the parotid and submandibular glands may be helpful in patients with motor neurone disease, though there is a small risk of causing transient worsening of the dysphagia.

A portable suction apparatus may be helpful. Difficulty in clearing secretions from the throat may be helped by the administration of a mucolytic agent such as carbocisteine or provision of a cough assist device. Some patients with bulbar dysfunction have periods where the mouth is too dry, particularly caused by nocturnal mouth breathing. In this circumstance an artificial saliva spray or gel may be helpful.

Any of the neuromuscular diseases that can lead to severe dysphagia can also cause respiratory failure, so that some patients will also require tracheostomy. Although a cuffed tracheostomy tube offers protection against the consequences of aspiration, it may exacerbate swallowing difficulties because of the restrictive effect it has on laryngeal movement and its impairment of coughing.

References

Adams CB, Kaye AH, Teddy PJ (1982). The treatment of trigeminal neuralgia by posterior fossa microsurgery. *J Neurol Neurosurg Psychiatry*, **45**, 1020–6.

Allen D, Dunn L (2004). Acyclovir or valaciclovir for Bell's palsy (idiopathic facial paralysis). *Cochrane Database Syst Rev*, CD001869.

Attal N, Cruccu G, Haanpaa M et al. (2006). EFNS guidelines on pharmacological treatment of neuropathic pain. *Eur J Neurol*, **13**, 1153–69.

Bademci G, Batay F, Yasargil MG (2006). "Triple cross" of the hypoglossal nerve and its microsurgical impact to entrapment disorders. *Minim Invasive Neurosurg*, **49**, 234–7.

Balamucki CJ, Stieber VW, Ellis TL et al. (2006). Does dose rate affect efficacy? The outcomes of 256 gamma knife surgery procedures for trigeminal neuralgia and other types of facial pain as they relate to the half-life of cobalt. *J Neurosurg*, **105**, 730–5.

Barker FG 2nd, Jannetta PJ, Bissonette DJ et al. (1996). The long-term outcome of microvascular decompression for trigeminal neuralgia. *N Engl J Med*, **334**, 1077–83.

Barker FG 2nd, Jannetta PJ, Bissonette DJ et al. (1995). Microvascular decompression for hemifacial spasm. *J Neurosurg*, **82**, 201–10.

Bauer CA, Coker NJ (1996). Update on facial nerve disorders. *Otolaryngol Clin North Am*, **29**, 445–54.

Baumgartner RW, Bogousslavsky J (2005). Clinical manifestations of carotid dissection. *Front Neurol Neurosci*, **20**, 70–6.

Ben-Menachem E (2001). Vagus nerve stimulation, side effects, and long-term safety. *J Clin Neurophysiol*, **18**, 415–8.

Blau JN, Kapadia R (1972). Idiopathic palsy of the recurrent laryngeal nerve: a transient cranial mononeuropathy. *BMJ*, **4**, 259–61.

Breeze R, Ignelzi RJ (1982). Microvascular decompression for trigeminal neuralgia. Results with special reference to the late recurrence rate. *J Neurosurg*, **57**, 487–90.

Burgess RC, Michaels L, Bale JF Jr et al. (1994). Polymerase chain reaction amplification of herpes simplex viral DNA from the geniculate ganglion of a patient with Bell's palsy. *Ann Otol Rhinol Laryngol*, **103**, 775–9.

Chalk C, Isaacs H (1990). Recurrent spontaneous accessory neuropathy. *J Neurol Neurosurg Psychiatry*, **53**, 621.

Chio A, Galletti R, Finocchiaro et al. (2004). Percutaneous radiological gastrostomy: a safe and effective method of nutritional tube placement in advanced ALS. *J Neurol Neurosurg Psychiatry*, **75**, 645–47.

Chole R, Patil R, Degwekar SS et al. (2007). Drug treatment of trigeminal neuralgia: a systematic review of the literature. *J Oral Maxillofac Surg*, **65**, 40–5.

Chowdhury SR, Chakraborty PP, Majumdar S et al. (2006). Spontaneous recovery in a case of Miller-Fisher syndrome presenting as polyneuritis cranialis. *N Z Med J*, **119**, U2059.

Christo PJ, Hobelmann G, Maine DN (2007). Post-herpetic neuralgia in older adults: evidence-based approaches to clinical management. *Drugs Aging*, **24**, 1–19.

Costa J, Espirito-Santo C, Borges A et al. (2005). Botulinum toxin type A therapy for hemifacial spasm. *Cochrane Database Syst Rev*, CD004899.

De Diego JI, Prim MP, De Sarria MJ et al. (1998). Idiopathic facial paralysis: a randomized, prospective, and controlled study using single-dose prednisone versus acyclovir three times daily. *Laryngoscope*, **108**, 573–5.

Eisen A, Bertrand G (1972). Isolated accessory nerve palsy of spontaneous origin. A clinical and electromyographic study. *Arch Neurol*, **27**, 496–502.

Elston JS (1986). Botulinum toxin treatment of hemifacial spasm. *J Neurol Neurosurg Psychiatry*, **49**, 827–9.

Fischbach F, Lehmann TN, Ricke J et al. (2003). Vascular compression in glossopharyngeal neuralgia: demonstration by high-resolution MRI at 3 tesla. *Neuroradiology*, **45**, 810–11.

Forster C, Brandt T, Hund E et al. (1996). Trigeminal sensory neuropathy in connective tissue disease: evidence for the site of the lesion. *Neurology*, **46**, 270–1.

Fotheringham J, Jackson K, Kersh R et al. (2005). Refeeding syndrome: life-threatening, underdiagnosed, but treatable. *QJM*, **98**, 318–9.

Furukawa T (1990). Numb chin syndrome in the elderly. *J Neurol Neurosurg Psychiatry*, **53**, 173.

Geha PY, Baliki MN, Chialvo DR et al. (2007). Brain activity for spontaneous pain of postherpetic neuralgia and its modulation by lidocaine patch therapy. *Pain*, **128**, 88–100.

Goins MR, Pitovski DZ (2004). Posttonsillectomy taste distortion: a significant complication. *Laryngoscope*, **114**, 1206–13.

Gorgulho AA, De Salles AA (2006). Impact of radiosurgery on the surgical treatment of trigeminal neuralgia. *Surg Neurol*, **66**, 350–6.

Gregory S, Siderowf A, Golaszewski AL et al. (2002). Gastrostomy insertion in ALS patients with low vital capacity: respiratory support and survival. *Neurology*, **58**, 485–7.

Gupta D, Vishwakarma SK (1987). Superior orbital fissure syndrome in trigemino-facial zoster. *J Laryngol Otol*, **101**, 975–7.

Hageman G, Ippel PF, Jansen EN et al. (1990). Familial, alternating Bell's palsy with dominant inheritance. *Eur Neurol*, **30**, 310–3.

Hagen NA, Stevens JC, Michet CJ Jr (1990). Trigeminal sensory neuropathy associated with connective tissue diseases. *Neurology*, **40**, 891–6.

Haines SJ, Jannetta PJ, Zorub DS (1980). Microvascular relations of the trigeminal nerve. An anatomical study with clinical correlation. *J Neurosurg*, **52**, 381–6.

Hamdy S, Mikulis DJ, Crawley A et al. (1999). Cortical activation during human volitional swallowing: an event-related fMRI study. *Am J Physiol*, **277**, G219–25.

Harding SP, Lipton JR, Wells JC (1987). Natural history of herpes zoster ophthalmicus: predictors of postherpetic neuralgia and ocular involvement. *Br J Ophthalmol*, **71**, 353–8.

Hato N, Matsumoto S, Kisaki H, Takahashi H et al. (2003). Efficacy of early treatment of Bell's palsy with oral acyclovir and prednisolone. *Otol Neurotol*, **24**, 948–51.

Hato N, Yamada H, Kohno H et al. (2007). Valacyclovir and prednisolone treatment for Bell's palsy: a multicenter, randomized, placebo-controlled study. *Otol Neurotol*, **28**, 408–13.

Hauser WA, Karnes WE, Annis J et al. (1971). Incidence and prognosis of Bell's palsy in the population of Rochester, Minnesota. *Mayo Clin Proc*, **46**, 258–64.

He L, Wu B, Zhou M (2006). Non-antiepileptic drugs for trigeminal neuralgia. *Cochrane Database Syst Rev 3*, CD004029.

Hemminki K, Li X, Sundquist K (2007). Familial risks for nerve, nerve root and plexus disorders in siblings based on hospitalisations in Sweden. *J Epidemiol Community Health*, **61**, 80–4.

Horton J, Means ED, Cunningham TJ et al. (1973). The numb chin in breast cancer. *J Neurol Neurosurg Psychiatry*, **36**, 211–6.

Illingworth RD, Porter DG, Jakubowski J. (1996). Hemifacial spasm: a prospective long-term follow up of 83 cases treated by microvascular decompression at two neurosurgical centres in the United Kingdom. *J Neurol Neurosurg Psychiatry*, **60**, 72–7.

Ishii K, Tamaoka A, Shoji S (1998). Dorsolateral pontine segmental infarction presenting as isolated trigeminal sensory neuropathy. *J Neurol Neurosurg Psychiatry*, **65**, 702.

Ito H, Ito H, Nakano S et al. (2007). Low-dose subcutaneous injection of botulinum toxin type A for facial synkinesis and hyperlacrimation. *Acta Neurol Scand*, **115**, 271–4.

Jacobson RR, Russell RW (1979). Glossopharyngeal neuralgia with cardiac arrhythmia: a rare but treatable cause of syncope. *BMJ*, **1**, 379–80.

Jannetta PJ (1977). Treatment of trigeminal neuralgia by suboccipital and transtentorial cranial operations. *Clin Neurosurg*, **24**, 538–49.

Juncos JL, Beal MF (1987). Idiopathic cranial polyneuropathy: A fifteen year experience. *Brain*, **110**, 197–211.

Kanai A, Suzuki A, Kobayashi M *et al.* (2006). Intranasal lidocaine 8 per cent spray for second-division trigeminal neuralgia. *Br J Anaesth*, **97**, 559–63.

Karibe H, Shirane R, Yoshimoto T (2004). Preoperative visualization of microvascular compression of cranial nerve IX using constructive interference in steady state magnetic resonance imaging in glossopharyngeal neuralgia. *J Clin Neurosci*, **11**, 679–81.

Kawasaki A, Borruat FX (2003). An unusual presentation of herpes zoster ophthalmicus: orbital myositis preceding vesicular eruption. *Am J Ophthalmol*, **136**, 574–5.

Kaye AH, Adams CB (1981). Hemifacial spasm: a long term follow-up of patients treated by posterior fossa surgery and facial nerve wrapping. *J Neurol Neurosurg Psychiatry*, **44**, 1100–3.

Keane JR (1996a). Cavernous sinus syndrome: an analysis of 151 cases. *Arch Neurol*, **53**, 967–71.

Keane JR (1996b). Twelfth-nerve palsy. Analysis of 100 cases. *Arch Neurol*, **53**, 561–6.

Khoo SG, Ullah I, Wallis F *et al.* (2007). Isolated hypoglossal nerve palsy: a harbinger of malignancy. *J Laryngol Otol*, **13**, 1–3.

Laha RK, Jannetta PJ (1977). Glossopharyngeal neuralgia. *J Neurosurg*, **47**, 316–20.

Langmore SE, Kasarskis EJ, Manca ML *et al.* (2006). Enteral tube feeding for amyotrophic lateral sclerosis/motor neuron disease. *Cochrane Database Syst Rev*, CD004030.

Lauria G, Majorana A, Borgna M *et al.* (2005). Trigeminal small-fiber sensory neuropathy causes burning mouth syndrome. *Pain*, **115**, 332–7.

Lecky BR, Hughes RA, Murray NM (1987). Trigeminal sensory neuropathy. A study of 22 cases. *Brain*, **110**, 1463–85.

Linnemann CC Jr, Alvira MM (1980). Pathogenesis of varicella-zoster angiitis in the CNS. *Arch Neurol*, **37**, 239–40.

Lossos A, Siegal T (1992). Numb chin syndrome in cancer patients: etiology, response to treatment, and prognostic significance. *Neurology*, **42**, 1181–4.

Massey EW, Moore J, Schold SC Jr (1981). Mental neuropathy from systemic cancer. *Neurology*, **31**, 1277–81.

Matthews WB (1993). Treatment of Bell's Palsy. In Matthews WB, Glaser GH, eds. *Recent Advances in Clinical Neurology 3*, Edinburgh, Churchill Livingstone.

McFarland HR (1976). Polyneuritis cranialis as the sole manifestation of Guillain Barre syndrome. *Mo Med*, **73**, 227–9.

Mehta J, Mahajan V, Khanna S (2002). Disseminated zoster with polyneuritis cranialis and motor radiculopathy. *Neurol India*, **50**, 228–9.

Miller RG, Rosenberg JA, Gelinas DF *et al.* (1999). Practice parameter: the care of the patient with amyotrophic lateral sclerosis (an evidence-based review): report of the quality standards subcommittee of the American academy of neurology: ALS practice parameters task force. *Neurology*, **52**, 1311–23.

Morgan M, Moffat M, Ritchie L *et al.* (1995). Is Bell's palsy a reactivation of varicella zoster virus? *J Infect*, **30**, 29–36.

Mori K, Iijima M, Koike H *et al.* (2005). The wide spectrum of clinical manifestations in Sjogren's syndrome-associated neuropathy. *Brain*, **128**, 2518–34.

Morosini A, Burke C, Emechete B (2003). Polyneuritis cranialis with contrast enhancement of cranial nerves on magnetic resonance imaging. *J Paediatr Child Health*, **39**, 69–72.

Murakami S, Honda N, Mizobuchi M *et al.* (1998). Rapid diagnosis of varicella zoster virus infection in acute facial palsy. *Neurology*, **51**, 1202–5.

Murakami S, Mizobuchi M, Nakashiro Y *et al.* (1996). Bell palsy and herpes simplex virus: identification of viral DNA in endoneurial fluid and muscle. *Ann Intern Med*, **124**, 27–30.

Osaki Y, Matsubayashi K, Okumiya K *et al.* (1995). Polyneuritis cranialis due to varicella-zoster virus in the absence of rash. *Neurology*, **45**, 2293.

Oxman MN, Levin MJ, Johnson GR *et al.* (2005). A vaccine to prevent herpes zoster and postherpetic neuralgia in older adults. *N Engl J Med*, **352**, 2271–84.

Ozenci M, Karaoguz R, Conkbayir C *et al.* (2003). Glossopharyngeal neuralgia with cardiac syncope treated by glossopharyngeal rhizotomy and microvascular decompression. *Europace*, **5**, 149–52.

Pearce JM (2006). Glossopharyngeal neuralgia. *Eur Neurol*, **55**, 49–52.

Phanthumchinda K, Sinsawaiwong S, Hemachudha T *et al.* (1997). Idiopathic hypertrophic cranial pachymeningitis: an unusual cause of subacute and chronic headache. *Headache*, **37**, 249–52.

Polo A, Manganotti P, Zanette G *et al.* (1992). Polyneuritis cranialis: clinical and electrophysiological findings. *J Neurol Neurosurg Psychiatry*, **55**, 398–400.

Prim MP, De Diego JI, Verdaguer JM *et al.* (2006). Neurological complications following functional neck dissection. *Eur Arch Otorhinolaryngol*, **263**, 473–6.

Regis J, Metellus P, Hayashi M *et al.* (2006). Prospective controlled trial of gamma knife surgery for essential trigeminal neuralgia. *J Neurosurg*, **104**, 913–24.

Resnick DK, Jannetta PJ, Bissonnette D *et al.* (1995). Microvascular decompression for glossopharyngeal neuralgia. *Neurosurgery*, **36**, 64–8.

Salinas RA, Alvarez G, Ferreira J (2004). Corticosteroids for Bell's palsy (idiopathic facial paralysis). *Cochrane Database Syst Rev*, CD001942.

Sampson JH, Grossi PM, Asaoka K *et al.* (2004). Microvascular decompression for glossopharyngeal neuralgia, long-term effectiveness and complication avoidance. *Neurosurgery*, **54**, 884–9; discussion 889–90.

Satoh T, Onoda K, Date I (2007). Fusion imaging of three-dimensional magnetic resonance cisternograms and angiograms for the assessment of microvascular decompression in patients with hemifacial spasms. *J Neurosurg*, **106**, 82–9.

Severson EA, Baratz KH, Hodge DO *et al.* (2003). Herpes zoster ophthalmicus in Olmsted county, Minnesota: have systemic antivirals made a difference? *Arch Ophthalmol*, **121**, 386–90.

Stieber VW, Bourland JD, Ellis TL (2005). Glossopharyngeal neuralgia treated with gamma knife surgery: treatment outcome and failure analysis. Case report. *J Neurosurg*, **102**(Suppl), 155–7.

Sweeney CJ, Gilden DH (2001). Ramsay Hunt syndrome. *J Neurol Neurosurg Psychiatry*, **71**, 149–54.

Sweeney PJ, Wilbourn AJ (1992). Spinal accessory (11th) nerve palsy following carotid endarterectomy. *Neurology*, **42**, 674–5.

Sweet WH, Wepsic JG (1974). Controlled thermocoagulation of trigeminal ganglion and rootlets for differential destruction of pain fibers. 1. Trigeminal neuralgia. *J Neurosurg*, **40**, 143–56.

Taha JM, Tew JM Jr (1996). Comparison of surgical treatments for trigeminal neuralgia: reevaluation of radiofrequency rhizotomy. *Neurosurgery*, **38**, 865–71.

Thornton FJ, Fotheringham T, Alexander M *et al.* (2002). Amyotrophic lateral sclerosis: enteral nutrition provision—endoscopic or radiologic gastrostomy? *Radiology*, **224**, 713–7.

Volpi A, Gross G, Hercogova J *et al.* (2005). Current management of herpes zoster: the European view. *Am J Clin Dermatol*, **6**, 317–25.

Watson CP (1995). The treatment of postherpetic neuralgia. *Neurology*, **45**, S58–60.

Wiles CM (1991). Neurogenic dysphagia. *J Neurol Neurosurg Psychiatry*, **54**, 1037–9.

Womack LW, Liesegang TJ (1983). Complications of herpes zoster ophthalmicus. *Arch Ophthalmol*, **101**, 42–5.

Yanagihara N, Yumoto E, Shibahara T (1988). Familial Bell's palsy: analysis of 25 families. *Ann Otol Rhinol Laryngol Suppl*, **137**, 8–10.

Young L (2006). Post-herpetic neuralgia: a review of advances in treatment and prevention. *J Drugs Dermatol*, **5**, 938–41.

CHAPTER 21

Polyneuropathy

Michael Donaghy

Contents

21.1　Diagnosis of polyneuropathy

Peripheral neuropathy has a multitude of causes, many of which can be diagnosed by careful clinical and electrophysiological evaluation. A fundamental distinction should be made between:

- *Polyneuropathy*, a generalized neuropathy affecting all peripheral nerve fibres.

- *Focal neuropathy*, which affects individual peripheral nerves either singly or multiply. Recognized causes of (multi-)focal peripheral neuropathy are listed in Table 22.1. In focal neuropathy the muscle wasting and weakness, reflex loss, and sensory disturbance are restricted to the territories of the affected peripheral nerve(s) or root(s). Occasionally widespread vasculitic involvement of the peripheral nervous system may produce the clinical picture of symmetrical polyneuropathy, rather than the multiple mononeuropathies more usually associated with vasculitis.

Clinical features. Typically polyneuropathy will cause the combination of distal limb muscle weakness, loss of tendon reflexes, and reduced distal limb sensation. There is variable involvement of the autonomic innervation, damage to which causes a dry, vasodilated foot or hand. Loss of tendon reflexes is a cardinal sign of polyneuropathy, often restricted to the ankle jerks in axonal degeneration, but involving more proximal reflexes in acquired demyelinating neuropathies which may involve more proximal segments or the nerve roots. Clinical features suggestive of demyelinating or conduction block polyneuropathy include:

- a relative lack of muscle wasting in relation to the degree of weakness because no denervation has occurred;

- weakness of proximal muscles as well as distal, because of nerve root involvement; and

Table 21.1 Causes of demyelinating polyneuropathy

Inherited:	Hereditary motor and sensory neuropathy Type 1 (Section 21.4.4)
	Hereditary motor and sensory neuropathy Type 3 (Dejerine–Sottas Disease) (Section 21.4.6)
	Hereditary neuropathy with liability to pressure palsies (Section 21.5)
	Adrenoleucodystrophy (Section 21.8.4)
	Krabbe's disease (Section 21.8.3)
	MNGIE (Section 21.7.6)
	Metachromatic leucodystrophy Section (21.8.2)
	Refsum disease (phytanic acid accumulation) (Section 21.8.1)
Acquired:	Guillain–Barré syndrome (Section 21.10.1)
	Chronic inflammatory demyelinating neuropathy (Section 21.11.2)
	Multifocal motor neuropathy with conduction block (Section 21.11.3)
	Pure motor demyelinating neuropathy (Section 21.11.4)
	Neuropathies associated with lymphoproliferative disorders (Section 21.12) or carcinoma (Section 21.13)
	Diphtheria (Section 21.14.4.)
Toxic:	Amiodarone neuropathy (Section 21.19.2)
	Perhexiline neuropathy
	Hexacarbon neuropathy (Section 21.21.7)

◆ disproportionate loss of joint position and vibration sensations compared to relative preservation of pain and temperature sensations which are carried by unmyelinated fibres.

Nerve conduction velocity measurements and electromyography (Section 3.5) should be used to distinguish between primarily demyelinating, axonal degeneration, or conduction block polyneuropathies (Fig. 21.1). Nerve conduction studies in demyelinating polyneuropathy show prolonged distal motor latencies, slowed motor conduction velocities, and prolonged F-wave latencies. Evidence of conduction block is particularly likely in acquired forms of demyelinating polyneuropathy. Demyelinating polyneuropathies generally hold a better prospect for recovery than axonal degeneration polyneuropathies although they tend to produce more severe clinical syndromes. Causes of demyelinating polyneuropathy are shown in Table 21.1.

Inherited polyneuropathy usually holds a poor prospect for recovery and other family members may require genetic counselling. Inherited polyneuropathy must be considered if there is a family history of neuropathy or foot deformity or parental consanguinity. Compared to acquired polyneuropathy, it tends to evolve slowly, and even marked degrees of weakness may not excite complaint by the patient. Positive sensory symptoms, such as paraesthesiae, spontaneous pains, or thermal sensations, suggest an acquired neuropathy, although they do occur in the polyneuropathy of the inherited metabolic disorders metachromatic leucodystrophy and Fabry disease. Often the suspicion of an inherited basis for a patient's polyneuropathy can only be confirmed by clinical or electrophysiological examination of relatives or molecular genetic studies.

Sensory neuropathies. Neuropathies purely or predominantly involving sensory fibres are shown in Table 21.2. Some purely sensory neuropathies spare large myelinated fibres, only affecting unmyelinated and small myelinated fibres. In such cases tendon reflexes are preserved and sensory nerve conduction is normal. However, there is usually associated dysfunction of unmyelinated autonomic fibres with postural hypotension and dry, red feet.

Motor neuropathies. Purely or predominantly motor neuropathies are listed in Table 21.3.

Autonomic neuropathies. Some disorders produce a purely or predominantly autonomic neuropathy (Table 21.4). In many the small unmyelinated sensory fibres are also affected, causing impaired pinprick and temperature sensations. Patients present with varying combinations of vasodilation and impaired sweating, sometimes with cardiovascular, gastrointestinal, micturition, or pupillomotor involvement (Freeman 2005).

Ageing. After the age of 65, an increasing proportion of asymptomatic people without neuropathic risk factors have unobtainable ankle jerks or loss of vibration sense from the feet, and a few have more extensive abnormalities, such as mild distal muscle weakness (Vrancken *et al.* 2006). Such features cannot be taken as sole evidence for peripheral nerve disease in the elderly and merely reflect the natural age-related loss of peripheral nerve axons (Jacobs and Love 1985).

21.2 Nerve biopsy

Biopsy of peripheral nerves makes a major diagnostic impact only in a small selected group of patients with peripheral neuropathy. It is of particular use in those patients suspected of suffering from the treatable conditions of vasculitic, sarcoid, or leprous neuropathy, or sensory perineuritis. It may establish the diagnosis in rare conditions such as giant axonal neuropathy or neuroaxonal dystrophy. Traditionally nerve biopsy has demonstrated amyloidosis, but rectal biopsy is similarly sensitive and less invasive. Hereditary neuropathy with liability to pressure palsies is diagnosable by nerve biopsy but this is being replaced by molecular genetic tests. Nerve biopsy usually has little role in establishing a diagnosis of acute or chronic inflammatory demyelinating polyneuropathy. Occasionally in these conditions, examination of teased nerve fibres may show segmental demyelination in suspected cases where motor nerve conduction velocities have been insufficiently slow to be diagnostic.

The nerve most commonly chosen for biopsy is the sural nerve at the ankle, less commonly the radial nerve at the wrist or the superficial peroneal nerve in the calf. The biopsy is carried out under local anaesthesia. Some favour fascicular, rather than full-thickness nerve biopsy, on the grounds that it leaves less residual sensory deficit. There is no difference in the degree of subsequent sensory loss or dysaesthetic pain comparing fascicular and full-thickness nerve biopsies. Full-thickness biopsy is advisable if vasculitis is suspected so as to permit inspection of the maximum number of epineurial blood vessels. The most common complications of nerve biopsy are failure of wound healing, or infection, particularly in patients receiving steroid therapy. Significant pain or paraesthesiae attributable to the biopsy occur in less than 10 per cent of patients after 1 year.

The specimen should be processed immediately by a laboratory experienced in peripheral nerve pathology. It should be divided to provide material for paraffin embedding, and for glutaraldehyde and osmium tetroxide fixation for single nerve fibre teasing, and for electron and light microscopy on 1-mm plastic-embedded sections; additional material may be frozen for immunofluorescent studies (Fig. 21.2). Morphometric analysis may be required to reveal subtle differences in the density of myelinated or unmyelinated fibres, or alterations in the fibre-size distribution; this process is time-consuming and technically demanding. Control morphometric values have been established for a wide age range (Jacobs and Love 1985) and the characteristic pathological changes in a wide variety of diseases (Richardson and De Girolami 1995) will be referred to in the following sections.

Quantitative assessment of immunohistochemically stained epidermal nerve fibres in small punch biopsies of skin is a valuable development in assessing small unmyelinated fibre involvement in neuropathies. This technique seems more sensitive than morphometry of sural nerve biopsies, is less technically complex, and detects unmyelinated fibre involvement in a wide range of polyneuropathies involving sensory fibres (Herrmann *et al.* 1999). Skin biopsy provides a repeatable tool for assessing progression of neuropathy or the effects of treatments. It can demonstrate loss of nerve fibres in those oft encountered patients with painful burning feet, in whom conventional clinical examination, nerve conduction studies, and sural nerve biopsy fail to reveal abnormalities (Periquet *et al.* 1999).

21.3 Treatment of polyneuropathy

21.3.1 General principles

Accurate diagnosis of the cause is essential to ensure that correct therapy is provided for certain conditions, for instance immunosuppression for vasculitic neuropathy and chronic inflammatory demyelinating neuropathy; intravenous immunoglobulin for multifocal motor neuropathy with conduction block; plasma exchange or intravenous immunoglobulin for the Guillain–Barré syndrome; cessation of toxic exposure to chemicals, drugs, or alcohol; vitamin replacement; or diet modification in Refsum disease. Hence, the same principles underlying treatment will be considered before detailed description of the different varieties of peripheral nerve disease.

Table 21.2 Causes of sensory peripheral neuropathy

Inherited:	Hereditary sensory and autonomic neuropathies (Section 21.6)
	Fabry disease (Section 21.8.5)
	Tangier disease (Section 21.8.7)
	Familial amyloid (Section 21.9.1)
Acquired:	Acute sensory polyneuritis (Section 21.18.10)
	Chronic idiopathic ataxic (sensory) neuropathy (Section 21.11.5)
	Sjögren's syndrome (Section 21.18.10)
	Primary amyloidosis (Section 21.9.2)
	Paraneoplastic sensory neuropathy (Section 21.13.1)
	Acquired immune deficiency syndrome (late) (Section 21.14.2)
	Leprosy (Section 21.14.1)
	Lyme disease (*Borrelia* infection) (Section 21.14.3)
	Sensory perineuritis (Section 21.16.1)
	Migrant sensory neuritis of Wartenberg (Section 21.16.2)
	Diabetes mellitus (Section 21.17.2)
	Vitamin B_{12} deficiency (Section 21.22.4)
	Vitamin E deficiency (Section 21.22.5)
Toxic:	Almitrine
	Cisplatin (Section 21.19.4)
	Didanosine (Section 21.19.7)
	Metronidazole (Section 21.19.14)
	Misonidazole

Table 21.3 Causes of motor peripheral neuropathy

Charcot-Marie-Tooth disease (some patients) (Section 21.4)
Porphyric neuropathy (Section 21.8.6)
Guillain–Barré syndrome (Section 21.10.1)
Acute motor axonal neuropathy (Section 21.10.2)
Multifocal motor neuropathy with conduction block (Section 21.11.3)
Pure motor demyelinating neuropathy (Section 21.11.4)
Diphtheria (Section 21.14.4)
Dapsone (Section 21.19.6)
Lead poisoning (Section 21.20.2)
Organophosphate poisoning (Section 21.21.9)

Physiotherapy is important to prevent muscle contractures and keep joints mobile, so that when regeneration of nerve fibres occurs, the limb may be in the best possible condition to profit by the return of nervous function. In the past, firm splinting was commonly employed in order to keep a paralysed muscle in a relaxed

Table 21.4 Clinically important autonomic involvement in peripheral neuropathy

Diabetic	(Section 21.17.5)
Amyloid deposition	Primary amyloidosis (Section 21.9.2)
	Familial amyloidosis (Section 21.9.1)
Hereditary sensory and autonomic neuropathy	Autosomal dominant sensory neuropathy (Section 21.6.1)
	Autosomal recessive sensory neuropathy (Section 21.6.2)
	Fabry disease (Section 21.8.5)
	Multiple symmetric lipomatosis (Section 21.7.4)
	Porphyria (Section 21.8.6)
	Familial dysautonomia (Riley-Day syndrome) (Section 21.6.3)
	Anhidrotic sensory neuropathy (Section 21.6.4)
	Small myelinated fibre deficiency (Section 21.6.5)
Idiopathic polyneuritis	Guillain–Barré syndrome (Section 21.10.1)
	Idiopathic autonomic neuropathy (Sections 21.10.4 and 21.11.7)
	Acute sensory neuropathy (Section 21.10.3)
	Adies pupil and the Ross syndrome (Section 13.3.4)
	Lambert–Eaton myasthenic syndrome (Section 24.10.2)
Paraneoplastic	(Section 21.11.7)
Toxic	Alcohol (Section 21.19.1)
	Amiodarone (Section 21.19.2)
	Cisplatin (Section 21.19.4)
	Heavy metals (Section 21.20)
	Perhexiline
	Seafood toxicity (Ciguatoxin) (Section 5.12.1)
	Taxol (Section 21.19.21)
	Vacor (Section 21.21.11)
	Vinca alkaloids (Section 21.19.24)
Infections	Trypanosoma cruzi (Chagas' disease)
	Leprosy (Section 21.14.1)
	Botulism (Sections 5.8.6 and 24.10.5)
	HIV (Section 21.14.2)
	Diphtheria (Section 21.14.4)

(from Donaghy M (2007). Autonomic dysfunction in peripheral nerve disease. Mathias C and Bannister R (eds). *Autonomic Failure* 5th Edition, OUP)

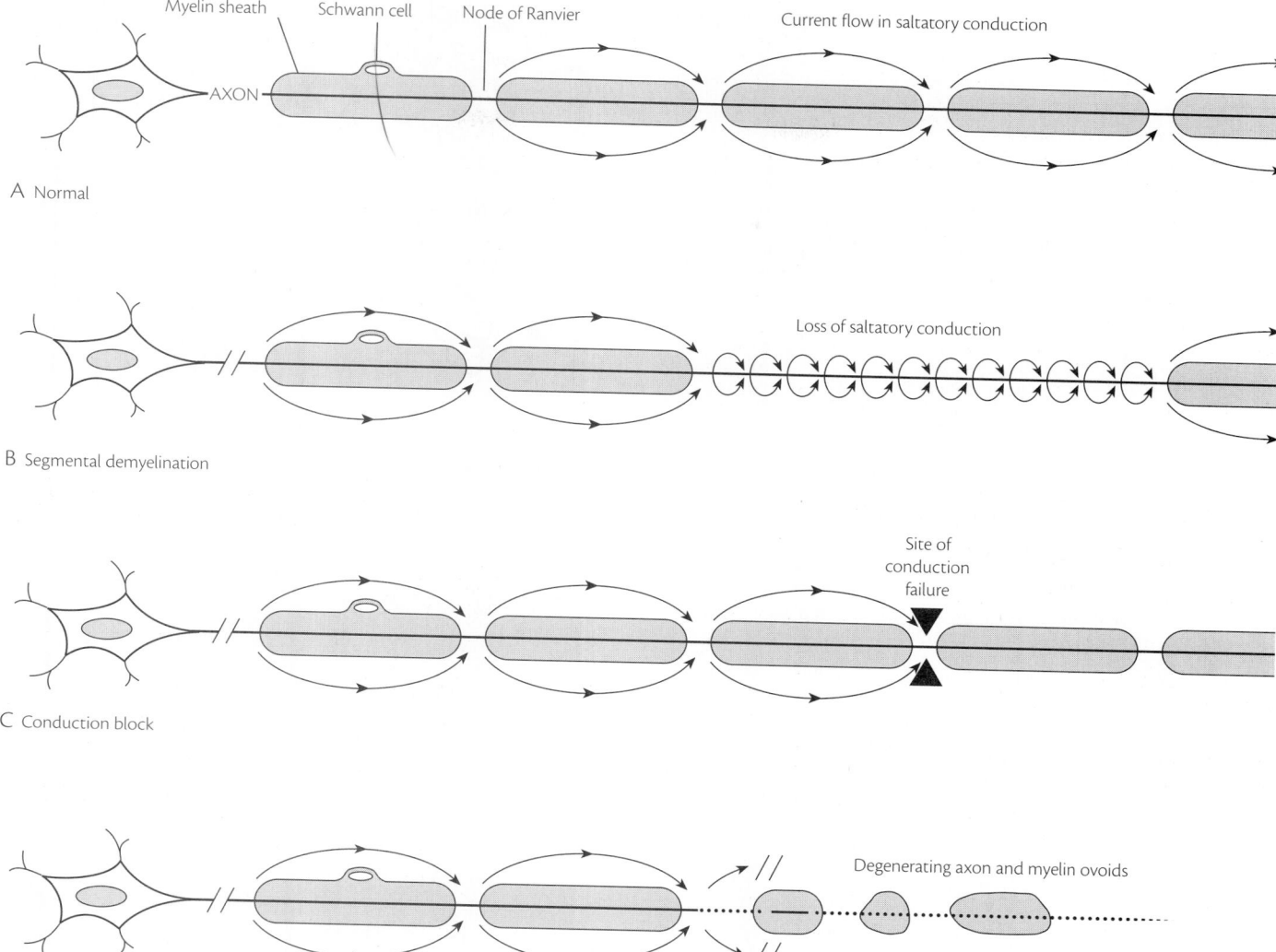

A Normal

B Segmental demyelination

C Conduction block

D Axonal degeneration

Fig. 21.1 Schematic illustration of axonal impulse conduction in A. normal axon, B. segmental demyelination after repopulation of the demyelinated segment of axon with sodium channels, C. conduction block due to a local blocking factor or acute segmental demyelination, and D. Wallerian axonal degeneration.

position and to preclude movement which was thought to promote contracture of antagonists. However splinting generally has little to commend it, as immobilized muscles tend to atrophy and become fibrotic more quickly.

21.3.2 Assessing recovery

The advent of effective therapies for disabling neuropathies brings the need for reliable assessment of recovery. This is useful both for formal clinical trials of treatment and for assessing effectiveness of a trial of treatment in an individual patient. Estimates of functional ability, such as walking, are more reliable and relevant than quantification of physical signs, such as strength of individual muscles. When assessing an individual's response to treatment, a set of measurements relevant to that patient's disability should be drawn up and measured pre- and post-treatment, paying particular attention to those which signify improvement of use in everyday life.

Various quantitative measures of the neurological examination can be undertaken. Muscle strength can be graded using the MRC scale:

Grade 0 = no contraction
Grade 1 = flicker of contraction
Grade 2 = active movement with gravity eliminated
Grade 3 = active movement against gravity
Grade 4 = active movement against gravity and resistance, and
Grade 5 = normal power.

However this MRC grading is relatively unreliable and unreproducible for the majority of neuropathies encountered in a civilian and non-surgical practice in which most weakness is either Grade 3 or 4. Nerve conduction studies can be quantified, particularly the velocity of motor conduction, the degree of block of motor conduction, or the amplitude of sensory nerve action potentials. However,

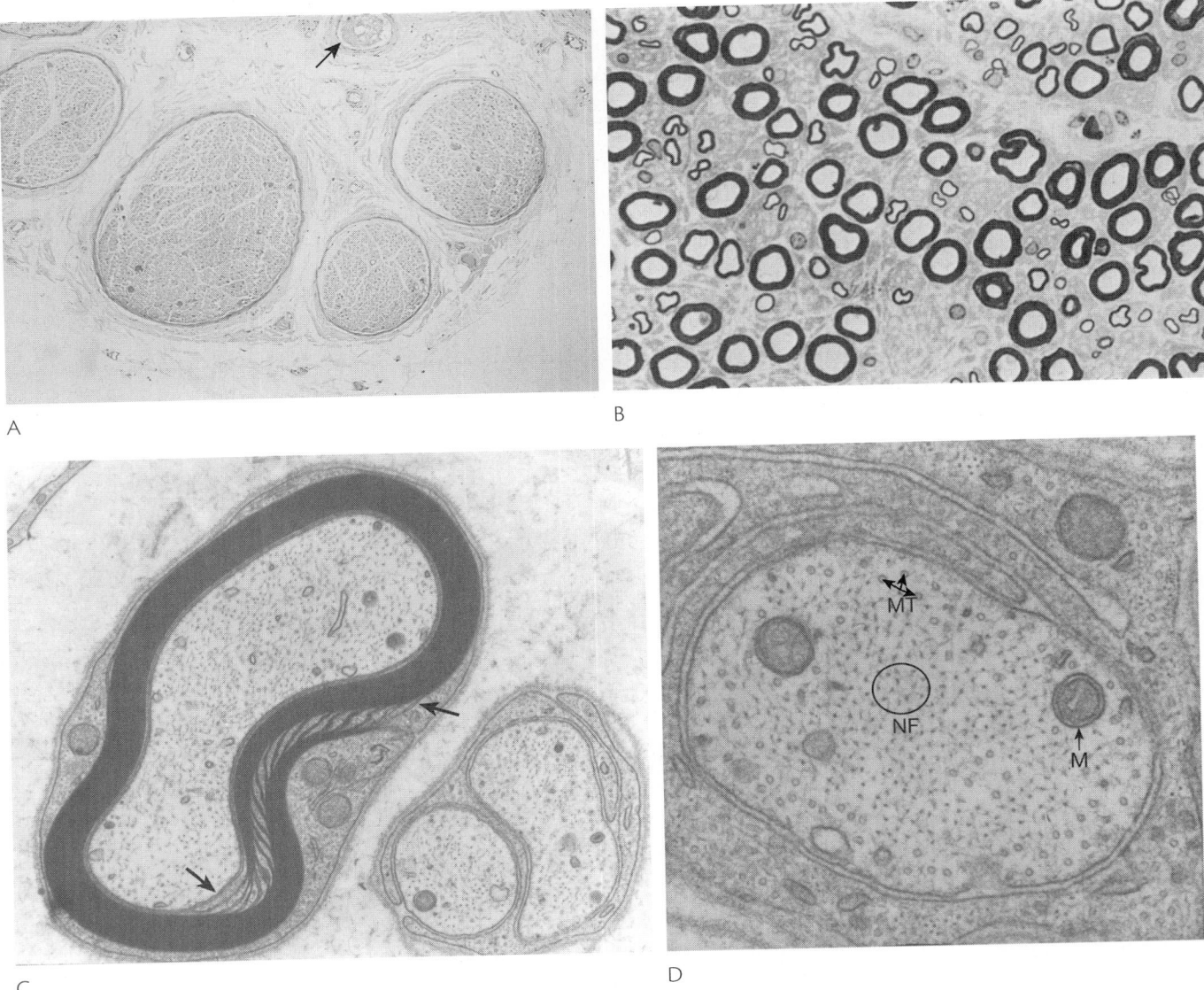

Fig. 21.2 Transverse sections of a peripheral nerve A. low power section of a complete nerve showing the fascicular organization and epineurial blood vessels (arrowed), (light microscopy), B. higher power showing the myelinated axons and endoneurial cells (light microscopy, semi-thin section, toluidine blue stain), C. electron microscopy of a myelinated fibre from a 3-month-old infant showing a Schmidt–Lanterman incisure in the myelin sheath (arrowed), D. electron microscopy of an unmyelinated nerve fibre showing the intra-axonal microtubules (MT), neurofilaments (NF), and mitochondria (M). (A. courtesy of Professor M Esiri, B. from Jacobs and Love 1985, C. and D. courtesy of Dr. R. King.)

such neurophysiological measures often show surprisingly little improvement even with clear-cut clinical improvement in conditions such as demyelinating neuropathy. Quantitative sensory testing devices have been developed, particularly for vibration and thermal sensation. Various clinical and electrophysiological parameters are integrated in the Total Neuropathy Score, shown in Table 21.5 (Cornblath *et al.* 1999).

In many circumstances a restricted set of measures can provide reliable evidence of improvement. For instance regaining the ability to heel-toe walk, stand on tiptoe, or perform Romberg's test provide clear-cut and reproducible evidence. Measurement of outstretched arm times, peg-sorting tasks, stair-climbing speed, or walking speed is also reliable unless you suspect that psychological factors may influence the patient's performance. It is useful for patients to monitor treatment against a variety of everyday tasks (Fig. 21.3).

Overall limb function can be quantified rapidly and reliably in neuropathies such as Guillain–Barré syndrome using a simple overall disability sum score, assessable by telephone if need be (Table 21.6) (Merkies *et al.* 2002). This is particularly valuable in monitoring recovery, response to treatment, or worsening in severity.

21.3.3 Immunomodulation

Immunomodulatory treatments are used successfully in a range of acquired neurological diseases including: acute idiopathic polyneuritis, chronic idiopathic polyneuropathies, neuropathies associated with lymphoproliferative disorders, vasculitic neuropathy, central nervous system vasculitic and collagen vascular disorders, multiple sclerosis, inflammatory myopathy, myasthenias, and cranial arteritis. Whilst detailed discussion of these immunomodulatory

Table 21.5 Total neuropathy score (reproduced from Cornblath *et al.* 1999, with permission)

Parameter	Score 0	1	2	3	4
Sensory symptoms	None	Symptoms limited to fingers or toes	Symptoms extend to ankle or wrist	Symptoms extend to knee or elbow	Symptoms above knees or elbows, or functionally disabling
Motor symptoms	None	Slight difficulty	Moderate difficulty	Require help/assistance	Paralysis
Autonomic symptoms, n	0	1	2	3	4 or 5
Pin sensibility	Normal	Reduced in fingers/toes	Reduced up to wrist/ankle	Reduced up to elbow/knee	Reduced to above elbow/knee
Vibration sensibility	Normal	Reduced in fingers/toes	Reduced up to wrist/ankle	Reduced up to elbow/knee	Reduced to above elbove/knee
Strength	Normal	Mild weakness	Moderate weakness	Severe weakness	Paralysis
Tendon reflexes	Normal	Ankle reflex reduced	Ankle reflex absent	Ankle reflex absent, others reduced	All reflexes absent
Vibration sensation (QST vibration)	Normal to 125% ULN	126–150% ULN	151–200% ULN	201–300% ULN	>300% ULN
Sural amplitude	Normal/reduced to <5% LLN	76–95% of LLN	51–75% of LLN	26–50% of LLN	0–25% of LLN
Peroneal amplitude	Normal/reduced to <5% LLN	76–95% of LLN	51–75% of LLN	26–50% of LLN	0–25% of LLN

QST = quantitative sensory test; ULN = limit of normal; LLN = lower limit of normal. Amplitudes refer to nerve conduction measurements.

treatments and their side effects lies outside the scope of this book, knowledge of important general principles is necessary for managing these diseases.

Steroids are given as oral prednisolone for chronic maintenance treatment, intravenous methylprednisolone for acute disorders, or dexamethasone for raised intracranial pressure. Consideration should be given to alternate day administration of prednisolone in neuromuscular disorders requiring long-term treatment. Increased susceptibility to infections, including opportunistic organisms, is a significant risk, particularly with long-term therapy. Steroid-induced diabetes mellitus, or exacerbation of previously controlled diabetes, may occur. Steroid myopathy may occur with long-term

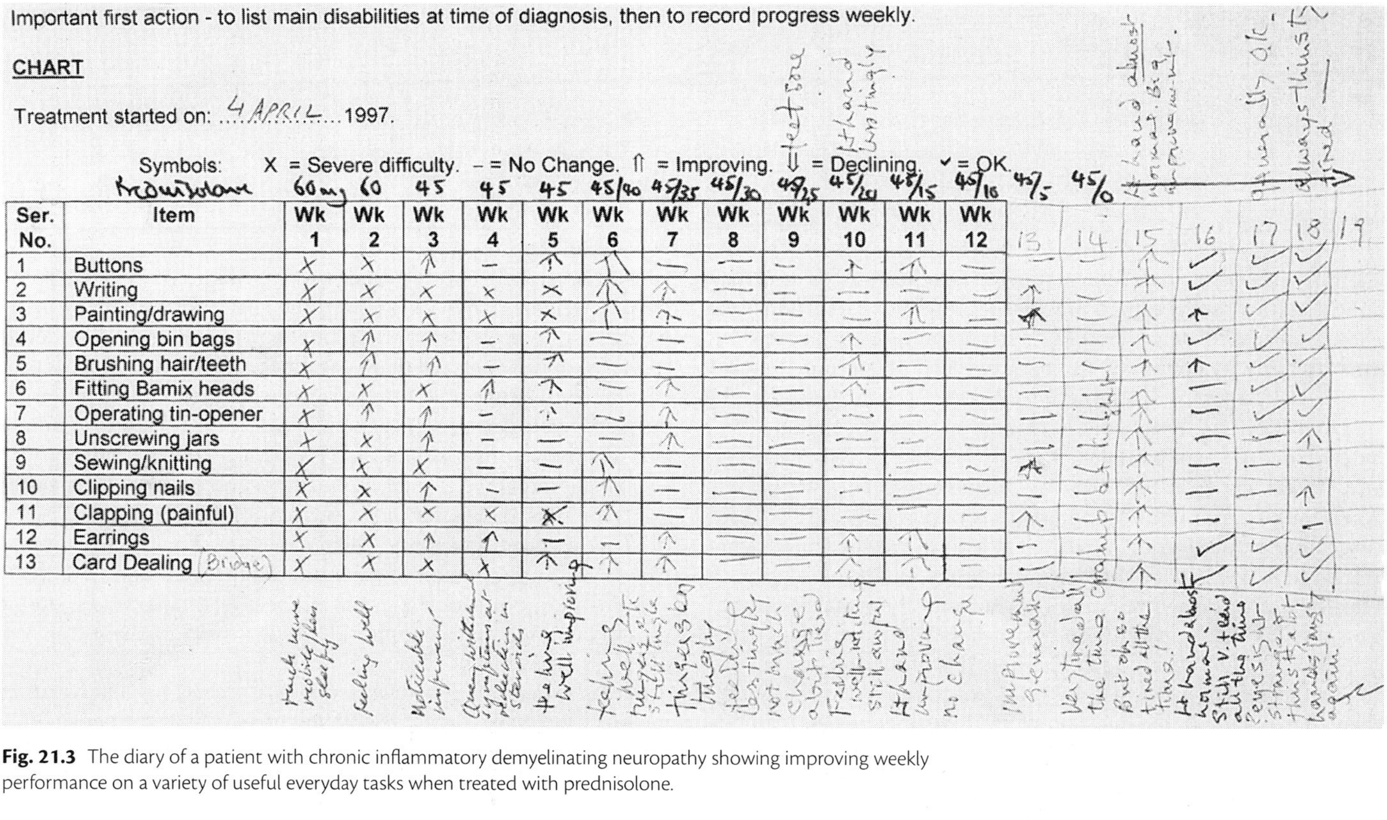

Fig. 21.3 The diary of a patient with chronic inflammatory demyelinating neuropathy showing improving weekly performance on a variety of useful everyday tasks when treated with prednisolone.

Table 21.6 The overall disability sum score (ODSS) (from Merkies *et al.* 2002)

Arm disability scale — function checklist	Not affected	Affected but not prevented	Prevented
Dressing upper part of body (excluding buttons/zips)	O	O	O
Washing and brushing hair	O	O	O
Turning a key in a lock	O	O	O
Using knife and fork (/spoon—applicable if the patient never uses knife and fork)	O	O	O
Doing/undoing buttons and zips	O	O	O

Arm grade

0 = Normal

1 = Minor symptoms or signs in one or both arms but not affecting any of the functions listed

2 = Moderate symptoms or signs in one or both arms affecting not preventing any of the functions listed

3 = Severe symptoms or signs in one or both arms preventing at least one but not all functions listed

4 = Severe symptoms or signs in both arms preventing all functions listed but some purposeful movements still possible

5 = Severe symptoms and signs in both arms preventing all purposeful movements

Leg disability scale—function checklist	No	Yes	Not applicable
Do you have any problem with your walking?	O	O	O
Do you use a walking aid?	O	O	O
How do you usually get around for about 10 metres?			
Without aid	O	O	O
With one stick or crutch or holding to someone's arm	O	O	O
With two sticks or crutches or one stick or crutch and holding to someone's arm	O	O	O
With a wheelchair	O	O	O
If you use a wheelchair, can you stand and walk a few steps with helps?	O	O	O
If you are restricted to bed most of the time, are you able to make some purposeful movements?	O	O	O

Leg grade

0 = Walking is not affected

1 = Walking is affected but does not look abnormal

2 = Walks independently but gait looks abnormal

3 = Usually uses unilateral support to walk 10 metres (25 feet) (stick, single crutch, one arm)

4 = Usually uses bilateral support to walk 10 metres (25 feet) (sticks, crutches, two arms)

5 = Usually uses wheelchair to travel 10 metres (25 feet)

6 = Restricted to wheelchair, unable to stand and walk few steps with help but able to make some purposeful leg movements

7 = Restricted to wheelchair or bed most of the day, preventing all purposeful movements of the legs (e.g. unable to reposition legs in bed)

Overall disability sum score = arm disability scale (range 0-5) + leg disability scale (range 0-7); overall range: 0 (no signs of disability) to 12 (maximum disability).

For the arm disability scale: allocate one arm grade only by completing the function checklist. Indicate whether each function is 'affected', 'affected but not prevented', or 'prevented'.

For the leg disability scale: allocate one leg grade only by completing the functional questions.

therapy, especially with fluorinated steroids such as dexamethasone. Cataract can be caused or worsened. Steroids can produce, or exacerbate, psychiatric conditions such as paranoia or depression. To offset the osteoporosis induced by chronic steroid therapy, prophylactic therapy with a biphosphorate or a calcitriol should be given from the outset, and hormone replacement therapy considered in post-menopausal women.

Azathioprine is often used as a steroid-sparing treatment when long-term immunosuppression is required. It too increases susceptibility to infection, mediated at least in part by leucopoenia related to dose-related bone marrow suppression. Hypersensitivity reactions, often involving abdominal pain and abnormal liver function tests, are not uncommon and necessitate permanent withdrawal. Regular full blood count and liver function testing are required; weekly for the first 4–8 weeks of therapy, and 3-monthly thereafter. Although the question of azathioprine causing lymphoproliferative disorders has been raised, especially in the setting of renal transplantation, there is no evidence of this complication when it is used in neurological disorders (Amato *et al.* 1993). Some female patients may wish to continue azathioprine during pregnancy so as to avoid deterioration in their neurological disorder. If so they should be advised that whilst there are occasional reports of chromosomal abnormalities and neonatal haematological disorders, the teratogenic risk is generally considered small to minimal and that the vast majority of such pregnancies end happily.

Cyclophosphamide is used in vasculitis and may be given either orally or pulsed intravenously. Susceptibility to infection is the chief early side effect and the full blood count should be monitored closely for dose-related bone marrow suppression. Simultaneous administration of Mesna helps avoid haemorrhagic cystitis, and it should be noted that cyclophosphamide greatly increases the risk of future bladder cancer.

Plasma exchange is generally carried out by centrifugal or filtration methods, generally for 5 days exchanging 50 ml plasma per kg body weight at each exchange. The rationale is removal of pathogenic antibodies in the plasma fraction, and replacement is usually with human albumin or gelatin solutions. In experienced hands, the technique is largely free of complications. Low-dose heparinization is used to prevent thrombosis and embolization from the indwelling venous catheter.

Intravenous immunoglobulin, IvIg, therapy is generally given in total doses of 2 g/kg body weight over between 2 and 5 days. Repeated administration every 6–10 weeks is required in the treatment of chronic neuropathies. Domiciliary administration may use different régimens, provides effective maintenance administration without hospitalization, and avoids loss of time from work or fluctuations in disease severity (Sewell *et al.* 1997). The precise mechanisms of IvIg's action in neurological disease remain unknown; immunomodulatory effects, anti-idiotypic antibodies, cytokine alterations, and direct effects on conduction block or remyelination are all possibilities (Stangel *et al.* 1999). Significant side effects are unusual, although up to 30 per cent may experience mild, self-limited reactions of headache, myalgia, fever, rash, or vasomotor reactions, which are generally controllable by varying the infusion rate, or by using antihistamines (Stangel *et al.* 2003). Such side effects seem commoner in IvIg-naïve patients. IgA-deficient patients may develop anaphylactic reactions. Rarely self-limited aseptic meningitis similar to that with OKT3 monoclonal antibodies, viscosity induced thromboembolic events, acute oliguric renal failure, or haemolytic anaemia have occurred. Naturally patients will be concerned about possible transmission of infections by IvIg. The solvent detergent step currently employed in purification inactivates HIV and hepatitis viruses. There is a theoretical risk of transmitting prion disease.

21.4 Charcot–Marie–Tooth disease

21.4.1 Range of disorders

Charcot–Marie–Tooth disease is the commonest cause of the peroneal muscular atrophy syndrome of distal leg muscle wasting and weakness, usually accompanied by pes cavus foot deformity (Figs 21.4 and 21.5). It is also known as peroneal muscular atrophy; hereditary motor and sensory neuropathy; hereditary hypertrophic

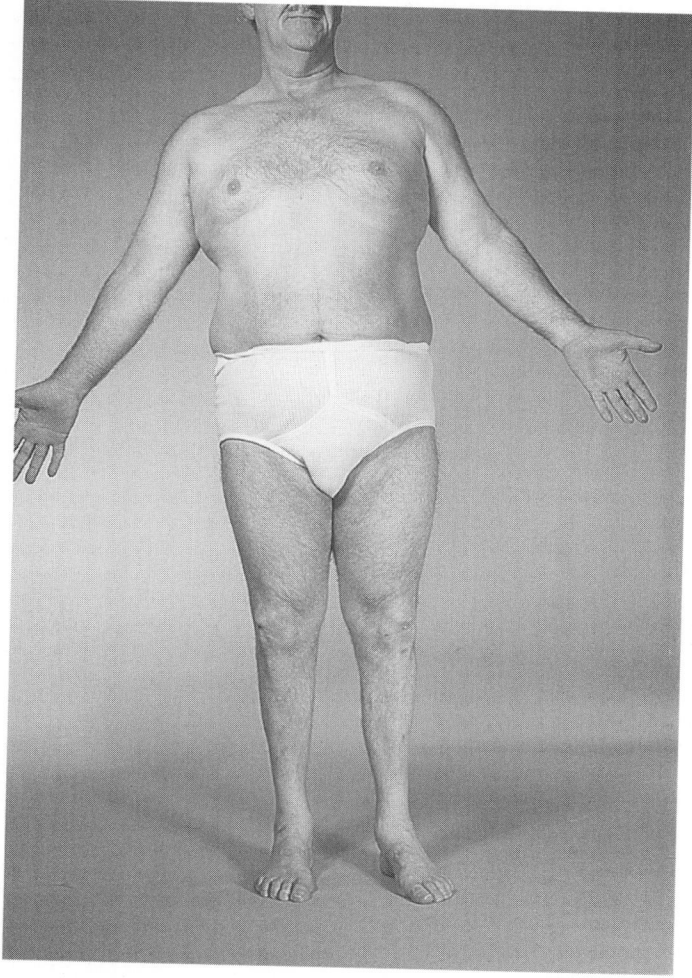

Fig. 21.5 Peroneal muscular atrophy in Charcot–Marie–Tooth disease showing the typical 'inverted champagne bottle' appearance of the legs due to muscle wasting below the knees.

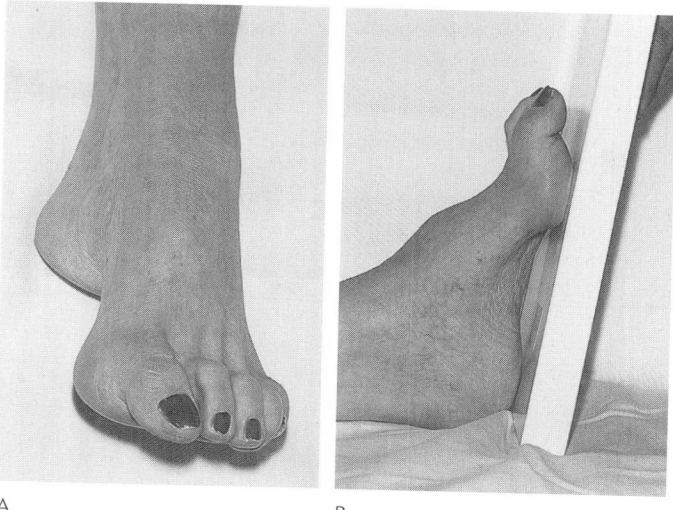

A   B

Fig. 21.4 Pes cavus in Charcot–Marie–Tooth disease. A. The hammer toe deformity, B. a defining feature of pes cavus is the ability to see daylight through the foot arch when the sole is placed against a flat surface.

neuropathy; Roussy–Lévy syndrome; Dejerine–Sottas disease. Charcot–Marie–Tooth death comprises a range of demyelinating and axonal loss neuropathies with various patterns of inheritance associated with numerous gene mutations.

Patients generally present in childhood or adolescence, but symptoms may become evident at any age from birth to senescence. Asymptomatic, yet affected, elderly relatives may be identified. The presenting symptoms are usually difficulty in walking or foot deformity. Positive sensory symptoms such as paraesthesiae make the diagnosis of Charcot–Marie–Tooth disease unlikely and should suggest an acquired neuropathy. Subacute deterioration in Charcot–Marie–Tooth disease, resembling superimposed chronic inflammatory demyelinating neuropathy, has occurred occasionally in association with MPZ and Cx32 mutations (Donaghy *et al.* 2000; Watanabe *et al.* 2002). Such patients may be steroid responsive and probably explain earlier reports of steroid-responsive inherited neuropathy. Occasionally Charcot–Marie–Tooth disease is associated with other neurological features such as spastic paraparesis, optic atrophy, pigmentary retinal degeneration, deafness, the dysmorphic Noonan syndrome, or the mental retardation

Table 21.7 Classification of Charcot–Marie–Tooth disease

Disease	Linkage	Gene	Typical features	Onset
1. DEMYELINATING (MCV <38 m/s) **Autosomal dominant (CMT1 or HMSN1)**				
CMT1A	17p11.2-12	Duplication PMP22 (rarely point mutation)	Typical CMT	Childhood- typically in 2nd decade
CMT1B	1q22-1q23	Point mutation P_o	Often more severe than CMT1A	Congenital to 2nd decade
CMT1C	16p13.1-p12.3	*LITAF/SIMPLE*	Typical CMT1	Childhood to 2nd decade
CMT1D	10q21-q22	Point mutation EGR2	Often more severe than CMT1A	Congenital or 1st decade
X-linked				
CMT1X	Xq13	Point mutation Cx32		
Severe dominant or recessive phenotypes: Dejerine-Sottas disease (DSD or HMSN-III)				
DSD-A (AD or AR)	17p11.2-12	Point mutation PMP22		
DSD-B (AD or AR)	1q22-q23	Point mutation P_o		
DSD-C (AD or AR)	10q21-q22	Point mutation EGR2		
DSD-D (AD)	8q23-q24			
Congenital hypomyelinating neuropathy (CHN)				
CHN-A (AD)	17p11.2-2-12	Point mutation PMP22		
CHN-B (AD)	1q22-q23	Point mutation P_o		
CHN-C (AD or AR)	10q21-22	Point mutation EGR2	Severe neuropathy, hypomyelination	
Hereditary neuropathy with liability to pressure palsies (HNPP)				
HNPP-A (AD)	17p11.2-12	Deletion PMP22		
Autosomal recessive (CMT4)				
CMT4A	8q13-21.1	*GDAP1*	Severe neuropathy. Vocal cord and diaphragm paralysis (some).	1st decade
CMT4B-1	11q22	*MTMR2*	Focally folded myelin. Severe. Facial and bulbar involvement.	Early childhood (~34 months)
CMT4B-2	11p15	*MTMR13*	Focally folded myelin, Tunisian families. Glaucoma.	1st or 2nd decade
CMT4C	5q23-q33	KIAA 1985	Resembles CMT1-, +scoliosis	1st or 2nd decade
CMT4D/HMSN-Lom	8q24	*NDRG1*	Progressive deafness, Bulgarian Gypsies (type Lom). Tongue atrophy.	1st decade
CCFDN	18q	CTDP1	Congenital cataracts, facial dysmorphism and neuropathy. Gypsies.	1st or 2nd decade (neuropathy)
CMT4F	19q13.1-13.3	Periaxin	Severe neuropathy. More sensory. Folded myelin.	Early childhood (12–24 months)
HMSN-Russe	10q22-q23		Severe form, Bulgarian Gypsies (type Russe)	1st or 2nd decade

Table 21.7 (Continued)

Disease	Linkage	Gene	Typical features	Onset
2. AXONAL (MCV ≥38 m/s)				
Autosomal dominant				
(CMT2 or HMSN-II)				
CMT2A	1p35-p36	KIF 1Bβ	Classical CMT2	
CMT2A	1p36.22	MFN2	Classical CMT2. More progressive. Optic atrophy.	Adult (mean ~20 years) 2nd decade; some asymptomatic 2nd or 3rd decade
CMT2B	3q13-q22	RAB7	Mainly sensory neuropathy, acral ulcerations (like HSAN)	
CMT2C	12q23-q24		Vocal cord and respiratory involvement	Adult
CMT2D	7p15	GARS	Predominant upper limbs and motor	2nd or 3rd decade
CMT2E	8p21	NF-L	Typical CMT (can resemble DI-CMT)	Childhood to 3rd decade
CMT2F	7q11	HSP27	Classical CMT + trophic changes	2nd or 3rd decade
CMT2G	12q12-q13.3		Classical CMT2	
CMT2L	12q24.23	HSP22	Classical CMT2	
CMT2	1q22-q23	Point mutation P₀	Hearing loss, pupillary dysfunction	4th or 5th decade
CMT2 (HMSNP)	3q13.1		CMT2. Proximal involvement.	
Autosomal recessive (AR-CMT2)				
AR-CMT2A	1q21.2-21.3	LMNA	Proximal. Rapid progression. Muscular dystrophy. Lipodystrophy. Cardiomyopathy.	
AR-CMT2B	19q13.3		Typical CMT2.	
AR-CMT2	8q21	GDAP1	Early onset. Vocal cord, diaphragm paralysis.	
X-linked (CMT2X)				
CMT2X	Xq24-q26		Deafness. Mental retardation.	
3. DOMINANT-INTERMEDIATE (DI-CMT)				
DI-CMTA	10q24.1-q25.1		Typical CMT, axonal and myelin pathology	1st decade
DI-CMTB	19p12-p13	DNM2	Typical CMT, axonal and myelin pathology	1st decade
DI-CMTC	1p34-p35	YARS	Typical CMT	

AD – autosomal dominant; **AR** – autosomal recessive; **CCFDN** – congenital cataracts, facial dysmorphism and neuropathy; **CHN** – congenital hypomyelinating neuropathy; **CMT** – Charcot-Marie-Tooth disease; **Cx32** – connexion 32; **CTDP1** – CTD phosphatase 1; **DNM21** – dynamin 2; **DSD** – Dejerine-Sottas disease; **EGR** – early growth response 2; **GARS** – glycyl-tRNA synthetase; **GDAP1** – ganglioside-induced differentiation associated protein 1; **HNPP** – hereditary neuropathy with liability to pressure palsy; **HMSN** – hereditary motor and sensory neuropathy; **HSAN** – Hereditary Sensory and Autonomic Neuropathy; **HSP** – small heat shock protein 22 or 27; **KIF 1Bβ** - microtubule motor KIF 1Bβ; **LITAF/SIMPLE** – lipopolysaccaride-induced tumour necrosis factor-α factor/for small integral membrane protein of the lysosome/late endosome; **LMNA** – lamin A/C; **MCV** – median nerve motor conduction velocity; **MTMR** – myotubularin-related protein; **NF-L** – neurofilament light gene; **NDRG1** – N-myc downstream regulated gene 1; **PMP22** – peripheral myelin protein 22; **P₀** – myelin protein zero; **RAB7** – GTP-ase late endosomal protein gene; **SBF2** – SET binding factor 2; **YARS** – tyrosyl-tRNA synthetase.
Hereditary neuropathies are classified by MIM (http://www.ncbi.nlm.Nib.gov/Omim/). (Adapted and updated from Reilly and Hanna 2002.)

associated with agenesis of the corpus callosum, also known as Andermann syndrome (Dupre *et al.* 2003).

Charcot–Marie–Tooth disease can be classified into Types 1–4 on clinical, electrophysiological, and genetic grounds. Many of the underlying molecular genetic abnormalities have been identified. The current classification consists of the autosomal dominant or X-linked *Type 1* demyelinating and *Type 2* axonal, with further subclassification based on inheritance pattern and gene identification. Currently, this classification mainly finds clinical utility in the electrophysiological differentiation of Types 1, demyelinating and 2, axonal forms of the disorder using a cut-off motor conduction velocity in the median nerve ≤ 38 m/s. However there are increasingly extensive subclassifications and overlap of these two types, often correlating with newly described mutations or linkage loci. *Type 3* refers to congenital hypomyelination neuropathy. Autosomal recessive forms are extremely rare with demyelinating forms being classified as Charcot–Marie–Tooth disease *Type 4*, and axonal forms simply described as autosomal recessive. Dominant intermediate

forms of Charcot–Marie–Tooth disease are unusual and defined by a range of motor nerve conduction velocities within the family that overlap those of Types 1 and 2.

Four gene abnormalities account for about two-thirds of all patients with Charcot–Marie–Tooth disease (Boerkoel *et al.* 2002). Over half of them have a reduplication of the gene for peripheral myelin protein 22, PMP22, the CMT1A duplication which produces the Type 1 demyelinating form. Five to ten per cent carry X-linked Connexin-32, Cx32 mutations which variably produce Type 2 axonal or milder degrees of Type 1 demyelinating neuropathy (Hattori *et al.* 2003). Another 5 per cent are due to various mutations of the myelin protein P₀ gene, MPZ, or the *PMP22* gene producing a mixture of Type 1 demyelinating and Type 2 axonal phenotypes.

Charcot–Marie–Tooth disease should be distinguished from the other common cause of the peroneal muscular atrophy syndrome: spinal muscular atrophy (Section 23.33). Such patients are less likely to have upper limb weakness, their tendon reflexes are relatively preserved, the sensory examination is normal, and the sensory nerve action potentials are normal (Harding and Thomas 1980a). Symptoms of distal spinal muscular atrophy usually start in childhood, although onset as late as 60 years has been recorded. Both autosomal dominant and recessive forms occur, the former tending to present earlier in life. The condition rarely produces severe disability.

21.4.2 Management

There is no drug therapy to stabilize or improve Charcot–Marie–Tooth disease, save in those occasional patients with superimposed chronic inflammatory demyelinating neuropathy who respond to immunomodulatory therapy. Footdrop may be overcome by ankle orthoses and physiotherapy may reduce tendon contractures, foot deformity, or scoliosis. There is a limited role for orthopaedic surgical correction of severe foot and ankle deformity. Genetic counselling should be undertaken. Prenatal diagnosis is possible in principle, and may have a role in offering prenatal diagnosis in potentially lethal or severely disabling infantile or childhood forms. However, in the vast majority of Charcot–Marie–Tooth disease patients, the disorder is only slowly progressive or stable, without ever causing overwhelming disability, and generally without significantly affecting life expectancy. Maternal Charcot–Marie–Tooth disease merits special obstetric attention since it carried increased risks of foetal presentation abnormalities, postpartum bleeding, and emergency caesarean section (Hoff et al. 2005).

21.4.3 Molecular genetic abnormalities

In total 37 identified genes or loci have been identified so far as responsible for the different forms of Charcot–Marie–Tooth disease (Table 21.7). Of these, 22 reflect demyelinating forms, 13 axonal, and 2 intermediate, with a range of autosomal dominant and recessive and sex-linked recessive transmission. The genetic causes of the severe childhood and congenital hypomyelinating forms are emerging. Confusingly, different mutations affecting the same gene can produce demyelinating, intermediate, or axonal forms of Charcot–Marie–Tooth disease—examples being Type 1B due to myelin protein P0 mutations, and GDAP1 mutations (Kuhlenbaumer et al. 2002; Reilly and Hanna 2002; Berciano and Combarros 2003). Some genetic abnormalities are used routinely for diagnostic testing: the 17p11.2 duplication, and mutations of the MPZ and Cx32 genes. Distinctive clinical and electrophysiological syndromes generally characterize the commoner mutations of these particular genes.

PMP-22. Seventy per cent of Charcot–Marie–Tooth disease Type 1, known as Type 1A, is associated with a duplication of the gene for peripheral myelin protein 22, *PMP22*, on chromosome 17. This occurs due to unequal crossover during meiosis, particularly on the father's side. This same myelin protein gene is affected by a mutation in the hypomyelinating *Trembler* mouse mutant (Timmerman et al. 1992) and is deleted in hereditary liability to pressure palsies (Section 21.5), a condition of hypomyelination (Lenssen et al. 1998). Thus increased copies of the *PMP22* gene are associated with decreased growth of the myelin spiral. Unsurprisingly the Charcot–Marie–Tooth disease Type 1A phenotype occurs in trisomy 17, Down's syndrome (Chance et al. 1992).

Myelin protein P0. This gene is an adhesive protein responsible for compaction of the myelin sheath. Abnormally wide spacing of the myelin lamellae results from some of its mutations (Gabreels-Festen et al. 1996). More than 80 point mutations of the MPZ cause a bewildering array of Charcot–Marie–Tooth disease Type 1B, demyelinating, Type 2, axonal, intermediate forms, congenital hypomyelinating and childhood onset forms, and a late onset form

of progressive demyelinating neuropathy with features resembling chronic inflammatory demyelinating polyneuropathy (Donaghy et al. 2000; Shy et al. 2004).

Connexin 32. X-linked forms of HMSN I are associated with more than 150 point mutations in the Connexin-32, Cx32, gene, which encodes the gap junctions occurring at the Schmidt–Lanterman incisures and the paranodal regions of the myelin sheath. These Cx32 mutations may interfere with rapid transport of small molecules between different myelin lamellae.

Many of the myriad genetic abnormalities underlying Charcot–Marie–Tooth disease affect functions likely to be fundamental to biology of the cells of many different tissues, rather than those specifically restricted to peripheral neurones or Schwann cells.

The demyelinating Type 1 phenotype of Charcot–Marie–Tooth disease can be associated with abnormalities of the various genes for the growth arrest protein peripheral myelin protein 22, PMP-22; early growth response element 2, EGR-2; ganglioside induced differentiation associated-protein-1, GDAP-1; myotubularin-related-proteins-2 and -13, MTMR-2; N-MYC-downstream-regulated-gene-1, NDRG-1; epithelial-growth-factor-related-protein-2, EGR-2; periaxin, PRX; a putative protein degradation gene LITAF/SIMPLE; with a novel SH-3/TPR domain protein; and the neurofilament light chain, NF-L (Warner et al. 1998; Kalaydjieva et al. 2000; Houlden et al. 2001a; Nelis et al. 2002; Takashima et al. 2002; Jordanova et al. 2003; Senderek et al. 2003; Street et al. 2003).

The inability to develop or maintain an axon in the axonal Type 2 form of Charcot–Marie–Tooth is founded in straightforward neurobiological logic when due to mutations in the genes for kinesin-motor-protein-1-B, K1F1B-beta, mitofusin 2, or neurofilament light chain, NFL (Fabrizi et al. 2007). These affect axonal transport of motor proteins and axonal structural intermediate filament proteins respectively (Zhao et al., 2001; Jordanova et al. 2003). It is less clear why the axonal Type 2 forms occur as a result of mutations in the small heat shock protein 27 gene, in the small GTP-ase late endosomal RAB-7 gene, or the lamin A/C gene, other mutations of which produce the Emery Dreyfuss muscular dystrophies, limb girdle muscular dystrophies, cardiomyopathy, and partial lipodystrophy (Chaouch et al. 2003; Tazir et al. 2004; Tang et al. 2005; Meggouh et al. 2006). It is intriguing to discover that broadly similar Charcot–Marie–Tooth phenotypes result from such an eclectic array of disturbances of cell biology.

Associated clinical findings occur in a small proportion of patients with Charcot–Marie–Tooth disease, and provide a clue as to the underlying genetic cause (Table 21.8).

21.4.4 Type 1: Demyelinating

This demyelinating neuropathy is the commonest form of Charcot–Marie–Tooth disease. It is also known as hereditary motor and sensory neuropathy Type 1, HMSN1. Motor nerve conduction in the median nerve is substantially slowed and nerve biopsy shows segmental demyelination, usually accompanied by hypertrophic 'onion-bulb' changes (Fig. 21.6). These concentric layers of Schwann-cell proliferation around axons represent previous cycles of recurrent demyelination followed by attempted remyelination. The inheritance is usually autosomal dominant with a high degree of intrafamilial concordance for the velocity of motor slowing in keeping with the underlying genetic heterogeneity. Seventy per cent of patients with the autosomal dominant form have reduplication of the *PMP22* gene, Type 1A (Thomas et al. 1997) and many of the remainder have point mutations of myelin protein P0, Type 1B (De Jonghe et al.

Table 21.8 Additional clinical abnormalities in Charcot—Marie—Tooth disease

Clinical abnormality	Type of Charcot–Marie–Tooth disease	Mutated gene
Prominent sensory loss/ sensory ataxia	CMT4F	PRX
	CMT2	P_0, RAB7
Late onset, pupillary anomalies and hearing loss	CMT2	P_0
Hearing loss	CMT1A	PMP22
	CMT4D	P_0, GJB1, NDRG1
	CMT4C	KIAA1985
Diplopia	CMT1D	EGR2
Vocal cord paresis	CMT2C	12q23-24
	AR-CMT2, CMT4A	GDAP1
Optic atrophy	CMT2A	MFN2
Upper limb predominance	CMT2D	GARS
CNS involvement: transient symptoms— ataxia, dysarthria, weakness with white matter MRI lesions	CMTX	GJB1
Agenesis of the corpus callosum and developmental delay	CMT/ACC	SLC12A6
Tremor		PMP22, P_0, GJB1
Skeletal deformities—severe scoliosis	CMT4C	KIAA1985
Glaucoma	CMT4B2	MTMR13/SBF2
Neutropenia	DI-CMT	DNM2

For key see Table 21.7 (Kindly provided by Dr. Yesim Parman, Istanbul Faculty of Medicine)

1999). Sex-linked recessive forms, sometimes mildly expressed in female carriers, are usually associated with Cx32 mutations, and may progressively deteriorate involving permanent axonal degeneration (Birouk *et al.* 1998). Autosomal recessive forms, Type 4, occur occasionally and generally produce more severe disability with even lower motor conduction velocities (Harding and Thomas 1980c).

Clinical features. Distal leg muscle atrophy and weakness, or pes cavus foot deformity (Fig. 21.4) are usually evident in the first or second decades of life. Leg weakness can become severe. Distal wasting may be such that the legs resemble inverted champagne bottles (Fig. 21.5). The ankle jerks are usually lost, but generalized areflexia occurs in only half the patients. Some hand weakness eventually occurs in most patients with Charcot–Marie–Tooth disease Type 1. A few develop tremor or ataxia of the limbs, the so-called 'Roussy–Lévy syndrome' which is not genetically distinct from Charcot–Marie–Tooth disease Type 1. All modalities of sensation may be impaired distally in the limbs. Occasional patients develop acrodystrophic changes secondary to severe sensory loss. Scoliosis, pupil abnormalities, or extensor plantar responses occasionally occur. Diaphragmatic weakness may cause dyspnoea or respiratory failure. Palpable nerve thickening, best detected at the great auricular nerve (Fig. 21.7) is found in about a quarter of the cases and is specific to the demyelinating forms of Charcot–Marie–Tooth disease (Harding and Thomas 1980b). Neurophysiological

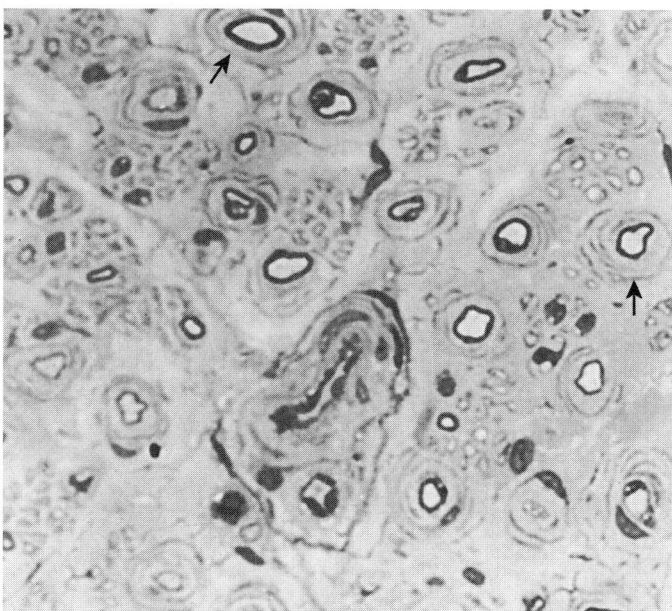

Fig. 21.6 Charcot–Marie–Tooth disease Type 1. Transverse section of a 1-mm araldite section of peripheral nerve stained with toluidine blue, showing multiple 'onion- bulbs' (arrows); sural nerve biopsy. (Courtesy of Dr. R. Madrid.)

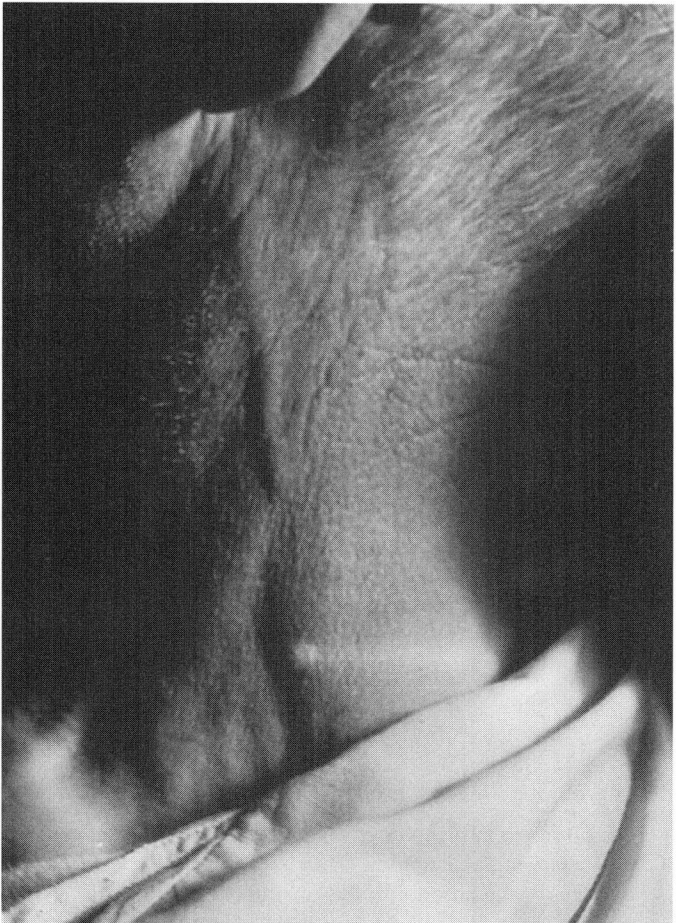

Fig. 21.7 Palpable enlargement of the greater auricular nerve in hereditary motor and sensory neuropathy Type 1. (Courtesy of the late Professor W.B. Matthews.)

examination of the patient and first-degree relatives is crucial for determining the genetic basis if screening for the usual molecular genetic mutations is negative.

Nerve conduction. The median nerve motor conduction velocity is 38 m/s or less, which distinguishes the condition from the neuronal form, Charcot–Marie–Tooth disease Type 2. Generalized motor slowing is already evident in infancy and early childhood (Garcia *et al.* 1998). Sensory nerve action potentials are absent or reduced in all patients, allowing distinction from distal spinal muscular atrophy (Harding and Thomas 1980b). Motor conduction velocities in Charcot–Marie–Tooth disease Type 1 rarely lie below 12 m/s, but if so, the diagnosis of Déjérine–Sottas disease or congenital hypomyelinating neuropathy should be considered (Ouvrier *et al.* 1987). Sural nerve biopsy shows hypertrophic onion-bulb changes (Fig. 21.6) and reduced myelinated fibre density. The spinalfluid protein is usually normal, but is occasionally elevated to 1 g/l or higher (Ouvrier *et al.* 1987).

Differential diagnosis. The most important differential diagnosis is from chronic inflammatory demyelinating polyneuropathy (Section 21.11.2) which is treatable. Chronic inflammatory demyelinating polyneuropathy too is associated with reduced motor conduction velocity and with segmental demyelination on nerve biopsy, and occasionally with onion-bulb changes. Pointers in favour of Charcot–Marie–Tooth disease Type 1 are pes cavus, palpable nerve thickening, preserved proximal limb muscle power, a spinal fluid protein of less than 0.8 g/l, a slow rate of progression of symptoms, a lack of positive sensory symptoms such as tingling, and onion-bulb changes on nerve biopsy. The diagnosis of Charcot–Marie–Tooth disease may be positively established by molecular genetic testing or by examining close relatives for signs of neuropathy, even if they are asymptomatic. Cases of Charcot–Marie–Tooth disease with ataxia, the Roussy–Lévy syndrome, should be distinguished from Refsum disease (Section 21.8.1) in which the serum phytanic acid is elevated, and from Friedreich's ataxia, in which nystagmus and extensor plantars are usual, inheritance is autosomal recessive, GAA trinucleotide expansion is present in intron 1 of the Frataxin gene is present and motor nerve conduction velocity is not markedly slowed.

21.4.5 **Type 2: Axonal**

Otherwise known as the neuronal form of hereditary motor and sensory neuropathy, Charcot–Marie–Tooth disease Type 2 reflects a reduction in the number of primary motor and sensory neurones. Type 2 is less common than Type 1. Motor nerve conduction velocities are higher than 38 m/s and inheritance is usually autosomal dominant (Harding and Thomas 1980b).

Syndromes. There are four clinical and genetically distinct syndromes of dominantly inherited Charcot–Marie–Tooth Type 2 although other variants are increasingly recognized:

- Type 2A, the commonest form, involves distal weakness and wasting with lesser degrees of sensory loss and areflexia starting in the second or third decade. It is a classical form of Charcot–Marie–Tooth disease.

- Type 2B shows a younger age of onset, often with foot ulceration and preserved ankle tendon reflexes, and may resemble hereditary sensory and autonomic neuropathy.

- Type 2C involves vocal cord or diaphragm paralysis in some members of the kinship.

- Type 2D characteristically produces more severe arm than leg muscle involvement.

A German kinship with Charcot–Marie–Tooth disease Type 2 has been described with sural nerve axonal swellings filled with neurofilaments (Vogel *et al.* 1985) but the dominant pattern of inheritance and their normal hair distinguish this family from giant axonal neuropathy (Section 21.7.1). Autosomal recessive forms are encountered occasionally, with an earlier onset of symptoms than the dominant form (Harding and Thomas 1980c). Rare X-linked dominant forms with childhood onset occur in which males are severely affected and females may suffer subclinical or mild disease; no male-to-male transmission occurs.

Clinical features. The clinical features of Charcot–Marie–Tooth disease Type 2 resemble those of Charcot–Marie–Tooth disease Type 1, but the onset of symptoms is generally later, usually in the second or third decade of life (Harding and Thomas 1980b). Onset can be delayed until old age. Patients most commonly present with difficulty in walking due to distal leg muscle weakness and wasting. Pes cavus foot deformity is less frequent than in Charcot–Marie–Tooth disease Type 1. The ankle jerks are usually absent. Hand weakness, tremor or ataxia of the arms, marked sensory loss, or generalized loss of tendon reflexes are less frequently encountered in Type 2 than in Type 1 (Harding and Thomas 1980b). Palpable nerve thickening does not occur in Type 2. Many patients with Charcot–Marie–Tooth Type 2 show little or no deterioration even when reassessed at intervals of many years and serious disability is uncommon. The median nerve motor conduction velocity is usually just within the normal range, and should not lie below 38–40 m/s. The median nerve sensory action potential is absent or reduced in amplitude (Harding and Thomas 1980b). The spinal fluid protein level is normal in Type 2. Nerve biopsies show axonal loss with little evidence of demyelination. Hypertrophic 'onion-bulb' changes are observed only rarely.

Differential diagnosis. Late onset Charcot–Marie–Tooth disease Type 2 should be distinguished from chronic idiopathic axonal polyneuropathy (Section 21.11.8). Sensory features predominate and progression occurs in this latter condition (Teunissen *et al.* 1997). The dominant inheritance should distinguish ataxic forms of Charcot–Marie–Tooth disease Type 2 from inherited causes of vitamin E deficiency (Section 21.22.5) and from Friedreich's ataxia, which has a poorer prognosis. Abnormal sensory nerve conduction distinguishes patients with predominantly motor forms of Charcot–Marie–Tooth disease Type 2 from distal spinal muscular atrophy.

21.4.6 **Infantile and childhood onset**

These heterogenous and relatively uncommon progressive sensorimotor demyelinating neuropathies start in infancy or early childhood and are inherited either autosomally recessively or dominantly (Lynch *et al.* 1997). They are also known as Déjérine–Sottas syndrome or congenital hypomyelination neuropathy. They are caused by various point mutations of the *PMP22*, *MPZ*, Periaxin, *GDAIP*, *MTMR2*, and *EGR2* genes (Parman *et al.* 2004) (Table 21.7). Affected children usually show delayed onset of walking and often develop ataxia and skeletal deformity. Palpable nerve thickening is common. One form of Charcot–Marie–Tooth disease Type 1, congenital hypomyelination neuropathy, presents at birth. It has a particularly poor prognosis; patients are unable to walk by their teens and may die at any age due to respiratory insufficiency. The prognosis is better if the onset is in childhood, but severe motor disability is usually evident by early adult life. Motor nerve conduction is extremely slow, to less than 12 m/s. This is lower than that generally measured in Charcot–Marie–Tooth disease Type 1 (Ouvrier *et al.* 1987). Spinal fluid protein is usually elevated to between 0.7 and 2.1 g/l. This can pose difficulties in distinction from the steroid-responsive chronic inflammatory demyelinating polyneuropathy of infancy and childhood (Section 21.11.2). In Charcot–Marie–Tooth disease Type 1 nerve biopsies show extensive onion-bulb formation and marked thinning of the myelin sheaths surrounding axons of all diameters (Ouvrier *et al.* 1987). Many axons are completely devoid of myelin sheaths in congenital hypomyelination neuropathy. The ataxic forms may cause severe disability and can be distinguished from early onset Friedreich's ataxia by their slow motor conduction velocities, and from Refsum disease by measurement of the serum phytanic acid level.

21.5 Hereditary neuropathy with liability to pressure palsies

Some families exhibit autosomal dominant inheritance of a tendency to develop mononeuropathies due to their nerves being unusually vulnerable to pressure or traction. This condition is known also as hereditary susceptibility, or liability, to pressure palsies, hereditary pressure sensitive neuropathy, or tomaculous neuropathy. Exposed nerves, such as the radial or lateral popliteal, are especially vulnerable. Painless brachial plexus lesions may result from sleeping in awkward postures or from the shoulder straps of heavy backpacks. Patients present with the motor and sensory features typical of the mononeuropathy in question and recovery

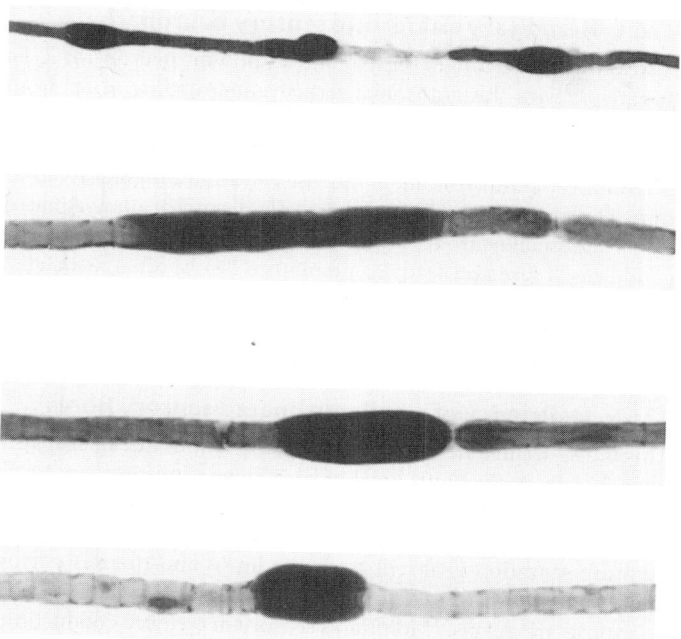

Fig. 21.8 Sausage-shaped myelin swellings on teased sural nerve fibres from a patient with hereditary pressure-sensitive ('tomaculous') neuropathy. (Reproduced from *Greenfield's Neuropathology* (5th edn.) by permission of Edward Arnold.)

occurs over days or weeks. Typically patients experience deadness or tingling of the fingertips after using scissors. Permanent disability may develop after recurrent episodes of paralysis affecting the same nerve. In many, nerve conduction studies show slight but generalized slowing of distal motor latencies or sensory nerve action potentials and minor slowing of motor conduction velocities (Andersson *et al.* 2000). This can be useful in detecting asymptomatic affected family members, or in raising suspicion of this hereditary neuropathy when investigating a seemingly uncomplicated peripheral nerve palsy. Teased fibres from nerve biopsies show characteristic 'tomaculous' (Latin: sausage) swellings due to redundant myelin loops as a result of overgrowth of the myelin spiral (Fig. 21.8). In approximately 80 per cent of the patients, there is a deletion of one *PMP22* gene at 17p11.2, and more pronounced evidence of background polyneuropathy (Lenssen *et al.* 1998). New mutations account for up to 5 per cent of all patients encountered with hereditary liability to pressure palsies.

21.6 Hereditary sensory and autonomic neuropathies

These inherited neuropathies reflect failure of development, or degeneration, of subpopulations of peripheral sensory and autonomic neurones. Prior to recognition that peripheral neuropathy underlay these disorders, they were termed as lumbo-sacral syringomyelia, inherited perforating ulcers or whitlows, acrodystrophic neuropathy, and congenital insensitivity to pain. Also some such patients were considered to suffer from congenital indifference or asymbolia to pain, a term which should be reserved for the rare situation in which there is lack of concern to a painful stimulus, yet one which is well received and in which the peripheral nerves are normal.

Lack of self-protection due to impaired pain appreciation leads to the development of a mutilating acropathy (Fig. 21.9), with ulceration and fissuring of the skin, long-bone fractures, Charcot joints, and digit amputation. The precise symptoms and signs of each neuropathy, and whether there are accompanying nerve conduction abnormalities, is determined by which subpopulation of sensory neurones is most affected. Patients with hereditary sensory and autonomic neuropathies should be instructed to avoid situations likely to cause thermal burns and trauma to the limbs, and to be assessed regularly by a chiropodist. They should be advised to ensure that their shoes are well-fitting and do not contain stones or sharp objects. Hereditary sensory and autonomic neuropathies should be distinguished from familial amyloidotic polyneuropathy (Section 21.9.1) which often affects sphincter control and sexual functioning.

There is no consensus for the classification of hereditary sensory and autonomic neuropathies. Here a descriptive classification (Donaghy *et al.* 1987) will be followed, indicating the corresponding numerical classification (Dyck *et al.* 1983) into Types I to V. A molecular genetic classification is emerging which subdivides some of the clinically recognized forms (Verpoorten *et al.* 2006) (Table 21.9).

21.6.1 Autosomal dominant sensory neuropathy

This is also known as hereditary sensory and autonomic neuropathy Type 1. Patients gradually lose all modalities of sensation from

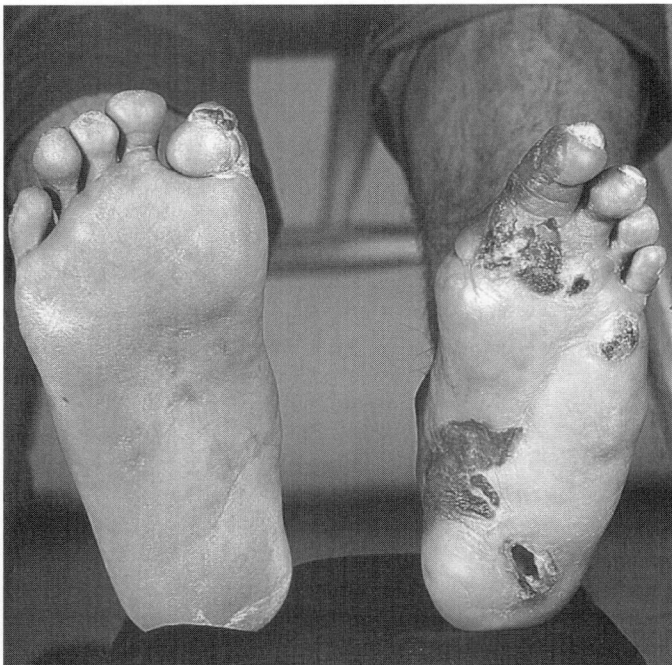

Fig. 21.9 Feet of a patient with hereditary sensory and autonomic neuropathy, showing chronic ulceration, loss of the left hallux, and shortening of the right hallux due to previous fracture.

the distal part of their limbs, particularly pain and temperature sensations. Spontaneous shooting pains in the legs may occur early in the disease. Foot ulcers, calluses, and foot deformity develop in many patients. Anhidrosis is not always present and may occur in the regions of impaired sensation. The ankle jerks are usually absent. Patients eventually develop mild distal muscle wasting and weakness, raising the question of overlap with more sensory forms of Charcot–Marie–Tooth disease. Sensory nerve action potentials are reduced or absent and motor conduction velocity is just within the normal range. Autopsy studies show loss of dorsal root ganglion neurones, with replacement by nodules of Nageotte indicating that sensory neurones have degenerated, rather than failing to develop. Mutations of the serine palmitoyltransferase long chain base subunit 1, *SPTLC1* gene underlie this form, reflecting a defect in sphingolipid biosynthesis. This dominantly inherited form can be associated rarely with deafness (Horoupian 1989).

Other forms of autosomal dominant sensory neuropathy are of early adult onset with severe sensory loss and autoamputations, involve cough and gastrooesophageal reflex, and overlap with Charcot–Marie–Tooth disease Type 2B (Section 21.4.5).

21.6.2 Autosomal recessive sensory neuropathy

Otherwise known as congenital sensory neuropathy, or hereditary sensory and autonomic neuropathy Type 2, this autosomal recessive condition presents in infancy or childhood with impairment of all modalities of sensation in the limbs and on the trunk. Marked distal limb mutilation occurs with whitlows, paronychia, plantar ulcers, painless long-bone fractures, and Charcot joints. Motor function is preserved. Tendon reflexes are usually lost. Anhidrosis of the hands and feet adversely affects skin texture and contributes

to ulceration. Sensory nerve action potentials are usually absent. The cutaneous sensory nerves are virtually devoid of myelinated fibres and depleted of unmyelinated fibres. The neuropathy seems to worsen progressively and patients progressively accumulate acrodystrophic changes. The prognosis is uncertain, but since published reports usually relate to patients under the age of 20, a shortened life span is inferred. Some are associated with mutations of hereditary sensory and autonomic neuropathy Type 2, *HSN2*, a gene of as yet unknown function. A similar form of hereditary sensory and autonomic neuropathy may be inherited as an X-linked recessive trait (Jestico *et al.* 1985). Another autosomal recessive form may be associated with spastic paraplegia (Cavanagh *et al.* 1979).

21.6.3 Familial dysautonomia

Otherwise known as the Riley–Day syndrome, or hereditary sensory and autonomic neuropathy Type 3, this rare congenital disorder is autosomally recessively inherited, principally by Ashkenazi Jews. Affected infants vomit, have impaired control of body temperature and blood pressure, sweat excessively, develop patchy skin blotching, have impaired tear formation, and are prone to pulmonary infection. Typically fungiform papillae are absent from the tongue. There is insensitivity to painful stimuli applied to the skin or eyes. The sensory deficit worsens with age, eventually involving kinaesthesia. There is areflexia and postural hypotension. Motor involvement eventually occurs. Nerve biopsy shows reduced numbers of unmyelinated fibres and a lack of large-diameter sensory myelinated fibres. Intradermal histamine injection fails to cause the erythematous flare of the triple response. About half of the patients die by the age of 20 due to renal or pulmonary failure, or sudden death (Axelrod 2004). The mutations affecting the IKBKAP gene lead to tissue specific expression of the mutant IkB kinase-associated protein.

21.6.4 Hereditary anhidrotic sensory neuropathy

Also known as hereditary sensory and autonomic neuropathy Type 4, patients with this autosomal recessive neuropathy suffer from congenital insensitivity to pain. They present in infancy with bouts of pyrexia, failure to thrive, retarded development, failure to respond to painful stimuli, anhidrosis, and mild mental retardation. The peripheral nerves are virtually devoid of unmyelinated axons and small neurones are absent from dorsal root ganglia. The condition is rare and leads to premature death, often associated with a bout of unexplained fever. It is due to mutations of the NTRK1 gene, a tyrosine kinase gene which encodes the high affinity nerve growth factor receptor (Indo 2002).

12.6.5 Deficiency of small myelinated sensory fibres

This is also termed hereditary sensory and autonomic neuropathy Type 5. A large Kashmiri kinship has shown autosomal recessive inheritance of a mutilating acropathy associated with bilateral corneal opacification due to neurotrophic keratitis. Pain and temperature sensation is absent from the limbs and there is patchy anhidrosis. Motor function, the tendon reflexes, and kinaesthetic sensations are normal. Motor and sensory nerve fibre conduction, which reflect the fastest conducting fibres, are normal. Sural nerve biopsy shows selective reduction of the smaller myelinated fibre population. An identical neuropathy has also occurred sporadically, but associated with normal corneas. This syndrome can reflect

Table 21.9 Classification of hereditary sensory and autonomic neuropathies

	HSAN classification	Locus	Gene	Clinical features	Inheritance	Onset
Autosomal dominant sensory neuropathy (Section 21.6.1)	Type I	9q22.2	SPTLC1	Distal sensory loss with pains	AD	Adult
				Deafness sometimes		
				Distal wasting		
				Mild autonomic involvement		
	Type I			Severe sensory loss and amputations	AD	Early adult
	Type IB	3p22-24		Additional cough and gastro-oesophageal reflux	AD	Adult
	CMT 2B	3q21.3	RAB7	Severe sensory loss, ulcers	AD	Adult
				Distal wasting		
Autosomal recessive sensory neuropathy (Section 21.6.2)	Type II	12p13.3	HSN2	Severe sensory loss legs and arms	AR	Childhood
	Type IIB			Whole body sensory loss	AR	Congenital
				Hypotonia		
Familial dysautonomia (Riley–Day syndrome) (Section 21.6.3)	Type III	9q31	IKBKAP	Severe autonomic dysfunction	AR	Congenital
				Loss of pain and temperature		
Hereditary anhidrotic sensory neuropathy (Section 21.6.4)	Type IV	1q21-22	NTRK1	Generalised anhidrosis	AR	Congenital
				Self-mutilation		
				Pain and temperature insensitivity		
				Mental retardation		
	Type IV		SCN9A	Channelopathy-associated insensitivity to pain	AR	Congenital
Deficiency of small myelinated sensory fibrres (Section 21.6.5)	Type V	1p13.1	NGFB	Loss of temperature and deep pain		Childhood
	Type V	1q21-22	NTRK1	Widespread anhidrosis		Congenital
				Distal loss of pain and temperature		

Key: AD – autosomal dominant; **AR** – autosomal recessive; **CMT 2B** – Charcot-Marie-Tooth disease Type 2B; **HSAN** – Hereditary Sensory and Autonomic Neuropathy; **HSN2** – Hereditary Sensory Neuropathy Type 2; **IKBKAP** – Inhibitor of kappa light polypeptide enhancer in B cells, kinase complex associated protein; **NTRK1** – Neurotrophic tyrosine kinase, receptor, type 1; **RAB7** – GTP-ase late endosomal protein gene; **SCN9A** – Sodium channel 9A; **SPTLC1** – Serine palmitoyltransferase long chain base subunit 1. (Adapted from Verpoorten et al 2006.)

mutations either of the neurotrophic tyrosine kinase receptor Type 1, N~TRK1 (Houlden *et al.* 2001b), or nerve growth factor β, NGFB, genes.

21.6.6 Congenital indifference to pain

True indifference to pain is rare. Most patients so described probably had hereditary sensory neuropathies with selective lack of small myelinated or unmyelinated sensory fibres. This was overlooked because there were no neuropathic abnormalities on examination and sensory nerve conduction was normal, given that large myelinated fibres are preserved in such neuropathies. One family has been described with dominantly inherited indifference to painful stimuli over the whole body and with normal morphometric examination of peripheral nerves (Landrieu *et al.* 1990).

21.7 Other inherited polyneuropathies

21.7.1 Giant axonal neuropathy

This rare progressive disorder of childhood involves both the peripheral and central nervous systems, causing peripheral neuropathy, ataxia, intellectual loss, and pyramidal tract dysfunc-

tion (Demir *et al.* 2005). It is characterized by accumulations of abnormally closely packed neurofilaments within swollen peripheral nerve axons (Fig. 21.10). The disorder is not restricted to neurofilaments and affects intermediate filament organization in all cell types, including glial fibrillary acidic protein in Schwann cells (Donaghy *et al.* 1988). Most patients have characteristic tightly curled hair, reflecting an abnormality of keratin, another intermediate filament protein. However, such hair is only variably present, even within affected kinships. The inheritance is autosomal recessive. The disorder is caused by mutations of the Gigaxon gene, GAN on chromosome 16q24.1 (Bruno *et al.* 2004; Demir *et al.* 2005). Patients present in childhood before dying from progressive disease in their second or third decade. Toxic neuropathy due to hexacarbons causes a similar histological picture, but the hair is normal, there is a history of exposure to glues or other solvents, and the age of onset is generally later (Section 21.21.7).

21.7.2 Neuropathy in cerebellar ataxia

Friedreich's ataxia (Section 39.4.1) affects the peripheral as well as the central nervous system. A stocking distribution of impaired light touch sensation spreads proximally with age. The Achilles

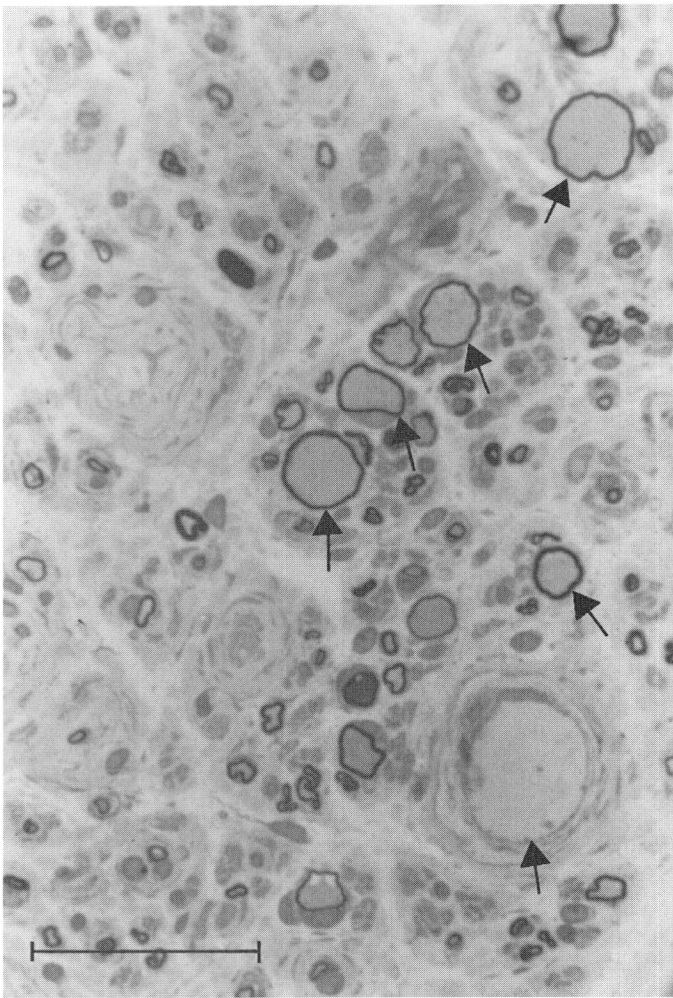

Fig. 21.10 Giant axonal neuropathy: semi-thin section of sural nerve showing numerous giant axonal swellings, some unmyelinated. Giant axonal profiles are shown by arrows; bar = 50 μm.

tendon reflexes are lost early, and vibration and joint-position sensations are impaired in the feet. Pes cavus may occur. The amplitude of sensory nerve action potentials is markedly reduced and sural nerve biopsy shows severe loss of the large myelinated fibres (Caruso *et al.* 1987). Despite the clinical evidence of progression, actively degenerating fibres are only relatively rarely encountered in biopsies, and the marked absence of large myelinated fibres is unlikely to be explicable by degeneration alone.

A large French kinship with dominantly inherited variable combinations of cerebellar ataxia and pure sensory neuropathy has been designated spinocerebellar ataxia Type 25, SCA25 (Stevanin *et al.* 2004).

21.7.3 Chédiak–Higashi syndrome

This rare autosomal recessive disease usually presents in infancy or childhood with febrile episodes, pyogenic infections, and partial oculocutaneous albinism. Death within the first two decades usually follows an accelerated phase with lymphoma-like proliferation. Peripheral blood granulocytes contain pathognomonic giant peroxidase-positive lysosomal granules. Patients develop features

of spinocerebellar degeneration, and peripheral neuropathy is particularly likely to occur during the accelerated phase. Prednisolone and vincristine may have some effect in controlling the neurological manifestations (Pettit and Berdal 1984). Although Chédiak–Higashi syndrome may be cured by allogeneic bone marrow transplantation in childhood, neurological abnormalities still develop in the third decade (Tardieu *et al.* 2005).

21.7.4 Multiple symmetric lipomatosis

Patients with this condition develop disfiguring multiple subcutaneous lipomata over the upper trunk and proximal arms also known as Madelung's disease (Fig. 21.11). The buttocks and legs are spared and are often relatively devoid of the usual subcutaneous fat.

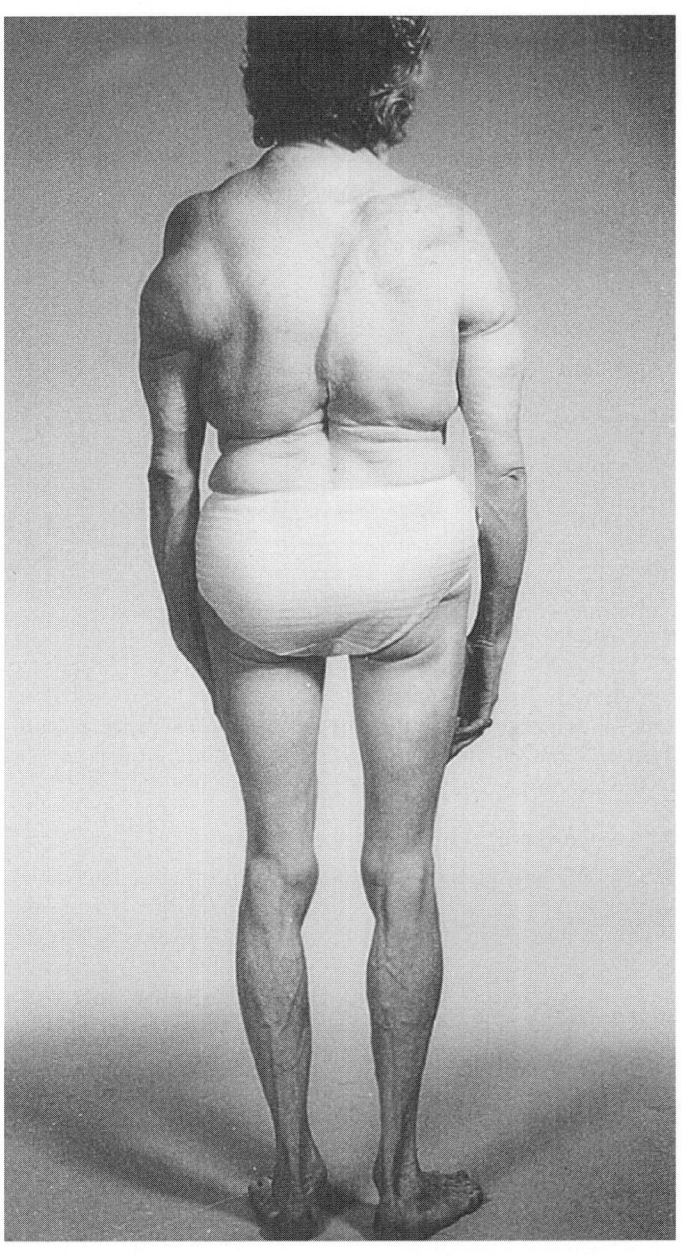

Fig. 21.11 Multiple symmetrical lipomatosis. Multiple lipmata are present on the neck, back, and upper arms, with scars from plastic surgery. Subcutaneous fat is sparse on the legs. (Reproduced with permission from Chalk *et al.* (1990.)

Up to 80 per cent of patients develop an axonal peripheral neuropathy, usually sensorimotor but occasionally with autonomic features. The neuropathy presents insidiously in middle age, is usually mild, and has a heterogenous basis. Excessive alcohol consumption was observed in many Italian patients, initially suggesting that alcoholism caused the neuropathy (Enzi *et al.* 1985). However, a large kinship has shown invariant association of peripheral neuropathy with multiple symmetrical lipomatosis, with possible autosomal recessive inheritance. Since no members of this kinship were alcoholic, the neuropathy of multiple symmetrical lipomatosis is likely to be genetically determined (Chalk *et al.* 1990). Other patients have abnormalities in Complex IV and multiple deletions of mitochondrial DNA (Klopstock *et al.* 1994).

21.7.5 Chorea-acanthocytosis

Muscle wasting and areflexia may occur in patients with limb chorea and orofacial dyskinesias associated with acanthocytes in peripheral blood. Nerve-conduction studies show an axonal degeneration neuropathy. The condition is often familial, either autosomal dominant involving mutations of the *CHAC* gene, or X-linked associated with the McLeod blood group variant gene XK (Danek *et al.* 2001; Saiki *et al.* 2003). It should be differentiated from the peripheral neuropathy and acanthocytosis which occurs in abetalipoproteinaemia, Bassen–Kornzweig disease, which is treatable with vitamin E (Section 21.8.8).

21.7.6 Mitochondrial disorders

Roughly a quarter of all patients with various forms of mitochondrial cytopathy (Section 24.6.3) have clinical features of mild sensorimotor neuropathy. Asymptomatic electrophysiological evidence of peripheral neuropathy is somewhat commoner (Yiannikas *et al.* 1986). Sural nerve biopsies show loss of large myelinated fibres, evidence of axonal degeneration, and the Schwann-cell cytoplasm may contain abnormal mitochondria with paracrystalline inclusions. There are particular associations with the **N**europathy **A**taxia **R**etinitis **P**igmentosa, or NARP syndrome, and with the MERRF and MELAS syndromes as well as point mutations of the mitochondrial genome (Bouillot *et al.* 2002).

Severe progressive polyneuropathy occurs in the distinctive syndrome of **M**itochondrial **N**euro**G**astro**i**ntestinal **E**ncephalomyopathy, or MNGIE. Supranuclear ophthalmoparesis and T_2-weighted brain MRI changes are usual accompaniments. Various autosomal recessively inherited mutations of the somatic gene encoding the thymidine phosphorylase gene, TP are responsible. The neuropathy may mimic Charcot–Marie–Tooth disease (Section 21.4) or chronic inflammatory demyelinating polyneuropathy (Section 21.11.2) which is unresponsive to immunomodulatory treatment (Bedlack *et al.* 2004).

21.7.7 Thermosensitive neuropathy

A French family has shown autosomal dominant inheritance of reversible episodes of ascending muscle weakness, paraesthesiae, and areflexia which seemed to be triggered by pyrexia over 38.5°C (Magy *et al.* 1997).

21.8 Inherited metabolic disorders causing polyneuropathy

This section addresses the peripheral-nerve manifestations of a number of diseases whose central nervous system manifestations are discussed in Chapter 10 and Section 37.7. In addition to those disorders to be discussed in detail, mild polyneuropathy has been described in lysosomal-storage diseases, Niemann–Pick, Gaucher, or GM gangliosidosis; hyperoxaluria; defective DNA repair, xeroderma pigmentosum, ataxia telangiectasia; and Cockayne's syndrome. Focal entrapment neuropathies may occur in the mucopolysaccharidoses.

21.8.1 Refsum disease

Refsum disease is also known as phytanic acid oxidase deficiency, heredopathia atactica polyneuritiformis, hereditary motor and sensory neuropathy Type IV, or phytanic acid accumulation. Demyelinating polyneuropathy is a central feature of Refsum disease, associated with retinitis pigmentosa, cerebellar ataxia, and a markedly raised spinal fluid protein. Night blindness is a common presenting symptom. Most patients also show hearing loss, anosmia, ichthyosis, skeletal abnormalities, and cardiomyopathy. The disease is rare, generally occurring in people of northern European racial stock (Refsum *et al.* 1984). A deficiency of phytanic acid α-hydroxylation due to mutations in the phytanoyl-CoA-hydrolase gene, *PHYH*, on chromosome 10p13, is inherited as an autosomal recessive trait. This results in a failure to oxidize exogenous phytol with resultant conversion to phytanic acid, which accumulates. The disorder is genetically heterogeneous, with a second gene, *PEX 7* on 6q22-24 causing a milder form of Refsum disease (van den Brink *et al.* 2003). The diagnosis is confirmed by demonstrating a high serum phytanic acid level. Phytates are derived chiefly from dietary dairy products and other animal fats; the chlorophyll of green vegetables is a less well-absorbed source.

Neuropathic symptoms first develop at any time from childhood to the third decade, and may be provoked by infections. Recurrent attacks of polyneuropathy have been described. The neuropathy is usually predominantly distal and motor, and generally preceded by night blindness. Severe disability may develop. Peripheral nerves may be palpably hypertrophied. Motor nerve conduction velocities are greatly slowed. Nerve biopsy confirms the demyelinating nature of the neuropathy, and may reveal hypertrophic changes and round paracrystalline inclusions within Schwann cells. The diagnosis of Refsum disease is of importance, since dietary restriction of phytate ingestion improves or stabilizes the neuropathy, with increased motor nerve conduction velocities. Plasma exchange also helps to lower the phytanic acid level, and is particularly valuable soon after diagnosis before dietary restriction becomes effective (Harari *et al.* 1991).

21.8.2 Metachromatic leucodystrophy

In this rare group of autosomal recessive diseases, also known as sulfatide lipidosis and arylsulfatase deficiency, a severe demyelinating sensorimotor neuropathy accompanies psychomotor retardation and seizures due to disease of the cerebral white matter. Motor nerve conduction velocity is substantially slowed with elevated spinal fluid protein. The different metachromatic leucodystrophies present variously in late infancy, childhood, or adulthood. Presentation with isolated peripheral neuropathy as late as the sixth decade is recorded (Comabella *et al.* 2001). Sulfatides accumulate in nervous tissue because of a range of mutations in the arylsulfatase A gene, ARSA located at 22q13, accounting for the varying age of onset and diverse clinical picture associated with this lysosomal storage disorder (Bertelli *et al.* 2006). The activity of the lysosomal enzymes arylsulphatase A and B are reduced in peripheral blood leucocytes. Metachromatic granules may be demonstrated in an early morning urine deposit. Nerve biopsy shows evidence of demyelination and remyelination, and the presence of metachromatic granules in Schwann-cell cytoplasm after staining with aniline dyes. Electron microscopy shows these to be zebra-like bodies, tuffstone, and prismatic inclusions (Martin *et al.* 1982). Bone marrow transplantation can be successful in ameliorating symptoms and slowing progression, particularly if performed sufficiently early in the disease (Peters and Steward 2003).

21.8.3 Krabbe disease

This rare autosomal recessive disease is also known as globoid cell leucodystrophy or galactosylceramide lipidosis. It is due to mutations of the galactosylceramide-beta-galactosidase gene, *GALC* on 14q24.3-32.1. It usually presents in infancy with psychomotor retardation (Husain *et al.* 2004). Demyelinating sensorimotor peripheral neuropathy generally develops later. Protruding ears may be a feature. A later onset form has been described. After an initial phase of hyperreflexia, which reflects central nervous system disease, patients become areflexic as the peripheral neuropathy develops. Motor nerve conduction velocity is substantially reduced. The diagnosis is confirmed by demonstrating reduced leucocyte galactosylceramide-beta-galactosidase levels.

21.8.4 Adrenoleucodystrophy

X-linked adrenoleucodystrophy (Section 37.6.2) is also known as adrenomyeloneuropathy. It causes adrenal insufficiency and neurological dysfunction associated with raised plasma levels of very-long-chain saturated fatty acids, such as hexacosanoic, pentacosanoic, and tetracosanoic acids. Different phenotypes occur with different ages of onset, within the same kinship, and even between monozygotic twins (Sobue *et al.* 1994). Over a hundred mutations in the *ALD* gene are identified, with little genotype–phenotype correlation (Dodd *et al.* 1997). The peripheral neuropathy is generally mild and causes little disability compared to the associated spastic and ataxic paraparesis. Nerve conduction may be slowed, but a mixed axonal and demyelinating picture is usual; female carriers may show electrophysiological evidence of neuropathy (van Geel *et al.* 1996).

21.8.5 Fabry's disease

This X-linked disorder is also known as angiokeratoma corporis diffusum or Anderson–Fabry disease and reflects α-galactosidase deficiency. It is due to mutations in the gene encoding the lysosomal hydrolase α-Gal A, at Xq22.1. This leads to accumulation of ceramide trihexoside in neural, renal, endothelial, and corneal tissues. It usually presents in childhood or early adult life with burning pain and paraesthesiae distally in the limbs or occasionally with a cramp-fasciculation syndrome in adulthood (Nance *et al.* 2006). Discrete crises of pain may be provoked by exercise or heat. These can be sufficiently severe to prevent or impede walking. Anhidrosis may occur. Most patients have angiokeratoma corporis diffusum, a characteristic crimson maculopapular rash in the 'bathing-trunks' area which may be overlooked on cursory examination (Fig. 21.12). Strokes, hypertension, renal failure, and corneal opacification are common (MacDermot *et al.* 2001). Muscle strength, kinaesthetic sensation, and nerve conduction studies are usually normal;

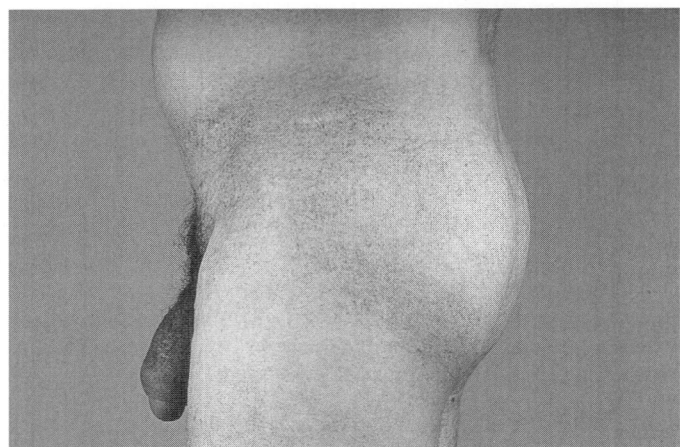

Fig. 21.12 Fabry disease. The typical crimson angiokeratoma corporis diffusum rash in the 'bathing-trunks' area.

thermal threshold testing is often abnormal (Luciano *et al.* 2002). Excess lysosomal storage of glycosphingolipids occurs in blood vessel walls, and selected populations of neurones, including ganglia. The diagnosis is confirmed by measuring blood leucocyte lysosomal α-galactosidase.

21.8.6 Porphyric neuropathy

Neurological symptoms occur in all three autosomal dominantly inherited hepatic porphyrias: acute intermittent, hereditary coproporphyria, and variegate porphyria. A fourth form is autosomal recessive, ALA dehydratase deficiency. All exhibit similar neuropsychiatric features during acute porphyric attacks. Only hereditary coproporphyria and variegate porphyria produce photosensitive skin lesions. Acute attacks may be diagnosed by measuring elevated levels of δ-aminolevulinic acid in the plasma, of porphobilinogen in the urine, and of faecal porphyrins. The underlying enzyme defects can be identified in cultured fibroblasts, the genes cloned, and disease-specific mutations mapped (Albers and Fink 2004).

Acute attacks are provoked by a wide variety of drugs, hormones, intercurrent infections, by reduced dietary carbohydrate intake or by lead intoxication (Section 21.20.2). Lead-exposed workers whose blood and urine lead levels lay within accepted safety limits may nonetheless develop wrist drop due to plumboporphyria (Dyer *et al.* 1993). During an acute attack, the urine characteristically turns red on standing due to oxidation of porphobilinogen. The first manifestation is usually abdominal pain. This is often associated with constipation, tachycardia, sweating, tremor, fever, and hypertension. These features reflect acute sympathetic outflow activity and may be associated with sudden death. They are followed by the neuropsychiatric features of insomnia, confusion, hallucinations, delusion, or depression. Seizures may occur. A dilutional hyponatraemia may contribute to the central nervous system dysfunction. Imaging evidence points to porphyrin precursor toxicity as the cause of acute neurological attacks, rather than the underlying haem deficiency (Solis *et al.* 2004).

A wide variety of neuromuscular manifestations have been described and occur in up to 40 per cent of porphyric attacks (Albers and Fink 2004). Acute weakness may start symmetrically in proximal muscles, preceded by myalgias and cramps. The onset may be restricted to only one limb for the first few hours or days. Weakness can take weeks to develop fully, and eventually may be most marked distally. The resultant tetraplegia is often maximal in the arms. Dysphagia, facial paralysis, and diplopia may occur in severe cases. Assisted ventilation may be required for diaphragm weakness. Tendon reflexes are absent or diminished in nearly all patients; characteristically the ankle jerks are most likely to be preserved. Paraesthesiae, dysaesthesiae, or numbness of the limbs may be present from the outset. Distal glove and stocking diminution of superficial sensation, and vibration sense abnormalities can be evident. It is unusual for the sensory disturbance to precede motor weakness. Some patients develop truncal sensory deficits, either in band-like distributions or in a bathing-trunks distribution. Postural hypotension and urinary retention can occur, reflecting autonomic neuropathy. Nerve conduction studies in established attacks are consistent with an axonal degeneration neuropathy. However, nerve conduction may be normal early in an attack, although mild myopathic features may be noted on electromyography of affected muscles. We still lack a cogent pathogenic explanation for the early weakness. The early proximal muscle involvement and relatively prompt recovery suggest a reversible myopathic or conduction block element, rather than axonal degeneration.

The treatment should aim to control pain with opiate analgesics. Sedation with phenothiazines is helpful. Epilepsy should be treated with low-dose clonazepam or sodium valproate and by correcting the underlying dilutional hyponatraemia. Other major anticonvulsants such as phenytoin, carbemazepine, or barbiturates are likely to worsen the severity of the underlying porphyric attack. The porphyric attack itself should be terminated by withdrawing precipitating drugs, by treating any underlying infections, by rehydration, and by administration of a high carbohydrate load either parenterally or orally. If the attack is severe or unresponsive to the preceding measures, haematin infusion may be effective. The neurological deficit recovers gradually over months although permanent deficits may result.

Identical neurological crises may occur in children with autosomal recessive hereditary tyrosinaemia (Mitchell et al. 1990) and in patients with lead poisoning. d-Aminolevulinic acid excretion is increased in both these conditions, as in porphyria.

21.8.7 Tangier disease

Plasma high-density lipoproteins are severely reduced in this rare autosomal recessive disorder which is also known as familial high-density lipoprotein deficiency or analphalipoproteinaemia. It is due to mutations in the ATP-binding cassette transporter 1 gene, *ABCA1*. Diagnostically there is absence or severe reduction in plasma HDL cholesterol and ApO A-1. Cholesterol esters accumulate in body tissues, producing cardiovascular disease, splenomegaly, and visibly swollen, orange-coloured pharyngeal tonsils. Most patients eventually develop peripheral neuropathy which can take either of two forms. Most commonly there is a relapsing and remitting multiple mononeuropathy which can be confused with leprosy (Sinha et al. 2004). Less often there is a slowly progressive symmetrical sensory neuropathy in which the distribution of the sensory disturbance resembles syringomyelia, with dissociated sensory loss affecting the upper limbs and wasting of the arm and facial muscles. These two contrasting peripheral neuropathies reflect severe demyelination and axonal degeneration, respectively (Zuchner et al. 2003). There is no specific treatment.

21.8.9 Abetalipoproteinaemia

This rare autosomal recessive disorder is also known as Bassen–Kornzweig disease syndrome or acanthocytosis. Associated with the acanthocytes in peripheral blood, there is fat malabsorption, retinitis pigmentosa, and a spinocerebellar degeneration resembling early onset Friedreich's ataxia. The associated axonal degeneration peripheral neuropathy produces areflexia and impaired vibration and joint-position sensation (Brin et al. 1986). It is due to deficiency of the microsomal triglyceride transfer protein (Wetterau et al. 1992). High-dose vitamin E therapy may prevent progression of the neurological symptoms and can produce improvement (Muller et al. 1985).

21.8.9 Cerebrotendinous xanthomatosis

Cholestanol accumulates in the tissues in this rare autosomal recessive disorder which is due to a block in hepatic bile acid synthesis resulting from mutations of the sterol 27-hydrolase gene, *CYP27*. A sensorimotor peripheral neuropathy is associated with juvenile cataracts, chronic diarrhoea, dementia, spastic paraparesis, cerebellar ataxia, and tendon xanthomas (Fig. 21.13) (Moghadasian et al. 2002). A high blood level of cholestanol confirms the diagnosis. Chenodeoxycholic acid therapy may produce some minor neurological improvement, including increased nerve conduction velocities. Presymptomatic detection of the mutation in at-risk family members allows presymptomatic treatment to try and stave off clinical disease.

21.9 Amyloid neuropathy

Amyloidotic polyneuropathy results from deposition within nerves of various non-branching, fibrillar proteins which all possess the crystallographic characteristic of forming a β-pleated sheet. Histologically, amyloid material is recognized by its property of staining with congo-red dye and exhibiting apple-green birefringence when viewed under polarizing light (Fig. 21.14).

Amyloidotic neuropathy occurs in two main groups of patients. The familial amyloidotic neuropathies reflect inherited substitutions of single amino acids in the proteins which are deposited: transthyretin, or pre-albumin, or less frequently apolipoprotein A-1 or gelsolin. Primary amyloidosis is due to tissue deposition of immunoglobulin light chains, usually derived from benign or

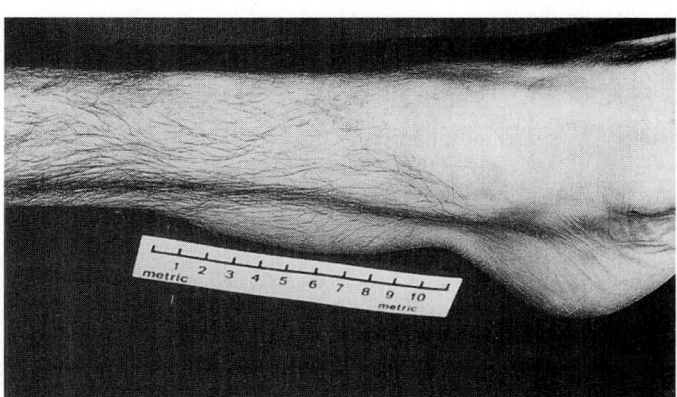

Fig. 21.13 Cerebrotendinous xanthomatosis. Achilles tendon xanthoma.

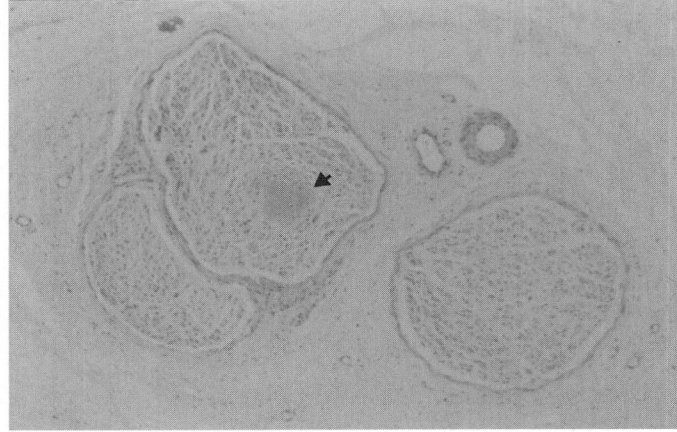

A

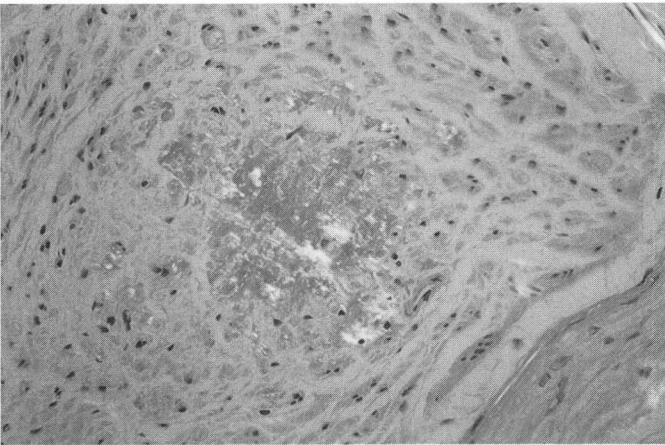

B

Fig. 21.14 Amyloid due to immunoglobulin light chain deposition in peripheral nerve. A. Eosinophilic deposit within a nerve fascicle, low power, haematoxylin, and eosin, B. Apple green birefringence of the amyloid deposit in polarized light, Congo red, higher power. (See Plate 28.)

malignant plasma cell tumours. It is rare for peripheral neuropathy, apart from carpal tunnel syndrome, to complicate the reactive or secondary amyloidoses, which are associated with circulating serum amyloid A protein, and which result from chronic inflammatory conditions.

21.9.1 Familial amyloidosis

Genetics. These neuropathies are inherited as autosomal dominant traits. Penetrance often varies within affected families. Onset of neuropathic symptoms is usually in the third to sixth decades. The original clinical and geographical classification into Types I–IV has been replaced by a molecular genetic classification (Hund *et al.* 2001). Molecular genetic analysis can now be applied to affected families for diagnostic, presymptomatic, and prenatal testing. Types I, originally Portuguese, and II, Indiana/Swiss, are due to the deposition of abnormal transthyretin within the body tissues, including the peripheral nerves. Well over 30 different point mutations have been identified in the transthyretin gene, *TTR*, on chromosome

18q11.2-q12.1 (Reilly *et al.* 1995; Plante-Bordeneuve *et al.* 1998). Of these the methionine 30, Met30 mutation is by far the commonest. Type III, Iowa, familial amyloidosis is due to deposition of mutant apolipoprotein A-1, resulting from various single base-pair substitutions in the gene on chromosome 11 (Nichols *et al.* 1988). Type IV, originally Finnish, is due to gelsolin gene mutations (Paunio *et al.* 1995; Conceicao *et al.* 2003).

Clinical features. Suspicion of hereditary amyloid neuropathy should always be raised by the combination of a small fibre neuropathy involving autonomic fibres coupled with cardiac disease. The polyneuropathy is similar in all these different types of familial amyloidosis. Numbness is usually associated with impaired pain and temperature sensations in the hands and feet. This may eventually lead to trophic ulceration. Spontaneous pains occur in the limbs. Areflexia develops. Weakness generally follows the sensory disturbance and affects distal muscles, particularly the small hand muscles and ankle dorsiflexors. Autonomic involvement affects the pupillary reactions to light, impairs gastrointestinal motility, and produces impotence and postural hypotension. The spinal fluid protein is often raised. Nerve conduction studies show an axonal degeneration neuropathy. Proteinuria, renal failure, and cardiac involvement contribute to premature death.

There are some noteworthy variations in the presentation of the different clinical forms of familial amyloidosis (Hund *et al.* 2001). In Type II the earliest symptom is usually a carpal tunnel syndrome, due to deposition of amyloid in the flexor retinaculum at the wrist. Renal and sphincter involvement do not occur in Type II amyloidosis. Type IV presents with lattice dystrophy of the cornea and patients may develop amyloid infiltration of the facial skin and involvement of the facial and auditory nerves.

Pathology. Nerve biopsies show amyloid deposition within nerve fascicles and around endoneurial blood vessels. At autopsy, widespread amyloid deposition is seen in the peripheral nervous system, affecting nerve plexuses, dorsal root and sympathetic ganglia. Leptomeningeal transthyretin amyloid deposits may have been symptomatic or noted in imaging. Multiple mechanisms probably contribute to the peripheral nerve damage. Amyloid deposits in the dorsal root ganglia may cause a sensory neuronopathy. Ischaemic changes may result from amyloid deposition around endoneurial blood vessels. Multifocal interruption of axons by small amyloid deposits along the course of a nerve may summate distally to produce a picture of diffuse fibre loss.

Treatment. Untreated, patients diagnosed with familial amyloid polyneuropathy survived about 10 years, although this varies geographically and in relation to different mutations. Liver transplantation has been introduced to treat transthyretin amyloid polyneuropathy on the grounds that the liver produces more than 90 per cent of this protein. Transplantation reduces mutated transthyretin in the blood and halts the progression of polyneuropathy with additional benefits for general health and gastrointestinal symptoms. There is little objective evidence of significant improvement in the neuropathy but further loss of myelinated axons from peripheral nerves largely ceases (Adams *et al.* 2000). Transplantation carries a significant mortality in amyloidosis, with only 60 per cent survival at 5 years (Parrilla *et al.* 1997). Poor prognosis is predicted by an already heavy load of amyloid deposition prior to transplantation, as revealed by symptomatic postural hypotension, urinary incontinence, or cardiac involvement. In principle, liver transplantation should be considered early in the disease so as to forestall amyloid deposition in the nerves and other organs.

21.9.2 Primary amyloidosis

Systemic amyloidosis due to immunoglobulin light-chain deposition is unusual before middle age. The peripheral nervous system is affected in about one-third of all cases (Duston *et al.* 1989). Various underlying lymphoproliferative disorders may be responsible for the monoclonal immunoglobulin production, ranging from malignant myeloma to benign paraproteinaemia. Serum paraproteinaemia and/or free light chains in the urine, known as Bence–Jones protein, may be detected. This form of amyloidosis is occasionally associated with hypernephroma.

Clinical features. Peripheral neuropathy is the presenting symptom in less than 10 per cent of patients with primary amyloidosis and these patients tend to have the longest survival. The initial symptoms tend to be sensory or, less frequently, autonomic. Impaired pain and temperature sensations and numbness affect the limbs. Spontaneous lancinating or burning dysaesthetic pains occur and may respond to carbamazepine therapy. Sometimes these lancinating pains are focal, for instance picking out a particular finger for a few weeks. Muscle weakness and areflexia occur later in the course of the neuropathy. Autonomic symptoms include postural hypotension, impotence, constipation, anhidrosis, and hypoactive pupils. Sensorimotor symptoms can be distributed asymmetrically, suggesting that individual peripheral nerves or dorsal root ganglia may be infiltrated to differing degrees by amyloid. Involvement of the ocular motor, trigeminal, or facial cranial nerves may be prominent (Traynor *et al.* 1991). Other features suggesting a diagnosis of primary amyloidosis include macroglossia, hepatosplenomegaly, proteinuria, or nephrotic syndrome, elevated serum alkaline phosphatase and paraproteinaemia (Park *et al.* 2003). Muscular stiffness, hypertrophy, and weakness occasionally complicate amyloid deposition in muscles. Nerve conduction studies show an axonal degeneration neuropathy. The spinal fluid protein is usually elevated. The diagnosis is confirmed by demonstrating amyloid deposition within biopsied peripheral nerves. However, it is simpler to establish the diagnosis of amyloidosis by rectal biopsy, which can be positive even when a nerve biopsy has not shown amyloid. The neuropathological features at autopsy resemble those of familial amyloidotic polyneuropathy (Section 21.9.1).

Treatment. The neuropathy progresses relentlessly. Eighty per cent of patients die within 3 years, usually due to associated renal or cardiac disease. Attempts at drug therapy have been generally unsuccessful, although there are recent glimmers of hope. Chemotherapy with melphalan and prednisolone does not alter perceptibly the downhill course of the neuropathy. However, trials have shown modestly enhanced survival in amyloidosis patients treated with melphalan and prednisolone compared to colchicine and retrospective analysis has shown that colchicine improves the median survival. Stabilisation or improvement of systemic manifestations of primary amyloidosis has followed autologous stem cell transplantation in selected patients; although the mortality is high in these with cardiovascular involvement (Mollee *et al.* 2004). However there has been no systematic study of how these more aggressive therapeutic approaches may benefit an existing polyneuropathy, or influence its emergence.

21.10 Acute idiopathic polyneuropathies and the Guillain–Barré syndrome

These conditions produce acute and diffuse demyelination or conduction block, or less frequently axonal degeneration affecting the spinal roots and peripheral nerves, and occasionally the cranial nerves. They are usually post-infective and recover spontaneously. The term Guillain–Barré syndrome includes two main entities now recognized as distinct: acute idiopathic demyelinating polyradiculoneuropathy (Section 21.10.1) and acute motor axonal neuropathy (Section 21.10.2).

21.10.1 Acute idiopathic demyelinating polyneuropathy

The term Guillain–Barré syndrome tends to be used interchangeably with acute idiopathic demyelinating polyneuropathy. Other names used for the condition have included: acute post-infective polyradiculoneuropathy, acute infectious polyneuritis, Landry–Guillain–Barré–Strohl syndrome, and post-infective polyneuritis.

Epidemiology. The Guillain–Barré syndrome is one of the commoner forms of polyneuropathy. Many cases were observed among troops during the 1914–18 war. The condition may occur in either sex, with slight male preponderance, and at any age, occasionally including infancy. The mean age of onset is around 40 but many series have shown a bimodal distribution with peaks in the third and sixth decades of life. There is no obvious seasonal clustering of cases. The crude average annual incidence rate varies in different countries from 0.6 to 1.9 per 100 000 people (Ropper *et al.* 1991; Chio *et al.* 2003). Familial occurrences do suggest some as yet unidentified genetic susceptibility factor (Geleijns *et al.* 2004).

Antecedent infections. Over half of Guillain–Barré syndrome patients experience symptoms of viral respiratory or gastrointestinal infections during the 1–3 weeks prior to the onset of neurological symptoms (Winer *et al.* 1988b). Serological studies have implicated a wide range of infective agents. Cytomegalovirus and *Campylobacter jejuni*, in approximately 30 per cent, are the commonest (Hadden *et al.* 2001). Epstein–Barr virus, *Mycoplasma pneumoniae*, human immunodeficiency virus, and childhood exanthems are also reported. The Guillain–Barré syndrome may accompany primary infection with HIV at a stage before viral antibodies are detectable in the serum; measurement of the p24 capsid antigen proving the underlying infection.

Cytomegalovirus and *Campylobacter* infections precipitate differing forms of Guillain–Barré syndrome. That associated with cytomegalovirus tends to occur in younger patients, with a high occurrence of respiratory muscle weakness, cranial nerve involvement, and significant sensory involvement (Visser *et al.* 1996). By contrast, *Campylobacter jejuni* infection is associated with preceding diarrhoeal illness in 70 per cent, a pure motor disorder (Section 21.10.2) is common, the electrophysiology often points to axonal dysfunction rather than demyelination, and recovery can be markedly slow. Forms of Guillain–Barré syndrome precipitated by both *Campylobacter* and cytomegalovirus may show delayed recovery compared to cases unassociated with these two infections (Visser *et al.* 1996).

When *Campylobacter jejuni* enteritis has precipitated Guillain–Barré syndrome, stool culture may be positive and serum IgM antibodies detected. Preceding *Campylobacter jejuni* infections can evoke Guillain–Barré syndrome even if there has been prompt treatment with antibiotics. Unusual forms of acute polyneuritis may occur following *Campylobacter* infection including variants with ophthalmoplegia. Different Penner serotypes of *Campylobacter* seem to provoke differing forms of acute polyneuritis, based upon studies in Japanese patients (Koga *et al.* 2005; Kimoto *et al.* 2006). Ganglioside epitopes on *Campylobacter* are thought to probe antibodies that cross-react with peripheral nerve glycolipids. Although some patients with Guillain–Barré syndrome, often the acute motor axonal variant, have anti-GM1 and anti-GD1$_A$ antibodies associated with infection by the HS:19 bacterial strain, this is by no means universal, and the association shows overlap with other clinical subtypes.

After immunization in 1976 of more than 40 million adults in the United States with swine influenza virus vaccine A/New Jersey/76 more than 500 cases of Guillain–Barré syndrome were reported in vaccinated individuals. It is estimated that this vaccine resulted in an excess incidence of one case of Guillain–Barré syndrome per 100 000 population, approximately doubling the normal incidence. No other causal relationship linking Guillain–Barré syndrome with vaccination by different strains of influenza virus has been shown. A prospective case-control study in England showed no significant excess of any form of vaccination during the 3 months preceding the Guillain–Barré syndrome (Winer *et al.* 1988a). The Guillain–Barré syndrome may be occasionally associated with underlying lymphoma, usually Hodgkin's disease. It can appear in patients already being treated with substantial doses of steroids, and is occasionally seen after renal transplantation from a cytomegalovirus positive donor, and after bone marrow or hepatic engraftment.

Immunopathogenesis. An autoimmune basis for the Guillain–Barré syndrome seems likely but remains unproven. Although antibodies to various gangliosides are described in Guillain–Barré syndrome, particularly following *Campylobacter* infection, it remains unclear whether these antibodies are pathogenic. Certainly no single antibody is ubiquitous for Guillain–Barré syndrome. Guillain–Barré syndrome bears a strong histological resemblance to experimental allergic neuritis, an acute monophasic disorder induced by immunization of experimental animals with peripheral nerve myelin proteins, particularly P2 and galactocerebroside. It is likely that diverse immunopathogenic mechanisms occur, including both antibody and cell-mediated immune mechanisms. Prominent neural inflammatory infiltrates can occur in both Guillain–Barré syndrome and experimental allergic neuritis.

Circumstantial support for autoantibody mediation of the neuropathy comes from the finding that plasma exchange shortens the duration of the disease. However, unlike most organ-specific autoimmune diseases, the Guillain–Barré syndrome shows no clear association with other autoimmune diseases or with major histocompatibility complex antigens. There is controversy as to whether demyelination or conduction block can be induced by injection of Guillain–Barré sera into animal nerves.

Pathology. The peripheral nerves in acute Guillain–Barré syndrome often show inflammatory cell infiltrate, with associated areas of demyelination, resembling experimental allergic neuritis. This inflammatory infiltrate is mainly perivascular and comprised of lymphocytes and macrophages. Electron microscopy shows that macrophages cause the myelin damage, and penetrate the

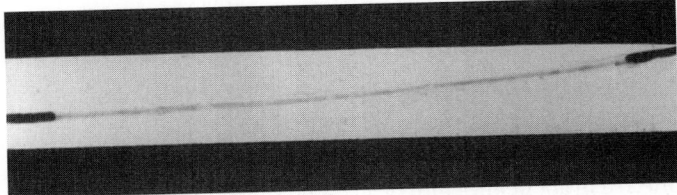

Fig. 21.15 Segmental demyelination: demyelinated internode of a teased surval-nerve fibre.

basement membrane around nerve fibres before stripping myelin sheaths off axons. Spinal nerve roots may be particularly affected, but changes are found at all levels of the peripheral nervous system. Teased peripheral sensory nerve fibre preparations may show marked segmental demyelination (Fig. 21.15). Some Wallerian degeneration may occur. Biopsy of the sural nerve may show surprisingly few abnormalities in comparison to the marked clinical severity of the neuropathy; this may reflect the distal and purely sensory nature of the sural nerve. Sensory nerve biopsy is generally unhelpful in establishing the diagnosis of Guillain–Barré syndrome; more typical demyelinative changes being present in motor nerves, which are not amenable to routine biopsy. Clinical criteria, spinal fluid protein elevation, and nerve conduction abnormalities remain the mainstay of diagnosis.

Occasional patients display motor–sensory axonal, rather than demyelinating, forms of Guillain–Barré syndrome. Opinion has been divided as to whether this represented secondary axonal degeneration induced by severe oedematous swelling of nerve roots, or whether it represented a primary attack on axonal antigens; evidence tends to support the latter (Lu *et al.* 2000). Characteristically, these patients have electrically inexcitable motor nerves early in their illness, electromyographic evidence of denervation within 2–5 weeks, marked muscle wasting, and protracted weakness with a generally poor recovery.

Clinical features

The neurological illness is preceded by symptoms of respiratory tract infection in approximately 40 per cent and gastrointestinal infection in less than 20 per cent in an English series; 8 per cent had undergone an operation in the preceding 3 months (Winer *et al.* 1988b). Neurological symptoms first develop 1–4 weeks after this infection. The Guillain–Barré syndrome produces a relatively symmetrical areflexic tetraparesis; the essential diagnostic criteria consist of progressive motor weakness of more than one limb, coupled with areflexia. Although sensory symptoms usually occur first, it is profound muscle weakness which is the main clinical feature in most patients once the disease is established.

Sensory features. In three-quarters of patients, the first neurological symptom is of paraesthesiae in the toes, less often in the fingers. Simultaneously, or soon afterwards, patients develop progressive limb weakness, often first noted as difficulty in walking. Despite the sensory nature of the initial symptoms, it is unusual for the eventual sensory loss to be particularly severe when compared to the profound motor loss. When sensory signs are present, they usually consist of impaired vibration and joint-position sensations. Half the patients experience pain which may be present from the outset and severe. It is generally maximal in the back and buttocks and may require short-term opiate analgesia. It usually resolves as recovery starts.

Motor features. Muscle weakness usually starts in the legs and ascends to the arms. Proximal muscle weakness may be prominent from the outset. The weakness is fairly symmetrical and usually involves the trunk musculature. It is unusual for the arms to be more severely weakened than the legs. Maximal weakness generally develops within 12–14 days of the onset of neurological symptoms. Although cessation of symptom progression within 4 weeks is often regarded as a necessary criterion for the diagnosis of Guillain–Barré syndrome, it is clear that in a small proportion of patients symptoms and signs continue to increase for up to 6 weeks from the onset. At the height of the disease, the majority of the patients are bed-bound and many of these have complete paralysis of all four limbs. Only 12 per cent remain able to walk throughout the illness. Those patients who become bed-bound and ventilator-dependent within 5 days tend to have the most prolonged disability and may develop severe permanent weakness. Significant sphincter dysfunction does not occur in Guillain–Barré syndrome, although urinary retention may result from abdominal wall weakness, particularly in patients with pre-existing urinary outflow tract obstruction.

Reflexes. Tendon reflexes are usually lost early in the disease. Total areflexia occurs in over 80 per cent of patients at some stage of the illness. The remainder usually lose their ankle jerks in isolation. Occasionally the tendon reflexes are preserved throughout the illness.

Cranial nerves. Approximately half the patients develop cranial-nerve palsies, usually in the wake of severe ascending limb weakness. Isolated unilateral or bilateral facial palsy is the commonest cranial-nerve lesion in Guillain–Barré syndrome. If weakness of the face is out of proportion to that of the limbs, Bannwarth's syndrome or Lyme disease (Section 21.14.3) should be considered, especially if the spinal fluid cell count is raised. Bulbar palsy and weakness of the muscles of mastication are the next commonest cranial-nerve abnormalities. With bulbar weakness there is a considerable risk of aspiration leading to acute respiratory failure or pneumonia, and endotracheal intubation should be performed if this seems likely to occur. Ocular palsy only occurs in about 10 per cent of patients, usually following severe limb and respiratory muscle weakness.

Breathing. Respiratory failure of sufficient severity to require assisted ventilation occurs in one-quarter of patients, although milder degrees of respiratory muscle involvement are much commoner. Patients with imminent respiratory failure may complain of orthopnoea and may be unable to complete more than a brief phrase of speech before pausing for breath. All patients with the evolving Guillain–Barré syndrome need to have their vital capacity and diaphragmatic movements assessed regularly so as to predict their requirement for assisted ventilation before a respiratory crisis occurs. Usually ventilation needs to be considered when the vital capacity falls below 1l in adults with otherwise normal lungs.

Autonomic dysfunction. This is common in the Guillain–Barré syndrome occurring in over 60 per cent (Zochodne *et al.* 1987). It contributes to the cardiac arrhythmias which are a leading cause of death, particularly in elderly patients. The presence of autonomic neuropathy cannot be predicted from the severity of the motor and sensory nerve abnormalities. Autonomic dysfunction may manifest either as excessive or as inadequate activity of the sympathetic or parasympathetic nervous systems. Wide fluctuations in blood pressure and heart rate, episodes of facial flushing, pupil abnormalities, patchy anhidrosis, paralytic ileus, or urinary retention may occur. Paroxysmal episodes of increased autonomic activity, causing hypertension, tachycardia, or facial flushing, are associated with a poor prognosis. They can be antecedents of sudden cardiac death in the Guillain–Barré syndrome. The pathophysiological basis of these varied autonomic manifestations is not known, but lymphocytic infiltrations of autonomic ganglia have been described.

Other neurological abnormalities. These occur occasionally in Guillain–Barré syndrome. Papilloedema occasionally develops. If so, it is sometimes associated with headache and raised spinal fluid pressure and tends to occur after a delay of some weeks. In some patients, it may reflect altered spinal fluid hydrodynamics resulting from the high protein content. However, other mechanisms must also be considered, since cases of papilloedema have been documented with normal spinal fluid protein levels. Optic neuritis and pyramidal tract signs are other rare manifestations which may point to a mild associated acute disseminated encephalomyelitis.

Relapsing forms. Recurrent Guillain–Barré syndrome occurs in up to 3 per cent, often after an interval of many years. The separate episodes may each be precipitated by new infections, such as recurrent cytomegalovirus exposure or two different infections such as respiratory syncytial virus and *C. jejuni*, or booster vaccinations with tetanus toxoid. Up to six separate episodes have been recorded, each in itself typical of Guillain–Barré syndrome. Relapsing Guillain–Barré syndrome can be distinguished from relapsing forms of chronic idiopathic demyelinating polyneuropathy by the rapidity of onset, the marked degree of recovery, normal CSF protein at the onset of an attack, the high incidence of preceding infections, and the lack of response to immunosuppressant drugs. The distinction from forms of chronic idiopathic demyelinating neuropathy of acute onset is particularly difficult, with late or multiple deteriorations being the chief distinguishing feature favouring the latter diagnosis (Ruts *et al.* 2005).

Regional variants. Some patients present without the ascending evolution of areflexic tetraparesis so typical of Guillain–Barré syndrome. However, their time course of deterioration and recovery, acellular spinal fluid with raised protein, and electrophysiological evidence of demyelination, coupled with some otherwise typical clinical features, make these likely to be regional variants of Guillain–Barré syndrome. Those which are recognized include: bifacial paresis, lateral rectus paresis plus paraesthesiae plus hyporeflexia, areflexic paraparesis, pharyngeal–cervical–brachial weakness, and Miller–Fisher syndrome coupled with weakness of bulbar or arm muscles (Ropper 1994). Isolated arm weakness may occur, either as a motor–sensory demyelination or as pure motor axonal involvement with anti-GM1 antibodies. No treatment trials have been undertaken for these rare variants, and it seems wise to treat them according to principles established for Guillain–Barré syndrome.

Investigation

CSF. In 80 per cent of cases the spinal fluid characteristically shows '*dissociation albumino-cytologique*' in which the protein content is elevated, often exceeding 2 g/l, with a normal cell count. Normal spinal fluid protein concentration is commonest when the spinal tap is performed during the first few days of neurological symptoms. This limits the diagnostic value of lumbar puncture in early cases. About 10 per cent of patients have a lymphocytic spinal fluid, which should raise consideration of Lyme disease or

HIV infection. Some patients develop a reduced serum sodium, possibly due to resetting of osmoreceptor responses.

Nerve conduction studies. These studies can be surprisingly normal early in the Guillain–Barré syndrome, despite severe paralysis. This reflects the purely radicular location of early demyelination or conduction block in many patients; conventional conduction studies merely measure motor conduction over the distal segments of peripheral nerves. Sometimes many peripheral nerves must be studied before diagnostic abnormalities are detected. Within the first 2 weeks, the commonest findings are of mildly prolonged distal motor latencies and of conduction block in which the amplitude of the compound muscle action potential progressively diminishes with more proximal sites of nerve stimulation. 'F'-waves may be absent or prolonged. As the disease progresses, the sensory nerve action potentials are usually lost, and motor slowing may become more evident distally. Permanent disability is predicted by electrical inexcitability of nerves early on and tends to be associated with electrophysiological evidence of axonal degeneration.

Differential diagnosis

The Guillain–Barré syndrome usually presents a distinctive clinical picture. The potential range of differential diagnosis of acutely evolving paralysis is enormous: spinal cord disease; neuromuscular transmission disorders; myopathy; vasculitic neuropathy; porphyria; malignant meningitis; infective neuropathies such as Diphtheria, Borreliosis, or Poliomyelitis; biological toxins such as tick paralysis or Botulism; drug and chemical toxins; metabolic abnormalities; critical illness polyneuropathy; and psychologically determined weakness (Ropper *et al.* 1991).

Acute spinal-cord lesions pose the commonest diagnostic difficulty and spinal MRI need to be undertaken in cases of doubt. However, the distinction is usually simple because of the extensor plantar responses, sensory level, prominent sphincter involvement, and the cellular spinal fluid encountered in acute ascending or transverse myelitis. It is rare for acute inflammatory myopathies to be confused with the Guillain–Barré syndrome. Pointers to primary muscle disease include the absence of sensory symptoms, preserved reflexes, normal spinal fluid protein, abnormal electromyogram, and raised serum creatine kinase levels.

Three rare acute neuropathies should be distinguished from the Guillain–Barré syndrome because they require different approaches to therapy. Borrelia infection causing Lyme disease or Bannwarth's syndrome (Section 21.14.3) is suggested by prominent unilateral or bilateral facial paralysis, radicular pain, and a cellular CSF. Porphyric polyneuropathy (Section 21.8.6) is associated with early neuropsychiatric abnormalities, abdominal pain, a purely motor syndrome, and preservation of the ankle jerks despite loss of the knee jerks. Diphtheritic polyneuropathy (Section 21.14.4) is now rare in Western countries, although resurgent in Eastern Europe, and should be considered in patients with descending demyelinating polyneuritis starting as bulbar palsy.

Treatment

Survival in the Guillain–Barré syndrome depends primarily upon meticulous attention to intensive care during the acute paralytic phase (Section 2.8) (Hughes *et al.* 2005). Feeding by naso-gastric tube should be instituted in those with bulbar dysfunction. Subcutaneous heparin and elastic stockings provide prophylaxis against deep venous thrombosis and pulmonary embolism. Vigilant electrocardiographic monitoring allows prompt recognition and treatment of cardiac arrhythmias which may be provoked by endotracheal suctioning or suxamethonium administration. Beta-blockers may be required for those with hypertensive crises. Patients with Guillain–Barré syndrome are particularly susceptible to hypotensive side effects of drugs, including thiopentone, frusemide, and morphine. Nursing care will prevent decubitus ulcers. Regular physiotherapy, and careful limb positioning will prevent muscle contractions in patients with prolonged paralysis. The gastrocnemius and soleus muscles are particularly prone to such contractures, which may lead to permanent walking disability even if muscle power returns.

Ventilation. Patients likely to deteriorate to the point of needing assisted ventilation should be alerted to this probability beforehand, whilst they can still ask questions, in a manner of calm planning. Endotracheal intubation and ventilation should be instituted without delay either if respiratory muscle failure is imminent or if paralysis of bulbar and laryngeal muscles places the patient at risk of choking. Assisted ventilation is usually required when the vital capacity has fallen to 15 ml/kg body weight; that is a vital capacity of approximately 1l for a 65 kg adult. Nasal endotracheal tubes are well tolerated by conscious patients and should be replaced by temporary tracheostomy if, as is usually the case, the period of ventilation is likely to exceed 1 week. Pulmonary atelectasis and infection are common in intubated patients and should be treated promptly with antibiotics and physiotherapy.

Steroids. Neither oral steroids nor intravenous high-dose steroids have a place in treating the Guillain–Barré syndrome (Guillain–Barré syndrome steroid trial group 1993). Addition of a 5-day course of 500 mg intravenous methylprednisolone to standard IvIg therapy does not improve 4-week outcome (van Koningsveld *et al.* 2004).

Plasma exchange (Section 21.3.3). This shortens the time taken for patients with Guillain–Barré syndrome to start to improve, to regain functional abilities such as walking, and reduces their requirement for assisted ventilation (Winer 2002). Plasma exchange enables the median patient to walk independently at 53 days compared to 85 days for controls, and allows 82 per cent to walk independently at 6 months compared to 71 per cent of controls. It is unclear whether plasma exchange improves survival or reduces the number of patients unable to walk at 1 year. Subgroup analysis suggests that those patients with acute motor axonal forms associated with diarrhoea and *Campylobacter* infection have a better outcome following IvIg than plasma exchange (Visser *et al.* 1999). To be maximally effective, plasma exchange needs to be started within the first week of neurological symptoms. It is unlikely to be effective if given after 2 weeks of neurological symptoms. Plasma exchange is recommended for those patients approaching inability to walk or with impairment of bulbar or respiratory function. Plasma-exchange schedules vary, but four or five 4-l exchanges using a continuous-flow technique, given on sequential days, are recommended. The plasma may be replaced by either albumin or fresh frozen plasma; the risk of non-A, non-B hepatitis being greater with the latter. About 10 per cent of patients treated by plasma exchange will subsequently undergo a mild relapse between 5 and 42 days later, which may be treated by a further course of plasma exchange. The factors determining poor outcome, such as advanced age or low compound muscle action potential amplitudes, appear to be the same for those receiving plasma exchange as for those receiving conservative therapy.

Intravenous immunoglobulin, IvIg, (Section 21.3.3). This treatment, given at 0.4 g/kg body weight/day for 5 days is at least equally effective to plasma exchange (Plasma exchange/Sandoglobulin

Guillain–Barré syndrome trial group 1997). IvIg has become the treatment of choice because it is immediately available, does not require cannulation of a major vessel, has fewer side effects than plasma exchange, and does not carry the same risks of exacerbating circulatory disturbances due to autonomic neuropathy. Also IvIg may be more effective than plasma exchange for the motor axonal subgroup resulting from diarrhoeal *Campylobacter* infections (Visser *et al.* 1999). There is concern that the easy availability of IvIg in district general hospitals may lead to Guillain–Barré syndrome being treated in intensive care units lacking expertise in the disease, with resultant increased death due to complications. As with plasma exchange, IvIg-treated patients may secondarily deteriorate within 2 weeks of treatment. It is unclear whether this simply reflects the natural history of underlying Guillain–Barré syndrome only temporarily modified by IvIg, or some specific IvIg effect, or indeed whether secondary deterioration is an indication for a second course of IvIg. Large scale trials of IvIg or plasma exchange have not been undertaken in children, but it seems logical to expect similar benefits to those seen in adults, with IvIg being preferable to plasma exchange, particularly given the problem of vascular access in small children.

Choice of immunotherapy. Plasma exchange and IvIg are equally effective. There is no additional benefit from combining the two or from giving plasma exchange and IvIg in sequence. Steroids have no place (Hughes *et al.* 2003). IvIg is the treatment of choice, particularly if it can be administered within 2 weeks of the first neurological symptoms. There is no evidence that plasma exchange or IvIg are at all effective if given more than 4 weeks after onset of neurological symptoms. Common sense suggests that the optimal benefit is to be gained by starting immunotherapy while the patient is still ambulant, although evidence for this is only available for plasma exchange.

Prognosis

Most patients with the Guillain–Barré syndrome will make a good spontaneous recovery if they receive competent supportive treatment. Even when general intensive care facilities are available, up to 10 per cent of patients may die in the acute phase of the disease. These patients are usually elderly and generally succumb to cardiac disease, pulmonary embolism, chest infection, or complications of intensive care or invasive procedures (Chio *et al.* 2003). The mortality is 4–5 per cent even for patients treated in specialist neurological units with plasma exchange or IvIg (Plasma Exchange/Sandoglobulin Guillain–Barré syndrome Trial Group 1997). Of the survivors, half make a full recovery but the others show some permanent residual symptoms and signs, usually weakness of distal leg muscles, absent ankle jerks, or distal sensory loss (Dornonville de la Cour and Jakobsen 2005). Even after IvIg or plasma exchange therapy, 16.5 per cent are unable to walk at 48 weeks (Plasma Exchange/Sandoglobulin Guillain–Barré syndrome Trial Group 1997). The factors predictive of poor outcome with slow recovery or permanent disability, include age over 60 years, a preceding diarrhoeal illness, development of severe paralysis within 5 days of the onset, respiratory failure requiring ventilation, and mean distal compound muscle action potentials of less than 20 per cent of normal.

21.10.2 Acute motor axonal neuropathy

Acute motor axonal neuropathy is a distinct subtype of Guillain–Barré syndrome which involves axonal degeneration or conduction block (Capasso *et al.* 2003) affecting motor fibres alone rather than the usual demyelination of both sensory and motor fibres. It was originally recognized in large summer epidemics in China, but is known to occur sporadically worldwide (Hafer-Macko *et al.* 1996). When compared to others with Guillain–Barré syndrome, acute motor axonal neuropathy patients have purely motor symptoms and signs, are more likely to have a more rapid evolution of limb weakness, plateau on average at 6 days compared to 9, have predominantly distal weakness, and are less likely to have cranial nerve involvement (Visser *et al.* 1996; Hiraga *et al.* 2003). Neurophysiology does not show the usual degree of motor slowing and prolongation of distal motor latencies, sensory nerve action potentials are generally preserved, motor nerves may be inexcitable, and electromyography often shows acute denervation changes. The differential diagnosis is similar to Guillain–Barré syndrome (Section 21.10.1) with particular consideration of poliomyelitis where that is still endemic. Subgroup analysis points to a better 6 months outcome if acute motor axonal neuropathy is treated with IvIg rather than plasma exchange (Visser *et al.* 1996; 1999). Acute motor axonal neuropathy is more likely to be associated with long-term or permanent disability. However many patients make substantial improvement in the early weeks after IvIg suggesting that reversible conduction block, rather than axonal degeneration, underlies much of the disability.

The pathogenesis of acute motor axonal neuropathy is of considerable interest given its clear association with preceding diarrhoeal illness caused by *Campylobacter jejuni* and the frequent development of anti-GMI gangliosides antibodies (Visser *et al.* 1996). Acute motor axonal neuropathy is particularly likely after infection with the Penner HS19 serotype of *C. jejuni*; *Campylobacter* lipopolysaccharides having ganglioside-like moieties raising the likelihood of antibodies cross-reacting with nerve (Koga *et al.* 2005; Kimoto *et al.* 2006). It is unknown whether some host susceptibility factor determines whether a *Campylobacter*-infected patient goes on to develop acute motor axonal neuropathy. This immunopathogenic mechanism is supported by demonstration of IgG and complement deposits on the axolemma at the nodes of Ranvier of motor fibres in fatal cases (Hafer-Macko *et al.* 1996).

21.10.3 Acute sensory neuropathy

Occasional patients with acute polyneuritis show profound limb sensory loss without weakness, particularly affecting joint, position, and vibration sensation, and with severe ataxia (Oh *et al.* 2001). Despite the lack of weakness, slowing of motor nerve conduction is usually demonstrable electrophysiologically. A high serum titre of anti-GD1b ganglioside antibody may be present (Pan *et al.* 2001). Patients show the same monophasic time course as for Guillain–Barré syndrome, and the same approach to treatment should be followed. Autopsies in such patients show lymphocytic infiltration and demyelination in the dorsal roots and sensory peripheral nerves.

This rare sensory form of polyneuritis should be distinguished from acute sensory neuronopathy, in which the limb sensory loss affects all modalities and often starts asymmetrically in the upper limbs, extends on to the trunk and face, there is no motor loss, recovery is unusual, and the pathology primarily involves loss of dorsal root ganglion neurones with lymphocytic infiltration (Hainfellner *et al.* 1996).

21.10.4 Acute autonomic neuropathy

Rarely patients present acutely with symptoms of autonomic neuropathy. Their symptoms are varied and reflect failure of the sympathetic and parasympathetic systems: postural hypotension, blurred vision, ptosis, pupillary abnormalities, dry mouth and eyes, anhidrosis, erectile failure, and constipation. Some patients may have varying degrees of associated thermal and pain sensation disturbances in the limbs. Spontaneous neuropathic pain may be prominent. Peripheral neuropathy may not be suspected initially because the sparing of larger myelinated fibres results in preserved tendon reflexes and normal sensory nerve action potentials (Suarez *et al.* 1994). Spontaneous recovery is the rule but is often incomplete. Prompt response to IvIg has been recorded (Mericle and Triggs 1997).

Within this diagnostic group are occasional patients with acute or subacute onset of a pure autonomic syndrome, without a sensory disturbance. Such patients not only have orthostatic hypotension, but may also show prominent cholinergic dysautonomia, reflected by dry eyes and mouth, abnormal pupil-light responses, upper gastrointestinal symptoms, and neurogenic bladder. These patients often have high levels of antibody to the ganglionic acetylcholine receptor of the α3 nicotinic type present in autonomic ganglia (Klein *et al.* 2003; Sandroni *et al.* 2004). It seems reasonable to consider early immunomodulatory therapy for such patients. Subacute presentations of this pure autonomic disorder can be paraneoplastic.

21.10.5 Miller Fisher syndrome

This distinctive syndrome comprises total external ophthalmoplegia, severe ataxia, and generalized tendon areflexia which all develop over a few days (Fisher 1956). The spinal fluid protein is elevated and patients recover over a matter of weeks. Some patients have combined features of Guillain–Barré and Miller Fisher syndromes in which the oculomotor disturbance and limb weakness occur within a few days of one another. Serial neurophysiological studies have shown evidence of peripheral nerve involvement in the Miller Fisher syndrome, with prolonged peripheral conduction in the blink reflex arc and subsequent recovery of motor nerve and 'F'-wave conduction velocities. The serum of over 90 per cent of patients contains antibodies against the GQ1b and GT1a gangliosides of both peripheral and central nervous systems (Willison and O'Hanlon 1999). Preceding infection with *Campylobacter jejuni* of the HS2 or HS4 serotypes is usual (Kimoto *et al.* 2006). The titre of this antibody tends to decline commensurate with clinical improvement. Some patients seem to respond promptly to either plasma exchange or intravenous immunoglobulin, but the overall impact of these treatments on eventual recovery is questioned (Mori *et al.* 2007).

There has been debate as to the existence of a central nervous system component to Miller Fisher syndrome. Indeed, some have considered the syndrome to be a form of brainstem encephalitis. Although brainstem encephalitis may present a similar clinical picture to the Miller Fisher syndrome, in addition it usually involves disturbed consciousness, extensor plantar responses, and MRI brain abnormalities, whilst tendon reflexes are preserved (Odaka *et al.* 2003). Some of these brainstem encephalitis patients have an axonal neuropathy too.

21.11 Chronic idiopathic polyneuropathies

21.11.1 The spectrum of disorders

Significant reversal of severe disability can be achieved with immunomodulatory treatment for many patients with this varied group of neuropathies. Although the sensorimotor demyelinating and axonal form is that most commonly encountered, the relative degrees of motor and sensory fibre involvement and the relative balance between demyelination, conduction block, and axonal degeneration, vary considerably in the different clinical subtypes. It is not known yet whether this reflects fundamentally different underlying pathogenic mechanisms, or whether the various clinical syndromes simply represent noteworthy peaks in a continuum. Clinical and electrophysiological distinction of these different syndromes is of practical importance because it influences the approach to treatment (Saperstein *et al.* 2001; Busby and Donaghy 2003). For instance steroids are usually highly effective in chronic inflammatory demyelinating sensorimotor polyneuropathy whereas they often cause deterioration, or at best are ineffective, in multifocal motor neuropathy with conduction block. As a general rule, intravenous immunoglobulin seems best effective when much of the disability is due to conduction block, whereas steroids and plasma exchange seem most effective when the disability is associated with histological demyelination as evidenced by slowed nerve conduction velocities.

21.11.2 Chronic inflammatory demyelinating polyneuropathy

This is a progressive, sometimes relapsing, steroid-dependent, demyelinating sensorimotor polyneuropathy primarily affecting the limbs. Usually it develops slowly, over months or years. Abrupt onset resembling Guillain–Barré syndrome can occur, yet with persistent symptoms (Mori *et al.* 2002). It is also known as chronic relapsing polyneuritis, chronic idiopathic demyelinating poly(radiculo)neuropathy, relapsing corticosteroid-dependent polyneuritis, or relapsing hypertrophic neuritis. Its recognition is of great importance because of the excellent response to immunomodulatory therapy in most patients.

Aetiology. The prevalence of chronic inflammatory demyelinating polyneuropathy increases with age from infancy to senescence with a mean of onset in the fifth decade. It is commoner in males. Accurate estimates of its incidence are not available, the overall prevalence is about 2 per 100 000, reaching 6.7/100 000 in the eighth decade in Australia (McLeod *et al.* 1999). Up to half the patients have a relapsing and remitting course, in which the initial deterioration can be rapid, resembling the Guillain–Barré syndrome. However, experience in Britain shows most patients to have stable or progressive neuropathy when one excludes fluctuation attributable to treatment changes. Unlike the Guillain–Barré syndrome, patients subsequently progress downhill over more than 2 months or undergo secondary deterioration some weeks after an initially satisfactory response to plasma exchange or intravenous immunoglobulin. The deterioration may be steady, or relapsing and remitting. In women, relapses are particularly associated with the third trimester of pregnancy or the immediate postpartum period. Up to a third of patients give a history of antecedent viral infection or vaccination. Serological evidence of previous cytomegalovirus infection is found in about half, although a directly causative relationship has not been established.

Various features point to an immunological mechanism for chronic inflammatory demyelinating polyneuropathy which might be considered as the chronic counterpart of Guillain–Barré syndrome. Nerve biopsies often show T-lymphocyte cell infiltrates, which may be slight, with early myelin stripping by macrophages. HLA antigen studies showed an increased frequency of the A3, B7, and DR2 antigens, and an association with specific GM haplotypes in a population of Australian patients (Feeney et al. 1990). An ubiquitous causative autoantibody has not been identified, although about 30 per cent have serum antibodies against myelin glycoprotein P_0 which are capable of inducing conduction block and demyelination on injection into rat sciatic nerve (Yan et al. 2001). It remains unknown whether there are other target antigens in the other patients, or the extent to which cell-mediated immunity may be important.

Pathology. The histological features in sural nerve biopsies are often indistinguishable from those of the Guillain–Barré syndrome. Teased fibres show segmental demyelination and thinly remyelinated internodes. Inflammatory infiltrates may be found in the endoneurium. Axonal loss may particularly affect large myelinated fibre populations. This range of abnormalities overlaps with those seen in chronic idiopathic axonal polyneuropathy (Section 21.11.8), limiting the diagnostic specificity of nerve biopsy from that condition (Bosboom et al. 2001). Perivascular macrophage clustering may be a particular marker of chronic inflammatory demyelinating neuropathy (Sommer et al. 2005). Rarely nerve biopsies show hypertrophic 'onion-bulb' formations, raising difficulties in distinguishing chronic forms from Charcot–Marie–Tooth disease Type I (Section 21.4.4). Although sural nerve biopsy is frequently undertaken in suspected chronic inflammatory demyelinating neuropathy, it rarely adds diagnostic information in patients with a characteristic clinical and electrophysiological picture with raised spinal fluid protein.

Clinical features

Three-quarters of patients present with a mixed sensorimotor neuropathy which is relatively symmetrical. Less commonly asymmetrical, or predominantly motor or sensory forms are encountered. Paraesthesiae are a common early feature and may be uncomfortable. Loss of vibration and joint-position senses is usually demonstrable and Rombergism is a common early symptom. Limb weakness is generally distributed both proximally and distally. A predominantly distal subtype also occurs, often in older men and associated with IgM paraproteinaemia: distal acquired demyelinating symmetric polyneuropathy (Mygland and Monstad 2003). Usually, all the reflexes are lost. The rate of deterioration varies but progression over more than 8 weeks is a distinguishing criterion from the Guillain–Barré syndrome (McCombe et al. 1987b). The cranial nerves are affected in about 15 per cent of patients, usually to a mild degree. Dysphagia, dysarthria, weakness of facial or masticatory muscles, and diplopia are the commonest cranial nerve manifestations. Respiratory failure is rare (Henderson et al. 2005). Papilloedema occasionally occurs. A coarse irregular action tremor may occur, seemingly unrelated to the mild degrees of proprioceptive loss or weakness, and resembles that seen in patients with paraproteinaemic neuropathy. This may reflect mismatch of muscle spindle afferent information from agonist and antagonist muscles due to severely slowed peripheral nerve conduction (Busby et al. 2003). Limb muscle weakness or sensory ataxia are the usual causes of significant disability.

Central nervous system involvement. Some patients with chronic inflammatory demyelinating polyneuropathy also have a clinical history of a relapsing multifocal central nervous system disorder. Cerebral magnetic resonance imaging may show periventricular plaques of demyelination, and evoked responses may be prolonged. Subclinical abnormalities of the central nervous system are present in a third to a half of patients. These findings pose questions of overlap with multiple sclerosis. They also raise the possibility that tremor in some patients with chronic inflammatory demyelinating polyneuropathy could be due to associated central nervous system involvement (Koller et al. 2005).

Nerve conduction studies. The mainstay of diagnosis is the demonstration of slowed motor nerve conduction, often with a degree of conduction block, in a patient with a chronically or subacutely progressive acquired peripheral neuropathy. Electrophysiological criteria have been proposed for the diagnosis of chronic inflammatory demyelinating polyneuropathy (Section 3.5.3): motor conduction velocities of less than 75 per cent of the lower limit of normal, distal motor latencies exceeding 130 per cent of the upper limit of normal, temporal dispersion, or conduction block following proximal stimulation and prolonged F-wave latencies. Sensory nerve action potentials are usually diminished or lost. Diagnostic difficulty may arise in patients, usually with early and mild disease, in whom the motor conduction velocity is insufficiently slow to be sure that the neuropathy is primarily demyelinating. In such patients, motor conduction velocities only just below the normal range are not uncommon.

CSF. This protein is elevated above 0.6 g/l in 50–85 per cent (Busby and Donaghy 2003).

Differential diagnosis

This most frequently causes difficulty in the distinction of chronic inflammatory demyelinating polyneuropathy from Charcot–Marie–Tooth disease Type I (Section 21.4.4), particularly if molecular genetic tests have been negative for the latter. Pointers favouring chronic inflammatory demyelinating polyneuropathy are a subacute rate of deterioration, relapsing-remitting progression of motor weakness, positive sensory symptoms such as paraesthesiae, raised spinal fluid protein, absence of a family history, and the absence of onion-bulb formations in a sural nerve biopsy. Motor nerve conduction studies tend to show multifocal slowing, conduction block, and dispersion of the distal compound muscle action potential in chronic inflammatory demyelinating polyneuropathy. By contrast the slowing is more uniform, without focal block, in Charcot–Marie–Tooth disease Type I. Associated deafness and pigmentary retinopathy should raise the possibility of Refsum disease, a rarely encountered possibility confirmable by blood phytanic acid measurement (Section 21.8.1). The mitochondrial disorder MNGIE (Section 21.7.6) should be considered in younger patients unresponsive to immunomodulatory treatment.

MRI-proven hypertrophy of cervical roots, brachial plexus, or the cauda equina may be noted in chronic inflammatory demyelinating polyneuropathy and Guillain–Barré syndrome (Duggins et al. 1999). Sometimes such MRI findings raise the question of a diffuse nerve root infiltrative process, but the presence of electrophysiologically proven demyelinating polyneuropathy is strong evidence against that, and should forestall nerve root biopsy. Similar cauda equine hypertrophy with leg neurological deficits occurs rarely in the absence of peripheral nerve conduction abnormalities (Burton *et al.* 2002). Hypertrophied roots and nerves may be more vulnerable to compression by stenosis of the lumbar spinal canal, in root exit foramina, and at common entrapment sites.

Chronic inflammatory demyelinating polyneuropathy is not usually a paraneoplastic phenomenon except in the sense of its common association with paraproteinaemias (Section 21.12.1). One should be suspicious of an underlying lymphoma, carcinoma (Section 21.13.2), or Castleman's disease (Section 21.12.3) in two circumstances. First when the neuropathy evolves relatively rapidly, and there are unusual features such as extensive cranial nerve involvement or neuropathic pain. Second when the patient relentlessly deteriorates despite immunosuppressant therapy.

Treatment

A proven diagnosis of chronic inflammatory demyelinating polyneuropathy means there is an excellent chance of recovery with immunomodulation therapy. Without treatment, chronic inflammatory demyelinating polyneuropathy is eventually fatal in up to 10 per cent of patients (Bouchard *et al.* 1999). Untreated, many of the remainder suffer protracted and serious disability. The degree of associated axonal loss may determine the chance of a good recovery, and may be lessened by prompt and early treatment.

Occasionally immunomodulatory treatment must be started in severe weakness to prevent further decline before it is clear whether the patient has Guillain–Barré syndrome, which would plateau by 4 weeks, subacute inflammatory demyelinating polyneuropathy progressing for up to 8 weeks, or chronic inflammatory demyelinating polyneuropathy, which should progress beyond 8 weeks. Usually it is preferable to give such patients a course of intravenous immunoglobulin or plasma exchange, rather than start steroids, since a secondary deterioration when the treatment effect wears off at 6–10 weeks will indicate that the underlying neuropathy continues to evolve, thus requiring more definitive long-term immunosuppressant therapy. An unusual intermediate form called subacute inflammatory demyelinating neuropathy evolving over 4–8 weeks is described (Oh *et al.* 2003).

Oral steroids. This therapy is the mainstay of treatment and often produces noteworthy improvements within 3 weeks. The results of therapy can be dramatic; bed-bound patients may regain almost normal motor function. Unfortunately not all patients respond to steroid therapy. It is elderly patients, or those with a significant degree of axonal degeneration, who tend to respond less well. Prednisolone administration schedules vary. An initial daily dosage of 60 mg is recommended, falling to 45 mg daily after 2 weeks, and converting to 45 mg on alternate days over the next 2–3 months. Steroid therapy may need to be continued for years and protection against osteoporosis should be prescribed (Section 21.3.3). Patients frequently relapse within a few months of withdrawing prednisolone or after reducing below the usual maintenance dose of 15–30 mg on alternate days.

Other immunosuppressant drugs are sometimes useful. Azathioprine is often added as a steroid-sparing agent, although there is no controlled evidence that it is beneficial in this condition. Nonetheless remission does seem to be maintainable by Azathioprine in some patients, particularly young women, who may relapse some months after this drug is stopped. Cyclosporin A can induce improvement in some steroid-resistant patients (Hodgkinson *et al.* 1990). Interferon-α 2A is effective in some patients resistant to other immunomodulatory therapy (Gorson *et al.* 1998), but conversely

the neuropathy has been reported to develop during treatment with interferons-α or -β and with tumour-necrosis-factor-α blockers (Koller *et al.* 2005; Richez *et al.* 2005). Anecdotal reports suggest responses to treatment-resistant chronic inflammatory demyelinating neuropathy to methotrexate or high dose cyclophosphamide (Brannagan *et al.* 2002; Fialho *et al.* 2006).

Plasma exchange (Section 21.3.3). This produces substantial improvement in 80 per cent of patients with either progressive or relapsing forms of chronic inflammatory demyelinating polyneuropathy (Hahn *et al.* 1996a). The neuropathy relapses some 4–10 weeks after a successful course of plasma exchange, and definitive long-term therapy should be commenced simultaneously unless repeated plasma exchange is envisaged.

Intravenous immunoglobulin, IvIg (Section 21.3.3). This produces significant improvement in about 65 per cent of chronic inflammatory demyelinating neuropathy patients (Hahn *et al.* 1996b). It can improve conduction block in peripheral nerves. The benefit of a 5-day course usually lasts 4–10 weeks, and the general indications are identical to plasma exchange. Because it is easier to administer, it provides a better option for maintenance therapy. It is similarly effective to oral prednisolone (Hughes *et al.* 2001).

Strategies. Strategies for immunomodulatory treatment vary in different clinical situations. Some patients have such mild forms of chronic inflammatory demyelinating polyneuropathy that the risks of treatment far outweigh the small benefits which could accrue. Usually a patient can be maintained on Prednisolone 15–30mg on alternate days, often in conjunction with Azathioprine. For inadequate responses, Immunoglobulin then plasma exchange should be tried, and in last resort Cyclosporin, Interferon-α 2A, Cyclophosphamide, or Methotrexate. Infantile and childhood chronic inflammatory demyelinating neuropathy responds to steroids or immunoglobulin. The elderly can be slow to begin what may be an ultimately useful response and plasma exchange or immunoglobulin should be considered with steroids from the outset. Withdrawal of immunomodulatory treatment is only likely to be a prolonged success in those unusual patients who fully remit, often children, adolescents, or young women; Azathioprine offers the chance of maintaining steroid-free remission once a good response has been obtained. Relapses on stopping or reducing therapy should be treated promptly since such patients seem to become less completely responsive due to accumulated axonal damage.

Given that steroids remain the mainstay of treatment in most patients, plasma exchange or IvIg are recommended in the following circumstances:

- those who fail to respond promptly or adequately to steroids;
- those in whom high initial steroid dosages pose contraindications, such as steroid-induced psychosis or brittle diabetes;
- to 'kick-start' an improvement in the elderly who are notoriously slow responders to steroids;
- if there is severe disability at the outset; and
- to reverse a relapse promptly so as to avoid the need for reinstituting very high steroid dosages.

Objective monitoring of therapy is important to judging its effectiveness (Section 21.3.2). Velocity of nerve conduction is of little help in monitoring the ongoing severity of any patient's neuropathy. Quantifiable foci of conduction block can provide useful guidance. Ultimately it is the clinical assessment of reliable parameters such as walking speeds, stair-climbing ability, manipulatory tasks such as buttons, Rombergism, and ability to stand on tiptoe or hop, which provides the best index of whether any patient's response to treatment is useful.

21.11.3 Multifocal motor neuropathy with conduction block

Many patients with multifocal motor neuropathies used to be diagnosed as suffering from benign forms of motor neurone disease solely affecting lower motor neurones. These patients may present at any age in adult life with symptoms that may have progressed slowly for 20 years or more.

Clinical picture. This varies immensely. Weakness is usually maximal distally, is often notably asymmetrical, and is more likely to start and predominate in the arms than the legs. In retrospect, often the first symptom has been inability to fully extend a single finger (Fig. 21.17), probably reflecting the onset of conduction block in a terminal branch of the posterior interosseous nerve (Slee *et al.* 2007). Muscle atrophy occurs with time. Occasionally a weakened muscle may be hypertrophied (Fig. 21.18). Myokymia or coarse fasciculations are observed sometimes in weakened muscles, and occasionally in remote muscles. Cranial nerve involvement can occur, affecting bulbar muscles, causing difficulty in differentiation from amyotrophic lateral sclerosis. Reflex loss is usually restricted to the affected muscles, although it can be more generalized. The critical physical sign pointing to conduction block is a muscle which is markedly weakened despite being unwasted.

Motor nerve conduction studies show varying combinations of multifocal motor conduction block, prolonged or absent F-waves, prolonged distal latencies, reduced motor nerve conduction velocities, or motor axonal loss with electromyographic evidence of denervation. The crucial electrodiagnostic feature is conduction block restricted to a nerve's motor fibres, at a site not vulnerable to compression (Fig. 21.19); unfortunately this often occurs in electrophysiologically inaccessible segments of an affected nerve, such as proximally. Pathophysiologically, these sites of conduction block seem associated with either depolarization or hyperpolarization (Kiernan *et al.* 2002; Priori *et al.* 2005). Despite inability to demonstrate conduction block electrophysiologically, some patients with

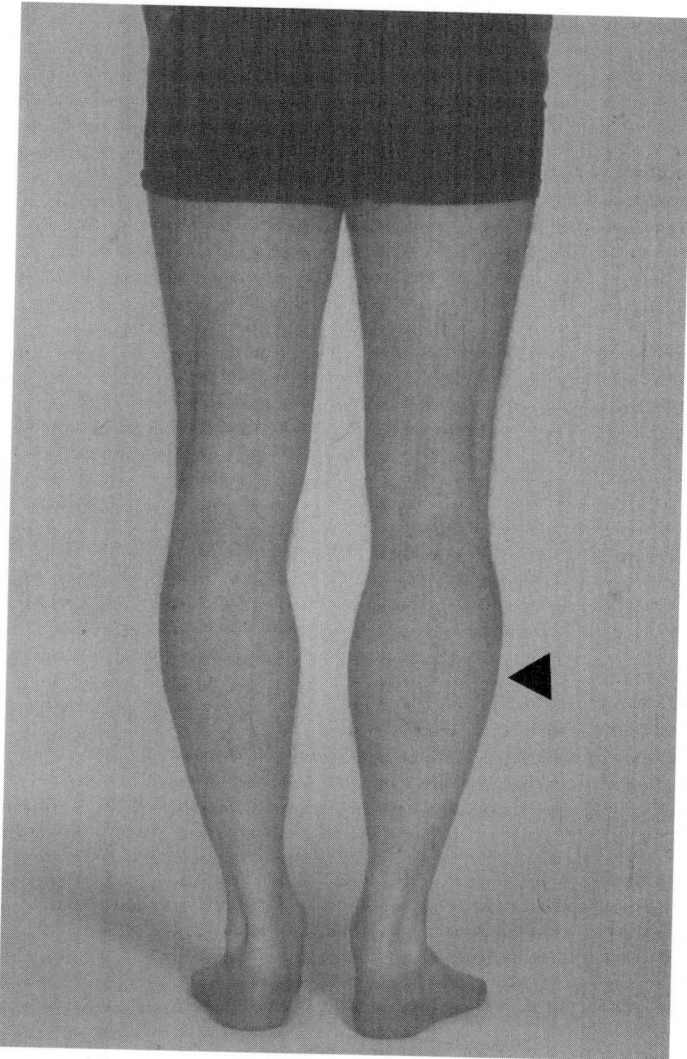

Fig. 21.17 Hypertrophy of the right, and weakened calf muscles (arrowed) in a patient with multifocal motor neuropathy with conduction block.

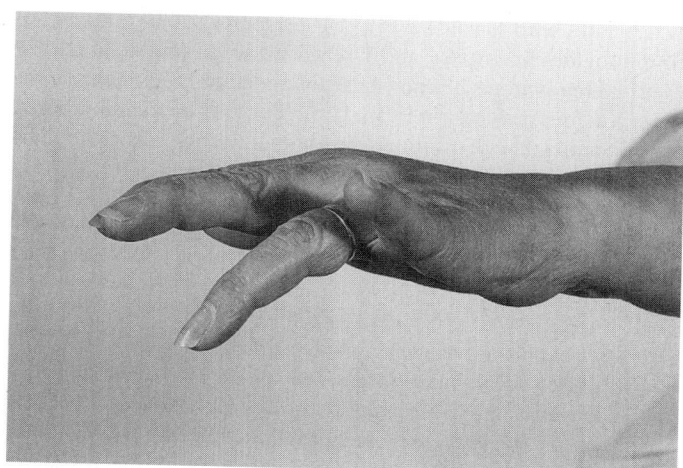

Fig. 21.16 Weakness of extension of a single finger: a common early symptom of multifocal motor neuropathy with conduction block.

the clinical phenotype typical of multifocal motor neuropathy respond well to IvIg (Delmont *et al.* 2006; Slee *et al.* 2007). Redefinition of the criteria for electrophysiological diagnosis of conduction block improves the diagnostic inclusion of IvIg responsive patients: removal of exclusions based on over-restrictive temporal dispersion, and allowing as little as 32 per cent reduction in compound muscle action potential following proximal stimulation (Ghosh *et al.* 2005a).

Serum antibodies to GM-1 gangliosides are present in approximately a third of patients with multifocal motor neuropathy with conduction block (Slee *et al.* 2007). It remains to be established whether anti-GM1 plays a pathogenic role in multifocal motor neuropathy (Willison and Yuki 2002). Interestingly neonatal motor neuropathy has been observed in a newborn from an α-GM1-antibody-positive mother with multifocal motor neuropathy (Attarian *et al.* 2004). It binds to nodes of Ranvier in peripheral nerves (Santoro *et al.* 1990). Focal deposition of immunoglobulins, and demyelination associated with inflammation, have been

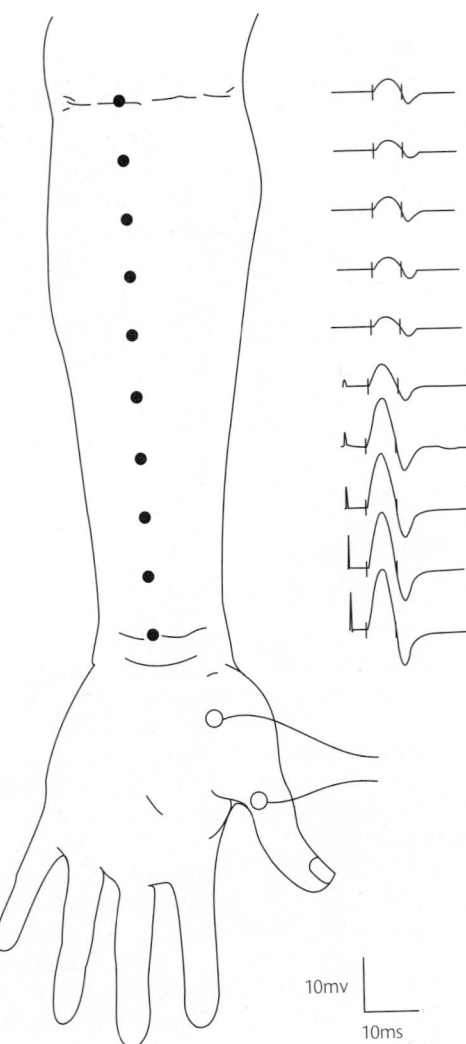

Fig. 21.18 Focal motor conduction block in a mid-forearm segment, demonstrated by inching the stimulating electrode along the median nerve in a patient with multifocal motor neuropathy with conduction block. Note the drop in compound muscle action potential amplitude with stimulation above mid-forearm. Sensory conduction was normal through this same segment. (Courtesy of Dr. M. Busby.)

10mv

10ms

observed in the motor roots (Oh *et al.* 1995). The spinal fluid is usually normal. Focal hypertrophy of nerves is often demonstrable, particularly in the brachial plexus (Beekman *et al.* 2005).

Sensory function. Usually there are no sensory symptoms and sensory nerve conduction is normal. A few patients report focal paraesthesiae but it is rare to demonstrate underlying abnormal sensory signs. Despite the purely motor features, minor involvement of sensory nerve fibres has been noted on biopsy.

Treatment. Untreated, multifocal motor neuropathy usually deteriorates steadily or in a stepwise fashion over many years. Occasionally it evolves subacutely causing severe disability within months. Some patients may stabilize with extremely minor degrees of motor involvement. It is uncertain whether true spontaneous remissions occur. Steroid treatment should be avoided since it is ineffective and often causes substantial motor deterioration (Busby and Donaghy 2003). Cyclophosphamide is effective but is not advisable as first-line therapy because of the serious side effect

profile (Section 21.3.3). It is best reserved for patients unresponsive to, or intolerant of, immunoglobulin, or in those rare instances where it is helpful as adjunctive therapy when the beneficial effect of immunoglobulin alone only lasts two or three weeks.

Intravenous immunoglobulin, IvIg, is the mainstay of treatment, often producing a clear clinical response within 36 h, usually maximal at 10–14 days, and wearing off at 6–12 weeks (Federico *et al.* 2000). The first treatment with immunoglobulin should be designed to determine whether the response sufficiently reverses disability to make regular treatment worthwhile; up to a third of patients show poor responses. Objective neurophysiological improvement in conduction block may occur with IvIg, more often showing improved temporal dispersion than changes in amplitude of compound muscle action potentials (Ghosh *et al.* 2005b). Self-infused home therapy at 2–3 weekly intervals is effective, time saving, and convenient, and it can be scheduled to avoid treatment-related fluctuations (Slee *et al.* 2007). The long-term benefits of IvIg are unknown and cases of continued downhill deterioration do occur (Van den Berg-Vos *et al.* 2002). However many patients continue responding well to IvIg for over a decade with little or no evidence of background deterioration of the disorder once a regular programme of maintenance therapy has been established which avoids the intermittent relapses which may allow axonal damage to accumulate. Interferon-β1a may be effective if IvIg or cyclophosphamide treatment have failed (Van den Berg-Vos *et al.* 2000a).

21.11.4 Pure motor demyelinating neuropathy

Occasionally patients are encountered with purely motor polyneuropathy which is symmetrical. Although this may involve more than legs, usually all four limbs are affected to a similar extent, particularly the distal muscles. The weakness lacks the asymmetry normally associated with multifocal motor neuropathy (Sabatelli *et al.* 2001; Busby and Donaghy 2003). It tends to present with deterioration over weeks to months rather than the very slow deterioration normally occurring in typical multifocal motor neuropathy with conduction block. Motor conduction studies show widespread slowing with variable degrees of conduction block. Anti-GM1 antibodies may be associated. The implications for choice of treatment underline the importance of differentiating pure motor demyelinating neuropathy from sensori-motor demyelinating neuropathy. Pure motor demyelinating neuropathy often deteriorates with steroids, whereas it responds well to intravenous immunoglobulin (Busby and Donaghy 2003). The likelihood of eventual remission is unknown, although natural remission can be observed after patients with young onset pass through adolescence. Even in patients with a similar clinical picture and anti-GM1 antibodies, where the electrophysiological picture reflects axonal degeneration rather than demyelination or conduction block, strength may improve over 6–24 weeks following cyclophosphamide and plasma exchange therapy (Pestronk *et al.* 1994).

21.11.5 Chronic ataxic polyneuropathy

The onset of chronic relapsing polyneuropathy can be preceded by ocular palsies occurring several weeks earlier (Donaghy and Earl 1985). These ocular palsies may be unilateral or bilateral, usually consisting of partial paralysis of ocular abduction. The polyneuropathy may be asymmetrical, markedly ataxic, worse in the arms, and include dysphagia. Nerve conduction studies may point to a primarily demyelinating disorder but the electrophysiological abnormality can be remarkably mild in relation to the clinical severity. Ataxic neuropathy and eye movement disorder is often associated with anti-GQ_{1B}, -GD_{1B}, -GD_3, or –GT_{1B} antibodies, IgM paraproteins, and cold agglutinins (Willison *et al.* 2001). Rombergism and pseudoathetosis (Fig. 21.16) are prominent and often outweigh the degree of demonstrable joint position sense loss. Paraesthesiae are a less common symptom than ataxia. This syndrome has become known by the acronym CANOMAD, **C**hronic **A**taxic **N**europathy with **O**phthalmoplegia, **M**-proteins, cold **A**gglutinins, and anti-**D**isialated ganglioside antibodies. Many patients do not exhibit the full syndrome at presentation, although many or all of the missing features appear with time; however the anti-GQ_{1B} antibody seems a reliable early marker. The eye movement disorder can be only intermittently symptomatic and manifest well after the initial neuropathic manifestations. Experience with immunomodulatory treatment shows that steroids are usually ineffective or may provoke deterioration whereas IvIg generally produces a response which is well maintained with long-term maintenance infusions (Busby and Donaghy 2003).

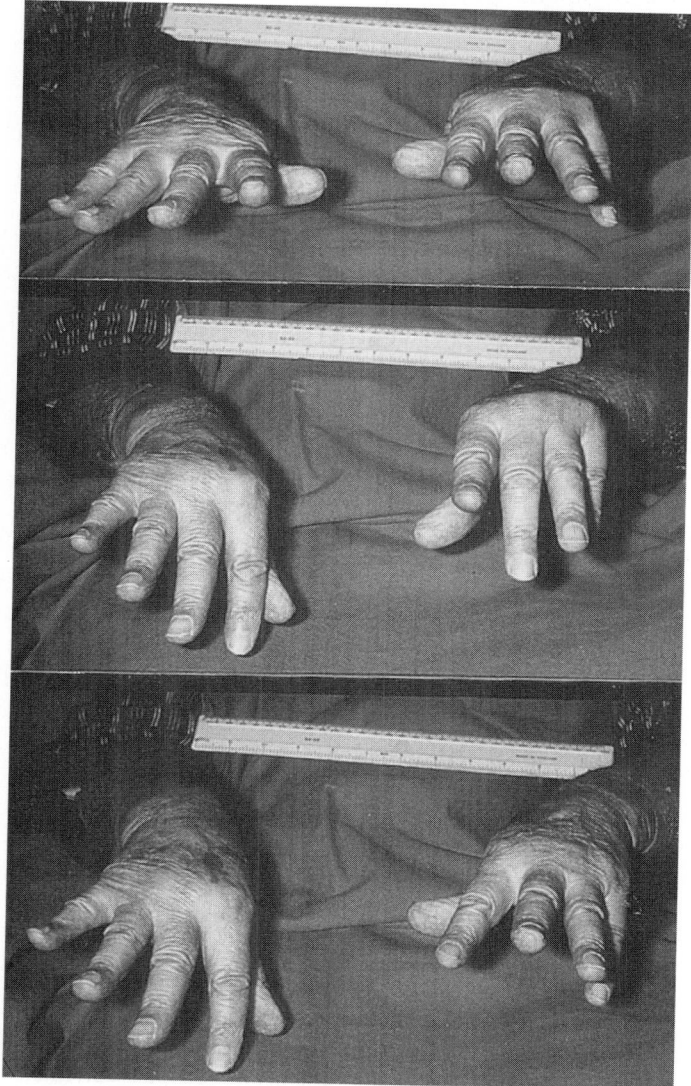

Fig. 21.19 Pseudoathetosis in a patient with sensory ataxic polyneuropathy. Frame intervals at 30 s.

Chronic ataxia may be due occasionally to a purely sensory form of chronic inflammatory demyelinating polyneuropathy affecting large myelinated kinaesthetic fibres. Patients present with limb ataxia, and sometimes numbness or pain, and are found to have profound loss of proprioceptive sensation which may even affect proximal joints. Muscle strength is normal and there is generalized areflexia. Despite this, motor nerve conduction is often slowed. Sensory nerve action potentials are absent. Demyelination may be present on sural nerve biopsy. The spinal fluid protein can be raised. The condition usually deteriorates progressively over months or years. Improvement may follow a trial of immunomodulatory therapy, including intravenous immunoglobulin (van Dijk *et al.* 1996).

Differential diagnosis of chronic sensory neuropathy. Chronic idiopathic neuropathy with purely sensory symptoms presents a difficult differential diagnostic problem. If associated with an eye movement disorder and anti-GQ_{1B} antibodies, the CANOMAD syndrome will be obvious. A syndrome of sensory ataxia with enlarged nerve roots on MRI, normal nerve conduction studies, somatosensory evoked potential abnormalities, and elevated CSF protein may respond to IvIg or steroids (Sinnreich *et al.* 2004). Paraneoplastic sensory neuropathy (Section 21.13.1) will be associated with small cell lung cancer or ovarian cancer, usually with subacute progression, involvement of all sensory fibre types, anti-Hu, antineuronal antibodies a mildly lymphocytic CSF, or other associated features of encephalomyelitis (Griffin *et al.* 1990; Graus *et al.* 1994). The sensory ganglionitis associated with Sjogren's syndrome (Section 21.18.10) may or may not have clear-cut symptoms of dry eyes and mouth, usually occurs in women, ataxia due to large fibre loss predominates, and there may be autonomic symptoms including Adie's pupil. An associated trigeminal neuropathy can occur, antinuclear antibody may be present, and the CSF is normal (Griffin *et al.* 1990; Sobue *et al.* 1993). A similar neurological picture may occur either acutely or chronically without Sjogren's syndrome (Griffin *et al.* 1990). Vitamin E deficiency (Section 21.22.5) can also produce a sensory ataxic neuropathy. Purely sensory presentations of chronic idiopathic axonal polyneuropathy (Section 21.11.8) also occur, usually in late adulthood, often with troublesome pain (Wolfe *et al.* 1999).

21.11.6 Multifocal motor and sensory neuropathy

Also known by the acronym MADSAMN, this rare condition is also referred to as multifocal motor and sensory demyelinating neuropathy or the Lewis–Sumner syndrome. Patients are usually middle aged with motor and sensory loss multifocally distributed and often predominantly in the arm (Van den Berg-Vos *et al.* 2000b; Busby and Donaghy 2003). It may resemble multifocal motor neuropathy (Section 21.11.3) except with prominent additional sensory involvement. Neurophysiologically there is evidence of multifocal conduction block and demyelination, and inflammatory demyelinating changes in biopsied nerves (Oh *et al.* 2005). Patients are usually either unresponsive to, or deteriorate, with steroids, whereas IvIg is effective much as for multifocal motor neuropathy.

21.11.7 Chronic autonomic neuropathy

Autonomic failure of gradual onset and slow progression is also termed pure autonomic failure, or idiopathic orthostatic hypotension. Such chronic autonomic neuropathies involve postural hypotension and erectile failure; altered control of micturition or sweating are common too. Such patients segregate into two groups (Klein *et al.* 2003). One involves high titres of antibodies to the ganglionic acetyl-choline-receptor. Associated cholinergic features of dry eyes and mouth, pupil abnormalities, neurogenic bladder, and gastrointestinal dysfunction. It represents the chronic counterpart of acute autonomic neuropathy (Section 21.10.4) and can be paraneoplastic. These patients may respond to immunomodulation. The second group are less well defined, but have low antibody titres with few symptoms of cholinergic failure. Their idiopathic autonomic failure resembles that seen in association with Parkinson's disease or multiple system atrophy, with prominent postural hypotension, often preceded by shoulder pain in a coat hanger distribution.

Presentation of hypotension in infancy, often coupled with hypotonia, hypothermia, and hypoglycaemia, and worsening during childhood and adolescence, should raise the possibility of dopamine-beta-hydroxylase deficiency. This very rare autosomal recessive disorder improves, with normalization of the low plasma noradrenaline levels, after administration of L-threo-dihydroxyphenyl serine, a precursor of noradrenaline (Senard and Rouet 2006).

21.11.8 Chronic idiopathic axonal polyneuropathy

This disorder usually starts in the sixth decade of life with clinical evidence of a mild sensorimotor, or less often a purely sensory, polyneuropathy, worse in the legs. It is a common polyneuropathy. All modalities of sensation may be impaired but paraesthesiae are uncommon. Neurophysiological and nerve biopsy studies point to axonal degeneration. Progression is slow, and eventual severe disability is rare. By definition an underlying cause is not discovered; a similar disorder in patients with paraproteinaemia tends to produce more severe arm involvement and worse disability (Notermans and Wokke 1996). Type 2 Charcot–Marie–Tooth disease (Section 21.4.5) can usually be differentiated by the positive family history, the predominance of motor involvement, the earlier age of onset, and the likelihood of pes cavus (Teunissen *et al.* 1997). The relationship to painful chronic cryptogenic sensory neuropathy with prominent pain (Wolfe *et al.* 1999) or to the painful burning foot syndrome (Periquet *et al.* 1999) is unclear. If abnormalities are limited to the legs, the potentially treatable lumbar canal stenosis syndrome should be sought by MRI. By definition, no cause can be found, although abnormal glucose tolerance tests are twice as common as in controls (Hughes *et al.* 2004; Hoffman-Snyder *et al.* 2006). There is no curative treatment for chronic idiopathic axonal polyneuropathy. A chronic relapsing axonal polyneuropathy with unusually severe motor involvement has responded promptly to intravenous immunoglobulin, presumably by reversal of widespread conduction failure (Katirji 1997) but this is not a likelihood in the majority of patients. Rapidly progressive axonal polyneuropathy, usually with subtle multifocal features, can occur occasionally in vasculitis (Section 21.15) (Vrancken *et al.* 2004).

21.12 Neuropathies associated with lymphoproliferative disorders

21.12.1 Benign paraproteinaemia

An increased incidence of peripheral neuropathy occurs in patients found to have monoclonal paraproteins on serum electrophoresis.

Such paraproteins only have an incidence of 0.1 per cent in the third decade of life, rising to 3 per cent in the eighth decade, yet they are found in 10 per cent of patients with idiopathic peripheral neuropathy (Latov 1995). An underlying haematological malignancy is detected in about 8 per cent, and subsequent malignant transformation occurs at less than 3/100 patient years (Eurelings *et al.* 2005). Diverse neuropathies are encountered. Only a proportion of paraproteinaemic proteins are likely to be directly causative of neuropathy. Thus in many cases the paraprotein is merely a coincidental finding and a trial of treatment should be considered along the usual lines for the idiopathic equivalent of that particular type of neuropathy (Busby and Donaghy 2003). A wide variety of peripheral neuropathies are encountered, mostly demyelinating. Amyloid neuropathy (Section 21.9.2) may result from immunoglobulin light-chain deposition in patients with paraproteinaemia. Vasculitic neuropathy (Section 21.15) is occasionally associated with cryoglobulins containing monoclonal rheumatoid factors.

Chronic inflammatory demyelinating polyneuropathy types. Demyelinating neuropathies indistinguishable from idiopathic chronic inflammatory demyelinating polyneuropathy are usually associated with IgG or IgA paraproteins and may respond well to immunosuppressant drugs, plasma exchange, or intravenous immunoglobulin. As a group, these paraprotein-associated chronic inflammatory demyelinating polyneuropathy-like neuropathies are more likely to progress slowly, cause less severe disability, and have prominent sensory involvement than idiopathic chronic inflammatory demyelating polyneuropathy (Simmons *et al.* 1995). Paraproteinaemia or other lymphoproliferative disorders can develop, or at least become evident, after the initial diagnosis of chronic demyelinating polyneuropathy. Slowly progressive demyelinating polyneuropathy occurs in about 5 per cent of patients with Waldenström's macroglobulinaemia, a disorder characterized by IgM hyperglobulinaemia, hyperviscosity, lymphadenopathy, hepatosplenomegaly, and lymphocytic infiltration of the bone marrow, and may antedate the systemic illness by some years.

Anti-myelin-associated glycoprotein activity. Some chronic sensorimotor demyelinating neuropathies are associated with IgM paraproteins possessing anti-myelin-associated glycoprotein, MAG, activity. Such patients usually develop sensory signs before motor, all go on to develop arm tremor and ataxia and usually stabilize at 2–5 years (Smith 1994). Nerve biopsies from such patients may show characteristic widely spaced myelin lamellae; similar morphological changes occurring in after passive transfer of the IgM paraprotein. Skin nerves show IgM on myelinated fibres, especially distally (Lombardi *et al.* 2005). The IgM paraproteins fix complement at the sites of separation of myelin lamellae. IgM paraproteinaemic demyelinating polyneuropathy with anti-MAG antibodies may show an unusually distal pattern of weakness (Katz *et al.* 2000). Unlike chronic inflammatory demyelinating polyneuropathy, such neuropathies normally respond poorly over the longer term to immunomodulation.

Anti-ganglioside activity. Purely motor neuropathies, often multifocal with conduction block, can be associated with paraproteins showing antibody activity against GM1 and GD1b gangliosides (Section 21.11.3). Predominantly sensory neuropathies due to combined axonal degeneration and demyelination are associated with paraproteins with antisulphatide activity (Ponsford *et al.* 2000). Chronic ataxic neuropathies associated with anti-GQ_{1B}

antibodies and intermittent ophthalmoplegia are often also associated with IgM paraproteinaemia, the CANOMAD syndrome (Section 21.11.5).

Anti-chondroitin sulphate activity. Predominantly sensory axonal degeneration neuropathy occasionally occurs in patients with IgM-k or IgM-λ paraproteins recognizing chondroitin sulphate. The first symptoms are usually peripheral numbness, paraesthesiae, or pain. Abnormalities of all modalities of sensation may be demonstrable. In some patients the disorder is associated with the skin condition, epidermolysis. In others there may be nerve thickening with features of focal entrapment neuropathy.

Treatment. The treatment of paraproteinaemic polyneuropathies can be difficult and relatively ineffective compared to idiopathic demyelinating neuropathies. Initially steroids, and plasma exchange or immunoglobulin, should be tried along the lines outlined for chronic inflammatory demyelinating polyneuropathy (Section 21.11.2). If this is insufficiently effective, the first decision concerns whether the patient has a sufficient degree of disability to warrant use of potentially dangerous chemotherapy. If so trials of the alkylating agents, cyclophosphamide or chlorambucil, fludarabine, cladribine, or Rituximab, can be considered.

21.12.2 Myelomatous neuropathy

Symptomatic neuropathies occur in about 5 per cent of patients with osteolytic multiple myeloma, although electrophysiological evidence of neuropathy may be present in up to 40 per cent of such patients. A wide range of neuropathies is encountered. The chronic demyelinating and amyloid neuropathies are probably a direct effect of paraproteins. A paraneoplastic sensory neuronopathy of the same type more usually associated with small-cell lung cancer can occur in myeloma. Conventional chemotherapy of the underlying myeloma has little effect on the amyloid neuropathy or the sensory neuronopathy. However, patients with demyelinating sensorimotor neuropathy can improve substantially with steroid therapy and plasma exchange given in addition to chemotherapy for their underlying myeloma. Pronounced improvement can occur after ablation therapy for a localized plasmacytoma.

Neuropathy is a common feature of osteosclerotic forms of myeloma, which are rare by comparison to osteolytic forms. The neuropathy associated with osteosclerotic myeloma, and the accompanying systemic illness, are identical to that seen in Castleman's disease, the POEMS syndrome, and the Crow–Fukase syndrome (Section 21.12.3). If solitary, the osteosclerotic lesion may be treated by localized irradiation or resection, leading to substantial improvement in the neuropathy over subsequent months.

21.12.3 Castleman's disease, POEMS syndrome

Progressively disabling, predominantly motor neuropathies may occur in association with elements of a characteristic syndrome: papilloedema, gynaecomastia, impotence, glucose intolerance, oedema, hepatosplenomegaly, and paraproteinaemia, usually IgA-λ. The skin changes are particularly characteristic and include diffuse cyanotic discolouration, poor capillary reperfusion after blanching, hypertrichosis, and diffuse non-dependent oedema. This constellation of features is particularly common in Japan where it is known as the Crow–Fukase syndrome. It is known otherwise by the acronymn POEMS syndrome, 'Polyneuropathy, Organomegaly, Edema, M band, and Skin changes'. Either osteosclerotic myeloma,

or angiofollicular lymph-node hyperplasia, Castleman's disease, may underlie this clinical syndrome. The neuropathy may be predominantly motor with severely reduced conduction velocities, or it may be sensorimotor with evidence of both demyelination and axonal loss. Roughly half show a good neurological response some months after the initiation of cyclophosphamide and prednisolone therapy, or melphelan and prednisolone, or high-dose cyclophosphamide and autologous blood stem cell transplantation (Jaccard *et al.* 2002). The remainder are relatively unresponsive, and in some a remorseless downhill progression occurs with eventual death despite chemotherapy for the underlying lymphoproliferative disorder.

21.12.4 Lymphomatous neuropathy

Five distinct types of polyneuropathy occur as an occasional remote accompaniment of lymphoma. The Guillain–Barré syndrome and chronic relapsing inflammatory demyelinating neuropathy, probably reflect disordered immune regulation, and should be treated according to standard principles (Sections 21.10.1 and 21.11.2) (Vallat *et al.* 1995). Paraneoplastic sensory neuronopathy occasionally complicates lymphoma, although this is a rare association in comparison to its incidence in small-cell carcinoma of the lung (Section 21.13.1). A subacute motor neuropathy may complicate Hodgkin's disease and other lymphomas and resolves spontaneously in most patients. Diffuse infiltration of nerves by non-Hodgkin's lymphoma can produce a progressive, painful, asymmetric polyneuropathy (van den Bent *et al.* 1999).

21.13 Carcinomatous neuropathy

Three types of peripheral neuropathy may occur as remote, or paraneoplastic effects of carcinoma: sensory neuronopathy, sensorimotor polyneuropathy, and, less frequently, vasculitic neuropathy, neuromyopathy, and autonomic neuropathy. Small-cell carcinoma of the lung is the commonest tumour to underlie paraneoplastic neuropathy. Carcinoma of the breast, ovary, or gastrointestinal tract, myeloma, or lymphoma, occur less frequently. Patients frequently present with neuropathy before experiencing symptoms from the underlying cancer itself. Although symptomatic paraneoplastic neuropathy is relatively uncommon, prospective studies in patients with lung and breast cancer reveal clinical evidence of polyneuropathy in up to 5 per cent, with subclinical electrophysiological abnormalities in another 20 per cent (Hughes *et al.* 1996).

21.13.1 Paraneoplastic sensory neuronopathy

This sensory neuropathy occurs more commonly in women, usually preceding tumour symptoms by 6–15 months and occasionally as long as 3 years (Section 38.4.3). It usually develops subacutely over a period of weeks before stabilizing spontaneously (Camdessanche *et al.* 2002). Less often it continues to deteriorate inexorably. This distinctive neuropathy may predominate in the arms and can be asymmetrical. Patients develop sensory ataxia due to loss of kinaesthetic sensation and may experience uncomfortable paraesthesiae. All modalities of sensation are impaired. The gait ataxia may prevent walking. Sensory loss can extend to the trunk and may contribute to impaired sphincter function. Muscle weakness does not occur because motor fibres are spared. The reflexes are usually lost. Occasionally such paraneoplastic sensory neuropathy runs a slowly progressive course without severe disability. Other forms of paraneoplastic encephalomyelitis frequently coexist, particularly limbic encephalitis (Section 38.4.2). A similar neuropathy occurs in Sjogren's syndrome (Section 21.18.10).

Nerve conduction studies show reduced or absent sensory nerve action potentials, while motor nerve conduction is normal. The spinal fluid is usually lymphocytic and proteinaceous. Neuropathologically there is profound loss of dorsal root ganglion neurones with lymphocytic infiltration; a 'dorsal root ganglionitis'. The serum or spinal fluid may contain high titres of an autoantibody directed against neuronal nucleoproteins of molecular weight 35–40 kDa, known as anti-Hu antibodies (Section 3.7.3). Anti-Hu antibody also occurs in patients with other forms of paraneoplastic encephalomyelitis and in sensorimotor neuropathy (Camdessanche et al. 2002). It may also be detected in low titre in some patients with small-cell lung cancer who do not have an associated neurological disorder. Detection of this anti-Hu antibody should always provoke careful search for an underlying tumour, which should be repeated after an interval if initially negative. Neither treatment nor removal of the underlying tumour, nor immunosuppression, is known to reverse the sensory neuronopathy. However, the underlying carcinoma should be carefully staged in case of the rare possibility of curative treatment.

21.13.2 Paraneoplastic sensorimotor neuropathy

The presence of muscle weakness distinguishes this from the purely sensory neuronopathy described above. Such neuropathies are a heterogeneous collection; they can be acute, subacute, or chronic in presentation, and primarily demyelinating or axonal in nature. They are not usually associated with other paraneoplastic neurological disorders and they can be associated with a wide range of underlying carcinomas. They may precede or follow tumour symptoms. Mild sensorimotor neuropathy may be detected in up to a quarter of patients with lung cancer (Hughes et al. 1996). Most usually the onset is subacute with limb weakness, sensory disturbance, and areflexia. Nerve conduction studies and electromyography may reveal axonal degeneration, in which case attempts at treatment with steroids are likely to be unsuccessful. Demyelinating neuropathies are occasionally encountered, although more commonly with underlying lymphoma than with carcinoma. Typical Guillain–Barré syndrome or chronic inflammatory relapsing demyelinating polyneuritis may occur, and the latter is often steroid-responsive. Nerve biopsies show variable combinations of axonal loss, segmental demyelination and remyelination, and perivascular lymphocytic infiltration.

21.13.3 Paraneoplastic vasculitic neuropathy

Mononeuritis multiplex due to vasculitis occasionally occurs with cancer of the prostate or lung, or lymphoma. Neuropathic symptoms may precede those due to the underlying tumour. Nerve and muscle biopsies allow histological diagnosis of microvasculitis. Cyclophosphamide therapy may lead to stabilization or improvement of the neuropathy (Oh et al. 1991).

12.14 Neuropathy due to infections

Peripheral neuropathy is a central clinical feature of some infections: leprosy, diphtheria, human immunodeficiency virus or HIV infection, borreliosis and herpes zoster. This section does not cover those peripheral neuropathies, such as the Guillain–Barré syndrome, which are infrequent and indirect manifestations of common infections with a wide variety of viruses and bacteria, all of which may share the common property of disturbing immune regulation or evoking antibodies which cross-react with nerve (Section 21.10). Demyelinating neuropathy is a rare accompaniment of Creutzfeldt–Jakob disease, both in the sporadic and inherited forms (Niewiadomska et al. 2002).

21.14.1 Leprosy

Aetiology. Leprosy or Hansen's disease, is due to infection of the skin, mucosal membranes, and peripheral nerves by *Mycobacterium leprae*, an acid-fast bacillus stainable by Ziehl–Neelsen's method. Infection is only likely after prolonged contact with patients suffering from bacillus-rich forms of the disease, especially if shed in nasal secretions. The skin is the commonest portal of entry. Leprosy is common in the Asian subcontinent but may be encountered anywhere in the world. In the Western world it is usually encountered in migrants from endemic areas.

Pathology. The histopathological and clinical picture varies widely in different individuals. This reflects different degrees of cell-mediated immunity. Three general forms may be distinguished within what is, in reality, a continuum which can be subclassified further (Jacobson and Krahenbuhl 1999). Patients with high immunity develop tuberculoid leprosy, which is not progressive, and is usually associated with a single granulomatous skin lesion containing few bacilli and which may involve an underlying peripheral nerve. Patients with low or absent immunity develop lepromatous leprosy in which copious bacilli multiply extensively in the cooler tissues in the body, with progressive and extensive involvement of skin and nerves. Most commonly patients manifest the intermediate or dimorphous forms which occupy the borderland between the tuberculoid and lepromatous varieties.

The exact pathogenetic mechanisms underlying nerve damage in leprosy are not clear. The advanced nerve damage in established tuberculoid leprosy may reflect the compressive and ischaemic consequences of the infiltrating cells forming the granuloma. Nerves in advanced lepromatous leprosy contain vast accumulations of bacilli which may disrupt nerve fibres by virtue of their sheer size. It is unlikely that early leprous neuropathy reflects primary infection of Schwann cells, because segmental demyelination is not prominent. In early leprous neuropathy, the bacilli are most prominent in macrophages and Remak cells, the supporting cells of unmyelinated fibres.

Clinical features. Early diagnosis is crucial since antibiotic therapy will prevent further irreversible nerve damage. The combination of skin and peripheral-nerve lesions is the hallmark of leprosy. There are three cardinal signs for clinical diagnosis:

- anaesthetic skin lesions;
- peripheral nerve enlargement (Fig. 21.20); and
- acid-fast bacilli on skin smear.

In *tuberculoid* forms, sharply demarcated, hairless, anaesthetic erythematous plaques, or hypopigmented macules, are associated with sensory and motor loss in the distribution of one or two damaged peripheral nerves. The nerve is often palpably enlarged. The greater auricular and superficial peroneal nerves are the most commonly affected. Occasionally tuberculoid leprosy affects nerves without an associated skin lesion, and detection of *M. leprae* by

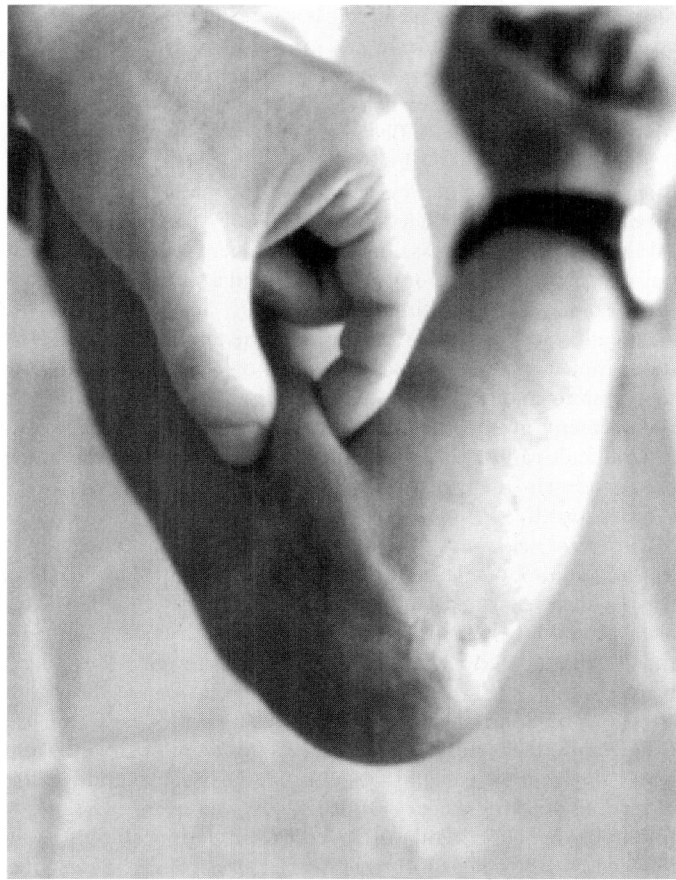

Fig. 21.20 A hugely palpable ulnar nerve in the upper arm in leprosy. Note that the nerve enlargement has led to suspected secondary compression in the cubital tunnel, hence the scar reflecting surgical release. (Courtesy of Dr. Colin McDougall.)

polymerase chain reaction in nerve biopsies aids diagnosis (Jardim *et al.* 2003).

In *lepromatous* leprosy, there is extensive skin involvement with erythematous macules, papules, or nodes. Skin thickening produces the characteristic leonine faces, with thickening of the nose and ear lobes and eventual perforation of the nasal septum. Nerves are diffusely and progressively involved, leading to mononeuritis multiplex.

In dimorphous or *intermediate* forms of leprosy, the skin and neuropathic changes lie between the tuberculoid and lepromatous forms and poorly defined hypopigmented skin lesions are characteristic. Sensory loss in leprosy tends to spare warm areas of the body, such as the palms, and preferentially affects the skin of cold areas. Patients with advanced leprous neuropathy develop profound pain and temperature loss, leading to acromutilation with trophic ulcers, Charcot joints and autoamputations.

Diagnosis. The clinical picture is usually characteristic, and failure to prove the diagnosis histologically should not deter the physician from advising drug therapy. The simplest method of proving the diagnosis of leprosy is to take skin biopsies or smears from both the centre and edge of a lesion, and to demonstrate acid-fast bacilli by Ziehl–Neelsen staining. Bacilli are most prominent within dermal nerves. Nerve biopsy is particularly valuable in

suspected cases without skin lesions and demonstrates bacilli or characteristic granulomatous reaction and inflammation (Chimelli *et al.* 1997). The lepromin skin test is only positive in tuberculoid forms.

Treatment. Leprosy is the world's commonest treatable neuropathy. Adequate early therapy prevents the development of disfiguring disability but will not allow recovery of nerves which are already severely damaged. The following chemotherapeutic regimens are currently recommended for adults (Jacobson and Krahenbuhl 1999):

◆ Borderline and lepromatous leprosy should be treated for a minimum of 2 years until skin scrapings and biopsies are negative for bacilli. Daily self-administration of dapsone (100 mg) and clofazimine (50 mg) orally should be accompanied by supervised administration once-monthly of clofazimine (300 mg) and rifampicin (600 mg).

◆ Patients with tuberculoid leprosy should receive daily dapsone (100 mg) with supervised monthly rifampicin (600 mg) for 6 months; single-dose combination therapy of rifampicin, ofloxacin, and minocycline, 'ROM', may also be effective.

Some physicians recommend that rifampicin be administered daily, rather than monthly, for both multi-bacillary and pauci-bacillary leprosy. Following chemotherapy, nerve grafting may restore sensation in patients with severe mononeuritic sensory loss causing acrodystrophic changes. Steroids are recommended to prevent treatment reactions in patients with hypersensitivity phenomena, such as erythema nodosum or iritis. Steroids are advised if a silent neuropathy develops after the initiation of chemotherapy, whether or not associated with systemic evidence of a reaction; such nerve fibre impairments are most likely in multibacillary forms of disease (Croft *et al.* 2000). Vasculitic neuropathy can develop years after effective treatment in nerves containing persisting leprosy antigen; steroid treatment is effective (Bowen *et al.* 2000).

21.14.2 Human immunodeficiency virus

A wide spectrum of peripheral neuropathy occurs in HIV infection and AIDS. It includes the Guillain–Barré syndrome and chronic inflammatory demyelinating neuropathy developing early in the course of the disease. Later in the disease, the symmetrical distal sensory neuropathy of AIDS must be differentiated from that caused by antiretroviral therapy (Section 21.19.7). A rapidly progressive multifocal motor and sensory polyradiculopathy due to cytomegalovirus or infiltrative lymphocytosis may occur (Gherardi *et al.* 1998). Necrotizing arteritic neuropathy can also occur in HIV-infected patients (Bradley and Verma 1996). Possible underlying HIV infection must be considered in a patient with undiagnosed polyneuropathy.

Guillain–Barré Syndrome. Typical Guillain–Barré syndrome (Section 21.10.1) occurs early in the course of HIV infection, often around the time of primary infection. HIV antibodies may not be present, and P24 antigen assays may be required to diagnose the infection. A clue to underlying HIV infection comes from finding a spinal fluid pleocytosis, generally 20–30 cells/mm^3. The usual treatment of the Guillain–Barré syndrome is recommended, including intravenous immunoglobulin administration, with particular care to avoid exposure to body fluids.

Chronic inflammatory demyelinating neuropathy. This regularly occurs in HIV-infected patients, often at a relatively early stage before the development of the acquired immunodeficiency syndrome. It should be treated in the usual way (Section 21.11.2). It should be recognized that steroid administration may augment the existing defect in cell-mediated immunity that occurs in HIV infection. Thus, immunoglobulin infusion may be particularly required in HIV-infected chronic inflammatory demyelinating polyneuropathy patients so as to minimize the use of immunosuppressive drugs.

Cytomegalovirus polyradiculoneuropathy. In patients with established HIV infection, cytomegalovirus causes a subacute polyradiculoneuropathy. Some patients may present with a sacral sensory loss and acute urinary retention, and progress to flaccid paraparesis within a few weeks (So and Olney 1994). Cytomegalovirus may be cultured from the spinal fluid and should be sought by polymerase chain reaction. Other patients develop a rapidly progressive multifocal sensory motor neuropathy affecting the limbs, in which dysaesthesiae and pain may be prominent from the outset. Cytomegalovirus may be detected by immunostaining within biopsied peripheral nerves or autopsied spinal-nerve roots. These nerves may contain gigantic cells with inclusions typical of cytomegalovirus infection. Without treatment, death soon follows the development of this neuropathy. Early therapy with ganciclovir or foscarnet can produce improvement.

Sensory polyneuropathy. A predominantly sensory, symmetrical polyneuropathy affects up to 30 per cent of patients with the acquired immunodeficiency syndrome. This neuropathy becomes increasingly common in the later stages of the illness and remains common despite the introduction of highly active antiretroviral therapy, or HAART (Simpson *et al.* 2003). The initial complaint is of painful paraesthesiae in the feet, and the ankle jerks are usually lost. Electrophysiology shows diminished or absent sensory nerve action potentials, without slowing of motor nerve conduction. The neuropathy progressively worsens in the feet. Lamotrigine is effective against the neuropathic pain (Simpson *et al.* 2003). It should be differentiated from the painful sensory neuropathy caused by ddI or ddC antiretroviral drugs (Section 21.19.7) which is likely to develop within months of starting the drug and then tends to deteriorate more rapidly.

21.14.3 Borreliosis

Infection with the tick-borne spirochaete *Borrelia burgdorferi* causes Lyme disease, a multisystem disorder comprising a characteristic expanding annular skin lesion called erythema chronicum migrans, oligoarthritis, carditis, meningoencephalitis, cranial neuritis, polyradiculopathy, and peripheral neuropathy (Halperin *et al.* 1996). Incomplete forms of the disease are frequent and should be recognized because of their impressive response to antibiotic therapy. The differential diagnosis of the neurological disorder includes tick-borne encephalitis, transmitted by the same tick bite (Logina *et al.* 2006). Neurological features occur in about 15 per cent, starting a few weeks to several months after the tick bite. Human infection usually follows tick bites during the summer. It is commonest in patients who have been in woodland areas populated by rodents, squirrels, or deer, the animal reservoirs for Borrelia. Infection is frequent in North America and mainland Europe, and also occurs less frequently in Britain, Australia, and Asia. The *Bannwarth syndrome* of lymphocytic meningoradiculitis is a form of Lyme disease, described in Europe before it was recognized that there was an underlying *Borrelia* infection.

Cranial neuritis. Facial palsy of acute onset, either unilateral or bilateral, commonly occurs in the early weeks of Borrelia infection. About a quarter of cases of Bell's palsy may be due to Borrelia infection in endemic areas (Halperin and Golightly 1992). The palsy is often incomplete and may be accompanied by subjective facial sensory disturbance. Facial nerve paralysis may be the only neurological feature of Lyme disease. It usually recovers without treatment but may be treated with oral antibiotics and a short course of oral prednisolone if seen within 24 h of onset.

Polyradiculopathy. Shooting pains in the territories of affected nerve roots are sometimes accompanied by reflex loss or sensorimotor abnormalities in the limbs. Sharp chest-wall pains reflect involvement of thoracic nerve roots. Half the patients with polyradiculopathy also have facial palsy. The polyradiculopathy of borreliosis may last some months, but usually resolves spontaneously.

Peripheral neuropathy. Mononeuritis multiplex, polyneuropathy, and acute brachial neuralgia may all occur in Lyme disease. Any of these neuropathies may accompany polyradiculitis and facial palsy. Mononeuritis multiplex and acute brachial plexus neuropathy tend to occur within the first few months of infection. Up to half of patients with untreated late Lyme disease develop a chronic polyneuropathy with intermittent limb paraesthesiae (Logigian and Steere 1992). Few neuropathic abnormalities are generally found on examination of such patients: mild distal glove- and-stocking sensory loss with preserved reflexes and motor function are the general rule. Nerve conduction studies show reduced and slowed sensory nerve action potentials, and, sometimes, increased distal motor latencies. Neurophysiological abnormalities are multifocal in nature, and there is no generalized motor slowing. Sural nerve biopsies may show mild axonal loss, with some perivascular lymphocytic infiltration; necrotizing vasculitis is not encountered.

Central nervous system involvement. This may be chronic and take a variety of forms resembling multiple sclerosis, other infective meningoencephalitides, stroke, or tumour (Oksi *et al.* 1996). MRI can show single or multiple enhancing brain lesions and pathologically these involve demyelination, lymphocytic blood vessel involvement, and detectable Borrelia DNA indicative of direct infection.

Investigations. The demonstration of elevated serum or spinal fluid IgM or IgG antibody titres to *Borrelia burgdorferi* has been the traditional diagnostic investigation. However antibody assays are of low diagnostic sensitivity early in infection, and do not discriminate between active and inactive infection later on. The spinal fluid contains a striking white cell pleocytosis of up to 700 cells/mm^3 in patients with polyradiculitis, but may be normal if isolated facial palsy or peripheral neuropathy are the only neurological features. Borrelia may be cultured from about 50 per cent of skin lesions but only 5 per cent of CSF specimens. Amplification of small amounts of Borrelial DNA in CSF or urine by polymerase chain reaction holds promise as the most specific means to prove infection, but current tests are negative in up to 50 per cent of patients depending upon the particular clinical manifestations and the stage of the disease (Schmidt 1997).

Treatment. The peripheral neuropathy or radiculopathy of Lyme disease responds well to intravenous Benzyl penicillin (2.4 g

6 hourly for 10 days), or ceftriaxone (2 g per day for 14 days), or doxycycline (Halperin *et al.* 1987).

21.14.4 Diphtheria

Polyneuropathy is the commonest and most important complication of diphtheria; the exotoxin of *Corynebacterium diphtheriae* has an affinity for peripheral nerves. Polyneuropathy is commoner in childhood rather than adult infections. Paralysis is more likely to follow severe local infections, which are usually faucial but may be extrafaucial. Antitoxin is given to reduce the incidence of paralysis, particularly in patients who can receive it early in the illness.

Palatal paralysis reflects the action of locally produced toxin upon the nerves to the bulbar musculature. Involvement of nerves by locally produced toxin accounts for localized paralysis following a cutaneous infection, the muscles paralysed being those supplied by the spinal segment from which the infected region is innervated. Localized neuropathy in one or more extremities was often seen after diphtheritic infection of limb wounds in the Middle East in the Second World War. Paralysis of accommodation, generalized polyneuropathy, and cardiac toxicity are due to blood-borne dissemination of the toxin to the ciliary muscles, peripheral nerves, and heart.

Diphtheria remains endemic in the Third World but is now rare in Western countries due to improved living conditions and childhood immunization programmes. Occasional outbreaks of clinical diphtheria do occur in previously immunized adults although the clinical severity is generally reduced. At least 10 per cent of adults vaccinated in childhood have insufficient residual immunity to protect against infection once Diphtheria returns to a population (Kjeldsen *et al.* 1985). This reduced immunity in adults resulted in a spectacular return of diphtheria to Russia and other eastern European countries after 1993 (Logina and Donaghy 1999; Piradov *et al.* 2001). Overall about 15 per cent of patients diagnosed with diphtheria develop polyneuropathy. Attenuated forms of diphtheritic polyneuropathy can occur in closed communities despite recent booster vaccination (Krumina *et al.* 2005).

Pathology. The primary lesion in the peripheral nerves is segmental demyelination accompanied by typical slowing of motor nerve conduction, which may persist for some time after clinical recovery. The neuropathic effects of the toxin are dose-dependent. Once bound to cells, the toxin becomes unavailable for inactivation by antitoxin.

Symptoms and signs. Paralysis of the palate is usually the earliest neurological symptom and appears a median of 10 days after the onset of localized throat Diphtheria, seen the typical palatal pseudomembrane. It is generally bilateral but may be unilateral. The voice becomes nasal, there is regurgitation of fluids through the nose on swallowing, and the larynx becomes paralysed, allowing inhalation and choking. The palatal reflex is usually lost. Twenty per cent develop ventilator-dependent respiratory failure. Improvement of bulbar symptoms occurs at median 30 days from onset. Secondary deterioration of bulbar function, sometimes enough to require ventilation for the first time, occurs in over a third at a median 40 days from initial onset (Logina and Donaghy 1999).

Paralysis of accommodation due to ciliary muscle involvement produces blurred vision for near objects. The pupillary reactions to light and on convergence are unimpaired. Paresis of the face or external ocular muscles may occur.

Generalized sensorimotor polyneuropathy affecting the limbs occurs in 90 per cent, at a median of 37 days from onset. It always occurs after the bulbar symptoms, and sometimes when the bulbar symptoms are already improving. About 50 per cent become unable to walk unaided; 30 per cent develop impaired bladder control. Blood pressure swings or cardiac arrhythmia reflect either autonomic neuropathy or cardiomyopathy (Logina and Donaghy 1999).

Diphtheritic hemiplegia is rare fortunately and is usually due to either embolism or thrombosis of a cerebral artery, or to acute post-infective encephalitis. Its effects are similar to those of other acquired forms of infantile hemiplegia. Meningism was once common in the acute stage with cervical rigidity or opisthotonos and rigidity of the limbs, so-called 'spasmodic diphtheria'. The CSF in such cases of presumed encephalopathy is usually normal in composition. Permanent bulbar palsy is a rare sequel.

Diagnosis. The chief differential diagnosis of diphtheria is from Guillain–Barré syndrome, which is an ascending, rather than descending, polyneuropathy (Section 21.10.1). Diphtheria is favoured by the high prevalence of bulbar and respiratory dysfunction at a time of little or no limb involvement, by evolution for longer than 4 weeks, by the preceding sore throat rather than catarrhal illness, and by the simultaneous involvement of other organs, particularly the heart (Logina and Donaghy 1999). The CSF protein tends to be elevated in both conditions. Throat cultures are positive in 98 per cent of diphtheria, and 8 per cent have the highly toxic 'bull-neck' form of the disease (Rakhmanova *et al.* 1996).

Prognosis. The prognosis of the paralysis is usually good now that endotracheal intubation prevents death due to bulbar or respiratory muscle failure. Nevertheless some limb symptoms persist in 80 per cent at 1 year, while 6 per cent are still unable to walk. Sixteen per cent of diphtheria patients die, but usually from cardiac or other organ involvement rather than paralysis (Logina and Donaghy 1999). Hemiplegia is a serious complication especially in children, as it may not only be fatal, but in patients who survive, recovery is usually incomplete, and epilepsy and dementia may follow.

Treatment. The treatment of diphtheria includes antibiotic therapy and injection of adequate doses of antitoxin as early as possible. Benzyl penicillin 1.2 g 6 hourly intravenously, should be given for 14 days, converting to oral penicillin when the patient can swallow normally. Erythromycin 500 mg four times daily is an alternative for patients with penicillin allergy. Diphtheria antitoxin should be given intravenously or intramuscularly; serum sickness may occur in up to 10 per cent of patients. It is unclear for how long after onset of diphtheria that antitoxin will be beneficial. In the absence of any formal clinical trials, retrospective evidence suggests little benefit on the incidence of paralysis or death if antitoxin is administered after the second day of the throat infection (Logina and Donaghy 1999). The general nursing care, and indications for assisted ventilation, are similar to those in Guillain–Barré syndrome (Section 21.10.1). Tracheostomy may be required early if paralysis of the pharynx or larynx leads to choking while feeding or drinking.

21.14.5 Herpes zoster

Herpes zoster is a reactivation of varicella-zoster virus which had been primarily acquired during chickenpox infection. Zoster is particularly likely in the elderly and the immunosuppressed, and may affect a fifth of all adults at some time in life. Few patients have more than one attack. Following primary chickenpox infection, the varicella-zoster virus becomes latent in the sensory ganglia and motor neurones. During zoster eruptions there is inflammation and haemorrhagic necrosis destroying neurones of the affected dorsal root ganglion, with a shingles eruption in the skin of the corresponding dermatome.

Clinical features. An attack of shingles is usually heralded by tingling in the dermatome or lancinating pains. These generally precede the visible rash by 2 or 3 days. Occasionally rash never develops (Fox *et al.* 2001). Erythematous macules and papules rapidly become vesicular, and the lesions accumulate over 3–5 days. Scabbing occurs 3–7 days later, and then dry by two weeks. The intensity of the vesicular eruption varies immensely, from a few vesicles only in mildly affected patients, to a dense oedematous rash covering the entirety of one or more dermatomes in more severely affected patients (Fig. 21.21). When zoster affects the ophthalmic division of the trigeminal nerve, conjunctivitis and keratitis can occur, associated with peri-orbital oedema. Electromyographically detectable motor involvement is associated with the skin eruption in 50 per cent of patients, and sometimes produces clinically evident muscular paralysis affecting the diaphragm, limb muscles, external ocular, or facial muscles (Haanpaa *et al.* 1997). Zoster eruptions in sacral dermatomes may produce paralysis of the bladder with haemorrhagic cystitis, and bowel ileus confusable with an acute abdomen. Clinical evidence of meningoencephalitis is uncommon. However, subclinical evidence of brain stem or spinal cord involvement at the relevant level is evident on MRI in over 50 per cent, and the CSF is lymphocytic, often with detectable varicella-zoster virus DNA in 60 per cent (Haanpaa *et al.* 1998).

Although the diagnosis of shingles can be confirmed by isolation of the varicella-zoster virus from vesicular fluid, or immediate demonstration by electron microscopy, the rash is usually sufficiently characteristic to allow unequivocal clinical diagnosis. However, if seeing a patient after resolution of a rash attributed to zoster, it is advisable to take a clear history concerning the radicular

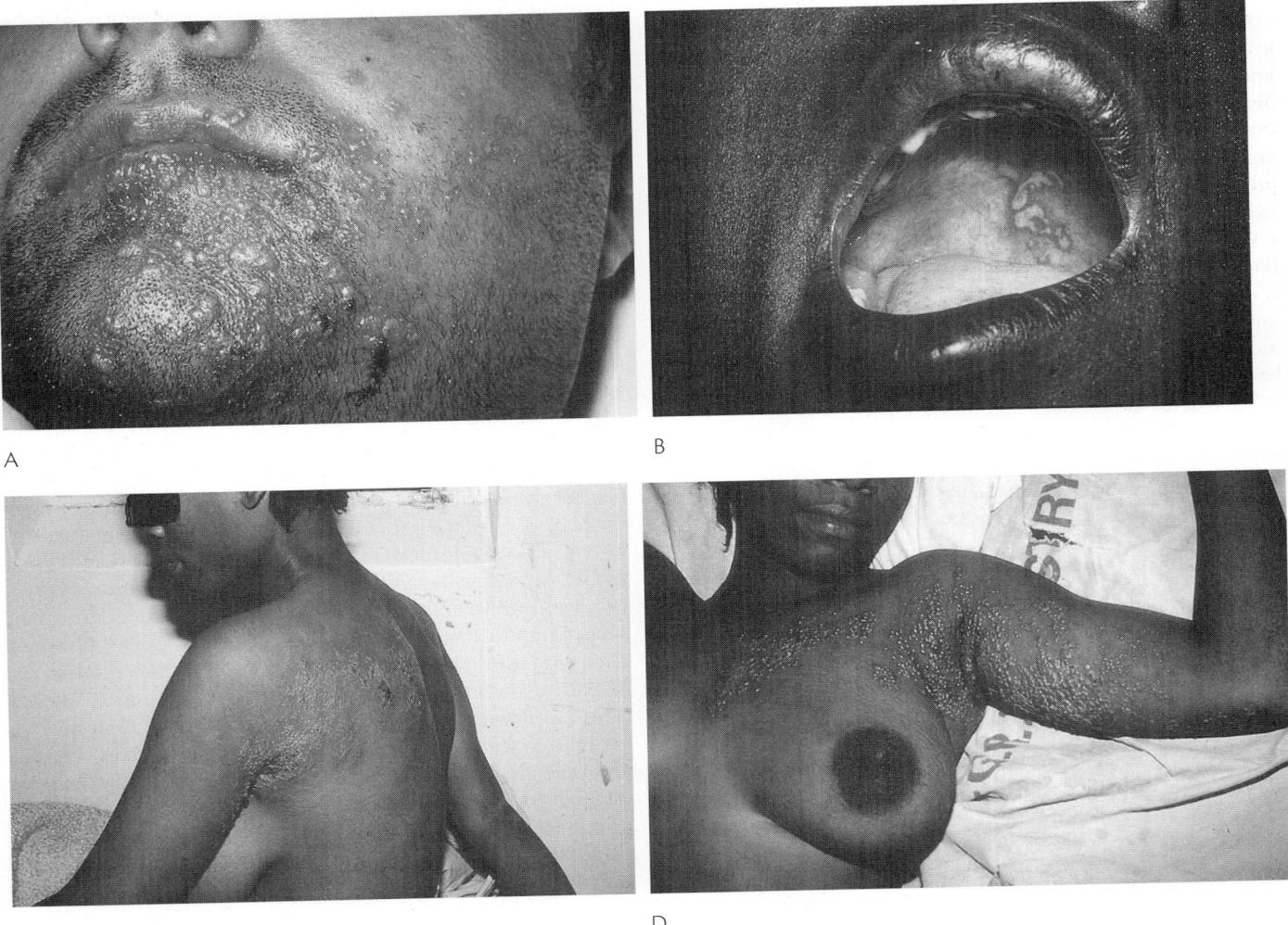

A

B

C

D

Fig. 21.21 Herpes zoster eruptions A. in the mandibular division of the trigeminal nerve; B. on the tongue and palate, the 'Ramsay Hunt syndrome'; C. T2 and T3 dermatomes posteriorly; and D. anteriorly. (Courtesy of Dr C Conlon.)

distribution and vesicular character of the eruption, to confirm the likely diagnosis when a patient is seen after resolution of the rash. Sometimes patients will self-diagnose other skin disorders as being attacks of zoster.

Treatment. The treatment of the acute attack should be with analgesics sufficiently potent to relieve pain. Anti-viral drugs, acyclovir, valaciclovir, or famciclovir, speed resolution of the acute eruption and may reduce the risk of prolonged pain. Oral anti-viral drugs are often prescribed for immunocompetent patients with uncomplicated zoster eruptions, although the extent of their value is unclear. Anti-viral drugs are clearly indicated in patients displaying clinical evidence of central nervous system involvement, or who are immunocompromised, and should be administered intravenously in more serious clinical situations (Cohen *et al.* 1999).

Post-herpetic neuralgia. This is the feared long-term complication of shingles. It is defined as pain persisting beyond 1 month. Quantification of sensory nerve fibres in skin biopsies from affected dermatomes show a more severe nerve fibre loss in those with post-herpetic neuralgia than in those without (Oaklander *et al.* 1998). Post-herpetic neuralgia is more likely in the elderly, those who have had a severe rash with significant pain, and those with ophthalmic involvement (Jung *et al.* 2004). Such patients merit anti-viral therapy during the acute zoster eruption so as to try and reduce the risk of subsequent post-herpetic neuralgia. Once established, post-herpetic neuralgia can be disturbingly resistant to local and systemic pain relieving measure. Amitriptyline may be beneficial, particularly if given from an early stage. Topical capsaicin ointment may relieve pain in some, but cause burning in others. Carbamazepine or gabapentin may be tried. Local measures may include topical lidocaine, regional nerve blocks, transcutaneous electrical stimulation, and acupuncture. Narcotic analgesics may be necessary if other measures fail.

21.15 Vasculopathic neuropathy

Ischaemia may produce focal damage to peripheral nerves, causing mononeuropathy or mononeuritis multiplex. This is most familiar in the context of necrotizing vasculitis or diabetic microvascular disease (Section 21.17.1) affecting the vasa nervorum. Patients with a possible diagnosis of vasculitic neuropathy should be investigated urgently so that treatment can be started to forestall further peripheral nerve damage.

21.15.1 Atherosclerosis and embolism

It is unusual for blockage of medium-sized arteries to cause obvious clinical features of neuropathy because of the rich longitudinal anastomosis of the peripheral nerve vasculature. Nonetheless, axonal loss is demonstrable histologically in nerves from limbs affected by chronic peripheral vascular disease (Nukada *et al.* 1982). In general, myelinated fibres seem more vulnerable to ischaemia than unmyelinated (Fujimura *et al.* 1991).

Acute embolic, thrombotic, or traumatic occlusion of a major artery to a limb may cause peripheral nerve dysfunction, but the neuropathic symptoms and signs are usually overshadowed by the prominent effects of associated acute ischaemia of skin and muscle. Prompt restoration of blood flow, for instance by embolectomy, may lead to full recovery of peripheral nerve function. Prolonged ischaemia leads to irreversible peripheral nerve damage which may

be associated with ischaemic contractures of muscles. Multifocal neuropathy resembling vasculitis may occur in the cholesterol emboli syndrome which may be precipitated by arterial catheterization procedures; muscle or nerve biopsy may demonstrate cholesterol clefts within small arteries (Bendixen *et al.* 1992). The creation of arteriovenous fistulae in the arm for haemodialysis can cause a distal axonal neuropathy, probably due to a vascular steal syndrome.

21.15.2 Non-systemic vasculitis

In some patients, necrotizing vasculitis is confined to the peripheral nervous system (Collins *et al.* 2003). Most patients with non-systemic vasculitic neuropathy develop multiple mononeuropathies affecting the limbs, thoracic roots, or the cranial nerves. Each mononeuropathy evolves over a few hours or days, producing muscle weakness, paraesthesiae, and pain, and global sensory disturbance in the territory of the affected nerve. Tendon reflexes may not be lost if the damage is restricted to the more distal segments of nerves. A minority of patients present with a symmetrical or asymmetrical distal polyneuropathy which may be sensorimotor or purely sensory. Thus, the potentially treatable condition of vasculitic neuropathy should be considered in patients with a distal axonal polyneuropathy which progresses quickly and for which no satisfactory diagnosis is apparent. Vasculitic neuropathy may occur in patients infected with HIV.

The ESR is elevated in a minority. The spinal fluid is usually normal. Nerve conduction studies show focal axonal loss and denervation of muscles. Occasionally conduction block may be demonstrated. Nerve or muscle biopsy will show that the walls of epineurial, perineurial, or muscular arteries are infiltrated by polymorphonuclear cells, and there is fibrinoid necrosis, destruction of the internal elastic lamina, and occlusion of the vessel lumen (Fig. 21.22). The yield of nerve biopsy is greatest if an electrophysiologically abnormal sensory nerve is chosen and biopsied to full thickness. Nerve biopsy is positive in over half of patients, and muscle biopsy improves the diagnosis of arteritis by a quarter (Vital *et al.* 2006). Thus the recommended diagnostic procedure is full thickness biopsy of a sural or superficial radial nerve if electrophysiologically normal, and if not a muscle biopsy.

There have been no controlled studies comparing different therapies for neuropathy in non-systemic or systemic vasculitis. Many consider that prednisolone alone provides adequate therapy for non-systemic vasculitis restricted to the peripheral nervous system. There are no clear guidelines as to the duration of steroid therapy, whether alternate-day steroids are effective as maintenance therapy, and whether relapses are common on cessation of steroid therapy. If steroids incompletely suppress the underlying vasculitis thereby allowing worsening of the neuropathy, cyclophosphamide therapy should be considered (Mathew *et al.* 2007). If the neuropathy is rapidly progressive and destructive, cyclophosphamide should be considered from the outset, given either orally or as intravenous pulses. Given the potential side effects of immunosuppressive drugs, it is desirable to obtain unequivocal histological proof of vasculitis before starting therapy. The benefits of therapy are first to prevent further peripheral nerve damage and second to allow the moderate degree of recovery of neurological function which occurs during the year after suppressing the vasculitis.

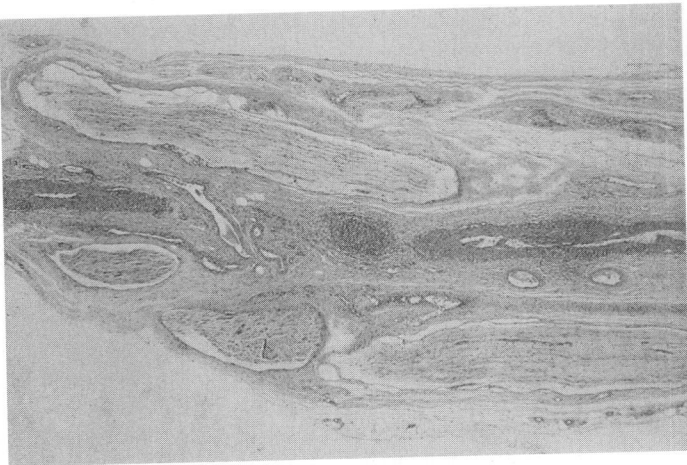
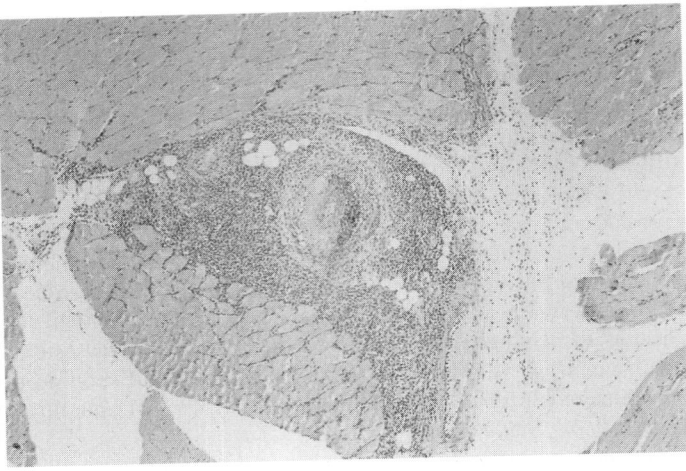

A   B

Fig. 21.22 Histological features in vasculitic neuropathy. A. Nerve biopsy, showing inflammatory cell infiltration of an epineural artery (longitudinal section, haematoxylin and eosin); B. muscle biopsy showing inflammatory cells surrounding and infiltrating the wall of a small artery, with occlusion of the lumen (haematoxylin and eosin).

21.15.3 Systemic vasculitis

The majority of patients with vasculitic neuropathy have underlying vasculitic involvement of systemic organs (Hawke *et al.* 1991). The neurological features resemble those of non-systemic vasculitic neuropathy, but often evolve more aggressively. The electrophysiological and nerve biopsy findings are identical to those in non-systemic vasculitic neuropathy (Section 21.15.2).

Neuropathy is a common feature of the systemic necrotizing vasculitides, which include polyarteritis nodosa, microscopic polyarteritis, Churg–Strauss syndrome, and Wegener's Granulomatosis (Hawke *et al.* 1991; Hattori *et al.* 1999; de Groot *et al.* 2001). The various patterns of neuropathy which have been identified in the systemic necrotizing vasculitides include mononeuritis multiplex, involvement of small cutaneous sensory nerves in the fingers or feet, symmetrical distal sensorimotor neuropathy, brachial plexopathy, and radiculopathy. Anti-neutrophil cytoplasmic antibodies may present in serum early on in Wegener's granulomatosis and polyarteritis nodosa. Often the choice of therapy is dictated by systemic manifestations such as renal involvement. Steroids alone are often effective in the Churg–Strauss syndrome, which is diagnosable by the distinctive eosinophilia and late-onset asthma. In patients with progressive vasculitic neuropathy, therapy should be instituted at the earliest opportunity to prevent the accumulation of irreversible nerve damage. Cyclophosphamide can produce dramatic remissions and cures in severe systemic necrotizing vasculitis to an extent that would be unlikely with steroids alone. However, considerable morbidity and mortality is associated with cyclophosphamide therapy, and it should be supervised by physicians familiar with the drug. It is recommended that cyclophosphamide should only be prescribed for patients with clear histological proof of systemic necrotizing vasculitis, or in whom the clinical syndrome is sufficiently distinctive to be beyond doubt. The rate of neurological relapses is greatly reduced by cyclophosphamide compared to prednisolone therapy, but probably at the cost of increased morbidity and mortality, attributable in part to cyclophosphamide side effects (Mathew *et al.* 2007).

Vasculitic neuropathy may occur in patients with nodular rheumatoid arthritis who may develop either mononeuritis multiplex, digital sensory neuropathy, or sensorimotor polyneuropathy. Vasculitic neuropathy also occurs in lupus erythematosus, Sjogren's syndrome, scleroderma, and in up to 14 per cent of patients with giant cell arteritis (Stefurak *et al.* 1999). Multifocal axonal polyneuropathy can complicate coeliac disease, although vasculitis is unproven as the pathogenic mechanism (Chin *et al.* 2006).

21.15.4 Cryoglobulinaemia

Neuropathy occurs in over 50 per cent of patients with essential mixed cryoglobulinaemia (Gemignani *et al.* 1992). Symmetrical sensorimotor polyneuropathy is commoner than mononeuritis multiplex. Nerve biopsies may show necrotizing vasculitis but diffuse endoneurial vessel damage and non-specific axonal loss is a commoner pathological finding. Most often the neuropathy produces painful dyaesthesias and sensory loss in a stocking distribution with prominent symptoms of restless legs, burnings, or formication (Gemignani *et al.* 1992). Prominent vasculitic purpura and Raynaud's phenomenon should suggest the possibility of cryoglobulinaemia. The diagnosis is proven by looking for immune precipitates in the serum of blood allowed to clot at 37°C. Plasma exchange, steroids, and cyclophosphamide have all been used therapeutically, with occasional success. Indeed, if a patient has been diagnosed with vasculitic neuropathy and continues to progress despite such treatment, the possibility of cryoglobulinaemia should be entertained. Hepatitis C infection commonly underlies essential mixed cryoglobulinaemia, and improvement in the neuropathy has been reported with interferon alpha therapy (Khella *et al.* 1995).

21.16 Sensory perineuritis and migrant sensory neuritis

21.16.1 Sensory perineuritis

This rare mononeuropathy causes pain and numbness in the territories of individual cutaneous nerves (Logigian *et al.* 1993).

Patients may present with severe pain in the feet induced by standing or walking. Tinel's sign may be produced by percussion along the course of affected nerves. Initially the disorder may be relapsing and remitting, but symmetrical distal sensory loss may eventually appear. Mixed motor and sensory nerve involvement has been described with perineuritis. Unlike the migrant sensory neuritis of Wartenberg, stretching of peripheral nerves does not produce electric shock sensations in sensory perineuritis. Biopsy of affected nerves shows a chronic inflammatory infiltrate in the perineurium surrounding some fascicles but not others. The differential diagnosis includes a rare purely sensory form of vasculitic neuropathy (Seo et al. 2004). Endoneurial blood vessels are spared. Similar histological findings have been described in the peripheral neuropathy of the Spanish toxic rapeseed oil syndrome. The disorder often responds to steroids.

21.16.2 Migrant sensory neuritis of Wartenberg

Patients with this disorder develop sudden pains in the territory of cutaneous nerves; the pain is induced by movements of a limb which stretch or distort the nerve. After repeated episodes of pain, cutaneous sensation may be lost in the nerve's territory for about 6 weeks. The relapsing and remitting nature of the sensory disturbance may initially suggest multiple sclerosis. Migrant sensory neuritis is probably commoner than generally appreciated and usually affects patients in middle life (Matthews and Esiri 1983). It has occurred in members of a family with dominantly inherited brachial plexus neuropathy (Thomas and Ormerod 1993). Nerve conduction studies may show diminished sensory nerve action potentials in affected nerves. The condition follows a benign course.

21.17 Diabetic neuropathy

21.17.1 Range of disorders

Diabetes mellitus is one of the commonest causes of disabling polyneuropathy. Two types of polyneuropathy are recognized and may coexist: symmetrical sensorimotor and autonomic. Various focal neuropathies occur, including diabetic proximal neuropathy, mononeuropathies of cranial and peripheral nerves, and truncal neuropathies (Watkins 1990; Said 1996). Two or more of these neuropathies commonly coexist within the same patient. Furthermore, many patients with established insulin-dependent diabetes mellitus have subclinical or electrophysiological evidence of sensory or autonomic polyneuropathy, despite being asymptomatic. In a population-based cohort study neuropathy was present in 66 per cent of insulin-dependent diabetics: polyneuropathy 54 per cent, carpal tunnel syndrome 11 per cent, visceral autonomic 7 per cent, and in 59 per cent of non-insulin-dependent diabetics: polyneuropathy 45 per cent, carpal tunnel syndrome 6 per cent, visceral autonomic 5 per cent. However, only about 20 per cent of diabetics have symptoms, and only 6 per cent of insulin-dependent, and 1 per cent of non-insulin-dependent diabetics had more severe forms of neuropathy (Dyck et al. 1993b). Neuropathy is significantly associated with diabetic retinopathy or nephropathy. Impaired glucose tolerance, without frank diabetes mellitus, is associated with axonal polyneuropathy predominantly affecting small fibres (Sumner et al. 2003).

21.17.2 Diabetic polyneuropathy

This is most prevalent in insulin-dependent diabetics of more than 20 years standing, and in those with hypertension or poor glycaemic control. Sensory fibres are mainly involved, often accompanied by a variable degree of autonomic neuropathy.

Clinical features. Sensory symptoms usually commence in the legs. The common initial symptoms are paraesthesiae, and burning or lancinating pains. The sensory loss usually reflects abnormalities of unmyelinated fibres with impaired pain and temperature sensations in a stocking distribution. The hands are also involved in more severe cases. In established cases vibration and joint-position sensations are impaired at the toes. These patients may have a sensory gait ataxia and a positive Romberg's sign. Tall diabetics are at greatest risk of sensory neuropathy, probably by virtue of their longer nerves. The ankle jerks are usually lost but generalized areflexia is less common. Involvement of autonomic fibres impairs sweating and prevents skin blood-flow regulation distally in the limbs, leading to a warm, dry foot with hard skin vulnerable to cracking. The combination of pain insensitivity and autonomic denervation predisposes the foot to skin ulceration. Neuropathic joints may develop. Motor loss is less common in diabetic polyneuropathy by comparison with the degree of sensory loss. Distal muscle weakness and wasting may be encountered in long-standing cases. However, marked motor involvement should provoke consideration of other possible contributing causes to the neuropathy apart from diabetes. Chronic inflammatory demyelinating polyneuropathy can occur in diabetics (Krendel et al. 1995). The diagnosis of sensory or autonomic polyneuropathy poses few difficulties in patients with recognized diabetes mellitus. The blood sugar should be checked in all patients presenting with sensory neuropathy since neuropathy may be the first symptom of diabetes.

Nerve conduction studies. These generally show diminished or absent sensory nerve action potentials with normal or only mildly impaired motor nerve conduction velocity. These electrophysiological findings occur in some diabetics before neuropathic symptoms develop. Electromyography often shows chronic denervation of distal muscles. Sensory nerve action potentials, which reflect conduction in large myelinated fibres, may be remarkably normal in those patients with selective loss of pain and temperature sensation.

Sural nerve biopsies. These usually show axonal loss and Wallerian degeneration affecting both myelinated and unmyelinated fibres. Less frequently diabetic nerves show segmental demyelination and remyelination, which is likely to be merely a secondary consequence of primary axonal atrophy (Said 1996). Painful forms of diabetic neuropathy show regenerative sprouting of unmyelinated fibres, possibly the pain results from abnormal discharges in the sprouts. Degeneration of the distal portions of dorsal column axons in the spinal cord is found at autopsy in patients with diabetic neuropathy. Thus both the central and peripheral branches of sensory neurones are vulnerable to hyperglycaemia. This observation, considered together with the greater vulnerability of sensory rather than motor neurones, suggests that it may be the perikarya of dorsal-root ganglion neurones rather than axons which are primarily affected by hyperglycaemia. Furthermore, because central nervous axonal branches of sensory neurones are unable to regenerate, full recovery of function is unlikely to result from treatments that merely promote peripheral nerve regeneration. Epineurial arteriolar walls may be thickened and endoneurial capillary lumens reduced in peripheral nerve, suggesting an ischaemic contribution to some cases of diabetic polyneuropathy. Morphometric examination of the entire length of nerves from elderly diabetics shows that multifocal ischaemic axonal loss may summate distally and contribute to the polyneuropathy (Dyck *et al.* 1986).

Pathophysiology. The abnormality of neuronal cell biology responsible for diabetic polyneuropathy is not known. It is likely that multiple pathogenetic mechanisms interact to varying degrees in producing a clinical picture of neuropathy which differs from patient to patient. Multifocal ischaemic neuropathy with distal summation of axonal loss may be a factor in elderly patients. However, it is likely that diffuse consequences of the metabolic disturbance are more prominent in causing the polyneuropathy of younger patients. Early trials of therapy were based on the notion that persistent hyperglycaemia may activate the polyol pathway in nerve causing sorbitol accumulation as a result of enhanced aldose reductase activity. The biochemical consequences of this for nerve conduction remain unproven and aldose reductase inhibitor therapy was unsuccessful in clinical trials. Insulin control does improve axonal excitability, with corresponding improvements in nerve conduction velocities (Kitano *et al.* 2004) but this is unlikely to be of direct relevance to the axonal degeneration of diabetic neuropathy. Accumulation of sugars can promote non-enzymatic glycosylation of peripheral-nerve proteins, probably altering their function and irreversibly cross-linking them by advanced glycosylation end-products (Ryle and Donaghy 1995). The slow phase of axonal transport of microtubule and neurofilament cytoskeletal proteins to the distal axon is reduced in experimental diabetes. This could alter the structural integrity of the distal axon and account for the axonal length-related neuropathy of diabetes (Medori *et al.* 1988). The APOE genotypes 3/4 and 4/4 increase the risk of polyneuropathy, by a degree equivalent to 15 extra years of age or duration of diabetes (Bedlack *et al.* 2003).

Treatment. Strict control of glycaemia provides the best hope of reducing the occurrence and progression of diabetic polyneuropathy (Diabetes Control and Complications Trial Research Group 1993). Optimal control of glycaemia improves vibration sensation in diabetic polyneuropathy. Restoration of glycaemia by pancreas transplantation halts the downhill progression of diabetic neuropathy and a minor degree of recovery is apparent 3.5 years after transplantation (Kennedy *et al.* 1990). It should be noted that protracted or recurrent hypoglycaemia due to over-zealous insulin treatment may cause a predominantly motor peripheral neuropathy, which is distal and symmetrically distributed, and may particularly affect the arms. A similar motor neuropathy may be seen in the hyperinsulinism of islet cell tumours. Treatment of established diabetics with the aldose reductase inhibitor sorbinil does not significantly improve the clinical or electrophysiological outcome over more than 3-year follow up (Sorbinil Retinopathy Trial Research Group 1993).

Acutely painful diabetic polyneuropathy. This can be severely disabling and difficult to treat. It usually improves after some months of strict glycaemic control. In newly diagnosed diabetics, pain can be precipitated or augmented by insulin therapy. Carbamazepine, phenytoin, gabapentin, or sodium valproate therapy may help to relieve shooting or stabbing pains (Kochar *et al.* 2004). Constant deep aching pain may respond to amitriptyline within a few days, but this drug may exacerbate a coexisting autonomic neuropathy causing urinary hesitancy or erectile failure. Skin care is essential so as to prevent chronic ulceration; cuts and abrasions should be treated promptly. Regular advice should be sought from a chiropodist, and the insides of shoes inspected daily for small stones and other irregularities.

21.17.3 Diabetic proximal neuropathy

This disorder ranges from the familiar extreme of acute asymmetrical painful proximal leg muscle weakness developing over a few days or weeks, to the less familiar extreme of symmetrical painless proximal muscle weakness developing over many weeks or months. Diabetic proximal neuropathy has been termed diabetic radiculoplexus neuropathy, diabetic myelopathy, polyradiculopathy, amyotrophy, lumbar plexopathy, mononeuropathy multiplex, femoral neuropathy, myopathy, or neuropathic cachexia. It is commonest in non-insulin-dependent diabetics, generally in their sixth or seventh decade. Previously unrecognized diabetes may present with proximal neuropathy. Proximal neuropathy is seldom accompanied by diabetic retinopathy or nephropathy.

Anterior thigh muscle pain is the usual first symptom. Proximal leg muscle weakness, mainly involving the quadriceps muscle, develops over the next few days or weeks. The knee jerks are lost in most patients. Despite the unilateral onset, bilateral weakness eventually occurs in over half of all patients. Occasionally the plantar responses are extensor, hence the original term 'diabetic myelopathy'. The neuropathy is usually accompanied by profound weight loss. Femoral nerve conduction is delayed. The spinal fluid protein is slightly elevated in most patients, indicating involvement of nerve roots.

Most patients improve neurologically after some months, and this is generally attributed to improved control of hyperglycaemia by insulin or oral hypoglycaemic agents (Coppack and Watkins 1991). Pain is the first symptom to resolve. The return of muscle power is usually substantial but is complete in only 20 per cent of patients. Recovery takes place over a period of 6–18 months. Up to one-fifth of patients may experience recurrence (Coppack and Watkins 1991). Although common sense requires that hyperglycaemia should be strictly treated in diabetic proximal neuropathy, there is no conclusive evidence that hypoglycaemic therapy promotes recovery over and above that which will occur spontaneously.

The pathogenetic mechanism responsible for diabetic proximal neuropathy remains unclear. The absence of associated diabetic retinopathy or glomerulopathy, the frequent bilaterality, and the relatively slow neurological deterioration in diabetic proximal neuropathy, have argued against vascular occlusion due to diabetic microangiopathy as the primary cause. Epineurial microvasculitis or inflammation is observed in some patients in biopsies of the intermediate cutaneous nerve of the thigh, a cutaneous branch of the femoral nerve (Said *et al.* 1994; Dyck *et al.* 1999). This raises the question of treating severe or enduring cases of diabetic proximal neuropathy with steroids or other immunomodulatory drugs (Krendel *et al.* 1995). The clinical features, pathology, and outcome of diabetic proximal neuropathy seem identical to those of non-diabetic radiculoplexus neuropathy (Section 22.6.1) (Dyck *et al.* 2001).

21.17.4 Diabetic truncal neuropathy

Attacks of truncal pain and sensory disturbance occur in diabetic patients. They may be recurrent, of variable severity, and may affect more than one thoracic nerve root territory. Truncal neuropathy typically affects non-insulin-dependent diabetics in their fifth to seventh decades. It is often accompanied by considerable weight loss, similar to that occurring in diabetic proximal neuropathy. The pain may not be strictly localized to the territory of a discrete dermatome. A sensory deficit is usually demonstrable on the trunk and can be restricted to the territory of a single anterior or posterior ramus. Abdominal protuberance may result from focal paralysis of abdominal wall muscles. Electromyography often shows denervation of paraspinal muscles. Skin nerve fibres are reduced in biopsies from affected compared to unaffected sensory territories (Lauria *et al.* 1998). The differential diagnosis includes multiple sclerosis and lesions affecting vertebral bones. Spontaneous recovery is usual but may take some months. Carbamazepine or amitriptyline can control the pain.

21.17.5 Diabetic autonomic neuropathy

Abnormal autonomic function is detectable in a sixth of all patients with insulin-dependent diabetes, although symptoms of autonomic peripheral neuropathy occur only in relatively few. Autonomic neuropathy usually coexists with a small fibre sensory peripheral neuropathy. The main symptoms are abnormal sweating or diarrhoea. Less frequently patients are troubled by postural hypotension, vomiting from gastroparesis, micturition difficulties, bladder infection due to atony, sexual impotence, and retrograde ejaculation. Symptomatic autonomic neuropathy predisposes patients to sudden death during anaesthesia, to cardiac arrhythmias, and it may reduce awareness of hypoglycaemia due to failure of catecholamine release. Although notably intermittent in severity, symptoms of autonomic disturbance tend to continue with little change in severity for many years. A wide variety of autonomic function abnormalities may be measured in diabetic patients (Said 1996). Dry warm feet, miosis, reduced pupil light reflexes, and ptosis may be observed. Iritis is associated with autonomic neuropathy in diabetics (Watkins 1990). The simplest reliable bedside tests consist of measuring postural hypotension, which reflects failure of sympathetic fibres, and measuring variability of the heart rate during deep breathing, this represents the sinus arrhythmia, in turn reflecting the parasympathetic innervation of the heart. Tight control of glycaemia by continuous subcutaneous insulin infusion or pancreatic transplantation produces only minor improvements in autonomic function (Kennedy *et al.* 1990). Diarrhoea can be particularly troublesome at night and may respond to codeine phosphate, clonidine, or one or two doses of tetracycline. Oral erythromycin can improve gastric emptying, possibly by mimicking the effects of motilin on gastrointestinal motility, and can be tried in patients disabled by serious vomiting due to diabetic gastroparesis (Janssens *et al.* 1990).

21.17.6 Diabetic mononeuropathy

Diabetics are particularly vulnerable to a wide range of mononeuropathies affecting peripheral or cranial nerves. Nerves vulnerable to compression are most commonly affected, such as the median in the carpal tunnel, the ulnar in the cubital groove, the radial at the humerus, the common peroneal at the fibular head, and the lateral cutaneous nerve of the thigh at the inguinal ligament. Painful oculomotor nerve palsies, often sparing the pupil, are common in older diabetics and resolve spontaneously. It is likely that pre-existing diabetic microvascular disease makes nerves unusually vulnerable to compression, although the utility of surgical decompression remains unproven (Chaudhry *et al.* 2006). These mononeuropathies may improve spontaneously, or with surgical release if at sites of compression. However, permanent residual abnormalities are common.

Some diabetics, usually elderly, develop a progressive and painful multifocal mononeuritis with necrotizing vasculitis found on nerve biopsy (Kelkar and Parry 2003; Said *et al.* 2003). Steroids seem effective in such patients.

21.17.7 Chronic inflammatory demyelinating polyneuropathy in diabetics

Diabetics seem to be ten-fold more vulnerable to chronic inflammatory demyelinating polyneuropathy (Section 21.11.2) than non-diabetics. Type 1, insulin-dependent, and Type 2, non-insulin-dependent, diabetics are equally vulnerable (Sharma *et al.* 2002a). This disorder can be differentiated from the normal diabetic polyneuropathy (Section 21.17.2) by the subacute development of weakness, including proximally, the large fibre sensory loss, and evidence of demyelination electrophysiologically (Uncini *et al.* 1999). The disorder is normally responsive to immunomodulation, including IvIg (Sharma *et al.* 2002b). When the occurrence of chronic inflammatory demyelinating polyneuropathy is suspected in a diabetic patient, it is advisable to give a trial of IvIg treatment (Section 21.3.3) to prove that the neuropathy is reversible. If so, long-term treatment should either continue with IvIg, or using oral prednisolone with adjustment of the patient's hypoglycaemic medication to offset the diabetogenic effect of steroids.

21.18 Neuropathy due to systemic medical disorders

21.18.1 Chronic renal failure

Clinical or electrophysiological evidence of polyneuropathy is detectable in over 50 per cent of patients with end-stage renal disease. Symptomatic polyneuropathy was much commoner before the era of widespread and earlier introduction of renal replacement therapy. Uraemic neuropathy develops very gradually and is uncommon if the glomerular filtration rate exceeds 10 ml/min. Although uraemia itself is responsible for the neuropathy in many patients with chronic renal failure, it should be recognized that neuropathy may be an independent feature of the underlying disease which has caused the chronic renal failure: diabetes mellitus, systemic vasculitis, myelomatosis, amyloidosis, and systemic lupus erythematosus. The neuropathies caused by these diseases are often focal or demyelinating in nature, unlike the axonal degeneration polyneuropathy of uraemia.

A restless leg syndrome is the commonest early symptom of uraemic polyneuropathy: crawling, pricking, and itching sensations occur at night. Burning paraesthesiae may develop. Muscle cramps and fatigability are followed by distal weakness and muscle atrophy. An autonomic neuropathy may cause sexual impotence and contribute to difficulties in intravascular fluid volume regulation, making postural hypotension a particular problem following fluid removal by dialysis. The earliest physical signs are loss of vibration sensation at the toes and absent ankle jerks. A mixed motor, autonomic, and multimodal sensory neuropathy eventually develops.

Nerve conduction studies reflect axonal degeneration of motor and sensory fibres. Nerve excitability is reduced; this axonal depolarization correlates with hyperkalaemia and improves with dialysis (Krishnan *et al.* 2006). Nerve biopsy shows loss of all sizes of nerve fibres, particularly large myelinated fibres. Less commonly, there is evidence of segmental demyelination or remyelination which may be secondary to axonal changes. Autopsy studies show degeneration of the dorsal columns of the spinal cord, representing the central axons of dorsal root ganglion cells. It is presumed that uraemic polyneuropathy results from the accumulation of neuro-toxic waste products, but the causative compounds have not been identified.

Renal replacement therapy normally prevents the neuropathy from deteriorating and may allow considerable recovery. The great-est improvement occurs in patients receiving renal transplants rather than those maintained by dialysis. Following successful renal transplantation, clinical and electrophysiological improvement starts after some months and continues slowly. However, full recovery of the neuropathy is unusual unless it was initially mild. A carpal tunnel syndrome may develop in patients receiving long-term dialysis, due to deposition of β_2-microglobulin amyloid in the flexor retinaculum. Acute or subacute neuropathy has occurred occasionally in patients treated by peritoneal dialysis. Such patients may have coexisting diabetes mellitus, the neuropathy has demyeli-nating features, and sometimes improves (Ropper 1993).

21.18.2 Hypothyroidism

Paraesthesiae, lancinating limb pains, or muscle cramps occur in half of all patients with established myxoedema. Sensory symptoms in the limbs may be the presenting feature of hypothyroidism. There are few neuropathic signs on examination; minor distal sen-sory changes are usually the only abnormalities. Distal spontane-ous and evoked pains either before or after thyroid replacement therapy, reflects a small fibre neuropathy or central sensitization (Ørstavik *et al.* 2006). Although characteristically slowly relaxing, the tendon reflexes are usually retained. Nerve biopsies may show segmental demyelination. The sensory symptoms may resolve on thyroid hormone replacement therapy. Carpal tunnel syndrome is common, and tarsal tunnel syndrome less common, in myxoede-ma. Acroparaesthesiae due to the former often resolve with thyroid hormone replacement therapy, but if not, surgical decompression may be required.

21.18.3 Acromegaly

Both polyneuropathy and the carpal tunnel syndrome are common in patients with acromegaly and are not thought to be due to the associated diabetes mellitus. The polyneuropathy is of insidious onset, causing distal paraesthesiae, depressed reflexes, distal muscle weakness and multimodal sensory disturbance. The peripheral nerves may be clinically enlarged. Nerve conduction is slightly slowed. Nerve biopsies show axonal loss with some demyelination and remyelination. The fascicular cross-sectional area is increased with accumulation of tissue subperineurially and endoneurially. It is unclear whether the neuropathy improves significantly with treatment of the underlying growth hormone excess.

Symptomatic carpal tunnel syndrome should be treated by surgical decompression.

21.18.4 Primary biliary cirrhosis

A distal sensory polyneuropathy may develop in patients with primary biliary cirrhosis. Sural nerve biopsy shows perineurial xanthomatous deposits distorting the normal architecture.

21.18.5 Systemic lupus erythematosus

A wide variety of peripheral neuropathies occur in systemic lupus erythematosus and should be treated according to principles already outlined. Some evidence of polyneuropathy occurs in 20 per cent, although often mild or even asymptomatic (Omdal *et al.* 1991). Demyelinating neuropathies of the acute Guillain–Barré type (Section 21.10.1) or steroid-responsive chronic inflammatory demyelinating polyneuropathy (Section 21.11.2) may occur. Chronic sensorimotor axonal degeneration neuropathies may be encountered. Focal neuropathy due to necrotizing arteritis, or car-pal tunnel syndrome may be the presenting feature of systemic lupus erythematosus (Stefurak *et al.* 1999). A small fibre sensory neuropathy may be demonstrated at skin biopsy by reduced num-bers of intradermal nerve fibres (Gøransson *et al.* 2006).

21.18.6 Sarcoidosis

A small proportion of patients with sarcoidosis have peripheral neuropathy. Neuropathy can be the presenting feature of sarcoido-sis. Cranial nerve palsies are most often encountered; these are often multiple, of variable severity, and particularly affecting the facial nerve. Mononeuropathy may affect any peripheral nerve, including the sensory nerves of the trunk. Sensorimotor polyneu-ropathy is less common and may take acute multifocal or purely sensory forms. Sensorimotor polyneuropathy may be associated with multiple small granulomas within biopsied nerves or muscles, with associated inflammatory or vasculitic features in some (Said *et al.* 2002). The response to steroids is usually good.

21.18.7 Eosinophilia–myalgia syndrome

This disorder followed months or years after ingestion of contami-nated l-tryptophan, a component of some body building food supplements. There was associated eosinophilia and brawny indu-ration of the skin. Some patients developed a painful inflammatory myopathy. Sensorimotor axonal degeneration neuropathies or multifocal neuropathies may occur. The combination of neuropa-thy and myopathy could result in respiratory failure requiring ven-tilation (Smith and Dyck 1990). A more chronic demyelinating neuropathy has also occurred (Freimer *et al.* 1992). The neuropa-thy of the eosinophilia–myalgia syndrome should be distinguished from other neuropathies associated with eosinophilia: hypereosi-nophilic syndrome (Section 21.18.8), necrotizing arteritis of the Churg–Strauss type (Section 21.15.3), and Hodgkin's disease (Section 21.12.4).

21.18.8 Hypereosinophilic syndrome

Symmetrical sensorimotor peripheral neuropathy due to axonal degeneration occurs in about one-tenth of patients with idiopathic hypereosinophilia (Monaco *et al.* 1988). Increased numbers of degranulated eosinophils are found in the blood. The systemic disorder may include a restrictive cardiomyopathy due to endomy-ocardial fibrosis.

21.18.9 Critical illness polyneuropathy

Sensorimotor polyneuropathy can develop in patients being ventilated for cardiorespiratory disease who develop multi-organ failure or sepsis. Prospective electrophysiological examination of patients with severe sepsis shows that abnormalities are common at the time of admission to intensive care, and predict subsequent development of critical illness neuropathy and myopathy (Khan *et al.* 2006). The compound muscle action potentials and sensory nerve action potentials are reduced in amplitude, and needle electrodes show evidence of limb muscle denervation. This neuropathy usually comes to light when patients fail to wean from the ventilator. The mortality in such patients is high, but those who recover neurologically do so over 3–6 months. This rapidity of recovery is faster than might be expected from a dense axonal degeneration polyneuropathy and suggests a degree of potentially reversible conduction failure. The disorder should be distinguished from Guillain–Barré syndrome by normal spinal fluid protein levels and the electrophysiological characteristic of axonal degeneration rather than demyelination. In a critically ill patient in the intensive care setting, the principal differential diagnosis is a critical illness myopathy, occurring most commonly in acute respiratory disorder such as asthma treated with non-depolarizing neuromuscular-blocking agents or high-dose steroids. Critical illness polyneuropathy and myopathy may coexist, and given that the creatine kinase level often remains normal, muscle biopsy is the only reliable way to diagnose the myopathy (Gutmann and Gutmann 1999).

21.18.10 Sjogren's syndrome

A sensory neuronopathy similar to the paraneoplastic disorder (Section 21.13.1) may occur in Sjogren's syndrome or the isolated sicca complex of dry eyes and dry mouth (Mori *et al.* 2005). This disorder can develop in patients without underlying cancer or Sjogren's syndrome and can be associated with Adie's pupil. The cases associated with Sjogren's syndrome usually have predominantly kinaesthetic sensory loss, may have neuropathic pain, dry eyes on Schirmer testing, a positive antinuclear factor, and they lack other features of paraneoplastic encephalomyelitis. Their sensory disturbance may stabilize or improve slightly. T_2-weighted MRI shows high signal intensity in the posterior columns cervically. No treatment is invariably effective in these sensory neuropathy syndromes, although there are reported benefits of IvIg in the ataxic neuropathy (Takahashi *et al.* 2003).

21.18.11 Gastrointestinal disease

Patients with inflammatory bowel disease can develop a range of demyelinating and axonal sensorimotor polyneuropathies and small fibre sensory neuropathies (Gondim *et al.* 2005). Immunomodulatory agents usually lead to improvement in those with demyelinating neuropathies. Approximately a quarter of patients with coeliac disease have a detectable chronic axonal neuropathy (Luostarinen *et al.* 2003). A range of neuropathies, usually axonal, is reported up to three times as frequently in patients with gluten sensitivity associated with antigliadin antibodies as in controls (Hadjivassiliou *et al.* 2006).

21.19 Drug-induced polyneuropathy

Proving that a drug causes peripheral neuropathy can be difficult unless prospective monitoring of patients has been carried out within a clinical trial, as is usually the case for cancer and HIV chemotherapy. Tinglings are a common medication side effect, but often reflect a physiological effect of the drug, rather than a degenerative peripheral neuropathy. Patients with pre-existing peripheral neuropathy, especially Charcot–Marie–Tooth disease seem unusually vulnerable to normally non-toxic doses of neuropathic drugs, such as Vincristine (Chaudhry *et al.* 2003; Weimer and Podwall 2006).

21.19.1 Alcohol

Alcoholics may develop either polyneuropathy or focal compressive peripheral nerve lesions. Pressure palsies often follow periods of stuporous immobility and generally affect the radial, ulnar, or common peroneal nerves. Polyneuropathy may develop insidiously in long-standing alcoholics. It usually presents with symmetrical burning dysaesthesiae or paraesthesiae distally, or with sensory ataxia. Physical signs of a symmetrical sensorimotor polyneuropathy are often restricted to the legs. Nerve conduction studies reflect axonal degeneration predominantly affecting sensory axons. Autonomic neuropathy may be present, and is associated with an increased risk of cardiovascular death. Alcoholic polyneuropathy may be caused either by direct toxicity of ethanol and its metabolites or by nutritional deficiency. Alcoholic neuropathy tends to be sensory dominant and slowly progressive, whereas thiamine deficiency produces marked motor involvement and subacute progression (Koike *et al.* 2003). Nutritional deficiency polyneuropathies are common in patients with associated Wernicke–Korsakoff syndrome (Section 34.5) and particularly reflect deficiency of vitamin B_1, thiamine. Ethanol, or its metabolites such as acetaldehyde, may have direct neurotoxic effects; this mechanism may be a particularly important cause of neuropathy in those alcoholics who are not malnourished.

Neuropathy is often ascribed erroneously to alcohol in any patients with moderately high consumption. In practice neuropathy is unlikely until a lifetime of alcohol consumption of 15 kg ethanol per kg body weight is reached; this equates for a 70 kg man to 300 ml whisky daily for 25 years (Monforte *et al.* 1995). Occasionally rapidly progressive polyneuropathy resembling Guillain–Barré syndrome, but without raised CSF protein or slowed nerve conduction, has occurred in alcoholics (Wohrle *et al.* 1998). Vitamin B_1 replacement therapy should be given to all patients with alcoholic polyneuropathy. Gradual improvement in the clinical and electrophysiological aspects of peripheral neuropathy occurs in those alcoholics who achieve long-term abstinence. Disulfiram, *Antabuse*, therapy may itself produce neuropathy in alcoholics (Section 21.19.8).

21.19.2 Amiodarone

This iodine-containing antiarrhythmic drug increases two-fold the occurrence of symmetrical distal sensorimotor polyneuropathy after prolonged administration (Vorperian *et al.* 1997). The neuropathy develops some months after starting treatment, can produce severe weakness, and is often associated with a raised spinal fluid protein level. Sural nerve biopsies show loss of myelinated fibres with lipid-laden lysosomes in Schwann cells; demyelination rather than axonal degeneration is thought to be the primary event (Pellissier *et al.* 1984). Approximately 70 per cent of patients taking 800 mg of amiodarone daily develop a reversible syndrome of tremor and ataxia which occurs independently of peripheral neuropathy.

221.19.3 Chloroquine

Chloroquine is used to treat malaria, amoebiasis, and chronic discoid lupus erythematosus. A combination of sensorimotor peripheral neuropathy with myopathy, or myopathy alone, may occur in patients taking doses of at least 500 mg daily for a year or more (Whisnant *et al.* 1963). The neuromyopathy improves steadily on stopping the drug.

21.19.4 Cisplatin

Cisplatin, carboplatin, and oxaloplatin are important drugs in treating ovarian, testicular, and bladder tumours. A predominantly sensory peripheral neuropathy, characterized by distal paraesthesiae, develops in almost all patients given a cumulative dose of 300–600 mg/m^2 cisplatin, often accompanied by Adriamycin (LoMonaco *et al.* 1992). Sensory ataxia may be severe. Sensory nerve conduction is abnormal, motor conduction is normal. Symptoms of neuropathy may start some weeks after the last dose of cisplatin has been administered, and may subsequently worsen for a few months. This phenomenon is known as coasting. Partial or complete recovery of the neuropathy may occur after cessation of cisplatin therapy, usually taking more than a year. Oxaloplatin differs in producing sensory symptoms within an hour of infusion and eventually almost all patients receiving a cumulative dosage of >540 mg/m^2 develop neuropathy (Quasthoff and Hartung 2002).

21.19.5 Colchicine

Mild neuromyopathy may be common in patients receiving colchicine as treatment for gout (Kuncl *et al.* 1987). Neuromyopathy is particularly likely to occur if there is associated mild chronic renal impairment. The myopathic element may involve severe proximal muscle weakness, electromyographic features resembling those of polymyositis, elevated serum creatine kinase levels, and electron microscopic evidence of accumulation of lysosomes and autophagic vacuoles in biopsies of proximal muscles. Creatine kinase levels return to normal within days of stopping colchicine, and proximal muscle strength improves over subsequent weeks. The neuropathic element is less pronounced than the myopathy, with distal limb sensory loss, tendon areflexia, and evidence of axonal degeneration on nerve conduction studies and sural nerve biopsies.

21.19.6 Dapsone

Motor neuropathy may complicate long-term therapy with 200–500 mg of dapsone daily (Gutmann *et al.* 1976). Such doses are generally used for the treatment of dermatological conditions, usually dermatitis herpetiformis, and dapsone neuropathy is less likely to complicate leprosy treatment. Weakness and wasting is most prominent distally in the limbs. Motor conduction velocities are normal or only slightly slowed and evoked muscle action potentials reduced. Tendon reflexes tend to be preserved although hypoactive. Muscle strength improves following withdrawal of dapsone.

21.19.7 Didanosine and antiretroviral drugs

Polyneuropathy occurs regularly in patients with HIV infection treated by the nucleoside analogue drugs didanosine, ddI, zalcidabine, ddC, lamuvidine, 3TC, stavudine, d4T, and fialuridine reverse transcriptase inhibitor, FIAV, and is a dose-limiting side effect (Dalakas 2001). This neuropathy is painful, purely or predominantly sensory, and often of explosive onset. The commonest signs are hyporeflexia of the ankle jerk, impaired pinprick and vibration sensation in the feet, and gait unsteadiness. The chance of neuropathy is dosage dependant, and it develops on average 8 weeks after starting high-dose therapy but develops later with contemporary lower dose therapy. After stopping the drug, the neuropathy may 'coast' with worsening symptoms, before stabilizing and improving. It can be difficult to differentiate this toxic neuropathy from the painful sensory neuropathy which occurs in patients with established HIV infection (Section 21.14.2). However these drug-toxic neuropathies differ in that they occur during the months after starting therapy, the painful symptoms usually evolve abruptly, and the hands are uninvolved. If in doubt, the drug should be withdrawn for some months to see if improvement occurs.

21.19.8 Disulfiram

An axonal degeneration sensorimotor neuropathy may develop after disulfiram, *Antabuse,* therapy for alcoholism. Symptoms improve after stopping the drug. Axonal neurofilament accumulations can be found in the sural nerve on electron microscopy. It is noteworthy that disulfiram is enzymatically converted to carbon disulphide which itself is known to cause a distal axonopathy with neurofilament accumulations (Section 21.21.2) (Ansbacher *et al.* 1982). Prospective studies suggest that peripheral nerve damage occurs at disulfiram doses of 250 mg/day, but not at 125 mg/day (Palliyath *et al.* 1990).

21.19.9 Ethambutol

Predominantly sensory neuropathy is an occasional consequence of treatment with the antituberculous drug ethambutol (Nair *et al.* 1980). Optic neuropathy is a commoner complication of long-term ethambutol administration.

21.19.10 Gold

Acute or subacute sensorimotor neuropathy can occur in patients some months after commencing gold therapy for rheumatoid arthritis (Katrak *et al.* 1980). Partial recovery occurs over the months following cessation of gold administration. Myokymia of limb muscles is a distinctive finding. Gold neuropathy should be distinguished from the mononeuritis multiplex that can occur in patients with aggressive rheumatoid disease with vasculitic features (Section 21.15.3).

21.19.11 Isoniazid

Patients with inherited slow drug-acetylation status may develop peripheral neuropathy when they receive long-term isoniazid therapy. Paraesthesiae and numbness are the initial symptoms and neuropathic pain may be prominent. Muscle weakness usually only appears in the later stages. Hyperalgesia and muscle cramping are distinctive features in many patients (Ochoa 1970). Isoniazid antagonizes the actions of vitamin B$_6$, or pyridoxine, and the neuropathy can be prevented by simultaneous administration of pyridoxine during isoniazid therapy. In patients who develop neuropathy, isoniazid therapy may be interrupted, vitamin B$_6$ given parenterally at 100–200 mg/day, and other antituberculous drugs continued. Variable degrees of improvement in the neuropathy follow these measures.

21.19.12 Leflunomide

A predominantly sensory and axonal degeneration polyneuropathy has been noted in patients treated with Leflunomide for rheumatoid arthritis (Bonnel and Graham 2004). The mean time of onset was after 6 months of therapy and early discontinuation led to some improvement.

21.19.13 Lithium

Occasional cases of peripheral neuropathy have been associated with lithium carbonate therapy for depression. Toxic levels of lithium were deemed responsible for a severe generalized sensorimotor neuropathy which shows electrophysiological and nerve biopsy evidence of axonal loss. Recovery can occur following drug cessation (Vanhooren et al. 1990).

21.19.14 Metronidazole

Sensory neuropathy may follow prolonged administration of the antibacterial drug metronidazole (Coxon and Pallis 1976). Paraesthesiae or numbness of the toes have usually been recorded only in patients who have received a total dose of at least 30 g. Sensory nerve action potentials are diminished in amplitude but of normal latency. Neuropathic symptoms resolve during the months after stopping the drug. Occasionally convulsions, encephalopathy, and cerebellar ataxia have been associated with metronidazole therapy.

21.19.15 Nitrofurantoin

An axonal degeneration sensory neuropathy may follow administration of large doses of the antibiotic, nitrofurantoin. Total dosages usually exceed 20 g, although neuropathy can occur after lower dosage in patients with impaired renal excretion (Lindholm 1967). Partial recovery may follow cessation of the drug.

21.19.16 Phenytoin

Peripheral neuropathy is demonstrable in up to 20 per cent of epileptic patients on long-term anticonvulsant therapy. It is usually relatively mild, involving sensory diminution in a stocking distribution and reduced ankle tendon reflexes. Although commonly attributed to phenytoin, the evidence relates this neuropathy to a wide range of anticonvulsants. It is commonest in patients receiving multiple drugs (Swift et al. 1981).

21.19.17 Pyridoxine

Sensory neuropathy with prominent ataxia may follow self-medication with megadoses of pyridoxine, vitamin B_6. The normal human daily pyridoxine requirement is approximately 0.004 g. Sensory neuropathy has usually followed daily oral ingestion of 2–6 g of pyridoxine or single massive parenteral doses (2 g/kg) in the treatment of mushroom poisoning (Albin et al. 1987). Neuropathy has also been recorded following long-term chronic consumption of lower doses (0.2 g daily) (Parry and Bredesen 1985). The rate of onset of symptoms is proportional to the magnitude of the daily dosage. The sensory neuropathy probably reflects damage to dorsal root ganglion neurones, with little potential for recovery.

21.19.18 Statins

A pharmacoepidemiologic database study reported an increased incidence of predominantly sensory and axonal polyneuropathy in patients receiving long-term statin treatment (Gaist et al. 2002). The neuropathy tends to be mild, with pain, paraesthesiae, numbness, and absent tendon reflexes in half of the cases.

21.19.19 Suramin

Either distal axonal, or subacute demyelinating, sensorimotor polyneuropathies have been noted in more than 80 per cent of patients receiving suramin doses sufficient to achieve plasma levels $\geq 350\ \mu g\ ml^{-1}$ during attempted treatment of hormone-refractory metastatic prostate cancer (Chaudhry et al. 1996).

21.19.20 Tacrolimus

Subacute demyelinating sensorimotor polyneuropathy has been noted 2 weeks to 6 months after starting tacrolimus, FK506, immunosuppression in transplant recipients (Bronster et al. 1995). It resembles chronic inflammatory demyelinating polyneuropathy and responds to plasma exchange or IvIg treatment. Recovery has been reported following substitution of tacrolimus by cyclosporin.

21.19.21 Taxol

The drugs docetaxel and paclitaxel, derived from yew tree needles, promote microtubule polymerization. They are valuable antitumour drugs but peripheral neuropathy is a dose-limiting side effect. Sensory neuropathy can occur in humans within days of high-dose paclitaxel therapy or more slowly at lower dosage regimens. Neuropathy seems more likely when the cumulative dose exceeds 600 mg/m^2 (Hilkens et al. 1996). Initially tingling and numbness affect the feet before spreading to the hands. Limb weakness can be present both distally and proximally. Nerve conduction studies show features of both axonal degeneration and demyelination. Symptoms and signs can progressively 'coast' after stopping taxols, but recovery usually starts by 8 weeks (New et al. 1996).

21.19.22 Thalidomide

Sensory neuropathy, with prominent paraesthesiae and muscle cramps, followed the use of thalidomide as a hypnotic in the 1960s before it was withdrawn for teratogenicity. Since its reintroduction for a variety of medical disorders, including graft versus host disease and erythema nodosum leprosum, thalidomide's capacity to cause a predominantly sensory neuropathy is regarded by many as being dose limiting. The neuropathy seems more likely with cumulative doses exceeding 20 g (Cavaletti et al. 2004).

21.19.23 Tumour necrosis factor inhibitors

Therapeutic monoclonal antibodies blocking tumour necrosis factor-α, enteracept and infliximab, have been associated with reports of chronic inflammatory demyelinating polyneuropathy and acute motor neuropathy with conduction block (Richez et al. 2005; Singer et al. 2004).

21.19.24 Vinca alkaloids

Vincristine and other vinca alkaloids disrupt microtubules and are used chiefly for treating lymphoma or leukaemia. The peripheral neuropathy initially develops after cumulative dosage of 4–19 mg/m^2 and causes paraesthesiae, areflexia, and mild autonomic symptoms. Modern drug administration schedules show early neuropathic symptoms to be more severe at higher dosage intensities, such as

1.33 mg/week (Verstappen *et al.* 2005). If the drug is continued, severe muscle weakness and sensory loss develop. Autonomic neuropathy may be severe, with paralytic ileus, features of acute abdomen, impotence, or postural hypotension. Laryngeal nerve palsies have been attributed to vincristine therapy. Nerve conduction studies point to a dying back axonal degeneration neuropathy (Quasthoff and Hartung 2002). Mild neurotoxicity inevitably occurs with therapeutically effective doses of vincristine. When patients develop numbness or mild manipulatory difficulties, it should be a warning to reduce or stop the drug. Vincristine should be stopped immediately if significant weakness or paralytic ileus develop. Functional recovery is usual if the drug is stopped before the advent of significant toxicity. Coasting, or off-therapy worsening, of the neuropathy occurs in over a quarter after stopping therapy (Verstappen *et al.* 2005). Permanently absent ankle tendon jerks are commonly noted in otherwise asymptomatic patients who have received courses of vincristine therapy. Patients with hereditary motor and sensory neuropathy may be unusually sensitive to vincristine neuropathy (Weimer and Podwall 2006).

21.20 Metal-poisoning polyneuropathy

This section considers peripheral neuropathies attributable to arsenic, lead, mercury, and thallium poisoning. Neuropathies due to therapy with cisplatin, lithium, and gold are considered in Section 21.19.

21.20.1 Arsenic

Neuropathy due to inorganic arsenic poisoning may develop insidiously in arsenic smelting workers. If it occurs acutely after single-dose poisonings, it develops 2–3 weeks later in victims who survive the initial shock and gastrointestinal disturbance of acute intoxication (Ratnaike 2003). Numbness and paraesthesiae are the initial symptoms, and abnormalities of vibration and position sensation are demonstrable. Distal leg muscle weakness may develop subsequently. The ankle tendon reflexes are invariably lost, those at the knee are sometimes lost, and the arm reflexes are generally preserved. White lines across the nails, Mee's lines, develop later. Sensory nerve action potentials are absent and motor conduction is mildly slowed initially before electrophysiological features of axonal degeneration supervene (Donofrio *et al.* 1987). Sural nerve biopsies show axonal degeneration. Slow improvement occurs over a period of years but permanent abnormalities are usual.

Chronic arsenical exposure in industrial workers may produce an asymptomatic sensorimotor neuropathy, detectable only by nerve conduction studies and unaccompanied by neuropathic signs on examination (Feldman *et al.* 1979). The arsenic level may be elevated in the blood, urine, or hair, depending upon the recency of exposure. Hyperkeratosis of the hands and feet may be noted. Gastrointestinal symptoms are not usually a feature of chronic arsenical poisoning. Chronic poisoning may cause anaemia with basophilic stippling of erythrocytes. Chelation therapy may be indicated, particularly in patients seen soon after the ingestion of a single dose.

21.20.2 Lead

Peripheral neuropathy due to inorganic lead poisoning usually occurs in metal smelting or battery manufacturing workers.

Organic lead intoxication has not been associated clearly to peripheral neuropathy. Lead neuropathy takes two forms, an acute or subacute predominantly motor disorder, and a chronic sensory disorder (Rubens *et al.* 2001). The blood lead level tends to reflect recent exposure.

Classically a purely motor peripheral neuropathy develops, particularly affecting much-used muscles, such as the wrist extensors of manual workers. Abdominal crampings are a common initial manifestation. Prominent wrist or foot drop are characteristic. Muscle weakness can be profound, causing respiratory failure. The degree of tendon reflex loss varies. Sensory loss is unusual. Children with lead neuropathy frequently have an associated encephalopathy (Section 5.7.3). Motor nerve conduction velocities may be slowed. Sural nerve biopsies show loss of the large myelinated axons with noteworthy paranodal demyelination. Lead interferes with porphyrin metabolism, which may explain the similarity between the symptoms of lead poisoning and those of porphyric neuropathy (Section 21.8.6): abdominal pain, motor neuropathy, and behavioural disturbance occur in both. In addition there is a rare inherited condition known as plumboporphyria, due to δ-aminolaevulinic acid dehydratase deficiency, in which porphyric neuropathy may be precipitated by occupational exposure to lead within accepted safety limits (Dyer *et al.* 1993).

Chronic lead intoxication can cause sensory polyneuropathy independently of its provocation of porphyria (Rubens *et al.* 2001). Distal sensory disturbance, reduced reflexes and autonomic vasomotor or sudomotor abnormalities are characteristic. Such patients do not have motor involvement. Delayed and reduced sensory nerve action potentials are found electrophysiologically. It is noteworthy that sensory features, which are absent in classical descriptions of lead neuropathy, are prominent in all other polyneuropathies caused by heavy metal poisoning. The free erythrocyte protoporphyrin level is the best guide to chronic lead exposure. Basophilic stippling of erythrocytes is seen on blood smear. Treatment involves identifying and removing the source of exposure, and using the chelating agents.

21.20.3 Mercury

Chronic exposure to inorganic or elemental mercury produces a mild peripheral neuropathy. Mild sensorimotor neuropathy has been noted following long-term exposure of industrial workers to inorganic mercury vapour and in dentists using mercury amalgam. Previously asymptomatic polyneuropathy may be demonstrable many years after occupational elemental mercury exposure (Albers *et al.* 1988). Elemental mercury poisoning occasionally resembles motor neurone disease (Adams *et al.* 1983).

Organic mercury poisoning typically causes the combination of paraesthesiae, sensory ataxia, and visual-field constriction (Section 5.7.5). Organic mercury poisoning has occurred after exposure to methyl mercury dust, after eating seafood which has accumulated methyl mercury from industrial effluent, Minamata disease, and after inadvertent consumption of seed grain treated with mercurial fungicides. Patients with mercurial neuropathy develop paraesthesiae in and around the mouth and in the fingers and the toes. Nerve conduction studies may be normal in patients with organic mercury toxicity, suggesting that the sensory loss can be due to central nervous system involvement. Extensive cerebellar involvement is evident at autopsy (Nierenberg *et al.* 1998).

The diagnosis of mercury poisoning may be confirmed by measuring blood, urine, and hair levels, which are differentially elevated depending upon the type and rate of exposure. The treatment involves identification and elimination of the source of exposure. Chelating agents such as dimercaprol or penicillamine are mainly effective against inorganic and elemental, rather than organic, mercury poisoning. It is unlikely that chelating agents remove mercury that is already bound to neural tissue.

21.20.4 Thallium

Polyneuropathy due to thallium normally follows suicidal or homicidal poisoning attempts with this tasteless, colourless rodenticide. High doses cause shock due to gastroenteritis and dehydration. If the victim survives, sensorimotor neuropathy becomes evident within a few days. Sensory symptoms occur first and consist of painful paraesthesiae, particularly affecting the feet (Kuo *et al.* 2005). The neuropathy may progress rapidly to involve the respiratory and bulbar muscles, thus resembling the Guillain–Barré syndrome. An associated autonomic neuropathy can result in tachycardia and hypertension. Central nervous system involvement occurs in severe poisoning, producing ataxia, optic neuropathy, confusional psychoses, and involuntary movements. Systemic features include dark pigmentation at the hair roots (Fig. 21.23) followed by alopecia, dry scaly skin, and Mee's lines on the nails. Neuropathological studies show axonal degeneration of fibres in peripheral nerves and dorsal columns. Electron microscopy reveals swollen axons containing large vacuoles and distended mitochondria. Although the neuropathy may eventually recover partially, permanent abnormalities are the rule unless the patient is seen early enough to undertake effective gastric decontamination. Chelating agents have not been effective. Haemoperfusion has been recommended for severe poisoning. Oral Berliner–Blue may promote faecal excretion of thallium.

21.21 Polyneuropathy due to industrial and agricultural chemicals

21.21.1 Acrylamide

Acrylamide monomer is catalytically polymerized in order to stabilize soil during mining and other earthworkings. The monomer is

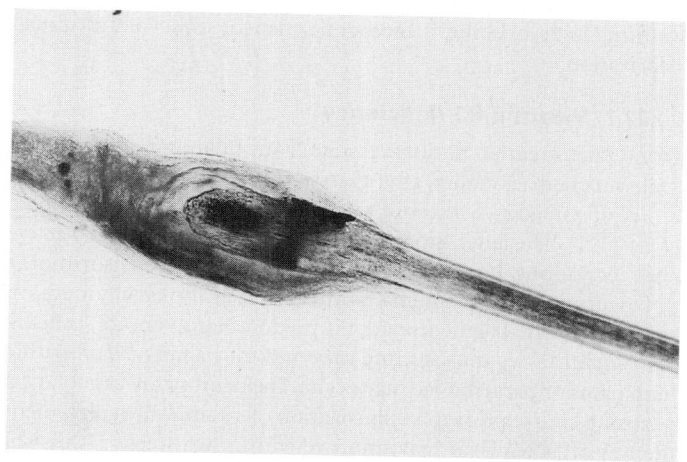

Fig. 21.23 Thallium poisoning: characteristic dark pigmentation at the root of a plucked hair. (Courtesy of Dr. M. Schwartz.)

neurotoxic. High-dose intoxication, as may occur after drinking contaminated well-water, causes a subacute encephalopathy followed some days later by signs of mild polyneuropathy (Igisu *et al.* 1975). Chronic low-dose intoxication generally occurs in construction workers following skin and inhalational exposure. Polyneuropathy may occur following exposure for as little as 4 weeks. The neuropathy involves both sensory and motor fibres. Positive sensory symptoms, such as paraesthesiae, are unusual. Diffuse areflexia is an early finding. Ataxia may be prominent. Contact dermatitis, blistering, and hyperhidrosis of the palms and soles may occur. Sensory nerve action potentials are small or absent and only a mild degree of motor slowing occurs. Sural nerve biopsy shows degeneration and regeneration of axons. Electron microscopy shows accumulations of disorganized neurofilaments in occasional axons, but giant axonal swellings are not seen (Davenport *et al.* 1976). In mild or subclinical cases, good recovery follows removal from exposure. Only partial recovery occurs from the more severe neuropathies.

21.21.2 Carbon disulphide

Peripheral neuropathy has been noted in workers using carbon disulphide in poorly ventilated conditions, in the rubber vulcanization or viscose rayon manufacturing industries. Distal sensorimotor loss and areflexia is usually restricted to the legs, but may involve the arms in severe cases. Electrophysiologically, the neuropathy is due to axonal degeneration (Vasilescu 1976). It is sometimes accompanied by encephalopathy with psychotic features.

21.21.3 Dimethylaminopropionitrile

Dimethylaminopropionitrile was used as a catalyst in the manufacture of polyurethane. Exposed workers developed an axonal degeneration sensorimotor neuropathy with noteworthy involvement of bladder control and sexual dysfunction. The initial symptoms were generally urinary hesitancy and impotence (Keogh *et al.* 1980).

21.21.4 Diethylene glycol

A rapidly ascending paralysis involving cranial muscles has been noted to start 8 days after diethylene glycol ingestion after initial manifestations of confusion and renal failure treated with haemodialysis (Rollins *et al.* 2002).

21.21.4 Ethylene oxide

Ethylene oxide gas is used both as a precursor for industrial chemicals and for sterilizing heat-sensitive devices used in healthcare. Industrial exposure to the gas has caused both encephalopathy and sensorimotor polyneuropathy (Gross *et al.* 1979).

21.21.6 Herbicides

Peripheral neuropathy has followed intense or repeated skin exposure to derivatives of the weed killer 2,4-D, 2,4-dichlorophenoxyacetic acid. Nausea, vomiting, and diarrhoea occur during the days immediately following exposure. First symptoms of peripheral neuropathy develop some days later and consist of painful paraesthesiae in the fingers and toes. Severe motor and sensory disability develop subsequently and recover incompletely. Once the neuropathy is established, motor nerve conduction velocities are moderately slowed (Goldstein *et al.* 1959).

21.21.7 Hexacarbons

n-Hexane and methyl *n*-butyl ketone 'MNBK' are used as solvents in glues and in flexographic printing. They are metabolized to the neurotoxic compound 2,5-hexanedione. Sensorimotor neuropathy has occurred in workers using such glues in shoe or furniture manufacture and following inhalational solvent abuse, 'glue-sniffing', (Altenkirch *et al.* 1977). The neurotoxic potency of hexacarbons is enhanced by simultaneous exposure to methylethylketone, which is often present in solvent mixtures but is not in itself neurotoxic. The peripheral neuropathy starts with numbness of the digits and may develop into severe symmetrical distal sensory and motor loss. Severe weakness may develop subacutely in 'glue-sniffers' and be misdiagnosed as the Guillain–Barré syndrome. Even after removal from exposure to the toxin, the neuropathy may continue to worsen for 2 or 3 months before stabilization and partial recovery take place. Electron microscopic examination of nerve biopsies shows giant axonal profiles swollen by accumulations of disorganized neurofilaments. Motor nerve conduction velocities are substantially slowed once the neuropathy is established. This may reflect the marked paranodal demyelination and myelin thinning that occur in relation to giant axonal change; since the neuropathy is not primarily demyelinating in nature.

21.21.8 Methylbromide

Sensorimotor peripheral neuropathy has been reported following chronic low-dose intoxication with methylbromide, a gas used as a fumigant and in fire extinguishers (Kantarjian and Shaheen 1963). Complete recovery of the neuropathy was reported within a year of removal from exposure.

21.21.9 Pesticides

Peripheral neuropathy has occurred following exposure to both organophosphorous and carbamate pesticides. In both cases, the symptoms of peripheral neuropathy develop after a delay of a few days or weeks following a single exposure. This delayed onset of neuropathy follows the earlier cholinergic phase of poisoning, in which acute paralysis, overwhelming bronchial secretions, bradycardia, and seizures may occur (Section 5.8.1).

In organophosphorous poisoning, symptoms of a predominantly motor neuropathy have been reported 1–3 weeks after acute exposure to the pesticides tricresylphosphate, mipafox, leptophos, trichlorphon, trichlornate, and methamidophos (Senanayake and Johnson 1982; Lotti *et al.* 1984). Not all organophosphates induce delayed peripheral neuropathy and it has been proposed that this capacity relates to their propensity to inhibit neurotoxic esterase (Lotti *et al.* 1984). It is uncertain whether neuropathy can follow chronic low-dose pesticide exposure, or whether it might follow acute intoxication by nerve gases intended for warfare. The initial symptoms of neuropathy consist of cramping in distal leg muscles accompanied by distal paraesthesiae and numbness. Progressive distal leg muscle weakness and hyporeflexia develop, followed by similar weakness of the arms. The severity varies, but severe quadriplegia may occur. Demonstrable sensory signs are mild or absent. Superimposed pyramidal tract abnormalities may be seen. Nerve conduction studies show denervation of muscles with little or no slowing of conduction velocity. Although the peripheral neuropathy recovers to some degree, the associated pyramidal tract abnormalities contribute to substantial long-term disability (Morgan and Penovich 1978).

Acute ingestion of carbamate pesticides may also produce delayed onset of peripheral neuropathy (Umehara *et al.* 1991). In comparison to organophosphorous poisoning, carbamate toxicity produces more prominent sensory signs and the degree of recovery is greater.

21.21.10 Trichloroethylene

The solvent and degreasing agent trichloroethylene can cause selective numbness of the facial skin, or polyneuritis cranialis in severe exposures (Feldman *et al.* 1992).

21.21.11 Vacor

This rodenticide is related to streptozotocin, a diabetogenic toxin. The acute onset of diabetes mellitus and severe autonomic failure, and a glove and stocking disturbance of pinprick sensation have been recorded after suicidal consumption of vacor (Pont *et al.* 1979). The autonomic failure is permanent with prominent disturbance of blood pressure and bladder control.

21.22 Vitamin deficiency polyneuropathy

21.22.1 The burning feet syndrome

Many nutritional neuropathies reflect the multiple vitamin deficiencies that occur during malnutrition due to starvation, chronic gastrointestinal disease, malnourished alcoholism, and contemporary bariatric surgery for morbid obesity (Thaisetthawatkul *et al.* 2004). Such patients often develop neurological illnesses which do not conform to the classic descriptions of dry beriberi due to thiamine deficiency or pellagra due to pyridoxine deficiency. Multiple vitamin deficiencies may produce the combination of predominantly sensory peripheral neuropathy with burning feet, amblyopia, sensorineural deafness, dizziness, myelopathy, and orogenital dermatitis, which is sometimes known as Strachan's syndrome. During the past century, this state has been described in malnourished native West Indians, jail inmates, sugar plantation labourers, and prisoners of war, particularly those held captive in the Far East during the Second World War (Cockerell and Ormerod 1993). Although the symptoms recover in some patients, they are permanent in many despite reinstitution of a balanced diet. The 'burning feet syndrome' is another permanent sequel of nutritional deprivation.

21.22.2 Vitamin B1 deficiency

Dry or neuropathic beriberi results from thiamine deficiency in malnourished alcoholics, after gastrectomy or in patients receiving a diet of milled rice without vitamin B supplementation (Koike *et al.* 2001). The latter group of patients are often physically active and consuming large amounts of carbohydrate. Sensorimotor polyneuropathy, leg oedema, and cardiomegaly usually develop simultaneously. The neuropathy is predominantly motor, initially affecting distal leg muscles, and may prevent walking. Neuropathic limb pains or paraesthesiae may occur. The neuropathy can follow a relapsing course prior to the institution of vitamin B_1 replacement therapy after which it may improve within two weeks (Ishibashi *et al.* 2003). Sural nerve biopsies from established cases show axonal degeneration predominantly affecting the larger myelinated fibres with a degree of secondary demyelination (Ohnishi *et al.* 1980). Motor nerve conduction is mildly slowed and sensory nerve action potentials diminished. Measurement of blood or urine thiamine levels is of limited value in making the diagnosis. The red cell transketolase activity is a more sensitive index, but does not distinguish between acute and chronic thiamine deficiency. After supplementation with vitamin B_1, strength and motor nerve conduction steadily improve and nerve biopsies show extensive regenerative activity.

21.22.3 Vitamin B6 deficiency

The full-blown syndrome of pellagra is only rarely encountered. It consists of a red-brown hyperkeratotic rash affecting exposed skin, gastrointestinal symptoms, neuropsychiatric features, and peripheral neuropathy. Approximately 50 per cent of pellagrins have sensorimotor peripheral neuropathy with noteworthy paraesthesiae, pain, and tenderness of distal leg muscles (Bomb *et al.* 1977). The peripheral neuropathy associated with isoniazid therapy is due to this drug's antagonism of vitamin B_6 (Section 21.19.11).

21.22.4 Vitamin B12 deficiency

Sensory peripheral neuropathy is occasionally encountered as the sole neurological manifestation of vitamin B_{12} deficiency, usually caused by underlying pernicious anaemia. It is normally overshadowed by the associated spinal-cord lesion known as subacute combined degeneration (Hemmer *et al.* 1998). The peripheral neuropathy contributes to paraesthesiae in the feet and distal loss of all modalities of sensation. The ankle jerks are absent and this physical sign correlates with the finding of reduced vitamin B_{12} levels in the elderly (Hin *et al.* 2006). Nerve conduction studies show diminished or absent sensory nerve action potentials. Motor nerve conduction studies generally show axonal neuropathy, but demyelinating features can occur. Symptoms improve little following initiation of vitamin B_{12} replacement injections (Saperstein *et al.* 2003).

21.22.5 Vitamin E deficiency

Long-standing vitamin E deficiency causes a sensory peripheral neuropathy associated with prominent ataxia, resembling a spinocerebellar degeneration (Chapter 39). Vitamin E deficiency occurs in patients with fat malabsorption due to cholestatic liver disease, short-bowel syndrome, or cystic fibrosis, and in abetalipoproteinaemia (Brin *et al.* 1986). Familial vitamin E deficiency is due to mutations of the α-Tocopherol transfer protein (Hentati *et al.* 1996). Chronic vitamin E deficiency can also cause a pigmentary retinopathy. Abnormalities of somatosensory-evoked potentials indicate an abnormality of central nervous system axons in the dorsal columns of the spinal cord. The diagnosis is proven by demonstrating a plasma tocopherol level reduced out of proportion to any reduction in plasma lipoprotein levels. The vitamin E content of the sural nerve is reduced, and this deficiency may precede the development of peripheral neuropathy (Traber *et al.* 1987). Vitamin E supplementation prevents further downhill progression of the neurological disorder (Sokol *et al.* 1985).

References

Adams CR, Ziegler DK, Lin JT (1983). Mercury intoxication simulating amyotrophic lateral sclerosis. *JAMA*, **250**, 642–3.

Adams D, Samuel D, Goulon-Goeau C *et al.* (2000). The course and prognostic factors of familial amyloid polyneuropathy after liver transplantation. *Brain*, **123**, 1495–504.

Albers JW, Fink JK (2004). Porphyric neuropathy. *Muscle Nerve*, **30**, 410–22.

Albers JW, Kallenbach LR, Fine LJ *et al.* (1988). Neurological abnormalities associated with remote occupational elemental mercury exposure. *Ann Neurol*, **24**, 651–9.

Albin RL, Albers JW, Greenberg HS *et al.* (1987). Acute sensory neuropathy-neuronopathy from pyridoxine overdose. *Neurology*, **37**, 1729–32.

Altenkirch H, Mager J, Stoltenburg G *et al.* (1977). Toxic polyneuropathies after sniffing a glue thinner. *J Neurol*, **214**, 137–52.

Amato MP, Pracucci G, Ponziani G *et al.* (1993). Long-term safety of azathioprine therapy in multiple sclerosis. *Neurology*, **43**, 831–3.

Andersson PB, Yuen E, Parko K *et al.* (2000). Electrodiagnostic features of hereditary neuropathy with liability to pressure palsies. *Neurology*, **54**, 40–4.

Ansbacher LE, Bosch EP, Cancilla PA (1982). Disulfiram neuropathy: a neurofilamentous distal axonopathy. *Neurology*, **32**, 424–8.

Attarian S, Azulay JP, Chabrol B *et al.* (2004). Neonatal lower motor neuron syndrome associated with maternal neuropathy with anti-GM1 IgG. *Neurology*, **63**, 379–81.

Axelrod FB (2004). Familial dysautonomia. *Muscle Nerve*, **29**, 352–63.

Bedlack RS, Edelman D, Gibbs JW 3rd *et al.* (2003). APOE genotype is a risk factor for neuropathy severity in diabetic patients. *Neurology*, **60**, 1022–4.

Bedlack RS, Vu T, Hammans S *et al.* (2004). MNGIE neuropathy: five cases mimicking chronic inflammatory demyelinating polyneuropathy. *Muscle Nerve*, **29**, 364–8.

Beekman R, van den Berg LH, Franssen H *et al.* (2005). Ultrasonography shows extensive nerve enlargements in multifocal motor neuropathy. *Neurology*, **65**, 305–7.

Bendixen BH, Younger DS, Hair LS *et al.* (1992). Cholesterol emboli neuropathy. *Neurology*, **42**, 428–30.

Berciano J, Combarros O (2003). Hereditary neuropathies. *Curr Opin Neurol*, **16**, 613–22.

Bertelli M, Gallo S, Buda A *et al.* (2006). Novel mutations in the arylsulfatase A gene in eight Italian families with metachromatic leukodystrophy. *J Clin Neurosci*, **13**, 443–8.

Birouk N, LeGuern E, Maisonobe T *et al.* (1998). X-linked Charcot-Marie-Tooth disease with connexin 32 mutations: clinical and electrophysiologic study. *Neurology*, **50**, 1074–82.

Boerkoel CF, Takashima H, Garcia CA *et al.* (2002). Charcot-Marie-Tooth disease and related neuropathies: mutation distribution and genotype-phenotype correlation. *Ann Neurol*, **51**, 190–201.

Bomb BS, Bedi HK, Bhatnagar LK (1977). Post-ischaemic paraesthesia in pellagrins. *J Neurol Neurosurg Psychiatry*, **40**, 265–7.

Bonnel RA, Graham DJ (2004). Peripheral neuropathy in patients treated with leflunomide. *Clin Pharmacol Ther*, **75**, 580–5.

Bosboom WM, van den Berg LH, Franssen H *et al.* (2001). Diagnostic value of sural nerve demyelination in chronic inflammatory demyelinating polyneuropathy. *Brain*, **124**, 2427–38.

Bouchard C, Lacroix C, Plante V *et al.* (1999). Clinicopathologic findings and prognosis of chronic inflammatory demyelinating polyneuropathy. *Neurology*, **52**, 498–503.

Bouillot S, Martin-Negrier ML, Vital A *et al.* (2002). Peripheral neuropathy associated with mitochondrial disorders: 8 cases and review of the literature. *J Peripher Nerv Syst*, **7**, 213–20.

Bowen JR, McDougall AC, Morris JH *et al.* (2000). Vasculitic neuropathy in a patient with inactive treated lepromatous leprosy. *J Neurol Neurosurg Psychiatry*, **68**, 496–500.

Bradley WG, Verma A (1996). Painful vasculitic neuropathy in HIV-1 infection: relief of pain with prednisone therapy. *Neurology*, **47**, 1446–51.

Brannagan TH 3rd, Pradhan A, Heiman-Patterson T *et al.* (2002). High-dose cyclophosphamide without stem-cell rescue for refractory CIDP. *Neurology*, **58**, 1856–8.

Brin MF, Pedley TA, Lovelace RE *et al.* (1986). Electrophysiologic features of abetalipoproteinemia: functional consequences of vitamin E deficiency. *Neurology*, **36**, 669–73.

Bronster DJ, Yonover P, Stein J *et al.* (1995). Demyelinating sensorimotor polyneuropathy after administration of FK506. *Transplantation*, **59**, 1066–8.

Bruno C, Bertini E, Federico A *et al.* (2004). Clinical and molecular findings in patients with giant axonal neuropathy (GAN). *Neurology*, **62**, 13–6.

Burton M, Anslow P, Gray W *et al.* (2002). Selective hypertrophy of the cauda equina nerve roots. *J Neurol*, **249**, 337–40.

Busby M, Donaghy M (2003). Chronic dysimmune neuropathy. A subclassification based upon the clinical features of 102 patients. *J Neurol*, **250**, 714–24.

Busby M, Nithi K, Mills K *et al.* (2003). The tremor associated with non-paraproteinaemic acquired demyelinating polyneuropathy—a case study. *J Neurol*, **250**, 486–7.

Camdessanche JP, Antoine JC, Honnorat J *et al.* (2002). Paraneoplastic peripheral neuropathy associated with anti-Hu antibodies. A clinical and electrophysiological study of 20 patients. *Brain*, **125**, 166–75.

Capasso M, Caporale CM, Pomilio F *et al.* (2003). Acute motor conduction block neuropathy. Another Guillain-Barre syndrome variant. *Neurology*, **61**, 617–22.

Caruso G, Santoro L, Perretti A *et al.* (1987). Friedreich's ataxia: electrophysiologic and histologic findings in patients and relatives. *Muscle Nerve*, **10**, 503–15.

Cavaletti G, Beronio A, Reni L *et al.* (2004). Thalidomide sensory neurotoxicity: a clinical and neurophysiologic study. *Neurology*, **62**, 2291–3.

Cavanagh NP, Eames RA, Galvin RJ *et al.* (1979). Hereditary sensory neuropathy with spastic paraplegia. *Brain*, **102**, 79–94.

Chalk CH, Mills KR, Jacobs JM *et al.* (1990). Familial multiple symmetric lipomatosis with peripheral neuropathy. *Neurology*, **40**, 1246–50.

Chance PF, Bird TD, Matsunami N *et al.* (1992). Trisomy 17p associated with Charcot-Marie-Tooth neuropathy type 1A phenotype: evidence for gene dosage as a mechanism in CMT1A. *Neurology*, **42**, 2295–9.

Chaouch M, Allal Y, De Sandre-Giovannoli A *et al.* (2003). The phenotypic manifestations of autosomal recessive axonal Charcot-Marie-Tooth due to a mutation in Lamin A/C gene. *Neuromuscul Disord*, **13**, 60–7.

Chaudhry V, Chaudhry M, Crawford TO *et al.* (2003). Toxic neuropathy in patients with pre-existing neuropathy. *Neurology*, **60**, 337–40.

Chaudhry V, Eisenberger MA, Sinibaldi VJ *et al.* (1996). A prospective study of suramin-induced peripheral neuropathy. *Brain*, **119**, 2039–52.

Chaudhry V, Stevens JC, Kincaid J *et al.* (2006). Practice Advisory: utility of surgical decompression for treatment of diabetic neuropathy: report of the Therapeutics and Technology Assessment Subcommittee of the American Academy of Neurology. *Neurology*, **66**, 1805–8.

Chimelli L, Freitas M, Nascimento O (1997). Value of nerve biopsy in the diagnosis and follow-up of leprosy: the role of vascular lesions and usefulness of nerve studies in the detection of persistent bacilli. *J Neurol*, **244**, 318–23.

Chin RL, Tseng VG, Green PH *et al.* (2006). Multifocal axonal polyneuropathy in celiac disease. *Neurology*, **66**, 1923–5.

Chio A, Cocito D, Leone M *et al.* (2003). Guillain-Barre syndrome: a prospective, population-based incidence and outcome survey. *Neurology*, **60**, 1146–50.

Cockerell OC, Ormerod IE (1993). Strachan's syndrome: variation on a theme. *J Neurol*, **240**, 315–8.

Cohen JI, Brunell PA, Straus SE *et al.* (1999). Recent advances in varicella-zoster virus infection. *Ann Intern Med*, **130**, 922–32.

Collins MP, Periquet MI, Mendell JR *et al.* (2003). Nonsystemic vasculitic neuropathy: insights from a clinical cohort. *Neurology*, **61**, 623–30.

Comabella M, Waye JS, Raguer N *et al.* (2001). Late-onset metachromatic leukodystrophy clinically presenting as isolated peripheral neuropathy: compound heterozygosity for the IVS2+1G-->A mutation and a newly identified missense mutation (Thr408Ile) in a Spanish family. *Ann Neurol*, **50**, 108–12.

Conceicao I, Sales-Luis ML, De Carvalho M *et al.* (2003). Gelsolin-related familial amyloidosis, Finnish type, in a Portuguese family: clinical and neurophysiological studies. *Muscle Nerve*, **28**, 715–21.

Coppack SW, Watkins PJ (1991). The natural history of diabetic femoral neuropathy. *QJM*, **79**, 307–13.

Cornblath DR, Chaudhry V, Carter K *et al.* (1999). Total neuropathy score: validation and reliability study. *Neurology*, **53**, 1660–4.

Coxon A, Pallis CA (1976). Metronidazole neuropathy. *J Neurol Neurosurg Psychiatry*, **39**, 403–5.

Croft RP, Nicholls PG, Steyerberg EW *et al.* (2000). A clinical prediction rule for nerve-function impairment in leprosy patients. *Lancet*, **355**, 1603–6.

Dalakas MC (2001). Peripheral neuropathy and antiretroviral drugs. *J Peripher Nerv Syst*, **6**, 14–20.

Danek A, Rubio JP, Rampoldi L *et al.* (2001). McLeod neuroacanthocytosis: genotype and phenotype. *Ann Neurol*, **50**, 755–64.

Davenport JG, Farrell DF, Sumi M (1976). "Giant axonal neuropathy" caused by industrial chemicals: neurofilamentous axonal masses in man. *Neurology*, **26**, 919–23.

de Groot K, Schmidt DK, Arlt AC *et al.* (2001). Standardized neurologic evaluations of 128 patients with Wegener granulomatosis. *Arch Neurol*, **58**, 1215–21.

De Jonghe P, Timmerman V, Ceuterick C *et al.* (1999). The Thr124Met mutation in the peripheral myelin protein zero (MPZ) gene is associated with a clinically distinct Charcot-Marie-Tooth phenotype. *Brain*, **122**, 281–90.

Delmont E, Azulay JP, Giorgi R *et al.* (2006). Multifocal motor neuropathy with and without conduction block: a single entity? *Neurology*, **67**, 592–6.

Demir E, Bomont P, Erdem S *et al.* (2005). Giant axonal neuropathy: clinical and genetic study in six cases. *J Neurol Neurosurg Psychiatry*, **76**, 825–32.

Diabetes control and Complications Trial Research Group (1993). The effect of intensive treatment of diabetes on the development and progression of long-term complications in insulin-dependent diabetes mellitus. *N Engl J Med*, **329**, 977–86.

Dodd A, Rowland SA, Hawkes SL *et al.* (1997). Mutations in the adrenoleukodystrophy gene. *Hum Mutat*, **9**, 500–11.

Donaghy M, Earl CJ (1985). Ocular palsy preceding chronic relapsing polyneuropathy by several weeks. *Ann Neurol*, **17**, 49–50.

Donaghy M, Brett EM, Ormerod IE *et al.* (1988). Giant axonal neuropathy: observations on a further patient. *J Neurol Neurosurg Psychiatry*, **51**, 991–4.

Donaghy M, Hakin RN, Bamford JM *et al.* (1987). Hereditary sensory neuropathy with neurotrophic keratitis. Description of an autosomal recessive disorder with a selective reduction of small myelinated nerve fibres and a discussion of the classification of the hereditary sensory neuropathies. *Brain*, **110**, 563–83.

Donaghy M, Sisodiya SM, Kennett R *et al.* (2000). Steroid responsive polyneuropathy in a family with a novel myelin protein zero mutation. *J Neurol Neurosurg Psychiatry*, **69**, 799–805.

Donofrio PD, Wilbourn AJ, Albers JW *et al.* (1987). Acute arsenic intoxication presenting as Guillain-Barre-like syndrome. *Muscle Nerve*, **10**, 114–20.

Dornonville de la Cour C, Jakobsen J (2005). Residual neuropathy in long-term population-based follow-up of Guillain-Barre syndrome. *Neurology*, **64**, 246–53.

Duggins AJ, McLeod JG, Pollard JD *et al.* (1999). Spinal root and plexus hypertrophy in chronic inflammatory demyelinating polyneuropathy. *Brain*, **122**, 1383–90.

Dupre N, Howard HC, Mathieu J et al. (2003). Hereditary motor and sensory neuropathy with agenesis of the corpus callosum. Ann Neurol, 54, 9–18.

Duston MA, Skinner M, Anderson J et al. (1989). Peripheral neuropathy as an early marker of AL amyloidosis. Arch Intern Med, 149, 358–60.

Dyck PJ, Kratz KM, Karnes JL et al. (1993b). The prevalence by staged severity of various types of diabetic neuropathy, retinopathy, and nephropathy in a population-based cohort: the Rochester Diabetic Neuropathy Study. Neurology, 43, 817–24.

Dyck PJ, Lais A, Karnes JL et al. (1986). Fiber loss is primary and multifocal in sural nerves in diabetic polyneuropathy. Ann Neurol, 19, 425–39.

Dyck PJ, Mellinger JF, Reagan TJ et al. (1983). Not 'indifference to pain' but varieties of hereditary sensory and autonomic neuropathy. Brain, 106, 373–90.

Dyck PJ, Norell JE, Dyck PJ (1999). Microvasculitis and ischemia in diabetic lumbosacral radiculoplexus neuropathy. Neurology, 53, 2113–21.

Dyck PJ, Norell JE, Dyck PJ (2001). Non-diabetic lumbosacral radiculoplexus neuropathy: natural history, outcome and comparison with the diabetic variety. Brain, 124, 1197–207.

Dyer J, Garrick DP, Inglis A et al. (1993). Plumboporphyria (ALAD deficiency) in a lead worker: a scenario for potential diagnostic confusion. Br J Ind Med, 50, 1119–21.

Enzi G, Angelini C, Negrin P et al. (1985). Sensory, motor, and autonomic neuropathy in patients with multiple symmetric lipomatosis. Medicine (Baltimore), 64, 388–93.

Eurelings M, Lokhorst HM, Kalmijn S et al. (2005). Malignant transformation in polyneuropathy associated with monoclonal gammopathy. Neurology, 64, 2079–84.

Fabrizi G, Cavallaro T, Angiari C et al. (2007). Charcot-Marie-Tooth disease type 2E, a disorder of the cytoskeleton. Brain, 130, 394–403.

Federico P, Zochodne DW, Hahn AF et al. (2000). Multifocal motor neuropathy improved by IVIg: randomized, double-blind, placebo-controlled study. Neurology, 55, 1256–62.

Feeney DJ, Pollard JD, McLeod JG et al. (1990). HLA antigens in chronic inflammatory demyelinating polyneuropathy. J Neurol Neurosurg Psychiatry, 53, 170–2.

Feldman RG, Niles C, Proctor SP et al. (1992). Blink reflex measurement of effects of trichloroethylene exposure on the trigeminal nerve. Muscle Nerve, 15, 490–5.

Feldman RG, Niles CA, Kelly-Hayes M et al. (1979). Peripheral neuropathy in arsenic smelter workers. Neurology, 29, 939–44.

Fialho D, Chan YC, Allen DC et al. (2006). Treatment of chronic inflammatory demyelinating polyradiculoneuropathy with methotrexate. J Neurol Neurosurg Psychiatry, 77, 544–7.

Fisher M (1956). An unusual variant of acute idiopathic polyneuritis (syndrome of ophthalmoplegia, ataxia and areflexia). N Engl J Med, 255, 57–65.

Fox RJ, Galetta SL, Mahalingam R et al. (2001). Acute, chronic, and recurrent varicella zoster virus neuropathy without zoster rash. Neurology, 57, 351–4.

Freeman R (2005). Autonomic peripheral neuropathy. Lancet, 365, 1259–70.

Freimer ML, Glass JD, Chaudhry V et al. (1992). Chronic demyelinating polyneuropathy associated with eosinophilia-myalgia syndrome. J Neurol Neurosurg Psychiatry, 55, 352–8.

Fujimura H, Lacroix C, Said G (1991). Vulnerability of nerve fibres to ischaemia. A quantitative light and electron microscope study. Brain, 114, 1929–42.

Gabreels-Festen AA, Hoogendijk JE, Meijerink PH et al. (1996). Two divergent types of nerve pathology in patients with different P0 mutations in Charcot-Marie-Tooth disease. Neurology, 47, 761–5.

Gaist D, Jeppesen U, Andersen M et al. (2002). Statins and risk of polyneuropathy: a case-control study. Neurology, 58, 1333–7.

Garcia A, Combarros O, Calleja J et al. (1998). Charcot-Marie-Tooth disease type 1A with 17p duplication in infancy and early childhood: a longitudinal clinical and electrophysiologic study. Neurology, 50, 1061–7.

Geleijns K, Brouwer BA, Jacobs BC et al. (2004). The occurrence of Guillain-Barre syndrome within families. Neurology, 63, 1747–50.

Gemignani F, Pavesi G, Fiocchi A et al. (1992). Peripheral neuropathy in essential mixed cryoglobulinaemia. J Neurol Neurosurg Psychiatry, 55, 116–20.

Gherardi RK, Chretien F, Delfau-Larue MH et al. (1998). Neuropathy in diffuse infiltrative lymphocytosis syndrome: an HIV neuropathy, not a lymphoma. Neurology, 50, 1041–4.

Ghosh A, Busby M, Kennett R et al. (2005a). A practical definition of conduction block in IvIg responsive multifocal motor neuropathy. J Neurol Neurosurg Psychiatry, 76, 1264–8.

Ghosh A, Virgincar A, Kennett R et al. (2005b). The effect of treatment upon temporal dispersion in IvIg responsive multifocal motor neuropathy. J Neurol Neurosurg Psychiatry, 76, 1269–72.

Goldstein NP, Jones PH, Brown JR (1959). Peripheral neuropathy after exposure to an ester of dichlorophenoxyacetic acid. JAMA, 171, 1306–9.

Gondim FA, Brannagan TH 3rd, Sander HW et al. (2005). Peripheral neuropathy in patients with inflammatory bowel disease. Brain, 128, 867–79.

Gøransson LG, Tjensvoll AB, Herigstad A et al. (2006). Small-diameter nerve fiber neuropathy in systemic lupus erythematosus. Arch Neurol, 63, 401–4.

Gorson KC, Ropper AH, Clark BD et al. (1998). Treatment of chronic inflammatory demyelinating polyneuropathy with interferon-alpha 2a. Neurology, 50, 84–7.

Graus F, Bonaventura I, Uchuya M et al. (1994). Indolent anti-Hu-associated paraneoplastic sensory neuropathy. Neurology, 44, 2258–61.

Griffin JW, Cornblath DR, Alexander E et al. (1990). Ataxic sensory neuropathy and dorsal root ganglionitis associated with Sjogren's syndrome. Ann Neurol, 27, 304–15.

Gross JA, Haas ML, Swift TR (1979). Ethylene oxide neurotoxicity: report of four cases and review of the literature. Neurology, 29, 978–83.

Guillain-Barre Syndrome Steroid Trial Group (1993). Double-blind trial of intravenous methylprednisolone in Guillain-Barre syndrome. Lancet, 341, 586–90.

Gutmann L, Gutmann L (1999). Critical illness neuropathy and myopathy. Arch Neurol, 56, 527–8.

Gutmann L, Martin JD, Welton W (1976). Dapsone motor neuropathy--an axonal disease. Neurology, 26, 514–6.

Haanpaa M, Dastidar P, Weinberg A et al. (1998). CSF and MRI findings in patients with acute herpes zoster. Neurology, 51, 1405–11.

Haanpaa M, Hakkinen V, Nurmikko T (1997). Motor involvement in acute herpes zoster. Muscle Nerve, 20, 1433–8.

Hadden RD, Karch H, Hartung HP et al. (2001). Preceding infections, immune factors, and outcome in Guillain-Barre syndrome. Neurology, 56, 758–65.

Hadjivassiliou M, Grunewald RA, Kandler RH et al. (2006). Neuropathy associated with gluten sensitivity. J Neurol Neurosurg Psychiatry, 77, 1262–6.

Hafer-Macko C, Hsieh ST, Li CY et al. (1996). Acute motor axonal neuropathy: an antibody-mediated attack on axolemma. Ann Neurol, 40, 635–44.

Hahn AF, Bolton CF, Pillay N et al. (1996a). Plasma-exchange therapy in chronic inflammatory demyelinating polyneuropathy. A double-blind, sham-controlled, cross-over study. Brain, 119, 1055–66.

Hahn AF, Bolton CF, Zochodne D et al. (1996b). Intravenous immunoglobulin treatment in chronic inflammatory demyelinating polyneuropathy. A double-blind, placebo-controlled, cross-over study. Brain, 119, 1067–77.

Hainfellner JA, Kristoferitsch W, Lassmann H *et al.* (1996). T-cell-mediated ganglionitis associated with acute sensory neuronopathy. *Ann Neurol*, **39**, 543–7.

Halperin JJ, Golightly M (1992). Lyme borreliosis in Bell's palsy. Long Island Neuroborreliosis Collaborative Study Group. *Neurology*, **42**, 1268–70.

Halperin JJ, Little BW, Coyle PK *et al.* (1987). Lyme disease: cause of a treatable peripheral neuropathy. *Neurology*, **37**, 1700–6.

Halperin JJ, Logigian EL, Finkel MF *et al.* (1996). Practice parameters for the diagnosis of patients with nervous system Lyme borreliosis (Lyme disease). Quality Standards Subcommittee of the American Academy of Neurology. *Neurology*, **46**, 619–27.

Harari D, Gibberd FB, Dick JP *et al.* (1991). Plasma exchange in the treatment of Refsum's disease (heredopathia atactica polyneuritiformis). *J Neurol Neurosurg Psychiatry*, **54**, 614–7.

Harding AE, Thomas PK (1980a). Hereditary distal spinal muscular atrophy. A report on 34 cases and a review of the literature. *J Neurol Sci*, **45**, 337–48.

Harding AE, Thomas PK (1980b). The clinical features of hereditary motor and sensory neuropathy types I and II. *Brain*, **103**, 259–80.

Harding AE, Thomas PK (1980c). Autosomal recessive forms of hereditary motor and sensory neuropathy. *J Neurol Neurosurg Psychiatry*, **43**, 669–78.

Hattori N, Ichimura M, Nagamatsu M *et al.* (1999). Clinicopathological features of Churg-Strauss syndrome-associated neuropathy. *Brain*, **122**, 427–39.

Hattori N, Yamamoto M, Yoshihara T *et al.* (2003). Demyelinating and axonal features of Charcot-Marie-Tooth disease with mutations of myelin-related proteins (PMP22, MPZ and Cx32): a clinicopathological study of 205 Japanese patients. *Brain*, **126**, 134–51.

Hawke SH, Davies L, Pamphlett R *et al.* (1991). Vasculitic neuropathy. A clinical and pathological study. *Brain*, **114**, 2175–90.

Hemmer B, Glocker FX, Schumacher M *et al.* (1998). Subacute combined degeneration: clinical, electrophysiological, and magnetic resonance imaging findings. *J Neurol Neurosurg Psychiatry*, **65**, 822–7.

Henderson RD, Sandroni P, Wijdicks EF (2005). Chronic inflammatory demyelinating polyneuropathy and respiratory failure. *J Neurol*, **252**, 1235–7.

Hentati A, Deng HX, Hung WY *et al.* (1996). Human alpha-tocopherol transfer protein: gene structure and mutations in familial vitamin E deficiency. *Ann Neurol*, **39**, 295–300.

Herrmann DN, Griffin JW, Hauer P *et al.* (1999). Epidermal nerve fiber density and sural nerve morphometry in peripheral neuropathies. *Neurology*, **53**, 1634–40.

Hilkens PH, Verweij J, Stoter G *et al.* (1996). Peripheral neurotoxicity induced by docetaxel. *Neurology*, **46**, 104–8.

Hin H, Clarke R, Sherliker P *et al.* (2006). Clinical relevance of low serum vitamin B12 concentrations in older people: the Banbury B12 study. *Age Ageing*, **35**, 416–22.

Hiraga A, Mori M, Ogawara K *et al.* (2003). Differences in patterns of progression in demyelinating and axonal Guillain-Barre syndromes. *Neurology*, **61**, 471–4.

Hodgkinson SJ, Pollard JD, McLeod JG (1990). Cyclosporin A in the treatment of chronic demyelinating polyradiculoneuropathy. *J Neurol Neurosurg Psychiatry*, **53**, 327–30.

Hoff JM, Gilhus NE, Daltveit AK (2005). Pregnancies and deliveries in patients with Charcot-Marie-Tooth disease. *Neurology*, **64**, 459–62.

Hoffman-Snyder C, Smith BE, Ross MA *et al.* (2006). Value of the oral glucose tolerance test in the evaluation of chronic idiopathic axonal polyneuropathy. *Arch Neurol*, **63**, 1075–9.

Houlden H, King RH, Wood NW, *et al.* (2001a). Mutations in the 5' region of the myotubularin-related protein 2 (MTMR2) gene in autosomal recessive hereditary neuropathy with focally folded myelin. *Brain*, **124**, 907–15.

Houlden H, King RH, Hashemi-Nejad A, *et al.* (2001b). A novel TRK A (NTRK1) mutation associated with hereditary sensory and autonomic neuropathy type V. *Anne Neurol*, 49:521–5.

Horoupian DS (1989). Hereditary sensory neuropathy with deafness: a familial multisystem atrophy. *Neurology*, **39**, 244–8.

Hughes R, Bensa S, Willison H *et al.* (2001). Randomized controlled trial of intravenous immunoglobulin versus oral prednisolone in chronic inflammatory demyelinating polyradiculoneuropathy. *Ann Neurol*, **50**, 195–201.

Hughes R, Sharrack B, Rubens R (1996). Carcinoma and the peripheral nervous system. *J Neurol*, **243**, 371–6.

Hughes RA, Umapathi T, Gray IA *et al.* (2004). A controlled investigation of the cause of chronic idiopathic axonal polyneuropathy. *Brain*, **127**, 1723–30.

Hughes RA, Wijdicks EF, Barohn R *et al.* (2003). Practice parameter: immunotherapy for Guillain-Barre syndrome: report of the Quality Standards Subcommittee of the American Academy of Neurology. *Neurology*, **61**, 736–40.

Hughes RA, Wijdicks EF, Benson E *et al.* (2005). Supportive care for patients with Guillain-Barre syndrome. *Arch Neurol*, **62**, 1194–8.

Hund E, Linke RP, Willig F *et al.* (2001). Transthyretin-associated neuropathic amyloidosis. Pathogenesis and treatment. *Neurology*, **56**, 431–5.

Husain AM, Altuwaijri M, Aldosari M (2004). Krabbe disease: neurophysiologic studies and MRI correlations. *Neurology*, **63**, 617–20.

Igisu H, Goto I, Kawamura Y *et al.* (1975). Acrylamide encephaloneuropathy due to well water pollution. *J Neurol Neurosurg Psychiatry*, **38**, 581–4.

Indo Y (2002). Genetics of congenital insensitivity to pain with anhidrosis (CIPA) or hereditary sensory and autonomic neuropathy type IV. Clinical, biological and molecular aspects of mutations in TRKA(NTRK1) gene encoding the receptor tyrosine kinase for nerve growth factor. *Clin Auton Res*, **12** (Suppl 1), I20–32.

Ishibashi S, Yokota T, Shiojiri T *et al.* (2003). Reversible acute axonal polyneuropathy associated with Wernicke-Korsakoff syndrome: impaired physiological nerve conduction due to thiamine deficiency? *J Neurol Neurosurg Psychiatry*, **74**, 674–6.

Jaccard A, Royer B, Bordessoule D *et al.* (2002). High-dose therapy and autologous blood stem cell transplantation in POEMS syndrome. *Blood*, **99**, 3057–9.

Jacobs JM, Love S (1985). Qualitative and quantitative morphology of human sural nerve at different ages. *Brain*, **108**, 897–924.

Jacobson RR, Krahenbuhl JL (1999). Leprosy. *Lancet*, **353**, 655–60.

Janssens J, Peeters TL, Vantrappen G *et al.* (1990). Improvement of gastric emptying in diabetic gastroparesis by erythromycin. Preliminary studies. *N Engl J Med*, **322**, 1028–31.

Jardim MR, Antunes SL, Santos AR *et al.* (2003). Criteria for diagnosis of pure neural leprosy. *J Neurol*, **250**, 806–9.

Jestico JV, Urry PA, Efphimiou J (1985). An hereditary sensory and autonomic neuropathy transmitted as an X-linked recessive trait. *J Neurol Neurosurg Psychiatry*, **48**, 1259–64.

Jordanova A, De Jonghe P, Boerkoel CF *et al.* (2003). Mutations in the neurofilament light chain gene (NEFL) cause early onset severe Charcot-Marie-Tooth disease. *Brain*, **126**, 590–7.

Jung BF, Johnson RW, Griffin DR *et al.* (2004). Risk factors for postherpetic neuralgia in patients with herpes zoster. *Neurology*, **62**, 1545–51.

Kalaydjieva L, Gresham D, Gooding R, *et al.* (2000). N-myc downstream-regulated gene 1 is mutated in hereditary motor and sensory neuropathy-Lom. *Am J Hum Genet*, **67**: 47–58.

Kantarjian AD, Shaheen AS (1963). Methyl bromide poisoning with nervous system manifestations resembling polyneuropathy. *Neurology*, **13**, 1054–8.

Katirji B (1997). Chronic relapsing axonal neuropathy responsive to intravenous immunoglobulin. *Neurology*, **48**, 1690–4.

Katrak SM, Pollock M, O'Brien CP *et al.* (1980). Clinical and morphological features of gold neuropathy. *Brain*, **103**, 671–93.

Katz JS, Saperstein DS, Gronseth G *et al.* (2000). Distal acquired demyelinating symmetric neuropathy. *Neurology*, **54**, 615–20.

Kelkar P, Parry GJ (2003). Mononeuritis multiplex in diabetes mellitus: evidence for underlying immune pathogenesis. *J Neurol Neurosurg Psychiatry*, **74**, 803–6.

Kennedy WR, Navarro X, Goetz FC *et al.* (1990). Effects of pancreatic transplantation on diabetic neuropathy. *N Engl J Med*, **322**, 1031–7.

Keogh JP, Pestronk A, Wertheimer D *et al.* (1980). An epidemic of urinary retention caused by dimethylaminopropionitrile. *JAMA*, **243**, 746–9.

Khan J, Harrison TB, Rich MM *et al.* (2006). Early development of critical illness myopathy and neuropathy in patients with severe sepsis. *Neurology*, **67**, 1421–5.

Khella SL, Frost S, Hermann GA *et al.* (1995). Hepatitis C infection, cryoglobulinemia, and vasculitic neuropathy. Treatment with interferon alfa: case report and literature review. *Neurology*, **45**, 407–11.

Kiernan MC, Guglielmi JM, Kaji R *et al.* (2002). Evidence for axonal membrane hyperpolarization in multifocal motor neuropathy with conduction block. *Brain*, **125**, 664–75.

Kimoto K, Koga M, Odaka M *et al.* (2006). Relationship of bacterial strains to clinical syndromes of Campylobacter-associated neuropathies. *Neurology*, **67**, 1837–43.

Kitano Y, Kuwabara S, Misawa S *et al.* (2004). The acute effects of glycemic control on axonal excitability in human diabetics. *Ann Neurol*, **56**, 462–7.

Kjeldsen K, Simonsen O, Heron I (1985). Immunity against diphtheria 25-30 years after primary vaccination in childhood. *Lancet*, **1**, 900–2.

Klein CM, Vernino S, Lennon VA *et al.* (2003). The spectrum of autoimmune autonomic neuropathies. *Ann Neurol*, **53**, 752–8.

Klopstock T, Naumann M, Schalke B *et al.* (1994). Multiple symmetric lipomatosis: abnormalities in complex IV and multiple deletions in mitochondrial DNA. *Neurology*, **44**, 862–6.

Kochar DK, Rawat N, Agrawal RP *et al.* (2004). Sodium valproate for painful diabetic neuropathy: a randomized double-blind placebo-controlled study. *QJM*, **97**, 33–8.

Koga M, Takahashi M, Masuda M *et al.* (2005). Campylobacter gene polymorphism as a determinant of clinical features of Guillain-Barre syndrome. *Neurology*, **65**, 1376–81.

Koike H, Iijima M, Sugiura M *et al.* (2003). Alcoholic neuropathy is clinicopathologically distinct from thiamine-deficiency neuropathy. *Ann Neurol*, **54**, 19–29.

Koike H, Misu K, Hattori N *et al.* (2001). Postgastrectomy polyneuropathy with thiamine deficiency. *J Neurol Neurosurg Psychiatry*, **71**, 357–62.

Koller H, Kieseier BC, Jander S *et al.* (2005). Chronic inflammatory demyelinating polyneuropathy. *N Engl J Med*, **352**, 1343–56.

Krendel DA, Costigan DA, Hopkins LC (1995). Successful treatment of neuropathies in patients with diabetes mellitus. *Arch Neurol*, **52**, 1053–61.

Krishnan AV, Phoon RK, Pussell BA *et al.* (2006). Sensory nerve excitability and neuropathy in end stage kidney disease. *J Neurol Neurosurg Psychiatry*, **77**, 548–51.

Krumina A, Logina I, Donaghy M *et al.* (2005). Diphtheria with polyneuropathy in a closed community despite receiving recent booster vaccination. *J Neurol Neurosurg Psychiatry*, **76**, 1555–7.

Kuhlenbaumer G, Young P, Hunermund G *et al.* (2002). Clinical features and molecular genetics of hereditary peripheral neuropathies. *J Neurol*, **249**, 1629–50.

Kuncl RW, Duncan G, Watson D *et al.* (1987). Colchicine myopathy and neuropathy. *N Engl J Med* **316**, 1562–8.

Kuo HC, Huang CC, Tsai YT *et al.* (2005). Acute painful neuropathy in thallium poisoning. *Neurology*, **65**, 302–4.

Landrieu P, Said G, Allaire C (1990). Dominantly transmitted congenital indifference to pain. *Ann Neurol*, **27**, 574–8.

Latov N (1995). Pathogenesis and therapy of neuropathies associated with monoclonal gammopathies. *Ann Neurol*, **37**, S32–42.

Lauria G, McArthur JC, Hauer PE *et al.* (1998). Neuropathological alterations in diabetic truncal neuropathy: evaluation by skin biopsy. *J Neurol Neurosurg Psychiatry*, **65**, 762–6.

Lenssen PP, Gabreels-Festen AA, Valentijn LJ *et al.* (1998). Hereditary neuropathy with liability to pressure palsies. Phenotypic differences between patients with the common deletion and a PMP22 frame shift mutation. *Brain*, **121**, 1451–8.

Lindholm T (1967). Electromyographic changes after nitrofurantoin (Furadantin) therapy in nonuremic patients. *Neurology*, **17**, 1017–20.

Logigian EL, Shefner JM, Frosch MP *et al.* (1993). Nonvasculitic, steroid-responsive mononeuritis multiplex. *Neurology*, **43**, 879–83.

Logigian EL, Steere AC (1992). Clinical and electrophysiologic findings in chronic neuropathy of Lyme disease. *Neurology*, **42**, 303–11.

Logina I, Donaghy M (1999). Diphtheritic polyneuropathy: a clinical study and comparison with Guillain-Barre syndrome. *J Neurol Neurosurg Psychiatry*, **67**, 433–8.

Logina I, Krumina A, Karelis G *et al.* (2006). Clinical features of double infection with tick-borne encephalitis and Lyme borreliosis transmitted by tick bite. *J Neurol Neurosurg Psychiatry*, **77**, 1350–3.

Lombardi R, Erne B, Lauria G *et al.* (2005). IgM deposits on skin nerves in anti-myelin-associated glycoprotein neuropathy. *Ann Neurol*, **57**, 180–7.

LoMonaco M, Milone M, Batocchi AP *et al.* (1992). Cisplatin neuropathy: clinical course and neurophysiological findings. *J Neurol*, **239**, 199–204.

Lotti M, Becker CE, Aminoff MJ (1984). Organophosphate polyneuropathy: pathogenesis and prevention. *Neurology*, **34**, 658–62.

Lu JL, Sheikh KA, Wu HS *et al.* (2000). Physiologic-pathologic correlation in Guillain-Barre syndrome in children. *Neurology*, **54**, 33–9.

Luciano CA, Russell JW, Banerjee TK *et al.* (2002). Physiological characterization of neuropathy in Fabry's disease. *Muscle Nerve*, **26**, 622–9.

Luostarinen L, Himanen SL, Luostarinen M *et al.* (2003). Neuromuscular and sensory disturbances in patients with well treated coeliac disease. *J Neurol Neurosurg Psychiatry*, **74**, 490–4.

Lynch DR, Hara H, Yum SW *et al.* (1997). Autosomal dominant transmission of Dejerine-Sottas disease (HMSN III). *Neurology*, **49**, 601–3.

MacDermot KD, Holmes A, Miners AH (2001). Anderson-Fabry disease: clinical manifestations and impact of disease in a cohort of 98 hemizygous males. *J Med Genet*, **38**, 750–60.

Magy L, Birouk N, Vallat JM *et al.* (1997). Hereditary thermosensitive neuropathy: an autosomal dominant disorder of the peripheral nervous system. *Neurology*, **48**, 1684–90.

Martin JJ, Ceuterick C, Mercelis R *et al.* (1982). Pathology of peripheral nerves in metachromatic leucodystrophy. A comparative study of ten cases. *J Neurol Sci*, **53**, 95–112.

Mathew L, Talbot K, Love S *et al.* (2007). Treatment of vasculitic peripheral neuropathy: a retrospective analysis of outcome. *QJM*, **100**, 41–51.

Matthews WB, Esiri M (1983). The migrant sensory neuritis of Wartenberg. *J Neurol Neurosurg Psychiatry*, **46**, 1–4.

McCombe PA, Pollard JD, McLeod JG (1987b). Chronic inflammatory demyelinating polyradiculoneuropathy. A clinical and electrophysiological study of 92 cases. *Brain*, **110**, 1617–30.

McLeod JG, Pollard JD, Macaskill P *et al.* (1999). Prevalence of chronic inflammatory demyelinating polyneuropathy in New South Wales, Australia. *Ann Neurol*, **46**, 910–3.

Medori R, Autilio-Gambetti L, Jenich H *et al.* (1988). Changes in axon size and slow axonal transport are related in experimental diabetic neuropathy. *Neurology*, **38**, 597–601.

Meggouh F, Bienfait HM, Weterman MA *et al.* (2006). Charcot-Marie-Tooth disease due to a de novo mutation of the RAB7 gene. *Neurology*, **67**, 1476–8.

Mericle RA, Triggs WJ (1997). Treatment of acute pandysautonomia with intravenous immunoglobulin. *J Neurol Neurosurg Psychiatry*, **62**, 529–31.

Merkies IS, Schmitz PI, van der Meche FG *et al.* (2002). Clinimetric evaluation of a new overall disability scale in immune mediated polyneuropathies. *J Neurol Neurosurg Psychiatry*, **72**, 596–601.

Mitchell G, Larochelle J, Lambert M *et al.* (1990). Neurologic crises in hereditary tyrosinemia. *N Engl J Med*, **322**, 432–7.

Moghadasian MH, Salen G, Frohlich JJ *et al.* (2002). Cerebrotendinous xanthomatosis: a rare disease with diverse manifestations. *Arch Neurol*, **59**, 527–9.

Mollee PN, Wechalekar AD, Pereira DL *et al.* (2004). Autologous stem cell transplantation in primary systemic amyloidosis: the impact of selection criteria on outcome. *Bone Marrow Transplant*, **33**, 271–7.

Monaco S, Lucci B, Laperchia N *et al.* (1988). Polyneuropathy in hypereosinophilic syndrome. *Neurology*, **38**, 494–6.

Monforte R, Estruch R, Valls-Sole J *et al.* (1995). Autonomic and peripheral neuropathies in patients with chronic alcoholism. A dose-related toxic effect of alcohol. *Arch Neurol*, **52**, 45–51.

Morgan JP, Penovich P (1978). Jamaica ginger paralysis. Forty-seven-year follow-up. *Arch Neurol*, **35**, 530–2.

Mori K, Hattori N, Sugiura M *et al.* (2002). Chronic inflammatory demyelinating polyneuropathy presenting with features of GBS. *Neurology*, **58**, 979–82.

Mori K, Iijima M, Koike H *et al.* (2005). The wide spectrum of clinical manifestations in Sjogren's syndrome-associated neuropathy. *Brain*, **128**, 2518–34.

Mori M, Kuwabara S, Fukutake T *et al.* (2007). Intravenous immunoglobulin therapy for Miller Fisher syndrome. *Neurology*, **68**, 1144–46.

Muller DP, Lloyd JK, Wolff OH (1985). The role of vitamin E in the treatment of the neurological features of abetalipoproteinaemia and other disorders of fat absorption. *J Inherit Metab Dis*, **8** (Suppl 1), 88–92.

Mygland A, Monstad P (2003). Chronic acquired demyelinating symmetric polyneuropathy classified by pattern of weakness. *Arch Neurol*, **60**, 260–4.

Nair VS, LeBrun M, Kass I (1980). Peripheral neuropathy associated with ethambutol. *Chest*, **77**, 98–100.

Nance CS, Klein CJ, Banikazemi M *et al.* (2006). Later-onset Fabry disease: an adult variant presenting with the cramp-fasciculation syndrome. *Arch Neurol*, **63**, 453–7.

Nelis E, Erdem S, Van Den Bergh PY, *et al.* (2002). Mutations in GDAP1: autosomal recessive CMT with demyelination and axonopathy. *Neurology*, **59**, 1865–72

New PZ, Jackson CE, Rinaldi D *et al.* (1996). Peripheral neuropathy secondary to docetaxel (Taxotere). *Neurology*, **46**, 108–11.

Nichols WC, Dwulet FE, Liepnieks J *et al.* (1988). Variant apolipoprotein AI as a major constituent of a human hereditary amyloid. *Biochem Biophys Res Commun*, **156**, 762–8.

Nierenberg DW, Nordgren RE, Chang MB *et al.* (1998). Delayed cerebellar disease and death after accidental exposure to dimethylmercury. *N Engl J Med*, **338**, 1672–6.

Niewiadomska M, Kulczycki J, Wochnik-Dyjas D *et al.* (2002). Impairment of the peripheral nervous system in Creutzfeldt-Jakob disease. *Arch Neurol*, **59**, 1430–6.

Notermans NC, Wokke JH (1996). Chronic idiopathic axonal polyneuropathy. *Muscle Nerve*, **19**, 1637–8.

Nukada H, Pollock M, Haas LF (1982). The clinical spectrum and morphology of type II hereditary sensory neuropathy. *Brain*, **105**, 647–65.

Oaklander AL, Romans K, Horasek S *et al.* (1998). Unilateral postherpetic neuralgia is associated with bilateral sensory neuron damage. *Ann Neurol*, **44**, 789–95.

Ochoa J (1970). Isoniazid neuropathy in man: quantitative electron microscope study. *Brain*, **93**, 831–50.

Odaka M, Yuki N, Yamada M *et al.* (2003). Bickerstaff's brainstem encephalitis: clinical features of 62 cases and a subgroup associated with Guillain-Barre syndrome. *Brain*, **126**, 2279–90.

Oh SJ, Claussen GC, Odabasi Z *et al.* (1995). Multifocal demyelinating motor neuropathy: pathologic evidence of 'inflammatory demyelinating polyradiculoneuropathy'. *Neurology*, **45**, 1828–32.

Oh SJ, Kurokawa K, de Almeida DF *et al.* (2003). Subacute inflammatory demyelinating polyneuropathy. *Neurology*, **61**, 1507–12.

Oh SJ, LaGanke C, Claussen GC (2001). Sensory Guillain-Barre syndrome. *Neurology*, **56**, 82–6.

Oh SJ, LaGanke C, Powers R *et al.* (2005). Multifocal motor sensory demyelinating neuropathy: inflammatory demyelinating polyradiculoneuropathy. *Neurology*, **65**, 1639–42.

Oh SJ, Slaughter R, Harrell L (1991). Paraneoplastic vasculitic neuropathy: a treatable neuropathy. *Muscle Nerve*, **14**, 152–6.

Ohnishi A, Tsuji S, Igisu H *et al.* (1980). Beriberi neuropathy. Morphometric study of sural nerve. *J Neurol Sci*, **45**, 177–90.

Oksi J, Kalimo H, Marttila RJ *et al.* (1996). Inflammatory brain changes in Lyme borreliosis. A report on three patients and review of literature. *Brain*, **119**, 2143–54.

Ørstavik K, Norheim I, Jorum E (2006). Pain and small-fiber neuropathy in patients with hypothyroidism. *Neurology*, **67**, 786–91.

Ouvrier RA, McLeod JG, Conchin TE (1987). The hypertrophic forms of hereditary motor and sensory neuropathy. A study of hypertrophic Charcot-Marie-Tooth disease (HMSN type I) and Dejerine-Sottas disease (HMSN type III) in childhood. *Brain*, **110**, 121–48.

Palliyath SK, Schwartz BD, Gant L (1990). Peripheral nerve functions in chronic alcoholic patients on disulfiram: a six month follow up. *J Neurol Neurosurg Psychiatry*, **53**, 227–30.

Pan CL, Yuki N, Koga M *et al.* (2001). Acute sensory ataxic neuropathy associated with monospecific anti-GD1b IgG antibody. *Neurology*, **57**, 1316–8.

Park MA, Mueller PS, Kyle RA *et al.* (2003). Primary (AL) hepatic amyloidosis: clinical features and natural history in 98 patients. *Medicine (Baltimore)*, **82**, 291–8.

Parman Y, Battaloglu E, Baris I *et al.* (2004). Clinicopathological and genetic study of early-onset demyelinating neuropathy. *Brain*, **127**, 2540–50.

Parrilla P, Ramirez P, Andreu LF *et al.* (1997). Long-term results of liver transplantation in familial amyloidotic polyneuropathy type I. *Transplantation*, **64**, 646–9.

Parry GJ, Bredesen DE (1985). Sensory neuropathy with low-dose pyridoxine. *Neurology*, **35**, 1466–8.

Paunio T, Sunada Y, Kiuru S *et al.* (1995). Haplotype analysis in gelsolin-related amyloidosis reveals independent origin of identical mutation (G654A) of gelsolin in Finland and Japan. *Hum Mutat*, **6**, 60–5.

Pellissier JF, Pouget J, Cros D *et al.* (1984). Peripheral neuropathy induced by amiodarone chlorhydrate. A clinicopathological study. *J Neurol Sci*, **63**, 251–66.

Periquet MI, Novak V, Collins MP *et al.* (1999). Painful sensory neuropathy: prospective evaluation using skin biopsy. *Neurology*, **53**, 1641–7.

Pestronk A, Lopate G, Kornberg AJ *et al.* (1994). Distal lower motor neuron syndrome with high-titer serum IgM anti-GM1 antibodies: improvement following immunotherapy with monthly plasma exchange and intravenous cyclophosphamide. *Neurology*, **44**, 2027–31.

Peters C, Steward CG (2003). Hematopoietic cell transplantation for inherited metabolic diseases: an overview of outcomes and practice guidelines. *Bone Marrow Transplant*, **31**, 229–39.

Pettit RE, Berdal KG (1984). Chediak-Higashi syndrome. Neurologic appearance. *Arch Neurol*, **41**, 1001–2.

Piradov MA, Pirogov VN, Popova LM et al. (2001). Diphtheritic polyneuropathy: clinical analysis of severe forms. Arch Neurol, **58**, 1438–42.

Plante-Bordeneuve V, Lalu T, Misrahi M et al. (1998). Genotypic-phenotypic variations in a series of 65 patients with familial amyloid polyneuropathy. Neurology, **51**, 708–14.

Plasma Exchange/Sandoglobulin Guillain-Barré Syndrome Trial Group (1997). Randomised trial of plasma exchange, intravenous immunoglobulin, and combined treatments in Guillain-Barré syndrome. Lancet, **349**, 225–30.

Ponsford S, Willison H, Veitch J et al. (2000). Long-term clinical and neurophysiological follow-up of patients with peripheral, neuropathy associated with benign monoclonal gammopathy. Muscle Nerve, **23**, 164–74.

Pont A, Rubino JM, Bishop D et al. (1979). Diabetes mellitus and neuropathy following Vacor ingestion in man. Arch Intern Med, **139**, 185–7.

Priori A, Bossi B, Ardolino G et al. (2005). Pathophysiological heterogeneity of conduction blocks in multifocal motor neuropathy. Brain, **128**, 1642–8.

Quasthoff S, Hartung HP (2002). Chemotherapy-induced peripheral neuropathy. J Neurol, **249**, 9–17.

Rakhmanova AG, Lumio J, Groundstroem K et al. (1996). Diphtheria outbreak in St. Petersburg: clinical characteristics of 1860 adult patients. Scand J Infect Dis, **28**, 37–40.

Ratnaike RN (2003). Acute and chronic arsenic toxicity. Postgrad Med J, **79**, 391–6.

Refsum S, Stokke O, Eldjarn L et al. (1984). Heredopathia atactica polyneuritisformis (Refsum disease). In Dyck PJ, Thomas PK, Lambert EH, Bunge RP, eds. Peripheral Neuropathy. WB Saunders, Philadelphia.

Reilly MM, Adams D, Booth DR et al. (1995). Transthyretin gene analysis in European patients with suspected familial amyloid polyneuropathy. Brain, **118**, 849–56.

Reilly MM, Hanna MG (2002). Genetic neuromuscular disease. J Neurol Neurosurg Psychiatry, **73 (Suppl 2)**, II12–21.

Richardson EP, De Girolami U (1995). Pathology of the Peripheral Nervous System. WB Saunders, Philadelphia.

Richez C, Blanco P, Lagueny A et al. (2005). Neuropathy resembling CIDP in patients receiving tumor necrosis factor-alpha blockers. Neurology, **64**, 1468–70.

Rollins YD, Filley CM, McNutt JT et al. (2002). Fulminant ascending paralysis as a delayed sequela of diethylene glycol (Sterno) ingestion. Neurology, **59**, 1460–3.

Ropper AH (1993). Accelerated neuropathy of renal failure. Arch Neurol, **50**, 536–9.

Ropper AH (1994). Further regional variants of acute immune polyneuropathy. Bifacial weakness or sixth nerve paresis with paresthesias, lumbar polyradiculopathy, and ataxia with pharyngeal-cervical-brachial weakness. Arch Neurol, **51**, 671–5.

Ropper AH, Wijdicks EF, Truax BT (1991). Guillain-Barre Syndrome. FA Davis, Philadelphia.

Rubens O, Logina I, Kravale I et al. (2001). Peripheral neuropathy in chronic occupational inorganic lead exposure: a clinical and electrophysiological study. J Neurol Neurosurg Psychiatry, **71**, 200–4.

Ruts L, van Koningsveld R, van Doorn PA (2005). Distinguishing acute-onset CIDP from Guillain-Barre syndrome with treatment related fluctuations. Neurology, **65**, 138–40.

Ryle C, Donaghy M (1995). Non-enzymatic glycation of peripheral nerve proteins in human diabetics. J Neurol Sci, **129**, 62–8.

Sabatelli M, Madia F, Mignogna T et al. (2001). Pure motor chronic inflammatory demyelinating polyneuropathy. J Neurol, **248**, 772–7.

Said G (1996). Diabetic neuropathy: an update. J Neurol, **243**, 431–40.

Said G, Goulon-Goeau C, Lacroix C et al. (1994). Nerve biopsy findings in different patterns of proximal diabetic neuropathy. Ann Neurol, **35**, 559–69.

Said G, Lacroix C, Lozeron P et al. (2003). Inflammatory vasculopathy in multifocal diabetic neuropathy. Brain, **126**, 376–85.

Said G, Lacroix C, Plante-Bordeneuve V et al. (2002). Nerve granulomas and vasculitis in sarcoid peripheral neuropathy: a clinicopathological study of 11 patients. Brain, **125**, 264–75.

Saiki S, Sakai K, Kitagawa Y et al. (2003). Mutation in the CHAC gene in a family of autosomal dominant chorea-acanthocytosis. Neurology, **61**, 1614–6.

Sandroni P, Vernino S, Klein CM et al. (2004). Idiopathic autonomic neuropathy: comparison of cases seropositive and seronegative for ganglionic acetylcholine receptor antibody. Arch Neurol, **61**, 44–8.

Santoro M, Thomas FP, Fink ME et al. (1990). IgM deposits at nodes of Ranvier in a patient with amyotrophic lateral sclerosis, anti-GM1 antibodies, and multifocal motor conduction block. Ann Neurol, **28**, 373–7.

Saperstein DS, Katz JS, Amato AA et al. (2001). Clinical spectrum of chronic acquired demyelinating polyneuropathies. Muscle Nerve, **24**, 311–24.

Saperstein DS, Wolfe GI, Gronseth GS et al. (2003). Challenges in the identification of cobalamin-deficiency polyneuropathy. Arch Neurol, **60**, 1296–301.

Schmidt BL (1997). PCR in laboratory diagnosis of human Borrelia burgdorferi infections. Clin Microbiol Rev, **10**, 185–201.

Senanayake N, Johnson MK (1982). Acute polyneuropathy after poisoning by a new organophosphate insecticide. N Engl J Med, **306**, 155–7.

Senard JM, Rouet P (2006). Dopamine beta-hydroxylase deficiency. Orphanet J Rare Disc, **1**, 7.

Senderek J, Bergmann C, Stendel C, et al. (2003). Mutations in a gene encoding a novel SH3/TPR domain protein cause autsomal recessive Charcot-Marie-Tooth type 4C neuropathy. Am J Hum Genet, **73**, 1106–19.

Seo JH, Ryan HF, Claussen GC et al. (2004). Sensory neuropathy in vasculitis: a clinical, pathologic, and electrophysiologic study. Neurology, **63**, 874–8.

Sewell WA, Brennan VM, Donaghy M et al. (1997). The use of self infused intravenous immunoglobulin home therapy in the treatment of acquired chronic demyelinating neuropathies. J Neurol Neurosurg Psychiatry, **63**, 106–9.

Sharma KR, Cross J, Ayyar DR et al. (2002b). Diabetic demyelinating polyneuropathy responsive to intravenous immunoglobulin therapy. Arch Neurol, **59**, 751–7.

Sharma KR, Cross J, Farronay O et al. (2002a). Demyelinating neuropathy in diabetes mellitus. Arch Neurol, **59**, 758–65.

Shy ME, Jani A, Krajewski K et al. (2004). Phenotypic clustering in MPZ mutations. Brain, **127**, 371–84.

Simmons Z, Albers JW, Bromberg MB et al. (1995). Long-term follow-up of patients with chronic inflammatory demyelinating polyradiculoneuropathy, without and with monoclonal gammopathy. Brain, **118**, 359–68.

Simpson DM, McArthur JC, Olney R et al. (2003). Lamotrigine for HIV-associated painful sensory neuropathies: a placebo-controlled trial. Neurology, **60**, 1508–14.

Singer OC, Otto B, Steinmetz H et al. (2004). Acute neuropathy with multiple conduction blocks after TNFalpha monoclonal antibody therapy. Neurology, **63**, 1754.

Sinha S, Mahadevan A, Lokesh L et al. (2004). Tangier disease--a diagnostic challenge in countries endemic for leprosy. J Neurol Neurosurg Psychiatry, **75**, 301–4.

Sinnreich M, Klein CJ, Daube JR et al. (2004). Chronic immune sensory polyradiculopathy: a possibly treatable sensory ataxia. Neurology, **63**, 1662–9.

Slee M, Selvan A, Donaghy M (2007). Multifocal motor neuropathy: the diagnostic spectrum and response to treatment. Neurology **69**, 1680–8.

Smith BE, Dyck PJ (1990). Peripheral neuropathy in the eosinophilia-myalgia syndrome associated with L-tryptophan ingestion. Neurology, **40**, 1035–40.

Smith IS (1994). The natural history of chronic demyelinating neuropathy associated with benign IgM paraproteinaemia. A clinical and neurophysiological study. *Brain*, **117**, 949–57.

So YT, Olney RK (1994). Acute lumbosacral polyradiculopathy in acquired immunodeficiency syndrome: experience in 23 patients. *Ann Neurol*, **35**, 53–8.

Sobue G, Ueno-Natsukari I, Okamoto H *et al.* (1994). Phenotypic heterogeneity of an adult form of adrenoleukodystrophy in monozygotic twins. *Ann Neurol*, **36**, 912–5.

Sobue G, Yasuda T, Kachi T *et al.* (1993). Chronic progressive sensory ataxic neuropathy: clinicopathological features of idiopathic and Sjogren's syndrome-associated cases. *J Neurol*, **240**, 1–7.

Sokol RJ, Guggenheim MA, Iannaccone ST *et al.* (1985). Improved neurologic function after long-term correction of vitamin E deficiency in children with chronic cholestasis. *N Engl J Med*, **313**, 1580–6.

Solis C, Martinez-Bermejo A, Naidich TP *et al.* (2004). Acute intermittent porphyria: studies of the severe homozygous dominant disease provides insights into the neurologic attacks in acute porphyrias. *Arch Neurol*, **61**, 1764–70.

Sommer C, Koch S, Lammens M *et al.* (2005). Macrophage clustering as a diagnostic marker in sural nerve biopsies of patients with CIDP. *Neurology*, **65**, 1924–9.

Sorbinil Retinopathy Trial Research Group (1993). The sorbinil retinopathy trial: neuropathy results. *Neurology*, **43**, 1141–9.

Stangel M, Kiefer R, Pette M *et al.* (2003). Side effects of intravenous immunoglobulins in neurological autoimmune disorders—a prospective study. *J Neurol*, **250**, 818–21.

Stangel M, Toyka KV, Gold R (1999). Mechanisms of high-dose intravenous immunoglobulins in demyelinating diseases. *Arch Neurol*, **56**, 661–3.

Stefurak TL, Midroni G, Bilbao JM (1999). Vasculitic polyradiculopathy in systemic lupus erythematosus. *J Neurol Neurosurg Psychiatry*, **66**, 658–61.

Stevanin G, Bouslam N, Thobois S *et al.* (2004). Spinocerebellar ataxia with sensory neuropathy (SCA25) maps to chromosome 2p. *Ann Neurol*, **55**, 97–104.

Street VA, Bennett CL, Goldy JD, *et al.* (2003). Mutation of a putative protein degradation gene LITAF/SIMPLE in Charcot-Marie-Tooth disease 1C. *Neurology*, **60**, 22–6.

Suarez GA, Fealey RD, Camilleri M *et al.* (1994). Idiopathic autonomic neuropathy: clinical, neurophysiologic, and follow-up studies on 27 patients. *Neurology*, **44**, 1675–82.

Sumner CJ, Sheth S, Griffin JW *et al.* (2003). The spectrum of neuropathy in diabetes and impaired glucose tolerance. *Neurology*, **60**, 108–11.

Swift TR, Gross JA, Ward LC *et al.* (1981). Peripheral neuropathy in epileptic patients. *Neurology*, **31**, 826–31.

Takahashi Y, Takata T, Hoshino M *et al.* (2003). Benefit of IVIG for long-standing ataxic sensory neuronopathy with Sjogren's syndrome. IV immunoglobulin. *Neurology*, **60**, 503–5.

Takashima H, Bokerkoel CF, John J, *et al.* (2002). Mutation of TDP1, encoding a topoisomerase l-dependent DNA damage repair enzyme, in spinocerebellar ataxia with axonal neuropathy. *Nat Genet*, **32**, 267–72.

Tang B, Liu X, Zhao G *et al.* (2005). Mutation analysis of the small heat shock protein 27 gene in Chinese patients with Charcot-Marie-Tooth disease. *Arch Neurol*, **62**, 1201–7.

Tardieu M, Lacroix C, Neven B *et al.* (2005). Progressive neurologic dysfunctions 20 years after allogeneic bone marrow transplantation for Chediak-Higashi syndrome. *Blood*, **106**, 40–2.

Tazir M, Azzedine H, Assami S *et al.* (2004). Phenotypic variability in autosomal recessive axonal Charcot-Marie-Tooth disease due to the R298C mutation in lamin A/C. *Brain*, **127**, 154–63.

Teunissen LL, Notermans NC, Franssen H *et al.* (1997). Differences between hereditary motor and sensory neuropathy type 2 and chronic idiopathic axonal neuropathy. A clinical and electrophysiological study. *Brain*, **120**, 955–62.

Thaisetthawatkul P, Collazo-Clavell ML, Sarr MG *et al.* (2004). A controlled study of peripheral neuropathy after bariatric surgery. *Neurology*, **63**, 1462–70.

Thomas PK, Marques W, Jr., Davis MB *et al.* (1997). The phenotypic manifestations of chromosome 17p11.2 duplication. *Brain*, **120**, 465–78.

Thomas PK, Ormerod IE (1993). Hereditary neuralgic amyotrophy associated with a relapsing multifocal sensory neuropathy. *J Neurol Neurosurg Psychiatry*, **56**, 107–9.

Timmerman V, Nelis E, Van Hul W *et al.* (1992). The peripheral myelin protein gene PMP-22 is contained within the Charcot-Marie-Tooth disease type 1A duplication. *Nat Genet*, **1**, 171–5.

Traber MG, Sokol RJ, Ringel SP *et al.* (1987). Lack of tocopherol in peripheral nerves of vitamin E-deficient patients with peripheral neuropathy. *N Engl J Med*, **317**, 262–5.

Traynor AE, Gertz MA, Kyle RA (1991). Cranial neuropathy associated with primary amyloidosis. *Ann Neurol*, **29**, 451–4.

Umehara F, Izumo S, Arimura K *et al.* (1991). Polyneuropathy induced by m-tolyl methyl carbamate intoxication. *J Neurol*, **238**, 47–8.

Uncini A, De Angelis MV, Di Muzio A *et al.* (1999). Chronic inflammatory demyelinating polyneuropathy in diabetics: motor conductions are important in the differential diagnosis with diabetic polyneuropathy. *Clin Neurophysiol*, **110**, 705–11.

Vallat JM, De Mascarel HA, Bordessoule D *et al.* (1995). Non-Hodgkin malignant lymphomas and peripheral neuropathies—13 cases. *Brain*, **118**, 1233–45.

van den Bent MJ, de Bruin HG, Bos GM *et al.* (1999). Negative sural nerve biopsy in neurolymphomatosis. *J Neurol*, **246**, 1159–63.

Van den Berg-Vos RM, Franssen H, Wokke JH *et al.* (2002). Multifocal motor neuropathy: long-term clinical and electrophysiological assessment of intravenous immunoglobulin maintenance treatment. *Brain*, **125**, 1875–86.

Van den Berg-Vos RM, Van den Berg LH, Franssen H *et al.* (2000a). Treatment of multifocal motor neuropathy with interferon-beta1A. *Neurology*, **54**, 1518–21.

Van den Berg-Vos RM, Van den Berg LH, Franssen H *et al.* (2000b). Multifocal inflammatory demyelinating neuropathy: a distinct clinical entity? *Neurology*, **54**, 26–32.

van den Brink DM, Brites P, Haasjes J *et al.* (2003). Identification of PEX7 as the second gene involved in Refsum disease. *Am J Hum Genet*, **72**, 471–7.

van Dijk GW, Notermans NC, Franssen H *et al.* (1996). Response to intravenous immunoglobulin treatment in chronic inflammatory demyelinating polyneuropathy with only sensory symptoms. *J Neurol*, **243**, 318–22.

van Geel BM, Koelman JH, Barth PG *et al.* (1996). Peripheral nerve abnormalities in adrenomyeloneuropathy: a clinical and electrodiagnostic study. *Neurology*, **46**, 112–8.

van Koningsveld R, Schmitz PI, Meche FG *et al.* (2004). Effect of methylprednisolone when added to standard treatment with intravenous immunoglobulin for Guillain-Barre syndrome: randomised trial. *Lancet*, **363**, 192–6.

Vanhooren G, Dehaene I, Van Zandycke M *et al.* (1990). Polyneuropathy in lithium intoxication. *Muscle Nerve*, **13**, 204–8.

Vasilescu C (1976). Sensory and motor coduction in chronic carbon disulphide poisoning. *Eur Neurol*, **14**, 447–57.

Verpoorten N, De Jonghe P, Timmerman V (2006). Disease mechanisms in hereditary sensory and autonomic neuropathies. *Neurobiol Dis*, **21**, 247–55.

Verstappen CC, Koeppen S, Heimans JJ *et al.* (2005). Dose-related vincristine-induced peripheral neuropathy with unexpected off-therapy worsening. *Neurology*, **64**, 1076–7.

Visser LH, Schmitz PI, Meulstee J *et al.* (1999). Prognostic factors of Guillain-Barre syndrome after intravenous immunoglobulin or plasma exchange. Dutch Guillain-Barre Study Group. *Neurology*, **53**, 598–604.

Visser LH, van der Meche FG, Meulstee J et al. (1996). Cytomegalovirus infection and Guillain-Barre syndrome: the clinical, electrophysiologic, and prognostic features. Dutch Guillain-Barre Study Group. *Neurology*, **47**, 668–73.

Vital C, Vital A, Canron MH et al. (2006). Combined nerve and muscle biopsy in the diagnosis of vasculitic neuropathy. A 16-year retrospective study of 202 cases. *J Peripher Nerv Syst*, **11**, 20–9.

Vogel P, Gabriel M, Goebel HH et al. (1985). Hereditary motor sensory neuropathy type II with neurofilament accumulation: new finding or new disorder? *Ann Neurol*, **17**, 455–61.

Vorperian VR, Havighurst TC, Miller S et al. (1997). Adverse effects of low dose amiodarone: a meta-analysis. *J Am Coll Cardiol*, **30**, 791–8.

Vrancken AF, Kalmijn S, Brugman F et al. (2006). The meaning of distal sensory loss and absent ankle reflexes in relation to age: a meta-analysis. *J Neurol*, **253**, 578–89.

Vrancken AF, Notermans NC, Jansen GH et al. (2004). Progressive idiopathic axonal neuropathy—a comparative clinical and histopathological study with vasculitic neuropathy. *J Neurol*, **251**, 269–78.

Warner LE, Mancias P, Butler IJ, et al. (1998). Mutations in the early growth response 2 (EGR2) gene are associated with herediatry myelinopathies. *Nat genet*, **18**, 382–4.

Watanabe M, Yamamoto N, Ohkoshi N et al. (2002). Corticosteroid-responsive asymmetric neuropathy with a myelin protein zero gene mutation. *Neurology*, **59**, 767–9.

Watkins PJ (1990). Natural history of the diabetic neuropathies. *QJM*, **77**, 1209–18.

Weimer LH, Podwall D (2006). Medication-induced exacerbation of neuropathy in Charcot Marie Tooth disease. *J Neurol Sci*, **242**, 47–54.

Wetterau JR, Aggerbeck LP, Bouma ME et al. (1992). Absence of microsomal triglyceride transfer protein in individuals with abetalipoproteinemia. *Science*, **258**, 999–1001.

Whisnant JP, Espinosa RE, Kierland RR et al. (1963). Chloroquine Neuromyopathy. *Mayo Clin Proc*, **38**, 501–13.

Willison HJ, O'Hanlon GM (1999). The immunopathogenesis of Miller Fisher syndrome. *J Neuroimmunol*, **100**, 3–12.

Willison HJ, O'Leary CP, Veitch J et al. (2001). The clinical and laboratory features of chronic sensory ataxic neuropathy with anti-disialosyl IgM antibodies. *Brain*, **124**, 1968–77.

Willison HJ, Yuki N (2002). Peripheral neuropathies and anti-glycolipid antibodies. *Brain*, **125**, 2591–625.

Winer JB (2002). Treatment of Guillain-Barre syndrome. *QJM*, **95**, 717–21.

Winer JB, Hughes RA, Anderson MJ et al. (1988b). A prospective study of acute idiopathic neuropathy. II. Antecedent events. *J Neurol Neurosurg Psychiatry*, **51**, 613–8.

Winer JB, Hughes RA, Osmond C (1988a). A prospective study of acute idiopathic neuropathy. I. Clinical features and their prognostic value. *J Neurol Neurosurg Psychiatry*, **51**, 605–12.

Wohrle JC, Spengos K, Steinke W et al. (1998). Alcohol-related acute axonal polyneuropathy: a differential diagnosis of Guillain-Barre syndrome. *Arch Neurol*, **55**, 1329–34.

Wolfe GI, Baker NS, Amato AA et al. (1999). Chronic cryptogenic sensory polyneuropathy: clinical and laboratory characteristics. *Arch Neurol*, **56**, 540–7.

Yan WX, Archelos JJ, Hartung HP et al. (2001). P0 protein is a target antigen in chronic inflammatory demyelinating polyradiculoneuropathy. *Ann Neurol*, **50**, 286–92.

Yiannikas C, McLeod JG, Pollard JD et al. (1986). Peripheral neuropathy associated with mitochondrial myopathy. *Ann Neurol*, **20**, 249–57.

Zhao C, Takita J, Tanaka Y et al. (2001). Charcot-Marie-Tooth disease type 2A caused by mutation in a microtubule motor KIF1Bbeta. *Cell*, **105**, 587–97.

Zochodne DW, Bolton CF, Wells GA et al. (1987). Critical illness polyneuropathy. A complication of sepsis and multiple organ failure. *Brain*, **110**, 819–41.

Zuchner S, Sperfeld AD, Senderek J et al. (2003). A novel nonsense mutation in the ABC1 gene causes a severe syringomyelia-like phenotype of Tangier disease. *Brain*, **126**, 920–7.

CHAPTER 22

Focal peripheral neuropathy

Michael Donaghy

Contents

22.1 Clinical diagnosis of focal neuropathy

22.1.1 Causes of focal neuropathy

Some causes of focal peripheral nerve damage are self-evident, such as involvement at sites of trauma, tissue necrosis, infiltration by tumour, or damage by radiotherapy (Table 22.1). Focal compressive and entrapment neuropathies are particularly valuable to identify in civilian practice, since recovery may follow relief of the compression. Leprosy is a common global cause of focal neuropathy, which involves prominent loss of pain sensation with secondary acromutilation, and requires early antibiotic treatment. Mononeuritis multiplex due to vasculitis requires prompt diagnosis and immunosuppressive treatment to limit the severity and extent

Table 22.1 Causes of focal peripheral neuropathy

Trauma (Section 22.2.1)		
Compression:	Entrapment (Section 22.2.3)	
	Nerve sheath tumours (Section 22.4.1)	
	Hereditary liability to pressure palsies (Section 21.5)	
	Mucopolysaccharidoses (Section 10.4.5)	
Infiltration:	Tumour (Section 22.4.2)	
	Leprosy (Section 21.14.1)	
	Sensory perineuritis (Section 21.16.1)	
	Sarcoidosis (Section 21.18.6)	
	Amyloidosis (Section 21.9)	
Ischaemic:	Vasculitis (Section 21.15)	
	Diabetes mellitus (Section 21.17)	
	Infarction (Section 21.15.1)	
	Irradiation (Section 22.4.3)	
Unknown mechanism:	Focal hypertrophic neuritis (Section 22.5.6)	
	Multifocal motor neuropathy (Section 21.11.3)	
	Tangier disease (Section 21.8.7)	
	Wartenberg's migrant sensory neuralgia (Section 21.16.2)	
	Neurofibromatosis type 2 (Section 11.2)	

of peripheral nerve damage. Various other medical conditions, both inherited and acquired, can present with focal neuropathy rather than polyneuropathy, the most common of which are diabetes mellitus and hereditary liability to pressure palsies. A purely motor focal presentation should raise the question of multifocal motor neuropathy with conduction block, which usually responds well to high-dose intravenous immunoglobulin infusions.

22.1.2 Differentiating root, plexus, and peripheral nerve lesions

A mononeuropathy is a lesion restricted to one single peripheral nerve, producing the characteristic motor, sensory, and reflex abnormalities distal to the site of the lesion. Multifocal neuropathy or mononeuritis multiplex are terms used to describe the co-existence of two or more separate mononeuropathies, most usually occurring in diabetes mellitus and vasculitis. When a patient harbours multiple mononeuropathies, the clinical picture may resemble a polyneuropathy, and the multifocal nature of the condition can be appreciated only from the history of separate onsets of symptoms in different peripheral nerve territories, and by careful motor and sensory examination to reveal differing densities of clinical involvement in adjacent nerve territories.

Clinical features are crucial to the often difficult distinction between lesions of the roots, plexuses, and peripheral nerves. A lesion of the brachial or lumbosacral plexus is suggested by a pattern of muscle weakness, reflex loss, and sensory disturbance which is not attributable to a lesion of a single spinal root nor to a single peripheral nerve. Not surprisingly, early on many plexus lesions are so anatomically restricted that they fail to satisfy this criterion. Furthermore, the roots, plexus, and peripheral nerves can all be involved simultaneously by pathological processes such as diabetes

mellitus, vasculitis, or tumour infiltration. Familiarity with the skin sensory territories and patterns of muscular innervation is crucial to the distinction between root lesions and peripheral nerve lesions (Fig. 22.1). Discrimination between plexus and multiple root lesions may be impossible on clinical grounds; however the involvement of autonomic nerve fibres in plexus lesions tends to produce a warm red and dry hand or foot. Furthermore, sensory nerve action potentials are preserved in root lesions because the dorsal root ganglion or the peripheral branches of the sensory axons are not affected. Proximal limb muscle weakness tends to be a feature of lesions involving the plexus or roots, rather than of lesions restricted to peripheral nerves.

22.1.3 Double crush lesions

Electrophysiologically proven compression of the median nerve in the carpal tunnel or of the ulnar nerve at the elbow are often associated with co-existing radiculopathy due to cervical spondylosis. This led to the proposal that proximal axonal compression might impair the distal axon's ability to resist otherwise subclinical compressive lesions. Thus a coexisting, but in itself subclinical, distal compression might become clinically obvious in the presence of a more proximal lesion. This notion of a 'double crush' syndrome gained popular acceptance despite a relative paucity of experimental support. However, in motor neurone disease, nerve conduction studies have shown that motor nerve fibres are not more vulnerable to focal compression at the elbow than are the healthy sensory nerve fibres (Chaudhry and Clawson 1997). Careful consideration of the various strands of evidence surrounding the double crush hypothesis concluded that it rarely, if ever, significantly determines clinical symptoms and signs (Wilbourn and Gilliatt 1997). Of course, that does not exclude simple summation of the separate deficits caused by clinically significant compressions of nerve fibres at two different levels, for instance a C6 radiculopathy combined with carpal tunnel syndrome. In such situations it can be difficult to decide which of the two lesions is the major determinant of clinical symptoms and thereby merits treatment.

22.1.4 Small hand muscle wasting

Wasting and weakness of intrinsic hand muscles is a common differential diagnostic problem which always raises the question of focal neuropathy affecting the median or ulnar nerves, but includes a wide range of other disorders (Table 22.2). These muscles are innervated by the anterior horn cells of the first thoracic segment of the spinal cord, with an occasional minor contribution from the eighth cervical. The causes of wasting, therefore, include lesions of the lower motor neurones at any point from this spinal segment to these muscles, together with certain other conditions in which primary muscular degeneration or secondary muscular wasting occurs. Electrophysiological assessment and, if necessary, magnetic resonance imaging of the cervical spinal cord and nerve roots, are of particular value in diagnosis.

Acute anterior horn cell lesions. Nowadays, acute anterior horn cell lesions are rare. The commonest causes are acute poliomyelitis and related viral infections. These are usually easily distinguished by the acute onset, the non-progressive nature of the subsequent weakness, and wasting and the absence of sensory loss. *Herpes zoster* is a rare cause but involves distinctive pain, rash, and sensory

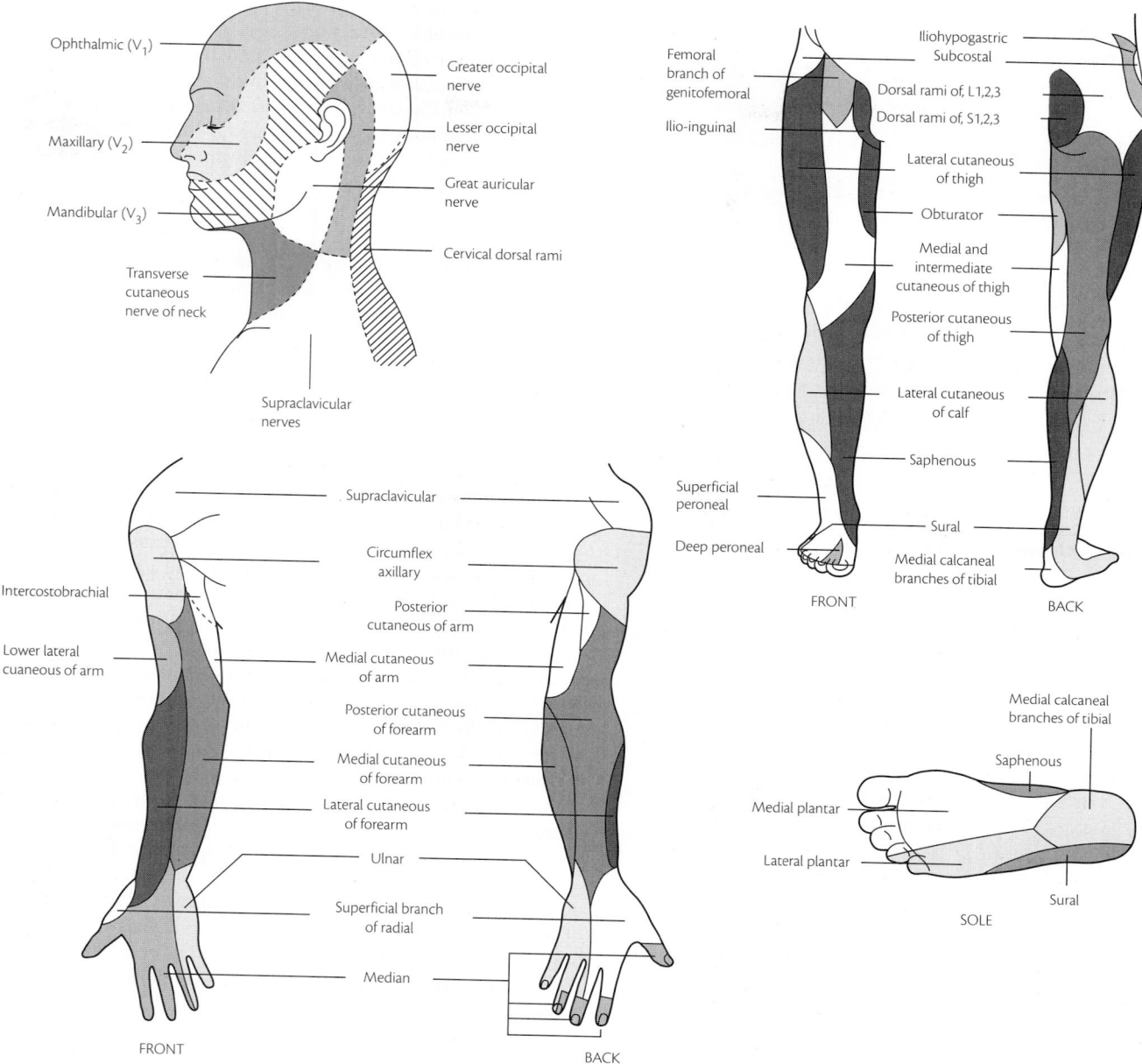

Fig. 22.1 Cutaneous territories supplied by spinal nerve roots and individual peripheral nerves.

disturbance. Acute postinfective polyradiculopathy, the Guillain–Barré syndrome, usually affects proximal limb muscles more severely than distal, at least at first, and like other forms of polyneuropathy it usually affects all four limbs. The acute motor axonal neuropathy variant of Guillain–Barré syndrome can produce small hand muscle weakness and wasting (Section 21.10). *Vascular lesions of the spinal cord* are a rare cause. Thrombosis of a branch of the anterior spinal artery can destroy anterior horn cells, but in such cases the corticospinal and spinothalamic tracts are usually damaged simultaneously. *Haematomyelia* or cord contusion following acute hyperextension injuries of the neck may also damage anterior horn cells in the cervical enlargement. Wasting is not usually confined to muscles innervated by the first thoracic segment and is generally associated with extensive sensory loss over the upper

limbs and with involvement of long ascending and descending tracts of the cord.

Slow-onset anterior horn cell lesions. The commonest chronic lesion is *amyotrophic lateral sclerosis*, which often begins with wasting of the small muscles in one or both hands. This condition is distinguished by its progressive course, the presence of fasciculation, and wasting of other muscle groups, the coexistence of corticospinal-tract degeneration, and the absence of sensory loss. Inherited *spinal muscular atrophy* of the distal type can begin in the small muscles of the hands, and often in the feet as well. Segmental spinal muscular atrophies often remain restricted to one arm. In *syringomyelia* wasting of the hand muscles is often an early symptom. The diagnosis depends upon the characteristic associated analgesia and thermo-anaesthesia, trophic lesions, and the frequent

Table 22.2 Causes of wasting of the small hand muscles

Mononeuropathy :	Median nerve (Abductor pollicis brevis) (Section 22.9.1)
	Ulnar nerve (Dorsal interosseous) (Section 22.9.3)
Polyneuropathy :	Chapter 21
Anterior Horn cell diseases :	Amyotrophic lateral sclerosis (Section 23.2)
	Poliomyelitis (Section 42.4.2)
	Distal spinal muscular atrophy (Section 23.3.3)
	Segmental spinal muscular atrophy (Section 23.3.4)
Spinal cord lesions :	Syringomyelia (Section 28.5.15)
	Ischaemia (Section 28.4.6)
	Spinal cord tumour (Section 28.5.6)
	Traumatic haematomyelia
	Multiple sclerosis (Section 37.5.3)
Spinal nerve root lesions :	Vertebral body collapse (Section 28.5.5)
	Infiltrating vertebral tumour (Section 28.5.5)
	Syphilitic meningomyelitis (Section 41.5.4)
	Trauma (Klumpke paralysis) (Section 22.5.2)
Central nervous system :	Parietal lobe lesions (Section 34.2.3)
	Foramen magnum lesions (Section 28.3.2)
Medial cord of brachial plexus :	Cervical rib (Section 22.5.3)
	Irradiation plexopathy (Section 22.4.3)
	Trauma (Section 22.5.2)
Muscular dystrophy :	Distal myopathy (Section 24.2.9)
	Myotonic dystrophy (Section 24.3)
Trophic disorders :	Chronic arthritis
	Shoulder-hand syndrome (Section 17.5.3)
	Ischaemic contracture (Section 22.2.2)

involvement of the corticospinal tracts. In *intrinsic tumour of the spinal cord* the signs of a progressive focal lesion at the cervical enlargement are usually accompanied by pain and evidence of involvement of long ascending and descending tracts. The ventral roots are occasionally involved in localized *syphilitic meningomyelitis*, or in *arachnoiditis* in which the cord substance usually also suffers. Such ventral root lesions can be distinguished from a lesion of the anterior horn cells only when the dorsal roots are also involved, giving root pain and impairment of sensation over the affected segmental cutaneous areas.

Lesions of the spinal nerves. A spinal nerve is formed from a fusion of the ventral and dorsal roots. For example, a lesion of the first thoracic nerve causes pain and often sensory loss in a radicular distribution along the ulnar border of the forearm, in addition to wasting of the small muscles of the hand. Although any spinal nerve may be compressed by a vertebral lesion, this is rarely the case for the first thoracic nerve since intervertebral disc disease or spondylosis are unusual at that level. Indeed, contrary to widespread misconception, TI root lesions are an unusual cause of

wasting of the small hand muscles, apart from occasions when presumed ischaemia in the first thoracic segment follows cord compression at a higher level. The first thoracic nerve can be compressed as a result of vertebral body collapse or extradural deposits due to malignancy. A traumatic lesion of the first thoracic spinal nerve is responsible for the *Dejerine–Klumpke type of birth palsy*. Lesions involving the first dorsal segment of the spinal cord, its ventral roots, and spinal nerve, usually cause loss of the cervical sympathetic innervation, as its pre-ganglionic fibres leave the cord at this level.

Lesions of the medial cord of the brachial plexus. Lesions of the medial cord of the plexus, for example the pressure of a *cervical rib*, cause wasting of some or all the muscles supplied by the ulnar nerve, including those in the forearm, in addition to the small hand muscles supplied by the median. The weakness and wasting may be confined to the hand. The distribution of pain and sensory loss often involves the eighth cervical and first thoracic segmental areas, that is, roughly, the supply of the ulnar nerve, together with part of the ulnar border of the forearm and arm.

Lesions of the median or ulnar nerves, and polyneuropathy. All lesions situated between the anterior horn cells of the first thoracic segment and the medial cord of the brachial plexus cause wasting of the small hand muscles. Distally to the medial cord of the plexus the innervation of these muscles is divided between the median and ulnar nerves. Lesions of these nerves are distinguished by the characteristic distribution of muscular wasting and sensory loss (Sections 22.9.1 and 22.9.3). Apart from localized lesions of these nerves, wasting in the hand may occur in various forms of *polyneuropathy* in which a glove pattern of sensory loss is usually present, with similar clinical features in the legs.

Central nervous system lesions. Small hand muscle wasting may occur in atrophy of the cervical spinal cord in advanced multiple sclerosis, but the underlying diagnosis will be obvious. Wasting of the contralateral body musculature, often starting in the thenar and hypothenar eminences, is a well-recognized although rare occurrence in parietal lobe lesions (Critchley 1953). Compression of the spinal cord above C4, or in the region of the foramen magnum occasionally leads to hand muscle wasting, possibly secondarily to vascular involvement (Symonds and Meadows 1937).

Muscular dystrophy. Wasting of the small hand muscles is found in some forms of *muscular dystrophy*, especially the *distal* type of *myopathy*. Less often it occurs in *myotonic dystrophy* in which muscles of the forearm are wasted but not so much those of the hands. The diagnosis depends upon the age of onset, the symmetrical character, distribution, and progressive course of the wasting. Fasciculation, sensory loss, or signs of involvement of the central nervous system will be absent. The familial nature of the disorder and the electrophysiological findings provide additional evidence for the muscular dystrophy.

Trophic disorders. Muscular wasting secondary to disuse in *arthritis* of the joints of the hand must not be overlooked. It is easily recognized on account of pain, swelling, and bony changes in the joints and particularly affects the dorsal interosseous muscles in rheumatoid arthritis. In the so-called *shoulder-hand syndrome*, or *algodystrophy*, pericapsulitis of the shoulder-joint is initially associated with painful swelling of the hand, with subsequent occurrence of atrophy of the small hand muscles and demineralization of bones: Sudeck's atrophy. *Ischaemic contracture* caused by major arterial blockage, for instance by fractures in the region of the elbow, leads to paralysis, wasting, and contracture of the muscles of the forearm and hand, with or without sensory loss.

22.2 Trauma, compression and ischaemia

22.2.1 Trauma

Traumatic nerve injury can produce differential damage to the various elements of a peripheral nerve. These have been classified in a manner which allows some prediction of possible recovery (Seddon 1944):

♦ *Neurotmesis* (Section 22.2.4) is complete anatomical division of the axons and connective tissue of a nerve; it includes complete transection.

♦ *Axonotmesis* (Section 22.2.5) refers to loss of continuity of the axons without disruption of their epineurial connective tissue sheath; Wallerian degeneration of the distal axon occurs over 3–5 days followed by axonal regeneration from the proximal stump at the rate of approximately 1mm/day. Axonotmesis can involve varying extents of endoneurial or perineurial disruption.

♦ *Neurapraxia* (Section 22.2.6) is a segmental block of conduction without axonal disruption; this usually involves paranodal or segmental demyelination, or mechanical damage to the myelin sheath, and conduction is restored following myelin repair.

22.2.2 Ischaemia

Nerve ischaemia and infarction due to narrowing or occlusion of vasa nervorum accounts for isolated cranial nerve lesions, usually of the third or sixth cranial nerve, in diabetes, and for mononeuritis multiplex or more diffuse polyneuropathy in diabetes and vasculitis. Large myelinated fibres are especially vulnerable (Fujimura *et al.* 1991). Ischaemic neuropathy is an important complication of atherosclerosis or embolic peripheral vascular disease (Nukada *et al.* 1996). Compression of a nerve for up to 20–30 min whilst unconscious or immobile, or tourniquet application above systolic pressure, produces paralysis and tingling which are immediately reversed by restoring nerve perfusion. More prolonged compression causes disruption of the myelin sheath by a combination of mechanical and ischaemic factors, and ultimately causes disruption of axons.

Ischaemic lesions involving both nerves and muscles may occur as a result of arterial occlusion or injury in closed limb fractures. An example is *Volkmann's ischaemic paralysis* of the arm and hand. Within a few hours of injury, painful sensory disturbance, paralysis, swelling, and cyanosis develop in the limb. Eventual fibrosis of muscles leaves the limb useless and neuropathic pain may ensue. The *anterior tibial syndrome* is a form of ischaemic paralysis of the anterior tibial muscles occurring after unaccustomed exertion. The muscles swell within their tight fascial compartment and may undergo partial or complete infarction. Early surgical decompression must be considered in both conditions to prevent permanent tissue damage, including infarction of the nerve.

22.2.3 Compression

Repeated or prolonged compression of a nerve causes a combination of ischaemia and mechanical deformation of the myelin sheath with local oedema. The initial neurapraxia progresses subsequently to axonotmesis. If the pressure is not relieved, perineurial fibrosis eventually develops and prevents regeneration. This is the sequence whereby lesions evolve in the neuropathies caused by herniated intervertebral disc, narrowed intervertebral foramen, cervical rib, median-nerve compression in the carpal tunnel, ulnar-nerve compression at the elbow, meralgia paraesthetica, and other so-called entrapment neuropathies.

22.2.4 Neurotmesis

Neurotmesis occurs as a result of open wounds, direct blunt injuries, severe traction upon a nerve during displaced fractures, and other forms of disruptive local damage such as misplaced injections. Retrograde degeneration occurs in the proximal stump for 2 or 3 cm and may be more marked after severe traction, or with associated sepsis. The severed distal axonal segment undergoes Wallerian degeneration over 3–5 days. Subsequently sprouting occurs from axons of the proximal stump. If damage to the connective tissue of the nerve is extensive, axonal regeneration towards the previous target may prove impossible resulting in neuroma formation on the proximal stump comprising nerve fibres and scar tissue. Complete division of a mixed peripheral nerve causes motor, sensory, vasomotor, sudomotor, and trophic manifestations corresponding anatomically to the territory supplied by the divided nerve.

Motor symptoms. Interruption of a nerve denervates its muscles causing flaccid paralysis, wasting developing after 2 weeks, and loss of the tendon reflex. Loss of muscle power in the affected nerve's territory may be partially offset by trick movements from the muscles innervated by an intact neighbouring nerve. Electromyography helps identify the denervated muscles from about 4 days after the injury. Nerve conduction in the nerve distal to the lesion slows soon after injury and is lost completely by 3–5 days.

Sensation. Division of a nerve causes complete loss of skin sensation only over the area exclusively supplied by the nerve, the *autonomous zone*. This is surrounded by an *intermediate zone*, which is the area of the nerve's skin territory overlapped by the supply of adjacent nerves. The autonomous and intermediate zones together constitute the *maximal zone* which is the full extent of the nerve's distribution. The cutaneous area over which light touch appreciation is lost is usually larger than that over which pinprick is lost. The appreciation and localization of pressure, of the pain induced by deep pressure, and the recognition of posture and passive movements at the joints, may be impaired as a result of nerve division. However these lost sensations are generally confined to a less extensive area than that which is anaesthetic to light touch.

Trophic change. Vasomotor and trophic disturbances which follow destruction of a motor or a mixed nerve are probably due, at least in part, to the interruption of efferent sympathetic fibres concerned in vasoconstriction. Such changes are most marked after injuries of the median, ulnar, and tibial division of sciatic nerves. After complete division the analgesic area of skin becomes warm and dry. Later, this skin becomes scaly and inelastic owing to retarded desquamation. The limb becomes oedematous when dependent. The analgesic area is liable to suffer injury, and heals slowly after damage, so that ulcers may develop. Growth of the nails and hair is usually slowed, although hypertrichosis occasionally occurs. Adhesions between tendons and their sheaths, and fibrous changes in the muscles and joints eventually set in, but can be offset by repeated passive movements of the joints from early on. Pericapsulitis of the shoulder joint, or 'frozen shoulder', can be very difficult to prevent.

22.2.5 Axonotmesis

This is the type of lesion produced experimentally by crushing a peripheral nerve with forceps, severing all or most of the axons but leaving intact the connective tissue sheath of the nerve. It may result from traction, advanced compression by entrapment, or direct blunt injuries such as fractures or dislocations. Wallerian degeneration occurs after 3–5 days in the axon distal to the injury. In acute self-limited cases axonal regeneration occurs at a rate of approximately 1 mm/day. A strongly positive Tinel sign over the lesion soon after injury, consisting of paraesthesiae evoked by percussing the nerve, indicates severance of axons, rather than neurapraxia. This positive Tinel sign will move peripherally along the nerve as axons regenerate distally, either through the intact nerve connective tissues in axonotmesis, or after successful nerve suture. Functional recovery is more rapid, complete, and accurate after axonotmesis than after suture repair of a complete division. The early clinical and electrophysiological effects are the same as those of neurotmesis.

22.2.6 Neurapraxia

Neurapraxia is a focal conduction block of intact axons produced by segmental myelin damage. Although nerve function is temporarily impaired or lost, recovery occurs too quickly to be explained by axonal regeneration. Neurapraxic lesions occur in various forms of nerve entrapment, compression, or traction provided the axons are not actually severed. Mechanical damage to the myelin sheath, which may simply consist of paranodal demyelination, seems to be the most important pathophysiological mechanism (Dawson *et al.* 1999). Although local ischaemia may play a role, it tends to produce axonal degeneration rather than demyelination. The following clinical features are typical of a neurapraxia:

- the loss of function is predominantly motor, there is little wasting, and the electromyography shows a reduced or absent interference pattern;
- sensory symptoms of numbness, tingling, or burning are common;
- sensory loss is often minimal particularly for modalities transmitted by unmyelinated fibres such as some aspects of touch, pain, and temperature sensation; by contrast, loss of position and vibration sensations transmitted by myelinated fibres are common;
- loss of sweating is unusual;
- nerve conduction distal to the lesion is preserved; an observation which rules out axonal degeneration if performed 7 or more days after an acute lesion.

Recovery is fairly rapid, usually beginning after a few days or weeks. It is usually complete within 9–12 weeks (Shyu *et al.* 1993) although occasionally, complete restoration of function may be delayed until 6 months. The recovery progresses irregularly and follows no anatomical order, but is always complete as long as the lesion was purely neurapraxic.

22.3 Surgical repair

The techniques of peripheral nerve surgery do not fall within the scope of this book, but it is important for neurologists to be aware of the general indications for surgical treatment and its potential. The purpose of surgical intervention after trauma consists of one or more of the following (Birch *et al.* 1998):

- To visualize whether a nerve has actually been severed or ruptured; this is most accurately determined within the first 3 days after injury.
- To restore continuity of a severed or ruptured nerve.
- To remove anything compressing or distorting a nerve, such as bone fragments or sutures.

Peripheral nerve repair, or neurorrhaphy, aims to approximate the severed ends so as to allow topographically accurate and unimpeded axonal regeneration. Primary repair is suitable for nerves which have been cleanly lacerated by sharp items, such as glass; this accounts for the majority of peacetime injuries. Traditional advice has been to treat contamination, infection, haematoma, or excessive instability of the adjacent skeleton before undertaking secondary suture of a nerve 3–4 weeks later. However, it is advocated increasingly that primary nerve repair should be undertaken at the same time as treating associated injuries as long as the patient's condition is stable, the surgical skill is available, and sepsis is not present. Direct surgical repair may restore the continuity of the epineurium with sutures, a suitable method for small nerves such as digital nerves with little non-neural tissue. Alternatively, it may employ the operating microscope to restore the continuity of groups of fascicles separately, which is more suitable for large mixed motor and sensory nerves, particularly for partial injuries. However this involves additional dissection which may offset the intended improvement in regeneration (Lee and Wolfe 2000).

After closed injuries with fractures, it can be difficult to tell whether the underlying nerve lesion is a neurotmesis, with no potential for regeneration, or an axonotmesis, which will recover. Various causes of nerve injury should be considered: tenting or entrapment of the nerve by sharp bone fragments, iatrogenic traction during manipulation, or by expanding haematoma (Ramachandran *et al.* 2006). Under these circumstances enough time should be allowed to permit regenerating nerve fibres to reach the most proximal muscle supplied by the nerve, calculating the rate of regeneration at 1 mm/day and allowing a slight margin. After this time has elapsed, if there is no recovery of function in that muscle, consideration should be given to exploration of the nerve. However it is unlikely that motor recovery will occur in adults if the operation is delayed for more than a year. Apart from surgical technique, the severity of the nerve injury, any associated arterial damage, the age of the patient, the site of the nerve injury, and the promptness of surgical repair are the most important determinants of outcome (Shergill *et al.* 2001). Children may regain normal motor and sensory functions; the elderly rarely do so. Considerable recovery often occurs after such repairs to distal sites, such as the ulnar nerve at the wrist, but is rare after repair to the brachial plexus.

For many years nerve grafting has been used to bridge large gaps in peripheral nerves. Grafting is of use when a segment of nerve has been lost and excessive tension would result from direct apposition of the cut ends, with resultant ischaemia. Autologous grafting material is obtained by sacrifice of nerves elsewhere, such as the sural nerve, sometimes with segments aligned in parallel to bridge the gap in a thicker nerve. Revascularization of the grafted segment is a concern, leading to the use of transposed vascularized nerve grafts in regions such as the brachial plexus (Lee and Wolfe 2000).

There are no rigorously established rules governing whether to operate on a peripheral nerve. Indications which commonly arise include the following (Birch *et al.* 1998):

♦ severe paralysis after a wound over the course of a nerve, or after a closed injury associated with tissue damage;

♦ a nerve lesion associated with a bone fracture which required early internal fixation;

♦ paralysis after closed traction injury of the brachial plexus;

♦ no beginnings of recovery from a neurapraxic lesion at 6 weeks, or from an axonotmesis after the calculated interval has elapsed;

♦ in some cases of entrapment neuropathy for instance due to haematoma, or if focal pain is occurring;

♦ surgical inspection of the nerve can be considered also when there is persistent pain long after injury or if there is an associated arterial lesion.

22.4 Peripheral nerve tumours and irradiation neuropathy

22.4.1 Primary tumours

Single neurofibromas and schwannomas may arise from nerve roots, from the brachial or lumbosacral plexuses, or from peripheral nerves. Multiple neurofibromas occur in von Recklinghausen's disease, the peripheral type of neurofibromatosis, Type I (Section 11.2). These tumours often produce no neurological deficit, merely being discovered on inspection or palpation (Fig. 22.2). When they develop in restricted spaces, such as intervertebral exit foramina, they cause progressive neurological dysfunction due to nerve compression. Nerve-sheath tumours hidden within the pelvis may attain an enormous size before detection; CT or MR imaging usually reveals them before palpation. Often, nerve-sheath tumours may be surgically removed with little or no increase in the pre-existing nerve damage. Occasionally focal peripheral nerve lesions are encountered in patients with Neurofibromatosis Type 2, the central form. In such patients no responsible neurofibroma is uncovered despite extensive MRI imaging of the relevant nerve, plexus, and roots, and the causative pathology is not known (Trivedi *et al.* 2000).

Nerve-sheath tumours can undergo malignant transformation, most commonly to neurofibrosarcoma. This is estimated to occur with a 10 per cent lifetime risk in Neurofibromatosis Type 1 and correlate principally with the presence of internal plexiform, rather than cutaneous, neurofibromas (Tucker *et al.* 2005). Such transformation produces a painfully enlarging tumour mass with dysfunction of the affected nerve or presents with metastases. Malignant nerve-sheath tumours can follow 2–30 years after therapeutic irradiation of the site, particularly in patients with a personal or family history of neurofibromatosis (Foley *et al.* 1980). Malignant nerve-sheath tumours have a poor prognosis even with aggressive treatment using chemotherapy and radiotherapy.

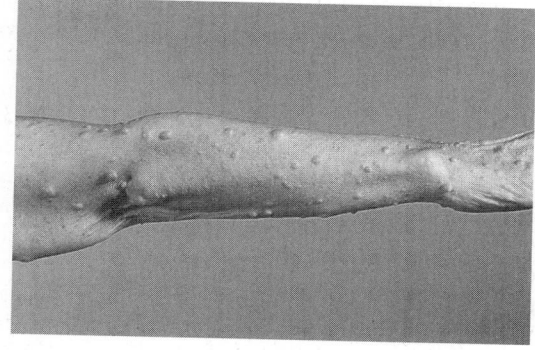

A

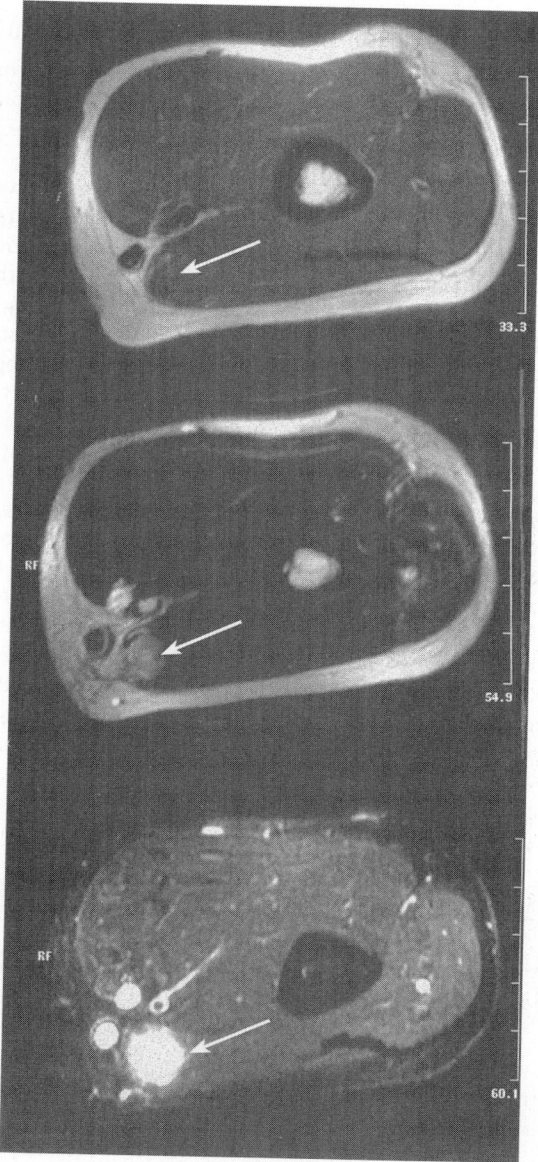

B

Fig. 22.2 A. Multiple neurofibromas on the arm of a patient with no neurological abnormalities. B. MRI of ulnar nerve neurofibroma (arrowed) 10 cm above elbow T1-weighted (top) T1 + Gd (middle) STIR (bottom) sequences.

Rarely, tumours arise from the neural elements themselves: neuroblastomas, ganglioneuroblastomas, and ganglioneurofibromas. Other rare tumours include the paraganglioma or chromaffinoma, which usually grows from the carotid body, glomus jugulare or the adrenal or retroperitoneal tissues, and the rare granular-cell tumour, which is usually solitary but rarely multifocal and occasionally malignant, and arises from the perineurium and may be derived from Schwann cells. Perineuroma, which causes a localized hypertrophic mononeuropathy, is a rare benign peripheral nerve tumour of perineurial cell origin (Bilbao *et al.* 1984).

Traumatic neuromas, or reactive pseudotumours, can cause pain. They can be excised with variable benefit to the patient. Morton's neuroma of the interdigital nerve of the foot is a distinctive example (Section 22.10.2).

22.4.2 Secondary tumour invading nerves or plexuses

Malignant tumours of non-neural tissue may invade peripheral nerves directly, causing profound pain and loss of neurological function. This commonly occurs in the brachial or lumbosacral plexus due to metastases or by direct extension of tumours which have arisen in nearby structures. Multifocal neuropathy occasionally occurs as a result of direct infiltration of nerves by large cell lymphoma, which can be of the angiotropic type (Levin and Lutz 1996). The clinical course of this lymphomatous neuropathy can be subacute with increasingly widespread nerve involvement, so that Guillain–Barré syndrome, or a paraneoplastic neuropathy are suspected unless an affected nerve is biopsied. Valuable improvement may follow chemotherapy.

Malignant infiltration of the *brachial plexus* usually results from carcinoma of the breast or bronchus, or malignant lymphoma (Fig. 22.3). Progressive sensorimotor dysfunction develops in one arm; over 80 per cent of patients experience severe pain: tumour infiltration may be palpable in the supraclavicular or axillary regions; and over 50 per cent of patients exhibit Horner's syndrome (Harper *et al.* 1989). *Pancoast's tumour* refers to infiltration of the lower brachial plexus and the lower cervical sympathetic chain by invasive carcinoma of the apex of the lung. If no tumour is palpable in a patient with painful plexopathy, CT or MRI will reveal the infiltrating tumour in most patients (Thyagarajan *et al.* 1995). Otherwise, diagnostic surgical exploration of the supraclavicular fossa may be required.

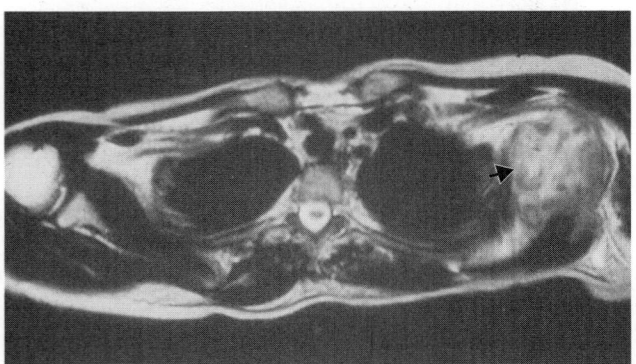

Fig. 22.3 MRI showing a non-Hodgkin's lymphoma in the peripheral portion of the brachial plexus adjacent to the shoulder joint (arrowed). (Courtesy of Dr. N. Moore.)

Table 22.3 Differential diagnosis between radiation and tumour plexopathy

	Radiation plexopathy	Tumour plexopathy
Initial symptom	Weakness	Pain
Distribution	Often bilateral	Unilateral
Site of weakness	Distal	Proximal
CT or MRI	Normal	Tumour mass
Myokymic discharges	50%	No

Malignant tumour infiltration of the *lumbosacral plexus* commonly results from direct spread from carcinoma of the colon, rectum, uterus, prostate, or ovary, and less frequently by metastatic spread from other tumours. Persistent severe unilateral local or radicular pain is followed by a predominantly proximal motor disturbance progressively worsening over weeks or months. Pelvic CT is abnormal in the majority of cases at initial presentation (Thomas *et al.* 1985). The mode of therapy for tumours invading nerve plexuses is determined by the known responsiveness of the tumour type in question. Pain relief may be achieved by local radiotherapy. The prognosis is poor, with over 80 per cent of patients dying within a year of diagnosis.

22.4.3 Irradiation plexopathy

Brachial plexopathy typically presents more than 6 months after supraclavicular irradiation treatment for breast cancer (Harper *et al.* 1989). Lumbosacral plexopathy presents more than 1 year after external or internal cavity irradiation of the pelvis for lymphoreticular, testicular, uterine, or ovarian malignancies. The threshold radiation dosages have not been clearly established. Slowly progressive arm or leg weakness develops, sometimes bilaterally, and there is relatively little pain compared to that experienced with tumour infiltration. In lumbosacral plexopathy, pelvic CT and MRI have enormously simplified the differential diagnosis between tumour infiltration and radiation-induced plexopathy (Thomas *et al.* 1985). Electromyography of weakened muscles shows low-frequency myokymic discharges in roughly half of patients with irradiation plexopathy. The differential diagnosis between radiation and tumour plexopathy is based on the features given in Table 22.3.

22.5 Brachial plexus lesions

22.5.1 Anatomy

The brachial plexus (Figs 22.4 and 22.5) is formed from the anterior primary divisions of the fifth, sixth, seventh, and eighth cervical (C5–C8) and the first thoracic (T1) spinal nerves. It sometimes receives a contribution from the fourth cervical (C4) or the second thoracic nerve (T2). Variations in the composition of the plexus are not uncommon. In the so-called *prefixed* type there is a contribution from C4; C5 is large and there may be no branch from T2. In the rare *postfixed* type there may be no branch from C4, and that from C5 is comparatively small, whereas the T2 contribution is quite substantial. The spinal segmental representation of muscles may be slightly higher or slightly lower than normal, according to whether the plexus is prefixed or postfixed. Intraplexus, rather than

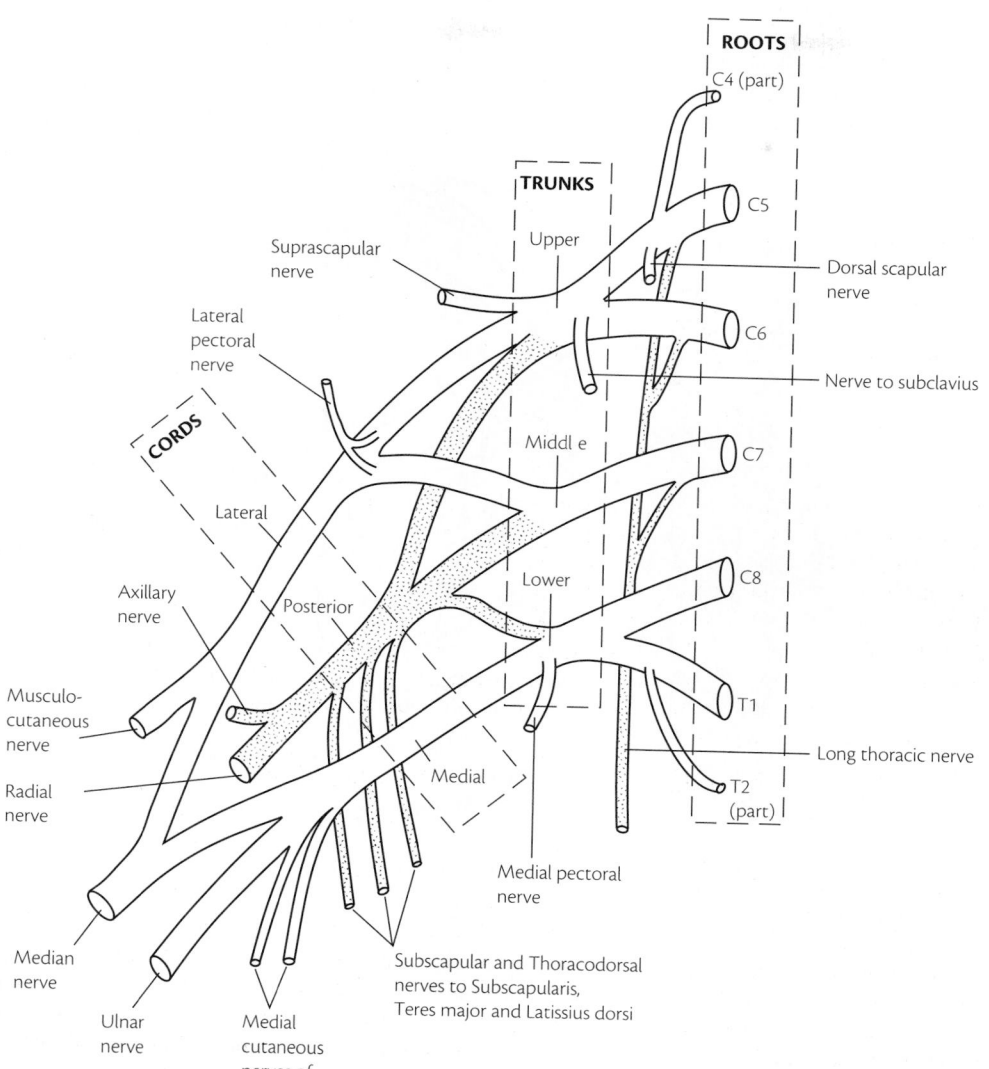

ROOTS
C4 (part)
C5
C6
C7
C8
T1
T2 (part)

TRUNKS
Upper
Middle
Lower

CORDS
Lateral
Posterior
Medial

Suprascapular nerve
Lateral pectoral nerve
Axillary nerve
Musculo-cutaneous nerve
Radial nerve
Median nerve
Ulnar nerve
Medial cutaneous nerves of arm and forearm
Medial pectoral nerve
Subscapular and Thoracodorsal nerves to Subscapularis, Teres major and Latissius dorsi
Dorsal scapular nerve
Nerve to subclavius
Long thoracic nerve

Fig. 22.4 Diagram showing how the trunks and cords of the brachial plexus are formed from the roots, and how they give rise to the peripheral nerves. (Modified from Mackinnon and Morris.)

segmental, variations are commonly encountered at surgical inspection.

The contributions to the plexus from the anterior primary divisions soon divide into anterior and posterior trunks, and from these its three cords are formed as shown diagrammatically in Fig. 22.4:

The lateral cord is formed by a union of anterior trunks of the C5, C6, and C7 nerves. From it arise the lateral pectoral and musculocutaneous nerves and the lateral head of the median nerve.

The medial or inner cord is formed by a combination of the anterior trunk of C8 with the contribution of the T1 to the plexus. It gives origin to the medial head of the median nerve, the ulnar nerve, the medial cutaneous nerves of the arm and forearm, and the medial pectoral nerve.

The posterior cord is formed by the union of the posterior trunks from the C5, C6, C7, C8, and sometimes the T1 nerves. It gives rise to the axillary and radial nerves, the two subscapular nerves, and the nerve to teres major.

Some proximal muscles are innervated by nerves derived from the plexus before formation of the three cords. The most important are:

♦ *The dorsal scapular nerve* (C5) supplies the levator scapulae and rhomboid muscles.

♦ *The long, or posterior, thoracic nerve* (C5, C6, C7) supplies the serratus anterior muscle.

♦ *The suprascapular nerve* (C5, C6) supplies the supraspinatus and infraspinatus muscles.

The brachial plexus is vulnerable to damage at many sites from many causes. Knowledge of its anatomical relationships is crucial to understanding these (Fig. 22.5). One or more spinal nerves may be involved by a lesion of the cervical spine, including congenital abnormality such as fusion of vertebrae in the Klippel–Feil syndrome, fracture dislocations, prolapsed intervertebral disc, spondylosis, malignant deposits, neurofibromas, and occasionally tuberculous, syphilitic, or fungal infection. The plexus itself may be injured by surgical, stab, or gunshot wounds, by

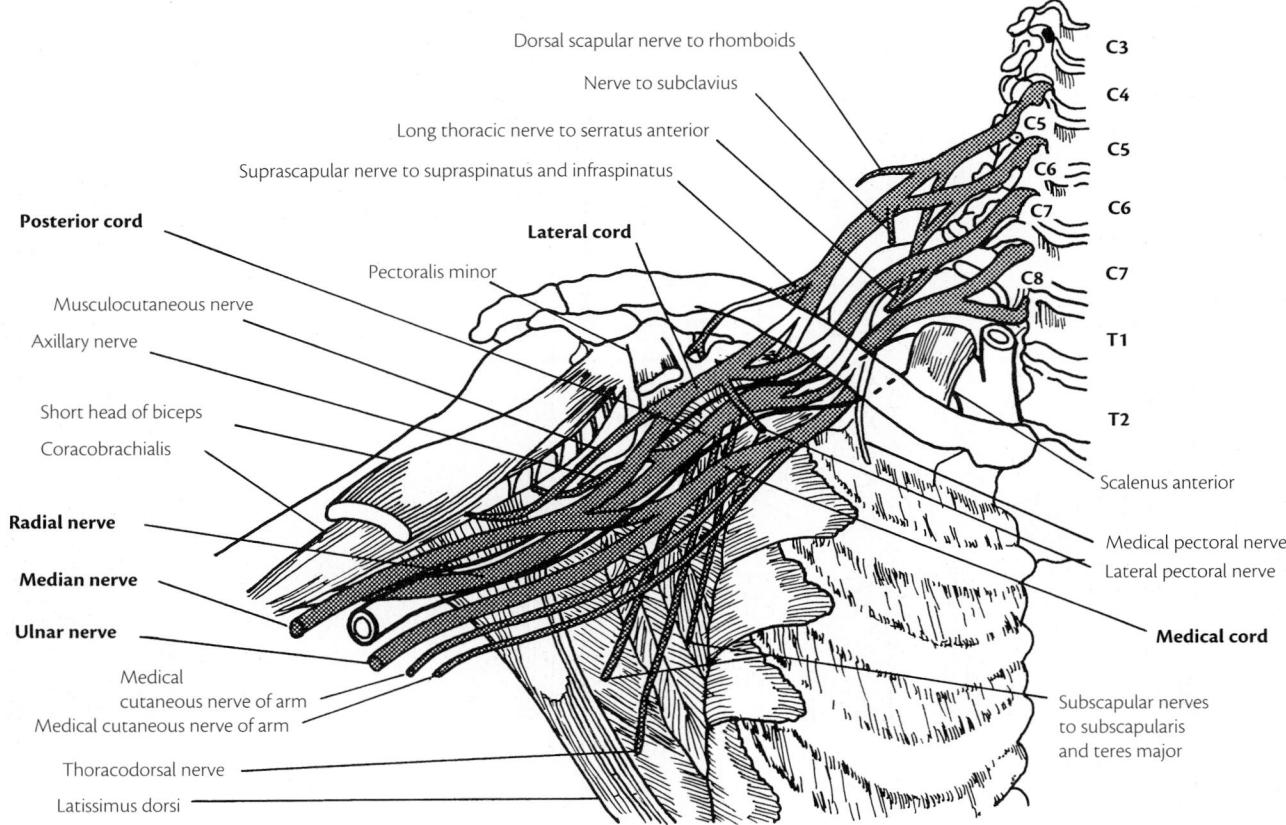

Fig. 22.5 The anatomical relationships of the brachial plexus and its derivative nerves. (From Aids to the Examination of the Peripheral Nervous System, fourth edition, Saunders Ltd, 2000.)

fracture of the clavicle, or by dislocation of the shoulder joint. Traction injuries of the upper plexus may follow forcible separation of the head and shoulder, and of the lower plexus by forced abduction of the arm. The lower plexus may be compressed by abnormalities such as cervical ribs in the thoracic outlet which defines the space between the clavicle and first rib. The plexus can be invaded by metastases, by malignant lymph nodes, or by an apical pulmonary neoplasm 'Pancoast's tumour' or it can be compressed by a neurofibroma. The character of the resultant motor and sensory disturbances depends upon the location of the lesion within the plexus. Complete lesions of the entire plexus usually follow traction injury.

Nerve conduction studies help discriminate between root, plexus, and peripheral nerve lesions. Sensory nerve action potentials are only lost if the lesion is distal to the dorsal-root ganglion, but their usefulness tends to be restricted to diagnosis of lesions affecting fibres derived from the C6, C7, and C8 roots. The axon response to histamine and the reflex vasodilatation to cold can also help in diagnosis; being absent when the lesion is distal to the dorsal-root ganglion and present when it is proximal. MRI or CT scanning may help localize compressive lesions of the nerve roots or plexus. CT myelography is essential for depicting the ventral and dorsal rootlets and their attachment to the spinal cord.

22.5.2 **Traumatic lesions of the brachial plexus**

Total plexus paralysis is rare. When the lesion is close to the vertebral column, all the muscles supplied by the plexus are paralysed and the cervical sympathetic may be involved too. When the plexus is involved at the level of the cords, the spinati, rhomboids, serratus anterior, pectorals, and cervical sympathetic supply may escape. Appreciation of light touch, pain, and temperature is lost over the forearm and hand, and over the outer surface of the arm in its lower two-thirds and sparing the T2 innervation over the upper inner arm and axilla. Joint position sense is lost from the fingers. All upper-limb tendon reflexes are absent.

Upper plexus paralysis or Erb's palsy. This is due to a lesion of the branch from C5 to the brachial plexus. Occasionally the C6 contribution is involved too. Upper plexus paralysis is usually the result of indirect violence, the nerve being torn by undue separation of the head away from the shoulder. It was once a common form of birth injury due to traction on the head when there was difficulty in delivering one shoulder (Fig. 22.6). It may occur in adults as a result of a fall on the shoulder forcing the head to one side, as on falling violently from a motor cycle. It occasionally follows general anaesthesia in patients whose arm has been held abducted and externally rotated. The muscles paralysed as a result of interruption of the C5 component are the biceps, deltoid, brachialis, supraspinatus, infraspinatus, and the rhomboids. When the C6 component is also involved there may be partial weakness of brachioradialis, serratus anterior, latissimus dorsi, triceps, pectoralis major, and extensor carpi radialis. The position of the limb is characteristic. It hangs at the side internally rotated at the shoulder, with the elbow extended and the forearm pronated in the 'waiter's tip' position. There is wasting of the paralysed muscles. Paralysis of

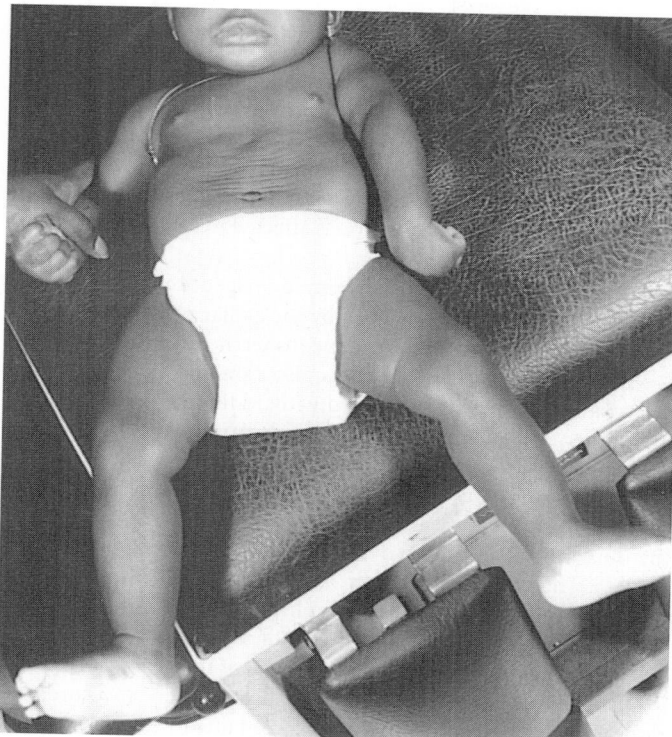

Fig. 22.6 Erbs Palsy due to an upper brachial plexus lesion on the left involving the C5, 6, and 7 roots. The arm hangs at the side internally rotated at the shoulder, with elbow extended and pronation of the forearm producing the so-called 'waiter's tip' posture. (Courtesy of Rolf Birch FRCS.)

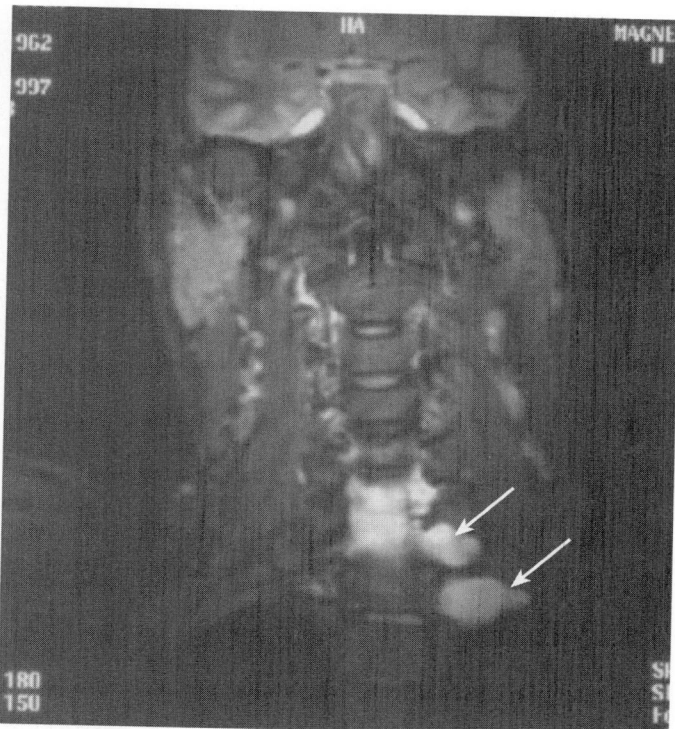

Fig. 22.7 T2-weighted MRI showing cerebrospinal fluid accumulation at the sites of traumatic avulsion of the C8 and T1 roots (arrowed) due to brachial plexus traction in a motor bike injury.

the deltoid renders abduction at the shoulder impossible. The elbow cannot be flexed because of paralysis of its flexors. External rotation at the shoulder is lost owing to paralysis of the spinati. Movements of the wrist and fingers are unaffected. The biceps and supinator jerks are lost. There is usually no sensory loss, except for a small area of anaesthesia and analgesia overlying the deltoid. The 'flare' response following local scratching of the skin is an axon reflex which is lost in lesions distal to the ganglia.

The principal determinant of a good outcome is a plexus lesion which is postganglionic rather than preganglionic, thereby affecting peripheral nerves rather than roots. Operative treatment can be beneficial and expert surgical intervention should be considered early on, especially when the upper plexus has been injured by a stab or gunshot wound. The earlier the repair, the more effective seems the relief from the neuropathic pain which can so impair functional recovery following adult brachial plexus injury (Kato *et al.* 2006). Surgery for closed traction injuries is not so successful if the lesion is shown to be proximal to the dorsal-root ganglia and involves root avulsion, but results are improving (Birch *et al.* 1998). Surgical transfer of all, or bundles from, an intact nerve can restore muscle contraction following preganglionic lesions, with lasting co-contraction being surprisingly unusual. The prognosis is poor if the 'flare' response is present, and particularly if myelography or MRI demonstrate a traumatic meningocele (Fig. 22.7) or a pattern of root-sleeve filling which indicates that the C5 root or spinal nerve has been torn from the spinal cord. Neurophysiological studies will not reveal axonal degeneration reliably until a week after the injury. The use of splints and physiotherapy is paramount

so as to prevent the development of irreversible muscle and joint contractures whilst awaiting neurological recovery. Aberrant reinnervation may give rise to synkinetic movements of shoulder girdle muscles and one-half of the diaphragm.

Lower plexus paralysis of the Klumpke type. The contribution of T1, and sometimes C8, to the brachial plexus may be torn as a result of traction on the abducted arm. Lower plexus paralysis may result from birth injury, from dropping falls during which the patient endeavours to clutch onto something overhead, or during sliding falls from motorcycles with the arm outstretched. The resulting paralysis and wasting involve all the small hand muscles, since these receive a T1 innervation. A claw-hand results from the unopposed action of the long flexors and extensors of the fingers (Fig. 22.8). When C8 is involved too there is wasting and weakness of the ulnar wrist flexors and the long finger flexors. Cutaneous anaesthesia and analgesia are present in a narrow zone along the ulnar border of the hand and for a variable distance up the forearm. Horner's syndrome due to associated cervical sympathetic paralysis is common when the more proximal T1 contribution to the plexus is damaged. The principles of investigation and treatment are similar to those from upper plexus lesions.

Birth lesions of the brachial plexus. Birth lesions of the infant's brachial plexus were common until the mid-20th century. They have declined dramatically since with recent estimates of less than 0.05 per cent for vertex deliveries. They are much commoner in breech deliveries, which often produce bilateral plexus injury (Birch *et al.* 1998). In contemporary British obstetric practice, the overall incidence is 1:2300 live births and sevenfold less frequent after

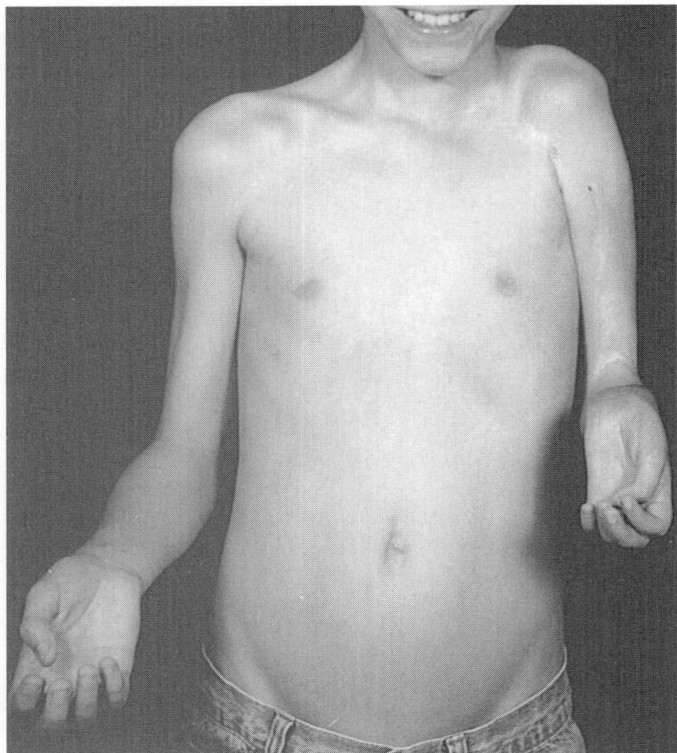

Fig. 22.8 Klumpke-type paralysis due to lower brachial plexus avulsion earlier in childhood. A marked disturbance of subsequent growth especially affects the clavicle, scapula, and humerus. Horner syndrome was present. Nerve transfer operations have restored some hand grasp. Nonetheless, the residual weakness is characteristic with paralysis of wrist flexion and a claw hand. (Courtesy of Rolfe Birch FRCS.)

caesarean section (Evans-Jones *et al.* 2003). Heavy babies born by the vertex are more at risk; in such cases shoulder dystocia is common and the upper trunk of the brachial plexus tends to be damaged by traction separating the head away from the shoulder. This produces the Erb type of paralysis. An associated diaphragm palsy due to phrenic nerve involvement is common. In contrast, breech deliveries tend to produce traction injuries of the lower brachial plexus, producing the Klumpke type of paralysis, often associated with a Horner syndrome. Occasionally the whole plexus is involved and this has been termed the Erb–Duchenne–Klumpke type. Overall by 6 months roughly half have recovered completely, whilst 2 per cent show no recovery (Evans-Jones *et al.* 2003). Full recovery is much more likely with the upper plexus lesions. Poor regeneration is subsequently associated with reduced limb bone growth in the affected arm. Up to 10 per cent have persisting disability either due to poor neural regeneration, or due to apraxia occurring because arm reinnervation occurred after the critical period for central motor programming (Brown *et al.* 2000). Chronic pain is rare compared to its frequency after similar lesions in adults, and sensory recovery is excellent even after operative treatment of severe injuries (Anand and Birch 2002). There is a growing tendency to advise surgical repair of the plexus when the neurophysiology shows severe axonal degeneration and there is a severe lower plexus injury, or when an upper plexus injury involves the phrenic nerve, or there is no recovery of proximal muscle power by 3 months postnatally (Birch *et al.* 1998, 2005). Associated posterior

dislocation of the shoulder can be successfully relocated at 1 year of age or later (Kambhampati *et al.* 2006).

Lesions of the cords of the plexus. The effects of lesions of the cords of the plexus can readily be deduced from Fig. 22.4:

The lateral cord. This is occasionally injured in dislocations of the shoulder. This causes paralysis of the biceps, coracobrachialis, and of all the muscles supplied by the median nerve, except those of the thenar eminence. Sensation is affected to a variable extent on the radial aspect of the forearm.

The posterior cord. This is rarely damaged, but if so there is paralysis of the muscles supplied by the axillary and radial nerves, and loss of sensation in their cutaneous territories.

Middle plexus paralysis. This is also rare and is equivalent to interruption of the posterior cord with additional paralysis of the latissimus dorsi, teres major, and subscapularis as a result of involvement of the thoracodorsal and subscapular nerves.

The medial or inner cord. It is most often injured by subcoracoid dislocation of the humerus. This paralyses the muscles supplied by the ulnar nerve, and those intrinsic hand muscles supplied by the median. The circumflex axillary nerve to deltoid is often damaged too. Sensory loss occurs along the ulnar border of the hand, forearm, and upper arm in the territories of the ulnar, medial cutaneous of the forearm, and medial cutaneous of the arm nerves.

Distal plexus lesions. The distal plexus, where the cords are resorted to form the major nerves, is contained within the medial brachial fascial compartment, which also contains the axillary artery and vein. A compartment syndrome due to haematoma or pseudoaneurism formation can follow percutaneous axillary vessel puncture (Tsao and Wilbourn 2003). This presents with symptoms most usually in the territories of the median and ulnar nerves, which run the longest courses through this compartment. Prompt recognition and surgical intervention may prevent nerve injury.

22.5.3 Thoracic outlet syndromes

Anatomy. The term 'thoracic outlet syndrome' refers to the symptoms and signs resulting from compression of the neurovascular bundle, consisting of brachial plexus, subclavian artery, and vein, by anomalies of the bones or soft tissues during its course between the neck and axilla. Usually thoracic outlet syndrome has a neurogenic presentation with symptoms of lower brachial plexus compression. By comparison syndromes due to subclavian arterial or venous compromise are relatively uncommon. The anatomical anomalies which have been associated with thoracic outlet syndromes are:

- an extra 'cervical' rib articulating with the seventh cervical vertebrae;

- a fibrous band in the same position as a cervical rib and emanating from an elongated C7 transverse process;

- healed clavicular or rib fractures with callus formation;

- abnormalities of the first thoracic rib;

- anomalies of positioning and insertion of the scalenus anterior or medius muscles;

- various other fibrous bands traversing the supraclavicular fossa (Wood *et al.* 1988).

It should be noted that such anomalies, particularly cervical ribs, are common in the general population, few of whom develop

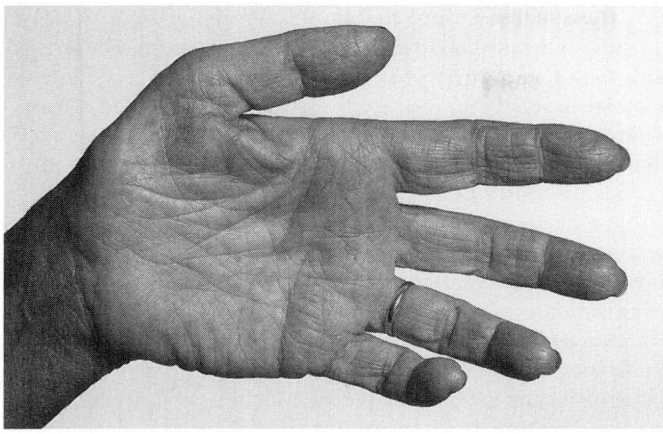

A

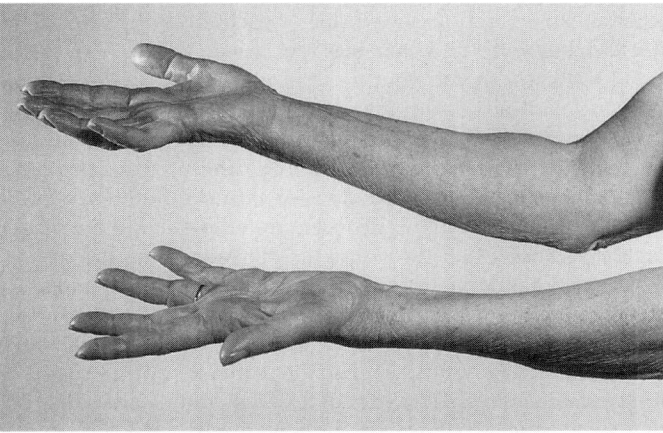

B

Fig. 22.9 Wasting of A. the thenar eminence and B. the forearm flexor compartment in a patient with a cervical rib and thoracic outlet syndrome.

studies confirming chronic post-ganglionic axonal loss and excluding a focal mononeuropathy (Gilliatt *et al.* 1970). Some such patients may have pain and sensory disturbance, predominantly in the ulnar forearm and sometimes in the ulnar side of the hand. These may be aggravated by use of the affected limb, particularly by carrying heavy objects or working with the arms above the head. Thoracic outlet syndrome is more common in women, with an age of onset ranging from the second to the eighth decade, peaking during the thirties.

Patients with definite thoracic outlet syndrome have harboured an asymptomatic cervical rib, or some other musculoskeletal anomaly, throughout life before eventually developing symptoms which gradually evolve into full-blown thoracic outlet syndrome over a period of years during adulthood. Thus, it is self-evident that intermediate or incomplete forms of thoracic outlet syndromes must exist as part of the natural history of thoracic outlet syndrome associated with cervical rib. Moreover, given the variable propensity for cervical ribs to cause thoracic outlet syndromes at all, it is likely that many patients will remain with intermediate forms for years, if not forever.

Unsurprisingly, such intermediate, or incomplete forms of thoracic outlet syndrome are harder to diagnose. Patients in this category should be regarded as having 'suspected' rather than 'definite' thoracic outlet syndrome. Sensory symptoms altered by arm usage, are particularly likely to occur in suspected thoracic outlet syndrome. Sensory loss in the T1 and C8 dermatomes, aggravation of pain and paraesthesia by using the affected arm, absence of nocturnal symptoms, and the occurrence of hand colour changes due to vascular involvement are common in this group. Wasting of the thenar eminence and the long finger flexors may not be evident, but these muscles are usually weakened if tested in a suitably sensitive manner to detect mild weakness. For instance, the examiner should use his own flexor digitorum profundus or dorsal interosseous muscles to test the patient's terminal interphalangeal joint flexion and finger abduction respectively (Figs 2.26, 2.12). Considerable diagnostic importance accrues from reproducing or precipitating sensory symptoms in the ulnar forearm by manually rolling the brachial plexus in the supraclavicular fossa from behind when the patient is standing with the arms hanging. Also, cervical ribs or other supraclavicular anomalies may be palpated by this manoeuvre. Tendon reflexes are usually preserved. Complaints of pain in the upper arm or shoulder are quite common in such patients, but are not a specific feature useful in diagnosis. Radial pulse obliteration by Wright's manoeuvre of hyperabduction and elevation of the arm or Adson's test of lateral rotation of the neck and hyperinflation of the chest is not very reliable as a diagnostic test because 'positive' results occur in many normal asymptomatic people. However if present only on the symptomatic side such vascular occlusion signs contribute to the level of diagnostic suspicion. In summary, three groups of patients who present are first those with advanced denervation of the small hand and forearm muscles without sensory symptoms, second the patient with a combination of motor and sensory symptoms, third patients with sensory symptoms alone.

Imaging. X-ray of the cervical spine and electrophysiology are the most useful investigations for supporting the diagnosis of thoracic outlet syndrome. Radiographic demonstration of a cervical rib contributes importantly to the diagnosis in a patient with a suitable clinical syndrome. Even if a patient is reputed to have had

thoracic outlet syndromes. It is hypothesized that variations in body segmentation may contribute to the likelihood of developing thoracic outlet syndrome. In an incompletely prefixed brachial plexus the contribution of the T1 spinal nerve to the lower trunk ascends angulating across the developing cervical rib. In incomplete postfixation of the brachial plexus the augmented T2 spinal nerve contribution ascends to join the lower trunk and angulates over the normal first thoracic rib. Pedants will note that the term 'thoracic outlet syndrome' is a misnomer because anatomically the thoracic outlet is bounded by the twelfth rib and closed by the diaphragm. More correct would be the term 'thoracic inlet syndrome' since it is this area of the supraclavicular fossa through which the brachial plexus passes. Nonetheless, the term 'thoracic outlet syndrome' has passed into general acceptance and it will continue to be used here.

The range of disorders. Controversy has surrounded the diagnosis of thoracic outlet syndromes. A definite diagnosis is relatively straightforward in the patient with weakness and wasting of the small hand muscles, particularly in the thenar eminence (Fig. 22.9A); sometimes with forearm flexor compartment wasting too (Fig. 22.9B); a radiographic cervical rib; and neurophysiological

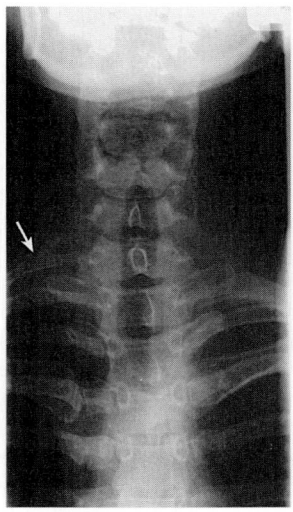
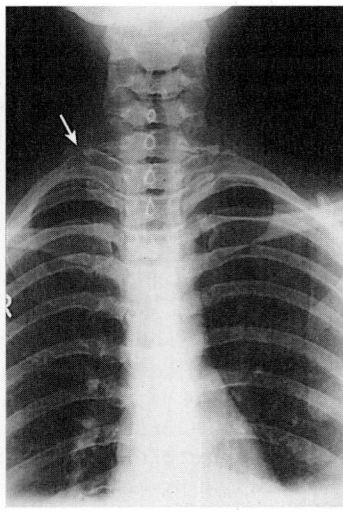

A   B

Fig. 22.10 Plain X-ray of the cervical spine in patients with thoracic outlet syndrome: A. due to right cervical rib (arrowed). Note this articulates with the C7 vertebra transverse process which is downturned in contrast to the upturned transverse processes of T1. B. An elongated downturned transverse process (arrowed) of the right C7 vertebra, found at operation to give rise to a fibrous band.

a 'normal' cervical spine X-ray, this should be inspected since sometimes the anomalous rib emanating from the C7 vertebral transverse process has been mistakenly identified as the 'normal' first thoracic rib. Apart from careful counting of the cervical vertebra, which can be difficult when the skull bones obscure the top of the neck, the key is to note that the seventh cervical vertebra has downturned transverse processes compared to the upturned transverse processes of the T1 vertebra (Fig. 22.10A). If the transverse process of C7 is unusually elongated and downturned, one should be suspicious that it is the origin of a radiographically invisible fibrous band (Fig 22.10B). In patients with suspected thoracic outlet syndrome, blinded analysis of volumetrically acquired magnetic resonance images of the brachial plexus detected deviation of the brachial plexus on about 80 per cent of symptomatic sides (Panegyres *et al.* 1993). MRI also detects instances of plexus distortion by post-traumatic callous of the first rib, by a hypertrophied serratus anterior muscle, or tumour compression.

Neurophysiological studies. These are important in diagnosing thoracic outlet syndrome. They exclude alternative diagnoses of common entrapment neuropathies such as carpal tunnel syndrome or ulnar nerve palsy. Electromyography shows denervation of forearm flexors and small hand muscles, and loss of sensory nerve action potentials confirms that the underlying lesion is post-ganglionic. Occasionally patients with suspected thoracic outlet syndrome will show normal nerve conduction studies. Whilst this should be regarded as evidence against brachial plexus compression, it should be noted that an ulnar sensory nerve action potential which lies within the normal range could have decreased from a higher premorbid level and should be compared with the potential in the unaffected arm. Also, it is usually the ulnar sensory nerve action potential, representing the C8 spinal nerve fibres, which is usually measured. However the less conventional

procedure of measuring the amplitude of the sensory action potential of the medial cutaneous nerve of the forearm, which represents the T1 spinal nerve, would be a more pertinent test for lower brachial plexus compression. Measurement of somatosensory-evoked potentials, or of motor conduction through the supraclavicular fossa following magnetic stimulation of the roots, has not proved to be diagnostically useful. No investigation has absolute diagnostic power for thoracic outlet syndrome, particularly in the milder suspected syndromes. The eventual diagnostic judgement must take into account both clinical features and investigations.

Management. The only definitive treatment for thoracic outlet syndrome is operative removal of the structure which is compressing or distorting the brachial plexus, usually a cervical rib or band. However, despite its extensive history, the surgical treatment of thoracic outlet syndrome is controversial for two reasons. First, the degree of certainty of the diagnosis has a crucial bearing upon whether an operation should be advised. Second, there is no consensus as to the most suitable type of operation.

Operation for thoracic outlet syndrome seems to be effective for relieving pain and sensory disturbance in 90 per cent of the cases. It improves muscle weakness in some 50 per cent of them, but produces no useful reinnervation of severely denervated hand muscles in full-blown and advanced cases of thoracic outlet syndrome (Donaghy *et al.* 1999). This raises an important question concerning the earliness of diagnosis and surgery so as to forestall permanent loss of hand muscle function. It is of little value to undertake an operation in a patient with severe wasting and weakness of hand muscles but without a troublesome sensory disturbance, because no benefit usually results. On the other hand, to operate early so as to forestall irreversible hand muscle denervation carries the risk that some patients will receive an unwarranted operation because the diagnosis of suspected thoracic outlet syndrome is incorrect. When advising upon possible surgery in suspected thoracic outlet syndrome, one should take account of whether the symptoms are disabling, and whether conservative methods aimed at altering posture and arm usage have failed. The patient should be cognizant of possible surgical complications, and, particularly if there is no cervical rib, the exploratory nature of such an operation. The risk of complications depends upon the experience of the surgeon and may be higher if the trans-axillary route is chosen. Noteworthy complications include traction injuries to the brachial plexus causing increased symptoms which usually resolve spontaneously, phrenic nerve injury resulting in diaphragm paralysis, complex regional pain syndrome, long thoracic nerve palsy, and cosmetic dissatisfaction with the supraclavicular scar.

Operations. Various operative approaches to decompressing the brachial plexus have been advocated for thoracic outlet syndrome. The anterior supraclavicular approach is probably the operation of choice since it provides the best exposure of the neurovascular bundle, cervical ribs, or fibrous bands, and can be used for first rib resection. Its disadvantages include the risk of damage to the long thoracic or phrenic nerves and the presence of a cosmetically undesirable scar. The trans-axillary route is popular, particularly in the United States, since it leaves a small hidden scar and requires little dissection, but it does involve strenuous abduction of the arm leaving the brachial plexus vulnerable to traction injury and haemostasis may prove difficult if there is intra-operative vascular damage. The infraclavicular approach is not commonly employed because it does not allow the anatomical abnormality to be visualized, despite providing excellent access to the anterior two-thirds of the first rib. It is particularly difficult to compare the effectiveness of these different operations because of the varying criteria which have been employed for pre-operative selection, and post-operative assessment (Donaghy *et al.* 1999).

Differential diagnosis. Thoracic outlet syndrome is generally distinguished from motor neurone disease by the presence of pain and sensory loss in the former, and by the more malignantly progressive and generalized muscle weakness of the latter. However, focal or segmental spinal muscular atrophies (Section 23.3.4) and multifocal motor neuropathy (Section 21.11.3) should be considered. In syringomyelia, wasting of the small hand muscles is associated with analgesia and thermo-anaesthesia, but the sensory loss is usually much more extensive than that associated with a cervical rib, deep tendon reflexes are lost, and signs of corticospinal-tract degeneration may be present (Section 28.5.15). Radiographic demonstration of a cervical rib must not be taken as sole proof that it is the cause of the patient's symptoms. Tumour of the lung apex will usually be visible radiographically. Other infiltrating tumours of the brachial plexus may be evident on CT scan or MRI. Entrapment of the median or ulnar nerves may be confused with cervical rib, but these diagnoses are established by the characteristic distribution of the motor and sensory symptoms produced by lesions of these nerves and the diagnostic nerve conduction abnormalities. However, some patients with thoracic outlet syndrome do have electrophysiological evidence of additional lesions of the median nerve at the wrist, or of the ulnar nerve at the elbow. For other causes of wasting in the hands, see Section 22.1.4.

22.5.4 Acute brachial neuritis

The essence of acute brachial neuritis consists of severe pain in the region of the shoulder, soon followed by weakness of upper limb muscles, usually those of the shoulder girdle. It is also known as shoulder girdle neuritis, neuralgic amyotrophy, cryptogenic brachial plexus neuropathy, or the Parsonage–Turner syndrome.

Pain is the initial symptom in most patients, often of sudden onset, generally severe and unilateral, and centred on the shoulder girdle (van Alfen *et al.* 2000). Only occasionally do patients develop muscle weakness without pain. This severe initial pain subsides within 4 weeks, and occasionally within hours, during which time muscle weakness develops on the same side. The weakness and subsequent muscle wasting usually affect the spinatus and deltoid muscles. Less frequently, the serratus anterior, trapezius, triceps, biceps, or diaphragmatic muscles may be affected. Forearm muscles are occasionally involved, most notably flexor pollicis longus causing weakness of thumb-tip flexion. Each of the above muscles may be involved in isolation, but multiple muscular involvement is more usual. Also involved may be the sternomastoid, laryngeal or neck extensor muscles, or the lumbosacral plexus. Any of these muscles may be involved in isolation. By comparison sensory symptoms are rarely prominent. However many patients experience paraesthesiae at the onset, or develop patches of mild skin sensory loss, most frequently over the shoulder in the territory of the circumflex axillary nerve.

The symptoms of acute brachial neuritis are usually unilateral. Occasionally bilateral involvement does occur, sometimes sequentially, although it is rarely symmetrical. Electrophysiological studies point to subclinical involvement of the asymptomatic limb's muscles in up to 25 per cent of patients. Careful clinical and electrophysiological studies show patchy involvement of upper-limb nerves, principally within the brachial plexus and its immediate branches, but sometimes affecting nerves more distally within the arm, such as the anterior interosseous nerve. The right arm is more frequently affected than the left. The prognosis is good for shoulder muscle involvement, but less good for forearm weakness. Good recovery of strength occurs in most patients by 3 years. However approximately half of them suffer prolonged milder pain or paresis with 22 per cent unable to work (van Alfen and van Engelen 2006). Recurrences occur in a quarter of the patients but are generally not as severe as the initial attack.

The estimated incidence is 2–3/100 000/year. It is commonly associated with preceding viral infections, exercise, surgery or trauma, pregnancy and the puerperium, or inoculations. The pathogenic mechanisms underlying acute brachial neuritis are unknown but it is thought to be immune-mediated. MRI may show patchy brachial plexus enhancement and biopsy shows florid T-lymphocyte infiltrates in plexus nerve trunks (Suarez *et al.* 1996).

The diagnosis is based on the clinical picture, and investigations are chiefly of help in excluding alternative diagnoses. The differential diagnosis includes hereditary recurrent brachial plexus neuropathy, vasculitic neuropathy, and prolapsed cervical intervertebral disc. Frequently accurate diagnosis has been delayed by an initial diagnosis of 'frozen shoulder'. No treatment is known to affect the long-term outcome; some consider that steroid or adrenocorticotropic hormone therapy diminishes the severity of pain in the early stages.

22.5.5 Hereditary neuralgic amyotrophy

A tendency to recurrent attacks of brachial neuritis may be inherited as an autosomal dominant trait. Both relapsing and remitting and chronic forms occur (van Alfen *et al.* 2000). Individual attacks are indistinguishable from those of acute brachial neuritis (van Alfen and van Engelen 2006) (Section 22.5.4). Complete recovery from discrete attacks is usual, although cumulative disability may develop with repeated attacks. Attacks may develop puerperally in genetically susceptible women. Apart from the family history, the features suggestive of a familial tendency to brachial neuritis include a prior history of unexplained vocal cord paralysis or, occasionally, of lumbosacral plexus lesions. Affected patients from some families display a distinctive facial asymmetry with close-set eyes, epicanthic folds, dwarfism, or cleft palate (Jeannet *et al.* 2001). Unlike hereditary liability to pressure palsies (Section 21.5), families with hereditary neuralgic amyotrophy do not show deletions, duplications, or point mutations of the *PMP-22* gene on chromosome 17, electrophysiology does not show a mild background polyneuropathy, and prominent pain occurs during attacks (Gouider *et al.* 1994). The two conditions are distinct genetic entities, and some families with hereditary neuralgic amyotrophy show linkage to chromosome 17q25, the locus containing various mutations of the Septin 9 gene (Kuhlenbaumer *et al.* 2005).

22.5.6 Chronic brachial plexopathy

Slowly progressive brachial plexus lesions are occasionally encountered in neurological practice. The majority involve a relatively painless progressive motor deficit with mild degrees of sensory loss and are usually unilateral. Generally, no firm diagnosis is established. However, some of these patients may suffer from *focal hypertrophic neuritis*, in which fusiform segmental enlargement of peripheral nerves, mimicking peripheral nerve tumours, may be palpated or visualized on CT (Cusimano *et al.* 1988). Chronic plexopathy due to tumour infiltration or irradiation should be distinguished in patients with known cancer (Section 22.4.3). Brachial plexopathy due to tumour infiltration is generally profoundly painful; it is particularly likely to occur in patients with lymphoma or carcinoma of the breast or lung, is rarely bilateral, and is associated with palpable supraclavicular masses or CT evidence of a mass within the brachial plexus. On the other hand, irradiation-induced brachial plexopathy is usually painless, is more likely to be bilateral, and is often associated with characteristic myokymic discharges on electromyography of affected muscles (Harper *et al.* 1989).

22.6 Lumbosacral plexus lesions

Most of the lumbosacral plexus lies protected within the bony ring of the pelvis, in contrast to the more exposed location of the brachial plexus. This shields the lumbosacral plexus from direct trauma, unless that is severe enough to fracture-dislocate the pelvic ring. However, tumours of the lumbosacral pelvis are relatively concealed from palpation compared to those of the brachial plexus. The introduction of CT and MRI of the retroperitoneum has greatly simplified the differential diagnosis of lumbosacral plexus lesions (Planner *et al.* 2006). A similar spectrum of disease affects both the lumbosacral and brachial plexuses, although idiopathic lumbosacral plexitis is encountered much less frequently than acute brachial neuralgia, its counterpart in the arm. Clinical differentiation of plexus lesions from those of roots or peripheral nerves is discussed in Section 22.1.2. It is important to recognize that lesions of the lumbosacral plexus, particularly when due to tumours or aneurysms, may induce radiating 'sciatica' pain, worsened by straight-leg-raising, leading to an initial diagnosis of a root lesion (Donaghy 2005). A number of major disease entities affecting the

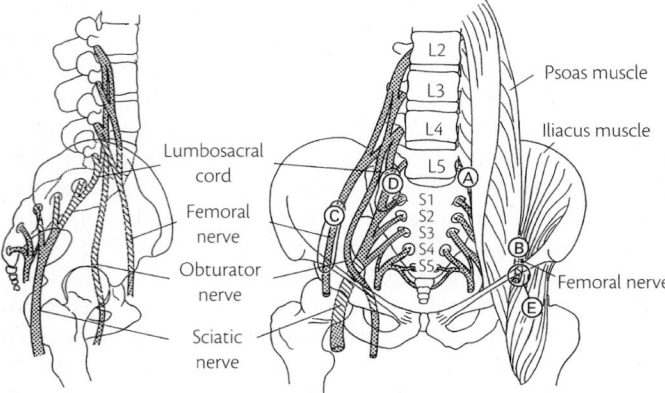

Fig. 22.11 Diagram of lumbosacral plexus showing the usual locations of lesions caused by different mechanical factors. A. Lumbosacral cord compression at the pelvic brim by the fetal head or mid-pelvic forceps causing maternal obstetric paralysis. B. Compression of the femoral nerve by surgical retractor blades in the gutter between the iliacus and psoas muscles. C. Compression of the femoral nerve by haematoma within the iliacus fascia. D. Lumbosacral trunk damage by fracture-dislocation of the sacroiliac joint. E. Angulation of the femoral nerve under the inguinal ligament during prolonged flexion and abduction of the hips in the lithotomy position.

lumbosacral plexus or the intrapelvic segments of nerves derived from it are discussed individually below, and the sites of some important lesions are shown in Fig. 22.11. Other important causes of lumbosacral plexus neuropathy are discussed elsewhere: *diabetic proximal neuropathy* (Section 21.17.3) and *vasculitic neuropathy* (Section 21.15), and tumour infiltration (Section 22.4.2).

22.6.1 Idiopathic lumbosacral plexopathy

This is the counterpart in the leg of acute brachial neuritis (Section 22.5.4). Also called lumbosacral radiculoplexus neuropathy, it seems indistinguishable from the proximal neuropathy of diabetics (Section 21.17.3) (Dyck *et al.* 2001). It may affect all age groups. Although uncommon, its frequency is probably underestimated because its very existence is under-recognized and because the symptoms may resemble those of prolapsed intravertebral disc. Patients experience an abrupt onset of unilateral severe pain in the anterior thigh if involvement of the lumbar plexus predominates, or in the buttock and posterior thigh if sacral plexus involvement is predominant. Muscle weakness is noted within 10 days of the onset of pain and may progressively worsen for some weeks before stabilizing. The pain tends to resolve as the weakness develops. The condition most commonly involves the upper portion of the plexus, producing an absent knee jerk, tenderness on palpation of a femoral nerve, and a positive femoral nerve stretch test. Symptoms of lower plexus involvement may closely resemble those of prolapsed intervertebral disc and the straight-leg-raising test may be positive. All patients experience muscle weakness in varying patterns but objective sensory disturbance occurs only in the minority. Sequential involvement of the other leg and weight loss are common. Spontaneous recovery occurs over months or years and is often incomplete, although the residual disability is rarely severe. It is not known whether steroid therapy influences the evolution of the weakness, or the extent of the subsequent recovery. Occasionally the syndrome is relapsing or progressive and immunosuppressant therapy or high dose intravenous immunoglobulin may be effective (Hollinger and Sturzenegger 2000). Leg nerve biopsies show endoneurial perivascular inflammatory infiltrates (Dyck *et al.* 2001). The differential diagnosis includes prolapsed intervertebral disc, which is not associated with such widespread weakness; vasculitic neuropathy, which more commonly affects more peripheral segments of nerves; diabetic proximal neuropathy; and tumour infiltration of the plexus.

22.6.2 Haemorrhagic lumbosacral plexopathy

There are two distinct syndromes of nerve compression by retroperitoneal haemorrhage. Large haematomas may diffusely compress the lumbar plexus within the psoas muscle, leading to weakness in both obturator and femoral nerve territories (Zarranz *et al.* 1981). More frequently, the intrapelvic portion of the femoral nerve is compressed by relatively smaller haematomas within the iliacus muscle and its indistensible fascia (Fig. 22.11C) (Tysvaer 1982). These syndromes usually occur during anticoagulant therapy but also in other bleeding diatheses, most notably haemophilia.

Femoral nerve compression by iliacus haematoma presents with pain in the groin or iliac fossa which radiates to the anteromedial thigh and medial calf. Hip extension exacerbates the pain and patients characteristically assume a hip-flexed posture. The quadriceps muscle is weak, the knee jerk depressed, and there is a variable degree of sensory loss in the femoral and long saphenous nerve territories. The haematoma may be palpable in the lower iliac fossa or visible at the groin. The condition should be distinguished from septic arthritis or haemarthrosis of the hip joint, both of which cause pain during passive rotation at the hip joint.

In contrast, lumbar plexus involvement in psoas muscle haematoma is not associated with pain on forced hip extension, the hip is not held flexed, no haematoma is palpable, and the thigh adductors are weakened, indicating involvement of the obturator nerve territory.

The diagnosis is confirmed by CT scan or MRI of the pelvis showing haematoma of the psoas or iliacus muscles, thereby differentiating the condition from pelvic abscess. Restoration of normal blood coagulation is the mainstay of therapy and usually allows good recovery. However, some patients with iliacus haematoma develop prolonged neurological disability. To avoid this early surgical drainage has been advocated to improve the chances of complete neurological recovery (Tysvaer 1982).

22.6.3 Trauma

In civilian practice, traction injury to the lumbosacral plexus is a common consequence of double vertical fracture-dislocation of the pelvic ring due to motor vehicle accidents (Stoehr 1978). The lumbosacral cord is the most usually affected portion of the plexus, because it is relatively fixed to the sacral ala (Fig. 22.11D), leading to weakness of muscles innervated by L5 and S1 (Ebraheim et al. 1997). Less frequently, the obturator or superior gluteal nerves or the L5, S1, S2, and S3 anterior roots may be involved. Lumbosacral plexus lesions are generally overlooked immediately following trauma because of the overwhelming nature of the other injuries. They are generally diagnosed only once it is recognized that recovery of limb movements is unusually delayed following pelvic fractures or fracture dislocations of the hip joint. Significant recovery of neurological function rarely occurs. There is little experience of surgical repair of lumbosacral plexus injuries.

22.6.4 Ischaemia

Aortoiliac vascular disease may produce neurological deficits localizable to the lumbosacral plexus. Diabetics undergoing renal transplantation seem especially vulnerable when the internal iliac artery is used to vascularize the graft. Intermittent claudication of buttock and leg muscles may be induced by exercise with pain, tingling, and motor deficits. This pain is particularly localized to the pelvis, and usually associated with stenosis of the internal iliac arteries. Successful treatment with interventional radiological endovascular therapy can be successful (Wohlgemuth and Stoehr 2002). Buttock claudication can also occur after endovascular stent-graft repairs of abdominal aortic aneurysms.

Buttock injections can cause ischaemia of the lumbosacral plexus as well as producing the more familiar problem of direct needle trauma to the sciatic nerve. Vasotoxic or crystalline drugs may be inadvertently introduced into the inferior gluteal artery in the medial aspect of the buttock, particularly if the aspiration test is omitted prior to injection. Toxic vasospasm, or vascular obstruction by crystals is thought to be responsible for such ischaemic lesions of the sciatic nerve or lumbar plexus (Stohr et al. 1980). In addition to neurological disturbance, the buttock skin develops a characteristic painful cyanosed swelling which may progress to gangrene, so-called 'embolia cutis medicamentosa'. Only partial

recovery of the neurological lesion occurs and persistent severe pain is common. Immediate administration of papaverine, heparin, and sympathetic blockade are recommended on suspicion of the diagnosis (Stohr et al. 1980), but these measures are of unproven value. Onset of painless lumbosacral plexopathy within 48 h has followed therapeutic injection of cisplatin or fluorouracil into the internal iliac artery (Castellanos et al. 1987).

22.6.5 Obstetric injuries

Postpartum footdrop occurs most commonly after labour which has been protracted, and involved cephalopelvic disproportion or a mid-pelvic forceps delivery. The lumbosacral cord is directly compressed at the pelvic brim, or as it overlies the sacro-iliac joint, most commonly by the infant's brow during an occiput anterior presentation (Fig. 22.11A). Weak ankle dorsiflexion and eversion are generally noticed only once the woman tries walking after delivery. Comprehensive neurophysiological studies localize the lesion to the expected site at the sacral ala where it is crossed by the lumbosacral trunk (L4, L5) before it joints the S1 root (Feasby et al. 1992). Complete spontaneous recovery within 3 months is usual. The recommended management of subsequent deliveries in women with a previous episode of peripheral pelvic nerve compression depends upon whether full recovery had occurred from the previous episode (Donaldson 1988). If there is residual neurological damage indicating axonal degeneration, elective caesarean section is advised. If full recovery had occurred, a trial of labour should be undertaken but the use of mid-pelvic forceps avoided. When an established lumbosacral plexus lesion develops during labour, it is recommended that labour should be allowed to continue normally since caesarean section is unlikely to reverse the neurological damage; however, mid-pelvic forceps should be avoided (Gonik et al. 1984).

22.6.6 Catamenial sciatica

The unusual developmental anomaly of implantation of endometriosis in the sciatic nerve at the sciatic notch may cause progressive sensorimotor sciatic nerve palsies in women of child-bearing age. These may be associated with perimenstrual pain in the buttock or posterior aspect of the thigh (Salazar-Grueso and Roos 1986). CT scan shows a contrast-enhancing mass in the sciatic notch. Pain may be relieved by danazol or progesterone, or by induction of an artificial menopause. Alternatively, the endometriosis may be removed microsurgically, particularly if fertility must be preserved (Salazar-Grueso and Roos 1986).

22.6.7 Surgical conditions

The intrapelvic portion of the femoral nerve may be damaged by laterally placed self-retaining retractor blades in the lower pelvis during abdominal hysterectomy or renal transplantation (Fig. 22.11B). The femoral nerve may also be damaged by prolonged and excessive angulation at the level of the inguinal ligament during operations under anaesthesia in the lithotomy position (Fig. 22.11E). Up to 70 per cent of recipients of hip prostheses display evidence of damage to the sciatic, femoral, or obturator nerves, although this is only a presenting symptom in a minority of about 1 per cent of the cases. Aneurysms of the iliac or hypogastric arteries may compress the lumbosacral plexus. In such cases, there is an abrupt onset of sciatic pain and impaired straight-leg-raising, suggestive of

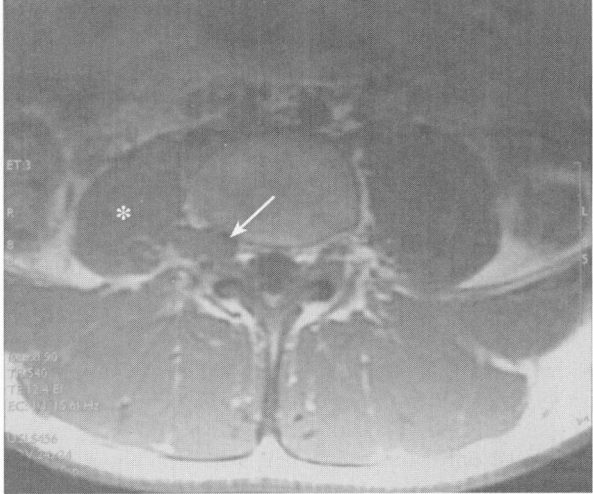

A

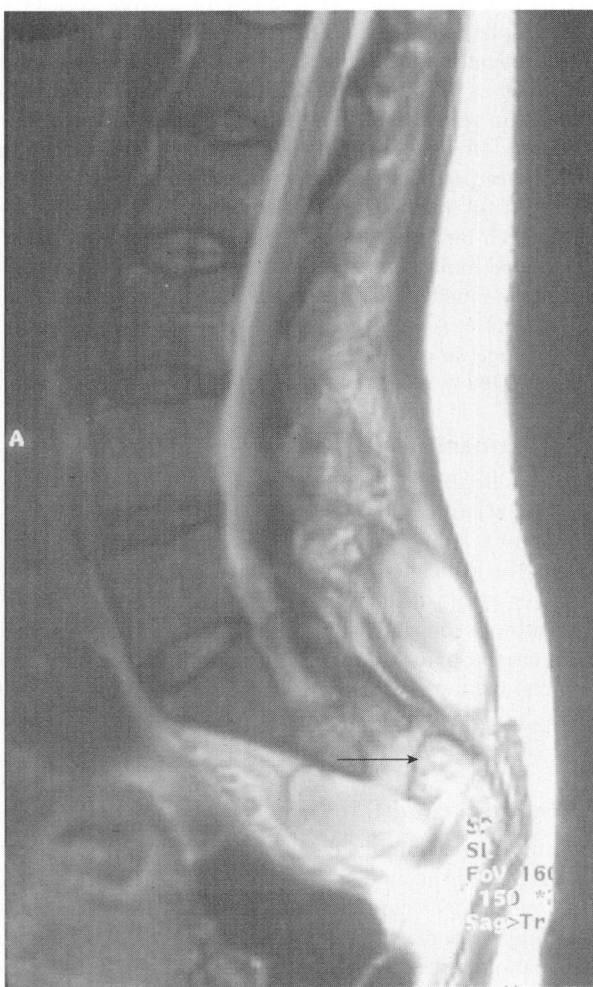

B

Fig. 22.12 MRI in the diagnosis of lumbosacral plexus lesions. A. Tumour (arrowed) involving the lumbar plexus on the postero medial border of the psoas muscle. Note the wasting of the psoas muscle on the affected side (*). B. A staphylococcal abscess in the Pouch of Douglas tracking posteriorly (arrowed) to the subcutaneous tissues in a patient with bilateral leg weakness and loss of sphincter control.

prolapsed intervertebral disc. On rectal examination a firm pulsatile mass is maybe palpable and MRI shows the aneurysm (Planner *et al.* 2006). MRI will detect tumours affecting the plexus (Fig. 22.12A). Patients with a pelvic abscess may develop leg weakness, which is often bilateral and associated with a variable severity of sensory symptoms, and sphincter involvement if the coccygeal plexus is also affected. The patient will usually be constitutionally unwell, with fever and leucocytosis, and MRI demonstrates the abscess (Fig. 22.12B).

22.7 Sacrococcygeal plexus lesions

Locally infiltrative tumours, principally derived from the rectum, prostate, and uterine cervix, may involve the sacrococcygeal plexus. These may involve the sacrum, with evidence of bone erosion on MRI. Pelvic abscess (Fig. 22.12B) or sacral osteomyelitis can also affect the sacrococcygeal plexus and will be revealed by MRI.

Sacral herpes zoster eruptions may lead to urinary retention or segmental paresis and are usually associated with mild lymphocytosis of the cerebrospinal fluid (Thomas and Howard 1972). Acyclovir therapy is generally recommended for this condition, although most patients do make complete or partial recoveries of motor function without this drug.

Acute ano-genital infection with the herpes simplex virus Type 2 may produce neuralgia, numbness, and paraesthesiae of the perineum, buttocks, and posterior thighs, followed by urinary retention, constipation, or erectile failure (Hemrika *et al.* 1986). Reduced anal tone, impaired anal and bulbocavernosus reflexes, and sacral dermatome sensory loss are usually encountered in such patients. Sacrococcygeal plexopathy due to primary genital herpes occurs most commonly in anally receptive homosexual males with herpetic proctitis. It is less frequent in women with herpetic vulvovaginitis and only rarely follows penile herpes in heterosexual males. Mild meningism and a spinal fluid lymphocytosis are usual in such patients. Herpes simplex virus type 2, rather than type 1, is the usual isolate. Acyclovir therapy is recommended. Even without acyclovir therapy, full recovery generally occurs; neurological symptoms generally last about 10 days and rarely more than 21 days. Urinary retention and ano-genital sensory loss have been reported during primary infection with HIV (Zeman and Donaghy 1991). This was associated with spinal-fluid lymphocytosis, and neurological recovery was only partial despite treatment with Zidovudine.

A post-partum lower sacral plexopathy can follow vaginal delivery (Ismael *et al.* 2000). Various combinations of persistent perineal sensory disturbance, micturition abnormalities, faecal incontinence, or anorgasmia occur. Direct trauma to sacral nerves is proposed given the association with multi-parity or mid-forceps rotation.

22.8 Phrenic and intercostal nerve lesions

22.8.1 Phrenic nerve paralysis

Dyspnoea, on exertion or lying flat, is the usual clinical symptom of phrenic nerve palsy. However this cause is usually overlooked because pulmonary and cardiac conditions come to mind initially. This orthopnoea is partially relieved by standing because descent of the diaphragm during inspiration is assisted by the weight of the liver attached to its underside. Unilateral diaphragmatic paralysis

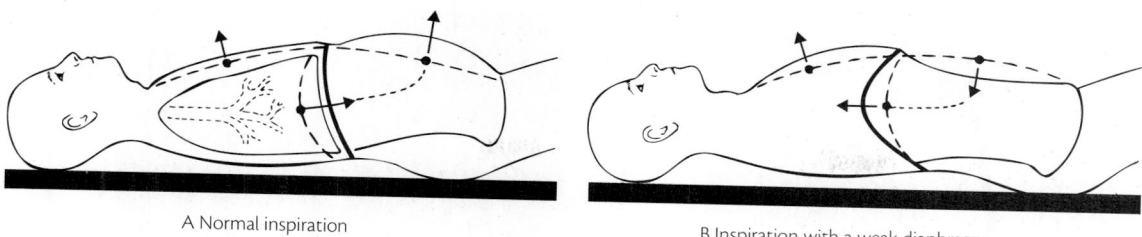

A Normal inspiration   B Inspiration with a weak diaphragm

Fig. 22.13 A. Diagram illustrating (left) during normal inspiration how the diaphragm moves downwards into the abdomen with the resultant outward movement of the abdominal wall, and (right) when the diaphragm is weak how the diaphragm is sucked upwards into the chest, with resultant indrawing of the upper abdominal wall, during inspiration.

may be symptomless whereas bilateral paralysis is a serious disorder. Diaphragm paralysis may be confirmed by clinical observation of paradoxical indrawing of the upper abdomen during inspiration with the patient supine (Fig. 22.13), by radiological screening of the diaphragm while the patient sniffs, by enhancement of the vital capacity when measured standing compared to lying, or by trans-diaphragmatic pressure measurement using balloons in the oesophagus and stomach.

The phrenic nerve derives mainly from the C3 and C4 roots, with an insignificant C5 contribution. It may be damaged at any stage in its course by trauma, by surgical procedures, or by tumours of the neck, mediastinum, or thorax. Conduction studies show that it is frequently involved in polyneuropathies (Newsom-Davis 1967). This is most familiar clinically in the context of respiratory failure due to the Guillain–Barré syndrome, in hereditary motor and sensory neuropathy, and in motor-neurone disease. Phrenic nerve involvement can occur in acute brachial neuritis, and syndromes of bilateral or recurrent isolated alternating phrenic nerve palsies are thought to be variants of acute brachial neuritis (Valls-Sole and Solans 2002). Diaphragm paralysis may also occur as an isolated phenomenon in patients with hereditary neuralgic amyotrophy (Section 22.5.5). Sometimes the cause of unilateral diaphragmatic paralysis cannot be established. The differential diagnosis of diaphragm paralysis is wide (Laroche *et al.* 1989) and includes lesions of the brainstem or spinal cord above the level of the phrenic outflow, traumatic lesions of the upper brachial plexus, disorders of neuromuscular transmission such as myasthenia gravis or the Lambert–Eaton myasthenic syndrome, and a wide range of primary muscle diseases, of which acid maltase deficiency most notably involves the diaphragm relatively selectively. An idiopathic bilateral isolated phrenic neuropathy occurs, often of relatively acute onset yet without pain or arm weakness, and unresponsive to IvIg (Lin *et al.* 2005). In about half of the patients, careful evaluation reveals more generalized evidence of neuralgic amyotrophy (Tsao *et al.* 2006).

22.8.2 Intercostal and truncal neuropathy

Intercostal neuralgia is a rare and ill-defined disorder which is diagnosed probably more often than it occurs. It is characterized by paroxysmal pain throughout the distribution of an intercostal nerve, frequently associated with cutaneous tenderness in the area it supplies, especially at the point of emergence of its lateral cutaneous branch. Before diagnosing this condition care must be taken to exclude the many other disorders which may be associated with similar pain due to compression of spinal dorsal roots by neoplasms of the spinal cord, nerve roots, or vertebrae or due to

inflammation, including arachnoiditis and syphilis. It may precede or follow an attack of herpes zoster.

Notalgia paraesthetica is a rare syndrome of localized burning pain over the scapula thought to be due to compression of the dorsal branches of the thoracic roots passing through the paraspinal muscles. A spinal nerve may be compressed as a result of localized collapse of the vertebral column, most often due to secondary carcinoma, trauma, or tuberculous caries. Spondylosis is often associated with root pains, which may also be produced by scoliosis. Pleurisy, both tuberculous and neoplastic, is sometimes mistakenly diagnosed as intercostal neuralgia, and the thorax is a common site of referred pain in visceral disease, especially diseases of the upper abdominal viscera including cholecystitis and carcinoma of the body or tail of the pancreas. Radicular pain or truncal neuropathy may also occur in diabetes mellitus or porphyria (Lauria *et al.* 1998; Albers and Fink 2004). Pain in the distribution of an intercostal nerve is also seen in some patients after thoracotomy and may be intractable. Truncal muscle involvement is not usually evident clinically, but flaccid bulging of a portion of the abdominal wall can be seen with lower thoracic root or intercostal nerve lesions.

Painful intercostal neuropathy should be treated with analgesics. Local anaesthetic injections may give temporary relief. If all else fails, the nerve may be injected with alcohol or phenol, care being taken that the needle does not penetrate the pleura. In occasional cases, surgical division of two or three intercostal nerves close to the spine has been tried but even after this operation pain may recur after a few months. Even posterior rhizotomy does not always afford permanent relief.

22.9 Upper limb mononeuropathy

22.9.1 The median nerve

Anatomy (Fig. 22.14). The median nerve is derived from C6, C7, C8, and T1 spinal nerves. It is formed by the union of two heads from the medial and lateral cords of the brachial plexus. It runs on the medial aspect of the upper arm and passes through the antecubital fossa anterior to the elbow joint. It usually passes between the two heads of the pronator teres in the upper forearm, before running down the anterior forearm deep to flexor digitorum superficialis. After passing under the transverse carpal ligament at the wrist, which forms the roof of the carpal tunnel, it enters the hand. There are no branches to muscles in the upper arm. In the forearm it supplies the following muscles, with branches given off in the order named: pronator teres (C6, C7), flexor carpi radialis (C6, C7), and palmaris longus. Also, via its anterior interosseous branch, it supplies flexor digitorum superficialis (C7, C8, T1), flexor

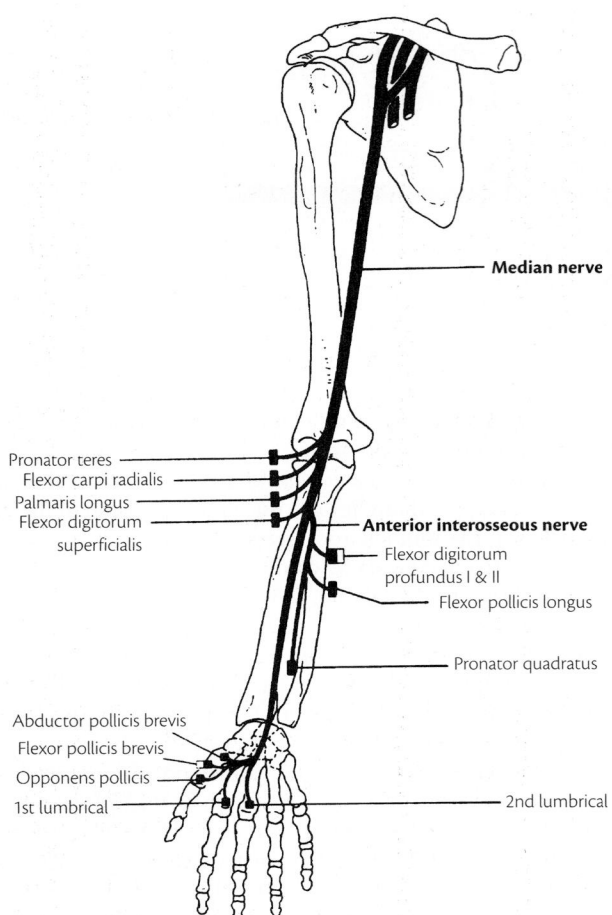

Fig. 22.14 Diagram of the median and anterior interosseous nerves showing the muscles they innervate. Note that the innervation of flexor digitorum profundus and flexor pollicis brevis is shared with the ulnar nerve.

digitorum profundus to the index and middle fingers (C7, C8), flexor pollicis longus (C7, **C8**), and pronator quadratus (C7, **C8**). In the hand the median nerve usually supplies the two radial lumbricals, opponens pollicis, abductor pollicis brevis, and the outer head of the flexor pollicis brevis, all of which are predominantly supplied by the T1 root with a small contribution from C8. Sometimes it supplies the first dorsal interosseous. Anomalies of median nerve innervation most commonly involve the thenar muscles; for instance opponens pollicis may receive pure ulnar nerve innervation. These generally reflect anastomoses between the ulnar and median nerves in the forearm. The most common of these is the Martin–Gruber anastomosis which occurs in up to 15 per cent of people, in which fibres destined for the thenar eminence lie in the ulnar nerve at the wrist but run in the median nerve at the elbow. This can be a source of clinical and electrophysiological confusion in carpal tunnel syndrome.

Trauma. The median nerve may be injured at any point of its course by stab or gunshot wounds. It is occasionally damaged by dislocation of the shoulder joint or by humerus fractures. The commonest acute traumatic lesion in civilian practice is a cut at the wrist, usually the result of the hand having been put through a window pane; in such cases the ulnar nerve may be damaged also. Painful lesions of the median nerve and/or of the medial and

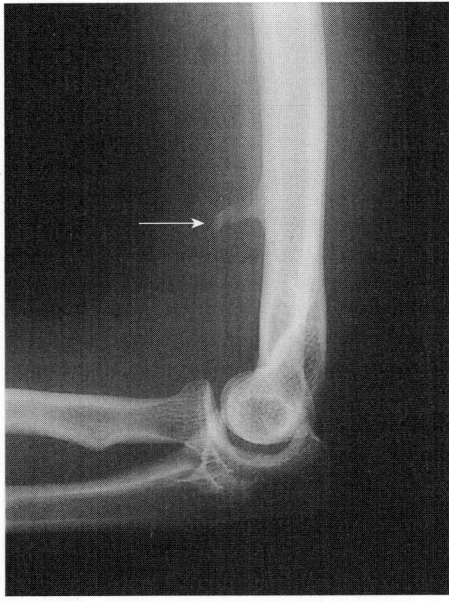

Fig. 22.15 Medial nerve compression by a ligament of Struthers. X-ray showing the bony spur (arrowed) on the shaft of the humerus from which the ligament originates before inserting into the medial epicondyle.

lateral cutaneous nerves of the forearm are an occasional complication of venepuncture in the antecubital fossa (Horowitz 1994), or of arterial cannulation or axillary block anaesthesia.

Proximal median neuropathy. This is uncommon and is normally associated with ulnar or radial palsy in cases of crutch paralysis, overlong use of surgical tourniquets, or compression during sleep or stupor. Other causes of high median nerve palsy include falls on the shoulder or arm, or hyperextension of the forearm. Not infrequently the cause remains obscure. Occasionally the median nerve is trapped by a *ligament of Struthers*, a fibrous band arising from a bony spur on the humerus about 5 cm above the medial epicondyle, to which it runs. Exploration and division of this band are indicated if its presence is suggested radiologically (Fig. 22.15).

In high median nerve lesions the radial flexor of the wrist is paralysed, so that when the wrist is flexed against resistance, the hand deviates to the ulnar side. The terminal phalanx of the thumb and the phalanges of the index finger cannot be flexed. There may be weakness of flexion of the phalanges of the remaining fingers, but not complete paralysis. Since the ulnar half of the flexor digitorum profundus is supplied by the ulnar nerve, the ring and little fingers are usually spared completely. Flexion at the metacarpophalangeal joints is carried out by the interossei and lumbricals, of which only

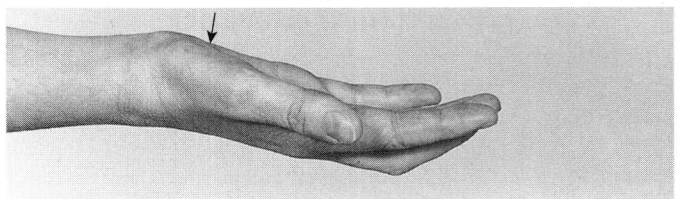

Fig. 22.16 Wasting of abductor pollicis brevis (arrowed) in a median nerve lesion. Note the concave appearance of the muscle at the base of the thumb along the first metacarpal bone when viewed from the side.

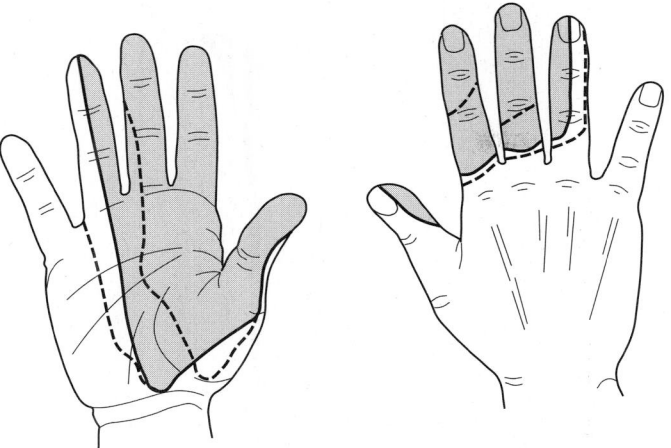

Fig. 22.17 The area of superficial sensory loss after a median nerve lesion above the wrist. This includes the territory supplied by the superficial palmar branch (——— usual boundary ﹘﹘﹘ variations). (Based on Head and Sherren 1905.)

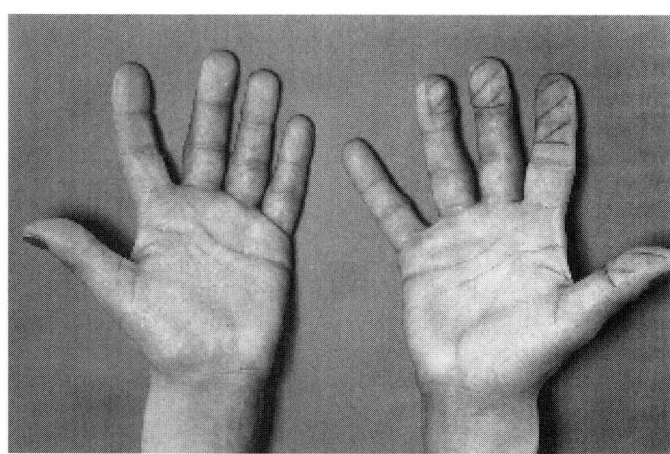

Fig. 22.18 The area of sensory loss in carpal tunnel syndrome. Sensation on the palms and proximal fingers in preserved because the superficial palmar branch of the median nerve does not traverse the carpal tunnel. These apparently masculine hands belong to a woman with acromegaly, the underlying cause of her carpal tunnel syndrome.

the two lumbricals for the index and middle fingers are innervated by the median. Paralysis of the muscles of the thenar eminence supplied by the median nerve leads to weakness of thumb opposition and abduction; this latter movement must be tested in a plane at right angles to the palm. In established lesions there is wasting of the paralysed muscles, especially conspicuous in the outer half of the thenar eminence, rendering the first metacarpal bone unduly prominent (Fig. 22.16). Superficial sensory loss following a median nerve lesion above the wrist is somewhat variable especially in regard to the appreciation of pinprick (Fig. 22.17).

The Pronator Syndrome. This presumed entrapment syndrome affecting the median nerve in the vicinity of pronator teres is held to be caused by hypertrophy of that muscle or by a variety of tendonous bands within pronator teres or the nearby origin of flexor digitorum superficialis. The characteristic symptom is aching and tenderness in the upper flexor forearm exacerbated by repetitive arm usage, especially in pronation. Sometimes there are paraesthesiae in the median nerve distribution but clear-cut weakness or wasting seems rare. Although electrophysiological studies are usually normal, they sometimes show denervation or loss of sensory nerve action potentials. The indications for, and results of, surgical exploration with a view to decompression are unclear (Dawson *et al.* 1999).

Median nerve compression at the wrist: Carpal tunnel syndrome. Compression of the median nerve in the carpal tunnel is common. Usually it is an idiopathic phenomenon spontaneously occurring generally in middle-aged women. It is often bilateral, although more symptomatic in the dominant hand. Often symptoms develop with repetitive hand movements. Usually the causes involve a reduction in the size of the carpal tunnel (Dawson *et al.* 1999). It occurs after fractures and arthritis involving the wrist joint, in tenosynovitis, in rheumatoid arthritis and gout, in association with ganglia, and in pyogenic infections of the hand. Endocrine disorders, such as myxoedema, acromegaly or diabetes mellitus, and pregnancy predispose to the condition. It may result from infiltrations with sarcoid or myeloma, or from amyloid deposition in and around the ligament in primary amyloidosis with light chain

deposition, familial amyloidosis with transthyretin deposition, and possibly in renal dialysis patients due to β-2 microglobulin deposition or arteriovenous fistulae for haemodialysis.

Pain and tingling occur in the hand and fingers typically awakening the patient at night or occurring on bunching up the hand for tasks such as writing. Many patients with carpal tunnel syndrome seem to think that tingling affects all their fingers, including the little finger; presumably this is because it can be so difficult to localize tingling and pain accurately. Carpal tunnel syndrome is the commonest cause of acroparaesthesiae. Often pain and paraesthesiae may be the only symptoms for many months or years. If cutaneous sensory loss develops on the digits it causes difficulty in handling small objects. Neurological examination is often completely normal. The commonest abnormal sign is some blunting of pinprick sensation on the tip of the index compared to the little finger. Less often abductor pollicis brevis is weak, or occasionally wasted. It should be noted that the area of sensory loss is limited to the distal palmar surface of the fingers (Fig. 22.18), which are innervated by the deep palmar branch of the median nerve, rather than its superficial palmar branch which arcs over the transverse carpal ligament. Tinel's sign of evoking tingling by percussing over the carpal tunnel, and Phalen's test of inducing paraesthesiae by flexing the wrist, are both unreliable.

Nerve conduction studies usually confirm the diagnosis reliably and should be performed whenever surgical release in envisaged. A prolonged distal motor latency on stimulating the nerve at the wrist and recording the muscle action potential from abductor pollicis brevis is diagnostic in moderate and severe carpal tunnel syndrome. The forearm motor conduction velocity will be normal. If a Martin–Gruber anastomosis between the median and ulnar nerves is present, the motor latency to abductor pollicis brevis will be preserved following proximal stimulation despite being prolonged after stimulation at the wrist. In mild cases, carpal tunnel syndrome may be suggested by reduction in the amplitude of sensory nerve action potentials when median nerve branches are stimulated in the fingers or palm, compared to the normality of sensory conduction in the ulnar nerve. Maximum sensitivity of

92 per cent comes from taking both distal motor latency to the second lumbrical muscle and palm-wrist sensory conduction abnormalities into account (Chang *et al.* 2002). MR imaging of the carpal tunnel produces nearly as high a sensitivity for diagnosing carpal tunnel syndrome, particularly if based on detecting median nerve signal change, but is much less specific (Jarvik *et al.* 2002).

Mild carpal tunnel symptoms can be treated conservatively, often with wrist splinting. If associated with pregnancy it will resolve after delivery and if with hypothyroidism, after endocrine correction. Indeed, prospective follow-up of untreated carpal tunnel syndrome shows that spontaneous resolution is most likely in hands with a short duration of symptoms, in younger people (Padua *et al.* 2001). Symptoms which constitute a considerable nuisance or disability by interfering with hand usage or sleep can be treated either by steroid injection of the carpal tunnel, or by surgical decompression. One injection of 20– 60 mg methylprednisolone into the carpal tunnel results in prolonged improvement in about half (Dammers *et al.* 2006). Surgical decompression is the only reliable long-term cure and is mandatory if conservative measures have failed or there is evidence of denervation. Many use surgery as primary treatment for carpal tunnel syndrome using a variety of approaches and techniques, including open and endoscopic surgery, under local anaesthesia (Dawson *et al.* 1999; Atroshi *et al.* 2006). Prospective comparison of surgical release and steroid injection shows better symptomatic and neurophysiological outcome at 20 weeks following surgery (Hui *et al.* 2005).

29.9.2 The anterior interosseous nerve

This nerve is purely motor, branching from the main trunk of the median nerve just after its emergence from pronator teres in the upper forearm. It supplies flexor pollicis longus (C7, **C8**), flexor digitorum profundus to the index and middle fingers (C7, **C8**) and pronator quadratus (C7, C8) (Fig. 22.14). Isolated lesions of this nerve, developing apparently spontaneously, are quite common, usually causing paralysis of flexion of the terminal phalanges of the thumb and index finger, which can cause disabling loss of pincer grip. It may be damaged by fractures and stab wounds around the elbow. Compression by fibrous bands and anomalous muscles occurs much as for the pronator syndrome affecting the median nerve. It is recognized increasingly that spontaneous isolated anterior interosseous nerve lesions, preceded by acute pain in the upper arm or forearm, probably represent a localized variant of acute brachial neuritis (Goulding and Schady 1993); it is well-recognized as a nerve to be particularly involved in more extensive forms of acute brachial neuritis (Section 22.5.4). Generally such patients start to improve by 10 months; virtually complete recovery takes up to 24 months if indeed it occurs. This raises doubts about the surgical practice of exploring the nerve for constricting fibrous bands if no recovery is evident by 3 months. Sometimes no cause can be established for acute and painless anterior interosseous nerve palsies. The diagnosis is suggested electrophysiologically by prolonged latency of motor conduction from the elbow to pronator quadratus. The median sensory nerve action potential is unaffected, in contrast to lesions involving the median nerve itself.

22.9.3 The ulnar nerve

Anatomy (Fig. 22.19). The ulnar nerve is derived from the C7, C8, and T1 spinal roots. It gives off no branches above the elbow, where it lies behind the medial epicondyle of the humerus before entering

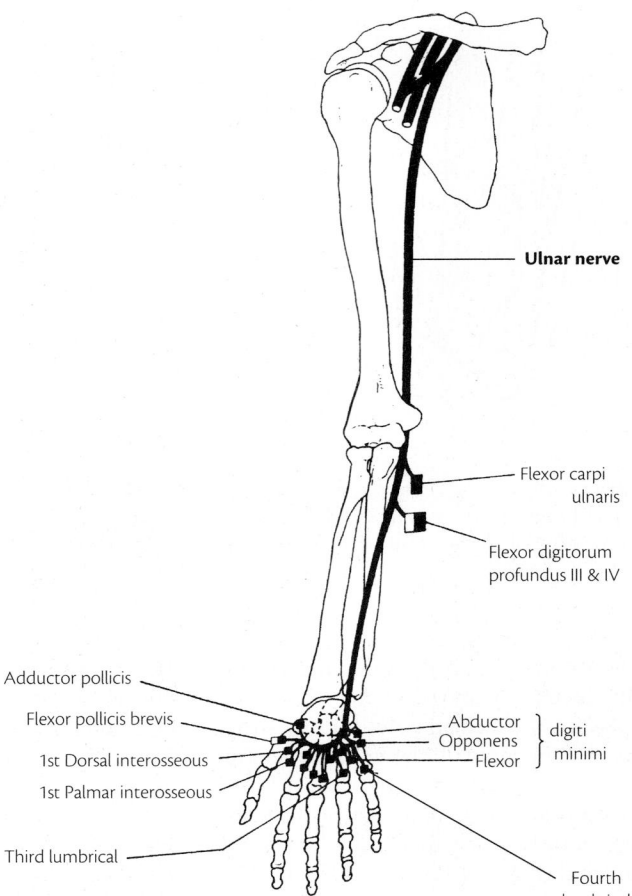

Fig. 22.19 Diagram of the ulnar nerve and its muscular innervation. Note that the innervation of flexor digitorum profundus and flexor pollicis brevis is shared with the anterior interosseous and median nerves respectively.

the cubital tunnel. In the forearm it supplies branches to flexor carpi ulnaris (C7, **C8**, T1) and the portion of flexor digitorum profundus (C7, **C8**) flexing the distal phalanx of the ring and little fingers. Although the nerves to these two muscles usually arise distal to the elbow, the anatomy varies enough to confound clinical localization; sometimes power in these muscles remains normal despite a proven ulnar nerve lesion at the elbow. The dorsal ulnar sensory branch, supplying the back of the hand, is given off a few centimetres above the wrist. In the hand the ulnar nerve usually supplies palmaris brevis, the muscles of the hypothenar eminence, the two medial lumbricals, the palmar and dorsal interossei, the transverse and oblique heads of the adductor pollicis, and the medial head of the flexor pollicis brevis, all of which predominantly receive a T1 root supply with a lesser C8 contribution. The first dorsal interosseous muscle is sometimes supplied by the median nerve. The ulnar nerve enters the hand through Guyon's canal, composed of ligaments between the hamate hook and pisiform bones, where it splits into a superficial terminal sensory branch and a deep motor branch.

Lesions above the elbow. These cause paralysis of all the muscles supplied by the ulnar nerve. Lesions of the ulnar nerve above the elbow are rare, but may be due to direct trauma or compression as for the median nerve (Section 22.9.1). Paralysis of flexor carpi

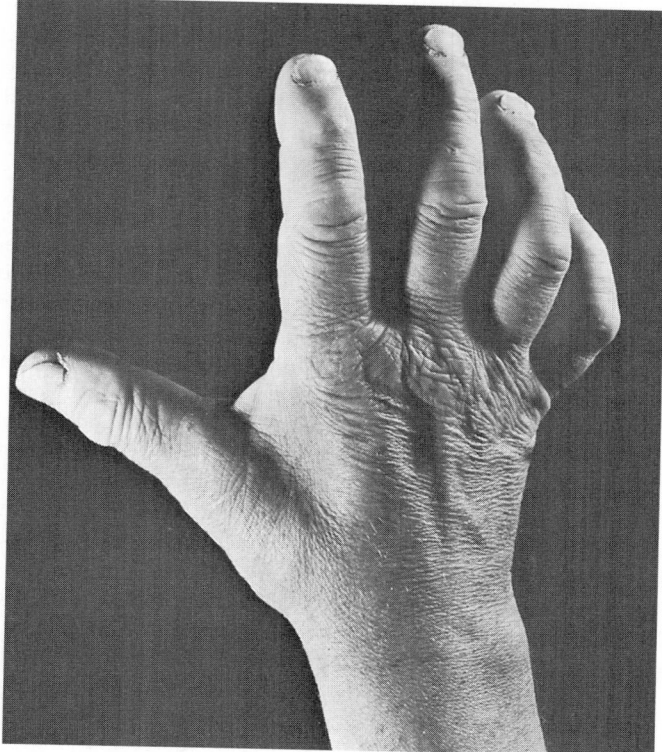

A

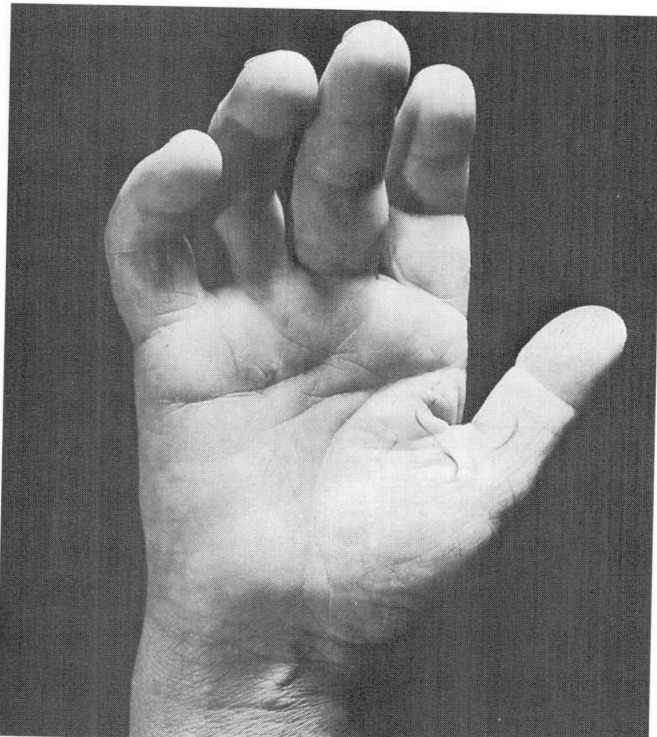

B

Fig. 22.20 An ulnar nerve palsy showing wasting of the first dorsal interosseous muscle, with the typical claw hand appearance due to hyperextension at the metacarpophalangeal joints of the little and ring fingers associated with inability to extend the interphalangeal joints.

ulnaris causes the hand to deviate to the radial side on flexion of the wrist against resistance. Paralysis of the ulnar half of the flexor digitorum profundus abolishes flexion of the little and ring fingers at the distal interphalangeal joints. Paralysis of the hypothenar eminence abolishes abduction of the little finger, and impairs flexion of this finger at the metacarpophalangeal joint. Paralysis of the interossei abolishes abduction and adduction of the fingers. When the interossei and lumbricals are paralysed, the fingers cannot be held with the metacarpophalangeal joints flexed and the interphalangeal joints extended. Paralysis of the transverse and oblique heads of the adductor pollicis weakens adduction of the thumb, most evident when the patient attempts to press the thumb flatly against the index finger. Wasting of the paralysed muscles is evident on the ulnar side of the front of the forearm, in the hypothenar eminence, the interosseous spaces, and the ulnar half of the thenar eminence. Paralysis of the small muscles of the hand causes 'claw-hand' (Fig. 22.20). This posture is produced by the unopposed action of the long finger extensor muscles in the presence of weakness of the lumbricals of the ring and little ringers.

After a lesion of the ulnar nerve at or above the elbow, the area of altered pinprick sensation varies, but usually covers the little finger, the ulnar border of the palm, and often the ulnar half of the ring finger (Fig. 22.21).

Lesions at the elbow. At the elbow the ulnar nerve may be affected by fractures and dislocations involving the lower end of the humerus and the elbow joint. Such injury to the nerve is usually immediate. Occasionally it becomes involved years later if such an injury has led to cubitus valgus, a 'tardy ulnar palsy'. Similarly the nerve may be damaged in the condylar groove by osteophytic outgrowths from arthritis of the elbow-joint, by a ganglion, or by a Charcot elbow joint. If the nerve dislocates from the condylar groove when the elbow flexes it may be damaged during anaesthesia or prolonged recumbency. The commonest site of chronic compression lies within the cubital tunnel just distal to the condylar groove, where the nerve is constricted beneath the aponeurotic origin of flexor carpi ulnaris (Fig. 22.22). The earliest symptoms are pain and paraesthesiae referred to the cutaneous distribution

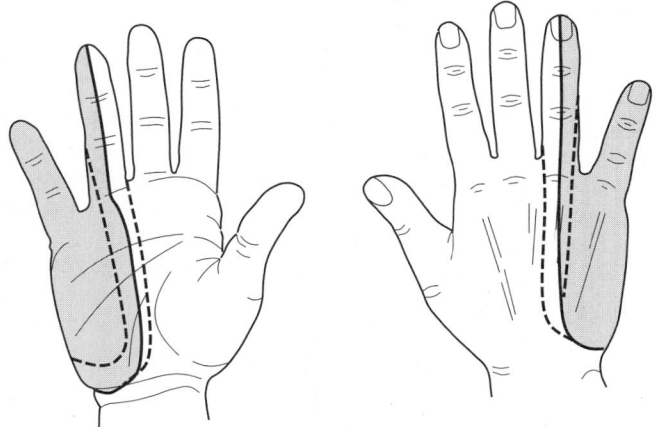

Fig. 22.21 The area of superficial sensory loss after an ulnar nerve lesion above the wrist. This territory includes the dorsal ulnar sensory branch, which supplies the back of the hand and fingers, and arises up to 10 cm proximal to the wrist; this territory is spared by a lesion of the ulnar nerve below the wrist (——— usual boundary ------- variations).

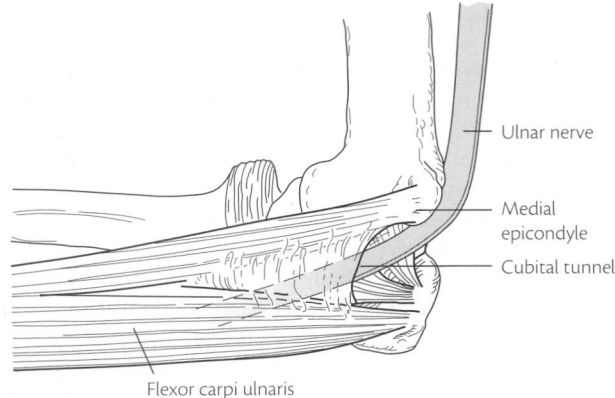

Fig. 22.22 Ulnar nerve compression at the elbow. Medial view of the nerve's passage through the condylar groove behind the medial epicondyle, where it can dislocate from the groove, or be compressed by bony abnormalities. The usual site of compression occurs in the cubital tunnel under the aponeurotic origin of flexor carpi ulnaris. (After Kincaid 1988.)

of the nerve, often first apparent only when the patient awakens in the morning after sleeping with the elbow flexed. Motor involvement is variable and often not symptomatic. It is surprising how rarely patients notice wasting of the first dorsal interosseous muscle, even if quite advanced.

Neither nerve conduction studies nor clinical examination can distinguish reliably between condylar groove and cubital tunnel compression. Electrophysiology merely shows motor conduction block or axonal degeneration localized to the segment of the ulnar nerve at the elbow and associated with a diminished sensory nerve action potential (Kincaid 1988). High resolution ultrasound may help in diagnosis and localization by showing nerve thickening (Beekman *et al.* 2004a). Management of compressive ulnar nerve lesions at the elbow has been confused (Collier and Burge 2001). The following approach is recommended (Collier and Burge 2001; Dawson et al. 1999). In general, a 3-month trial of conservative measures is advised so that surgery can be avoided if spontaneous recovery starts after neurapraxic lesions. Conservative therapy consists of avoiding resting on the elbows and padding the elbows. If the patient has a bony deformity of the condylar canal, and the neuropathy does not respond to conservative measures, it should be transposed anteriorly. If the nerve recurrently dislocates from the condylar groove, and the neuropathy fails to respond to elbow padding and avoiding leaning on the elbows, medial epicondylectomy should be considered. In the most common situation of presumed idiopathic compression in the cubital tunnel, the simplest operation is to decompress the aponeurotic roof of the tunnel, which minimizes the risk of operative nerve damage (Arle and Zager 2000). Medial epicondylectomy or anterior transposition of the nerve can be considered later if the neuropathy continues to progress, or at the time of surgical inspection if other abnormalities are found unexpectedly. Surgical treatment improves outcome compared to conservative management (Beekman *et al.* 2004b).

Lesions at the wrist. Damage at the wrist proximal to Guyon's canal produces paralysis which is confined to those small muscles of the hand supplied by the nerve. Flexor carpi ulnaris and the ulnar half of flexor digitorum profundus escape. Sensory loss is as shown in Fig. 22.21 except that the territory may be spared on the

dorsum of the hand and fingers which is supplied by the dorsal ulnar sensory branch. Lesions in and around Guyon's canal produce a bewildering array of sensory and motor deficits depending upon the combinations of involvement of the superficial terminal sensory branch and the deep palmar motor branch before and after it subdivides to the hypothenar and thenar muscles. When only the deep palmar branch is involved well distal to Guyon's canal, sensory loss does not occur and the hypothenar muscles escape, weakness being most marked in the first dorsal interosseous muscle.

At the wrist the ulnar nerve may be injured by cuts, and the median nerve may be simultaneously involved. The entire palmar branch of the ulnar nerve, including its superficial sensory component, is occasionally compressed or injured on the anterior aspect of the wrist in Guyon's canal. A pressure neuropathy of the deep palmar branch of the ulnar nerve sometimes occurs in individuals whose occupation or recreation involves prolonged or recurrent pressure upon the outer part of the palm, for instance in cyclists (Capitani and Beer 2002). Rarely benign tumours or ganglia may compress the main trunk or the deep branch. Measurement of the sensory nerve action potential from the little finger, and comparison of the distal motor latencies to abductor digiti minimi and the first dorsal interosseous, and incorporating mid-palmar stimulation sites, are helpful in pin-pointing the precise level of ulnar

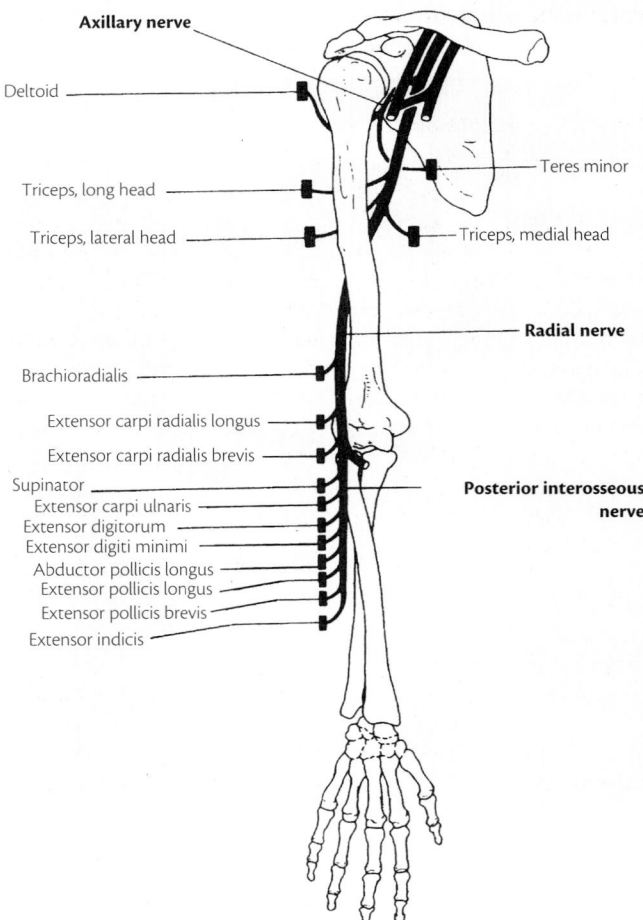

Fig. 22.23 Diagram of the radial, posterior interosseous, and axillary nerves showing the muscles they innervate.

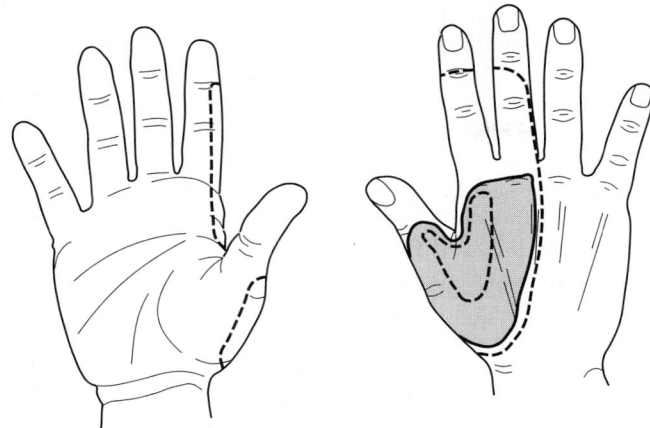

Fig. 22.24 The area of superficial sensory loss in a radial nerve lesion below the origin of the posterior cutaneous nerve of the forearm in the axilla. This skin territory corresponds to the area innervated by the superficial radial nerve (———— usual boundary -------- variations).

nerve lesions to the lower forearm, wrist, or hand (Cowdery *et al.* 2002). The choice of treatment for ulnar nerve lesions in the wrist and hand depends upon the cause. For instance, lesions of the deep palmar branch that are due to repetitive palm trauma require a change in the use of the hand. Small ganglia or tumours may prove visible on magnetic resonance imaging of the palm.

22.9.4 The radial nerve

Anatomy (Fig. 22.23). The radial nerve continues from the posterior cord of the brachial plexus and is derived from the C5, C6, C7, C8 spinal nerves. It innervates the following muscles in the upper arm where it winds posteriorly in the spiral groove of the humerus deep to the triceps muscle: triceps (C6, C7, C8), brachioradialis (C5, C6), extensor carpi radialis longus (C5, C6). In the lateral aspect of the upper forearm it divides into the superficial radial nerve, which is purely sensory, and the posterior interosseous nerve, which is purely motor. The posterior interosseous nerve innervates supinator (C6, C7), extensor digitorum (**C7**, C8), extensor carpi ularis (**C7**, C8), and the three extensors of the thumb (C7, C8).

The radial nerve above the origin of the posterior cutaneous nerve of the forearm in the axilla carries sensation from the lower half of the radial aspect of the arm, from the middle of the posterior aspect of the forearm, and also from a variable area on the dorsum of the hand. This area on the hand corresponds to the territory supplied by the superficial radial nerve (Fig. 22.24).

Lesions. Compression of the radial nerve in or above the axilla by crutches or during stupor causes paralysis and wasting of all the muscles it supplies. There is wrist and finger drop and the triceps weakness localizes the high level of this lesion. The triceps and brachioradialis reflexes are lost and the sensory loss involves both posterior cutaneous nerve of the forearm and superficial radial nerve territories.

The nerve is most vulnerable to damage in the upper arm in the spiral groove of the humerus (Mondelli *et al.* 2005). The triceps power and reflex will be preserved and the main feature is wrist and finger drop accompanied by loss of the brachioradialis reflex. Such

lesions are most familiar as a 'Saturday night paralysis' in which the nerve is compressed when the patient, often intoxicated, falls asleep with the arm extended over the arm of a chair. Following a pressure palsy of the radial nerve, sensory loss is variable and may be absent but is found usually on the dorsum of the hand between the thumb and index finger. Less frequently, radial nerve lesions of the upper arm follow blunt trauma, misplaced deep intramuscular injections, fractures of the humerus, and careless positioning of the arm during general anaesthesia.

Radial nerve conduction studies may be helpful in determining the site of the lesion (Mondelli *et al.* 2005). Following compressive lesions, recovery can be anticipated in 9–12 weeks if nerve conduction studies show a neurapraxic lesion with preservation of the radial sensory action potential, no denervation changes on sampling the brachioradialis or the forearm extensors, and evidence of focal conduction block on stimulating the radial nerve in the upper arm (Shyu *et al.* 1993). A light cockup splint may be used to maintain extension of the wrist, thereby allowing the hand to be used while awaiting recovery. The prognosis of lesions of the radial nerve is often good, as most are simple neurapraxias. Even after complete division and suture, signs of returning muscular function are usually evident within 8 months, according to the level of the lesion.

The purely sensory superficial radial nerve may be compressed during an aberrant route through the forearm extensor musculature, or by radius fractures, tight bindings on the wrists, or ruptured synovial effusions from the elbow joint. The nerve lies superficially, especially in the distal forearm, and often tenderness or a Tinel sign can be evoked at the site of the lesion. The area of sensory loss is shown above (Fig. 22.24). The sensory nerve action potential is absent or small (Mondelli *et al.* 2005).

22.9.5 The posterior interosseous nerve

This purely motor nerve arises from the radial nerve at the elbow and passes through the supinator muscle in the arcade of Frohse. Lesions result in weakness of supinator (C6, C7), extensor digitorum (**C7**, C8), extensor carpi ulnaris (**C7**, C8), and the three extensors of the thumb (C7, C8) (Fig. 22.23). This results in wrist and finger drop without sensory loss. The terminal motor branches are separate, and long, so that individual extensor muscles may be paralysed differentially or in isolation (Ay *et al.* 2005). Differential finger drop is a common early symptom of multifocal motor neuropathy (Section 21.11.3). The posterior interosseous nerve may be compressed by various masses such as lipomas, or by abnormalities arising from the elbow joint, or by fibrous anomalies of the arcade of Frohse as it passes through the supinator muscle adjacent to the lateral epicondyle. Fractures and dislocations of the head of the radius are common causes. Surgical exploration should be considered if there is no sign of recovery after 9–12 weeks; occasionally fibrous compressive bands will be found. Multifocal motor neuropathy should be considered in patients with spontaneously occurring, painless, posterior interosseous nerve palsy.

Repetitive pronation-supination of the forearm, such as using a screwdriver, may be followed by local aching which is difficult to differentiate from lateral epicondylitis. Some patients with such 'resistant tennis elbow' may be suffering from compression of the posterior interosseous nerve but the very existence of such a condition is unclear.

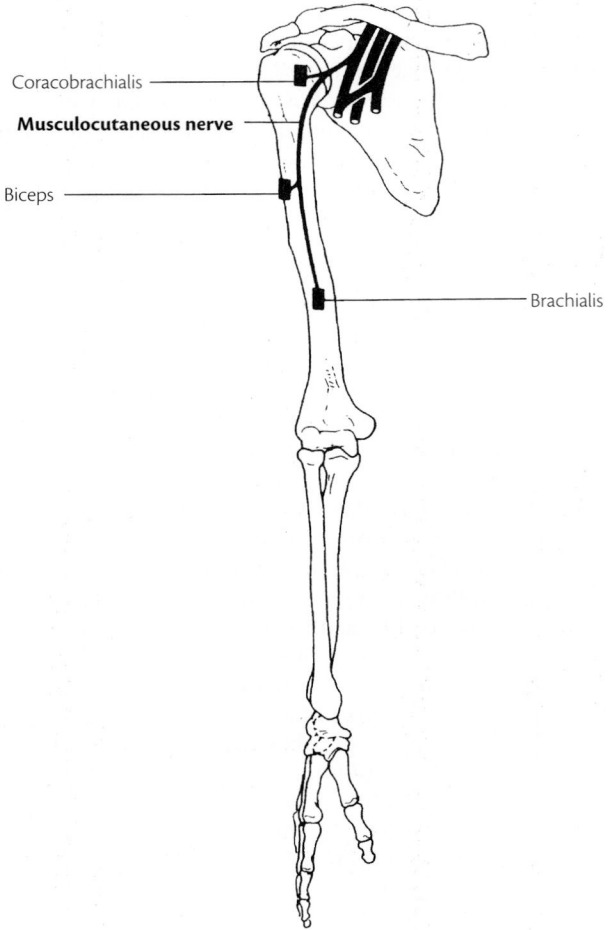

Coracobrachialis —

Musculocutaneous nerve —

Biceps —

— Brachialis

Fig. 22.25 Diagram of the musculoskeletal nerve showing the muscles it innervates.

22.9.6 The musculocutaneous nerve

The musculocutaneous nerve is a branch of the lateral cord of the brachial plexus, its fibres being derived from the C5 and C6 spinal nerves (Fig. 22.25). It supplies the biceps and part of the brachialis, the two principal flexors of the elbow. Its sensory distribution is to the radial border of the forearm as low as the carpometacarpal joint of the thumb, a territory also known as the lateral cutaneous nerve of the forearm (Fig. 22.1). The musculocutaneous nerve may be absent in less than 10 per cent of people, with its function taken over by branches of the median nerve (Prasada Rao and Chaudhary 2001).

Division of the musculocutaneous nerve causes weakness of flexion of the elbow-joint, though some flexion can still be carried out by brachioradialis and that part of the brachialis which is innervated by the radial nerve. Sensation is impaired over the radial forearm. The musculocutaneous nerve is rarely injured alone, but may be damaged by upper brachial plexus injury, dislocation of the head of the humerus, by penetrating wounds, by clavicle fractures, or traction injuries. Temporary paralysis of the biceps occasionally occurs after compression anteriorly on the shoulder. The differential diagnosis is from a C6 root lesion, in which case the power and tendon reflex of the brachioradialis would be lost too and the sensory disturbance would extend below the wrist to the thumb and index finger.

22.9.7 The circumflex axillary nerve

The axillary nerve arises from the C5 and C6 spinal nerves and the posterior cord of the brachial plexus (Fig. 22.23). It innervates the deltoid muscle (**C5**, C6) and its cutaneous branch supplies sensation to an area extending from the acromion process to as far as half way down the outer aspect of the upper arm (Fig. 22.1). Injury to the axillary nerve causes wasting and weakness of the deltoid with paralysis of abduction at the shoulder and numbness in the sensory territory (Fig. 22.26). In clinical practice the actual area of sensory loss is often no more than a small area near the insertion of the deltoid. The axillary nerve may be involved by injuries in the region of the neck of the humerus including dislocation of the shoulder joint, fractures of the upper humerus, and deep intramuscular injections into the deltoid. It may be injured by attempts to reposition a dislocated shoulder, and checking the sensory territory is advised beforehand. It is the nerve most often involved in acute brachial neuritis (Section 22.5.4), in which case there is usually severe and persistent pain in the shoulder region for several hours or even a few weeks before the paralysis is noted. Other non-traumatic causes are rare, but it does occur in volleyball players (Paladini *et al.* 1996). In cases due to trauma or

A

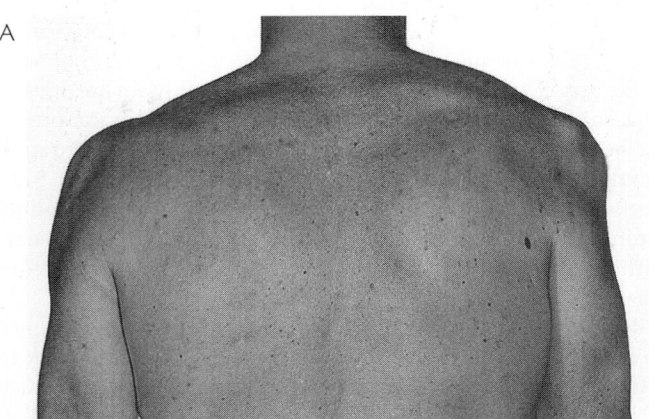

B

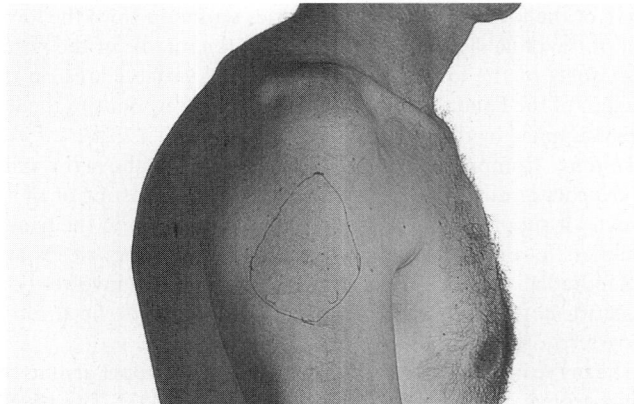

Fig. 22.26 An axillary (circumflex) nerve lesion, showing A. wasting of the right deltoid muscle and B. the area of sensory loss on the upper outer arm.

traction, operative inspection of the nerve should be considered if there is no evidence of recovery within a few months (Kline and Kim 2003).

22.9.8 The suprascapular nerve

This nerve is derived from the upper trunk of the brachial plexus and supplies the infraspinatus (**C5**, C6) and supraspinatus (**C5**, C6) muscles which respectively externally rotate, and initiate abduction at, the shoulder. There is no sensory component and weakness, if noticed at all, presents with inability to initiate shoulder abduction when the arm is hanging at the side, or with difficulty in externally rotating the arm whilst attempting to write across the page. Isolated traumatic lesions of this nerve are rare, though it is occasionally damaged as a sequel of scapular fracture. The nerve may suffer entrapment by ganglion cysts or a hypertrophied inferior transverse scapular ligament as it passes through the suprascapular foramen in sportsmen such as fencers. This may be an unrecognized source of shoulder pain following injury. Operative inspection and compression should be considered in traumatic or suspected compression which fails to recover spontaneously (Kim *et al.* 2005). The spinatus muscles are commonly involved in acute brachial neuritis, often in conjunction with deltoid (Section 22.5.4).

22.9.9 The long thoracic nerve

The long thoracic nerve is derived from the C5, C6, and C7 spinal roots before the formation of the brachial plexus. It supplies the serratus anterior muscle and has no sensory territory. Serratus anterior fixes the scapula to the chest wall when the arm is pushed forward in front of the body such as when doing press-ups. Weakness of the muscle causes characteristic winging of the scapula during such movements (Fig. 22.27). Although this weakness may be noted on shoulder movements, equally often the shoulder blade protrusion is first noticed in the bathroom mirror or by others on the beach, or when its protrusion interferes with squeezing through tight spaces. The long thoracic nerve is injured alone most frequently as a result of direct pressure upon the shoulder during carrying, strenuous shoulder exercise, from blows to the shoulder, and during surgical procedures in the axilla. Commonly it is involved in acute brachial neuritis in which case other muscles are usually involved too (Section 22.5.4). It is occasionally involved

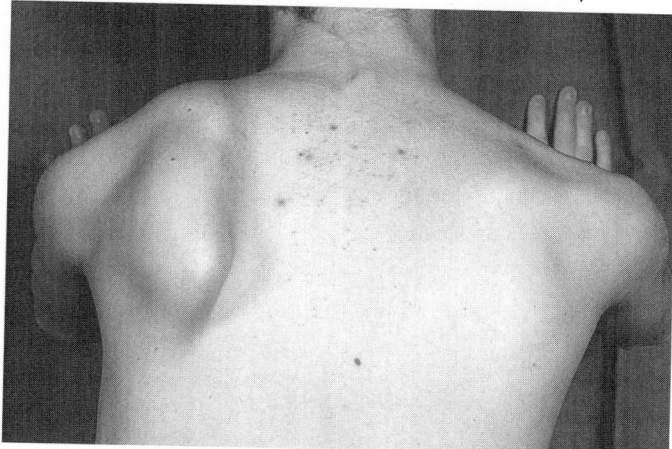

Fig. 22.27 A long thoracic nerve lesion causing winging of the left scapula when the outstretched arm is pushed forwards against the wall.

in inflammation secondary to apical pleurisy. Autosomal dominantly inherited familial long thoracic nerve palsy is probably a form of hereditary neuralgic amyotrophy (Section 25.5.5) (Phillips 1986). Conduction velocity can be measured along the long thoracic nerve and assists in assessing possible lesions. Idiopathic palsies may recover well but traumatic lesions rarely do so (Friedenberg *et al.* 2002). Then the question arises as to whether to undertake orthopaedic surgical fixation of the lower angle of the scapula to the ribcage. However, fixation reduces the overall range of shoulder movements and is rarely permanently effective. Dynamic repairs are preferred, such as transposing the insertion of part of pectoralis major onto the lower scapula.

22.10 Lower limb mononeuropathy

22.10.1 The sciatic nerve

Anatomy (Fig. 22.28). The sciatic nerve is derived from the sacral plexus, which is formed by a fusion of the ventral primary divisions of the L4 and L5 and the S1, S2, and S3 spinal nerves. The nerve is composed of two trunks, a medial, consisting of the ventral divisions of L4, L5, S1, S2, S3, which is destined to form the tibial nerve and a lateral, consisting of the dorsal divisions of L4, L5, S1, S2,

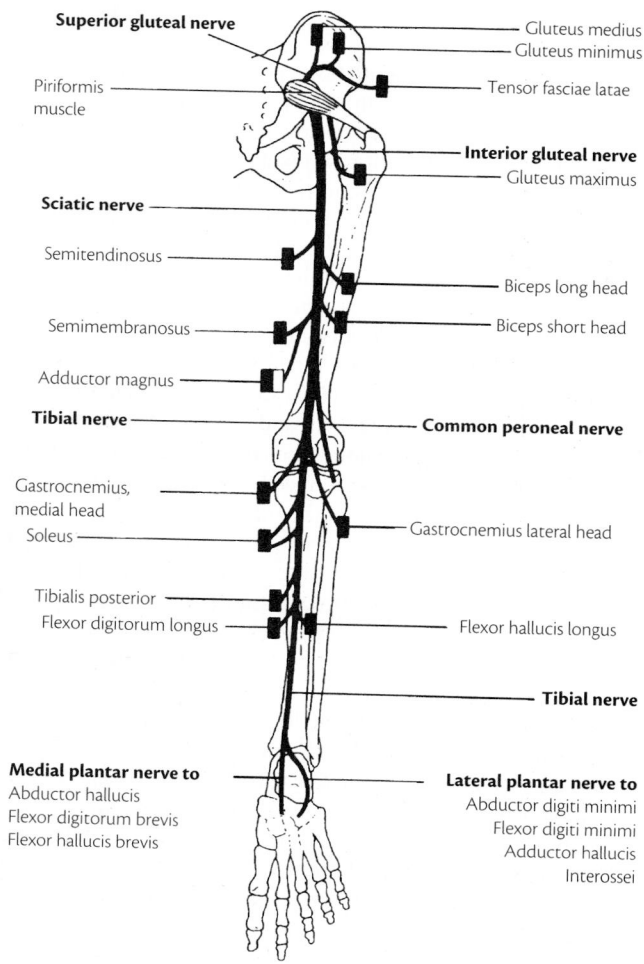

Fig. 22.28 Diagram of the sciatic, tibial, gluteal, and plantar nerves showing the muscles they innervate.

which gives rise to the common peroneal, or lateral popliteal, nerve. These two trunks, although bound together by connective tissue, are distinct structures from the moment of their formation in the sacral plexus. This separation of the tibial and peroneal divisions of the nerve means that buttock trauma, such as misplaced injections, may affect either division separately, more usually the more superficially located common peroneal. The sciatic nerve leaves the pelvis by passing through the great sciatic notch in company with the superior gluteal nerve to gluteus medius and minimus (**L4**, **L5**, S1), inferior gluteal nerve to gluteus maximus (**L5**, **S1**, S2), and the posterior cutaneous nerve of the thigh. It enters the buttock running behind the hip joint where it usually lies deep to the piriformis muscle. However, anatomical variations occur in this relationship to piriformis, ignorance of which can lead to damage during hip joint surgery; the two trunks can pass on separate sides of this muscle, through the muscle, or anterior to it. Then it enters the back of the thigh, lying midway between the great trochanter of the femur and the ischial tuberosity, before descending deep to the hamstring muscles. The sciatic nerve terminates at a variable point between the sciatic notch and the popliteal fossa by separating into the common peroneal and tibial nerves. The sciatic nerve supplies the following muscles in the thigh: semitendinosus, semimembranosus, and biceps (all L5, **S1**, S2), and part of the adductor magnus. Through its terminal branches of the post tibial (Section 22.10.2) and common peroneal nerves (Section 22.10.3) the sciatic nerve supplies all muscles below the knee.

Effects of lesions. After complete interruption of the sciatic nerve there is paralysis of flexion of the knee, which is carried out by the hamstrings, and of all the muscles below the knee. Foot-drop occurs as a result of paralysis of the anterior tibial group of muscles and the peronei. The patient can stand and walk, but drags the toes of the affected foot and is unable to stand on his toes or heel on the paralysed side. The skin sensory territory of the sciatic nerve lies entirely below the knee on the lateral half of the calf and dorsum of the foot, representing the common peroneal territory, and the sole and outer border of the foot, the posterior tibial territory (Fig. 22.1). In sciatic nerve lesions the saphenous nerve territory is spared since this is a branch of the femoral nerve which supplies the medial aspects of the calf and the instep. The ankle-jerk and plantar reflex are lost, but the knee jerk is retained. Vasomotor and trophic changes are usually conspicuous after complete division: oedema, dry skin, loss of foot sweating, and sometimes ulcers on the sole.

Causes of lesions. The sciatic nerve, or one of its trunks, is usually damaged as a result of fractures and dislocations of the hip joint, pelvis, or femur, penetrating wounds of the buttock and thigh, and hip joint replacement surgery (Section 22.6.7) (Yuen *et al.* 1994). Sciatic nerve pressure palsies in the buttock can occur during coma, anaesthesia, enforced recumbency, or meditation. A noteworthy cause of a sciatic nerve lesion is a misplaced injection given too far medially in the buttock rather than in the upper outer quadrant. In such cases, it is usually the common peroneal division of the nerve which is damaged because it lies more superficially. The question of entrapment by the piriformis muscle is debated; this *'pyriformis syndrome'* is said to involve buttock pain radiating down the leg like sciatica, but it is rarely associated with significant clinical or electrophysiological deficits and the validity or frequency of the syndrome remain unclear. The sciatic nerve may be compressed within the pelvis by neoplasms, or by deposits of endometriosis at the sciatic notch (Section 22.6.6). Complete division of the nerve is rare. Radiographs of the pelvis and hip joint are necessary in cases resulting from trauma. CT or MRI will show lesions within the lower pelvis or sciatic notch. Lack of denervation of paraspinal muscles on electromyography will help distinguish sciatic nerve lesions from root compression. Nerve conduction studies are generally unhelpful in localizing the precise site of a proximal lesion (Dawson *et al.* 1999).

Treatment of sciatic and peroneal nerve lesions. Surgical treatment is rarely undertaken although nerve inspection and grafting may improve outcome, particularly for injuries of the tibial division (Kim *et al.* 2004). It is important to prevent contracture of the Achilles tendon and the foot should be splinted in dorsiflexion day and night and the ankle moved through its full range passively. Recovery is always slow after suturing a completely divided nerve. In compressive or traction lesions return of voluntary power is rarely complete, cannot be expected over less than 12–18 months, and usually takes 2–3 years (Yuen *et al.* 1994).

22.10.2 The tibial nerve

Anatomy (Fig. 22.28). The tibial nerve is an end branch of the sciatic nerve, from which it separates at any point between the sciatic notch and the popliteal fossa. Then it travels deep to the gastrocnemius muscle and enters the foot via the 'tarsal tunnel' roofed by the flexor retinaculum on the medial aspect of the ankle (Fig. 22.29). It then divides into the calcaneal sensory branches and medial and lateral plantar nerves which supply sensation to the sole of the foot and innervate the intrinsic foot muscles. The tibial nerve supplies gastrocnemius and soleus (S1, S2), tibialis posterior (L4, L5), flexor digitorum longus (L5, **S1**, S2), and flexor hallucis longus (L5, **S1**, S2). It gives rise to the medial and lateral plantar nerves supplying the small muscles of the foot (S1, S2).

Effects of lesions. After division of the tibial nerve above gastrocnemius the calf and sole muscles are paralysed and wasted and the foot assumes a position of talipes calcaneovalgus. The ankle-jerk is lost, and the plantar reflex may also be inelicitable. Skin sensation is lost over the sole, including the plantar aspect of the toes and the dorsal aspect of their terminal phalanges (Fig. 22.1).

Causes of lesions. The tibial nerve can be subject to trauma or ischaemia, compressed in the popliteal fossa by knee joint cysts, arterial aneurisms, and nerve sheath tumours or trapped by the tendonous origin of soleus (Mastaglia 2000; Drees *et al.* 2002).

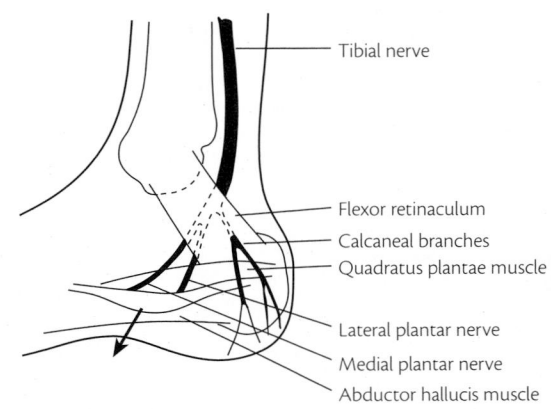

Fig. 22.29 The tarsal tunnel through which the tibial nerve enters the foot. (From Stewart 1993.)

Rarely the posterior tibial nerve may be compressed at the ankle in the tarsal tunnel (Fig. 22.29). This gives rise to burning pain in the sole of the foot and toes and paraesthesiae or sensory loss over almost the entire sole of the foot, termed '*posterior' tarsal tunnel syndrome*. This syndrome is sometimes a true entrapment by tendon analogous to carpal tunnel syndrome. More often it seems to result from tight shoes or plaster casts, rheumatoid arthritis, swellings or tenosynositis in the tarsal tunnel, and has been described in hypothyroidism and acromegaly (Stewart 1993). Comparison of motor latency and of sensory nerve action potentials in the medial and lateral plantar nerves both help in diagnosis, but are somewhat unreliable (Dawson *et al.* 1999; Patel *et al.* 2005).

Occasionally interdigital nerves may be compressed by the adjacent metatarsal heads as they enter the medial aspect of the sole of the foot. This produces pain locally with numbness and tingling in adjacent toes, worsened by walking or squeezing the forefoot. It usually occurs in the third interspace and is sometimes associated with fusiform neuroma formation, a so-called *Morton's neuralgia*. Surgical excision of this neuroma may cure the pain (Siu and Chandran 2005).

22.10.3 The common peroneal nerve

Anatomy (Fig. 22.30). The common peroneal, or lateral popliteal, nerve is an end branch of the sciatic nerve from which it separates anywhere between the sciatic notch and the popliteal fossa. It should not be forgotten that pure common peroneal nerve palsies can result from lesions high in the thigh or buttock. Then it winds round the neck of the fibula bone to enter the anterior tibial compartment by passing through the 'fibular tunnel' in the superficial head of peroneus longus. It divides into the superficial and deep peroneal nerves. The superficial nerve supplies the peroneal muscles (L5, S1) and then supplies sensation to the skin of the lower lateral calf and dorsum of the foot (Fig. 22.1). The deep peroneal nerve runs deep in the anterior tibial compartment, supplying tibialis anterior (**L4**, L5), extensor digitorum longus (**L5**, S1), extensor hallucis longus (**L5**, S1), and extensor digitorum brevis (**L5**, S1), and its terminal branch supplies the skin between the first and second toes and the adjacent dorsum of the foot (Fig. 22.1).

Effects of lesions. After division of the common peroneal nerve there is paralysis of dorsiflexion of the foot and toes and of eversion of the foot; foot-drop results. Inversion is lost when the foot is dorsiflexed, but weak inversion is possible in plantar-flexion. When the nerve is divided above the point of origin of the superficial peroneal nerve, sensation is impaired over the dorsum of the foot, including the first two toes, and over the antero-lateral aspect of the lower half of the calf. When the lesion lies below the origin of the superficial peroneal nerve, sensory loss is restricted to the first two toes and adjacent dorsum of the foot.

Causes of lesions. The common peroneal nerve may be injured by penetrating wounds around the knee joint, or by upper fibular fractures. Selective deep peroneal branch injury can occur with arthroscopic knee surgery. External compression is common, causes including tight bandages or plaster casts applied to the knee, pressure during sleep, and habitual sitting with crossed legs. Emaciated patients are particularly vulnerable, especially those with cancer (Koehler *et al.* 1997). The popliteal fossa should be palpated carefully for bursae, cysts, or tumours; MRI can reveal such lesions compressing the nerve (Iverson 2005) (Fig. 22.31). MRI may reveal peroneal intraneural ganglia potentially amenable to surgery (Spinner *et al.* 2003). Prolonged kneeling or squatting

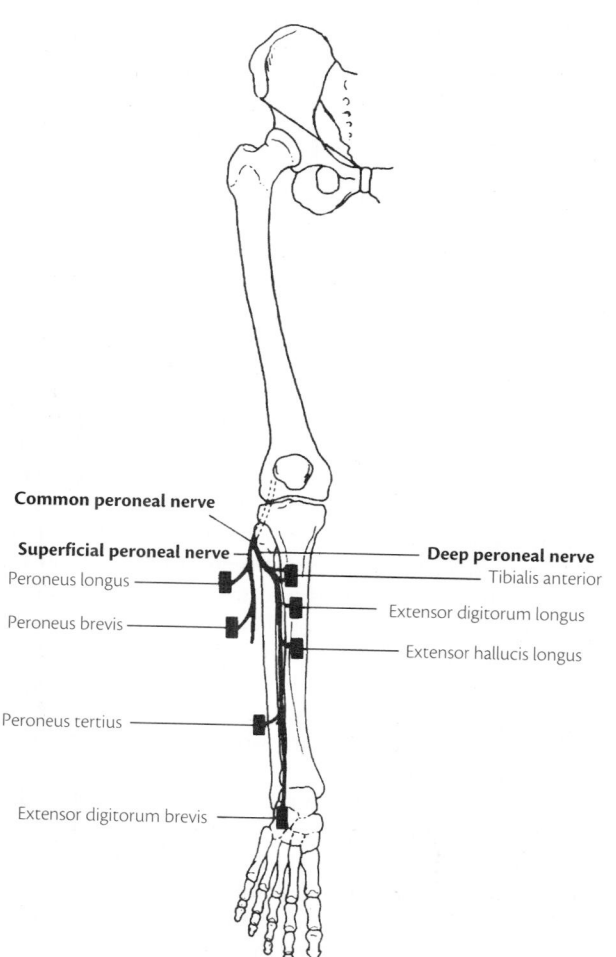

Fig. 22.30 Diagram of the common peroneal nerve showing the muscles it innervates.

Common peroneal nerve

Superficial peroneal nerve — **Deep peroneal nerve**

Peroneus longus — Tibialis anterior

Peroneus brevis — Extensor digitorum longus

— Extensor hallucis longus

Peroneus tertius

Extensor digitorum brevis

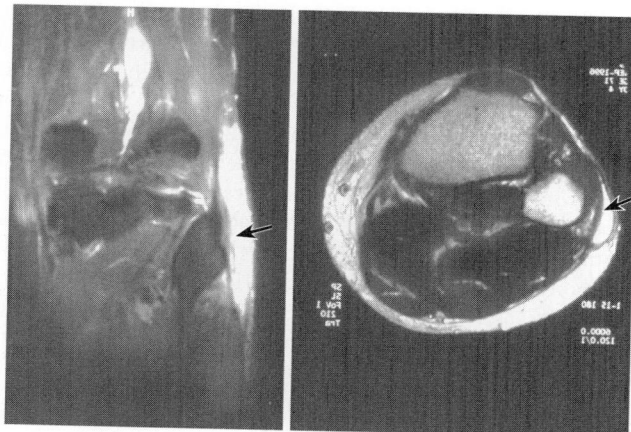

Fig. 22.31 MRI of the knee longitudinally (left) and transversely (right) in a patient with common peroneal nerve palsy caused by a biceps tendon bursa (arrowed).

can compress the nerve, for instance in slaters working on roofs with one leg flexed and lying under the other with its outer aspect against the roof surface. Occasionally the nerve is entrapped by a tight fibrous band in the fibular tunnel, relieved by surgery. In cases of entrapment or compression neuropathy, the muscles which the nerve innervates do not always suffer equally. The peronei are usually less severely affected than the anterior tibial group, and the area of sensory loss is often less than that found after complete division (Sourkes and Stewart 1991). Rarely the deep peroneal nerve is selectively compressed by oedema in the anterior compartment syndrome following trauma, ischaemia, or exercise of the lower leg. Terminal sensory branches of the superficial peroneal nerve may be entrapped where they pierce the fascia above the ankle becoming symptomatic after minor ankle trauma (Styf 1989).

Spontaneously occurring common peroneal nerve palsies of uncertain cause are quite common. They often recover in 2–3 months with avoidance of leg crossing and kneeling. If not, MRI may reveal bursae, cysts, ganglia, or tumours. Sometimes surgical inspection of the nerve at the fibular neck is required in those for whom there was no convincing explanation for the mononeuropathy, and in whom no spontaneous recovery occurs by 3 months. For it is only at operative inspection that entrapment by fibrous band can be identified. The diagnosis of a common peroneal palsy can be confirmed by measurements of nerve conduction velocity along the nerve, which may show conduction block across the head of the fibula, particularly if a more proximal stimulation point is used and recordings made from a range of muscles (Sourkes and Stewart 1991). Slowing of sensory conduction across that segment of the nerve localized the lesion accurately in 64 per cent of a series of 47 patients (Singh *et al.* 1974). Sometimes the amplitude of the compound muscle action potential recorded by surface electrodes over the extensor digitorum brevis muscle during supramaximal stimulation of the nerve may be larger on stimulation at the knee than at the ankle. This is due to an anomalous branch of the superficial peroneal nerve, the accessory deep peroneal nerve, which passes alongside the peroneus brevis muscle and behind the lateral malleolus, and supplies the lateral part of the extensor digitorum brevis (Dessi *et al.* 1992).

22.10.4 The sural nerve

The sural nerve normally arises from the tibial nerve in the upper calf. It attains a superficial route half way down the calf and runs lateral to the Achilles' tendon, where it usually receives a contribution from the common peroneal nerve. It passes behind the lateral malleolus to enter the foot and supply sensation to the outer border of the foot from the heel to the 4th and 5th toes (Fig. 22.1). Lesions cause numbness and paraesthesiae in this area. There is no motor component. The commonest lesion is total or partial transection for the purposes of diagnostic nerve biopsy, and uncomfortable dysaesthesiae may result (Section 21.2). Otherwise sural mononeuropathy is rare, but described following local trauma, surgery, lacerations, and external compression at the ankle (Reisin *et al.* 1994). Lesions can result from compression by a Baker's cyst at the knee.

22.10.5 The femoral nerve

Anatomy (Fig. 22.32). The femoral nerve is derived from the lumbar plexus in the psoas major muscle, which it innervates,

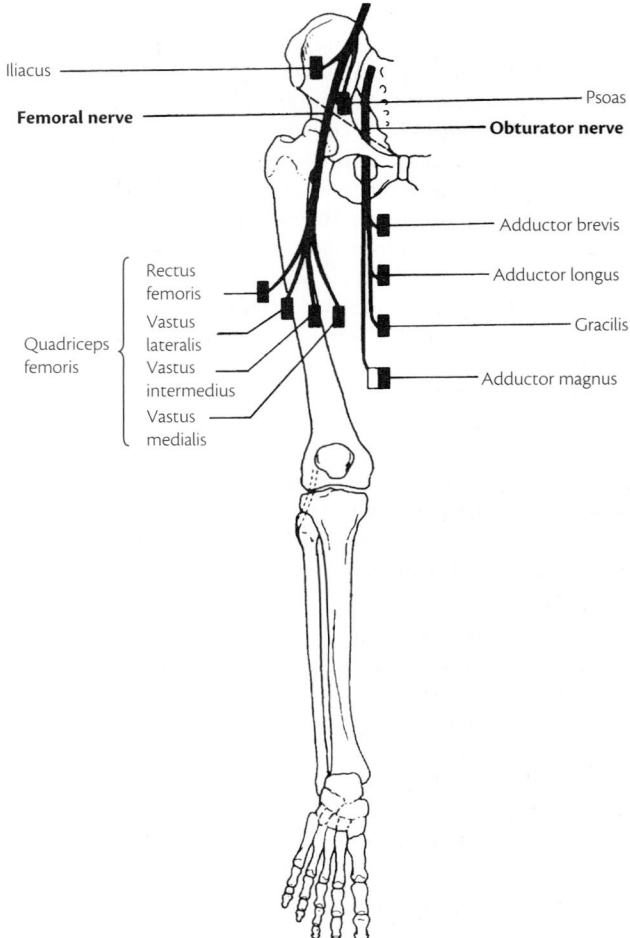

Fig. 22.32 Diagram of the femoral and obturator nerves showing the muscles they supply.

arising from the dorsal parts of the L2, L3, and L4 spinal nerves, posterior to the obturator nerve. After passing through the pelvis where it is vulnerable to compression in the gutter between iliacus and psoas (Section 22.6.7), it enters the femoral triangle of the thigh beneath the inguinal ligament, lateral to the femoral sheath and vessels. In the abdomen it sends a branch to iliacus. In the femoral triangle it divides into terminal branches which include the supply to the quadriceps (L2, **L3**, **L4**). It gives articular branches to the hip and knee joints. The intermediate and medial cutaneous branches supply the anterior and medial aspects of the lower two-thirds of the thigh (Fig. 22.1). The saphenous nerve supplies sensation to the inner aspect of the calf and instep (Fig. 22.1).

Effects of lesions. Proximal lesions of the femoral nerve produce slight weakness of hip flexion due to paralysis of iliacus, but the principal motor disturbance is weakness of knee extension owing to paralysis of quadriceps. In consequence the leg gives way in walking and climbing stairs and arising from sitting is difficult. The knee jerk is lost. Sensation is lost over the inner calf area innervated by the saphenous branch of the nerve. Usually the clinical picture is unmistakable. It is distinguished from a lumbar plexus lesion by the preservation of thigh adductor power supplied by the obturator nerve. Causalgia may occur in the distribution of the saphenous nerve after partial lesions.

Causes of lesions. The intrapelvic femoral nerve may be damaged by pelvic surgery, psoas abscess, pelvic neoplasia, or iliacus haematoma (Section 22.6.2). It may be injured in fractures of the pelvis or of the femur, or by hip dislocation or hip replacement. Lesions of this nerve are rarely seen as a result of penetrating wounds of the thigh, as the proximity of the femoral artery renders most such injuries rapidly fatal. Femoral neuropathy may be induced by the lithotomy position. The commonest lesion is diabetic proximal neuropathy (Section 21.17.3) and idiopathic lumbosacral plexopathy may occur also (Section 22.6.1). Measurement of femoral nerve conduction velocity and terminal latency are valuable in localization by narrowing down muscular involvement to the femoral nerve distribution, but rarely help localize the level of the nerve injury. CT scan or MRI is essential for the diagnosis of intrapelvic femoral nerve lesions.

22.10.6 The saphenous nerve

This long and superficial sensory branch of the femoral nerve is usually damaged by penetrating injuries or surgery, particularly varicose vein ligation, occasionally it is compressed by stirrups at the inner knee during surgery or childbirth. It supplies sensation to the medial calf and instep (Fig. 22.1). An entrapment syndrome at the exit from Hunter's canal, some 10 cm proximal to the medial femoral condyle, has been postulated. However there is generally insufficiently clear evidence of sensory loss to make a definite diagnosis of entrapment mononeuropathy (Stewart 1993).

22.10.7 The lateral cutaneous nerve of the thigh

The lateral cutaneous nerve of the thigh is purely sensory and derived from the dorsal divisions of the L2 and L3 spinal nerves. After passing through psoas major and winding round the lateral wall of the pelvis, it enters the thigh beneath the lateral end of the inguinal ligament. It emerges superficially from the fascia lata of the thigh about 10 cm distal to the anterior superior iliac spine. Then it divides into an anterior and a posterior branch which carry sensation from the lateral and anterolateral aspects of the thigh from the buttock almost down to the knee (Fig. 22.1). Lesions are usually thought to be due to entrapment and kinking where the nerve passes through the inguinal ligament. This is commonest in obese or pregnant patients, who experience uncomfortable paraesthesiae in the lateral thigh particularly on walking, a syndrome known as '*meralgia paraesthetica*'. Numbness is demonstrable in the nerve's territory. Evidence of subclinical entrapment at the inguinal ligament is a common autopsy finding (Jefferson and Eames 1979). Meralgia paraesthetica often resolves spontaneously. When pain is more troublesome, repeated infiltration of local anaesthetic around the lateral half of the inguinal ligament may relieve symptoms. If this is unsuccessful, surgical neurolysis just medial to the anterior superior iliac spine reduces or cures symptoms in most (Siu and Chandran 2005). Occasionally intrapelvic causes underlie isolated lesions of the lateral cutaneous nerve of the thigh; these are similar to causes of intrapelvic femoral nerve lesions and some will be revealed by pelvic MRI. They may present with similar symptoms to lesions at the inguinal ligament, although pain induced by walking is less likely.

22.10.8 The obturator nerve

Anatomy (Fig. 22.32). The obturator nerve is derived from the L2, L3, and L4 spinal nerves by branches anterior to those forming the femoral nerve. The union of these roots occurs in the psoas muscle and the nerve emerges from the pelvis by the obturator foramen to supply the thigh adductors (**L2, L3**, L4). It gives branches to the knee and hip joints. Its skin territory is variable, generally comprising a small area on the upper medial thigh (Fig. 22.1).

Lesions. Injury to the obturator nerve causes paralysis of the thigh adductors except for the flexor fibres of the adductor magnus, which are innervated by the sciatic. Sensory loss is often not detected. Isolated lesions are rare but can occur with hip and pelvic fractures or hip joint surgery, obstructed labour, femoral arterial procedures, abdominal or pelvic surgery, or pelvic cancers (Rogers et al. 1993; Sorenson *et al.* 2002). Occasionally the nerve may be compressed in the obturator canal by an obturator hernia or as a result of osteitis pubis following genito-urinary surgery. If the cause remains undiagnosed, CT or MRI of the pelvis is necessary to detect an occult tumour. In obturator neuropathy, the knee jerk and quadriceps power are normal, thereby excluding a lumbar root lesion or plexopathy.

22.11 Perineal and groin mononeuropathy

22.11.1 The ilioinguinal nerve

This nerve, derived from the L1 and L2 roots, passes round the abdominal wall in the region of the pelvic rim and through the inguinal canal to provide sensation to a narrow band of skin across the upper thigh, inguinal region, iliac crest, and the base of the scrotum, or the labium majoris (Fig. 22.1). It contributes motor fibres to the lower abdominal muscles. It is occasionally affected by direct injury, by herniorrhaphy repair, or by other surgical and invasive procedures on the lower abdomen or iliac crest (Stulz and Pfeiffer 1982). Entrapment in abdominal wall muscles near the anterior superior iliac spine has been described. Pain in the groin, sometimes causing the patient to adopt a flexed posture, is the usual manifestation, and injection of local anaesthetic around the nerve or division of it usually affords relief (Lee and Dellon 2000; Stewart 1993).

22.11.2 The genitofemoral nerve

The genitofemoral nerve derives from the L1 and L2 spinal roots and is formed from the lumbar plexus in the psoas muscle. It runs down that muscle to separate into the genital and femoral branches. The genital branch supplies the scrotal cremaster muscle and sensation to the scrotum or labium majoris and the upper medial thigh. The femoral branch supplies skin in the femoral triangle on the upper anterior thigh. This nerve may be entrapped by intra-abdominal adhesions, especially after appendicectomy, or may be damaged by injury to the groin (Nakano 1978). The genital branch can be involved by inguinal hernia surgery. Genitofemoral nerve lesions also can follow the wearing of excessively tight jeans (O'Brien 1979). There is pain in the internal inguinal ring, relieved by hip flexion, and there may be sensory impairment in the skin of the femoral triangle. The cremasteric contraction reflex evoked by scratching the inner upper thigh may be lost on the affected side. Surgical decompression or division is needed occasionally.

22.11.3 The pudendal nerve

The pudendal nerve is derived from the S2, S3, and S4 spinal nerve contributions to the sacral plexus. It leaves the pelvis through the sciatic notch, then passes anteriorly into the perineum. It supplies sensation to the scrotum, perineal skin, bladder, rectum, lower vagina, and labium, and innervates the erectile tissue of the clitoris or penis. Its dorsal nerve of the penis or clitoris transmits erotic sensations from those organs. Surgical trauma and mass lesions within the pelvis can damage the pudendal nerve. Once outside the pelvis it can be compressed by prolonged bicycle riding using a hard seat, or by the perineal posts used during hip joint surgery to allow traction for reducing hip fractures (Stewart 1993). Lesions seem more commonly symptomatic in men who complain of perineal pain and numbness of half of the penis, sometimes with mild erectile failure.

References

Albers JW, Fink JK (2004). Porphyric neuropathy. *Muscle Nerve*, **30**, 410–22.

Anand P, Birch R (2002). Restoration of sensory function and lack of long-term chronic pain syndromes after brachial plexus injury in human neonates. *Brain*, **125**, 113–22.

Arle JE, Zager EL (2000). Surgical treatment of common entrapment neuropathies in the upper limbs. *Muscle Nerve*, **23**, 1160–74.

Atroshi I, Larsson GU, Ornstein E et al. (2006). Outcomes of endoscopic surgery compared with open surgery for carpal tunnel syndrome among employed patients: randomised controlled trial. *BMJ*, **332**:1473.

Ay S, Apaydin N, Acar H et al. (2005). Anatomic pattern of the terminal branches of posterior interosseous nerve. *Clin Anat*, **18**, 290–5.

Beekman R, Schoemaker MC, Van Der Plas JP et al. (2004a). Diagnostic value of high-resolution sonography in ulnar neuropathy at the elbow. *Neurology*, **62**, 767–73.

Beekman R, Wokke JH, Schoemaker MC et al. (2004b). Ulnar neuropathy at the elbow: follow-up and prognostic factors determining outcome. *Neurology*, **63**, 1675–80.

Bilbao JM, Khoury NJ, Hudson AR et al. (1984). Perineurioma (localized hypertrophic neuropathy). *Arch Pathol Lab Med*, **108**, 557–60.

Birch R, Ahad N, Kono H et al. (2005). Repair of obstetric brachial plexus palsy: results in 100 children. *J Bone Joint Surg Br*, **87**, 1089–95.

Birch R, Bonney G, Wynn Parry C (1998). *Surgical Disorders of the Peripheral Nerves*. Churchill Livingstone, Edinburgh.

Brown T, Cupido C, Scarfone H et al. (2000). Developmental apraxia arising from neonatal brachial plexus palsy. *Neurology*, **55**, 24–30.

Capitani D, Beer S (2002). Handlebar palsy-a compression syndrome of the deep terminal (motor) branch of the ulnar nerve in biking. *J Neurol*, **249**, 1441–5.

Castellanos AM, Glass JP, Yung WK (1987). Regional nerve injury after intra-arterial chemotherapy. *Neurology*, **37**, 834–7.

Chang MH, Wei SJ, Chiang HL et al. (2002). Comparison of motor conduction techniques in the diagnosis of carpal tunnel syndrome. *Neurology*, **58**, 1603–7.

Chaudhry V, Clawson LL (1997). Entrapment of motor nerves in motor neuron disease: does double crush occur? *J Neurol Neurosurg Psychiatry*, **62**, 71–6.

Collier A, Burge P (2001). (iii) Management of mechanical neuropathy of the ulnar nerve at the elbow. *Curr Orthop*, **15**, 256–63.

Cowdery SR, Preston DC, Herrmann DN et al. (2002). Electrodiagnosis of ulnar neuropathy at the wrist: conduction block versus traditional tests. *Neurology*, **59**, 420–7.

Critchley M (1953). *The Parietal Lobes*. Edward Arnold, London.

Cusimano MD, Bilbao JM, Cohen SM (1988). Hypertrophic brachial plexus neuritis: a pathological study of two cases. *Ann Neurol*, **24**, 615–22.

Dammers JW, Roos Y, Veering MM et al. (2006). Injection with methylprednisolone in patients with the carpal tunnel syndrome: A randomised double blind trial testing three different doses. *J Neurol*, **253**, 574–7.

Dawson D, Hallett M, Wilbourn AJ (1999). *Entrapment Neuropathies*. Lippincott-Raven, Philadelphia.

Dessi F, Durand G, Hoffmann JJ (1992). The accessory deep peroneal nerve: a pitfall for the electromyographer. *J Neurol Neurosurg Psychiatry*, **55**, 214–5.

Donaghy M (2005). Lumbosacral plexus lesions. In Dyck PJ, Thomas P, eds. *Peripheral Neuropathy*, Elsevier Saunders, Philadelphia.

Donaghy M, Matkovic Z, Morris P (1999). Surgery for suspected neurogenic thoracic outlet syndromes: a follow up study. *J Neurol Neurosurg Psychiatry*, **67**, 602–6.

Donaldson J (1988). *Neurology of Pregnancy*. WB Saunders, London.

Drees C, Wilbourn AJ, Stevens GH (2002). Main trunk tibial neuropathies. *Neurology*, **59**, 1082–4.

Dyck PJ, Norell JE, Dyck PJ (2001). Non-diabetic lumbosacral radiculoplexus neuropathy: natural history, outcome and comparison with the diabetic variety. *Brain*, **124**, 1197–207.

Ebraheim NA, Lu J, Biyani A et al. (1997). The relationship of lumbosacral plexus to the sacrum and the sacroiliac joint. *Am J Orthop*, **26**, 105–10.

Evans-Jones G, Kay SP, Weindling AM et al. (2003). Congenital brachial palsy: incidence, causes, and outcome in the United Kingdom and Republic of Ireland. *Arch Dis Child Fetal Neonatal Ed*, **88**, F185–9.

Feasby TE, Burton SR, Hahn AF (1992). Obstetrical lumbosacral plexus injury. *Muscle Nerve*, **15**, 937–40.

Foley KM, Woodruff JM, Ellis FT et al. (1980). Radiation-induced malignant and atypical peripheral nerve sheath tumors. *Ann Neurol*, **7**, 311–8.

Friedenberg SM, Zimprich T, Harper CM (2002). The natural history of long thoracic and spinal accessory neuropathies. *Muscle Nerve*, **25**, 535–9.

Fujimura H, Lacroix C, Said G (1991). Vulnerability of nerve fibres to ischaemia. A quantitative light and electron microscope study. *Brain*, **114**, 1929–42.

Gilliatt RW, Le Quesne PM, Logue V et al. (1970). Wasting of the hand associated with a cervical rib or band. *J Neurol Neurosurg Psychiatry*, **33**, 615–24.

Gonik B, Stringer CA, Cotton DB et al. (1984). Intrapartum maternal lumbosacral plexopathy. *Obstet Gynecol*, **63**, 45S–46S.

Gouider R, LeGuern E, Emile J et al. (1994). Hereditary neuralgic amyotrophy and hereditary neuropathy with liability to pressure palsies: two distinct clinical, electrophysiologic, and genetic entities. *Neurology*, **44**, 2250–2.

Goulding PJ, Schady W (1993). Favourable outcome in non-traumatic anterior interosseous nerve lesions. *J Neurol*, **240**, 83–6.

Harper CM, Jr., Thomas JE, Cascino TL et al. (1989). Distinction between neoplastic and radiation-induced brachial plexopathy, with emphasis on the role of EMG. *Neurology*, **39**, 502–6.

Hemrika DJ, Schutte MF, Bleker OP (1986). Elsberg syndrome: a neurologic basis for acute urinary retention in patients with genital herpes. *Obstet Gynecol*, **68**, 37S–39S.

Hollinger P, Sturzenegger M (2000). Chronic progressive primary lumbosacral plexus neuritis: MRI findings and response to immunoglobulin therapy. *J Neurol*, **247**, 143–5.

Horowitz SH (1994). Peripheral nerve injury and causalgia secondary to routine venipuncture. *Neurology*, **44**, 962–4.

Hui AC, Wong S, Leung CH et al. (2005). A randomized controlled trial of surgery vs steroid injection for carpal tunnel syndrome. *Neurology*, **64**, 2074–8.

Ismael SS, Amarenco G, Bayle B et al. (2000). Postpartum lumbosacral plexopathy limited to autonomic and perineal manifestations: clinical and electrophysiological study of 19 patients. *J Neurol Neurosurg Psychiatry*, **68**, 771–3.

Iverson DJ (2005). MRI detection of cysts of the knee causing common peroneal neuropathy. *Neurology*, **65**, 1829–31.

Jarvik JG, Yuen E, Haynor DR *et al.* (2002). MR nerve imaging in a prospective cohort of patients with suspected carpal tunnel syndrome. *Neurology*, **58**, 1597–602.

Jeannet PY, Watts GD, Bird TD *et al.* (2001). Craniofacial and cutaneous findings expand the phenotype of hereditary neuralgic amyotrophy. *Neurology*, **57**, 1963–8.

Jefferson D, Eames RA (1979). Subclinical entrapment of the lateral femoral cutaneous nerve: an autopsy study. *Muscle Nerve*, **2**, 145–54.

Kambhampati SB, Birch R, Cobiella C *et al.* (2006). Posterior subluxation and dislocation of the shoulder in obstetric brachial plexus palsy. *J Bone Joint Surg Br*, **88**, 213–9.

Kato N, Htut M, Taggart M *et al.* (2006). The effects of operative delay on the relief of neuropathic pain after injury to the brachial plexus: a review of 148 cases. *J Bone Joint Surg Br*, **88**, 756–9.

Kim DH, Murovic JA, Tiel R *et al.* (2004). Management and outcomes in 353 surgically treated sciatic nerve lesions. *J Neurosurg*, **101**, 8–17.

Kim DH, Murovic JA, Tiel RL *et al.* (2005). Management and outcomes of 42 surgical suprascapular nerve injuries and entrapments. *Neurosurgery*, **57**, 120–7; discussion 120–7.

Kincaid JC (1988). AAEE minimonograph #31: the electrodiagnosis of ulnar neuropathy at the elbow. *Muscle Nerve*, **11**, 1005–15.

Kline DG, Kim DH (2003). Axillary nerve repair in 99 patients with 101 stretch injuries. *J Neurosurg*, **99**, 630–6.

Koehler PJ, Buscher M, Rozeman CA *et al.* (1997). Peroneal nerve neuropathy in cancer patients: a paraneoplastic syndrome? *J Neurol*, **244**, 328–32.

Kuhlenbaumer G, Hannibal MC, Nelis E *et al.* (2005). Mutations in SEPT9 cause hereditary neuralgic amyotrophy. *Nat Genet*, **37**, 1044–6.

Laroche CM, Mier AK, Spiro SG *et al.* (1989). Respiratory muscle weakness in the Lambert-Eaton myasthenic syndrome. *Thorax*, **44**, 913–8.

Lauria G, McArthur JC, Hauer PE *et al.* (1998). Neuropathological alterations in diabetic truncal neuropathy: evaluation by skin biopsy. *J Neurol Neurosurg Psychiatry*, **65**, 762–6.

Lee CH, Dellon AL (2000). Surgical management of groin pain of neural origin. *J Am Coll Surg*, **191**, 137–42.

Lee S, Wolfe S (2000). Peripheral nerve injury and repair. *J Am Acad Orthop Surg*, **8**, 243–52.

Levin KH, Lutz G (1996). Angiotropic large-cell lymphoma with peripheral nerve and skeletal muscle involvement: early diagnosis and treatment. *Neurology*, **47**, 1009–11.

Lin PT, Andersson PB, Distad BJ *et al.* (2005). Bilateral isolated phrenic neuropathy causing painless bilateral diaphragmatic paralysis. *Neurology*, **65**, 1499–501.

Mackinnon P and Morris J (1994). *Oxford Textbook of Functional Anatomy*, Vol 1. Oxford University Press, Oxford.

Mastaglia FL (2000). Tibial nerve entrapment in the popliteal fossa. *Muscle Nerve*, **23**, 1883–6.

Mondelli M, Morana P, Ballerini M *et al.* (2005). Mononeuropathies of the radial nerve: clinical and neurographic findings in 91 consecutive cases. *J Electromyogr Kinesiol*, **15**, 377–83.

Nakano KK (1978). The entrapment neuropathies. *Muscle Nerve*, **1**, 264–79.

Newsom-Davis J (1967). Phrenic nerve conduction in man. *J Neurol Neurosurg Psychiatry*, **30**, 420–6.

Nukada H, van Rij AM, Packer SG *et al.* (1996). Pathology of acute and chronic ischaemic neuropathy in atherosclerotic peripheral vascular disease. *Brain*, **119**, 1449–60.

O'Brien MD (1979). Genitofemoral neuropathy. *Br Med J*, **1**, 1052.

Padua L, Aprile I, Saponara C *et al.* (2001). Multiperspective assessment of peripheral nerve involvement in diabetic patients. *Eur Neurol*, **45**, 214–21.

Paladini D, Dellantonio R, Cinti A *et al.* (1996). Axillary neuropathy in volleyball players: report of two cases and literature review. *J Neurol Neurosurg Psychiatry*, **60**, 345–7.

Panegyres PK, Moore N, Gibson R *et al.* (1993). Thoracic outlet syndromes and magnetic resonance imaging. *Brain*, **116**, 823–41.

Patel AT, Gaines K, Malamut R *et al.* (2005). Usefulness of electrodiagnostic techniques in the evaluation of suspected tarsal tunnel syndrome: an evidence-based review. *Muscle Nerve*, **32**, 236–40.

Phillips LH 2nd (1986). Familial long thoracic nerve palsy: a manifestation of brachial plexus neuropathy. *Neurology*, **36**, 1251–3.

Planner AC, Donaghy M, Moore NR (2006). Causes of lumbosacral plexopathy. *Clin Radiol*, **61**, 987–95.

Prasada Rao PV, Chaudhary SC (2001). Absence of musculocutaneous nerve: two case reports. *Clin Anat*, **14**, 31–5.

Ramachandran M, Birch R, Eastwood DM (2006). Clinical outcome of nerve injuries associated with supracondylar fractures of the humerus in children: the experience of a specialist referral centre. *J Bone Joint Surg Br*, **88**, 90–4.

Reisin R, Pardal A, Ruggieri V *et al.* (1994). Sural neuropathy due to external pressure: report of three cases. *Neurology*, **44**, 2408–9.

Rogers LR, Borkowski GP, Albers JW *et al.* (1993). Obturator mononeuropathy caused by pelvic cancer: six cases. *Neurology*, **43**, 1489–92.

Salazar-Grueso E, Roos R (1986). Sciatic endometriosis: a treatable sensorimotor mononeuropathy. *Neurology*, **36**, 1360–3.

Seddon H (1944). Three types of nerve injury. *Brain*, **66**, 237–88.

Shergill G, Bonney G, Munshi P *et al.* (2001). The radial and posterior interosseous nerves. Results fo 260 repairs. *J Bone Joint Surg Br*, **83**, 646–9.

Shyu WC, Lin JC, Chang MK *et al.* (1993). Compressive radial nerve palsy induced by military shooting training: clinical and electrophysiological study. *J Neurol Neurosurg Psychiatry*, **56**, 890–3.

Singh N, Behse F, Buchthal F (1974). Electrophysical study of peroneal palsy. *J Neurol Neurosurg Psychiatry*, **37**, 1202–13.

Siu TL, Chandran KN (2005). Neurolysis for meralgia paresthetica: an operative series of 45 cases. *Surg Neurol*, **63**, 19–23; discussion 23.

Sorenson EJ, Chen JJ, Daube JR (2002). Obturator neuropathy: causes and outcome. *Muscle Nerve*, **25**, 605–7.

Sourkes M, Stewart JD (1991). Common peroneal neuropathy: A study of selective motor and sensory involvement. *Neurology*, **41**, 1029–33.

Spinner RJ, Atkinson JL, Scheithauer BW *et al.* (2003). Peroneal intraneural ganglia: the importance of the articular branch. Clinical series. *J Neurosurg*, **99**, 319–29.

Stewart JD (1993). Compression and entrapment neuropathies. In Dyck PJ, Thomas P, Griffin J, Low P, Poduslo J, eds. *Peripheral Neuropathy*. WB Saunders, Philadelphia.

Stoehr M (1978). Traumatic and postoperative lesions of the lumbosacral plexus. *Arch Neurol*, **35**, 757–60.

Stohr M, Dichgans J, Dorstelmann (1980). Ischaemic neuropathy of the lumbosacral plexus following intragluteal injection. *J Neurol Neurosurg Psychiatry*, **43**, 489–94.

Stulz P, Pfeiffer KM (1982). Peripheral nerve injuries resulting from common surgical procedures in the lower portion of the abdomen. *Arch Surg*, **117**, 324–7.

Styf J (1989). Entrapment of the superficial peroneal nerve. Diagnosis and results of decompression. *J Bone Joint Surg Br*, **71**, 131–5.

Suarez GA, Giannini C, Bosch EP *et al.* (1996). Immune brachial plexus neuropathy: suggestive evidence for an inflammatory-immune pathogenesis. *Neurology*, **46**, 559–61.

Symonds C, Meadows S (1937). Compression of the spinal cord in the neighbourhood of the foramen magnum. *Brain*, **60**, 52–84.

Thomas JE, Cascino TL, Earle JD (1985). Differential diagnosis between radiation and tumor plexopathy of the pelvis. *Neurology*, **35**, 1–7.

Thomas JE, Howard FM Jr. (1972). Segmental zoster paresis--a disease profile. *Neurology*, **22**, 459–66.

Thyagarajan D, Cascino T, Harms G (1995). Magnetic resonance imaging in brachial plexopathy of cancer. *Neurology*, **45**, 421–7.

Trivedi R, Byrne J, Huson SM *et al.* (2000). Focal amyotrophy in neurofibromatosis 2. *J Neurol Neurosurg Psychiatry*, **69**, 257–61.

Tsao BE, Ostrovskiy DA, Wilbourn AJ *et al.* (2006). Phrenic neuropathy due to neuralgic amyotrophy. *Neurology*, **66**, 1582–4.

Tsao BE, Wilbourn AJ (2003). The medial brachial fascial compartment syndrome following axillary arteriography. *Neurology*, **61**, 1037–41.

Tucker T, Wolkenstein P, Revuz J *et al.* (2005). Association between benign and malignant peripheral nerve sheath tumors in NF1. *Neurology*, **65**, 205–11.

Tysvaer AT (1982). Computerized tomography and surgical treatment of femoral compression neuropathy. Report of two cases. *J Neurosurg*, **57**, 137–9.

Valls-Sole J, Solans M (2002). Idiopathic bilateral diaphragmatic paralysis. *Muscle Nerve*, **25**, 619–23.

van Alfen N, van Engelen BG (2006). The clinical spectrum of neuralgic amyotrophy in 246 cases. *Brain*, **129**, 438–50.

van Alfen N, van Engelen BG, Reinders JW *et al.* (2000). The natural history of hereditary neuralgic amyotrophy in the Dutch population: two distinct types? *Brain*, **123**, 718–23.

Wilbourn AJ, Gilliatt RW (1997). Double-crush syndrome: a critical analysis. *Neurology*, **49**, 21–9.

Wohlgemuth WA, Stoehr M (2002). Percutaneous arterial interventional treatment of exercise-induced neurogenic intermittent claudication due to ischaemia of the lumbosacral plexus. *J Neurol*, **249**, 988–92.

Wood VE, Twito R, Verska JM (1988). Thoracic outlet syndrome. The results of first rib resection in 100 patients. *Orthop Clin North Am*, **19**, 131–46.

Yuen EC, Olney RK, So YT (1994). Sciatic neuropathy: clinical and prognostic features in 73 patients. *Neurology*, **44**, 1669–74.

Zarranz JJ, Simon R, Salisachs P (1981). Acute anticoagulant-induced compressive lumbar plexus neuropathy. A clinico-pathological study. *Eur Neurol*, **20**, 469–72.

Zeman A, Donaghy M (1991). Acute infection with human immunodeficiency virus presenting with neurogenic urinary retention. *Genitourin Med*, **67**, 345–7.

CHAPTER 23

The motor neurone disorders

Pamela Shaw

Contents

23.1 Introduction

23.1.1 Classification of motor neurone disorders

The motor neurone diseases are a group of disorders in which there is selective loss of function of upper and/or lower motor neurones in the motor cortex, brainstem, and spinal cord resulting in impairment in the nervous system control of voluntary movement. The term 'motor neurone disease', often abbreviated to 'MND', is used differently in different countries. In the United Kingdom it is used as an umbrella term to cover the related group of neurodegenerative disorders including amyotrophic lateral sclerosis, the commonest variant, as well as progressive muscular atrophy, primary lateral sclerosis, and progressive bulbar palsy. However, in many other countries amyotrophi\c lateral sclerosis, referred to as ALS, has been adopted as the umbrella term for this group of clinical variants of motor system degeneration. There is a tendency now internationally to use the ALS/MND abbreviation to cover this group of conditions. Careful diagnosis within the motor neurone diseases is essential for advising about prognosis, potential genetic implications, and for identifying those with acquired lower motor neurone syndromes who may benefit for the administration of immunomodulatory therapy. Table 23.1 classifies the motor neurone disorders according to whether the upper motor neurone, the lower motor neurone, or both groups of cells are affected.

23.1.2 Differential diagnosis

In patients with motor neurone diseases, precise differential diagnosis requires clinical and neurophysiological assessment as to whether the disorder involves the upper motor neurones, the lower motor neurones, or both. Consideration should then be given to factors such as age of onset, rapidity of progression, evidence of an inherited disorder, and the anatomical distribution of the clinical features to determine the likely diagnosis. As a general rule,

Table 23.1 Classification of the motor neurone disorders

◆ **Combined upper and lower motor neurone disorders**

Amyotrophic lateral sclerosis (ALS)
    Familial adult onset
    Familial juvenile onset
    Sporadic
    ALS-plus syndromes
        ALS with frontotemporal dementia
        Western Pacific ALS–Parkinsonism–dementia complex

◆ **Upper motor neurone disorders**

Primary lateral sclerosis
The hereditary spastic paraplegias
Neurolathyrism
Konzo

◆ **Lower motor neurone disorders**
Hereditary

The spinal muscular atrophies (SMAs)
- Proximal autosomal recessive SMA of childhood (associated with SMN mutations)
  Type 1 Werdnig Hoffman disease
  Type II Intermediate form
  Type III Wohlfart–Kugelberg–Welander disease
  Type IV Adult onset
- Acute infantile forms of SMA not associated with SMN mutations
  SMA with pontocerebellar hypoplasia
  SMA and arthrogryphosis +/- bone fractures
  X-linked SMA with arthogryphosis
  Lethal congenital contracture syndromes types 1 and 2
- Autosomal dominant proximal SMA
- Distal spinal muscular atrophy /hereditary motor neuronopathy
  SMA with respiratory distress (SMARD)
  Autosomal recessive distal SMA
  Distal SMA with upper limb predominance
  Distal SMA with lower limb predominance
  Distal SMA with vocal cord paralysis
  Congenital non-progressive SMA involving the lower limbs
  Scapuloperoneal SMA
Kennedy's disease (X-linked spinobulbar neuronopathy)
Hexosaminidase deficiency (GM2 gangliosidosis)

Acquired

Monomelic focal and segmental spinal muscular atrophies
Multifocal motor neuropathies
Acute motor axonal neuropathy
Post-polio syndrome
Post-irradiation syndrome

Infective disorders

Acute poliomyelitis
West Nile fever
Other viral infections eg enterovirus 71 and rabies
HIV associated motor neurone disorder
Lyme disease
Tick borne encephalitis
Creutzfeld–Jacob disease (amyotrophic forms)

◆ **Disorders of the bulbar motor system**

Kennedy's disease (X-linked bulbospinal neuronopathy)
Brown–Vialetto–van Laere syndrome
Fazio Londe disease

◆ **Toxic disorders of the motor neuron**

Neurolathyrism
Konzo
Heavy metal toxicity (lead, mercury)
Western Pacific ALS–Parkinsonism–dementia complex
Post-irradiation motor neurone injury

◆ **Disorders of motor neurone overactivity**

Neuromyotonia
Stiff person syndrome

◆ **Miscellaneous motor neurone disorders**

Endocrinopathies e.g. hyperthyroidism, hyperparathyroidism, hypoglycaemia
Copper deficiency syndrome
Benign cramp-fasciculation syndrome

---

the sensory system, sphincter control, and cognitive function are usually preserved in motor neurone diseases.

The clinical features of lower motor neurone involvement are muscle wasting, fasciculations, and flaccid weakness. Tendon reflexes are usually preserved until the muscle denervation is severe. In amyotrophic lateral sclerosis, the upper motor neurone component tends to preserve the reflexes and the preservation of reflexes in a limb with severe muscular atrophy may be an important clue to the diagnosis. Motor neurone diseases will usually give rise to significant denervation atrophy of weakened muscles. If muscle bulk is relatively preserved in a weak muscle it should raise the possibility of conduction block rather than denervation, as found in multifocal motor neuropathy with conduction block. Fasciculations are visible flickering movements within the muscle belly which are insufficient to produce movement at the joint and are most commonly observed in the large proximal muscle such as deltoid and quadriceps which have large motor units. Fasciculations which are not visible clinically may be detected by electromyography. Fasciculations can only be regarded as indicative of motor neurone disease if associated with evidence of denervation by clinical or electrophysiological assessment.

In patients with suspected motor neurone disease, nerve conduction studies will rule out sensorimotor polyneuropathy; pure motor demyelinating neuropathies or multifocal motor neuropathy with conduction block. Maximal motor conduction velocity is often reduced in nerves supplying denervated muscles as a consequence of degeneration of large motor axons. However, in amyotrophic lateral sclerosis the motor conduction velocity rarely falls below 80 per cent of the lower limit of normal and F waves or distal motor latencies rarely exceed 1.25 times the upper limit of normal. Results outside these limits should raise the possibility of a primary demyelinating motor neuropathy. Electromyography helps to distinguish denervation from myopathy and may also detect subclinical denervation in limbs which are clinically normal in patients with amyotrophic lateral sclerosis. Muscle biopsy may occasionally be required to rule out myopathy, particularly in patients with atypical features or slowly progressive proximal weakness.

Signs of upper motor neurone involvement in motor neuron diseases include increased muscle tone, clonus, pyramidal distribution weakness, and extensor plantar responses. In many patients with amyotrophic lateral sclerosis the presence of clonus and extensor plantar responses is often obscured by the profound denervation changes in the distal lower limb muscles, thereby obscuring clinical confirmation of upper motor neurone involvement. Changes in threshold and parameters of central motor conduction following transcranial magnetic stimulation of the motor cortex may be seen in patients with amyotrophic lateral sclerosis, but the changes observed are complex and may vary with the stage or anatomy of the disease and are insufficiently sensitive to provide a robust and useful tool for detecting subclinical involvement of the upper motor neurone.

The group of patients with amyotrophic lateral sclerosis who do appear to have a pure lower motor neurone disorder without clear clinical signs of upper motor neurone involvement, present a diagnostic challenge. In this situation it is important to exclude treatable lower motor neurone disorders such as multifocal motor neuropathy with conduction block and a trial of immunomodulatory therapy may be indicated where diagnostic uncertainty persists. The passage of time may resolve the diagnostic uncertainty as patients with the pure lower motor neurone variant of amyotrophic lateral sclerosis, known as progressive muscular atrophy, will typically show a brisk rate of disease progression.

The commonest diagnostic problem is to distinguish amyotrophic lateral sclerosis from other motor neurone disorders carrying a better prognosis, particularly if signs of upper motor neurone involvement are lacking. Multifocal motor neuropathy with conduction block usually develops over many years, tends to present with asymmetrical involvement of the upper limbs, may present with difficulty in extending one or more fingers, is associated with electrophysiological evidence of motor nerve conduction slowing or block, and the patient may have antiganglioside antibodies detectable in blood (Section 21.11.3). Patients with post-polio syndrome tend to deteriorate rather slowly by comparison with amyotrophic lateral sclerosis and present with slight further loss of neuromuscular function some decades after an earlier attack of acute poliomyelitis (Section 23.3.7). Kennedy's disease should be suspected in patients who appear to have amyotrophic lateral sclerosis with bulbar involvement, but who do not deteriorate as rapidly as expected (Section 23.5.1). In men with pure lower motor neurone disorders the presence of gynaecomastia should always be looked for. Other diagnostic alerts for Kennedy's disease include the presence of mentalis contractions or fasciculations and clinical or neurophysiological evidence of sensory nerve involvement.

Muscle fasciculations are common in the normal population and are frequently observed in the calf muscles after exercise. Patients with benign fasciculations who come to neurological attention are frequently individuals with medical knowledge who fear the development of motor neurone disease, or those with a family history of amyotrophic lateral sclerosis. Benign fasciculations are most commonly observed in the calf muscles, are not associated with any other abnormal clinical signs and electromyography shows no evidence of denervation.

23.2 Amyotrophic lateral sclerosis/motor neurone disease

Historical descriptions of amyotrophic lateral sclerosis, or ALS, were reported as early as 1824 by Charles Bell and others (Goldblatt 1968; Tyler and Sheffner 1991). However, Charcot was the first to describe the condition in detail, based on careful clinico-pathological correlation and he coined the term amyotrophic lateral sclerosis (Charcot and Joffroy 1869). Motor neurone disease, or MND, as used by Brain in the first edition of this textbook (Brain 1933), is the name used for the disease most commonly in the United Kingdom and represents an umbrella term encompassing amyotrophic lateral sclerosis and three other disorders that are considered its clinical variants: primary lateral sclerosis, progressive muscular atrophy, and progressive bulbar palsy. In the USA, amyotrophic lateral sclerosis is sometimes known as Lou Gehrig's disease after the famous Yankee baseball player who developed the disease at the peak of his sporting career and died in 1941. Amyotrophic lateral sclerosis is sometimes referred to as ALS/MND.

Amyotrophic lateral sclerosis is a neurodegenerative disorder that causes progressive injury and cell death of lower motor neurones within the brainstem and spinal cord and upper motor neurones in the motor cortex. The disease has an incidence of about 2 per 100 000 and a prevalence of 6–8 /100 000 (Chancellor and Warlow 1992; Traynor et al. 1999). There are approximately 5000 individuals with amyotrophic lateral sclerosis at any one time in the UK. The incidence is fairly uniform throughout the world, with the exception of a few high incidence foci, for example, on the Kii peninsula of Japan and the Western Pacific island of Guam. The disease predominantly affects middle-aged and elderly individuals, with a mean age of onset of 55 years, though younger individuals can also be affected. For reasons that are not understood, men are affected more commonly than women, with a male/female ratio of approximately 1.6/1. Motor neurone disease is sporadic in 90 per cent of cases, but in 5–10 per cent of cases the disease is familial, usually with an autosomal dominant mode of inheritance.

23.2.1 Clinical features

Amyotrophic lateral sclerosis is characterized and defined by the presence of clinical features reflecting degeneration of upper motor neurones of the cerebral cortex, and lower motor neurones of the brainstem and spinal cord. Lower motor neurone degeneration causes weakness, atrophy, and fasciculation of the limb and bulbar musculature. Muscle cramps are a common symptom, and patients may be aware of fasciculation as twitching or flickering movements of their muscles. Clinical signs resulting from degeneration of the upper motor neurones include the incongruous presence of active or brisk tendon reflexes in a wasted limb, increased muscle tone, and sometimes the presence of Hoffman's or Babinski's signs. Upper motor neurone dysfunction within the bulbar territory causes pseudobulbar palsy, where the snout and jaw reflexes may be exaggerated, and increased tone within the bulbar muscles may cause slowing of repetitive movements of the tongue as well as strained effortful speech. Emotional lability, producing difficulty in controlling episodes of laughing or crying, often accompanies upper motor neurone signs in the bulbar region.

In amyotrophic lateral sclerosis the disease often starts focally and asymmetrically in the upper limb, lower limb, or bulbar territories.

This is followed by progressive spread of injury to contiguous groups of motor neurones, so that the patient's clinical features often show an anatomically logical progression. By the end of the disease course most patients will have features of upper and lower motor neurone dysfunction affecting all four limbs and the bulbar musculature.

Upper and lower limb features

The first clinical problem is present in the upper or lower limbs in approximately 75 per cent of patients. At first asymmetry or unilaterality of symptoms and signs is common and the first noticeable problem for the patient commonly involves the distal limb muscles. Affected individuals may notice weakness, wasting or clumsiness of one hand, or simply difficulty with everyday actions such as turning a key or lifting a heavy object. In the lower limbs, foot drop or a tendency to trip over the toes of one foot is a common presenting symptom. The gait may become slowed or clumsy due to weakness and spasticity of one or both lower limbs. The patient may be aware of limb muscle cramps which can sometimes precede other clinical features of amyotrophic lateral sclerosis by months or years. In addition, muscle fasciculation may be noticed by the patient, particularly arising from the large proximal limb muscles such as biceps, triceps, pectoralis major, and quadriceps. As the condition progresses, patients often develop a characteristic pattern of limb muscle weakness. In the upper limbs, the intrinsic hand muscles, particularly the thenar group, tend to be affected severely and early, whereas other muscle groups such as triceps and the finger flexors are relatively spared until late in the disease course. In the lower limbs, the pattern of weakness is often in a pyramidal distribution, with flexors weaker than extensors, early weakness of hip flexion and ankle dorsiflexion and often more severe weakness of the distal muscles. Examination of the patient will frequently reveal a combination of upper and lower motor neurone features with proximal fasciculation, muscle wasting, particularly distally (Fig. 23.1), the characteristic pattern of weakness described above, but in addition tone may be mildly increased, the reflexes are brisk and extensor plantar responses may be present.

Bulbar features

Bulbar symptoms are the presenting clinical problem in approximately 25 per cent of patients with amyotrophic lateral sclerosis and this presentation is particularly common in elderly women. The onset of amyotrophic lateral sclerosis in the bulbar regions has a less favourable prognosis than limb-onset disease (Chancellor *et al.* 1993; del Aguila *et al.* 2003). The first problem is usually slurring of speech which initially may only be apparent when the individual is tired at the end of the day, after prolonged use of the voice, or after alcohol. The dysarthria may rarely be accompanied by dysphonia. The patient may notice particular difficulty with specific sounds such as 's' and may notice other features of bulbar muscle weakness such as difficulty in pursing the lips or whistling. The onset of dysarthria is often initially attributed to a 'minor stroke', but it soon becomes apparent that the speech difficulty is a progressive problem. Patients with amyotrophic lateral sclerosis often have a mixed spastic/flaccid dysarthria. The speech develops a tight, strangled quality due to the upper motor neurone component, but superimposed on this, the flaccid lower motor neurone weakness of the palate and nasopharynx give the speech a nasal quality. Examination of the patient will often reveal a combination of upper and lower motor neurone bulbar signs with weakness of the facial

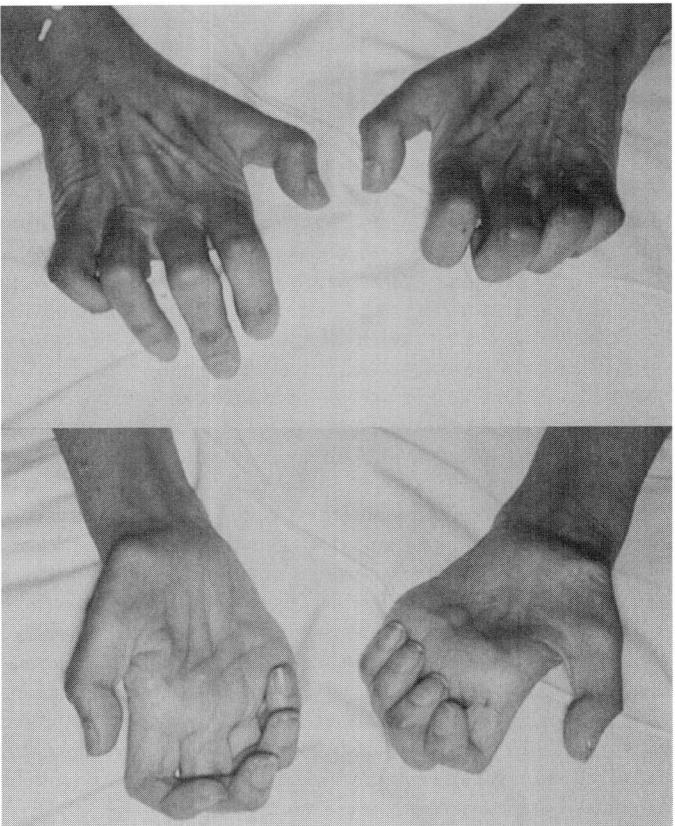

Fig. 23.1 Wasting of the intrinsic hand muscles in a patient with amyotrophic lateral sclerosis.

muscles, a spastic, weak, wasted, and fasciculating tongue and a brisk jaw jerk. As the disease progresses the patient may develop weakness of the palate and muscles of mastication. Occasionally forcible jaw clenching is a troublesome symptom causing trauma to the mouth. Dysphagia becomes a problem later, often several weeks or months after the onset of the speech problem. Initially dysphagia tends to be more pronounced for liquids compared to solids. Gradually certain food items such as lettuce, foods with a crumbly texture, or items with a sharp taste become difficult to swallow and may cause choking episodes. Mealtimes may become prolonged and arduous, with frequent episodes of coughing. When dysphagia reaches a significant level, the patient will have a problem with excess saliva pooling in the mouth. If accompanied by lower facial weakness, the patient will suffer from drooling which causes considerable social embarrassment. As the bulbar dysfunction becomes increasingly severe, the patient may develop complete anarthria. Severe dysphagia causes the patient to be at risk of aspiration and chest infection.

Other clinical features

Weight loss. This is a common feature of amyotrophic lateral sclerosis. Multiple factors may contribute to this, with loss of muscle bulk due to amyotrophy, difficulty maintaining nutrition in the face of dysphagia, and loss of appetite resulting from reactive anxiety and depression or from immobility. Severe weight loss and nutritional deficiencies may themselves exacerbate muscle weakness.

Neck muscle weakness. This is common later in the course of disease, causing difficulty holding the head upright, the 'dropped head syndrome'. This problem commonly causes pain and increases difficulties with swallowing and communication.

Respiratory muscle weakness. Occasionally, onset of the disease in motor neurones innervating the respiratory muscles causes breathlessness as a presenting symptom (Nightingale *et al.* 1982; De Carvalho *et al.* 1996). More commonly, respiratory failure develops insidiously during the course of the disease, causing dyspnoea and orthopnoea. Significant diaphragmatic weakness may be apparent from paradoxical movement of the abdominal wall during inspiration and from a marked decline in the forced vital capacity when measured in the supine compared to the upright position. Symptoms of nocturnal carbon dioxide retention may develop, including morning headaches, anorexia, and daytime somnolence.

Extramotor features. Amyotrophic lateral sclerosis is traditionally considered to be a pure motor disorder, which spares sensation, cognition, and autonomic function. However, there is increasing evidence that, despite the relative vulnerability of motor neurones to the degenerative process, other types of neurones are also affected. Thus, amyotrophic lateral sclerosis is now regarded as a multisystem neurodegenerative disorder in which the earliest and most severe degeneration tends to involve motor neurones. In most patients, the evolution of motor dysfunction is lethal before the development of overt signs of central nervous system pathology in other regions, although occasional patients may spontaneously develop a severe multisystem degeneration (Machida *et al.* 1999). Patients whose survival is prolonged by ventilatory support may develop widespread features of extramotor system involvement.

Cognitive impairment. Frontotemporal dementia (Section 34.6.5) encompasses a range of clinical features that result from atrophy of the frontal and anterior temporal regions of the brain. In the frontal variant of frontotemporal dementia, there is prominent personality change and conduct disorder, and affected individuals often become disinhibited or apathetic with emotional blunting, and loss of insight. Behaviour tends to become ritualized or stereotypical, and eating behaviour often becomes abnormal. Visuospatial function and memory are relatively preserved. Two syndromes of progressive language dysfunction are seen with frontotemporal dementia. Primary progressive aphasia, which may progress to mutism, is characterized by effortful speech production with phonological and grammatical errors, and word retrieval difficulties. Patients may also develop semantic dementia, with impairment of naming and word comprehension but preservation of fluent, grammatical speech output (Neary *et al.* 1998; McKhann *et al.* 2001; Hodges *et al.* 2004). There is overlap between these clinical syndromes, as patients with either syndrome may later develop deterioration in personality and behaviour (Bak *et al.* 2001), while language abnormalities may be found in patients with frontotemporal dementia (Neary *et al.* 1990; Strong *et al.* 1996).

Approximately 5 per cent of patients with motor neurone disease will develop overt features of frontotemporal dementia (Hudson 1981). Cognitive dysfunction may precede, follow, or coincide with the features of motor dysfunction. The most common symptoms seen in frontotemporal dementia in association with motor neurone disease are progressive deterioration in personality and behaviour (Ringholz *et al.* 2005), but primary progressive aphasia is also described.

Motor neurone disease patients without overt dementia may show more subtle features of frontal lobe dysfunction. This problem may be under-recognized in the normal clinic setting, partly because of the difficulty in assessing cognitive function in patients with bulbar dysfunction and severe motor deficits. Detailed neuropsychological assessment demonstrates that up to 50 per cent of patients with sporadic motor neurone disease develop features of frontal lobe dysfunction (Ringholz *et al.* 2005), There is evidence indicating that cognitive impairment may be more common in patients with primary lateral sclerosis or predominant upper motor neurone signs (Caselli *et al.* 1995), and in those with prominent bulbar involvement (Schreiber *et al.* 2005).

Parkinsonism. This is described in cases of sporadic motor neurone disease, and occurs more frequently in cases with accompanying dementia (Qureshi *et al.* 1996). The prevalence of Parkinsonism in motor neurone disease may be underestimated, as extrapyramidal features can be masked in patients with muscle weakness, wasting, and spasticity. Subclinical defects in dopaminergic transmission have been demonstrated by positron emission tomography scanning in patients with sporadic motor neurone disease, without clinical evidence of an extrapyramidal disorder (Takahashi *et al.* 1993).

Disorders of eye movement. Supranuclear ophthalmoplegia, including limitation of ocular movement, slow ocular movement, and spasmodic gaze fixation, have been described in motor neurone disease. Although the majority of these patients develop these features after a disease course extended by invasive ventilation, they can occasionally be seen in 'natural disease' (Hayashi *et al.* 1989; Mizutani *et al.* 1990). Difficulty initiating eyelid opening, with preserved reflex eyelid movements, often referred to as 'eyelid apraxia', is also seen in patients with motor neurone disease, and may be associated with supranuclear vertical gaze impairment (Abe *et al.* 1995).

Sensory impairment. Patients with amyotrophic lateral sclerosis not infrequently complain of non-specific sensory phenomena, such as tingling in the fingers. Sensory examination is usually normal, but nerve conduction studies and somatosensory-evoked potentials reveal the presence of abnormalities in peripheral and central sensory pathways in up to 60 per cent of patients (Bosch *et al.* 1985; Shefner *et al.* 1991). There are isolated case reports of more profound sensory involvement in association with amyotrophic lateral sclerosis (Wakabayashi *et al.* 1998).

Selective sparing of specific motor neurone groups. The eye movements tend to be spared in motor neurone disease. Even in advanced disease, when patients would otherwise be 'locked in', often limited communication can be retained by movements of the eyes. Similarly, the strength of the pelvic floor muscles is relatively preserved, so that patients with motor neurone disease usually remain continent throughout the course of the disease. These selectively spared muscle groups reflect the fact that motor neurones in the oculomotor nuclei of the brainstem and Onuf's nucleus in the sacral spinal cord are less vulnerable to the pathological process in motor neurone disease compared to those innervating limb and bulbar musculature.

Clinical variants of amyotrophic lateral sclerosis

There are several clinical variants that describe the predominant clinical features of the patient at the time of presentation. As the disease progresses, however, the majority of patients will develop features of upper and lower motor neurone degeneration affecting the limbs and bulbar musculature, the most common variant of amyotrophic lateral sclerosis.

Progressive muscular atrophy. Approximately 5–10 per cent of patients with motor neurone disease will present with clinical features reflecting only degeneration of lower motor neurone groups in the spinal cord, in the absence of any evidence of upper motor neurone pathology. Some of these patients will later develop brisk reflexes or extensor plantar responses. Clinical examination may not reliably detect upper motor neurone involvement late in the disease, when there is severe global weakness and amyotrophy. It is noteworthy that in those patients who develop no clinical upper motor neurone signs during life, approximately 50 per cent will have pathological evidence of corticospinal tract pathology at autopsy (Ince *et al.* 2003). This suggests that progressive muscular atrophy is part of the same pathological spectrum of disease as amyotrophic lateral sclerosis.

Primary lateral sclerosis. Primary lateral sclerosis, first described by Erb (1875), is a degenerative disorder of the upper motor neurone pathways causing spasticity of the limb and bulbar muscles. It is usually a sporadic disorder of insidious onset, and commonly starts in the fifth decade or later as a spastic paraparesis. Bulbar or upper limb onset has also been described. The course is gradually progressive, and although patients ultimately develop a severe spastic spinobulbar paresis, survival is usually prolonged compared to patients with classical amyotrophic lateral sclerosis (Pringle *et al.* 1992). Many patients with primary lateral sclerosis will survive more than 10 to 15 years after symptom onset.

Not all case series of primary lateral sclerosis have adequately excluded alternative diagnoses, particularly prior to the availability of modern imaging techniques. Pringle and colleagues described a series of 8 patients with progressive symmetrical spinobulbar spasticity in whom alternative diagnoses were carefully excluded, and proposed diagnostic criteria for primary lateral sclerosis (Pringle *et al.* 1992). Although the cardinal feature of primary lateral sclerosis is upper motor neurone involvement, in most case series described, there is usually evidence clinically or electrophysiologically of some lower motor neurone dysfunction, which may only develop after several years (Kuipers-Upmeijer *et al.* 2001; Le Forestier *et al.* 2001). The evolution of primary lateral sclerosis into amyotrophic lateral sclerosis has been described, with the onset of generalized amyotrophy developing after many years of a slowly progressive degeneration of the upper motor neurone system, (Bruyn *et al.* 1995), and primary lateral sclerosis may occur as a phenotypic manifestation of familial amyotrophic lateral sclerosis. These findings strongly indicate that primary lateral sclerosis is part of the spectrum of amyotrophic lateral sclerosis, with predominant but not exclusive degeneration of upper motor neurones.

Progressive bulbar palsy. Bulbar onset motor neurone disease occurs in approximately 25 percent of patients and is most common in women of middle to elderly age. Progressive bulbar palsy usually progresses to involve the limbs, although clinical and electrophysiological abnormalities affecting the limbs may not be found at the time of presentation.

Segmental variants. Several variants of amyotrophic lateral sclerosis have been described, in which the disease follows a more segmental pattern than is typical in classical disease. The flail arm syndrome, progressive amyotrophic diplegia, may occur in up to 10 per cent of patients with motor neurone disease (Fig. 23.2) (Gamez *et al.* 1999; Katz *et al.* 1999). It is much more common in men than

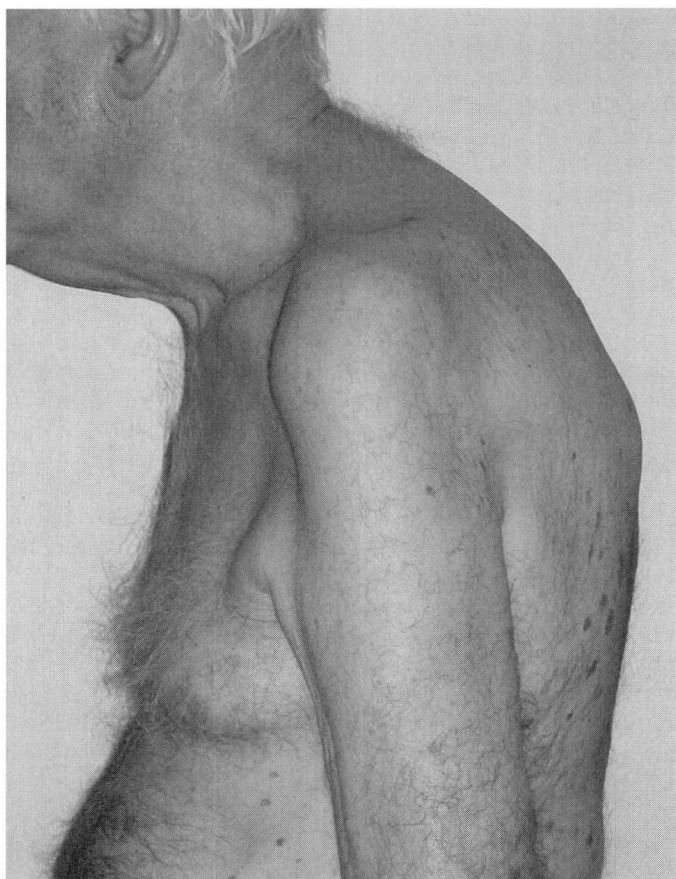

Fig. 23.2 The flail arm syndrome variant of amyotrophic lateral sclerosis.

in women, and has a longer median survival than classical amyotrophic lateral sclerosis (Hu *et al.* 1998). It is characterized by profound symmetrical weakness and wasting of the upper limbs with hyporeflexia. Although signs of pyramidal tract involvement may be seen in the lower limbs, there is little or no functional impairment of the bulbar muscles or legs at presentation, and these regions of the motor system are usually involved only late in the disease course. A similar focal presentation in the lower limbs with progressive paraparesis is recognized. Other forms of segmental motor neurone degeneration are described in Section 23.3.4.

Madras form. A specific subtype of early onset sporadic amyotrophic lateral sclerosis was identified in Southern India in 1970 (Meenakshisundaram *et al.* 1970; Saha *et al.* 1997; Gourie-Devi and Nalini 2003b). The phenotype is that of young onset amyotrophic lateral sclerosis with bulbar and limb involvement. Lower motor neurone features tend to predominate, but upper motor neurone signs are present and include brisk tendon reflexes and extensor plantar responses. The onset is usually in the second or third decades and the disorder is more common in males. Typical early features comprise distal upper limb weakness with wasting, dysarthria, dysphagia, and progressive sensorineural hearing loss. Progressive deafness is found in more than 50 per cent of affected individuals and is an important feature distinguishing Madras amyotrophic lateral sclerosis from other forms with early sporadic onset. Spasticity and distal leg weakness frequently develop and the disease course is one of progressive disability but with long-term survival. Electromyography studies show chronic partial denervation. Pathological findings from one case have been described (Shankar *et al.* 2000). There was depletion of lower motor neurones from the spinal cord and brainstem; neuronal loss and gliosis in the cochlear nucleus; demyelination and axonal loss in the cochlear nerve, and degeneration of the corticospinal tracts with accompanying gliosis. There was no description of whether ubiquitinated inclusion bodies were present in surviving motor neurones.

Disease course, prognosis, and death

Symptoms and signs are often focal at the time the patient with amyotrophic lateral sclerosis first presents to medical attention, though neurophysiological evidence of motor neurone injury may be more widespread. In the majority of cases there is an inexorable progression of the pathological process, often seemingly to contiguous groups of motor neurones. At the end stage of the disease, the patient will usually have significant motor dysfunction affecting all four limbs and the bulbar muscles as well as compromise of respiratory function. The patient may become entirely dependent upon nursing care, bulbar dysfunction may progress to anarthria, and enteral feeding may be required due to severe dysphagia.

Table 23.2 Summary of revised El Escorial research diagnostic criteria for amyotrophic lateral sclerosis (ALS)

The diagnosis of amyotrophic lateral sclerosis requires:

(A:1) Evidence of LMN degeneration by clinical, electrophysiological, or neuropathological examination;

(A:2) Evidence of UMN degeneration by clinical examination, and

(A:3) Progressive spread of symptoms or signs within a region or to other regions, as determined by history or examination,

Together with

(B) The absence of:

(B:1) Electrophysiological and pathological evidence of other disease that might explain the signs of LMN and/or UMN degeneration, and

(B:2) Neuroimaging evidence of other disease processes that might explain the observed clinical and electrophysiological signs

---

Within these principles, one can diagnose:

Definite ALS

♦ UMN signs and LMN signs in three regions

Probable ALS

♦ UMN signs and LMN signs in two regions with at least some UMN signs rostral to LMN signs

Probable ALS—Laboratory supported

♦ UMN signs in one or more regions and LMN signs defined by EMG in at least two regions

Possible ALS

♦ UMN signs and LMN signs in one region (together), or

♦ UMN signs in two or more regions

♦ UMN and LMN signs in two regions with no UMN signs rostral to LMN signs

---

UMN: (upper motor neurone) signs: clonus, Babinski sign, absent abdominal skin reflexes, hypertonia, loss of dexterity.

LMN: (lower motor neurone) signs: atrophy, weakness. If only fasciculation: search with EMG (electromyography) for active denervation.

Regions reflect segmental neuronal pools: bulbar, cervical, thoracic, and lumbosacral.

Certain clinical features are associated with a worse prognosis for the clinical course of the disease including: older age at onset of symptoms; early compromise of respiratory function; bulbar onset of symptoms and more rapid presentation of the patient to medical attention (Chancellor *et al.* 1993; del Aguila *et al.* 2003; Millul *et al.* 2005). The mean survival from symptom onset is approximately 3 years, although the rate of progression varies between individuals. Rapid variants of disease may progress to death within a few months, but approximately 4 per cent of patients will survive for more than 10 years (Turner *et al.* 2003). The usual cause of death is progressive respiratory failure, which may be accompanied by bronchopneumonia.

23.2.2 Diagnostic criteria

The El Escorial Diagnostic criteria for amyotrophic lateral sclerosis were agreed as an international consensus to facilitate therapeutic trials and multinational research collaborations (Brooks 1994). These criteria were subsequently revised and the key features of the revised criteria are shown in Table 23.2 (Brooks *et al* 2000). In essence there must be a combination of upper and lower motor neurone signs, evidence of progression over at least 6 months, and other conditions that may mimic amyotrophic lateral sclerosis must be excluded by appropriate investigations. These criteria are stricter than the burden of proof for the diagnosis of amyotrophic lateral sclerosis usually applied in clinical practice and indeed some individuals die from the effects of amyotrophic lateral sclerosis without ever reaching the classification criteria required for definite diagnosis (Traynor *et al.* 2000b). Clinicians should be cautious about applying these criteria in clinical practice as patients may be confused and less able to come to terms with their illness by being told that they have probable or possible amyotrophic lateral sclerosis rather than a more definite diagnosis. Nevertheless, the criteria provide a structured approach to the evaluation of patients with amyotrophic lateral sclerosis and facilitate the inclusion of uniform populations of patients in clinical research studies.

23.2.3 Differential diagnosis and investigation

The main conditions to be considered in the differential diagnosis of amyotrophic lateral sclerosis are listed in Table 23.3. Diagnostic errors are not uncommon in amyotrophic lateral sclerosis and an important aspect of follow-up care is to review the diagnosis in patients whose symptoms or disease course have atypical features. Of crucial importance is not to miss the diagnosis of potentially treatable disorders or those with a more benign prognosis. In a population-based survey from Scotland (Davenport *et al.* 1996) amyotrophic lateral sclerosis was misdiagnosed in 8 per cent of 552 patients included on the register. In an Irish population-based study which reported 437 patients' referrals diagnosed initially as amyotrophic lateral sclerosis, an alternative diagnosis was found in 7 per cent (Traynor *et al.* 2000a). Most of the misdiagnosed cases had lower motor neurone syndromes: 22 per cent had multifocal motor neuropathy with conduction block (Sections 23.3.6 and 21.11.3) and 13 per cent had Kennedy's disease (Section 23.5.1). Other forms of motor neuropathy should also be considered including acute motor axonal neuropathy (Section 21.10.2), a variant of Guillain–Barré syndrome (Griffin *et al.* 1996; Léger and Salachas 2001) and acute porphyric neuropathy (Section 21.8.6). Myasthenia gravis (Section 24.10.1), particularly with prominent

Table 23.3 Differential diagnosis of amyotrophic lateral sclerosis/motor neurone disease

- Other acquired motor neuropathies including multifocal motor neuropathy with conduction block, acute motor axonal neuropathy, and acute porphyric neuropathy.
- Kennedy's disease
- Myasthenia gravis
- Benign fasciculation syndromes
- Peripheral nerve hyperexcitability with myokymia
- Focal mechanical lesions of the spinal cord or brainstem e.g. multilevel spondylotic radiculomyelopathy, syringomyelia, tumours of base of skull, spinal cord, cauda equina/conus medullaris
- Hereditary spastic paraplegia
- Adult onset spinal muscular atrophy and hereditary motor neuropathies
- Myopathies including inclusion body myositis and polymyositis
- Multisystem neurodegenerative disorders e.g. frontotemporal dementia; multisystem atrophy; progressive supranuclear palsy; corticobasal degeneration; spinocerebellar atrophies (SCA 6 and SCA3)
- Infective disorders eg HTLVI; HIV; Lyme disease
- Post-polio syndrome
- Paraneoplastic syndromes
- Metabolic disorders e.g. hyperthyroidism and hyperparathyroidism

bulbar involvement, may occasionally mimic features of amyotrophic lateral sclerosis, though upper motor neurone clinical signs will be absent.

Benign fasciculation. This is relatively common, affecting approximately 1 per cent of the population, and often causes great anxiety, particularly in individuals with medical knowledge. The syndrome comprises chronic muscle fasciculation, most commonly in the calf muscles, which may be accompanied by cramps and is often worsened by exercise. Fasciculation may be detected by electromyography, but is not accompanied by evidence of progressive denervation (Blexrud *et al.* 1993). One study describing 121 patients with a diagnosis of benign fasciculation, 33 per cent of them health workers, seen at the Mayo clinic, reported no cases developing amyotrophic lateral sclerosis during a prolonged period of follow-up (Blexrud *et al.* 1993). However, individuals presenting with fasciculations and cramps may very rarely go on to develop amyotrophic lateral sclerosis (de Carvalho and Swash 2004). Muscle cramps and twitching in the presence or absence of weakness may occur in peripheral nerve hyper-excitability with myokymia (Hart *et al.* 2002; Gutmann and Gutmann 2004; Lagueny 2005). Myokymia usually consists of rippling continuous muscle contractions, rather than the discrete random twitching typically seen with fasciculation.

Focal structural lesions. These lesions of the brainstem or spinal cord may sometimes be confused with amyotrophic lateral sclerosis, particularly multilevel spondylotic radiculomyelopathy which may produce upper motor neurone signs accompanied by lower motor neurone signs in the upper and lower limbs. Other structural lesions which may occasionally be mistaken for amyotrophic lateral sclerosis include mass lesions at the skull base; slowly growing tumours of the cervical spinal cord, cauda equina or conus medullaris, and syringomyelia. Many of these patients will have pain and sensory disturbance as well as motor dysfunction. Other more peripheral lesions such as infiltration or compression of the lower roots of the brachial plexus by a Pancoast tumour (Section 22.4.2) or a cervical rib (Section 22.5.3) will also usually be accompanied by pain and sensory features. Occasionally patients with infiltrative lesions of the base of tongue may mimic the features of bulbar onset amyotrophic lateral sclerosis and the accompanying pain is an important diagnostic clue. In the presence of these structural lesions electrophysiological and imaging studies will usually lead to the correct diagnosis.

Hereditary spastic paraplegia variants (Section 23.4.2). These may have lower motor neurone features as well as the typical slowly progressive upper motor neurone disorder primarily affecting the lower limbs. There is clinical overlap between hereditary spastic paraplegia and the primary lateral sclerosis variant of motor neurone disease (Brugman *et al.* 2005; Strong and Gordon 2005). Patients with the Troyer syndrome, SPG 20, and Silver syndrome, SPG 17, typically have both upper and lower motor neurone features, but a much more slowly progressive clinical course than typical amyotrophic lateral sclerosis. Adult onset forms of spinal muscular atrophy and hereditary motor neuronopathy can pose diagnostic difficulties. Individuals with apparently sporadic lower motor neurone syndromes who survive more than 4 years from the onset of symptoms may turn out to have relatively benign forms of late onset spinal muscular atrophy (van den Berg-Vos *et al.* 2003a,b) (Section 23.3). Clues to the diagnosis of hereditary motor neuronopathy or spinal muscular atrophy include slow progression, family history, pes cavus, and other features such as vocal cord involvement.

Inclusion body myositis (Section 24.7.3) may mimic the progressive muscular atrophy variant. This is one of the most common types of myopathy presenting in those over 50 years of age (Griggs *et al.* 1995; Dalakas 2006). In one series of 70 patients with inclusion body myositis, 13 per cent had originally received a diagnosis of amyotrophic lateral sclerosis (Dabby *et al.* 2001). Important clues to the diagnosis of inclusion body myositis include early and prominent weakness of finger flexors and quadriceps muscles, which tend only to become weak late in the course of amyotrophic lateral sclerosis. Dysphagia may occur. Re-examination of patients both clinically and by electrophysiology where there is diagnostic uncertainty is very important and muscle biopsy may sometimes be indicated. Polymyositis may sometimes be misdiagnosed as amyotrophic lateral sclerosis. For example one reported case of biopsy-proven polymyositis was reported to show clinical features including dysphagia, possible upper motor neurone signs, normal creatine kinase level, and fasciculations on electromyography, causing understandable diagnostic uncertainty (Ryan *et al.* 2003).

Multisystem degenerative disorders. Disorders in which features of upper and lower motor neurone damage may occur include the frontotemporal dementia syndromes; subtypes of spinocerebellar atrophy including SCA6 and SCA3 (Section 39.8) (Ohara *et al.* 2002; Seilhean *et al.* 2004), multiple system atrophy (Section 40.3.8), progressive supranuclear palsy (Section 40.3.9), and corticobasal degeneration (Section 40.3.10) (Neary *et al.* 1990; Strong *et al.* 2003; Kertesz *et al.* 2005; Mott *et al.* 2005). These multisystem disorders all have prominent features of involvement of the central nervous system outside the upper and lower motor neurones including involvement of basal ganglia, the cerebellum, and cortical structures, so that diagnostic confusion with amyotrophic lateral sclerosis is uncommon.

Infective. Infective and post-infective disorders occasionally need to be considered in the differential diagnosis of amyotrophic lateral sclerosis. HTLV1 associated myelopathy (Sections 28.5.8; 42.4.1) typically produces a slowly progressive spastic paraparesis with early involvement of the bladder. It has been reported to occasionally be associated with amyotrophic lateral sclerosis-like features, with wasting and fasciculation of the tongue and limbs and electrophysiological evidence of widespread denervation (Matsuzaki *et al.* 2000; Silva *et al.* 2005). Patients have been reported with an amyotrophic lateral sclerosis-like disorder in the presence of HIV infection and some patients have apparently improved neurologically with highly active anti-retroviral treatment, 'HARRT' (Moulignier *et al.* 2001; Calza *et al.* 2004). Often the motor disorder occurring in the context of HIV is not entirely typical of amyotrophic lateral sclerosis, with a younger age of onset and a more rapidly progressive disease. The post-polio progressive muscular atrophy syndrome (Section 23.3.7) may sometimes be mistaken for amyotrophic lateral sclerosis, but its clinical course is much more slowly progressive. It may affect up to one-third of individuals who suffered acute paralytic poliomyelitis and is usually regarded as a syndrome of motor decompensation related to ageing (Ramlow *et al.* 1992; Ragonese *et al.* 2005). Lyme disease (Section 42.5.2) has been described as causing a progressive motor neurone disorder which may improve following the administration of antibiotic therapy (Hemmer *et al.* 1997). Lyme disease is not established as a cause of amyotrophic lateral sclerosis, but it is reasonable to consider the condition as a potential mimicking syndrome in individuals with exposure to tick bites, or when the CSF contains oligoclonal immunoglobulin bands. It is noteworthy that antibodies to *Borrelia burgdorferi* are commonly encountered in people living in endemic areas (Halperin *et al.* 1990). Shoulder girdle amyotrophy is a feature of infection by the tick borne encephalitis virus (Logina *et al.* 2006).

Malignancy. Most association studies have failed to demonstrate a convincing link between amyotrophic lateral sclerosis and malignant disease (Section 38.4.4) (Jokelainen 1976; Evans *et al.* 1990; Rosenfeld and Posner 1991; Freedman *et al.* 2005). However, some reports have suggested a possible association with several types of haematological malignancy and melanoma (Younger *et al.* 1990; Rowland *et al.* 1995; Gordon *et al.* 1997; Freedman *et al.* 2005). An amyotrophic lateral sclerosis-like syndrome has been reported in association with anti-Hu antibodies, and breast cancer may be linked to a primary lateral sclerosis like disorder without anti-Hu antibodies (Forsyth *et al.* 1997). Predominantly motor neuropathies or neuronopathies can be associated with anti-Hu or anti-Yo antibodies (Khwaja *et al.* 1998; Graus *et al.* 2001). Overall, a causative association between amyotrophic lateral sclerosis and cancer is not generally regarded as proven, although there are occasional reports of remission of motor dysfunction following therapy for malignant disease.

Metabolic disorders. These disorders occasionally mimic the features of amyotrophic lateral sclerosis. Hyperthyroidism may present with muscle weakness, wasting, and fasciculation (Rosati *et al.* 1980; Chotmongkol 1999). Some patients have superimposed upper motor neurone signs, a combination resulting in a clinical picture similar to that of amyotrophic lateral sclerosis (Fisher *et al.* 1985; Shaw *et al.* 1988). Treatment of the hyperthyroid state usually results in complete or near complete recovery of the upper motor neurone signs. Hyperparathyroidism may present with weakness and brisk reflexes and a causal relationship between hyperparathyroidism has been suggested (Patten and Pages 1984), but not established with certainty (Jackson *et al.* 1998).

Investigation

The diagnosis of amyotrophic lateral sclerosis is essentially clinical and there is no specific diagnostic test. Neurophysiological evaluation and MRI of the brain and spine are the most useful investigations. Sensory nerve conduction is usually normal and motor nerve conduction velocity is also normal unless there has been severe depletion of large diameter motor axons. It is important that conduction block is carefully excluded and this may require evaluation of proximal nerve segments. Electromyography is valuable in demonstrating neurogenic changes that cannot be explained by a single nerve, root, or plexus lesion (Mills 2003). Assessment of the thoracic paraspinal muscles can be particularly valuable in differentiating amyotrophic lateral sclerosis from multilevel spondylotic radiculomyelopathy (Kuncl *et al.* 1988). Imaging is useful to exlude the presence of structural disorders. A variety of blood tests may be helpful in distinguishing the amyotrophic lateral sclerosis mimic syndromes outlined above. Muscle biopsy is only indicated in rare or atypical cases where diagnostic uncertainty persists in the light of the initial investigation results.

23.2.4 Pathology

Motor neurone disease has been considered as traditionally a pure motor disorder, but the selectivity of the disease process for the motor system is now recognized to be relative rather than absolute. Careful clinical and pathological studies have revealed involvement in extra-motor parts of the central nervous system such as changes in other long tracts, including sensory and spinocerebellar pathways, and in neuronal groups such as substantia nigra neurones and dentate granule cells in the hippocampus. The description of ubiquitinated intraneuronal inclusions in 1988 highlighted a common molecular pathology in motor neurone disease (Leigh *et al.* 1988; Lowe *et al.* 1989) and has provided evidence of widespread involvement of extramotor regions of the central nervous system. Furthermore, the ubiquitinated inclusion is described in a number of related disorders, including primary lateral sclerosis, progressive muscular atrophy, and frontotemporal dementia, supporting the hypothesis that these conditions represent a clinicopathological spectrum of the same disease process. Thus, motor neurone disease is now regarded as a multi-system disease in which the motor neurones tend to be affected earliest and most severely (Ince *et al.* 1998a).

Gross pathological changes. The gross pathological changes of amyotrophic lateral sclerosis consist of atrophy of the cerebral precentral gyrus, and shrinkage, sclerosis, and pallor of the lateral and anterior corticospinal tracts tracts of the spinal cord. Thinning of the hypoglossal nerves and anterior spinal roots may be observed, and there is atrophy of the somatic musculature.

Lower motor neurone pathology. Motor neurone disease patients will typically have lost 50 per cent of the lower motor neurones in the limb enlargement areas of the spinal cord at autopsy (Ince 2000). Many of the remaining lower motor neurones show atrophic and basophilic changes that are likely to represent part of the spectrum of an apoptosis, programmed cell death pathway (Martin 1999). The depletion of lower motor neurones is accompanied by diffuse astrocytic gliosis in the spinal grey matter. There is relative preservation of motor neurones in the nucleus of Onufrowitz, Onuf's nucleus, in the sacral spinal cord (Mannen *et al.* 1977), which innervates skeletal muscles of the pelvic floor, and in the cranial motor nuclei of the oculomotor, trochlear, and abducens

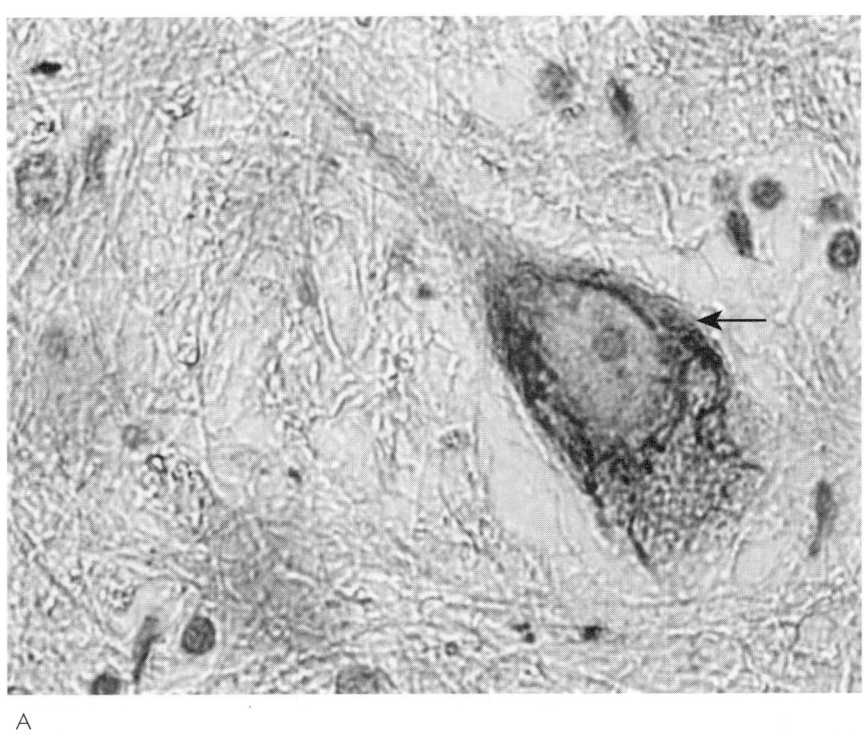

A

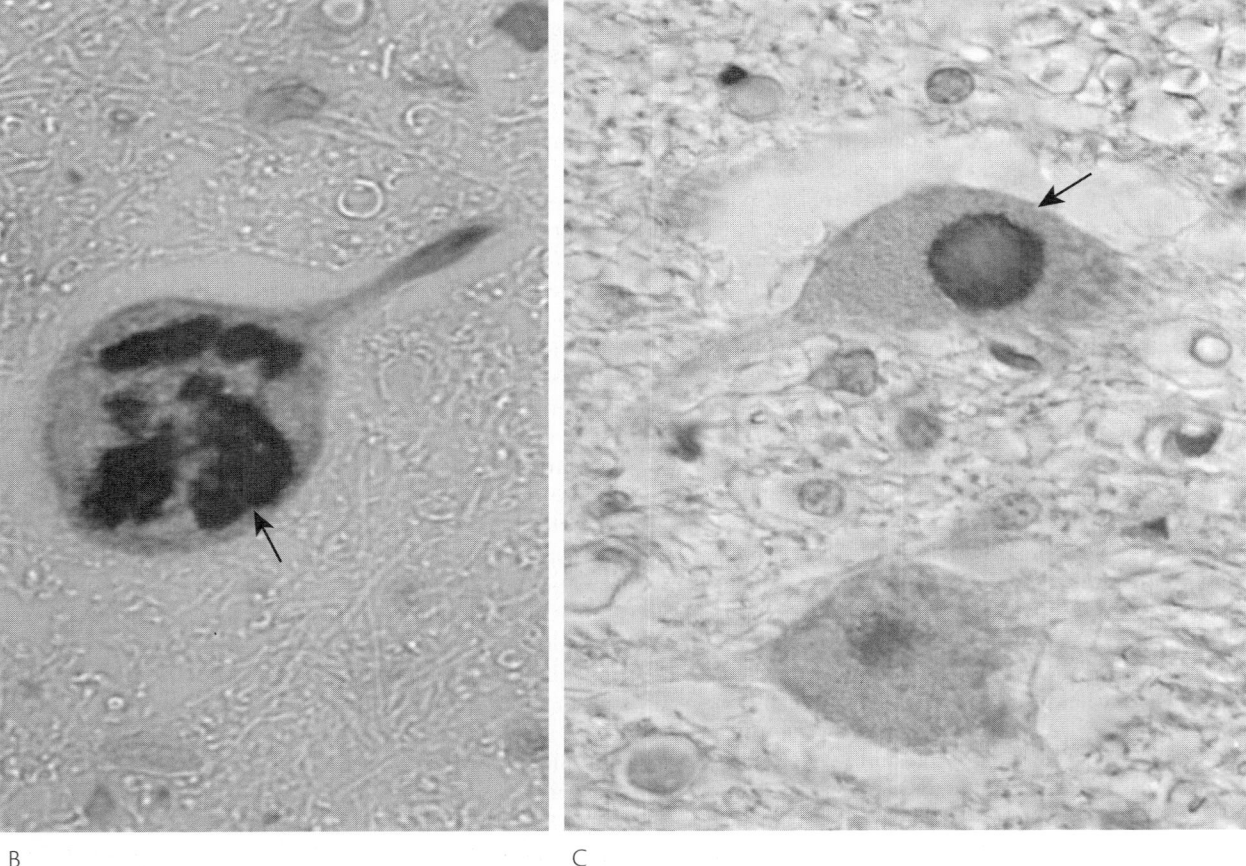

B   C

Fig. 23.3 Inclusion bodies within spinal cord motor neurones in amyotrophic lateral sclerosis. A. Skein like ubiquitinated inclusion; B. Hyaline conglomerate neurofilament inclusion body in a patient with superoxide dismutase related familial amyotrophic lateral sclerosis; C. Compact ubiquitinated inclusion; A,C immunohistochemistry for ubiquitin; B immunohistochemistry for neurofilament: all × 40 obj. Arrows in A, B, and C indicate the relevant inclusion.

nerves which control eye movements. The selective resistance of these two groups of motor neurones in motor neurone disease is unexplained.

A cardinal feature of lower motor neurone pathology in motor neurone disease is the presence of inclusion bodies within the soma and proximal dendrites. *Ubiquitinated inclusions* are the most frequent neuronal lesion, and are found in virtually 100 per cent of cases (Ince *et al.* 2003; Piao *et al.* 2003). These inclusions show a range of appearances including thread-like profiles; skeins (Fig. 23.3A) of varying compactness and more compact spherical bodies (Fig. 23.3C) (Ince *et al.* 1998a). Ubiquitin is a small highly conserved protein which becomes covalently bound to intracellular proteins targeted for disposal. The resulting tagged protein is degraded in the proteolytic channel of the 26S proteasome. Ubiquitin immunoreactivity is a prominent feature of the neuropathology of many neurodegenerative diseases, where it is found as a component of various inclusion bodies. Whereas in several other neurodegenerative diseases, the protein to which ubiquitin binds is known, for instance tau or α-synuclein ubiquitin-positive inclusions, in motor neurone disease the protein substrate for ubiquitination is not currently known with certainty, though TD43 represents an important candidate (Neumann *et al.* 2006).

Bunina bodies, first described in 1962 are eosinophilic inclusions present in the soma of lower motor neurones (Bunina 1962). They are found in 86 per cent of amyotrophic lateral sclerosis cases (Piao *et al.* 2003), and are specific to the disease. They have been shown to be immunoreactive to the proteinase inhibitor cystatin C, but their nature, origin, and significance remain unclear (Okamoto *et al.* 1993).

Neurofilament conglomerate inclusions are large inclusion bodies detectable with silver stains and with antibodies directed at both the phosphorylated and non-phosphorylated forms of heavy- and medium-chain neurofilament proteins (Ince *et al.* 1998b). Motor neurones containing these prominent inclusion bodies lose the normal neurofilament cytoskeleton from the remainder of the cell body. Neurofilament conglomerate inclusions are seen only infrequently in sporadic cases of motor neurone disease and are most commonly described in familial motor neurone disease caused by specific mutations in the SOD1 gene (Fig. 23.3B) (Rouleau *et al.* 1996; Ince *et al.* 1998). In normal motor neurones, neurofilament proteins within the cell body are predominantly non-phosphorylated, and there is progressive phosphorylation of neurofilaments in the axonal compartment. Several reports have documented a diffuse increase in neurofilament phosphorylation within the soma of spinal motor neurones in motor neurone disease (Munoz *et al.* 1988; Sobue *et al.* 1990). Swellings termed *spheroids* have been described in the axons of lower motor neurones in motor neurone disease. These are composed of abnormally orientated accumulations of neurofilaments, and are presumed to represent focal abnormalities of axonal cytoskeletal regulation, with failure of axonal transport (Ince 2000). Axonal spheroids are not however, disease specific, and are present in normal subjects, although the numbers of spheroids may be increased in motor neurone disease. Spheroids have been described in the cell body of spinal motor neurones (Hirano *et al.* 1967), and by immunocytochmistry appear similar to neurofilament conglomerate inclusions suggesting that the same dysregulation of cytoskeletal function may underlie both types of cellular lesion.

Upper motor neurone pathology. A major feature of the pathology of motor neurone disease is axonal loss within the descending pyramidal motor pathway, associated with secondary myelin pallor and gliosis of the corticospinal tracts. This pathology gives rise to the 'lateral sclerosis' within the spinal cord originally described by Charcot. Myelin pallor of the corticospinal tract is most prominent in the cervical cord and medullary pyramids, and in many cases is not demonstrable above the level of the medulla. This finding suggests that upper motor neurone changes in motor neurone disease arise from a dying back axonopathy, with distal denervation due to axonal loss preceding degeneration of the cell bodies of the upper motor neurones and loss of pyramidal cells from the cortex (Ince 2000).

In the motor cortex, pathological changes are highly variable, even in patients with well-established signs of upper motor neurone involvement clinically. Severely affected cases will usually show a reduction in the population of giant pyramidal neurones, Betz cells, in the motor cortex, either due to loss of these neurones, or a reduction in their size so that they are indistinguishable from smaller neighbouring pyramidal cells. Intracellular inclusions, including ubiquitinated inclusions, have not been convincingly described in Betz cells. The motor cortex may show astrocytic gliosis of varying severity in the grey matter and underlying subcortical white matter. Evidence of microglial activation, can be detected with immunocytchemical markers in the corticospinal tract within the brain and spinal cord (Troost *et al.* 1990; Ince *et al.* 2003).

Sensory and cerebellar pathways. Despite the paucity of sensory signs clinically, the ascending sensory pathways of the dorsal column are commonly affected in amyotrophic lateral sclerosis. In the Japanese literature, the prevalence of dorsal column pallor in familial disease is emphasized, with the concept of 'familial amyotrophic lateral sclerosis with posterior column involvement'. However, more recent work suggests that this change is detectable at autopsy in up to 50 per cent of all sporadic amyotrophic lateral sclerosis cases (Ince *et al.* 2003). There is also evidence of degenerative changes in peripheral sensory nerves, and loss of large afferent nerve fibres (Kawamura *et al.* 1981). Cerebellar involvement in amyotrophic lateral sclerosis is not usually recognized clinically, although late involvement of these pathways could go undetected in the presence of severe weakness. Pathologically, degeneration of the spinocerebellar pathway, reflected by cell loss from the thoracic nucleus of Clarke, and by pallor of ascending spinocerebellar pathways, is a relatively frequent finding in sporadic motor neurone disease (Brownell *et al.* 1970).

Muscle pathology. Loss of lower motor neurones result in denervation and atrophy of skeletal muscle. Histologically denervated muscle shows clusters of angular atrophic fibres and fibre-type grouping which results from serial denervation and reinnervation arising from collateral sprouting of axons of surviving motor neurones within muscle (Ince 2000).

Primary lateral sclerosis and progressive muscular atrophy. The pathological features of several patients with primary lateral sclerosis have been described (Pringle *et al.* 1992; Watanabe *et al.* 1997). The most prominent features described are atrophy of the prefrontal gyrus, with loss of Betz cell somata, and pallor and atrophy of the lateral corticospinal tracts. In cases where ubiquitin immunochemistry was undertaken, classical ubiquitinated inclusions were seen in a small number of lower motor neurones, and gliosis of the spinal ventral horns has also been described (Pringle *et al.* 1992; Watanabe *et al.* 1997). Ubiquitinated inclusion bodies in the cerebral cortex in a distribution typically found in amyotrophic lateral sclerosis-dementia cases has also been described (Kawashima *et al.* 1998), suggesting that primary lateral sclerosis may merge into the spectrum of amyotrophic lateral sclerosis dementia.

A pathological study of 14 patients with the clinical phenotype of progressive muscular atrophy demonstrated typical ubiquitinated inclusions in the spinal cord or bulbar motor neurone groups in most cases (Ince *et al.* 2003). Interestingly, 50 per cent of this series were also shown to have evidence of corticospinal tract involvement demonstrated by immunostaining for active microglia and macrophages.

The pathological findings of occasional abnormalities of the lower motor neurones in primary lateral sclerosis, and of corticospinal tract involvement in progressive muscular atrophy support the clinical evidence of overlap between these conditions and amyotrophic lateral sclerosis (Rowland 1999). Ubiquitinated inclusions are the characteristic pathological feature of amyotrophic lateral sclerosis, primary lateral sclerosis, and progressive muscular atrophy suggesting that these disorders share common pathophysiological mechanisms (Mackenzie and Feldman 2005). Primary lateral sclerosis and progressive muscular atrophy should be considered part of the spectrum of amyotrophic lateral sclerosis.

Pathological features of prolonged disease. Patients whose survival is prolonged by invasive ventilation may develop clinical and pathological features of involvement of less vulnerable motor neurone groups and eventually widespread involvement of the central nervous system (Hayashi and Kato 1989; Mizutami *et al.* 1990; Sasaki *et al.* 1992). These patients may lose voluntary control of the ocular muscles, and become totally 'locked-in'. They also lose control of external sphincters, and develop decubitus ulcers, features

not typically seen in motor neurone disease. Electroencephalograms performed at later stages of disease show diffuse slowing (Hayashi *et al.* 1989) and MR imaging demonstrates progressive cerebral atrophy, including the frontal and temporal lobes, precentral gyrus, postcentral gyrus, anterior cingulate gyrus, and corpus callosum (Kato *et al.* 1993).

Pathology of cognitive impairment. Patients with motor neurone disease and dementia have both the characteristic motor system ubiquitinated inclusions, together with cerebral pathology, which consists of small globular ubiquitinated inclusions within the dentate granule cells (Fig. 23.4), and a variable component of neocortical ubiquitinated neurites and small neuronal ubiquitinated inclusions (Munoz *et al.* 2003; Mackenzie and Feldman 2003). Other limbic structures, including the amygdala and parahippocampal gyrus may also show ubiquitinated inclusion pathology. Autopsy series indicate that 20–50 per cent of non-demented motor neurone disease patients have similar cerebral pathology, although the severity and distribution tends to be less extensive (Wilson *et al.* 2001; Mackenzie *et al.* 2003; Mackenzie and Feldman 2005). The degree of overlap in pathological findings, neuropsychological deficits, and imaging studies in classical motor neurone disease without dementia, motor neurone disease inclusion dementia, and amyotrophic lateral sclerosis-frontotemporal dementia suggest that they represent a spectrum of clinical disease with a common pathological substrate.

23.2.5 Disease pathogenesis

The primary pathogenetic processes underlying amyotrophic lateral sclerosis are multifactorial and the precise mechanisms underlying selective cell death in the disease are at present

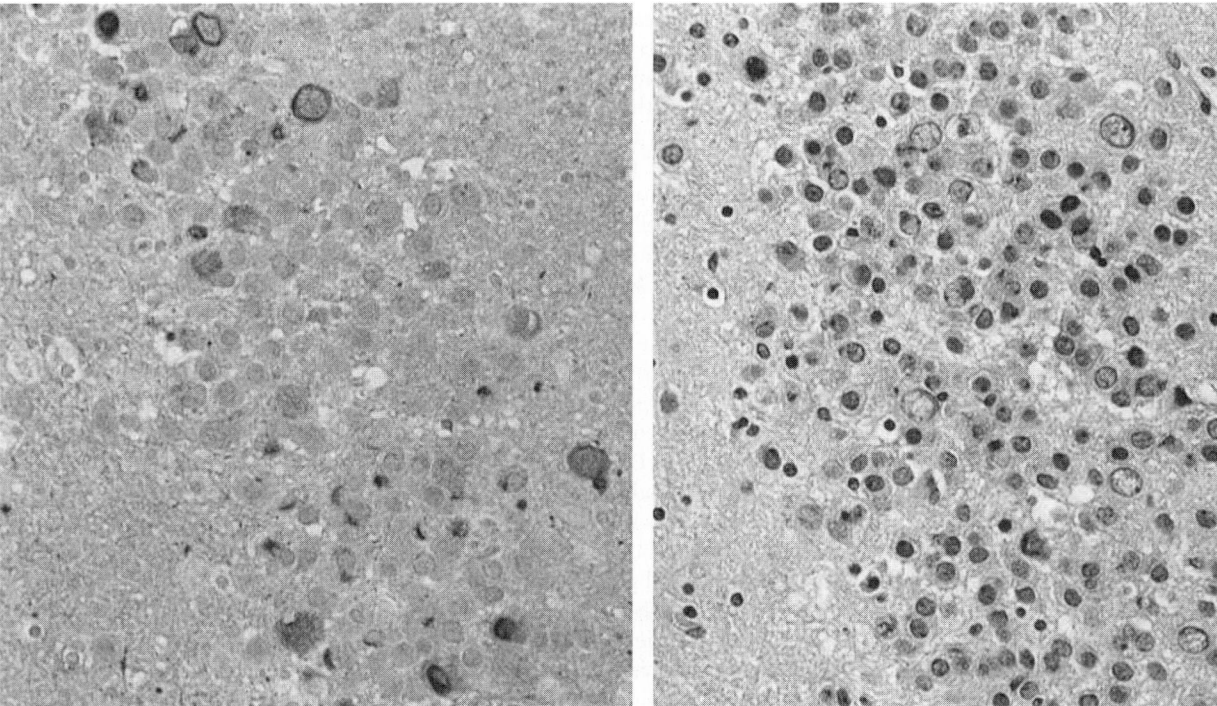

Fig. 23.4 Ubiquitinated inclusions in the dentate granule cells of the hippocampus in a patient with amyotrophic lateral sclerosis dementia. Immunohistochemsitry for ubiquitin: × 20 obj.

incompletely understood. Current understanding of the neurodegenerative process in amyotrophic lateral sclerosis suggests that there may be a complex interplay between multiple mechanisms including genetic factors, oxidative stress, excitotoxicity, and protein aggregation as well as damage to critical cellular processes, including axonal transport and organelles such as mitochondria (Bruijn and Cleveland 1996; Bruijn et al. 2004; Shaw 2005). Recently there has been growing interest in the role played in motor neurone injury by neighbouring non-neuronal glial cells and in dysfunction of particular molecular signalling pathways. The relative importance of these different pathways may well vary in different subgroups of patients, and a very important task for the future is to further define the subgroups of amyotrophic lateral sclerosis. Evidence has also accumulated that the final process of motor neurone death is likely to occur via a caspase-dependent programmed cell death pathway resembling apoptosis.

Genetics of motor neurone disease

Patients with amyotrophic lateral sclerosis report a family history of the disease in about 5–10 per cent of cases. The most common pattern of inheritance is autosomal dominant, with complete penetrance, although recessive or X-linked inheritance occurs in some pedigrees. Linkage studies in amyotrophic lateral sclerosis are rendered difficult by age-dependent onset in adults, short disease duration, heterogeneity of presentation, and misdiagnosis. Linkage to 12 different chromosomal loci has been established in familial motor neurone disease, and for 6 of these, the underlying genetic defect has been identified (Table 23.4).

ALS 1: Copper–zinc superoxide dismutase, SOD1. Familial motor neurone disease with adult onset is clinically indistinguishable from sporadic amyotrophic lateral sclerosis in individual cases. Twenty per cent of families with autosomal dominant motor neurone disease show mutations in the gene on chromosome 21q22.1 which encodes the free radical scavenging enzyme superoxide dismutase, referred to as SOD1 (Rosen et al. 1993). More than 100 mutations have now been identified, the majority of which are missense mutations. SOD1 is a ubiquitously expressed metalloenzyme whose major function is to convert intracellular superoxide free radicals to hydrogen peroxide. SOD1 is an abundant protein in the central nervous system, accounting for about 1 per cent of brain protein. SOD1 was initially thought to be confined to the cytosolic compartment of cells but it is now recognized that a small proportion of the protein is located in the intermembrane space of mitochondria (Okado-Matsumoto and Fridovich 2001). The reasons why motor neurones are especially vulnerable to injury in the presence of SOD1 mutations, are not yet clear. Despite 13 years of intensive research effort, the pathways leading to the cell death of motor neurones in the presence of SOD1 mutations have not been fully identified, although there is a convincing body of evidence that the mutant SOD1 protein exerts its detrimental effects through a toxic gain of function rather than a loss of function. Most of our current level of understanding of disease mechanisms in amyotrophic lateral sclerosis has come from the study of the effects of SOD1 mutations but, even in this defined genetic subgroups of disease, the pathways to neurodegeneration appear to be complex and multifactorial.

There is considerable variation in disease phenotype in terms of age of onset and rate of disease progression in human SOD1 related motor neurone disease. It is apparent that the clinical phenotype must be modified by other genetic and/or environmental factors.

There has been much interest in the *D90A SOD1* mutation, which has a dominant inheritance in some genetic backgrounds, but is recessively inherited with two mutated copies of the gene required to cause disease, in Scandinavian populations, implying a co-inherited protective factor (Andersen et al. 1996).

Mutant *SOD1* transgenic mice have been genetically engineered and develop a disease which clinically and pathologically resembles human motor neurone disease. The most extensively studied are *SOD1 G93A*, *SOD1 G37R*, and *SOD1 G85R* (Bruijn et al. 1996). Transgenic rats, carrying *G93A* or *H46R SOD1* also develop a motor neurone disease phenotype (Nagai et al. 2001). In addition cellular models of *SOD1* related motor neurone disease have been generated which have helped to elucidate cellular mechanisms of disease (Pasinelli et al. 1998; Cookson et al. 2002). The toxic gain of function of mutant *SOD1* has not yet been fully defined, but there are several pathophysiological processes which may be involved, including oxidative stress, mitochondrial dysfunction, excitotoxicity, protein aggregation, and inflammation. These mechanisms are not mutually exclusive and it is possible that all of these factors play a role in the development of motor neurone injury.

ALS 2: Alsin. In 2001 alsin was identified as the causative gene for an autosomal recessive form of juvenile amyotrophic lateral sclerosis linked to chromosome 2q33 (Hadano et al. 2001; Yang et al. 2001). Mutations in alsin can also cause a motor neurone degenerative disorder with a predominant upper motor neurone phenotype: juvenile recessive primary lateral sclerosis; infantile onset ascending hereditary spastic paralysis (Eymard-Pierre et al. 2002); or autosomal recessive complicated hereditary spastic paraplegia (Gros-Louis et al. 2003).

ALS 2 encodes a 184 KDa protein which contains three putative guanine nucleotide exchange factor domains and is alternatively spliced to generate a short and a long transcript. These factors are known to activate small GTPase proteins by stimulating the release of GDP in exchange for GTP. Given the conserved guanine nucleotide exchange factor domains of ALS2, it is predicted to function as an activator of particular small GTPases. The small GTPases control a range of important cellular processes and function as binary switches—alternating between inactive GDP-bound and active GTP-bound states. The alsin protein is widely expressed, but enriched within the central nervous system, where it is localized to the cytoplasmic face of endosomal membranes (Yamanaka et al. 2003). The functions of alsin are still being investigated, but to date it has been shown to function as an activator of the small GTPase protein Rab5 (Otomo et al. 2003; Kunita et al. 2004). This implies that alsin is important in endosomal dynamics and the working hypothesis is that alsin normally regulates trafficking of signalling molecules important for proper development and/or maintenance of health of motor neurones. ALS2 knockout mice have been generated but the motor system phenotype so far arising appears to be very mild (Kris et al. 2003).

ALS 4: Senataxin The ALS4 locus linked to chromosome 9q34 was originally identified in a single large pedigree with juvenile onset, autosomal dominant amyotrophic lateral sclerosis. The disease course in this family was indolent and did not reduce life expectancy. Three different missense mutations, L3095, R2136H, and T3I, in three families with this subtype of motor neurone disease were identified (Chen et al. 2004). The *SETX* gene encodes senataxin, a large 302.8KD protein of unknown function. Much of the protein has no homology with other known proteins but there is one DNA/RNA helicase domain. DNA/RNA helicase proteins are known to have roles in processes such as repair, replication, recombination or transcription of DNA and RNA processing, RNA transcript stability, and the initiation of translation. Recessive loss of function mutations in *SETX* are associated with ataxia-oculomotor apraxia type 2 (Moreira *et al.* 2004). It is predicted that the different phenotype of dominantly inherited ALS4 is likely to be caused by a toxic gain of function of the mutated senataxin protein.

ALS 8: VAPB. Nishimura and co-workers described a novel missense mutation, P565, in the *VAPB* gene for vesicle-associated membrane protein/synaptobrevin associated membrane protein, located on chromosome 20q13.3, in a Brazilian family with ALS8, an autosomal dominant slowly progressive disorder characterized by fasciculation, cramps, and postural tremor (Nishimura *et al.* 2004). They subsequently found the same mutation in six further families with different clinical phenotypes, including late onset spinal muscular atrophy and classical rapidly progressive amyotrophic lateral sclerosis. Vesicle associated proteins are intracellular membrane proteins that can associate with microtubules and have been shown to function in membrane transport. The VAPB protein has three identifiable structural domains. The first 150 residues form an MSP domain conserved between all members of this protein family; the central region contains an amphipathic helical structure predicted to form a coiled/coil protein–protein interaction motif and at the carboxy terminus is a hydrophobic region that acts as a membrane anchor. Preliminary cell biological studies have indicated that the wild-type VAPB protein localizes predominantly to the endoplasmic reticulum. The P56S mutation disrupts the subcellular distribution and induces the formation of intracellular protein aggregates (Nishimura *et al.* 2004).

Dynactin mutation. A mutation, G595, substitution in the gene encoding the P150 subunit of dynactin, DCTN1, has been identified in a single family with a slowly progressive lower motor neurone degenerative disorder (Puls *et al.* 2003). The described family had an unusual phenotype, presenting in early adulthood with respiratory difficulties due to vocal cord paralysis, progressive facial weakness, and weakness and atrophy of the hands and later development of neurogenic changes distally in the lower limbs. Mutations of the p150 subunit of DCTN1 were subsequently identified in patients with familial amyotrophic lateral sclerosis and frontotemporal dementia, and one case of apparently sporadic amyotrophic lateral sclerosis (Munch *et al.* 2004; Munch *et al.* 2005).

The dynactin–protein complex is required for dynein mediated retrograde axonal transport of vesicles and organelles along the microtubule system. The amino acid change caused by the mutation is predicted to distort the folding of the microtubule binding domain of dynactin. Overexpression of the P50 subunit of dynactin has been shown to disrupt the function of this protein complex and causes late onset progressive motor neurone degeneration in genetically engineered mice (La Monte *et al.* 2002).

Other amyotrophic lateral sclerosis loci. The genes for several other subtypes of amyotrophic lateral sclerosis remain to be identified. Three separate families have shown linkage to chromosome 16. Amyotrophic lateral sclerosis with fronto-temporal dementia has been mapped to a 17-cM interval chromosome 9q21 (Hosler *et al.* 2000; Ostojic *et al.* 2003), and one Swedish family with a similar phenotype without linkage to the chromosome 9 locus has recently been identified, suggesting genetic heterogeneity for this subtype of disease. Motor neurone degeneration may occur in patients with fronto-temporal dementia and Parkinson's disease associated with mutations in the microtubule associated protein tau MAPT (Clark *et al.* 1998). The mutant tau protein forms filamentous inclusions and insoluble aggregates that are associated with neurodegeneration. Some patients with familial fronto-temporal dementia, Parkinsonism, and amyotrophic lateral sclerosis do not have identified mutations in tau suggesting that further genes causing this triad of features remain to be identified (Kowalska *et al.* 2003).

Possible genetic risk factors in sporadic amyotrophic lateral sclerosis. There have been reports of genetic variants found in individuals with apparently sporadic disease (summarized in Table 23.4) (Kunst 2004; Shaw 2005).

Oxidative stress

Cellular injury by free radical species is a major potential cause of the age-related deterioration in neuronal function which occurs in neurodegenerative diseases. There has been particular interest in the role of oxidative stress in amyotrophic lateral sclerosis given that *SOD1* mutations, which encode a key cellular anti-oxidant defence protein, underlie approximately 20 per cent of familial cases. Studies of CSF and human post-mortem central nervous system tissue have demonstrated the presence of biochemical changes which represent the effects of oxidative stress and these changes are more pronounced in amyotrophic lateral sclerosis cases compared to controls (Shaw *et al.* 1995b; Ferrante *et al.* 1997; Smith *et al.* 1998; Tohgi *et al.* 1999). Fibroblasts cultured from the skin of patients with both familial and sporadic amyotrophic lateral sclerosis show increased sensitivity to oxidative insults compared to those from control cases (Aguirre *et al.* 1998).

In relation to the toxic gain of function of the mutant superoxide dismutase 1 protein, oxidative damage and/or metal mishandling have been strongly implicated. The main hypotheses have been that mutations alter the structure of this protein, allowing greater access of abnormal substrates to the active copper site of the dimeric enzyme, resulting in the production of damaging free radical species including peroxynitrite and hydroxyl radicals. Nitration of tyrosine residues on cellular proteins by peroxynitrite can have damaging consequences (Beckman *et al.* 1993). Some *SOD1* mutations render the protein more likely to form a zinc deficient variant (Crow *et al.* 1997), which in turn makes the copper site more accessible to abnormal substrates. *In vitro* studies have demonstrated that zinc deficient superoxide dismutase 1 causes peroxynitrite dependent cell death (Estevez *et al.* 1999). However, a body of experimental work has raised questions as to whether the toxicity of mutant superoxide dismutase 1 can be explained by copper-dependent oxidative mechanisms. For example, superoxide dismutase 1 that has been manipulated not to bind copper by mutating the four histidine residues for copper binding still causes motor neurone disease in transgenic mice. Also, knock out of the gene encoding the copper chaperone protein normally required for insertion of copper into the enzyme, has no effect on the disease phenotype in *SOD1* transgenic mice (Subramaniam *et al.* 2002). It seems plausible that mutant superoxide dismutase 1 may induce oxidative stress by a mechanism beyond its own catalytic activity and transcriptional repression of anti-oxidant response genes under the control of the transcription factor NRF2 is one potential pathway of interest (Kirby *et al.* 2005).

Table 23.4 Genetics of familial amyotrophic lateral sclerosis (ALS) and genetic association factors in sporadic amyotrophic lateral sclerosis

	Age of onset	Inheritance	Chromosome	Gene
FAMILIAL ALS				
ALS1	Adult	Dominant	21q22.1	SOD1
ALS2	Juvenile	Recessive	2q33	ALSIN
ALS3	Adult	Dominant	18q21	
ALS4	Juvenile	Dominant	9q34	Senataxin
ALS5	Juvenile	Recessive	15q15.1-q21.1	
ALS6	Adult	Dominant	16q12	
ALS7	Adult	Dominant	20ptel-p13	
ALS8	Adult	Dominant	20q13.33	VAPB
ALS-X	Adult	Dominant	Xp11-q12	
ALS-DEMENTIA				
ALS-FTD	Adult	Dominant	9q21-22	
ALS-FTD	Adult	Dominant	9p21.3-13.3	
ALS-PD-dementia	Adult	Dominant	17q21-22	MAPT
MND-ID	Adult	Dominant	17q21-22	not tau related
DLDH and ALS	Adult	Dominant	3	CHMP2B
SPORADIC ALS GENETIC ASSOCIATIONS	**Variant**		**Chromosome**	**Gene**
	6 del / 2 ins		22q12	Neurofilament heavy KSP region
	1 del		Mitochondria	Cytochrome c oxidase
	Point mutation		Mitochondria	tRNA (Ile)
	1 del		14q11-q12	AP endonuclease
	Splicing		11p13	EAAT2
	Promotor SNPs		6p12	VEGF
	Copy number		5q13	SMN
	E4 isoform		19q13	Apo E
	Point mutations		14q11.2	Angiogenin
	Point mutations		3p12.1	CHMP2B

Excitotoxicity

Glutamate is the major excitatory transmitter in the human central nervous system and there is great complexity in the molecular structure of the repertoire of receptors for this neurotransmitter system. Excitotoxicity is the term coined for neuronal injury induced by excessive stimulation of glutamate receptors, by mechanisms which include derangement of intracellular calcium homeostasis, and excessive free radical production. Motor neurones appear particularly susceptible to toxicity via activation of cell surface AMPA receptors (Carriedo et al. 1996). Glutamatergic toxicity has been implicated as a contributory factor to motor neurone injury in amyotrophic lateral sclerosis (Heath and Shaw 2005). The key findings are that the expression and function of the major glial glutamate re-uptake transporter protein EAAT$_2$ may be impaired in the central nervous system of motor neurone disease patients and that CSF, and therefore central nervous system extracellular fluid, levels of glutamate appear to be abnormally elevated at least in a proportion of motor neurone disease patients (Rothstein et al. 1995; Shaw et al. 1995a; Fray et al. 1998; Spreux-Varoquaux et al. 2002). Excitotoxicity has provided a potential mechanistic link between SOD1 mutant mediated motor neurone disease and the sporadic form of the disease. The presence of mutant superoxide dismutase 1 increases the sensitivity of motor neurones to glutamate toxicity (Kruman et al. 1999); causes alteration in AMPA receptor subunit expression (Spalloni et al. 2004); as well as reduced expression of the major glutamate re-uptake transporter EAAT$_2$

(Bendotti et al. 2001). Whether as a primary or a propagating process, it appears that glutamate toxicity plays a contributory role to the injury of motor neurones in amyotrophic lateral sclerosis. This is supported by the finding that anti-glutamate therapy with riluzole has some effect, albeit modest, in prolonging survival in human amyotrophic lateral sclerosis patients, and in SOD1 mutant mouse models (Lacomblez et al. 1996; Gurney et al. 1996).

Mitochondrial dysfunction

Mitochondria serve multiple important intracellular functions including generation of intracellular ATP, buffering of intracellular calcium, generation of intracellular free radicals, and involvement in the initiation of apoptotic cell death. Age-related deterioration in mitochondrial function may contribute to the development of late-onset neurodegenerative diseases. There is a body of evidence emerging from investigation of human material and cellular and animal models indicating that mitochondrial dysfunction may contribute to motor neurone injury in amyotrophic lateral sclerosis and this has been reviewed (Beal 2000; Menzies et al. 2002b).

The key evidence for mitochondrial dysfunction in human amyotrophic lateral sclerosis includes:

- alteration in the morphology of mitochondria in hepatocytes, muscle and motor neurones;
- increased mitochondrial volume and calcium levels within motor axon terminals in muscle biopsies from sporadic amyotrophic lateral sclerosis cases (Siklos et al. 1996);

- reduced complex IV activity in spinal motor neurons in sporadic amyotrophic lateral sclerosis (Borthwick *et al*. 1999);

- high frequency of mitochondrial DNA mutations in motor cortex tissue in sporadic amyotrophic lateral sclerosis (Dhaliwal *et al*. 2000);

- Multiple mutations and decreased mitochondrial DNA in muscle and spinal cord in sporadic motor neurone disease (Wiedemann *et al*. 2002); and

- amyotrophic lateral sclerosis like phenotypes in several patients with genetic mutations affecting mitochondrial proteins (Comi *et al*. 1998; Borthwick *et al*. 2006).

In cellular models of superoxide dismutase 1 related motor neurone disease, expression of a *G93A SOD1* mutant results in the development of abnormally swollen mitochondria which display impaired activity of complexes II and IV of the mitochondrial respiratory chain, impaired cellular bioenergetic status, and alteration in the mitochondrial proteome (Menzies *et al*. 2002a; Wood-Allum *et al*. 2006). Molecular targeting of mutant superoxide dismutase 1 to the mitochondria but not to the nucleus or endoplasmic reticulum leads to activation of the apoptosis cascade and cell death (Takeuchi *et al*. 2002). In some strains of mutant *SOD1* transgenic mice mitochondrial vacuolation within motor neurones is an early feature of the pathology (Wong *et al*. 1995). Whereas superoxide dismutase 1 was previously considered to be an exclusively cytosolic protein, it is now recognized also to reside in the intermembrane space of mitochondria. Superoxide dismutase 1 has been shown to accumulate in vacuolated mitochondria in mutant *SOD1* mice. It has been demonstrated that the activities of several complexes of the mitochondrial respiratory chain are reduced prior to disease onset and that these changes increase with age. Oxidative damage to mitochondrial protein and lipids and decreased ATP synthesis have been reported at the onset of the murine disease. Translocation of cytochrome C, an initiator of apoptosis, from the mitochondria to the cytosol has been demonstrated during disease progression in the mice. Recent studies have reported that mutant superoxide dismutase 1 is selectively and aberrantly recruited to the cytoplasmic face of mitochondria in spinal cord tissue from mutant *SOD1* transgenic mice and that the anti-apoptotic protein Bcl2 may be entrapped within large protein aggregates of superoxide dismutase 1 within spinal cord tissue, which may result in reduced availability of this protein to regulate apoptosis (Liu *et al*. 2004; Pasinelli *et al*. 2004).

Therapeutic effects of compounds which modulate mitochondrial function have begun to be investigated in *SOD1* transgenic mouse models. Creatine buffers energy levels within the cell, maintains ATP levels, and stabilizes mitochondrial creatine kinase which inhibits opening of the mitochondrial permeability transition pore. Administration of creatine to G93A transgenic mice improved motor function and extended survival in a dose-dependent manner, as well as causing a decrease in biochemical indices of oxidative damage in the spinal cord (Klivenyi *et al*. 1999). Minocycline, a tetracycline derivative which inhibits microglial activation and blocks release of cytochrome c from mitochondria, also slows disease in mutant *SOD1* mice (Zhu *et al*. 2002).

Cytoskeleton and axonal transport

Motor neurones, which in the human nervous system may have axons up to 1 m in length, are highly reliant on an efficient intracellular transport system with anterograde and retrograde components. It is interesting that in *SOD1* mutant mice, axonal transport is demonstrably impaired several months before clinical disease onset (Wiliamson *et al*. 1999). The kinesin complex of proteins are important molecular motors for anterograde axonal transport on the microtubule system. Mutations of genes encoding several kinesin proteins have been shown to cause several types of motor neurone degeneration including a hereditary spastic paraplegia, SPG10, and type 2A Charcot–Marie–Tooth disease, though have not yet been associated with amyotrophic lateral sclerosis. The dynein–dynactin complex is the important motor for retrograde transport on the microtubule system, returning components such as multivesicular bodies and neurotrophic factors back to the perikaryon. Mutations in dynein and the dynactin complex which is an activator of cytoplasmic dynein, cause progressive motor neurone disease in mice (Lamonte *et al*. 2002; Hafezparast *et al*. 2003). Mutations in the P150 subunit of dynactin may cause a form of motor neurone disease in human subjects (Puls *et al*. 2003).

Neurofilament proteins form a major component of the cytoskeleton of neurones and important functions include maintenance of cell shape and axonal calibre, as well as axonal transport. Neurofilament subunits are assembled in the motor neurone cell body, and transported down the axon by slow axonal transport. Accumulation and abnormal assembly of neurofilaments are common pathological hallmarks of amyotrophic lateral sclerosis. Ubiquitinated inclusions with compact or Lewy body like morphology within surviving motor neurones in amyotrophic lateral sclerosis may show immunoreactivity for neurofilament epitopes. In some cases of *SOD1* mutation related amyotrophic lateral sclerosis, large argyrophilic neurofilament conglomerate inclusions have been observed in the cell bodies and axons of motor neurones (Ince *et al*. 1998). Approximately 1 per cent of sporadic amyotrophic lateral sclerosis cases have deletions of insertions in the KSP repeat region of the neurofilament heavy chain gene, *NFH* (Figlewicz *et al*. 1994; Tomkins *et al*. 1998). Pathological changes within motor neurones develop in mice overexpressing NF-light or -heavy subunits, or in mice expressing mutations in the *NFL* gene. Transgenic mice which carry *SOD1* mutations, also show alterations in neurofilament organization, with the development of neurofilament spheroids, as well as reduced neurofilament protein and decreased transport rate in the ventral root axons. Genetic manipulations to alter the expression of neurofilament proteins have been shown to alter the disease course in *SOD1* transgenic mice. Increased expression of *NFH*, resulting in trapping most neurofilaments within the cell body, robustly improves the disease course, by as much as 6 months in mutant *SOD1* mice (Couillard-Despres *et al*. 1998).

Protein aggregation

Misfolding of proteins with the formation of intracellular aggregates is a key feature of multiple neurodegenerative diseases. There is continuing debate as to whether such aggregated proteins play an important role in disease pathogenesis, whether they represent harmless by-standers, or whether they could be beneficial to the cell by sequestration of toxic abnormal proteins. In the *SOD1* transgenic mouse model of familial amyotrophic lateral sclerosis, the mutant superoxide dismutase 1 protein forms conspicuous cytoplasmic inclusions in motor neurones and sometimes in astrocytes, which develop before the onset of motor dysfunction.

Several hypotheses have been put forward to explain how mutant superoxide dismutase 1 aggregates could generate cellular toxicity:

◆ Sequestration of other proteins required for normal motor neurone function. Several additional proteins have been shown to be present in superoxide dismutase 1 aggregates including: CCS, the copper chaperone for SOD1, ubiquitin neurofilaments, glial fibrillary acidic protein, two neuronal glutamate transporters, BCl2 and proteins involved in chaperone and proteosome functions.

◆ By repeatedly misfolding, the superoxide dismutase 1 aggregates may reduce the availability of chaperone proteins required for the folding and function of other essential intracellular proteins.

◆ The superoxide dismutase 1 mutant protein aggregates may reduce proteasome activity needed for normal protein turnover.

◆ Inhibition of the function of specific organelles such as mitochondria by aggregation on or within these organelles.

Overexpression of chaperone proteins has been shown to reduce mutant superoxide dismutase 1 aggregation and enhances the survival and function of motor neurones in culture (Takeuchi *et al.* 2002). In addition, arimoclomol, a drug which up-regulates the expression of heat shock proteins increases the life span of *G93A SOD1* mice by 22 per cent (Kieran *et al.* 2004). Clearly protein aggregates, which can be identified by ubiquitin immunostaining, are a feature of sporadic as well as familial motor neurone disease. Superoxide dismutase 1 containing aggregates are not a characteristic feature of sporadic motor neurone disease and determining the nature of the protein inclusions in the sporadic disease is a very important research goal. A recent report has identified TDP-43, the TAR-DNA-binding protein of 43kDa, as a component of ubiquitinated inclusions in ALS (Neumann *et al.* 2006). TDP-43 is known to be a nuclear factor which plays a role in the regulation of transcription and alternative splicing. Further investigation is required to determine the role played by this protein in the pathogenesis of amyotrophic lateral sclerosis.

Inflammatory cascades and the role of non-neuronal cells

There has been much recent interest in the possibility that non-neuronal cells, including activated microglia and astrocytes, may contribute to the pathogenesis and/or propagation of the disease process in amyotrophic lateral sclerosis. Several studies in genetically engineered mouse models have suggested that expression of mutant superoxide dismutase 1 in neurones alone is insufficient to cause motor neurone degeneration and that involvement of non-neuronal cells in the vicinity of motor neurones may be required. Chimeric mice have been produced which have both normal and mutant *SOD1* expressing cells (Clement *et al.* 2003). Motor neurones expressing mutant superoxide dismutase 1 can escape disease if surrounded by a sufficient number of normal non-neuronal cells. Conversely normal motor neurones surrounded by mutant superoxide dismutase 1 containing non-neuronal cells, developed signs of cellular injury, with the development of ubiquitinated protein deposits. Thus, the mutant enzyme may cause neurotoxicity indirectly by perturbing the function of non-neuronal cells such as microglia (Boillee *et al.* 2006). Microglia play a critical role as resident immunocompetent and phagocytic cells within the central nervous system. Activation is associated with transformation to phagocytic cells capable of releasing potentially cytotoxic molecules including reactive oxygen species, nitric oxide, proteases, and pro-inflammatory cytokines (Gonzales-Scarano and Baltuch 1999). Given this, there is little doubt that activated microglia can inflict significant damage on neurones but their role is complex and they are capable of stimulating neuroprotective as well as neurotoxic effects. Proliferation of activated microglia is a prominent histological feature in the spinal ventral horn both in mutant *SOD1* trangenic mice and in human amyotrophic lateral sclerosis. In the mice, microglial activation is present before the onset of significant motor neurone loss or clinical signs of disease. Various inflammatory cytokines or enzymes are upregulated in the spinal cord or CSF of amyotrophic lateral sclerosis patients: 1L-6, 1L-1β, cyclo-oxygenase 2, and prostaglandin E2 or, in the spinal cord of mutant *SOD1* mice: 1L-1β, TNF-α, cyclo-oxygenase 2, and prostaglandin E2 (Almer *et al.* 2002; Hensley *et al.* 2002; Tikka *et al.* 2002). Microglia appear to mediate the toxicity to neurones in culture of CSF from patients with motor neurone disease by releasing factors which enhance glutamate toxicity. Minocycline, which inhibits microglial activation ameliorates disease progression in mutant *SOD1* mice (Zhu *et al.* 2002).

There is a tendency in amyotrophic lateral sclerosis for the disease to start focally and to spread to contiguous groups of motor neurones (Brooks *et al.* 1995). It would be very relevant to identify molecules that contribute to this propagation and clearly those released from activated microglia would be plausible candidates.

Apoptosis

Apoptosis describes the controlled removal of cells by an energy-dependent cell death programme. Key molecules which regulate apoptosis include: the caspase family of proteolytic enzymes which, when activated by cleavage, orchestrate cell destruction by digesting several intracellular targets including structural and regulatory proteins; the Bcl2 family of oncoproteins where the balance and subcellular distribution between pro- and anti-apoptotic members is crucial in regulating cell survival or destruction; and the apoptosis inhibitor family of proteins which suppress apoptosis by preventing proteolytic activation of specific caspases. Several pathways triggering caspase activation have been identified including: release of pro-apoptotic factors such as cytochrome c from mitochondria; activation of cell surface ligand receptor systems of the tumour necrosis factor family including Fas-Fas ligand; and stress to the endoplasmic reticulum with activation of caspase 12.

There is evidence that motor neurones may die in amyotrophic lateral sclerosis according to a programmed cell death pathway resembling apoptosis (Guegan and Przedborski 2003; Sathasivam and Shaw 2005). Key evidence from human post-mortem studies includes:

◆ structural morphology of degenerating motor neurones compatible with the apoptosis as well as internucleosomal DNA fragmentation detected by TUNEL staining;

◆ increased expression of specific apoptosis related molecules for instance Le^y antigen and prostate apoptosis response-4 protein in spinal cord;

◆ alteration in the balance of expression and subcellular compartmental localization of pro- and anti-apoptotic members of the Bcl2 family in a direction favouring apoptosis; and

◆ significant increases in the activities of caspases 1 and 3 in the spinal cord.

Investigation of cellular models of *SOD1* mutation related amyotrophic lateral sclerosis has shown that motor neuronal cells expressing mutant *SOD1* are more likely to die by apoptosis when oxidatively stressed (Cookson *et al.* 2002). In addition, under unstressed basal culture conditions, these mutant superoxide dismutase 1 containing cells appear to be 'primed' for cell death by expressing early molecular markers of apoptosis (Sathasivam *et al.* 2005). In the mutant *SOD1* transgenic mouse model, there is evidence of DNA laddering, increased expression, and activation of caspase 1 and caspase 3 in the spinal cord of symptomatic mice, and alterations in the balance of key members of the Bcl2 protein family in a direction favouring apoptosis (Li *et al.* 2000). Cross breeding experiments between G93A *SOD1* transgenic mice and mice genetically engineered to over-express anti-apopototic molecules results in amelioration of the murine disease. The administration of caspase inhibitors has a partial neuroprotective effect in cellular models (Sathasivam *et al.* 2005) and intraventricular administration of a broad spectrum caspase inhibitor to mutant *SOD1* mice prolongs life span by approximately 20 per cent (Li *et al.* 2000).

Cellular features of motor neurones predisposing to neurodegeneration

One of the unsolved mysteries in neurodegenerative diseases, including amyotrophic lateral sclerosis, is the selective vulnerability of particular neuronal groups to the neurodegenerative process. Superoxide dismutase 1 is a ubiquitously distributed anti-oxidant defence protein, yet when the protein is mutated, it is the motor neurones which are most susceptible to injury. Certain cell specific features of motor neurones may predispose to age-related degeneration (Shaw and Eggett 2000; Durham *et al.* 2003). Key features are likely to include the cell size of motor neurones which has downstream consequences for intracellular transport, energy metabolism, and neurofilament content. The neurones vulnerable to injury in motor neurone disease have particular sensitivity to glutamatergic toxicity via AMPA receptor activation and differ from most other neuronal groups in expressing a high preponderance of calcium permeable AMPA receptors, lacking the GluR2 subunit (Williams *et al.* 1997). Motor neurones also have a relative lack of expression of calcium buffering proteins (Ince *et al.* 1993) and appear to have a high threshold for mounting a protective heat shock response. (Durham 2003) Recent studies suggest that the properties of mitochondria from the spinal cord may differ from those of mitochondria from other tissues (Sullivan *et al.* 2004).

23.2.6 Epidemiology

The incidence of motor neurone disease reported from recent epidemiological studies ranges from 1 and 3 per 100 000, with point prevalence rates of 6–8 per 100 000 (Chancellor and Warlow 1992; Traynor *et al.* 1999). Most of the studies have been conducted from developed countries and relatively little is known of the incidence and prevalence in developing countries, or in specific racial or ethnic groups. Pockets of high incidence are described amongst the Chamorro indigenous population of the Western Pacific island of Guam, on the Kii peninsula of Japan and amongst the Auyu and Jakai people of Irian Jaya (Plato *et al.* 2003; Kuzuhara and Kokubo 2005). The explanation for the strikingly increased incidence and prevalence of motor neurone disease in these geographical foci remains uncertain. Although the prevalence remains high in Guam compared to typical populations in western countries, there has been a substantial decrease over the last half century.

Some studies have indicated that the overall age-related incidence of motor neurone disease has increased over several decades (Lilienfeld *et al.* 1989; Maasilta *et al.* 2001). However, it is unclear whether this is due to demographic factors, better ascertainment, or to changed exposure to unknown environmental risk factors. A gradual increase in the prevalence of motor neurone disease is expected given the changing age structure of the population and with the introduction of therapies which extend survival of patients.

The incidence of motor neurone disease increases with age, being very low under the age of 40 and peaking at approximately 75 years of age. The reported mean age of onset in sporadic motor neurone disease varies between 55 and 65 years in most studies, with a range varying between the third decade and the ninth decade (Jokelainen 1976; Kurtzke 1991). The mean age of onset in patients with familial motor neurone disease is about a decade earlier. Occasionally patients with classical motor neurone disease present in the second or third decade of life. Sporadic motor neurone disease is commoner in men than women, with a male/female ratio of around 1.6:1. This ratio approaches 1:1 in familial motor neurone disease. Women are relatively over-presented in older age groups, although the standardized age-related incidence is greater in elderly men. Bulbar onset is also more common in older patients, especially in older women (Haverkamp *et al.* 1995; Forbes *et al.* 2004). In most large clinic-based or population-based series, 5–10 per cent of cases are classified as familial.

Risk factors

The only risk factors proven to have an association with the development of motor neurone disease are gender, a positive family history, and increasing age. At the present time no environmental risk factors are regarded as of proven causative significance. Environmental factors which have been reported to increase the risk of developing motor neurone disease include trauma, physical activity, participation in athletic pursuits, dietary habits, alcohol consumption, cigarette smoking, residence in rural rather than urban areas, and working in certain occupations, for example the leather industry or electrical work. Using an evidence-based medicine approach, Armon (2003) concluded that smoking is probably associated with amyotrophic lateral sclerosis, but that the evidence in favour of other reported environmental factors was not strong. Recent studies have reported an increased risk of developing motor neurone disease in military personnel, in airline pilots, and in Italian professional football players (Horner *et al.* 2003; Chio *et al.* 2005; Weisskopf *et al.* 2005). Further investigation is required to substantiate these findings and to address the underlying causes of the reported occupational associations.

23.2.7 Management

There are two main goals in the management of motor neurone disease. The first is the alleviation of symptoms which occur during the course of the disease to maintain quality of life. The second is the administration of neuroprotective therapy to slow the progression of motor neurone injury and neurodegeneration.

Supportive care and symptom control

The management of motor neurone disease poses considerable ethical, logistical, and educational problems. Ethical issues are involved in aspects of management including the use of artificial methods for maintaining nutrition, ventilatory support, the use of neuroprotective drugs to slow disease progression, and the use of opiate medication in the terminal phase of the disease. The logistical and educational problems arise from the relative rarity of motor neurone disease and the fact that many health care professionals have little experience in dealing with the rapidly progressive weakness and bulbar and respiratory failure which may occur during the course of the disease. The coordinated action of multiple health care professionals within a multidisciplinary team can lessen the difficulties experienced by patients and by their families.

Symptomatic therapy aimed at alleviating the distressing symptoms which often arise during the course of motor neurone disease can do much to improve the quality of life for the patient. Detailed discussion of all of these therapies is beyond the scope of this chapter. Some of the common symptoms which may develop, and their symptomatic therapies are highlighted in Table 23.5. Many of the symptomatic therapies currently recommended by clinicians have not been assessed in rigorous controlled trials. The evidence base for some of these therapies was reviewed several years ago by an American Academy of Neurology task-force (Miller *et al.* 1999). A few areas of progress in the symptomatic management of motor neurone disease will be highlighted.

Specialist multidisciplinary clinics. There is an increasing tendency in the United Kingdom and worldwide for patients with motor neurone disease to be managed in specialist clinics. This allows the coordination of an experienced multi-disciplinary team, that includes the neurologist, specialist nursing staff, physiotherapist, occupational therapist, speech therapist, dietician, social worker, and orthotist. Input may also be required from other specialist teams including respiratory medicine, gastroenterology, and palliative care. Considerable support is also provided by patient associations such as the Motor Neurone Disease Association.

Nutritional management. Weight loss is universal in motor neurone disease patients and may be due to dysphagia, loss of muscle mass, or anorexia. Weight loss, malnutrition, and dehydration can aggravate muscle weakness and shorten lifespan, whilst frequent choking spells can make mealtimes intolerable. Malnutrition is an independent prognostic factor for survival in amyotrophic lateral sclerosis with an almost 8-fold increased risk of death in patients who are malnourished (Desport *et al.* 1999). The management of the nutritional status of motor neurone disease patients has improved in recent years. When the patient first begins to develop swallowing problems a few simple measures can be helpful. Dysphagia may be increased by anxiety and the social embarrassment resulting from slowness in eating, dribbling, and choking. Patients should be encouraged to eat in as relaxed and comfortable an environment as possible. Sucking ice before meals may decrease choking spells. Attention to food consistency is important, and the family should receive advice from a dietician. If the patient is continuing to lose weight, then nutritional supplements of liquid or semi-solid consistency may be helpful and a range of these should be tried to ascertain which preparations are most palatable for the individual patient. More active therapeutic intervention for the dysphagia should be considered when the following problems are apparent:

♦ continuing weight loss of more than 10–20 per cent of the normal body weight despite the above measures;

♦ dehydration;

♦ aspiration with resultant respiratory infection; and

♦ meal times have become too prolonged and tiring or intolerable due to frequent choking spells.

For long-term enteral feeding percutaneous endoscopic gastrostomy, or PEG, is the procedure of choice. The placement of a percutaneous endoscopic gastrostomy is a relatively straight-forward procedure which can be performed under a local anaesthetic. After placement many patients report great relief and increased well-being, though as yet no large scale quality of life studies have been conducted. Percutaneous endoscopic gastrostomy feeding in motor neurone disease has been the subject of a recent Cochrane review (Langmore *et al.* 2006), and represents one of the major advances in symptomatic care for patients, leading to weight stabilization and adequate nutrional and fluid intake, although a survival benefit has yet not been convincingly shown. The need for percutaneous endoscopic gastrostomy feeding should be anticipated as the risks of the procedure are higher once the patient's forced vital capacity falls below 50 per cent (Miller *et al.* 1999). In some patients, technical difficulties may be experienced in the insertion of a percutaneous endoscopic gastrostomy tube and in this situation, a radiologically guided method may be used (Thornton *et al.* 2002). In patients with a low vital capacity, the use of non-invasive positive pressure ventilation during percutaneous endoscopic gastrostomy insertion has been shown to improve tolerance and safety of the procedure (Gregory *et al.* 2002). A full strength feeding regimen providing 1500–2000 calories per day can be introduced over 48 h following feeding tube insertion. Feeding can be managed by syringe boluses at normal meal-times, or by continuous infusion by pump which can be given overnight. Some patients defer consenting to undergo percutaneous endoscopic gastrostomy insertion until an advanced disease stage is reached. In these patients, as well as attention to respiratory status, vigilance is required for the possibility of the re-feeding syndrome as a post-operative complication (Fotheringham *et al.* 2005).

Respiratory support. Respiratory muscle weakness develops insidiously during the course of motor neurone disease, causing dyspnoea, orthopnoea, and symptoms of carbon dioxide retention, which include daytime somnolence, morning headaches, and lack of restorative sleep, with frequent waking. The management of the respiratory complications include the following general measures. Attention should be given to the detection and prevention of aspiration pneumonia. Antibiotic therapy should be used at the first indication of a chest infection. When the patient has difficulty in clearing secretions from the chest, chest physiotherapy and postural drainage should be used if possible. The provision of a suction machine and the prescription of a mucolytic agent such as carbocisteine to reduce the viscosity of secretions may also be helpful. Patients will breathe more comfortably during sleep if placed in a semi-upright position. When patients experience bouts of severe dyspnoea, accompanied by extreme anxiety or panic, a small dose of lorazepam may be useful, 0.5–1 mg sublingually. If breathlessness causes distress during the later stages of the disease, the use of small amounts of morphine will be useful. Further depression of respiration can usually be avoided if the initial dose is small and increments are gradual.

Assisted ventilation, coupled with appropriate nutritional support, could theoretically extend the patient's life indefinitely, and the implications of initiating such respiratory support must be clearly thought through and discussed for each patient. There are

Table 23.5 Symptomatic therapies in amyotrophic lateral sclerosis

SYMPTOM	TREATMENT
Muscle weakness and fatigue	— Physiotherapy to prevent joint stiffness and muscle contractures — Appliances to maintain mobility and independence such as walking aids, wheel chairs, ankle-foot orthoses, head supports, mobile arm supports bathroom aids, etc. — Acetylcholinesterase inhibitors (e.g. pyridostigmine) can cause a short-term improvement in fatigue in some patients, but are not used routinely
Fasciculations, cramps, spasticity	— Spasmolytic agents (baclofen, tizanidine): Dose must be carefully titrated as loss of tone can worsen mobility — Quinine sulphate for cramps — Low-dose diazepam for cramps or fasciculations
Sialorrhoea (drooling) and difficulty in clearing secretions	— Hyoscine transdermal patches, amitriptyline, or atropine — Intrasalivary gland injection of botulinum toxin — Portable suction devices — Low-dose parotid irradiation may be considered if drug treatment is not successful. — Carbocisteine reduces viscosity of secretions
Pseudobulbar affect	— Responds well to amitriptyline or selective serotonin reuptake inhibitors (SSRIs)
Depression and anxiety	— Tricyclic antidepressants or SSRIs — Psychological counselling
Insomnia	— Treatment should be directed at the cause of insomnia. Common causes in motor neurone disease are respiratory insufficiency, anxiety, depression, muscle cramps, and inability to change position — Sedatives should be administered with care in patients with respiratory compromise
Constipation	— Review medications (analgesics and anticholinergics worsen constipation) and ensure adequate fluid intake — Bulk-forming or osmotic laxatives, glycerol suppositories
Musculoskeletal pain	— Non-steroidal anti-inflammatory agents and physiotherapy — More potent analgesic agents may be required in the later stages
Dysarthria	— Simple strategies to improve communication can be taught by a speech therapist. When these become ineffective, a variety of communcation aids are available, such as a light-writer
Dysphagia	— Attention to food consistency, nutritional supplements. — Gastrostomy tube insertion—radiologically guided or percutaneous endoscopic gastrostomy
Dyspnoea	— Influenza prevention vaccinations — Antibiotics for chest infection — Attention to sleeping position — Sublingual lorazepam for choking, stridor — Non-invasive ventilation — In some countries tracheostomy and invasive ventilation may be considered — Cough assist devices
Terminal care	— Assistance from palliative care team — Analgesics and anxiolytic agents in incremental doses as necessary to relieve distress.

considerable international differences in the use of assisted ventilation to manage respiratory failure in amyotrophic lateral sclerosis. The progressive nature of the condition has acted as a deterrent, in some countries, for the active management of respiratory dysfunction in many patients, particularly when other motor disabilities are extensive. Full 24-h intermittent positive pressure ventilation via a tracheostomy, is an option that is chosen only rarely by fully informed patients. The costs of tracheostomy ventilation, in terms both of financial resources and the caregiver support required, are substantial. Non-invasive intermittent positive pressure ventilation via a mask (Fig. 23.5) is a practical option for respiratory support. Non-invasive positive pressure ventilation used overnight has been

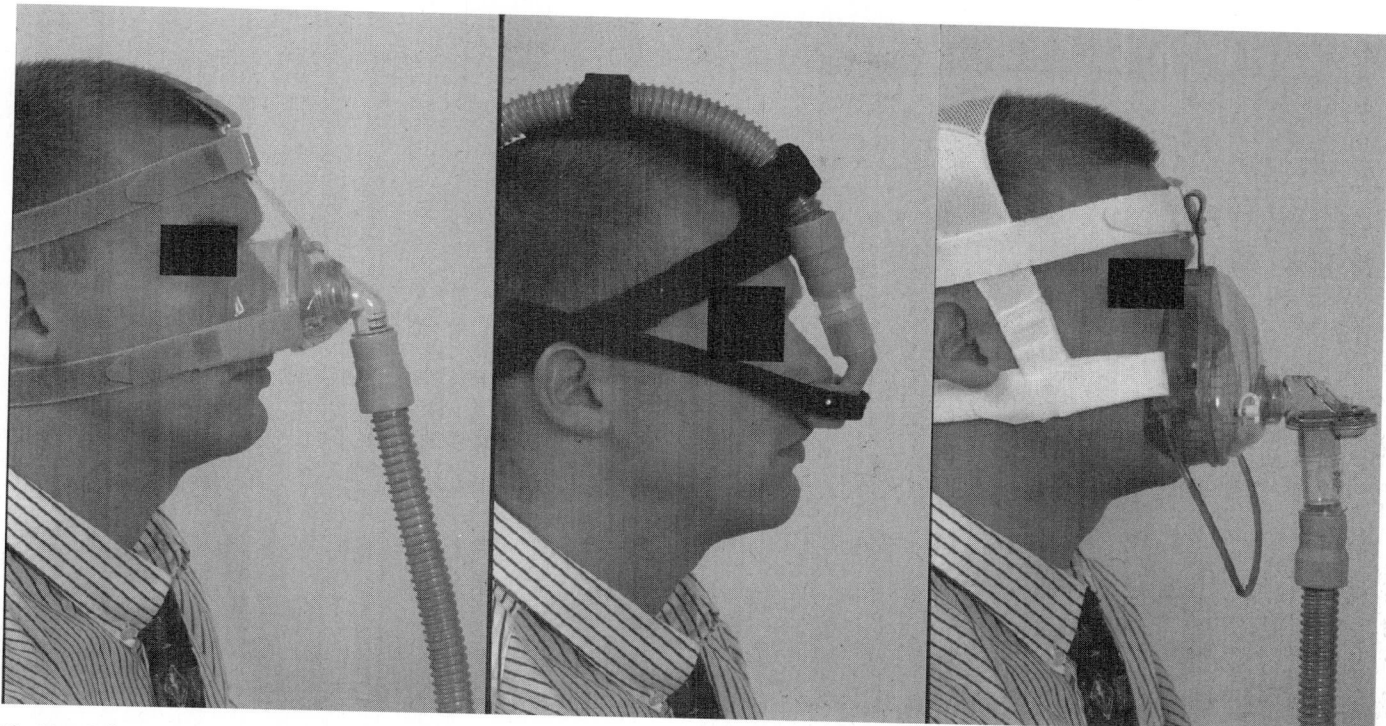

Fig. 23.5 Different types of mask which can be used in non-invasive ventilation, including nasal mask, a mask that avoids pressure over the bridge of the nose and a full facial mask.

shown to alleviate symptoms of chronic hypoventilation and to significantly improve several measures of quality of life (Lyall *et al.* 2001; Bourke *et al.* 2003). In the United Kingdom only a small proportion of patients with amyotrophic lateral sclerosis are treated with non-invasive positive pressure ventilation, and there is marked variation in clinical practice (Bourke *et al.* 2002). This may partly be due to regional variation in the availability of non-invasive positive pressure ventilation. The early signs and symptoms of hypoventilation are also subtle and easily overlooked.

Recent work has demonstrated that the criteria most predictive of symptomatic benefit from non-invasive positive pressure ventilation in patients with motor neurone disease are: orthopnoea, daytime hypercapnia, nocturnal oxygen desaturation, together with relatively preserved bulbar function (Bourke *et al.* 2003). A recently published randomized, controlled trial of non-invasive ventilation in motor neurone disease has shown that in patients without severe bulbar dysfunction, survival is significantly extended with a median survival benefit of approximately 7 months, with maintenance of and improvement in multiple quality of life measures (Bourke *et al.* 2006). The survival benefit from non-invasive positive pressure ventilation in this group is much greater than from currently available neuroprotective therapy. Patients with severe bulbar dysfunction have more difficulty in tolerating non-invasive positive pressure ventilation for hours at a time and in this group there was some improvement in sleep-related symptoms but no demonstrable survival benefit.

End of life care. In United Kingdom practice the opening up of hospice places for patients with amyotrophic lateral sclerosis has greatly improved the quality of care and support for patients in the later stages of the disease. If motor neurone disease patients are not

ventilated, they will almost always die in their sleep from hypercapnic coma. In the terminal phases of illness the aim of treatment is to ensure that the patient is comfortable, and opiate and anxiolytic medication should be used as required to alleviate discomfort or distress.

Neuroprotective therapy

There is no therapy currently available which has a dramatic effect in slowing disease progression in amyotrophic lateral sclerosis. However, some small steps have been made towards this ultimate goal in recent years. An understanding of the molecular pathways that lead to motor neurone death (Section 23.2.4), is needed in order to target therapeutic strategies. These insights into the mechanisms of neuronal degeneration have led to the development of a number of compounds which protect neurones in cell culture and in animal models of motor neurone disease. Over 50 potential neuroprotective agents have been tested in clinical trials which have been extensively reviewed elsewhere (Meininger *et al.* 2000; Turner and Leigh 2003). The larger recent trials, and their theoretical and experimental basis are summarized in Table 23.6.

Riluzole. Has been shown to significantly slow disease progression and is the only neuroprotective agent licensed for use in motor neurone disease. It is a sodium channel blocker whose primary mechanism of action is to reduce excitotoxicity through inhibition of glutamate release. Also it has been shown to have several other potentially neuroprotective effects. Two double-blind placebo-controlled trials of riluzole have been carried out in more than 1100 patients (Bensimon *et al.* 1994; Lacomblez *et al.* 1996). A Cochrane review of riluzole therapy in motor neurone disease concluded that there is a statistically significant, although modest, effect in prolonging survival by approximately 3 months (Miller *et al.* 2002).

Table 23.6 Recent clinical trials in amyotrophic lateral sclerosis targetting different postulated disease mechanisms

	Therapeutic agent	Rationale	Result of clinical trial in ALS patients
Excitotoxicity	Riluzole	Sodium channel blocker that inhibits presynaptic glutamate release. Slowed disease progression in *SOD1* mouse model of amyotrophic.	Modest significant survival benefit
	Branched chain amino acids	Activate glutamate dehydrogenase to reduce glutamate levels	No significant benefit.
	Gabapentin	Reduces glutamate activity. Slowed disease progression in *SOD1* mutant mouse models of motor neurone disease.	Trend towards slowing of disease progression in pilot trial not duplicated in further trial.
	Topiramate	Reduces glutamate activity. Protects against motor neurone degeneration *in vitro*.	No significant benefit
Oxidative stress	Vitamin E (α-tocopherol)	Supplementation of the diet of *SOD1* transgenic mice with vitamin E delayed onset of symptoms and slowed disease progression	No survival benefit. Significantly more patients remained in a milder disease state after 12 months of treatment.
	N-acetylcysteine	*N*-acetylcysteine is a precursor of the antioxidant glutathione	No significant difference in survival or disease progression in an under-powered trial.
Neurotrophic factors	CNTF (subcutaneous)	These neurotrophic factors promote survival of motor neurones *in vitro* and arrest disease progression in the *wobbler* mouse model of motor neuron disease.	No benefit shown, detrimental effect at higher doses
	BDNF (subcutaneous and intrathecal)		No significant benefit. Intrathecal trial terminated early due to increased incidence of adverse events in the treated group (unpublished)
	IGF-1 (subcutaneous)	Promotes motor neurone survival in several models of neuronal injury	Significant slowing of disease progression in a US trial not duplicated in a European study. Cochrane review concluded IGF-1 use could not be recommended.
Mitochondrial dysfunction	Creatine	Phosphocreatine allows the rephosphorylation of ADP to ATP. Oral creatine supplementation may improve cellular energy deficits, and prolongs survival in *SOD1* transgenic mice.	No significant benefit
Miscellaneous compounds	Xaliproden	Oral neurotrophic agent. Neurotrophic effects in animal models of neurodegeneration	No significant benefit
	Pentoxifylline	A phosphodiesterase inhibitor already used in the treatment of peripheral vascular disease. Identified as a potential therapeutic target for amyotrophic lateral sclerosis through screening in transgenic mice.	No significant benefit
	Ono-2506	An inhibitor of astrocyte activation. Neuroprotective effects in cell culture and animal models of neuronal injury.	Overall phase II trial showed no significant benefit, but post-hoc analysis indicated possible benefit in patient subgroup early in the disease course
	Novartis TCH346	Prevents neuronal apoptosis	No significant benefit

No clear effect on muscle strength was demonstrated, and neither trial evaluated quality of life. In view of the high cost to benefit ratio, there has been controversy about the use of riluzole worldwide. In the United Kingdom, its use is recommended by the National Institute for Clinical Excellence, which estimated the cost of therapy to be £34 000 to £43 500 per quality-adjusted life year, or QALY. Riluzole therapy is relatively expensive and the average survival benefit to be expected is modest, but against this must be weighed the arguments that the patient population requiring the drug is relatively small, and that these patients are facing a lethal disease for which no other therapy is available. At presentit is unknown whether the modest overall therapeutic effect ofriluzole conceals individual good responders and non-responders.

Other neuroprotective agents. Several compounds that appeared to protect neurones from degeneration in cell culture and animal models have had disappointing results in human clinical trials. There are two possible explanations for this. First, the models used may not accurately reproduce human disease or the testing in the models may be insufficiently rigorous. Several drugs effective in SOD1 transgenic mice have not been beneficial in human trials, including gabapentin, creatine, topiramate, and vitamin E. Potential explanations for failure of translation into effective human neuroprotective therapies include starting therapy presymptomatically in mice, and deficiencies in the design of mouse trials including failure to take into account gender and litter effects. Also problems with the design and methodology of human clinical trials in the past could mask a modest clinical benefit. To improve the design and implementation of clinical trials in motor neurone disease, the World Federation of Neurology published consensus guidelines in 1998 (Miller *et al.* 1999).

Future developments. The next few years are likely to see further progress in defining the molecular mechanisms of cell death underlying the neurodegenerative process in amyotrophic lateral sclerosis. It can be anticipated that further genetic mutations associated with familial disease and the contribution of genetic factors to the sporadic form of the disease will be identified. The continuing use and refinement of cellular and animal models will allow researchers to understand the sequential molecular events leading to motor neurone cell death and to evaluate new neuroprotective strategies.

Several potential neuroprotective agents are being evaluated or will shortly be entered into clinical trials in amyotrophic lateral sclerosis. These include minocycline, arimoclomol, glatiramer acetate, co-enzymeQ10, celecoxib, and ONO-2506. Automated laboratory assays of neurodegeneration can rapidly screen thousands of chemicals to identify lead compounds for drug development. The traditional reluctance of pharmaceutical companies to invest heavily in rarer diseases has recently been addressed by the emergence of non-profit-making biotech companies and academic institutions using high-throughput drug screening to identify compounds of interest. Future neuroprotective therapy for patients with amyotrophic lateral sclerosis may well involve a 'cocktail' of pharmacological agents aimed at different mechanisms contributing to the biochemical cascade of cell injury. Gene therapy approaches using viral vectors pseudotyped to ensure retrograde transport within motor neurone axons following intramuscular injection have shown great promise in murine models of motor neurone disease (Azzouz *et al.* 2004) and phase 1 human trials are expected in the near future. Cell replacement therapy using stem cells poses particular difficulties in relation to amyotrophic lateral sclerosis, not least because of the long axonal processes required for the normal function of motor neurones and the inhibitory signals preventing the effective growth and correct synaptic alignment of newly generated motor neuronal axons. Perhaps the earliest promise of cell replacement therapy for motor neurone disease will be to try to create a supportive environment of non-neuronal cells in the vicinity of motor neurones to achieve neuroprotection of existing differentiated motor neurones. Measures to improve supportive care for patients and their families will continue to develop including the judicious use of ventilatory support. As clinical and scientific developments allow the prospect of increasing the duration of the disease and prolonging survival of patients with motor neurone disease, very careful attention needs to be paid to the quality of life of afflicted individuals.

23.3 Disorders of the lower motor neurone

23.3.1 Spinal muscular atrophy

The term spinal muscular atrophy encompasses a group of genetically determined pure lower motor neurone disorders in which degeneration of the anterior horn cells leads to progressive, symmetrical muscle weakness, and wasting, with sparing of sensation, and absence of pyramidal tract involvement. As the bulbar musculature may be affected, and motor neurone degeneration is therefore not confined to the spinal cord, an alternative term 'hereditary motor neuronopathy' was proposed, and both terms are currently in use. There are difficulties with the definition and classification of spinal muscular atrophy, which are gradually being resolved as the underlying genetic defects are identified. By definition spinal muscular atrophy is genetically determined, but adult patients sometimes present with what appears to be a sporadic form, and a genetic basis is therefore unproven. The neurodegenerative process in spinal muscular atrophy has a predilection for motor neurones, but in some families, pyramidal tract and sensory involvement are also seen. Thus the condition shows some overlap with other neurodegenerative diseases affecting the central and peripheral motor systems.

Proximal recessive spinal muscular atrophy of childhood

This is the most common type of spinal muscular atrophy. It is inherited as an autosomal recessive disorder and is one of the most common lethal childhood autosomal recessive diseases. Patients develop predominantly proximal limb weakness, with relative sparing of the facial muscles and the diaphragm. Proximal spinal muscular atrophy is divided into subtypes, according to severity and age of onset (Zerres and Rudnik-Schoneborn 1995):

- *Type I spinal muscular atrophy*, or *Werdnig–Hoffman disease*, presents with severe generalized muscle weakness and hypotonia at birth, or by the age of 6 months. Affected children never sit or walk, and usually die from respiratory insufficiency within the first 2 years of life.

- *Type II spinal muscular atrophy* is an intermediate form with onset of muscle weakness before the age of 18 months, and patients can sit, but are never able to walk unaided, and survival is usually limited to adolescence.

- *Type III spinal muscular atrophy*, or *Wohlfart–Kugelberg–Welander disease*, presents after the age of 18 months. Patients gain the ability to stand and walk, but often become wheelchair-dependent in adolescent or adult life though life expectancy is normal.

- Onset of recessive proximal spinal muscular atrophy has been described in adulthood, and this is sometimes designated *type IV spinal muscular atrophy*.

Genetics of chromosome 5q13 linked spinal muscular atrophy

Linkage analysis revealed that the three subtypes of proximal recessive spinal muscular atrophy mapped to chromosome 5q13. This region of chromosome 5 is complex, and characterized by low copy repeats, which may account for instability of this region, and trigger frequent deletions or gene conversions. Within the spinal muscular atrophy critical region, there is a 500 kB inverted duplication, with four genes present in at least two copies, telomeric and centromeric: the survival motor neuron gene, *SMN*, the neuronal apoptosis inhibitory protein gene, *NAIP*, the gene encoding

BTF2p44, a subunit of RNA polymerase II involved in transcription, and a putative RNA binding protein, H4F5. Homozygous deletion of the telomeric copies of all four genes have been described in patients with spinal muscular atrophy, but the frequency of gene deletions is much higher for *SMN1* than for the other 3 genes and it is now well established mutations affecting *SMN1* cause 5q13-linked spinal muscular atrophy. In a series of 525 patients with classical spinal muscular atrophy, 96 per cent were linked to chromosome 5q13, and all of these showed mutations in *SMN1* (Wirth *et al.* 2000). Furthermore, mice possess only one survival motor neuron gene, *Smn*, loss of which is embryonically lethal. However, $Smn^{-/-}$;*SMN2* mice that carry one or two copies of human *SMN2* develop a phenotype of motor neuron degeneration that resembles spinal muscular atrophy in type (Monani *et al.* 2000).

Duplication of the *SMN* gene occurred more than 5 million years ago, before the separation of human and chimpanzee lineages and subsequent sequence divergence in *Homo sapiens* has led to 5 base pair differences between *SMN1*, and its centromeric homolog *SMN2*. The primary gene sequences of *SMN1* and *SMN2* predict identical proteins, but the translationally silent change in exon 7 of *SMN2* decreases the activity of an exonic splicing enhancer, leading to skipping of exon 7 and a truncated protein in 80 per cent of the transcript produced by *SMN2*. Homozygous absence of *SMN2*, found in about 5 per cent of controls, has no clinical phenotype. The majority of 5q13-linked spinal muscular atrophy patients show homozygous absence of *SMN1* exon 7. This may occur through gene deletion, often a large deletion that includes the whole gene, or several genes within the critical region for spinal muscular atrophy. Alternatively, *SMN1* may be replaced by a copy of *SMN2* during DNA replication, a process known as gene conversion.

Multiple more subtle intragenic mutations have been described in spinal muscular atrophy, the majority of which produce a truncated protein, either through splice site mutations that disrupt exon 7, or through nonsense or frameshift mutations which introduce a stop codon. Missense mutations show an interesting pattern of clustering which provides some insight into the functional domains of the SMN protein. A tyrosine–glycine rich sequence at the C-terminal of SMN encompasses five of the described missense mutations, and is a highly conserved sequence, identical in yeasts and nematodes. This region has homology to RNA interacting proteins. A further cluster of mutations is seen in the central region of *SMN*, which constitutes a so-called Tudor domain, an evolutionarily conserved sequence of unknown function found in many eukaryotic proteins.

Genetic modifying factors in spinal muscular atrophy
SMN2 produces only 20 per cent of full-length protein. This is insufficient to rescue the phenotype in homozygous deletion of *SMN1*, but there is strong evidence that *SMN2* is a disease-modifying gene for spinal muscular atrophy. A molecular basis for the wide variation in the severity of the phenotype resides in the fact that homozygous absence of *SMN1* can be due to gene deletion or to conversion to *SMN2*. A patient may have anything from 1 to 4 copies of *SMN2*, leading to a progressive increase in the amount of full-length protein. Both the copy number of *SMN2* and the protein levels of SMN have been shown to correlate with severity of disease phenotype (Feldkotter *et al.* 2002; Lefebvre *et al.* 1997). However, the copy number of *SMN2* in types I, II, and III spinal muscular

atrophy overlaps, therefore this alone cannot expain the phenotypic variation of the disease, and other disease-modifying factors must exist.

The majority of type I spinal muscular atrophy patients display large scale 5q13 deletions, removing *SMN1* and adjacent microsatellite markers, whereas type III patients tend to have small deletions or gene conversions affecting only *SMN1* (Rodrigues *et al.* 1996). This suggests that a spinal muscular-atrophy-modifying locus distinct from *SMN1* lies in the 5q13 interval. *NAIP* is a good candidate as it functions as a negative regulator of apoptosis. *NAIP* deletion occurs in 45 per cent of type I spinal muscular atrophy patients, and 18 per cent of type II and III patients, therefore loss of *NAIP* may lead to a more severe disease phenotype (Roy *et al.* 1995). *H4F5* lies closer to *SMN1* than any other known gene. Homozgous deletions of this gene have been found in 90 per cent of type I spinal muscular atrophy cases, and it has also been proposed as a disease-modifying gene. However, deletions of both *NAIP* and *H4F5* may simply reflect larger chromosomal deletions that involve both *SMN1* and *SMN2*, and their role in the pathogenesis is not confirmed.

The SMN protein
SMN produces a protein of 294 amino acids that is widely expressed. In spinal muscular atrophy patients, the level of the SMN protein is only moderately reduced in muscle and lymphoblasts, but is reduced 100-fold in the spinal cord of type I patients (Coovert *et al.* 1997). SMN protein shows diffuse cytoplasmic expression, but within the nucleus, is clustered in suborganelles called 'gems', for 'gemini of coiled bodies'. These are similar in size and number to, and often associated with Cajal or coiled bodies, which are known to have a role in mRNA metabolism. The SMN protein oligomerizes, and associates with six proteins named Gemins, to form the SMN complex. Self-association occurs through an oligomerization domain in exon 6, and appears to be essential for its activity (Lorson *et al.* 1998).

The SMN complex interacts with several proteins, many of which are involved in RNA metabolism. These are ubiquitous cellular processes, which would indicate that the clinical features of spinal muscular atrophy may be caused by a particular susceptibility of lower motor neurones to defects in RNA handling. Alternatively, SMN may have as yet unidentified functions that are specific to the motor neurone. The list of proteins reported to interact with SMN also includes several which are not involved in RNA metabolism, including profilin, the FUSE binding protein, ZRP1, and p53. The functional significance of these interactions is currently unknown.

Within the cytoplasm, the SMN complex has an important role in the assembly of spliceosomal small nuclear ribonucleoproteins, snRNPs (Pellizzoni *et al.* 1998; Yong *et al.* 2004). SMN has also been shown to have a function in pre-mRNA splicing in the nucleus. SMN mutations found in patients with spinal muscular atrophy cause deficiencies in splicing regeneration activity and interactions with other key proteins, but it is uncertain whether the function of SMN in pre-mRNA splicing explains the motor neurone specific pathology in spinal muscular atrophy. The SMN complex also interacts with viral transcriptional activators, and with RNA polymerase II, pol II, which physically and functionally couples transcription, splicing, and polyadenylation, an association mediated by RNA helicase A, RHA. Expression of a dominant-negative mutant of SMN causes accumulation of pol II and RHA in the nucleus, and inhibits transcription *in vivo*, suggesting a role for SMN in transcriptional regulation (Pellizzoni *et al.* 2001).

Motor neurone-specific functions of SMN. SMN has been shown to interact with two heteronuclear ribonucleoproteins, hnRNP-R and hnRNP-Q, which are considered to play important roles in mRNA editing, transport, and splicing. HnRNP-R is predominantly located in the axons of motor neurones, where it colocalizes with SMN, a finding which led to the confirmation of a motor-neurone-specific function of SMN (Rossoll *et al.* 2002). In zebrafish, knock-down of SMN protein levels in the developing embryo, caused pathfinding defects specific to the motor axon (McWhorter *et al.* 2003). Similarly, primary motor neurones cultured from a transgenic spinal muscular atrophy mouse model show reduced axon growth, while overexpression of SMN and hnRNP-R in cultured neuronal cells promoted neurite outgrowth (Rossoll *et al.* 2003). These findings raise the possibility that SMN is involved in transport of mRNA molecules in the axons of motor neurones.

Differential diagnosis of spinal muscular atrophy

Other disorders may present in infancy with hypotonia and a pattern of weakness identical to Werdnig–Hoffman disease, which are distinguished by associated features. The aetiological relationship of these disorders to classical spinal muscular atrophy has been clarified by testing for *SMN* mutations:

- *Spinal muscular atrophy with pontocerebellar hypoplasia* presents with neonatal hypotonia, nystagmoid eye movements, cortical blindness, and mental retardation. Affected patients do not have mutations in *SMN* (Rudnik-Schoneborn *et al.* 2003).

- *Spinal muscular atrophy and arthrogryposis.* Arthrogryposis, congenital joint contractures, is caused by decreased foetal movements *in utero* that can occur in the context of several underlying problems, including neuropathies, myopathies, and oligohydramnios. Some infants with 5q13-linked spinal muscular atrophy have arthrogryposis, the presence which was previously regarded as an exclusion criterion for spinal muscular atrophy.

- *Spinal muscular atrophy with arthrogryphosis and bone fractures* is characterized by a pattern of weakness indistinguishable from spinal muscular atrophy type 1, and congenital long-bone fractures (Kelly *et al.* 1999). It is genetically distinct from *SMN*-related spinal muscular atrophy, and may be autosomal recessive or X-linked recessive.

- *X-linked spinal muscular atrophy/arthrogryphosis* is linked to the short arm of chromosome X. Affected infants have congenital joint contractures, facial dysmorphia, chest deformities, hypotonia, and areflexia, and electromyographic studies and muscle biopsy consistent with loss of lower motor neurones (Kobayashi *et al.* 1995).

- *Lethal congenital contracture syndrome 1* is a disorder of multiple congenital contractures with neuropathological changes resembling those of spinal muscular atrophy, restricted to Finland. The disorder is fatal in the third trimester, and has associated features of intrauterine growth retardation, foetal hydrops, and facial abnormalities. Linkage has been established to chromosome 9p34 (Makela-Bengs *et al.* 1998).

- *Lethal congenital contracture syndrome 2.* A similar syndrome, affecting Israeli Bedouins, is distinguished by additional craniofacial and ocular findings, lack of hydrops, multiple pterygia,

fractures, and bladder abnormalities. Linkage to both 5q13 and 9p34 have been excluded (Landau and Mishori-Dery *et al.* 2003).

23.3.2 Autosomal dominant proximal spinal muscular atrophy

A missense mutation, P56S, in the vesicle-associated membrane protein, *VAPB* gene on chromosome 20q13.3 was identified in seven kindreds, with three different phenotypes of autosomal dominant motor neurone disease (Nishimura *et al.* 2004). Some patients had an atypical slowly progressive form of amyotrophic lateral sclerosis with tremor, ALS 8, some had typical rapidly progressive familial amyotrophic lateral sclerosis, while others presented with adult onset proximal spinal muscular atrophy, Finkel Type. *VAPB* is discussed further in Section 23.2.5.

23.3.3 Distal spinal muscular atrophy/hereditary motor neuronopathy

The distal distal spinal muscular atrophies or the distal hereditary motor neuronopathies, or dHMN, are a more diverse group of disorders. They commonly manifest as a peroneal muscular atrophy syndrome, distinct from the Charcot–Marie–Tooth syndrome (Section 21.4), which causes a similar pattern of weakness, by the lack of sensory involvement. The clinical picture is one of progressive weakness of the toes and feet which may be associated with foot deformity and which extends over time to involve the distal upper limb muscles. Some patients have unusual or additional features, including predominant involvement of the hands, vocal cord paralysis, diaphragm paralysis, and pyramidal tract signs. Harding proposed seven subtypes of distal hereditary motor neuronopathy (Harding 1993; European CMT Consortium 1998), on the basis of genetic and clinical criteria. This classification was reviewed by the European Charcot–Marie–Tooth Consortium in 1998, but this is constantly evolving with the identification of novel clinical and genetic entities, the finding that previously delineated phenotypes show genetic heterogeneity, and that single gene disorders can vary widely in phenotype (Irobi *et al.* 2004).

Spinal muscular atrophy with respiratory distress

In a series of 200 patients with infantile-onset spinal muscular atrophy, Rudnik-Schoneborn *et al.* (1996) found that approximately 1 per cent had diaphragmatic weakness and did not have deletions of the *SMN* gene on chromosome 5q. This subtype of infantile spinal muscular atrophy, known by the acronym SMARD, is characterized by severe breathing difficulties and limb weakness which is predominantly distal. SMARD1 is caused by mutations in the gene encoding immunoglobulin mu-binding protein 2, *IGHMBP2*, which has RNA helicase activity, and is involved in pre-mRNA processing (Grohmann *et al.* 2001). Some affected infants have evidence of sensory and autonomic nerve involvement as well as the motor deficit. Mutations in the homologous murine gene, *ighmbp2* are responsible for spinal muscular atrophy in the 'neuromuscular degeneration', nmd mouse, which shows close phenotypic resemblance to SMARD1 (Cox *et al.* 1998).

Autosomal recessive distal spinal muscular atrophy

This disorder causes slowly progressive distal limb weakness and has a very variable age of onset. In the most severely affected patients, there is evidence of diaphragmatic involvement, with a decrease in vital capacity and elevation of the hemidiaphragms on chest radiographs. In a large consanguineous Lebanese pedigree, genetic linkage has been mapped to chromosome 11q though mutations in *IGHMBP2* have been excluded. A distinct form of autosomal recessive distal spinal muscular atrophy, identified in families from the Jerash region of Jordan, has been mapped to chromosome 9p21.1-p12.

Distal spinal muscular atrophy with upper limb predominance

Pedigrees with distal spinal muscular atrophy with upper limb predominance, designated dHMN-V, often include affected individuals with evidence of mild sensory disturbance or upper motor neurone features. The disorder was linked to chromosome 7p in a large Bulgarian family (Christodoulou *et al.* 1995). Subsequently a Mongolian family was identified in which Charcot–Marie–Tooth disease type 2D and distal spinal muscular atrophy with upper limb predominance segregated in the same kindred. All affected members had weakness and wasting of the intrinsic muscles of the hands. Those with no sensory deficit and peroneal muscle weakness had a diagnosis of distal spinal muscular atrophy, while in other affected individuals Charcot–Marie–Tooth disease was diagnosed on the basis of a glove and stocking sensory loss. Both disorders in this family were linked to the same region of chromosome 7p. Mutations were subsequently uncovered in the gene encoding glycyl tRNA synthetase, *GARS*, in Charcot–Marie–Tooth disease type 2D families, in families with dHMN-V, and in the family described with both disorders (Antonellis *et al.* 2003). Identification of the underlying genetic disorder has therefore confirmed that Charcot–Marie–Tooth Type 2D and distal spinal muscular atrophy with upper limb predominance are phenotypic variants of a single disorder. GARS is a member of the family of aminoacyl tRNA sythetases involved in diverse cellular processes, including charging tRNAs with their appropriate amino acids. This defect would be expected to affect every glycine-containing protein, and it is at present unknown why this gene defect specifically impairs the function of neurones.

Linkage to chromosome 7p15 was excluded in a large Austrian family with dHMN-V. In this family and several others, heterozygous mutations were identified in the Berardinelli–Seip congenital lipodystrophy gene, (Windpassinger *et al.* 2004). The gene encodes seipin, an integral membrane protein of the endoplasmic reticulum, and null mutations cause Berardinelli–Seip congenital lipodystrophy. *N88S* and *S90L* mutations affect glycosylation of seipin, resulting in the formation of intracellular aggregates. The same mutations were also found to be associated with the Silver syndrome variant of hereditary spastic paraplegia (Section 23.4.2). This is a further example of the same genetic defect causing two quite distinctive phenotypes. The clinical features associated with *BSCL2* mutations have recently been broadened to include individuals who have lower limb predominant distal amyotrophy in the absence of pyramidal tract signs, and some with a combination of spasticity and severe amyotrophy in the lower limbs (Irobi *et al.* 2004).

Distal spinal muscular atrophy with lower limb predominance

Designated dHMN-II, this is usually inherited as an autosomal dominant trait, but sporadic cases are frequently described, which may reflect late-onset recessive disease, non-genetic aetiology, or new mutations. One large Belgian pedigree was linked to chromosome 12q24 and subsequently heterozygous mutations were found in the gene encoding the small heat shock 22 kDa protein 8, *HSPB8/HSP22*, in four families (Irobi *et al.* 2004c). Missense mutations

have also been identified in the interacting partner of HSP22, heat shock 27kDa protein 1, *HSPB1/HSP27*, both in patients with dHMN-II and in pedigrees with Charcot–Marie–Tooth Type 2F (Evgrafov *et al.* 2004). Most of the *HSP22* and *HSP27* mutations disrupt a conserved αcrystallin domain in these proteins and cell biological studies have shown that *HSP22* mutants show greater binding to HSP27, resulting in the formation of intracellular aggregates (Irobi *et al.* 2004a). HSP27 is also involved in the organization of the neurofilament network, important for the maintenance of the axonal cytoskeleton and for axonal transport, and mutant HSP27 perturbs neurofilament assembly (Evgrafov *et al.* 2004).

Distal spinal muscular atrophy with vocal cord paralysis

Distal spinal muscular atrophy with vocal cord paralysis, designated dHMN-VII, characterized by distal limb weakness and dysphonia, has been described to belinked to chromosome 2q14 in two large Welsh families with common ancestry (McEntagart *et al.* 2001). Three disorders have been described which show similar clinical features to dHMN-VII: one family with distal spinal muscular atrophy and vocal cord paralysis also had sensorineural hearing loss, (Bolthauser *et al.* 1989) while Charcot–Marie–Tooth disease type IIC is characterized by progressive vocal cord paralysis, distal limb weakness, and sensory loss (Donaghy and Kennett 1999). The gene alterations underlying these two disorders are currently unknown. A third disorder, presenting with breathing difficulty due to vocal cord paralysis, progressive facial weakness, and weakness and atrophy of the hands, has been found to be caused by a missense mutation in the gene encoding dynactin, *DCTN1*, on chromosome 2p13. This mutation is predicted to distort the folding of the dynactin microtubule binding domain, and lead to dysfunction of dynactin-mediated retrograde axonal transport (Puls *et al.* 2003).

Congenital non-progressive spinal muscular atrophy affecting the lower limbs

This relatively benign disorder, described in pedigrees from Holland and Canada, has been shown by linkage analysis to map to chromosome 12q23-24 (van der Vleuten *et al.* 1998). Again, the disorder is genetically heterogeneous, as in one family linkage to this region has been excluded.

Scapuloperoneal spinal muscular atrophy

Muscle weakness and wasting in a scapuloperoneal distribution may be myopathic or neurogenic in origin. Several large pedigrees with neurophysiological evidence of denervation have been described. In one pedigree, linkage has been mapped to chromosome 12q24.1-12q24.31 (Isozumi *et al.* 1996) which is close to the region of chromosome 12 to which a myopathic scapuloperoneal syndrome is linked (Wilhelmsen *et al.* 1996). Interestingly many of the families described with scapuloperoneal spinal muscular atrophy include individuals with both myopathic and neurogenic changes in affected muscles. Congenital non-progressive spinal muscular atrophy affecting the lower limbs maps to the same region as scapuloperoneal spinal muscular atrophy, but it has not yet been established whether these disorders are allelic.

23.3.4 Monomelic focal and segmental spinal muscular atrophies

Monomelic amyotrophy has been chiefly reported from Japan and India. Hiroyama *et al.* (1959) reported unilateral atrophy of the upper limb and gave it the term 'juvenile muscular atrophy of unilateral upper extremity'. Later a report from India described cases of atrophy of the muscles of one lower limb and described it as the 'wasted leg syndrome' (Prabhakar *et al.* 1981). Gourie-Devi *et al.* (1984) suggested the term monomelic amyotrophy to cover both of these entities. Monomelic amyotrophy has been reported to account for 8–29 per cent of all motor neuron diseases in the series reported from India (Gourie-Devi *et al.* 1984; Saha *et al.* 1997).

Monomelic amyotrophy is a benign variant of motor neurone disease that predominantly affects young men. Lower motor neurone features of wasting and weakness are usually confined to one upper or less commonly the lower limb, without involvement of other components of the nervous system. Patients with the upper limb variant often report tremulousness of the fingers. In the upper limb, the muscles innervated by the C7 to T1 spinal segments i.e. the intrinsic hand muscles and flexors and extensors of the wrists and fingers, tend to be most severely affected. Relative sparing of the brachioradialis muscle amongst the surrounding atrophic forearm muscles is a characteristic feature of this condition (Hirayama *et al.* 1963). The disorder is usually sporadic, although rarely it may be familial (Nalini *et al.* 2004). In the lower limb variant, the muscle atrophy most commonly involves both proximal and distal muscles, but some patients have predominant involvement of either the thigh or lower leg musculature (Prabhakar *et al.* 1981; Gourie-Devi *et al.* 1984). Muscle cramps and fasciculations are observed in 20–30 per cent of patients and unilateral pes cavus may be a presenting feature in the lower limb variant.

Some patients develop similar symptoms in the contralateral limb, but the disorder usually remains strikingly asymmetrical. In most cases, the onset is insidious, with a slow progression over 2–4 years, followed by a plateau phase, although further weakness may evolve for up to 8 years (Peiris *et al.* 1989; Gourie-Devi and Nalini 2003a). The tendon reflexes are depressed. Seldom, if ever, does weakness spread to involve other parts of the body, and upper motor neurone signs are absent. It does not evolve to amyotrophic lateral sclerosis.

A neurogenic pattern on electromyography, and histological evidence of neurogenic atrophy of muscle indicate anterior horn cell pathology. Pathological studies in two patients, who died of coincidental causes, showed atrophy of the affected region of the spinal cord with severe, asymmetrical, bilateral loss of anterior horn motor neurones (Hirayama *et al.* 1987; Araki *et al.* 1989). Imaging studies have shown evidence of forward displacement of the dural sac during neck flexion in patients with Hirayama syndrome, and this has led to the proposal that the condition may be caused by intermittent compression and ischaemia of the cervical spinal cord. However, other reports have failed to confirm this (Schroder *et al.* 1999; Willeit *et al.* 2001) and thus the pathogenic mechanisms underlying Hirayama syndrome and other forms of monomelic amyotrophy remain uncertain. Mutations in the *SMN* and *SOD1* genes have been excluded. A mitochondrial DNA mutation has been reported from Italy in one patient with monomelic amyotrophy and sensorineural hearing loss (Fetoni *et al.* 2004).

23.3.5 Hexosaminidase deficiency

The GM2 gangliosidoses are a group of recessively inherited disorders in which deficiency of the lysosomal enzyme, β hexosaminidase A, leads to abnormal intracellular accumulation of lipids in neurons and glia (Section 10.4.2). Tay-Sachs disease, the classical infantile

form of GM2 gangliosidosis associated with mutations in both alleles of the *HEXA* gene, is the most frequent form of the disorder and is particularly found amongst Ashkenazi Jews. Affected children, after a few months of normal development, start to regress, lose head control and become hypotonic and apathetic, with the development of seizures and cortical blindness. Examination will frequently reveal a large head, a macular cherry red spot, spasticity, quadriparesis, and an exaggerated startle response. The disorder is usually fatal before the age of 5–6 years. GM2 gangliosidosis can also occur as a later onset disease, arbitrarily divided into juvenile, early adult, and late adult forms. The late-onset forms can cause a disorder of the upper and lower motor neurons. However, there is usually evidence clinically of a multisystem disorder with involvement of the cerebellum and its connections, the autonomic nervous system, and extrapyramidal manifestatons such as Parkinsonism, dystonia, or choreoathetosis. Patients may also have evidence of peripheral neuropathy and develop features of dementia or psychosis. Pathologically the most characteristic change is the presence of swollen ballooned neurons with storage material in lysosomes and characteristic membranous cytoplasmic inclusion bodies. Meganeurites are formed which are associated with aberrant dendritic, neuritic, and synaptic growth.

Several reports have described the phenotype of motor neurone disease in adult hexosaminidase deficiency (Kaback *et al.* 1978; Johnson *et al.* 1982; Mitsumoto *et al.* 1985). Gudesblatt and colleagues described 52 patients who had atypical amyotrophic lateral sclerosis and 4 of these had partial hexosaminidase A deficiency (Gudesblatt *et al.* 1988). None of 50 patients with typical amyotrophic lateral sclerosis had abnormal HexA activity and similar findings were reported by Drory and colleagues (Drory *et al.* 2003). Most cases with a motor system disorder in the context of HexA deficiency have additional neurological manifestations and a pure amyotrophic lateral sclerosis phenotype is extremely rare (Karni *et al.* 1988; Parboosingh *et al.* 1997). Muscle weakness tends to develop insidiously during the second decade of life. Weakness is usually first apparent in the proximal muscles and may be associated with muscle fasciculation. Atrophy of the intrinsic hand muscles can also be seen as an early feature. Upper motor neurone features may be present or absent. As described above, the patients on close examination or on follow up over time will often have features indicative of a progressive multisystem neurodegenerative disorder.

23.3.6 Multifocal motor neuropathy with conduction block

Multifocal motor neuropathy (Section 21.11.3) is an acquired disorder which is important to distinguish from amyotrophic lateral sclerosis because it has a much more benign prognosis and is potentially amenable to therapy with intravenous immunoglobulin (Slee *et al.* 2007). It is characterized by slowly progressive, asymmetrical weakness which develops gradually or with stepwise progression over several years. Initially weakness may not be accompanied by significant muscle wasting. The age of onset is usually between 20 and 50 years and the disorder is more common in men. Weakness is more common in the upper limbs compared to the legs and is usually distal. Common initial symptoms include wrist drop and weakness of grip. Weakness is often more pronounced than would be expected for the degree of muscle wasting which is a clinical clue to the presence of conduction block. However, atrophy may become pronounced in patients with long duration disease. Muscle cramps and fasciculations are reported by two-thirds of patients. Multifocal motor neuropathy is commonly initially misdiagnosed as amyotrophic lateral sclerosis (Traynor *et al.* 2000). The slowly progressive disease course, the absence of upper motor neurone signs, and the presence of conduction block on neurophysiological examination, which may only be found after repeated testing, are important clinical clues pointing to the correct diagnosis. Raised titres of anti-GM1 ganglioside antibodies in serum may also be a diagnostic clue, though unfortunately the presence of these antibodies is not specific for multifocal motor neuropathy (Taylor *et al.* 1996).

23.3.7 Post-polio progressive muscular atrophy

Acute poliomyelitis is described in detail in Chapter 42.4.2. Some patients who recover partially or fully from acute poliomyelitic weakness develop a new syndrome of progressive motor deficit many years after the original illness. This post-polio syndrome may affect one-third of patients who have had acute paralytic poliomyelitis. The newly reported fatigue, pain, cramps, wasting, weakness, and functional deterioration develop in areas overtly or subclinically affected during the acute attack (Trojan and Cashman 2005). Typically, several decades elapse between the acute paralysis and the onset of post-polio syndrome, with the peak incidence 20–25 years after the original illness. Often no objective change in muscle strength can be detected on manual muscle testing over several years in patients who nevertheless complain of progressive weakness or of progressive difficulty carrying out activities of daily living. Post-polio syndrome must be distinguished from the non-specific symptoms of joint instability; nerve, root, or plexus compression; and increasing scoliosis which may be late secondary effects resulting from the original weakness. A recent systematic review of the relevant literature concluded that at present conclusions cannot be drawn regarding the functional course or prognostic factors in late-onset polio sequelae and that further work, including long-term follow-up studies of unselected patient populations are needed (Stolwijk-Swuste *et al.* 2005). The cause has not been completely elucidated, but it is likely to be due to distal axonal degeneration of enlarged post-poliomyelitis motor units. Care must be taken to identify coincidental neurological disorders and orthopaedic complications of longstanding weakness, limb dysfunction, and spinal deformity. Although currently there is no specific disease-modifying treatment for post-polio syndrome, an interdisciplinary management programme can be useful in controlling symptoms. There is some evidence that supervised aerobic muscle training and introduction of non-invasive ventilation for patients with respiratory impairment may be helpful measures (Farbu *et al.* 2006). A recent randomized controlled trial of intravenous immunoglobulin showed slight improvement in some parameters including muscle strength and SF-36 subscale vitality score, but no significant change in overall quality of life or pain (Gonzalez *et al.* 2006).

23.4 Disorders of the upper motor neurone

23.4.1 Primary lateral sclerosis

Primary lateral sclerosis is considered as part of the spectrum of amyotrophic lateral sclerosis (Section 23.2.1).

23.4.2 Hereditary spastic paraplegia

Hereditary spastic paraplegia, first described in the 1880s, is a group of hereditary neurodegenerative or neurodevelopmental diseases which affects approximately 1 in 10 000 individuals. The main feature of the clinical phenotype is progressive lower limb spasticity due to degeneration of the corticospinal tracts within the spinal cord. Hereditary spastic paraplegia is most commonly inherited as an autosomal dominant trait, but autosomal recessive and X-linked recessive forms also exist. The hereditary spastic paraplegia phenotype may be 'pure' in which the spastic paraparesis occurs in isolation or 'complicated' where the spastic paraparesis is one component of a much more complex neurological and/or systemic disorder. Hereditary spastic paraplegia shows extreme genetic heterogeneity. To date more than 30 genetic loci have been identified and genes have been identified at 14 of these (Table 23.7). The recent discovery of multiple genes is rapidly shaping new concepts of the cellular mechanisms of degeneration of the long axons of the corticospinal tract in hereditary spastic paraplegia. It is apparent that motor neurons provide an extreme example of the potential difficulties for a cell in trafficking, transport, and energy metabolism, and that the longest axons of the central nervous system may be specifically vulnerable to several distinct biochemical perturbations.

Clinical features

Pure hereditary spastic paraplegia shows progressive spastic paraparesis as the major feature, but other clinical features may be observed including bladder disturbance, mild distal muscle wasting, pes cavus, dorsal column dysfunction, and loss of ankle reflexes. The most common presenting symptom in pure forms is gait disturbance, the patient often complaining of lower limb stiffness, balance difficulties, or of a tendency to fall. In young children, delayed motor milestones or a tendency to walk on the toes, may be the first indications of a problem. The major features apparent on neurological examination consist of lower limb spasticity, hyperreflexia, and extensor plantar responses which may be accompanied by a mild pyramidal distribution weakness. A characteristic feature of hereditary spastic paraplegia is that the patient's disability usually arises from prominent spasticity and any accompanying muscle weakness is often very mild.

There is considerable variation in age at onset and severity of the spastic paraparesis even within affected members of the same pedigree, suggesting that other genetic or environmental factors may impact on the phenotype observed. The reported age at onset of pure hereditary spastic paraplegia ranges from infancy to the eighth decade (Harding 1981). Several studies have indicated that an early age of onset, <35 years, tends to be associated with relatively slow disease progression, the majority of individuals retaining the ability to walk even in elderly life. In contrast late onset, >35 years, tends to show more rapid disease progression, and many patients become non-ambulant in the seventh and eighth decades of life (Harding 1981). Hereditary spastic paraplegia does not, in general, reduce life expectancy. Approximately 25–30 per cent of patients have a subclinical phenotype in which symptoms which are so mild that the condition may not be revealed without a neurological examination.

Complicated hereditary spastic paraplegia shows the spastic paraparesis as only one component of a much more complex disorder with additional clinical features (Table 23.8). Some of the complicated hereditary spastic paraplegia phenotypes are extremely

Table 23.7 Genetic classification of hereditary spastic paraplegia

Genome database Designation	Chromosome	Inheritance	Phenotype	Genetic Defect
SPG1	Xq28	X-linked	Complicated	L1CAM
SPG2	Xq22	X-linked	Both	PLP
SPG3	14q11.2	AD	Pure	Atlastin
SPG4	2p22	AD	Both	Spastin
SPG5	8p12-q13	AR	Pure	
SPG6	15q11.1	AD	Pure	NIPA1
SPG7	16q24.3	AR	Both	Paraplegin
SPG8	8q24	AD	Pure	KIAA0196
SPG9	10q23.3-24.2	AD	Complicated	
SPG10	12q13	AD	Pure	KIF5A
SPG11	15q13-15	AR	Both	
SPG12	19q13	AD	Pure	
SPG13	2q24-q34	AD	Pure	HSP60
SPG14	3q27-q28	AR	Complicated	
SPG15	14q	AR	Complicated	
SPG16	Xq11.2	X-linked	Pure	
SPG17	11q12-q14	AD	Complicated	Seipin
SPG18	Pending			
SPG19	9q33-q34	AD	Pure	
SPG20	13q12.3	AR	Complicated	Spartin
SPG21	15q22.31	AR	Complicated	Maspardin
SPG22	Pending			
SPG23	1q24-q32	AR	Complicated	
SPG24	13q14	AR	Complicated	
SPG25	6q23.3-q24.1	AR	Complicated	
SPG26	12p11.1-12q14	AR		
SPG27	10q22.1-q24.1	AR		
SPG28	14q21.3-Q22.3	AR	Pure/distal sensory loss	
SPG29	1p21.1-1p27.1	AR	Complicated	
SPG30	2q37.3	AR	Pure	
SPG31	2p12	AD	Pure	REEP1
SPG32				
SPG33	10q24.2	AD	Pure	ZFYVE27

Key: AD = autosomal dominant

AR = autosomal recessive

rare, having been described only in single families. Certain complicated phenotypes show characteristic clinical features associated with particular hereditary spastic paraplegia loci. For example, hereditary spastic paraplegia associated with cognitive impairment is most commonly described in patients with spastin, SPG4, mutations (White *et al.* 2000; Webb *et al.* 1998). In patients with the SPG9 subtype of hereditary spastic paraplegia there is a very distinctive clinical phenotype with the development of cataracts, severe gastroesophageal reflux, and an axonal neuropathy superimposed on the spastic paraparesis (Seri *et al.* 1999). Pigmentary macular degeneration is an additional feature observed in patients with hereditary spastic paraplegia linked to the SPG15 locus (Hughes *et al.* 2001). Peripheral neuropathy has been described in patients with hereditary spastic paraplegia linked to the SPG11 and SPG14 loci (Vazza *et al.* 2000; Mostacciuolo *et al.* 2000). In the Silver variant of hereditary spastic paraplegia, SPG17, wasting and weakness of the hands, is a striking feature (Patel *et al.* 2001). Patients with Troyer syndrome, SPG20, have dysarthria, distal amyotrophy, short

stature, and developmental delay in addition to spastic paraparesis (Patel *et al.* 2002).

Epilepsy has been described as a feature complicating hereditary spastic paraplegia, including in families with spastin mutations (Gigli *et al.* 1993; Yih *et al.* 1993). There does not appear to be an association with particular types of seizures and the epilepsy may occur before or after the onset of the spastic paraparesis.

Dementia and cognitive impairment have been reported in complicated hereditary spastic paraplegia pedigrees (White *et al.* 2000) and have also been described as an isolated accompaniment to spastic paraparesis in both autosomal dominant and recessive families (Cross and McKusick 1967; Webb *et al.* 1998; Pridmore *et al.* 1995). The cognitive abnormalities observed, for example impairments of attention, perceptual speed, visuomotor coordination, and forgetfulness, are in keeping with a sub-cortical type of dementia and features suggesting major cortical involvement, such as dysphasia, agnosia, and dyscalculia, are usually absent. Several families with dementia complicating HSP have been described with spastin mutations, SPG4, (Webb *et al.* 1998; White *et al.* 2000). Affected individuals in SPG4 families may have subclinical cognitive impairment detectable by neuropsychological evaluation (Byrne *et al.* 2000).

Distal amyotrophy is one of the commonest additional features seen in patients with hereditary spastic paraplegia. Several pathologies can give rise to this amyotrophy including lower motor neurone loss, axonal neuropathy, and central axonopathy. Several characteristic syndromes of hereditary spastic paraplegia with amyotrophy have been described:

- The commonest type with amyotrophy is also known as peroneal muscular atrophy with pyramidal features or *hereditary motor and sensory neuropathy type V* (Harding and Thomas 1984). It is usually transmitted as an autosomal dominant trait and develops in the second decade of life or later. Affected individuals have amyotrophy associated with axonal motor and sensory neuropathy as well as features of spastic paraparesis.

- *Silver syndrome* consists of autosomal dominant hereditary spastic paraplegia complicated by striking amyotrophy of the hands (Silver 1966). There is variation in age at onset, ranging from childhood to late adult, and in severity. Some Silver syndrome pedigrees are linked to the SPG17 locus, but there is clearly genetic heterogeneity as other families are not linked to this site.

- *Troyer syndrome* is an autosomal recessive form of hereditary spastic paraplegia complicated with amyotrophy of the hands and feet, pseudobulbar palsy, choreoathetosis, short stature, and mental retardation (Cross and McKusick 1967b; Auer-Grumbach *et al.* 1999). It was originally described as a childhood onset disorder in an Old Order Amish population in the USA. More recently, it has been suggested that Troyer syndrome should be broadened to include families with later onset, lack of movement disorder, atrophy or partial agenesis of the corpus callosum, and non-Amish origin. Troyer syndrome is linked to SPG20 and the gene encodes a protein named spartin (Patel *et al.* 2002).

- *Charlevoix–Saguenay syndrome*, described in families from Quebec, appears similar to Troyer syndrome, but with the additional clinical feature of ataxia (Bouchard *et al.* 1978).

- A further rare phenotype of recessive hereditary spastic paraplegia complicated by amyotrophy is described resembling juvenile

Table 23.8 Additional clinical features in complicated hereditary spastic paraplegia (HSP)

Clinical feature	HSP subtype	Additional comments
Amyotrophy	Peroneal muscular atrophy	Amyotrophy associated with an axonal sensory and motor neuropathy (AD)
	Silver syndrome	Severe wasting of the small muscles of the hand with sparing of the lower limb musculature. Linked to SPG17 (AD)
	Troyer syndrome	Distal wasting in the limbs with delayed development, spastic quadraparesis, pseudobulbar palsy, choreathetosis, and short stature. Linked to SPG20 (AR).
	Charlevoix–Saguenay syndrome	Similar to Troyer syndrome with additional ataxia, described in Quebec (AR)
	Resembling juvenile FALS	Childhood onset (AR)
Cardiac defects	-	Associated with mental retardation
Cerebellar signs	-	Dysarthria with a mild upper limb ataxia
Deafness	Sensori-neural	X-linked
Dementia	Subcortical or cortical pattern	Dementia can occur in isolation with HSP, when it tends to be of the subcortical type, or be part of a much more complex phenotype (AR and AD). Linkage to SPG4 locus in a number of families
Endocrine dysfunction	Kallmann's syndrome	Hypogonadotrophic hypogonadism and anosmia
Epilepsy	-	Various epileptic seizure types have been descibed incuding; absence, simple/complex partial, atonic, grand mal, and myoconic
Extrapyramidal Signs	Choreoathetosis Dystonia and rigidity Mast syndrome	Dementia, dysarthria, and athetosis in Amish people with onset in 2nd decade (AR)
Hyperekplexia	-	Neonatal hypertonia and an exaggerated startle response (AD).
Icthyosis	Sjögren–Larsson syndrome	Also with mental retardation and occasionally a pigmentary macular degeneration (AR).
Retinal changes	Optic atrophy Retinal degeneration Kjellin syndrome	Pigmentation seen in SPG15. Dysarthria, upper limb ataxia, dementia, retinal degeneration +/– amyotrophy (AR).
Sensory neuropathy	Asymptomatic Childhood onset Adult onset	Sensory neuropathy detected only on clinical examination. With painless ulcers and deformities secondary to neuropathic bone resorption. Trophic skin changes and foot ulcers.
Others	SPG 1	Mental retardation, aphasia, a shuffling gait, and adducted thumbs. Caused by mutations in *L1CAM* gene (X-linked).
	SPG 9	Bilateral cataracts, gastroesophageal reflux, and amyotrophy

Key: AD = autosomal dominant, AR = autosomal recessive, FALS = familial amyotrophic lateral sclerosis.

onset familial amyotrophic lateral sclerosis. In these pedigrees a childhood onset spastic paraparesis is observed with prominent wasting of the distal musculature. Neurophysiological examination shows electromyographic changes in keeping with lower motor neurone degeneration (Bruyn *et al.* 1993).

Sensory neuropathy of variable severity with onset in childhood or adult life is also a common additional feature in complicated hereditary spastic paraplegia. Severely affected patients may develop chronic painless cutaneous ulcers and neuropathic bone resorption occurring in early life, but in some patients the sensory neuropathy may be subclinical and only detected with neurophysiological testing (Schady and Smith 1994).

Pathology

Detailed neuropathological findings have been reported in relatively few cases of hereditary spastic paraplegia. Therefore the extent to which the known clinical and genetic heterogeneity is reflected in pathological heterogeneity remains to be defined. In addition, because genetic characterization has only recently emerged, there are few pathological reports on genetically characterized cases. The core neuropathological features of hereditary spastic paraplegia were first described by Strümpell and confirmed in a series of subsequent reports (Sack *et al.* 1978; Strumpell 1886). The spinal cord shows pallor of the lateral and frequently also the anterior corticospinal tracts, with loss of axons and myelin

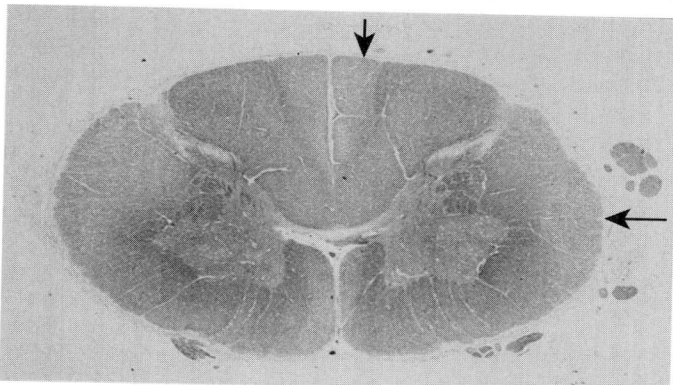

Fig. 23.6 Spinal cord pathology in hereditary spastic paraplegia showing myelin pallor of the lateral corticospinal tracts (arrow) and the medial part of the posterior columns (arrow).

which most markedly affects the longest descending axons in the lumbosacral region (Fig. 23.6). There is also pallor of the dorsal columns, particularly the medial fibres within the fasciculus gracilis. Involvement of the spinocerebellar tracts is described in approximately 50 per cent of cases. Depletion of Betz cells from the motor cortex is reported in some cases, but anterior horn cells in the spinal cord have usually been reported as appearing normal. Degeneration has occasionally been described in neurones of the dorsal nucleus of Clarke. Characteristically there is more severe involvement of the distal part of the corticospinal tracts and dorsal columns so that the most severe changes in these fibre pathways are observed in the lumbar and cervical cord respectively. This has led to the hypothesis that in hereditary spastic paraplegia the neurodegeneration occurs as a 'dying back' axonopathy, affecting the most distal part of these long axons first.

A recent study described the distribution of spastin in the normal human central nervous system, as well as the molecular pathology of three cases of hereditary spastic paraplegia with defined mutations in spastin (Wharton *et al.* 2003). Spastin was shown to be a neuronal protein widely distributed within the central nervous system. Within motor neurones, spastin was predominantly expressed in the cytoplasm of the cell body, with some extension of staining into the proximal neurites and axons. Interestingly, although not prominently affected clinically in SPG4, the spinal cord lower motor neurones showed evidence of cytopathology, with the presence of hyaline inclusion bodies some of which stained for β-tubulin; and variable loss of expression of n-phosphorylated neurofilament protein, β-tubulin, spastin, and mitochondria from the cell bodies, suggesting altered partitioning of cytoskeletal components and organelles. In addition all three cases showed evidence of tau pathology, with neurofibrillary tangles, neuropil threads, and glial tau pathology.

In many neuropathological reports of pure hereditary spastic paraplegia cerebral structures, with the exception of Betz cells, are said to be uninvolved. However subclinical involvement of other parts of the central nervous system is increasingly recognized and in particular there is emerging recognition of cognitive impairment is some families with hereditary spastic paraplegia. The neuropathological substrate of such cognitive impairment currently remains poorly defined, though isolated case reports have been described (Ferrer *et al.* 1995; White *et al.* 2000). In autosomal recessive forms of hereditary spastic paraplegia associated with mutation in the paraplegin gene, SPG7, muscle biopsies have shown ragged red fibres (Casari *et al.* 1998). Scattered muscle fibres show negative histochemical reaction for cytochrome oxidase, but preserved or elevated succinate dehydrogenase activity and peripheral accumulation of mitochondria. These changes, which are typical of oxidative phosphorylation defects in muscle, support a role for mitochondrial dysfunction in the pathogenesis of paraplegin mutation-associated hereditary spastic paraplegia. Skeletal muscle has not been widely surveyed in other types of hereditary spastic paraplegia, though it has been reported that muscle biopsies from some hereditary spastic paraplegia patients in whom spastin and paraplegin mutations had been excluded showed biochemical evidence of impairment in mitochondrial respiratory chain function (McDermott *et al.* 2003b).

Differential diagnosis and investigation

Molecular genetic testing is increasingly used in the diagnosis. Although 14 causative genes have now been identified, testing for mutations in these genes is not necessary routinely available to the clinicians involved in patient care. Therefore, at present hereditary spastic paraplegia remains a diagnosis of exclusion and conditions which should be considered in the differential diagnosis are included in Table 23.9. It is clearly important to exclude diagnoses in which there is a treatable cause for the spastic paraparesis including structural lesions of the spinal cord, vitamin B_{12} deficiency, multiple sclerosis, or dopa responsive dystonia. It is also important to exclude other motor system disorders such as familial amyotrophic lateral sclerosis, in which the clinical course and prognosis are significantly different from those to be expected in patients with hereditary spastic paraplegia.

MRI in hereditary spastic paraplegia may show a degree of spinal cord atrophy, but no other structural abnormalities of the central nervous system are usually apparent. Mild to moderate atrophy of the corpus callosum has been reported and several of these families have shown linkage to the SPG11 locus (Casali *et al.* 2004). There are occasional reports of atrophy of other intracranial structures and of cerebral hemisphere white matter lesions. Nerve conduction studies and electromyography are normal in the majority of cases of pure hereditary spastic paraplegia. Central motor conduction times have been reported to show either unrecordable, or delayed responses to the lower limbs, and usually normal values for the upper limbs (Pelosi *et al.* 1991; Schady *et al.* 1991). Somatosensory evoked potentials from the lower limbs have been reported to be small or absent (Pelosi *et al.* 1991; Aalfs *et al.* 1993).

The most useful molecular genetic test is screening for mutations in the spastin, SGP4 gene, which will be detected in approximately 40 per cent of patients with autosomal dominant hereditary spastic paraplegia. Such mutation screening is, however, costly and time consuming as mutations have been identified scattered throughout the spastin gene and therefore all 17 exons need to be analysed, at least in the index cases (Lindsey *et al.* 2000; McDermott *et al.* 2006). Mutations in atlastin, SPG3A, will be detected in another 10–15 per cent of families with autosomal dominant hereditary spastic paraplegia (Zhao *et al.* 2001) and the diagnostic yield is likely to be highest when targeted to patients with an age of onset of symptoms within the first decade of life.

Table 23.9 Differential diagnosis of hereditary spastic paraplegia

◆ Multiple sclerosis
◆ Cerebral palsy
◆ Spondylotic spinal disease
◆ Amyotrophic lateral sclerosis/motor neurone disease
◆ Other structural spinal cord disorders e.g. tumours, arteriovenous malformation
◆ Arnold–Chiari malformation
◆ Adrenoleukodystrophy/ adrenomyeloneuropathy
◆ Leukodystrophies e.g. Krabbe disease, metochromatic leukodystrophy
◆ Dopa responsive dystonia
◆ Vitamin B12 deficiency / subacute combined degeneration of the cord
◆ Abetalipoproteinaemia
◆ Spinocereballar ataxias
◆ HTLV-1 infection, tropical spastic paraparesis
◆ Neurosyphilis
◆ Neurolathyrism
◆ Vitamin E deficiency
◆ Arginase deficiency

Genetics and molecular mechanisms

To date 33 genetic loci associated with subtypes of hereditary spastic paraplegia have been identified and the 14 causative genes have been identified (Table 23.7). These causative genetic mutations may converge to cause disruption in several different cellular processes which may underlie hereditary spastic paraplegia and related disorders.

Autosomal dominant hereditary spastic paraplegia. To date 12 loci for autosomal dominant forms have been identified and the causative gene determined at 9 of these loci. The genes at SPG3A, SPG4, SPG6, SPG8, SPG10, SPG13, SPG17, SPG 31, and SPG 33 have been identified as *atlastin, spastin, NIPA1, KIAA0196* or strumpellin, the kinesin heavy chain *gene KIF5A*, heat shock protein 60 (*hsp60*), seipin, *REEP1*, and *ZFYVE27* respectively (Hazan *et al.* 1999; Zhao *et al.* 2001; Hansen *et al.* 2002; Reid *et al.* 2002; Rainier *et al.* 2003; Windpassinger *et al.* 2004; Mannan *et al.* 2006; Zuchner *et al.* 2006; Valdmanis *et al.* 2007;). The genes at the remaining loci SPG8, SPG9, SPG12, and SPG19 are as yet unknown.

◆ *Atlastin related hereditary spastic paraplegia; SPG3A.* Hereditary spastic paraplegia linked to SPG3A on chromsome14q11.2 accounts for approximately 10 per cent of autosomal dominant cases (Zhao *et al.* 2001). Affected patients tend to have a pure phenotype, a young age of onset, and the presence of distal amyotrophy in the upper and lower limbs. Chronic neurogenic change on electromyography is another possible distinctive feature. Affected patients are sometimes misdiagnosed as having cerebral palsy. The *SPG3A* gene consists of 14 exons and encodes a protein of 588 amino acids which has conserved motifs for GTPase binding and hydrolysis and is structurally homologous to guanylate binding protein 1, a member of the dynamin family of large GTPases (Zhao *et al.* 2001). Its predicted structure indicates that it is likely to be an integral membrane protein, with two transmembrane domains (Zhu *et al.* 2003). Dynamins are a group of proteins known to be involved in vesicle trafficking, including the formation of clathrin coated vesicles from the plasma membrane and receptor mediated endocytosis; Golgi membrane dynamics; the maintenance and distribution of mitochondria and associate with cytoskeletal components including actin and the microtubule network. These functions are clearly important for neurotransmission, the action of neurotrophic factors, and axonal transport. The atlastin protein is localized most abundantly in the central nervous system, where it is enriched in pyramidal neurons, particularly in layer V of the cortex, including the motor cortex, and in the hippocampus (Zhu *et al.* 2003). At the subcellular level, it has recently been shown to co-localize with the Golgi apparatus (Zhu *et al.* 2003).

Most of the mutations in atlastin described to date are missense mutations which have been located in exons 4, 7, and 8. Of these only one, 650G>A, disrupts the GTPase motif directly (Muglia *et al.* 2002). It has been speculated that the other missense mutations may exert their pathogenic effect by introducing an altered secondary protein structure which disrupts GTPase activity by disturbing multimerization or protein–protein interactions. It will be of interest in the future to explore the effects of atlastin mutations on the structure and function of the Golgi apparatus, axonal transport and growth, as well as parameters such as endocyosis and vesicle trafficking, as perturbations of these functions could underlie the development of the distal axonopathy of hereditary spastic paraplegia.

◆ *Spastin related hereditary spastic paraplegia; SPG4.* Spastin is made up of 17 exons mapping to chromosome 2p21-p22, and encodes a ubiquitously expressed 616 amino acid protein. SPG4 is the most frequent form of autosomal dominant hereditary spastic paraplegia, accounting for approximately 40 per cent of cases. Spastin related hereditary spastic paraplegia does not appear to have a distinctive clinical phenotype compared to other pure autosomal dominant types. In addition, there is no clear correlation between the type of spastin mutation identified and observed phenotype.

The age at onset for spastin related hereditary spastic paraplegia is highly variable as is the severity of the phenotype. Most affected individuals remain ambulant, but the clinical extremes of the phenotype range from asymptomatic patients with detectable lower limb pyramidal signs found only when clinically examined, about a quarter, to a minority of severely affected patients who become chairbound or bedridden. There is some evidence that disease progression and severity is worse in those with late onset disease and that the frequency of features such as paresis, amyotrophy, dorsal column involvement, and urinary disturbance increase with disease duration (Fonknechten *et al.* 2000). The majority of hereditary spastic paraplegia associated with spastin mutation is of a pure phenotype, but descriptions of more complicated phenotypes have emerged recently. Dementia or less severe disturbances of cognition complicating SPG4 has been described in several families and active progression of cognitive deterioration over a 3-year period in older patients with spastin related hereditary spastic paraplegia has been demonstrated (McGonagle *et al.* 2004). Epilepsy associated with spastin related hereditary spastic paraplegia has also been reported in several families (Mead *et al.* 2001) There are isolated reports of a variety of additional features complicating SPG4 including restless legs, myoclonus, atypical seizures, dysarthria, erectile dysfunction, severe constipation, ileus, and faecal incontinence.

Spastin shares homology with a large family of proteins known as the ATPases Associated with diverse cellular Activities, or AAA, which are involved in a range of cellular processes including cell cycle regulation, gene expression, organelle biogenesis, vesicle mediated protein transport, and as molecular chaperones co-operating in the assembly, function, and disassembly of protein complexes (Patel and Latterich 1998). All these AAA proteins share a common functional domain known as the AAA cassette which contains highly conserved motifs including Walker A, Walker B, and the AAA minimal consensus domain. Spastin has been shown to possess a microtubule interacting and trafficking domain and also has nuclear localization sequences (Ciccarelli et al. 2003). Spastin has particular homology with two other proteins, katanin and SKD1. Katanin is a microtubule severing protein and is involved in the dynamic regulation of the microtubule cytoskeleton throughout the cell cycle and SKD1 is an endosomal morphology and trafficking protein. There is evidence from cellular models generated in several laboratories that spastin may also be involved in regulating the microtubule cytoskeleton (Errico et al. 2002; McDermott et al. 2003a). Microtubules are dynamic polymers of alpha and beta tubulin. They perform essential roles, forming the mitotic spindle in dividing cells, acting as tracks for transport of various cellular cargoes including membranous organelles such as mitochondria and provide essential cytoskeletal support in all cells. Evidence to date suggests that spastin, like katanin, acts as a microtubule severing protein. It seems that missense mutations in spastin may lead to entrapment of the protein in a microtubule bound state. Thus, current knowledge indicates that spastin is involved in regulating microtubule dynamics and rearrangement both in proliferating and in post-mitotic cells. It is thought likely that spastin influences microtubule dynamics in growth cones, thus regulating the stability of axons and axonal transport. Long axons such as those of the corticospinal tracts are likely to be most dependent on fine regulation of the microtubule transport system by the action of spastin.

Of the mutations in the spastin gene published to date, approximately 11 per cent are nonsense mutations, 26 per cent frameshift with consequent premature termination codon, 28 per cent missense, and 35 per cent are splice site mutations. Recently, it has been reported that partial spastin deletions are relatively common and may account for approximately 18 per cent of cases of autosomal dominant hereditary spastic paraplegia (Beetz et al. 2006; Beetz et al. 2007). There is no particular 'hot spot' for mutation within the gene, with mutations throughout the length of the gene, making mutation screening a lengthy process. Most of the mutations are predicted to have a detrimental effect on the conserved AAA cassette. The broad mutational spectrum observed in spastin related hereditary spastic paraplegia initially suggested that axonal injury might be caused by a loss of function/ haploinsufficiency effect. However, the demonstration of altered microtubule regulation in cells overexpressing missense spastin mutations suggests the possibility that a dominant negative pathogenic mechanism may be involved, at least in individuals with missense mutations. Spastin mutants with a missense mutation in the AAA cassette appear to bind constitutively to microtubules through the intact N terminal microtubule binding domain and could easily prevent the normal action of wild type spastin or otherwise block transport along the microtubule

system. It is possible that more than one pathogenic mechanism may be involved, depending on the type of mutation. However, it seems likely that the pathogenesis of spastin related hereditary spastic paraplegia may be due to impairment of the fine regulation of the microtubule cytoskeleton in long axons.

- *Non-imprinted in Prader–Willi/ Angelman loci 1, or NIPA1 related hereditary spastic paraplegia; SPG6* is located within the 15q11 chromosomal region which is deleted in Prader-Willi and Angelman syndromes. Two families with autosomal dominant hereditary spastic paraplegia were initially described with the same T45R mutation in the NIPA1 gene (Rainier et al. 2003). This mutation disrupts an interspecies conserved amino acid. The function of NIPA1 was previously unknown, but it is strongly expressed in neuronal tissues and structural analysis of the encoded protein predicts that it possesses 9 transmembrane domains, suggesting that it may function as a membrane transporter or receptor (Rainier et al. 2003). It is possible therefore that the disease mechanism may involve altered signal transduction or small molecule transport. Recent evidence indicates that the NIPA1 protein functions as a magnesium transporter (Goytain et al. 2006). Individuals with Prader–Willi or Angelman syndromes with deletions including NIPA1 do not develop features of hereditary spastic paraplegia which suggests that the disease mechanism likely involves a dominant negative or gain of function rather than haploinsufficiency.

- *Strumpellin or KIAA0196 related hereditary spastic paraplegia; SPG8.* Three mutations have been identified in families linked to the SPG8 locus (Valdmanis et al. 2007). The function of the encoded 1159 amino acid protein strumpellin is currently relatively unknown.

- *KIF5A related hereditary spastic paraplegia; SPG10.* The kinesin heavy chain proteins KIF5A, KIF5B, and KIF5C are part of a multisubunit complex, kinesin-1, that acts as a microtubule motor involved in anterograde fast axonal transport. KIF5A has an exclusively neuronal expression, enriched in motor neurones (Hirokawa 1998). The heavy chain proteins represent the force producing subunit of kinesin, possessing a motor domain that interacts with the microtubule track and hydrolyses ATP. A single family has been described with pure, autosomal dominant hereditary spastic paraplegia associated with a missense mutation at an invariate asparagine residue, N256S, in the motor domain of the neuronal kinesin heavy chain protein KIF5A (Reid et al. 2002). The mutation is predicted to cause loss of function or a dominant negative effect on the neuronal kinesin I motor and is predicted to cause abnormal fast axonal transport of cargoes vital for the distal axon. A homologous mutation in Drosophila larvae results in 'organelle jams' within axons of motor neurones (Hurd et al. 1996). Recently a second autosomal dominant hereditary spastic paraplegia pedigree was described with a missense mutation, R280C, at an invariant arginine residue in exon 10, located in a region of the protein involved in microtubule binding activity (Fichera et al. 2004).

- *Heat shock protein 60 related hereditary spastic paraplegia; SPG13.* The gene at the SPG13 locus has been identified as the mitochondrial chaperonin, heat shock protein 60. One family with a V72I substitution and autosomal dominant pure hereditary spastic paraplegia was described in 2002 (Hansen et al. 2002).

The use of a complementation assay showed that wild type heat shock protein 60 and its co-chaperone heat shock protein 10, but not V72I heat shock protein 60 could support the growth of *E. coli* in which the homologous bacterial genes had been deleted (Hansen *et al.* 2002). Heat shock protein 60 is a mitochondrial protein which is upregulated for example in conditions of cellular stress, for example in the presence of an accumulation of unfolded proteins within the mitochondrial matrix (Zhao *et al.* 2002).

- *Seipin related hereditary spastic parapalegia or Silver syndrome, SPG17.* Silver syndrome is a form of autosomal dominant hereditary spastic paraplegia in which spastic paraparesis is accompanied by atrophy of the hand muscles, especially the thenar muscles and sometimes the distal lower limb muscles (Silver 1966). Neurophysiological studies indicate lower motor neurone or motor root involvement in addition to the corticospinal tract dysfunction (Warner *et al.* 2004). Following evidence of linkage of some families to chromosome 11q12-q14, heterozygous missense mutations in the Berardinelli–Seip congenital lipodystrophy type 2 gene, *BSCL2*, were found in two families, one with Silver syndrome and one with hereditary distal motor neuropathy (Windpassinger *et al.* 2004). The encoded protein has been termed seipin and the two mutations result in amino acid substitutions, N88S and S90L, in an N-glycosylation motif. Seipin is an integral membrane protein of the endoplasmic reticulum and the disruption of glycosylation caused by the missense mutations appears to result in protein aggregation. The clinical phenotype of this genetic disorder has recently been expanded to include patients with predominantly lower limb amyotrophy and also to some patients without signs of corticospinal tract pathology (Irobi *et al.* 2004b). It had previously been demonstrated that null mutations affecting seipin caused autosomal recessive Berardinelli–Seip congenital lipodystrophy type 2, with the phenotype of near absence of adipose tissue from early childhood and severe insulin resistance. Currently it is not understood how mutant forms of seipin cause two such distinct disease states (Agarwal and Garg 2004).

- *Receptor expression-enhancing protein 1 related hereditary spastic paraplegia; SPG31.* Recently mutations were identified in six autosomal dominant hereditary spastic paraplegia families linked to the SPG 31 locus on chromosome 2p12 (Zuchner *et al.* 2006). This represented 6.5 per cent of index cases with hereditary spastic paraplegia. Receptor expression-enhancing protein 1 is a widely expressed protein which localizes to mitochondria, although its precise function has not yet been elucidated.

- *ZFYVE27 related hereditary spastic paraplegia; SPG33.* A recent report described a single German family with a mutation in *ZFYVE27* (Mannan *et al.* 2006). The encoded protein is a novel member of the FYVE-finger family of proteins and is a specific spastin-binding protein.

- *Other autosomal dominant hereditary spastic paraplegia loci.* The genes associated with SPG9, SGG12, and SPG19 have not yet been identified. Spastic paraparesis at the SPG9 has been described in one family in which the phenotype is complicated by bilateral cataracts, gastro-oesophageal reflux with vomiting, and amyotrophy (Seri *et al.* 1999). A further family has been described with cataracts, learning difficulties, and skeletal abnormalities (Slavotinek *et al.* 1996).

Autosomal recessive hereditary spastic paraplegia. Autosomal recessive hereditary spastic paraplegia is also clinically and genetically heterogeneous and is uncommon compared to the dominantly inherited forms. Autosomal recessive pedigrees have been linked to 15 genetic loci and include both pure and complicated phenotypes. The genes underlying three subtypes have been identified to date: SPG7, paraplegin (Casari *et al.* 1998); SPG20/Troyer syndrome, spartin (Patel *et al.* 2002); and SPG21, maspardin (Simpson *et al.* 2003). The genes associated with the other 12 autosomal recessive loci have not yet been identified.

- *Paraplegin related hereditary spastic paraplegia; SPG7.* Patients with paraplegin mutations may have a pure or a complicated phenotype. In complicated cases the spastic paraparesis may be accompanied by a variety of other features including optic atrophy, cortical and cerebellar atrophy, dysphagia, dysarthria, distal amyotrophy, or sensorimotor neuropathy (Casari *et al.* 1998; McDermott *et al.* 2001). The *SPG7* gene comprises 17 exons and the encoded protein, paraplegin, is a ubiquitously expressed nuclear encoded mitochondrial metalloprotease which is a member of the ATPase associated with diverse cellular activities or AAA, protein family. In yeast, Afg3p and Rca1p form a high molecular weight hetero-oligomeric complex in the inner mitochondrial membrane. From studies of homologous proteins in yeast, it was predicted that paraplegin would function by forming multimeric complexes which have proteolytic and chaperone-like functions in the mitochondria, essential for the normal assembly and turnover of respiratory chain complexes. Recent studies employing a paraplegin knockout mouse have shown that the first detectable abnormality is the appearance at 4 months of age of abnormal enlarged mitochondria in synaptic terminals in the grey matter of the lumbar spinal cord (Ferreirinha *et al.* 2004). These mitochondrial abnormalities correlate in time with the onset of a significant functional motor deficit and precede signs of axonal degeneration by several months. Clearly, mitochondrial dysfunction within axonal terminals could lead to problems with ATP generation, regulation of calcium homeostasis, and in free radical metabolism. The paraplegin knockout mice also develop accumulations of neurofilaments and organelles in swollen axons indicating abnormalities in anterograde transport. Retrograde transport is delayed in symptomatic mice and this could clearly contribute to the axonal degeneration by affecting the transport of mitochondria, the trafficking of endosomes, and the internalization of neurotrophic factors. Work with cellular models has demonstrated that paraplegin co-assembles with a homologous protein AFG3L2 into a high molecular weight complex within the inner mitochondrial membrane (Atorino *et al.* 2003). Using fibroblasts from *SPG7* patients it was demonstrated that this complex is defective in the presence of paraplegin mutations, with a resultant decrease in complex 1 activity of the mitochondrial respiratory chain and increased sensitivity of the cells to oxidative stress. Several families have so far been described with paraplegin mutations. All the mutations described in some way affect the conserved AAA domain, either truncating the protein or by causing an in-frame deletion. Patients may show evidence of mitochondrial dysfunction on histochemical analysis of muscle (Casari *et al.* 1998; McDermott *et al.* 2001).

- *Spartin-related hereditary spastic paraplegia-Troyer syndrome; SPG 20.* Troyer syndrome, named after the family in which it was first identified, was originally described in 1967 in an Old Order Amish community (Cross and McKusick 1967b), a genetically isolated population with high prevalence of autosomal recessive disorders. The cardinal clinical features include spastic paraparesis, pseudobulbar palsy, distal amyotrophy, mild developmental delay, and subtle skeletal abnormalities including short stature. Imaging studies may show cerebral white matter abnormalities and thinning of the corpus callosum (Proukakis *et al.* 2004). In 2002, a frameshift mutation, 1110delA, in SPG20 encoding spartin was found to be the genetic cause of this disorder (Patel *et al.* 2002). The *SPG20* gene on chromosome 13q12.3 comprises nine exons and encodes a 72.7 kDa protein of 666 amino acids named spartin, which is ubiquitously expressed in adult tissues. Spartin shares homology with the other proteins, SNX15, VPS4, and Skd1, known to be involved in endosomal morphology and dynamics and protein trafficking, suggesting that spastin may serve a similar role. Spartin also shares homology with the N terminal region of spastin which forms the microtubule interacting and trafficking domain thought to be responsible for binding to microtubules. Thus, evidence to date indicates that spartin may function in the control of endosomal trafficking and/or microtubule dynamics (Crosby and Proukakis 2002). A very recent report described that spartin localizes to mitochondria (Lu *et al.* 2006).

- *Maspardin related hereditary spastic paraplegia—Mast syndrome; SPG 21*. Mast syndrome is an autosomal recessive, complicated form of HSP accompanied by dementia, which is also found at relatively high frequency among the Old Order Amish community (Cross and McKusick 1967a). Clinical features may include developmental delay, pseudobulbar palsy, cerebellar and extrapyramidal dysfunction, as well as thinning of the corpus callosum, cerebral atrophy, and white matter demyelination on MR imaging of the brain. 14 affected patients were found to be homozygous for a single base pair, 601insA, insertion in the gene encoding the acid-cluster protein of 33 kDa, *ACP33*, which has been renamed maspardin (Simpson *et al.* 2003). This is a frameshift mutation which causes a truncated protein product which is likely to cause a loss of function effect. One study had reported the subcellular distribution of maspardin in a T cell-line (Zeitlman *et al.* 2001) where it localized to vesicles involved in the early endosome recycling pathway and to acidic organelles. Further work is required to define the localization and function of maspardin within neurons but current evidence indicates that maspardin may be involved in protein sorting and trafficking and that defective trafficking might be important in the pathogenesis of Mast syndrome.

X-linked forms of hereditary spastic paraplegia. These forms are relatively rare and are largely seen in paediatric practice. There are three X-linked loci: SPG1, SPG2, and SPG16. The genes involved at the SPG1 and two loci have been known for some time and the molecular mechanisms of disease are relatively well understood. The gene at SPG16 is not yet known.

- *L1 cell adhesion molecule related hereditary spastic paraplegia; SPG1*. SPG1 is a rare developmental disorder of varying severity and patients usually have a complex phenotype with mental retardation and congenital musculoskeletal abnormalities, such as the absence of extensor hallucis longus, in addition to spastic paraparesis. Mutations in the L1 cell adhesion molecule gene,

L1CAM, are also found in X-linked hydrocephalus, X-linked agenesis of the corpus callosum and the MASA syndrome of mental retardation, aphasia, shuffling gait and adducted thumbs. The diseases are now considered to be part of a clinical syndrome with the acronym CRASH: corpus callosum hypoplasia, retardation, adducted thumbs, spastic paraplegia, and hydrocephalus (Fransen *et al.* 1995). L1CAM is a cell surface, membrane associated glycoprotein, and a member of the immunoglobulin superfamily of proteins, which is expressed predominantly within neurons and Schwann cells. It is a protein with complex extracellular and intracellular interactions and has important functions in neuronal adhesion, axonal outgrowth, and pathfinding during central nervous system development. L1CAM plays an essential role in the correct formation of the corticospinal tract (Dahme *et al.* 1997; Demyanenko *et al.* 1999). In knockout mice the normal anatomy of the corticospinal tracts is disrupted with failure of decussation of corticospinal axons across the midline at the level of the pyramids (Castellani *et al.* 2000). L1CAM is a component of the Sema3A receptor complex and mice deficient in L1CAM fail to respond to directive signalling from Sema3A.

- *Proteolipid protein related hereditary spastic paraplegia; SPG2*. The proteolipid protein 1 gene, *PLP*, encodes proteoipid protein 1 and its minor DM20 isoform. Both pure and complicated hereditary spastic paraplegia phenotypes may be observed in patients with mutations in proteolipid protein (Bonneau *et al.* 1993). Patients with a complicated phenotype may have a cerebellar syndrome, mental retardation, and optic atrophy in addition to spastic paraparesis. Mutations in the proteolipid protein gene also cause Pelizaeus–Merzbacher disease, a severe dysmyelinating syndrome characterized by nystagmus, ataxia, spasticity, abnormal movements, optic atrophy, and microcephaly, usually resulting in death during adolescence (Section 10.2.10). The phenotype of disease caused by mutations in the proteolipid protein gene can be considered as a continuous spectrum with milder SPG2 at one end and the more severe Pelizaeus–Merzbacher disease at the other. Proteolipid protein is the major myelin protein of the central nervous system, accounting for approximately 50 per cent of total myelin protein in the adult brain. DM20 is an alternatively spliced form, which lacks 35 amino acid residues. The function of the proteolipid protein and DM20 proteins has not been established with certainty, but they are considered likely to have an important function in stabilizing the structure of central nervous system myelin by forming the intraperiod line. It is possible that the two protein isoforms have different functions. The DM20 isoform appears earlier during development and is thought to play a role in glial cell development, while the proteolipid protein isoform is expressed later and may play a role in myelin assembly and maintenance. Mutations in the *PLP* gene that do not affect the the DM20 isoform are associated with the milder Pelizaeus–Merzbacher disease phenotype or SPG2. Conversely mutations which reduce the level of the DM20 isoform are associated with the more severe Pelizaeus–Merzbacher disease phenotypes (Griffiths *et al.* 1998a). Plp knockout mice develop normal myelin sheaths despite lack of PLP/DM20, but subsequently develop a severe axonopathy (Griffiths *et al.* 1998b). It was recently demonstrated that absence of PLP/DM20 in oligodendrocytes results in early impairment of fast axonal transport and multifocal accumulations of membranous organelles, in a mouse model with a null mutation in *PLP* (Edgar *et al.* 2004). Axonal degenerative changes were found to be concentrated at the distal regions of long axons.

Treatment of hereditary spastic paraplegia

So far, increased knowledge of the molecular basis of hereditary spastic paraplegia has had little impact on clinical practice. At present symptomatic treatments can be used to alleviate spasticity, bladder dysfunction, and the lower limb discomfort which can be troublesome features of hereditary spastic paraplegia. Neuroprotective therapies are not yet available to ameliorate the distal axonopathy.

23.5 Disorders of the bulbar motor system

23.5.1 Kennedy's disease

Kennedy's disease, or spinobulbar muscular atrophy is an X-linked degenerative disorder of lower motor neurones (Kennedy *et al.* 1968). Initial symptoms consist of hand tremor, fasciculations, and muscle cramps, followed by progressive weakness and atrophy of limb and bulbar musculature (Sinnreich and Klein 2002; Lee *et al.* 2005). Limb muscle weakness is predominantly proximal in distribution and mainly affects the lower limbs. There are no clinical signs of upper motor neurone dysfunction and tendon reflexes are reduced or absent. Weakness of the lower facial and tongue muscles causes dysarthria and jaw weakness may cause the mouth to hang open. Pharyngeal involvement causing dysphagia, and respiratory muscle weakness causing breathlessness, are less common. Perioral fasciculations causing 'quivering' of the chin is a characteristic feature. Patients frequently have evidence of mild androgen insensitivity with gynaecomastia, testicular atrophy, oligospermia, and erectile dysfunction. The disease course is much more slowly progressive compared to amyotrophic lateral sclerosis and typically life expectancy is not reduced, though a proportion of patients may die from the effects of respiratory muscle weakness (Atsuta *et al.* 2006). Patients may become wheelchair dependent over 2–3 decades, although some remain ambulatory until late in life. Sensory symptoms are uncommon and mild. However there is evidence that neurodegeneration is not exclusive to motor neurones. Mild distal sensory loss is frequently present in the lower limbs, and sensory nerve conduction studies show decreased or absent sensory nerve potentials in keeping with depletion of dorsal root ganglion cells demonstrated pathologically (Ferrante and Wilbourn 1997). Heterozygous female carriers of spinobulbar musclar atrophy may show mild clinical manifestations of the disease such as muscle cramps, mild facial weakness, and fasciculations, and neurophysiological studies showing mild changes of chronic denervation (Sobue *et al.* 1993).

Kennedy's disease is often misdiagnosed initially as amyotrophic lateral sclerosis. Though rare, it is important not to miss the diagnosis because of the genetic implications for the family and the comparatively more benign disease course. The diagnosis should be considered in any male patient with a phenotype of a pure lower motor neurone disorder, particularly when the disease course is relatively indolent, gynaecomastia is present, or when there is evidence of a mild concomitant sensory neuropathy. DNA examination for the CAG repeat expansion in exon 1 of the androgen receptor gene is diagnostic.

Genetics and disease pathogenesis

Spinobulbar muscular atrophy was linked to chromosome Xq12-21 in 1986 (Fischbeck *et al.* 1986). The disease-causing mutation was reported in 1991 identifying expansion of a trinucleotide CAG repeat in the androgen receptor gene from a normal length of 17–26 repeats, to a disease associated length of 40–52 repeats (La Spada *et al.* 1991). This CAG repeat region encodes a long tract of glutamine residues, beginning at amino acid 58. Disease severity and age of onset show some correlation with the size of the repeat expansion (Doyu *et al.* 1992). Patients with minimally expanded polyglutamine tracts in the androgen receptor gene have been reported to have unusual clinical manifestations, including essential tremor (Kaneko *et al.* 1993a), hypertrophic cardiomyopathy (Kaneko *et al.* 1993b), and very late onset of muscular weakness (Doyu *et al.* 1993). Spinobulbar muscular atrophy is one of the triplet repeat expansion disorders. In all these disorders, neurodegeneration occurs when the poly(Q) tract reaches a critical length of about 40 repeats and there tends to be a similar age of onset and rate of progression of disease, which correlate with the length of the poly(Q) tract (Zoghbi and Orr 2000). All are progressive neurological disorders that affect only a subset of neurones, despite ubiquitous expression of the expanded protein.

The androgen receptor is a nuclear receptor in the steroid receptor superfamily. The protein, encoded by eight exons, contains three functional domains: a carboxy-terminal hormone-binding domain, a DNA-binding domain, and an amino-terminal transactivation domain, which contains the polyglutamine tract. The androgen receptor is phosphorylated and bound to heat shock proteins in the cytoplasm. On ligand binding, it is transported to the nucleus, where it binds to DNA, and acts as a transcription factor. Expansion of the polyglutamine tract results in reduced target gene transactivation which may account for the features of decreased androgen sensitivity (Lieberman *et al.* 2002). However, complete loss of androgen receptor function, in individuals with testicular feminization syndrome, does not lead to motor neurone degeneration. Reduction in androgen receptor function does not therefore appear to be the major mechanism leading to motor neurone degeneration in spinobulbar muscular atrophy, and it is thought that neurodegeneration occurs through a toxic gain of function of the altered protein. There is evidence that polyglutamine repeats themselves are neurotoxic whilst the protein in which it is expressed is likely to modify the specific pattern of neurodegeneration seen in each trinucleotide repeat disorder (Ordway *et al.* 1997).

A transgenic mouse model of Kennedy's disease has been developed with a full-length androgen receptor and very long repeat expansions, which phenotypically reproduces spinobulbar muscular atrophy (Katsuno *et al.* 2002). There is evidence that toxicity of the mutant androgen receptor is ligand dependent, as it only occurs with the higher androgen levels present in males. Castration of the transgenic male mice reduces symptom severity, whilst treatment of transgenic females with exogenous testosterone worsens their phenotype (Katsuno *et al.* 2002). Leuprorelin, an LHRH agonist that reduces testosterone release from the testis, inhibits nuclear accumulation of mutant androgen receptor protein, resulting in rescue of the motor dysfunction in male transgenic mice (Katsuno *et al.* 2004). Human clinical trials of leuprorelin in spinobulbar muscular atrophy patients have commenced.

Several potential mechanisms for the toxic gain of function of the expanded androgen receptor protein have been proposed.

Misfolding of the polyQ expanded proteins in spinobulbar muscular atrophy and other polyglutamine diseases is thought to lead to formation of aggregates, which are seen as nuclear inclusions. Nuclear inclusions containing the amino-terminal epitopes of the mutant androgen receptor are found within motor neurones and certain non-neuronal tissues in patients with spinobulbar muscular atrophy (Li *et al.* 1998) and in cell culture and animal models of the disease. The role of protein aggregation in the pathogenesis of neurodegeneration is, however, uncertain. The presence of ubiquitin, proteasomes, and molecular chaperones within nuclear inclusions in spinobulbar muscular atrophy implicates protein misfolding in disease pathogenesis. There is evidence that polyglutamine proteins may directly impair the ubiquitin–proteasome system, possibly by exceeding the capacity of proteasomes or sequestering molecular chaperones, which would then be unavailable to degrade other substrates (Bence *et al.* 2001). Ubiquitin-dependent proteolysis has a central role in regulating fundamental cellular events, such as cell division and apoptosis, and represents a potential mechanism linking protein aggregation to cellular dysregulation and cell death.

Nuclear localization of mutant androgen receptor appears to be a requirement for neurotoxicity (Katsuno *et al.* 2002). Nuclear accumulation of mutant polyglutamines may cause toxicity through interaction with transcription factors, with resultant disruption of gene transcription. Down-regulation of gene expression seems to be an early event in polyglutamine disease pathogenesis. In the case of spinobulbar muscular atrophy, interest has focussed on the CREB binding protein, CBP, which is sequestered by the mutant polyglutamine (McCampbell *et al.* 2000). CBP is a transcriptional coactivator that possesses histone acetyltransferase activity. Hyperacetylation of histones marks transcriptionally active regions of chromatin. As CBP is present in a functionally limiting level in cells, its sequestration by mutant polyglutamine could have major effects on gene expression. In neurons, CBP is a key component of neurotrophic factor signalling pathways, including that of vascular endothelial growth factor, VEGF. Transgenic mice that overexpress mutant human androgen receptor have reduced spinal cord VEGF[164] protein levels, even in presymptomatic stages, and polyglutamine-expanded androgen receptor has been shown to interfere with CBP-mediated transcription of VEGF (Sopher *et al.* 2004). Reversal of the histone acetylation defect caused by CBP sequestration, using histone deacetylase inhibitor drugs has been shown to reduce polyglutamine-associated cell death in both *Drosophila* and mouse models of Huntington's disease (Steffan *et al.* 2001; McCampbell *et al.* 2001).

The mutant androgen receptor protein may also adversely affect axonal transport. A motor neurone cell line transfected with mutant androgen receptor, develops protein aggregates which affect the distribution of the motor protein kinesin, and consequently the distribution of mitochondria. This suggests that protein aggregation within motor neurones in Kennedy's disease disrupts kinesin-mediated fast axonal transport (Piccioni *et al.* 2002).

23.5.2 Brown–Vialetto–van Laere syndrome

The Brown–Vialetto–van Laere syndrome is a rare disorder of unknown aetiology characterized by progressive weakness of bulbar muscle groups associated with sensorineural deafness (Brucher *et al.* 1981; Hawkins *et al.* 1990; Sathasivam *et al.* 2000). In most cases progressive deafness is the initial symptom followed by involvement of the lower motor cranial nerves (Fig. 23.7). Lower

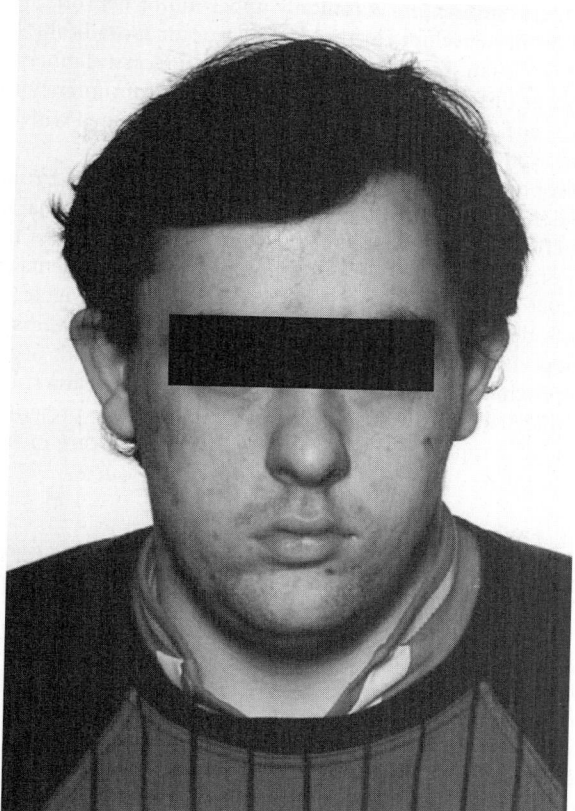

A

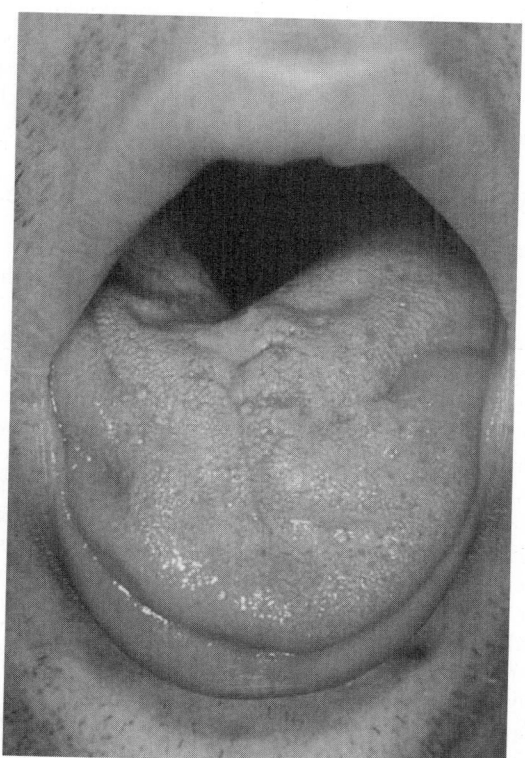

B

Fig. 23.7 A patient with Brown–Vialetto–van Laere syndrome showing bilateral facial weakness (A) and wasting of the tongue (B).

motor neurone and less commonly upper motor neurone signs in the limbs may develop. The disorder may occur sporadically, but in some cases is familial, usually with autosomal recessive inheritance. The age of onset of the first symptom varies from infancy to the third decade. A period of worsening may be associated with intercurrent events such as an infection.

Neurophysiological examination reveals changes of chronic or active denervation in limb and bulbar muscles with normal nerve conduction velocities. Detectable hyperintensity of brainstem nuclei on MRI has been recently described in some patients (Koul *et al.* 2006). The clinical course of the condition is variable; some patients deteriorate progressively from the outset, others have abrupt periods of deterioration interspersed with periods of stability. Approximately half of patients reported in the literature survive more than 10 years. Management of patients at the present time consists of symptomatic treatment and supportive care including gastrostomy feeding and respiratory support measures.

23.5.3 Fazio–Londe disease

Fazio–Londe disease is a rare, and presumed to be inherited either autosomal recessively or dominantly, form of progressive motor neurone degeneration with onset in infancy (McShane *et al.* 1992). The disorder develops in children within the first 5 years of life. Stridor is a prominent early symptom, often leading to an initial diagnosis of croup. Progressive bulbar paralysis then develops, with dysarthria and dysphagia. Typically, over the next 2 years the limb muscles become involved and respiratory failure develops resulting in death. Occasional cases of later onset and longer survival have been described. Fazio–Londe syndrome and Brown–Vialetto–van Laere syndrome may overlap, as shown in one family in which four siblings born to consanguineous parents were variably affected by pontobulbar weakness, deafness, and limb weakness (Dipti *et al.* 2005). Two siblings with the phenotype of Fazio–Londe disease have been described who had a mitochondrial respiratory chain defect (Roeleveld-Versteegh *et al.* 2004).

23.6 Toxic disorders of the motor neurone

23.6.1 Neurolathyrism and konzo

The development of motor system disorders has been associated with dietary dependence on food plants with neurotoxic potential, notably the grass pea or chickling pea *Lathyrus sativus* and cassava *Manihot esculenta* Crantz, in specific geographical regions of the world. Dietary reliance on the grass pea or on insufficiently processed cassava has resulted in outbreaks of neurolathyrism, and konzo or neurocassivism, respectively. These are self-limiting neurodegenerative disorders confined to the upper motor neurone system. Investigation of epidemics of lathyrism and konzo has indicated that individual susceptibility to the toxic effects of these plants varies with gender, age, nutritional state, level of physical exercise, as well as the toxin content of the plants, methods of preparation of staple food, and duration of dietary exposure. Epidemics of both disorders tend to occur when adverse environmental conditions force heavy dietary reliance on the grass pea and cassava, which are plants relatively resistant to drought and pest attack. Despite the clinical similarities of the two neurotoxic disorders, outbreaks of these diseases have occurred in two distinct non-overlapping geographical areas of the world. Lathyrism outbreaks have occurred in the Indian subcontinent, China, and in Ethiopia and Eritrea on the Horn of Africa as well as parts of Europe. By contrast konzo has been reported mainly from sub-Saharan Africa (Rosling and Tylleskar 1995; Rosling and Tylleskar 2000).

The common clinical picture in both disorders is of a symmetrical spastic paraparesis. In the prodromal phase subjects affected by lathyrism or konzo complain of leg stiffness, weakness, and cramping. Tremor is common and acute reversible sensory symptoms are often reported. Bladder involvement is not common. In mild cases spasticity of the legs is only revealed when the subject is asked to run; in severe cases the individual may become bedridden. In time, severely affected patients may show disuse muscle atrophy. Sensory examination, mentation, cerebellar and cranial nerve function are usually normal. Konzo, the cassava related disorder, is potentially more serious and may cause a spastic tetraparesis as well as pseudobulbar signs (Tshala-Katumbay *et al.* 2001). Severe cases of konzo may also develop visual impairment and bilateral optic neuropathy (Mwanza *et al.* 2003). The term konzo means 'tied legs' in the Kiyaka language. There is evidence for another cassava associated neurological disorder which occurs in older subjects with a chronic high intake of incompletely detoxified cassava (Osuntokum 1981). This consists of a slowly evolving ataxic neuropathy with or without evident upper motor neurone signs, which may be accompanied by visual and sensorineural hearing deficits.

The two diseases show differential patterns of gender susceptibility. Lathyrism tends to affect young males more commonly and more severely than females (Spencer *et al.* 1984). Konzo appears to affect males and females similarly, while women of child bearing age are more susceptible than men of the same age (Tylleskar *et al.* 1993). Excessive physical activity is often reported at the outset of disease and may be a predisposing factor that stresses the motor system and promotes susceptibility of the cortical motor neurones.

Electrophysiological examination in both disorders indicates marked dysfunction of the pyramidal tracts with absent responses in the lower limbs following magnetic or transcranial magnetic stimulation of the motor cortex or prolonged central motor conduction times (Hugon *et al.* 1990, Tshala Katumbay *et al.* 2002). Most subjects have normal peripheral motor and sensory conduction velocities and electromyography. Somatosensory evoked response studies indicate that subclinical involvement of sensory pathways is relatively common. A study using MRI in two Tanzanian patients with konzo revealed no abnormalities (Tylleskar *et al.* 1993). Imaging studies are not available for patients with lathyrism. Pathological studies are sparse or lacking. A study of the brain in one individual who developed lathyrism 31 years prior to his death revealed loss and shrinkage of pyramidal neurons in the upper part of the precentral gyrus (Filiminoff 1926). A few studies of the spinal cord in cases of lathyrism have shown predominantly distal and symmetrical degeneration of lateral and ventral corticospinal tracts, sometimes accompanied by distal degeneration of the spinocerebellar and gracile tracts (Hirano *et al.* 1976; Striefler *et al.* 1977).

The molecular mechanisms underlying these two neurotoxic disorders have still to be elucidated. β-N-oxalylamino-L-alanine, or BOAA, appears to be the main neurotoxic compound in the grass pea and is considered the likely cause of human lathyrism. The mechanism of neurotoxicity appears to be excitotoxic, involving excessive glutamatergic neuronal stimulation. BOAA is a potent agonist at the AMPA subtype of glutamate receptor and an inhibitor of glutamate uptake (Spencer 1999) and recent evidence indicates that activation of group I metabotropic glutamate receptors may also play a part in the toxic effects (Kusama-Eguchi et al. 2004). Other studies have provided evidence that BOAA may disrupt mitochondrial function by triggering glutathione loss and protein thiol oxidation (Ravindranath 2002). At present, there is no robust animal model of lathyrism that develops features of cortical motor neurone degeneration.

In konzo the disease pathogenesis is not properly understood and again there is no animal model of the condition. Epidemiological studies have consistently shown a link with dietary intake of inadequately processed bitter cassava, combined with a low protein intake (Rosling and Tylleskar 2000). Bitter varieties of cassava contain large amounts of cyanogenic glucosides, linamarin, and lotaustralin, and the levels of these poisons depend on the prevalent environmental conditions. Weeks or months of dependency on incompletely detoxified cassava, combined with low intake of proteins which normally provide the source of sulphur amino acids to convert cyanide to thiocyanate, leads to outbreaks of konzo. The mechanism by which cyanogenic toxins damage the upper motor neurone system has not been established with certainty: hypotheses include mitochondrial inhibition; modulation of the glutamatergic neurotransmitter system (Spencer 1999); and carbamoylation of proteins in the nervous system (Mellado et al. 1982). The cyanide metabolite cyanate is considered the likely aetiological factor in the cassava induced ataxic (myelo)neuropathy (Kogure et al. 1975).

There is no effective treatment for these persistently disabling disorders. Once the acute or subacute onset of the disease has passed, the disability remains unchanged and irreversible. Little benefit has been reported from centrally acting spasmolytic agents. The role of therapies such as botulinum toxin injection and physiotherapy to reduce muscle spasm and prevent joint contractures has not been reported. However, both lathyrism and konzo can be prevented by education, modifying food preparation, or changing dietary practice, and international networks have been established to aim at preventing outbreaks of both lathyrism and konzo.

23.6.2 Western Pacific amyotrophic lateral sclerosis-Parkinsonism-dementia complex

In the 1950s a high incidence geographical focus of amyotrophic lateral sclerosis was identified on the Western Pacific island of Guam, where the indigenous Chamorro people had a rate of the disease of approximately 50–100 times that of Western populations (Waring et al. 2004). Subsequently a unique neurodegenerative condition amongst the same population, which was given the term Parkinsonism–dementia complex was described (Hirano et al. 1961). Affected individuals developed severe dementia, and rigid akinetic Parkinsonism with marked postural deformities. Hyperreflexia and muscle atrophy affecting mainly the distal extremities were frequently detected. Familial clustering of both disorders was observed. The occurrence of Parkinsonism–dementia complex and amyotrophic lateral sclerosis in a geographically isolated population in similar age groups, and with overlap in symptomatology and familial aggregation led to the suspicion that the two diseases may have a common cause (Murakami 1999). Pathologically, Guamanian amyotrophic lateral sclerosis–Parkinsonism–dementia complex appears to be a tauopathy, characterized by the formation of neurofibrillary tangles. Variation in the anatomical distribution of pathology accounts for the varying clinical presentations, with tangles predominantly seen in the motor system in Guamanian amyotrophic lateral sclerosis (Ince and Codd 2005).

The cause of Parkinsonism–dementia complex and Guamanian amyotrophic lateral sclerosis is unknown. A genetic cause has not been found despite apparent familial clustering of the diseases. Several observations support an environmental aetiology, including the fact that Filipino immigrants are susceptible to the condition, and that there is a trend for increasing age of onset and decreasing disease incidence over time (Galasko et al. 2002). The leading hypothesis relates to ingestion of an environmental toxin present in the seed of the false sago palm, Cycas circinalis (Ince and Codd 2005). Beta-methylaminoalanine, BMAA, is one of several chemicals with neurotoxic properties present in the seeds of cycad plants which are used by the native Chamorros for food and medicine. BMAA can selectively injure motor neurones via AMPA/kainate receptor activation followed by increased intracellular calcium and free radical generation (Rao et al. 2006). BMAA toxicity seemed unlikely as a cause of Guamanian amyotrophic lateral sclerosis when studies showed that the Chamorro practice of flour preparation from cycad seeds largely removed the toxin and that neurotoxic effects in primates were only observed following massive exposure to BMAA. However, the cycad hypothesis has emerged recently based on a new understanding of Chamorro food practices, a cyanobacterial origin of BMAA in cycad tissue and a possible mechanism for biomagnification of this neurotoxin in Guam ecosystem (Ince and Codd 2005). Free living cyanobacteria produce high levels of BMAA in the root system of cycads and the toxin is then concentrated in the seed tissues. The traditional Chamorro diet includes the prized delicacy of the fruit bat, which feeds on cycad seed components and reportedly bioaccumulates BMAA (Cox et al. 2003). Plant and animal proteins provide a previously unrecognized reservoir for the slow release of the BMAA toxin. Patients with amyotrophic lateral sclerosis–Parkinsonism–dementia complex have been reported to show elevated levels of BMAA in post-mortem brain tissue (Cox et al. 2003). Further investigation of this aetiological hypothesis and indeed of the potential role of cyanobacterial toxicity in sporadic neurodegenerative diseases is ongoing.

23.6.3 Post-irradiation motor disorders

Approximately 5 per cent of patients undergoing radiotherapy for breast cancer develop brachial plexopathy, but in addition to motor problems, sensory change is usually prominent (Olsen et al. 1993; Jaeckle 2004). Neurophysiological examination may reveal myokymia and motor nerve conduction block (Esteban and Traba 1993). Delayed progressive bulbar dysfunction with myokymia has also been described in some patients following radiotherapy (Glenn and Ross 2000). Post-irradiation myelopathy can produce a relatively pure lower motor neurone syndrome (Lamy et al. 1991).

A predominantly motor disorder can develop in the lower limbs following radiotherapy involving exposure of the lumbar spinal canal as part of therapy for testicular tumours, lymphoma, or other neoplasms (Section 5.9.5). The latency between radiotherapy and onset of symptoms varies between a few months and several decades. The lower limb weakness is often distal and asymmetrical and associated with loss of reflexes. Wasting may not be apparent even in weak muscles, indicating that conduction block may contribute to weakness, although this is difficult to demonstrate. In most patients the leg weakness gradually progresses and may eventually result in severe disability, though there may be periods of stabilization of symptoms. Patients may eventually develop sensory symptoms, but objective sensory signs are usually absent and sensory nerve action potentials are normal (Bowen *et al.* 1996). Sphincter and sexual function are usually relatively well preserved initially. Attempted therapeutic intervention with steroids or intravenous immunoglobulin has not shown any convincing benefit. MRI with gadolinium may show nodular enhancement of the conus and cauda equina (Bowen *et al.* 1996; Hsia *et al.* 2003). The CSF protein may be normal or elevated. Myokymia may be noted on electromyography. Microscopic examination has shown a vasculopathy of the proximal spinal nerve roots, with preservation of motor neurone cell bodies and spinal cord architecture (Bowen *et al.* 1996). With improved safety of radiotherapy regimes and the substitution of chemotherapy for radiotherapy in the treatment of testicular cancer, it is expected that the incidence of this clinical problem will decrease over time

23.7 Disorders of motor neurone overactivity

23.7.1 Neuromyotonia

Clinicians use several terms, including myokymia, neuromyotonia, Isaac's syndrome, and cramp fasciculation syndrome to describe the motor manifestations of peripheral nerve hyperexcitability (Section 24.10.4). Neuromyotonia is a rare syndrome of spontaneous and continuous muscle fibre activity reflecting a peripheral nerve hyperexcitability disorder (Maddison 2006). This results in stiffness and cramping of muscles, often worse following exercise, though muscle power is frequently normal. Neuromyotonia can occur as a feature of a heterogeneous group of underlying disorders. The majority of cases are acquired and some of these are associated with an underlying peripheral neuropathy, myasthenia gravis, or the administration of penicillamine (Hart *et al.* 2002). Up to 20 per cent of cases may be associated with an underlying malignancy, usually thymoma or small cell lung cancer (Caress *et al.* 1997). Neuromyotonia can also occasionally be found in patients with hereditary neuropathies (Hahn *et al.* 1991). In acquired neuromyotonia, voltage-gated potassium channel, or VGKC, antibodies indicative of an auto-immune pathogenesis may be found in approximately 40 per cent of patients (Hart *et al.* 1997). There is evidence that these antibodies may reduce K^+ currents by cross-linking of potassium channels (Tomimitsu *et al.* 2004).

Neurophysiological assessment allows neuromyotonia to be distinguished from true myotonia. Electromyography features include spontaneous, continuous, irregularly occurring doublet or multiplet single motor unit, or partial motor unit, high intraburst frequency of 30–300 Hz (Fig. 23.8). Patients do not have detectable central neurophysiological abnormalities and motor cortex excitability parameters, and central motor conduction times are normal (Maddison *et al.* 2006). The pathogenic abnormality is thought to lie within the terminal branches of the peripheral motor nerves. Symptoms of neuromyotonia may improve substantially with carbamazepine or phenytoin therapy and patients may respond to intravenous immunoglobulin therapy or plasma exchange.

The differential diagnosis of neuromyotonia includes true myotonic disorders, myokymia which is restricted to muscle innervated by injured motor axons, the stiff person syndrome (Section 23.7.2), and the benign cramp fasciculation syndrome.

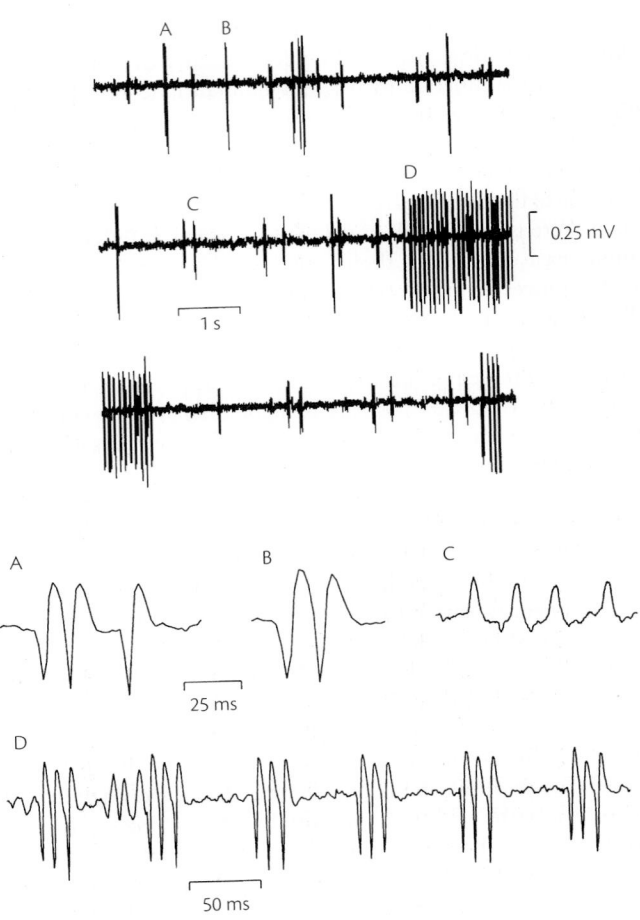

Fig. 23.8 Neuromyotonia. The upper traces are of a 25-s continuous needle electromyography recording from medial gastrocnemius muscle in a patient with acquired neuromyotonia. Sections of this are shown at faster time–bases below. Motor units are seen to fire spontaneously and irregularly as doublets (B), triplets (A) and multiplets (C), with intraburst frequencies of up to 120 Hz. The prolonged discharge seen in the middle of the recording consists of rapidly firing triplets of more than one motor unit (D). (Courtesy of Dr Paul Maddison.)

23.7.2 Stiff person syndrome

Stiff person syndrome is a rare distressing neurological disorder first described in 1956 (Moersch and Woltman 1956) (Sections 38.3.4 and 40.10.3). The symptoms often begin with tightness or stiffness of the trunk muscles which will usually spread to involve the proximal muscles of the limbs. Persistent contraction of the axial muscles leads to truncal rigidity, a characteristic lumbar hyperlordosis, and restriction of movement of the hips and spine. Superimposed on this background rigidity, the patient will often be affected by painful muscle spasms which may be precipitated by sudden movement, noise, tactile stimulation, or emotional upset. Specific phobia is a frequent non-motor symptom of stiff person syndrome (Henningsen and Meinck 2003). Neurophysiological examination reveals motor unit firing at rest simultaneously from agonist and antagonist muscles. The spontaneous muscle activity is of central rather than peripheral origin, related to release of polysynaptic spinal and brainstem reflexes (Thompson 2001). Up to 80 per cent of patients with classical stiff person syndrome have autotantibodies in the serum and CSF to glutamic acid decarboxy-lase, GAD, a rate limiting enzyme responsible for the synthesis of gamma-aminobutyric acid, GABA, and concentrated in GABAergic nerve terminals and in pancreatic β cells (Rakocevic *et al.* 2004). A high proportion of patients have coexisting diabetes mellitus or other autoimmune diseases. More recently antibodies to GABA A-receptor-associated-protein have been found in up to 70 per cent of patients with stiff person syndrome, with experimental evidence that this may inhibit the surface expression of the GABA A receptor and the normal functioning of GABAergic neurotransmission (Raju *et al.* 2006). In patients with classical stiff person syndrome there is a paucity of neuropathological abnormalities, suggesting that functional impairment of inhibitory neurotransmitter circuit-ry is sufficient to cause disease without overt structural damage. Imaging studies are usually normal in classical stiff person syn-drome, but magnetic resonance spectroscopy has shown reduced levels of GABA in specific brain regions (Levy *et al.* 2005) and tran-scranial magnetic stimulation has revealed significantly enhanced motor cortex excitability (Koerner *et al.* 2004).

Several variants of the stiff person syndrome have been described (Shaw 1999). Progressive encephalomyelitis with rigidity is usually a paraneoplastic disorder which may present with similar clinical features. However, the clinical course is usually relentlessly progres-sive, often resulting in death within a few months. Histopathological examination reveals widespread perivascular lymphocytic cuffing and infiltration, with neuronal loss in the lower brainstem and spinal cord, especially in the spinal grey matter (Whitely *et al.* 1976). A second variant is the jerking stiff person syndrome, in which myoclonic jerking is a prominent additional feature (Leigh *et al.* 1980). A third variant, in which antibodies to the synaptic vesicle protein amphiphysin I are detected, has been described in associa-tion with breast cancer (Folli *et al.* 1993). Barker and colleagues drew attention to a fourth clinical sub-type, the stiff limb syndrome (Barker *et al.* 1998). Affected individuals present with rigidity, abnormal fixed posturing, and painful spasms usually of the distal lower limb and sparing the trunk. Brainstem abnormalities and sphincter dysfunction may develop later in the disease course. This focal variant is less likely to be associated with detectable autoanti-bodies, or to respond well to therapy and is more likely to lead to loss of ambulation. Recently a series of patients with stiff person syndrome and elevated anti-GAD antibodies in the CSF has been reported who also had cerebellar disease, gait ataxia, dysarthria, and oculomotor abnormalities (Rakocevic *et al.* 2006).

The rarity of stiff person syndrome limits the feasibility of controlled clinical trials, but the available evidence indicates that drugs which increase cortical and spinal inhibition or enhance GABA such as benzodiazepines, baclofen, valproate, gabapentin, tiagabine, and immunomodulatory therapies such as prednisolone, intravenous immunolglobulin, and plasmapheresis are effective therapies.

References

Aalfs CM, Koelman JH, Posthumus Meyjes FE *et al.* (1993). Posterior tibial and sural nerve somatosensory evoked potentials: a study in spastic paraparesis and spinal cord lesions. *Electroencephalogr Clin Neurophysiol*, **89**, 437–41.

Abe K, Fujimura H, Tatsumi, C *et al.* (1995). Eyelid "apraxia" in patients with motor neuron disease. *J Neurol Neurosurg Psychiatry*, **59**, 629–32.

Agarwal AK, Garg A (2004). Sepin: a mysterious protein. *Trends Mol Med*, **10**, 440–4.

Aguirre T, Van Den Bosch L, Goetschalckx K *et al.* (1998). Increased sensitivity of fibroblasts from amyotrophic lateral sclerosis patients to oxidative stress. *Ann Neurol*, **43**, 452–7.

Almer G, Teismann P, Stevic Z *et al.* (2002). Increased levels of the pro-inflammatory prostaglandin PGE2 in CSF from ALS patients. *Neurology*, **58**, 1277–9.

Andersen PM, Forsgren L, Binzer, M *et al.* (1996). Autosomal recessive adult-onset amyotrophic lateral sclerosis associated with homozygosity for Asp90Ala CuZn-superoxide dismutase mutation. A clinical and genealogical study of 36 patients. *Brain*, **119**, 1153–72.

Antonellis A, Ellsworth RE, Sambuughin N *et al.* (2003). Glycyl tRNA synthetase mutations in Charcot-Marie-Tooth disease type 2D and distal spinal muscular atrophy type V. *Am J Hum Genet*, **72**, 1293–9.

Araki K, Ueda Y, Michinaka C *et al.* (1989). An autopsy case of juvenile muscular atrophy of unilateral upper extremity (Hirayama's disease). *Nippon Naika Gakkai Zasshi*, **78**, 674–5.

Armon C (2003). An evidence-based medicine approach to the evaluation of the role of exogenous risk factors in sporadic amyotrophic lateral sclerosis. *Neuroepidemiology*, **22**, 217–28.

Atorino L, Silvestri L, Koppen M *et al.* (2003). Loss of m-AAA protease in mitochondria causes complex I deficiency and increased sensitivity to oxidative stress in hereditary spastic paraplegia. *J Cell Biol*, **163**, 777–87.

Atsuta N, Watanabe H, Ito M *et al.* (2006). Natural history of spinal and bulbar muscular atrophy (SBMA): a study of 223 Japanese patients. *Brain*, **129**, 1446–55.

Auer-Grumbach M, Fazekas F, Radner H *et al.* (1999). Troyer syndrome: a combination of central brain abnormality and motor neuron disease? *J Neurol*, **246**, 556–61.

Azzouz M, Ralph GS, Storkebaum E *et al.* (2004). VEGF delivery with retrogradely transported lentivector prolongs survival in a mouse ALS model. *Nature*, **429**, 413–7.

Bak TH, O'Donovan DG, Xuereb JH *et al.* (2001). Selective impairment of verb processing associated with pathological changes in Brodmann areas 44 and 45 in the motor neurone disease-dementia-aphasia syndrome. *Brain*, **124**, 103–20.

Barker RA, Revesz T, Thom M *et al.* (1998). Review of 23 patients affected by the stiff man syndrome: clinical subdivision into stiff trunk (man) syndrome, stiff limb syndrome, and progressive encephalomyelitis with rigidity. *J Neurol Neurosurg Psychiatry*, **65**, 633–40.

Beal MF (2000). Mitochondria and the pathogenesis of ALS. *Brain*, **123**, 1291–2.

Beckman JS, Carson M, Smith CD *et al.* (1993). ALS, SOD and peroxynitrite. *Nature*, **364**, 584.

Beetz C, Nygren AOH, Schickel J *et al.* (2006). High frequency of partial SPAST deletions in autosomal dominant hereditary spastic paraplegia. *Neurology*, **67**, 1926–30.

Beetz C, Zucnner S, Ashley-Koch A *et al.* (2007). Linkage to a known gene but no mutation identified: comprehensive re-analysis of SPG4 HSP pedigrees reveals large deletions as the sole cause. *Human Mutation*, **28**, 739–40.

Bence NF, Sampat RM, Kopito RR (2001). Impairment of the ubiquitin-proteasome system by protein aggregation. *Science*, **292**, 1552–5.

Bendotti C, Tortarolo M, Suchak SK *et al.* (2001). Transgenic SOD1 G93A mice develop reduced GLT-1 in spinal cord without alterations in cerebrospinal fluid glutamate levels. *J Neurochem*, **79**, 737–46.

Bensimon G, Lacomblez L, Meininger V (1994). A controlled trial of riluzole in amyotrophic lateral sclerosis. ALS/Riluzole Study Group. *N Engl J Med*, **330**, 585–91.

Blexrud MD, Windebank AJ, Daube JR (1993). Long-term follow-up of 121 patients with benign fasciculations. *Ann Neurol*, **34**, 622–5.

Boillee S, Yamanaka K, Lobsiger C. S. *et al.* (2006). Onset and progression in inherited ALS determined by motor neurons and microglia. *Science*, **312**, 1389–92.

Boltshauser E, Lang W, Spillmann T *et al* (1989). Hereditary distal muscular atrophy with vocal cord paralysis and sensorineural hearing loss: a dominant form of spinal muscular atrophy? *J Med Genet*, **26**, 105–8.

Bonneau D, Rozet JM, Bulteau C *et al.* (1993). X linked spastic paraplegia (SPG2): clinical heterogeneity at a single gene locus. *J Med Genet*, **30**, 381–4.

Borthwick GM, Johnson MA, Ince PG et al. (1999). Mitochondrial enzyme activity in amyotrophic lateral sclerosis: implications for the role of mitochondria in neuronal cell death. Ann Neurol, 46, 787–90.

Borthwick GM, Taylor RW, Walls TJ et al. (2006). Motor neuron disease in a patient with a mitochondrial tRNAIle mutation. Ann Neurol, 59, 570–4.

Bosch EP, Yamada T, Kimura J (1985). Somatosensory evoked potentials in motor neuron disease. Muscle Nerve, 8, 556–62.

Bouchard JP, Barbeau A, Bouchard R et al. (1978). Autosomal recessive spastic ataxia of Charlevoix-Saguenay. Can J Neurol Sci, 5, 61–9.

Bourke SC, Bullock RE, Williams TL et al. (2003). Noninvasive ventilation in ALS: indications and effect on quality of life. Neurology, 61, 171–7.

Bourke SC, Tomlinson M, Williams TL et al. (2006). Effects of non-invasive ventilation on survival and quality of life in patients with amyotrophic lateral sclerosis: a randomised controlled trial. Lancet Neurol, 5, 140–7.

Bourke SC, Williams TL, Bullock RE et al. (2002). Non-invasive ventilation in motor neuron disease: current UK practice. Amyotroph Lateral Scler Other Motor Neuron Disord, 3, 145–9.

Bowen J, Gregory R, Squier M et al. (1996). The post-irradiation lower motor neuron syndrome neuronopathy or radiculopathy? Brain, 119, 1429–39.

Brain WR (1933). Diseases of the Nervous System. Oxford University Press, London, 1st edition, p. 801.

Brooks BR (1994). El Escorial World Federation of Neurology criteria for the diagnosis of amyotrophic lateral sclerosis. Subcommittee on motor neuron diseases/amyotrophic lateral sclerosis of the World Federation of Neurology Research Group on neuromuscular diseases and the El Escorial "Clinical limits of amyotrophic lateral sclerosis" workshop contributors. J Neurol Sci, 124 (Suppl), 96–107.

Brooks BR, Miller RG, Swash M, Munsat TL (2000). El Escorial revisited: revised criteria for the diagnosis of amyotrophic lateral sclerosis. Amyotroph Lateral Scler Other Motor Neuron Disord, 1, 293–9.

Brooks BR, Shodis KA, Lewis DH et al. (1995). Natural history of amyotrophic lateral sclerosis. Quantification of symptoms, signs, strength, and function. Adv Neurol, 68, 163–84.

Brownell B, Oppenheimer DR, Hughes JT (1970). The central nervous system in motor neurone disease. J Neurol Neurosurg Psychiatry, 33, 338–57.

Brucher JM, Dom R, Lombaert A et al. (1981). Progressive pontobulbar palsy with deafness: clinical and pathological study of two cases. Arch Neurol, 38, 186–90.

Brugman F, Wokke JH, Scheffer H et al. (2005). Spastin mutations in sporadic adult-onset upper motor neuron syndromes. Ann Neurol, 58, 865–9.

Bruijn LI, Cleveland DW (1996). Mechanisms of selective motor neuron death in ALS: insights from transgenic mouse models of motor neuron disease. Neuropathol Appl Neurobiol, 22, 373–87.

Bruijn LI, Miller TM, Cleveland DW (2004). Unraveling the mechanisms involved in motor neuron degeneration in ALS. Annu Rev Neurosci, 27, 723–49.

Bruyn RP, Koelman JH, Troost D et al. (1995). Motor neuron disease (amyotrophic lateral sclerosis) arising from longstanding primary lateral sclerosis. J Neurol Neurosurg Psychiatry, 58, 742–4.

Bruyn RP, Scheltens P, Lycklama a Nijeholt J et al. (1993). Autosomal recessive paraparesis with amyotrophy of the hands and feet. Acta Neurol Scand, 87, 443–5.

Bunina TL (1962). On intracellular inclusions in familial amyotrophic lateral sclerosis. Zh Nevropatol Psikhiatr Im S S Korsakova, 62, 1293–9.

Byrne PC, Mc Monagle P, Webb S et al. (2000). Age-related cognitive decline in hereditary spastic paraparesis linked to chromosome 2p. Neurology, 54, 1510–7.

Calza L, Manfredi R, Freo E et al. (2004). Transient reversal of HIV-associated motor neuron disease following the introduction of highly active antiretroviral therapy. J Chemother, 16, 98–101.

Caress JB, Abend WK, Preston DC et al. (1997). A case of Hodgkin's lymphoma producing neuromyotonia. Neurology, 49, 258–9.

Carriedo SG, Yin HZ, Weiss JH (1996). Motor neurons are selectively vulnerable to AMPA/kainate receptor-mediated injury in vitro. J Neurosci, 16, 4069–79.

Casali C, Valente EM, Bertini E et al. (2004). Clinical and genetic studies in hereditary spastic paraplegia with thin corpus callosum. Neurology, 62, 262–8.

Casari G, De Fusco M, Ciarmatori S et al. (1998). Spastic paraplegia and OXPHOS impairment caused by mutations in paraplegin, a nuclear-encoded mitochondrial metalloprotease. Cell, 93, 973–83.

Caselli RJ, Smith BE, Osborne D (1995). Primary lateral sclerosis: a neuropsychological study. Neurology, 45, 2005–9.

Castellani V, Chedotal A, Schachner M et al. (2000). Analysis of the L1-deficient mouse phenotype reveals cross-talk between Sema3A and L1 signaling pathways in axonal guidance. Neuron, 27, 237–49.

Chancellor AM, Slattery JM, Fraser H et al. (1993). The prognosis of adult-onset motor neuron disease: a prospective study based on the Scottish Motor Neuron Disease Register. J Neurol, 240, 339–46.

Chancellor AM, Warlow CP (1992). Adult onset motor neuron disease: worldwide mortality, incidence and distribution since 1950. J Neurol Neurosurg Psychiatry, 55, 1106–15.

Charcot J-M, Joffroy A (1869). Deux cas d'atrophie musculaire progressive avec lesions de la substance grise et de faisceaux anterolateraux de la moelle epiniere. Arch Physiol Norm Pathol, 3, 744–57.

Chen YZ, Bennett CL, Huynh HM et al. (2004). DNA/RNA helicase gene mutations in a form of juvenile amyotrophic lateral sclerosis (ALS4). Am J Hum Genet, 74, 1128–35.

Chio A, Benzi G, Dossena M et al. (2005). Severely increased risk of amyotrophic lateral sclerosis among Italian professional football players. Brain, 128, 472–6.

Chotmongkol V (1999). Amyotrophic lateral sclerosis syndrome and hyperthyroidism: report of 4 patients. J Med Assoc Thai, 82, 615–8.

Christodoulou K, Kyriakides T, Hristova AH et al. (1995). Mapping of a distal form of spinal muscular atrophy with upper limb predominance to chromosome 7p. Hum Mol Genet, 4, 1629–32.

Ciccarelli FD, Proukakis C, Patel H et al. (2003). The identification of a conserved domain in both spartin and spastin, mutated in hereditary spastic paraplegia. Genomics, 81, 437–41.

Clark LN, Poorkaj P, Wszolek Z et al. (1998). Pathogenic implications of mutations in the tau gene in pallido-ponto-nigral degeneration and related neurodegenerative disorders linked to chromosome 17. Proc Natl Acad Sci USA, 95, 13103–7.

Clement AM, Nguyen MD, Roberts EA et al. (2003). Wild-type nonneuronal cells extend survival of SOD1 mutant motor neurons in ALS mice. Science, 302, 113–7.

Comi GP, Bordoni A, Salani S et al. (1998). Cytochrome c oxidase subunit I microdeletion in a patient with motor neuron disease. Ann Neurol, 43, 110–6.

Consortium. Report (1998). 53rd ENMC International Workshop on Classification and Diagnostic Guidelines for Charcot-Marie-Tooth Type 2 and Distal Hereditary Motor Neuropathy. Naarden, The Netherlands.

Cookson MR, Menzies FM, Manning P et al. (2002). Cu/Zn superoxide dismutase (SOD1) mutations associated with familial amyotrophic lateral sclerosis (ALS) affect cellular free radical release in the presence of oxidative stress. Amyotroph Lateral Scler Other Motor Neuron Disord, 3, 75–85.

Coovert DD, Le TT, McAndrew PE et al. (1997). The survival motor neuron protein in spinal muscular atrophy. Hum Mol Genet, 6, 1205–14.

Couillard-Despres S, Zhu Q, Wong PC et al. (1998). Protective effect of neurofilament heavy gene overexpression in motor neuron disease induced by mutant superoxide dismutase. Proc Natl Acad Sci USA, 95, 9626–30.

Cox GA, Mahaffey CL, Frankel WN (1998). Identification of the mouse neuromuscular degeneration gene and mapping of a second site suppressor allele. Neuron, 21, 1327–37.

Cox PA, Banack SA, Murch SJ (2003). Biomagnification of cyanobacterial neurotoxins and neurodegenerative disease among the Chamorro people of Guam. Proc Natl Acad Sci USA, 100, 13380–3.

Crosby AH, Proukakis C (2002). Is the transportation highway the right road for hereditary spastic paraplegia? Am J Hum Genet, 71, 1009–16.

Cross HE, McKusick VA (1967a). The mast syndrome. A recessively inherited form of presenile dementia with motor disturbances. Arch Neurol, 16, 1–13.

Cross HE, McKusick VA (1967b). The Troyer syndrome. A recessive form of spastic paraplegia with distal muscle wasting. Arch Neurol, 16, 473–85.

Crow JP, Sampson JB, Zhuang Y et al. (1997). Decreased zinc affinity of amyotrophic lateral sclerosis-associated superoxide dismutase mutants leads to enhanced catalysis of tyrosine nitration by peroxynitrite. J Neurochem, 69, 1936–44.

Dabby R, Lange DJ, Trojaborg W et al. (2001). Inclusion body myositis mimicking motor neuron disease.Arch Neurol 58, 1253–6.

Dahme M, Bartsch U, Martini, R. et al. (1997). Disruption of the mouse L1 gene leads to malformations of the nervous system. Nat Genet, 17, 346–9.

Dalakas MC (2006). Sporadic inclusion body myositis--diagnosis, pathogenesis and therapeutic strategies. Nat Clin Pract Neurol, 2, 437–47.

Davenport RJ, Swingler RJ, Chancellor AM et al. (1996). Avoiding false positive diagnoses of motor neuron disease: lessons from the Scottish Motor Neuron Disease Register. J Neurol Neurosurg Psychiatry, 60, 147–51.

de Carvalho M, Matias T, Coelho F et al. (1996). Motor neuron disease presenting with respiratory failure. J Neurol Sci, 139, 117–22.

de Carvalho M, Swash M (2004). Cramps, muscle pain, and fasciculations: not always benign? Neurology, 63, 721–3.

del Aguila MA, Longstreth WT Jr, McGuire V et al. (2003). Prognosis in amyotrophic lateral sclerosis: a population-based study. Neurology, 60, 813–9.

Demyanenko GP, Tsai AY, Maness PF (1999). Abnormalities in neuronal process extension, hippocampal development, and the ventricular system of L1 knockout mice. J Neurosci, 19, 4907–20.

Desport JC, Preux PM, Truong TC et al. (1999). Nutritional status is a prognostic factor for survival in ALS patients. Neurology, 53, 1059–63.

Dhaliwal GK, Grewal RP (2000). Mitochondrial DNA deletion mutation levels are elevated in ALS brains. Neuroreport, 11, 2507–9.

Dipti S, Childs AM, Livingston JH et al. (2005). Brown-Vialetto-Van Laere syndrome; variability in age at onset and disease progression highlighting the phenotypic overlap with Fazio-Londe disease. Brain Dev, 27, 443–6.

Donaghy M, Kennett R (1999). Varying occurrence of vocal cord paralysis in a family with autosomal dominant hereditary motor and sensory neuropathy. J Neurol, 246, 552–5.

Doyu M, Sobue G, Mitsuma T et al. (1993). Very late onset X-linked recessive bulbospinal neuronopathy: mild clinical features and a mild increase in the size of tandem CAG repeat in androgen receptor gene. J Neurol Neurosurg Psychiatry, 56, 832–3.

Doyu M, Sobue G, Mukai E et al. (1992). Severity of X-linked recessive bulbospinal neuronopathy correlates with size of the tandem CAG repeat in androgen receptor gene. Ann Neurol, 32, 707–10.

Drory VE, Birnbaum M, Peleg L et al. (2003). Hexosaminidase A deficiency is an uncommon cause of a syndrome mimicking amyotrophic lateral sclerosis. Muscle Nerve, 28, 109–12.

Durham HD (2003). Factors underlying the selective vulnerability of motor neurons to neurodegeneration. Blue books of practical neurology: motor neuron disorders.:379–400.

Edgar JM, McLaughlin M, Yool D et al. (2004). Oligodendroglial modulation of fast axonal transport in a mouse model of hereditary spastic paraplegia. J Cell Biol, 166, 121–31.

Erb W (1875). Uber einen wenig bekannten spinalen symptomenkomplex. Berl Klin Wochenschr, 12, 357–9.

Errico A, Ballabio A, Rugarli EI (2002). Spastin, the protein involved in autosomal dominant hereditary spastic paraparesis, is involved in microtubule dynamics. Hum Mol Genet, 11, 153–63.

Esteban A, Traba A (1993). Fasciculation-myokymic activity and prolonged nerve conduction block. A physiopathological relationship in radiation-induced brachial plexopathy. Electroencephalogr Clin Neurophysiol, 89, 382–91.

Estevez AG, Crow JP, Sampson JB et al. (1999). Induction of nitric oxide-dependent apoptosis in motor neurons by zinc-deficient superoxide dismutase. Science, 286, 2498–500.

Evans BK, Fagan C, Arnold T (1990). Paraneoplastic motor neuron disease and renal cell carcinoma: improvement after nephrectomy. Neurology, 40, 960–2.

Evgrafov OV, Mersiyanova I, Irobi J et al. (2004). Mutant small heat-shock protein 27 causes axonal Charcot-Marie-Tooth disease and distal hereditary motor neuropathy. Nat Genet, 36, 602–6.

Eymard-Pierre E, Lesca G, Dollet S et al. (2002). Infantile-onset ascending hereditary spastic paralysis is associated with mutations in the alsin gene. Am J Hum Genet, 71, 518–27.

Farbu E, Gilhus NE, Barnes MP et al. (2006). EFNS guideline on diagnosis and management of post-polio syndrome. Report of an EFNS task force. Eur J Neurol, 13, 795–801.

Feldkotter M, Schwarzer V, Wirth R et al. (2002). Quantitative analyses of SMN1 and SMN2 based on real-time lightCycler PCR: fast and highly reliable carrier testing and prediction of severity of spinal muscular atrophy. Am J Hum Genet, 70, 358–68.

Ferrante MA, Wilbourn AJ (1997). The characteristic electrodiagnostic features of Kennedy's disease. Muscle Nerve, 20, 323–9.

Ferrante RJ, Browne SE, Shinobu LA et al. (1997). Evidence of increased oxidative damage in both sporadic and familial amyotrophic lateral sclerosis. J Neurochem, 69, 2064–74.

Ferreirinha F, Quattrini A, Pirozzi M et al. (2004). Axonal degeneration in paraplegin-deficient mice is associated with abnormal mitochondria and impairment of axonal transport. J Clin Invest, 113, 231–42.

Ferrer I, Olive M, Rivera R et al. (1995). Hereditary spastic paraparesis with dementia, amyotrophy and peripheral neuropathy. Neuropath Appl Neurobiol, 21, 255–61.

Fetoni V, Briem E, Carrara F et al. (2004). Monomelic amyotrophy associated with the 7472insC mutation in the mtDNA tRNASer(UCN) gene. Neuromuscul Disord, 14, 723–6.

Fichera M, Lo Giudice M, Flaco M et al. (2004). Evidence of kinesin heavy chain (KIF5A) involvement in pure hereditary spastic paraplegia. Neurology, 63, 1108–10.

Figlewicz DA, Krizus A, Martinoli MG et al. (1994). Variants of the heavy neurofilament subunit are associated with the development of amyotrophic lateral sclerosis. Hum Mol Genet, 3, 1757–61.

Filiminoff IN (1926). Zur pathologisch-anatomischen Charakteristik des Lathyrismus. Z Gesamte Neurol Psychiatr, 105, 76–92.

Fischbeck KH, Ionasescu V, Ritter AW et al. (1986). Localization of the gene for X-linked spinal muscular atrophy. Neurology, 36, 1595–8.

Fisher M, Mateer JE, Ullrich I, Gutrecht JA (1985). Pyramidal tract deficits and polyneuropathy in hyperthyroidism, Combination clinically mimicking amyotrophic lateral sclerosis. Am J Med, 78, 1041–4.

Folli F, Solimena M, Cofiell R et al. (1993). Autoantibodies to a 128-kd synaptic protein in three women with the stiff-man syndrome and breast cancer. N Engl J Med, 328, 546–51.

Fonknechten N, Mavel D, Byrne P et al. (2000). Spectrum of SPG4 mutations in autosomal dominant spastic paraplegia. Hum Mol Genet, 9, 637–44.

Forbes RB, Colville S, Swingler RJ (2004). The epidemiology of amyotrophic lateral sclerosis (ALS/MND) in people aged 80 or over. Age Ageing, 33, 131–4.

Forsyth PA, Dalmau J, Graus F et al. (1997). Motor neuron syndromes in cancer patients. Ann Neurol, 41, 722–30.

Fotheringham J, Jackson K, Kersh R, Gariballa SE (2005). Refeeding syndrome: life-threatening, underdiagnosed, but treatable. QJM, 98, 318–9.

Fransen E, Lemmon V, Van Camp G et al. (1995). CRASH syndrome: clinical spectrum of corpus callosum hypoplasia, retardation, adducted thumbs, spastic paraparesis and hydrocephalus due to mutations in one single gene, L1. Eur J Hum Genet, 3, 273–84.

Fray AE, Ince PG, Banner SJ et al. (1998). The expression of the glial glutamate transporter protein EAAT2 in motor neuron disease: an immunohistochemical study. Eur J Neurosci, 10, 2481–9.

Freedman DM, Travis LB, Gridley G et al. (2005). Amyotrophic lateral sclerosis mortality in 1.9 million US cancer survivors. Neuroepidemiology, 25, 176–80.

Galasko D, Salmon DP, Craig UK et al. (2002). Clinical features and changing patterns of neurodegenerative disorders on Guam, 1997-2000. Neurology, 58, 90–7.

Gamez J, Cervera C, Codina A. (1999). Flail arm syndrome or Vulpian-Bernhart's form of amyotrophic lateral sclerosis. J Neurol Neurosurg Psychiatry, 67, 258.

Gigli GL, Diomedi M, Bernardi G et al. (1993). Spastic paraplegia, epilepsy, and mental retardation in several members of a family: a novel genetic disorder. Am J Med Genet, 45, 711–6.

Glenn SA, Ross MA (2000). Delayed radiation-induced bulbar palsy mimicking ALS. *Muscle Nerve*, **23**, 814–7.

Goldblatt D (1968). Motor neuron disease: historical introduction. In *Motor Neuron Diseases*, Norris FH, Kurland LT (eds). New York: Grune and Stratton, 1968, pp. 3–11.

Gonzalez-Scarano F, Baltuch G (1999). Microglia as mediators of inflammatory and degenerative diseases. *Annu Rev Neurosci*, **22**, 219–40.

Gonzalez H, Sunnerhagen KS, Sjoberg I et al. (2006). Intravenous immunoglobulin for post-polio syndrome: a randomised controlled trial. *Lancet Neurol*, **5**, 493–500.

Gordon PH, Rowland LP, Younger DS et al. (1997). Lymphoproliferative disorders and motor neuron disease: an update. *Neurology*, **48**, 1671–8.

Gourie-Devi M, Nalini A (2003a). Long-term follow-up of 44 patients with brachial monomelic amyotrophy. *Acta Neurol Scand*, **107**, 215–20.

Gourie-Devi M, Nalini A (2003b). Madras motor neuron disease variant, clinical features of seven patients. *J Neurol Sci*, **209**, 13–7.

Gourie-Devi M, Suresh TG, Shankar SK (1984). Monomelic amyotrophy. *Arch Neurol*, **41**, 388–94.

Goytain A, Hines R, El-Husseini A et al. (2006). NIP1A (SPG6), the basis for autosomal dominant form of hereditary spastic paraplegia, encodes a functional Mg2+ transporter. *J Biol Chem* (In Press).

Graus F, Keime-Guibert F, Rene R et al. (2001). Anti-Hu-associated paraneoplastic encephalomyelitis: analysis of 200 patients. *Brain*, **124**, 1138–48.

Gregory S, Siderowf A, Golaszewski AL et al. (2002). Gastrostomy insertion in ALS patients with low vital capacity: respiratory support and survival. *Neurology*, **58**, 485–7.

Griffin JW, Li CY, Ho TW et al. (1996). Pathology of the motor-sensory axonal Guillain-Barre syndrome. *Ann Neurol*, **39**, 17–28.

Griffiths I, Klugmann M, Anderson T et al. (1998a). Current concepts of PLP and its role in the nervous system. *Microsc Res Tech*, **41**, 344–58.

Griffiths I, Klugmann M, Anderson T et al. (1998b). Axonal swellings and degeneration in mice lacking the major proteolipid of myelin. *Science*, **280**, 1610–3.

Griggs RC, Askanas V, DiMauro S et al. (1995). Inclusion body myositis and myopathies. *Ann Neurol*, **38**, 705–13.

Grohmann K, Schuelke M, Diers A et al. (2001). Mutations in the gene encoding immunoglobulin mu-binding protein 2 cause spinal muscular atrophy with respiratory distress type 1. *Nat Genet*, **29**, 75–7.

Gros-Louis F, Meijer IA, Hand CK et al. (2003). An ALS2 gene mutation causes hereditary spastic paraplegia in a Pakistani kindred. *Ann Neurol*, **53**, 144–5.

Gudesblatt M, Ludman MD, Cohen JA et al. (1988). Hexosaminidase A activity and amyotrophic lateral sclerosis. *Muscle Nerve*, **11**, 227–30.

Guegan C, Przedborski S (2003). Programmed cell death in amyotrophic lateral sclerosis. *J Clin Invest*, **111**, 153–61.

Gurney ME, Cutting FB, Zhai P et al. (1996). Benefit of vitamin E, riluzole, and gabapentin in a transgenic model of familial amyotrophic lateral sclerosis. *Ann Neurol*, **39**, 147–57.

Gutmann L, Gutmann L (2004). Myokymia and neuromyotonia 2004. *J Neurol*, **251**, 138–42.

Hadano S, Hand CK, Osuga H et al. (2001). A gene encoding a putative GTPase regulator is mutated in familial amyotrophic lateral sclerosis 2. *Nat Genet*, **29**, 166–73.

Hafezparast M, Klocke R, Ruhrberg C et al. (2003). Mutations in dynein link motor neuron degeneration to defects in retrograde transport. *Science*, **300**, 808–12.

Hahn AF, Parkes AW, Bolton CF et al. (1991). Neuromyotonia in hereditary motor neuropathy. *J Neurol Neurosurg Psychiatry*, **54**, 230–5.

Halperin JJ, Kaplan GP, Brazinsky S et al. (1990). Immunologic reactivity against Borrelia burgdorferi in patients with motor neuron disease. *Arch Neurol*, **47**, 586–94.

Hansen JJ, Durr A, Cournu-Rebeix I et al. (2002). Hereditary spastic paraplegia SPG13 is associated with a mutation in the gene encoding the mitochondrial chaperonin Hsp60. *Am J Hum Genet*, **70**, 1328–32.

Harding AE (1981). Hereditary "pure" spastic paraplegia: a clinical and genetic study of 22 families. *J Neurol Neurosurg Psychiatry*, **44**, 871–83.

Harding AE (1993). Inherited neuronal atrophy and degeneration predominantly of lower motor neurones. *Peripheral Neuropathy*, **2**, 1051–64.

Harding AE, Thomas PK (1984). Peroneal muscular atrophy with pyramidal features. *J Neurol Neurosurg Psychiatry*, **47**, 168–72.

Hart IK, Maddison P, Newsom-Davis J et al. (2002). Phenotypic variants of autoimmune peripheral nerve hyperexcitability. *Brain*, **125**, 1887–95.

Hart IK, Waters C, Vincent A et al. (1997). Autoantibodies detected to expressed K+ channels are implicated in neuromyotonia. *Ann Neurol*, **41**, 238–46.

Haverkamp LJ, Appel V, Appel SH (1995). Natural history of amyotrophic lateral sclerosis in a database population. Validation of a scoring system and a model for survival prediction. *Brain*, **118**, 707–19.

Hawkins SA, Nevin NC, Harding AE (1990). Pontobulbar palsy and neurosensory deafness (Brown-Vialetto-Van Laere syndrome) with possible autosomal dominant inheritance. *J Med Genet*, **27**, 176–9.

Hayashi H, Kato S (1989). Total manifestations of amyotrophic lateral sclerosis. ALS in the totally locked-in state. *J Neurol Sci*, **93**, 19–35.

Hazan J, Fonknechten N, Mavel D et al. (1999). Spastin, a new AAA protein, is altered in the most frequent form of autosomal dominant spastic paraplegia. *Nat Genet*, **23**, 296–303.

Heath P, Shaw PJ (2005). Excitotoxicity in amyotrophic lateral sclerosis. In *Amyotrophic Lateral Sclerosis*, H. Mitsumoto, S. Przedborski, P. Gordon, M. DelBene (eds). Marcel Dekker Inc, New York, pp. 297–336.

Hemmer B, Glocker FX, Kaiser R et al. (1997). Generalised motor neuron disease as an unusual manifestation of Borrelia burgdorferi infection. *J Neurol Neurosurg Psychiatry*, **63**, 257–8.

Henningsen P, Meinck HM (2003). Specific phobia is a frequent non-motor feature in stiff man syndrome. *J Neurol Neurosurg Psychiatry*, **74**, 462–5.

Hensley K, Floyd RA, Gordon B et al. (2002). Temporal patterns of cytokine and apoptosis-related gene expression in spinal cords of the G93A-SOD1 mouse model of amyotrophic lateral sclerosis. *J Neurochem*, **82**, 365–74.

Hirano A, Kurland LT, Krooth RS et al. (1961). Parkinsonism-dementia complex, an endemic disease on the island of Guam. I. Clinical features. *Brain*, **84**, 642–61.

Hirano A, Kurland LT, Sayre GP (1967). Familial amyotrophic lateral sclerosis. A subgroup characterized by posterior and spinocerebellar tract involvement and hyaline inclusions in the anterior horn cells. *Arch Neurol*, **16**, 232–43.

Hirano A, Llena JF, Streifler M et al. (1976). Anterior horn cell changes in a case of neurolathyrism. *Acta Neuropathol*, **35**, 277–83.

Hirayama K, Tokokura Y, Tsubaki T (1959). Juvenile muscular atrophy of unilateral upper extremity–a new clinical entity. *Psychiatry Neurology*, **61**, 2190–97.

Hirayama K, Tomonaga M, Kitano K et al. (1987). Focal cervical poliopathy causing juvenile muscular atrophy of distal upper extremity: a pathological study. *J Neurol Neurosurg Psychiatry*, **50**, 285–90.

Hirayama K, Tsubaki T, Toyokura Y et al. (1963). Juvenile muscular atrophy of unilateral upper extremity. *Neurology*, **13**, 373–80.

Hirokawa N (1998). Kinesin and dynein superfamily proteins and the mechanism of organelle transport. *Science*, **279**, 519–26.

Hodges JR, Davies RR, Xuereb JH et al. (2004). Clinicopathological correlates in frontotemporal dementia. *Ann Neurol*, **56**, 399–406.

Horner RD, Kamins KG, Feussner JR et al. (2003). Occurrence of amyotrophic lateral sclerosis among Gulf War veterans. *Neurology*, **61**, 742–9.

Hosler BA, Siddique T, Sapp PC et al. (2000). Linkage of familial amyotrophic lateral sclerosis with frontotemporal dementia to chromosome 9q21-q22. *JAMA*, **284**, 1664–9.

Hsia AW, Katz JS, Hancock SL, Peterson K (2003). Post-irradiation polyradiculopathy mimics leptomeningeal tumor on MRI. *Neurology*, **60**, 1694–6.

Hu MT, Ellis CM, Al-Chalabi A et al. (1998). Flail arm syndrome: a distinctive variant of amyotrophic lateral sclerosis. *J Neurol Neurosurg Psychiatry*, **65**, 950–1.

Hudson AJ (1981). Amyotrophic lateral sclerosis and its association with dementia, parkinsonism and other neurological disorders: a review. *Brain*, **104**, 217–47.

Hughes CA, Byrne PC, Webb S *et al.* (2001). SPG15, a new locus for autosomal recessive complicated HSP on chromosome 14q. *Neurology*, **56**, 1230–3.

Hugon J, Ludolph A, Gimenez-Roldan S *et al.* (1990). Electrophysiological evaluation of human lathyrism—results in Bangladesh and Spain. In *ALS New Advances in Toxicology and Epidemiology*, Clifford Rose F, Norris F (eds). Smith-Gordon, London, pp. 49–56.

Hurd DD, Saxton WM (1996). Kinesin mutations cause motor neuron disease phenotypes by disrupting fast axonal transport in Drosophila. *Genetics*, **144**, 1075–85.

Ince P. (2000). Neuropathology. *Amyotrophic Lateral Sclerosis*, 83–12.

Ince P, Stout N, Shaw P *et al.* (1993). Parvalbumin and calbindin D-28k in the human motor system and in motor neuron disease. *Neuropathol Appl Neurobiol*, **19**, 291–9.

Ince PG, Codd GA (2005). Return of the cycad hypothesis - does the amyotrophic lateral sclerosis/parkinsonism dementia complex (ALS/PDC) of Guam have new implications for global health? *Neuropathol Appl Neurobiol*, **31**, 345–53.

Ince PG, Evans J, Knopp M *et al.* (2003). Corticospinal tract degeneration in the progressive muscular atrophy variant of ALS. *Neurology*, **60**, 1252–8.

Ince PG, Lowe J, Shaw PJ (1998a). Amyotrophic lateral sclerosis: current issues in classification, pathogenesis and molecular pathology. *Neuropathol Appl Neurobiol*, **24**, 104–17.

Ince PG, Tomkins J, Slade JY *et al.* (1998b). Amyotrophic lateral sclerosis associated with genetic abnormalities in the gene encoding Cu/Zn superoxide dismutase: molecular pathology of five new cases, and comparison with previous reports and 73 sporadic cases of ALS. *J Neuropathol Exp Neurol*, **57**, 895–904.

Irobi J, De Jonghe P, Timmerman V (2004a). Molecular genetics of distal hereditary motor neuropathies. *Hum Mol Genet*, **13** Spec No 2, R195–202.

Irobi J, Van den Bergh P, Merlini L *et al.* (2004b). The phenotype of motor neuropathies associated with BSCL2 mutations is broader than Silver syndrome and distal HMN type V. *Brain*, **127**, 2124–30.

Irobi J, Van Impe K, Seeman P *et al.* (2004c). Hot-spot residue in small heat-shock protein 22 causes distal motor neuropathy. *Nat Genet*, **36**, 597–601.

Isozumi K, DeLong R, Kaplan J *et al.* (1996). Linkage of scapuloperoneal spinal muscular atrophy to chromosome 12q24.1-q24.31. *Hum Mol Genet*, **5**, 1377–82.

Jackson CE, Amato AA, Bryan WW *et al.* (1998). Primary hyperparathyroidism and ALS: is there a relation? *Neurology*, **50**, 1795–9.

Jaeckle KA (2004). Neurological manifestations of neoplastic and radiation-induced plexopathies. *Semin Neurol*, **24**, 385–93.

Johnson WG, Wigger HJ, Karp HR *et al.* (1982). Juvenile spinal muscular atrophy: a new hexosaminidase deficiency phenotype. *Ann Neurol*, **11**, 11–6.

Jokelainen M (1976). The epidemiology of amyotrophic lateral sclerosis in Finland. A study based on the death certificates of 421 patients. *J Neurol Sci*, **29**, 55–63.

Kaback M, Miles J, Jaffe M (1978). Hexosaminidase A deficiency in early adulthood: A new type of GM2-gangliosidosis. *Am J Hum Genet*, **30**, 30–31A.

Kaneko K, Igarashi S, Miyatake T *et al.* (1993a). 'Essential tremor' and CAG repeats in the androgen receptor gene. *Neurology*, **43**, 1618–9.

Kaneko K, Igarashi S, Miyatake T *et al.* (1993b). Hypertrophic cardiomyopathy and increased number of CAG repeats in the androgen receptor gene. *Am Heart J*, **126**, 248–9.

Karni A, Navon R, Sadeh M (1988). Hexosaminidase A deficiency manifesting as spinal muscular atrophy of late onset. *Ann Neurol*, **24**, 451–3.

Kato S, Hayashi H, Yagishita A (1993). Involvement of the frontotemporal lobe and limbic system in amyotrophic lateral sclerosis: as assessed by serial computed tomography and magnetic resonance imaging. *J Neurol Sci*, **116**, 52–8.

Katsuno M, Adachi H, Kume A *et al.* (2002). Testosterone reduction prevents phenotypic expression in a transgenic mouse model of spinal and bulbar muscular atrophy. *Neuron*, **35**, 843–54.

Katsuno M, Adachi H, Tanaka F *et al.* (2004). Spinal and bulbar muscular atrophy: ligand-dependent pathogenesis and therapeutic perspectives. *J Mol Med*, **82**, 298–307.

Katz JS, Wolfe GI, Andersson PB *et al.* (1999). Brachial amyotrophic diplegia: a slowly progressive motor neuron disorder. *Neurology*, **53**, 1071–6.

Kawamura Y, Dyck PJ, Shimono M *et al.* (1981). Morphometric comparison of the vulnerability of peripheral motor and sensory neurons in amyotrophic lateral sclerosis. *J Neuropathol Exp Neurol*, **40**, 667–75.

Kawashima T, Kikuchi H, Takita M *et al.* (1998). Skein-like inclusions in the neostriatum from a case of amyotrophic lateral sclerosis with dementia. *Acta Neuropathol*, **96**, 541–5.

Kelly TE, Amoroso K, Ferre M *et al.* (1999). Spinal muscular atrophy variant with congenital fractures. *Am J Med Genet*, **87**, 65–8.

Kennedy WR, Alter M, Sung JH (1968). Progressive proximal spinal and bulbar muscular atrophy of late onset. A sex-linked recessive trait. *Neurology*, **18**, 671–80.

Kertesz A, McMonagle P, Blair M *et al.* (2005). The evolution and pathology of frontotemporal dementia. *Brain*, **128**, 1996–2005.

Khwaja S, Sripathi N, Ahmad BK *et al.* (1998). Paraneoplastic motor neuron disease with type 1 Purkinje cell antibodies. *Muscle Nerve*, **21**, 943–5.

Kieran D, Kalmar B, Dick JP *et al.* (2004). Treatment with arimoclomol, a coinducer of heat shock proteins, delays disease progression in ALS mice. *Nat Med*, **10**, 402–5.

Kirby J, Heath PR, Allen S *et al.* (2005). Mutant SOD1 alters the motor neurone transcriptome: implications for familial amyotrophic lateral sclerosis. *Brain*, **128**, 1686–706.

Klivenyi P, Ferrante RJ, Matthews RT *et al.* (1999). Neuroprotective effects of creatine in a transgenic animal model of amyotrophic lateral sclerosis. *Nat Med*, **5**, 347–50.

Kobayashi H, Baumbach L, Matise TC *et al.* (1995). A gene for a severe lethal form of X-linked arthrogryposis (X-linked infantile spinal muscular atrophy) maps to human chromosome Xp11.3-q11.2. *Hum Mol Genet*, **4**, 1213–6.

Koerner C, Wieland B, Richter W *et al.* (2004). Stiff-person syndromes: motor cortex hyperexcitability correlates with anti-GAD autoimmunity. *Neurology*, **62**, 1357–62.

Kogure K, Busto R, Cassel J *et al.* (1975). Effects of high-dose cyanate upon cerebral energy metabolism of the rat. *Pharmacology*, **13**, 391–400.

Koul R, Jain R, Chacko A *et al.* (2006). Pontobulbar palsy and neurosensory deafness (Brown-Vialetto-van Laere syndrome) with hyperintense brainstem nuclei on magnetic resonance imaging: new finding in three siblings. *J Child Neurol*, **21**, 523–5.

Kowalska A, Konagaya M, Sakai M *et al.* (2003). Familial amyotrophic lateral sclerosis and parkinsonism-dementia complex--tauopathy without mutations in the tau gene? *Folia Neuropathol*, **41**, 59–64.

Kris J, Millecamps S, Zhu Q (2003). Creation of a mouse model for juvenile amuotroph latgeral sclerosis. *Amyotroph Lateral Scler Other Motor Neuron Disord*, **4**, 11.

Kruman II, Pedersen WA, Springer JE *et al.* (1999). ALS-linked Cu/Zn-SOD mutation increases vulnerability of motor neurons to excitotoxicity by a mechanism involving increased oxidative stress and perturbed calcium homeostasis. *Exp Neurol*, **160**, 28–39.

Kuipers-Upmeijer J, de Jager AE, Hew JM (2001). Primary lateral sclerosis: clinical, neurophysiological, and magnetic resonance findings. *J Neurol Neurosurg Psychiatry*, **71**, 615–20.

Kuncl RW, Cornblath DR, Griffin JW (1988). Assessment of thoracic paraspinal muscles in the diagnosis of ALS. *Muscle Nerve*, **11**, 484–92.

Kunita R, Otomo A, Mizumura H *et al.* (2004). Homo-oligomerization of ALS2 through its unique carboxyl-terminal regions is essential for the ALS2-associated Rab5 guanine nucleotide exchange activity and its regulatory function on endosome trafficking. *J Biol Chem*, **279**, 38626–35.

Kunst CB (2004). Complex genetics of amyotrophic lateral sclerosis. *Am J Hum Genet*, **75**, 933–47.

Kurtzke JF (1991). Risk factors in amyotrophic lateral sclerosis. *Adv Neurol*, **56**, 245–70.

Kusama-Eguchi K, Kusama T, Suda A *et al.* (2004). Partial involvement of group I metabotropic glutamate receptors in the neurotoxicity of 3-N-Oxalyl-L-2,3-diaminoprpanoic acid (L-beta-ODAP). *Biol Pharm Bull*, **27**, 1052–8.

Kuzuhara S, Kokubo Y (2005). Atypical parkinsonism of Japan: amyotrophic lateral sclerosis-parkinsonism-dementia complex of the Kii peninsula of Japan (Muro disease): an update. *Mov Disord*, **20**, S108–13.

La Spada AR, Wilson EM *et al.* (1991). Androgen receptor gene mutations in X-linked spinal and bulbar muscular atrophy. *Nature*, **352**, 77–9.

Lacomblez L, Bensimon G, Leigh PN (1996). Dose-ranging study of riluzole in amyotrophic lateral sclerosis. Amyotrophic Lateral Sclerosis/Riluzole Study Group II. *Lancet*, **347**, 1425–31.

Lagueny A (2005). Cramp-fasciculation syndrome. *Rev Neurol*, **161**, 1260–6.

LaMonte BH, Wallace KE, Holloway BA *et al.* (2002). Disruption of dynein/dynactin inhibits axonal transport in motor neurons causing late-onset progressive degeneration. *Neuron*, **34**, 715–27.

Lamy C, Mas JL, Varet B, Ziegler M (1991). Postradiation lower motor neuron syndrome presenting as monomelic amyotrophy. *J Neurol Neurosurg Psychiatry*, **54**, 648–9.

Landau D, Mishori-Dery A (2003). A new autosomal recessive congenital contractural syndrome in an Israeli Bedouin kindred. *Am J Med Genet A*, **117**, 37–40.

Langmore SE, Kasarskis EJ, Manca ML *et al.* (2006). Enteral tube feeding for amyotrophic lateral sclerosis/motor neuron disease. *Cochrane Database Syst Rev*, 4, CD004030.

Le Forestier N, Maisonobe T, Piquard A *et al.* (2001). Does primary lateral sclerosis exist? A study of 20 patients and a review of the literature. *Brain*, **124**, 1989–99.

Lee JH, Shin JH, Park KP *et al.* (2005). Phenotypic variability in Kennedy's disease: implication of the early diagnostic features. *Acta Neurol Scand*, **112**, 57–63.

Lefebvre S, Burlet P, Liu Q *et al.* (1997). Correlation between severity and SMN protein level in spinal muscular atrophy. *Nat Genet*, **16**, 265–9.

Leger JM, Salachas F (2001). Diagnosis of motor neuropathy. *Eur J Neurol*, **8**, 201–8.

Leigh PN, Anderton BH, Dodson A *et al.* (1988). Ubiquitin deposits in anterior horn cells in motor neurone disease. *Neurosci Lett*, **93**, 197–203.

Leigh PN, Rothwell JC, Traub M *et al.* (1980). A patient with reflex myoclonus and muscle rigidity: "jerking stiff-man syndrome" *J Neurol Neurosurg Psychiatry*, **43**, 1125–31.

Levy LM, Levy-Reis I, Fujii M *et al.* (2005). Brain gamma-aminobutyric acid changes in stiff-person syndrome. *Arch Neurol*, **62**, 970–4.

Li M, Miwa S, Kobayashi Y *et al.* (1998). Nuclear inclusions of the androgen receptor protein in spinal and bulbar muscular atrophy. *Ann Neurol*, **44**, 249–54.

Li M, Ona VO, Chen M *et al.* (2000). Functional role and therapeutic implications of neuronal caspase-1 and -3 in a mouse model of traumatic spinal cord injury. *Neuroscience*, **99**, 333–42.

Lieberman AP, Harmison G, Strand AD *et al.* (2002). Altered transcriptional regulation in cells expressing the expanded polyglutamine androgen receptor. *Hum Mol Genet*, **11**, 1967–76.

Lilienfeld DE, Chan E, Ehland J *et al.* (1989). Rising mortality from motoneuron disease in the USA, 1962-84. *Lancet*, **1**, 710–3.

Lindsey JC, Lusher ME, McDermott CJ *et al.* (2000). Mutation analysis of the spastin gene (SPG4) in patients with hereditary spastic paraparesis. *J Med Genet*, **37**, 759–65.

Liu J, Lillo C, Jonsson PA *et al.* (2004). Toxicity of familial ALS-linked SOD1 mutants from selective recruitment to spinal mitochondria. *Neuron*, **43**, 5–17.

Logina I, Krumina A, Karelis G *et al.* (2006). Clinical features of double infection with encephalitis (TBE) and Lyme Correliosis (LB) transmitted with tick bite. *J Neurol Neurosurg and Psychiatry*, **77**, 1350–3.

Lorson CL, Strasswimmer J, Yao, J. M. *et al.* (1998). SMN oligomerization defect correlates with spinal muscular atrophy severity. *Nat Genet*, **19**, 63–6.

Lowe J, Aldridge F, Lennox G *et al.* (1989). Inclusion bodies in motor cortex and brainstem of patients with motor neurone disease are detected by immunocytochemical localisation of ubiquitin. *Neurosci Lett*, **105**, 7–13.

Lu J, Rashid F, Byrne PC (2006). The hereditary spastic paraplegia protein spartin localises to mitochondria. *J Neurochem*, **98**, 1908–19.

Lyall RA, Donaldson N, Fleming T *et al.* (2001). A prospective study of quality of life in ALS patients treated with noninvasive ventilation. *Neurology*, **57**, 153–6.

Maasilta P, Jokelainen M, Loytonen M *et al.* (2001). Mortality from amyotrophic lateral sclerosis in Finland, 1986-1995. *Acta Neurol Scand*, **104**, 232–5.

Machida Y, Tsuchiya K, Anno M *et al.* (1999). Sporadic amyotrophic lateral sclerosis with multiple system degeneration: a report of an autopsy case without respirator administration. *Acta Neuropathol*, **98**, 512–5.

Mackenzie IR, Feldman H (2003). The relationship between extramotor ubiquitin-immunoreactive neuronal inclusions and dementia in motor neuron disease. *Acta Neuropathol*, **105**, 98–102.

Mackenzie IR, Feldman HH (2005). Ubiquitin immunohistochemistry suggests classic motor neuron disease, motor neuron disease with dementia, and frontotemporal dementia of the motor neuron disease type represent a clinicopathologic spectrum. *J Neuropathol Exp Neurol*, **64**, 730–9.

Maddison P (2006). Neuromyotonia. *Clin Neurophysiol*, **117**, 2118–27.

Maddison P, Mills KR, Newsom-Davis J (2006). Clinical electrophysiological characterization of the acquired neuromyotonia phenotype of autoimmune peripheral nerve hyperexcitability. *Muscle Nerve*, **33**, 801–8.

Makela-Bengs P, Jarvinen N, Vuopala K *et al.* (1998). Assignment of the disease locus for lethal congenital contracture syndrome to a restricted region of chromosome 9q34, by genome scan using five affected individuals. *Am J Hum Genet*, **63**, 506–16.

Mannan AU, Krawen P, Sauter SM *et al.* (2006). ZFYVE27 (SPG33), a novel spastin-binding protein, is mutated in hereditary spastic paraplegia. *Am J Hum Genet*, **79**, 351–7.

Mannen T, Iwata M, Toyokura Y *et al.* (1977). Preservation of a certain motoneurone group of the sacral cord in amyotrophic lateral sclerosis: its clinical significance. *J Neurol Neurosurg Psychiatry*, **40**, 464–9.

Martin LJ (1999). Neuronal death in amyotrophic lateral sclerosis is apoptosis: possible contribution of a programmed cell death mechanism. *J Neuropathol Exp Neurol*, **58**, 459–71.

Matsuzaki T, Nakagawa M, Nagai, M. *et al.* (2000). HTLV-I-associated myelopathy (HAM)/tropical spastic paraparesis (TSP) with amyotrophic lateral sclerosis-like manifestations. *J Neurovirol*, **6**, 544–8.

McCampbell A, Taye AA, Whitty L *et al.* (2001). Histone deacetylase inhibitors reduce polyglutamine toxicity. *Proc Natl Acad Sci USA*, **98**, 15179–84.

McCampbell A, Taylor JP, Taye AA *et al.* (2000). CREB-binding protein sequestration by expanded polyglutamine. *Hum Mol Genet*, **9**, 2197–202.

McDermott CJ, Burness CE, Kirby J *et al.* (2006). Clinical features of hereditary spastic paraplegia due to spastin mutation. *Neurology*, **67**, 45–51.

McDermott CJ, Dayaratne RK, Tomkins J *et al.* (2001). Paraplegin gene analysis in hereditary spastic paraparesis (HSP) pedigrees in northeast England. *Neurology*, **56**, 467–71.

McDermott CJ, Grierson AJ, Wood JD *et al.* (2003a). Hereditary spastic paraparesis: evidence of disrupted intracellular transport associated with spastin mutation. *Ann Neurol*, **54**, 748–59.

McDermott CJ, Taylor RW, Hayes C *et al.* (2003b). Investigation of mitochondrial function in hereditary spastic paraparesis. *NeuroReport*, **14**, 485–8.

McEntagart M, Norton N, Williams H *et al.* (2001). Localization of the gene for distal hereditary motor neuronopathy VII (dHMN-VII) to chromosome 2q14. *Am J Hum Genet*, **68**, 1270–6.

McKhann GM, Albert MS, Grossman M *et al.* (2001). Clinical and pathological diagnosis of frontotemporal dementia: report of the Work Group on frontotemporal dementia and pick's disease. *Arch Neurol*, **58**, 1803–9.

McMonagle P, Byrne, P, Hutchinson, M. (2004). Further evidence of dementia in SPG4-linked autosomal dominant hereditary spastic paraplegia. *Neurology*, **62**, 407–10.

McShane MA, Boyd S, Harding B *et al.* (1992). Progressive bulbar paralysis of childhood. A reappraisal of Fazio-Londe disease. *Brain*, **115**, 1889–900.

McWhorter ML, Monani UR, Burghes AH *et al.* (2003). Knockdown of the survival motor neuron (Smn) protein in zebrafish causes defects in motor axon outgrowth and pathfinding. *J Cell Biol*, **162**, 919–31.

Mead SH, Proukakis C, Wood N et al. (2001). A large family with hereditary spastic paraparesis due to a frame shift mutation of the spastin (SPG4) gene: association with multiple sclerosis in two affected siblings and epilepsy in other affected family members. J Neurol Neurosurg Psychiatry, 71, 788–91.

Meenakshisundaram E, Jagannathan K, Ramamurthi B (1970). Clinical pattern of motor neuron disease seen in younger age groups in Madras. Neurol India, 18 (Suppl 1), 109.

Meininger V, Salachas F (2000). Review of clinical trials. Amyotroph Lateral Scler: pp. 389–405.

Mellado W, Slebe JC, Maccioni RB (1982). Tubulin carbamoylation. Functional amino groups in microtubule assembly. Biochem J, 203, 675–81.

Menzies FM, Cookson MR, Taylor RW et al. (2002a). Mitochondrial dysfunction in a cell culture model of familial amyotrophic lateral sclerosis. Brain, 125, 1522–33.

Menzies FM, Ince PG, Shaw PJ (2002b). Mitochondrial involvement in amyotrophic lateral sclerosis. Neurochem Int, 40, 543–51.

Miller RG, Mitchell JD, Lyon M et al. (2002). Riluzole for amyotrophic lateral sclerosis (ALS)/motor neuron disease (MND). Cochrane Database Syst Rev, 2, CD001447.

Miller RG, Rosenberg JA, Gelinas DF et al. (1999). Practice parameter: the care of the patient with amyotrophic lateral sclerosis (an evidence-based review): report of the Quality Standards Subcommittee of the American Academy of Neurology: ALS Practice Parameters Task Force. Neurology, 52, 1311–23.

Mills KR (2003). Neurophysiological investigations of motor neuron disorders. Blue Books of Practical Neurology: Motor Neuron Disorders: pp 57–72.

Millul A, Beghi E, Logroscino G et al. (2005). Survival of patients with amyotrophic lateral sclerosis in a population-based registry. Neuroepidemiology, 25, 114–9.

Mitsumoto H, Sliman RJ, Schafer IA et al. (1985). Motor neuron disease and adult hexosaminidase A deficiency in two families: evidence for multisystem degeneration. Ann Neurol, 17, 378–85.

Mizutani T, Aki M, Shiozawa R et al. (1990). Development of ophthalmoplegia in amyotrophic lateral sclerosis during long-term use of respirators. J Neurol Sci, 99, 311–9.

Moersch FP, Woltman HW (1956). Progressive fluctuating muscular rigidity and spasm ("stiff-man" syndrome); report of a case and some observations in 13 other cases. Mayo Clin Proc, 31, 421–7.

Monani UR, Sendtner M, Coovert DD et al. (2000). The human centromeric survival motor neuron gene (SMN2) rescues embryonic lethality in Smn(-/-) mice and results in a mouse with spinal muscular atrophy. Hum Mol Genet, 9, 333–9.

Moreira MC, Klur S, Watanabe M et al. (2004). Senataxin, the ortholog of a yeast RNA helicase, is mutant in ataxia-ocular apraxia 2. Nat Genet, 36, 225–7.

Mostacciuolo ML, Rampoldi L, Righetti E et al. (2000). Hereditary spastic paraplegia associated with peripheral neuropathy: a distinct clinical and genetic entity. Neuromuscul Disord, 10, 497–502.

Mott RT, Dickson DW, Trojanowski JQ et al. (2005). Neuropathologic, biochemical, and molecular characterization of the frontotemporal dementias. J Neuropathol Exp Neurol, 64, 420–8.

Moulignier A, Moulonguet A, Pialoux G et al. (2001). Reversible ALS-like disorder in HIV infection. Neurology, 57, 995–1001.

Muglia M, Magariello A, Nicoletti G et al. (2002). Further evidence that SPG3A gene mutations cause autosomal dominant hereditary spastic paraplegia. Ann Neurol, 51, 794–5.

Munch C, Rosenbohm A, Sperfeld AD et al. (2005). Heterozygous R1101K mutation of the DCTN1 gene in a family with ALS and FTD. Ann Neurol, 58, 777–80.

Munch C, Sedlmeier R, Meyer T et al. (2004). Point mutations of the p150 subunit of dynactin (DCTN1) gene in ALS. Neurology, 63, 724–6.

Munoz DG, Dickson DW, Bergeron C et al. (2003). The neuropathology and biochemistry of frontotemporal dementia. Ann Neurol, 54, S24–8.

Munoz DG, Greene C, Perl DP et al. (1988). Accumulation of phosphorylated neurofilaments in anterior horn motoneurons of amyotrophic lateral sclerosis patients. J Neuropathol Exp Neurol, 47, 9–18.

Murakami N (1999). (Parkinsonism-dementia complex on Guam). Ryoikibetsu Shokogun Shirizu, (27 Pt 2), 57–60.

Mwanza JC, Tshala-Katumbay D, Kayembe DL et al. (2003). Neuro-ophthalmologic findings in konzo, an upper motor neuron disorder in Africa. Eur J Ophthalmol, 13, 383–9.

Nagai M, Aoki M, Miyoshi I et al. (2001). Rats expressing human cytosolic copper-zinc superoxide dismutase transgenes with amyotrophic lateral sclerosis: associated mutations develop motor neuron disease. J Neurosci, 21, 9246–54.

Nalini A, Lokesh L, Ratnavalli E (2004). Familial monomelic amyotrophy: a case report from India. J Neurol Sci, 220, 95–8.

Neary D, Snowden JS, Gustafson L et al. (1998). Frontotemporal lobar degeneration: a consensus on clinical diagnostic criteria. Neurology, 51, 1546–54.

Neary D, Snowden JS, Mann DM et al. (1990). Frontal lobe dementia and motor neuron disease. J Neurol Neurosurg Psychiatry, 53, 23–32.

Neumann M, Sampathu DM, Kwong LK et al. (2006). Ubiquitinated TDP-43 in frontotemporal lobar degeneration and amyotrophic lateral sclerosis. Science, 314, 130–3.

Nightingale S, Bates D, Bateman DE et al. (1982). Enigmatic dyspnoea: an unusual presentation of motor-neurone disease. Lancet, 1, 933–5.

Nishimura AL, Mitne-Neto M, Silva HC et al. (2004). A mutation in the vesicle-trafficking protein VAPB causes late-onset spinal muscular atrophy and amyotrophic lateral sclerosis. Am J Hum Genet, 75, 822–31.

Ohara S, Iwahashi T, Oide T et al. (2002). Spinocerebellar ataxia type 6 with motor neuron loss: a follow-up autopsy report. J Neurol, 249, 633–5.

Okado-Matsumoto A, Fridovich I (2001). Subcellular distribution of superoxide dismutases (SOD) in rat liver: Cu,Zn-SOD in mitochondria. J Biol Chem, 276, 38388–93.

Okamoto K, Hirai S, Amari M et al. (1993). Bunina bodies in amyotrophic lateral sclerosis immunostained with rabbit anti-cystatin C serum. Neurosci Lett, 162, 125–8.

Olsen NK, Pfeiffer P, Johannsen et al. (1993). Radiation-induced brachial plexopathy: neurological follow-up in 161 recurrence-free breast cancer patients. Int J Radiat Oncol Biol Phys, 26, 43–9.

Ordway JM, Tallaksen-Greene S, Gutekunst CA et al. (1997). Ectopically expressed CAG repeats cause intranuclear inclusions and a progressive late onset neurological phenotype in the mouse. Cell, 91, 753–63.

Ostojic J, Axelman K, Lannfelt L et al. (2003). No evidence of linkage to chromosome 9q21-22 in a Swedish family with frontotemporal dementia and amyotrophic lateral sclerosis. Neurosci Lett, 340, 245–7.

Osuntokun BO (1981). Cassava diet, chronic cyanide intoxication and neuropathy in the Nigerian Africans. World Rev Nutr Diet, 36, 141–73.

Otomo A, Hadano S, Okada, T. et al. (2003). ALS2, a novel guanine nucleotide exchange factor for the small GTPase Rab5, is implicated in endosomal dynamics. Hum Mol Genet, 12, 1671–87.

Parboosingh JS, Figlewicz DA, Krizus, A. et al. (1997). Spinobulbar muscular atrophy can mimic ALS: the importance of genetic testing in male patients with atypical ALS. Neurology, 49, 568–72.

Pasinelli P, Belford ME, Lennon, N. et al. (2004). Amyotrophic lateral sclerosis-associated SOD1 mutant proteins bind and aggregate with Bcl-2 in spinal cord mitochondria. Neuron, 43, 19–30.

Pasinelli P, Borchelt DR, Houseweart MK et al. (1998). Caspase-1 is activated in neural cells and tissue with amyotrophic lateral sclerosis-associated mutations in copper-zinc superoxide dismutase. Proc Natl Acad Sci USA, 95, 15763–8.

Patel H, Cross H, Proukakis C et al. (2002). SPG20 is mutated in Troyer syndrome, an hereditary spastic paraplegia. Nat Genet, 31, 347–8.

Patel H, Hart PE, Warner TT et al. (2001). The Silver syndrome variant of hereditary spastic paraplegia maps to chromosome 11q12-q14, with evidence for genetic heterogeneity within this subtype. Am J Hum Genet, 69, 209–15.

Patel S, Latterich, M. (1998). The AAA team: related ATPases with diverse functions. Trends Cell Biol, 8, 65–71.

Patten BM, Pages M (1984). Severe neurological disease associated with hyperparathyroidism. Ann Neurol, 15, 453–6.

Peiris JB, Seneviratne KN, Wickremasinghe HR et al. (1989). Non familial juvenile distal spinal muscular atrophy of upper extremity. J Neurol Neurosurg Psychiatry, 52, 314–9.

Pellizzoni L, Charroux B, Rappsilber J *et al.* (2001). A functional interaction between the survival motor neuron complex and RNA polymerase II. *J Cell Biol*, **152**, 75–85.

Pellizzoni L, Kataoka N, Charroux B *et al.* (1998). A novel function for SMN, the spinal muscular atrophy disease gene product, in pre-mRNA splicing. *Cell*, **95**, 615–24.

Pelosi L, Lanzillo B, Perretti A *et al.* (1991). Motor and somatosensory evoked potentials in hereditary spastic paraplegia. *J Neurol Neurosurg Psychiatry*, **54**, 1099–102.

Piao YS, Wakabayashi K, Kakita A *et al.* (2003). Neuropathology with clinical correlations of sporadic amyotrophic lateral sclerosis: 102 autopsy cases examined between 1962 and 2000. *Brain Pathol*, **13**, 10–22.

Piccioni F, Pinton P, Simeoni S *et al.* (2002). Androgen receptor with elongated polyglutamine tract forms aggregates that alter axonal trafficking and mitochondrial distribution in motor neuronal processes. *Faseb J*, **16**, 1418–20.

Plato CC, Garruto RM, Galasko D *et al.* (2003). Amyotrophic lateral sclerosis and parkinsonism-dementia complex of Guam: changing incidence rates during the past 60 years. *Am J Epidemiol*, **157**, 149–57.

Prabhakar S, Chopra JS, Banerjee AK *et al.* (1981). Wasted leg syndrome: a clinical, electrophysiological and histopathological study. *Clin Neurol Neurosurg*, **83**, 19–28.

Pridmore S, Rao, G, Abusah, P. (1995). Hereditary spastic paraplegia with dementia. *Aust N Z J Psychiatry*, **29**, 678–82.

Pringle CE, Hudson AJ, Munoz DG *et al.* (1992). Primary lateral sclerosis. Clinical features, neuropathology and diagnostic criteria. *Brain*, **115**, 495–520.

Proukakis C, Cross H, Patel H *et al.* (2004). Troyer syndrome revisited. A clinical and radiological study of a complicated hereditary spastic paraplegia. *J Neurol*, **251**, 1105–10.

Puls I, Jonnakuty C, LaMonte BH *et al.* (2003). Mutant dynactin in motor neuron disease. *Nat Genet*, **33**, 455–6.

Qureshi AI, Wilmot G, Dihenia B *et al.* (1996). Motor neuron disease with parkinsonism. *Arch Neurol*, **53**, 987–91.

Ragonese P, Fierro B, Salemi G *et al.* (2005). Prevalence and risk factors of post-polio syndrome in a cohort of polio survivors. *J Neurol Sci*, **236**, 31–5.

Rainier S, Chai JH, Tokarz D *et al.* (2003). NIPA1 gene mutations cause autosomal dominant hereditary spastic paraplegia (SPG6). *Am J Hum Genet*, **73**, 967–71.

Raju R, Rakocevic G, Chen Z *et al.* (2006). Autoimmunity to GABAA-receptor-associated protein in stiff-person syndrome. *Brain*, **67**, 1068–70.

Rakocevic G, Raju, R, Dalakas MC (2004). Anti-glutamic acid decarboxylase antibodies in the serum and cerebrospinal fluid of patients with stiff-person syndrome: correlation with clinical severity. *Arch Neurol*, **61**, 902–4.

Rakocevic G, Raju R, Semino-Mora C *et al.* (2006). Stiff person syndrome with cerebellar disease and high-titer anti-GAD antibodies. *Neurology*, **67**, 1068–70.

Ramlow J, Alexander M, LaPorte R *et al.* (1992). Epidemiology of the post-polio syndrome. *Am J Epidemiol*, **136**, 769–86.

Rao SD, Banack SA, Cox PA *et al.* (2006). BMAA selectively injures motor neurons via AMPA/kainate receptor activation. *Exp Neurol*, **201**, 244–52.

Ravindranath V (2002). Neurolathyrism: mitochondrial dysfunction in excitotoxicity mediated by L-beta-oxalyl aminoalanine. *Neurochem Int*, **40**, 505–9.

Reid E, Kloos M, Ashley-Koch A *et al.* (2002). A kinesin heavy chain (KIF5A) mutation in hereditary spastic paraplegia (SPG10). *Am J Hum Genet*, **71**, 1189–94.

Ringholz GM, Appel SH, Bradshaw M *et al.* (2005). Prevalence and patterns of cognitive impairment in sporadic ALS. *Neurology*, **65**, 586–90.

Rodrigues NR, Owen N, Talbot K *et al.* (1996). Gene deletions in spinal muscular atrophy. *J Med Genet*, **33**, 93–6.

Roeleveld-Versteegh AB, Braun KP, Smeitink JA *et al.* (2004). Mitochondrial respiratory chain disease presenting as progressive bulbar paralysis of childhood. *J Inherit Metab Dis*, **27**, 281–3.

Rosati G, Aiello I, Tola R *et al.* (1980). Amyotrophic lateral sclerosis associated with thyrotoxicosis. *Arch Neurol*, **37**, 530–1.

Rosen DR, Siddique T, Patterson D *et al.* (1993). Mutations in Cu/Zn superoxide dismutase gene are associated with familial amyotrophic lateral sclerosis. *Nature*, **362**, 59–62.

Rosenfeld MR, Posner JB (1991). Paraneoplastic motor neuron disease. *Adv Neurol*, **56**, 445–59.

Rosling H, Tylleskar T (1995). Konzo. *Tropical Neurology*, 353–64.

Rosling H, Tylleskar T (2000). Cassava. *Exp Clin Neurotoxicol*, 338–43.

Rossoll W, Jablonka S, Andreassi C *et al.* (2003). Smn, the spinal muscular atrophy-determining gene product, modulates axon growth and localization of beta-actin mRNA in growth cones of motoneurons. *J Cell Biol*, **163**, 801–12.

Rossoll W, Kroning AK, Ohndorf UM *et al.* (2002). Specific interaction of Smn, the spinal muscular atrophy determining gene product, with hnRNP-R and gry-rbp/hnRNP-Q: a role for Smn in RNA processing in motor axons? *Hum Mol Genet*, **11**, 93–105.

Rothstein JD, Van Kammen M, Levey AI *et al.* (1995). Selective loss of glial glutamate transporter GLT-1 in amyotrophic lateral sclerosis. *Ann Neurol*, **38**, 73–84.

Rouleau GA, Clark AW, Rooke K *et al.* (1996). SOD1 mutation is associated with accumulation of neurofilaments in amyotrophic lateral sclerosis. *Ann Neurol*, **39**, 128–31.

Rowland LP (1999). Primary lateral sclerosis: disease, syndrome, both or neither? *J Neurol Sci*, **170**, 1–4.

Rowland LP, Sherman WL, Hays AP *et al.* (1995). Autopsy-proven amyotrophic lateral sclerosis, Waldenstrom's macroglobulinemia, and antibodies to sulfated glucuronic acid paragloboside. *Neurology*, **45**, 827–9.

Roy N, Mahadevan MS, McLean M *et al.* (1995). The gene for neuronal apoptosis inhibitory protein is partially deleted in individuals with spinal muscular atrophy. *Cell*, **80**, 167–78.

Rudnik-Schoneborn S, Forkert R, Hahnen E *et al.* (1996). Clinical spectrum and diagnostic criteria of infantile spinal muscular atrophy: further delineation on the basis of SMN gene deletion findings. *Neuropediatrics*, **27**, 8–15.

Rudnik-Schoneborn S, Sztriha L, Aithala GR *et al.* (2003). Extended phenotype of pontocerebellar hypoplasia with infantile spinal muscular atrophy. *Am J Med Genet A*, **117**, 10–7.

Ryan A, Nor AM, Costigan D *et al.* (2003). Polymyositis masquerading as motor neuron disease. *Arch Neurol*, **60**, 1001–3.

Sack GH, Huether CA, Garg N (1978). Familial spastic paraplegia-clinical and pathologic studies in a large kindred. *Johns Hopkins Med J*, **143**, 117–21.

Saha SP, Das SK, Gangopadhyay PK *et al.* (1997). Pattern of motor neurone disease in eastern India. *Acta Neurol Scand*, **96**, 14–21.

Sasaki S, Tsutsumi Y, Yamane K *et al.* (1992). Sporadic amyotrophic lateral sclerosis with extensive neurological involvement. *Acta Neuropathol*, **84**, 211–5.

Sathasivam S, Grierson AJ, Shaw PJ (2005). Characterization of the caspase cascade in a cell culture model of SOD1-related familial amyotrophic lateral sclerosis: expression, activation and therapeutic effects of inhibition. *Neuropathol Appl Neurobiol*, **31**, 467–85.

Sathasivam S, O'Sullivan S, Nicolson A *et al.* (2000). Brown-Vialetto-Van Laere syndrome: case report and literature review. *Amyotroph Lateral Scler Other Motor Neuron Disord*, **1**, 277–81.

Sathasivam S, Shaw PJ (2005). Apoptosis in amyotrophic lateral sclerosis--what is the evidence? *Lancet Neurol*, **4**, 500–9.

Schady W, Dick JP, Sheard A Crampton S (1991). Central motor conduction studies in hereditary spastic paraplegia. *J Neurol Neurosurg Psychiatry*, **54**, 775–9.

Schady W, Smith CM (1994). Sensory neuropathy in hereditary spastic paraplegia. *J Neurol Neurosurg Psychiatry*, **57**, 693–8.

Schreiber H, Gaigalat T, Wiedemuth-Catrinescu U *et al.* (2005). Cognitive function in bulbar- and spinal-onset amyotrophic lateral sclerosis. A longitudinal study in 52 patients. *J Neurol*, **252**, 772–81.

Schroder R, Keller E, Flacke S *et al.* (1999). MRI findings in Hirayama's disease: flexion-induced cervical myelopathy or intrinsic motor neuron disease? *J Neurol*, **246**, 1069–74.

Seilhean D, Takahashi J, El Hachimi KH *et al.* (2004). Amyotrophic lateral sclerosis with neuronal intranuclear protein inclusions. *Acta Neuropathol*, **108**, 81–7.

Seri M, Cusano R, Forabosco P et al. (1999). Genetic mapping to 10q23.3-q24.2, in a large Italian pedigree, of a new syndrome showing bilateral cataracts, gastroesophageal reflux, and spastic paraparesis with amyotrophy. Am J Hum Genet, 64, 586–93.

Shankar SK, Gourie-Devi M, Shankar, L. et al. (2000). Pathology of Madras type of motor neuron disease (MMND)--a histological and immunohistochemical study. Acta Neuropathol, 99, 428–34.

Shaw PJ (1999). Stiff-man syndrome and its variants. Lancet, 353, 86–7.

Shaw PJ (2005). Molecular and cellular pathways of neurodegeneration in motor neurone disease. J Neurol Neurosurg Psychiatry, 76, 1046–57.

Shaw PJ, Bates D, Kendall-Taylor P (1988). Hyperthyroidism presenting as pyramidal tract disease. BMJ, 297, 1395–6.

Shaw PJ, Eggett CJ (2000). Molecular factors underlying selective vulnerability of motor neurons to neurodegeneration in amyotrophic lateral sclerosis. J Neurol, 247, I17–27.

Shaw PJ, Forrest V, Ince PG et al. (1995a). CSF and plasma amino acid levels in motor neuron disease: elevation of CSF glutamate in a subset of patients. Neurodegeneration, 4, 209–16.

Shaw PJ, Ince PG, Falkous G et al. (1995b). Oxidative damage to protein in sporadic motor neuron disease spinal cord. Ann Neurol, 38, 691–5.

Shefner JM, Tyler HR, Krarup C (1991). Abnormalities in the sensory action potential in patients with amyotrophic lateral sclerosis. Muscle Nerve, 14, 1242–6.

Siklos L, Engelhardt J, Harati Y et al. (1996). Ultrastructural evidence for altered calcium in motor nerve terminals in amyotropic lateral sclerosis. Ann Neurol, 39, 203–16.

Silva MT, Leite AC, Alamy AH et al. (2005). ALS syndrome in HTLV-I infection. Neurology, 65, 1332–3.

Silver JR (1966). Familial spastic paraplegia with amyotrophy of the hands. Ann Hum Genet, 30, 69–75.

Simpson MA, Cross H, Proukakis C et al. (2003). Maspardin is mutated in mast syndrome, a complicated form of hereditary spastic paraplegia associated with dementia. Am J Hum Genet, 73, 1147–56.

Sinnreich M, Klein CJ (2002). Bulbospinal muscular atrophy: Kennedy's disease. Arch Neurol, 61, 1324–6.

Slavotinek AM, Pike M, Mills K et al. (1996). Cataracts, motor system disorder, short stature, learning difficulties, and skeletal abnormalities: a new syndrome? Am J Med Genet, 62, 42–7.

Slee M, Selvan A, Donaghy M (2007). Multifocal motor neuropathy: the diagnostic spectrum and response to treatment. Neurology, 67, 1680–87.

Smith RG, Henry YK, Mattson MP et al. (1998). Presence of 4-hydroxynonenal in cerebrospinal fluid of patients with sporadic amyotrophic lateral sclerosis. Ann Neurol, 44, 696–9.

Sobue G, Doyu M, Kachi T et al. (1993). Subclinical phenotypic expressions in heterozygous females of X-linked recessive bulbospinal neuronopathy. J Neurol Sci, 117, 74–8.

Sobue G, Hashizume Y, Yasuda T et al. (1990). Phosphorylated high molecular weight neurofilament protein in lower motor neurons in amyotrophic lateral sclerosis and other neurodegenerative diseases involving ventral horn cells. Acta Neuropathol, 79, 402–8.

Sopher BL, Thomas PS Jr, LaFevre-Bernt MA et al. (2004). Androgen receptor YAC transgenic mice recapitulate SBMA motor neuronopathy and implicate VEGF164 in the motor neuron degeneration. Neuron, 41, 687–99.

Spalloni A, Albo F, Ferrari F et al. (2004). Cu/Zn-superoxide dismutase (GLY93-->ALA) mutation alters AMPA receptor subunit expression and function and potentiates kainate-mediated toxicity in motor neurons in culture. Neurobiol Dis, 15, 340–50.

Spencer PS (1999). Food toxins, ampa receptors, and motor neuron diseases. Drug Metab Rev, 31, 561–87.

Spencer PS, Schaumburg HH, Cohn DF et al. (1984). Lathyrism: a useful model of primary lateral sclerosis. Research Progress in Motor Neuron Disease, 312–27.

Spreux-Varoquaux O, Bensimon G, Lacomblez L et al. (2002). Glutamate levels in cerebrospinal fluid in amyotrophic lateral sclerosis: a reappraisal using a new HPLC method with coulometric detection in a large cohort of patients. J Neurol Sci, 193, 73–8.

Steffan JS, Bodai L, Pallos J et al. (2001). Histone deacetylase inhibitors arrest polyglutamine-dependent neurodegeneration in Drosophila. Nature, 413, 739–43.

Stolwijk-Swuste JM, Beelen A, Lankhorst GJ et al. (2005). The course of functional status and muscle strength in patients with late-onset sequelae of poliomyelitis: a systematic review. Arch Phys Med Rehabil, 86, 1693–701.

Striefler M, Cohn DF, Hirano A et al. (1977). The central nervous system in a case of neurolathyrism. Neurology, 27, 1176–8.

Strong MJ, Gordon PH (2005). Primary lateral sclerosis, hereditary spastic paraplegia and amyotrophic lateral sclerosis: discrete entities or spectrum? Amyotroph Lateral Scler Other Motor Neuron Disord, 6, 8–16.

Strong MJ, Grace GM, Orange JB et al. (1996). Cognition, language, and speech in amyotrophic lateral sclerosis: a review. J Clin Exp Neuropsychol, 18, 291–303.

Strong MJ, Lomen-Hoerth C, Caselli RJ et al. (2003). Cognitive impairment, frontotemporal dementia, and the motor neuron diseases. Ann Neurol, 54, S20–3.

Strumpell A (1886). Ueber eine bestimmte Form der primaren combinierten Systemerkrankung des Ruckenmarks. Arch Psychiatr Nervenkr, 17, 17–38.

Subramaniam JR, Lyons WE, Liu J et al. (2002). Mutant SOD1 causes motor neuron disease independent of copper chaperone-mediated copper loading. Nat Neurosci, 5, 301–7.

Sullivan PG, Rabchevsky AG, Keller JN et al. (2004). Intrinsic differences in brain and spinal cord mitochondria: Implication for therapeutic interventions. J Comp Neurol, 474, 524–34.

Takahashi H, Snow BJ, Bhatt MH et al. (1993). Evidence for a dopaminergic deficit in sporadic amyotrophic lateral sclerosis on positron emission scanning. Lancet, 342, 1016–9.

Takeuchi H, Kobayashi Y, Ishigaki S et al. (2002). Mitochondrial localization of mutant superoxide dismutase 1 triggers caspase-dependent cell death in a cellular model of familial amyotrophic lateral sclerosis. J Biol Chem, 277, 50966–72.

Taylor BV, Gross L, Windebank AJ (1996). The sensitivity and specificity of anti-GM1 antibody testing. Neurology, 47, 951–5.

Thompson PD (2001). The stiff-man syndrome and related disorders. Parkinsonism Relat Disord, 8, 147–53.

Thornton FJ, Fotheringham T, Alexander M et al. (2002). Amyotrophic lateral sclerosis: enteral nutrition provision--endoscopic or radiologic gastrostomy? Radiology, 224, 713–7.

Tikka TM, Vartiainen NE, Goldsteins G et al. (2002). Minocycline prevents neurotoxicity induced by cerebrospinal fluid from patients with motor neurone disease. Brain, 125, 722–31.

Tohgi H, Abe T, Yamazaki K et al. (1999). Remarkable increase in cerebrospinal fluid 3-nitrotyrosine in patients with sporadic amyotrophic lateral sclerosis. Ann Neurol, 46, 129–31.

Tomimitsu H, Arimura K, Nagado, T. et al. (2004). Mechanism of action of voltage-gated K+ channel antibodies in acquired neuromyotonia. Ann Neurol, 56, 440–4.

Tomkins J, Usher PA, Slade JY et al. (1998). Novel insertion in the KSP region of the neurofilament heavy gene in amyotrophic lateral sclerosis (ALS). NeuroReport, 9, 3967–70.

Traynor BJ, Codd MB, Corr B et al. (1999). Incidence and prevalence of ALS in Ireland, 1995-1997: a population-based study. Neurology, 52, 504–9.

Traynor BJ, Codd MB, Corr, B. et al. (2000a). Amyotrophic lateral sclerosis mimic syndromes: a population-based study. Arch Neurol, 57, 109–13.

Traynor BJ, Codd MB, Corr B et al. (2000b). Clinical features of amyotrophic lateral sclerosis according to the El Escorial and Airlie House diagnostic criteria: A population-based study. Arch Neurol, 57, 1171–6.

Trojan DA, Cashman NR (2005). Post-poliomyelitis syndrome. Muscle Nerve, 31, 6–19.

Troost D, Van den Oord JJ, Vianney de Jong JM (1990). Immunohistochemical characterization of the inflammatory infiltrate in amyotrophic lateral sclerosis. Neuropathol Appl Neurobiol, 16, 401–10.

Tshala-Katumbay D, Eeg-Olofsson KE, Kazadi-Kayembe T et al. (2002). Analysis of motor pathway involvement in konzo using transcranial electrical and magnetic stimulation. Muscle Nerve, 25, 230–5.

Tshala-Katumbay D, Eeg-Olofsson KE, Tylleskar T *et al.* (2001). Impairments, disabilities and handicap pattern in konzo--a non-progressive spastic para/tetraparesis of acute onset. *Disabil Rehabil*, **23**, 731–6.

Turner MR, Leigh PN (2003). Disease modifying therapies in motor neurone disorders: the present position and potential future developments. *Motor Neuron Disorders*, **28**, 497–545.

Turner MR, Parton MJ, Shaw CE *et al.* (2003). Prolonged survival in motor neuron disease: a descriptive study of the King's database 1990-2002. *J Neurol Neurosurg Psychiatry*, **74**, 995–7.

Tyler HR, Sheffner J (1991). Amyotrophic lateral sclerosis. *Handbook of Clinical Neurology*, 169–215.

Tylleskar T, Howlett WP, Rwiza HT *et al.* (1993). Konzo: a distinct disease entity with selective upper motor neuron damage. *J Neurol Neurosurg Psychiatry*, **56**, 638–43.

Valdmanis PN, Meijer IA, Reynolds A *et al.* (2007). Mutations in the KIAA0196 gene at the SPG8 locus cause hereditary spastic paraplegia. *Am J Hum Genet*, **80**, 152–61.

Van Den Berg-Vos RM, Van Den Berg LH, Visser J *et al.* (2003a). The spectrum of lower motor neuron syndromes. *J Neurol*, **250**, 1279–92.

van den Berg-Vos RM, Visser J, Franssen H *et al.* (2003b). Sporadic lower motor neuron disease with adult onset: classification of subtypes. *Brain*, **126**, 1036–47.

van der Vleuten AJ, van Ravenswaaij-Arts CM, Frijns CJ *et al.* (1998). Localisation of the gene for a dominant congenital spinal muscular atrophy predominantly affecting the lower limbs to chromosome 12q23-q24. *Eur J Hum Genet*, **6**, 376–82.

Van Es HW, Van den Berg LH, Franssen H *et al.* (1997). Magnetic resonance imaging of the brachial plexus in patients with multifocal motor neuropathy. *Neurology*, **48**, 1218–24.

Warner TT, Patel H, Proukakis C *et al.* (2004). A clinical, genetic and candidate gene study of Silver syndrome, a complicated form of hereditary spastic paraplegia. *J Neurol*, **251**, 1068–74.

Watanabe R, Iino M, Honda M (1997). Primary lateral sclerosis. *Neuropathology*, **17**, 220–4.

Webb S, Coleman D, Byrne P *et al.* (1998). Autosomal dominant hereditary spastic paraparesis with cognitive loss linked to chromosome 2p. *Brain*, **121**, 601–9.

Weisskopf MG, O'Reilly EJ, McCullough ML *et al.* (2005). Prospective study of military service and mortality from ALS. *Neurology*, **64**, 32–7.

Wharton SB, McDermott CJ, Grierson AJ *et al.* (2003). The cellular and molecular pathology of the motor system in hereditary spastic paraparesis due to mutation of the spastin gene. *J Neuropathol Exp Neurol*, **62**, 1166–77.

White KD, Ince PG, Lusher M *et al.* (2000). Clinical and pathologic findings in hereditary spastic paraparesis with spastin mutation. *Neurology*, **55**, 89–94.

Whiteley AM, Swash M, Urich H (1976). Progressive encephalomyelitis with rigidity. *Brain*, **99**, 27–42.

Wiedemann FR, Manfredi G, Mawrin C *et al.* (2002). Mitochondrial DNA and respiratory chain function in spinal cords of ALS patients. *J Neurochem*, **80**, 616–25.

Wilhelmsen KC, Blake DM, Lynch T *et al.* (1996). Chromosome 12-linked autosomal dominant scapuloperoneal muscular dystrophy. *Ann Neurol*, **39**, 507–20.

Willeit J, Kiechl S, Kiechl-Kohlendorfer U *et al.* (2001). Juvenile asymmetric segmental spinal muscular atrophy (Hirayama's disease): three cases without evidence of "flexion myelopathy". *Acta Neurol Scand*, **104**, 320–2.

Williams TL, Day NC, Ince PG *et al.* (1997). Calcium-permeable alpha-amino-3-hydroxy-5-methyl-4-isoxazole propionic acid receptors: a molecular determinant of selective vulnerability in amyotrophic lateral sclerosis. *Ann Neurol*, **42**, 200–7.

Williamson TL, Cleveland DW (1999). Slowing of axonal transport is a very early event in the toxicity of ALS-linked SOD1 mutants to motor neurons. *Nat Neurosci*, **2**, 50–6.

Wilson CM, Grace GM, Munoz DG *et al.* (2001). Cognitive impairment in sporadic ALS: a pathologic continuum underlying a multisystem disorder. *Neurology*, **57**, 651–7.

Windpassinger C, Auer-Grumbach M, Irobi J *et al.* (2004). Heterozygous missense mutations in BSCL2 are associated with distal hereditary motor neuropathy and Silver syndrome. *Nat Genet*, **36**, 271–6.

Wirth B. (2000). An update on the mutation spectrum of the survival motor neuron gene (SMN1) in autosomal recessive spinal muscular atrophy (SMA). *Hum Mutat*, **15**, 228–37.

Wong PC, Pardo CA, Borchelt DR *et al.* (1995). An adverse property of a familial ALS-linked SOD1 mutation causes motor neuron disease characterized by vacuolar degeneration of mitochondria. *Neuron*, **14**, 1105–16.

Wood-Allum CA, Barber SC, Kirby J *et al.* (2006). Impairment of mitochondrial anti-oxidant defence in SOD1-related motor neuron injury and amelerioration by ebselen. *Brain*, **129**, 1693–709.

Yamanaka K, Vande Velde C, Eymard-Pierre E *et al.* (2003). Unstable mutants in the peripheral endosomal membrane component ALS2 cause early-onset motor neuron disease. *Proc Natl Acad Sci USA*, **100**, 16041–6.

Yang Y, Hentati A, Deng HX *et al.* (2001). The gene encoding alsin, a protein with three guanine-nucleotide exchange factor domains, is mutated in a form of recessive amyotrophic lateral sclerosis. *Nat Genet*, **29**, 160–5.

Yih JS, Wang SJ, Su MS *et al.* (1993). Hereditary spastic paraplegia associated with epilepsy, mental retardation and hearing impairment. *Paraplegia*, **31**, 408–11.

Yong J, Wan L, Dreyfuss G (2004). Why do cells need an assembly machine for RNA-protein complexes? *Trends Cell Biol*, **14**, 226–32.

Younger DS, Rowland LP, Latov N *et al.* (1990). Motor neuron disease and amyotrophic lateral sclerosis: relation of high CSF protein content to paraproteinemia and clinical syndromes. *Neurology*, **40**, 595–9.

Zeitlman L, Sirim P, Kremmer E, Kolanus W (2001). Cloning of ACP33 as a novel intracellular ligand of CD4. *J Biol Chem*, **276**, 9123–32.

Zerres K, Rudnik-Schoneborn S (1995). Natural history in proximal spinal muscular atrophy. Clinical analysis of 445 patients and suggestions for a modification of existing classifications. *Arch Neurol*, **52**, 518–23.

Zhao Q, Wang J, Levichkin IV *et al.* (2002). A mitochondrial specific stress response in mammalian cells. *Embo J*, **21**, 4411–9.

Zhao X, Alvarado D, Rainier S *et al.* (2001). Mutations in a newly identified GTPase gene cause autosomal dominant hereditary spastic paraplegia. *Nat Genet*, **29**, 326–31.

Zhu PP, Patterson A, Lavoie B *et al.* (2003). Cellular localization, oligomerization, and membrane association of the hereditary spastic paraplegia 3A (SPG3A) protein atlastin. *J Biol Chem*, **278**, 49063–71.

Zhu S, Stavrovskaya IG, Drozda M *et al.* (2002). Minocycline inhibits cytochrome c release and delays progression of amyotrophic lateral sclerosis in mice. *Nature*, **417**, 74–8.

Zoghbi HY, Orr HT (2000). Glutamine repeats and neurodegeneration. *Annu Rev Neurosci*, **23**, 217–47.

Zuchner S, Wang G, Tran-Viet K-N *et al.* (2006). Mutations in the novel mitochondrial protein REEP1 cause hereditary spastic paraplegia type 31. *Am J Hum Genet*, **79**, 365–9.

CHAPTER 24

Muscle diseases

David Hilton-Jones

Contents

24.1 Introduction

This chapter is concerned with those disorders in which the primary pathological process affects skeletal muscle, for which in everyday clinical practice the term myopathy is a convenient shorthand. However, it must be stressed that diseases of the motor nerves and neuromuscular junction can produce an identical clinical picture to several of the myopathies, and this will be emphasized many times throughout the chapter when considering differential diagnosis. Indeed sometimes, despite one's best efforts, one is left uncertain as to whether the primary disease process is in the nerves or muscles—it may be that in some conditions the disease process directly affects both nerves and muscles. The intimate relationship, both structural and functional, between nerves and the muscles they innervate means that disease of one may have a profound effect on the other—the most striking example is the change that occurs to skeletal muscle fibre-type distribution in denervation.

Huge advances have been made in our understanding of neuromuscular disorders, for instance over the last two decades diseases of nerve, neuromuscular junction, and skeletal muscle. As yet not all of these have been translated into effective therapy. Myasthenia gravis and many of the idiopathic inflammatory myopathies are treatable by relatively crude immunosuppressive regimens, but eventually it is to be hoped that more detailed knowledge of their immunopathogenesis will lead to specific, and safer, therapies. The genetic and molecular bases of several forms of muscular dystrophy are now known. This aids diagnosis in individual cases and should in the future lead on to specific therapies, either through genetic engineering methods or by strategies that replace a defective protein or ameliorate the effects of the accumulation of a non-functional protein. Diseases associated with abnormal ion channel function, the so-called channelopathies, are appearing throughout neurology, and the myopathies provide several interesting examples. Primary metabolic disorders, although relatively rare, are a topic of considerable interest, none more so than the mitochondrial cytopathies.

What has come as great surprise is the phenotypic variability associated with some genetic defects. The same mutation in different family members may present in different fashion. For instance caveolin mutations may present as asymptomatic elevation of serum creatine kinase, as a limb–girdle dystrophy, as rippling muscle disease, or as a distal myopathy. Otherwise, different mutations with the same gene may give rise to very varied phenotypes, for instance lamin A/C mutations presenting as Emery–Dreifuss syndrome, limb–girdle muscular dystrophy, isolated cardiomyopathy, progeria, skeletal malformations, Charcot–Marie–Tooth disease, or partial lipodystrophy. This has complicated classification of inherited neuromuscular disorders with either purely clinical or purely molecular classifications not proving entirely satisfactory for practical use.

At present, no specific therapies are available for the majority of the myopathies, particularly those that are likely to present to neurological clinics. This increases rather than decreases the burden on the clinician. An enormous amount can be done to help the patient and family and in many countries specialist clinics are now available that aim to provide an integrated approach to management. In addition, for many of these disorders genetic issues are of enormous importance, but may be overlooked by the non-specialist. If a case of an X-linked disorder is diagnosed then it is a serious oversight not to offer counselling to appropriate female family members. In the case of myotonic dystrophy, failure to identify an asymptomatic female carrier of the gene, who then goes on to have a congenitally affected child, is tragic.

This introductory section will cover aspects of the structure and function of the neuromuscular system, and the clinical and laboratory approach to diagnosis. Thereafter, individual disorders will be discussed with the extent of coverage reflecting the commonness or otherwise of the disorder, rather than the degree of research interest.

24.1.1 Structure and function

It is not necessary to have an in-depth knowledge of the anatomy, physiology, and biochemistry of the neuromuscular system in order to understand, diagnose, and treat most of the conditions to be discussed in this chapter. However, some understanding of the basic processes involved is undoubtedly advantageous. This section covers aspects of structure and function. Biochemical processes are discussed in Section 24.6.

From the practising clinician's point of view, the neuromuscular system can be considered to have three major components: the *lower motor neurone*, the *neuromuscular junction*, and *muscle fibres*. The cell bodies of the *lower motor neurones* are situated in the brainstem motor nuclei and in the grey matter of the anterior horns of the spinal cord, hence their alternative name, anterior horn cells. The motor neurone axons travel in the cranial nerves and the spinal ventral roots. The latter contribute to the brachial and lumbosacral plexuses from which the peripheral nerves arise, which are generally mixed nerves in that they also convey sensory fibres. In terms of differential diagnosis, disorders of anterior horn cells and motor neurones, such as the spinal muscular atrophies, are of considerable importance because clinically they may closely mimic myopathies. Thus, we now know that some patients previously diagnosed as having Becker muscular dystrophy in fact had chronic spinal muscular atrophy.

A more detailed description of the structure and functioning of the *neuromuscular junction* is given in Section 24.10. In brief, the arrival of a nerve impulse opens voltage-gated calcium channels in the nerve terminal. This leads to an influx of calcium ions which triggers the release of many quanta of acetylcholine, ACh. The ACh diffuses across the synaptic cleft to interact with ACh receptors on the post-synaptic muscle membrane. This leads to an influx of cations, mainly sodium, which causes transient depolarization of the muscle-fibre membrane, the end-plate potential, which in turn activates voltage-gated sodium channels producing an action potential in the muscle fibre. Neuromuscular junction disorders, like anterior horn cell disorders discussed above, are not associated with sensory symptomatology and may also mimic conditions in which the primary pathology is at muscle fibre level.

Skeletal muscle is composed of numerous *muscle fibres* grouped together in fasciculi (Fig. 24.1). Each fibre is a multinucleate cell containing numerous myofibrils, the contractile proteins, sarcoplasm, equivalent to the cytoplasm of mononuclear cells, mitochondria, and the sarcotubular system (Fig. 24.2). The muscle fibre membrane is called the sarcolemma and in normal muscle the nuclei lie just beneath it. In clinical practice we recognize diseases that affect the muscle fibre membrane, including several types of muscular dystrophy and the channelopathies, biochemical processes occurring in the sarcoplasm and mitochondria, such as the primary metabolic

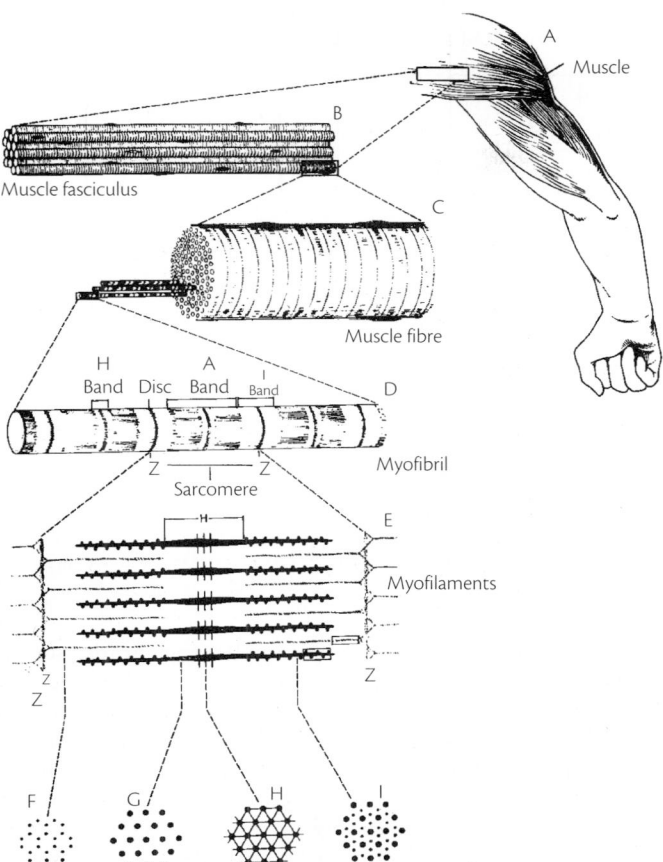

Fig. 24.1 Histological and molecular structure of skeletal muscle. (Reproduced from Patton *et al.* (1976) and modified from Bloom and Fawcett (1970).)

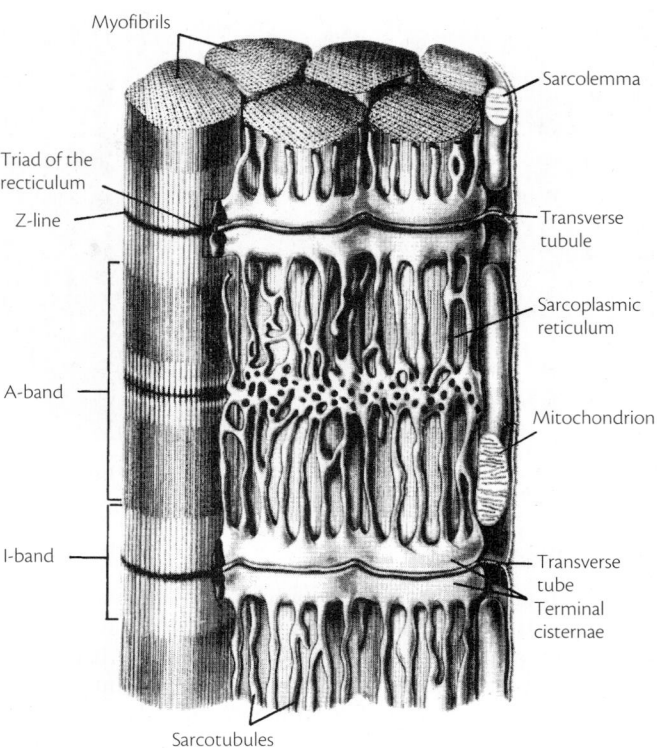

Fig. 24.2 Muscle structure, showing relation of endoplasmic reticulum and of transverse tubular system to fibrils. (Reproduced from Patton *et al.* (1976) and previously published in Bloom and Fawcett (1970).)

myopathies and mitochondrial cytopathies, and some forms of muscular dystrophy, the sarcotubular system including Brody's syndrome and tubular aggregate myopathy, the microvascular system as in dermatomyositis, inflammatory or immune processes that destroy muscle fibres such as polymyositis as well as conditions which involve the interstitial tissues.

The muscle fibre action potential is propagated deep into each fibre by the transverse tubular system (Fig. 24.2). The T-tubules are intimately associated with the sarcoplasmic reticulum and depolarization of this leads to release of calcium ions which in turn activate the contractile proteins actin, myosin, and others; an intensely energy-consuming process fuelled by adenosine triphosphate, ATP. Relaxation is achieved through re-uptake of calcium ions into the sarcoplasmic reticulum, which is also an energy-dependent process mediated by sarcoplasmic reticulum Ca^{++}-ATPase.

The *motor unit* is an important concept physiologically and when interpreting the results of clinical neurophysiological studies and muscle biopsy. Each muscle fibre is innervated by only one axon. But, each motor neurone undergoes terminal axonal branching and innervates many muscle fibres. These fibres are scattered randomly throughout the muscle, lying between fibres supplied by other motor neurones. Another important, and related, concept is that of *fibre types*. In many animals two types of muscles are recognized, red and white. The red muscle is composed of type 1 muscle fibres that are rich in myoglobin, giving the red colour, and mitochondria, are relatively slow to contract, but resistant

to fatigue. White muscle, on the other hand, is composed of type 2 fibres which have fewer mitochondria and greater glycolytic capacity. It has fast-twitch characteristics and fatigues relatively rapidly. Human muscles in health are composed of a mixture of both fibre types which gives the characteristic chequer-board appearance to a cross-sectional muscle biopsy stained to differentiate between fibre types (Fig. 24.3). A number of histochemical reactions applied to muscle biopsy specimens allow differentiation between different fibre types. The most important technique is myofibrillar ATP-ase staining. Preincubation in acid or alkaline buffers allows distinction between type 1 and 2 fibres. In fact, various other subtypes can be identified, notably type 2 into types 2A, B, and C. Identification of these is of limited diagnostic value in certain pathological states. Considering again the *motor unit*, there is now overwhelming evidence that the innervating neurone determines the fibre type. Thus, we should think in terms of type 1 motor neurones, which will make any muscle fibre they innervate have the characteristics of type 1 muscle fibres, as described above and similarly for type 2 motor neurones. Returning to the chequer-board appearance (Fig. 24.3), the distribution of the different fibre types is due to the random innervation of fibres scattered throughout the muscle by individual motor neurones. In pathological conditions that are associated with denervation and subsequent reinnervation this random pattern changes. Surviving axons sprout and innervate several adjacent previously denervated fibres, converting them all to the same fibre type. The resulting appearance on a cross-sectional muscle biopsy is of fibre-type grouping (Fig. 24.4)

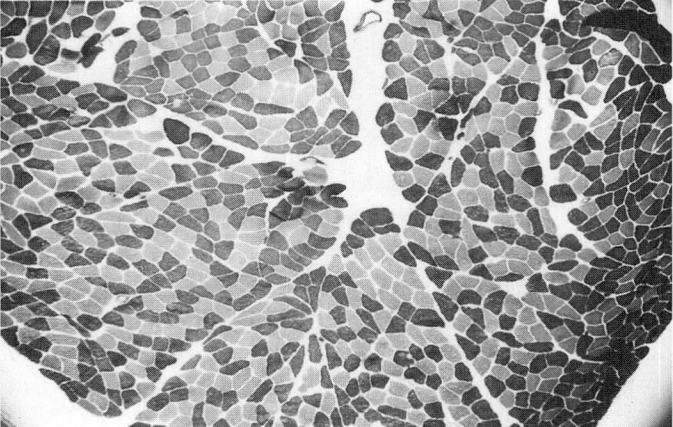

Fig. 24.3 Normal muscle fibre-type distribution. ATPase stain at pH 9.6. Type I fibres pale, type II fibres pale. (Courtesy of Dr Waney Squier.)

24.1.2 Classification

There is no universally accepted approach to the classification of neuromuscular disorders. Important recent contributions in terms of comprehensive cataloguing of disorders include ICD10-NA, the World Health Organisation's 'Application of the International Classification of Diseases to Neurology', and 'Classification of Neuromuscular Disorders' (Rowland and McLeod 1994). These comprehensive listings are valuable for coding and other purposes, but in everyday clinical practice what is needed is a simpler classification that helps in focusing investigations, and also in focusing the clinician's mind.

The first major division is between acquired and inherited myopathies (Table 24.1). Sub-classification within each of the categories listed will be discussed in the appropriate section. As a general observation, at the time of writing, no specific therapy is available for any of the inherited conditions listed in Table 24.1, although symptomatic treatment can be very helpful in some, notably the channelopathies of primary periodic paralyses and myotonia congenita, and some forms of congenital myasthenia. Genetic counselling, and in some cases the provision of an antenatal diagnostic service, is of major importance. The particular importance of recognizing acquired myopathies is that many of them are treatable.

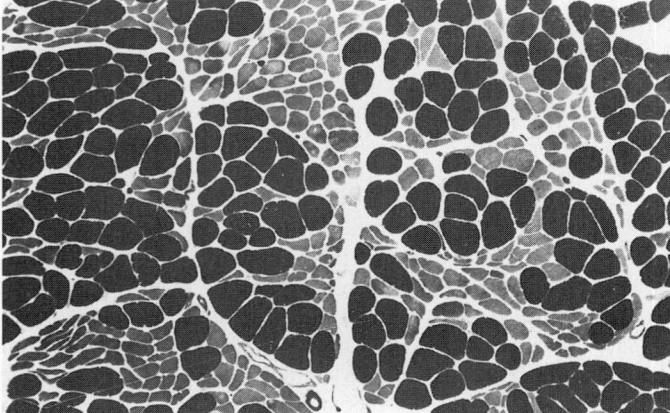

Fig. 24.4 Fibre-type grouping in denervation/reinnervation. Compare with Fig. 24.3 (Courtesy of Dr Waney Squier.)

Table 24.1 Inherited and acquired myopathies

Inherited
Muscular dystrophies
Myotonic dystrophy
Congenital myopathies
Channelopathies
Primary metabolic disorders
Congenital myasthenic syndromes*

Acquired
Drug and toxin induced
Endocrine
Secondary metabolic
Inflammatory
Paraneoplastic
Myasthenia gravis*
Lambert–Eaton myasthenic syndrome*

*Arguably, these disorders might be classified as channelopathies with there being a primary, inherited, ion channel abnormality in some of the congenital myasthenic syndromes and secondary, acquired, ion channel dysfunction or destruction in myasthenia gravis and Lambert–Eaton syndrome.

This may be by correcting the primary underlying disorder, such as an endocrinopathy or metabolic abnormality, by removal of a causative drug or toxin, or by immune modulation as with the idiopathic inflammatory myopathies, myasthenia gravis, or Lambert–Eaton myasthenic syndrome. It bears stressing that at the bedside even the most experienced neuromuscular clinician may not be able to distinguish between a form of muscular dystrophy, a chronic inflammatory myopathy, and a metabolic disorder—misdiagnosis may have far-reaching consequences not just for the patient but also for family members in the present and future generations.

24.1.3 Clinical approach

Some disorders are so characteristic that a confident specific diagnosis may be made at the bedside. Further investigation can then be limited to a confirmatory test. For example, classical type 1 myotonic dystrophy may be confirmed by demonstration of the specific gene abnormality. Even if clinical assessment does not give a specific diagnosis, it will often point to the general nature of the problem, differentiating between a dystrophy, inflammatory myopathy, or metabolic disorder, thus guiding the course of further investigations. This section reviews certain aspects of the history and physical examination as they relate to myopathic disorders, and is followed by further sections detailing particular investigative methods. Muscle has a limited range of responses to a wide range of disease, so the symptoms and signs associated with myopathies are relatively few.

Weakness is the commonest presenting symptom of muscle disease. The distribution, date of onset, and rate of progression must be determined. Fluctuating weakness is seen in myasthenic syndromes, and in some of the channelopathies and metabolic myopathies.

Considering the cranial musculature variable ptosis and diplopia are characteristic of myasthenia gravis. Constant ptosis is seen in mitochondrial chronic progressive external ophthalmoplegia, myotonic dystrophy, oculopharyngeal muscular dystrophy, and several congenital myopathies. Limitation of eye movements, with or without diplopia, is seen in mitochondrial chronic progressive external ophthalmoplegia, myasthenic syndromes, oculopharyngeal muscular dystrophy, and thyroid ophthalmopathy. Constant or fluctuating ptosis and diplopia may be seen with thyroid ophthalmopathy.

Facial weakness in myopathic disorders is usually bilateral and symmetrical, and mild weakness is easily missed. Classical symptoms include difficulty whistling, sucking through a straw, and blowing up balloons. Facial weakness can be striking in myasthenia gravis, myotonic dystrophy, and facioscapulohumeral muscular dystrophy. If present at all, it is usually mild in other dystrophies and inflammatory myopathies. Bulbar weakness, causing dysarthria and dysphagia, is common in myasthenia gravis. Dysphagia may be the presenting symptom in oculopharyngeal muscular dystrophy. It is seen in severe cases of idiopathic inflammatory myopathy and occasionally as a relatively early feature in inclusion body myositis.

Selective involvement of limb muscles may provide a powerful clue as to the diagnosis. In most of the acquired myopathies (Table 24.1), the selectivity is in the form of proximal greater than distal weakness. A similar distribution may be seen in certain neuropathies, particularly the inflammatory demyelinating polyradiculoneuropathies, and in myasthenic syndromes. It is with the dystrophies, including myotonic dystrophy, that the most remarkable selectivity may be seen, and as yet no satisfactory explanation for this phenomenon has been advanced. Individual disorders are discussed later. Particularly striking examples include facioscapulohumeral muscular dystrophy, oculopharyngeal muscular dystrophy, and myotonic dystrophy. In facioscapulohumeral dystrophy the scapular and biceps and triceps humeral muscles may be profoundly weak and wasted, but sitting between them deltoid appears normal. A few disorders characteristically have distal weakness as an early feature, the commonest examples being myotonic dystrophy and inclusion body myositis.

Respiratory muscle weakness may be seen in the later stages of many neuromuscular disorders and is often the major factor contributing to the patient's demise, as in Duchenne muscular dystrophy, or the rigid spine syndrome. Importantly, respiratory failure may be the presenting feature of a number of disorders, most notably adult-onset acid maltase deficiency and motor neurone disease. The earliest symptoms of respiratory insufficiency include nightmares, early-morning headache, and excessive day-time sleepiness. Only later does the patient complain of breathlessness on exertion and orthopnoea. Bedside testing should include vital capacity measurement, erect and supine, and looking for paradoxical abdominal movement indicative of diaphragmatic weakness; on inspiration, with the patient lying flat, the upper abdomen will be drawn inwards.

Pain is perhaps the next commonest symptom in a neuromuscular clinic. It is worth noting that if pain is the sole complaint and physical examination is normal, then often a specific diagnosis is not achieved. Pain may be very localized or be widespread (Table 24.2), or may be exercise-induced (Table 24.3).

Patients may mean one of several things when they complain of muscle *cramps*. Ordinary cramps are neurogenic in origin; motor nerve hyperactivity causes painful muscle contraction, and electromyography shows high-frequency motor unit discharges. They are usually benign but predisposing factors include metabolic disturbances and hypothyroidism. Pathological states associated with neurogenic hyperactivity include neuromyotonia and stiff-man syndrome. However, cramp may also be used to describe problems arising at muscle fibre level. *Myotonia* is caused by recurrent muscle fibre membrane depolarization. The patient complains of delayed relaxation of grip, or jaw stiffness when chewing. The most common cause is myotonic dystrophy, but the phenomenon may be much more severe in the rarer condition of myotonia congenita. Grip and percussion myotonia are readily demonstrated at the bedside (Fig. 24.5). Electrically silent *contractures* are a feature of some metabolic disorders, notably glycogenoses such as myophosphorylase deficiency, McArdle's disease, and the extremely rare conditions of Brody's syndrome and rippling muscle disease.

The term *contracture* is also used to describe the shortening of muscles and associated inability to passively stretch them caused by progressive fibrosis of the muscle. This is seen as a late feature in

Table 24.2 Disorders causing localized or generalized muscle pain

Localized
Trauma
Ischaemia
Infection
Bacterial
Parasitic
Acute alcoholic myopathy
Some glycogenoses e.g. McArdle's disease
Inflammation sarcoidosis
Eosinophilic fasciitis
Neuralgic amyotrophy
Generalized
Acute idiopathic inflammatory myopathies
Infections
Viral
Toxoplasmosis
Drug-induced myopathies
Metabolic myopathies
Metabolic bone disease
Hypothyroid myopathy
Carnitine palmitoyltransferase deficiency
Polymyalgia rheumatica
Connective tissue disorders
Guillain–Barré syndrome
Porphyria

Table 24.3 Causes of exercise-induced muscle pain

Metabolic myopathies
Glycogenoses
Mitochondrial cytopathies
Carnitine palmitoyltransferase deficiency
Muscular dystrophies
Duchenne
Becker
Tubular aggregate myopathy
Dermatomyositis
Ischaemia with claudication

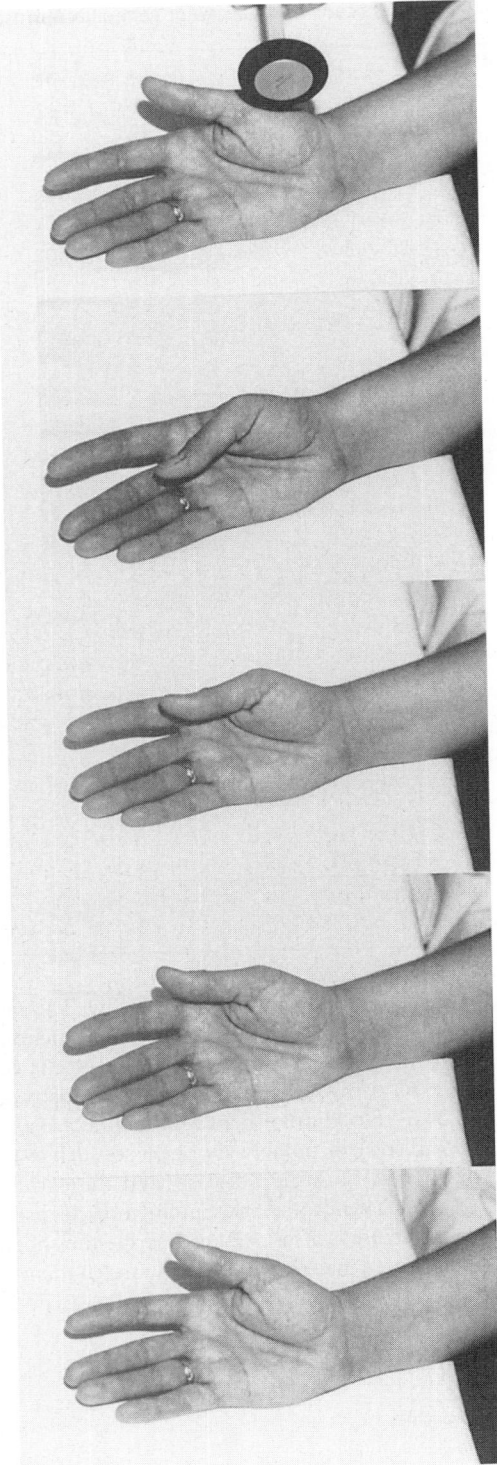

Fig. 24.5 Percussion myotonia. Sequential photographs taken at approximately 3 second intervals, after a sharp tap to the thenar eminence.

it tends to parallel it. Muscle *pseudohypertrophy*, which is probably due to fatty and fibrous tissue infiltration, is seen in Duchenne and Becker muscular dystrophy and less commonly in autosomal limb–girdle dystrophies. True *hypertrophy*, a form of work-hypertrophy, is seen in myotonia congenita and neuromyotonia.

Myoglobinuria is noted by the patient as passage of dark urine and is due to extensive muscle damage that allows the release of myoglobin into the bloodstream. It poses a risk of renal failure from acute tubular necrosis. Commoner causes are noted in Table 24.4.

Tendon reflexes tend to be preserved in myopathic disorders in contrast to neurogenic conditions in which they are depressed or lost at an early stage. In the Lambert–Eaton myasthenic syndrome the reflexes are reduced or absent, but show striking post-contraction potentiation.

The history taking and examination must extend beyond the neuromuscular system. Important aspects in the history include family history, current or recent drug and toxin exposure including anaesthetics, and past medical history. A systematic review of all other body systems is important as many myopathies are part of a multisystem disorder, or a purely myopathic condition may lead to secondary effects elsewhere, such as the symptoms associated with respiratory failure. Similarly, physical examination must include a fairly detailed general medical assessment, not forgetting respiratory function, as well as assessment of the central nervous system, and peripheral nervous system for evidence of sensory nerve dysfunction. Cardiomyopathy and dysrhythmias, which may be either sub-clinical or which may dominate the clinical picture, are seen in association with a number of skeletal myopathies (Table 24.5). Central nervous system involvement is common in mitochondrial cytopathies, including pigmentary retinopathy, ophthalmoplegia, deafness, cognitive impairment, or movement disorder. Sensory symptoms and signs point away from a disorder affecting the anterior

Table 24.4 Causes of myoglobinuria

Intensive exercise in normal individuals	
Inherited myopathies	
Metabolic	Glycogenoses, lipid storage disorders, malignant hyperthermia:
Dystrophic	Duchenne and Becker muscular dystrophy:
Acquired myopathies	
Inflammatory	(Dermatomyositis and polymyositis)
Infections	(Viral and bacterial)
Ischaemia	
Trauma	
Crush injury	
Electric shock	
Status epilepticus	
Drugs	
Opiates	
Clofibrate	
Statins	
Toxins	
Alcohol	
Animal venoms	
Others	
Neuroleptic malignant syndrome	
Metabolic disorders	
Heat stroke	

many myopathies and neurogenic disorders but is also an important early symptom and sign in Emery–Dreifuss muscular dystrophy, Bethlem myopathy, and rigid spine syndrome.

Wasting is not a common presenting symptom. In myopathies weakness tends to precede wasting, whereas in neurogenic disorders

Table 24.5 Skeletal myopathies associated with heart involvement

Dysrhythmias
Myotonic dystrophy
Emery–Dreifuss muscular dystrophy
Mitochondrial cytopathies

Cardiomyopathy
Duchenne and Becker muscular dystrophy
Emery–Dreifuss muscular dystrophy
Limb-girdle muscular dystrophy
Dermatomyositis
Debranching enzyme deficiency
Alcoholism
Endocrine myopathies

horn cell, neuromuscular junction, or muscle, but it should be noted that compressive neuropathies are common and a secondary consequence of immobility particularly of the ulnar nerve at the elbow and common peroneal nerve at the knee.

24.1.4 Biochemical investigations

The most familiar biochemical test for investigation of a suspected myopathy is estimation of the serum *creatine kinase* level. Muscle damage, from whatever cause, leads to release of creatine kinase into the bloodstream. The specific skeletal muscle isoform is creatine kinase MM, in contrast to cardiac muscle creatine kinase MB, but generally only the total creatine kinase activity is measured. Serum creatine kinase is always elevated in Duchenne, Becker, and Emery–Dreifuss muscular dystrophy. It is usually elevated in inflammatory myopathies, glycogenoses, congenital muscular dystrophy, autosomal limb–girdle muscular dystrophies, and Kennedy's syndrome, a neurogenic disorder. It is often normal in myotonic dystrophy, facioscapulohumeral muscular dystrophy, mitochondrial cytopathies, and drug-induced myopathies, especially steroid-myopathy. An elevated creatine kinase in an apparently normal individual is a teasing clinical problem with several causes (Table 24.6).

In mitochondrial cytopathies serum and spinal fluid *lactate* and *pyruvate* levels may be elevated. Serum levels are measured in exercise tests, described below. In disorders of lipid metabolism, blood and urine *carnitine* and *acyl-carnitine* levels and ratios may be altered.

Exercise tests are used to investigate suspected metabolic myopathies. The *forearm exercise test* is used to screen for disorders of glycogenolysis and glycolysis and biochemical considerations are

Table 24.6 Causes of an elevated serum creatine kinase (CK) level in 'normal' individuals

Hypothyroidism

Female carriers of Duchenne/Becker muscular dystrophy gene

Male with Becker muscular dystrophy mutation but no weakness

Presymptomatic muscular dystrophies e.g. dysferlinopathy

Caveolinopathy

Susceptibility to malignant hyperthermia

Drug-induced sub-clinical myopathy especially statins

Strenuous physical exercise

discussed further in Section 24.6.1. In brief, venous blood from the forearm is assayed after vigorous use of the flexor muscles, typically by repeatedly squeezing a rubber bulb. The normal response is a rise in lactate and ammonia levels. The former is blunted or absent, and the latter accentuated, with defects of the glycogenolytic/glycolytic pathway.

Aerobic exercise is most easily performed on a static bicycle. In mitochondrial disorders there is an excessive rise in serum lactate and in the lactate/pyruvate ratio. This is mainly a research procedure.

Phosphorus magnetic resonance spectroscopy is a powerful tool for investigating aspects of muscle energy metabolism *in vivo* but availability is very limited. It allows measurement of the phosphorus-containing molecules ATP, phosphocreatine, phosphomonoesters, and inorganic phosphate in muscle, all of which play a vital role in normal energetic processes. Intracellular pH can also be determined. It can replace, and indeed give additional information over, the forearm and aerobic exercise protocols described above.

Prolonged fasting may be used to assess disorders of lipid metabolism but is potentially very hazardous and its use should be restricted to specialist units.

For most of the primary metabolic myopathies the diagnosis is proven by *enzyme assay*, most frequently on a sample of muscle frozen and stored at −70°C at the time of biopsy. In some conditions assay may be on blood cells, urine, cultured fibroblasts, or liver biopsy.

24.1.5 Neurophysiological studies

These are discussed in Section 3.5 but a brief review of their value in studying suspected myopathies is appropriate here. Nerve conduction studies are an essential part of the evaluation. As noted, primary neurogenic disorders may clinically mimic a myopathy. Nerve conduction is normal in diseases that affect only muscle, but may be abnormal in conditions, such as the mitochondrial cytopathies, in which nerve and muscle pathology may coexist. Similar comments apply to somatosensory evoked potential studies.

The most useful technique for studying primary muscle disease is concentric needle electrode electromyography, with a single fibre modification of this technique, being a major tool for the study of neuromuscular junction disorders. Four particular features can be documented.

Insertional activity describes the electrical response seen when the electromyography needle first enters the muscle or is subsequently moved and is probably the result of muscle fibre damage. Increased activity is typical of denervation but may also be seen in inflammatory myopathies. Decreased activity is seen in atrophic and fibrotic muscle and during attacks of periodic paralysis.

Spontaneous activity can be seen once any insertional activity has settled. Fibrillation potentials and positive sharp waves are typical of denervation but are also seen in inflammatory myopathies and active muscular dystrophies. Fasciculation potentials indicate anterior horn cell disease but may rarely be seen in thyrotoxic myopathy. Myotonic discharges are most frequently encountered in myotonic dystrophy and the less common condition of myotonia congenita, more rarely in acid maltase deficiency and hypothyroidism.

Motor unit potentials may be analysed in terms of amplitude, duration, number of phases, and the stability of the waveform. The characteristic appearance in myopathic disorders is of polyphasic potentials with reduced amplitude and duration. Instability of the waveform has the same significance as increased jitter and is seen in neuromuscular junction disorders, denervation, and some inflammatory myopathies.

The *interference pattern* is more complex and develops at a lower force of contraction in myopathic disorders. This reflects early recruitment of the surviving muscle fibres.

In *single-fibre electromyography* a special electrode is positioned in such a way that it detects the potentials from two muscle fibres, each innervated by a terminal branch of the same axon. In response to nerve stimulation or voluntary contraction the two fibres fire nearly, but not quite, simultaneously. The slight difference between the potentials is called the interpotential difference, and reflects the difference in conduction times of the two terminal pathways. *Jitter* is measured as the mean consecutive difference of successive interpotential intervals. The main source of jitter is neuromuscular transmission, and thus it is increased in myasthenic disorders. It is also increased in mitochondrial cytopathies and in denervation.

Repetitive nerve stimulation and measurement of the evoked compound muscle action potential is also useful for the study of neuromuscular transmission. Myasthenia gravis is associated with a decremental response at 3 Hz stimulation, whereas in the Lambert–Eaton syndrome the resting compound muscle action potential is small but increases in amplitude following brief voluntary contraction or with repetitive stimulation at 20 Hz.

24.1.6 Imaging methods

There is considerable variability in the use of muscle imaging, some units not offering it at all, others placing great value in it. Potential information that can be obtained includes assessing the distribution and severity of muscle involvement, assessment of disease progress, and selection of a suitable muscle, or region within one muscle, for biopsy. With respect to the latter, ultrasound has been fairly widely used, particularly in paediatric practice and by those who prefer needle to open biopsy. Distribution of muscle involvement may sometimes give a powerful clue towards the diagnosis. Thus, MRI demonstration of involvement of the forearm flexor muscles before clinical evidence of finger flexion weakness is typical of inclusion body myositis (Sekul *et al.* 1997).

24.1.7 Muscle biopsy

Although muscle biopsy is of considerable importance in the investigation of myopathies it must be remembered that muscle has a limited repertoire of responses to a wide range of pathological insults and that the appearances in a small biopsy specimen may not be representative of the underlying disease process. In other words, the biopsy findings must be interpreted within the context of the clinical evaluation and the results from other tests. This section discusses the biopsy procedure, specimen handling, and preparation, and a brief review of major pathological features. There are some excellent monographs on the subject of muscle biopsy (Dubowitz and Sewry 2007).

In general, the muscle selected for biopsy should be moderately weak. Very weak muscles often just show end-stage changes of fibrosis and fat replacement. It is important that normal statistics for the chosen muscle are available; for example, certain muscles normally show predominance of a particular fibre type. For this reason only a limited number of muscles are normally used for

biopsy, the commonest being deltoid and vastus lateralis component of quadriceps. It is largely a matter of personal preference and experience as to whether the muscle sample is obtained through an open approach or by the use of a biopsy needle. Both are performed under local anaesthesia, except sometimes in small children. Needle biopsy leaves a smaller scar but less material is obtained; this can cause problems with orientation and if pathological changes are patchy, as they often are in inflammatory myopathies, they may be missed.

The specimen for light microscopy is orientated under a dissecting microscope and snap frozen by immersion in cooled isopentane. Sections of 6 μm thickness are cut in a cryostat and mounted on cover slips, ready for staining. A panel of routine histological and histochemical stains are applied to each specimen (Table 24.7). Additional histochemical stains may be indicated, such as myophosphorylase staining in suspected McArdle's disease. Immunocytochemistry is a particularly valuable technique for the study of muscular dystrophies and idiopathic inflammatory myopathies. The sections are incubated with an antibody directed against a specific antigen, and the antibody is visualized by use of a coloured tag. Specific examples are considered when discussing individual disorders.

Electron microscopy is of limited value in routine practice. Notable exceptions include the demonstration of filamentous inclusions in inclusion-body myositis and the characterization of ultrastructural abnormalities in several of the congenital myopathies.

Major pathological changes. In *denervation* features include small angular fibres and fibre-type grouping, reflecting reinnervation. *Muscular dystrophies* of limb–girdle and Xp21 Duchenne and Becker type show an increase in the normal variability of fibre size, necrosis, regenerating fibres, fibre splitting, increased numbers of central nuclei, and an increase in fibrous tissue. In myotonic dystrophy and facioscapulohumeral muscular dystrophy the biopsy is often normal or shows only non-specific features and biopsy does not form part of their routine assessment. Immunocytochemistry shows dystrophin to be absent in Duchenne and reduced in Becker muscular dystrophy. In the autosomal recessive limb–girdle dystrophies one or more of the sarcoglycans may be reduced or absent. The *idiopathic inflammatory myopathies* are characterized by the presence of inflammatory infiltrates, composed mainly of lymphocytes, necrotic fibres and, in dermatomyositis, capillary loss, infarcts, and perifascicular atrophy. Immunocytochemistry permits identification of lymphocyte sub-types and the demonstration of

Table 24.7 Routine muscle biopsy dyes and stains, and their uses

Tissue dyes	
Haematoxylin & eosin	Histology
Modified Gomori trichrome	Histology
Periodic acid Schiff	Demonstrating excess glycogen
Oil red O	Demonstrating excess lipid
Histochemistry	
ATPase (at pH 4.3, 4.6 & 9.4)	Fibre typing
NADH	Myofibrillar architecture
Succinate dehydrogenase	Mitochondrial distribution
Cytochrome oxidase	Mitochondrial function
Acid phosphatase	Lysosomal activity

Table 24.8 Gene mutations in myopathic disorders

	Disorders
Large deletion	Duchenne and Becker dystrophy ~70% patients
Small deletions & point mutations	Duchenne and Becker dystrophy ~30% patients
Deletion	Mitochondrial cytopathies mitochondrial DNA
Point mutations	Many metabolic myopathies
	Limb-girdle dystrophies e.g. sarcoglycanopathies
	X-linked Emery–Dreifuss syndrome
	Channelopathies
	Myotonia congenita
	Congenital myasthenic syndromes
	Mitochondrial cytopathies mitochondrial DNA
Nucleotide repeat expansions	Myotonic dystrophies
	Oculopharyngeal muscular dystrophy
Deletion of repeat units	Facioscapulohumeral muscular dystrophy

major histocompatibility complex type I antigen expression on muscle fibres. Many *congenital myopathies* are associated with ultra-structural abnormalities, either affecting fibre architecture, as in central core disease or in the form of abnormal structures or accumulations within fibres, as in nemaline myopathy. In *metabolic myopathies* the appearances are often unremarkable. Excess glycogen and lipid accumulation may be seen in the glycogenoses and lipid metabolism disorders respectively, but are not invariable. In *mitochondrial cytopathies* the classical finding is of ragged-red fibres seen on Gomori's modified trichrome stain and reflecting aggregations of mitochondria, but these are not always present. NADH and succinate dehydrogenase staining is often abnormal and there may be cytochrome oxidase negative fibres. Electron microscopy is not essential, but shows accumulations of structurally abnormal mitochondria which may contain inclusions.

24.1.8 Molecular genetic studies

The rate of discovery in this area is so rapid that even specialists in the neuromuscular field have difficulty keeping their knowledge up to date. Web databases are vital such as http://www.musclegenetable.org. The traditional approach to gene identification was by working backwards from the mutated protein. That approach is not possible when the gene product is not known. Positional cloning, also known as 'reverse genetics', allows us to move in the opposite direction. By studying polymorphic markers in large informative families the chromosomal position of the relevant gene can be identified, and then finer mapping techniques can be used to find its exact position. Expressed transcripts from that area are assessed for possible candidate genes. Once the gene has been identified the protein product can be deduced.

At present, the ability to identify a specific gene defect aids precise diagnosis and carrier detection, allows accurate genetic counselling and pre-natal diagnosis, and may sometimes indicate the likely severity of the phenotype. In the future, knowledge of the gene defect will presumably be essential if genetic engineering

approaches to therapy are to be considered. Alternatively, a detailed understanding of the structure and function of the protein product may lead to therapeutic approaches by biochemical means. The types of mutation seen in the commoner inherited myopathies are shown in Table 24.15.

24.1.9 Genetic counselling

This is a major factor when considering inherited neuromuscular disorders, but one that is sometimes neglected. Whoever provides counselling must have a detailed understanding not only of the genetic issues but also of the disease itself. The affected individual will want to know not only the statistical risk of their offspring being affected but also the possible severity in those that are. This is a straightforward matter for those conditions that follow simple Mendelian rules but is very much more difficult for those that do not, such as the trinucleotide repeat expansion disorders and mitochondrial cytopathies. These issues are discussed further when considering individual disorders. Asymptomatic relatives of affected individuals provide a particular challenge. The counselling issues for women in families carrying an X-linked disorder, usually Duchenne or Becker, are well understood. For autosomal dominant disorders a relative may be asymptomatic because they do not carry the relevant gene. Alternatively, they may carry the gene but not show its manifestations because they are too young to do so, or because there is variability of expression for that disorder, known as reduced penetrance. Gene testing in asymptomatic individuals needs considerable care and thought. Carrier and pre-symptomatic testing is generally not performed in childhood. It can never be assumed that a relative is unaffected on hearsay evidence, but their are difficulties in seeing individuals who believe themselves to be normal and then demonstrating abnormalities, relating to the disease in question, on examination. On the other hand, it is a tragedy if an asymptomatic woman in a family known to suffer from myotonic dystrophy, gives birth to a child with the severe congenital form of the disease when, given appropriate counselling, they might have opted for gene testing, prenatal diagnosis, and selective termination of pregnancy. Increasingly, pre-implantation genetic diagnosis techniques are becoming available which although emotionally challenging in their own right remove the traumas associated with selective termination of pregnancy.

24.2 Muscular dystrophies and related disorders

An early, and still generally useful, definition of these conditions is that they are primary, inherited, progressive, degenerative disorders of muscle. In practice, a few are non-progressive and in some conditions degenerative changes are slight or absent. Reclassification may be appropriate when the molecular basis of each form has been determined, but to date such new evidence has if anything confused rather than illuminated due to the unexpected breadth of genotype–phenotype variability.

Much experimental evidence had suggested that the muscle fibre membrane might be the major site of pathology in the dystrophies and this appeared to be borne out when dystrophin was identified; its absence causes Duchenne muscular dystrophy but if it is present in reduced quantity or in a truncated form the phenotype is that of Becker dystrophy. Subsequently it was shown that absence of membrane associated proteins, sarcoglycans, functionally related to dystrophin is responsible for some forms of limb–girdle dystrophy. However, we now know that some dystrophies are due to defects of:

- sarcoplasmic proteins: calpain in one form of limb–girdle dystrophy or a protein kinase in myotonic dystrophy type 1;
- nuclear membrane proteins: emerin and lamin A/C in Emery–Dreifuss syndrome;
- myofibrillar proteins: myotilin in a form of limb–girdle muscular dystrophy.

Since genetic counselling is such an important aspect of management, a classification based on the Mendelian pattern of inheritance combined with phenotypic description is particularly useful in everyday practice (Table 24.9). Despite its name, myotonic dystrophy is usually considered separately (Section 24.3).

24.2.1 Dystrophinopathies

This term encompasses those conditions in which there is a primary abnormality of dystrophin. The clinical phenotype of Duchenne dystrophy is stereotyped and had been well characterized long before the discovery of dystrophin. Becker dystrophy was known to have a more variable phenotype, but quite how variable has only become apparent over the last decade. At its most severe Becker dystrophy is indistinguishable from Duchenne. A less severe form presents in late-middle age. In yet another form weakness may be absent and the picture is that of cramps and episodic rhabdomyolysis and myoglobinuria.

Dystrophin and associated proteins

After dystrophin came the identification of other related proteins (Fig. 24.6). Although the precise function of this protein complex remains much debated the most favoured opinion is that it acts to stabilize the inherently fragile muscle fibre membrane during the rigours of contraction and relaxation. The complex links intracellular actin to proteins in the extracellular matrix.

Duchenne muscular dystrophy

This is the commonest form of muscular dystrophy and being an X-linked recessive disorder principally affects boys. Rarely, girls are affected, severely, as a result of chromosomal abnormalities, including Turner's syndrome or translocations. Up to 10 per cent of female carriers may show evidence of muscle disease, usually weakness or hypertrophy, or rarely cardiomyopathy, and prior to the identification of

Table 24.9 Classification of the muscular dystrophies

X-linked
Duchenne
Becker
Emery-Dreifuss

Autosomal dominant
Facioscapulohumeral
Oculopharyngeal
Scapuloperoneal
Limb-girdle (~10%)
Distal myopathies

Autosomal recessive
Limb-girdle (~90%)
Congenital
Distal myopathies

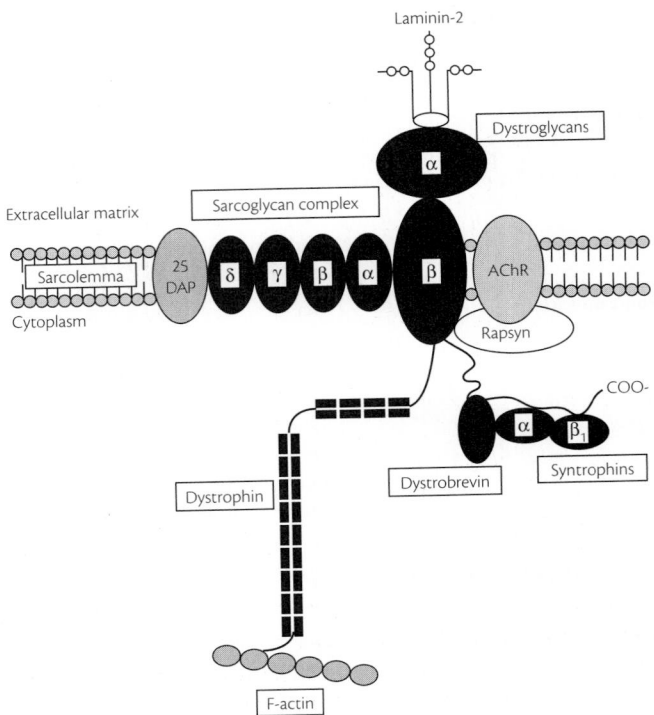

Fig. 24.6 The dystrophin associated protein complex. (Courtesy of Rosie Fisher.)

dystrophin were often diagnosed as having limb–girdle dystrophy; the basis is skewed X-inactivation (Yoshioka *et al.* 1998).

The incidence is 20 to 30/100,000 live-born males. One-third are known to have a previous family history, one-third are new mutations, and one-third are born to an unknowing carrier. These figures emphasize the difficulties associated with genetic counselling for this disorder.

Clinical features. The principal clinical picture is of progressive weakness, affecting limbs and respiratory muscles, but important additional features include muscle hypertrophy, speech delay, intellectual impairment, and cardiac involvement. These boys appear normal at birth and in the first year of life. Walking is clumsy and delayed, with about one-half still not walking at 18 months. They do not achieve the ability to run, hop and jump. As the lower limb weakness progresses the gait becomes more and more waddling and the abdomen protuberant because of increasing lumbar lordosis. Climbing stairs is difficult and the child rises from the floor in a characteristic fashion, Gower's manoeuvre, first turning prone and then placing their hands on their knees and thighs to push themselves erect. Symptomatic shoulder-girdle weakness develops within 2–3 years but is evident on examination much earlier. Ambulation is lost at about 10 years of age. During adolescence the weakness progresses, spinal deformity increases, and contractures, which first involved the hip flexors and Achilles tendons, progress rapidly particularly affecting elbows and knees. Deformity may prevent comfortable seating in a wheelchair and should be a major target for the physiotherapist. Residual power in finger and wrist flexor muscles allows the boy to use a keyboard and to control an electric wheelchair. Mild facial weakness is evident in the late stage. Particularly in the early-stages, muscle hypertrophy, and later pseudohypertrophy due to fatty replacement of muscle, develops in most boys. It is most evident in the calves, quadriceps, and muscles of mastication. It is not pathognomonic and, for example, may also be seen in spinal muscular atrophy. Overall, the increasing weakness is associated with progressive muscular atrophy and loss of tendon reflexes.

Respiratory muscle weakness develops early, as shown by forced vital capacity measurement, and respiratory function is later further compromised by the spinal deformity. The cough becomes weak, leading to aspiration and increased risk of chest infection. Ventilatory insufficiency leads to excessive day-time sleepiness, poor appetite, early morning headache, and, commonly, a fear of sleeping. Untreated, ventilatory failure used to lead to death around the age of 20 years, but the use of non-invasive ventilation has substantially prolonged survival.

Cardiac involvement is an inevitable feature of the disease, but is uncommonly symptomatic, perhaps because of the boys' limited physical activity and thus demand on the heart. The electrocardiogram changes early, with a characteristic appearance of tall R waves in the right precordial leads and deep narrow Q waves in the left precordial and limb leads. Echocardiographic changes appear later and are usually not gross. A small proportion of boys develop symptomatic cardiac failure, which is occasionally the cause of death. Increasingly, presymptomatic treatment with diuretics, ACE inhibitors and β-blockers is being considered and may be important given the prolonged survival associated with the introduction of non-invasive ventilation.

Delayed speech development is a common but under-recognized feature and the wise clinician presented with a boy with this symptom checks the serum creatine kinase level even in the absence of other abnormality. As a group, the average intelligence quotient is reduced to about 80, but many are of normal or above-average intelligence.

Diagnosis. A highly confident diagnosis can be made on the basis of the clinical picture and the finding of a very high serum creatine kinase level. Up to the age of 5 years, including at birth, the creatine kinase activity is typically in the order of 100-times the upper limit of normal, and is never within the normal range. As the disease progresses and muscle mass is lost the creatine kinase activity falls. A definitive diagnosis depends upon demonstration of an underlying gene defect. In about two-thirds this consists of a large deletion, easily detected in the laboratory. In the other one-third the gene defect is currently not readily demonstrated due to technical and cost considerations, but it may consist of a small deletion or duplication or of a point mutation. In these patients the diagnosis is based upon demonstration, by immunocytochemistry and Western blotting, of absence or severe deficiency of dystrophin in a muscle biopsy specimen (Fig. 24.7). The classical histological features of muscular dystrophy are present, consisting of increased variability of fibre size, rounding of fibres, fibre splitting, increased numbers of central nuclei, hyaline fibres, muscle fibre necrosis, phagocytosis and regeneration, and replacement by fat and fibrous tissue (Fig. 24.8).

Management. Genetic counselling issues have already been discussed (Section 24.1.9). It is now accepted that steroids prolong ambulation, which has several advantages including beneficial effects with respect to spinal deformity and ventilatory function. However, side-effects are significant and further trials are indicated (Manzur *et al.* 2004).

There is huge interest in the potential for specific therapy given our understanding of the basic underlying molecular mechanism but practical realization is likely to be some time away. Therapeutic options include correcting or ameliorating the basic genetic defect, for instance by use of antisense oligonucleotides, upregulation of associated proteins such as utrophin, and myoblast transfer (Wells 2006).

Overall, current management involves physical approaches and dealing with the profound psychological problems associated with a relentlessly progressive debilitating and eventually fatal disorder.

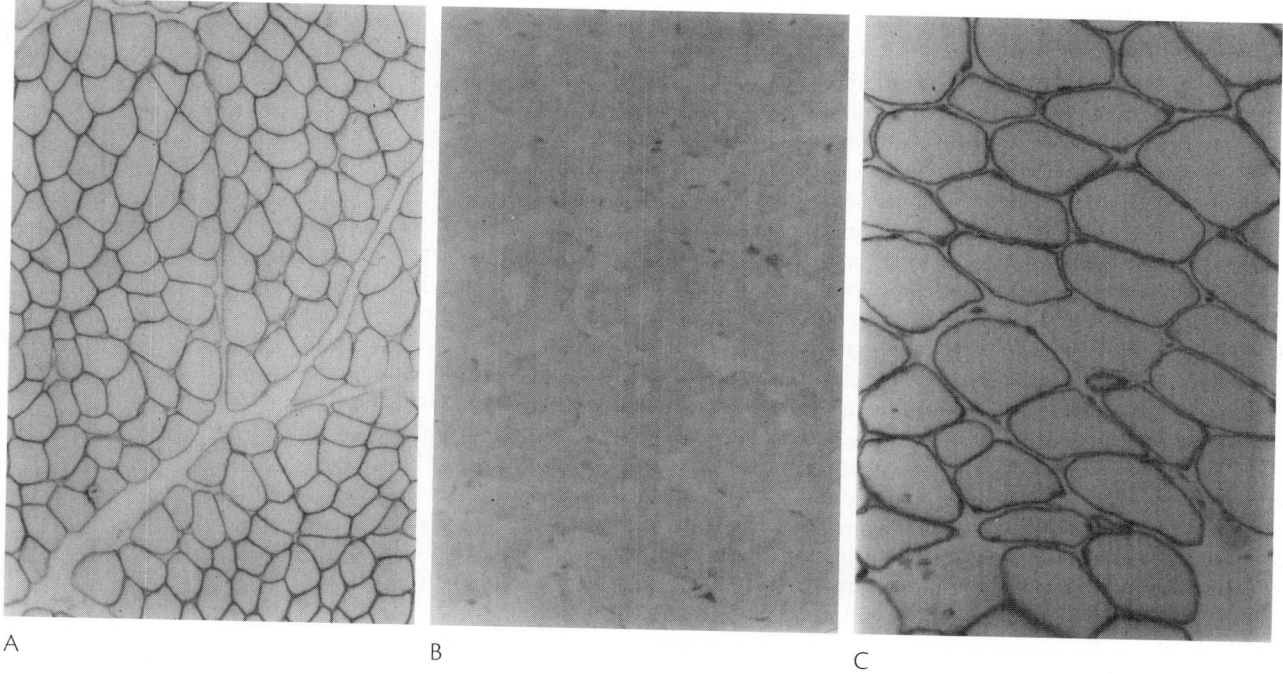

A   B   C

Fig. 24.7 Dystrophin staining. (A) Normal muscle—note the regular staining around the periphery of each muscle fibre. (B) Duchenne dystrophy—note complete absence of staining. (C) Becker dystrophy—note irregular staining and single central fibre with almost absent staining. (Courtesy of Dr Waney Squier.)

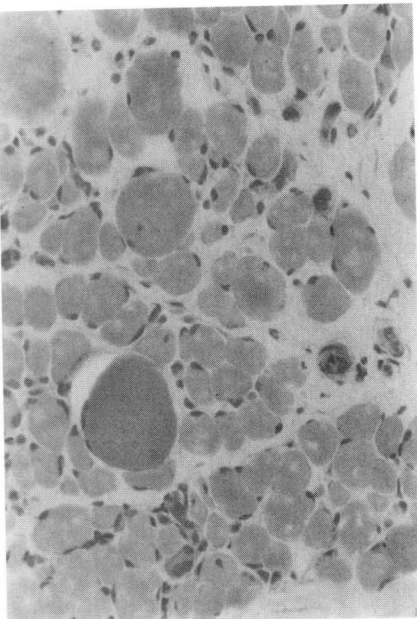

Fig. 24.8 Duchenne muscular dystrophy. Haematoxylin and eosin stain. Note the increased variability in fibre size, roundness of fibres, hypercontracted/hyaline fibre, and increased connective tissue. (Courtesy of Dr Waney Squier.)

Regular exercise is important and can continue even after the boy becomes wheelchair bound. Bed rest for intercurrent illnesses must be avoided because it is associated with rapid and irreversible further muscle deterioration. Contractures can be delayed, and to some extent reversed, by regular passive exercises—these should be taught to the family by a suitably experienced physiotherapist. Additional techniques include splinting and the use of callipers and other walking aids. The role of surgery for contractures is complicated and controversial. Any such surgery should never be performed without very detailed assessment by appropriately experienced physician, physiotherapist, and orthotist, in conjunction with a similarly experienced surgeon. This should be done in very close collaboration with the patient and family. Sometimes, although such procedures may prolong walking the boy decides that he would rather opt for wheelchair existence.

Once wheelchair-bound, spinal curvature rapidly increases. It can be helped by very careful attention to the seating arrangement and the use of a spinal orthosis jacket. Spinal fusion, by one of several procedures, may be of value but by the time it is required there are substantial risks from anaesthesia and surgery relating to the respiratory muscle and cardiac involvement. Rhabdomyolysis and a malignant hyperthermia-like reaction are additional complications. As described above, close liaison between all interested parties is required.

Occupational therapy has a major role to play and constant reassessment, as the disease progresses and needs change, is required. Dietary advice is needed, to avoid obesity. Emotional support comes from many sources and the care advisor system developed by the Muscular Dystrophy Campaign in Britain is an excellent example of how this can be provided. There are numerous issues relating to schooling, most will go to a mainstream school, and the potential for social isolation. In the early stages the boy should be encouraged towards hobbies and pastimes that he will be able to continue when his disability increases. Parental guilt and depression are common.

Ventilatory failure is inevitable and, if untreated, leads to death around the age of 20 years. The introduction of non-invasive nocturnal ventilation has had a dramatic effect on survival as well as improving quality of life (Eagle *et al.* 2002).

Becker muscular dystrophy

The prevalence of this condition is about 2.5/100,000. Originally delineated clinically as a milder phenotype of Duchenne muscular dystrophy we now know that its basis is either a reduced amount of dystrophin or the presence of dystrophin of lower than normal molecular weight, and that the phenotype varies in severity from a condition indistinguishable from Duchenne, to a late-adult onset proximal myopathy, and can also present as a disorder characterized by cramps, myalgia, and myoglobinuria without weakness. At a genetic level, Becker muscular dystrophy is usually associated with an in-frame deletion as opposed to Duchenne in which the deletion is out-of-frame.

In typical Becker muscular dystrophy weakness is first evident between the ages of 5 and 20 years. A rather common history is that of an adolescent who has been at the bottom of the class for several years with respect to sporting success, particularly in running. As the lower limb weakness progresses activities such as climbing stairs and rising from a chair become increasingly difficult. Upper limb weakness may not be symptomatic for many years but is often evident earlier on examination. Other common early features are exercise-induced calf cramps, toe-walking, and calf hypertrophy (Fig. 24.9). The distribution of muscle weakness parallels that seen in Duchenne, with early involvement of hip flexors and extensors, quadriceps, pectoralis major, latissimus dorsi, and brachioradialis. The age of loss of ambulation is very variable and many remain ambulant, albeit with an ungainly gait, into late-middle age.

Cardiac involvement is underestimated (Hoogerwaard *et al.* 1997). At presentation patients should have an electrocardiogram and echocardiogram. These should be repeated periodically, at intervals depending on previous findings and symptomatology. Rapid progression, from being asymptomatic to death from cardiac failure, may occur. Optimal management remains controversial but may involve the use of diuretics, β-blockers, ACE inhibitors, and rarely, cardiac transplantation.

Rarer cases of very late onset, 60 years and later, often presenting with quadriceps weakness, have been described. Another characteristic phenotype is of exercise-induced myalgia and cramps, sometimes with episodes of rhabdomyolysis (Samaha and Quinlan 1996) which may initially suggest a metabolic myopathy.

Diagnosis. The principles are as given for diagnosing Duchenne muscular dystrophy. In the one-third of patients who do not have a readily demonstrable gene abnormality, confirmation is easy if Western blotting shows dystrophin of reduced molecular weight or a severe reduction in quantity. Lesser reductions may be more difficult to interpret and there is overlap with the sarcoglycanopathies, limb–girdle dystrophies, in which there may be secondary reduction of dystrophin. Immunocytochemistry is very suggestive if it shows patchy dystrophin staining (Fig. 24.7) but can never alone exclude the diagnosis. As in Duchenne the serum creatine kinase is invariably elevated in early stages. Electromyography shows changes consistent with primary muscle disease with the additional presence of fibrillation potentials and positive sharp waves.

Management. Surgical intervention is rarely required. Contractures are a late feature, following loss of ambulation. The general

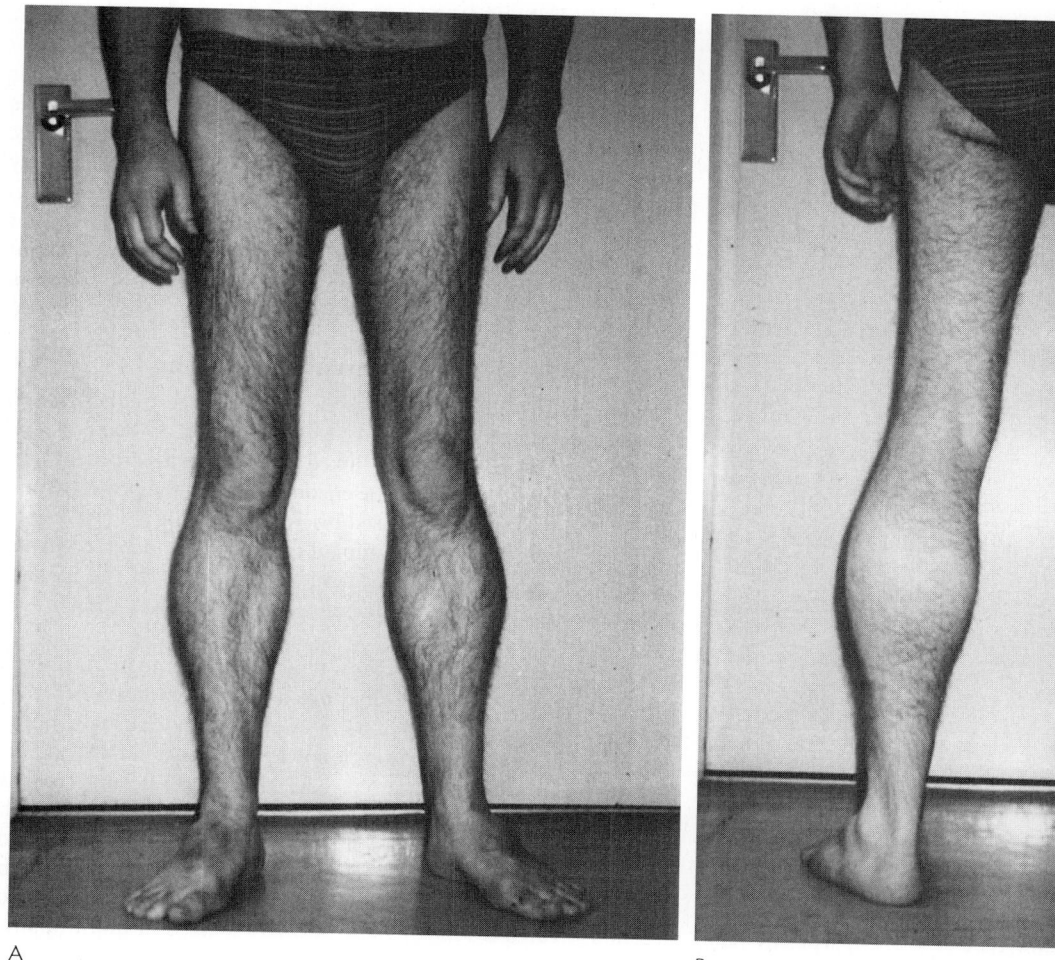

A   B

Fig. 24.9 Becker muscular dystrophy. Note the quadriceps wasting (A) and asymmetrically enlarged calves (B).

principles of management are as discussed for Duchenne muscular dystrophy. Particular attention needs to be paid to the heart and to ventilatory function. Respiratory failure and its associated symptoms occur only in late-stages, long after ambulation has been lost. Nocturnal positive pressure ventilation with a mask affords symptomatic relief.

Female carriers

Up to 10 per cent of female carriers of dystrophin gene mutations may show evidence of muscle disease, weakness or hypertrophy, and prior to the identification of dystrophin were often diagnosed as having limb–girdle dystrophy; the basis is skewed X-inactivation (Yoshioka *et al.* 1998). It has been known for some time that occasional carriers present with cardiomyopathy, even in the absence of skeletal muscle involvement, and recent studies have shown that subclinical cardiomyopathy is relatively common (Grain *et al.* 2001). It is recommended that all carriers undergo cardiological assessment at diagnosis and at intervals thereafter (Bushby *et al.* 2003).

24.2.2 Limb–girdle muscular dystrophies

The term limb–girdle muscular dystrophy acquired widespread usage in the 1950s and has been the subject of much debate ever since. It was used to separate a group of disorders that had in common

limb–girdle weakness and absence of early facial weakness, from the X-linked dystrophies and autosomal dominant facioscapulohumeral muscular dystrophy. Autosomal dominant, autosomal recessive, and sporadic forms were included. Some patients so classified were subsequently shown to have Becker muscular dystrophy, to be manifesting female carriers of the Duchenne/Becker gene abnormality, to have spinal muscular atrophy, or one of a variety of metabolic myopathies, such as acid maltase deficiency. Recent genetic and molecular studies have allowed us to re-evaluate the whole concept of limb–girdle muscular dystrophy (Laval and Bushby 2004; Straub and Bushby 2006) (Table 24.10). It is currently possible to reach a molecular genetic diagnosis in about 60 per cent of patients presenting with this type of clinical syndrome—implying that there are still many more genes to be identified. Recessive forms account for about 90 per cent, dominant for 10 per cent, of the limb–girdle dystrophies.

After the discovery of dystrophin, the dystrophin-associated proteins were identified (Fig. 24.6) and these seemed a likely candidate for other forms of muscular dystrophy, as subsequently proved to be the case. The concept that all muscular dystrophies might relate to an abnormality of a sarcolemma-associated protein proved to be too simplistic.

Table 24.10 Classification of the limb–girdle muscular dystrophies

	Gene location	Protein	Gene
Autosomal dominant			
LGMD 1A	5q31	Myotilin	MYOT
LGMD 1B	1q11–q21	Lamin A/C	LMNA
LGMD 1C	3p25	Caveolin 3	CAV3
Autosomal recessive			
LGMD 2A	15q15.1–q21.1	Calpain 3	CAPN3
LGMD 2B	2p12–14	Dysferlin	DYSF
LGMD 2C	13q12	γ-sarcoglycan	SGCG
LGMD 2D	17q21	α-sarcoglycan	SGCA
LGMD 2E	4q12	β-sarcoglycan	SGCB
LGMD 2F	5q33–q34	δ-sarcoglycan	SGCD
LGMD 2G	17q11–q12	Telethonin	TCAP
LGMD 2H	9q33.2	TRIM32	TRIM32
LGMD 2I	19q13.33	Fukutin-related protein	FKRP
LGMD 2J	2q31	Titin	TTN
LGMD 2K	9q34.1	POMT-1	POMT1

Of the recessive limb–girdle muscular dystrophies the commonest, in most Caucasian populations, are types 2A, calpainopathy, and 2I, fukutin-related protein. Early muscle wasting and contractures are features of 2A, whereas 2I may importantly be associated with cardiomyopathy and early ventilatory impairment, including ventilatory failure when still ambulant. Type 2I is phenotypically very similar to Becker muscular dystrophy and must always be considered in the differential diagnosis. Type 2B, dysferlinopathy, is allelic with Miyoshi distal myopathy, a condition that presents in late-adolescence with gastrocnemius wasting and weakness, associated with a high serum creatine kinase level. A patient presenting with Miyoshi eventually 'evolves' into a typical limb–girdle muscular dystrophy pattern of weakness. The sarcoglycanopathies, types 2C, D, E, and F, are most commonly seen in populations where consanguinity is common and can include an early-onset severe dystrophy mimicking Duchenne. Titin mutations when present on both alleles cause limb–girdle muscular dystrophy 2J, whereas a heterozygous mutation on just one allele presents as a late-onset distal myopathy.

The genes associated with three forms of dominant limb–girdle muscular dystrophy have been identified. Each is associated with a wide range of phenotypic manifestations. Myotilin mutations may present as limb–girdle muscular dystrophy 1A, or as myofibrillar myopathy associated with distal weakness and sometimes cardiomyopathy (Olive *et al.* 2005). Lamin A/C mutations not only present as limb–girdle muscular dystrophy 1B but also as Emery–Dreifuss syndrome, cardiomyopathy, lipodystrophy, neuropathy, progeria, and skeletal abnormalities; this group of conditions is referred to as the laminopathies (Rankin and Ellard 2006; Woodman *et al.* 2004). Caveolin mutations cause not only limb–girdle muscular dystrophy 1C but also rippling muscle disease, asymptomatic elevation of serum creatine kinase, and distal myopathy (Woodman *et al.* 2004).

Diagnosis. This still lies mainly within the preserve of specialist and research laboratories. Loss of any one component of the dystrophin-associated protein complex leads to secondary loss, or at least lack of proper localization, of the other components; this complicates interpretation of immunocytochemical techniques. A combination of immunocytochemical labelling of muscle biopsy sections, together with western blotting, is invaluable in helping to identify the defective or absent protein. Absolute confirmation of the diagnosis requires DNA analysis of the appropriate gene, but that is time consuming and expensive given the lack of common mutations and large size of many of the genes.

24.2.3 Emery–Dreifuss muscular dystrophy

This highly distinctive syndrome is characterized by the triad of early contractures, humero-peroneal distribution of weakness, and cardiac conduction defects. X-linked recessive forms are associated with mutations in the *STA* gene encoding coding the protein emerin; autosomal forms are caused by mutations in the *LMNA* gene encoding lamin A/C, with dominant forms being much more common than recessive. Both proteins are localized to the nuclear membrane. Other laminopathies associated with *LMNA* mutations are discussed above.

Clinical features. Contractures typically affect the neck, elbows, and ankles (Fig. 24.10). They may pre-date any demonstrable weakness. Limited neck flexion and elbow extension are usually noted first, from about 2 years of age, followed by toe-walking. Weakness is initially in a humero-peroneal distribution, but later spreads to involve the peri-scapular muscles and proximal lower limb muscles. Severe weakness is uncommon and most men remain mobile. Cardiac involvement is usually in the form of conduction abnormalities, ranging from first degree to complete heart block, and if untreated is a major cause of death. Dilated cardiomyopathy is less common.

Female carriers are usually asymptomatic but rarely may develop severe cardiac conduction problems in the absence of skeletal muscle involvement. Therefore, screening of at-risk relatives and periodic review is advisable.

Diagnosis. The serum creatine kinase may be normal or elevated. Electromyography shows changes indicating primary muscle disease and light microscopy of skeletal muscle shows dystrophic changes. In the X-linked form, immunocytochemistry shows absence of emerin in skeletal muscle and skin, and skin biopsy may be used to identify female carriers. Numerous mutations have been identified within the emerin gene. In the autosomal form, Immunocytochemistry is unhelpful and the diagnosis rests on *LMNA* mutation analysis.

Management. By far the most important aspect is identification and management of cardiac involvement. There is a high incidence of sudden death (Pinelli *et al.* 1987). Regular electrocardiographic monitoring is required and the patient should be made aware of the need to report immediately symptoms such as palpitation and dizziness. There is a strong argument in favour of the use of implantable defibrillators (Meune *et al.* 2006). The progression of contractures may be reduced by physiotherapy and Achilles tenotomies are often helpful.

24.2.4 Facioscapulohumeral muscular dystrophy

This is another condition in which recent genetic advances have led to a better understanding of the extent of phenotypic variability

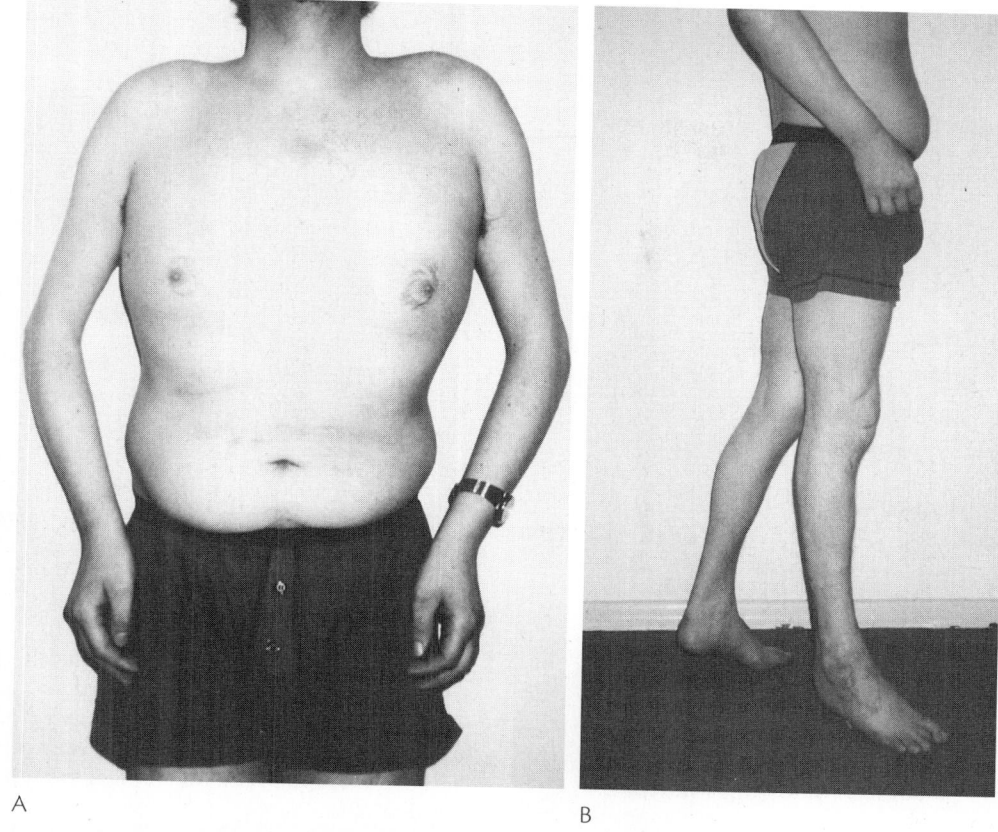

Fig. 24.10 Emery-Dreifuss syndrome. Note the humeral muscle wasting and elbow contractures (A), and the Achilles tendon contractures causing toe-walking (B).

(Tawil *et al.* 1998). It is inherited as an autosomal disorder, with a prevalence of about 5/100 000, and gains its name from the characteristic early pattern of muscle involvement. New mutations are not uncommon. The relevant gene and gene product are unknown. The disease is associated with deletion of an integral number of 3.3 kb repeats in the subtelomeric region of chromosome 4q35. The current view is that this rearrangement has a position-effect on one or more genes proximal to the repeat (Tawil and Van Der Maarel 2006).

Clinical features. Onset is typically in the second decade. Uncommonly, onset may be in the first year of life; the parent typically has a much milder form of the disease and this probably reflects somatic mosaicism. Penetrance is high but not complete and clinical assessment alone may fail to identify gene carriers. Females tend to be less severely affected than males.

The first symptoms, although often not recognized by the patient, relate to facial weakness; poor whistle, difficulty sucking on a straw and blowing-up balloons. At presentation, 90 per cent have demonstrable facial weakness (Fig. 24.11).

As the name implies, there is selective weakness and wasting of muscles around the shoulder girdle, affecting serratus anterior, rhomboids, and the lower part of trapezius, the scapular fixator muscles, and pectoralis major, but with preservation of deltoid. Often scapular winging is noted before weakness (Fig. 24.12). Weakness causes the patient difficulty abducting their arms and performing tasks above shoulder height. On examination, the inability to abduct the arms to more than about 60° and the elevation

of the scapulae due to the pull of the unaffected deltoid produces a highly characteristic appearance (Fig. 24.13). Unusually for a myopathy, the picture may be quite markedly asymmetric (Fig. 24.14).

Weakness of tibialis anterior is evident in most patients at presentation and rarely foot-drop is a presenting symptom. Similarly, upper arm weakness (biceps more than triceps) may be demonstrated. Padberg (1998) has described seven stages in the progression of the disease (Table 24.11); progression from one stage to the next may take decades and abortive cases occur. About 20 per cent of patients over the age of 50 years require a wheelchair. Spinal and pelvic girdle weakness may lead to marked lumbar lordosis. The rare early-onset form is associated with severe facial weakness which may be misdiagnosed as Möbius syndrome (Fig. 24.15).

Mild sensorineural deafness can be detected in many adults but is rarely symptomatic. It may be more evident in early-onset cases. Retinal vascular disease, Coats' disease, may be demonstrated by fluorescein angiography but routine eye examination is normal and visual symptoms very rare. Pain is common and under-recognized (Bushby *et al.* 1998). It is largely mechanical in origin.

The extra-ocular and pharyngeal musculature are not involved. Apart from occasional tightening of the Achilles tendons, contractures are not a major feature.

Diagnosis. This is by DNA analysis. If these are negative neurophysiological studies and muscle biopsy may be needed to exclude other disorders that may mimic the condition, although these are extremely rare: polymyositis, mitochondrial cytopathy, and possibly neurogenic disorders. The serum creatine kinase is often normal.

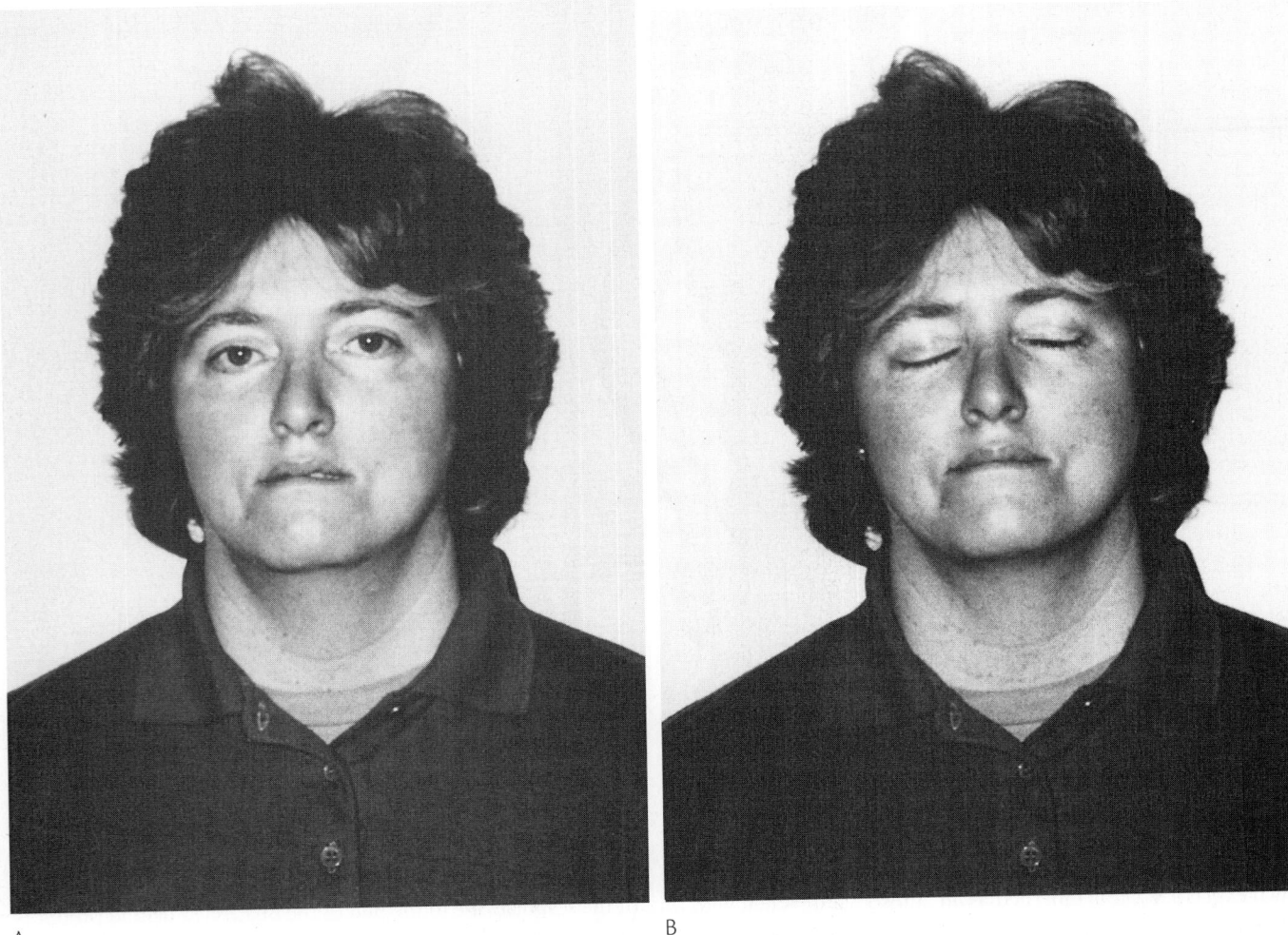

A   B

Fig. 24.11 Facioscapulohumeral muscular dystrophy. The facial weakness may not be striking because it is symmetrical (A). On attempted forceful eye closure there is incomplete burying of the eye-lashes (B).

Electromyographic findings are those associated with primary muscle disease. Depending on the muscle selected, biopsy findings are sometimes minimal. In a clinically affected muscle there are dystrophic changes and, a subject of much debate, inflammatory infiltrates.

Management. No drugs have been demonstrated to be unequivocally effective and none can be recommended. Pain management is important and usually revolves around physiotherapy and analgesics/anti-inflammatory drugs. Scapula fixation surgery may be helpful, perhaps especially in younger patients with marked scapular instability but preserved biceps and triceps function.

24.2.5 Scapuloperoneal syndrome

The clinical features of this rare syndrome are indicated by the title; periscapular wasting and weakness with associated scapular winging, and wasting and weakness of tibialis anterior causing weakness of ankle dorsiflexion. In other words, the clinical features resemble the phenotype of facioscapulohumeral muscular dystrophy without the facial weakness. Many cases represent a *forme fruste* of facioscapulohumeral dystrophy and have a typical DNA deletion, but some do not (Felice *et al.* 2000). A similar phenotype may be

seen in Emery–Dreifuss syndrome and probably in several autosomal dystrophies. Neurogenic cases have been described, representing a form of spinal muscular atrophy, but few have been subjected to modern diagnostic methods.

24.2.6 Oculopharyngeal muscular dystrophy

Striking features of this condition include its late-onset, typically in the fifth or sixth decade, and the highly selective pattern of muscle involvement. It is caused by a trinucleotide repeat expansion in the *PABPN1* gene on chromosome 14 (Brais *et al.* 1998). In the common autosomal dominant form one of the alleles is expanded from the normal 6 repeats, to 8 or more. In a rarer autosomal recessive form both alleles contain 7 repeats. The differential diagnosis includes myasthenia gravis, mitochondrial cytopathy, and thyroid ophthalmopathy.

Clinical features. The first feature is usually ptosis, often asymmetricslly, less often dysphagia. As the disease progresses the ptosis may obscure vision (Fig. 24.16). Slight external ophthalmoplegia may develop but is rarely severe and diplopia is uncommon. There is often striking over-activity of frontalis in an attempt to overcome the ptosis.

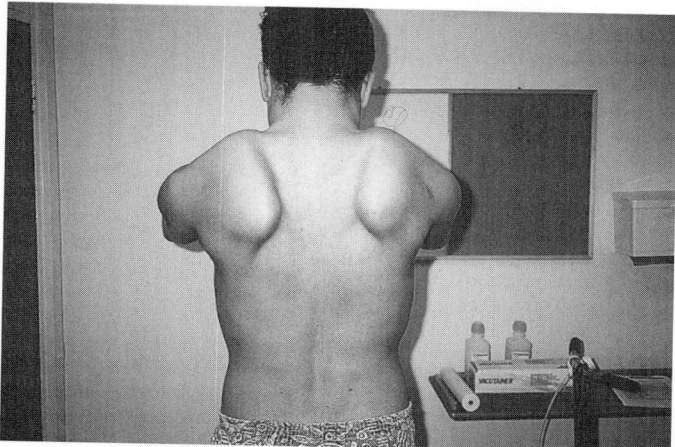

Fig. 24.12 Facioscapulohumeral muscular dystrophy. Classical scapular winging when the patient pushes against the wall with straight arms.

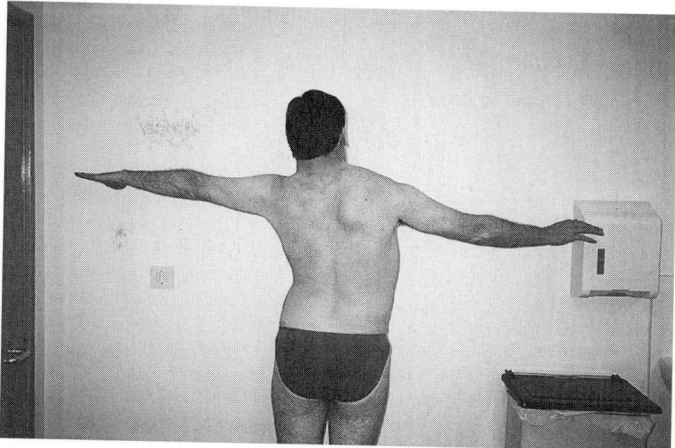

Fig. 24.14 In facioscapulohumeral muscular dystrophy the weakness is often asymetrical. Compare with Fig. 24.13

Dysphagia follows within a few years, initially for solids and later for fluids. Nasal regurgitation may occur. At this stage mild facial weakness is usually evident. Although not included in the name, in later stages proximal limb–girdle weakness may develop. This is usually mild and affects the shoulder–girdle more often than the pelvic–girdle.

Diagnosis. Diagnosis is based on DNA analysis. Although muscle biopsy shows characteristic finding of 8.5 nm intranuclear tubular filments, it is not required for diagnosis.

Management. If the ptosis interferes with vision, then corrective surgery is helpful (Fig. 24.16). Dysphagia is managed as for other neuromuscular disorders that impair swallowing. Options include advice from a speech therapist, cricopharyngeal myotomy, botulinum, and gastrostomy.

24.2.7 Congenital muscular dystrophy

These autosomal recessive disorders are characterized by congenital or early-infantile onset of weakness with variable central nervous system involvement including ocular. A common molecular theme is that they are associated with defects affecting proteins in the extracellular matrix (Schessl *et al.* 2006).

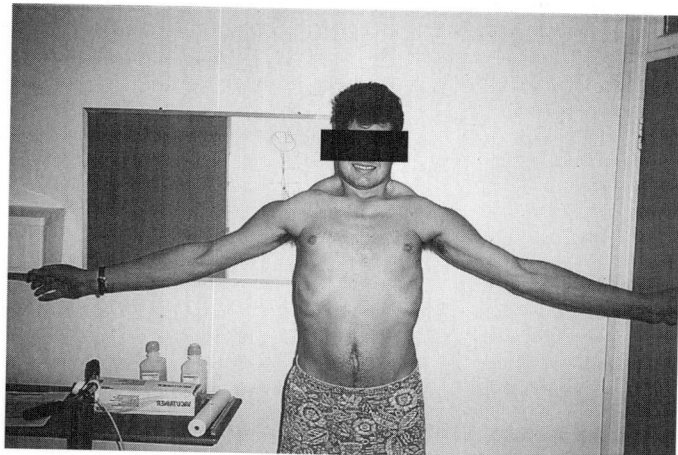

Fig. 24.13 Facioscapulohumeral muscular dystrophy. The patient is attempting to elevate his hands above shoulder height, but is unable to do so.

One of the first to be delineated at a molecular level was merosin, laminin α2, deficiency. These children fail to achieve independent ambulation, epilepsy is common, and although mental development is normal they invariably have white matter abnormalities on MRI.

An important emerging theme is that of disorders of α-dystroglycan O-mannosyl-linked glycosylation (Muntoni *et al.* 2004). Proteins involved in this process, and abnormalities of which have been linked to congenital dystrophies, include POMT1, POMT2, POMGnT1, Fukutin, Fukutin-related protein or FKRP, and LARGE. Some are associated with gross central nervous system and eye involvement. Recognized phenotypes include Fukuyama muscular dystrophy, Walker–Warburg syndrome and muscle–eye–brain disease.

Disorders affecting the collagen genes *COL6A1*, *COL6A2*, and *COL6A3* are associated with either Ullrich congenital muscular dystrophy or Bethlem myopathy (Section 24.2.8). They are sometimes referred to as the collagenopathies (Lampe and Bushby 2005). Ullrich is usually inherited in autosomal recessive fashion and is associated with distal joint laxity and proximal contractures, and with abnormalities of the skin texture and healing properties.

As an exception to the theme of extracellular matrix dysfunction, rigid spine muscular dystrophy is caused by mutations in the *SEPN1* gene which encodes a selenoprotein in the endoplasmic reticulum. As the name indicates, spinal rigidity is a characteristic feature. Many of these patients remain ambulant, but they are at considerable risk of developing ventilatory failure. Mutations in the same gene cause multiminicore disease. Many other

Table 24.11 The stages of facioscapulohumeral muscular dystrophy (after, Padberg, 1998)

Stage 0	Facial weakness
Stage 1	Periscapular weakness and wasting (scapular winging)
Stage 2	Tibialis anterior weakness
Stage 3	Pelvic girdle weakness
Stage 4	Difficulty on stairs and rising from a chair
Stage 5	Wheelchair outdoors
Stage 6	Wheelchair indoors

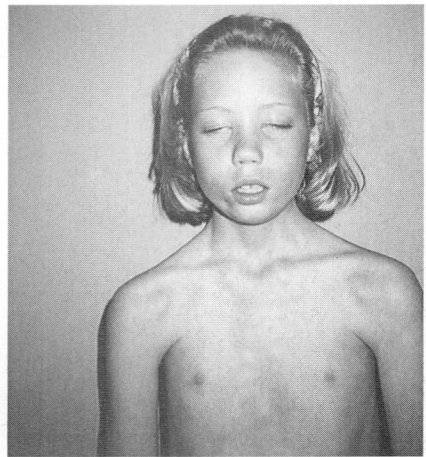

A

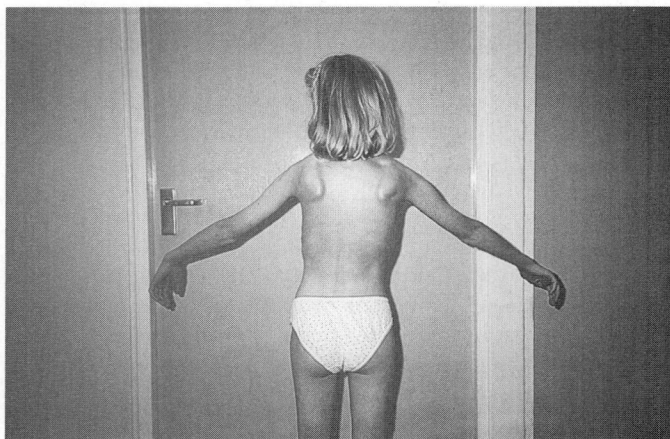

B

Fig. 24.15 Facioscapulohumeral muscular dystrophy. Severe, early-onset form. Note the profound facial weakness (A) and inability to fully abduct at the shoulders (B).

phenotypes have been described but their molecular basis remains uncertain.

24.2.8 Bethlem myopathy

This rare condition is due to a mutation of one of the collagen VI genes and with Ullrich congenital muscular dystrophy falls into the category of collagenopathy. It is inherited as an autosomal dominant condition. The age of onset varies from infancy to adolescence. Proximal lower limb weakness leads to a Gower's sign which tends to disappear in adolescence and reappear in early adult life. Thereafter there is slow progression and most patients remain ambulant. In the upper limbs there is mild proximal weakness. There is generalized slimness of the muscles. A characteristic feature is the development of contractures involving the interphalangeal joints, elbows, ankles, and often other joints. The serum creatine kinase may be slightly elevated and electromyography shows myopathic changes. Diagnosis is established by DNA analysis. The differential diagnosis includes laminopathy.

24.2.9 Distal myopathies

Like the congenital muscular dystrophies the distal myopathies are currently the subject of intensive research, genetic discovery, and

A

B

Fig. 24.16 Oculopharyngeal muscular dystrophy. Before (A) and after (B) ptosis correction surgery.

re-classification (Mastaglia *et al.* 2005). Their basic feature is early-involvement of distal muscles, more often in the lower than the upper limbs, although later widespread weakness may occur. They must be distinguished from other conditions in which there is substantial distal weakness including Charcot–Marie–Tooth disease, distal spinal muscular atrophy, myotonic dystrophy, inclusion body myositis and scapuloperoneal syndrome.

The first type to be clearly defined was the autosomal dominant, late-onset, distal myopathy of Welander. Initially described in Sweden it has now been reported in other countries, but usually in people with Swedish ancestors. Weakness is first noted in the hands in the 4th to 6th decade and appears later in the distal lower limbs. It progresses slowly but remains confined to the distal muscles. Although the relevant gene has not been identified, all patients have a common shared haplotype indicating common ancestry (von Tell *et al.* 2002).

Two autosomal dominant late-onset forms starting in the anterior compartment of the lower legs have been defined and are sometimes referred to as tibial muscular dystrophy. The Markesbery–Griggs form, caused by mutations in *ZASP*, later causes distal upper limb weakness. The Udd form is mostly but not entirely confined to Finns and is caused by mutations in the *TTN* gene encoding titin; homozygous mutations of the gene are associated with an early onset form of limb–girdle muscular dystrophy type 2J. An autosomal dominant late-onset form affecting the posterior compartment of the lower limb is associated with mutations in the *MYOT* gene encoding myotilin and is thus allelic with LGMD 1A.

Autosomal dominant childhood or early adult life onset distal myopathy affecting the anterior compartment of the lower limb, Laing distal myopathy, is caused by mutations in the *MYH7* gene.

Nonaka described an autosomal recessive form of distal myopathy in Japan with anterior compartment weakness developing in early adult life with additional involvement of iliopsoas and neck flexor muscles, and the presence of rimmed vacuoles on muscle biopsy. It was subsequently shown that this disorder is allelic with hereditary inclusion body myopathy, quadriceps sparing myopathy, initially described in Iranian Jews. This phenotype has now been reported worldwide and is due to mutation in the *GNE* gene.

Miyoshi and others described an early-onset, autosomal recessive, myopathy affecting particularly gastrocnemius, and causing difficulty climbing stairs and hopping. Onset is typically in late adolescence. Serum creatine kinase levels are always substantially elevated. It is now known that this is a phenotypic variant of limb–girdle muscular dystrophy type 2B, and the cause is a mutation of the *DYSF* gene encoding dysferlin.

24.3 **The myotonic dystrophies**

It is now recognized that there are two separate conditions meriting the title myotonic dystrophy. Although they share many clinical similarities, and probably have a common molecular basis, they are distinct disorders. Most clinicians will be familiar with myotonic dystrophy type 1, also referred to as Steinert's disease and first described about a century ago. Myotonic dystrophy type 2 is also referred to as proximal myotonic myopathy, proximal myotonic dystrophy and as Ricker's disease the latter reflecting the seminal contribution to its delineation (Ricker *et al.* 1995).

Myotonic dystrophy type 1 is by far the commonest inherited myopathy seen in adult life. The exact prevalence is uncertain but is about 12/100 000. There remains considerable uncertainty as to the prevalence of myotonic dystrophy type 2. In some countries,

for instance Germany, it appears to be about as common as myotonic dystrophy type 1, whereas in others, such as the United Kingdom, it appears to be rare. Some of that apparent variability may reflect ascertainment.

The molecular basis of each is an unstable nucleotide repeat expansion; in myotonic dystrophy type 1 a CTG trinucleotide repeat in the 3′-untranslated region of the *DMPK* gene, and in myotonic dystrophy type 2 a CCTG tetranucleotide repeat in intron 1 of the *ZNF9* gene. It is believed that these untranscribed abnormalities lead to disruption of mRNA metabolism which affects the function of many other genes in part through aberrant splicing (Machuca-Tzili *et al.* 2005). For example, aberrant splicing of the chloride channel gene leads to myotonia and of the insulin receptor gene to insulin resistance.

These mutations are unstable in mitosis and meiosis. The former means that the size of the mutation increases in different tissues during the lifetime of the individual. Given that the mitotic rate differs between different tissues, the size of the mutation also differs, not just at any one time point but also during life. The instability in meiosis means that the ova or sperm of affected individuals tend to carry a larger expansion than in the parent tissue. Disease onset therefore tends to be earlier and the eventual severity greater, in the offspring, a phenomenon known as anticipation.

They are both multisystemic disorders with clinically important considerations relating to involvement of the heart, particularly conduction abnormalities, the respiratory system resulting in ventilatory failure and infections, and brain.

24.3.1 **Myotonic dystrophy type 1**

Clinically, myotonic dystrophy type 1 can for practical clinical purposes be considered to present as one of four main phenotypes, in order of severity:

◆ a severe congenital form;

◆ an early-childhood onset form;

◆ the classical adult form, the commonest presentation;

◆ a late-onset form that may be asymptomatic and show no abnormal signs.

These reflect the size of the underlying CTGn repeat expansion Anticipation is often striking. Although the severe congenital form is only seen if the mother is the transmitting parent (Fig. 24.17), it is wrong to think that male transmission is not associated with anticipation. Indeed, for smaller repeat sizes, expansion tends to be greater with male transmission, but not into the range associated with congenital disease.

Clinical features. The *classical adult form* typically presents in late-adolescence or early adult life. It is common to see asymptomatic individuals, for example picked up during family screening, who, despite their lack of symptoms, have characteristic findings on examination. The distribution of muscle involvement is highly characteristic and difficult to confuse with other myopathies. The characteristic facial appearance (Fig. 24.17) is the result of ptosis and wasting and weakness of the facial muscles and muscles of mastication, together with premature balding, which is more commonly seen in males. In the limbs there is distal weakness, affecting the hands and wrists more than the feet and ankles. Patients complain of difficulties with activities such as unscrewing the lid of a bottle and wringing out a cloth. Tripping, due to foot-drop, develops later. As the disease progresses the weakness spreads proximally and in

Fig. 24.17 Myotonic dystrophy. Anticipation—the disease is more severe in her two sons.

Table 24.12 Non-skeletal muscle manifestations of myotonic dystrophy

Heart	Conduction problems
Eyes	Cataracts
Central Nervous System	Mental retardation
	Excessive day-time sleepiness
Endocrine	Testicular atrophy
	Reduced fertility
	Male-pattern baldness
Smooth muscle	Dysphagia and aspiration
	"Irritable-bowel-like" symptoms
	Constipation
	Faecal soiling
	Incoordinate uterine contractions
Peripheral nerve	Rarely symptomatic
Pilomatrixomas	

late-middle age a small proportion of patients become wheelchair-bound, although most retain some independent ambulation. Also characteristic is the presence of wasting and weakness of sterno-mastoid and other neck flexor muscles; patients have difficulty lifting their head from the pillow and in a car their head is thrown backwards when their partner accelerates too rapidly.

Myotonia describes delayed muscle relaxation following voluntary contraction or percussion. It is due to repetitive discharge of the muscle fibre membrane and is rarely absent in myotonic dystrophy even though the patient might not be aware of it. If symptomatic, the patient may complain of stiffness of the hands causing difficulty relaxing the grip, or of stiffness of the tongue or jaw when chewing and swallowing. Contrary to some reports, it is not always worse in cold weather and inability to let go after shaking the examiners hand is seen only in severe cases. It is best demonstrated by showing slowness of grip relaxation or by percussion of the thenar eminence or extensor digitorum communis muscle (Fig. 24.5). 'Warm-up' occurs so that the myotonia becomes less evident with repeated contractions. Myotonia diminishes as weakness advances.

Respiratory muscle, including diaphragm, weakness develops in later stages and together with aspiration relating to dysphagia contributes to chest infections which are a common terminal event. Anaesthesia is particularly hazardous, with delayed recovery of spontaneous ventilation. Hypoventilation, with associated clinical features, may occur but excessive day-time sleepiness is usually due to other mechanisms.

Myotonic dystrophy is a multi-system disease with important manifestations outside skeletal muscle (Table 24.12). Foremost amongst these is cardiac involvement (Bushby *et al.* 2003). This takes the form of conduction problems and rhythm disturbances, typically atrial flutter or fibrillation, less commonly ventricular arrhythmias, and is undoubtedly a major cause of sudden death in these patients. The patient should have an annual electrocardiogram and 24 hour electrocardiographic monitoring if new abnormalities appear or symptoms such as pre-syncope, or palpitations develop. In many, but not all, patients the electrocardiographic changes evolve from normality, to first-degree block, to minor intraventricular conduction abnormalities, left anterior hemiblock, and then to higher forms of block. In a few, sudden death occurs even when a recent electrocardiogram

was normal or there is a pacemaker *in situ*. Anaesthesia is particularly hazardous, with the added risk of respiratory failure.

Excessive day-time sleepiness is present in about three-quarters and may be very disabling. It is rarely due to hypoventilation and appears to be a primary cerebral problem. Smooth-muscle involvement is also under-diagnosed. In adults irritable-bowel-like symptoms and constipation are common. Megacolon and pseudo-obstruction occurs rarely.

Cataracts are common. The earliest changes, which develop in early adult life, are of polychromatic dots in the anterior and posterior subcapsular regions. Late-stage cataracts are, at the bedside, indistinguishable from ordinary senile cataracts.

As a group, intelligence quotient tends to be lower than average, but many individuals are of normal intelligence.

The *late-onset* form, associated with only a small CTG trinucleotide expansion, may be asymptomatic. Commonly, the patient develops cataracts at a relatively young age, and myotonic dystrophy should be considered in all patients who develop early cataracts. Myopathic features are absent or limited to mild facial and hand weakness. There may be premature balding.

The *congenital form* is the most dramatic. The absence of myotonia and the fact that the mother may be asymptomatic means that the diagnosis is sometimes missed. This form is associated with a very large CTG trinucleotide expansion and the prognosis is poor (Reardon *et al.* 1993). Many affected foetuses abort spontaneously. Third trimester polyhydramnios is common. Talipes may be present. At birth there is marked hypotonia, respiratory insufficiency, and feeding difficulties. The facial appearance at birth, with a tented-upper lip reflecting muscle weakness, is characteristic. The feeding and respiratory problems settle and the hypotonia gradually resolves. Motor milestones are somewhat delayed and there is invariably mental retardation which eventually results in the need for special schooling. Fecal soiling is often a major problem. Dysarthria is often marked. As the child reaches adolescence the features of the classical adult form start to become evident, including myotonia.

There is also often delay in diagnosing the *childhood-onset form* of the disease unless the condition is already known in the family. By definition, and in contrast to the congenital form, there are no problems in the perinatal period. Motor milestones may be slightly delayed. Problems become more apparent around the start of schooling with evidence of cognitive delay and poor language development. Dysarthria is common. The children complain of fatigue, and slowness of activities is often striking. Although often missed, facial weakness is almost invariable, together with weakness of neck flexion.

Diagnosis. The diagnosis is established on the basis of DNA analysis and no other diagnostic tests are required. The differential diagnosis for the classical adult form is not wide, but includes myotonic dystrophy type 2.

Management. There are many strands to successful management. Genetic counselling not only for the patient but also for at-risk relatives is vitally important. It must be remembered that a female carrying the gene may be asymptomatic but still be at risk of having a child with the severe congenital form of the disease. It is negligent to fail to offer such people counselling. Antenatal diagnosis, based on chorionic villus sampling is readily available.

Regular electrocardiograms are important and the patient must be instructed to report symptoms such as dizzyness, fainting and palpitation. Ambulatory 24 hour electrocardiographic monitoring should be performed if the electrocardiogram changes or symptoms develop. A proportion of patients will eventually require a pacemaker, but the timing of pacemaker insertion, and whether or not implantable defibrillators should be used, is controversial.

Chest infections may be reduced by influenza and pneumococcus vaccinations. Hypoventilation related symptoms are relatively uncommon but must be considered and if appropriate treated with non-invasive ventilation, although the experience of many experts is that this is often poorly tolerated. The patient must be aware of the cardiac and respiratory hazards of anaesthesia and many carry a warning card or wear a bracelet or medallion.

Cataracts are treated surgically when vision is sufficiently impaired. Bowel complaints are treated symptomatically. Excessive daytime sleepiness can be very disabling. Taking a nap after meals and at other chosen times may limit inappropriate periods of sleep. In a small minority it is associated with hypoventilation and fragmentation of sleep, in which case non-invasive ventilation may be of value. Modafinil helps some patients, sometimes dramatically. A trial of treatment is justified in all patients with troublesome sleepiness (Talbot *et al.* 2003).

Myotonia rarely requires treatment. Drugs that are effective, including mexilitine, procainamide, and phenytoin, are theoretically contraindicated because of their potential action on cardiac conduction, although in practice few problems have been encountered. The effects of weakness may be in part ameliorated by appropriate advice from an occupational therapist and physiotherapist and the use of orthoses. Speech therapy may help swallowing problems and the dysarthria which is often a major problem for children with congenital myotonic dystrophy.

24.3.2 Myotonic dystrophy type 2

Whilst there are many apparent clinical similarities between myotonic dystrophy types 1 and 2, there are also enough differences to mean that confusion between the two disorders is unlikely (Day *et al.* 2003). Because the phenotype is rather less distinctive than in myotonic dystrophy type 1, the possibility of myotonic dystrophy type 2 might not be considered after bedside evaluation and one pointer may be the identification of myotonia by electromyography. As in myotonic dystrophy type 1, a suggestive family history may be lacking. The following comments emphasize the differences between the two conditions.

Despite the instability of the mutation, anticipation is less evident and a congenital onset form of myotonic dystrophy type 2 has not been described. In one large series, median age at onset was 48 years (Day *et al.* 2003). Central nervous system involvement is probably absent in type 2, including excessive daytime sleepiness. Bowel involvement, which is common in myotonic dystrophy type 1, appears to be absent in type 2. Cataracts are seen in both. Cardiac involvement is less in myotonic dystrophy type 2, but insulin resistance and frank diabetes more common than in type 1.

The distribution and pattern of myotonia appears to be very similar in myotonic dystrophy types 1 and 2, but overall seems to be rather less striking in type 2, at least at a clinical level. In myotonic dystrophy type 2 myotonia was clinically evident in ~75 per cent of patients, and present on electromyography in ~90 per cent (Day *et al.* 2003). In myotonic dystrophy type 2 the muscle weakness is more marked proximally than distally, whereas in type 1 the distal muscles, particularly finger flexors, are invariably much weaker than proximal muscles. In myotonic dystrophy type 2 the hip flexor and extensor muscles are most prominently involved. Many authors have commented on day-to-day fluctuations of strength in type 2. In addition, proximal muscle, especially thigh, discomfort and pain is frequently reported in DM2. Ventilatory muscle weakness has not been described in myotonic dystrophy type 2.

The diagnosis of myotonic dystrophy type 2 is by *ZNF9* gene mutation analysis. Occasionally, the diagnosis may first be suspected following electromyography showing myotonia or muscle biopsy revealing characteristic features including pyknotic nuclear clumps and denervation-like changes, but neither is necessary to establish the diagnosis.

Management guidelines are similar to those for myotonic dystrophy type 1 but cardiorespiratory problems are less evident and genetic counselling issues less complex.

24.4 Congenital myopathies

Although many of these conditions present early in life with hypotonia, the term congenital is not entirely accurate as some are not evident at birth and some may not present until adulthood. This is yet another area in which genetic and molecular developments are currently leading to reclassification and new understanding (Goebel 2005). The common features of these disorders include frequent presentation as a 'floppy infant', morphological changes such as high-arched palate, long face, and skeletal deformity, generalized muscle slimness and weakness, slow or non-progression, normal or only slightly elevated serum creatine kinase, and specific structural changes within muscle, indeed they are sometimes called ultrastructural myopathies. They may be sporadic, autosomal recessive or dominant, or X-linked, and some individual disorders may show more than one pattern of inheritance. There is an important association between central core disease and malignant hyperthermia.

24.4.1 Central core disease

Presentation is most often in infancy but asymptomatic adult cases are recognized. Inheritance is autosomal dominant and most cases are caused by mutations in the *RYR1* gene encoding the ryanodine receptor. Mutations in the *RYR1* gene are also associated with malignant hyperthermia, multiminicore disease, and core/rod disease

(Robinson *et al.* 2006). The weakness is usually mild and is either generalized, also affecting the face, or affects mainly proximal lower limb muscles. Scoliosis and contractures are occasional features. Respiratory failure is unusual. Progression, if it occurs, is slow. There is a complex association with malignant hyperthermia. About one-third of patients with central core disease are malignant hyperthermia susceptible. Conversely, most patients with malignant hyperthermia have normal muscle histology. The pathological features include type I fibre predominance and centrally placed cores, best seen using the NADH-TR reaction, which run the length of affected fibres.

24.4.2 Nemaline myopathy

The major pathological feature of this condition is the presence of nemaline rods, which are of uncertain constitution in the subsarcolemmal region where they appear red using the Gomori trichrome method. Genetic forms (autosomal dominant or recessive, or sporadic cases) are associated with mutations in *TPM2*, *TPM3*, *ACTA1*, *NEB*, *TNNT1*, encoding respectively for tropomyosin 2 and 3, α-actin, nebulin, and slow troponin T.

Most cases present at birth as floppy infants, but later onset can occur. There is facial and proximal weakness. The dysmorphic features of a long face, high-arched palate and chest deformity reflect intra-uterine and congenital weakness. Respiratory failure, probably indicating diaphragmatic involvement, is relatively common and respiratory function must be monitored throughout life. It can be managed by nocturnal mask positive pressure assisted ventilation. Typically, progression of weakness is very slow.

Nemaline rods may also be seen as a non-specific pathological finding in a variety of late-onset acquired myopathies including bent-spine and dropped head syndrome, hypothyroidism, in association with ventilatory failure, and associated with monoclonal gammopathy (Chahin *et al.* 2005).

24.4.3 Congenital fibre-type disproportion

The characteristic pathological feature is disproportion between the sizes of type I and type II muscle fibres. Normally these are approximately equal in size but in this condition, due to type II fibre hypertrophy, the difference exceeds 25 per cent.

The typical presentation is as a floppy infant. Contractures and congenital dislocation of the hip are common. There is a wide range of severity but marked weakness is uncommon. There is no further progression, and indeed often improvement, after two years of age. Sporadic, X-linked recessive, and autosomal recessive and dominant patterns of inheritance have been reported.

The disorder is clearly genetically heterogeneous. Recently, causative mutations have been identified in *ACTA1*, encoding α-actin, and *SEPN1* encoding Selenoprotein N1 genes. *ACTA1* mutations are also associated with nemaline myopathy, and *SEPN1* mutations with multiminicore disease and rigid spine muscular dystrophy.

24.4.4 Multiminicore disease

Pathologically this disorder is characterized by multiple minicores within muscle fibres. These are areas of myofibrillar disruption and are devoid of mitochondria, so are revealed by oxidative enzyme stains for NADH, COX, and SDH.

This is usually a relatively mild condition with onset of generalized weakness in early infancy with very slow or no progression. As in nemaline myopathy diaphragmatic weakness and respiratory failure can occur. There is clinical overlap with rigid spine muscular dystrophy, and it is now known that these disorders are allelic, and the underlying cause is mutations within the *SEPN1* gene encoding selenoprotein N1. *RYR1* mutations, typically associated with central core disease, can also cause multiminicore disease.

24.4.5 Centronuclear myopathy

This condition is also known as myotubular myopathy because it was thought that the abnormal fibres, with central nuclei, resembled myotubes and that the cause might be arrested maturation. That is now doubted and the term centronuclear myopathy is preferred.

The cause of the severe X-linked form is a mutation in the *MTM1* gene encoding myotubularin. There is often a history of miscarriages and neonatal male deaths. Those males born alive have severe weakness and respiratory insufficiency and most die in early infancy.

Autosomal dominant forms are of lesser severity. Onset is usually in infancy or early childhood, rarely in early adult life. Weakness is predominantly proximal and some have facial weakness, ptosis, and external ophthalmoplegia. Progression is usually slow or absent. There is genetic heterogeneity but recently it has been shown that some cases are due to mutations in the *DNM2* gene, encoding dynamin 2, which can also be associated with Charcot–Marie–Tooth disease (Bitoun *et al.* 2005).

24.5 Non-dystrophic myotonias and periodic paralysis

Many of the conditions to be discussed in this section are caused by an inherited defect of a membrane-bound ion channel. Arguably, they could have been included in a section entitled 'Channelopathies' which, in the neuromuscular field, also might have included malignant hyperthermia and various congenital myasthenic syndromes, as well as the acquired conditions of myasthenia gravis, Lambert–Eaton syndrome, and neuromyotonia. Considering neurology as a whole, other channelopathies include epilepsy, migraine, and certain ataxias. Thus, the channelopathies are a clinically disparate group but represent a newly defined class of disorders which are currently the focus of considerable research activity (Hanna 2006). They may be subdivided on the basis of whether the involved channel is voltage-dependent/gated as for the sodium, calcium, and chloride channelopathies, and indirectly the ryanodine receptor, or ligand-dependent, as for acetylcholine receptor.

The term *non-dystrophic-myotonias* serves to distinguish the conditions accompanied by myotonia described in this section from the myotonic dystrophies (Section 24.3), which in turn are now considered separate from the muscular dystrophies (Section 24.2).

The periodic paralyses are characterized by episodic weakness. Often there is a change in the serum potassium level during an attack, and this led to the original classification of hyper-, hypo-, and normo-kalaemic forms. Primary forms are inherited channelopathies and are now classified upon the basis of the ion channel involved, whereas secondary forms are due to systemic metabolic disturbance. Primary hyperkalaemic periodic paralysis may be accompanied by myotonia.

24.5.1 Myotonia congenita

Although first described by Thomsen, who himself had the disease, as an autosomal dominant disorder, it was shown later by Becker that it is very much more frequently an autosomal recessive condition. Different mutations of the same gene, CLCN1, are responsible for each. The chloride channel has an important role in membrane repolarization.

Clinical features. The recessive and dominant form are very similar, but with the recessive type tending to be more severe. The characteristic feature is generalized myotonia, first evident in childhood, without persistent weakness. The myotonia eases with repetitive contractions, so-called 'warm-up'. In the cranial nerve territory myotonia can cause difficulty chewing, and in early childhood there may be striking slowness of relaxation of the periorbital muscles, looking a little like blepharospasm, when crying. Grip myotonia causes problems similar to those seen in myotonic dystrophy but without the additional problem of weakness. In the lower limbs the stiffness may cause the patient to fall as they start to walk and this is most evident after a period of rest; one patient regularly fell after having been standing-to-attention and then ordered to march, to the irritation of the Sergeant-Major! Transient weakness, lasting several seconds only, may be seen in the recessive form, for example when first making a forceful fist. In some patients there is striking muscle hypertrophy (Fig. 24.18) which represents a form of work-hypertrophy consequent upon the myotonia.

At the bedside, myotonia is easily demonstrated on muscle percussion and as slowness of relaxation of grip. Facial muscle myotonia may be seen as lid-lag, demonstrated by the patient looking upwards for several seconds, and then suddenly down towards the ground. The upper lids lag behind the downward eye movement, or show delayed relaxation following forceful eye-closure (Fig. 24.19).

Diagnosis. Electromyography shows generalized myotonia. Exercise protocols and cooling during electromyography may help to identify the likely genetic cause of the myotonia and direct further investigation (Fournier *et al.* 2006). Serum creatine kinase is normal or mildly elevated. The diagnosis is confirmed by DNA studies, but they are not yet widely available. The differential diagnosis, which does not usually cause many difficulties, includes the myotonic dystrophies and sodium channelopathies.

Management. Many patients cope without treatment. For those that require relief from the myotonia, mexiletine is the drug of choice.

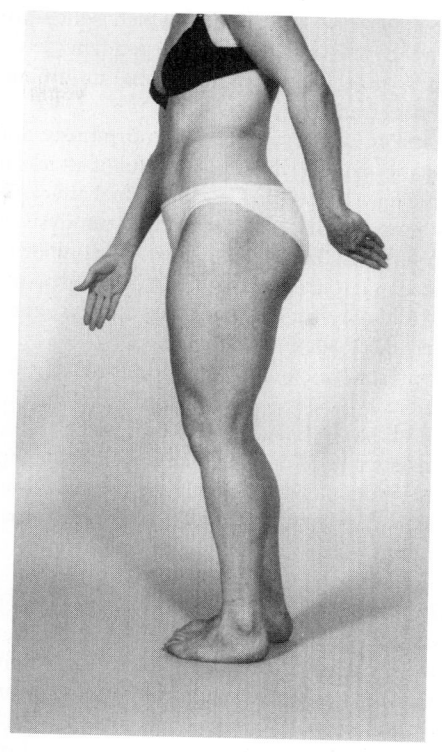

Fig. 24.18 Myotonia congenita. Note the muscle hypertrophy.

24.5.2 **Primary hypokalaemic periodic paralysis**

This form of periodic paralysis is usually caused by one of several mutations affecting the *CACNA1S* gene, which encodes the dihydropyridine receptor. Less commonly it is associated with mutations in the *SCN4A* gene encoding the sodium channel, which more typically cause hyperkalaemic periodic paralysis. Skeletal muscle has two functionally linked types of calcium channel related to the T-tubular/sarcoplasmic reticular system, which are involved in excitation–contraction coupling. The dihydropyridine receptor is an L-type voltage-gated calcium channel located in the T-tubular membrane.

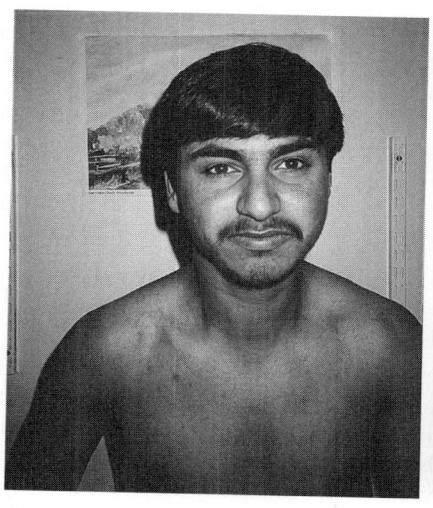 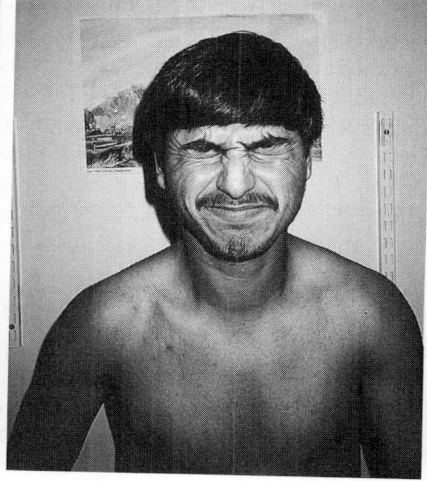 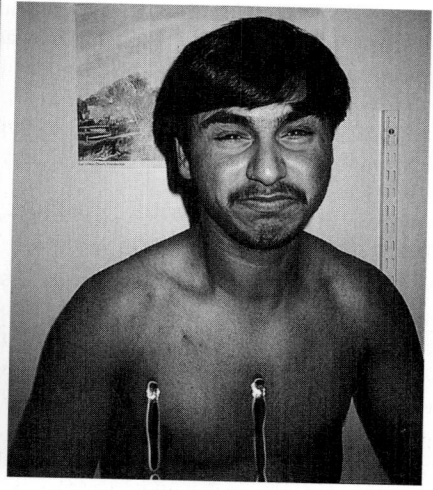

A   B   C

Fig. 24.19 Myotonia congenita. At rest (A), during forceful eye-closure (B), and 5 seconds after relaxing (C).

It is coupled to the ryanodine receptor, which is the second calcium channel and which itself is not voltage dependent.

The disease is inherited as an autosomal dominant condition with reduced penetrance in women.

Clinical features. The severity varies enormously. Some carriers are asymptomatic, some have only a handful of attacks in their lifetime, and some have daily episodes. Individual attacks may involve one or a few muscles, or cause generalized paralysis, but even in these cases, respiratory failure is unusual. The bulbar musculature is usually spared and it is easy to see how somebody complaining of total inability to move, but still able to speak, move their head, and to breathe, may be labelled as hysterical.

Most cases present in childhood or adolescence, a few not until the third decade. Typically, the patient wakes in the morning with the paralysis and the weakness resolves during the course of the day. Rarely, attacks last several days. Precipitants include high carbohydrate food and emotional upset. Gentle exercise may abort an attack.

Many patients develop a slowly progressive proximal myopathy and this is largely independent of the occurrence of paralytic attacks.

Diagnosis. Increasingly, this will become DNA-based. Demonstration of a low serum potassium level during a spontaneous attack is very helpful. Provocative tests are rarely indicated. Exercise-testing during electromyography can be informative and guide further molecular diagnosis (Fournier *et al.* 2004). Muscle biopsy is not part of the routine diagnostic process, but in patients with permanent weakness it may show a vacuolar myopathy.

Management. Patients should be advised to avoid high-carbohydrate meals and excessive physical exercise. Infrequent mild attacks may not require treatment. More troublesome attacks usually respond to oral potassium chloride, up to 10 g, given as an unsweetened aqueous solution. For prevention of frequent attacks, acetazolamide is the drug of choice. Low doses, 125 mg on alternate days, may be sufficient. There is no specific treatment for the permanent myopathy.

24.5.3 Thyrotoxic periodic paralysis

This is an important differential diagnosis from primary hypokalaemic periodic paralysis. It is sporadic, is commoner in males, and the majority of cases occur in Orientals (Kung 2006). Features of thyrotoxicosis may be absent. The clinical and biochemical features, and precipitating factors to paralytic attacks, are indistinguishable from those associated with primary hypokalaemic periodic paralysis, with the exception that the onset is typically in adult life. In a male over the age of 21 years developing periodic paralysis, the thyrotoxic form is far more likely than the primary form.

24.5.4 Primary hyperkalaemic periodic paralysis and related disorders

Three different conditions are known to be associated with mutations affecting the skeletal muscle sodium channel gene, *SCN4A*; hyperkalaemic periodic paralysis, paramyotonia congenita, and potassium-aggravated myotonia. Different clinical expressions may be seen within members of the same family. Inheritance is autosomal dominant, usually with full penetrance.

Hyperkalaemic periodic paralysis

The clinical presentation is similar to the hypokalaemic, or calcium channel, form with episodic weakness in childhood and attacks often occurring before breakfast. They are often fairly brief and may be accompanied by muscle aching. Precipitants include cold, fasting, rest after exercise, emotional stress, pregnancy, alcohol, and potassium-loading. Patients may find out for themselves that intake of a carbohydrate, such as a sweet drink, may abort an attack. The attacks tend to reduce in frequency in later life but, independent of acute episodes, a progressive proximal myopathy may develop.

Emphasizing the relationship to the other two conditions, myotonia may be evident clinically, most often affecting the facial musculature, or electromyographically, and there may be features of paramyotonia.

The diagnostic approach is similar to that for the hypokalaemic form. Electromyography may be helpful and, increasingly DNA-based diagnosis is available. An attack can be precipitated by potassium-loading but this is not without hazard and should not be part of routine assessment. The form previously classified as normokalaemic, on the basis of no change in serum potassium level during an attack, is now known to be caused by sodium-channel gene mutations.

Treatment consists of avoiding precipitating factors, frequent carbohydrate-rich meals and drug therapy if the attacks are frequent and disabling. Low-dose thiazide diuretics are first choice but acetazolamide also works, although with greater risk of side-effects. Inhalation of a β-adrenergic stimulant, such as salbutamol, may abort attacks.

Paramyotonia congenita

The overlap with hyperkalaemic periodic paralysis was noted above. The characteristic feature of this condition is paramyotonia, which describes myotonia precipitated and exacerbated by exercise and cold. In other forms of myotonia repeated use of a muscle leads to reduction in myotonia, the 'warm-up' phenomenon, but in this condition the reverse applies. The facial and distal upper limb muscles are most affected. There is marked cold-sensitivity, which not only exacerbates the myotonia but may also induce weakness which can last for hours. Onset is in early infancy. Later, some patients develop paralytic attacks.

Diagnosis is based on the clinical picture, demonstration of cold-exacerbated myotonia, and DNA studies. Treatment is often not required, but mexiletine is effective.

Potassium-aggravated myotonia

This is currently the preferred term for disorders previously labelled as myotonia fluctuans, myotonia permanens, and acetazolamide-responsive myotonia. Features include fluctuating myotonia, sometimes severe, which is exacerbated by exercise and, as the name implies, potassium loading, but not by cold, and the absence of weakness. New mutations, affecting the sodium channel gene, appear to be common. Myotonia congenita is an important differential diagnosis.

24.5.5 Andersen–Tawil syndrome

The three major clinical elements are potassium-sensitive periodic paralysis, the cardiac dysrhythmia of bidirectional ventricular tachycardia, and dysmorphic features affecting face and hands. Partial manifestations, including long-QT syndrome, are common and the full phenotypic range is yet to be defined (Davies *et al.* 2005). The molecular basis is mutation in the *KCNJ2* gene encoding the inward rectifier potassium channel Kir2.1.

24.5.6 **Secondary periodic paralyses**

Any disease or drug which changes the total body potassium level can induce weakness, which is usually persistent but may be periodic. Weakness is common if the serum potassium level drops below 2.5 mmol/l. In severe cases the respiratory muscles may be involved and rhabdomyolysis can occur. Generalized weakness may develop with serum levels above 6 mmol/l, but is rarely severe and the main risk is cardiac dysfunction. Weakness responds rapidly to correction of the serum potassium level.

24.5.7 **Schwartz–Jampel syndrome**

The major clinical features of this rare autosomal recessive syndrome are muscle stiffness, particularly affecting the face and giving a characteristic appearance with blepharophimosis and a tight appearance around the mouth, short stature, and chondrodysplasia. Although the muscle stiffness is often stated to be myotonia, there remains debate as to the exact underlying neurophysiological basis. It is caused by mutations in the *HSPG2* gene, encoding the basement membrane protein perlecan (Stum *et al.* 2006).

24.6 **Primary metabolic myopathies**

This section is concerned with primary, genetically determined, metabolic disorders of skeletal muscle. Many of these are multisystem disorders, reflecting the widespread distribution of some of the enzymes involved, whereas others are confined to skeletal muscle. In many, the basic problem is disruption of normal energy-generating processes. In some, the accumulation of lipid and glycogen may disrupt muscle-fibre function. A brief overview of normal energetic processes is followed by discussion of individual disorders. In everyday clinical practice the mitochondrial cytopathies are the most frequently encountered of the metabolic myopathies.

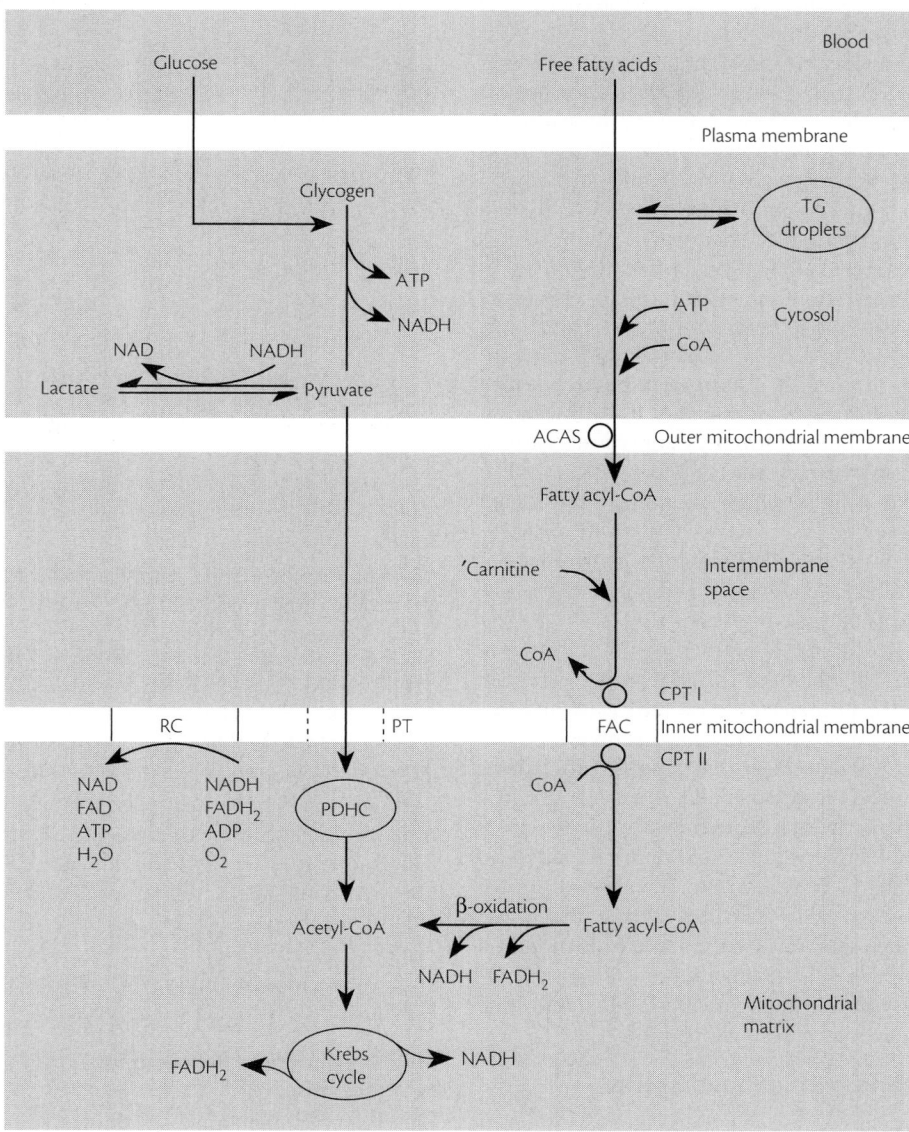

Fig. 24.20 Major metabolic pathways in skeletal muscle. ACAS, acyl-CoA synthetase; ADP, adenosine diphosphate; ATP, adenosine riphosphate; CoA, coenzyme A; CPT, carnitine palmitoyltransferase; FAC, fatty acyl carnitine; FAD, flavine adenine dinucleotide; FADH$_2$, reduced FAD; mm, mitochondrial membrane; NAD, nicotinamide adenine dinucleotide; NADH, reduced NAD; PDH, pyruvate dehydrogenase complex; PT, pyruvate translocase; RC, respiratory chain; TG, triglyceride. (Reproduced from The Oxford Textbook of Medicine, Oxford University Press.)

Normal energetic processes

Although individual pathways of energy metabolism, and the relationships between them, are complex, they can be simplified to allow ready understanding of the metabolic myopathies (Fig. 24.20). Adenosine triphosphate, ATP, is the major immediate energy source. It provides the energy for essential housekeeping metabolic functions as well as providing energy for contraction. Resting muscle uses very little energy, but vigorous exercise can increase ATP utilization by more than two orders of magnitude. Muscle is well adapted to meet this demand. ATP can be re-synthesized through three major pathways; the creatine kinase reaction, glycolysis, and oxidative phosphorylation (Fig. 24.20). The relative contribution from each depends upon the state of nutrition and, more importantly, the level and duration of exercise.

At rest the major fuel source is circulating free fatty acids. These are transported into mitochondria, across the inner mitochondrial membrane by the carnitine transport system, into the matrix, and then they undergo β-oxidation. The latter generates the energy-rich electron carriers of reduced nicotinamide adenine dinucleotide, NADH, and reduced flavine adenine dinucleotide, FADH$_2$. Transfer of their electrons to molecular oxygen in the electron transport respiratory chain releases energy, via the pumping of protons, for the generation of ATP, the process being known as oxidative phosphorylation. Circulating glucose entering muscle may undergo glycolysis, with the product pyruvate then being further metabolized within mitochondria, but the contribution to energy generation is small and much of the glucose is stored as glycogen for future use.

During early strenuous exercise the oxidative pathways cannot cope with the demand for ATP generation because the resting blood flow provides inadequate delivery of oxygen and substrate. The creatine kinase reaction provides a short-term buffer but then glycogenolysis becomes critical. This produces ATP, but much less efficiently than oxidative metabolism, NADH, and pyruvate. Because of the relative lack of oxygen pyruvate cannot undergo further oxidative metabolism. The increasing NADH/NAD ratio would inhibit glycolysis, which is prevented by the reduction of pyruvate to lactate. This explains the lactic acidosis seen in disorders of oxidative metabolism such as the mitochondrial cytopathies, and the failure to generate lactate in disorders of glycogenolysis and glycolysis.

As exercise continues, muscle blood flow increases, the respiratory rate rises, and fatty acids are mobilized from fat stores. Glycogen stores are depleted. Thus, in sustained exercise, fatty acids and oxidative metabolism again become the main energy source.

Observations in clinical practice bear out these simplifications. Disorders of glycogen and glucose metabolism typically produce symptoms in early, particularly intense, exercise, but if activity is sustained a 'second-wind' develops due to the return to fatty acid metabolism. Defective fatty acid metabolism, insufficient to produce symptoms at rest, is exposed by sustained exercise and fasting which limits carbohydrate metabolism in muscle. Mitochondrial disorders affecting respiratory chain function are often symptomatic at rest, and multisystem in their manifestations, reflecting the critical central role of oxidative phosphorylation.

24.6.1 Disorders of carbohydrate metabolism

The major steps in glycogenolysis are shown in Fig. 24.21 together with those enzymes known to be associated with metabolic myopathies. Acid maltase deficiency differs from the other glycogenoses in that it involves a pathway not directly involved in energy generation.

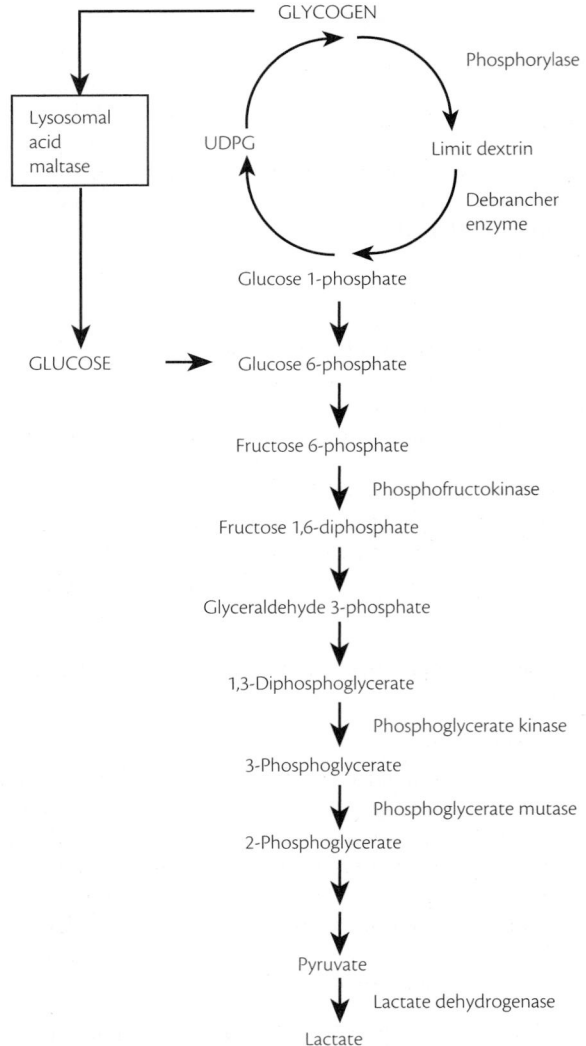

Fig. 24.21 Glycogenloysis and glycolysis. (Reproduced from The Oxford Textbook of Medicine, Oxford University Press.)

These are autosomal recessive disorders, with the exception of phosphoglycerate kinase deficiency which is X-linked. Although the genetic basis of many of these conditions has been identified, routine DNA-based diagnosis is not yet readily available, nor is it necessary in everyday practice. Diagnosis is based on metabolic investigations. The serum creatine kinase level is usually elevated. In many, forearm exercise is associated with impaired venous lactate production. This information, and more, can be obtained by phosphorus magnetic resonance spectroscopy studies, but these are not widely available. In some cases enzyme deficiency may be demonstrated by histochemical methods applied to a muscle biopsy specimen. Muscle biopsy typically shows glycogen accumulation. Diagnosis is confirmed by enzyme assay.

Myophosphorylase deficiency or Type V glycogenosis, McArdle's disease

The classical presentation is with exercise intolerance in childhood. Exercise induces pain and stiffness. More extreme activity may produce electrically silent cramps and myoglobinuria. Patients often discover the 'second-wind' phenomenon for themselves.

Progressive proximal weakness often develops in middle age and may be the mode of presentation in later-onset cases, who in retrospect often have a history of mild exercise intolerance. Myophosphorylase is a muscle-specific enzyme and so clinical manifestations are restricted to skeletal muscle.

Debranching enzyme deficiency or Type III, Cori-Forbes disease

The commonest form presents in early infancy with hypoglycaemia, seizures, failure to thrive, and hepatomegaly. Myopathic features may be slight. During childhood the liver problems tend to settle but the myopathy progresses and in adult life a significant cardiomyopathy may develop. Less commonly presentation is in adulthood with a progressive proximal, less frequently distal, myopathy. A history of a protuberant abdomen in childhood and mild exercise intolerance may be obtained. Overall, exercise intolerance is less prominent than in phosphorylase deficiency.

Phosphofructokinase deficiency or Type VII, Tarui's disease

This is much rarer than the conditions described above, with less than 50 cases having been described. The myopathic features are similar to myophosphorylase deficiency. Most patients have laboratory evidence of increased haemolysis, reflected by increased bilirubin and raised reticulocyte count, hyperuricaemia, although the clinical accompaniments of jaundice, gallstones, and gout are uncommon.

Other glycogenoses

Deficiencies of phosphoglycerate kinase, phosphoglycerate mutase, and lactate dehydrogenase cause exercise intolerance with myalgia and myoglobinuria. Only a handful of cases have been described. Branching enzyme is required for glycogen synthesis. Deficiency usually causes progressive hepatosplenomegaly, with death by 4 years if untreated by transplantation, and over one-half have a myopathy. Very rare presentations with cardiac and/or skeletal myopathy, but without liver disease, have been described.

Acid maltase deficiency or Type II, Pompe's disease

This condition differs in many ways from the other glycogenoses. The functional role of the enzyme is not clear but it is not apparently involved in energy-generating pathways and deficiency is not associated with exercise intolerance. Three main clinical forms are recognized; infantile, childhood, and adult. The severity relates to the amount of residual enzyme activity. The infantile form was the type described by Pompe and is a multisystem disorder causing progressive generalized weakness, failure to thrive, organomegaly including tongue enlargement, and death by two years due to cardiac or respiratory failure. The childhood form presents later, but by age 15 years. There is a progressive proximal myopathy and later cardiomyopathy. Death occurs in the second or third decade from cardiorespiratory failure.

The adult form, which may not present until the seventh decade, is usually restricted to skeletal muscle. It usually presents as a proximal myopathy, clinically indistinguishable from many other causes of 'limb–girdle syndrome'. Diaphragmatic involvement is common and up to one-third of patients present with respiratory failure, although in retrospect there may have been symptoms of limb weakness, which is usually evident on examination. With nocturnal ventilatory support many patients can lead active lives for many years before progressive limb weakness causes increasing disability. Muscle biopsy shows glycogen accumulation in vacuoles of lysosomal origin. Diagnosis is confirmed by enzyme assay, either on the muscle biopsy specimen or cultured fibroblasts. Lymphocytes also show glycogen storage and an air-dried blood film treated with periodic acid Schiff reagent provides a quick and reliable diagnostic test.

Enzyme replacement therapy has been shown to be effective in the severe infantile form (Klinge *et al.* 2005). Anecdotal reports suggest some benefit in adult-onset cases but the results of a formal trial are still awaited.

24.6.2 Disorders of lipid metabolism

These rare conditions are sometimes referred to as lipid storage myopathies but this term obscures the fact that in many cases lipid accumulation, as shown by fat stains applied to muscle biopsy samples, is not evident. Multisystem features may be more striking than myopathy. Fatty acids entering muscle are activated by acyl-CoA synthetase at the outer mitochondrial membrane (Fig. 24.20). The fatty acyl-CoA is converted to an acyl-carnitine, which is then transported across the inner mitochondrial membrane by the carnitine-dependent transporter system involving carnitine palmitoyltransferases I and II. In the mitochondrial matrix fatty acyl-CoA is reformed and then undergoes β-oxidation, as described above. Lipid disorders may involve the carnitine system or β-oxidation.

The importance of fatty acid oxidation to muscle energy metabolism has already been discussed. In addition, fatty acids are the substrate for hepatic ketogenesis; ketone bodies are a major auxiliary fuel source for the central nervous system. Further, fatty acids are required for hepatic gluconeogenesis. These latter two observations explain the multi-organ involvement that may occur with these disorders.

Carnitine deficiency

Secondary carnitine deficiency is common. Its causes include many defects of intermediary metabolism, notably defects of fatty acid β-oxidation, mitochondrial cytopathies, and haemodialysis. Primary carnitine deficiency is a rare autosomal recessive disorder presenting in early life with hypoketotic hypoglycemia, or later with skeletal myopathy or cardiomyopathy. It is due to mutations in the *SLC22A5* gene, encoding the carnitine transporter OCTN2. The disease responds to dietary carnitine supplementation.

Carnitine palmitoyltransferase deficiency

Although rare, this is the commonest disorder of fatty acid metabolism causing a myopathy. It presents with myalgia, rhabdomyolysis, and myoglobinuria, precipitated by long-sustained exercise, classically on an army route march or by similar exercise, or by fasting. During an attack muscle biopsy may show some lipid accumulation but often does not. Screening is by tandem mass spectrometry on a blood sample taken after an overnight fast. Specific diagnosis is by enzyme assay, usually on cultured fibroblasts. Numerous mutations in the *CPT II* gene have been identified (Isackson *et al.* 2006).

β-oxidation disorders

The investigation of these conditions is complex (Tyni *et al.* 2002). The commonest is medium-chain acyl-CoA dehydrogenase deficiency and this usually presents in childhood with Reye-like episodes involving encephalopathy and hypoketotic hypoglycaemia, and sudden death. Rare cases present later with recurrent rhabdomyolysis. A common point mutation allows DNA diagnosis in about 90 per cent of patients. Many other disorders of β-oxidation have been described but all are rare. They usually present in infancy with multisystem disease but myopathy may predominate. Correct diagnosis is important in determining therapy. Fasting must be avoided. Diet supplementation with medium-chain fatty acids may help in cases where the metabolic block affects enzymes responsible for long-chain fatty acid metabolism. Carnitine deficiency can be corrected by dietary supplementation.

24.6.3 Mitochondrial cytopathies

Although it was myopathy that first led to the identification of this disease group, they are now recognized to be multisytem disorders affecting not only the central and peripheral nervous system but also every other organ system in the body (Section 10.5). They were initially defined at a morphological level; light and electron microscopy demonstrated accumulations of mitochondria which often showed structural abnormalities (Fig. 24.22). We now know that morphological changes such as ragged-red fibres are not always present, or are more subtle, such as some fibres showing deficient cytochrome oxidase staining. The next stage involved isolation of mitochondria from such patients and investigation of metabolic pathways, particularly the respiratory chain. It soon became clear that there was a poor correlation between the phenotype and identification of a particular metabolic defect. In 1988 it was shown that some cases were associated with an abnormality of the mitochondrial genome and subsequent studies have shown a broad correlation between genotype and phenotype. Recently, the importance of nuclear DNA mutations causing autosomal patterns of inheritance have been more fully appreciated (DiMauro 2006). Within this somewhat recherché field there is continuing debate as to whether mitochondrial disorders should be lumped together or split into specific syndromes; in practice both approaches can be useful. This section will consider basic aspects of mitochondrial

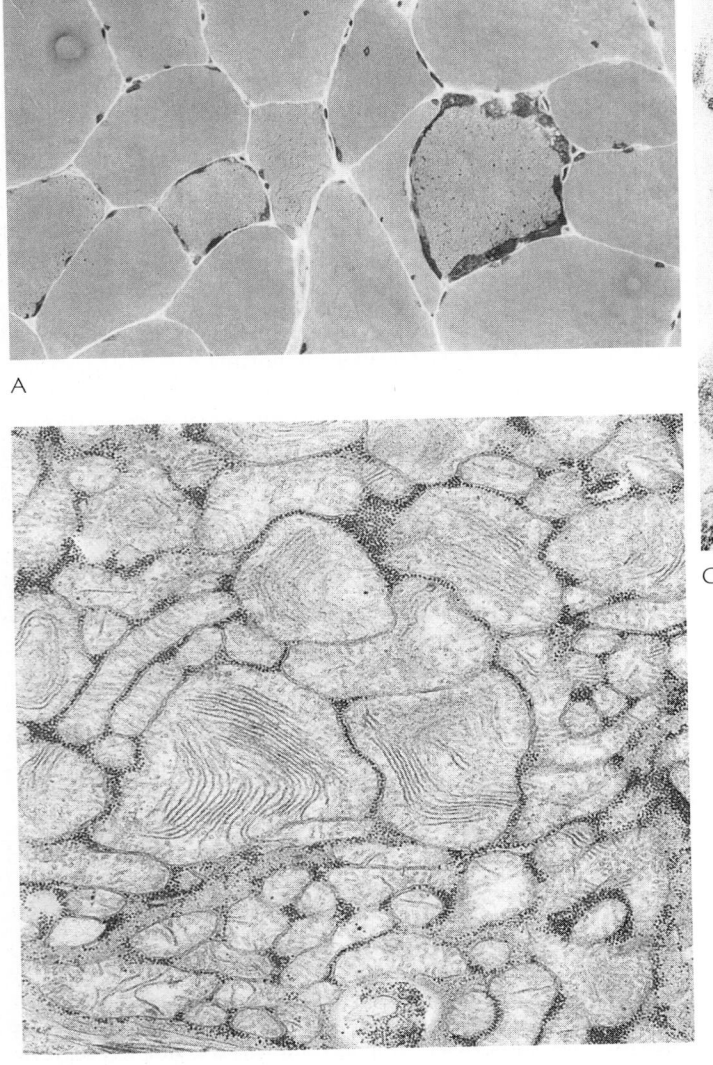

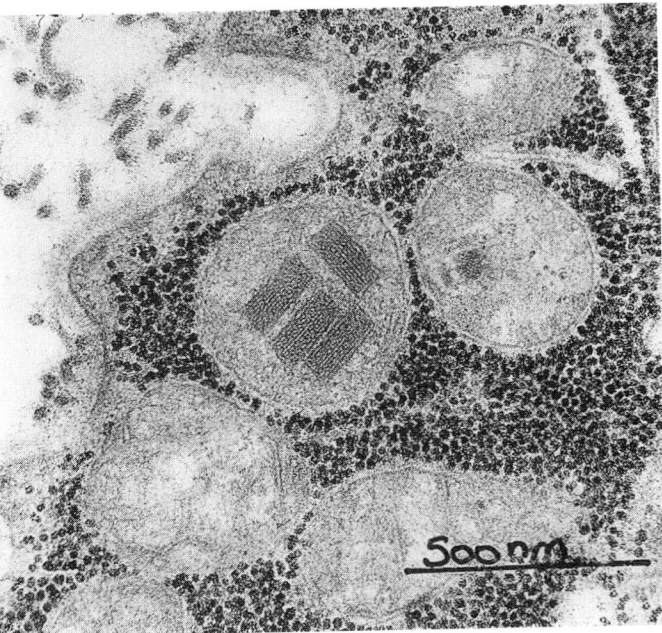

Fig. 24.22 Mitochondrial cytopathy. Modified Gomori trichrome stain showing ragged-red fibres (A). Electron microscopy—morphologically abnormal mitochondria (B). Electron microscopy—paracrystalline inclusions (C).

biology, genetics, clinical features, characteristic syndromes, diagnosis, and treatment. It will only deal with conditions in which muscle involvement is usually present and will thus exclude conditions such as Leber's hereditary optic neuropathy and non-neurological disorders such as Pearson syndrome (Section 10).

Mitochondrial biology

Many of the biochemical functions of mitochondria have been discussed above. The respiratory chain, vital to oxidative phosphorylation, consists of five complexes. Each of these is composed of many subunits. Most of these are encoded by nuclear DNA but some of the subunits in four of the five complexes are encoded by mitochondrial DNA.

Mitochondrial DNA (mtDNA) is a 16.5 kb circular double strand of DNA. A number of properties relating to mtDNA are important when considering mitochondrial diseases:

* It is exclusively maternally inherited.

* It encodes some of the subunits of four of the five respiratory chain complexes, 13 subunits in total, as well as 22 tRNAs and two ribosomal RNAs.

* Each mitochondrion contains several copies of mtDNA and normally these are all the same known as homoplasmy.

* In diseases with mtDNA mutations there is a mixture of both normal wild-type and mutated mtDNA; heteroplasmy.

* There is a threshold effect so symptoms are only manifest if the mutant:wild-type ratio is above a certain level. The critical ratio may vary from organ to organ depending upon the metabolic demands of that tissue.

* Nuclear genes influence mitochondrial biogenesis and thus nuclear genetic abnormalities, which will be autosomally inherited, may cause secondary mtDNA abnormalities.

Genetics

An understanding of this is vital, particularly with respect to genetic counselling. Many cases are sporadic, some show classical maternal inheritance, and some show an autosomal pattern of inheritance. Different types of mtDNA mutation tend to be associated with particular inheritance patterns. Mutation types include *major rearrangements* with deletions, typically a length of about 5 kb is deleted or, less commonly, duplications of a stretch of mtDNA. In most such cases there is a single type of mutation. There will thus be heteroplasmy with two populations: wild-type and mutant. Patients with these types of mutation are generally sporadic and their offspring at little risk. Clinically, such patients most frequently have chronic progressive external ophthalmoplegia.

The next commonest type of mutation is a *point mutation* in mtDNA, typically affecting a tRNA. Such mutations are maternally inherited and are often associated with multisystem disease, particularly with central nervous system involvement.

Some patients show *variable deletions*, meaning that there are several populations of mutated mtDNA each with a different-sized deletion. The usual clinical presentation is with chronic progressive external ophthalmoplegia, and inheritance is autosomal dominant, less commonly recessive. With *mitochondrial depletion* there is a reduced amount of wild-type mtDNA. Inheritance is autosomal dominant or recessive. Many nuclear genes associated with these autosomally inherited mitochondrial cytopathies have been identified including *POLG, ANT1* and *C10orf2* or *Twinkle* (Naimi *et al.* 2006).

Table 24.13 Organ involvement in mitochondrial cytopathies

Organ	Involvement
Muscle	Chronic external ophthalmoplegia
	Proximal myopathy
	Exercise intolerance
Eyes	Pigmentary retinopathy
	Optic atrophy
	Cataracts
Brain	Encephalopathy
	Stroke-like episodes
	Epilepsy
	Dementia
	Extra-pyramidal/movement disorders
	Ataxia
	Deafness
	Leigh's syndrome
	Raised CSF protein
Peripheral nerve	Neuropathy; typically sub-clinical and axonal
Heart	Cardiac conduction problems
	Cardiomyopathy
Gut	Hypomotility/pseudo-obstruction
Liver	Failure
Endocrine	Diabetes mellitus
	Hypoparathyroidism
	Short stature
Kidney	Fanconi syndrome
Blood	Sideroblastic anaemia
	Folate deficiency
Skin	Multiple lipomatosis

Clinical features

The principal clinical features of the mitochondrial cytopathies are shown in Table 24.13. Such a list makes it clear why these disorders are so often considered in differential diagnosis. The most commonly encountered and significant clinical features include chronic progressive external ophthalmoplegia, asymptomatic retinal pigmentation, proximal myopathy, epilepsy, encephalopathy, deafness, ataxia, cardiac conduction abnormalities, short stature, and diabetes. A number of characteristic syndromes are discussed below but it must be remembered that presentation of mitochondrial cytopathy may be with any one, or combination, of the features listed in Table 24.13.

Characteristic syndromes

There are three relatively stereotyped syndromes which present little diagnostic difficulty.

Chronic progressive external ophthalmoplegia. This is the most frequently encountered phenotype and is seen at all ages. Ptosis and progressive limitation of eye movements develop over many years (Fig. 24.23). Diplopia is uncommon. There is often an asymptomatic peripheral pigmentary retinopathy. In adult life it may occur in isolation, with deafness, with proximal limb weakness, and with diabetes. In childhood the combination of chronic progressive

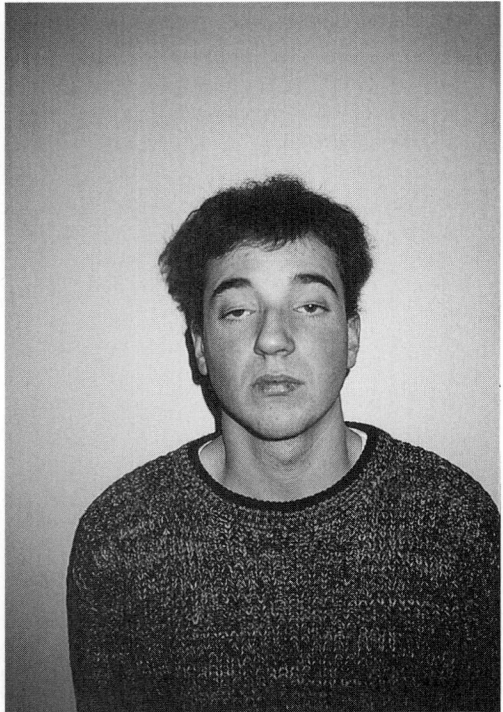

A

B

Fig. 24.23 Mitochondrial chronic progressive external ophthalmoplegia. Note the ptosis (A), and failure of eye movement on attempting to look fully to the right (B).

external ophthalmoplegia with, variably, retinopathy, heart block, ataxia and elevated CSF protein is referred to as Kearns–Sayre syndrome. Some use this eponym loosely to describe all cases of chronic progressive external ophthalmoplegia. Many cases are sporadic and muscle mtDNA shows a single deletion. Autosomal dominant, and less frequently recessive, chronic progressive external ophthalmoplegia is associated with multiple mtDNA deletions and nuclear gene defects as described above. The important differential diagnoses in adult life include myasthenia gravis, oculopharyngeal muscular dystrophy, and thyroid ophthalmopathy.

MELAS syndrome. This acronym is derived from *m*itochondrial myopathy, *e*ncephalopathy, *l*actic *a*cidosis, and *s*troke-like episodes. It is most frequently related to a point mutation, at base point *3243—tRNA*$^{Leu(UUR)}$A→G, but the same phenotype may be seen with other point mutations, and conversely the 3243 mutation may cause other phenotypes including chronic progressive external opthalmoplegia. Presentation is in childhood with stroke-like episodes, often affecting the occipital cortex. Initially recovery from such episodes may be good but with time major deficits develop. Epilepsy and encephalopathic episodes are common. Exercise may precipitate lactic acidosis with associated systemic upset. Other clinical features may be present, including short stature and deafness. MELAS is inherited maternally but penetrance appears to be low and it is relatively uncommon to see two people in the same family with the full syndrome, but restricted features in many family members, such as deafness and diabetes, are common.

MERRF syndrome. *M*yoclonus *e*pilepsy develops in late-adolescence or early adult life and later may be accompanied by generalized tonic-clonic convulsions. Dementia and ataxia are common associated features. Muscle biopsy shows *r*agged *r*ed *f*ibres but symptomatic myopathy is unusual. It is most frequently associated with the *8344-tRNA*^{Lys} A→G mutation, but as with MELAS there is genetic and phenotypic heterogeneity. Inheritance is maternal but also as with MELAS many individuals carrying the mutation are asymptomatic or oligosymptomatic.

Diagnosis

Many laboratories now offer mtDNA analysis for major rearrangements involving deletions and duplications for the commoner point mutations, and this may be all that is required if the phenotype is one of the three characteristic syndromes described above. At present, very few laboratories offer a nuclear gene diagnostic service. A practical point to note is that point mutations can be identified in a blood sample whereas deletions require a sample of muscle. In other situations investigation is complex and requires specialist facilities including exercise testing, magnetic resonance spectroscopy, and biochemical studies on extracted mitochondria.

Treatment

This is very much in its infancy and arguably no specific therapies yet exist (Smith *et al.* 2004). Anecdotal reports of benefit from vitamins, co-factors, and co-enzyme Q have not been substantiated by formal trials. Prednisolone seems to help some children with MELAS. Exercise regimes may help. Careful attention must be paid to potentially treatable complications such as heart involvement, diabetes, and epilepsy.

24.6.4 **Myoglobinuria and rhabdomyolysis**

Myoglobin is a haem protein which transfers oxygen from the muscle fibre membrane to mitochondria. Membrane damage leads to release of the protein into the bloodstream, myoglobinaemia and excretion into the urine, myoglobinuria, causing darkening of the urine from pale brown to black. Myoglobinuria is accompanied by massive elevation of the serum creatine kinase level. The principal danger is acute tubular necrosis, and forced alkaline diuresis is often advised to reduce the risk of this, although it is of unproven benefit. Some of the principal causes of myoglobinuria are listed in Table 24.4.

24.6.5 Malignant hyperthermia

The relationship between malignant hyperthermia and central core disease was mentioned in Section 24.4.1. The central event in this condition is disturbed calcium homoeostasis in the sarcoplasmic reticulum; sudden influx of calcium causes hypermetabolism and muscle contracture. Attacks are triggered by succinylcholine and, more potently, volatile inhalational anaesthetic agents, notably halothane. The incidence is about 1 in 50 000 anaesthetics, and is probably the commonest cause of death during anaesthesia. The mortality rate is greatly reduced by prompt recognition and treatment with intravenous dantrolene. Affected individuals may show persistent elevation of the serum creatine kinase and malignant hyperthermia is one cause of idiopathic hyperthermia.

The clinical features are of rigidity, which may be localized typically to the masseter, or generalized, accompanied by rapidly increasing body temperature and tachycardia. Metabolically there is acidosis, elevation of the serum creatine kinase, and myoglobinuria.

Whether or not associated with central core disease, malignant hyperthermia is inherited as an autosomal dominant disorder. There is clear evidence of genetic heterogeneity. In pigs the equivalent condition, which is inherited as an autosomal recessive, is invariably related to a mutation in the ryanodine receptor gene *RYR1*. In man, many patients have *RYR1* mutations but some families have been linked to other loci, including *CACNAIS*. The *RYR1* mutations are numerous and spread throughout the gene. DNA-based diagnosis is not yet feasible in routine practice.

Susceptible individuals may be assessed by *in vitro* contracture testing; a muscle biopsy sample is exposed to caffeine, halothane, and ryanodine. Malignant hyperthermia-susceptible individuals show a lower contractile threshold. False-positive and false-negative results may occur. Many clinicians do not use this test but simply advise all at-risk individuals to assume that they are susceptible and to carry documentation/pendants/bracelets to that effect.

24.7 Inflammatory myopathies

Inflammatory myopathies are defined by the presence of inflammatory infiltrates within muscle. There are many unrelated causes and there is no entirely satisfactory classification (Table 24.14). The commonest are the idiopathic inflammatory myopathies. Dermatomyositis and polymyositis are particularly important, partly because they can mimic many other muscle disorders but mainly because they are treatable. Inclusion body myositis is the commonest acquired myopathy in older patients. Despite the presence of inflammatory infiltrates it is doubtful that it is a true inflammatory myopathy, and unlike the other two conditions mentioned it rarely responds significantly to immunosuppression. However, it is conveniently retained in this category because of clinical similarities with dermatomyositis and polymyositis. The incidence of dermatomyositis and polymyositis is roughly similar and in the order of 1/100,000. Pure polymyositis is rare, but myositis associated with connective tissue disease is relatively common and the myositis is not always clinically significant. This section will concentrate almost exclusively on the idiopathic inflammatory myopathies. Until fairly recently, dermatomyositis and polymyositis were assumed to have a similar pathogenetic basis with the main clinical difference being the presence or absence of skin involvement. There

Table 24.14 The inflammatory myopathies

Idiopathic
- Dermatomyositis
- Polymyositis
- Inclusion body myositis

Associated with collagen vascular diseases
- Systemic lupus erythematosus
- Mixed connective tissue disease
- Scleroderma
- Sjögren's syndrome
- Rheumatoid arthritis

Infective
- Viral: Coxsackie, Epstein–Barr, adenovirus, influenza, HIV, HTLV I
- Parasitic
- Bacterial
- Fungal

Miscellaneous
- Eosinophilic myositis
- Associated with vasculitis
- Granulomatous
- Orbital myositis
- Graft v host disease

is now overwhelming evidence that despite their clinical similarities they have different pathogeneses and a few specific clinical differences (Dalakas 2004). Their immunopathogenesis is reviewed briefly below. As yet, this information has not translated into specific immunotherapies. Although inclusion body myositis shares pathological similarities with polymyositis it is probable that these are secondary to some as yet undefined primary nuclear degenerative process.

24.7.1 Immunopathogenesis

Accumulated evidence indicates that dermatomyositis is a humorally mediated autoimmune disorder. Complement-dependent attack leads to destruction of capillaries, in muscle and skin. In muscle the resulting microangiopathy leads to the characteristic pathological features of infarction and perifascicular atrophy (Fig. 24.24). Whether it is deposition of circulating immune complexes or the binding of an antibody to an endothelial antigen which triggers the lytic complement pathway is unknown. Immunocytochemical studies, which now form part of the routine assessment of muscle biopsies from patients with these disorders, show deposition of the complement C5b-9 membrane attack complex in capillaries. Inflammatory infiltrates are predominantly perivascular, with a predominance of B-lymphocytes over T-lymphocytes, and a high CD4/CD8 ratio.

Conversely, polymyositis relates to cell-mediated immunity. A characteristic feature is partial invasion of non-necrotic muscle fibres by CD8+ cytotoxic T-cells. Inflammatory infiltrates tend to be within fascicles (Fig. 24.25), and T-cells predominate over B-cells. Invaded and non-invaded fibres show major histocompatability complex (MHC) class I expression.

Dermatomyositis and polymyositis are associated with a number of autoantibodies, the pathological significance of which remains uncertain. Some are particularly seen in so-called overlap cases in which myositis coexists with a connective tissue disorder such as Sjögren's syndrome, scleroderma, systemic lupus erythematosus,

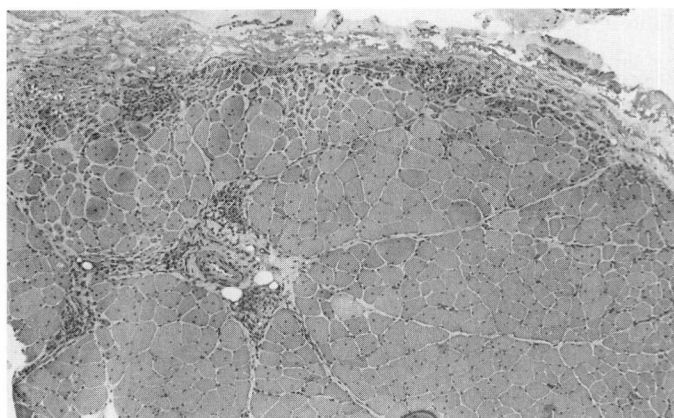

Fig. 24.24 Dermatomyositis. Note the perifascicular atrophy and perivascular inflammation. (Courtesy of Dr Waney Squier.)

CREST syndrome, and mixed connective tissue disease. A number of myositis-specific antibodies have been described, one of which, anti-Jo, correlates strongly with the presence of interstitial lung disease. Anti-Jo and other antisynthetase antibodies are also associated with arthritis, Raynaud's, and 'mechanic's hands' with thickening and cracking of the skin of the hands and fingers. Apart from anti-Jo, these antibodies are not yet used widely in clinical practice.

The immunopathological features of inclusion body myositis are very similar to those of polymyositis but additional strands of evidence suggest that these may be secondary. One view is that the primary disease process leads to destruction of myonuclei. Accumulation of abnormal proteins similar to those seen in Alzheimer's disease, such as amyloid-β precursor protein, has led to the suggestion that inclusion body myositis is primarily a myodegenerative disease (Askanas and Engel 2006).

24.7.2 Dermatomyositis and polymyositis

There is considerable clinical overlap between these conditions, and between polymyositis and inclusion body myositis, and also potential for confusion at a pathological level. The major clinical and pathological features in dermatomyositis, polymyositis and inclusion body myositis are summarized in Table 24.15.

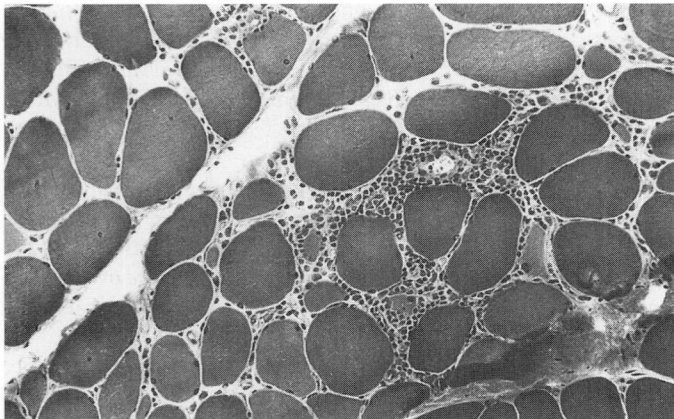

Fig. 24.25 Polymyositis. Note that the inflammatory infiltration is endomysial. (Courtesy of Dr Waney Squier.)

Clinical features

Rash is present in about 90 per cent of patients with dermatomyositis. In its absence the diagnosis can still be made on pathological grounds. The commonest appearances are of erythema of the face and exposed parts of the upper chest with red/purple discolouration of the skin over the knuckles and dilatation of the nail-bed capillaries. The rash is photosensitive, hence the distribution. Characteristic, but less common, is the violaceous/purple discolouration of the eyelids. Rash may precede or follow the onset of muscle weakness, which in some cases may be trivial. In dermatomyositis the weakness comes on sub-acutely, over a matter of weeks, whereas in polymyositis the course is much more protracted and weakness may have been evolving for a year or more before the patient seeks help. With acute onset the muscles are often painful, tender, and swollen but otherwise significant discomfort in the idiopathic inflammatory myopathies is uncommon. The weakness is proximal. When severe, the respiratory and bulbar muscles may also be involved.

Extra-muscular manifestations are common in dermatomyositis and include Raynaud's and arthralgia. Cardiac involvement, with conduction and contractile abnormalities, is one cause of death. In both dermatomyositis and polymyositis interstitial pulmonary fibrosis may occur and is associated with the presence of anti-Jo antibody.

Up to 20 per cent of patients with dermatomyositis, more in the older population, have an associated malignancy, the detection of which may precede or follow the diagnosis of dermatomyositis. Appropriate screening includes chest, abdominal, and pelvic imaging, mammography, rectal and vaginal examination, and basic blood tests. Consideration should be given to repeating these after one year.

Diagnosis

Standard 'inflammatory' markers such as ESR and CRP are often normal and must not be relied upon. Serum creatine kinase is often but not always elevated. Generally it is high in acute cases and often normal or only minimally elevated in chronic cases. Electromyography typically shows spontaneous activity, in the form of fibrillation potentials and positive sharp waves, and a myopathic pattern of motor unit potentials. The gold-standard for diagnosis is muscle biopsy (Table 24.15 and Figs 24.24 and 24.25). The pathological changes are patchy and a single biopsy, particularly if small, may be normal and have to be repeated.

Treatment

Despite a lack of controlled trials, prednisolone is accepted as the initial treatment of choice, typically at a dose of 1 mg/kg body weight. This is frequently combined with azathioprine at 2.5 mg/kg body weight or methotrexate up to 25 mg/week as a 'steroid-sparing' agent. Some clinicians automatically use combined treatment from the outset, others only later if it is clear that the patient is going to need a relatively high dose of prednisolone. After a month or so of high-dose daily prednisolone treatment, and if the serum creatine kinase has fallen into the normal range, the dose of prednisolone is gradually reduced on an alternate day basis. The rate of reduction depends upon the clinical response and, to some extent, the serum creatine kinase. It is much easier to determine the response in patients who presented acutely than in those with grumbling onset of disease. Particularly in this latter group azathioprine or methotrexate may be useful as it is very difficult to judge the appropriate dose of prednisolone, with the danger of giving too

Table 24.15 Major clinical and pathological features in the idiopathic inflammatory myopathies

	Dermatomyositis	Polymyositis	Inclusion body myositis
Clinical features			
Sex	F > M	M = F	M >> F
Age of onset	Any	20 yrs +	50 yrs +
Onset	Subacute/acute	Chronic	Chronic
Distribution of weakness	Proximal	Proximal	Proximal + distal + asymmetric (Typically quadriceps + finger flexors)
Muscle pain/swelling	In acute cases	No	No
Skin involvement	Often	No	No
Raynaud's, arthralgia	Frequent	Infrequent	No
Dysphagia	In severe cases	Infrequent	Occasional
Association with cancer	Up to 20%	Probably not	No
Cardiac involvement	Yes	No	No
Interstitial lung disease	Associated with anti-Jo	Associated with anti-Jo	No
Pathological features			
Scattered necrotic fibres	-/+	++	++
Infarcts	++	-	-
Scattered atrophic fibres	+	++	++
Perifascicular atrophy	++	-	-
Zonal myofibrillar loss	++	-	-
Capillary loss	++	-	-
Rimmed vacuoles	-	-	++
15nm filaments	-	-	++
Partial invasion	-	++	++
Perivascular inflammation	++	-	-
Endomysial inflammation	+	++	++

little, and losing benefit, or too much and inducing a secondary steroid-myopathy.

Alternatives to azathioprine and methotrexate include ciclosporin, mycophenolate mofetil and cyclophosphamide. Intravenous immunoglobulin is effective in dermatomyositis, but not demonstrably superior to cheaper drug regimes. Plasma exchange has been less well assessed but theoretically may be helpful in dermatomyositis. None of these treatments have been evaluated adequately in controlled trials.

Appropriate advice should be given with respect to steroid-induced osteoporosis prophylaxis. Adequate nutrition is important, as is encouragement of physical activity. Skin rashes respond to topical steroids and are helped by sun-blocking creams.

Prognosis is difficult to determine accurately in a given individual. Poor prognostic factors include advanced age, long-standing weakness at time of initiation of therapy, associated malignancy, and

lung and heart involvement. Most patients require treatment for at least two years, many for five years, and some remain treatment-dependent. Most patients show some response to therapy. In younger patients with a short history one generally hopes to return them to normal or near-normal strength. In an older patient with long-standing weakness one may prevent progression but gain little improvement in strength.

24.7.3 Inclusion body myositis

In its most characteristic form this is a readily recognisable condition. The diagnosis is sometimes first made when a patient diagnosed as having polymyositis fails to respond to therapy. As noted, its pathogenesis is still much argued.

Clinical features

Onset is unusual before the age of 50 years. It is much commoner in men. Rarely, otherwise typical cases of sporadic inclusion body myositis are seen to be familial. This is not the same as hereditary inclusion body myositis which occurs in younger patients and in which different pathological features are seen. Progression of weakness is slow. Typically the quadriceps are involved early and inclusion body myositis is the commonest cause of 'isolated quadriceps myopathy'. Also involved early and highly selectively are the finger flexor muscles, causing profound weakness of grip. Together, these two features are virtually pathognomonic. Overall, and in contrast to dermatomyositis and polymyositis, distal weakness is common and there is often asymmetry between the two sides. Dysphagia is usually a late feature, but rarely a presenting symptom. Extra-muscular clinical features are absent, but associations with connective tissue diseases and autoantibodies have been noted.

Diagnosis

Serum creatine kinase is normal or modestly elevated. Electromyography shows features similar to those seen in dermatomyositis and polymyositis but in addition 'neurogenic' changes in the form of large amplitude long duration motor unit potentials are often seen. Diagnostic confirmation is by muscle biopsy (Table 24.15), particularly by the demonstration of 15 nm intranuclear and cytoplasmic filaments or congo red positive inclusions.

Treatment

Immunosuppression is generally ineffective or at best shows only a very modest transient benefit. It is arguable whether a trial of such therapy is justified. The disease progresses very slowly, but in some patients it does become profoundly disabling. Orthoses at wrists, knees, and ankles may be helpful.

24.7.4 Other inflammatory myopathies

Associations with other autoimmune/connective tissue diseases have been discussed. Sometimes the term 'overlap syndrome' is used to describe such cases—this simply reflects our incomplete understanding of the pathogenesis of these disorders.

There are important associations between AIDS and inflammatory myopathy (Authier and Gherardi 2006). The commonest is a polymyositis-like condition virtually indistinguishable from idiopathic polymyositis clinically and pathologically. Retrovirus is present in interstitial cells but does not appear to invade muscle fibres. Patients with AIDS are at risk of opportunistic muscle infections, such as microsporidia or toxoplasmosis. Zidovudine can cause mitochondrial depletion and a myopathy in which myalgia

Table 24.16 Thyroid-associated muscle disorders

Hypothyroid myopathies
Thyrotoxic myopathy
Thyroid ophthalmopathy, Graves disease
Thyrotoxic periodic paralysis
Myasthenia gravis

predominates over weakness. HTLV-I can also cause a myositis either alone or in association with tropical spastic paraparesis. It can also be associated with an inclusion body myositis-like picture.

Granulomata in muscle are probably not that uncommon in sarcoidosis, but clinically evident myopathy is rare.

Worldwide, infective myopathies are common, but are mostly seen in underdeveloped parts and are closely related to economic and sanitary conditions (Crum-Cianflone 2006). Geographical restrictions are becoming less evident with increased air travel. Parasitic infections are common, and include toxoplasmosis, trypanosomiasis, malaria, cysticercosis, echinococcosis or hydatidosis, and trichinellosis or trichinosis. These involve complex life-cycles with intermediate and definitive hosts and much debate as to optimal approaches to treatment.

24.8 Endocrine myopathies

Weakness, usually proximal and affecting the pelvic girdle earlier than the shoulder girdle, is a common feature of many endocrinopathies. It resolves upon correction of the underlying disorder. Muscle biopsy shows non-specific type II muscle fibre atrophy. The commonest is glucocorticoid excess in the form of iatrogenic steroid myopathy but the clinical features parallel those of Cushing's disease. The most frequently encountered primary endocrine myopathies are those associated with thyroid disease.

24.8.1 Steroid myopathy

Some two-thirds of cases of Cushing's syndrome are due to an ACTH-producing pituitary adenoma, one-sixth due to ectopic ACTH production by a tumour, and one-sixth due to a cortisol-secreting tumour of the adrenal cortex. Weakness is present in about 60 per cent of patients with Cushing's syndrome. Iatrogenic steroid myopathy is most frequently seen in association with the use of 9-α fluorinated steroids, which include dexamethasone, triamcinolone, and betamethasone. Topical application can also cause myopathy. Individuals vary in their susceptibility to developing myopathy, and women are more at risk than men.

Clinical features

Weakness, which is often accompanied by myalgia, starts in a pelvifemoral distribution and later spreads to the trunk and shoulder girdle. Atrophy is common. It is rare to develop myopathy without other features of glucocorticoid excess being apparent.

Diagnosis

This is predominantly clinical. Serum creatine kinase is usually normal, electro-myography is often normal or may show non-specific myopathic features, and muscle biopsy shows type II fibre atrophy. An occasional clinical dilemma is distinguishing between increasing weakness due to reactivation of disease and that due to

steroids in a patient being treated for an inflammatory myopathy. electromyography, serum creatine kinase, and muscle biopsy may help, but sometimes all one can do is alter the dose of steroids and monitor the clinical response.

Treatment

Stopping steroids, or correcting the underlying endocrine disorder, is usually followed by full recovery of the myopathy. If steroid therapy cannot be stopped, then a non-fluorinated drug, such as prednisolone, should be used at the lowest possible dosage and preferably on an alternate-day regime.

24.8.2 Acute steroid myopathy

This iatrogenic condition merits separate mention. It was first clearly delineated in patients with asthma being given a combination of high-dose parenteral steroids and a neuromuscular blocking agent, but has now been described in patient receiving steroids alone. Weakness, which may be profound and associated with respiratory failure, develops acutely or subacutely. Recovery occurs over 6 to 12 months. The pathological features are of small angular fibres and selective loss of myosin filaments.

24.8.3 Thyroid-associated myopathies

The muscle disorders associated with thyroid disease show the greatest clinical variability of all of the endocrine myopathies (Table 24.16). In myasthenia gravis there is an increased incidence of autoimmune thyroid disease and thyroid function should form part of the routine assessment of such patients.

Hypothyroidism

Although weakness is common it is less striking than in hyperthyroidism and is very rarely the presenting symptom. In both it is proximal in distribution. A combination of hypothyroidism, muscle hypertrophy and weakness, and slowness of movement, is sometimes referred to as the Kocher–Debré–Semelaigne syndrome in childhood and Hoffman's syndrome in adulthood. The serum creatine kinase is almost invariably elevated. Occult hypothyroidism is an important cause of otherwise idiopathic elevation of serum creatine kinase and of aspartate and alanine transaminases; the latter sometimes leads to a spurious search for liver disease. Muscle features recover after restoring a euthyroid state.

Hyperthyroidism

Weakness is present in about one-half of thyrotoxic patients, usually becoming apparent shortly after the onset of other thyrotoxic symptoms, but in up to 10 per cent may be the presenting feature. The shoulder girdle may be affected before the pelvic girdle and in some patients early distal weakness is apparent. Atrophy is usually slight. Severe weakness raises the possibility of coexistent myasthenia gravis. The serum creatine kinase is typically normal. Recovery of the myopathy follows successful treatment of the endocrine disorder.

Thyroid ophthalmopathy, Graves' disease

Although most frequently associated with hyperthyroidism, this can occur in hypothyroid and, clinically most challenging, euthyroid patients. Eyelid lag and retraction are common in hyperthyroidism. Thyroid ophthalmopathy refers to those patients in whom the ocular involvement is much more severe. Conjunctival injection and swelling is followed by diplopia and proptosis due to enlargement of the extra-ocular muscles and orbital soft tissues.

In later stages vision is further threatened by papilloedema and optic nerve compression. All of these changes may be unilateral. An important clinical subgroup is those patients with diplopia only, in whom the diagnosis is often missed.

Diagnosis is easy if there is obvious clinical or laboratory evidence of thyroid disease. If the T4, T3, and TSH are normal, further investigations include antimicrosomal and antithyroglobulin antibodies and a TRH stimulation test. Thyroid-stimulating immunoglobulins are present in many patients. Indirectly, the diagnosis may be supported by demonstrating extra-ocular muscle swelling by ultrasound, CT, or MRI.

The patient should be made euthyroid. Lid lag and retraction may respond to topical application of guanethedine 10 per cent. For severe disease treatment options include surgical decompression of the orbit, orbital irradiation, and high-dose prednisolone.

Thyrotoxic periodic paralysis

Clinically, this resembles hypokalaemic periodic paralysis. Attacks typically start in early adult life. The incidence is very much higher in Orientals than other races and there is a very strong male predominance. Attacks cease when the underlying thyroid disorder is corrected.

24.8.4 Disorders of vitamin D and calcium metabolism

There is a complex relationship between vitamin D metabolism, calcium and phosphate homoeostasis, and parathyroid hormone activity. In clinical practice the most commonly encountered disorders include osteomalacia and metabolic bone disease associated with renal failure.

Osteomalacia

Pelvic girdle weakness is the presenting symptom in about one-third of patients. However, bone pain tends to dominate the clinical picture with particular involvement of the pelvis, femora, and ribs. The gait is typically waddling and Gower's manoeuvre may be present. The diagnosis of osteomalacia can usually be confirmed by demonstrating reduced bone density, low blood calcium and phosphate levels, and elevated serum alkaline phosphatase activity. Bone pain responds quickly to treatment but weakness may take much longer to recover.

Renal failure

The term renal osteodystrophy refers to the disturbed vitamin D metabolism and secondary hyperparathyroidism complicating chronic uraemia. It is often associated with pelvic girdle weakness which responds favourably to dialysis, transplantation, or vitamin D therapy. Dialysis osteodystrophy was due to aluminium toxicity arising from high levels of the metal in water used to prepare the dialysate.

Parathyroid gland dysfunction

Symptomatic weakness is uncommon in hypo and hyperparathyroidism.

24.8.5 Other endocrine myopathies

Weakness, usually predominantly proximal, may also be seen in acromegaly, hypopituitarism, ACTH excess, primary aldosteronism, and Addison's disease.

24.9 Toxic, nutritional, and drug-induced myopathies

The most important toxic myopathies are those associated with ethanol consumption. Although malnutrition causes muscle wasting, specific nutritional myopathies are rare. Drug-induced myopathies are of considerable importance and are probably under-recognized.

24.9.1 Ethanol-related myopathies

Up to two-thirds of chronic alcoholics show evidence of proximal weakness and muscle wasting, most evident around the pelvic girdle (Section 5.2.10). It is painless and rarely severe. It improves with abstinence. Clinically the picture may be complicated by coexistent alcoholic neuropathy.

More dramatic is acute alcoholic myopathy which typically develops in a chronic alcoholic following a binge. It may be restricted to a single muscle or be generalized. Cramps precede muscle swelling, which may be dramatic, and weakness. In the lower limb the picture may mimic deep vein thrombosis. Swelling may induce a compartment syndrome necessitating surgery. There is extensive muscle fibre breakdown, rhabdomyolysis, with gross elevation of the serum creatine kinase, myoglobinuria, posing a risk of renal failure, and sometimes hyperkalaemia. Recovery, which may be incomplete, occurs over several weeks.

24.9.2 Other toxic myopathies
Eosinophilia–myalgia syndrome

This condition was due to a contaminant called 'peak E' in a batch of L-tryptophan made by a single manufacturer. Early features included myalgia, skin rash, dyspnoea, arthralgia, fever, and weight loss. Several months later some patients developed scleroderma-like skin infiltration, persistent myalgia, and weakness.

Toxic oil syndrome

This condition, which affected many thousands of people in Spain in the early 1980s, bears close similarities to the eosinophilia–myalgia syndrome. It was due to ingestion of an illegally imported, reprocessed, denatured rapeseed oil. Features included myalgia, eosinophilia, respiratory distress, scleroderma-like skin changes, wasting, and weakness.

Snake venoms

Many are myotoxic (Section 5.12.5).

24.9.3 Nutritional myopathies

Osteomalacia is discussed above. Vitamin E deficiency may be seen in chronic cholestasis, malabsorption syndromes, and abetalipoproteinaemia. Myopathy is one feature of a multi-faceted neurological disorder associated with vitamin E deficiency (Sections 21.22.5; 39.7.4).

Nutritional carnitine deficiency is uncommon but may be caused by prolonged parenteral nutrition. It is occasionally seen in patients on renal dialysis and in pregnancy. It causes proximal weakness.

24.9.4 Statin toxicity

This merits specific comment because muscle symptoms are a relatively common side effect of this class of drugs which are widely, and increasingly frequently, prescribed and in some countries can now be purchased without a prescription (Christopher-Stine 2006).

Much of the literature surrounding myotoxic side effects is very poor, coming from specialists with little or no expertise in the field of muscle disease. This partly explains the confusion in the literature with respect to the term myositis which has been used to describe statin-induced muscle symptoms associated with a raised serum creatine kinase but, extraordinarily, without any pathological evidence of an inflammatory process in the muscle. Myositis, if it occurs at all, is rare and current evidence suggests that myotoxicity is a class effect, relating to inhibition of metabolic pathways. Statins inhibit 3-hydroxy 3-methylglutaryl coenzyme A, HMG-CoA, reductase. This pathway is involved in coenzyme Q_{10} synthesis and it has been proposed, but remains unproven, that low levels of CoQ might explain the myotoxicity.

The features of statin myotoxicity range from asymptomatic elevation of serum creatine kinase, to myalgia usually with elevation of creatine kinase, to a severe necrotizing but not inflammatory myopathy with rhabdomyolysis. The latter is rare, probably about 0.1 per million prescriptions, but potentially life-threatening.

Most commonly, symptoms and serum creatine kinase elevation settle rapidly on discontinuation of the drug. Persistent abnormality often turns out to be due to a pre-existing myopathy that had not been recognized prior to starting the drug. This may also explain many of the case reports of a previously unrecognized myopathy being 'revealed' by the statin. Alternatively, and perhaps particularly in the case of metabolic myopathies, a minor pre-existing metabolic disorder insufficient to cause symptoms may be exacerbated and become symptomatic as a result of the additional metabolic insult caused by the statin (Vladutiu *et al.* 2006).

The risk of statin myotoxicity is increased substantially by the combined use of fibrates or ciclosporin, or by other drugs and agents, such as grapefruit juice, that inhibit cytochrome p450. However, the benefits of statin therapy are such that there is considerable anxiety about depriving patients of their use. In high-risk patients it is still reasonable to use a statin, but starting at a low dose and monitoring symptoms and serum creatine kinase closely. Some statins may be less myotoxic than others, such as pravastain, which is not metabolized by the cytochrome p450 system, but none can be considered to be risk-free.

24.9.5　Other drug-induced myopathies

The mechanisms of drug-induced myotoxicity are many and varied and, in many cases, poorly understood (Mastaglia 2006). They include direct toxicity, immune mechanisms and electrolyte disturbance. Some also have a neurotoxic effect. For practical purposes a classification based on the clinical presentation is most useful (Table 24.17). This table by no means includes all drugs reported to cause myotoxicity, but lists those most commonly implicated.

24.10　Disorders of the neuromuscular junction

The clinical conditions relating to dysfunction at the neuromuscular junction are myasthenia gravis, which is by far the commonest, the Lambert–Eaton myasthenic syndrome, congenital myasthenic syndromes, and acquired neuromyotonia or Isaacs' syndrome. Each of these relates to disturbed ion channel function. This is immune-mediated in myasthenia gravis, the Lambert–Eaton syndrome and neuromyotonia, and due to a genetic defect in the congenital myasthenic syndromes. The relevant major functional components of the neuromuscular junction are illustrated schematically in Fig. 24.26. In myasthenia gravis, the antibodies are directed against the acetylcholine receptors, in Lambert–Eaton syndrome, against the voltage-gated calcium channels, VGCC, and in neuromyotonia, against the voltage-gated potassium channels, VGKC.

Two classes of ion channel are involved: the voltage-gated sodium, potassium, and calcium channels, and the ligand-gated acetylcholine receptors. Depolarization of the nerve terminal membrane, which depends on voltage-gated sodium channels, VGNC, opens the calcium channels, VGCC; influx of calcium triggers release of quanta of acetylcholine. Acetylcholine binds to the α-subunits of the receptors on the muscle fibre membrane, which allows influx of sodium ions; this generates the end-plate potential which in turn activates muscle fibre membrane voltage-gated sodium channels to create the action potential. This subsequently triggers muscle contraction. Acetylcholine in the synaptic cleft is destroyed by acetylcholinesterase. Nerve terminal repolarization, and closure of the calcium channels are dependent upon inactivation of the voltage-gated sodium channels and opening of the voltage-gated potassium channels. The acetylcholine receptor is a pentameric structure, with a central ion channel, composed of four subunits and has foetal, $\alpha_2,\beta,\delta,\gamma$, and adult, $\alpha_2,\beta,\delta,\varepsilon$, forms (Fig. 24.26).

Table 24.17 Drug-induced myopathies

Focal myopathy	Intramuscular injections
Acute/subacute painful myopathy	Cholesterol-lowering drugs
	Opiates
	ε-aminocaproic acid
	Amiodarone
	Cyclosporin
	Emetine
	β-blockers
	Zidovudine
	Vincristine
Acute rhabdomyolysis	Opiates
	Cocaine
	Amphetamines
	Phencyclidine
	Stains
Chronic painless myopathy	Corticosteroids
	Chloroquine
	Colchicine
	Perhexiline
	Amiodarone
Hypokalaemia	Diuretics
	Purgatives
	Amphotericin B
	Carbenoxolone
	Liquorice
Inflammatory myopathy	D-penicillamine
	Procainamide

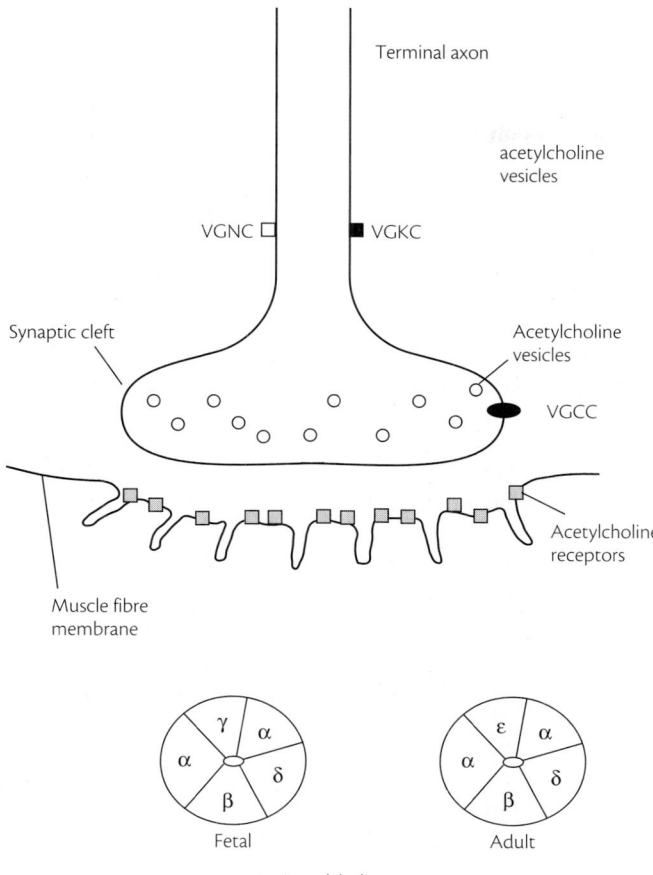

Fig. 24.26 Major functional features of the neuromuscular junction (see Section 24.10).

24.10.1 **Myasthenia gravis**

The prevalence of myasthenia gravis is in the order of 5 to 10 per 100 000 population. The major clinical features are of weakness and, characteristically, excessive fatiguability. It is associated with an increased incidence of other autoimmune diseases, notably thyroid dysfunction, rheumatoid arthritis, and other connective tissue disorders. Penicillamine may induce myasthenia which is clinically indistinguishable from the idiopathic variety and in which anti-acetylcholine receptor, anti-AChR, antibodies are present. It resolves following drug withdrawal.

Myasthenia may present at any age. Five major groups of patients can be identified:

- Onset in early adult life; female preponderance; anti-AChR antibody titres are high; the thymus shows medullary hyperplasia.

- Later onset; male preponderance; anti-AChR antibody titres tend to be low; the thymus is atrophic.

- Later onset; equal sex ratio; intermediate levels of anti-AChR antibodies; thymoma present.

- Any age; absence of anti-AChR antibodies, but presence of anti-muscle specific kinase, MuSK, antibodies; predominantly oculo-bulbar-respiratory involvement with minor limb involvement; more resistant to therapy than anti-AChR antibody myasthenia gravis; thymus normal/atrophic.

- Any age; seronegative myasthenia; male predominance; anti-AChR antibodies not detectable by standard assay; thymus atrophy; often restricted to ocular involvement.

As will be noted, the presence or absence of anti-AChR antibodies and the state of the thymus gland influence therapeutic approaches.

Neonatal myasthenia describes the condition of transient weakness in the newborn due to transplacental transfer of anti-AChR antibody from an affected mother. Symptoms resolve within a few weeks.

Clinical features

Although any muscle may be involved, those most commonly affected at presentation are the extraocular muscles, causing ptosis and diplopia. Facial and bulbar muscle weakness causes dysarthria and dysphagia. Proximal limb weakness exceeds distal. Respiratory muscle weakness severe enough to cause ventilatory embarrassment usually develops only in the presence of substantial limb weakness. Selective weakness of the neck extensors is a rather common feature in older, typically male, patients and causes head-drop. The most characteristic feature of myasthenia gravis is excessive fatiguability. This may be demonstrated on examination, for example showing the development of ptosis on attempted sustained up-gaze, but is evident in the history. Weakness varies during the course of the day and during activity, and is eased by rest. Dysarthria increases the longer the patient speaks. Weakness is exacerbated by emotion, heat, menstruation, and intercurrent infection.

In a small proportion of patients the disease remains confined to the extra-ocular muscles. If the disease has not spread beyond the eyes within 2 years of onset it is unlikely to do so later. Patients with ocular myasthenia often do not have anti-AChR antibodies. When left untreated, the condition gradually spreads to involve more muscles and when severe may cause respiratory failure. The spontaneous remission rate is low.

Recently, a specific subgroup of patients with anti-MuSK antibodies has been identified (Evoli *et al.* 2003). These patients tend to be female and have pronounced oculo-bulbar involvement and frequent respiratory crises. Tongue and facial muscle atrophy, and persisting weakness, are common (Farrugia *et al.* 2006). They are more resistant to anticholinesterase and immunosuppressant drugs than those patients with anti-AChR antibodies.

Diagnosis

The diagnosis is confirmed by demonstrating the presence of either anti-AChR or anti-MuSK antibodies. False-positive results are exceedingly rare. Sensitivity is limited to about 50 per cent in purely ocular myasthenia and about 90 per cent in generalized myasthenia.

If antibodies are not detected, the next most useful investigation is electromyography. The traditional, and simplest, test is to stimulate a nerve at 3Hz and to measure the compound muscle action potential. A decrement of >10 per cent when comparing the fifth to the first response is abnormal. More sensitive, but technically more difficult, is single fibre electromyography (Section 3.5.6). Studies of orbicularis oculi are particularly valuable in cases of ocular myasthenia.

The thymus is imaged by CT or MRI. Most patients with thymoma have anti-striated muscle antibodies in their serum.

Thyroid function should also be assessed.

Although well-known, discussion of the edrophonium, Tensilon®, test is left until last for several reasons. The principle is that edrophonium, a short-acting anticholinesterase, given intravenously produces a transient improvement in strength in patients with myasthenia gravis. Although this is generally true it must be noted that:

- false-negative results may occur;
- false-positive results undoubtedly occur and are seen in myasthenic syndromes, motor neurone disease, mitochondrial myopathies including chronic external ophthalmoplegia, which is a major differential diagnosis of ocular myasthenia, and other neuromuscular disorders;
- rarely, cardiorespiratory collapse may occur, particularly if atropine pretreatment has not been given to forestall vagal bradycardia.

For these reasons, this test should only be done by experienced personnel, when indicated during the diagnostic process. Generally, it is not required and is most likely to be helpful in ocular myasthenia gravis in which detectable antibodies may be absent and single fibre electro-myography, even of orbicularis oculi, may be normal.

Treatment

If a thymoma is present it should be removed because it may be locally invasive. If removal is impossible or incomplete, radiotherapy and chemotherapy should be considered. However, this will not benefit the myasthenia.

Most patients will show some response to the anticholinesterase drug pyridostigmine given orally in a dose of up to 60 mg six times daily, but this will satisfactorily control only relatively mild disease.

In patients with purely ocular myasthenia, patients over the age of about 50 years with generalized disease, and patients without anti-AChR antibodies, the next approach to treatment is alternate-day prednisolone. This is introduced slowly, to prevent acute deterioration, up to a dose of 1.5 mg/kg body weight every alternate day. Following remission the dose is gradually reduced and the minimum effective dose determined. Azathioprine at a dose of 2.5 mg/kg body weight/day is frequently added as a steroid-sparing agent and it may be possible eventually to withdraw prednisolone and to maintain the patient on azathioprine alone. The maximum benefit of azathioprine is not seen for at least 6 months. Regular monitoring of liver function is required. For patients intolerant of azathioprine, other useful, but potentially more toxic immunosuppressants include methotrexate, mycophenolate mofetil, cyclosporin, cyclophosphamide, and rituximab.

In patients under the age of about 50 years with generalized myasthenia and anti-AChR antibodies, thymectomy may induce remission in about one-quarter and lead to improvement in about one-half of patients. Considerable doubt remains about the risk to benefit ratio, and an international thymectomy trial is underway. Morbidity from trans-sternal thymectomy is low in specialized hands and when patients are given optimal medical management prior to surgery. Such medical management includes the immunosuppressant regimes described above and the use of plasma exchange and intravenous immunoglobulins. These latter two treatments are equally efficacious and give symptomatic improvement for up to six weeks. They are useful during the initiation of prednisolone treatment, which in itself can cause transient worsening of the myasthenia, as well as prior to thymectomy. Immunosuppressant therapy continues following thymectomy but is gradually tapered as the patient's condition permits.

Prednisolone therapy is likely to be long-term and appropriate consideration must be given to potential side-effects, particularly the risk of osteoporosis.

Immunopathogenesis

Although numerous important observations relating to immune system function in myasthenia gravis have been made, many fundamental questions remain unanswered. Anti-AChR antibodies are of IgG class, although in cases that are seronegative using the standard assay there is evidence that IgM class antibodies may be involved. Antibodies, which bind to the α-subunit, cause acetylcholine receptor dysfunction in one of several ways: complement-mediated lysis; modulation of receptor turnover; and receptor blockade. Antibody titres correlate with disease activity in an individual patient, but not between patients. The mechanism by which anti-MuSK antibodies leads to myasthenia is unclear (Conti-Fine et al. 2006)

The thymus in seropositive patients typically shows medullary hyperplasia. Anti-AChR antibody is produced by thymic cells, AChR-reactive T-cells are present, and myoid cells within the thymus express AChR.

It is hoped that when the afferent and efferent limbs of the immune pathways are better understood that it will be possible to develop more selective immunotherapies.

24.10.2 Lambert–Eaton myasthenic syndrome

This is a pre-synaptic disorder in which the quantal release of acetylcholine is reduced as a result of antibodies directed against voltage-gated calcium channels. It is very much less common than myasthenia gravis. In about 60 per cent of cases it is associated with small cell lung cancer and up to 3 per cent of patients with this type of tumour may develop Lambert–Eaton myasthenic syndrome, although the diagnosis is probably not infrequently missed.

Clinical features

Presentation is usually with gait disturbance. This may be attributed directly to weakness but sometimes the description may be less specific, such as one patient who said 'my legs get in the way when I am walking'. Others refer to 'walking through treacle'. There may be mild ptosis but extra-ocular signs are much less evident than in myasthenia gravis. Autonomic features are common and include impotence, in males, and dryness of the mouth. On examination, strength and tendon reflexes may augment transiently following forceful voluntary contraction of the relevant muscle.

Diagnosis

The characteristic electromyographic finding is of a small compound muscle action potential that shows marked increase in amplitude following voluntary contraction or high frequency nerve stimulation, (Section 3.5.6). Anti-voltage-gated calcium channel antibodies can be detected in most patients. An underlying small cell lung cancer must be sought at presentation and, in smokers, at intervals thereafter.

Treatment

Tumour treatment often improves the neurological disorder. Symptomatic relief is obtained from 3,4-diaminopyridine which blocks voltage-gated potassium channels and thus delays nerve repolarization. In the absence of tumour, treatment with prednisolone and azathioprine, as for myasthenia gravis, often induces remission. Plasma exchange and intravenous immunoglobulin provide transient benefit and supplement other forms of treatment.

24.10.3 Congenital myasthenic syndromes

This is an extremely rare group of conditions with an overall prevalence in the order of 1/1 000 000. They are genetically determined, non-autoimmune, disorders. The common features are onset in infancy, fatiguable weakness, and a decremental response to repetitive nerve stimulation. There is as yet no entirely satisfactory clinical or genetic system of classification. The commonest are post-synaptic disorders in which there is a mutation affecting one of the AChR sub-units, commonly the ε sub-unit, which alters the kinetics of the receptor. Recently, early- and late-onset syndromes have been associated with mutations affecting the end-plate protein rapsyn, which has a role in acetylcholine receptor clustering (Burke *et al.* 2003). Inherited, childhood-onset, limb–girdle myasthenia with absence of oculo-bulbar features has been associated with mutations affecting DOK-7, another protein important for maintaining normal neuromuscular junction structure (Beeson *et al.* 2006). Also recognized are pre-synaptic disorders and congenital endplate acetylcholinesterase deficiency. Apart from the post-synaptic slow-channel syndrome, which is dominant, these various disorders are autosomal recessive conditions.

Therapeutic options are limited. Anticholinesterase drugs offer symptomatic benefit, except in patients with end-plate acetylcholinesterase deficiency. Immune therapies, such as those used in myasthenia gravis are of no value.

24.10.4 Acquired neuromyotonia

Previously known as Isaacs' syndrome of continuous muscle fibre activity, this rare condition is characterized by hyperexcitability of nerves leading to muscle twitching or myokymia, cramps, paraesthesia, and sweating, (Section 23.7.1). Electromyography shows high-frequency bursts of continuous motor unit discharges, or multiplet discharges. It may be seen in association with a number of acquired disorders including myasthenia gravis, demyelinating polyneuropathy, or carcinoma and inherited neuropathies or spinal muscular atrophy. Some acquired cases are autoimmune in origin and relate to the presence of antibodies directed against voltage-gated potassium channels. In these patients immunosuppression with prednisolone and azathioprine may be of benefit and in severe cases plasma exchange may give transient benefit. Most patients gain symptomatic benefit from lamotrigine, carbamazepine, or phenytoin.

24.10.5 Botulism

Botulism is caused by a powerful neurotoxin produced by the anaerobic organism *Clostridium botulinum*. The toxin binds to specific acceptors on the cholinergic nerve terminal, is internalized via endocytosis, and then cleaves one of the proteins involved in exocytosis. The net result is paralysis and parasympathetic blockade due to inhibition of calcium-mediated release of acetylcholine. The commonest forms are infant botulism, seen in the first six months of life and due to over-colonization of the gastrointestinal tract by *Cl. botulinum*, and food-borne botulism caused by ingestion of toxin in tainted food. Wound botulism is rare.

24.11 Miscellaneous muscle disorders

This section deals with those conditions which do not readily fit into any of the aforementioned categories, or in which there is debate as to whether or not there is primary involvement of skeletal muscle.

24.11.1 Chronic fatigue syndrome

Of the many symptoms associated with this syndrome, the most prominent are fatigue, myalgia, impaired concentration, poor memory, sleep disturbance, and emotional lability. Functional studies have failed to demonstrate clear evidence of a peripheral disorder to explain the patients' perceived neuromuscular difficulties. Early claims for persistent viral infection proved to be unfounded, but at least in some cases enteroviral RNA may be detected in muscle. Metabolic studies in muscle have given conflicting results. Despite some reports suggesting otherwise, muscle biopsy is usually either normal or shows changes reflecting reduced physical activity. It is probable that in the vast majority of patients with chronic fatigue syndrome that there is no primary disorder affecting peripheral nerves, the neuromuscular junction, or skeletal muscle. The significance of gene expression profiling studies is as yet unclear (Fang *et al.* 2006).

24.11.2 Polymyalgia rheumatica

It is doubtful if there is primary muscle involvement in this condition but it is important in the present context because it may be confused with the idiopathic inflammatory myopathies. In practice, there should rarely be difficulty distinguishing between them. Polymyalgia rheumatica is characterized by pain and stiffness often early in the morning, around the neck and shoulder girdle, less frequently around the pelvic girdle, and the absence of weakness, although this may seem to be present because of the discomfort. It is rare before the age of 55 years and the ESR is almost invariably elevated. There is a prompt clinical response to prednisolone. Some patients go on to develop giant cell arteritis (Sections 18.7.1; 36.2.8). The aetiology of these two conditions is unclear but there is evidence of vascular involvement and much of the pain and stiffness may relate to chronic synovitis.

Electromyography is usually normal, as is the serum creatine kinase. Muscle biopsy may show type II fibre atrophy, probably reflecting disuse.

24.11.3 Neoplastic disorders

Primary muscle tumours are rare and their classification has caused difficulty (Newton *et al.* 1995). Embryonal rhabdomyomas in early childhood affect the head and neck region and the genito-urinary tract, 'sarcoma botryoides'. In adults they typically involve the head, neck, and throat region. Cardiac rhabdomyomas are seen in tuberous sclerosis. In adults, rhabdomyosarcomas arise in the legs. They are highly aggressive, metastasize early, and respond poorly to radiotherapy.

Apart from the result of local invasion such as breast carcinoma invading pectoralis, secondary tumours in skeletal muscle are rare. Other tumours that may arise in muscle, but not from the muscle itself, include angiomata, dermoid tumours, and neurofibromata.

24.11.4 Paraneoplastic disorders

The most important, and best defined, paraneoplastic disorders affecting skeletal muscle and the neuromuscular junction are dermatomyositis (Sections 24.7.2; 38.4.7) and the Lambert–Eaton myasthenic syndrome (Section 24.10.2). The term carcinomatous neuromyopathy is widely used to describe a syndrome of symmetrical proximal weakness, often with depressed tendon reflexes. Many such cases are probably neurogenic in origin but such patients are, understandably, not always investigated in detail. Endocrine gland tumours causing increased or decreased hormone production may produce weakness, as may tumours that cause metabolic derangement. Weakness in association with cancer may be due to the rather non-specific effect of a number of factors including cachexia, malnutrition, infection, and inactivity.

24.11.5 Myositis ossificans

This term is used to describe two unrelated conditions in neither of which is muscle inflammation, myositis, a major feature.

Localized myositis ossificans

This is an acquired disorder and in many, but not all, cases it arises as a result of trauma, either minor repeated trauma or a single substantial injury. Repeated trauma in sportsmen or women is a well-recognized cause. Initially there is localized swelling and tenderness, which is followed by bone formation. Small lesions may resolve spontaneously. Larger areas of calcification, impairing movement, may need to be resected.

Fibrodysplasia ossificans progressiva

This is an autosomal dominant condition often caused by a new mutation. There has been recent debate as to the significance of mutations in the noggin gene (Fontaine *et al.* 2005). A congenital shortening malformation of the great toe is present almost invariably, with the thumb being affected less frequently. Endochondral ossification of skeletal muscles occurs in a specific order and the patients are described as developing a second skeleton. It causes profound immobility and most patients are wheelchair bound by the third decade of life. No treatment has been shown to be of benefit.

24.11.6 Compartment syndromes

Some muscles are contained within semirigid fibroosseous compartments, the most important examples being the anterior tibial compartment and the volar compartment of the forearm. If the muscles within these compartments swell the pressure rises rapidly. The causes of swelling include:

- ischaemia due to arterial compression due to displaced fracture, tourniquet pressure, clamping during surgery or haematoma;

- direct trauma;

- drugs that induce rhabdomyolysis such as alcohol or heroin.

The rising pressure further impedes capillary blood flow, and thus a vicious circle of increasing ischaemia develops. Nerves within the compartment become ischaemic, and if the pressure is sufficient infarction may occur.

The clinical features are of pain and swelling, and sensory and motor involvement relating to the peripheral nerves compressed within the compartment. Extensive muscle necrosis may lead to myoglobinuria sufficient to cause renal failure, a crush syndrome.

Clinical assessment may be aided by pressure measurements using a wick-catheter inserted into the compartment. Treatment is surgical by subcutaneous fasciotomy.

The contracture that may arise as a result of fibrosis of the damaged muscle is known as Volkmann's ischaemic contracture.

A chronic form of compartment syndrome is also recognized, and the commonest example involves the anterior tibial compartment in athletes, less frequently in more sedentary individuals. Local pain develops on exercise and resolves with rest. The symptoms may be shown to parallel an increase in pressure within the compartment during exercise. The problem may respond to prolonged rest but in some patients fasciotomy is required.

References

Askanas V, Engel WK (2006). Inclusion-body myositis: a myodegenerative conformational disorder associated with Abeta, protein misfolding, and proteasome inhibition. *Neurology*, **66**, S39–48.

Authier FJ, Gherardi RK (2006). Muscular complications of human immunodeficiency virus (HIV) infection in the era of effective anti-retroviral therapy. *Rev Neurol*, **162**, 71–81.

Beeson D, Higuchi O, Palace J *et al.* (2006). Dok-7 mutations underlie a neuromuscular junction synaptopathy. *Science*, **313**, 1975–8.

Bitoun M, Maugenre S, Jeannet PY *et al.* (2005). Mutations in dynamin 2 cause dominant centronuclear myopathy. *Nat Genet*, **37**, 1207–9.

Brais B, Bouchard JP, Xie YG *et al.* (1998). Short GCG expansions in the PABP2 gene cause oculopharyngeal muscular dystrophy. *Nat Genet*, **18**, 164–7.

Burke G, Cossins J, Maxwell S *et al.* (2003). Rapsyn mutations in hereditary myasthenia: Distinct early- and late-onset phenotypes. *Neurology*, **61**, 826–8.

Bushby K, Muntoni F, Bourke JP (2003). 107th ENMC international workshop: the management of cardiac involvement in muscular dystrophy and myotonic dystrophy. 7th-9th June 2002, Naarden, The Netherlands. *Neuromuscul Disord*, **13**, 166–72.

Bushby KM, Pollitt C, Johnson MA *et al.* (1998). Muscle pain as a prominent feature of facioscapulohumeral muscular dystrophy (FSHD): four illustrative case reports. *Neuromuscul Disord*, **8**, 574–9.

Chahin N, Selcen D, Engel AG (2005). Sporadic late onset nemaline myopathy. *Neurology*, **65**, 1158–64.

Christopher-Stine L (2006). Statin myopathy: an update. *Curr Opin Rheumatol*, **18**, 647–53.

Conti-Fine BM, Milani M, Kaminski HJ (2006). Myasthenia gravis: Past, present and future. *Eur J Clin Invest*, **116**, 2843–54.

Crum-Cianflone NF (2006). Infectious myositis. *Best Pract Res Clin Rheumatol*, **20**, 1083–97.

Dalakas MC (2004). Inflammatory disorders of muscle: progress in polymyositis, dermatomyositis and inclusion body myositis. *Curr Opin Neurol*, **17**, 561–7.

Davies NP, Imbrici P, Fialho D *et al.* (2005). Andersen-Tawil syndrome: New potassium channel mutations and possible phenotypic variation. *Neurology*, **65**, 1083–9.

Day JW, Ricker K, Jacobsen JF *et al.* (2003). Myotonic dystrophy type 2: molecular, diagnostic and clinical spectrum. *Neurology*, **60**, 657–64.

DiMauro S (2006). Mitochondrial myopathies. *Curr Opin Rheumatol*, **18**, 636–41.

Dubowitz V, Sewry C (2007). *Muscle Biopsy. A Practical Approach*. 3rd Edn Elsevier, Amsterdam.

Eagle M, Baudouin SV, Chandler C *et al.* (2002). Survival in Duchenne muscular dystrophy: improvements in life expectancy since 1967 and the impact of home nocturnal ventilation. *Neuromuscul Disord*, **12**, 926–9.

Evoli A, Tonali PA, Padua L *et al.* (2003). Clinical correlates with anti-MuSK antibodies in generalized seronegative myasthenia gravis. *Brain*, **126**, 2304–11.

Fang H, Xie Q, Boneva R *et al.* (2006). Gene expression profile exploration of a large dataset on chronic fatigue syndrome. *Pharmacogenomics*, **7**, 429–40.

Farrugia ME, Robson MD, Clover L et al. (2006). MRI and clinical studies of facial and bulbar muscle involvement in MuSK antibody-associated myasthenia gravis. Brain, 129, 1481–92.

Felice KJ, North WA, Moore SA et al. (2000). FSH dystrophy 4q35 deletion in patients presenting with facial-sparing scapular myopathy. Neurology, 54, 1927–31.

Fontaine K, Semonin O, Legarde JP et al. (2005). A new mutation of the noggin gene in a French Fibrodysplasia ossificans progressiva (FOP) family. Genet Couns, 16, 149–54.

Fournier E, Arzel M, Sternberg D et al. (2004). Electromyography guides toward subgroups of mutations in muscle channelopathies. Ann Neurol, 56, 650–61.

Fournier E, Viala K, Gervais H et al. (2006). Cold extends electromyography distinction between ion channel mutations causing myotonia. Ann Neurol, 60, 356–65.

Goebel HH (2005). Congenital myopathies in the new millennium. J Child Neurol, 20, 94–101.

Grain L, Cortina-Borja M, Forfar C et al. (2001). Cardiac abnormalities and skeletal muscle weakness in carriers of Duchenne and Becker muscular dystrophies and controls. Neuromuscul Disord, 11, 186–91.

Hanna MG (2006). Genetic neurological channelopathies. Nat Clin Pract Neurol, 2, 252–63.

Hoogerwaard EM, de-Voogt WG, Wilde AA et al. (1997). Evolution of cardiac abnormalities in Becker muscular dystrophy over a 13-year period. J Neurol, 244, 657–63.

Isackson PJ, Bennett MJ, Vladutiu GD (2006). Identification of 16 new disease-causing mutations in the CPT2 gene resulting in carnitine palmitoyltransferase II deficiency. Mol Genet Metab, 89, 323–31.

Klinge L, Straub V, Neudorf U et al. (2005). Safety and efficacy of recombinant acid alpha-glucosidase (rhGAA) in patients with classical infantile Pompe disease: results of a phase II clinical trial. Neuromuscul Disord, 15, 24–31.

Kung AW (2006). Clinical review: Thyrotoxic periodic paralysis: a diagnostic challenge. J Clin Endocrinol Metab, 91, 2490–5.

Lampe AK, Bushby KM (2005). Collagen VI related muscle disorders. J Med Genet, 42, 673–85.

Laval SH, Bushby KM (2004). Limb–girdle muscular dystrophies—from genetics to molecular pathology. Neuropathol Appl Neurobiol, 30, 91–105.

Machuca-Tzili L, Brook D, Hilton-Jones D (2005). Clinical and molecular aspects of the myotonic dystrophies: A review. Muscle and Nerve, 32, 1–18.

Manzur AY, Kuntzer T, Pike M et al. (2004). Glucocorticoid corticosteroids for Duchenne muscular dystrophy. Cochrane Database Syst Rev, CD003725.

Mastaglia FL (2006). Drug induced myopathies. Pract Neurol, 6, 4–13.

Mastaglia FL, Lamont PJ, Laing NG (2005). Distal myopathies. Curr Opin Neurol, 18, 504–10.

Meune C, Van Berlo JH, Anselme F et al. (2006). Primary prevention of sudden death in patients with lamin A/C gene mutations. N Engl J Med, 354, 209–10.

Muntoni F, Brockington M, Torelli S et al. (2004). Defective glycosylation in congenital muscular dystrophies. Curr Opin Neurol, 17, 205–9.

Naimi M, Bannwarth S, Procaccio V et al. (2006). Molecular analysis of ANT1, TWINKLE and POLG in patients with multiple deletions or depletion of mitochondrial DNA by a dHPLC-based assay. Eur J Hum Genet, 14, 917–22.

Newton WA Jr., Gehan EA, Webber BL et al. (1995). Classification of rhabdomyosarcomas and related sarcomas. Pathologic aspects and proposal for a new classification—an Intergroup Rhabdomyosarcoma Study. Cancer, 76, 1073–85.

Olive M, Goldfarb LG, Shatunov A et al. (2005). Myotilinopathy: refining the clinical and myopathological phenotype. Brain, 128, 2315–26.

Padberg GW (1998). In Neuromuscular Disorders:Clinical and Molecular Genetics, Emery AEH, ed. pp. 105–21. J. Wiley and sons, Chichester.

Pinelli G, Dominici P, Merlini L et al. (1987). Cardiologic evaluation in a family with Emery-Dreifuss muscular dystrophy. G Ital Cardiol, 17, 589–93.

Rankin J, Ellard S (2006). The laminopathies: a clinical review. Clin Genet, 70, 261–74.

Reardon W, Newcombe R, Fenton I et al. (1993). The natural history of congenital myotonic dystrophy: mortality and long term clinical aspects. Arch Dis Child, 68, 177–81.

Ricker K, Koch MC, Lehmann-Horn F et al. (1995). Proximal myotonic myopathy. Clinical features of a multisystem disorder similar to myotonic dystrophy. Arch Neurol, 52, 25–31.

Robinson R, Carpenter D, Shaw MA et al. (2006). Mutations in RYR1 in malignant hyperthermia and central core disease. Hum Mutat, 27, 977–89.

Rowland LP, McLeod JG (1994). Classification of neuromuscular disorders. J Neurol Sci, 124 (Suppl), 109–30.

Samaha FJ, Quinlan JG (1996). Myalgia and cramps: dystrophinopathy with wide-ranging laboratory findings. J Child Neurol, 11, 21–4.

Schessl J, Zou Y, Bonnemann CG (2006). Congenital muscular dystrophies and the extracellular matrix. Semin Pediatr Neurol, 13, 80–9.

Sekul EA, Chow C, Dalakas MC (1997). Magnetic resonance imaging of the forearm as a diagnostic aid in patients with sporadic inclusion body myositis. Neurology, 48, 863–6.

Smith PM, Ross GF, Taylor RW et al. (2004). Strategies for treating disorders of the mitochondrial genome. Biochim Biophys Acta, 1659, 232–9.

Straub V, Bushby K (2006). The childhood limb–girdle muscular dystrophies. Semin Pediatr Neurol, 13, 104–14.

Stum M, Davoine CS, Vicart S et al. (2006). Spectrum of HSPG2 (Perlecan) mutations in patients with Schwartz-Jampel syndrome. Hum Mutat, 27, 1082–91.

Talbot K, Stradling J, Crosby J et al. (2003). Reduction in excess daytime sleepiness by modafinil in patients with myotonic dystrophy. Neuromuscul Disord, 13, 357–64.

Tawil R, Figlewicz DA, Griggs RC et al. (1998). Facioscapulohumeral dystrophy: a distinct regional myopathy with a novel molecular pathogenesis. FSH Consortium. Ann Neurol, 43, 279–82.

Tawil R, Van Der Maarel SM (2006). Facioscapulohumeral muscular dystrophy. Muscle Nerve, 34, 1–15.

Tyni T, Pourfarzam M, Turnbull DM (2002). Analysis of mitochondrial fatty acid oxidation intermediates by tandem mass spectrometry from intact mitochondria prepared from homogenates of cultured fibroblasts, skeletal muscle cells, and fresh muscle. Pediatr Res, 52, 64–70.

Vladutiu GD, Simmons Z, Isackson PJ et al. (2006). Genetic risk factors associated with lipid-lowering drug-induced myopathies. Muscle Nerve, 34, 153–62.

von Tell D, Somer H, Udd B et al. (2002). Welander distal myopathy outside the Swedish population: phenotype and genotype. Neuromuscul Disord, 12, 544–7.

Wells DJ (2006). Therapeutic restoration of dystrophin expression in Duchenne muscular dystrophy. J Muscle Res Cell Motil, 27, 387–98.

Woodman SE, Sotgia F, Galbiati F et al. (2004). Caveolinopathies: mutations in caveolin-3 cause four distinct autosomal dominant muscle diseases. Neurology, 62, 538–43.

Yoshioka M, Yorifuji T, Mituyoshi I (1998). Skewed X inactivation in manifesting carriers of Duchenne muscular dystrophy. Clin Genet, 53, 102–7.

SECTION 6

Structural disease affecting brain, spinal cord, and nerve roots

CHAPTER 25

Head injury

Ian Whittle

Contents

25.1 Epidemiology of head injury

Head injury or traumatic brain injury is a ubiquitous phenomenon in all societies and affects up to 2 per cent of the population per year (Bullock *et al.* 2006). Although the causes of head injury and its distribution within populations vary, it can have devastating consequences both for the patient and family (Tagliaferri *et al.* 2006). In some countries severe traumatic brain injury is the commonest cause of death in people under 40 years (Lee *et al.* 2006), and it is estimated that the sequelae of head injury cost societies billions of dollars per year. Understanding of the pathophysiology, diagnosis, and management have all improved dramatically in the last few decades (Steudel *et al.* 2005). However within western society, perhaps one of the greatest benefits has been the reduction in severe craniocerebral injuries following motor vehicle accidents. This has arisen because of increased safety in car design, seat-belt legislation, the introduction of air-bags, enforcement of speed limits, and the societal conformity to drink-driving legislation. For instance, because of these changes, in the last 15 years the number of severe head injuries managed in the Clinical Neuroscience unit in Edinburgh has decreased by around 66 per cent. Unfortunately in some developing countries one legacy of increased traffic, particularly of motor cycles, is an epidemic of head injuries amongst young adults (Lee *et al.* 2006). With the number of severe head injuries declining in many countries the challenge will be to provide better care for patients with minor head injury, about 10 times more common than severe injury (Steudel *et al.* 2005)

Ageing patients who tend to fall over, falls associated with increased alcohol consumption, and domestic or social assaults probably now contribute to the majority of head injuries (Flanagan *et al.* 2005; Steudel *et al.* 2005; Tagliaferri *et al.* 2006). Sporting injuries are fortunately uncommon as a cause of severe craniocerebral injury, although horse riding accidents can sometimes be devastating particularly in teenage girls. In some countries injuries from hand guns and other missiles are common (Aryan *et al.* 2005), but in European countries many such injuries are self-inflicted. Prompt management of intracranial haematoma, which occurs in 25–45 per cent of severe head injuries, 3–12 per cent of moderate injuries, and 0.2 per cent of minor injuries, and the rehabilitation of patients with head injury are now important areas in clinical neuroscience (Flanagan *et al.* 2005; Bullock *et al.* 2006b, c).

25.2 Primary craniocerebral injury

25.2.1 Scalp injuries

The scalp and underlying soft tissues, particularly the temporalis and occipitofrontalis muscles, galea, and periostium are the first tissue layers subjected to insult during traumatic craniocerebral injury. Certain types of injury lacerate the scalp and underlying soft tissues whereas blunt impact injuries lead to crush or contusional injury of the scalp and underlying muscles with scalp haematoma or bruising. In children this can lead to a subgaleal haematoma large enough to cause anaemia (Anton *et al.* 1999). Since the scalp is extremely well supplied by blood vessels which lie just superficial to the galeal layers, profuse blood loss may occur if pressure is not applied to the wound or bleeding vessels ligated (Turnage *et al.* 2000). If the superficial temporal artery or occipital artery is damaged there may be false aneurysm development or even traumatic arteriovenous fistulae (Amirjamshidi *et al.* 2000; Cadamy *et al.* 2003; Aquilina *et al.* 2005). Because of the rich blood supply even extensive scalp wounds generally heal very well and quickly.

Another important factor about scalp injuries is the possibility of an underlying compound depressed skull fracture. Often the fracture does not lie directly underneath the laceration because of tissue distortions generated by the forces applied during the insult. A high index of suspicion, careful evaluation of the mechanism of injury, careful clinical examination, and interpretation of either skull X-ray or CT scan are required to forestall later intracranial and extracranial infection (Shokunbi *et al.* 2000).

25.2.2 Skull fractures

Skull fractures are termed open or compound, which underly a scalp laceration, or closed, in which there is no scalp laceration. They may be linear, comminuted with the bone broken into fragments or depressed. The presence of a skull fracture is an important pointer to the likelihood of significant primary or secondary brain injury, especially if accompanied by a depressed Glasgow coma score (Table 25.1). A patient without a fracture and a Glasgow coma score of 15 has a risk of intracranial haematoma of 1:8000. If the patient is in coma with no fracture the risk rises to around 1:30, and if there is a fracture the risk of intracranial haematoma is 1:4. Similar patterns are seen in children with head injury (Teasdale *et al.* 1990). These associations are not surprising given the considerable force required to fracture the skull.

Depending on the amount of force delivered to the cranium there may be diastasis or separation of the sutures (Fig. 25.1c). Diastasis is usually seen in the context of a severe head injury that runs to either the coronal or saggital sutures. Diastasis of the sutures is usually associated with dural tearing and large subgaleal haematomas due to periostial elevation by extravasation of blood from the fracture site. An underlying major venous sinus may be involved and cause haematoma accumulation.

Table 25.1 The estimated number of patients attending hospital in the United Kingdom with head injury/year/million total population and the absolute risk of traumatic intracranial haematoma (from Teasdale *et al.* 1990)

	Adults		Children	
	Number attending	Risk(l:n)	Number attending	Risk(l:n)
No fracture				
Conscious	10700	7866	8100	12599
Impaired consciousness	630	180	280	580
Coma	75	27	26	65
Fracture				
Conscious	10	4	110	157
Impaired consciousness	49	5.1	19	25
Coma	46	3.6	12	12

Depressed fractures are caused by focal impact injury, for instance hammer blows, the edge of a bottle, or the end of a baseball bat (Fig. 25.1). Such fractures reflect relatively focussed high-impact energy transfer to the cranium. The central point of impact is often maximally depressed and the inner and outer tables of the bone may separate with the inner table being indriven. This may tear the dura and cause an underlying brain laceration or contusion. The extent of depression is related to the thickness of the skull, the region of the skull traumatized, and the amount of force conveyed.

25.2.3 Concussion

Transient loss of consciousness or concussion is a common feature of even minor head injury. It is a common occurrence in many contact sports. The pathophysiology has recently been reviewed extensively (McCrory and Berkovic 2001; Shaw 2002; Anderson *et al.* 2006) and the major theories critically evaluated. These invoke vascular, reticular, centripetal, pontine cholinergic, and convulsive hypotheses. It was concluded that only the convulsive theory is readily compatible with the experimental neurophysiological data from EEG and evoked potentials and could provide a totally viable explanation for the clinical features of concussion. According to the convulsive theory part of impact force to the cranium transmits a rotational and deaccelerative force to the cerebrum causing a collision with the inner skull vault, that leads to transient de-afferentation of the cortex. This functional deafferentation of the cortex is a consequence of diffuse mechanically induced depolarization and synchronized discharge of cortical neurons. Such a mechanism fits well with the relatively high incidence of concussive convulsions of 1 in 70, or what has been considered early post-traumatic epilepsy (McCrory and Berkovic 1998). These concussive convulsions are not associated with later onset epilepsy (Perron *et al.* 2001).

A convulsive theory can also explain traumatic amnesia, autonomic disturbances, and the miscellaneous collection of symptoms of the post-concussion syndrome more adequately than any of its rivals (Shaw 2002). Concussion *per se* does not appear to be related to any single recognizable neuroradiological or neuropathological

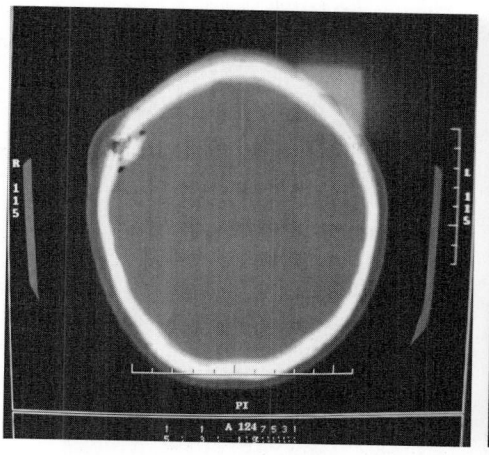
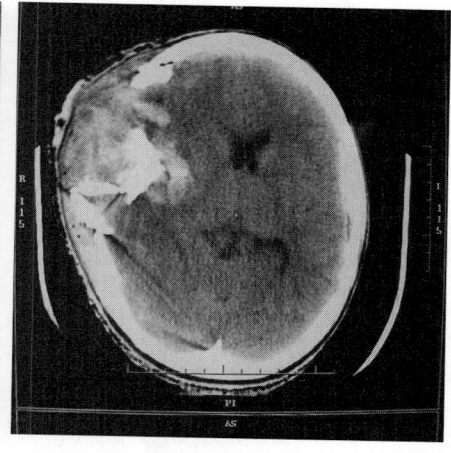

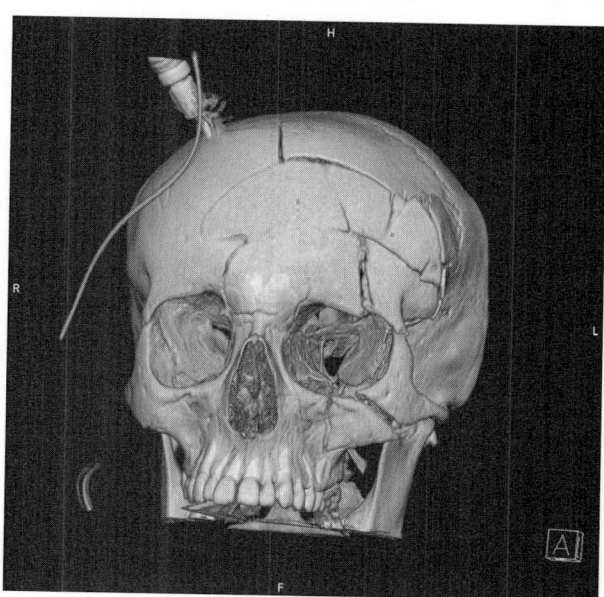

Fig. 25.1 Skull fractures of varying severity shown by a range of CT scans. (A) A compound depressed vault fracture following a hammer blow to the head. As well as the depressed fragment several pockets of intracranial air can be seen adjacent to the fracture. (B) A severe compound depressed fracture with a large area of cranial vault disruption and a large intracranial haematoma underlying the injury. (C) is a 3 dimensional reconstruction of a complex craniofacial fracture. The left zygoma, orbital rim, frontal and temporal bones are clearly fractured. The intracranial pressure monitor is reconstructed on the contralateral side of the head.

entity although repeated episodes of concussion, as in boxers, can lead to chronic changes in the brain (Roberts *et al.* 1990).

25.2.4 Traumatic subarachnoid haemorrhage

Traumatic subarachnoid haemorrhage is a common finding on CT scan following moderate or severe head injury. It is associated with a poorer outcome than in similar patients with idiopathic subarachnoid haemorrhage (Mattioli *et al.* 2003; Hanlon *et al.* 2005). The poorer outcome seems to be associated with volume of initial haemorrhage, admission Glasgow coma score, and associated brain lesions (Chieregato *et al.* 2005). In many patients traumatic subarachnoid haemorrhage on an initial scan is followed by evolution of other brain lesions (Mattioli *et al.* 2003; Fainardi *et al.* 2004).

From the clinical perspective traumatic subarachnoid haemorrhage may mimic spontaneous subarachnoid haemorrhage, since not infrequently patients suffering aneurysmal rupture may fall and hit their heads at the time of rupture. Although spontaneous subarachnoid haemorrhage classically has a basal distribution and is heavier around the aneurysm, in some cases a high index of suspicion is required and CT-angiography is warranted.

25.2.5 Acute extradural haematoma

This is a collection of blood between the inner table of bone and endostial dural, and is usually a complication of a skull fracture. The dura is sheared from the inner table of bone with associated tears in meningeal vessels, and bleeding from diploic veins or adjacent venous sinus. Classically acute extradural haematoma tends to occur in the middle fossa following a temporal fracture and tearing of either the trunk or a branch of the middle meningeal artery. However, extradural haematoma can occur in any location, and in 31 per cent of surgical cases no bleeding source can be found (Bullock *et al.* 2006a). It is readily diagnosable on CT scanning because of its concave shape (Fig. 25.2A). An extradural haematoma occurs more readily in children and young adults since the dura is only loosely attached onto the inner table of the skull. With advancing age over > 50 years the dura frequently ossifies onto the

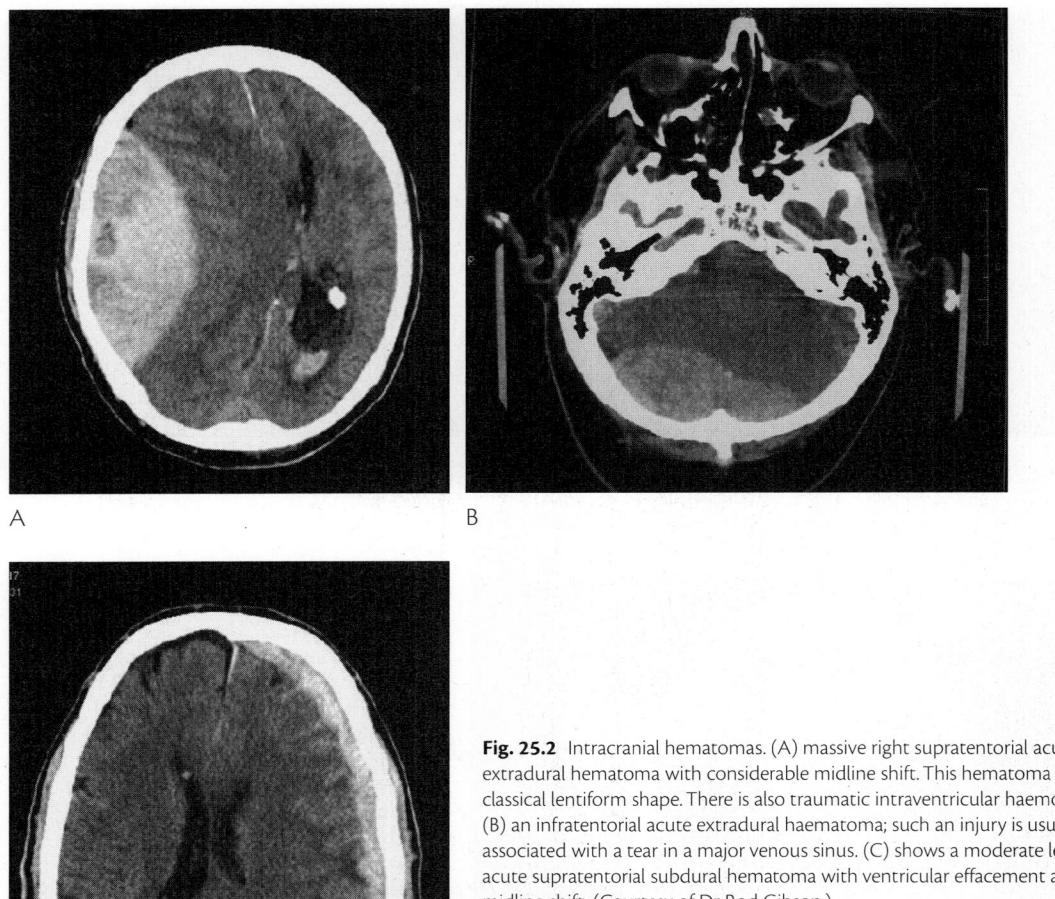

Fig. 25.2 Intracranial hematomas. (A) massive right supratentorial acute extradural hematoma with considerable midline shift. This hematoma has the classical lentiform shape. There is also traumatic intraventricular haemorrhage. (B) an infratentorial acute extradural haematoma; such an injury is usually associated with a tear in a major venous sinus. (C) shows a moderate left-sided acute supratentorial subdural hematoma with ventricular effacement and midline shift. (Courtesy of Dr Rod Gibson.)

inner table thus making development of the extradural plane by a haematoma much less likely (Bullock *et al.* 2006a).

Since extradural haematoma complicates skull fracture in many patients, the primary brain injury may be minimal. In such cases there is often a 'lucid' interval before enlargement of the haematoma leads to deterioration in the Glasgow coma score. In a recent review, lucid intervals were documented in 47 per cent of patients with extradural haematoma (Bullock *et al.* 2006b). In head-injured patients presenting with coma, with a Glasgow coma score < 8, around 10 per cent have an extradural haematoma that requires evacuation (Bullock *et al.* 2006a).

25.2.6 **Acute subdural haematoma**

This is a more common intracranial haematoma than extradural haematoma. It is due to either laceration of vessels, especially the small cerebral veins, on either the brain surface or bridging the subdural space, or traumatic 'bursting' of a cerebral contusion into the subdural space. It is not uncommon following falls in the elderly and inebriated patients, assaults, and motor vehicle accidents

(Bullock *et al.* 2006b). Diagnosis is made by CT, which shows a haematoma that is concave on its inner surface and between the brain and skull (Fig. 25.2B,C). Morbidity and mortality is often high because of the severity of associated primary or secondary brain injury. Traumatic subarachnoid haemorrhage, which occurs in 14–25 per cent of acute subdural haematomas, extradural haematoma, occuring in 6–14 per cent, brain contusions, and significant extracranial injuries all compound management and lead to poor outcome (Bullock *et al.* 2006a). A recent review states cumulative mortalities of between 40 and 60 per cent for all patients with acute subdural haematoma requiring surgery and 57–68 per cent for patients in coma subsequently requiring evacuation (Bullock *et al.* 2006a).

Spontaneous acute subdural haematoma may also occur due to rupture of intracranial aneurysms into the subdural space (O'Sullivan *et al.* 1994), in patients with coagulation abnormalities or on anticoagulant therapy (Depreitere *et al.* 2003), after weight lifting, or use of anabolic steroids or cocaine (Alaraj *et al.* 2005).

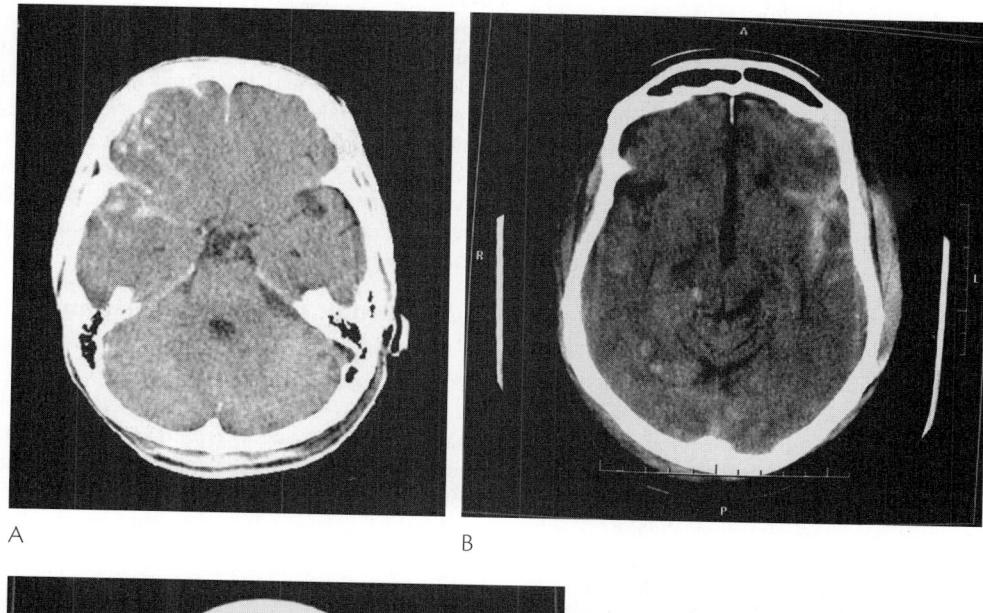

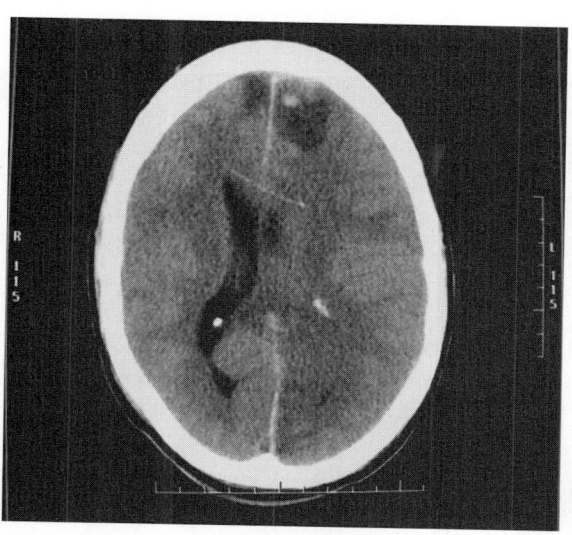

Fig. 25.3 Cerebral contusions shown by (A) relatively localized right frontal and anterior temporal contusions following a fall on the back of the head. (B) Right parietal contusions and left traumatic subarachnoid haemorrhage within the sylvian fissure in an elderly alcoholic following a simple fall. (C) Severe brain swelling complicating bifrontal and left temporal contusions 4 days after injury. There is left lateral ventricular effacement, and midline shift to the right. There is also early infarction in the territory supplied by the left posterior cerebral artery due to uncal herniation.

25.2.7 **Intracerebral haematoma and brain contusions**

Craniocerebral trauma can cause focal intracerebral haematoma. This may be a focus or foci of brain contusions, or *ab initio* a substantial intracerebral haematoma (Fig. 25.3). Brain contusions are common in the orbito-frontal and temporal polar regions (Ribas *et al.* 1992). These lesions are found in between 8–35 per cent of craniocerebral trauma patients with the incidence being higher in patients with severe injury (Bullock *et al.* 2006c). Some patients with this type of injury may present in initially good Glasgow coma score but be drowsy. Brain contusions commonly evolve both with increasing perilesional oedema and enlargement of their haemorrhagic component (Yamaki *et al.* 1990; Lobato *et al.* 1997). The latter mechanism may give rise to delayed traumatic intracerebral haematoma (Young *et al.* 1984). In other patients such delayed haematomas may occur in regions of the brain that were normal on the original CT scan (Gentleman *et al.* 1989).

As the intracerebral haematoma and perilesional oedematous brain enlarges, an initially well patient may deteriorate dramatically, intracranial pressure may rise precipitously, and brain herniation may occur (Fig. 25.3C). Studies of regional cerebral blood flow in traumatic intracerebral haematoma have shown the haemorrhagic core is ischaemic with variable flows in the surrounding oedematous brain, resulting in ischaemia in around 25 per cent (Chieregato *et al.* 2004). The patterns of blood flow also varied with time, with initial ischaemia, recovery for 2–4 days, and then a further ischaemic phase. Contusions can cause longer-term, severe cognitive and focal sensorimotor deficits. Surgery for contusions and or traumatic intracerebral haematoma accounts for about 20 per cent of trauma surgery (Bullock *et al.* 2006c).

25.2.8 **Diffuse axonal injury**

This type of injury is caused by severe unrestrained rotational head movements. It is common after high speed motor vehicle accidents. The neuropathology of diffuse axonal injury is complex. Understanding of the neuropathology has evolved over the last 20 years as clinicopathological correlation and advances in immunohistochemistry of axonal proteins has led to more detailed clinical and experimental study (Graham and Gennarelli 1997;

Graham *et al.* 2000). It was initially considered to be a mechanical phenomenon that resulted in focal or diffuse shearing of axons from the neuronal bodies. However neuropathological studies have shown that rapid stretch of axons can damage the internal axonal cytoskeleton resulting in a loss of elasticity and impairment of axoplasmic transport (Smith *et al.* 2003). A subsequent cascade of intracellular events centres around impaired axonal transport. There is disruption and impaction of neurofilaments lead to axonal swelling, with either discrete bulb formation or elongated varicosities that accumulate and impede transported proteins (Smith *et al.* 2003; Marmarou *et al.* 2006). Calcium entry into damaged axons is thought to initiate further damage by the activation of proteases. Ultimately, the swollen axons may become disconnected and degenerate. This process varies in different axonal groups and the response pattern is time dependent (Chung *et al.* 2005).

In patients sustaining widespread diffuse axonal injury, the Glasgow coma score is usually low from the moment of injury and the intracranial pressure is often normal. Paradoxically the CT scan may also appear normal, or there may be only small punctate brain contusions often located in the centrum ovale, dorsolateral brain stem, and corpus callosum. Recently diffusion transfer-MRI of patients with mild diffuse axonal injury has shown changes in mean diffusivity, and fractional anisotropy can detect regions of damage (Inglese *et al.* 2005). Other MR techniques such as gradient-recalled echo and turbo proton echo-planar spectroscopic imaging MRI sequences also can detect large numbers of diffuse axonal injury type lesions (Giugni *et al.* 2005). Because of the diffuse nature of the brain injury severe neurological and cognitive deficits are common in survivors.

25.2.9 Chronic subdural haematoma

An increasingly common problem with the ageing population is chronic subdural haematoma. It can occur after relatively minor head trauma particularly in the elderly since the subdural space may be capacious due to craniocerebral disproportion caused by cerebral atrophy within a fixed diameter cranium. Elderly patients, on aspirin or anticoagulants are particularly predisposed to chronic subdural haematoma. It is an extracerebral collection, which varies in viscosity from a crank-case oil-type consistency, derived from break down of the blood clot, to a fluid which in colour and

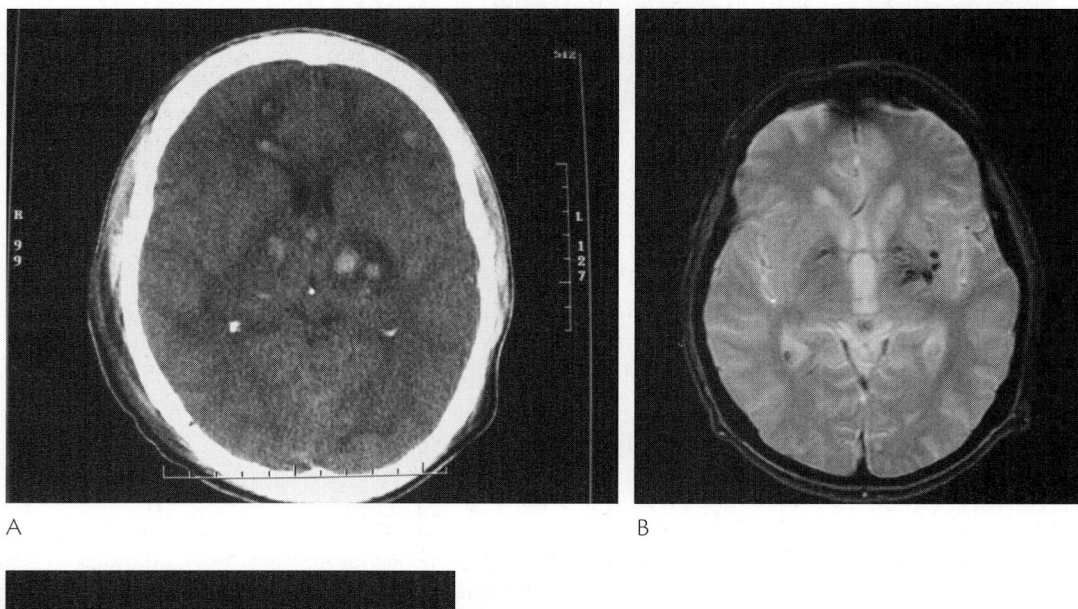

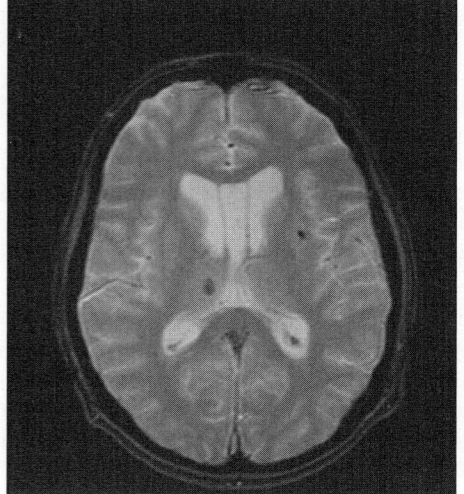

Fig. 25.4 Diffuse axonal injury. (A) This CT scan shows characteristic petechial-type haemorrhages deep within the brain associated with severe rotational head injuries and diffuse axonal injury. (B) and (C) show MRI T2-weighted gradient echo sequences of a 65-year-old man who suffered a severe head injury after a fall from a mountain bike. The corresponding CT scan was almost normal. The MR scan shows scattered petechial haemorrhages in the left globus pallidus region (B) and throughout the brain (C). This man initially had a severe extrapyramidal syndrome before going on to making an excellent recovery. (Courtesy of Dr David Summers.)

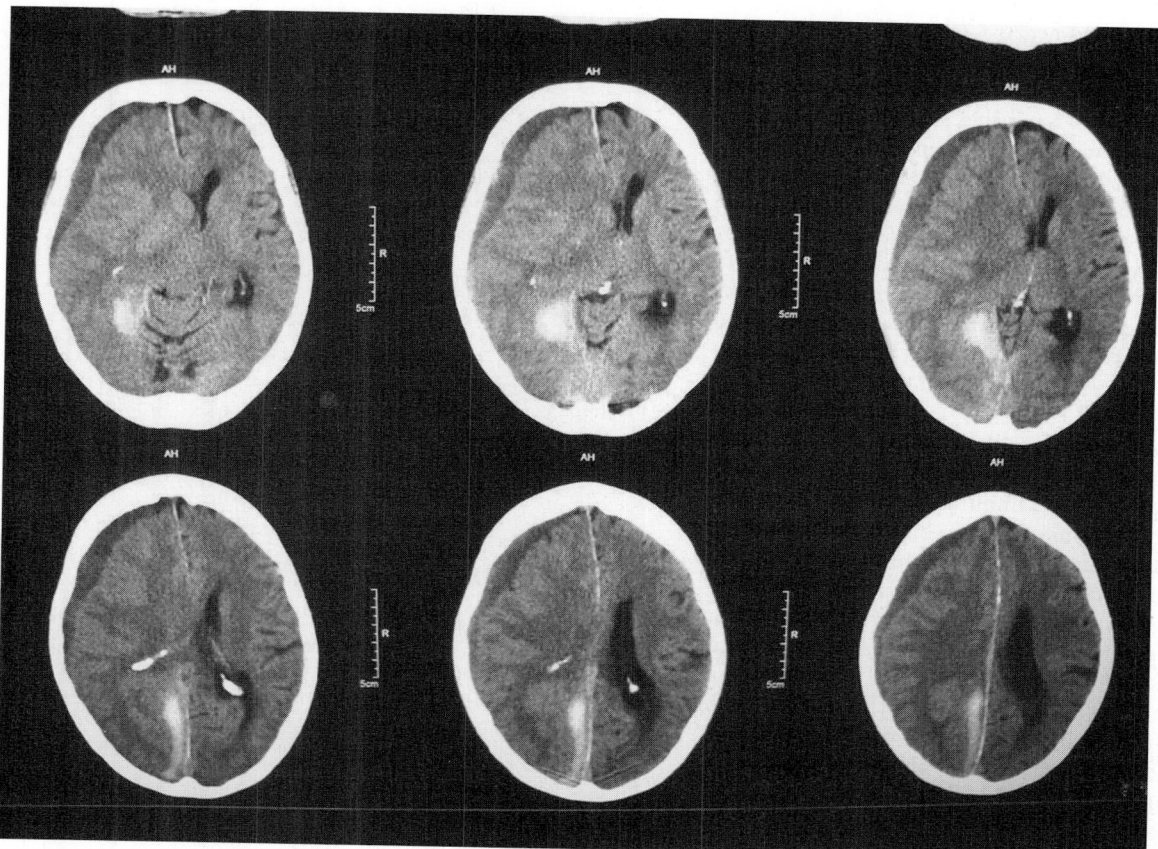

Fig. 25.5 Chronic right fronto-parietal subdural haematoma shown by a series of CT scans. This is causing local mass effect with ventricular effacement and midline shift to the left. There is also a more recent acute parafalcine subdural haematoma occipitally on the right. Collections of mixed density are a feature of chronic subdural haematoma.

consistency resembles blood-stained, brownish CSF. Not infrequently there is a mixture of blood clot, resolving brownish watery fluid, and xanthochromic CSF in loculations within a large chronic subdural haematoma. These components can be readily seen on CT scans (Fig. 25.5).

The pathogenesis of chronic subdural haematoma has been studied in some detail. There is now convincing evidence that the haematoma membrane and fluid release pro-angiogenic factors such as vascular endothelial growth factor, the pro-angiogenic factors angiopoietin 1 and 2, and tie, their receptor (Vaquero *et al.* 2002; Hohenstein *et al.* 2005). Many chronic subdural haematomas resolve without any treatment whilst those in other patients go on to become larger and larger with evidence of fresh and recurrent bleeding. The pathophysiological basis of these phenomena are not well understood.

Because the subdural collection can accumulate slowly, there may be significant midline shift, sometimes with very few signs and symptoms. Such haematomas can either mimic a wide range of clinical neurological syndromes or be asymptomatic. Some of these clinical syndromes are probably related to regional ischaemia caused by the subdural haematoma, rather than raised intracranial pressure (Inao *et al.* 2001). Many elderly patients are found to have incidental chronic subdural haematoma, or haematomas, when CT scanned for other reasons. In the majority of cases when symptoms

are minimal conservative management is warranted. Repeat imaging often shows spontaneous resolution of the haematoma.

25.2.10 Traumatic brain oedema

The nature of the oedema associated with traumatic brain injury has been long debated. For many years it was considered vasogenic. However more recent experimental and clinical findings suggest it is cellular or cytotoxic in origin (Marmarou *et al.* 2006) This is an important finding since therapies directed at traumatic brain oedema are vital in controlling raised intracranial pressure (Sections 25.3.1 and 26.3).

25.3 Secondary craniocerebral injury

25.3.1 Raised intracranial pressure

The brain is enclosed within the skull which is a rigid bony container. Intracranial pressure therefore depends on the relative volumes of intracranial blood, CSF, and brain parenchyma. Intracranial pressure also fluctuates in response to changes in intrathoracic pressure, being increased by coughing, defecation, or cardiac pulsation. These transient increases do no harm. In a normal supine adult intracranial pressure is the same as the CSF pressure obtained at lumbar puncture, around 5–18 cm water or 4–15 mm mercury). In patients with an acute traumatic intracranial mass lesion consisting of haemorrhage and oedema, the additional mass volume is initially compensated for by a reduction in cerebral blood volume and CSF volume. However, a critical point is soon reached where no further compensation is possible, and any additional volume insult will lead to exponential rises in intracranial pressure. Eventually a final plateau pressure is reached when intracranial pressure is close to mean arterial pressure (Lofgren *et al.* 1973; Steiner *et al.* 2006) Generalized or localized increases in intracranial pressure may lead to ventricular effacement and midline shift, marked displacement of intracranial structures with brain herniation syndromes. There may also be either focal or generalized impairment of brain perfusion.

25.3.2 Cerebral perfusion pressure

The cerebral perfusion pressure, CPP, is the sum of the mean arterial pressure, MAP, less the intracranial pressure, ICP: thus CPP = MAP-ICP. Progressive rises in intracranial pressure lead to increases in mean arterial pressure and reflex bradycardia. However, if there is a severe and sustained elevation of intracranial pressure autoregulation will be ineffective and cerebral perfusion may be focally or generally compromised, leading to cerebral ischaemia and infarction. The relative contributions of raised intracranial pressure and hypotension to the impaired cerebral perfusion pressure may be important for directing therapy (Marmarou *et al.* 2005), and systemic vascular hypotension can produce profound secondary insult to the brain (Chesnut *et al.* 1993). Focal brain ischaemia may complicate the brain herniation syndromes. A cerebral perfusion pressure of > 60–70 mmHg is generally required to sustain adequate cerebral perfusion (Chan *et al.* 1992; Steiner *et al.* 2006). Although children and young adults can tolerate lower levels, profound and prolonged reduction of cerebral perfusion pressure often produce devastating brain hypoxia and ischaemia (Chesnut *et al.* 1993; Kirkness *et al.* 2005).

25.3.3 Brain herniation syndromes

Within the cranial cavity there are three compartments, two supratentorial and one infratentorial, that are separated by the falx and tentorium (Fig. 25.6). The supratentorial compartments communicate beneath the falx, and both communicate with the infratentorial compartment through the tentorial hiatus. The infratentorial compartment also communicates with the spinal compartment through the foramen magnum. Because the brain has a jelly-like consistency it is easily compressed by haematomas. As a result brain displacement caused by a traumatic mass lesion may lead to shifts between compartments (Miller *et al.* 1997). These give rise to either subtle or obvious clinical syndromes.

- *Subfalcine herniation of the cingulate gyrus.* With a unilateral supratentorial traumatic mass the ipsilateral cingulate gyrus may herniate beneath the free edge of the falx (Miller *et al.* 1997). The anterior cerebral artery may be compressed sufficiently to cause medial hemispheric infarction. Usually there are no specific clinical signs except deteriorating conscious level due to hemispheric compression.

- *Transtentorial uncal herniation.* With large ipsilateral supratentorial intracranial traumatic mass lesions the uncal medial part of the temporal lobe is pushed down through the tentorial notch to become wedged between the tentorial edge and the midbrain (Miller *et al.* 1997). This compresses the midbrain, cerebral aqueduct, and ipsilateral posterior cerebral artery which curves

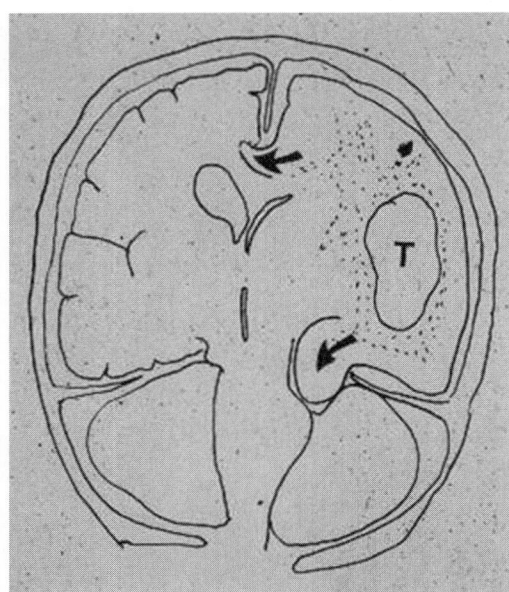

Fig. 25.6 Diagram showing the intracranial compartment, the tentorium cerebelli and falx cerebri, and both subfalcine and transtentorial herniations (arrowed). Herniation syndromes arise because of lacal mass effect and pressure gradients. If T is considered a mass of any etiology, for instanc e tumour, haematoma, abscess, it may cause perilesional oedema, ventricular effacement, and midline shift. If the mass effect is large enough herniation can occur beneath the falx or across the tentorium.

over the tentorial edge in its path posteriorly to the occipital region. With progressive displacement the contralateral cerebral peduncle may be pushed against the tentorial edge. The phenomenon of uncal herniation leads to a classical syndrome of midbrain and oculomotor dysfunction. If this is not recognized or cannot be effectively treated punctuate midbrain haemorrhages, posterior cerebral artery territory infarction, and contralateral ventricular dilatation may occur. The clinical features of an uncal herniation, most often due to a traumatic intracranial haematoma, are:

- Glasgow coma score falls;

- Motor responses become asymmetrical;

- Ipsilateral pupil dilation with loss of light reactivity;

- Blood pressure rises;

- Bradycardia;

- Respiratory rate slowing with apnoea; and

- Posterior cerebral artery infarction in some survivors.

- *Bilateral uncal herniation.* Bilateral uncal herniation, also termed central herniation, can occur with severe bilateral hemispheric swelling. This is common following myocardial infarction or other major cardiac events that render the brain diffusely ischaemic and hypoxic.

- *Foraminal herniation of the tonsils.* Mass lesions within the posterior cranial fossa, which is relatively small in volume compared to the supratentorial compartments, can cause herniation of the cerebellar tonsils and medulla downwards through the foramen magnum (Miller *et al.* 1997). The cerebellar impaction leads to medullary compression, which in turn can lead to a dramatic decrease in the Glasgow coma score, acute hypertension, bilateral extensor responses, and bilateral fixed dilated pupils followed by sudden respiratory arrest. A similar syndrome may occur following removal of CSF by lumbar puncture in patients with raised intracranial pressure due to a posterior fossa tumour.

25.3.4 Pyrexia, seizures, and hypoxia

Pyrexia, seizures, and hypoxia complicate the course of many patients with head injury (Chesnut et al. 1993). Hypoxia can lead to devastating global brain injury and may be caused by an associated pneumothorax, adult respiratory distress syndrome or 'shock lung', pulmonary oedema, ventilator-associated respiratory infections, or aspiration pneumonitis. Measurement of the difference between arterial and venous blood oxygenation levels through a jugular bulb catheter gives an estimate of the metabolic status of the brain, and can be used to guide therapies (de Deyne et al. 1998). Pyrexia and seizures increase the metabolic demands of the brain often at times of inadequate perfusion reserve, disordered regulation of regional cerebral blood flow, or impaired cerebral perfusion pressure. The absolute temperature level, and its duration, required to cause secondary hyperthermic neuronal damage is not known (Childs et al. 2006).

25.4 Clinical features, assessment, and management

Neurological assessment and management of a head-injured patient form a continuum. Early assessment determines initial management, and evolution of the management strategy depends upon subsequent assessments of clinical, investigative, and monitoring parameters. This allows early recognition of potential complications of the primary head injury, as they emerge, with subsequent introduction of therapies that will minimize potential secondary cerebral insults such as hypotension, ischaemia, hypoxia, pyrexia, seizures, infection, and raised intracranial pressure. The introduction of telemedicine, using either video conferencing or image transfer technology, means that remote advice can be provided for many patients about assessment and emergency management relatively soon after their primary trauma assessment (Wong et al. 2006).

25.4.1 Primary clinical screening and Glasgow coma score

Patients with head injury often have extracranial injury. It is important to remember that head injury alone *never* causes hypovolaemic shock. Indeed an isolated severe head injury may cause systemic vascular hypertension and bradycardia, the *Cushing response*. A priority in treatment is the management of associated systemic complications such as severe chest injury, intra-abdominal haemorrhage, complex limb or pelvic fractures, or other injuries causing major blood volume loss. All these phenomena will lead to secondary cerebral ischaemia and hypoxia with resultant brain damage.

The Glasgow coma scale, often known as the 'GCS', is a measure of conscious level (Table 25.2). The maximum score is 15 and the minimum is 3. It sums three component domains of eye opening, limb motor, and verbal responses.

The Glasgow coma score can be used in patients of all ages and is reproducible between different observers (Teasdale et al. 1978).

The *best verbal* component of the scale needs to be adjusted to take into account the age of children, particularly under five years (Reilly et al. 1988) (Table 25.3). The best post-resuscitation Glasgow coma score is used to classify severity of head injury. The severity of craniocerebral injury is classified as:

◆ mild, Glasgow coma score from 15 to 13;

◆ moderate from 12 to 9; and

◆ severe, 8 or less.

Change in Glasgow coma score over time is very informative and trend of change is probably more important than an absolute reading at any one point in time. Introduction of the Glasgow coma score has greatly facilitated the classification and objective management of head-injured patients. It is used internationally and has underpinned major advances in head injury by its precise categorization of patient cohorts.

Coma is defined as a Glasgow coma score of 8 or less. In general, trauma patients with a Glasgow coma score of 8 or less are intubated and ventilated as part of their primary management. This is to allow adequate oxygenation and ventilation to prevent hypoxia, to avoid aspiration pneumonitis by protecting the airway, and to allow hyperventilation, which reduces the $PaCO_2$ thereby lowering intracranial pressure through cerebral vasoconstriction

The Glasgow coma score is not a substitute for neurological examination in head- injured patients but merely an important adjunct to the examination. Neurological examination should routinely include an assessment of pupillary size, eye movements, and reaction. This information together with the Glasgow coma score has a significant prognostic value on patient outcomes (Jennett et al. 1979).

The head should be examined for signs of a basal skull fracture. *Battle's sign* of ecchymosis or bruising over the mastoid process is indicative of a middle fossa skull fracture. Racoon eyes, with severe bilateral perorbital oedema and bruising, is indicative of a possible anterior fossa skull fracture. The maxillofacial skeleton should be

Table 25.2 The Glasgow coma scale (Teasdale and Jennett 1974)

Eye-opening response
4 Spontaneous
3 To speech
2 To painful stimulus
1 None
Best motor response in upper limbs
6 Obeys commands
5 Localizes
4 Withdraws (normal flexion)
3 Flexes abnormally (spastic flexion)
2 Extends
1 None
Verbal response
5 Oriented
4 Confused
3 Inappropriate words
2 Incomprehensive sounds
1 None

Table 25.3 Adelaide modification of the Glasgow coma scale for children (Simpson and Reilly 1982)

Eye response	
4 Spontaneous	
3 To speech	
2 To pain	
1 None	
Best motor response in upper limbs	Highest score appropriate for age
6 Obeys commands	>2 years
5 Localizes to pain	6 months–2 years
4 Normal flexion to pain	>6 months
3 Spastic flexion to pain	<6 months
2 Extension to pain	<6 months
1 None	
Best verbal response	
5 Oriented to place	>5 years
4 Words	>12 months
3 Vocal sounds	>6 months
2 Cries	<6 months
1 None	

assessed for Le Forte type fractures, that involve the maxillary and zygomatic bones. The nose and ears should be checked for CSF leaks, although these can be difficult to ascertain in the presence of significant bleeding from these regions. The scalp should be surveyed for lacerations and penetrating injuries. Neurological examination of the limbs will give a guide to focal brain injury, spinal injury, or peripheral nerve injury. In patients in coma who had been paralysed and ventilated before transfer, an assessment of the level of muscle paralysis is vital as a prelude to interpreting brain stem reflexes.

As with all injured patients, management commences with establishing airway, breathing, and circulation. The neck should be immobilized until a cervical spine injury has been excluded. The Glasgow coma score should be documented on arrival and following resuscitation, and the findings of a neurological survey recorded. Many patients with head injury are under the effects of alcohol and other drugs that affect the conscious level. If in doubt, assume that depressed consciousness is due to brain injury. Continued monitoring of conscious level over time by means of Glasgow coma score is a key aspect of ongoing assessment and management.

Patients with minor or moderate head injury, not requiring surgery, can be managed in a general neurosurgical ward and treated symptomatically. This generally involves observations and administration of analgesia, antiemetics, and intravenous fluids. Guidelines about investigation and management of such patients has been published by the Scottish Intercollegiate Guidelines Network (SIGN Guideline no 46, 2000. www.Sign.ac.uk; Royal College of Surgeons, 2005) and the United Kingdom's National Institute for Clinical Excellence (NICE) in 2004 (www.NICE.org.uk)

25.4.2 Neuroradiological investigation

Following resuscitation, cardiorespiratory stabilization, and prioritization if there are multiple injuries, a head CT scan is performed to determine the status of the brain parenchyma and presence of any intracranial haematoma. CT is the investigation of choice following head injury, and certainly should be done if there is a skull fracture on a skull radiograph. Additional indications for CT in minor head-injured patients include failure to reach Glasgow coma scale = 15 within 2 h, suspected basal skull fracture, more than two episodes of vomiting and age > 65 years (Stiell *et al.* 2001). It provides ready visualization of intracranial haematoma, brain contusions, depressed bone fragments, intracranial air and associated maxillofacial fractures (Fig. 25.1). Information from initial brain scanning may also forewarn the clinician about potential complications such as the likelihood of worsening intracranial post-traumatic brain oedema, delayed CSF fistulae, and requirements for rehabilitation.

The findings on initial brain scan provide, in combination with the Glasgow coma score and pupillary status, good prognostic value in determining outcome following head injury (Marshall *et al.* 1992; Wardlaw *et al.* 2002) (Table 25.4). Factors associated with a poorer outcome are midline shift > 1.5 cm, effacement of the perimesencephalic cistern, traumatic subarachnoid haemorrhage, and diffuse brain swelling (Eisenberg *et al.* 1990). Findings on CT are used also to guide initial surgical therapies. Indications for intracranial clot evacuation are > 5 mm midline shift, substantial haematoma volume, compression of the basal cisterns, or compound cranial wounds need to be surgically explored. If the CT scan shows minimal apparent brain injury, but the patient is in coma an intracranial pressure monitor is inserted for postoperative or elective intracranial pressure monitoring. Many patients' brain injuries evolve with time and this can be seen on serial scanning (Oertel *et al.* 2002).

25.4.3 Surgery for intracranial haematoma and contusions

Intracranial haematoma commonly results from neurotrauma. Indications for surgery in this group of patients include decreased Glasgow coma scale and midline shift of 5 mm or more, persisting headache emanating from small extradural haematoma, and persisting raised intracranial pressure despite multimodality intensive care management. The decision to operate is also influenced by the patient's age, any comorbidities, the haematoma size, and any pupil changes (Bullock *et al.* 2006a–c). Preoperatively the platelet count, blood coagulation parameters, and haemoglobin need to be checked. Blood should be available if transfusion is likely to be required during surgery.

Table 25.4 Diagnostic categories for severe head injury based upon CT scanning (Marshall *et al.* 1991)

Diffuse injury I	No visible pathology seen CT scan
Diffuse injury II	Cisterns are present with shift 0–5mm; No high- or mixed-density lesion >25 ml; May include bone fragments and foreign bodies
Diffuse injury III (swelling)	Cisterns compressed or absent; Shift 0–5mm; No high-or mixed-density lesion >25ml
Diffuse injury IV (shift)	Shift >5 mm; No high- mixed-density lesion >25 ml
Evacuated mass lesion	Any lesion evacuated surgically
Non- evacuated mass lesion	High- or mixed-density lesion >25 ml not evacuated surgically

Acute subdural and extradural haematomas are usually diagnosed by brain imaging shortly after the head injury in which case surgery becomes, in many cases, an integral part of primary management so as to optimize intracranial pressure. Craniotomy is commonly performed for acute subdural or extradural haematomas and for large haemorrhagic contusions or traumatic intracerebral haematoma. The precise surgical technique varies with osteoplastic flaps, free bone flaps, replacement or non-replacement of the bone flap (Bullock *et al.* 2006a, b). From a technical viewpoint these procedures are relatively straightforward. Particular complications may arise during surgery due to lacerated or torn major venous sinuses, reactive brain oedema with brain herniation through the craniotomy after removal of the blood clot, bleeding from inaccessible regions, and haemodynamic instability. Ideally the surgery should be performed as soon as possible, although some patients may require transfer to a neurosurgical centre. Despite such delay, outcome is generally better if the surgery is performed at a neurosurgical centre (Bullock *et al.* 2006a). The timing of surgery for an acute extradural haematoma is most important in those patients in coma, a better outcome being associated with shorter duration of pupil inequality (Haselsberger *et al.* 1988; Sakas *et al.* 1995; Bullock *et al.* 2006a).

◆ *Acute extradural haematoma.* The indication for surgery is a haematoma volume > 30 ml regardless of Glasgow coma score, whereas patients with haematomas of smaller volume, < 5 mm midline shift, and Glasgow coma score > 8 may be observed closely (Bullock *et al.* 2006a). In many patients the longer term outcome after surgery is often excellent, however mortality rates are 10 per cent for adults and 5 per cent in children (Bullock *et al.* 2006a). Clearly however if the extradural haematoma is merely one component of a more severe craniocerebral injury, and between 30 and 50 per cent have associated intracranial lesions, the outcome may be less satisfactory (Bullock *et al.* 2006a). Results for surgical evacuation of extradural haematoma are influenced adversely by a low initial Glasgow coma score, advancing age, and the presence of traumatic subarachnoid haemorrhage, and having intracranial pressure > 35 mmHg during the course of their management (Lobato *et al.* 1983; van den Brink *et al.* 1999; Bullock *et al.* 2006a). The importances of preoperative midline shift, haematoma volume, haematoma location, and mixed-density of the blood clot on CT scan are unclear (Bullock *et al.* 2006a). Some studies suggest worse outcome with increased haematoma volume (Rivas *et al.* 1988; Lee *et al.* 1998) whereas others did not (van den Brink *et al.* 1999).

◆ *Acute subdural haematoma.* Indications for surgery are similar to the parameters for acute extradural haematomas with > 5 mm midline shift, and >10 mm haematoma thickness being the primary indicators (Bullock *et al.* 2006b). Similar factors of age, initial Glasgow coma score, and pupil asymmetry generally influence outcome following surgical evacuation of a subdural haematoma. However the overall outcome is much worse for subdural than extradural haematomas. This is partly because many patients with acute subdural haematomas have suffered more severe craniocerebral injuries, with associated traumatic subarachnoid haemorrhage, acute extradural haematoma, and brain contusions, than those with extradural haematomas. Furthermore the age range of patients with acute subdural haematoma tends to be slightly older, with a mean of 31 to 47 years compared to

20–30 years for patients with an extradural haematoma (Bullock *et al.* 2006a, b) and their initial Glasgow coma score is lower, with between 37 and 80 per cent having an initial Glasgow coma score < 9 (Bullock *et al.* 2006b). As a result many of these patients, despite evacuation of the haematoma, sustain lengthy periods of raised intracranial pressure as well as medical comorbidities that may impair longer term outcomes. Up until the mid-1980s approximately 50 per cent of patients with acute subdural haematomas would die with considerable morbidity in the survivors. Although recent figures show improvement, considerable morbidity remains in the survivors (see Section 25.2.6).

◆ *Brain contusions.* The surgical indications in patients with brain contusions are Glasgow coma score < 8, persistently raised intracranial pressure despite maximal multimodality management in intensive care, midline shift > 5 mm, and contusion size > 20 ml (Bullock *et al.* 2006c). In patients with parenchymal haematomas and contusions, complicated by severe brain swelling and uncontrolled intracranial pressure either contusionectomy or decompressive surgery (Section 25.4.4) should be considered (Bullock *et al.* 2006c). An enigma of neurotrauma is that some patients may only be confused and agitated despite severe contusions on initial CT scan. These patients can be managed conservatively but need to be observed closely since many contusions enlarge with time after head injury (Yamaki T *et al.* 1990; Lobato *et al.* 1997). Many contusions are multiple and tend to lie predominantly in the inferior frontal and anterior temporal lobe. Removal of such brain areas, even if contused, may further compromise cognitive function in survivors. Bad prognostic variables include older age, lower Glasgow comma score, absent papillary responses, respiratory difficulties, and compression of the basal cisterns (Bullock *et al.* 2006c).

25.4.4 Refractory post-traumatic intracranial hypertension

With early surgical evacuation of intracranial haematomas comes a cohort of patients who survive the i nitial craniocerebral insult only to develop medically refractory, severe raised intracranial pressure during their stay in intensive care. In most of these patients it is therefore routine to perform post-operative intracranial pressure monitoring. A range of commercial systems are now available for invasive intracranial pressure monitoring (Steiner *et al.* 2006). These systems provide important information about the intracranial biodynamics and are important guides to therapy. In many patients who develop post-operative or primary refractory raised intracranial pressure, there is no role for removal of brain tissue. Therefore a procedure involving a bony decompression of the skull with opening of the dura has been advocated to control the intracranial pressure (Jiang *et al.* 2005; Aarabi *et al.* 2006; Timofeev *et al.* 2006).

This type of surgery involves a large cranial incision to remove two-thirds of the anterior hemispheric skull vault. In some patients it may require bifrontal craniotomy depending on the precise anatomy of the brain swelling and decompression desired. Such procedures are generally associated with a rapid reduction in raised intracranial pressure, mean 10 mmHg (Aarabi *et al.* 2006), improvement in cerebral oxygenation, and in cerebral blood flow (Reithmeier *et al.* 2005). However, the clinical outcome is often highly unpredictable, but approximately 70 per cent of patient lives are saved, 50 per cent have a favourable outcome with good recovery or moderate disability, 20 per cent remain severely disabled, and the incidence of vegetative outcomes is low, ranging up to 14 per cent (Jiang *et al.* 2005; Aarabi *et al.* 2006: Timofeev *et al.* 2006). Improvement seems unrelated to the improvement in brain-monitored procedures immediately in the post-decompression period, patient age, pupillary response, and amount of brain shift (Aarabi *et al.* 2006). Complications included hydrocephalus in 10 per cent, haemorrhagic swelling ipsilateral to the craniectomy site in 16 per cent, and subdural hygroma (Aarabi *et al.* 2006).

There has been one small clinical trial in children that suggests decompression is beneficial (Taylor *et al.* 2001). Because this procedure is rather mechanically based and aspects such as timing of surgery, extent of craniectomy, and extent of durotomy remain undefined, two randomized controlled trials, DECRAN and ICP rescue, are currently being undertaken in patients who require such procedures. These are clearly required since, if many of these patients survive in a severely disabled state, the benefit of such an operation would be questionable (Sahuquillo and Arikan 2006).

25.4.5 Cranial reconstruction and repair of CSF fistulae

Most *skull fractures* are simple, closed linear injuries and require no specific treatment. The management of other skull fractures depends upon the type, location, and complexity, the extent of depressed fragments, and whether the wound is clean or contaminated. A closed depressed fracture can be left without surgical treatment unless there is a significant cosmetic deformity, since the incidence of infection is very low. With an open injury there is a high incidence of infection if the wound is not cleansed and debrided and the depressed fragments elevated (Shokunbi *et al.* 2000). Not infrequently there are foreign bodies in-driven to the depressed bone. Common findings are pieces of scalp, hair, clothing, and soil contamination, particularly if the injury is caused by a blast. Any retained foreign body in the brain will become infected eventually to cause a brain abscess and a high incidence of epilepsy. One of the principles learnt from war surgery is that compound depressed fractures need to be cleaned, debrided, and elevated and the dura needs primary repair to decrease the likelihood of such catastrophic infective complications (Simpson *et al.* 2004). The scalp should be closed under no tension. These basic principles have been applied in all recent wars, although with the advent of CT scanning and antibiotics lesser procedures have been advocated by some, albeit with a higher complication rate (Carey 2003).

Basal skull fractures may be complicated by dural tears and CSF leaks through the nose or ear. Generally otorrhoea stops spontaneously, or may require temporary lumbar CSF drainage, and rarely requires surgical repair. Cranionasal fistula with CSF rhinorrhoea can become chronic and predispose a patient to meningitis. These fistulae may occur in the setting of a complex craniofacial injury or linear fracture through the cribriform plate of the ethmoid bone, sphenoid bone, or frontal sinus region. If there is a persistent aerocele or CSF leak then the focus of bone and dural damage should be identified and craniotomy undertaken to formally repair the anterior fossa dura. This may require a synthetic or autologous graft material being tacked over the dural deficit as well as tissue glues to hold the graft in place whilst healing occurs. Such surgery is usually a delayed undertaking after the acute phase of injury. Anosmia is a common complication of either the primary injury or the secondary repair process (Section 16.1.4).

In patients with *cranionasal CSF fistula* there is no role for prophylactic antibiotics since these merely increase the pathogenicity of infecting organisms (Eljamel 1993; Choi *et al.* 1996). Generally patients developing meningitis following traumatic cranionasal fistulae have a pneumococcal infection with low morbidity and mortality if the meningitis is recognized and treated (Section 41.5.1). Despite apparent satisfactory repair of the dura and skull base, CSF fistulae may be recurrent in some patients and identification of the egress site can be very difficult. These patients need to be counselled about the potential for meningitis and the need to present urgently following onset of suggestive symptoms.

A particular complication of skull fracture in children is termed a *'growing' skull fracture*. This occurs when the dura is torn by the fracture and arachnoid herniates between the skull fragments and leads over some time to erosion of the bony edges (Muhonen *et al.* 1995). The patients present with a pulsatile scalp lump, and craniotomy and dural repair is required.

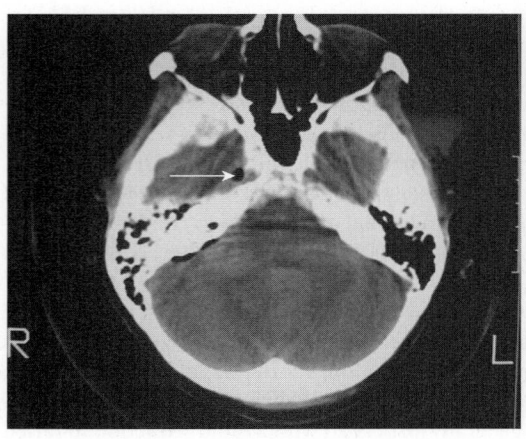
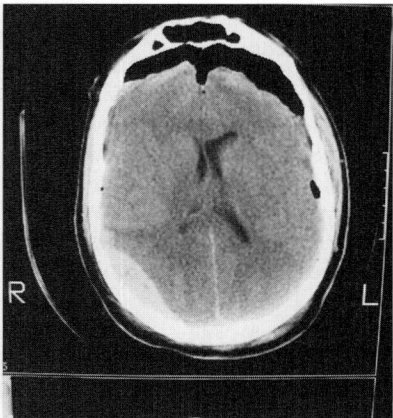

A   B

Fig. 25.7 Intracranial air following neurotrauma shown by CT scans. (A) shows bubbles of air around the right petrous bone and lateral to the sphenoid bone (arrow) following a right middle fossa fracture. (B) shows a large bifrontal aerocoele following an anterior cranial fossa fracture. There is also a small right posterior parietal extradural haematoma.

25.4.6 Surgery for chronic subdural haematoma

Surgical treatment, when indicated, is by drainage of the collection. This can be done through burrholes or craniotomy, with intra-operative irrigation of the sub-dural space, with or without postoperative drainage of the subdural space (Weigel *et al.* 2003; Cenic *et al.* 2005).Outcome is partly related to patient medical co-morbidity, which is frequently high, and the presenting functional status and Glasgow coma score (Gelabert-Gonzalez *et al.* 2005). Requirement for re-drainage, and clinical recurrence is a problem after all types of primary surgery and approximately 15 per cent of patients require more than one operation. There has been little change in outcomes in the last 20 years for patients with chronic subdural haematoma, and there is no consensus as to what is the best operation (Weigel *et al.* 2003; Cenic *et al.* 2005). Part of the difficulty with many patients with chronic subdural haematoma is failure of the brain to re-expand quickly following evacuation of the haematoma. This predisposes to the sub-dural space re-filling either acutely of chronically with fresh haematoma.

25.5 Intensive care and medical management

25.5.1 The role of steroids

Since steroids have a role in stabilizing cell membranes, maintaining the integrity of the blood brain barrier, and altering cytokine release, they have been used empirically in the management of head injuries in many centres. A meta-analysis of the potential benefit of steroids following neurotrauma could not confirm either a beneficial or a harmful effect (Alderson *et al.* 1997). As a result, a very large international multicentre randomized placebo-controlled study was performed, CRASH. The final results at 6 months after head injury of this study suggested that in the population analysed the risk of death was higher in the corticosteroid group than in the placebo group at 25.7 per cent against 22.3 per cent; relative risk 1.15, (95 per cent CI 1.07–1.24; $p=0.0001$), as was the risk of death or severe disability 38.1 per cent vs. 36.3 per cent; odds ratio 1.05; (0.99–1.10) (Edwards *et al.* 2005). This has shown that there is no role for steroids in the management of head-injured patients. Additionally it highlights the need for very large clinical trials of any medical intervention in head injury since therapeutic benefits may be small. Underpowering may well explain the failure of many compounds to benefit head-injured patients in previous trials (Ghajar *et al.* 2004).

25.5.2 Multimodality intensive care

Some patients following head injury are in coma from onset with a Glasgow coma score < 9, others have multi-trauma, and some patients may require special monitoring following neurosurgery. The latter categories of patients are usually managed in an intensive care setting. The pathophysiology of traumatic brain injury evolves over several days and the principal aim of management is to limit secondary damage due to ischaemia, and brain herniation caused by either raised intracranial pressure or systemic hypotension, and hypoxia. The concept of primary brain injury and secondary brain insults developed during the 1980s led to a comprehensive restructuring of management of patients with head injury (Chesnut *et al.* 1993; Signorini *et al.* 1999). The basic principles of head injury management became based upon stabilizing the brain environment by maintaining normal or near normal intracranial pressure, maintaining cerebral brain perfusion pressure, preventing hypoxia of $PaO_2 < 60$ mmHg and oxygen saturation < 90 per cent, avoiding hyperglycaemia, and attempting to maintain a normal core body temperature.

Guidelines concerning the optimal parameters for the general management of such patients have recently been published (Brain Trauma Foundation, 2000 and 2003; NICE, 2004: and SIGN), although it is generally admitted that the level of evidence to support various monitoring and treatment strategies is sub-optimal (Steiner *et al.* 2006). Paradoxically one recent retrospective study from two intensive care units in the Netherlands showed no benefit from multimonitoring directed intracranial pressure/cerebral perfusion pressure treatments and increases in both treatment intensity and respirator days (Cremer *et al.* 2005).

Intracranial pressure is often severely elevated following neurotrauma, because of oedema, intracranial haematoma, contusions, engorgement of the brain vasculature, hydrocephalus, or even infection (Section 26.5). In adults a sustained intracranial pressure that exceeds 25 mmHg is associated with a poorer outcome, whereas the critical thresholds for children are not clearly defined but probably lie between 15 and 18 mmHg depending on age (Steiner *et al.* 2006). Severely brain-injured patients are therefore kept paralysed, sedated, and ventilated and arterial and intracranial pressure monitored. In specialist supraregional neurointensive care units, jugular venous oxygen saturation, SjO_2, brain tissue oxygen, $tipO_2$, brain temperature, brain microdialysis to measure extracellular metabolites, and EEG are also often monitored. (Haitsma *et al.* 2002; Kett-White *et al.* 2002; Steiner *et al.* 2006) Such parameters give useful data about the levels of hyperventilation required, the effects of mannitol and barbiturates in reducing intracranial pressure, and the effects of raising systemic arterial blood pressure to maintain mean arterial pressure > 90 mmHg. These are essential research tools and in addition they can indicate unexpected or unpredictable adverse effects of various therapies (Steiner *et al.* 2006).

Cerebral blood flow is sometimes directly related to mean arterial pressure after head injury due to loss of autoregulation (Steiner *et al.* 2006). A cerebral perfusion pressure of > 60–70 mmHg is generally required to sustain adequate brain oxygenation. Elevation of cerebral perfusion pressure, by the use of inotropes to raise mean arterial pressure, from 32 to 67 mmHg significantly improved $tipO_2$ by 62 per cent. Such manoeuvres can help avoid global or regional ischaemia (Kiening *et al.* 1997). Aggressive attempts to maintain cerebral perfusion pressure > 70 mmHg are associated with an increased risk of adult respiratory distress and are not associated with significant increases in brain tissue oxygen (Kiening *et al.* 1997; Brain Trauma Foundation, 2003). Although children and young adults can tolerate lower levels of cerebral perfusion pressure, the functional consequence of prolonged lowering of cerebral perfusion pressure is an unfavourable neurological outcome (Kiening *et al.* 1997; Czosnyka *et al.* 2006).

There are numerous ways of monitoring intracranial pressure with differing merits (Steiner *et al.* 2006). Different patterns of waveform have been described (Section 26.2). Waveform analysis can give information about cerebrospinal compensatory reserve and possible responses to hyperventilation (Czosnyka *et al.* 2006; Steiner *et al.* 2006). Methods for treating intracranial pressure greater than 25 mmHg include hyperventilation of patients to maintain low carbon dioxide levels of around 25 mmHg whilst maintaining oxygenation and thus preventing hypoxia, the use of osmotic agents such as mannitol and furosemide and more recently the use of hypertonic saline (Ogden *et al.* 2005). The benefit of mannitol has recently been subject to a meta-analysis (Wakai *et al.* 2007) In the acute management of comatose patients with severe head injury, the administration of high-dose mannitol resulted in reduced mortality with a relative risk of 0.56 (95 per cent CI 0.39 to 0.79) and reduced death and severe disability, relative risk of 0.58 (95 per cent CI 0.47 to 0.72) when compared with conventional-dose mannitol. Two trials however suggested 7.2 per cent hypertonic saline reduces intracranial pressure more effectively than either 15 per cent or 20 per cent mannitol (Battison *et al.* 2005; Harutjunyan *et al.* 2005) and is associated with better outcome than mannitol, relative risk for death = 1.25 (95 per cent CI 0.47 to 3.33) (Wakai *et al.* 2007).

Agents such as barbiturates which are very effective in depressing brain metabolism and lowering intracranial pressure also decrease systemic arterial pressure A Cochrane Review concluded that there is no evidence that barbiturate therapy in patients with acute severe head injury improved outcome, probably because it resulted in a fall in blood pressure in 25 per cent of treated patients (Roberts 2000). Mannitol therapy for raised intracranial pressure seems to produce a lower mortality when compared to pentobarbital treatment (Wakai *et al.* 2007). Despite the known problems of systemic hypotension, it is still not known whether it is better to maintain cerebral perfusion pressure with a high systemic arterial pressure, or to have a lower intracranial pressure and lower systemic arterial pressure with the same cerebral perfusion pressure (Grande *et al.* 2002; Marmarou *et al.* 2005; Steiner *et al.* 2006).

In Douglas Miller's pioneering studies of secondary insults following head injury despite measurement of multiple modalities including intracranial pressure, mean arterial pressure, PaO_2, and temperature, it was only raised intracranial pressure that was associated statistically with a poorer outcome (Signorini *et al.* 1999). However pyrexia may be associated with a poorer outcome because hypothermia has on occasions been neuroprotective following head injury (Strachan *et al.* 1989). Attempts at brain cooling or systemic hypothermia in an effort to control pyrexia problems have not led to satisfactory results in randomized clinical controlled trials. This is largely because hypothermia can lead to increased chest infections and sepsis which counteract the beneficial effects of the hypothermia on the brain. Meta-analysis of trials of hypothermia in head injury (Henderson *et al.* 2003), concluded that iatrogenic hypothermia may confer a marginal benefit in neurological outcome, but there does not appear to be clear evidence of lower mortality rates.

Monitoring multiple variables in the brain of patients in neurointensive care units is now made possible by the ready availability of intraparenchymal and intraventricular brain monitoring probes. Brain temperature, and brain tissue CO_2, oxygen, pH, and bicarbonate can all be measured using multi-parametric probes (Zauner *et al.* 1997; Jaeger *et al.* 2005; Steiner *et al.* 2006). Local tissue metabolism can be measured using intraparenchymal microdialysis from which biochemical parameters such as glucose level, lactate, pyruvate, amino acid, and monoamines can be studied. (Zauner *et al.* 1997; Hlateky *et al.* 2002) The jugular bulb oxygen level is also routinely measured. Such multimodality monitoring coupled with intensive medical and nursing management has reduced in

mortality and poor outcomes following severe craniocerebral injury (Lu *et al.* 2005). Excluding those patients who are admitted with a Glasgow coma score of 3 and bilaterally fixed or dilated pupils, around 27 per cent of patients with severe traumatic brain injury will die currently with the rest making variable degrees of recovery (Marshall *et al.* 1998; Lu *et al.* 2005).

25.5.3 Post-traumatic epilepsy and the role of anti-convulsants

Seizures complicate the management of a large number of patients following craniocerebral trauma. Depending on the series reviewed whether civilian or military, post-traumatic epilepsy may occur in between 4 and 53 per cent of patients (Salazar *et al.* 1985; Annegers *et al.* 1998; Englander *et al.* 2003; Frey 2003). The patients predisposed to seizures in the first week after injury are those with intracranial haematoma, particularly cerebral contusions, intracerebral haemorrhage, and acute subdural haematoma, those with penetrating craniocerebral injuries from missiles, younger patients, lower Glasgow coma score on admission, and chronic alcoholism (Salazar *et al.* 1985; Annegers *et al.* 1998; Englander *et al.* 2003; Frey 2003;). As many as 86 per cent of patients experiencing one seizure after traumatic brain injury will have a second in the next 2 years, but the risk of seizures falls off with time after injury. A seizure in the first week predisposes to later onset seizures. Other risk factors include age greater than 65 years, and amnesia for more than one day (Annegers *et al.* 1998; Frey 2003).

Prophylactic use of anticonvulsants in the intensive care setting has recently been justified by a meta-analysis of randomized controlled trial which showed a lower incidence of seizures in the first week following neurotrauma in those patients given anticonvulsants compared to those given placebo (Schierhout *et al.* 2001). Despite the adverse pathophysiological consequences caused by early post-traumatic seizures such as raised intracranial pressure, altered metabolic requirements of the brain, and increases in cerebral blood flow, the meta-analysis showed no reduction in mortality, late onset seizures, or neurological disability.

Later onset seizures, at more than one week after injury, are not decreased by prophylactic use of anticonvulsant therapy (Schierhout *et al.* 2001; Beghi 2003). Given that there are specific and general side effects associated with a range of anticonvulsants, most neuro-clinicians managing head-injured patients would now use anticonvulsants only following a first seizure. The duration of anticonvulsant therapy after the acute phase of head injury is highly conjectural and at this stage there are no good guidelines (Beghi 2003; Frey 2003).

25.5.4 Rehabilitation after brain injury

Following the acute period of head injury management in intensive care and the neurosurgical ward, many patients will remain with impairments and disability that require prolonged periods of neuro-rehabilitation, now a sub-specialty in its own right. The aims of neuro-rehabilitation are to optimize the environment for clinical recovery (Levy *et al.* 2005), retrain patients in various neuromotor and cognitive skills, and to identify and address any potential problems when the patient is restored to the community (Salazar *et al.* 2000; Sarajuuri *et al.* 2005; van den Broek 2005). It is generally considered that brain plasticity has restored function to an optimal level after about 2 years but many patients go on recovering, albeit less spectacularly, for a longer period.

25.6 The consequences of craniocerebral injury

25.6.1 The post-traumatic syndrome

The vast majority of patients who sustain a concussional head injury have a mild head injury with normal CT brain imaging. However many such patients experience elements of a post-concussional syndrome or post-traumatic syndrome. The major components of this syndrome are dizziness, headaches, cognitive dysfunction, disturbances of sleep, fatigue, and blank spells (Anderson *et al.* 2006; Hellawell *et al.* 1990). Most of these symptoms resolve spontaneously after 2–12 weeks, and early intervention by a specialist service reduces social restriction and symptomatology (Wade *et al.* 1998; King 2003). Post-traumatic headache often has features of both tension headache and migraine and mostly resolves within 6 months, but can be problematic for much longer, and may be aggravated by analgesic misuse (Section 18.6.9) (Lane and Arciniegas 2002).

Although there may be no findings on CT imaging in the vast majority, several findings suggest an organic cause for these post-traumatic symptoms. These include diffuse microscopic axonal injury, MRI abnormalities in the white matter, abnormalities in regional cerebral blood flow, subtle abnormalities on EEG and brain stem evoked responses, and abnormal glucose metabolism in certain regions of the brain (King 2003). The syndrome is also worse in those with a pre-existing psychopathological disorder, and in those seeking compensation. Treatment is directed at the symptomatology, using simple analgesics or amytriptiline for headache, benzodiazepine for acute anxiety, antidepressants for depression, or psychostimulants for fatigue and cognitive slowing (Anderson *et al.* 2006). In patients with chronic post-traumatic syndrome, management strategies require correct evaluation of the possible cause of symptoms; however this is often difficult because organic and psychological symptoms become intimately related (Ruff 2005). Management may also be compounded if litigation is involved.

25.6.2 Persistent cognitive and behavioural changes

Although many patients survive severe neurotrauma, and will have no physical deficit, a large proportion will have cognitive or behavioural impairments to varying degrees. In part these deficits may relate to the focal nature of the head injury as well as more generalized brain damage. Factors influencing the extent of post-traumatic cognitive dysfunction include Glasgow coma score at presentation, age, presence of traumatic subarachnoid haemorrhage, and the presence of intracranial haematomas (Halliwell *et al.* 1999; Hanlon *et al.* 2005). Problems with short-term memory, executive functions, judgement, insight, planning, impulse control mood disorders, and social interactions can bedevil the lives of patients who have otherwise made good sensory motor recoveries from the head injury (Halliwell *et al.* 1999: Wilson *et al.* 2000). Aspects of these dysfunctions can be picked up on the Glasgow outcome score and Extended Glasgow outcome score (Wilson *et al.* 2000). Problems with lack of social restraint and impulse control can produce behavioural difficulties which compromise rehabilitation and prevent integration of the patient back into the community. In some patients this involved unmasking of subtle pre-existing personality traits, whilst in others it may involve complete change in the patient's personality and behaviour. Irritability and aggression are common features of the more severely damaged patient. Unfortunately in many cases these latter features can lead to social isolation and a predilection to further head injury because of loss of insight and loss of restraint on anti-social behaviour.

The neuropathological substrates of cognitive and behavioural dysfunction after traumatic brain injury vary enormously. Some patients may have no obvious abnormalities on CT scan yet diffusion tensor MRI can reveal white matter disruption that correlates with their cognitive dysfunction (Nakayama *et al.* 2006). At the other extreme there may be focal or multifocal gliosis or post-traumatic lateral ventricular dilatation, due to white matter and cortical volume loss, which correlates well with cognitive dysfunction (Wilde *et al.* 2006). Hydrocephalus is found in approximately 40 per cent of patients after severe head injury and 27 per cent after moderate head injury (Poca *et al.* 2005). Post-traumatic hydrocephalus may not be due simply to just ex vaccuo dilatation, that is ventricular enlargement due to loss of brain parenchymal volume. However intracranial pressure monitoring and studies of CSF outflow resistance reveal a variable pattern of relationships

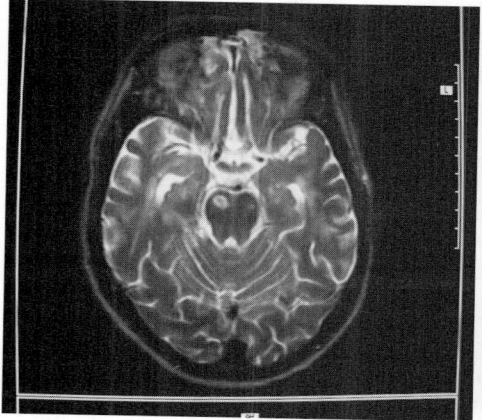
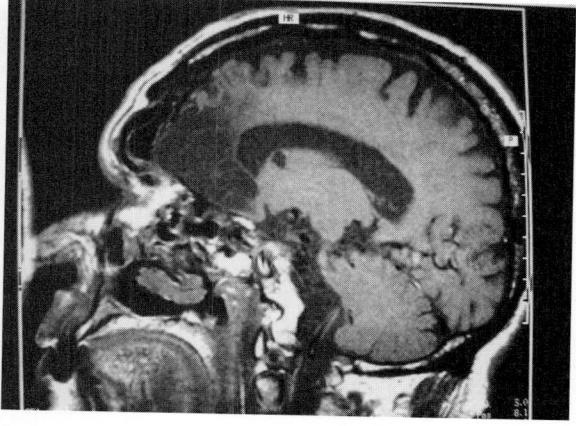

A   B

Fig. 25.8 Post-traumatic encephalomalacia shown by MRI. (A) shows a T2-weighted image showing diffuse white matter loss, thinning of the cortical mantle, enlargement of the temporal horn of the lateral ventricles, and two foci of signal change in the lower midbrain. (B) T1-weighted MRI shows a large area of frontal encephalomalacia following contusional brain injury. There is also marked gyral atrophy throughout the brain.

between the hydrocephalus, intracranial pressure measurements, and outflow resistance (Marmarou *et al.* 1996). Seventy-seven per cent of patients with post-traumatic hydrocephalus have normal intracranial pressure, whilst it is raised in 23 per cent. In 20 per cent of those having hydrocephalus and normal intracranial pressure CSF outflow resistance is raised suggesting a component of communicating hydrocephalus. (Marmarou *et al.* 1996).

25.6.3 Post-traumatic states of limited awareness

Some patients survive severe brain injuries but are left severely disabled. Previously many of these patients were classified as vegetative (Section 33.6) (Bates 2005). Recently positron emission tomography and other investigational studies have revealed that there is a spectrum of underlying conditions (Laureys *et al.* 2004; Owen *et al.* 2006), that include the 'locked-in' syndrome (Section 33.5), a minimally conscious state, and persistent vegetative state (Section 33.6). Such conditions are particularly characterized by problems in arousal and awareness. Emergence from coma is usual after 3–4 weeks when arousal occurs, but recovery of awareness may not follow. Recent judicial decisions which allowed decisions to limit treatment by withdrawing life-sustaining therapy have highlighted the difficulty of managing such patients (McLean 2005). Many patients can exist in a vegetative condition for many years and occasionally some show either spontaneous late improvement (Avesani *et al.* 2006) or improvement after deep brain stimulation (Yamamoto and Katayama 2005). This latter intervention, and the performance of research on patients unable to give consent raises several ethical issues, but such research is essential if the condition is to be scientifically studied, experts trained to manage patients with such conditions, and rational management plans to be developed (Laureys *et al.* 2004; Bates 2005; McLean 2005; Pickard *et al.* 2005).

25.6.4 Glasgow outcome score

As the Glasgow coma scale helped categorize patients into severity of head injury, the Glasgow outcome score, referred to as the GOS, and Extended Glasgow outcome score, GOSE, have helped to categorize patients' outcomes. These are particularly used in clinical trials and objective evaluation of head injury management

Table 25.5 The original Glasgow outcome scale and Extended Glasgow outcome scale, with some variations. The Extended Glasgow outcome scale divides the outcomes denoted as severe disability, moderate disability, and good recovery into two separate categories (modified from Jennett and Teasdale, 1981)

Extended scale		Original scale	Contracted scales		
Dead		Dead	Dead		
Vegetative		Vegetative			Dead or vegetative
Degree of disability:	5		Dependent		
	4	Severely disabled			Severely disabled
	3				
	2	Moderately disabled		Independent	Independent
	1		Independent		
	0	Good recovery			
Total categories:	8	5	3		

(Table 25.5) (Teasdale *et al.* 1998). The Glasgow outcome score is generally dichotomized into good outcome, with good recovery, minimal or moderate disability, and poor outcome consisting of severe disability or death. In a recent clinical trial the severe disability group was dichotomized also into those who could have some semi-independent life and those who were totally dependent. The Glasgow outcome score is very important in comparing outcome from various groups in a standardized fashion and can be reliably obtained from telephone interviews (Wilson *et al.* 2000; Pettigrew *et al.* 2003).

25.6.5 Predisposition to later cerebral neurodegeneration

Although 'punch-drunkenness' has long been recognized as a chronic post-traumatic encephalopathy, more recent interest has focussed on how trauma may interact with genetic factors to predispose to premature brain atrophy or degeneration in normal individuals. The mechanisms underlying post-traumatic Parkinson's disease and early onset dementia after neurotrauma are of considerable interest (Section 40.3.7). One study of boxers suggested that possession of one *APOE4* allele predisposed to greater brain dysfunction (Jordan *et al.* 1997), however studies of a range of patients in Canada (Chamelian *et al.* 2004), the United Kingdom (Johnson *et al.* 2006), and Zulus in South Africa (Nathoo *et al.* 2003) failed to find any correlation between *APOE4*, *IL-1α*, and *IL-1β* allelic status and worsen outcome following brain injury. Numerous studies looking at putative mechanisms of neurodegeneration and neuroprotection have been studied in transgenic mice. APOE and tau proteins may have central importance because of their roles in neuronal integrity and known relationship to dementias (Section 34.6) (Brecht *et al.* 2004). It is likely that in the next decades these, and other, molecular biological studies will provide insight into the genetic factors that interact with clinical parameters to influence prognosis after traumatic brain injury.

References

Aarabi B, Hesdorffer DC, Ahn ES *et al.* (2006). Outcome following decompressive craniectomy for malignant swelling due to severe head injury. *J Neurosurg*, **104**, 469–79.

Alaraj AM, Chamoun RB, Dahdaleh NS *et al.* (2005). Spontaneous subdural haematoma in anabolic steroids dependent weight lifters: reports of two cases and review of literature. *Acta Neurochir*, **147**, 85–7; discussion 87–8.

Alderson P, Roberts I (1997). Corticosteroids in acute traumatic brain injury: systematic review of randomised controlled trials. *BMJ*, **28**, 314(7098), 1855–9.

Amirjamshidi A, Abbassioun K, Rahmat H (2000). Traumatic aneurysms and arteriovenous fistulas of the extracranial vessels in war injuries. *Surg Neurol*, **53**, 136–45.

Anderson T, Heitger M, Macleod AD (2006). Concussion and mild head injury. *Pract Neurol*, **6**, 342–57.

Annegers JF, Hauser WA, Coan SP *et al.* (1998). A population-based study of seizures after traumatic brain injuries. *N Engl J Med*, **338**, 20–4.

Anton J, Pineda V, Martin C *et al.* (1999). Posttraumatic subgaleal haematoma: a case report and review of the literature. *Pediatr Emerg Care*, **15**, 347–9.

Aquilina K, Carty F, Keohane C *et al.* (2005). Pseudoaneurysm of the occipital artery: an unusual cause of persisting headache after minor head injury. *Ir Med J*, **98**, 215–7.

Aryan HE, Jandial R, Bennett RL et al. (2005). Gunshot wounds to the head: gang- and non-gang-related injuries and outcomes. *Brain Inj*, **19**, 505–10.

Avesani R, Gambini MG, Albertini G (2006). The vegetative state: a report of two cases with a long-term follow-up. *Brain Inj*, **20**, 333–8.

Bates D (2005). The vegetative state and the Royal College of Physicians guidance. *Neuropsychol Rehabil*, **15**, 175–83.

Battison C, Andrews PJ, Graham C et al. (2005). Randomized, controlled trial on the effect of a 20% mannitol solution and a 7.5% saline/6% dextran solution on increased intracranial pressure after brain injury. *Crit Care Med*, **33**, 196–202; discussion 257–8.

Beghi E (2003). Overview of studies to prevent posttraumatic epilepsy. *Epilepsia*, **44** (Suppl 10), 21–6.

Brain Trauma Foundation (2003).

Brecht WJ, Harris FM, Chang S (2004). Neuron-specific apolipoprotein e4 proteolysis is associated with increased tau phosphorylation in brains of transgenic mice. *J Neurosci*, **24**, 2527–34.

Bullock MR, Chesnut R, Ghajar J et al. (2006a). Guidelines for the surgical management of traumatic brain injury author group: Introduction. *Neurosurgery*, **58**, S2-1–S2-3.

Bullock MR, Chesnut R, Ghajar J et al. (2006b). Surgical management of traumatic brain injury author group. Surgical management of acute epidural hematomas. *Neurosurgery*, **58**, S7–15.

Bullock MR, Chesnut R, Ghajar J et al. (2006c). Surgical management of traumatic brain injury author group. Surgical management of acute subdural hematomas. *Neurosurgery*, **58**, S16–24.

Cadamy AJ, McNaughton GW, Helliwell R (2003). Traumatic pseudoaneurysms of the superficial temporal artery. *Eur J Emerg Med*, **10**, 236–7.

Carey ME (2003). The treatment of wartime brain wounds: traditional versus minimal debridement. *Surg Neurol*, **60**, 112–19.

Cenic A, Bhandari M, Reddy K (2005). Management of chronic subdural haematoma: a national survey and literature review. *Can J Neurol Sci*, **32**, 501–6.

Chamelian L, Reis M, Feinstein A (2004). Six-month recovery from mild to moderate traumatic brain injury: the role of APOE-epsilon4 allele. *Brain*, **127**, 2621–8. [Epub 2004 October 20].

Chan KH, Miller JD, Dearden NM (1992). The effect of changes in cerebral perfusion pressure upon middle cerebral artery blood flow velocity and jugular bulb venous oxygen saturation after severe brain injury. *J Neurosurg*, **77**, 55–61.

Chesnut RM, Marshall LF, Klauber MR (1993). The role of secondary brain injury in determining outcome from severe head injury. *J Trauma*, **34**, 216–22.

Chieregato A, Fainardi E, Morselli-Labate AM et al. (2005). Factors associated with neurological outcome and lesion progression in traumatic subarachnoid hemorrhage patients. *Neurosurgery*, **56**, 671–80; discussion 671–80.

Chieregato A, Fainardi E, Servadei F et al. (2004). Centrifugal distribution of regional cerebral blood flow and its time course in traumatic intracerebral hematomas. *J Neurotrauma*, **21**, 655–66.

Childs C, Vail A, Leach P et al. (2006). Brain temperature and outcome after severe traumatic brain injury. *Neurocrit Care*, **5**, 10–14.

Choi D, Spann R (1996). Traumatic cerebrospinal fluid leakage: risk factors and the use of prophylactic antibiotics. *Br J Neurosurg*, **10**, 571–5.

Chung RS, Staal JA, McCormack GH et al. (2005). Mild axonal stretch injury in vitro induces a progressive series of neurofilament alterations ultimately leading to delayed axotomy. *J Neurotrauma*, **22**, 1081–91

Cremer OL, van Dijk GW, van Wensen E (2005). Effect of intracranial pressure monitoring and targeted intensive care on functional outcome after severe head injury. *Crit Care Med*, **33**, 2207–13.

Czosnyka M, Hutchinson PJ, Balestreri M (2006). Monitoring and interpretation of intracranial pressure after head injury. *Acta Neurochir Suppl*, **96**, 114–18.

De Deyne C, Van Aken J, Decruyenaere J et al. (1998). Jugular bulb oximetry: review on a cerebral monitoring technique. *Acta Anaesthesiol Belg*, **49**, 21–31.

Depreitere B, Van Calenbergh F, van Loon J (2003). A clinical comparison of non-traumatic acute subdural haematomas either related to coagulopathy or of arterial origin without coagulopathy. *Acta Neurochir*, **145**, 541–6; discussion 546.

Edwards P, Arango M, Balica L et al. CRASH trial collaborators (2005). Final results of MRC CRASH, a randomised placebo-controlled trial of intravenous corticosteroid in adults with head injury-outcomes at 6 months. *Lancet*, **365**, 1957–9.

Eisenberg HM, Gary HE Jr, Aldrich EF et al. (1990). Initial CT findings in 753 patients with severe head injury. A report from the NIH Traumatic Coma Data Bank. *J Neurosurg*, **73**, 688–98.

Eljamel MS (1993). Antibiotic prophylaxis in unrepaired CSF fistulae. *Br J Neurosurg*, **7**, 501–5.

Englander J, Bushnik T, Duong TT et al. (2003). Analyzing risk factors for late posttraumatic seizures: a prospective, multicenter investigation. *Arch Phys Med Rehabil*, **84**, 365–73.

Fainardi E, Chieregato A, Antonelli V et al. (2004). Time course of CT evolution in traumatic subarachnoid haemorrhage: a study of 141 patients. *Acta Neurochir*, **146**, 257–63; discussion 263.

Flanagan SR, Hibbard MR, Gordon WA (2005). The impact of age on traumatic brain injury. *Phys Med Rehabil Clin N Am*, **16**, 163–77.

Frey LC (2003). Epidemiology of posttraumatic epilepsy: a critical review. *Epilepsia* **44** (Suppl 10), 11–17.

Gelabert-Gonzalez M, Iglesias-Pais M, Garcia-Allut A (2005). Chronic subdural haematoma: surgical treatment and outcome in 1000 cases. *Clin Neurol Neurosurg*, **107**, 223–9.

Gentleman D, Nath F, Macpherson P (1989). Diagnosis and management of delayed traumatic intracerebral haematomas. *Br J Neurosurg*, **3**, 367–372.

Ghajar J, Hesdorffer DC (2004). Steroids CRASH out of head-injury treatment. *Lancet Neurol*, **3**, 708.

Giugni E, Sabatini U, Hagberg GE et al. (2005). Fast detection of diffuse axonal damage in severe traumatic brain injury: comparison of gradient-recalled echo and turbo proton echo-planar spectroscopic imaging MRI sequences. *Am J Neuroradiol*, **26**, 1140–8.

Graham DI, Gennarelli TA (1997). Trauma. In Graham DI, Lantos PL, eds. *Greenfield's Neuropathology*, pp. 197–248. Arnold, London.

Graham DI, McIntosh TK, Maxwell WL et al. (2000). Recent advances in neurotrauma. *J Neuropathol Exp Neurol*, **59**, 641–51.

Grande PO, Asgeirsson B, Nordstrom CH (2002). Volume-targeted therapy of increased intracranial pressure: the Lund concept unifies surgical and non-surgical treatments. *Acta Anaesthesiol Scand*, **46**, 929–41.

Haitsma IK, Maas AI (2002). Advanced monitoring in the intensive care unit: brain tissue oxygen tension. *Curr Opin Crit Care*, **8**, 115–20.

Hanlon RE, Demery JA, Kuczen C et al. (2005). Effect of traumatic subarachnoid haemorrhage on neuropsychological profiles and vocational outcome following moderate or severe traumatic brain injury. *Brain Inj*, **19**, 257–62.

Harutjunyan L, Holz C, Rieger A et al. (2005). Efficiency of 7.2% hypertonic saline hydroxyethyl starch 200/0.5 versus mannitol 15% in the treatment of increased intracranial pressure in neurosurgical patients - a randomized clinical trial [ISRCTN62699180]. *Crit Care*, **9**(5), R530–40. [Epub 2005 August 9].

Haselsberger K, Pucher R, Auer L (1988). Prognosis after acute subdural or epidural haemorrage. *Acta Neurochir*, **90**, 111–6.

Hellawell D, Taylor R, Pentland B. Cognitive and psychosocial outcome following moderate or severe traumatic brain injury. (1999) *Brain Injury*, **13**, 489–504.

Henderson WR, Dhingra VK, Chittock DR (2003). Hypothermia in the management of traumatic brain injury. A systematic review and meta-analysis. *Intensive Care Med*, **29**, 1637–44. [Epub 2003 August 12].

Hlatky R, Robertson CS (2002). Multimodality monitoring in severe head injury. *Curr Opin Anaesthesiol*, **15**, 489–93.

Hohenstein A, Erber R, Schilling L et al. (2005). Increased mRNA expression of VEGF within the haematoma and imbalance of angiopoietin-1 and -2 mRNA within the neomembranes of chronic subdural haematoma. *J Neurotrauma*, **22**, 518–28.

Inao S, Kawai T, Kabeya R et al. (2001). Relation between brain displacement and local cerebral blood flow in patients with chronic subdural haematoma. *J Neurol Neurosurg Psychiatry*, **71**, 741–6.

Inglese M, Makani S, Johnson G et al. (2005). Diffuse axonal injury in mild traumatic brain injury: a diffusion tensor imaging study. *J Neurosurg*, **103**, 298–303.

Jaeger M, Soehle M, Meixensberger J (2005). Brain tissue oxygen (PtiO2): a clinical comparison of two monitoring devices. *Acta Neurochir Suppl*, **95**, 79–81.

Jennett B, Teasdale G (1981). *Management of Head Injuries*, pp. 306. FA Davis Coy, Philadelphia.

Jennett B, Teasdale G, Braakman R et al. (1979). Prognosis of patients with severe head injury. *Neurosurgery*, **4**, 283–9.

Jiang JY, Xu W, Li WP et al. (2005). Efficacy of standard trauma craniectomy for refractory intracranial hypertension with severe traumatic brain injury: a multicenter, prospective, randomized controlled study. *J Neurotrauma*, **22**, 623–8.

Johnson VE, Murray L, Raghupathi R et al. (2006). No evidence for the presence of apolipoprotein epsilon4, interleukin-1alpha allele 2 and interleukin-1beta allele 2 cause an increase in programmed cell death following traumatic brain injury in humans. *Clin Neuropathol*, **6**, 255–64.

Jordan BD, Relkin NR, Ravdin LD et al. (1997). Apolipoprotein E epsilon4 associated with chronic traumatic brain injury in boxing. *JAMA*, **278**, 136–40.

Kett-White R, Hutchinson PJ, Czosnyka M et al. (2002). Multi-modal monitoring of acute brain injury. *Adv Tech Stand Neurosurg*, **27**, 87–134.

Kiening KL, Hartl R, Unterberg AW et al. (1997). Brain tissue pO2-monitoring in comatose patients: implications for therapy. *Neurol Res*, **19**, 233–40.

King NS (2003). Post-concussion syndrome: clarity amid the controversy? *Br J Psychiatry*, **183**, 276–8.

Kirkness CJ, Burr RL, Cain KC (2005). Relationship of cerebral perfusion pressure levels to outcome in traumatic brain injury. *Acta Neurochir Suppl*, **95**, 13–6.

Lane JC, Arciniegas DB (2002). Post-traumatic headache. *Curr Treat Options Neurol*, **4**, 89–104.

Laureys S, Owen AM, Schiff ND (2004). Brain function in coma, vegetative state, and related disorders. *Lancet Neurol*, **3**, 537–46.

Lee KK, Seow WT, Ng I (2006). Demographical profiles of adult severe traumatic brain injury patients: implications for healthcare planning. *Singapore Med J*, **47**, 31–6.

Lee E, Hung Y, Wang L et al. (1998). Factors influencing the functional outcome of patients with acute epidural hematomas. Analysis of 200 patients undergoing surgery. *J Trauma*, **45**, 946–52.

Levy M, Berson A, Cook T et al. (2005). Treatment of agitation following traumatic brain injury: a review of the literature. *NeuroRehabilitation*, **20**, 279–306.

Lobato R, Cordobes F, Rivas J (1983). Outcome from severe head injury related to the type of intracranial lesion. A computerized tomography study. *J Neurosurg*, **59**, 762–74.

Lobato RD, Gomez PA, Alday R (1997). Sequential computerized tomography changes and related final outcome in severe head injury patients. *Acta Neurochir*, **139**, 385–91.

Lofgren J, von Essen C, Zwetnow NN (1973). The pressure-volume curve of the cerebrospinal fluid space in dogs. *Acta Neurol*, **49**, 557–74.

Lu J, Marmarou A, Choi S et al. (2005). Impact and Abic Study Group: Mortality from traumatic brain injury. *Acta Neurochir Suppl*, **95**, 281–5.

McLean SA (2005). Permanent vegetative state: the legal position. *Neuropsychol Rehabil*, **15**, 237–50.

McCrory PR, Berkovic SF (1998). Concussive convulsions. Incidence in sport and treatment recommendations. *Sports Med*, **25**, 131–6.

McCrory PR, Berkovic SF (2001). Concussion: the history of clinical and pathophysiological concepts and misconceptions. *Neurology*, **57**, 2283–9.

Marmarou CR, Povlishock JT (2006). Administration of the immunophilin ligand FK506 differentially attenuates neurofilament compaction and impaired axonal transport in injured axons following diffuse traumatic brain injury. *Exp Neurol*, **197**, 353–62.

Marmarou A, Saad A, Aygok G (2005). Contribution of raised ICP and hypotension to CPP reduction in severe brain injury: correlation to outcome. *Acta Neurochir*, **95**, 277–80.

Marmarou A, Foda MA, Bandoh K (1996). Posttraumatic ventriculomegaly: hydrocephalus or atrophy? A new approach for diagnosis using CSF dynamics. *J Neurosurg*, **85**, 1026–35.

Marmarou A, Signoretti S, Fatouros PP et al. (2006). Predominance of cellular edema in traumatic brain swelling in patients with severe head injuries. *J Neurosurg*, **104**, 720–30.

Marshall LF, Maas AI, Marshall SB (1998). A multicenter trial on the efficacy of using tirilazad mesylate in cases of head injury. *J Neurosurg*, **89**, 519–25

Marshall LF, Marshall SB, Klauber MR et al. (1992). The diagnosis of head injury requires a classification based on computed axial tomography. *J Neurotrauma*, **9** (Suppl 1), S287–92.

Mattioli C, Beretta L, Gerevini S et al. (2003). Traumatic subarachnoid hemorrhage on the computerized tomography scan obtained at admission: a multicenter assessment of the accuracy of diagnosis and the potential impact on patient outcome. *J Neurosurg*, **98**, 37–42.

Miller JD, Ironside JW (1997). Raised intracranial pressure, oedema and hydrocephalus. In Graham DI, Lantos P, eds. *Greenfield's Neuropathology*, 6th Edition, pp. 157–95. Arnold, London.

Muhonen MG, Piper JG, Menezes AH (1995). Pathogenesis and treatment of growing skull fractures. *Surg Neurol*, **43**, 367–72; discussion 372–3.

Nakayama N, Okumura A, Shinoda J et al. (2006). Evidence for white matter disruption in traumatic brain injury without macroscopic lesions. *J Neurol Neurosurg Psychiatry*, **77**, 850–5. [Epub 2006 March 30].

Nathoo N, Chetry R, van Dellen JR et al. (2003). Apolipoprotein E polymorphism and outcome after closed traumatic brain injury: influence of ethnic and regional differences. *J Neurosurg*, **98**, 302–6.

Oertel M, Kelly DF, McArthur D (2002). Progressive hemorrhage after head trauma: predictors and consequences of the evolving injury. *J Neurosurg*, **96**, 109–16.

Ogden AT, Mayer SA, Connolly ES Jr (2005). Hyperosmolar agents in neurosurgical practice: the evolving role of hypertonic saline. *Neurosurgery*, **57**, 207–15; discussion 207–15.

O'Sullivan MG, Whyman M, Steers JW et al. (1994). Acute subdural haematoma secondary to ruptured intracranial aneurysm: diagnosis and management. *Br J Neurosurg*, **8**, 439–45.

Owen AM, Coleman MR, Boly M (2006). Detecting awareness in the vegetative state. *Science*, **313**, 1402.

Perron AD, Brady WJ, Huff JS (2001). Concussive convulsions: emergency department assessment and management of a frequently misunderstood entity. *Acad Emerg Med*, **8**, 296–8.

Pettigrew LE, Wilson JT, Teasdale GM (2003). Reliability of ratings on the Glasgow Outcome Scales from in-person and telephone structured interviews. *J Head Trauma Rehabil*, **18**, 252–8.

Pickard JD, Coleman MR, Czosnyka M (2005). Hydrocephalus, ventriculomegaly and the vegetative state: a review. *Neuropsychol Rehabil*, **15**, 224–36.

Poca MA, Sahuquillo J, Mataro M et al. (2005). Ventricular enlargement after moderate or severe head injury: a frequent and neglected problem. *J Neurotrauma*, **22**, 1303–10.

Reilly PL, Simpson DA, Sprod R *et al.* (1988). Assessing the conscious level in infants and young children: a paediatric version of the Glasgow Coma Scale. *Childs Nerv Syst*, **4**, 30–3.

Reithmeier T, Lohr M, Pakos P (2005). Relevance of ICP and ptiO(2) for indication and timing of decompressive craniectomy in patients with malignant brain edema. *Acta Neurochir*, **147**, 947–52.

Ribas GC, Jane JA (1992). Traumatic contusions and intracerebral hematomas. *J Neurotrauma*, **9**, S265–78.

Rivas J, Lobato R, Sarabia R *et al.* (1988). Extradural hematoma: analysis of factors influencing the courses of 161 patients. *Neurosurgery*, **23**, 44–51.

Roberts I (2000). Barbiturates for acute traumatic brain injury. *Cochrane Database Syst Rev* (2) CD 000033.

Roberts GW, Allsop D, Bruton C (1990). The occult aftermath of boxing. *J Neurol Neurosurg Psychiatry*, **53**, 373–8.

Royal College of Surgeons of England Trauma Committee (2005). The Royal College of Surgeons of England: a position paper on the acute management of patients with head injury. *Ann R Coll Surg Engl*, **87**, 323–5.

Ruff R (2005). Two decades of advances in understanding of mild traumatic brain injury. *J Head Trauma Rehabil*, **20**, 5–18.

Sahuquillo J, Arikan F (2006). Decompressive craniectomy for the treatment of refractory high intracranial pressure in traumatic brain injury. *Cochrane Database Syst Rev*, **25**, CD003983.

Sakas D, Bullock M, Teasdale G (1995). One-year outcome following craniotomy for traumatic hematoma in patients with fixed dilated pupils. *J Neurosurg*, **82**, 961–5.

Salazar AM, Jabbari B, Vance SC *et al.* (1985). Epilepsy after penetrating head injury. I. Clinical correlates: a report of the Vietnam Head Injury Study. *Neurology*, **35**, 1406–14.

Salazar AM, Warden DL, Schwab K *et al.* (2000). Cognitive rehabilitation for traumatic brain injury: a randomized trial. Defense and Veterans Head Injury Program (DVHIP) Study Group. *JAMA*, **283**, 3075–81.

Sarajuuri JM, Kaipio ML, Koskinen SK *et al.* (2005). Outcome of a comprehensive neurorehabilitation program for patients with traumatic brain injury. *Arch Phys Med Rehabil*, **86**, 2296–302

Schierhout G, Roberts I (2001). Anti-epileptic drugs for preventing seizures following acute traumatic brain injury. *Cochrane Database Syst Rev*, (4), CD000173.

Shaw NA (2002). The neurophysiology of concussion. *Prog Neurobiol*, **67**, 281–344.

Shokunbi MT, Komolafe EO, Malomo AO *et al.* (2000). Scalp closure without fracture elevation does not reduce the risk of infection in patients with compound depressed skull fractures. *Afr J Med Med Sci*, **29**, 293–6.

Signorini DF, Andrews PJ, Jones PA *et al.* (1999). Adding insult to injury: the prognostic value of early secondary insults for survival after traumatic brain injury. *J Neurol Neurosurg Psychiatry*, **66**, 26–31.

Simpson DA, David DJ - Herbert Moran Memorial Lecture (2004). World War I: the genesis of craniomaxillofacial surgery? *ANZ J Surg*, **74**, 71–7.

Smith DH, Meaney DF, Shull WH (2003). Diffuse axonal injury in head trauma. *J Head Trauma Rehabil*, **18**, 307–16.

Steiner LA, Andrews PJD (2006). Monitoring the injured brain: ICP and CBF. *Brit J Anesthesia*, **97**, 26–38.

Steudel WI, Cortbus F, Schwerdtfeger K (2005). Epidemiology and prevention of fatal head injuries in Germany--trends and the impact of the reunification. *Acta Neurochir*, **147**, 231–42; discussion 242.

Stiell IG, Wells GA, Vandemheen K *et al.* (2001). The Canadian CT Head Rule for patients with minor head injury. *Lancet*, **357**, 1391–6.

Strachan RD, Whittle IR, Miller JD (1989). Hypothermia and severe head injury. *Brain Inj*, **3**, 51–5.

Tagliaferri F, Compagnone C, Korsic M *et al.* (2006). A systematic review of brain injury epidemiology in Europe. *Acta Neurochir*, **148**, 255–68.

Taylor A, Butt W, Rosenfeld J (2001). A randomized trial of very early decompressive craniectomy in children with traumatic brain injury and sustained intracranial hypertension. *Childs Nerv Syst*, **17**, 154–62.

Teasdale G, Knill-Jones R, van der Sande J (1978). Observer variability in assessing impaired consciousness and coma. *J Neurol Neurosurg Psychiatry*, **41**, 603–10.

Teasdale GM, Murray G, Anderson E *et al.* (1990). Risks of acute traumatic intracranial haematoma in children and adults: implications for managing head injuries. *BMJ*, **300**, 363–7.

Teasdale GM, Pettigrew LE, Wilson JT *et al.* (1998). Analyzing outcome of treatment of severe head injury: a review and update on advancing the use of the Glasgow Outcome Scale. *J Neurotrauma*, **15**, 587–97.

Timofeev I, Kirkpatrick PJ, Corteen E *et al.* (2006). Decompressive craniectomy in traumatic brain injury: outcome following protocol-driven therapy. *Acta Neurochir*, **96**, 11–16.

Turnage B, Maull KI (2000). Scalp laceration: an obvious 'occult' cause of shock. *South Med J*, **93**, 265–6.

van den Brink WA, Zwienenberg M, Zandee SM (1999). The prognostic importance of the volume of traumatic epidural and subdural haematomas revisited. *Acta Neurochir*, **141**, 509–14.

van den Broek MD (2005). Why does neurorehabilitation fail? *Head Trauma Rehabil*, **20**, 464–73.

Vaquero J, Zurita M, Cincu R (2002). Vascular endothelial growth-permeability factor in granulation tissue of chronic subdural haematomas. *Acta Neurochir*, **144**, 343–6; discussion 347.

Wade DT, King NS, Wenden FJ *et al.* (1998). Routine follow up after head injury: a second randomised controlled trial. *J Neurol Neurosurg Psychiatry*, **65**, 177–83.

Wakai A, Roberts I, Schierhout G (2007). Mannitol for acute traumatic brain injury. *Cochrane Database Syst Rev*, **24**, CD001049.

Wardlaw JM, Easton VJ, Statham P (2002). Which CT features help predict outcome after head injury? *J Neurol Neurosurg Psychiatry*, **72**, 188–92; discussion 151.

Weigel R, Schmiedek P, Krauss JK (2003). Outcome of contemporary surgery for chronic subdural haematoma: evidence based review. *J Neurol Neurosurg Psychiatry*, **74**, 937–43.

Wilde EA, Bigler ED, Pedroza C *et al.* (2006). Post-traumatic amnesia predicts long-term cerebral atrophy in traumatic brain injury. *Brain Inj*, **20**, 695–9.

Wilson JT, Pettigrew LE, Teasdale GM (2000). Emotional and cognitive consequences of head injury in relation to the Glasgow outcome scale. *J Neurol Neurosurg Psychiatry*, **69**, 204–9.

Wong HT, Poon WS, Jacobs P (2006). The comparative impact of video consultation on emergency neurosurgical referrals. *Neurosurgery*, **59**, 607–13; discussion 607–13.

Yamaki T, Hirakawa K, Ueguchi T *et al.* (1990). Chronological evaluation of acute traumatic intracerebral haematoma. *Acta Neurochir*, **103**, 112–5.

Yamamoto T, Katayama Y (2005). Deep brain stimulation therapy for the vegetative state. *Neuropsychol Rehabil*, **15**, 406–13.

Young HA, Gleave JR, Schmidek HH *et al.* (1984). Delayed traumatic intracerebral hematoma: report of 15 cases operatively treated. *Neurosurgery*, **14**, 22–5.

Zauner A, Doppenberg EM, Woodward JJ *et al.* (1997). Continuous monitoring of cerebral substrate delivery and clearance: initial experience in 24 patients with severe acute brain injuries. *Neurosurgery*, **41**, 1082–91; discussion 1091–3.

CHAPTER 26

Raised intracranial pressure, cerebral oedema, and hydrocephalus

Ian Whittle

Contents

26.1 Introduction

The brain is protected by the cranial skeleton. Within the intracranial compartment are also cerebrospinal fluid, CSF, and the blood contained within the brain vessels. These intracranial components are in dynamic equilibrium due to the pulsations of the heart and the respiratory regulated return of venous blood from the brain. Normally the mean arterial blood pressure, systemic venous pressure, and brain volume are regulated to maintain physiological values for intracranial pressure, ICP. There are a range of very common disorder such as stroke, and much less common, such as idiopathic intracranial hypertension, that are associated with major disturbances of intracranial pressure dynamics. In some of these the contribution to pathophysiology is relatively minor whereas in others it may be substantial and be a major contributory factor to morbidity or even death.

Intracranial pressure can be disordered because of brain oedema, disturbances in CSF flow, mass lesions, and vascular engorgement of the brain. Each of these may have variable causes and there may be interactions between mechanisms. In this chapter the normal regulation of intracranial pressure is outlined and some common disease states in clinical neurological practice that are characterized by either primary or secondary problems in intracranial pressure dynamics described.

26.2 Intracranial pressure–volume dynamics

26.2.1 Brain tissue water and volume

The adult human brain weighs about 1500 g and contains about 70 per cent water. The water content of white matter, 68 per cent, is less than that of cortex, 73 per cent. The intracellular compartment comprises 1100–1300 ml, the extracellular space 100–150 ml, the intracranial CSF volume 75–100 ml, and cerebral blood volume 75–100 ml (Papadopoulus *et al.* 2004). Others suggest that the normal brain extracellular space amounts to 12–19 per cent of total brain volume (Go 1997). The volumes of the various compartments are controlled by local mechanisms such as the blood–brain barrier, the choroid plexus, endogenous brain water production, the arachnoid villi, and systemic factors such as vasopressin, atriopeptides, renin-angiotensin, and aldosterone (Go 1997; Kimelburg 2004). Recently subcellular mechanisms of water transport into and out of the brain have been characterized and the important role of tight junction proteins in the blood–brain barrier and the perivascular, periependymal aquaporin family defined (Papadopoulus *et al.* 2002, 2004; Kimelburg 2004).

26.2.2 The blood–brain barrier

The blood–brain barrier is a barrier, which limits the egress of water and impedes efflux of most ions and other compounds from the vascular compartment to brain extracellular space (Ballabh *et al.* 2004; Kimelburg 2004; Papadopoulus *et al.* 2004). Its role is to maintain a homeostatic environment for neurons to function effectively and to exclude potentially toxic substances. The cellular elements of the blood–brain barrier are: endothelial cells, astrocytic end-feet or foot-processes, and pericytes. The endothelium and pericytes have an abluminal basement membrane. Within the structural components of the blood–brain barrier lie a range of transporter and receptor systems that control transmembrane and transluminal physiology.

The blood–brain barrier endothelial cells are characterized by the absence of fenestrations, extensive tight junctions, and sparse pinocytic vesicular transport. Endothelial cells interconnected by transmembrane proteins comprise the tight junction. Tight junctions consist of three integral membrane proteins, claudin, occludin, and junction adhesion molecules, and also a number of cytoplasmic accessory proteins such as zona occludens-1, 2, and 3, and cingulin. These cytoplasmic proteins link membrane proteins to the cytoskeleton protein actin in order to maintain the structural and functional integrity of the endothelium (Papadoupulus *et al.* 2004).

Pericytes appear to play a key role in angiogenesis, structural integrity, and differentiation of the vessel, and formation of endothelial tight junction. In an *in vitro* culture model pericytes stabilized the capillary-like structure formed by endothelial cells co-cultured with astrocytes by preventing apoptosis of the endothelium. Astrocytic end-feet tightly ensheath the pericytes and endothelium vessel wall and release trophic factors that are critical for the induction and maintenance of the blood–brain barrier and expression of a range of enzymes (Janzer and Raff 1987; Hayashi *et al.* 1997).

The predominant aquaporin protein in the central nervous system, aquaporin 4 is a major pathway for osmotically driven water transport and is found within the peri-capillary astrocytic foot processes, the external and sub-ependymal glial limiting membrane, and within ependymal cells (Papadoupulus *et al.* 2004). It facilitates movement of water into and out of the brain since disruption of the protein can significantly alter patterns of brain oedema and water clearance (Amiry-Moghaddam *et al.* 2004). Experiments in aquaporin 4 knock out mice have shown either an increased resistance to brain oedema or a particular susceptibility depending on the experimental model used (Papadopoulos *et al.* 2004).

Although the blood–brain barrier is permeable to water, muscle capillaries are 1000 times more permeable. The bulk or convective flow of both water (*J*) and compounds across the blood–brain barrier is determined by a range of biophysical parameters such as the hydrostatic pressure in the capillary (P_{cap}) and brain (P_{tiss}), hydraulic conductivity or coefficient of ultrafiltration (*L*) across the blood–brain barrier, the osmotic pressure in the capillary (Π_{cap}) and brain tissue (Π_{tiss}), and σ the reflection coefficient of solutes

$$J = L\,(P_{cap} - P_{tissue})\,\Pi + \sigma\,(\Pi_{cap} - \Pi_{tissue})$$

In the brain σ is essentially 1 for both ions and proteins since the tight junctions make the blood vessels impermeable. Under normal circumstances there is net pressure driving water from the brain extracellular space to the capillaries (Kimelburg 2004).

Molecules and proteins cross the blood–brain barrier in several ways. These are by: (1) passive diffusion, particularly for lipid-soluble substances; (2) facilitative and energy-dependent receptor-mediated transporters e.g. transferrin receptor, low density lipoprotien-receptor; (3) carrier-mediated transporters which provide essential brain nutrients e.g. GLUT-1 for glucose, CAT-1 for basic amino acids, LAT-1 for neutral amino acids, EAAT-1 for acidic amino acids; and (4) absorptive trancytosis e.g. albumin. Additionally there are important mechanisms that efflux compounds from the endothelial membrane and basolateral membrane. Such well-described proteins are P-glycoprotein (MDR-1), BCRP, OAT, and OCT families.

Passive diffusion, or extraction percentage (*E*), of a compound across the blood–brain barrier is dependent upon relationships between the capillary surface (*S*) area and capillary permeability (*P*) and cerebral blood flow (CBF).

$$\log\,(1 - E) = P.S/CBF$$

Thus pathophysiological changes in the blood–brain barrier, blood osmolality, dysregulated cerebral blood flow, or increased capillary pressure will influence the permeability of the blood–brain barrier as well as passive diffusion of water, ions, proteins, and other compounds into the brain (Go 1997).

26.2.3 CSF formation and flow

The CSF is formed largely by the choroid plexus in the lateral ventricles, third ventricle and fourth ventricle. The endothelium of the choroid plexus does not have a blood–brain barrier, therefore fluid leaks into the epithelial cells. These have an asymmetrical distribution of carrier systems that result in net transport of Na+, Cl–, and HCO_3 ions with subsequent osmotic H_2O movement into the ventricle (Johnston and Teo 2000; Kimelburg 2004). CSF is an ultrafiltrate of blood and has a low sodium content, a glucose level of approximately two-thirds of blood level, and very low protein content. Formation of CSF is at a pressure of 11 mmHg (Rosenborg *et al.* 1997) and at a rate of approximately 0.3 ml per minute. There are many factors (Johnston and Teo 2000) influencing CSF formation that include endogenous factors such as CSF pressure, hypoxia, pH, hypoglycaemia, and various therapeutic agents such as acetozolamide, frusemide, amiloride, omeprazole or glucocorticoids. Although CSF formation may be constant between pressures of -10–20 cm H_2O there is evidence that it decreases at higher pressures and is dependent on sympathetic stimulation (Johnston and Teo 2000). Some contribution to the CSF volume comes form both brain production of water from aerobic metabolism of glucose, circa 30 ml/day, and movement of fluid from the extracellular space of the brain, through the ependyma and glia limitans into the ventricular and sub-arachnoid spaces (Papodopoulus *et al.* 2004). The precise contribution of each source to total CSF production is not well understood (Fig. 26.1) (Johnston *et al.* 2000).

CSF flow is from the lateral ventricles through the foramen of Munro into the anterior third ventricular region through the third ventricle, down the aqueduct of Sylvius in the fourth ventricular region, and thence from the fourth ventricle through the foramina of Luchka and Magendie into the cisterna magna and pontomedullary cisterns (Fig. 26.1). From there it flows caudally into the spinal

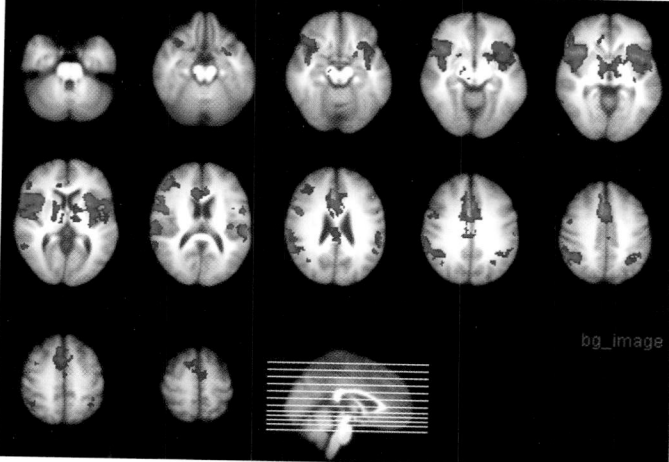

Plate 1 Functional MRI, or fMRI, showing areas of brain activity, as represented by increased blood flow, when a painful heat stimulus is applied to the limb of a patient with rheumatoid arthritis. (Courtesy of Professor I Tracey.) (See also Fig. 3.7.)

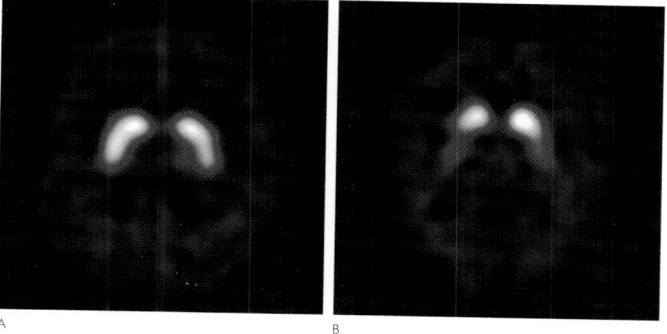

Plate 2 DAT scan of patient with Parkinson's disease compared with patient with essential tremor. The findings in normal patients or patients with essential tremor show normal signal intensity and isotope uptake in the basal ganglia (A), whereas in Parkinson's disease there is reduced uptake (B), which is often asymmetrically, in unilateral Parkinson's disease. (Courtesy of Dr D Grosset.) (See also Fig. 3.8.)

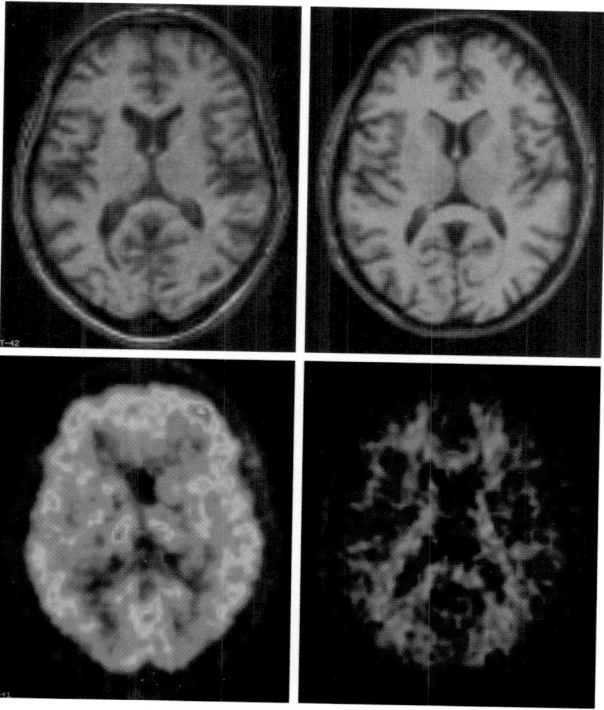

Plate 3 Positron emission tomography, PET, in a 56-year-old patient with Alzheimer's disease (left) and a 55-year-old normal control. Top: MRI brain. Bottom: 11C-PIB PET scan for amyloid. (Courtesy of Professor D Brooks.) (See also Fig. 3.9.)

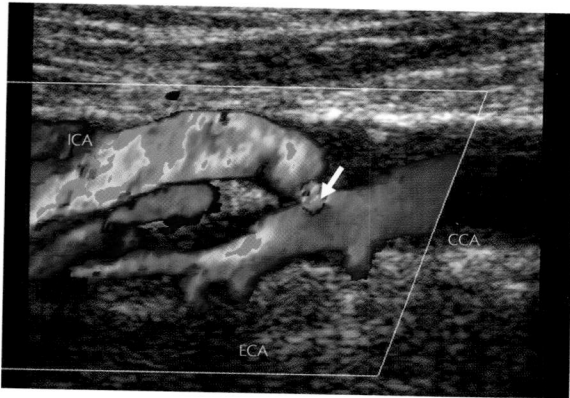

Plate 4 A colour flow Doppler ultrasound of the carotid bifurcation showing a plaque (arrow) at the origin of the internal carotid artery and the resultant stenosis. (See also Fig. 3.12.)

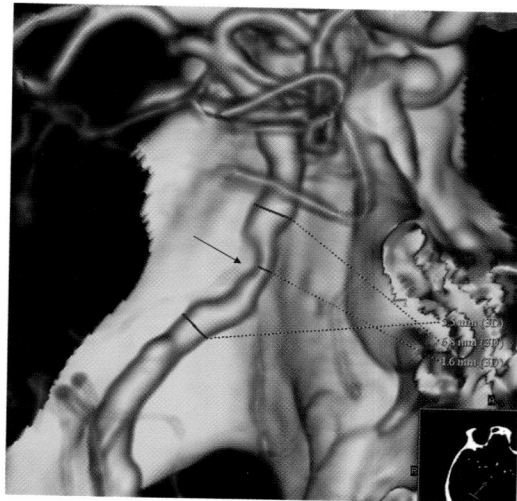

Plate 5 A CT angiogram three-dimensional reconstruction showing a stenosis (arrow) of the distal vertebral artery. (See also Fig. 3.13.)

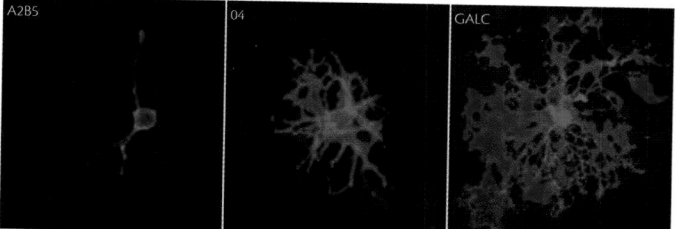

Plate 6 Stages in maturation of the oligodendrocyte lineage from progenitor (A2B5+) to precursor (O4 +) and mature myelinating cell (GalC +). (See also Fig. 8.3).

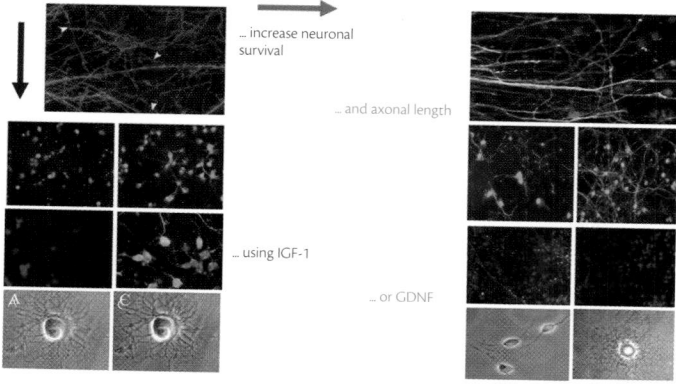

Plate 7 Medium conditioned by cells of the oligodendrocyte lineage supports neuronal survival and axonal growth acting through defined growth factors. (See also Fig. 8.4.)

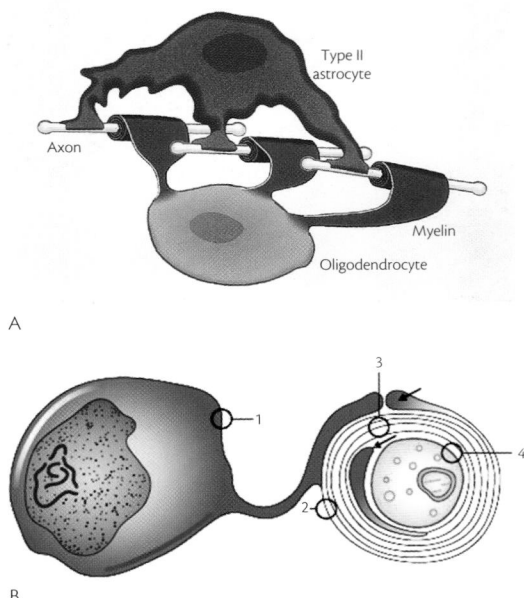

A

B

Plate 8 A. The cellular architecture of myelinated axons. B. Schematic drawing of the relationship between an oligodendrocyte and the internodal myelin of a nerve fibre: 1, the oligodendrocyte membrane; 2, the myelin surface; 3, compact myelin; 4, the myelin / axon interface. (Reproduced with permission from McAlpine's Multiple Sclerosis 4th edition 2005.) (See also Fig. 8.5.)

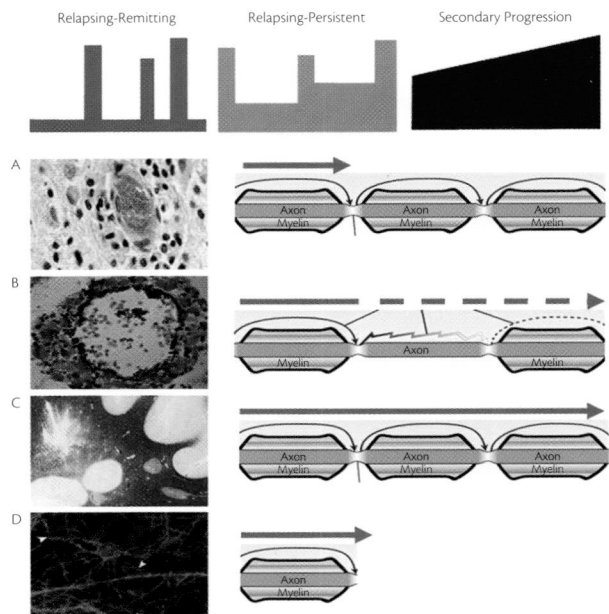

Plate 10 Cartoon to show the stages of injury to myelinated axons and their expression as the clinical course of multiple sclerosis. A. Inflammation causes transient conduction block, with recovery. B. Persistent inflammation leads to acute axonal injury and demyelination. C. Remyelination is associated with recovery of structure and function. D. Persistent demyelination is associated with chronic axonal loss and astrocytosis. (See also Fig. 8.7.)

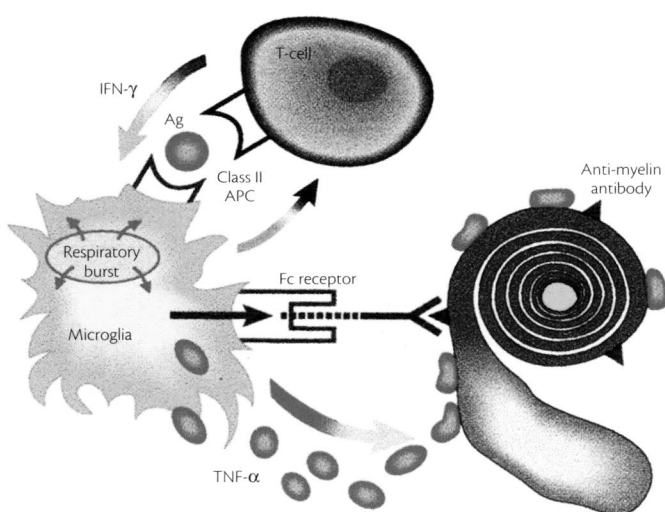

Plate 9 Scheme to represent interactions between activated T cells, microglia, and the oligodendrocyte. (Reproduced with permission from McAlpine's Multiple Sclerosis 4th edition 2005.) (See also Fig. 8.6 B.)

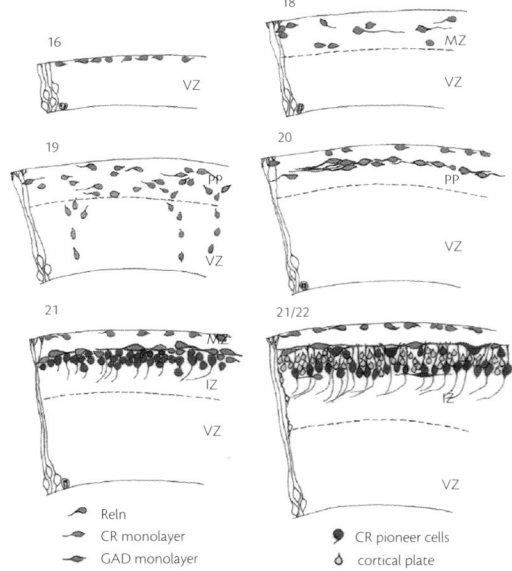

Plate 11 Diagrammatic representation of the developmental events proposed by Meyers and her colleagues for the early development of the human neocortex. All figures were drawn at the same magnitude with the aid of a camera lucida. *Blue*, Reln; *light green*, early calretinin (CR) immunoreactive neurons; *red*, glutamic acid decarboxylase (GAD) immunoreactive neurons; *dark green*, CR immunoreactive 'pioneer cells'; yellow, cortical plate neurons. The first Reln immunoreactive neurons appear at Carnegie stage 16 (5 weeks post-conceptual age) and increase in number from Carnigie stages 17 to 19 (6 to 6.5 weeks post-conceptual age). The first CR immunoreactive neurons appear at Carnigie stage 19 (6.5 weeks post-conceptual age) in what could now be called the preplate. GAD immunoreactive neurons, first appear at stage 20 (7 weeks post-conceptual age). Concurrently, Reln immunoreactive neurons settle in the subpial compartment. At stage 21 (7 weeks post-conceptual age), the pioneer cells send the first corticofugal fibres. The preplate is split apart into a minor superficial component and a large deep component, the subplate, through the first cohorts of the cortical plate, at stages 21 and 22 (8 weeks post-conceptual age). *IZ*, intermediate zone; *MZ*, marginal zone; *PP*, preplate; *VZ*, ventricular zone. (From Meyer *et al.* 2000.) (See also Fig. 9.4.)

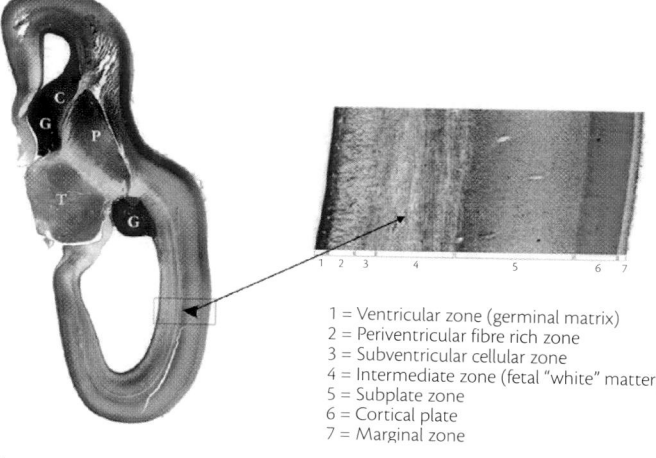

1 = Ventricular zone (germinal matrix)
2 = Periventricular fibre rich zone
3 = Subventricular cellular zone
4 = Intermediate zone (fetal "white" matter)
5 = Subplate zone
6 = Cortical plate
7 = Marginal zone

Plate 12 Horizontal section through the brain in an 18-week post-conceptual age foetus. Low power and high power view stained with cresyl violet. C, caudate nucleus; G, Ganglionic eminence; P, putamen; T, thalamus. (Adapted from Kostovic et al. 2002.) (See also Fig. 9.5.)

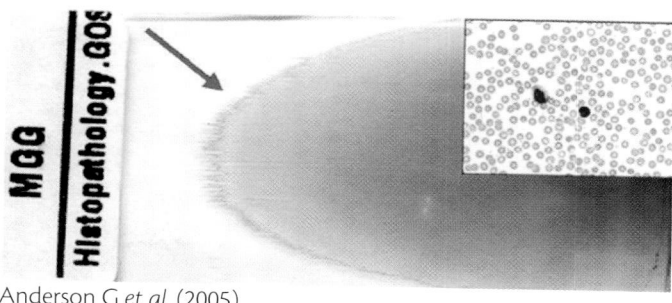

Anderson G *et al.* (2005)

Plate 14 Vacuolated lymphocytes. (A) Low power image of a blood film, illustrating the correct area in which to look for the presence of vacuolated lymphocytes (arrow), and (inset) high power photomicrograph demonstrating a small vacuolated lymphocyte and monocyte (May–Grunwald–Giemsa staining; original magnification, ×400). (See also Fig. 10.2 A.)

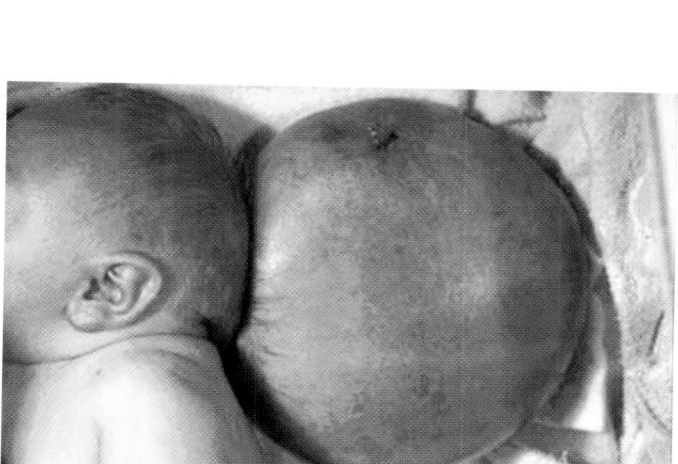

A

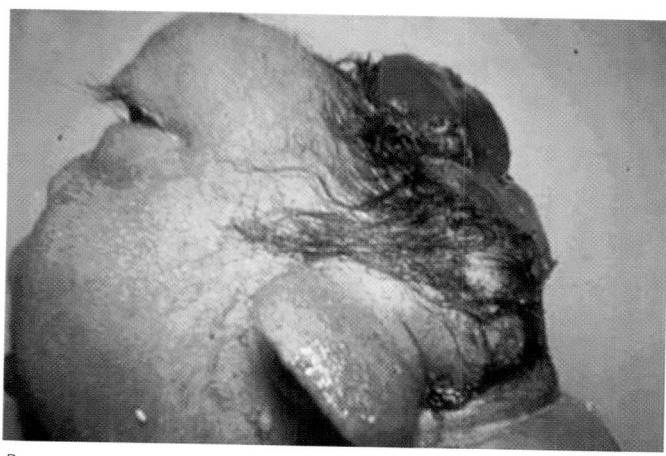

B

Plate 13 (A) Neonate with an occipital encephalocele; (B) Neonate with anencephaly. (Reprinted with permission of the Department of Pathology, Virginia Commonwealth University and the VCU Health System.) (See also Fig. 9.9.)

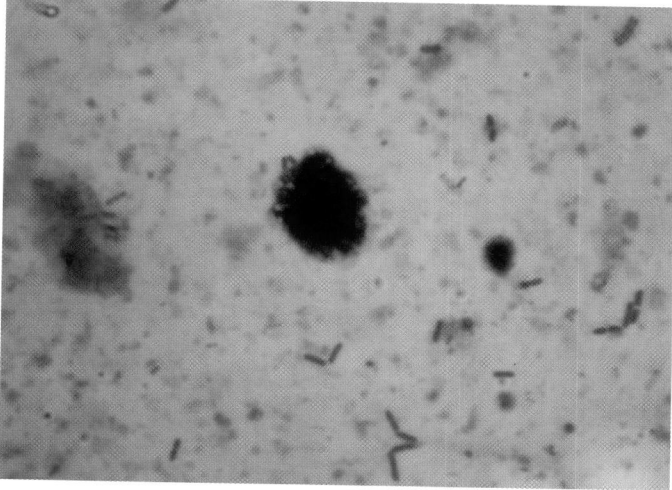

Plate 15 Late onset metachromatic leukodystrophy. T2-weighted MRI (A) axial and (B) coronal showing diffusely altered cerebral white matter signal. (Courtesy of Dr P. Anslow). (C) Urinary deposit stained with toluidine blue showing a shed epithelial cell containing brown metachromatic material. (Courtesy Dr W. Squier.) (See also Fig. 10.4 C.)

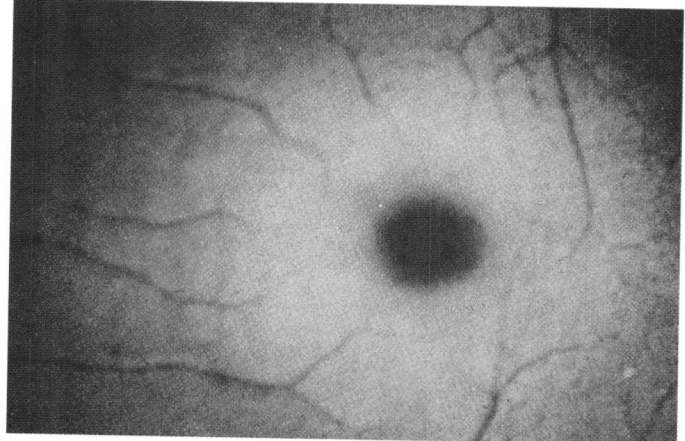

Plate 16 Cherry red spot at the macula on fundoscopy. Gaucher's disease showing a cherry red macular spot with surrounding retinal pallor. (See also Fig. 10.9 A.)

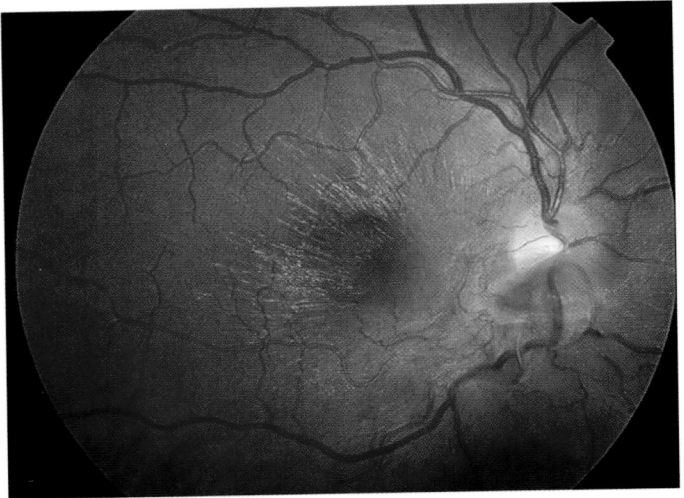

A

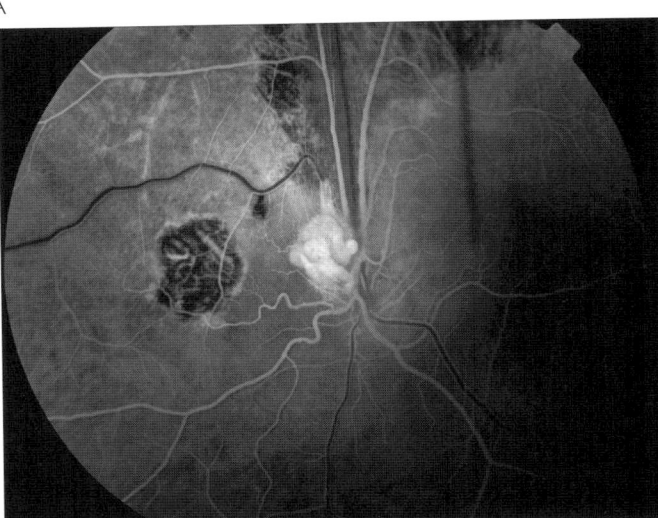

B

Plate 17 Von-Hippel–Lindau Disease. A. Haemangioma adjacent to the optic disc with secondary exudates in the macular region. B. Fluorescein angiogram showing two retinal haemangiomas, the one on the left treated. (Courtesy of Dr E. Snodgrass.) (See also Fig. 11.6.)

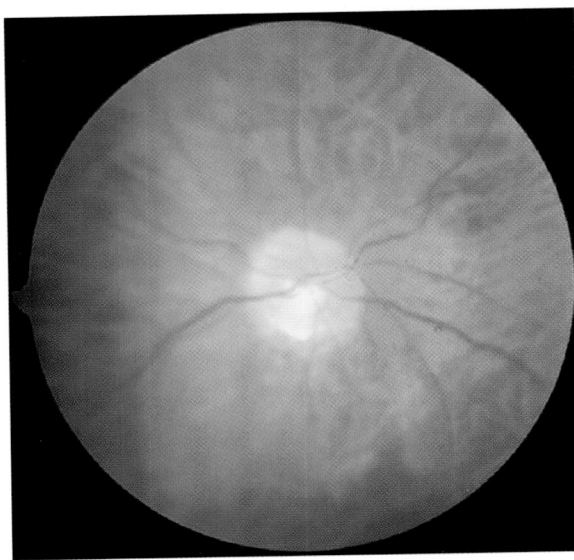

Plate 19 Central retinal artery occlusion. (See also Fig. 12.4.)

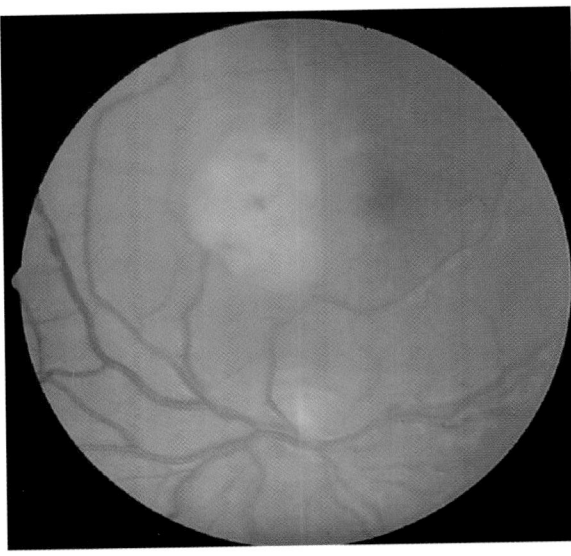

Plate 20 Fundus showing retinal haematoma associated with tuberose sclerosis. (See also Fig. 12.5.)

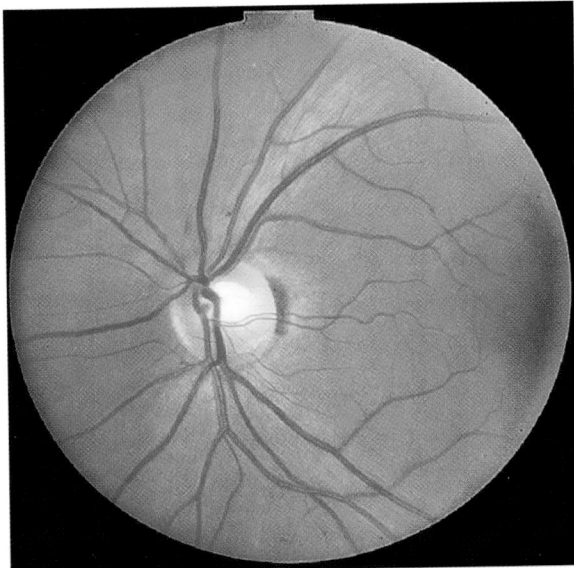

Plate 18 A normal fundus and optic disc. (See also Fig. 12.3.)

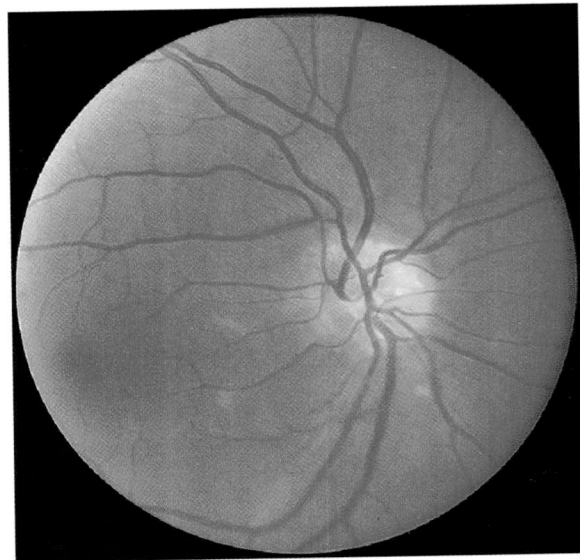

Plate 21 Pseudopapilloedema showing optic nerve head drusen. (See also Fig. 12.6.)

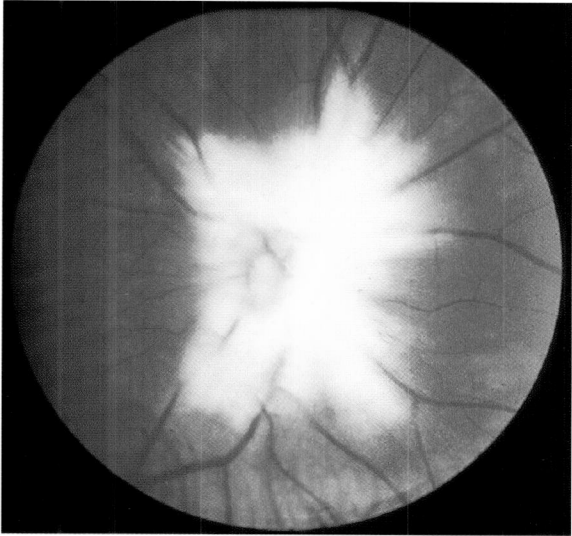

Plate 22 Myelinated retinal nerve fibres. (See also Fig. 12.7.)

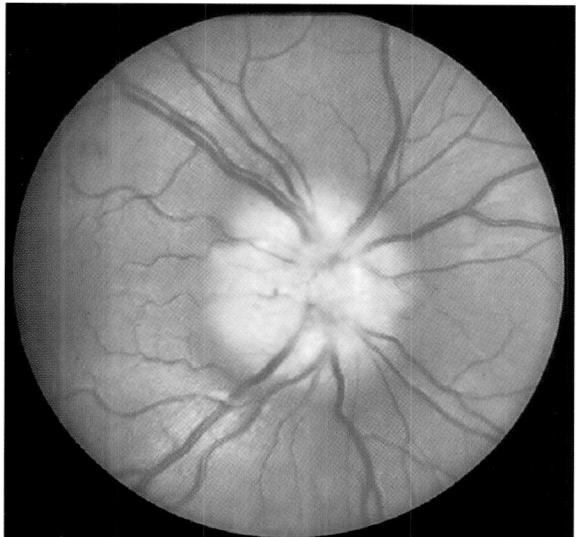

Plate 23 Papilloedema due to raised intracranial pressure. (See also Fig. 12.8.)

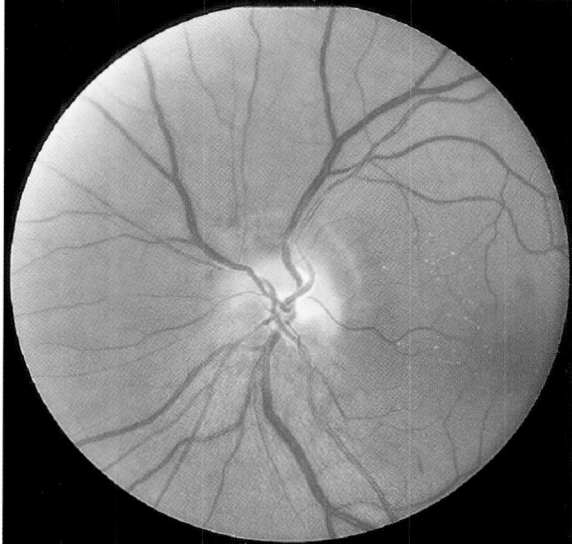

Plate 24 Anterior ischaemic optic neuropathy. (See also Fig. 12.9.)

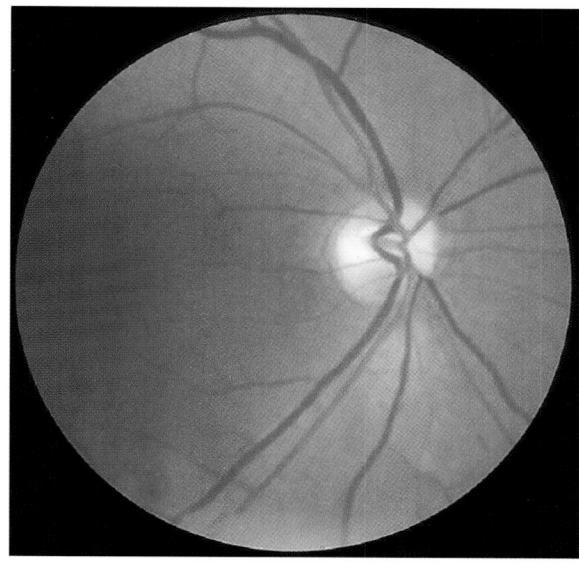

Plate 25 Optic atrophy. (See also Fig. 12.10.)

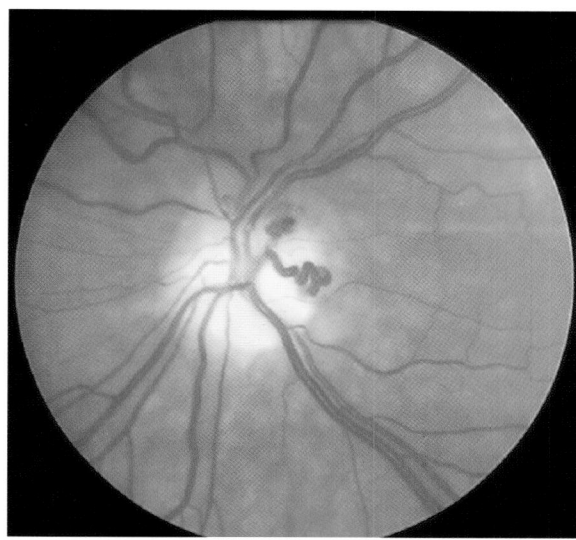

Plate 26 Optociliary shunts due to an optic nerve sheath meningioma. (See also Fig. 12.11.)

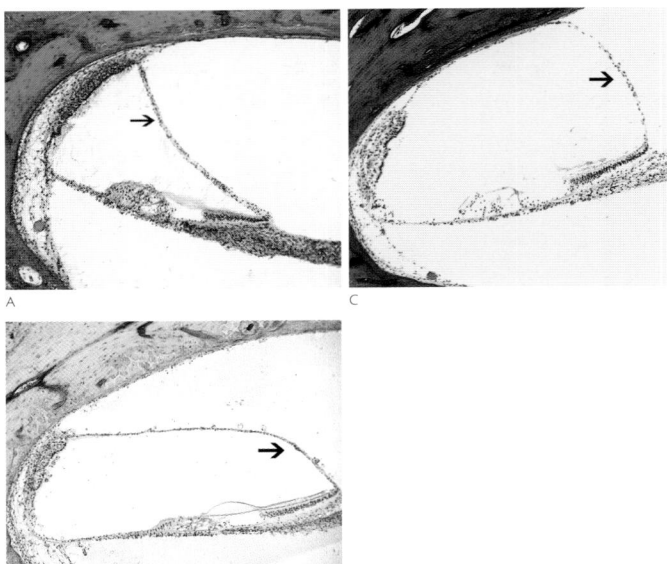

Plate 27 Histological sections through the cochlea showing (A) normal anatomy of apical turn, (B) mild endolymphatic hydrops with bulging of Reissner's membrane and (C) severe hydrops with Reissner's membrane blown up and contacting the top of scala tympani, with destruction of the organ of Corti in a case of Menière disease. Arrows points to Reissner's membrane. (Courtesy of Professor Leslie Michaels.) (See also Fig. 15.14.)

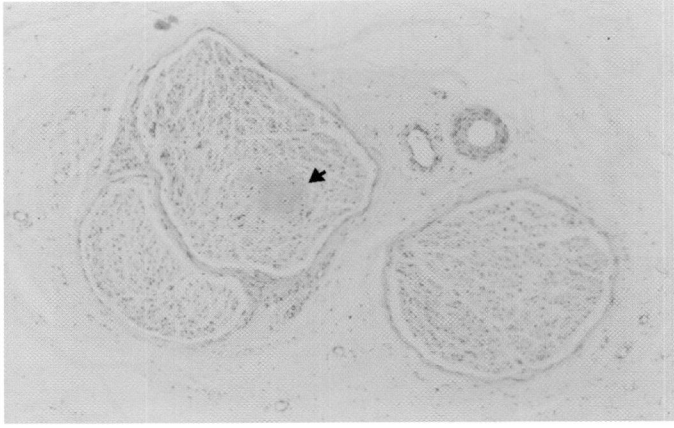

A

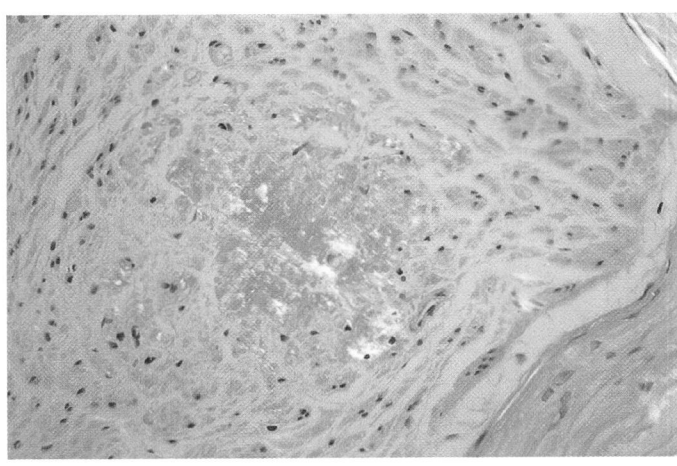

B

Plate 28 Amyloid due to immunoglobulin light chain deposition in peripheral nerve. A. Eosinophilic deposit within a nerve fascicle, low power, haematoxylin, and eosin, B. Apple green birefringence of the amyloid deposit in polarized light, Congo red, higher power. (See also Fig. 21.14.)

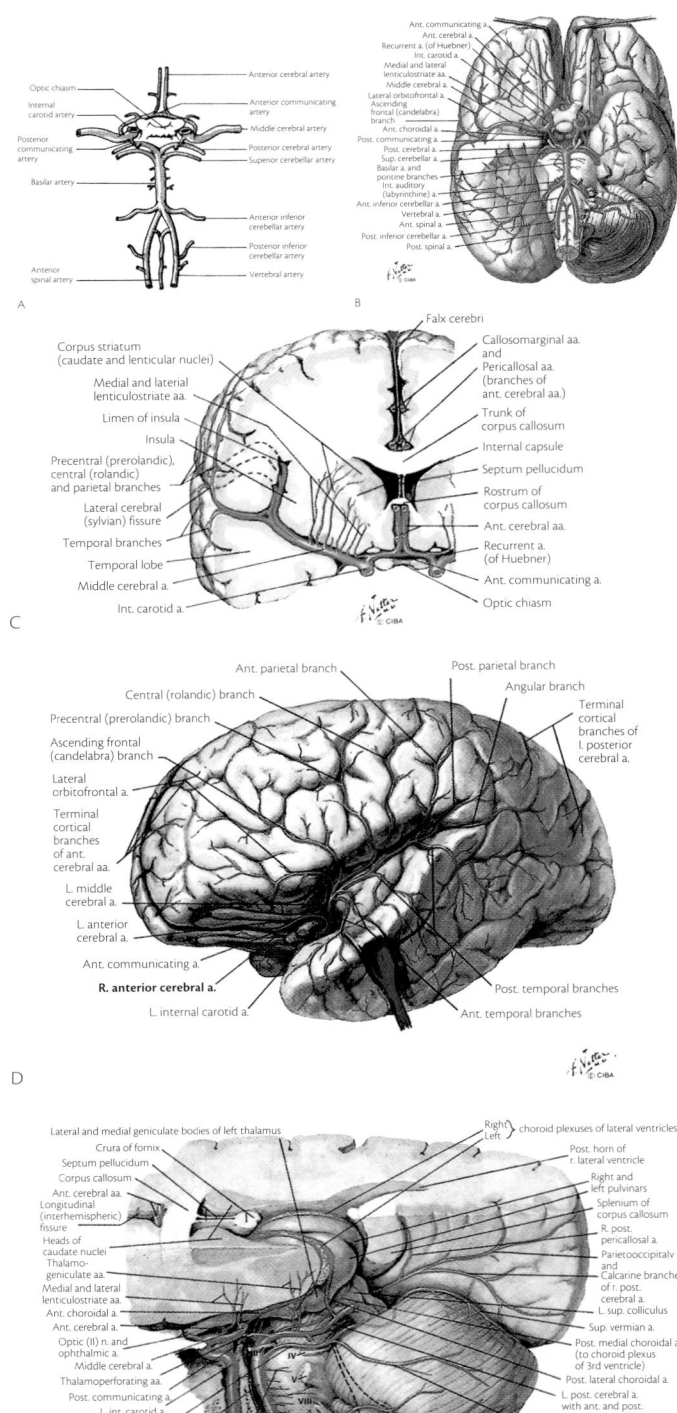

Plate 29 The Cerebral Arterial Circulation. (A) The circle of Willis at the base of the brain as seen from below. There is considerable anatomical variation and this figure represents but one rather typical arrangement. (B) A view of the base of the brain to show the circle of Willis, vertebrobasilar arterial system, middle and anterior cerebral arteries. (C) A coronal section of the brain through the optic chiasm to show the penetrating lenticulostriate branches of the main stem of the middle cerebral artery. (D) A lateral view of the brain to show the cortical branches of the middle cerebral artery. (E) Lateral view of the brain to show the vertebral, basilar and posterior cerebral arteries and their branches in the posterior fossa. (Figures 35.8(B)–(E) reprinted with permission from Clinical Symposia, illustrated by Frank N. Netter, MD, copyright 1990 CIBA-GEIGY Corporation. All rights reserved.) (See also Fig. 35.8.)

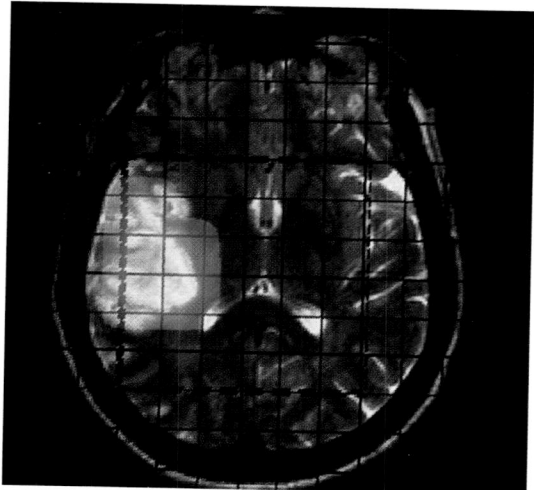

Plate 30 A metabolic 'map' of lactate in the brain by MR spectroscopy, superimposed on a structural MR scan. There is an obvious excess of lactate in the infarcted area in the distribution of a branch of the right middle cerebral artery. (Courtesy of Dr Joanna Wardlaw.) (See Fig. 35.13 B.)

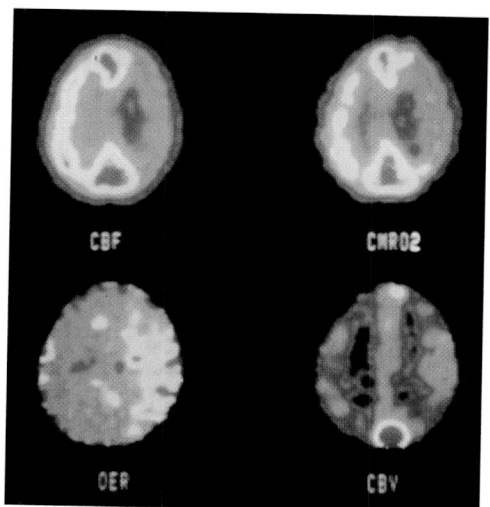

(i)

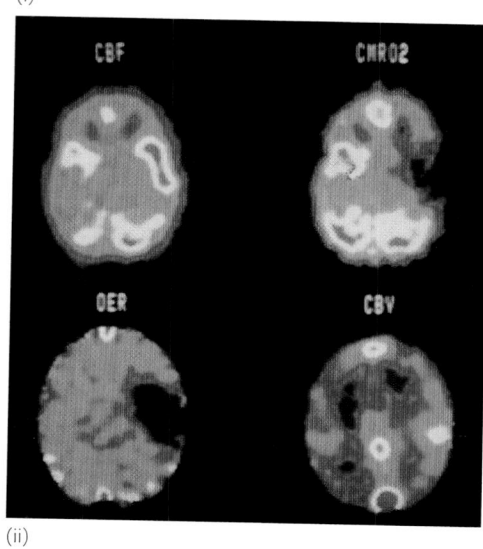

(ii)

Plate 31 Positron emission tomography (PET) scan 90 minutes after acute left middle cerebral artery occlusion: (i) cerebral blood flow (CBF) is low, cerebral metabolic rate for oxygen ($CMRO_2$) is low, but the oxygen extraction ratio (OER) (fraction) is maximal (i.e. ischaemia). (ii) PET scan a week later to show increased CBF and cerebral blood volume (CBV), despite low $CMRO_2$ and OER (i.e. absolute luxury perfusion). (Scans kindly provided by Professor Richard Frackowiak.) (See also Fig. 35.14 B.)

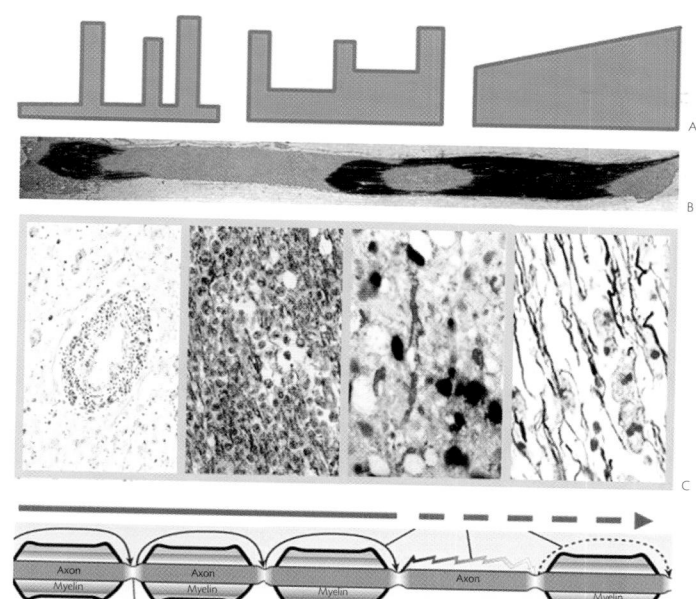

Plate 32 The pathogenesis of multiple sclerosis. A. The clinical course is depicted. B. Macroscopic areas of demyelination in the optic nerve. C. The panels show perivascular infiltration by activated lymphocytes, microglial removal of myelin debris, remyelination, and axonal injury. D. Interruption in saltatory conduction. (See also Fig. 37.1.)

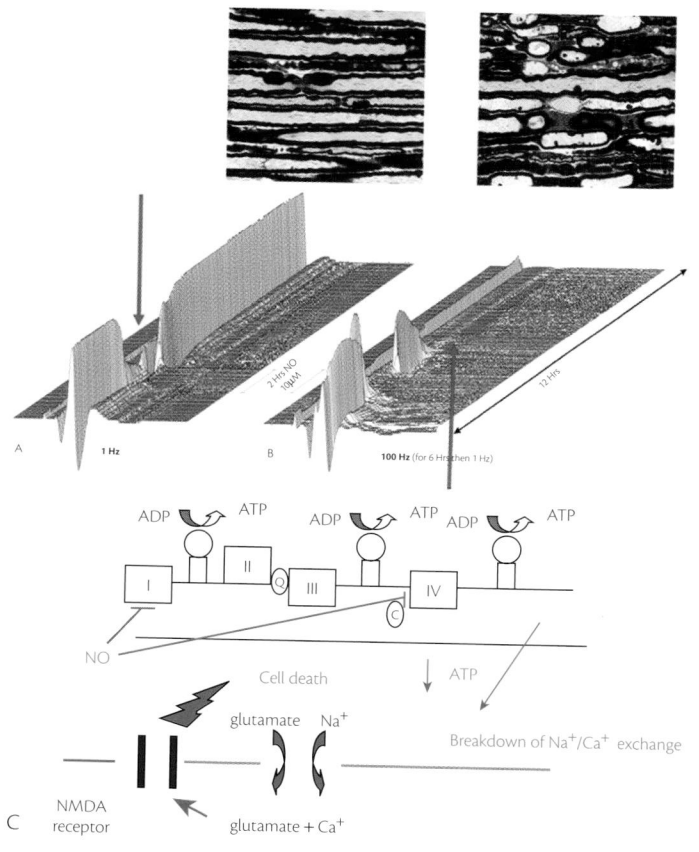

Plate 33 The effect of a nitric oxide donor, spermine NONOate, on conduction along axons passing through a demyelinating lesion in the rat spinal cord based on averaging 128 successive records over a 4-h recording. A. Transient failure of conduction in the rat spinal cord following application of nitric oxide donors without histological damage. B. Persistent failure of conduction in the rat spinal cord following prolonged exposure to donors of nitric oxide and prolonged electrical stimulation with evidence for glial and axonal injury. Reproduced from Smith *et al.* (2001). C. Cartoon to show the combination of energy failure due to mitochondrial damage with consequential excitotoxicity resulting from failure of normal ion exchange mechanisms acting across the 'axonal' membrane. (See also Fig. 37.3.)

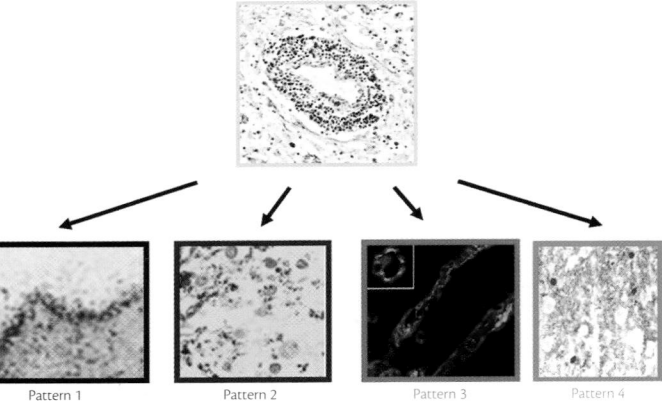

Pattern 1   Pattern 2   Pattern 3   Pattern 4

Plate 34 Heterogeneity in multiple sclerosis plaques. Focal perivascular lymphocytic infiltration is associated with four patterns of further tissue injury. (See also Fig. 37.4.)

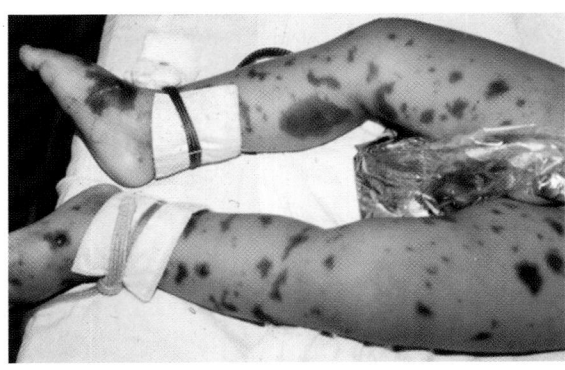

A

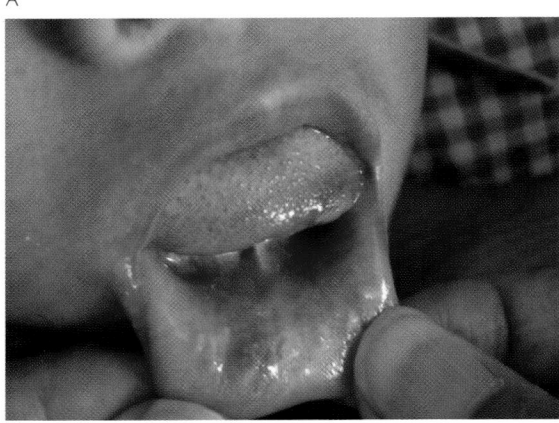

B

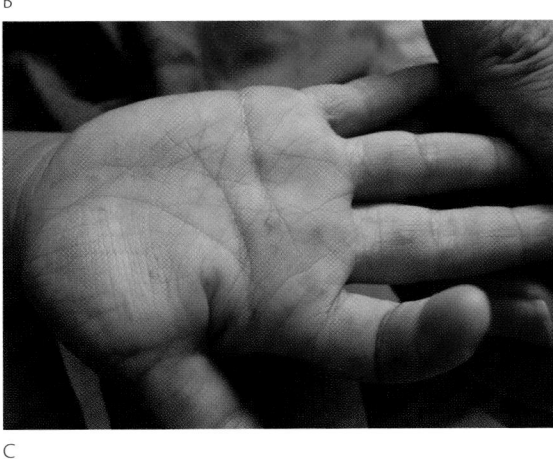

C

Plate 35 Exanthema and enanthemas in meningitis. Meningococcal disease purpuric rash (a). Enterovirus 71, hand, foot, and mouth disease with ulcers on the tip of the tongue and lower lip (b) and palm vesicles (c). (Photographs by T Solomon.) (See also Fig. 41.2.)

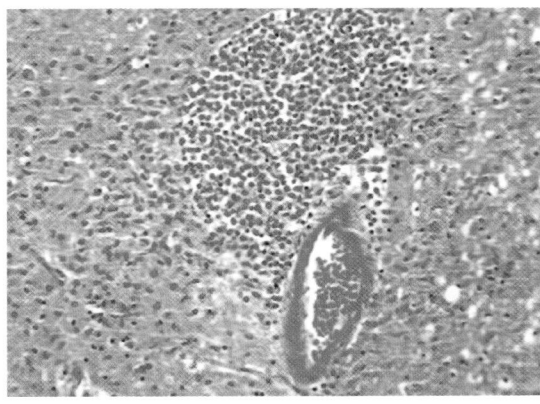

A

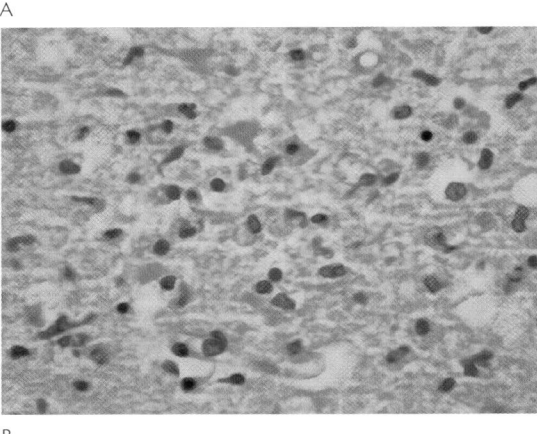

B

Plate 36 Histopathological picture of the temporal cortex of a man who died from herpes simplex virus encephalitis. (A) Intense perivascular inflammatory infiltrate consisting of activated microglia, macrophages and lymphocytes; haematoxylin, and eosin staining ×20. (B) High power view showing microglia and dead neurons with nuclear dissolution (karyolysis) and hypereosinophilia within the cytoplasm retaining the original pyramidal contour, haematoxylin and eosin staining ×40. (Pictures courtesy of Dr Daniel Crooks and Solomon *et al.* 2007.) (See also Fig. 42.1.)

A

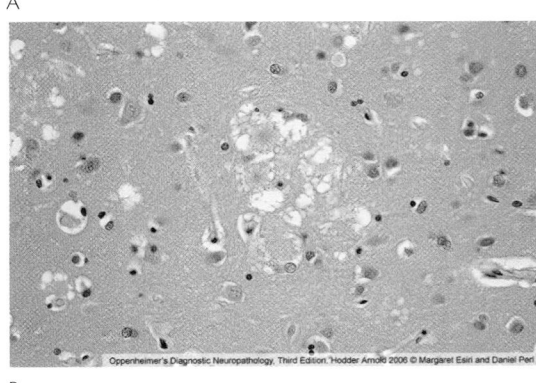

B

Plate 37 Neuropathology of sporadic Creutzfeldt–Jakob disease. (A) Low power view showing cerebral cortical spongiform change. (B) Higher power view showing a florid amyloid plaque. (Reproduced from Esiri and Perl 2006.) (See also Fig. 42.13.)

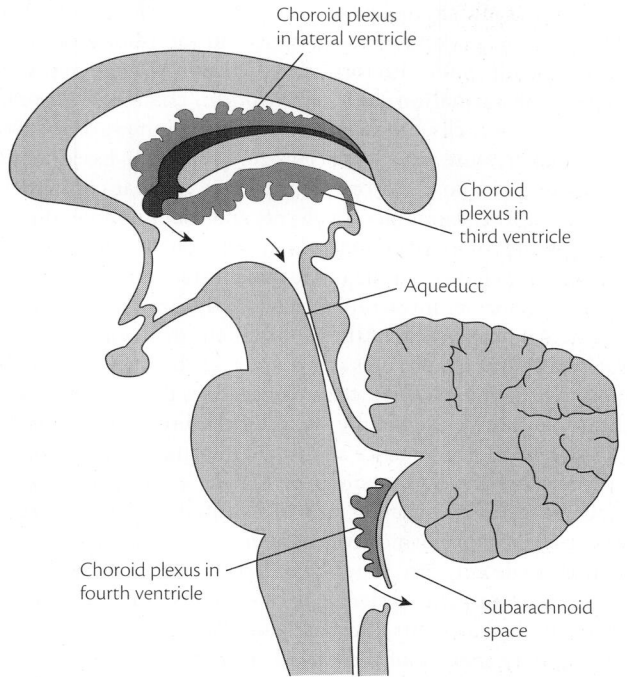

Fig. 26.1 CSF flow through the brain. Diagrammatic midline section through the brainstem and diencephalic region showing the disposition of the lateral, third, and fourth ventricular choroid plexi, and flow of CSF from the lateral ventricle, through the Foramen of Monro into the third ventricle (arrows). Then down the aq`ueduct into the fourth ventricle and egressing from there vis the foramen of Magendi. (From Brodal, The Central Nervous System, OUP 1998.)

subarachnoid space and also rostrally into the basal subarachnoid cisterns. The CSF then passes over the surface of the cerebral hemispheres and is reabsorbed in the arachnoid granulations adjacent to the superior sagittal sinus (Fig. 26.2). Perfusion times are around 2 h to traverse the entire brain surface and 1–1.5 h to reach the lumbar cistern (Greitz *et al.* 1996). Absorption into the arachnoid granulations normally occurs at a pressure of 5 mmHg (Kimelburg 2004), however the precise physiological mechanism remains unclear (Johnston and Teo 2000). Other outflow routes for CSF have been recognized along the sheaths of major blood vessels and cranial nerves (Greitz *et al.* 1996; Johnston and Teo 2000; Abbott 2004). There is also evidence for connections between the CSF and nasal lymphatic vessels (Johnston *et al.* 2004); although this pathway is important in sheep, its significance in humans is not known.

26.2.4 Cerebral blood volume

The cerebral blood volume is composed of the arterial inflow, blood in the capillaries and in the venous vessels. Cerebral blood flow is generally independent of mean arterial pressure (Fig. 26.3), but is closely regulated by arterial $PaCO_2$ systemically and locally by regional factors such as release of endothelin and nitrous oxide. Many of the veins in the brain have a large capacity for engorgement therefore cerebral blood volume can vary quite considerably. The rate of venous outflow is controlled by intrathoracic pressure, patency of the major venous cranial sinuses, and hydrostatic pressure. Under normal circumstances, cerebral blood volume totals about 150 ml with average cerebral blood volumes in grey matter being 5.5 ml blood per 100 ml brain and 1.4 ml blood per 100 ml brain for white matter. Quantitative determination of cerebral blood

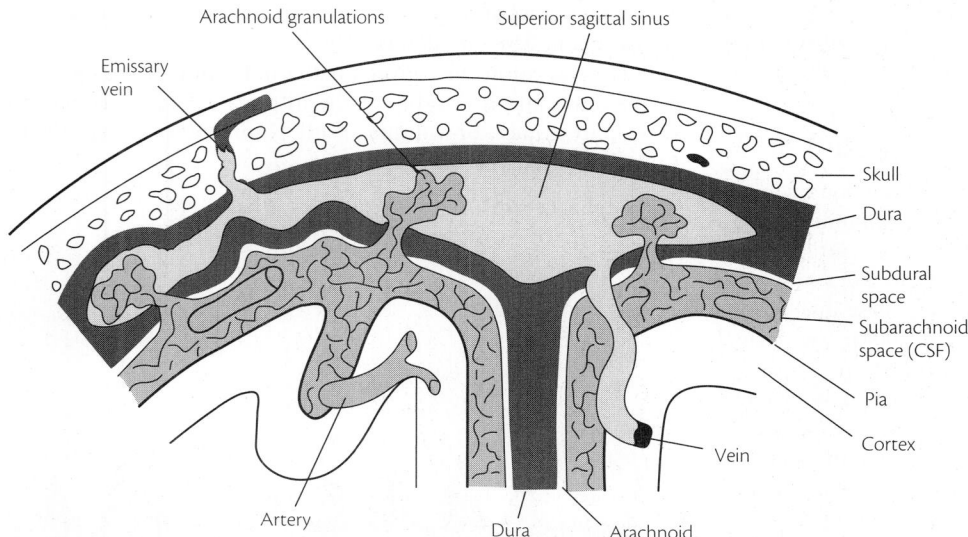

Fig. 26.2 Resorption of cerebrospinal fluid. Diagrammatic coronal section of the skull, the superior sagittal sinus, attached falx cerebri, CSF space, and arachnoid granulations. The CSF passes from the basal subarachnoid cisterns over the cerebral convexities to the paramedian arachnoid granulations where it enters the venous sinus. Thrombosis of the venous sinuses, or stenoses in the outflow pathways can lead to raised venous pressure and impaired CSF re-absorption. (From Brodal, The Central Nervous System, OUP 1998.)

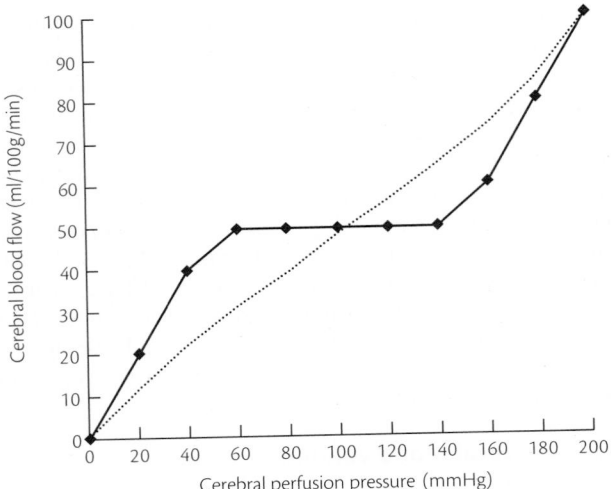

Fig. 26.3 Autoregulation of cerebral blood flow. Under normal circumstances cerebral blood flow remains constant over a range of cerebral perfusion pressure from 50 to 150 mmHg. (solid line + black squares). However when normal autoregulation is lost, cerebral blood flow becomes passively dependent on cerebral perfusion pressure (dotted line). In these circumstances vascular hypotension can lead to either focal or global ischemic brain damage.

volume is important in a range of neurological conditions and this can now be done using T1 vascular space occupancy MRI following gadolinium-DTPA infusion. (Lu *et al.* 2005). Factors that alter arterial inflow to the brain or cerebral venous outflow can cause considerable volumetric fluctuations in cerebral volume. These volume changes can occur acutely, for instance during a Valsalva manoeuvre, or chronically. Raised cerebral venous pressure will have an effect on intracranial pressure dynamics, and this has been implicated in the pathogenesis of idiopathic intracranial hypertension (Section 26.5.6). Previously engorgement of the cerebral vasculature was considered to be a major factor in brain swelling associated with neurotrauma; however it is now considered that cytotoxic brain oedema is the predominant factor causing brain swelling after neurotrauma (Marmarou 2004).

26.2.5 The physiology of intracranial pressure

The cranium contains the brain, CSF, and cerebral blood volume. Since the cranium in adults is a closed box of fixed volume there is a dynamic relationship caused by interactions between cardiac pulsations, arterial cerebral blood flow, venous outflow, the compliance or stiffness of the brain, and the CSF volume. These interactions produce intracranial pressure (Czosnyka *et al.* 2004). The normal intracranial pressure in adults is between 5 and 15 mmHg and is lower in children. Under normal circumstances however, intracranial pressure may transiently rise to 50 or 60 mmHg. Such physiological circumstances occur when performing Valsalva-like manoeuvres. This inhibits venous return from the brain to the heart because of the raised intrathoracic pressure. As a result the cerebral blood volume increases. Therefore, whilst weightlifting or straining to defecate, intracranial pressure may shoot up but when the action is finished it has a reciprocal decrease back to normal pressure. Interactions between the cerebral blood volume, CSF volume, brain volume, and mass lesion volume are fundamental to intracranial pressure dynamics.

The normal intracranial pressure wave form reflects oscillations from the cardiac systole/diastole cycle and the respiratory cycle, (Czosnyka *et al.* 2004). Factors affecting brain volume and water content, CSF formation and reabsorption, changes in cerebral blood volume as well as 'mass' insults will contribute variably to the intracranial pressure wave form (Figs 26.4 and 26.5). Essentially an increase in the volume of one of the three normal components consisting CSF, cerebral blood volume, and brain or an additional fourth volume component, such as a mass lesion, which will require compensatory changes in the other three intracranial components to maintain normal intracranial pressure dynamics. This was the fundamental proposition of the Munro–Kelly doctrine.

An isolated intracranial pressure reading, as obtained from lumbar puncture or tapping a CSF access device, does not, however, really indicate the pressure dynamic status of the brain (Johnston and Teo 2000). In clinical practice additional information about intracranial pressure can be measured by placing a monitoring device into the brain, into the lateral ventricle, or into the subdural space. This is routinely performed for many patients in neurosurgery in whom there is concern about raised intracranial pressure. The commercially available systems are reliable, robust, and generally have small pressure drifts after insertion. Data can be downloaded for computers analysis of the periods of monitoring into various waveforms, pressure thresholds, and time characteristic of the intracranial pressure levels. The problem of how to quantify measured abnormalities and how to weigh the significance of abnormalities such as raised intracranial pressure, level of hypertension, or duration of pressure rise, still remains a challenge despite use of computerized programs (Johnston and Teo 2000; Czosnyka *et al.* 2004). Current developments are investigating the utility of non-invasive measurment of intracranial pressure using MRI. This technique is in its infancy but could be of clinical benefit if it is validated (Raksin *et al.* 2003). Once again the challenge of interpreting the intracranial pressure data remains.

Since many variables can affect intracranial pressure clinicians have sought ways of evaluating the hydrodynamic, rather than static, status of the CSF circulation. The 'pressure volume curve', PVI (d*P*/d*V*) and compliance gives a better indication of the tightness of the brain and the ability of the intracranial compartment to withstand a mass or volume insult (Johnston and Teo 2000). Compliance is defined as:

$$\text{Compliance} = \Delta V/\Delta P = \Delta V/(P\text{max} - P\text{base}).$$

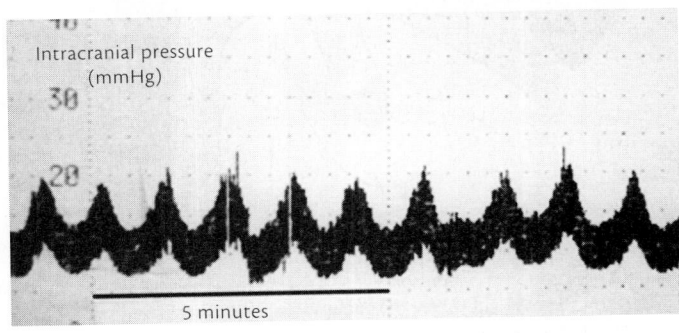

Fig. 26.4 Intracranial pressure trace showing classical B-waves peaking at an intracranial pressure of 20 mmHg. These are symmetrical, one per minute pressure oscillations related to cerebral blood flow. (Courtesy of Dr Y. H. Yau.)

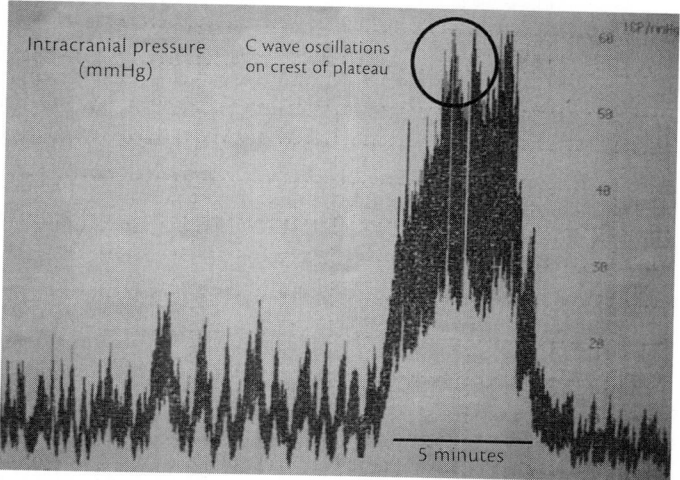

Fig. 26.5 Intracranial pressure waveform recorded following surgery for an intracranial tumour. There are baseline oscillations from 0 to 10 mmHg, with occasional C waves of 20–25 mmHg. There is also a characteristic A-wave or plateau wave that lasts 5 min with a pressure peak of 50–60 mmHg. Pressure waves of this magnitude can result in brain herniations or alterations of conscious state. (Courtesy of Dr YH Yau.)

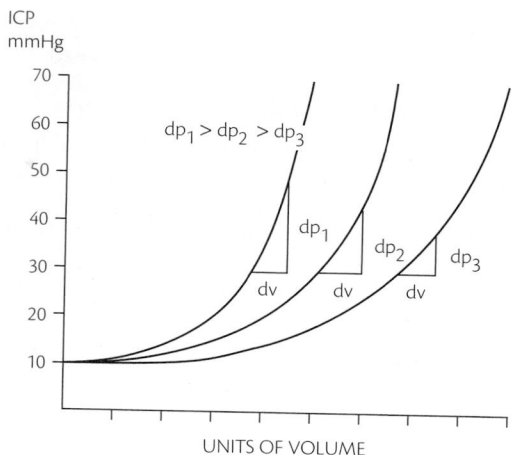

Fig. 26.6 Three pressure–volume curves showing the changes in intracranial pressure that can occur from a baseline pressure following a volume insult such as intracranial haematoma. The curves could represent responses of three different individuals such as a healthy young person (dp_1/dv), middle-aged person (dp_2/dv), or elderly person with brain atrophy (dp_3/dv) subject to the same 'volume insult'. For each given unit of volume insult (dv) the pressure response is variable (dp) because of differences in brain compliance.

Where V is the volume of fluid injected, and P the maximal and baseline intracranial pressure. Figure 26.6 demonstrates this concept, in that for a given starting intracranial pressure the three patients will have different pressure–volume curves and hence different clinical responses to an identical intracranial mass insult such as a haematoma. Conceptually the three figures can be thought of as a young person with a healthy brain and small CSF volume, a middle-aged person with a mildly atrophic brain and slight increase in CSF volume, and an elderly person with brain atrophy and hydrocephalus. Similar patterns of pressure–volume curves could be obtained from the same person subjected to a mass lesion that is administered in equal volumetric aliquots but at variable rates. Thus the first curve would represent relatively quick formation of a mass lesion, the second a slower formation, and the third the slowest formation.

Additional information about CSF disequilibrium and hydrodynamic status can be obtained from CSF infusion methods (Johnston and Teo 2000; Czosnyka *et al.* 2004). These techniques involve 'challenging' the CSF pathways with either a constant rate/volume infusion, or constant pressure infusion technique and then measuring the CSF pressure responses (Marmarou *et al.* 2005b). Although providing information about CSF outflow resistance (R_o), there is often a large variability in reported results between centres, and a potential for iatrogenic CSF infection, thus making clinical application problematic in a patient with possible symptomatic chronic hydrocephalus (Morgan *et al.* 1991; Marmarou *et al.* 2005b).

26.3 Cerebral oedema

26.3.1 Origins and classification

Brain 'swelling' has been recognized as a neuropathological entity for centuries (Bell 1983) However its pathogenesis long remained unclear and was subject to considerable debate. Although brain oedema was recognized as an increase in its tissue volume due to increase in brain water content, it can be categorized as arising from either

◆ a defect in the permeability of brain capillary system vasogenic brain oedema;

◆ or dysfunction of cell membranes cytotoxic brain oedema; (Klatzo 1967).

The former is largely extracellular, whereas cytotoxic brain oedema is predominantly intracellular. Additional categorizations have been of interstitial brain oedema due to transependymal spread of CSF in hydrocephalus (Fishman 1976), and hydrostatic and hypoosmotic oedema (Miller 1979). Hydrostatic oedema occurs with failure of cerebrovascular resistance resulting in transudation of an ultrafiltrate into the extracellular space from engorged capillaries. Since the human brain does not have a lymphatic system, water accumulates relatively easily within the brain parenchyma. Clearance of the excess fluid therefore depends very much on convection of the fluid through the white matter (Abbott 2004) to the subependymal area then transependymal displacement into the ventricular CSF and through the glia limitans into the subarachnoid space.

Brain oedema and swelling plays a crucial role in the pathophysiology of a range of neurological disorders including stroke, neurotrauma, intracranial tumours, hepatic encephalopathy, and intracranial infections (Go 1997; Marmarou 2004). Recent experimental and clinical studies have however shown that both cytotoxic and vasogenic oedema may complicate either head injury or stroke; however the relative time course of oedema subtype predominance varies (Unterberg *et al.* 2004). These, and other studies have used MRI diffusion weighted imaging and derived values such as apparent diffusion co-efficient to measure brain water non-invasively (Marmarou 2004; Ranjan *et al.* 2005; Pasco *et al.* 2006). Raised intracranial pressure associated with the primary brain disorder and coupled with secondary brain oedema is a major contributor to the morbidity and mortality associated with these conditions.

26.3.2 Vasogenic brain oedema

Vasogenic brain oedema is caused by either increased blood–brain barrier permeability to water or increased permeability of an abnormal capillary bed within the brain, such as neovascularity associated with an intracranial tumour. The permeability of the normal blood–brain barrier is increased by local or generalized infective processes, for instance brain abscess, subdural empyema, infective encephalitis, and thrombophlebitis. In each of these processes there may be changes in integrin expression on the cerebral endothelium leading to local leukocyte adherence, activation of microglia, changes in the release of nitric oxide, and alterations in local superoxide dismutase (Koedel et al. 1995). As a result of this inflammatory response there is cytokine release which disrupts tight junction adherence proteins and alters endothelial cell biophysics so that water permeability into the brain is increased (Mathur et al. 1992, Koedel et al. 1995; Davies 2002; Sanni et al. 2004; Stenzel et al. 2005).

In infective or neoplastic disorders the net result of 'leaky' capillaries is an increase in water transfer from the vascular compartment to the extracellular space of the brain (Davies 2002). Although cerebral blood flow is greater in the cortex than white matter, vasogenic brain oedema fluid is predominantly a white matter phenomenon because its tissue hydraulic resistance is considerably lower than the cortex. Water tends to flow by bulk convection along the white matter tracts of the cerebral hemisphere, producing a characteristic MRI or CT appearance (Fig. 26.7). White matter

water content may increase from 68 to 75 per cent depending on the cause and the proximity of the lesion and can be measured non-invasively using T2 weighted MRI. As well as local brain oedema, there is obliteration of adjacent cerebral sulci, effacement of the ventricular system, and possibly midline shift, which depends on the extent of the oedema. The extent of the oedema, and the rate of progression of the underlying pathological process influence the degree of elevation of intracranial pressure and the decrease in the local brain compliance.

Peritumoural oedema is predominantly vasogenic in origin, arising from a neoplastic capillary bed devoid of tight junctions, a continuous endothelium, or the other characteristic features of the blood–brain barrier (Davies 2002; Papadopoulos et al. 2004). Since mature astrocytes secrete trophic factors for the blood–brain barrier formation (Janzer et al. 1987) the degree of increased permeability in neoplasia is related to the extent of de-differentiation of neoplastic astrocytes (Davies 2002). Thus WHO grade I and II tumours generally have little peritumoural oedema, whereas the more malignant grade III and IV tumours are characterized by the oedema. Peritumoural oedema has been recognized for many years on CT scans and modern MRI techniques allow quantification of many parameters related to oedema pathogenesis. Dynamic contrast-enhanced MRI, DCE-MRI, can be used to obtain measurements of tumour permeability by applying a pharmacokinetic model to a dynamic series of T_1-weighted MR images acquired following intravenous injection of a contrast agent (Barbier et al. 2001). Similarly, dynamic susceptibility contrast MRI, DSC-MRI, can be used to obtain measurements of cerebral perfusion, again by tracking a bolus of contrast agent, using a T_2^*-weighted sequence

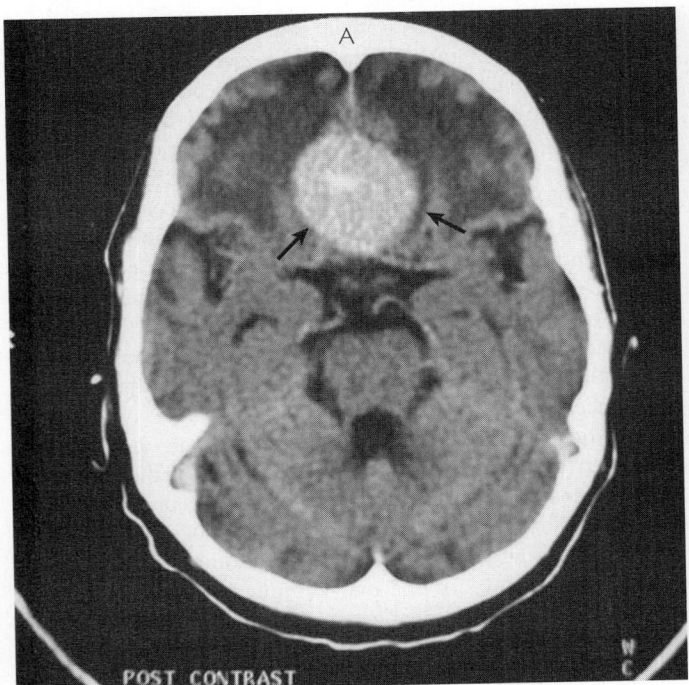

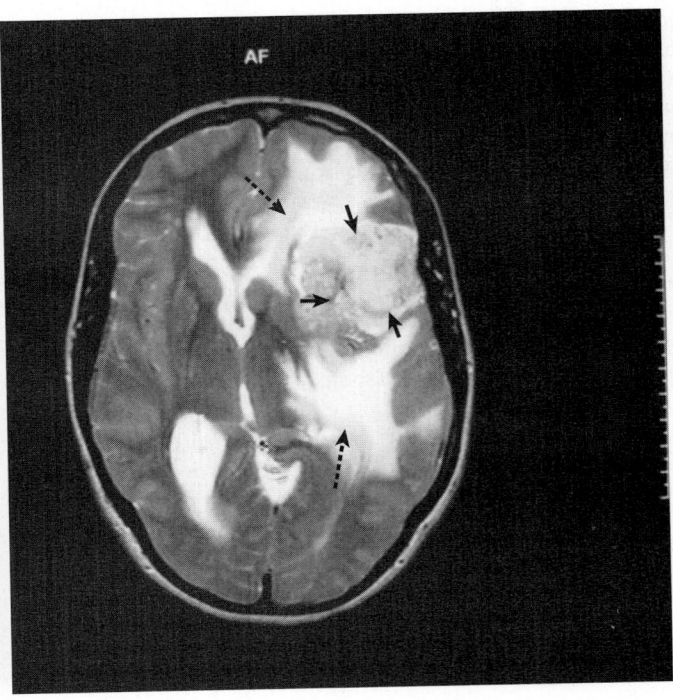

A

B

Fig. 26.7 Vasogenic perilesional edema around menigiomas: A. Shows a midline frontobasal meningioma (arrows) on CT scan with bilateral frontal lobe edema; B. Shows a T_2-weighted MRI of a sphenoid wing meningioma (solid arrows) with profound midline shift due to oedema in the frontal and parietal regions (broken arrows).

(Kamman *et al.* 1988). The longitudinal relaxation time T_1 has been shown in excised tissue (Fatouros *et al.* 1991) and human brain (MacFall *et al.* 1987) to be related to the tissue water content and can be measured with MRI using many different techniques such as inversion recovery (Wang *et al.* 1987) or multiple flip angle gradient echo acquisitions (Basser *et al.* 1994). Finally, diffusion tensor MRI, DT-MRI, enables quantification of the water diffusion tensor in the brain (Basser *et al.* 1994) from which can be derived the mean diffusivity (D_{av}) which is thought to provide an overall measure of the mobility of water in the extracellular space (Basser and Pierpaoli 1996).

Both experimental and clinical observations support the concept that brain oedema per se is not detrimental to neurological dysfunction. As long as brain perfusion is adequate there is no neocortical dysfunction. The causes of brain dysfunction associated with vasogenic brain oedema relate primarily to the underlying cause of the oedema. Peritumoral brain oedema and perifocal infective brain oedema may be associated with brain hypofunction causing focal neurological deficits and/or focal hyperfunction causing seizures. In each case the mechanisms are complex (Whittle 1992; Beaumont and Whittle 2000). Extravasated proteins, which increase in quantity and molecular weight with worsening of blood–brain barrier permeability, toxic compounds released by inflammatory cells, organisms, and neoplastic cells all contribute to the pathophysiology.

26.3.3 Cytotoxic brain oedema

This type of brain oedema is characterized by swelling of the cells in the brain. As a result of this intracellular compartmental swelling the extracellular space decreases. With this type of oedema the blood–brain barrier remains intact. If cellular swelling is not controlled the cell wall bursts resulting in neuronal or glial loss. This process is caused by various pathological processes, including energy depletion and increased permeability of the cell membrane or toxins which interfere with sodium–potassium membrane pumps (Unterberg *et al.* 2004). As a result there is an unbalanced influx of osmotically active solutes, especially Na^+, with passive influx of water. The original neuropathological description of this syndrome was generated by triethyltin poisoning and it was this model which was used to characterize the neuropathological and ultrastructural changes. As a clinical entity cytotoxic oedema mainly results from various toxins, cerebral ischemia (Fig. 26.8),

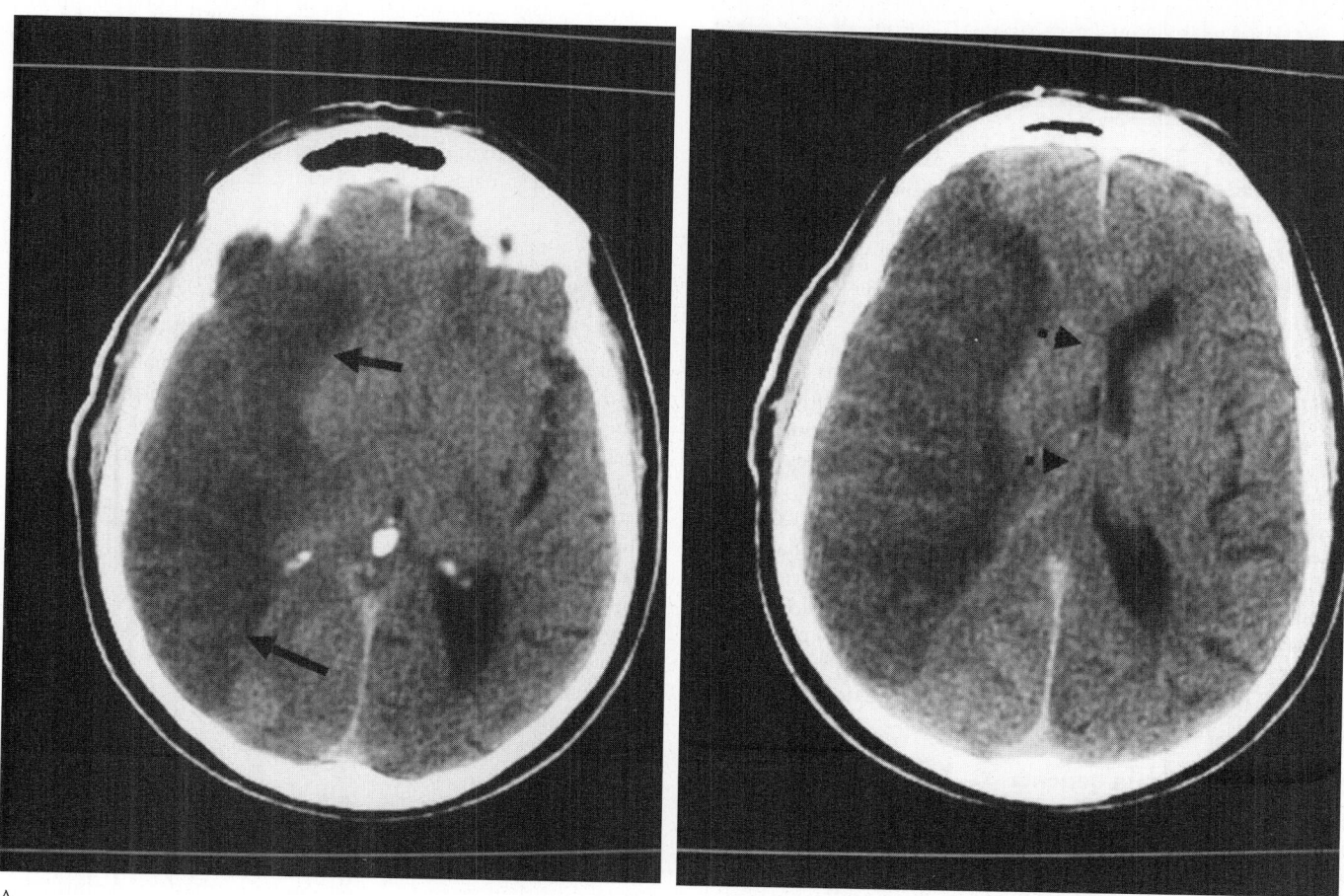

A   B

Fig. 26.8 Two CT scans showing a large infarct in the middle cerebral artery territory 24 h after thrombotic occlusion. A. shows the extent of the infarct, and increased water content due to cytotoxic edema in the infracted tissue (arrows). B. Shows the significant mass effect and midline shift (arrows). Sometimes decompressive craniectomy is required to prevent fatal brain herniation.

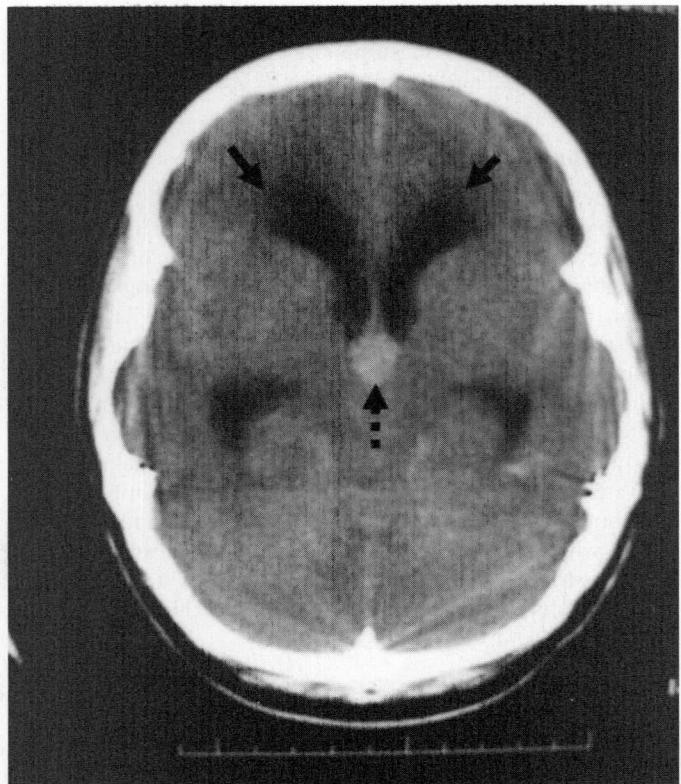

Fig. 26.9 CT scan showing bilateral ventricular hydrocephalus due to a colloid cyst (broken arrow) obstructing the foramina of Monro. Interstitial oedema, which appears as low signal density, is clearly seen around the frontal horns of the lateral ventricles (solid arrows).

neurotrauma, and hepatic encephalopathy (Ranjan *et al.* 2005; Pasco *et al.* 2006).

26.3.4 Interstitial brain oedema

This phenomenon is classically seen in patients with relatively acute obstructive hydrocephalus. As a result of the obstruction the ventricular volume increases and then the intraventricular CSF pressure rises with a resultant increase in the pressure gradient across the ependyma–brain interface (Johnston *et al.* 2000). As a consequence CSF passes into the subependymal extracellular space. The fluid is then driven through the white matter by convection to the sulci or basal subarachnoid spaces (Abbott 2004). Both the hydrocephalus and interstitial brain oedema may cause major perturbation of intracranial pressure dynamics. The effects of the interstitial oedema on brain structure and function are poorly understood (Johnston and Teo 2000).

26.3.5 Osmotic brain oedema

This type of increase in brain water occurs due to significant reductions in the osmotic coefficient of blood. It is classically seen with iatrogenic or self-inflicted water intoxication. As a result the osmotic pressure of the blood drops and the transcapillary gradient increases, with water passing into the brain. Profound bilateral diffuse brain oedema can result, with death. It may also occur if cerebral tissue is hyperosmolar as after cerebral ischemia or following neurotrauma (Kimelburg 2004).

26.3.6 Brain oedema in diabetes and liver failure

Cerebral oedema can complicate the management and clinical course of patients with diabetic ketoacidosis. This phenomenon is almost uniquely limited to children and occurs in about 1 per cent of those with diabetic ketoacidosis. In most patients the brain swelling subsides without sequelae after treatment of the diabetes (Levitsky 2004). Patients predisposed to this form of cerebral oedema often have severe dehydration, hyponatraemia, severe acidosis, hypocapnoea, and bicarbonate therapy. There are some contradictory features between animal models and patients with hyperglycaemia and cerebral oedema. A recent study using MRI suggested the oedema was principally vasogenic rather than cytotoxic (Glaser *et al.* 2004). These complex mechanisms remain far from understood.

Another metabolic disorder that can be complicated by raised intracranial pressure is acute liver failure. Hyperammonaemia of >200 μmol/l is believed to be a primary cause and this, together with disorders of metabolism of glutamine and alanine, can lead to an osmotically driven brain oedema (Tofteng *et al.* 2006). A diffusion-weighted MRI study suggested that the brain oedema in acute liver failure was predominantly cytotoxic in origin (Ranjan *et al.* 2005). In a large cohort analysis of 332 patients with acute liver failure and severe encephalopathy, the likelihood of liver transplantation, a surrogate marker for the severity of hepatic dysfunction, was directly associated with the use of intracranial pressure monitoring. These patients with intracranial pressure monitoring received a high proportion of vasopressor and intracranial pressure-related medication. In patients receiving liver transplants, the 30-day survival period was similar in both those monitored, who presumably had higher intracranial pressure, and those unmonitored, who presumably had normal intracranial pressure (Vaquero *et al.* 2005).

26.4 Hydrocephalus

26.4.1 Obstructive and communicating hydrocephalus

Hydrocephalus is defined as an increase in intracranial CSF volume. This may be caused by atrophy of the brain with compensatory ex-vacuo increase in ventricular volume, known as asymptomatic ventriculomegaly. In most cases such ventriculomegaly is not associated with signs or symptoms since there is no hydrodynamic disturbance of CSF flow or reabsorption. Abnormalities in CSF formation, flow, or absorption can lead to dilatation of the ventricular system and cause symptomatic hydrocephalus. This results from deranged intracranial pressure hydrodynamics due to imbalance between CSF formation and absorption. There are still many unknown factors regulating CSF formation and flow, and the challenges are to devise better methods of clinical measurement for these parameters so as to gain better insights into normal physiology and hence pathophysiology (Section 26.2.3) (Johnston and Teo 2000; Czosnyka *et al.* 2004).

From the clinical perspective excess formation of CSF is extremely rare and indeed has been only documented in several cases of choroid plexus papilloma. The vast majority of cases of hydrocephalus are due to disturbances of CSF flow. Some of the CSF pathways, in particular the foramina of Munro and the aqueduct of Sylvius, are narrow or obstruct easily (Figs 26.9 and 26.10). *Obstructive hydrocephalus* refers to hydrocephalus in which there is a blockage of CSF flow within the ventricular system. *Communicating hydrocephalus*

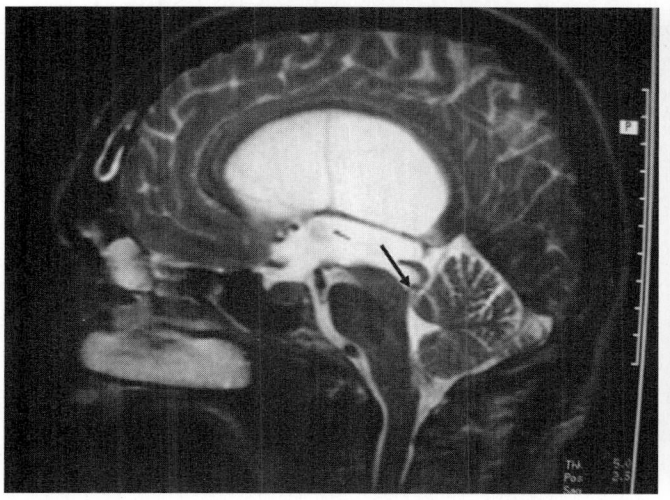

A

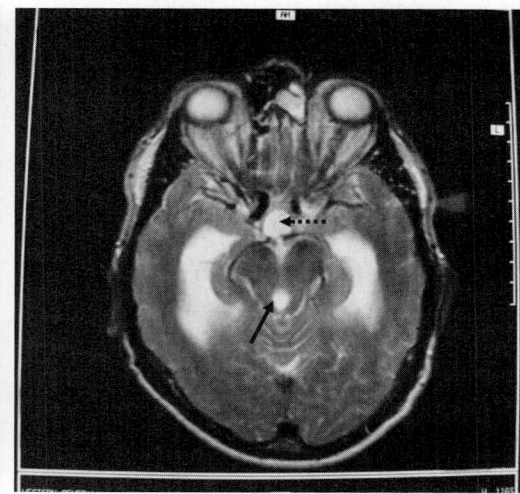

B

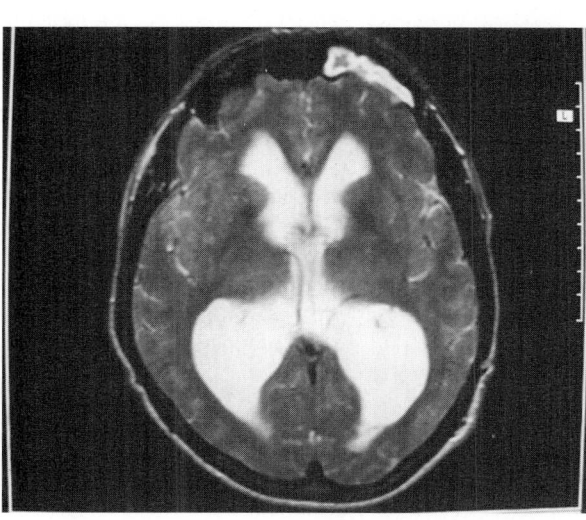

C

Fig. 26.10 Obstructive hydrocephalus due to aqueduct stenosis in a 35-year-old man shown in a series of three T_2-weighted MRI scans. A. Sagittal view showing stenosis of the distal aqueduct, with dilatation of the proximal aqueduct (arrow), third and lateral ventricles, whilst the fourth ventricle is normal in size. The corpus callosum in thinned. B. Axial view showing dilatation of the rostral aqueduct (arrow) and ballooning of the floor of the third ventricle into the interpeduncular CSF cistern region (broken arrow). The temporal horns of the lateral ventricles are also grossly dilated. C. Shows gross enlargement of the third and lateral ventricles.

refers to situations in which the CSF exits the fourth ventricle but there is a failure of either reabsorption of CSF at the arachnoid granulations or a failure of the CSF to reach the structures (Fig. 26.11).

26.4.2 **Common causes of hydrocephalus**

In children, developmental anomalies of the brain, such as Chiari II malformation, Dandy–Walker syndrome, and congenital aqueduct stenosis are frequent causes (Sections 9.2.6; 9.3.2). Obstructive hydrocephalus may be caused by a tumour of the lateral ventricle, a neurocytoma, a tumour of the foramen of Munro region, such as colloid cyst (Fig. 26.9) or subependymal giant cell astrocytoma, compression of the aqueduct by a tectal plate tumour, or fourth ventricular outflow obstruction by a cerebellar or brainstem tumour (Table 26.1). Communicating hydrocephalus is common after adult subarachnoid haemorrhage, tuberculous meningitis, and neonatal intracranial or intraventricular haemorrhage. Neuroimaging with CT or MRI is diagnostic of both hydrocephalus and the likely cause, but usually tells us little about CSF hydrodyamics (Johnston and Teo 2000).

26.4.3 **Clinical features**

Hydrocephalus may produce a range of clinical features from the subtle to the extremely serious. Communicating hydrocephalus may present with mild symptoms and signs of cognitive decline, gait disturbance, and headache in various permutations. This commonly follows apparent recovery from subarachnoid haemorrhage, basal meningitis, or head injury. Obstructive hydrocephalus generally presents with a more acute syndrome, characterized by progressively worsening headaches, psychomotor slowing, visual disturbance, and not infrequently papilloedema. If obstructive hydrocephalus is severe and acute it can cause extreme intracranial pressure waves which result in hydrocephalic 'attacks'. During these episodes, which most commonly occur in patients with ventriculo-peritoneal shunt malfunction or colloid cysts (Fig. 26.9), the patient usually lapses into transient coma, and develops extensor posturing of the upper and lower limbs, often with dilated pupils. If not recognized and treated these attacks can cause sudden death. Rarer symptom complexes due to hydrocephalus include a Parkinsonian-like syndrome with akinetic mutism and catatonia

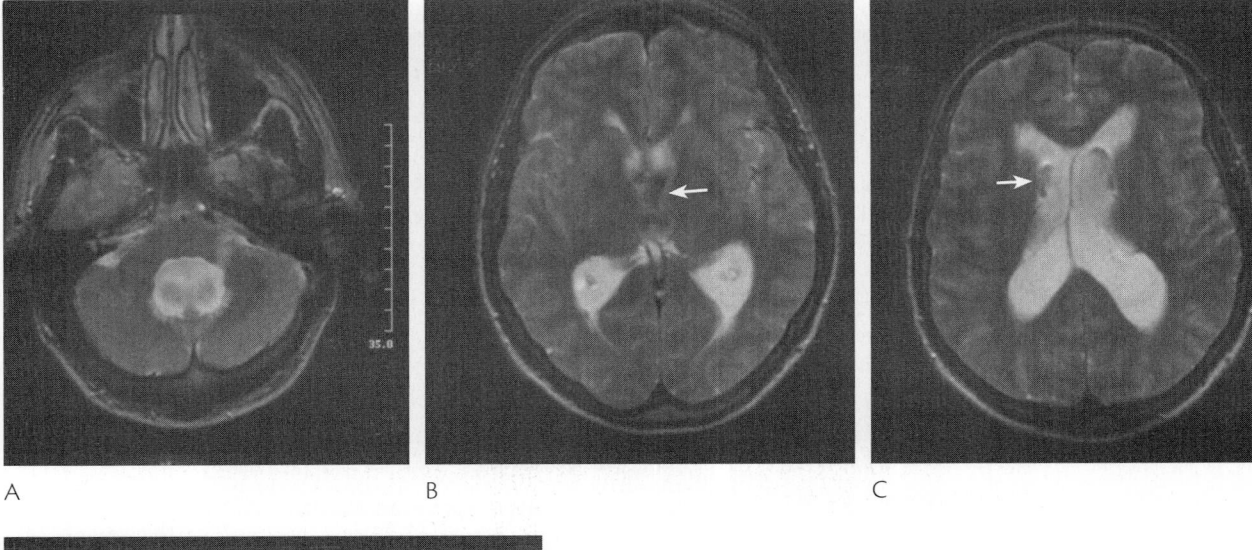

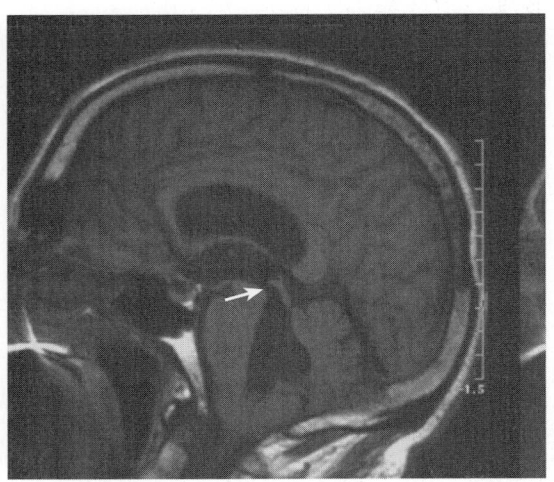

Fig. 26.11 Communicating or 'normal-pressure' hydrocephalus in a 50-year-old person. A. Enlargement of the fourth ventricle with turbulent CSF flow (axial T_2-weighted MRI). B. Enlargement of the lateral and third ventricles, with turbulent CSF in the third ventricle (arrow) (axial T_2-weighted MRI). C. Enlargement of the lateral ventricles with turbulent flow related to the interventricular foramina (arrow) (axial T_2-weighted MRI). D. Distension of the third and fourth ventricles, and the cerebral aqueduct (arrow) (midline sagittal T_1-weighted MRI). Note that MRI of communicating hydrocephalus shows distension of the cerebral aqueduct and fourth ventricle, and sometimes turbulent, flowing CSF. These are not features seen in the obstructive hydrocephalus due to aqueduct stenosis shown by Fig. 26.10. (Courtesy of Dr P Anslow.)

(Zeidler *et al.* 1998; Chu *et al.* 2006). In children prior to closure of the skull features, hydrocephalus may cause enlargement of the head, with or without 'sunsetting' eyes (Fig. 26.12). If associated with tonsillar herniations, there may be severe neck pain and torticollis.

26.4.4 Arrested or compensated hydrocephalus

Some patients have apparently asymptomatic chronic hydrocephalus or become shunt independent. They may have had a shunt during childhood but without subsequent revision for 15–20 years. This unusual group of patients is often termed 'arrested hydrocephalus'. Many have an enlarged cranium. Most of these patients have developed compensatory mechanisms for CSF absorption, but have little reserve. Several series have however demonstrated that many of these patients have abnormal intracranial pressure wave forms when monitored overnight (Johnston *et al.* 1983; Yau *et al.* 2002). Therefore if they suffer a minor head injury with traumatic subarachnoid haemorrhage, they may decompensate acutely and require insertion of a shunt system. Experimental and clinical studies suggest these patients suffer progressive brain damage due to white matter and axonal damage (Del Bigio *et al.* 2003). Therefore careful clinical

monitoring of motor and cognitive functioning is required (Johnston and Teo 2000). Cognitive improvement after third ventriculostomy has been described (Burtscher *et al.* 2003).

26.4.5 Chronic hydrocephalus and 'normal pressure', hydrocephalus

Chronic hydrocephalus is very common in the elderly. It poses the question of whether it merely represents a compensatory ventricular dilatation either due to atrophy, or alternatively to another intracranial disease such as subarachnoid haemorrhage or meningitis, or whether it reflects a pathophysiological response to increased CSF outflow resistance. The latter situation has been defined as *idiopathic normal pressure hydrocephalus*. This involves a syndrome of progressive gait impairment, cognitive decline, and urinary incontinence as the classic diagnostic triad. However since ventricular enlargement is very common in the elderly precise diagnosis is extremely difficult because many such patients have cognitive decline due to Alzheimer's disease or multi-infarct dementia, gait disturbances due to hip disorders or Parkinson's disease, and urinary incontinence due to prostatic or gynaecological problems (Maramarou *et al.* 2005a,b; Relkin *et al.* 2005).

Table 26.1 Some causes of obstructive and communicating hydrocephalus

Obstructive hydrocephalus

Lesion location	*Pathology*
Lateral Ventricle	Intraventricular meningioma Neurocytoma Choroid plexus papilloma
Foramen of Monro	Colloid cyst Subependymal giant cell astrocytoma
Peri-Aqueductal	Aqueduct stenosis (congenital or acquired) Pineal region tumours Tectal glioma
Fourth Ventricular	Chiari II malformation, Dandy–Walker malformation, Primitive neuroectodermal tumour, Ependymoma, Astrocytoma

Communicating hydrocephalus

Idiopathic symptomatic adult hydrocephalus

Post subarachnoid haemorrhage

Post infective (Tuberculous, bacterial)

Raised CSF protein

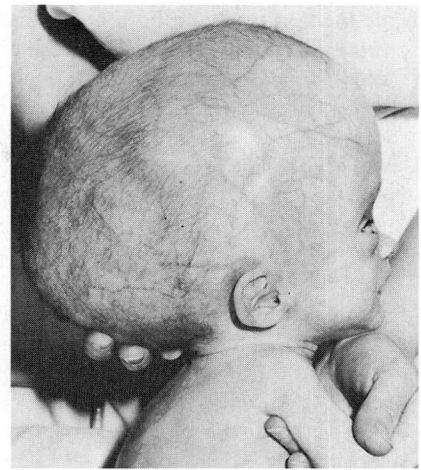

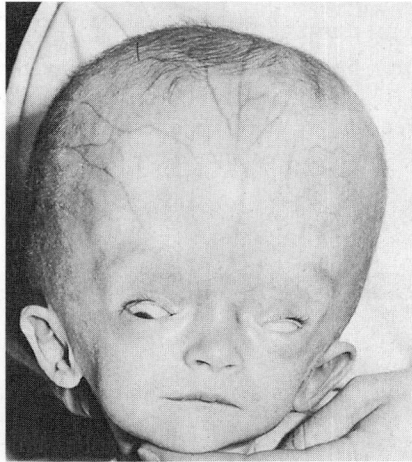

Fig. 26.12 Infantile obstructive hydrocephalus causing gross enlargement of the head associated with lumbar meningomyelocele and an Arnold–Chiari malformation (courtesy of Mr L P Lassman). Fortunately with improved neonatal diagnosis and paediatric care such presentations in Western society are now very rare.

Investigation of the patient with a suspected normal pressure hydrocephalus syndrome includes lumbar puncture, neuropsychological profiling, as well as CSF drainage tests, intracranial pressure monitoring, and measurements of CSF outflow resistance. Unfortunately none of these tests are diagnostic (Marmarou *et al.* 2005b). Establishing the diagnosis is, however, important since there can be a very high complication rate following inadvertent shunting in patients without precise diagnosis. Unfortunately even in those patients with either a complete clinical triad or those thoroughly investigated using intracranial pressure dynamic testing, shunting only improves patients in between 31 and 90 per cent (Bergsneider *et al.* 2005). The variable response to CSF diversion in this cohort of patients has given rise to the concept of a shunt non-responsive version of the disorder (Relkin *et al.* 2005).

Lumbar puncture in a group of patients with suspected normal pressure hydrocephalus revealed opening pressures of 15+/−4.5 cm water. This is only slightly higher than opening pressure in a range of normal volunteers 12.2+/−3.4 cm water (Relkin *et al.* 2005). If the CSF protein is raised it may suggest CSF outflow obstruction as a cause of the hydrocephalus. Transient high pressure B waves are detectable during prolonged intraventricular monitoring adults with symptomatic idiopathic normal pressure hydrocephalus (Graff-Radford *et al.* 1989). However their frequency and the period of monitoring time in which they occur has not been adequately defined for diagnostic purposes. A recent study evaluating intraventricular compliance in this group of patients also failed to reveal definitive thresholds (Yau *et al.* 2002). The role of intracranial pressure monitoring in these patients is to define those with symptomatic hydrocephalus that is due to causes other than suspected normal pressure hydrocephalus. There is currently no good quality evidence, better than Class III, to support continuous intracranial pressure monitoring to determine the frequency of A or B waves (Marmarou *et al.* 2005b).

Removal of CSF via a 'tap test' has been used since the first descriptions of normal pressure hydrocephalus (Hakim and Adams 1965). It is common practice to remove 40–50 ml of CSF and then re-evaluate the patient. The difficulty lies is determining what constitutes a significant improvement following the tap test, for instance walking speeds or mini-mental test scores. Recent studies suggest that the tap test has good positive predictive value (Malm *et al.* 1995; Walchenbach *et al.* 2002). However specificity is low. Prolonged external lumbar drainage has also been evaluated in patients being considered for shunting for normal pressure hydrocephalus. This test is felt to be superior to CSF tap testing in both predictive value and sensitivity but hospital admission is required and complications may result. If a lumbar catheter is to be placed then CSF outflow resistance can also be evaluated. Several centres have performed studies of CSF outflow resistance (R_0) in normal pressure hydrocephalus patients; however absolute values are dependent on the method used and both results and thresholds seem centre specific (Marmarou *et al.* 2005b). Generally it is felt that CSF R_0 is a more accurate test than a CSF tap test.

Neuropsychological investigations invariably show a subcortical dementing process with slowing of thought, inattentiveness, and apathy (Devito *et al.* 2005). In the early stages executive function and behavioural and personality changes may also be apparent. The neuroanatomical basis for these manifestations is unclear but is thought to relate to the frontostriatal system. The differential diagnosis on neuropsychological testing alone is wide because of the overlap with other common neurodegenerative disorders such as Alzheimer's disease. Because of the difficulties of standardizing prognostic evaluation of patients with idiopathic normal pressure hydrocephalus, a multicentre study has been suggested to answer many of the difficult questions in this area (Relkin *et al.* 2005).

26.4.6 Surgical management

The treatment of hydrocephalus requires the underlying cause to be addressed. If hydrocephalus is obstructive, and the primary cause is treated but symptoms and hydrocephalus persist then CSF diversion is indicated. The commonest surgical treatment is CSF diversion using a pressure-regulated or flow-regulated valve system, a ventriculo-peritoneal or a ventriculo-atrial shunt (Bergschneider *et al.* 2005) (Fig. 26.13). Programmable shunt systems are now available also that allow alteration of the CSF drainage pressure in the valve system (Drake *et al.* 2000). CSF shunting restores absorption of CSF but in an extra-cranial compartment. CSF shunting has revolutionized treatment of hydrocephalus in the last 4 decades. However it is a treatment bedevilled by failure and by other complications such as infections, low-pressure syndromes, brain displacements, and extracerebral collections (Drake *et al.* 2000; Johnston and Teo 2000; Kulkarni *et al.* 2001; Kestle *et al.* 2003). This remains the case despite considerable advances in shunt technology, endoscopic surgery, and brain imaging. Paradoxically overall shunt effectiveness and complication profiles are similar despite the different varieties of commercially available modern shunts (Drake *et al.* 2000; Kestle *et al.* 2003). CSF shunting in adult hydrocephalus can cause particular problems if the 'hydrocephalus' is in fact just related to age-related brain atrophy and subsequent ventriculomegaly, rather than representing a CSF hydrodynamic disturbance (Bergschneider *et al.* 2005). Such problems can be common in the elderly patient with suspected normal pressure hydrocephalus (Maramarou *et al.* 2005b).

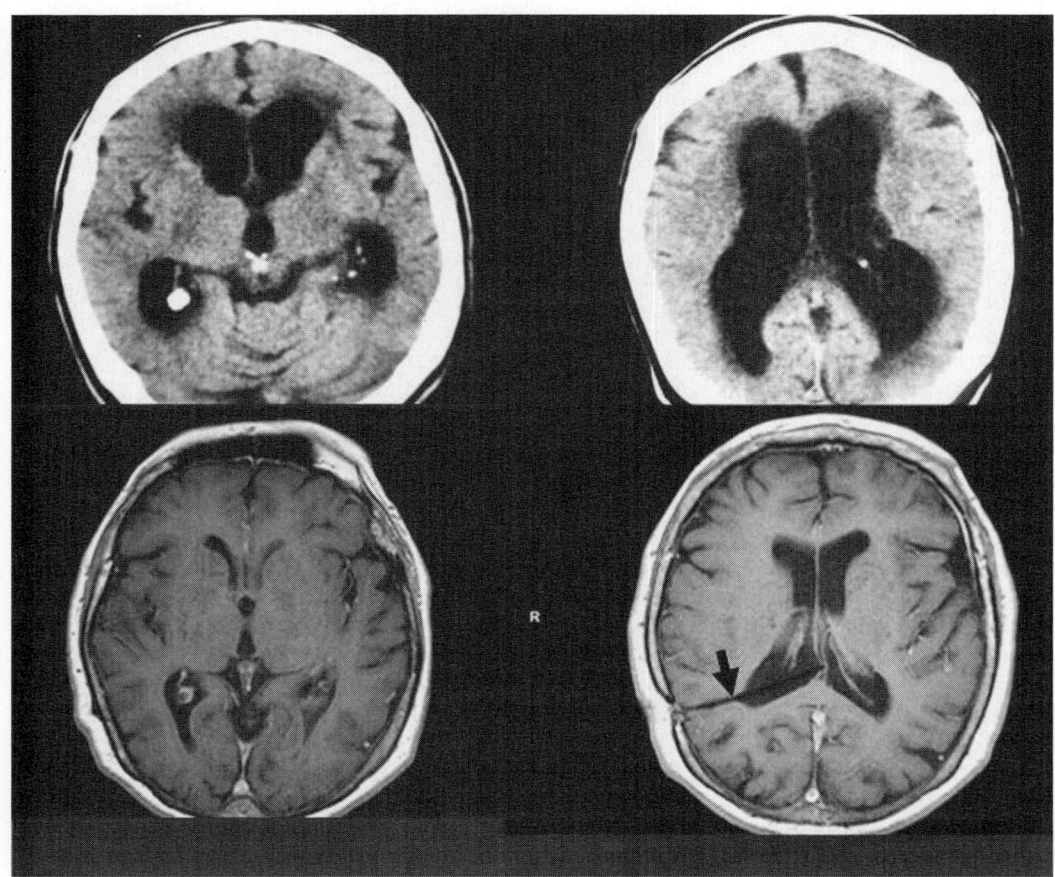

Fig. 26.13 Communicating hydrocephalus. Top: Series of two preoperative CT scans in an elderly adult who presented with ataxia, cognitive decline, and incontinence that progressed to a state of akinetic mutism with rigidity and catatonia. Bottom: Following ventriculoperitoneal shunting the communicating hydrocephalus resolved on MRI scan with complete recovery of the patient. The ventricular catheter can clearly be seen in the lateral ventricle in the right MRI (arrow).

The recent development of fine calibre endoscopes has led to increasing use of third ventriculostomy for congenital aqueduct stenosis, and tectal or pineal region tumours (Li *et al.* 2005). Re-establishing normal pathways of CSF flow avoids some of the difficulties of shunt insertion, and can lead to good long-term results particularly in patients older than 6 months (Kadrian *et al.* 2005; Li *et al.* 2005). However in a clinical trial in children the complication rates, and failure of the endoscopic procedure were very similar to CSF shunt insertion (Tuli *et al.* 1999).

26.5 Raised intracranial pressure in clinical practice

26.5.1 Management: general principles

Problems with intracranial pressure dynamics are a common feature of a range of clinical conditions in both adult and paediatric neurological practice (Table 26.2). Problems arise because of either mass lesions causing brain herniation syndromes, or from the intracranial pressure being either focally or diffusely raised so that cerebral perfusion is compromised (Section 25.3.2). 'Mass lesion' in this context can refer to either a true mass such as an intracranial haematoma or brain tumour, to an area of oedematous brain or to

Table 26.2 Some common and important clinical causes of raised intracranial pressure

Traumatic brain injury
Major cerebral arterial thrombosis
Primary intracranial haemorrhage
Primary and secondary brain tumours
Intracranial infection
Obstructive hydrocephalus
Communicating hydrocephalus
Idiopathic intracranial hypertension
Sagittal sinus thrombosis
Hypoxia and toxins
Liver failure

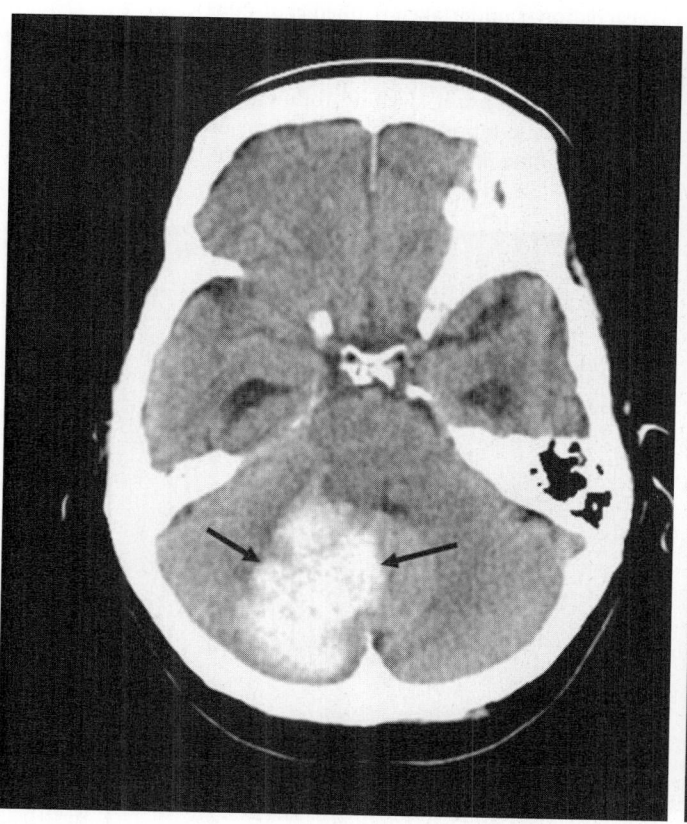

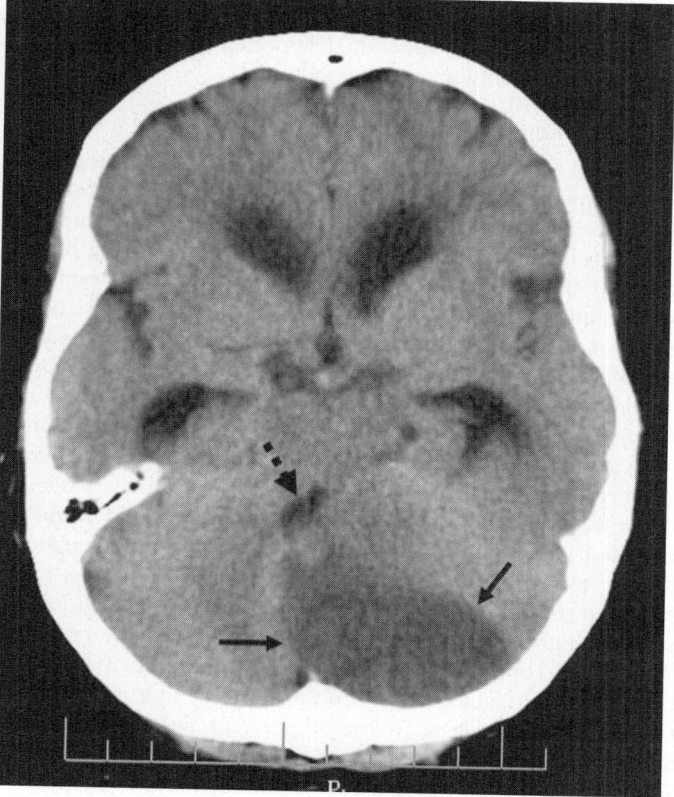

A   B

Fig. 26.14 The spectrum of 'stroke' in the posterior cranial fossa for which the role of surgery remains uncertain. CT scan (A) shows a spontaneous lateral and paramedian cerebellar hematoma (arrows) in a hypertensive patient. CT scan (B) shows a cerebellar infarction (arrows) with swelling, compression and distortion of the fourth ventricle (broken arrow). In both these situations the pressure in the posterior fossa is raised and intracranial pressure may also be aggravated by secondary obstructive hydrocephalus which is obvious in (B). The role of hematoma removal and decompressive posterior fossa craniectomy remains uncertain. (Images courtesy of Dr R Gibson and Dr D Summers.)

an increase in CSF volume. As noted above, the rate of increase in size of the 'mass lesion', patient age, co-existing primary pathologies such as systemic hypertension or diabetes, and baseline brain compliance will determine the intracranial pressure reponse to the disease process (Section 26.2.5). The most acute and life-threatening effects of raised intracranial pressure occur following neurotrauma with the development of intracranial haemorrhage and secondary brain oedema. Medical conditions with life threateningly raised intracranial pressure occur due to primary intracerebral haemorrhage, and brain swelling following embolic or thrombotic stroke. In patients with acutely raised intracranial pressure problems causing a significant impairment of conscious state, with a Glasgow Coma Score < 9, the principles of management of acute life-threatening intracranial hypertension include (Section 25.5):

◆ intubation and ventilation for airway protection and oxygenation;

◆ controlled hyperventilation;

◆ intravenous mannitol;

◆ removal of mass lesions;

◆ drainage of any hydrocephalus;

◆ medical management of hypertension or hypotension;

◆ control of seizures (Mayer *et al.* 2005b).

Different problems in management are posed by the more chronically raised intracranial pressure caused by intracranial neoplasms, infections, idiopathic intracranial hypertension or idiopathic intracranial hypertension, and liver failure.

26.5.2 Stroke: surgical options

Following a large primary, supratentorial intracerebral haemorrhage or large volume ischemic brain infarction, such as following a middle cerebral artery thrombosis, raised intracranial pressure and brain herniation syndromes are a common cause of death and morbidity (Fig. 26.8). Removal of a haematoma initially seemed a logical approach to primary management. However a very large recent randomized controlled trial showed that there is no overall benefit from surgical decompression of a primary intracerebral haematoma which is not due to arteriovenous malformation or aneurysm (Mendelow *et al.* 2005). In patients with brain herniation following aneurysmal rupture and an associated large intracerebral haemorrhage, prognosis is generally poor. Most of these patients will be World Federation of Neurosurgery sub-arachnoid haemorrhage grade 4 (a Glasgow Coma Score between 9 and 12) or grade 5 (a score of 8 or less). Although removal of the haematoma may lead to control of immediate intracranial pressure-related problems, the high morbidity and mortality result from the primary aneurysmal rupture, delayed cerebral ischaemia, associated metabolic disturbances, and hydrocephalus. The mortality for the patients in this group is around 50 per cent, although younger patients tend to do better.

The prognosis for intracerebral haematoma following arteriovenous malformation rupture is much better. This may be in part because the pressure generating the haematoma is non-arterial and therefore the primary brain injury associated with the haematoma is significantly less than with a ruptured aneurysm or primary intracerebral haematoma. Additionally many of the patients tend to be much younger and therefore more able to withstand the primary and secondary brain insult. The World Federation of

Neurosurgery grade of subarachnoid haemorrhage of the patient at admission seems to be the primary prognostic determinant (Lawton *et al.* 2005). Patients in grade 1–3 (wit a Glosgow Coma Score >13) do better than those in grade 4 or 5. Paradoxically patients presenting with an intracerebral haemorrhage from a ruptured arteriovenous malformation did better after surgical resection than those patients with arteriovenous malformations that had not ruptured. This finding almost certainly reflects the rapid onset of clinical deficits following the malformation rupture and formation of an intracerebral haematoma, and their resolution following haematoma removal together with removal of the arteriovenous malformation. By contrast those patients electively operated on for an arteriovenous malformation are in better condition preoperatively but subject to the direct morbidity of the surgical process (Lawton *et al.* 2005).

The benefit of removal of primary intracerebellar haemorrhages is contentious (Fig. 26.14). There has been no appropriately powered randomized controlled trial addressing this issue, but several prospective studies have suggested prognostic factors predictive of outcome. In general, patients in good neurological condition (Glasgow coma score 13–15) with small haematomas of <3 cm diameter do well with conservative treatment. Those with large haematomas >3 cm, including those in good neurological status, and all comatose patients, generally do worse whether or not surgical evacuation is undertaken (Cohen *et al.* 2002).

Recently there has been a vogue for performing decompressive hemicraniectomy in an attempt to save the life of patients with severe secondary brain swelling following ischaemic infarction such as middle cerebral artery occlusion or internal carotid artery occlusion (Uhl *et al.* 2004; Kilincer *et al.* 2005). Often the clinical

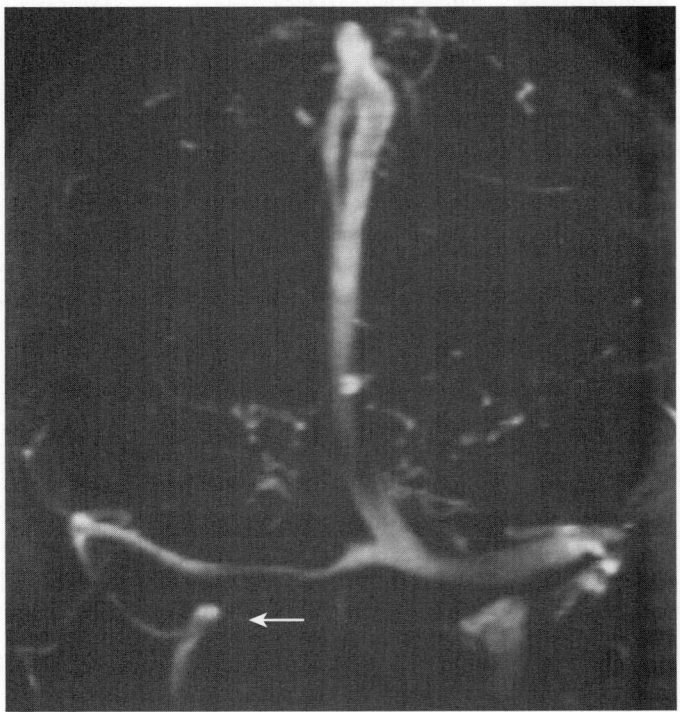

Fig. 26.15 MR Venogram showing stenosis (arrow) of the transverse sinus. The contribution of such abnormalities to idiopathic intracranial hypertension is not well understood.

condition is not too bad upon presentation in such patients but with swelling of the brain during the first few days there is progressive loss of consciousness with signs of uncal herniation (Fig. 26.8). Surgical removal of this swollen brain would be an unattractive and radical step given that some of this brain might regain function later. Accordingly trials are addressing the question of decompressive surgery whereby a large area of the cranium is removed, together with opening the dura, to allow the brain to decompress externally. Preliminary results from clinical series suggest that such surgery may save lives (Uhl *et al.* 2004; Kilincer *et al.* 2005; Malm *et al.* 2006). More recently data from three European randomized controlled trials of decompressive surgery after stroke, DECIMAL, DESTINY, HAMLET, were pooled to demonstrate a beneficial effect of craniectomy if perfomed with 48 h of stroke onset (Vahedi *et al.* 2007). It is likely therefore that this type of surgery will become part of acute management of some stroke patients.

26.5.3 Stroke: medical management

Primary intracerebral haemorrhage accounts for about 15 per cent of strokes and has a 1 month case fatality of around 42 per cent (Dennis 2003). Given this high incidence and morbidity associated with it there have been studies looking at how to minimize morbidity using medical therapeutics, particularly in view of the limitations of surgery. The major cause of additional morbidity and secondary deterioration in these patients is due to rebleeding. Studies have focussed on reducing rebleeding rate with one therapy of interest being recombinant activated factor VIIa. This helps stop bleeding and has been evaluated in multitrauma situations. In a randomized double blind placebo controlled trial of 399 patients with spontaneous intracerebral haematoma, treatment with recombinant factor VIIa in various dosages within 4 h of intracerebral haemorrhage onset limited growth of the haematoma by about 50 per cent (Mayer *et al.* 2005a). This was associated with a 38 per cent reduction in mortality versus placebo, and significantly improved functional outcomes at 90 days. In view of these promising findings a further phase 3 trial is currently being conducted to evaluate the effects of early haemostatic therapy on haematoma evolution.

As well as controlling intracranial pressure, maintenance of cerebral perfusion pressure is important in many patients with primary or secondary intracerebral haemorrhage. The difficulty lies in balancing the beneficial effects of a lowering mean arterial pressure to reduce rebleeding rate, against the potentially detrimental effects of impaired cerebral perfusion pressure. The American Stroke Association guidelines recommend that mean arterial pressure be maintained at or below 130 mmHg for patients with intracerebral haematoma and a history of hypertension (Mayer *et al.* 2005b). As with intracerebral haemorrhage following neurotrauma it is recommended that cerebral perfusion pressure should be maintained above 70 mmHg. Drugs such as labetolol, esmolol, nicardipine, enalapril, thenodopam can be used depending on specific indications and contraindications. A Cochrane review reveals that there is no evidence that corticosteroids are of either benefit or hazard after intracerebral haematoma, either spontaneous or following aneurysmal rupture (Feigin *et al.* 2005).

26.5.4 Brain tumours: medical options

Many primary and secondary brain tumours present with raised intracranial pressure. The contributing factors consist of the size of the neoplastic mass, the extent of peritumoural oedema, and the

location of the tumour if it is also causing obstructive hydrocephalus. The use of glucocorticoids in the management of tumour-associated peritumoural oedema has been established since the early 1960s (Kaal *et al.* 2004). In several large non-randomized trials there has been significant reduction of morbidity and mortality following intracranial surgery for neoplasms in those patients prescribed pre-operative corticosteroid. The post-operative course was also much smoother. The precise mode of action of glucocorticoids is not well understood (Whittle 1992; Wick *et al.* 2004). Both in experimental and human glioma measurements of capillary permeability show decreases after steroids (Swaroop *et al.* 2001). Suppression of vascular endothelial growth factor is probably important (Heiss *et al.* 1996).

Although patients often improved dramatically within 4–24 h of steroid therapy often there was no associated decrease in the measured intracranial pressure to correlate with the clinical improvement (Miller *et al.* 1977). However, brain compliance does rise which accords with later observations of brain peritumoural water content being significantly reduced over the first 72 h after steroid therapy (Sinha *et al.* 2004). Such reductions were seen with malignant glioma, meningioma, and metastasis despite considerable variation in the mechanisms and quantity of brain oedema between these tumours. Positron emission tomography studies suggest that cerebral blood volume also falls after steroids; however recent MRI studies have shown no significant change in either cerebral blood flow or cerebral blood volume (Bastin *et al.* 2006).

In addition to the value of steroids in treating symptoms of severe raised intracranial pressure such as severe headache, nausea, and vomiting associated with tumour-associated brain oedema, they also improve peritumoural brain function (Whittle 1992). Although the neuropathology of peritumoural brain infiltration is now better understood, the molecular mechanisms whereby glucocorticoids rectify brain function remain obscure as do those pertaining to decreased permeability in neoplastic capillary beds (Goel *et al.* 2003).

The precise dose of steroid to be administered is also contentious. Traditionally new patients have been started on 16 mg dexamethase divided into four equal daily doses. However a clinical trial showed that there is very little difference in the efficacy of 4 mg/day versus 16 mg/day (Vecht *et al.* 1994). Another small trial suggested twice daily steroid dosage to be as effective as four times daily (Weisman *et al.* 1991). Since most intracranial tumours cannot be treated definitively many patients are placed on chronic daily steroid dosage which is usually titrated against symptoms. Thus they may be given steroids during courses of radiotherapy, chemotherapy, and in the terminal course of their disease. The side-effects associated with long-term steroid therapy have been well described and include a moon face, development of acneform rashes, hyperglycaemia, proximal myopathy, peptic ulceration, glycosuria, uncomfortable oral candidiasis, and pruritus due to anal candida. Occasional patients do not respond to steroids and may die rapidly of fulminating brain oedema and tumour progression (Sinha *et al.* 2003).

26.5.5 Brain tumours: surgical options

The surgical management of intracranial tumours causing mass effect has been to excise them. This mechanical approach leads to a reduction in intracranial pressure in the cases of both malignant glioma and of meningioma (Whittle *et al.* 2004; Mitchell *et al.* 2005).

Table 26.3 Some factors associated with or predisposing to idiopathic intracranial hypertension (modified from Ball and Clarke 2006)

Female gender
Obesity
Polycystic ovary syndrome
Medications
- Oral contraceptives
- Corticosteroid withdrawal
- Antibiotics (Tetracycline, nalidixic acid)
- Nonsteroidal anti-inflammatory drugs
- Vitamin A imbalances
- Lithium
Thrombophilic tendency
Stenosis of the major venous sinus

The mass lesion, and associated neoplastic capillary bed, is removed to relieve both the mass responsible for the raised intracranial pressure and the source of the associated vasogenic brain oedema. The extent to which this reduces symptoms and signs varies and is very dependent upon the pathology of the mass lesion (Whittle *et al.* 1998). Several series suggest that craniotomy and tumour removal for malignant glioma ameliorate signs in about 19–32 per cent of patients; however most (58–76 per cent) are unchanged and a significant percentage (9–26 per cent) get worse (Whittle 2003). An attractive alternative therapy to surgery for patients with lesions less than 3 cm in diameter and significant co-morbidity is stereotactic radiosurgery. This therapy uses focussed radiotherapy to 'fry' lesions in a single treatment fraction. It is particularly useful in patients with brain metastases, especially those who have more than one lesion or who have co-morbidities making surgery particularly hazardous (Vecht 2003).

26.5.6 Idiopathic intracranial hypertension

Idiopathic intracranial hypertension is also known as *benign intracranial hypertension* or *pseudotumour cerebri*. The syndrome is characterized by raised intracranial pressure, papilloedema, and the absence of any focal neurological deficit except sometimes a sixth cranial nerve palsy. Since it can lead to visual failure it is not necessarily benign. The pathophysiologically is poorly understood leading to problems with the nomenclature (Johnston and Teo 2000). Idiopathic intracranial hypertension is often associated with obesity, menstrual irregularities, and female gender as well as a range of relatively uncommon abnormalities of the vascular haemostasis. A similar syndrome can be precipitated by a range of therapeutic agents, especially oral contraceptives and tetracyclines (Ball and Clarke 2006). There is evidence from neuroradiological and coagulation studies that thrombophilia may be a common predisposing factor in idiopatic intracranial hypertension (Dunkley and Johnston 2004).

Clinically the major features are headache and papilloedema due to the raised intracranial pressure. There may be associated visual obscuration, enlargement of the blind spot, and diplopia. Neuroradiological investigations reveal normal ventricular size. Lumbar puncture usually reveals opening CSF pressure of >25 cm of water. CSF composition is normal. Although the syndrome has been recognized for many years the aetiology remains poorly under-

stood (Johnston and Teo 2000; Ball and Clarke 2006). Recent MRI studies have discounted significant brain oedema in these patients (Bastin *et al.* 2003) although high resolution MR venography consistently shows abnormalities in the transverse sinuses when compared to normals (Higgins *et al.* 2006) (Fig. 26.15). The role of these venous stenoses remains the subject of some controversy, although some patients have been treated by stenting of the stenoses (Bono *et al.* 2005; Owler *et al.* 2005).

The mainstay of treatment for idiopathic intracranial hypertension patients is to avoid and/or correct any predisposing factors (Ball and Clarke 2006). An idiopathic intracranial hypertension-like syndrome can be found with a range of medical disorders (Table 26.3). Since many patients are obese, weight reduction should be attempted, although often unsuccessful. There have been infrequent reports of symptomatic relief of headache in small numbers of patients following major weight reduction (Ball and Clarke 2006). Serial lumbar punctures have been employed to transiently decrease intracranial pressure. In some patients a few lumbar punctures seem to provide long-lasting relief from the syndrome. Other patients gain only transient relief from lumbar puncture. They may require medical treatment with acetazolomide, a carbonic anhydrase inhibitor. Although acetazolomide may decrease CSF formation by 60 per cent, its main role probably lies in treating the milder cases of idiopathic intracranial hypertension (Johnston and Teo 2000). Diuretics have also been tried together with corticosteroids although the latter tend to exacerbate any weight gain. The relative failure of agents regarded as decreasing CSF production may reflect an altered balance of CSF formation, between the choroid plexus and extra-choroid sites, in some conditions with raised intracranial pressure (Johnston and Teo 2000).

CSF diversionary procedures such as lumbo-peritoneal shunting and ventriculo-peritoneal shunting have been used for patients with severe visual symptoms and intractable unremitting headaches (Ball and Clarke 2006). Both of these surgical procedures pose technical difficulties in such patients leading to a high shunt complication rate; revision is often required. Additionally lumbo-peritoneal shunting can cause iatrogenic tonsillar herniation that mimics a Chiari I malformation (Johnston and Teo 1998). Severe papilloedema can be treated using optic nerve sheath fenestration or even subtemporal decompression (Johnston and Teo 2000; Ball and Clarke 2006).

Unfortunately a Cochrane review has revealed that there is very little evidence base for these various therapies in the literature (Lueck *et al.* 2005; Ball and Clarke 2006). Most therapies are used on an *ad hoc* basis with the aim of symptomatic control. More recent studies with venous stenting have described variable success but their long-term outcome remains to be determined (Owler *et al.* 2005; Bono *et al.* 2005).

References

Abbott A (2004). Evidence for bulk flow of brain interstitial fluid: significance for physiology and pathology. *Neurochem Int*, **45**, 545–52.

Amiry-Moghaddam M, Frydenlund DS, Ottersen OP (2004). Anchoring of aquaporin-4 in brain: molecular mechanisms and implications for the physiology and pathophysiology of water transport. *Neuroscience*, **129**, 999–1010.

Ball AK, Clarke CE (2006). Idiopathic intracranial hypertension. *Lancet Neurol*, **5**, 433–42.

Ballabh P, Braun A, Nedergaard M (2004). The blood-brain barrier: an overview. Structure, regulation and clinical implications. *Neurobiol Dis*, **16**, 1–13.

Barbier EL, Lamalle L, Descorps M (2001). Methodology of brain perfusion imaging. *J Magn Reson Imaging*, **13**, 496–520.

Basser PJ, Mattiello J, Le Bihan D (1994). Estimation of the effective self-diffusion tensor from the NMR spin echo. *J Magn Reson B*, **103**, 247–54.

Basser PJ, Pierpaoli C (1996). Microstructural and physiological features of tissues elucidated by quantitative diffusion tensor MRI. *J Magn Reson B*, **111**, 209–19.

Bastin ME, Sinha S, Farrall AJ *et al.* (2003). Diffuse brain oedema in idiopathic intracranial hypertension: a quantitative magnetic resonance imaging study. *J Neurol Neurosurg Psychiatry*, **74**, 1693–6.

Bastin ME, Carpenter TK, Armitage PA *et al.* (2006). Effects of dexamethasone on cerebral perfusion and water diffusion in patients with high-grade glioma. *Am J Neuroradiol*, **27**, 402–8.

Beaumont A, Whittle IR (2000). The pathophysiology of tumour associated epilepsy. *Acta Neurochirurg*, **142**, 1–15.

Bell BA (1983). A history of the study of brain edema. *Neurosurgery*, **13**, 724–8.

Bergsneider M, Black PM, Klinge P *et al.* (2005). Surgical management of idiopathic normal-pressure hydrocephalus: *Neurosurgery*, **57**, S29–39; discussion ii–v.

Black P (1980). Idiopathic normal pressure hydrocephalus: Results of shunting in 62 patients. *J Neurosurg*, **52**, 371–7.

Bono F, Giliberto C, Mastrandrea C *et al.* (2005). Transverse sinus stenoses persist after normalization of the CSF pressure in IIH. *Neurology*, **65**, 1090–3.

Burtscher J, Bartha L, Twerdy K *et al.* (2003). Effect of endoscopic third ventriculostomy on neuropsychological outcome in late onset idiopathic aqueduct stenosis: a prospective study. *J Neurol Neurosurg Psychiatry*, **74**, 222–5.

Chu W, Zeidler M, Whittle IR (2006). Ventriculomegaly or symptomatic hydrocephalus? Beware of atypical neurological syndromes. *Age Ageing*, **35**, 319.

Cohen ZR, Ram Z, Knoller N *et al.* (2002). Management and outcome of non-traumatic cerebellar haemorrhage. *Cerebrovasc Dis*, **14**, 207–13.

Czosnyka M, Czosnyka Z, Momjian S *et al.* (2004). Cerebrospinal fluid dynamics. *Physiolog Measurement*, **25**, R51–76.

Davies DC (2002). Blood-brain barrier breakdown in septic encephalopathy and brain tumours. *J Anat*, **200**, 639–46.

Dennis MS (2003). Outcome after brain haemorrhage. *Cerebrovasc Dis*, **16**, 9–13.

Del Bigio MR, Wilson MJ, Enno T (2003). Chronic hydrocephalus in rats and humans: white matter loss and behaviour changes. *Ann Neurol*, **53**, 337–46.

Devito EE, Pickard JD, Salmond CH *et al.* (2005). The neuropsychology of normal pressure hydrocephalus (NPH). *Br J Neurosurg*, **19**, 217–24.

Drake JM, Kestle JR, Tuli S (2000). CSF shunts 50 years on--past, present and future. *Childs Nerv Syst*, **16**, 800–4.

Dunkley S, Johnston IH (2004). Thrombophilia as a common predisposing factor in pseudotumour cerebri. *Blood*, **103**, 1972–73.

Fatouros PP, Marmarou A, Kraft KA *et al.* (1991). In vivo brain water determination by T1 measurements: effect of total water content, hydration fraction, and field strength. *Magn Reson Med*, **17**, 402–13.

Feigin VL, Anderson N, Rinkel GJ (2005). Corticosteroids for aneurysmal subarachnoid haemorrhage and primary intracerebral haemorrhage. *Cochrane Database Syst Rev*, **20**, CD004583.

Fishman RA (1976). Brain edema. *NEJM*, **293**, 705–11.

Glaser NS, Wootton-Gorges SL, Marcin JP *et al.* (2004). Mechanism of cerebral edema in children with diabetic ketoacidosis. *J Pediatr*, **145**, 164–71.

Go KG (1997). The normal and pathological physiology of brain water. *Adv Tech Stand Neurosurg*, **23**, 47–142.

Goel S, Wharton SB, Brett LP *et al.* (2003). Morphological changes and stress responses in neurons in cerebral cortex infiltrated by diffuse astrocytoma. *Neuropathology*, **23**, 262–70.

Graff-Radford N, Godersky J, Jones M (1989). Variables predicting surgical outcome in symptomatic hydrocephalus in the elderly. *Neurology*, **39**, 1601–4.

Greitz D, Hannerz J (1996). A proposed model of cerebrospinal fluid circulation: observations with radionuclide cisternography. *Am J Neuroradiol*, **17**, 431–8.

Hakim S and Adams R (1965). The special clinical problem of symptomatic hydocephalus with normal cerebrospinal fluid pressure: Observations on cerebrospinal fluid dynamics. *J Neurol Sci* **2**, 307–327.

Hayashi Y, Nomura M, Yamagishi S *et al.* (1997). Induction of various blood-brain barrier properties in non-neural endothelial cells by close apposition to co-cultured astrocytes. *Glia*, **19**, 13–26.

Heiss JD, Papavassiliou E, Merrill MJ *et al.* (1996). Mechanism of dexamethasone suppression of brain tumor-associated vascular permeability in rats. Involvement of the glucocorticoid receptor and vascular permeability factor. *J Clin Invest*, **98**, 1400–8.

Janzer RC and Raff MC (1987). Astrocytes induce blood-brain barrier properties in endothelial cells. *Nature*, **325**, 253–7.

Johnston IH, Teo C (2000). Disorders of CSF hydrodynamics. *Childs Nerv Syst*, **16**, 776–99.

Johnston I, Jacobson E, Besser M (1998). The acquired Chiari malformation and syringomyelia following spinal CSF drainage: a study of incidence and management. *Acta Neurochir*, **140**, 417–27; discussion 427–8.

Johnston M, Zakharov A, Papaiconomou C *et al.* (2004). Evidence of connections between cerebrospinal fluid and nasal lymphatic vessels in humans, non-human primates and other mammalian species. *Cerebrospinal Fluid Res*, **1**, 2.

Kaal EC, Vecht CJ (2004). The management of brain edema in brain tumors. *Curr Opin Oncol*, **16**, 593–600.

Kadrian D, van Gelder J, Florida D *et al.* (2005). Long-term reliability of endoscopic third ventriculostomy. *Neurosurgery*, **56**, 1271–8; discussion 1278.

Kamman RL, Go KG, Brouwer W *et al.* (1988). Nuclear magnetic resonance relaxation in experimental brain edema: effects of water concentration, protein concentration, and temperature. *Magn Reson Med*, **6**, 265–74.

Kestle JR, Drake JM, Cochrane DD *et al.* (2003). Insertion trial participants. Lack of benefit of endoscopic ventriculoperitoneal shunt insertion: a multicenter randomized trial. *J Neurosurg*, **98**, 284–90.

Kilincer C, Asil T, Utku U *et al.* (2005). Factors affecting the outcome of decompressive craniectomy for large hemispheric infarctions: a prospective cohort study. *Acta Neurochir*, **147**, 587–94; discussion 594.

Kimelburg HK (2004). Water homeostasis in the brain: basic concepts. *Neuroscience*, **129**, 851–60.

Klatzo I (1967). Presidential address. Neuropathological aspects of brain edema. *J Neuropath Exp Neurol*, **26**, 1–14.

Koedel U, Bernatowicz A, Paul R *et al.* (1995). Experimental pneumococcal meningitis: cerebrovascular alterations, brain edema, and meningeal inflammation are linked to the production of nitric oxide. *Ann Neurol*, **37**, 313–23.

Kulkarni AV, Rabin D, Lamberti-Pasculli M *et al.* (2001). Repeat cerebrospinal fluid shunt infection in children. *Pediatr Neurosurg*, **35**, 66–71.

Lawton MT, Du R, Tran MN *et al.* (2005). Effect of presenting hemorrhage on outcome after microsurgical resection of brain arteriovenous malformations. *Neurosurgery*, **56**, 485–93; discussion 485–93.

Levitsky LL (2004). Symptomatic cerebral edema in diabetic ketoacidosis: the mechanism is clarified but still far from clear. *J Pediatr*, **145**, 149–50.

Li KW, Nelson C, Suk I *et al.* (2005). Neuroendoscopy: past, present, future. *Neurosurgical focus*, **19**, E1.

Lu H, Law M, Johnson G *et al.* (2005). Novel approach to the measurement of absolute cerebral blood volume using vascular-space-occupancy magnetic resonance imaging. *Magn Reson Med*, **54**, 1403–11.

Lueck C, McIlwaine G (2005). Interventions for idiopathic intracranial hypertension. *Cochrane Database Syst Rev*, **20**, CD003434.

MacFall JR, Wehrli FW, Breger RK *et al.* (1987). Methodology for the measurement and analysis of relaxation times in proton imaging. *Magn Reson Imaging*, **5**, 209–20.

Malm M, Kristensen B, Karlsson T *et al.* (1995). The predictive value of cerebrospina fluid dynamic tests in patients with the idiopathic adult hydrocephalus syndrome. *Arch Neurol*, **52**, 783–9.

Malm J, Bergenheim AT, Enblad P *et al.* (2006). The Swedish malignant middle cerebral artery infarction study: long-term results from a prospective study of hemicraniectomy combined with standardized neurointensive care. *Acta Neurol Scand*, **113**, 25–30.

Maramarou A (2004). The pathophysiology of brain edema and elevated intracranial pressure. *Cleve Clin J Med*, **71**, S6–8.

Marmarou A, Bergsneider M, Relkin N *et al.* (2005a). Development of guidelines for idiopathic normal-pressure hydrocephalus: introduction. *Neurosurgery*, **57**, S2-1–S2-3.

Marmarou A, Bergsneider M, Klinge P *et al.* (2005b). The value of supplemental prognostic tests for the preoperative assessment of idiopathic normal-pressure hydrocephalus. *Neurosurgery*, **57**, S2-17–S2-28.

Mathur A, Khanna N, Chaturvedi UC (1992). Breakdown of blood-brain barrier by virus-induced cytokine during Japanese encephalitis virus infection. *Int J Exp Pathol*, **73**, 603–11.

Mayer SA, Brun NC, Begtrup K *et al.* (2005a). Recombinant activated factor VII intracerebral hemorrhage trial investigators. Recombinant activated factor VII for acute intracerebral hemorrhage. *N Engl J Med*, **352**, 777–85

Mayer SA, Rincon F (2005b). Treatment of intracerebral haemorrhage. *Lancet Neurol*, **4**, 662–72.

Mendelow AD, Gregson BA, Fernandes HM *et al.* (2005). Early surgery versus initial conservative treatment in patients with spontaneous supratentorial intracerebral haematomas in the International Surgical Trial in Intracerebral Haemorrhage (STICH): a randomised trial. *Lancet*, **365**, 387–97.

Miller JD, Sakalas R, Ward JD *et al.* (1977). Methylprednisolone treatment in patients with brain tumors. *Neurosurgery*, **1**, 114–7.

Miller JD (1979). The management of cerebral oedema. *Br J Hosp Med*, **21**, 152–65.

Mitchell P, Ellison DW, Mendelow AD (2005). Surgery for malignant gliomas: mechanistic reasoning and slippery statistics. *Lancet Neurol*, **4**, 413–22.

Morgan MK, Johnston IH, Spittaler PJ (1991). A ventricular infusion technique for the evaluation of treated and untreated hydrocephalus. *Neurosurgery*, **29**, 832–6; discussion 836–7.

Owler BK, Parker G, Halmagyi GM *et al.* (2005). Cranial venous outflow obstruction and pseudotumor Cerebri syndrome. *Adv Tech Stand Neurosurg*, **30**, 107–74.

Papadopoulos MC, Saadoun S, Binder DK *et al.* (2004). Molecular mechanisms of brain tumor edema. *Neuroscience*, **129**, 1011–20.

Papadopoulos MC, Krishna S, Verkman AS (2002). Aquaporin water channels and brain edema. *Mt Sinai J Med*, **69**, 242–8.

Pasco A, Minassian AT, Chapon C *et al.* (2006). Dynamics of cerebral edema and the apparent diffusion coefficient of water changes in patients with severe traumatic brain injury. A prospective MRI study. *Eur Radiol*, **17** [Epub ahead of print]

Raksin PB, Alperin N, Sivaramakrishnan A *et al.* (2003). Noninvasive intracranial compliance and pressure based on dynamic magnetic resonance imaging of blood flow and cerebrospinal fluid flow: review of principles, implementation, and other noninvasive approaches. *Neurosurg Focus*, **15**, 14:e4.

Ranjan P, Mishra AM, Kale R *et al.* (2005). Cytotoxic edema is responsible for raised intracranial pressure in fulminant hepatic failure: in vivo demonstration using diffusion-weighted MRI in human subjects. *Metab Brain Dis*, **20**, 181–92.

Relkin N, Marmarou A, Klinge P *et al.* (2005). Diagnosing idiopathic normal-pressure hydrocephalus. *Neurosurgery*, **57**, S2-4–S2-16.

Sanni LA, Jarra W, Li C *et al.* (2004). Cerebral edema and cerebral hemorrhages in interleukin-10-deficient mice infected with Plasmodium chabaudi. *Infect Immun*, **72**, 3054–8.

Sinha S, Bastin ME, Whittle IR (2003). Rapid clinical deterioration in a patient with multi-focal glioma despite corticosteroid therapy: a quantitative MRI study. *Br J Neurosurg*, **17**, 537–40; discussion 540.

Sinha S, Bastin ME, Wardlaw JM *et al.* (2004). Effects of dexamethasone on peritumoural oedematous brain: a DT-MRI study. *J Neurol Neurosurg Psychiatry*, **75**, 1632–5.

Stenzel W, Dahm J, Sanchez-Ruiz M *et al.* (2005). Regulation of the inflammatory response to Staphylococcus aureus-induced brain abscess by interleukin-10. *J Neuropathol Exp Neurol*, **64**, 1046–57.

Swaroop GR, Holmes MC, Kelly PAT *et al.* (2001). The effects of dexamethasone of tumour blood flow, capillary permeability and iNOS in experimental glioma. *J Clin Neuroscience*, **8**, 35–9.

Tofteng F, Hauerberg J, Hansen BA *et al.* (2006). Persistent arterial hyperammonemia increases the concentration of glutamine and alanine in the brain and correlates with intracranial pressure in patients with fulminant hepatic failure. *J Cereb Blood Flow Metab*, **26**, 21–7.

Tuli S, Alshail E, Drake J (1999). Third ventriculostomy versus cerebrospinal fluid shunt as a first procedure in pediatric hydrocephalus. *Pediatr Neurosurg*, **30**, 11–5.

Uhl E, Kreth FW, Elias B *et al.* (2004). Outcome and prognostic factors of hemicraniectomy for space occupying cerebral infarction. *J Neurol Neurosurg Psychiatry*, **75**, 270–4.

Unterberg AW, Stover J, Kress B *et al.* (2004). Edema and brain trauma. *Neuroscience*, **129**, 1021–9.

Vahedi K, Hofmeijer J, Juettler E *et al.* (2007). Early decompressive surgery in malignant infarction of the middle cerebral artery: a pooled analysis of three randomised controlled trials. *Lancet Neurol*, **6**, 215–22.

Vaquero J, Fontana RJ, Larson AM *et al.* (2005). Complications and use of intracranial pressure monitoring in patients with acute liver failure and severe encephalopathy. *Liver Transpl* **11**, 1581–9.

Vecht CJ, Hovestadt A, Verbiest HB *et al.* (1994). Dose-effect relationship of dexamethasone on Karnofsky performance in metastatic brain tumors: a randomized study of doses of 4, 8, and 16 mg per day. *Neurology*, **44**, 675–80.

Vecht C (2003). Management of brain metastases. In Williams C, ed. *Evidence Based Oncology*, pp. 574–8. BMA publishing, London.

Walchenbach R, Geiger E, Thomeer R *et al.* (2002). The value of temporary external lumbar CSF drainage in predicting the outcome of shunting on normal pressure hydrocephalus. *J Neurol Neurosurg Psychiatry*, **72**, 503–6.

Wang HZ, Riederer SJ, Lee JN (1987). Optimizing the precision in T1 relaxation estimation using limited flip angles. *Magn Reson Med*, **5**, 399–416.

Weissman DE, Janjan NA, Erickson B *et al.* (1991). Twice-daily tapering dexamethasone treatment during cranial radiation for newly diagnosed brain metastases. *J Neurooncol*, **11**, 235–9.

Whittle IR (1992). The origins and management of peritumoural brain dysfunction. *Neurosurg Quart*, **2**, 174–99.

Whittle IR, Thomson AM, Taylor R (1998). The effects of resective surgery for left sided intracranial tumours on language function: a prospective study. *Lancet*, **351**, 1014–8.

Whittle I R, Collie D, Smith C *et al.* (2004). Meningiomas. *Lancet*, **363**, 1535–43.

Whittle IR (2003). How effective is surgery for malignant glioma? In Williams C, ed. *Evidence Based Oncology*, pp. 565–7 BMA publishing, London.

Wick W, Kuker W (2004). Brain edema in neurooncology: radiological assessment and management. *Onkologie*, **27**, 261–6.

Yau YH, Piper IR, Contant CF *et al.* (2002). Clinical experience in the use of the Spiegelberg automated compliance device in the assessment of patients with hydrocephalus. *Acta Neurochir Suppl*, **81**, 171–2.

Zeidler M, Dorman PJ, Ferguson IT *et al.* (1998). Parkinsonism associated with obstructive hydrocephalus due to idiopathic aqueductal stenosis. Short report. *J Neuro Neurosurg Psychiatry*, **64**, 657–9.

CHAPTER 27

Tumours of the brain and skull

Robert Grant

Contents

27.1 Introduction and terminology

Over the last 10 years, there there have been several important advances in cell biology, molecular genetics, and targeted therapies in neuro-oncology. Improved neurosurgical techniques such as frameless stereotaxy, awake craniotomy, and intra-operative MRI, safer methods of directing radiotherapy, new chemotherapy approaches, and novel modalities of therapy provide optimism that there will eventually be some improvements in treatment-related morbidity and survival. There has also been an increasing change from individual clinician decision making to decision making by multidisciplinary teams of neurosurgeons, neurologists, clinical oncologists, neuropathologists, neuroradiologists, and specialist nurses with the aim of improving decision making, management planning across specialties, communication, and enrolment in suitable clinical trials. In addition, Good Clinical Practice guidelines, an international ethical and scientific quality standard for designing, conducting, recording, and reporting trials, increases the onus and responsibilities on clinical investigators to perform trials to the highest standard and to have the trials externally monitored, and the trial conduct and results audited. While these obligatory and statutory responsibilities are labour intensive and time consuming, they should improve the quality of trials by limiting the possibility of unintentional bias or fraud. Improving the recording of serious adverse event reporting through trial quality assurance and quality control procedures will help ensure that a balanced view of the effects of a drug or procedure is identified earlier than in the past. It will be interesting to see how research develops over the next decade.

Tumours of the brain and skull can be divided by site and relationship to the cranial fossae:

♦ extracranial but involving neural tissue; head and neck lesions

Table 27.1 World Health Organization typing of central nervous system tumours

1.	**Tumours of neuroepithelial tissue**
1.1	***Astrocytic tumours***
1.1.1	Astrocytoma
1.1.2	Anaplastic (malignant) astrocytoma
1.1.3	Glioblastoma
1.1.4	Pilocytic astrocytoma
1.1.5	Pleomorphic xanthoastrocytoma
1.1.6	Subependymal giant cell astrocytoma
1.2	***Oligodendroglial tumours***
1.2.1	Oligodendroglioma
1.2.2	Anaplastic (malignant)
1.3	***Ependymal tumours***
1.3.1	Ependymoma
1.3.2	Anaplastic (malignant)
1.3.3	Myxopapillary ependymoma
1.3.4	Subependymoma
1.4	***Mixed gliomas***
1.4.1	Oligo-astrocytoma
1.4.2	Anaplastic (malignant)
1.4.3	Other
1.5	***Choroid plexus tumours***
1.5.1	Choroid plexus papilloma
1.5.2	Choroid plexus carcinoma
1.6	***Neuroepithelial tumours***
1.6.1	Astroblastoma
1.6.2	Polar spongioblastoma
1.6.3	Gliomatosis cerebri
1.7	***Neuronal and mixed neuronal-glial tumours***
1.7.1	Gangliocytoma
1.7.2	Dysplastic gangliocytoma of cerebellum
1.7.3	Desmoplastic neuroepithelial tumour
1.7.4	Dysembryoplastic neuroepithelial tumour
1.7.5	Ganglioglioma
1.7.6	Anaplastic (malignant) ganglioglioma
1.7.7	Central neurocytoma
1.7.8	Paraganglioma of the filum terminale
1.7.9	Olfactory neuroblastoma
1.8	***Pineal parenchymal tumours***
1.8.1	Pineocytoma
1.8.2	Pineoblastoma
1.8.3	Mixed/transitional pineal tumours
1.9	***Embryonal tumours***
1.9.1	Medulloepithelioma
1.9.2	Neuroblastoma
1.9.3	Ependymoblastoma
1.9.4	Primitive neuroectodermal tumours
2.	**Tumours of cranial and spinal nerves**
2.1	***Schwannoma (neurinoma)***
2.2	***Neurofibroma***
2.3	***Malignant peripheral nerve sheath tumour***

3.	**Tumours of the meninges**
3.1	***Tumours of meningothelial cells***
3.1.1	Meningioma
3.1.2	Atypical meningioma
3.1.3	Papillary meningioma
3.1.4	Anaplastic (malignant) meningioma
3.2	***Mesenchymal, non-meningothelial tumours***
Benign neoplasms	
3.2.1	Osteocartilaginous tumours
3.2.2	Lipoma
3.2.3	Fibrous histiocytoma
3.2.4	Others
3.2.5	Haemangiopericytoma
Malignant neoplasms	
3.2.6	Chondrosarcoma
3.2.7	Malignant fibrous histiocytoma
3.2.8	Rhabdomyosarcoma
3.2.9	Meningeal sarcomatosis
3.2.10	Others
3.3	***Primary melanocytic lesions***
3.3.1	Diffuse melanosis
3.3.2	Melanocytoma
3.3.3	Malignant melanoma
3.4	***Tumours of uncertain histogenesis***
3.4.1	Haemangioblastoma
4.	**Lymphoma & haemopoietic tumours**
4.1	***Malignant lymphoma***
4.2	***Plasmocytoma***
4.3	***Granulocytic sarcoma***
4.4	***Other***
5.	**Germ cell tumours**
5.1	***Germinoma***
5.2	***Embryonal carcinoma***
5.3	***Yolk sac tumour***
5.4	***Choriocarcinoma***
5.5	***Teratoma***
5.6	***Mixed germ cell tumour***
6.	**Cysts and tumour-like lesions**
7.	**Tumours of the sellar region**
7.1	***Pituitary adenoma***
7.2	***Pituitary carcinoma***
7.3	***Craniopharyngioma***
8.	**Local extensions of regional tumours**
8.1	***Paraganglioma***
8.2	***Chordoma***
8.3	***Chondroma/chondrosarcoma***
8.4	***Carcinoma***
9.	**Metastatic tumours**
10.	**Unclassified tumour**

- intracranial but extracerebral;
  - anterior cranial fossa, e.g. olfactory groove,
  - middle cranial fossa, e.g. pituitary region and sphenoid wing,
  - posterior cranial fossa, e.g. cerebello-pontine angle, and craniocervical junction;
- within the brain substance, e.g. primary or secondary intracerebral tumours.

Histological grading is crucial to the understanding of brain tumours and their eventual management. The World Health Organisation classification (Table 27.1) (Kleihues *et al.* 1993a)

recognizes that tumours develop via different pathways and that the behaviour of some grades of malignancy are quite distinct. The classification also allows standardization of reporting for epidemiological studies. The World Health Organization have revised the grading system of central nervous system tumours to take into account recently defined histological entities such as pleomorphic xanthoastrocytoma and dysembryoplastic neuroepithelial tumours (Table 27.2). Identification of the type of tumour and predictions of the prognosis and response to treatment can be refined by genotyping methods of identifying genetic changes common in certain tumour types, for instance Ch 1p and 19q loss or methyl guanine methyl transferase, MGMT, promoter methylation status.

Table 27.2 World Health Organization grading system of central nervous system tumours on a malignancy scale where Grade IV signifies greater malignancy

Tumour Group	Tumour Type	Grade I	Grade II	Grade III	Grade IV
Astrocytic tumours	Subependymal giant cell	*			
	Pilocytic	*			
	Low grade		*		
	Pleomorphic xanthoastrocytoma		*	*	
	Anaplastic			*	
	Glioblastoma				*
Oligodendrogliomas	Low grade		*		
	Anaplastic			*	
Oligo-astrocytomas	Low grade		*		
	Anaplastic			*	
Ependymal tumours	Subependymoma	*			
	Myxopapillary	*			
	Low grade		*		
	Anaplastic			*	
Choroid plexus	Papilloma	*			
	Carcinoma			*	*
Neuronal/Glial cell	Gangliocytoma	*			
	Ganglioglioma	*	*		
	Desmoplastic infantile ganglioglioma	*			
	Dysembryoplastic neuroepithelial	*			
	Central neurocytoma	*			
Pineal tumours	Pineocytoma		*		
	Pineocytoma/pineoblastoma		*	*	
	Pineoblastoma				*
Embryonal tumours	Medulloblastoma				*
	Other primitive neuroectodermal tumours				*
	Medulloepithelioma				*
	Neuroblastoma				*
	Ependymomblastoma				*
Cranial/spinal nerve	Schwannoma	*			
	Malignant peripheral nerve sheath			*	*
Meningeal tumours	Meningioma	*			
	Atypical meningioma		*		
	Papillary meningioma		*	*	
	Haemangiopericytoma		*	*	
	Anaplastic meningioma			*	

27.2 Epidemiology of brain tumours

27.2.1 Mortality

Brain tumours are the second most common cause of death from neurological disease, surpassed only by stroke. Neuroepithelial brain tumours are the fifth most common cause of death from malignancy under the age of 65 years. If other intrinsic brain tumours, including metastases and primary central nervous system lymphomas are considered, it is very likely that intracranial tumours are one of the three most common causes of death from cancer in the working population.

27.2.2 Incidence and prevalence

Information from cancer registry statistics show that neuroepithelial tumours are the most common solid malignancy in children, the fourth most common in the under-45 age group, and the eighth most common under the age of 65 years. However, cancer registries frequently do not reliably record incidence data on all primary intracranial tumours, particularly meningiomas, acoustic neuromas, pituitary adenomas, and primary central nervous system lymphoma (Counsell *et al.* 1997).

Meningiomas, acoustic neuromas, and pituitary tumours account for almost 25 per cent of all intracranial tumours (Table 27.3), and there is evidence that more tumours are being identified, although this may reflect the wider availability of brain imaging. Primary central nervous system lymphomas associated with AIDS, have also increased. Brain metastases affect 20 per cent of patients with cancer and account for over 50 per cent of all

intrinsic brain tumours. The incidence of intracranial metastases may also be increasing, as the brain will also act as a 'sanctuary site' for metastases as treatment of the systemic components of the tumour improves.

The incidence of all intracranial tumours is between 21 to 30 cases per 100 000 population each year (Counsell *et al.* 1996; Counsell and Grant 1998; Pobereskin and Chadduck 2000) (Table 27.3). Assuming no demographic or regional differences, one would expect approximately 200 000 new cases of intracranial tumour each year in Europe and 100 000 new cases each year in North America. Over the last 50 years the incidence of intracranial tumours in developed countries has increased (McKinney 2004). The increase is predominantly in the elderly population and may be due to better case ascertainment and improved health care for the elderly and the shifting demographics to a more elderly population (Radhakrishnan *et al.* 1995). Studies produce a wide range of incidences for neuroepithelial tumours, 2.5 to 9.1/10^5/year, but studies with good case ascertainment are at the higher figure (Kaye *et al.* 1993; D'Alessandro *et al.* 1995). The age specific incidence of primary brain tumours has a small peak in early childhood, 3.5/10^5 cases/year, dips in the 15 to 24 years age range, at 2.9/10^5/year, and then progressively climbs to its highest incidence in the 65 to 74 years age range of 24/10^5/year before falling off slightly to 9.6/10^5/year in the >85 years age range (Counsell *et al.* 1996). Cerebral metastases are rare in the under 25 years age group, at 1/10^5/year, but increase gradually to 53.7/10^5/year in the 65 to 74 years (CI 41.6–68.2/10^5/year.) age group. Meningeal tumours are also very rare under 25 years of age, at 0.4/10^5/year and increase steadily to a peak incidence of 9.0/10^5/year in the age range 75 to 85 (CI 3.6–18.5/10^5/year). A meta-analysis of incidence studies, suggests that the differences could be accounted for by methods of identifying patients, inclusion criteria for what represented primary intracranial tumours, and the scanning techniques used to identify cases. Methodological guidelines for future incidence studies in brain tumours have been suggested, in order to standardize future incidence studies to allow meaningful analysis (Counsell and Grant 1998).

Medulloblastomas and ependymomas occur most frequently in childhood or early adolescence. Grade 2 astrocytomas and oligodendroglioma in early adults aged 20–40 and grade 3 and 4 astrocytomas, the malignant gliomas, most commonly in the >40 age range although they can occur at any age.

27.2.3 Referral pattern

Patients with intracranial tumours can present to a variety of hospital-based clinicians or primary-care physicians as a result of increased availability of direct access CT scanning (Grant *et al.* 1996). Following imaging diagnosis, further specialist referral is almost always the rule. Referral occurs at a time of great uncertainty for the patient and family, and close collaboration between clinicians is essential to reduce anxiety. A clear explanation to the patient and family and a simple well constructed management plan must be given, which should be consistent between specialists. Recently, multidisciplinary team meetings have being advocated to discuss individual patients and advise on the best management plan. These meetings are often tumour site specific, so for pituitary tumours they involve endocrinologist, neurosurgeons, clinical oncologist, ophthalmologist, and specialist nurse (NICE: http://www.nice.org.uk/page.aspx?o=282278). Communication with

Table 27.3 Incident cases of different intracranial tumours diagnosed in SE Scotland: and expected United Kingdom frequency, assuming similar demographics to SE Scotland

		Crude Incidence (per 10^5) (95% CI)	Expected UK cases/year
1.	Neuroepithelial	8.2 (6.8–9.8)	4641
2.	Tumours of cranial and spinal nerves	0.7 (0.3–1.2)	396
3.	Tumours of the meninges	3.0 (2.6–4.0)	1698
4.	Lymphoma and haemopoietic tumours	0.7 (0.3–1.2)	396
5.	Germ cell tumours	0.1 (0.0–0.4)	57
6.	Cysts and tumour-like lesions	0.1 (0.0–0.5)	57
7.	Tumours of the sellar region	2.5 (1.7–3.3)	1415
8.	Local extensions from regional tumours	0.1 (0.0–0.4)	57
9.	Metastatic tumours	14.3 (12.4–16.3)	8093
10.	Unclassified	0.5 (0.1–0.9)	283
	Total intra-cranial tumours	30.2	17 093

patient, family, and general practitioner should be unambiguous and supportive (Davies and Hopkins 1997a).

27.2.4 Geographical, racial, and social influences

Males are 1.1 to 1.7 times more likely than females to develop primary intracerebral brain tumours (Counsell et al. 1996) and females are about 2.2 times more likely to develop meningiomas (Counsell and Grant 1998). There do not appear to be any specific geographical factors that influence the incidence of primary brain tumours (McKinney 2004, Wrensch et al. 2002). Some studies report that white Americans have a higher incidence of brain tumours than blacks, Hispanics, and Japanese Americans; however, others have only shown a higher incidence in black children than whites (Bunin 1987) and have not found a higher incidence in whites compared with Hispanics.

The level of affluence is associated with the likelihood of developing neuroepithelial brain tumours in adults and children. The incidence of neuroepithelial tumours is approximately 2.3 times greater in areas of SE Scotland with the greatest affluence than with the least affluent (McLoone 1993). The relationship between deprivation category and incidence of neuroepithelial tumours is linear (Counsell et al. 1996). The incidence of metastases was about twice as high in the least affluent groups than the most affluent (X_2 trend 6.18, $p=0.01$). The relationship between neuroepithelial tumours and social deprivation is very unlikely to be due to inequality of access to care in the health service because of the National Health Service and because the inverse relationship is seen with cerebral metastases.

27.2.5 Risk factors

Inherited cancer syndromes and very rare cases due to therapeutic radiation are the only causative factors that have been unequivocally identified (IARC 2003). The strongest possible risk factor is therapeutic cranial irradiation. Therapeutic irradiation for scalp ringworm was associated with a 4-fold increase in the risk of brain tumour (Ron et al. 1988). A Japanese study of atomic bomb survivors failed to demonstrate any increased risk of brain tumours in children exposed to radiation in utero (Kato et al. 1989). Certain chemicals such as polycyclic aromatic hydrocarbons and nitroso compounds can induce brain tumours in experimental animals, but their role in human brain tumours remain unproven (Maekawa and Mitsumori 1990). Vinyl chloride can cause brain tumours in rats and a Swedish study showed that workers in a vinyl chloride plant were twice as likely to die from brain tumours (Hagmar et al. 1990). There is no good evidence that cellular telephones are associated with brain tumours. Dietary and other chemical exposures, maternal birth characteristics, head trauma, infections, vaccinations and various medications, including anti-nauseants during pregnancy, have been suggested as risk factors, but none have shown strong enough correlations to confirm a link (Wrensch et al. 2002). Immunodeficiency states are a strong risk factor for developing primary central nervous system lymphoma.

27.2.6 Genetic Factors

Many different genes have been associated with familial cancer predisposition and have associations with central nervous system tumours (Table 27.4). There is a genetic predisposition to brain tumours in patients with the family cancer syndrome (Li and Fraumeni 1969). Li–Fraumeni syndrome has been linked to a mutation in p53 on chromosome 17p13 (Malkin et al. 1990).

Table 27.4 Genes responsible for familial predisposition to cancer and development of astrocytomas

Genes responsible for familial predisposition to cancer		
Chromosome region	Gene/locus symbol	Familial cancer
3p14	RCC1	von Hippel–Lindau disease
9q22.3	NBCCS	Gorlin syndrome
9q34	TSC1	Tuberous sclerosis
11q22–q23	ATM	Ataxia telangectasia
13q14	RB1	Retinoblastoma
16p13.3	TSC2	Tuberous sclerosis
17p13	p53	Li–Fraumeni syndrome
17q11	NF1	Neurofibromatosis type 1
22q	NF2	Neurofibromatosis type 2

Genes responsible for development of astrocytomas		
	Gene locus	Protein action
Growth factor receptors		
	7p12.3	EGFR
	4q12	PDGFR-A-R
Signal transduction		
	10q23.3	PTEN/MMAC
Retinoblastoma pathway (restriction point G1)		
	9p21	CDKN 2A/B
	7q21	CDK6
	11q13	CCND1
	6p21	CCND3
	13q14.1-14.2	RB1
p53 pathway		
	9p21	P14^{ARF}
	12q14.3-12q15	MDM2
	1p32	MDM4
	17p13.1	TP53

p53 mutations have been frequently found in all grades of astrocytoma and represent an early event in malignant transformation of astrocytic cells (Ohgaki et al. 1993). Glioblastomas with p53 mutations generally have arisen from a low grade astrocytoma and are often called 'secondary' glioblastomas. Glioblastomas that have arisen de novo, i.e. not preceded by a low grade glioma, generally do not have a p53 mutation and are called 'primary' glioblastomas.

In hereditary syndromes with linked gene mutations such as tuberous sclerosis at 9q32–34, about 5 per cent of cases will develop cerebral gliomas. Tuberous sclerosis is an autosomal dominant condition and genes that are involved are TSC1 gene on Ch 9q34 and TSC2 gene on Ch 16p13. The incidence is thought to be between 1 and 2/10,000.

Neurofibromatosis type 1, another autosomal dominant condition with near complete penetrance, is related to mutations at Ch 17q11–22 and is associated with cerebral or optic nerve glioma in between 4 and 45 per cent of patients. In neurofibromatosis type 2, with mutations at Ch 22q12, schwannomas, ependymomas, and meningiomas are found more frequently than in the general population. Although neurofibromatosis type 1 is an autosomal dominant disorder with near complete penetrance, about half of neurofibromatosis type 1 cases are new mutations. Neurofibramatosis type 2 is also autosomal dominant.

Von Hippel–Lindau disease is an autosomal dominant condition characterized by the development of haemangioblastomas of the central nervous system and other sites. It is due to a defect in the tumour suppressor gene on 3p25–26. Up to 50 per cent of patients develop cerebellar haemangioblastomas and 15 per cent develop spinal haemangioblastomas.

Other rare syndromes such as Gorlin's syndrome, Turcot's syndrome, and Gardner's syndrome are associated with brain tumours. Gorlin's syndrome is an autosomal dominant condition with naevoid basal cell carcinoma and is associated with medulloblastoma. Turcot's syndrome and familial polyposis is associated with an increased incidence of glioma and Gardner's syndrome with medulloblastoma.

27.3 Biological basis

Our knowledge of cellular and molecular biology has expanded exponentially over the last 10 years (Rasheed *et al.* 1999). The developments have been used to confirm the histological nature of tumours, determine the proliferating potential of malignant cells and provide insight into the control mechanisms behind tumour cell proliferation and tumour suppression. Three classes of genes have been implicated in the pathogenesis of cancer: oncogenes, tumour suppressor genes, and mismatch repair genes. Oncogenes are derived from normal cellular genes, or proto-oncogenes, which function as growth stimulators. However, if activated, uncontrolled cellular proliferation occurs. Tumour suppressor genes normally restrain cellular proliferation but when inactivated by gene mutation, will result in uncontrolled cellular proliferation. Knudsen proposed the 'Two hit hypothesis' which suggested that the loss of both alleles of the tumour suppressor gene was necessary before a cancer could develop. Patients with a germline mutation in one of these alleles require a somatic change in the other allele before a tumour develops. The third class of gene important to the development of tumours are the mismatch repair genes. The products of these genes identify mismatches that occur at point mutations in the genome and repair these. If there is a problem with failure to identify mismatches, gene repair and control are affected. Carcinogenesis may be initiated by a mutation or alteration in certain genes, for instance deletion or rearrangement of DNA, or by several different changes in several genes. The mutation can be random, inherited, or induced by radiation, chemicals, or other insult.

27.3.1 Growth factors

Growth factors are usually under the control of proto-oncogenes and tumour suppressor genes. Several proto-oncogenes have been associated with the development of malignant gliomas, such as *c-erb B-1, c-sis, c-myc, ras, c-fos,* and *ros* oncogenes. *Erb-b* encodes for epidermal growth factor receptor, EGFR, and c-sis encodes for platelet derived growth factor, PDGF. The most common growth factor in malignant glioma is EGFR. EGFR is over-expressed in about 40 to 60 per cent of malignant gliomas but expression is uncommon in astrocytoma. This suggests that the EGFR occurs as a late feature of deregulation of cell division and not an initiating factor of tumourogenesis. Amplification or over-expression of EGFR is associated with resistance to chemotherapy-induced apoptosis. There is, however, not a close relationship between over-expression of EGFR and survival in tumours of the same grade. PDGF receptor amplifications are common in low-grade astrocytomas, suggesting that actroption of PDGF may be an early event in pathogenesis of malignant astrocytoma. Several other growth factors such as transforming growth factors, TGF-alpha and TGF-beta, fibroblast growth factor, and insulin-like growth factor are also over-expressed in some glioblastomas. TGF-alpha is a polypeptide growth factor that has sequence homology for EGF and binds to EGFR. TGF-alpha is frequently amplified in anaplastic astrocytoma and glioblastoma multiforme and may act as a growth factor ligand for EGFR and form an autocrine growth loop leading to proliferation of glioma cells. The amplification and over-expression of TGF-alpha might be an earlier event in a gradual process of tumourigenesis. TGF-beta stimulates astrocytoma cells to migrate and invade (Yamada *et al.* 1995).

The Bcl-2 protein family may contribute to impaired ability of glioma cells to undergo apoptosis. Bcl-2 controls apoptosis through release of cytochrome C from mitochondria and activation of caspases, which initiate the cell death process.

27.3.2 Tumour suppressor factors

Several different chromosomes have been reported as having possible loci for tumour suppressor genes. Low grade brain tumours frequently demonstrate loss of genetic material on chromosomes 6, 13, 17p, or 22. *P53* gene is the most well known tumour suppressor genes and is situated on chromosome 17p13.1. Mutations of the *p53* gene are implicated in many cancers. In gliomas, dysfunction of key components of the apoptotic pathways, prevent apoptosis of damaged cells and contribute to tumour development and malignant progression. In response to DNA damage, wild-type p53 induces G1-S cell cycle arrest and either mediates DNA repair or initiates apoptosis if the damage is too great to repair. Inactivation of p53 is an early event in glioma development and it promotes genomic instability by allowing highly damaged cells to survive and accumulate further mutations. As the grade of malignancy increases, other allelic changes are seen in addition. Allelic loss of Ch 19q is found in 46 per cent of grade 3 gliomas, but only 11 per cent of grade 2 tumours (von Deimling *et al.* 1994). The most common chromosomal abnormality, present only in glioblastoma, is loss of chromosome 10. A region on chromosome 10q23.3 called *PTEN* acts as a tumour suppressor gene. *PTEN* alterations occur with a high frequency in primary glioblastoma, usually in elderly patients where the tumour arises *de novo*, but also occur with a low frequency in secondary glioblastoma, usually in young patients where the glioblastoma develops from a low grade to a high grade tumour. In *primitive neuro-ectodermal tumours*, PNET-medulloblastomas and PNET variants, deletions of the short arm of chromosome 17p, or duplication of the long arm 17q occur in a high percentage of cases and there is frequently also loss of information on chromosome 10, 11, and 22. Seventy per cent of meningiomas have loss on chromosome 22. Despite the large number of potential tumour suppressor genes, there are no specific losses of genetic material that are unique to a particular histological group of tumours and no specific deletion or other genetic alteration is found in all tumours.

27.4 Clinical features and imaging diagnosis

27.4.1 Extracranial tumours affecting neural tissue

Patients with tumours of the head and neck most commonly present with headache, lower cranial neuropathies, or facial pain. Where the involvement is in the nasopharynx, one should consider nasopharyngeal carcinoma, adenoid cystic carcinoma and metastatic tumours.

Where there is involvement of the carotid body region, *glomus jugulare* should be considered. Patients with glomus jugulare tumours commonly present with lower cranial nerve palsies, such as dysphonia, dysphagia with wasted tongue, or weak palate (Section 20.7). If extensive, a glomus jugulare can also cause pulsatile tinnitus, headaches, and hearing loss. They are usually large when eventually discovered. Plain X-rays, with a submento-vertex view, will best demonstrate the enlargement of the jugular foramen. CT scanning shows the strongly enhancing tumour mass with erosion of the adjacent bone. MRI will demonstrate the 'salt and pepper' appearance caused by the flow voids within the tumour. Gadolinium enhanced coronal MRI scan is particularly useful to delineate the extent of the tumour and the relationship to the brain stem.

Where there is occipital headache, and lower cranial neuropathies, tumour involvement of the skull base should be considered. In the midline, chordoma, chondroma, and chondrosarcoma are all possible. Imaging of the head and neck by CT scan or MRI may demonstrate a lesion with or without soft tissue involvement. CT scanning is superior when bony involvement is present, as in skull osteomas, Paget's disease and fibrous dysplasia, and MRI with its multiplanar capabilities is superior for soft tissue visualization, as in nasopharyngeal carcinoma, metastases, and glomus jugulare tumours. Contrast enhancement will better delineate blood vessels from surrounding soft tissue structures and otorhinolaryngological assessment, angiography, and simple blood tests, such as alkaline phosphatase and myeloma screen, may also be helpful.

27.4.2 Intracranial extracerebral lesions

Anterior cranial fossa

Patients with olfactory groove meningiomas usually present late when the tumour is large enough to cause headache, seizures, or personality changes. Anosmia is a rare complaint in the absence of other symptoms.

Middle cranial fossa including pituitary region

Pituitary tumours may present because of endocrine effects, hormone excess or hypopituitarism, or mass effect. Women with prolactinomas commonly present with amenorrhoea and galactorrhoea and are referred to gynaecologists. Men with prolactinomas usually present later than women and may complain of headache, reduced sexual function and visual field defects. The presence of acromegaly or steroid excess will point to a diagnosis of a growth hormone or ACTH secreting macroadenoma respectively and should stimulate a request for imaging of the pituitary region. Plain lateral skull X-rays may show a 'double floor' or erosion of the sella tursica in the presence of a pituitary macroadenoma. Skull X-ray, however, may be entirely normal in macroadenoma and is always normal in microadenomas and is insufficient to exclude any intracranial tumour. Investigations for hormone secreting pituitary tumour include prolactin levels, growth hormone levels, and

serum or urinary cortisols plus imaging of the pituitary gland by multiplanar contrast enhanced MRI, or contrast enhanced coronal CT with fine cuts through the pituitary gland (Fig. 27.1).

Craniopharyngiomas commonly present with symptoms and signs of mild hypopituitarism or diabetes insipidus. Almost 90 per cent of men complain of impotence, while most women complain of amenorrhoea. Children present with short stature and 40 per cent of patients will be hypothyroid at presentation while 25 per cent have adrenal insufficiency. Fifty per cent of patients will have diabetes insipidus and headache. Craniopharyngiomas are usually a complex combination of cysts and solid tumour with calcification (Fig. 27.2). There is generally no surrounding oedema in the brain. Craniopharyngiomas can be difficult to differentiate from dermoid or epidermoid tumours or Rathke's pouch cysts, but craniopharyngiomas generally have more complex cysts than epidermoids, thicker irregular walls, and more calcification on CT scan. Hypopituitarism is also found in 20 per cent of patients with epidermoid and dermoid cysts in the suprasellar or parasellar areas (Fonari *et al.* 1990).

Patients with *visual field loss* due to optic nerve or chiasm compression will be referred to ophthalmologists, physicians, or neurologists. The visual impairment may be due to pathology in the nerve, for instance glioma, or pressure on the nerve from a tumour, such as pituitary tumour, craniopharyngioma, meningioma, or metastasis, or a cyst such as dermoid, epidermoid, or Rathke's pouch. Symptoms will lead to imaging of the anterior visual pathway. Multiplanar gadolinium enhanced MRI is the investigation of choice.

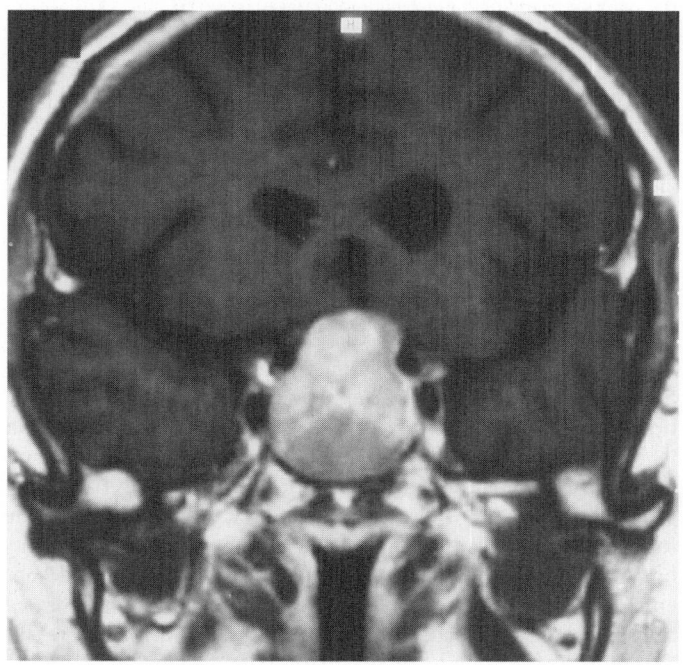

Fig. 27.1 Non-functioning radiological grade IV pituitary macroadenoma with suprasellar extension in a male who was referred following his third minor road traffic accident in 3 months where he hit parked cars on his left side. Examination revealed a classical bitemporal homonymous hemianopia and optic nerve pallor. He was found to have panhypopituitarism. He was treated with hydrocortisone, thyroxine, and DDAVP and a transphenoidal hypophysectomy was performed with almost complete resection. He elected to have a 'wait and watch' policy rather than pituitary radiotherapy. Coronal Gadolinium enhanced MR scan.

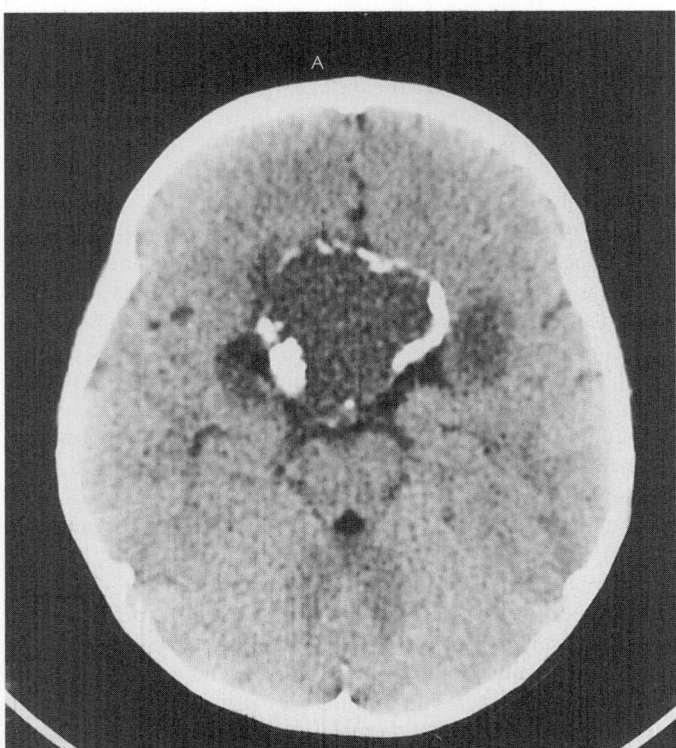

Fig. 27.2 Craniopharyngioma with cysts and calcification. Axial non-contrast CT scan. As in this case, it can be difficult to distinguish from an exophytic intracerebral tumour such as an oligodendroglioma. Further imaging in the coronal plane by MRI with gadolinium contrast can be helpful.

The coronal images will provide useful information about expansion of the optic nerves consistent with an optic nerve glioma or meningioma, and about the parasellar region and the relationship with any extrinsic pressure on the optic chiasm. Pituitary macroadenomas usually cause expansion of the pituitary fossa and displace the optic chiasm upwards, producing a bitemporal field loss which starts in the superior temporal quadrants. The visual field defects with pituitary macroadenomas will vary depending on whether the optic chiasm is prefixed or postfixed. Craniopharyngiomas expand downwards from the hypothalamus and cause a bitemporal hemianopia most frequently involving the inferior temporal quadrants, but can also cause a variety of visual field defects depending on where the tumour presses on the visual apparatus. Epidermoid and dermoid cyst generally appear on CT scanning as well-circumscribed lesions with low density between that of CSF and brain due to cholesterol or keratin granules. The wall of epidermoid cysts may be thinly calcified, and since the contents are avascular they do not enhance with contrast. Dermoids are more heterogeneous, have a thicker wall, rarely enhance, and more commonly demonstrate calcification. There is generally no surrounding oedema in the brain. On T1-weighted MRI sequences, epidermoids exhibit a variable signal, white when the lipid content is high or black if lipid content is low. Classically, epidermoids have low signal on T1-weighted images and very high signal on T2-weighted images. Dermoids give high signal on T1-weighted images in areas containing fat. They give variable signal where there is a combination of fat, muscle, and bone and may be mistaken for a craniopharyngioma or mixed germ cell

tumour, or teratoma. Mixed germ cell tumours are more heterogeneous than germ cell tumours, or germinomas, because they contain a variety of tissues including bone, cartilage, hair, and fatty tissue. Enhancement following contrast is common in germ cell tumours. Tumours such as hypothalamic astrocytomas and oligodendrogliomas can also extend downwards to cause chiasmal or optic nerve compression and hypopituitarism. These tumours are usually solid with areas of calcification but can also sometimes be exophytic. Rathke's cleft cysts are simple intrasellar cysts containing CSF. Meningiomas are usually easily differentiated from cysts and other tumours, however, en plaque meningioma of the optic nerve may be difficult to visualize even with gadolinium enhanced MRI and should always be considered as a potential diagnosis in patients with progressive optic nerve disease even in the absence of a clear mass lesion on MRI.

Patients who present with peri-orbital pain, ocular muscle paralysis, or ptosis may have tumours in the orbit, often metastases or lymphoma, lacrimal gland carcinoma, or tumours of the sphenoid wing, usually meningioma, carcinoma, dermoid, epidermoid, large pituitary tumours, or craniopharyngioma. Differential diagnosis will depend on the speed of onset of symptoms and the imaging appearance.

Posterior fossa

If the presenting complaint is facial numbness or weakness, deafness, tinnitus, or vertigo, patients are likely to be sent by their family practitioners to see a physician or otorhinolaryngologist. The differential diagnosis includes acoustic neuroma, meningioma, haemangioblastoma, dermoid, epidermoid, lipoma, and metastasis.

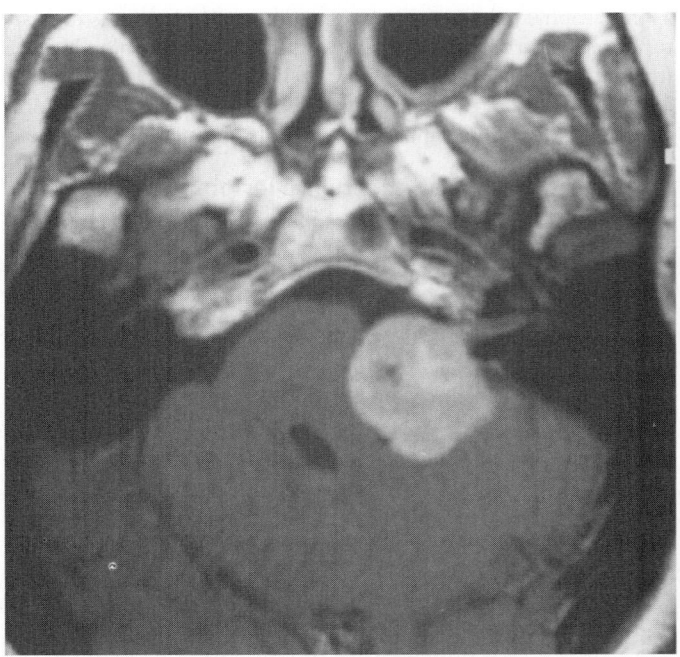

Fig. 27.3 Acoustic neuroma. Axial gadolinium enhanced MRI scan showing distortion of the pons and some midline shift. This 53-year-old woman was fit but presented with moderate unilateral deafness and slight unsteadiness. Examination revealed unilateral sensorineural deafness, horizontal jerk nystagmus, and a diminished corneal reflex. There were no signs of neurofibromatosis. She had complete surgical resection and was left with unilateral profound deafness, partial lower motor neurone facial weakness, and an anaesthetic cornea.

The most common *cerebellopontine angle tumour* is an acoustic neuroma (Fig. 27.3). Patients most commonly present with deafness, tinnitus, or vertigo. Patients with meningiomas less commonly have acoustic nerve symptoms and more commonly present with other cranial nerve involvement, especially facial numbness and facial weakness, however, differentiation on clinical grounds is unreliable. MRI is most sensitive imaging technique to delineate lesions of the middle or posterior cranial fossae. Acoustic neuromas usually expand the acoustic nerve and may cause expansion of the internal auditory meatus. Small tumours enhance uniformly and are usually easy to distinguish from other tumours; however if acoustic neuromas are very large it may be difficult to identify the origin of the tumour and distinguish it from a meningioma. Meningiomas strongly enhance uniformly on CT reflecting the vascularity of these tumours, but necrosis, cysts, and calcification can alter the signal characteristics on MRI making differentiation from dermoids or even haemangioblastomas rather difficult. Cholesteatomas and epidermoids can commonly be differentiated from meningiomas and acoustic neuromas by their relative lack of enhancement.

If the lower cranial nerves are involved it is imperative to get good imaging of the base of the skull, neural exit foramena and extracranial soft tissues in the neck.

27.4.3 Intracerebral lesions

Headache, memory or personality changes, and seizures are the most common initial symptoms in patients with primary intracerebral tumours. However, patients are commonly referred to hospital only when focal symptoms or signs become obvious, such as seizures, hemiparesis papilloedema, dysphasia, or hemianopia. Hemiparesis or hemisensory loss are the most common symptoms at hospital presentation (Table 27.5). Nearly all patients who have weakness or numbness complain of these symptoms, thus directing the clinician to the abnormality on examination. Only 7 per cent of patients with malignant glioma complain of visual symptoms, yet over 20 per cent have signs of visual field loss and 23 per cent have papilloedema; therefore, careful examination of the visual fields and fundi is important in anyone complaining of headaches or symptoms suggestive of disturbance of higher mental function. Most commonly, the upper motor neurone weakness is mild initially and affects fine manipulation first and mild progressive lower limb weakness, involving hip flexion, knee flexion, and ankle dorsiflexion (Fig. 27.4). Clinical follow up using quick sensitive simple tests such as the timed nine hole peg test, timed 10 m walk, and a test of memory and grading of dysphasia are usually

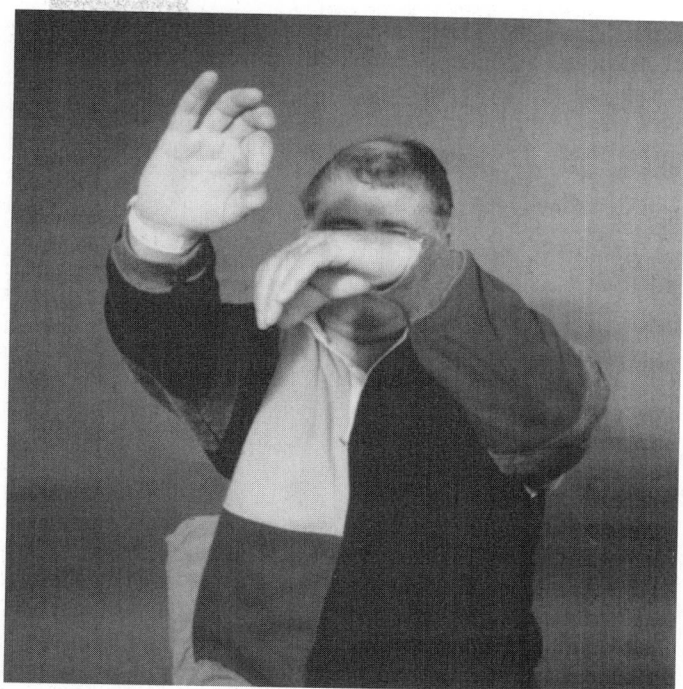

Fig. 27.4 Patient with a right frontal astrocytoma, treated by radiotherapy, with mild to moderate distal weakness of his left hand and leg which lead to problems with manipulation and speed of walking. Response to treatment was followed clinically using the timed nine hole peg test and the timed 10 m walk.

sufficient to assess clinical response to treatment (Grant *et al.* 1994; Clyde *et al.* 1998). The Barthel Activities of Daily Living Index (Section 6.5.2) may be useful in elderly patients or in patients with metastases where the weakness is severe, but is insensitive and the 'ceiling effect' precludes its use in trials of glioma in general and does not record cognitive disability or dysphasia. The Karnofsky Performance scale is useful for grading patients for entry into studies, but in practice is usually used a 3 point grading scale (≥70, 60–50, <50) rather than an 11 point scale (100, 90, 80, …, 10, 0). It can be used to follow individual patients although intra-observer and inter-observer errors limit its usefulness.

Stroke-like onset or collapse with coma, occurs in about 5 per cent of patients with intracerebral tumours and is most commonly related to haemorrhage into a malignant glioma or metastasis. Stroke-like presentations and subacute presentations with cognitive deficits,

Table 27.5 Common examination abnormalities found at first hospital presentation in patients with intracerebral tumours

	High grade glioma (%)	Low grade glioma (%)	Other primary (%)	Metastasis (%)	All tumours (%)
Nil	13.2	44.7	5.6	12.4	14.4
Hemiparesis/ hemisensory	53.8	23.4	16.7	45.4	42.7
Cognitive/ personality	32.1	17.0	25.0	26.9	26.9
Papilloedema	23.6	14.9	22.2	9.2	14.4
Dysphasia	20.7	8.5	11.1	13.2	14.4
Hemianopia	18.9	6.4	11.1	6.0	9.6

visual field disorders, or dysphasia are more common in the elderly and most commonly suggest a poor prognosis.

Late onset epilepsy, defined as first seizure after age 18, is a common presentation in patients who have low grade gliomas and meningiomas. It has been estimated that between 3 per cent and 10 per cent of patients with late onset epilepsy have an underlying tumour of some form. Seizures are the first presenting symptom in 54 per cent of low grade gliomas, 50 per cent of anaplastic astrocytomas, 26 per cent of meningiomas, 19 per cent of glioblastomas, 15 per cent of metastases and 11 per cent of primary central nervous system lymphomas. Over a follow-up period of 3 years, the prevalence of seizures rises to 70 per cent in low grade glioma, 56 per cent in anaplastic astrocytoma, 48 per cent in glioblastoma, 44 per cent in meningioma, 39 per cent in primary central nervous system lymphoma, and 31 per cent in metastases. Tumour associated epilepsy is focal in approximately 50 per cent of patients, partial epilepsy with secondary generalization in 25 per cent, and tonic–clonic seizures without warning in 25 per cent of patients.

Children are more likely to have posterior fossa or deep thalamic region tumours and present with cerebellar symptoms; symptoms of raised intracranial pressure are more frequent.

Early identification of patients with intra-cerebral tumours is important but very difficult. Various attempts to draw up referral guidelines for patients with intracerebral tumours have been devised (http://www.nice.org.uk/page.aspx?o=cg027niceguideline), which include a timescale by which such patients should be seen and scanned.

CT and MR brain scanning has dramatically improved the management of patients with brain tumours, but diagnostic interpretation is not without its difficulties. The three levels of diagnosis are: is it a tumour; if so, what type of tumour is it; and if it is a glioma, what grade of glioma is it?

- **Is it a tumour?** Neuroradiologists will correctly predict an intracerebral tumour on a CT scan in about 90 to 95 per cent of cases. However, approximately 10 per cent of patients will have had a previous CT or MRI scan that has been reported as either normal or an alternative pathology, such as stroke (Okamoto *et al.* 2004a and b). In these cases MRI will usually demonstrate an abnormality more clearly, but the aetiology of the lesion may not be clear. Even in the best centres 5 to 10 per cent of CT scans reported by a radiologist as having an intracerebral tumour will later be found to have non-malignant pathologies. The differential diagnosis of a non-contrast enhancing lesion, with standard doses of contrast, includes demyelination, encephalitis, infarct, post-traumatic and non-specific changes. The differential diagnosis in patients with contrast enhancing lesions includes demyelination, arterio-venous malformation, haemorrhagic stroke, and cerebral abscess. In some patients who present with a stroke-like onset, it may not be evident that the haemorrhage has occurred into an existing mass lesion. The common tumours to present with intra-tumoural haemorrhage are glioblastoma, metastatic lung cancer, melanoma, and choriocarcinoma. The 'open ring' imaging sign after contrast MR imaging is considered to be a relatively specific sign of a demyelinating lesion that mimics a brain tumour (Masdeu *et al.* 2000).

- **What type of tumour is it?** Errors in reporting of CT or MRI are even more common when attempts are made to predict the type of malignancy. The main areas of difficulty are where tumours have an exophytic extension with involvement of the meninges, intense contrast enhancement or sometimes calcification of meningeal/vascular origin, as for meningioma/haemangiopericytoma, or of glial origin such as glioblastoma or oligodendroglioma. In these cases it may be very difficult to say whether the tumour is extracerebral and invading the brain or intrinsic and becoming exophytic. It has been estimated that 5 per cent of brain images reported as single or multiple metastases by experienced neuroradiologists will in fact turn out to be primary brain tumours, either glioma or primary central nervous system lymphoma. In one study of single brain metastasis, 11 per cent of patients with known systemic malignancy with a solitary brain lesion thought on CT imaging to be a metastasis turned out to have a different pathology; in some cases the pathology was not a tumour at all (Patchell *et al.* 1990). MRI scanning may be more discriminatory. Primary central nervous system lymphoma can be unifocal in 60 per cent or multifocal in 40 per cent. Cells are densely packed and generally homogenously enhance and thus are commonly mistaken for metastases.

- **If it is a glioma, what grade of glioma is it?** There are some characteristics on imaging that are more common in a particular histology, but no single characteristic is specific (Table 27.6). Astrocytomas and oligodendrogliomas are commonly homogeneous and may be cystic or show areas of calcification and usually do not enhance. By contrast anaplastic astrocytoma and glioblastoma multiforme are generally heterogeneous with cysts or necrosis, commonly demonstrate shift of midline structures with significant oedema and show contrast enhancement. Algorithms have been suggested based on contrast enhancement, space occupation, cyst formation, necrosis, and oedema to help predict the grade of malignancy; however these only predict about 60 per cent of cases correctly. In a recent study, at a time when CT and MRI were readily available, 45 per cent of patients who were suspected of having an astrocytoma had an anaplastic astrocytoma and 5 per cent had a non-malignant histology following biopsy (Kondziolka *et al.* 1993). Pilocytic astrocytomas, subependymal giant cell astrocytoma, myxopapillary ependymoma, and desmoplastic neuro-epithelial tumours can show contrast enhancement and may be misdiagnosed as malignant gliomas or metastases.

Histopathological correlation has proved that enhancing areas on post-contrast CT and MRI scans correspond to densely cellular, hypervascular tissue of viable tumour (Fig. 27.5) (Burger *et al.* 1983; Whelan *et al.* 1988). A consistent finding from stereotactic biopsy studies is of a variable zone of microscopic tumour infiltration outside the enhancing area, extending at least as far as the abnormal signal on the T2-weighted images. There may be even greater extension of isolated tumour cells beyond these radiologically defined boundaries (Burger *et al.* 1988; Kelly *et al.* 1987).

The small uniform cells of a primary central nervous system lymphoma are densely packed together. These tumours are commonly more radio-dense than surrounding brain on non-contrast enhanced CT scan, giving the appearance that some contrast has been given (Fig. 27.6). When contrast is given they can enhance intensely. Although characteristic, these features are not specific for primary central nervous system lymphoma and can be seen in other tumours comprising of tightly packed cells, as in some metastases, some small cell malignant gliomas, and medulloblastomas.

Table 27.6 Imaging characteristics and pointers towards diagnosis of common intrinsic tumours of the brain

	Pilocytic astrocytoma	Astrocytoma	Oligodendroglioma	Medulloblastoma	Malignant glioma	Metastasis	Primary CNS lymphoma	Germ cell tumour
Peak age	10–30 yr	20–40 yr	20–50 yr	1–20 yr	40–70 yr	50–80 yr	50–70 yr	10–30 yr
Site	Usually cerebral or midline cerebellar	Adult cerebral; child cerebellar	Frontotemporal	Cerebellum, IV ventricle	Cerebral	Anywhere	Periventricular, anywhere	Pineal, suprasellar
Single/multiple	Single	Single	Single	Single/CSF	95% single	33% single	60% single CSF>20% Vitreous>20%	90% single CSF>20%
Usual characteristics								
Borders	Well demarcated	Diffuse/infiltrating	Well defined	Well defined	Serpiginous	Well defined	Indistinct	Distinct
Cysts	Common	Uncommon	Uncommon	Uncommon	Occ. necrotic	Occasionally	Nil	Occasionally
Calcification	10–40%	5–10%	50%	Nil	<5%	<3%	Nil	10–15%
Peritumoral oedema	Nil or mild	Mild	Mild	Moderate	Moderate	Moderate/severe	Mild/severe	Nil/mild
Mass effect	Nil or mild	Mild	Mild/moderate	Moderate	Moderate	Moderate/severe	Mild/severe	Nil/mild
Investigations								
CT	Low density Enhancing nodule	Low density No enhancement	Low density No enhancement	Occ. high density Enhancement	Low density Border enhances	Low density Uniform/border	Occ. high density Uniform/border	Low density Uniform
T₁-MRI	Iso/hyperintense	Hypointense	Hypointense	Hypointense	Hypointense	Hypointense	Hyper/isointense	Iso/hypointense
T₁ Gad-MRI	Enhancing nodule	No enhancement	No enhancement	Enhancement	Border enhances	Uniform/border	Uniform/border	Uniform
T₂-MRI	Hyperintense	Hyperintense	Hyperintense	Hyperintense	Hyperintense	Hyperintense	Hyperintense	Hyperintense
Other helpful tests	–	–	–	MRI spine CSF	–	Chest/abdo. CT Tumour markers	Slit-lamp (eyes) CSF	CSF Tumour markers
Differential diagnosis	Abscess Malig. glioma Ganglioglioma Craniopharyngioma Germinoma	Multiple sclerosis Infarct Oligodendroglioma Malig. glioma	Meningioma AVM Astrocytoma Craniopharyngioma Malig. glioma	Astrocytoma Malig. glioma Lymphoma Germinoma	Abscess Stroke Metastasis Oligodendroglioma Lymphoma	Stroke Abscess Malig. glioma Lymphoma	Multiple sclerosis Sarcoid Metastasis Malig. glioma Toxoplasmosis	Meningioma PNET Metastasis Malig. glioma Metastasis

abdo., Abdominal; AVM, arterio-venous malformation; CNS, central nervous system; CSF, cerebrospinal fluid; Gad, gadolinium; Malig, malignant; Occ., occasional(ly); PNET, primitive neuro-ectodermal tumour.

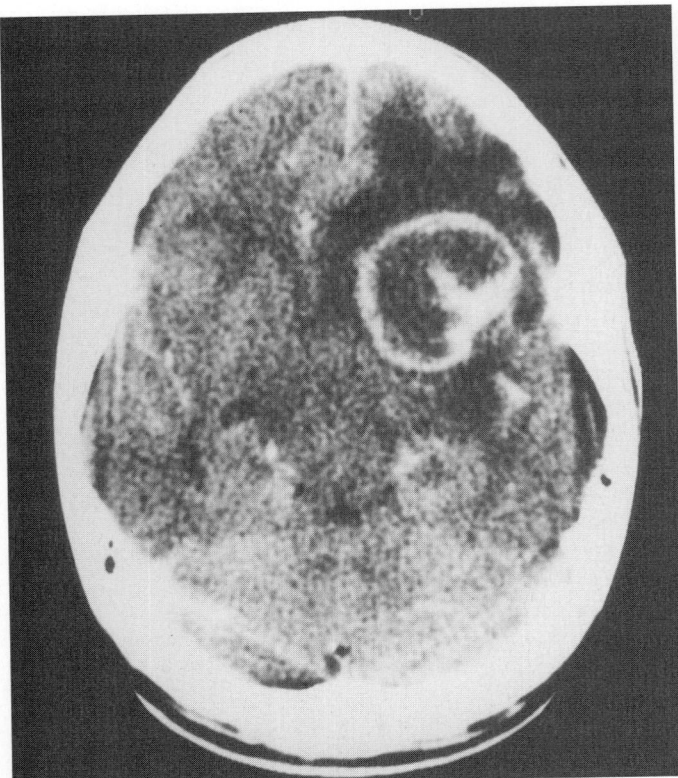

Fig. 27.5 Glioblastoma multiforme. Axial contrast enhanced CT scan. Differential diagnosis would include metastasis or even abscess, such as Toxoplasmosis, depending on the clinical story. Histological confirmation by stereotactic biopsy or resection is usually advisable. Freehand biopsy has a high risk-to-benefit ratio and should be avoided.

Frequently in AIDS patients, primary central nervous system lymphomas can demonstrate ring enhancement similar to an abscess. Primary central nervous system lymphomas are commonly located in a periventricular distribution deep in the white matter and approximately two-thirds occur in the cerebral hemispheres and one-third infratentorially.

The pineal gland lies between the splenium of the corpus callosum above and the superior colliculus below. Numerous tumour types can arise in the pineal region: germ cell tumours, pineocytomas, gliomas, metastases, cysts. Patients commonly present with Parinaud's syndrome with limitation of upgaze, convergence, and impaired pupillary reaction to light and accommodation. Eyelid retraction, Collier's sign, or ptosis can also occur. Some patients will have diabetes insipidus and if there is a beta-human chorionic gonadotrophin, beta HCG, secreting germinoma precocious puberty can occur. CT or MRI scan with gadolinium will demonstrate the lesion which is generally causing an obstructive hydrocephalus. The most common tumours in this area are teratomas or germinomas. Pineocytomas are less common and arise from the pineal parenchymal cells. Gliomas can arise in the pineal region and metastases to the pineal region also occur. In addition to these solid tumours, pineal region cysts can occur and can be simple, filled with CSF, or can be epidermoid or dermoid cysts. Germinomas are usually solid, enhance uniformly, and may be surrounded by calcium. Teratomas also enhance but are heterogenous with multiloculated cysts. Non-germinomatous germ cell tumours

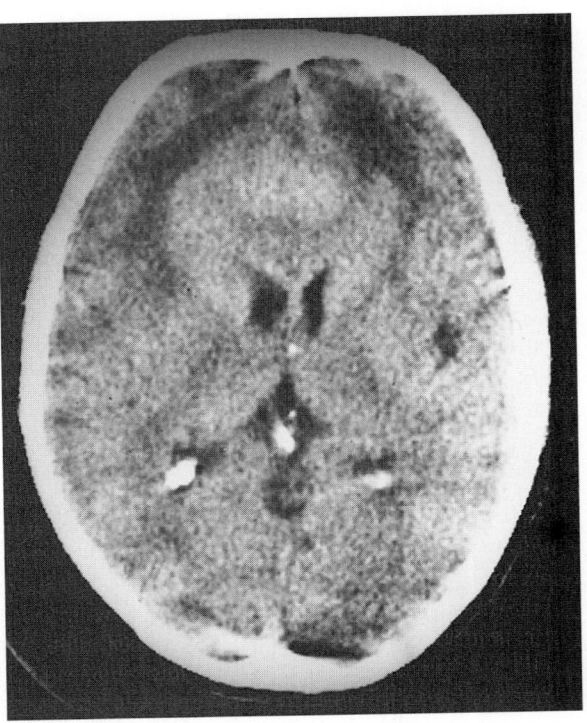

A

B

Fig. 27.6 Primary central nervous system lymphoma. A. Axial non-contrast CT scan demonstrating periventricular and corpus callosal tumour with pseudo-enhancement due to densely packed small cells. Pseudo-enhancement is occasionally found with other small cell tumours, such as medulloblastoma and small cell lung cancer. B. Profound uniform contrast enhancement.

are heterogenous and may enhance irregularly or not at all. Choriocarcinoma bleeds frequently and intratumoural haemorrhage or bleeding into a cyst may occur. Pineocytomas have low signal on T1-weighted MRI, high signal on T2-weighted images and enhance uniformly, but there can be calcification within the tumour, or haemorrhage if there are any pineoblastoma elements. Pineal gliomas can vary on imaging depending on whether they are low or high grade tumours. Tumours at this site quite commonly seed to the spinal canal, therefore full spinal MR imaging is also important. Serum and CSF, obtained in the absence of hydrocephalus, analysis for germ cell markers can be valuable. Malignant teratoma, germinoma with syncytioblastic cells, embryonal carcinoma and, endodermal sinus tumours may have elevated levels of alphafetoprotein. Choriocarcinoma and embryonal carcinoma may have elevated levels of beta HCG, whereas only germinomas and germinomas with syncytiotrophoblastic cells have elevated levels of placental alkaline phosphatase. Non germ cell tumours are negative for all these markers. CSF histology is usually negative. In the presence of these hormones, the most likely diagnosis is a teratoma or choriocarcinoma. If the CSF is negative for these markers, then biopsy of the pineal tumour is recommended. Shunting or fenestration of the ventricles may be necessary prior to definitive operation or at the time of operation. Transventricular biopsies run the risk of bleeding from the veins surrounding the anterior pineal region. Resection of the tumour from behind requires retraction of the occipital lobe and is frequently associated with temporary or permanent visual field defects.

27.5 Importance of histological confirmation

It is wise not to be overconfident about the suspected type or grade of a tumour when discussing the situation with the patient and relative prior to biopsy or resection, for all the reasons mentioned above regarding diagnostic certainty. The main classification scheme used to type and grade tumours is the World Health Organization classification (Table 27.1). Management strategies and prognosis are considerably changed depending on the tumour type and tumour grade. For example the diagnostic workup and management of certain types of tumours, such as primary central nervous system lymphoma, medulloblastoma, germinoma are completely different from that for others such as glioma, meningioma, or metastasis. Some tumour types are very sensitive to radiation, for instance germinoma, lymphoma, whilst others, such as glioma, teratoma, are strongly resistant to radiation. Some tumours are chemosensitive, for instance lymphoma, germinoma, and anaplastic oligodendroglioma, while others are relatively chemoresistant, such as glioblastoma or meningioma.

The main histological characteristics used to grade primary malignancy are: increased cellularity, increased number, and atypical appearance of mitotic activity, pleomorphism or anaplasia, proliferation of the vascular endothelium, and tissue necrosis. Grading systems such as Dumas–Duport, have been validated by correlation between the level of anaplasia and survival after diagnosis (Kim *et al.* 1991).

Previously, a key diagnostic criterion for glioblastoma multiforme was the presence or absence of necrosis, but the revised World Health Organization system, WHO-2, and the Dumas–Duport systems do not require necrosis for classification as glioblastoma multiforme (Kleihues *et al.* 1993b; Dumas–Duport

et al. 1988). Necrosis did predict a shorter survival, however, the magnitude of the survival difference between patients with and without necrosis was small, with median survival 10.9 months versus 12.5 months respectively. This supports the WHO-2 classification of endothelial proliferation without necrosis as sufficient to make a diagnosis of glioblastoma multiforme (Barker *et al.* 1996). Most centres use the WHO system or occasionally the Dumas–Duport system.

The heterogeneous nature of some tumours, means that insufficient sampling of the tumour can lead to undergrading of tumours and result in inappropriate treatment planning. The use of image directed stereotactic techniques has led to a 95 per cent diagnostic success rate, even in small tumours in sites with difficult access, and allows several samples from representative areas of the tumour to be taken so that grading is more reliable (Bernstein and Parrent 1994). Interobserver concordance is also variable in these heterogeneous tumours: glioblastoma multiforme–89 per cent; anaplastic astrocytoma–77 per cent; oligoastrocytoma–48 per cent; oligodendroglioma–39 per cent (Aldape *et al.* 2000). Genotyping of these tumours will better define their nature.

27.6 Head, neck, and skull tumours

Tumours of the head and neck can be divided based on their position in relation to the skull and, if biopsied, by their cell type. Extra-cranial tumours of the head and neck are usually due to carcinoma of the paranasal sinuses, nasopharynx, oral cavity, or oropharynx or are a result of tumours arising from local structures, such as blood vessels in the case of glomus jugulare, or bone in chordoma or chondroma. These tumours commonly produce cranial neuropathies.

27.6.1 Carcinoma

Head and neck cancer usually affects patients in the fifth and sixth decades. The most common primary tumours are carcinoma of:

- the nasal cavity and paranasal sinuses,
- nasopharynx,
- oral cavity,
- oropharynx.

Squamous cell carcinoma is the most common cell type. Risk factors are smoking and possibly alcohol. Up to 10 per cent of patients with head and neck tumours have second malignancies which also tend to be cancers associated with tobacco or alcohol, originating in lung, oesophagus, or stomach.

The nasal cavity and paranasal sinuses

Tumours of the nasal cavity and sinuses have an incidence of $<1/10^5$/year, affect men twice as often as women and usually occur in patients over 60 years. Higher incidences are found in Japan and South Africa. Occupational factors are important such as nickel or chromium exposure, radium and isopropyl alcohol. The frequency of nasal and paranasal cancers is higher in furniture, shoe, and textile industries (Roush 1979). Tumours in this region are commonly squamous cell carcinomas, but esthesioneuroblastomas and adenoid cystic tumours can also be aggressive and invade the cranial cavity. Tumours are frequently far advanced by the time of diagnosis, because the symptoms of nasal blockage or discharge are frequently ignored. Extension to the orbit may cause diplopia and proptosis and extension upwards may result in direct invasion of the cribriform plate and frontal lobes with anosmia and headache. Surgery usually involves a combined approach from a skull base team comprising neurosurgeon, otorhinolaryngologist, and faciomaxillary surgeon. Factors that make surgical resection difficult or sometimes impossible include the involvement of the base of skull, nasopharynx, or sphenoid sinus. In inoperable patients, local disease can be controlled in a substantial proportion of patients using radiation therapy.

Nasopharynx

Nasopharyngeal carcinoma is endemic in China and North Africa. Dietary factors such as nitrosomines in salt-cured food and viral factors, particularly Epstein–Barr virus, have been implicated in the malignant transformation into nasopharyngeal cancer. These tumours usually occur predominantly in 40 to 60 year olds and males are twice as likely to be affected. Nasopharyngeal carcinoma most commonly present as a lump in the neck, but can also cause nasal blockage and deafness from blockage of the eustachian tube or invade the skull base or cavernous sinus causing cranial nerve palsies. Surgical biopsy confirms the diagnosis and radiation therapy is the treatment of choice. At recurrence, surgery has a limited role. Chemotherapy can be helpful to treat widespread metastatic disease, which occurs in a high proportion of patients with this tumour. Cisplatinum, methotrexate, and epirubicin all have some palliative effect. Five year survival rates of 50 to 60 per cent can be achieved with combinations of these treatment modalities.

Oral cavity

Tumours of the oral cavity are typically squamous and a close link with smoking has been identified. Alcohol may also be implicated in the aetiology. Diagnosis is usually earlier than other head and neck tumours, and neural involvement is less common. These tumours can however cause the 'numb-chin syndrome' due to involvement of the mental nerve, a branch of the mandibular division of the trigeminal nerve (Section 20.1.6).

Oropharynx

Tumours of the oropharynx usually occur in patients older than 50 years and are 4-fold commoner in males. Smoking and alcohol are the most common risk factors. Presentation is commonly quite late and lymph node spread and neural involvement are frequently found. Bulbar palsy or isolated involvement of the lower cranial nerves are not uncommon. Surgery and radiation therapy are the mainstays of treatment and survival depends on the site, extent, and staging of the tumour. Laryngeal and hypopharyngeal carcinomas cause local pain or referred pain to the ear and lymphadenopathy in the neck. Dysarthria and swallowing difficulties and recurrent laryngeal nerve palsy can also occur.

In general, if patients with head and neck tumours have early disease, radical resection and post-operative radiotherapy can be curative, although the cosmetic and functional result of treatment may affect quality of life. In advanced disease, where cure is not possible, it may be best to treat the primary site by radiation and any loco-regional lymph nodes by surgical resection and radiotherapy. The 5 year survival for nasal and sinus carcinoma is approximately 60 per cent, but depends on the extent of primary disease

at presentation. Chemotherapy in the initial treatment regime can lead to significant tumour regression in 60 to 90 per cent of cases. A systematic review of concomitant radiotherapy and chemotherapy in patients with locally advanced head and neck cancer revealed an overall survival benefit (OR=0.62; 95 per cent CI: 0.52–0.74; $p < 0.00001$; RR=0.83, RD=11 per cent) although combined treatment produced more adverse events than radiation therapy alone (Browman *et al.* 2001). Meta-analyses of individual patient data from 31 randomized controlled trials of neoadjuvant chemotherapy for patients with locally invasive head and neck cancer and meta-analyses of adjuvant chemotherapy revealed no significant survival advantage over loco-regional treatment alone (Pignon *et al.* 2000).

27.6.2 Chordoma

Chordomas are very rare slow growing tumours that arise from remnants of the notocord and generally occur in the midline in the region of the clivus at the base of the skull. They are locally invasive and usually present late because of the non-descript nature of the chronic headache or neck pain. They may present with intermittent diplopia, facial numbness if the upper clivus is involved, or as a nasopharyngeal mass with multiple lower cranial neuropathy, affecting glossopharyngeal, vagus, accessory, and hypoglossal nerves if the lower clivus is involved. Most clivus chordomas produce destruction of the clivus and have extradural extension. CT scan best demonstrates bone destruction but MRI better demonstrates the tumour margins and soft tissue structures and blood vessels. MRI demonstrates low signal intensity on T1-weighted scans. Chordomas rarely calcify or enhance. MRI demonstrates low signal intensity on T1-weighted scans. Imaging alone cannot adequately differentiate between chordoma, chondroid chordoma, and chondrosarcoma; however, chondrosarcomas usually arise along the petro-occipital fissure and more commonly calcify. Occasionally chondrosarcomas can arise in the midline in the region of the clivus.

Because of the approximation of the tumour to sensitive neural and vascular tissues, complete surgical removal is not usually possible but surgery is required to make the diagnosis and to reduce bulk disease. Care should be taken to keep the dura intact when performing trans-sinus surgery because of the risk of post-operative meningitis; however, since 50 per cent of these tumours have already breached the dura by the time of diagnosis, careful post-operative packing with fat and muscle grafts is essential. Radical resection of anterior skull base tumours has improved (Lawton *et al.* 1995) in the hands of teams of surgeons specializing in skull base surgery, but the advice is to follow surgical removal with local radiation therapy. There are no randomized studies of the effect of radiation and conventional photon irradiation has shown no dose–response relationship (Tai *et al.* 1995). Despite this there have been advances in radiosurgical methods, using gamma knife, linear accelerator, and proton beam that have been purported to give better local control and spare surrounding tissues (Schulz-Ertner *et al.* 2002; Krishnan *et al.* 2005). Chemotherapy, even with multi-agent chemotherapy is only rarely effective (Scimeca *et al.* 1996). The natural history of chordoma is minimally affected by surgery and radiotherapy, although symptom control can be achieved adequately. Five year survival rates range from 10 to 58 per cent in different series, but most series suggest that <50 per cent of patients are alive at 5 years. Systemic metastases occur in between 10 and 30 per cent.

27.6.3 Glomus tumours

Glomus tumours arise from the glomus jugulare and can spread medially to involve the middle ear or skull base and therefore can present with hoarseness and dysphagia due to lower cranial nerve palsies (Section 20.7.3). In some cases these tumours can cause pulsatile tinnitus and deafness and the tumour can be seen behind the tympanic membrane. CT or MRI scan is usually sufficient to make the diagnosis, but angiography is also valuable to define the blood supply to the tumour and to embolise the tumour preoperatively. Surgery can be hazardous and usually involves a skull base team. Surgery requires mastoid and a sub-occipital craniectomy and can be complicated by meningitis or facial nerve palsy. The place of radiation therapy is debatable. The tumour is benign, but complete resection is often not possible. Local control can be achieved by radiation therapy alone in approximately 90 per cent of patients with inoperable glomus tumours (Springate and Weichselbaum 1990).

27.6.4 Fibrous dysplasia

Fibrous dysplasia is a benign fibrous process that can involve the skull vault or the base of the skull. The condition can present in isolation or as part of the McCune–Albright syndrome of fibrous dysplasia, cafe au lait spots, and endocrinopathy comprising precocious puberty, thyrotoxicosis, primary hyperparathyroidism, or hyerprolactinaemia. McCune–Albright syndrome is due to postzygotic somatic mutations in the gene encoding G alpha s proteins, *GNAS1* (Ringel *et al.* 1996). Pathologically, the bone is replaced by fibrous tissue composed of collagen and fibroblasts. It can be difficult to distinguish fibrous dysplasia from an ossifying fibroma. Headaches or cranial nerve involvement are the common presenting symptoms. Craniobasal fibrous dysplasia can produce progressive visual loss due to extradural optic nerve compression if it affects the bones of the orbit or conductive and sensorineural hearing loss if the disease affects the temporal bone. A canal cholesteatoma is found in 40 per cent of patients with temporal bone involvement by fibrous dysplasia. Surgical resection is usually the treatment of choice. Radiation therapy may also be beneficial in relieving symptoms. Malignant transformation to a fibrosarcoma is rare but can occur.

27.6.5 Skull osteomas

Cranial osteomas are common and not usually symptomatic. Occasionally because of their size or site they can cause cosmetic disfigurement, headache, cranial neuropathies, or seizures. Rarely they are associated with Gardner's syndrome. A new classification for cranial osteomas has been suggested (Haddad *et al.* 1997). The classification divides osteomas into parenchymal, dural, skull base, and skull vault, with the latter being divided into enostotic and exostotic variants. CT scan is the imaging technique of choice. Indications for surgery include progressive ophthalmoplegia, neurological signs, and significant cosmetic deformity.

27.6.6 Paget's disease

Cranial Paget's disease, osteitis deformans, can be monoostotic or polyostotic. Paget's disease involves the cranial vault and temporal bones in 30 to 40 per cent of cases. It is characterized by excessive and disorganized bone formation and resorption. Patients may present with headaches, hearing loss, tinnitus and vertigo, or hemifacial spasm. The bone alkaline phosphatase is almost always elevated and is a useful marker, but the best marker to follow cranial involvement with Paget's is the serum propeptide carboxyterminal of type 1 procollagen for new bone formation and the urinary C-terminal telopeptide of type 1 collagen for bone resorption (Alvarez *et al.* 1997). Calcitonin and etidronate and newer bisphosphonates, such as pamidronate, alendronate can help control the turnover of new bone and help control pain and prevent neurological or orthopaedic complications (Selby *et al.* 2002).

27.7 Intracranial extracerebral tumours

Intracranial extracerebral tumours arise from pituitary gland, a pituitary adenoma, meninges, a meningioma, nerves, a neuroma, or neuroepithelial remnants, a dermoid or epidermoid, and are usually benign. Metastases directly to the dural space do occur, especially with breast cancer and lymphoma, but frequently they are a late complication in patients with widespread systemic involvement or as part of a malignant meningitic process where they can produce mass lesions.

27.7.1 Pituitary tumours

Pituitary tumours can arise from any cell in the anterior pituitary gland, account for 10 per cent of intracranial tumours and have an incidence of 2 to 3/100 000/year. They are most commonly benign adenomas, although pituitary carcinomas and metastases to the pituitary do occur. Tumours can produce symptoms and signs through hypersecretion of hormones or related to the mass effect of the tumour on the optic nerves or structures adjacent to the pituitary gland. Tumours generally have to be >1cm before they cause any symptoms of neural or vascular compression. When the tumour expands the pituitary gland it can cause hypofunction of certain hormones and produce hypothyroidism, amenorrhoea, or Addison's disease. Involvement of the neurohypophysis can result rarely in diabetes insipidus. When the tumour extends laterally it can involve the cavernous sinus and result in cranial neuropathies. Haemorrhage into the pituitary gland can cause pituitary apoplexy.

Hormone secreting adenomas

The most common type of pituitary adenoma is the microadenoma. Microadenomas by definition are less than 10 mm in diameter, do not expand the pituitary fossa and usually secrete prolactin. About 70 per cent of prolactin secreting tumours are microadenomas.

- *Prolactin secreting microadenomas* are more common in women who present earlier because of secondary amenorrhoea, infertility, or galactorrhoea. Prolactinomas in men probably present at a later stage and are more commonly macroadenomas; headache, loss of libido, and visual failure or visual field defect are the usual symptoms. Prolactinomas commonly cause of hyperprolactinaemia with blood levels of >100 to 200 mg/l, although prolactinomas cannot be excluded at lower prolactin levels. Occasionally prolactin levels of >100 mg/l are found in idiopathic hyperprolactinaemia and this may lead to diagnostic uncertainty. Other causes of hyperprolactinaemia include drugs, hypothyroidism, and renal failure. To differentiate between a prolactinoma and hyperprolactinaemia from other causes, thyrotrophin response to a dopamine receptor antagonist may be used, since only prolactinomas may have an increased response. Non-functioning tumours have an incidence of 1/100 000/year and slowly expand causing upward displacement of the optic chiasm and a characteristic bitemporal hemianopia with headache and sometimes panhypopituitarism. Mortality associated with hypopituitarism due to non-functioning tumours or their treatment is increased 2 to 3-fold. Most clinically non-functioning adenomas express gonadotrophin hormone subunits *in vitro* or occasionally *in vivo*. By the time of diagnosis by clinical, radiological, and hormonal studies symptoms have usually been present for some years.

◆ *Cushing's disease* can occur and is associated with elevated adrenocorticotrophin hormone. The clinical features of central obesity, 'moon' facies, buffalo hump, abdominal striae, and hypertension, usually make the condition obvious. Some patients do not have the classical appearance but instead complain of depression and lethargy.

◆ *Thyrotrophin, TSH, secreting* pituitary adenomas are commonly macroadenomas and are usually associated with thyrotoxicosis. Occasionally, longstanding primary hypothyroidism can result in pituitary hyperplasia with elevated TSH levels and enlargement of the pituitary gland. TSH secreting pituitary adenomas will require surgical intervention.

◆ *Growth hormone secreting* adenomas are rare but present with giantism during puberty and acromegaly after fusion of the epiphyses.

Pituitary metastases

Pituitary metastases are infrequent and usually occur in the context of known systemic cancer with metastatic spread to other sites, but occasionally can be the only manifestation of metastatic spread. Most cases present with headache or diabetes insipidus. Panhypopituitarism and visual field impairment is demonstrated in about 25 per cent (Sioutos *et al.* 1996).

Radiological classification

Pituitary tumours can be radiologically classified based on their size and growth characteristics. Grade 0 is where there are no imaging abnormalities, grade I is where there are minor changes in the pituitary but the tumour is <1 cm, grade II is where there is diffuse enlargement but no focal sellar destruction, grade III is where there is local invasion of the sella and grade IV is where there is extensive destruction of the sella. Further subclassification will depend on the extent and direction of supra or parasellar extension. Coronal and sagittal MRI with gadolinium contrast gives high definition information about the pituitary, parapituitary region, and adjacent soft tissues; however it is not as good as CT at demonstrating bone erosion. High resolution contrast enhanced coronal CT scan with 1.5 mm contiguous slices will also provide information about the homogeneity of the pituitary gland and its relationship to surrounding structures. CT of a microadenoma will characteristically show low density within the gland; however, small cysts in the pituitary occur in normal people and therefore the diagnosis of microprolactinoma should not be made on radiological grounds alone. Characterization of the tumour by MRI plus or minus MR angiography is usually sufficient to plan surgery (Fig. 27.1). Formal arterial angiography is not usually required.

Management Recommendations for service provision and guidelines for management have been developed (Royal College of Physicians 1997).

Dopamine agonists There is general agreement that bromocriptine is the treatment of choice in prolactin secreting adenomas. Bromocriptine is a dopamine agonist that directly stimulates specific pituitary cell membrane dopamine D2 receptors and inhibits prolactin synthesis and secretion. Treatment will usually cause a reduction in size of any macroadenoma and reduce or normalize prolactin levels in blood. There have been reports of macroadenomas enlarging despite treatment with bromocriptine and close review of visual acuity and fields is recommended. If patients are unable to tolerate bromocriptine because of side effects, other dopamine agonists can be tried, including pergolide, cabergoline, or lisuride.

Pregnancy During pregnancy there can be enlargement of the normal pituitary gland by up to 70 per cent and therefore special care has to be taken during pregnancy in patients with pituitary tumours, especially macroadenomas. During pregnancy bromocriptine can be discontinued if the tumour is a microadenoma. Ideally, it is advisable to wait till any macroadenoma has been adequately treated before a planned pregnancy is attempted. Bromocriptine should be restarted after pregnancy or during pregnancy if there is neurological deterioration.

Surgery Prolactin secreting macroadenomas can also be treated by surgical resection, if there is serious visual compromise or side effects with medical treatment. Visual improvement occurs in 80 per cent of cases but prolactin levels frequently do not reduce to normal. The recurrence rate of macroadenomas after surgery ranges from 25 per cent to 75 per cent, therefore radiotherapy after surgical resection may be necessary if drug therapy is not possible. Neurosurgery or radiotherapy is sometimes recommended before pregnancy particularly for macroadenomas that have not responded to bromocriptine.

Non-functioning adenomas are usually macroadenomas. Drugs are ineffective at reducing the tumour size. Surgery is the treatment of choice and most adenomas can be dealt with via a transphenoidal approach. Surgery will confirm the diagnosis and relieve compression by the tumour on surrounding structures. About 75 per cent of patients with visual field defects will have some recovery of visual fields post-operatively and headache is relieved in over 90 per cent (Ebersold *et al.* 1986; Comtois *et al.* 1991). Post-operative complications occur in <10 per cent of cases and include diabetes inspidus, CSF leak, transient hyponatraemia, meningitis, and sinusitis. Diabetes insipidus in the immediate post-operative period occurs in 10 to 20 per cent of patients but is permanent in only 2 to 5 per cent. It is uncommon to achieve complete resection in macroadenomas because most have invaded the dura or surrounding structures. Post-operative imaging can be performed at 2 to 3 months and close follow up by MRI will usually identify tumour regrowth and determine which patients may benefit from radiotherapy. Some centres advocate radiation therapy to any residual tumour, but the frequency of radiation induced complications should be considered especially in younger patients.

Pituitary Cushing's syndrome can be difficult to diagnose. High dose dexamethasone suppression test, using 2 mg dexamethasone 6 hourly for 48 hours, with suppression of the morning serum cortisol to <50 per cent basal value, points to pituitary-dependent disease. Some centres recommend a single high dose overnight test of 8 mg or an intravenous test of 1 mg/hour for 7 hours. If there is good endocrine evidence of a pituitary ACTH secreting adenoma, transphenoidal surgery will result in cure in over 80 per cent and a low recurrence rate of 5 per cent. Plasma ACTH will be elevated, whereas it would be unmeasurable in adrenal disease. Hypokalaemia favours ectopic ACTH secretion. On the corticotrophin-releasing hormone test with bilateral inferior petrosal sampling an exaggerated serum cortisol response over basal suggests pituitary-dependent disease. Pituitary and adrenal MRI and chest imaging may identify an ectopic source. These tumours are frequently microadenomas and transphenoidal surgery is the treatment of choice. 15 to 40 per cent of patients with ACTH-secreting macroadenomas have a poor response to surgery and post-operative radiotherapy, and adrenal blocking drugs or bilateral adrenalectomy are commonly required. Patients who have received pituitary radiation therapy are at long-term risk of hypopituitarism and require continuing endocrine assessment (Brada *et al*. 1993).

Growth hormone secreting adenomas can be treated with somatostatin analogues or bromocriptine where surgery is contraindicated or unsuccessful (Barkan *et al*. 1988). Bromocriptine can also lower growth hormone levels in up to 75 per cent of cases, but normal growth hormone levels are only achieved in 10 to 20 per cent, treatment must be lifelong, and there may not be any shrinkage of macroadenomas. Surgery or radiotherapy is usually required. The growth hormone response to transphenoidal resection is not as good as the endocrine response to surgery for ACTH-secreting tumours. The aim is to achieve basal GH levels of <5 mu/l, suppressing to <2 mu/l after glucose loading. This will not be achieved in almost 50 per cent after trans-sphenoidal surgery. Post-operative radiotherapy is frequently required. Radiotherapy is moderately effective and may be used as the primary treatment where surgery is contraindicated or as an adjunct to surgery. It may take many years to achieve normal growth hormone levels and usually a combination of surgery and radiotherapy or radiotherapy and drugs are required (Bloom *et al*. 1984).

27.7.2 Meningeal tumours

Meningiomas are benign slow growing intracranial extracerebral tumours that account for 15 to 20 per cent of all intracranial tumours. Only 25 per cent of meningiomas are symptomatic at presentation. The frequency of meningiomas increases with age (Radhakrishnan *et al*. 1995). Women are more than twice as commonly affected. Risk factors include gender, previous ionising radiation, and neurofibromatosis type 2. The World Health Organization histological classification and grading system selects characteristics that predict an aggressive behaviour of the tumour and the risk of early recurrence (Kleihues *et al*. 1993b). There are many histological sub-types of meningioma and these tumours can be graded as benign meningiomas, atpical meningiomas or malignant meningiomas depending on the degree of anaplasia, cellular atypia, mitoses, and necrosis. Proliferation indices identified by immunohistochemical methods on pathological specimens may also predict an aggressive nature. Meningiomas express progesterone and oestrogen receptors and receptors for platelet derived growth factor (Black *et al*. 1996).

Eighty five per cent of meningiomas are supratentorial. The most common sites are over the convexities of the skull, the falx, or tentorium followed by the sphenoid ridge, suprasellar areas, and olfactory groove. CT and MRI will demonstrate a well-demarcated enhancing lesion with a dural base that may involve or displace adjacent nerves, or produce significant oedema in adjacent brain (Fig. 27.7). Angiography is not usually necessary; however, it may be helpful to identify the feeding vessels and allow immediate pre-operative embolization to reduce the potential for severe haemorrhage from the tumour during operation. If the pathologist is not informed that pre-operative embolization has been performed an incorrect diagnosis of malignant meningioma may be made because of the necrosis in the specimen. In symptomatic meningiomas with brain oedema, dexamethasone 2 to 4 mg t.i.d. will usually produce speedy relief of symptoms. Although most meningiomas are benign tumours, operation is not always straightforward. The extent of surgical excision is graded by Simpson's grading of extent of surgical resection (Simpson 1957). Asymptomatic meningiomas, especially in the elderly, are best left alone. Large symptomatic meningiomas will usually require surgery, but pre-operative embolization may reduce the vascularity of the tumour and make surgery easier. Surgical mortality can be as high as 14 per cent and the 10 year survival can range from 43 to 77 per cent. Convexity, parasagittal, lateral sphenoid, and olfactory groove meningiomas can usually be resected completely with low morbidity. Suprasellar, cavernous sinus, clivus, tentorial, and posterior fossa meningiomas are more difficult, although improved surgical techniques have resulted in more radical resection. Morbidity is much higher in these sites and there can be a high recurrence rate. It has been estimated that the 10 year risk of recurrence is 9 to 20 per cent where the surgeon feels there has been a complete resection and 18 to 50 per cent recurrence where subtotal resection has been performed. Meningiomas at the base of the skull involving the sphenoid ridge may require a joint surgical approach by both facio-maxillary and neurosurgeons.

In symptomatic meningiomas in the elderly or at sites that increase operative risk, stereotactic radiation therapy or radiosurgery, as sole treatment, may reduce the size of the tumour or slow the

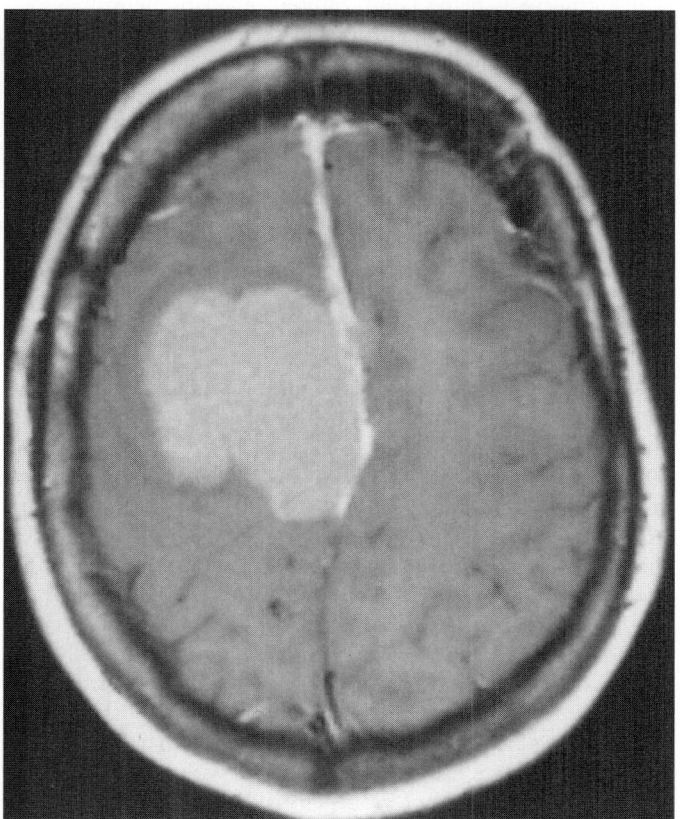

Fig. 27.7 Meningioma. Axial gadolinium enhanced MR scan.

growth rate. Radiosurgery is usually only considered for relatively small tumours of less than 3 cm in diameter that do not impinge on structures such as the pituitary gland or optic nerves, abut the ventricles, where operation would be hazardous or following subtotal resection. In these selected situations, 96 per cent of patients with benign meningiomas, 76 per cent with atypical meningiomas and 19 per cent with malignant meningiomas will have disease control at 5 years (Friedman *et al.* 2005). High dose fractionation regimes can be associated with 50 to 60 per cent short-term complications and 10 to 20 per cent permanent complications.

There is no role for hormonal therapy or chemotherapy in the management of benign meningioma. In recurrent, atypical or malignant meningioma, where repeated surgery is hazardous and maximal radiotherapy has been previously given, drug therapies have been tried. Antiprogesterones have been used with some apparent success in some patients with meningioma (Black 1993). Anti-oestrogens such as tamoxifen 40 mg/m^2 twice daily for 4 days followed by 10 mg twice daily, may produce a reduction in size of the tumour in 15 per cent of patients (Goodwin *et al.* 1995). Hydroxyurea has also been tried with limited success. In general, chemotherapies and not very effective and the balance of risk: benefit has not been proven.

27.7.3 Neural tumours

Cerebellopontine angle tumours

Acoustic neuromas, schwannomas, account for approximately 80 per cent of extra-axial lesions in the region of the cerebellopontine angle and 4 to 10 per cent of intracranial tumours overall (Mahaley *et al.* 1990; Grant *et al.* 1996). Other tumours in the region of the cerebellopontine angle include meningiomas in 10 per cent, primary cholesteatoma in 5 to 10 per cent, glomus jugulare tumours in 1 per cent, facial or trigeminal neuroma in 1 to 2 per cent, and metastasis in 1 to 5 per cent. The incidence of acoustic neuroma is approximately 1/100 000/yr. The acoustic nerve is the most common site for a neuroma in 85 per cent, although they can also arise from the trigeminal nerve in 1 to 8 per cent, facial nerve in 0.5 to 1 per cent and spinal roots in 10 to 15 per cent. Schwannomas arise from the junction between the peripheral schwann cell nerve sheath and the central glial nerve sheath. Ninety five per cent of acoustic neuromas are sporadic and 5 per cent are dominantly inherited as part of neurofibromatosis type 1 or type 2. Karyotype analysis in sporadic schwannoma may be normal. The most common abnormality is monosomy of chromosome 22, and there may be deletions in the long arm of chromosome 22q. The *NF2* gene has been isolated on chromosome 22q to a 6-Mb region of the q12 band of the long arm of chromosome 22 and it is highly likely that the relevant area on this chromosome involves a tumour suppressor gene. Mutations to the *NF2* gene are also likely to be an important step in the pathogenesis of sporadic unilateral acoustic neuroma and have been found in 40–70 per cent of cases.

Acoustic neuroma

Up to 85 to 90 per cent of acoustic neuromas arise from the vestibular branch of the nerve. Acoustic neuromas present with unilateral slowly progressive hearing loss in 95 per cent, sometimes associated with non-specific unsteadiness in 77 per cent, tinnitus in 71 per cent or vertigo. Commonly the sensorineural deafness occurring in 90 per cent is associated with facial sensory loss in 50 per cent or facial weakness in 10 per cent. Hydrocephalus due to obstruction of CSF pathways can lead to raised intracranial pressure. MRI is the scanning procedure of choice.

The tumours are very slow growing. Almost half of the tumours do not grow perceptibly over a 5 year follow up. Of those that do enlarge, 75 to 80 per cent of cases the growth rate is only 1 to 2 mm/year (Bederson *et al.* 1991). There may therefore be a case for conservative management with careful follow-up rather than intervention, especially in the elderly or those in poor general health, where the tumour is small or if there is contralateral deafness and retained hearing in the affected ear. Growth rate apparently does not correlate with the age of the patient, the size of the tumour or the duration of symptoms (Bederson *et al.* 1991).

Surgery

Surgical management has a high morbidity, particularly in patients with retained hearing. A suboccipital approach has the advantage of possibly retaining any existing hearing, but may require cerebellar traction and can cause post-operative headaches and cerebellar symptoms. The translabyrinthine approach has the advantage of requiring little in the way of cerebellar retraction, and because the surgery is largely extradural, the complications of meningitis and headache are less; however hearing is always lost post-operatively. The middle fossa approach is useful for small tumours and may spare hearing; however there are increased complication rates from facial nerve paresis and possibly sequelae such as seizures or dysphasia as a result of temporal lobe traction. Monitoring brainstem auditory evoked potential intra-operatively can significantly decrease the morbidity of surgery, especially when trying to preserve hearing. Prolongation of the latency of wave V of the brainstem auditory evoked potential is usually an early sign that the acoustic nerve is being compromised. Acoustic neuromas may require a joint surgical approach by both ear, nose, and throat surgeons and neurosurgeons. Surgery should aim to remove the tumour completely and to preserve facial nerve function and where possible preserve hearing. Neurosurgery for acoustic neuroma has a post-operative mortality of approximately 5 per cent. Mortality is related to the size of the tumour and age of patient (Hardy *et al.* 1989). These percentages are heavily influenced by selection of patients, experience of the surgeon, and possibly the surgical approach. Mortality using the translabyrinthine approach is probably not significantly different from the suboccipital or middle fossa approaches when patient characteristics are taken into account. There is general consensus that small tumours are better approached by a translabyrinthine approach whereas large tumours are best approached suboccipitally or by middle fossa approach. Translabyrinthine surgery results in complete hearing loss but this is not a problem if hearing is already lost pre-operatively and there may be slightly more chance of preserving facial nerve function. Anatomical preservation of the facial nerve can be achieved in about 80 to 90 per cent of cases, but anatomical preservation is not always associated with good facial nerve function. It is exceptionally rare for post-operative hearing to be better than pre-operative hearing using either the suboccipital or middle fossa approaches. Where hearing preservation is the main aim, for instance if there is already contralateral deafness and the affected ear has maintained hearing, a suboccipital approach has advantages. In selected centres it has been suggested that with experienced neurosurgeons and the aid of brainstem auditory evoked potentials, complete resection of the tumour can be accomplished with preservation of hearing in 50 per cent of patients with tumours smaller than 2 cm and in >80 per cent of patients who have a tumour of <1cm diameter (Post *et al.* 1995). However, a recent review shows that only a few patients have truly normal hearing after surgery (Sanna *et al.* 1995). Delayed deterioration in hearing years after successful operation is well recognized in up to 50 per cent of patients, although the cause remains uncertain (Shelton *et al.* 1990; Ogunrinde *et al.* 1994). Attempts at maintaining hearing by minimizing resection can be complicated by recurrence of the tumour.

Radiotherapy

In some cases where hearing is preserved and the tumour is small, stereotactic radiosurgery or stereotactic radiotherapy using a linear accelerator can be effective. Stereotactic radiosurgery uses a single fraction of high dose but small volume radiation to the tumour. Conformal beam stereotactic radiotherapy using a linear accelerator and fractionating the treatment over several days and reducing the dose of each fraction, has potential advantages in that radiation-induced neural side effects are reduced by reducing the fraction size. Radiation therapy is not usually advised for tumours of >3 cm because of the increased risks of central nervous system side effects. The aims of radiation therapy are to prevent growth of acoustic neuroma and retain neurological function. Tumours <3 cm in diameter show shrinkage or 'stabilization' in 97 per cent at 5 years after stereotactic radiosurgery (Flickinger *et al.* 2001). This apparent success has to be compared with the natural history of acoustic neuroma. One study of conservative management demonstrated that 71 per cent of acoustic neuromas do not enlarge over 3.4 years (Deen *et al.* 1996). Short-term complications from stereotactic radiosurgery or stereotactic radiotherapy using a linear accelerator are infrequent; however it will be some years before one can fully ascertain the effect of radiation on the acoustic nerve and surrounding structures, particularly in patients with normal pre-radiation hearing. Hearing is very likely to become impaired with time and only 50 per cent of patients with preserved hearing following radiation therapy will maintain this at 6 months and only 45 per cent at 1 to 2 years. Two years after stereotactic radiosurgery, preserved facial nerve function is achieved in 90 per cent and trigeminal nerve function in 75 per cent of cases who had no deficit immediately post-radiotherapy. In the long term, cases of radiation related cancer in the treatment field have been described. The results of conformal beam stereotactic radiotherapy are as good as stereotactic radiosurgery but longer term side effects appear to be less, probably reflecting the fractionation schedule and reduction in fraction size. Where surgery is contra-indicated because of poor health or poor risk: benefit ratio, radiosurgery or conformal stereotactic radiotherapy should be considered the treatment of choice, and there is adequate evidence that these treatments have a therapeutic role. The role of radiotherapy in the treatment of small acoustic neuromas remains controversial.

27.8 Primary intracerebral tumours

27.8.1 Gliomas

Cerebral gliomas are the most common primary intrinsic brain tumours. Gliomas are locally invasive and even after apparently successful macroscopic resection they recur at the same site in 95 per cent of cases. They spread outwith the central nervous system in less than 1 per cent, although it has been estimated that CSF spread occurs in up to 5 per cent.

For practical purposes gliomas can be divided into low grade gliomas, World Health Organization or Dumas–Duport grade 1 and 2, and high grade gliomas, World Health Organization or Dumas–Duport grade 3 and 4.

Low grade gliomas

Low grade gliomas account for approximately 20 per cent of all cerebral gliomas. Symptoms will depend on the site of the tumour. Prognosis depends on age at diagnosis, length of pre-operative symptoms, epilepsy, and extent of resection (Piepmeier *et al.* 1996; Salcman 1995). World Health Organization grade 1 gliomas include rare entities like pilocytic astrocytoma, pleomorphic xanthoastrocytoma, and subependymal giant cell astrocytoma and are potentially curable if they can be completely resected. Post-operative radiotherapy is not required. If there is suboptimal resection, these tumours are so slow growing that it is unlikely that early cranial radiotherapy has any advantage over a wait and watch policy, particularly since radiation induced side effects increase with the passage of time and there can be extended survival even in the group with subtotal resection. In symptomatic cases where the risks of resective surgery are considered too great, radiation therapy can produce long-term symptomatic benefit. Approximately 80 per cent of patients with pilocytic astrocytomas are alive at 15 years (Shaw *et al.* 1997; Shaw 1995).

World Health Organization grade 2 gliomas, fibrillary or protplasmic astrocytomas, oligoastrocytomas, and oligodendrogliomas, commonly present with seizures without neurological deficit. In adult patients with low grade glioma, older age, astrocytoma histology, presence of neurologic deficits before surgery, largest tumour diameter > 6 cm and tumour crossing the midline were important adverse prognostic factors for survival. These factors can be used to identify low-risk and high-risk patients (Pignatti *et al.* 2002). Patients under the age of 40 years have a median survival of 8 years compared with 5.5 years for patients aged between 40 and 50 years and 1.6 years if older than 50 years. If patients have a good performance status, the median survival is 7.4 years compared with 1.6 years if they have a poor performance status (Eyre *et al.* 1993). Most tumours are situated in the frontotemporal regions and are frequently diffuse, extending throughout a lobe at presentation (Figs 27.8, 27.9, and 27.10). Clearly the diffuse infiltrating astrocytomas cannot be resected. For more focal low grade astrocytomas in non-eloquent areas, resection may be feasible. There is uncertainty of how best to treat patients with low grade gliomas. It is probably advisable to biopsy lesions suggestive of low grade glioma because there can be foci of higher grade despite the lack of enhancement. However, if seizures are the only symptom, a wait and watch policy is preferred by some clinicians and patients. The growth rate of low grade gliomas is very slow and has been estimated as 4.1 mm/year (Fig. 27.11). There are no randomized controlled trials of resection versus biopsy in low grade glioma and it is unlikely that such a study would see completion because of the excellent survival with astrocytomas, with 46 per cent five year survival and oligodendroglioma 73 per cent five year survival (Shaw *et al.* 1997; Shaw 1995). Radiation can reduce the size of a tumour but because of the good survival, this has to be balanced with the

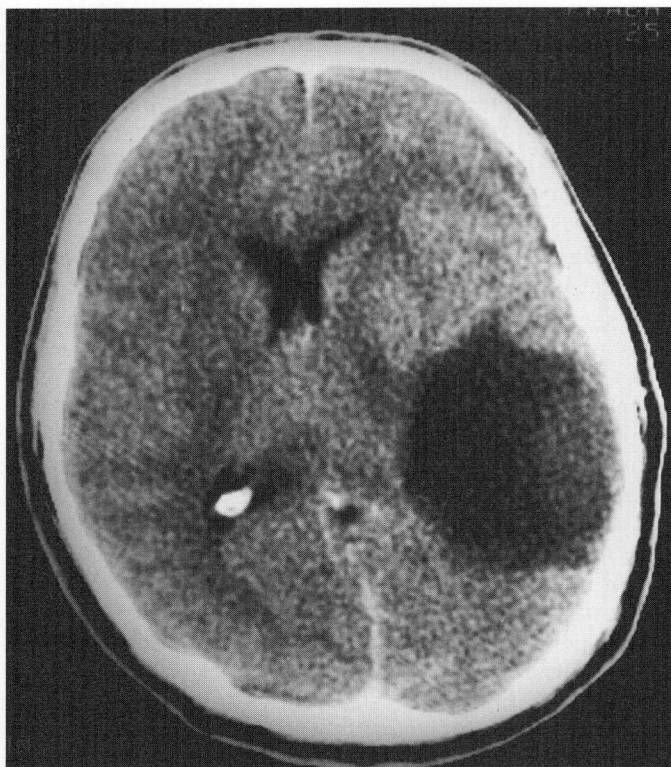

Fig. 27.8 Fibrillary astrocytoma demonstrated by stereotactic biopsy of a 32-year-old male with late onset epilepsy. Axial CT scan with contrast. The patient elected not to enter the EORTC low grade study and adopted a 'watch and wait' policy.

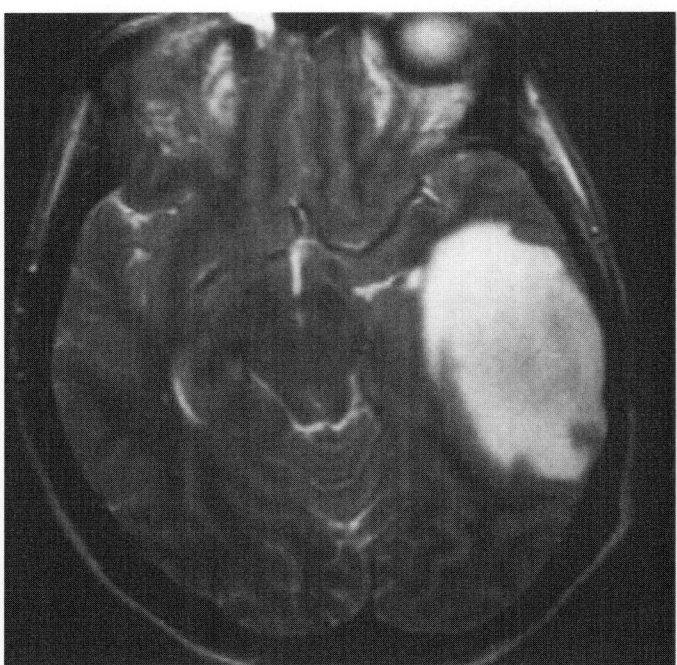

Fig. 27.9 Fibrillary astrocytoma. Axial T2-weighted MRI of same patient as Fig. 27.8, but 2 years later. Seizures remained reasonably controlled but he developed occasional predominantly left sided throbbing headaches, without diurnal variation or any other features. MRI shows slight but further space occupying effect with displacement of the middle cerebral artery anteriorly. Interestingly migraine with or without aura develops in a proportion of patients with brain tumours of any type, irrespective of whether there is mass effect on the scan. The aura commonly corresponds to the site of the tumour. It responds usually to standard antimigrainous drugs but occasionally steroids are useful. This patient was given steroids and was then treated with radiation 60Gy in 30 fractions over 6 weeks.

increased likelihood of developing delayed radiation-induced toxicity. A randomized controlled trial of early versus delayed radiotherapy for low grade gliomas accrued 314 patients. Median progression free survival was 5.3 years with early radiotherapy compared to 3.4 years for the wait-and-see group; however, overall survival was similar in both groups at 7.4 years compared to 7.2 years respectively. In the control group, 65 per cent received radiotherapy at progression. The study did not examine late complications of radiotherapy or quality of life (Karim *et al.* 2002; van den Bent 2005). A randomized study of radiation versus radiation plus Choroethyl-Cyclohexyl-Nitrosourea, CCNU, demonstrated more radiological responses in the radiation alone group, at 79 per cent versus 54 per cent, failed to demonstrate any survival advantage with adjuvant chemotherapy and there were significant haematological toxicities in the chemotherapy treated group (Eyre *et al.* 1993). More recently, PCV chemotherapy with Procarbazine, CCNU, and Vincristine or Temozolomide have been advocated prior to radiation therapy as this can occasionally lead to reduction in tumour size or symptomatic improvement in low grade glioma; however, the role of chemotherapy has yet to be proven. At the time of recurrence most tumours have changed to higher grades of malignancy and chemotherapy may have something to offer (Muller *et al.* 1977).

High grade glioma

High grade gliomas: anaplastic astrocytoma, glioblastoma multiforme, and anaplastic oligodendroglioma, account for approxi-

mately 80 per cent of all cerebral gliomas. Prognosis depends on age, grade, performance status, and possibly, the extent of remaining disease after surgery. Age is the most important independent prognostic variable at presentation followed by grade. A recursive partitioning analysis is available from three Radiation Therapy Oncology Group trials (Curran *et al.* 1993). Patients with poor performance status are less likely to be offered treatment and the patients with poor performance status invariably also have shorter survival, even accounting for age and grade. Site of tumour and volume of tumour on pre-operative CT or MRI does not seem to be prognostically important. Some studies suggest that the amount of peritumoural oedema, extensive contrast enhancement or the volume of tumour remaining on a scan performed at 48 to 72 hours post-operatively are associated with poor outcome (Hammoud *et al.* 1996; Muller *et al.* 1977; Piepmeier *et al.* 1996). Care must be taken when interpreting enhanced images with respect to timing of injection of contrast and the time of performance of the scan. Tumours can significantly alter in their enhancement on dynamic scanning depending on the delay between contrast injection and performance of imaging (Fig. 27.12).

Surgery It is not certain whether the type of surgery performed, such as biopsy versus resection, has any survival benefit other than in patients who are on the verge of coning or who have hydrocephalus. Sterotactic biopsy can undergrade a glioma, because of inadequate sampling which in turn may result in patients with stereotactic biopsy anaplastic astrocytomas appearing to do worse than

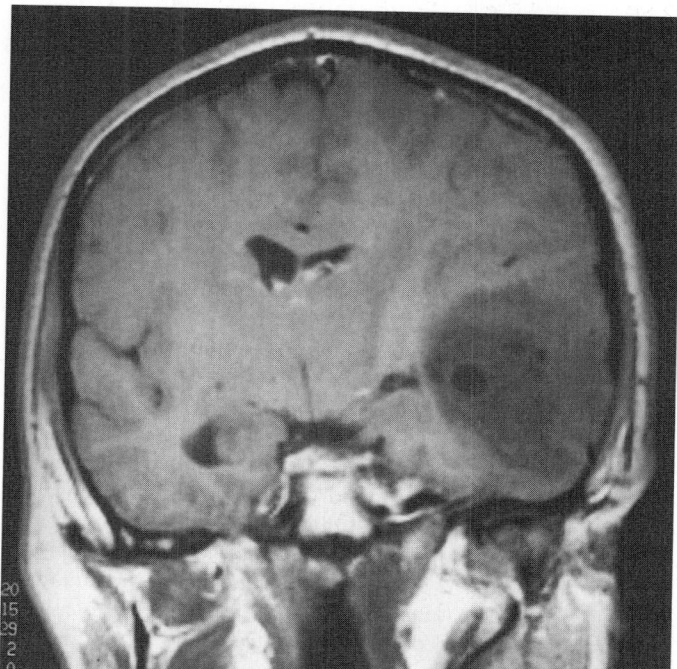

Fig. 27.10 Fibrillary astrocytoma. Coronal gadolinium enhanced MR scan of same patient as in Figs 27.8 and 27.9, at the same time as Fig. 27.9. This demonstrates the mass effect with displacement of the temporal horn medially and upwards and the lack of contrast enhancement. The tumour probably remains low grade since the patient remains well 2 years after this MR scan with occasional seizures and infrequent 'migraine headaches'. Recent scans reveal little change other than slightly less mass effect.

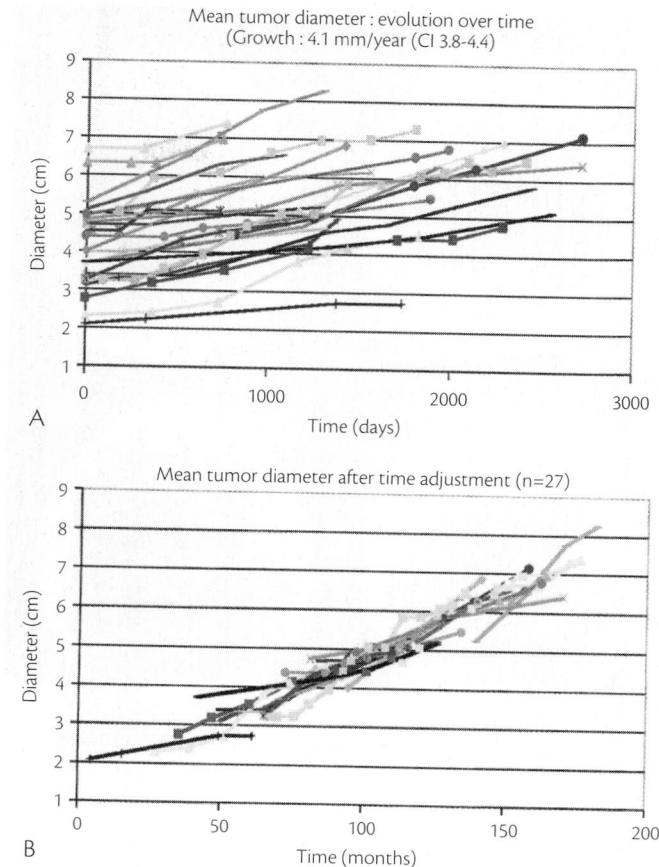

Fig. 27.11 Growth rate of low grade gliomas estimated from serial MR scans. (Courtesy of Professor J-Y Delattre.)

those where the diagnosis has been based on an extensive resection, purely because of grading error (Glantz *et al.* 1991). One or two prospective studies have collected data before or after resection and show that about one third improve after resection, 58 to 76 per cent stay the same and 9 to 26 per cent deteriorate (Fadul 1988; Sawaya 1998). There are several phase 2 studies and a recursive partitioning analysis that support the view that extent of resection influences survival (Simpson *et al.* 1993; Schiff and Shaffrey 2003). These however are affected by a number of confounding variables, including size, site, age, further treatment, and lack of a comparable control group, and it is not possible to say with certainty that resective surgery extends median survival. A Cochrane systematic review (Hart *et al.* 2007) identified only one randomized controlled trial of biopsy versus resection in patients with radiologically 'obvious' glioblastoma multiforme in patients older than 65 years (Vuorinen *et al.* 2003). This study of 30 patients randomized to either maximum resection or stereotactic biopsy reported that 23 per cent had diagnoses other than glioblastoma, consisting of stroke, metastasis, lymphoma, or uncertainty, leaving 13 in the biopsy group and 10 in the resection group who were referred for radiotherapy. Data was available in 18 of these patients; the overall median survival time was 21 weeks, with resection survival of 24 weeks versus biopsy of 12 weeks (*p* <0.035). There was no significant difference in the time to deterioration between the groups; the amount of radiotherapy received had an effect on survival (*p* <0.001) (Vuorinin *et al.* 2003). This study was underpowered to give any firm conclusion to whether resection prolongs survival.

If biopsy is done this should be by stereotactic technique rather than freehand, because of the higher complication rate associated with the latter. New operative adjuncts, including awake craniotomy, functional MRI, and intra-operative MRI have been used in selected cases in highly specialized centres where the tumour is in an eloquent area. One small randomized controlled trial of neuro-navigation did not show any survival advantage or reduction in death or disability (Willems *et al.* 2006). The goal of resective neurosurgery should be as complete resection as possible along its macroscopic boundaries. If achieved without complications, this provides reliable histological diagnosis, potentially improves the patient's neurological status and may make the tumour more sensitive for additional therapies, such as chemotherapy (Salcman 1987; Shapiro *et al.* 1989). The degree of tumour removal in most studies has been determined by the intra-operative perception of the neurosurgeon. With the increasing availability of neuro-imaging, it has become clear that the surgeon's opinion at the time of operation of what represents a total resection bears little resemblance to the post-operative MRI appearances. Intra-operative imaging is only performed in a few centres. 5-Aminolevulinic acid, Gliolan™, is a chemical that is taken up preferentially by brain tumour tissue and fluoresces under blue light. It has been used to assist the surgeon in identifying remaining tumour at surgery and so may help improve the extent of resection of gliomas. A randomized controlled trial in selected patients with peripherally situated gliomas which

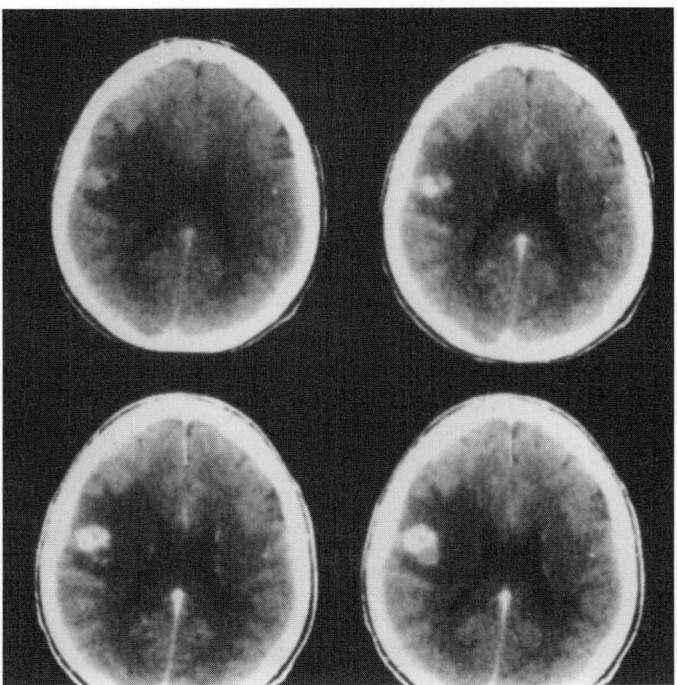

Fig. 27.12 Right hemisphere tumour. CT scan with contrast taken at different 5 minute intervals after contrast administration. (Courtesy of Dr P Warnke.)

demonstrated that it was possible to resect more, it took longer for the tumour to return on imaging, but it did not affect the overall survival. There was a slightly higher early neurological complication rate in the group given 5-Amino-levaliric acid (Stummer *et al.* 2006).

Focal complications following surgery include haematoma, abscess, stroke, and seizures, whilst systemic complications include thromboembolism and pneumonia. Morbidity rates range from 11 to 32 per cent with post-mortality rates ranging from 0 to 20 per cent. Variations in reporting these complications and how long after surgery the mortality is recorded will account for some of the larger variations. The complication rates are lower for biopsy than for resection. Post-operative enhancement on CT scan performed before the 5th post-operative day reflects residual tumour (Cairncross *et al.* 1985; Jeffries *et al.* 1981). Examination of serial post-operative MR scans has demonstrated that post-operative imaging during days 1 to 3 after resection of a high grade glioma avoids artefacts due to post-operative enhancement and the delineation of tumour was vastly superior to post-operative CT (Albert *et al.* 1994). MRI studies have suggested that post-operative residual tumour was a more important prognostic variable than age or performance status and the incidence of tumour recurrence is directly related to the volume of residual tumour after initial resection (Albert *et al.* 1994; Berger 1995). However, these results must be interpreted with caution since the selection of patients for aggressive resection based on tumour location and demarcation from surrounding normal tissue may introduce selection bias.

Radiotherapy. Malignant glioma is one of the most aggressive tumours in man. It rarely spreads outwith the central nervous system, is highly radioresistant, and has a predilection to loco-regional recurrence. Each of these three factors has lead to particular approaches to primary treatment and management of 'recurrence'. There is good randomized controlled evidence from the early 1970s and 1980s that radiation therapy improves survival in patient with high grade gliomas (Walker *et al.* 1978, 1980). Radiation therapy increases the median survival from 4 to 5 months to about 9 months. A randomized controlled trial has demonstrated that 60 Gy in 30 fractions over 6 weeks was superior to 45 Gy in 20 fractions over 4 weeks and resulted in a prolongation of survival by 3 months in the group treated with 60 Gy (Bleehen *et al.* 1991). The current standard practice is to give 60 Gy in 30 fractions over 6 weeks. Radiation therapy is usually directed at the area of the enhancing tumour plus at least 2 cm margin of peritumoural oedema (Chang *et al.* 1983; Halperin *et al.* 1996). Focal radiation with 40 Gy in 20 fractions over 4 weeks to the tumour and peritumoural oedema is usually followed by a further 20 Gy boost to the enhancing tumour and 1 to 2 cm margin over 2 weeks. The wide margins are because tumour cells can be found 2 cm or more from the apparent radiological boundary of the tumour in areas that simply look 'oedematous' on CT or MRI and most studies demonstrate that relapse occurs within 2 cm of the enhancing rim of the tumour in 80 per cent of cases (Halperin *et al.* 1989; Wallner *et al.* 1989). Boosting the radiation dose to the centre of the tumour is now standard practice in most centres. Dose escalation of radiation using conformal beam therapy or stereotactic radiation aims to maximally treat the centre of the tumour and spare normal tissue outwith the 2 cm margin to reduce long-term morbidity from radiation damage.

Hyperfractionation, multiple small fractions per day to a higher overall dose, and 'acceleration' consisting of multiple treatments per day with the same dose but reducing the overall treatment time, do not improve survival. Focal therapies including interstitial brachytherapy, neutron therapy, and particle pions have not revealed a significant survival advantage in small randomized controlled trials (Laperriere *et al.* 1998; Duncan *et al.* 1986; Pickles *et al.* 1997). The effect of radiation is greater in those under 60 years. In the over 70 age group the effect of radiation therapy remains controversial, but one recently reported randomized controlled trial in patients >70 years of age with anaplastic gliomas or glioblastoma who had a Karnofsky performance score of >70 found a 8 week improvement in progression free survival and 11 weeks median overall survival from standard radiotherapy consisting of 50 Gy in 1.8 Gy fractions, when compared with standard care (Keime-Guibert *et al.* 2005). The 1 year survival was 13.8 per cent with radiotherapy compared with 2.6 per cent in those managed conservatively. A Phase III study that compared the addition or stereotactic radiotherapy to conventional radiotherapy plus BCNU with conventional radiotherapy plus BCNU alone for supratentorial malignant glioma failed to demonstrate any survival advantage or changes in patterns of failure with the addition of stereotactic radiotherapy (Souhami *et al.* 2004).

Chemotherapy. Numerous chemotherapy agents, such as Nitrosoureas, Procarbazine, or Platinum derivatives, have been shown to reduce tumour size in about one-third of patients in phase II studies (Mahaley 1991). The beneficial effect of chemotherapy has to be balanced with potentially serious side effects. The risk: benefit ratio will depend on individual patient- and tumour-related factors. Younger patients, patients with anaplastic astrocytoma of grade 3, and patients with pure or mixed oligodendroglioma, are more likely to harbour chemosensitive tumours. Tumour-related genotypic factors such as chromosomal deletions on chromosomes 1p and 19q in oligodendroglioma and DNA repair enzyme inactivation by methylation at the promotor site of Methylguanine-DNA methyltransferase, MGMT, in glioblastoma. Approximately 50 per cent of malignant gliomas have an inactivated MGMT and these patients are associated with better prognosis and response to treatment with alkylating agents such as BCNU or Temozolomide, whereas tumours with unmethylated MGMT have a poorer prognosis and are very unlikely to respond to chemotherapy (Hegi *et al.* 2005).

◆ *Is chemotherapy better than supportive care?* A randomized controlled trial demonstrated that chemotherapy with a nitrosourea extends survival beyond what can be expected from supportive care, but the difference in median survival was only 4.5 weeks in the valid study group (Walker *et al.* 1978). Both groups received the similar corticosteroid use during the trial, about two weeks on average. Chemotherapy produced an 8 per cent absolute increase in survivors at 6 months and a 10.6 per cent absolute increase in survivors at one year. This represented twice as many survivors at 6 months and four times as many at 1 year with chemotherapy (Walker *et al.* 1978, 1980).

◆ *Is chemotherapy better than supportive care plus regular steroids?* A randomized controlled trial has looked at methylprednisolone given for one week each month, compared to chemotherapy with procarbazine or BCNU (Green *et al.* 1983). In this study chemotherapy with BCNU or procarbazine was superior to methylprednisolone; however the survival difference was only really noticeable 12 months after treatment and treatment with procarbazine only extended survival by a median of 2 weeks for the total randomized population or 7 weeks for the 'valid study group'. With BCNU median survival increased by 9 weeks for the total randomized population or 10 weeks for the 'valid study group'.

◆ *What is the likelihood of responding to chemotherapy?* The likelihood of response is related to the age of the patient at the start of chemotherapy and grade of tumour (Nelson *et al.* 1988). Nitrosoureas produce a partial response in almost 40 per cent of patients with high grade glioma under the age of 40, 17 per cent of patients between 40 and 59 years but only 5 per cent of patients ≥60 years (Grant *et al.* 1995). Median survival from the time of starting chemotherapy is approximately 43 weeks in younger patients of <60 years and 24 weeks in patients over the age of 60 years. Differences in response rate, time to progression, and survival related to age persist following adjustment for grade of tumour. The risk of myelosuppressive complications requiring admission to hospital is approximately 16 per cent in patients <60 years of age but 35 per cent in patients >60 years. Patients ≥60 years are therefore at a greater risk from chemotherapy and have less chance of a response or prolonged survival (Grant *et al.* 1995). It would seem reasonable to consider chemotherapy either adjuvantly or on recurrence in patients with high grade glioma who are young but careful thought should be given before suggesting chemotherapy to patients over the age of 60 years (Nelson *et al.* 1988).

◆ *Is the imaging response to chemotherapy related to time to progression or survival?* Although one might reasonably think that magnitude of response would be related to duration of response or survival, this has only been demonstrated in patients with anaplastic oligodendroglioma who have achieved a >90 per cent

imaging response (Cairncross and Eisenhauer 1995). The magnitude of tumour response has not been demonstrated to be related to duration of response or survival in high grade glioma (Grant *et al.* 1997). It has been estimated that therapy must be effective in >70 per cent of patients before one would see a significant effect on survival. Imaging may show little change although there is profound clinical change. Figure 27.13 shows an extreme example of this, where clearly the patient 'progressed' with no perceptible change in imaging. Response must take into account clinical, imaging, and steroid information. In addition, some patients appear to respond quickly to chemotherapy on imaging but then progress rapidly despite chemotherapy, presumably as a result of acquired resistance. Other patients respond slowly to chemotherapy but have more prolonged responses despite discontinuation of chemotherapy (Grant *et al.* 1997). In patients who do respond to chemotherapy, speed of response is not associated with duration of response. However, in one study where serial measures of tumour volume following chemotherapy measured, likelihood of achieving a response associated with the size of tumour in glioblastoma multiforme with only small volume tumours having a response (Grant *et al.* 2002).

◆ *Does chemotherapy increase survival?* The main aim of chemotherapy is to prevent disease progression and extend survival. A individual patient data meta-analysis of 12 randomized controlled trials of chemotherapy after surgery and radiotherapy in high grade glioma demonstrated a 2 month median survival benefit in patients treated with chemotherapy from 10 months to 12 months, an increase in 1 year survival from 40 to 46 per cent and an increase in 2 year survival from 10 to 15 per cent (Hazard ratio 0.85, 95 per cent CI 0.78–0.92), equating to a 15 per cent risk reduction of death (Glioma Meta-analysis Trial Group 2002). This small median survival advantage has to be balanced by the side effect profile. In addition, very few people in any of these randomized controlled trails were aged >70. An ad hoc analysis in those

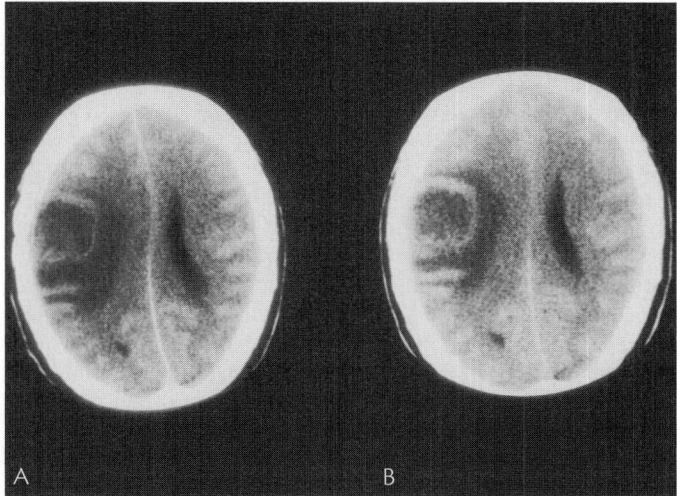

Fig. 27.13 Right parietal glioblastoma in a 54-year-old male. A. CT scan with contrast when asymptomatic following radiation and chemotherapy. B. CT scan with contrast taken 1 week later when he presented to the emergency department with headache and 'central cone'. Despite steroids and mannitol he died within 40 minutes of the CT scan. The two scans show virtually no change in the degree of enhancement or space occupation.

involved in the metaanalysis who were aged >70 demonstrated a non-significant trend to have a poorer survival with chemotherapy than without (personal communication). Two recent studies have confirmed the findings of this meta-analysis. The first has shown that BCNU impregnated wafers, Gliadel™, inserted at the time of initial resective surgery improve survival by 2 months compared with placebo wafers (Westphal *et al.* 2003). The BCNU impregnated wafer study had strict selection criteria and it has been estimated that only 25 per cent of patients with glioblastoma multiforme would be suitable. A randomized controlled trial of concomitant and adjuvant Temozolomide in addition to standard radiation therapy versus standard radiation therapy in patients with glioblastoma has demonstrated a 2.5 month survival benefit and 25 per cent 2 year survival (Stupp *et al.* 2005).

◆ *Is combination chemotherapy superior to single agent nitrosourea?* Most phase III studies have not shown any survival advantage with PCV combination chemotherapy versus a single agent nitrosourea such as BCNU (Shapiro *et al.* 1989; Mahaley *et al.* 1987; Levin *et al.* 1985). In a subgroup analysis of a randomized study, a doubling of time to progression and survival at the 50th and 25th percentile was demonstrated in patients with anaplastic astrocytoma, a Karnofsky score of ≥70, who had completed radiotherapy and >x1 course of chemotherapy (Levin *et al.* 1990). This was an interesting exploratory analysis, which may be important, but it was a post hoc analysis and may simply reflect a chance finding. In addition, high dose oral procarbazine is probably every bit as effective as nitrosoureas and has been shown to produce a significant number of partial responses in patients who have progressed while taking nitrosureas (Green *et al.* 1983; Newton *et al.* 1990). It has not been demonstrated that combination therapy with PCV is superior to sequential single agent chemotherapy, such as BCNU followed by high dose procarbazine. It remains uncertain whether oral Temozolomide is superior to adjuvant chemotherapy with PCV and this is the subject of a further randomized controlled trial.

When considering the lack of long-term efficacy from chemotherapy, and methods to overcome this, one has to examine primary mechanisms for drug resistance. Firstly, it must be possible to deliver the drug to the site of the tumour in sufficient concentration to have an effect. Secondly, drug must be able to pass through the blood brain barrier at the 'advancing' edge of the tumour, which may not be well vascularized. Chemotherapy penetration at the necrotic centre or enhancing nidus, will not prevent tumour progression at the margin. Thirdly, intracellular mechanisms of drug resistance, such as lack of intracellular drug activation, drug inactivation for instance by glutathione-s-transferases, DNA enzyme repair for instance by O^6-alkylguanine-DNA-alkyltransferase and active removal of chemotherapy from the cell by P-glycoprotein, an energy-dependent drug efflux pump mediated through multidrug resistance gene overexpression, must be overcome. Increasing drug delivery by intra-arterial chemotherapy, can reduce systemic effects of chemotherapy while maintaining the local concentration of chemotherapy to the tumour, however, the potential benefits are limited by local toxicity to the brain with encephalopathy or seizures and to the eye causing optic neuropathy. Randomized studies have shown no benefit in terms of survival over systemic chemotherapy (Mahaley *et al.* 1986; Green *et al.* 1989). Chemotherapy with BCNU and Cisplatin given prior to radiation was found to be

no better than adjuvant chemotherapy and had more toxic side effects (Grossman *et al.* 2003). Most studies therefore seem to support the view that chemotherapy can be useful whether given adjuvantly or at the time of clinical or radiological relapse after radiation therapy, although no one has achieved the 25 per cent 2 year survival achieved with concomitant and adjuvant Temozolomide compared with only 10 per cent in the control arm (Walker *et al.* 1978; EORTC Brain Tumor Group 1981; Stupp *et al.* 2005). This in part may be due to more stringent entry criteria. Temozolomide is less effective in biopsied cases and cases with moderate disability, a World Health Organization performance status of 2 or above. The long-term complication rate, in terms of radiation-induced leucoencephalopathy, of the radiation therapy plus concomitant and adjuvantly Temozolomide group remains uncertain.

Other therapies. Gene therapy techniques are currently being used to try and target elements in the tumour gene. Many genes have been implicated in the development and growth of gliomas. Despite very promising phase II studies and tremendous enthusiasm, a phase III trial of gene therapy with Herpes Simplex Virus-tk and radiation therapy did not show any advantage over radiation therapy alone (Rainov 2000). The gene therapy was injected directly into the tumour resection boundaries after maximal tumour resection and this was followed by radiotherapy. The failure may have in part been due to a very poor rate of transfection. There are other methods using adenoviral vectors with improved gene transfer efficiency of up to 10 per cent and stronger transgene expression in the tumour, which in one small pilot randomized studies suggest a benefit (Immonen *et al.* 2004). This study however was underpowered and the active treatment group had a lower median age and fewer glioblastoma multiforme patients.

The main emerging approaches to treatment of malignant glioma involve 'Targeted therapies'. These treatments target genetic and cellular alterations in cell signalling that are more commonly found in tumour cells such as amplification of oncogenes such as epidermal growth factor receptors on the cell surface or mutations of tumour suppressor genes, for instance, the phosphatase and tensin homologue, PTEN. Blood vessels can also act as a target such as vascular endothelial growth factor, VEGF, and platelet derived growth factor, PDGF, which are over-expressed. EGFR, VEGF, and PDGF also activate pathways which result in up-regulation of the anti-apoptotic protein Bcl-xl which help cells survive. These factors lead to proliferation of tumour cells, stimulate angiogenesis, and reduce the ability for cells to become apoptotic. Proliferating cells probably express different receptors to invading tumour cells. Cells in the process of invading over-express integrin cell surface adhesion molecules and matrix metalloproteinases that degrade extracellular proteins and allow the tumour cells to migrate and invade into surrounding brain.

There are no tumour specific antigens, but there are tumour associated antigens. These can act as targets for antibodies which may down-regulate or stop the action of oncogenes or growth factors, or be used to attach conjugated toxins, radiotherapy, or chemotherapy. These tumour associated antigens can be targeted by monoclonal antibodies such as Cetuximab which can block EGFR; EGFR tyrosine kinase inhibitors such as gefitinib, erlotinib, or imatinib; conjugated toxins such as pseudomonas exotoxins conjugated to transforming growth factor alpha or interleukin-13; anti-angiogenic agents, such as thalidomide and anti-invasive agents such as

Marimistat or cilengitide. There are also some novel cell growth and migration inhibitors such as Accutane. These all act at a molecular or biological target level. Cetuximab binds to EGFR and prevents ligand binding, blocks ligand-induced tyrosine kinase activation and receptors become internalized. It induces apoptosis and inhibits angiogenesis *in vitro* and has shown some success in phase II clinical trials. I^{131} labelled murine anti-tenascin monoclonal antibody 81C6, Neuradiab™ injected directly into the tumour rim after resection appears to improve survival in phase II studies when compared with historical controls. Monoclonal antibodies conjugated with isotopes such as Yttrium or pseudomonas exotoxin (Precise) have also been used in small phase II studies and are currently in phase III studies. Transferrin receptors are expressed tumour associated antigens and transferring conjugated with diphtheria toxin, Transmid™, is in phase III studies. Tyrosine kinase inhibitors such as erlotinib have been used in malignant glioma and have produced a 20 per cent response in selected groups. These small molecules are now being increasingly used in combination with standard therapies such as radiotherapy and chemotherapy. Because these small molecules are likely to be broken down systemically, most studies involve direct injection at the time of surgical resection or infusions into the tumour cavity using convection enhanced delivery technologies within the sealed tumour cavity given under pressure to allow diffusion of the small molecule to 2 cm from the cavity. Thalidomide inhibits integrin receptors and inhibits neovascularization. It is a drug notorious for its teratogenic effects in the foetus causing in particular phocomelia, and is contraindicated in women of childbearing age who have a chance of pregnancy. Matrix metalloproeinases such as Marimistat have been used in malignant glioma along with Temozolomide, but can cause joint pain in up to 50 per cent.

New small molecule therapies may be associated with hitherto unexpected side effects. Many treatments will only be possible if the patient is fit enough for resection. They may only be associated with inhibition of tumour progression, rather than causing tumour regression. New protocol designs and clinical response measures may be necessary and the likely way forward will be combinations of these therapies accompanied by standard treatments. The treatments are expensive. It seems unlikely that one particular small molecule or single target will lead to cure since only a percentage of tumours express these antigens and not all tumour cells express a specific target. They need to be proven in large well designed randomized controlled trials.

Oligodendrogliomas

Although malignant glioma continues to have a poor prognosis, anaplastic oligodendroglioma can be responsive to chemotherapy for prolonged periods, and chemotherapy prior to radiation therapy may prove to be more effective. Studies suggest that PCV chemotherapy produces a 75 per cent response rate, with 17 per cent stable disease and only 8 per cent have progressive disease (Macdonald *et al.* 1990; Kritis *et al.* 1993; Macdonald 1994). In this moderately chemoresponsive tumour, where a large proportion of patients will respond, even those >60 years, there is some evidence that patients who have a 'major response' defined as >90 per cent reduction in contrast enhancing area, have better survival than those with partial response or stable disease (Cairncross and Eisenhauer 1995; Macdonald 1994). As mentioned above, tumours with Ch 1p and Ch 19q deletions seem to do particularly well with chemotherapy. A preliminary report of a randomized controlled trial of pre-radiotherapy PCV versus radiotherapy followed by chemotherapy as necessary demonstrated that pre-radiotherapy PCV does not impart a survival advantage for histologically defined anaplastic oligodendrogliomas and anaplastic oligoastrocytomas, but may prolong progression-free survival at the expense of greater acute toxicity (Cairncross *et al.* 2004).

27.8.2 Medulloblastomas

Medulloblastoma is the most common childhood central nervous system tumour, but is a relatively rare tumour in adults. About 25 per cent of all medulloblastomas occur in adults, aged >16 years. The cell of origin of this tumour remains uncertain but is probably of embryonic origin with the possibility of taking different lines of differentiation, otherwise known as **p**rimitive **n**euroectodermal **t**umour or PNET. The most common genotypic abnormality, in 30–45 per cent of cases, is the loss of genetic material from chromosomal arm 17p (Ellison 2002). This site is apparently the location of a suppressor gene, the removal of which allows for the expression of the tumour. Medulloblastomas usually arise in the midline and are most commonly found in the cerebellum, approximately 95 per cent in children. Presentation is usually either related to a cerebellar ataxia, or raised intracranial pressure related to obstructive hydrocephalus. Cranial nerve presentations, especially diplopia, also occur. The tumour has low signal on T1-weighted MRI and high signal on T2-weighted scans and there is usually gadolinium enhancement on T1-weighted images.

Primitive neuroectodermal tumours are malignant tumours with a high propensity to disseminate throughout the CSF. Staging of the tumour is therefore important, but can wait till after the definitive surgical procedure. Because of the possible contamination of CSF at the time of surgery it is probably best to defer CSF analysis till 10 to 14 days after the operation; however the MRI of the spine to look for gadolinium enhancing nodules can be performed any time after surgery. Thirty to forty per cent of children will have CSF dissemination although this may be greater in children under 5 years of age (Deutsch 1988). The bad prognostic features for medulloblastoma appear to be age <3 years, CSF dissemination, and possibly extent of resection (Evans *et al.* 1990). Current treatment guidelines are largely determined by clinically based prognostic factors, the most important of which are tumour location and the extent of tumour spread. Although the cure rate for high-risk primitive neuroectodermal tumours has improved, the irreversible sequelae of craniospinal axis radiation treatment in patients who survive have motivated researchers to investigate more fully which patients can safely receive less treatment (Jakacki 2005).

The management of acute obstructive hydrocephalus is controversial. Some surgeons suggest steroids and tumour resection, whereas others suggest steroids and external ventricular drainage or steroids and ventriculoperitoneal shunting prior to a definitive operation. Surgery for medulloblastoma is performed in the prone position, to avoid the risks of air embolus and pneumocephalus or systemic hypotension when operating in the seated position. A repeat MRI scan at 48 hours may be helpful for assessing the extent of tumour resection.

Radiotherapy to the tumour improves survival in children with medulloblastoma. In addition there is good evidence that craniospinal irradiation reduces the risk of recurrence from CSF dissemination. A randomized prospective trial demonstrated improved disease control using 36 Gy of craniospinal irradiation with a posterior fossa boost to 54 Gy compared with 23.4 Gy with a posterior fossa boost to 54 Gy in 30 fractions (Deutsch *et al.* 1996). This study was prematurely closed after an interim analysis at 16 months demonstrated a significant number of relapses in the low dose craniospinal irradiation group. Craniospinal irradiation also improves the number of 10 year survivors (Landverg *et al.* 1980; Castro-Vita *et al.* 1980). Radiation therapy to the craniospinal axis can have serious long-term toxicities: neuropsychological, neuro-endocrine with growth retardation and hypothyroidism, bone marrow depression, and second neoplasms in long-term survivors particularly of thyroid and other central nervous system neoplasms. In attempts to try and reduce these toxicities and to try to improve survival, chemotherapy has been used. Medulloblastoma is a relatively chemosensitive tumour. Randomized studies have compared radiation with radiation and CCNU, vincristine chemotherapy. The overall survival was 53 per cent at 5 years and 45 per cent at 10 years. Although an early analysis demonstrated a significantly better disease free survival with the addition of chemotherapy, there were subsequently late relapses in the chemotherapy group and no statistically significant effect on disease free survival was apparent. On post-hoc subgroup analyses, chemotherapy was thought to benefit patients with subtotal resection, brainstem involvement, and more extensive disease at presentation (Tait *et al.* 1990). A further randomized controlled study of the same agents in North America came to the same conclusions but also noted a survival advantage in patients with extensive disease in a subgroup analysis (Evans *et al.* 1990).

Currently, there is a vogue for chemotherapy in addition to craniospinal radiotherapy with a boost to the primary site for patients with a high risk of progression, in an attempt to improve survival and reduce the dose of craniospinal irradiation (Cohen *et al.* 1996). Approximately 10 to 20 per cent of patients will eventually have metastatic spread outwith the central nervous system, especially to lungs and bone.

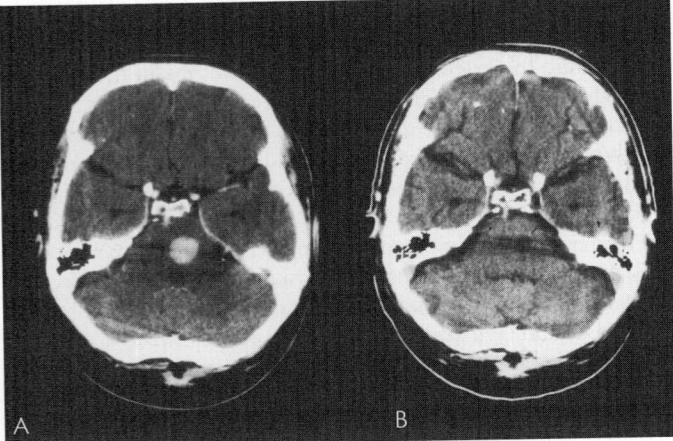

Fig. 27.14 Primary central nervous system lymphoma involving the pons. A. Axial CT scan with contrast. B. Same patient admitted for biopsy one week after starting steroids showing a 'disappearing tumour'. Steroids were discontinued and the tumour 'returned' on imaging when rescanned 3 weeks later. Biopsy confirmed primary central nervous system lymphoma.

With treatment, approximately 33 to 66 per cent of children are alive at 5 years and 25 to 50 per cent are alive at 10 years, depending on the patient selection for the survival analyses (Bloom *et al.* 1991). Younger age at diagnosis was the most prominent risk factor for neurocognitive deficits among survivors, despite reductions in craniospinal irradiation dosing and efforts to limit the boost volume. Young patients show substantial problems with the development of reading skills and intelligence (Mulhern *et al.* 2005).

27.8.3 Primary central nervous system lymphomas

Primary central nervous system lymphoma accounts for approximately 1 per cent of all primary intracerebral tumours and is almost always B-cell in origin. They occur in higher frequencies in patients with some form of immunosuppression. Approximately 2 to 6 per cent of patients with AIDS and 0.5 to 1 per cent of patients following transplantation will develop a primary central nervous system lymphoma. Primary central nervous system lymphoma is also found in association with Wiskott Aldrich syndrome, systemic lupus, idiopathic thrombocytopenic purpura, Sjogren's, and sarcoidosis (Ling *et al.* 1997; Remick *et al.* 1990). Despite the association with immunosuppressive states, primary central nervous system lymphoma more commonly occurs sporadically in the immunocompetent. Epstein–Barr virus can be detected in all AIDS-related primary central nervous system lymphomas raising the possibility of a direct virus related cause (MacMahon *et al.* 1991; Guterman *et al.* 1996). The incidence was increasing in the 1990s but has stabilized or may be falling (Kadan-Lottick *et al.* 2002). Primary central nervous system lymphoma is classified as an extranodal lymphoma, stage 1_E, and it most commonly presents as single or multiple contrast enhancing space occupying lesion(s) within the brain or as cells in the eye, 'vitreous lymphoma', or CSF, 'leptomeningeal lymphoma'. Primary central nervous system lymphoma is most commonly confined to central nervous system or eye and even at post-mortem only 10 per cent of patients are found to have disease outwith the central nervous system. The tumour is commonly in a periventricular site and this may explain the high incidence of CSF involvement. Approximately 20 per cent of patients will have CSF involvement at the time of presentation and a further 20 per cent will have had or develop a uveitis with 'floaters' and progressive loss of vision due to lymphoma of the vitreous (Hochberg *et al.* 1988). The diagnosis may be suspected if the CT or MR scan is typical, but biopsy is essential. There is no evidence that resection is superior to stereotactic biopsy. The tumours frequently lose their enhancement when steroids are given and it is not uncommon for the tumour to 'disappear' on CT scan after steroids which makes stereotactic biopsy difficult (Fig. 27.14). These tumours have an increased propensity to bleed when biopsied compared with other intracerebral tumours. Staging of primary central nervous system lymphoma is important. The International Primary Central Nervous System Lymphoma Collaborative Group has published guidelines for standardized baseline evaluation in newly diagnosed cases (Abrey *et al.* 2005). These include centralized pathology review and immunotyping; neurological, ophthalmological, slit lamp, and cognitive examination; HIV serology, lactate dehydrogenase, CSF cytology, flow cytometry, and 24 hour urine for creatinine clearance; contrast enhanced MRI of brain, CT scan of chest and abdomen, bone marrow biopsy, and testicular ultrasound in elderly males. Slit lamp examination may identify uveitis due to vitreous involvement, retinal detachment or haemorrhages, optic neuropathy or papilloedema. Pathologists may have difficulty identifying lymphomatous cells from chronic inflammatory cells and both populations can co-exist. The CSF protein is usually increased and one third have CSF glucose levels below the lower limit of the accepted range. CT or MR scan of the chest and abdomen and bone marrow biopsy is almost always negative (Hochberg and Miller 1988).

Corticosteroids are cytotoxic to lymphocytes and up to 40 per cent show a significant initial response, and occasional prolonged remissions with steroids have been reported. Patients younger than 60 years treated with >40 Gy to the whole brain and a boost to 50 Gy to the tumour bed along with chemotherapy appear to have a better survival with a median of around 2 to 3 years (Reni et al. 1997). Escalating the boost to 60 Gy does not appear to produce any survival advantage (Nelson et al. 1992). Chemotherapy used to be kept in reserve for relapsed primary central nervous system lymphoma; however, now chemotherapy with high dose methotrexate has been found to be effective and has good CSF penetration (Lachance et al. 1994; Glass et al. 1994), the trend is to give chemotherapy and radiotherapy in patients <60 years. With the combined regimes 50 per cent respond and 2 year survival rates in up to 70 per cent of selected patients (Ferreri et al. 2003). Some authorities suggest chemotherapy only in patients >60 years, because of high frequency of neurotoxicity with radiation and chemotherapy (Batchelor et al. 2003). Others consider that chemotherapy-related toxicity in the elderly is so frequent and prolonged remissions so few that treatment with chemotherapy is not justified, especially in patients with other co-morbidities. These various protocols have not been compared in randomized controlled trials.

Patients older than 60 years given chemo-radiation have a very high incidence of treatment-related neurotoxicity with dementia, ataxia, and incontinence with a median time to onset of about a year (Abrey et al. 1998). Neurotoxicity is much less common in patients under 60 years with a frequency of 30 per cent during 96 months of follow-up. Intrathecal methotrexate used to be the treatment of choice for CSF disease, but this treatment is no longer required if high doses intravenous methotrexate are given. Ocular disease can be treated symptomatically by steroids and frequently is misdiagnosed as non-lymphomatous related uveitis. Steroids usually only have a temporary effect and radiation to the posterior two-thirds of the eye or chemotherapy with high dose cytosine arabinoside or methotrexate can be useful.

Radiation therapy plus or minus chemotherapy certainly produces a clinical and radiological response in most cases and the median survival is in the region of 2 to 3 years in non-immuno-compromised patients. The median survival of AIDS patients with primary central nervous system lymphoma, treated with radiation plus or minus chemotherapy, is approximately 3 months.

27.8.4 Pineal region tumours

Pineal region tumours can arise from a number of cell types. The most common tumours in this area are teratomas or germinomas. Pineocytomas are less common and arise from the pineal parenchymal cells and gliomas can also arise in the pineal region. Occasionally metastases can spread to the pineal region. In addition to these solid tumours, pineal region cysts can occur and can be simple, filled with CSF or can be epidermoid or dermoid cysts. It can be

difficult to distinguish with any certainty by imaging characteristics whether a pineal region tumour is a pineal glioma, an ependymoma, a pineocytoma, germinoma, or teratoma. Serum and CSF analysis can be helpful. If germ cell markers such as alphafetoprotein and beta human chorionic gonadotropin are positive, biopsy of the lesion is not required. Elevated germ cell markers indicate a malignant germ cell tumour. These patients can be treated with radiation therapy and chemotherapy and followed by measuring the tumour markers. If the lesion enlarges but markers reduce, surgical debulking may be necessary (Lee et al. 1995). In the absence of tumour markers, surgical confirmation is strongly advised. Hydrocephalus may have to be dealt with first, either by external drainage or placement of a ventriculo-peritoneal shunt. In the presence of mild hydrocephalus, some surgeons prefer performing a definitive resection and placing a ventricular drain at the time of surgery, which can be later clamped and removed or changed into a ventriculo-peritoneal shunt. Other authors have suggested performing a third ventricular ventriculostomy (Goodman 1993).

If the frozen section reveals germinoma, then there is no need to proceed to complete resection because this tumour is so radiosensitive and chemosensitive. Cure rates for germinoma exceed 90 per cent at 10 years. If the biopsy confirms a benign pineal region tumour, such as dermoid, epidermoid, pilocytic astrocytoma, ependymoma, or pineocytoma, maximum resection should be attempted and is generally the only treatment necessary. If the biopsy demonstrates a malignant pineal region tumour, such as a non-germinomatous germ cell tumour, choriocarcinoma, embryonal cell carcinoma, immature teratoma, endodermal sinus tumour, or malignant glioma, then maximum resection is probably advisable, especially if the patient is young and has a good performance status. Complications of surgery include disorders of eye movement, ataxia, and cognitive problems. If a supratentorial approach to the tumour is taken, there is a higher incidence of visual field defects and hemiparesis. Surgery for pineal region tumours is associated with a morbidity of 12 per cent and mortality rate up to 8 per cent (Bruce and Stein 1993). Preoperative or 12 day post-operative spinal MRI may identify spinal seeding from the pineal region tumour. Standard radiation schedules for germinoma consist of 40 Gy to the whole brain and a boost to 55 Gy to the pineal region. Radiation therapy dosages of <50 Gy are associated with increased risk of recurrence (Schild et al. 1993). The need for craniospinal irradiation is uncertain, but most centres only suggest this for documented CSF metastases. Germinomas are also chemosensitive and chemotheraphy has been advocated in young patients in order to delay radiation or possibly reduce the dosage of radiation given. Germinomas and non-germinomatous germ cell tumours are sensitive to cisplatinum, plus or minus etoposide, and to cyclophosphamide, and pre-radiation chemotherapy is now being advocated by some centres (Patel et al. 1992). Pineoblastomas respond only partially to radiation and less well to chemotherapy. Various chemotherapy regimes have been tried with only limited success.

27.8.5 Craniopharyngiomas

Craniopharyngiomas are benign tumours that usually present in childhood or early adulthood. They arise from the embryological remnants of Rathke's pouch in the suprasellar area or pituitary region. Symptoms usually present in adolescence or early adulthood. In children the most common symptom is growth failure and adults present with sexual dysfunction in men and amenorrhoea in women. The tumour can present with hypopituitarism in 25 to 40 per cent, diabetes insipidus in 50 per cent, visual failure from pressure on the optic chiasm or optic nerves in 40 to 70 per cent, raised intracranial pressure from hydrocephalus from obstruction of the third ventricle in 20 to 40 per cent, or with personality and memory problems. The tumour is usually a mixture of cysts and solid components, where the cysts contain thick fluid like engine oil containing cholesterol crystals. Skull X-ray may demonstrate calcification in the suprasellar region, but MRI scan is the most valuable investigation (Fig. 27.2). The sagittal and coronal scans provide invaluable information to the surgeon. Differential diagnosis includes meningioma, optic nerve glioma, teratoma, dermoid or epidermoid cyst, metastasis or sarcoidosis.

After correction of any endocrinopathy, definitive operation can be performed in relative safety. The mainstay of treatment is resection of the tumour, although this has to be tempered by its tendency to be adherent to surrounding structures. Total resection is frequently impractical and attempts at aggressive resection have resulted in high morbidity and mortality of up to 20 per cent and recurrence rates of 30 to 40 per cent (Yasergil et al. 1990; Weiss et al. 1989; Wen et al. 1989). Most authors suggest safe subtotal resection and either post-operative radiotherapy to the residual disease or radiotherapy at the time of recurrence, depending on the age of the patient (Weiss et al. 1989; Wen et al. 1989). Some still suggest attempted complete removal as the best approach (Yasergil et al. 1990). There are no randomized studies of safe subtotal resection +/− radiation therapy versus maximum possible resection +/− radiation therapy and no randomized studies of early radiation versus delayed radiation. If radiation is given after resection, the usual advised dose is at least 54 Gy. At recurrence, it is not uncommon to get enlargement of one of the cysts. Cystic recurrences can be treated by placement of a reservoir and aspiration of the cyst intermittently (Gutin et al. 1980). A different approach is to instil ^{32}P a beta emitting isotope with limited penetrance. This has been reported to result in cyst regression in >80 per cent of cases with good symptomatic relief (Pollock et al. 1995). Twenty year survival for craniopharyngioma approaches 60 per cent, but recurrence is common and morbidity is significant (Regine et al. 1992). Craniopharyngiomas probably have a better prognosis when diagnosed in adults than when diagnosed in childhood.

27.9 **Metastatic intracerebral tumours**

Improved treatment of systemic malignancies has lead to an increase in the frequency of brain metastases, possibly because the brain may act as a 'sanctuary site' for cancer cells during systemic chemotherapy (Greig et al. 1990). The incidence of intracerebral metastases is approximately $14/10^5$/year; therefore each year in the United Kingdom one would expect approximately 8000 new cases. Metastases account for 45 per cent of all intracranial tumours and 60 per cent of intracerebral tumours. The frequency of brain metastases varies depending on the primary tumour but ranges from 12–35 per cent of all cancer patients (Posner et al. 1978; Galicich et al. 1996). Brain metastasis are generally a late manifestation of cancer and systemic metastases frequently co-exist, however, 36 per cent of patients do not have a past history of cancer at initial presentation. It has been estimated that only 19 per cent of patients do not have metastases at other sites at the time of presentation, but this was based on a cancer hospital population (Cairncross

et al. 1980). The frequency of isolated brain metastases is likely to be higher in a general hospital population study. Lung cancer, cancer of unknown origin, breast cancer and melanoma account for 90 per cent of brain metastases (Grant et al. 1996). It is very unusual for patients with breast or gastrointestinal tract malignancies to present with brain metastases as the initial manifestation of cancer. If there is no history of malignancy at presentation with brain metastases the primary site is most commonly lung, in 55 per cent, or the primary tumour is not identified prior to death in 40 per cent (Grant et al. 1996).

Seizures are the presenting symptom in about 16 per cent and will eventually occur in up to 40 per cent of patients at some stage (Posner 1980). A randomized controlled trial has failed to show any advantage to prophylactic prescription of anticonvulsants in patients with brain metastases (Glantz et al. 1996). Anticonvulsants should not be given prophylactically in patients with metastases. Anticonvulsants are helpful in controlling frequency of seizures. Patients with cancer are at an increased risk of developing deep venous thrombosis, because of immobility and possibly as a result of hypercoagulation related to cancer (Dhami et al. 1993). This risk may be increased if hemiparesis results from a brain metastasis. Patients at risk of deep vein thrombosis with multiple risk factors such as hemiparesis, cancer, or operation should be managed with elastic stockings and subcutaneous heparin (Monreal et al. 1996). A literature review of patients with cancer who had either prophylactic therapy for deep vein thrombosis or pulmonary embolus showed that prophylactic anticoagulation or vena caval filter did not improve quality adjusted life expectancy. However, anticoagulant therapy provided a 9 per cent gain in quality adjusted life expectancy for patients with acute deep vein thrombosis and a 16 per cent gain for patients who survived a pulmonary embolus, and vena caval filter yielded 11 and 18 per cent gains respectively (Sarasin et al. 1993).

Steriods Symptomatic management of brain metastases with steroids for headache, focal neurological deficit, or cognitive problems is very effective in the short term in reducing the effect of brain oedema. The mechanism of action of steroids remains uncertain. Steroids do reduce the amount of oedema around the tumour and repair a leaky blood brain barrier, but the symptomatic relief within 6–24 hours after starting antedates any obvious change on CT scan or MRI. Approximately 70 per cent of patients improve with steroids. A reasonable starting dose would be 4 mg of dexamethasone intravenously four times a day, then changing to oral medication and altering the timing to give the last dose before 6 pm, because insomnia is a particularly common side effect of treatment. Steroids usually reach their maximal effect by 7 days and a gradual reduction in dose is advised then whether or not there has been an improvement in neurological deficit. Steroids almost certainly will provide a slight survival benefit, although this has never been proven in any randomized controlled trial and probably will never be. Their long-term use however is limited by the growth of the tumour, and the systemic side effects of long-term treatment.

Prognosis In patients with cerebral metastases will depend on the age; Karnofsky performance score ≥70; control at primary site and systemic metastatic spread. These factors have been used in recursive partitioning analysis to form classes with quite different survival (Gasper et al. 1997) (Fig. 27.15). Patients with single metastasis, Karnofsky performance score ≥70, aged <65 years, with controlled primary and no extracranial disease have a median

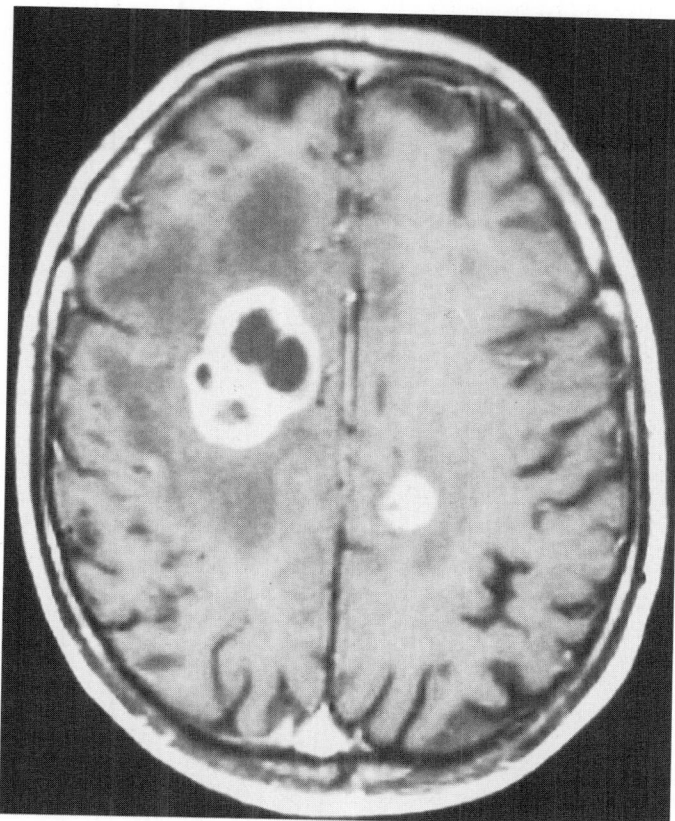

Fig. 27.15 Multiple cerebral metastases. Axial gadolinium enhanced MR scan.

survival of 13.5 months compared with only 6 months if the brain metastases are multiple and <2.3 months if the performance score is < 70.

Untreated patients usually die in a median of 4 weeks. There is evidence that prophalactic radiotherapy in small cell lung cancer prevents development of brain metastases and extends survival as well without producing cognitive problems if the fraction size is kept below 3 Gy. In situations where brain metastases are present and long-term survival and therefore long-term cognitive damage is likely to be reduced, radiotherapy is associated with median survivals of 3 to 6 months. Case selection undoubtedly plays a large part in this; nevertheless, palliative whole brain radiotherapy is the treatment of choice for most patients with brain metastases, since most are multiple, in non-resectable sites, and most patients have systemic spread elsewhere. The most effective dose of radiotherapy is widely debated. There does not appear to be a difference between 20 Gy over one week; 30 Gy in 2 weeks and 50 Gy over 4 weeks (Berk 1995).

27.9.1 Single metastases

Single brain metastasis refers to a single metastasis from a systemic tumour irrespective of the extent of spread to other organs. The term 'solitary brain metastasis' refers to the brain being the only site of systemic spread. Solitary brain metastasis is uncommon but has a better prognosis, if the brain disease can be controlled. Approximately 30 to 40 per cent of cerebral metastases are single (Delattre *et al.* 1988). Metastases from colon, kidney, and breast are more frequently single than metastases from lung or melanoma,

but because lung cancer has a higher incidence than colon or kidney, lung cancer remains the most likely cause.

Surgery There is no evidence that operation for a single brain metastasis extends survival in patients with active cancer at other sites. Nevertheless, most patients with known cancer and a presumed single brain metastasis should be considered for resection because of the higher radiological diagnostic error rate for single brain lesions. In one study of patients with known systemic cancer and a CT brain scan suggestive of single metastasis, 11 per cent were found to have a different histological diagnosis, frequently non-malignant (Patchell *et al.* 1990). If patients are being considered for surgical resection of a single brain metastasis, it is usually advisable to perform an MR brain scan, to determine whether there are actually multiple micro-metastases not identified on CT scanning (Fig. 27.16). The diagnostic accuracy approaches 95 per cent. Surgical resection of a cerebral metastasis is feasible in selected patients. A surgically accessible lesion can be defined as one that is superficial, that is close to the brain surface or abutting a fissure or sulcus, and can be operated on with minimal parenchymal resection. This type of metastasis can frequently be resected even in eloquent areas of the brain. Surgery has the benefit of removing the lesion, reducing the need for long term steroids, potentially improving quality of life and providing a small survival gain in certain situations. Three randomized controlled studies have examined the place of resection of a single brain metastasis in patients with stable disease elsewhere. Two have demonstrated that resection improves survival to 40 weeks versus 15 weeks and to 10 months versus 6 months respectively (Patchell *et al.* 1990; Vecht *et al.* 1993). The third failed to demonstrate any difference (Mintz *et al.* 1996). A Cochrane Review and meta-analysis did not demonstrate a statistically significant difference in survival between the two treatments Hazard Ratio 0.74 (95 per cent CI 0.39 to 1.40, $p=0.35$). It did suggest a possible improvement in functional independent survival with surgery (Hart and Grant 2005). The duration of functional independence was better in the surgically treated group and there were fewer deaths from neurological disease. The post-operative mortality rate was 4 per cent in each group in one study where biopsy and radiation was compared with resection and radiation (Patchell *et al.* 1990) and 9 per cent in the second study in the group treated by resection and radiation versus 0 per cent in the group who received radiation therapy only (Vecht *et al.* 1993). Is there a need for whole brain radiotherapy after surgical resection? One small trial of post-surgical radiotherapy versus no post-surgical radiotherapy demonstrated a reduction in radiological recurrence with whole brain radiotherapy and fewer cases of neurological deterioration; however, there was no difference in functional independence or survival (Patchell *et al.* 1998). Recursive partitioning analysis class 1 patients may best be managed by delaying whole brain radiotherapy, since they are at the greater risk of delayed radiation related side effects.

Radiotherapy The majority of patients with single brain metastasis will not be suitable for surgery, due to the tumour being inaccessible, systemic disease, or other health-related factors, and radiation remains the accepted palliative treatment for most patients. There is no doubt that certain patients can have a symptomatic and imaging response to radiation. These patients usually are younger than 60 years, have a good Karnofsky performance score of >70, radiosensitive tumours and controlled primary tumour, and metastatic disease confined to the brain. Failure of radiation

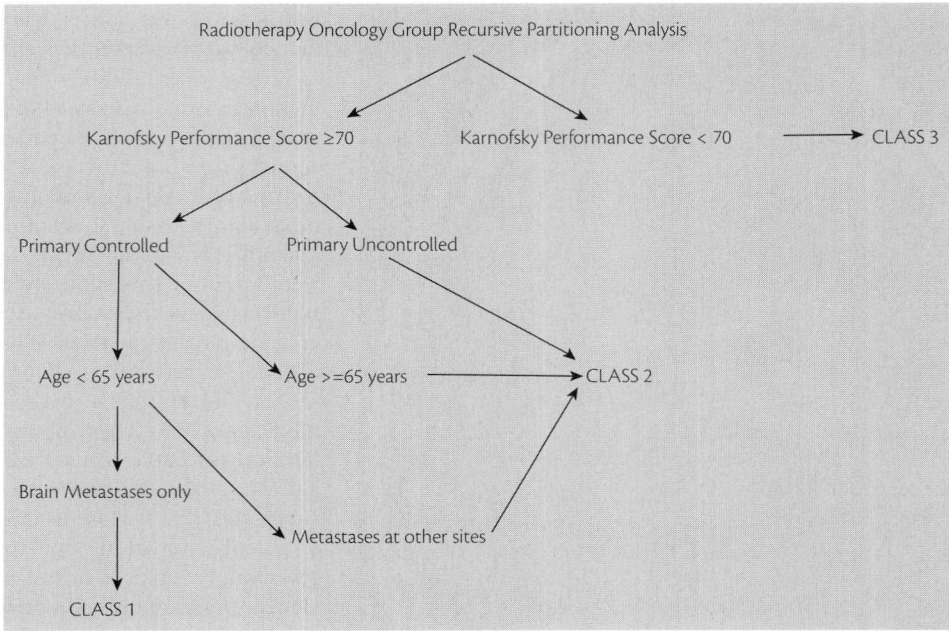

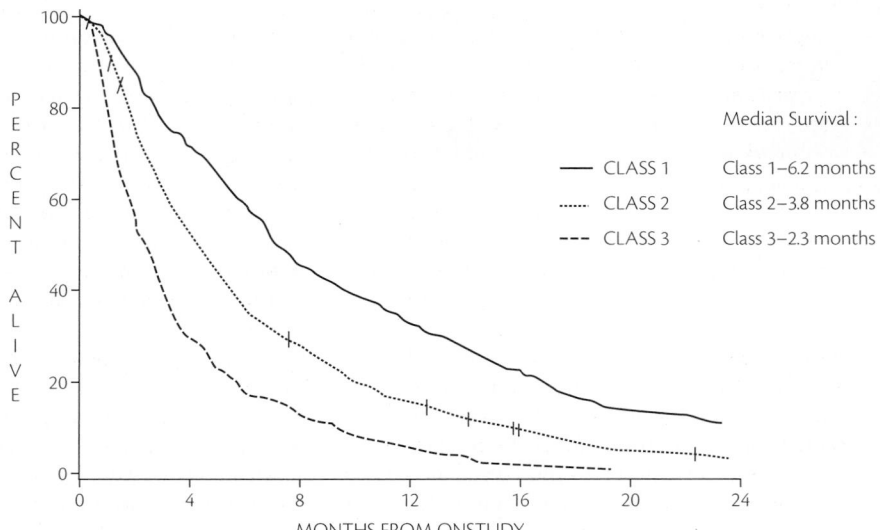

Fig. 27.16 Predictive Classification of Survival in patients with cerebral metastases. (After Gasper *et al.* 2000.)

to have any clinical or imaging response is more commonly seen elderly patients and patients with a Karnofsky performance scale <70 (Deiner-West *et al.* 1989). The optimal dose fractionation schedule for treatment of brain metastases remains uncertain and varies widely from 20 Gy given over 1 week to 50 Gy over 4 weeks. Conventional whole brain radiation therapy is thought to increase median survival in patients with brain metastasis by 3 to 6 months (Cairncross *et al.* 1980); however, this is based on retrospective non-randomized matched controlled series and is probably optimistic.

Technical advances in radiotherapy have re-opened the debate about the value of surgery for single brain metastasis. Stereotactic radiotherapy using a linear accelerator with capability for three dimensional conformal external radiation and a non-invasive removable frame, which allows fractionated treatments, or stereotactic radiosurgery using multiple cobalt-60 sources and a fixed rigid surgically attached stereotactic frame, the 'gamma knife' for

single session treatment may be as effective as surgical resection for single brain metastasis. Radiosurgery is high dose single fraction external irradiation of a stereotactically well defined target. For technical reasons the metastasis must be <3 cm in diameter and ideally should not border the ventricles, brainstem, or cranial nerves. Radiosurgery is an option for treatment in patients with single metastasis who are unfit for surgery or have a metastasis in a surgically inaccessible site. Highly selected series have demonstrated local control in 80 per cent of treated cases and an incidence of radiation necrosis of approximately 5 to 10 per cent (Flickinger *et al.* 1994; Alexander *et al.* 1995). Stereotactic radiosurgery using the conventional LINAC system or gamma knife has probably superseded the use of interstitial brachytherapy, which requires the placement of radioactive implants into the bed of the tumour after surgical resection or by stereotactic implantation and has a high incidence of radionecrosis. Because of the possibilities of treating single brain metastasis with surgical resection or stereotactic radiosurgery, there is now debate whether there is any need to treat the whole brain, if they are truly single. Randomized controlled trials to examine surgical resection versus radiosurgery alone are underway. A recent study of patients with between 1 and 3 non-resectable brain metastases treated with whole brain radiation therapy with or without stereotactic radiosurgery boost demonstrated no increase in the overall survival. Overall survival was 6.5 months with stereotactic radiosurgery boost compared to 5.7 months for radiotherapy alone. Post-hoc subgroup analyses of patients with a single metastasis demonstrated a better survival with the addition of stereotactic radiosurgery of 6.5 months versus 4.9 months. There was no significant difference in the cause of death (Andrews *et al.* 2004).

27.9.2 Multiple metastases

Management of young patients with multiple brain metastases is different from patients with single metastasis. Firstly, there is less chance of misdiagnosis on CT or MR scan (Fig. 27.16); secondly, there is less chance of having two or three metastases at surgically resectable sites, and thirdly, there is usually less opportunity for stereotactic radiotherapy, because the metastases are multiple. Conventional whole brain radiation without histological confirmation is almost always the management of choice. In patients with multiple brain metastases but stable systemic disease, there is no evidence that surgery and cranial radiation is superior to radiation alone. There are highly selected reported cases of good symptom control and extended survival from operation on two or three brain metastases at surgically accessible sites although such cases are few and far between. However, the effectiveness of surgery for multiple cranial metastases remains highly debatable and survival is probably poorer than patients with single brain metastasis (Hazuka *et al.* 1993). The series suggesting an advantage to surgery for multiple brain metastases were poorly matched, naturally highly selective, did not compare with patients treated with whole brain radiotherapy alone, and it is difficult to determine whether resections were complete or not (Bindal *et al.* 1993; Hazuka *et al.* 1993).

Systemic chemotherapy in a very selected patient group with potentially chemoresponsive tumours, such as breast, small cell lung or germ cell tumours, may be offered and may improve systemic disease, but usually the blood brain barrier will limit the efficacy in patients with brain metastases (Kristjansen *et al.* 1988; Boogerd *et al.*

Table 27.7 Causes of malignant meningitis

Primary central nervous system tumours	Systemic tumours	Haematological malignancies
Overall 1–32%	Overall 4–15%	Overall 5–15%
Medulloblastoma, 30–50%	Breast, 12–34%	Acute lymphocytic leukaemia, 40%
Ependymoma, 10–20%	Lung, 10–26%	Acute myelocytic leukaemia, 7%
Glioblastoma, 1–5%	Melanoma, 17–25%	Lymphoma, 7–30%
Primary central nervous system lymphoma, 20–30%	Gastrointestinal tract, 4–14%	
Oligodendroglioma, 5%	Unknown, 1–7%	

1992). Attempts to overcome the effect of the blood brain barrier by using fat-soluble chemotherapeutic agents or by giving a bradykinin analogue such as RMP-7, to increase the permeability of the blood brain barrier and allow chemotherapy such as carboplatin to cross the blood brain barrier have been tried with very limited success.

Management of elderly patients with multiple brain metastases and active systemic disease is palliative with aims being symptomatic control and supportive care. There is no good evidence that radiation extends survival in elderly, disabled patients with active primary disease. Predicted survival of this group without treatment is in the region of 2 months

27.10 Malignant meningitis

Malignant meningitis is defined as diffuse or widespread multifocal neoplastic involvement of the subarachnoid space. It can be due to spread from primary central nervous system tumours, metastatic spread from systemic malignancies, or due to haematological malignancies (Table 27.7) (Grossman *et al.* 1991; Walker 1991; Recht 1991). The pathogenesis is probably multifactorial. Haematogenous spread via the choroid plexus is considered to be the most common route of spread especially for haematogenous malignancies, although rupture of cerebral metastases or spread along perivascular spaces of perforating vessels is very likely in cases related to primary central nervous system malignancies and a percentage of cases with intraparenchymal metastases. In one study 50 per cent of patients with malignant meningitis from solid systemic malignancies had previously had intraparenchymal metastases (Grant *et al.* 1994). A further but less common possibility is spread from deposits in the subdural space or associated with epidural spinal cord compression and spread along the nerve roots. The dura is thick and acts as a strong physical barrier to direct spread but about 5 per cent of patients with epidural spinal cord compression have co-existing malignant meningitis.

Primary central nervous system malignancies that abut the ventricles or lie close to the surface of the brain are most likely to spread to the CSF. CSF spread occurs in 30 to 50 per cent of cases with medulloblastoma, 10 to 20 per cent of cases with ependymoma and 1 to 5 per cent of glioblastoma. This CSF spread is most commonly asymptomatic but 'dropped' metastases especially from tumours in the posterior fossa, such as ependymoma and medulloblastoma, can result in a cauda equina syndrome or spinal cord compression. Imaging of the spinal canal is an important investigation to consider prior to planning radiation therapy or further management.

Malignant meningitis is seen in 4 to 8 per cent of autopsied cases dying with systemic cancer. Approximately 5 to 10 per cent of patients with breast or lung primaries will develop malignant meningitis and these are the two most common primary sites in most series. Nevertheless the occurrence of malignant meningitis is higher in rare malignancies such as melanoma, in 10 to 15 per cent, and systemic lymphoma, especially non-Hodgkins lymphoma at 30 per cent. In haematological malignancies such as acute lymphocytic leukaemia, 40 per cent of patients have malignant cells in the CSF and in acute myelocytic leukaemia 7 per cent of patients have CSF involvement (Walker 1991).

Pathology Macroscopically, there is opacification of the meninges usually at the base of the brain but also over the convexities or cauda equina region. The pathological changes are similar to infective meningitis. Diffuse or multifocal tumour infiltrates occur with

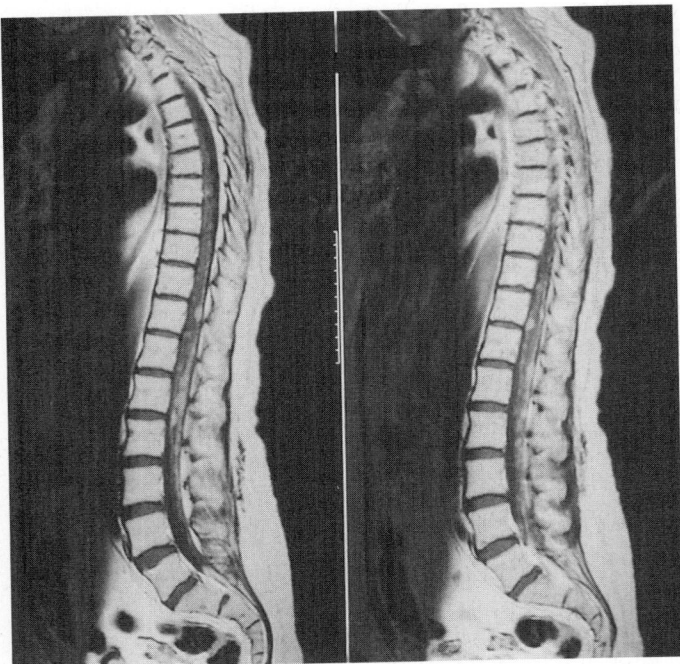

Fig. 27.17 Carcinomatous meningitis. Sagittal gadolinium enhanced MR scan of spinal canal leptomeningeal metastases.

reactive fibrosis and lymphocytosis. It is not uncommon for the reactive lymphocytosis found in the CSF to cause confusion between systemic malignancy with reactive lymphocytosis, lymphoma, or even infective meningitis. These infiltrates are commonly at the ventral surface of the brain and in the cerebral and cerebellar sulci. Tumour can encase the basal meningeal vessels causing ischaemia and infarction of the perforating vessels or encase the perineurium of the cranial or spinal nerves causing ischaemia and then degeneration. Occasionally tumour will invade nerves.

Clinical features The pathophysiology of malignant meningitis can be predicted from the pathology. Hydrocephalus resulting from obliteration of the foramina of Magendi and Lushka occurs rarely but slowed egress of CSF via the arachnoid villi can produce a communicating hydrocephalus with raised intracranial pressure. Interference with the blood supply to the parenchyma causes infarction. Metabolic competition between tumour and nerves may be the reason for the gradual onset of cranial neuropathy or radiculopathy although direct invasion of the nerves and parenchyma undoubtedly also occurs. Malignant meningitis is usually a late complication of cancer and often presents at the same time as advancing disease at other sites. Patients will often have already had intraparenchymal disease. The diagnosis of malignant meningitis should be considered particularly in patients with neurological symptoms or signs affecting a combination of cranial nerve, spinal root, and cerebral cortex (Grossman and Moynihan 1991; Walker 1991; Recht 1991). It has been estimated that 50 per cent of patients have mild memory impairment at diagnosis and dementia occurs in 30 per cent. Headache is a presenting feature in 40 per cent and usually comes on gradually but becomes increasingly severe and intractable. The characteristics of the headache may be those of raised intracranial pressure or meningeal or vascular headache. Focal or generalized seizures occur in 5 to 10 per cent of cases. Neurological signs are frequently asymptomatic, for instance,

absent reflexes, mild weakness, or subtle sensory signs. The single most common feature is a cranial neuropathy occurring in 80 per cent. The extraocular muscles are most commonly affected, in 75 per cent, followed by the facial nerve in 47 per cent or acoustic nerve in 40 per cent. The optic nerve is involved in about 38 per cent with papilloedema in 19 per cent, and reduced visual acuity in 17 per cent. Spinal root disease is the presenting symptom in about 25 per cent of cases and can be associated with back pain, limb pain, numbness, or weakness. Commonly there is a mixture of upper and lower limb spinal root symptoms to be present at the same time. On examination at the time of diagnosis 80 per cent have weakness in one or more roots and 60 per cent will have absent reflexes at some level. Occasionally cauda equina symptoms occur with sensory deficits from L1 downwards.

Investigations depend on the site of involvement and the differential diagnosis. The two most useful investigations are an imaging investigation at the clinically determined level, gadolinium enhanced MRI scan (Fig. 27.17) or myelography and CSF analysis. T1-weighted MRI with gadolinium enhancement will reveal an abnormality in 30 to 70 per cent of cases, depending on the case series reported. CSF analysis will reveal an elevated intracranial pressure of >160 mm CSF in 45 to 65 per cent of cases. CSF may also reveal an elevated protein of >0.5g/l in 81 to 89 per cent, low CSF glucose of <2.5mmol/l) in 31–41 per cent and 'lymphocytosis' of >5 cells/mm³ but negative cultures in 54 to 72 per cent (Wasserstrom *et al.* 1982). It is important to send at least 5 ml of fresh CSF quickly to the laboratory or the likelihood of obtaining a diagnostic sample may be impaired. The first CSF sample will be positive for malignant cells in about 54 per cent of cases. This is substantially higher in diffuse cellular malignant meningitis, in 75 per cent, than multifocal nodular meningeal disease, in 38 per cent. Repeated lumbar puncture will identify a further 30 per cent of cases missed on the first. If two lumbar punctures do not demonstrate malignant cells, the likelihood of a positive cytological diagnosis reduces dramatically, with only 1 per cent subsequently being positive (Wasserstrom *et al.* 1982). In the presence of repeated negative cytological specimens, the diagnosis can sometimes be made by cisternal puncture, in 2 per cent, or from sampling from ventricular CSF if there is a shunt or a ventricular access device, in 2 per cent. Approximately 10 per cent of cases will persistently have negative cytological CSF examinations, even in the presence of multi nodular deposits in the subarachnoid space. It is highly likely that the natural history, management plan, and response to treatment will be different in diffuse highly cellular malignant meningitis compared with the predominantly multi-nodular form. It may be possible to improve on the diagnostic accuracy and specificity of tumour type by using immunohistological staining methods on cytospun preparations of CSF. Epithelial membrane antigen, cytokeratins CAM 5.2, prostate specific antigen, and thyroglobulin can confirm the diagnosis and give clues to site of the primary if previously unknown. B and T cell markers may be supportive of lymphoma, HMB 45 is a relatively specific marker for melanoma and placental alkaline phosphatase may confirm germinoma. Glial fibrillary acidic protein positive staining of cells can demonstrate that the cells are of glial origin. Alphafetoprotein estimation and human chorionic gonadotrophin may support a diagnosis of teratoma or choriocarcinoma.

Management of malignant meningitis is controversial. In most cases CSF disease is a late pre-terminal manifestation of widespread disease. Treatment will depend on tumour type, with lymphoma and breast more responsive than melanoma and lung; extent of tumour spread systemically; whether the symptomatic site is cranial or spinal; raised intracranial pressure with hydrocephalus whether or not shunted; previous treatment; effectiveness of systemic chemotherapy; and meningeal deposits versus diffuse CSF pleocytosis. Most commonly, symptom control is all that can be reasonably offered using steroids, anticonvulsants and analgesia. If there is hydrocephalus, this can be shunted. It is uncertain how effective intra-reservoir chemotherapy treatment is in the presence of shunted hydrocephalus and the frequency of encephalopathy in this treatment group is higher. If there are symptomatic solid leptomeningeal metastases, radiation therapy occasionally stops progression or helps symptoms although relief is usually short-lived. If there are no solid leptomeningeal metastases, systemic disease is potentially treatable, and the patient is not severely impaired, then treatment of the CSF may be worthwhile. This is usually best given after placement of an Ommaya reservoir, since the distribution of chemotherapy is probably better than using the lumbar route (Shapiro *et al.* 1975). Nevertheless, placement of a reservoir and intra-reservoir treatment is not without complications. From a personal review of the literature, technical problems with placement occur in 6.5 to 28 per cent of cases, infection in 4.9 to 50 per cent, toxic complications of treatment in 1.7 to 20 per cent, and the mortality in reported series range from 0.5 to 8.3 per cent.

The only drugs routinely used are, methotrexate, cytosine arabinoside, or thio-tepa. In practice, if the systemic tumour is not sensitive to these agents then the CSF disease will not be sensitive. If CSF disease comes under control and malignant cells disappear from the CSF, the methotrexate can be reduced to once a week or discontinued, and clinical and CSF follow-up will determine whether further treatment is necessary. Intrathecal methotrexate should be given with preservative free saline and treatment with folinic acid, should be started at 15 mg orally every 12 hours on the day of treatment and for the following 24 hours. Folinic acid reverses the peripheral side effects and can prevent mucositis and marrow suppression. Methotrexate should be withheld if the white blood cell count falls below 3000/mm^3 or platelets below 100 000/mm^3. If leucoencephalopathy develops, methotrexate should probably be replaced by cytosine arabinoside 40 mg intraventricularly. There are approximately 5 per cent toxic deaths from treatment and 15 per cent develop confusion, disorientation, headache, nausea, or vomiting within 48 hours of methotrexate treatment although this usually resolves after 48 hours. Arachnoiditis and transverse myelitis can occur with methotrexate. The late complications of treatment with methotrexate include leucoencephalopathy and a necrotising encephalopathy. In certain situations, systemic chemotherapy may be valuable in treating the CSF disease or extending survival, for instance with high dose intravenous methotrexate for primary central nervous system lymphoma or breast carcinoma (Grant *et al.* 1994; Ackland *et al.* 1987). Liposomal cytosine arabinoside has been used in a controlled trail against soluble cytosine arabinoside and has been shown to have a better cytological response rate of 71 per cent versus 15 per cent (Glantz *et al.* 1999). Intrathecal topotecan has also been used recently (Blaney *et al.* 2003).

Prognosis For treated malignant meningitis this varies from series to series and one suspects that this is an area where 'gearing' of results and publication bias plays a large role in the apparent effectiveness of treatment. CSF becomes negative for malignant cells in approximately 40 per cent of cases. In general 25 per cent of patients have symptomatic improvement, 50 per cent remain stable for short periods of several weeks and 25 per cent progressively decline and die in 6 weeks. Periods of stability and improvement range from 1 week to 2.5 years with a median of 3 months. Median survival for untreated malignant meningitis is approximately 4 to 6 weeks. With 'aggressive' treatment one-third are dead in 6 weeks, median survival ranges from 9 to 24 weeks and 10 per cent survive >1 year. In selected cases where there is no systemic disease, two-thirds of patients will remain stable or improve, with median survival of 10 months, and 20 per cent are alive at 1 year (Kim and Glantz 2001).

27.11 Complications of treatment

27.11.1 Surgery

The complications of surgery for brain tumours will depend on whether the tumour is sited intracerebrally, intracranial extracerebrally, or in the neck or skull. Complications will also depend on surrounding important structures such as nerves, endocrine or vascular structures, the radiological appearance of the tumour regarding size, uniformity, or necrotic areas, tumour pathology, the experience of the neurosurgeon, and the state of general health of the patient.

It is unlikely that the frequency of side effects recorded in the literature reflects the day to day frequency of side effects in general neurosurgery units (Maurice-Williams 1997). The complications of neurosurgery can be divided into non-surgical and surgical complications.

Non-surgical complications

Patients with cancer are at an increased risk of deep venous thrombosis and pulmonary embolus. The additional risks of a surgical operation with bedrest and possible intra-tumoural bleeding make peri-operative management difficult. Deep venous thrombosis and pulmonary embolus pose a serious risk and prophylaxis with elasticated stockings and subcutaneous heparin are usually indicated (Frim *et al.* 1992). Electrolyte disturbances secondary to diabetes insipidus or syndrome of inappropriate antidiuretic hormone secretion can lead to a stormy peri-operative course. Patients may be systemically unwell as a result of malignancy or super-added chest or urinary tract infections. Endocrine deficiencies should be treated prior to neurosurgery on the pituitary gland and close attention paid to any peri-operative endocrine complications.

Surgical complications

Post-operative haematomas at the operative site occur in approximately 5 per cent of patients. This frequency is probably higher in patients with malignant melanoma, choriocarcinoma, lung carcinoma, glioblastoma and lymphoma. This higher risk of bleeding may influence decision on whether stereotactic biopsy or an open procedure is performed in some cases. In operations for an intracerebral tumour, a transient neurological deficit will occur in approximately 10 per cent of patients post-operatively, slightly less with stereotactic operations. There is recovery of the neurological deficit in approximately 50 per cent of cases. There is also a risk of seizures as a result of operation in those who have no prior history of seizures and sometimes a flurry of seizures occur in the post-operative period. Intra-operative stroke will occur in less than 1 per cent depending on selection. Tumours in the region of the sylvian fissure are probably best operated on by an open procedure, because of the moderately high risk of damaging of the branches of the middle cerebral artery with catastrophic results. Post-operative infective meningitis and cerebral abscess are rare complications of craniotomy but still occur. Post-operative hydrocephalus is also uncommon but occurs particularly in patients undergoing posterior fossa surgery, where hydrocephalus is probably due to post-operative brain swelling or contamination of the CSF by blood or debris. Operations in the region of the temporal lobe can result in significant memory deficits which are sometimes not appreciated because pre-operative cognitive assessments are commonly not performed as post-operative bedside assessments more commonly concentrate on focal weakness or sensory impairments. Operations on the head and neck may damage cranial nerves, such as the facial nerve in parotid surgery, infra-orbital nerve in maxillary surgery, palatal and vocal cord paralysis due to damage to branches of the vagus in radical neck dissections or ipsilateral Horner's syndrome.

Mortality from craniotomies for malignant glioma or metastasis approach 5 per cent, and for stereotactic neurosurgery are approximately 1 per cent, depending on the selection of patients and the experience of the surgeon (Cabantog and Bernstein 1994).

27.11.2 Radiation Therapy

The toxic side effects of cranial radiation can be divided into local effects and central nervous system effects (Section 5.9). The side effects or radiation therapy depend on the dose fractionation schedule used, the natural history of the underlying disease and the likelihood of get a radiotherapeutic response.

Local effects

Some people will feel nauseated about 30 minutes to 1 hour after treatments, and find that small meals with a low fat content are usually preferable to a large lunch or dinner. Patients will develop alopecia, but the degree and likelihood of recovery will depend on the dose and fractionation schedule used. Hair loss starts about two or three weeks into treatment and maximum regrowth has occurred by 6 months. Frequently the hair that returns in the irradiated area is fine and curly and may be of a slightly different colour. Skin can become dusky red, dry and itchy about three weeks into treatment and slight deafness can occur due to wax build up. Most people feel tired and sleepy at the end of a course of radiation and some feel sick.

Some years after cranial irradiation there may be further local neuroendocrine or neural complications. Pituitary failure can occur if the pituitary has received a moderately large dosage directly for pituitary adenoma, or is in the treatment field as in fronto-temporal low grade glioma. Radiation usually affects the prolactin and sex hormones first, with an increase in prolactin levels and decrease in FH and LSH, causing problems with periods or infertility, then the thyroid stimulating hormone falls producing secondary hypothyroidism. If the optic nerve is in the treatment field, one commonly finds an afferent pupillary defect with optic neuropathy which is usually asymptomatic or only produces mild visual acuity disturbance. Years after temporal lobe or posterior fossa irradiation one may find mild sensorineural hearing loss.

Central nervous system effects

The most serious central nervous system complications of radiation to the nervous system are: acute encephalopathy; subacute or early delayed demyelination; delayed cerebral radiation necrosis; and chronic leucoencephalopathy.

Acute encephalopathy is rare, but comes on usually within 24 hours of cranial irradiation. Symptoms consist of headache, nausea vomiting, fever or worsening of neurological deficits. Occa-sionally, swelling causes cerebral herniation. The likelihood of developing the acute encephalopathy is related to dose and whether the patient is pre-treated with steroids prior to radiation. It can be difficult in some patients with brain tumours to know whether the deterioration is attributable to radiation or to progression of the underlying tumour. The treatment is steroids and consider reducing the fraction size of radiation.

Early delayed reaction is common and is seen 4 weeks to 3 months after completion of cranial radiation. In patients with cerebral tumours the symptoms are indistinguishable from those of tumour progression, except there is usually commonly a feeling of excessive tiredness and nausea. In cases who have died with early delayed subacute radiation reaction and who have had a post-mortem, there are changes in the white matter of the brain stem or cerebrum of demyelination, similar to those of multiple sclerosis. The treatment is to re-institute steroids for a period of 4 to 8 weeks and the gradually reduce them and discontinue.

Delayed cerebral radiation necrosis is infrequent and can start months or years after cerebral irradiation. In patients who have cerebral tumours, the clinical and radiological appearances mimic tumour recurrence. MRI cannot adequately distinguish active tumour from radiation necrosis. Positron emission tomography and single photon emission tomography can sometimes give an indication of whether there is increased radioisotope uptake consistent with active tumour or reduced uptake related to an avascular mass consistent with radiation necrosis, but neither technique is infallible and the only sure way to find out is to resect the necrotic mass. In patients with malignant glioma, it is common to see areas of necrosis within the tumour consistent with radiation damage and other areas of active tumour. Radiation necrosis is characterized by fibrinoid necrosis, luminal narrowing or occlusion, medial fibrosis and adventitial proliferation in small arteries. There may also be bizarre, multinucleated astrocytes and foci of necrosis. The necrosis is thought to be due to ischaemia secondary to changes in the small and medium vessels. Other hypotheses for the necrosis are that the radiation-induced changes in the glia produce demyelination and white matter damage or that the radiation causes release of cytokines into surrounding brain which results in tissue damage. Necrosis as a result of tumour progression does not have the same degree of small vessel occlusive and fibrotic changes but has significant endothelial proliferation.

Chronic leucoencephalopathy is usually only found in long-term survivors of cranial irradiation. Ten per cent of patients who survive for more than 1 year after radiotherapy for cerebral metastases will develop cognitive problems. Relatives notice that the patient may lack motivation, there is psychomotor retardation, memory impairment, and ataxia or apraxia of gait. As time passes there may be urinary incontinence, marked dementia, inability to walk due to apraxia or ataxia and cortical myoclonus. High dose and large fractionation schedules are thought to be associated with a higher incidence of radiation-induced leucoencephalopathy (DeAngelis *et al.* 1989; Klein *et al.* 2002). Recent studies however have suggested that radiotherapy in glioma is not the main reason for cognitive deficits (Taphoorn and Klein 2004). The tumour itself and other medical treatments also contribute. The CT or MR scan shows diffuse white matter changes in the cortical white matter and ventricular dilatation or generalized atrophy. The clinical picture is similar to normal pressure hydrocephalus or basal ganglia disease. Patients do not improve with lumbar puncture or ventricular shunting.

27.11.3 Chemotherapy

Side effects of chemotherapy may be a property of the mode of delivery or of the agent itself. Modes of delivery of chemotherapy include direct chemotherapy into the tumour bed at the time of surgery, intra-arterial chemotherapy where the tumour is confined to an area supplied by one artery, usually the internal carotid, or systemic chemotherapy either intravenously or orally. Intra-arterial chemotherapy requires arterial catheterization and commonly a general anaesthetic since there can be severe pain in the distribution supplied by the sensory fibres within the artery. This is a direct toxic effect on the artery. In addition even with supra-ophthalmic instillation of chemotherapy, there can be significant optic nerve toxicity due to turbulence and back flow along the ophthalmic artery. This can result in unilateral visual loss or ocular necrosis. Intravenous chemotherapy must be given cautiously in a fast flowing arm vein. If chemotherapy gets into the soft tissues of the arm it can result in a severe local thrombophlebitis.

The other effects of systemic chemotherapy relate to the toxic effect of the individual drugs. Nitrosoureas such as BCNU cause nausea that starts about two hours after starting an infusion that may persist for 24 to 48 hours. Facial flushing or dizziness may occur during infusion which is rate dependent and resolves on stopping the drug. Bone marrow suppression occurs with almost all agents and is maximal 4 to 6 weeks after receiving the drug and usually settles by 8 weeks. Risks of infection, bleeding and tiredness are greatest around 4 weeks after treatment. Lung toxicity usually starts after a total dose of 1 g of BCNU. A restrictive ventilatory defect is found and it is valuable to monitor vital capacity regularly in patients who receive >1 g total dose. Renal function should also be monitored. Procarbazine can also cause haematological and gastrointestinal symptoms, but in addition can cause flu-like symptoms, rash, and neurological symptoms of ataxia, headaches, paraesthesia, and dizziness. Procarbazine can cause hypertensive crisis and severe gastrointestinal symptoms if given with antidepressants, alcohol, or tyramine rich foods such as cheese or bananas. Vincristine causes neurotoxicity with neuropathy, myopathy, and autonomic disturbance, gastrointestinal symptoms, and sometimes alopecia. Haematological toxicity is mild.

Steroid use can result in weight gain, oedema, electrolyte problems, diabetes, osteoporosis, thinning of skin, and predisposition to infections and peptic ulcers. Cisplatinum derivatives cause neurotoxicity with peripheral neuropathy or deafness, renal toxicity, and bone marrow suppression. High dose cytosine arabinoside and 5 fluorouracil can cause reversible cerebellar ataxia, in addition to bone marrow suppression and gastrointestinal and liver toxicity. Alopecia and infertility can occur with any of the drugs. Methotrexate can cause bone marrow suppression, mucositis, and rarely pneumonitis. It is nephrotoxic and hepatotoxic and can cause an encephalopathy if given in high dose. Toxic effects to intrathecal methotrexate are mentioned in Section 27.10.

27.12 Recovery and rehabilitation

Recovery and rehabilitation from a diagnosis of brain tumour or from medical treatment of a brain tumour may involve many specialists and support services. The first step is often coming to terms with the diagnosis and this can be eased by accurate, understandable medical information about the disease and its treatment options. This is best done by a doctor experienced in managing patients with brain tumours (Richards 1990). The Royal College of Physicians have produced guidelines regarding breaking bad news to the patients with malignant glioma and their relatives. (Davis *et al.* 1997). Written information can often be a helpful reminder to patients and relatives, particularly where there are cognitive difficulties. There are also helpful general informational leaflets about different sorts of brain tumours from charitable organizations. Occasionally, patients with brain tumours will have disabling anxiety or will become clinically depressed and may require special counselling or antidepressant medication. Neuropsychological symptoms may be related to the disease, for instance seizures or fronto-limbic involvement with tumour, treatment with anticonvulsants, steroids, radiation therapy or chemotherapy, or fear of the future including loss of health, independence, work, family position, or relationships. Neurocognitive support from psychologists, psychiatrists, support groups, and self help groups may all play a role in neuropsychological recovery by providing information, enhancing personal control and teaching coping mechanisms.

Physical rehabilitation involves improvement of neurological impairment by steroids, anticonvulsants, painkillers, or speech therapy, coping with physical disability using walking aids, eating aids, adaptation of the home and by maximizing physical independence thus reducing handicap by encouraging re-integration into home, work, and past-times where feasible, and minimizing unnecessary hospital contact. In the early post-operative period, physical rehabilitation usually progresses alongside medical therapy in hospital, however, there may be a feeling of active treatment for the tumour grinding to a halt after radiation and this is paradoxically sometimes a period of despair and anxiety while patients await 'what's next'. Patients should be made aware of possible early delayed effects, to allay the fear of early recurrence and should have a target directed plan for recovery which includes their own programme for rehabilitation and for periods of rest.

References

Abrey LE, Bachelor TT, Ferreri AJM *et al.* (2005). Report of an international workshop to standardise baseline evaluation and response criteria for primary central nervous system lymphoma. *J Clin Oncol*, **23**, 5034–43.

Abrey LE, DeAngelis LE, Yahalom J (1998). Long term survival in primary CNS Lymphoma. *J Clin Oncol*, **16**, 859–63.

Ackland S, Schilsky R (1987). Review article; High dose methotrexate: A critical reappraisal. *J Clin Oncol*, **5**, 2017–31.

Albert FK, Forsting M, Sartor K *et al.* (1994). Early post-operative magnetic resonance imaging after resection of malignant glioma: Objective evaluation of residual tumor and its influence on regrowth and prognosis. *Neurosurgery*, **34**, 45–61.

Aldape K, Simmons ML, Davis RL *et al.* (2000). Discrepancies in diagnoses of neuroepithelial neoplasms: the San Francisco Bay Area Adult Glioma Study. *Cancer*, **88**, 2342–9.

Alexander E, Moriarty TM, Davis RB *et al.* (1995). Sterotactic radiosurgery for the definitive noninvasive treatment of brain metastases. *J Natl Cancer Inst*, **87**, 34–40.

Alvarez L, Peris P, Pons F *et al.* (1997). Relationship between biochemical markers of bone turnover and bone scintigraphic indices in assessment of Paget's disease activity. *Arth Rheum*, **40**, 461–8.

Andrews DW, Scott CB, Sperduto PW *et al.* (2004). Whole brain radiation therapy with or without steotactic radiosurgery boost for patients with one to three brain metastases: phase III results of RTOG 9508 randomised trial. *Lancet*, **363**, 1665–72.

Barkan, AL, Lloyd RV, Chandler WF *et al*. (1988). Pre-operative treatment of acromegaly with long acting sandostatin: shrinkage of invasive pituitary macroadenomas and improved surgical remission rate. *J Clin Endocrinol Metab*, **67**, 1040–8.

Barker FG, Davis RL, Chang SM *et al*. (1996). Necrosis as a prognostic factor in glioblastoma multiforme. *Cancer*, **77**, 1161–6.

Batchelor TT, Carson K, O'Neill A *et al*. (2003). NABTT CNS Consortium: The treatment of central nervous system lymphoma (PCNSL) with methotrexate and deferred radiotherapy—NABTT 96-07. *J Clin Oncol*, **21**, 1044–9.

Bederson JB, von Ammon K, Wichmann WW *et al*. (1991). Conservative treatment of patients with vestibular tumors. *Neurosurgery*, **28**, 646–51.

Berger MS (1995). Role of surgery in diagnosis and management. In Appuzzo MLJ, ed. *Benign Cerebral Glioma*, pp. 293–307. Park Rtdge, AANS.

Berk L (1995). An overview of radiotherapy trials for the treatment of brain metastases. *Oncology*, **9**, 1205–19.

Bernstein M, Parrent AG (1994). Complications of CT-guided stereotactic biopsy of intra-axial brain lesions. *J Neurosurg*, **81**, 165–8.

Bindal RK, Sawaya R, Leavens Me *et al*. (1993). Surgical treatment of multiple brain metastases. *J Neurosurg*, **79**, 210–6.

Black PM (1993). Meningiomas. *Neurosurgery*, **32**, 643–57.

Black PM, Carroll R, Zhang J (1996). The molecular biology of hormone and growth factor receptors in meningiomas. *Acta Neurochir (Suppl)*, **65**, 50–3.

Blaney SM, Heideman R, Berg S *et al*. (2003). Phase 1 clinical trial of intrathecal topotecan in patients with neoplastic meningitis. *J Clin Oncol*, **21**, 143–7.

Bleehen NM, Stenning SP (1991). A Medical Research Council trial of two radiotherapy doses in the treatment of grades 3 and 4 astrocytoma. The Medical Research Council Brain Tumour Working Party. *Br J Cancer*, **64**, 769–74.

Bloom B, Kramer S (1994) Conventional radiation therapy in the management of acromegaly. In *Secretory Tumors of the Pituitary Gland*.

Bloom HJG, Glees J, Bell J (1991). The treatment and long term prognosis of children with intracranial tumors: a study of 610 cases, 1950–1981. *Int J Radiat Oncol Biol Phys*, **18**, 723–45.

Boogerd W, Dalesio O, Bais EM *et al*. (1992). Response of brain metastases from breast cancer to systemic chemotherapy. *Cancer*, **69**, 972–80.

Brada M, Rajan B, Traish D *et al*. (1993). The long term efficacy of conservative surgery and radiotherapy in the control of pituitary adenomas. *Clin Endocrinol*, **38**, 571–8.

Browman GP, Hodson DI, Mackenzie RJ *et al*. (2001). Choosing a concomitant chemotherapy and radiotherapy regimen for squamous cell head and neck cancer: A systematic review of the published literature with subgroup analysis. *Head Neck*, **23**, 579–89.

Bruce JN, Stein BM (1993). Complications of surgery for pineal region tumors. In Post KD, Friedman ED, McCormick PC, eds. *Postoperative Complications in Intracranial Neurosurgery*, pp. 74–86. Thieme Medical Publishers, New York.

Bunin G (1987). Racial patterns of childhood brain cancer by histologic type. *J Natl Cancer Inst*, **78**, 875–80.

Burger PC, Dubois PJ, Schold SC *et al*. (1983). Computerised tomographic and pathologic studies of untreated, quiescent, and recurrent glioblastoma multiforme. *J Neurosurg*, **58**, 159–69.

Burger PC, Heinz ER, Shibata T *et al*. (1988). Topographic anatomy and CT correlations in untreated glioblastoma multiforme. *J Neurosurg*, **68**, 698–704.

Cabantog AM, Bernstein M (1994). Complications of first craniotomy for intr-axial brain tumour. *Can J Neurol Sci*, **21**, 213–8.

Cairncross JG, Eisenhauer EA (1995). Response and control: lessons from oligodendroglioma. *J Clin Oncol*, **13**, 2475.

Cairncross JG, Kim J-H, Posner JB (1980). Radiation therapy for brain metastases. *Ann Neurol*, **7**, 529–41.

Cairncross JG, Pexman JHW, Rathbone MP *et al*. (1985). Post-operative contrast enhancement in patients with brain tumor. *Ann Neurol*, **17**, 570–2.

Cairncross G, Seiferheld W, Shaw E *et al*. (2004). An intergroup randomized controlled clinical trial (RCT) of chemotherapy plus radiation (RT) versus RT alone for pure and mixed anaplastic oligodendrogliomas: Initial report of RTOG 94-02. *J Clin Oncol*; 2004 ASCO Annual Meeting Proceedings (Post-Meeting Edition). Vol 22, No 14S (July 15 Supplement), 1500.

Castro-Vita H, Salazar OM, Scarantino C *et al*. (1980). Medulloblastomas. *Rev Int Radiol*, **5**, 77–82.

Chang CH, Horton J, Schoenfeld D (1983). Comparison of post-operative radiotherapy and chemotherapy in the multidisciplinary management of malignant gliomas. *Cancer*, **52**, 977–1007.

Clyde Z, Chataway J, Slattery J *et al*. (1998). Significant change in tests of neurological impairment in patients with brain tumours. *J Neuro-Oncol*, **39**, 81–90.

Cohen BH, Packer RJ (1996). Chemotherapy for medulloblastomas and primitive neuroectodermal tumors. *J Neuro-Oncol*, **29**, 55–68.

Comtois R, Beauregard H, Somma M *et al*. (1991). The clinical and endocrine outcome to trans-sphenoidal microsurgery of non functioning pituitary adenomas. *Cancer*, **68**, 860–6.

Counsell CE, Collie DA, Grant R (1996). Incidence of intracranial tumours in Lothian Region of Scotland, 1989-90. *J Neurol Neurosurg Psychiatry*, **61**, 143–50.

Counsell C, Collie D, Grant R (1997). Limitations of using a cancer registry to identify incident primary intracranial tumours. *J Neurol Neurosurg Psychiatry*, **63**, 94–7.

Counsell C, Grant R (1998). Incidence studies of primary and secondary intracranial tumours: a systematic review of their methodology and results. *J Neuro-Oncol*, **37**, 241–50.

Curran WJ, Scott CB, Horton J *et al*. (1993). Recursive partitioning analysis of prognostic factors in three Radiation Therapy Oncology Group malignant glioma trials. *J Natl Cancer Inst*, **85**, 704–10.

D'Alessandro G, Di Giovanni M, Iannizzi L *et al*. (1995). Epidemiology of primary intracranial tumors in the Valle d'Aosta (Italy) during the 6-year period 1986-1991. *Neuroepidemiology*, **14**, 139–46.

Davis E, Hopkins A, on behalf of a Working Group (1997). Good Practice in the management of adults with malignant cerebral glioma: clinical guidelines. *Br J Neurosurg*, **11**, 318–30.

DeAngelis LM, Delattre JY, Posner JB (1989). Radiation induced dementia in patients cured of brain metastases. *Neurology*, **39**, 789–96.

Deen HG, Ebersold MJ, Harner SG *et al*. (1996). Conservative management of acoustic neuroma: an outcome study. *Neurosurgery*, **39**, 260–6.

Deiner-West M, Dobbins TW, Phillips TL *et al*. (1989). Identification of an optimal subgroup for treatment evaluation of patients with brain metastases using RTOG study 7916. *Int J Radiat Oncol Biol Phys*, **16**, 669–73.

Delattre JY, Krol G, Thaler HT *et al*. (1988). Distribution of brain metastases. *Arch Neurol*, **45**, 741–4.

Deutsch M (1988). Medulloblastoma: staging and treatment outcome. *Int J Radiat Oncol Biol Phys*, **14**, 1103–7.

Deutsch M, Thomas PR, Krischer J *et al*. (1996). Results of a prospective randomized trial comparing standard dose neuraxis irradiation (3,600 cGy/20) with reduced neuraxis irradiation (2,340 cGy/13) in patients with low-stage medulloblastoma. A combined Children's Cancer Group-Pediatric Oncology Group Study. *Pediatr Neurosurg*, **24**, 167–76.

Dhami MS, Bona RD (1993). Thrombosis in cancer patients. *J Postgrad Med*, **93**, 131–3, 137–40.

Dumas-Duport C, Scheithauer B, O'Fallon J *et al*. (1988). Grading of astrocytomas. A simple and reproducible method. *Cancer*, **62**, 2152–65.

Duncan W, McLelland J, Jack WJ *et al*. (1986). Report of a randomized pilot study of the treatment of patients with supratentorial gliomas using neutron irradiation. *Br J Radiol*, **59**, 373–7.

Ebersold MJ, Quast LM, Laws ER *et al*. (1986). Long term results in transsphenoidal removal of non functioning pituitary adenomas. *J Neurosurg*, **64**, 713–9.

Ellison D (2002). Classifying the medulloblastoma: insights from morphology and molecular genetics. *Neuropathol Appl Neurobiol*, **28**, 257–82.

EORTC Brain Tumor Group (1981). Evaluation of CCNU, VM-26 plus CCNU, and procarbazine in supratentorial brain gliomas. *J Neurosurg*, **55**, 27–31.

Evans A, Jenkin D, Sposto R *et al.* (1990). The treatment of medulloblastoma: results of a prospective randomized trial of radiation with and without CCNU, vincristine and prednisone. *J Neurosurg*, **72**, 575–82.

Eyre HJ, Crowley JJ, Townsend JJ *et al.* (1993). A randomized trial of radiotherapy versus radiotherapy plus CCNU for incompletely resected low grade gliomas: a Southwest Oncology Group study. *J Neurosurg*, **78**, 909–14.

Fadul C, Wood J, Thaler H *et al.* (1988). Morbidity and mortality of craniotomy for excision of supratentorial glioma. *Neurology*, **38**, 1374–9.

Ferreri AJM, Abrey LE, Blay J-Y *et al.* (2003). Management of primary central nervous system lymphoma: A summary statement from the 8th International Conference on Malignant Lymphoma. *J Clin Oncol*, **21**, 2407–14.

Flickinger JC, Kondziolka D, Niranjan AS *et al.* (2001). Results of acoustic neuroma radiosurgery: An analysis of five years' experience using current methods. *J Neurosurg*, **94**, 1–6.

Flickinger JC, Kondziolka D, Lunsford LD *et al.* (1994). A multi-institutional experience with steroetactic radiosurgery for solitary brain metastasis. *Int J Radiat Oncol Biol Phys*, **28**, 797–802.

Fonari M, Solero CL, Lasio G *et al.* (1990). Surgical treatment of intracranial dermoid and epidermoid cysts in children. *Childs Nerv Syst*, **6**, 66–70.

Friedman WA, Murad GJ, Bradshaw P *et al.* (2005). Linear accelerator surgery for meningiomas. *J Neurosurg*, **103**, 206–9.

Frim DM, Barker FG, Poletti CE *et al.* (1992). Post-operative low-dose heparin decreases thromboembolic complications in neurosurgical patients. *Neurosurgery*, **30**, 830–3.

Galicich JH, Arbit E, Wronski M (1996). Metastatic brain tumours. 807-12. In Wilkins RH, Rengachary SS, eds. *Neurosurgery*, 2nd edition, pp. 807–12. McGraw-Hill, New York.

Gasper L, Scott C, Rotman M *et al.* (1997). Recursive Partitioning Analysis (RPA) of prognostic factors in three Radiation Therapy Oncology Group (RTOG) brain metastases trials. *Int J Radiat Oncol Biol Phys*, **37**, 745–51.

Glantz MJ, Berger PC, Herndon JE 2nd *et al.* (1991). Influence of the type of surgery on the histologic diagnosis in patients with anaplastic gliomas. *Neurology*, **41**, 1741–4.

Glantz MJ, Cole BF, Friedberg MH *et al.* (1996). A randomised, blinded, placebo controlled trial of divalproex sodium prophalaxis in adults with newly diagnosed brain tumors. *Neurology*, **46**, 985–91.

Glantz MJ, LaFolette S, Jaeckle KA *et al.* (1999). Randomized trial of slow release versus standard formulation of cytarabine for the intrathecal treatment of lymphomatous meningitis. *J Clin Oncol*, **17**, 3110–6.

Glass J, Gruber ML, Cher L *et al.* (1994). Pre-irradiation methotrexate chemotherapy of primary central nervous system lymphoma: long-term outcome. *J Neurosurg*, **81**, 188–95.

Glioma Meta-analysisTrial Group (2002). Chemotherapy for high grade glioma. *Cochrane Database Syst Rev*, CD003913(3).

Goodman R (1993). Magnetic resonance imaging-directed stereotactic endoscopic third ventriculostomy. *Neurosurg*, **32**, 1043–7.

Goodwin JW, Crowley J, Eyre HJ *et al.* (1995). A phase II evaluation of tamoxifen in unresectable or refractory meningiomas: a south west oncology group study. *J Neuro-Oncol*, **15**, 75–7.

Grant R, Slattery J, Gragor A *et al.* (1994a). Recording neurological impairment in clinical trials og glioma. *J Neuro-Oncol*, **19**, 37–49.

Grant R, Liang BC, Page MS *et al.* (1995). Age influences chemotherapy response in astrocytomas. *Neurology*, **45**, 929–33.

Grant R, Liang BC, Slattery J *et al.* (1997). Chemotherapy response criteria in malignant glioma. *Neurology*, **48**, 1336–40.

Grant R, Naylor B, Greenberg HS *et al.* (1994b). Clinical outcome in aggressively treated meningeal carcinomatosis. *Arch Neurol*, **51**, 457–61.

Grant R, Walker M, Hadley D *et al.* (2002). Imaging response to chemotherapy with RMO-7 and carboplatin in malignant glioma: size matters but speed does not. *J Neurooncol*, **57**, 241–5.

Grant R, Whittle IR, Collie D *et al.* (1996). Referral pattern and management of patients with malignant brain tumours in South East Scotland. *Health Bull*, **54**, 212–22.

Green SB, Byar DP, Walker MD *et al.* (1983). Comparisons of carmustine, procarbazine and high dose methylprednisolone as additions to surgery and radiotherapy for the treatment of malignant glioma. *Cancer Treat Rep*, **67**, 121–32.

Green SB, Shapiro WR, Burger PC *et al.* (1989). Randomized comparison of intra-arterial (IA) cisplatinum and intravenous (IV) PCNU for the treatment of primary brain tumours (BTCG study 8420A). *Proc Am Soc Clin Oncol*, **8**, 26.

Greig NH, Ries LG, Yancik R *et al.* (1990). Increasing annual incidence of primary malignant brain tumors in the elderly. *J Natl Cancer Inst*, **82**, 1621–4.

Grossman S and Moynihan T (1991). Neoplastic meningitis. *Neurol Clin*, **9**, 843–56.

Grossman SA, O'Neill A, Grunnet M *et al.* (2003). Phase III study comparing three cycles of infusional carmustine and cisplatin followed by radiation therapy with radiation therapy and concurrent carmustine in patients with newly diagnosed supratentorial glioblastoma multiforme: Eastern Cooperative Oncology Group trial 2394. *J Clin Oncol*, 21, 1485–91.

Guterman KS, Hair LS, Morgello S (1996). Epstein-Barr virus and AIDS-related primary central nervous system lymphoma. Viral detection by immunohistochemistry, RNA in situ hybridization, and polymerase chain reaction. *Clin Neuropath*, **15**, 79–86.

Gutin PH, Klemme WM, Lagger RL *et al.* (1980). Management of the unresectable cystic craniopharyngioma by aspiration through an Ommaya reservoir drainage system. *J Neurosurg*, **52**, 36–40.

Haddad FS, Haddad GF, Zaatari G (1997). Cranial osteomas: their classification and management. Report on a giant osteoma and review of the literature. *Surg Neurol*, **48**, 143–7.

Hagmar L, Akesson B, Nielsen J *et al.* (1990). Mortality and cancer morbidity in workers exposed to low levels of vinyl chloride monomer at a polyvinyl chloride processing plant. *Am J Ind Med*, **17**, 553–65.

Halperin EC, Bentel G, Heinz ER *et al.* (1989). Radiation therapy treatment planning in supratentorial glioblastoma multiforme: An analysis based on post mortem topographic anatomy with CT correlations. *Int J Radiat Oncol Biol Phys*, **17**, 1347–50.

Halperin EC, Herndon J, Schold SC *et al.* (1996). A phase III randomized prospective trial of external beam radiotherapy, mitomycin C, carmustine, and 6-mercaptopurine for the treatment of adults with anaplastic glioma of the brain. CNS Cancer Consortium. *Int J Radiat Oncol Biol Phys*, **34**, 793–802.

Hammoud MA, Sawaya R, Shi W *et al.* (1996). Prognostic significance of pre-operative MRI scans in glioblastoma multiforme. *J Neuro-Oncol*, **27**, 65–73.

Hardy DG, Macfarlane R, Baguley D *et al.* (1989). Surgery for acoustic neuroma: an analysis of 100 translabyrinthine operations. *J Neurosurg*, **71**, 799–804.

Hart MG, Grant R - Cochrane Review (2005). Surgical resection and whole brain radiation therapy versus brain radiation therapy alone for single brain metastasis. *Cochrane database.*

Hart MG, Grant R, Metcalfe SE. Cochrane Review (2007). Biopsy versus resection for high grade glioma. *Cochrane Database Syst Rev*, CD002034.

Hazuka MB, Burleson W Stroud DN *et al.* (1993). Multiple brain metastases are associated with poor survival in patients treated with surgery and radiotherapy. *J Clin Oncol*, **11**, 369–73.

Hegi ME, Diserens AC, Gorlia T *et al.* (2005). MGMT Gene Silencing and benefit from Temozolomide in Glioblastoma. *N Engl J Med*, **352**, 997–1003.

Hochberg FH, Miller DH (1988). Primary central nervous system lymphoma. *J Neurosurg*, **68**, 835–53.

IARC (2003). Tumours of the nervous system. In Stewart BW, Kleihues P, eds. *World Cancer Report*, IARC Press, Lyon.

Immonen A, Vapalahti M, Tyynela K *et al.* (2004). AdvHSV-tk gene therapy with intravenous ganciclovir improves survival in human glioma: a randomised controlled study. *Mol Ther*, **10**, 967–72.

Jakacki RI (2005). Treatment strategies for high-risk medulloblastoma and supratentorial primitive neuroectodermal tumors. Review of the literature. *J Neurosurg*, **102**, 44–52.

Jeffries BF, Kishore PRS, Singh KS *et al.* (1981). Contrast enhancement in the post-operative brain. *Radiology*, **139**, 409–13.

Kadan-Lottick NS, Skluzarek MC *et al.* (2002). Decreasing incidence of primary central nervous system lymphoma. *Cancer*, **95**, 193–202.

Karim AB, Afra D, Cornu P *et al.* (2002). Randomized trial on the efficacy of radiotherapy for cerebral low-grade glioma in the adult: European Organization for Research and Treatment of Cancer Study 22845 with the Medical Research Council study BRO4: an interim analysis. *Int J Radiat Oncol Biol Phys*, **52**, 316.

Kato H, Yoshimoto Y, Schull WJ (1989). Risk of cancer among children exposed to atomic bomb irradiation in utero: a review. In Napalkov NP, Rice JM, Tomatis L, Yamasaki H, eds. *Perinatal and Mutational Carcinogenesis*, pp. 365–74. International Agency for Research on Cancer, Lyon.

Kaye AH, Giles GG, Gonzales M (1993). Primary central nervous system tumors in Australia: a profile of clinical practice from the Australian brain tumor registry. *Aus N Z J Surg*, **63**, 33–8.

Keime-Guibert Chinot O, Taillandier L, Cartalat-Carel S *et al.* (2005). Phase III study comparing radiotherapy with supportive care in older patients with newly diagnosed anaplastic astrocytomas or Glioblastoma multiforme (GBM). An ANOCEF group trial. Abst. World Fed *Neuro-Oncol*, 2005 (Edinurgh).

Kelly PJ, Dumas-Duport C, Kispert DB *et al.* (1987). Imaging based stereotaxic serial biopsies in untreated intracranial glial neoplasms. *J Neurosurg*, **66**, 865–74.

Kim L, Glantz MJ (2001). Neoplastic meningitis. *Curr Treat Options Oncol*, **2**, 517–27.

Kim TS, Halliday AL, Hedley-White ET *et al.* (1991). Correlates of survival and the Dumas-Duport grading system for astrocytomas. *J Neurosurg*, **74**, 27–37.

Kleihues P, Burger PC, Scheithauer BW (1993a). In World Health Organisation. *Histological Typing of Tumours of the Central Nervous System*, 2nd edition. Springer Verlag, Berlin.

Kleihues P, Burger PC, Scheithauer BW (1993b). The new WHO classification of brain tumours. *Brain Pathol*, **3**, 255–68.

Klein M, Heimans JJ, Aaronson NK *et al.* (2002). Effect of radiotherapy and other treatment related factors on mid term to long term cognitive sequelae in low grade gliomas: a comparative study. *Lancet*, **360**, 1361–8.

Kondziolka D, Lunsford LD, Martinez AJ (1993). Unreliability of contemporary neurodiagnostic imaging in evaluating suspected adult supratentorial (low grade) astrocytoma. *J Neurosurg*, **79**, 533–6.

Krishnan S, Foote RL, Brown PD, Pollock BE *et al.* (2005). Radiosurgery for cranial base chordomas and chondrosarcomas. *Neurosurgery*, **56**, 777–84.

Kristjansen PE, Hansen HH (1988). Brain metastases from small cell lung cancer treated with combination chemotherapy. *Eur J Cancer Clin Oncol*, **24**, 545–9.

Kritis AP, Yung WKA, Bruner J *et al.* (1993). The treatment of anaplastic oligodendrogliomas and mixed gliomas. *Neurosurgery*, **32**, 365–70.

Lachance DH, Brizel DM, Gockerman JP *et al.* (1994). Cyclophosphamide, doxorubicin, vincristine and prednisone for primary central nervous system lymphoma: short duration response and multifocal intracerebral recurrence preceeding radiotherapy. *Neurology*, **44**, 1721–7.

Landverg TG, Lindgren ML, Cavalin-Stahl EK *et al.* (1980). Improvements in the radiotherapy of medulloblastoma 1946–1975. *Cancer*, **45**, 670–8.

Laperriere NJ, Leung PM, McKenzie S *et al.* (1998). Randomized study of brachytherapy in the initial management of patients with malignant astrocytoma. *Int J Rad Onc Biol Phys*, 41, 1005–11.

Lawton MT, Hamilton MG, Beals SP *et al.* (1995). Radical resection of anterior skull base tumors. (Review). *Clin Neurosurg*, **42**, 43–70.

Lee A, Chan G, Fung C *et al.* (1995). Paradoxical response of a pineal immature teratoma to combination chemotherapy. *Med Pediatr Oncol*, **24**, 53–7.

Levin VA, Silver P, Hannigan J *et al.* (1990). Superiority of post radiotherapy adjuvant chemotherapy with CCNU, procarbazine, and vincristine (PCV) over BCNU for anaplastic gliomas: NCOG 6G61 Final report. *Int J Radiat Oncol Biol Phys*, **18**, 321–4.

Levin VA, Wara WM, Davis RL *et al.* (1985). Phase III comparison of chemotherapy with BCNU and the combination of procarbazine, CCNU, and vincristine administered after radiation therapy with hydroxyurea to patients with malignant gliomas. *J Neurosurg*, **63**, 218–23.

Li FP, Fraumeni JF Jr (1969). Soft tissue sarcoma, breast cancer and other neoplasms. A family syndrome? *Ann Intern Med*, **71**, 747–52.

Ling SM, Roach M, Larson DA *et al.* (1997). Radiotherapy of primary central nervous system lymphoma in patients with and without human immunodeficiency virus. Ten years experience at the University of California San Francisco. *Cancer*, **73**, 2570–82.

Macdonald DR (1994). Low grade gliomas, mixed gliomas and oligodendrogliomas. *Semin Oncol*, **21**, 236–48.

Macdonald DR, Gaspar LE, Cairncross JG (1990). Successful chemotherapy for newly diagnosed aggressive oligodendroglioma. *Ann Neurol*, **27**, 573–4.

MacMahon EME, Glass JD, Harris NL *et al.* (1991). Epstein-Barr virus in AIDS-related primary central nervous system lymphoma. *Lancet*, **338**, 969–73.

Maekawa A, Mitsumori K (1990). Spontaneous occurrence and chemical induction of neurogenic tumours in rats — influence of host factors and specificity of chemical structure. *Crit Rev Toxicol*, **20**, 287–310.

Mahaley MS (1991). Neuro-oncology Index and review (adult brain tumours): Radiotherapy, chemotherapy, immunotherapy, photodynamic therapy. *J Neuro-Oncol*, **11**, 85–147.

Mahaley MS, Mettlin C, Natarajan N *et al.* (1990). Analysis of patterns of care of brain tumor patients in the United States: A study of the brain tumor section of the AANS and CNS, and the commission on cancer of the ACS. *Clin Neurosurg*, **36**, 347–52.

Mahaley MS, Whaley RA, Blue M *et al.* (1986). Central neurotoxicity following intracarotid BCNU chemotherapy for malignant glioma. *J Neuro-Oncol*, **3**, 297–314.

Mahaley MS, Whaley RA, Krigman MR *et al.* (1987). Randomized phase III trial of single versus multiple chemotherapeutic treatment following surgery and during radiotherapy for patients with anaplastic gliomas. *Surg Neurol*, **27**, 430–2.

Malkin D, Li FP, Strong LC *et al.* (1990). Germ line p53 mutations in a familial syndrome of breast cancer, sarcoma and and other neoplasms. *Science*, **250**, 1233–8.

Masdeu JC, Quinto C, Olivera C *et al.* (2000). Open-ring imaging sign: highly specific for atypical brain demyelination. *Neurology*, **54**, 1427–33.

Mathiesen T, Kihlstrom L, Karlsson B *et al.* (2003). Potential complications following radiotherapy for meningiomas. *Surg Neurol*, **60**, 193–8.

Maurice-Williams RS (1997). The notes in the cupboard: the question of intellectual honesty in neurosurgery. *Br J Neurosurg*, **11**, 277–9.

Metcalfe SE, Grant R (2001). Biopsy versus resection of malignant glioma. *Cochrane Database Syst Rev*, CD002034.

McKinney P (2004). Brain Tumours: incidence, survival and aetiology. *J Neurol Neurosurg Psychiatry*, **75**, 12–7.

McLoone P (1993). Cartstairs codes for Scottish postcode sectors from the 1991 census. Cartstairs Codes for Scottish Postcode Sectors from the 1991 Census. Public Health Research Unit, University of Glasgow, Glasgow.

Mintz AH, Kestle J, Rathbone MP *et al.* (1996). A randomized trial to assess the efficacy of surgery in addition to radiotherapy in patients with a single intracerebral metastasis. *Cancer*, **78**, 1470–6.

Monreal M, Alastrue A, Rull M *et al.* (1996). Upper extremity deep venous thrombosis in cancer patients with venous access devices — prophalaxis with a low molecular weight heparin (Fragmin). *Thromb Haemost*, **75**, 251–3.

Mulhern RK, Palmer SL, Merchant TE *et al.* (2005). Neurocognitive Consequences of Risk-Adapted Therapy for Childhood Medulloblastoma. *J Clin Oncol*, **23**, 5511–9.

Muller W, Afra D, Schroder R (1977). Supratentorial recurrences of gliomas: Morphological studies in relation to time intervals with astrocytomas. *Acta Neurochir*, **37**, 75–91.

Nelson DF, Diener-West M, Horton J *et al.* (1988). Combined modality approach to treatment of malignant gliomas: Re-evaluation of RTOG 7401/ECOG 1374 with long term follow-up: A joint study of the Radiation Therapy Oncology Group and Eastern Cooperative Oncology Group. *NCI monogr*, **6**, 279–84.

Nelson DF, Martz KL, Bonner H *et al.* (1992). Non Hodgkin's lymphoma of the brain: Can high dose, large volume radiation therapy improve survival? Report of a prospective trial by the Radiation Therapy Oncology Group (RTOG): RTOG 8315. *Int J Radiat Oncol Biol Phys*, **23**, 9–17.

Newton H, Junck L, Bromberg J *et al.* (1990). Procarbazine chemotherapy in the treatment of recurrent malignant astrocytomas after radiation and nitrosourea failure. *Neurology*, **40**, 1743–6.

NICE. National Institute of Clinical Excellence: Improving outcomes in brain and other CNS tumours. Guidance manual. http://www.nice.org.uk/page.aspx?o=282278

NICE National Institute of Clinical Excellence: Referral for suspected cancer. http://www.nice.org.uk/page.aspx?o=cg027niceguideline

Ogunrinde OK, Lunsford LD, Flickinger JC *et al.* (1994). Stereotactic radiosurgery for acoustic nerve tumors in patients with useful preoperative hearing: results at 2-year follow-up examination. *J Neurosurg*, **80**, 1011–7.

Ohgaki H, Eibl RH, Schwab M *et al.* (1993). Mutations of the p53 tumor supressor gene in neoplasms of the central nervous system. *Mol Carcinog*, **8**, 74–80.

Okamoto K, Furasawa, Ishikawa K *et al.* (2004a). Mimics of brain tumor: part I. *Radiat Med*, **22**, 63–76.

Okamoto K, Furasawa, Ishikawa K *et al.* (2004b). Mimics of brain tumor: part II. *Radiat Med*, **22**, 135–42.

Patchell RA, Tibbs PA, Regine WF *et al.* (1998). Postoperative radiotherapy in the treatment of single metastases to the brain: a randomised trial. *JAMA*, **280**, 1485–9.

Patchell RA, Tibbs PA, Walsh JW *et al.* (1990). A randomized trial of surgery in the treatment of single metastases to the brain. *N Engl J Med*, **322**, 494–500.

Patel SR, Buckner JC, Smithson WA *et al.* (1992). Cisplatinum based chemotherapy in primary central nervous system germ cell tumours. *J Neuro-Oncol*, **12**, 47–52.

Piepmeier J, Christopher S, Spencer D *et al.* (1996). Variations in the natural history and survival of patients with supratentorial low grade astrocytomas. *Neurosurgery*, **38**, 872–9.

Pickles T, Goodman GB, Rheaume DE *et al.* (1997). Pion radiation for high grade astrocytoma: results of a randomized study. *Int J radiat Oncol, Biol Phys*, **37**, 491–7.

Pignatti F, van den Bent MJ, Curran D *et al.* (2002). Prognostic factors for survival in adult patients with cerebral low-grade glioma. *J Clin Oncol*, **20**, 2076–84.

Pignon JP, Bourhis J, Domenge C *et al.* (2000). Chemotherapy added to locoregional treatment for head and neck squamous cell carcinoma: three metaanalyses of updated individual data. *Lancet*, 949–55.

Pobereskin LH, Chadduck JB (2000). Incidence of brain tumours ib two English counties: a population based study. *J Neurol Neurosurg Psychiatry*, **69**, 464–71.

Pollock BE, Lunsford LD, Kondziolka D *et al.* (1995). Phosporus-32 intracavity irradiation of cystic craniopharyngiomas: current technique and long term results. *Int J Radiat Oncol Biol Phys*, **33**, 1944–52.

Posner JB, Chernik NL (1978). Intracranial metastases from systemic cancer. *Adv Neurol*, **19**, 579–87.

Posner JB (1980). 189-207. Clinical manifestations of brain metastasis. In Weiss L, Gilbert HA, Posner JB, eds. *Brain Metastasis*, pp 189–207. GK Hall, Boston.

Post KD, Eisenberg MB, Catalano PJ (1995). Hearing preservation in vestibular schwannoma surgery: What factors affect outcome? *J Neurosurg*, **83**, 191–6.

Radhakrishnan K, Mokri B, Parisi JE *et al.* (1995). The trends in incidence of primary brain tumors in the population of Rochester, Minnesota. *Ann Neurol*, **37**, 67–73.

Rainov NG (2000). A phase III clinical evaluation of herpes simplex virus type 1 thymidine kinase and ganciclovir gene therapy as an adjuvant to surgical resection and radiation in adults with previously untreated glioblastoma multiforme. *Hum Gene Ther*, **20**, 2389–401.

Rasheed BA, Wiltshire RH, Bigner RN *et al.* (1999). Molecular pathogenesis of malignant gliomas. *Curr Opin Oncol*, **11**, 162–67.

Recht LD (1991). Neurologic complications of systemic lymphoma. *Neurol Clin*, **9**, 1001–15.

Regine WF, Kramer S (1992). Pediatric craniopharyngiomas: Long term results of combined treatment with surgery and radiation. *Int J Radiat Oncol Biol Phys*, **24**, 611–7.

Remick SC, Diamond C, Migliozzi JA *et al.* (1990). Primary central nervous system lymphoma in patients with and without the acquired immune deficiency syndrome: a retrospective analysis and review of the literature. *Medicine*, **69**, 345–60.

Reni M, Ferreri AJ, Garancini MP *et al.* (1997). Therapeutic management of primary central nervous system lymphoma in immunocompetent patients: results of a critical review of the literature. *Ann Oncol*, **8**, 227–34.

Richards T (1990). Chasms in communication. *BMJ*, **301**, 1407–8.

Ringel MD, Schwindinger WF, Levine MA (1996). Clinical implications of genetic defects in G proteins. The molecular basis of McCune-Albright syndrome and Albright hereditary osteodystrophy. *Medicine*, **75**, 171–84.

Ron E, Modan B, Boice JD Jr *et al.* (1988). Tumours of the brain and nervous system after radiotherapy in childhood. *N Eng J Med*, **319**, 1033–9.

Roush GC (1979). Epidemiology of cancer of the nose and paranasal sinuses. *Head Neck Surg*, **2**, 3–11.

Royal College of Physicians (1997). Pituitary tumours: Recommendations for service provision and guidelines for management of patients. Consensus statement of a working party. *Pubs R Coll Physicians Lond.*

Salcman M (1995). The natural history of low grade gliomas. In Apuzzo MLJ, ed. *Benign Cerebral Glioma*, pp 213–29. AANS, Park Ridge.

Salcman M (1987). Surgical decision making for malignant brain tumours. *Clin Neurosurg*, **35**, 285–313.

Sanna M, Karmarkar S, Landolfi M (1995). Hearing preservation in vestibular schwannoma surgery: fact or fantasy? *J Laryngol Otol*, **109**, 374–80.

Sarasin FP, Eckman MH (1993). Management and prevention of thromboembolic events in patients with cancer-related hypercoagulable states: a risky business. *J Gen Intern Med*, **8**, 476–86.

Schiff D, Shaffrey ME (2003). Role of resection for newly diagnosed malignant gliomas. *Exp Rev Anticancer Ther*, **3**, 621–30.

Schild SE, Scheithauer BW, Schomberg PJ *et al.* (1993). Pineal parenchymal tumors: clinical, pathologic, and therapeutic aspects. *Cancer*, **72**, 870–80.

Schulz-Ertner D, Haberer T, Jakel O *et al.* (2002). Radiotherapy for cordomas and low grade chondrosarcomas of the skull base with carbon ions. *Int Radiat Oncol Biol Phys*, **53**, 36–42.

Scimeca PG, James-Herry AG, Black KS *et al.* (1996). Chemotherapy treatment of malignant chordoma in children. *J Pediatr Hematol Oncol*, **18**, 237–40.

Selby PL, Davie MW, Ralston SH *et al.* (2002). Guidelines on the management of Paget's disease of bone. *Bone*, **31**, 366–77.

Shapiro WR, Green SB, Burger PC *et al.* (1989). Randomized trial of three chemotherapy regimens and two radiotherapy regimens in postoperative treatment of malignant glioma. BTSG Trial 8001. *J Neurosurg*, **71**, 1–9.

Shapiro WR, Young DF, Mehta BM (1975). Methotrexate: distribution in cerebrospinal fluid after intravenous, ventricular and lumbar injections. *N Engl J Med*, **293**, 161–6.

Shaw EG, Scheithauer BW, O'Fallon JR (1997). Supratentorial gliomas: a comparative study by grade and histologic type. *J Neuro-Oncol*, **31**, 273–8.

Shaw E (1995). The low grade glioma debate: evidence defending the position of early radiotherapy. *Neurosurg Clin*, **42**, 488–94.

Shelton C, Hitselberger WE, House WF *et al.* (1990). Hearing preservation after acoustic tumor removal: long term results. *Laryngoscope*, **100**, 115–9.

Simpson D (1957). The recurrence of intrathecal meningioma after surgical treatment. *J Neurol Neurosurg Psychiatry*, **20**, 22–39.

Simpson JR, Horton J Scott C et al. (1993). Influence of location and extent of surgical resection on survival of patients with glioblastoma multiforme; results of three consecutive Radiation Therapy Oncology Group (RTOG) clinical trials. *Int J Radiat Biol Phys*, **26**, 239–44.

Sioutos P, Yen V, Arbit E (1996). Pituitary gland metastases. *Ann Surg Oncol*, **3**, 94–9.

Souhami L, Seiferheld W, Brachman D et al. (2004). Randomized Prospective Comparison Of Stereotactic Radiosurgery (SRS) followed by conventional Radiotherapy (RT) With BCNU To RT With BCNU alone for selected patients with Supratentorial Glioblastoma Multiforme (GBM): Report Of RTOG 93-05 Protocol. *Int J Radiat Oncol Biol Phys*, **60**, 853–60.

Springate SC, Weichselbaum RR (1990). Radiation or surgery for chemodectoma of the temporal bone: a review of local control and complications. *Head Neck*, **12**, 303–7.

Stummer W, Pichmeier U, Meinel T et al. (2006). Fluorescence-guided surgery with 5-aminolevulinic acid for resection of malignant glioma: a randomised controlled multicentre phase III trial. *Lancet Oncol*, **7**, 392–401.

Stupp R, Mason WP, van den Bent MJ (2005). Radiotherapy plus concomitant and adjuvant Temozolomide for Glioblastoma. *N Engl J Med*, **352**, 987–96.

Tai PT, Craighead P, Bagdon F (1995). Optimization of radiotherapy for patients with cranial chordoma. A review of dose response ratios for photon techniques. *Cancer*, **75**, 749–56.

Tait DM, Thornton-Jones H, Bloom HJ et al. (1990). Adjuvant chemotherapy for medulloblastoma: the first multicentre controlled trial of the International Society of Paediatric Oncology (SIOP 1). *Eur J Cancer*, **26**, 464–9.

Taphoorn MJ, Klein M (2004). Cognitive deficits in adult patients with brain tumours. *Lancet Neurol*, **3**, 159–68.

Van den Bent MJ, Afra D, de Witte O et al. (2005). Long-term efficacy of early versus delayed radiotherapy for low-grade astrocytoma and oligodendroglioma in adults: the EORTC 22845 randomised trial. *Lancet*, **366**, 985–9.

Vecht CJ, Haaxma-Reiche H, Noordijk EM et al. (1993). Treatment of single brain metastasis: radiotherapy alone or combined with neurosurgery? *Ann Neurol*, **33**, 583–90.

von Deimling A, Bender KB, Jahnke R et al. (1994). Loci associated with malignant progression in astrocytomas: a candidate on chromosome 19q. *Cancer Res*, **54**, 1397–401.

Vuorinen V, Hinkka S, Farkkila M et al. (2003). Debulking or biopsy of malignant glioma in elderly people—a randomised study. *Acta Neurochir*, **145**, 5–10.

Walker RW (1991). Neurological complications of leukaemia. *Neurol Clin*, **4**, 989–99.

Walker MD, Alexander E Jr, Hunt WE et al. (1978). Evaluation of BCNU and/or radiotherapy in the treatment of anaplastic gliomas: a co-operative clinical trial. *J Neurosurg*, **49**, 333–43.

Walker MD, Green SB, Byar DP et al. (1980). Randomized comparisons of radiotherapy and nitrosoureas for the treatment of malignant glioma after surgery. *N Engl J Med*, **303**, 1323–9.

Wallner KE, Galicich JH, Krol G et al. (1989). Patterns of failure following treatment for glioblastoma multiforme and anaplastic astrocytoma. *Int J Radiat Oncol Biol Phys*, **16**, 1405–9.

Wasserstrom W, Glass J, Posner J (1982). Diagnosis and treatment of leptomeningeal metastases from solid tumours. *Cancer*, **49**, 759–72.

Weiss M, Sutton L, Marcial V et al. (1989). The role of radiation therapy in the management of childhood craniopharyngioma. *Int J Radiat Oncol Biol Phys*, **17**, 1313–21.

Wen BC, Hussey DH, Staples J et al. (1989). A comparison of roles of surgery and radiation therapy in the management of craniopharyngiomas. *Int J Radiat Oncol Biol Phys*, **16**, 17–24.

Westphal M, Hilt DC, Bortey E et al. (2003). A phase 3 trial of local chemotherapy with biodegradable carmustine (BCNU) wafers (Gliadel wafers) in patients with primary malignant glioma. *Neuro-Oncol*, **5**, 79–88.

Whelan HT, Clanton JA, Wilson RE et al. (1988). Comparisons of CT and MRI brain tumor imaging using a canine glioma model. *Pediatr Neurol*, **4**, 279–83.

Willems PW, Taphoorn MJ, Burger H et al. (2006). Effectiveness of neuronavigation in resecting solitary intracerebral contrast enhancing tumors: a randomized controlled trial. *J Neurosurg*, **104**, 360–8.

Wrensch MR, Minn Y, Chew T et al. (2002). Epidemiology of primary brain tumors: current concepts and review of the literature. *Neuro-Oncol*, **4**, 278–99.

Yamada N, Kato M, Yamashita H et al. (1995). Enhanced expression of transforming growth factor-beta and its type I and type II receptors in human glioblastoma. *Int J Cancer*, **62**, 386–92.

Yasergil MG, Curic M, Kis M et al. (1990). Total removal of craniopharyngiomas. Approaches and long term results in 144 patients. *J Neurosurg*, **73**, 3–11.

CHAPTER 28

Spinal cord disorders

David Bates

Contents

28.1 Introduction

This and Chapter 29 address non-traumatic, pathological disorders involving the spinal cord, cauda equina, and nerve roots within the vertebral canal. The systems are anatomically close, lying within the vertebral canal, so there is an introductory section on the anatomy of these nerve structures. The innervation of the bladder, rectum, corpus cavernosum, and seminal vesicles is reviewed in Chapter 29, but pathological disorders outside the vertebral canal, which affect sphincter function, are discussed in Section 22.7.

28.1.1 Differential diagnosis

Non-traumatic spinal cord disease may be caused by compression due to tumour, infection or haematoma, inflammation, infection or post-infection, metabolic disturbances, infarction, and degeneration. The diagnosis is often made easier by the clinical assessment: the patient's age, the speed of onset of the disease, severity of the deficits, the pattern of motor and sensory involvement, and presence of pain and sphincter symptoms are all important in making an assessment of the site and likely nature of the spinal disease.

Investigations are obligatory to confirm a diagnosis and to direct therapy. MRI is the most useful investigation. It has largely replaced myelography which should now only be considered in patients with indwelling cardiac pacing wires. Additional investigations including examination of the cerebrospinal fluid, evoked potentials, and specific blood tests may be required and the value of plain X-rays, CT scan, and, in some instances, angiography should not be overlooked.

The remainder of this chapter will consider specific disorders, identifying pathology, clinical presentation, investigation, and management. Acute and chronic conditions are considered separately and those affecting the cauda equina, spinal root, and sphincters are considered in the Chapter 29.

28.2 Clinical anatomy of the spinal cord

28.2.1 The spinal cord and long tracts

The spinal cord extends from the foramen magnum, where it joins the medulla oblongata, to the level of the first or second lumbar vertebra; it is within the vertebral canal throughout is length. It is oval, flattened in the antero-posterior axis, and has two enlargements, one in the cervical and one in the lumbar region, corresponding to the outflow of nerves to the upper and lower limbs. At its lower end it terminates in the conus medullaris from the end of which the filum terminale continues downwards to the posterior surface of the coccyx. The surface of the spinal cord shows several longitudinal grooves; there is a deep anterior fissure and a shallower posterior medium sulcus and on the lateral aspect two sulci, termed antero-lateral and postero-lateral. From these lateral sulci a series of root filaments emerge anteriorly and posteriorly on each side. At intervals several filaments from the postero-lateral sulcus unite to form a dorsal root upon which is situated the dorsal root ganglion; similarly, those from the antero-lateral sulcus unite to form a ventral root. The dorsal and ventral roots are paired, they join just distal to the dorsal root ganglion and form the spinal nerve which exits the canal through the intervertebral foramen.

The spinal cord is organized into segments, one corresponding to each pair of spinal nerves. There are eight cervical, twelve dorsal or thoracic, five lumbar, five sacral, and one coccygeal segment. The spinal cord ends at the first or second lumbar vertebra and all the nerves below the first lumbar descend to their respective foramina in a bundle of nerves which resembles a horse's tail and is called the cauda equina.

The spinal cord, like the brain, is surrounded by three meninges: the pia mater, a fibrous membrane, forms the immediate covering of the cord and fine septa penetrate into the cord substance; the arachnoid mater is a delicate, transparent membrane lying superficially to the pia mater from which it is separated by the sub-arachnoid space, containing cerebrospinal fluid, CSF, and being bridged by trabeculae; outside the arachnoid is the dura mater which lines the vertebral canal, from which it is separated by the epidural space containing fat and the internal vertebral venous plexus. The arachnoid extends to the second sacral vertebra, the dura to the third sacral vertebra. The spinal cord is suspended within the dural sheath by a series of ligamenta denticulata, which extend laterally from each side of the cord terminating in tooth-like attachments to the inner aspect of the dura.

On transverse section the spinal cord is divided into the central grey matter, which has an H-shape, and the peripheral white matter. The grey matter consists of ganglion cells and nerve fibres and the white matter of ascending and descending fibres with their myelin sheaths.

The white matter, which consists of longitudinal bundles of nerve fibres, is divided into three columns on each side. The anterior column lies between the anterior fissure and the anterior horn of grey matter, the lateral column lies laterally to the grey matter between the ventral and dorsal roots, which is between the antero-lateral and postero-lateral sulci, and the posterior column lies between the posterior medium septum and the posterior horn of grey matter.

The anterior column contains ascending and crossed fibres in the ventral spino-thalamic tract, together with descending fibres in the olivo-spinal, vestibulo-spinal, tecto-spinal, and ventral cortico-spinal tracts. The lateral column contains the major descending motor pathway, the lateral cortico-spinal tract, with the smaller descending rubro-spinal tract and the ascending and crossed lateral spino-thalamic tract, and the ventral and dorsal spino-cerebellar tracts. The dorsal column contains the ascending, uncrossed gracile and cuneate fascicles carrying axons sub-serving proprioceptive and vibration sensation to their first synapse in the respective nuclei

in the brainstem. In the lateral cortico-spinal tract the descending motor neurones destined for the lumbo-sacral segments run laterally to those destined for the cervical segment and a similar layered relationship exists for ascending fibres in the crossed spino-thalamic tracts. However, in the posterior columns fibres from the lower limbs lie more medial than those ascending from the upper limbs so the dorsal column enlarges progressively from the lower thoracic to the cervical region.

Descending sympathetic autonomic fibres travel in the intermedio-lateral columns of the grey matter on each side of the thoracic cord and in the upper lumbar segments. Parasympathetic neurones lie in the anterior horns of the grey matter of the sacral segments.

28.2.2 **Segmental organization**

The fact that the spinal cord ends at the first or second lumbar vertebra explains why the segmental organization of the cord is anatomically higher in each of its levels than the spinal nerves. In the cervical region the cord segment is 1–2 vertebrae above the level of the exiting nerve root, in the thoracic region 3–4 vertebrae higher and even more in the lumbo-sacral and coccygeal segments. The H-shaped mass of grey matter within the spinal cord consists of anterior and posterior horns on each side, united by a grey commissure in the middle of which lies the central canal. The anterior horns contain ganglion cells, the axons of which enter the anterior roots and form the lower motor neurone running to the effector muscle. The alpha motor neurones which innervate skeletal muscles are larger than the gamma motor neurones which run to the muscle spindles. The motor neurones are arranged in groups transversely and columns longitudinally and these specific groups and columns in the cervical and lumbar enlargements innervate specific muscle groups and individual muscles via the relevant spinal nerves. The total number of lower motor neurones in the anterior horns of the spinal cord is remarkably consistent, both from side to side and between individuals (Tomlinson *et al.* 1973). The nerve cell groups in the posterior horn of the grey matter comprise the substantia gelatinosa which are the cell bodies of axons entering through the dorsal roots and which synapse with other neurones whose axons cross the mid-line to ascend in the contralateral spinothalamic tracts. Other cells in the dorsal funicular group, or nucleus proprius, may have similar function whereas the neurones of the nucleus dorsalis synapse and send axons into the dorsal spino-cerebellar tracts.

At each segmental level the entering dorsal root sends some axons to synapse in the dorsal horn but others, which have their cell bodies in the dorsal root ganglion, remain ipsilateral and ascend in the posterior funiculus.

At each segmental level the dorsal, sensory nerve root with its appendage, the dorsal root ganglion, separates from the ventral motor root just inside the intervertebral foramen, the two uniting distally to form the individual spinal nerve root and, in the cervical and lumbo-sacral levels become part of the nerve plexus.

28.2.3 **Blood supply of the spinal cord**

Arterial

The spinal cord is richly supplied with blood. There are two posterior spinal arteries, each derived from the corresponding vertebral or posterior inferior cerebellar artery at the level of the foramen magnum. These two posterior vessels traverse the length of the spinal cord lying just in front of, or just behind, the dorsal nerve roots. There is a single anterior spinal artery formed by the union of a

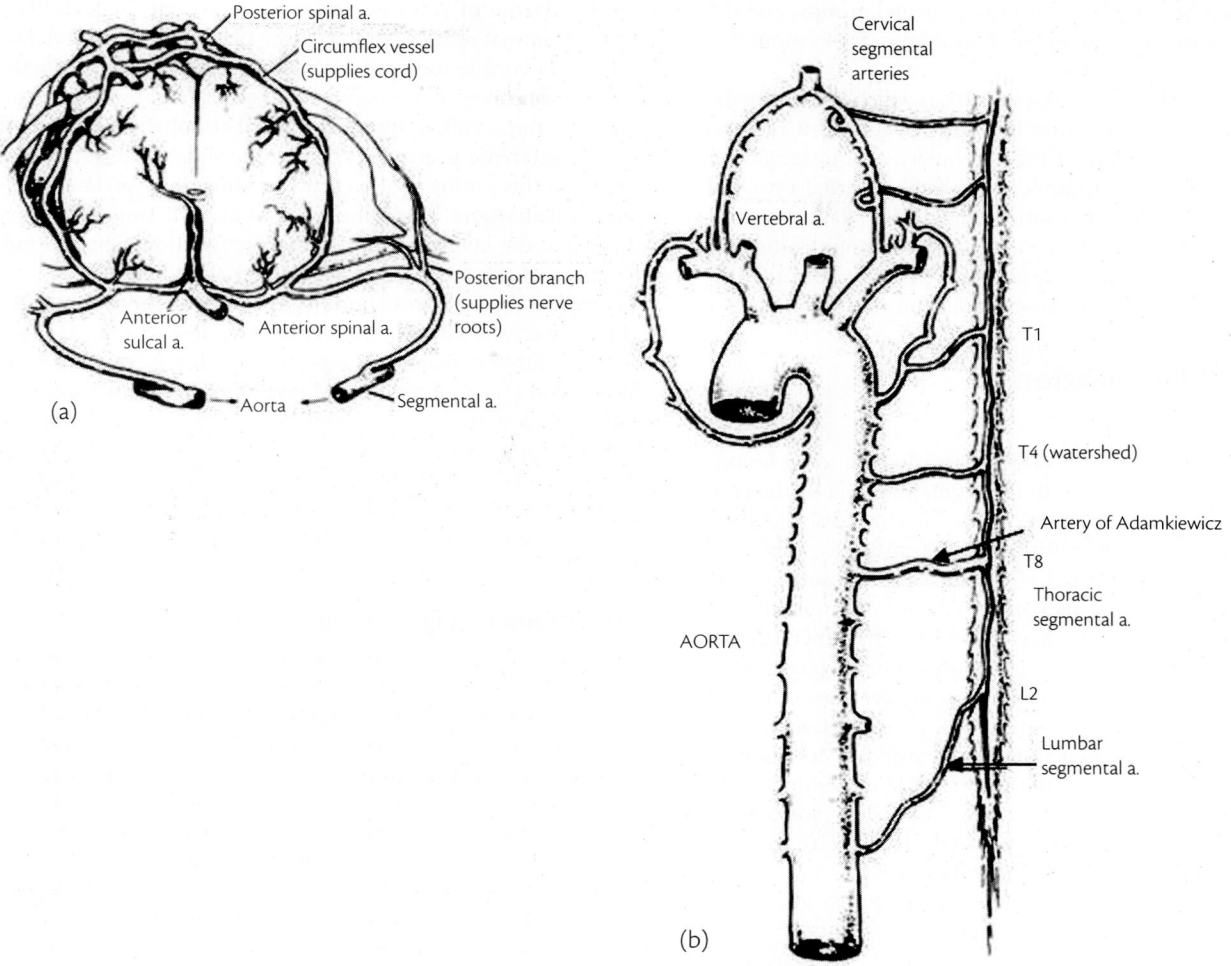

Fig. 28.1 Arterial supply of spinal cord and nerve roots (a) cross section (b) longitudinal diagram showing the origin of the anterior spinal artery. (From Henson and Parsons (1967).)

branch from each vertebral artery which descends throughout the length of the spinal cord in the anterior median fissure. The spinal arteries are reinforced at each intervertebral foramen by segmental arteries derived from the vertebral costo-cervical trunk, intercostal and lumbar arteries. Two of these segmental arteries tend to be larger than the rest and of greater importance, one in the lower cervical region, commonly at C6, and one, the great anterior radicular artery of Adamkiewicz, which usually enters the spinal cord between the T5 and T8 segments on the left side (Fig. 28.1).

At each segmental level the cord is surrounded by small circumferential vessels that run over the surface, form anastomoses between the anterior and posterior spinal vessels and send horizontal branches into the cord to supply the white matter and part of the posterior horns of the grey matter.

The direction of blood flow in the anterior spinal artery may not be the same throughout its length (Bolton 1939). Flow can be in both rostral and caudal directions, is variable from one individual to another, and may be modified by vascular disease. The measurement of spinal cord blood flow in animals has confirmed that both motor and sensory activity at a spinal segmental level is associated with temporary vasodilatation and increased blood flow in the relevant portion of the cord, cauda equina, and nerve roots (Blau and

Rushworth 1958). The anterior spinal artery supplies all but the posterior part of the posterior columns and the posterior horns, which are supplied by the posterior spinal arteries. Occlusion of the anterior spinal artery therefore results in massive infarction of the spinal cord whereas damage to the posterior spinal arteries is less significant.

Descending branches from the spinal arteries supply the cauda equina and the vessels in the lowest segments of the cord and the roots and cauda equina receive tributaries from the ilio-lumbar and lateral sacral branches of the internal iliac arteries.

Venous drainage

The spinal veins derived from the spinal cord substance terminate in a plexus in the pia mater where there are six tortuous, often plexiform longitudinal channels, one along the anterior median fissure, a second along the posterior median sulcus, and two others situated on either side, one pair just behind and the other just in front of the line of attachment of the ventral and dorsal nerve roots. These six vessels communicate freely with one another and above pass into the corresponding veins of the medulla oblongata and drain into the intracranial venous sinuses. The posterior half of the cord is drained by posterior medullary veins; the anterior medullary group has one lateral and two medial groups, and the anastomotic pattern may explain the clinical features of venous infarction of the cord (Hughes 1971).

There is an internal vertebral venous plexus which lies within the vertebral canal between the dura and the vertebra, this receives tributaries from the spinal cord and is arranged as four longitudinal veins. These vessels communicate with the external vertebral venous plexus, via the foramina and, though this is relatively unimportant in the normal situation. However thrombophlebitis may reach the spinal veins by this route and back pressure from abdominal and thoracic contents can cause venous thrombosis.

28.3 Clinical localization

28.3.1 General features

This section considers the general features of spinal cord lesions and specific findings according to the segmental level of the cord lesion. The signs and symptoms of spinal cord disorders are dependent upon the level, longitudinal and transverse extent, and pathological nature of the underlying cause. Because there are many potentially treatable causes of spinal cord dysfunction, and because delays in diagnosis may have an adverse effect on outcome, a thorough knowledge of the clinical manifestations is essential.

The classical features of spinal cord disease involve sensory, motor, and sphincter manifestations, but partial, or early, lesions may involve only one or two of these functions. Symptoms described by patients include weakness below the level of the lesion, with difficulties in walking and/or in upper limb function. The gait may be described as weak, stiff, a tendency to drag the legs or unsteady and there may be symptoms of painful muscle spasms or cramps. Sensory symptoms include numbness, tingling, pins and needles, hypersensitivity, burning sensations, altered sensation of temperature, and tight band like feelings or the sensation of swelling below the lesion. When a cord lesion is complete loss of all voluntary movement and sensation will occur below it. Pain may occur at the level of the lesion, particularly when there is spinal cord compression from extrinsic disease involving the vertebral column. The most common sphincter disturbances, due to spinal cord disease are urgency, frequency, and incontinence of urine. Hesitancy or urinary retention is less common except in acute transverse spinal cord lesions where retention, as part of 'spinal shock' is the rule. Constipation is common and faecal incontinence rare. In males erectile dysfunction is a common symptom and other autonomic changes, such as postural hypotension, excessive sweating, and vasomotor disturbances occur below the site of the lesion; Horner's syndrome may occur with lesions in the cervical spinal cord.

Clinical symptoms and signs can be divided into those occurring at the level of the lesion and those below the level of the lesion due to interruption of the long tracts. At the level of the lesion there may be lower motor neurone signs with focal muscle wasting, fasciculation, and hypo or areflexia due to damage to anterior horn cells. This pattern of motor involvement will be focal and segmental. Radicular pain or dermatomal sensory loss may occur due to damage to sensory nerve roots. Damage to the long tracts is usually the most serious consequence of spinal cord disease. Damage to the lateral and anterior columns causes upper motor neurone signs below the level of the lesion with pyramidal pattern of weakness, being greater in the anti-gravity muscles, or the extensors of the arms and flexors of the legs, spasticity and hyper-reflexia with absent abdominal reflexes and extensor plantar responses. Acute, severe spinal cord lesions produce a flaccid paraplegia below the level of the lesion with a temporary phase of hypotonia and areflexia, sometimes called 'spinal shock', before the appearance of more characteristic upper motor neurone signs. The differential diagnosis in this setting is of an acute inflammatory polyneuropathy or Guillain–Barré Syndrome (Section 21.10.1) but the sensory findings and the involvement of sphincter function, which is rare in polyneuropathy, are helpful in making the distinction.

A complete cord syndrome results in loss of all sensory modalities below the level of the lesion. Partial syndromes may produce variable findings; damage to the posterior columns causes loss of joint position sense, vibration, and two-point discrimination ipsilateral to the lesion, with a positive Romberg's sign and an ataxic gait. In the upper limb pseudo-athetoid movements may occur when the hands are held out-stretched with the eyes closed. When damage affects the spino-thalamic pathways there is contralateral loss of pain and temperature sensation below the level of the lesion.

28.3.2 Foramen magnum syndrome

Lesions at the level of the foramen magnum are characterized by upper motor neurone weakness and sensory loss affecting any modality below the level of the cranium. Rarely there may be deficits affecting the lowest cranial nerves. The only manifestation may be a high spinal cord syndrome but commonly neck stiffness and pain radiating into a shoulder, occipital headache with weakness of the upper or lower extremities, numbness of the hands or arms, and clumsiness may be described. It is common to find an upper motor neurone monoparesis, hemiparesis, or quadriparesis, and the loss of sensation may involve all modalities though it can be 'cape like' and as high as the second cervical dermatome. There may be pseudo-athetosis of the fingers due to loss of joint position sense and there can be proximal atrophy of muscles in the upper extremities. Sometimes 'electric shock'-like sensation is described by the patient, similar to Lhermitte's symptom. Rarely, down-beat nystagmus will be found and more rarely involvement of the twelfth cranial nerve. Classically symptoms may begin in one limb then spread in a 'clockwise' or 'counter-clockwise' direction to the ipsilateral lower limb, contralateral lower limb, then contralateral upper limb, or to the contralateral upper limb, contralateral lower limb, then ipsilateral lower limb.

The differential diagnosis is often between a foramen magnum and upper cervical cord tumour. The importance of the syndrome is that it may be due to benign tumours such as a meningioma or neurofibroma, although severe pain raises the possibility of metastases (Fig. 28.2). Arnold Chiari malformations may occasionally present in this way.

28.3.3 Cervical spinal cord lesions

Upper cervical spine

Neck pain is common, there will be both sensory and motor problems affecting the upper and lower limbs, there will be tetraparesis or tetraplegia and, if the lesion is above the level of C4, the diaphragm may be affected, compromising respiration when lying. Sensory impairment will involve all four limbs and the trunk. With extrinsic lesions pressing upon the cord or with inflammation in

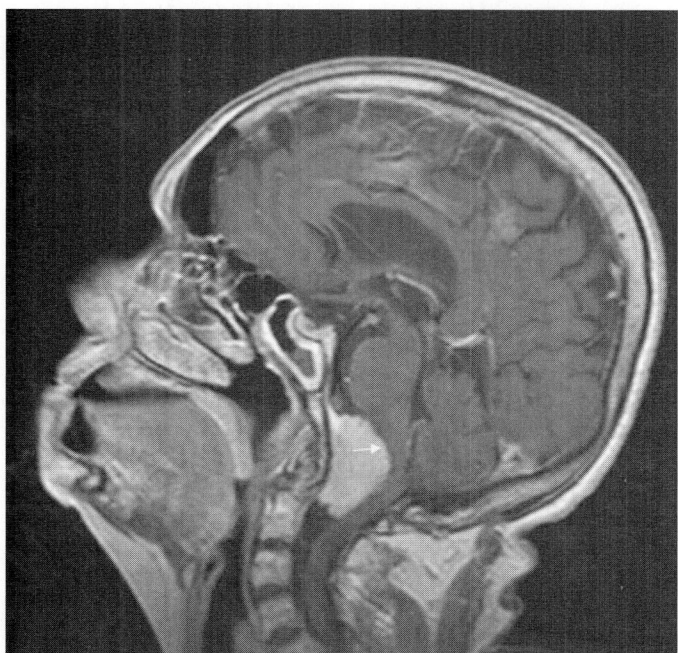

Fig. 28.2 Foramen magnum meningioma (arrow) compressing the medulla and cervical spinal cord. T1-weighted MRI with Gd enhancement in sagittal image.

the cord reflexes will be brisk with extensor plantar responses, but if the lesion is central within the cervical spinal cord, as in syringomyelia, upper limb reflexes may be depressed and associated with a suspended sensory disturbance.

Mid-cervical spine

Damage between the third and sixth cervical vertebra will affect the fifth, sixth, and seventh cervical nerve roots. There will be atrophy and weakness of muscles innervated by those roots, particularly the rhomboids, deltoids, spinati, biceps, brachioradialis, and triceps. There will be spastic paralysis of the lower limbs and of the C8 and T1 innervated muscles in the upper limbs. Biceps, brachioradialis, and triceps reflexes will be diminished or lost and there may be 'inversion' of reflexes; striking the biceps tendon resulting in flexion of the fingers. C8 reflexes will be brisk. Sensory symptoms occur in the lower limbs, trunk, and medial aspects of the upper limbs, including the ring and little fingers.

Lower cervical and upper thoracic spine

When damage is at the level of the lower cervical spine the eighth cervical and first thoracic nerve roots are likely to be involved. There will be relatively normal power above the level of the lesion but weakness and atrophy of the flexors of the wrists and fingers and of the small muscles of the hand. There may be a Horner's syndrome, though this is more commonly seen with lesions affecting the first thoracic nerve root peripherally. Tendon reflexes in the upper limbs will be retained though the finger jerk may be absent. There will be spastic paralysis of the trunk and lower limbs, sparing of the bladder, and sensory abnormality in the lower limbs, trunk, and medial arm, probably involving the medial forearm and digits.

28.3.4 Thoracic spinal cord lesions

Upper and mid thoracic spine

Damage in the upper thoracic spine will manifest as atrophic paralysis confined to the intercostal muscles innervated by the segments involved. Movements of the diaphragm will be normal but there will be spastic paralysis of the muscles of the abdomen and lower limbs and a sensory level demonstrable on the trunk.

The lower thoracic spine

When damage occurs below the mid-thoracic spine cord involvement is likely to be below the 10th thoracic segment. Sensory loss will reach the level of the umbilicus, the lower abdominal recti will be paralysed, and when the patient lying on a bed is asked to raise the head from the bed, the umbilicus should be drawn upwards, Beevor's sign. The upper abdominal reflexes will be preserved, the lower abdominal reflexes lost, and there should be spastic paralysis of the lower limbs.

If the lesion occurs lower, affecting the 12th thoracic and first lumbar segments the abdominal muscles will be spared, the abdominal reflexes will be intact, but the cremasteric reflexes will be diminished or lost and there will be spastic paralysis of the lower limbs.

When damage occurs at the lowest part of the thoracic spine, involvement of the bottom of the spinal cord and conus medullaris is likely. If the third and fourth lumbar segments are involved hip flexion will remain intact, but there will be wasting and weakness of quadriceps and hip adductors with reduction or loss of knee jerks, but a spastic paralysis below that level. The ankle jerks will be brisk and the plantar responses extensor. Sensory loss will be present below the level of the knees, spreading onto the posterior calves, thighs, buttocks, and perineum.

When the lesion affects the first and second sacral segments then flexion of the hip, adduction of the thigh and extension of the knee and dorsiflexion of the foot will be preserved. There will be atrophic paralysis of the intrinsic muscles of the foot and calf with weakness of knee flexion and hip abduction and extension. The knee jerks will be present, the ankle jerks absent, and the plantar responses mute. The anal and bulbo-cavernosus reflexes will be retained and sensory loss will involve the buttocks and perineum, extending down the posterior aspect of the lower limbs to the soles of the feet.

When the third and fourth sacral segments are involved the large bowel and bladder are paralysed with retention of urine and constipation, due to the uninhibited action of the sympathetic nerve supply. The external sphincters are paralysed and the anal and bulbo-cavernosus reflexes are lost. There is usually sensory loss in the perineum and in the buttocks in a 'saddle' distribution. Power in the lower limbs and the reflexes are normal.

28.3.5 Motor symptoms

Motor symptoms due to damage to the spinal cord are of two types. At the level of the cord injury there will be damage to the anterior horn cells and anterior roots and, depending upon the length of the lesion, lower motor neurone signs will be evident at segmental levels. The spinal cord or nerve root damage may be lateralized indicating the site of the lesion and will involve flaccid weakness, ultimately muscle wasting and the loss of relevant deep tendon reflexes.

Below the level of the spinal cord lesion there may initially, in severe lesions, be an acute flaccid paresis due to 'spinal shock' but within a matter of hours to days the signs of spastic weakness will emerge with increased tone, brisk reflexes, and a pyramidal distribution of the motor weakness affecting, predominantly, the flexors of the lower limbs and the extensors of the upper limbs. The most helpful signs of spinal cord damage are the plantar responses. When there has been spinal injury and the plantar responses can be demonstrated to be extensor then, because the spinal cord ends at second or third lumbar vertebra, the damage is almost invariably in the cervico-dorsal spine and radiology of the lumbar spine is unlikely to be relevant.

In the cervical spine the combination of radicular and lower motor neurone signs with upper motor signs below the level of the lesion may be enhanced by the finding of 'inversion' of reflexes. Most commonly this occurs where damage at C5/6 root level also affects the spinal cord, rendering hyperactive the anterior horn cell pool below that level. Striking the C5/6 reflex with a tendon hammer results in absence of the C5/6 reflex, but demonstrates enhancement of the C8 reflex in finger flexion and, on rare occasions, even triceps causing elbow extension. This combination of upper and lower motor lesions in the cervical spinal cord is very suggestive of cervical spondylosis causing a radiculo-myelopathy.

Ultimately there may be the development of contractures in a chronically spastic limb and trophic change in the colour and appearance of the skin.

28.3.6 Sensory symptoms

Although motor symptoms are most important in identifying the site of the lesion sensory symptoms are helpful in identifying its completeness. When a spinal cord lesion is total all sensory modalities below the level of the lesion will be affected. If the lesion is partial affecting the posterior aspect of the cord, then there may only be loss of proprioception and if partial affecting one or other side of the cord, then crossed anaesthesia with ipsilateral proprioceptive loss as in the Brown–Sequard syndrome can be identified.

In keeping with the symptoms and signs of motor disturbance there may be acute pain in a radicular distribution at the site of the lesion with loss of sensory modalities below it. Occasionally spinal cord lesions cause hyperpathia, or increased sensitivity, below the level of the lesion, most classically after trauma or ischaemia.

In addition to the immediate symptoms and signs of sensory disturbance with a spinal cord lesion there may be the development of trophic changes in the skin and ultimately ulceration due to anaesthesia together with the finding of Charcot joints.

28.3.7 Autonomic disturbances

Lesions in the cervical region may be associated with Horner's syndrome and those affecting the T1 root may result in loss of sweating in the ipsilateral hand. Occasionally central cord lesions, such as syringomyelia, may present with hyperhydrosis due to sympathetic malfunction in one or both upper limbs.

Vasomotor changes are common below the level of a spinal cord injury. Those affecting the cervical or dorsal spine and damaging the sympathetic nervous outflow may result in disturbances of bowel motility, vasomotor changes and postural hypotension, and damage to bladder and bowel control, most commonly causing urgency and frequency of micturition and constipation. Impotence is a common symptom of spinal injury in the male.

28.4 Acute cord syndromes

28.4.1 Spinal shock

When there is major injury to the cervical or upper thoracic spinal cord the clinical syndrome of 'spinal shock' may occur. It is characterized by loss of motor, sensory and autonomic function below the level of the lesion. The more severe and the higher the spinal cord injury the greater the duration and severity of the 'spinal shock'.

Most commonly patients have flaccid paralysis with loss of cutaneous and deep tendon reflexes and anaesthesia to all sensory modalities below the level of the injury. There may also be systemic hypotension, cutaneous hyperaemia, and bradycardia as a result of autonomic dysfunction and unopposed vagal tone. Whether the syndrome occurs due to vascular hypoperfusion of the spinal cord after trauma, or is due to abnormal neurotransmitter production is uncertain.

Clinically problems are caused by the difficulty in assessing the nature of the spinal injury and the uncertainty about the duration of the syndrome. Usually 'spinal shock' resolves within hours and weakness or numbness which persists is likely to be due to physical cord injury. There are, however, reports of both absent reflexes and autonomic changes lasting for days then resolving, particularly with high cervical lesions. In practical terms it is important that clinicians do not mistake 'spinal shock' for hypovolaemic shock. The latter is due to fluid loss and responds to volume repletion, the former responds to sympathomimetic agents, rather than to volume replacement.

28.4.2 Complete versus incomplete lesions

Complete transverse cord lesion

Though best exemplified by traumatic cord injury, acute inflammatory or ischaemic disease can cause identical effects to transection. The level of the cord damage is apparent and below it all sensory modalities are lost, bladder and bowel function is affected, and there is total paralysis with the eventual development of increased tone and hyper-reflexia below the level of the lesion. There may also be significant postural hypotension and, if the lesion is high, quadriplegia and possible loss of diaphragmatic control (Fig. 28.3).

Central cord syndrome

Damage in the mid to low cervical region is common following a fall in an elderly person with hyperextension of the neck which was previously the site of cervical spondylosis. It may also be seen with acute inflammatory lesions and acute ischaemic lesions. It is characterized by weakness that is more marked in the arms than in the legs, there is patchy sensory loss with dysaesthesiae and a-reflexia in the upper limbs. The preferred investigation is MRI of the cervical spine and, in those cases due to trauma, prognosis depends upon evidence of ischaemia or haematoma within the spinal cord. Central lesions due to inflammatory transverse myelopathies are frequently necrotic and prognosis is guarded, the use of steroids is recommended, and plasmapheresis may be considered.

Anterior cord syndrome

Damage or disease affecting the anterior part of the spinal cord is most commonly seen in ischaemic anterior spinal artery occlusion or trauma. It may follow large disc herniations and fractures with retro-pulsed bone fragments. It is characterized by complete paralysis below the level of the lesion with hypalgesia at the level of the injury and loss of spinothalamic sensation beneath.

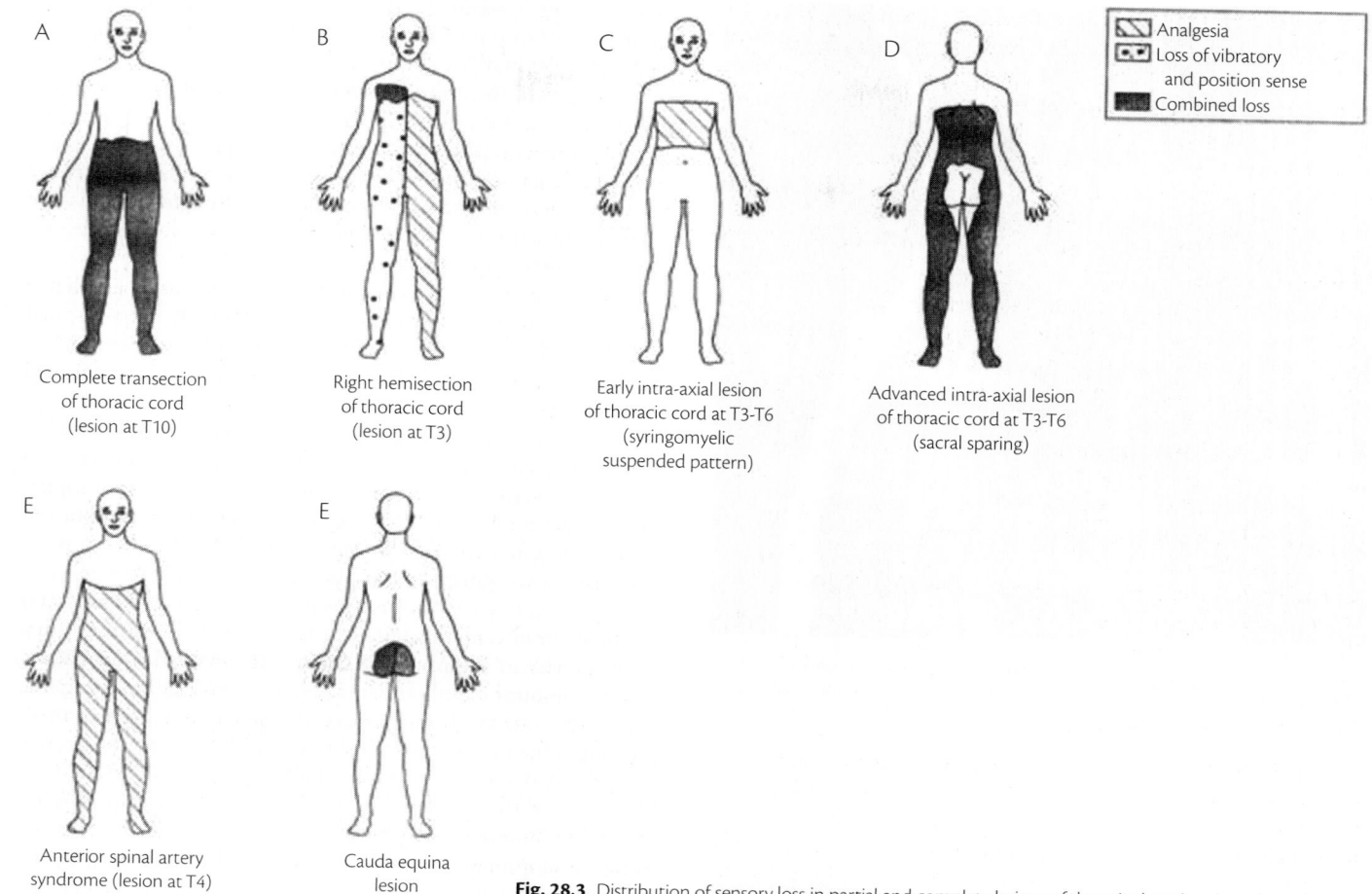

A Complete transection of thoracic cord (lesion at T10)

B Right hemisection of thoracic cord (lesion at T3)

C Early intra-axial lesion of thoracic cord at T3-T6 (syringomyelic suspended pattern)

D Advanced intra-axial lesion of thoracic cord at T3-T6 (sacral sparing)

Analgesia
Loss of vibratory and position sense
Combined loss

E Anterior spinal artery syndrome (lesion at T4)

E Cauda equina lesion

Fig. 28.3 Distribution of sensory loss in partial and complete lesions of the spinal cord and cauda equina.

Posterior column sensation is preserved due to the separate blood supply from the posterior spinal arteries.

Prognosis for recovery of motor function is poor, though some sensory improvement can occur. Surgical intervention may be required, interventional angiography has little to offer.

Posterior cord syndrome

Inflammatory disease of the central nervous system, particularly multiple sclerosis, may cause focal damage to the posterior columns, often towards their lateral elements, in the cervical spine. This can result in lack of proprioception and, in multiple sclerosis, the bilateral 'useless hands of Oppenheim' may result where there is sensory ataxia of both hands. If a posterior cord syndrome is more extensive, following infarction of the posterior spinal arteries, then there may be sparing of pinprick and thermal sensation below the level of the lesion but there will be paralysis and total loss of proprioception below that level. Posterior cord syndromes may also occur with vitamin B12 deficiency.

Brown–Sequard syndrome

Defined as hemi-section of the cord this may follow trauma but is more commonly seen with inflammatory demyelinating disease as in multiple sclerosis. It is occasionally seen with tumours where one half of a spinal cord is damaged. The patient has ipsilateral loss of motor control and posterior column function below the level of the injury, but contra lateral loss of pain and temperature sensation, usually from one or two dermatomes below the level of the proprioceptive loss, due to the crossing fibres (Section 2.5.3, Fig. 2.54). This loss of sensory function may be called 'dissociated', that is with 'posterior column' and 'spinothalamic' loss on opposite sides of the body. There is variable loss of sphincter function and the prognosis for recovery is usually good.

28.4.3 Trauma and compression

One of the most common clinical situations faced by the neurologist results from acute or chronic spinal cord injury. Often it results in permanent neurological deficit. There are few areas of neurology in which delayed or missed diagnosis, poor initial management, and lack of timely investigation, have more significant effects upon disability (Table 28.1).

Injury

Approximately half of spinal cord injury occurs in the cervical region, the remainder is distributed in the thoracic, lumbar, and sacral regions. In developed countries road traffic accidents and high-risk sports have now overtaken work-related and domestic accidents as the leading cause of spinal cord injury. Pre-hospital

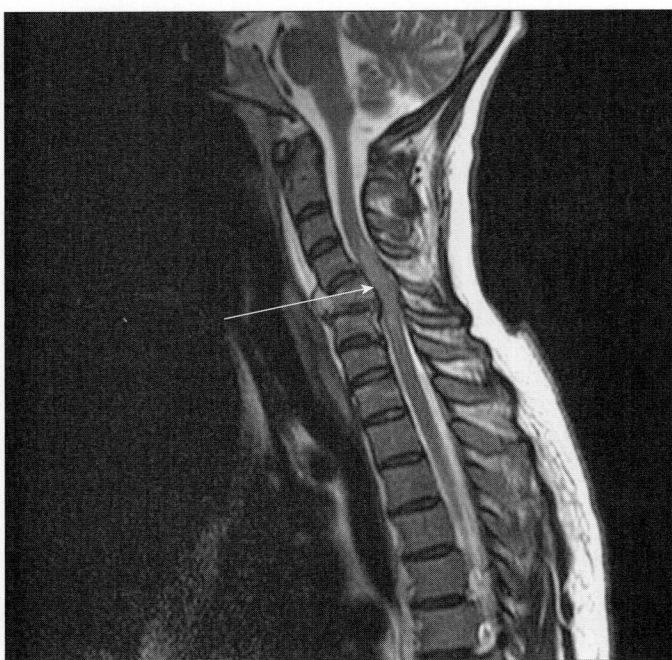

Fig. 28.4 Acute flexion injury to cervical spine causing crush fracture of C5 and cord compression (arrowed). There is high signal in the cord indicating oedema.

Table 28.1.   Causes of acute myelopathy

Spinal cord compression
Intervertebral disc prolapse
Subdural/epidural haematoma
Spinal epidural abscess
Vertebral fracture or dislocation

Inflammatory and demyelinating
Multiple sclerosis
Transverse myelitis
Acute necrotic myelopathy
Devic's neuromyelitis optica
Sarcoidosis

Ischaemia
Anterior spinal artery occlusion
Fibrocartilaginous embolism
Dissecting aortic aneurism
Decompression sickness
Haematomyelia

Infective
Spinal-cord abscess
Viral myelitis
Schistosomiasis
Brucellosis
Syphilis

emergency care is essential and immobilization of the cervical spine is a major part of out of hospital management following injury in which the spine may have been damaged.

The severity of the injury varies considerably. At one end of the spectrum are mild and transient phenomena such as 'stingers' which are recognized by most professional sportsmen and 'transient paraplegia' seen in contact sports and with falls. The other end of the spectrum includes immediate irreversible paralysis due to spinal cord transection or compression with vascular injury resulting in infarction (Fig. 28.4).

Spinal cord injury, even when due to trauma, is usually associated with pre-existing conditions, such as a congenitally narrow spinal canal, cervical spondylosis, spondylolisthesis, or osteopenia, all of which make trauma more likely to result in significant spinal cord injury.

Spinal cord injury frequently occurs in the setting of multisystem injury and only 20 per cent of patients have isolated cord injury, the great majority also sustaining traumatic brain injury, haemo-pneumothorax, or fractures of long bones. It is important that the emergency room physician and neurologist called to assess such a patient recognize the effect of multiple traumas in worsening spinal cord injury and the converse effects of spinal cord injury creating hypotension and shock thereby worsening more diffuse injury.

The severity of the injury to the spinal cord varies depending upon the amount of sub-arachnoid space surrounding the spinal cord. There is relatively more space available at the foramen magnum and in the upper cervical spine, at which level fractures rarely injure the cord, except in the case of atlanto-axial dislocation, than in the mid to the lower cervical canal where the subarachnoid space is narrower, anterior and posterior dimensions may be reduced by the development of cervical spondylosis and folding of the ligamentum flavum and more minor injuries result in major trauma to the cord.

Lower in the spine there is gradually increasing spinal arachnoid space and consequently injury to the lower spinal cord, conus medullaris, and cauda equina is relatively less common. Once again the pre-existence of lumbar spondylosis, disc disease, or canal narrowing is likely to worsen outcome from trauma.

Acute compression myelopathies

Acute spinal cord compression is a medical and surgical emergency and differentiation from non-compressive myelopathy is vital. When there is compression of the spinal cord by a disc protrusion or epidural haematoma the prognosis for recovery is inversely related to the time taken to establish the diagnosis and relieve the compression. Prognosis is also affected by the severity of the compression, pre-existing spinal conditions, and the age of the patient.

Intervertebral disc prolapse

Prolapse of an intervertebral disc in the cervical region, or in the dorsal region can impinge upon and compress the spinal cord. Lesions in the lumbar region may affect the conus medullaris but are more likely to compress nerve roots.

Most disc prolapses occur laterally and in the cervical region result in root symptoms and signs, extending into an upper limb. Central disc prolapse in the cervical spine does occur, most commonly in the situation of acute hyperextension of the neck, as in falling forwards; thoracic disc protrusions can be central and compress the cord which is 'bowstringed' over the dorsal kyphosis.

Direct and acute central disc prolapse can cause a transverse cord syndrome with a motor and sensory level but asymmetrical prolapse may cause a partial Brown Sequard syndrome which may vary in severity.

The investigation of choice is MRI with both sagittal and axial views; the MRI scan can also reveal the degree of damage within the cord as high signal areas on T2-weighted images in the cord parenchyma (Fig. 28.5).

Acute central disc protrusion of a cervical or thoracic intervertebral disc causing spinal cord symptoms and signs is a surgical emergency. The neck should be immobilized, surgical opinion sought, and decompression performed when indicated.

Subdural or epidural haematoma

Bleeding within the spinal canal but outside the dura may occur following blunt trauma or after surgical procedures, such as lumbar puncture in people who are anticoagulated or thrombocytopenic. For reasons described above haematomata are more likely to cause pressure when in the cervical or dorsal region than in the lumbar region and should always be considered when a sub-acute deterioration occurs following blunt or open trauma.

Spontaneous epidural haematoma is a rare condition which is usually seen in the thoracic region and can be due to a vascular malformation (Hernandez 1982). It is a surgical emergency. The site of the haematoma will be identified by the presence of pain and discomfort and the level of spinal injury identified clinically. MR scanning is the investigation of choice and haematomas that are more than a few days old produce a characteristic high signal on T2 MR images due to formation of methaemoglobin.

Spinal epidural abscess

Spinal epidural abscess is potentially a neurosurgical emergency; if untreated irreversible paraplegia may develop due to cord compression and consequent ischaemia. The most common infectious agent is *Staphylococcus aureus* and the most common association is within discitis or infection of the relatively avascular intervertebral disc. Occasionally infection may occur with anaerobes, streptococci, and gram negative bacilli. There are two main sources of infection into the spinal epidural space, the most common is via bacteraemia or septicaemia, sometimes following an innocuous skin infection and at other times associated with endocarditis or profound septicaemia. Increasingly, haematogenous spread to the intervertebral discs is seen in intravenous drug abusers. The organism almost certainly spreads via the blood stream to the vascular endplates of the vertebra, then infiltrating the avascular intervertebral disc and rarely begins directly within the epidural space. The second source of epidural infection is following a direct and open procedure on the spine. It may occur after surgery, local injection and, rarely, the performance of lumbar puncture or epidural anaesthesia. Although open trauma to the spine is a potential cause of epidural abscess there is no evidence that closed trauma causes discitis or epidural infection.

The patient usually presents with fever and localized back pain, radicular pain at the level of infection may occur and is followed by the development of a transverse cord syndrome with sensory, motor, and sphincter deficits below the level of the lesion. The clinical features may occur sub-acutely and both fever and leucocytosis may be absent. An elevated ESR and focal tenderness to palpation of the spinal column at the level of infection may be helpful clues but the physician must have a high level of suspicion in all patients who have had septicaemia and those who have undergone recent instrumentation to the spine.

MRI is the investigation of choice and will demonstrate the extradural compressive lesion and frequently demonstrate abnormalities in the adjacent vertebra and intervertebral disc. Gadolinium enhancement of the infection may be demonstrated (Fig. 28.6). The CSF, if taken from a site below the lesion, will usually reveal a mild pleocytosis, slightly elevated protein, and a normal glucose, the typical pattern for a para-meningeal focus of infection. Lumbar puncture is not, however, a necessary investigation in this situation, rarely gives useful information about the nature of the infection, and may potentially be dangerous if the abscess is causing complete occlusion of the spinal canal.

MRI diagnosis should be followed by neurosurgical intervention with laminectomy and abscess drainage when indicated. Intravenous antibiotics are required and should be started as soon as the epidural mass is demonstrated. Blood cultures should be taken and may reveal the cause of the infection.

The prognosis is variable, being better when there is rapid diagnosis and surgical intervention where necessary and worse when there is a delay in diagnosis or surgery allowing pressure upon the cord to result in infarction of the tissue. Rarely a subdural abscess may develop, often in association with local infective lesions. Its presentation and management is the same as for extradural abscess.

When spinal infections, such as vertebral osteomyelitis, discitis, and non-compressive epidural or paraspinal abscesses are identified radiologically but not causing significant neurological dysfunction and compression they can usually be managed without surgery. Appropriate antibiotic therapy is all that is required with monitoring to ensure that, should signs of instability, progressive bone destruction, or neurological dysfunction arise, surgery can be performed urgently.

Penetrating spinal cord injury

Increasingly, acute injury to the spinal cord is seen with missile injury, particularly gunshot and stab wounds. These are both surgical emergencies, with missile injuries foreign material must be removed and potential CSF leaks corrected. Where there is complete or partial transection of the cord from a missile injury it will be irreparable. Antibiotic treatment is required.

Stab wounds to the spine are usually deflected to one side and may result in a Brown–Sequard syndrome with dural laceration and CSF leak. Once again they are surgical emergencies; require appropriate imaging in a neurosurgical centre and the provision of antibiotics prior to operation.

28.4.4 Inflammatory lesions

Acute inflammatory lesions affecting the spinal cord are demyelinating, post-infective or vasculitic. The distinction between the different types may not be immediately apparent but some attempt at differentiation is important because the treatment may differ. In all cases the investigation of choice is MRI of the affected spinal area and extensive scanning of the whole spine can be important, both diagnostically and in excluding other lesions.

Multiple sclerosis

The spinal cord is involved pathologically and clinically at some stage of the disease in almost all patients (Section 37.5). In about

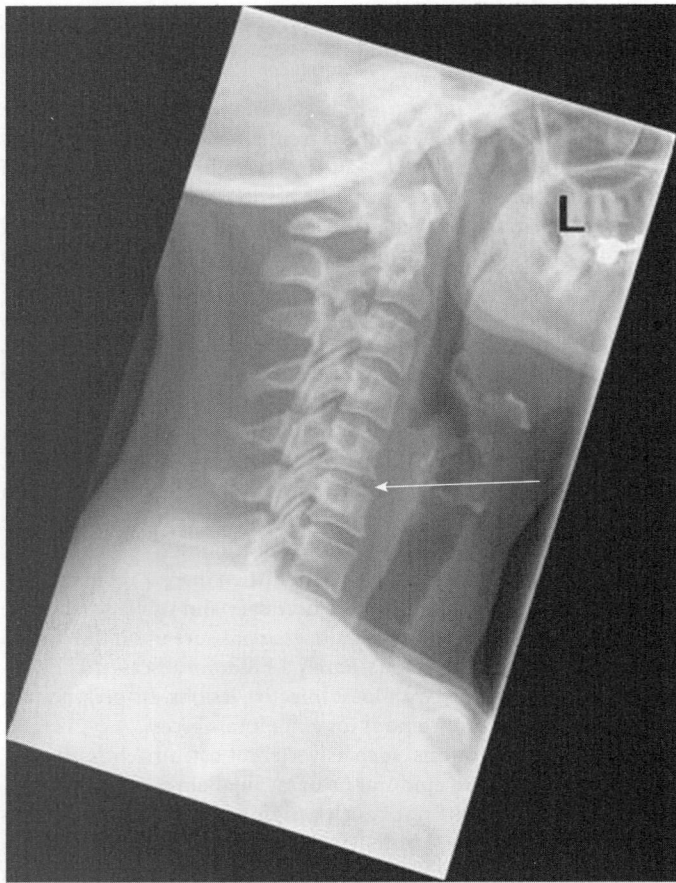

A

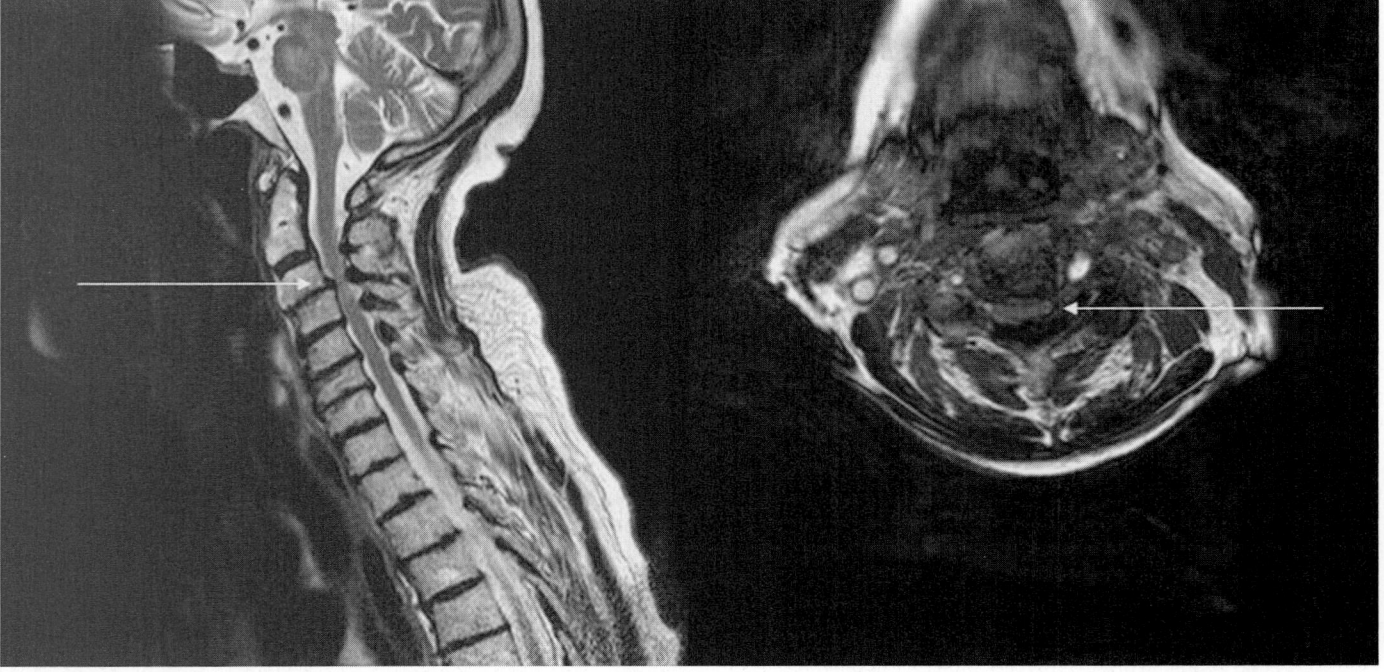

B

Fig. 28.5 Acute Cervical Disc Prolapse. A. X-ray of cervical spine showing loss of lordosis and disc space narrowing at C5/6 (arrow). B. T2-weighted sagittal and axial MRI showing severe cervical spondylosis with retrolisthesis and disc prolapse at C3/4 causing compressive myelopathy with signal change in the spinal cord.

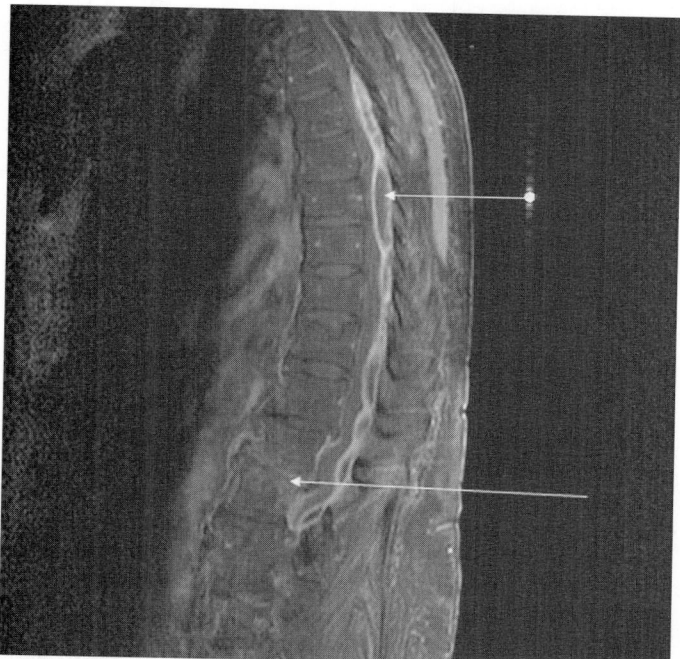

Fig. 28.6 T1-Gd enhanced MRI of epidural abscess in mid dorsal spine compressing the spinal cord (top arrow). Loss of disc space and bone change at site of discitis (bottom arrow).

50 per cent of patients the initial presentation is of a spinal cord disturbance. Sensory, motor, and sphincter symptoms may all occur in any combination, though typically the symptoms indicate only partial disturbance of spinal cord function and sometimes only explicable on the basis of more than one lesion. Focal demyelinating lesions due to multiple sclerosis almost always occupy only part of the cross section of the spinal cord (Oppenheim 1978) (Fig. 28.7).

Common sensory symptoms include Lhermitte's symptom tingling or electrical shock-like sensations spreading down the back and into the limbs on neck flexion. This is a common manifestation of demyelinating lesions involving the posterior columns in the cervical cord and may occur in isolation or in association with other cord symptoms. It is rarely seen in association with cervical spondylosis or other structural cervical pathology and usually indicates inflammatory disease in the cervical cord. Other sensations, which may be unilateral or bilateral, involve only the lower limbs or only the upper limbs, or be associated with a sensory level on the torso, include tingling, numbness, dysaesthesiae, loss of temperature sensation, or a tight 'elastic band like' sensation. A common sensory symptom in the upper limbs is loss of proprioception in both hands, 'the useless hand of Oppenheim' which indicates two postero-lateral cervical spine lesions and is almost diagnostic of multiple sclerosis.

Motor symptoms include limb weakness and spasms, cramps and the sensation of heaviness, unsteadiness on walking and dragging

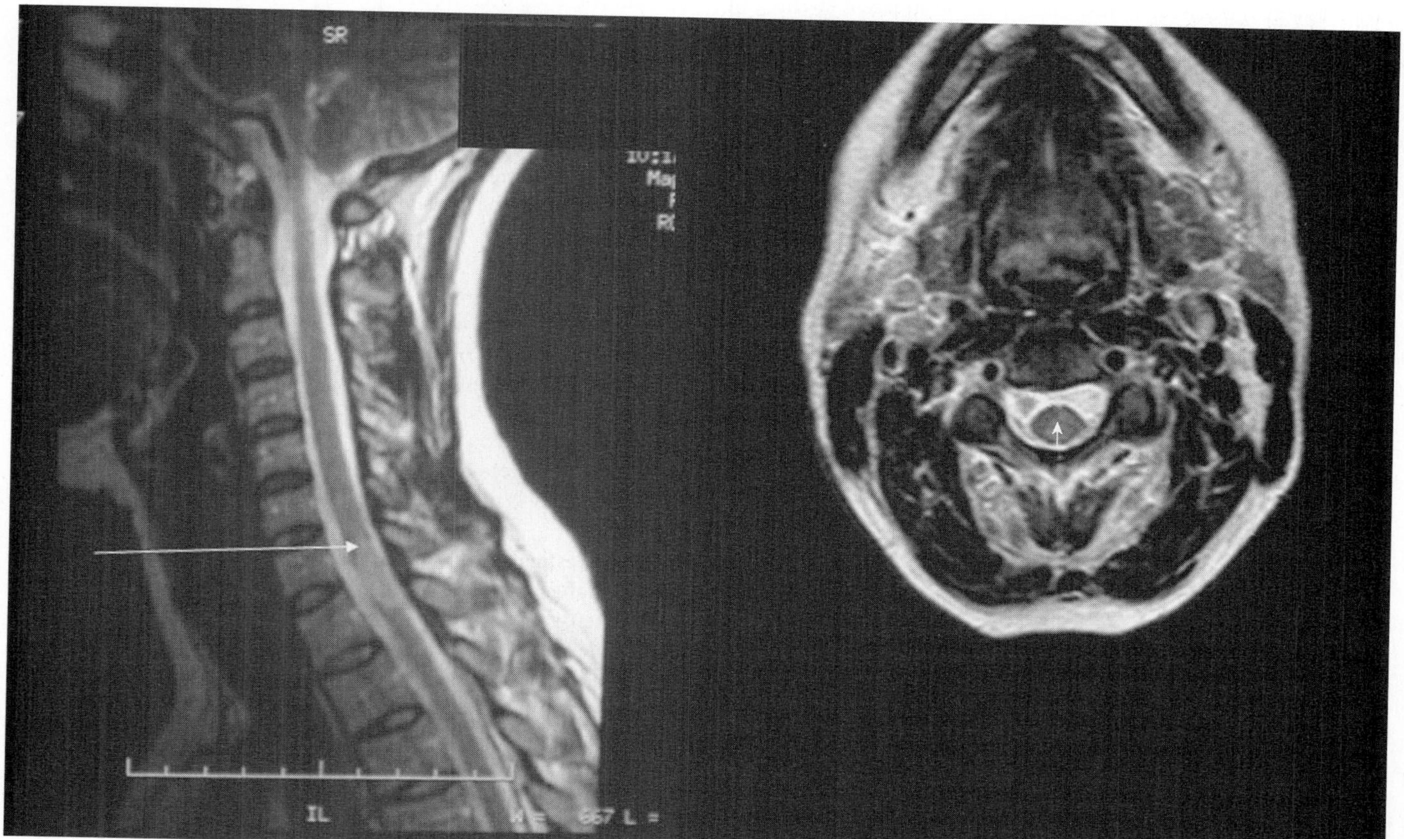

Fig. 28.7 Myelitis in multiple sclerosis. T2-weighted MRI of short partial lesion (arrowed) of inflammatory demyelination in central posterior cervical cord due to multiple sclerosis.

or stiffness of the lower limbs. They too may be unilateral or bilateral and symmetrical or asymmetrical.

Sphincter disturbance is most commonly manifest as increased frequency and urgency of micturition and occasional incontinence. Constipation may occur and, rarely, faecal urgency or incontinence. Impotence is a common symptom of cord demyelination in males.

The pattern of evolution of the initial attack in multiple sclerosis is usually over a few days, then stabilizing for a few weeks and resolving gradually. Examination may be surprisingly normal, though frequently there will be signs such as extensor plantar responses, brisk deep tendon reflexes, loss of vibration sensation, and absence of the abdominal reflexes. A Brown–Sequard syndrome, deafferentation of one or both hands, or the signs of a partial spinal cord lesion affecting both lower limbs and with a segmental level on the torso is most common. There may be pyramidal weakness of muscles as a monoparesis, hemiparesis, paraparesis, tri-or tetra-paresis associated with the expected reflex changes.

Although the partial nature of the spinal cord involvement and other associated features, such as retrobulbar neuritis, an internuclear ophthalmoplegia, or a relevant past history may indicate the probability of the diagnosis, the investigation of the patient seen with an acute spinal episode must include an urgent spinal MRI, both to confirm the presence of demyelinating lesions and exclude other pathology, particularly compression. Multiple sclerosis most commonly causes small areas of high signal on T2-weighted images of the spinal cord, usually less than one segment in length and almost always less than three segments. There may be apparent swelling of the cord and gadolinium enhancement can be demonstrated.

Sometimes more than one focal cord lesion can be identified which may indicate dissemination of lesions in space. When this is seen it is advisable to obtain additional MRI brain since, in about 50 per cent of cases, this will reveal cerebral white matter lesions typical of multiple sclerosis. Such lesions are prognostically important, their presence indicating a substantial risk of the development of clinically definite multiple sclerosis within 5–10 years (O'Riordan *et al.* 1998) and fulfilling some of the requirements for the new diagnostic criteria of multiple sclerosis (Polman *et al.* 2005).

The indications for further investigation including evoked potentials, CSF examination, and repeated MRI in an attempt to confirm the diagnosis of multiple sclerosis and determine the indications for the use of disease modifying therapy are discussed under multiple sclerosis (Sections 37.5.5; 37.5.8).

Recurrent acute spinal cord relapses are common during the course of multiple sclerosis and, when the diagnosis is secure, do not indicate the need for further MRI. Nonetheless, the occurrence of a complete transverse cord lesion or the association with other symptoms such as pain, fever, or local tenderness in the spine indicates the need for further imaging to exclude the possibility of a second pathology.

Some patients with multiple sclerosis will not show recovery from the original episode of spinal disturbance and may progress to primary progressive multiple sclerosis. In this situation repeat MR imaging is appropriate, examination of CSF to search for oligoclonal bands is obligatory, and evoked potentials may be helpful in establishing the diagnosis (McDonald *et al.* 2001).

The initial treatment of a partial spinal cord lesion in multiple sclerosis is with high dose steroids in the form of methyl prednisolone given in a dose of 0.5–1g intravenously, daily for 3–5 days.

If symptoms become persistent then therapy with the spasmolytics baclofen, tizanidine, dantrolene, or gabapentin should be considered and where painful dysaesthesiae or paraesthesiae persist the use of gabapentin, pregabalin, amitriptyline, or carbamazepine can be considered. Where there is sustained urgency of micturition a bladder relaxant such as oxybutinin or detrusitol may be considered, after assessing residual urine by ultrasound, and ultimately the use of intravesical botulinum toxin. Sildenafil can be considered for erectile dysfunction in males.

Patients who show frequent relapses affecting the spine and relapsing remitting multiple sclerosis, but retain ambulation, should be considered for long-term treatment with disease modifying therapy, including the beta-interferons and glatiramer acetate.

Transverse myelitis

Transverse myelitis may occur as a post-infectious inflammatory process causing acute or sub-acute complete spinal cord dysfunction (Section 37.4.3). Most cases are preceded by an infectious illness and are thought to have an immuno-pathogenic basis. Rarely transverse myelitis may follow vaccination and in a proportion of cases no prior event can be identified.

Some people with transverse myelitis are shown to have high levels of autoantibodies or collagen vascular disorders, particularly systemic lupus erythematosis, the primary antiphospholipid syndrome and Sjogren's syndrome. In these situations the pathological basis may be inflammatory or vasculitic.

In the majority of cases transverse myelitis is a monophasic disorder, but there are occasional reports, particularly in non-Caucasian populations, where multiple, recurrent episodes occur.

In the most common setting of post-infectious transverse myelitis, the preceding infection may be identified clinically or serologically, usually varicella, Epstein–Barr virus, mycoplasma, campylobacter, but there may simply have been a non-specific upper respiratory tract or gastrointestinal tract infection. People of all ages can be affected and the presentation is seen in childhood and the elderly.

The onset may be with discomfort and pain in the spine, but this symptom is usually not prominent. Weakness and paraesthesiae develop in the lower limbs and early urinary retention or incontinence is common. Symptoms evolve rapidly, typically reaching their peak within hours to days. The most common areas affected are the cervical or thoracic spinal cord and the upper level of symptoms and signs vary accordingly. Occasionally, even when a complete transverse lesion has developed clinically continuing disease activity is manifest by an ascending level of motor and sensory loss.

The investigation of choice is MRI which reveals swelling and signal change, high signal on T2-weighted images, which is typically extensive over more than three cord segments and involving the whole antero-posterior diameter of the spinal cord on the sagittal images (Fig. 28.8). In the acute phase there is patchy gadolinium enhancement.

CSF examination usually shows a moderate mono-nuclear pleocytosis, though may be a-cellular, the protein level is moderately elevated but the glucose is normal. Oligoclonal bands are sometimes present but may disappear with time. Those cases which have been examined pathologically show perivascular inflammation, sometimes with demyelination and in more severe cases with widespread inflammation and necrosis.

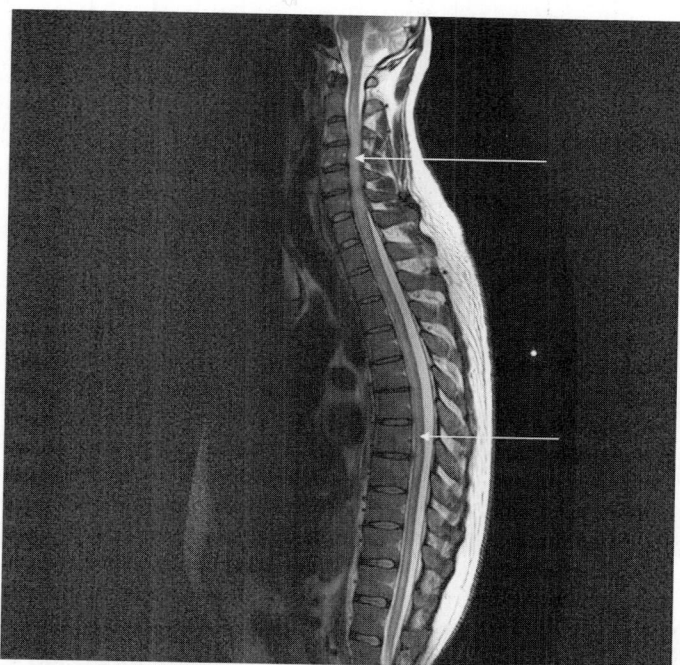

Fig. 28.8 Transverse myelitis or neuromyelitis optica. T2-weighted MRI of longitudinally extensive inflammatory lesion in spinal cord (as occurring between arrows).

Brain MRI occasionally shows disseminated white matter lesions suggesting a more diffuse asymptomatic acute demyelinating encephalo-myelitis-like picture but more often brain imaging is normal. It is imperative that further microbiological investigations, including polymerase chain reaction, studies, are undertaken to identify infection and serology performed to identify collagen vascular disease or autoantibodies.

In the acute stage therapy is usually a short course of high dose corticosteroids with intravenous methyl prednisolone 0.5–1 g daily for 3–5 days with or without oral steroids thereafter. There is some evidence that plasma exchange and intravenous immunoglobulin may help but the role of these therapies remains uncertain. The prognosis is variable, some people make an excellent recovery despite complete paraplegia in the acute stage, but others are left with a permanent transection of the spinal cord.

In those cases where acute transverse myelopathy has been shown to be associated with systemic lupus erythematosis or other autoimmune conditions, vigorous immunosuppression with high dose steroids and cytotoxic agents may be employed. It should be recognized that such therapy is predominantly directed to the prevention of further lesions and is unlikely to reverse any established neurological deficit. In those cases thought to have a primary antiphospholipid syndrome or where a coagulopathy is thought to be contributing to the spinal cord disorder, anticoagulation with heparin then with oral anticoagulants and the use of anti-platelet agents should be considered.

Acute necrotic myelopathy

This is probably a variant of transverse myelitis. It is rarer and characterized by the development of acute complete transverse myelopathy followed by permanent severe disability with flaccid paraplegia or quadriplegia, areflexia, and an atonic bladder. Pathologically there is widespread necrotic change over a number of segments of the spinal cord involving both white and grey matter. It may be seen as the severe end of the spectrum of post-infectious or autoantibody related myelopathy, but it has also been reported in people who have been successfully treated for pulmonary tuberculosis.

Devic's neuromyelitis optica, neuromyelitis optica, optico-spinal multiple sclerosis

The nosological status of Devic's neuromyelitis optica is controversial, partially because of difference in definitions (Section 37.4.4). Originally, Devic's disease was applied to a severe, almost complete extensive transverse myelitis in association with severe and permanent optic neuritis, the events occurring within months of one another.

More recently neuromyelitis optica has been applied to conditions with inflammatory lesions in the spinal cord and optic nerves, typically sparing the brain, and with longitudinal extensive lesions in the cord, pleocytosis in CSF, frequent autoantibodies, and negative oligoclonal bands (Wingerchuck 2007). A significant proportion of these patients have been demonstrated to have a specific neuromyelitis optica IgG which is believed to be directed against the aquaporin-4 epitope on water channels distributed throughout the axial portion of the nervous system (Lennon *et al.* 2005) (Fig. 28.9).

Uncertainty persists as to whether the non-Caucasian variant of inflammatory demyelination optico-spinal multiple sclerosis, which in many cases resembles neuromyelitis optica, but which may have lesions within the cerebrum, is the same disease.

In terms of the spinal cord these conditions cause central and almost complete transverse myelitis which is longer than three segments and associated with the presence of systemic autoantibodies. The prognosis for recovery from neuromyelitis optica is poor; there is often permanent paraparesis and visual loss. The suggested therapies of plasmaphoresis and cytotoxic agents with long-term steroids have been partially effective, the humoral antibody detected in a proportion of these patients now raises the question of treatment with the specific anti-CD20 monoclonal antibody rituximab, though any potential efficacy awaits demonstration in a controlled, therapeutic trial.

Pathologically lesions in Devic's disease differ from those of inflammatory demyelination, including hyalinization of blood vessels, extensive necrotic change, and the presence of eosinophils. It is assumed that immunopathological mechanisms are likely to be important and the recently discovered antibody may be causative, though further experiments to establish transmission are awaited.

28.4.5 Vascular disorders

Spinal cord ischaemia and infarction may occur with embolism, occlusion, or dissection of the anterior spinal artery due to atheroma, aortic disease, or a period of hypotension. Rarely may it complicate aortography or angiography. Posterior spinal artery occlusion has been described after the intrathecal use of astringents such as phenol. When ischaemia occurs the neurones and central spinal grey matter are usually more vulnerable than the long white matter tracts.

Ischaemia and infarction

Occlusion of the anterior spinal artery causes infarction of the anterior and lateral columns of the cord. When occlusion occurs in

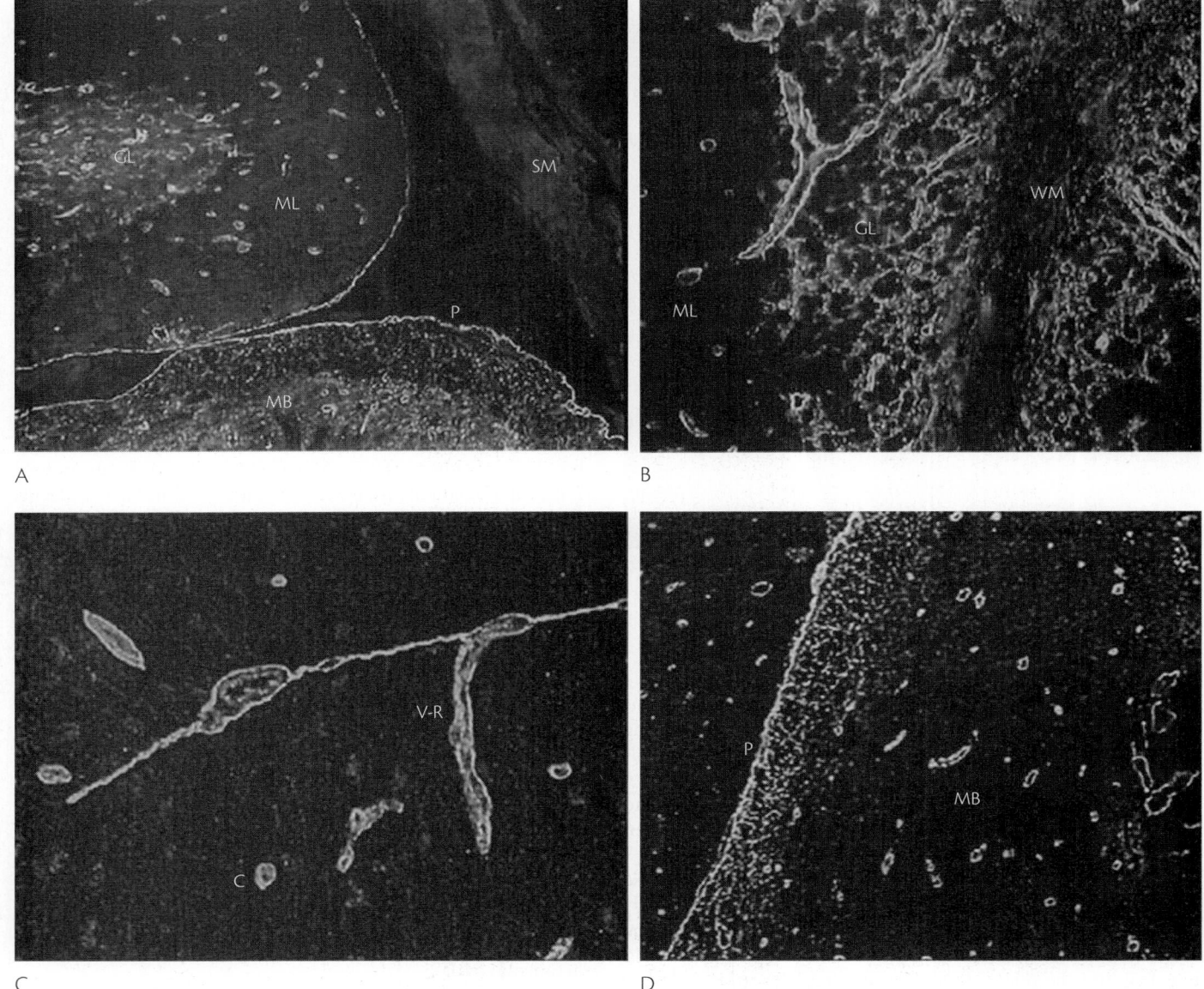

A

B

C

D

Fig. 28.9 Neuromyelitis optica Immuno-fluorescent pattern of bound NMO-Iga (aquaporin & antibody) in mouse central nervous system.
A. Showing linear staining of juxtaposed pial membranes (P) of the cerebellar cortex and mid brain (MB) and their micro vessels. The adjacent gut smooth muscle and vessels (SM) is not stained.
B. There is prominent microvascular staining in the cerebellar molecular layer (ML), granular layer (GL), and white matter (WM).
C. Shows linear staining in the cerebellar cortex including the pia, pial lining of Virchow-Robin (V-R) spaces and on microvessels, including capillaries.
D. Staining of the sub-pia of the mid-brain (MB). (From Lennon et al. 2004.)

the cervical region there is infarction from the fourth cervical to the third thoracic segments (Spiller 1909). Clinically, there is neck pain with paraesthesiae extending into the upper limbs followed by a flaccid paralysis of both arms with loss of pain and temperature sensation below the level of the lesion but with preservation of light touch, vibration, and joint position sense due to the integrity of the posterior columns. Initially there is a flaccid paralysis of the lower limbs, spinal shock, but, if the patient survives, spastic weakness of the lower limbs develops with increased reflexes and extensor plantar responses. There is usually retention of urine and faeces in the early stages but automatic bladder and bowel control may be achieved. In severe cases paralysis remains complete

and prognosis is poor but when infarction is less extensive or the ischaemia is reversible the lower limbs may show a variable degree of recovery.

In the thoracic region anterior spinal artery occlusion is usually a complication of dissecting aneurysm of the aorta, though it may follow emboli from an atheromatous plaque, vascular surgical procedures, or a drop in perfusion pressure. It is seen following cardiac arrest and as a rare complication of angiography. Rarer causes include the inadvertent injection of contrast into the thyrocervical trunk during cerebral angiography, emboli of fibro-cartilage from intervertebral disc degeneration entering the bone marrow, atrial myxoma, Sickle cell anaemia, and non-compressive Paget's disease causing a spinal artery steal phenomenon due to arteriovenous connections (Herzberg and Bayliss 1980).

Fibro-cartilaginous embolism often occurs in the healthy athlete in association with pain in the back followed by the rapid development of a severe transverse myelopathy or the syndrome of anterior spinal artery occlusion. There may be a history of exertion or back injury during sporting activities a day or so before the syndrome begins. Such cases have been shown to be due to occlusion of small vessels within the spinal cord with fibro-cartilage, consequent infarction of the cord in association with local herniation of nucleus pulposus of an intervertebral disc into an adjacent vertebral body. It is thought that the pressure of this herniation forces disc material into vessels within the bone marrow causing subsequent embolization locally (Tosi *et al.* 1996).

With the more common anterior spinal cord infarction due to dissecting aortic aneurysm there is usually pain in the abdomen and back followed by total permanent flaccid paralysis of the lower limbs, sphincter paralysis, and loss of pain and temperature sensation to a level at about the umbilicus, the 10th thoracic segment of the cord. There is usually some preservation of light touch and vibration and joint position sense in the classical anterior spinal artery syndrome.

Thrombosis of posterior spinal arteries, though more rare, causes loss of proprioception, vibration sensation and light touch and venous infarction may occur commonly in the presence of the dural spinal arteriovenous malformation (Fig. 28.10).

When a complete anterior spinal artery occlusion occurs the clinical syndrome is usually apparent, but partial infarction or ischaemia of the cord may be more difficult to recognize when due to occlusion of a posterior spinal artery or a feeding radicular vessel. There may then be weakness, sensory impairment, and discomfort restricted to one limb, or asymmetrically in both lower limbs and

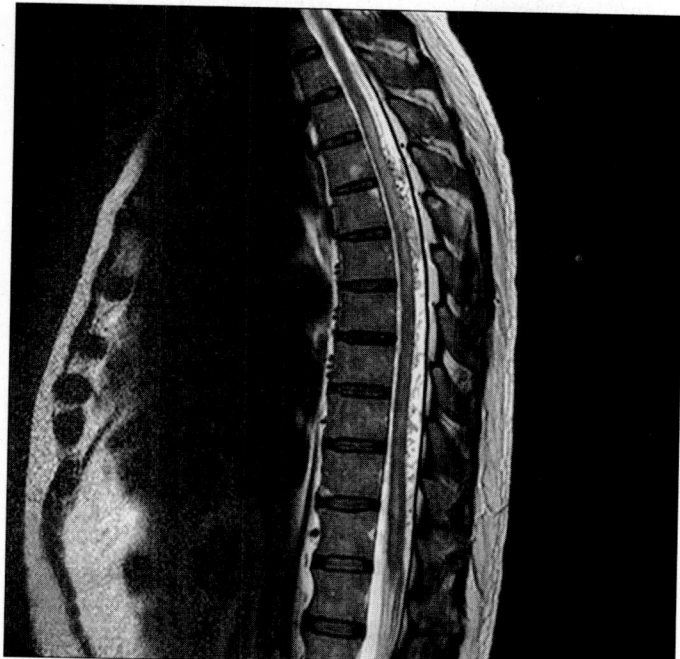

Fig. 28.10 Long ischaemic lesion throughout the dorsal spinal cord causing high signal on MRI. The abnormal signal in the CSF posterior to the cord is suggestive of a dural arteriovenous malformation.

there may be considerable recovery after such a localized infarction. Recurrent transient episodes of spinal ischaemia, like cerebral transient ischaemic attacks, are extremely uncommon, but may be important prognostic symptoms preceding infarction.

Collagen vascular disorders such as systemic lupus erythematosis or polyarteritis nodosa may cause infarction of the spinal cord and emboli may occur from bacterial endocarditis or as part of decompression sickness, Caisson disease (Section 5.10.4). This is usually seen in underwater divers when, with decompression, nitrogen bubbles form in spinal vessels, usually in the upper thoracic cord. Symptoms may be mild and transient but can be devastating with a complete transverse myelopathy. This is a medical emergency and the patient should be treated with immediate recompression in a hyperbaric chamber; when the acute deficit is severe prognosis is variable but remarkable recovery may be seen with recompression.

In general the diagnosis of spinal cord infarction is made from the clinical features, but should always be supported by spinal MRI. This is important to rule out alternative, compressive pathology and in most instances will demonstrate the area of the infarction. Infarction is seen as a high signal region within the cord on T2-weighted images, sometimes associated with swelling, and extending over a variable number of segments. The lesion may be seen predominantly in the distribution of the anterior spinal artery and there may be changes in adjacent vertebra due to ischaemia.

Management of spinal cord infarction includes identification of the cause, search for sources of embolization, and exclusion of a coagulation disorder. Anticoagulation or interventional angiography may be indicated with venous infarction secondary to thrombosis in an arteriovenous malformation.

Haematomyelia

Haemorrhage into the spinal cord occurs much less commonly than bleeding into the brain. It is usually the result of trauma but non-traumatic causes include arteriovenous vascular malformations, bleeding into an intramedullary metastasis, a primary coagulation disorder, or the abuse of anti-coagulants. Very rarely no underlying cause can be found.

The presentation is of a sudden spinal cord syndrome, usually with pain at the level of the haemorrhage, followed by the development of a transverse cord syndrome with loss of motor and sensory function below the level of the lesion. The clinical symptoms are indistinguishable from those due to epidural or sub-dural haemorrhage and from acute infarction of the spinal cord, apart from the distribution of symptoms.

The diagnosis must be made with MRI, the cord will be swollen at the level of the haemorrhage and there will be signal loss on a T2-weighted sequence MRI scan due to the presence of deoxyhaemoglobin. Surgical evacuation of a discrete haematoma is generally advised, though the benefits are unproven.

28.4.6 Infections

Direct infection of the spinal cord itself is rare. By contrast bacterial infections commonly invade the tissues around the cord. Viral infections are more common and protozoal and parasitic infections occur.

Spinal cord abscess

Intramedullary pyogenic spinal abscess is a rare disorder which requires early diagnosis and treatment. The source of infection is usually blood borne, from bacteraemia or septicaemia, though it may occur from a local skin infection with contiguous spread through local tissues to the spinal cord (Fig. 28.11). The infecting organism is usually *Staphylococcus aureus*. The presentation is with fever, elevated peripheral white cell count, focal spinal pain, and a rapidly developing transverse cord lesion. Rarely the presentation may be sub-acute and there may not be systemic manifestations of infection.

Investigation is by MRI which will show a high signal intramedullary lesion on T2-weighted images with gadolinium enhancement, classically in a 'ring' pattern. The radiological appearances are often indistinguishable from primary intraspinal tumour, intraspinal metastasis, or an acute demyelinating lesion. CSF examination will show mild pleocytosis, moderate elevation of protein, and normal glucose. The diagnosis depends upon a high index of clinical suspicion in patients who have had bacteraemia or septicaemia, or those with local surgical procedures close to the spine. The need for surgery is important to achieve the diagnosis and to drain the abscess. High dose intravenous antibiotics should be given as soon as the diagnosis is suspected.

Viral myelitis

Viral infections of the spinal cord include poliomyelitis, coxsackie viruses, herpes zoster, herpes simplex, HIV, Epstein–Barr virus, cytomegalovirus, and HTLV1 (Section 42.4). The latter produces a characteristic chronic evolving paraparesis and is considered in the section on chronic myelopathies. All of the viruses which can cause acute infective myelitis may be associated with a post-infective myelitis which is likely to be due to immune mediated responses triggered by the viral antigen. Both poliomyelitis and coxsackie viruses produce an acute inflammatory meningomyelitis with cord involvement, predominantly affecting the anterior horn cells, leading to patchy multi-focal or sometimes extensive muscle

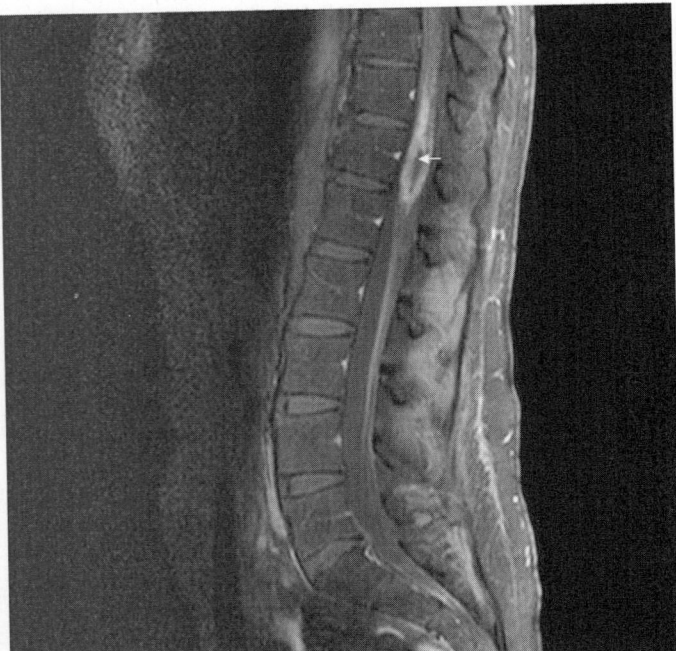

Fig. 28.11 Spinal cord abscess at level of conus (arrow). Gd enhanced T1-weighted MRI.

weakness and wasting. There is corresponding reflex loss but no sensory disturbance. In the acute phase the CSF contains a mononuclear pleocytosis with elevation of protein. Varicella zoster virus causes a sensory dermatomal effect, often with marked pain due to predominant involvement of the dorsal root ganglia. The associated dermatomal vesicular rash will normally establish the diagnosis. More extensive involvement of the adjacent spinal cord or roots can lead to features of myelopathy and segmental myotomal defects with weakness, wasting, and areflexia.

A transverse myelitis due to viral infection, such as those described, is rare but is indistinguishable from a post-infectious transverse myelitis. Such infections are most likely to occur in immuno-compromised states, including people with HIV infection (Section 43.3.6). The diagnosis is confirmed by the isolation of the viral antigen from CSF by polymerase chain reaction; high dose acyclovir is indicated for cord infections due to herpes simplex or zoster.

In patients with AIDS, a vacuolar myelopathy sometimes develops (Section 43.3.6). The tempo is sub-acute with symptoms typically evolving over weeks. Motor, sensory, and sphincter abnormalities all appear, sometimes asymmetrical and predominantly in the legs, though the arms may also be involved. Pathologically, there is vacuolation in the spinal cord white matter which is most marked in the thoracic region. This manifestation is seen in patients with AIDS, rather than asymptomatic HIV positive individuals, and therefore Highly Active Anti-Retroviral agents, HAART, should delay or prevent its occurrence.

Parasitic infestations

Schistosomiasis is an important cause of spinal cord syndromes in Africa, South America, and the Far East (Section 43.2.10). The most common cause of acute myelitis is *Schistosoma mansonii*, but infection may also occur with *S. haematobium* or *S. japonicum*. The thoraco-lumbar cord is the most likely area of involvement, either by a localized granulomatous process or due to ova in the arteries and veins causing ischaemic damage (Fig. 28.12). An acute necrotic myelitis has also been reported (Queiroz *et al.* 1979). MRI reveals a high signal intrinsic cord lesion on T2-weighted images and gadolinium enhancement on T1-weighted images. The spinal cord may be swollen and the CSF may contain pleocytosis with an elevated protein. The diagnosis may be made by the identification of schistosoma ova in the faeces or tissues or by demonstrating a positive serological response to shistosomes in the CSF. Treatment is with praziquantel and should prevent progression of the disease but existing neurological deficits will not reverse.

28.4.7 Granulomatous disease

Sarcoidosis is the only common granulomatous disease to affect the spinal cord. Intramedullary masses of granuloma may present with an acute, sub-acute, or chronic myelopathy. Lepto-meningeal involvement with sarcoid can also occur and gadolinium enhanced MRI, the investigation of choice, may show granulomatous tissue enhancing within the spinal cord and highlighting the meninges and roots.

The granuloma can occur at any level in the spinal cord and may be associated with involvement of the optic nerve, facial nerve, auditory nerves, hypothalamus, brainstem, and hemispheres. Most patients have evidence of sarcoidosis outside the nervous system. The serum angiotensin converting enzyme tends to be positive and enzyme levels can be detected in CSF, though they are not essential

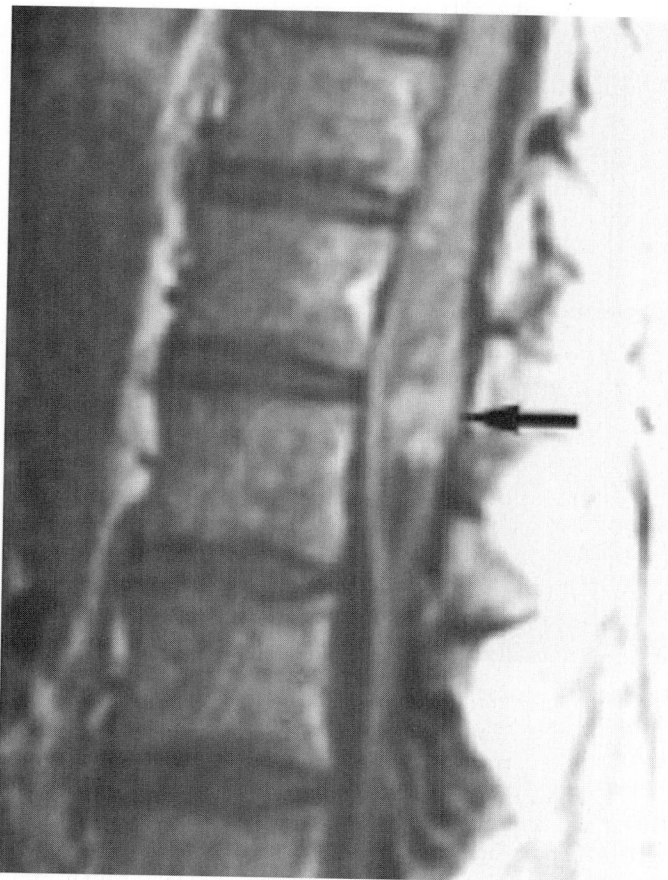

Fig. 28.12 Schistosomiasis. Gd enhanced T1-weighted MRI at the level of the conus shows swelling with areas of low and high signal.

for diagnosis. CSF will normally contain a pleocytosis and a high protein.

The treatment of acute sarcoid granulomata in the nervous system is with corticosteroids, initially in high doses. Long-term maintenance of steroids may be required to prevent a relapse of cord or central nervous system disease and long-term maintenance steroids can be coupled with immunosuppressive therapy to reduce the steroid effect.

28.4.8 Miscellaneous acute myelopathies

Brucellosis causes meningitis and meningo-encephalitis and may cause intrinsic cord lesions (Section 42.5.11). It can present with an acute myelopathy when diagnosis is established by MRI and positive Brucella serology in the CSF.

Neurosyphilis is another cause of acute or chronic spinal cord syndromes and may be identified by blood and CSF serology (Section 42.5.1).

Toxic myelopathies are rare, but the use of clioquinol in excessive doses causes sub-acute myelo-optic neuropathy which can present with an acute spinal cord syndrome.

28.5 Chronic and evolving cord syndromes

Chronic lesions affecting the spinal cord may be due to either compressive or non-compressive lesions (Table 28.2) with contrasting symptomatology. Compression of the spinal cord is classically associated with pain at the site of the pressure, the slow evolution of motor and sensory disturbances which may be a symmetrical and associated with lower motor neurone signs and lower sensory neurone symptoms at the site of the lesion and the relative sparing of bladder and bowel until late in the disease. Intrinsic spinal cord lesions, or non-compressive disease, tend to affect bladder and bowel early, be relatively painless, and not associated with lower motor neurone signs at the site of the lesion.

28.5.1 Chronic spinal cord compression

Chronic compression of the spinal cord occurs with diseases of the vertebral column and from other lesions pressing upon the spinal cord. The two most common vertebral column conditions leading to chronic spinal compression are firstly spondylosis together with protrusion of intervertebral discs and the formation of osteophytes reducing size of the spinal canal, and secondly, metastases from distant carcinoma. Less frequent causes include neoplasms arising from the vertebra, such as sarcoma, myeloma, osteoma, cordoma, and haemangioma; vertebral infections due to tuberculosis, staphylococcal osteomyelitis or sphylitic osteitis; the osteitis deformans of Paget's disease; and, rarely, achondroplasia, mucopolysacaridoses, and severe kyphoscoliosis. Bone marrow swelling in thalassaemia can occasionally compress the spinal cord and aneurysms may erode vertebra and cause pressure upon the cord. Atlanto-axial subluxation, particularly in rheumatoid arthritis, or spondylolisthesis at any level in the spine may narrow the spinal canal.

Other causes of chronic compression include extra-dural abscess following systemic infection or vertebral osteitis, arachnoiditis due to syphilis, tuberculosis, sarcoidosis or other granulomatous processes, infiltration of the meninges with leukaemia or lymphoma or other en-plaque malignancy. Developmental arachnoid cysts and parasitic cysts such as hydatid and cystercercosis are rare causes of chronic compression. Intra and extra-medullarly tumours may compress the cord chronically and dural herniation of the spinal cord through a spontaneous or created defect in the meninges can result in a progressing cord deficit.

28.5.2 Cranio-vertebral junction anomalies

Lesions in the region of the foramen magnum, most commonly a neurofibroma or meningioma, usually produce a slowly evolving quadriparesis with a characteristic cluster of associated physical signs. These include pain in the neck and occipital region with radiation of pain and sensory loss in the distribution of the C2 to C4 dermatomes. This may be accompanied by sensory loss in the first division of the trigeminal nerve which may be central and extend gradually due to involvement of the descending spinal tract of the fifth nerve in its intra-medullary spinal course. There may also be contralateral loss of pain and temperature sensation, stiffness of the neck muscles, and the head may be held rigidly. A Horner's syndrome ipsilateral to the lesion may occur and there may be the diagnostic sign of down-beat nystagmus made worse on lateral gaze. Involvement of the 11th and 12th cranial nerves can cause weakness of trapezius and the tongue. The weakness which gradually evolves in the arms and legs is often asymmetrical and may extend from one arm to the contralateral arm, leg, then ipsilateral leg in a circular way. It is important to remember that lesions at the level of the foramen magnum can interfere with the phrenic nerve and cause problems with respiration, particularly in the recumbent position.

Table 28.2. Causes of chronic spinal cord compression

Diseases of the vertebral column

Common

    Metastatic carcinoma

    Cervical spondylosis/disc protrusion

    Traumatic fracture dislocation

Less common

    Primary vertebral neoplasms

    Sarcoma

    Myeloma

    Osteoma

    Chordoma

    Haemangioma

    Vertebral infections

    Tuberculosis

    Staphylococcal osteomyelitis

    Syphilitic osteitis

    Craniocervical junction abnormalities

    Rheumatoid atlantoaxial subluxation/pannus

    Paget's disease

Rare

    Achondroplasia

    Mucopolysaccharidosis

    Juvenile osteochondritis

    Thalassaemia

Other causes within the spinal canal

    Extradural abscess

    Metastatic infection

    Vertebral osteitis

    Arachnoiditis

    Syphilis

    Tuberculosis

    Sarcoidosis

    Meningeal infiltration

    Lymphoma

    Leukaemia

    Arachnoidal cysts

    Parasitic cysts

    Hydatid

    Cysticercos

    Extramedullary tumours

    Intramedullary tumours

    Dural herniation of the spinal cord

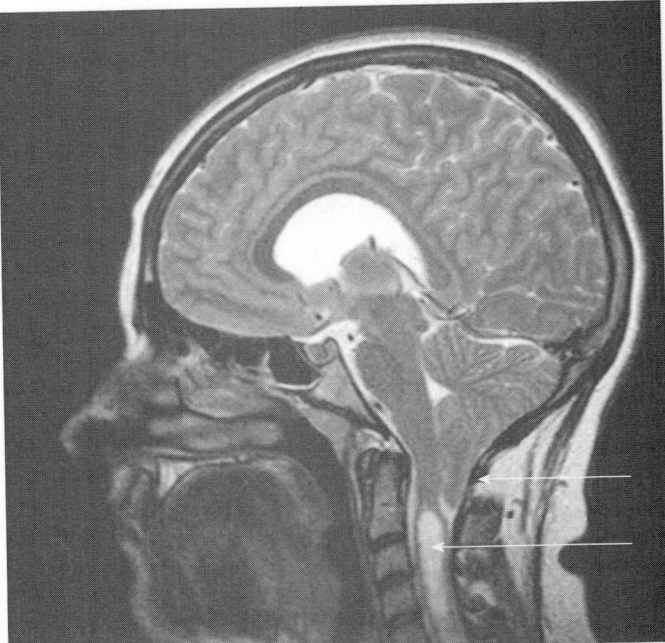

Fig. 28.13 Syringomyelia. Arnold Chiari malformation with descent of the cerebellar tonsils (upper arrow) and a cervical syrinx (lower arrow). Sagittal T2-weighted MRI.

lateral and central, combined with osteophyte formation on the vertebral bodies and soft tissue changes in the paravertebral tissue frequently result in compression of the cervical cord with or without root involvement. This slowly progressive degenerative process is 'cervical spondylosis' and becomes increasingly common with age. The widespread use of MRI scan has confirmed a high prevalence in the normal population, albeit usually without symptoms. Spondylotic changes are most common in the mid to lower cervical spine with the maximal frequency and severity of involvement at C5/6 (Fig. 28.14).

The effect of cervical spondylosis upon the spinal cord is complex. There is direct compression, the protruding disc or osteophyte may interfere with the blood supply, and tethering of the cord by the ligamentum denticulata and spinal roots may cause narrowing of the intervertebral foramina allowing ordinary neck movements to produce cumulative trauma. The ligamentum flavum, lying posteriorly, may become corrugate. The result is a condition of patchy degeneration in the cervical cord, termed 'cervical myelopathy' (Wilkinson 1973).

The clinical picture of cervical spondylosis may be indistinguishable from the progressive spastic paraplegia seen in multiple sclerosis and, since the former is so common, it is not unusual to find the two disorders co-existing. The presence of radicular signs in the upper limbs is most consistent with cervical spondylosis; sometimes extensive lower motor neurone signs in the arms, including wasting, fasciculation, and areflexia, may raise the possibility of motor neurone disease.

Though neurophysiology may help in determining radicular disease from anterior horn cell disease, MRI is the diagnostic investigation of choice. It must be remembered that many healthy older adults will have moderate spondylotic changes with mild cord indentation or compression and that it is only when the cord

Such lesions are commonly due to structural abnormality at the foramen magnum. MRI will differentiate between neurofibroma or meningioma and an Arnold Chiari malformation (Fig. 28.13) and also demonstrate anterior atlanto-axial sub-luxation in rheumatoid arthritis, psoriatic arthropathy, and the mucopolysaccharoidoses.

28.5.3 Cervical spondylosis

The most common cause of progressing cervical myelopathy in adult life is cervical spondylosis. Chronic disc protrusions both

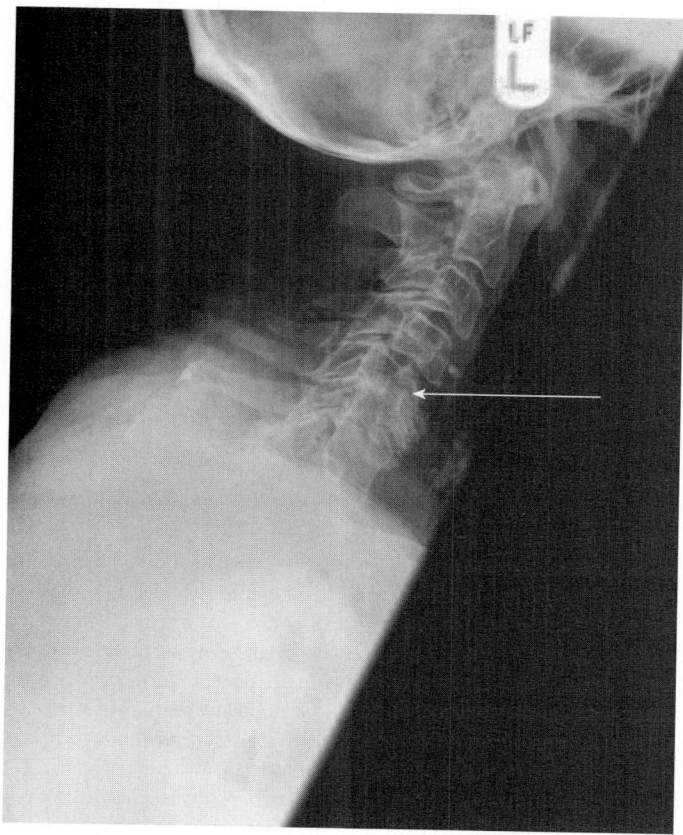

A

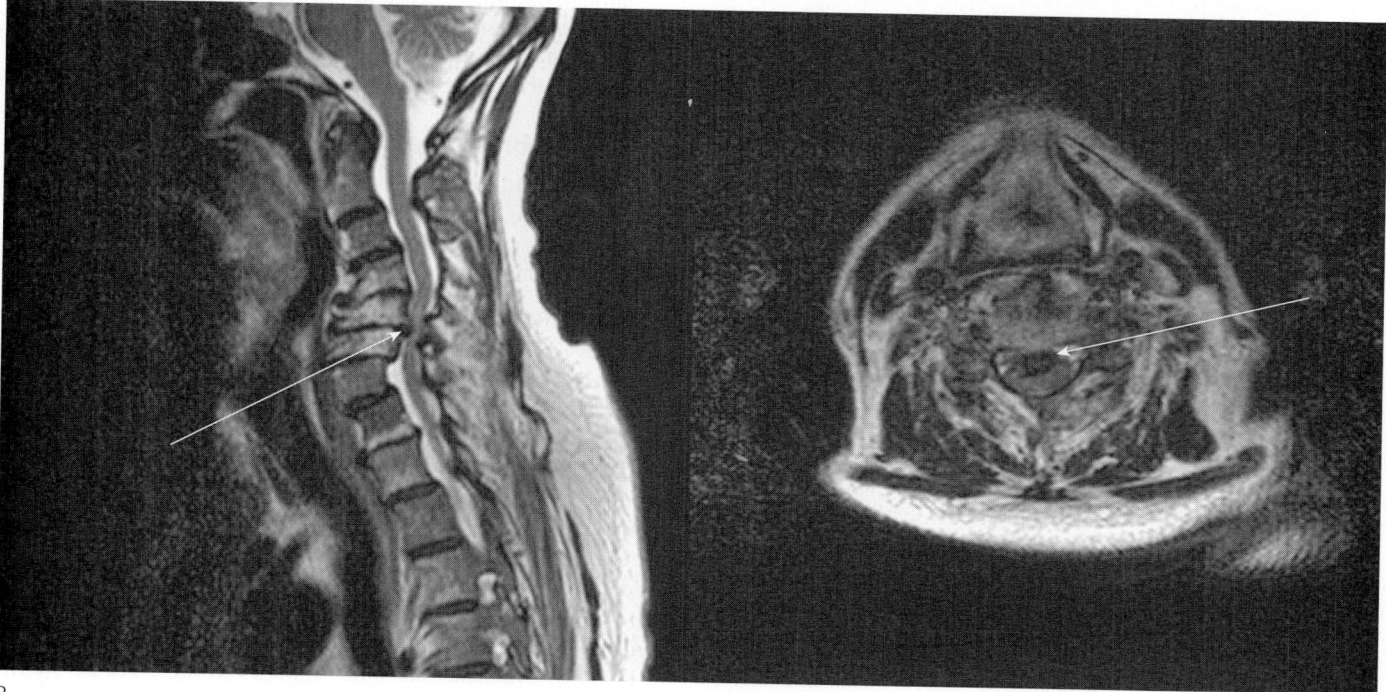

B

Fig. 28.14 Cervical Spondylitic Myelopathy (A). X-ray of cervical spine showing severe spondylosis with ankylosis at C5/6/7 (arrow) and narrowing of the spinal canal. (B). T2-weighted MRI sagittal (left) and axial (right) showing severe cervical spondylosis with retrolisthesis at C3/4 causing myelopathy with signal change in the spinal cord.

compression is moderate or marked, sometimes associated with signal change within the cord, that it is to be regarded as the sole cause of significant myelopathic functional deficits. The presence of high signal in the cord on T2-weighted images at the level of compression indicates structural damage to the cord parenchyma by the compression and associated ischaemia. The older investigation of myelography is now only necessary when MRI is contraindicated as in those with cardiac pacemakers, intracranial aneurysm clips, or orbital metal fragments.

The natural history of cervical myelopathy due to cervical spondylosis is of a fluctuating condition in which there are periods of deterioration interwoven with periods of stability and even mild improvement. Pain in the neck radiating to the arms fluctuates and though some people have severe disability which increases remorselessly, most are left with a varying degree of residual disability. In some cases immobilization of the neck in a hard or soft collar has been said to arrest clinical progression, but immobilization is less effective in this condition than in acute cervical disc prolapse.

Surgical decompression should always be considered when spinal cord damage is progressing, particularly when there is evidence of severe compression radiologically. When there is multi-level disease or significant canal stenosis laminectomy may be the operation of choice, where there is a single or two adjacent disc levels anterior removal of the disc or discs with either spinal fusion or replacement of disc material should be considered (Cloward 1980).

28.5.4 Other bony lesions

Atlanto-axial subluxation

Separation of the odontoid process of the axis, sometimes seen as a congenital abnormality or as the result of trauma, rheumatoid arthritis, psoriatic arthropathy or mucopolysaccharoidoses may permit abnormal movement of the atlas on the axis leading, in time, sometimes gradually after years, sometimes more acutely, to a myelopathy due to compression of the cord (Stevens *et al.* 1971). The sub luxation can be demonstrated on CT myelography or by MRI and can be detected clinically by movement of the head on the neck in the Sharp and Purser test. The preferred treatment is occipital cervical fusion, though immobilization in a position of flexion may be considered. Operative decompression of the upper cervical cord can be complicated by haematomyelia.

Spondylolisthesis

Though most commonly seen in the lumbar spine as a cause of lumbo-sacral radiculopathy, spondylolisthesis can also occur in the cervical spine, resulting in reduction of the antero-posterior diameter of the spinal canal and consequent compression of the spinal cord. The signs are similar to those of cervical spondylosis, the investigation identical, and the treatment involves fusion of the relevant vertebra.

Paget's disease

Paget's disease, or osteitis deformans, is a primary bone disease in which there is both resorption of normal bone and formation of abnormal new bone (Section 27.6.6). Its cause is uncertain; diagnosis is established on the basis of characteristic radiographic findings together with an elevated serum alkaline phosphatase. Neurological symptoms result from bony overgrowth in the skull and deformity of the skull shape, together with deformity and overgrowth in the vertebral column leading to compression of neural structures.

Occasionally the Paget's bone may show sarcomatous change and the highly vascular abnormal bone may also reduce blood supply to adjacent neural structures by a 'steal' phenomenon.

Spinal cord compression may occur at any level but is most common in the upper thoracic region. In keeping with extrinsic compression pain is prominent, fractures and dislocations may occur, and myelopathies can be seen without compression and are assumed to be ischaemic and vascular. When cord compression is demonstrated treatment with calcitonin, mythramycin, and diphosphonates have been shown to reduce pain and improve neurological function. Rarely surgical decompression may be indicated.

Tuberculous spinal osteitis

This is now rare in developed countries, but used to be seen in children and young adults. It was thought due to the consumption of unpasteurized milk and was frequently due to the bovine tubercle bacillus. In the developing world tuberculous spinal infection is still a common problem, it most usually affects the thoracic spinal cord, the infective process begins in the vertebral body, spreads across the disc to adjacent bodies, leading to their collapse and the formation of an angular deformity of the spine. The deformity as such is rarely the cause of compression of the cord which is more usually affected by an extra-dural tuberculous abscess or the development of tuberculous meningomyelitis. In addition to compression of the cord interference with the blood supply of adjacent segments either by compression of radicular arteries or by tuberculous endarteritis is an important factor in producing myelopathy and paraplegia.

Conservative treatment with anti-tuberculous chemotherapy rather than surgical decompression, often gives remarkably good outcome even in severe cases.

A similar, but much less common osteitis may occur with syphilis, spinal compression, and cord ischaemia is created in a similar pathological way and the response to appropriate conservative therapy is also excellent.

28.5.5 Vertebral tumours

A variety of primary and secondary tumours and swellings of the spinal bones can result in chronic compression. These are commoner than intrinsic tumours of the spinal cord or soft tissue swellings related to the meninges as the cause of chronic compression.

Vertebral metastases

Secondary carcinoma is the most common vertebral neoplasm. It is rare in those under the age of 35 years and the commonest primary is derived from lung, breast, thyroid, or prostate. Less commonly primaries from the uterus, stomach, kidney, and large bowel may metastasize to spine. The vertebral metastasis is usually blood borne, but sometimes when the spinal lesion is at the same segmental level as the primary growth peri-neural lymphatic drainage is believed to be the cause. Carcinomatous deposits eat into and erode the spongy portions of the vertebral body which then collapses (Fig. 28.15). The spinal cord may be compressed either as the result of the ensuing spinal deformity or by an extradural extension of the growth. Usually the spinal roots are compressed before the cord so that the typical local and radicular pain indicating an extrinsic lesion may be present for some time before the development of the cord syndrome.

When multiple vertebrae are involved treatment may be palliative and in such circumstances adequate analgesia with opiates and

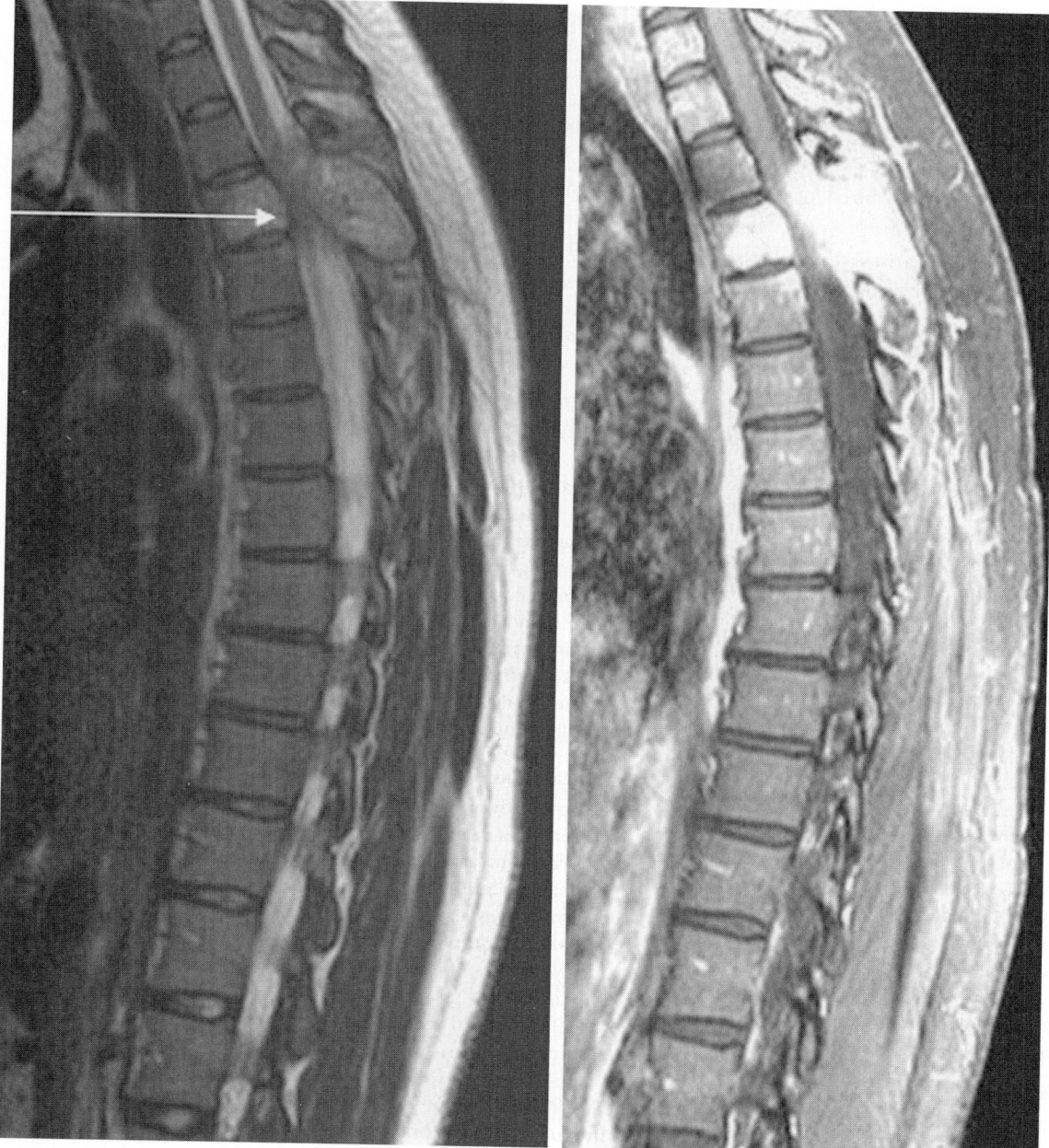

Fig. 28.15 Vertebral metastasis compressing the dorsal spinal cord. T2-weighted and T1-Gd enhanced MRI.

the use of radiotherapy should be considered. Chemotherapy appropriate to the particular form of cancer may be indicated and in many cases of acute compression a single level emergency laminectomy, decompression, and bone replacement is considered. The prime indication for surgery used to be to relieve pressure and obtain a surgical biopsy prior to radiotherapy and chemotherapy but modern techniques allow some degree of reconstitution of the spine prior to radiotherapy and chemotherapy.

The decision about the method of treatment of the metastasis depends upon its nature, the management of the primary tumour, and the evidence for more metastases. Radiotherapy sometimes has remarkable effect on pain relief and in some cord compression relief is achieved with a combination of steroids and radiotherapy. Powerful analgesics and, in some cases, surgical pain relief with cordotomy, stereotactic thalamotomy, or spinal cord stimulation may be required.

Other vertebral neoplasms

◆ *Sarcomas* arising in a vertebra or invading the spinal column from adjacent tissues may cause chronic spinal cord compression. Cavernous haemangiomata are rare tumours within the vertebra and even more rarely enlarge to cause spinal cord or nerve root compression. The radiological changes with prominent vertical striations or a honeycomb pattern in the bone are typical and patchy high signal is seen on unenhanced T1-weighted images within enhancement after gadolinium.

- *Myeloma* may occur as solitary tumour or simultaneously in several vertebral bodies. When present as a solitary lesion it may cause cord compression and biopsy is required for identification and to determine therapy.

- *Osteomas* are rare tumours, usually arising from the posterior part of a vertebral body and tending to compress the spinal cord from anteriorly.

- *Chondromas* are usually not true tumours but merely intervertebral disc protrusions associated with spondylosis.

- *Chordomas*, which arise from remnants of the notocord, most commonly develop in the clivus in the cranium or in the sacro-coccyxgeal region but may rarely be seen in the cervical or thoracic areas and cause cord compression.

- When *lymphoma* or *leukaemia* spreads to the spine it usually infiltrates the dura mater extensively but occasionally invades the tissues of the cord and can have a space occupying effect with symptoms and signs and nerve root compression. Such vertebral neoplasms blend into the spectrum of malignant meningitis which occurs with many carcinomas, affects the cranial nerves and nerve roots, may be associated with spinal cord disease, and is usually diagnosed with CSF examination, including cytology and gadolinium enhanced MRI scans which show thickening and enhancement of nerve roots.

28.5.6 Intraspinal tumours

Tumours within the spine are conveniently divided into extra-dural and intra-dural. The latter can further be divided into those arising outside the spinal cord, extra-medullary tumours, and those lying within the cord, intra-medullary tumours. About 20 per cent of spinal tumours are extra-dural, 60 per cent intra-dural but extra-medullary, and 20 per cent intra-medullary. Extra-dural tumours, those due to involvement of the vertebra and paraspinal tissues, are commonly secondary and have been described above.

Extramedullary, intradural tumours

The most common extra-medullary, intra-dural tumours are meningiomas and neurofibromas (Alter 1975). Meningiomas are more common in the middle-aged female. In the spine extra-medullary, intra-dural tumours are about three times as common as intra-medullary tumours.

- *Neurofibromas* usually arise from the spinal roots, more frequently the posterior than the anterior. They may be single or multiple and may or may not be associated with neurofibromatosis. Exceptionally an extra-medullary neurofibroma may grow through the intervertebral foramen adopting a 'dumbbell' shape. The extra-spinal portion may be palpable at the side of the vertebra. Neurofibromas may develop at any level of the spinal canal and occur equally in the sexes (Fig. 28.16).

- *Meningiomas* arise from the arachnoid covering the roots or the spinal cord. They are almost always in the thoracic region and affect females much more frequently than males (Fig. 28.17). Sarcomatous changes in spinal meningiomas and primary extra-medullary sarcomas occur but are rare. *Lipomas* are seen most

commonly in association with occult spina bifida and spinal dysraphism. Other developmental anomalies which may mimic growths in causing cord compression include dorsal neuroenteric cysts, a residuum of the yolk sac, and congenital extra-dural cysts, intra-spinal meningoceles.

- *Epidermoid cysts* may develop intraspinally several years after lumbar puncture due to implantation of fragments of epidermis into the spinal canal. They are more commonly associated with spinal dysraphism and are constitutional.

- *Cordomas* are rare malignant tumours arising from a remnant of the notocord and, in the spine, are almost invariably in the sacro-coccyxgeal region. They may occur rarely in the cervical or thoracic region.

- *Dermoid cysts* and other forms of teratoma may also be found within the spinal canal.

Intra-medullary tumours

Intra-medullary spinal tumours may mimic any of the cerebral gliomas (Kernohan *et al.* 1931). Ependymomas are most common, accounting for more than 40 per cent of tumours (Fig. 29.1). The next most common intra-medullary tumour is an astrocytoma, but in children medulloblastoma are found and at all ages, oligodendrogliomata, ganglioneuromata, and haemangioblastomata may be seen. Intra-medullary metastases from carcinoma are rare. Leukaemic and lymphomatous deposits may occur in the cord, as can tuberculomas.

Clinical Features

Metastases from primary spinal neoplasm are rare but ependymoma of the filum terminale has been reported to metastasize into the body. By contrast intra and extra-medullary spinal deposits from intracranial gliomas, particularly medulloblastomata, are not uncommon, especially in children.

Apart from the meningioma, which is much more common in elderly females, other spinal tumours show no sex bias. Tumours may develop at any age, though the majority occur between the ages of 20 and 60 years. Meningiomas are rare in childhood as are all spinal tumours, compared to those found within the cerebrum.

The thoracic cord is the most common site for extra-dural and extra-medullary tumours, approximately 2/3 of all extra-medullary tumours are situated in the dorsal or dorso-lateral aspect of the cord and approximately 1/3 in the ventral or ventro-lateral aspects.

Intrinsic tumours of the spinal cord tend to be painless and present with early involvement of bladder and bowel. Extrinsic tumours tend to cause local pain, local lower motor neurone and lower sensory neurone signs, and are late to involve the bladder and bowel.

MRI is the investigation of choice for the diagnosis of all forms of spinal tumour. Neurofibromas are usually hyperintense on T2-weighted images, meningiomas may be isointense on T1- and T2-weighted images and show prominent gadolinium enhancement. Lipomas produce a characteristic high signal on T1-weighted images and low on T2-weighted, and intra-medullary tumours are usually associated with swelling of the cord and with gadolinium enhancement.

Extra-medullary, intra-dural tumours, which are usually benign and cause cord compression resulting in a progressing spastic paraplegia, should be surgically removed. In general, a good prognosis can be predicted if the diagnosis is made promptly and

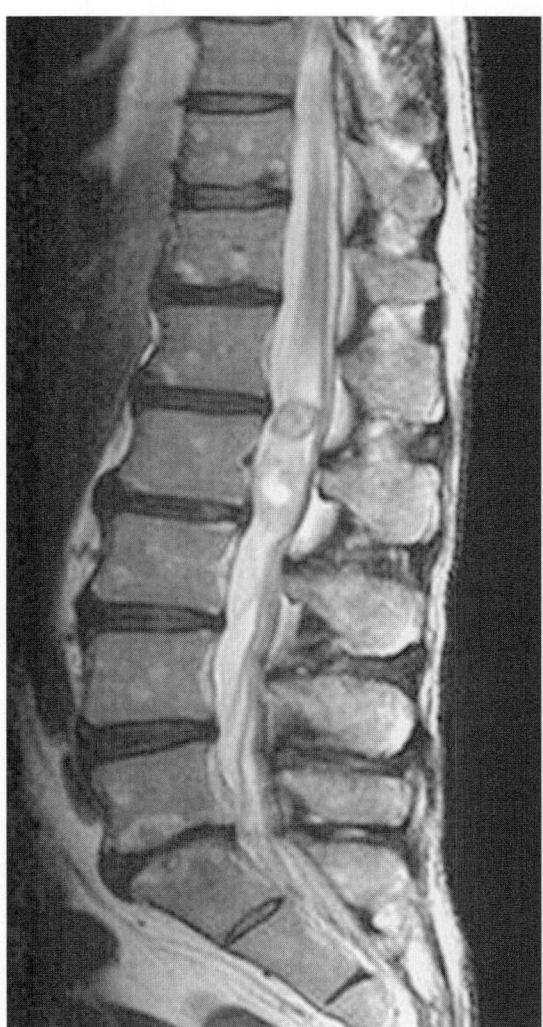
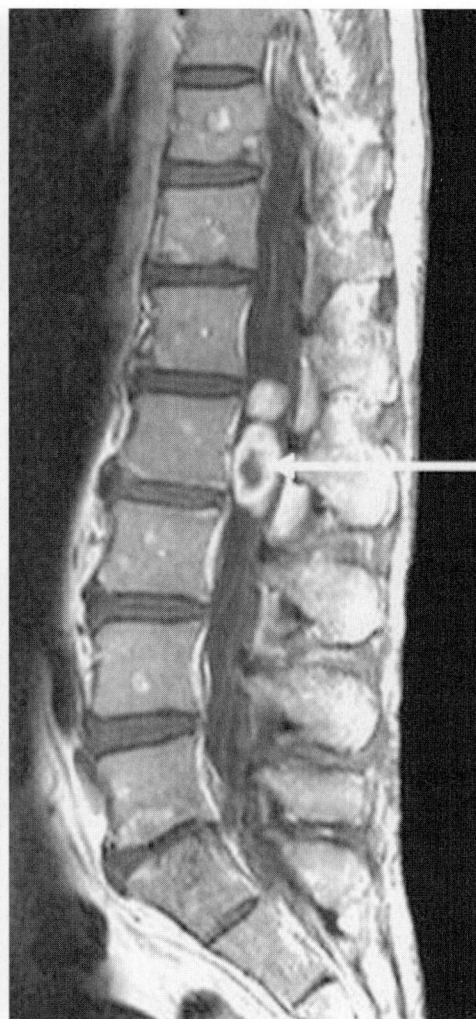

Fig. 28.16 Cystic neurofibroma within the lumbar theca. T2-weighted (left) and T1-Gd enhanced (right) MRI.

the pre-surgical deficit is not severe. Intra-medullary gliomas or ependymomas may be indolent with a history evolving over years or decades. Surgical debulking of ependymomas may be possible in some instances and some cases of intramedullary glioma may benefit from radiotherapy and chemotherapy.

28.5.7 Meningeal abnormalities

Arachnoid cysts

Arachnoid cysts, which are presumed to be developmental in origin, differ from the neuroenteric cysts in not being associated with spina bifida. They are an occasional cause of cord compression, most commonly in children, adolescents, and young adults. They are extrinsic and therefore associated with radicular pain, together with signs of spinal cord dysfunction developing in a step-like manner. Most are in the dorsal region posterior to the cord and communicate with the sub-arachnoid space by a narrow orifice which used to be demonstrated on myelography (Fig. 28.19).

Nowadays the lesions are diagnosed by MRI which shows the presence of a cyst-like structure with the signal characteristics of CSF. They are particularly common in Marfan's syndrome and ankylosing spondylosis and may be found in the sacral canal.

Surgical decompression and excision of the cyst is indicated, alternatively the cyst may be marsupialized.

Cord herniation through a dural tear

Rarely the meninges may be torn as the result of trauma or surgical intervention. Even more rarely a tear may appear in the dura where there is no prior history of trauma or operation. Such a dural defect is most commonly seen in the upper or mid-thoracic region and usually lies on the ventral aspect of the cord. There may be herniation of the spinal cord through the dural defect into the extra-dural space (Fig. 28.19). The clinical presentation is with a progressive thoracic cord syndrome with motor and sensory features, often asymmetric and not causing complete paraplegia. MRI reveals a characteristic abnormality with the extra-dural herniation being shown separate from the normal spinal cord (Housmann and Moseley 1996). Surgical treatment aims to return the cord to its correct location and close the dural defect; the possibility of ischaemic injury to the cord must not be overlooked.

28.5.8 Infective causes

Although many viruses cause acute myelitis chronic and progressing myelopathy is seen only with retroviral infection.

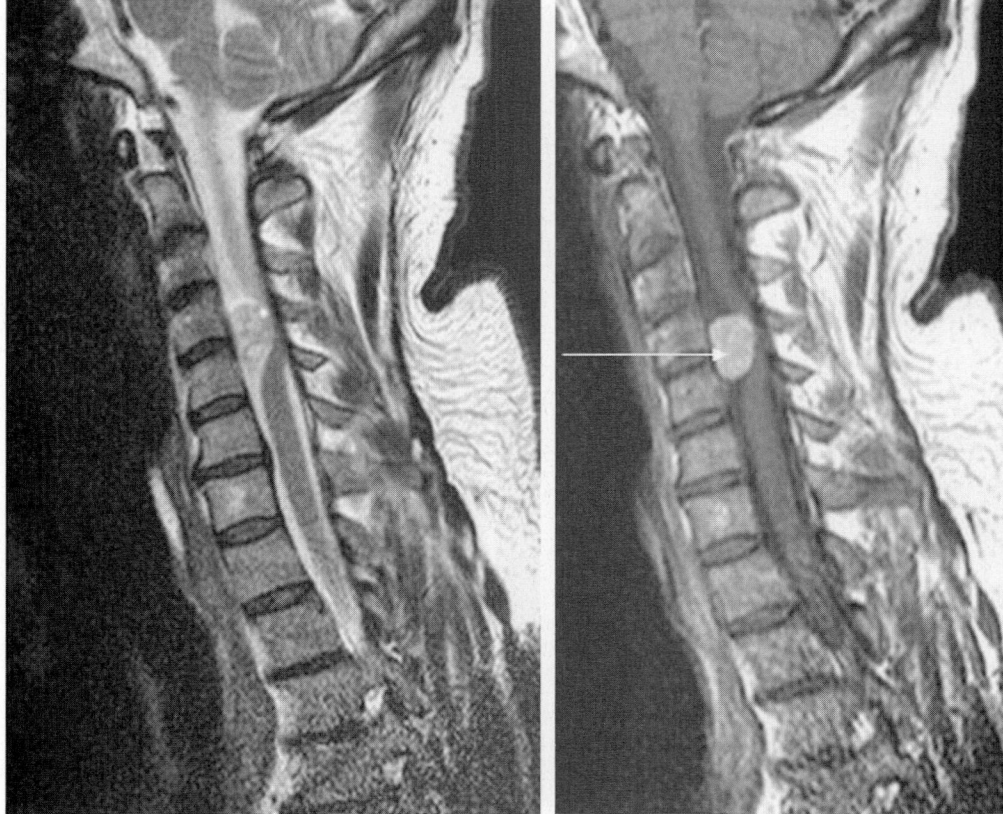

Fig. 28.17 Anterior cervical meningioma compressing the spinal cord. T2-weighted (left) and T1-Gd enhanced (right) MRI.

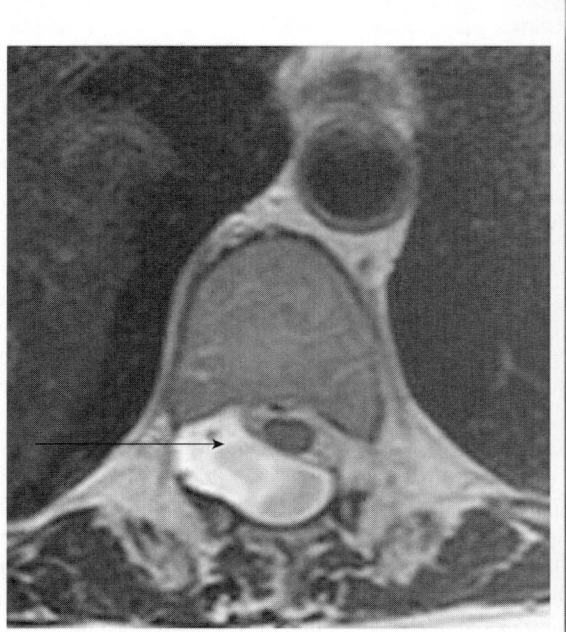

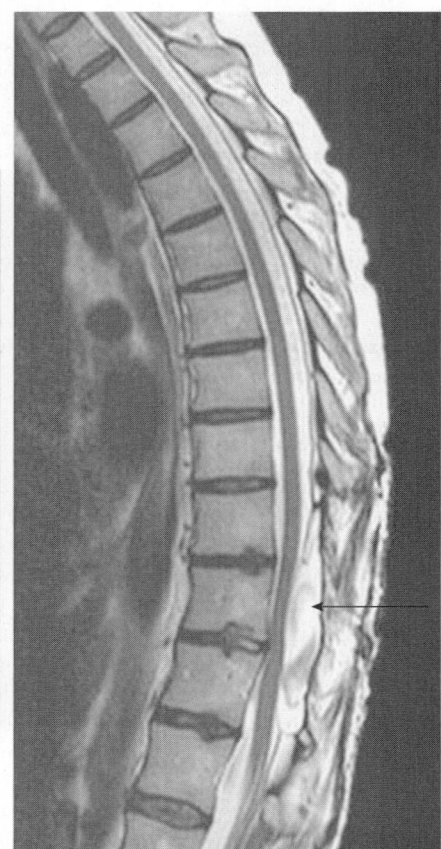

Fig. 28.18 Dorsal arachnoid cyst compressing the spinal cord. T2-weighted MRI.

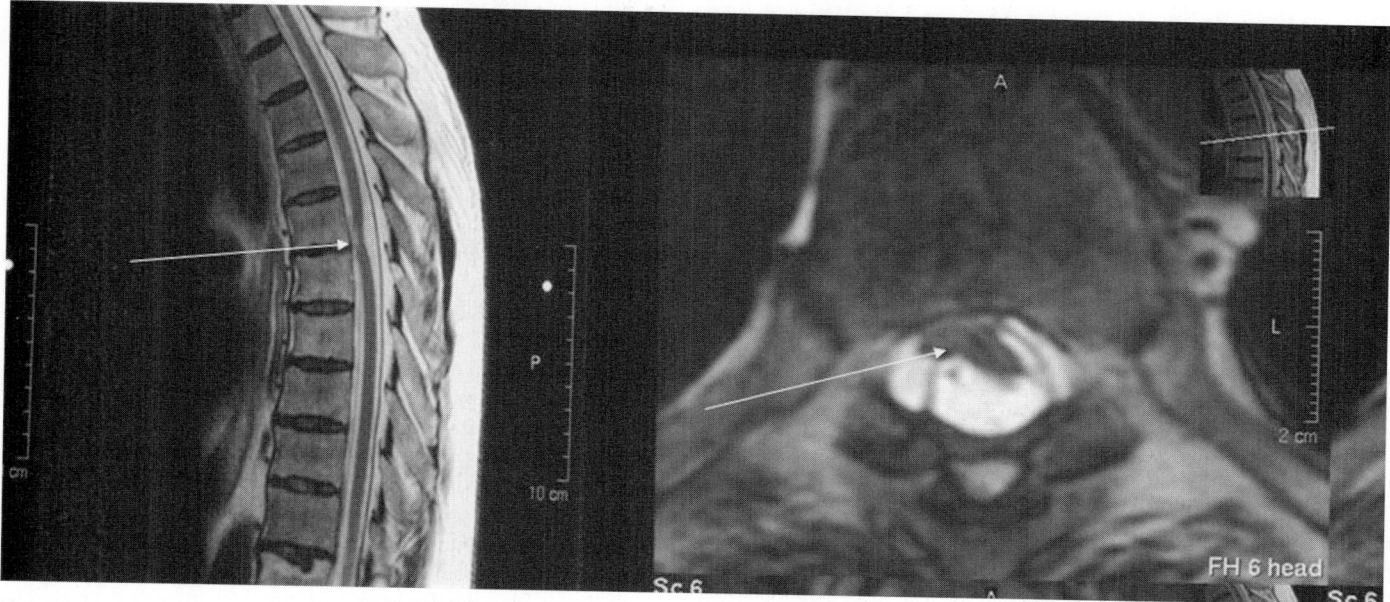

Fig. 28.19 Herniation of spinal cord through a dural tear. T2-weighted MRI.

HIV infection

In the late stages of infection with HIV1, frequently in association with AIDS–dementia complex, patients with AIDS may develop a slowly progressing myelopathy combined with a neuropathy (Section 43.3.6). There may be a combination of a paraparesis with increased tone, peripheral sensory loss, sphincter disturbance, and impotence. Most commonly the chronic myelopathy affects the thoracic spinal cord and MRI scan may show atrophy of the cord. Pathologically there is evidence of vacuolar change within the cord, though the pathogenesis of this change is uncertain. Highly active anti-retroviral therapy, HAART, does not prevent disease progression, suggesting that the myelopathy is not due to direct infection, but rather to indirect 'bystander' effects of the immune condition. There is some suggestion that the mechanism may be analogous to that seen in vitamin B12 deficiency. Therapy with l-methionine has been attempted.

HTLV-1-associated myelopathy

Infection with human T-lymphotrophic virus type 1, HTLV-1, is endemic in Japan and the Caribbean. It causes a progressive paraparesis which clinically resembles primary progressive multiple sclerosis (Section 42.4.1). When such a progressing non-compressive myelopathy occurs in a patient of a non-Caucasian origin HTLV-1 infection should be considered (Cruickshank *et al.* 1989). Although most common in the Caribbean, southern United States, southern Japan, South America, and Africa it has been reported in Afro-Caribbean migrants in the UK and may occur many years after migration.

The clinical picture is that of a gradually progressive spinal cord syndrome evolving over years with increasing paraparesis, spasticity, brisk reflexes in the lower limbs, and extensor plantar responses. Disturbance of sphincter control occurs early, in keeping with intrinsic disease of the cord, and is prominent. Paraesthesiae and dysaesthetic pain is prominent in some patients, often radiating from the buttocks downwards to the feet and sometimes described as aching, tingling, burning, and sharp. It may be made worse by

activity. It is rarely accompanied by low back pain but objective sensory signs are usually absent; there may be reduction in pin prick sensation, light touch sensation, and vibration in the distal part of the legs.

Symptoms and signs in the upper limbs are uncommon, the brain is not involved. MRI will show spinal cord atrophy, most marked in the thoracic region. Cerebral MRI may show mild abnormalities in the white matter which are non-specific and usually age related.

The CSF often contains a mild mononuclear pleocytosis of between 5 and 50 cells per cubic millimetre with a normal glucose and normal or slightly increased protein, but with oligoclonal bands that are not matched in the serum. All patients have antibodies to HTLV1 in the CSF and serum.

Pathologically the disease has been shown to be due to perivascular and meningeal inflammation, areas of demyelination, gliosis, and necrosis. The central grey matter, posterior columns, and cortico-spinal tracts are preferentially involved and it has been suggested that it is due to HTLV1 infection of lymphocytes causing an aberrant auto-immune reaction that damages the cord. The pathological changes may be mediated by HTLV1 specific cytotoxic T-cells rather than due to direct viral injury. These immune cells appear to become activated in response to interactions with retroviral ENV and TAX proteins.

At present no anti-viral agent effectively controls HTLV1. The most effective therapies seem to involve the use of immunomodulating agents, such as steroids and interferon alpha, or plasmapheresis. Future therapies are likely to target the pathogenetic effects of HTLV1 reactive T-cells.

HTLV-2 associated myelopathy

There is a separate and antigenically distinct human T-lymphotropic virus type 2, HTLV-2, which is seen increasingly in intravenous drug users. It has been suggested that the myelopathy in this situation is usually due to a combined infection, but there has recently been the suggestion that spastic paraparesis, hyper-reflexia, spastic bladder and atrophic myelopathy can be seen in people infected with HTLV-2 but not HTLV-1. Both agents can be identified by appropriate polymerase chain reaction amplification of viral DNA in the CSF.

28.5.9 Toxic and deficiency conditions

Copper deficiency

Although the haematological effects of copper deficiency have been recognized for years, only relatively recently has copper deficiency been recognized as a cause of myelopathy, with a spastic gait and prominent sensory ataxia, central nervous system demyelination, peripheral neuropathy, and optic neuritis. The cause of the copper deficiency may not be apparent, though most patients have undergone gastric surgery. The picture mimics the myeloneuropathy of vitamin B12 deficiency (Section 28.5.10) and there can be an associated myelodysplasia which resolves with copper supplementation, and the neurological deficits stabilize but do not recover (Kumar *et al.* 2005).

Lathyrism

Lathyrism is a chronic toxic nutritional disease caused by long-term ingestion of large quantities of flour made from the drought resistant chickling pea, *Lathyrus sativus* (Section 23.6.1). During times of famine in some regions of Africa and India when there is a shortage of wheat and other grains the diet involves the regular ingestion of chickling peas for months.

A characteristic sub-acute or chronic spinal cord syndrome develops with weakness and spasticity of the lower limbs, pins and needles, and numbness in the legs. There may be urinary urgency or incontinence and erectile dysfunction. Tremor and other involuntary movements can occur in the upper limbs.

Pathologically there is degeneration of the white matter tracts in the spinal cord and the responsible neurotoxin is beta-*N*-oxyalylamino-l-alanine, BOAA, which is a glutamate receptor agonist. It causes increased intracellular levels of reactive oxygen species and subsequent impairment of mitochondrial oxidative phosphorylation (Spencer *et al.* 1986). The condition is irreversible, though the toxin should of course be removed; it tends to be self-limiting and does not appear to affect life expectancy.

Konzo

The chronic dietary ingestion of a neurotoxin derived from flour made from short-soaked cassava roots is endemic in eastern Africa (Section 23.6.1). The clinical syndrome is very similar to lathyrism, causing a spastic paraparesis with hyper-reflexia and the complaints of cramps and dysaesthesiae. The upper limbs are less affected.

The condition is thought due to the liberation of cyanohydrins from the flour which are metabolized to thiocyanate which in turn stimulates the AMPA glutamate receptor sub-type causing excitotoxic neuronal injury.

28.5.10 Sub-acute combined degeneration of the spinal cord

Vitamin B12 deficiency causes a characteristic sub-acute myelopathic syndrome associated with large fibre peripheral neuropathy, known as sub-acute combined degeneration of the spinal cord. It is rare, but it its importance lies in its response to treatment which

must be instituted quickly to minimize the extent of permanent neurological deficit.

Vitamin B12 myelopathy usually occurs in middle life, the average age of onset is 50 years. It may begin in the 20s or as late as the 70s and affects the sexes equally. Familial occurrence is rare and although most cases of B12 myelopathy are associated with megaloblastic anaemia the association is not inevitable: anaemia may be slight or absent in spite of severe spinal degeneration, and only 10–15 per cent of patients with pernicious anaemia suffer the neurological disorder.

Symptoms of confusion, depression, and dementia have been reported due to B12 deficiency with normal blood and bone marrow findings. Some cases of tobacco–alcohol amblyopia have been linked to traces of cyanide in tobacco smoke which interferes with the utilization of vitamin B12; thus hydroxycobalamine but not cyanocobalamine is given to treat this condition. Vitamin B12 deficiency is usually associated with gastric achylia causing a lack of intrinsic factor, secreted by the normal stomach to facilitate the absorption of vitamin B12. The only function of instrinsic factor is to make possible the absorption of vitamin B12 in the terminal ileum via specific receptors on ileal muscosal cells. The absence of intrinsic factor is often associated with circulating serum antibodies to gastric parietal cells which are the source of the intrinsic factor, and strongly suggest that pernicious anaemia is an autoimmune disease.

There is considerable evidence that methyl group transfer, necessary in metabolism of myelin, requires the presence of both vitamin B12 and methyl tetrahydrofolic acid via the methyalanine synthetase reaction which is dependent upon B12; methylalanine may possibly have a protective effect.

The commonest cause of vitamin B12 deficiency is absence of the instrinsic factor, but impaired absorption of the intrinsic factor B12 complex may also occur in coeliac disease, small bowel lymphoma, or after resection of the terminal ileum. True dietary deficiency of vitamin B12 is uncommon but occurs in some vegans, a strict vegetarian group who do not eat any animal products. Lack of intrinsic factor is commonly due to autoimmune disease, but may also follow partial or total gastrectomy, small bowel disease, and diverticulosis or fisutulae of the small intestine as in Crohn's disease. Malabsorption will also occur if the intrinsic factor is biologically inert, due to pancreatic disease, or the effects of drugs, particularly colchicine. Chronic addiction to nitrous oxide inhalation can also produce a myelo-neuropathy indistinguishable from that due to B12 deficiency (So and Simon 1991).

When a megaloblastic anaemia is diagnosed it may be due either to folic acid deficiency or vitamin B12 deficiency. If folic acid is given alone for the anaemia this may aggravate the neurological symptoms of vitamin B12 deficiency. Megaloblastic anaemia has also been reported in infancy due to hereditary transcobalamine deficiency even though the serum B12 was normal. Improvement in this situation may follow administration of large doses of parenteral B12 (Thomas *et al.* 1982).

Pathology

Macroscopic changes in the nervous system are few, though there may be slight cerebral atrophy and thinning of the posterior columns of the spinal cord. When the spinal cord is sectioned there is demyelination in the posterior and lateral columns. Histologically there is focal demyelination scattered throughout the white matter associated with the accumulation of lipid filled macrophage and gemistocytic astrocytes. The changes are most striking in the heavily myelinated fibres of the posterior columns and lateral columns (Fig. 28.20). There is secondary degeneration of the long tracts, again most marked in the posterior columns and cortico-spinal tracts. In the most severely affected areas, both myelin sheaths and axon cylinders are lost leaving vacuolated spaces separated by a fine glial meshwork. Rarely similar areas of degeneration may be found in the cerebral white matter, with degenerative changes, especially in association fibres. In the peripheral nerves there can be loss of the larger myelinated fibres and evidence of axonal degeneration. Segmental demyelination of the peripheral nerve is also seen.

Systemic pathological changes associated with pernicious anaemia include glossitis, hyperplasia of the bone marrow in the long bones, slight enlargement of the spleen, excess iron in the reticulo-endothelial system, and atrophy of all coats of the stomach wall.

Neurological manifestations

The clinical presentation is due to the combined features of posterior column, cortico-spinal tract, and peripheral nerve degeneration. Involvement of the optic nerves and brain may also occur. The onset of symptoms is usually gradual but can be remarkably rapid. Initial symptoms consist of paraesthesiae with tingling sensations often in the fingers before the toes: the involvement of upper limbs before lower limbs in an apparently peripheral neuropathic process should always raise the possibility of sub-acute combined degeneration of the spinal cord. There may be sensations of numbness, coldness, and tightness with sharp stabbing pains occasionally reported and the sensation of the extremities being swollen or encased in tight bandages or constricting bands. The paraesthesiae may spread from the feet and legs onto the trunk, giving a sense of constriction around the chest or abdomen.

Motor symptoms occur later, initially being reported as a sensation of tiredness on walking, then a feeling of instability, particularly when walking in the dark or standing with eyes closed, a liability to stumble, and ultimately weakness of the legs.

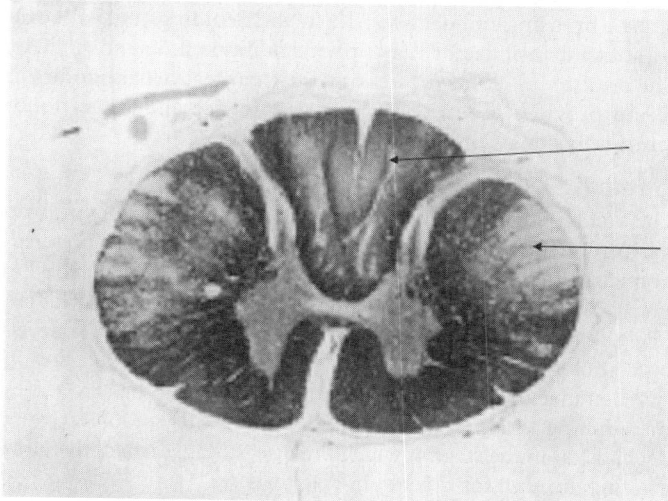

Fig. 28.20 Vitamin B12 deficiency causing sub-acute degeneration of the spinal cord. Transverse section at C3 showing loss of myelin staining of the dorsal columns and corticospinal tracts (arrowed).

Objective sensory changes are usually present, predominantly affecting posterior column sensation. Postural sense and appreciation of passive movement and vibration sense are impaired in the feet and later in the fingers. Cutaneous sensation to light touch, pin prick, heat and cold is impaired in the periphery causing the characteristic 'glove and stocking' distribution of superficial sensory loss. The calves may be tender to pressure and the proximal border of the anaesthesia may ascend.

In some cases weakness and stiffness, in others sensory ataxia predominates in the lower limbs, but objectively both weakness and sensory ataxia are usually demonstrable in all four limbs, being most severe in the lower limbs where there is a positive Romberg's sign. There may be peripheral muscle wasting as the peripheral neuropathy progresses (Section 21.22.4).

Tendon reflexes vary considerably; in about 50 per cent of cases the ankle jerks are absent when the patient is first seen, the knee jerks are lost less commonly and in some instances both may be increased. The plantar reflexes are initially flexor but become extensor in the great majority of patients. When degeneration is confined to the posterior columns sensory ataxia is the predominant feature. Conversely spastic paraparesis may be the sole presentation in some and in others the signs of peripheral neuropathy predominate.

Sphincter disturbance tends to occur late, initially causing hesitancy or precipitancy of micturition and later retention or incontinence. Impotence may occur early.

Bilateral primary optic atrophy, with some visual impairment, is seen in about 5 per cent of cases and can be the presenting feature with central scotoma. Nystagmus is relatively common, the pupils may be small but react normally, and otherwise the cranial nerves are normal. Dysarthria is rarely identified.

Mental changes are common and their importance has been stressed, they may be present before anaemia or signs of spinal cord disease. There may be a mild dementia without impaired memory and intellectual capacity or a confusional psychosis with disorientation and paranoid tendencies, Korsakoff's syndrome. Alternatively the mental disorder may be affective, manifesting as irritability or severe depression.

The EEG may show diffuse activity but returns to normal after appropriate treatment. The CSF is normal. MRI scans show high signal in the spinal cord white matter (Fig. 28.21), particularly in the posterior columns on axial scans and there may be diffuse signal abnormalities in the cerebral white matter. These imaging changes in the cord and white matter of the brain have shown striking resolution when B12 therapy is instituted (Hemmer *et al.* 1998).

Associated findings

Gastric achlorhydria is constant in pernicious anaemia but free acid may be present in the gastric juice when the deficiency is due to low dietary intake or malabsorption. There is usually macrocytic anaemia with a high mean cell volume, megalocytes or even megaloblasts in the circulating blood, poikilocytosis, anisocytosis, polychromatophillia, and leucopenia with a relative lymphocytosis may all occur. Even when the peripheral blood count is normal the bone marrow may be abnormal.

Glossitis is common but can be slight or absent when the anaemia is mild. Other symptoms, which may be apparent when the anaemia is severe, include dyspnoea, the characteristic lemon tint

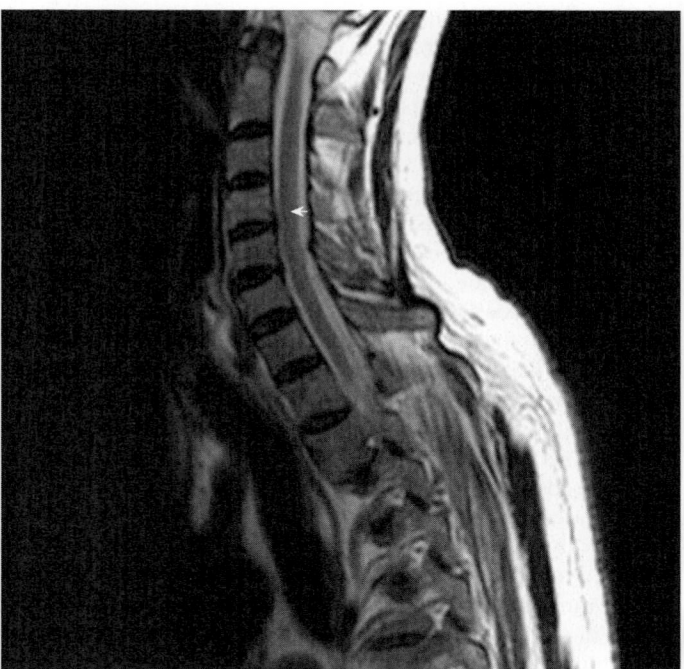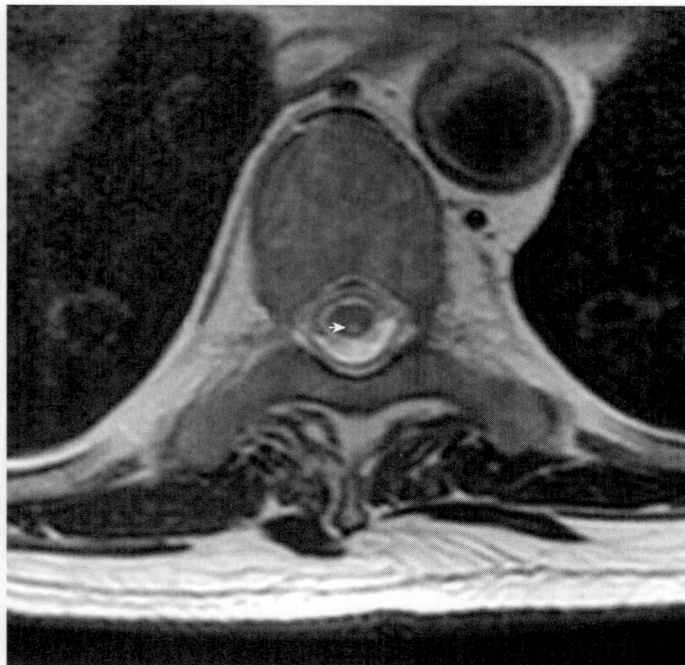

Fig. 28.21 Sub-acute combined degeneration due to B12 deficiency. High signal on T2-weighted MRI in posterior columns of cervical spinal cord.

of the skin and sclera, cardiac dilatation, haemic murmurs over the heart, and oedema most marked in the lower limbs. The spleen may be palpable and gastrointestinal symptoms are common, especially anorexia, flatulence, and diarrhoea, particularly when the deficiency is due to intestinal disease.

Differential diagnosis

The neurological picture must be distinguished from tabes dorsalis, multiple sclerosis, spinal cord compression, other intrinsic myelopathies, and other causes of polyneuropathy. *Neuro-syphilis* is usually identified by the presence of reflex irido-plegia and appropriate serological tests in blood, CSF, or both. *Multiple sclerosis* often has evidence of more diffuse disease with both optic disc pallor and nystagmus and the ankle jerks are usually exaggerated. The difficulty in distinction is most likely to occur in those patients with a slowly progressive spastic paraplegia, but the course of multiple sclerosis is usually more chronic, there should not be anaemia, and the serum B12 level should be normal. The MRI appearance of the conditions is different. *Spinal cord compression* can cause an ataxic paraplegia of gradual onset. There is usually a well-defined upper level of motor disability and sensory loss, a finding which is rare in B12 deficiency. MRI is of crucial importance in detecting compressive lesions and though cervical spondylotic myelopathy may produce signs closely resembling those of B12 myeloneuropathy, it should be distinguished by MRI scan. Both conditions may co-exist and vitamin B12 levels should be assessed if there is doubt.

Other peripheral neuropathies are relatively easy to distinguish from B12 deficiency because of the associated symptoms and signs of spinal cord disease. If neuropathy is the sole or the predominant manifestation the distinction may be entirely dependent upon estimation of the serum B12 and other serologic tests. There may be a co-existing deficiency of other B vitamins, but usually in B12 deficiency the sensory manifestations are more severe, relatively painless, and the electrophysiological findings are predominantly those of an axonal neuropathy with partial denervation. Visual evoked potentials may show significant conduction delay.

Whenever B12 deficiency is suspected on neurological grounds, a blood count should be performed and serum B12 should be estimated. Some phenothiazines, notably chlorpromazine, may interfere with the estimation of serum B12 and give falsely low levels and the assay may be further complicated by the presence of other myeloproliferative or hepatic disorders. Investigation of absorption of B12 by using radioactive B12 in the Schilling test with and without intrinsic factor may provide additional diagnostic help. The presence of elevated titres of gastric parietal cell antibodies in serum provides useful confirmatory evidence and when doubt remains marrow biopsy may be indicated.

Prognosis

The average survival of patients with pernicious anaemia prior to the introduction of liver treatment was about 2 years. The prescription of hydroxycobalamine given parenterally restores the blood to normal and the health of the patient should be maintained indefinitely. Such patients, with adequate treatment, should never develop neurological complications.

When myeloneuropathy is already present it can be arrested by introducing vitamin B12 therapy, but the degree of recovery depends upon the stage at which the disease has reached, therefore creating the need for urgency in diagnosis.

Although peripheral nerves can sometimes regenerate this is not possible in the spinal cord, though some remyelination can occur as shown by MRI. The patient may therefore expect considerable improvement in the symptoms of polyneuropathy with loss of pins and needles and dysaesthesiae, improvement in the glove and stocking anaesthesia, and return of muscle strength but the extensor plantar responses and spastic weakness of the lower limbs may persist. Even when the disease has been arrested by treatment intercurrent infection may lead to exacerbation.

Therapy

Vitamin B12 must be given intramuscularly; oral treatment requires very large doses and even when intrinsic factor is given the results are inconsistent. Treatment should be begun with 1000 µg of vitamin B12 given every 2 or 3 days for five doses to restore the tissue stores. After this 1000 µg should be given weekly for 6 months, then 1000 µg per month is usually sufficient. The dose may need to be increased if infection or renal insufficiency develops. Vitamin B12 must be given for the rest of the patient's life.

Folic acid is not only ineffective in treating vitamin B12 deficiency but may be deleterious as the administration of a folate load can produce a secondary B12 deficiency with exacerbation of neurological symptoms. Where there are severe residual disabilities appropriate symptomatic management will be needed with physiotherapy and anti-spasticity medication.

28.5.11 Vascular disorders

Arterio-venous malformations

Dural arteriovenous malformations are acquired and increase in prevalence with age. They are most often found over the surface of the spinal cord and are less often intramedullary. The characteristic clinical presentation is in late middle age or older, though rarely they may occur in young adults. Males are affected far more often than females. Most dural arteriovenous malformations are found in the mid to lower thoracic cord, sometimes extending to the conus and less often more rostrally in the cervical area. Dural arteriovenous malformations extend over a number of segments and macroscopically appear as large tortuous veins lying on the dorsal aspect of the cord. The feeding vessel is usually one of the radicular arteries, commonly the artery of Adamkiewicz, or one of the other dorsal branches arising from the aorta and feeding the spinal artery system.

The clinical presentation is of a late middle-aged male who develops a slowly progressive thoracic cord syndrome. There may be a history of exercise intolerance with weakness and sensory disturbance in the lower limbs, followed by difficulties with sphincter control and impotence. Examination reveals upper motor neurone signs in the lower limbs, but in addition there may be some lower motor neurone signs involving the highest lumbar segments, such as wasting, fasciculation, and weakness of quadriceps with reduced or absent knee responses. This is thought to represent the commonly found caudal extension of the malformation into the conus, causing the combination of upper and lower motor neurone signs. Auscultation of the spine may detect the rare finding of a spinal bruit.

Although the aetiology of dural arteriovenous malformations is uncertain they are seen following venous thrombosis in paravertebral venous sinuses and, since they are acquired, probably relate to the development of back pressure into the venous system. It is rare to identify an underlying coagulopathy in these patients.

Although the diagnosis used to require supine myelography and spinal angiography recent advances in MRI have superseded these investigations. High resolution MRI now reveals a characteristic, diagnostic combination of abnormalities in almost all cases and spinal angiography is rarely required for diagnostic purposes.

MRI typically reveals increased signal with swelling of the cord on T2-weighted images, usually involving several segments in the lower thoracic region. There may be patchy gadolinium enhancement and the pathognomic finding is of multiple, serpiginous, small signal voids closely applied to the dorsal surface of the cord at the same level as the intrinsic cord signal abnormality. These are thought to represent the dilated and tortuous dural veins (Fig. 28.22). MR angiography or rapid MR imaging after a bolus of gadolinium may show the lesion more directly. Spinal angiography is required to determine the level of the fistula and the suitability of the lesion for embolization.

The natural history of dural arteriovenous malformations is one of gradual progression in clinical defects, typically over a number of years. Acute decompensation does occur, probably due to venous infarction, but it is less common. Haemorrhage is a very rare complication. Because of the poor prognosis therapeutic intervention is recommended once a neurological deficit has been identified. The main aim of active treatment is to prevent further deterioration, although occasionally there may be improvement in existing symptoms. Most cases are considered suitable for embolization, an important criterion being that there should be adequate alternative arterial supply to the cord separate from the artery which is to be catheterized. In experienced hands the results of embolization are very good with partial or complete obliteration of the malformation. There is a risk of inducing cord infarction acutely and recurrences of the malformation may occur. Surgical extirpation of the malformation can be achieved in some cases but is now only used when embolization is unsuccessful or unsuitable because it carries a higher morbidity. In some cases where an underlying coagulation disorder is identified or if there has been venous infarction in the cord due to thrombosis, anticoagulation may be considered. This inevitably increases the risk of haemorrhage from the malformation.

Cavernous angioma

Although more common within the cerebrum there are reports of isolated cavernomas or cavernous angiomas being identified as intramedullary lesions within the cervical, dorsal, and lumbar spine. They present a variety of symptoms, depending upon their site and are not usually susceptible to surgery. They may rarely bleed and cause an increase in symptoms.

28.5.12 System degenerations

Both inherited and acquired conditions occur which appear to cause degeneration of long tracts and neurones within the spinal cord.

Hereditary spastic paraplegia

There are several different types of hereditary spastic paraplegia presenting clinically with different ages of onset and different modes of inheritance (Section 23.4.2). Many different genetic loci have been identified and there is no single genetic test for this condition.

Hereditary spastic paraplegia tends to present as a chronic syndrome with almost purely motor involvement, though there is occasional sphincter disturbance. Both autosomal dominant and recessive forms exist; the former is more common in young adults, the latter in childhood.

In the absence of a positive family history it is difficult to differentiate hereditary spastic paraplegia from other progressive,

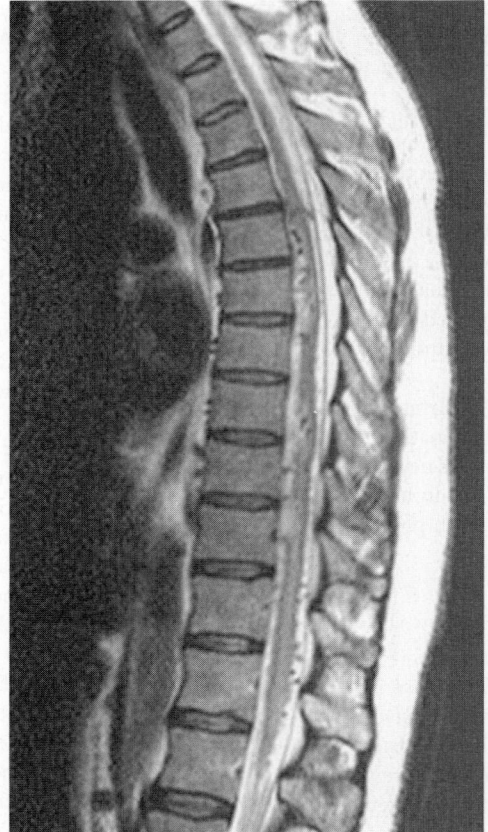
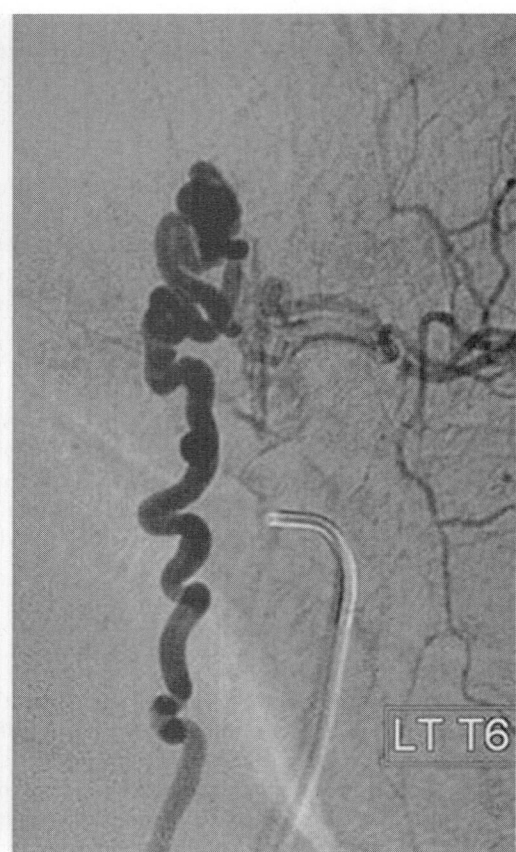

Fig. 28.22 Dural arterio-venous malformation (left) T2-weighted MRI showing spinal cord infarction and intraspinal vessel flow voids. Angiogram (right) showing feeding vessels prior to occlusion.

predominantly or entirely upper motor neurone syndromes, including the primary progressive form of multiple sclerosis and amyotrophic lateral sclerosis. One useful feature is that in hereditary spastic paraplegia there is great disproportion between the spasticity, which is severe, and the weakness, which is minimal. In addition, the gait is very spastic, often with a marked lumbar lordosis, but power in the lower limb muscles is remarkably retained. Reflexes are pathologically brisk, plantar responses are extensor, and there is commonly pes cavus.

In hereditary spastic paraplegia the CSF is a cellular and there are no oligoclonal bands, but there may be a mild elevation in protein. MRI of the spinal cord shows no focal signal change, unlike multiple sclerosis, and the cord size may be normal or show diffuse atrophy. Brain imaging is normal, again unlike multiple sclerosis, but in the absence of specific genetic markers the diagnosis remains one of exclusion.

Genetic counselling should be offered to the patient and family, the course is usually one of slow progression, though patients typically remain ambulant. Spasticity may be helped by the use of baclofen, tizanidine, or dantrolene sodium; intrathecal baclofen may be considered. Physiotherapy should be offered and, where necessary, advice about bladder control provided.

Adrenomyeloneuropathy

This X-linked perioxysomal disorder is most commonly seen as a progressive cerebral syndrome in boys (Sections 10.2.4; 37.7.2). It has a uniformly poor prognosis. Some individuals present with a slowly progressive myelopathy during adult life and this milder clinical picture may occur in carriers of the X-linked abnormality.

Onset is usually in young adulthood, between the ages of 20 and 40 years. Males are more frequently affected, though a progressive spastic paraparesis can be seen in heterozygous females. The patient slowly develops a spastic paraparesis over a number of years; together with upper motor neurone signs, there may be loss of ankle jerks, indicating co-existing peripheral nerve involvement, and sensory manifestations including paraesthesiae, loss of sensation of light touch, pain, vibration, and joint position. The lower limbs are predominantly involved, but more generalized central nervous system involvement may occasionally be indicated by the presence of cerebellar signs or mild changes in cognition.

MRI of the spinal cord is normal, apart from mild atrophy, most marked in the thoracic region. Brain MRI is normal in some patients, but in others there are symmetrical cerebral white matter high signal abnormalities on T2-weighted images which are most often seen in the posterior regions around the trigone and occipital horns. They may become more extensive involving the cerebellar peduncles as well.

Some patients have co-existing Addison's disease with skin hyperpigmentation, hypotension, a low serum cortisol, and an abnormal synacthen test. The diagnostic investigation is measurement in the plasma of very long chain fatty acids. The biochemical abnormality of fatty acids has been amended with dietary supplementation of erucic acid, but patients with established symptoms have not been convincingly demonstrated to improve. An adult presentation of progressive spastic paraparesis with cerebral MRI white matter abnormalities has been described in Krabbe's globoid-cell leukodystrophy (Section 37.7.1).

Motor neurone diseases

Although the combination of upper and lower motor neurone signs in the absence of sensory disturbance is classical in the diagnosis of motor neurone disease some forms of the illness, which predominantly affect one or other cell type, can cause difficulties in diagnosis (Section 23.2).

Amyotrophic lateral sclerosis. Amyotrophic lateral sclerosis, the usual form of motor neurone disease, manifests with both upper and lower motor neurone features in several limbs. In some instances there may be involvement of the bulbar musculature and, once the full clinical picture of upper and lower motor neurone involvement is apparent, there is little difficulty with diagnosis.

A few patients, however, present with an initial phase in which the symptoms and signs are largely, if not entirely, attributable to upper motor neurone involvement. In this situation the differential diagnosis includes other causes of progressive spastic paraplegia, especially multiple sclerosis, cervical spondylosis, which may co-exist, and primary lateral sclerosis. The presence of sensory symptoms and signs clearly points towards an alternative diagnosis, but these are not always present in either multiple sclerosis or cervical spondylosis.

In general the tempo of amyotrophic lateral sclerosis is more rapid than that for patients with progressive spastic paraparesis due to multiple sclerosis or cervical spondylosis and the degree of muscle weakness is greater.

Additional investigations often help to clarify the diagnosis. In multiple sclerosis MRI will demonstrate lesions in the cerebral white matter and spinal cord, the CSF contains oligoclonal bands and in some patients the visual evoked potentials are delayed. In amyotrophic lateral sclerosis brain and axial spinal cord images may show symmetrical high signal in the corticospinal tracts on T2-weighted images in about 60 per cent of patients (Fig. 28.23), a characteristic finding when present and entirely different from the multi-focal and asymmetrical lesions found in multiple sclerosis. The CSF may show an elevated protein level but there are no oligoclonal bands and the visual evoked potentials are normal. Although there may be no clinical signs of lower motor neurone involvement in the early stages of amyotrophic lateral sclerosis, electromyography may reveal evidence of denervation and serial studies over several months may show anterior horn cell involvement becoming more widespread.

Electromyography may also show denervation in cervical spondylosis, but this is restricted to upper limb muscles and shows little spread in myotomal involvement on follow-up. MRI is required to demonstrate cervical spondylosis, but it should be remembered that a degree of spondylosis will frequently co-exist in middle-aged patients who develop either multiple sclerosis or amyotrophic lateral sclerosis. Careful judgement of all available clinical and investigative data is needed to determine the relative importance of the spondylosis and the underlying neurological disease in contributing to the patient's clinical state.

Particular problems may arise in patients who have a combination of cervical and lumbar spondylosis where denervation may be found

in both upper and lower limbs together with upper motor neurone signs from the cervical myelopathy. In this situation the absence of sensory involvement and the evidence of progression over time on electromyography will be diagnostic of motor neurone disease.

The management of amyotrophic lateral sclerosis is discussed elsewhere (Section 23.2).

Primary lateral sclerosis. This rare progressive syndrome is generally regarded as a form of motor neurone disease (Section 23.2.1). Even after many years the clinical syndrome is entirely confined to the upper motor neurone and there is no evidence of denervation.

The presentation is usually in middle age with a slowly progressive spastic paraparesis. Sensory features are absent and the tempo is slower than that in amyotrophic lateral sclerosis, patients may survive for 10–15 years. Symptoms typically involve lower limb function first with weakness, stiffness, and difficulty in walking, but the upper limbs and bulbar muscles may all become involved. In the fully developed syndrome there is a spastic quadriparesis with pseudo-bulbar palsy.

Cognition remains intact and eye movements are unaffected. Bladder dysfunction is rare and late. MRI does not reveal the signal change in the cortico-spinal tract seen in amyotrophic lateral sclerosis but sagittal T1-weighted brain images may reveal focal atrophy of the pre-central gyrus (Pringle *et al.* 1992). Motor evoked potentials reveal a prolonged central motor conduction time and CSF is normal, except for mildly elevated protein.

Symptomatic treatment of spasticity is indicated, but there is no therapy known to modify the underlying course of the disease. Riluzole has not been shown to be effective in this condition.

28.5.13 Inflammatory disorders

Multiple sclerosis

In about 5–10 per cent of patients a progressive spinal paraplegia is the presenting manifestation of multiple sclerosis (Section 37.5).

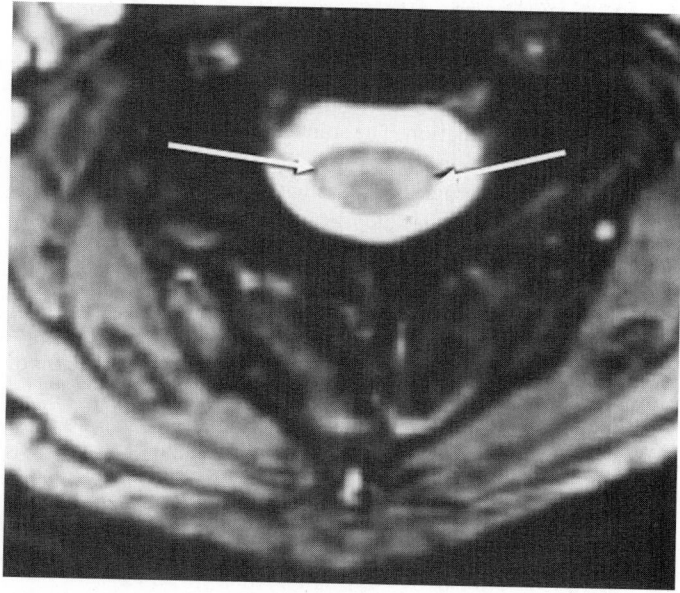

Fig. 28.23 Motor neurone disease. Axial T2-weighted MRI of upper cervical cord showing bilateral high signal in the lateral columns consistent with corticospinal tract degeneration.

This situation in which multiple sclerosis presents as a slowly progressive neurological disorder is called primary progressive multiple sclerosis. A small number of patients in this clinical subgroup have other progressive symptoms, such as optic neuropathy, cerebellar ataxia, and dementia, but a progressive myelopathy is by far the most common syndrome.

The age of onset of primary progressive multiple sclerosis is older than the more common relapsing remitting form of the disease and patients over 50 years may present. In addition, unlike relapsing remitting disease, where there is a 2:1 predominance of females, primary progressive disease is equal in gender prevalence. The syndrome may be purely motor or combined with sensory deficits of which the most common are paraesthesiae, dysaesthesiae, and loss of vibration sensation in the lower limbs. Sphincter dysfunction is frequent in keeping with an intrinsic cord lesion and the most common symptoms are urgency and incontinence with, in males, erectile dysfunction.

Some patients may demonstrate clinical features indicative of disease above the level of the foramen magnum, optic atrophy, internuclear ophthalmoplegia, or ataxia, but the majority have symptoms confined to the spinal cord. MRI shows some cerebral white matter abnormalities in most patients but usually few lesions compared to relapsing remitting disease. Although many patients with relapsing remitting multiple sclerosis may go on to develop a progressing spinal syndrome, known as secondary progressive multiple sclerosis, they have more significant changes on MRI in the brain and spinal cord.

The real value of MRI in this condition is to exclude compression and sometimes to demonstrate intrinsic demyelinating cord lesions. Most patients have oligoclonal immunoglobulin bands in the CSF, but not in the serum and frequently abnormalities of visual or brainstem evoked responses.

The pathogenesis of progressive spastic paraplegia in multiple sclerosis includes both focal and diffuse changes. There are multiple foci of demyelination throughout the spinal cord, predominantly involving the white matter tracts and sometimes extending into the central grey matter. There is also a variable degree of axonal loss which probably contributes to the progressive, irreversible defect. Diffuse microscopic changes may also be seen in normal appearing white matter in the brain and spinal cord with areas of gliosis or peri-vascular inflammation. The spinal cord is frequently atrophic, which may be shown on MRI.

The treatment of progressive spastic paraparesis due to multiple sclerosis is symptomatic. Baclofen, tizanidine, and dantrolene sodium may all alleviate spasticity and flexor spasms and the advice of a multi-disciplinary neuro-rehabilitation team will be indicated when there are significant functional difficulties. Severe spasticity may sometimes be helped by the use of intrathecal baclofen delivered via a subcutaneous reservoir and pump and the use of botulinum toxin to relieve spasm is also considered.

In those patients in whom there is a sudden rapid decline a short course of high dose cortico-steroids may be considered, but there is no convincing evidence of efficacy and there is no evidence to show that the new disease modifying treatments which have an effect in relapsing remitting disease have any role to play in the management of primary progressive disease. Immunosuppression with agents such as azathioprine, mitoxantrone, and methotrexate has sometimes been used, but there is no evidence of efficacy from controlled clinical trials.

Paraneoplastic myelopathy

Myelopathy is an uncommon manifestation of paraneoplastic disease neurologically. Paraneoplastic syndromes involving the cerebellum, brainstem, limbic system, and dorsal root ganglia are established and are much more common than myelopathy (Section 38.4). A sub-acute progressive cord syndrome has been described in association with malignancy; the clinical presentation is with motor, sensory, and sphincter defects evolving over days to weeks and associated with small cell carcinoma of the lung or lymphoma. MRI of the spinal cord may be normal or there may be gadolinium enhancement over several segments, a situation not unlike that found in transverse myelitis (Mokri *et al.* 1998).

CSF shows a mild mononuclear pleocytosis, often with protein elevation and occasionally with oligoclonal bands. Treatment involves the extirpation of the primary malignancy and the use of immunosuppressants on the grounds that most paraneoplastic syndromes have an immunopathological basis. Usually, however, the latter approach is ineffective unless the primary tumour can be effectively removed.

Progressive encephalomyelitis with rigidity

There is a rare syndrome with increasing limb tone causing rigidity, myoclonic jerking, and deep tendon hyper-reflexia evolving over months. Spinal MRI is normal but CSF reveals a mono-nuclear pleocytosis, elevated protein, and oligoclonal bands. Pathological changes have been demonstrated within the cervical region where there is loss of inter-neurones and some patients have been demonstrated to have antibodies to glutamic acid decarboxylase, GAD, having features in common with the Stiff Person syndrome (Section 40.10.3). This may occasionally be a paraneoplastic manifestation and treatment involves drugs to reduce myoclonus and rigidity together with the use of immunosuppression and plasmapheresis.

28.5.14 Radiation myelopathy

Radiation injury to the central nervous system can result in a delayed pathological process in which there is vascular hyalinization and occlusion. Degeneration of fibre tracts and necrosis involving both white and grey matter may occur. In a small proportion of patients who have undergone radiation to the chest or neck a delayed myelopathy has been reported where the spinal cord was thought included in the radiation field. At post-mortem the vascular changes cause cord ischaemia and this is thought to be the major mechanism.

Three types of spinal cord syndrome have been described following radiation (Section 5.9.1). The first is a transient and benign disorder manifesting limb paraesthesiae and a positive Lhermitte's symptom, usually appearing some 2–6 months after radiotherapy and possibly associated with swelling of the cervical spinal cord on MRI. Fortunately these symptoms subside spontaneously and the true pathological basis is unclear.

The more common and more serious problem is the development of a steadily progressive myelopathy which appears after a longer delay. The most common latent period is between 12–18 months, but symptoms have been reported to occur more than 5 years after treatment. Either the thoracic or cervical cord can be affected depending upon the level of the preceding radiotherapy. Symptoms tend to develop gradually with paraesthesiae, dysaesthesiae, weakness and stiffness of the lower and occasionally the lower limbs. Sphincter disturbance develops. The initial symptoms and signs may be asymmetrical but there is a progression over weeks to months to a more or less complete paraplegia with sensory loss below the level of the lesion. Pain is infrequent as a symptom.

The third and rarest form of post-radiation syndrome occurs at the site of radiation and causes a segmental lower motor neurone disorder with weakness, wasting, and reflex loss suggesting degeneration of the anterior horn cells in the irradiated part of the cord.

MRI in progressive radiation myelopathy reveals high signal on T2-weighted images often with swelling of the cord at the level of the irradiation. In the early clinical stages there may be gadolinium enhancement and the radiological changes are indistinguishable from other inflammatory or neoplastic lesions within the cord. The clue lies in the correspondence of the lesion level with that of previous radiation and the presence of radiation induced changes in adjacent vertebral bodies is a valuable diagnostic feature. CSF may be normal, though there can be a minor increase in protein.

It is suggested that radiation myelopathy will not occur where the total dose is less than 6000 cGy given over 30–70 days with a daily fraction of not more than 200 cGy and a weekly fraction of not more than 900 cGy (Cagan *et al.* 1980). Radiation myelopathy should therefore be avoidable. Although treatment with corticosteroids and anticoagulation has been tried there is no proof of efficacy for any intervention after radiation.

28.5.15 Syringomyelia

Syringomyelia is a chronic disorder characterized pathologically by the presence of a long cavity, or syrinx, surrounded by gliosis, situated in the central part of the spinal cord and sometimes extending up into the medulla, syringobulbia. The principal clinical features are cutaneous analgesia and thermoanaesthesia often with preservation of light touch and postural sensibility, but with muscular wasting and trophic changes. The upper limbs are most commonly affected and associated with symptoms of cortico-spinal tract dysfunction in the lower limbs. The condition was first described by Ollivier in 1824.

Aetiology and pathogenesis

For many years it was thought that syringomyelia was due to a congenital abnormality, perhaps causing abnormal closure of the central canal of the spinal cord in the embryo. Others thought that the condition was degenerative of unknown cause. It is now apparent that 'communicating syringomyelia' is the more common variety and is associated with congenital anomalies and other lesions in the neighbourhood of the foramen magnum. These include the Arnold Chiari type I anomaly which consists of congenital extension of the cerebellar tonsils below the foramen magnum, cranio-vertebral development abnormalities with or without occult hydrocephalus, and basal arachnoiditis (Barnett *et al.* 1973). It is suggested that such abnormalities, as well as the Dandy–Walker syndrome of closure of the foramina of Magendie–Luschka prevent, possibly intermittently, the egress of CSF from the fourth ventricle into the sub-arachnoid space. As a result pressure waves of fluid are forced down into the central canal of the cord which thus becomes dilated, known as hydromyelia.

This view is generally accepted, though opinions differ as to the exact nature of the hydrodynamic mechanisms involved. The fact that a syringomyelic cavity is sometimes found alongside an apparently normal spinal canal can be accounted for by the fact that with dilatation of the canal its ependymal lining quickly disap-

pears and diverticula may form which dissect outwards, downwards, or upwards, alongside the canal in the central grey matter. In most large series the Chiari type I anomaly has been the most common congenital anomaly found, but basal arachnoiditis, developing as a sequel to previous trauma, sub-arachnoid haemorrhage, or meningitis, or no evident cause, accounts for about a quarter of cases. Arachnoiditis produced by cisternal injection of kaolin in experimental animals has been shown to produce experimental syringomyelia.

It has been suggested that perinatal trauma may either produce the cerebellar tonsillar ectopia or induce syringomyelia in the presence of such a development anomaly. It is also clear that primary cerebellar ectopia can be present without causing syringomyelia but with other neurological signs, such as hydrocephalus, paraparesis, or a cerebellar syndrome. True communicating syringomyelia has also been described as a complication of glioma in the mid-brain and upper cord.

In non-communicating syringomyelia, by contrast, the condition is more often due to or associated with spinal injury with or without paraplegia, spinal arachnoiditis, or spinal tumour. In these cases the cavity may develop in the thoracic or lumbar cord first. Indeed except in cases of spina bifida, with which hydromyelia may be associated, the discovery of a lumbar syrinx in a patient without history of injury should always raise the possibility of a spinal glioma or ependymoma, though intramedullary metastasis or extramedullary tumours are also seen. In cases of traumatic paraplegia or arachnoiditis the cavities usually ascend from the site of the lesion, but in upper cervical lesions downwards cavitation may be found. The cavitation has been attributed to a combination of factors including venous obstruction, exudation of protein, and ischaemia. Oedema may be another factor.

The prevalence of syringomyelia is 5–10 per 100 000. The pathological condition is probably more common since the widespread availability of MR scanning has identified some individuals with asymptomatic syringes which are usually small. The condition has been described in more than one member of a family and other congenital malformations, including spina bifida, have been found in families with affected members. It is more common in males than females and symptoms can appear at any age between 10 and 60 years but usually present in young adult life.

Pathology

The typical pathological changes are most frequently found in the lower cervical and upper thoracic regions of the cord. Extension to the medulla is common and rarely the process may reach the pons or even as high as the internal capsule. Thoraco-lumbar and lumbo-sacral syringomyelia are rare and usually due to true hydromyelia associated with developmental anomalies, although ascending cavitation following traumatic transverse lesions of the cord or in association with cord tumours may occur.

The affected region of the cord may be enlarged, mainly in the transverse plane and in rare cases the enlargement is sufficient to cause erosion of the bones in the spinal canal, or at least widening of the anterior–posterior diameter (Fig. 28.13). Transverse section of the cord reveals a cavity surrounded by a zone of translucent gelatinous material which, microscopically, contains glial cells and fibres. The protein content of the fluid in the cavity is raised and there is little difference between the 'communicating' and 'non-communicating' syringomyelia.

The expanding cavity and surrounding gliosis affecting the less resistant grey matter in the centre of the spinal cord, more severely than the dense white matter, allows the cavity to invade the anterior horns of the grey matter causing atrophy of anterior horn cells, degeneration of their axons in the ventral roots and peripheral nerves, and consequent wasting of muscles. The reflexes are lost. Extension to the brainstem, syringobulbia, usually occurs first in the postero-lateral medulla near the spinal nucleus of the trigeminal nerve and the nucleus ambiguous, so that the earliest signs of brainstem dysfunction are usually due to the involvement of these nuclei. Compression of the long ascending and descending tracts of the cord or brainstem occurs later, giving secondary degeneration most marked in the cortico-spinal tracts then in the spino-thalamic tracts, and last in the posterior columns. Haemorrhage into a syringomyelic cavity is one uncommon form of haematomyelia.

Presentation

The symptoms of syringomyelia are readily interpreted as due to a progressive lesion in the central region of the spinal cord. The onset is usually insidious, but rarely may develop rapidly and sometimes in association with an episode of coughing, sneezing, or straining. Wasting and weakness of the small muscles of the hands are common early symptoms and the patient may notice loss of feeling in the hands or recognize suffering painless injuries. Sometimes the development of scoliosis in childhood is the clue to the condition.

Sensory symptoms and signs

The commonest site and presentation is with an elongated cavity in the central grey matter extending longitudinally through several segments in the lower cervical and upper thoracic cord. The lesion may be predominantly unilateral, therefore interrupting desiccating sensory fibres on one side of the cord derived from several consecutive dorsal roots. These are the fibres which conduct impulses concerned with the appreciation of pain, heat and cold, and those forms of sensibility are therefore affected whilst others travelling in the donal columns are preserved.

This 'dissociated' sensory loss was described originally by Charcot and usually appears first along the ulnar border of the hand, forearm, and arm, and on the upper part of the chest and back on one side in a half cape distribution. There is a horizontal lower border on the chest wall and the sensory disturbance ends at the mid-line.

Sometimes sensory loss may be impaired in a glove distribution and when the lesion extends centrally the area of dissociated sensory loss becomes bilateral. As the lesion extends upwards and downwards in the cord so the area of sensory impairment extends to the radial sides of the upper limbs and neck, and downwards over the thorax exhibiting a distribution 'en cuirasse' (Fig 28.23).

When the lesion reaches the upper cervical segments it begins to involve the spinal tract and nucleus of the trigeminal nerve, receiving fibres concerned with the appreciation of pain, heat and cold on the face. There is extension of the area of dissociated sensory loss in a concentric 'onion skinning' manner from behind forwards onto the face, sparing the sensation at the tip of the nose and the upper lip.

The progression of the spinal lesion later causes compression of the lateral spinothalamic tracts on one or both sides, leading to loss of appreciation of pain, heat and cold over the lower parts of the body. There may be an area of normal sensibility over the abdomen intervening between the area of thoracic anaesthesiae due to inter-ruption of the decussating fibres and sensory loss in the lower limbs due to compression of the spino-thalamic tracts. When the spino-thalamic tract is compressed at the level of the medulla appreciation of pain, heat and cold is impaired or lost over the whole of the contra lateral side of the body.

The posterior columns are usually the last of the sensory pathways to suffer, but late in evolution there may be loss of appreciation of posture, passive movement, and vibration. This is most likely to involve the lower limbs and there may be extensive anaesthesia.

Thermoanaesthesia can be detected by the patient since hot water no longer feels hot over the affected parts and analgesia exposes the patient to injuries, especially burns on the fingers which, being painless, are not noticed. Spontaneous pains, though by no means invariable, are sometimes troublesome and the patient may describe burning, aching, or shooting sensations which can resemble the lightning pains of tabes dorsalis. More commonly the pain is continuous and burning in nature and may present on the face or an upper limb.

If the lesion begins in the thoraco-lumbar or lumbo-sacral region of the cord the dissociated loss will have an appropriate distribution. In non-communicating cases, secondary to spinal cord trauma, or other lesions, an ascending sensory level after months or years should suggest the presence of an ascending syrinx.

Motor symptoms and signs

The earliest motor manifestations are usually muscular weakness and wasting, due to compression or destruction of the anterior horn cells. Since the lesion usually begins in the cervico-thoracic cord muscular wasting usually appears in the small hand muscles. It may be bilateral or unilateral, but as the lesion extends the wasting spreads to involve hands, forearms, and arms. The shoulder girdles and upper intercostal muscles may be involved. The wasting and weakness is never as marked as that seen in motor neurone disease and fasciculation is uncommon.

Contractures may develop particularly in the hand and forearm muscles and as the lesion extends to the postero-lateral medulla there may be involvement of the nucleus ambiguous causing paresis of the soft palate, pharynx, and vocal cord occasionally giving rise to laryngeal stridor and common causing dysphagia.

The other motor functions of the cranial nerves are rarely affected, though paralysis of the mandibular muscles, facial muscles, and lateral rectus has been recorded. The tongue is commonly involved and nystagmus is often seen. The nystagmus is variable, sometimes being phasic and present in lateral gaze but may be dissociated, and vertical nystagmus on up-gaze is reported. Paralysis of the ocular sympathetic on one or both sides may cause Horner's syndrome, though the reaction to light should be preserved.

Compression of the cortico-spinal tracts in the spinal cord causes weakness with spasticity and extensor plantar responses in the later stages. There is rarely loss of significant power in the lower limbs, but tendon reflexes are exaggerated in the lower limbs, though diminished or lost in the upper limbs. This loss of reflexes is due to interruption of the reflex arc as the syrinx extends into the lateral grey matter. The sphincters are rarely affected since the anterior columns tend to be spared.

Trophic symptoms and signs

Trophic symptoms may sometimes be very marked. True hypertrophy involving all tissues can be present in one limb, one half of the body, or even the tongue. Loss of sweating or excessive sweating may occur, usually over the face and hands. Excessive sweating may be spontaneous or excited when the patient takes hot or highly seasoned food. Twenty per cent of patients have osteoarthropathy, Charcot's joints, the shoulders, elbows, and cervical spine are most often affected, less often the joints of the hands, the temporomandibular joints, the sterno-clavicular joints, and the acromio-clavicular joints. Atrophy and decalcification of bones around joints with erosion of surfaces and subsequent bony destruction are seen radiographically. These joint changes are anaesthetic and painless. The affected joint is often enlarged and movement evokes loud crepitus but is painless. The long bones may be brittle.

Skin changes, including cyanosis, hyperkeratosis and thickening of the sub-cutaneous tissues causing a swelling of the fingers, described as 'la main succulente' (Fig. 28.24). The analgesia renders the patient exceptionally liable to repeated minor injuries and healing is slow. Ulceration, infection, and necrosis of bone is not uncommon and gangrene may be seen. The scars of former injuries are usually evident on the palmar surface of the fingers.

Syringobulbia

The medulla may be involved by upward extension from the cord, or may be the primary site of the disorder when the onset of symptoms may be sudden or gradual. Trigeminal pain, vertigo,

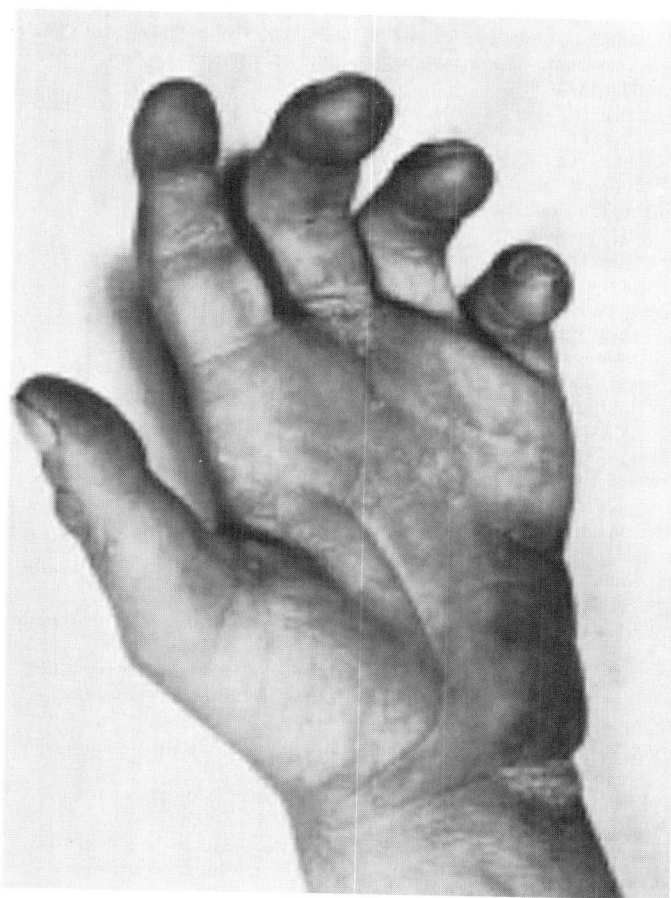

Fig. 28.24 The 'main succulent' of syringomyelia with painless cuts, clawing of the fingers, and autonomic changes.

facial, palatal, or laryngeal palsy, or wasting of the tongue may all be seen.

Morvan's syndrome

This title is applied to patients in whom there is progressive loss of pain sensation, ulceration, loss of soft tissue, and resorption of the phalanges with muscular atrophy, not only in the hands, but sometimes in the feet with perforating ulcers. Whilst such changes in the hands rarely occur in syringomyelia, a similar syndrome may occur in leprosy when all four extremities are involved. The most common cause is hereditary sensory neuropathy (Section 21.6).

Associated abnormalities

Many developmental anomalies have been described in association with syringomyelia occurring either in affected individuals or in members of their families. Deformities of the sternum, kyphoscoliosis, asymmetry of breasts, increase in the ratio between arm and body length, acrocyanosis of the hands, curved fingers, enuresis, and anomalies of the hair and ears, particularly with low hairline and short neck are identified. There may also be cervical rib, spina bifida, basilar impression of the skull, fusion of the cervical vertebrae in the Klippel–Feil syndrome with shortening of the neck, hydrocephalus, and pes cavus. Light brown pigmentation, either in the form of spots or diffuse sheets may be seen in segmental distribution, commonly over the shoulders.

Imaging

MRI is the diagnostic investigation of choice. It is now widely available and has enabled a rapid and non-invasive diagnosis to be made earlier than in the past. T2-weighted saggital and axial spin echo images reveal the low signal central cavity in the spinal cord, the longitudinal extent of which is highly variable (Fig. 28.25). Cord expansion is usually apparent. When the syrinx is associated with a Chiari malformation the latter is also readily demonstrated on saggital T1-weighted images which include the level of the foramen magnum. Where the differential diagnosis includes intrinsic spinal cord tumour, gadolinium enhancement may identify enhancing tumour tissue. Myelography followed by immediate and delayed CT scanning, the previous 'gold standard' diagnostic investigation has now been rendered obsolete, except where MRI is contraindicated.

Other investigations

The CSF shows no abnormality unless the cavity is large enough to cause a block when the protein content of the fluid is raised. Single fibre electromyographic studies have shown a relatively consistent pattern of involvement of the cervical anterior horn cells.

Diagnosis

There is little difficulty in making the diagnosis of syringomyelia clinically in advanced cases. The combination of wasting and trophic lesions of the hands with dissociated sensory loss and long tract dysfunction in the lower limbs is distinctive (Section 22.1.4). A clinically based diagnosis is more difficult in the early stages, but MRI scan allows early and accurate diagnosis in almost all cases. There is a wide differential diagnosis:

◆ *Intramedullary tumour of the spinal cord*, especially ependymoma, may simulate the condition clinically. As a rule it progresses more rapidly and blockage of the sub-arachnoid space with resulting CSF changes soon occurs. The same is true of *extramedullary spinal tumours* where pain is usually a more prominent symptom. *Haematomyelia* may produce similar signs, but develops more acutely.

◆ *Cervical spondylosis* may cause wasting of the proximal upper limbs and paraesthesiae in the hands with spastic weakness of the lower limbs, but it does not cause dissociated sensory loss.

◆ *Motor neurone disease* may simulate syringomyelia when it begins with wasting of the small muscles of the hand, but sensory loss of absent and muscle wasting develops more rapidly with fasciculation almost constant.

◆ *Thoracic outlet syndrome* due to cervical rib may cause difficulties in diagnosis (Section 22.5.3). Since the two can co-exist the distinction may be difficult. Pain in the ulnar border of the hand and forearm is more common with cervical rib and sensory disturbance is usually more evident with syringomyelia.

◆ *Hereditary motor sensory neuropathy*, or *Charcot Marie Tooth disease* (Section 21.4), is distinguished from syringomyelia by the fact that muscle wasting usually appears first in the lower limbs and there is a distal glove and stocking sensory loss to all modalities (Section 21.4).

◆ The trophic symptoms of *Raynaud's disease* may simulate syringomyelia in the hands, but there is no dissociated sensory loss and the classical blanching of the fingers is diagnostic.

◆ *Syringobulbia* presents diagnostic difficulty when it occurs alone and must be distinguished from other medullary lesions. Thrombosis of the posterior–inferior cerebellar atrophy may produce a similar sensory loss but is distinguished by its acute onset and vertigo. Tumours of the medulla may simulate syringobulbia and require differentiation with MRI scan. Progressive bulbar palsy is distinguished by the lack of sensory disturbance and eye signs and the diagnosis of basilar impression, which may closely simulate and be associated with syringomyelia, can be distinguished only on imaging.

Prognosis

The course of syringomyelia, if untreated, is progressive, though progress is often slow and arrest may occur sometimes lasting for years. A sudden intensification of symptoms may occur after coughing, straining, or minor trauma or be caused by haemorrhage in a syringomyelic cavity. Exceptionally, distension of the spinal cord may become so marked as to produce a complete transverse myelopathy leading to paraplegia.

Treatment

Symptomatic treatment for pain and spasticity may be required, protection of analgesic areas and earlier treatment of cutaneous lesions in order to promote healing are essential. In non-communicating secondary to spinal tumour or arachnoiditis, laminectomy with partial or complete removal of the causal tumour decompression and drainage of arachnoid cysts or the tumour itself with division of fibrous bands have all been helpful.

When ascending cavitation follows a complete traumatic transverse lesion of the cord the process may be arrested by total excision of a segment of spinal cord at and below the level of the injury and with incomplete post-traumatic myelopathy and syrinx, syringostomy, which is shunting of the cavity, may relieve pain.

In those cases with an associated Chiari malformation decompression of the foramen magnum and upper cervical cord is sometimes performed. The results are variable but there may be relief of head and neck pain and sometimes reduction in long tract signs. Less often is there a beneficial effect on the segmental sensory and motor deficits. The response may be better if surgery is performed early in the course of the disease and the value of syringostomy in cases of communicating syringomyelia is uncertain.

References

Alter M (1975). Statistical aspects of spinal cord tumours. In Vinken PJ, Bruyn GW, eds. *Handbook of Clinical Neurology*, Vol. 19, North Holland, Amsterdam.

Barnett HJM, Foster JB, Hudgson P (1973). *Syringomyelia in Major Problems in Neurology Series* Ed JN Walton, Saunders, London.

Blau JN, Rushworth G (1958). Observations on the blood vessels of the spinal cord and their response to motor activity. *Brain*, **81**, 354–63.

Bolton D (1939). The blood supply of the spinal cord. *J Neurol Psychiatry*, **2**, 137–48.

Cagan RA, Woollin M, Gilbert HA *et al.* (1980). Comparison of the tolerance of the brain and spinal cord to injury by radiation. In Gilbert HA, Cagan RA, eds. *Radiation Damage to the Nervous System*, Ravenpress, New York.

Cloward RB (1980). Acute cervical spine injuries CIBA, Clinical Symposia **31**, 1.

Cruickshank JK, Rudge P, Dalgleish AG *et al.* (1989). Tropical spastic paraparesis and human T-cell lymphotrophic virus type I in the United Kingdom. *Brain*, **112**, 1057–90.

Hemmer B, Glocker FX, Schumacher M *et al.* (1998). Sub-acute combined degneration: clinical, electrophysiological and magnetic resonance imaging findings. *J Neurol Neurosurg, Psychiatry*, **65**, 822–7.

Henson RA, Parson M (1967). Ischaemic lesions of the spinal cord: an illustrated review. *QJM*, **36**, 205–22.

Hernandez D, Vinuela F, Feasby TE (1982). Recurrent paraplegia with total recovery from spontaneous spinal epidural haematoma. *Ann Neurol*, **11**, 623–4.

Herzberg L, Bayliss E (1980). Spinal cord syndrome due to non-compressive Paget's disease. *Lancet*, **ii**, 13–5.

Housmann ON, Moseley IE (1996). Idiopathic dural herniation of the thoracic spinal cord. *Neuroradiology*, **39**, 503–10.

Hughes JT (1971). Venous infarction of the spinal cord. *Neurology*, **21**, 794–800.

Kernohan JW, Woltman HW, Adson AW (1931). Intra-medullary tumours of the spinal cord. *Arch Neurol Psychiatry*, **25**, 679–701.

Kumar N, Elliott MA, Hoyer JD *et al.* (2005). Myelodisplasia, myeloneuropathy and copper deficiency. *Mayo Clin Proc*, **80**, 943–6.

Lennon VA, Kryzer TJ, Pittock SJ *et al.* (2005). IgG marker of optic-spinal multiple sclerosis binds to the aquaporin-4 water channel. *J Exp Med*, **202**, 473–7.

Lennon VA, Wingerchuk DM, Kryzer TJ *et al.* (2004). A serum autoantibody marker of neuromyelitis optica: distinction from multiple sclerosis. *Lancet*, **364**, 2106–12.

McDonald WI, Compston A, Adan G *et al.* (2001). Recommended diagnostic criteria for multiple sclerosis: guidelines from the international panel on the diagnosis of multiple sclerosis. *Ann Neurol*, **50**, 121–7.

Mokri B, Weinshenker BG, Goudreu JL *et al.* (1998). Long-tract myelopathy: a novel paraneoplastic syndrome. *Ann Neurol*, **44**, 486–8.

O'Riordan J I, Thompson AJ, Kingsley DPE *et al.* (1998). The prognostic value of brain MRI in clinically isolated syndromes of the central nervous system. A 10 year follow-up. *Brain*, **121**, 495–503.

Oppenheim DR (1978). The cervical cord in multiple sclerosis. *Neuropath Appl Neurobiol*, **4**, 151–62.

Polman CH, Wolinski JS, Reingold SC *et al.* (2005). Multiple sclerosis diagnostic criteria: 3 years later. *Multiple Sclerosis*, **11**, 5–12.

Pringle CE, Hudson AJ, Munoz DG *et al.* (1992). Primary lateral sclerosis. Clinical features. Neuropathology and diagnostic criteria. *Brain*, **115**, 495–520.

Queiroz LdeS, Nucci A, Facure NO *et al.* (1979). Massive spinal necrosis in schistosomiasis. *Arch Neurol*, **36**, 517–9.

So YT, Simon RP (1991). Deficiency diseases of the nervous system. In Bradley WG, Daroff RB, Fenichel GM *et al.* eds. *Neurology in Clinical Practice*, Butterworth Heinmann, Boston.

Spencer DS, Roy DN, Ludolph A *et al.* (1986). Lathyrism: evidence for role of the neuro-excitatory amino-acid BOAA. *Lancet*, ii, 1066–7.

Spiller WG (1909). Thrombosis of the cervical anterior median spinal artery: syphilitic acute anterior polio-myelitis. *J Nerve Ment Dis*, **36**, 601–13.

Stevens JM, Cartlidge NEF, Saunders M *et al.* (1971). Atlanto-axial sub-luxation and cervical myelopathy in rheumatoid arthritis. *QJM*, **40**, 391–408.

Thomas PK, Hoffbrand AV, Smith AS (1982). Neurological involvement in hereditary transcobalamine 2 deficiency. *J Neurol Neurosurg Psychiatry*, **45**, 74–7.

Tomlinson BE, Irving D, Rebeiz JJ (1973). Total number of limb motor neurones in the human lumbo-sacral cord and an analysis of the accuracy of various sampling procedures. *J Neurol Sci*, **20**, 313–27.

Tosi L, Rigoli G, Beltramello A (1996). Fibrocartilaginous embolism of the spinal cord: a clinical and pathological consideration. *J Neurol Neurosurg Psychiatry*, **60**, 55–60.

Wilkinson M (1973). *Cervical Spondylosis*, 2nd edition, Heinemann, London.

Wingerchuck DM (2007). Diagnosis and treatment of neuromyelitis optica. *Neurologist*, **13**, 2–11.

CHAPTER 29

Cauda equina lesions, radiculopathies, and sphincter disorders

David Bates

Contents

Pathological processes involving the spinal roots and cauda equina present with symptoms of lower motor neurone and first order sensory neurone damage. Pain is a common, though not inevitable, symptom. Pathological processes may be acute, as with a prolapsed intervertebral disc or chronic and extend over many years, as with spondylotic bony changes or structural diseases such as spondylolisthesis. The cauda equina carries innervation to the bladder, rectum, corpus cavernosum, and seminal vesicles and damage commonly presents with sphincter disturbance and impotence. In general the nerve roots throughout the spine and cauda equina are more resistant to injury and pathological processes than the spinal cord; rapid diagnosis and surgical intervention where indicated, may improve outcome considerably.

29.1 Cauda equina and conus medullaris lesions

Disease and damage affecting the spinal roots characteristically cause lower motor neurone signs and symptoms, together with sensory involvement, commonly pain or numbness. There is loss of reflexes and may be objective trophic change in the skin involved. It should be relatively easy to establish the level of a single radiculopathy based upon the sensory, motor, and reflex changes. It is more difficult to assign the cause and localization when several adjacent roots are involved in one process.

29.1.1 Anatomy

The cauda equina normally begins at the level of the disc space between the first and second lumbar vertebra where it separates from the conus medullaris. It comprises the lumbar, sacral, and coccygeal nerve roots from the second lumbar root downwards and the fibres lie posteriorly within the lumbo-sacral theca, moving forwards segment by segment to exit the intervertebral foramina. The constitution of the nerve roots is precisely the same as in the rest of the spine, though the roots of the lumbar, sacral, and coccygeal nerves descend with an increasing degree of obliquity to their respective exits. The roots of the lower lumbar and upper sacral nerves are the largest of all nerve roots and their axons the most numerous. The roots of the coccygeal nerve are the smallest.

The second, third, and fourth sacral nerves carry visceral nerve fibres which join the parasympathetic part of the autonomic nervous system running directly into the pelvic plexus.

At each intervertebral foramen the spinal nerves have the intervertebral discs lying anteriorly with the synovial facet joints lying posteriorly. Each nerve is accompanied by a radicular artery and there is a plexus of small veins around it.

29.1.2 Conus medullaris syndromes

Conus medullaris damage is often seen in combination with lesions of the cauda equina, and the precise site of disease may be difficult to identify when several lumbo-sacral levels are involved. Logically the involvement of the conus medullaris should cause upper motor neurone symptoms and signs, but commonly presents with sphincteric disturbances, saddle anaesthesia over the third to fifth sacral dermatomes and impotence, without abnormality of the legs. There may be a combination of loss of reflexes in the lower limbs with extensor plantar responses or the apparently paradoxical finding of some reflexes being increased and others reduced or lost.

Lesions of the cauda equina and conus medullaris may cause similar symptoms and signs, including local, referred and radicular pain, sphincter disturbance, loss of buttock and leg sensation, and leg weakness.

Processes confined to the conus medullaris, as opposed to those which also involve the cauda equina, usually have less severe, but bilateral, pain, motor findings are mild but symmetrical and sensory findings, confined to the saddle distribution, bilateral and symmetrical. There is early involvement of the sphincters and male sexual function and the process tends to evolve suddenly. There may be loss of ankle jerks and the plantar responses may be extensor.

29.1.3 Cauda equina syndromes

It may be impossible to distinguish a lesion arising in the conus medullaris from one in the cauda equina. In general lesions of the cauda equina are more likely to cause pain which may be severe and asymmetrical, significant motor weakness is more likely, often with visible fasciculation in the affected muscles. Sensory loss is variable and may be asymmetrical. Both knee and ankle tendon reflexes may be lost. The plantar responses will not be extensor. Sphincter disturbances occur less commonly and later, and impairment of male sexual function is usually less severe.

The clinical picture is variable depending upon the site and causation of the cauda equina damage.

Compression by neoplasm

Ependymoma or neurofibroma, and more rarely chordoma, lymphoma, meningioma or secondary neoplasm, may cause compression of the cauda equina (Fernside and Adams 1978) (Fig. 29.1). When spina bifida occulta is present an associated lipoma may cause compression and there can also be fibrous bands or chronic

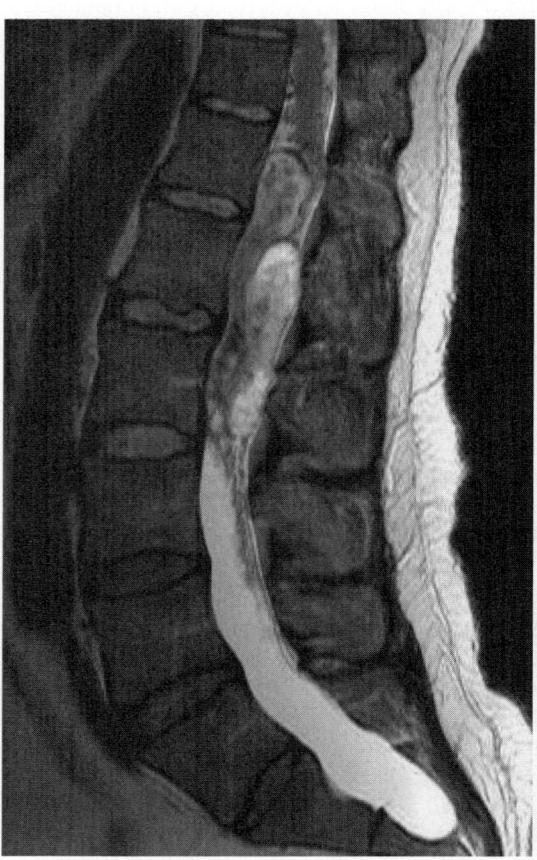
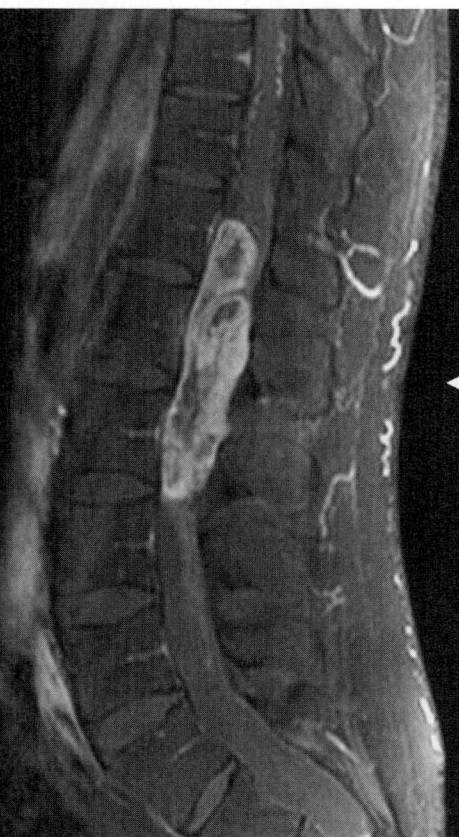

Fig. 29.1 Ependymoma of the conus medullaris causing compression and a progressing cauda equina syndrome. T2-weighted and T1-Gd-enhanced MRI.

arachnoiditis associated with the congenital abnormality. More rarely perineural or Tarloff, cysts lying on the posterior sacral nerve roots can cause compression but they are most commonly asymptomatic. Ependymomas and neurofibromas are approximately equal in frequency and are both seen more often in men than women.

Clinical distinction between a tumour arising in the cauda equina from one arising in the conus medullaris and compressing the cauda equina may be impossible. A small tumour may cause symptoms limited to one or two roots on one side for many years, whereas a large growth may involve the whole of the cauda equina and become symptomatic relatively acutely. Anatomically the lower roots are more likely to be affected than the upper since they have a longer intrathecal course and can be implicated by tumours at any level from the first lumbar vertebra into the sacrum.

Pain is the first symptom in most cases of cauda equina tumour. It is usually felt in the lumbar or sacral regions as a dull aching pain and may be exacerbated by jerky movements, coughing, or sneezing. Subsequently there may be referral of pain into one or both lower limbs, in the distribution of a dermatome, though pain may rarely be referred to the bladder, rectum, or testis.

Since the lower motor neurones are affected motor symptoms are most characteristically those of weakness and wasting in the distribution of the affected nerve. Most commonly paralysis of muscles below the knee and of the hamstrings and glutei are seen due to their innervations from the sacral roots. In such cases the ankle jerks are depressed or absent, the plantar responses may be absent but the knee jerks are commonly intact.

The sensory loss is dependent upon the particular nerve roots involved. When the lowest sacral roots are involved there is a saddle-shaped area of anaesthesia and analgesia which may not be symmetrical in the perineum, and on the buttocks and back of the thighs.

Higher lesions may affect the L5 and S1 dermatomes over the foot and lateral aspect of the leg. When the very lowest sacral segments are involved the patient may describe altered sensation in the urethra but usually bladder sensation remains intact so that the patient is aware of the desire to micturate.

At some stage in the compressive process sphincter control will become affected, but commonly only late. Compression of the second, third, and fourth ventral and dorsal sacral roots interrupts the reflex arc subserving evacuation of the bladder and rectum. There is therefore usually retention of urine and constipation of faeces due to the unopposed contraction of the internal sphincters, even though the external sphincters are paralysed. There is erectile dysfunction in the male and loss of the anal and bulbo-cavernous reflexes.

Trophic changes in the lower limbs occur late but include cyanosis and coldness, the feet tend to become oedematous and trophic ulceration may occur which is slow to heal. There may be Charcot joints and painless bone infection in the feet.

The investigation and management of tumours of the cauda equina is identical to that described for spinal cord tumours (Section 28.5.6). Imaging is with MRI, often with gadolinium enhancement and a surgical opinion is obligatory.

Acute central disc protrusion

A central intervertebral disc protrusion, most usually at the L4/5 or L5/S1 levels, can cause acute compression of the cauda equina (Fig. 29.2). Urinary retention is a prominent early symptom, pain may not be severe and sensory examination characteristically reveals a saddle-shaped sensory loss over the buttocks and perineum. This condition is a neurosurgical emergency and if suspected urgent MRI should be undertaken. The prognosis for recovery of bladder function is directly related to the delay between symptom

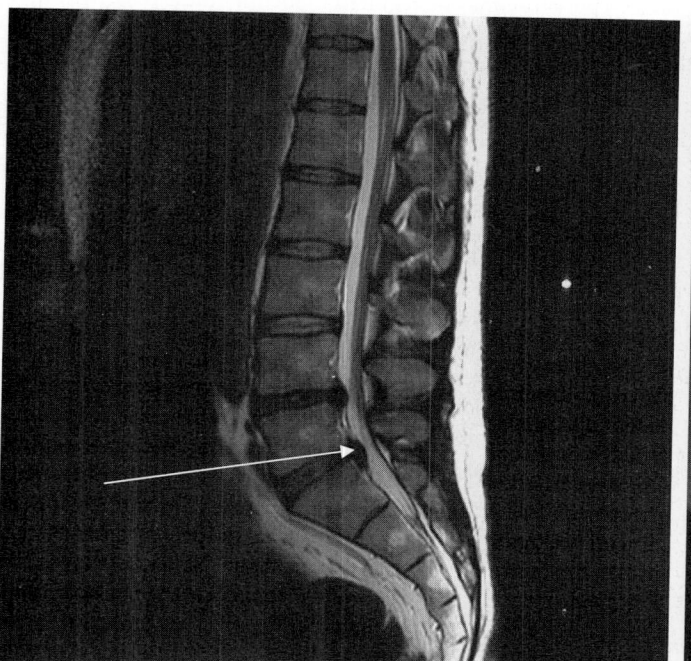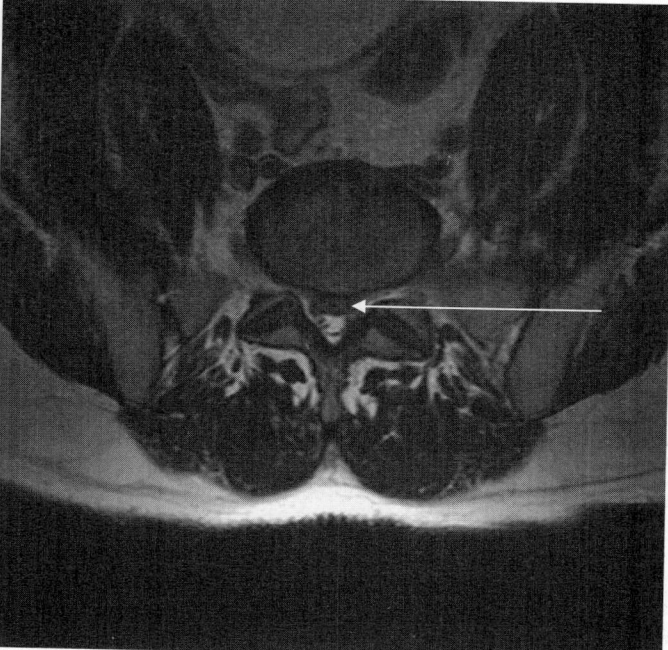

Fig. 29.2 Acute central disc prolapse at the L5/S1 level causing cauda equina syndrome. T2-weighted MRI.

onset and surgical decompression. Painful retention of urine is common in the older male but, when it occurs in the young male or a female, should immediately give rise to consideration of this syndrome. It should be remembered that retention of urine in a patient with back pain, though often due to the use of narcotic analgesics, makes obligatory imaging of the lumbo-sacral spine to exclude a compressive lesion. Acute urinary retention in the patient with a suspected ruptured disc or other possible intrathecal mass lesion calls for emergency neuroradiological investigation. Spinal imaging is obligatory if patients present with weak or flaccid lower extremities but at least partially preserved sensation, but absent knee or ankle jerks and with asymmetrical sensory loss or saddle anaesthesia and particularly with urinary retention. In the absence of urinary retention the chance of having a cauda equina syndrome is low (O'Laoire *et al.* 1981).

Spinal canal stenosis

Chronic degenerative changes, such as disc protrusion and osteophytic hypertrophy of the lumbar vertebra and facet joints can lead to narrowing or stenosis of the lumbar canal. This is more likely if there is pre-existing congenital narrowing of the canal. The condition is most commonly seen in the older population, men more than women and may also be related to arteriopathic changes in the aorta and spinal vessels. The most striking symptom of lumbar canal stenosis is poor exertional tolerance with the development of weakness, pain, or numbness in the lower limbs when standing or walking. It does not usually occur with equivalent exercise cycling. The patient will characteristically describe an improvement in symptoms if the spine is flexed forwards and by sitting, though lying may occasionally cause the symptom. The phenomenon is also called claudication of the cauda equina. It can be confused with, and may be coincident with, vascular claudication symptoms in the lower limbs due to peripheral vascular disease. The classical feature of claudication may only be present in two-thirds of patients.

MRI will show degenerative vertebral column changes with narrowing of the lumbo-sacral canal. Most of these patients do not have objective signs of nerve root dysfunction and many remain stable for years without developing progressive neurological deficit. These patients can be managed with mild analgesics and advice; those with intractable pain or progressive neurological deficit require surgical intervention (Goh *et al.* 2004). If the lesion can be shown to be local and related to disc disease then a single level discectomy might be performed. More commonly decompressive laminectomy is required, especially if there is a progressing neurological deficit.

Other factors which may contribute to lumbar canal stenosis include the presence of degenerative osteophytes, spondylolisthesis, facet joint hypertrophy, thickening of the ligamentum flavum, and disc herniation. The patient usually has some low back pain and the distinction from vascular intermittent claudication can usually be made by the fact that stopping walking and standing still does not improve symptoms in lumbar canal stenosis but causes their resolution in true intermittent claudication.

Non-compressive disease of the cauda equina

Chronic, focal, or diffuse inflammation of the spinal theca can cause neurological symptoms due to inflammation, adhesion, and distortion of the nerve roots. Such chronic spinal arachnoiditis or chronic adhesive arachnoiditis may occur at any level in the spinal cord, but is most commonly symptomatic in the lumbo-sacral region

causing a cauda equina syndrome (Fig. 29.3). The cause of adhesive arachnoiditis is not always identifiable. It used to be seen following meningo-vascular syphilis or after staphylococcus or tuberculous infection of the vertebra. Meningococcal, pneumococcal, and viral meningitides have all been reported as potentially causative and some cases follow spinal surgery. In the past myelography with oil-based dyes was thought responsible for some cases and intra-thecal or epidural chemical usage as in spinal anaesthesia has been suggested to be causative (Sections 5.5.5 and 5.5.6). In most cases no cause is identified despite full investigation, although trauma and the presence of a lumbar intervertebral disc prolapse have been considered causative. There are rare examples of familial arachnoiditis.

Arachnoiditis probably exerts its principal pathological effect by interference with the blood supply to the roots of the cauda equina. Rarely may it be associated with pressure effects due to the formation of arachnoid cysts. MRI characteristically shows loss of the normal homogenous high signal of CSF on T2-weighted images with clumping of the nerve roots and distortion of their normal smooth passage through the theca.

The symptoms include local and radicular pain, paraesthesiae, and, less commonly, weakness, wasting of muscles, and sphincter dysfunction. The spinal fluid tends to be abnormal with an increase in the protein level and a moderate mononuclear pleocytosis. There is no effective therapy for arachnoiditis and although surgical debridement has been attempted it is usually unsuccessful and there is no proof of efficacy for epidural or intrathecal steroids. Most treatment is aimed at symptomatic control of pain and improvement of sphincter function.

Malignant meningitis

Infiltration of the meninges in the lower lumbar theca with secondary neoplasm or lymphoma is not uncommon. It is usually painful, progressive, and reveals typical abnormalities in the CSF with high protein and the presence of malignant cells. Repeated lumbar puncture may be required to identify these cells. MRI may show enhancement of the nerve roots and occasionally deposits of tumour may be seen as swellings on the roots (Fig. 29.4).

Post-radiation damage

A predominantly delayed and motor disorder resulting in weakness and wasting of the lower limbs may occur between 3 and 25 years after irradiation of para-aortic nodes, and therefore the distal spinal cord and cauda equina, in the treatment of testicular, bladder, or ovarian tumours (Section 5.9.5). MRI may show gadolinium enhancement of the cauda equina and pathology appears to be of a radiation-induced vasculopathy of the proximal spinal roots with relative preservation of the neuronal cell bodies and spinal cord. This condition is likely to be due to a motor radiculopathy affecting the irradiated portion of the cauda equina proximal to the dorsal root ganglia. In keeping with the information in relation to radiation-induced myelopathy the threshold exposure appears to be greater than 40 Gy in those patients who develop this relentless deterioration of post-irradiation lumbo-sacral radiculopathy (Bowen *et al.* 1996). More modern control of the radiation dosage seems to avoid the subsequent development of the syndrome.

29.1.4 Management of acute lesions

Compression

Acute compression of the cauda equina often requires emergency neurosurgical decompression. The causes of acute compression of

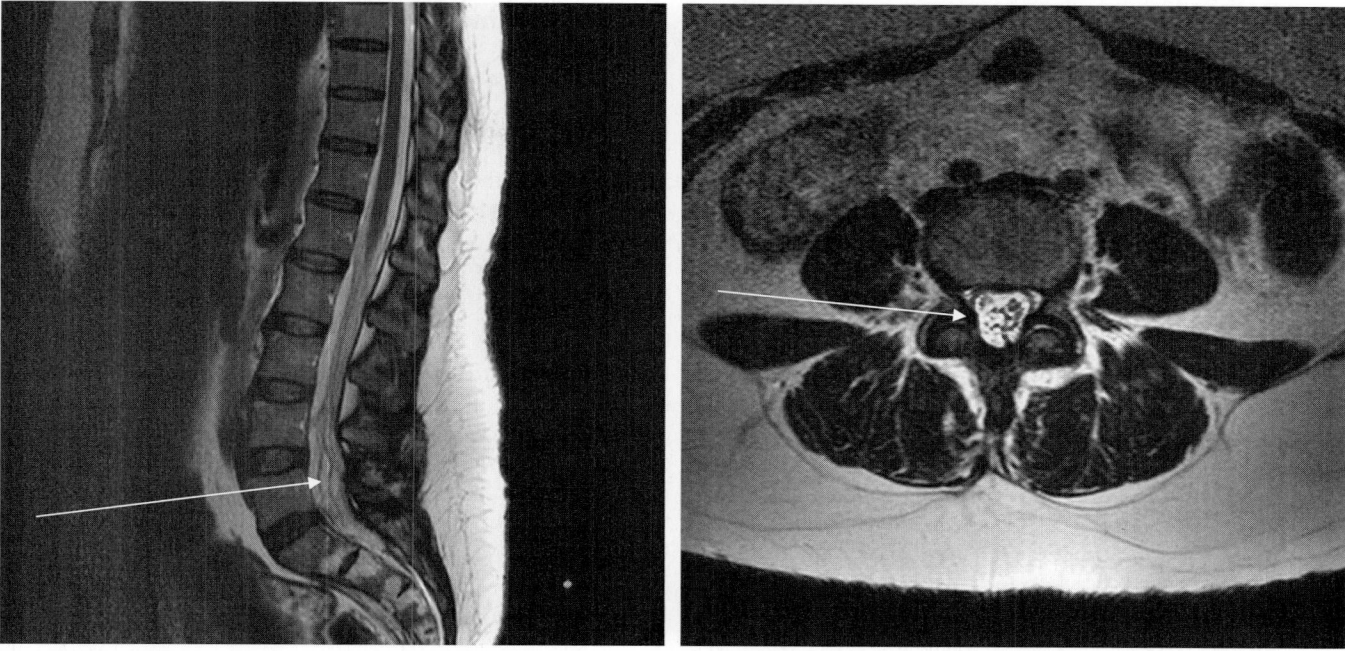

Fig. 29.3 Arachnoiditis causing clumping of tortuous nerve roots in the cauda equina. T2-weighted MRI of lumbar spine.

the cauda equina are similar to those of spinal cord compression: an acute disc extrusion, most usually at the L4/5 or L5/S1 interspace due to a central disc protrusion; sub-dural and epidural haematomas; sub-dural and epidural abscesses, and vertebral crush fractures following trauma (Fig. 29.5).

The potential for recovery of function is better for the cauda equina than for the spinal cord even with a profound deficit and there is therefore every reason to consider decompression as rapidly as possible because there is a direct correlation between time to decompression and functional recovery.

Ischaemia

When an acute cauda equina syndrome is identified but imaging is normal, one of the possible implications is that there has been ischaemia to the nerve roots in the cauda equina. This is commonly

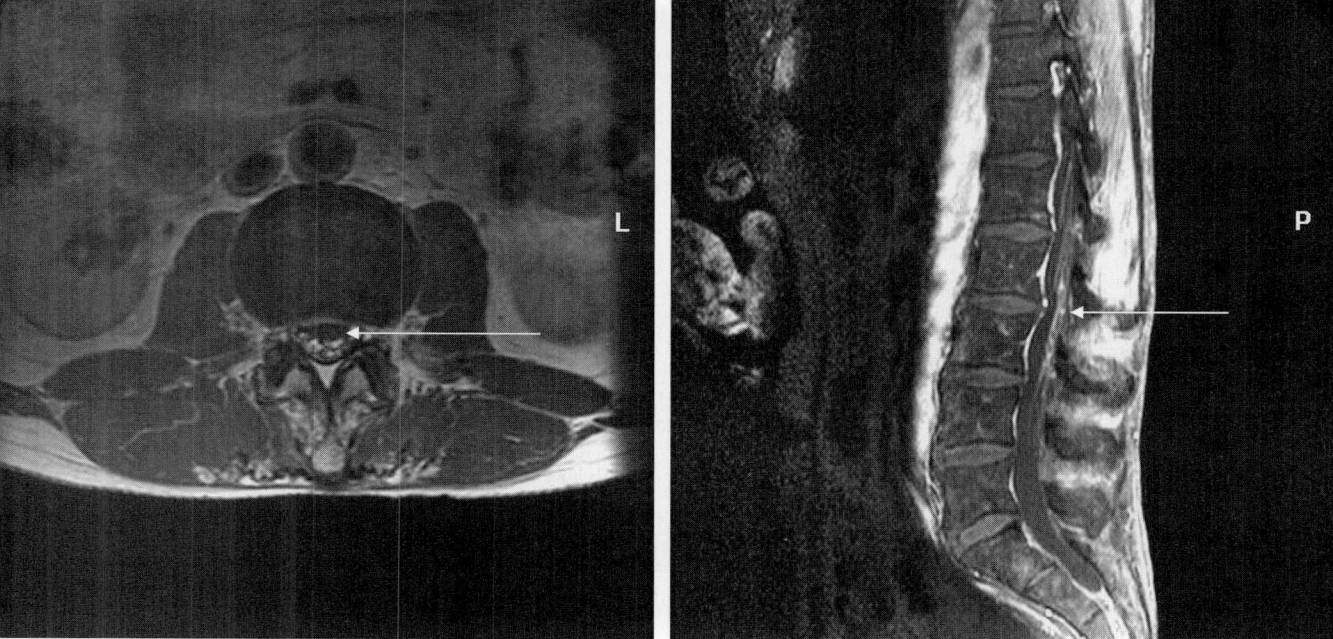

Fig. 29.4 Malignant meningitis of cauda equina revealed by enhancing nodules on the nerve roots on T1-Gd-enhanced MRI.

associated with atheroma of the terminal aorta or small vessel disease as in diabetes mellitus and vasculitis affecting the spinal vessels. It is usually a diagnosis of exclusion.

Inflammation

The presentation of inflammatory disease of the cauda equina is usually subacute over hours to days. A patchy and asymmetrical cauda equina syndrome develops, imaging excludes a structural lesion and may rarely show high signal within the conus or gadolinium enhancement of the nerve roots. The CSF should identify an inflammatory reaction and oligoclonal bands may be positive. Immunomodulatory therapies may be tried but have no proven role.

Infection

The acute or subacute evolution of a cauda equina syndrome with normal MRI or one that shows enhancement of nerve roots with gadolinium may be seen in virus infections, particularly with herpes virus. There may be the diagnostic rash within the distribution of the sacral nerve roots or simply an inflammatory reaction within the CSF. Polymerase chain reaction for viral DNA in the CSF may be diagnostic. A particularly severe form may be due to cytomegalovirus in HIV-infected patients (Section 21.14.2). Antiviral drugs should be chosen accordingly.

Post-infective

Cauda equina syndromes may occasionally be seen as part of a post-infective radiculopathy. The diagnosis is by exclusion, though neurophysiology may help by identifying demyelinating lesions in the lumbo-sacral nerve roots. There is no therapy of proven benefit.

29.1.5 Management of chronic lesions

Compression

The evolution of a cauda equina syndrome over months or years raises the possibility of an intrathecal tumour, such as an ependymoma or neurofibroma. Meningioma is less common and lipoma in association with spina bifida occulta should be considered. Less commonly dermoids and secondary tumours may present in the same way. The investigation of the gradual evolution of a chronic cauda equina syndrome is with MR imaging. Surgical decompression should be considered.

Ischaemia

In association with lumbar canal stenosis and the symptom of intermittent claudication of the cauda equina there may be the gradual evolution of a cauda equina syndrome, presumed to occur on the basis of progressing ischaemia. MRI will confirm canal stenosis, or occasionally the presence of spondylolisthesis, which may then be surgically approached in an attempt to arrest the progression of the chronic cauda equina syndrome.

Inflammation

Chronic inflammatory processes affecting the cauda equina are rare. Granulomatous disease such as sarcoid may occasionally be seen and the evolution of arachnoiditis following surgical intervention, meningitis, or intrathecal chemicals should always be considered. The typical appearance on MRI is of clumping of the nerve roots within the cauda equina and this, together with the CSF findings, is usually diagnostic. There is little effective therapy, steroids are frequently tried and may be helpful in granulomatous causes.

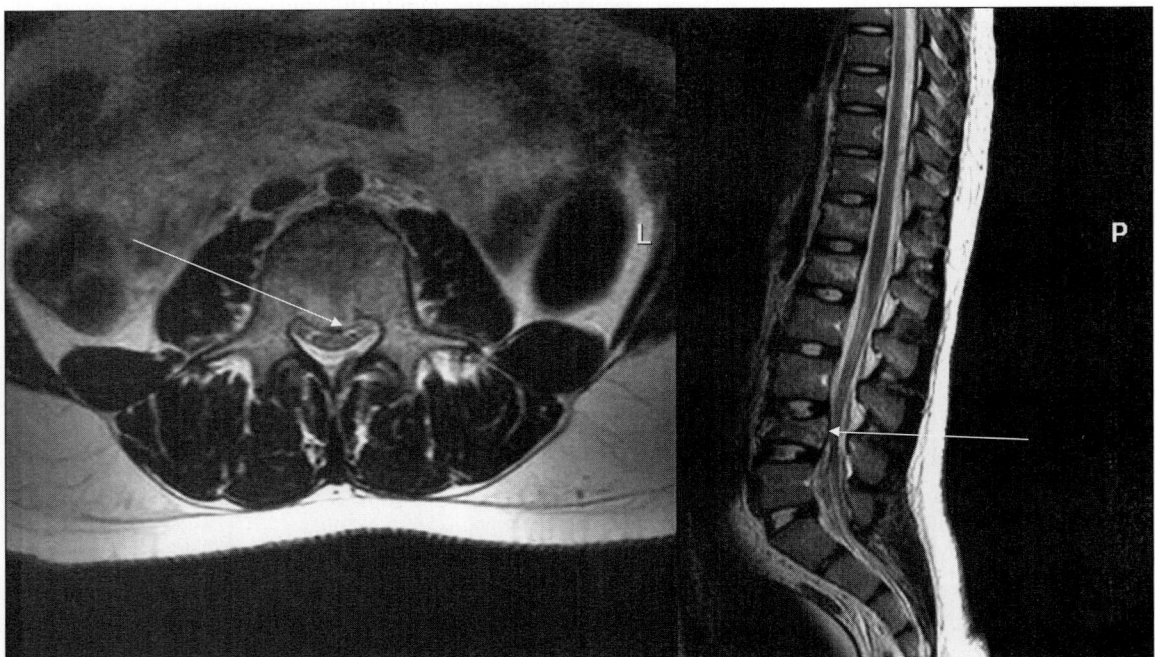

Fig. 29.5 Traumatic crush fracture of the L4 vertebra causing compression of cauda equina. T2-weighted axial and sagittal MRI.

Table 29.1 The Symptoms and signs of radiculopathies

Cervical root lesions

C1–3 Sensory loss is found over the back (C2) and side (C3) of the neck; motor supply to a number of neck muscles is interrupted but is not usually clinically apparent.

C4 There is sensory loss in a cape distribution between the side of the neck and the top of the shoulder. Unilateral lesions cause weakness of rhomboids. Bilateral lesions of C2–4 may cause bilateral diaphragm paralysis, manifested as dyspnoea and a reduced vital capacity, especially pronounced on lying flat.

C5 Sensory loss extends from the outer surface of the shoulder down into the lateral arm and forearm. There is weakness and wasting of deltoids, spinati, and pectorals, with loss of the pectoral reflex.

C6 Sensory loss involves the distal lateral forearm and hand, the thumb, and sometimes the index finger. There is weakness and wasting of biceps brachii, brachioradialis, forearm flexors, and pronator teres. The biceps and brachioradialis jerks are absent. The triceps jerk may also be reduced or absent.

C7 Sensory loss involves the index, middle, and ring fingers, and a strip in the middle of the hand both on the palmar and dorsal surface. There is weakness and wasting of triceps, wrist, and sometimes finger extensors and supinator. The triceps jerk is absent.

C8 Sensory loss involves the little finger, and medial aspect of the hand and forearm. There is wasting and weakness of intrinsic hand muscles, especially interossei, and hypothenar eminence, and sometimes weakness of finger extension. The finger flexion reflex is lost.

Thoracic root lesions

T1 Sensory loss involves the medial aspect of the proximal forearm and upper arm. There is weakness and wasting of the muscles of the thenar eminence.

T2–T12 There is a band of sensory loss unilaterally at a level on the trunk according to each nerve. It is helpful to remember that there is an interface of the C4 and T2 dermatomes at the upper chest level, the nipple is at approximately T4, and the umbilicus is at approximately T10. Although the muscle supply to intercostal muscles or abdominal wall muscles may be interrupted, there are usually no clinical motor signs of a unilateral lesion, except for a loss of abdominal reflex at the appropriate level (T9–10 for upper abdominal reflexes, T11–12 for lower abdominal reflexes).

Lumbar root lesions

L1 Sensory loss is found in the groin and upper buttock, and there is weakness of iliopsoas.

L2 Sensory loss occurs over the proximal anterior and medial thigh. There is weakness of hip flexion and adduction.

L3 Sensation is lost over distal anterior and medial thigh and medial aspect of the knee. There is weakness of hip flexors and may be mild weakness of quadriceps.

L4 Sensory loss occurs over the anterior and medial shin. There is weakness and wasting of quadriceps, and some weakness of tibialis anterior and ankle inversion. The knee jerk is absent.

L5 The loss of sensation is most apparent on the dorsum and inner aspect of the foot. There is weakness and wasting of the anterior tibial muscles, leading to footdrop and weakness of toe dorsiflexion and eversion. Hip abduction is also weak due to involvement of gluteal muscles.

Sacral root lesions

S1 Sensory loss occurs on the sole and outer aspects of the foot, as well as the calf. There is weakness of plantar flexion of the foot and toes. The ankle jerk is absent. Hip abduction and extension may be weak.

S2–S5 The main findings are of sensory loss, involving the back of the thigh (S2) the buttocks (S3–5) in a saddle-shaped distribution, and the perineum (S5). Bladder retention is the rule with bilateral lesions involving multiple sacral roots in the cauda equina.

Tumours

Characteristically the gradual evolution of a painful cauda equina syndrome often with the finding on MRI of small nodules visible on the sacral nerve roots raises the possibility of a malignant radiculopathy. It is most commonly seen with primary breast or lung tumours, though lymphomas can also present in this way, diagnosis is confirmed by CSF cytology or, rarely, meningeal biopsy. Definitive treatment is usually impossible.

Paraneoplastic

There are rare syndromes of the gradual evolution of cauda equina syndromes in association with known primary tumour in which no direct infiltration can be demonstrated (Section 38.4). It is assumed that in this situation there is a paraneoplastic involvement of the lumbo-sacral nerve roots. No immunomodulatory treatment is known to help.

29.2 Radiculopathies

The distribution of symptoms and signs of a radiculopathy depend on the particular nerve root involved (Table 29.1) (Fig. 29.6).

29.2.1 Acute cervical disc prolapse

MRI demonstrates that the nucleus pulposus of cervical intervertebral discs commonly shows degeneration or dehydration. Such a disc may then rupture through the posterior annulus into the spinal canal. Such a prolapse is commonly lateral, resulting in compression

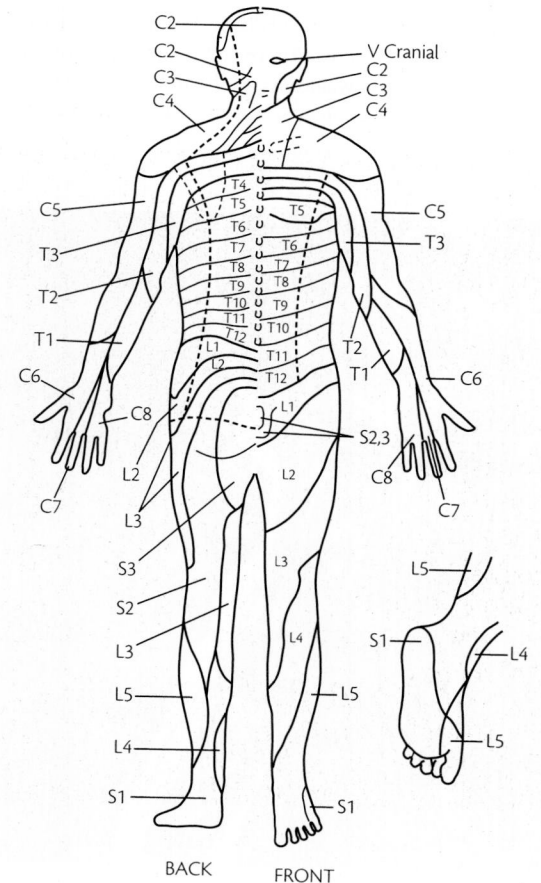

Fig. 29.6 The dermatomes supplied by individual spinal roots.

of nerve roots or more rarely central when it may compress the cervical spinal cord. Cervical radiculopathy is commonly acute and, though sometimes related to trauma, is most commonly seen without any identifiable preceding traumatic event. Disc herniation is more common in people under the age of 45 years. In older patients radicular symptoms are more likely to be related to foraminal stenosis due to degenerative change.

In the cervical spine disc prolapse is most common in the mid to lower cervical region, at C4/5, C5/6, or C6/7, and may result in radiculopathy or myelopathy alone, or in combination as a radiculomyelopathy.

Acute disc prolapse most commonly causes intense pain in the neck extending into the arm in the distribution of the relevant dermatome. Headache may occur and the neck and root pain is made worse by neck movement, coughing, sneezing, or the Valsalva manoeuvre. Neck rotation may increase pain and the neck is often held rigidly with evidence of paravertebral spasm.

Pain may be the only manifestation of the acute disc prolapse but in other instances there may be associated sensory and motor symptoms and signs identifying the root lesion. The most common sites for disc herniation in the cervical region are at C4/5, C5/6, and C6/7 and the character and site of the neurological manifestations depend upon the particular root involved. There may be dysaesthesiae and numbness in the distribution of the C5, C6, or C7 roots. Higher root levels are rarely involved and lower roots are usually spared. The relevant tendon reflex will be depressed and there will be weakness in the distribution of muscles innervated by the root.

Disc herniation at C4/5 affects the C5 root. It tends to cause pain, pins and needles, and altered sensation over the shoulder and down to the forearm. There may be weakness of the deltoid, spinati, pectoralis, and sometimes of the biceps, muscles. The pectoralis and

sometimes the biceps reflexes may be lost. Spread of the biceps reflex to the finger flexors, an increased triceps reflex, or an inverted biceps (Section 2.3.2) reflex suggests that there is also a myelopathy.

At the level of C5/6 prolapse the C6 nerve root is involved. Pins and needles extend to the thumb and distal forearm with weakness typically in the brachio-radialis, and biceps muscles. The biceps and brachio-radialis tendon reflexes may be diminished or inverted.

Lesions at C6/7 compress the C7 nerve root and cause paraesthesiae extending, typically, into the middle finger with weakness of the triceps muscle and extensors of wrist and fingers and a diminished triceps tendon reflex.

Plain cervical spine X-ray may show narrowing of the relevant intervertebral disc. The optimum investigation is MRI which will reveal disc prolapse, seen most clearly on T2-weighted images. The high signal cuff of CSF surrounding the nerve root, as it passes into the exit foramen can be seen to be reduced or obliterated and the soft tissue mass of the protruding disc is visible (Fig. 29.7). On occasion it may be difficult to identify the difference between an acutely protruded disc and a more chronic osteophyte encroachment, in which case CT imaging can be diagnostic. Neurophysiology may be useful in difficult diagnostic cases, but electromyographic denervation in muscles is unlikely to be shown for several weeks.

When pain is the predominant symptom most disc protrusions can be treated conservatively and will improve within 4–8 weeks. Non-steroidal anti-inflammatory agents are commonly given, occasionally with the use of a cervical collar and physical therapy. Large prolapses with clear radicular defects usually require surgical intervention which may be needed urgently. If patients have intractable weakness or pain which does not improve then nerve root decompression should be considered. A randomized controlled clinical trial is underway comparing operative versus conservative therapy. A central disc prolapse causing a myelopathy should be

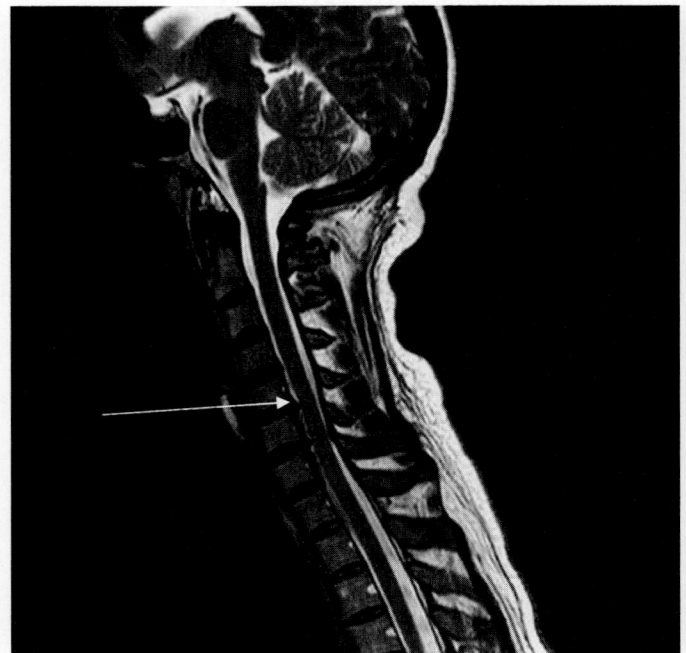
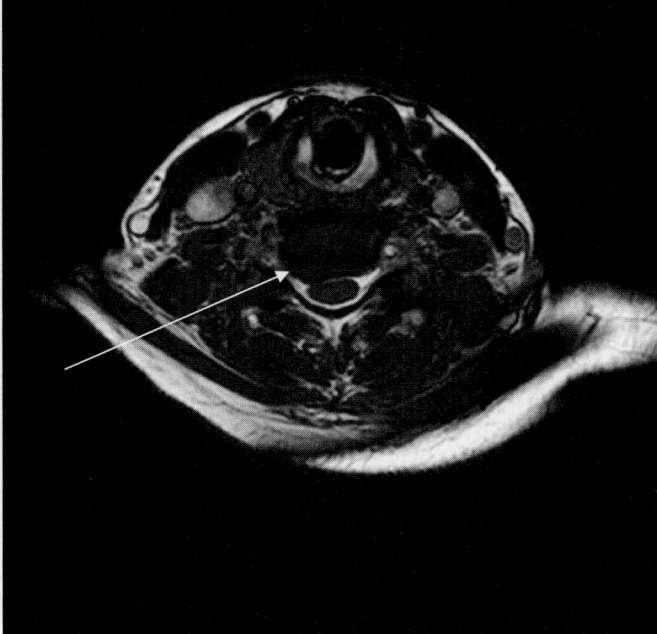

Fig. 29.7 Acute right-sided prolapsed disc at C5/6 compromising the C6 root exit foramen. T2-weighted sagittal and axial MRI.

managed by anterior decompression with preservation of the disc space, but a lateral disc prolapse can often be treated successfully by posterior foraminal decompression.

29.2.2 Spondylotic cervical radiculopathy

The process of spondylosis in the spine increases with aging to a greater or lesser extent in all individuals. The cervical spinal column has 37 joints that are in motion throughout life and degenerative osteoarthritis and spondylosis are ubiquitous with increasing age. Patients who have dystonia or other cervical movement disorders may be predisposed to premature cervical spine degeneration. Those with congenitally narrow canals are more liable to the development of a myelopathy.

Plain cervical X-rays commonly show loss of height of disc spaces with the formation of osteophytes on the vertebral body above and below the narrowed disc (Fig. 28.14A). As the intervertebral disc spaces narrow, osteophytic development at the bone margins encroaches upon the intervertebral foramina. Degenerative changes in the synovial facet joints cause similar stenosis. Radiculopathy occurs when the spondylotic process encroaches upon the nerve root canal and the combination of spinal pain, radicular discomfort, dysaesthesiae, numbness and weakness within the distribution of a cervical root or roots may indicate the presence of cervical spondylosis. Compression of the cervical cord particularly occurs when there is a significant narrowing of the cervical spinal canal. A combination of radiculopathy and myelopathy, known as radiculomyelopathy, is characteristic of cervical spondylosis.

The spondylotic process most often involves the fifth, sixth, and seventh cervical vertebrae and the roots most often compromised are C6 and C7. Pain and sensory loss in the corresponding dermatome is associated with segmental weakness, wasting, and reflex loss in the myotome of the root concerned. With impingement upon the spinal cord there may be spreading or inversion of reflexes to lower levels, indicating radiculopathic damage and hyperexcitability of anterior horn cells lower in the spinal cord.

The appearances of cervical spondylosis are demonstrable on plain X-ray but MRI is needed to identify the severity of entrapment (Fig. 28.14B+C). CT scan may help distinguish between entrapment by soft tissue and by bone.

Cervical spondylotic radiculopathy is usually treated conservatively with adequate analgesia, physiotherapy, and possibly the use of a cervical spinal collar. This is frequently sufficient to allow resolution of symptoms, reduce pain, and may improve power. In general, surgery is not helpful in reducing pain and discomfort in the neck or back, though it may be effective in reducing symptoms of radiculopathy when the nerve root canal may be enlarged by an anterior or posterior approach. The anterior approach is preferred because the posterior approach, with a partial laminectomy, allows the diseased foramen to remain mobile with a higher risk of continued pain.

The commonest indication for surgery to the cervical spine is a progressing cervical spondylotic myelopathy. An anterior approach at one or two levels can be considered or more major surgery performed by posterior approach.

29.2.3 Other causes of cervical radiculopathy

Many processes involving the meninges, including inflammation, infection, and infiltration, may involve nerve roots in the cervical spine. Infections, such as tuberculosis and syphilis were hitherto common causes of upper limb radiculopathy. Now borrelia, cytomegalovirus particularly in immunocompromised patients, and other herpes viruses may all cause painful radiculopathy. Granulomatous conditions, such as sarcoidosis, and neoplastic lesions causing malignant meningitis, commonly seen with breast or bronchogenic neoplasms, lymphoma, or leukaemia may also cause slowly progressive cervical radiculopathy.

The symptoms are those of a cervical radiculopathy affecting multiple root levels. Slow progression raises the possibility of inflammatory or neoplastic disease, commonly with depressed reflexes and frequently with severe pain which is worse during the night.

MRI is the investigation of choice and gadolinium-enhanced studies usually demonstrate diffuse patchy or multi-focal, nodular meningeal enhancement, raising the possibility of carcinomatous meningitis. CSF examination characteristically shows an increase in mononuclear white cells and a high protein, the glucose occasionally being reduced. Appropriate microbiological investigations should be performed, including polymerase chain reaction detection of viral genome. Cytology may help identify malignant cells. Quite frequently more than one lumbar puncture is required to detect the cytological abnormality.

Very rarely, when other diagnostic tests are negative, meningeal biopsy may be performed to look for the source of the condition. It should be directed towards an area of meninges displaying gadolinium enhancement on MRI.

29.2.4 Thoracic nerve root lesions

Thoracic disc entrapment

Though intervertebral discs in the thoracic region do prolapse, they tend to do so posteriorly rather than laterally and are most likely to cause impingement upon the dorsal spinal cord. Very rarely a lateral prolapse, often associated with trauma, may cause entrapment of a thoracic root and the symptom is then of intense pain in the distribution of the dermatome supplied by that that nerve root.

Thoracic spondylosis

Degenerative changes are less common in the thoracic spine; osteophytes may develop on the anterior or posterior aspects of the vertebral bodies but rarely cause clinical radiculopathy. Thoracic disc herniation may be seen on MRI but is usually asymptomatic. Thoracic myelopathy due to disc herniation is estimated to have an annual incidence of approximately one case per million and is most common in middle life.

Non-compressive causes of dorsal radiculopathy

Intercostal neuralgia, or diabetic truncal neuropathy (Section 21.17.4), is seen in diabetes mellitus where it is thought due to localized vasculitis of an individual spinal root. Herpes zoster infection not uncommonly involves the dorsal root ganglia in the thoracic region (Section 21.14.5). The presentation of shingles within the distribution of a nerve root causes a vesicular rash often followed by intractable pain.

29.2.5 Lumbo-sacral disc degeneration and prolapse

One of the most common of all symptoms is that of low back pain radiating into a leg. It is usually the result of a lumbo-sacral radiculopathy due to intervertebral disc degeneration in the spine followed by disc prolapse. Some degree of disc degeneration in the lumbosacral region is almost universal with ageing. Degenerative changes in the lumbar intervertebral discs, including dehydration and narrowing, become increasingly common with age. They are seen most frequently at L4/5 and L5/S1 levels. It follows that L5 or S1 radiculopathies are those most commonly seen in clinical practice. The process whereby the nucleus pulposus of the disc becomes extruded through a weakened annulus fibrosus is identical to that in the cervical region. The most common direction of protrusion is posterolateral causing compression of the nerve root with nerve root symptoms and signs. Less commonly a large central disc protrusion may cause an acute cauda equina syndrome (Section 29.1.3) with little in the way of radiated pain (Fig. 29.8).

The most common symptom of an intervertebral disc syndrome is low back pain, which is typically chronic, variable, and fluctuating. Low back pain due to disc prolapse may be difficult to distinguish from other non-specific musculo-skeletal disorders. Pain in the back can develop as a sudden, severe pain due to acute disc prolapse. This may be precipitated by lifting, twisting, or turning and is typically associated with sciatica where the pain extends down the back of the leg. The pain is made worse by bending, walking, coughing, sneezing, and even sitting. The pain from an L5 or S1 radiculopathy radiates from the buttock, down the back or side of the thigh and leg to the ankle or foot and may be associated with pins and needles or numbness in a similar distribution. Similar sciatica pain occurs occasionally in lumbosacral plexus lesions, in which case lumbar spine imaging will be normal (Section 22.6).

Examination characteristically reveals loss of the normal lumbar lordosis often with paravertebral muscle spasm. Straight leg raising is limited due to sciatic pain in the case of lower lumbar disc prolapse. Hip extension is restricted with the rarer upper lumbar root compression. The motor and sensory signs depend upon the root or roots involved. They are commonly unilateral and most typically consist of depression of the knee jerk with loss of the adductor reflex subserved by the L4 root associated with loss of sensation on the medial aspect of the shin, weakness of dorsiflexion of the great toe, and altered sensation on the outer aspect of the shin innervated by the L5 root, or absence of the ankle jerk subserved by the S1 root.

An L5 root lesion, typically due to an L4/5 disc herniation causes pain radiating into the medial foot and the great toe with pins and needles on the medial aspect of the foot and weakness in the extensor hallucis longus, the ankle dorsiflexors, and the peroneal muscles.

An S1 root compression, usually due to an L5/S1 disc herniation, commonly results in lateral foot pain with pins and needles on the outer aspect of the foot, depression or absence of the ankle jerk, and weakness of the peroneal muscles and sometimes ankle flexors.

Rarely an L4 root lesion occurs, due to protrusion of the L3/L4 disc, causing pain radiating into the medial shin, the medial hamstring reflex is depressed and there may be altered sensation over the L4 dermatome.

MRI is the preferred diagnostic investigation with sagittal and axial views demonstrating the site and direction of the disc prolapse. On sagittal images there is usually loss of high signal on T2-weighted images within the disc implying dehydration and reduction of the height of the intervertebral disc space. The axial images show the posterolateral protrusion with narrowing of the nerve root in the intervertebral foramen (Fig. 29.9). Similar views can be achieved with CT scanning, but MRI provides far better soft tissue discrimination and better delineation of the cauda equina to exclude a more central lesion. The value of CT is to distinguish between soft disc and bony osteophyte. Neurophysiology may aid in localizing the lesion and is particularly helpful in separating monoradiculopathy from other conditions such as multiple radiculopathies, plexopathies, and peripheral neuropathies.

When pain is the major symptom, conservative treatment with bed rest and analgesia frequently leads to remission, often within 4–6 weeks. Bed rest, followed by limited activity, physiotherapy, and the use of non-steroidal anti-inflammatory agents are usually sufficient. The decision on whether or not to offer surgical treatment must take into account the severity of the signs and the length of the history. Acute motor weakness, such as a foot drop due to an L5 root lesion, is usually treated surgically, often with good results in the short term. Patients who lack objective neurological signs of nerve root dysfunction are least likely to benefit from lumbar disc surgery. Currently trials are underway to assess the indications for surgery and the natural history of conservative management

A not uncommon long-term complication following surgery is fibrosis at the site of the operation which can lead to recurrent radicular pain and neurological signs. Repeat gadolinium-enhanced MRI assists in distinguishing a recurrent disc prolapse from this

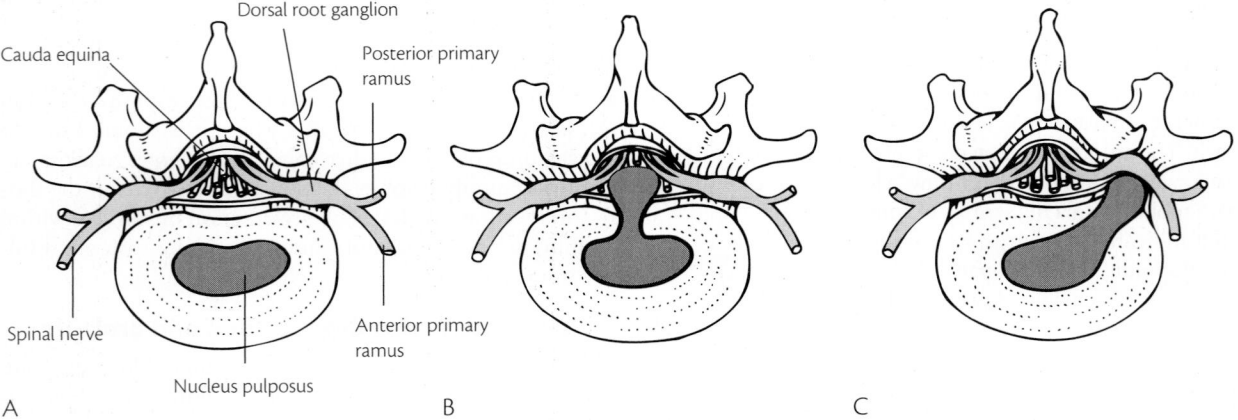

Fig. 29.8 A. The normal relationship between the lumbar intervertebral discs and the nerve roots. B. Central and C. lateral prolapse of a lumbar intervertebral disc, causing compression of (B), the cauda equina, and (C) the exiting nerve root, respectively. (From Donaghy M (1997), *Neurology*, Oxford University Press.)

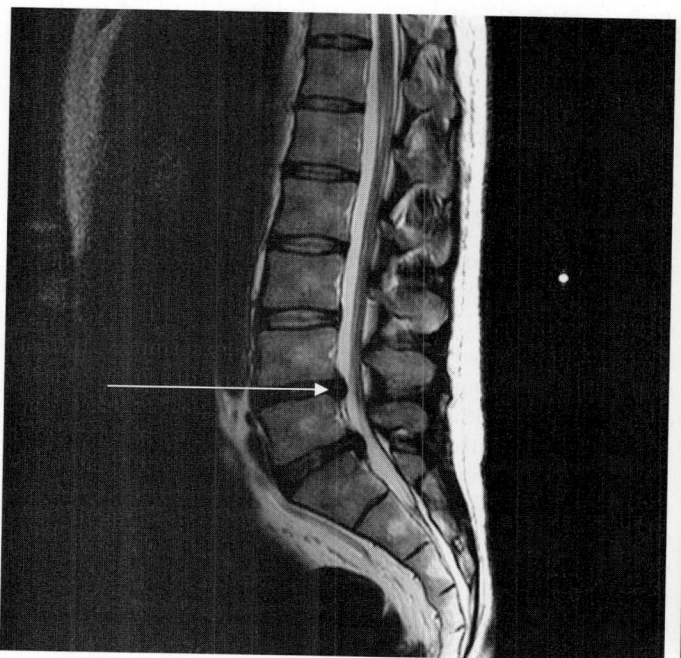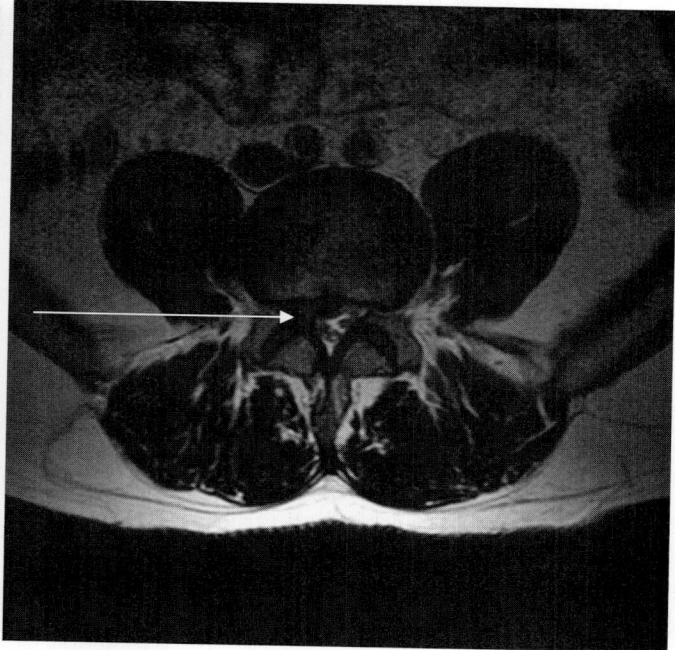

Fig. 29.9 Lateral lumbar disc prolapse causing compression of right L5 nerve root. T2-weighted MRI.

fibrotic reaction which is seen as an area of enhanced tissue in the region of the nerve root and canal. Post-surgical fibrosis is treated symptomatically, not by further surgery.

29.2.6 Lumbar spondylosis, spondylolisthesis, and spondylitis

More than 80 per cent of people are said to experience episodes of acute low back pain which recur and about 5 per cent of people report chronic low back pain throughout life. Radiological evidence of osteoarthritis and lumbar spondylosis increases with age and is universal by the age of 60 years. The presence or absence of these findings does not correlate with symptoms of back pain; routine radiology of the lumbar spine is not to be encouraged. When a potential structural problem is considered in the lumbar region the investigation of choice is MRI.

Spondylolysis is most common in the lumbar region, being reported in up to 5 per cent of individuals. It is usually asymptomatic, but when spondylolisthesis occurs, in 1–2 per cent of the population, the spinal canal is narrowed and symptoms may arise.

Spondylolisthesis describes slippage of one vertebra upon the next, commonly L4 on L5 or L5 on S1. It can be due to congenital anomalies of the facet joints (Fig. 29.10). In this situation the upper vertebra of the pair moves forwards resulting in pain and ultimately neurological symptoms as the result of trapping nerve roots and consequent ischaemia. The resulting syndrome can be one of an isolated lower lumbar radiculopathy or of a more widespread cauda equina syndrome. In such cases imaging reveals the abnormality and surgical procedures with stabilization of the vertebra may be necessary to relieve neural compression.

Inflammatory spondyloarthropathies, including ankylosing spondylitis, Reiter's syndrome, psoriatic arthritis, and that seen with inflammatory bowel disease tend not to cause neurological complications until spinal disease is advanced with rigidity and kyphosis. They may be associated with atlanto-axial subluxation, vertebral fractures, and disc disease. In the lumbo-sacral spine they cause spinal canal stenosis. Ankylosing spondylitis is associated with posterior lumbosacral diverticula which distinguish this form of cauda equina syndrome from other chronic spinal arachnoiditis and may respond to surgical decompression.

29.2.7 Lumbosacral radiculitis or infiltration

Many inflammatory conditions involve the lumbo-sacral nerve roots. Herpes simplex type II infection can cause an acute painful syndrome often mimicking a disc prolapse but with normal MRI scan (Section 22.7). This form of herpes virus, which causes genital herpes, can be associated with recurrent aseptic meningitis or with an acute lumbo-sacral radiculopathy. Manifestations include urinary retention, weakness of the distal lower limb muscles with loss of ankle or knee jerks, and sensory loss mainly in the sacral or lower lumbar distribution. Neurological episodes can occur in association with, or separate from an episode of genital herpes and CSF examination shows a mononuclear pleocytosis with a positive polymerase chain reaction for herpes simplex type II virus DNA. Treatment is with intravenous acyclovir.

Cytomegalovirus infection is a recognized cause of lumbar radiculopathy in patients with AIDS and can be treated with gancyclovir and foscarnet (Section 21.14.2). Its occurrence is less common with Highly Active Anti-Retroviral Therapy, HAART, for AIDS.

Chronic inflammatory demyelinating polyneuropathy may be associated with enlargement of nerve roots in the cauda equina (Section 21.11.2). The swollen nerve roots can be seen on MRI, often with gadolinium enhancement, and represents onion bulb formation caused by Schwann cell mediated remyelination. This swelling of nerves may, rarely, produce clinical evidence of a cauda equina syndrome. Treatment is as for the underlying polyneuropathy.

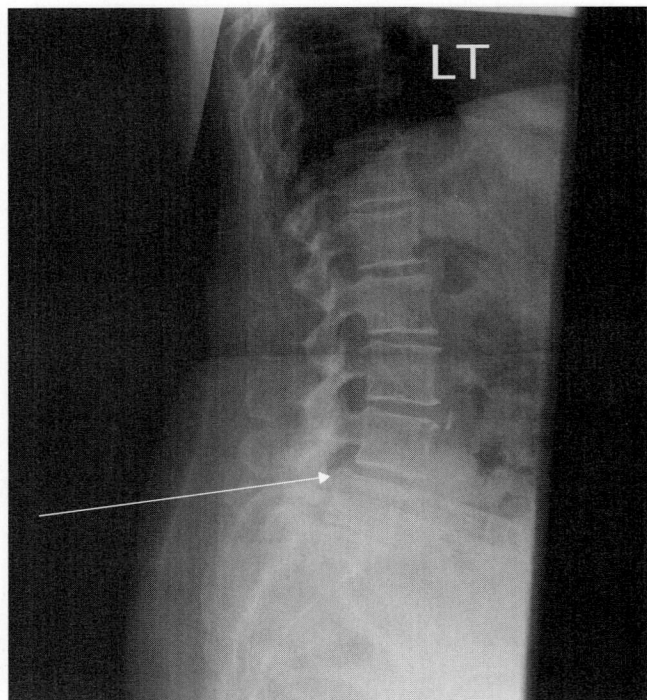

Fig. 29.10 Spondylolisthesis of L4 anteriorly on L5 (arrowed). Plain radiograph.

An isolated hypertrophy of the cauda equina, without evidence of peripheral neuropathy, is an occasional cause of leg dysfunction and has been associated with bronchial adenoma (Burton *et al.* 2002).

Very rarely in Charcot Marie Tooth disease Type III (Section 21.4.6) there may be enlarged lumbosacral nerve roots resulting in a similar cauda equina syndrome.

The lumbosacral roots are liable to involvement with granulomatous conditions such as sarcoid. Meningeal infiltration and compression by enlarging granulomas may cause a cauda equina syndrome and diagnosis, in the absence of systemic disease is difficult and may require biopsy. Neither blood nor CSF angiotensin converting enzyme levels are definitive in diagnosis. Steroid therapy is effective acutely but its long-term benefit is less established and the use of cytotoxic agents to act as steroid sparing agents should be considered.

Single or multiple root lesions, often painful, can be due to meningeal infiltration by neoplasms, such as lymphoma, leukaemia, and carcinoma especially from breast and lung. Diagnosis may be suggested by MRI enhancement and nodules on the roots and confirmed by cytology of the CSF.

29.2.8 Sacral nerve root lesions

Compression

With the exception of the S1 root, which can be entrapped by an L5/S1 disc prolapse, the lower sacral nerve roots are unlikely to be entrapped by disc prolapses. However they are potentially injured when a direct posterior prolapse impinges upon the cauda equina. It is more common for the lower sacral roots to be affected by processes involving the sacrum, such as cordoma, or by the presence of an ependymoma or a lipoma within the spinal theca.

Non-compressive lesions

The sacral spinal roots are vulnerable to processes of infiltration, inflammation, and infection. Their function can also be impaired by ischaemia, both with large vessel disease, as in Leriche syndrome, and in the smaller vessel diseases, as in diabetes mellitus.

29.3 Disorders of micturition

29.3.1 Innervation of the bladder

The parasympathetic nerve supply from the second, third, and fourth sacral nerves, the nervi erigentes, join the vesical plexus. From this plexus both sympathetic and parasympathetic components containing both motor and afferent fibres extend to the bladder. The parasympathetic nerves carry motor fibres to the detrusor muscle, a non-striated muscle of three poorly distinguished layers which empties the bladder, and inhibitory fibres to the sphincter vesicae. The sympathetic fibres, which originate from the lower thoracic and upper lumbar segments of the cord, supply inhibitory fibres to the detrusor and motor fibres to the sphincter vesicae, but their predominant role is vasomotor. Detailed reviews are available concerning the physiology of micturition (Pearman and England 1976) and the pathophysiology of incontinence (Swash 1985).

When the parasympathetic nerve is stimulated the longitudinal fibres of the detrusor pull up the bladder neck, open it, and the circular fibres exert pressure on the bladder contents. There are functional alpha and beta adrenergic sympathetic fibres innervating the muscle of the bladder wall and stimulation of the latter allows the bladder to fill, stimulation of the former, which are more common in the neck of the bladder, cause the internal sphincter to contract. Thus, phenoxybenzamine, by blocking the alpha receptors opens the bladder neck and, by its role in blocking muscarinic cholinergic receptors, increases functional bladder capacity. The sympathetic nervous system is less important in the physiology of bladder control in humans than the parasympathetic system (Brindley 1986, 1988).

In infancy, bladder evacuation occurs reflexly. The reflex arc runs through the sacral cord segments, via the cauda equina. Control over bladder evacuation, as the child grows, is associated with increasing ability to inhibit this evacuation reflex. Control of the inhibitory impulse is achieved by the sympathetic system which allows closure of the sphincter and inhibits contraction of the detrusor muscles. Also it becomes possible to overcome this inhibition voluntarily, thereby initiating the act of micturition, which then completes reflexly.

Three nerve mechanisms control bladder function:

◆ the sacral reflex arc for evacuation

◆ the inhibitory influence of the sympathetic

◆ voluntary control which overcomes the latter and initiates micturition. This voluntary control is exerted by the actions of the muscles of the pelvic floor and the striated muscle sphincters of the urethra innervated by the pudendal nerve. These help to maintain continence. When these relax, and the abdominal wall muscles are voluntarily contracted, intra-abdominal pressure increases and the act of micturition is assisted. Therefore any weakness of these muscles worsens failure of autonomic nervous control.

Bladder sensation, allowing the feeling of fullness and the desire to micturate, is sensed by afferent impulses in the second and third sacral nerve roots via the spino-thalamic tracts. Urethral pain travels in the same pathway, but urethral touch and pressure are transmitted via the posterior columns of the spinal cord.

The cortical centre for bladder sensation lies in the post-central gyrus at the vertex of the cerebral hemisphere. The corresponding area of the pre-central gyrus is thought to be the site of origin of the motor impulses which initiate the act of micturition. Parasagittal lesions affecting this region of the cortex bilaterally may cause urinary retention. Unilateral, or more often bilateral, lesions in the superior frontal gyrus may give rise to urgency and frequency of micturition and incontinence, or sometimes retention, due to the sensation giving rise to the desire to micturate being diminished or absent. A micturition control centre probably exists in the pons in humans and other mammalian species (Griffiths *et al.* 1990; Blok *et al.* 1997) which may explain why lesions in this region are occasionally associated with urinary hesitancy or retention (Fig. 29.11).

Voluntary initiation of micturition usually occurs in response to an awareness of bladder distension. The descending motor pathway concerned with voluntary bladder evacuation lies in the lateral columns of the spinal cord at the mid-position.

29.3.2 Disturbances of bladder control

The bladder performs two functions: storage and emptying. Neural programmes for each of these functions, which probably exist respectively in the dorsal tegmentum of the pons and the supra-pontine influence of the frontal lobes, switch from one state to the other (Morrison *et al.* 2002). Most of the time the bladder is in storage mode and the decision to void is determined by the perceived state of bladder fullness with the timing influenced by assessment of a socially suitable moment.

During the storage phase the sympathetic nervous system inhibits the smooth muscle of the detrusor. Voluntary pundendal nerve activation can also be employed to close the striated muscle of the urethral sphincter and pelvic floor. Inhibition of the parasympathetic system prevents detrusor contraction. Reversal of this reciprocal activation–inhibition of the sphincter and detrusor muscles allows voiding to begin. Relaxation of the striated muscle of the sphincter is followed seconds later by contraction of the smooth muscle of the detrusor, with the result that the bladder empties.

Positron emission tomography studies have shown that during filling of the bladder there is increasing activity in the peri-aqueductal grey matter in the pons and bilaterally in the frontal lobes. (Blok *et al.* 1997) The anterior regions of the frontal lobes are critical for bladder control. Cerebral causes of bladder disturbance cause inappropriate and unaware incontinence, nocturnal enuresis, and may cause retention of urine, but they can also result in urgency and frequency of micturition and urge incontinence. The process of micturition is co-ordinated normally in this situation since the damage lies in the higher control processes.

Lesions in the spinal cord are probably the most common causes of neurogenic bladder dysfunction. Spinal cord disease is characterized by an abnormally overactive, small capacity bladder that produces the symptoms of urgency and frequency of micturition. Urge incontinence is likely when there are associated motor difficulties and poor neural drive to the detrusor muscle during attempts to void. Detrusor–sphincter dyssynergia explains the incomplete bladder emptying and the presence of a post-micturition residual which predisposes to urinary tract infection. Since bladder innervation in the spinal cord is more caudal than the lower limbs, any form of spinal cord disease that causes bladder dysfunction is likely to produce clinical signs in the lower limbs unless the lesion is limited to the conus medullaris.

A lesion affecting the conus medullaris is typified by the combination of upper and lower motor neurone signs in the legs, together with urinary symptoms. Typically there is asymmetrical wasting of the calves and intrinsic muscles of the feet, but the prominent bladder symptoms and extensor plantar responses point to the site of the lesion lying in the conus medullaris, rather than the cauda equina. The bladder disorder may be a mixture of detrusor hyper-reflexia and incomplete emptying.

Damage to the cauda equina leaves the detrusor decentralized rather than denervated because the post-ganglionic parasympathetic innervation is unaffected. This may explain why the nature of bladder dysfunction after a cauda equina lesion is unpredictable, ranging from spontaneous evacuation to detrusor hyperreflexia.

29.3.3 Cerebral lesions

Neural pathways involved in the voluntary initiation of micturition can be interrupted above the level of the spinal cord resulting in retention of urine, usually in association with severe bilateral corticospinal tract signs. Bilateral lesions involving the pre-central cortex can also cause retention of urine. Damage to this part of the cortex or its pathways almost certainly accounts for the retention of urine or urge incontinence, which occur as symptoms of frontal lobe damage due to intracranial tumour, anterior communicating artery aneurysm, penetrating brain wounds, and after prefrontal leucotomy.

In people who develop urinary incontinence following stroke or diffuse cerebral disorders including Alzheimer's disease and other degenerative diseases the causes are probably multi-factorial. In the dementia seen with normal pressure hydrocephalus (Section 26.4.5), in which incontinence is a cardinal feature, improvement in urodynamic function may be seen within hours of lumbar puncture.

When primary nocturnal enuresis persists in otherwise normal children it is probably due to delay in developing inhibition of reflex bladder evacuation. When secondary nocturnal enuresis occurs the child has acquired abnormal conditioned reflexes or psychological disturbances whereby bladder evacuation continues to occur during sleep. Childhood enuresis in childhood is rarely due to spinal cord or cauda equina lesions, such as spina bifida occulta, but such causes should always be excluded.

When a cerebral cause of bladder disturbance is suspected imaging of the cerebrum is indicated to identify structural pathology.

29.3.4 Spinal cord lesions

The assessment and investigation of bladder dysfunction due to disease within the spinal canal depends on obtaining an adequate history. Because of the length of the spinal cord symptoms are most commonly upper motor neurone in type, with urgency and frequency of micturition being most common. With damage to the cauda equina the symptom is more likely to be painless retention with overflow incontinence.

Partial lesions of the spinal cord above the conus medullaris can affect inhibitory fibres destined for the sympathetic outflow

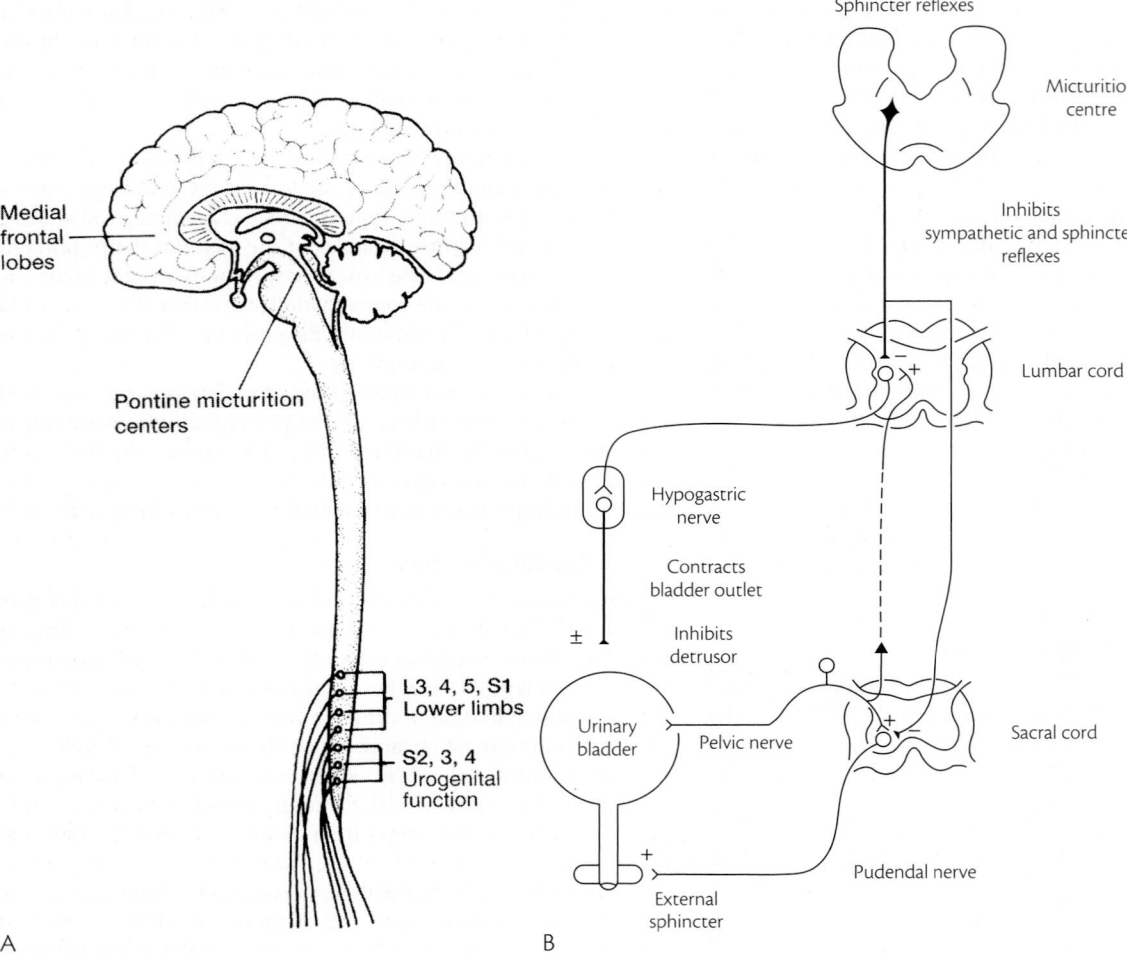

A

B

Fig. 29.11 A. Levels of innervation of the urinary bladder; the cortex, pons, and lumbo-sacral outflow. Any spinal cord lesion causing neurological symptoms and signs in the lower limbs will be likely to affect bladder control. It would be unusual for spinal pathology to affect bladder control without causing a neurological deficit in the lower limbs unless it was confined to the conus. B. Diagram showing detrusor–sphincter reflexes. During the storage of urine, distention of the bladder produces low-level vesical afferent firing, which in turn stimulates (1) the sympathetic outflow to the bladder outlet (base and urethra) and (2) pudendal outflow to the external urethral sphincter. These responses occur by spinal reflex pathways and represent 'guarding reflexes', which promote continence. Sympathetic firing also inhibits detrusor muscle and transmission in bladder ganglia. At the initiation of micturition, intense vesical afferent activity activates the brainstem micturition centre, which inhibits the spinal guarding reflexes. (From Bannister and Mathias, Autonomic Failure 4th Edn, Oxford University Press.)

or descending fibres concerned in the voluntary initiation of micturition. When the former occurs the patient has difficulty in retaining urine and micturition is precipitated. This symptom, called urgency, is common in most spinal cord diseases and seen early in intrinsic lesions affecting the spinal cord, such as multiple sclerosis and later in extrinsic compressive lesions, such as spinal neurofibroma and meningioma.

More severe, but incomplete, lesions of the cord tend to impair voluntary control over micturition. This allows urinary retention to occur because of the uninhibited action of the sympathetic nervous system. When both voluntary control and sympathetic inhibition occur together there is a combination of retention and urgency or urge incontinence. The combination causes frequent and precipitate micturition but with an increase in the amount of residual urine remaining in the bladder immediately after

micturition, which may be demonstrated by bladder ultrasound examination. This too is a common finding in intrinsic cord lesions, such as multiple sclerosis, and the pathophysiological abnormality of bladder control is known as detrusor–sphincter-dyssynergia.

When conduction through the spinal cord is completely interrupted by transection or a severe transverse lesion above the level of the conus there is retention of urine during the phase of spinal shock. This is followed by enhancement of reflex activity in the distal portion of the spinal cord causing reflex evacuation of the bladder mediated by the sacral reflex arc. When this reflex evacuation develops it may be facilitated by stimuli applied to the sacral cutaneous areas. If the spinal cord damage affects the sacral segments of the cord the bladder can be atonic, presumably due to destruction of the sacral reflex arc in the conus medullaris.

Classical examples of spinal cord induced bladder disturbance are seen after traumatic paraplegia, in multiple sclerosis, in which the incidence of bladder dysfunction is estimated to be 75 per cent, transverse myelitis, tropical spastic paraparesis, and arteriovenous malformations of the spinal cord. The identification of a neurogenic bladder due to spinal cord disease makes obligatory imaging of the cervico-dorsal spinal cord to identify the cause and exclude surgically correctable lesions.

29.3.5 Sacral reflex arc lesions

Accumulation of urine causes bladder distension which stimulates afferent activity in the hypogastric and pelvic nerve fibres. Continence depends upon discharges arising in Onuf's nucleus whose efferent fibres in the pudendal nerve maintain contraction of the external urethral sphincter and inhibit transmission at postganglionic parasympathetic neurones innervating the detrusor. This ensures that urethral pressure exceeds the pressure of detrusor contraction and allows the bladder to continue to fill.

Voiding occurs when there is inhibition of sympathetic pudendal nerve firing. This relaxes the sphincter and allows cholinergically mediated parasympathetic contraction of the detrusor muscle. The main function of the sympathetic system is therefore to maintain continence by closure of the sphincter by alpha adrenergic receptor stimulation and to inhibit the detrusor via beta-adrenergic receptor stimulation. Continence requires selective facilitation and inhibition of these various spinal reflexes.

The sacral reflex arc mediates evacuation of the bladder. Interruption of the arc usually results in retention of urine due to the unopposed action of the sympathetic nervous system. Lesions of the conus medullaris interrupt the central fibres of the reflex. Lesions of the cauda equina affecting the second, third, and fourth sacral nerves interrupt both the afferent and efferent pathways. All these conditions result in retention of urine. Common causes include diabetic neuropathy and pelvic surgery; rarer causes include amyloid neuropathies and immune neuropathies. In tabes dorsalis the reflex arc is interrupted on the afferent side due to degeneration of the afferent neurones.

Incomplete lesions of the conus medullaris or cauda equina are associated with reflex evacuation of the bladder. With complete lesions of the cauda equina it is more common for the bladder to remain atonic, with dilatation allowing accumulation of large volumes of urine which may be followed by overflow, dribbling incontinence. Such bladder distension is usually painless and may not be recognized by the patient. These symptoms may occur with spina bifida and a tethered spinal cord.

29.3.6 Sphincter myotonia

Painful urinary retention in young women without overt neurological disease or signs is rare. It can occur in association with abnormal electromyography in the urethral sphincter. Myotonic discharges are seen with complex repetitive discharges and decelerating bursts (Fowler 1999). Originally thought to be psychogenic, the aetiology of this syndrome remains uncertain. The condition manifests only in pre-menopausal women and has been associated with polycystic ovaries; thus it is possible that the myotonia relates to a hormonal abnormality.

Clinical neurological examination, imaging, and formal neurological assessment may be normal with the result of patients being misdiagnosed. The voiding difficulty may persist for months or years and may require clean intermittent self-catheterization. Recently sacral nerve stimulation has been found to be helpful in some patients.

29.3.7 Autonomic disturbances

Bladder dysfunction may arise as a result of autonomic disturbance. Nocturia is a common symptom in primary autonomic failure (Sections 21.10.4 and 21.11.7), together with symptoms of frequency, urgency, incontinence, and retention. The bladder may be atonic on cystometrography. Frequently there will be associated symptoms such as constipation and sexual disturbance.

Neurogenic bladder symptoms are not necessarily indicative of lesions affecting the spinal cord and cauda equina. In Parkinson's disease bladder symptoms usually occur late in the course of the disease and, in men, are likely to be due to or associated with prostatic outflow obstruction.

When urinary symptoms occur early in the course of a disorder involving relatively mild extrapyramidal features, the diagnosis of multiple system atrophy should be considered (Section 40.3.8). Indeed the onset of urogenital symptoms in this disorder may precede overt neurological involvement by many years. Multiple system atrophy probably affects the central nervous system at several sites which are important for bladder control, thus explaining why urinary complaints occur early and can be so severe. The neuronal loss in the pons causes detrusor hyperactivity coupled with incomplete bladder emptying resulting from loss of parasympathetic innervation to the detrusor due to neuronal degeneration in the intermedial lateral columns of the spinal cord. In addition, loss of neurones in Onuf's nucleus results in denervation of the urethral sphincter. When bladder symptoms occur as part of the earliest evidence of a system degeneration it is often necessary to image the lumbo-sacral spine to exclude structural causes before the full nature of the disease becomes apparent.

29.3.8 Assessment and investigation

A simple and formal approach to investigation of the neurogenic bladder has been outlined (Fowler 1996).

Urological investigations usually take secondary importance to spinal or cerebral imaging which is necessary to exclude structural lesions that might require neurosurgical intervention. When bladder dysfunction is attributable to lower motor neurone disturbances, the initial assessment should be MRI of the lumbosacral spine. Intrapelvic lesions can damage bladder nerve pathways particularly post-operatively or as a result of neoplasia. Most spinal cord syndromes which affect bladder function result in urgency and frequency of micturition. Localizing the site of the spinal cord lesion depends upon other associated clinical features, such as pain, a sensory level, and motor signs with the underlying lesion being located by MRI.

Ultrasound of the bladder establishes its size and post-micturition ultrasound establishes the amount of residual urine. Formal urological assessment is rarely indicated but includes cystoscopy, urethroscopy, and cystometrography to establish the nature of the neurogenic problems.

Electromyography plays a part in assessing activity within the urethral sphincters. It may help identify denervation of the nerve fibres from Onuf's nucleus, as part of a system degeneration or local conus abnormality and also to identify the rare syndrome of sphincter myotonia which increases activity in the urethral sphincters in young women (Section 29.3.6).

29.4 Disturbances of bowel function

29.4.1 The rectum: neuroanatomy and physiology

The nerve supply of the rectum is similar to that of the bladder, having inhibitory sympathetic and stimulatory parasympathetic nervous system components. The internal sphincter, which normally maintains continence is innervated by the sympathetic nervous system and relaxes in response to rectal dilatation. Voluntary control of the bowel is only exerted by the external sphincter and there is no voluntary inhibitory system. The puborectalis muscle, which maintains an acute anorectal angle, is most important in achieving continence and is effective even when the internal and external sphincters are incompetent.

As with the bladder, the lower bowel usually functions in a storage mode. Continence is maintained by the combination of the acute ano-rectal angle, internal anal sphincter tone, and voluntary contraction of the external anal sphincter and pelvic floor. The process of defaecation involves a series of neurological events that begin in response to the sensation of a full rectum. When this sensation is perceived, and the moment judged to be suitable, defaecation is initiated by raising intra-abdominal pressure and allowing descent of the pelvic floor and sphincter relaxation.

Defaecation can be initiated as a voluntary act but the sites of the higher level processes involved are unknown and their pathways through the brainstem and spinal cord to Onuf's nucleus are incompletely mapped but assumed to be similar to those subserving micturition.

29.4.2 Bowel disorders

Bowel function is rarely affected by lesions of the frontal lobes, although the neurological control of anorectal motility probably involves centres within these lobes. Little information is available from functional imaging studies and it is rare to see specific problems with bowel function in patients with frontal lobe lesions (Silverman *et al.* 1997). Control of faecal continence may be lost in diffuse degenerative disease of the brain, especially when the frontal lobes are involved as in Alzheimer's disease.

Spinal cord disorders commonly affect bowel function, usually resulting in constipation. As with micturition, intrinsic spinal cord lesions such as multiple sclerosis and syringomyelia tend to cause early bowel symptoms. By contrast extrinsic compressive lesions cause late development of symptoms. Loss of both rectal sensation and the desire to defaecate means that bowel emptying must be induced at convenient times by cutaneous stimulation, the use of suppositories or enemas, or by manual evacuation. Occasionally diversion procedures are required (Norton and Henry 1999). Rarely patients with spinal cord disease may develop faecal urgency and incontinence, particularly in partial lesions of the spinal cord, especially multiple sclerosis. Such symptoms may be helped by the use of alpha adrenergic agonists.

When the sacral innervation of the rectum has been damaged, most commonly following pelvic surgery or trauma, automatic activity is mediated by the enteric plexus in the bowel wall. This produces contraction of the rectum and relaxation of the anal sphincter in response to a rise of tension within the rectum. This reflex activity is more pronounced when the sacral innervation remains intact, as after complete division of the spinal cord above the sacral enlargement. This autonomic system is relatively inefficient due to the limited force of rectal contraction. Thus there is a tendency for all disturbances of rectal innervation to result in constipation. Reflex defaecation may occur and, in the presence of an intact sacral reflex arc, can be provoked by cutaneous stimuli applied to the sacral dermatomes, a trick employed by some paraplegic patients.

In most patients with complete spinal cord or cauda equina lesions, satisfactory control of the bowels may be achieved by means of twice weekly enemas or suppositories, or by manual evacuation of faeces.

29.5 Disturbances of sexual function

29.5.1 Innervation of the sexual organs

The seminal vesicles are innervated from the inferior hypogastric plexus, a combination of sympathetic fibres from the superior hypogastric plexus and parasympathetic fibres from the pelvic splanchnic nerves. The corpora cavernosa of the penis are innervated from the hypogastric plexus and parasympathetic nerves from the S2 spinal segment. Stimulation of the hypogastric plexus results in penile erection and seminal emission (Brindley 1988). The commonest symptom with spinal cord lesions is for loss of such function, though rarely partial lesions may result in over-activity of the hypogastric plexus with the occurrence of priaprism.

In the female the utero-vaginal plexus arises from the inferior hypogastric plexus and travels with vessels to ramify in the myometrium and endometrium of the uterus and cervix. The sympathetic nerves cause uterine contraction and vasoconstriction and the parasympathetic uterine inhibition and vasodilatation. However their actions are complicated by the major effect of circulating hormones. Intact sensory pathways are essential for normal sexual function and gratification.

29.5.2 Sexual disorders

Physiological sexual responses in men and women have been divided into four phases: excitement, plateau, orgasm, and resolution (Masters and Johnson 1970). Much remains to be discovered about cortical control of sexual function, though it is recognized that cerebral processes determine libido and desire. The ability to sustain a sexual response is determined by spinal autonomic reflexes.

It is recognized that sexual dysfunction is more common in patients with frontal and temporal lobe lesions, given their predominant role in libido. Sexual dysfunction is not uncommon after head injury, particularly where there is cognitive damage. Although hypersexual behaviour may occur after frontal lobe damage, it is more common to have a reduction in sexual behaviour (O'Carroll *et al.* 1991) (Section 34.10).

In the male high cervical cord lesions cause loss of psychogenic erections although spontaneous or reflex erections may still occur. In low spinal canal lesions, particularly those involving the cauda equina, there may be erectile failure.

In partial spinal lesions, as occur commonly in multiple sclerosis, more than 60 per cent of men have erectile dysfunction. Lesions in the anterior spinal tracts cause loss of orgasmic sensation in both sexes and all types of lesion are associated with a difficulty in sustaining erection. Sildenafil has a role in the management of such symptoms and mechanical devices may also be considered.

In women with multiple sclerosis sexual dysfunction is common, affecting more than 50 per cent. Libido is most affected. Additional problems are posed by lower limb spasticity, loss of pelvic sensation, genital dysaesthesiae, and fear that incontinence will occur during intercourse.

References

Blok BF, Willemsem AT, Hostege G (1997). A PET study on brain control of micturition in humans. *Brain*, **120**, 111–21.

Bowen J, Gregory R, Squier M *et al.* (1996). The post irradiation lower motor neurone syndrome. *Brain*, **119**, 1429–39.

Brindley GS (1986). Sacral root and hypogastric plexus stimulators and what these models tell us about autonomic actions on the bladder and urethra. *Clin Sci*, **70** (Suppl 14), 41s–4s.

Brindley GS (1988). The actions of parasympathetic and sympathetic nerves in human micturition, erection and seminal emission and their restoration in paraplegic patients by implanted electrical stimulation. *Proc R Soc Med*, **235**, 111–6.

Burton M, Anslow P, Gray W *et al.* (2002). Selective hypertrophy of the cauda equina nerve roots. *J Neurol*, **249**, 337–40.

Fearnside MR, Adams CBT (1978). Tumours of the cauda equina. *J Neurol Neurosurg Psychiatry*, **41**, 24–31.

Fowler CJ (1996). Investigations of the neurogenic bladder. *J Neurol Neurosurg Psychiatry*, **60**, 6–13.

Fowler CJ (1999). Neurological disorders of micturition and their treatment. *Brain*, **122**, 1213–31.

Goh JK, Kalifa W, Anslow P *et al.* (2004). The clinical syndrome associated with lumbar canal stenosis. *Eur Neurol*, **52**, 242–9.

Griffiths D, Holstege C, deWall H *et al.* (1990). Control and coordination of bladder and uretheral function in the brain stem of the cat. *Neurourol Urodyn*, **9**, 63–82.

Masters WH, Johnson VE (1970). *Human Sexual Inadequacy*. Little, Brown, Boston.

Morrison J, deGroat W, Downie J *et al.* (2002). Neurophysiology and neuropharmacology. In Abrams P, Cardozo L, Khoury S, Wein W, eds. *Incontinence 2nd International Consultation on Incontinence*, Health Publication, Plymouth.

Norton C, Henry M (1999). Investigation and treatment of bowel problems. In Fowler CJ, ed. *Neurology of Bladder, Bowel and Sexual Dysfunction*. Butterworth-Heinemann, Boston.

O'Carroll R, Woodrow J, Maroun F (1991). Psychosexual and psychosocial sequelae of closed head injury. *Brain Inj*, **5**, 303–13.

O'Laoire SA, Crockard HA, Thomas DG (1981). Prognosis for sphincter recovery after operation for cauda equina compression owing to lumbar disc prolapse. *BMJ*, **282**, 1852–4.

Pearman JW, England EJ (1976). The urinary tract. In Vinken PJ, Bruyn GW, eds. *Handbook of Clinical Neurology*, Vol. 26, pp. 409. North Holland, Amsterdam.

Silverman D, Munakata J, Ennes H *et al.* (1997). Regional cerebral activity in normal and pathological perception of visceral pain. *Gastroenterology*, **112**, 64–72.

Swash M (1985). New concepts of incontinence. *BMJ*, **290**, 4–5.

Seizures and alterations of consciousness and thought

CHAPTER 30

Seizures and related disorders in children

Helen Cross

Contents

30.1   Introduction

Epilepsy is most prevalent at each end of the age spectrum—the very young and the very old. Up to 1 per cent of the childhood population will have active epilepsy at any time. Of these 60–70 per cent will be controlled on medication or enter into spontaneous remission, however the remainder will continue to have seizures despite the range of treatment available. There will be associated comorbidity of learning and behaviour difficulties in a significant proportion, and these may take precedence in management over the seizures themselves. Careful evaluation of each individual child with regard to the possible diagnosis and associated comorbidities is required in all children presenting with recurrent paroxysmal episodes in order to optimize management.

30.2   The differential diagnosis of paroxysmal disorders in childhood

The most common cause of poor response to treatment in epilepsy is misdiagnosis. About 20–40 per cent of children arriving in a tertiary epilepsy clinic for further assessment do not have epilepsy, but rather conditions involving paroxysmal episodes that are not epileptic in origin (Uldall *et al.* 2006). Common causes of misdiagnosis include misconceptions of what is important in the history. Too much weight may be put on factors other than the events themselves such as family history, or abnormal neurological examination findings. Or one may be unaware of the range of conditions that may present in childhood that are not epileptic. Epileptic seizures are defined as changes in movement or behaviour that occur as the result of a primary change in the electrical activity of the brain. Numerous other types of event may involve such changes but as secondary phenomena to a non-epileptic primary event (National Institute of Clinical Excellence 2004).

30.2.1 Syncope and related disorders

Any event that reduces the supply of oxygenated blood to the brain will result in a loss of consciousness; and if prolonged, secondary epileptic phenomena may occur as a result of the hypoxia, causing a hypoxic seizure. In childhood the most common causes are breath-holding attacks, or reflex anoxic seizures. These typically occur in young children, 6 months–5 years, in response to a noxious stimulus. Children may either have a prolonged expiratory apnoea, causing cyanosis and hence a cyanotic breath holding attack, or a reflex asystole causing 'pallid' attacks. The description of events is key to diagnosis; most children grow out of the tendency and often there is a family history of such events. Vasovagal attacks do occur but in the older population—typical syncope is rare under 10 years, and in such children a cardiac cause must be considered. In addition atypical features should prompt a cardiac evaluation such as occurrence on exercise, or a prolonged event. This aside it is good practice for an electrocardiogram to be performed on all new presentations of collapse (Scottish Intercollegiate Guidelines Network 2005).

Hyperekplexia is an autosomal dominant disorder, the result of a mutation in the glycine receptor gene, where children present from birth with excessive startle, resulting in tonic spasm that can be very profound and result in hypoxic seizures. The history with the supportive examination finding of startle to nose tapping will suggest the diagnosis. Although attacks may become less severe with age, in infancy they can be life threatening.

30.2.2 Tics, cataplexy, and movement disorders

The most common phenomena to mimic an epileptic episode are tics (Section 40.6.1). Simple motor tics are common in young children, usually involve the upper body or face, and are short lived in their natural history. Many may also appear to be familial. Only a minority go on to develop Gilles de la Tourette's syndrome (Section 40.6.3) with vocal as well as motor tics. In the very young, paroxysmal abnormal eye movements can be the presenting feature of alternating hemiplegia, the intermittent hemiplegia itself only becoming apparent beyond the first year of life. Cataplexy (Section 32.3.1), with drops from loss of muscle tone triggered by emotion, may be misdiagnosed as epileptic drop attacks. Abnormal upper body movements seen associated with feeding may be related to oesophageal reflux, Sandifer's syndrome, and may be seen particularly in the neurologically abnormal or developmentally delayed child. The movements are thought to be induced by an attempted change in intrathoracic pressure to relieve pain from oesophagitis. Other paroxysmal episodes that may be misdiagnosed include benign paroxysmal vertigo (Section 15.3.2) and benign paroxysmal torticollis (Section 40.4.1). These conditions can come on before the age of five, with obvious torticollis or with paroxysmal vertiginous episodes of which the child will be aware, and lead to an unwillingness to move or unsteadiness on their feet. The episodes usually last minutes to hours; the conditions resolve with age. Paroxysmal movement disorders and dystonias may also be misdiagnosed as epileptic, and indeed anticonvulsant treatment may

be effective in some. Therefore a response to anticonvulsants should not be used as a diagnostic tool.

30.2.3 Behavioural and psychiatric conditions

Various behavioural phenomena may be misdiagnosed as epileptic attacks. The most common referral to the neurology clinic will be of day dreaming being questioned as absence attacks; in the vast majority simple day dreaming or non attention will be the diagnosis. Key features confirming this are a situational occurrence and the ability to be distracted from attacks, sometimes by physical touch. In very young children self-gratification behaviour can cause concern as the children can appear distracted, unaware of their surroundings and become flushed. They may also become distressed if distracted. Such events are very common and not abnormal although the diagnosis can be distressing to parents. Toddlers may have distal movements associated with excitement that are completely benign called 'overflow movements' or shuddering attacks. Stereotypies and ritualistic behaviours can cause concern but are common in the learning disabled and autistic population. Episodic rage is commonly referred for assessment as being possibly epileptic in origin but is almost never directly the result of epileptic discharge, although it may be seen as a postictal phenomenon. Pseudoseizures are seen in children, rarely under 10 years, and commonly coexist with real epileptic seizures. Their diagnosis requires careful assessment and evaluation, particularly looking for a cause, although sexual abuse is only likely to account for a small number.

30.2.4 Parasomnias

A wide range of sleep phenomena (Section 32.4) exist in normal children; a parent will only become aware of this when forced to sleep with a young child. Benign sleep myoclonus is commonly misdiagnosed as epilepsy in the first 6 months of life leading to children being overtreated with antiepileptic medication. The key differentiation is that jerks only occur during sleep in developmentally normal children. Older children and adults may also experience myoclonic jerks as a normal phenomenon on going off to sleep (Section 32.4.1). Night terrors may occur at the same time each night; children awake apparently terrified and unresponsive, and will appear distressed for several minutes before settling back to sleep. These may be aborted by waking the child at a similar hour each night prior to the time of the likely attack.

30.3 Diagnosis of epilepsy in childhood

30.3.1 Definitions and differences from adults

Arguably there is little difference in the definition of epilepsy between adults and children: it is a condition whereby the individual is prone to recurrent epileptic seizures. By definition this means at least two epileptic seizures. In certain circumstances, particularly certain syndromes, the diagnosis of epilepsy may be made after a single seizure and a diagnostic EEG abnormality. This is probably of most relevance when considering when to treat after a single seizure if the risk of further seizures is judged to be high.

30.3.2 Epidemiology

The incidence of epilepsy in the first year of life is higher than at any other age during childhood (Doose *et al.* 1983; Hauser 1993; Braathen and Theorell 1995; Olafsson *et al.* 1996). Age-specific incidence rates for the first year of life documented in studies to date show great variability from 80 to 256/100 000. Differences in geographical study areas and methodology may be responsible for such variation. Retrospective study designs and studies where records of a single centre providing services to a catchment area are the only source of cases are associated with possible under-ascertainment and bias. There is evidence that the incidence of epilepsy in childhood has been declining over the past decades (Hauser *et al.* 1993). The availability of diagnostic criteria in later time periods may have resulted in reduced inclusion of non-epileptic patients. A recent prospective community-based study has estimated an incidence in children under 2 years of life of 62.3 (95 per cent CI: 47.4–81.9)/100 000 (Eltze *et al.* 2007). Neonatal seizures may be more prevalent, with incidence figures of 70 to 270/100 000, and also higher in preterm infants at 58 to 132/1000. Incidence figures fall for the remainder of childhood, but still remain higher than in adolescence.

Population data suggest the majority of children who present with epilepsy in childhood have a good prognosis, both for remission as well as control of seizures. Studies show that two-thirds of children will be seizure free long term (Sillanpaa and Schmidt 2006). Furthermore population studies suggest the majority do well. However there are select populations in whom the risk for comorbidity and long-term seizures remain high.

Mortality is related to the risk of associated systemic disease, as well as accidents. However there is also a risk of sudden unexpected death in epilepsy amongst those who continue to have convulsive seizures, and those with associated learning and physical disability are most at risk (Forsgren *et al.* 2005).

30.3.3 The role of investigation

The history is the key to the diagnosis of epileptic seizures, and hence of epilepsy. It is of particular value if obtained from an eye witness which is usually the parent in the case of children. However if the event has happened outside the home such as in school further information may need to be sought and results of investigations may assume greater importance in support of the diagnosis. An EEG should be performed in any child with suspected epilepsy; in most this will be following the second seizure. The most important role of the EEG is not to confirm the diagnosis of epilepsy, unless of course an event is captured during the EEG recording, but to enable a syndrome diagnosis. The child's EEG requires skilled interpretation; developmental changes seen with age may be misinterpreted if reviewed by an individual not experienced in paediatric EEG. In a

small number of children an EEG may be considered after a first seizure; for example in a child with a history of a single sleep seizure involving facial or bulbar features where benign rolandic epilepsy may be suspected, or where a teenager presents with an initial tonic clonic seizure and generalized spike wave seen on EEG would suggest an idiopathic generalized epilepsy. If the EEG is normal and yet the diagnosis is still suspected then a sleep EEG should be performed. Specialist EEG investigations such as ambulatory EEG and video EEG telemetry (Section 3.3.5) would only be performed in specific circumstances, perhaps where diagnosis remains in question or where localization of seizure onset is to be defined as part of presurgical evaluation.

Unlike in adults, neuroimaging is not routinely performed in all children presenting with epilepsy. Usually imaging will not help in diagnosis, but may point to the aetiology. MRI is the imaging of choice, although in rare circumstances CT scan may reveal the aetiology promptly in an acute presentation of seizures. MRI should be considered in all children presenting with epilepsy under the age of 2 years, in all those presenting with a focal epilepsy where a benign syndrome cannot be confidently diagnosed and in those whose seizures continue despite a trial of two anticonvulsant medications (NICE 2004). The role of other investigation remains in determining an aetiology, and will be driven by the presentation (Section 30.6.3).

30.3.4 Underlying conditions presenting with seizures

It is important to consider the underlying cause of an epilepsy. There are few neurometabolic or neurodegenerative conditions that present with seizures as a first isolated symptom. These are most common within the first year of life, and are seen with decreasing frequency with age. It is important to diagnose these in view of their different prognoses and the possible relevance to antenatal diagnosis in future pregnancies. The conditions to be considered are outlined in Table 30.1 (see also Table 10.2 VI 1–16).

All children presenting within the first year require evaluation as to whether there is an underlying neurometabolic condition. In an older child it is the presence of associated symptomatology, whether a fluctuation in neurological status or neurological decline that may point to whether further evaluation is warranted. It is often difficult to decide whether a neurological deterioration is related to the epilepsy, or due to an underlying causative process of both the epilepsy and the decline. This may take time to evaluate, so as to determine whether there is any degree of fluctuation related to

Table 30.1 Neurometabolic and neurodegenerative conditions presenting with seizures

Infancy	1–5 years	5–10 years	Adolescence and adulthood
Metabolic:	Mitochondrial cytopathy	Subacute sclerosing	Pogressive myoclonic epilepsies of:
Non-ketotic hyperglycinaemia	Homocysteinuria	panencephalitis	Lafora body type
D-glyceric aciduria	Rett syndrome	HIV	Unverricht–Lundberg
Hyperammonaemia	Late infantile neuronal	Progressive neuronal	Sialidoses
Biotinidase deficiency	ceroid-lipofuscinosis	degeneration of childhood	Progressive neuronal degeneration of childhood
Sulphite Oxidase deficiency	Gaucher's type III	Wilson's disease	
Late infantile neuronal ceroid-lipofuscinosis		Niemann Pick type C	
Menkes syndrome			
Krabbe disease			
Tay Sachs			
Peroxisomal disorders			

seizures suggestive of epileptic encephalopathy (Section 30.5.2) as opposed to the steady deterioration with loss of skills that would suggest a neurodegenerative process. The emergence of neurological signs generally, although not exclusively (Neville and Boyd 1995; Dale and Cross 1999), suggests an underlying progressive pathology. Investigations should be targeted according to the condition suspected, and may include CSF studies, biopsies, or further genetic analysis.

Mitochondrial disease

Certain mitochondrial diseases have seizures as part of their clinical phenotype. Some patterns are well recognized such as MELAS, **m**yoclonic **e**pilepsy, **l**actic **a**cidosis, and **s**troke-like episodes or MERRF, **m**yoclonic **e**pilepsy with **r**agged **r**ed **f**ibres (Sections 10.5 and 24.6.3). Other less-defined disorders present with evidence of multisystem involvement with epileptic seizures and developmental regression. Key investigative marker are raised plasma and CSF lactate.

Alper's disease, or progressive neuronal degeneration of childhood, with associated hepatic involvement may result from mitochondrial gene mutations particularly of the nuclear polymerase gamma gene, *POLG* (Section 10.5.6) (Ferrari *et al.* 2005; Horvath *et al.* 2006). Lactic acidosis may be absent. Children with Alper's disease may present with an aggressive form of epilepsy, with frequent status epilepticus, epilepsia partialis continua, and developmental regression. The EEG is suggestive with spikes or polyspikes, often unilaterally and prominent posteriorly, in association with slow waves and a disorganized background (Boyd *et al.* 1986). Visual evoked potentials are also grossly abnormal with normal electroretinogram. Signs of liver impairment may be seen only relatively late in the disease and may be determined only on biopsy. The condition is invariably fatal; some previously reported cases of valproate-induced liver failure may have been unmanifesting cases of Alper's.

Niemann Pick disease type C

This condition may present insidiously with onset of seizures in early to mid-childhood, and cognitive regression that may only be apparent sometime after the diagnosis of epilepsy (Section 10.4.4). A movement disorder ensues, with extrapyramidal characteristics. The condition has two hallmarks, cataplexy which may be misdiagnosed as epileptic drop attacks, and failure of upward gaze. It is a progressive autosomal recessive disease, characterized by late accumulation of multiple lipid molecules in association with abnormal tubulovesicular trafficking (Patterson and Platt 2004). At the cellular level, the disorder is characterized by accumulation of unesterified cholesterol and glycolipids in the lysosomal/late endosomal system. Approximatively 95 per cent of patients have mutations in the *NPC1* gene at 18q11 which encodes a large membrane glycoprotein primarily located at late endosomes. The remainder have mutations in the *NPC2* gene at 14q24.3 which encodes a small soluble lysosomal protein with cholesterol-binding properties (Vanier and Millat 2003). The gene product *NPC1* protein is not suitable for transduction therapies, and gene replacement or repair is to yet be practicable for Niemann Pick disease type C or related disorders. Treatment is symptomatic, and death occurs usually before the third decade.

Neuronal ceroid lipofuscinosis

Late infantile neuronal ceroid lipofuscinosis often presents with seizures as the first manifestation around the third year

of life; developmental plateau then follows and evidence of a progressive neurodegenerative disease manifests (Section 10.3.2). Although generalized tonic-clonic seizures may be the first manifestation, myoclonus becomes prominent. Blindness eventually develops due to retinal atrophy. This is in contrast to the early juvenile form where visual failure is likely to be the first manifestation and seizures occur relatively late in the course. The EEG in late infantile neuronal ceroid lipofuscinosis can be suggestive by demonstrating prominent posterior spikes on slow photic stimulation (Fig. 30.1). Further neurophysiological studies show enlarged visual evoked potentials and a reduced or absent electroretinogram.

For the majority of families affected by one of these diseases, a biochemical and/or genetic diagnosis can be achieved (Williams *et al.* 2006). Classical late infantile neuronal ceroid lipofuscinosis is caused by mutations in the *CLN2* gene leading to tripeptidyl-peptidase 1 deficiency. Early juvenile neuronal ceroid lipofuscinosis is caused by mutations in *CLN3* gene. Other variants may result from mutations in other *CLN* genes.

Other progressive myoclonic epilepsies

Children may manifest with myoclonus as a significant component of the epilepsy syndrome. In many of early onset this may be part of a static encephalopathy or one of the syndromes discussed below (Section 30.5). A small number of children, particularly those presenting in later childhood, may be manifesting a progressive myoclonic epilepsy. In some presenting in teenage years, such as those with Unverricht–Lundberg disease, initially the diagnosis of an idiopathic generalized epilepsy such as juvenile myoclonic epilepsy may have been made. However there is often keenness to make such a diagnosis, however atypical the presentation, in view of the more benign nature of the disease. Certainly a consideration of a progressive myoclonic epilepsy should be raised if there is lack of response to medication, and where myoclonus is a prominent feature to which there is no diurnal variation.

Unverricht–Lundberg disease. This may be subtle in its presentation, confused at the outset with an idiopathic generalized epilepsy. First described in 1891 by Unverricht in Estonia and later Lundborg in Sweden (1903) the diagnosis is confirmed by the finding of an increased number of dodecamer repeats in the promoter region of the *EPM1* gene at 21q22.3 which encodes cystatin B (Lalioti *et al.* 1997). This inhibits the papain family of cysteine proteases, involved in the initiation of apoptosis. Onset is also in the teenage years. Cognitive decline is not inevitable and if present may be another pointer to a progressive disease in the absence of frequent seizures such as Lafora body disease. Where there may be suspicion, a simple screen is provided by somatosensory potentials performed in conjunction with the EEG, whence an abnormally large somatosensory evoked potential would indicate that further evaluation is warranted (Fig. 30.2). The nature of further investigation would of course depend on timing of the clinical presentation (Table 30.1). Myoclonus becomes the main problem with a pseudoataxia.

Lafora body disease. This presents in the teenage years but may be earlier. It is a myoclonic epilepsy with episodes of generalized status epilepticus. Some however may present with apparent focal occipital lobe seizures. Progressive cognitive decline ensues quite rapidly, with only a short survival. Up to 80 per cent are caused by mutations in the *EPM2A* gene at 6q24 which encodes laforin

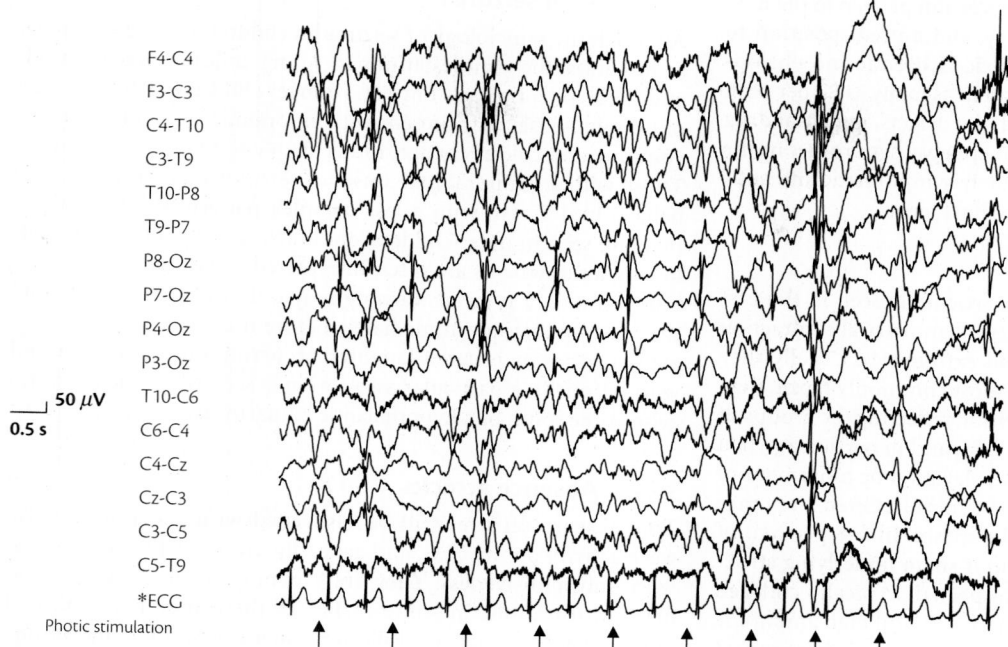

Fig. 30.1 Neuronal ceroid lipofuscinosis. Enhanced response posteriorly on EEG to photic stimulation suggestive of the late infantile form.

(Serratosa *et al.* 1995, 1999). Diagnosis may be made by mutation analysis although screening is by skin biopsy from an area where apocrine sweat glands are present, usually the axilla, to look for Lafora bodies, consisting of excessive abnormal glycogen with excessively long linear peripheral chains.

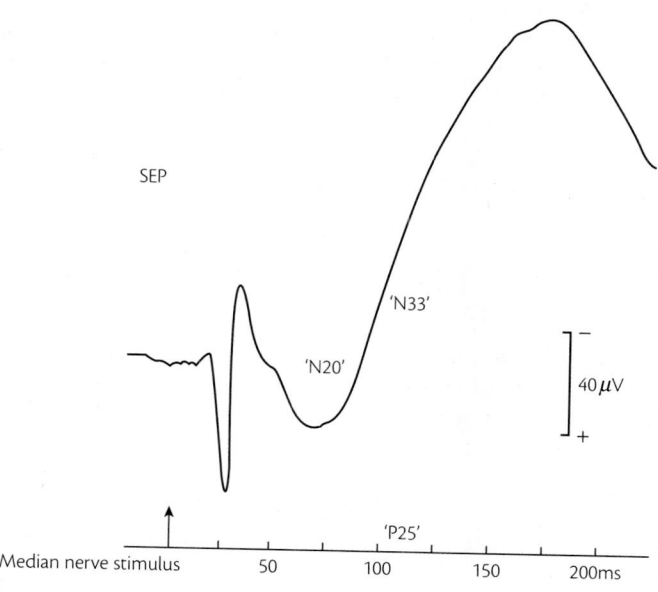

Fig. 30.2 Unverricht–Lundberg disease. Exaggerated somatosensory potential suggestive of this progressive myoclonic epilepsy. Normal SEP amplitudes—P25 <8.6 μV, N33 <8.4 μV, P25-N33 <34 μV.

30.4 Seizure types in children

30.4.1 Neonatal seizures

Neonatal seizures may take many forms, but may be subtle. Such seizures make up to 50 per cent of the total, with clonic seizures making up the majority of the remainder. Rarely myoclonic or tonic seizures may be seen. Generalized tonic clonic seizures are virtually never seen (Pitt and Pressler 2005).

Subtle seizures may involve abstruse clinical signs including ocular manifestations, mouthing, sucking, autonomic symptoms, posturing, or pedalling, which may not have been interpreted as seizural in origin. In addition the differential diagnosis of abnormal movements in neonates is important; non epileptic jerky movements or jitteriness are common, particularly when there is metabolic imbalance such as hypocalcaemia. Apnoea is common in the premature infant but very rarely in isolation the result of seizure activity. Benign neonatal sleep myoclonus may be misinterpreted as seizural in origin by the inexperienced; the clue is jerks only occurring during sleep in an otherwise normal baby. Video EEG monitoring of at-risk babies in neonatal units has shown unrecognized electrical seizures to be common. Furthermore treatment with antiepileptic drugs may lead to resolution of clinical signs despite persistence of electrical seizures. The prognostic significance of this electrical-clinical dissociation is unknown, as the electrical seizure is more likely to reflect aetiology.

Interpretation of the neonatal EEG requires experience and expertise in view of the natural changes in EEG characteristics over time. EEG features of neonatal seizures include rhythmic activity, with sudden distinct onset and ending, over a minimal duration of 10 s. Focal origin with spread, electrodecremental events, and periodic discharges are also highly suspicious of seizure discharges

(Mizrahi and Claney 2000). A burst suppression pattern to the EEG will be indicative of cerebral pathology, and a predisposition to seizures. Seizures will be related to hypoxic-ischaemic encephalopathy, cerebral haemorrhage, or infarction in as many as 80 per cent. Other causes include meningitis, metabolic defects, maternal drug withdrawal, and only rarely a neonatal epilepsy syndrome. In the vast majority therefore, seizures are acutely symptomatic and antiepileptic drugs are only required in the short term.

30.4.2 Infantile spasms

Spasms are a particular seizure type, particularly seen in the first year of life. They involve a sudden bilateral symmetrical contraction of the muscles of the neck, trunk, and extremities. Usually they occur repetitively in clusters. They are most commonly flexor, flexor-extensor, or more rarely purely extensor, reflecting the groups of muscles involved. Their intensity may vary, and in particular when manifesting subtly, for instance as head elevation or movement of only one limb, they may be overlooked clinically. Often a cry may be heard coincidentally or just after the spasm. In 6–8 per cent of patients spasms may be unilateral; and if so an underlying structural lesion of the brain should be suspected. Typically children presenting with infantile spasms will have a typical high amplitude disorganized slow waves interspersed with spikes and sharp waves, termed hypsarrhythmia (Fig. 30.3), but this does not always occur. Hypsarrhythmia is seen in about 40–70 per cent and thus an abnormal but not truly hypsarrhythmic EEG does not exclude the diagnosis of infantile spasms.

30.4.3 Focal seizures

Although the semeiology of seizures in children may be distinctive as in adults, the manifestations in young children may be subtle and their epileptic basis unclear. Younger children with discharges of occipital onset may present with autonomic symptoms, such as vomiting and colour change, rather than overt visual phenomena. Those with a temporal lobe onset of seizures may simply exhibit behavioural arrest, with a less complex pattern of automatisms than are seen in older children or adults; sucking infants or swallowing automatisms are often interpreted as normal movements. Furthermore, a preictal warning may only manifest as the child seeking out an adult immediately before the seizure. Gelastic seizures present as isolated laughter occurring out of context and without mirth; a particular association is seen when such seizures arise as the result of a hypothalamic hamartoma.

30.4.4 Absence attacks

True absence seizures will be associated with generalized spike wave activity on the EEG, and generally involve behavioural arrest with unresponsiveness. They are generally abrupt in onset, and brief in duration. In some instances there may be associated phenomenology such as eyelid movement, occasional mouthing, or some loss of tone. They are most often misdiagnosed as day dreaming, but more commonly daydreaming is misinterpreted as absence seizures. It should be borne in mind that behavioural arrest in children may be the sole manifestation of a focal seizure,

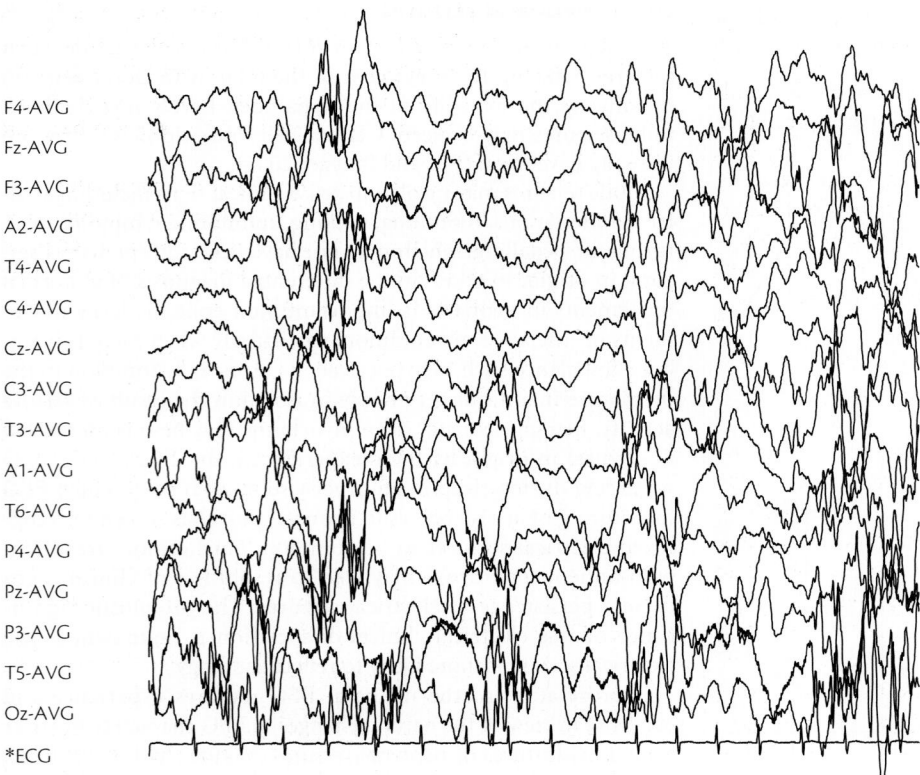

Fig. 30.3 Infantile spasms. Highly abnormal disorganized high amplitude spike and slow wave on EEG characteristic of hypsarrhythmia.

leading to difficulty in distinguishing whether attacks reflect generalized absence or focal discharges unless defined by EEG.

30.4.5 Myoclonus

Myoclonic attacks are typically seen as brief jerks of the limbs, or head nods, representing jerky movements of antagonistic muscle groups associated with spike wave discharges on the EEG, in which the spike corresponds to the myoclonic jerk. In some patients the phenomenon of negative myoclonus occurs in which the discharge itself is not associated with any muscle contraction for 50–400 ms, and then followed by a jerk as the EEG returns to normal.

30.4.6 Atonic seizures

Seen almost exclusively in children, atonic seizures are most likely to cause injury. Being of abrupt onset, and with a sudden generalized loss of tone, they cause the child to drop to the floor with a high risk of traumatic injury.

30.4.7 Tonic seizures

Tonic seizures involve sudden sustained stiffening of the whole body, associated with generalized fast activity on EEG. During sleep they may be extremely subtle, but if occurring whilst awake they may cause a drop to the floor whose considerable force leads to injury. Although the hallmark of certain distinctive epilepsy syndromes, such as Lennox–Gastaut (Section 30.5.5), such tonic seizures may be induced by medications including carbamazepine and phenytoin. Parents may have noted a characteristic noise, usually the result of forced expiration as muscles stiffen at the onset. Trembling movements may occur toward the end of tonic seizures leading to confusion with generalized tonic-clonic attacks.

30.4.8 Generalized tonic-clonic seizures

Characteristically these seizures involve stiffness, followed by clonic jerking of the limbs. In children this clonic jerking may be subtle, and simply considered to be a twitching, particularly of the face. In general tonic-clonic seizures last longer than tonic seizures. Seizures of focal onset in children may also progress to become secondarily generalized, a more common occurrence in children aged under two. In such cases the child may experience a warning prior to the manifestation of generalized stiffening, or may exhibit a prior attack typical of a focal seizure.

30.5 Epilepsy syndromes in childhood

30.5.1 The value of classification

The International League Against Epilepsy first proposed a classification of epilepsy, primarily by seizure type (1981) and subsequently revised according to epilepsy syndromes (Proposal for revised classification of epilepsies and epileptic syndromes 1989). A syndrome is defined as a cluster of symptoms and signs that define a unique epileptic condition. By definition this must include more than the seizure phenomena alone, and may include information from other investigations including as EEG and, increasingly, MRI. Many syndromes are age related, with many of the distinctive syndromes presenting in childhood, with characteristic seizure types and specific EEG features.

There are both advantages and disadvantages to using such a classification. Advantages include the ability to predict prognosis,

which is helpful to outline to the parents at the outset. In addition, there may be generally accepted optimal pharmacological approaches to specific syndromes, defining particular drug responses that may be more favourable, or indeed drugs that should be avoided (National Institute for Clinical Excellence 2004). Disadvantages include the possible operational difficulties in making the diagnosis of specific syndromes, and the academic controversy that surrounds the delineation of certain syndromes.

This original classification subdivided the epilepsies into those which were localization-related and those which were generalized, with further categories for those 'special syndromes' which were neither. This classification also depended on the terminologies of idiopathic, where there was presumed to be no underlying brain pathology, or symptomatic, where there was underlying brain pathology, and of cryptogenic, where underlying brain pathology was presumed but unproven. Although there has been criticism in such classification, studies have shown its broad utility in classifying children at presentation in all but 12 per cent of cases, and that such diagnoses remain durable over time (Berg et al. 2000). Nonetheless some of the terminology is recognized as confusing, for instance the distinction between symptomatic and cryptogenic. Furthermore, with time the list of syndromes has lengthened dramatically yet without clarifying the definition between the well-accepted syndromes and those which are less well defined.

Recognizing the problems of the existing classification, the International League against Epilepsy proposed a further classification in 2001 (Engel et al. 2001). This proposes an axis system, that starts with the diagnosis of an epileptic seizure, moves to that of epilepsy, and then to the syndromic diagnosis (Table 30.2). It disposes of the terminologies of simple or complex as well as cryptogenic, which it proposes replacing by presumed symptomatic, and suggests the term focal instead of partial. A full list of existing syndromes is provided (www.ILAE-epilepsy.org). It also enables aetiology to be included in the classification, as well as any comorbidity. Such a classification is easier to understand, and can be used as a basis for teaching basic diagnostic principles.

The majority of the syndromes of epilepsy are age related, and therefore are diagnosed in childhood. Diagnosing a syndrome has considerable advantages; it enables a prognosis to be outlined from the start so expectations can be outlined, and may define suitable management, and particularly treatment.

30.5.2 The concept of 'epileptic encephalopathy'

An epileptic encephalopathy is defined as a condition in which the epileptiform abnormalities themselves are believed to contribute to the progressive disturbance in cerebral function (Engel et al. 2001). It occurs relatively commonly in early-onset epilepsies, one study

Table 30.2 The International League against Epilepsy classification of epilepsies proposed in 2001. A full list of syndromes is available at www.ILAE-epilepsy.org

Axis 1	Is it a seizure?
Axis 2	Is it epilepsy?
Axis 3	If epilepsy, what epilepsy syndrome?
Axis 4	What is the underlying aetiology?
Axis 5	Are there additional impairments?

estimated that 39 per cent of 504 consecutive children who had their first epileptic seizure were between the ages of 28 days and 3 years (Dalla Bernardina *et al.* 1983). By definition, it is presumed that the pathophysiological mechanism is ongoing epileptic activity at a critical point of brain development, in view of which it has to be thought of as potentially, at least in part, reversible. Accordingly optimal if not complete seizure control should therefore be the aim in all young children with epilepsy. The syndromes listed within the International League against Epilepsy classification as epileptic encephalopathies include Infantile Spasms or West Syndrome, severe myoclonic epilepsy of infancy (Section 30.5.4), Lennox–Gastaut syndrome, Continuous Spike Wave of Slow Sleep, and Landau–Kleffner syndrome (Section 30.5.5). However any epilepsy presenting with aggressive seizure onset and cognitive decline, in the absence of a progressive aetiology, could be defined as an epileptic encephalopathy, including myoclonic astatic epilepsy and some symptomatic focal epilepsies.

30.5.3 Syndromes presenting neonatally

The majority of seizures presenting in the neonatal period are acutely symptomatic of underlying disorders (Section 30.4.1). In addition there are a small number of rare syndromes that commence in the neonatal period and which demand recognition.

Benign idiopathic neonatal convulsions

Otherwise known as fifth day fits, this disorder presents between day 4 and 6 in up to 90 per cent of cases. Seizures may be frequent, involve clonic movements with or without apnoea, and resolve spontaneously within a few days. Typically children will present with seizures of unknown origin and therefore be fully evaluated for an acute symptomatic cause; EEG has been reported to show a characteristic feature 'theta pointu alternant' or may be normal. The long-term outcome is debated; in one study 'abnormalities were reported in the long term in up to half suggesting not such a benign condition (Pryor *et al.* 1981; North *et al.* 1989). However in many studies follow-up is only of short duration and outcome cannot truly be evaluated (Plouin and Anderson 2005).

Benign familial neonatal convulsions

These have a slightly earlier onset on day 2–3. These children usually have a family history with evidence of autosomal dominant inheritance. Linkage has been demonstrated to potassium channel genes *KCNQ2* and *KCNQ3*. The EEG is often normal. The prevalence at 11 per cent of later epilepsy in this group is higher than expected amongst the normal population in limited studies (Plouin and Anderson 2005).

Pyridoxine-dependent seizures

These are a rare but treatable subgroup of neonatal seizures (Section 10.7.1 and Table 10.2, V2). Accordingly any child presenting with seizures within the first year of life should be considered for a trial of pyridoxine. In general this syndrome presents with generalized seizures soon after birth with associated encephalopathy involving hyperalertness, irritability, systemic features of thermal dysregulation, respiratory distress, and sometimes metabolic acidosis. Although traditionally diagnosed after a trial and ultimately trial withdrawal of pyridoxine the metabolic defect is now known to be a deficiency of brain α-aminoadipate semialdehyde dehydrogenase. Urinary α-aminoadipate semialdehyde may be diagnostic, and mutations of the aldehyde dehydrogenase, *ALDH, 7A1* gene on

chromosome 5q31 encoding antiquitin, or α-aminoadipic semialdehyde dehydrogenase, have recently been discovered as a major cause of pyridoxine-dependent epilepsy (Mills *et al.* 2006).

30.5.4 Syndromes presenting in infancy

Febrile convulsions

Febrile seizures are listed within the epilepsy classification, although by definition are single events in the majority. They may be defined as 'an event in infancy or childhood, usually occurring between 3 months and 5 years of age, associated with fever but without evidence of intracranial infection or defined cause' (National Institute of Health 1981). Ninety per cent occur before 3 years of age, the majority during the second year with a peak incidence between 18 and 24 months. The majority occur in isolation, lasting less than 10 min, and with no evidence of focality. Some children develop 'complex' febrile seizures, defined as a seizure lasting longer than 10 min, evidence of focality, or 2 or more seizures within 24 h or within the same febrile illness. There has been much discussion as to the relationship of early febrile seizures to the later development of epilepsy, particularly around the relationship of prolonged febrile seizures to the development of hippocampal sclerosis and temporal lobe epilepsy. The risk of developing epilepsy remains higher amongst those who have a complex febrile seizure (Verity et al. 1985). Adults coming to surgery for epilepsy from hippocampal sclerosis also have a higher prevalence of a history of febrile convulsions in childhood (Section 31.11.2). There is also evidence from animal studies and possibly from imaging, that this group run the risk of hippocampal damage (Liu *et al.* 1995; Scott *et al.* 2003). However, the genetic determinants remain complex, and other syndromes have been discovered where seizures are linked to fever. There is a risk of recurrence of febrile seizures in 30–40 per cent, a higher risk seen in those with early age of onset, family history, and a shorter duration of illness prior to the first seizure. There is no evidence that prophylactic medication prevents or alters outcome from febrile seizures, or that investigation outside looking for a source of infection is helpful. The latter is particularly important in children presenting under 12 months of age as intracranial infection may be masked. The main aim of management remains temperature control. A supply of emergency rescue medication may be indicated if there is a history of a prolonged seizure or a seizure recurrence during the same illness.

Benign partial epilepsy of infancy

At present this entity is usually diagnosed on retrospective review of the history, the response to anticonvulsants, and the developmental progress (Watanabe *et al.* 1990, 1993). The true prevalence of this epilepsy remains unknown since most children presenting in this age group are likely to have seizures that respond poorly to anticonvulsants and have a poor prognosis with regard to neurodevelopmental outcome. It is recognized that this category includes a small number of children who presented with focal seizures, that responded promptly to anticonvulsants, have a good prognosis for long-term seizure remission, and good developmental progress. Many have a familial component.

Neonatal or early myoclonic encephalopathy and early infantile epileptic encephalopathy or Ohtahara syndrome

Neonatal or early myoclonic epileptic encephalopathy and early infantile epileptic encephalopathy or Ohtahara syndrome are rare, present in the neonatal period or early infancy, and are both associated with a burst-suppression pattern on the EEG. Early myoclonic epileptic encephalopathy is characterized by fragmentary and erratic myoclonias as well as partial seizures. The main seizure types at the outset of early infantile epileptic encephalopathy are tonic spasms and partial seizures. Structural brain abnormalities appear to be more common in early infantile epileptic encephalopathy, whereas early myoclonic epileptic encephalopathy may be a manifestation of inborn errors of metabolism (Ohtahara and Yamatogi 2003). In both prognosis is poor with regard to seizure control and neurodevelopment although probably more so in early myoclonic epileptic encephalopathy with at least 50 per cent mortality before 12 months. Some cases of early infantile epileptic encephalopathy show a characteristic evolution with age to West and Lennox–Gestaut syndromes (Section 30.5.5) suggesting that early infantile epileptic encephalopathy is an age-dependent epileptic encephalopathy.

Possible metabolic causes of epilepsy are cited in Tables 10.2 V and 30.1 and Section 10.7.1. Two are of particular note:

- *Biotinidase deficiency* is a treatable cause which should not be overlooked. Biotinidase is required for the use of biotin (Section 10.7.1). An autosomal recessively inherited deficiency of this enzyme results in complex changes, due to the secondary effects upon other biotin-dependent enzymes. Thus biotinidase deficiency will manifest as the consequences of biotin deficiency, the extent of which will depend on the duration of the condition prior to its recognition. Children present with cutaneous and neurological symptoms (Bartlett 2001). The most common neurological symptoms at presentation are seizures and alopecia, dermatitis, and hypotonia may be associated. In children who present with seizures in the first year of life with unclear aetiology, a trial of biotin should be considered until the results of biotinidase assay become available.

- *Sulphite oxidase deficiency.* This may be an isolated enzyme deficiency or the clinically identical molybdenum cofactor deficiency which combines the deficiencies of sulphite oxidase and xanthine dehydrogenase. It is a rare inborn error of metabolism associated with lens dislocation and a severe neurological picture, including seizures, developmental delay, and abnormalities of muscle tone and microcephaly (Johnson and Wadman 1995) (Section 10.7.1). In particular there is a high incidence of neonatal seizures (Slot *et al.* 1993). Later onset may occur with considerable phenotypic variability (Hughes *et al.* 1998). Screening for the disorder is by testing for sulphite in a fresh urine sample.

Migrating partial seizures of infancy

This syndrome starts in the first year of life and has an extremely poor prognosis for control of seizures and developmental outcome. It is likely to have multiple aetiologies. The initial presentation suggests an early onset of focal epilepsy, relatively resistant to anticonvulsant medication. As the first year evolves, these seizures of initially focal onset show evidence of 'migration' to the contralateral hemisphere intra-ictally, as judged by clinical or EEG criteria (Fig. 30.4). Other features are recurrent status epilepticus, and proneness to seizures with intercurrent illness. The prognosis is extremely poor both for developmental outcome and mortality, with three of the original series of 14 not surviving (Coppola *et al.* 1995).

West syndrome and infantile spasms

The triad of infantile spasms, hypsarrhythmia, and developmental plateau is termed West Syndrome, originally described by the paediatrician Charles West in his own son in the 19th century. However the presentation of typical spasms despite the absence of true hypsarrhythmia does not exclude the diagnosis and such children should receive the same treatment relatively urgently. The described EEG pattern is a supposed interictal pattern observed mainly in the awake state. Typical hypsarrhythmia is present mainly during the early stages of the disorder and preceding the clinical manifestations by weeks. However a profoundly abnormal EEG resembling hypsaarrhythmia (Fig. 30.3) may be seen with some global developmental malformations such as lissencephaly (Section 9.2.5) and is unlikely to change with treatment.

Given the highly disorganized nature of the EEG it is not surprising that the affected child's awareness is likely to be impaired. Developmental slowdown may precede the spasms, and is an almost invariable finding at their onset, occurring in 68–85 per cent (Matsumoto *et al.* 1981; Riikonen 1984). Autistic withdrawal is not uncommon at onset and persists in a high proportion of children. The prognosis is strongly determined by the underlying aetiology and some degree of learning impairment is seen in up to 90 per cent, with only 18 per cent achieving a longitudinal intelligence quotient of >51 (Riikonen 1982). The evidence suggests that a better prognosis is related to earlier treatment of spasms, and in particular a shorter duration of hypsarrhythmia (Rener-Primec *et al.* 2006).

Diagnosing infantile spasms is important because it justifies early aggressive treatment. The treatment of choice remains between vigabatrin and steroids. Protocols differ in dosing regimes as well as in the duration of treatment courses depending on geographical location and availability of these drugs. Vigabatrin is not licensed in the USA, natural ACTH has been replaced by synthetic ACTH in some European countries and Japan, and different types of oral steroids are in use. Despite a large number of published trials investigating efficacy and tolerability of vigabatrin, glucocorticosteroids, ACTH, and other agents for the treatment of infantile spasms, only limited conclusions can be drawn due to differences in design and treatment protocols as well as poor overall quality of most randomized studies (Hancock *et al.* 2003; Mackay *et al.* 2004). Vigabatrin remains the medication of choice if infantile spasms results from tuberous sclerosis (Section 11.1) (Hancock *et al.* 2003). A recent multicentre randomized controlled trial compared the relative effectiveness of Tetracosactide, a synthetic ACTH, and Prednisolone versus vigabatrin for the short-term treatment, of cryptogenic and symptomatic infantile spasms excluding patients with tuberous sclerosis (Lux *et al.* 2004). It was observed that 78 per cent of patients enrolled to Prednisolone and Tracosactide ceased spasms by day 14 of treatment versus 54 per cent on vigabatrin (mean difference 19 per cent, 95 per cent CI 1–36 per cent, $P=.045$). After a year there was no difference between the treatment groups with respect to proportion of patients with absence of spasms, 75 per cent, and seizure-free patients, 57 per cent (Lux *et al.* 2005). More robust evidence is required to make a clear recommendation on the best treatment of infantile spasms. The choice remains between steroids and vigabatrin as first-line therapy and this may be influenced by the aetiology, and preference of the physician and parents.

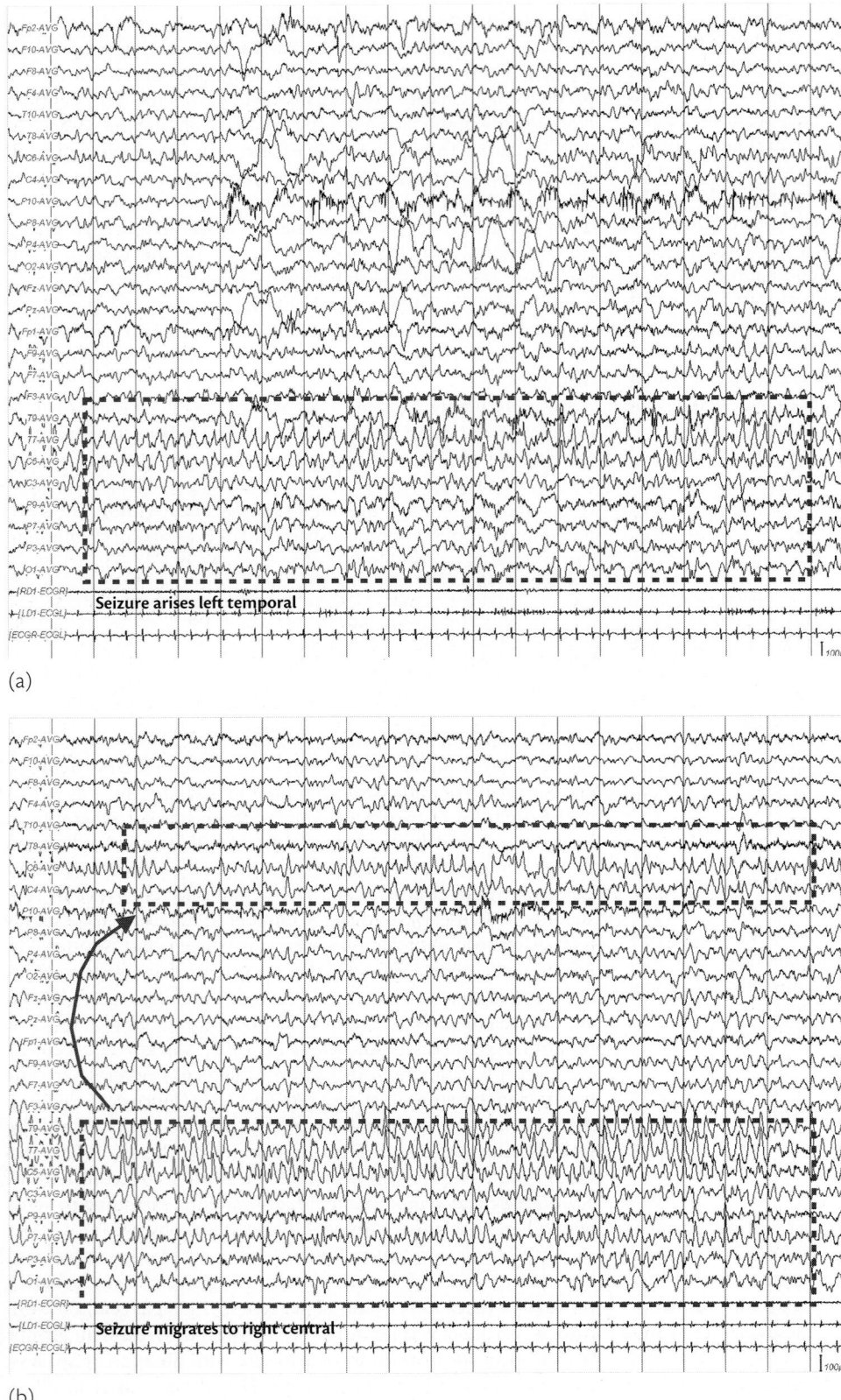

(a)

(b)

Fig. 30.4 Migrating partial seizures of infancy. Suggestive EEG in a 14-month-old child. The seizure appears to start in the left temporal region (a) but switches to the contralateral side intraictally (b).

Severe myoclonic epilepsy of infancy, Dravets syndrome

This has become a well-recognized clinical syndrome shown to be associated with a sodium channel gene mutation, *SCN1A*, on chromosome 2 (Harkin *et al.* 2007). A typical course will include presentation with prolonged and probably febrile lateralized seizures in the first year with normal early development. This is followed in the second decade by the development of other seizure types, including myoclonic jerks, as well as developmental plateauing (Dravet *et al.* 1992). The early EEG may be normal; 40 per cent show some degree of photosensitivity that may resolve. Sometimes the condition starts relatively later, or has other atypical features, and has been termed severe myoclonic epilepsy borderline (Harkin *et al.* 2007). Up to 80 per cent have *SCN1A* gene mutations. Phenotypic correlation with specific mutations has been attempted and some of the variability may reflect mosaicism (Marini *et al.* 2006). Although seizures may be resistant to medication, and children may continue to be troubled by recurrent status epilepticus, particular responsiveness has been demonstrated to sodium valproate, clobazam, stiripentol (Chiron *et al.* 2000), and topiramate (Ceulemans *et al.* 2004). Certain medications, namely lamotrigine, need to be avoided as they may exacerbate seizures (Guerrini *et al.* 1998).

Generalized epilepsy with febrile seizures plus

This syndrome is characterized by heterogeneous epilepsies including febrile seizures and mild to severe generalized epilepsies (Scheffer and Berkovic 1997). The most common phenotypes are febrile seizures and febrile seizures plus, in which seizures persist beyond 6 years of age. The most severe phenotypes include febrile seizures plus with absences, myoclonus, or atonic seizures. The genetics are important as they link febrile seizures with epilepsy. Mutations in sodium channels genes, *SCN1A* and *SCN1B*, and GABA$_A$ receptor subunit genes, *GABRG2* and *GABRD*, have been described in different pedigrees (Scheffer and Berkovic 2003).

Symptomatic focal epilepsy

Children with structural lesions of the brain may present with devastating onset of focal epilepsy in infancy, sometimes associated with an apparent epileptic encephalopathy. All children presenting with epilepsy in association with a unilateral structural lesion should be referred to an epilepsy surgery programme for evaluation at an early stage given the likelihood they will not respond to anticonvulsant medication; surgery should be considered early on to maximize their developmental potential (Cross *et al.* 2006). Furthermore, children presenting apparent focal epilepsy without an aetiology evident on MRI should be evaluated in detail given that lesions may be difficult to detect on initial imaging, or even become less apparent with normal myelination of the brain (Eltze *et al.* 2005).

30.5.5 Syndromes presenting in early childhood

Benign epilepsy with occipital spikes

Although initially identified as an early-onset epilepsy with occipital spikes or occipital paroxysms, with or without 'fixation-off sensitivity', it is now recognized that it is more akin to an 'autonomic' epilepsy, and is commonly called Panayiotopoulos syndrome (Panayiotopoulos *et al.* 1993). Children may present with only a single or with recurrent seizures whose semeiology is typically a change of awareness, and vomiting with or without other autonomic manifestations such as flushing or pallor. The age of onset is usually 3–6 years. Such episodes may be prolonged and perhaps misdiagnosed as a non-epileptic encephalopathy. Occipital spikes may be seen on EEG in up to 40 per cent. The prognosis is excellent with most children having only 1–3 seizures, with remission occurring 1–2 years from onset. Thus the need for treatment needs to be carefully considered, although most will respond promptly to carbamazepine. Brain MRI is advisable as structural occipital lesions may present with similar seizure semeiology.

Myoclonic astatic epilepsy

This syndrome is thought to be an idiopathic epilepsy (Doose *et al.* 1970). It is characterized by the onset of generalized tonic clonic seizures and later myoclonic, astatic, and myoclonic-astatic seizures causing drop attacks. Seizures may occur very frequently, and children may be prone to periods of nonconvulsive status epilepticus. Although not classed as one of the epileptic encephalopathies, affected children may be obtunded during periods of very frequent seizures. Although seizures may be frequent and difficult to control in the early stages, this eases after 1–3 years. Those affected by frequent drop attacks may have been classified previously as having Lennox–Gastaut Syndrome. The prognosis for long-term cognitive outcome remains relatively better than for the other early-onset seizure syndromes with 75 per cent achieving the normal range (Guerrini *et al.* 2002). Poor prognostic indicators include frequent episodes of nonconvulsive status epilepticus or the development of tonic seizures. The anticonvulsants most likely to be of benefit include sodium valproate, with or without lamotrigine, benzodiazepines, ethosuximide (Doose *et al.* 1987), and levetiracetam (Labate *et al.* 2006). There is also anecdotal evidence that these children may be particularly responsive to the ketogenic diet (Section 30.7.4) (Oguni *et al.* 2002) and this should receive early consideration.

Lennox–Gastaut syndrome

One of the epileptic encephalopathies, this syndrome may present first in early childhood with multiple seizure types, or may evolve in a child who has previously presented with West syndrome (Section 30.5.4). It was originally described by Gastaut as an epileptic encephalopathy with diffuse spike and wave complexes and multiple types of attack, including tonic seizures (Gastaut *et al.* 1966). Although the term has been broadened often to describe difficult epilepsies of childhood including drop attacks, it should be defined by a triad of symptoms including multiple seizure types, slow spike wave complexes on EEG, and impairment of cognitive function. Tonic seizures occur in a high percentage of patients with slow spike and wave complexes (Gastaut *et al.* 1966; Chevrie and Aicardi 1972) and their documented presence is a prerequisite for a diagnosis of Lennox–Gastaut syndrome. Atypical absence and atonic or tonic falls are also seen in the majority. Generalized paroxysmal fast activity is the second EEG hallmark of Lennox-Gastaut syndrome, with or without seizures, and a period of sleep recording may be required to reveal their presence. Effective treatment options for the multiple seizures and comorbidities of Lennox–Gastaut syndrome remain limited and long-term prognosis is poor for many patients. The newer agents topiramate, lamotrigine, and felbamate have been shown to be effective in randomized controlled trials (Hancock and Cross 2003).

Landau–Kleffner syndrome and continuous spike wave of slow sleep

These two syndromes, although often listed together are not synonymous. Continuous spike wave of slow sleep is a syndrome characterized by neuropsychological and behavioural change temporally related to the development of near continuous spike wave activity on EEG during slow wave sleep. It was first described as 'subclinical status epilepticus' induced by sleep in children (Patry *et al.* 1971). Although the terms continuous spike wave of slow sleep and electrical status epilepticus of slow sleep are used synonymously, some investigators have proposed that electrical status epilepticus of slow sleep should be used to designate the EEG abnormalities and continuous spike wave of slow sleep syndrome for the combined electroclinical picture (Galanopolou *et al.* 2000). Initially a proportion of 85 per cent of slow wave sleep needed to be occupied by spike and wave activity as a criterion of diagnosis, but several authors now accept a lower proportion. Clinical associations include global or selective neuropsychological impairments such as acquired aphasia, or dyspraxia. Motor features include ataxia, dystonia, or unilateral deficits. The seizure types vary with focal and or apparently generalized seizures, tonic clonic seizures, absences, partial motor seizures, complex partial seizures, or epileptic falls; however tonic seizures are never seen.

Landau–Kleffner syndrome is the condition of acquired epileptic aphasia. It is characterized by acquired language regression associated with some seizures, but more importantly an epileptiform EEG pattern, often bitemporal but in some cases typical of electrical status epilepticus of slow sleep. Whilst the clinical presentation is diverse, the treatment may remain similar, with steroids and benzodiazepines being most likely to be beneficial. Lamotrigine and levetiracetam have been reported to be effective. The prognosis regarding both seizures and the EEG abnormality remains good in both conditions, although the cognitive and language is relatively unpredictable and may relate to early treatment success.

30.5.6 Syndromes presenting in mid-childhood

Benign epilepsy with centrotemporal spikes

This is the most common epilepsy presenting in childhood. Typically children present aged between 5 and 10 years with nocturnal seizures, either focally involving the mouth and face, or generalized. This benign condition carries an excellent prognosis, with all children reported to enter remission by 14 years of age (Loiseau *et al.* 1988). Some have deemed treatment unnecessary in view of the guaranteed remission. Despite the good prognosis for remission, there is still the risk of recurrent generalized seizures, which carry the risk of morbidity, and 10–20 per cent experience seizures exclusively from the awake state (Lerman 1985). In addition some report that benign epilepsy with centrotemporal spikes may not be so benign because of associated cognitive and behavioural effects which may be related to the frequency of seizures, therefore representing an epileptiform phenomenon (Metz Lutz *et al.* 1999). Thus the decision to treat will depend on the degree to which the seizures are interfering with the child's life. The treating physician should decide whether treatment is desired together with the family and child bearing in mind that there is likely to be an excellent response to antiepileptic drugs and that these will not be needed long term. The medication of choice is carbamazepine with prompt seizure control in 80 per cent of patients.

A small group of patients presenting similarly show some atypical features. A progressive course with cognitive regression is associated with progression to electrical status epilepticus of slow sleep. Caution is required when initiating carbamazepine in such patients as it may exacerbate electrical status epilepticus of slow sleep. Nonetheless, the vast majority of children presenting with the features of benign epilepsy with centrotemporal spikes are likely to have a benign course, and any unexpected deterioration on treatment requires further evaluation by sleep EEG and a change of medication should be considered. Some consider that if an initial sleep EEG recording has demonstrated very frequent discharges and there is concern about cognition, an alternative medication such as valproate is indicated.

Idiopathic childhood occipital epilepsy, Gastaut type

This has a later mean age of onset than the benign epilepsy with occipital spikes of Panayiotopoulos (Section 30.5.5). The seizures almost always involve visual phenomena at the outset, which may be well described and drawn by the child. The seizures may occur in isolation, or proceed to secondary generalization with hemiclonic seizures in about 40 per cent. One-third of the original series of patients had a severe postictal headache, at times accompanied by nausea or vomiting (Gastaut *et al.* 1992). Therefore the condition may be confused with migraine. The interictal EEG demonstrates unilateral or bilateral paroxysmal posterior spike waves on eye closure 'fixation off sensitivity', as does the benign epilepsy with occipital spikes in younger children (Section 30.5.5). Full remission by late teenage occurs in the majority.

Childhood absence epilepsy

This syndrome accounts for about 8 per cent of epilepsy in school age children (Cavazzuti 1980). Children present with brief episodes of behavioural arrest, of which the EEG correlate is a 3 Hz spike and wave (Fig. 30.5). Attacks are unprovoked, and occur in any situation. They may be provoked in clinic or during EEG recordings by hyperventilation. The syndrome is almost certainly genetically determined, with 75 per cent concordance for monozygotic twins (Hauser and Anderson 1986). The prognosis is excellent with up to 80 per cent achieving long-term remission. The children with the best outlook are those with absences commencing early in childhood, whose absences are not associated with myoclonus and who do not have tonic-clonic seizures preceding or coinciding with the onset of absences. Medications that may be effective include sodium valproate, lamotrigine, and ethosuximide. The absence attacks may be exacerbated by medications such as carbamazepine and phenytoin.

Juvenile absence epilepsy

Children present with a later onset of absences than childhood absence epilepsy, but with a slightly faster frequency of slow wave activity (Section 31.5.1).

Other absence epilepsy syndromes

Myoclonic absence epilepsy. This may be confused at presentation with childhood absence epilepsy, especially since the EEG abnormality is also likely to be a 3 Hz spike-wave discharge. However attacks are characterized by rhythmic jerks predominantly of the upper limbs that also occur at 3 per second. Small longitudinal series suggest that learning difficulty is common in this group, even prior to presentation. The prognosis for seizure control is guarded; around 50 per cent go into remission but this is less likely if generalized tonic clonic seizures occur (Tassinari *et al.* 1995). The response to anticonvulsant medication is variable; good response has been seen with valproate or ethosuximide (Tassinari *et al.* 1995)

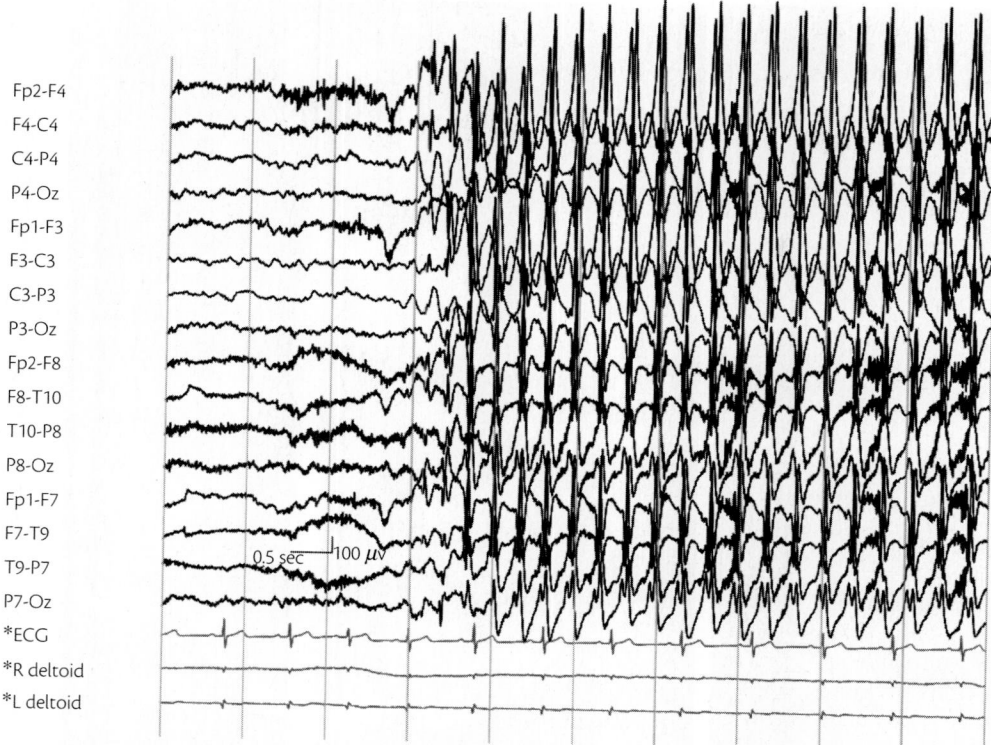

Fig. 30.5 Childhood absence epilepsy. EEG showing the characteristic 3 Hz spike and wave discharge.

or with either combined with lamotrigine (Manonmani and Wallace 1994).

Eyelid myoclonus associated with absences. Also known as eyelid myoclonia with absences, this was first reported by Jeavons (1977), with subsequent extensive description by Giannakodimos and Panayiotopoulos (1996). The onset is in childhood and the seizures begin with rhythmic fast brief jerking of the eyelids and upward jerking of the eyeballs, followed by brief mild absence. The attacks occur mainly on closing the eyes. The EEG shows 3–6 Hz generalized polyspike and wave, and all patients are photosensitive. Most patients experience occasional tonic-clonic seizures on sleep deprivation. Treatment is as for other idiopathic generalized epilepsies using sodium valproate either alone or in combination with ethosuximide or clonazepam. Although the photosensitivity may disappear with age, the eyelid myoclonia persists and remains resistant to treatment even if absences appear controlled.

Perioral myoclonia with absence. Myoclonus of the lower face may be associated with absence attacks. This involves rhythmic contractions causing protrusion of the lips and twitching of the corners of the mouth associated with a mild absence (Panyiotopoulos *et al.* 1995). These seizures may build up to absence status and then to a tonic-clonic seizure. Although similar medications should be trialled as in childhood absence epilepsy, this syndrome may not respond well to treatment.

Rasmussen's encephalitis

Rasmussen syndrome is a progressive condition that most commonly presents between 5 and 10 years, although may occur outside these age limits, even in adulthood. Classically children present with a focal epilepsy that becomes increasingly difficult to treat. Commonly this evolves to a hemiepilepsy, 60 per cent experiencing epilepsia partialis continua at some time in their clinical course. This often heralds an increasing hemiparesis. Hemiatrophy is seen of the contralateral cerebral hemisphere (Bien *et al.* 2005) (Fig. 30.6).

The term 'encephalitis' originates from the original surgical studies by Rasmussen who on performing hemispherectomy on some of these individuals noted that they had pathological evidence for a chronic encephalitis of no apparent aetiology. The evidence now points to a causative autoimmune process. Original studies suggested humoural immunity with a role for GluR3 antibodies, but these have been shown to be non-specific and cellular immunity involving T-cell dysfunction is now considered likely (Li *et al.* 1997; Bauer *et al.* 2002; Watson *et al.* 2004). The mechanism by which such immune disorders produce a uni-hemisphere abnormality remains unexplained.

Despite variable responses to steroids and intravenous immunoglobulin (Hart *et al.* 1994), the only curative treatment available remains hemispherectomy. Although this abolishes seizures it necessarily results in hemiparesis and hemianopia. Detailed presurgical assessment is required for such patients to enable optimal planning of any surgical procedure, particularly given the risk of cognitive decline and the influence of cerebral dominance. The decision about when to operate needs to balance the development of cognitive decline against the risk of functional deficits should surgery be undertaken. The key determinant is language function, although some degree of transfer and recovery may be seen even after relatively late onset disease and surgery (Boatman *et al.* 1999).

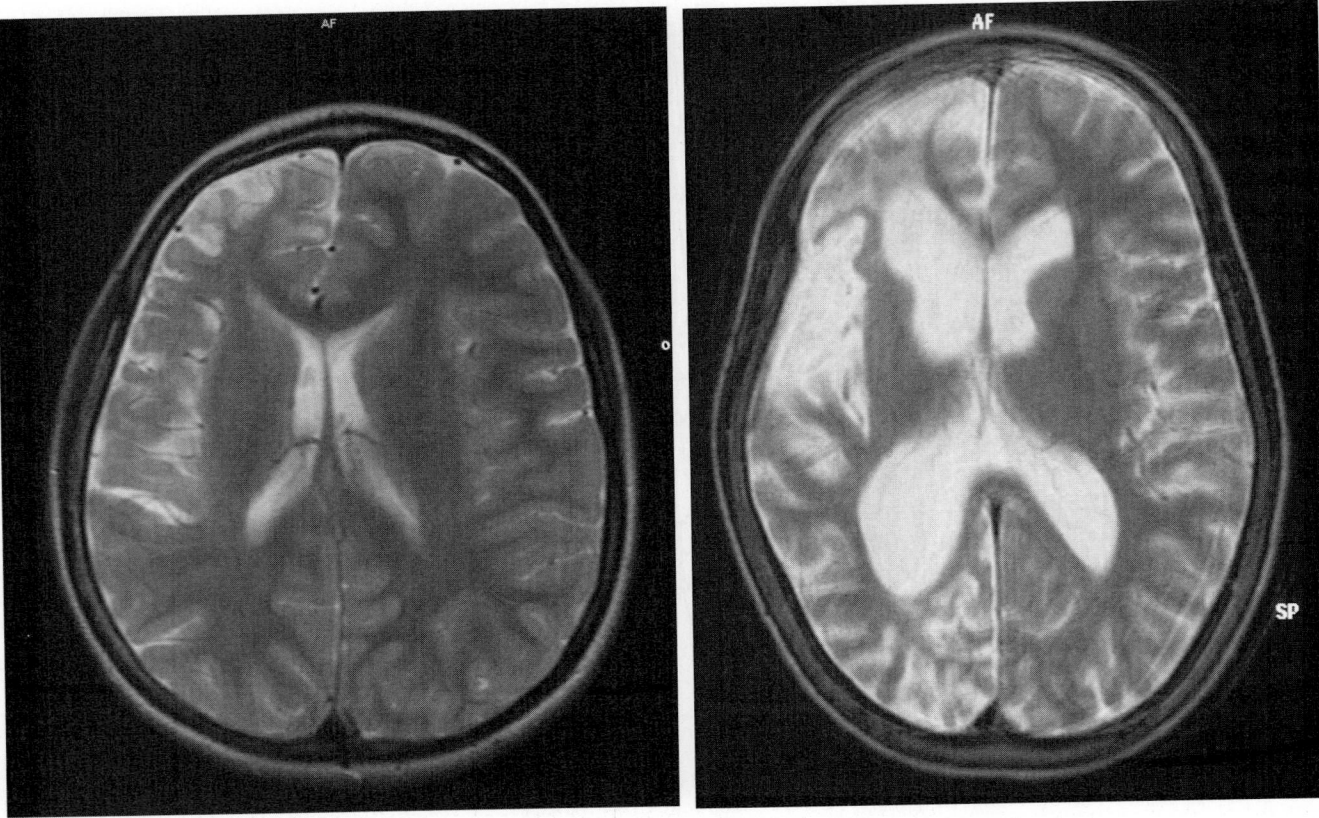

Age 8 years   Age 12 years

Fig. 30.6 Rasmussen's syndrome. Progressive right cerebral atrophy demonstrated on MR scans 4 years apart in a child with progressive left focal motor seizures and hemiparesis.

30.5.7 **Syndromes presenting in late childhood**

Many of the idiopathic generalized epilepsies persisting into adulthood may originally present in mid- to late childhood. Juvenile absence epilepsy (Section 31.5.1) is likely to have a slightly later onset than childhood absence epilepsy (Section 30.5.6), and the spike wave activity is of slightly faster frequency at 4–5 Hz. The prognostic implication is that although individuals may respond promptly to anticonvulsants, they are unlikely to wean successfully from the medication. Furthermore some patients may present as the onset of juvenile myoclonic epilepsy (Section 31.5.1), with the myoclonus and generalized tonic-clonic seizures only manifesting at a later stage.

30.5.8 **Epilepsy syndromes into adulthood**

It is estimated that 65 per cent of epilepsy that starts in childhood persists into adulthood. There are particular syndromes where this is likely. Of the idiopathic group this includes the later onset idiopathic generalized epilepsies, as well as the benign epilepsies with occipital paroxysms and the Lennox–Gastaut syndrome (Section 30.5.5). In absence epilepsies it can be difficult to tease out those risk factors that may make an individual less likely to grow out of epilepsy. The defining features, particular those that may indicate poor prognosis are outlined in Table 30.3. These include a very

early or a later age of onset, prolonged duration of attacks, and occurrence of generalized tonic-clonic convulsions.

30.6 **The causes of childhood epilepsy**

30.6.1 **Genetics of the epilepsies**

Studies over the past 10 years have revealed a central role for genetically determined ion channel abnormalities in the pathophysiology of the idiopathic epilepsies. However many of these epilepsies follow complex inheritance and do not have single gene association. Epilepsies with monogenic inheritance are uncommon and are associated with mutations in genes that encode subunits of voltage-gated and ligand-gated channels (Table 30.4). Whereas in some inheritance appears autosomal dominant, in others inheritance is more complex and genotype–phenotype correlation is not clear. Nonetheless, the discovery of such genetic associations has improved our understanding of these epilepsies, and sometimes contributed to management. Understanding will increase in the future although at present many genetic studies still remain only available on a research basis.

It is imperative to determine family history on evaluating a child initially presenting with epilepsy since the genetics can provide an important clue to the diagnosis. However the exact nature of the

Table 30.3 Differential diagnosis of childhood absence epilepsies

	Childhood absence epilepsy (Section 30.5.6)	Juvenile absence epilepsy (Section 31.5.1)	Absence with eyelid myoclonus (Section 30.5.6)	Myoclonic absence epilepsy (Section 30.5.6)
Age of onset (usual)	3–12 years (6–7 years)	7–16 years (10–12 years)	2–5 years	11 months–12 years (7 years)
AS duration	<10 s	>15 s	3–6 s	10–60 s
AS frequency	Many/day	1–10/day	Many/day	Many/day
Other seizures	Rare 40% GTCS	Common GTCS/myoclonus	Common GTCS (photosensitive)	Myoclonus with absence (3 Hz)
EEG	3 Hz spike-wave	3.5–4 Hz spike-wave	3–6 Hz spike-wave	3 Hz spike-wave
Prognosis	6% persist	Lifelong	Lifelong	>50% persist > 10 years

GTCS = Generalized tonic-clonic seizures

AS = Absence seizure

reported episodes is just as important and may be described by older members of the family.

30.6.2 Chromosomal disorders

There are many genetically determined disorders of which epilepsy is part of the symptom presentation. In children with chromosomal anomalies there is increasing recognition that epilepsy is a frequent and significant part of the clinical problem. Furthermore, underlying chromosomal anomalies are found in an increasing number of children where epilepsy had been the manifesting feature of an otherwise undetermined disorder. It is likely that loss of or abnormal functioning of genetic loci on involved chromosomes involves mechanisms leading to increased seizure susceptibility, cortical excitability, or changes in neurotransmitter functions. Alternatively epilepsy may originate in structural brain abnormalities occurring in association with the chromosomal abnormality.

Although ranging in clinical severity epilepsies occurring in the setting of chromosomal anomalies are generally difficult to treat. In a few, such as Angelman syndrome, the epilepsy or the EEG pattern is characteristic or very suggestive of that chromosomal anomaly. However in the majority of disorders this is not so although our understanding of karyotype–phenotype relationships continues to develop. Increasing knowledge of the EEG patterns and epilepsies associated with particular chromosomal disorders may contribute to diagnosis. Also the presence of a chromosomal anomaly may suggest target genes responsible for seizure pathogenesis.

Angelman syndrome

Angelman syndrome occurs in 1 of 2000–12 000 in the general population and may account for up to 6 per cent of cases of severe mental retardation and epilepsy. The genetic abnormality in the majority of cases is a large deletion of the maternally inherited chromosome 15q11-13. Yet deletion of this same region of the paternally inherited chromosome also results in Prader–Willi syndrome in which seizures are uncommon (Magenis *et al.* 1990). Epilepsy is very common in Angelman syndrome, reportedly occurring in up to 90 per cent, with onset usually between 18 and 24 months of age, often with convulsions and fever (Viani *et al.* 1995). Tonic-clonic, atypical absences, complex partial, myoclonic, tonic, and atonic have all been reported, often occurring in clusters. They may become quiescent in later childhood and only emerge again in adulthood (Laan *et al.* 1996). Status epilepticus of generalized tonic-clonic, absence, and myoclonic forms all occur. In particular nonconvulsive status whether myoclonic or atypical absence, may go unrecognized in the context of the associated developmental delay despite treatment potentially leading to significant benefit (Viani *et al.* 1995) The EEG is characteristic in Angelman syndrome (Fig. 30.7):

◆ Most commonly early symmetrical persistent rhythmic activity at 4–6 per second not associated with drowsiness (Laan *et al.* 1996);

◆ More persistent posterior sharp theta with small spikes presenting into adulthood, either seen spontaneously or brought on by eye closure (Boyd *et al.* 1988; Kette *et al.* 2003);

◆ Anterior rhythmic activity is most frequently reported. Variants of this 'delta pattern' have been characterized: hypsarrhythmic-like; ill-defined slow spike and wave variant; triphasic-like variant and slow variant (Kette *et al.* 2003). Persistence of anterior rhythmic slow triphasic waves is noted in adulthood (Laan *et al.* 1996).

Table 30.4 Genetically determined epilepsies due to mutations of ion channel genes

	Mode of inheritance	Type/channel	Gene
Benign neonatal familial convulsions	Autosomal dominant	Ion/Potassium	KNCQ2/3
Benign familial neonatal infantile seizures	Autosomal dominant	Ion/sodium	SCN2A
Severe myoclonic epilepsy of infancy	Complex	Ion/sodium	SCN1A
Generalized epilepsy with febrile seizures plus	Complex	Ion/Sodium Ligand/GABA	SCN1A SCN1B GABRG2
Autosomal dominant frontal lobe epilepsy	Autosomal dominant	Ligand/Nicotinic Acetylcholine receptor	CHRNA4

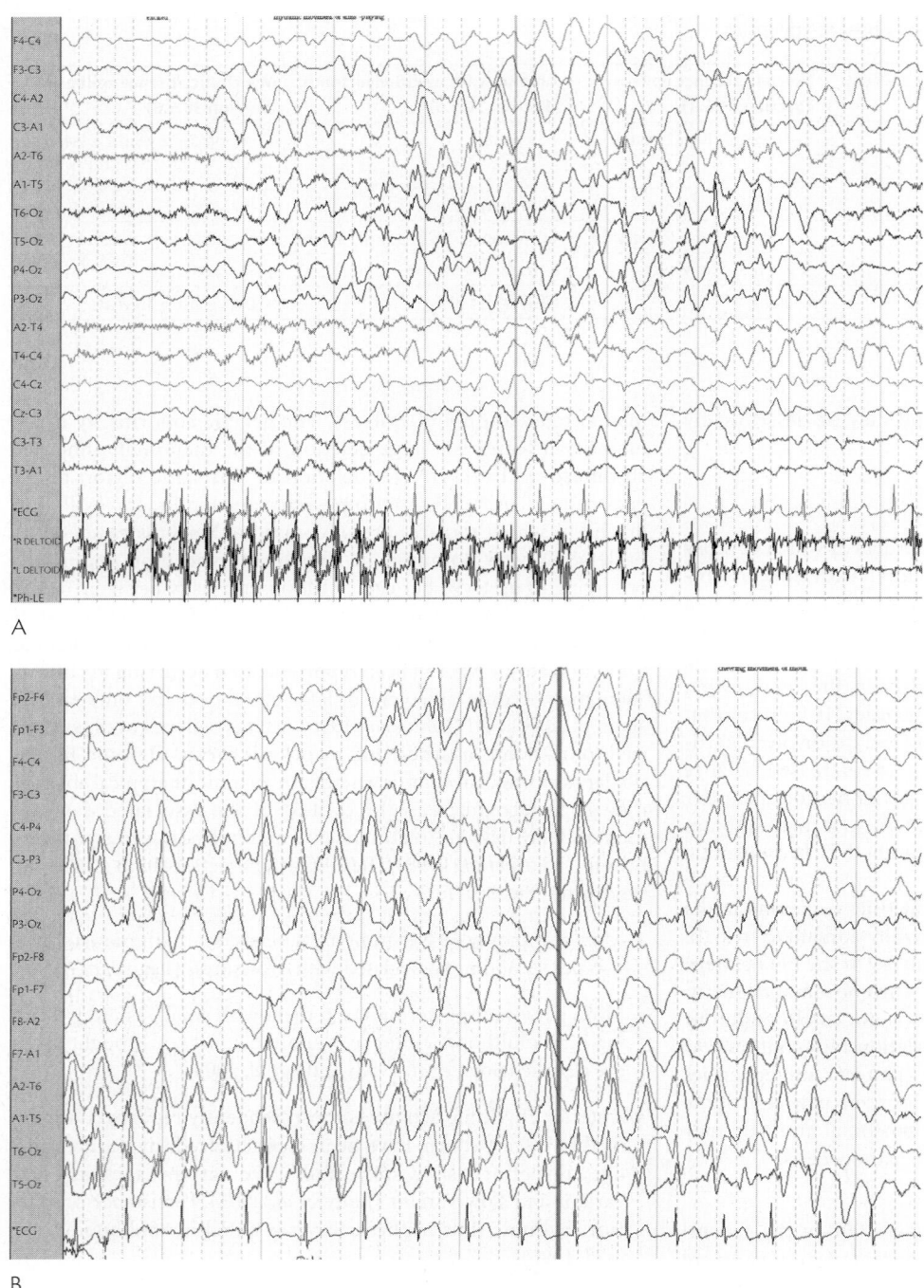

Fig. 30.7 Angelman's syndrome. Characteristic findings on EEG. (A) Persistent rhythmic 4–6/s activities often reaching more than 200 mV not associated with drowsiness. (B) Prolonged runs of rhythmic 2–3/s activity sometimes associated with discharges forming ill-defined spike wave complexes.

These EEG features may be present at different times in the same patient and suggest the diagnosis of Angelman syndrome before emergence of the full clinical phenotype. The presence of all three EEG features in the setting of developmental delay provides strong evidence for Angelman syndrome.

Cortical myoclonus is particularly well described as responding to piracetam (Guerrini *et al.* 1996). Initially the seizures are particularly resistant to treatment, but this improves in later childhood. Sodium valproate has been noted to be useful either with or without a benzodiazepine (Viani *et al.* 1995).

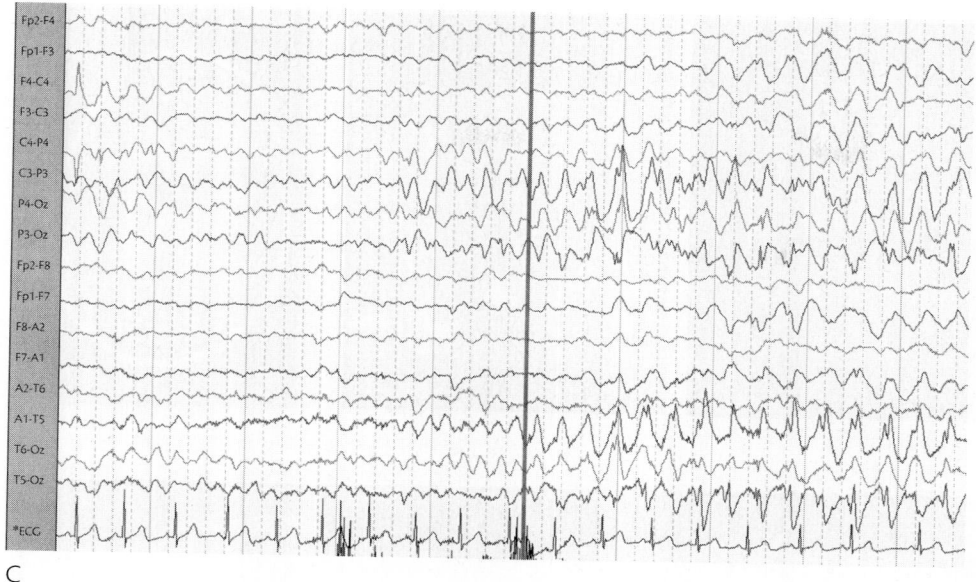

C

Fig. 30.7 (Continued) (C) Spikes mixed with 3–4/s components reaching 200 μV, mainly posteriorly, facilitated by eye closure.

Topiramate and phenobarbitone may also be effective (Franz *et al.* 2000).

Miller–Dieker syndrome

This syndrome includes lissencephaly (Section 9.2.5) in association with a characteristic facies of prominent forehead, bitemporal hollowing, a short nose, and protuberant upper lip. It results from deletions or rearrangements of 17p13.3, involving the *LIS-1* gene. The EEG may show widespread high amplitude spike and slow wave discharges reminiscent of hypsarrhythmia. Later the EEG may show the widespread fast activity commonly seen in association with such brain malformations. As a result of the lissencephaly, the epilepsy will present early and may prove very resistant to medication.

Ring chromosome abnormalities

There are increasing reports of epilepsies associated with ring chromosome abnormalities. In the majority these are not associated with the dysmorphic features suggestive of a karyotype abnormality.

The ring abnormality of *chromosome 20* is that most commonly reported. In 50 per cent of such children epilepsy presents before the age of 6 years and characteristically the seizures have a focal nature with visual symptoms, nocturnal tonic seizures, or arousals with a frontal semeiology often being reported. The patients are prone to nonconvulsive status epilepticus. The EEG may manifest as continuous bifrontal rhythmic theta/delta waves with accompanying spikes or sharp waves or may show continuous diffuse abnormalities (Ville *et al.* 2006). Typically the response to medication is poor. Furthermore the rate of behavioural disorder is high, with poor attention and concentration, impulsivity, disinhibition, obsessive behaviours, and aggressive outbursts being common. In addition, patients may acquire cognitive difficulties over time.

Patients with ring *chromosome 14* present with epilepsy of earlier onset, severe to profound learning difficulties, speech impairment, microcephaly, and dysmorphism, particularly with ocular abnormalties (Schmidt *et al.* 1981).

30.6.3 Structural causes

Malformations of cortical development, ischaemic lesions, and tumours remain the most common structural causes of epilepsy in childhood. MRI has greatly enhanced the sensitivity of detecting these abnormalities, leading to further insights into the causes of childhood epilepsy. Brain malformations pose a high risk of drug-resistant epilepsy (Fig. 30.8). Although the prevalence of malformations is low in a newly diagnosed population, it rises as drug-resistant populations are reviewed and ultimately is highest amongst those being assessed for surgery. Classifications of malformations have been founded on neuroimaging and histological criteria, and more recently there has been an increasing contribution from genetics with the discovery of causal mutations. A summary of the most recent proposed classification scheme is summarized in Table 30.5 (Barkovich *et al.* 2005). The full range of disorders of brain segmentation, of neuronal and glial proliferation, and of neuronal migration are discussed in Sections 9.2.3, 9.2.4, and 9.2.5 respectively.

30.7 Management of epilepsy in childhood

30.7.1 General principles

At the initial diagnosis it is important to discuss all aspects of diagnosis and management. This includes basic principles of management, such as advice on safety in the event of the child having a seizure. Children should be able to participate in as many activities as possible, although with precautions. Climbing is likely to be dangerous and therefore not advised although other activities are possible with precautions—for example swimming under one-to-one supervision or cycling provided not in busy traffic. In an older child one should consider discussing lifestyle issues

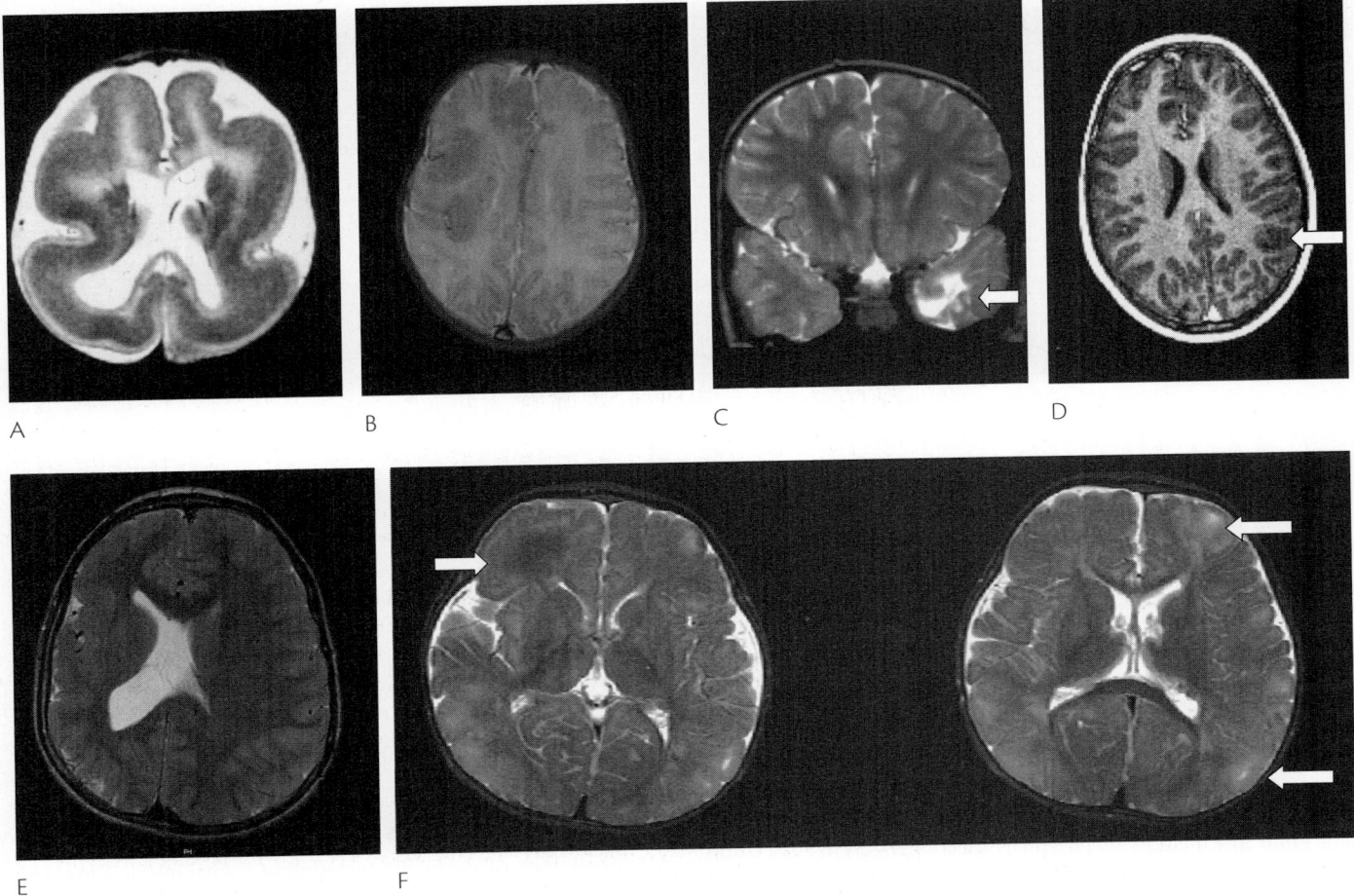

Fig. 30.8 Cerebral malformations on MRI in children presenting with epilepsy. (A) Lissencephaly showing no cortical sulcal pattern; (B) Mulitilobar dysplasia; (C) Dysembryoplastic neuroepithelial tumour of the temporal lobe; (D) Focal cortical dysplasia (arrowed); (E) Right hemipolymicrogyria; and (F) Multiple tubers of tuberous sclerosis (arrowed).

such as eventual driving, alcohol consumption, contraception, and conception.

30.7.2 When to consider treatment

In the majority of children medication is likely to be advised after the diagnosis of epilepsy has been made, that is after the second seizure. The choice of medication and its timing will depend on the particular diagnosis of the epilepsy, and requires full discussion with the family. In certain circumstances treatment may be postponed since certain idiopathic syndromes may not warrant immediate treatment. For example, benign epilepsy with occipital spikes (Section 30.5.5) may present with single episode of status epilepticus, and no further seizures thereafter. Children with benign epilepsy with centrotemporal spikes may have infrequent seizures, although the designation of 'benign' refers to the prompt response to treatment if prescribed and the tendency to grow out of seizures; nonetheless, the seizures themselves may well cause morbidity. A full discussion should balance the risks of the epilepsy against those of the treatment and should involve parents, any other carers, and often the child so as to decide on the advisability of starting treatment.

In other circumstances treatment should be considered early on if the risks of recurrence are considered high after a single seizure, for instance where there is an associated neurological abnormality,

or where a diagnosis of idiopathic generalized epilepsy is suspected on EEG.

30.7.3 Pharmacological management

When a decision has been made to start treatment, the most suitable anticonvulsant drug needs to be considered. This depends particularly on the type of epilepsy, or preferably the epilepsy syndrome, which has been diagnosed. Both conventional and newer antiepileptic drugs are used in children; their licensing position will depend on country and the availability of efficacy and safety data which has been obtained in children. Drug dosage needs to be adjusted in relation to the specific pharmacokinetic characteristics in young children, such as slower gastrointestinal absorption, higher volumes of distribution, and shorter clearance periods. In particular the reduced clearance rates can lengthen with age, and further dose reduction may be required to avoid the risk of toxicity as the child gets older. In any situation, general principles of antiepileptic drug introduction remain the same, whether a drug is required, if so which drug may be best for the child and the introduction of that drug slowly, with subsequent titration according to clinical response. The initial decision requires choice of which antiepileptic drug is indicated and to introduce this in a gradual titration to the optimal dose. If this fails, an alternative drug should be introduced

Table 30.5 Classification for malformations of cortical development (adapted from Barkovich *et al.* 2005)

Embryological Event	Examples
Failure of ventral induction	Holoprosencephaly
I. Abnormal neuronal / glial proliferation or apoptosis	
A. Decreased proliferation / Increased apoptosis	Microcephaly
B. Increased proliferation	Megalancephaly
C. Abnormal proliferation (abnormal cell types):	
Non-neoplastic	Tuberous sclerosis
	Cortical dysplasia with balloon cells
	Hemimegalencephaly
Neoplastic	Dysembryoblastic neuroepithelial tumour
	Ganglioglioma
	Gangliocytoma
II Abnormal neuronal migration	
A. Lissencephaly / Subcortical band heterotopia spectrum	*LIS 1, DCX, ARX, RELN* mutations
B. Cobblestone complex	Walker–Warburg
C. Heterotopia	Fukyama muscular dystrophy
	Subependymal (periventricular)
	Subcortical
	Marginal glioneural
III Abnormal cortical organization, including late neuronal migration	
A. Polymicrogyria and schizencephaly	Unilateral polymicrogyria
	Bilateral perisylvian syndrome
B. Cortical dysplasia without balloon cells	
C. Microdysgenesis	
Unclassified	
A. Secondary to inborn errors of metabolism	Mitochondrial and pyruvate metabolic disorders
	Peroxisomal disorders
B. Other unclassified malformations	Sublobar dysplasia

slowly, and if effective, the initial medication should be slowly withdrawn. Polytherapy should be avoided; although two medications may be synergistic in their action there is little evidence to suggest more than two medications will improve efficacy. However, occasionally a small minority of children require more than two drugs for optimal seizure control. Side effects may be problematic and their possibility should be discussed prior to drug introduction; the concerns of the parents should be considered. It is important to remember that seizures may be aggravated by certain medications, such as the exacerbation of electrical status epilepticus of slow wave sleep by carbamazepine (Section 30.5.5).

The aim of treatment is always to achieve seizure control with minimal if any side effects. If full seizure control cannot be achieved it may be worth considering slow withdrawal of medication so as to achieve optimal rather than complete seizure control. Of course, in some circumstances seizure control may not be the primary area of concern of parents and the coexistent comorbidities may demand greater attention (Section 30.7.7).

30.7.4 The ketogenic diet

The ketogenic diet has been used in the treatment of epilepsy for almost one hundred years. An osteopath in the early 20th century, prior to the introduction of anticonvulsant medication, discovered that starving people with epilepsy of all but water led to remission of seizures. However, such extremes of treatment cannot be advised. In 1921 Wilder suggested the use of a diet to mimic the metabolic effects of starvation; whereas in starvation the body breaks down body fat to produce ketones, a similar effect can be achieved by giving fat within the diet as the main energy source (Wilder 1921). Cohort studies have established that this can be effective although it lost favour when anticonvulsant drugs were introduced. Yet with time it became apparent that not all respond to anticonvulsant drugs and that the diet can be extremely effective for selected individuals. But the target population remains to be clearly delineated.

The classical ketogenic diet has provided a main fat source of long chain fat and is designed on the basis of a fat to carbohydrate ratio, including protein, of 3 or 4:1. Many of the studies in the early years report use of this diet (Freeman *et al.* 1998). Possible side effects of the diet provoked concern, and a more palatable way of giving the diet was proposed in which the main energy source was provided by medium chain triglyceride. It remained low in carbohydrate (Huttenlocher *et al.* 1971). A recent randomized controlled trial revealed a ketogenic diet to be effective over and above no change in treatment and no difference in efficacy between the classical and medium chain triglyceride ketogenic diets (Neal *et al.* 2008). In these circumstances parental choice or the child's wishes may confound interpretation of efficacy; this needs further examination.

Children may be considered for a ketogenic diet if they fail to respond to anticonvulsant medication. The diet requires a high degree of dietetic advice, for meal calculation, as well as considering the commitment by the parents and the child. There is no clear evidence that any age group responds over and above any other (Freeman *et al.* 1998). Although there is little experience in adults; the ketogenic diet poses different compliance problems for teenagers in view of the protein restriction. In addition there is no clear evidence as to which syndromes or seizure types may respond best, although anecdotal evidence suggests particular benefit in myoclonic astatic epilepsy, and reports in younger children suggest a particular benefit in infantile spasms (Kossoff *et al.* 2002; Nordli *et al.* 2002).

30.7.5 Status epilepticus

Status epilepticus is defined as recurrent repetitive seizures without regaining full awareness in between seizures or as continuous seizure activity for more than 30 min. Aggressive treatment of prolonged convulsive seizures is justified in view of the likelihood of increased morbidity in seizures lasting longer than 30 min. Treatment is determined by protocols, with defined times after which poor response to treatment requires intensive care (Appleton *et al.* 2000) (Fig. 30.9). Emergency treatment of seizures in the community has been enhanced by the availability of rescue medication in the form of rectal diazepam, or more recently buccal midazolam (Scott *et al.* 1999). However once in hospital intravenous protocols should be used; the most common cause of admission to the intensive care unit is overmedication with benzodiazepines (Chen *et al.* 2004). Initial treatment should be with a benzodiazepine, but

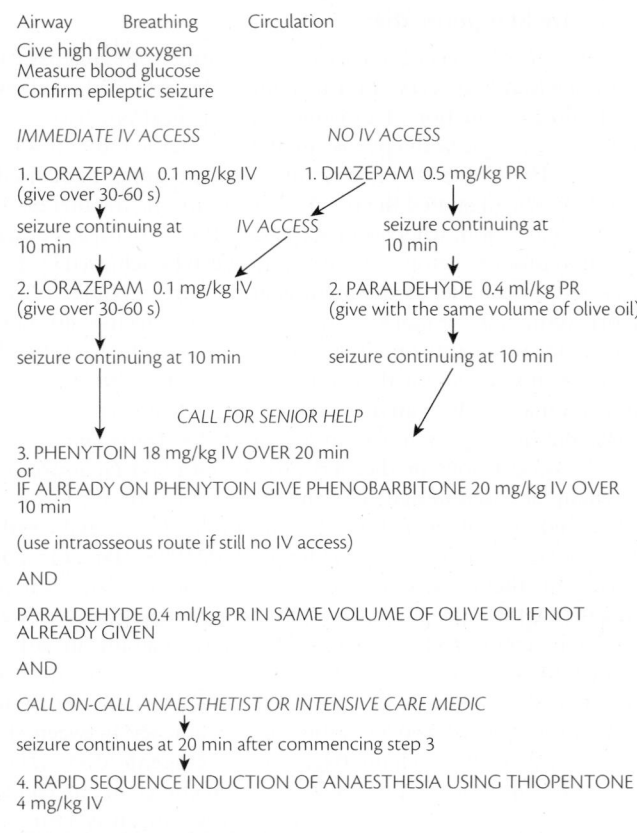

Airway   Breathing   Circulation
Give high flow oxygen
Measure blood glucose
Confirm epileptic seizure

IMMEDIATE IV ACCESS   *NO IV ACCESS*

1. LORAZEPAM 0.1 mg/kg IV   1. DIAZEPAM 0.5 mg/kg PR
(give over 30–60 s)

seizure continuing at   *IV ACCESS*   seizure continuing at
10 min   10 min

2. LORAZEPAM 0.1 mg/kg IV   2. PARALDEHYDE 0.4 ml/kg PR
(give over 30–60 s)   (give with the same volume of olive oil)

seizure continuing at 10 min   seizure continuing at 10 min

CALL FOR SENIOR HELP

3. PHENYTOIN 18 mg/kg IV OVER 20 min
or
IF ALREADY ON PHENYTOIN GIVE PHENOBARBITONE 20 mg/kg IV OVER
10 min

(use intraosseous route if still no IV access)

AND

PARALDEHYDE 0.4 ml/kg PR IN SAME VOLUME OF OLIVE OIL IF NOT
ALREADY GIVEN

AND

CALL ON-CALL ANAESTHETIST OR INTENSIVE CARE MEDIC

seizure continues at 20 min after commencing step 3

4. RAPID SEQUENCE INDUCTION OF ANAESTHESIA USING THIOPENTONE
4 mg/kg IV

TRANSFER TO INTENSIVE CARE UNIT

Fig. 30.9 Treatment of status epilepticus. Recommended protocol as agreed by National Institute of Clinical Excellent in 2004 (Appleton *et al.* 2000). Key: IV— intravenous. PR—per rectum.

following the protocol to load promptly with phenytoin if there is no response to the benzodiazepine; repeated dosage with benzodiazepines is undesirable.

A distinction should be made between convulsive and nonconvulsive status. Nonconvulsive status is a change in clinical state or behaviour, in association with change in EEG from the usual baseline. When occurring within syndromes where it may be more prevalent, such as Lennox–Gastaut Syndrome, there is only slim evidence that similarly aggressive treatment to that used in convulsive status is justified; indeed further change from baseline may be difficult to determine. Nevertheless it is important to be vigilant as to the possible occurrence of nonconvulsive status in view of the possible impact on cognition and learning.

30.7.6 The role of surgery

Children with symptomatic focal epilepsy which is unresponsive to medication should be considered early on for curative surgical resection. Traditionally there has been a reluctance to consider children as candidates for surgery due to the perception of uncertain seizure prognosis, and the relatively invasive nature of the presurgical evaluation. However, with improvements in syndromic diagnosis, and the definition of aetiology allowed by modern imaging, along with advances in neurosurgery and neuroanaesthesia, risks have been minimized and surgery has become available to a wider range of children with epilepsy.

Like adults, children may be considered either for resective surgery, or in certain circumstances for functional procedures in which tissue is not removed but function modified (Section 31.11). In children the range of potential procedures is wide, particularly the resections, and these differ proportionally from adult practice. Similar principles apply, particularly that seizures should be proven to arise from one functionally silent area of the brain, of variable size. Hemispherectomy or multilobar resection account for about two-thirds of procedures undertaken in children (Harvey *et al.* 2008). Moreover children with early-onset catastrophic epilepsy of focal origin, and those with early-onset focal epilepsy associated with certain syndromes, such as Sturge–Weber syndrome (Section 11.3) or tuberous sclerosis, (Section 11.1) should be considered for surgery early in the natural history of their epilepsy (Cross *et al.* 2006).

Freedom from seizures is the primary aim of paediatric epilepsy surgery. However secondary aims may be of equal, if not of primary, importance to some families and require careful consideration with counselling prior to surgery. Seizure outcome from focal resection relates to the completeness of the resection, whether determined structurally (Edwards *et al.* 2000) or electrographically (Paolicchi *et al.* 2000). Wider procedures such as hemidisconnection primarily relate to the underlying pathology, with lower rates of seizure freedom seen for developmental malformations, particularly hemimegalancephaly (Section 9.2.4) (Devlin *et al.* 2003). Developmental and behavioural outcomes are often of concern to parents; an argument in favour of early surgery has often been that of improved developmental outcome. Yet objective support for this is scarce. Much of the data on neurodevelopmental progress appears to show a maintained intelligence quotient trajectory against peers (Devlin *et al.* 2003; Freitag and Tuxhorn 2005). Small prospective series suggest that surgery performed under the age of 1 year is likely to improve developmental trajectories than surgery performed later (Loddenkemper *et al.* 2007). This is particularly important given that almost 50 per cent of children coming to surgery have seizure onset under 12 months of age (Harvey *et al.* 2008). Little data is available on behavioural outcomes. Rates of behavioural difficulty are high in this population, but studies show that little prediction can be made as to possible improvement or deterioration after surgery (Devlin *et al.* 2003; McLellan *et al.* 2005).

Vagal nerve stimulation (Section 31.11.5) has a place in the management of children with drug-resistant epilepsy, although its exact role remains unclear. Data overall suggest that in children whose epilepsy is not associated with an identifiable underlying lesion, and who are not resective surgical candidates, an improvement of at least 50 per cent is likely to be seen in 40–45 per cent (Wheless and Maggio 2002; Uthman *et al.* 2004). Freedom from seizures cannot be a goal of this procedure and realistic expectations about outcome must be discussed with families. Some early data on small numbers of children suggested accumulated benefit over time but this has not been replicated in larger series. Benefits beyond reduced seizure frequency may also be seen such as reduced seizure duration or increased awareness.

30.7.7 The comorbidities of childhood epilepsy

Integral to the management of children with epilepsy is the recognition and management of coexisting learning or behaviour difficulties. The occurrence of learning difficulty in children with particularly early-onset epilepsy is high, although the pathophysiological explanation remains unclear in many. In those where it precedes the onset of the epilepsy it is likely to be related to the underlying aetiology. In others where developmental plateau or regression occurs at the onset or during the course of the epilepsy, it is presumed to be related to the underlying epileptic activity, an epileptic encephalopathy (Section 30.5.2). In many it is likely to be a combination of factors, with possible contributions also from medication side effects. It is important to recognize such difficulties in view of the likely impact on schooling and possibly behaviour.

The rate of behavioural difficulty seen in children with epilepsy is greater than in children with other non-neurological chronic diseases (Davies *et al.* 2003). There is also data to suggest such behaviour difficulties antedate seizure presentation (Austin *et al.* 2001). Whether such difficulties reflect part of the underlying aetiology to the epilepsy itself or represent sub-clinical seizure activity often remains unclear. Higher rates of behaviour disorder are seen in children with 'complicated' as opposed to 'pure' epilepsy (Davies *et al.* 2003), suggesting a major contribution from the underlying aetiology. Other compounding factors include seizure occurrence, other ongoing electrical activity, medication, and psychosocial influences. There is no reason why such children should not be assessed as any other child with behaviour disorder, and behaviour modification programmes tried prior to consideration of medication.

References

Appleton R, Choonara I, Martland T *et al.* (2000). The treatment of convulsive status epilepticus in children. The status epilepticus working party, members of the status epilepticus working party. *Arch Dis Child*, **83**, 415–9.

Austin JK, Harezlak J, Dunn DW (2001). Behavior problems in children before recognized seizures. *Pediatrics*, **107**, 115–22.

Barkovich AJ, Kuzniecky RI, Jackson GD *et al.* (2005). A developmental and genetic classification for malformations of cortical development. *Neurology* **65**, 1873–87.

Bartlett K (2001) Biotinidase deficiency in vitamin responsive conditions. In Baxter P, ed. *Paediatric Neurology*, pp. 1–11. Mackeith Press, London.

Bauer J, Bien CG, Lassmann H (2002). Rasmussen's encephalitis: a role for autoimmune cytotoxic Tlymphocytes. *Curr Opin Neurol*, **15**, 197–200.

Berg AT, Shinnar S, Levy SR *et al.* (2000). How well can epilepsy syndromes be identified at diagnosis? A reassessment 2 years after initial diagnosis. *Epilepsia*, **41**, 1269–75.

Bien CG, Granata T, Antozzi C *et al.* (2005). Pathogenesis, diagnosis and treatment of Rasmussen encephalitis. *Brain*, **128**, 454–71.

Boatman D, Freeman J, Vining E *et al.* (1999). Language recovery after left hemispherectomy in children with late onset seizures. *Ann Neurol*, **46**, 579–86.

Boyd SG, Harden A, Egger J *et al.* (1986). Progressive neuronal degeneration of childhood with liver disease ('Alpers' disease'): characteristic neurophysiological features. *Neuropediatrics*, **17**, 75–80.

Boyd SG, Harden A, Patton MA (1988). The EEG in early diagnosis of the Angelman (happy puppet) syndrome. *Eur J Pediatr*, **147**, 508–13.

Braathen G, Theorell K (1995). A general hospital population of childhood epilepsy. *Acta Paediatr*, **84**, 1143–6.

Cavazzuti GB (1980). Epidemiology of different types of epilepsy in school age children of Modena, Italy. *Epilepsia*, **21**, 57–62.

Ceulemans B, Boel M, Claes L *et al.* (2004). Severe myoclonic epilepsy in infancy: toward an optimal treatment. *J Child Neurol*, **19**, 516–21.

Chevrie JJ, Aicardi J (1972). Childhood epileptic encephalopathy with slow spike-wave. A statistical study of 80 cases. *Epilepsia*, **13**, 259–71.

Chin RF, Verhulst L, Neville BG *et al.* (2004). Inappropriate emergency management of status epilepticus in children contributes to the need for intensive care. *JNNP*, **75**, 1584–8.

Chiron C, Marchand MC, Tran A *et al.* (2000). Stiripentol in severe myoclonic epilepsy in infancy: a randomised placebo-controlled syndrome-dedicated trial. *Lancet*, **356**, 1638–42.

Commission on Classification and Terminology of the International League against Epilepsy (1989). Proposal for revised classification of epilepsies and epileptic syndromes. *Epilepsia*, **30**, 389–99.

Coppola G, Plouin P, Chiron C *et al.* (1995). Migrating partial seizures in infancy: a malignant disorder with developmental arrest. *Epilepsia*, **36**, 1017–24.

Cross JH, Jayakar P, Nordli D *et al.* (2006). Proposed criteria for referral and evaluation of children for epilepsy surgery: recommendations of the Subcommission for Paediatric Epilepsy Surgery. *Epilepsia*, **47**, 952–9.

Dale RC, Cross JH (1999). Ictal hemiparesis. *Dev Med Child Neuro*, **41**, 344–7.

Dalla Bernardina B, Colamaria V, Capovilla G *et al.* (1983). Nosological classification of epilepsies in the first three years of life. *Prog Clin Biol Res*, **124**, 165–83.

Davies S, Goodman R, Heyman I (2003). A population survey of mental health problems in children with epilepsy. *Dev Med Child Neurol*, **4**, 292–5.

Devlin AM, Cross JH, Harkness W *et al.* (2003). Clinical outcomes of hemispherectomy for epilepsy in childhood and adolescence. *Brain*, **126**, 556–66.

Doose H, Baier WK (1987). Epilepsy with primarily generalized myoclonic-astatic seizures: a genetically determined disease. *Eur J Pediatri*, **146**, 550–4.

Doose H, Sitepu B (1983). Childhood epilepsy in a German city. *Neuropediatrics*, **14**, 220–4.

Doose H, Gerken H, Leonhardt R *et al.* (1970). Centrencephalic myoclonic-astatic petit mal. Clinical and genetic investigation. *Neuropadiatrie*, **2**, 59–78.

Dravet C, Bureau M, Guerrini R, *et al.* (1992). Severe myoclonic epilepsy in infants. In Roger J, Bureau M, Dravet C, *et al.* eds *Epileptic syndromes in infancy, childhood and adolescence* 2nd ed. John Libbey, London.

Edwards JC, Wyllie E, Ruggieri PM *et al.* (2000). Seizure outcome after surgery for epilepsy due to malformation of cortical development. *Neurology*, **55**, 1110–4.

Eltze C, Chong WK, Harding B *et al.* (2005). Focal cortical dysplasia in infants; some MRI lesions almost disappear with maturation of myelination. *Epilepsia*, **46**, 1988–22.

Eltze C, Scott RC, Chin RFM *et al.* (2007). Epilepsy in infancy study: first results of a collaborative population-based study in north London. *Dev Med Child Neurol*, **49**, 17.

Engel J Jr (2001). A proposed diagnostic scheme for people with epileptic seizures and with epilepsy: report of the ILAE Task Force on Classification and Terminology. *Epilepsia*, **42**, 796–803.

Ferrari G, Lamantea E, Donati A *et al.* (2005) Infantile hepatocerebral syndromes associated with mutations in the mitochondrial DNA polymerase-gammaA. *Brain*, **128**, 723–31.

Forsgren L, Hauser WA, Olafsson E *et al.* (2005). Mortality of epilepsy in developed countries: a review. *Epilepsia*, **46**, 18–27.

Franz DN, Glauser TA, Tudor C *et al.* (2000). Topiramate therapy of epilepsy associated with Angelman's syndrome. *Neurology*, **54**, 1185–8.

Freeman JM, Vining EPG, Pillas DJ (1998). The efficacy of the ketogenic diet -1998: a prospective evaluation of intervention in 150 children. *Pediatrics*, **102**, 1358–63.

Freitag H, Tuxhorn I (2005). Cognitive funciton in preschool children after epilepsy surgery: rationale for early intervention. *Epilepsia*, **46**, 561–7.

Galanopolou AS, Bojko A, Lado F *et al.* (2000). The spectrum of neuropsychiatric abnormalities associated with electrical status epilepticus in sleep. *Brain Dev*, **22**, 279–95.

Gastaut H, Broughton R (1972). *Epileptic Seizures*, pp. 37–47. CC Thomas, Springfield, Illinois.

Gastaut H, Roger J, Soulayrol R *et al.* (1966) Childhood epileptic encephalopathy with diffuse slow spike-waves (otherwise known as 'petit mal variant') or Lennox Syndrome. *Epilepsia*, **7**, 139–79.

Gastaut H, Roger J, Bureau M (1992). Benign epilepsy of childhood with occipital paroxysms. Update. In Roger J, Bureau M, Dravet C *et al.* eds. *Epileptic syndromes in infancy, childhood and adolescence* 2nd ed. John Libbey, London.

Giannakodimos S, Panayiotopoulos CP (1996). Eyelid myoclonia with absences in adults: a clinical and video-EEG study. *Epilepsia*, **37**, 36–44.

Guerrini R, De Lorey TM, Bonanni P *et al.* (1996). Cortical myoclonus in Angelman syndrome. *Ann Neurol*, **40**, 39–48.

Guerrini R, Dravet C, Genton P *et al.* (1998). Lamotrigine and seizure aggravation in severe myoclonic epilepsy. *Epilepsia*, **39**, 508–12.

Hancock E, Cross H (2003). Treatment of Lennox-Gastaut syndrome (Cochrane Review). In: *The Cochrane Library, Issue 3*. John Wiley and Sons Ltd, Chichester.

Hancock E, Osborne J, Milner P (2003). Treatment of infantile spasms. *Cochrane Database Syst Rev* CD001770.

Harkin LA, McMahon JM, Iona X *et al.* (2007) The spectrum of SCN1A related infantile epileptic *Encephalopathies*, **130**, 843–52.

Hart YM, Cortez M, Andermann F *et al.* (1994). Medical treatment of Rasmussen's syndrome (chronic encephalitis and epilepsy): effect of high-dose steroids or immunoglobulins in 19 patients. *Neurology*, **44**, 1030–6.

Harvey AS, Cross JH, Shinnar S *et al.* (2008). Seizure syndromes, aetiologies and procedures in paediatric epilepsy: a 2004 International Survey. *Epilepsia*, **49**, 146–155.

Hauser WA, Anderson VE (1986). Genetics of Epilepsy. In Pedley TA, Meldrum BS, eds. *Recent Advances in Epilepsy*, pp. 21–36. Churchill Livingstone, Edinburgh.

Hauser WA, Annegers JF, Kurland LT (1993). Incidence of epilepsy and unprovoked seizures in Rochester, Minnesota: 1935-1984. *Epilepsia*, **34**, 453–68.

Horvath R, Hudson G, Ferrari G *et al.* (2006) Phenotypic spectrum associated with mutations of the mitochondrial polymerase gamma gene. *Brain*, **129**, 1674–84.

Hughes EF, Fairbanks L, Simmonds HA *et al.* (1998). Molybdenum cofactor deficiency-phenotypic variability in a family with late onset variant. *Dev Med Child Neurol*, **40**, 57–61.

Huttenlocher PR, Wilbourn AJ, Signore JM (1971). Medium-chain triglycerides as a therapy for intractable childhood epilepsy. *Neurology*, **21**, 1097–103.

Kette D, Joaquina QA, Rosi MG *et al.* (2003). Angelman syndrome: difficulties in EEG pattern recognition and possible misinterpretations. *Epilepsia*, **44**, 1051–63.

Kossoff EH, Pyzik PL, McGrogan JR *et al.* (2002). Efficacy of the ketogenic diet for infantile spasms. *Pediatrics*, **109**, 780–3.

Jeavons PM (1977). Nosological problems of myoclenic epilepsies in childhood and adolescence. *Dev Med Child Neurol*, **19**, 3–6.

Johnson JL, Wadman SK (1995) Molybdenum cofactor deficiency and isolated sulfte oxidase deficiency. In Scriver CR, Beaudet AL, Sly WS, Valle D, eds. *The Metabolic and Molecular Bases of Inherited Disease*, 7th edition, pp. 2271–83. McGraw-Hill, New York.

Laan LA, den Boer AT, Hennekam RC *et al.* (1996). Angelman syndrome in adulthood. *Am J Med Genet*, **66**, 356–60.

Labate A, Colosimo E, Gambardella A *et al.* (2006). Levetiracetam in patients with generalised epilepsy and myoclonic seizures: an open label study. *Seizure*, **15**, 214–8.

Lalioti M, Scott HS, Buresi C *et al.* (1997). Dodecamer repeat in cystatin B in progressive myoclonic epilepsy (EPM1). *Nature*, **356**, 847–51.

Lerman P (1985). Benign partial epilepsy with centrotemporal spikes In Roger J, Dravet C, Bureau M *et al.*, eds. *Epileptic Syndromes in Infancy, Childhood & Adolescence*, pp. 50–158. John Libbey Eurotext, London.

Li Y, Uccelli A, Laxer KD *et al.* (1997). Local-clonal expansion of infiltrating T lymphocytes in chronic encephalitis of Rasmussen. *J Immunol*, **158**, 1428–37.

Liu Z, Mikati M, Holmes G (1995) Mesial temporal sclerosis: pathogenesis and significance. *Pediatr Neurol*, **12**, 5–1.

Loddenkemper T, Holland KD, Stanford LD *et al.* (2007). Developmental outcome after epilepsy surgery in infancy. *Pediatrics*, **119**, 930–5.

Loiseau P, Duche B, Cordova S *et al.* (1988). Prognosis of benign childhood epilepsy with centrotemporal spikes: a followup study of 168 patients. *Epilepsia*, **29**, 229–35.

Lux AL, Edwards SW, Hancock E *et al.* (2004). The United Kingdom Infantile Spasms Study comparing vigabatrin with prednisolone or tetracosactide at 14 days: a multicentre, randomised controlled trial. *Lancet*, **364**, 1773–8.

Lux AL, Edwards SW, Hancock E *et al.* (2005). The United Kingdom Infantile Spasms Study (UKISS) comparing hormone treatment with vigabatrin on developmental and epilepsy outcomes to age 14 months: a multicentre randomised trial. *Lancet Neurol*, **4**, 712–7.

Mackay MT, Weiss SK, Webber T *et al.* (2004). Practice parameter: medical treatment of infantile spasms: report of the American Academy of Neurology and the Child Neurology Society. *Neurology*, **62**, 1668–81.

Magenis RE, Toth-Fejel S, Allen LJ *et al.* (1990). Comparison of the 15q deletions in Prader-Willi and Angelman syndromes: specific regions, extent of deletions, parental origin, and clinical consequences. *Am J Med Genet*, **35**, 333–49.

Manonmani V, Wallace SJ. (1994) Epilepsy with myoclonic absences *Arch Dis Child*, **70**, 288–90.

Marini C, Mei D, Cross JH *et al.* (2006) Mosaic SCN1A mutation in familial severe myoclonic epilepsy of infancy. *Epilepsia*, **47**, 1737–40.

Matsamoto A, Watanabe K, Negro T *et al.* (1981) Infantile Spasms: etiologic factors, clinical aspects and long term prognosis in 200 cases. *Eur J Pediatr*, **135**, 239–44.

McLellan A, Davies S, Heyman I *et al.* (2005). Psychopathology in children with epilepsy before and after temporal lobe resection. *Dev Med Child Neurol*, **47**, 666–72.

Metz Lutz MN, Kleitz C, de Saint Martin *et al.* (1999). Cognitive development in benign focal epilepsies of childhood. *Dev Neurosci*, **21**, 182–90.

Mills PB, Struys E, Jakobs C *et al.* (2006). Mutations in antiquitin in individuals with pyridoxine-dependent seizures. *Nat Med*, **12**, 307–9.

Mizrahi EM, Clancy RR (2000). Neonatal seizures: early onset seizure syndromes and their consequences for development. *Mental Retardation and Dev Disabilities Research Reviews*, **6**, 229–41.

National Institute for Clinical Excellence (NICE) (2004). The epilepsies: diagnosis and management of the epilepsies in children and young people in primary and secondary care. Clinical Guideline 20, October.

National Institutes of Health (1981). Consensus development conference on febrile seizures. Proceedings. *Epilepsia*, **2**, 377–81.

Neal EG, Chaffe H, Schwartz R *et al.* (2008). The ketogenic diet in the treatment of epilepsy: a randomised, controlled trial (NCT00564915) *Lancet neurology*, **7**, 500–6.

Neville BGR, Boyd SG (1995). Selective epileptic gait disorder. *J Neurol Neurosurg Psychiatry*, **58**, 371–3.

NICE guideline (2004). The Epilepsies: diagnosis and management of the epilepsies in children and young people in primary and secondary care. www.nice.org.uk/CG020.

Nordli D, Kuroda MM, Carroll J *et al.* (2002). Experience with the ketogenic diet in infants. *Pediatrics*, **108**, 129–133.

North KN, Storey GN, Henderson-Smart DJ (1989). Fifth day fits in the newborn. *Aust Paediatr J*, **25**, 284–7.

Oguni H, Tanaka T, Hayashi K *et al.* (2002) Treatment and long-term prognosis of myoclonic-astatic epilepsy of early childhood. *Neuropediatrics*, **33**, 122–33.

Ohtahara S, Yamatogi Y (2003). Epileptic encephalopathies in early infancy with suppression-burst. *J Clin Neurophysiol*, **20**, 398–407.

Olafsson E, Hauser WA, Ludvigsson P *et al.* (1996). Incidence of epilepsy in rural Iceland: a population-based study. *Epilepsia*, **37**, 951–5.

Panayiotopoulos CP (1993). Benign childhood partial epilepsies: benign childhood seizure susceptibility syndromes. *J Neurol Neurosurg Psychiatry*, **56**, 2–5.

Panayiotopoulos CP, Ferrie CD, Giannakodimos S *et al.* (1995). Perioral myoclonia with absences. In Duncan JS, Panayiotopoulos CP, eds. *Typical Absences and Related Syndromes*, pp. 221–30. Churchill Communication, Europe.

Paolicchi JM, Jayakar P, Dean P *et al.* (2000). Predictors of outcome in pediatric epilepsy surgery. *Neurology*, **54**, 642–7.

Patry G, Lyagoubi S, Tassinari CA (1971). Subclinical 'electrical status epilepsticus' induced by sleep in children. *Arch Neurol*, **24**, 242–52.

Patterson MC, Platt F (2004). Therapy of Niemann-Pick disease, type C. *Biochim Biophys Acta*, **1685**, 77–82.

Plouin P, Anderson VE (2005). Benign familial and non familial neonatal seizures. In Roger J, Bureau M, Dravet C *et al.*, eds. *Epileptic Syndromes in Infancy, Childhood and Adolescence*, pp.3–15. John Libbey, Eurotext.

Pryor DS, Don N, Macourt DC (1981). Fifth day fits: a syndrome of neonatal convulsions. *Arch Dis Child*, **56**, 753–8.

Putt M, Pressler R (2005) Neurophysiological testing in the newborn and infant. *Early Human Development*, **81**, 939–46.

Rener-Primec Z, Stare J, Neubauer D (2006). The risk of lower mental outcome in infantile spasms increases after three weeks of hypsarrhythmia duration. *Epilepsia*, **47**, 2202–5.

Riikonen R (1982). A long-term follow-up study of 214 children with the syndrome of infantile spasms. *Neuropediatrics*, **13**, 14–23.

Riikonen R (1984). Infantile Spasms: modern practical aspects. *Acta Pediatr Scand*, **73**, 1–12.

Scheffer IE, Berkovic SF (1997). Generalised epilepsy with febrile seizures plus. A genetic disorder with heterogeneous clinical phenotypes. *Brain*, **120**, 479–90.

Scheffer IE, Berkovic SF (2003). The genetics of human epilepsy. *Trends Pharmacol Sci*, **24**, 428–33.

Schmidt R, Eviatar L, Wong M *et al.* (1981). Ring chromosome 14: a distinct clinical entity. *J Med Genet*, **18**, 304–7.

Schwartz RH, Eaton J, Bower BD *et al.* (1989) ketogenic diets in the treatment of epilepsy: short-term clinical effects. *Dev Med Child Neurol*, **31**, 145–51.

Scott RC, Besag FM, Neville BGR (1999). Buccal midazolam and rectal diazepam for treatment of prolonged seizures in childhood and adolescence: a randomised trial. *Lancet*, **353**, 623–6.

Scott R, King MD, Gadian DG *et al.* (2003). Hippocampal abnormalities after prolonged febrile convulsion: a longitudinal MRI study. *Brain*, **126**, 2551–7.

Scottish Intercollegiate Guidelines Network (SIGN) (2005). Diagnosis and management of epilepsies in children and young people.

Serratosa JM, Delgado-Escueta AV, Posada I *et al.* (1995). The gene for progressive myoclonic epilepsy of the Lafora type maps to chromosome 6q. *Hum Mol Genet*, **4**, 1657–63.

Serratosa JM, Gomez-Garre P, Gallardo MA *et al.* (1999) A novel protein tyrosine phosphatase gene is mutated in progressive myoclonic epilepsy of the Lafora type (EMP2). *Hum Mol Genet*, **8**, 345–52.

Sillanpaa M, Schmidt D (2006). Prognosis of seizure recurrence after stopping antiepileptic drugs in seizure-free patients: a long-term population-based study of childhood-onset epilepsy. *Epilepsy Behav*, **8**, 713–9.

Slot HN, Overweg-Plandsoen WC, Bakker HD *et al.* (1993). Molybdenum cofactor deficiency: an easily missed cause of neonatal convulsions *Neuropediatrics*, **24**, 139–42.

Tassinari CA, Michelucci R, Rubboli G *et al.* (1995). Myoclonic absence epilepsy. In Duncan JS, Panayiotopoulos CP, eds. *Typical Absences and Related Syndromes*, pp. 187–95. Churchill Communication Europe, London.

Uldall P, Alving J, Hansen LK *et al.* (2006). The misdiagnosis of epilepsy in children admitted to a tertiary epilepsy centre with paroxysmal events. *Arch Dis Child*, **91**, 219–21.

Uthman BM, Reichl AM, Dean JC *et al.* (2004). Effectiveness of vagus nerve stimulation in epilepsy patients: a 12 year observation. *Neurology*, **63**, 1124–6.

Vanier MT, Millat G (2003). Niemann Pick Disease Type C. *Clin Genet*, **64**, 269–81.

Verity CM, Butler NR, Golding J (1985). Febrile convulsions in a national cohort followed up from birth: prevalence and recurrence in the first five years of life. *BMJ*, **116**, 329–37.

Viani F, Romeo A, Viri M *et al.* (1995). Seizure and EEG patterns in Angelman's syndrome. *J Child Neurol*, **10**, 467–71.

Ville D, Kaminska A, Bahi-Buisson N *et al.* (2006). Early pattern of epilepsy in the ring chromosome 20 syndrome. *Epilepsia*, **47**, 543–9.

Watanabe K, Negoro T, Aso K (1993). Benign partial epilepsy with secondarily generalized seizures in infancy. *Epilepsia*, **34**, 635–8.

Watanabe K, Yamamoto N, Negoro T *et al.* (1990). Benign infantile epilepsy with complex partial seizures. *J Clin Neurophysiol*, **7**, 409–16.

Watson R, Jiang Y, Bermudez I *et al.* (2004). Absence of antibodies to glutamate receptor type 3 (GluR3) in Rasmussen encephalitis. *Neurology*, **63**, 43–50.

Wheless JW, Maggio V (2002). Vagus nerve stimulation therapy in patients younger than 18 years. *Neurology*, **59**, s21–5.

Wilder RM (1921). The effects of ketonemia on the course of epilepsy. *Mayo Clinic Proceedings*, **2**, 307–8.

Williams RE, Aberg L, Autti T *et al.* (2006). Diagnosis of the neuronal ceroid lipofuscinoses: an update. *Biochim Biophys Acta*, **1762**, 865–72.

CHAPTER 31

Seizures, epilepsy, and other episodic disorders in adults

David Chadwick

Contents

31.1 The differential diagnosis of seizures

31.1.1 Definition of seizures and epilepsy

Epilepsy, or more correctly a seizure, is most easily defined in physiological terms, being 'the name for occasional sudden, excessive, rapid, and local discharges of grey matter' (Jackson 1873). It is more difficult to offer a comprehensive clinical definition of epileptic seizures and epilepsy because of the varied clinical manifestations produced by cerebral neuronal discharge. However, an epileptic seizure can be defined as an intermittent and stereotyped disturbance of consciousness, behaviour, emotion, motor function, or sensation that on clinical grounds is believed to result from cortical neuronal discharge. Epilepsy can then be defined as a condition in which seizures recur, usually spontaneously.

The differential diagnosis of epilepsy is large because of the enormous range of symptoms that can occur during seizures. Inevitably, the differential diagnosis for tonic-clonic seizures is very different from that for simple partial seizures with autonomic symptoms. The most common clinical problem is the differential diagnosis from other causes of transient loss of consciousness associated with collapse, the commonest cause of which is syncope.

31.1.2 Syncope

Syncope defined as a sudden loss of consciousness associated with the inability to maintain postural tone, followed by spontaneous recovery. It is common and results from a wide variety of causes (Table 31.1). It occurs across the age range, with an incidence of 4.7/1000 in women in the third decade of life rising to 11.1/1000 in the eighth decade and has a varied prognosis. In the elderly it is associated with significant morbidity and mortality (Soteriades et al. 2002).

Neurally mediated syncope

The commonest cause of syncope is vaso-vagal in young people who have no serious underlying pathology. It is precipitated by unpleasant sights or pain, standing for prolonged periods, or after exposure to heat, hunger, dehydration, and alcohol excess. It is posture dependent: symptoms can start while sitting, but loss of consciousness usually occurs when the individual stands. The mechanisms leading to fainting are an abnormal autonomic response consisting of vasodilation and increased vagal tone. Pre-syncopal symptoms appear before hypotension and bradycardia and persist after the blood pressure returns to normal. The subject usually has a feeling of warmth with a dry mouth and a desire for fresh air or a drink of water. Nausea can develop quickly along with deep, sighing respiration, blurring of vision with spots in front of the eyes and loss of colour vision, noises in the ears, vertigo, and depersonalization. It should be noted that the majority of these symptoms do not occur in seizures.

The onset of these symptoms is usually gradual and eye witnesses comment on pallor and sweating. The subject will collapse if they remain standing. The collapse may be rigid or flaccid and some form of clonic or other positive motor phenomena are common, 'convulsive syncope', (Lempert et al. 1994) raising the spectre of epilepsy for the inexperienced diagnostician. The EEG shows only generalized slow activity at this time, without any epileptiform features. It is controversial as to whether true secondary epileptic seizures can occur in as a result of prolonged cerebral hypoperfusion because of syncope.

Table 31.1 Causes of syncope

Neurally mediated reflex syncope
Vasovagal syncope
Carotid sinus syncope
Situational syncopes:
micturition syncope
cough/sneeze syncope
valsalva
swallow syncope
Glossopharyngeal neuralgia
Orthostatic Hypotension
Syndromes of autonomic failure (Section 2.7.3)
Drug induced
Hypovolaemia
Cardiac syncope
Dysrhythmias (heart block, tachycardias, etc.):
Sinus node dysfunction
Atriventricular conduction system disease
Paroxysmal tachycardia
Inherited syndromes:
long QT syndrome
brugada syndrome
Drug-induced
Structural cardiac or cardiopulmonary disease:
Obstructive valvular disease (particularly aortic stenosis)
Myocardial infarction
Atrial myxoma
Obstructive Cardiomyopathy
Shunts
Pulmonary hypertension
Cerebrovascular disease
Steal syndromes

Loss of consciousness is brief and on recovery the subject is usually nauseated and tremulous with continued pallor and sweating. Crucially the syncopal individual recalls recovery at the site of their collapse, while someone with a tonic-clonic seizure has their first recall on the way to hospital. They may, however, feel washed out and unwell for some time afterwards.

Carotid sinus syncope

Whilst compression of the carotid sinus in young people rarely causes any symptoms, in the elderly, particularly those with heart disease, it can cause bradycardia and syncope often presenting as 'falls'. Carotid sinus hypersensitivity may be diagnosed when 5–10 s of carotid sinus massage results in a 50mmHg or greater drop in systolic blood pressure or a 3 s or greater ventricular pause (Brignole et al. 2004). Most commonly this is due to reflex vagal inhibition of the heart. However, more rarely it may be caused by a vasopressor effect leading to a fall in blood pressure independent of heart rate. Very rarely, pressure on one carotid can lead to almost immediate loss of consciousness if there is a grossly stenotic or occluded contralateral artery so that ipsilateral carotid compression in effect leads to a standstill in much of the cerebral circulation.

Carotid sinus syncope is most commonly seen in patients with arteriosclerosis and hypertension but may sometimes occur with local neoplastic disease in the neck or aneurysmal dilatation of the sinus. The usual precipitant in all cases is a sudden turn of the head inducing dizziness and fainting.

Pulmonary disorders and syncope

Cough syncope is the commonest respiratory cause of syncope. It tends to be seen in middle-aged smokers, usually male, who are usually overweight and have chronic obstructive airways disease. Fainting is precipitated by a paroxysm of continuous coughing. The individual experiences light-headedness followed by unconsciousness. Recovery is usually quick. The prolonged bout of coughing elevates intrathoracic pressure and thereby impedes venous return and cardiac output. It is doubtful that people with normal respiratory function can cough for a sufficiently long period to induce cough syncope (Pederson *et al.* 1966). More rarely, cough syncope, as well as cough headache, have been associated with cerebellar ectopia and syringomyelia.

Primary or secondary pulmonary hypertension may be associated with syncope because of a failure of the right ventricle to increase output on demand. Syncope may therefore occur during effort in a fashion similar to that seen with aortic stenosis (Ross 1988), with Eisenmenger syndrome and with pulmonary embolus (Soloff and Rodman 1967).

Oesophageal syncope

Syncope may occur as a reflex response to sensory stimuli within the territories of the glossopharyngeal or vagus nerves. Guberman and Catching (1986) described 29 patients with syncope associated with swallowing. Almost all patients had some form of oesophageal disorder or cardiac disease, such as ischaemic heart disease or heart block. Rarely cancer of the head and neck can be associated with syncope occurring spontaneously or on swallowing.

Syncope in the period after eating may not be uncommon, particularly in the elderly or in patients with autonomic failure. Syncope can occur as an accompaniment of glossopharyngeal neuralgia (Riley *et al.* 1942) (Sections 19.2.2; 20.3.3). The pain is precipitated by stimulation, or movement of the oropharynx during chewing, swallowing, or coughing, and precedes syncope. Carbamazepine can be used, although microvascular decompression has also been successful (Tsuboi *et al.* 1985).

Pelvic syncope

Syncope during and immediately after micturition is not uncommon. It seems to occur in two different groups of people: young healthy men and older people of either sex who have concurrent medical problems (Kapoor *et al.* 1985). In the first group, syncope usually occurs at the end of voiding after the patient has got out of bed in the middle of the night. It seems to be predisposed to by sleep deprivation, hunger, and intercurrent infection. Bladder distension may be one means of reflexly precipitating fainting which can also occur following the decompression of a painfully distended bladder. The role of the Valsalva manoeuvre during voiding is uncertain.

Syncope may on occasions occur with defaecation in older patients (Pathy 1978). Other pelvic examinations and interventions such as prostatic examination and insertion of intra-uterine devices are associated occasionally with syncopal episodes.

Cardiac syncope

Cardiac syncope may differ from vasovagal syncope as its onset is often rapid and without warning other than palpitation (Ross 1988). It should be suspected in patients with increasing age, vascular risk factors, and known cardiac disease. On rare occasions life threatening dysrhythmias are seen in young people with no previous cardiac history. These patients may undergo neurological referral. Both the congenital long QT syndrome (Schwartz 1985)

and Brugada syndrome (Antzelevitch *et al.* 2002) result from cardiac channelopathies and should be suspected when syncope, collapse, or atypical convulsions are precipitated by exercise, or occur during rest or sleep, or when there is a family history of sudden death. In these circumstances, a routine electrocardiogram should be examined for evidence of prolonged QTc and the characteristic ST segment elevation in leads V1-3 with apparent right bundle branch block, respectively.

Orthostatic hypotension

Syncope in syndromes of autonomic failure develops slowly over several minutes. A large number of central and peripheral nervous system disorders are associated with symptoms of autonomic failure in which syncope is prominent (Sections 2.7.3 and 21.11.7). Syncope occurs on standing, beginning with a feeling of light-headedness which progresses slowly. Visual obscurations may occur with other central symptoms of cerebral hypo-perfusion but patients show an absence of sweating, bradycardia or pallor.

Investigation of possible syncope and blackouts

For the great majority of subjects a well-taken history and eye witness account will allow confident clinical differentiation between simple syncope and seizures. Difficulties may arise in cases of atypical or convulsive syncope where, no single clinical feature will categorically allow a differentiation. The major differences are summarized in Table 31.2. It is of paramount importance to recognize that while seizures characterized by sudden collapse, loss of consciousness, and rapid recovery, but without significant positive motor phenomena, are seen commencing in childhood, they rarely begin for the first time in adults.

Neurologists will see a number of patients where differentiation between syncope and seizures is difficult. For this reason they should be familiar with guidelines on the investigation of syncope (Brignole *et al.* 2004) and be able to recognize serious cardiac conditions requiring intervention. When syncope occurs with suspected or certain heart disease, cardiac evaluation will include echocardiography, stress testing, and prolonged electrocardiographic monitoring. If severe or recurrent syncope occurs in the absence of heart disease, investigation is required with electrocardiogram, tilt testing, carotid massage, and if negative, prolonged electrocardiographic monitoring. Guidelines for epilepsy draw

Table 31.2 The differences between vaso-vagal syncope and seizures characterized by collapse

	Syncope	Seizures
Posture	Upright	Any posture
Pallor and sweating	Invariable	Uncommon
Onset	Gradual	Sudden/aura
Injury	Unusual	Not uncommon
Convulsive jerks	Not uncommon	common
Incontinence	Rare	Common
Unconsciousness	Seconds	Minutes
Recovery	Rapid	Often slow
Postictal confusion	Rare	Common
Frequency	Infrequent	May be frequent
Precipitating factors	Crowded places Lack of food Unpleasant circumstances	Rare

Table 31.3 Electrocardiographic abnormalities suggesting an arrhythmic syncope (Brignole *et al.* 2004)

- Bifascicular block (defined as either left bundle branch block or right bundle branch block combined with left anterior or left posterior fascicular block)
- Other intraventricular conduction abnormalities (QRS duration >0.12 sec)
- Mobitz I second degree atrioventricular block
- Asymptomatic sinus bradycardia (<50 bpm), sinoatrial block or sinus pause >3 s in the absence of negatively chronotropic medications
- Pre-excited QRS complexes
- Prolonged QT interval
- Right bundle branch block pattern with ST-elevation in leads V1–V3 (Brugada syndrome)
- Negative T waves in right precordial leads, epsilon waves and ventricular late potentials suggestive of arrhythmogenic right ventricular dysplasia
- Q waves suggesting myocardial infarction

Table 31.4 The differences between epileptic seizures and pseudoseizures

	Epileptic seizure	Pseudoseizure
Onset	Sudden	May be gradual
Retained consciousness in prolonged seizure	Very rare	Common
Pelvic thrusting	Rare	Common
Flailing, thrashing, asynchronous limb movements	Rare	Common
Rolling movements	Rare	Common
Cyanosis	Common	Unusual
Tongue biting and other injury	Common	Less common
Stereotyped attacks	Usual	Uncommon
Duration	Seconds or minutes	Often many minutes
Gaze aversion	Rare	Common
Resistance to passive limb movement or eye opening	Unusual	Common
Prevention of hand falling on to face	Unusual	Common
Induced by suggestion	Rarely	Often
Postictal drowsiness or confusion	Usual	Often absent
Ictal EEG abnormality	Almost always (except with simple and some complex partial seizures)	Almost never
Postictal EEG abnormal (after seizure with impairment of consciousness)	Usually	Rarely

attention to the need for standard electrocardiogram recordings in all subjects with a diagnosis of epilepsy or suspected epilepsy (NICE 2004). The electrocardiographic abnormalities which suggest an arrhythmic cause of syncope are listed in Table 31.3.

In considering the differentiation between syncope and seizures it has to be recognized that seizures can themselves be associated with significant cardiac dysrhythmia (Rugg-Gunn *et al.* 2004). Some complex partial seizures are described as 'temporal lobe syncope' (Delgado-Escueta *et al.* 1982). In these, sudden falls occurred without warning, but were followed by amnesia and gradual recovery. They are, however, exceptionally rare as a new seizure type in adults, and often accompanied by other seizure types in keeping with a frontal onset.

31.1.3 Psychogenic alteration in consciousness

Psychogenic non-epileptic attacks

The misdiagnosis of epilepsy is common. Psychogenic non-epileptic attacks, or pseudoseizures, cause the greatest diagnostic difficulty, most commonly mimicking tonic-clonic seizures, or more rarely minor convulsive seizures and syncope (Groppel *et al.* 2000) (Section 4.7.12). Their incidence in the community is difficult to ascertain but could represent 4 per cent of all cases of epilepsy (Sigurdardottir and Olafsson 1998). The incidence increases dramatically as more selected and apparently refractory populations are examined. Thus up to 20 per cent of referrals to tertiary centres are misdiagnosed, and up to 50 per cent of cases of drug-refractory status epilepticus are due to non-epileptic attacks (Howell *et al.* 1989). Their recognition is important because of the potentially tragic consequences of both inappropriate long-term anti-epileptic drug treatment (Smith *et al.* 1999) and intensive care unit admission (Reuber *et al.* 2004).

The diagnosis of pseudoseizures may be difficult, particularly when they occur in patients who also have a history of true epileptic seizures. Eye witness accounts and videotapes of events can raise suspicions, but no single clinical feature differentiates non-epileptic attacks from epilepsy, although a number of factors may be of value (Table 31.4). The most useful clinical features of attacks are resistance to eye opening in pseudoseizures and the absence of pupillary dilatation which is invariably present in tonic-clonic seizures.

The temporal pattern of attacks should, however, alert the clinician to the possibility of non-epileptic attacks. They are refractory to anticonvulsant therapy, in contrast to epilepsy where convulsive seizures in particular are likely to be well controlled. Failure to control apparent tonic-clonic seizures in a patient who develops attacks after the first decade of life, in whom there is no identifiable cerebral disease and in whom interictal EEG recordings have never shown significant epileptiform abnormalities is almost pathonemonic of non-epileptic attacks. With modern anti-epileptic drugs very few patients continue to have frequent tonic-clonic seizures and those who do have preceding severe cerebral insults resulting in intellectual and neurological impairments. In short they have 'bad brains'; evidence of which is strikingly absent in the majority of people most with non-epileptic attacks.

Some positive features allow identification of subjects at risk. Non-epileptic attacks most commonly occur in women, with onset most commonly in the second or third decades of life (Roy 1979; Howell *et al.* 1989). Individuals often have a significant history of self-poisoning and self-injury, and previous episodes of unexplained physical symptoms. They come from dysfunctional families, and there is often a history of physical and sexual abuse (Francis and Baker 1999; Reuber and Elger 2003).

The management of non-epileptic attacks is more difficult than the diagnosis. Symptomatic events will usually need to be recorded by videotelemetry and be shown to be characteristic of the usual attacks. An open discussion of the non-epileptic nature is necessary after which anti-epileptic drugs can be withdrawn, usually with advice that this alone will often result in an improvement in attacks. The psychological and psychiatric background needs to be explored by people with experience of the area and an on going programme of treatment and support instituted. The prognosis is variable, and other non-physical symptoms commonly occur in the long term (Reuber *et al.* 2003). There is controversy about the incidence of epilepsy in people with non-epileptic attacks. It is probably low but those with both problems are exceptionally difficult to manage.

Hyperventilation

Hyperventilation is common and the bulk of cases may go unrecognized but some may be misdiagnosed as epilepsy (Riley 1982). Common manifestations include dizziness, detachment, blurred vision, tingling, muscle spasm, tetany, palpitation, dyspnoea and chest pain, heartburn, epigastric pain, muscle cramps, and fatigue (Section 4.3.4). Some form of alteration in consciousness is common and up to 15 per cent of patients may lose consciousness during attacks (Pincus 1978). The wide variety of symptoms experienced by patients with hyperventilation will most commonly be confused with complex partial seizures. The most useful factors, which allow differentiation are that hyperventilation attacks are commonly precipitated by stressful circumstances and that they lack a stereotype nature with different types of symptoms referable to different systems occurring on different occasions. A simple diagnostic test involves asking the patient to re-breathe from a paper bag held over the mouth and nose during attacks induced by voluntary hyperventilation.

Panic attacks

Panic attacks can be mistaken for complex partial seizures. They often include hyperventilation but also commonly encompass abdominal discomfort, a choking feeling, fear, autonomic symptoms, and sometimes even loss of consciousness. The episodes, however, are usually clearly precipitated and often more prolonged than seizures. Panic attacks are most likely to occur in association with phobic anxiety states and patients usually have considerable insight into the nature of their attacks.

Rage outbursts

It is controversial whether violent behaviour is common in people with epilepsy and complex partial seizures (van Elst *et al.* 2000). It is common for individuals who describe sudden outbursts of violent behaviour with minimal provocation, often with some associated patchy amnesia, to be referred for neurological evaluation with a presumptive diagnosis of a seizure disorder. It is striking that individuals are invariably young men from deprived backgrounds who have often themselves been abused. Violence occurs in response to minimal stimuli and is often directed against a specific family member. This is clearly distinct from epilepsy and the term 'episodic dyscontrol' has been used (Maletzky 1973). Strict guidelines must be applied before ever accepting that aggressive or violent behaviour is part of a seizure disorder (Treiman and DelgadoEscueta 1983). An epileptic basis to such attacks should only be accepted where definite seizures occur at other times and where violent behaviour is a consistent feature of that individual's seizures. Violence can only be accepted as seizure related where it is brief and poorly directed.

Fugue states

States of psychogenic wandering are prolonged, usually with sudden recovery of awareness (Section 4.7.12). It is often impossible to obtain a clear account of behaviour during such attacks but this usually appears to have been quite normal. Subjects have a dense amnesia for the period of time concerned and individuals usually have an associated depression and the need to escape from some stressful life situation (Stengel 1943). Such episodes may be confused with complex partial status or other forms of non-convulsive status epilepticus (Mayeux *et al.* 1979), but here there is usually evidence of abnormal behaviour during the amnesia, which is briefer and more frequent (Kopelman *et al.* 1994), and more akin to transient global amnesia (Section 31.1.5).

31.1.4 Focal cerebral ischaemia

On rare occasions it is difficult to differentiate between focal seizures and focal ischaemia due to either migraine or thrombo-embolic disease.

Transient ischaemic attacks will rarely be confused with seizures because they develop more slowly and last longer. They are virtually never accompanied by altered consciousness, and motor and sensory phenomena that occur are almost uniformly negative. Whilst rarely focal seizures may rarely be accompanied by predominantly negative motor or sensory problems (Lesser *et al.* 1987), the greatest difficulties occur in rare haemodynamic transient ischaemic attacks in which weakness may be accompanied by some shaking and tremor (Yanagihara *et al.* 1985).

Migraine and epilepsy occur in the same individual more commonly than would be expected by chance. Loss of consciousness is not uncommon in migraine though usually it takes the form of syncope associated with nausea and hypotension (Risser 1985). The complex relationship between migraine and epilepsy has been reviewed (Haut *et al.* 2006). On occasions migrainous episodes may induce frank seizures and arteriovenous malformations may cause both seizures and migraine-like phenomena. It does seem that migraine and benign Rolandic and occipital epilepsies commonly co-exist, as they may in some mitochondrial disorders such as the MELAS syndrome of Mitochondrial Encephalopathy, Lactic Acidosis, and Stroke-like episodes (Section 35.4.8).

31.1.5 Transient global amnesia

The syndrome of transient global amnesia describes an abrupt onset of amnesia usually accompanied by repetitive questioning, in an individual who remains alert and communicative (Fisher and Adams 1958). The amnesia lasts for hours and attacks recur only rarely. The aetiology of this syndrome remains controversial but it is highly likely that it has different causes which may include thrombo-embolic disease, migraine, and epilepsy (Hodges and Warlow 1990). Up to 7 per cent of patients with transient global amnesia are likely to have an epileptic basis to their attacks. These can usually be identified by attacks that are brief and which recur over a short period of time. They are common on wakening, and individuals may have some partial recall. Sometimes such individuals will also describe some features at the start of the attacks, which would support focal seizure onset, for example, olfactory hallucination. Other types of simple or complex partial seizures occur, although transient amnesia can be the sole manifestation of seizures (Zeman *et al.* 1998).

31.1.6 Sleep phenomena

A number of sleep phenomena may be confused with seizures (Section 32.4). *Hypnic jerks* are usually single jerks that occur in the very early stages of sleep (Section 32.4.1). They usually lead to arousal and may be accompanied by a feeling of falling, a cry, or some other kind of brief sensory disturbance. *Periodic movements of sleep* are rhythmic and repetitive leg movements more commonly seen in later life (Section 32.4.5). They usually consist of jerking movements, usually dorsiflexion of the foot and extension of the toes that can occur repetitively during non-rapid eye movement sleep (Coleman *et al.* 1980). *Sleep walking* is a form of automatic behaviour occurring during deep non-rapid eye movement sleep and is much more common in children than in adults (Section 32.4.2). It usually ceases by the mid-teens. In adults it must be distinguished from post-ictal automatism following sleep seizures or from complex partial seizures resulting in automatic behaviour (Pedley and Guilleminault 1977).

Abnormalities of sleep can also lead to confusion with seizures because of daytime disturbances. *Narcolepsy* should not lead to significant confusion (Section 32.3.1). However, *cataplexy* (Section 32.3.1) in which there may be sudden collapse with loss of postural tone triggered by emotion or startle or loud noise could potentially be confused with atonic drop attacks. However, the age of onset of such episodes should preclude real confusion; atonic seizures most commonly occur during the first decade of life whereas symptoms of the narcoleptic syndrome rarely begin before the second or third decades of life. Some subjects with narcolepsy can also exhibit automatic behaviour when they appear to be only half awake. The individual appears drowsy and absent minded, though may be capable of carrying on relatively complex tasks for which they are subsequently amnesic.

Sleep apnoea, which is most commonly obstructive in nature and associated with obesity and night-time snoring leads to day-time drowsiness (Section 32.3.4). Occasionally episodes of day-time sleepiness may be associated with respiratory obstruction and jerks that can lead to referral with the suggestion of a seizure disorder.

31.1.7 Movement disorders

Some unusual movement disorders may on occasion cause confusion with seizures. Non-epileptic myoclonus must be differentiated from epileptic myoclonus (Section 40.7). *Paroxysmal kinesogenic choreoathetosis* is a rare disorder in which short-lasting tonic spasms with writhing movements, usually affecting an arm or a leg occur (Kertesz 1967) (Section 40.4.7). The onset is usually in adolescence and there is a strong male predominance, and often a family history. The attacks are precipitated by sudden movement after a period of rest and are often preceded by a peculiar sensation in the limb before the movement commences. They can occur frequently but respond readily to antiepileptic drugs, though there is no evidence to suggest that they are epileptic in nature. The syndrome is so striking that it should not easily be confused with a seizure disorder although rarely some patients with movement-induced seizures and abnormal ictal EEGs have been described. A non-kinesigenic form is described with very similar movements but usually of much longer duration (Mount and Reback 1940). This is familial with onset in infancy or childhood and there is no male predominance. Tonic spasms induced by movement can occur in multiple sclerosis as may paroxysmal episodes of dysarthria and ataxia (Section 37.5.3).

31.1.8 Metabolic events

A number of metabolic disturbances may result in acute symptomatic seizures (Section 31.7). Hypoglycaemia is unusual in that it

may be recurrent and give rise to diagnostic confusion with epilepsy. Hypoglycaemia is most commonly seen in diabetics receiving insulin or oral hypoglycaemic agents. Diabetics may be particularly sensitive to hypoglycaemia and can experience symptoms at higher blood glucose levels than non-diabetic subjects. Marks (1981) describes different types of neuroglycopenia, the commonest of which is acute. As blood sugar falls, pallor, sweating, and tachycardia develop associated with confusion, collapse, and occasionally coma. True seizures may occur during the course of hypoglycaemia further complicating diagnosis. Hypoglycaemia must always be considered in the differential diagnosis of epilepsy in a diabetic population but perhaps the greatest diagnostic difficulty will arise in the rare cases of insulin-secreting tumours in non-diabetic patients. Here, hypoglycaemia is most likely to occur during the course of the night. Less common is a syndrome of 'subacute neuroglycopenia' characteristic of insulinoma. Subjective symptoms are not marked but there is mild confusion and clumsiness. There may be disinhibition, suggesting intoxication with ataxia and slurred speech. This confused behaviour of hypoglycaemia may be difficult to differentiate from complex partial seizures.

31.1.9 Other miscellaneous events

A negative motor phenomenon that may sometimes be confused with epilepsy is the syndrome of cryptogenic or benign drop attacks of middle-aged women (Stevens and Matthews 1973) (Section 2.6.5). These result in a sudden fall, usually onto the knees, without any clouding of consciousness. The episodes tend to be infrequent and the outcome seems to be quite benign. It must be emphasized that while atonic, tonic, and myoclonic seizures can cause brief episodes of falling with rapid recovery in children and occasionally in adolescence, such attacks do not commence in adult life.

McCrory (1997) drew attention to *concussive convulsions* as a non-epileptic phenomenon in Australian rules football. Convulsions began within 2 sec of the head injury and usually consisted of a brief tonic phase followed by bilateral myoclonic jerks. Some versive head movements and asymmetric posturing were seen in some individuals. No convulsion lasted for more than 150 sec and none of the players had any behavioural or neuropsychological features that suggested anything other than a mild concussion. While these events have been widely assumed to represent a form of post-traumatic epileptic seizure, there is now considerable evidence to the contrary. Follow-up of subjects with such events by Jennett indicated that immediate convulsions were not a predictor of late post-traumatic epilepsy, in contradistinction to other seizures occurring in the first post-traumatic week, and that on the whole they tended to be associated with relatively mild concussive head injury (Jennett 1975). Similar events can occur with head injuries during other sports such as horse racing (Chadwick 1997). It is important that they are recognized as non-epileptic in order that inappropriate restrictions are avoided.

31.2 Mechanisms of seizures and epilepsy

In man spikes and sharp waves are the electroencephalographic hallmarks of interictal recordings of many patients with epilepsy. Such activity appears to be due to a hypersynchronization of electrical activity within an abnormal pool of neurones, and they are rarely seen in the EEGs of non-epileptic patients (Section 31.9.1). Simplistically, epileptic seizures occur when excitatory influences in the cerebral hemispheres outweigh inhibitory influences. Study of the basic mechanisms of the human epilepsies is, of course, fraught with ethical and practical difficulty. Thus much knowledge has been accumulated from a number of animal models of seizures and epilepsy. While the direct relevance to the human epilepsies remains in some doubt, it seems likely that knowledge gained in this way will be highly informative.

There is considerable evidence that the fundamental building block of most seizure disorders is the paroxysmal depolarization shift and an associated high frequency burst firing of neurones (Prince 1978). This phenomenon is one that can be observed in isolated cells, within simple neuronal circuits, in animal models of epilepsy, and indeed in human post-operative material (Fig. 31.1). Paroxysmal depolarization shift may occur normally as part of the spontaneous activity of CA3 hippocampal pyramidal neurones (Wong and Prince 1981), and in pyramidal cells of layers 4 and 5 of the neocortex (Gutnick *et al.* 1982). While paroxysmal depolarization shift and burst firing can be intrinsic properties of neurones, the propagation and synchronization of this kind of activity to produce either interictal spikes or seizures, requires a contribution from neuronal circuits, which may themselves exhibit abnormalities predisposing neurones within them to behave in an abnormal fashion. The concept of paroxysmal depolarization shift and burst firing underlies not only to our understanding of the basic mechanisms underlying epilepsy, but also to the mechanisms of and targets for anti-epileptic drug action (White 1997).

31.2.1 Molecular and cellular factors

A number of factors control neuronal excitability. These include voltage-gated ion channels, neurotransmitter ligand activated ion channels, neuromodulators, and second messenger systems. Ligand-gated ion channels are responsible for communication between cells while voltage-gated channels determine how inhibitory and excitatory influences are integrated in a way that determines the propagation of impulses to other neurones.

Voltage-gated channels

Neuronal membranes are usually polarized to a potential of -90 mV by the activity of Na^+–K^+–ATPase transporter systems. Voltage-gated ion channels are membrane-spanning proteins composed of different sub-units that when open, permit the passage of ions. Openings may be transient or persistent depending on the nature of the channel. Most channels will open on depolarization of the membrane, but some open when the membrane is hyperpolarized. Channels may exhibit a number of different states (Fig. 31.2).

Voltage-gated sodium channels are intimately involved in the propagation of action potentials, the rapid upstroke being due to an opening of fast transient channels at about –60 mV. In addition there is also a persistent component of the current that may be of greater relevance to epilepsy. Toxins that prolong sodium channel opening cause burst firing and seizures (Mantegazza *et al.* 1998). Phenytoin and carbamazepine are able to block repetitive firing of neurones effect on voltage-gated sodium channels that is both use-dependent and voltage-dependent; sodium valproate has a similar action also (Fig. 31.3). For a number of reasons, this mechanism of action may be seen as having ideal properties. It will tend to block the pathological neuronal activity of repetitive firing with relatively little effect on more physiological patterns of activity, and would also be predicted to be highly effective in preventing the spread of seizure activity. The importance of voltage-gated sodium channels in human epilepsy is further emphasized by the finding of molecular abnormalities in families with generalized epilepsy and febrile seizures and patients with severe myoclonic epilepsy of infancy (Escayg *et al.* 2000; Claes *et al.* 2001).

Voltage-dependent calcium channels contribute to dendritic spikes, slow somatic depolarizations, and associated burst discharges, and by doing so, trigger neurotransmitter release. Six sub-classes of calcium channels are known to exist: L, N, T, P, Q, and R. T channels have a low threshold of activation at around –70 mV. They inactivate relatively rapidly. These channels are found in

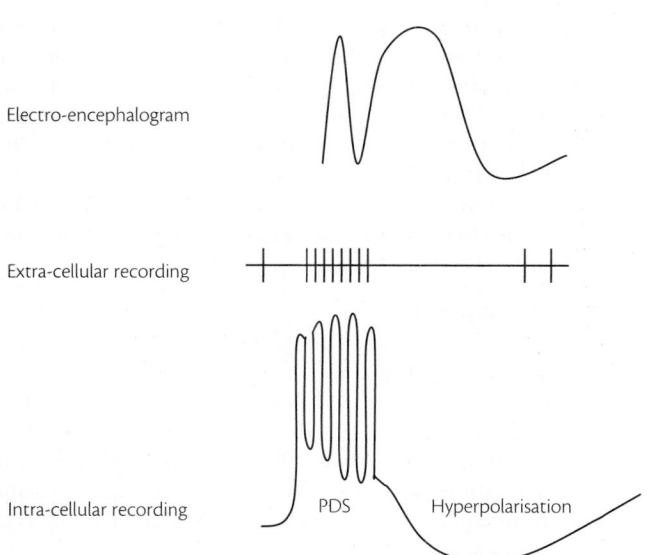

Fig. 31.1 Epileptiform activity in the EEG and its relationship to intracellular events. Surface spikes are generated by synchronous paroxysmal depolarizations (PDS) in large groups of neurones.

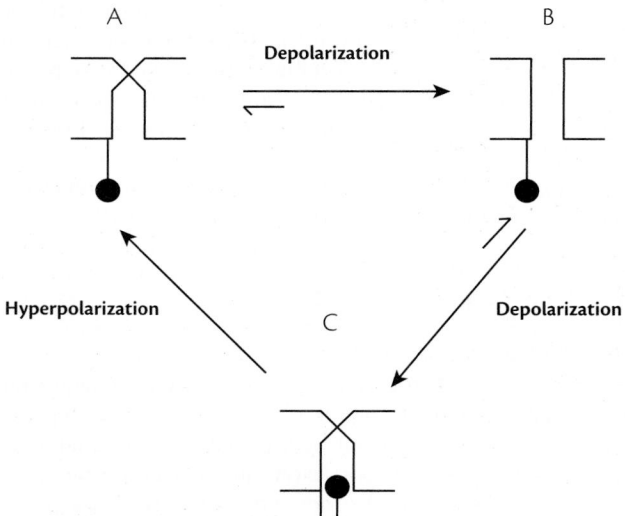

Fig. 31.2 Different states in which the voltage sensitive Na+ channel may exist.
A—Closed but can be activated.
B—Open allowing entry of Na+ ions.
C—Closed and inactivatable, a state which prevents repetitive firing and is made more probable by anti-epileptic drugs such as phenytoin etc.

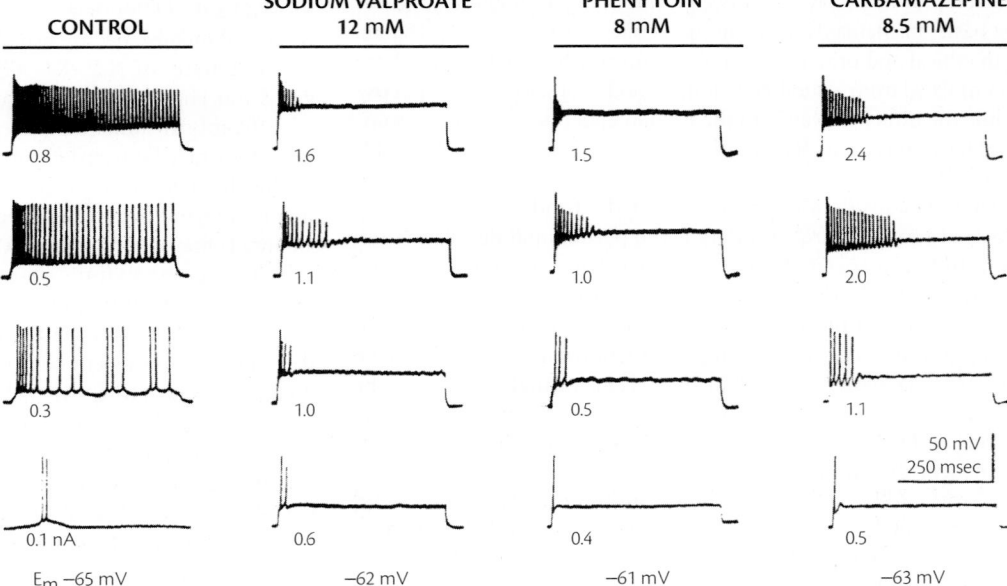

Fig. 31.3 The effects of increasing depolarizing current pulses on intracellular membrane potentials of spinal neurones in culture exposed to different anticonvulsants. With increasing current there is more prolonged depolarization and burst firing in control medium. Sodium valproate, phenytoin, and carbamazepine all block repetitive firing, the blockade increasing with increasing currents applied. (Adapted from Macdonald (1994) with permission.)

high concentrations in thalamic neurones and play an important role in the generation of generalized spike wave discharges. They appear sensitive to anti-absence drugs such as ethosuximide, methadione derivatives, and sodium valproate (Coulter *et al.* 1990). Over-expression of low-threshold Ca^+ channels appears to be implicated in animal models of absence and generalized spike wave (Burgess and Noebels 1999). A mutation of *CACNA1A* has been found in a patient with absence epilepsy (Jouvenceau *et al.* 2001).

Voltage-gated potassium channels are very diverse in their nature. Delayed rectifying potassium currents, I_k, activate at potentials more positive than –40 mV and seem to contribute to spike repolarization. Fast transient currents activate at –45 mV to –60 mV and they appear to play an important role in the regulation of repetitive firing by prolonging after-spike hyperpolarization and slowing down firing rate. Inward rectifying potassium currents activate in response to hyperpolarization, and help regulate resting membrane potentials. Potassium currents can be blocked with tetraethylammonium and 4-aminopyridine (Jones and Heinemann 1987).

Mutations of *KCNQ2* and *KCNQ3* encoding K^+ channel subunits can lead to benign neonatal convulsions (Biervert *et al.* 1998; Singh *et al.* 1998). The potentiation of voltage-dependent potassium currents is currently being explored as a potential target for new anti-epileptic drugs.

Inhibitory neurotransmission

γ-aminobutyric acid, GABA, is the major inhibitory neurotransmitter in the forebrain, being present at approximately 30 per cent of all synapses in the central nervous system. The majority of GABAergic neurones are short axon interneurones forming local inhibitory loops. Three distinct types of GABA receptor are recognized. $GABA_A$ and $GABA_C$ receptors are receptors linked to chloride ion channels. Metabotropic $GABA_B$ receptors may be pre- or post-synaptic and are coupled to calcium or potassium ion channels via GTP proteins (Bowery 1993; Macdonald and Olsen 1994). $GABA_A$ receptors are large molecular weight proteins that

contain a number of binding sites, not only for GABA but also for barbiturates, benzodiazepines, picrotoxin, and anaesthetic steroids. Binding of GABA leads to an opening of the chloride and potassium ion channels and resultant hyperpolarization, known as inhibitory post-synaptic potentials. The receptor has been cloned and five different subunit families isolated. Receptors constructed from different combinations of sub-units appear to exhibit differing pharmacological properties, though binding sites for GABA and barbiturates appear highly conserved.

Drugs such as allylgylcine, which prevents the synthesis of GABA, picrotoxin, and bicuculline which block GABA receptors, as well as penicillin, are all potent convulsant agents. The activity of benzodiazepines and barbiturates at $GABA_A$ receptor appears responsible for their anti-epileptic activity (White 1997). Vigabatrin, a rationally developed anti-epileptic drug, is a suicidal inhibitor of GABA-transaminase, the enzyme responsible for GABA metabolism. Tiagabine potentiates GABAergic activity by blocking the re-uptake of GABA into neurones and glia. Mutations of the *GABRG2* gene encoding for $GABA_A$ subunits have been found in generalized epilepsy and febrile seizures (Wallace *et al.* 1998; Baulac *et al.* 2001). On the other hand, enhanced GABAergic inhibition via $GABA_B$ receptors appears to worsen absence seizures in humans and experimental animals (Section 31.2.2).

Excitatory neurotransmission

The major excitatory neurotransmitters are the amino acids L-glutamate and L-aspartate. They exert their synaptic influences by interacting with a number of different types of receptors, which are identified because of specificity for binding different molecules. Binding at the DL-α-amino-3-hydroxy-5-methyl-isoxazolepropionic acid, AMPA, receptor makes its associated ion channel permeable to both sodium and potassium. It desensitizes rapidly (Tang *et al.* 1989) and is probably responsible for the majority of rapid excitatory neurotransmission. Four glutamate receptors genes, GluRs1-4, have now been cloned and appear to encode sub-units that can express the known electrophysiology and pharmacology of the

AMPA receptor. The kainate receptor is also coupled to a channel permeable to sodium and potassium. It does not, however, appear to desensitize at the same rate as the AMPA receptor. Both these channels differ strikingly from the *N*-methyl-D-aspartate, NMDA, receptor. This appears to be a much more complex receptor site that has an absolute requirement for the presence of a co-agonist, glycine, in order to result in channel opening (Johnson and Ascher 1987). Magnesium, zinc, polyamines, and steroids can also modulate the site. When membranes are hyperpolarized the channel is blocked by magnesium, a blockade that is reversed when the membrane depolarizes. Opening of the channel allows the entry of both sodium and calcium. This acts as an amplification mechanism that leads to prolonged activation of already excited neurones and associated burst firing (Williamson and Wheal 1992). Calcium entry may also ultimately result in excitotoxicity and cell death. Like the AMPA and kainate receptor, the NMDA receptor exists as a number of sub-families.

Excitatory amino acids are also able to interact with metabotropic receptors that activate second messenger systems to influence biochemical pathways and ion channels. These receptors are found both pre-synaptically and post-synaptically. Activation usually results in pre-synaptic inhibition and post-synaptic excitation. These receptors may have an important role in supporting epileptic activity (Arvanov *et al.* 1995).

Acetylcholine receptors

The role of nicotinic acetylcholine receptors in the central nervous system excitabailty is poorly understood, though they may serve as ligand-gated sodium channels. Mutation of a gene, *CHRNA4*, which encodes for the β2 sub-unit of the receptor has been associated with autosomal dominant nocturnal frontal lobe epilepsy (Steinlein *et al.* 1995).

31.2.2 **Epileptic activity in neuronal systems**

While molecular changes may predispose to burst firing of neurons, synchronization of such activity, necessary for seizures also requires the involvement of neuronal circuits. Here our knowledge of the way in which circuits operate to cause or suppress seizures is fragmentary. We have little indication of why a fixed alteration in a gene-product may give rise to an intermittent and paroxysmal disturbance. While we may be able to identify the causes of epilepsy in many more patients at molecular or lesional levels, what are the changes within the brain that result in clustering of seizures on 2 or 3 days, but not on others in a month? This remains the fundamental clinical question in epilepsy.

Focal epileptogenesis

The most studied and clinically relevant model of focal seizures and epileptogenesis is the hippocampus and hippocampal sclerosis, the most common form of focal epilepsy in man. Mesial temporal structures are interconnected by a reverberating loop involving enterorhinal cortex, dentate gyrus, CA3, CA1, enterorhinal cortex (Fig. 31.4). Normal spontaneous activity of CA3 pyramidal neurones consists of paroxysmal depolarization shifts and associated burst firing of the cell body and apical dendrites (Wong and Prince 1979; 1981). Function of the hippocampal brain slice can be studied in material from normal animals exposed to a variety of chemical manipulations as well as in slices from animals expressing a chronic model of epilepsy (Traub and Jefferys 1998). In normal brain material, bursting activity in CA3 neurones has about a 30 per cent chance of evoking bursts in connected neurones. This probability increases when the pyramidal cell is excited by many inputs simultaneously.

This system is also subject to plasticity in response to a number of stimuli or to damage. Three phenomena; cell loss, mossy fibre sprouting, and neurogenesis may all enhance the potential for

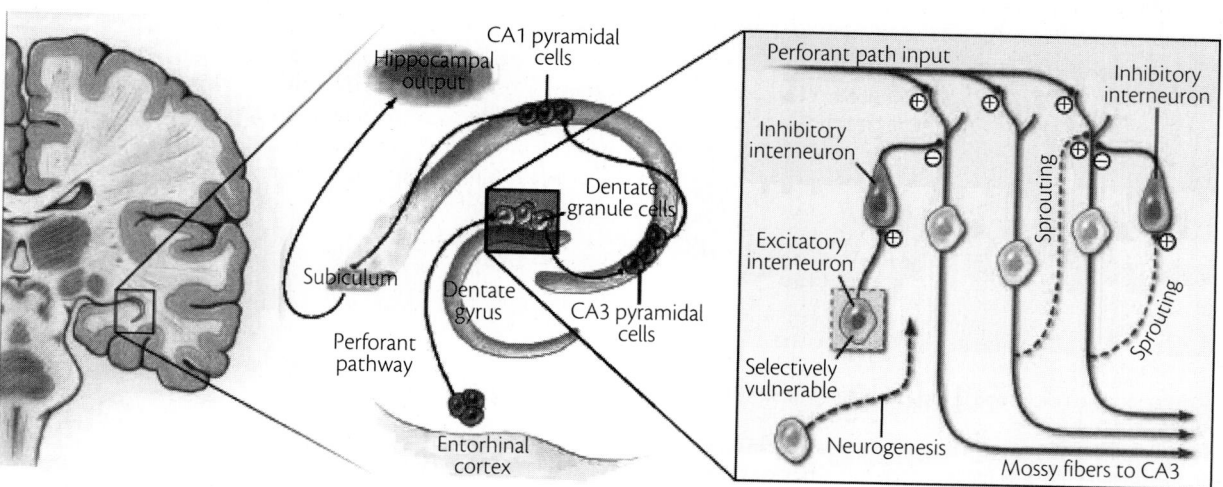

Fig. 31.4 Hippocampal sclerosis is the most common identified pathological feature in cases of mesial temporal-lobe epilepsy. Normally, input to the hippocampus comes from the entorhinal cortex to the dentate granule cells through the perforant path. Dentate granule cells project to the CA3 sector as the first step in the hippocampal output pathway. A close-up of the dentate granule-cell layer reveals several morphologic changes characteristic of hippocampal sclerosis that may play a part in epileptogenesis. Newly sprouted mossy fibres from dentate granule cells can synapse on dendrites of neighbouring dentate granule cells, resulting in a recurrent excitatory circuit. They can also sprout onto inhibitory interneurons. Excitation interneurons, which normally activate inhibitory interneurons, may be selectively vulnerable to brain insults. Finally, neurogenesis of new dentate granule cells continues into adult life, and these neurons may integrate themselves into abnormal circuits. (Reproduced from Chang and Lowenstein 2003 with permission.)

seizure generation while resulting in pathological change that can lead to hippocampal sclerosis. In the kindling model of epilepsy, repeated sub-convulsive electrical stimulation, usually to the amygdala, leads to increasing after-discharge and ultimately to behavioural seizures (Goddard *et al.* 1969). Repeated focal applications of convulsant agents can lead to a similar phenomenon. It appears that limbic structures are particularly sensitive to the development of kindling when compared to neocortex, a situation that is reflected in man, where the temporal lobe is by far the most common site of seizure onset (Section 31.4.2).

At a pathological level, kindled animals show evidence of neuronal loss in the hippocampus accompanied by sprouting of the mossy fibre axons of the dentate granule cells (Sutula *et al.* 1988). It seems that sprouting probably requires the death of neurones and that the sprouting fibres take up synaptic sites that are thereby vacated. Similar changes are seen after experimental damage from status epilepticus (Tauck and Nadler 1985). If new synaptic connections are made to other excitatory cells, this would represent a process potentially contributing to the hyperexcitability of the

kindled brain. More recently there has been evidence of post-natal neurogenesis occurring in the dentate gyrus through life (Eriksson *et al.* 1998). In the pilocarpine model of chronic epilepsy, seizures can induce neurogenesis, neurones being abnormally integrated into existing circuits (Parent and Lowenstein 1997). These changes therefore provide a potential basis for the clinical predisposition of the mesial temporal structures to produce seizures and a plausible mechanism for some of the progressive changes that may be seen as part of drug-resistant epilepsies. Full proof of the concept is however, lacking.

Generalized epilepsies

Here the sudden onset, the bilateral synchrony, and in the case of simple absences and myoclonus, the non-evolving pattern of electrical activity clearly suggests some very generalized disturbance of neuronal activity in both hemispheres. Historically, there have been two schools of thought that would explain such a phenomenon; that it is a primary cortical process or that there is an unspecified 'centrencephalic' system in which structures of the upper brainstem

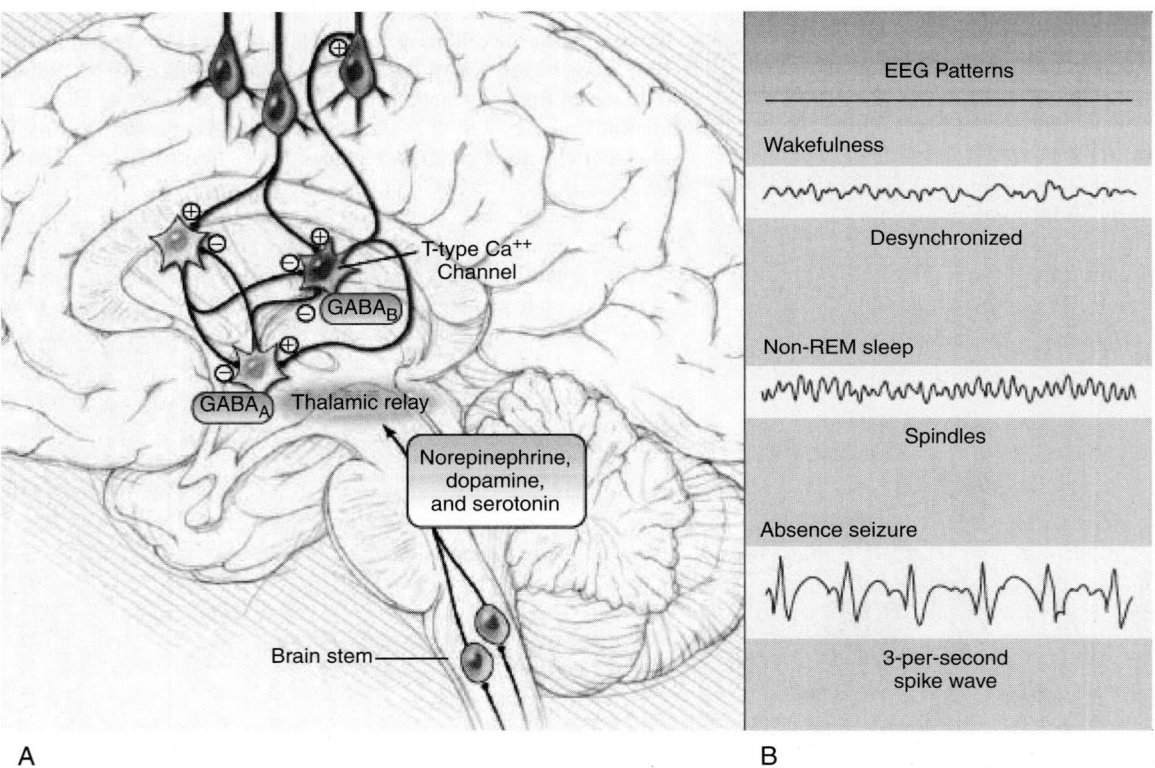

A
B

Fig. 31.5 The normal thalamocortical circuit and EEG patterns during wakefulness, non-rapid eye movement (non-REM) sleep, and absence seizures.

Panel A shows the normal thalamocortical circuit. Thalamic relay neurons can activate the cortical pyramidal neurons in either a tonic mode or a burst mode, the latter made possible by T-type calcium channels. The mode of thalamocortical activation is controlled largely by input from the thalamic reticular neurons, which hyperpolarize the relay neurons through γ-aminobutyric acid type B ($GABA_B$) receptors and are themselves inhibited by neighbouring reticular neurons through activation of GABA type A ($GABA_A$) receptors. Cortical pyramidal neurons activate the thalamic reticular neurons in a feed-forward loop. Ascending noradrenergic, serotonergic, and dopaminergic inputs from brainstem structures appear to modulate this circuit.

Panel B shows EEG patterns of wakefulness, non-REM sleep, and absence seizures. During wakefulness, the cortex is activated by the thalamus in a tonic mode, allowing for processing of external sensory inputs. This results in a desynchronized appearance of the EEG. During non-REM sleep, the cortex is activated in a burst mode, resulting in the EEG appearance of rhythmic sleep spindles. During an absence seizure, the normal thalamocortical circuit becomes dysfunctional, allowing burst activation of the cortex to occur during wakefulness, which results in the EEG appearance of rhythmic spike-wave discharges and interrupts responsiveness to external stimuli. (Reproduced from Chang and Lowenstein 2003 with permission.)

and thalamus are responsible for generating the spike wave discharge and driving a cortical synchrony. Gloor (1968) has pointed out that these two hypotheses are not mutually exclusive and has developed a 'generalized cortico-reticular' hypothesis.

The cellular substrate for these phenomena is now well-understood (Fig. 31.5). It is dependent on a thalamocortical circuit that includes the nucleus reticularis thalami. The circuits involve excitatory glutaminergic synapses and inhibitory gabergic synapses. The behaviour of thalamic neurones and the circuit is largely determined by the presence of a high density of calcium T chanels, which results in Ca^{2+}/K^{+}-dependent burst firing. These in turn can give rise to strong inhibitory postsynaptic potentials mediated by $GABA_B$ receptors in thalamocortical relay neurones.

This circuitry is important in activating cortical neurones during sleep–waking cycles. Tonic activity in the relay neurones occurs during wakefulness and rapid eye movement sleep, but they fire in the burst mode during non-rapid eye movement sleep. In the awake animal, thalamic neurones are maintained at a resting potential of approximately –50 mV because of the effects of normal afferent activity of brainstem-activating systems. In this state, calcium T channels are not activated. During drowsiness and sleep, however, thalamic neurones hyperpolarize and begin to exhibit typical repetitive burst firing that contributes to sleep spindles in the EEG (Steriade et al. 1993). The classical anti-absence drug ethosuximide acts by causing a voltage-dependent blockade of T-type calcium currents a property shared by valproate. Hyperpolarization and burst firing is greatly facilitated by GABAergic activity via $GABA_B$ mechanisms. Thus, $GABA_B$ receptor agonists such as baclofen, exacerbate absence and $GABA_B$ antagonists have anti-absence properties in animal models (Hosford et al. 1992; Liu et al. 1992). These phenomena probably also explain the effects of vigabatrin in exacerbating absence seizures.

31.3 Epidemiology of seizures and epilepsy

Despite problems with differing definitions of epilepsy and case ascertainment methods, there is remarkable agreement about the epidemiology of epilepsy in different populations in the developed world (Sander et al. 1990; Sander 2003) Incidence rates vary in the range of 20–55/100 000 per year whereas the prevalence for active epilepsy is in the range of 4–10/1000. Age-specific incidence, prevalence, and cumulative incidence are described for a population in Rochester, Minnesota (Fig. 31.6) and replicated in other populations. It can be seen that the incidence of epilepsy is highest at the extremes of life but that there are significant differences between the cumulative incidence and prevalence of epilepsy, indicating that the majority of patients who develop epilepsy do not suffer from a chronic disorder. The cumulative incidence of epilepsy by the age of 70 may be as high as 2–3 per cent of the population. There is evidence that the incidence in children may be falling with time (Section 30.3.2), although it may be rising in the elderly. Incidence and prevalence, is higher in third world countries than the developed world, with higher rates usually found in rural as opposed to urban communities.

Most studies suggest a slightly higher incidence and prevalence in men than women, with approximately two-thirds of cases being partial epilepsies and approximately one-third generalized. Epidemiological studies identify a wide range of risk factors for the development of epilepsy (Fig. 31.7).

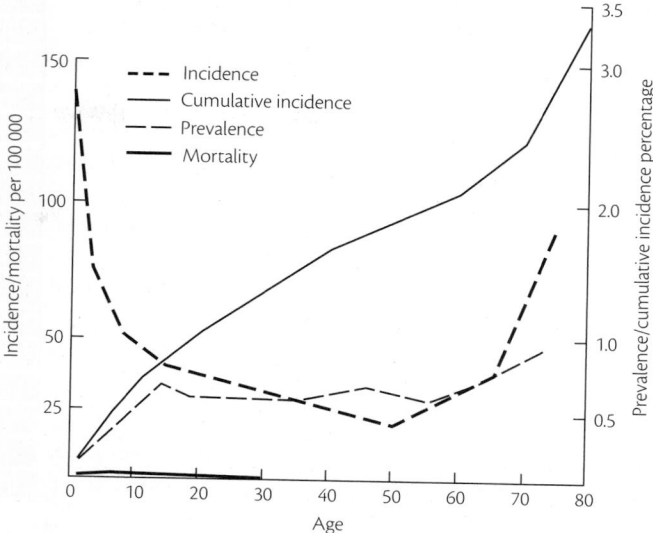

Fig. 31.6 Age-specific incidence, prevalence and cumulative incidence rates for epilepsy in Rochester, Minnesota 1935–74. (Reproduced from Anderson et al. 1986 with permission.)

31.3.1 The prognosis of epilepsy

Community rather than hospital-based studies (Annegers et al. 1979b) and the National General Practice Survey of Epilepsy (Cockerell et al. 1997) are the best source of information in this area. Four hundred and fifty seven patients identified in Rochester, with a history of two or more non-febrile seizures were followed for at least 5 years, and in the case of 141 for 20 years. The probability of being in a remission lasting for 5 years or more was 61 per cent at 10 years, and as high as 70 per cent at 20 years (Fig. 31.8). Similarly in the general practice survey 68 per cent of patients achieved a 5-year remission by 9 years of follow-up. Further support for such high rates of remission is obtained from studies of patients followed prospectively from diagnosis and the commencement of therapy, which show that between 50 and 77 per cent of such patients are 'controlled' depending on how control is defined (Reynolds 1987).

The converse of this information is that 20–30 per cent of patients with epilepsy never achieve remissions; they have a refractory epilepsy that is associated with psychosocial handicap (Jacoby et al. 1996). Relative few patients switch between seizure and seizure-free states, indicating that epilepsy is bimodal in its outcome. The question then arises as to what determines prognosis?

The type of epilepsy or epilepsy syndrome is of considerable importance. Thus, juvenile myoclonic epilepsy is life long, but benign Rolandic epilepsy never recurs during adult life. However, many children and adults with epilepsy cannot be classified by epilepsy syndrome and even within syndromes there may be considerable variation in outcome. In West's syndrome, follow-up for up to 35 years showed that 30 per cent had died, but 24 per cent survived with normal intelligence and 35 per cent were seizure-free (Riikonen 1996). Partial epilepsies have a poorer prognosis than generalized epilepsy (Cockerell et al. 1997).

The age of onset of epilepsy is perhaps one of the most important factors affecting outcome. The commencement of seizures within the first year of life, when it is usually symptomatic of cerebral

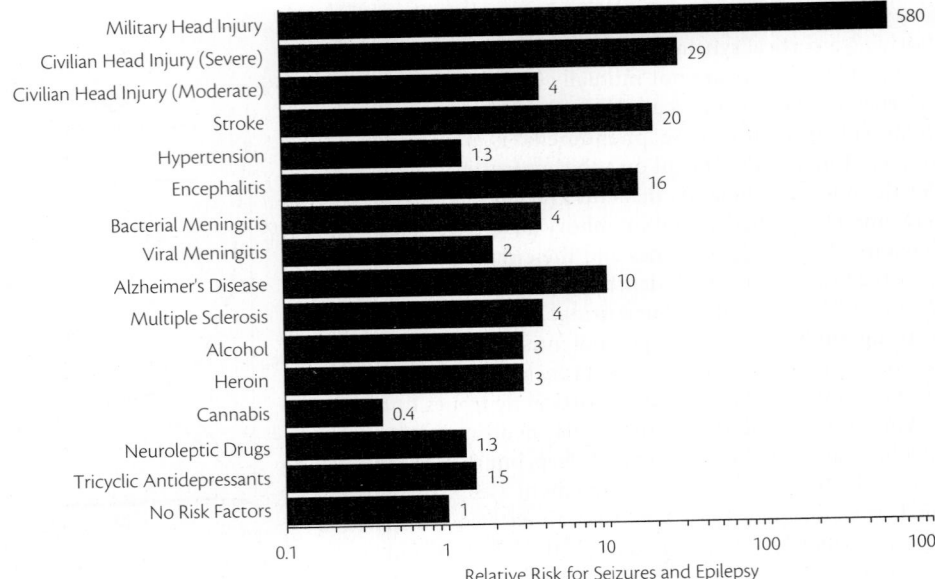

Fig. 31.7 Relative risks for specific risk factors predisposing to epilepsy. (Adapted from Hauser & Hesdorffer 1990 with permission.)

pathology and indicative of one of the malignant childhood epilepsies, carries a particularly adverse prognosis (Sofijanov 1982). In childhood the risk of intractability falls strongly with each additional year of age of onset (Berg *et al.* 1996). The same effect of age is seen in studies which include adults, though the effect is weaker.

Whatever the age of onset, the long-term outcome of epilepsy can usually be predicted around the time of diagnosis. The outcome is worse for the more seizures that occur before diagnosis (Cockerell 1997). Annegers *et al.* (1979a) showed that most patients who achieve remission do so early during the course of treatment. With continuing seizures it becomes progressively less likely that an individual patient will enter remission. Thus, there is a plateau in the number of patients in remission 15–20 years after the onset of epilepsy.

Most studies indicate that symptomatic epilepsy, with an identified cause, carries a poorer prognosis and individuals with associated neurological, cognitive, behavioural, and psychiatric impairment also fare worse (Hauser and Hesdorffer, 1990).

31.3.2 Mortality from epilepsy

All available studies show an increased mortality ratio for epilepsy of between 2 and 3 times the expected (Cockerell 1997). A number of factors appear to contribute to this excess mortality. The greatest excess occurs in the early years of life and is more obvious in men than women. The risk of mortality is greatest in the early years following diagnosis, and is highest for patients with tonic-clonic seizures, seizures that recur frequently, and for those with remote symptomatic epilepsy.

Some of the excess mortality seems to be associated with the underlying aetiology of an epilepsy rather than the occurrence of seizures themselves. The association of epilepsy with cerebrovascular disease in later life probably accounts for the excess mortality associated with the diagnosis of epilepsy in the elderly. The association with mental handicap and cerebral palsy in younger age groups seems to be of considerable importance. Both neoplasms and arteriovenous malformations contribute to the excess mortality from symptomatic epilepsies.

Some controversy surrounds the role of sudden unexpected death in people with epilepsy. A population-based incidence study (Ficker *et al.* 1998) indicates that the incidence of sudden unexplained death was over 20 times higher in people with epilepsy than in the community as a whole. The overall incidence of sudden death in this study in people with epilepsy was 0.35 per 1000 patient years. Thus, it remains an extremely rare event. The major risk factors for sudden unexpected deaths appear to be young age, early onset of seizures, the presence of tonic-clonic seizures, the male sex, and being in bed. While status epilepticus

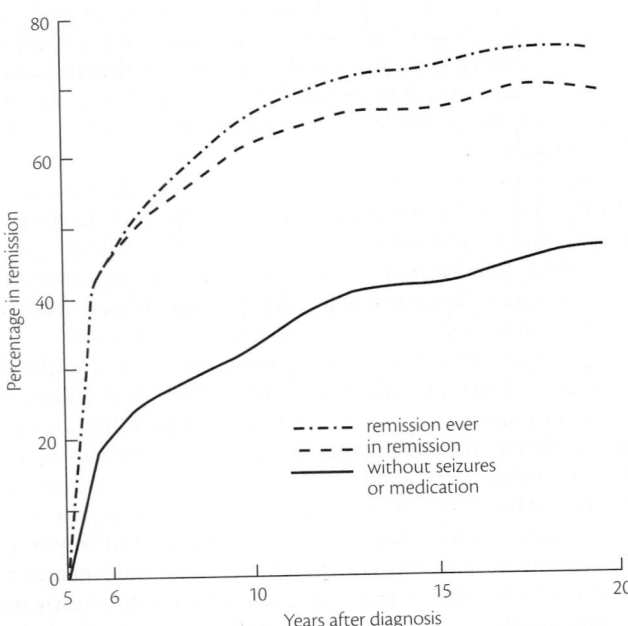

Fig. 31.8 Prognosis for 5-year remission of epilepsy following diagnosis in a cohort from Rochester, Minnesota. (From Annegers *et al.* 1979b.)

continues to be associated with mortality its rarity means that it does not contribute significantly to the excess mortality associated with epilepsy.

Accident is a not uncommon cause of death in epilepsy, accounting for up to 10 per cent of all deaths, drowning being responsible for the great majority (Blisard and McFeeley 1988). Accidents are also a common cause of injury. In a community-based study in the United Kingdom, 24 per cent of those with active epilepsy-reported head injuries within the previous year, and 16 per cent burns (Buck et al. 1997).

31.4 Classification of seizures

The great heterogeneity of clinical phenomena that can be associated with seizure discharge necessitates some system of classification. An international classification of epileptic seizures was proposed in 1981 (Table 31.5) (Commission on classification and terminology of the International League Against Epilepsy 1981). This classification broadly divides seizures into *focal or partial seizures*, which begin locally, and which may spread or evolve into secondary generalized tonic-clonic seizures, and *generalized seizures* in which the onset is sudden and in which both cerebral hemispheres are involved in the discharge from a very early point in the seizure. The classification makes use of both clinical and electroencephalographic information. However, similar seizures may occur at different ages and have very different implications. Conversely, a patient may experience differing seizures during the course of their life so that a classification of different epileptic syndromes based on seizure types occurring within the syndrome, age of onset, and aetiology will also be of vital importance in the management of patients with epilepsy. A classification of epilepsy syndromes is presented in Table 31.6 (Commission on

Table 31.5 Classification of seizures (Commission on classification and terminology of the International League against Epilepsy, 1981)

Partial seizures (seizures beginning locally)

Simple (consciousness not impaired)

  with motor symptoms

  with somatosensory or special sensory symptoms

  with autonomic symptoms

  with psychic symptoms

Complex (with impairment of consciousness)

  beginning as simple partial seizures (-progressing to complex seizure) impairment of consciousness at onset

  impairment of consciousness only

  with automatism

Partial seizures becoming secondarily generalized

Generalized seizures

Absence seizures

  simple (petit mal)

  complex

Myoclonic seizures

Clonic seizures

Tonic seizures

Tonic-clonic seizures

Atonic seizures

Classification and Terminology of the International League Against Epilepsy 1989b). There has been further discussion about the adequacy of these classifications (Engel 2001), but they have not been replaced. It is must be recognized that aetiological classification may be as important as classification based on clinical phenomenology.

31.4.1 Simple partial seizures

Classical neurological teaching has suggested that motor and sensory phenomena may be used to infer precise localizing value. This has increasingly been questioned as more detailed intracranial recording has become available to correlate with observed clinical phenomena. Thus, the clinical phenomena of seizures reflect not only the site of origin, but also the structures through which seizure discharge spreads during the seizure.

Simple motor seizures

Simple partial seizures with motor signs may give rise to clonic or tonic movements involving any part of the body. Seizures most often involve the face or hand area, because of the disproportionate amount of the motor cortex occupied by the somatotopic representation of these parts of the body. Whilst clonic seizures involving an arm or a leg may be taken as a reasonably satisfactory indication of seizures involving the motor strip, movements of the eye or facial muscles around the eye can be produced by occipital discharge and clonic movements of the mouth, tongue, or pharynx by temporal discharge (Lesser et al. 1987). True Jacksonian 'march' with a slow spread of clonic activity from one muscle group to another is uncommon, but when seen does seem to imply relatively specific localization to the pre-central gyrus.

It is evident that tonic motor seizures resulting in version or dystonic posturing of limbs have considerably less localizing value than was previously understood. They may arise from involvement of wider areas of cortex that include the pre-motor region and supplementary motor area. Versive movement, however, can occur in seizures arising in temporal, parietal, and occipital lobes as well as in generalized seizures. Ipsilateral head turning can be as common as contralateral head turning (Ochs et al. 1984). However, version occurs earlier in seizures with frontal onset than in those with temporal onset.

Negative motor phenomena may occur during simple partial seizures. Speech arrest may be the most common manifestation. Whilst speech arrest with a preserved ability to understand speech suggests a seizure in the dominant inferior frontal gyrus, less specific forms of speech arrest can occur with seizure onset in the supplementary motor areas of the dominant or non-dominant hemisphere (Geier et al. 1977). Very rarely, inhibition of movement has been described as part of simple partial seizures. Post-ictal paralysis, or Todd's paralysis, seems specific for seizures involving the contralateral motor strip at some point. Versive seizures tend not to be followed by such paralysis.

Simple sensory seizures

These are rare and occur in no more than 2 per cent of patients with epilepsy. Somatosensory seizures usually arise in the post-central area, but almost always spread to the pre-motor strip at an early stage so that motor phenomena dominate the seizure semiology. They most commonly affect the face or hand and may occasionally spread as do Jacksonian seizures. Sensations are usually those of paraesthesiae or numbness. Rarely, disturbing or painful sensations can occur. Post-ictal numbness similar to Todd's phenomena may occur.

Primitive visual symptoms including spots, flashes of light, or patterns in one visual field are most commonly associated with occipital seizures. On occasion occipital seizures may produce visual symptoms that involve both half visual fields. More complex visual hallucinations are less likely to have an occipital onset.

Whilst non-specific 'dizziness' is often described by patients as part of a simple partial seizure, it usually seems that this term is used because of difficulties in describing complex sensory disturbances. True vertigo must be exceptionally uncommon. Buzzing, hissing, whistling, and ringing noises can be experienced most commonly with involvement of the lateral parts of the temporal lobe.

Olfactory and gustatory symptoms are commonly associated with medial temporal involvement or involvement of the frontal orbital regions. Smells and tastes are usually unpleasant but may be difficult to characterize further than this.

Visceral symptoms are a common component of simple partial seizures. They are most often associated with involvement of limbic structures of the temporal and frontal lobes. Most common is an epigastric sensation, sometimes described as butterflies or nausea which tends to rise characteristically into the throat or mouth. More rarely stomach pain, belching, or even vomiting can occur. Autonomic symptoms can include pallor, flushing and sweating, pupillary dilatation, and increases in heart rate. Involuntary micturition or defaecation whilst not uncommon in seizures associated with loss of consciousness is extremely rare in simple partial seizures.

Complicated psychic symptoms are not uncommon during simple partial seizures. Again they usually indicate involvement of mesial temporal or frontal limbic structures. Psychic symptoms are often associated with other olfactory, gustatory, or autonomic disturbances. Dysmnesic symptoms are perhaps the most common with sensations of familiarity, known as *déjà vu*, or strangeness, *jamais vu*. Memory flashbacks or playbacks may occur and may merge into rather non-specific symptoms of dream-like states, unreality and depersonalization on the one hand, and more formed illusions and hallucinations combining visual and auditory aspects on the other. A variety of perceptual changes may occur during seizures with objects appearing larger or smaller or changing in shape or being perseverated (Gloor *et al.* 1982). Emotional experiences are often described, the most frequent being intense fear. Pleasurable sensations are much rarer but laughter can occur. Anger or rage is extremely rare as a true ictal disturbance.

31.4.2 Complex partial seizures

Complex partial seizures in adults are of considerable importance. They are the predominant seizure type in approximately 40 per cent of patients with epilepsy (Juul-Jensen and Foldspang 1983). They tend to be more resistant to anti-epileptic drug treatment with at best only 50 per cent of patients attaining long-term remissions. Patients with this seizure type commonly present additional psychological or psychiatric handicap, which greatly adds to the complexity of the management problems that they present. Additionally, it is in this group of patients that a surgical approach to the treatment of epilepsy is likely to be most successful.

The international classification of seizures emphasizes that complex partial seizures must include some impairment of consciousness and amnesia. They may or may not be preceded by symptoms of a simple partial seizure and they may or may not be associated with automatism. The site of origin for complex partial seizures is most commonly in the medial temporal lobes in 70 per cent of cases, and the frontal lobes in 20 per cent. The onset of the complex partial seizure occurs when the seizure discharge spreads to involve the limbic system bilaterally. It is important to emphasize that complex partial seizures are not identical or synonymous with temporal lobe seizures and that there is no absolute correlation between the phenomenology of complex partial seizures and their site of origin.

The most typical form of temporal lobe complex partial seizure will start with a visceral aura of a simple partial seizure. There is then an arrest of activity and a motionless stare followed by a phase of stereotyped automatism. Such automatisms occur in a similar fashion in most of an individual patient's seizures and most commonly consist of lip smacking, chewing, or swallowing, or picking at clothes or fidgeting with objects. Walking or running or verbal automatisms are less frequent. Dystonic asymmetrical posturing is common. Stereotyped automatisms are usually followed by a phase of confusion associated with reactive automatisms. In these, the patient may continue with a previous activity or begin some form of activity that may be considerably influenced by the patient's immediate environment. Restraint during this phase of a seizure may sometimes give rise to reactive violent behaviour. It is probable that the majority of such typical complex partial seizures arise from the hippocampus, but by no means all do so. Patients with this seizure type often have well localized anterior temporal spikes and sharp waves that may be unilateral or bilateral.

Complex partial seizures arising from the frontal regions are more likely to exhibit some of the following phenomena: they are usually frequent, brief, and less commonly followed by post-ictal confusion, the onset of seizures is without warning and automatisms begin immediately without a preceding motionless stare, automatisms tend to be bilateral and include thrashing, rolling, kicking, or bicycling movements, sexual automatism with pelvic thrusting may be more common with seizures of frontal onset. There is a marked predominance of sleep seizures and complex partial status seems to be more common with frontal seizures (Williamson *et al.* 1985; Quesney 1986). Inter-ictally the EEG may be unremarkable or show apparently generalized epileptiform abnormalities.

31.4.3 Partial seizures with secondary generalization

Secondarily generalized seizures are relatively uncommon and may only contribute approximately 9 per cent of seizures in adults compared to approximately 16 per cent in children (Gastaut *et al.* 1975). Furthermore, they are particularly amenable to anti-epileptic drug treatment and the majority of patients developing partial or focal epilepsy in adult life have very few secondarily generalized seizures even though their partial seizures remain a management problem. In adults, secondary generalized seizures not infrequently occur during sleep. Late-onset tonic-clonic seizures occurring during sleep must be regarded as having a focal onset until proved otherwise.

31.4.4 Generalized seizures

Whilst typical absence, myoclonus and generalized tonic-clonic seizures may begin in adolescence and early adult life, other forms of generalized seizures such as atypical absence, tonic and atonic seizures are features of age-related childhood epilepsies. While they may persist into adult life, they rarely if ever commence in adult life. However, many children with severe childhood epilepsies tend to develop more typical partial seizures (particularly complex partial seizures) or may expect remission when they enter adult life (Huttenlocher and Hapke 1990).

31.4.5 Status epilepticus

Almost all seizures are self-limiting events, though the mechanisms leading to termination of seizures are poorly understood. Status epilepticus can be defined as two or more seizures occurring without full recovery of function, or as more continuous seizure activity for 30 min or more. While it is true that there are as many types of status epilepticus as there are types of seizures, it is increasingly apparent that status has a dynamic of its own that goes beyond the semiology of seizures themselves. At a simplified level, status may be classified into generalized convulsive status epilepticus, non-convulsive status epilepticus including complex partial and absence status, and simple partial status.

Generalized convulsive status epilepticus is most common. Best estimates of its incidence lie in the range of 200–300/1 000 000 (Shorvon 1994). It probably makes up about 70 per cent of all cases of status and is associated with the highest morbidity and mortality. Mortality rates for generalized convulsive status epilepticus remain high with most series demonstrating rates of 10–20 per cent. It is more common in children than in adults.

Generalized convulsive status epilepticus may have many different causes. The number of cases with a previous history of chronic epilepsy may be falling and less than 5 per cent of people with epilepsy ever experience generalized convulsive status epilepticus. In this group, discontinuation of anti-epileptic drugs is by far the most common provocative factor. Other causes of status in those without epilepsy include drug abuse particularly of alcohol, metabolic disorders, cerebral hypoxia, stroke, trauma, tumour, and neurological infection. The outcome is largely determined by the aetiology and by the speed at which effective treatment is instituted.

At onset of generalized convulsive status epilepticus, typical and relatively discrete tonic-clonic seizures may be observed. However, with progression, convulsive activity commonly becomes less marked and may eventually be confined to myoclonic jerks of the head, eyes, and face. At this stage, the usual EEG characteristics may be lost and more periodic activity is seen (Treiman *et al.* 1990). This has been called subtle convulsive status and may represent a terminal state. It is most frequently described in series that include large numbers of cases of status following cardiorespiratory arrest during anoxic coma, and it remains somewhat controversial whether it really represents an anoxic rather than a seizure-related state.

Early in generalized convulsive status epilepticus there is immediate catecholamine release resulting in tachycardia, hypertension, and hyperglycaemia. Pulmonary oedema may occur as may a combined respiratory and metabolic acidosis. Hyperpyrexia commonly occurs and appears important in determining poor outcomes. This, combined with blood and CSF leucocytosis, can suggest the presence of infection when they are, in fact, the result of the status. Late in status epilepticus the blood pressure and heart rate begin to fall, hypoglycaemia may occur, and renal failure may be caused by rhabdomyolysis.

Few conditions should be confused with generalized convulsive status epilepticus though it is not unusual for recurrent pseudoseizures to be managed by the inexperienced as true status epilepticus (Howell *et al.* 1989). Acute encephalopathies due to hypoxia, uraemia or other metabolic causes, which are associated with multifocal myoclonus may also on occasions be confused with status epilepticus.

Complex partial status epilepticus is of uncertain incidence, largely because it is under-diagnosed (Shorvon 1994). Probably one-third of cases occur without a previous history of epilepsy and often start with one or two discrete tonic-clonic seizures. It may well be that the condition is more common in subjects with frontal than temporal lobe epilepsy. The typical picture is of a twilight state with varying degrees of confusion. Typical automatisms of complex partial seizures may or may not be seen. The EEG is crucial for a diagnosis but a variety of changes occur from localized seizure discharges through periodic lateralized epileptiform discharges to generalized patterns of disturbance. Episodes vary in length, some examples lasting months are documented (Shorvon 1994).

The differential diagnosis of complex partial status epilepticus is exceptionally wide and includes a variety of psychiatric disorders as well as any unexplained confusional state or alteration of consciousness. There may be surprisingly little morbidity to even prolonged episodes of complex partial status epilepticus and there are very few case reports of significant memory impairment following such episodes.

Simple partial status epilepticus is probably even rarer and commonly takes a somatomotor form. In adults, stroke, tumour, and non-ketotic hypoglycaemia are probably the most common causes.

31.5 Epilepsy syndromes in adult life

A systematic classification of epilepsy syndromes has been proposed as have modifications of these (Commission on classification and terminology of the International League Against Epilepsy 1989a; Engel 1998; Soteriades 2002). Most epilepsy syndromes are age-related and occur in childhood, whilst the majority of the epilepsies of adult life are symptomatic partial or focal epilepsies. For this reason the classification has less relevance to adults, in whom accurate localization of the onset of partial seizures and determination of their aetiology is of greater importance. The classification has been subject to considerable criticism as many epidemiological studies fail to classify syndromes in the majority of patients with epilepsy. Some reference does, however, need to be made to those specific epilepsy syndromes where seizures have a predilection to continue into adult life.

31.5.1 Idiopathic generalized epilepsies

These are a relatively well-defined group of syndromes with age-specific onsets and generalized spike wave abnormality in the EEG. They are primarily genetic though the mode of inheritance is likely to be complex and heterogeneous. They account for at least 15–20 per cent of human epilepsy (Jallon and Latour 2005). Often they are of childhood onset but may persist into adulthood (Sections 30.4 and 30.5).

Childhood absence epilepsy

This is discussed in Section 30.4.4. In those with absence persisting into adult life there may be features suggesting alternative diagnoses such as eye-lid or peri-oral myoclonus with absence (Panayiotopoulos *et al.* 1992; 1994; 1995), or that the absences were part of a juvenile myoclonic or absence epilepsy. When tonic-clonic seizures occur in subjects with this syndrome in adult life, they occur without an aura and have a tendency to occur within 1–2 h of wakening.

Table 31.6 International classification of epilepsies, epileptic syndromes, and related seizure disorders (Commission on Antiepileptic Drugs of the International League against Epilepsy, 1989)

1. Localization-related (focal, local, partial)

Idiopathic (primary)

1.1   Benign childhood epilepsy with centro-temporal spikes (Section 30.5.6)

Childhood epilepsy with occipital paroxysms (Section 30.5.6)

Primary reading epilepsy (Section 31.6.1)

Symptomatic (secondary)

1.2   Temporal lobe epilepsies

Frontal lobe epilepsies

Parietal lobe epilepsies

Occipital lobe epilepsies

Chronic progressive epilepsia partialis continua of childhood

Syndromes characterized by seizures with specific modes of precipitation

Cryptogenic

1.3   Defined by:

Seizure type (see Table 31.5)

Clinical features

Aetiology

Anatomical localization

2. Generalized

2.1   Benign neonatal familial convulsions (Section 30.5.3)

Benign neonatal convulsions (Section 30.5.3)

Benign myoclonic epilepsy in infancy (Section 30.5.4)

Childhood absence epilepsy (pyknolepsy) (Section 30.5.6)

Juvenile myoclonic epilepsy (impulsive petit mal) (Section 31.5.1)

Epilepsies with grand mal seizures on wakening

Other generalized idiopathic epilepsies

Epilepsies with seizures precipitated by specific modes of activation

Cryptogenic symptomatic

2.2.   West's syndrome (infantile spasms. Blitz–Nick–Salaam Krampfe) (Section 30.5.4)

Lennox–Gastaut syndrome (Section 30.5.5)

Epilepsy with myoclonic-astatic seizures (Section 30.5.5)

Epilepsy with myoclonic absences (Section 30.5.6)

2.3.1   Non-specific aetiology

Early myoclonic encephalopathy (Section 30.5.4)

Early infantile epileptic encephalopathy with suppression bursts (Section 30.5.4)

2.3.2   Specific syndromes

Epileptic seizures may complicate many disease states

3. Undetermined epilepsies

With both generalized and focal seizures

3.1   Neonatal seizures (Section 30.4.1)

Severe myoclonic epilepsy in infancy (Section 30.5.4)

Epilepsy with continuous spike-wave during slow-wave sleep (Section 30.5.5)

Acquired epileptic aphasia (Landau–Kleffner syndrome) (Section 30.5.5)

Other undetermined epilepsies

3.2   Without unequivocal generalized or focal features

4. Special syndromes

Situation-related seizures

4.1   Febrile convulsions (Section 31.5.4)

Isolated seizures or isolated status epilepticus

Seizures occurring only when there is an acute or toxic event due to factors such as alcohol, drugs, eclampsia, non-ketotic hyperglycaemia

---

Sodium valproate is undoubtedly the drug of choice for the treatment of persisting seizures in adults with this syndrome.

Juvenile absence epilepsy

This syndrome is less common than childhood absence epilepsy and may account for 20 per cent of patients with typical absence. Absence seizures begin between 7 and 16 but tend to be much less frequent than those of childhood absence epilepsy, while a greater proportion, 80 per cent, have tonic-clonic seizures. Seventy-five per cent of these latter seizures occur on awakening. Myoclonic seizures can also occur in this syndrome, perhaps in as many as 16 per cent of patients (Wolf and Inoue 1984). The EEG shows a high incidence of poly-spike and wave and photosensitivity. Valproate probably remains the drug of choice with remission in 85 per cent of patients In distinction from childhood absence epilepsy, the syndrome is probably lifelong (Bouma *et al.* 1996).

Other absence epilepsies

A number of other described absence epilepsy syndromes exist that present in childhood, but are likely to persist into adult life. These are discussed elsewhere (Section 30.4.4).

Juvenile myoclonic epilepsy

This is a common epilepsy syndrome, which was first comprehensively described by Janz and Christian (1957). It is a genetic disorder and up to 25 per cent of patients have a family history. In some pedigrees the gene encoding for this disorder has been shown by linkage analysis to lie on the short arm of the 6th chromosome (Bai *et al.* 2002), while in others linkage to chromosome 15 has been found (Taske *et al.* 2002) suggesting that genetic heterogeneity exists.

It accounts for approximately 5–10 per cent of the epilepsies. Seizures can begin between 8 and 26, but 80 per cent commence between the ages of 12 and 18 (Section 30.4.5). Myoclonic jerks are usually symmetric and mostly affect the upper limbs. On occasion lateralized myoclonus can occur mimicking focal motor seizures. Myoclonus can on occasions be associated with absence or typical absence seizures can occur independently. Seizures most commonly occur after wakening but there may also be a second peak in seizure susceptibility during the evening. Sleep deprivation appears to be a potent provocative factor.

Ninety per cent of patients have generalized tonic-clonic seizures and 10 per cent of patients have absence seizures. It is usually the tonic-clonic seizures that precipitate medical referral and patients may not recognize the important association with jerks which may have preceded the first tonic-clonic seizure by some time. In identifying this syndrome it is important to ask specifically for a history of myoclonus.

The EEG typically shows polyspike and spike wave activity at a more rapid rate than the classic 3 cycle per second spike wave. Photosensitivity is extremely common and is usually identified in about 30 per cent of patients, though this phenomenon may frequently be blocked by treatment with valproate (Goosses 1984).

Valproate is the drug of choice in this syndrome. Remission may be uncommon if patients are treated with other anti-epileptic drugs (Delgado-Escueta and Enrile-Bacsal 1984) and carbamazepine and vigabatrin may exacerbate myoclonus. Remissions of this syndrome are drug-dependant and 90 per cent of patients with it will relapse if drugs are withdrawn.

Tonic-clonic seizures on awakening

Tonic-clonic, or clonic-tonic-clonic, seizures predominate in this syndrome although the presence of occasional myoclonus or absence does not preclude the diagnosis. Janz (1962) examined the timing of tonic-clonic seizures in 2825 patients. Thirty-three per cent had tonic-clonic seizures on awakening compared to 44 per cent occurring during sleep and 23 per cent occurring at random. There was a strong association between tonic-clonic seizures occurring on wakening and generalized spike-wave activity in the EEG. The syndrome has the widest range of age at onset of the idiopathic generalized epilepsies, from 5 to 30 years, with peak onset at 17–20 years.

Precipitating factors include sleep deprivation, sudden arousal, and alcohol intake but catamenial exacerbations are also prominent. Avoidance of precipitating factors is important in management and valproate is the drug of choice for this syndrome.

31.5.2 Symptomatic generalized epilepsies

These childhood epilepsies frequently carry a malign prognosis and are associated with a significant mortality (Section 30.5). However, many children with such epilepsies will survive into the adult age range. They frequently exhibit multiple handicaps. In adult life there is often a change in the nature of seizures with myoclonic, tonic, atonic, and atypical absence seizures becoming less frequent. More typical simple partial and complex partial seizures become evident. The characteristic of these seizures is often that they appear multi-focal from a clinical and electroencephalographic point of view. Whilst such patients may continue to have a severe epilepsy during adult life, seizure frequency tends to be less than in childhood.

31.5.3 Idiopathic partial epilepsies

Some idiopathic age-related syndromes, such as benign Rolandic epilepsy, are not seen in adult life as they have an early age of onset, a uniformly excellent prognosis, and do not continue into adult life. However, the range of genetically determined partial epilepsies seen in adults is increasing.

Autosomal dominant nocturnal frontal lobe epilepsy has been described in a number of families from Australia, United Kingdom, and Canada (Scheffer *et al.* 1995). Many patients previously diagnosed as having paroxysmal nocturnal dystonia probably have this seizure disorder. Seizures begin in childhood and persist into adult life. They occur during sleep, often in clusters and are typical frontal lobe seizures that are brief, associated with vocalization, and include varieties of motor activity including thrashing, tonic stiffening, and clonic jerks. Occasional secondarily generalized tonic-clonic seizures occur. The interictal EEG is usually normal as can be ictal EEG recording. Most patients are responsive to carbamazepine. Some families show linkage to markers on chromosome 20q. A mis-sense mutation has been demonstrated in the gene encoding the α-4 sub-unit of the nicotinic acetylcholine receptor (Steinlein *et al.* 1995). Other families do not show this linkage and there may be genetic heterogeneity.

Other genetically determined partial epilepsies probably occur in adult life. Berkovic *et al.* (1994) described 19 monozygotic twins concordant for a temporal lobe epilepsy with a very benign course. Similar cases have been reported in non-twin families. The age of onset was anywhere between 10 and 60 years. Seizures tended to be brief simple or complex partial temporal lobe seizures, which were very easily controlled with carbamazepine or phenytoin. Ottman *et al.* reported a family with characteristic temporal lobe seizures associated with an auditory aura. There was evidence of linkage to chromosome 10q with mutations in *LGI1*, the function of which is unclear (Ottman *et al.* 2004).

31.5.4 Cryptogenic partial epilepsy

Mesial temporal lobe epilepsy

The syndrome of mesial temporal lobe epilepsy remains best described as a cryptogenic epilepsy in spite of increasing understanding of its aetiological and pathological mechanisms. There is no good epidemiological data about its incidence and prevalence, but information from surgical series would suggest that it probably is a major contributor to the total number of cases of temporal lobe epilepsy. Up to two-thirds of cases have a history of prolonged febrile convulsions in childhood (Williamson *et al.* 1993) and in others there may be a history of trauma or infection. A case-controlled study has suggested that complicated febrile seizures during childhood may be the aetiological factor in up to 20 per cent of all patients of complex partial seizures (Rocca *et al.* 1987). The pathology shows neuronal loss and gliosis in the hippocampus, known as hippocampal sclerosis, associated with synaptic reorganization resulting in functional hypersynchronization and hyperexcitability. This pathological change can be seen following seizures in experimental animals (Sutula *et al.* 1988) and on occasions is also seen in man associated with hamartomas and cortical dysplasias.

The most striking clinical feature of mesial temporal lobe epilepsy is its progressive nature. Typical temporal lobe seizures may be noted in as early as the first decade of life. However, when they begin they are often simple partial seizures that are brief, and may continue for years before the diagnosis is made. When recognized in childhood, seizures are often suppressed only to return in adolescence or early adult life (French *et al.* 1993). By this time, the seizures are more severe and commonly are typical temporal lobe complex partial seizures and some episodes of secondary generalization start to appear. Ninety per cent of subjects describe an aura, most commonly visceral sensations, olfactory hallucination, and memory disturbance. Subsequently there is often a progression to complex partial seizures beginning with motor arrest and staring, oro-alimentary automatisms, and fumbling and picking automatisms of the hands. Versive symptoms may occur, either early or late in the seizures. There is also evidence of a progressive disturbance of cognitive function, particularly memory (Elger *et al.* 2004). which can be associated with increased severity of MRI changes with the duration of disease (Salmenpera *et al.* 1998).

Investigations may show a normal inter-ictal EEG or alternatively characteristic anterior temporal sharp waves which are bilaterally independent in up to a third of patients. High resolution T1-weighted MR imaging shows hippocampal atrophy that is most commonly unilateral but can be bilateral (Fig. 31.9). T2-weighted MRI and FLAIR sequences show high signal in the affected hippocampus.

Whilst anti-epileptic drug treatment reduces or abolishes secondarily generalized tonic-clonic seizures, the complex partial

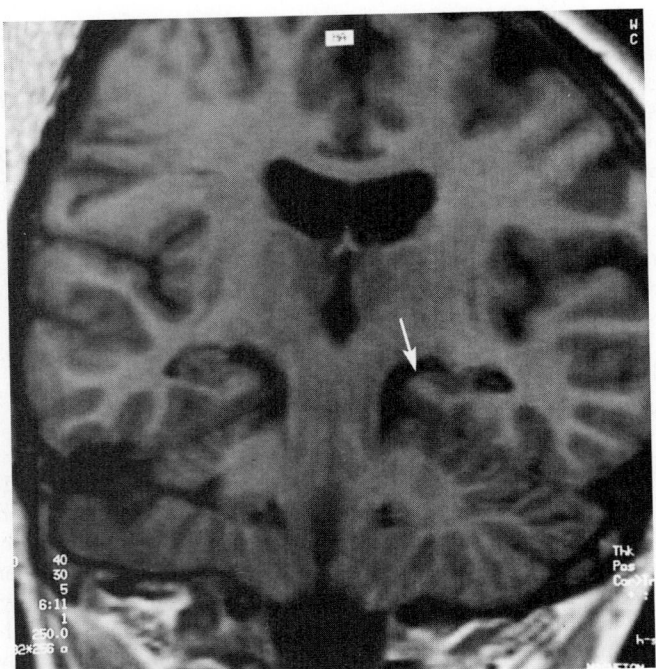

Fig. 31.9 Coronal T$_1$-weighted images, showing marked left mesial temporal sclerosis and atrophy (arrow).

seizures tend to be refractory and many patients exhibit considerable psychosocial disadvantage. Surgical treatment will usually abolish disabling seizures in about 60–70 per cent of subjects (Section 31.11.2). For this reason, it is important that the syndrome is recognized early so that the impact of the disorder can be minimized.

Rasmussen's encephalitis

This syndrome is fortunately rare and although usually seen in children (Section 30.5.4), it can present in adults. It comprises epilepsia partialis continua, slowly progressive hemiplegia, and intellectual impairment with progressive atrophy of one cerebral hemisphere. Histological examination reveals a low grade inflammatory response (Rasmussen *et al.* 1958). While a variety of viruses have been suggested as an aetiological agent, there are no consistent findings in this area. Recently some cases have demonstrated antibodies to GluR3, glutamate receptor, and immunosuppression can be helpful (Leach *et al.* 1999). Many cases require hemispherectomy for seizure control (Section 31.11).

31.6 Factors precipitating seizures

A number of factors may appear to precipitate seizures in susceptible individuals. These may be classified as either specific sensory stimuli or actions, causing reflexly induced seizures, or non-specific precipitants.

31.6.1 Reflexly induced seizures

Whilst the term 'reflex epilepsy' has been widely applied, two-thirds of patients who have reflexly induced seizures also have apparently spontaneous seizures occurring at other times.

Photosensitive epilepsy is the most common reflex epilepsy, accounting for 0.5–8 per cent of patients with epilepsy, but a much greater proportion of those with idiopathic generalized epilepsy

(Jallon and Latour 2005). The crudest visual stimulus to evoke seizures is flicker or flash, a factor that is made use of in the routine recording of most EEGs. This form of sensitivity is most common in childhood and juvenile absence epilepsy and juvenile myoclonic epilepsy and some forms of progressive myoclonic epilepsy. It is a much rarer accompaniment of symptomatic generalized epilepsies and is seen rarely in partial epilepsy. Flicker stimuli may be produced in the environment by television and video games, stroboscopic illumination or sunlight passing through trees or railings or other regularly spaced objects. Most individuals are maximally sensitive between 15 and 20 Hz. Maximum sensitivity often occurs just after the eyes are closed or when the eyes are open. Stimulation of one eye only reduces sensitivity. The greater the proportion of the visual field taken up by stimuli and the greater the luminance then the greater the potential for photosensitivity. In many individuals patterned flash is a potent stimulus.

The electrophysiological correlate of photosensitivity is the photoconvulsive response. This is most commonly seen in females during adolescence and may disappear in adult life. A photoconvulsive response consists of bilaterally synchronous spike-wave activity, which persists for a second or more after the cessation of the flash stimulus. It must be clearly differentiated from photic following and photomyoclonic responses, which have no significant association with photically induced seizures.

About a third of flash-sensitive patients exhibit sensitivity to patterns, the most potent of which are strong stripes, and 70 per cent do so if the pattern oscillates (Wilkins and Lindsay 1985). The most important practical implication of pattern sensitivity is television epilepsy. A television picture is created by variations in the brightness of a spot that scans the screen repeatedly from left to right. The pattern that is generated in this way is similar to a vibrating pattern, which is a very potent stimulus to pattern-sensitive individuals.

Seizures can be prevented in susceptible individuals by maintaining a satisfactory distance from the television set and using the remote control to adjust the picture. More complex preventative methods involve viewing the screen through polarized spectacles so as to produce only monocular stimulation (Wilkins and Lindsay 1985). Most anti-epileptic drugs block photic sensitivity. Valproate is the treatment of choice.

Particular interest has been aroused by the occurrence of seizures with computer games (Fish *et al.* 1994). The incidence has been estimated as approximately 1.5 per 100 000 of the population between 7 and 19 years. Most subjects show sensitivity to flicker or pattern, but in some cases it may be absent, suggesting a contribution from more complex visual or cognitive stimuli.

Primary reading epilepsy can be viewed as a form of visually induced epilepsy. The characteristic seizures in this disorder are myoclonic jerks of the jaw, which may proceed to tonic-clonic seizures. Both focal and paroxysmal EEG abnormalities have been described in this condition (Wilkins and Lindsay 1985). A number of different mechanisms may be involved in producing seizures with reading. In some patients, the lines of print may act as patterns, in others eye movements may provoke the seizures. In some patients neither pattern sensitivity nor eye movements appear important and in these comprehension of the written material may be the important provocative factor.

Sudden noise or other startle may give rise to seizures, particularly in mentally handicapped patients (Anderman and Andermann 1986). More complex auditory stimuli can also provoke seizures in *musicogenic epilepsy* (Poskanzer *et al.* 1962), where seizures are usually complex partial in type. A variety of other complex reflex epilepsies have been described including *eating epilepsy* which gives rise to complex partial seizures and *writing epilepsy* producing jerking of the writing hand. Other cognitive functions such as arithmetic and listening to spoken language can evoke seizures in rare cases.

Touch or muscle stretch may occasionally provoke seizures although more typically evoking myoclonic jerking in patients with progressive myoclonic or post-hypoxic myoclonus. Immersion in hot or cold water can act as seizure precipitant, particularly in the Indian sub-continent (Gururaj and Satishchandra 1992). Rarely seizures may be provoked by movement though this is much less common than paroxysmal choreathetosis.

31.6.2 Non-specific precipitants

Sleep epilepsy. The sleep–waking cycle can have a profound influence on the occurrence of seizures in susceptible individuals. Many patients have tonic-clonic seizures only during sleep (Gibberd and Bateson 1974; D'Alessandro *et al.* 2004). A number of epilepsy syndromes show a predilection for seizures during non-rapid eye movement sleep including the idiopathic partial epilepsies, electrical status epilepticus during slow wave sleep, and temporal and frontal lobe epilepsies, with greater likelihood of secondary generalization (Bazil and Walczak 1997). Sleep may enhance focal epileptogenic discharges and tonic-clonic seizures limited to sleep in the adult should usually be regarded as having a partial onset until proved otherwise. When a pattern of only sleep seizures has been established, the risk of waking seizures is low at 13 per cent over 6 years (D'Alessandro *et al.* 2004).

Awakening seizures. Seizures occurring shortly after wakening are common in the idiopathic generalized epilepsies: juvenile myoclonic epilepsy, tonic-clonic seizures on awakening, and childhood and juvenile absence epilepsy. These epilepsies are particularly sensitive to sleep deprivation or to sudden rousing from deep sleep. The symptomatic generalized epilepsies are characterized by occurrence of seizures that are independent of the sleep–waking cycle.

Catamenial epilepsy. Many women with epilepsy are subject to catamenial exacerbation of seizures although it is uncommon to see women who only have seizures corresponding to such a pattern. The time of greatest susceptibility seems to be in the few days preceding the onset of menstruation. There is evidence of oestrogens being potentially epileptogenic and progestogens potentially anticonvulsant (Scharfman and MacLusky 2006). In spite of this, regulation of periods using oral contraceptive preparations has not shown any benefit in suppressing seizures. Where periods are regular the prescription of a benzodiazepine such as clobazam for a number of days around the period of maximum risk can be beneficial (Feeley *et al.* 1982). The effects of pregnancy on seizure control in women with epilepsy are unpredictable. In some individuals seizures may increase in frequency, in some they may decrease but in the majority there is no significant change in seizures frequency.

31.7 Acute symptomatic seizures in adults

The great majority of seizures and epilepsies developing in adult life will be regarded as symptomatic, a view reinforced by the advent of modern MRI. It may be useful to divide causes of seizures and epilepsy in adult life into:

◆ *acute symptomatic seizures* occurring in response to systemic illness or cerebral insult,

◆ *remote symptomatic epilepsies* in which epilepsy develops as a result of persisting cerebral lesion or damage (Section 31.8).

Some aetiologies such as head injury, stroke, and intracranial infections may cause both acute symptomatic seizures and remote symptomatic epilepsy. The presence of the one is not necessarily associated with the other. Sander *et al.* (1990) found that the commonest remote symptomatic causes of epilepsy were vascular disease in 15 per cent and tumour in 6 per cent. Remote symptomatic epilepsy was commonest in the elderly where vascular disease accounted for 49 per cent of cases. Tumour causes only 1 per cent of epilepsy below 30 years of age, but accounts for 19 per cent of cases between 50 and 59 years. Trauma caused 3 per cent of cases, infection 2 per cent. Acute symptomatic seizures occurred in 15 per cent, and alcohol was the commonest single cause at 6 per cent, its highest incidence of 27 per cent occurring between 30 and 39 years of age.

31.7.1 Causes of acute symptomatic seizures

Acute symptomatic seizures occur commonly: of 1758 patients admitted to an intensive care unit, 217 exhibited neurological complications, 61 of whom had seizures (Bleck *et al.* 1993). When they occur as a result of systemic disorders they are associated with an acute encephalopathy and most commonly seizures are tonic-clonic. Patients may also exhibit tremor, asterixis, and multifocal myoclonus. The risk of tonic-clonic seizures is more determined by the rate of a metabolic change than the absolute levels reached. They are particularly common in the elderly, where they may account for as much as 77 per cent of the incidence of seizures (Loiseau *et al.* 1990). Focal seizures can be seen in association with acute cerebral insults when they are usually recognized as focal motor seizures, sometimes as epilepsia partialis continua. It is suggested that complex partial seizures rarely occur as acute symptomatic seizures but such seizures could be difficult to recognize when accompanied by an acute confusional state or coma.

Whilst acute symptomatic seizures may be suppressed by short-acting anti-epileptic drugs such as benzodiazepines they do not usually require longer term anti-epileptic drug treatment. Acute symptomatic seizures are often resistant to benzodiazepines and other anti-epileptic drugs requiring correction of the underlying metabolic abnormality to suppress seizures.

31.7.2 Disorders of fluid and electrolyte balance

Hypernatraemia may occur in gastroenteritis, fever, sweating, burns, and diabetes, and due to gross fluid restriction or excessive salt intake. It is usually defined as a serum sodium concentration above 145 mEq/l. Much higher concentrations can be tolerated in chronic hypernatraemia, with seizures occurring more commonly during the phase of correction. Altered consciousness is common and focal or generalized tonic-clonic seizures occur most commonly in patients who are also uraemic or acidotic or in whom there is acute elevation towards 160 mEq/l (Riggs 2002).

Hyponatraemia is more common than hypernatraemia, occurring with congestive cardiac failure, liver disease, nephrotic syndrome, water-overload, diuretic misuse, as well as with renal disease and syndromes of inappropriate antidiuretic hormone secretion which can result from neurological disorders. Tonic-clonic seizures occur and occasionally status epilepticus. Again the rate of change in serum sodium concentration is of prime importance, with encephalopathy and seizures seen with rapid falls to 115 mEq/l or less (Adrogue and Madias 2000). Hyponatraemia is associated with a high mortality and too rapid a correction by the over-enthusiastic use of hypertonic saline has been associated with the occurrence of central pontine myelinolysis (Section 37.6).

Hypocalcaemia may be seen in hypoparathyroidism, vitamin D deficiency, acute pancreatitis, and pseudohypoparathyroidism. Up to 70 per cent of patients with hypoparathyroidism may have seizures (Messing and Simon 1986) associated with tetany, altered consciousness, and abnormal behaviour and dyskinesia. Both tonic-clonic and focal motor seizures are described.

Hypercalcaemia most commonly occurs in disseminated malignant disease and hyperparathyroidism. It results in weakness, drowsiness, and confusion, but seizures are uncommon. When they occur they may be generalized or focal motor seizures (Riggs 2002).

Hypomagnesaemia may be seen in inflammatory bowel disease, bowel resection, and other malabsorption syndromes, and is often associated with other electrolyte disturbance. Neurological syndromes may be seen with levels below 1.0 mmol/l. The clinical state, with startle, tremor, and myoclonus and Chvostek's sign, may be indistinguishable from hypocalcaemia, although tetany may be less common. Hypomagnesaemia should be considered if seizures continue in a treated hypocalcaemic patient.

31.7.3 Metabolic disorders

Diabetes. Seizures seem particularly common in association with hyperosmolar non-ketotic hyperglycamia (Venna and Sabin 1981). They may occur in up to a quarter of such patients and simple partial motor seizures account for approximately 80 per cent of such seizures. Epilepsy partialis continua can occur. Such seizures may be very resistant to anti-epileptic drug treatment but seem to respond rapidly to the correction of the hyperglycaemia. By contrast, seizures seem to be very rare in ketoacidotic diabetic coma (Messing and Simon 1986).

Hypoglycaemia. This is usually seen in diabetic patients using insulin or hypoglycaemic drugs. It occurs more rarely with insulinoma, other neoplasms, or severe liver disease. Seizures, usually tonic-clonic seizures without an aura, may occur in 7 per cent of patients (Malouf and Brust 1985).

Thyroid disease. Seizures appear extremely uncommon in hyperthyroidism. They do occur occasionally as a feature of Hashimoto's encephalopathy (Section 38.3.1) that can precede other manifestations of thyrotoxicosis. Myoclonic seizures may occur along with tonic-clonic seizures (Ghika-Schmid *et al.* 1996). Seizures seem more common in patients with hypothyroidism and may occur in as many as a quarter of patients with myxoedema coma (Jellinek 1962). Patients do not seem to be at risk of continued seizures after the underlying thyroid abnormality has been corrected.

Porphyria. Seizures may occur in approximately 15 per cent of patients during episodes of acute intermittent porphyria and may be a presenting feature (Section 21.8.6). Drug control of seizures can be a significant management problem as phenytoin, barbiturates, and carbamazepine can all induce attacks of porphyria. Benzodiazepines may be used with caution but it has also been suggested that magnesium sulphate may also be effective in controlling seizures. Of the newer anti-epileptic drugs, pregabalin and levetiracetam may be safe.

Liver disease. Seizures are a feature of acute hepatic failure but not of chronic hepatic dysfunction. Seizures may be focal but are more commonly tonic-clonic seizures often preceded by multifocal myoclonus. Neurological features of encephalopathy and seizures are common following liver transplantation (Saner *et al.* 2006).

Renal failure. Acute uraemic encephalopathy commonly presents with motor excitability including tremors, asterixis, multifocal myoclonus, chorea, and dystonia. Convulsions occur in as many as a third of patients; most are tonic-clonic seizures but focal motor seizures and epilepsia partialis continua can occur (De Deyn *et al.* 1992). Seizures have also been reported during dialysis, in the dialysis disequilibrium syndrome, and as part of dialysis encephalopathy, a sub-acute progressive disorder in which speech disorders, dementia, and myoclonus are prominent.

31.7.4 Drug-related seizures

Drugs, and particularly alcohol, are a common cause of seizures (Messing *et al.* 1984) and many different drugs have been associated with seizures (Table 31.7). The Boston collaborative drugs surveillance programme (1972) reported 26 cases of drug-induced convulsions in approximately 33 000 in-patients, an incidence of 0.08 per cent. Most commonly implicated were penicillin, hypoglycaemic drugs, lignocaine, and psychotropic agents. Messing *et al.* (1984) reviewed case records of over 3000 patients presenting with seizures and found that they were drug-related in 1.7 per cent, the most common drugs involved being isoniazid, psychotropic drugs,

Table 31.7 Drugs associated with seizures

Anaesthetics	Antibiotics	Antipsychotic agents
Propofol	Benzylpenicillin	Chlorpromazine
Halothane	Carbenicillin	Lithium
Ketamine	Oxacillin	Clozapine
	Ampicillin	
Analeptics	Cycloserine	**Radiographic contrast media**
Nikethamide	Isoniazid	
Aminophylline	Nalidixic acid	Meglumine carbamate
Amphetamines	Ciprofloxacin and Fluoroquinalones	Meglumine iothalamate
Ephedrine		Metrizamide
Analgesics		**Miscellaneous**
Cocaine	**Anti-epileptic drugs**	D-Penicillamine
Pethidine	Phenobarbitone	Baclofen
Dextropropoxyphene	Phenytoin	Hyperbaric oxygen
Merperidine	Ethosuximide	Folate
	Vigabatrin	Piperazine
	Carbamazepine	Cyclosporin
Antidysrhythmics		Interferon
Disopyramide	**Antidepressants**	Mefloquine
Lignocaine	Tricyclics	
	Bupropion	
	Mianserin	
	Maprotiline	

bronchodilators, hypoglycaemic agents, lignocaine, and penicillin. The majority of seizures are tonic-clonic but whilst most begin without an aura, there was a simple partial motor onset in nine patients.

Seizures may be provoked in two ways: there may be specific neural excitatory effects for some drugs or alternatively there may be non-specific effects resulting from high doses of drugs often administered during self-poisoning. Most drug-induced seizures are dose-related and particular care must be exercised when drugs are administered parenterally or intrathecally. Patients with renal or hepatic failure may be at risk because of inability to metabolize potentially convulsant drugs and individuals with a previous history of epilepsy or pre-existing brain disease may be particularly at risk. The subject has been recently reviewed (Ruffmann *et al.* 2006).

Antibiotics. Penicillin has potent epileptogenicity in animals and has been widely used as a model for both focal and generalized epilepsies. It acts as a GABA antagonist and may also bind to benzodiazepine receptors (Curtis *et al.* 1972; Antoniadis *et al.* 1980). Benzylpenicillin is probably the most potent antibiotic in causing seizures but ampicillin and cephalosporins carry some risk. Newer quinalone antibiotics may similarly interfere with GABAergic mechanisms. Isoniazid may cause seizures because of its action in antagonizing pyridoxine, a co-enzyme required for the synthesis of GABA (Blakemore 1980).

Psychotropic drugs. Tricyclic antidepressants are particularly likely to cause myoclonus and convulsions when taken in overdose but seizures may occur in up to 1 per cent of patients taking therapeutic dosages (Lowry and Dunner 1980). Amitriptyline is probably the antidepressant with the highest risk, but monoamine oxidase inhibitors and selective serotonin uptake inhibitors appear relatively safe (Dailey and Naritoku 1996). Antipsychotic usage may be complicated by seizures. Phenothiazines, particularly chlorpromazine, are associated with a 1–2 per cent incidence of seizures (Logothetis 1967). Of the atypical antipsychotics, clozapine may have a high risk of seizures (Pacia and Devinsky 1994) but pimozide and sulpiride seem relatively safe. Lithium toxicity may be associated with seizures (Section 5.5.3).

Analeptic drugs. Most central nervous system stimulant drugs are capable of causing seizures and problems most commonly arise with theophylline and its derivatives which significantly lower seizure threshold possibly by elevating cyclic GMP levels in brain (Walker 1981).

Drugs of abuse. Cocaine, amphetamines, narcotics, and phencyclidine have been associated with seizures (Holland *et al.* 1993) (Section 5.3). They have stimulant actions on the central nervous system and lower seizure threshold. Cocaine presents the most frequent problem with tonic-clonic seizures occurring in 5 per cent of emergency department attendances with acute toxicity (Pascual-Leone *et al.* 1990). Organic solvents have been reported to cause epilepsy (Jacobsen *et al.* 1994). It should be noted that seizures are rarely seen with narcotic abuse.

Transplantation and immunosuppressants. Seizures can occur in up to 20 per cent of children undergoing renal transplantation (McEnery *et al.* 1989), but only 4 per cent of liver transplants (Wijdicks *et al.* 1996). Cyclosporin and some other immunosuppressant drugs give rise to seizures (Section 5.6.3). Inevitably, drug effects may interact with other factors such as metabolic disturbance and infection. Cyclosporin-induced seizures may occur in 1–2 per cent of renal transplants and 5 per cent of bone marrow transplants (Wijdicks *et al.* 1995).

Withdrawal seizures. The withdrawal of chronically administered sedative drugs which show tolerance is a well-recognized cause of seizures and may occur with alcohol, barbiturates, and benzodiazepines. The best studied of withdrawal seizures are those that occur with alcohol and which may be a part of the delirium tremens syndrome (Section 5.2.1). The abuse of alcohol is an important cause of seizures in the community and it must be considered in adults developing tonic-clonic seizures for the first time (Hillbom 1980). The risk is clearly related to the dose of alcohol consumed (Lechtenberg and Worner 1992). It seems that abrupt, absolute, or relative withdrawal of alcohol is most commonly responsible for causing seizures. Clustering of seizures occurs between 7 and 48 h after the withdrawal of alcohol. Sixty per cent of patients have more than 1 seizure, but status epilepticus occurs in less than 5 per cent of patients. During the withdrawal period photo-myoclonic and photo-convulsive responses may be seen in the EEG. On occasions seizures can occur in patients whilst intoxicated with alcohol.

Non-compliance. Non-compliance with anti-epileptic drug medication is well recognized to be a common cause of seizures in people with epilepsy and is an important cause of status epilepticus (Aminoff and Simon 1980). Abuse of barbiturates and subsequent withdrawal in non-epileptics has been an important cause of seizures. This effect is dose-related and seizures occur with other withdrawal symptoms such as insomnia, tremor, anorexia, and autonomic over-activity. The EEG of patients undergoing barbiturate withdrawal shows features of both photo-myoclonic and photo-convulsant responses. Benzodiazepines can also be associated with withdrawal seizures.

31.8 Causes of remote symptomatic epilepsy

It is well recognized that a number of cerebral insults and pathologies predispose to the development of epilepsy.

31.8.1 Hypoxic ischaemic cerebral insults

A static encephalopathy, manifesting as learning disability and motor handicap present from birth, is commonly associated with seizure disorders. As many as 50 per cent of individuals with both mental handicap and cerebral palsy have seizures (Hauser *et al.* 1987). Some but by no means all may have been caused by cerebral hypoxia. The more severe the mental and physical handicap the higher the risk of epilepsy (Edebol-Tysk 1989). The great majority will develop seizures early in life but up to 15 per cent of patients may have a seizure disorder that starts after the age of 15 (Forsgren *et al.* 1990). Whereas in childhood this symptomatic epilepsy commonly involves generalized seizures including myoclonus, tonic and atonic seizures, and infantile spasms, there is an increasing predominance of partial seizures and secondary generalized tonic-clonic seizures as individuals mature. In this population the outcome for epilepsy is poor, only a third achieving seizure remissions of a year or more and a third having at least one seizure per month. Early brain damage is one of the strongest factors predicting a poor outcome of epilepsy.

Generalized cerebral hypoxia during adult life seems much less likely to result in seizures. Acute hypoxia, when severe, is commonly associated with convulsions and multifocal myoclonus

(Wardrope *et al.* 1991). In patients with post-hypoxic coma following an anoxic insult, seizures seem to be much rarer and when they occur they may be associated with an adverse prognosis (Bates *et al.* 1977).

31.8.2 Head injury

The relationship between head injury, acute symptomatic seizures occurring within the first week of injury, and late post-traumatic epilepsy represents perhaps the best studied of all causes of epilepsy. Head injury is a well-known cause of epilepsy and most patients developing epilepsy will recall the occurrence of a minor head injury sometime before the development of seizures. However, it is clear that only specific types of head injury carry a significant risk of post-traumatic epilepsy.

Although perhaps 2 per cent of all concussive head injuries result in epilepsy (Annegers *et al.* 1980), head trauma was the cause of seizures in 3 per cent of patients registered in the National General Practitioners Survey of Epilepsy (Sander *et al.* 1990). There can be no doubt that the more severe the head injury, the higher the risk of post-traumatic epilepsy (Annegers *et al.* 1975b, 1998).

Missile injuries and epilepsy. Brain injuries caused by missiles provide a well-defined and relatively homogenous group of injuries that are fortunately rare in civilian life. They have been fully studied in cohorts of patients from the First World War through to the Vietnam War. Overall, it would seem that 50 per cent of patients surviving such injuries will develop post-traumatic epilepsy and that the relative risk of developing epilepsy will initially be 580 times higher than a general age-matched population during the first year, falling to 25 times higher after 10 years (Salazar *et al.* 1985). The risk of epilepsy correlates with the severity of the injury, the amount of tissue loss, and the severity of neurological impairment. Infection and abscess formation probably exert an important influence on the incidence of epilepsy. Seizures during the first year seems to predict the duration and frequency of subsequent epilepsy.

Blunt injuries. The most satisfactory unselected population of patients developing post-traumatic epilepsy has been studied by Annegers *et al.* (1998). The relationship of the severity of head injury to subsequent risk of epilepsy over 15–20 years was:

◆ Mild head injuries, as defined by loss of consciousness or amnesia of less than 30 min carry an increased incidence ratio for epilepsy of 1.5 (95 per cent CI 1–2.2);

◆ moderate head injuries defined by loss of consciousness or amnesia of between 30 min and 24 h or a skull fracture carry an incidence ratio of 2.9 (95 per cent CI 1.9–4.1) for the development of epilepsy;

◆ severe head injuries, with loss of consciousness or amnesia in excess of 24 h, or subdural haematoma or brain contusion, are accompanied by a considerably increased risk with a standardized incidence ratio of 17.0 (95 per cent CI 12.3–23.6) for the development of epilepsy.

Subdural haematoma and brain contusion were the factors that greatly increased this risk. Within the neurosurgical population studied by Jennett (1975), an increased risk of epilepsy was determined by the presence of a compound depressed fracture, intracranial haemorrhage, or the occurrence of early acute symptomatic seizures within the first post-traumatic week. In those few subjects with all three risk factors, the likelihood of post-traumatic epilepsy could be as high as 50–80 per cent.

Annegers *et al.* have further clarified the importance of early, as opposed to immediate, seizures in predicting late post-traumatic epilepsy by showing that early seizures have no independent effect on the risk of late post-traumatic epilepsy, but merely act as a marker of injuries of sufficient severity to cause late epilepsy.

Concussive convulsions. Jennett recognized that seizures occurring immediately on impact do not seem to carry any subsequent excess risk of epilepsy and indeed characterize relatively mild concussive injuries. McCrory *et al.* (1997) studied such immediate 'concussive convulsions' in Australian sportsmen and came to the same conclusions. They questioned whether these events are seizures rather than a form of acute temporary decerebration.

Post-traumatic anti-epileptic drugs. A systematic review of 10 randomized controlled trials showed there was consistent evidence that the early treatment with anti-epileptic drugs, most commonly phenytoin but also carbamazepine and phenobarbitone, resulted in a relative risk for early seizures of 0.34 (95 per cent CI 0.21-–0.54). Treatment of 100 patients would result in 10 patients avoiding seizures during the first week, who would otherwise have had them. However, the studies did not provide evidence that this benefit was accompanied by any reduction in mortality and there is no evidence that a reduction in early seizures is accompanied by a reduction in late post-traumatic epilepsy (Schierhout and Roberts 1998). Current evidence from intervention studies and community-based epidemiology would indicate that early seizures act simply as a marker for more severe head injuries rather than an important part of a pathogenic process that leads from severe head injury to either late post-traumatic epilepsy or death. More recently, studies of the potential effects of valproate prophylaxis have been undertaken (Temkin *et al.* 1999). Slightly under 15 per cent of patients had late seizures when treated with phenytoin compared to approximately 20 per cent of patients receiving valproate treatment, a relative risk of 1.4 (95 per cent CI 0.8–2.4) and the study was stopped because of excess deaths in the valproate-treated group. Thus, there is reasonable evidence that immediate administration of anti-epileptic drugs reduces the incidence of early seizures, and that phenytoin has the best evidence to support its use in this role. This effect will, however, carry with it a small risk of acute idiosyncratic reactions, as long as phenytoin is only used during the first post-traumatic week. There seems no justification for the routine use of anti-epileptic drugs in patients with severe head injuries.

Post-craniotomy seizures. Craniotomy can be viewed as a type of head injury, although the pathology underlying its necessity contributes significantly to risk. The overall incidence of seizures occurring after supratentorial craniotomy has been estimated at 17 per cent during a follow-up period of at least 5 years (Foy *et al.* 1981a). The incidence varied from 3 to 92 per cent depending on the condition for which the craniotomy was undertaken.

Approximately one-fifth of patients undergoing aneurysm surgery develop post-operative seizures (North *et al.* 1983). The incidence varies according to the site of the aneurysm; 7.5 per cent from internal carotid aneurysm, 21 per cent from anterior communicating aneurysm, and 39 per cent for a middle cerebral artery aneurysm (Foy *et al.* 1981a). Additional factors influencing the incidence may be the presence of an intracerebral haematoma, cortical damage, splitting the silvian fissure, cerebral swelling and peri-operative aneurysmal rupture, and the length of surgery

(Foy *et al.* 1981b). That surgery is responsible for at least part of the risk of epilepsy associated with aneurysm is suggested by the halving of risk in conservatively managed survivors following aneurysmal subarachnoid haemorrhage and in those treated by endovascular coiling (Molyneux *et al.* 2005). Arteriovenous malformations and spontaneous intracerebral haematoma from other causes carry risks of epilepsy of 50 and 20 per cent respectively and surgical treatment does seem to be an additional risk factor for these conditions (Crawford *et al.* 1986). Cavernous angiomas probably represent 10–20 per cent of all central nervous system vascular malformations, and seizures are the only symptom in up to 70 per cent (Farmer *et al.* 1988). The MR changes, with high T2 signal core and halo of low signal due to haemosiderin, suggests that many may bleed asymptomatically. Here surgery seems highly effective in seizure control.

The incidence of new seizures following meningioma surgery is of the order of 20 per cent (North *et al.* 1983). The incidence is higher for parasagittal lesions than for convexity or basal tumours. Some 44 per cent of patients who have pre-operative seizures do not have any further seizures post-operatively. Surgery for supratentorial abscess carries a very high risk. With sufficiently long follow-up virtually all patients develop epilepsy. Ventricular shunting procedures can be associated with a 24 per cent risk of seizures and multiple shunt revisions and shunt infections significantly increase the risks (Copeland *et al.* 1982).

Thirty-seven per cent of all patients who experience post-operative seizures do so within the first week and 40 per cent of this group continue to have later seizures. By the time 1 year has elapsed 77 per cent of those who will develop seizure disorders will have done so and by 2 years, 92 per cent will have had their first seizure (Foy *et al.* 1981b).

The possible effects of prophylaxis in high risk patients has been studied by a number of authors (Temkin *et al.* 1998). There is no evidence that phenytoin or carbamazepine significantly reduce the incidence of post-craniotomy seizures in high-risk groups of patients, nor do they seem to effect the likelihood of persistence of the seizure disorder over a period of time. This use of prophylactic anti-epileptic drugs can be associated with a high incidence for adverse effects, particularly drug-induced rash (Chadwick *et al.* 1984). For this reason prophylactic treatment does not seem to be justified.

31.8.3 Intracranial tumours

The relationship between intracranial tumours and epilepsy is well established. Consequently there is considerable pressure to image patients presenting with epilepsy. In fact, brain tumours are responsible for late-onset epilepsy in only about 10 per cent of cases. They were found in 6 per cent of patients registered with the National General Practitioner Survey of Epilepsy (Sander *et al.* 1990). The incidence of tumours rises steeply where seizures are clearly focal in nature (Sumi and Teasdall 1963). Tumours of the frontal, parietal, and occipital lobes seem to carry the highest risk of epilepsy (Mauguiere and Courjon 1978).

In general, about 40 per cent of those with seizures due to tumour have seizures as the first manifestation. The interval between the first seizure, the diagnosis of the tumour, and the development of further neurological problems is commonly prolonged (Smith *et al.* 1991). This reflects the fact that the majority of tumours that present only with epilepsy tend to be benign. Indeed, presentation

with an initial symptom of epilepsy is one of the most powerful prognostic factors indicating a good prognosis (Smith *et al.* 1991). Such intracerebral tumours are most commonly shown to be low density, non-enhancing lesions on initial CT scans and to be relatively low grade astrocytomas or oligodendrogliomas. The incidence of seizures with cerebral metastases is lower, probably about 20 per cent at the time of presentation (Cohen *et al.* 1988). The prognosis for tumour-associated epilepsies is poor. Only 11 of 164 patients achieved a 1-year remission of epilepsy with anti-epileptic drug treatment and 50 per cent of patients with tumour epilepsies in adult life die within 4 years. However, 20–30 per cent show prolonged survival (Smith *et al.* 1991).

There is a considerable dilemma about the management of intracerebral tumours. Meningiomas and other well-defined intracerebral lesions should be treated surgically where this is practical and where the patient is not old or infirm. Seizures will be suppressed in about 40 per cent of patients with meningiomas, and up to 80 per cent of other patients undergoing lesionectomy of other discreet tumours (Cascino *et al.* 1990). In contrast many neurologists find it difficult to recommend aggressive treatment with biopsy or tumour debulking, and radiotherapy in a patient with a low-grade infiltrative tumour whose only symptom is epilepsy and in whom there is a good prospect of good quality survival for many years. In such cases the tumour is rarely fully resectable and the risk-benefit for early surgical treatment remains uncertain (Karim *et al.* 1996; Shaw *et al.* 2002) (Section 27.8.1).

Cerebral tumours are another area in which prophylactic treatment with anti-epileptic drugs has been advocated for those even before the first seizure. Trials have to date failed to show any benefit in preventing the first or later seizures (Weaver *et al.* 1995; Sirven *et al.* 2004). The studies available are small and do not exclude clinically important benefits. The striking factor was the low risk of *de novo* seizures in the studies in treated and untreated patients. Given that anti-epileptic drugs have a considerable propensity to interact with chemotherapeutic agents and steroids, and that tumour-related epilepsies tend to be drug resistant, there seems no justification for this approach.

31.8.4 Cerebrovascular disease

Cerebrovascular disease and stroke become an increasingly common cause of epilepsy in the later years of life (Loiseau *et al.* 1990). It has been estimated that cerebrovascular disease may account for 15 per cent of new cases of epilepsy (Sander *et al.* 1990) and more than 50 per cent of new cases in the elderly (Camilo and Goldstein 2004). A community-based study of stroke showed a 5-year actuarial risk of seizures of 11.5 per cent, over 30 times that expected. The risk was highest after subarachnoid haemorrhage, 30 per cent, and primary intracerebral haematoma, in 25 per cent, while the risk for ischaemic stroke of 9 per cent was restricted to survivors of large anterior circulation stokes. Seizures within 24 h of stroke onset were a risk factor for late seizures, which occurred in 35 per cent of such patients (Burn *et al.* 1997). Risk factors for late epilepsy include cortical involvement, stroke severity, and haemorrhagic stroke (Camilo and Goldstein 2004). However, asymptomatic carotid occlusion (Cocito *et al.* 1982) and asymptomatic cerebral infarction (Shorvon *et al.* 1984) may be found in patients presenting with epilepsy in later life and seizures may also precede a stroke (Burn *et al.* 1997; Cleary *et al.* 2004).

Cerebral venous sinus thrombosis is increasingly recognized as an important type of stroke, in younger people (Section 35.15.1). Seizures commonly complicate the acute illness, but the long-term prognosis appears good, with few survivors developing epilepsy. In a series of 77 patients 28 had acute symptomatic seizures, but only four had late seizures and epilepsy, all of whom had acute seizures (Preter *et al.* 1996).

Arteritic disorders can be accompanied by seizures as part of stroke-like syndromes or acute encephalopathies (Sections 36.2 and 36.3). Anywhere between 17 and 50 per cent of patients with systemic lupus erythematosis and central nervous system involvement have seizures, Seizures may also complicate, to a lesser degree, involvement in polyarteritis nodosa, Behcet's disease, and mixed connective tissue disease (Shannon and Goetz 1989). Seizures can occur in hypertensive encephalopathy and in subacute bacterial endocarditis.

A rare cerebrovascular cause of seizures is the hyper-reperfusion syndrome occurring after carotid endarterectomy where seizures can occur on the first post-operative day. It is doubtful that this has a significant risk for late epilepsy (Nielsen *et al.* 1995).

31.8.5 Neurological infections

A wide range of viral, bacterial, opportunistic, and parasitic infestations can be associated with seizures. Infections accounted for 3 per cent of seizure disorders in the epidemiological study in Rochester, Minnosota (Hauser and Kurland 1975). The 20-year risk of developing unprovoked seizures following common central nervous system infections was 6.8 per cent, almost 7 times the expected rate. Increased incidence of seizures was highest during the 5 years after a neurological infection but continued to be elevated for as long as 15 years (Annegers *et al.* 1988). The risk was highest, at 22 per cent, for patients with a viral encephalitis associated with acute symptomatic seizures, and 10 per cent for patients with viral encephalitis without early seizures. For bacterial meningitis associated with early seizures the risk was 13 per cent, but only 2.4 per cent for those without early seizures. Aseptic meningitis does not pose an increased risk of seizures.

Seizures commonly occur during acute viral encephalitis, particularly with herpes simplex encephalitis when the seizures are frequently focal in nature. Pre-natal infection with cytomegalovirus, rubella, and herpes can produce retardation associated with late epilepsy (Forsgren *et al.* 1990). Seizures can also occur with subacute measles encephalitis and subacute rubella encephalitis as well as with 'slow virus infections' including subacute sclerosing panencephalitis and Creutzfeld–Jacob disease, but in both these conditions myoclonus tends to dominante the picture. Infection with HIV can be associated with seizures not only because of an increased risk of opportunistic infections, but also because the direct neurotropic effects of the virus (Wong *et al.* 1990). Of 100 patients who were HIV positive and had seizures, 45 had evidence of opportunistic infection or central nervous system lymphoma, 24 evidence of the AIDS–dementia complex, but 23 had no other identifiable cause for their seizures (Holtzman *et al.* 1989).

Bacterial infections causing meningitis are occasionally associated with seizures, particularly if complicated by cortical vein or sinus thrombosis or cerebral abscess. Tuberculous meningitis may present with seizures. In the Indian subcontinent tuberculomas may be a common cause of epilepsy associated with disappearing ring-enhancing CT lesions (Goulatia *et al.* 1987). Other causes of chronic meningitis are not infrequently associated with seizures,

for instance cryptococcus, and candida. Perhaps the commonest infestation associated with seizures is cysticercosis or which may account for 50 per cent of incident cases in adults in developing countries (Medina *et al.* 1990) (Section 43.2.9). Others include schistosomiasis, hydatid, malaria, and toxoplasmia (Bittencourt *et al.* 1988).

31.8.6 Malformations of cortical development

The introduction of modern high resolution MRI has shown that malformations of cortical development are important and common causes of epilepsy (Section 9.2). A series from Norway found that 13 of 303 patients had such developmental anomalies (Brodtkorb *et al.* 1992), while 16 of 222 patients with temporal lobe epilepsy had developmental abnormalities in another series (Lehericy *et al.* 1995). Malformations of cortical development may be generalized, when they are usually associated with developmental delay and a static encephalopathy from birth as well as a severe epilepsy, or they may be focal in nature with seizures as the only symptom. Such focal abnormalities may account for 15–20 per cent of the adult population with intractable epilepsy (Kuzniecky and Jackson 1997). Many have a genetic basis and some form a part of other well-recognized symptom complexes such as the neurocutaneous syndromes, for example tuberose sclerosis (Section 11.1). They may also be environmental and related to infection, toxins, and radiation.

Focal cortical dysplasia is usually frontal or temporal with MRI findings of abnormal gyral thickening often associated with abnormalities of the underlying white matter with blurring of the grey–white matter interface (Fig. 31.10). On occasions, the abnormality may be 'transmantle' extending from the ventricle to the cortical surface. The changes may be difficult to differentiate from those of tuberose sclerosis. Seizures usually present in the first decade and are often refractory to drug treatment. Surgical resection offers benefits in many cases.

The heterotopias are, by definition, normal cells present in an abnormal location (Section 9.2.5). *Subcortical band heterotopia* is seen in women and consists of MR changes with a circumferential

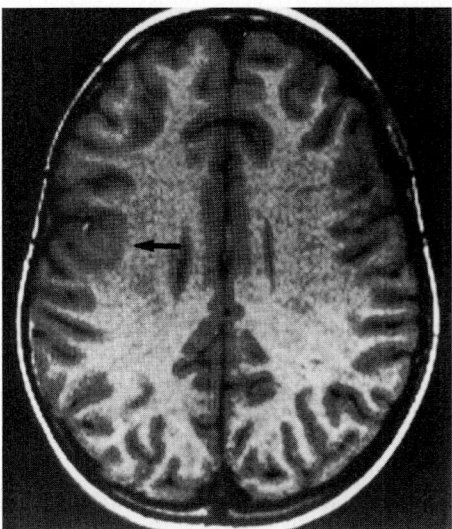

Fig. 31.10 Focal cortical dysplasia shown in axial MRI. (Reproduced from Kuzniecky and Jackson 1997 with permission.)

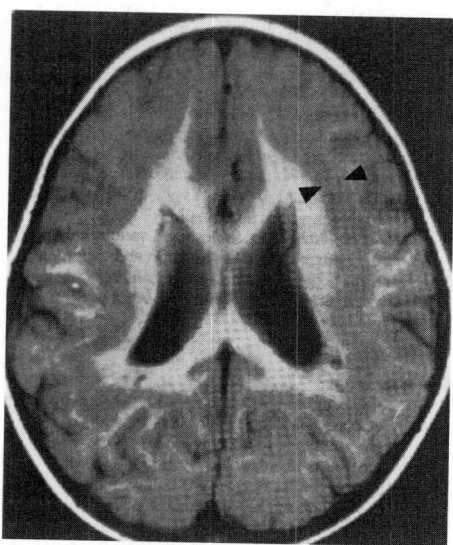

Fig. 31.11 Subcortical-band heterotopia on MRI T1-weighted images. (Reproduced from Kuzniecky and Jackson 1997 with permission.)

band of subcortical grey matter (Fig. 31.11). Lisencephaly results in an abnormally smooth cortical surface and may be incompatible with survival. Both conditions can be seen within the same family and segregate with familial epilepsy. Mutations in the *DCX* gene on the X chromosome and *LIS1* gene on chromosome 17 have been implicated in the disorders (Sisodiya 2004).

Periventricular or subependymal nodular heterotopia is probably the commonest developmental disorders seen in patients with epilepsy. MR appearances are of multiple smooth nodules of grey matter lining the lateral ventricles, which may be associated with other focal subcortical heterotopias (Fig. 31.12). On rare occasions, focal subcortical heterotopia can be seen in the absence of subependymal heterotopia. Most patients have relatively normal development but perhaps 80 per cent of patients with the disorder have epilepsy. When familial there is evidence of more women being affected, with fewer surviving males. The condition can be caused by mutations of *FLNA* on Xq28, encoding for filamin A, a protein required for neuronal migration.

Polymicrogyria refers to the presence of an area with many abnormally small gyri with an irregular cortical surface. When diffuse, it can be associated with severe developmental delay, with localized epilepsy. *Schizencephaly* describes the presence of grey matter lined clefts that extend from the cortical surface to the ependymal lining and may represent a form of polymicrogyria. The condition can be unilateral or bilateral. Developmental delay and contralateral hemiparesis are common.

The importance of the recognition and accurate diagnosis of cortical dysplasia lies in the potential for surgical treatment in some patients. Cortical dysplasias of the temporal lobe seem to have by far the best surgical outcome with up to 50 per cent being seizure-free (Kuzniecky and Jackson 1997). Unfortunately, the majority of cortical dysplasias are extra-temporal and here lower seizure-free rates are seen when localized resection of focal dysplasias is attempted.

Dysembrioplastic neuro-epithelial tumours are present in many patients diagnosed as having low-grade temporal gliomas or hamartomas in early series (Daumas-Duport *et al.* 1988). They can be reasonably described as malformations of cortical development due to abnormal proliferation or apoptosis. The lesions are usually seen within areas of dysplastic cortex and seem to have a predilection for the temporal lobes though they may also occur in the frontal lobes. They may make up between 5 and 10 per cent of pathology found in temporal lobectomy series and virtually never show any form of malignant transformation.

31.8.7 Other causes of symptomatic seizures and epilepsy

Neurodegenerative disorders can be associated with epilepsy. In Alzheimer's disease seizures occur usually late in the illness in up to 15 per cent of patients (Romanelli *et al.* 1990) (Section 34.6.2). The relative risk of epilepsy for autopsy-proven causes of Alzheimer's disease is around 10 (Hauser *et al.* 1986). Myoclonus is also evident, particularly in patients with familial Alzheimer's disease and with Alzheimer's change complicating Down's syndrome. In contrast, seizures appear rare in frontotemporal degenerations (Section 34.6.4).

An increased incidence of seizures has been noted in association with multiple sclerosis, at 3–6 times the expected rate (Poser and Brinar 2003) (Section 37.5.3). It may be that seizures are particularly likely to occur as an acute symptomatic phenomenon related to clinical relapse and plaque formation. More chronic epilepsy can be seen in severely disabled patients with frontal lobe syndromes and numerous frontal subcortical plaques (Moreau *et al.* 1998).

31.9 The diagnosis of epilepsy

The diagnosis of epilepsy in the adult is essentially clinical, and based on a detailed description of events experienced by the patient before, during, and after a seizure. An eye witness account is particularly important. In view of the social and economic implications, diagnostic errors must be avoided at all costs. Thus, the first rule about diagnosing epilepsy is never to make the diagnosis without incontrovertible clinical evidence. If there is any doubt, the

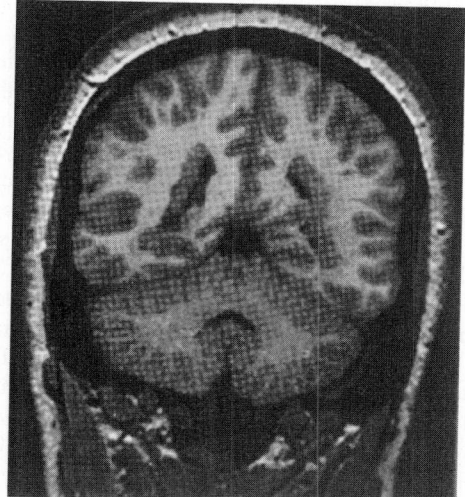

Fig. 31.12 Periventricular nodular heterotopia on MRI T$_1$-weighted images. (Reproduced from Kuzniecky and Jackson 1997 with permission.)

clinician should resist the temptation to attach a label and should rely on the passage of time and the further description of symptomatic events to reach a firm conclusion. Few people with epilepsy will come to harm from a delay in diagnosis whereas a false-positive diagnosis is always gravely damaging. In patients with possible syncope there is an imperative to reach a clear diagnosis as this group have excess morbidity and mortality (Section 31.1.2).

However, it is not enough simply to decide that a patient's attacks are epileptic in nature. Other considerations must be addressed in the diagnostic process.

◆ Are the seizures merely acute symptomatic seizures requiring treatment only of the underlying condition (Section 31.7), or are they part of an epilepsy?

◆ If the seizures are thought to be part of an epilepsy, an adequate classification of seizures and of the epilepsy syndrome must be attempted because of the important prognostic, therapeutic, and aetiological implications.

◆ Wherever possible, a cause should be identified because this might require treatment in its own right, and to fully inform the patient about his or her condition.

While the differentiation of seizures from other episodic events is made on clinical grounds, investigations have particular importance in answering the subsequent diagnostic questions.

31.9.1 **The electroencephalogram**

The EEG may rarely provide information that adds weight to the clinical diagnosis, but more importantly aids the classification of epilepsy. Routine inter-ictal EEG recording is one of the most abused investigations in clinical medicine and wrong interpretation of it use results in great human suffering. The diagnostic value of an inter-ictal EEG is widely misunderstood (Binnie 1997). Often the EEG is requested either to exclude or to prove a diagnosis of epilepsy— something that can seldom, if ever, be done. Erroneous interpretation of the EEG is probably the commonest reason for non-epileptic events being diagnosed as seizures. Particular problems are caused by misinterpretation of non-epileptiform sharp transients, such as 6 and 14 per second positive spikes and benign epileptiform transients of sleep, and responses to hyperventilation and photic stimulation.

In a population of patients with definite epilepsy from a tertiary referral centre, 30 per cent exhibited epileptiform activity in every routine inter-ictal record, while 11 per cent never did (Ajmone-Marsan and Zivin 1970). Sleep recording may increase sensitivity; epileptiform abnormalities occurred in 63 per cent of subjects without discharges in an intial awake record. When it is recognized that simple partial and brief complex partial seizures can occur without detectable changes at scalp electrodes in ictal records, the sensitivity of standard inter-ictal EEG recording will always be poor. It must be remembered that these figures apply to populations with clinically definite epilepsy, and considerably overestimate the value of the EEG in populations for whom there is diagnostic uncertainty. Interictal epileptiform features were found in 39 per cent of adults with a first presentation of seizures or epilepsy, many of whom had recordings within 24 h of a first seizure (King *et al.* 1998). This rate was again increased by subsequent sleep-deprived recording. The specificity of the inter-ictal EEG is best demonstrated by series that screened military personnel (Gregory *et al.* 1993).

In a total of 21 000 individuals epileptiform abnormalities were found in only 2.4 per 1000.

The inter-ictal EEG is potentially important in two clinical settings. First, it may be difficult to differentiate absence seizures from complex partial seizures. In patients with seizures occurring without an aura that are characterized by a brief period of absence with or without automatism. The finding of generalized spike wave or focal spike activity, respectively, will clarify the diagnosis. This differentiation has important implications for treatment and prognosis. Second, in patients with tonic-clonic seizures without an aura, especially when these occur during sleep, the EEG can again differentiate between generalized epilepsies characterized by generalized spike wave and seizures with a focal onset in which there may be localized abnormalities.

Ictal EEG, using ambulatory or more satisfactorily videotelemetry techniques, is important in distinguishing epileptic seizures from non-epileptic attacks. However, movement and other artefacts may complicate interpretation. Identifying tonic-clonic seizures should, however, present no problems because of the post-ictal changes. Differentiating between non-convulsive psychogenic episodes and complex partial seizures, particularly those of frontal origin can still be difficult.

31.9.2 **Neurological imaging**

Over the past decade the role of neurological imaging in the diagnosis and management of epilepsy has changed considerably. On the one hand CT scanning is universally available, but MR scanning has become enormously sophisticated and capable of demonstrating many abnormalities not previously recognized by CT.

MRI is more sensitive for most of the cerebral pathologies associated with chronic epilepsy, with the exception of calcification which it does not demonstrate well. Mesial temporal sclerosis, low-grade neoplasia, vascular lesions, particularly caveromas, and developmental abnormalities are all likely to be missed on CT, but readily demonstrated by MRI. In the elderly, however, MRI often shows a high incidence of lesions, whose clinical relevance to the seizures may be uncertain. In a series of 300 children and adults presenting with first seizures, MRI identified epileptogenic lesions in 17 per cent of 154 patients with definite partial epilepsy and 18 per cent with unclassified epilepsy, but was normal in all patients with clinical and EEG evidence of a generalized epilepsy syndrome (King *et al.* 1998). MRI was more sensitive than CT in the early detection of causative lesions. Indeed, CT detected only half the 17 tumours found, although it was uncertain how many of these would have been optimally treated by surgery if diagnosed.

At present a reasonable approach is to undertake imaging at the point of diagnosis in all adults except those who can be identified clinically and neurophysiologically as having one of the syndromes of idiopathic generalized epilepsy. While MRI is preferable, CT scanning is effective in identifying the more aggressive tumours causing epilepsy, but could miss more indolent gliomas. MRI is indicated in all patients with epilepsy who appear refractory to pharmacological treatment irrespective of previous imaging studies. Here MRI should be tailored to the individual, but will usually include high definition T1-weighted thin-slice scans, often with hippocampal volumetry, T2-weighted and FLAIR images.

Other technologies that produce 'functional' images such as positron emission tomography, PET, single photon emission computerized tomography, SPECT, and functional MRI and spectroscopy are largely experimental and to date have usually been used in the assessment of patients for surgical treatment of their epilepsy (Sections 3.1.5 and 3.1.6).

31.10 Pharmacological treatment of epilepsy

31.10.1 General principles

At a time when there is a sudden and dramatic increase in the number and choice of drugs to treat epilepsy, it is perhaps important to begin by considering some broad principles that need to be applied to the treatment of an individual patient:

- *Certainty of diagnosis.* The diagnosis of seizures or epilepsy should be secure. There is little or no place for a therapeutic trial, when the diagnosis is uncertain. Acute symptomatic seizures must be differentiated from seizures occurring spontaneously as a part of epilepsy. The former will rarely need anything other than acute treatment of seizures together with a treatment of the underlying cause, as in alcohol withdrawal or acute metabolic disorders.

- *Deciding when to start treatment.* Initiating or changing anti-epileptic drug therapy needs full and adequate discussion with the patient who should be made fully aware of the aims of treatment, the benefits, and potential adverse effects. Many decisions to be made in treatment of epilepsy are not clear cut, and require balanced judgement. The patient's personal circumstances and views are crucial to ensuring compliance with regimes. In many circumstances the doctor should be a provider of relevant information rather than a decision-taker. Compliance is a major issue in the long-term management of epilepsy and poor compliance may not identify a 'bad' patient, so much as defining a poor doctor-patient relationship involving an inadequately informed patient.

- *Purpose of treatment.* The ultimate aim of treating epilepsy will be no seizures and no drugs. Unfortunately, this is not readily achievable for many patients with epilepsy who have a chronic disorder. The first step in treating epilepsy will always be to choose the minimum effective dose of an optimally effective anti-epileptic drug. In practice, this means initiating treatment, and slowly increasing this dose if and when further seizures occur. This monotherapy approach, using a single anti-epileptic drug will usually be successful in 50–70 per cent of new patients presenting with epilepsy. Alternative monotherapies, or combined treatments involving polytherapy will only be necessary in the minority with more severe epilepsies. In this group of patients a law of diminishing returns applies. Briefly stated, the longer the seizures remain poorly controlled the less likelihood of remission of epilepsy (Annegers *et al.* 1979a). This has two consequences. First, often some agreement will need to be reached with the patient as to an acceptable compromise between a reduced seizure frequency and the severity of unwanted side-effects of anti-epileptic drugs. Second, non-pharmacological treatments such as surgery may demand serious consideration at a relatively early stage.

- *Choice of treatment.* In choosing between different drugs, judgements must address the relative efficacy of a drug, and its tolerability and safety profile, for an individual patient. These factors will contribute to the overall effectiveness of an anti-epileptic drug. Comparative judgements of efficacy, tolerability, and effectiveness are best based on the results of appropriate randomized clinical trials (Marson and Chadwick 1996). In addition to these fundamental principles, it is helpful if anti-epileptic drugs are simple for patients to use, needing no more than twice daily dosing without requiring regular anti-epileptic blood level monitoring.

Most comparative studies of anti-epileptic drugs identify the spectrum of adverse effects of anti-epileptic drugs as the factor contributing most to determining their relative effectiveness. It has proved difficult to detect significant differences in efficacy outcomes in comparative monotherapy studies, because of inadequate design and powering of studies (Glauser *et al.* 2006). Anti-epileptic drugs possess dose-related adverse effects, largely affecting the central nervous system, as well as idiosyncratic side effects. In addition, because of the long periods of time for which they may be taken, these drugs have also been associated with chronic toxicity, as well as teratogenicity as they may be taken through the child-bearing years. All these issues need to be taken into account in choosing drug treatment.

31.10.2 Starting therapy

In the past anti-epileptic treatment has been advocated before seizures occur. Such prophylactic treatment has been recommended for patients with a high prospective risk of epilepsy after head injury and craniotomy for various neurosurgical conditions. Because no clear evidence exists that anti-epileptic treatment is effective in preventing late epilepsy (Section 31.8.2) (Temkin *et al.* 1998), it seems better to delay treatment until seizures have occurred rather than to adopt a policy of treatment of all those at risk— particularly as there may be a high incidence of side effects with prophylactic treatment and of poor compliance.

It is also common for clinicians to see patients with a first tonic-clonic seizure, infrequent such seizures, or seizures with minor symptomatology for whom there would be uncertainty about commencing treatment. Such patients have been studied in two randomized controlled trials (Marson *et al.* 2005; Leone *et al.* 2006). These provide evidence that early treatment fails to modify the longer term prognosis which is universally good, but that it can reduce the incidence of seizures in the shorter term. The chief factors affecting the risk of seizure recurrence in the short term are increasing numbers of seizures, an abnormal EEG, and the presence of a history of, or current evidence of underlying brain disease of which seizures are symptomatic. The presence of these factors make it possible to identify a small number of higher risk patients for whom treatment may be worthwhile (Kim *et al.* 2006).

When two or more unprovoked seizures have occurred within a short interval, anti-epileptic therapy is usually indicated. Problems do, however, arise in defining a short interval. Most would include periods of 6 months to 1 year within the definition. Even where seizures occur close together, the identification of specific precipitating factors may make it more important to counsel patients than to commence drug therapy. The most common examples are photically induced and alcohol-withdrawal seizures in adults.

31.10.3 Choice of drug

There is agreement that patients with newly diagnosed epilepsy should be treated with a single drug. The key issue in the choice of this first drug at diagnosis is an accurate and adequate diagnosis of seizure type and, if possible, epilepsy syndrome. By no means all

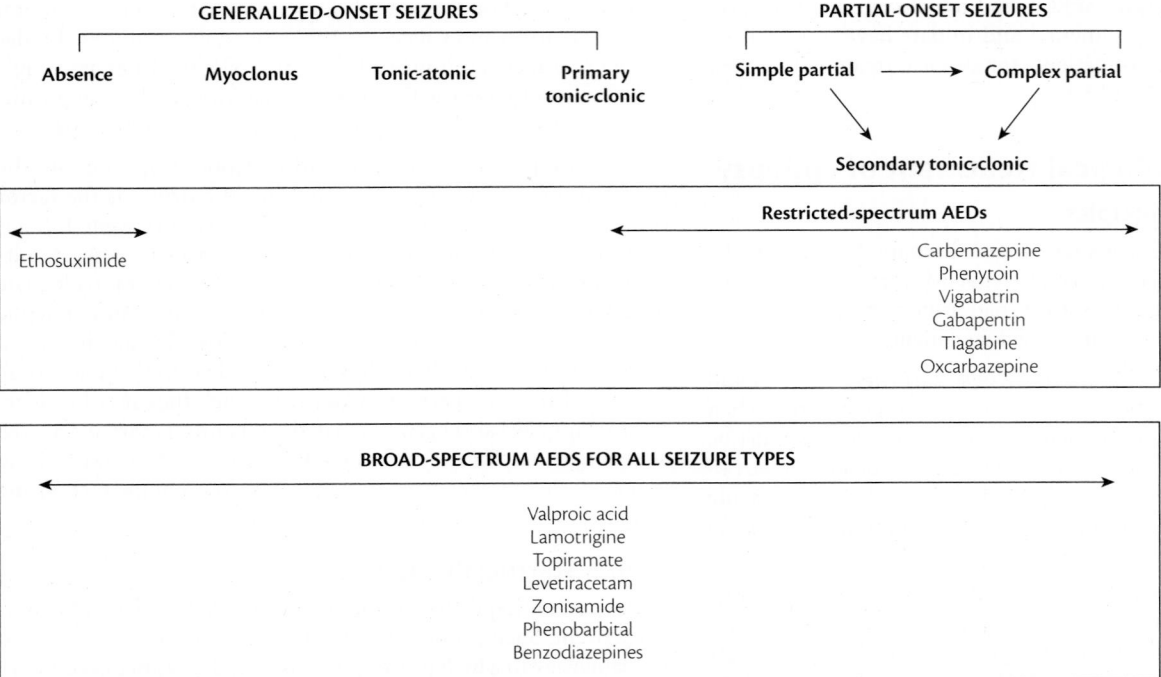

Fig. 31.13 Spectrum of efficacy of anti-epileptic drugs (AEDs). (Reproduced from McCorry *et al.* 2004 with permission.)

drugs are effective against all seizure types. The spectrum of efficacy of drugs is represented graphically in Fig. 31.13. It is particularly important to avoid the use of drugs that may themselves exacerbate seizures. Hence, there is evidence that both carbamazepine and vigabatrin (Perucca *et al.* 1998) may exacerbate absence and myoclonic seizures in the generalized epilepsy syndromes. It is here that syndromic classification of epilepsy becomes most important as a drug should be chosen that would be effective against all seizure types known to occur in that syndrome, rather than only those that have occurred in an individual patient to date.

It is now becoming possible to associate particular anti-epileptic drug mechanisms with effects against different seizure types and different adverse effects. Inevitably, many anti-epileptic drugs, may have multiple mechanisms of action, some of which are not fully understood. Table 31.8 summarizes different mechanisms of action for anti-epileptic drugs (Perucca 2005). Several drugs modify ionic sodium conductance across membranes, binding to ion channels in order to maintain them in an inactivated state thereby blocking repetitive neuronal firing. Drugs that possess this property include phenytoin, carbamazepine, lamotrigine, oxcarbazepine, topiramate, valproate, and zonisamide. All are effective in preventing partial seizures and both generalized and secondarily generalized tonic-clonic seizures. Most of these drugs, with the possible exception of valproate, can be associated with dose-related neurotoxicity syndromes that include ataxia, nystagmus, and diplopia. The second direct membrane effect is displayed by drugs such as ethosuximide and possibly valproate which modify slow or calcium-T currents in the thalamus. This mechanism seems particularly effective against spike wave mechanisms and absence seizures. These same calcium conductances can be enhanced by gabergic inputs to the thalamus

via $GABA_B$ receptors (Crunelli and Leresche 1991), which may explain the exacerbation of absences by vigabatrin.

A number of anti-epileptic drugs exert their properties through modulation of the $GABA_A$ receptor/chloride ionophore. Thus, both benzodiazepines and barbiturates bind close to this site to increase chloride-ion conductance and maintain membrane hyperpolarization. Newer drugs such as vigabatrin and tiagabine may have more direct effects in prolonging the synaptic action of GABA. To date evidence suggests that these drugs are effective against partial and secondary generalized seizures but may exacerbate spike wave epilepsies. They are less likely to cause sedation and ataxia but may have a higher risk of psychiatric disorder including depression. Drugs, which interfere with excitatory neurotransmission via glutamate and aspartate receptors, may yet prove valuable anti-epileptic drugs. Some of the anti-epileptic properties of felbamate may be due to its ability to interfere with the action of glycine in facilitating glutaminergic activity, and a number of drugs with potential glutaminergic activity have entered clinical trial programmes with varying success.

Actions, common side-effects and indications for use are briefly summarized in Table 31.9. Further details of the pharmacokinetics of these drugs are also presented later in the chapter.

Partial, localization-related epilepsies

Currently, a large number of drugs can be considered for treating patients with cryptogenic and symptomatic partial epilepsies. These include both older and newer drugs, including those with a spectrum of efficacy limited to the partial epilepsies and those with a broader spectrum of effects (see Fig.13.13). Evidence-based guidelines have indicated that carbamazepine is for most patients the optimal first choice drug (NICE 2004). This guideline will

Table 31.8 Mechanism of action of anti-epileptic drugs

	Blockade of Na^{++} channels	Blockade of Ca^{++} channels	GABA$_A$ receptor potentiation	Increased synaptic GABA	Other actions
Older Drugs					
Carbamazepine	++	+(L)			
Valproate	+	+ (T)	?	+	
Phenytoin	++				
Barbiturates	+		++	+	
Ethosuximide		++ (T)			
Benzodiazepines			++		
Newer drugs					
Vigabatrin				++	
Lamotrigine	++	++ (N,P/Q,R,T)		+	K$^+$ currents
Gabapentin		++(N,P/Q)		+	
Pregabalin		++(N,P/Q)			
Felbamate	++		+	+	NMDA receptors
Topiramate	++	+(L)	+		AMPA/Kainate receptors Carbonic anhydrase inhibition
Tiagabine				++	
Oxcarbazepine	++	+(N,P)			
Levetiracetam		+(N)	+		Modulation of SV2A protein, K$^+$ currents
Zonisamide	++	+ +(N,P,T)	+		Carbonic anhydrase inhibition

require updating following the outcome of a large study comparing carbamazepine with newer anti-epileptic drugs. This showed that lamotrigine was superior to carbamazepine for time to treatment failure, being better tolerated and of equivalent efficacy to carbamazepine (Marson *et al.* 2007b) and was cost effective. Gabapentin and topirimate were inferior to lamotrigine, but there was uncertainty about the place of oxcarbazepine. The study did not include pregabalin, zonisamide, or levetiracetam for which there is some limited evidence that its effectiveness may be similar to carbamazepine (Brodie *et al.* 2007). The newer drugs are likely to remain second choice montherapies or will be used as adjunctive therapy with one of the first-choice sodium channel drugs, lamotrigine or carbamazepine or oxcarbazepine.

Generalized epilepsies

The management and treatment of the cryptogenic and symptomatic generalized epilepsies is more relevant to childhood epilepsy and will not be discussed further here (Section 30.7.3). The choice of drug therapy in the idiopathic generalized epilepsies is, however, a major issue as collectively these may represent between 20 and 30 per cent of all human epilepsies. Evidence-based guidelines have indicated that valproate is for most patients the optimal first choice drug (NICE 2004). This guideline will require updating following the outcome of a large study comparing valproate with lamotrigine and topiramate (Marson *et al.* 2007a). This shows that valproate remains the most clinically effective drug for this group of epilepsies, topiramate being poorly tolerated and lamotrigine lacking efficacy compared to valproate. These results create particular difficulties in decision making for women in their child-bearing years, for

whom valproate may present problems during pregnancy (Section 31.10.5).

Unclassified epilepsy

Decisions about starting anti-epileptic drug treatment often have to be made in the face of some uncertainty concerning a syndromic classification. While the clinician may be certain that seizures have occurred, there may be insufficient information available from a few poorly witnessed events to provide a definite syndrome diagnosis. Common situations in which this occurs include the patient with witnessed tonic-clonic seizures during sleep, and infrequent day-time trance-like episodes reflecting absence or complex partial seizures. Where this uncertainty exists it is relatively unusual for the EEG or other investigations to provide definitive information. In these circumstances, a broad spectrum anti-epileptic drug, such as valproate is to be preferred (Marson *et al.* 2007a).

31.10.4 Adverse effects of anti-epileptic drugs

When prescribing anti-epileptic drugs it is important that patients are counselled about the possible adverse effects and their significance.

Acute dose-related toxicity

Most anti-epileptic drugs including phenytoin, carbamazepine, lamotrigine, barbiturates, and benzodiazepines, give rise to a nonspecific encephalopathy associated with high blood concentrations. Patients exhibit sedation and nystagmus and, with increasing blood levels, ataxia, dysarthria, and ultimately confusion and drowsiness.

Table 31.9 Efficacy and toxicity of anti-epileptic drugs

	Structure	Indications	Contra-indications	Dosage in adults	Optimal plasma levels	Adverse Effects			
						Dose related	Idiosyncratic	Chronic	Teratogenicity
Carbamazepine		Drug of choice: Partial epilepsy	Idiopathic generalized epilepsy	600–1600 mg/day with gradual introduction because of autoinduction	4–10 µg/ml, but very variable upper limit to tolerability	Dizziness, diplopia, and unsteadiness	Rash and acute hypersensitivity reactions Aplastic anaemia (1:200 000)	Few that are well documented (hyponatraemia and neutropenia)	Spina bifida in 1% of pregnancies
Clobazam		Second choice drug: Probable broad spectrum. Useful for treating clusters of seizures		20–60 mg/day Therapeutic and adverse effects may show tolerance		Drowsiness and sedation, but less than other benzodiazepines			
Ethosuximide		Second choice drug: Absence persisting into adult life	Partial epilepsy and generalized tonic-clonic seizures	0.5–2.0 g/day	40–100 µg/ml	Nausea, drowsiness, and dizziness	Rash and acute hypersensitivity reactions. SLE-like syndromes		Little information
Felbamate		Occasional use: Lennox–Gastaut syndrome		1200–4800 mg/day		Insomnia and gastrointestinal intolerance	Aplastic anaemia (1:3000–5000) Hepatic failure	Weight loss	
Gabapentin		Second choice drug: Partial epilepsies		900 mg–4.8 g/day		Drowsiness, ataxia, and sedation	None known	Weight gain	
Lamotrigine		First choice drug: Broad spectrum for partial epilepsy and possibly generalized syndromes		100–800 mg/day		Diplopia, dizziness, and sedation	Rash and acute hypersensitivity reactions (particularly with valproate co-medication)		

Drug	Indication	Dose	Therapeutic level	Dose-related side effects	Idiosyncratic effects	Chronic/teratogenic effects
Lorazepam	First choice drug: status epilepticus / Drug of choice: status epilepticus	0.1 mg/kg				
Oxcarbazepine	Drug of choice: for partial epilepsy—broadly comparable with carbamazepine / Idiopathic generalized epilepsy	600–3000 mg/day	50–150 µmol/l	Dizziness, diplopia, and unsteadiness, but less frequent than carbamazepine	Rash, but less frequent than carbamazepine. 25 per cent of patients sensitive to Carbamazepine will also be sensitive to Oxcarbazepine	Hyponatraemia
Phenobarbitone	Occasional use in partial and generalized epilepsies (excepting absence) and status	60–200 mg/day	15–35 µg/ml but limits often modified by tolerance	Drowsiness, sedation and unsteadiness, adverse effects on cognition and behaviour	Rash	Tolerance and habituation. Dupuytrens contracture and connective tissue disorders. Hare-lip/cleft palate, and cardiological abnormalities
Phenytoin	Second choice drug: Partial epilepsy and generalized tonic-clonic seizures	200–600 mg/day	10–20 µg/ml. Monitoring is indicated whenever there is poor control of seizures or side effects.	Drowsiness, ataxia, and dysarthria. Rarely abnormal movements	Rash and acute hypersensitivity reactions	Gum hypertrophy, coarsening of facial features, hirsuitism, and acne. SLE-like syndromes. Hare-lip/cleft palate & cardiological abnormalities
Levetiratcetam	Second choice drug for partial and generalized seizures	750–3000 mg/day		Fatigue, somnolence, dizziness		Little information
Pregabalin	Second choice drug: Partial epilepsies	75–600 mg/day		Drowsiness, ataxia, and sedation	None known	Weight gain

(continued)

Table 31.9 (Continued) Efficacy and toxicity of anti-epileptic drugs

	Structure	Indications	Contra-indications	Dosage in adults	Optimal plasma levels	Adverse Effects			
						Dose related	Idiosyncratic	Chronic	Teratogenicity
Primidone		Rarely used: Probable efficacy as phenobarbitone		500–1500 mg/day	As phenobarbitone to which it is metabolized	Drowsiness, sedation, and unsteadiness, adverse effects on cognition and behaviour	Rash	Tolerance and habituation. Dupuytrens contracture and connective tissue disorders	Hare-lip/cleft palate and cardiological abnormalities
Tiagabine		Second choice drug: Partial epilepsy	Idiopathic generalized epilepsy	15–60 mg/day		Dizziness, depression, tremor. May be exacerbation of partial seizures at higher doses.			
Topiramate		Second choice: Broad spectrum		100–800 mg/day		Sedation, cognitive difficulty.		Renal calculi	
Valproate (Sodium)		First choice broad spectrum drug: may be less effective in partial epilepsy than carbamazepine		1–3 g/day	Of no value	Tremor, irritability, and occasional confusion.	Gastric intolerance. Hepatotoxicity (rare in adults). Pancreatitis	Weight gain, alopecia, insulin intolerance, polycystic ovarian syndrome	Spina bifida in 2–3 per cent of pregnancies. Foetal valproate syndrome
Vigabatrin		Final choice drug for partial epilepsies. May be useful in adult survivors of West's syndrome	Idiopathic generalized epilepsy	1.5–6.0 g/day	Of no value	Depression	Psychosis	Weight gain, Visual field constriction.	

In some instances seizure frequency may increase with high blood levels and occasionally involuntary movements are seen, particularly with phenytoin. Phenytoin is especially likely to result in dose-related toxicity because of its unusual pharmacokinetics (Section 31.10.8). Carbamazepine may cause similar symptoms if the dose is not built up slowly, because of its ability to auto-induce liver microsomal enzymes. Valproate does not appear to be associated with this typical syndrome of neurotoxicity, but some patients with high blood levels may exhibit restlessness and irritability, sometimes with a frank confusion state, sometimes with elevated blood ammonia concentrations. Postural tremor is a more common dose-related adverse effect.

All anti-epileptic drugs can have adverse effects on cognitive function and behaviour with increasing dose and blood concentrations, although there is little evidence that they are common in patients using therapeutic doses as monotherapy (Loring et al. 1994). While carbamazepine and valproate may have smaller risks of this type than barbiturates and phenytoin, the newer generation of drugs may be better tolerated (McCorry et al. 2004).

Drug interactions may increase the risk of dose-related toxicity. Thus valproate may greatly prolong the half-life of lamotrigine, making dosage reduction necessary during co-medication. Similarly, withdrawal of enzyme-inducing drugs can give rise to similar problems.

Acute idiosyncratic toxicity

Most anti-epileptic drugs, particularly phenytoin, carbamazepine, and lamotrigine, may cause a delayed hypersensitivity reaction consisting of maculopapular erythematous eruption which, in more severe cases, may be associated with fever, lymphadenopathy, and hepatitis. The incidence of allergic skin reaction with phenytoin may be as high as 10 per cent and with carbamazepine up to 15 per cent with rapid introduction of the drug (Chadwick et al. 1984). Lamotrigine is also associated with similar problems, requiring slow titration particular with co-medication with valproate. The incidence of serious reactions such as Stevens Johnson syndrome or acute epidermal necrolysis is much lower.

Marrow aplasia is a rare complication of carbamazepine, but more common with felbamate (Kaufman et al. 1997). Reports of fatal cases of liver failure in association with valproate therapy largely concern children under the age of 2 years who are often multiply handicapped and receiving many different anti-epileptic drugs. It may be that they have an underlying error of metabolism that predisposes them to liver failure (Dreifuss et al. 1987). The potential for rare idiosyncratic side-effects of levetiracetam, topiramate, and gabapentin is currently uncertain, but probably small.

Chronic toxicity

Anti-epileptic drugs are unusual in that they may be administered to patients over a long period as treatment for chronic epilepsy. This may lead to the development of a wide variety of syndromes of chronic toxicity (Table 31.9). A number of factors seem to predispose to the development of these disorders, including polypharmacy, dosage, and duration of therapy. Whilst valproate and carbamazepine may have fewer chronic toxic effects than barbiturates and phenytoin, the delay in recognizing quite common chronic toxic effects with the older agents should warn us that continued vigilance is needed in the use of the newer anti-epileptic drugs. Some patients exposed to long-term treatment with vigabatrin

have developed severe irreversible concentric visual field loss (Eke et al. 1997). Quantitative visual field assessment can reveal asymptomatic, usually nasal, visual field constriction, associated with electroretinographic changes in keeping with retinal cone system dysfunction (Krauss et al. 1998), in larger numbers of patients treated with vigabatrin. Further follow-up of 32 patients continuing monotherapy with vigabatrin and 19 patients continuing carbamazepine from the randomized controlled trial reported by Kalviainen et al. (1998) showed that 41 per cent of vigabatrin-treated patients had visual field constriction, compared to no carbamazepine treated patients, indicating a causal relationship between long-term vigabatrin exposure and visual field loss.

There is some concern about the incidence of weight gain on anti-epileptic drugs. Valproate, gabapentin, and pregabalin are perhaps most troublesome in this respect, although topiramate and zonisamide may be associated with weight loss.

The polycystic ovary syndrome of amenorrhea, oligomenorrhea, abnormal cycle intervals, or menometrorrhagia, and signs of hyperandrogenism, including hirsuitism, acne, and alopecia is more common in women with epilepsy. Obesity is a common association. The symptomatic syndrome is much less common than the finding of polycystic ovaries on ultrasound. A number of studies have shown that polycystic ovary syndrome is over-represented among women with epilepsy, but there is controversy as to whether this is due to drug effects (Bilo et al. 2001; Isojarvi et al. 2001). Epilepsy itself may be related to polycystic ovary syndrome via epileptiform discharges disrupting normal hypothalamic function.

There is evidence that populations of people with epilepsy have a higher prevalence of osteopenia and or osteoporosis than would be expected. Enzyme-inducing anti-epileptic drugs may contribute to this as may reduce mobility in people with epilepsy.

31.10.5 Pregnancy and anti-epileptic drugs

All anti-epileptic drugs must be regarded as potentially teratogenic, making adequate preconception counselling essential. It is uncertain how this risk compares to that of seizures during pregnancy but the optimal policy is to suppress seizures with the lowest effective dose of a single anti-epileptic drug, whenever possible. Older drugs such as phenytoin and barbiturates, may increase the risk of major foetal malformation two- to three-fold, the most common malformations being hare-lip, cleft palate, and cardiovascular anomalies. The risks are higher with polytherapy than with monotherapy.

With the advent of pregnancy registers there is now greater certainty about the incidence of malformations. The United Kingdom registry provides data on over 3600 pregnancies. Of commonly used monotherapies, carbamazepine was associated with an overall malformation rate of 2.2 per cent, lamotrigine 3.2 per cent, and valproate 6.2 per cent, compared to a rate of 3.5 per cent in women with epilepsy who had not taken drugs during pregnancy (Morrow and Craig 2003). These are average figures and there is good evidence that risk increases with dose. It may be that the dose–risk relationship is particularly steep for doses of lamotrigine above 200 mg/day. There is insufficient evidence on the safety of gabapentin, topiramate, and levetiracetam at present to indicate that they are safe, although the data for levetiracetam appears promising.

There is an association between neural tube defects and exposure to valproate and carbamazepine (Rosa 1990; Samren et al. 1997).

A clear dose effect was evident for valproate with a threshold at 1 g/day. Early screening for neural tube defects, using ultrasound and amniocentesis, and testing for alpha-fetoprotein, therefore seems to be indicated in women becoming pregnant while taking these drugs. It is now good practice to prescribe folate supplements to sexually active women of child-bearing age taking anti-epileptic drugs.

In addition to major abnormality, dysmorphic syndromes have been described with a number of anti-epileptic drugs. Phenytoin may be associated with epicanthic folds, hypertelorism, broad flat nasal bridges, and distal digital hypoplasia (Hanson *et al.* 1976). Valproate has been associated with inferior epicanthic folds, flat nasal bridges with upturned nasal tips, a shallow philtrum, and down-turned mouths (DiLiberti *et al.* 1984). These abnormalities can be associated with radial ray dysplasia. Carbamazepine has been associated with similar features with microcephaly in addition (Jones *et al.* 1989). The main concern with all these syndromes is the possible association with growth retardation and developmental delay.

A large retrospective study (Adab *et al.* 2004) has shown that valproate may be strongly associated with impairments of verbal intelligence and a dysmorphic syndrome in children. The risks begin with daily doses over 800 mg/day and over 40 per cent of exposed children had a low or exceptionally low verbal intelligence quotient and consequent educational and behavioural difficulties. There is further consistent data from smaller studies (Gaily *et al.* 2004) highlighting the risk for valproate and the safety of carbamazepine. However, the effect of frequent tonic-clonic seizures during a pregnancy on verbal intelligence may be of similar degree to that of valproate exposure, complicating decision making for women with idiopathic generalized epilepsy.

31.10.6 Long-term anti-epileptic drug therapy

The majority of patients developing epilepsy achieve a long-lasting remission soon after the start of therapy. For these patients drug withdrawal may be considered after 2, 3, or more years (Section 31.10.9). Some 20 per cent of patients developing epilepsy have a chronic disorder, never completely controlled by drugs. Of patients who are not controlled, but comply with, maximal tolerated doses of a single anti-epileptic drug, about 11 per cent may respond to an alternative monotherapy, while 40 per cent become seizure-free if a first failure is due to intolerance (Kwan and Brodie 2000). Smaller numbers who graduate to polytherapy can become seizure-free. However, a policy of polytherapy inevitably increases the risks of dose-related, idiosyncratic, and chronic toxicities. The law of diminishing returns applies. Thus, for this group of patients a realistic aim may not be complete remission of seizures but the compromise of a reduced seizure frequency involving less severe seizures, to be achieved with one drug, or at the most two.

Some patients may continue to have seizures but not be disabled by them. For instance, they may have very infrequent seizures, seizures that are symptomatically minor or confined to sleep. In such patients there is usually little to be gained from alternative drugs or additional drugs, assuming that a single drug is being used which is suitable for the seizure type and epilepsy syndrome.

Patients who continue to be disabled by the occurrence of seizures despite treatment with a single drug in optimal dosage demand further careful consideration. In particular it is important to consider whether there are factors that would explain an unsatisfactory response to therapy, such as unidentified structural pathology, the presence of complex partial seizures, or poor compliance. If this is not the case, then it is important to review the diagnosis: a common reason for failure of therapy is that the patient does not have epilepsy or that not all reported symptoms are epileptic in nature.

Where none of these conditions apply, it may be reasonable to try alternative drugs as monotherapy, and then to undertake a trial of the addition of a second drug. However, this demands careful discussion with the patient and the understanding that the second drug will be withdrawn in the absence of a satisfactory sustained response.

31.10.7 Refractory epilepsy and rational polytherapy

The single randomized controlled trial that has examined policies of alternative monotherapy versus add-on therapy in patients failing a single anti-epileptic drug for poor efficacy, produced inconclusive results (Beghi *et al.* 2003). However, placebo controlled studies of new anti-epileptic drugs demonstrate that new anti-epileptic drugs in partial epilepsy increase the odds of a reduction in seizure frequency of 50 per cent or by two to five times compared to placebo (McCorry *et al.* 2004). Empirically, all those experienced in treating epileptic patients recognize that most with refractory epilepsy will receive combinations of therapy and that it will be extremely difficult to reduce therapy to achieve treatment with a single drug. Thus, at present the weight of evidence suggests that drug combinations can possess greater efficacy in patients failing on monotherapy. What remains uncertain is the degree of benefit and the extent to which it may be offset by increased risk of adverse events.

There are many examples of polytherapy causing dose-related neurotoxicity: sulthiame inhibiting the metabolism of phenytoin; or valproate to inhibiting the metabolism of lamotrigine, both resulting in symptoms of intoxication. However, the ability of patients to tolerate carbamazepine and lamotrigine may be strongly influenced by whether or not they are taking simultaneously other drugs with actions on sodium conductances. Thus, the incidence of ataxia and diplopia in placebo controlled add-on studies of lamotrigine is strikingly higher than that for GABAergic drugs such as vigabatrin and tiagabine when used in similar trials (Marson *et al.* 1997).

The impact of polytherapy on the incidence of chronic toxicity and teratogenicity in patients with epilepsy is more difficult to assess. There is a consensus that chronic toxicity is more commonly seen in patients exposed to long-term polytherapy and that the incidence of teratogenicity rises strikingly with the number of anti-epileptic drugs administered during pregnancy. Thus, administration of three drugs during a pregnancy may be associated with anything up to a 50 per cent incidence of major foetal malformations (Nicene *et al.* 1980).

Consideration of the above evidence leads to two conclusions. First there is an urgent need for pragmatic clinical trials to examine the benefits of combination therapy. Secondly that when polytherapy is used it should embrace a number of principles of rational polytherapy:

◆ It is best to combine anti-epileptic drugs with different mechanisms of action than to prescribe combinations of anti-epileptic drugs that have similar mechanisms of action. Otherwise the additional efficacy will be limited, but the incidence of adverse events would be expected to be multiplied.

- It is best to select anti-epileptic drugs with relatively little potential for pharmacokinetic interaction.
- Patients treated with polytherapy demand more intensive monitoring, both clinically and with anti-epileptic drug levels.

What remains unknown is whether certain combinations of drugs may provide particular benefits in effectiveness. There are no satisfactory studies in this area. It may be that for partial epilepsies for which the first-line treatment is a sodium channel blocking drug this may be effectively combined with clobazam or levetiracetam.

31.10.8 Monitoring anti-epileptic drugs

Pharmacokinetic data (Table 31.10) defines drug absorption, distribution, metabolism, and elimination. One clinical application of pharmacokinetics is therapeutic drug monitoring in serum or plasma.

Phenytoin has a non-linear relation between the dose and the serum concentration (Richens and Dunlop 1975). This results in a narrow therapeutic window, and monitoring is necessary to avoid neurotoxicity in patients, whose dosage is being increased. The

Table 31.10 Pharmacokinetics of anti-epileptic drugs

	Absorption (Time (hours) to peak plasma conc. after oral dose)	Protein-binding (per cent)	Active metabolites	Metabolism (half life in hours)	Important interactions
Carbamazepine	4–24	75	10,11-epoxide	8–30	Enzyme inducer reducing blood levels of phenytoin, barbiturates, lamotrigine, topiramate, tiagabine, and oral contraceptives. Its own metabolism shows auto-induction and is induced by other enzyme-inducing anti-epileptic drugs.
Clobazam	1–3	90	N-desmethyl	10–50	
Ethosuximide	2–6	–		40–70	
Felbamate	2–6	25		12–24	Reduces carbamazepine levels, but increases epoxide. Increases phenytoin, valproate, and phenobarbitone levels. Its levels are reduced by enzyme-inducing anti-epileptic drugs but slightly increased by valproate.
Gabapentin	2–3	–		5–7	Not metabolized and no interactions.
Lamotrigine	2–3	50		12–48	Is not an enzyme inducer, but it may reduce oestrogen levels. Its metabolism may be induced by other anti-epileptic drugs enzyme inducers, and inhibited by valproate
Levetiracetam	1–2	<10		5–8	Not metabolized and no interactions
Oxcarbazepine	2–6	40		8–10	Less enzyme induction than carbamazepine, but usual precautions with oral contraceptives.
Phenobarbitone	1–6	45		50–160	Enzyme inducer reducing blood levels of phenytoin, carbamazepine, lamotrigine, topiramate, tiagabine, and oral contraceptives. Its own metabolism shows auto-induction and is induced by other enzyme-inducing anti-epileptic drugs.
Phenytoin	4–12	90		9–140	Enzyme inducer reducing blood levels of carbamazepine, barbiturates, lamotrigine, topiramate, tiagabine, and oral contraceptives. Its own metabolism shows auto-induction and is induced by other enzyme inducing anti-epileptic drugs.
Pregabalin	1	–		6	Not metabolized and no interactions.
Primidone	2–5	20	Phenobarbitone Phenylethyl-malonamide	4–12	Enzyme inducer reducing blood levels of phenytoin, carbamazepine, lamotrigine, topiramate, tiagabine, and oral contraceptives. Its own metabolism shows auto-induction and is induced by other enzyme-inducing anti-epileptic drugs.
Tiagabine	1–2	95		4–9	Is not an enzyme inducer, but its metabolism is induced by other anti-epileptic drugs enzyme inducers.
Topiramate	1.5–4	15		12–24	Is not an enzyme inducer, but its metabolism may be induced by other anti-epileptic drugs enzyme inducers.
Valproate (Sodium)	1–4	90		8–20	Enzyme inhibitor of lamotrigine metabolism.
Vigabatrin	1–2	–		5–7	Phenytoin levels may fall, but mechanism is uncertain.
Zonisamide	2–5	40–50		60	Lower levels with enzyme inducers.

concept of the 'therapeutic' or 'optimal' range for phenytoin has been extended to other anti-epileptic drugs, and many laboratories now routinely estimate serum concentrations of drugs other than phenytoin.

A single measurement will give a good approximation of the steady state concentration for drugs with long half-lives such as phenytoin or phenobarbitone, but not for drugs with short half-lives. Measurements of sodium valproate concentrations from specimens taken at random during the day are impossible to interpret, as they may represent peak, trough, or intermediate concentrations. However collecting early morning specimens for measuring troughs is rarely practicable.

Even when concentrations of free drugs and their metabolites in the blood are known, important pharmacodynamic considerations may alter the relationship between the blood concentration and therapeutic effect. Thus for sodium valproate the onset of action is slower and longer lasting than can be explained by the pharmacokinetics of the drug (Rowan *et al.* 1979). Similarly, tolerance to the neurotoxic and therapeutic effects of benzodiazepines and barbiturate drugs is not explained by pharmacokinetic changes and must be due to drug–receptor interaction.

There are further fundamental biological reasons for doubting the value of routine monitoring of blood concentrations of anti-epileptic drugs. The upper limit of a therapeutic range may be defined as the concentration of the drug at which toxic effects are likely to appear. The most consistent relationship between the serum concentration and toxic effect is for phenytoin, but even with this drug some patients may tolerate, and indeed require, serum concentrations above 20 μg/ml. For sodium valproate, phenobarbitone, and carbamazepine there is a wide variation in individual tolerance of serum concentrations.

The lower limit of the therapeutic range is even more difficult to define, and most patients have epilepsy that is controlled by anti-epileptic serum concentrations well below the optimal range. Unquestioning acceptance of therapeutic ranges creates problems: patients with satisfactory control of seizures and low blood concentrations of drugs may have their doses needlessly increased, and patients who tolerate and need high blood concentrations may have their doses reduced.

Routine blood level monitoring is a valuable aid in the management of certain categories of patients:

- those receiving phenytoin or multiple drug treatment in whom dosage adjustment is necessary because of dose-related toxicity or poor seizure control;

- mentally retarded patients in whom the assessment of toxicity may be difficult;

- patients with renal or hepatic disease;

- patients who may not be complying with treatment.

31.10.9 Withdrawing anti-epileptic drugs

Both population and cohort studies have demonstrated that 70–80 per cent of patients diagnosed and treated for epilepsy will attain long-term remission exceeding 2 years. The decision to start a trial of drug withdrawal should be made by the patient after appropriate advice. This needs to cover difficult topics, which include an individual's risk of relapse on withdrawing treatment, and indeed, on continued treatment, the eventual timing of any trial of withdrawal and the longer term outlook if seizures recur.

The possibility of a further driving ban may make those dependent on driving rather reluctant to try withdrawing medication.

Most commonly, periods of 2 years or more of treatment are considered necessary before consideration of withdrawal. It is usually suggested that longer seizure-free periods result in a lower risk of recurrence. This is most likely to be due to selection bias, patients who relapse while still on medication after shorter periods of time being excluded.

Recent randomized controlled trials have examined policies of differing lengths of treatment prior to stopping medication. Children who entered remission within 2 months of starting treatment were randomized to stop medication after 6 or 12 months (Peters *et al.* 1998). Six months after the first follow-up, 22 per cent still on anti-epileptic drugs had relapsed despite treatment compared to 37 per cent who had been withdrawn from their drugs. However, by 24 months after randomization the risk of relapse was 49 and 48 per cent respectively. The profile of the relationship between the time-elapsed seizure-free and the risk of seizures in the following year on treatment is illustrated in Fig. 31.14 (Medical Research Council Anti-epileptic Drug Withdrawal Study Group 1991). This is derived from two groups of patients; those patients who had a recurrence of seizures during the course of the study, all of whom reverted to treatment after such a seizure, and all those patients randomized to continue treatment at the outset of the study. The risk of a seizure whilst on treatment in the next year is about 50 per cent falling to approximately 20 per cent after 1 year seizure-free. By 4–5 years after a seizure, the risk in the next year falls to about 10 per cent. The risk for seizures after this time changes relatively little so that a policy of considering discontinuation of anti-epileptic drugs after 2–5 years in adults seems reasonable.

Some epilepsy syndromes are very clearly associated with particular levels of risk of relapse after stopping treatment. Absence epilepsy has an uncertain prognosis for remission. Although in the short term, most become seizure-free on treatment, about 25 per cent relapse when medications are withdrawn. Juvenile myoclonic epilepsy has an excellent response to treatment, but relapses occur in almost all patients when medications are stopped. Most studies find a favourable prognosis for epilepsy with onset in childhood. Studies including both childhood and adolescent onset epilepsy usually find a substantially increased risk of relapse in those with

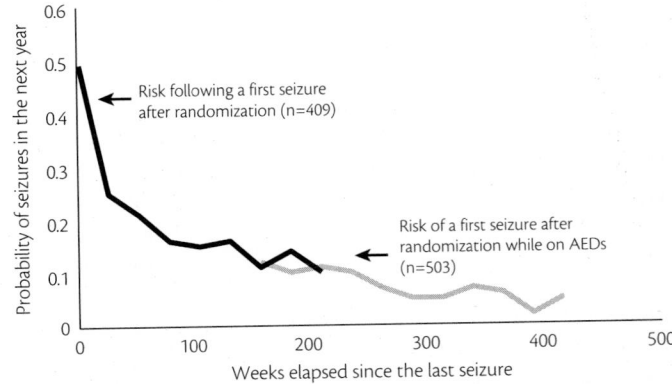

Fig. 31.14 The probability of a further seizure in a following year, while on drug therapy, and its relationship to the passage of time. Data is derived from the Medical Research Council (1991) study of antiepileptic drug (AED) withdrawal. Early risk is calculated from those patients who experienced a seizure following randomization, and who were thereafter treated. Later risk is calculated from the group of patients randomized to continued therapy, whose median duration of remission was 3 years.

adolescent onset (Berg and Shinnar 1994). Individuals with an identifiable aetiology associated with their epilepsy, who have remote symptomatic epilepsy (Section 31.8), are less likely to enter remission than those with idiopathic or cryptogenic epilepsy. Once in remission, they are about 50 per cent more likely to relapse if medication is stopped.

There is considerable controversy over the value of the EEG in predicting the prognosis for relapse after stopping treatment. Overall, data suggests that the EEG is of greater prognostic significance in children than in adults.

Most evidence indicates that the majority of patients who relapse when medication is stopped will regain acceptable control when treatment is re-introduced. If a decision is taken to withdraw treatment, clear advice should be offered about the speed of withdrawal. From a practical point of view, it seems reasonable to taper most regimes gradually over a 2–3-month period. For most adults a seizure recurrence will require the prompt re-institution of the anti-epileptic drug regime that was previously successful.

31.10.10 Managing status epilepticus

Status epilepticus is a medical emergency because of the mortality and morbidity that can result both from the systemic complications and from the continuing epileptic activity which itself may result in neuronal damage and loss. The satisfactory management of convulsive status epilepticus demands aggressive treatment with rapidly acting anti-epileptic drugs aimed at abolishing motor and electrical evidence of status, as well as all the necessary cardiovascular, respiratory, and metabolic support (Meierkord et al. 2006). This will not be easily available outside an intensive therapy setting. An approach to management is outlined in Fig. 31.15. Early suppression of seizures must be combined with adequate investigation to disclose the cause of status and institution or maintenance of satisfactory longer term anticonvulsant therapy.

The choice of anti-epileptic drug lies between those available for intravenous use, which enter the brain rapidly and have an immediate mode of action. Thus benzodiazepines, barbiturates, and phenytoin can all be considered. Data comparing different drugs is sparse and derived from small numbers of patients, with the exception of a large study comparing phenobarbitone, lorazepam, phenytoin, and diazepam plus phenytoin (Treiman et al. 1998). The only regime shown to be different was phenytoin alone, which was less successful than lorazepam. Certainly lorazepam should now be preferred to diazepam in view of its longer action. It should be administered as a loading dose of 0.1 mg/kg. It is still worthwhile combining this with phenytoin (20 mg/kg) as the drug allows a seamless transfer to oral administration at a future date.

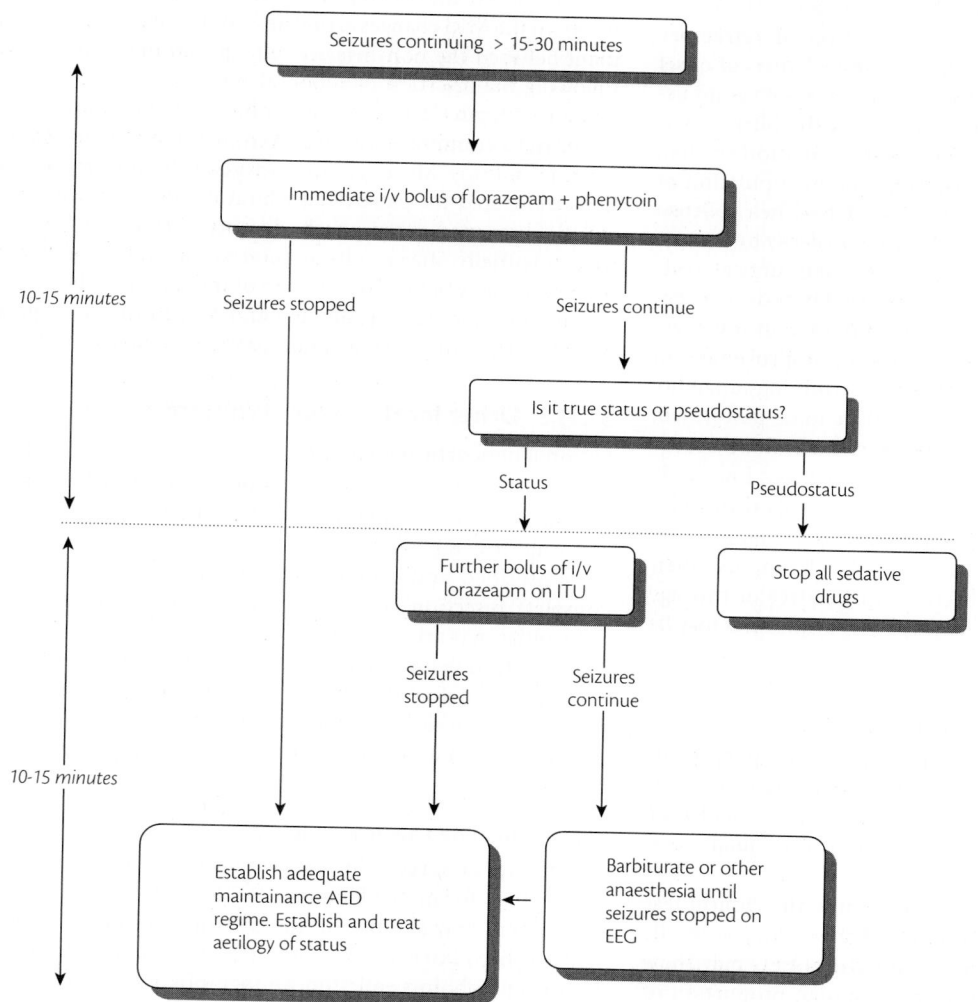

Fig. 31.15 Flow diagram of the management of status epilepticus in adults. AED: antiepileptic drug ICU: intensive care unit

31.11 **Surgical treatment of epilepsy**

31.11.1 **Introduction**

Whilst the surgical treatment of epilepsy was pioneered in the United Kingdom over 100 years ago it has never been made widely available to patients with epilepsy. Yet, the increasing sophistication of EEG investigation, neurological imaging, and neuropsychology mean that this form of treatment can be highly successful in large numbers of patients.

The axiom of surgical treatment requires:

- either the accurate identification of a localized site of seizure onset with the aim of a curative resection,

- or the disconnection of epileptogenic zones so as to interrupt seizure spread by means of a palliative procedure such as callosotomy or multiple pial resections.

Inevitably excisions of epileptogenic lesions and zones will also involve some degree of interruption of their connections. Of the varying procedures undertaken at centres worldwide, 68 per cent of operations involve some form of temporal lobe surgery whilst extra-temporal cortical excisions accounted for 24 per cent of operations. Two per cent were hemispherectomies and 6 per cent corpus callosotomy (Engel 1986).

To be considered for any of these procedures patients need to demonstrate a history of medically refractory epilepsy. There may be some controversy about a precise definition of refractory epilepsy but this will usually be established within 2 years of onset if optimal doses of suitable anti-epileptic drugs have been administered singly or in combination. Most patients with epilepsy syndromes of poor prognosis exhibit these at presentation so that, thereafter, there can be little optimism that the manipulation of drug therapy is likely to radically alter the outcome of their epilepsy. Patients should be sufficiently disabled by their epilepsy to warrant the risks of the necessary pre-surgical evaluation and surgical treatment, currently approximately 0.5 per cent mortality and 5 per cent morbidity. There should be a high probability that an improvement in seizure control will lead to a significant improvement in the individual's quality of life. Other factors determining suitability of treatment will include the type of procedure to be performed. Logic dictates that the above criteria may be relaxed where neurological imaging shows the presence of a lesion such as a low-grade tumour that would demand surgical treatment in its own right. The timing of surgical treatment is important, given that it is rare for individuals over the age of 30 years to radically improve their psychosocial status even if they become seizure-free at this age (Crandall 1987). A number of different epilepsy syndromes may be considered for surgery.

31.11.2 **Mesial temporal lobe epilepsy**

There is no doubt that patients with mesial temporal epilepsy due to hippocampal sclerosis can expect excellent results of surgery, with 60–70 per cent of patients becoming seizure-free (Wiebe *et al.* 2001). Other foreign tissue lesions in this region have equally good outcomes after surgery.

A detailed clinical history will usually identify the clinical features of this syndrome (Section 31.5.4), and the history of febrile seizures is always a helpful marker. The inter-ictal EEG may show anterior temporal spikes or sharp waves in a high proportion of cases. Modern high quality MRI has revolutionized investigation with its ability to demonstrate hippocampal atrophy with T1 volumetry, abnormal T2 signal, and FLAIR. In specialist centres functional imaging using positron emission tomography or single photon emission tomography may also be of value. Epileptogenic zones are characteristically hypometabolic and hypoperfused during the interictal state but may become hypermetabolic and hyperperfused ictally, or immediately post-ictally (Fish 1989). The role of techniques such as MR spectroscopy and functional imaging remain to be defined.

Neuro-psychological assessment is essential. An overall intelligence quotient of less than 70 would tend to indicate diffuse cerebral damage thereby reducing the likelihood of a good outcome to surgery. Particular evaluation of verbal and visual memory can be important in lateralizing deficits that are likely to be ipsilateral to seizure onset. Testing after intracarotid amytal is helpful in further defining memory deficits related to each temporal lobe and lateralizing speech. Where a well lateralized memory deficit is found, the likelihood of a good outcome is high. Amytal testing is also important in ensuring that memory function in the temporal lobe contralateral to that on which surgery is being considered is adequate to sustain memory post-operatively. If there is agreement between ictal semiology, MRI, inter-ictal EEG, and neuro-psychology, no further investigations may be necessary, although some centres undertake ictal recording in all cases.

Invasive EEG monitoring is now only necessary in those patients in whom the MRI changes are bilateral or in whom there is disagreement between the non-invasive tests. Recordings can be obtained following the insertion of subdural electrodes and stereotactically implanted depth electrodes or a combination of the two.

The most common surgical procedure is the en bloc anteriotemporal lobectomy. More recently amygdalo-hippocampectomy has been pioneered for patients in whom it can be shown that there is a definite medial temporal onset to seizures. This procedure may potentially offer results as good as classical en bloc resection with potentially better preservation of memory, but the resection is potentially more technically demanding and the two operations have not been compared in a randomized controlled trial.

31.11.3 **Other focal epilepsy syndromes**

Lesional neocortical epilepsy

Partial epilepsies caused by tumour, vascular malformation, or localized cortical dysplasia constitute another, not uncommon, surgically treatable group of epilepsies. Where lesions are well circumscribed and remote from functionally important areas, complete resection will be associated with a high success rate. Difficulties may arise from lesions close to eloquent areas for which more detailed neurophysiological and functional mapping may be necessary to define the extent of the epileptogenic zone. This is particularly the case with focal cortical dysplasia where the extent of the lesion may be difficult to define purely from structural imaging.

Non-lesional neocortical epilepsy

The results of surgery in this group of patients are significantly poorer than for mesial temporal and lesional neocortical epilepsy, but surgery may still be considered where seizure semiology and well-localized inter-ictal spiking suggest a satisfactory localization. Ictal single photon emission tomography may have a particular value in the investigation of patients and allow satisfactory planning of subdural or intracerebral ictal recording. Surgery will usually consist of extensive resection away from eloquent areas. In eloquent areas multiple sub-pial transections may provide palliation.

Hemispheric epilepsy syndromes

Hemispherectomy may be suitable for patients with intractable epilepsy and an infantile hemiplegia with a useless hand and hemianopia. Such epilepsy may occur with Sturge–Weber syndrome, Rasmussen's encephalitis, and childhood stroke. Overall, 70–80 per cent of patients become seizure-free following this operation and behavioural abnormalities can also improve. The operation, however, fell into disrepute as up to 25 per cent of those undergoing hemispherectomy developed delayed complications. Most suffered from recurrent subdural haemorrhage from the subdural membrane lining the hemispherectomy cavity. A modified hemispherectomy procedure to eliminate the large extra-dural space and to isolate the ventricular system from the subdural cavity reduces these complications. Probably hemispherectomy should be restored now to its previous role in children with infantile hemiplegia and epilepsy and also in those rare children and adults with chronic progressive focal encephalitis (Hart *et al.* 1994; Rasmussen *et al.* 1958).

Other partial epilepsy syndromes

Gelastic seizures associated with hypothalamic hamartoma are rare. The success of surgical treatment is controversial (Cascino *et al.* 1993; Valdueza *et al.* 1994).

31.11.4 Generalized epilepsies

Section of the corpus callosum and hippocampal commissure is an accepted palliative procedure for uncontrolled secondarily generalized seizures (Spencer *et al.* 1987). The procedure seeks to prevent the generalization of seizures, particularly those that generalize rapidly resulting in falls from tonic and atonic seizures. Early procedures were often complicated by ventriculitis, meningitis, and hydrocephalus, by more severe and frequent focal seizures immediately post-operatively, and by a characteristic disconnection syndrome of mutism, apraxia of the non-dominant limbs, agnosia, apathy, confusion, and infantile behaviour. Refinements of the surgical procedure and the introduction of anterior and two-stage operations have reduced the morbidity. In some series up to 80 per cent of patients have had a complete cessation of generalized seizures with falls, although about 25 per cent may have more intense partial seizures than previously.

The selection criteria for corpus callosotomy are more poorly defined than for other surgical procedures. The operation will be considered most commonly in children and adolescents with very severe epilepsy involving a generalized or multifocal origin to seizures or with seizures of sudden onset resulting in falls.

31.11.5 Vagal nerve stimulation

A commercial device to stimulate the vagal nerve is now available (neuroCybernectic prosthesis; Cyberonics, Houston TX, USA). The device consists of a pulse generator, a bipolar lead to stimulate the nerve, and a programming wand and software with handheld magnets capable of switching the stimulator on or off. The device is implanted to stimulate the left vagal nerve, since stimulation on this side is less likely to cause cardiac effects. Two randomized controlled trials of the device have been undertaken in patients

with refractory complex-partial seizures (The Vagus Nerve Stimulation Study Group 1995; Handforth *et al.* 1998). Patients were randomly assigned high stimulation with frequent on/off cycles and stimulus intensity adjusted to the highest comfortable level or low stimulation at low frequency and intensity consistent with the patients being aware of the physical symptoms of vagal stimulation. In the low-stimulation groups, mean reduction in seizure frequency was 6 and 15 per cent in the two studies respectively compared with 25 and 28 per cent in the high-stimulation groups. The proportion of patients who had a 50 per cent reduction in seizures from baseline was 13 and 15 per cent for the low-stimulation groups and 31 and 23 per cent for the high-stimulation groups.

Calculation of the number needed to treat shows that stimulators would have to be implanted in 5–12 patients to see one of them have a 50 per cent reduction in seizures if the low-stimulation group is a true 'placebo' response, or four to seven stimulators implanted assuming a 10 per cent placebo response rate, which is the average for that in new anti-epileptic drug studies. In both clinical trials, patients reported a tolerable level of increased cough, voice alteration, and throat discomfort. There is great uncertainty about the efficacy of vagal nerve stimulation in other types of epilepsy (Chadwick 2001).

References

Adab N, Jacoby A, Smith D *et al.* (2001). Additional educational needs in children born to mothers with epilepsy. *J Neurol Neurosurg Psychiatry*, **70**, 15–21.

Adab N, Kini U, Vinten J *et al.* (2004). The longer term outcome of children born to mothers with epilepsy. *J Neurol Neurosurg Psychiatry*, **75**, 1575–83.

Adrogue HJ, Madias NE (2000). Hyponatremia. *N Engl J Med*, **342**, 1581–9.

Ajmone-Marsan C, Zivin LS (1970). Factors related to the occurrence of typical paroxysmal abnormalities in the EEG records of epileptic patients. *Epilepsia*, **11**, 361–8.

Aminoff MJ, Simon RP (1980). Status epilepticus: causes, clinical features and consequences in 98 patients. *Am J Med*, **69**, 657–66.

Anderman F, Andermann E (1986). Excessive startle syndromes: Startle disease, jumping and startle epilepsy. In Fahn S, Marsden CD, Van Woest M, eds. *Advances in Neurology*, pp. 321–38. Raven Press, New York.

Anderson VE, Hauser WA, Rich SS (1986). Genetic heterogeneity in the epilepsies. In Delgado-Escuata AV, Ward AA, Woodbury DM, Porter RJ, eds. *Advances in Neurology*. Raven Press, New York.

Annegers JF, Hauser WA, Elverback LR (1979a). Remission of seizures and relapse in patients with epilepsy. *Epilepsia*, **20**, 729–37.

Annegers JF, Hauser WA, Elverback LR (1979b). Remission of seizures and relapse in patients with epilepsy. *Epilepsia*, **20**, 729–37.

Annegers JF, Grabow JD, Groover RV *et al.* (1980). Seizures after head trauma: A population study. *Neurology*, **30**, 683–9.

Annegers JF, Hauser WA, Beghi E *et al.* (1988). The risk of unprovoked seizures after encephalitis and meningitis. *Neurology*, **38**, 1407–10.

Annegers JF, Hauser WA, Coan SP *et al.* (1998). A population based study of seizures after traumatic brain injuries. *N Engl J Med*, **338**, 20–4.

Antoniadis A, Mueller WE, Wollert U (1980). Benzodiazepine receptor interactions may be involved in the neurotoxicity of various penicillin derivatives. *Ann Neurol*, **8**, 71–3.

Antzelevitch C, Brugada P, Brugada J *et al.* (2002). Brugada syndrome: a decade of progress. *Circ Res*, **91**, 1114–8.

Arvanov V, Holmes K, Keele N *et al.* (1995). The functional role of metabotropic glutamate receptors in epileptiform activity induced by 4-aminopyridine in the rat amygdala slice. *Brain Res*, **669**, 140–4.

Bai D, Alonso ME, Medina MT *et al.* (2002). Juvenile myoclonic epilepsy: linkage to chromosome 6p12 in Mexico families. *Am J Med Genet*, **113**, 268–74.

Bates D, Caronna JJ, Cartlidge NEF *et al.* (1977). A prospective study of nontraumatic coma: methods and results in 310 patients. *Ann Neurol*, **2**, 211–20.

Baulac S, Huberfeld G, Gourfinkel-An I *et al.* (2001). First genetic evidence of GABA(A) receptor dysfunction in epilepsy: a mutation in the gamma2-subunit gene. *Nat Genet*, **28**, 46–8.

Bazil CW, Walczak TS (1997). Effects of sleep and sleep stage on epileptic and nonepileptic seizures. *Epilepsia*, **38**, 56–62.

Beghi E, Gatti G, Tonini C *et al.* (2003). Adjunctive therapy versus alternative monotherapy in patients with partial epilepsy failing on a single drug: a multicentre, randomised, pragmatic controlled trial. *Epilepsy Res*, **57**, 1–13.

Berg AT, Shinnar S (1994). Relapse following discontinuation of antiepileptic drugs: a meta- analysis. *Neurology*, **44**, 601–8.

Berg, AT, Levy SR, Novotny EJ *et al.* (1996). Predictors of intractable epilepsy in childhood: a case-control study. *Epilepsia*, **37**, 24–30.

Berkovic SF, Howell RA, Hopper JL (1994). Familial temporal lobe epilepsy: a new syndrome with adolescent/adult onset and a benign course. In Wolf P, ed. *Epileptic Seizures and Syndromes*, pp. 257–63. John Libbey, London.

Biervert C, Schroeder BC, Kubisch C *et al.* (1998). A potassium channel mutation in neonatal human epilepsy. *Science*, **279**, 403–6.

Bilo L, Meo R, Valentino R *et al.* (2001). Characterization of reproductive endocrine disorders in women with epilepsy. *J Clin Endocrinol Metab*, **86**, 2950–6.

Binnie CD (1997). The Electroencephalogram: advances and pitfalls. In Porter RJ, Chadwick D, eds. *The Epilepsies 2*, pp. 111–40. Butterworth-Heinemann, Boston.

Bittencourt PRM, Gracia CM, Lorenzana P (1988). Epilepsy and parasitosis of the central nervous system. In Pedley TA, Meldrum BS, eds. *Recent Advances in Epilepsy*, pp. 123–60. Raven Press, New York.

Blakemore WF (1980). Isoniazid. In Spencer PS, Schaumburg HH, eds. *Experimental and Clinical Neurotoxicology*, pp. 476–89. Williams and Wilkins, Baltimore.

Bleck TP, Smith MC, Pierre-Louis SJ-C *et al.* (1993). Neurologic complications of critical medical illnesses. *Crit Care Med*, **21**, 98–103.

Blisard KS, McFeeley PJ (1988). The spectrum of neuropathological findings in deaths associated with seizure disorders. *J Forensic Sci*, **22**, 910–4.

Boston Collaborative Drug Surveilllance Program (1972). Drug-induced convulsions. *Lancet*, **2**, 677–9.

Bouma PA, Westendorp RG, van Dijk JG *et al.* (1996). The outcome of absence epilepsy: a meta-analysis. *Neurology*, **47**, 802–8.

Bowery NG (1993). GABAB receptor pharmacology. *Annu Rev Pharmacol Toxicol*, **33**, 109–47.

Brignole M, Alboni P, Benditt DG *et al.* (2004). Guidelines on management (diagnosis and treatment) of syncope--update 2004. *Europace*, **6**, 467–537.

Brodie MJ, Perucca E, Ryvlin P *et al.* (2007). Comparison of levetiracetam and controlled-release carbamazepine in newly diagnosed epilepsy. *Neurology*, **68**, 402–8.

Brodtkorb E, Nilsen G, Smevik O (1992). Epilepsy and anomalies of neuronal migration: MRI and clinical aspects. *Acta Neurol Scand*, **86**, 24–32.

Buck D, Baker GA, Jacoby A *et al.* (1997). W. Patients' experiences of injury as a result of epilepsy. *Epilepsia*, **38**, 439–44.

Burgess DL, Noebels JL (1999). Single gene defects in mice: the role of voltage-dependent calcium channels in absence models. *Epilepsy Res*, **36**, 111–22.

Burn J, Dennis M, Bamford J *et al.* (1997). Epileptic seizures after a first stroke: the Oxfordshire community stroke project. *BMJ*, **315**, 1582–7.

Camilo O, Goldstein LB (2004). Seizures and epilepsy after ischemic stroke. *Stroke*, **35**, 1769–75.

Cascino GD, Kelly PJ, Hirschorn KA *et al.* (1990). Stereotactic resection of intra-axial cerebral lesions in partial epilepsy. *Mayo Clin Proc*, **65**, 1053–60.

Cascino GD, Andermann F, Berkovic SF *et al.* (1993). Gelastic seizures and hypothalamic hamartomas: evaluation of patients undergoing chronic intracranial EEG monitoring and outcome of surgical treatment. *Neurology*, **43**, 747–50.

Cavazzuti GB (1980). Epidemiology of different types of epilepsy in schoolage children of Modena, Italy. *Epilepsia*, **21**, 57–62.

Chadwick D (2001). Vagal-nerve stimulation for epilepsy. *Lancet*, **357**, 1726–7.

Chadwick D, Shaw MD, Foy P *et al.* (1984). anticonvulsant concentrations and the risk of drug induced skin eruptions. *J Neurol Neurosurg Psychiatry*, **47**, 642–4. 84.

Chadwick DW (1997). Wrong diagnosis may deprive people of their livelihood. *BMJ*, **314**, 1283.

Chang BS, Lowenstein DH (2003). Epilepsy. *N Engl J Med*, **349**, 1257–66.

Claes L, Del-Favero J, Ceulemans B *et al.* (2001). De novo mutations in the sodium-channel gene SCN1A cause severe myoclonic epilepsy of infancy. *Am J Hum Genet*, **68**, 1327–32.

Cleary P, Shorvon S, Tallis R (2004). Late-onset seizures as a predictor of subsequent stroke. *Lancet*, **363**, 1184–6.

Cocito L, Favle E, Reni L (1982). Epileptic seizures in cerebral arterial occlusive disease. *Stroke*, **13**, 189–95.

Cockerell OC, Johnson AL, Sander JWAS *et al.* (1997). Prognosis of epilepsy: A review and further analysis of the first nine years of the British National General Practice Study of Epilepsy, a prospective population-based study. *Epilepsia*, **38**, 31–46.

Cohen N, Strauss G, Lew R *et al.* (1988). Should prophylactic anticonvulsants be administered to patients with newly diagnosed cerebral metastases. A retrospective analysis. *J Clin Oncol*, **6**, 1621–4.

Coleman RM, Pollak CP, Weitzman ED (1980). Periodic movements in sleep (nocturnal myoclonus): relation to sleep disorders. *Ann Neurol*, **8**, 416–21.

Commission on Antiepileptic Drugs of the International League against Epilepsy (1989). Guidelines for clinical evaluation of antiepiletic drugs. *Epilepsia*, **30**, 400–8.

Commission on classification and terminology of the International League Against Epilepsy (1981). Proposal for revised clinical and electroencephalographic classification of epileptic seizures. *Epilepsia*, **22**, 489–501.

Commission on Classification and Terminology of the International League Against Epilepsy (1989a). Proposal for revised classification of epilepsies and epileptic syndromes. *Epilepsia*, **30**, 389–99.

Commission on Classification and Terminology of the International League Against Epilepsy (1989b). Proposal for revised classification of epilepsies and epileptic syndromes. *Epilepsia*, **30**, 389–99.

Copeland GP, Foy P, Shaw MDM (1982). The incidence of epilepsy after ventricular shunting operations. *Surg Neurol*, **17**, 279–81.

Coulter DA, Huguenard JR, Prince DA (1990). Differential effects of petit mal anticonvulsants on thalamic neurones: calcium current reduction. *Br J Pharmacol*, **100**, 800–6.

Crandall PH (1987). Cortical resections. In Engel J, ed. *Surgical Treatment of Epilepsies*, pp. 377–404. Raven Press: New York.

Crawford PM, West CR, Shaw MD *et al.* (1986). Cerebral arteriovenous malformations and epilepsy: factors in the development of epilepsy. *Epilepsia*, **27**, 270–5.

Crunelli V, Leresche N (1991). A role for GABAB receptors in excitation and inhibition of thalamocortical cells. *Trends Neurol Sci*, **14**, 16–21.

Curtis DR, Game CJA, Johnston GAR *et al.* (1972). Convulsant action of penicillin. *Brain Res*, **43**, 242–5.

D'Alessandro R, Guarino M, Greco G *et al.* (2004). Risk of seizures while awake in pure sleep epilepsies: a prospective study. *Neurology*, **62**, 254–7.

Dailey JW, Naritoku DK (1996). Antidepressants and seizures: clinical anecdotes overshadow neuroscience. *Biochem Pharmacol*, **52**, 1323–9.

Daumas-Duport C, Scheithauer BW, Chodkiewicz JP *et al.* (1988). Dysembryoplastic neuroepithelial tumor: a surgically curable tumor of young patients with intractable partial seizures. *Neurosurgery*, **23**, 545–56.

De Deyn PP, Saxena VK, Abts H *et al.* (1992). Clinical and pathphysiological aspects of neurological complications in renal failure. *Acta Neurol Belg*, **92**, 191–206.

Delgado-Escueta AV, Bascal FE, Treiman DM (1982). Complex partial seizures on closed circuit television and EEG: a study of 691 attacks in 79 patients. *Ann Neurol*, **57**, 292–300.

Delgado-Escueta AV, Enrile-Bacsal F (1984). Juvenile myoclonic epilepsy of Janz. *Neurology*, **34**, 285–94.

DiLiberti JH, Farndon PA, Dennis NR *et al.* (1984). The fetal valproate syndrome. *Am J Genet*, **19**, 473–81.

Dreifuss FE, Santilli N, Langer DH *et al.* (1987). Valproic acid fatalities: a retrospective review. *Neurology*, **37**, 379–85.

Edebol-Tysk K (1989). Epidemiology of spastic tetraplegic cerebral palsy in Sweden. I. Impairments and disabilities. *Neuropediatrics*, **20**, 41–5.

Eke T, Talbot JF, Lawden MC (1997). Severe persistent visual field constriction associated with vigabatrin. *BMJ*, **314**, 180–1.

Elger CE, Helmstaedter C, Kurthen M (2004). Chronic epilepsy and cognition. *Lancet Neurol*, **3**, 663–72.

Engel J (1986). Outcome with respect to epileptic seizures. In Engel J, ed. *Surgical Treatment of the Epilepsies*, pp. 553–72. Raven Press, New York.

Engel J Jr (1998). Classifications of the international league against epilepsy: time for reappraisal. *Epilepsia*, **39**, 1014–7.

Engel J Jr (2001). Classification of epileptic disorders. *Epilepsia*, **42**, 316.

Eriksson PS, Perfilieva E, Bjork-Eriksson T *et al.* (1998). Neurogenesis in the adult human hippocampus. *Nat Med*, **4**, 1313–7.

Escayg A, MacDonald BT, Meisler MH *et al.* (2000). Mutations of SCN1A, encoding a neuronal sodium channel, in two families with GEFS+2. *Nat Genet*, **24**, 343–5.

Farmer J-P, Cosgrove JR, Villemure J-G *et al.* (1988). Intracerebral cavernous angiomas. *Neurology*, **38**, 1699–704.

Feeley M, Calvert R, Gibson J (1982). Clobazam in catamenial epilepsy: a model for evaluating anticonvulsants. *Lancet*, **ii**, 71.

Ficker DM, So EL, Shen WK *et al.* (1998). Population-based study of the incidence of sudden unexplained death in epilepsy. *Neurology*, **51**, 1270–4.

Fish D (1989). CT and PET in drug resistant epilepsy. In Trimble MR, ed. *Chronic Epilepsy, its Progress and Management*, pp. 59–72. Wiley, Chichester.

Fish DR, Quirk JA, Smith SJM *et al.* (1994). *National Survey of Photosensitivity and Seizures Induced by Electronic Screen Games*. Department of Trade and Industry, London.

Fisher CM, Adams RD (1958). Transient global amnesia. *Trans Am Neurol Assoc*, **83**, 143–6.

Forsgren L, Edvinsson SO, Blomquist HK *et al.* (1990). Epilepsy in a population of mentally retarded children and adults. *Epilepsy Res*, **6**, 234–48.

Foy PM, Copeland GP, Shaw MDM (1981a). The incidence of postoperative seizures. *Acta Neurochir*, **55**, 253–64.

Foy PM, Copeland GP, Shaw MDM (1981b). The natural history of postoperative seizures. *Acta Neurochir*, **57**, 15–22.

Francis P, Baker GA (1999). Non-epileptic attack disorder (NEAD): a comprehensive review. *Seizure*, **8**, 53–61.

French JA, Williamson PD, Thadani VM *et al.* (1993). Characteristics of medial temporal lobe epilepsy I. Results of history and physical examination. *Ann Neurol*, **34**, 774–80.

Gaily E, Kantola-Sorsa E, Hiilesmaa V *et al.* (2004). Normal intelligence in children with prenatal exposure to carbamazepine. *Neurology*, **62**, 28–32.

Gastaut H, Gastaut JL, Gonclaves e Silva GE *et al.* (1975). Relative frequency of different types of epilepsy: a study employing the classification of the International League Against Epilepsy. *Epilepsia*, **16**, 457–61.

Geier S, Bancaud J, Talairach J *et al.* (1977). The seizures of frontal lobe epilepsy. *Neurology*, **27**, 951–8.

Ghika-Schmid F, Ghika J, Regli F *et al.* (1996). Hashimoto's myoclonic encephalopathy: an undiagnosed treatable condition? *Mov Disord*, **11**, 555–62.

Giannakodimos S, Panayiotopoulos CP (1996). Eyelid myoclonia with absences: a clinical and video-EEG study in adults. *Epilepsia*, **37**, 36–44.

Gibberd FB, Bateson MC (1974). Sleep epilepsy: its pattern and prognosis. *BMJ*, **2**, 403–5.

Glauser T, Ben-Menachem E, Bourgeois B *et al.* (2006). ILAE treatment guidelines: evidence-based analysis of antiepileptic drug efficacy and effectiveness as initial monotherapy for epileptic seizures and syndromes. *Epilepsia*, **47**, 1094–120.

Gloor P (1968). Generalized corticoreticular epilepsies. Some considerations on the pathophysiology of generalized bilaterally synchronous spike and wave discharge. *Epilepsia*, **9**, 249–63.

Gloor P, Olivier A, Quesney LF *et al.* (1982). The role of the limbic system in experential phenomena of temporal lobe epilepsy. *Ann Neurol*, **12**, 129–44.

Goddard GV, McIntyre DC, Leech CK (1969). A permanent change in brain function resulting form daily electrical stimulation. *Exp Neurol*, **25**, 295–330.

Goosses R (1984). Die beziehung der fotosensibilitat zu den verschiedenen epileptischen syndromen. West Berlin.

Goulatia RK, Verma A, Mishra NK *et al.* (1987). Disappearing CT lesion in epilepsy. *Epilepsia*, **28**, 523–7.

Gregory RP, Oates T, Merry RTG (1993). Electroencephalogram epileptiform abnormalities in candidates for aircrew training. *Electroencephalogr Clin Neurophysiol*, **86**, 75–81.

Groppel G, Kapitany T, Baumgartner C (2000). Cluster analysis of clinical seizure semiology of psychogenic nonepileptic seizures. *Epilepsia*, **41**, 610–4.

Guberman A, Catching J (1986). Swallow syncope. *Can J Neurol Sci*, **13**, 267–9.

Gururaj G, Satishchandra P (1992). Correlates of hot water epilepsy in rural south India: a descriptive study. *Neuroepidemiology*, **11**, 173–9.

Gutnick MJ, Connors BW, Prince DA (1982). Mechanisms of neocortical epileptogenesis in vitro. *J Neurophysiol*, **48**, 1321–35.

Handforth A, DeGiorgio CM, Schachter SC *et al* (1998). Vagus nerve stimulation therapy for partial-onset seizures: a randomized active-control trial. *Neurology*, **51**, 48–55.

Hanson JW, Myrianthopoulos NC, Harvey MAS *et al* (1976). Risks to the offspring of women treated with hydantoin anticonvulsants, with emphasis on the fetal hydantoin syndrome. *J Pediatrics*, **89**, 662–8.

Harding GFA, Jeavons PM (1995). *Photosensitive Epilepsy*. Cambridge University Press, Cambridge, UK.

Hart YM, Cortez M, Andermann F *et al.* (1994). Medical treatment of Rasmussen's Syndrome (chronic encephalitis and epilepsy): effect of high dose steroids and immunoglobulins in 19 patients. *Neurology*, **44**, 1030–6.

Hauser WA, Anderson VE (1986). Genetics of epilepsy. In Pedley TA, Meldrum BS, eds. *Recent Advances in Epilepsy*, Vol. 3, pp. 21–36. Churchill Livingstone, Edinburgh.

Hauser WA, Hesdorffer DC (1990). *Epilepsy: Frequency, Causes and Consequences*. Demos Publications, New York.

Hauser WA, Kurland LT (1975). The epidemiology of epilepsy in Rochester, Minnesota, 1935 through 1967. *Epilepsia*, **16**, 166–82.

Hauser WA, Morris ML, Heston LL *et al.* (1986). Seizures and myoclonus in patients with Alzheimer's disease. *Neurology*, **36**, 1226–30.

Hauser WA, Shinnar S, Cohen H (1987). Clinical predictors of epilepsy among children with cerebral palsy and/or mental retardation. *Neurology*, **37**, 150–65.

Haut SR, Bigal ME, Lipton RB (2006). Chronic disorders with episodic manifestations: focus on epilepsy and migraine. *Lancet Neurol*, **5**, 148–57.

Hillbom ME (1980). Occurrence of cerebral seizures provoked by alcohol abuse. *Epilepsia*, **21**, 459–66.

Hodges JR, Warlow CP (1990). Syndromes of transient amnesia: towards a classification. A study of 153 cases. *J Neurol Neurosurg Psychiatry*, **53**, 834–43.

Holland RW, Marx JA, Earnest MP *et al.* (1993). Grand mal seizures temporally related to cocaine use: clinical and diagnostic features. *Emerg Med*, **22**, 758.

Holtzman D, Kaku D, So Y (1989). New-onset seizures associated with HIV infection: causation and clinical features in 100 cases. *Am J Med*, **87**, 173–7.

Hosford DA, Clark S, Cao Z *et al* (1992). The role of GABAB receptor activation in absence seizures of lethargic (lh/lh) mice. *Science*, **257**, 398–400.

Howell SJL, Owen L, Chadwick DW (1989). Pseudostatus Epilepticus. *QJM*, **266**, 507–19.

Huttenlocher PR, Hapke RJ (1990). A followup study of intractable seizures in childhood. *Ann Neurol*, **28**, 699–705.

Isojarvi JI, Tauboll E, Tapanainen JS *et al.* (2001). On the association between valproate and polycystic ovary syndrome: a response and an alternative view. *Epilepsia*, **42**, 305–10.

Jackson JH (1873). On the anatomical, physiological and pathological investigation of epilepsies. West Riding Lunatic Asylum Medical Reports, p. 3.

Jacobsen M, Baelum J, Bonde JP (1994). Temporal epileptic seizures and occupational exposure to solvents. *Occup Environ Med*.

Jacoby A, Baker GA, Steen N *et al.* (1996). The clinical course of epilepsy and its psychosocial correlates: findings from a U.K. Community study. *Epilepsia*, **37**, 148–61.

Jallon P, Latour P (2005). Epidemiology of idiopathic generalized epilepsies. *Epilepsia*, **46** (**Suppl 9**), 10–4.

Janz D (1962). The grand-mal epilepsies and the sleeping-waking cycle. *Epilepsia*, **3**, 69–109.

Janz D, Christian W (1957). Inpulsiv-petit mal. *J Neurol*, **176**, 346–86.

Jellinek EH (1962). Fits, faints, coma and dementia in myxoedema. *Lancet*, **2**, 1010–2.

Jennett B (1975). *Epilepsy after Nonmissile Head Injuries*. Heinemann Medical Books, London.

Johnson JW, Ascher P (1987). Glycine potentiates the NMDA response in cultured mouse brain neurons. *Nature*, **325**, 529–31.

Jones KL, Lacro RV, Johnson KA (1989). Pattern of malformations in the children of women treate with carbamazepine during pregnancy. *N Engl J Med*, **320**, 1661–6.

Jones RSG, Heinemann U (1987). Pre- and post-synaptic K+ and Ca²+ fluxes in area CA1 of the rat hippocampus in vitro: effects of Ni²+, TEA and 4-AP. *Exp Brain Research*, **68**, 205–9.

Jouvenceau A, Eunson LH, Spauschus A *et al.* (2001). Human epilepsy associated with dysfunction of the brain P/Q-type calcium channel. *Lancet*, **358**, 801–7.

Juul-Jensen P, Foldspang A. Natural history of epileptic seizures. *Epilepsia*, **24**, 297–312.

Kalviainen R, Nousiainen I, Mantyjarvi M *et al.* (1998). Initial vigabatrin montherapy is associated with increased risk of visual field constriction. *Epilepsia*, **39**, 72.

Kapoor WN, Peterson JR, Karpf M (1985). Micturition syncope: a reappraisal. *JAMA* **253**, 796–8.

Karim AB, Maat B, Hatlevoll R *et al.* (1996). A randomized trial on dose-response in radiation therapy of low-grade cerebral glioma: European Organization for Research and Treatment of Cancer (EORTC) Study 22844. *Int J Radiat Oncol Biol Phys*, **36**, 549–56.

Kaufman DW, Kelly JP, Anderson T *et al.* (1997). Evaluation of case reports of aplastic anemia among patients treated with felbamate. *Epilepsia*, **38**, 1265–9.

Kertesz A (1967). Paroxysmal kinesigenic choreoathetosis. *Neurology*, **17**, 680–90.

Kim LG, Johnson A, Marson AG *et al.* (2006). Predicting the risk of seizure recurrence after single seizures and early epilepsy. *Lancet Neurol*, **5**, 317–22.

King MA, Newton MR, Jackson GD *et al.* (1998). Epileptology of the first-seizure presentation: a clinical, electroencephalographic, and megnetic resonance imaging study of 300 consecutive patients. *Lancet*, **352**, 1007–11.

Kopelman MD, Panayiotopoulos CP, Lewis P (1994). Transient epileptic amnesia differentiated from psychogenic "fugue": neuropsychological, EEG, and PET findings. *J Neurol Neurosurg Psychiatry*, **57**, 1002–4.

Krauss GL, Johnson MA, Miller NR (1998). Vigabatrin-associated retinal cone system dysfunction: electroretinogram and ophthalmologic findings. *Neurology*, **50**, 614–8.

Kuzniecky RI, Jackson GD (1997). Developmental disorders. In Engel J Jr, Pedley TA, eds. *Epilepsy: A Comprehensive Textbook*, pp. 2517–32.Lippincott-Raven, New York.

Kwan P, Brodie MJ (2000). Early identification of refractory epilepsy. *N Engl J Med*, **342**, 314–9.

Leach JP, Chadwick DW, Miles JB *et al.* (1999). Improvement in adult-onset Rasmussen's encephalitis with long-term immunomodulatory therapy. *Neurology*, **52**, 738–42.

Lechtenberg R, Worner TM (1992). Seizure incidence enhancement with increasing alcohol intake. *Ann NY Acad Sci*, **654**, 474–6.

Lehericy S, Dormont D, Semah F *et al.* (1995). Developmental abnormalities of the medial temporal lobe in patients with temporal lobe epilepsy. *Am J Neuroradiol*, **16**, 617–23.

Lempert T, Bauer M, Schmidt D (1994). Syncope: a videometric analysis of 56 episodes of transient cerebral hypoxia. *Ann Neurol*, **36**, 233–7.

Leone MA, Solari A, Beghi E (2006). Treatment of the first tonic-clonic seizure does not affect long-term remission of epilepsy. *Neurology*, **67**, 2227–9.

Lesser R P, Luders H, Dinner D S *et al.* (1987). Simple partial seizures in epilepsy. In Luders H, Lesser RP, eds. *Electroclinical Syndromes*, pp. 223–78. SpringerVerlag, London.

Liu Z, Vergnes M, Depaulis A *et al.* (1992). Involvement of intrathalamic GABAB neurotransmission in the control of absence seizures in the rat. *Neuroscience*, **48**, 87–93.

Logothetis J (1967). Spontaneous epileptic seizure and electroencephalographic changes in the course of phenothiazine therapy. *Neurology*, **17**, 869–77.

Loiseau J, Loiseau P, Duche B *et al.* (1990). A survey of epileptic disorders in southwest France: seizures in elderly patients. *Ann Neurol*, **27**, 232–7.

Loring DW, Meador KJ, Thompson WD (1994). Neurodevelopmental effects of phenytoin and carbamazepine. *JAMA*, **272**, 850–1.

Lowry MR, Dunner FJ (1980). Seizures during tricyclic therapy. *Am J Psychiatry*, **137**, 1461–2.

Lund M (1952). Epilepsy associated with intracranial tumours. *Acta Psych Neurol Scand*, Suppl 81.

Macdonald RL, Olsen RW (1994). GABAA receptor channels. *Annu Rev Neurosci*, **7**, 569–602.

Maletzky BM (1973). The episodic dyscontrol syndrome. *Dis Ner Sys*, **34**, 178–85.

Malouf R, Brust JCM (1985). Hypoglycaemia: Causes, neurological manifestations and outcome. *Ann Neurol*, **17**, 421–30.

Mantegazza M, Franceschetti S, Avanzini G (1998). Anemone toxin (ATX II)-induced increase in persistent sodium current: effects on the firing properties of rat neocortical pyramidal neurones. *J Physiol*, **507**, 105–16.

Marks V (1981). Symptomatology. In Marks V, Rose FC, eds. *Hypoglycaemia*, pp. 458–63. Blackwell Scientific Publications, Oxford.

Marson A, Jacoby A, Johnson A *et al.* (2005). Immediate versus deffered antiepileptic drug treatment for early epilepsy and single seizures: a randomised controlled trial. *Lancet*, **365**, 2007–13.

Marson AG, Al-Kharusi AM, Alwaidh M *et al.* (2007a). The SANAD study of effectiveness of valproate, lamotrigine, or topiramate for generalised and unclassifiable epilepsy: an unblinded randomised controlled trial. *Lancet*, **369**, 1016–26.

Marson AG, Al-Kharusi AM, Alwaidh M *et al.* (2007b). The SANAD study of effectiveness of carbamazepine, gabapentin, lamotrigine, oxcarbazepine, or topiramate for treatment of partial epilepsy: an unblinded randomised controlled trial. *Lancet*, **369**, 1000–15.

Marson AG, Chadwick D (1996). Comparing antiepileptic drugs. *Curr Opin Neurol*, **9**, 103–6.

Marson AG, Kadir ZA, Hutton JL *et al.* (1997). The new antiepileptic drugs: a systematic review of their efficacy and tolerability. *Epilepsia*, **38**, 859–80.

Mauguiere F, Courjon J (1978). Somatosensory epilepsy: a review of 127 cases. *Brain*, **101**, 307–32.

Mayeux R, Alexander M P, Benson D F *et al.* (1979). Poriomania. *Neurology*, **29**, 1616–9.

McCorry D, Chadwick D, Marson A (2004). Current drug treatment of epilepsy in adults. *Lancet Neurol*, **3**, 729–35.

McCrory PR, Bladin PF, Berkovic SF (1997). Retrospective study of concussive convulsions in elite Australian rules and rugby league footballers: phenomenology, aetiology, and outcome. *BMJ*, **314**, 171–4.

McEnery PT, Nathan J, Bates SR (1989). Convulsion in children undergoing renal transplantation. *J Paediatr*, **115**, 532–6.

Medical Research Council Antiepileptic Drug Withdrawal Study Group (1991). Randomized study of antiepileptic drug withdrawal in patients in remission. *Lancet*, **337**.

Medina MT, Rosas E, Rubio-Donnadieu F *et al.* (1990). Neurocysticercosis as the main cause of late-onset epilepsy in Mexico. *Arch Intern Med*, **150**, 325–7.

Meierkord H, Boon P, Engelsen B *et al.* (2006). EFNS guideline on the management of status epilepticus. *Eur J Neurol*, **13**, 445–50.

Messing RO, Simon RP (1986). Seizures as a manifestation of systemic disease. *Neurologic Clinics*, **4**, 563–84.

Messing RO, Closson RG, Simon RP (1984). Drug-induced seizures: A 10 year experience. *Neurology*, **34**, 1582–6.

Molyneux AJ, Kerr RS, Yu LM *et al.* (2005). International subarachnoid aneurysm trial (ISAT) of neurosurgical clipping versus endovascular coiling in 2143 patients with ruptured intracranial aneurysms: a randomised comparison of effects on survival, dependency, seizures, rebleeding, subgroups, and aneurysm occlusion. *Lancet*, **366**, 809–17.

Moots PL, Maciunas RJ, Eisert DR *et al.* (1995). The course of seizure disorders in patients with malignant gliomas. *Arch Neurol*, **52**, 717–24.

Moreau Th, Sochurova D, Lemesle M *et al.* (1998). Epilepsy in patients with multiple sclerosis: radiologocal-clinical correlations. *Epilepsia*, **39**, 893–6.

Morrow J and Craig J (2003). Anti-epileptic drugs in pregnancy: current safety and other issues. *Expert Opin Pharmacother*, **4**, 445–56.

Mount LA, Reback J (1940). Familial paroxysmal choreoathetosis: preliminary report on a hitherto undescribed clinical syndrome. *Arch Neurol*, **44**, 841–6.

Nakane Y, Okuma T, Takashi R *et al.* (1980). Multi-institutional study on the teratogenicity and fetal toxicity of antiepileptic drugs. A report of a collaborative study group in Japan. *Epilepsia*, **21**, 663–80.

NICE (National Institute for Clinical Excellence) (2004). The epilepsies – the diagnosis and management of the epilepsies in adults and children in primary and secondary care. Clinical Guideline 20. (http://guidance.nice.org.uk/CG20/niceguidance/pdf/English).

Nielsen TG, Sillesen H, Schroeder TV (1995). Seizures following carotid endarterectomy in patients with severely compromised cerebral circulation. *Eur J Vasc Endovasc Surg*, **9**, 53–7.

North JB, Penhall RK, Hanieh A *et al.* (1983). Phenytoin and postoperative epilepsy: a double blind study. *J Neurosurg*, **58**, 672–7.

Ochs R, Bloor P, Quesney F, Ives J *et al.* (1984). Does headturning during a seizure have lateralising or localising significance? *Neurology*, **34**, 884–90.

Ottman R, Winawer MR, Kalachikov S *et al.* (2004). LGI1 mutations in autosomal dominant partial epilepsy with auditory features. *Neurology*, **62**, 1120–6.

Pacia SV, Devinsky O (1994). Clozapine-related seizures: experience with 5,629 patients. *Neurology*, **44**, 2247–9.

Panayiotopoulos CP, Chroni E, Daskalopoulos C *et al.* (1992). Typical absence seizures in adults: clinical, EEG, video-EEG findings and diagnostic/syndromic considerations. *J Neurol Neurosurg Psychiatry*, **55**, 1002–8.

Panayiotopoulos CP, Ferrie CD, Giannakodimos S *et al.* (1994). Perioral myoclonia with absences: a new syndrome? In Wolf P, ed. *Epileptic Seizures and Syndromes*, pp. 143–53. John Libbey, London.

Panayiotopoulos CP, Ferrie CD, Giannakodimos S *et al.* (1995). Perioral myoclonia with absences. In Buncan JS, Panayiotopoulos CP, eds. *Typical Absences and Related Epileptic Syndromes*, pp. 221–30. Churchill Livingstone, London.

Parent JM, Lowenstein DH (1997). Mossy fiber reorganization in the epileptic hippocampus. *Curr Opin Neurol*, **10**, 103–9.

Pascual-Leone A, Dhuna A, Altafullah I *et al.* (1990). Cocaine-induced seizures. *Neurology*, **40**, 404–7.

Pathy MS (1978). Defaecation syncope. *Age and Ageing*, **7**, 233–6.

Pederson A, Sandoe E, Hvidberg E *et al.* (1966). Studies on the mechanism of tussive syncope. *Acta Medica Scandinavica*, **179**, 653–61.

Pedley TA, Guilleminault C (1977). Episodic nocturnal wanderings responsive to anticonvulsant drug therapy. *Ann Neurol*, **2**, 30–5.

Perucca E (2005). An introduction to antiepileptic drugs. *Epilepsia*, **46 (Suppl 4)**, 31–7.

Perucca E, Gram L, Avanzini G *et al.* (1998). Antiepileptic drugs as a cause of worsening of seizures. *Epilepsia*, **39**, 5–17.

Peters ACB, Brouwer OF, Geerts AT *et al* (1998). Randomised prospective study of early discontinuation of antiepileptic drugs in children with epilepsy: Dutch study of epilepsy in childhood. *Neurology*, **50**, 724–30.

Pincus JH (1978). Disorders of conscious awareness: hyperventilation syndrome. *Br J Hosp Med*, **19**, 312–3.

Poser CM, Brinar VV (2003). Epilepsy and multiple sclerosis. *Epilepsy Behav*, **4**, 6–12.

Poskanzer DC, Brown AE, Miller H (1962). Musicogenic epilepsy caused only be a discrete frequency band of church bells. *Brain*, **85**, 77–92.

Preter M, Tzourio CT, Ameri A *et al.* (1996). Long-term prognosis in cerebral venouis thrombosis. *Stroke*, **27**, 243–6.

Prince DA (1978). Neurophysiology of epilepsy. *Annu Rev Neurosci*, **1**, 395–415.

Quesney LF (1986). Seizures of frontal lobe origin. In Pedley TA, Meldrum BS, eds. *Recent Advances in Epilepsy*, pp. 81–110. Churchill Livingstone, Edinburgh.

Rasmussen T, Olszewski J, LloydSmith D (1958). Focal seizures due to localized chronic encephalitis. *Neurology*, **8**, 435–48.

Reuber M, Baker GA, Gill R *et al.* (2004). Failure to recognize psychogenic nonepileptic seizures may cause death. *Neurology*, **62**, 834–5.

Reuber M, Elger CE (2003). Psychogenic nonepileptic seizures: review and update. *Epilepsy Behav*, **4**, 205–16.

Reuber M, Pukrop R, Bauer J *et al.* (2003). Outcome in psychogenic nonepileptic seizures: 1 to 10-year follow-up in 164 patients. *Ann Neurol*, **53**, 305–11.

Reynolds EH (1987). Early treatment and prognosis of epilepsy. *Epilepsia*, **28**, 97–106.

Richens A, Dunlop A (1975). Serum phenytoin levels in the management of epilepsy. *Lancet*, **ii**, 247–9.

Riggs JE (2002). Neurologic manifestations of electrolyte disturbances. *Neurol Clin*, **20**, 227–39, vii.

Riikonen R (1996). Long-term outcome of West Syndrome: a study of adults with a history of infantile spasms. *Epilepsia*, **37**, 367–72.

Riley HA, German WJ, Wortis H *et al.* (1942). Glossopharyngeal neuralgia initiating or associated with cardiac arrest. *Trans Am Neurol Assoc*, **68**, 28–30.

Riley TL (1982). Syncope and hyperventilation. In Riley TL, Roy A, eds. *Pseudoseizures*, pp. 34–61. Williams & Wilkins.

Risser WL (1985). Syncope in adolescents. *Am Fam Physician*, **32**, 117–23.

Rocca WA, Sharbrough FW, Hauser WA *et al.* (1987). Risk factors for complex partial seizures: a population-based case control study. *Ann Neurol*, **21**, 22–31.

Romanelli MF, Morris JC, Ashkin K *et al.* (1990). Advanced Alzheimer's disease is a risk factor for lateonset seizures. *Arch Neurol*, **47**, 847–50.

Rosa FH (1990). Spina bifida in maternal carbamazepine exposure cohort data. *Teratology*, **41**, 587–8.

Ross RT (1988). *Syncope*. W B Saunders, London.

Rowan AJ, Binnie CD, Warfield CA *et al.* (1979). The delayed effect of sodium valproate on the photoconvulsive response in man. *Epilepsia*, **20**, 61–8.

Roy A (1979). Hysterical seizures. *Arch Neurol*, **36**, 447–54.

Ruffmann C, Bogliun G, Beghi E (2006). Epileptogenic drugs: a systematic review. *Expert Rev Neurother*, **6**: 575–89.

Rugg-Gunn FJ, Simister RJ, Squirrell M *et al.* (2004). Cardiac arrhythmias in focal epilepsy: a prospective long-term study. *Lancet*, **364**, 2212–9.

Salazar AM, Jabbari B, Vance SC *et al.* (1985). Epilepsy after penetrating head injury: I. Clinical correlates. *Neurology*, **35**, 1406–14.

Salmenpera T, Kalviainen R, Partanen K *et al.* (1998). Hippocampal damage caused by seizures in temporal lobe epilepsy [letter] [see comments]. *Lancet*, **351**, 35.

Samren EB, van Duijn CM, Koch S *et al.* (1997). Maternal use of antiepileptic drugs and the risk of major congenital malformations: a joint European study of human teratogenesis associated with maternal epilepsy. *Epilepsia*, **38**, 981–90.

Sander JW (2003). The epidemiology of epilepsy revisited. *Curr Opin Neurol*, **16**, 165–70.

Sander JWAS, Hart YM, Johnson AL *et al.* (1990). National general practice study of epilepsy: newly diagnosed epileptic seizures in a general population. *Lancet*, **336**, 1267–71.

Saner F, Gu Y, Minouchehr S *et al* (2006). Neurological complications after cadaveric and living donor liver transplantation. *J Neurol*, **253**, 612–7.

Scharfman HE, MacLusky NJ (2006). The influence of gonadal hormones on neuronal excitability, seizures, and epilepsy in the female. *Epilepsia*, **47**, 1423–40.

Scheffer IE, Berkovic SF (1997). Generalized epilepsy with febrile seizures plus. A genetic disorder with heterogeneous clinical phenotypes. *Brain*, **120**, 479–90.

Scheffer IE, Bhatia KP, Lopes-Cendes I *et al.* (1995). Autosomal dominant nocturnal frontal lobe epilepsy—a distinctive clinical disorder. *Brain*, **118**, 61.

Schierhout G, Roberts I (1998). Prophylactic antiepileptic agents after head injury: a systematic review. *J Neurol Neurosurg Psychiatry*, **64**, 108–12.

Schwartz PJ (1985). Idiopathic long QT syndrome: progress and questions. *Am Heart J*, **109**, 399–411.

Shannon M, Goetz G (1989). Connective tissue diseases and the nervous system. In Aminoff M, ed. *Neurology and General Medicine*, pp. 389–412. Churchill Livingstone, New York.

Shaw E, Arusell R, Scheithauer B *et al.* (2002). Prospective randomized trial of low- versus high-dose radiation therapy in adults with supratentorial low-grade glioma: initial report of a North Central Cancer Treatment Group/Radiation Therapy Oncology Group/Eastern Cooperative Oncology Group study. *J Clin Oncol*, **20**, 2267–76.

Shorvon S (1994). *Satus Epilepticus: Its Clinical Features and Treatment in Children and Adults*. Cambridge University Press, Cambridge.

Shorvon SD, Gilliat RW, Cox TCS *et al.* (1984). Evidence of vascular disease from CT scanning in late onset epilepsy. *J Neurol Neurosurg Psychiatry*, **47**, 225–30.

Sigurdardottir KR, Olafsson E (1998). Incidence of psychogenic seizures in adults: a population-based study in Iceland. *Epilepsia*, **39**, 749–52.

Singh NA, Charlier C, Stauffer D *et al.* (1998). A novel potassium channel gene, KCNQ2, is mutated in an inherited epilepsy of newborns. *Nat Genet*, **18**, 25–9.

Sirven JI, Wingerchuk DM, Drazkowski JF *et al.* (2004). Seizure prophylaxis in patients with brain tumors: a meta-analysis. *Mayo Clin Proc*, **79**, 1489–94.

Sisodiya SM (2004). Malformations of cortical development: burdens and insights from important causes of human epilepsy. *Lancet Neurol*, **3**, 29–38.

Smith D, Defalla BA, Chadwick DW (1999). The misdiagnosis of epilepsy and the management of refractory epilepsy in a specialist clinic. *QJM*, **92**, 15–23.

Smith DF, Hutton JL, Sandemann D *et al.* (1991). The prognosis of primary intracerebral tumours presenting with epilepsy: the outcome of medical and surgical management [see comments]. *J Neurol Neurosurg Psychiatry*, **54**, 915–20.

Sofijanov NG (1982). Clinical evolution and prognosis of childhood epilepsies. *Epilepsia*, **23**, 61–9.

Soloff LA, Rodman T (1967). Acute pulmonary embolism. II. Clinical features. *Am Heart J*, **74**, 710–24.

Soteriades ES, Evans JC, Larson MG *et al.* (2002). Incidence and prognosis of syncope. *N Engl J Med*, **347**, 878–85.

Spencer SS, Gates JR, Reeves AR *et al.* (1987). Corpus callosum section. In Engel J, ed. *Surgical Treatment of the Epilepsies*, pp. 425–44. Raven Press.

Steinlein OK, Mulley JC, Propping P *et al.* (1995). A missense mutation in the neuronal nicotinic receptor a subunit is associated with autosomal dominant nocturnal frontal lobe epilepsy. *Nat Genet*, **11**, 201.

Stengel E (1943). Further studies on pathological wandering (fugue with the impulse to wander). *J Ment Sci*, **89**, 224–41.

Steriade M, McCormick DA, Sejnowski TJ (1993). Thalamocortical oscillations in the sleeping and aroused brain. *Science*, **262**, 679–85.

Stevens DL, Matthews WB (1973). Cryptogenic drop attacks. An affliction of women. *Br Med J*, **1**, 439–42.

Sumi SM, Teasdall RD (1963). Focal seizures. A review of 150 cases. *Neurology*, **13**, 582–86.

Sutula T, He XX, Cavazos J *et al.* (1988). Synaptic reorganization in the hippocampus induced by abnormal functional activity. *Science*, **239**, 1147–50.

Tang C-M, Dichter M, Morad M (1989). Quisqualate activates a rapidly inactivating high conductance ionic channel in hippocampal neurons. *Science*, **243**, 1474–7.

Taske NL, Williamson MP, Makoff A *et al.* (2002). Evaluation of the positional candidate gene CHRNA7 at the juvenile myoclonic epilepsy locus (EJM2) on chromosome 15q13-14. *Epilepsy Res*, **49**, 157–72.

Tauck DL, Nadler JV (1985). Evidence of functional mossy fiber sprouting in hippocampal formation of kainic acid-treated rats. *J Neurosci*, **5**, 1016–22.

Temkin NR, Dikmen SS, Winn HR (1998). Clinical trials for seizure prevention. *Adv Neurol*, **76**, 179–88.

Temkin NR, Dikmen SS, Anderson GD *et al.* (1999). Valproate therapy for prevention of posttraumatic seizures: a randomized trial. *J Neurosurg*, **91**, 593–600.

The Vagus Nerve Stimulation Study Group (1995). A randomized controlled trial of chronic vagus nerve stimulation for treatment of medically intractable seizures. *Neurology*, **45**, 224–30.

Traub RD, Jefferys JGR (1998). Epilepsy in vitro: electrophysiology and computer modeling. In Engel J, Pedley TA, eds. *Epilepsy: A Comprehensive Textbook*, pp. 405–18. Lippincott-Raven, New York.

Treiman DM, Meyers PD, Walton NY *et al.* (1998). A comparison of four treatments for generalized convulsive status epilepticus. Veterans Affairs Status Epilepticus Cooperative Study Group. *N Engl J Med*, **339**, 792–8.

Treiman DM, Walton NY, Kendrick C (1990). A progressive sequence of electroencephalographic changes during generalised convulsive status epilepticus. *Epilepsy Res*, **5**, 49–60.

Treiman DN, DelgadoEscueta AV (1983). Violence in epilepsy: a critical review. In Pedley TA, Meldrum BS, eds. *Recent Advances in Epilepsy*, pp. 179–209. Churchill Livingstone, Edinburgh.

Tsuboi M, Suzuki K, Nagao S *et al.* (1985). Glossopharyngeal neuralgia with cardiac syncope. A case successfully treated with microvascular decompression. *Surg Neurol*, **24**, 279–83.

Valdueza JM, Cristante L, Dammann O *et al.* (1994). Hypothalamic hamartomas with special reference to gelastic epilepsy and surgery. *Neurosurgery*, **34**, 949–58.

van Elst LT, Woermann FG, Lemieux L *et al.* (2000). Affective aggression in patients with temporal lobe epilepsy: a quantitative MRI study of the amygdala. *Brain*, **123**, 234–43.

Venna N, Sabin TD (1981). Tonic focal seizures in nonketotic hyperglycaemia of diabetes mellitus. *Arch Neurol*, **38**, 512–4.

Walker JE (1981). Effect of aminophylline on seizure thresholds and brain regional cyclic nucleotides in the rat. *Exp Neurol*, **74**, 299–304.

Wallace RH, Wang DW, Singh R *et al.* (1998). Febrile seizures and generalized epilepsy associated with a mutation in the Na+-channel beta1 subunit gene SCN1B. *Nat Genet*, **19**, 366–70.

Wardrope J, Ryan F, Clark G *et al.* (1991). The Hillsborough tragedy. *BMJ*, **303**, 1381–5.

Weaver S, Forsyth P, Fulton D *et al.* (1995). A prospective randomised study of prophylactic anticonvulsants (AC) in patients with primary brain tumors (PBT) or metastatic brain tumors (MBT) and without prior seizures (Sz): a preliminary analysis of 67 patients. *Neurology*, **45**, A263.

White HS (1997). Mechanisms of antiepileptic drugs. In Porter RJaCD, ed. *The Epilepsies 2*. Butterworth-Heinemann, pp. 1–30.

Wiebe S, Blume WT, Girvin JP *et al.* (2001). A randomized, controlled trial of surgery for temporal-lobe epilepsy. *N Engl J Med*, **345**, 311–8.

Wijdicks EF, Wiesner RH, Krom RA (1995). Neurotoxicity in liver transplant recipients with cyclosporine immunosuppression. *Neurology*, **45**, 1962–4.

Wijdicks EFM, Plevak DJ, Wiesner RH *et al.* (1996). Causes and outcome of seizures in liver transplant recipients. *Neurology*, **47**, 1523–5.

Wilkins A, Lindsay F (1985). Common forms of reflex epilepsy: physiological mechanisms and techniques for treatment. In Pedley TA MBS, ed. *Recent Advances in Epilepsy*. Churchill-Livingstone, pp 239–72.

Williamson PD, French JA, Thadani VM *et al.* (1993). Characteristics of medial temporal lobe epilepsy II. Interictal and ictal scalp electroencephalography, neuropsychological testing, neuroimaging, surgical results and pathology. *Ann Neurol*, **34**, 781–7.

Williamson PD, Spencer DD, Spencer SS *et al.* (1985). Complex partial seizures of frontal lobe origin. *Ann Neurol*, **18**, 497–504.

Williamson R, Wheal HV (1992). The contribution of AMPA and NMDA receptors to graded bursting activity in the hippocampal CA1 region in an acute in vitro model of epilepsy. *Epilepsy Res*, **12**, 179–88.

Wolf P, Inoue Y (1984). Therapeutic response of absence seizures in patients of an epilepsy clinic for adolescents and adults. *J Neurol*, **231**, 225–9.

Wong MC, Suite NDA, Labar DR (1990). Seizures in human immunodeficiency virus infection. *Neurology*, **47**, 640–2.

Wong RKS, Prince DA (1979). Dendritic mechanisms underlying penicillininduced epileptiform activity. *Science*, **204**, 1228–31.

Wong RKS, Prince DA (1981). Afterpotential generation in hippocampal pyramidal cells. *J Neurophysiol*, **45**, 86–97.

Yanagihara T, Piepgras DG, Klass DW (1985). Repetitive involuntary movements associated with episodic cerebral ischaemia. *Ann Neurol*, **18**, 244–50.

Zeman, AZ, Boniface SJ, Hodges JR (1998). Transient epileptic amnesia: a description of the clinical and neuropsychological features in 10 cases and a review of the literature. *J Neurol Neurosurg Psychiatry*, **64**, 435–43. 98.

CHAPTER 32

Sleep disorders

Paul Reading

Contents

Despite major advances in our understanding of its neurobiology, sleep remains an enigma. Its true function and even the amount needed for optimum brain performance remain uncertain (Frank 2006). However, the need to sleep is imperative, reflecting the fact that sleepiness, like hunger and thirst, is a true drive state. Sleepiness can only be satiated by sleep itself. Moreover, severely disordered sleep can profoundly affect cognition, mental health, and physical well-being.

Although sleep medicine has a traditionally low profile in neurology teaching and practice, sleep-related phenomena are frequently associated with numerous neurological disorders. Conversely, sleep problems can adversely affect familiar conditions such as headache and epilepsy. Furthermore, in large surveys, sleep-related symptoms are undoubtedly common with 25 per cent of the population reporting problems that significantly and regularly impact on daily activities.

32.1 Normal sleep

The traditional view that sleep is largely a passive or necessarily restful process has largely been superseded. By contrast, rather than simply reflecting the absence of wakefulness, sleep is actively orchestrated with a highly reproducible and complex internal architecture (Harris 2005). Sleep is divided into non-rapid eye movement, NREM, stages I to IV, and rapid eye movement, REM, states on the basis of physiological parameters (Fig. 32.1). Light non-rapid eye movement sleep of stages I and II comprises about 55 per cent of total sleep time whereas deep non-rapid eye movement, stages III and IV, sleep fills 25 per cent. The remaining 20 per cent is occupied by the curious state of REM sleep in which vivid dreams predominate.

A typical distribution of healthy adult sleep is shown in Fig. 32.2. Four or five rapid eye movement cycles occur through the night. It should be recognized that occasional arousals from nocturnal sleep may be regarded as normal and that seemingly random body movements or shuffles in position occur regularly throughout the night often at shifts of sleep stage. In rapid eye movement sleep, however, despite high levels of cerebral metabolic activity that loosely correspond to dream mentation, general motor activity is profoundly suppressed and any observable movements are confined to occasional minor jerks.

The precise mechanisms that control shifts between wake and the various sleep stages are poorly characterized although a current theme proposes the existences of multiple sleep switches (Saper *et al.* 2001). In essence, discrete brainstem centres for wake, non-rapid eye movement, and rapid eye movement sleep interact by mutually inhibitory pathways, creating so-called 'flip-flop' switches. With these, transitions between wake and sleep states are made both efficiently and quickly. Switches between non-rapid eye movement and rapid eye movement sleep states may also occur by a similar process (Lu *et al.* 2006).

Sleep generally deteriorates with age and determining the limits of normal sleep can be difficult. Most classifications of disordered

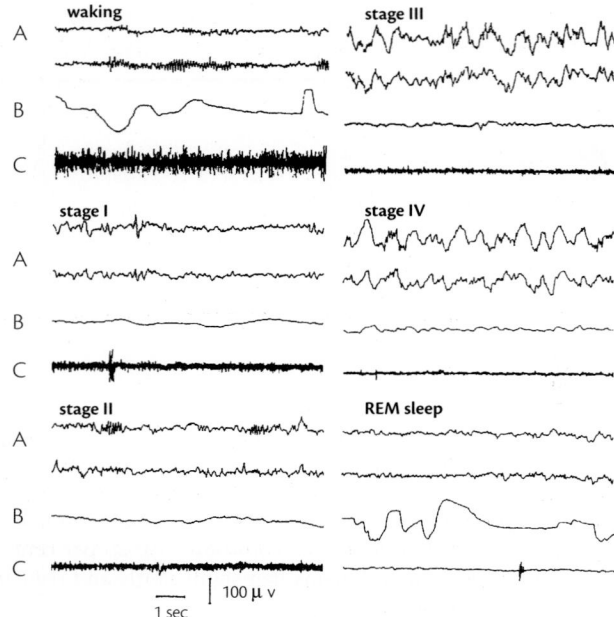

Fig. 32.1 Summary of the parameters defining the four stages of non-rapid eye movement sleep and rapid eye movement sleep compared to wakefulness. A. shows the EEG either centrally (upper trace) or occipitally (lower trace); B. is the oculogram revealing eye movements; C. is electromyographic activity recorded from the chin.

In stage II non-rapid eye movement sleep, sleep spindles are seen on the EEG. In stage IV sleep, large amplitude delta frequency waves comprise more than 50 per cent of the EEG.

In rapid eye movement sleep, the EEG is low amplitude and desynchronized. By definition, eye movements are prominent. Apart from minor myoclonic activity, the EMG trace is silent.

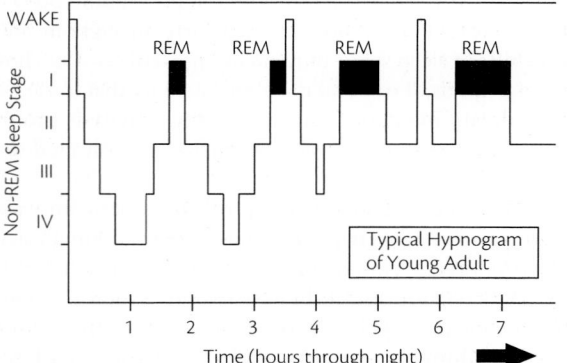

Fig. 32.2 A typical hypnogram of a young adult showing four cycles of non-rapid eye movement (non-REM) and rapid eye movement (REM) sleep. Two brief awakenings are shown which can be considered normal. The proportion of deep non-rapid eye movement sleep (stages III and IV) is highest in the first few hours of sleep whereas rapid eye movement sleep predominates towards the end of the sleep period.

sleep are symptom-based and the recently revised International Classification of Sleep Disorders (ICSD-2) recognizes eight categories (American Academy of Sleep Medicine 2005):

- Insomnias;
- Sleep-related breathing disorders;
- Hypersomnias of central origin;
- Circadian rhythm sleep disorders;
- Parasomnias;
- Sleep-related movement disorders;
- Isolated symptoms, normal variants, and unresolved issues; and
- Other sleep disorders.

32.2 Insomnia

Chronic insomnia is loosely defined as the perception of inadequate sleep for a period of more than 4 weeks. Such inability to fall asleep or maintain continuous sleep is a common symptom and has a number of causes both intrinsic and extrinsic (Table 32.1). It is rare for organic cerebral pathology to underlie primary insomnia and persistently maladaptive attitudes or behaviours are usually responsible. An index event or illness can often be identified. The common forms of primary insomnia are probably best treated by behavioural modification including a combination of cognitive behaviour therapy and relaxation techniques (Morin *et al.* 1999). The intermittent use of short-acting hypnotics may be helpful although long-term drug treatment is rarely beneficial.

If symptoms of inadequate sleep date back to childhood, the term 'idiopathic insomnia' is sometimes used. Although its neurobiology remains obscure, at some level, this disorder probably reflects a constitutionally impaired sleep drive such that the normal homeostatic pressure to sleep is inadequate.

The interplay between psychological distress and chronic insomnia is complex with each element potentially fuelling the other. Psychiatric input to treat any significant mood disorder can therefore be helpful in attempting to resolve sleep-related symptoms.

32.3 Hypersomnia

Significant excessive daytime sleepiness is reported by 5 per cent of the population and is most often due to poor quality or diminished overnight sleep (Table 32.2). It is important to distinguish true sleepiness or drowsiness from fatigue and lethargy which often have different causes. Within the abnormally sleepy population, approximately 2 per cent have a primary sleep disorder in which the most striking complaint is an inability to stay awake appropriately despite the desire to do so.

32.3.1 Narcolepsy

Introduction and clinical features

Narcolepsy is not a rare disorder with an estimated prevalence of 1 per 2000 in Caucasian populations. However, differences in case ascertainment and the availability of sleep services have led to considerable variance in reported rates worldwide (Silber *et al.* 2002). Moreover, there is undoubtedly a spectrum of severity and many mildly affected individuals are undiagnosed, misdiagnosed, or only

Table 32.1 Some causes of primary and secondary insomnia

	Causes	Comments
Primary insomnia	*Intrinsic sleep disorders*	
	Psychophysiological insomnia	Sometimes called 'conditioned insomnia'
	Paradoxical insomnia	Formerly called 'sleep–wake misperception'
	Idiopathic insomnia	History dates back to childhood
	Extrinsic sleep disorders	
	Poor sleep hygiene	
	Environmental sleep disorder	Examples include sleep disordered bed partners or pets interfering with sleep; usually results in daytime sleepiness
	Altitude insomnia	Mild hypoxaemia produces poor sleep because of unstable respiratory control overnight
	Drug-dependent insomnia	Hypnotics, stimulants, or alcohol may be responsible
Secondary insomnia	*Neurological conditions*	
	Restless legs syndrome (Section 40.11.1)	An important and treatable cause of insomnia
	Parkinson's disease (Section 40.3.1)	Sleep fragmentation can be an integral part of the condition
	Morvan's syndrome (Sections 23.7.1; 24.10.4)	A rare paraneoplastic or autoimmune syndrome with neuromuscular hyperexcitability and severe insomnia as cardinal features
	Fatal familial insomnia (Section 42.9.8)	A very rare familial prion disease with significant thalamic pathology as the presumed substrate for severe insomnia
	Medical disorders	
	Asthma	
	Gastro-oesophageal reflux	An important and often overlooked diagnosis
	Chronic pain syndromes including fibromyalgia	A high percentage of light non-rapid eye movement sleep often seen
	Psychiatric causes	
	Secondary to medication	
	Mood disorders including anxiety, depression, and mania	

Table 32.2 Excessive day time sleepiness. Some extrinsic and intrinsic causes

Intrinsic causes	Primary causes
	Narcolepsy
	Idiopathic hypersomnia
	Kleine–Levin syndrome (intermittent sleepiness)
	Causes secondary to a chronic disorder
	Sleep-disordered breathing
	Restless legs syndrome and periodic limb movement disorder
	Parkinson's disease
	Multiple sclerosis
	Head injuries
	Encephalitis
Extrinsic causes	Sleep deprivation or insufficient sleep
	Drug-related hypersomnia
	Environmental sleep disorder
	Shift work sleep disorder
	Jet lag

correctly diagnosed after many years of symptoms (Kryger *et al.* 2002). It most often starts in adolescence with a second minor peak in early middle age and symptoms are life-long. Most sufferers develop coping strategies to minimize the impact of the syndrome although this may lead to social isolation.

Narcolepsy is important not least because it is usually disabling, influencing every aspect of daily living (Dodel *et al.* 2004). Many narcoleptics feel a sense of underachievement partly because treatment is frequently either delayed or only partially effective. A perceived lack of medical interest in the disease together with the adverse effects on schooling, careers, and relationships understandably produces frustration. Secondary mood disorders are seen in many patients.

Rather than reflecting true hypersomnolence over a 24-h period, narcolepsy is best viewed as a primary disorder of sleep–wake regulation (Fig. 32.3) with an inability to stay awake for more than a few hours as the cardinal symptom (Dauvilliers *et al.* 2007). Indeed, excessive daytime sleepiness not explained by another disorder remains an essential diagnostic criterion. Many subjects describe sudden and irresistible urges to sleep, often in public or inappropriate situations, invariably worse if they are unoccupied or bored. Short naps lasting minutes can often be restorative, however, in contrast to most other sleep disorders causing hypersomnolence. A minority of sufferers are relatively unaffected by excessive daytime sleepiness and other features of the syndrome predominate.

Cataplexy is a curious phenomenon highly specific to narcolepsy and present to varying degrees in two-thirds of patients (Overeem *et al.* 2001). It is particularly important to identify typical cataplexy

Typical 24-hour hypnograms in a control and an untreated narcoleptic patient

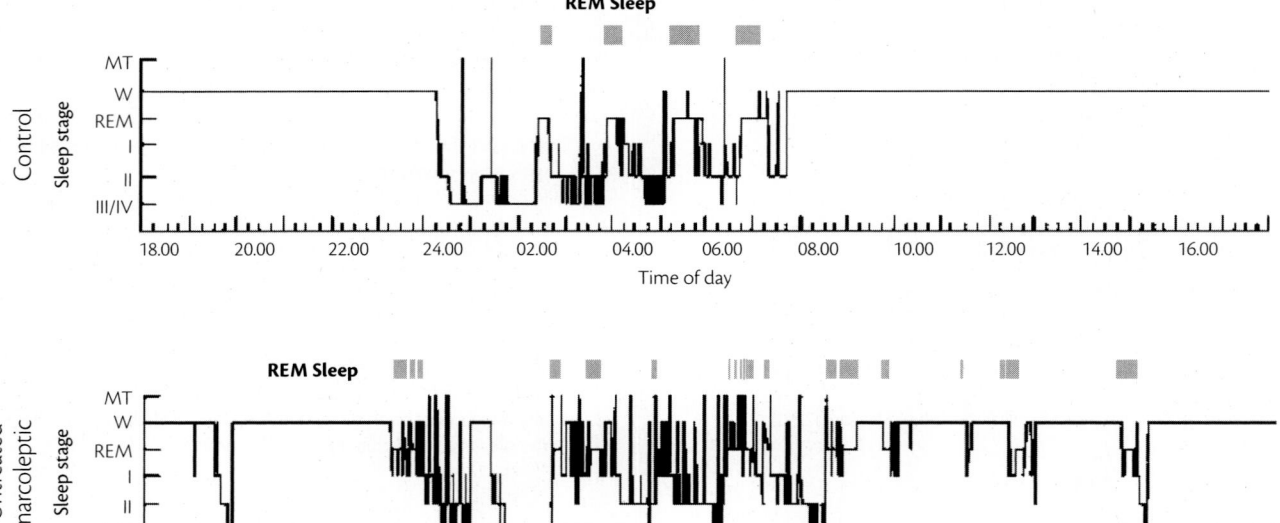

Fig. 32.3 Comparisons of typical hypnograms over 24 h in a control and an untreated narcoleptic. In the narcoleptic trace, there is severe disruption of the usual pattern with numerous daytime naps containing rapid eye movement (REM) sleep. Overnight, the usual sleep architecture is disorganized in the narcoleptic with several awakenings and associated movement. W– wake, MT– significant movements.

since its presence in a subject with excessive daytime sleepiness is considered diagnostic of narcolepsy. Full-blown episodes reflect an intrusion of profound muscle paralysis that descends over a few seconds from head to the lower limbs, often causing collapse to the floor. Identifiable triggers usually have an emotional content. Laughter and other positive emotions such as pleasant surprise are the most common precipitants although frustration and anger can also reliably provoke episodes. In some subjects, the mere thought or anticipation of an emotional event can cause unpredictable collapse. Reassuringly, attacks are rare in dangerous situations and most sufferers only report cataplexy when relatively relaxed in familiar environments with friends. It is therefore very uncommon for physicians to witness episodes, making a reliable history crucial for confident diagnosis. Importantly, partial or focal attacks are common and can be subtle, perhaps confined to the jaw or neck. Occasionally, an inability to tell the punchline of a joke due to a stuttering dysarthria may be the only manifestation. Facial twitching or head bobbing is very common as an episode commences and can lead to diagnostic confusion. Crucially, awareness is preserved in cataplexy although in rare instances when attacks last more than a minute or so, dream-like intrusions and altered consciousness may intercede. Severely affected individuals may have over 20 attacks a day often reporting that the amusement or frustration induced by the cataplectic episodes themselves can prolong the period of weakness.

It is widely thought that cataplexy occurs because rapid eye movement sleep paralysis intrudes into the wakeful state.

Indeed, as in rapid eye movement sleep, a subject is rendered temporarily areflexic during an episode as a result of descending inhibitory neural impulses from lower brainstem centres directly onto motor neurons. Some evidence suggests that this phenomenon may occur to a minor degree during emotion in control subjects, adding credibility to the adage 'going weak with laughter' (Overeem *et al.* 1999).

Sleep paralysis and hallucinations around sleep–wake transition are the other two components of the narcoleptic 'tetrad' first described over fifty years ago (Yoss and Daly 1957). Only 25 per cent of subjects have all four elements, however, and the presence of these other symptoms, particularly sleep paralysis, is not specific to narcolepsy. *Sleep paralysis* is usually frightening primarily because of an inability to take deep breaths voluntarily. Most episodes occur at the point of waking although narcoleptics typically also report episodes at sleep onset. Accompanying sensations of being crushed with or without vivid visual hallucinations may add to the distress of the episodes. Like cataplexy, this phenomenon reflects the intrusion of rapid eye movement sleep elements into the wakeful or drowsy state.

Hallucinations occurring at sleep onset, hypnagogic, or as the subject wakes, hypnopompic, are usually visual and can be both vivid and disturbing, especially in children. They represent fragments of dream mentation intruding into the conscious state, reinforcing the notion that narcoleptics cannot adequately maintain consistent boundaries between states of wakefulness and sleep.

When questioned, the majority of narcoleptics have fragmented nocturnal sleep. Although this may be due to the intrusion of a parasomnia or obstructed breathing, both of which are commoner in narcoleptics, the primary problem is one of sleep regulation and maintenance. The notion that narcoleptics have problems sleeping at night is counterintuitive to some but is an important addition to the original descriptions of the syndrome.

Other symptoms are commonly reported in narcolepsy. In addition to obvious naps, most narcoleptics will experience numerous 'micro-sleeps' through the day in which awareness during routine activities is compromised for a few seconds. The resulting lapses lead to automatic and inappropriate behaviours with worrying consequences for complex and potentially dangerous tasks such as driving. Although difficult to characterize, many narcoleptics also report significant problems with memory and concentration as a result of their sleep–wake difficulties. Furthermore, increasing evidence suggests abnormalities of appetite, particularly at night, with cravings for sweet foods. Moderate obesity is commonly seen in childhood narcolepsy and may have a metabolic explanation since there is no clear link with excessive food intake (Kotagal *et al.* 2004). Indeed, some evidence suggests that overweight narcoleptics eat less than normal controls.

Pathogenesis and diagnosis

Since the discovery in 1984 that Japanese narcoleptics were extremely likely to carry the human leucocyte antigen, HLA, haplotype DR2 (Juji *et al.* 1984), an autoimmune basis for the syndrome has been thought likely. The predisposing antigen has since been established as DQB1*0602. This is present in over 90 per cent of narcoleptics with cataplexy and around 50 per cent of those without cataplexy, compared to a frequency of 20 per cent in control populations. Of interest, homozygosity for DQB1*0602 appears to confer an even greater risk for the syndrome. However, there remains no direct evidence for autoimmunity either in the form of serum markers or CSF abnormalities.

A major breakthrough in understanding the neurobiology of narcolepsy occurred in 1999 when two groups independently demonstrated abnormalities of a recently described neuropeptide, hypocretin, also called orexin, in separate animal models. The well-established autosomal recessive Doberman model was shown to have dysfunctional hypocretin receptors (Lin *et al.* 1999) whilst a mouse hypocretin knockout model developed convincing clinical features of narcolepsy with cataplexy (Chemilli *et al.* 1999). Subsequently, it has been demonstrated that CSF hypocretin is virtually absent both in sporadic canine models and also human narcoleptics with cataplexy (Ripley *et al.* 2001). Indeed, a CSF hypocretin level of less than 110 pg/ml is now considered diagnostic.

Post-mortem evidence has confirmed that pathology in narcoleptic brains is confined specifically to hypocretin neurons (Thannickal *et al.* 2000). Confusingly, however, in rare familial narcolepsy and in sporadic cases without typical cataplexy, hypocretin levels can be preserved, implying there is more than one pathogenetic mechanism for certain forms of the syndrome.

Following the unexpected involvement of the hypocretin system in human narcolepsy, it has been intensely studied in intact animals. Around 30 000 neurons containing the peptide are confined to the lateral hypothalamus but innervate all the arousal systems in the brain. Levels of hypocretin rise towards the end of the waking day, especially in the presence of peptide hunger signals or if the subject is expecting food (Saper 2006). Activity of hypocretin neurons, therefore, appears to stabilize a state of wakefulness when the organism needs to be alert. In narcolepsy, their absence leads to inappropriate switches between sleep and wakefulness. Moreover, transitions between behavioural states may be incomplete, explaining the intrusion of rapid eye movement sleep phenomena such as paralysis into wakefulness. The precise mechanism by which emotional stimuli, in particular, trigger cataplexy, however, remains elusive.

If typical cataplexy is absent and CSF hypocretin levels cannot be easily be measured, a positive diagnosis of narcolepsy can be made following a multiple sleep latency test. This test measures the propensity for a subject to fall sleep by recording the average length of time to reach light sleep in a conducive environment over four or five nap opportunities between 9 am and 3 pm. If the mean sleep latency is less than 8 min and rapid eye movement sleep is achieved within 15 min on at least two occasions, the criteria for narcolepsy are fulfilled. Reliable results depend on ensuring a reasonable night's sleep preceding the investigation. The multiple sleep latency test also requires a strict protocol to avoid false negative results.

Secondary narcolepsy

Narcoleptic symptoms including cataplexy have been reported in a number of neurological conditions (Nishino and Kanbayashi 2005). Given the recent advances in the understanding the neurobiology of the primary syndrome, it is not surprising that pathology in the region of the hypothalamus such as tumours around the third ventricle can lead to secondary narcolepsy, presumably by depletion of hypocretin-containing neurons. However, the mechanism of severe sleepiness or sleep–wake dysregulation after head injury or as components of other conditions such as multiple sclerosis and Parkinson's disease can be difficult to explain.

The various subtypes of narcolepsy are shown in Table 32.3.

Table 32.3 Subtypes of narcolepsy and associated features.

	Narcolepsy with cataplexy (sporadic)	Narcolepsy without cataplexy (sporadic)	Familial narcolepsy	Secondary (symptomatic) narcolepsy
REM sleep reached within 15 min on two or more occasions in MSLT	85%	100% (by definition)	uncertain	75%
HLA DQB1*0602 positivity	85–93%	35–56%	65–79%	Uncertain
Presence of low or undetectable CSF Hcrt-1 levels	>90%	14%	38%	Variable; reported instances of very low levels in individual cases
Proposed or presumed pathogenesis	Autoimmune destruction of Hcrt synthesizing neurons	Partial Hcrt deficiency; Unknown mechanism in many	Multiple genotypes; Hcrt system very rarely involved directly	Damage to Hcrt containing neurons in the lateral hypothalamus

Key : REM: rapid eye movement; MSLT: multiple sleep latency test; HLA: human leucocyte antigen; CSF: cerebrospinal fluid; Hcrt-1: hypocretin-1

Treatment

Advice on lifestyle helps a proportion of narcoleptics. Planned naps, especially after meals, may improve wakefulness. Furthermore, the avoidance of large meals rich in refined carbohydrates is reportedly beneficial to some.

The majority of narcoleptics, however, benefit from medication to improve daytime wakefulness (Table 32.4) although few are normalized (Mignot and Nishino 2005). Modafinil is the most widely used wake-promoting agent that has partly replaced traditional psychostimulants. Its mechanism of action remains obscure although a direct effect on arousal centres in the hypothalamus is postulated. It has no definite positive effect on cataplexy. Side-effects are rare and include headache or gastrointestinal upset. Interactions with the oral contraceptive pill and uncertainty over safety in pregnancy may limit its use in young women.

In severe sleepiness or if modafinil is unsuccessful, central stimulants with a predominantly dopaminergic action such as dexamfetamine are helpful, especially if used flexibly. Despite prescriber concerns, it is rare for psychological addiction to occur in narcolepsy although tolerance may require increasing doses with time. At a practical level, cardiovascular side-effects such as hypertension are relatively rare but necessitate caution in the elderly. Given the different mechanisms of action, a combination of modafinil and a psychostimulant is appropriate. Additional use of caffeine and setting aside time for planned naps may reduce the need for medication.

About a half of narcoleptic subjects also require specific treatment for cataplexy. Although the evidence base is small, most anti-depressants will suppress cataplexy by increasing cerebral monoaminergic activity and inhibiting the tendency to enter rapid eye movement sleep. The side-effect profile of most anti-depressant drugs, particularly the tricyclics, may limit their usefulness in cataplexy. A new approach for troublesome cataplexy is to use sodium oxybate although emerging trial evidence suggests this drug helps daytime sleepiness as well. It is a liquid preparation taken before bedtime and, due to its short half-life, once during the night, if the subject is awake. After several weeks of therapy, the effects on cataplexy are striking with almost 90 per cent of attacks abolished (Xyrem International Study Group 2005). Inadvertent daytime naps, objective and subjective measures of daytime sleepiness also improve. The agent appears to work, in part, by inducing deep restorative sleep early in the night such that the sleep drive is effectively dissipated by the following morning. The drug should be used with extreme caution in any patient living alone or with young children in case confusional episodes from deep sleep are provoked. However, if disturbed nocturnal sleep is a major symptom, it appears a logical treatment given that standard benzodiazepine hypnotic agents rarely induce refreshing sleep in narcolepsy.

Following the recent findings that most narcoleptics are deficient of the neuropeptide, hypocretin, a future goal will be to develop replacement therapy. Indeed, if hypocretin levels are increased in animal models by intracerebral infusions, there appear to be clinical effects. In humans, the development of an oral analogue that will penetrate the blood–brain barrier is a current pharmacological goal.

32.3.2 Idiopathic hypersomnia

Idiopathic hypersomnia is a diagnosis of exclusion most often made when excessively sleepy patients do not fulfil the criteria for narcolepsy. Depending on precise definitions, it is probably 10 times less common than narcolepsy. Classical cases report difficulty waking in the morning followed by prolonged unrefreshing

Table 32.4 Commonly used drug treatments for the narcoleptic syndrome

	Total 24-h dose range	Comments
Excessive daytime sleepiness		
Modafinil	200–600 mg	Different mechanism of action to traditional psychostimulants
Dexamfetamine	5–60 mg	Tolerance can develop but dependence rare
Methylphenidate	10–80 mg	Similar to amphetamine but possibly smoother action; long-acting preparation available
Sodium oxybate	4.5–9 g	Taken through the night; may act synergistically with daytime stimulants
Cataplexy		
Venlafaxine	75–225 mg	Possibly the anti-depressant with most anti-cataplectic properties
Clomipramine	10–150 mg	Potent but side-effects often limit use
Fluoxetine	20–44 mg	Appropriate for mild cataplexy; few side effects
Sodium oxybate	4.5–9 g	Taken at night; up to 90% of attacks may be abolished after 4 weeks of treatment
Disturbed nocturnal sleep		
Clonazepam	0.5–2 mg	Sleep continuity improved but sleep quality not usually refreshing; intermittent rather than continuous use advisable
Sodium oxybate	4.5–9 g	Deep non-rapid eye movement increased; overall sleep quality improved

daytime naps despite long and deep nocturnal sleep. Low mood and frequent automatic behaviours are commonly reported (Bassetti and Aldrich 1997). However, no specific narcoleptic features such as cataplexy are present and CSF hypocretin levels are generally normal. Sleep investigations should confirm a shortened daytime sleep latency of less than 8 min preceded by normal, yet prolonged, nocturnal sleep. A new category of idiopathic hypersomnia without prolonged overnight sleep has been proposed although this is controversial and distinction from atypical or monosymptomatic narcolepsy can be difficult.

As in narcolepsy, although usually with less satisfactory results, the treatment of idiopathic hypersomnia consists of modafinil alone or in combination with traditional psychostimulants.

32.3.3 Kleine–Levin syndrome

Kleine–Levin syndrome is a rare and poorly characterized sleep disorder most commonly seen in adolescents (Arnulf *et al.* 2005a). The primary feature is periodic hypersomnia lasting days to weeks, recurring at intervals of weeks to months. During symptomatic periods, the subject is generally drowsy and usually displays abnormal behaviours. These include simple irritability, hallucinations, hypersexuality, and abnormal appetite producing hyperphagia. Investigations are generally unhelpful although an excess of rapid eye movement sleep is occasionally seen during episodes. Intermittent hypothalamic dysfunction is a plausible, if speculative, mechanism to explain the symptom complex. Secondary causes are very rare and reportedly include a wide variety of neurological conditions such as multiple sclerosis and Prader–Willi syndrome.

Treatments are empirical and usually ineffective although the syndrome tends to resolve spontaneously after several years. Amphetamine may help during episodes and lithium may be used as a prophylactic agent.

32.3.4 Sleep apnoea

A nocturnal apnoea is defined as a cessation of breathing for 10 s or more and is most often secondary to an obstruction in the soft palate. Obstructive sleep apnoea is the commonest cause of severe daytime sleepiness and usually occurs on the background of severe snoring. Central adiposity in males is the commonest risk factor although enlarged tonsils and a receding jaw can also predispose to the problem (Stradling and Crosby 1991). In deep non-rapid eye movement and rapid eye movement sleep, either the hypoxia generated by the breathing difficulty or the effort made by respiratory control centres to overcome the obstruction will partially arouse the subject, severely disrupting sleep.

Apart from causing daytime sleepiness, it is thought obstructive sleep apnoea has significant consequences for general health and may be an independent risk factor for vascular disease, diabetes, and hypertension (Shamsuzzaman 2003). Although neurologists will not usually be involved in treating obstructive sleep apnoea, it is important to recognize the syndrome in any patient and to refer to a sleep service for assessment. The best treatment is usually continuous positive airways pressure, or CPAP, delivered by a nasal mask. Mandibular advancement devices, palatal surgery, wake-promoting drugs, and even tracheostomy may be considered in individual cases.

Central sleep apnoea may occur as an isolated phenomenon secondary to brainstem pathology or as part of severe obstructive sleep apnoea, especially in morbidly obese subjects. It is probably under-recognized in conditions with autonomic dysfunction such as multiple system atrophy. It is a complex problem to treat especially since drug therapies rarely help to increase respiratory drive.

32.4 Parasomnias and sleep-related movement disorders

Parasomnias can simply be defined as intermittent undesirable events arising from sleep that are not epileptic in nature. The spectrum is large, ranging from visual imagery at sleep onset to complex motor behaviours, occasionally with violent components. Family members and bed partners are usually more concerned than the subjects themselves who often remain oblivious to any nocturnal disturbance. Parasomnias are generally classified according to the sleep stage from which they arise although some parasomnias are not 'state dependent'.

A simple yet valid scheme to explain most parasomnias is shown in Fig. 32.4. The brain normally exists in three distinct and mutually exclusive states, namely wakefulness, non-rapid eye movement sleep, and rapid eye movement sleep. Normally, switches occur seamlessly and relatively quickly between these states through the sleep–wake cycle. The majority of parasomnias result from abnormal state transitions such that elements of one state intrude into the boundary of another. A subject can be considered 'caught' for a variable period of time in a separate abnormal state somewhere in between wake and sleep.

With the exception of certain rapid eye movement sleep disorders, the neuroanatomical basis of parasomnias remains entirely obscure. The high prevalence of familial aggregation suggests genetic factors and predominance in childhood implies a maturational component, particularly in non-rapid eye movement parasomnias.

32.4.1 Parasomnias at the sleep–wake transition

It is almost a universal experience to have occasional unpleasant sensations of falling through space at the point of sleep onset with resulting brief muscular contractions. In some subjects these *hypnic jerks* can regularly interfere with sleep onset and recur through the night. In others there are accompanying explosive *sensory phenomena*, sometimes with severe head pain as a component. Treatments with short-acting hypnotic agents may be justified in severe cases.

More complex and prolonged phenomena comprising a variety of *rhythmical movements* also tend to occur during extreme drowsiness just before sleep. Head banging is the commonest manifestation in children. The problem tends to resolve with time although can persist into adulthood and disturb bed partners. Various patterns of movement are seen with the head, neck, and trunk most commonly involved. The view that the movements are semi-voluntary, as part of a sleep-inducing habit, does not concord with the observation that the phenomenon arises only from deep or even rapid eye movement sleep in a minority of subjects (Stepanova *et al.* 2005).

32.4.2 Non-rapid eye movement parasomnias

Non-rapid eye movement parasomnias are characterized by sudden but partial arousals from deep sleep, usually stage 4, resulting in behaviours for which the subject usually has no subsequent

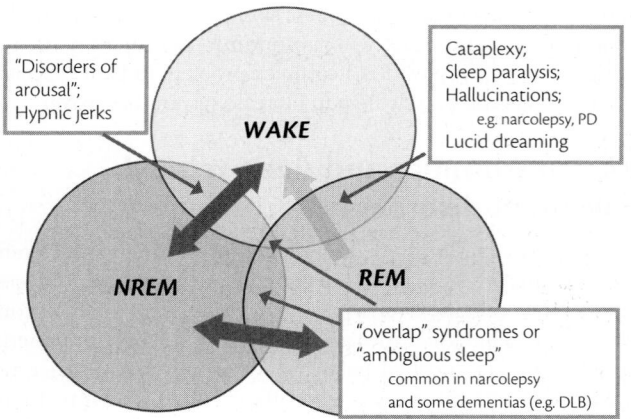

Fig. 32.4 A graphical demonstration depicting the normal transitions between the mutually exclusive states of wakefulness, non-rapid eye movement (NREM), and rapid eye movement (REM) sleep. The switch from rapid eye movement sleep to wake can occur directly and would normally lead to a dream experience. Most parasomnias occur because of abnormal or inefficient state transitions. Sleep walking and related phenomena occur when a subject arouses incompletely from deep non-rapid eye movement sleep. Hypnic jerks may occur when the brain fails to fall asleep in its entirety. The narcoleptic symptoms of sleep paralysis, cataplexy, and hallucinations at sleep–wake transition occur when elements of rapid eye movement sleep intrude into wakefulness. Parkinsonian hallucinations probably represent rapid eye movement sleep imagery occurring in the drowsy wakeful state. Some subjects report the ability to control their dreams, lucid dreaming, which can be considered as wakeful consciousness intruding into the rapid eye movement sleep state. In some narcoleptics or in severe dementia, it can be very difficult to stage sleep accurately and 'overlap' syndromes producing ambiguous sleep can occur. Key: PD – Parkinson's disease; DLB – dementia with Lewy bodies.

recollection. Based on clinical features, sleep-walking, confusional arousals, and night terrors are recognized as three separate phenomena all due to abnormal arousal from deep sleep. Within this notional spectrum, however, there may be considerable overlap and the type of episodes may change with age.

In *sleepwalking*, the subject will typically leave the bedroom and may well engage in complex behaviours such as cooking and eating. Communication is possible at a basic level although it is usually clear to observers that the subject is not fully alert or responsive. Concerns often arise when there are attempts to leave the house or if there are any violent elements to the episodes.

Confusional arousals refer to brief episodes of disorientation in which the subject may sit up in bed and survey the environment before returning to sleep.

Night terrors are dramatic episodes often lasting several minutes in which the subject suddenly arouses from sleep typically with a loud scream and extreme agitation. Motor and autonomic indications of extreme fear are usually alarming to parents and observers.

All these arousal disorders tend to occur within an hour of sleep when non-rapid eye movement sleep is at its deepest. It is rare for events to recur through the night. If there is any recall, it is usually vague and related to a non-specific fear or urge to leave the bedroom in the case of night terrors. Particularly deep sleep following a period of deprivation or induced by drugs including alcohol may increase the likelihood of events. General stress, changes in schedule, or sleeping in a new environment are further recognized precipitants.

Non-rapid eye movement parasomnias are common in the first decade of life, affecting at least 6 per cent of children on a regular basis (Mason and Pack 2007). Persistence into adulthood occurs in around 15 per cent of these. A confident distinction between nocturnal epilepsy and parasomnias can usually be made from clinical features alone although investigations and video analysis may be required in certain cases. Particularly in adults, overnight investigations may reveal an additional sleep disorder such as sleep apnoea or periodic leg movements that may partially arouse the subject and help trigger a parasomnia.

It is rarely necessary to treat non-rapid eye movement parasomnias with medication, especially in children. However, if disturbances are frequent or likely to cause danger, short courses of benzodiazepines such as low-dose clonazepam before bed are usually helpful. In the absence of any substantial evidence, antidepressants such as paroxetine are also used to good effect, presumably by effects on sleep architecture.

32.4.3 Rapid eye movement sleep parasomnias

Rapid eye movement parasomnias include nightmares, rapid eye movement sleep paralysis, and rapid eye movement sleep behaviour disorder. Given the propensity for rapid eye movement sleep to occur late in the night, these parasomnias are typically reported between 3 am and 6 am., in contrast to the earlier occurrences of arousal disorders from non-rapid eye movement sleep.

Nightmares represent arousals from unpleasant dreams and are universal experiences. However, up to 4 per cent of adults have frequent or intrusive nightmares often in the context of psychological stress or substance abuse. Nightmares with recurring themes are a hallmark of post-traumatic stress disorder. Certain drugs such as beta-blockers can trigger nightmares as may the sudden withdrawal of anti-depressant agents that normally suppress rapid eye movement sleep.

Symptoms of *sleep paralysis* seen in around 40 per cent of narcoleptics can also occur as an isolated phenomenon, occasional with a familial pattern. As in narcolepsy, the profound paralysis is usually disturbing. Typically, prolonged episodes can be aborted by a tactile stimulus from a bed partner. If treatment is thought necessary, tricyclic antidepressants are usually helpful

An increasingly recognized rapid eye movement parasomnia occurs when abnormal motor activity intrudes into rapid eye movement sleep, reflecting a fault in the normal mechanisms that render dreaming subjects completely atonic. So-called rapid eye movement *sleep behaviour disorder* was first formally described in 1986 (Schenck *et al.* 1986). It is predominantly an affliction of middle-aged or elderly males and has an intimate relation to several neurodegenerative diseases, particularly Parkinsonism. Over 65 per cent of subjects free of any movement disorder during wakefulness at the onset of symptoms will develop Parkinson's disease within 10 years of follow-up (Schenk *et al.* 2003). The nocturnal episodes are brief and generally explosive, usually involving the upper limbs. There is often an apparently aggressive intent but injuries to bed partners are incidental and violence is rarely directed. In mild cases, episodes are confined to vocalization or swearing with little observable movement. If awoken during an event, dream recall is the norm although most remain oblivious to their behaviours if their sleep remains continuous. Intriguingly, pleasant dreams or those with a sexual content are very rare whereas reports of being chased by aggressors or attacked by animals are typical themes (Olson *et al.* 2000).

It has been proposed that when rapid eye movement sleep behaviour disorder is part of an established neurodegenerative syndrome it may indicate underlying synuclein pathology given its predominance in Parkinsonian syndromes. However, if overnight studies are performed, rapid eye movement sleep without atonia and even full-blown rapid eye movement sleep behaviour disorder are also common in other pathologies including tauopathies such as progressive supranuclear palsy (Arnulf *et al.* 2005b).

It is often necessary to treat rapid eye movement sleep behaviour disorder on a long-term basis to prevent injury either to the sufferer or the bed partner. Clonazepam in a dose range 0.25–2 mg is usually effective and melatonin has been used as a second-line agent. If there are suspicions of an additional breathing-related sleep disorder, overnight investigations are warranted as clonazepam may worsen obstructive sleep apnoea, for example.

32.4.4 Other parasomnias

There exist a number of rare parasomnias that can arise from any sleep stage. *Nocturnal catathrenia* is characterized by high-pitched monotonous groans that occur on expiration after deep inspirations. The events last around 10 s and invariably disturb bed partners, recurring in clusters through the night, even in rapid eye movement sleep (Vetrugno *et al.* 2001). No consistently effective treatments have been reported.

Bruxism refers to teeth grinding and is common. In sleep, it may affect up to 8 per cent of the population to varying degrees and be an important cause of sleep disruption especially if associated with another sleep disorder (Lavigne and Montplaisir 1994). It occurs at any age and in any sleep stage. In neurological patients, it may trigger facial pain and precipitate migraine attacks or other chronic headache syndromes. If not suspected clinically, it is often picked up as 1-Hz interferences on overnight EEG recordings. Dental occlusal appliances are usually helpful.

Nocturnal enuresis may be considered as a parasomnia in some. It can occur during any sleep stage and is commoner in children with other parasomnias. It can be difficult to treat although behavioural techniques including scheduled awakenings are usually effective. Drugs such as tricyclics or desmopressin may be needed in resistant cases.

32.4.5 Periodic leg movements of sleep

Periodic leg movements of sleep are characterized by stereotyped leg movements occurring in clusters every 30 s or so throughout sleep, especially in the light non-rapid eye movement stages. The movements themselves tend to be fairly slow, evolving over 1–5 s and typically involving both legs although one or the other may predominate. An episode tends to start with great toe extension and spreads to include ankle dorsiflexion, followed by knee and hip flexion in severe cases. It is relatively rare for subjects to be aware of the leg movements although bed partners may complain. The phenomenon increases dramatically with age and is strongly associated with restless legs syndrome (Section 40.11.1).

If periodic leg movements of sleep are demonstrated after overnight investigation, it can be difficult to gauge their clinical significance, especially if there are no associated EEG arousals (Silber 2001). Further complications may arise if there are other reasons for fragmented sleep such as obstructive sleep apnoea, in which case periodic leg movements of sleep may be triggered as a secondary epiphenomenon.

Treatments for restless legs syndrome also ameliorate periodic leg movements of sleep. Dopamine agonists are usually effective although it is difficult to predict in advance whether any response will be clinically meaningful.

32.5 Circadian rhythm disorders

If both quality and quantity of sleep are normal over 24 h but a subject is unable to sleep or stay awake at the desired or expected time, a circadian rhythm disorder may be diagnosed. Most commonly this problem has a clear extrinsic cause such as shift work or jet travel but in some situations there is almost certainly dysfunction of the internal clock mechanism. Behavioural or motivational factors may contribute to the generation of highly irregular sleep–wake patterns especially in younger subjects.

In mammals, the primary biological clock is sited in an area of the hypothalamus called the suprachiasmatic nucleus. The mechanism of the clock at a sub-cellular level has been extensively researched and appears very similar across all animal species, including humans (Turek 2004). In strict isolation with no external cues, the periodicity of the human clock is around 24.3 h. In real life this rhythm is entrained precisely to 24 h primarily by light cues acting on retinal cells that contain a newly discovered retinal pigment, melanopsin (Hattar *et al.* 2003). A retinal tract to the hypothalamus allows this information to influence the clock mechanism. Subjects blind from birth frequently report difficulty in adapting to a conventional sleep–wake cycle because their internal clocks run a little 'slower' than average without light entrainment. Very rarely, sighted individuals also suffer from a similar non-24 h sleep–wake disorder although the precise mechanism remains obscure.

32.5.1 Delayed sleep phase syndrome

Subjects diagnosed with delayed sleep phase syndrome can be considered as extreme 'night owls' such that they are simply unable to sleep before 2 am or later (Lu and Zee 2006). The main concern is usually the subsequent inability to wake effectively for school or work. It is important to exclude significant mood disorder as a driver for the abnormal cycle. Similarly, delayed sleep phase syndrome would not be diagnosed in those who simply prefer the solitude of night and avoid daytime interactions. Sleep diaries and wrist actigraphy can help confirm the diagnosis which mostly affects adolescents with a prevalence estimated at 1 per cent. Sufferers and their families are very commonly frustrated by this sleep disorder and the relative lack of its recognition.

Treatment is difficult and starts with a strict schedule and general sleep hygiene measures. Melatonin taken around 2 h before desired sleep onset time may help with sleep onset although long-term use of hypnotics is usually unsuccessful. Phototherapy from a light box on waking may also help 'reset' the internal clock.

32.5.2 Advanced sleep phase syndrome

This is an extremely rare disorder but of interest because a familial form has been identified and the relevant gene analysed (Toh *et al.* 2001). The point mutation occurs in a period gene, *hPer2*, such that the circadian sleep–wake period is 23.3 h. This results in subjects sleeping and waking at least 4 h earlier than expected. Other indications of disturbed circadian rhythm include melatonin secretion and core temperature.

Humans also generally show 'phase advance' with increasing age. Common experience suggests that many elderly subjects will fall asleep in the evening and wake early in the day, especially in institutions where this may be encouraged as part of a convenient regime.

32.5.3 Shift work sleep disorder

An increasing number of people are employed in jobs requiring shift work in a variety of patterns. Rotating shifts, in particular, do not allow circadian rhythms to adapt and frequently lead to difficulties either in staying awake for employment or sleeping effectively during daylight hours. Of potential concern are the secondary effects of sleep deprivation on cognitive performance in tasks demanding sustained attention or decision making, especially in occupations involving heavy industry or transportation. Most shift workers find it increasingly difficult to adapt their sleep–wake cycle as they age. Moreover, additional sleep problems such as obstructive sleep apnoea may worsen the situation.

If shift work is causing significant symptoms and cannot be avoided, treatment is a challenging area if simple sleep hygiene advice fails to help. Planned naps may be beneficial and shift patterns that rotate by delaying work time rather than advancing it are generally easier to cope with. Regular medication is controversial with concerns over dependency, especially with regard to hypnotic agents. Regular caffeine may be used and wake-promoting drugs such as modafinil have been licensed in severe shift work sleep

disorder. However, the concept of shift work sleep disorder as a problem requiring drug treatment lies uncomfortably with many physicians.

32.6 Assessing sleep disorders

In assessing a patient with a sleep disorder, the importance of a detailed history from the subject and ideally a bed partner or close family member cannot be over-emphasized. Together with a sleep diary, when appropriate, the majority of diagnoses can be made with moderate confidence on history alone. With important exceptions such as sleep apnoea, where quantification of the problem is important, it is relatively rare for investigations to add useful diagnostic information. Clearly, if a reliable history is not available in the case of a subject sleeping alone, for example, tests can be invaluable. The availability of sleep facilities varies dramatically throughout the world, often dependent on how the tests are financed. The following section is based on a United Kingdom perspective where sleep medicine is relatively under-resourced.

32.6.1 Insomnia

Where insomnia is an isolated symptom, overnight tests are rarely useful. However, in subjects complaining of extremely reduced overnight sleep, surrogate monitoring of sleep using wrist actigraphy may be useful in demonstrating paradoxical insomnia in which there is a misperception of the amount of sleep obtained.

An algorithm for assessing chronic insomnia is shown in Fig. 32.5. Chronic insomnia also associated with daytime sleepiness and frequent naps is likely to have a secondary identifiable cause.

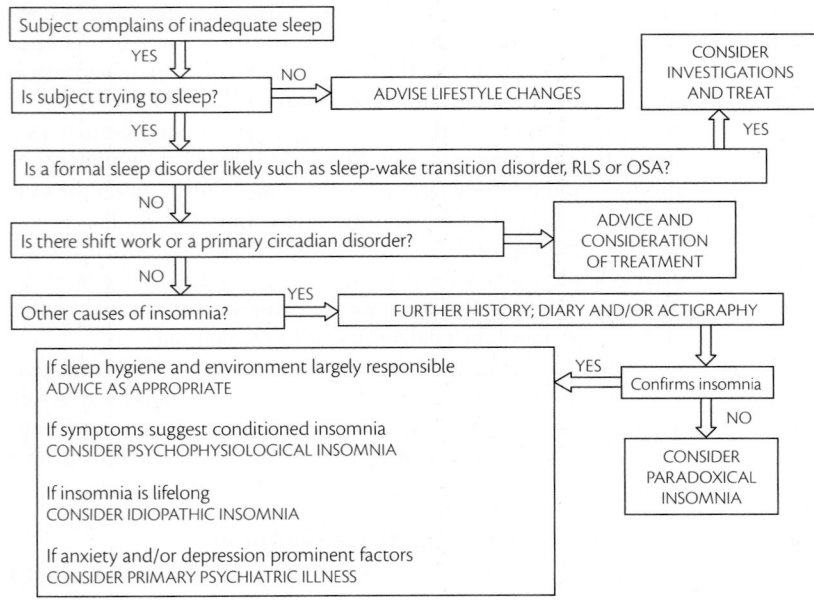

Fig. 32.5 Algorithm for the assessment of a subject with insomnia. Key: RLS – restless legs syndrome; OSA – obstructive sleep apnoea.

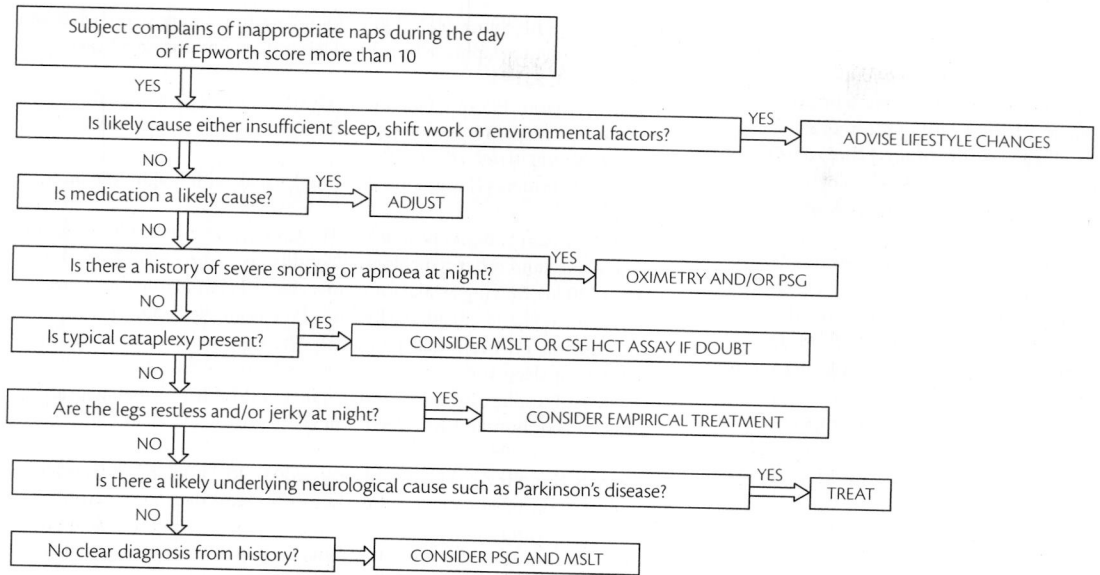

Fig. 32.6 Algorithm for the assessment of excessive daytime sleepiness. Most authorities regard a score of 10 or over on the subjective Epworth sleep scale as significant. Oximetry can usually be performed overnight in the patient's home with a finger monitor.

Key: PSG – polysomnography; MSLT – multiple sleep latency test; HCT – hypocretin, also called orexin.

32.6.2 Excessive daytime sleepiness

If excessive daytime sleepiness is the primary complaint, it is normally possible to identify an underlying cause, even if the answer is simply insufficient overnight sleep. Care should be taken in establishing that sleepiness per se is the symptom of concern and not lethargy or fatigue which are more likely to have psychological or motivational substrates.

An algorithm for assessing a sleepy subject is shown in Fig. 32.6.

32.6.3 Parasomnias

Non-rapid eye movement parasomnias are difficult to investigate and reliance on a good history is central to a confident diagnosis. Capturing an event on overnight recording is rare and investigations on asymptomatic nights are usually unremarkable. Particularly in adults, an additional sleep disorder may sometimes be precipitating a parasomnia. If so, it is advisable to perform overnight investigations to detect arousals secondary to apnoeas or leg movements, for example. Differentiating non-rapid eye movement parasomnias from nocturnal epilepsy can be difficult and video analysis, ideally of several episodes, can be crucial for diagnosis. The provision of video recorders to patients' families in order to capture events at home may be more productive and cost-effective than formal overnight recording in a hospital setting. Table 32.5 outlines some differentiating features between non-rapid eye movement parasomnias and nocturnal epilepsy.

Table 32.5 A summary of some key differences between non-rapid eye movement parasomnias and nocturnal epilepsy

	Non-rapid eye movement sleep arousal disorder	Nocturnal (frontal lobe) epilepsy
Age at onset	Early childhood	Adolescence or later
Positive family history	80–90%	<40%
Number per month	1–3	Usually >10
Number per night	1	Usually several
Semiology	Complex and non-stereotyped	Stereotyped
Duration	Minutes	Seconds
Timing	First third of night	Random
Sleep stage	Non-rapid eye movement stage III or IV	Most often stage II
Ictal EEG	High amplitude delta activity	Epileptic activity rarely seen
Triggers	Commonly identified	Rare
Natural history	Spontaneous remission	Persistent

References

American Academy of Sleep Medicine (2005). *The International Classification of Sleep Disorders*, 2nd edition. AASM, Westchester.

Arnulf I, Zeitzer JM, File J *et al*. (2005a). Kleine-Levin syndrome: a systematic review of 186 cases in the literature. *Brain*, **128**, 2763–76.

Arnulf I, Merino-Andreu M, Bloch F *et al*. (2005b). REM sleep behavior disorder and REM without atonia in patients with progressive supranuclear palsy. *Sleep*, **28**, 349–54.

Bassetti C, Aldrich MS (1997). Idiopathic hypersomnia. A series of 42 patients. *Brain*, **120**, 1423–35.

Chemelli RM, Willie JT, Sinton CM *et al*. (1999). Narcolepsy in orexin knockout mice: molecular genetics of sleep regulation. *Cell*, **98**, 437–51.

Dauvilliers Y, Arnulf I, Mignot E (2007). Narcolepsy with cataplexy. *Lancet*, **369**, 499–511.

Dodel R, Peter H, Walbert T *et al*. (2004). The socioeconomic impact of narcolepsy. *Sleep*, **27**, 1123–8.

Frank G (2006). The mystery of sleep function: current perspectives and future directions. *Rev Neurosci*, **17**, 375–92.

Harris CD (2005). Neurophysiology of sleep and wakefulness. *Respir Care Clin*, **11**, 567–86.

Hattar S, Lucas RJ, Mrosovsky N *et al*. (2003). Melanopsin and rod-cone photoreceptive systems account for all major accessory visual functions in mice. *Nature*, **424**, 75–81.

Juji T, Satake M, Honda Y *et al*. (1984). HLA antigens in Japanese patients with narcolepsy: all the patients were DR2 positive. *Tissue Antigens*, **24**, 316–9.

Kotagal S, Krahn LE, Slocumb N (2004). A putative link between childhood narcolepsy and obesity. *Sleep Med*, **5**, 147–50.

Kryger MH, Walid R, Manfreda J (2002). Diagnoses received by narcolepsy patients in the year prior to diagnosis by a sleep specialist. *Sleep*, **25**, 36–41.

Lavigne GJ, Montplaisir JY (1994) Restless legs syndrome and sleep bruxism: prevalence and association among Canadians. *Sleep*, **17**, 739–43.

Lin L, Faraco J, Li R *et al*. (1999). The sleep disorder canine narcolepsy is caused by a mutation in the hypocretin (orexin) receptor 2 gene. *Cell*, **98**, 365–76.

Lu BS, Zee PC (2006). Circadian rhythm sleep disorders. *Chest*, **130**, 1915–23.

Lu J, Sherman D, Devor M *et al*. (2006). A putative flip-flop switch for control of REM sleep. *Nature*, **441**, 589–94.

Mason TBA, Pack AI (2007). Paediatric parasomnias. *Sleep*, **30**, 141–51.

Mignot E, Nishino S (2005). Emerging therapies in narcolepsy-cataplexy. *Sleep*, **28**, 754–63.

Morin C, Hauri PJ, Espie CS *et al*. (1999). Nonpharmacologic treatment of chronic insomnia. An American Academy of Sleep Medicine review. *Sleep*, **22**, 1134–56.

Nishino S, Kanbayashi T (2005). Symptomatic narcolepsy, cataplexy and hypersomnia, and their implications in the hypothalamic hypocretin/orexin system. *Sleep Med Rev*, **9**, 269–310.

Olson EJ, Boeve BF, Silber MH (2000). Rapid eye movement sleep behaviour disorder: demographic, clinical and laboratory findings in 93 cases. *Brain*, **123**, 331–9.

Overeem S, Mignot E, van Dijk JG *et al*. (2001). Narcolepsy: clinical features, new pathophysiologic insights, and future perspectives. *J Clin Neurophysiol*, **18**, 78–105.

Overeem S, Lammers GJ, van Dijk JG (1999). Weak with laughter. *Lancet*, **354**, 838.

Ripley B, Overeem S, Fujiki N *et al*. (2001). CSF hypocretin/orexin levels in narcolepsy and other neurological conditions. *Neurology*, **57**, 2253–8.

Saper CB (2006). Staying awake for dinner: hypothalamic integration of sleep, feeding and circadian rhythms. *Prog Brain Res*, **153**, 43–52.

Saper CB, Chou TC, Scammell TE (2001). The sleep switch: hypothalamic control of sleep and wakefulness. *Trends Neurosci*, **24**, 726–31.

Schenck CH, Bundlie SR, Ettinger MG *et al*. (1986). Chronic behavioral disorders of human REM sleep: a new category of parasomnia. *Sleep*, **9**, 263–308.

Schenck CS, Bundlie SR, Mahowald MW (2003). REM behavior disorder (RBD): delayed emergence of parkinsonism and/or dementia in 65 per cent of older men initially diagnosed with idiopathic RBD, and an analysis of the minimum tonic and/or phasic electromyographic abnormalities found during REM sleep. *Sleep*, **26**, 316.

Shamsuzzaman AS, Gersh BJ, Somers SK (2003). Obstructive sleep apnea: implications for cardiac and vascular disease. *JAMA*, **290**, 1906–14.

Silber MH (2001). Controversies in sleep medicine: periodic limb movements. *Sleep Med*, **2**, 367–9.

Silber MH, Krahn LE, Olson EJ (2002). Diagnosing narcolepsy: validity and reliability of new diagnostic criteria. *Sleep Med*, **3**, 109–13.

Stepanova I, Nevsimalova S, Hanusova J (2005). Rhythmic movement disorder in sleep persisting into childhood and adulthood. *Sleep*, **28**, 851–7.

Stradling JR, Crosby JH (1991). Predictors and prevalence of obstructive sleep apnoea and snoring in middle-aged men. *Thorax*, **46**, 85–90.

Thannickal TC, Moore RY, Nienhuis R *et al*. (2000). Reduced number of hypocretin neurons in human narcolepsy. *Neuron*, **27**, 469–74.

Toh K, Jones CR, He Y *et al*. (2001). An hPer2 phosphorylation site mutation in familial advanced sleep phase syndrome. *Science*, **291**, 1040–3.

Turek FW (2004). Circadian rhythms: from the bench to the bedside and falling asleep. *Sleep*, **27**, 1600–2.

Vetrugno R, Provini F, Plazzi G *et al*. (2001). Catathrenia (nocturnal groaning): a new type of parasomnia. *Neurology*, **56**, 681–3.

Xyrem International Study Group (2005). Further evidence supporting the use of sodium oxybate for the treatment of cataplexy: a double-blind, placebo-controlled study in 228 patients. *Sleep Med*, **6**, 415–21.

Yoss RE, Daly DD (1957). Criteria for the diagnosis of the narcoleptic syndrome. *Proc Staff Meet Mayo*, **32**, 320–8.

CHAPTER 33

Coma

Martin Rossor

Contents

33.1 Introduction

33.1.1 Neural basis of consciousness

We all have an intuitive sense of what is meant by consciousness generally and what it is to be conscious ourselves. However since, in a sense, consciousness is a primary element in experience it cannot be readily defined in terms of anything else.

Although a simple definition of consciousness remains elusive, a number of components, relevant neurologically, can be considered to contribute; wakefulness, perceptual awareness and concept of self, and of experience of awareness (Zeman 2001). The neurology of sleep and awake states is the best understood and is covered in Chapter 32. Awareness of visual and auditory stimuli enrich our consciousness. However, the example of blindsight in patients and in monkeys (Cowey and Stoerig 1995), in which the subject will respond correctly to visual stimuli but without being aware, suggests that particular areas of cortex are required for the awareness of a sensory stimulus. Awareness of self, for example recognition in a mirror, which emerges in children at around 18 months and may be evident in chimpanzees, might also be considered an important component. Whether this concept of self survives in patients with dementia, in which 'consciousness' is considered to be present, is unclear.

These studies would suggest that, with improved understanding, consciousness as a neurobiological phenomenon, is likely to be fractionated into a number of discrete components. Nevertheless, at present a distinction of value to the clinical neurologist is that between the content of consciousness and the state of consciousness itself. The content of consciousness depends upon the activities of the cerebral cortex, the thalamus and their interrelationship; lesions of these structures will diminish the content of consciousness without as a rule changing the state of consciousness as such.

By contrast the ascending reticular activating system, which extends from the lower border of the pons to the ventromedial thalamus, profoundly influences the state of consciousness or arousal. The cells of origin of this system occupy a paramedian area in the brainstem, extending from the lower part of the pons to a rostral level which includes the posterior hypothalamus, the thalamic intralaminar nuclei, and the septal area (Fig. 33.1). The landmark studies of Magoun and his collaborators demonstrated that there were two broad inputs to the cerebral cortex, namely those which alter the activity of the greater part or the whole of the cerebral cortex, and those which activate very specific cortical projection areas, such as the visual or the primary sensory cortices. Destruction of the reticular system does not interfere with the action of the sensory impulses on a specific projection area, but it eliminates the tonic impulses from the hypothalamic-reticular system to the cortex as a whole. The reticular activating system is now seen as a complex system which subsumes noradrenergic and cholinergic projections to the cortex. Drugs which tend to produce unconsciousness, such as anaesthetics and hypnotics, selectively depress the ascending reticular activating system, while those which cause wakefulness have the opposite, facilitatory effect.

Further observations which support the distinction between content and state of consciousness is the observation that sleep is a separate active physiological process and not merely a feature of arousal. Moreover, sleeping and waking can occur in man even after total bilateral destruction of the cerebral hemispheres as in the vegetative state (Section 33.6.1). Thus to cause coma, defined as a state of unconsciousness in which the eyes are closed and sleep–wake cycles absent (Plum and Posner 1980), a lesion of the cerebral hemispheres

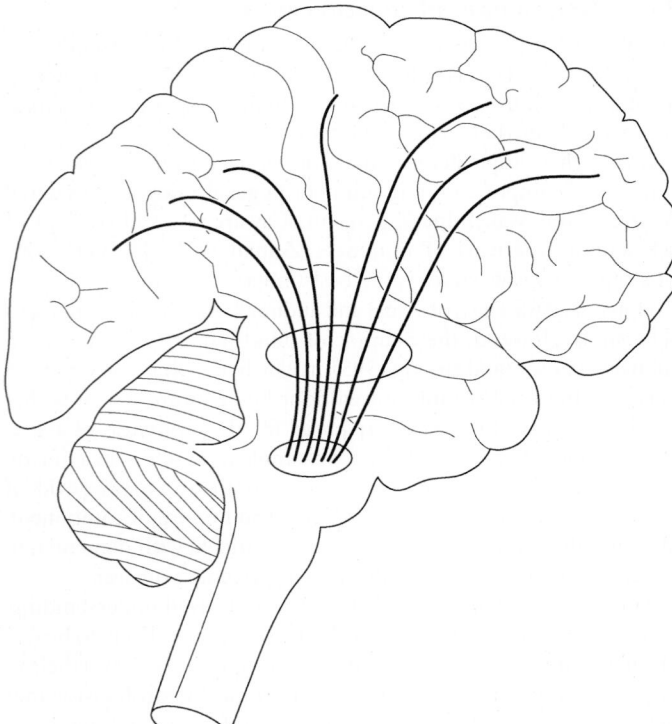

Fig. 33.1 The reticular activating system extends from the lower third of pons to thalamus. Lesions disrupting this system as well as extensive lesions of cerebral cortex will lead to coma.

needs to be extensive and bilateral. Lesions of the brainstem must be above the lower third of the pons and destroy both sides of the paramedian reticulum.

Many pathological processes can thus be responsible for stupor or coma, for example, head injury, tumour, vascular and inflammatory lesions, and most commonly toxic and metabolic states which usually lead to unconsciousness primarily through their effect upon the brainstem. In the series of Plum and Posner (1980) of 500 cases of stupor or coma, initially of unknown aetiology, 101 proved to be due to supratentorial lesions probably producing their effects by indirect action upon the brainstem, 65 to subtentorial lesions, and 326 to diffuse or metabolic brain dysfunction, while there were eight cases of psychiatric 'coma'.

33.1.2 Terminology

Between full consciousness and pathological complete unconsciousness or coma, there is a continuum of severity. In addition, there exists many states which differ not only in degree, but also in quality and especially in the nature of impairment of consciousness and by implication the content of such consciousness as remains (Giacino 1997). In an attempt to distinguish states of impaired consciousness that differ in degree, a number of terms have been introduced. An early definition of *coma* required that the patient could not be aroused by any stimulus, however vigorous and painful. *Semicoma* was then defined as complete loss of consciousness with a response only at the reflex level, while less severe degrees of impairment of consciousness were entitled severe, moderate, and mild confusion. The term semicoma is now obsolete. *Stupor* has been applied to the situation wherein the subject can only be aroused by vigorous and continuous external stimulation; as previously defined, the patient in coma elicits no response to any external stimulus or inner need (Plum and Posner 1980). More recently the term '*minimally conscious state*' has been introduced for the patient who demonstrates inconsistent but reproducible evidence of awareness of the environment or self (Giacino *et al.* 2002).

The terms clouding of consciousness, obtundation, and lethargy all imply mild disturbances of arousal, but are too imprecise to be of clinical value. The use of terms other than coma and stupor to indicate the degree of impairment of consciousness is beset with difficulties and more important is the use of coma scales, such as the Glasgow Coma Scale (Section 25.4.1), which indicates the severity of coma using a number of easily identifiable behavioural features (Teasdale and Jennett 1974; Teasdale *et al.* 1978).

Patients who survive coma, in other words in a state of unresponsiveness to external stimuli with eyes closed, do not remain in this state for more than 2–3 weeks, but rather develop a persistent unresponsive state in which sleep–wake cycles return. This is seen after severe brain injury and the implication is that brainstem function returns with sleep–wake cycles, eye opening in response to verbal stimuli and normal respiratory control. However, they show no apparent understanding or behavioural responses that would allow one to infer that they are truly conscious. There are no verbal responses and no discrete localizing motor responses. The term proposed by Teasdale and Jennett (1974) is the *vegetative state*, which has now replaced the terms 'coma vigil' and 'apallic syndrome'.

Cairns introduced the term *akinetic mutism* (Cairns *et al.* 1941), to describe the features of a patient with an epidermoid cyst of the third ventricle which resembled sleep, in being associated with general muscular relaxation, but differed from sleep in that, although the patient's eyes remained apparently alert to moving objects, strong afferent stimuli were necessary to achieve arousal. The state has been described as one of 'motionless, mindless wakefulness' and despite the patient's immobility, there are few signs of damage to descending motor pathways. Most cases are caused by bilateral lesions of the orbitomesial frontal cortex, the limbic system including the cingulate or the diencephalic reticular system (Nemeth *et al.* 1986; Marin 1990). Akinetic mutism overlaps with the minimally conscious state and the term is rarely used now (Anonymous 1995).

Also different is the so-called de-efferented state or *'locked-in' syndrome* in which the patient is fully aware of his surroundings being conscious and alert, but usually tetraplegic, aphonic, and anarthric so that he can communicate only through blinking or by vertical eye movements (Section 33.5). It is therefore most important to distinguish this state in which the patient can hear, respond, and indeed, display complex ideas despite extensive paralysis, from akinetic mutism and the chronic vegetative state. Functional neuroimaging has compared cerebral metabolism in these states. The lowest, comparable to that seen with general anaesthesia, is seen in coma and the vegetative state with preservation in the locked-in state (Laureys *et al.* 2004).

33.1.3 Approaches to differential diagnosis

General examination. A patient in coma requires most detailed and systematic examination since the clue to the cause of the unconsciousness may lie in any organ system. On arrival in an accident and emergency department, however, immediate attention will need to be given to suitable resuscitation measures. Thus the airway will need to be protected and if necessary intubation and assisted respiration instituted. Attention then needs to be given to the circulation and on establishing intravenous access blood should be withdrawn, stored, and estimated for glucose, other biochemical parameters, and possible toxicological screening. Attention is then directed towards the assessment of the patient in terms of the severity of the coma and to diagnostic evaluation. Although the patient is unable to provide a history, it is essential that this is not neglected. All possible information should be gathered, particularly about the mode of onset which can be obtained from relatives, paramedics, ambulance personnel, or bystanders. Previous medical history including epilepsy, diabetes, and drug history is of great importance and the patient's general practitioner may need to be contacted. Clues may also be obtained from the patient's clothing, or handbag.

The general examination is as important as the neurological examination and careful examination for rigors or trauma require complete exposure and 'log roll' to examine the back. Needle marks should also be specifically looked for.

A careful neurological examination will inevitably need to await stabilization of the patient. If head trauma is suspected the examination must await adequate stabilization of the neck. Documentation of the severity of coma is essential for subsequent management. The Glasgow Coma Scale (Section 25.4.1; Table 33.1) provides a widely understood scale of proven value and should be assessed at the outset. Following this, particular attention should be paid to brainstem and motor functions.

Pupils. The pupils must be assessed for size, any inequality, and the reaction to bright light. Particular care should be taken with small pupils or the pinpoint pupils of pontine haemorrhage, and to assess the light response, a magnifying glass may be necessary. An important general rule is that coma due to metabolic disturbance is associated with preservation of the pupillary light response and most metabolic encephalopathies give small pupils with preserved light reflex. Drugs such as atropine and cerebral anoxia tend to dilate the pupils and opiates will constrict them. Structural lesions are more commonly associated with pupillary asymmetry and with loss of light reflex. Midbrain tectal lesions give round, regular, medium-sized pupils which do not react to light but may show hippus; nuclear midbrain lesions also as a rule give medium-sized pupils, fixed to all stimuli, which are often irregular and unequal. A third-nerve lesion distal to the nucleus gives a fixed, dilated pupil on the side of the lesion. Tegmental lesions in the pons give bilaterally small pupils which in pontine haemorrhage may be pinpoint, although reactive. A lateral medullary lesion can give an ipsilateral Horner's syndrome, while the pupil on the side of an occluded carotid artery causing cerebral infarction is often small.

Ocular movements. The position of the eyes at rest and the presence of spontaneous eye movement should be assessed. Then the reflex responses to oculocephalic and oculovestibular manoeuvres determined. In the unconscious patient with diffuse cerebral disturbance but intact brain stem function, slow roving eye movements can be observed. A frontal lobe lesion may cause deviation of the eyes towards the side of the lesion, while conversely a lateral pontine lesion can cause conjugate deviation to the opposite side. Conjugate deviation downwards indicates a midbrain lesion and disconjugate ocular deviation will also indicate a structural brainstem lesion.

The oculocephalic or doll's head response can be tested by rotating the head from side to side and observing the position of the eyes. If the eyes move conjugately in the opposite direction to that of head movement, the response is positive and indicates an intact pons. However, cervical trauma will need to have been excluded before undertaking this test and the oculovestibular responses may be more reliable. The oculovestibular responses are tested by the instillation of ice cold water into the external auditory meatus having confirmed that there is no tympanic rupture. A normal response in a conscious patient is the development of nystagmus with the quick phase away from the stimulated side. This requires intact cerebropontine connections. In an unconscious patient with an intact brainstem, a tonic response occurs without the correcting fast phase of nystagmus with the tonic movement towards the stimulated side. With both the oculocephalic and oculovestibular responses, disconjugate movement of the eyes, for example that occurring with damage to the medial longitudinal fasciculus causing an internuclear ophthalmoplegia may be observed. Complete absence of these brainstem ocular reflexes indicates severe brainstem involvement. The corneal response is normally well preserved until late.

Respiration. *Cheyne–Stokes respiration*, in which hyperpnoea alternates with apnoea, is commonly found in comatose patients but is relatively non-specific. Rapid regular respiration is also common in comatose patients (Leigh and Shaw 1976) and is often found with pneumonia or acidosis. *Central neurogenic hyperventilation* (Plum and Swanson 1958) has been described in patients with dysfunction of the brainstem tegmentum. It is a rare syndrome comprising elevated arterial oxygen tension, pO_2, decreased arterial carbon dioxide tension, pCO_2, and respiratory alkalosis in the absence of any evidence of pulmonary disease. Most such patients have brainstem tumours, and most, but not all are in coma (Rodriguez *et al.* 1982). Sometimes the condition may complicate hepatic encephalopathy (Plum 1982). Brainstem lesions may also give apneustic breathing with a pause at full inspiration (Plum and Alvord 1964) or ataxic, irregular respiration with random deep and shallow breaths; a pattern which is seen particularly with medullary lesions (Plum and Swanson 1958).

Motor function. With the examination of the motor system, particular attention should be directed towards asymmetry of tone or movement. The plantar responses are usually extensor, but asymmetry is again important. The tendon reflexes are less useful. The motor response to painful stimuli should be assessed carefully and forms part of the Glasgow Coma Scale (Table 33.1).

The most commonly used painful stimuli are supraorbital nerve pressure and nailbed pressure. Rubbing of the sternum should be avoided as this can cause bruising and distress the relatives. Patients may localize or exhibit a variety of responses. Flexion of the upper limb with extension of the lower limb, decorticate response, and extension of the upper and lower limb, the decerebrate response, may be observed. Again asymmetry is important. In general, extensor responses in a decerebrate pattern indicate a more severe disturbance and prognosis.

Head and neck. The head should be examined carefully for evidence of injury and the skull should be palpated for depressed fractures. The ears and nose should be examined for haemorrhage and leakage of CSF. Examination of the fundi may demonstrate papilloedema or subhyaloid or retinal haemorrhages. In the presence of trauma to the head, associated trauma to the neck should be assumed until proven otherwise. If established as safe to do so, the cervical spine should be gently flexed and a positive Kernig's sign sought, which may indicate a meningitis or subarachnoid haemorrhage. Neck stiffness though may occur with raised intracranial pressure and incipient tonsillar herniation.

Investigation of coma. At presentation blood will be taken for determination of glucose, electrolytes, liver function, calcium, osmolality, and blood gases. Blood should also be stored for a subsequent drug and toxin screen if needed. Following the clinical examination, a broad distinction between a metabolic cause with preserved pupillary responses or structural cause of coma is likely to have been established. Although most patients with coma will require CT or MR imaging, and indeed all with persisting coma, clearly this is of greater urgency when a structural lesion is suspected. In the absence of focal signs, but with evidence of meningitis, a lumbar puncture may need to precede scanning as a matter of clinical urgency. In other situations lumbar puncture should be delayed until after the brain scan because of the risk of precipitating a pressure cone secondary to a cerebral mass lesion. All patients will require a chest X-ray and ECG and more detailed investigations of systemic disease will be directed by the clinical examination. The EEG in the diagnosis of coma can be of value, but it is relatively limited. It is of value in identifying the occasional patient with sub-clinical status epilepticus and clearly of value in assessing the patient who has been admitted following an unsuspected seizure. Fast activity is commonly found with drug overdose and slow wave abnormalities with metabolic and anoxic coma. An isoelectric EEG may occur with drug induced comas, but otherwise indicates severe cerebral damage.

33.1.4 Measurement of coma

It is most important that the level of coma should be documented from the earliest opportunity. Subsequent record will indicate whether the level of consciousness is improving; if so decisions concerning further investigation and therapy are less urgent. The Glasgow Coma Scale provides an easy clinical assessment and is now widely used (Teasdale and Jennett 1974) (Table 33.1). It is also incorporated into several intensive care unit scoring systems such as the Acute Physiology and Chronic Health Evaluation, APACHE II (Knaus *et al.* 1985). There is a broad association between the Glasgow Coma Scale and mortality across the whole range of scores validating its role as an index of the severity of brain injury (Gennarelli *et al.* 1994).

33.1.5 Management of the unconscious patient

The management of the unconscious patient will consist of treatment of the underlying cause where possible, and the maintenance of the normal physiology in terms of respiration, circulation, and nutrition whilst the patient is unconscious. Recent improvements in intensive care have been so dramatic, with improved technology, that patients in a comatose state can be maintained in a condition of adequate health.

In the short term, the unconscious patient should be nursed on their side without a pillow. Attention will clearly need to be paid to the airway requiring, as a minimum, an oral airway, although usually patients will require intubation and if prolonged coma, tracheostomy. The unconscious patient will have retention or incontinence of urine and will require catheterization. Intravenous fluid is necessary and if coma persists, then adequate nutrition is required. Disturbances of electrolytes, particularly sodium, are common in the intensive care situation and need scrupulous monitoring.

33.1.6 Prognosis in coma

In general, coma carries a serious prognosis. Nevertheless, this is dependent to a large extent on the underlying cause. Coma due to depressant drugs carries an excellent prognosis provided that resuscitative and supportive measures are available and no anoxia has

Table 33.1 The Glasgow Coma Scale

	Score
Eye-opening	
Nil	1
To pain	2
To voice	3
Spontaneously (with blinking)	4
Motor response	
Nil	1
Extension	2
Flexion	3
Withdrawal	4
Localizing	5
Voluntary	6
Verbal response	
Nil	1
Groans	2
Words (expletive)	3
Disorientated	4
Orientated	5
Total	15

Derived from Teasdale and Jennett (1974).

been sustained. Thus coma from depressant drugs even persisting for days can be associated with a full and complete recovery. In general, metabolic causes, apart from anoxia, carry a better prognosis than structural lesions and head injury. A number of factors contribute to predicting the outcome and considerable work has recently been directed towards identifying these. Many studies have addressed the outcome of coma and show broad agreement. Early studies tended to differ in the outcome measures used, but following the lead of Jennett and Bond (1975), five simple outcome measures are widely used:

- good recovery indicates patients who return to their normal life;

- moderate disability indicates patients who achieve independence in daily living, but do not resume their previous level of function;

- severe disability refers to patients who regain some cognitive functions but are dependent on others for daily support;

- in the vegetative state (Section 33.6.1) patients awaken but give no sign of cognitive awareness; and

- with no recovery patients remain in coma until death.

In general, the length of coma is of poor prognostic significance, as is increasing age. Whether the underlying cause is metabolic or structural, the brainstem reflexes early in the coma are an important predictor of outcome. For example, in a series of head injuries, if the pupils were fixed at 24 h, 91 per cent of 1000 patients died and only 4 per cent made satisfactory recovery (Jennett et al. 1979). A similar pattern is found with non-traumatic coma. Thus only 2 per cent of patients made a moderate or good recovery with absent brainstem reflexes at 1 day, and there were no recoveries other than to a vegetative state, or severe disability if reflexes were absent at 3 days (Levy et al. 1981).

Neurologists are often asked to give a prognosis on patients arriving in intensive care units following cardiopulmonary arrest. In general the absence of pupillary light and corneal reflexes 6 hours after the onset of coma is very unlikely to be associated with survival (Snyder et al. 1981). The persisting vegetative state carries uniformly poor prognosis, although a partial return to the minimally conscious state can occur. The prognosis for anoxic vegetative states is worse than for cerebral trauma. The probability of a return of awareness is less then 1 per cent after 3 months in anoxic cases and 12 months in traumatic cases (The Multisociety Task force on PVS 1994). However, rare cases of late recovery have been reported (Rosenberg et al. 1977; Shuttleworth 1983).

33.2 Metabolic causes of coma

The diagnosis of metabolic coma is one not to be missed due to the potential reversibility provided that the metabolic disturbances are rectified early. There are no specific pointers to a metabolic cause for coma although in general the pupils are reactive albeit small, and seizure activity often complicates the picture. The metabolic causes of coma are legion (Kunze 2002).

33.2.1 Renal coma

Uremic coma may occur in acute or chronic renal failure, but with the more widespread use of haemodialysis and peritoneal dialysis, unsuspected renal failure as a cause of coma is encountered only rarely (Brouns and DeDeyn 2004). It more commonly enters the

differential diagnosis where there are a number of potential causes of coma. The metabolic changes produced in renal failure are complex, and a raised blood urea cannot alone be responsible for the loss of consciousness, although the blood urea concentration does provide a useful index of severity of the renal failure. There is usually metabolic acidosis, accompanied by complex electrolyte disturbances. Water intoxication, due to fluid retention, with a serum osmolality of less than 260 mOsm/l, is a factor in some cases. As is general with cerebral insults, the speed of metabolic disturbance may dictate the severity of the disturbance of consciousness. Although the clinical features are similar with both acute and chronic renal failure, acute metabolic disturbance may result in an impairment of consciousness that would not occur with the more gradual change encountered in chronic renal failure. Headache, vomiting, dyspnoea, mental confusion, drowsiness or restlessness, and insomnia are early symptoms and later muscular twitchings, asterixis, myoclonus, and generalized convulsions are likely to precede the coma. The raised blood urea or creatinine establishes the diagnosis but differentiation from hypertensive encephalopathy, often accompanied by signs of renal impairment, can be difficult (Bolton and Young 1990).

Patients undergoing dialysis may develop iatrogenic causes of impaired consciousness. The *dialysis disequilibrium syndrome*, which is more common in children and during rapid changes in blood solutes, is a temporary self-limiting disorder, but which can be fatal. Animal studies indicate that rapid osmotic shift of water into the brain is the main problem which is correctable by careful control. The dialysis disequilibrium syndrome is usually accompanied by headache, nausea, vomiting and restlessness before drowsiness, and marked somnolence. It can occur during or just after dialysis treatment, but usually resolves in 1 or 2 days.

Whereas the dialysis disequilibrium syndrome is self-limiting, *dialysis encephalopathy* results in progressive dysarthria, mental changes and progression to seizures, myoclonus, asterixis, and focal neurological signs. Terminally, there may be coma. Characteristically, the EEG reveals paroxysmal bursts of irregular generalized spike and wave activity. The dialysis encepholopathy or dialysis dementia syndrome has been attributed to the neurotoxic effects of aluminium, arising both from the use of aluminium-containing antacids and a high aluminium content in the water. Dialysis dementia reached its peak prevalence in the mid 1970s before preventive action was taken. However, rare cases are still occasionally seen.

Patients undergoing renal transplantation may experience confusion and seizures as part of an acute graft rejection—a *rejection encephalopathy syndrome* (Section 5.6). This is usually seen within the first 3 months of transplantation and responds to increased immunosuppression.

33.2.2 Hepatic coma

Acute liver failure can lead to rapid onset of coma in a jaundiced patient and the diagnosis is usually obvious. By contrast, patients with chronic cirrhosis, who are not jaundiced but have portosystemic shunting or portocaval surgical shunts, may experience decompensation with episodes of confusion leading to coma. Portosystemic encephalopathy can also rarely occur in patients without liver disease. Such patients are also particularly vulnerable to precipitation of hepatic coma by gastrointestinal haemorrhage, infection, the use of certain diuretics, sedatives and analgesics, general anaesthesia, and the ingestion of high protein food or ammonium compounds. Hepatic coma is usually of subacute onset, although it can be sudden, with an initial confusional state often accompanied by bilateral asterixis or flapping tremor. Asterixis, considered a negative myoclonus jerk, results in sudden loss of a maintained posture. It can be readily elicited by asking the subject to maintain extension at the wrist; asterixis will be seen as a sudden loss of posture (Fig. 33.2). As coma supervenes, there is often decerebrate and/or decorticate posturing with extensor plantar responses. The diagnosis rests upon the presence of physical signs of liver disease, including hepatic foetor and biochemical evidence of disturbed liver function. The EEG is characteristically abnormal with paroxysms of bilaterally synchronous slow waves in the delta range or with occasional triphasic waves.

The disturbance of consciousness has traditionally been considered to be due to raised ammonia, and in acute hepatic failure ammonia concentrations are markedly increased. By contrast, ammonia levels are more variable in the chronic hepatic encephalopathy with decompensation. Increased GABA transmission may play a role as patients are particularly susceptible to benzodiazepines. The management of severe acute hepatic encephalopathy is aimed at control of cerebral oedema guided by intracranial pressure monitoring, until a liver transplant is available. Management of the decompensation in chronic encephalopathy requires the identification of precipitating factors such as drugs and infection. Clearance of protein from the gut by dietary control, lactulose, and antibiotics such as neomycin are widely used, although there is a paucity of data on efficacy (Shawcross and Jalan 2005).

33.2.3 Pancreatic encephalopathy

The biochemical cause of pancreatic encephalopathy is unclear. It can rarely occur with acute pancreatitis, but is more commonly seen as episodic stupor or coma in chronic relapsing pancreatitis. The impairment of consciousness usually begins between the second and fifth day, is characterized by an acute confusional state with hallucinations, followed by focal or generalized convulsions, and there may be corticospinal-tract dysfunction (Menza and Murray 1989).

33.2.4 Salt and water imbalance

Since sodium is the principal serum cation, hypo-osmolality is largely equivalent to hyponatraemia (Kumar and Berl 1998). It is an important cause of confusional states and, if severe, stupor and coma. Low sodium is usually due to impaired water excretion attributable to renal insufficiency, chronic sodium depletion which can often be diuretic induced, or to the syndrome of inappropriate antidiuretic hormone (ADH) secretion (Adrogué and Madias 2000). This is often seen in neurological practice for example with infections, head injury, a variety of drugs, and following surgery, although the latter may be associated with the syndrome of cerebral salt wasting in which there is polyuria and natriuresis (Singh *et al.* 2002). Lesions within the region of the hypothalamus are also an important neurological cause. Very rarely some psychiatrically disturbed patients may be compulsive water drinkers giving rise to hyponatraemia.

The speed of change in sodium concentration is critical as to whether neurological features emerge in *hyponatraemia*. Certainly the diagnosis should be considered with sodium concentrations below 120 mmol/l but even concentrations up to 128 mmo/l may be implicated if there has been a sudden fall. Conversely, patients with chronic sodium depletion may survive concentrations down to 110 mmol/l with little to find neurologically. Hyponatraemia is associated with seizures and can be associated with focal neurological deficits. Although hyponatraemia with neurological disturbance represents a medical emergency, sudden correction can give rise to central pontine myelinolysis (Section 37.6). Thus correction should be done by water restriction in the first instance and hypertonic saline used only when severe and symptomatic aiming not to exceed an increase of 8 mmol/l/24 h (Adrogue and Madias 2000).

Hypernatraemia, defined as sodium above 145 MEq/l, may also cause confusional states and lethargy but less often coma. It may be seen in children with severe diarrhoea and is common in patients with coma who are unable to drink normally or when parenteral fluids have been inadequate. It is seen in adults with diabetes insipidus, although the increased thirst protects such patients from the severe neurological consequences. Very rarely, it may be seen in neurological disease with diencephalic involvement when there is

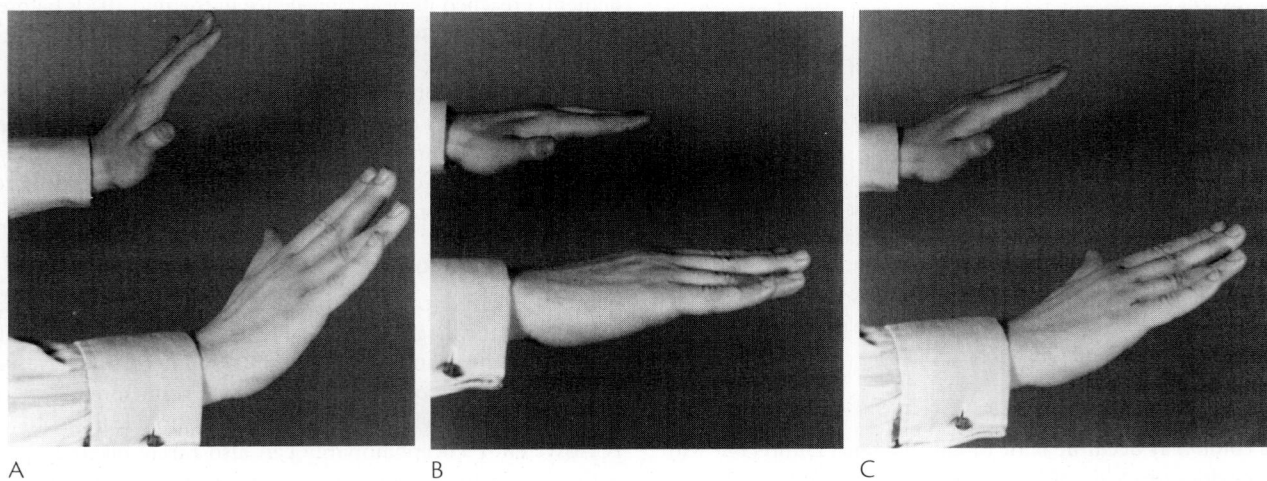

A   B   C

Fig. 33.2 Asterixis is best elicited by asking the patient to extend the wrists with the arms outstretched (A). Asterixis is seen as a sudden loss of posture with momentary flexion (B) and (C).

an impaired thirst response. This has been reported with tumours and with infections (Maxwell *et al.* 1987).

33.2.5 Inborn errors of metabolism

A vast array of hereditary disorders of metabolism can result in episodic stupor or coma as well as occurring as a terminal feature. The majority are very rare and usually present in childhood (Table 10.2). Associated biochemical disturbance can provide clues for example, inherited disorders of the urea cycle resulting in hyperammonaemia in children. Porphyria can present with disturbance of consciousness in adults particularly after the administration of a variety of drugs. Mitochondrial disorders can also present for the first time in adults as an encephalopathy (Howard *et al.* 1995).

33.2.6 Disturbed glucose metabolism

Hyperglycaemia. Diabetic ketoacidosis is usually associated with Type 1 diabetes. The blood sugar is usually above 40 mmol/l together with ketonuria. There is often a secondary lactic acidosis which needs to be distinguished from the lactic acidosis that can follow severe anoxia, methyl alcohol ingestion or administration of paraldehyde. Patients are dehydrated and there is rapid shallow respiration and occasionally acetone can be detected on the breath. Although patients may be drowsy, coma is rare. By contrast, the hyperglycaemic non-ketotic state is associated with Type II diabetes and is more commonly seen in the elderly. Coma is more common than with ketoacidosis, as are focal signs and seizures, and the onset is gradual over days. There is profound cellular dehydration and patients are at risk of developing cerebral venous thrombosis which may contribute to the disturbance of consciousness. In addition to diabetes, hyperglycaemic non-ketotic coma may be induced by drugs, acute pancreatitis, burns, and heat stroke.

Hypoglycaemia. By contrast to diabetic coma, hypoglycaemic coma is of much more rapid onset. Symptoms appear with blood sugars of less than 2.5 mmol/l; initially autonomic disturbance with sweating and pallor, and then symptoms of inattention and irritability before progressing to stupor, coma, and frequent seizures. Some patients may present with a focal onset for example with a hemiparesis (Malouf and Brust 1985). Plantar responses are frequently found to be extensor. Patients may be hypothermic. The diagnosis is relatively straightforward if the patient is known to be taking insulin. Spontaneous hypoglycaemia may be more difficult to diagnose and patients with insulinomas are usually diagnosed late. There may be a long history of intermittent symptoms and the history often suggests that these occur in relation to fasting or exercise. Hypoglycaemia may also be associated with hepatic disease, alcohol, cerebral malaria, hypopituitarism, and Addison's disease. Treatment is with glucose and glucagon (Kearney and Dang 2007) which should be administered together with thiamine, if there is any doubt as to the diagnosis, to avoid exacerbating Wernicke encephalopathy. Unless treated promptly, hypoglycaemia results in irreversible brain damage. Cerebellar Purkinje cells, cerebral cortex, and particularly the hippocampus and basal ganglia are particularly affected in a pattern similar to anoxia. Dementia and a cerebellar ataxia are the clinical sequelae of inadequately treated hypoglycaemia.

33.2.7 Endocrine causes of coma

Pituitary failure. This is a rare cause of coma and is the result of a number of factors which include hypoglycaemia, hypotension, hypothermia, and impaired adrenocortical function. There is usually a history of fatigue, occasionally depression, and loss of libido, but because of the slow onset of symptoms these are often missed. Patients with hypopituitarism are very sensitive to infections and to sedative drugs which are often the precipitation of impaired consciousness. Acute onset of hypopituitarism occurs with haemorrhagic infarction in pre-existing tumours, 'pituitary apoplexy'; patients present with impaired consciousness, meningism, and opthalmoplegia.

Hypothyroidism commonly causes mental symptoms together with headaches, poor concentration and apathy; this is frequently diagnosed as depression. With progression there is increasing somnolence and, as with hypopituitarism, patients become sensitive to drugs and infections. These and cold weather, particularly in the elderly, may precipitate myxoedemic coma. Myxoedema coma has a high mortality; it is associated with hypoglycaemia and hyponatraemia and there may be additional adrenal insufficiency (Rodriguez *et al.* 2004). Diagnosis may be missed if not thought of and hypothermia requires a low reading thermometer to detect.

Hyperthyroidism. This can cause mild mental symptoms usually of anxiety, restlessness, and reduced attention. Very rarely, patients may develop a 'thyroid storm' with agitated delirium which can progress to coma. Such patients may have an associated bulbar paralysis (Newcomer *et al.* 1983). By contrast some patients, particularly the elderly, may develop an apathetic form of thyrotoxicosis with depression leading to apathy, confusion, and coma without any signs of hypermetabolism (Thomas *et al.* 1970). As with myxoedema coma, the diagnosis is easily missed if not considered.

Adrenocortical failure. Mental changes are common in Addison's disease and secondary hypoadrenalism. Undiagnosed Addison's disease is frequently associated with behavioural changes and fatigue. Intercurrent infection or trauma may precipitate coma and associated metabolic disturbances of hypotension, hypoglycaemia, and dehydration. Tendon reflexes are often absent and raised intracranial pressure with papilloedema can occur (Jefferson 1956). Acute adrenal failure due to meningococcal septicaemia, the Friedrichsen–Waterhouse syndrome, is now rarely seen as a cause of sudden coma in infants. Acute adrenal failure due to HIV infection however, can occur.

33.2.8 Disturbed calcium and magnesium metabolism

Hypercalcaemia. This is an important cause of mental confusion and apathy, often with headache. If severe, this can progress to stupor and even coma (Gatewood *et al.* 1975). Causes of hypercalcaemia are numerous and are probably seen most commonly with metastatic bone disease including multiple myeloma (Bushinsky and Monk 1998).

Hypocalcaemia. This primarily affects the peripheral nervous system with tetany and sensory disturbance. Only very rarely will it cause disturbance of consciousness although it can be associated with intracranial hypertension and papilloedema (Grant 1953).

Hypomagnesemia. It occurs with inadequate intake and is usually seen in patients with prolonged parenteral feeding. It is often overshadowed by other metabolic disturbances including hypocalcaemia but can give rise to a similar clinical picture.

Hypermagnesemia. This can occur due to renal insufficiency although rarely at a clinically significant level. Overzealous replacement of magnesium and its use clinically, for example, in the treatment of eclampsia, can give rise to magnesium intoxication with major central nervous system depression.

33.2.9 Drugs

Poisoning, drug abuse, and alcohol intoxication are the major causes of coma, accounting for up to 30 per cent of those presenting through Accident and Emergency departments. Despite its frequency, the majority, perhaps as high as 80 per cent, require only simple observation in their management.

Self-poisoning

The most commonly encountered drugs in suicide attempts are benzodiazepines, paracetamol, and antidepressants. Narcotic overdoses can occur, particularly commonly with purchase of street supplies of heroin of variable purity. Characteristic pinpoint pupils and shallow respirations provide the diagnosis and may be aided by obvious needle marks. Some addicts, however, use mydriatics to mask the pupillary changes. The coma is easily reversible with naloxone, but in view of its short half-life, this may need to be continued as an infusion. Solvent abuse and glue sniffing should be considered in the undiagnosed patient with coma. Drugs may also result in disturbed consciousness due to secondary metabolic derangement, for example, the acidosis with ethylene glycol and with carbon monoxide poisoning (Balzan *et al.* 1996). Blood for subsequent toxicology is important although it is rarely available to influence acute management; in view of the short half-life of many drugs, urine samples should also be collected.

Psychiatric drugs

Many of the drugs used in psychiatry may cause coma in toxic doses and occasionally as an idiosynchratic response. Tricyclic/antidepressants are often taken in suicide attempts. Cardiac dysrhythmias are common and may require temporary pacing. Neuroleptics can be associated with the neuroleptic malignant syndrome (Section 5.5.2) and in modest doses can result in drowsiness and even coma in patients with dementia with Lewy bodies (Section 34.6.3). The selective serotonin re-uptake inhibitors can be associated with the serotonin syndrome especially in the setting of polypharmacy (Boyer and Shannon 2005).

Alcohol intoxication

Coma due to alcohol intoxication is often apparent from the history, flushed face, rapid pulse, and low blood pressure. This can be aided by the smell of alcohol on the breath, but this should never be assumed as the cause of coma. Due to the widespread social use of alcohol, this is a common finding in patients with coincidental illness. Moreover, patients who are intoxicated are at increased risk of hypothermia and of head injury, either of which can be the primary cause of coma. At low plasma concentrations of alcohol, mental changes are common and at higher levels, coma ensues. In non-alcoholics, levels of 450 mg/dl may prove fatal.

33.3 Cerebrovascular causes of coma

Cerebrovascular disease is a frequent cause of coma. A common mechanism is the impairment of perfusion of the reticular activating system as occurs with hypotension, brainstem herniation due to parenchymal haemorrhage or swelling from infarct, or more rarely, extensive brainstem infarction.

33.3.1 Subarachnoid haemorrhage

Loss of consciousness is common with subarachnoid haemorrhage and only about half of patients recover from the initial effects of the haemorrhage. Contributing causes to coma are the acute rise in intracranial pressure and later, the emergence of vasospasm. The diagnosis is made on the history and in patients in whom there is loss of consciousness, blood is normally apparent on neuroimaging (Section 35.16). Secondary metabolic effects may also contribute and hyponatraemia is a common finding. This may occur with renal salt loss and volume contraction, so-called 'cerebral salt wasting'. It is proposed that this is due to increased release of brain natriuretic peptide, causing increase in urine volume and sodium excretion (Berendes *et al.* 1997).

33.3.2 Parenchymal haemorrhage

A large parenchymal haemorrhage may cause a rapid decline in consciousness either from rupture into the ventricles or from subsequent herniation and brainstem compression. Pontine haemorrhages are commonly associated with rapid onset of coma which carries a poor prognosis (Balci *et al.* 2005). Of particular importance is the diagnosis of a cerebellar haemorrhage or of a cerebellar infarct with subsequent oedema and direct brainstem compression. Diagnosis and early decompression can be lifesaving.

33.3.3 Hypotension

The critical blood flow in man required to maintain effective cerebral activity is about 20 ml/100g/min and any fall below this rapidly leads to cerebral insufficiency. The causes are legion and include syncope in younger patients and commonly cardiac disease in older patients.

33.3.4 Hypertensive encephalopathy

Hypertensive encephalopathy is now less common with better control of blood pressure. Patients can present with impaired consciousness and a grossly raised blood pressure with papilloedema. Posterior cortical involvement with visual deficits preceding coma and posterior cortical changes on MRI are characteristic. Brainstem or cerebellar swelling with coma in the absence of cerebral involvement may be seen (Chang and Keane 1999). The blood pressure rise needed to precipitate encephalopathy is variable and depends on the rapidity. In general a diastolic of at least 130 mmHg is needed but in a previously normotensive individual smaller rises may be sufficient especially in pregnant women. In chronic hypertension a diastolic of 150 mmHg may be sustained without symptoms.

33.3.5 Eclampsia

Eclampsia presents in the second half of pregnancy and represents a failure of autoregulation with raised blood pressure. Since the upper limit of autoregulation is directly related to the individuals normal blood pressure, the level at which eclampsia occurs may be very variable; conventionally a diastolic of 90 mmHg or a rise of 15 mmHg from previous level is required for the diagnosis. Neuropathologically there are ring haemorrhages around occluded small vessels with fibrinoid deposits. The clinical characteristics are those of seizures, cortical blindness, and coma. Management is the control of convulsions and raised blood pressure. Parental magnesium is commonly employed but this may give rise to hypermagnesemia.

33.3.6 Cerebral venous thrombosis

Thrombosis of the deep venous system including the internal cerebral vein tends to be more common in women and can be associated with a rapid decline in level of consciousness following headache, nausea, and vomiting (Crawford *et al.* 1995) (Section 35.15).

33.4 Other causes of coma

33.4.1 Hypoxia

Cerebral hypoxia or lack of oxygen to the brain has devastating effects. It occurs in a whole variety of settings including high altitude, suffocation, drowning, carbon monoxide poisoning, cardiac arrest, and pulmonary disease, with or without ventilatory failure. Traditionally hypoxia or anoxia has been classified as anoxic anoxia where there is low arterial oxygen, for example with high altitude or pulmonary disease; anaemic anoxia due to decreased haemoglobin, and ischemic anoxia due to impaired cerebral blood flow. Since hypoxia and ischemia frequently overlap, and the terms are sometimes used interchangeably in clinical practice, the term hypoxic-ischemic encephalopathy (Section 5.4.2) is often used.

The cerebral dysfunction arising from hypoxia depends upon its rate of development. Acute hypoxia is more devastating and patients may maintain function with chronically lowered oxygen tensions and relatively modest symptoms, which if they were to occur acutely would result in neurological dysfunction. Arterial oxygen tensions of 5.0–6.5 kPa are associated with a decline in cognitive function and at pressures below 4.0 kPa consciousness is usually lost. Anoxia following cardiac and respiratory arrest are common causes of coma in patients in intensive care units. Recovery from cardiorespiratory arrest with associated coma is generally poor. In a study of 210 patients followed up at 1 year, only 13 per cent had regained independent function (Levy *et al.* 1981). Individuals without a pupillary light reflex on day 1 never regained independence (Levy *et al.* 1985). The absence of pupillary light reflexes were confirmed as an indicator of poor prognosis together with the absence of motor responses to pain on day 3 in a systematic review of 33 outcome studies (Zandbergen *et al.* 1998). Rarely a delayed post anoxic encephalopathy may occur following a period of recovery. The patient deteriorates with progressive impairment of consciousness and motor disturbance after a period of some 5–10 days. This is associated with widespread leukoencepalopathy.

Accidental and intentional carbon monoxide poisoning is an important cause of hypoxic coma (Section 5.4.1). Neither the early clinical features nor the concentration of carboxyhaemoglobin predict outcome reliably. As with postae cardiac arrest late delayed sequelae are seen. Cerebral malaria is an important cause of hypoxic ischemic encephalopathy in Africa (Section 42.6.1).

33.4.2 Hypercapnia

Ventilatory failure and pulmonary disease lead to hypercapnia in association with the hypoxia, together with respiratory acidosis. Individuals may also develop hypercapnia with scuba diving. Clinical features include asterixis, myoclonus, conjunctival injection, and papilloedema. There is increasing impairment of consciousness and ultimately coma with symptoms generally appearing with $PaCO_2$ greater than 10 kPa. Patients with chronic bronchitis often have markedly impaired blood gases but lack dyspnoea. Such patients are at risk because their resistance to hypercapnia means that they require a hypoxic drive to maintain ventilation. The injudicious administration of oxygen can result in marked deterioration with resulting coma.

33.4.3 Disturbance of thermoregulation

Hyperthermia

Loss of thermoregulation giving rise to hyperthermia or heat stroke (Section 5.10.1) is a clinical emergency. This may be encountered after prolonged exertion in a hot environment for example, in endurance sports such as cycle racing and marathon running. The initial rise in body temperature with profuse sweating is followed by hyperpyrexia, an abrupt cessation of sweating, and then the rapid onset of neurological dysfunction. Ataxia is an early feature usually seen at temperatures of 41°C and above, followed by increasing obtundation, coma, convulsion, and death. This may be exacerbated by certain drugs and can be seen with 'ecstasy' abuse, in individuals with prolonged dancing and loss of thirst reaction. Tetanus, pontine haemorrhage and lesions in the floor of the 3rd ventricle may also give rise to hyperpyrexia as may neuroleptics in the neuroleptic malignant syndrome and malignant hyperpyrexia with anaesthetic. Survivors of heat stroke may be left with permanent neurological sequelae of paraperesis or cerebellar ataxia (Yaqub and Al-Deeb 1998).

Hypothermia

Hypothermia may also give rise to impairment of consciousness. This may be seen with hypopituitarism and hypothyroidism, and also induced by certain drugs, commonly seen previously with chlorpromazine. However, there is always a risk of hypothermia in the elderly with inadequately heated rooms and this may be exacerbated by immobility, for example in those with Parkinson's disease. It will also occur in patients exposed to low temperature environments or cold water immersion. In these situations the history is clear. There is usually generalized rigidity and muscle fasciculation but shivering usually stops below 32°C. Hypoxia and hypercapnea are common. Gradual warming is necessary and may require peritoneal dialysis with warm fluids. The diagnosis is important to establish and requires a low reading rectal thermometer.

Spontaneous periodic hypothermia, or Shapiro's syndrome, is a rare syndrome of recurrent hypothermia often associated with agenesis of the corpus callosum, polydipsia, polyuria, and hyponatraemia (Mooradian *et al.* 1984). Some cases of callosal agenesis have also been described with periodic hyperthermia, 'reverse Shapiro's syndrome'.

33.4.4 Raised intracranial pressure

Raised intracranial pressure *per se*, as occurs with benign intracranial hypertension does not cause coma. Rather mass effects such as tumours, abscesses, haemorrhage, subdural and extradural haematomata will cause impairment of consciousness due to distortion of the reticular activating system (Section 26.5). This is the classical feature of brainstem herniation. The clinical development and pattern depends on a number of features such as normal variation in the tentorial aperture, site of lesion, and the speed of development and rapid deterioration may relate to vessel occlusion. Lesions located deeply, laterally, or in the temporal lobes will have a greater tendency to herniation and loss of consciousness than those located at a distance such as the frontal and occipital lobes (Andrews *et al.* 1988). The speed of development of a mass is also important. Slowly growing tumours may achieve a substantial size and distortion of cerebral structure without impairment of consciousness by contrast to small rapidly expanding lesions.

Two patterns of downward herniation of the brain are commonly recognized although typically these may coincide (Section 26.5). *Central herniation* involves downward displacement of the upper brainstem which can be demonstrated on MRI (Wijdicks and Miller 1997). By contrast with *uncal herniation* the medial temporal lobe herniates through the tentorium. In the former, small pupils are followed by midpoint pupils and irregular respiration gives way to hyperventilation as coma deepens. With uncal herniation there is classically a unilateral dilated pupil due to compression of the third nerve and asymmetric motor signs. As coma deepens the opposite pupil loses the light reflex and may constrict briefly before enlarging (Ropper 1990). Less common is upward transtentorial herniation with lesions in the posterior fossa, and often accompanied by tonsillar herniation through the foramen magnum (Cuneo *et al.* 1979). Cingulate herniation involves displacement of the cingulate gyrus under the falx and is often followed by central herniation.

33.4.5 **Head injury**

Trauma is the leading cause of death below the age of 45 and head injury accounts for half of all trauma deaths (Section 25.1). It is also a major cause of patients presenting with coma. A history is usually available and if not, signs of injury such as bruising of the scalp or skull fracture lead one to the diagnosis. Many head injured patients are intoxicated but alcohol does not significantly alter the coma scale in such patients (Stuke *et al.* 2007). However, as with assuming that alcohol on the breath provides a direct clue to a cause of coma, evidence of head injury need not necessarily imply that this is the cause. Other causes of impaired consciousness, for example epileptic seizure, may have resulted in a subsequent head injury. Multiple fractures may also result in cerebral fat embolism.

Impact damage can be diffuse or focal. Rotational forces of the brain cause surface cortical contusions and even lacerations. These are most obvious frontotemporally because of the irregular sphenoidal wing and orbital roof. Tearing of veins may give rise to subdural bleeding. This is less important than the damage to nerve fibres. Diffuse axonal injury is now seen as the major consequence of head injury and associated coma (Adams *et al.* 1982). Mild degrees of axonal injury also occur with concussion and brief loss of consciousness. Experimental primate studies show a direct relationship between the extent of diffuse axonal injury and the length of coma (Gennarelli *et al.* 1982). Nearly all patients who survive head trauma in a persistent vegetative state have MRI evidence of diffuse axonal injury (Kampfl *et al.* 1998). The typical features of diffuse axonal injury on CT are small punctate haemorrhages in the white matter, especially corpus callosum, and the basal ganglia. MRI is more sensitive, albeit less practical, and new sequences such as diffusion tensor imaging may add further precision (Huisman *et al.* 2004).

Secondary damage can occur from parenchymal haemorrhage, brain swelling due to oedema and vascular dilatation, all of which will lead to raised intracranial pressure. This will reduce perfusion pressure which can be accentuated by systemic hypoxia and blood loss. The specific instances of subdural and extradural haematomata which may cause impairment of consciousness following apparent recovery are important to diagnose as they are readily surgically treatable.

33.4.6 **Infections**

Systemic infections may result in coma as an event secondary to metabolic and vascular disturbance or seizure activity. However direct infections of the central nervous system as with meningitis and encephalitis can all be associated with coma. When meningitis causes coma, the onset of symptoms is usually subacute, before losing consciousness the patient complains of intense headache, associated with fever and neck stiffness. Meningococcal meningitis, however, may be fearfully rapid in onset (Section 41.5.2). The diagnosis is confirmed by identifying the characteristic changes in the CSF, from which it may be possible to isolate the causative organism. Prompt treatment of acute meningitis is imperative and may of necessity precede diagnostic confirmation. The onset of encephalitis is also usually subacute, and often associated with fever or seizures although herpes simplex encephalitis may be explosive at onset, leading to coma within a matter of hours (Section 42.3.1). Treatment with acyclovir by necessity precedes definitive diagnosis. The acute parainfectious encepalopathies, acute disseminated encephalomyelitis (Section 37.4.1), and acute hemorrhagic leucoencephalopathy commonly lead to coma.

A number of parasitic infections may give rise to coma of which by far the most important is cerebral malaria (Section 42.6.1). Two to ten per cent of cases of infection with *Plasmodium falciparum* are associated with cerebral malaria which carries a 25 per cent mortality rate. There is an acute profound mental obtundation or psychosis leading to coma with extensor plantar responses. Although the CSF may show increased protein, characteristically there is no pleocytosis. Patients often develop hypoglycaemia and lactic acidosis which may contribute to the coma.

Independently of direct infections of the central nervous system, septic patients commonly develop an encephalopathy (Section 42.5.13). In some patients this can be severe with a prolonged coma. Lumbar puncture in such patients is usually normal or only associated with a mildly elevated protein level. The EEG is usually abnormal, ranging from diffuse theta through to triphasic waves and suppression or burst-suppression. Whilst there is high mortality there is the potential for complete reversibility and so the presence of coma should not prevent an aggressive approach to management of such patients including, for example, haemodialysis to deal with acute renal failure (Bolton *et al.* 1993). The aetiology of septic encephalopathy is probably multifactorial and includes microabscesses, reduced cerebral blood flow, cerebral oedema, and disruption of the blood brain barrier from the action of inflammatory mediators (Papadopulos *et al.* 2000; Davies *et al.* 2006).

33.4.7 **Miscellaneous causes**

Seizures are a common cause of coma with a period of unconsciousness following a single generalized seizure commonly being between 30 and 60 min. Following status epilepticus, there may be a prolonged period of coma. If the history is unavailable, clues may be provided from trauma to the tongue or inside of the mouth. Seizures secondary to metabolic disturbances may have a longer period of coma.

Basilar migraine is often associated with some impairment of consciousness and familial hemiplegic migraine can be accompanied by prolonged coma (Section 18.2.1).

Extensive white matter disease may also result in impaired consciousness, for example progressive multifocal leucoencephalopathy

(Section 42.3.14) and severe end-stage multiple sclerosis. *Prion disease* may lead to coma over a short period of 6–8 weeks but this is following a progressive course of widespread neurological disturbance (Section 42.8).

33.4.8 Psychogenic unresponsiveness

In psychogenic unresponsiveness the patient, though apparently unconscious, (Plum and Posner 1980) usually shows some response to external stimuli (Section 4.8.6). For example, an attempt to elicit the corneal reflex may cause a vigorous contraction of the orbicularis oculi. Marked resistance to passive movement of the limbs may be present, and signs of organic disease are absent. The caloric and optokinetic responses and the EEG are all generally normal. It is not possible to mimic the roving eye movements of coma, nor is it possible to mimic the slow closure that is seen after the eyelids are raised by the examiner. However, the diagnosis of catatonic stupor as a cause of psychogenic unresponsiveness is more difficult, as in such cases the EEG may be abnormal. Catatonia, in which the limbs maintain a posture passively imposed by the examiner, may be a helpful sign in such cases.

33.5 The locked in state

The de-efferented state or 'locked in' syndrome occurs with lesions of the medulla or anterior pons with sparing of the tegmentum and thus preservation of consciousness. It is an important diagnosis to consider in the differential diagnosis of coma particularly for the unwary. It is most commonly seen with basilar artery infarcts (Patterson and Grabois 1986). The patient, despite being tetraplegic and anarthric, is fully aware of their surroundings, a tragic experience eloquently described by JD Bauby in 'The Diving Bell and the Butterfly' (Bauby 1997). Functional imaging shows much higher glucose metabolism compared with the vegetative state (Levy *et al.* 1987). The EEG shows a reactive alpha rhythm consistent with consciousness and event related potentials are normal (Onofrj *et al.* 1997). So called alpha coma, in which an alpha rhythm is found in association with coma, is seen usually with brainstem infarcts, but is usually unresponsive to stimuli, although rarely this can occur with drug intoxication when the alpha rhythm may be more responsive (Carroll and Mastaglia 1979). In the classical locked in syndrome, there is preservation of vertical eye movements which can provide a means of communication. Total locked in syndrome may occur with lesions of the cerebral peduncles in which there is a complete ophthalmoplegia (Karp and Hurtig 1974). Although most commonly occurring in association with basilar territory infarct, it can occur with demyelination as in central pontine myelinolysis and a similar syndrome can occur with peripheral disorders with severe polyneuropathies, myasthenia, and neuromuscular blocking agents. Even with infarcts, patients can make good recoveries and even limited recovery can enable patients to return home (Smith and Delargy 2005).

33.6 Chronic disorders of consciousness

Many patients in coma will either recover in the first one or two days or succumb to the cerebral insult. Patients who survive in coma, who are in a state of unresponsiveness to external stimuli with eyes closed, do not remain in this state for more than 2–3 weeks but develop a persistent unresponsive state in which sleep–wake cycles

return. There is, however, no evidence of consciousness; they are awake but not aware. Jennett and Plum (1972) proposed the term 'persistent or chronic vegetative state'. These are patients in whom brainstem function recovers but there is no cerebral function. The term 'persistent' implies permanence and so there has been a trend towards use of the term 'vegetative state' to separate the clinical syndrome from any prognostic implication (Bernat 2006).

Some patients will emerge from coma with profound cerebral dysfunction but with some responsiveness to stimuli. Similarly patients in a vegetative state may regain some function with time. Accordingly, the term 'minimally conscious state' has been introduced to define this syndrome (Giacino *et al.* 2002).

33.6.1 Vegetative state

By definition, the vegetative state is a clinical condition of complete unawareness of the self and the environment, which is accompanied by sleep–wake cycles and complete or partial preservation of hypothalamic and brainstem autonomic functions but bladder and bowel incontinence. There should be no evidence of sustained, reproducible, purposeful, or voluntary behavioural responses to visual, auditory, tactile, or noxious stimuli and no evidence of language comprehension or expression. The persistent vegetative state has been defined as one that is present 1 month after acute traumatic or non-traumatic brain injury or lasting for at least 1 month in patients with degenerative or metabolic disorders or developmental malformations (The Multi-Society Task Force on PVS 1994). Others have defined the vegetative state as a 'clinical condition of unawareness of self and environment in which the patient breathes spontaneously, has a stable circulation, and shows cycles of eye closure and eye opening which may simulate sleep and waking' (Royal College of Physicians 2003). They went on to define the continuing vegetative state as being when the vegetative state continues for more than 4 weeks. It then becomes increasingly unlikely that the condition is part of the recovery phase from coma and the diagnosis for continuing vegetative state can be made. The permanent vegetative state can be made when the diagnosis of irreversibility has been established with a high degree of clinical certainty. It is considered a reasonable diagnosis when a patient has been in a vegetative state following head injury for more than 12 months or following other causes of brain damage for more than 6 months. As discussed above there is an increasing use of the term 'vegetative state' rather than permanent vegetative state.

Extensive damage to the cerebral cortex thalamus and the connecting fibres but with sparing of the brainstem and hypothalamus is essentially the neuropathological correlate of the vegetative state (Kinney and Samuel 1994). In general, traumatic brain injury, the commonest cause of the vegetative state, has prominent white matter involvement due to diffuse axonal injury (Section 33.4.5). Hypoxic-ischemic insults cause greater damage to the cortical and thalamic neurons, and laminar necrosis in the cerebral cortex is often seen. Profound thalamic damage with intact brainstem was found in the case of Karen Quinlan, who suffered a cardio-pulmonary arrest in 1975 and died 10 years later (Kinney *et al.* 1994).

The vegetative state demands considerable skill for diagnosis and requires reassessment over a period of time to ensure that indeed there is no purposeful response to external stimuli. The diagnosis may be supported by functional imaging studies. In patients who have remained in a vegetative state for prolonged periods of time, the cerebral metabolism drops to below 50 per cent of the normal

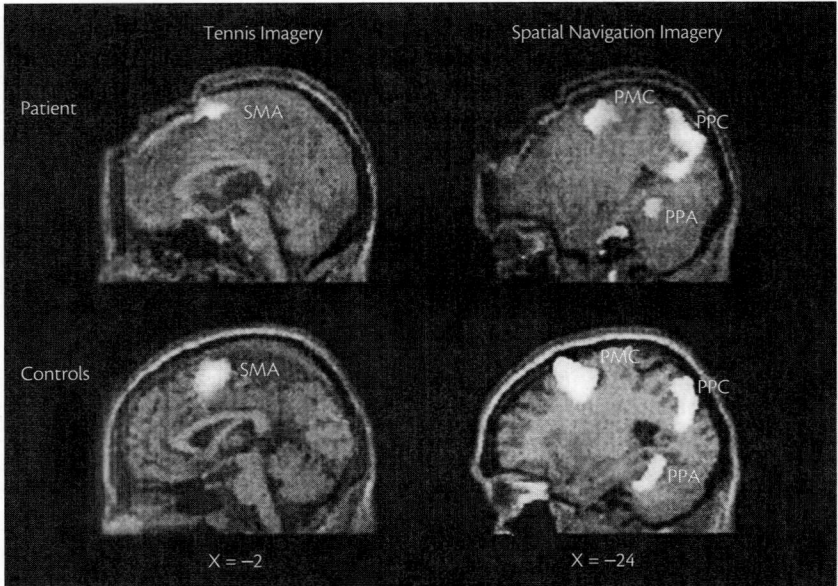

Fig. 33.3 Functional MRI of a patient with vegetative state (top) compared to controls (bottom) showing supplementary motor area (SMA) activity when told to imagine playing a game of tennis (left). On being asked to imagine moving around the house (right) patient and controls showed similar activity in the parahippocampal gyrus (PPA), posterior parietal lobe (PPC), and lateral premotor cortex (PMC). (From Owen *et al.* 2006 with permission.)

range. Cortical activation studies using auditory, visual, and somatosensory stimuli in general show activation of primary sensory areas but not subsequent activation of the polymodal secondary association areas which are believed to subserve the function of awareness (Laureys *et al.* 2004).

33.6.2 Management and prognosis

The management of the vegetative state has attracted considerable debate (Howard and Miller 1995). Until the diagnosis of the vegetative state has been made beyond doubt, full nursing support and nutrition, usually by gastrostomy, should be provided. In patients in whom an undoubted diagnosis has been made, recovery of consciousness for a post-traumatic persistent vegetative state is unlikely after 12 months and for a non-traumatic persistent vegetative state, extremely rare after 3 months. The lifespan in persistent vegetative state is reduced and for most patients, life expectancy ranges between 2 and 5 years although survival beyond 10 years is reported. When a diagnosis has been established, an application may be made, often involving application to the court and depending upon national guidelines, for withdrawal of tube feeding.

However, it is clearly difficult to make categorical statements that there is no awareness and certainly in the early months of the vegetative state patients may be moving towards the minimally conscious state. In a study of 40 patients with a referral diagnosis of vegetative state, it was found that 43 per cent did not fulfil the criteria. Most of the misdiagnosed patients were blind or severely visually impaired (Andrews *et al.* 1996)

33.6.3 Minimally conscious state

The term 'minimally conscious state' was introduced to cover those patients in whom there is profound impairment of cerebral function, on recovery from coma or after recovery from a period of the vegetative state. The key feature is that, although patients may be poorly responsive, there is evidence of awareness, for example by following simple commands, gestural or verbal responses to yes/no questions; reaching for objects, sustained visual pursuit to moving stimuli, and smiling or crying to emotional stimuli but not to neutral stimuli (Giacino *et al.* 2002). The neuropathology in the minimally conscious state is generally less severe than that seen in the vegetative state with less thalamic involvement and in traumatic cases less diffuse axonal injury (Jennett *et al.* 2001).

Assessment of patients with a minimally conscious state, and its distinction from the vegetative state, is fraught with difficulty. Observation over prolonged periods of time and repeated assessments are important. Observations of nursing personnel and family members are also valuable. The disorders of consciousness scale, DOCS, has been developed for the assessment of awareness in severely impaired patients (Pape *et al.* 2005a, b).

Activation studies in patients in the minimally conscious state do show activation of secondary association areas in additional to primary sensory areas. Some patients in the vegetative state have also shown more extensive activation. Owen *et al.* (2006) reported a 23-year-old woman following severe traumatic brain injury who, given instructions to 'imagine playing tennis' showed activation of the supplementary motor area similar to that seen in controls (Fig. 33.3). By contrast when asked to imagine walking through her home, activation was seen in the parahippocampal gyrus and parietal cortex. Similarly, two patients in a vegetative state showed temporal association area activation in response to their name (Di *et al.* 2007). This was similar to four patients in the minimally conscious state. Whether such reports will challenge the concept of complete lack of awareness in the vegetative state or, more likely, help to identify patients who are recovering some awareness and moving to the minimally conscious state, remains to be seen.

33.7 Brain death

This problem has assumed considerable importance in recent years, first because of the increasing difficulty of deciding whether it is justifiable to maintain life indefinitely with artificial support in patients with severe brain damage, and second because of the difficult question of deciding when a cerebral lesion is irreversible and death is imminent, so that viable organs for donation may be removed. It is now widely accepted that brain death denotes death of the individual and is a generally accepted legal definition.

33.7.1 Definition

Many different criteria have been developed, but all are broadly similar (Halevy and Brody 1993). The diagnosis of brain death generally equates to functional death of the brainstem, and when brainstem death has occurred, there is no possible chance of recovery (Pallis 1982; Wijdicks 2001). Although there are some national variations, most criteria include the same conditions for diagnosing brain death and employ similar clinical bedside tests for confirming brainstem death. The American Academy of Neurology published practical guidelines to determine brain death in 1995 (Quality Standards SubCommittee 1995). Guidance is also provided on the interval for re-evaluation depending on age of patients. The following are the UK criteria (Conference of Medical Royal Colleges and their Faculties in the United Kingdom 1976).

33.7.2 Criteria for diagnosis

Conditions for considering diagnosis of brain death.

All of the following should co-exist:

◆ The patient must be deeply comatosed. There should be no suspicion that this is due to central nervous system depressant drugs, primary hypothermia must have been excluded and any metabolic or endocrine contribution or cause of the coma carefully assessed.

◆ The patient must be maintained on a ventilator because spontaneous respiration had previously become inadequate or ceased altogether. Neuromuscular blocking agents must have been excluded as a cause of respiratory failure. The failure of neuromuscular transmission from other causes may need to be excluded by use of a nerve stimulator.

◆ There should be no doubt that the patient's condition is due to irremediable structural brain damage and the diagnosis of the underlying cause of brain death should have been fully established.

Tests for confirming brain death

All brain stem reflexes should be absent as follows:

◆ The pupils are fixed in diameter and do not respond to sharp changes in the intensity of incident light.

◆ There is no corneal reflex.

◆ The vestibulo-ocular reflexes are absent. These are absent when no eye movement occurs during or after the slow infusion by syringe of 20 ml of ice cold water into each external auditory meatus in turn, clear access to the tympanic membrane having been established by direct inspection. This test may be contraindicated on one or other side by local trauma.

◆ No motor responses within the cranial nerve distribution can be elicited by adequate stimulation of any somatic area.

◆ There is no gag reflex or reflex response to bronchial stimulation by a suction catheter passed down the trachea.

◆ No respiratory movements occur when the patient is disconnected from the mechanical ventilator for long enough to ensure that the arterial carbon dioxide tension rises above the threshold for stimulating respiration—that is the $paCO_2$ must normally reach 6.7 kPa, or 50 mmHg. Hypoxia during disconnection should be prevented by delivering oxygen at 6 l per minute through a catheter into the trachea. Many patients in intensive care units are ventilated to maintain relative hypocapnoea and so ventilation may need to be adjusted before apnoea testing to allow the pCO_2 to rise slowly beforehand.

It is well established that spinal cord function can persist in the absence of brainstem function thus tendon reflexes may be preserved and a number of spinal reflexes observed especially during apnoea testing or tracheal suction (Spittler et al. 2000). One dramatic and potentially distressing response is the *Lazarus sign* in which the arms may elevate slowly and the trunk flex. None of these spinal reflexes indicate preservation of brain, as opposed to spinal function.

The tests should be repeated by two independent clinicians to ensure that there is no observer error. The timing of the two tests is not fixed and will depend to some extent on the underlying cause and the age of the patient, particular care being taken with infants. It may be as long as 24 or even 48 h. It is important that experienced clinicians, one of whom is a consultant, perform the tests.

The criteria have been found reliable in clinical practice, for example of 1003 survivors of severe head injury, none would have been suspected of being brainstem dead on the above criteria (Jennett et al. 1981).

33.8 Coma in children

The basic principles of the approach to coma in children are the same as for adults (Kirkham 2001). However, the Glasgow Coma Scale (Teasdale and Jennett 1974), originally intended for use in adults with closed head injury, cannot be applied without modification in the very young, principally because the motor and verbal components require adjustment for age. A number of special coma scales have since been devised, the Paediatric Coma Scale having the highest interobserver reliability (Simpson et al. 1991). To overcome the difficulty of assessing verbal ability in the intubated patient, a grimace score has been proposed as a surrogate measure, also with moderately good interobserver reliability (Kirkpatrick 1997). A fundamental problem in assessing coma in children is that, whilst a Glasgow Coma Scale score of 8 or less in adults signifies a high probability of intracranial hypertension and need for intensive care, no such correlative data are available in children (Gemke and Tasker 1998). This makes evaluation of interventions and prognosis difficult.

33.8.1 Causes

The major causes of coma in children are accidental and non-accidental injury, infection, hypoxia, and epilepsy (Seshia et al. 1977, 1983) although the aetiology remained unknown in 14 per cent of one series (Wong et al. 2001).

Inborn errors of metabolism as causes of coma are relatively more common in young children, for example those arising from urea cycle disorders, disorders of glycogen storage, pyruvate metabolism, respiratory chain disorders, and medium chain acyl-CoA dehydrogenase deficiency, only rarely seen in adult neurology (Tables 10.1, 10.2). Inborn errors of metabolism may also mimic the features of birth asphyxia in neonates.

Reye syndrome is now very rarely seen in Western countries since the association with aspirin use has been recognized and prevented. Occasional cases are associated with varicella and influenza B infection. The initial stage starts 3–6 days after the appearance of the varicella rash and may progress rapidly from vomiting to increasing coma, metabolic disturbance, liver failure, and cerebral oedema with herniation.

Hypertonic dehydration, usually secondary to gastroenteritis in small children may cause subdural haematomas or venous sinus thrombosis as water is drawn out of the brain. Over rapid correction of dehydration may precipitate cerebral oedema and seizures.

Head injury may be caused by non-accidental means, and a history of trauma is not always available. It should be suspected if there are signs of other injuries (Caffey 1972), fractures of differing ages, or if the account of the injury is discrepant to that expected from the clinical examination. Coma precipitated by trauma may develop after a longer lucid interval than in adults. In infants, the relative elasticity of the skull, along with the presence of open fontanelles and unfused sutures can lead to a different pattern of intracranial pathology with haematomas, dural and tentorial tears, and oedema (Zimmerman and Bilaniuk 1981). It is important to recognize that child abuse may also result in coma through asphyxia or poisoning.

Cardiovascular causes of coma are rare in children outside the context of cyanotic congenital heart disease complicated by cerebral venous thrombosis or abscess, or post circulatory bypass cardiac surgery. It may be difficult clinically to decide whether a young child who has for example had complex cardiac surgery or cardiorespiratory arrest and is requiring intensive care is in coma or not. Intensive clinical monitoring is required using EEG if necessary. Clinical evaluation of the paralysed sedated child is impossible and EEG monitoring becomes particularly important.

33.8.2 Prognosis

There are relatively fewer data on the prognosis of coma in children than for adults. In non-traumatic coma, infants have a higher mortality of 44 per cent compared to older children, 24 per cent (Seshia *et al.* 1983; Johnston and Seshia 1984). All patients who died had internal or external ophthalmoplegia or problems regulating body temperature. Children with anoxic injuries, for instance due to cardiac arrest, have the highest mortality. Duration of coma best predicts disability in survivors (Johnston and Seshia 1984). Slow, poorly reactive EEG patterns at day 10 of coma also predict neurological impairment (Bricolo *et al.* 1978). In general, children with traumatic head injury fare better than adults with the opportunity for substantial recovery (Berger *et al.* 1985).

In young children the potential for apparent 'recovery' includes the process of normal development. It is mistrading to use the term chronic vegetative state in very young children. Certainly it should not be used under the age of 1 year. In older children although the adult definitions appear to apply in the short to medium term, there is a very common tendency for some recovery of awareness and responsiveness albeit with very severe disability. Modern neonatal care allows the survival of some very severely cerebrally damaged babies, in whom the brainstem is sufficiently robust to support vegetative functions.

In general the criteria for brain death are as in adults but with caution being exercised in neonates. It is suggested that up to 2 months of age there should be two examinations and it may be helpful to obtain two EEGs 48 h apart (Ashwal and Schneider 1989). That up to 1 year these two examinations and EEGs can be 24 h apart and that over a year, the usual prior criteria and two examinations 12–24 h apart may suffice. Under a year some form of angiographic demonstration of lack of cerebral blood flow may provide a useful adjunct and in another study, use of middle cerebral artery pulsed Doppler ultrasound was found to be extremely helpful as an adjunct to this diagnosis (Kirkham *et al.* 1987).

References

Anonymous (1995). Recommendations for use of uniform nomenclature pertinent to patients with severe alterations in consciousness. American Congress of Rehabilitation Medicine [published erratum appears in *Arch Phys Med Rehabil*, **76**(4), 397] [see comments]. *Arch Phys Med Rehabil*, **76**, 205–9.

Adams JH, Graham DI, Murray LS *et al.* (1982). Diffuse axonal injury due to nonmissile head injury in humans: an analysis of 45 cases. *Ann Neurol*, **12**, 557–63.

Adrogue HJ, Madias NE (2000). Hyponatremia. *N Engl J Med*, **342**, 1581–9.

Andrews BT, Chiles BW, Olsen WL *et al.* (1988). The effect of intracerebral hematoma location on the risk of brain-stem compression and on clinical outcome. *J Neurosurg*, **69**, 518–22.

Andrews K, Murphy L, Munday R *et al.* (1996). Misdiagnosis of the vegetative state: retrospective study in a rehabilitation unit [see comments]. *BMJ*, **313**, 13–6.

Ashwal S, Schneider S (1989). Brain death in the newborn. *Pediatrics*, **84**, 429–37.

Balci K, Asil T, Kerimoglu M *et al.* (2005). Clinical and neuroradiological predictors of mortality in patients with primary pontine hemorrhage. *Clin Neurol Neurosurg*, **108**, 36–9.

Balzan MV, Agius G, Galea DA (1996). Carbon monoxide poisoning: easy to treat but difficult to recognise [see comments]. *Postgrad Med J*, **72**, 470–3.

Bauby J-D (1997). *The Diving-Bell & the Butterfly*. Fourth Estate, London.

Berendes E, Walter M, Cullen P *et al.* (1997). Secretion of brain natriuretic peptide in patients with aneurysmal subarachnoid haemorrhage. *Lancet*, **349**, 245–9.

Berger MS, Pitts LH, Lovely M *et al.* (1985). Outcome from severe head injury in children and adolescents. *J Neurosurg*, **62**, 194–9.

Bernat JL (2006). Chronic disorders of consciousness. *Lancet*, **8**, 367, 1181–92. Review. Erratum in: *Lancet*, 2006 **24**, 367, 2060.

Bolton CF, Young GB (1990). *Neurological Complications of Renal Disease*. Butterworth-Heinemann, Stoneham, MA.

Bolton CF, Young GB, Zochodne DW (1993). The neurological complications of sepsis. *Ann Neurol*, **33**, 94–100.

Boyer EW, Shannon M. (2005). The serotonin syndrome. *N Engl J Med*, **17**, 352(11), 1112–20.

Bricolo A, Turazzi S, Faccioli F *et al.* (1978). Clinical application of compressed spectral array in long-term EEG monitoring of comatose patients. *Electroencephalogr Clin Neurophysiol*, **45**, 211–25.

Brouns R, De Deyn PP (2004). Neurological complications in renal failure: a review. *Clin Neurol Neurosurg*, **107**, 1–16.

Bushinsky DA, Monk RD (1998). Calcium. *Lancet*, **352**, 306–11.

Caffey J (1972). On the theory and practice of shaking infants. Its potential residual effects of permanent brain damage and mental retardation. *Am J Dis Child*, **124**, 161–9.

Cairns H, Oldfield RC, Pennybacker JB *et al.* (1941). Akinetic mutism with an epidermoid cyst of the 3rd ventricle. *Brain*, **64**, 273–90.

Carroll WM, Mastaglia FL (1979). Alpha and beta coma in drug intoxication uncomplicated by cerebral hypoxia. *Electroencephalogr Clin Neurophysiol*, **46**, 95–105.

Chang GY, Keane JR (1999). Hypertensive brainstem encephalopathy: three cases presenting with severe brainstem edema. *Neurology*, **53**, 652–4.

Conference of Medical Royal Colleges and their Faculties in the United Kingdom (1976). Diagnosis of brain death. Statement issued by the honorary secretary of the Conference of Medical Royal Colleges and their Faculties in the United Kingdom on 11 October 1976. *BMJ*, **2**, 1187–8.

Cowey A, Stoerig P (1995). Blindsight in monkeys [see comments]. *Nature*, **373**, 247–9.

Crawford SC, Digre KB, Palmer CA *et al.* (1995). Thrombosis of the deep venous drainage of the brain in adults. Analysis of seven cases with review of the literature. *Arch Neurol*, **52**, 1101–8.

Cuneo RA, Caronna JJ, Pitts L *et al.* (1979). Upward transtentorial herniation: seven cases and a literature review. *Arch Neurol*, **36**, 618–23.

Davies NW, Sharief MK, Howard RS (2006). Infection-associated encephalopathies: their investigation, diagnosis, and treatment. *J Neurol*, **253**, 833–45.

Di HB, Yu SM, Weng XC *et al.* (2007). Cerebral response to patient's own name in the vegetative and minimally conscious states. *Neurology*, **68**, 895–9.

Gatewood JW, Organ CH Jr, Mead BT (1975). Mental changes associated with hyperparathyroidism. *Am J Psychiatry*, **132**, 129–32.

Gemke RJ, Tasker RC (1998). Clinical assessment of acute coma in children. *Lancet*, **351**, 926–7.

Gennarelli TA, Thibault LE, Adams JH *et al.* (1982). Diffuse axonal injury and traumatic coma in the primate. *Ann Neurol*, **12**, 564–74.

Gennarelli TA, Champion HR, Copes WS *et al.* (1994). Comparison of mortality, morbidity, and severity of 59, 713 head injured patients with 114,447 patients with extracranial injuries. *J Trauma*, **37**, 962–8.

Giacino JT (1997). Disorders of consciousness: differential diagnosis and neuropathologic features. *Semin Neurol*, **17**, 105–11.

Giacino JT, Ashwal S, Childs N *et al.* (2002). The minimally conscious state: definition and diagnostic criteria. *Neurology*, **58**, 349–53. Review.

Grant DK (1953). Papilloedema and fits in hypoparathyroidism. *Q J Med*, **22**, 243.

Halevy A, Brody B (1993). Brain death: reconciling definitions, criteria, and tests. *Ann Intern Med*, **119**, 519–25.

Howard RS, Miller DH (1995). The persistent vegetative state. *Br Med J*, **310**, 341–2.

Howard RS, Russell S, Losseff N *et al.* (1995). Management of mitochondrial disease on an intensive care unit. *QJM*, **88**, 197–207.

Huisman TA, Schwamm LH, Schaefer PW *et al.* (2004). Diffusion tensor imaging as potential biomarker of white matter injury in diffuse axonal injury. *Am J Neuroradiol*, **25**, 370–6.

Jefferson A (1956). Clinical correlation between encephalopathy and papilloedema in Addison's disease. *J Neurol Neurosurg Psychiatry*, **19**, 21.

Jennett B, Bond M (1975). Assessment of outcome after severe brain damage. *Lancet*, **1**, 480–4.

Jennett B, Gleave J, Wilson P (1981). Brain death in three neurosurgical units. *BMJ Clin Res Ed*, **282**, 533–9.

Jennett B, Plum F (1972). Persistent vegetative state after brain damage. A syndrome in search of a name. *Lancet*, **1**, 734–7.

Jennett B, Teasdale G, Braakman R *et al.* (1979). Prognosis of patients with severe head injury. *Neurosurgery*, **4**, 283–9.

Jennett B, Adams JH, Murray LS *et al.* (2001). Neuropathology in vegetative and severely disabled patients after head injury. *Neurology*, **27**(56), 486–90.

Johnston B, Seshia SS (1984). Prediction of outcome in non-traumatic coma in childhood. *Acta Neurol Scand*, **69**, 417–27.

Kampfl A, Franz G, Aichner F *et al.* (1998). The persistent vegetative state after closed head injury: clinical and magnetic resonance imaging findings in 42 patients. *J Neurosurg*, **88**, 809–16.

Karp JS, Hurtig HI (1974). 'Locked-in' state with bilateral midbrain infarcts. *Arch Neurol*, **30**, 176–8.

Kearney T, Dang C (2007). Diabetic and endocrine emergencies. *Postgrad Med J*, **83**, 79–86. Review.

Kinney HC, Korein J, Panigrahy A *et al.* (1994). Neuropathological findings in the brain of Karen Ann Quinlan. The role of the thalamus in the persistent vegetative state [see comments]. *N Engl J Med*, **330**, 1469–75.

Kinney HC, Samuels MA (1994). Neuropathology of the persistent vegetative state. A review. *J Neuropathol Exp Neurol*, **53**, 548–58. Review.

Kirkham FJ, Levin SD, Padayachee TS *et al.* (1987). Transcranial pulsed Doppler ultrasound findings in brain stem death. *J Neurol Neurosurg Psychiatry*, **50**, 1504–13.

Kirkham FJ (2001). Non-traumatic coma in children. *Arch Dis Child*, **85**, 303–12.

Kirkpatrick PJ (1997). On guidelines for the management of the severe head injury. *J Neurol Neurosurg Psychiatry*, **62**, 109–11.

Knaus WA, Draper EA, Wagner DP *et al.* (1985). APACHE II: a severity of disease classification system. *Crit Care Med*, **13**, 818–29.

Kumar S, Berl T (1998). Sodium. *Lancet*, **352**, 220–8.

Kunze K (2002). Metabolic encephalopathies. *J Neurol*, **249**, 1150–9. Review.

Laureys S, Owen AM, Schiff ND (2004). Brain function in coma, vegetative state, and related disorders. *Lancet Neurol*, **3**, 537–46. Review.

Leigh RJ, Shaw DA (1976). Rapid regular respiration in unconscious patients. *Arch Neurol*, **33**, 356–61.

Levy DE, Bates D, Caronna JJ *et al.* (1981). Prognosis in nontraumatic coma. *Ann Intern Med*, **94**, 293–301.

Levy DE, Caronna JJ, Singer BH *et al.* (1985). Predicting outcome from hypoxic-ischemic coma. *JAMA*, **8**, 253(10):1420–6.

Levy DE, Sidtis JJ, Rottenberg DA *et al.* (1987). Differences in cerebral blood flow and glucose utilization in vegetative versus locked-in patients. *Ann Neurol*, **22**, 673–82.

Malouf R, Brust JC (1985) Hypoglycemia: causes, neurological manifestations, and outcome. *Ann Neurol*, **17**, 421–30.

Marin RS (1990). Differential diagnosis and classification of apathy. *Am J Psychiatry*, **147**, 22–30.

Maxwell MH, Kleeman CR, Narins RG (1987). *Clinical Disorders of Fluid and Electrolyte Metabolism*, 4th edition. McGraw Hill, New York.

Menza MA, Murray GB (1989). Pancreatic encephalopathy. *Biol Psychiatry*, **25**, 781–4.

Mooradian AD, Morley GK, McGeachie R *et al.* (1984). Spontaneous periodic hypothermia. *Neurology*, **34**, 79–82.

Multisociety Task Force on PVS (1994).

Nemeth G, Hegedus K, Molnar L (1986). Akinetic mutism and locked-in syndrome: the functional-anatomical basis for their differentiation. *Funct Neurol*, **1**, 128–39.

Newcomer J, Haire W, Hartman CR (1983). Coma and thyrotoxicosis. *Ann Neurol*, **14**, 689–90.

Onofrj M, Thomas A, Paci C *et al.* (1997). Event related potentials recorded in patients with locked-in syndrome. *J Neurol Neurosurg Psychiatry*, **63**, 759–64.

Owen AM, Coleman MR, Boly M *et al.* (2006). Detecting awareness in the vegetative state. *Science*, **313**, 1402.

Pallis C (1982). ABC of brain stem death. From brain death to brain stem death. *BMJ Clin Res Ed*, **285**, 1487–90.

Papadopoulos MC, Davies DC, Moss RF *et al.* (2000). Pathophysiology of septic encephalopathy: a review. *Crit Care Med*, **28**, 3019–24.

Pape TL, Heinemann AW, Kelly JP *et al.* (2005a). A measure of neurobehavioral functioning after coma. Part I: Theory, reliability, and validity of Disorders of Consciousness Scale. *J Rehabil Res Dev*, **42**, 1–17.

Pape TL, Senno RG, Guernon A *et al.* (2005b). A measure of neurobehavioral functioning after coma. Part II: Clinical and scientific implementation. *J Rehabil Res Dev*, **42**, 19–27.

Patterson JR, Grabois M (1986). Locked-in syndrome: a review of 139 cases. *Stroke*, **17**, 758–64.

Plum F (1982). Mechanisms of 'central' hyperventilation. *Ann Neurol*, **11**, 636.

Plum F and Alvord EC Jr (1964). Apneustic breathing in man. *ArchNeurol Psychiatry*, **10**, 101.

Plum F, Posner JB (1980). *Stupor and Coma*, 3rd edition. Davis, Philadelphia.

Plum F, Swanson AG (1958). Abnormalities in the central regulation of respiration in acute and convalescent poliomyelitis. *Arch Neurol Psychiatry*, **80**, 267.

Quality Standards Subcommittee of the American Academy of Neurology (1995) Practice parameters for determining brain death in adults (summary statement). Neurology **45**, 1012–4.

Rodriguez I, Fluiters E, Perez-Mendez LF *et al.* (2004). Factors associated with mortality of patients with myxoedema coma: prospective study in 11 cases treated in a single institution. *J Endocrinol*, **180**, 347–50.

Rodriguez M, Baele PL, Marsh HM *et al.* (1982). Central neurogenic hyperventilation in an awake patient with brainstem astrocytoma. *Ann Neurol*, **11**, 625–8.

Ropper AH (1990). The opposite pupil in herniation. *Neurology*, **40**, 1707–9.

Rosenberg GA, Johnson SF, Brenner RP (1977). Recovery of cognition after prolonged vegetative state. *Ann Neurol*, **2**, 167.

Royal College of Physicians (2003). *The Vegetative State: Guidance On Diagnosis And Management*. [Report of a working party]. Royal College of Physicians, London.

Seshia SS, Johnston B, Kasian G (1983). Non-traumatic coma in childhood: clinical variables in prediction of outcome. *Dev Med Child Neurol*, **25**, 493–501.

Seshia SS, Seshia MM, Sachdeva RK (1977). Coma in childhood. *Dev Med Child Neurol*, **19**, 614–28.

Shawcross D, Jalan R (2005). Dispelling myths in the treatment of hepatic encephalopathy. *Lancet*, **365**, 431–3.

Shuttleworth E (1983). Recovery to social and economic independence from prolonged postanoxic vegetative state. *Neurology*, **33**, 372–4.

Simpson DA, Cockington RA, Hanieh A *et al.* (1991). Head injuries in infants and young children: the value of the Paediatric Coma Scale. Review of literature and report on a study. *Childs Nerv Syst*, **7**, 183–90.

Singh S, Bohn D, Carlotti AP *et al.* (2002). Cerebral salt wasting: truths, fallacies, theories, and challenges. *Crit Care Med*, **30**, 2575–9. Review.

Smith E, Delargy M. (2005). Locked-in syndrome. *BMJ*, **330**, 406–9.

Snyder BD, Gumnit RJ, Leppik IE *et al.* (1981). Neurologic prognosis after cardiopulmonary arrest: IV. Brainstem reflexes. *Neurology*, **31**, 1092–7.

Spittler JF, Wortmann D, von During M *et al.* (2000). Phenomenological diversity of spinal reflexes in brain death. *Eur J Neurol*, **7**, 315–21.

Stuke L, Diaz-Arrastia R, Gentilello LM *et al.* (2007). Effect of alcohol on Glasgow coma scale in head-injured patients. *Ann Surg*, **245**, 651–5.

Teasdale G, Jennett B (1974). Assessment of coma and impaired consciousness. A practical scale. *Lancet*, **2**, 81–4.

Teasdale G, Knill JR, van-der Sande SJ (1978). Observer variability in assessing impaired consciousness and coma. *J Neurol Neurosurg Psychiatry*, **41**, 603–10.

The Multi-Society Task Force on PVS (1994). Medical aspects of the persistent vegetative state (1). *N Engl J Med*, **330**, 1499–508.

Thomas FB, Mazzaferri EL, Skillman TG (1970). Apathetic thyrotoxicosis: A distinctive clinical and laboratory entity. *Ann Intern Med*, **72**, 679–85.

Wijdicks EF, Miller GM (1997). MR imaging of progressive downward herniation of the diencephalon. *Neurology*, **48**, 1456–9.

Wijdicks EF (2001). The diagnosis of brain death. *N Engl J Med*, **344**, 1215–21. Review.

Wong CP, Forsyth RJ, Kelly TP *et al.* (2001). Incidence, aetiology, and outcome of non-traumatic coma: a population based study. *Arch Dis Child*, **84**, 193–9.

Yaqub B, Al Deeb S (1998). Heat strokes: aetiopathogenesis, neurological characteristics, treatment and outcome. *J Neurol Sci*, **156**, 144–51.

Zandbergen EG, de Haan RJ, Stoutenbeek CP *et al.* (1998). Systematic review of early prediction of poor outcome in anoxic-ischaemic coma. *Lancet*, **5**, 1808–12.

Zeman A (2001). Consciousness. *Brain*, **124**, 1263–89.

Zimmerman RA, Bilaniuk LT (1981). Computed tomography in pediatric head trauma. *J Neuroradiol*, **8**, 257–71.

CHAPTER 34

Neuropsychological disorders, dementia, and behavioural neurology

Martin Rossor

Contents

34.1 Introduction

34.1.1 General principles

The diseases which disrupt the cerebral cortex and its subcortical connections result in a wide variety of clinical features. These include the classical syndromes of higher cortical dysfunction such as the dysphasias, dyspraxias, amnesias, and agnosias together with a wide variety of behavioural and emotional disturbances. Such disorders frequently overlap with the clinical disciplines of clinical psychology and psychiatry. Historically there has been a broad split between those diseases which are seen by neurologists and those that are seen by psychiatrists. To some extent the distinction reflects the different clinical approaches employed; neurologists concentrate on the generality of disease caused by lesions in defined areas, whereas psychiatrists often deal with diseases that show a greater interaction with the individuals own personal history and place in society (Lishman 1987). In this chapter disturbances of higher cortical function, the dementias, and behavioural aspects of neurological lesions are discussed. Awareness of the occasional presentation of psychiatric disease to the neurologists is important and further details are available in textbooks of psychiatry. A review of clinical syndromes referable to identified areas of the cerebral cortex, is followed by a functional approach which discusses the main neuropsychological syndromes. The more generalized cognitive impairment seen with the dementias such as Alzheimer's disease, dementia with Lewy bodies, and the frontotemporal lobar degenerations are then reviewed followed by areas of neuropsychiatric overlap.

34.1.2 Anatomy and physiology of the cerebral cortex

The human cerebral cortex consists of six laminae comprising in total some 28×10^9 neurones and approximately the same number of glial cells: from the external surface, these are the molecular layer, the external granular, the external pyramidal, the internal granular, the internal pyramidal, and the multiform or fusiform. The interconnection of the neurons comprises a staggering 10^{12} synapses. Despite this dramatic development, the basic structural organization of the cerebral cortex in modular terms is the same across species. The basis of the modular organization is the minicolumn representing some 80–100 neurons connected vertically and within each minicolumn are all of the major cortical neuronal types. Two broad categories of neuronal cell types can be distinguished, the large pyramidal cells which are the origin of the main outflow tracts and which utilize glutamate as the main neurotransmitter. The smaller non-pyramidal cells have predominantly local connections and primarily utilize the inhibitory amino acid gamma aminobutyric acid together with a variety of coexistent neuropeptides. There is an increasing sub-categorization of the small non-pyramidal cells and an increasing understanding of the intrinsic connectivity of these cells within the minicolumns. Cortical columns are minicolumns bound together by horizontal connections over a short range and which share physiological features based in part upon shared input and output characteristics. Layers 2 and 3 project to other cortical areas and layers 5 and 6 are primarily sub-cortical projections. Columns vary only from 300 to 600 microns in diameter across a very wide range of species. The expansion of the human brain is due to the increase of the total number of modular units rather than a difference in their size (Rakic 1995).

The organization of the cerebral cortex, in terms of the functional characteristics of the cortical columns, was established using single cell recording for the somatic sensory cortex by Mountcastle and the visual cortex by Hubel and Wiesel (Mountcastle 1997). The response characteristics of a somatic sensory column depend both upon modality and topography of the receptive fields. Similarly in the visual cortex, columns can be defined by various properties of increasing complexity from ocularity and place through to orientation. This modular organization in anatomical and functional terms accords with the general view from cognitive psychology of modular processing (Fodor 1983). As the modular processing becomes more complex, the defining characteristics of the cortical column become the inputs from other columns, the outputs of which represent a further level of cortical processing. This achieves greatest complexity in the association cortices or homotypical cortical areas.

There is regional specialization within the cerebral cortex which is reflected to some extent in architectonic differences, such as that originally identified by Brodmann (1909). However, particularly at the higher level of cortical processing, as represented by the association areas, cortical operations are also distributed. Evidence for distributed networks is provided both by functional brain imaging (Ramnani et al. 2004) and by anatomical evidence of enormous convergent and reciprocal connections, for example those between the parietal and frontal cortex (Goldman 1988).

34.2 Clinical syndromes associated with specific areas of the cerebral cortex

A number of clinical syndromes are recognized as characteristic of lesions in specific areas of the cerebral cortex. The syndromes largely, but not invariably, involve loss of function, although in some instances loss of inhibition may present as release phenomena. The descriptions of many of these clinical syndromes were derived from patients with discrete lesions due to ischaemic infarcts or tumours and formerly required follow-up to post mortem to determine the location. Modern neuroimaging has considerably improved the power of such studies.

It should be emphasized, however, that the observed associations are between clinical syndromes and brain areas and do not necessarily locate function to a specific brain region. Although a specific function may be lost following damage at a given site, the function itself is more likely to depend upon successful integration of a neural network. The particular area would be part of such a network and assumed to be of central importance.

In addition to the location of clinical syndromes to specific cortical areas, some syndromes may be better interpreted as representing disconnections of one area from another. The concept of disconnection syndromes was originally postulated by Dejerine and other early neurologists, and re-explored in considerable detail by Geschwind (1965). Some of the clinical syndromes arising from damage to the corpus callosum are most easily explained in terms of disconnection.

Functional imaging has contributed further to our knowledge of localization of function. Baseline measures of blood flow or cerebral glucose metabolism with positron emission tomography, PET scanning, can identify areas of reduced basal function when abnormalities on structural imaging may not be readily apparent. Activation studies can provide information on changes in cerebral blood flow, or deoxygenation of haemoglobin, and thus regions which are activated during a specific cognitive task relative to another. Such functional imaging studies have revealed that activation is often widely distributed, indicating that distributed networks are involved in such cognitive tasks. However, as with structure function relationships, caution has to be exercised in their interpretation. Whilst lesion studies show areas which may be necessary for a particular function, they can only localize deficits and it cannot be inferred that the specific function occurs in that area. They will show areas and structures that are necessary but not sufficient for that function. Similarly, activation may occur in areas that are not necessarily essential to a function. For example, activation studies of language may show areas of increased blood flow in the right hemisphere and yet lesion studies indicate that dysphasia rarely occurs with lesions in the same area.

The most widely used activation studies involve subtraction paradigms (Petersen et al. 1988). A baseline task is compared statistically with an activation task that engages the cognitive component of interest. When the baseline data are subtracted then areas that are activated are believed to relate directly to the particular cognitive component under study. More complicated models have been used to deal, for example, with language function where it may be difficult to identify an appropriate baseline task. In these paradigms, referred to as cognitive conjunctions, areas of common activation rather than areas of different activation are sought (Price and Friston 1997). Functional imaging has provided valuable insights into the distributive networks involved in many cognitive processes such as language (Gabrieli 1998; Price 1998). It does however remain a research tool and as yet has had limited impact on routine neurological management.

34.2.1 The frontal lobes

The frontal lobes lie rostral to the central sulcus and superior to the Sylvian fissure. It is in humans that the frontal lobes show the greatest development compared with other primates. Within the frontal lobes are the primary motor cortices located within the precentral area together with the supplementary motor areas and the frontal eye fields. The dominant frontal lobe encompasses Broca's area and the adjacent area of the motor cortex is involved with the motor control of the oropharynx, lesions of which result in impairment of articulation and phonation. It is, however, the prefrontal cortex, Brodmann's areas 9, 10, 11, 12 and 45, 46, and 47 which are particularly developed in humans and yet have a less clearly defined function.

The frontal cortex has widespread connections with other areas of the brain. The pyramidal cells of the motor cortex form the major fronto-striatal outflow tract. Similarly there are extensive projections from subcortical structures into the frontal cortex, notably the dopaminergic, noradrenergic, and cholinergic cortical projection systems.

Early studies with experimental frontal lobe lesions in non-human primates revealed impaired performance on a number of tasks which suggested perseverative responses and difficulty in switching between preferred modes of response. These difficulties with switching cognitive sets were explored further using the Wisconsin card-sorting test (Milner 1963). Patients tend to perseverate on these tasks and yet on other tasks such as cognitive estimates patients may be quite impulsive and unable to monitor their performance.

The combination of perseverative responses, lack of initiative, and impulsivity have been brought together in the hypothesis of a supervisory attentional system for the frontal lobes (Norman and Shallice 1980). In this model the frontal lobes have an important role in both selecting appropriate behavioural responses and inhibiting inappropriate ones. This can explain the paradoxical combination of both aspontaneity and lethargy together with impulsivity even within the same patient. Breakdown in such a system results in markedly impaired social behaviour and adaptability, and yet formal testing on intelligence may often be spared (Shallice 1982). One of the most distinctive features of the frontal lobe syndrome is a change in personality, most commonly towards disinhibition. The effect upon personality of massive bifrontal lesions was well demonstrated by the celebrated case of Phineas Gage who in 1848 had a crowbar driven through the front of his skull. He was described as 'fitful, irreverent, indulging at times in the greatest profanity . . . . manifesting but little deference for his fellows, impatient of restraint or advice when it conflicts with his desires, at times pertinaciously obstinate, yet capricious and vacillatory'. The disinhibited behaviour may result in childish excitement, 'moria', or joking and pathological punning, 'Witzelsucht'; there may in addition be sexual indiscretions and exhibitionism. Alternatively, the syndrome may present with lack of initiative and profound psychomotor slowing with inability to persist with a task, abulia. This may be accompanied, particularly with bifrontal lesions, by urinary incontinence which can occasionally be seen with unilateral lesions. This incontinence is commonly associated with lack of concern and social awareness which is a useful clinical clue since this type of incontinence is rarely found with generalized dementing conditions such as Alzheimer's disease until late in the disease.

The behavioural disturbance can be striking and precede changes on formal tests of frontal lobe function by many years. A distinction has been drawn between dorsolateral frontal lesions which may be associated with cognitive decline and apathy and with orbitomedial with prominent behavioural change (Devinsky et al. 1995; Blair and Cipolotti 2000). Often the changes in behaviour are most apparent to the spouse who may feel that they are married to somebody entirely different. Some of the behavioural features in frontal lobe lesions resemble the impairment of social interactions found in autistic spectrum disorder (Section 9.6.4).

Clinical examination may be less revealing than the history and often patients perform relatively well on formal tests of intelligence. Specific tasks, however, such as the Wisconsin card-sorting test, the Weigl sorting test, cognitive estimates, verbal fluency, and bimanual motor tasks will show impairment but even this may be patchy and inconsistent. The patient's appearance, however, may be clearly abnormal, appearing unkempt, unwashed, and lacking all spontaneity. Neurological examination may reveal primitive reflexes such as the rooting reflexes, both tactile and oral, and sucking reflexes if severe. Grasp reflexes are easily elicited by running the hand across the palm and may be elicited from the foot (Seyffarth and Denny-Brown 1948). The grasp reflex and instinctive grasp reaction can be seen as part of a more generalized magnetic behaviour, elicited as utilization behaviour (Lhermitte 1983; Shallice et al. 1989). This is a striking example of environmental dependency in which presentation of an object will elicit a behavioural response regardless of whether or not it is appropriate. Patients offered spectacles, for example, may place them on their nose followed by further pairs, until three or four are stacked upon one another, the inappropriateness of this completely eluding the patient.

As lesions extend more posteriorly in the dominant frontal lobe there may be an associated non-fluent anterior dysphasia and impairment of speech production. Lesions in relation to the orbital surface may result in unilateral visual failure and anosmia. The latter will rarely be found unless specifically sought and is a characteristic feature of olfactory groove meningiomata (Fig. 34.1). Prominent changes in tone may be found on examination which increases in response to the stimulus, so-called paratonia or Gegenhalten. More posteriorly placed lesions may result in mild pyramidal signs but most striking are bifrontal medial lesions causing prominent gait impairment with truncal instability, often referred to as gait apraxia (Meyer and Barron 1960). With the availability of CT and MR scanning, frontal lobe tumours are far more readily diagnosed. However, with the gradual onset of personality changes, many cases are still missed until late. In general, prominent frontal lobe syndromes are seen more commonly with tumours and degenerative disease than with vascular disease (Bogousslavsky 1994), with the exception of anterior cerebral artery infarcts which commonly result in abulia (Kumral et al. 2002).

34.2.2 The temporal lobes

The Sylvian fissure separates the temporal lobe from the frontal lobe and rostral part of the parietal lobe. There is, however, no clear boundary between the posterior temporal lobe and the parietal and occipital lobes. The temporal lobes have easily discernible gyri, the superior, middle, and inferior temporal gyri, and also the parahippocampal and hippocampal convolutions. The hippocampus demonstrates a three-layered cortical structure in contrast to the

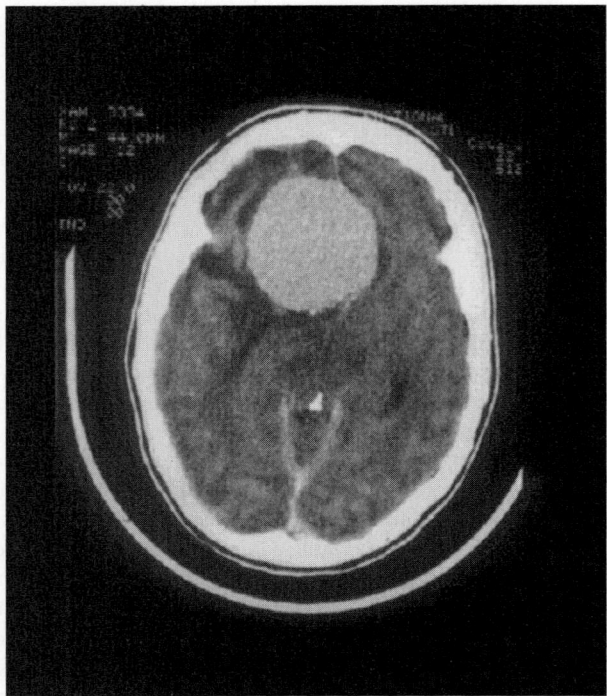

Fig. 34.1 CT scan reveals an olfactory groove meningioma in a patient with anosmia and a frontal lobe syndrome.

six-layered neocortex. Heschl's gyrus in the Sylvian fissure represents the primary auditory receptive area and fibres terminating here do so in a tonotopic arrangement. Deep within the temporal lobe is the amygdala.

Discrete lesions of the anterior temporal poles may be clinically silent and in general, lesions of the non-dominant temporal lobe are less obvious clinically; sometimes the only clue on neurological examination being a superior quadrantic visual field defect or behavioural change. However, bitemporal lesions and those of the dominant temporal lobe may result in profound functional impairments. The predominant lesion of the dominant temporal lobe is that of a language impairment, classically a Wernicke's aphasia (Section 34.4.4) or rarely a pure word deafness or auditory verbal agnosia.

Unilateral lesions of the dominant temporal lobe can be shown to result in impairment of memory for verbal material by contrast to impaired visual memory with non-dominant lesions (Milner 1971). Unilateral lesions of Heschl's gyrus rarely result in deafness but careful testing with binaural testing reveals subtle abnormalities in the area contralateral to the lesions. The inability to recognize faces, prosopagnosia, is usually seen with bilateral lesions but well described with unilateral non-dominant temporoparietal lesions (Warrington and James 1967). Patients with prosopagnosia may have a variety of deficits ranging from an inability to recognize the face through to loss of familiarity and inability to match faces. This can sometimes be so severe that patients may be incapable of recognizing members of their own family but can immediately do so on hearing their voice.

Vary rarely lesions of the temporal lobe can result in a true auditory agnosia. Simple perception of sound and pure tone is intact but the interpretation of complex sounds is severely impaired. In pure

auditory agnosia recognition of all noises, the sound of a bell, dogs barking, running water is lost (Hecaen 1962). More common is the loss of appreciation of music, or amusia seen with right temporal lobe lesions. The patient's appreciation of melody, timbre, and rhythm all tend to be impaired and they may also have difficulty in musical recognition (Stewart *et al.* 2006).

Bilateral temporal lobe lesions are far more devastating. These cases are rare and occur after herpes encephalitis or in the later stages of frontotemporal degeneration (Section 34.6.4). They are normally associated with a dysphasia but in addition profound memory impairments (Section 34.4.7) (Milner 1958; Milner *et al.* 1968). Bilateral removal of the temporal lobes in the monkey produces a striking behavioural state referred to as the *Kluver–Bucy syndrome*. The monkeys show increased exploratory behaviour in which they will examine objects by oral and manual manipulation with apparent inability to recognize them visually. They are usually placid but with hypersexuality. Similar behaviours in humans have been described with herpes encephalitis and with frontotemporal degenerations (Cummings and Duchen 1981; Lilly *et al.* 1983). The frontotemporal degenerations encompass a variety of neuropathologies, including Pick's disease. They are often associated with obsessional behaviour, change in eating habits, hyperreligiosity, and both hypo and hypersexuality. Reduced empathy and diminished emotional responses are characteristic. Some of these features may reflect involvement of the amygdala (Section 34.6.4). The rare lipoid proteinosis *Urbach–Wiethe disease* can result in progressive bilateral degeneration of the amygdalae. Studies in such patients reveal that the appreciation of fear can be particularly impaired but patients do develop deficits in both negative and positive emotional processing (Siebert *et al.* 2003)

A variety of episodic symptoms may be found with temporal lobe lesions which are usually on an epileptic basis and range from auditory hallucinations to disruption of time perception and disturbances of sexual behaviour. In addition, however, chronic bilateral lesions, particularly those of the medial temporal lobes may cause profound disintegration of personality and behaviour. In its extreme form patients react to any stimulus by excessive rage with screaming, biting, and spitting. This distressing clinical picture is seen most commonly in survivors of herpes encephalitis.

34.2.3 The parietal lobes

The parietal lobes lie behind the central sulcus and above the Sylvian fissure but the posterior boundaries with the occipital and temporal lobes are not clearly defined. Immediately behind the central sulcus is the primary sensory cortex which is delineated posteriorly by the post central sulcus. The superior temporal sulcus curves upwards posteriorly into the inferior parietal lobule in relation to the angular gyrus. This is adjacent to the posterior extremity of the Sylvian fissure, which also curves up into the inferior parietal lobule in relation to the supramarginal gyrus. The parietal lobes are well developed in man and continue to develop until about the seventh year of life. There are extensive connections with other association areas.

Lesions within the parietal lobes can present with an enormous variety of disturbances of higher cortical function, some of which are quite dramatic in their presentation. Dominant parietal lobe lesions are often associated with dysphasia. This may be predominantly a motor or sensory dysphasia depending upon the anteroposterior location. Inferior parietal lesions are associated with deficits of verbal short-term memory as assessed by digit span. More posterior lesions are associated with dyslexia, dysgraphia, and dyscalculia, and ideational apraxia is a consistent feature of dominant parietal lobe lesions. *Gerstmann's syndrome* refers to the association of finger agnosia, dyscalculia, right left disorientation, and agraphia. Although this is commonly seen with lesions of the dominant angular gyrus (Gerstmann 1940), this particular clustering of deficits is no more common than other patterns. Lesions of either parietal lobe may result in visuospatial disturbance. The integration of visual information with spatial information in the posterior parietal lobe can give rise to visuomotor or optic ataxia (Rondot *et al.* 1977), in which patients have difficulty touching an object under visual guidance. This forms a major component of *Balint's syndrome* (Hecaen and Ajuriaguerra 1954), in which patients also have an inability to direct gaze into the peripheral field despite full eye movements. Both parietal lobes are involved with selective attention (Mesulam 1981). Patients with non-dominant parietal lesions have marked impairment of selective attention which will affect both right- and left-sided space but most commonly left-sided space. This may be so severe that they will ignore the left side of the body resulting in problems in dressing and in shaving only one-half of their face. If this is associated with a left hemiplegia they may be unaware of the deficit, anosognosia, and indeed deny that the paralysed limb has anything to do with them.

On examination, patients with parietal lobe lesions may show the obvious neglect of the left side with an arm hung loosely at the side or out of the sleeve of their jacket with the left face unshaven and associated pyramidal signs. Field defects are most commonly inferior quadrantanopias. Gaze impersistence, an inability to sustain lateral gaze on testing eye movements, is seen quite commonly and can be frustrating for the examiner. Cortical sensory loss may be found in the contralateral limbs and is most readily picked up by testing two-point discrimination, which is impaired together with astereognosis, that is inability to recognize objects by their shape, or alternatively inability to recognize figures written on the hand, agraphesthesia. Neglect can be unmasked by simultaneous tactile or visual stimuli, when the patient may only recognize one of the two simultaneously presented stimuli, ignoring that contralateral to the lesion. The sensory testing may be difficult with patients showing considerable variability and being easily fatigued (Critchley 1953). Simple bedside tests of visuospatial function which can be helpful include drawing a cube or clock face, when patients will often ignore features to the left hand side. So-called dressing apraxia can be assessed by watching the patient put on clothing and the task made more difficult by inverting one sleeve.

34.2.4 The occipital lobes

The occipital lobes are separated on the medial surface from the parietal lobes by the parieto-occipital fissure but there are no clearly defined margins between the parietal and temporal lobes on the lateral surface. The occipital lobes subsume the termination of the visual pathways, and the primary visual cortex, Brodmann area 17, lies on either side of the calcarine fissure which runs from the occipital pole to the splenium of the corpus callosum. It has the histological characteristic that its fourth layer is divided into two granular cell layers by a thickened band of heavily myelinated fibres, the external band of Baillarger. This band is visible to the naked eye, hence the name of striate cortex. The classical findings with lesions of the occipital cortex are homonymous visual field defects; a homonymous hemianopia when confined to one occipital lobe. Bilateral lesions may cause altitudinal defects, since the termination of the optic radiation is topographically arranged with lower retinal fibres terminating in the cortex below the calcarine fissure. Superior quadrantic defects are found with inferior lesions and vice versa.

With extensive bilateral lesions a variety of abnormalities are found ranging from complete cortical blindness, through to subtle visual disturbances. With complete cortical blindness the pupillary responses are preserved as is visual imagery in dreams, but cortical evoked responses and the alpha rhythm on the EEG are both lost. Strikingly, patients with complete cortical blindness may develop a visual anosognosia, *Anton's syndrome*, with denial of their loss of sight. These patients may walk around as if they can see but will bump into objects and often explain their difficulties by complaining about the light or their loss of glasses. With partial recovery from cortical blindness or with lesions involving the visual association areas there may be a variety of disturbances of higher visual processing. These may fractionate into distinct syndromes such as visual disorientation in which the ability to locate objects within the visual field is considerably impaired and patients may be effectively blind. Other syndromes involve selective loss of colour, achromatopsia, and very rarely impairment of movement perception (Zihl *et al.* 1983; Shipp *et al.* 1994). Rarely patients may develop genuine visual object agnosia if the lesion involves the occipito-temporal areas. These patients are unable to recognize objects by sight but can do so on palpation or by sound. The lesions are usually bilateral but may be found with dominant lesions. These syndromes of visual disorientation and visual agnosia reflect the dorsal 'where' and ventral 'what' processing streams respectively. The dorsal stream links visual information with spatial information in the parietal lobe. The ventral stream links the visual information with semantic stores in the temporal lobe.

Patients may complain of a variety of visual hallucinations and illusions, more commonly with bilateral or non-dominant lesions. Many of these are associated with epileptic phenomena, such as elementary unformed hallucinations and flashes of light, colours or geometric forms. They may be seen within the setting of a hemianopic disturbance. Striking visual illusions may occur with metamorphopsias and marked changes in shape. In addition polyopia, or multiple images, may be seen or the striking disturbance of palinopsia in which perserveration of visual images occurs and colour may spread outside the geometric confines of the object.

34.3 Subcortical syndromes

Although the cerebral cortex is viewed as the main seat of cognitive behaviour, it is becoming increasingly recognized that damage to subcortical structures can give rise to profound behavioural and cognitive disturbance. In many instances, these may mimic deficits arising from lesions within the cerebral cortex, for example the dysphasia and dyspraxia which may be seen with dominant thalamic lesions (Graff *et al.* 1984). These may share characteristics with cortical lesions by virtue of the extensive neural interconnections; an interpretation supported by observed changes in metabolism within connected areas of the cerebral cortex demonstrated on single photon emission tomography scanning subsequent to subcortical infarcts (Perani *et al.* 1987). Many of the behavioural disturbances are discussed elsewhere, but some structures are particularly notable. The importance of the amygdalae in behaviour has already been referred to and rare instances of a Kluver–Bucy syndrome have been described (Cummings and Duchen 1981). In addition, bilateral damage to the amygdalae can result in disturbance of memory as well as disturbed social behaviour (Tranel and Hyman 1990) and impaired emotional facial recognition (Adolphs *et al.* 1999); the latter may contribute to some of the social behavioural disturbances in frontotemporal degeneration (Lavenu *et al.* 1999).

Disease of the basal ganglia can give rise to prominent cognitive and behavioural dysfunction, most commonly observed with massive destructions following haemorrhage, infarcts, or intoxications. Bilateral thalamic infarcts are dominated by disturbances of attention, but fluctuating aphasia is characteristic of left-sided lesions, whereas right thalamic infarcts can result in the 'left neglect' syndromes which mimic non-dominant parietal lesions (Castaigne *et al.* 1981; Graff *et al.* 1984). Thalamic haemorrhages tend to be more dramatic in their presentation and usually cause a relatively fluent aphasia with marked fluctuation and superimposed hypophonia (Luria 1977). Caudate lesions are more commonly associated with behavioural disturbances, usually abulia, than with motor syndromes (Bhatia and Marsden 1994). A syndrome of dysphasia with dysarthria and orofacial dyspraxia can occur with dominant head of caudate infarcts (Naeser *et al.* 1982), and dysphasic syndromes can also occur with lesions of the adjacent white matter or internal capsule (Damasio *et al.* 1982). Apraxia is very rare with lesions confined to basal ganglia but can be seen more commonly with lesions of the thalamus (Pramstaller and Marsden 1996). Bilateral lesions of the basal ganglia can result in profound behavioural disturbance. Marked apathy, similar to a frontal lobe syndrome, can occur with bilateral lesions of the globus pallidus, so-called pure psychic akinesia (Laplane *et al.* 1984). Bilateral infarction of the head of caudate has resulted in severely aggressive and criminal behaviour (Richfield *et al.* 1987). Occlusions of the basilar artery at the bifurcation, the so-called 'top of the basilar' syndrome, result in complex disorders of eye movement with convergent spasm, retraction nystagmus, and skew deviation. It is commonly accompanied by memory disturbance, an agitated confusional state with prominent visual hallucinations, so-called peduncular hallucinosis (Caplan 1980).

In addition, isolated brainstem lesions can result in cognitive impairment predominantly executive with sparing of posterior cortical function and memory, presumed to be due to involvement of ascending forebrain projections (Garrard *et al.* 2002). A dysexecutive syndrome can also be seen with cerebellar lesions and include visuospatial and memory deficits—the cerebellar cognitive affective syndrome. More posterior lobar and vermal lesions are reported to have more prominent behavioural changes with blunting of affect or disinhibition (Schmahmann and Sherman 1998)

34.4 Neuropsychological syndromes

In the preceding section clusters of neuropsychological deficits which are characteristic of damage to particular areas of cerebral cortex were described. Often relatively pure deficits may present to neurologists and more detailed understanding of the precise nature of the functional impairment, over and above the localizing significance, can be helpful in diagnosis and management.

34.4.1 Disorders of perception and the agnosias

Lissauer introduced the terms apperceptive and associative 'mind blindness' to distinguish between patients whose abilities to perceive and discriminate an object are impaired and those patients who are unable to recognize the object having correctly perceived it. Subsequently Freud introduced the term agnosia which subsequently replaced the term 'mind blindness'.

Visual agnosias have been most widely studied and Lissauer's original analysis has subsequently been proved useful in the analysis of neurological patients. In order to perceive an object a number of features must be analysed and processed such as shape, colour, location within space and movement; these may each be selectively damaged. Shape discrimination can be assessed by asking the patient to discriminate between rectangles and squares of increasing similarity (Efron 1968). Preservation of shape and location but with loss of colour, or achromatopsia, is rare (Zeki 1990) and is usually seen with bilateral damage to the fusiform and lingual gyri. There is commonly an associated superior altitudinal field defect (Meadows 1974a). Impairment of visual location, visual disorientation (Holmes 1918), can cause great impairment and render the patient functionally blind. Finally patients have very rarely been described with inability to detect movement indicating a further dimension of early visual processing (Zihl *et al.* 1983).

These abnormalities of early visual processing are most commonly seen with bilateral occipital and occipito-parietal lesions but can be seen with unilateral lesions in which case the deficit is found in the contralateral field of vision. In these instances the functional impairment may be less prominent and patients less likely to present with a specific history. These disorders of visual processing are usually due to ischaemic lesions, but visual disorientation can be found in degenerative lesions of the occipital and parietal cortex, sometimes referred to as posterior cortical atrophy, which can be a presenting feature of Alzheimer's disease (Benson *et al.* 1988).

In *apperceptive agnosia* there is impairment of the generation of a structured percept of an object despite adequate initial processing of shape, colour, and location. This impairment can be demonstrated in patients who have difficulty coping with perceptually difficult visual stimuli, such as incomplete line drawings, overlapping line drawings (De *et al.* 1969), fragmented letters, and unusual views (Warrington and Taylor 1973). Unlike the retinotopic organization of early visual processing, the entire visual field is involved in patients with apperceptive agnosia, and the minimal lesion is usually found in the posterior non-dominant parietal lobe.

Associative visual agnosia is very rare (Hecaen and Angelergues 1963). Patients can describe an object well and copy drawings of it precisely but it has no meaning for them. They are, however, able to recognize the object immediately through other sensory channels. Patients with visual agnosia usually have bilateral lesions but these can occur with unilateral left posterior parietal lesions. Classically, visual agnosia has been interpreted as a disconnection of the percept from a central meaning system but an alternative interpretation is the existence of modality-specific meaning systems and visual agnosia would thus be seen as a loss of the specific meaning system associated with the visual domain i.e. a visual semantic memory impairment (McCarthy and Warrington 1990).

Within the overall category of visual agnosias some defects have been singled out for particular consideration because they present as striking clinical deficits, examples include the inability to recognize colour, colour agnosia, and inability to recognize faces, prosopagnosia. *Colour agnosia* (Hecaen and Albert 1978), implies intact colour perception and semantic knowledge of colour, but the majority of cases appear to have an associated impairment of colour naming. Faces present a perceptually difficult visual task and clearly problems in visual processing and perception will result in difficulty in face recognition. However, some patients present with a relatively selective impairment of facial recognition, *prosopagnosia*. Such cases can present a striking clinical picture in that patients may be unable to recognize those very close to them but can immediately do so when they hear them speak or by looking for other clues in their dress or mannerisms. Prosopagnosia is usually associated with right occipito-temporal lesions and commonly associated with a left homonymous superior quadrantanopia (Meadows 1974b). Prosopagnosia can be analysed in a similar way to object recognition, employing tests to assess perceptual analysis such as matching pictures of faces and matching facial expressions. Some patients may have normal performance on face matching tasks but be quite unable to recognize familiar faces (De 1986). Others may show intact face matching but be unable to match facial expression (Etcoff 1984). These theoretical aspects of prosopagnosia have generated a number of information processing models of face recognition (Bruce and Young 1986).

Agnosias in other sensory domains have been far less well studied and have a less secure theoretical basis. Following Lissauers original terminology these have also been divided into apperceptive and associative agnosias (Vignolo 1982). *Cortical deafness* can occur with bilateral temporal lesions which result in deficits of discrimination, temporal sequencing and spatial localization of sound. *Pure word deafness* as an isolated agnosia for speech sounds (Goldstein 1974) usually overlaps with Wernicke's aphasia. *Auditory agnosia* is extremely rare and refers to patients with intact hearing and intact language comprehension but who are unable to recognize meaningful non-verbal sounds. *Amusia*, an inability to produce and/or appreciate music, is seen most commonly with non-dominant hemisphere lesions (Brust 1980), although the reading and writing of music notation may be seen with left hemisphere lesions.

Tactile agnosia is even less secure as a distinct syndrome. Patients with parietal lesions will often have difficulty with the appreciation of size, texture, and shape of objects held in the hand, and *astereognosis*, which is more strictly referred to as stereoanasthesia or stereohypoasthesia. These patients will have impairment of two-point discrimination and sometimes subtle proprioceptive changes. Strictly speaking astereognosis should exist when shape and discrimination is intact as evidenced by the patients' description, but recognition impaired. Astereognosis, however, may occur as a disconnection syndrome in patients with callosal lesions (Geschwind and Kaplan 1962). In such patients objects can be recognized in the right hand but not in the left, they can, however, be correctly identified from a visual array. This is interpreted as disconnection of the sensory information from the right parietal cortex reaching the left hemisphere language area. Although rarely diagnosed, patients are also reported with olfactory and gustatory agnosias.

34.4.2 Disorders of spatial awareness and the body image

We are aware of the existence of our bodies, their position in space, and the relation of their parts to one another because we receive data through numerous sensory channels; these include vision, cutaneous sensibility, and proprioceptive information from the muscles, joints, and labyrinths. The somatic impulses pass via the ventral nucleus of the thalamus to the supramarginal gyrus which is thus concerned with awareness of the opposite side of the body. This concept of the body in consciousness is known as the body-image or body-schema.

Symptoms of disorders of the body-image may be positive or negative. The chief positive symptom is the phantom, an illusion of the persistence of a part of the body lost by amputation such as a phantom limb, or an illusory awareness of a part from which sensation has been lost through interruption of afferent pathways. *Phantom limbs* after amputation may be painless or painful (Riddoch 1941). The painless phantom soon becomes less obtrusive, and gradually shortens eventually to disappear. Painful phantoms may persist indefinitely and cause much distress.

Impairment of spatial sense together with neglect is most commonly seen following right parietal lesions and can present a dramatic clinical picture. Such patients frequently manifest spatial disorientation both for external space and for the body-image. These patients often have an associated left hemiparesis, again indicating a relationship with the non-dominant parietal lobe (Benson *et al.* 1976). Patients with right hemisphere lesions exhibit neglect that is most obvious to the left side of the body and left space and these patients may not shave the left side of their face, may eat food only on the right side of the dinner plate, and may demonstrate neglect dyslexia. Most striking is inattention to left-sided deficits, such as left hemiparesis, so called *anosognosia* (Babinski 1914). In a less severe degree, patients may recognize their left hemiparesis but are unconcerned, a complication which can make rehabilitation extremely difficult (Denes *et al.* 1982).

There are two main theories of spatial neglect in association with lesions of the non-dominant parietal lobe. In the one it is proposed that there is a central representation of space which is damaged; in the other it is proposed that there is a defect of selective attention, a function for which the right hemisphere is dominant. The widely quoted evidence for the former is the report of two patients who were asked to imagine and describe a plaza in Milan, first when facing the cathedral and then from the perspective of the cathedral itself. The patients' descriptions ignored the left side on both occasions but resulted in a complete description when both imagined pictures were combined (Bisiach and Luzzati 1978).

34.4.3 Apraxia

Apraxia may be defined as the inability to carry out a purposive movement, the nature of which the patient understands, in the absence of severe motor weakness or paralysis, sensory loss or ataxia. For example, a patient who is asked to protrude his tongue is unable to do so on request, though he may carry out inappropriate movements such as opening his mouth; yet a moment later he spontaneously protrudes his tongue to lick his lips. Apraxia may involve any movement which is normally initiated voluntarily; movements of the eyes, face, muscles of articulation, chewing and swallowing, manipulation of objects, gestures with the upper limb, walking or sitting down. Apraxia is seen most commonly with left hemisphere lesions and is then found in association with dysphasia.

The terminology of apraxia is particularly confusing since it includes a number of conditions that are not genuinely apraxic and additionally employs terms which are derived from the theoretical framework developed by Liepmann at the turn of the century. He defined three types of apraxia, namely limb kinetic, ideomotor, and ideational. These were based upon a theoretical neuroanatomical model similar to those which have been developed for speech. *Limb-kinetic apraxia* represents a loss of dexterity particularly of hand movements and affecting all classes of movement. This can be the most difficult to identify in the presence of basal ganglia and pyramidal dysfunction (see Leiguarda and Marsden 2000; Zadikoff and Lang 2005). *Ideomotor dyspraxia* refers to the poor performance of a motor act in response to a verbal command; the patients know what to do but not how to achieve it. It has been interpreted in terms of disconnection although the most secure disconnection model for ideomotor dyspraxia is seen in patients with lesions of the anterior corpus callosum who have difficulty with performing tasks to command using the left hand, i.e. the 'praxis centre' in the left hemisphere is disconnected from the right motor cortex which controls the left hand (Geschwind 1965). This is sometimes referred to as callosal apraxia and is commonly seen with anterior cerebral artery infarcts.

Ideational apraxia was defined by Liepmann as an inability to perform a sequential motor act, even though each could be carried out separately. However, this is relatively non-specific and may be a feature of frontal lobe lesions reflecting a difficulty of programming rather than of dyspraxia. Ideational apraxia has been defined as an impairment in manipulation of actual objects (De Renzi *et al.* 1968). Subsequently, it was postulated that these patients have an agnosia for tool usage (De Renzi and Lucchelli 1988) and in this context is associated with lesions of the posterior dominant parietal lobe. Patients with clinically obvious apraxia are relatively rare but severely disabled. They have considerable difficulty with using a knife and fork and many other common tasks which require manual dexterity and they will often look at their hands in a bemused way. By contrast, patients who demonstrate ideomotor dyspraxia at the bedside may not be functionally disabled, representing merely a feature of a specific neurological examination. Assessment at the bedside includes dexterity, pantomime of transitive such as 'show me how to use a screwdriver', and intransitive such as 'show me how to wave goodbye' movements, use of real objects, sequence of movements, and copying hand postures; the latter exploring visuospatial motor encoding.

In addition to the syndromes described above, which relate to manual dexterity, other body part dyspraxias have been described. A dissociation between limb apraxia and an *axial dyspraxia*, demonstrated by difficulty with adopting truncal postures such as a boxer's stance, has been described. Patients with *gait apraxia* show severe impairment of walking and often of standing with additional truncal instability (Meyer and Barron 1960). *Orofacial dyspraxia* is found with dominant frontal lesions and often associated with a cortical dysarthria (Nathan 1947); such patients show a characteristic inability to make oral movements to command but with sparing of eye movements. When asked to cough they will frequently repeat the word 'cough', they will be observed however to carry out such motor acts spontaneously. This may be seen on a degenerative basis (Tyrrell *et al.* 1991).

Dressing apraxia and constructional apraxia reflect the difficulties with spatial encoding rather than motor programming and illustrate some of the confusion of terminology. *Dressing apraxia* is associated with non-dominant parietal lobe lesions and is normally seen in a context of spatial impairment and left-sided neglect.

Constructional apraxia (Kleist 1922) refers to a disorder of the spatial disposition of an action and is illustrated at the bedside by inability to copy a cube or to make a simple arrangement of matches. This was originally identified as a disconnection syndrome between spatial analysis and voluntary action. This remains to be proven and apparent constructional apraxia may be due to more than one defect. Patients with left hemisphere lesions may make dyspraxic errors and commonly they will retain the spatial organization but simplify a diagram, whereas those with right hemisphere lesions show impairment of the spatial organization and will often neglect the left side (Arrigoni and De Renzi 1964; Warrington *et al.* 1966).

34.4.4 Dysphasia

Since so much of the complexity of human behaviour depends upon language, impairment in this domain often presents early and with striking features. Historically the language disorders were also the first to be associated with precise focal brain lesions. The terms dysphasia and aphasia are often used interchangeably. American usage favours dysphasia for developmental or congenital language disorders, reserving aphasia for the acquired disorders of language. In Europe dysphasia has been applied in the strict sense of a partial acquired language disorder with aphasia referring to complete absence of language. This, however, is rarely adhered to strictly.

Aphasias are disturbances of language and not simply motor speech dysfunctions, thus a patient with aphasia will also have impairment of other aspects of language such as writing. The term *aphemia* is often used for patients with impaired speech but with intact writing. Dysarthrias refer to impaired speech sound production and dysphonia to local disturbances of the larynx and pharynx. Different terms have been used to describe the clinical spectrum of language disturbance. Many of these arose out of the early descriptions of neurologists which were based on a variety of theoretical constructs, for example conduction, transcortical motor and transcortical sensory aphasias were terms introduced to describe syndromes based upon theoretical models of language developed by Wernicke and Lichtheim. These models implied precise localization of function with fibre tracts connecting them. These were assumed to have an anatomical basis. Subsequently considerable advances were made in analysing individual components of speech comprehension and production using information- processing models. The theoretical models and diagrams in these instances relate to individual components of the process without implying any anatomical correlates. However, in some instances the information processing models can be correlated with the earlier theoretical and neurological models (Shallice 1981a). With improved techniques of neuro-imaging and particularly the opportunities of functional neuroimaging with activation paradigms on positron emission tomography scanning, considerable advances in the anatomical correlates of the individual processes in language can be anticipated.

A broad distinction between fluent and non fluent aphasias is followed here as this can provide a useful starting point for the clinician (Goodglass *et al.* 1964; Benson 1967) and within this framework the broad distinction between disturbances of production and comprehension are considered. Disturbances of reading, the dyslexias; of writing, the dysgraphias; and of calculation, the dyscalculias, are often associated but are considered separately.

Many of the clinical language syndromes can be identified by simple bedside testing although detailed neuropsychological assessment is required both for quantitation and careful dissection of the individual components of language failure. Examination should include careful observation of the patient's spontaneous speech which can be assisted by the use of picture description. The fluency should be noted and the occurrence of paraphasias documented. These may be phonemic paraphasias in which one or two of the syllables of the word are mistaken or substituted, for example 'stable' for 'table', a pattern more commonly found with anteriorlesions. By contrast semantic paraphasias, the substitution of semantically similar words, for example 'chair' for 'table', are found more commonly with posterior lesions. Comprehension can be tested at the single word and at the sentence level and there are a variety of neuropsychological tests available for this such as the Peabody test. For bedside testing, word comprehension can be assessed by giving the patient verbal instructions but care has to be exercised both in patients in whom intellect may be impaired and in patients with dyspraxia, if the task is motor dependent. Individual word comprehension can be tested by verbal definition and confrontational naming, which can also assess word retrieval. Repetition can provide evidence of dissociations between errors in spontaneous speech and repetition, and serves to distinguish the clinical syndromes of conduction aphasias and the transcortical aphasias. Finally both reading and writing should be assessed.

Aphasias are most commonly encountered in the setting of strokes and neoplasms producing focal lesions, and patients will often have associated neurological signs such as hemiparesis or visual field defects. More difficult to assess are patients with aphasia as part of a more generalized cognitive impairment, such as occurs with degenerative dementias. In addition patients with psychosis or with acute confusional states can create problems, although the language process itself is preserved if carefully observed. Thus paraphasic errors are rare in the psychosis although the rare schizophrenic 'word salad' can create diagnostic difficulty. However other evidence of disturbed behaviour is apparent, whereas the patient with jargon aphasia as part of a Wernicke's dysphasia is seen to behave normally in the realm of non-language behaviour. The mute patient presents a particular diagnostic challenge. Patients may be mute because of a severe dysphasia, but more commonly are anarthric in which case writing will be preserved. Alternatively mutism may be associated with disturbances in attention such as occurs with frontal and subfrontal disease and may be seen in akinetic mutism.

Clearly any assessment of a patient requires knowledge of handedness. The vast majority of people have language represented in the left hemisphere. Very rarely, less than 1 per cent of right-handed individuals may develop aphasia with right hemisphere lesions, termed crossed aphasia in dextrals (Zangwill 1979). In approximately half of patients who are left-handed language also resides in the left hemisphere.

Non-fluent and Broca's aphasia

In 1861 Broca described the case of Monsieur Leborgne who had sustained a stroke with damage to the left inferior frontal gyrus and underlying white matter. The patient was initially mute and was then left with a severe non-fluent aphasia. Broca's original terminology for this, aphemia, has subsequently been confined to patients with impairment of speech but preservation of writing, usually with lesions of the dominant inferior motor cortex. The term aphasia, introduced by Trousseau, supplanted the term aphemia

(Schiff *et al.* 1983). The striking feature of Broca's aphasia is that the speech is non-fluent being both slow and reduced in output and patients are often mute initially. The content of the speech may be impaired with frequent phonemic errors and is usually agrammatic with the omission of prepositions, adjectives, and adverbs. Repetition and confrontation naming are normally impaired although the patients are often helped by cueing. Writing is faulty both in morphology and in terms of spelling and grammar.

One of the clinically distinguishing features of Broca's aphasia is that the impairment is largely confined to language expression with relative preservation of comprehension. Indeed auditory comprehension of individual words is very well preserved, although performance on tests used to explore sentence comprehension is usually impaired (Goodglass and Kaplan 1972); similar subtle impairment of comprehension can be found in reading.

The traditional Broca's area, established in the original cases, is the posterior part of the inferior frontal gyrus. An associated ideomotor apraxia of the non-dominant hand may be found depending upon the extent of subcortical damage. Some patients will show quite striking dysarthria and associated orofacial dyspraxia.

Two distinct clinical patterns of Broca's aphasia have been described (Mohr *et al.* 1978), depending on the size of the lesion. Lesions confined to Broca's area and subcortical white matter, are usually due to embolic strokes in the anterior branches of the left middle cerebral artery, and are associated with rapid recovery of expressive speech. Patients with occlusions of the middle cerebral artery, sparing the territory of the inferior division which supplies the temporal lobe, or patients with occlusion of the internal carotid, have a more widespread lesion which renders them globally aphasic initially, and often unable to comprehend, together with a dense hemiparesis. However, over the subsequent months of recovery, comprehension improves and a residual Broca's aphasia remains.

The majority of patients with Broca's aphasia have suffered strokes, but the syndrome may also be found with tumours, although in a less pure form. Some cases of selective left hemisphere degeneration may also present with a non-fluent speech, as in primary progressive aphasia (Mesulam 1982), and some patients with degeneration of the frontal lobe, such as occurs in Pick's disease, may develop a striking orofacial dyspraxia with speech impairment (Tyrrell *et al.* 1991).

Lichtheim had originally proposed a model of cortical concept centres for words which were connected with the motor centre or word sound centre by transcortical pathways. Theoretical syndromes based on this model and involving disconnections between the centres were postulated for conduction, transcortical motor and transcortical sensory aphasias. In *transcortical motor aphasia* patients make frequent errors of speech production with a very low output and frequent phonemic paraphasias. Repetition is, however, intact. Transcortical motor aphasia is seen most commonly with anterior cerebral artery lesions with an associated ideomotor apraxia of the left hand and a right hemiparesis affecting the leg more than the arm.

By contrast with transcortical motor aphasia patients with *conduction aphasia* have a profound impairment of repetition with frequent phonemic paraphasias. This was explained as a disconnection between intact comprehension and intact motor centres which would allow for relatively normal spontaneous speech. Anatomically, this was attributed to lesions of the arcuate fasciculus which indeed are often associated with conduction aphasias.

Speech production requires the correct selection and ordering of phonemes, impairment of which gives rise to the characteristic phonemic errors seen with a conduction aphasia. A pattern of deficit has also been recognized due to impairment of the motor coordination required for phoneme production, so called kinetic speech production impairment. These are often associated with lesions in the inferior precentral gyrus (Lecours 1976). These disturbances of phoneme selection and expression can be distinguished from dysarthrias in which there are characteristic impairments in swallowing and generation of meaningless sounds which require the same motor apparatus.

Nominal dysphasia

Intact speech is dependent upon appropriate word retrieval. Patients with impaired word retrieval may be relatively fluent but their speech appears empty with frequent circumlocutions. Clinically this can be tested by confrontational naming when abnormalities may be apparent, particularly for low frequency words, that are less obvious in spontaneous speech. Word finding difficulties are found quite frequently in neurological practice, but can occur as a relatively isolated finding in patients with left temporal lobe lesions. The study of word retrieval has proved to be a fertile area for theoretical modelling and a number of important clinical observations have been made. First, naming may be modality specific, for example, patients may be unable to name when presented with the object visually but are able to do so when presented to touch, so called *optic aphasia*. In addition, patients have been described in whom a tactile naming impairment is confined to the left hand, and such cases are most easily interpreted as disconnection syndromes involving lesions of the corpus collosum (Geschwind and Kaplan 1962). Another distinction has been drawn between patients whose failure at naming is consistent, and patients in whom the failure varies between different testing sessions and appears to be sensitive to the precise timing of confrontation. Different names are failed on different occasions and an object may be correctly named, but not if immediately shown again. The latter has been interpreted as an impairment of access to the word store, so called *semantic access dysphasia* (Warrington and McCarthy 1983). This syndrome of impaired access can be contrasted with patients in whom confrontational naming is impaired because they have lost their verbal semantic memory; in these instances the same words tend not to be named and patients can quite often verbalize their loss of comprehension.

One of the striking clinical features of word finding difficulty is the phenomenon of category specificity, namely that certain categories of words are more impaired than others. In some instances, the specificity is so striking as to have been recognized as a distinct syndrome, for example, colour anomia in patients who are unable to name but can adequately match colours; letter naming or letter anomia, and body part naming, or autotopagnosia. Many additional dissociations have been demonstrated, for example, between action naming and object naming, living and inanimate objects etc. These *category-specific dysphasias* provide important theoretical insights into language organization and have been proposed to depend upon the association of the word with various attributes at the time meaning is acquired. Thus, for example, words which are associated with a strong visual component, of which colour would be the most striking, can be contrasted with those, such as tools and manipulable objects, which would depend on a major

proprioceptive input when the word acquires its meaning (Warrington and McCarthy 1987). Such an account would accord well with current concepts of parallel distributed processing across neural networks. These examples within the verbal domain reflect how concepts and meaning may be organized across all domains and deficits in verbal, visual, and possibly other sensory domains can be seen in semantic dementia (Section 34.6.4).

Wernicke's aphasia

Shortly after Broca's description, Karl Wernicke outlined the features of a fluent aphasia which in many respects provided the clinical counterpart of a Broca's aphasia. The striking feature is that such patients speak fluently but the speech is often empty of meaning with frequent semantic paraphasias. At times the paraphasias may be profound with frequent neologisms, so called jargon aphasia. Comprehension is invariably impaired as is reading. Writing reflects the language impairment with frequent semantic paraphasias and spelling errors. Repetition, as with Broca's aphasia is impaired. Wernicke's aphasia is commonly due to a vascular lesion and the acute onset of a jargon aphasia is usually due to an embolus to the inferior division of the middle cerebral artery. The area involved is the posterior superior temporal gyrus, often extending into the inferior parietal area. By contrast to Broca's aphasia, Wernicke's aphasia is often unaccompanied by neurological deficit on examination which to the unwary can lead to faulty diagnosis of a psychotic disturbance. A right superior quadrananopia may be found if carefully sought.

Intact comprehension requires intact speech perception and patients may be seen with so called *pure word deafness*. A number of these patients, with bilateral temporal lobe lesions, have been shown to have impairment of auditory temporal acuity (Auerbach *et al.* 1982). However, some patients have impairment of phoneme discrimination which can be associated with left temporal lobe lesions.

As the counterpart to transcortical motor aphasia, Lichtheim proposed the syndrome of *transcortical sensory aphasia* in which patients are able to repeat, and whose speech is fluent, but is associated with impaired comprehension and paraphasias. The model postulated that the word recognition centre was intact, as shown by an intact repetition, but dissociation from a central meaning system resulted in impaired comprehension. This syndrome is now interpreted as an impairment of verbal semantic memory. It can be seen with posterior border zone infarcts but more commonly in frontotemporal degeneration where it is the major feature of the clinical syndrome of semantic dementia (Section 34.6.4). Rarely patients may be seen with isolation of the speech area who are neither able to speak nor to understand but are able to repeat (Geschwind *et al.* 1968).

Dysprosody

Patients with Broca's aphasia often have impairment of the normal rhythm of speech, so called dysprosody. Sometimes this may be so pronounced as to be referred to as the 'foreign accent syndrome' (Blumstein *et al.* 1987). Impairment of the emotional expression and comprehension of speech is often found with right hemisphere lesions and attempts have been made to seek comparable dysphasia syndromes to those found with left hemisphere lesions (Ross 1981).

Subcortical aphasias

Dysphasias are classically associated with cortical lesions but subcortical damage is increasingly being recognized as a cause of dysphasia. To some extent these mimic the classical dysphasia syndromes depending upon their location, for example, anterior lesions resulting in a non-fluent aphasia. In many instances reduced metabolism can be demonstrated on positron emission tomography scanning in the appropriate cortical area and are presumed to reflect impaired projection systems. Other syndromes have been described, for example, ischaemic lesions of the head of the caudate nucleus and associated internal capsule present with striking orofacial dyspraxia, dysarthria, and non-fluent dysphasia (Naeser *et al.* 1982). Left thalamic haemorrhages frequently give rise to an aphasia, often with fluctuations in arousal with concomitant fluctuation in language function.

34.4.5. Disorders of speech production and dysarthria

Dysarthria as a disorder of articulation does not involve any disturbance in the proper construction and use of words. In the dysarthric patient, symbolic verbal formulation is normal: only the mechanism of verbal sound production is faulty. When so severely affected that the patient is totally unable to articulate, it is referred to as anarthria. *Dysphonia* is applied to local disturbance of the larynx and may also render the patient mute. As well as structural abnormalities of the larynx, impaired innervation of laryngeal muscles can lead to dysphonia. A 'bovine cough' is a simple clinical sign of inability to close the larynx.

The articulatory muscles on each side appear to be innervated by both cerebral hemispheres. Hence a unilateral corticospinal lesion, for example in the internal capsule, may cause temporary but not permanent dysarthria. However, an extensive unilateral lesion involving the motor cortex may cause persistent dysarthria, especially when the dominant hemisphere is involved, in this case the dysarthria is often associated with some degree of Broca's aphasia. Dysarthria is consistently produced, however, by bilateral corticospinal lesions, due, for example, to congenital diplegia, vascular lesions of both internal capsules, degeneration of both corticospinal tracts, as in motor neurone disease, and lesions such as tumours involving both corticospinal tracts together in the midbrain. With such lesions, the articulatory muscles are weak and spastic and the tongue appears smaller, firmer, and less mobile than normal. The jaw-jerk and the palatal and pharyngeal reflexes are exaggerated. Speech is slurred and often explosive, production of consonants, especially labials and dentals, being severely affected. Spastic dysarthria is usually associated with dysphagia and often with impairment of voluntary control over emotional expression, the syndrome of 'pseudobulbar palsy'.

With lesions of the corpus striatum, articulation is impaired, partly, at least, as a result of muscular rigidity. Thus in Wilson's disease and in Parkinsonism, articulation is slow and slurred owing to immobility of the lips and tongue and the pitch of the voice is monotonous. In the dystonias and Huntington's disease, dysarthria is common; indeed in severe cases speech may be unintelligible. In these diseases irregular respiration may also contribute to the dysarthria.

The co-ordination of articulation suffers severely when the cerebellar vermis is damaged and also when lesions involve the cerebellar connections in the brainstem. Speech in such cases is often explosive with slurring and undue separation of individual syllables; scanning or syllabic speech. Ataxic dysarthria of this character is seen after acute cerebellar lesions and in multiple sclerosis and the hereditary ataxias.

Lower motor neurone lesions cause wasting and weakness, and often fasciculation, of the muscles of articulation; a true bulbar palsy. In the early stages, the pronunciation of labials suffers most. Later, progressive weakness of the tongue impairs the production of dentals and gutterals, and weakness of the soft palate gives the voice a nasal quality. There is often associated dysphonia and finally total anarthria. Motor neurone disease is the commonest cause, but paresis of the bulbar muscles may also be seen in syringobulbia, bulbar poliomyelitis, cranial polyneuritis, and brainstem tumours.

Combinations of these varieties of dysarthria are common; for example, in multiple sclerosis the articulatory muscles may be both spastic and ataxic and in motor neurone disease, a combination of upper and lower motor neurone lesions may be present.

Diseases of the muscles, such as myasthenia gravis, polymyositis, and muscular dystrophy involving facial muscles, lead to a dysarthria similar to that resulting from lesions of the lower motor neurones. In myasthenia, fatigability may cause increasing slurring if the patient is asked to count aloud. In the myotonias, impaired muscular relaxation may add a spastic quality to the speech.

Palilalia

Palilalia is a rare disorder of speech which, as its name implies (Gk: palin, again; lalein, to chatter), is characterized by repetition of a phrase which the patient reiterates with increasing rapidity. Palilalia most frequently occurs in post-encephalitic Parkinsonism, in general paresis, frontotemporal degenerations, and in pseudobulbar palsy due to vascular lesions. In echolalia the patient repeats or echoes words or brief phrases spoken by the examiner. It is often seen in frontotemporal degenerations and in autism.

34.4.6 Agraphia, alexia, and acalculia

The majority of aphasic syndromes are associated with impairment of writing or dysgraphia. This is such a frequent association clinically that sparing of writing with impaired speech usually indicates that the speech impairment is due to a disruption of speech production rather than a pure dysphasic syndrome. Similarly, some patients may demonstrate dysgraphia with intact speech. *Agraphias* have been broadly divided into those which affect the processes of spelling and those which affect writing. The latter can be seen as a particular type of ideational dyspraxia, although patients are described in whom praxis is otherwise preserved. Disorders of spelling as such have also been subdivided and have allowed the generation of theoretical models similar to the informational processing models for reading (Shallice 1981b). Dysgraphias are typically associated with posterior dominant parietal lesions.

Acquired disorders of reading, or *dyslexias*, are commonly found with dysphasias but can present in isolation. These were originally classified by Dejerine into two broad groups, those with and those without dysgraphia. In clinical terms this distinction has stood the test of time. *Dyslexia with dysgraphia* is commonly seen with lesions of the left angular gyrus. This type of dyslexia is often seen together with agraphia, acalculia, and anomia and has been described as a distinct syndrome by Gerstmann, although this is only an observed clustering and these are not dependent on each other (Benton 1961). In the syndrome of *alexia without agraphia*, patients are unable to read but are able to write even though they cannot read their own writing. This is often found with colour anomia and patients usually have a right hemianopia. It is most commonly seen with lesions of the left parieto-occipital area, often involving the splenium of the corpus callosum. The classical interpretation of this syndrome has been that of a disconnection of visual information in the intact left field from the left angular gyrus. However, this does not readily explain a striking feature of these patients, which is the ability to read using a letter by letter strategy. Information-processing models have now largely replaced the neurological models of reading (Marshall and Newcombe 1966). Using these models, alexia without agraphia can be interpreted as a word form dyslexia supported by the fact that these patients are more impaired on reading script than print (Warrington and Shallice 1980). It is argued that following initial word form recognition, analysis may proceed either by a phonological route or by a sight vocabulary route. For further discussion of the information processing models see McCarthy and Warrington (1990).

Impairment of arithmetic ability is relatively common and is seen frequently with aphasia. However, it can be recognized as a selective lesion and most commonly is found with left parietal lesions. *Dyscalculia* commonly arises as a consequence of dyslexia or dysgraphia and so-called spatial acalculia due to visuospatial disorganization when written arithmetic calculations are performed.

34.4.7 Disorders of memory

Memory is the ability to store and subsequently retrieve past experience and is central to many cognitive functions, and the maintenance of an autobiographical memory is central to personal identity.

It is clear that memory is not a unitary function and there is a profusion of different terms. It is usual, however, to draw a distinction between short term or primary, and long term or secondary memory. Neurologists often use 'short-term' memory to describe memory for recent autobiographical events but it is preferable to confine this term to the concept of 'immediate' memory, developed by neuropsychologists. We use short-term or immediate memory to remember a telephone number, and is tested at the bedside using the digit span. A normal person can usually retain a maximum of seven or eight digits with rapid forgetting over some 30s unless rehearsed. Although less commonly tested at the bedside, patients may also have a selective impairment of short-term visual memory. The original simple model that memory involved entry into the short-term store before consolidation into a long-term store 'secondary memory' is no longer tenable as patients with impaired digit span may have normal learning and secondary memory. Current theories of the role of primary memory range from a component of the working memory model of Baddeley (1986), to involvement in language fluency or as a safety backup resource (Shallice and Warrington 1970). From the neurologist's point of view, reduced digit span is commonly seen with impairment of attention but as an isolated finding can be related to lesions of the inferior dominant parietal lobe where it often occurs with dyscalculia.

Disabling memory impairments arise when individuals lose the ability to maintain an autobiographical memory. This can occur in a variety of clinical situations such as dementia and confusional states, but it is the patients with otherwise intact cognition who have provided the main basis for study. Such patients typically have two components to their impairment, an *anterograde amnesia*, which is a deficit in acquiring new memories following the illness, and a *retrograde amnesia*, a loss of recall for events prior to the illness.

The patient HM, who had bilateral medial temporal lobe resections extending back some 8 cm from the temporal poles, has been intensively studied and provided important insights into amnesia. These studies demonstrate a profound impairment of both verbal and non-verbal learning with intact immediate recall as evidenced by normal digit span (Scoville and Milner 1957; Milner *et al.* 1968). Subsequent similar cases of amnesia have followed this pattern. The status of remote memories has been less secure. So-called Ribot's 'law' states that there is a direct relationship between the strength of a memory and its recency, i.e. old memories being better preserved, and indeed this is often observed at the bedside. However there are problems of interpretation since it is difficult to match the saliency of the remembered events and the apparent preservation of old memories probably just reflects a small stock of overlearned anecdotes. The preservation of these overlearned episodes are likely to engage semantic memory systems as evidenced by the better recall of recent as opposed to distant events in semantic dementia. Despite the profound memory loss in HM and other similar cases, there is preservation of certain types of learning, for example improved performance on the recognition of fragmented letters can be demonstrated and, most strikingly, a retained ability to acquire new motor skills often without recollection of having done so.

In the case of HM, impaired learning of both verbal and visual material was found, but these can be selectively impaired. Group studies indicate that patients with left hemisphere damage have impaired verbal memory and the converse for non-verbal or visual memory. Often these material-specific memory impairments do not present clinically, but rather are found on specific testing. On occasions, however, some do present to the neurologist, the most striking being a topographical memory impairment in patients who are unable to recall familiar routes or buildings (Patterson and Zangwill 1945).

Studies of amnesia had previously focussed on failure at various putative stages of memory in terms of input, storage, and retrieval. Impairments of input and consolidation suggested that the strength and endurance of a memory depended on the extent of processing and thus consolidation (Craik and Lockhart 1972). This interpretation of amnesia in terms of impaired storage had argued that consolidation and retrieval mechanisms were intact but that there was an increased rate of forgetting. More recently, however, attention has focussed on the dissociations between preserved and impaired memory functions, observations which cannot easily be accommodated within a simple unitary model of memory with consolidation and retrieval models. Cohen and Squire (1980) contrasted procedural learning, which reflects 'knowing how', and declarative knowledge, the 'knowing that' memory. This describes the situation of patients acquiring new motor skills often without explicit knowledge of having done so. One patient for example, was able to learn a new piano tune without any recollection of having done so, and replay if prompted with the initial bars (Starr and Phillips 1970). However, the dimension of declarative memory does not easily explain the deficits seen in patients such as HM in whom memory for words, part of declarative knowledge, is well preserved. Tulving (1973) drew attention to the difference between episodic memory, such as for day-to-day events, and semantic memory. Other approaches have looked at the dynamic processes involved rather than observed dichotomies, for example the processes involved in implicit and explicit learning, which are those variably dependent upon the degree of conscious recall.

Causes of amnesia

A large variety of diseases may be associated with amnesia (Kopelman 2002). Most commonly memory impairment is seen with confusional states or dementia, but in these instances the amnesia is part of a wider spectrum of cognitive impairment (see below). Diseases affecting the medial temporal lobes and other structures on the limbic circuit may cause amnesia. Thus midline tumours in the region of the third ventricle, such as craniopharyngiomas, colloid cysts, massive pituitary tumours (Williams and Pennybacker 1954), thalamic gliomas, and tumours of the splenium of the corpus collosum, believed to be due to involvement of the fornix (Rudge and Warrington 1991) can all cause severe amnesia. Inflammatory disorders such as sarcoidosis and other granulomatous lesions in the same areas and limbic encephalitis as a paraneoplastic phenomenon (Henson and Urich 1982) are also associated with amnesia. A profound memory impairment is also commonly seen with herpes encephalitis due to the selective involvement of the medial temporal cortex. Vascular events of the posterior cerebral artery which supplies medial temporal lobe and hippocampus can cause amnesia (Benson et al. 1974). This often occurs with bilateral damage in association with basilar artery syndromes, but can occur with unilateral cerebral artery occlusions particularly in the elderly, which may be due to pre-existing contralateral hippocampal damage.

The Korsakoff syndrome

This is a striking amnesia which usually follows a Wernicke's encephalopathy (Section 34.5). Patients present with a profound amnesia and are quite unable to remember events even within the last half an hour but may be shown to have implicit learning, for example of motor skills. Other tests of cognitive function are well preserved in the pure form of Korsakoff syndrome but in clinical practice a spectrum may be seen with the more generalized cognitive impairment of alcoholic dementia at one end and patients with a pure Korsakoff syndrome at the other (Cutting 1978). It is, however, the striking contrast between the profound amnesia and the relatively minor additional cognitive defects that characterizes Korsakoff's syndrome.

Transient loss of memory

This is also a common clinical problem with both anterograde and retrograde amnesia, the latter shrinking on recovery leaving a gap in memory for the period of anterograde memory impairment subsequent to the onset. Temporary memory loss may occur with a variety of conditions which result in either generalized cerebral dysfunction or with selective disturbance of the medial temporal lobe and diencephalon, such as in temporal lobe epilepsy. Cerebral tumours may cause episodic memory loss which may also be on an epileptic basis (Lisak and Zimmerman 1977). Head injury and drugs, especially alcohol (Goodwin et al. 1969) and benzodiazepines, are common causes. Transient memory impairment may also occur as a feature of transient ischaemic episodes in posterior cerebral territory. A picture resembling transient global amnesia may also occur with migraine but this is usually with a clinical history of previous migraine attacks and the episodes are normally followed by headache (Caplan et al. 1981). Psychogenic amnesia and hysterical fugue states, a not infrequent topic of newspaper stories, are discussed below.

Transient epileptic amnesia

In addition to the transient memory loss associated directly with seizures, patients with transient epileptic amnesia suffer recurrent episodes of amnesia and report gaps in their autobiographical memory for which they have no recall (Zeman et al. 1998) A sleep EEG may be required to demonstrate temporal lobe epileptiform discharges. Patients with transient epileptic amnesia have been shown to have an accelerated rate of forgetting (Manes et al. 2005).

Transient global amnesia

This is a striking syndrome affecting the middle aged and the elderly. The history is obtained of sudden onset of impairment of episodic or autobiographical memory. The attacks are frequently reported to have been triggered by sexual intercourse or sudden cold; winter bathing in the elderly being a classic history (Fisher 1982). The attacks last from 1 to 24 h and the patients may repeatedly ask questions and appear anxious but otherwise able to drive home and there have been reports of virtuoso musical performances during an attack. Testing during an attack reveals profound loss of episodic memory with a variable retrograde component. Investigations are usually normal (Quinette et al. 2006) and a full recovery can be anticipated, although some patients are left with mild memory impairment (Mazzucchi et al. 1980). Recurrences are rare but do occur in a small proportion (Shuping et al. 1979). A vascular cause may underlie the syndrome of transient global amnesia (Fisher and Adams 1958), although supportive evidence has been difficult to establish and epidemiological studies have suggested a closer link to migraine (Hodges and Warlow 1990). Venous ischaemia has been suggested as a mechanism due to observed triggers which can be associated with a Valsalva manoeuvre (Winbeck et al. 2005).

Paramnesias

Confabulation, the production of memories without any basis in real events, can accompany amnesia and is most commonly observed in Korsakoff's syndrome. However, confabulation can be seen in patients with adequate performance on routine memory testing and is therefore more appropriately considered as a paramnesia (Stuss et al. 1978). Two classes of confabulation have been distinguished (Berlyne 1972). Momentary or provoked confabulation occurs in response to questions and requests for specific information that the patient might reasonably be expected to know. More florid is fantastic confabulation in which the patient will spontaneously produce bizarre accounts often far in excess of what might be required in response to the situation. Confabulation is found most commonly in association with frontal lobe lesions, particularly in medial frontal lobe lesions (Stuss et al. 1978; Damasio et al. 1985).

Reduplicative paramnesia (Pick 1903) describes a behavioural disturbance in which the patient transposes or reduplicates places. For example, he might claim that his home is not his normal home but a very similar building in which he is staying or alternatively that the hospital in which he has been admitted is next door. Such patients usually have bifrontal or right frontal lesions (Kapur et al. 1988). A similar syndrome is the *Capgras syndrome* in which patients refuse to recognize people, often close relatives who are familiar to them, claiming them to be impostors. Capgras' syndrome is usually seen in association with right hemisphere lesions (Alexander et al. 1979). Patients with the *Fregoli syndrome* believe that persons are all the same but adopting disguises. Reduplicative paramnesias are often seen in dementia with Lewy bodies.

34.5 **Wernicke–Korsakoff syndrome**

In 1887, Korsakoff reported a syndrome of mental changes, predominantly of memory impairment in association with polyneuropathy. The tendency to confabulate was noted, as was the link with alcohol abuse, but it was also reported with typhoid fever and prolonged vomiting. Six years earlier, Wernicke had described an acute illness of ataxia, ophthalmoplegia, polyneuropathy, and a confusional state. It was later observed that these syndromes often occurred sequentially in the same patient. Now the majority of cases are associated with alcohol abuse and thiamine deficiency plays a key role in Wernicke's encephalopathy (Section 5.2.3).

Wernicke's encephalopathy presents with an acute or subacute confusional state together with oculomotor disturbance and ataxia. The confusional state involves inattention and disorientation but can progress to drowsiness and even to coma and death if the underlying thiamine deficiency is not recognized and treated. The examination may reveal nystagmus, VI nerve palsies, conjugate gaze palsies, and ataxia. Criteria reflecting the classic triad of encephalopathy, oculomotor disturbance and ataxia have been developed (Caine *et al.* 1997). The occurrence of any two of the following: dietary deficiency, oculomotor palsies, cerebellar dysfunction, and altered mental state, can provide high diagnostic accuracy.

Korsakoff's syndrome emerges as the Wernicke's encephalopathy resolves and the confusional state clears to reveal a profound amnesia with both an inability to recall recent events and to learn new facts (Kopelman 1995). It may emerge gradually without a prominent preceding encephalopathic stage. The key component is the impairment of memory out of proportion to other cognitive domains. The deficit relates to event or episodic memory with sparing of semantic memory, and implicit learning can be demonstrated. Classically, patients also confabulate and will easily confuse the temporal sequence of memories that are recalled; this is believed to relate to additional frontal damage and indeed, many patients lack insight and initiative indicative of a frontal syndrome. However, more widespread cognitive deficits may be seen along with a clinical picture which merges with 'alcohol dementia' (Cutting 1978). Some have suggested that Korsakoff's syndrome and alcohol dementia are all part of the same spectrum but memory impairment of the former does set it apart as does involvement of diencephalic structures. Bilateral haemorrhagic lesions in the areas of the third and fourth ventricle and aqueduct are characteristic of Wernicke's encephalopathy and the diencephalic lesions are critical to the emergent amnesia of Korsakoff's syndrome. The original proposal that damage to the dorsal medial nucleus of the thalamus was the minimal lesion (Victor *et al.* 1989), has been considered too specific and anterior and medial thalamic and mammillary body involvement are also implicated. Thiamine deficiency is believed to underlie the pathogenesis of the disorder and explains its occurrence in other disorders such as prolonged vomiting, nutritional deficiency, and hyperemesis gravidarum. In patients who abuse alcohol, there is considerable variability in susceptibility to Wernicke–Korsakoff syndrome suggesting not only subtle differences in diet but also potential genetic factors.

The key to the treatment is to make the diagnosis which should be considered in all patients presenting acutely with unexplained

cognitive impairment or coma. Treatment is with intravenous thiamine, at least 100 mg daily. In the acute situation it is important to give thiamine before intravenous glucose as a sudden glucose load may exhaust the thiamine required as a cofactor in the pentose phosphate shunt. (Heye *et al.* 1994).

34.6 **The Dementias**

34.6.1 **Approaches to a differential diagnosis**

The term dementia refers to the clinical syndrome of impairment in multiple domains of cognitive function, which must include impairment of episodic memory, in a patient who remains alert with normal arousal. This serves to distinguish patients with dementia from those with confusional states or delirium, who have abnormal arousal, and from those with a single discrete cognitive deficit such as a dysphasia. This definition has been used to create operational criteria, such as those of the Diagnostic and Statistical Manual of the American Psychiatric Association, DSM-IV. An important feature of these definitions is that the dementia must be sufficiently severe to interfere with social and occupational function. There is clearly no simple cut-off from normality and whether this criterion is met will depend upon premorbid function and on the patient's social and occupational demands. This is particularly relevant to the elderly in whom some decline of cognitive function, particularly memory, is commonplace. Attempts have been made to distinguish patients with a progressive dementia, normally due to Alzheimer's disease, from less severe memory impairment associated with ageing. The terms benign senescent forgetfulness, age-associated memory impairment, and age-associated cognitive decline have been used but it is clear that many patients presenting in late life with memory complaints will be in the early stages of Alzheimer's disease even if not fulfilling the formal criteria of dementia. The term *mild cognitive impairment* was introduced to reflect this group of patients who had a memory impairment but were not demented (Petersen *et al.* 2001). Patients with this amnestic mild cognitive impairment progress to fulfil criteria for Alzheimer's disease at a rate of 10–15 per cent per year. With the recognition that mild cognitive impairment may be heterogeneous with non-amnestic presentations and multiple domain involvement, the concept has been expanded beyond only memory (Winblad *et al.* 2004; Jicha and Petersen 2007).

The definition of dementia is a broad one and it is not surprising, therefore, that the syndrome can be associated with a large variety of diseases (Table 34.1). All of these may be associated with widespread cognitive impairment but the clinical patterns vary and can lead to characteristic features. A broad distinction has been drawn between subcortical and cortical dementias (Albert *et al.* 1974; Cummings 1986). The term *subcortical dementia* was originally applied to the cognitive deficits seen in progressive supranuclear palsy and Huntington's disease, and is characterized by a marked slowness in cognition with additional impairments of motivation and attention. Indeed if the patient is allowed time then performance on routine neuropsychological testing may improve, but performance in everyday life remains severely compromised. The diseases that are most commonly found with this type of dementia are those affecting the basal ganglia and frontal connections. Dysphasia, dyspraxia, and agnosia are not prominent in these patients, by contrast to those with *cortical dementia*, in whom the

Table 34.1 Illustrative causes of adult-onset dementia

Primary cerebral degenerations	Cerebral Infections and Inflammatory Disorders	Cerebrovascular disease	Metabolic and toxic causes
Alzheimer's disease	Neurosyphilis	Multiple cortical infarcts	These generally present as confusional states
Down's syndrome and Alzheimer histopathology	Viral encephalitis esp Herpes simplex	Multiple lacunar infarcts	Hypothyroidism
Pick's disease	HIV infection		Hyper- and hypocalcemia
Progressive supranuclear palsy	Progressive multifocal leucoencephalopathy	Binswanger's disease	Hypoglycemia
Corticobasal degeneration	Subacute sclerosing panencephalitis	Congophilic angiopathy including hereditary	Hypo- and hypernatremia
Frontal lobe degeneration	Subacute rubella encephalitis	Thrombotic thrombocytopoenic purpura	Uremia
Dementia with motor neuron disease inclusions	Viral, bacterial, and fungal meningitides	Subacute diencephalic angioencephalopathy	Dialysis dementia
Parkinsonism–dementia complex of Guam	Whipples disease	Subdural haematoma	Chronic hepatic encephalopathy
Parkinson's disease and dementia	Behcet's syndrome	Giant aneurysms	Hashimoto's encephalopathy
Dementia with Lewy bodies	Disseminated encephalomyelitis	Arteriovenous malformations	Wernicke–Korsakoff syndrome
	Multiple sclerosis	Hyperviscosity syndromes	Alcoholic dementia
Thalamic degeneration	Sarcoidosis	Dutch amyloid angiopathy	Marchiafava–Bignami disease
Calcification of the basal ganglia with neurofibrillarytangles		Cerebral autosomal dominant arteriopathy with subcortical infarcts and leucoencephalopathy (CADASIL)	Hypoxia
Huntington's disease			Drugs, poisons, heavy metals
Spinocerebellar degenerations		Cranial arteritis	Vitamin B_{12} deficiency
Progressive myoclonic epilepsy		Cerebral arteritides inc polyarteritis nodosa, systemic lupus erythematosis, thromboangiitis obliterans, and granulomatous angiitis	Pellagra
			Malabsorption syndrome and celiac encephalopathy

Prion dementias	Neoplasms	Inherited metabolic and storage disorders	Miscellaneous
Sporadic and familial Creutzfeldt–Jakob Disease	Meningiomas	Porphyria	Aqueduct stenosis
Gerstmann–Straussler–Schenke Syndrome	Gliomas especially callosal	Wilson's disease	Normal pressure hydrocephalus
Kuru	Parapituitary tumours	Mitochondrial cytopathies	Open and closed head injuries
Iatrogenic Creutzfeldt–Jakob Disease	Pineal and midbrain tumours	Kuf's disease	Dementia pugilistica
Variant Creutzfeldt–Jakob disease	Cerebral lymphoma	Metachromatic and adrenoleucodystrophies	Post cerebral irradiation
Familial fatal insomnia	Cerebral metastases	Membranous lipodystrophy	
	Carcinomatous meningitis	Cerebrotendinous xanthomatosis	
	Paraneoplastic syndromes including limbic encephalitis		

memory impairment is generally not improved by cues, and the speed of cognition is relatively normal. The prototypic cortical dementia is Alzheimer's disease and the characteristic clinical pattern reflects damage to the cortical association areas (Table 34.2).

Of the many diseases listed in Table 34.1 some are very rare, and in others dementia is only an occasional clinical feature. The cognitive impairment may vary from minor deficits found only on specific testing to a more prominent dementia. Vascular dementia, for example, subsumes a large variety of different pathological entities, with a variable cognitive profile.

Only a minority of patients with dementia are found to have a treatable disease in published series, the most common being the subcortical dementia of depressive illness, but other treatable diseases include benign neoplasms, infections, vitamin deficiencies, and normal pressure hydrocephalus. However, the proportion of patients with reversible dementia seen by neurologists has lessened with earlier and better assessment (Clarfield 2003). In view of the profound consequences of dementia and the emerging treatment possibilities, a comprehensive approach to investigation and early diagnosis is justified, although clearly the extent of investigations will be determined by the clinical features and age of the patient. Both European (Waldemar *et al.* 2007) and US (Doody *et al.* 2001) guidelines on investigation and management of patients with dementia have been published. It would be inappropriate to undertake intensive investigation in a very elderly patient with multiple systemic illnesses and cognitive impairment; by contrast, the young

Table 34.2 Cortical and subcortical dementias
Derived in part from Cummings (1986)

	Subcortical dementia	Cortical dementia
Severity	Mild to moderate	More severe earlier in course
Speed of cognition	Slow	Normal but frequent errors
Neuropsychology	Memory impairment, recall aided by cues	More severe memory impairment unaided by cues. Dysphasia, dyspraxia, agnosia
Mood	Apathy Depression	Depression less common
Motor abnormalities	Extrapyramidal Dysarthria	Common *Gegenhalten*
Neuropathology	Prominent changes in striatum and thalamus	Prominent changes in cortical association areas

or atypical patient requires exhaustive investigation. All patients presenting with dementia should undergo a careful neuropsychological assessment. This will establish whether the patient does indeed have a generalized cognitive impairment and will help to detail the pattern of deficits. Routine biochemical and haematological determinations, which should include an ESR, thyroid function, syphilis and Borrelia serology, and autoantibody screen, will identify many of the potentially remediable causes. Vitamin B_{12} is often found to be low but rarely implicated as a cause of dementia. At-risk individuals with a subcortical dementia may require HIV testing. Chest X-ray and ECG complete the general assessment which will also serve to identify comorbid illnesses which may contribute to the cognitive impairment. Other blood tests, for example, genotyping, will depend upon the clinical features.

All patients should have neuroimaging, either CT scanning or MRI. Tumours and hydrocephalus can be excluded by both imaging modalities but MRI has the additional advantage of more detailed imaging of white matter disease in demyelination and vascular disease. Increasingly, MRI can also provide information on regional atrophy, such as the selective atrophy of hippocampi in Alzheimer's disease, which is of value in differential diagnosis (Jack *et al.* 1999). Serial MRI may also demonstrate rates of atrophy in degenerative disease outside the normal range (Fox and Schott 2004). Positron emission tomography, PET, can also identify regional deficits of blood flow and cerebral metabolism in degenerative dementia, using either 15-oxygen or fluorodeoxyglucose scanning. Thus, a characteristic feature of Alzheimer's disease is posterior biparietal bitemporal hypometabolism (Frackowiak *et al.* 1981) which can be contrasted with frontal deficits in the fronto-temporal degenerations (Silverman *et al.* 2001). The use of single photon emission tomography is more readily available and although it lacks the quantitative information provided by positron emission tomography scanning can also contribute to the identification of regional deficits, most useful in frontal degeneration where atrophy may be difficult to identify on structural images. The EEG may identify patients with subclinical seizure activity and the characteristic changes of the spongiform encephalopathies; in addition it can assist in the distinction of patients with Alzheimer's

disease, with early slow wave changes, from patients with fronto-temporal degenerations in whom the EEG is relatively well preserved (Chan *et al.* 2004). Examination of the cerebrospinal fluid is important in patients in whom one suspects an inflammatory or infective cause and increasingly CSF protein markers are useful such as $A\beta_{1-42}$ and tau in Alzheimer's disease (Blennow 2003). A relatively specific protein marker, P14-3-3 is found in cerebrospinal fluid in Creutzfeldt–Jakob disease (Lemstra 2000). Tissue biopsy, of muscle to detect mitochondrial cytopathy, of skin for Kufs disease, or of tonsil for variant Creutzfeldt–Jakob disease (Hill *et al.* 1999) may be diagnostic. In rare instances meningeal and cerebral tissue biopsy will be necessary, and may be the only way to establish a diagnosis of isolated cerebral vasculitis (Warren *et al.* 2005).

The majority of the diseases causing dementia which are listed in Table 34.1 are discussed elsewhere. In general, most cause a largely subcortical picture. In many of these instances there will be other clinical clues. The primary degenerative dementias give rise to a dementia with few other neurological findings in the early stages and these are dealt with in detail here.

34.6.2 Alzheimer's disease

In 1906, Alois Alzheimer described a 51-year-old lady with dementia and senile plaques and neurofibrillary tangles at autopsy. Originally Alzheimer's disease was viewed as a rare pre-senile dementia. However, clinicopathological studies in the 1960s demonstrated an overall relationship between dementia and the presence of senile plaques in both young and elderly demented patients, (Blessed *et al.* 1968) and the view emerged that Alzheimer's disease and senile dementia of the Alzheimer type was a single disease. Nevertheless, within the broad group of Alzheimer's disease cases, clinical and pathological heterogeneity can be observed. This originally focused on the age at onset with subtle distinctions being made between early and late onset disease. At a neuropathological level, distinctions have been drawn between cases that consist predominantly of neurofibrillary tangles and those that consist predominantly of senile plaques. However, the most robust biological categorization relates to cases with a clear family history. Up to 40 per cent have a family history of an affected first degree relative (Farrer *et al.* 1990); rarely there is a clear autosomal dominant history. Pathogenic mutations in three different genes have been identified in this group, namely *presenilin 1* and *2* and amyloid precursor protein, *APP*, genes which account for the majority of the autosomal dominant familial Alzheimer's disease cases described. Inheritance of an E_4 allele of the apolipoprotein E gene is associated with an increased risk of developing Alzheimer's disease (Corder *et al.* 1993).

It is now recognized that Alzheimer's disease is the major cause of dementia. Epidemiological studies indicate a doubling of the prevalence of dementia with each decade above 65 years to a prevalence approaching 50 per cent in those aged 85 and above (Evans *et al.* 1989). Approximately 70 per cent of cases of dementia will be due to Alzheimer's disease, either alone or in combination with vascular disease. Moreover vascular risk factors per se are also associated with a higher incident rate of Alzheimer's disease in longitudinal studies.

Clinical features

Alzheimer's disease is a disorder of middle and late life. Early onset cases are described in the fourth and fifth decade but these are rare and almost exclusively familial. The clinical features of familial

Alzheimer's disease associated with mutations in the *APP* and *presenilin* genes are broadly similar to sporadic disease apart from the age at onset. However, cases with mutation at *APP* 692 have more amyloid angiopathy with cerebral haemorrhages and thus, share similarities to hereditary cerebral haemorrhage with amyloidosis of the Dutch type due to mutations at *APP* 693, and some *preenilin 1* mutations are associated with a spastic paraparesis (Crook *et al.* 1998).

The classical presentation of Alzheimer's disease is with memory impairment. In some patients there may be a relatively prolonged course with isolated memory deficits until late into the disease (Hodges 2006). Patients who fulfil the criteria for mild cognitive impairment (Section 34.6.1) with an isolated memory impairment often progress to Alzheimer's disease confirmed at autopsy or found to have neurofibrillary tangles suggestive of early Alzheimer's disease (Petersen *et al.* 2006). The memory impairment primarily affects episodic and autobiographical memory, the patient forgets appointments, and mislays objects. Procedural memory and learning may be relatively preserved and, as with Korsakoff's syndrome, patients may demonstrate procedural learning without apparent parallel declarative learning. It is often stated that remote memory is preserved but if specifically investigated remote memory is also disrupted. Language impairment can occur relatively early and emergence of verbal semantic memory impairment implicates involvement of temporal lobe structures, extending beyond the hippocampus. Alzheimer patients frequently complain of difficulty with people's names which may reflect early impairment in the semantic memory system (Thompson *et al.* 2002). Dyspraxia is generally a late feature, although ideomotor dyspraxia is often found if specifically sought. Visuospatial and visuoperceptual deficits are also prominent and in some patients may be the presenting feature. Patients are quite often unaware of their cognitive deficits, often being brought to the attention of doctors by their relatives. The denial or anosognosia is not related to severity (Feher *et al.* 1991).

In addition to the classical presentation with impairment of episodic memory, some patients present with prominent frontal or language impairment (Galton *et al.* 2000). Patients with a biparietal presentation can be difficult to distinguish from corticobasal degeneration. Posterior cortical atrophy was a term used to describe patients with a variety of posterior parietal and parieto occipital features (Benson *et al.* 1988). The majority of such patients do have Alzheimer's disease and are sometimes referred to as the visual variant; early on dorsal visual processing deficits are the most prominent (McMonagle *et al.* 2006). However, some patients can present similarly with dementia with Lewy bodies (Section 34.6.3), Creutzfeldt–Jakob disease (Section 34.6.5), and corticobasal degeneration (Section 40.3.10).

In addition to the cognitive deficits, patients with Alzheimer's disease develop a number of neuropsychiatric features. Depression is very common and to a minor degree occurs in the majority of patients (Wragg *et al.* 1989) but needs to be distinguished from apathy. Psychosis is also common as the disease progresses, often with delusions of theft which can be difficult to manage for family members. Hallucinations do occur but if prominent should raise the possibility of dementia with Lewy bodies (Section 34.6.3). Agitation is also common as the disease progresses and is associated with increased burden of neurofibrillary tangles in the orbitofrontal and cingulate cortex (Tekin *et al.* 2001).

The general neurological examination is relatively normal in Alzheimer's disease at presentation, although motor abnormalities of extrapyramidal type commonly emerge as the disease progresses. Many patients with additional bradykinesia are found to have dementia with Lewy bodies. Primitive reflexes such as the instinctive grasp reaction, rooting, and sucking occur late. Generalized seizures occur in 10–20 per cent over the total course of the disease and myoclonus is relatively common in familial Alzheimer's disease.

Structural neuroimaging characteristically shows medial temporal lobe atrophy which can be seen on CT scan or more specifically, hippocampal atrophy on MRI (Scheltens *et al.* 2002) (Fig. 34.2). Functional imaging with single photon emission tomography or positron emission tomography reveals a posterior biparietal bitemporal pattern of hypometabolism. EEG shows slowing and loss of alpha rhythm relatively early in the disease. Routine CSF examination is normal but $A\beta_{1-42}$ concentration is reduced and both total tau and phosphotau increased. The combination of the two results in high sensitivity and specificity in younger Alzheimer's disease patients at least (Schoonenboom *et al.* 2004). The criteria for diagnosis of Alzheimer's disease provide levels of probable and possible diagnosis; definite Alzheimer's disease is reserved for those in whom histological confirmation of plaques and tangles is available (McKhann *et al.* 1984). Clinical diagnosis has steadily improved and, using these criteria, high sensitivity of 80–90 per cent but lower specificity of around 70 per cent are commonly reported (Knopman *et al.* 2001).

Neuropathology

The histopathological hallmarks of the disease are neurofibrillary tangles and senile plaques. Neurofibrillary tangles are intraneuronal and found predominantly in the allocortex and temporoparietal neocortex (Fig. 34.3). There is a predilection for pyramidal cells to be involved, particularly in layers three and five of the neocortex, the CA1 layer of the hippocampus, subiculum, and layers two and five of the entorhinal cortex. Braak and Braak quantified regional tangle formation in normal elderly and mild to severe Alzheimer's disease cases and suggested a staging system with progression of the disease from entorhinal cortex and hippocampus to neocortex (Braak and Braak 1991). This progression can be visualized *in vivo* by mapping the progressive atrophy in patients with familial Alzheimer's disease using MRI (Scahill *et al.* 2002). Neurofibrillary tangles consist of paired helical filaments which can be seen under electron microscopy (Kidd 1963), with a filament diameter of 10 nm wound in a double helix with a periodicity of 160 nm. Neurofibrillary tangles are also found within the dystrophic neurites of senile plaques. The major component of the paired helical filament has been shown to be the microtubule-associated protein tau which is abnormally phosphorylated (Lee *et al.* 1991). All six tau isoforms are deposited in Alzheimer's disease (Goedert *et al.* 1992) which distinguishes the tau pathology from that found in progressive supranuclear palsy, Pick's disease, and the other tauopathies.

Senile plaques are also found predominantly in neocortical association areas and consist of glial processes, abnormal nerve endings or dystrophic neuritis, and a central core of β-amyloid; they vary between 25 and 200 microns in diameter. β-amyloid is also deposited in cerebral blood vessels. The β-amyloid protein has been isolated and shown to be derived from a much larger transmembrane molecule, the amyloid precursor protein (Kang *et al.* 1987). Diffuse plaques are not associated with dystrophic neurites

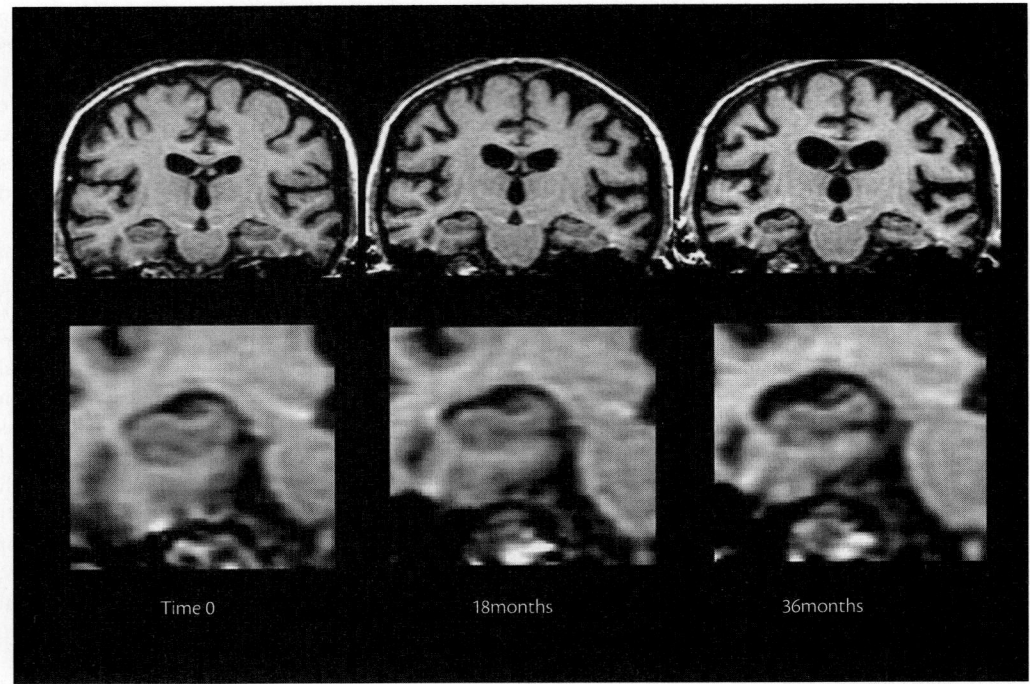

Fig. 34.2 Progressive hippocampal atrophy in a patient with Alzheimer's disease. Coronal T$_1$ weighted MRI.

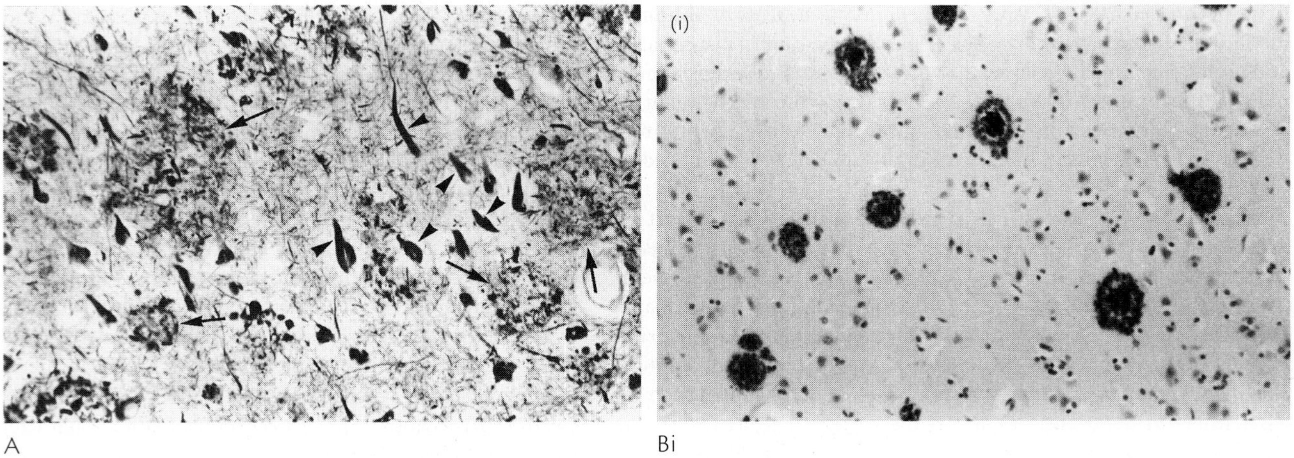

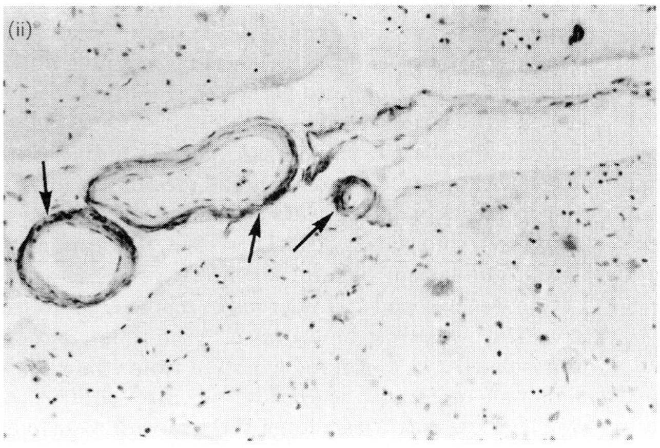

Time 0   18months   36months

Fig. 34.3 Alzheimer's disease. A. Senile plaques (arrows) and neurofibrillary tangles (arrowheads) in the hippocampus. Modified Bielschowsky silver impregnation. ×300. B. Positive immunostaining of (i) plaques (×250) and (ii) blood vessels (×250), with an antibody to βA$_4$ protein. Avidin–biotin complex method. (Courtesy of Professor BH Anderton, PL Lantos and Mr A Brady.)

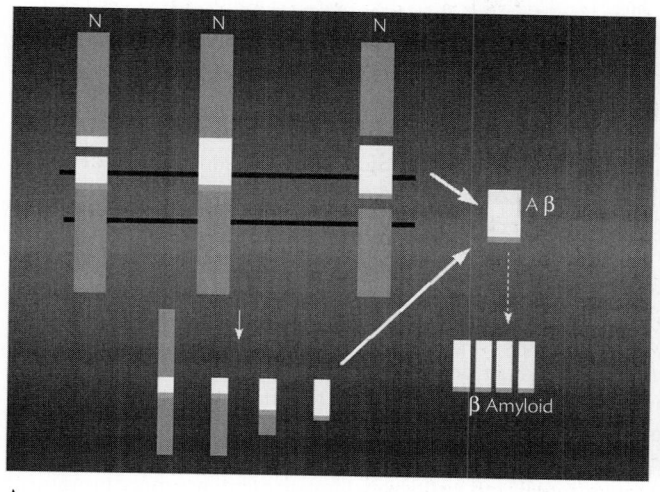
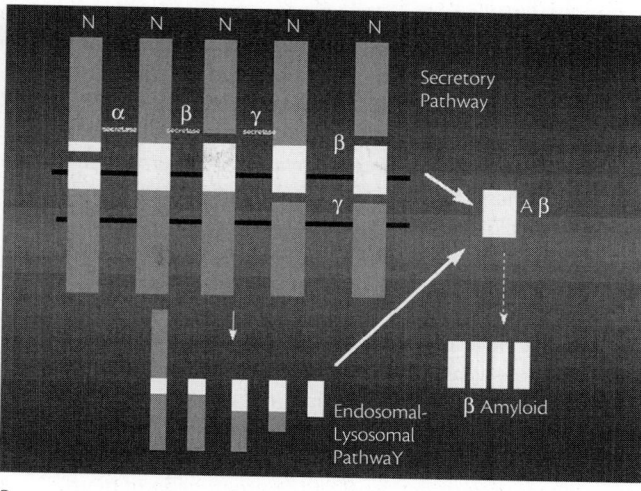

Fig. 34.4 Amyloid precursor protein (APP) metabolism. A. Normal metabolism of APP to release soluble $A\beta_{1-40}$. B. Metabolism of APP with release of $A\beta_{1-42}$ leads to increased formation of β-amyloid.

and are believed to precede the classical mature plaque. The central amyloid core consists predominantly of the $A\beta_{1-42}$ species which may act as a nidus for subsequent fibrillary amyloid deposition (Iwatsubo *et al.* 1995).

Neuronal cell loss is also maximal in the hippocampus and association areas of the neocortex. Cell loss also occurs in subcortical nuclei which include the amygdala and the origins of the subcortical projection systems, the nucleus basalis of Meynert, the nucleus raphe, and the locus coeruleus. The damage to the nucleus basalis and septal nuclei results in the cholinergic deficit, a consistent but not specific feature of Alzheimer's disease. Replacement of the cholinergic deficit is the basis of current symptomatic treatment. There is an overall association between cell loss and the histological features of plaques and tangles and it is assumed that cells which contain neurofibrillary tangles are degenerating. Other histological changes found predominantly in the hippocampus include eosinophilic Hirano bodies and vacuolar changes in cytoplasm referred to as granulovacuolar degeneration. The severity of the histological changes of senile plaques and neurofibrillary tangles show an overall association with severity of dementia. They are not, however, specific to Alzheimer's disease and neurofibrillary tangles in the hippocampus are found commonly in normal old age, as are limited numbers of senile plaques throughout the cortex. This has led to quantitative criteria for diagnosis based on neuritic plaques, the Consortium to Establish a Registry of Alzheimer's Disease criteria now largely superseded by a combination with the Braak score of neurofibrillary tangle distribution, 'the National Institute on Aging—Reagan Criteria' (Newell *et al.* 1999).

The cause of Alzheimer's disease is still not established but considerable advances have been made which identify a central role for amyloid deposition. This is best understood in patients with autosomal dominant familial Alzheimer's disease. The rare mutations in the *APP* gene either increase the total amount of β-amyloid produced or alter the processing to favour amyloid deposition. The *APP* 670-671 mutation increases the total amount of β-amyloid peptide produced from the precursor protein and is similar in this

regard to Down's syndrome, where there is an increased amount of amyloid produced due to a gene dosage effect arising from the trisomy 21. The amyloid precursor protein is normally cleaved at putative β- and γ-secretase sites to release the 40 amino acid Aβ peptide whose physiological function is unclear. It can be measured as a soluble peptide in both plasma and CSF but will form fibrillary aggregates of β-amyloid in senile plaques. Mutations at *APP* 717, close to the γ-secretase site, result in a subtle increase in the proportion of molecules extended at the C terminus $A\beta_{1-42}$ (Fig. 34.4). $A\beta_{1-42}$ has a greater propensity to fibril formation and can act as a nidus for subsequent $A\beta_{1-40}$ deposition. The familial Alzheimer's disease cases associated with mutations in the *presenilin 1* and *2* genes (Cruts and Van Broeckhoven 1998), also result in altered metabolism of APP resulting in a relative increase in $A\beta_{1-42}$; and presenilin 1 is a component of the γ-secretase. *ApoE4*, also known to be an important genetic risk factor, is associated with enhanced amyloid deposition and may stabilize fibril formation.

These lines of evidence all support a central role for amyloid deposition and the 'amyloid cascade' hypothesis (Hardy and Higgins 1992). This model predicts that the neuronal loss and neurofibrillary tangle formation are secondary events, but as yet, it is not established how exactly neurodegeneration is triggered but protofibrils found early in the amyloid aggregation process are neurotoxic. Recent evidence for this central role is provided by transgenic mouse models arising from over expression of a human mutated *APP* gene, which result in senile plaque formation and can be prevented by immunization with $A\beta_{1-42}$ (Schenk *et al.* 1999). The secondary role for neurofibrillary tangles in this pathogenic cascade does not diminish their potential importance as a final common pathway.

Treatment

Advice, support, and a sensible prognosis for the carer should be available and information can be obtained from national Alzheimer's Disease Societies and Alzheimer's Disease International. Treatment of comorbid disease such as infections and heart failure are important as these may cause further deterioration in cognition and behaviour. Patients with dementia are very sensitive to the cognitive side effects of drugs (McKeith and Cummings 2005). Agitation and delusional symptoms may need to be treated but neuroleptic medication should be kept to an absolute minimum as there may be an increased risk of cerebrovascular adverse events. Atypical antipsychotics may be better but are still reported to be associated with a small increase in mortality (Schneider *et al.* 2005). Depression is a common early feature of Alzheimer's disease and a trial of a selective serotonin reuptake inhibitor is indicated if there is a clinical suspicion of depressive symptomology. Tricyclics should be avoided because of anticholinergic side effects. Patients who hold a driving licence should inform the national driving authority at the time of diagnosis according to national guidelines but the decision of when to stop driving will depend upon their cognitive function (Waldemar *et al.* 2007).

There is no current treatment to affect the progression of neurodegeneration in Alzheimer's disease but, following the effects of immunotherapy in transgenic models (Schenk *et al.* 1999), a number of trials of active and passive immunization are underway. One active immunization study had to be stopped due to development of meningoencephalitis. However, patients subsequently coming to autopsy had apparent clearance of amyloid (Nicoll *et al.* 2003). Inhibition of cholinesterase activity in the brain will enhance levels of acetylcholine and thus ameliorate the cholinergic deficit arising from degeneration of the ascending projections from basal forebrain to hippocampus and the neocortex. Donepezil, rivastigmine, and galantamine all have similar symptomatic effects which are modest but improvement in cognitive measures and activities of daily living is seen in 20–25 per cent of patients with mild to moderate Alzheimer's disease over and above a placebo response (Cochrane database of systematic reviews: 2006). There is no evidence of any effect on disease progression. Memantine, a non-competitive *N*-methyl-D-aspartate, or NMDA, receptor agonist represents the second class of licenced drugs for Alzheimer's disease, and is currently approved for moderate to severe disease. It may be combined with cholinesterase inhibitors (Tariot *et al.* 2004)

34.6.3 Dementia with Lewy bodies and Parkinson's disease dementia

Terminological confusion has surrounded the group of patients with Parkinsonian features and dementia. It has been known for a long time that patients with Parkinson's disease may develop cognitive impairment and such patients are found to have Lewy bodies, the pathological hallmark of Parkinson's disease, throughout the cerebral cortex. With systematic autopsy studies, such cases have emerged as being common, perhaps constituting as many as 15–20 per cent in some series of older patients (Rahkonen *et al.* 2003). Series based on clinical diagnosis would suggest a similar proportion. However, there are overlaps in terms of both the clinical and pathological features with Alzheimer's disease, for example, senile plaques found in dementia with Lewy bodies and Lewy bodies are found in classical Alzheimer's disease, making assessment difficult. However, it is now sufficiently clear that there is a distinct clinical picture with management implications to justify its own nosological status (McKeith *et al.* 2004). Dementia with Lewy bodies is now the preferred term, but this condition has previously been, and still is referred to as cortical Lewy body disease, Lewy

body dementia, senile dementia of the Lewy body type, and Lewy body variant of Alzheimer's disease.

Cognitive impairment emerges in the majority of patients with Parkinson's disease with time, typically after about 10 years, and if severe justifies the term Parkinson's disease dementia. At the end stage dementia with Lewy bodies and Parkinson's disease dementia will be indistinguishable. Conventionally dementia with Lewy bodies is reserved for cases in which the Parkinsonian syndrome follows or coincides with the cognitive syndrome. For research studies a maximum of 1 year before emergence of the dementia is allowable.

The characteristic clinical picture is that of a cognitive impairment which precedes or follows closely a symmetrical Parkinsonian syndrome although dementia with Lewy bodies may be unaccompanied by extrapyramidal failures. Tremor is rare compared with classical brainstem Lewy body Parkinson's disease. The cognitive impairment is similar to Alzheimer's disease except that executive, visuoperceptual, and visuospatial impairments are more prominent and the impairment of episodic memory less severe. The additional striking feature is marked fluctuation of cognitive function which can appear as a confusional state with impairment of attention. Reduplicative paramnesias and hallucinations complete the picture. In between these periods of worsening cognition, caregivers will often report episodes of lucidity. Hallucinations also occur late in Alzheimer's disease, however, if present at an early and mild stage, this is a strong indicator of an underlying diagnosis of dementia with Lewy bodies (Ballard *et al.* 1999). The features were formalized into criteria in which in addition to the central feature of dementia two out of three core features of a Parkinsonian syndrome, fluctuations and hallucinations were required to be present. These criteria have been updated (McKeith *et al.* 2005) with added suggestive features of a rapid eye movement sleep behaviour disorder, severe neuroleptic sensitivity, and low dopamine transporter uptake by single photon emission tomography or positron emission tomography imaging. If one or more suggestive features occur with one core feature, then a diagnosis of probable dementia with Lewy bodies can be made. In some patients, the disorder is of an apparent subacute onset with rapid deterioration and myoclonus may occur prompting a diagnosis of Creutzfeldt–Jakob disease.

The EEG usually shows slowing. MRI may show atrophy with relative greater involvement of the parahippocampal gyrus than hippocampus in contrast to Alzheimer's disease (O'Brien *et al.* 1997). Functional imaging shows the biparietal bitemporal pattern of Alzheimer's disease, but in addition, more prominent occipital changes which may reflect the visuoperceptual problems and hallucinations (Albin *et al.* 1996). Dopamine transporter imaging, which is reduced in Parkinson's disease and dementia with Lewy bodies, is spared in Alzheimer's disease and so may prove a useful diagnostic marker (O'Brien *et al.* 2004).

Histologically the Lewy body in dementia with Lewy bodies is identical to that found in brainstem Lewy body Parkinson's disease, consisting of eosinophilic inclusions. Alpha-synuclein is the major component of Lewy bodies (Spillantini *et al.* 1997) and immunohistochemistry not only shows the widespread distribution of Lewy bodies but also neuritic change in affected cells, so-called Lewy neurites (Fig. 34.5). Cases of Lewy body pathology can be classified according to whether brainstem, limbic/basal forebrain, or neocortical areas are involved. Limbic and neocortical involvement are the predominant patterns in dementia with Lewy bodies. Many cases have associated β-amyloid plaques but typically without the extensive

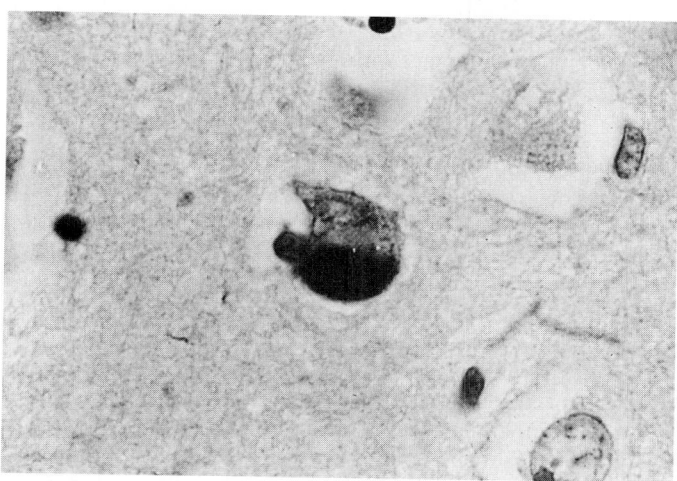

Fig. 34.5 A Lewy body in the insular cortex. Alpha-synuclein immunohistochemistry, magnification ×900. (Courtesy of Professor BH Anderton and Dr T Revesz.)

tau-positive neuritic change found in Alzheimer's disease. Many cases, however, will have additional neurofibrillary tangles and as the tangle load increases the probability that the pathological features are associated with a dementia with Lewy bodies phenotype is reduced (McKeith *et al.* 2004).

Management is difficult (McKeith *et al.* 2005). Treatment of the Parkinsonian syndrome may worsen the confusional state. Patients are exquisitely sensitive to neuroleptics which not only precipitate a profound extrapyramidal syndrome but also result in worsening cognition. There is some evidence that the atypical neuroleptics such as olanzepine and risperidone may be safer, but even here, there is a risk of worsening the clinical state and in general, all such drugs should be avoided if at all possible. Autopsy studies have demonstrated a cholinergic deficit comparable to or even greater than that found in Alzheimer's disease, and cholinersterase inhibitors are of value. A trial of rivastigmine in patients with Parkinson's disease dementia demonstrated cognitive improvement comparable to that seen in Alzheimer's disease (Emre *et al.* 2004).

34.6.4 Frontotemporal degenerations

The frontotemporal degenerations are a group of disorders which are considered together as they all share the characteristic feature of a degenerative process affecting the frontal and/or temporal lobes (Snowden *et al.* 1996). The anatomical distribution determines the clinical features which may often be asymmetric. These disorders are characterized by disturbances of language, speech production, frontal dysexecutive, and behavioural features. Clinical descriptions of the prototypic syndromes, frontotemporal dementia, also referred to as frontal variant, progressive non-fluent aphasia, and semantic dementia have been published (Neary *et al.* 1998), but other presentations, for example, primary progressive prosopagnosia (Tyrrell *et al.* 1990) also occur. The neuropathological processes underlying these diseases are variable and include Pick's disease, non-specific frontal lobe degeneration, hereditary tauopathies, corticobasal degeneration, and ubiquitin-positive tau-negative motor neurone disease type inclusions; Alzheimer's disease and prion disease can also present with the features of a frontal syndrome.

The difficulty in predicting the underlying neuropathology from the clinical features demands a clinical descriptive approach to the frontotemporal degenerations before considering the potential diseases defined neuropathologically. The nosology surrounding these disorders is confused further by the terminology relating to Pick's disease. Originally, Pick described cases of focal atrophy with the clinical features of dysphasia and/or a frontal lobe syndrome. This purely descriptive clinical terminology has been championed by some who suggest the term *Pick syndrome* (Kertesz and Munoz 1998) but in parallel, the development started by Alzheimer, who observed the swollen Pick cells and argentophilic inclusions, Pick bodies, has defined a particular type of neuropathology. More recently, our understanding of the molecular pathology of Pick's disease has further defined its nosological status.

Asymmetrical cortical degeneration is another term that was introduced by Caselli and Jack (1992) to refer to the group of slowly progressive focal degenerations. In a sense, all degenerative dementias start as a focal syndrome, even including Alzheimer's disease, which initially is that of a memory impairment. Most of these focal cortical degenerations are subsumed within the frontotemporal degenerative group but primary progressive apraxia, often associated with corticobasal degeneration and posterior cortical atrophy (Benson *et al.* 1988), usually associated with Alzheimer's disease, fall outside the general rubric of the frontotemporal degenerations.

Typically, the frontotemporal degenerations show structural or functional imaging changes indicative of frontotemporal degeneration. The EEG is characteristically normal by contrast to Alzheimer's disease and CSF examination is usually unremarkable.

Management of the frontotemporal degenerations can be very difficult with no available treatment to stop the progression. A randomized trial of trazodone has shown some benefit on the behavioural disturbance (Lebert *et al.* 2004).

Frontotemporal dementia

Although the term frontotemporal dementia has been retained as one of the prototypic clinical syndromes of frontotemporal degeneration (Neary *et al.* 1998), the clinical features are those of a frontal syndrome although the disease process does with time extend into the temporal lobes. An alternative term widely used is frontal variant of frontotemporal degeneration. It is the behavioural change as opposed to language or speech impairment, which characterize this group of patients. Social skills deteriorate early with difficulties at work, and the change in personality can be particularly distressing for the spouse. Some patients become disinhibited and overactive, reflecting predominant involvement of the orbitomedial frontal cortex, whilst others become apathetic, reflecting dorsolateral involvement. Patients who are at first overactive and at times aggressive, may become quieter as the disease progresses. Loss of empathy is common and distressing to the family (Gregory *et al.* 2002).

Other behavioural features include changes in food preference, usually towards sweet foods, ritualistic behaviours, and daily routines. Preservation of other cognitive skills means that the patient can go off walking in a stereotyped route, without ever getting lost. Dramatic failures on tests sensitive to frontal lobe function are apparent, although some patients may do well even on these tests, but fail on those of social cognition (Blair and Cipolotti 2000). Utilization behaviour may be observed. Speech in some gradually reduces with the features of a dynamic aphasia.

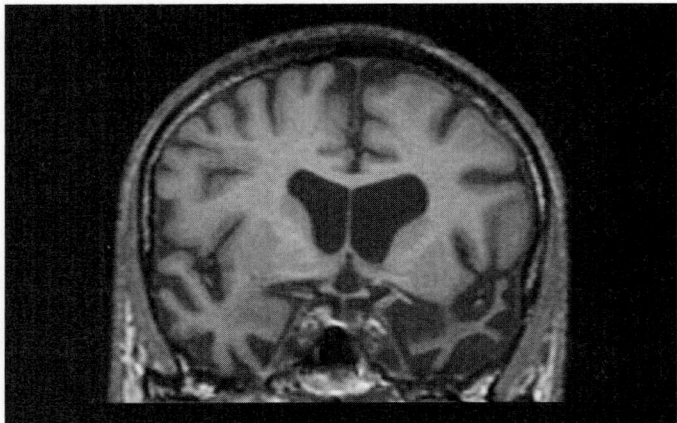

Fig. 34.6 A patient with semantic dementia showing left temporal lobe atrophy on Coronal T$_1$-weighted MR image.

Some patients develop a frontotemporal degeneration in association with motor neurone disease. Characteristically, this involves the anterior horn cells as opposed to long tracts and fasciculation is seen characteristically in the proximal upper limb (Mitsuyama *et al.* 1979).

Progressive non-fluent aphasia

A non-fluent aphasia with relative preservation of comprehension results in non-fluent and effortful speech, although writing early on may be preserved (Grossman and Ash 2004). Comprehension is relatively preserved even as the disease progresses to mutism. At this stage, patients may travel alone or even drive a car without difficulty. Social skills are preserved early but with disease progression, behavioural changes do emerge. Structural imaging shows left perisylvian atrophy which can also be seen on functional imaging. (Nestor *et al.* 2003)

Semantic dementia

The third commonly encountered syndrome is referred to as semantic dementia and describes patients whose presenting feature is that of a verbal semantic memory impairment (Snowden *et al.* 1989; Hodges *et al.* 1992). Speech is fluent and on first encounter the disorder may be missed. However, more detailed examination soon reveals the naming and comprehension deficit. Speech production is spared but it does become progressively more empty and communication more difficult. A surface dyslexia is common, representing a dependence on a phonological as opposed to lexical route to reading; it is characterized by difficulty with irregular words, tending to 'regularize' their pronunciation. Thus to pronounce 'pint' to rhyme with 'stint' would be a regularization error. As the disease progresses, the semantic memory deficit may extend into the visual domain, and rarely this can be the presenting feature, with the appearance of a visual associative agnosia. Behavioural changes with altered eating and ritualistic stereotypes occur. The disease starts with language impairment, reflecting left temporal lobe involvement, but with involvement of the right temporal lobe, patients can develop a prosopagnosia and an inability to recognize emotional expressions, impairing further their ability to communicate. MRI and functional imaging reveal asymmetric left anterior temporal lobe atrophy involving the anterior hippocampus

(Chan *et al.* 2001) (Fig. 34.6). In some patients, right temporal lobe atrophy predominates with more florid behavioural changes and in some, prosopagnosia is the presenting feature (Rosen *et al.* 2005).

Neuropathology of the frontotemporal degenerations

A variety of neuropathological changes are associated with the frontotemporal degenerations (Forman *et al.* 2006). Although particular syndromes tend to be associated with certain pathologies, there is overlap such that the underlying disease process cannot at present be predicted reliably from the clinical presentation alone. Moreover, it is not yet clear whether the apparent neuropathological entities do indeed represent distinct nosological entities or merely a spectrum of patterns of neuronal degeneration. Recent advances in molecular pathology are helping to clarify this complex group of disorders.

Tau-inclusions. In Pick's disease, there is striking asymmetric focal atrophy usually with a distinct border. Affected gyri may be very thin with a 'knife-edged' appearance. The anterior temporal and frontal lobes are predominantly involved with the superior temporal gyrus characteristically spared, particularly posteriorly. Astrocytic gliosis is variable. The hallmark lesion is the presence of tau-positive, ubiquitin-positive argentophilic inclusions known as Pick bodies, which are widespread and typically prominent in the dentate gyrus; there may in addition, be ballooned neurons or Pick cells. The balloon cells are tau positive and Aβ crystallin positive, and can be identified in a number of cases of frontotemporal degeneration as well as in typical cases of corticobasal degeneration (Binetti *et al.* 1998). Recent studies have demonstrated that the tau inclusions consist only of tau isoforms which contain the three repeat microtubule-binding domains, 'three repeat tau'; this now distinguishes Pick's disease from the other tauopathies (Delacourte *et al.* 1996). A wider neuropathological substrate has been attributed to Pick's disease in the past, but it is now considered a rare predominantly sporadic disorder. Pick inclusions or the features of corticobasal degenerations are a common finding at autopsy in patients with progressive non-fluent aphasia.

Hereditary tauopathies. Families with apparent autosomal dominant inheritance of frontotemporal degeneration have long been recognized. Many, but by no means all, are now linked to mutations in the *tau* gene (Poorkaj *et al.* 1998; Hutton *et al.* 1998). Neuropathologically these are associated with glial tau inclusions, tau positive ballooned neurons, 'Pick cells', and atypical Pick bodies. The clinical phenotype is variable and often includes an extrapyramidal syndrome and amyotrophy. Before the discovery of *tau* mutations, these cases were referred to as 'frontotemporal dementia with Parkinsonism linked to chromosome 17' (Foster *et al.* 1997).

Ubiquitin inclusions. Frontotemporal degeneration with motor neurone disease shows similar changes in the frontal lobe to those described under frontal lobe degeneration but with additional loss of motor neurones in the spinal cord. In addition, ubiquitin positive, tau negative inclusion bodies are found particularly in layer 2 of the frontal and temporal cortex. With improved immunohistochemistry an increasing number of cases are found to have ubiquitin positive, tau negative inclusions, which is a common finding in patients presenting with semantic dementia (Rossor *et al.* 2000; Davies *et al.* 2005). A proportion of cases of ubiquitin positive inclusions are familial and have characteristic intranuclear deposits, in some of these families mutations in the progranulin gene are found (Baker *et al.* 2006; Cruts *et al.* 2006). A large Danish family, with frontotemporal degeneration and ubiquitin inclusions, is associated with a mutation in *CHMP2b*, part of the ESCRT pathway involved in protein degradation (Skibinski *et al.* 2005).

Dementia lacking distinctive histology. Frontal lobe degeneration of the non-Alzheimer type (Brun 1993) lacks inclusion bodies or other hallmark features, and has also been referred to as 'dementia lacking distinctive histology', (Knopman *et al.* 1990) and non-specific dementia (Kim *et al.* 1981). Atrophy tends to be symmetrical in contrast to the asymmetry of Pick's disease, but can be severe, affecting the frontal lobes. There is microvacuolation or mild spongiosis and astrocytic gliosis, particularly in cortical laminae 1–3. There is no abnormal tau immunoreactivity or inclusions, but some cases have been reclassified as ubiquitin positive inclusions. Some cases of frontotemporal degeneration are found to have neurofilament inclusions (Josephs *et al.* 2003) and rarely, cases with a phenotype of frontal lobe degeneration are reported in whom the underlying histopathology is that of Alzheimer's disease or prion disease.

34.6.5 Prion diseases

The prion diseases (Section 42.8) also referred to as the spongiform encephalopathies, comprise Creutzfeldt–Jakob disease, variant Creutzfeldt–Jakob disease, Gerstmann Straussler–Schenker syndrome, Kuru and familial fatal insomnia; diseases that are now grouped together because they share a common disease mechanism involving aberrant protein folding (Prusiner *et al.* 1998). The novel disease mechanism which results in a disorder that can be both hereditary and transmissible, together with a threat of an epidemic of variant Creutzfeldt–Jakob disease consequent upon the bovine spongiform encephalopathy crisis in Europe, has focused considerable attention on these diseases despite their rarity.

The transmission of *Kuru*, the spongiform encephalopathy found amongst the Fore highlanders of Papua New Guinea, to non-human primates by Gajdusek led to the concept of a 'slow virus' although the transmissible agent remained elusive. Ultimately this was shown to be a protein devoid of nucleic acid—the prion protein PrP (Prusiner 1991). The transmissible prion protein, PrPSc, is derived from a normal cellular protein, PrPc, by post translational modification resulting in a high beta sheet content; the mechanism of the subsequent cellular degeneration is not established. The abnormal isoform of the protein, PrPSc, has the ability to induce aberrant folding of the host protein, hence the transmission which can occur either through cannibalism as in Kuru, or iatrogenically as with growth hormone derived from cadaveric pituitary glands and surgical interventions. Species differences in the PrP sequence make transmission of the disease across species inefficient. This species barrier can occur experimentally and is now believed to have occurred with variant Creutzfeldt–Jakob disease following the bovine spongiform encephalopathy epidemic. Mutations in the PrP gene are believed to facilitate the aberrant protein folding and underlie familial Creutzfeldt–Jakob disease, familial fatal insomnia and Gerstmann Straussler–Schenker syndrome. A common methionine/valine polymorphism at PrP 129 is a genetic risk factor for the disease (Palmer *et al.* 1991).

Prion diseases are multisystem central nervous system disorders with variable degrees of dementia, cerebellar, pyramidal and extrapyramidal features. *Creutzfeldt–Jakob disease* occurs world-wide with an annual incidence of about one per million. It is invariably fatal.

The classical triad of dementia, myoclonus, and abnormal EEG with periodic or pseudoperiodic complexes is seen typically in patients between the ages of 50 and 70 years. It has a subacute onset; rarely the disease can be extremely rapid with death within 2 months whereas in others there is a slower progression of the disease over 1–2 years (Brown *et al.* 1986). The different clinical phenotypes in terms of progression have been related to different isoforms of the aberrant prion protein, PrPSc (Parchi *et al.* 1996). About 10 per cent of cases of prion disease are familial with autosomal dominant transmission due to mutations in the *PrP* gene. The phenotype can be varied, some mimicking Alzheimer's disease and others Huntington's disease; cases with the 144 base-pair insert tend to have a slow progression with a variable phenotype (Collinge *et al.* 1992). Patients with a prominent cerebellar component were originally described as *Gerstmann Straussler–Schenker syndrome* and *familial fatal insomnia* (Lugaresi *et al.* 1986) is characterized by insomnia with loss or dramatic reduction of slow wave and rapid eye movement sleep, together with autonomic disturbance.

Iatrogenic cases have been associated with corneal grafting, dura mater grafts, and in-depth electrode recording. The majority of iatrogenic cases have been associated with cadaveric pituitary derived growth hormone treatment, a practice which was discontinued in the mid-1980s. These cases have more cerebellar features and less dementia (Fradkin *et al.* 1991). *Variant Creutzfeldt–Jakob disease* has emerged in the United Kingdom and France in the last 10 years and is believed to have resulted from ingestion of contaminated food from bovine spongiform encephalopathy infected cattle and recently by blood transfusions from individuals harbouring the disease (Llewelyn *et al.* 2004). Recent incidence cases of Kuru suggest that incubation periods in excess of 40 years can occur with human to human prion disease transmission (Collinge *et al.* 2006). Variant Creutzfeldt–Jakob disease is associated with a characteristic PrPSc isoform (Collinge *et al.* 1996). The cases of variant Creutzfeldt–Jakob disease are younger than classical Creutzfeldt–Jakob disease, some even in their teens, with early depression, anxiety, and dysaesthesiae. Cerebellar and basal ganglia features are more prominent than cognitive impairment early in the disorder (Will *et al.* 2000).

The characteristic histopathology is the spongiform change, although this is variable; it is associated with neuronal intracytoplasmic vacuolation together with astrocytosis and gliosis. The abnormal prion protein, PrPSc, can be demonstrated on immunohistochemistry and can form plaques, especially in the cerebellum in Gerstmann Straussler–Schenker syndrome. Variant Creutzfeldt–Jakob disease is also associated with PrP plaques reminiscent of Kuru.

Blood tests are usually normal in prion disease, although there can be mildly abnormal liver function tests. PrP genotyping will identify mutations in the familial cases and cases of variant Creutzfeldt–Jakob disease reported to date have all been methionine homozygous at PrP 129. MRI may show increased signal in the basal ganglia and cortex in sporadic Creutzfeldt–Jakob disease, most readily seen with diffusion weighted imaging (Tschampa *et al.* 2005). In variant Creutzfeldt–Jakob disease a characteristic increase in MRI signal in the posterior thalamus, the pulvinar sign, is best seen with FLAIR images (Zeidler *et al.* 2000). Cerebrospinal fluid is unremarkable although P14-3-3, which probably reflects rapid neuronal disintegration, is found in most classical Creutzfeldt–Jakob disease cases. The EEG may be normal or non-specific early in the disease but in classical Creutzfeldt–Jakob disease pseudoperiodic or periodic complexes are seen. Pharyngeal tonsillar biopsy in variant Creutzfeldt–Jakob disease can reveal the specific PrP^{Sc} isoform (Hill *et al.* 1999).

At present there is no specific treatment for the prion diseases (Trevitt and Collinge 2006). The PrP^{Sc} is highly resistant to degradation and thus instruments cannot be routinely sterilized. Neurosurgical instruments must be quarantined and if the diagnosis confirmed, destroyed. There is no need to barrier nurse patients but disposable instruments should be used for invasive procedures, and all samples clearly indicate the suspected diagnosis.

34.6.6 Vascular cognitive impairment

Impairment of blood supply to the brain used to be considered to be the main cause of dementia in the elderly until it was recognized that such a mechanism is rarely, if ever, implicated. Multiple small strokes, referred to as *multi-infarct dementia*, were subsequently identified to be an important mechanism both clinically and at autopsy (Hachinski *et al.* 1974). In most early neuropathological and clinical series, vascular dementia accounted for some 10–20 per cent of dementia cases alone, but importantly is a major concomitant of Alzheimer's disease in the elderly so called mixed dementia (Neuropathology group 2001). Moreover, up to 25 per cent of patients 3 months after a stroke were considered to have dementia using Diagnostic and Statistical Manual of the American Psychiatric Association, DSM-IV, criteria and up to 60 per cent had cognitive impairment (Pohjasvaara *et al.* 1997). The incidence of vascular dementia may be falling with better management of vascular risk factors. It is also believed that many cases of dementia with Lewy bodies were previously diagnosed clinically as vascular dementia. If cases of dementia where there is a vascular component are considered, then there is no doubt that vascular disease is a major cause or contributor to cognitive impairment (Hachinski and Bowler 1993). The term vascular dementia replaced multi-infarct dementia as it reflected the considerable heterogeneity of the condition and includes cases due to haemorrhage, small lacunar infarcts, large cortical infarcts, and vasculitides. However, the dependence of the term on the dementia criteria meant that only late cases would be included and the term vascular cognitive impairment is increasingly used (O'Brien *et al.* 2003).

Clinical criteria for the diagnosis of vascular dementia have been dominated by the development of criteria for Alzheimer's disease. Thus, memory remains as a key component and yet may be relatively less important in vascular dementia (Bowler *et al.* 1997) in which executive dysfunction is prominent. Early criteria assumed that stepwise deterioration and motor abnormalities would be characteristic and from this was developed the Hachinski score. Patients with a score of 4 or less were considered likely to be degenerative by contrast to those with a score of 7 or more who were thought to have a multi-infarct dementia. This remains a useful guide and series have been pathologically verified (Moroney *et al.* 1997). Criteria (Roman 1998) have been developed which require the appearance of cognitive impairment within 3 months of a stroke or sudden onset and fluctuation of cognitive impairment. In view of the potential contribution of focal neuropsychological deficits from a discrete stroke, the cognitive criteria for dementia are that, there should be memory impairment plus at least two other affected domains. There should also be relevant vascular changes on imaging which are thought to be directly related.

However, very different proportions of cases are diagnosed as vascular dementia depending upon the use of different consensus criteria. (Wetterling *et al.* 1996).

Various vascular pathologies are associated with vascular cognitive impairment: single discrete cortical infarcts, multiple infarcts or multi-infarct dementia, subcortical arteriosclerotic encephalopathy or Binswanger disease, hypoperfusion dementia, and haemorrhages. In reality, these may overlap.

Single discrete infarcts, for example, in right middle cerebral and posterior cerebral artery territories and thalamic infarcts, can present with a picture suggestive of dementia. Much more common however, are accumulation of deficits from multiple individual cortical and/or subcortical infarcts. Men are more commonly affected than women and there is usually a vascular history particularly of hypertension. There is a gradual accumulation of cognitive deficits with episodes of confusion or focal neurology but the onset may be insidious with gradual progression in the absence of stepwise deterioration. If there are mainly subcortical infarcts, patients tend to have a subcortical pattern of cognitive deficit with cognitive slowing and additional motor features. Some may develop an extrapyramidal syndrome and in others a pseudobulbar palsy can be prominent with pathological laughing and crying. Neuropathologically, multiple small subcortical infarcts appear to be more important in vascular dementia than single large infarcts (Esiri *et al.* 1997).

Subcortical arteriosclerotic encephalopathy: Binswanger's disease

Binswanger originally described eight cases of periventricular demyelination and dementia. This was considered a rarity until the advent of neuroimaging and many patients with white matter changes on scanning acquired this diagnosis. Clinically, it is very similar to patients with multiple subcortical infarcts namely frontal and subcortical cognitive features, dysarthria, and pseudobulbar palsy. Gait impairment may occur early and is characterized by a wide-based shuffling gait in contrast to the narrower base seen in Parkinson's disease (Thompson and Marsden 1987). Criteria have been suggested for the diagnosis of Binswanger's disease (Bennett *et al.* 1990).

Much confusion has arisen from attempts to diagnose Binswanger's disease from neuroimaging. Non-specific periventricular white matter abnormalities are common both in patients with dementia and in the non-demented elderly and the term leukoaraiosis has been proposed (Hachinski *et al.* 1987). Leukoaraiosis appears as low attenuation on CT scan particularly around the frontal and occipital horns and as increased signal on T2-weighted MRI (Fig. 34.7). Neuropathologically, there is demyelination, gliosis, and hyalinosis with fibrinoid necrosis of small blood vessels similar to that seen in hypertension. Minor degrees of white matter disease are also seen in pure Alzheimer's disease.

Increasingly the term subcortical vascular dementia or subcortical vascular cognitive impairment is used to encompass both 'Binswanger's disease' and the lacunar state as these normally co-exist. The subcortical cognitive impairment reflects the anatomical distribution of disease (Gold *et al.* 2005)

Treatment of vascular cognitive impairment is primarily that of management of vascular disease risk factors such as hypertension, smoking, diabetes, carotid stenosis, and heart disease. There have been few control trials of management of risk factors and its effect on cognition but, treatment of isolated systolic hypertension in the

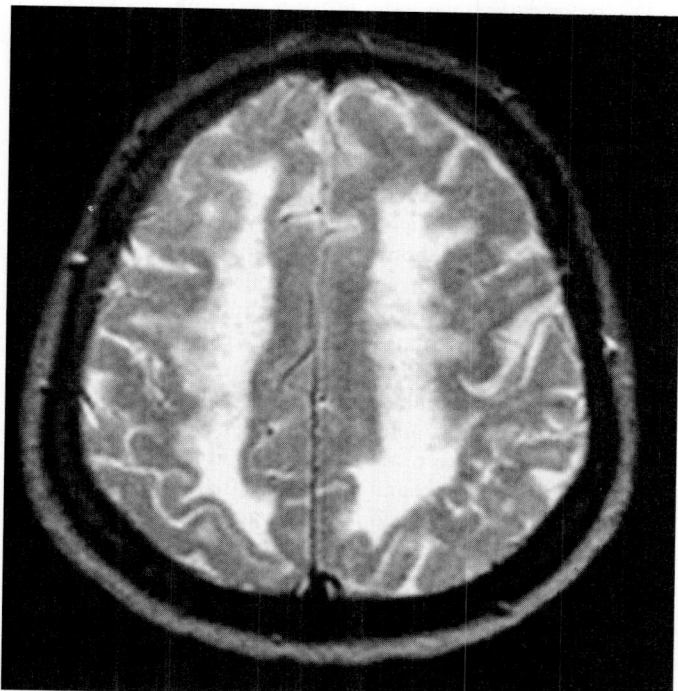

Fig. 34.7 Binswanger's disease. Increased signal on T_2-weighted MRI in a patient with hypertension and small vessel disease resulting in cognitive impairment.

elderly may reduce the incidence of dementia (Forette *et al.* 1998). Statins may be less important as hypertension is a greater risk factor than hypercholesterolaemia for small vessel disease (Shepherd *et al.* 2002). Cholinesterase inhibitors are reported to be of some benefit but it is not clear how much may relate to coexistent Alzheimer's disease (Bowler *et al.* 2003).

Other causes of vascular dementia

Significant cognitive impairment, sufficient to justify the criteria of dementia, can occur after subarachnoid haemorrhage, subdural haematomas, and global ischaemia following cardiac arrest with laminar necrosis and hippocampal cell loss, 'hypoperfusion dementia' (Sulkova 1987). A variety of vasculitides can also be associated with the early development of cognitive impairment and even present as a dementia; these include systemic lupus erythematosus (Section 36.3.1) and primary cerebral angiitis (Section 36.2.2) which is usually accompanied by headaches. Sneddon's syndrome (Rebollo *et al.* 1983) is the association of livedo reticularis with cerebrovascular disease and can present with cognitive impairment. A number, but not all, are associated with anticardiolipin antibodies.

The rare cases of *hereditary cerebral amyloidosis* both of the Icelandic, the Dutch, and the Flemish type, can be associated with cognitive impairment although the salient clinical feature is that of recurrent cerebral haemorrhage. Cerebral autosomal dominant arteriopathy with subcortical infarcts and leukoencephalopathy, known as *CADASIL* (Dichgans *et al.* 1998) is characterized by recurrent subcortical ischaemic events with the subsequent development of a pseudobulbar palsy and cognitive impairment. Early symptoms include migraine-like headache and psychiatric disturbance. The MRI scan shows a striking leukoencephalopathy in addition to multiple small infarcts with characteristic signal changes in the temporal pole and external capsule. This condition, which is increasingly recognized is linked to mutations in the Notch 3 gene (Joutel *et al.* 1996). Most mutations are found in exons 3 and 4. Notch 3 deposition can be demonstrated in the smooth muscle of the arteriole wall, and skin and muscle biopsies can be highly specific although sensitivity is low (Markus *et al.* 2002).

34.6.7 Miscellaneous causes of dementia

Dementia-plus syndromes

Since cognition is so easily disrupted by diseases affecting the cortex and its subcortical connections, cognitive impairment or dementia is very common in neurological practice. In the majority of cases the other clinical features provide the diagnostic clues. These dementia-plus syndromes are numerous and include Huntington's disease, some spinocerebellar ataxias, and a variety of inherited metabolic disorders, such as metachromatic leucodystrophy, Kuf's disease, lysosomal storage disorders and mitochondrial cytopathies. Normal pressure hydrocephalus can present as the classic triad of dementia, incontinence, and gait disturbance, but has tended to be over-diagnosed (Section 26.4.5). The cognitive impairment is very much that of a subcortical impairment with cognitive slowing. Patients with prominent cognitive impairment are more likely to have a coexisting degenerative disease and rarely, if ever, respond to shunting. Multiple sclerosis patients will often develop cognitive impairment and in some this can be a prominent and even presenting feature.

Infections

Syphilis is the classic infection which can be associated with dementia (Section 42.5.1). However, late cases of *Borrelia burgdorferi*, Lyme disease, can also be associated with dementia (Krupp *et al.* 1991) (Section 42.5.2) and Whipple's disease (Section 42.5.8) can rarely involve the cerebral cortex with cognitive impairment which responds to antibiotic treatment (Singer 1998). The other major infection associated with dementia is HIV encephalopathy and dementia can be a presenting feature (Janssen *et al.* 1992) (Section 43.3.5). The clinical features are those of a subcortical dementia: minor degrees of cognitive impairment insufficient to fulfil criteria for dementia are termed 'HIV-associated minor cognitive/motor disorder'. Only a small proportion of such patients progress to frank dementia. There is evidence that dementia is becoming less common with antiretroviral treatment but the prevalence of minor cognitive impairment has remained unchanged, possibly because of prolonged survival (Cysique *et al.* 2004). Progressive multifocal leucoencephalopathy (Section 42.3.14) is frequently associated with cognitive impairment and occurs in a variety of immunosuppressed patients as well as those with HIV.

Other causes

One of the reasons for neuroimaging of patients with dementia is to identify tumours, the treatment of which can result in cognitive improvement. However, with malignant tumours irradiation itself may give rise to late cognitive impairment (Keime-Guibert *et al.* 1998) (Section 5.9.2). The paraneoplastic phenomenon of limbic encephalitis, usually associated with carcinoma of the bronchus, can result in a memory impairment that precedes diagnosis of the tumour by a number of years and can mimic Alzheimer's disease (Section 38.4.2). However, examination of the CSF usually reveals oligoclonal bands and occasionally pleocytosis (Bakheit *et al.* 1990). Non-paraneoplastic limbic encephalitis has recently been associated with antibodies to the voltage-gated potassium channels (Vincent *et al.* 2004; Thieben *et al.* 2004). Drugs, particularly barbiturates, can lead to cognitive slowing as can heavy metals, such as lead, arsenic, manganese, and mercury. Workers in the felt hat industry who were exposed to mercury frequently developed a confusional state with cognitive impairment, hence the term 'mad as a hatter' (Section 5.7.5). Dementia pugilistica, arising from recurrent head injury, particularly in boxers, is associated with tangles and presents as a cognitive impairment with dysarthria and an extrapyramidal syndrome.

A variety of rare degenerative dementias are gradually being delineated although some, such as argyrophilic grain dementia (Braak *et al.* 1989) are really only diagnosed at autopsy. Kosaka (1994) described a series of patients with basal ganglia calcification on neuroimaging with neurofibrillary tangles but no senile plaques. The rare *Worster–Drought syndrome*, a familial disorder with dementia and spastic paraparesis is now recognized as a novel amyloidosis due to a stop codon mutation in the *BRI* gene (Vidal *et al.* 1999). The disease is now also referred to as familial British dementia. The MRI reveal deep white matter changes, lacunar infarcts, and prominent callosal atrophy (Mead *et al.* 2000). Patients with a frontotemporal dementia, Paget's disease, and inclusion body myositis have mutations in the valosin-containing protein which is involved in ubiquitination and degradation of proteins (Schroder *et al.* 2005).

34.7 Acute confusional states and delirium

Acute confusional states are extremely common, especially in the elderly, and may occur in up to 25 per cent of those admitted to medical and surgical wards (Taylor and Lewis 1993). It is very common in intensive care units (McNicoll *et al.* 2003), and after heart bypass (Santos *et al.* 2004). By contrast to dementia, confusional states are usually short-lived and potentially reversible with prominent impairments of attention and arousal (Lipowski 1990). The term 'acute confusional state' tends to be most commonly used in Europe and the term 'delirium' to be used in the US. The terms toxic psychosis, acute brain syndrome, acute encephalopathy, and transient cognitive disorder are all synonymous (Lipowski 1990). Patients are likely to be elderly, although the very young are also vulnerable. The condition is often seen within the setting of systemic disease or multiple drug therapy. Disruption of attention, a loss of the ability to focus on specific stimuli, is a key feature. This consists of impairment of selective attention with the patient unable to attend to the examiner and being continually distracted by irrelevant environmental stimuli (Geschwind 1982; Inouye 1990). Arousal may be impaired with somnolence and descent into stupor, or alternatively agitation may occur. However, attention is impaired even if arousal is enhanced. These disturbances of selective attention and arousal result in marked disorientation in place and time, particularly of a sense of time of day, but as with dementia personal identity is normally preserved. The sleep–wake cycle is disrupted with patients being awake at night leading to the familiar feature of the patient wandering into the street in their night clothes. Memory is impaired, in particular short-term memory as assessed by the digit span, is greatly disrupted by the impairment of attention. This is not, however, specific to confusional states.

There are frequent paramnesias such as reduplicative paramnesia and the Capgras syndrome (Section 34.4.7). Misperceptions, particularly with visual hallucinations are prominent, whereas auditory hallucinations should raise the suspicion of a functional psychosis. Patients may have bizarre and very frightening hallucinations often with image distortion, known as metamorphopsia, and these can occur particularly with drugs and toxins. Language is relatively preserved although the content is abnormal. Errors may relate to the disrupted thought processes and errors on naming may reflect visual misperceptions or reflect paramnesic errors; these disturbances on confrontational naming have been referred to as non-aphasic misnaming (Weinstein and Kahn 1952). As might be predicted, judgement is grossly impaired, particularly since anosognosia is the rule rather than the exception and may result in patients being combative and aggressive when being examined or if any attempt is made to restrain them. Mood lability is common and fluctuations in the cognitive state are frequent, such that the patient may be seen in a lucid interval when examined. The onset of a confusional state is acute or subacute over hours or days, in contrast to the common dementias, and fluctuation characteristic.

In addition to the cognitive features, clinical examination commonly reveals autonomic disturbance with tachycardia, hypertension, sweating, fever, and tachypnoea. Asterixis, myoclonus, increased tone, and carphologia, or plucking at the bedclothes, are often found. There may be evidence of systemic disease such as cardiac failure, respiratory failure or infections and these should be carefully sought.

Common difficulties with differential diagnosis include the functional psychoses (Lishman 1987). In general, patients with functional psychoses are orientated in time and place and not overtly dysphasic, although there may be a bizarre content. An abrupt onset in an older patient without a psychiatric history would be most unusual for functional psychosis. Similarly, any clouding of consciousness or disorientation suggests an acute confusional state, as does the predominance of visual over auditory hallucinations.

The distinction from dementia may be difficult with chronic confusional states. Moreover, the two often co-exist; acute confusional states may evolve into dementia, and patients with dementia are very prone to acute confusional states due to infections or drugs. Dementia with Lewy bodies (Section 34.6.3) may also present with features of a confusional state. Patients with dementia may also develop confusion with reduced or unfamiliar environmental cues, so called 'sundowning' at night. In general, when a patient presents with cognitive impairment suggestive of an acute confusional state with rapid onset and fluctuation, the more aggressive should be the investigation for reversible underlying causes.

The disturbance of arousal in acute confusional states can be attributed to dysfunction of the reticular activating system, and patients will have associated disturbances of the sleep–wake cycle and may deteriorate into stupor or even coma. Abnormalities of arousal, however, are not necessary to make a diagnosis of an acute confusional state in those instances in which impaired attention is prominent. Attention requires both disregard of irrelevant environmental stimuli and sustained attention on relevant stimuli. The association cerebral cortices provide the anatomical substrate of selective attention (Mesulam 1981) with particular involvement of the non-dominant parietal lobe as evidenced by the occurrence of acute confusional states with right middle cerebral artery infarcts

(Mesulam *et al.* 1976). A disturbance of selective attention may also occur with frontal lobe lesions.

The disturbances that cause acute confusional states tend to be generalized and it has been proposed that the selective vulnerability of the systems that subserve arousal and selective attention is due to their polysynaptic characteristics. It has also been suggested that the ascending cholinergic system may be particularly vulnerable because of the prominent confusional state that can occur with anticholinergic drugs (Tune *et al.* 1981; Han *et al.* 2001).

Causes

The causes of acute confusional states are legion. Acquired metabolic disturbances, particularly hypoxia and hypoglycaemia are common and require prompt treatment. Thiamine deficiency should always be considered in patients who are admitted to accident and emergency departments without a clear history, since the administration of glucose can precipitate a Wernicke's encephalopathy. Drugs are a very common cause, particularly in patients on multiple drug regimens (Carter *et al.* 1996). Anticholinergics, anti Parkinsonian, and benzodiazepine drugs, as causes of confusional states are commonly encountered in neurological patients (Foy *et al.* 1995; Jain 2001). Recreational drugs including alcohol, cocaine, 'crack', amphetamines and LSD are all well recognized causes. Acute confusional states commonly occur following surgery and are often due to a combination of drugs, hypoxia, electrolyte disturbance, and infection but may also be exacerbated by fragmented sleep. Confusional states are common after cataract surgery in the elderly and may relate to sensory deprivation (Summers and Reich 1979). Approximately 30 per cent of patients undergoing open heart surgery or coronary artery bypass grafting develop confusional states (Smith and Dimsdale 1989): recognition of a confusional state in the setting of an intensive care unit can be diagnostically challenging. Cerebrovascular disease can cause acute confusional states, particularly right middle cerebral artery infarcts (Mesulam *et al.* 1976), but also occur with posterior cerebral artery infarcts (Medina *et al.* 1974). Migraine attacks may rarely be associated with a confusional state, particularly in children. Epilepsy is a common cause both during the seizure and as a post-ictal phenomenon.

Investigation

Investigation should be thorough and will in part be directed by the general medical examination. Not only will this include screening of blood sugar, electrolytes, liver function tests, and a search for infection, but where necessary a drug screen. Neuroimaging, EEG to detect diffuse slowing or low voltage fast activity, and examination of CSF may all be necessary.

Management and prognosis

Management depends upon treatment of the underlying cause and symptomatic measures should be directed towards the maintenance of constant environmental stimuli so as to avoid over and under-stimulation; a night-light may be valuable (Meagher *et al.* 2001). Drug therapy should be kept to an absolute minimum, but patients may rarely require a neuroleptic for marked agitation, atypical antipsychotics with low anticholinegic properties are preferable.

Although in general the prognosis is good, delirium can be life-threatening and is an important determinant of institutionalization (Cole 2004). Many elderly patients may have underlying degenerative disease which renders them vulnerable to confusional states, and in whom a stable cognitive deficit may emerge as the confusional state clears (Levkoff *et al.* 1992; Jackson *et al.* 2004).

34.8 Hallucinosis

Hallucinations are a common accompaniment of confusional states, but they can also occur in a variety of other disorders. A *hallucination* may be defined as a sensory perception occurring without an external stimulus. An *illusion* is defined as a misinterpretation of an external stimulus and can occur in normal people, particularly with fatigue. A *delusion*, by contrast, is an idea or thought, such as a false concept of persecution, which has no substance in fact; in contrast to visual and auditory hallucinations, it is purely a thought process with no sensory content. Hallucinations and delusions may occur together in various toxic or confusional states and in psychotic illnesses, such as schizophrenia.

The principal circumstances in which hallucinations may occur are:

◆ in dreaming and the hypnagogic state;

◆ in pathological disorders of sleep;

◆ as a result of disease of the peripheral sense organs;

◆ focal disturbance of the central nervous system and in neurodegenerative diseases;

◆ drug-induced hallucinations; and

◆ in certain psychoses.

Hallucinations are not uncommon in states of drowsiness and will occur in hypnagogic states, when falling asleep, and hypnopompic, on awakening from sleep, but to an exaggerated extent in patients suffering from narcolepsy. Visual hallucinations can occur particularly in the elderly with reduced visual acuity and can occur with lesions anywhere along the visual pathway, *Charles Bonnet syndrome* (Santhouse *et al.* 2000). Visual hallucinations are a core feature of dementia with Lewy bodies. Extracampine hallucinations in which there is a sense of a person outside the visual field also occur commonly in dementia with Lewy bodies (Chan and Rossor 2002). Auditory hallucinations, including music, may also occur following acquired deafness, and again may be more common in the elderly (Stewart *et al.* 2006). Hallucinations involving various sensory modalities, together with perceptual illusions and other disorders of consciousness are particularly liable to occur as a result of lesions of the temporal lobes. These may frequently occur as epileptic phenomena. The perceptual illusions include disordered visual perception, for example macropsia or micropsia, and a similar alteration in auditory perception, feelings of unreality of the self or the surroundings, and disturbances of awareness of the body. Visual hallucinations have been described in which the individual feels that he is observing his own body from outside his physical self; this unusual phenomenon has some affinity with sensations of intense depersonalization or unreality. Visual hallucinations also sometimes occur as a result of epileptic discharge arising in the posterior part of the temporal lobe or in the parieto-occipital region. Agitated delirium and visual impairment may, for example, result from medial temporo-occipital infarction (Medina *et al.* 1977). L'hermitte described peduncular hallucinosis with lesions of the upper part of the brainstem which was interpreted as a dissociation of the state of sleep. Hallucinations may also result from brainstem compression (Dunn *et al.* 1983) and elementary auditory hallucinations may occur with pontine lesions, so-called 'pontine auditory hallucinosis'.

A variety of drugs may result in hallucinations, mescaline and lysergic acid, LSD, being notorious. However, withdrawal of alcohol can result in hallucinosis, which may become permanent particularly in established alcoholics. Hallucinations are an important diagnostic feature of the psychoses. They may occur with severe affective disorders, such as depression in which they are often associated with morbid features. Auditory hallucinations occurring in clear consciousness which are mood incongruent are very suggestive of schizophrenia. It is characteristic that such auditory hallucinations involve argument about the patient in the third person.

34.9 Psychiatric disorders in neurological practice

34.9.1 Mood disturbance

Affective disturbances commonly involve the neurologist, either because patients with a primary diagnosis of depression present with neurological symptoms, because the patient develops a depressive reaction to their neurological disability, or, less commonly, a depressive illness arises directly as a result of central nervous system disease.

A *depressive illness* is the commonest psychiatric diagnosis to be found in neurological practice (Kirk and Saunders 1977). The commonest presentations are headache, dizziness, and memory impairment. A careful history may unearth depressive symptomatology, such as persistent dysphoria, loss of appetite and libido, and disturbed sleep patterns with early morning waking. Suicidal ideation should be specifically sought. However, patients may not volunteer or may minimize mood symptoms, focusing instead on somatic complaints. Depressive pseudodementia (Caine 1981) is an important diagnosis since it is the main cause of a reversible dementia. The term 'pseudodementia' is not ideal, as these patients do have impaired cognition and probably share some of the neurochemical disturbances found in other dementias, namely deficits in the ascending monoaminergic projections to cerebral cortex. The clinical features are those of a subcortical dementia with marked slowing of cognition. The patients often complain of their memory disturbance and are clearly distressed by it. Patients give many 'don't know' answers to questions, in contrast to patients with Alzheimer's disease who will give incorrect or circumlocutary refutable answers. Effortful, as opposed to implicit memory function, is particularly affected. As a general rule, patients who spontaneously complain of memory loss should always raise the possibility of a depressive illness. Depression is, however, common early in Alzheimer's disease and requires appropriate treatment.

A depressed mood as a reaction to neurological disability is commonplace and can be difficult to distinguish from persistent depression which may be more intimately related to the neurological disturbance itself. Diseases of the basal ganglia and connections to frontal lobe are particularly liable to cause depressive illness. Thus, frontal meningiomata were commonly misinterpreted as depressive illness particularly in the pre-scanning days. Depression is a common accompaniment of Parkinson's disease that is not relieved by improvement in motor function following drug treatment (Mindham *et al.* 1976; Santamaria *et al.* 1986); this may indicate an overlap in the underlying biochemical disturbance of monoamine systems. Cerebrovascular disease is a frequent cause of mood changes with up to a third of patients developing depression after stroke. It is important for management since many patients may become depressed some months after their stroke and appropriate follow-up is essential (Wade *et al.* 1987). Some studies have suggested a high prevalence of depression with dominant hemisphere infarcts, particularly with more anteriorly located lesions (Starkstein and Robinson 1993). This has not been confirmed in other studies, although the high prevalence of depression is recognized (House *et al.* 1990). Hypomania and euphoria are much less common, but can occur with cerebral infarcts and with Huntington's disease. It is often stated that patients with multiple sclerosis develop an euphoria and although this may be observed, depressive illness is still the major mood change (Schiffer 1987).

34.9.2 Anxiety

Anxiety is defined in the Diagnostic and Statistical Manual of the American Psychiatric Association, DSM-IV, as 'The apprehensive anticipation of future danger or misfortune accompanied by a feeling of dysphoria or somatic complaints of tension. The focus of anticipated danger may be internal or external'. Both DSM-IV and ICD-10 classify anxiety disorders into a number of categories; the most relevant to neurological practice are generalized anxiety disorder, panic disorder, specific phobias, and post traumatic stress disorder. As with depression, patients may develop anxiety in response to their neurological illness and, in addition, patients with anxiety may present to the neurologist by virtue of their symptoms. Generalized anxiety may result in a number of symptoms of nervousness, fatigue, and loss of concentration. These are generally non-specific symptoms and the key features are the additional acute attacks of anxiety. These attacks are associated with nervousness and increasing panic which if severe, as in phobic-anxiety attacks, may result in intense sensations of depersonalization and fear of impending death. There may be a sense of being smothered together with dyspnoea and frequently palpitations, both of which may result in referrals to a cardiologist. Nausea, urinary frequency, and vertigo during attacks are common. Hyperventilation, by virtue of the reduction in pCO_2, will itself contribute to the sense of vertigo and result in paraesthesiae classically affecting the fingers and circumoral region. If sufficiently severe, this may result in tetany (Section 4.3.4).

Panic attacks may occur in relation to obvious precipitating factors particularly in relation to specific phobias, such as open spaces or heights. They are quite frequent following trauma and caffeine may precipitate anxiety in normal people and panic attacks in those who are susceptible. They may, however, occur in apparently unprovoked situations. Diagnostic confusion occurs with epilepsy, labyrinthine disturbance, essential tremor, and cognitive impairment. Anxiety and panic attacks usually present in patients in their twenties and are rare as an initial presentation after the age of 40. Forced hyperventilation for a period of 2–3 min may reproduce many of the symptoms aiding diagnosis, but care should be taken in the interpretation since many of the symptoms of paraesthesiae, vertigo, and derealization can be precipitated as a normal concomitant of forced hyperventilation.

Post traumatic stress disorder is usually associated with both anxiety and depression. Key diagnostic features are flashbacks to the original trauma which may be triggered by relevant stimuli. Patients characteristically avoid situations which might be associated with the original trauma. Such symptoms are common immediately after a traumatic event but are considered abnormal if persisting for more than a month.

34.9.3 Obsessive–compulsive disorders

Obsessions are recurrent, intrusive thoughts or images that are often repulsive to the individual. Compulsions are repetitive and stereotyped behaviours that are found in association with obsessions and are usually performed according to certain rules and may be used to neutralize obsessional thoughts. Commonest amongst the obsessive–compulsive disorders are ritual washing and cleaning, and ritual checking. Pure obsessive disorder with stereotyped thoughts, which are often of an illicit sexual or sacrilegious nature, are less common. A close association of obsessive–compulsive disorders with neurological disease has been recognized since the development of obsessive–compulsive behaviour was observed in association with post-encephalitic Parkinsonism, especially with oculogyric crises (Section 40.3.6). There is also a high prevalence of obsessive–compulsive disorder in Gilles de la Tourette syndrome (Robertson *et al.* 1988) (Section 40.6.3) in patients with Sydenham's chorea (Swedo *et al.* 1989) (Section 40.5.7) and in paediatric autoimmune neuropsychiatric disorders associated with streptococcal infections (Swedo *et al.* 1998) and the Kleine–Levin syndrome (Arnulf *et al.* 2005) (Sections 32.3.3; 34.11). Neurological abnormalities are frequently found on examination in patients with obsessional slowness, and include speech and gait abnormalities, cogwheel rigidity and tics, together with frontal neuropsychological deficits (Hymas *et al.* 1991). This suggests dysfunction within the fronto-striatal connections, a view supported by focal hypometabolism on 15-oxygen positron emission tomography scanning in the orbital frontal, pre-motor, and mid-frontal cortex (Sawle *et al.* 1991).

34.10 Disorders of sexual behaviour

Disturbances of sexual behaviour, as opposed to disturbance of the mechanics of sexual activity arising from disease to the spinal cord and peripheral nerves (Section 29.5.2), is seen predominantly with diseases of the basal ganglia and frontal and temporal lobes. Sexual imagery may form part of the obsessive–compulsive disorder and disturbed behaviour may arise in association with post-encephalitic Parkinson's disease. Similarly, cases of increased libido and paraphilia have been found in patients on treatment for Parkinson's disease (Quinn *et al.* 1983). Disturbances in sexual behaviour in patients with frontal lobe lesions probably reflect disinhibition. Lesions of the temporal lobes are frequently associated with disturbances in sexual behaviour, found most commonly as part of the spectrum of temporal lobe epilepsy. Reduced libido with impotence as an inter-ictal phenomenon is common, but hypersexuality can occur as an immediate post-ictal phenomenon (Blumer 1970). In addition to hyper- or hyposexuality, the development of paraphilias can occur with temporal lobe lesions or temporal lobe epilepsy. The most famous case was that described by Mitchell *et al.* (1954) of a 38-year-old man who had experienced pleasure in looking at, and imagining, safety pins, since adolescence. Increasingly this would be associated with sexual arousal and on occasions would result in the precipitation of temporal lobe seizures. As the attacks became more frequent, he developed impotence. Post-ictally, he would occasionally dress in his wife's clothing. He was found to have a left temporal lobe focus with increased EEG activity on looking at safety pins. He underwent left temporal lobectomy and at operation, gliosis was found. Following surgery, there was a resolution of his attacks, restoration of potency, and cessation of the paraphilia.

Sexual disinhibition can occur in all the common dementias (Alagiakrishnan *et al.* 2005) but is particularly common with frontotemporal dementia. This reflects the involvement of medial basal frontal cortex and amygdalae, structures commonly implicated in acquired hypersexuality or paraphilias (Miller *et al.* 1986). Malpositioned ventricular catheters affecting the septal area can also lead to hypersexuality (Gorman and Cummings 1992).

Finally, side effects of a variety of drugs in neurological practice may alter sexual behaviour. Neuroleptics, antihypertensives and anticonvulsants may reduce libido. Dopaminergic drugs can be associated with increased libido and rarely with emergence of paraphilias.

34.11 Eating disorders

Neuroendocrine control of feeding behaviour is carefully balanced and an increasing number of neurotransmitters, hormones, and receptors are now known to be involved, including neuropeptide Y, PYY, corticotrophin releasing factor, leptins, ghrelin, and insulin (Woods *et al.* 1998). These are primarily under hypothalamic control and eating disorders are commonly found with hypothalamic lesions. Anorexia and bulimia nervosa, classically considered psychiatric disorders, may relate to hypothalamic disturbance. These disorders are seen primarily in adolescent and young women. *Anorexia nervosa* is associated with a distorted body image and the fear of gaining weight and individuals have increased physical activity, reduced caloric intake and at times, dramatic weight loss. *Bulimia nervosa* involves episodic gorging followed by self-induced vomiting, laxative, and diuretic abuse (Yager and Andersen 2005).

Changes in eating behaviour can also be seen with the frontotemporal degenerations (Section 34.6.4) and a variety of drugs in neurological practice, for example, sodium valproate and steroids can lead to enhanced appetite.

The Kleine–Levin syndrome

This is rare and originally described in males, but can rarely be seen in females (Section 32.3.3). It typically starts in adolescence, often following infection or head injury, and may resolve with time. It is characterized by episodes, typically lasting from 4 to 7 days, of hypersomnia, hypersexuality, hyperphagia, and altered mood (Asnulf *et al.* 2005). The hyperphagia may herald the onset of an attack with the individual eating raw and cooked food with a voracious appetite (Critchley 1962), the same clinical syndrome has been reported with a localized diencephalic encephalitis.

Sleep-related eating disorder

This was first reported in 1955 as 'night eating syndrome' (Stunkard *et al.* 1955) and is linked to somnambulism. Patients develop nocturnal hyperphagia, insomnia, and subsequent obesity. It is more common in women than men and there is commonly partial or complete amnesia for the event. Low-dose pramipexole (Provini *et al.* 2005) and topiramate have been reported to be beneficial (Winkelman *et al.* 2003).

References

Adolphs R, Tranel D, Hamann S et al. (1999). Recognition of facial emotion in nine individuals with bilateral amygdala damage. *Neuropsychologia*, **37**, 1111–7.

Alagiakrishnan K, McCracken P, Feldman H (2006). Treating vascular risk factors and maintaining vascular health: is this the way towards successful cognitive ageing and preventing cognitive decline? *Postgrad Med J*, **82**, 101–5.

Albert ML, Feldman RG, Willis AL (1974). The 'subcortical dementia' of progressive supranuclear palsy. *J Neurol Neurosurg Psychiatry*, **37**, 121–30.

Albin RL, Minoshima S, D'Amato CJ et al. (1996). Fluoro-deoxyglucose positron emission tomography in diffuse Lewy body disease. *Neurology*, **47**, 462–6.

Alexander MP, Stuss DT, Benson DF (1979). Capgras syndrome: a reduplicative phenomenon. *Neurology*, **29**, 334–9.

Arrigoni G, De Renzi E (1964). Constructional apraxia and hemispheric locus of lesion. *Cortex*, **1**, 170.

Arnulf I, Zeitzer JM, File J et al. (2005). Kleine-Levin syndrome: a systematic review of 186 cases in the literature. *Brain*, **128**, 2763–76.

Auerbach SH, Allard T, Naeser M et al. (1982). Pure word deafness. Analysis of a case with bilateral lesions and a defect at the prephonemic level. *Brain*, **105**, 271–300.

Babinski J (1914). Contribution a l'étude des trouble mentaux dans l'hemiplégie organique cérébral (anosognosie). *Rev Neurol*, **27**, 845–52.

Baddeley A, Bressi, Della Sala S et al. (1986). Dementia and working memory. *Quart J Exp Psychiatry*, **38**, 603–18.

Bakheit AMO, Kennedy PGE, Behan PO (1990). Paraneoplastic limbic encephalitis: clinico-pathological correlations. *J Neurol Neurosurg Psychiatry*, **53**, 1084–8.

Ballard C, Holmes C, McKeith I et al. (1999). Psychiatric morbidity in dementia with Lewy bodies: a prospective clinical and neuropathological comparative study with Alzheimer's disease. *Am J Psychiatry*, **156**, 1039–45.

Baker M, Mackenzie IR, Pickering-Brown SM et al. (2006). Mutations in progranulin cause tau-negative frontotemporal dementia linked to chromosome 17. *Nature*, **442**, 916–9.

Bennett DA, Wilson RS, Gilley DW et al. (1990). Clinical diagnosis of Binswanger's disease. *J Neurol Neurosurg Psychiatry*, **53**, 961–5.

Benson DF (1967). Fluency in aphasia: Correlation with radioactive scan localisation. *Cortex*, **3**, 373–94.

Benson DF, Marsden CD, Meadows JC (1974). The amnesic syndrome of posterior cerebral artery occlusion. *Acta Neurol Scand*, **50**, 133–45.

Benson DF, Gardner H, Meadows JC (1976). Reduplicative paramnesia. *Neurology*, **26**, 147–51.

Benson F, Davis J, Snyder BD (1988). Posterior Cortical atrophy. *Arch Neurol*, **45**, 789–93.

Benton AL (1961). The fiction of the 'Gerstmann Syndrome'. *J Neurol Neurosurg Psychiatry*, **24**, 176–81.

Berlyne N (1972). Confabulation. *Br J Psychiatry*, **120**, 31–9.

Bhatia KP, Marsden CD (1994). The behavioural and motor consequences of focal lesions of the basal ganglia in man. *Brain*, **117**, 859–76.

Binetti G, Growdon JH, Vonsattel JP (1998). Pick's disease. In Growdon JH, Rossor MN, eds. *The Dementias*, pp. 7–44. Butterworth-Heinemann, Newton.

Bisiach E, Luzzatti C (1978). Unilateral neglect of representational space. *Cortex*, **14**, 129–33.

Blair RJR, Cipolotti L (2000). Impaired social response reversal: A case of 'acquired sociopathy'. *Brain*, **123**, 1122–41.

Blennow K, Hampel H (2003). CSF markers for incipient Alzheimer's disease. *Lancet Neurol*, **2**, 605–13.

Blessed G, Tomlinson B, Roth M (1968). The association between quantitative measures of Dementia and of senile change in the cerebral grey matter of elderly subjects. *Br J Psychiatry*, **114**, 797–811.

Blumer D (1970). Hypersexual episodes in temporal lobe epilepsy. *Am J Psychiatry*, **126**, 1099–106.

Blumstein SE, Alexander MP, Ryalls JH et al. (1987). On the nature of the foreign accent syndrome: A case study. *Brain Lang*, **31**, 215–44.

Bogousslavsky J (1994). Frontal stroke syndromes. *J Eur Neurol*, **34**, 306–15.

Bowler JV (2003). Acetylcholinesterase inhibitors for vascular dementia and Alzheimer's disease combined with cerebrovascular disease. *Stroke*, **34**, 584–6.

Bowler JV, Eliasziw M, Steenhuis R et al. (1997). Comparative evolution of Alzheimer disease, vascular dementia, and mixed dementia. *Arch Neurol*, **54**, 697–703.

Braak H, Braak E (1991). Neuropathological staging of Alzheimer-related changes. *Acta Neuropathol*, **82**, 239–59.

Braak H, Braak E, Bohl J et al. (1989). Alzheimers-disease—amyloid plaques in the cerebellum. *J Neurol Sci*, **93**, 277–87.

Brodmann K (1909). *Vergleichende Lokalisationslehre der Grosshirnrinde in ihren Prizipien dargestellt auf Grund des Zellenbauer*. J.A. Basth, Leipzig.

Brown P, Cathala F, Castaigne P et al. (1986). Creutzfeldt-Jakob disease: clinical analysis of a consecutive series of 230 neuropathologically verified cases. *Ann Neurol*, **20**, 597–602.

Bruce V, Young A (1986). Understanding face recognition. *Br J Psychol*, **77**, 305–27.

Brun A (1993). Frontal-lobe degeneration of non-Alzheimer type revisited. *Dementia*, **4**, 126–31.

Brust JC (1980). Music and language: musical alexia and agraphia. *Brain*, **103**, 367–92.

Caine ED (1981). Pseudodementia. Current concepts and future directions. *Arch Gen Psychiatry*, **38**, 1359–64.

Caine D, Halliday GM, Kril JJ et al. (1997). Operational criteria for the classification of chronic alcoholics: identification of Wernicke's encephalopathy. *J Neurol Neurosurg Psychiatry*, **62**, 51–60.

Caplan LR (1980). 'Top of the basilar' syndrome. *Neurology*, **30**, 72–9.

Caplan L, Chedru F, Lhermitte F et al. (1981). Transient global amnesia and migraine. *Neurology*, **31**, 1167–70.

Carter GL, Dawson AH, Lopert R (1996). Drug-induced delirium. Incidence, management and prevention. *Drug Saf*, **15**, 291–301.

Caselli RJ, Jack CR (1992). Asymmetric cortical degeneration syndromes—a proposed clinical classification. *Arch Neurol*, **49**, 770–80.

Castaigne P, Lhermitte F, Buge A et al. (1981). Paramedian thalamic and midbrain infarct: clinical and neuropathological study. *Ann Neurol*, **10**, 127–48.

Chan D, Rossor MN (2002). '-but who is that on the other side of you?' Extracampine hallucinations revisited. *Lancet*, **360**, 2064–6.

Chan D, Fox NC, Scahill RI et al. (2001). Patterns of temporal lobe atrophy in semantic dementia and Alzheimer's disease. *Ann Neurol*, **49**, 433–42.

Chan D, Walters RJ, Sampson EL et al. (2004). EEG abnormalities in frontotemporal lobar degeneration. *Neurology*, **62**, 1628–30.

Clarfield AM (2003). The decreasing prevalence of reversible dementias: an updated meta-analysis. *Arch Intern Med*, **163**, 2219–29.

Cohen NJ, Squire LR (1980). Preserved learning and retention of pattern analysing skill in amnesia: Dissociation of 'knowing how' and 'knowing that'. *Science*, **210**, 207–10.

Cole MG (2004). Delirium in elderly patients. *Am J Geriatr Psychiatry*, **12**, 7–21.

Collinge J, Brown J, Hardy J et al. (1992). Inherited prion disease with 144 base pair gene insertion. 2. Clinical and pathological features. *Brain*, **115**, 687–710.

Collinge J, Sidle KCL, Meads J et al. (1996). Molecular analysis of prion strain variation and the etiology of new variant cjd. *Nature*, **383**, 685–90.

Collinge J, Whitfield J, McKintosh E et al. (2006). Kuru in the 21st century-an acquired human prion disease with very long incubation periods. *Lancet*, **367**, 2068–74.

Corder EH, Saunders AM, Strittmatter WJ et al. (1993). Gene dose of apolipoprotein E type 4 allele and the risk of Alzheimer's disease in late onset families. *Science*, **261**, 921–3.

Craik FIM, Lockhart RS (1972). Levels of processing: a framework for memory research. *J Verb Learning Verb Behaviour*, **11**, 671–84.

Critchley M (1953). *The Parietal Lobes*. Arnold, London.

Critchley M (1962). Periodic hypersomnia and megaphagia in adolescent males. *Brain*, **85**, 627–56.

Crook R, Verkkoniemi A, Perez-Tur J et al. (1998). A variant of Alzheimer's disease with spastic paraparesis and unusual plaques due to deletion of exon 9 of presenilin 1. *Nat Med*, **4**, 452–5.

Cruts M, Van Broeckhoven C (1998). Molecular genetics of Alzheimer's disease. In Growdon JH, Rossor MN, eds. *The Dementias*, pp. 155–70. Butterworth-Heinemann, Boston.

Cruts M, Gijselinck I, van der Zee J *et al.* (2006). Null mutations in progranulin cause ubiquitin-positive frontotemporal dementia linked to chromosome 17q21. *Nature*, **24**, 920–4.

Cummings JL (1986). Subcortical dementia. Neuropsychology, neuropsychiatry, and pathophysiology. [Review]. *Br J Psychiatry*, **149**, 682–97.

Cummings JL, Duchen LW (1981). Kluver-Bucy syndrome in Pick disease: Clinical and pathologic correlations. *Neurology*, **31**, 1415–22.

Cutting J (1978). The relationship between Korsakoff's syndrome and 'alcoholic dementia'. *Br J Psychiatry*, **132**, 240–51.

Cysique LA, Maruff P, Brew BJ (2004). Prevalence and pattern of neuropsychological impairment in human immunodeficiency virus-infected/acquired immunodeficiency syndrome (HIV/AIDS) patients across pre- and post-highly active antiretroviral therapy eras: a combined study of two cohorts. *J Neurovirol*, **10**, 350–7.

Damasio AR, Damasio H, Rizzo M *et al.* (1982). Aphasia with nonhemorrhagic lesions in the basal ganglia and internal capsule. *Arch Neurol*, **39**, 15–24.

Damasio AR, Graff RN, Eslinger PJ *et al.* (1985). Amnesia following basal forebrain lesions. *Arch Neurol*, **42**, 263–71.

Davies RR, Hodges JR, Kril JJ *et al.* (2005). The pathological basis of semantic dementia. *Brain*, **128**, 1984–95.

De Renzi E, Lucchelli F (1988). Ideational apraxia. *Brain*, **111**, 1173–88.

De Renzi E, Pieczuro AC, Vignolo LA (1968). Ideational apraxia: a quantitative study. *Neuropsychologia*, **6**, 41–52.

De RE (1986). Prosopagnosia in two patients with CT scan evidence of damage confined to the right hemisphere. *Neuropsychologia*, **24**, 385–9.

De RE, Scotti G, Spinnler H (1969). Perceptual and associative disorders of visual recognition. Relationship to the side of the cerebral lesion. *Neurology*, **19**, 634–42.

Delacourte A, Robitaille Y, Sergeant N *et al.* (1996). Specific pathological Tau protein variants characterize Pick's disease. *J Neuropathol Exp Neurol*, **55**, 159–68.

Denes G, Semenza C, Stoppa E *et al.* (1982). Unilateral spatial neglect and recovery from hemiplegia: a follow-up study. *Brain*, **105**, 543–52.

Devinsky O, Morrell MJ, Vogt BA (1995). Contributions of anterior cingulate cortex to behaviour. *Brain*, **118**, 279–306.

Dichgans M, Mayer M, Uttner I *et al.* (1998). The phenotypic spectrum of CADASIL: clinical findings in 102 cases. *Ann Neurol*, **44**, 731–9.

Doody RS, Stevens JC, Beck C *et al.* (2001). Practice parameter: management of dementia (an evidence-based review). Report of the quality standards subcommittee of the American academy of neurology. *Neurology*, **56**, 1154–66.

Dunn DW, Weisberg LA, Nadell J (1983). Peduncular hallucinations caused by brainstem compression. *Neurology*, **33**, 1360–1.

Efron R (1968). What is perception? In Cohen RS, Wartofsky M, eds. *Boston Studies in the Philosophy of Science*, pp. 137. Humanities Press, New York.

Emre M, Aarsland D, Albanese A *et al.* (2004). Rivastigmine for dementia associated with Parkinson's disease. *N Engl J Med*, **351**, 2509–18.

Esiri MM, Wilcock GK, Morris JH (1997). Neuropathological assessment of the lesions of significance in vascular dementia. *J Neurol Neurosurg Psychiatry*, **63**, 749–53.

Etcoff NL (1984). Selective attention to facial identity and facial emotion. *Neuropsychologia*, **22**, 281–95.

Evans DA, Funkenstein HH, Albert MS *et al.* (1989). Prevalence of Alzheimer's disease in a community population of older persons. Higher than previously reported. *JAMA*, **262**, 2551–6.

Farrer LA, Myers RH, Cupples LA *et al.* (1990). Transmission and age-at-onset patterns in familial Alzheimer's disease: evidence for heterogeneity. *Neurology*, **40**, 395–403.

Feher E, Mahurin R, Inbody S *et al.* (1991). Anosognosia in Alzheimer's disease. *Neuropsychiatry Neuropsychol Behav Neurol*, **4**, 136–46.

Fisher CM (1982). Transient global amnesia. Precipitating activities and other observations. *Arch Neurol*, **39**, 605–8.

Fisher CM, Adams RD (1958). Transient global amnesia. *Trans Am Neurol Assoc*, **83**, 143–6.

Fodor J (1983). *The Modularity of Mind*. MIT Press, Cambridge.

Forette F, Seux ML, Staessen JA *et al.* (1998). Prevention of dementia in randomised double-blind placebo-controlled Systolic Hypertension in Europe (Syst-Eur) trial. *Lancet*, **352**, 1347–51.

Forman MS, Farmer J, Johnson JK (2006). Frontotemporal dementia: clinicopathological correlations. *Ann Neurol*, **59**, 952–62.

Foster NL, Wilhelmsen K, Sima AA *et al.* (1997). Frontotemporal dementia and parkinsonism linked to chromosome 17: a consensus conference. Conference Participants. *Ann Neurol*, **41**, 706–15.

Fox NC, Schott JM (2004). Imaging cerebral atrophy: normal ageing to Alzheimer's disease. *Lancet*, **363**, 392–4.

Foy A, O'Connell D, Henry D *et al.* (1995). Benzodiazepine use as a cause of cognitive impairment in elderly hospital inpatients. *J Gerontol A Biol Sci Med Sci*, **50**, M99–106.

Frackowiak RJ, Pozzilli C, Legg NJ *et al.* (1981). Regional cerebral oxygen supply and utilization in dementia. A clinical and physiological study with oxygen-15 and positron tomography. *Brain*, **104**, 753–78.

Fradkin JE, Schonberger LB, Mills JL *et al.* (1991). Creutzfeldt-Jakob disease in pituitary growth hormone recipients in the United States. *JAMA*, **265**, 880–4.

Gabrieli JD (1998). Cognitive neuroscience of human memory. *Annu Rev Psychol*, **49**, 87–115.

Galton CJ, Patterson K, Xuereb JH *et al.* (2000). Atypical and typical presentations of Alzheimer's disease: a clinical, neuropsychological, neuroimaging and pathological study of 13 cases. *Brain*, **123**, 484–98.

Garrard P, Bradshaw D, Jager HR *et al.* (2002). Cognitive dysfunction after isolated brain stem insult. An underdiagnosed cause of long term morbidity. *J Neurol Neurosurg Psychiatry*, **73**, 191–4.

Gerstmann J (1940). Syndrome of finger agnosia, disorientation for right and left: agraphia and acalculia. *Arch Neurol Psychiatry*, **44**, 398–408.

Geschwind N (1965). Disconnexion syndromes in animals and man. *Brain*, **88**, 585–641.

Geschwind N (1982). Disorders of attention: a frontier in neuropsychology. *Philos Trans R Soc Lond B Biol Sci*, **298**, 173–85.

Geschwind N, Kaplan E (1962). A human cerebral deconnection syndrome. *Neurology*, **12**, 675–85.

Geschwind N, Quadfasel FA, Segarra JM (1968). Isolation of the speech area. *Neuropsychologia*, **6**, 327–40.

Goedert M, Spillantini MG, Cairns NJ *et al.* (1992). Tau-proteins of alzheimer paired helical filaments—abnormal phosphorylation of all 6 brain isoforms. *Neuron*, **8**, 159–68.

Gold G, Kovari E, Herrmann FR *et al.* (2005). Cognitive consequences of thalamic, basal ganglia, and deep white matter lacunes in brain aging and dementia. *Stroke*, **36**, 1184–8.

Goldman RP (1988). Topography of cognition: parallel distributed networks in primate association cortex. *Annu Rev Neurosci*, **11**, 137–56.

Goldstein M (1974). Auditory agnosia for speech ('pure-word deafness'): A historical review with current implications. *Brain Lang*, **1**, 195–204.

Goodglass H, Kaplan E (1972). *The Assessment of Aphasia and Related Disorders*. Lea and Febiger, Philadelphia.

Goodglass H, Quadfasel FA, Timberlake WH (1964). Phrase length and the type of severity of aphasia. *Cortex*, **1**, 133–53.

Goodwin DW, Crane JB, Guze SB (1969). Alcoholic 'blackouts': a review and clinical study of 100 alcoholics. *Am J Psychiatry*, **126**, 191–8.

Gorman DG, Cummings JL (1992). Hypersexuality following septal injury. *Arch Neurol*, **49**(3), 308–10.

Graff RN, Eslinger PJ, Damasio AR *et al.* (1984). Nonhemorrhagic infarction of the thalamus: behavioral, anatomic, and physiologic correlates. *Neurology*, **34**, 14–23.

Gregory C, Lough S, Stone V *et al.* (2002). Theory of mind in patients with frontal variant frontotemporal dementia and Alzheimer's disease: theoretical and practical implications. *Brain*, **125**, 752–64.

Grossman M, Ash S (2004). Primary progressive aphasia: a review. *Neurocase*, **10**, 3–18.

Hachinski VC, Bowler JV (1993). Vascular dementia [letter; comment]. *Neurology*, **43**, 2159–60.

Hachinski V, Lassen N, Marshall J (1974). Multi-infarct dementia. A cause of mental deterioration in the elderly. *Lancet*, **27**, 207–9.

Hachinski VC, Potter P, Merskey H (1987). Leuko-araiosis. *Arch Neurol*, **44**, 21–3.

Han L, McCusker J, Cole M *et al.* (2001). Use of medications with anticholinergic effect predicts clinical severity of delirium symptoms in older medical inpatients. *Arch Intern Med*, **161**, 1099–105

Hardy JA, Higgins GA (1992). Alzheimer's disease: the amyloid cascade hypothesis. *Science*, **256**, 184–5.

Hecaen H (1962). Clinical symptomatology in right and left hemispheric lesions. In Mountcastle VB, ed. *Interhemispheric Relations and Cerebral Dominance*, pp. 215. John Hopkins, Baltimore.

Hecaen H, Ajuriaguerra J (1954). Balint's syndrome (psychic paralysis of visual fixation) and its minor forms. *Brain*, **77**, 373–400.

Hecaen H, Angelergues R (1963). *Le cécité Psychique*. Masson, Paris.

Hecaen H, Albert ML (1978). *Human Neuropsychology*. John Wiley & Sons, New York.

Henson RA, Urich H (1982). *Cancer and the Nervous System*. Blackwell, Oxford.

Heye N, Terstegge K, Şirtl C *et al.* (1994). Wernicke's encephalopathy—causes to consider. *Intensive Care Med*, **20**, 282–6.

Hill AF, Butterworth RJ, Joiner S *et al.* (1999). Investigation of variant Creutzfeldt-Jakob disease and other human prion diseases with tonsil biopsy samples. *Lancet*, **353**, 183–9.

Hodges JR (2006) Alzheimer's centennial legacy: origins, landmarks and the current status of knowledge concerning cognitive aspects. *Brain*, **129** (Pt 11), 2811–22.

Hodges JR, Warlow CP (1990). The aetiology of transient global amnesia. A case-control study of 114 cases with prospective follow-up. *Brain*, **113**, 639–57.

Hodges JR, Patterson K, Oxbury S *et al.* (1992). Semantic dementia. Progressive fluent aphasia with temporal lobe atrophy. *Brain*, **115**, 1783–806.

Holmes G (1918). Disturbances of visual orientation. *Br J Opthalmal*, **2**, 449–68.

House A, Dennis M, Warlow C *et al.* (1990). Mood disorders after stroke and their relation to lesion location. A CT scan study. *Brain*, **113**, 1113–29.

Hutton M, Lendon C, Rizzu P *et al.* (1998). Association of missense and 5'-splice-site mutations in tau with the inherited dementia FTDP-17. *Nature*, **393**, 702–5.

Hymas N, Lees A, Bolton D *et al.* (1991). The neurology of obsessional slowness. *Brain*, **114**, 2203–33.

Inouye SK, van Dyck CH, Alessi CA *et al.* (1990). Clarifying confusion: the confusion assessment method. A new method for detection of delirium. *Ann Intern Med*, **113**, 941–8.

Iwatsubo T, Mann DM, Odaka A *et al.* (1995). Amyloid β Protein (Aβ) Deposition: Aβ42(43) Precedes Aβ40 in Down Syndrome. *Ann Neurol*, **37**, 294–9.

Jack CR, Petersen RC, Xu YC *et al.* (1999). Prediction of AD with MRI-based hippocampal volume in mild cognitive impairment. *Neurology*, **52**, 1397–403.

Jackson JC, Gordon SM, Hart RP *et al.* (2004). The association between delirium and cognitive decline: a review of the empirical literature. *Neuropsychol Rev*, **14**, 87–98.

Jain KK (2001). *Drug-induced Neurological Disorders*, 2nd edition. Seattle-gottingen, Hogrefe & Huber.

Janssen RS, Nwanyanwu OC, Selik RM *et al.* (1992). Epidemiology of human-immunodeficiency-virus encephalopathy in the United States. *Neurology*, **42**, 1472–6.

Jicha GA, Petersen RC (2007). *Mild Cognitive Impairment*. (Blue Books of Practical Neurology) The Dementias 2.

Josephs KA, Holton JL, Rossor MN *et al.* (2003). Neurofilament inclusion body disease: a new proteinopathy? *Brain*, **126**, 2291–303.

Joutel A, Corpechot C, Ducros A *et al.* (1996). Notch3 mutations in cadasil, a hereditary adult-onset condition causing stroke and dementia. *Nature*, **383**, 707–10.

Kang J, Lemaire H-G, Unterbeck A *et al.* (1987). The precursor of Alzheimer's disease amyloid A4 protein resembles a cell-surface receptor. *Nature*, **325**, 733–7.

Kapur N, Turner A, King C (1988). Reduplicative paramnesia: possible anatomical and neuropsychological mechanisms. *J Neurol Neurosurg Psychiatry*, **51**, 579–81.

Keime-Guibert F, Napolitano M, Delattre JY (1998). Neurological complications of radiotherapy and chemotherapy. *J Neurol*, **245**, 695–708.

Kertesz A, Munoz D (1998). Pick's disease, frontotemporal dementia, and Pick complex: emerging concepts. *Arch Neurol*, **55**, 302–4.

Kidd M (1963). Paired helical filaments in electronmicroscopy in Alzheimer's disease. *Nature*, **197**, 192–3.

Kim RC, Collins GH, Parisi JE *et al.* (1981). Familial dementia of adult onset with pathological findings of a nonspecific nature. *Brain*, **104**, 61–78.

Kirk C, Saunders M (1977). Primary psychiatric illness in a neurological out-patient department in North East England. An assessment of symptomatology. *Acta Psychiatr Scand*, **56**, 294–302.

Kleist K (1922). Die psychomotorischen Storungen und ihr Verhaltnis zu den Motilitatsstorungen bei der Stammganglien. *Mschr Pychiat Neurol*, **52**, 253–302.

Knopman DS, Mastri AR, Frey WH *et al.* (1990). Dementia lacking distinctive histologic features: a common non-Alzheimer degenerative dementia. *Neurology*, **40**, 251–6.

Knopman DS, DeKosky ST, Cummings JL *et al.* (2001). Practice parameter: diagnosis of dementia (an evidence-based review). Report of the quality standards subcommittee of the American academy of neurology. *Neurology*, **56**, 1143–53.

Kopelman MD (2002). Disorders of memory. *Brain*, **125**, 2152–90.

Kosaka K (1994). Diffuse neurofibrillary tangles with calcification: a new presenile dementia. *J Neurol Neurosurg Psychiatry*, **57**, 594–6.

Krupp LB, Masur D, Schwartz J *et al.* (1991). Cognitive functioning in late Lyme borreliosis. *Arch Neurol*, **48**, 1125–9.

Kumral E, Bayulkem G, Dvyapan D *et al.* (2002). Spectrum of anterior cerebral artery territory infarction: clinical and MRI findings. *Eur J Neurol*, **9**, 615–24.

Laplane D, Baulac M, Widlocher D *et al.* (1984). Pure psychic akinesia with bilateral lesions of basal ganglia. *J Neurol Neurosurg Psychiatry*, **47**, 377–85.

Lavenu I, Pasquier F, Lebert F *et al.* (1999). Perception of emotion in frontotemporal dementia and Alzheimer disease. *Alzheimer Dis Assoc Disord*, **13**, 96–101.

Lebert F, Stekke W, Hasenbroekx C *et al.* (2004). Frontotemporal dementia: a randomised, controlled trial with trazodone. *Dement Geriatr Cogn Disord*, **17**, 355–9.

Lecours AR (1976). The 'Pure Form' of the phonetic disintegration syndrome (pure anarthria); anatomo-clinical report of a historical case. *Brain Lang*, **3**, 88–113.

Lee VMY, Balin BJ, Otvos L *et al.* (1991). A68—a major subunit of paired helical filaments and derivatized forms of normal-tau. *Science*, **251**, 675–8.

Leiguarda RC, Marsden CD (2000). Limb apraxias: higher-order disorders of sensorimotor integration. *Brain*, **123**, 860–79.

Lemstra AW, van Meegen MT, Vreyling JP *et al.* (2000). 14-3-3 testing in diagnosing Creutzfeldt-Jakob disease: a prospective study in 112 patients. *Neurology*, **55**, 514–6.

Levkoff SE, Evans DA, Liptzin B *et al.* (1992). Delirium. The occurrence and persistence of symptoms among elderly hospitalized patients. *Arch Intern Med*, **152**, 334–40.

Lhermitte F (1983). 'Utilization behaviour' and its relation to lesions of the frontal lobes. *Brain*, **106**, 237–55.

Lilly R, Cummings JL, Benson DF *et al.* (1983). The human Kluver-Bucy syndrome. *Neurology*, **33**, 1141–5.

Lipowski ZJ (1990). *Acute Confusional States*, 2nd edition. Oxford University Press, New York.

Lisak RP, Zimmerman RA (1977). Transient global amnesia due to a dominant hemisphere tumor. *Arch Neurol*, **34**, 317–8.

Lishman WA (1987). *Organic Psychiatry, the Psychological Consequences of Cerebral Disorder*, 2nd edition. Blackwell Scientific, Oxford.

Llewelyn CA, Hewitt PE, Knight RS *et al.* (2004). Possible transmission of variant Creutzfeldt-Jakob disease by blood transfusion. *Lancet*, **363**, 417–21.

Lugaresi E, Medori R, Montagna P *et al.* (1986). Fatal familial insomnia and dysautonomia with selective degeneration of thalamic nuclei. *N Engl J Med*, **315**, 997–1003.

Luria AR (1977). On quasi-aphasic speech disturbances in lesions of the deep structures of the brain. *Brain Lang*, **4**, 432–59.

Manes F, Graham KS, Zeman A *et al.* (2005). Autobiographical amnesia and accelerated forgetting in transient epileptic amneseia. *J Neurol Neurosurg Psychiatry*, **76**, 1387–91.

Markus HS, Martin RJ, Simpson MA *et al.* (2002). Diagnostic strategies in CADASIL. *Neurology*, **59**, 1134–8.

Marshall JC, Newcombe F (1966). Syntactic and semantic errors in paralexia. *Neuropsychologia*, **4**, 169–76.

Mazzucchi A, Moretti G, Caffarra P et al. (1980). Neuropsychological functions in the follow-up of transient global amnesia. *Brain*, **103**, 161–78.

McCarthy RA, Warrington EK (1990). *Cognitive Neuropsychology. A Clinical Introduction*. Academic Press, London.

McKeith I, Cummings J (2005). Behavioural changes and psychological symptoms in dementia disorders. *Lancet Neurol*, **4**, 735–42.

McKeith I, Mintzer J, Aarsland D et al. (2004). Dementia with Lewy bodies. *Lancet Neurol*, **3**, 19–28.

McKeith IG, Dickson DW, Lowe J et al. (2005). Diagnosis and management of dementia with Lewy bodies: third report of the DLB Consortium. *Neurology*, **27**, 1863–72.

McKhann G, Drachman D, Folstein M et al. (1984). Clinical diagnosis of Alzheimer's Disease: Report of the NINCDS–ADRDA work group under the auspices of Department of health and human services task force on Alzheimer's disease. *Neurology*, **34**, 939–44.

McMonagle P, Deering F, Berliner Y et al. (2006). The cognitive profile of posterior cortical atrophy. *Neurology*, **66**(3), 331–8.

McNicoll L, Pisani MA, Zhang Y et al. (2003). Delirium in the intensive care unit: occurrence and clinical course in older patients. *J Am Geriatr Soc*, **51**, 591–8.

Mead S, James-Galton M, Revesz T et al. (2000). Familial British dementia with amyloid angiopathy: early clinical, neuropsychological and imaging findings. *Brain*, **123**, 975–91.

Meadows JC (1974a). Disturbed perception of colours associated with localized cerebral lesions. *Brain*, **97**, 615–32.

Meadows JC (1974b). The anatomical basis of prosopagnosia. *J Neurol Neurosurg Psychiatry*, **37**, 489–501.

Meagher DJ (2001). Delirium: optimising management. *BMJ*, **322**, 144–9.

Medina JL, Chokroverty S, Rubino FA (1977). Syndrome of agitated delirium and visual impairment: a manifestation of medial temporo-occipital infarction. *J Neurol Neurosurg Psychiatry*, **40**, 861–4.

Medina JL, Rubino FA, Ross E (1974). Agitated delirium caused by infarctions of the hippocampal formation and fusiform and lingual gyri: a case report. *Neurology*, **24**, 1181–3.

Mesulam MM (1981). A cortical network for directed attention and unilateral neglect. *Ann Neurol*, **10**, 309–25.

Mesulam MM. (1982) Slowly progressive aphasia without generalized dementia. *Ann Neurol*, **11**(6), 592–8.

Mesulam MM, Waxman SG, Geschwind N et al. (1976). Acute confusional states with right middle cerebral artery infarctions. *J Neurol Neurosurg Psychiatry*, **39**, 84–9.

Meyer JS, Barron DW (1960). Apraxia of gait: a clinico-physiological study. *Brain*, **83**, 261–84.

Miller BL, Cummings JL, McIntyre H et al. (1986) Hypersexuality or altered sexual preference following brain injury. *J Neurol Neurosurg Psychiatry*, **49**(8), 867–73.

Milner B (1958). Psychological defects produced by temporal lobe excision. *Res Publ Ass Nerv Ment Dis*, **36**, 244–57.

Milner B (1963). Effects of different brain lesions on card sorting. *Arch Neurol*, **9**, 90.

Milner B (1971). Interhemispheric differences in the localization of psychological processes in man. *Br Med Bull*, **27**, 272–7.

Milner B, Corkin S, Teuber HL (1968). Further analysis of the hippocampal amnesic syndrome: 14 year follow up of HM. *Neuropsychologia*, **6**, 215.

Mindham RH, Marsden CD, Parkes JD (1976). Psychiatric symptoms during l-dopa therapy for Parkinson's disease and their relationship to physical disability. *Psychol Med*, **6**, 23–33.

Mitchell W, Falconer MA, Hill D (1954). Epilepsy with fetishism relieved by temporal lobectomy. *Lancet*, **2**, 626–30.

Mitsuyama Y, Takamiya S (1979). Presenile dementia with motor neuron disease in Japan. A new entity? *Arch Neurol*, **36**, 592–3.

Mohr JP, Pessin MS, Finkelstein S et al. (1978). Broca aphasia: pathologic and clinical. *Neurology*, **28**, 311–24.

Moroney JT, Bagiella E, Desmond DW et al. (1997). Meta-analysis of the Hachinski Ischemic Score in pathologically verified dementias. *Neurology*, **49**, 1096–105.

Mountcastle VB (1997). The columnar organization of the neocortex. *Brain*, **120**, 701–22.

Naeser MA, Alexander MP, Helm EN et al. (1982). Aphasia with predominantly subcortical lesion sites: description of three capsular/putaminal aphasia syndromes. *Arch Neurol*, **39**, 2–14.

Nathan PW (1987). Facial apraxia and apraxic dysarthria. *Brain*, **70**, 449–78.

Neary D, Snowden JS, Gustafson L et al. (1998). Frontotemporal lobar degeneration: a consensus on clinical diagnostic criteria. *Neurology*, **51**, 1546–54.

Nestor PJ, Graham NL, Fryer TD et al. (2003). Progressive non-fluent aphasia is associated with hypometabolism centred on the left anterior insula. *Brain*, **126**, 2406–18.

Neuropathology Group. Medical Research Council Cognitive Function and Aging Study (2001). Pathological correlates of late-onset dementia in a multicentre, community-based population in England and Wales. Neuropathology Group of the Medical Research Council Cognitive Function and Ageing Study (MRC CFAS). *Lancet*, **357**, 169–75.

Newell KL, Hyman BT, Growdon JH, Hedley-Whyte ET. (1999) Application of the National Institute on Aging (NIA)-Reagan Institute criteria for the neuropathological diagnosis of Alzheimer disease. *J Neuropathol Exp Neurol*, **58**(11), 1147–55.

Nicoll JA, Wilkinson D, Holmes C et al. (2003). Neuropathology of human Alzheimer disease after immunization with amyloid-beta peptide: a case report. *Nat Med*, **9**, 448–52.

Norman DA, Shallice T (1980). *Attention to action: willed and automatic control of behaviour*. 99, Centre for Human Information Processing, University of California, San Diego.

O'Brien JT, Desmond P, Ames D et al. (1997). Magnetic resonance imaging correlates of memory impairment in the healthy elderly: Association with medial temporal lobe atrophy but not white matter lesions. *Int J Geriatr Psychiatry*, **374**, 369–74.

O'Brien JT, Erkinjuntti T, Reisberg B et al. (2003). Vascular cognitive impairment. *Lancet Neurol*, **2**, 89–98.

O'Brien JT, Colloby S, Fenwick J et al. (2004). Dopamine transporter loss visualized with FP-CIT SPECT in the differential diagnosis of dementia with Lewy bodies. *Arch Neurol*, **61**, 919–25.

Palmer MS, Dryden AJ, Hughes JT et al. (1991). Homozygous prion protein genotype predisposes to sporadic Creutzfeldt-Jakob disease. *Nature*, **352**, 340–2.

Parchi P, Castellani R, Capellari S et al. (1996). Molecular basis of phenotypic variability in sporadic Creutzfeldt-Jakob disease. *Ann Neurol*, **39**, 767–78.

Paterson A, Zangwill OL (1945). A case of topographical disorientation associated with a unilateral cerebral lesion. *Brain*, **68**, 188–212.

Perani D, Vallar G, Cappa S et al. (1987). Aphasia and neglect after subcortical stroke. A clinical/cerebral perfusion correlation study. *Brain*, **110**, 1211–29.

Petersen SE, Fox PT, Posner MI et al. (1988). Positron emission tomographic studies of the cortical anatomy of single-word processing. *Nature*, **331**, 585–9.

Petersen RC, Stevens JC, Ganguli M et al. (2001). Practice parameter: early detection of dementia: mild cognitive impairment (an evidence-based review). Report of the quality standards subcommittee of the American academy of neurology. *Neurology*, **56**, 1133–42.

Petersen RC, Parisi JE, Dickson DW et al. (2006) Neuropathologic features of amnestic mild cognitive impairment. *Arch Neurol*, **63**(5), 665–72.

Pick A (1903). On reduplicative paramnesia. *Brain*, **26**, 260–7.

Pohjasvaara T, Erkinjuntti T, Vataja R et al. (1997). Dementia three months after stroke. Baseline frequency and effect of different definitions of dementia in the Helsinki Stroke Aging Memory Study (SAM) cohort. *Stroke*, **28**, 785–92.

Poorkaj P, Bird TD, Wijsman E et al. (1998). Tau is a candidate gene for chromosome 17 frontotemporal dementia. *Ann Neurol*, **43**, 815–25.

Pramstaller PP, Marsden CD (1996). The basal ganglia and apraxia. *Brain*, **119**, 319–40.

Price CJ (1998). The functional anatomy of word comprehension and production. *Trends Cogn Sci*, **2**, 281–8.

Price CJ, Friston KJ (1997). Cognitive conjunction: a new approach to brain activation experiments. *Neuroimage*, **5**, 261–70.

Provini F, Albani F, Vetrugno R et al. (2005). A pilot double-blind placebo-controlled trial of low-dose pramipexole in sleep-related eating disorder. Eur J Neurol, 12, 432–6.

Prusiner SB (1991). Molecular biology of prion diseases. Science, 252, 1515–22.

Prusiner SB, Scott MR, DeArmond SJ et al. (1998). Prion protein biology. Cell, 93, 337–48.

Quinette P, Guillery-Girard B, Dayan J et al. (2006). What does transient global amnesia really mean? Review of the literature and thorough study of 142 cases. Brain, 129, 1640–58.

Quinn NP, Toone B, Lang AE et al. (1983). Dopa dose-dependent sexual deviation. Br J Psychiatry, 142, 296–8.

Rahkonen T, Eloniemi-Sulkava U, Rissanen S et al. (2003). Dementia with Lewy bodies according to the consensus criteria in a general population aged 75 years or older. J Neurol Neurosurg Psychiatry, 74, 720–4.

Rakic P (1995). A small step for the cell, a giant leap for mankind: a hypothesis of neocortical expansion during evolution. Trends Neurosci, 18, 383–8.

Ramnani N, Behrens TE, Penny W et al. (2005). New approaches for exploring anatomical and functional connectivity in the human brain. Biol Psychiatry, 56, 613–9.

Rebollo M, Val JF, Garijo F et al. (1983). Livedo reticularis and cerebrovascular lesions (Sneddon's syndrome). Clinical, radiological and pathological features in eight cases. Brain, 106, 965–79.

Richfield EK, Twyman R, Berent S (1987). Neurological syndrome following bilateral damage to the head of the caudate nuclei. Ann Neurol, 22, 768–71.

Riddoch G (1941). Phantom limbs and body shape. Brain, 64, 197–222.

Robertson MM, Trimble MR, Lees AJ (1988). The psychopathology of the Gilles de la Tourette syndrome. A phenomenological analysis. Br J Psychiatry, 152, 383–90.

Roman GC (1998). Diagnostic criteria in cerebrovascular diseases and vascular dementia. Eur J Neurol, 5, S3–8.

Rondot P, de Recondo J, Dumas JL (1977). Visuomotor ataxia. Brain, 100(2), 355–76.

Rosen HJ, Allison SC, Schauer GF et al. (2005). Neuroanatomical correlates of behavioural disorders in dementia. Brain, 128, 2612–25.

Ross ED (1981). The aprosodias. Functional-anatomic organization of the affective components of language in the right hemisphere. Arch Neurol, 38, 561–9.

Rossor MN, Revesz T, Lantos PL, Warrington EK (2000). Semantic dementia with ubiquitin-positive tau-negative inclusion bodies. Brain, 123, 267–76.

Rudge P, Warrington EK (1991). Selective impairment of memory and visual perception in splenial tumours. Brain, 114, 349–60.

Santamaria J, Tolosa E, Valles A (1986). Parkinson's disease with depression: a possible subgroup of idiopathic parkinsonism. Neurology, 36, 1130–3.

Santhouse AM, Howard RJ, ffytche DH (2000). Visual hallucinatory syndromes and the anatomy of the visual brain. Brain, 123, 2055–64.

Santos FS, Velasco IT, Fraguas R Jr (2004). Risk factors for delirium in the elderly after coronary artery bypass graft surgery. Int Psychogeriatr, 16, 175–93.

Sawle GV, Hymas NF, Lees AJ et al. (1991). Obsessional slowness. Functional studies with positron emission tomography. Brain, 114, 2191–202.

Scahill RI, Schott JM, Stevens JM et al. (2002) Mapping the evolution of regional atrophy in Alzheimer's disease: unbiased analysis of fluid-registered serial MRI. Proc Natl Acad Sci USA, 99, 4703–7.

Scheltens P, Fox N, Barkhof F et al. (2002). Structural magnetic resonance imaging in the practical assessment of dementia: beyond exclusion. Lancet Neurol, 1, 13–21.

Schenk D, Barbour R, Dunn W et al. (1999). Immunization with amyloid-beta attenuates Alzheimer disease-like pathology in the PDAPP mouse. Nature, 400, 173–7.

Schiff HB, Alexander MP, Naeser MA et al. (1983). Aphemia. Clinical-anatomic correlations. Arch Neurol, 40, 720–7.

Schiffer RB (1987). The spectrum of depression in multiple sclerosis. An approach for clinical management. Arch Neurol, 44, 596–9.

Schmahmann JD, Sherman JC (1998). The cerebellar cognitive affective syndrome. Brain, 121, 561–79.

Schneider LS, Dagerman KS, Insel P (2005). Risk of death with atypical antipsychotic drug treatment for dementia: meta-analysis of randomized placebo-controlled trials. JAMA, 294, 1934–43.

Schoonenboom NS, Pijnenburg YA, Mulder C et al. (2004). Amyloid beta(1-42) and phosphorylated tau in CSF as markers for early-onset alzheimer's disease. Neurology, 62, 1580–4.

Schroder R, Watts GD, Mehta SG et al. (2005). Mutant valosin-containing protein causes a novel type of frontotemporal dementia. Ann Neurol, 57, 457–61.

Scoville WB, Milner B (1957). Loss of recent memory after bilateral hippocampal lesions. J Neurol Neurosurg Psychiatry, 20, 11–21.

Seyffarth H, Denny-Brown D (1948). The grasp reflex and the instinctive grasp reaction. Brain, 71, 109–83.

Shallice T (1981a). From Neuropsychology to Mental Structure. Cambridge University Press, New York.

Shallice T (1981b) Phonological agraphia and the lexical route in writing. Brain, 104, 413–29.

Shallice T (1982). Specific impairments of planning. Philos Trans R Soc Lond B Biol Sci, 298, 199–209.

Shallice T, Burgess PW, Schon F et al. (1989). The origins of utilization behaviour. Brain, 112, 1587–98.

Shallice T, Warrington EK (1970). Independent functioning of verbal memory stores: a neuropsychological study. Q J Exp Psychol, 22, 261–73.

Shepherd J, Blauw GJ, Murphy MB et al. (2002). Pravastatin in elderly individuals at risk of vascular disease (PROSPER): a randomised controlled trial. Lancet, 360, 1623–30.

Shipp S, de Jong BM, Zihl J et al. (1994) The brain activity related to residual motion vision in a patient with bilateral lesions of V5. Brain, 117 (Pt 5), 1023–38.

Shuping JR, Toole JF, Rollinson RD (1979). Transient global amnesia. Trans Am Neurol Assoc, 104, 183–6.

Siebert M, Markowitsch HJ Bartel P (2003). Amygdala, affect and cognition: evidence from 10 patients with Urbach-Wiethe disease. Brain, 126, 2627–37.

Silverman DH, Small GW, Chang CY et al. (2001). Positron emission tomography in evaluation of dementia: Regional brain metabolism and long-term outcome. JAMA, 286, 2120–7.

Singer R (1998). Diagnosis and treatment of Whipple's disease. Drugs, 55, 699–704.

Skibinski G, Parkinson NJ, Brown JM et al. (2005). Mutations in the endosomal ESCRTIII-complex subunit CHMP2B in frontotemporal dementia. Nat Genet, 37, 806–8.

Smith LW, Dimsdale JE (1989). Postcardiotomy delirium: conclusions after 25 years? Am J Psychiatry, 146, 452–8.

Snowden JS, Goulding PJ, Neary D (1989). Semantic dementia: a form of circumscribed cerebral atrophy. Behav Neurol, 2, 167–82.

Snowden JS, Neary D, Mann DMA (1996). Fronto-temporal Lobar Degeneration. Churchill Livingstone, Edinburgh.

Spillantini MG, Schmidt ML, Lee VM et al. (1997). Alpha-synuclein in Lewy bodies [letter]. Nature, 388, 839–40.

Starkstein SE, Robinson RG (1993). Depression in cerebrovascular disease. In Starkstein SE, Robinson RG, eds. Depression in Neurological Disease, pp. 28–49. John Hopkins University Press, Baltimore.

Starr A, Phillips L (1970) Verbal and motor memory in the amnestic syndrome. Neuropsychologia, 8, 75–88.

Stewart L, von Kriegstein K, Warren JD et al. (2006). Music and the brain: disorders of musical listening. Brain, 129, 2533–53.

Stunkard AJ, Grace WJ, Wolff HG (1955). Night-eating syndrome; pattern of food intake among certain obese patients. Am J Med, 19, 78–86.

Stuss DT, Alexander MP, Lieberman A et al. (1978). An extraordinary form of confabulation. Neurology, 28, 1166–72.

Sulkava R, Erkinjuntti T (1987). Vascular dementia due to cardiac arrhythmias and systemic hypotension. Acta Neurol Scand, 76, 123–8.

Summers WK, Reich TC (1979). Delirium after cataract surgery: review and two cases. Am J Psychiatry, 136, 386–91.

Swedo SE, Leonard HL, Garvey M et al. (1998). Pediatric autoimmune neuropsychiatric disorders associated with streptococcal infections: clinical description of the first 50 cases. Am J Psychiatry, 155, 264–71.

Swedo SE, Rapoport JL, Cheslow DL et al. (1989). High prevalence of obsessive-compulsive symptoms in patients with Sydenham's chorea. Am J Psychiatry, 146, 246–9.

Tariot PN, Profenno LA, Ismail MS (2004) Efficacy of atypical antipsychotics in elderly patients with dementia. J Clin Psychiatry, 65 (Suppl 11), 11–5. Review.

Taylor D, Lewis S (1993). Delirium. *J Neurol Neurosurg Psychiatry*, **56**, 742–51.

Tekin S, Mega MS, Masterman DM *et al.* (2001). Orbitofrontal and anterior cingulate cortex neurofibrillary tangle burden is associated with agitation in Alzheimer disease. *Ann Neurol*, **49**, 355–61.

Thieben MJ, Lennon VA, Boeve BF *et al.* (2004). Potentially reversible autoimmune limbic encephalitis with neuronal potassium channel antibody. *Neurology*, **62**, 1177–82.

Thompson PD, Marsden CD (1987). Gait disorder of subcortical arteriosclerotic encephalopathy: Binswanger's disease. *Mov Disord*, **2**, 1–8.

Thompson SA, Graham KS, Patterson K *et al.* (2002). Is knowledge of famous people disproportionately impaired in patients with early and questionable Alzheimer's disease? *Neuropsychology*, **16**, 344–58.

Tranel D, Hyman BT (1990). Neuropsychological correlates of bilateral amydala damage. *Arch Neurol*, **47**, 349–55.

Trevitt CR, Collinge J (2006). A systematic review of prion therapeutics in experimental models. *Brain*, **129**, 2241–65.

Trimble MR (2000). Behaviour and Personality Disturbances. In Bradley WG, Daroff RB, Fenichel GM, Marsden CD, eds. *Neurology in Clinical Practice: Principles of Diagnosis and Management*, 3rd edition, pp. 89–104. Butterworth-Heinemann, Boston.

Tschampa HJ, Kallenberg K, Urbach H *et al.* (2005). MRI in the diagnosis of sporadic Creutzfeldt-Jakob disease: a study on inter-observer agreement. *Brain*, **128**, 2026–33.

Tulving E (1973). Episodic and semantic memory. In Tulving E, Donaldson W, eds. *Organization of Memory*, pp. 382. Academic Press, New York.

Tune LE, Damlouji NF, Holland A *et al.* (1981). Association of postoperative delirium with raised serum levels of anticholinergic drugs. *Lancet*, **2**, 651–3.

Tyrrell PJ, Kartsounis LD, Frackowiak RS *et al.* (1991). Progressive loss of speech output and orofacial dyspraxia associated with frontal lobe hypometabolism [see comments]. *J Neurol Neurosurg Psychiatry*, **54**, 351–7.

Tyrrell PJ, Warrington EK, Frackowiak RSJ *et al.* (1990). Progressive degeneration of the right temporal-lobe studied with positron emission tomography. *J Neurol Neurosurg Psychiatry*, **53**, 1046–50.

Victor M, Adams RD, Collinge GH (1989). *The Wernicke-Korsakoff's Syndrome and Related Disorders due to Alcoholism and Malnutrition.* FA Davis IS Davis, Philadelphia.

Vidal R, Frangione B, Rostagno A *et al.* (1999). A stop-codon mutation in the BRI gene associated with familial British dementia. *Nature*, **399**, 776–81.

Vignolo L (1982). Auditory agnosia. *Philos Trans R Soc Lond B Biol Sci*, **298**, 49–57.

Vincent A, Buckley C, Schott JM *et al.* (2004). Potassium channel antibody-associated encephalopathy: a potentially immunotherapy-responsive form of limbic encephalitis. *Brain*, **127**, 701–12.

Wade DT, Legh SJ, Hewer RA (1987). Depressed mood after stroke. A community study of its frequency. *Br J Psychiatry*, **151**, 200–5.

Waldemar G, Dubois B, Emre M *et al.* (2007) Recommendations of the diagnosis and management of Alzheimer's disease and other disorders associated with dementia: EFNS guideline. *Eur J Neurol*, **14**, e1–26.

Warren JD, Schott JM, Fox NC *et al.* (2005). Brain biopsy in dementia. *Brain*, **128**, 2016–25.

Warrington EK, James M and Kinsbourne M (1966). Drawing disability in relation to laterality of cerebral lesion. *Brain*, **89**, 53–82.

Warrington EK, Taylor AM (1973). The contribution of the right parietal lobe to object recognition. *Cortex*, **9**, 152–64.

Warrington EK, Shallice T (1980). Word-form dyslexia. *Brain*, **103**, 99–112.

Warrington EK, McCarthy R (1983). Category specific access dysphasia. *Brain*, **106**, 859–78.

Warrington EK, McCarthy RA (1987). Categories of knowledge. Further fractionations and an attempted integration. *Brain*, **110**, 1273–96.

Weinstein EA, Kahn RL (1952). Non-aphasic misnaming (paraphasia) in organic brain disease. *Arch Neurol Psychiatry*, **62**, 72–9.

Wetterling T, Kanitz RD, Borgis KJ (1996). Comparison of different diagnostic criteria for vascular dementia (ADDTC, DSM-IV, ICD-10, NINDS-AIREN). *Stroke*, **27**, 30–6.

Will RG, Zeidler M, Stewart GE *et al.* (2000). Diagnosis of new variant Creutzfeldt-Jakob disease. *Ann Neurol*, **47**, 575–82.

Williams M, Pennybacker J (1954). Memory disturbance in third ventricle tumours. *J Neurol Neurosurg Psychiatry*, **17**, 115–23.

Winbeck K, Etgen T, von Einsiedel HG *et al.* (2005). DWI in transient global amnesia and TIA: proposal for an ischaemic origin of TGA. *J Neurol Neurosurg Psychiatry*, **76**, 438–41.

Winblad B, Palmer K, Kivipelto M *et al.* (2004). Mild cognitive impairment—beyond controversies, towards a consensus: report of the International Working Group on Mild Cognitive Impairment. *J Intern Med*, **256**, 240–6.

Winkelman JW (2003) Treatment of nocturnal eating syndrome and sleep-related eating disorder with topiramate. *Sleep Med*, **4(3)**, 243–6.

Woods SC, Seeley RJ, Porte D *et al.* (1998). Signals that regulate food intake and energy homeostasis. *Science*, **280**, 1378–83.

Wragg RE, Jeste DV (1989). Overview of depression and psychosis in Alzheimer's disease. *Am J Psychiatry*, **146**, 577–87.

Yager J, Andersen AE (2005). Clinical practice. Anorexia nervosa. *N Engl J Med*, **353**, 1481–8.

Zadikoff C, Lang AE (2005) Apraxia in movement disorders. *Brain*, **128** (Pt 7), 1480–97.

Zadikoff C, Lang AE, Leiguarda RC *et al.* (2000). Limb apraxias: higher-order disorders of sensorimotor integration. *Brain*, **123**, 860–79.

Zangwill OL (1979). Two cases of crossed aphasia in dextrals. *Neuropsychologia*, **17**, 167–72.

Zeidler M, Sellar RJ, Collie DA *et al.* (2000). The pulvinar sign on magnetic resonance imaging in variant Creutzfeldt-Jakob disease. *Lancet*, **355**, 1412–8.

Zeki S (1990). A century of cerebral achromatopsia. *Brain*, **113**, 1721–77.

Zeman AZ, Boniface SJ, Hodges JR (1998). Transient epileptic amnesia: a description of the clinical and neuropsychological features in 10 cases and a review of the literature. *J Neurol Neurosurg Psychiatry*, **64**, 435–43.

Zihl J, von Cramon CD, Mai N (1983). Selective disturbance of movement vision after bilateral brain damage. *Brain*, **106**, 313–40.

SECTION 8

Vascular, demyelinating, inflammatory and degenerative disorders of the central nervous system

CHAPTER 35

Cerebrovascular diseases

Peter Rothwell

35.1 Introduction

This chapter is concerned with those diseases of the cerebral and ocular circulation that cause ischaemia or infarction of the brain and eye or spontaneous haemorrhage into or around the brain. The main clinical manifestations of these diseases are transient ischaemic attack and stroke.

A transient ischaemic attack, often referred to as a 'TIA' is an acute loss of focal brain or monocular function with symptoms lasting less than 24 h and which is thought to be due to inadequate cerebral or ocular blood supply as a result of arterial thrombosis, low flow, or embolism associated with arterial, cardiac, or haematological disease.

A stroke is rapidly developing clinical symptoms and/or signs of focal, and at times global (applied to patients in deep coma and to those with subarachnoid haemorrhage) loss of brain function, with symptoms lasting more than 24 h or leading to death, with no apparent cause other than that of vascular origin (Hatano 1976).

Separating transient ischaemic attack from stroke on the basis of this arbitrary 24-h time limit is helpful because the differential diagnosis of transient ischaemic attack differs from stroke. However, there is no qualitative difference between a transient ischaemic attack and an ischaemic stroke: anything that causes a transient ischaemic attack may, if more severe or prolonged, cause an ischaemic stroke (Sempere *et al.* 1998). Therefore, the investigation of a patient with a transient ischaemic attack and a mild ischaemic stroke is identical, and the long-term management to reduce the risk of further vascular events is similar.

Stroke is a very common condition. Each year there are about one million strokes in the European Union (Sudlow and Warlow 1997) making it by far the most common neurological disorder

(MacDonald *et al* 2000). About 25 per cent of men and 20 per cent of women can expect to suffer a stroke if they live to be 85 years old (Bonita 1992) and stroke is the second most common cause of death worldwide (Murray and Lopez 1996). However, mortality data underestimate the true burden of stroke, since in contrast to coronary heart disease and cancer, the major burden of stroke is chronic disability rather than death (Wolfe 2000). Brain diseases, of which stroke comprises a large proportion, cause 23 per cent of healthy years lost and around 50 per cent of years of life lived with disability in Europe (Olesen and Leonardi 2003). About a third of stroke survivors are functionally dependent at 1 year and stroke is the commonest cause of neurological disability in the developed world (Murray and Lopez 1996, MacDonald *et al.* 2000). Stroke also causes secondary medical problems including dementia, depression, epilepsy, falls, and fractures. In the United Kingdom, the costs of stroke are estimated to be nearly twice those of coronary heart disease (British Heart Foundation Statistics Database 1998, Rothwell 2001b) accounting for about 6 per cent of total National Health Service and Social Services expenditure (Rothwell 2001b). As the population ages over the next two decades, the total stroke rate will probably increase unless there are substantial decreases in age and sex-specific incidence (Rothwell *et al.* 2004a). Stroke deaths are projected to increase from 4.5 million worldwide in 1990 to 7.7 million in 2020 when stroke will account for 6.2 per cent of the total burden of illness (Bonita 1992; Sudlow and Warlow 1997; Menken *et al.* 2000).

About 85 per cent of all first ever strokes are ischaemic, 10 per cent are due to primary intracerebral haemorrhage and about 5 per cent are due to subarachnoid haemorrhage (Rothwell *et al.* 2004a) (Fig. 35.1). Within ischaemic stroke, 25 per cent are caused by large artery disease, 25 per cent by small vessel disease, 20 per cent by cardiac embolism, 5 per cent by other rarer causes, and the remaining 25 per cent are of undetermined aetiology. Ischaemic stroke may also be classified by anatomical location using clinical features into total anterior circulation stroke; partial anterior circulation stroke, lacunar stroke, and posterior circulation stroke. This is of some help in identifying the likely underlying pathology and gives information as to prognosis.

35.2 Epidemiology of stroke

Understanding of the epidemiology of stroke has lagged that of coronary heart disease partly because of a lack of research funding for stroke (Rothwell 2001b; Pendlebury *et al.* 2004; Pendlebury 2007) and partly because stroke is a much more heterogeneous disorder. Separate assessment of the different stroke subtypes should ideally be made in epidemiological studies of stroke but this was often not possible in early studies because of a lack of brain and vascular imaging and remains problematic today because of the frequent difficulty in ascribing a cause for a given stroke even when imaging is available.

35.2.1 Mortality

Stroke mortality rises rapidly with age (Rothwell *et al.* 2005a). The age-standardized death rate attributed to stroke varies 6-fold between developed countries while very little is known about the developing world (Inzitari *et al.* 1995; Connor *et al.* 2007). Particularly high reported rates of stroke occur in eastern Europe, and Japan, and particularly low rates in certain parts of North

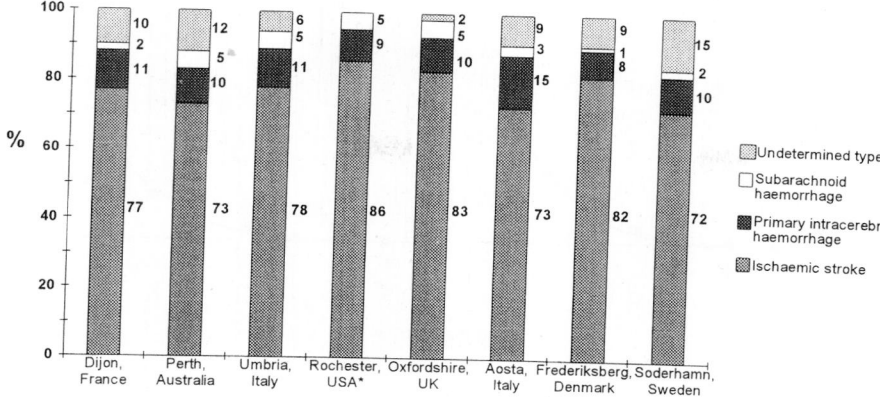

Fig. 35.1 The percentage distribution of the three pathological types of first-ever-in-a-lifetime stroke from age 45 to 84 years in eight comparable community-based studies in the 1980s and 1990s. The numbers by the columns refer to percentages (with permission from Sudlow and Warlow 1997.)

America and some, western European countries (Feigin *et al.* 2003). The reasons for these differences are unclear but one possibility is that the stroke subtypes more likely to be fatal, particularly intracranial haemorrhage or cardioembolic stroke, are more frequent in countries with high stroke mortality.

35.2.2 Incidence, prevalence, and time trends

The incidence of new cases of first ever stroke can only be reliably assessed in prospective population-based studies (Sudlow and Warlow 1996; Feigin *et al.* 2003; Rothwell *et al.* 2004a) since hospital-based studies are subject to referral bias. One of the most comprehensive population-based studies of stroke incidence is the Oxford Vascular Study, OXVASC, which has near-complete case ascertainment, irrespective of age (Coull *et al.* 2004). Previous studies such as the MONICA project and the Framingham study, had an age cut off at 65 or 75 years or relied on voluntary participation. The OXVASC study showed that the incidence of stroke in the United Kingdom, including subarachnoid haemorrhage, is approximately 2.3/1000/year (Rothwell *et al.* 2005a), with about a quarter of events occuring under the age of 65 and about a half above the age of 75 (Fig. 35.2). The incidence of cerebrovascular events in OXVASC was similar to that of acute coronary vascular events in the same population during the same period (Fig. 35.3), with a similar age distribution (Rothwell *et al.* 2005a). Stroke prevalence in the United Kingdom is about 5/1000 population but the exact figure depends on the population age and sex structure, incidence, and survival which may vary by time and place. The prevalence is around 50/1000 in men and 25/1000 in women aged 65 to 74 (Wyller *et al.* 1994; Geddes *et al.* 1996; Bots *et al.* 1996).

A reduction in stroke incidence over the last two decades would be expected, given that randomized trials have shown several interventions to be effective in the primary and secondary prevention of stroke. Indeed, it has been estimated that full implementation of currently available preventive strategies could reduce stroke incidence by as much as 50–80 per cent (Murray *et al.* 2003; Wald and Law 2003). Stroke mortality rates certainly declined from the 1950s to 1980s in North America and Western Europe (Bonita *et al.* 1990; Thom 1993), but this decline has since levelled-off. Although apparent trends in stroke mortality are very difficult to interpret because of changes over time in death certification practices and case-fatality, stroke incidence also appeared to decline in the 1960s and 1970s in the United States, Asia, and Europe (McGowern *et al.* 1992; Kodama 1993; Tunstall-Pedoe *et al.* 1994;

Numminen *et al.* 1996). However, the majority of subsequent studies during the 1980s and 1990s, when effective preventive treatments had become more widely available, have shown either no change (Wolf *et al.* 1992; Bonita *et al.* 1993; Stegmayr *et al.* 1994) or more commonly an increase in age and sex-adjusted incidence (Johansson *et al.* 2000, Medin *et al.* 2004).

The most recent studies of time trends in stroke incidence do suggest that age-specific incidence is now finally falling (Sarti *et al.* 2003; Rothwell *et al.* 2004a; Anderson *et al.* 2005; Hardie *et al.* 2005). Between 1981–84 and 2002–2004 a 40 per cent reduction in the incidence of fatal and disabling stroke was found in Oxfordshire, United Kingdom (Rothwell *et al.* 2004a), although this was accompanied by an apparent rise in the incidence of transient ischaemic attack and minor stroke particularly in the oldest old (Fig. 35.4). The reasons for the decline in incidence of major stroke in recent years are unclear but may be related to a decline in the prevalence of causative risk factors or to treatment of risk factors such as hypertension and elevated cholesterol.

35.2.3 Racial and social factors

There are racial and social differences in susceptibility to stroke (Forouhi and Sattar 2006) and in the incidence of the various stroke subtypes. Some of these racial differences are in part caused by differences in risk factor prevalence: hypertension and diabetes are more common in blacks and coronary heart disease is more common in whites, for example (Sacco 2001). Other differences are not properly understood, such as the much higher proportion of stroke due to intracerebral haemorrhage in south east Asia and the far east than in western countries (Sudlow and Warlow 1996; Feigin *et al.* 2003).

Black populations

The burden of stroke is higher in blacks than whites for both ischaemic, particularly small vessel stroke, and haemorrhagic stroke (Woo *et al.* 1999; Schneider *et al.* 2004; Pandey and Gorelick 2005; White *et al.* 2005; Wolfe *et al.* 2006a,b). This pattern may be related in part to a higher prevalence of hypertension and diabetes (Gillum 1999; Sacco 2001). Intracranial large artery occlusive vascular disease also appears to be more common in blacks compared with Caucasians (Sacco *et al.* 1995; Wityk *et al.* 1996; Lynch and Gorelick 2000).

Maori and Pacific Islands

People in New Zealand have a higher stroke incidence than Europeans, perhaps due to differences in risk factors and health-related behaviours (Bonita *et al.* 1997; Feigin *et al.* 2006).

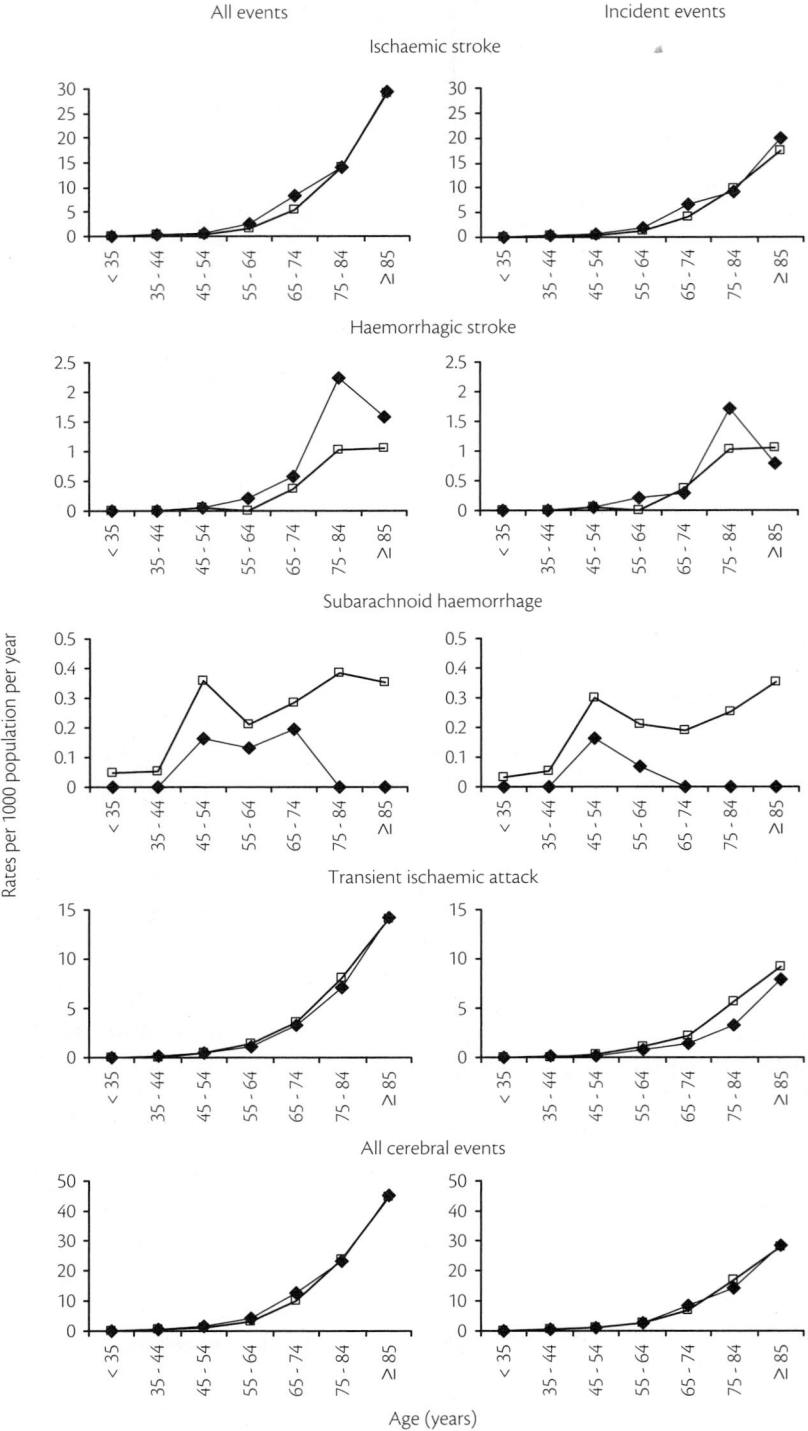

Fig. 35.2 Age-specific rates of all events and incident events for different types of acute cerebrovascular events in men (diamonds) and women (open squares) in Oxfordshire from 2002–2005. (Rothwell *et al.* 2005a.)

Japanese and Chinese populations

Stroke, particularly primary intracerebral haemorrhage is more common in Japan and China than in Western countries (Huang *et al.* 1990) and this is accompanied by less extracranial but more intracranial arterial disease (Feldmann *et al.* 1990; Leung *et al.* 1993; Sacco *et al.* 1995; Wityk *et al.* 1996).

South Asian populations

People of south Asian origin in the United Kingdom have a high prevalence of coronary heart disease and stroke, central obesity, as evidenced by high waist-to-hip ratio, insulin resistance, non-insulin dependent diabetes and hypertension (Cappuccio 1997; Kain *et al.* 2002; Bhopal *et al.* 2005). This increase in vascular risk

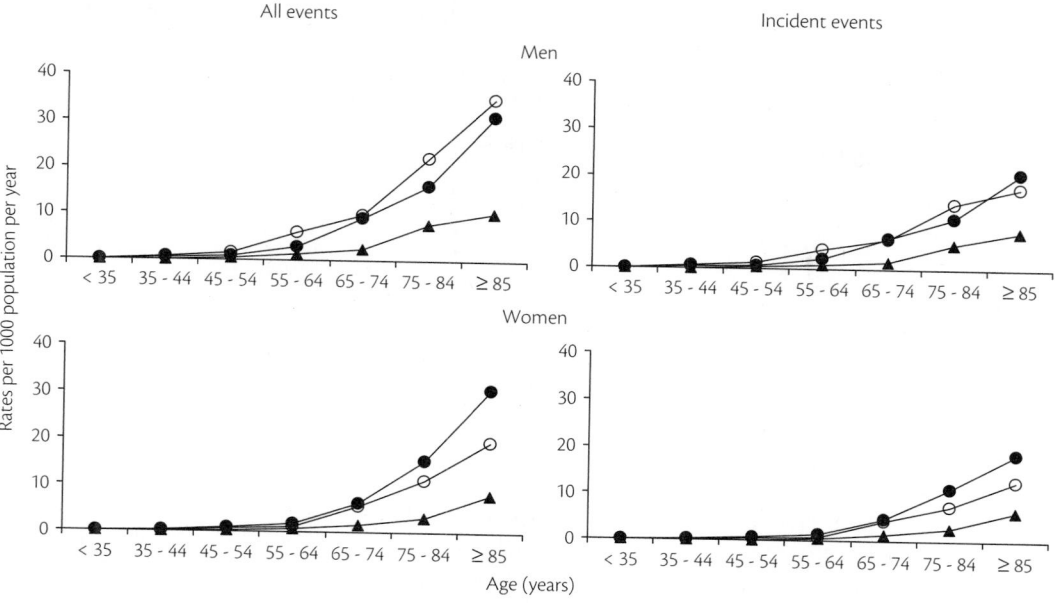

Fig. 35.3 Age-specific rates of all events and of incident events for stroke (i.e. not including transient ischaemic attack; closed circles), myocardial infarction, and sudden cardiac death combined (i.e. not including unstable angina; open circles), and acute peripheral vascular events (triangles) in men and women in Oxfordshire from 2002 to 2005. (Rothwell *et al.* 2005a.)

seems to be due in part to a genetic susceptibility, such as high serum lipoprotein A levels, potentiated by dietary and life-style induced changes in lipid levels.

Deprivation

In the United Kingdom, both stroke incidence and poor outcome after stroke are greater in areas of socio-economic disadvantage (Kaplan and Keil 1993; Avendaño *et al.* 2004). This is partly because poverty is associated with adverse health behaviours and risk factors such as smoking (Hart *et al.* 2000a). There is also evidence that poor maternal and infant health is associated with increased mortality from stroke in later life (Barker 1995; Martyn *et al.* 1996). However, the adverse effect of socio-economic deprivation also appears to be cumulative throughout life (Davey Smith *et al.* 1997; Hart *et al.* 2000b).

35.2.4 Seasonal and diurnal variation

In most studies, both stroke mortality and hospital admission rates are higher in winter than summer (Douglas *et al.* 1991; Pan *et al.* 1995; Feigin and Wiebers 1997). This seasonal variation might be explained by the complications of stroke being more likely to occur in the winter, for instance pneumonia, and cannot be simply assumed to reflect stroke incidence. Where incidence has been measured in the community there is little seasonal variation, at least in temperate climates, although primary intracerebral haemorrhage is somewhat more likely in the winter months and on cold days (Rothwell *et al.* 1996; Jakovljevic *et al.* 1996).

Stroke occurs most frequently in the hour or two after waking in the morning, but whether this applies to all subtypes of stroke is difficult to say because of the relatively small proportion of intracranial haemorrhages in most studies (Kelly-Hayes *et al.* 1995; Elliott 1998). Subarachnoid haemorrhage is very unlikely to occur during sleep and, in general, most likely during strenuous activities (Wroe *et al.* 1992).

35.2.5 Risk factors

There are many more data on risk factors for acute coronary events than for ischaemic stroke (Bhatia and Rothwell 2005) because of more intensive investigation in routine clinical practice and because heart disease receives much higher levels of research funding than stroke (Rothwell 2001b; Pendlebury 2007).

The main risk factors for stroke are listed in Table 35.1. Although there are probably few qualitative differences there are quantitative differences between risk factors for ischaemic stroke and coronary heart disease. Smoking, raised plasma cholesterol and male sex are stronger risk factors for myocardial infarction, while hypertension is a stronger risk factor for stroke. The tendency for epidemiologists to lump all types of stroke together might explain some of the quantitative differences between stroke and coronary heart disease risk factors. For example, it is possible that strokes associated with large artery disease have a more similar risk factor profile to coronary heart disease than cardioembolic stroke. To date, differences in risk factor relationships to the different stroke subtypes are unclear (Schulz *et al.* 2004, Jackson and Sudlow 2005).

Age

Age is the strongest risk factor for ischaemic stroke of all subtypes and for primary intracerebral haemorrhage, but is less important for subarachnoid haemorrhage (Bamford *et al.* 1990a; Rothwell *et al.* 2005a). Overall stroke incidence at age 75–84 is about 25 times higher than at age 45–54 (Fig. 35.2).

Sex and sex hormones

Ischaemic stroke is slightly more common in men than in women (Fig. 35.2) although the male excess is less marked than in coronary heart disease and peripheral arterial disease (Rothwell *et al.* 2005a). This excess of vascular events in men has been attributed to differences in endogenous sex hormones, although there is little

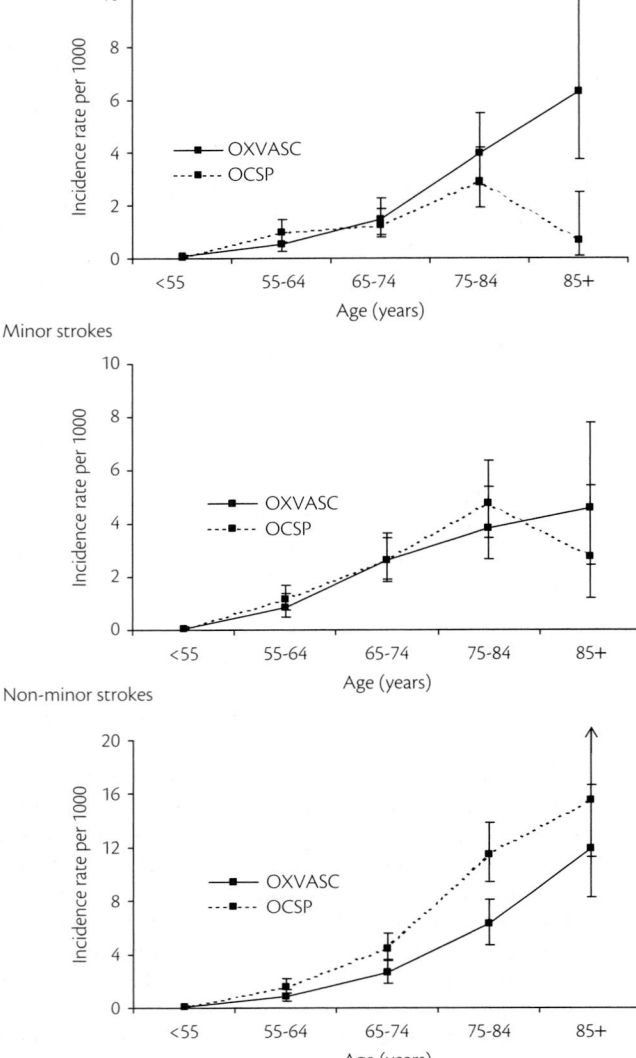

Minor strokes

Non-minor strokes

Fig. 35.4 The age-specific incidence of transient ischaemic attack, minor stroke and major disabling stroke in Oxfordshire in 1981–84 (OCSP) and 2002–04 (OXVASC). (Rothwell *et al.* 2004a.)

Table 35.1 Risk factors for ischaemic stroke (not necessarily 'causal' or independent)

Factors associated with an increased risk of vascular disease
Age
Male sex
Increasing blood pressure
Cigarette smoking
Blood lipids*
Diabetes mellitus
Increasing plasma fibrinogen
Raised factor VII coagulant activity*
Raised tissue plasminogen activator antigen*
Low blood fibrinolytic activity*
Raised von Willebrand factor
Raised haematocrit
Atrial fibrillation
Sex hormones
Excess alcohol consumption
Obesity and diet*
Physical inactivity
Raised white blood cell count
Recent infection*
Hyperhomocyst(e)inaemia
Snoring*
Corneal arcus*
Psychological factors*
Low serum albumin*
Diagonal earlobe crease*
Impaired ventilatory function*
Family history of stroke
Social deprivation
Evidence of pre-existing vascular disease
Myocardial infarction/angina
Cardiac failure
Left ventricular hypertrophy
Peripheral vascular disease
Cervical arterial bruit and stenosis
Transient ischaemic attacks

* Somewhat uncertain association with stroke.

direct evidence to support this hypothesis. Exogenous high-dose oestrogen given to elderly men with prostatic cancer increases their risk of vascular deaths (Byar and Corle 1988) and use of hormone replacement therapy in women after the menopause is associated with an increased risk of acute coronary sydrome, stroke, and venous thromboembolism (Farquhar *et al.* 2005; Gabriel *et al.* 2005). The equalization of vascular risk in elderly males and females is thus probably not explained by the natural menopause, although bilateral oophorectomy without oestrogen replacement in premenopausal women doubles the risk of vascular events (van der Schouw *et al.* 1996). Oral contraceptives increase the risk of ischaemic stroke, although the risk is less for haemorrhagic stroke, but the increased risk is small for low-dose oestrogen preparations unless there are associated factors such as migraine or cigarette smoking (Chang *et al.* 1999; Bousser and Kittner 2000; Donaghy *et al.* 2002).

Blood pressure

Increasing blood pressure is strongly associated with subsequent stroke risk (Gil-Nunez and Vivancos-Mora 2005; Goldstein and Hankey 2006). The relationship between usual diastolic blood pressure and stroke is 'log-linear' throughout the normal range, with no evidence of a threshold below which the risk becomes stable (Rodgers *et al.* 1996) (Fig. 35.5). Stroke incidence about doubles with each 7.5 mmHg increase in usual diastolic blood pressure in Western populations and with each 5.0 mmHg in Japanese and Chinese populations (MacMahon *et al.* 1990; Eastern Stroke and Coronary Heart Disease Collaborative Research Group 1998; Lewington *et al.* 2002). The strength of the association between blood pressure and stroke is attenuated with increasing age, although the absolute risk of stroke in the elderly is far higher than in the young (Lewington *et al.* 2002). Nevertheless, hypertension is still a risk factor in the very elderly, although it is weaker because stroke may be associated with low blood pressure due to cardiac failure and other co-morbid conditions (Birns *et al.* 2005). Moreover, in patients with bilateral severe carotid stenosis or

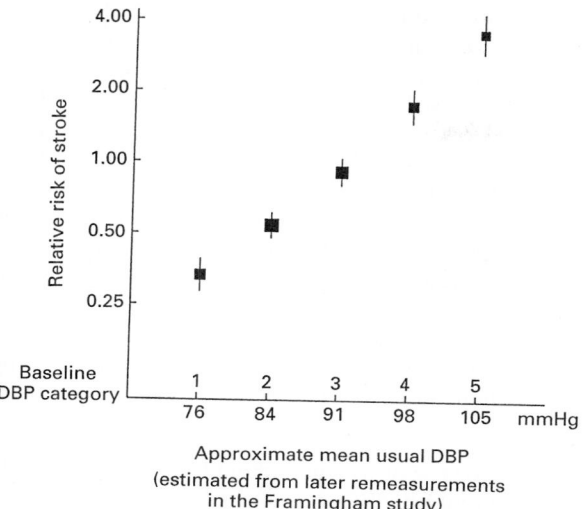

Fig. 35.5 The relative risk of stroke related to the usual diastolic blood pressure (DBP) at baseline from an overview of seven prospective studies in mostly western populations. The error bars are 95 per cent confidence intervals. (Courtesy of MacMahon *et al.* 1990.)

occlusion, stroke risk is higher at low blood pressures suggesting that aggressive blood pressure lowering may be harmful in this group (Rothwell *et al.* 2003c). The relationship with systolic blood pressure is similar and possibly stronger than with diastolic blood pressure (Keli *et al.* 1992) and even 'isolated' systolic hypertension is associated with increased risk (Sagie *et al.* 1993; Petrovitch *et al.* 1995; Lewington *et al.* 2002). Hypertension is more common in stroke patients from black populations as compared to whites (Sacco 2001).

Cigarette smoking

Cigarette smoking is associated with stroke with a relative risk about 1.5: there is a dose–response relationship, males and females are equally affected and the association seems to become weaker in the elderly (Hankey 1999). The evidence for a link between cigarette smoking and primary intracerebral haemorrhage is less clear (Vessey *et al.* 2003). Smoking has been related to the extent of carotid disease on arterial angiography (Homer *et al.* 1991), on ultrasound (Fine-Edelstein *et al.* 1994; Howard *et al.* 1998), and in identical twins discordant for smoking (Haapanen *et al.* 1989).

Diabetes mellitus

Diabetes doubles the risk of ischaemic stroke (Tuomilehto and Rastenyte 1999; Wannamethee *et al.* 1999; Rothwell 2005) and the risk of fatal stroke is higher in those with a higher HbA1c at diagnosis (Stevens *et al.* 2004). There is a high prevalence of diabetes in stroke patients from black populations (Sacco 2001). It should be noted that any studies linking diabetes with stroke mortality data will exaggerate the association, because diabetics who have a stroke are more likely to die of it than non-diabetics (Jorgensen *et al.* 1994). Randomized trials have shown that diabetic treatment reduces the risk of microvascular complications of diabetes and decreases progression of carotid intima media thickness but not does not necessarily reduce the incidence of stroke and other macrovascular events (UK Prospective Diabetes Study (UKPDS) Group 1998; The Diabetes Control and Complications

Trial/Epidemiology of Diabetes Interventions and Complications Research Group 2003).

Blood lipids

Increasing levels of total plasma cholesterol and low-density lipoprotein cholesterol, and to a lesser extent decreasing levels of high-density lipoprotein cholesterol, are strong risk factors for coronary heart disease, whereas blood triglyceride levels are less predictive. Reduction of plasma cholesterol by 1mmol/l reduces the relative risk of coronary events by at least a third (Lewington and Clarke 2005), with little diminution of benefit in the elderly (Huxley *et al.* 2002; Clarke *et al.* 2002; Baigent *et al.* 2005). The relationship between blood lipids and stroke is much weaker but there is some evidence that cholesterol is negatively associated with intracranial haemorrhage which obscures the weak positive association with ischaemic stroke in studies where stroke subtype is not characterized (Prospective Studies Collaboration 1995; Eastern Stroke and Coronary Heart Disease Collaborative Research Group 1998; Koren-Morag *et al.* 2002; Zhang *et al.* 2003).

Data from the Heart Protection Study of cholesterol lowering in patients with known vascular disease or diabetes have shown that such therapy reduced the risk of stroke on follow up but did not show a reduction in recurrent stroke (Heart Protection Study Collaborative Group 2002; Collins *et al.* 2004) possibly because of lack of differentiation of stroke subtype or the fact that patients were at low risk of stroke recurrence since the incident strokes occurred on average 4.6 years before the study onset. However, more recently, the Stroke Prevention by Aggressive Reduction in Cholesterol Levels, SPARCL, trial showed that atorvastatin in patients who had had a stroke or transient ischaemic attack within 1–6 months before study entry, did reduce overall stroke risk (Amarenco *et al.* 2006), although there was a significant increase in risk of intracerebral haemorrhage on statin treatment. Interestingly, the same trend had been found in the Heart Protection Study in the 3280 patients with previous stroke or transient ischaemic attack (Collins *et al.* 2004), in whom simvastatin 40 mg also increased the risk of haemorrhagic stroke. Thus, the randomized evidence does suggest that there is a causal association between plasma LDL cholesterol and risk of ischaemic stroke, but more work is required to determine the cause of the increase in risk of haemorrhagic stroke.

Atrial fibrillation

Non-rheumatic atrial fibrillation is by far the most common cause of cardioembolic stroke but cannot cause more than one-sixth of all ischaemic strokes since it is present in about this proportion of ischaemic stroke patients (Sandercock *et al.* 1992; Schulz and Rothwell 2003; Rothwell *et al.* 2004a) except in the *very* elderly, where its prevalence is highest (Wolf *et al.* 1991a). The average absolute risk of stroke in unanticoagulated non-rheumatic atrial fibrillation patients without prior stroke is about 4 per cent per year, 5–6 times greater than in those in sinus rhythm (Wolf *et al.* 1991b; Hart *et al.* 1999; Lip 2005); the risk is much higher again in patients with rheumatic atrial fibrillation.

The stroke risk associated with atrial fibrillation in an individual patient is higher in the presence of a previous embolic event, increasing age, hypertension, diabetes, left ventricular dysfunction, and an enlarged left atrium (Stroke Prevention in Atrial Fibrillation Investigators 1992, 1995; Atrial Fibrillation Investigators 1994, 1998; Di Pasquale *et al.* 1995; Lip and Boos 2006). At least 10 similar stroke risk stratification schemes for atrial fibrillation have been published (Lip and Boos 2006; Nattel and Opie 2006). The best validated of these, the CHADS2 score, awards 1 point each for **C**ongestive heart failure, **H**ypertension, **A**ge ≥ 75 years, and **D**iabetes mellitus and 2 points for prior **S**troke or transient ischaemic attack. Patients with a CHADS2 score of 0 have a stroke risk of 0.5 per cent per year whilst those with a score of 6 have a yearly stroke risk of 15 per cent or more.

Paroxysmal atrial fibrillation carries the same stroke risk as persistent atrial fibrillation (Lip and Hee 2001; Saxonhouse and Curtis 2003) and should be treated similarly. There is no evidence that conversion to sinus rhthym followed by pharmacotherapy to try and maintain such rhythm is superior to rate control in terms of mortality and stroke risk (Segal et al. 2001; Blackshear and Safford 2003; Hart et al. 2003).

Some of the association between atrial fibrillation and stroke must be coincidental because atrial fibrillation can be caused by coronary and hypertensive heart disease, both of which may be associated with atheromatous disease or primary intracerebral haemorrhage. Although anticoagulation markedly reduces the risk of first or recurrent stroke, this is not necessarily evidence for causality because this treatment may be working in other ways, such as by inhibiting artery-to-artery embolism although trials of warfarin in secondary prevention of stroke in sinus rhythm have shown no benefit over aspirin.

Cardioembolism

Apart from atrial fibrillation, there are many other causes of cardioembolic stroke including prosthetic heart valves and patent foramen ovale.

Obesity and the metabolic syndrome

Any relationship between obesity and stroke is likely to be confounded by the positive association of obesity with hypertension, diabetes, hypercholesterolaemia, and lack of exercise, and the negative association with smoking and concurrent illness. Nevertheless, stroke is more common in the obese and abdominal obesity appears to be an independent predictor of stroke (Suk et al. 2003). The constellation of metabolic abnormalities including central obesity, decreased high density lipoprotein, elevated triglycerides, elevated blood pressure, and impaired glucose tolerance is known as the metabolic syndrome and is associated with a 3-fold increase risk of type 2 diabetes and a 2-fold increase in cardiovascular risk (Eckel et al. 2005; Grundy et al. 2005).

Metabolic syndrome is thought to be the main driver for the modern day epidemic of diabetes and vascular disease. As well as primary prevention of acute vascular events in patients with the metabolic syndrome (Eckel et al. 2005; Grundy et al. 2005), an additional aim should be prevention of progression to frank diabetes. Both lifestyle modification with diet and exercise (Tuomilehto et al. 2001; Diabetes Prevention Program Research Group 2002), and angiotensin converting enzyme, or ACE, inhibitors or angiotensin antagonists (Yusuf et al. 2000; Dahlof et al. 2002; Julius et al. 2004) have been shown to be effective in reducing progression to diabetes in patients with the metabolic syndrome.

Diet

Relating various dietary constituents to the risk of vascular disease is difficult since observational data are likely to be biased. As noted above, there is good evidence that dietary and lifestyle modification can improve vascular risk in patients with the metabolic syndrome but randomized trials of dietary interventions including fish oil supplementation and vitamin supplementation have generally been disappointing (Steinberg 1995; Orencia et al. 1996; Stephens 1997; GISSI-Prevenzione Investigators 1999; Yusuf et al. 2000; Hooper et al. 2000b; Leppala et al. 2000; Heart Protection Study Collaborative Group 2002).

Reduction in salt intake reduces blood pressure in both normotensive and hypertensive individuals (He and MacGregor 2004) and there is evidence that adding salt to food increases the risk of cerebral haemorrhage (Jamrozik et al. 1994). In contrast, a high intake of potassium may reduce stroke risk (Khaw and Barrett-Connor 1987; Whelton et al. 1997). It remains unclear whether reducing sodium intake lowers stroke risk.

Exercise

A systematic review of 23 studies found that moderate and high levels of physical activity are associated with reduced risk of all stroke (Lee et al. 2003). This reduced risk is thought to be related to lower body weight, blood pressure, blood viscosity, fibrinogen concentration, and better lipid profiles.

Alcohol

A systematic review of 35 observational studies indicated that heavy alcohol consumption of >5 units/day increases the relative risk of stroke by 1.6 (95 per cent CI 1.4–1.9) particularly haemorrhagic stroke (RR 2.18, 95 per cent CI 1.5–3.2) (Reynolds et al. 2003). Moderate consumption of alcohol appears to lower stroke risk by comparison to abstention (Reynolds et al. 2003; Elkind et al. 2006). Whether 'binge' drinking is associated with stroke is uncertain (Hillbom et al. 1999) but it is possible that irregular drinking carries a higher risk (Mazzaglia et al. 2001).

Haemostatic variables

Despite much effort, very few consistent associations have been found between coagulation parameters, fibrinolytic activity, platelet behaviour, and risk of stroke (Markus and Hambley 1998; Sacco 2001). Although there is a relationship between increasing plasma fibrinogen and stroke (Rothwell et al. 2004c; Danesh et al. 2005), it is attenuated by adjusting for cigarette smoking and other confounding variables such as infections and social class (Brunner et al. 1996; Lowe et al. 1997). Raised plasma factor VII coagulant activity, raised tissue plasminogen activator antigen, low blood fibrinolytic activity, and raised von Willebrand factor are risk factors for coronary heart disease and perhaps also for stroke (Meade et al. 1993; Qizilbash et al. 1997; Macko et al. 1999).

Haematocrit

Although cerebral blood flow is strongly related to haematocrit, any effect of increasing haematocrit on risk of stroke, or type of stroke, is weak and confounded by cigarette smoking, blood pressure, and plasma fibrinogen (Welin et al. 1987). However, raised haematocrit does seem to be associated with an increased case-fatality in ischaemic stroke (Allport et al. 2005).

Infections and inflammation

Both chronic and acute infection have been implicated in the development and stability of atheromatous plaques (Danesh et al. 1997). However, observational epidemiological studies of infections in general (Grau et al. 1998), and of chronic dental infection (Beck et al. 1996), together with serological evidence of specific infectious agents, such as *Chlamydia pneumoniae*, *Helicobacter pylori*, and cytomegalovirus, have not shown convincing evidence of a relationship with stroke or coronary heart disease (Markus and Mendall 1998; Danesh *et al.* 1999, 2000; Fagerberg *et al.* 1999; Glader *et al.* 1999; Strachan *et al.* 1999; Ngeh *et al.* 2003; Danesh *et al.* 2003). An early randomized trial to eliminate chlamydia was far too small to be reliable (Gurfinkel *et al.* 1997) but further antibiotic intervention trials are ongoing (Danesh 2005; Jespersen *et al.* 2006).

Homocysteinaemia

There is an association between the rare inborn recessive condition of homocystinuria and arterial and venous thrombosis and observational data linking coronary heart disease, stroke, and venous thromboembolism with increasing plasma homocysteine (Wald *et al.* 2002, 2004). This led to trials of olic acid and pyridoxine supplementation to lower homocysteine levels (Hankey 2002; Hankey and Eikelboom 2005). Results from such trials have so far been disappointing: the Vitamin Intervention for Stroke Prevention Study, or VISP, and NORVIT (Toole *et al.* 2004; Bonaa *et al.* 2006) trials showed no treatment effect on recurrent stroke, coronary events, or deaths. Preliminary results from the VITATOPS trial have shown no evidence of reduced levels of inflammation, endothelial dysfunction, or the hypercoagulability postulated to be increased by elevated homocysteine levels in patients with previous transient ischaemic attack or stroke treated with folic acid, vitamin B12, and vitamin B6 (Dusitanond *et al.* 2005). However, a recent systematic review of all randomized trials of homocysteine lowering does suggest a modest reduction in stroke risk (Wang *et al.* 2007).

Non-stroke vascular disease

Coronary heart disease is associated with ischaemic stroke in post-mortem (Stemmermann *et al.* 1984), twin (Brass *et al.* 1996), case–control (Feigin *et al.* 1998), and cohort studies (Harmsen *et al.* 1990; Shaper *et al.* 1991; Wolf *et al.* 1991b; Touze *et al.* 2006) as are electrocardiographic abnormalities, cardiac failure, left ventricular hypertrophy, claudication, and asymptomatic peripheral vascular disease (Hankey *et al.* 2006; Leys *et al.* 2006).

Abdominal aortic aneurysms occur in about 10–20 per cent of patients with cerebrovascular disease but it is not known if people with aneurysms have more stroke or other vascular events, compared with people without aneurysms (Hollander *et al.* 2003; Leys *et al.* 2006).

Other possible associations

Innumerable other risk factors have been linked with coronary heart disease, and to a lesser extent stroke, but data are sparse and there is probably a lot of confounding.

35.2.6 Hereditary disorders

A few strokes are clearly 'familial' with a simple Mendelian pattern of inheritance of the underlying cause (Table 35.2). Some of these genetic causes of stroke are described below:

Cerebral autosomal dominant arteriopathy with subcortical infarcts and leucoencephalopathy

Known by the acronym CADASIL, this is an autosomal dominant syndrome characterized by recurrent small vessel ischaemic stroke in middle age and sub-cortical dementia with pseudobulbar palsy usually in the absence of vascular risk factors (Singhal *et al.* 2004).

Table 35.2 Some causes of stroke which can be 'familial' (including intracranial venous thrombosis and intracranial haemorrhage)

Vascular anomalies	Intracranial vascular malformation Saccular aneurysm Hereditary haemorrhagic telangiectasia
Connective tissue anomalies	Ehlers–Danlos syndrome Pseudoxanthoma elasticum Marfan's syndrome Fibromuscular dysplasia Polycystic kidney disease Mitral leaflet prolapse
Haematological diseases	Haemophilia and other coagulation factor deficiencies Sickle-cell disease/trait Antithrombin III deficiency Protein C deficiency Activated protein C resistance Protein S deficiency Plasminogen abnormality/deficiency Dysfibrinogenaemia
Others	Familial hypercholesterolaemia Cerebral amyloid angiopathy (Icelandic and Dutch forms) Neurofibromatosis Tuberous sclerosis Homocysteinaemia Fabry's disease Migraine Cardiac myxoma Cardiomyopathy Von Hippel–Lindau disease Mitochondrial cytopathy Cerebral autosomal dominant arteriopathy with subcortical infarcts and leucoencephalopathy (CADASIL)

Mood disturbance and migraine with aura often precede the strokes but there is considerable phenotypic variation (Dichgans *et al.* 1998). Death usually occurs in the sixth or seventh decade. Brain MRI is always abnormal in symptomatic patients, and often in asymptomatic subjects too, and shows widespread focal, diffuse, and confluent white matter changes (Fig. 35.6), particularly in the periventricular and subcortical regions (Chabriat *et al.* 1995, 1998; Hutchinson *et al.* 1995; Dichgans *et al.* 1998). Changes at the temporal poles, the external capsule, and the corpus callosum are characteristic.

The disease has been reported from many parts of the world and there are now more than 500 families described. The prevalence of genetically proven disease in the West of Scotland has been reported to be 1.98 per 100 000 adults and the probable mutation prevalence was estimated to be 4.14 (3.04–5.53) per 100 000 adults (Razvi *et al.* 2005).

The underlying small vessel arteriopathy is distinct from arteriosclerotic and amyloid angiopathy and can be found in skin and muscle biopsies as well as in the leptomeningeal and perforating arteries of the brain (Jung *et al.* 1995). There is concentric thickening of the arterial walls with extensive deposition of eosinphilic granular material in the media and internal elastic lamina (Fig. 35.6).

The genetic locus is on chromosome 19q12 (Tournier-Lasserve *et al.* 1993). The deleterious mutations in the human equivalent of

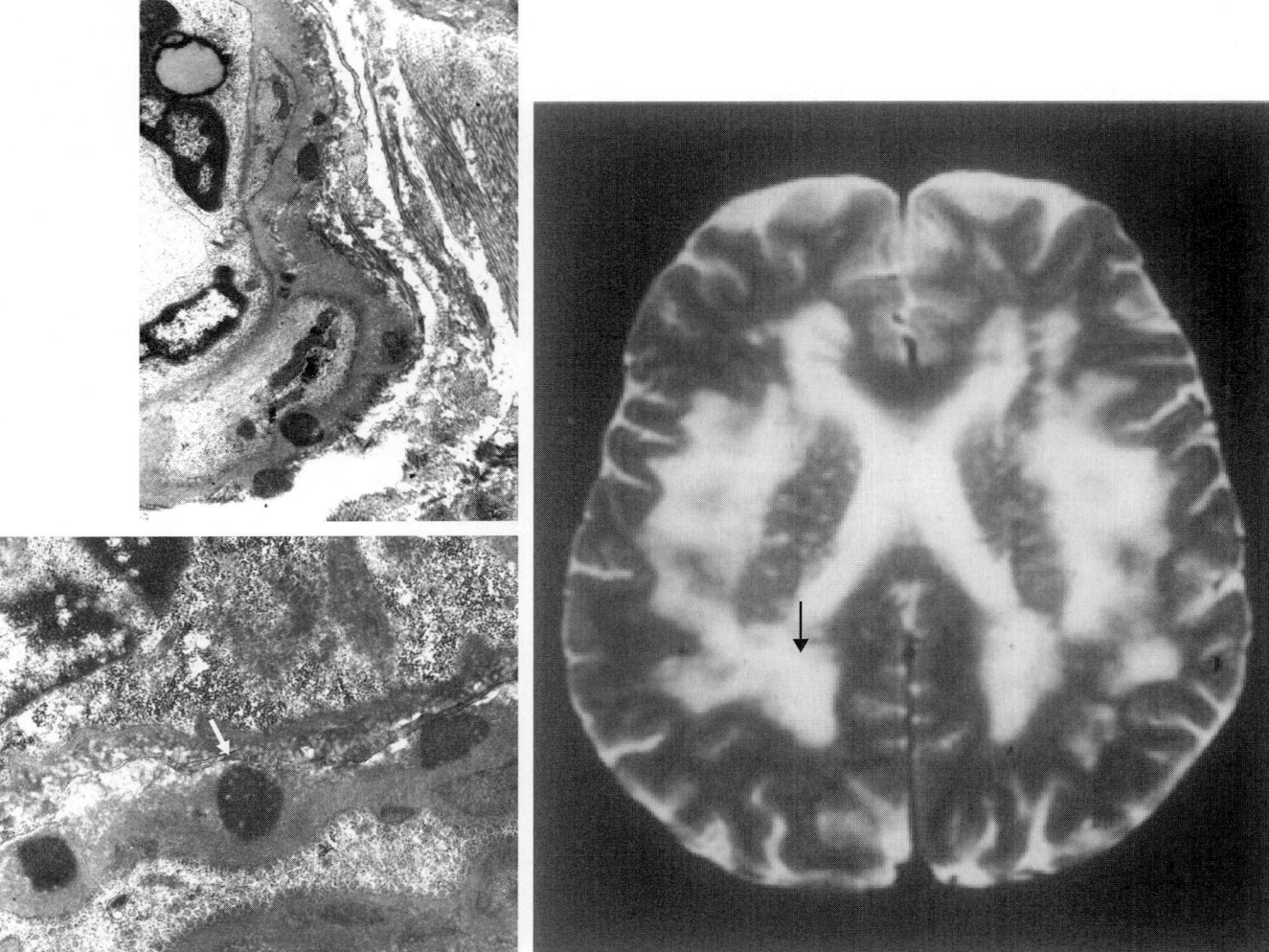

Fig. 35.6 Investigations from a patient with cerebral autosomal dominant arteriopathy with subcortical infarcts and leucoenceophalopathy, CADASIL: Electron micrographs (left) of muscle showing thickening of the basal lamina of the small blood vessels and the presence of dense eosinophilic material (arrow); MR brain imaging (right) showing widespread leukoairaosis.

the mouse *Notch 3* gene were found to be causative for CADASIL (Joutel *et al.* 1996; Dichgans *et al.* 1996). The *Notch 3* gene is involved in mediating signal transduction between neighbouring cells. In contrast to other *Notch* genes, which are ubiquitously expressed, *Notch 3* is mainly expressed in vascular smooth muscle cells. Arteries from transgenic mice show a diminished flow-induced dilation and the pressure-induced myogenic tone is significantly increased in their arteries compared with that in wild-type mice (Dubroca *et al.* 2005). Around 70 per cent of the hitherto characterized mutations cluster in exons 3 and 4 (Joutel *et al.* 1997). So far, there appears to be no genotype–phenotype correlation. The vast majority of known patients are members of affected families, but *de novo* mutations have been reported and so CADASIL can affect patients without a family history (Joutel *et al.* 2000).

The diagnosis should be considered in patients under 70 years of age who present with symptoms of subcortical cerebrovascular disease especially if they occur without classical risk factors but with an appropriate family history. The characteristic changes on

MRI, which is the most useful and sensitive screening tool, are apparent in all symptomatic and a large number of yet asymptomatic patients. The diagnosis is confirmed in the majority of patients by screening for mutations, particularly in exons 3 and 4. The search for rarer mutations however, is very costly and not usually done except for research. A useful alternative is to search for granular osmiophilic material in skin biopsies, although they can also be normal (Ebke *et al.* 1997).

Hereditary dyslipidaemias

Hereditary dyslipidaemias such as familial hypercholesterolaemia, type II and type IV hyperlipidaemia, and Tangier diease predispose to premature large vessel atherosclerosis and hence stroke (Meschia 2003; Hutter *et al.* 2004).

Connective tissue disease

Several inherited connective tissue disorders predispose to arterial dissection and other vascular abnormalities including aneurysms and vasoocclusive disease.

Fabry's disease

Fabry's disease is an X-linked disorder causing deficiency of α-galactosidase, which leads to an accumulation of glycosphingolipids in vascular endothelial and other cells. The associated cerebrovascular disorders are mainly ischaemic stroke, but intracerebral haemorrhage and subarachnoid haemorrhage can also occur. Around two-thirds of the infarcts involve the vertebrobasilar territory. Strokes usually occur from the third decade onward. Systemic nonvascular features of the phenotype include angiokeratomata, painful acroparaesthesiae, and renal failure (Section 21.8.5). The frequency of previously unknown Fabry's disease in patients with cryptogenic stroke under the age of 55 has been reported to be up to 5 per cent (Rolfs et al. 2005).

Cardiac disorders

Hypertrophic cardiomyopathy is frequently autosomal dominant with incomplete penetrance. Mutations in several genes encoding structural muscle proteins have been found (Franz et al. 2001). About 20 per cent of cases of dilated cardiomyopathy are familial with autosomal dominant, recessive, X-linked, and mitochondrial inheritance is seen (Franz et al. 2001). Most cardiomyopathies predispose to arrhythmia but there are also a number of primary cardiac arrhythmias including the long QT syndromes of Jervell, Lange, and Nielsen which are autosomal-recessive, and the Romano Ward which is autosomal dominant. A variety of mutations in sodium and potassium channels have been implicated (Viskin and Long 1999). There are also reports of familial atrial fibrillation (Brugada et al. 1997). Atrial myxoma (Section 35.4.3) is associated with a high risk of embolization and stroke and can be familial, particularly in younger patents and in men (Carney 1985).

Mitochondrial disorders

Various mitochondrial disorders may be associated with stroke-like episodes and cardiomyopathy (see Section 35.4.8).

Haematological disorders

Various genetic blood disorders including sickle cell anaemia and familial thrombophilias (see Section 35.4.7) are associated with stroke.

35.2.7 Genetic risk factors for 'sporadic' stroke

The contribution of genetic factors to stroke risk in populations has been difficult to establish, as it has for coronary heart disease (Alberts 1991) and is made more difficult by the fact that stroke is a heterogeneous clinical syndrome with numerous underlying pathologies and that many of the risk factors for stroke have strong genetic components. Reliable interpretation of published family history studies is undermined by major heterogeneity, insufficient detail, and potential publication and reporting bias, with much stronger associations in small studies and methodologically less rigorous studies (Flossmann et al. 2004). Few studies consider the number of affected and unaffected relatives, or phenotyped strokes in detail and the majority of studies do not adjust associations for intermediate phenotypes (Flossmann et al. 2005). No twin study and only a minority of family history studies have differentiated between ischaemic and haemorrhagic stroke in the proband. There are very few data on the influence of family history on stroke severity and no data on stroke recovery. Generally, genetic influences are stronger in patients with a relatively early age of stroke-onset, and

a family history of stroke confers a higher risk of stroke if onset was below 70 years of age (Schulz et al. 2004).

Based on the assumption that at least some of the risk of apparently sporadic stroke is genetic, large numbers of studies have now been done in an attempt to identify the genes involved (Dichgans 2007). However, it seems likely that the genetic component of stroke risk is modest and that many, indeed probably hundreds, of genes are involved, each one contributing only a small increased risk. Studies so far have generally not been large enough to detect reliably the sort of small effects that might realistically be expected and have had other methodological limitations, including poor choice of controls in case-control studies, inadequate distinction between the different pathological types and subtypes of stroke for which genetic influences may differ, failure to replicate positive results in an independent and adequately sized study, and testing of multiple genetic or subgroup hypotheses with no adjustment of p-values for declaring statistical significance (Dichgans and Markus 2005; Sudlow et al. 2006).

Candidate gene studies

Most genetic studies so far have been candidate gene studies, in which the frequency of different genotypes at a specific locus or loci within a gene or genes thought likely to be in some way connected with stroke risk are compared between stroke cases and stroke-free controls. Candidate genes have generally been selected on the basis of their known or presumed involvement in the control of factors or pathways likely to influence stroke risk: blood pressure, lipid metabolism, inflammation, coagulation, homocysteine metabolism, and so on (Hassan and Markus, 2000; Casas et al. 2004). Rigorous meta-analyses of candidate gene studies, both in stroke and other vascular diseases such as coronary heart disease, have highlighted various methodological problems, particularly the inadequate size of studies (Keavney et al. 2000; Wheeler et al. 2004; Sudlow et al. 2006). Large numbers of candidate gene studies have together identified a handful which, on the basis of results from meta-analyses, seem likely to influence risk of ischaemic stroke modestly. These genes include those encoding factor V Leiden, methylenetatrahydrofolate reductase, prothrombin and angiotensin converting enzyme (Casas et al. 2004; Casas et al. 2006).

Linkage studies

As yet there have been far fewer stroke genetics studies that use more traditional genetic study designs, based on collecting information and DNA from related individuals with and without the disease of interest. This is at least partly because family members of stroke patients are often no longer alive, and so obtaining information and samples for DNA extraction from large enough numbers of relatives is challenging (Hassan et al. 2002). The Icelandic deCODE group identified two candidate genes for ischaemic stroke, encoding the enzymes phosphodiesterase-4D and arachidonate 5-lipoxygenase-activating protein ALOX5AP (Gulcher et al. 2005). There is still some debate, however, about their influence in non-Icelandic populations, since their effect on ischaemic stroke risk has been confirmed in only a few replication studies (Rosand et al. 2006; Gulcher et al. 2006).

Whole genome association studies

The combination of technological developments allowing rapid genotyping at multiple loci, the attraction of non-hypothesis driven genetic studies, and the recognized limitations of traditional linkage approaches have led to an increasing interest in genome-wide association studies, where multiple polymorphisms across the genome are genotyped and compared in cases and controls, looking for loci where significant differences may suggest genetic influences on disease risk. Only one small, uninformative preliminary study in stroke has been published so far (Matarin *et al.* 2007).

Association with intermediate phenotypes

Genetic studies have started to emerge of so-called intermediate phenotypes, markers of predisposition to stroke or other vascular diseases, which can be measured in large numbers of subjects both with and without vascular risk factors or disease. These intermediate phenotypes include carotid intima media thickness and leukoaraiosis as measured or graded on CT or MR brain scans, and both linkage and candidate gene approaches have been used (Humphries and Morgan 2004; Dichgans and Markus 2005).

35.3 **Blood supply to the brain**

The brain is only 2 per cent of total body weight, but at rest it receives 20 per cent of the cardiac output of blood and consumes about 20 per cent of the total inspired oxygen. This rich blood supply is delivered by the two internal carotid and two vertebral arteries (Fig. 35.7), which anastomose at the base of the brain to form the circle of Willis (Fig. 35.8). The carotid arteries supply the anterior, and the vertebrobasilar arterial system supplies the posterior, portions of the brain.

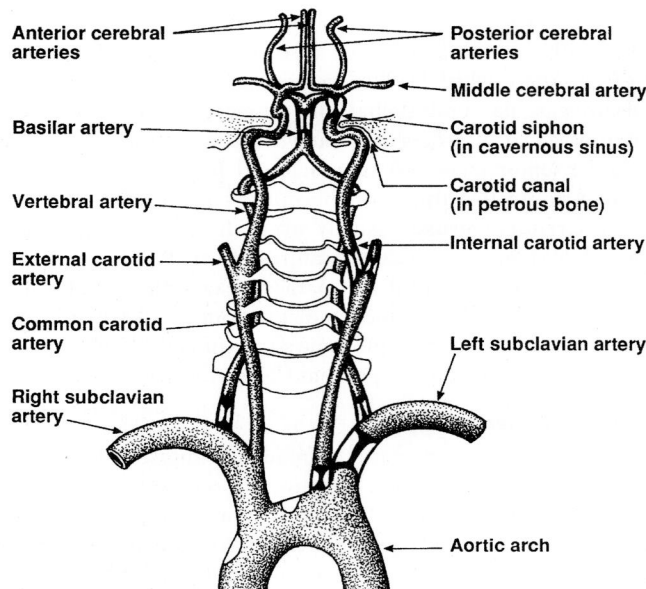

Fig. 35.7 The anatomy of the arterial circulation to the brain and eye. White indentations into the arterial lumen represent sites at which atherothrombosis is particularly common.

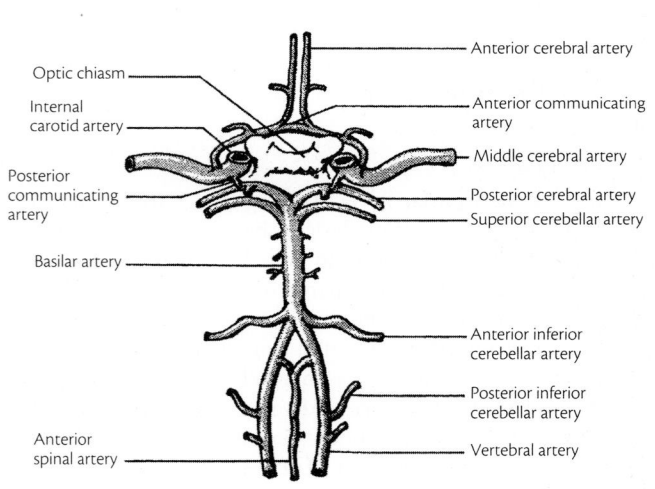

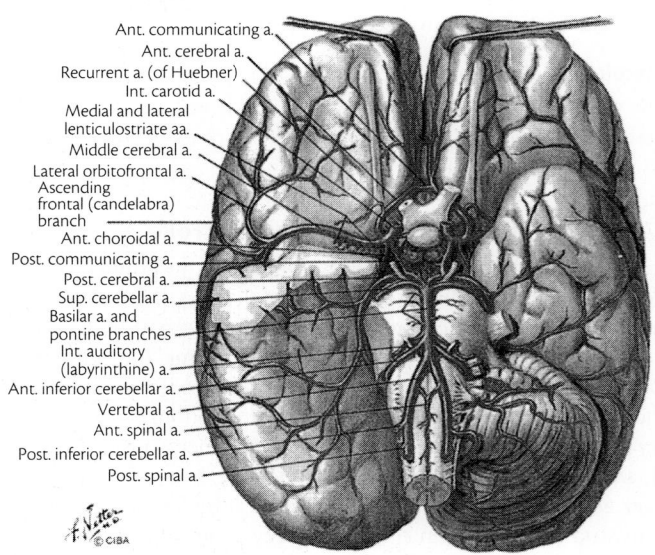

Fig. 35.8 The Cerebral Arterial Circulation. (A) The circle of Willis at the base of the brain as seen from below. There is considerable anatomical variation and this figure represents but one rather typical arrangement. (B) A view of the base of the brain to show the circle of Willis, vertebrobasilar arterial system, middle and anterior cerebral arteries. (See Plate 29.)

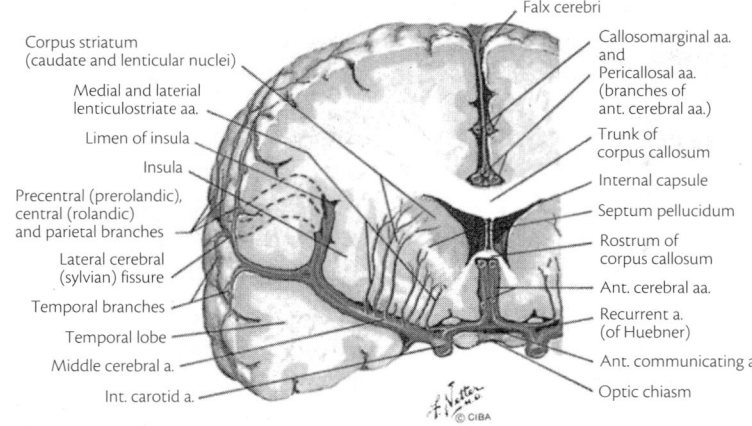

Falx cerebri

Corpus striatum (caudate and lenticular nuclei)

Medial and lateral lenticulostriate aa.

Limen of insula

Insula

Precentral (prerolandic), central (rolandic) and parietal branches

Lateral cerebral (sylvian) fissure

Temporal branches

Temporal lobe

Middle cerebral a.

Int. carotid a.

Callosomarginal aa. and Pericallosal aa. (branches of ant. cerebral aa.)

Trunk of corpus callosum

Internal capsule

Septum pellucidum

Rostrum of corpus callosum

Ant. cerebral aa.

Recurrent a. (of Huebner)

Ant. communicating a.

Optic chiasm

C

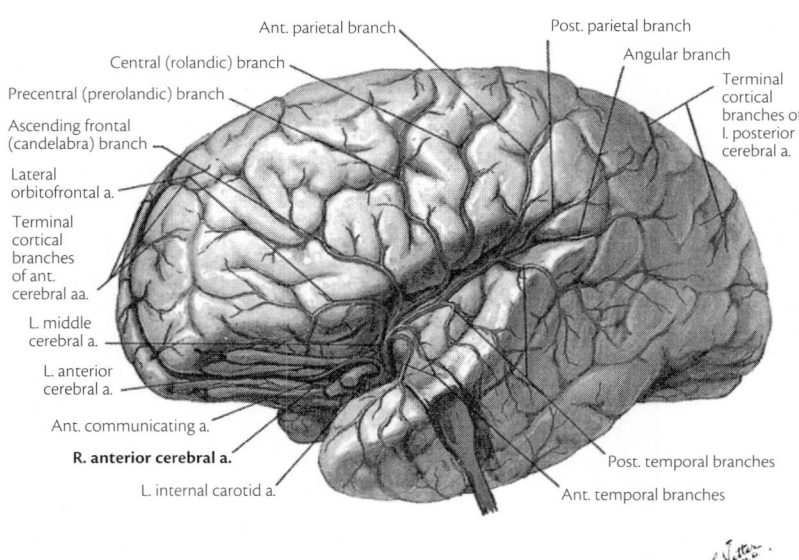

Ant. parietal branch

Central (rolandic) branch

Precentral (prerolandic) branch

Ascending frontal (candelabra) branch

Lateral orbitofrontal a.

Terminal cortical branches of ant. cerebral aa.

L. middle cerebral a.

L. anterior cerebral a.

Ant. communicating a.

R. anterior cerebral a.

L. internal carotid a.

Post. parietal branch

Angular branch

Terminal cortical branches of l. posterior cerebral a.

Post. temporal branches

Ant. temporal branches

D

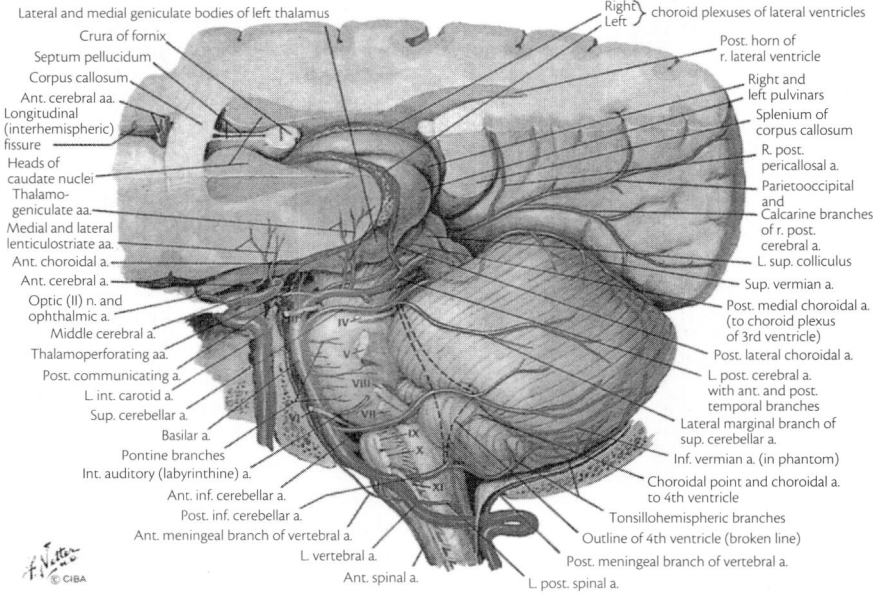

Lateral and medial geniculate bodies of left thalamus

Crura of fornix

Septum pellucidum

Corpus callosum

Ant. cerebral aa.

Longitudinal (interhemispheric) fissure

Heads of caudate nuclei

Thalamo-geniculate aa.

Medial and lateral lenticulostriate aa.

Ant. choroidal a.

Ant. cerebral a.

Optic (II) n. and ophthalmic a.

Middle cerebral a.

Thalamoperforating aa.

Post. communicating a.

L. int. carotid a.

Sup. cerebellar a.

Basilar a.

Pontine branches

Int. auditory (labyrinthine) a.

Ant. inf. cerebellar a.

Post. inf. cerebellar a.

Ant. meningeal branch of vertebral a.

L. vertebral a.

Ant. spinal a.

Right / Left } choroid plexuses of lateral ventricles

Post. horn of r. lateral ventricle

Right and left pulvinars

Splenium of corpus callosum

R. post. pericallosal a.

Parietooccipital and Calcarine branches of r. post. cerebral a.

L. sup. colliculus

Sup. vermian a.

Post. medial choroidal a. (to choroid plexus of 3rd ventricle)

Post. lateral choroidal a.

L. post. cerebral a. with ant. and post. temporal branches

Lateral marginal branch of sup. cerebellar a.

Inf. vermian a. (in phantom)

Choroidal point and choroidal a. to 4th ventricle

Tonsillohemispheric branches

Outline of 4th ventricle (broken line)

Post. meningeal branch of vertebral a.

L. post. spinal a.

E

Fig. 35.8 (C) A coronal section of the brain through the optic chiasm to show the penetrating lenticulostriate branches of the main stem of the middle cerebral artery. (D) A lateral view of the brain to show the cortical branches of the middle cerebral artery. (E) Lateral view of the brain to show the vertebral, basilar and posterior cerebral arteries and their branches in the posterior fossa. (Figures 35.8(B)–(E) reprinted with permission from Clinical Symposia, illustrated by Frank N. Netter, MD, copyright 1990 CIBA-GEIGY Corporation. All rights reserved.) (See Plate 29.)

35.3.1 Anatomy of the cerebral circulation

The detailed anatomy of the cerebral circulation is well described by Sheldon (1981). There is individual variation, particularly in the branches of the main cerebral arteries and thus in the territories of supply of the various major arteries, which may be asymmetrical and even change with time, depending in part on the availability of functional collaterals (van der Zwan *et al.* 1992)(Fig. 35.9). Well-recognized developmental anomalies include:

◆ marked inequality in size of the two vertebral arteries;

◆ the left vertebral artery arising directly from the aorta;

◆ the right common carotid artery arising from the aortic arch;

◆ a combined origin of the left common carotid and innominate arteries;

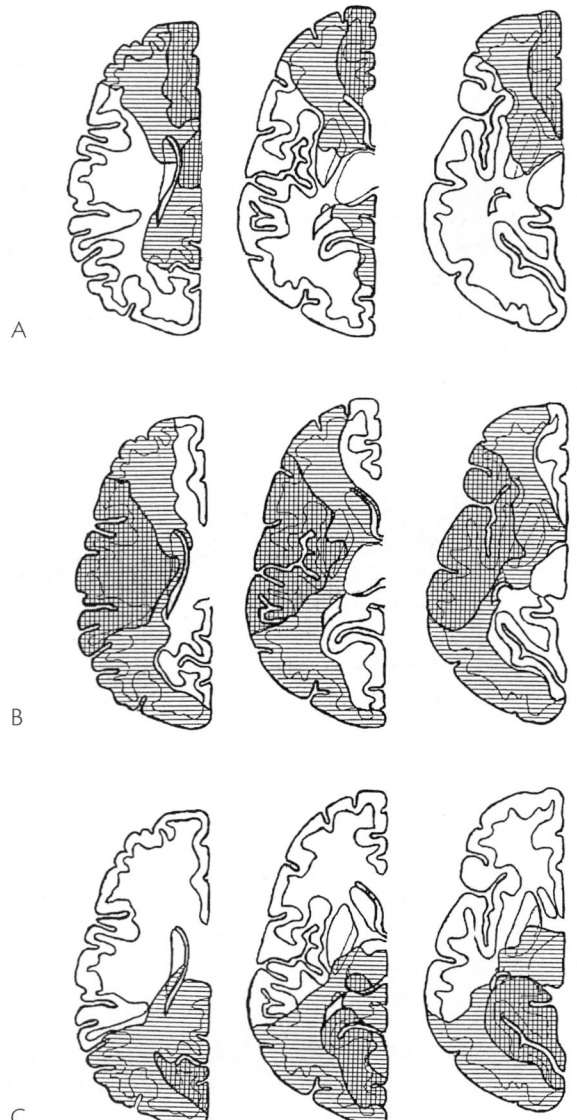

Fig. 35.9 Cerebral arterial territories. Horizontal sections of the brain to show areas supplied by the anterior (A), middle (B), and posterior (C) cerebral arteries. The vertical lines represent the smallest and the horizontal lines the largest areas of supply in various individuals. (With permission from Dr Albert van der Zwan, University of Utrecht.)

◆ hypoplasia or absence of one, or less often both, posterior communicating artery(ies);

◆ hypoplasia or absence of the anterior communicating artery;

◆ hypoplasia or absence of the proximal part of one anterior cerebral artery so that blood flow to both anterior cerebral arteries comes from one internal carotid artery;

◆ a persistent trigeminal artery joining the internal carotid artery to the basilar artery; and

◆ a paired or fenestrated basilar artery.

The **internal carotid artery** starts as the carotid sinus at the bifurcation of the **common carotid artery** at the level of the thyroid cartilage. It runs up the neck, without any branches, to the base of the skull where it passes through the foramen lacerum to enter the carotid canal of the petrous bone. It then runs through the cavernous sinus in an S-shaped curve, the carotid siphon, pierces the dura and exits just medial to the anterior clinoid process, and then bifurcates into the anterior cerebral artery and the larger middle cerebral artery.

The **external carotid artery** also starts at the bifurcation. Branches supply the jaw, face, scalp, neck, and meninges via the superficial temporal, facial, and occipital arteries.

The **ophthalmic artery** is the first major branch of the internal carotid artery and arises in the cavernous sinus. It passes through the optic foramen to supply the eye and other structures in the orbit.

The **posterior communicating artery** is the next artery to arise from the internal carotid artery and passes back to join the first part of the posterior cerebral artery, so contributing to the circle of Willis. Tiny branches supply the adjacent optic chiasm, optic tract, hypothalamus, thalamus, and midbrain.

The **anterior choroidal artery** arises from the last section of the internal carotid artery, just beyond the posterior communicating artery origin, and supplies the optic tract, internal capsule, medial parts of the basal ganglia, the medial part of the temporal lobe, thalamus, lateral geniculate body, proximal optic radiation, and midbrain. Occasionally it arises from the proximal middle cerebral artery or posterior communicating artery. Minor twiglets from the distal internal carotid artery contribute blood to the pituitary gland, optic chiasm, and nearby structures, including the meninges.

The **anterior cerebral artery** passes horizontally and medially to enter the interhemispheric fissure, anastomoses with its counterpart of the opposite side via the **anterior communicating artery**, curves up around the genu of the corpus callosum, and supplies the anterior and medial parts of the cerebral hemisphere. Small branches also supply parts of the optic nerve and chiasm, hypothalamus, anterior basal ganglia, and internal capsule.

The **middle cerebral artery** enters the Sylvian fissure and divides into 2–4 branches which supply the lateral parts of the cerebral hemisphere. From its main trunk a medial and lateral group of tiny lenticulostriate arteries and arterioles pass upwards to penetrate the base of the brain and supply the basal ganglia and internal capule (Marinkovic *et al.* 1985). Some of these small penetrating vessels extend up into the white matter of the corona adiata in the centrum semiovale towards the small medullary perforating branches of the cortical arteries coming down from above.

The **vertebral artery** arises from the proximal subclavian artery and ascends to pass through the transverse foramina of the sixth to second cervical vertebrae, giving off small muscular branches on the way. It then passes posteriorly around the articular process of the atlas to enter the skull through the foramen magnum. It unites with the opposite vertebral artery on the ventral surface of the brainstem at the pontomedullary junction to form the basilar artery. Branches to the meninges arise at the foramen magnum. The vertebral artery gives rise to the anterior and posterior spinal arteries, the posterior inferior cerebellar artery to inferior vermis and inferior and posterior surfaces of the cerebellar hemispheres and brainstem, and small penetrating arteries to the medulla.

The **basilar artery** ascends ventral to the pons to the ponto-midbrain junction in the interpeduncular cistern where it divides into the two posterior cerebral arteries. Numerous small branches penetrate the brainstem and cerebellum. The basilar artery also gives rise to the anterior inferior cerebellar artery, which supplies the rostral cerebellum, brainstem, inner ear, and the superior cerebellar artery supplying the brainstem, superior half of the cerebellar hemisphere, vermis, and dentate nucleus.

The **posterior cerebral artery** encircles the midbrain close to the oculomotor nerve at the level of the tentorium and supplies the inferior part of the temporal lobe, and the occipital lobe (Marinkovic *et al.* 1987). Many small perforating arteries arise from the proximal portion of the posterior cerebral artery to supply the midbrain, thalamus, hypothalamus, and geniculate bodies. Sometimes a single perforating artery supplies the medial part of each thalamus, or both sides of the midbrain. In about 15 per cent of individuals the posterior cerebral artery is a direct continuation of the posterior communicating artery, its main blood supply then coming from the internal carotid artery rather than the basilar artery.

The meninges are supplied by branches of the external carotid artery, internal carotid artery, and vertebral arteries. The most prominent branches from the external carotid artery are the middle meningeal artery and tributaries of the ascending pharyngeal and occipital arteries. Most of the branches from the internal carotid artery arise near the cavernous sinus and from the ophthalmic artery in the orbit. Branches from the vertebral artery arise at the foramen magnum. There are numerous meningeal anastomoses between these small arteries.

The scalp is supplied by branches of the external carotid artery, particularly the superficial temporal, occipital, and posterior auricular arteries. Above the orbit there is a contribution from terminal branches of the ophthalmic artery. There is a rich anastomotic network between the various arteries of the scalp.

35.3.2 Collateral blood supply to the brain

This topic is best described by Liebeskind (2003). Normally the internal carotid artery provides blood to the anterior two-thirds of the ipsilateral cerebral hemisphere and the posterior circulation is supplied by the vertebral, basilar, and posterior cerebral arteries. Collateral channels may develop in response to occlusion of one or more of the intracerebral vessels, particularly if flow limitation is gradual rather than sudden. Unlike the normal cerebral blood supply, the functional capacity of the collateral blood supply to respond to changes in perfusion pressure is limited. Collateral blood flow may develop via:

◆ The *Circle of Willis*, which is formed by the proximal part of the two anterior cerebral arteries connected by the anterior commu-

nicating artery, and the proximal part of the two posterior cerebral arteries, which are connected to the distal internal carotid arteries by the posterior communicating arteries. However, about 50 per cent of circles have one or more hypoplastic or absent segments, usually one of the communicating arteries, and atheroma may limit the potential for collateral flow (Fig. 35.8).

◆ *Around the orbit* branches of the external carotid artery anastomose with branches of the ophthalmic artery if the internal carotid artery is severely stenosed or obstructed. Collateral flow from the external carotid artery into the orbit then passes retrogradely through the ophthalmic artery to fill the carotid siphon, middle cerebral artery, and anterior cerebral artery. Sometimes flow may even reach the posterior cerebral artery and vertebrobasilar system.

◆ *Muscular branches of the vertebral artery* in the neck, distal to a vertebral obstruction, may receive blood retrogradely from occipital and ascending pharyngeal branches of the external carotid artery, or from the deep and ascending cervical arteries. Also, anastomoses can develop between branches of the subclavian artery and external carotid artery when the common carotid artery is obstructed.

◆ *Leptomeningeal anastomoses* on the surface of the brain may develop between cortical branches of the anterior, middle, and posterior cerebral arteries and, to a lesser extent, between pial branches of the cerebellar arteries.

◆ *Dural anastomoses* can develop between meningeal branches of the internal carotid artery, external carotid artery, and vertebral arteries. Occasionally small dural anastomoses develop between cortical leptomeningeal and dural arteries.

◆ *Parenchymal anastomoses* occasionally develop in the precapillary bed of the perforating arteries at the base of the brain supplying the basal ganglia.

◆ *Anterior choroidal artery*, a branch of the internal carotid artery, can anastomose with the posterior choroidal artery, a branch of the posterior cerebral artery.

35.3.3 Venous drainage

The venous anatomy is very variable. In general, venous blood flows centrally via the deep cerebral veins, and peripherally via the superficial cerebral veins into the dural venous sinuses, which lie between the endosteal outer and meningeal inner layer of the dura, which drain into the internal jugular veins (Stam 2005) (Fig. 35.10). The cerebral veins are thin-walled, have no valves, and the blood flow is often in the same direction as in neighbouring arteries. There are numerous venous connections between the cerebral veins and dural sinuses, as well as with the venous system of the meninges, skull, scalp, and nasal sinuses, so facilitating the propagation of thrombus or spread of infection between these vessels.

35.3.4 The regulation of cerebral blood flow

It is not practical to measure cerebral blood flow in routine practice but increasing knowledge of cerebral blood flow regulation, and the relationship between cerebral blood flow and cerebral metabolism, has had a major influence on the understanding of the pathophysiology of impaired perfusion reserve and of acute ischaemic stroke (Frackowiak 1986; Marchal *et al.* 1996a; Baron 2001; Rutgers *et al.* 2004).

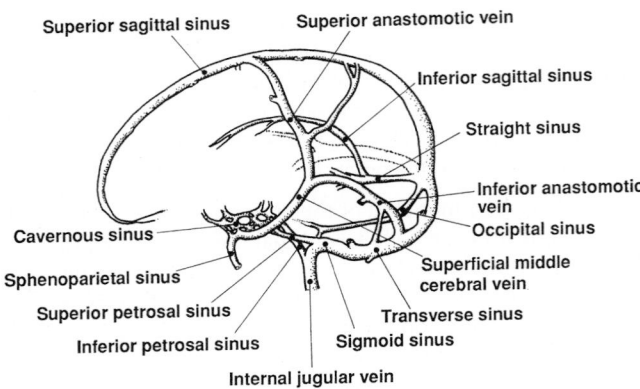

Superior sagittal sinus
Superior anastomotic vein
Inferior sagittal sinus
Straight sinus
Inferior anastomotic vein
Cavernous sinus
Occipital sinus
Sphenoparietal sinus
Superficial middle cerebral vein
Superior petrosal sinus
Transverse sinus
Inferior petrosal sinus
Sigmoid sinus
Internal jugular vein

Fig. 35.10 Diagram of the venous drainage of the brain. (Modified with permission from F.A. Davis and Co.)

Cerebral blood flow in normal man is about 50 ml/100 g of brain/min. Using regional techniques such as positron emission tomography (Frackowiak *et al.* 1980), it has been shown that cerebral blood flow, cerebral blood volume, and cerebral energy metabolism measured as cerebral metabolic rate of oxygen, CMRO2, or of glucose, CMRglu, are all coupled, and higher in grey than in white matter. This means that the oxygen extraction fraction is similar (approximately one-third) throughout the brain (Leenders *et al.* 1990). Therefore, in normal resting human brain, cerebral blood volume is a reliable reflection of function or CMRO2. There is a gradual fall of cerebral blood flow, cerebral blood volume, CMRglu, and CMRO2 with age, but they remain coupled so the oxygen extraction fraction remains more or less constant (Blesa *et al.* 1997).

Cerebral blood flow and blood gas tensions

Cerebral blood flow is very susceptible to small changes in PaCO2: an acute rise of 1 mmHg causes an immediate increase in cerebral blood flow of about 5 per cent due to dilatation of cerebral resistance vessels. In chronic respiratory failure, however, adaptation occurs so that cerebral blood flow is normal despite the hypercapnia. Modest changes in arterial oxygen tension do not affect cerebral blood flow, but when the PaO2 falls below about 50 mmHg, and oxygen saturation starts falling, there is a fall in cerebral vascular resistance and cerebral blood flow rises (Brown *et al.* 1985). Increasing PaO2 above the normal level has little effect on cerebral blood flow.

Cerebral blood flow and brain function

Increasing regional functional activity of the brain, for instance in the motor cortex contralateral to voluntary hand movements, increases regional metabolic activity in the same area (Lassen *et al.* 1977; Geisler *et al.* 2006). The increasing CMRO2 and CMRglu are achieved not by increasing oxygen extraction fraction or the glucose extraction fraction, but by rapid local vasodilatation over seconds of the cerebral resistance vessels, increase in cerebral blood volume and, therefore, in cerebral blood flow. Conversely, low functional and metabolic demand, as occurs in a cerebral infarction, are associated with a low cerebral blood flow.

Cerebral blood flow, perfusion pressure, and autoregulation

Cerebral blood flow depends on cerebral perfusion pressure and cerebrovascular resistance. The perfusion pressure is the difference between systemic arterial pressure at the base of the brain when in

the recumbent position and the venous pressure at exit from the subarachnoid space, the latter being approximated by the intracranial pressure. Cerebral perfusion pressure divided by cerebral blood flow gives the cerebrovascular resistance. In normal man, cerebral blood flow remains almost constant when the mean systemic blood pressure is between about 50 and 170 mmHg which, under normal circumstances when the intracranial venous pressure is negligible, is the same as the cerebral perfusion pressure (Fig. 35.11). This homeostatic mechanism to maintain a constant cerebral blood flow in the face of changes in cerebral perfusion pressure is known as *autoregulation* (Reed and Devous 1985; Powers 1993). Autoregulation is less effective in the elderly, so that postural hypotension is more likely to be symptomatic (Wollner *et al.* 1979; Parry *et al.* 2006).

Within the autoregulatory range, as cerebral perfusion pressure falls there is, within seconds, vasodilatation of the small cerebral resistance vessels, a fall in cerebrovascular resistance, a rise in cerebral blood volume, and therefore cerebral blood flow remains constant (Fig. 35.12) (Aaslid *et al.* 1989). If vasodilatation is maximal and cerebral perfusion pressure continues to fall due to a drop in systemic blood pressure or an increase in intracranial pressure, cerebral blood flow starts to decline as the cerebral perfusion reserve is exhausted. However, metabolic activity is maintained by increasing oxygen extraction fraction: this is 'misery' perfusion or oligaemia (Fig. 35.12). Eventually the oxygen extraction fraction can rise no more, and with further cerebral perfusion pressure reduction, metabolic activity as measured by CMRO2 starts to fall, and metabolism becomes limited by perfusion. This is what is normally meant by ischaemia; the perfusion reserve is exhausted and flow is inadequate to meet the metabolic demands of the tissues. At around this point the patient becomes symptomatic with non-focal features such as faintness if the whole brain is involved, or focal features such as hemiparesis if only part of the brain is involved.

If the perfusion pressure rises above the autoregulatory range, where compensatory vasoconstriction and cerebral perfusion pressure are maximal, then hyperaemia occurs followed by vasogenic oedema, raised intracranial pressure, and the clinical syndrome of hypertensive encephalopathy.

Cerebral perfusion reserve

It follows from the above that the ratio of cerebral blood flow to cerebral blood volume is a measure of *cerebral perfusion reserve* (Schumann *et al.* 1998). Below about 6.0, even if cerebral blood flow is still normal, vasodilatation and cerebral blood volume are

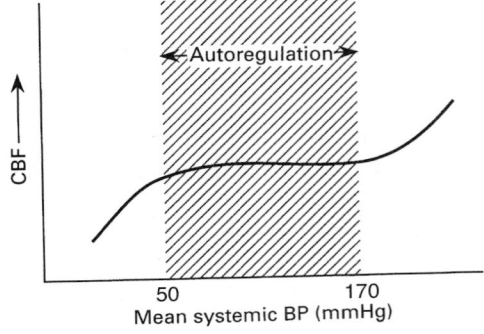

Fig. 35.11 A diagram to illustrate autoregulation of cerebral blood flow (CBF) in normal man. Between a mean systemic blood pressure (BP) of about 50 and 170 mmHg cerebral blood flow remains roughly constant.

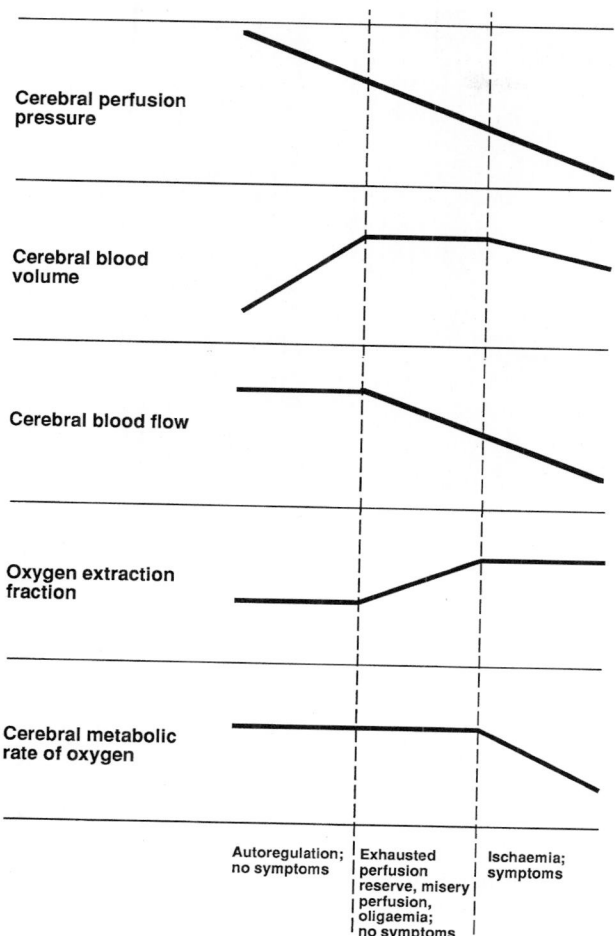

Fig. 35.12 A model to demonstrate the compensatory mechanisms which maintain cerebral metabolic activity as cerebral perfusion pressure falls.

maximal and the reserve is exhausted, as shown by a rising oxygen extraction fraction on positron emission tomography scanning.

Chronically impaired perfusion reserve tends to occur when one or both internal carotid arteries are stenosed by at least 50 per cent of the luminal diameter (Brice *et al.* 1964; DeWeese *et al.* 1970; Schroeder 1988), or are occluded, *and* the collateral circulation is inadequate (Powers *et al.* 1987; Kluytmans *et al.* 1999). Under these circumstances of maximal vasodilatation, the brain is vulnerable to any further fall in cerebral perfusion pressure and cerebral metabolism is beginning to become impaired with the appearance of structural abnormalities on MRI (van der Grond *et al.* 1996; Isaka *et al.* 1997; Derdeyn *et al.* 1999).

Indirect assessment of perfusion reserve can be achieved by using transcranial Doppler ultrasound, single photon emission computed tomography, positron emission tomography, dynamic CT or functional MRI to measure cerebral blood flow response to hypercapnia during CO_2 inhalation, breath holding, or after intravenous acetazolamide, a carbonic anhydrase inhibtor (Arigoni *et al.* 2000; Kikuchi *et al.* 2001; Shiogai *et al.* 2003; Shiino *et al.* 2003). However, there is uncertainty about how these various tests should be standardized and how to define 'normality', given a continuous variable is being measured. It should be noted that indirect methods of measuring perfusion reserve are inaccurate when the

normal relationships between cerebral blood flow, cerebral blood volume, oxygen extraction fraction, and vascular reactivity break down, as they may well do in newly ischaemic or infarcted brain.

Impaired perfusion reserve is associated with an increased likelihood of recurrent stroke (Yamauchi *et al.* 1996), prior ischaemic events in patients with carotid occlusion (Derdeyn 1999; 2005), presence of silent brain infarction, and increased likelihood of need for carotid shunting in carotid endarterectomy (Kim *et al.* 2000). Extra-cranial to intracranial bypass surgery has been shown to improve cerebral perfusion reserve in patients with large vessel occlusive disease and is currently being assessed (Adams *et al.* 2001; Grubb *et al.* 2003) as a treatment for secondary prevention of stroke in patients with carotid occlusion and reduced cerebral perfusion reserve in whom the risk of ipsilateral stroke on medical treatment is high (Grubb and Powers 2001). Medical treatment with angiotensin converting enzyme inhibitors has also been shown to increase cerebral perfusion reserve in patients with previous minor stroke (Hatazawa *et al.* 2004).

Cerebral blood flow, hypertension, and stroke

In chronically hypertensive patients, the autoregulatory range is shifted upwards so that cerebral blood flow starts falling and ischaemic symptoms occur at a higher systemic blood pressure than normal (Strandgaard and Paulson 1992) but autoregulation appears otherwise to be maintained (Traon *et al.* 2002) except in malignant hypertension (Immink *et al.* 2004). The upward shift of autoregulation appears to return towards normal when hypertension is treated. Conversely, hypertensive encephalopathy is more likely to occur in acute hypertension when the upper limit of autoregulation is still normal, such as occurs in eclampsia.

Autoregulation is impaired, or abolished, in damaged areas of brain so that cerebral blood flow becomes 'pressure passive' and follows perfusion pressure. Both static and dynamic cerebral autoregulation are impaired in patients with stroke (Strandgaard *et al.* 1992; Eames *et al.* 2002; Georgiadis *et al.* 2002; Novak *et al.* 2003) but treatment of hypertension with angiotensin converting enzyme inhibitors and angiotensin II receptor antagonists subacutely after stroke appears to lower systemic pressure without compromising cerebral blood flow (Paulson and Waldemar 1990; Moriwaki *et al.* 2004; Nazir *et al.* 2004).

35.3.5 Pathophysiology of acute cerebral ischaemia
The consequences of ischaemia

The brain normally derives its energy from the oxidative metabolism of glucose. Because there are negligible stores of glucose in the brain, when cerebral blood flow falls and the brain becomes ischaemic, a series of neurophysiological and functional changes occur at various thresholds of flow before cell death, or infarction occurs (Fig. 35.13). The degree of cell damage depends not only on the depth of ischaemia but also its duration and the availability of collateral circulation (Liebeskind 2003; Zemke *et al.* 2004; Harukuni and Bhardwaj 2006). Different mechanisms are responsible for reversible loss of cellular function, and for irreversible cell death, and there are also differences between the mechanisms that cause death of neurones, glia, and endothelial cells, and perhaps between white matter and grey matter.

When cerebral blood flow falls below about 20 ml/100 g brain/min, the oxygen extraction fraction becomes maximal and CMRO2 begins to fall (Fig. 35.12), resulting in ischaemia (Wise *et al.* 1983).

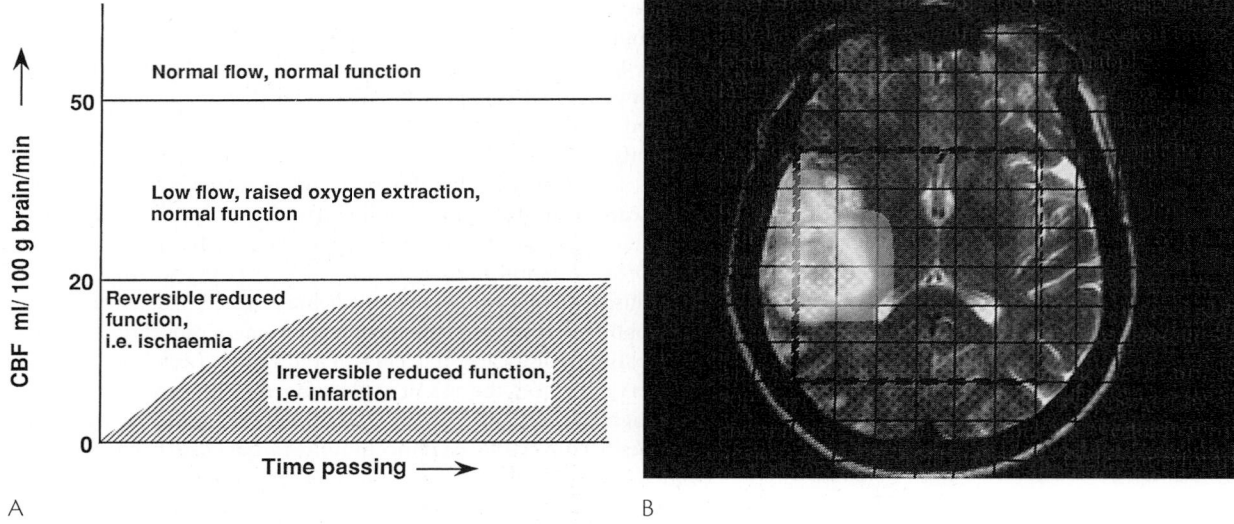

Fig. 35.13 (A) The thresholds of cerebral ischaemia at different levels of cerebral blood flow (CBF) in relation to the duration of ischaemia. (B) A metabolic 'map' of lactate in the brain by MR spectroscopy, superimposed on a structural MR scan. There is an obvious excess of lactate in the infarcted area in the distribution of a branch of the right middle cerebral artery. (Courtesy of Dr Joanna Wardlaw.) (See Plate 30.)

The EEG flattens, evoked responses disappear, and neurological signs appear. In fact, a high oxygen extraction fraction is only seen early after acute ischaemic stroke, in the first day or so. If flow is restored, functional recovery is still possible (Fig. 35.13). With increasing ischaemia, impaired protein synthesis is the earliest detectable metabolic change followed by the inefficient anaerobic metabolism of glucose which causes a rise in lactate production, fall in intra and extra cellular pH, and impaired phosphocreatine and ATP synthesis, and so energy failure. The energy-dependent functions of the cell membranes, including ion pumps, become progressively affected; water, sodium, and chloride enter cells as the beginnings of cytotoxic oedema, together with cytotoxic calcium, and potassium leaks out. Cellular transport mechanisms and neurotransm

itter systems fail, neurotoxic excitatory transmitters are released from neurones and free oxygen radicals, nitric oxide, and lipid peroxides are formed damaging cells further. (Back *et al.* 2004; Warner *et al.* 2004; Zemke *et al.* 2004).

An acute inflammatory response with the migration of neutrophils and then of monocytes and macrophages into the ischaemic area, with activation of microglia, is another possible mechanism of neuronal death. (Zheng and Yenari 2004). At flows below about 10 ml/100 g brain/min infarction occurs, and even if flow is restored, function does not recover. Later on apoptosis rather than necrosis may be responsible for neuronal death (Zhang *et al.* 2004). With MR spectroscopy it is possible to track some of these metabolic changes in man with high levels of lactate indicating anaerobic metabolism and low *N*-acetyl aspartate, a neuronal marker, being found in large infarcts with a poor prognosis (Graham *et al.* 1995; Federico *et al.* 1998).

At the stage of infarction, CMRO2 is low and cerebral blood flow is either appropriately low as well, with a normal oxygen extraction fraction representing pure metabolic depression, or the oxygen extraction fraction is low indicating that cerebral blood flow is in excess of the low metabolic demands of the infarcted tissue, luxury perfusion (Fig. 35.14). In absolute luxury perfusion, cerebral blood flow is increased causing hyperperfusion, while in relative luxury perfusion it is normal or decreased.

The ischaemic penumbra

Around acutely infarcted brain there is an ischaemic penumbra (Astrup *et al.* 1981). Here the blood flow is low, function depressed, and the oxygen extraction fraction high. In other words there is viable tissue with misery perfusion where the needs of the tissue are not being met. The tissue may die or recover, depending on the speed and extent of restoration of blood flow. This concept opens up the possibility of a therapeutic time window during which restoration of flow or neuronal protection from ischaemic damage might prevent both immediate cell death and the recruitment of neurons for apoptosis. Other brain areas may show relative or absolute hyperaemia due to good collateral flow, reperfusion after an occluded artery has been reopened, inflammation, and vasodilatation in response to hypercapnia. Here oxygen extraction fraction is low and there is luxury perfusion indicating that flow is in excess of metabolic requirements, perhaps because the tissue has been irreversibly damaged.

Recently positron emission tomography studies have demonstrated, in man, that about one-third of the ultimately infarcted tissue on late CT is in areas where, within hours of stroke onset, there had been potentially viable 'penumbral' tissue (Marchal *et al.* 1996a). However, it is still not clear for how long this penumbral region persists in a potentially viable state, although time periods as long as 18 h have been suggested and it appears that some recovery is possible if flow is restored (Lassen *et al.* 1991; Furlan *et al.* 1996). Accurate information about the ischaemic penumbra, and any areas of luxury perfusion, requires positron emission tomography, which is not practical in the routine management of acute ischaemic stroke. More recently MR diffusion weighted and perfusion imaging (Baird *et al.* 1997; Barber *et al.* 1998a), and dynamic CT have been proposed to delineate the penumbra but there is uncertainty over the exact interpretation of these techniques (Guadagno *et al.* 2004).

The therapeutic time window and reperfusion

Spontaneous recanalization, at least of middle cerebral artery occlusion, occurs in up to two-thirds of patients within a week of stroke onset, many in the first 48 h (Fieschi *et al.* 1989; Kaps *et al.* 1992; Zanette *et al.* 1995; Arnold *et al.* 2005). In general, the CT

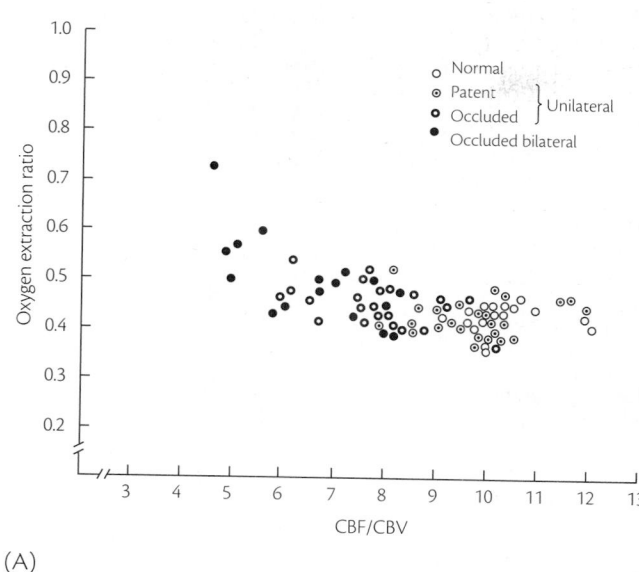

(A)

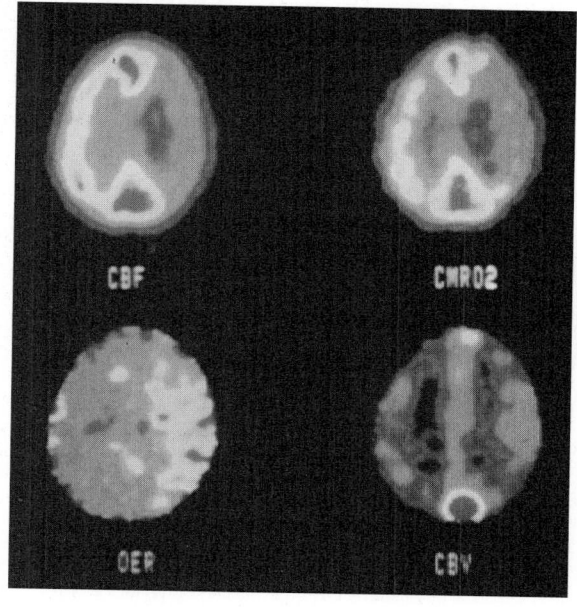

(Bi)

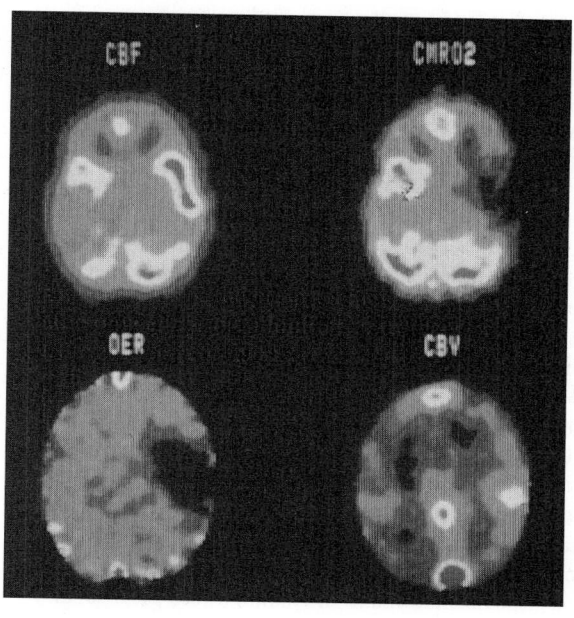

(Bii)

Fig. 35.14 (A) Relation between oxygen extraction ratio (or fraction) and cerebral blood flow divided by cerebral blood volume (CBF/CBV) in each of the 82 middle-cerebral artery regions from 32 patients with patent or occluded carotid arteries, and nine normal subjects (with permission from Gibbs *et al.* 1984). (B) Positron emission tomography (PET) scan 90 minutes after acute left middle cerebral artery occlusion: (i) cerebral blood flow (CBF) is low, cerebral metabolic rate for oxygen (CMR0$_2$) is low, but the oxygen extraction ratio (OER) (fraction) is maximal (i.e. ischaemia). (ii) PET scan a week later to show increased CBF and cerebral blood volume (CBV), despite low CMR0$_2$ and OER (i.e. absolute luxury perfusion). (Scans kindly provided by Professor Richard Frackowiak.) (See Plate 31.)

and functional outcomes are both better with recanalization and reperfusion, and even with early hyperperfusion, than if the middle cerebral artery remains occluded (Wardlaw *et al.* 1993; Marchal *et al.* 1996b; Barber *et al.* 1998b).

Ischaemic cerebral oedema

Cerebral ischaemia causes not only reversible and then irreversible loss of brain function, but also cerebral oedema (Symon *et al.* 1979; Hossman 1983). Ischaemic oedema is partly 'cytotoxic' and partly 'vasogenic'. Cytotoxic oedema starts early, within minutes of stroke onset, and affects the grey more than the white matter where damaged cell membranes allow intra-cellular water to accumulate. Vasogenic oedema, which starts rather later, within hours of stroke onset, affects more the white matter where the damaged blood–brain barrier allows plasma constituents to enter the extracellular

space. Ischaemic cerebral oedema reaches its maximum in 2–4 days and then subsides over a week or two.

Cerebral oedema not only increases local hydrostatic pressure and compromises blood flow further but also causes mass effect, brain shift, and eventually brain herniation (Fig. 35.15). Death in the first week after cerebral infarction is often due to these mass effects.

Diaschisis

Acute or chronic cerebral injury may cause effects in remote areas of brain (Meyer *et al.* 1993), so called 'diaschisis', by reducing neuronal inputs and metabolic activity: in the contralateral cerebellum and ipsilateral internal capsule, thalamus and basal ganglia after cortical lesions, in the ipsilateral cortex following internal capsule and thalamic lesions, and in the contralateral hemisphere. The functional consequences of diaschisis are not clear (Bowler *et al.* 1995).

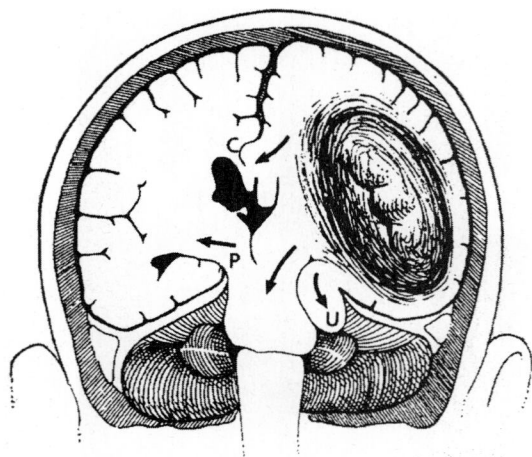

Fig. 35.15 A large oedematous cerebral infarct in the distribution of the middle cerebral artery causes herniation of the cingulate gyrus (c) under the falx cerebri; of the ispsilateral uncus (u) under the tentorium to compress the oculomotor nerve, posterior cerebral artery and brainstem; and of the contralateral cerebral peduncle (p) to cause ipsilateral hemiparesis. (With permission from Plum and Posner, 1985.)

35.3.6 Pathophysiology of acute intracerebral haemorrhage

The events following intracerebral haemorrhage have been most intensively studied for the most common type: rupture of one or more deep perforating arteries. The extravasated blood causes disruption of white matter tracts and irreversible damage to neurones in the deep nuclei or the cortex. The resultant increase in intracranial pressure may threaten other parts of the brain, particularly when the intracranial pressure reaches levels of the same order of magnitude as the arterial pressure, bringing the cerebral perfusion pressure close to zero (Rosand *et al.* 2002). Direct mechanical compression of the brain tissue surrounding the haematoma and, to some extent, vasoconstrictor and pro-inflammatory substances in extravasated blood also lead to impaired blood supply (Castillo *et al.* 2002; Butcher *et al.* 2004). Cellular ischaemia leads to further swelling from oedema (Gebel *et al.* 2002; Siddique *et al.* 2002), which is initially cytotoxic and later vasogenic.

Hydrocephalus may be an additional space-occupying factor. This complication is especially likely to occur with cerebellar haematomas, but a large haematoma in the region of the basal ganglia may also cause enlargement of the ventricular system, by rupture into the third ventricle, or through dilatation of the opposite lateral ventricle, with midline shift and obstruction of the third ventricle, whilst the ipsilateral ventricle is compressed. The zone of ischaemia around the haematoma may swell through systemic factors such as hypotension or hypoxia. Often there is also loss of cerebral autoregulation in the vasculature supplying the region of the haematoma. Some perifocal ischaemic damage occurs at the time of bleeding and cannot be prevented, but it is uncertain whether the vicious cycle of ongoing ischaemia causing steadily increasing pressure can be interrupted in its early stages.

35.4 Causes of transient ischaemic attack and ischaemic stroke

The causes of cerebral ischaemia are listed in Table 35.3. Rare conditions are proportionately more common in young stroke

Table 35.3 The causes of cerebral ischaemia and infarction

Arterial wall disorders
Atherothromboembolism
Intracranial small vessel disease (lipohyalinosis, arteriolosclerosis, microatheroma)
Trauma (Table 35.5)
Dissection (Table 35.6)
Fibromuscular dysplasia
Congenital arterial anomalies
Moyamoya syndrome
Embolism from arterial aneurysms
Inflammatory vascular diseases (Table 35.7)
Leukoaraiosis
Irradiation
Infections
Embolism from the heart (Table 35.4)
Haematological disorders (Table 35.8)
Miscellaneous conditions
Pregnancy/puerperium
Oral contraceptives and other female sex hormones
Drug abuse
Cancer
Perioperative
Migraine
Inflammatory bowel disease
Chronic meningitis
Homocystinaemia
Fabry's disease
Mitochondrial cytopathy
Hypoglycaemia
Fibrocartilaginous embolism
Snake bite
Fat embolism
Epidermal naevus syndrome
Susac's syndrome
Nephrotic syndrome
Cerebral autosomal dominant arteriopathy with subcortical infarcts and leucoencephalopathy (CADASIL)

patients, in whom degenerative arterial disease is unusual, but can still be a cause of stroke in the elderly. Venous infarction is considered later (Section 35.15).

Epidemiological studies have shown that about 25 per cent of ischaemic strokes are caused by identifiable atherothromboembolism from large artery disease, 25 per cent by small vessel disease, 20 per cent are cardioembolic, about 5 per cent are caused by rarities, and the remainder are of undetermined aetiology (Schulz and Rothwell 2003).

35.4.1 Large vessel disease and atherothromboembolism

Atheroma seems to be an almost inevitable accompaniment of ageing, at least in developed countries. Atherosclerosis is a multifocal disease affecting large- and medium-sized arteries particularly where there is branching, tortuosity, or confluence of vessels (Fig. 35.7). Turbulence caused by changes in blood flow direction is thought to contribute to endothelial damage and ultimately plaque formation. Atheroma begins in childhood as fatty streaks possibly in response to endothelial injury and over many years arterial smooth muscle cells proliferate and the intima is invaded by macrophages, fibrosis occurs, and cholesterol is deposited to form fibrolipid plaques (Ross 1999; Goldschmidt-Clermont 2005; Gotto 2005). Individuals with atheroma in one artery usually have widespread vascular disease making them at high risk of ischaemic heart disease, stroke, and claudication (Mitchell and Schwartz 1962; Rothwell 2001a) particularly among white males who often have accompanying hypercholesterolaemia. There appear to be important racial differences in the distribution of atheroma and race is an independent predictor of lesion location. White males tend to develop atheroma in the extracranial cerebral vessels, aorta, and coronary arteries whereas intracranial large vessel disease appears to be relatively more common in black, hispanic, and Asian populations (Feldmann et al. 1990; Leung et al. 1993; Sacco et al. 1995; Wityk et al. 1996; White et al. 2005) and tends to affect younger patients and those with type I diabetes mellitus (Sacco et al. 1995). Some but not all sources report that women have more intracranial disease than men.

Pathological, angiographic, and ultrasonic studies show that the most common extracranial sites for atheroma are the aortic arch, the proximal subclavian arteries, the carotid bifurcation (Fig. 35.16), and the vertebral artery origins (Fig. 35.17). Plaques in the subclavian arteries frequently extend into the origin of the

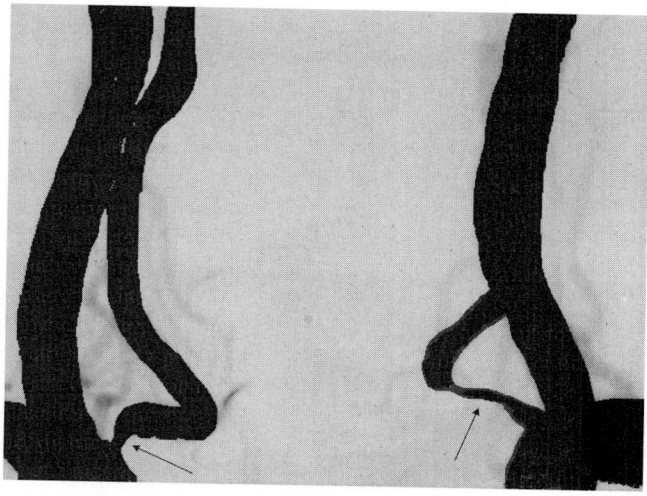

Fig. 35.17 Bilateral vertebral artery stenosis (arrows). Digitally subtracted arterial angiogram of the origins of both common carotid and vertebral arteries.

vertebral arteries and similar plaques may occasionally occur at the origin of the innominate arteries. Frequently, the second portion of the vertebral artery as it passes through the transverse foramen is also affected but the atheroma, which tends to form a ladder-like arrangement opposite cervical discs and osteophytes, does not normally restrict the lumen size significantly.

Intracranial arteries are morphologically different from extracranial arteries in having no external elastic membrane, fewer elastic fibres in the media and adventitia, and a thinner intimal layer. The major sites for atheroma formation in the anterior circulation are the carotid siphon, the proximal middle cerebral artery, and the anterior cerebral artery around the anterior communicating artery origin. In the posterior circulation, the intracranial vertebral arteries are often affected just after they penetrate the dura (Fig. 35.18) and distally near the basilar artery origin. Plaques are also found in the proximal basilar artery and also prior to the origin of the posterior cerebral arteries. The mid-basilar segment may be affected around the origins of the cerebellar arteries (Fig. 35.19). Occlusion of a branch artery at its origin by disease in the parent vessel seems to occur more commonly in the posterior circulation, 'basilar branch occlusion', than in the anterior circulation where occlusion of the small perforating arteries is usually caused by intrinsic small vessel disease.

Atheromatous medium-sized arteries at the base of the brain, particularly the vertebral and basilar arteries, may become affected by dolichoectasia. The arteries are widened, tortuous, and elongated, and may be visualized on MRI or if the walls are calcified on CT. Dolichoectasia is usually found in elderly patients with hypertension and diabetes and may cause stroke through embolization of thrombus or by occlusion of small branch arteries. In younger patients it should raise the possibility of Fabry disease.

Stroke mechanisms due to large artery atherosclerosis

The various patterns of arterial occlusion commonly seen in routine practice are shown in Fig. 35.20. Each of these types of stroke can be caused by large artery atherosclerosis. There are four principal mechanisms by which atherosclerotic lesions may cause ischaemic stroke:

◆ Thrombi may form on lesions and cause local occlusion.

◆ Embolization of plaque debris or thrombus may block a more distal vessel. Emboli follow the prevailing direction of flow in

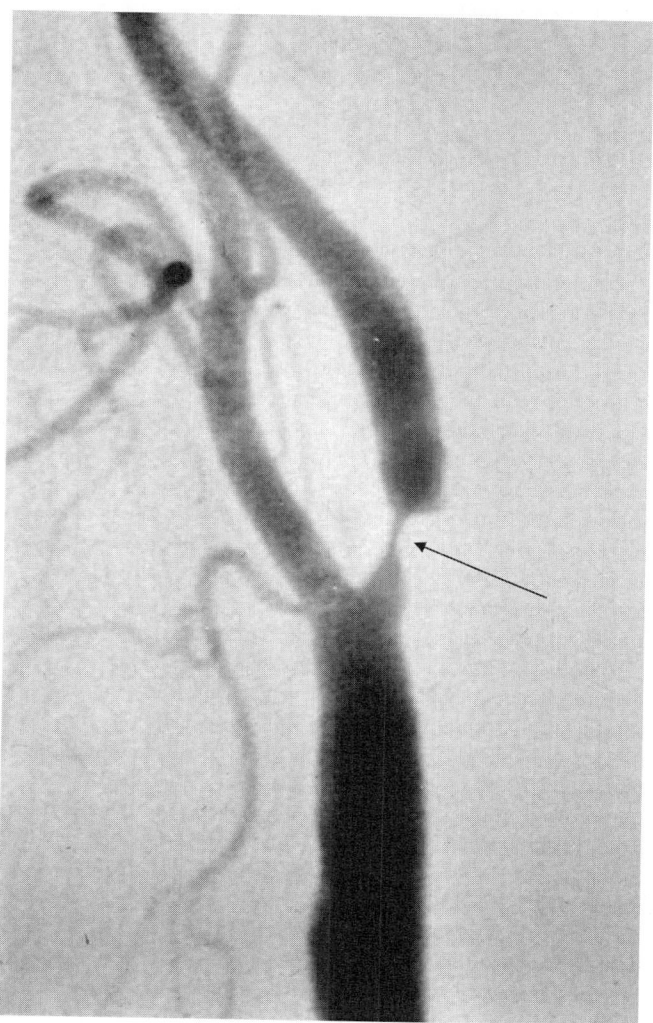

Fig. 35.16 Severe stenosis of the internal carotid artery (arrow). Digitally subtracted arterial angiogram.

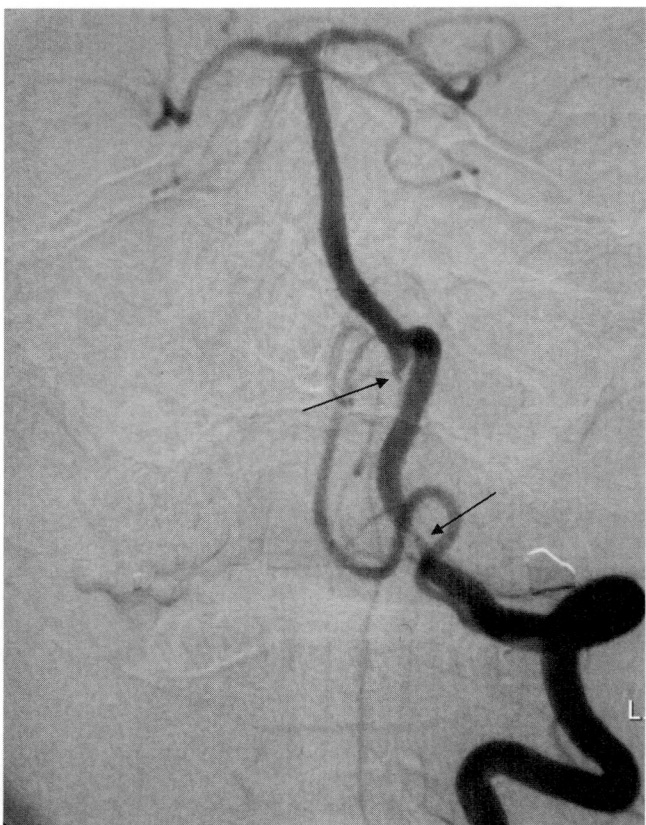

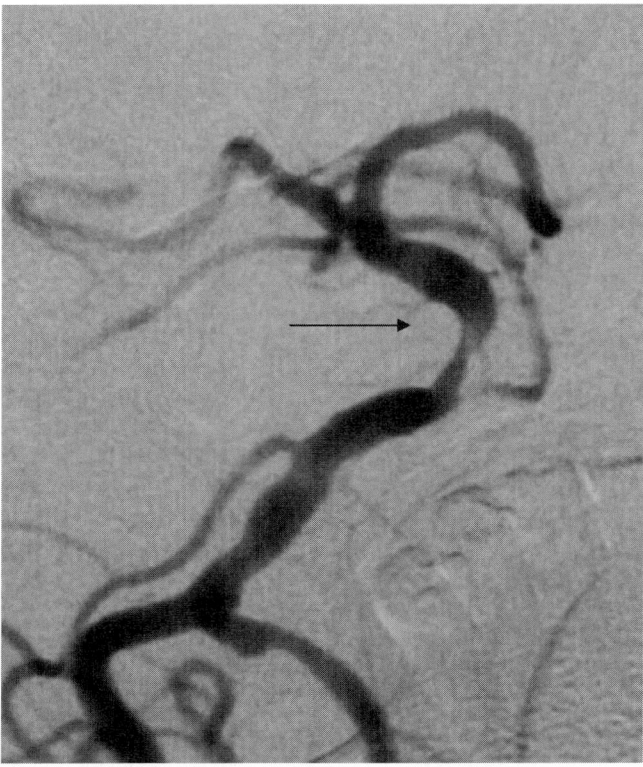

Fig. 35.19 Stenosis (arrow) of an ectatic basilar artery. Digitally subtracted arterial angiogram.

Fig. 35.18 Occlusion of the distal right vertebral artery (arrow) and a tight stenosis of the left vertebral artery (arrow). Digitally subtracted arterial angiogram.

a vessel thus most emboli from the internal carotid arteries will travel to the retina or the anterior two-thirds of the ipsilateral cerebral hemisphere. However, in patients with vascular disease, abnormal flow patterns may arise owing to vessel occlusion and collateral flow. For example, infarction may occur ipsilateral to a chronically occluded internal carotid artery as emboli from the contralateral internal carotid pass via the anterior communicating artery. Emboli are usually the cause of obstruction of the anterior circulation intracranial vessels (Lhermitte, Gautier, Derouesne 1970; Ogata *et al.* 1994) at least in white males in whom intracranial disease is relatively rare.

◆ Small vessel origins may be occluded by growth of plaque in the parent vessel, such as in the basilar artery or proximal middle cerebral artery.

◆ Severe reduction in the diameter of the vessel lumen caused by plaque growth may lead to hypoperfusion and infarction of distal 'borderzone' brain regions where blood supply is poorest.

Approximately 90 per cent of atherothromboembolic strokes in whites are caused by atheroma in the extracranial vessels whereas intracranial disease appears to be equally important in blacks and Hispanics (Sacco *et al.* 1995; Wityk *et al.* 1996). Atheromatous disease in the ascending aorta and the aortic arch is increasingly recognized as a source of cerebral emboli and an independent risk factor for ischemic stroke *in vivo* (Amarenco *et al.* 1994; Jones *et al.* 1995; Heinzlef *et al.* 1997; MacLeod *et al.* 2004).

Plaque activation and stroke risk

Atheromatous plaques are typically slow-growing or quiescent for long periods but may suddenly develop fissures or ulcers (Fig. 35.21). Activated plaques trigger platelet aggregation, thrombus formation (Viles-Gonzalez *et al.* 2004; Redgrave *et al.* 2006), and embolism. In keeping with the concept of acute intermittent activation of plaques, the likelihood of stroke in patients with cerebral atheromatous disease varies with time, being highest in the few days after a transient ischaemic attack or stroke (Coull *et al.* 2004; Rothwell and Warlow 2005) and large artery strokes are particularly likely to recur early (Lovett *et al.* 2004). Furthermore, emboli are more often detected with transcranial Doppler sonography if carotid stenosis is recently symptomatic (Dittrich *et al.* 2006; Markus 2006) and the rate of Doppler-detected emboli in the middle cerebral artery tends to decline with time after stroke (Kaposzta *et al.* 1999).

Plaque irregularity or ulceration, which is best visualized on catheter angiography (Fig. 35.22) but can sometimes be seen on contrast-enhanced MR angiography (Fig. 35.23), is independently associated with increased stroke risk, probably because irregularity represents plaque ulceration and instability with thrombosis and so likely complicating embolism (Molloy and Markus 1999; Rothwell *et al.* 2000a, Lovett *et al.* 2004). There is also evidence to suggest that ulcerated carotid plaques are more likely than smooth plaques to be associated with vascular events in other territories such as the coronary arteries (Rothwell *et al.* 2000b) suggesting that plaque activation is a systemic phenomenon. The trigger for activation is not known but infective, inflammatory, or genetic mechanisms have been proposed.

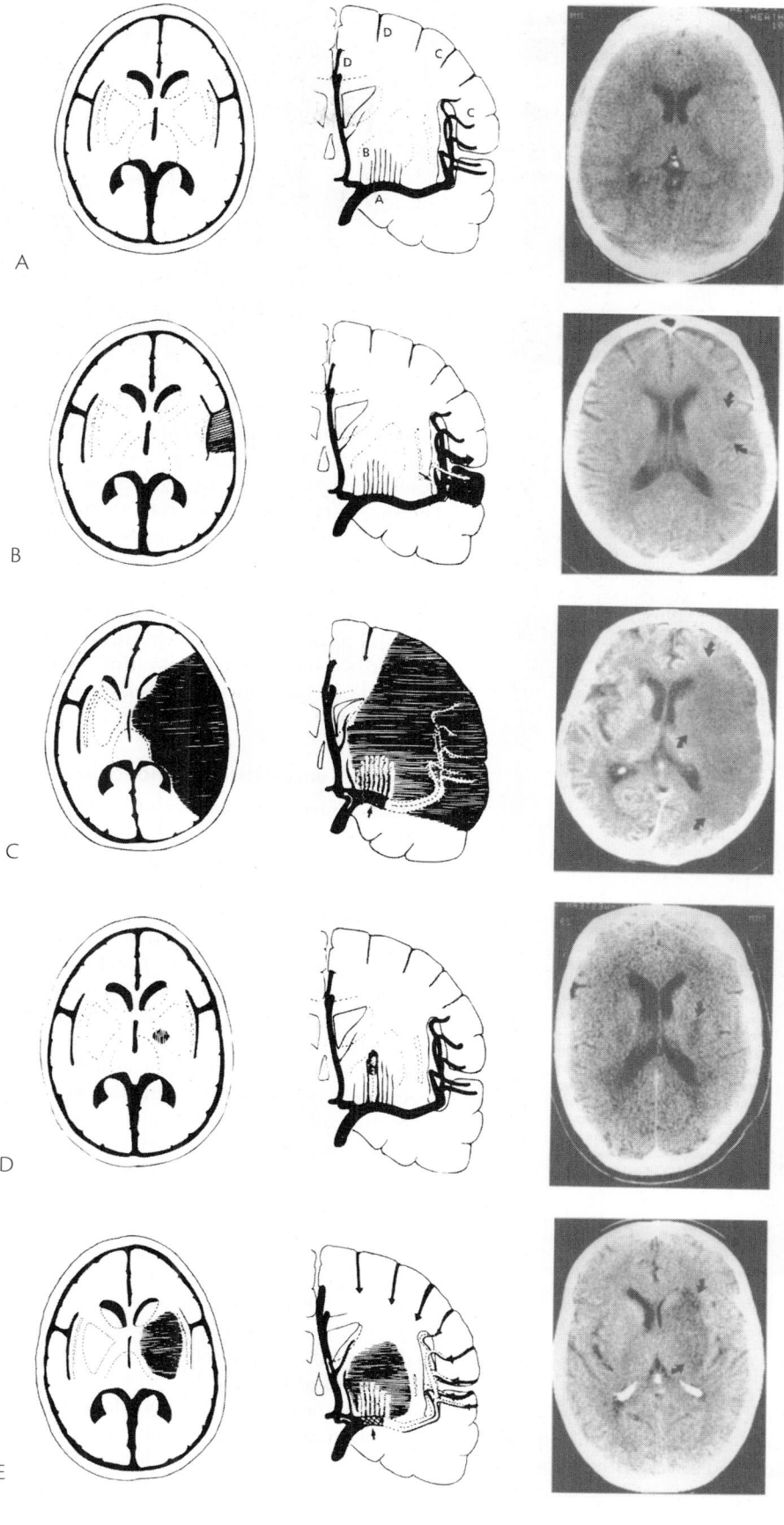

Fig. 35.20 Various patterns of arterial occlusion causing different types of ischaemic stroke. Left-hand column, diagram of axial CT brain scan through the level of the basal ganglia; middle column, diagram of the middle cerebral artery and anterior cerebral arteries on a coronal brain section; right-hand column, corresponding CT brain scan. A, main trunk of middle cerebral artery; B, lenticulostriate perforating branches of the middle cerebral artery; C, cortical branches of the middle cerebral artery; D, cortical branches of the anterior cerebral arteries.

(A) Normal arterial anatomy and CT scan;

(B) occlusion (usually embolic—arrow—from heart, aorta, or internal carotid artery) of a cortical branch of the middle cerebral artery and restricted cortical infarct on CT (arrows);

(C) occlusion (usually embolus—arrow—as in (b) above) of middle cerebral artery trunk to cause infarction of entire middle cerebral artery territory (arrows);

(D) occlusion of a single lenticulostriate artery to cause a lacunar infarct (arrow); note that the patient has an old lacunar infarct in the opposite hemisphere;

(E) occlusion of the middle cerebral artery trunk with good cortical collaterals from the anterior and posterior cerebral arteries to cause a striatocapsular infarct (arrows).

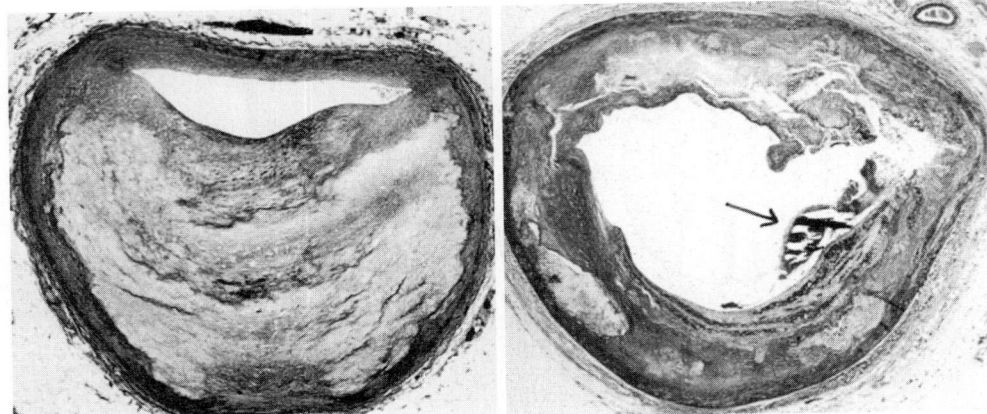

Fig. 35.21 Atheromatous plaque. Histological sections showing a thick fibrous cap (left) and a ruptured plaque (right) with adherent thrombus (arrow).

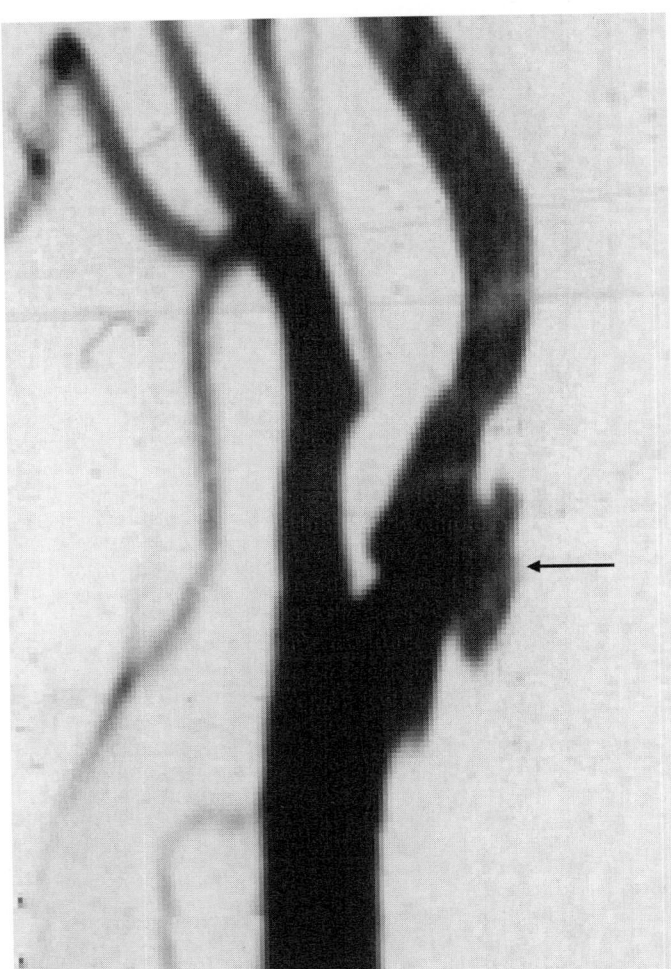

Fig. 35.22 Ulcerated plaque at the carotid bifurcation. Digitally subtracted arterial angiogram with contrast seen within the plaque (arrow).

Cholesterol embolization syndrome

This rare disorder is thought to be caused by rupture of atheromatous plaques in elderly people with widespread disease, either spontaneously or as a complication of instrumentation of large atheromatous arteries, and possibly of anticoagulation or thrombolysis.

Cholesterol debris is released and embolizes to the microcirculation of organs including the brain and spinal cord. Hours or days after instrumentation or surgery there is the subacute onset of a syndrome very similar to systemic vasculitis or infective endocarditis, with malaise, fever, abdominal pain, proteinuria and renal failure, stroke-like episodes, drowsiness, confusion, skin petechiae, splinter haemorrhages, livedo reticularis, cyanosis of fingers and toes, raised erythrocyte sedimentation rate, neutrophil leucocytosis, and eosinophilia. The diagnosis is made by finding cholesterol debris in the microcirculation of biopsy material, usually from the kidney but sometimes from skin or muscle (Cross 1991; Rhodes 1996).

35.4.2 Small vessel disease and leukoariosis

The small penetrating arteries of the brain, less than about 500 microns in diameter, are not supported by a good collateral circulation. These arteries include the lenticulostriate branches of the middle cerebral artery, the thalamoperforating branches of the proximal posterior cerebral artery, and the perforating arteries to the brainstem. Therefore, occlusion is likely to cause infarction, albeit in a small restricted area of brain. Such 'lacunar' infarcts comprise about one-quarter of first ischaemic strokes (Bamford *et al.* 1987; Sempere *et al.* 1998; Schulz and Rothwell 2003). Because the case fatality is low, about 1 per cent, there are few pathological data, but it does seem that these small arteries are much less likely to be occluded by emboli either from the heart or from extracranial sites of atherothrombosis compared with the trunk or cortical branches of the middle cerebral artery (Tegeler *et al.* 1991; Boiten *et al.* 1996; Gan *et al.* 1997). Furthermore, ischaemic lacunar strokes are less often associated with middle cerebral artery emboli detected with transcranial Doppler (Koennecke *et al.* 1998).

Although not universally accepted, it is generally thought that these small perforating arteries are occluded by thrombus complicating not atheroma but a distinct small vessel arteriopathy 'hyaline arteriosclerosis' or 'simple small vessel disease' (Lammie *et al.* 1997). This is an almost universal change in the small arteries and arterioles of the aged brain particularly in the presence of hypertension or diabetes. The muscle and elastin in the arterial wall are replaced by collagen, there is subintimal hyalinization, the wall becomes thickened and the lumen narrowed, and the vessel becomes tortuous (Fig. 35.24). In complex small vessel disease, there is more aggressive disorgabization of the small vessel walls

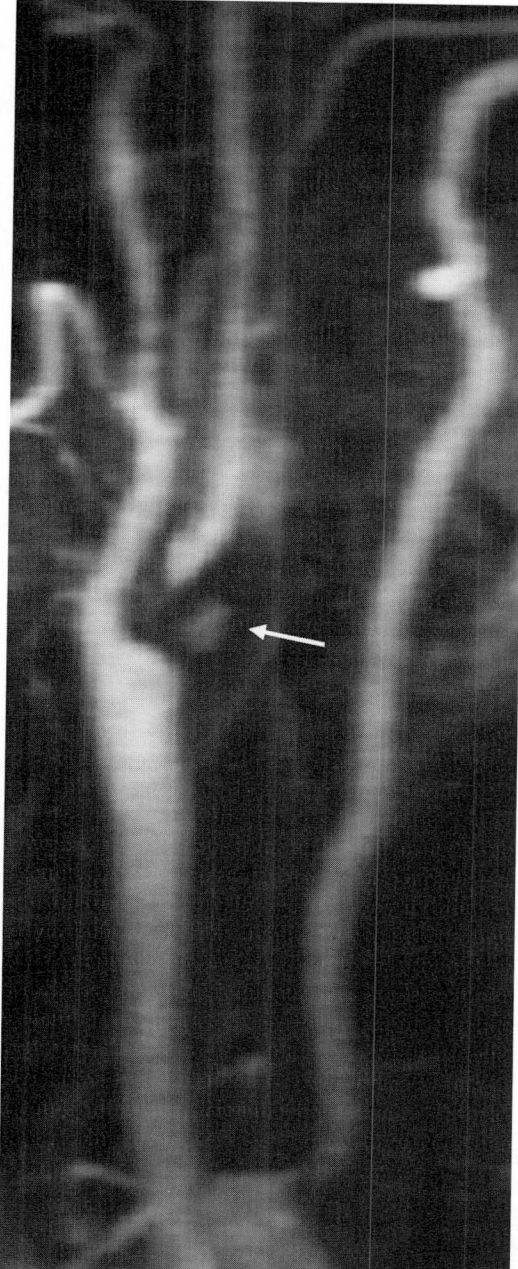

Fig. 35.23 Severe carotid stenosis due to an ulcerated plaque. Contrast-enhanced magnetic resonance angiogram with contrast seen within the plaque (arrow).

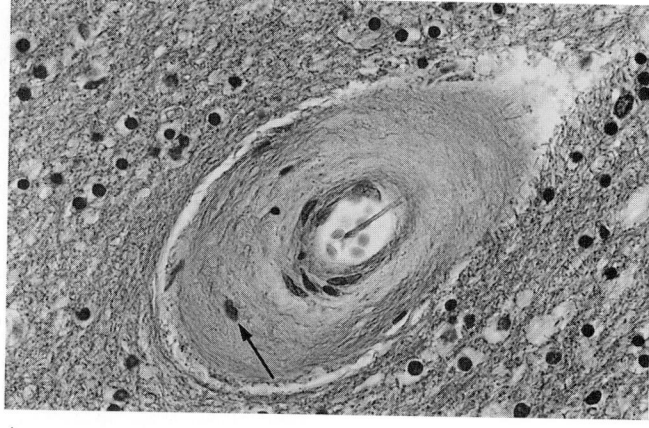

A

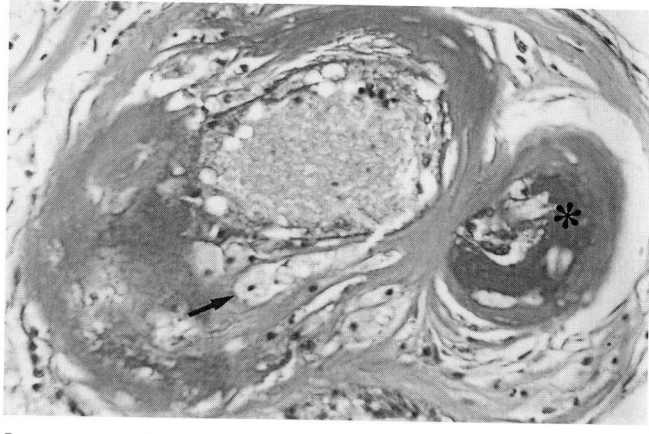

B

Fig. 35.24 Penetrating lenticulostriate artery branches in the putamen. Photomicrograph illustrating two distinctive patterns of vessel pathology: (A) Concentric hyaline wall thickening with a few remaining vascular smooth muscle cell nuclei (arrow). The lumen remains patent. Such 'simple' small vessel disease is an almost invariable feature of elderly brains, most prominent in hypertensives and diabetics. Haematoxylin and eosin: ×420. (B) A complex, disorganized vessel segment showing an asymmetric destructive process with focal fibrinoid material (asterix) and mural foam cells (arrow). The lumen is visible cut in two planes of section. This 'complex' vessel lesion corresponds to what CM Fisher termed 'lipohyalinosis', and in this case the lesion was adjacent to, and presumably the cause of, a right striatocapsular lacunar infarct. Haematoxylin and eosin: ×210. (Courtesy of Dr Alistair Lammie.

accompanied by foam cell infiltration. Whether or not simple and complex small vessel diseases are related is unclear.

The current view is that both 'complex' small vessel disease and atheroma at or near the origin of the small perforating vessels arising from the major cerebral arteries cause most of the small deep infarcts responsible for lacunar ischaemic strokes which make up about one-quarter of symptomatic cerebral ischaemic events (Bamford *et al.* 1987; Schulz and Rothwell 2003). However, this hypothesis is not universally accepted (Millikan and Futrell 1990) since there is little direct post-mortem evidence of occlusion of these vessels leading to lacunar infarcts. Certainly, at least some small infarcts in the brainstem and internal capsule can be due to atheroma at the mouth of the small penetrating vessels spreading from atheroma of the larger parent artery (Fisher and Caplan 1971; Fisher 1979). It is also conceivable that this small vessel arteriopathy can lead to small, deep haemorrhages as well as lacunar infarcts (Labovitz *et al.* 2007); indeed, both types of stroke often coincide (Samuelsson *et al.* 1996; Kwa *et al.* 1998).

Leukoariosis

There are a number of alternative terms for leukoariosis: Binswanger's disease, chronic progressive subcortical encephalopathy, subcortical arteriosclerotic encephalopathy, periventricular leucoencephalopathy. This reflects the confusion between the

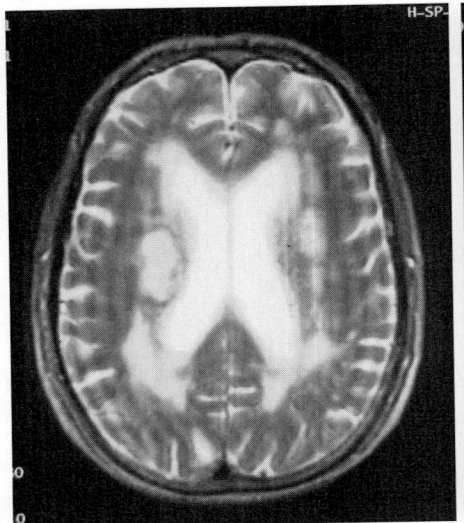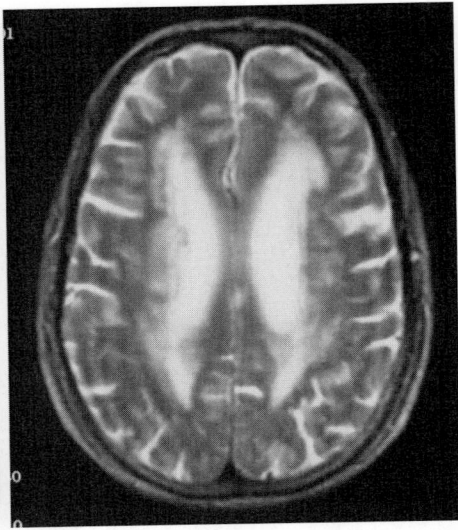

Fig. 35.25 Advanced periventricular leukoaraiosis. MR brain scan showing the typical periventricular abnormalities.

clinical, radiological, and pathological literature (Munoz 2006). On CT, there is roughly symmetrical but irregular periventricular hypodensity, with or without ventricular dilatation and focal white matter hypodensities. This is better seen as high signal on T2-weighted MR images (Fig. 35.25). This periventricular radiological appearance is due to demyelination, axonal loss, and gliosis thought to occur as a consequence of diffuse rather than focal ischaemia in the distribution of the long perforating arteries from the pial surface of the brain. Vascular occlusion has not been seen (Caplan 1995; Pantoni and Garcia 1997).

Leuokoariosis is frequent in the normal elderly but is seen particularly in those with hypertension, dementia, or stroke but not with increasing carotid stenosis severity (Bots *et al.* 1993; Adachi *et al.* 1997; Munoz 2006). It is also common in patients with cerebral amyloid angiopathy. It is a risk factor for ischaemic, particularly lacunar, and haemorrhagic stroke (Inzitari 2003) and is associated with increased bleeding risk on anticoagulants (Gorter 1999). It seems likely that the association between leukoariosis and stroke occurs because hypertension causes both pathological syndromes in the same individual: leukoaraiosis is not itself the cause of the stroke.

35.4.3 Embolism from the heart

Approximately 20 per cent of ischaemic stroke is cardioembolic. There are a large number of potential cardiac sources of embolism (Table 35.4) but it may be difficult to be certain whether an identified putative embolic source is the cause of a stroke, particularly if there are alternative causes such as coexistent large artery disease, or if the stroke is lacunar and unlikely to be caused by cardiac embolism.

Atrial fibrillation is an important cause which has already been discussed (Section 35.2.5).

Coronary artery disease

Overall, there is about a 5-fold relative excess risk of stroke in the first few days and weeks after myocardial infarction but the

Table 35.4 Cardiac sources of embolism

Paradoxical emoblism from the venous system
Atrial septal defect
Ventricular septal defect
Patent foramen ovale
Pulmonary arteriovenous fistula
Left atrium
Atrial fibrillation
Sinoatrial disease
Myxoma
Inter-atrial septal aneurysm
Mitral valve
Rheumatic stenosis or regurgitation
Infective endocarditis
Non-bacterial thrombotic or marantic endocarditis
Prosthetic valve
Mitral annulus calcification
Libman-Sacks endocarditis
Papillary fibroelastoma
Left ventricular mural thrombus
Acute myocardial infarction
Left ventricular aneurysm
Cardiomyopathy
Myxoma
Blunt chest injury
Mechanical artificial heart
Aortic valve
Rheumatic stenosis or regurgitation
Infective endocarditis
Non-bacterial thrombotic or marantic endocarditis
Prosthetic valve
Calcification and/or sclerosis
Syphilis
Congenital cardiac disorders, particularly with right to left shunt
Cardiac surgery, catheterisation, angioplasty
Others: primary oxalosis, hydatid cyst

absolute risk of clinically evident systemic embolism is well under 5 per cent (Dutta *et al.* 2006a). The risk of embolism is higher in anterior infarcts, large infarcts, and the presence of a dyskinetic wall segment. Some post-myocardial infarction strokes may be due to hypotension and boundary zone infarction, atrial fibrillation with left atrial thrombus, paradoxical embolism, coronary and aortic instrumentation, while others are primarily haemorrhagic as a consequence of antithrombotic and thrombolytic drugs. Rarely, the same non-atheromatous disorder can cause both ischaemic stroke and acute Myocardial Infarction: giant cell arteritis, aortic arch dissection, or infective endocarditis. The long-term risk of stroke after acute myocardial infarction is about 1.5 per cent per annum, 8 per cent in 5 years (Martin *et al.* 1993; Loh *et al.* 1997).

Mitral leaflet prolapse

Mitral leaflet prolapse is a common incidental finding. It can be complicated by gross mitral regurgitation, infective endocarditis, atrial fibrillation and left atrial thrombus, and thus embolism to the brain. However there is no excess risk of first or recurrent stroke in patients with uncomplicated mitral leaflet prolapse (Orencia *et al.* 1995a, b).

Calcification of the aortic and mitral valves

Calcification, and possibly sclerosis, of the aortic and mitral valves may be a cause of embolism of calcific or complicating thrombotic material. However, these degenerative disorders of heart valves are so common, particularly in the elderly, that it has been very difficult to associate them causally with stroke (Boon *et al.* 1996).

Paradoxical embolism and patent foramen ovale

Post-mortem examples have established that paradoxical embolism can occur from venous thrombi through the right to the left side of the heart. Emboli may pass through a patent foramen ovale, which is found in about one-quarter of healthy people, or an atrial septal defect, or a ventriculoseptal defect (Gautier *et al.* 1991; Jeanrenaud and Kappenberger 1991; Cabanes *et al.* 1993). There is an increased incidence of patent foramen ovale in patients with cryptogenic stroke (Mas *et al.* 2001; Lamy *et al.* 2002) but the risk of recurrent stroke in patients with a patent foramen ovale is low and so routine endovascular closure cannot be recommended (Mas 2003; Amarenco 2005; Homma and Sacco 2005; Kizer and Devereux 2005; Messe *et al.* 2005) particularly since there is evidence of a continuing risk of stroke after closure of patent foramen ovale (Wahl *et al.* 2001).

Atrial septal aneurysm

Atrial septal aneurysm is an echocardiographic finding in some normal people. The combination of atrial septal aneurysm and patent foramen ovale was thought to carry a higher stroke risk than patent foramen ovale alone (Mas *et al.* 2001; Lamy *et al.* 2002), with a reported risk of recurrent stroke in such patients as high as 15 per cent (Mas *et al.* 2001) but more recent data from a larger study have cast doubt on this observation (CODICE Study Group 2006). Interestingly, there appears to be an association between patent foramen ovale and migraine, particularly migraine with aura, and this is particularly strong where there is coexistent atrial septal aneurysm. There are anecdotal reports of improvement in migraine symptoms following patent foramen ovale closure (Holmes 2004; Diener *et al.* 2005).

Infective endocarditis

About one-fifth of patients with infective endocarditis have an ischaemic stroke or transient ischaemic attack as a result of embolism of valvular vegetations (Section 42.5.9). Cerebrovascular symptoms usually occur before the infection has been controlled and may be the presenting feature, although often the patient has already been hospitalized for other symptoms (Hart *et al.* 1990; Salgado 1991). Haemorrhagic transformation of an infarct, occurs in 20 to 40 per cent. Primarily haemorrhagic strokes, intracerebral or, rarely, subarachnoid, are more commonly caused by pyogenic vasculitis and vessel wall necrosis than by mycotic aneurysms which can be single or multiple and most often affect the distal branches of the middle cerebral artery (Masuda *et al.* 1992; Krapf *et al.* 1999). These aneurysms tend to resolve with time and cerebral angiography to detect unruptured aneurysms with a view to surgery is unnecessary (van der Meulen *et al.* 1992).

Early institution of the correct antibiotic therapy is the most effective way to prevent thromboembolism in infective endocarditis, the risks of which are highest in the first 24–48 hours after diagnosis. Anticoagulation should not be given to patients with native valve or bioprosthetic valve endocarditis because of the risk of intracerebral haemorrhage from mycotic aneurysms and arteritis and the reduction in embolism risk with antibiotic therapy. For patients with mechanical valves who are on long-term anticoagulation at the time of developing infective endocarditis, the correct management is unclear. Other neurological complications of infective endocarditis include: meningitis, diffuse encephalopathy, acute mononeuropathy, cerebral abscess, discitis, and headache (Jones and Siekert 1989; Kanter and Hart 1991).

Fever, cardiac murmur, and vegetations are not invariably present in patients with infective endocarditis. Thus blood cultures are indicated in unexplained stroke particularly if there is raised erythrocyte sedimentation rate, mild anaemia, neutrophil leukocytosis, and disturbed liver function, or a history of intravenous drug abuse. The CSF can be normal, but $>\cdot 100$ polymorphs/mm^3 is said to suggest endocarditis although similar counts have been described in intracerebral haemorrhage and in haemorrhagic transformation of an infarct, but not in ischaemic stroke (Powers 1986).

Non-bacterial thrombotic or marantic endocarditis

Small, friable, and sterile vegetations made of fibrin and platelets can be found on the heart valves of patients with cancer, and in the antiphospholoipid antibody syndrome systemic lupus erythematosis, and possibly protein C deficiency. Thrombotic emboli from such vegetations can be demonstrated using trans-oesophageal echocardiography and are frequently seen in patients with cancer and cerebral ischaemia (Dutta *et al.* 2006b).

Prosthetic heart valves

Prosthetic valves, particularly mechanical ones, are associated with thrombosis and embolism, and infective endocarditis. There seems to be little difference in this respect between the different mechanical valves, but those in the mitral position are most prone to thrombosis. The overall risk of clinically evident embolism is 1–2 per cent per annum on anticoagulants (Vongpatanasin *et al.* 1996).

Cardiac myxomas

Cardiac myxomas are rare, occasionally familial, and arise in any heart chamber but 75 per cent are found in the left atrium.

Tumour material, or complicating thrombus, may embolize and often there are features of intracardiac obstruction (dyspnoea, cardiac failure, syncope) and constitutional upset (malaise, weight loss, fever, rash, arthralgia, myalgia, anaemia, raised ESR, hyper-gammaglobulinaemia) (Ekinci Donnan 2004). Myxomatous emboli impacted in cerebral arteries may cause aneurysmal dilatation with subsequent intracerebral or subarachnoid haemorrhage (Sabolek et al. 2005).

Dilating cardiomyopathies

Cardiomyopathies may be complicated by intracardiac thrombus but associated embolic stroke is rare.

Cardiac surgery

Cardiac surgery is complicated by stroke or retinal/optic nerve infarction in about 2 per cent of cases, the risk being greater for valve than for coronary artery surgery (Newman et al. 2006) but early post-operative confusion is much more common and cognitive deficits may persist for some weeks (Newman et al. 2006). Possible mechanisms include embolization during or after surgery, hypotension, cholesterol embolization, simultaneous carotid endarterectomy, thrombosis asssociated with heparin-induced thrombocytopenia, and intracranial haemorrhage due to anticoagulation or thrombocytopenia.

Sinoatrial disease

Sinoatrial disease or the sick sinus syndrome is associated with systemic embolism, particularly if there is bradycardia alternating with tachycardia, or atrial fibrillation (Bathen et al. 1978).

Instrumentation of the coronary arteries and aorta

Such procedures may dislodge valvular or atheromatous debris causing neurological complications (Ayas and Wijdicks 1995) and the cholesterol embolization syndrome.

35.4.4 Arterial dissection and trauma

Trauma to the arteries supplying the brain is probably a more common cause of stroke than is generally realized, partly because of the wide variety of potential causes of injury (Table 35.5). The most common consequence of trauma that can lead to stroke is arterial dissection.

Arterial dissection

Arterial dissection is a common cause of ischaemic stroke and transient ischaemic attack in young adults. Sometimes there is a predisposing cause (Schievink 2001; Rubinstein et al. 2005) (Table 35.6) but often there is no explanation. The artery may become occluded by the wall haematoma itself, thrombosis and embolism may complicate occlusive or non-occlusive dissections, and aneurysmal bulging of the weakened wall may occur (O'Connell et al. 1985). Arterial rupture is unusual.

There are a number of diagnostic pointers to cervical dissection which should be looked for in the history and examination:

- Potential neck injury;
- Pain. In the neck, side of the head, face or eye ipsilateral to internal carotid artery dissection. Pain at the back of the head and neck, usually unilaterally, for vertebral dissection.
- Horner's syndrome as a result of damage to sympathetic nerves around the internal carotid artery. This occurs in up to 50 per cent.

Table 35.5 Causes of injury of the arteries supplying the brain

Penetrating injury
Missile wounds
Neck laceration
Neck surgery
Tonsillectomy
Oral trauma
Catheter angiography
Jugular vein cannulation

Non-penetrating injury
Blow to the neck
Carotid compression tests
Attempted strangulation
Neck injury: fracture, subluxation, or dislocation
Sudden neck movements: whiplash injury, 'head-banging', ceiling painting, head injury, head turning, or minor falls
Yoga
Neck manipulation
Labour
Tonic-clonic seizure
Vomiting
Bronchoscopy
Atlanto-axial dislocation
Occipito-atlantal instability
Fractured base of skull
Cervical rib
Fractured clavicle

- Self-audible bruit which may be described as tinnitus owing to dissection adjacent to the base of the skull. This occurs in about 30 per cent.
- Ipsilateral palsy of a cranial nerve. This occurs in about 10 per cent, often affecting the XII or another lower cranial nerve, rarely the III nerve.
- Cervical root lesions have been reported in association with vertebral artery dissections from pressure or ischaemia.

The presence of cranial neuropathy may result in a misdiagnosis of brainstem stroke. Cranial nerve palsies may result from local pressure from the false internal carotid artery lumen or thromboembolism or haemodynamic compromise to the blood supply of the nerve. The III nerve receives its blood supply from the ophthalmic artery, branches of the internal carotid or the posterior cerebral artery, and thus may become ischaemic after carotid dissection. However, this is very rare.

Table 35.6 Causes of dissection of the extra- and intracranial arteries

Traumatic
Penetrating injury
Non-penetrating injury

Spontaneous
Fibromuscular dysplasia
Cystic medial necrosis
Marfan's syndrome
Ehlers–Danlos syndrome
Pseudoxanthoma elasticum
Inflammatory arterial disease
Infective arterial disease, e.g. syphilis

The features listed above may precede the onset of cerebral ischaemia by hours or days, and relevant points in the history may thus not be volunteered spontaneously by the patient. Alternatively, diagnostic pointers to dissection may be absent altogether and thus the diagnosis becomes one of exclusion or confirmation by imaging. Cervical arterial dissection generally has a benign prognosis, with a 95 per cent 10-year survival, although the risk of stroke is high during the first few days and weeks. Secondary aneurysm formation can cause symptoms due to local pressure but does not appear to increase the risk of thromboembolism.

The incidence of diagnosed internal carotid artery dissection is around 1–4 per 100 000 per year. Vertebral dissection is a little less common. The actual incidence of dissections is likely to be considerably higher, but the diagnosis is often missed, particularly in older patients. Usually only one artery is involved but in about 10 per cent of cases multiple arteries may be affected simultaneously or in close succession. Recurrence rates are low at around 1 per cent per annum except in familial cases of arterial dissection or hereditary connective tissue disorder where rates are higher (Leys *et al.* 1995).

On angiography there is usually a long, tapered, narrow, or occluded segment, perhaps with an intimal flap, double lumen, or intraluminal thrombus, and sometimes an associated aneurysm. Intracranial arterial occlusion, presumably embolic, may be seen. Carotid dissection can often be strongly suspected on Duplex (Sturzenegger *et al.* 1993; 1995; Flis *et al.* 2007), but the most sensitive and specific imaging evidence of both carotid and vertebral dissection comes from a combination of axial MRI through the lesion, to show the acute haematoma in the arterial wall, with MR angiography (Auer *et al.* 1998; Flis *et al.* 2007) (Fig. 35.26).

Intracranial arterial dissection

Intracranial dissection is much rarer. It may present with subarachnoid haemorrhage due to rupture of a pseudo-aneurysm, as well as

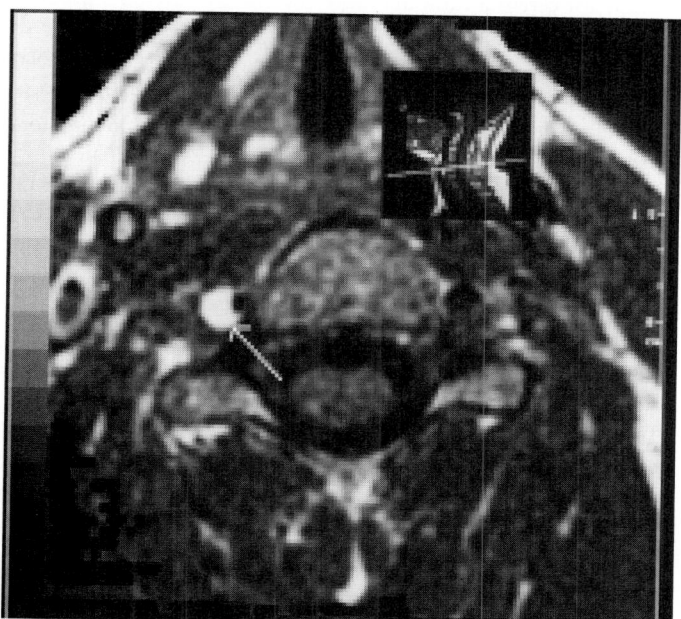

Fig. 35.26 Arterial dissection. High-signal in the wall of the right internal carotid artery (arrow) indicating a recent dissection. MRI.

with ischaemic stroke, and is less often diagnosed during life (Farrell *et al.* 1985; de Bray *et al.* 1997; Chaves *et al.* 2002).

Aortic arch dissection

Aortic arch dissection can cause profound hypotension with global, and sometimes boundary zone, cerebral ischaemia, or focal cerebral ischaemia if the dissection spreads up one of the neck arteries. Clues to this diagnosis are anterior chest or interscapular pain, along with diminished, unequal, or absent arterial pulses in the arms or neck, a normal electrocardiogram unlike acute myocardial infarction, acute aortic regurgitation, and pericardial effusion.

Trauma

Penetrating and non-penetrating neck injuries are more likely to damage the carotid than the better protected vertebral artery. The vertebral artery appears to be more vulnerable to rotational and hyperextension injuries of the neck, particularly at the level of the atlas and axis. Laceration, dissection, and intimal tears may be complicated by thrombosis and then embolism and, therefore, ischaemic stroke at the time of, or some days or even weeks after the injury. Later stroke may be a consequence of the formation of a traumatic aneurysm, arteriovenous fistula, or a fistula between the carotid and vertebral arteries (Davis and Zimmerman 1983).

The subclavian artery can be damaged by a fractured clavicle or a cervical rib, with later embolization up the vertebral arteries or even up the right common carotid artery (Prior *et al.* 1979).

35.4.5 Rare arterial disorders

Although probably relatively rare, there are a large number of arterial disorders of anomalies that can cause ischaemic stroke (Tables 35.2 and 35.3). The frequency of many of these disorders has undoubtedly been underestimated due to underinvestigation, particularly the tendency to confine vascular imaging to a snapshot of a small region around the carotid bifurcations.

Fibromuscular dysplasia

Fibromuscular dysplasia is a rare segmental disorder more common in females affecting small- and medium-sized arteries (Slovut and Olin 2004). It is most common in the renal arteries, causing hypertension. The mid-cervical portion of the internal carotid artery is the most frequently affected artery to the brain but the vertebral arteries at the level of the first two cervical vertebrae may also be involved. Intracranial pathology is exceptional (Arunodaya *et al.* 1997). Histologically there is fibrosis and thickening of the arterial wall alternating with atrophy, giving a typical angiographic appearance of a 'string of beads' (Fig. 35.27). Fibromuscular dysplasia is associated with intracranial saccular aneurysms and arteriovenous malformations and dissection. Fibromuscular dysplasia of some arteries to the brain is found in up to 1 per cent of routine post-mortems so that any association with cerebral ischaemia or infarction may be no more than coincidence. Occasionally, however, it may be complicated by thrombosis and embolism. The natural history is unknown.

Congenital arterial anomalies

Occasionally the carotid arteries are hypoplastic or absent, and kinking, acute angulation, tortuosity, and looping of the internal carotid artery may be seen on angiograms (Metz *et al.* 1961). Such appearances can be due to atheroma, fibromuscular dysplasia, or congenital abnormality. There is a tendency to regard anomalies in

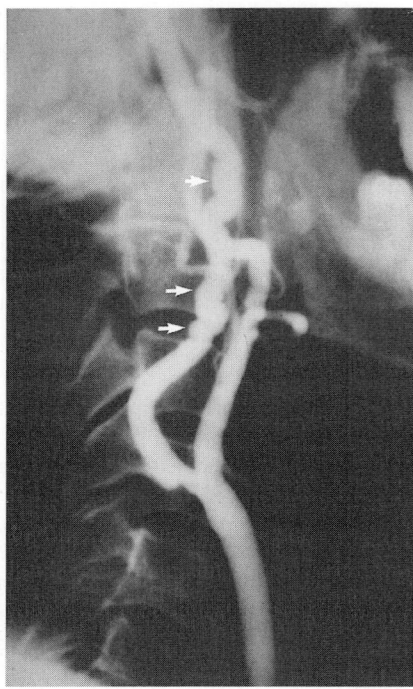

Fig. 35.27 Fibromuscular dysplasia. Lateral view of a selective carotid angiogram showing the typical 'string of beads' appearance (arrows).

children and young adults as 'congenital' and those in the middle-aged and elderly as 'atherosclerotic'.

Congenital carotid loops

Carotid loops may be associated with aneurysm formation and rarely embolism, endothelial damage, and thrombosis, and exceptionally with focal ischaemia on head movement (Sarkari *et al.* 1970; Desai and Toole 1975). Rarely, these loops may cause hypoglossal nerve lesions or pulsatile tinnitus.

Some inherited disorders of connective tissue (Table 35.2) can present with, or be complicated by, arterial dissection or even rupture, intra and extra cranial aneurysm formation, caroticocavernous fistula, and mitral leaflet prolapse: the Ehlers-Danlos syndrome (North *et al.* 1995), pseudoxanthoma elasticum (Mayer *et al.* 1994), and Marfan's syndrome (Bowen *et al.* 1987; Schievink *et al.* 1994).

Moyamoya syndrome

In Japanese 'moyamoya' means 'puff of smoke' and describes the characteristic radiological appearance of the fine anastomotic collaterals that develop from the perforating and pial arteries at the base of the brain, the orbital and ethmoidal branches of the external carotid artery, and leptomeningeal and transdural vessels, in response to severe stenosis or occlusion of one, or both, distal internal carotid arteries (Yonekawa and Kahn 2003). The circle of Willis and the proximal cerebral and basilar arteries may also be involved.

Moyamoya seems to be mainly confined to the Japanese and other Asians, and in most cases the cause is unknown (Bruno *et al.* 1988; Chiu *et al.* 1998). Some cases are familial (Kitahara *et al.* 1979), others appear to be due to a generalized fibrous disorder of arteries (Aoyagi *et al.* 1996), and a few may be due to a congenital

hypoplastic anomaly affecting arteries at the base of the brain, or associated with Down's syndrome (Cramer *et al.* 1996). The syndrome may present in infancy with recurrent episodes of cerebral ischaemia and infarction, mental retardation, headache, epileptic seizures, and occasionally involuntary movements. In adults, subarachnoid or primary intracerebral haemorrhage are also common owing to rupture of collateral vessels. There have also been a few reports of associated intracranial aneurysms (Iwama *et al.* 1997) and also cerebral arteriovenous malformations.

Embolism from intra and extra cranial arterial aneurysms

Embolism from thrombus within the cavity of an aneurysm is rare and is difficult to prove in an individual case in which there may be other potential sources of embolization. Intracranial aneurysms more commonly present with rupture and subarachnoid haemorrhage whereas internal carotid artery aneurysms tend to cause pressure symptoms including a pulsatile and sometimes painful mass in the neck or pharynx, ipsilateral Horner's syndrome, or compression of the lower cranial nerves. Extracranial vertebral artery aneurysms may cause pain in the neck and arm, a mass, spinal cord compression, and upper limb ischaemia (Catala *et al.* 1993).

Irradiation

Excessive irradiation of the head and neck can damage intra- and extracranial arteries, both large and small. Within the radiation field, months or years later a localized, stenotic, and sometimes apparently atheromatous, lesion may become symptomatic. There can be considerable fibrosis of the arterial wall and even aneurysm formation (Zuber *et al.* 1993; Griewing *et al.* 1995; Bitzer and Topka 1995; O'Connor and Mayberg 2000). The most common causes of this large artery variant are radiotherapy to the neck following laryngeal cancinoma, which leads to disease around the carotid bifurcation, and radiotherapy to pituitary tumours, which leads most commonly to disease in the basilar artery. Management is uncertain and the prognosis variable. Patients who have had radiotherapy for cerebral tumours more commonly develop a progressive small vessel vasculopathy.

35.4.6 Inflammatory vascular disease

There are a number of acute, subacute, and chronic inflammatory 'vasculitic' disorders of the arterial or venous wall (Table 35.7). These disorders may be associated with ischaemic stroke, intracranial haemorrhage, intracranial venous thrombosis (Section 35.15), or by a generalized encephalopathy. Angiographic appearances can be diagnostic, particularly in larger artery vasculitis, but angiography is not particularly sensitive and the abnormailities or non-specific and so diagnosis is often made on the basis of the clinical syndrome. Contrast-enhanced high-resolution MRI can be useful in showing contrast uptake in the walls of thickened and inflamed large cerebral arteries.

Giant cell arteritis

Giant cell arteritis is the most common vasculitic cause of stroke, and is associated particularly with posterior circulation ischaemia (Nesher 2000; Ronthal *et al.* 2003; Eberhardt and Dhadly 2007). Medium and large arteries are affected, especially branches of the external carotid artery, the ophthalmic artery, and the vertebral artery. The patients are elderly, with the diagnosis being rare under age 60 years. Malaise, polymyalgia, and other systemic symptoms

Table 35.7 Inflammatory vascular diseases causing stroke

Giant cell arteritis (Section 36.2.8)

Takayasu's disease (Section 36.2.9)

Systematic lupus erythematosus (Section 36.3.1)

Antiphospholipid antibody syndrome

Primary systematic vasculitis (Section 36.2.3)

Rheumatoid disease (Section 36.3.2)

Sjogren's syndrome (Section 36.3.3)

Behcet's disease (Section 36.5.1)

Relapsing polychondritis

Progressive systemic sclerosis (Section 36.3.4)

Sarcoid angiitis (Section 36.4)

Primary vasculitis of the central nervous system (Section 36.2.2)

Malignant atrophic papulosis

Acute posterior multifocal placoid pigment epitheliopathy

Buerger's disease

are frequently present. The ESR is usually raised, often to over 100 mm in the first hour (see Section 36.2.8).

Takayasu's disease

Takayasu's disease is a chronic vasculitis, histologically identical to giant cell arteritis but affecting only the aorta and large arteries arising from it, mainly in young Oriental women (Seko 2007) (Section 36.2.9). Systemic features are common: malaise, weight loss, arthralgia, fever. Treatment is with immunosuppression, which may result in remission, and occasionally surgical procedures may be undertaken (Liang and Hoffman 2005). The neurological complications reflect progressive narrowing and eventual occlusion of the large arteries in the neck: claudication of the jaw muscles, ischaemic oculopathy, syncope, seizures, confusion, boundary zone infarction, and rather rarely focal ischaemic stroke or TIAs (Hoffman et al. 2000). In addition, there may be ischaemia of the arms, and of the kidneys to cause hypertension, as well as ischaemic necrosis of the lips, nasal septum, and palate. Other causes of a similar aortic arch syndrome include advanced atheroma, giant cell arteritis, syphilis, subintimal fibrosis, arterial dissection, trauma, and coarctation.

Systemic lupus erythematosus

Systemic lupus erythematosus is more likely to cause a subacute or chronic generalized encephalopathy than symptomatic focal ischaemia (Mills 1994; Moore 1997; Jennekens and Kater 2002a, b; D'Cruz et al. 2007) (Section 36.3.1). The underlying vascular pathology where present, appears to be intimal proliferation rather than a vasculitis. The extracranial arteries are largely unaffected, but embolism from heart valve vegetations is quite common, particularly when there are circulating antiphospholipid antibodies (Mitsias and Levine 1994; Roldan et al. 1996). Intracranial venous thrombosis is rare (Vidailhet et al. 1990). In some patients with little clinical evidence of systemic lupus erythematosus, there is prominent livedo reticularis which, when associated with stroke, is referred to as *Sneddon's syndrome*, in which antiphospholipid antibodies are particularly common (Stockhammer et al. 1993, Kalashnikova et al. 1994; Boesch et al. 2003; Hilton and footitt 2003).

Antiphospholipid syndrome

Antiphospholipid syndrome is a constellation of various recurrent clinical events as well as specific immunological features: arterial and venous thrombosis including recurrent ischaemic stroke or transient ischaemic attack and intracranial venous thrombosis, migraine-like episodes, recurrent miscarriage, livedo reticularis, cardiac valvular vegetations, thrombocytopenia, false positive syphilis serology and persistently raised (> 20 units) circulating IgG anticardiolipin antibodies, and/or the circulating lupus anticoagulant, usually detected by prolongation of the activated partial thromboplastin time (Katzav et al. 2003; Brey 2005; Sanna et al. 2005; Lim et al. 2006; Merrill 2007). The antiphospholipid antibodies are found in some normal people, and in systemic lupus erythematosus, hence an isolated finding of a raised antibody level in a patient with stroke is of uncertain significance. Antibodies are not uncommonly present after acute stroke, but only where they remain on re-testing after 6 weeks should the diagnosis of antiphospholipid syndrome be made, particularly if other clinical features are lacking.

Primary systemic vasculitis

Primary systemic vasculitis is a group of related disorders including polyarteritis nodosa, Wegener's granulomatosis, the Churg–Strauss syndrome, and various hypersensitivity vasculitides (Section 36.2.3). They are very rare and have cerebrovascular consequences, similar to those of systemic lupus erythematosus (Futrell 1995; Savage et al. 1997; Ferro 1998). Stroke is usually lacunar (Reichart 2000) and there is often associated haematuria, eosinophilia, and circulating anti-neutrophil cytoplasmic antibodies.

Rheumatoid disease

Rheumatoid disease is rarely complicated by a systemic vasculitis (Section 36.3.2), which can involve the brain (Genta et al. 2006). Occasionally atlanto-axial dislocation causes symptomatic vertebral artery compression (Howell and Molyneux 1988).

Sjogren's syndrome

Sjogren's syndrome is occasionally complicated by systemic vasculitis causing focal cerebral ischaemia, global encephalopathy, and aseptic meningitis (Hietaharju et al. 1993; Bragoni et al. 1994; Delalande et al. 2004) (Section 36.3.3).

Behcet's disease

Neurological involvement in Behcet's disease (Section 36.5.1) may be subclassified into two major forms: a vascular-inflammatory process with focal or multifocal parenchymal involvement; and cerebral venous sinus thrombosis and intracranial hypertension. The vasculitis and meningitis may affect cerebral arteries, particularly in the posterior circulation, to cause ischaemic stroke, and possibly intracranial haemorrhage (Farah et al. 1998; Krespi et al. 2001; Siva et al. 2004; Borhani Haghighi et al. 2005).

Relapsing polychondritis

Relapsing polychondritis may be complicated by a generalized encephalopathy, stroke-like episodes, and ischaemic optic neuropathy as a result of systemic vasculitis (Stewart et al. 1988; Hsu et al. 2006).

Progressive systemic sclerosis

Progressive systemic sclerosis (Section 36.3.4) is hardly ever complicated directly by stroke, although a carotid and cerebral vasculopathy has been described (Heron *et al.* 1998; Lucivero *et al.* 2004).

Sarcoid angiitis

Sarcoid (Section 36.4) affects the cerebral vessels only rarely, usually causing a generalized encephalopathy rather than focal features due to ischaemia or haemorrhage (Zajicek 2000; Gullapalli and Phillips 2004; Spencer *et al.* 2005).

Primary vasculitis of the central nervous system

Primary vasculitis or isolated angiitis of the central nervous system is a very rare disorder that affects leptomeningeal, cortical, and sometimes spinal cord blood vessels (Section 36.2.2). It is 'isolated' in the sense that it is confined to the central nervous system, although histologically it is similar to sarcoid angiitis and it can occur in association with herpes zoster, HIV, and other infections. The course is subacute, often leading to death in weeks or months, with mental confusion and impairment, headache, vomiting, stroke-like episodes, and myelopathy. Systemic symptoms are very uncommon. Diagnosis is only really possible from meningeal/cortical biopsy (Hankey GJ 1991; Vollmer *et al.* 1993; MacLaren *et al.* 2005).

Malignant atrophic papulosis

Malignant atrophic papulosis, or Dego's disease, is a very rare syndrome consisting of crops of painless pinkish papules on the trunk and limbs that heal as distinctive circular porcelain-white scars. It may be complicated by ischaemic lesions in the gut, brain, spinal cord, and nerve roots due to endothelial proliferation in small arteries (Sotrel *et al.* 1983; Subbiah *et al.* 1996).

Acute posterior multifocal placoid pigment epitheliopathy

Acute posterior multifocal placoid pigment epitheliopathy is a rare and usually benign and self-limiting chorioretinal disorder with rapidly deteriorating central vision. However it can be complicated by systemic vasculitis, aseptic meningitis, and stroke (Comu *et al.* 1996; de Vries *et al.* 2006).

Buerger's disease

Buerger's disease, thromboangitis obliterans, is a rare inflammatory disorder of small- and medium-sized arteries and veins, chiefly of the limbs and almost never of the cerebral circulation, much more common in men than women and mainly seen in smokers (Calguneri *et al.* 2004; Olin and Shih 2006).

35.4.7 Haematological disorders

A number of haematological disorders may occasionally cause ischaemic stroke and transient ischaemic attack (Tatlisumak and Fisher 1996; Arboix and Besses 1997; Markus and Hambley 1998; Matijevic and Wu 2006) (Table 35.8).

Polycythaemia

Polycythaemia is usually defined as a haematocrit above 0.50 in males and 0.47 in females. Polycythaemia rubra vera or primary polycythaemia, a myeloproliferative disorder, may be complicated by transient ischaemic strokes, ischaemic stroke, or intracranial venous thrombosis (Silverstein *et al.* 1962; Pearson and Wetherley-Mein 1978; Markus and Hambley 1998). Ischaemic complications may occur because the platelet count is raised and platelet activity enhanced, or because of increased whole-blood viscosity. Paradoxically there may also be a haemostatic defect as a result of defective platelet function, so causing intracranial haemorrhage.

Table 35.8 Haematological disorders causing ischaemic stroke

Polycythaemias
Essential thrombocythaemia
Leukaemia
Sickle-cell disease/trait and other haemoglobinopathies
Iron deficiency anaemia
Paraproteinaemias
Paroxysmal nocturnal haemoglobinuria
Thrombotic thrombocytopenic purpura
Disseminated intravascular coagulation
Thrombophilias and other causes of 'hypercoagulability'

Secondary polycythaemia is due to a raised red cell mass, and may result from chronic hypoxia, smoking, congenital cyanotic heart disease, renal tumour, or cerebellar haemangioblastoma. Probably it is also a risk factor for stroke.

Essential thrombocythaemia

Essential thrombocythaemia, or idiopathic primary thrombocytosis, is another myeloproliferative disorder in which the platelet count is raised, usually to over $1000 \times 10^9/l$, but sometimes only to $500 \times 10^9/l$. Secondary thrombocytosis occurs in malignancy, splenectomy, hyposplenism, surgery, trauma, haemorrhage, iron deficiency, infections, polycythaemia rubra vera, myelofibrosis, and the leukaemias. There is a tendency for arterial and venous thrombosis and, paradoxically, intracranial haemorrhage because the platelets are haemostatically defective (Arboix *et al.* 1995; Harrison *et al.* 1998; Mosso *et al.* 2004; Ogata *et al.* 2005).

Leukaemia and lymphoma

Leukaemia and lymphoma may cause intracranial haemorrhage particularly in acute myeloid leukaemia, most commonly in association with acute disseminated intravascular coagulation although other haemostatic defects or central nervous system infiltration may be responsible (Rogers 2003; Glass 2006). The haemorrhage is often fulminant with bleeding usually occurring in the brain or subdural compartment, and rarely in the subarachnoid space. Occasionally, cerebral venous thrombosis or arterial occlusion may occur. *Malignant angioendotheliosis*, an intravascular lymphoma, is a very rare cause of stroke-like episodes fading into a progressive global encephalopathy (Chapin *et al.* 1995; Zukerman 2006).

Sickle-cell disease and other haemoglobinopathies

Sickle-cell disease and rarely other haemoglobinopathies may be complicated by ischaemic stroke or, sometimes, intracranial haemorrhage (Razvi and Bone 2006; Switzer *et al.* 2006). Patients are usually homozygote children although sometimes a sickle-cell crisis in an adult heterozygote, provoked by hypoxia, can be responsible. Small and large arteries, as well as veins, develop a fibrous vasculopathy and are occluded by thrombi as a result of the abnormally rigid red blood cells and raised whole blood viscosity, thrombocytosis, and impaired fibrinolytic activity.

Iron deficiency anaemia

If severe, iron deficiency anaemia causes non-specific neurological symptoms which are presumably hypoxic in origin, such as faintness, poor concentration, giddiness, tiredness, and general weakness. Just occasionally, transient ischaemic attacks and even ischaemic stroke seem to be provoked by profound anaemia, but usually only in association with severe extracranial occlusive arterial disease, or thrombocytosis (Akins *et al.* 1996; Keung and Owen 2004).

Paraproteinaemias

Multiple myeloma and macroglobulinaemia, are associated with anaemia because of defective erythropoesis and this causes non-specific neurological symptoms. A haemostatic defect due to reduced platelet number and perhaps reactivity as a result of complicating uraemia, may cause intracranial haemorrhage. However, most of the 'cerebral' features of these patients can be explained by the 'hyperviscosity syndrome' which is characterized by headache, ataxia, diplopia, dysarthria, lethargy, drowsiness, poor concentration, visual blurring, and deafness. The same syndrome can be seen in primary polycythaemia or leukaemias. Arterial or venous cerebral infarction may occur and at post-mortem the microcirculation is occluded with acidophilic material thought to be precipitates of the abnormal proteins (Davies-Jones 1995). It is exceptional for patients with neurological involvement not to have a raised ESR.

Paroxysmal nocturnal haemoglobinuria

Paroxysmal nocturnal haemoglobinuria is a very rare acquired disorder in which haemopoetic stem cells become peculiarly sensitive to complement-mediated lysis. Venous, and perhaps arterial, thrombosis occurs in the brain and elsewhere. Almost always patients are anaemic at neurological presentation and there may be a history of dark urine, evidence of haemolysis, and a low platelet and granulocyte count (Al-Hakim *et al.* 1993; Audebert *et al.* 2005).

Thrombotic thrombocytopenic purpura

Thrombotic thrombocytopenic purpura is a rare acute or subacute disease in adults, rather similar to the haemolytic uraemic syndrome in children. Haemorrhagic infarcts due to platelet microthrombi occur in many organs, and in the brain they may cause stroke-like episodes (Matijevic and Wu 2006). More commonly, the presentation is with a global encephalopathy on the background of systemic malaise, fever, skin purpura, renal failure, haematuria, and proteinuria. The blood film shows thrombocytopenia, haemolytic anaemia, and fragmented red cells. The differential diagnosis includes systemic lupus erythematosus, infective endocarditis, idiopathic thrombocytopenia, heparin induced throbocytopenia with thrombosis, non-bacterial thrombotic endocarditis, and disseminated intravascular coagulation.

Disseminated intravascular coagulation

Widespread haemorrhagic brain infarcts and intracranial haemorrhages tend to cause an acute or subacute global encephalopathy rather than stroke-like episodes. The diagnosis is confirmed by a low platelet count, low plasma fibrinogen, and raised fibrin degradation products and D-dimer.

Thrombophilias

Thrombophilias and other causes of 'hypercoagulability' are very rare causes of stroke (Matijevic and Wu 2006). Antithrombin III deficiency, protein S deficiency, protein C deficiency, activated protein C resistance due to factor V Leiden mutation, and plasminogen abnormality or deficiency can all cause peripheral and intracranial venous thrombosis. This is usually recurrent and often

a family history can be obtained. Also these disorders may cause arterial thrombosis, although paradoxical embolism must always be considered as an explanation. However, such deficiencies may be incidental and it is important to exclude other causes of stroke.

35.4.8 Miscellaneous rare causes

There are many other proven but relatively rare causes of cerebral ischaemia or infarction, some of the more important of which are listed in Table 35.3 and discussed below.

Pregnancy and the puerperium

Pregnancy is complicated by stroke in about 10 per 100 000 deliveries in developed countries, about twice the background rate (Turan and Stern 2004; Helms and Kittner 2005; Hender *et al.* 2006). The risks of ischaemic stroke, intracerebral hemorrhage, and subarachnoid hemorrhage are not increased in the 9 months of gestation except for a high risk in the 2 days prior and 1 day postpartum. The remaining 6 weeks postpartum also have an increased risk of ischaemic stroke and intracerebral hemorrhage, though less than the peripartum period. Although there are some rare causes of stroke specific to pregnancy and the postpartum period, eclampsia, cardiomyopathy, postpartum cerebral venous thrombosis, and, possibly, paradoxical embolism warrant special consideration. The diagnostic and therapeutic approaches to stroke during pregnancy and the postpartum period are similar to the approaches in the nonpregnant woman with some minor modifications based on consideration of the welfare of the foetus. There is a theoretical risk of magnetic resonance imaging exposure during the first and second trimester but the benefit to the mother of reliable diagnosis will often outweigh the risk.

Causes and mechanisms particularly relevant to pregnancy include intracranial venous thrombosis, arterial dissection during labour, acute middle cerebral or other large artery occlusion, low-flow infarction and disseminated intravascular coagulation complicating eclampsia, vasoconstriction secondary to drugs, infective endocarditis, peripartum cardiomyopathy, sickle cell crisis, and intracranial haemorrhage(s) due to eclampsia, anticoagulants, rupture of a pre-existing aneurysm, or vascular malformation. Metastases of choriocarcinoma can present with stroke-like episodes and on CT look remarkably like primary intracerebral haemorrhages. However, many cases remain cryptogenic and the risk of recurrent stroke in a future pregnancy is unknown.

Migraine

Numerous studies have shown that migraine is an independent risk factor for stroke both during, and remote from, the migraine attack (Chang *et al.* 1999; Donaghy *et al.* 2002; Bousser and Welch 2005; Tietjen 2005). A cerebral infarction can occur during a migraine attack, cerebral ischaemia can induce migrainous symptoms. The term 'migrainous stroke' should be reserved for a persisting focal neurological deficit that starts during a typical migrainous aura, with or without headache, and that mimics the symptomatology of previously experienced auras (Bousser *et al.* 1985). Such migrainous strokes usually cause a homonymous hemianopia or focal sensory deficit without persisting disability, and do not appear to recur very often (Hoekstra-van Dalen *et al.* 1996).

Sometimes arterial occlusion is demonstrated by angiography and the cause is postulated to be *in situ* thrombosis complicating vasospasm. No provoking factors are known. Other possible causes of stroke in the context of headache must be considered: carotid dissection, antiphospholipid antibody syndrome, mitochondrial cytopathy, ruptured vascular malformation, cerebral autosomal dominant arteriopathy with subcortical infarcts and leukoencephalopathy. Migraine auras without headache may be confused with transient ischaemic attack (Section 35.6.2).

Epidemiological studies suggest the existence of close but complex relationships between estrogens, migraine, and stroke in women before menopause (Chang *et al.* 1999; Donaghy *et al.* 2002; Bousser 2004). Migraine, particularly without aura, is strongly influenced by estrogens as illustrated by the frequency of onset at puberty, of menstrual migraine, and of improvement during pregnancy. Migraine, particularly with aura, is a risk factor for ischaemic stroke with a relative risk of 3, further increased by tobacco smoking and oral contraceptive use (Chang *et al.* 1999; Donaghy *et al.* 2002). The pathophysiological mechanism underlying these close relationships remains unknown. In practice, given the very low absolute risk of stroke in young women, there is no systematic contraindication to oral contraceptive use in young female migraineurs but rather a firm recommendation for no smoking and for the use of low-estrogen-content pills or progestogens only, particularly in migraine with aura. Both ischaemic stroke and migraine can be consequences of many underlying vascular disorders. Elevated blood pressure, an important stroke risk factor, is less common in migraineurs.

Acquired antiphospholipid antibodies, not clearly a cause of migraine per se, may increase the risk of infarction in migraineurs. Hereditary conditions, including cerebral autosomal dominant arteriopathy with sub-cortical infarcts and leukoencephalopathy, also known by the acronym CADASIL, mitochondrial myopathy, encephalopathy, lactic acidosis and stroke, often abbreviated to MELAS and hereditary haemorrhagic telangiectasia, appear to predispose to both migraine and stroke.

Purported mechanisms for migraine-associated stroke include:

◆ involvement of the vasculature, including vasospasm, arterial dissection, and small vessel arteriopathy;

◆ hypercoagulability, involving elevated von Willebrand Factor, or platelet activation; and

◆ elevated risk of cardioembolism through patent foramen ovale, or from atrial septal aneurysm.

Triptans and ergotamines, used to treat acute migraine attacks, appear to be safe in low-risk populations, but should be avoided in haemiplegic migraine, basilar migraine, and in patients with prior cerebral or cardiac ischaemia.

Drug abuse

The use of certain recreational drugs shows a marked temporal association with the onset of both haemorrhagic and ischaemic strokes, often within minutes to 1 h after the use of the drug (Neimann *et al.* 2000; O'Connor *et al.* 2005). Acute, severe elevation of blood pressure, cardiac dysrhythmias, cerebral vasospasm, and embolization due to foreign material injected with the diluents can cause stroke with a close temporal association with taking drugs such as cocaine, amphetamines, and opiates (Section 5.3). Stroke can also result from less acute pathologies, such as vasculitis and embolization due to infective endocarditis or dilated cardiomyopathy. Rupture of aneurysms and arteriovenous malformations have been detected in up to half of the patients with haemorrhagic stroke due to cocaine abuse. Cocaine causes cerebral infarction, intracerebral and subarachnoid haemorrhage, usually within hours of use (Section 5.3.2). Possible aetiological mechanisms include an acute rise in blood pressure, an underlying vascular malformation or aneurysm, cardiac arrhythmia, cardiomyopathy and cerebral embolism, and cerebral vasoconstriction.

Amphetamines can cause a small vessel vasculopathy leading to intracerebral haemorrhage or infarction (Heye and Hankey 1996) (Section 5.3.4). Other sympathomimetic drugs such as ephedrine, phenylpropanolamine, fenfluramine, and phentermine may cause stroke by similar mechanisms, as can 'ecstasy' (Wen *et al.* 1997). Intravenous drug users are at high risk of endocarditis, often due to Staphylococcus aureus.

Cancer

Cancer may cause stroke (Rogers 2003) through:

◆ embolism of non-infected heart valve vegetations: non-bacterial thrombotic, or marantic, endocarditis;

◆ infection: fungi, herpes zoster, bacterial endocarditis;

◆ tumour emboli;

◆ haemorrhage into primary tumours: malignant astrocytoma, oligodendroglioma, medulloblastoma, haemangioblastoma;

◆ metastases: melanoma, germ cell tumours, choriocarcinoma, lung, hypernephroma;

◆ haemostatic failure: leukaemias, etc, hyperviscosity syndrome or 'hypercoagulability';

◆ disseminated intravascular coagulation; and

◆ intracranial venous thrombosis. Irradiation damage or neoplastic compression or invasion of neck arteries are both unusual causes of ischaemic stroke.

Peri-operative stroke

Peri-operative stroke complicates under 2 per cent of non-cardiac surgical procedures. It can be due to hypotension and boundary zone infarction, trauma to and dissection of neck arteries, paradoxical embolism, fat embolism, infective endocarditis, myocardial infarction, atrial fibrillation, and a haemostatic defect caused by antithrombotic drugs or disseminated intravascular coagulation. It is more common in patients with previous strokes, other manifestations of vascular disease, and chronic obstructive lung disease (Limburg *et al.* 1998). Simultaneous carotid endarterectomy and coronary bypass grafting is associated with 10–15 per cent risk of death, stroke, or myocardial infarction and the risk is higher in those with bilateral as opposed to unilateral carotid disease (Naylor *et al.* 2003a, b).

Chronic meningitis

Meningitis caused by tuberculous, syphilitic, and fungal infections may involve the arteries at the base of the brain, or the perforating arteries, and so be complicated by ischaemic stroke and intracranial haemorrhage. Very occasionally acute local infections such as tonsillitis, pharyngitis, lymphadenitis can cause inflammation and secondary thrombosis in the carotid artery in the neck. Otitis media or mastoiditis may cause dural sinus thrombosis. Cerebral arterial and venous thrombosis may result from bacterial meningitis, ophthalmic herpes zoster, chicken pox, leptospirosis, HIV infection, cat scratch disease, neurotrichinosis, and possibly borreliosis.

Inflammatory bowel disease

Both ulcerative and Crohn's colitis may occasionally be complicated by intracranial venous thrombosis, arterial occlusion, and intra-cerebral haemorrhage (Lossos *et al.* 1995). Mechanisms include thrombocytosis, hypercoagulability, immobility and paradoxical embolism, vasculitis, and dehydration. The bowel disease is not necessarily severe at the time of the stroke. Coeliac disease can also be complicated by a cerebral vasculitis but this often presents with an encephalopathy rather than a stroke (Mumford *et al.* 1996).

Homocystinuria

Homocystinuria is an autosomal recessive inborn error of metabolism, is complicated by cerebral arterial or venous thrombosis (Schimke *et al.* 1965; Visy *et al.* 1991; Rubba *et al.* 1994). Heterozygotes may have an increased risk of vascular disease.

Fabry's disease

Fabry's disease is occasionally complicated by ischaemic stroke, particularly in the vertebrobasilar territory, but usually not until after other more common features are well established (Meschia *et al.* 2005; Razvi and Bone 2006; Rolfs *et al.* 2005) (Section 21.8.5).

Mitochondrial cytopathy

Mitochondrial cytopathy may present with stroke-like episodes often complicated by epilepsy and encephalopathy. CT scanning often shows hypodensities, particularly in the occipital regions, and calcification of the basal ganglia (Fig. 35.28). MRI shows T2-weighted hyperintensities in the temporo-occipito-parietal regions which do not correspond to classical vascular territories (Fig. 35.29) and which may disappear on subsequent scans. There are often other clinical features, such as calcification of the basal ganglia, migraine, episodic vomiting, short stature, sensorineural deafness, diabetes, and learning disability (Section 10.5). A particular syndrome is known as MELAS: Mitochondrial Encephalopathy, Lactic Acidosis and Stroke-like episodes. The blood and CSF lactate are usually raised, most patients have an abnormal muscle biopsy, and diagnosis can often be made by detection of the relevant genetic mutations.

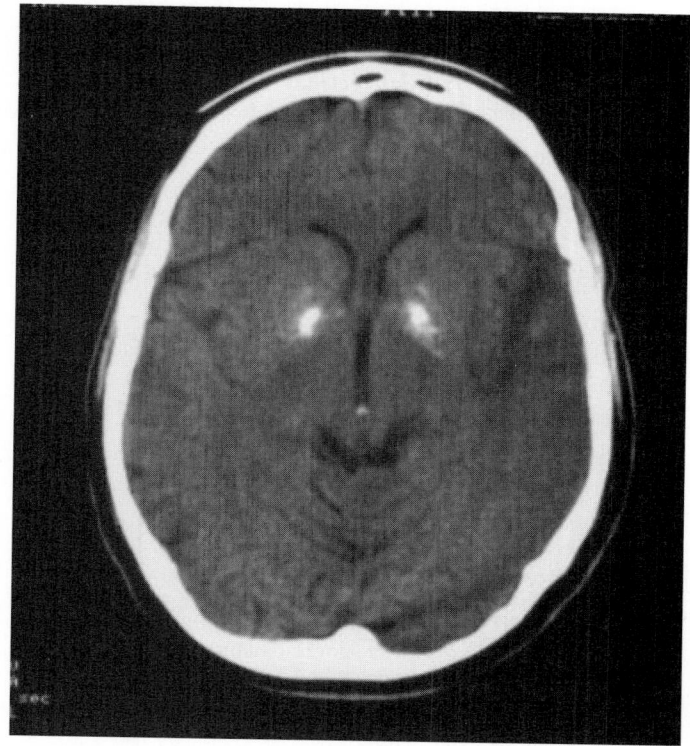

Fig. 32.28 Mitochondrial cytopathy. CT brain scan of a 40-year-old man showing calcification of the basal ganglia and hypodensity in the left temporal lobe.

Hypoglycaemia

Hypoglycaemic drugs, and rarely an insulinoma, are a well-recognized but rare cause of transient focal neurological episodes, particularly right hemiplegia and aphasia masquerading as transient ischaemic attacks. These episodes tend to occur on waking in the morning, or after exercise. By the time the patient is seen, the blood glucose may well have returned to normal. Persisting focal deficits seem to be unusual (Malouf and Brust 1985; Wallis *et al.* 1985; Service 1995; Shanmugam *et al.* 1997).

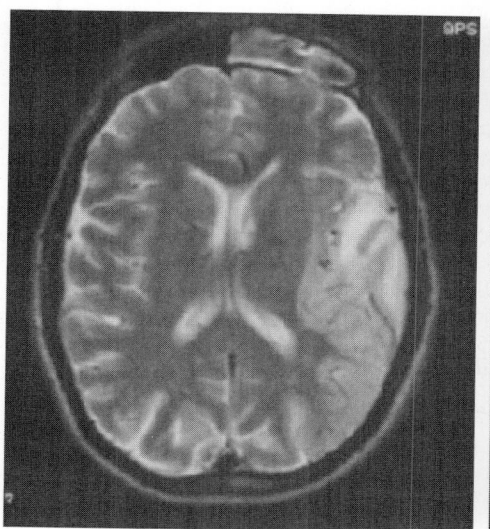
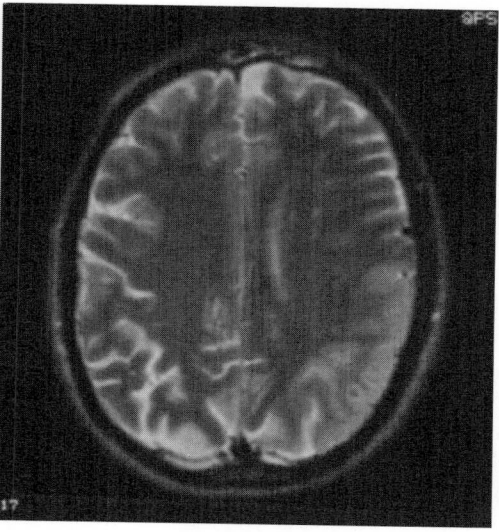

Fig. 35.29 Mitochondrial cytopathy. T2-weighted MR brain images from a patient showing primarily cortical hyperintensity in the left temporo-parietal region not typical of middle cerebral artery branch infarction.

Hypercalcaemia and Hyponatraemia

Hypercalcaemia (Longo and Witherspoon 1980) and, hyponatraemia (Ruby and Burton 1977; Berkovic *et al.* 1984) have been reported to cause transient ischaemic attack-like episodes.

Fibrocartilaginous embolism

Fibrocartilaginous embolism is a rare and curious disorder where fibrocartilaginous emboli, presumably from degenerative intervertebral disc material, are found in various organs, the spinal cord more often than the brain (Freyaldenhoven *et al.* 2001).

Snake bite

Injection of venom may cause intracranial haemorrhage as a consequence of defibrination and other haemostatic defects, and rarely ischaemic stroke (Bashir and Jinkins 1985) (Section 5.12.4).

Fat embolism

Fat embolism, which usually occurs following long bone fracture or surgery, most commonly causes a global encephalopathy, but on occasion there may be focal features, presumably reflecting local ischaemia (Jacobson *et al.* 1986; van Oostenbrugge *et al.* 1996).

Epidermal naevus syndrome

Epidermal naevus syndrome, a sporadic neurocutaneous disorder, can be complicated by stroke (Dobyns and Garg 1991).

Susac's syndrome

Susac's syndrome is a rare triad of branch retinal artery occlusions, hearing loss and microangiopathy of the brain causing a subacute encephalopathy, almost always in women (Gross *et al.* 2004).

35.5 Causes of spontaneous intracranial haemorrhage

Spontaneous intracranial haemorrhage occurs:

♦ primarily within the brain: primary intracerebral haemorrhage;

♦ into the subarachnoid space: subarachnoid haemorrhage (discussed in Section 35.16);

♦ sometimes into the ventricles: intraventricular haemorrhage;

♦ into the subdural space: subdural haemorrhage.

The exact site of origin of bleeding may be unclear: a saccular aneurysm may rupture into the brain as well as into the subarachnoid space, or disruption of a small perforating artery may cause intraventricular haemorrhage as well as a basal ganglia haematoma. Even at post mortem there may be uncertainty because the source of the haemorrhage may well have been destroyed, particularly if small, for instance a tiny intracranial vascular malformation. The causes of intracranial haemorrhage are much the same, whatever the primary site of the bleeding, although their relative frequency varies with the site (Table 35.9).

35.5.1 Primary intracerebral haemorrhage

Primary intracerebral haemorrhage is more common than subarachnoid haemorrhage, and its incidence increases with age (Fig. 35.2). It is probably more frequent in south east Asian, Japanese, and Chinese populations than in Caucasians. It is usually caused by intracranial small vessel disease associated with hypertension, cerebral amyloid angiopathy, or intracranial vascular malformations (Sutherland and Auer 2006). Less common causes

Table 35.9 Causes of spontaneous intracranial haemorrhage

Hypertension; chronic or acute
Aneurysms
 Saccular
 Atheromatous
 Mycotic
 Myxomatous
 Dissecting
Cerebral amyloid angiopathy (Section 35.5.4)
Intracranial vascular malformations (Sections 9.4.1and 35.5.4)
 Arteriovenous; cerebral or dural
 Venous
 Cavernous
 Telangiectasis
Haemostatic failure
 Haemophilia and other coagulation disorders
 Thrombocytopenia
 Thrombotic thrombocytopenic purpura
 Anticoagulation
 Therapeutic thrombolysis
 Antiplatelet drugs
 Polycythaemia rubra vera
 Essential thrombocythaemia
 Paraproteinaemias
 Disseminated intravascular coagulation
 Renal failure
 Liver failure
 Snake bite (Section 5.12.4)
Inflammatory vascular disease (Section 35.4.6)
Haemorrhagic transformation of cerebral infarction, venous more often than arterial
Intracranial venous thrombosis (Section 38.15)
Sickle-cell disease/trait
Moyamoya syndrome (Section 35.4.5)
Carotid endarterectomy (Section 35.14.4)
Posterior fossa and other intracranial surgery
Delayed post-traumatic 'spat-apoplexie'
Alcoholic binge
Wernicke's encephalopathy (Section 5.2.3)
Vascular tumours
 Melanoma
 Choriocarcinoma
 Malignant astrocytoma (Section 27.8.1)
 Oligodendroglioma (Section 27.8.1)
 Medulloblastoma (Section 27.8.2)
 Haemangioblastoma (Section 11.4.2)
 Choroid plexus papilloma
 Hypernephroma
 Endometrial carcinoma
 Bronchogenic carcinoma
Drug abuse (Section 5.3.1)
Infections
 Herpes simplex (Section 42.4.1)
 Leptospirosis (Section 42.5.3)
 Anthrax (Section 5.8.6)
 Chronic meningitis
Scorpion bite (Section 5.12.5)
Silastic dural substitute

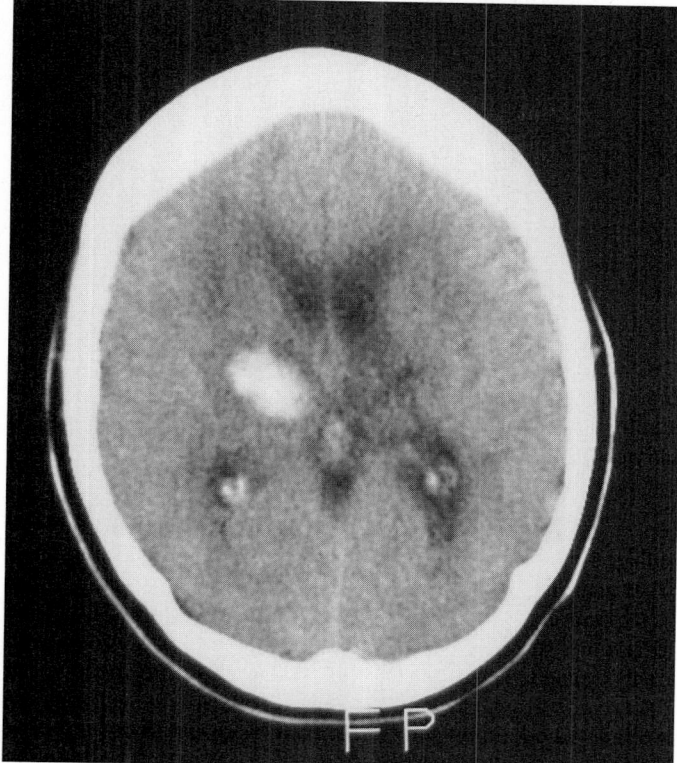

Fig. 35.30 A typical deep 'hypertensive' primary intracerebral haemorrhage. CT brain scan.

include saccular aneurysms, haemostatic defects, particularly those induced by anticoagulation or therapeutic thrombolysis, and possibly antiplatelet drugs, infective endocarditis, cerebral vasculitis, and drug abuse (Neiman *et al*. 2000; O'Connor *et al*. 2005).

The site of primary intracerebral haemorrhage provides information as to the cause: 'hypertensive' haemorrhages (Fig. 35.30) tend to occur in the basal ganglia, thalamus, and pons, while lobar haemorrhages are more often due to cerebral amyloid angiopathy, vascular malformations, and haemostatic failure (Dickinson 2001; Smith and Eichler 2006; Sutherland and Auer 2006). Multiple haemorrhages suggest certain specific causes (Table 35.10). Rarely, primary intracerebral haemorrhage is familial (Table 35.2).

The haematoma continues to expand after stroke onset, frequently causing further deterioration (Brott *et al*. 1997; Leira *et al*. 2004). Some brainstem haemorrhages evolve subacutely, particularly

Table 35.10 Causes of multiple spontaneous intracerebral haemorrhages

Cerebral amyloid angiopathy
Metastases
Haemostatic defect
Thrombolytic drugs
Intracranial venous thrombosis
Inflammatory vascular disease
Intracranial vascular malformations
Malignant hypertension
Eclampsia
Multiple haemorrhagic infarcts (usually embolic from the heart)
Occult head injury
Drug abuse

those caused by a vascular malformation (O'Laoire *et al*. 1982; Howard 1986). Any large haematoma may cause brain shift, transtentorial herniation, brainstem compression, and raised intracranial pressure. Haematomas in the posterior fossa are also particularly likely to cause obstructive hydrocephalus. Rupture into the ventricles, or on to the surface of the brain is common causing blood to appear in the subarachnoid space.

35.5.2 Primary intraventricular haemorrhage

Primary intraventricular haemorrhage is very unusual, except in premature babies. In adults, a cause is not always found. Some may be due to a vascular malformation in the ventricular wall (Gates *et al*. 1986; Darby *et al*. 1988). The clinical features are so similar to subarachnoid haemorrhage that it can only be differentiated on CT or at post-mortem.

35.5.3 Subdural haemorrhage

Subdural haemorrhage is usually traumatic rather than spontaneous, although trauma can be either very minor or unremembered in elderly patients and alcoholics (Section 25.2.6). Rupture causes include:

- rupture of a vascular malformation in the dura;
- rupture of a peripheral aneurysm, mycotic more likely than saccular;
- a haemostatic defect, particularly therapeutic anticoagulation;
- a superficial cerebral tumour can be responsible; and
- as a very rare complication of lumbar puncture or spontaneous intracranial hypotension (Section 18.6.7).

Often no convincing cause is found. Acute subdural haemorrhage appears hyperintense on CT brain scan whereas chronic subdural haematomas appear hypodense (Fig. 35.31). Haematomas of intermediate age, approximately 4–6 weeks, are often isodense to grey matter on CT and can be overlooked.

35.5.4 Specific causes of intracerebral haemorrhage

Some of the many potential causes of intracerbral haemorrhage (Table 35.9) are considered in more detail below.

Chronic hypertension

Hypertension causes thickening and disruption of the walls of the small arteries which perforate the base of the brain, particularly the lenticulostriate arteries in the region of the basal ganglia. It is rupture of these abnormal vessels which is thought to cause hypertensive primary intracerebral haemorrhage, although it is almost impossible to prove a cause and effect relationship in individuals because the haemorrhage destroys the exact site of the bleeding (Takebayashi and Kaneko 1983). In practice, the clinical diagnosis of 'hypertensive' primary intracerebral haemorrhage is based on the lack of any alternative explanation in a patient known to have had hypertension, or who clearly has evidence of hypertensive organ damage. However, it seems very likely that other factors, such as cerebral amyloid angiopathy, may interact with hypertension to cause primary intracerebral haemorrhage in a particular individual.

Cerebral amyloid angiopathy

Cerebral amyloid angiopathy is an organ-specific form of amyloid deposition in small and medium arteries, and less commonly veins, of the cerebral cortex and meninges, particularly in the elderly.

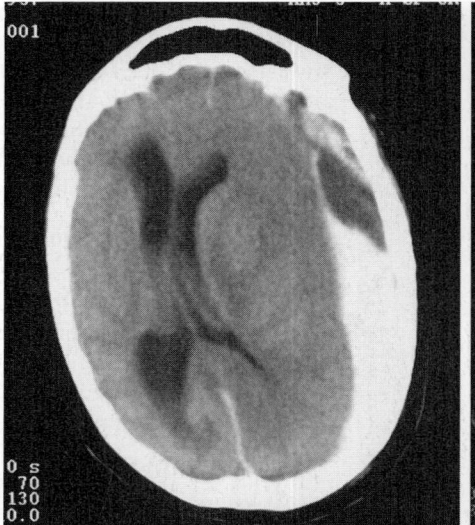
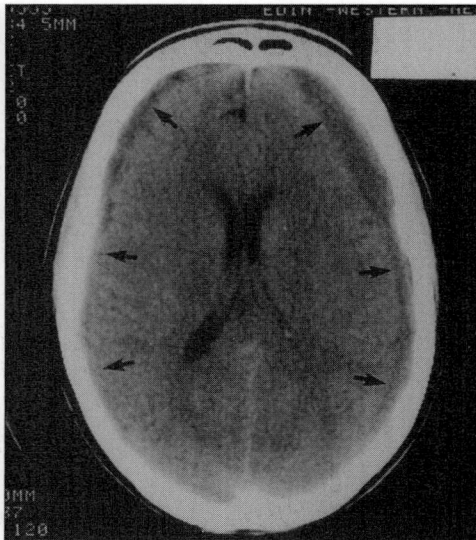

Fig. 35.31 Acute hyperintense left subdural haematoma with mass effect (left) and chronic bilateral hypodense subdural haematoma (arrows) (right). CT brain scans.

Frequently, there is additional subcortical small vessel disease and demyelination. Cerebral amyloid angiopathy can be associated with Alzheimer's disease, Down's syndrome, cerebral vasculitis, cerebral irradiation, and dementia pugilistica. It is thought to be the cause of lobar haemorrhages in the elderly, which are often multiple and recurrent, and possibly of subarachnoid haemorrhage (Smith and Eichler 2006; Thanvi and Robinson 2006). Patients, with or without lobar haemorrhage, may have progressive dementia and a history of minor stroke-like episodes and transient ischaemic attacks, and even focal epileptic seizures. These phenomena are possibly due to small primary intracerebral haemorrhages, known as microbleeds, which are undetectable except by gradient-echo MRI (Fig. 35.32) (Zhang-Nunes *et al.* 2006; Maia *et al.* 2007). Dominantly inherited forms of cerebral amyloid angiopathy cause primary intracerebral haemorrhage in young adults in Iceland (Jensson *et al.* 1987) and in middle-aged adults in Holland (Bornebroek *et al.* 1997), and a syndrome of progressive dementia, ataxia, and spasticity-associated sometimes with stroke in the United Kingdom (Plant *et al.* 1990).

Intracranial vascular malformations

Intracranial vascular malformations are uncommon, probably congenital, and sometimes familial (Byrne 2005). Those in the dura, draining into the sinuses rather than cerebral veins, can also be caused by skull fracture, craniotomy, or dural sinus thrombosis The overall intracranial vascular malformations detection rate is about three per 100 000 population per annum and the prevalence is about 20 per 100 000 (Brown *et al.* 1996). There are four main types:

Arteriovenous malformations

Arteriovenous malformations present most commonly and consist of an abnormal fistulous connection(s) between one or more hypertrophied feeding arteries and dilated draining veins (Fig. 35.33) (Clatterbuck *et al.* 2005). The blood supply is derived from one cerebral artery or, more often, several, sometimes with a contribution from branches of the external carotid artery. Arteriovenous

malformations vary from a few millimetres to several centimetres in diameter. About 15 per cent are associated with aneurysms on their feeding arteries. Some grow during life but a few shrink or even disappear, and some are multiple. These fistulae occur in or on the brain, or in the dura of the intracranial sinuses.

Arteriovenous malformations can present at almost any age with:

◆ partial or secondarily generalized epileptic seizures;

◆ haemorrhage which is more often intracerebral than subarachnoid or subdural;

◆ as a mass lesion;

◆ with a carotico-cavernous fistula due to a dural arteriovenous malformation;

◆ transient ischaemic attack-like episodes;

◆ with a self-audible bruit;

◆ with the syndrome of benign intracranial hypertension due to increased pressure in cerebral draining veins or sinuses; particularly if a dural arteriovenous malformation is near the transverse/sigmoid sinus and petrous bone; and

◆ with high output cardiac failure in neonates and infants.

Headache, although common, is not by itself diagnostically helpful and may well be a coincidence. Rarely, a bruit can be heard over the skull or orbits. A brainstem arteriovenous malformation can present like multiple sclerosis, with fluctuating symptoms and signs of brainstem dysfunction, perhaps due to recurrent haemorrhage.

CT scan may show calcification and rather non-specific hypo- or hyperdensity, while an enhanced scan is likely to show the dilated vessels of large malformations. MRI is more sensitive, showing evidence of old haemorrhage and vascular flow voids (Fig. 35.34). Angiography is the definitive investigation but even this can be normal with small malformations.

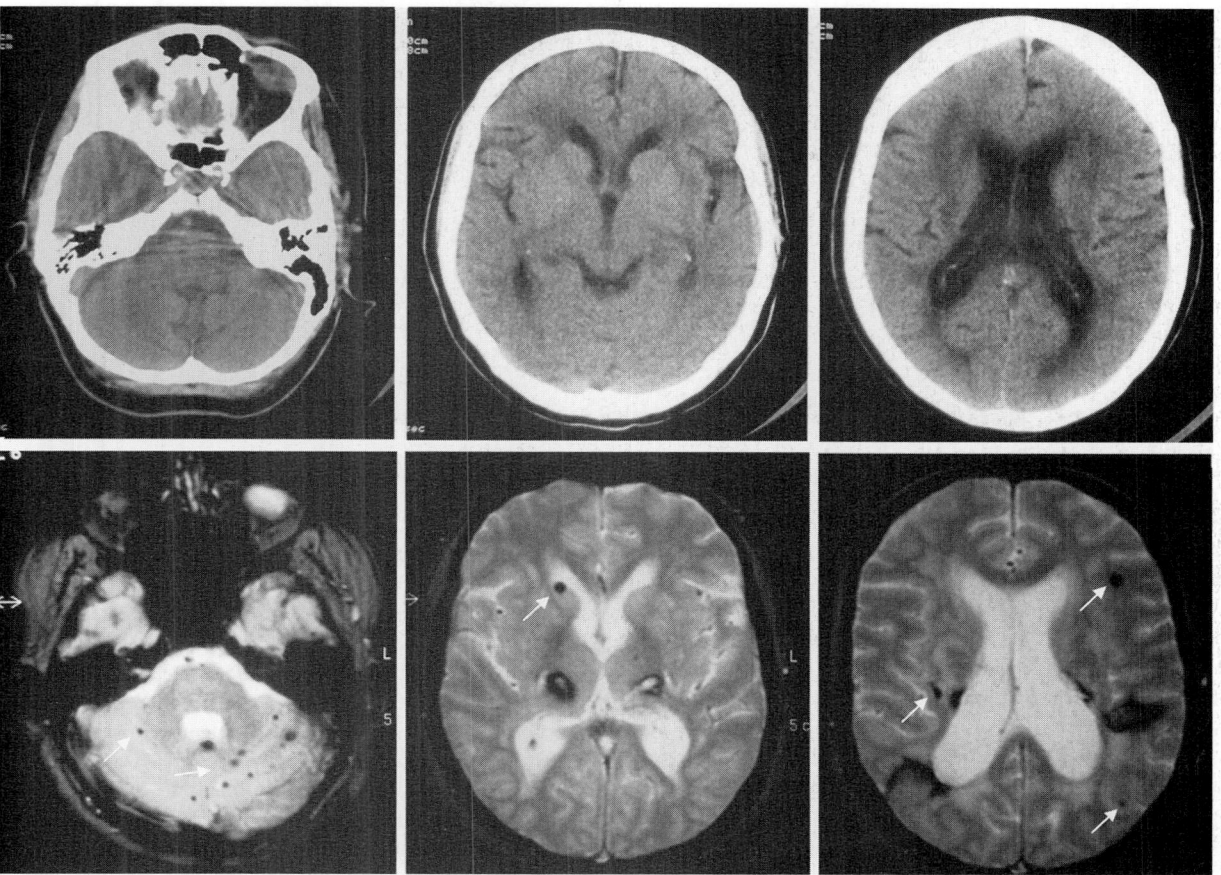

Fig. 35.32 Cerebral amyloid angiopathy. CT brain scans (top) from a patient showing only leukoaraiosis, whereas gradient echo MR brain imaging performed on the same day (below) shows evidence several previous intracerebral haemorrhages and multiple microbleeds (arrows).

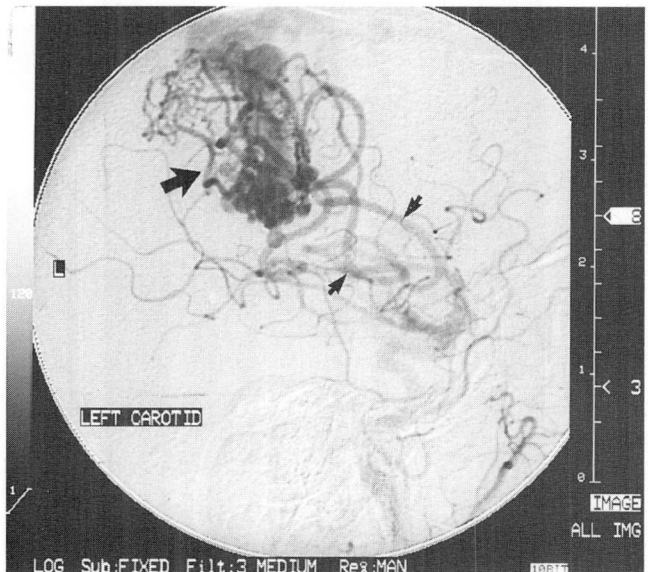

Fig. 35.33 Large arteriovenous malformation (arrow). Intracranial view of selective carotid angiogram showing supply by branches of the middle cerebral artery (small arrows).

Venous malformations

Venous malformations consist of collections of venous channels and a large draining vein. Most are asymptomatic but they can possibly present with haemorrhage into the ventricles, or seizures. On contrast CT, the draining vein may appear as a linear enhancing streak (Fig. 35.35), but a flow void on MRI is more sensitive. The definitive diagnosis is made on the venous phase of a cerebral angiogram.

Cavernous malformations

Cavernous malformations, or cavernomas, are sharply circumscribed collections of thin-walled sinusoidal vessels lined with a single layer of endothelium without intervening brain parenchyma or identifiable mature vessel wall elements. They are sometimes multiple and occasionally familial (Labauge *et al.* 2007). Most are asymptomatic and picked up incidentally on MRI. They can present with seizures, recurrent subacute brainstem syndromes, a mass lesion, and less commonly with haemorrhage. The angiogram is usually normal but CT can show a hypo- or hyperdense area that may enhance, perhaps with calcification, usually without surrounding oedema or mass effect. MRI reveals sharply circumscribed lesions, typically with evidence of haemosiderin as a result of old but asymptomatic haemorrhage (Fig. 35.36). However, they cannot

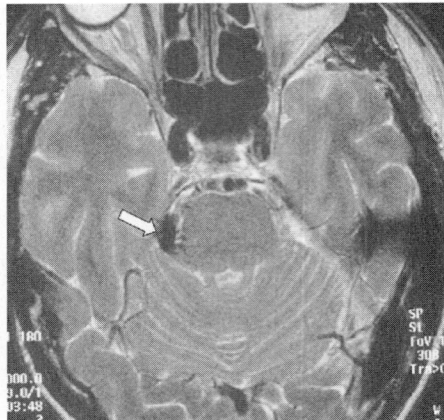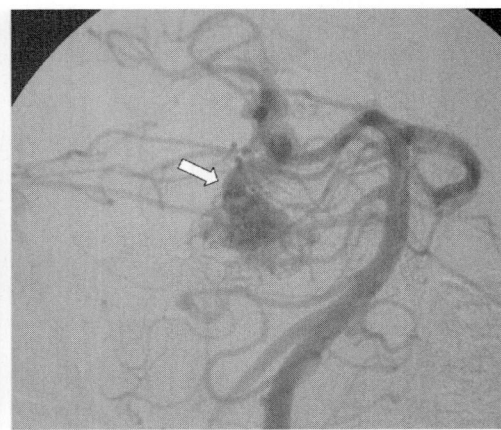

Fig. 35.34 Dural arteriovenous malformation (arrows) at the right cerebellopontine angle causing tinnitus. T2-weighted MRI (left) and cerebral angiogram (right).

always be distinguished from small arteriovenous malformations (Kattapong *et al.* 1995).

Telangiectasias

Telangiectasias are collections of dilated capillaries which are usually of no clinical significance (Milandre *et al.* 1987). They may be associated with hereditary haemorrhagic telangiectasia, the Osler–Weber–Rendu syndrome, but this is more likely to be associated with neurological complications from a pulmonary arteriovenous malformations with right-to-left shunting, such as cerebral hypoxia, brain abscess, paradoxical and septic embolism, or from an associated intracranial arteriovenous malformations or aneurysm (McDonald *et al.* 1998).

Carotico-cavernous fistula

Carotico-cavernous fistula is an abnormal connection between the carotid arterial system and the cavernous sinus (Section 13.4.5). It may occur spontaneously, especially in the elderly, or as a result of a ruptured dural arteriovenous malformations or intracavernous internal carotid artery aneurysm, closed or penetrating head injury, Ehlers–Danlos syndrome or pseudoxanthoma elasticum. With high flow direct fistulae from the internal carotid artery itself, the onset is dramatic with unilateral pulsating exophthalmos and an orbital bruit, often audible to the patient. In addition, there may be orbital pain, papilloedema, dilated conjunctival veins and chemosis, glaucoma, monocular visual loss, and involvement of the third, fourth, sixth, and the first and perhaps the second sensory division of the trigeminal nerve. The ophthalmoplegia may also be due to hypoxia and swelling within the extra ocular muscles. Dural fistulae present more insidiously because the blood flow is lower from small meningeal branches of the internal or external carotid arteries in the cavernous sinus. If there is no spontaneous resolution, it may be possible to obliterate the fistula with balloon catheterization.

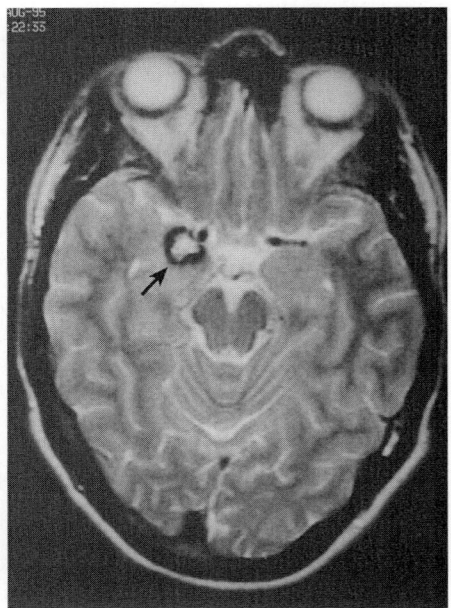

Fig. 35.35 The large draining vein of a venous malformation (arrow). Enhanced CT scan.

Fig. 35.36 Cavernous malformation (arrow). T2-weighted MR brain scan. The 'black' halo representing old haemorrhage is very characteristic.

35.6 Clinical features and differential diagnosis of transient ischaemic attacks

35.6.1 Symptoms and signs

The causes of a transient ischaemic attack, often abbreviated to TIA, are the same as the causes of stroke in general. However, the differential diagnosis of transient ischaemic attack is different owing to the transient nature of the symptoms. Brain imaging is used to rule out the occasional structural lesion causing 'transient focal neurological attacks' (Table 35.11) but may also contribute prognostic information and may allow definite localization of the site of ischaemia and thus the targeting of further investigation and secondary preventive treatments.

Symptoms and ischaemic territory

Symptoms are of sudden onset and are 'focal' indicating a disturbance in a particular area of brain, or in one eye (Flemming *et al.* 2004; Sherman 2004). Motor symptoms are the most common: weakness, clumsiness, or heaviness usually on just one side of the body (Table 35.12). Unilateral sensory symptoms are described as numbness, tingling, or deadness. Speech may be dysphasic, dysarthric, or both. Transient monocular blindness, or amaurosis fugax, affects the upper or lower half of vision, or all the vision of one eye, and is often described like a 'blind or shutter' coming down from above, or up from below. However, transient monocular ischaemia can also cause partial visual loss, such as blurring or dimming. Transient monocular blindness must be distinguished from transient homonymous hemianopia, although this can be difficult even when the patient is a very good historian.

Simultaneous bilateral transient motor or sensory loss are almost always due to brainstem ischaemia. Sudden simultaneous bilateral blindness in elderly patients usually indicates bilateral occipital ischaemia. Vertigo, diplopia, dysphagia, unsteadiness, tinnitus, amnesia, drop attacks, and dysarthria may be caused by global cerebral ischaemia, or by non-vascular causes such as motor neurone disease or myaesthenia in the case of dysarthria. If these symptoms occur in isolation, the diagnosis of transient ischaemic attack should only be considered after exclusion of other possibilities (Gomez *et al.* 1996). Global symptoms such as a reduced level of

Table 35.11 Causes of transient focal neurological attacks

Focal cerebral ischaemia: a 'transient ischaemic attack'
Migraine with aura (Section 18.2.1)
Partial epileptic seizures (Section 31.4)
Structural intracranial lesions
Tumour
Chronic subdural haematoma
Vascular malformation
Giant aneurysm
Multiple sclerosis (Section 37.5.3)
Labyrinthine disorders: Meniere's disease or benign positonal vertigo
Peripheral nerve or root lesion
Metabolic
Hypoglycaemia
Hyperglycaemia
Hypercalcaemia
Hyponatraemia
Psychological

Table 35.12 Clinical features and vascular distribution of transient ischaemic attacks

Symptoms	Vascular distribution		Proportion[†]
	Carotid	Vertebrobasilar	%
Unilateral weakness, heaviness, or clumsiness	+	+	50
Unilateral sensory symptoms	+	+	35
Dysarthria*	+	+	23
Transient monocular blindness	+	−	18
Dysphasia	+	(+)	18
Unsteadiness/ataxia*	(+)	+	12
Bilateral simultaneous blindness	−	+	7
Vertigo*	−	+	5
Homonymous hemianopia	(+)	+	5
Diplopia	−	+	5
Bilateral motor loss	−	+	4
Dysphagia*	(+)	+	1
Crossed sensory and motor loss	−	+	1

* In general if these symptoms are *isolated* it is best not to diagnose definite TIA

[†] The proportion of TIA patients with various symptoms from the Oxfordshire Community Stroke Project many patients had more than one symptom such as weakness *and* numbness, etc. (unpublished)

consciousness are almost never caused by a transient ischaemic attack. They can only be accepted as resulting from a transient ischaemic attack if there are additional focal symptoms that are unlikely to be epileptic or syncopal. Occasionally a blackout and the consequent fall can cause a transient ischaemic attack or stroke.

If more than one body part is involved, the symptoms usually start simultaneously in all parts, persist for a while and then gradually wear off over a few minutes, particularly in the case of amaurosis fugax, or an hour or so. If a patient still has symptoms more than an hour after the onset, the chances are that complete recovery will take more than 24 h. A mild headache accompanying the neurological symptoms is quite common, usually ipsilateral to the affected carotid territory, but most commonly in posterior circulation transient ischaemic attacks. If cerebral symptoms last less than a minute, particularly if they are 'sensory', the diagnosis of transient ischaemic attack is difficult to sustain. In contrast, symptoms of retinal ischaemia may be very short lived.

The symptoms of a transient ischaemic attack enable categorisation of attacks by arterial territory affected: carotid in about 80 per cent or vertebrobasilar in 20 per cent. This has important implications for further investigation and secondary prevention. Such categorization may be straightforward where there are definite cortical symptoms such as dysphasia or brainstem symptoms such as diplopia. However, because the motor and sensory pathways are supplied by both vascular systems at different points in their course, it is not always possible to distinguish which territory is involved (Table 35.12). Ischaemia in the territory of supply of the deep perforating arteries may be suspected if the patient has a transient lacunar syndrome and no positive evidence of cortical involvement such as dysphasia.

Mechanisms of ischaemia

Most transient ischaemic attacks are probably caused by arterial occlusion (Bogousslavsky *et al.* 1986). Rarely they may be secondary to low flow distal to a severely stenosed or occluded artery in the neck following a drop in blood pressure: as after antihypertensive medication or vasodilators, after standing or sitting up quickly, after a heavy meal or a hot bath, on exercise, or during cardiac arrhythmia (Caplan and Sergay 1976; Ruff *et al.* 1981; Ross Russell and Page 1983; Kamata *et al.* 1994). Such low-flow transient ischaemic attacks may be atypical: symptoms may take some minutes to develop, there may be irregular shaking or dystonic posturing of the arm or leg contralateral to the cerebral ischaemia, or there is monocular or binocular visual blurring, dimming, fragmentation, or bleaching, often just in bright light (Hess *et al.* 1991; Schulz and Rothwell 2002). Symptoms of focal brainstem ischaemia due to intermittent obstruction of a vertebral artery by cervical osteophytes are rare, presumably because collateral blood flow to the brainstem is usually sufficient.

'Subclavian steal' is caused by retrograde flow in the vertebral artery. It is a common angiographic and ultrasound finding when there is stenosis or occlusion of the subclavian artery proximal to the vertebral artery origin, particularly on the left, or of the innominate artery. When the ipsilateral arm is exercised, the increased blood flow to meet the metabolic demand may be enough to 'steal' more blood down the vertebral artery, away from the brainstem into the axillary artery. If there is poor collateral blood flow to the brainstem, then symptoms may occur, but this is very rare. The subclavian disease is almost always severe enough to be detectable by unequal radial pulses and blood pressures, and often there is a supraclavicular bruit (Fig. 35.37) (Cho *et al.* 2007).

Signs

Owing to their brief duration, patients are rarely examined during a transient ischaemic attack at a time when focal neurological signs might indicate the site of the lesion, although this now occurs more frequently with the advent of acute stroke services

and thrombolysis. However, non-neurological signs may help elucidate the cause of the attacks. A focal arterial bruit over the carotid bifurcation (Fig. 35.37) is predictive of internal carotid artery stenosis, although if very tight the stenosis or occlusion may not cause a bruit at all (Hankey and Warlow 1990). Bruits may occur in normal arteries, particularly in women (Hennerici *et al.* 1981; Schulz and Rothwell 2001). Atheromatous carotid stenosis is usually too proximal at the carotid bifurcation for patients to hear their own bruit: self-audible bruits are more likely to be due to distal internal carotid artery stenosis due to dissection or fibromuscular dysplasia, a dural arteriovenous fistula near the petrous temporal bone, a glomus tumour, a caroticocavernous fistula, raised intracranial pressure, or heightened awareness of their normal pulse (Thie *et al.* 1993; Waldvogel *et al.* 1998). Absent or unequal common carotid or superficial temporal pulses suggests common carotid or external carotid disease respectively.

Fibrin-platelet emboli are occasionally observed moving through the retinal circulation. More often cholesterol emboli are seen at arteriolar branching points as glittering yellow bodies that reflect the ophthalmoscope light, but usually do not obstruct blood flow. Such findings strongly suggest proximal atheroma (Wijman *et al.* 2004). 'Calcific' retinal emboli look solid, white, and non-reflective, they tend to impact at the optic disc edge and suggest embolism from aortic or mitral valve calcification (Ramakrishna *et al.* 2005). Later, embolized vessels may become white and 'sheathed' but this appearance eventually disappears. Large artery disease elsewhere, suggested by unequal arm blood pressure/pulses, known as coronary artery disease, claudication, femoral bruits, or absent foot pulses, suggests that transient ischaemic attacks are due to atherothromboembolism of the extracranial arteries, while a cardiac valvular lesion may suggest embolism from the heart.

35.6.2 Differential diagnosis

Transient ischaemic attacks are but one cause of 'transient focal neurological attacks' (Table 35.11) and 'transient monocular blindness' (Table 35.13). There is no diagnostic test to confirm a transient ischaemic attack although the advent of new imaging

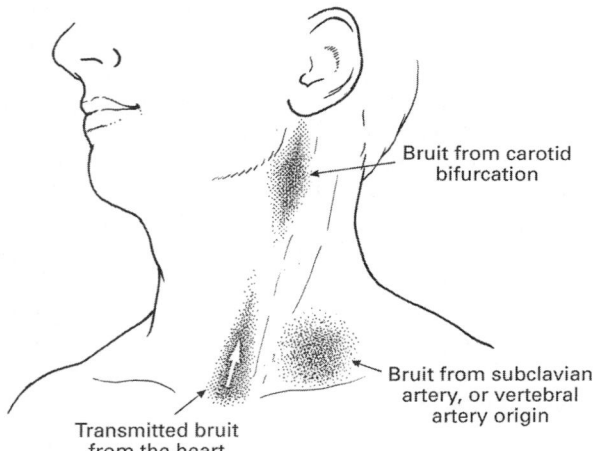

Fig. 35.37 The localization and implications of cervical bruits in various places. Note that a bruit arising from the carotid bifurcation is high up under the angle of the jaw. Supraclavicular bruits can be due to either subclavian or vertebral artery origin stenosis.

Table 35.13 Causes of transient monocular blindness or 'amaurosis fugax'

Ischaemia: a 'transient ischaemic attack'
Glaucoma
Uhthoff's phenomenon in retrobulbar neuritis (Section 37.4.2)
Raised intracranial pressure with papilloedema (Section 12.4.3)
Retinal haemorrhage
Retinal venous thrombosis (Section 12.3.3)
Retinal detachment
Macular degeneration (Section 12.3.2)
Intra-orbital tumour
Carotico-cavernous fistula (Section 13.4.5)
Retinal migraine
Intracranial dural malformation (Section 35.5.4)
Paraneoplastic retinopathy (Section 38.4.6)
Reversible diabetic cataract
Ureitis–glaucoma–hyphaema syndrome

techniques, particularly diffusion-weighted MRI (Section 35.8.3), has allowed the diagnosis to be made with near certainty in some patients. Some conditions are particularly frequently misdiagnosed as transient ischaemic attacks and will be discussed below.

Migraine with aura

Migraine is not a major diagnostic problem if the aura is associated with a headache with or without nausea and vomiting or if a known migraineur develops a typical aura without headache. However, occasionally migraine auras start in middle age and if there is no headache, they can be confused with transient ischaemic attacks. The time course of the symptoms is the key to distinguishing between transient ischaemic attack and migraine aura: migrainous auras start slowly, spread and intensify over several minutes, and usually fade in 20–30 min (Dennis and Warlow 1992). The symptoms tend to begin in one domain, particularly vision, fade and move on to another, such as language, and tend to be positive involving flashing lights or tingling rather than the negative symptoms typical of a transient ischaemic attack such as weakness, visual loss, or numbness. However, it is important to note that a progressive or stuttering pattern of symptom onset, positive visual phenomena, and headache are also compatible with vertebrobasilar transient ischaemic attack.

Epilepsy

Epilepsy is not a diagnostic problem unless the seizures are partial. Partial sensory seizures tend to cause positive symptoms such as tingling, and symptoms 'march' across a hand or foot, and up the limb in around a minute and may eventually be accompanied by focal motor seizures or secondary generalization. Sudden speech arrest seems to be more often epileptic, and not necessarily arising in the dominant hemisphere, than due to ischaemia which is more likely to cause dysphasic speech (Cascino et al. 1991). Transient inhibitory seizures may mimic the focal motor weakness of transient ischaemic attack, but are most unusual (Kaplan 1993).

Intracranial structural lesions

Occasionally intracranial structural lesions such as subdural haematoma, or tumour may cause transient ischaemic attacks, or at least episodes resembling transient ischaemic attack. Of course such lesions may be an incidental presence. Compression of an intracranial artery is perhaps an explanation for those patients with a space-occupying lesion, although focal seizures misdiagnosed as transient ischaemic attacks are another possibility. Intracranial vascular malformations might cause local steal of blood, thereby causing a transient ischaemic attack, or perhaps cause focal epileptic attacks which mimic transient ischaemic attacks.

Transient global amnesia

Transient global amnesia is a characteristic but uncommon clinical syndrome (Section 31.1.5), usually occurring in the middle-aged or elderly (Hodges and Warlow 1990a, b; Quinette et al. 2006). The onset is sudden with severe anterograde amnesia usually accompanied by retrograde amnesia. The attack lasts several hours, after which the patient recovers the ability to lay down new memories and recall old ones, but never has recall of the period of the attack itself. During the attack the patient is fully conscious, has no loss of personal identity, looks normal if a little subdued and bewildered, and has no other symptoms apart perhaps from some headache and nausea. The patient can perform normal everyday activities, even driving, but typically asks the same question repetitively because of the anterograde amnesia. A witness account is required to differentiate such attacks from hysterical fugues, alcoholic amnesic states, or complex partial seizures. Emotional upset, Valsalva manoevres, defaecation, sexual intercourse, and other physical exertions, often outdoor activities in a cold day, may be precipitants (Quinette et al. 2006). In most cases, the prognosis is excellent and attacks do not usually recur.

The aetiology of transient global amnesia is unclear. Various mechanisms have been proposed including temporary metabolic abnormality in the medial temporal lobes, venous hypertension, and ischaemia (Bettermann 2006; Roach 2006; Menendez Gonzalez and Rivera 2006). Sometimes a diagnosis of epilepsy, usually of complex partial type, becomes apparent subsequently. This is particularly likely if the transient global amnesia had been short lived, for less than an hour, had occurred on wakening, and had recurred early (Zeman et al. 1998). Recent developments in neuroimaging have confirmed that medial temporal lobe changes accompany transient global amnesia (Sander and Sander 2005). The presence of diffusion-weighted MRI abnormality (Sedlaczek et al. 2004; Sander and Sander 2005) has led to the proposal that transient global amnesia may be an ischaemic phenomenon. However such changes are not diagnostic of ischaemia and can occur following seizures.

Other neurological disorders

Occasionally, non-structural neurological disorders such as motor neuron disease or myasthaenia gravis may present with transient symptoms of sudden onset. These are particularly likely to consist of dysarthria or dysphagia in elderly patients. Unless the diagnosis is made initially, subsequent deterioration is often put down to 'recurrent stroke'.

Cryptogenic drop attacks

These affect middle-aged and elderly women, almost only when walking rather than just standing or sitting (Section 2.6.5). Without warning, the patient falls to the ground. There is no loss of consciousness or leg weakness. The attacks may recur but then disappear as mysteriously as they came. There is usually no known cause, although carotid sinus syncope and orthostatic hypotension are possibilities (Dey et al. 1996). There appear to be no serious prognostic implications. Sudden weakness of both legs can occur in brainstem ischaemia and, rarely, if both anterior cerebral arteries are supplied from the same stenosed internal carotid artery. Bilateral motor, sensory, or visual impairments can also be due to bihemispheric boundary zone ischaemia distal to severe carotid disease (Sloan and Haley 1990). Finally, spinal cord 'TIAs' do occur but are even rarer than spinal cord infarction (Cheshire et al. 1996). Cataplexy is almost invariably precipitated by excitement or emotion and seldom causes the patient to fall over (Section 32.3.1).

Psychogenic attacks

Psychogenic attacks are usually situational, for instance occurring in open spaces. Suggestive features include young age of less than 50, lack of vascular risk factors, symptoms affecting the left non-dominant side (Rothwell 1994), hyperventilation, other medically unexplained symptoms, or non-organic motor or sensory signs.

35.7 Clinical features of acute stroke

35.7.1 Diagnosis of stroke

The diagnosis of stroke is relatively straightforward if there is focal brain dysfunction of sudden onset or which was first present on waking. There may be some progression over the first few minutes or hours, particularly in posterior circulation stroke. Usually the deficit stabilizes by 12–24 h and, if the patient survives, recovery starts within a few days in most cases. Various scoring systems or similar strategies have been developed to aid clinicians in identifying stroke (Nor *et al.* 2005; Hand *et al.* 2006a), but these are not infallible and the potential differential diagnoses (Table 35.14) must always be borne in mind.

If the history is consistent with stroke, there is only a 5 per cent chance of a CT or MR brain scan showing an intracranial mass lesion rather than the expected changes consistent with stroke (Sandercock *et al.* 1985). This risk is higher when the speed of onset is uncertain. Features indicative of an intracranial tumour include recent headaches, seizures, papilloedema, a worsening deficit over days or weeks, and the presence of a primary tumour elsewhere. Chronic subdural haematoma is suggested by prior head injury, more drowsiness, confusion and headache out of proportion to the severity of the neurological deficit, a fluctuating course, use of anticoagulants, or chronic alcohol abuse. Other diagnoses are usually obvious: multiple sclerosis occurs at a younger age, peripheral nerve or root lesion are accompanied by clinical signs and or pain, post-seizure hemiparesis is suggested by the history, metabolic encephalopathy by global rather than focal neurological features, somatization and hysteria by young age and inconsistent signs, encephalitis by fever, clinical symptoms and signs and a diffusely abnormal EEG, and intracranial abscess by fever and a predisposing cause such as sinusitis or a congenital heart lesion (Norris and Hachinski 1982).

Occasionally head injury causing intracerebral haemorrhage can be missed whilst haemorrhagic stroke may cause a fall and subsequent head injury, thus the sequence of events may be unclear (Berlit *et al.* 1991). Ischaemic stroke following head injury may be due to neck artery dissection. Residual signs from an old stroke may become more pronounced with intercurrent illness or post seizure.

Table 35.14 Differential diagnosis of acute stroke

Intracranial tumour: glioma, meningioma etc.
Subdural haematoma
Epileptic seizure
Metabolic/toxic encephalopathy: hypoglycaemia, hepatic failure, alcohol intoxication
Cerebral abscess
Viral encephalitis
Hypertensive encephalopathy
Multiple sclerosis
Head injury
Peripheral nerve lesion
Psychogenic: somatization, hysteria
Creutzfeldt–Jakob disease

35.7.2 Determining the site of the lesion

Determining the location of the stroke provides useful prognostic information, helps to establish the cause, and enables selection of suitable investigation and treatment.

Using only symptoms and signs, stroke patients can be divided into four main clinical syndromes (Bamford *et al.* 1991; Mead *et al.* 1999):

◆ total anterior circulation syndrome;

◆ partial anterior circulation syndrome;

◆ lacunar syndrome; and

◆ posterior circulation syndrome.

These categories provide information on early prognosis, residual disability, and risk of recurrence.

Stroke localization using clinical data is not infallible: in about one-quarter of cases where a recent lesion is visible on brain imaging, it is not in the expected place (Mead *et al.* 1999). For example, although most pure motor strokes are caused by a lacunar infarct as a result of small vessel disease, in a few cases the CT or MR scan shows striatocapsular infarction caused by middle cerebral artery occlusion with good cortical collaterals (Fig. 35.20). Thus, patients may need to be reclassified on the basis of the brain imaging results.

Total anterior circulation syndrome

A large haematoma in one cerebral hemisphere, or an infarct affecting a large proportion of the middle cerebral artery territory, causes a characteristic clinical syndrome: contralateral hemiparesis; with or without a sensory deficit; involving the whole of at least two of the three body areas, the face, upper limb or lower limb; a homonymous visual field defect; and a cortical deficit consisting of dysphasia, neglect, or visuospatial problems. Cognitive or visual field defects may have to be assumed in drowsy patients. Deviation of the eyes towards the affected hemisphere is common but recovers in a few days. A large haematoma may cause midline shift, transtentorial herniation, and coma within 24 h. By contrast, these changes take 2 or 3 days to evolve with large infarcts as cerebral oedema develops.

Total anterior circulation infarcts are usually due to the acute occlusion of the internal carotid artery or embolic occlusion of the proximal middle cerebral artery from a cardiac or proximal arterial source (Caplan 1993; Lindgren *et al.* 1994; Wardlaw *et al.* 1996, Georgiadis *et al.* 2004). Sometimes the cortex is relatively spared due to good pial collaterals or rapid recanalization of the occluded artery, and infarction is largely subcortical in the distribution of several lenticulostriate arteries. This may be seen as a characteristic area of '*striatocapsular infarction*' on brain imaging (Fig. 35.38). This clinical syndrome is not as severe as a total anterior circulation syndrome with less cognitive deficit, and often without homonymous hemianopia (Nicolai *et al.* 1996).

Partial anterior circulation syndrome

A lobar haemorrhage, or a cortical infarct, causes a more restricted clinical syndrome consisting of only two of the three components of the total anterior circulation syndrome, or isolated higher cortical dysfunction such as dysphasia, a predominantly proprioceptive deficit in one limb, a motor/sensory deficit restricted to one body area or part of one body area (Bassetti *et al.* 1993; Aerden

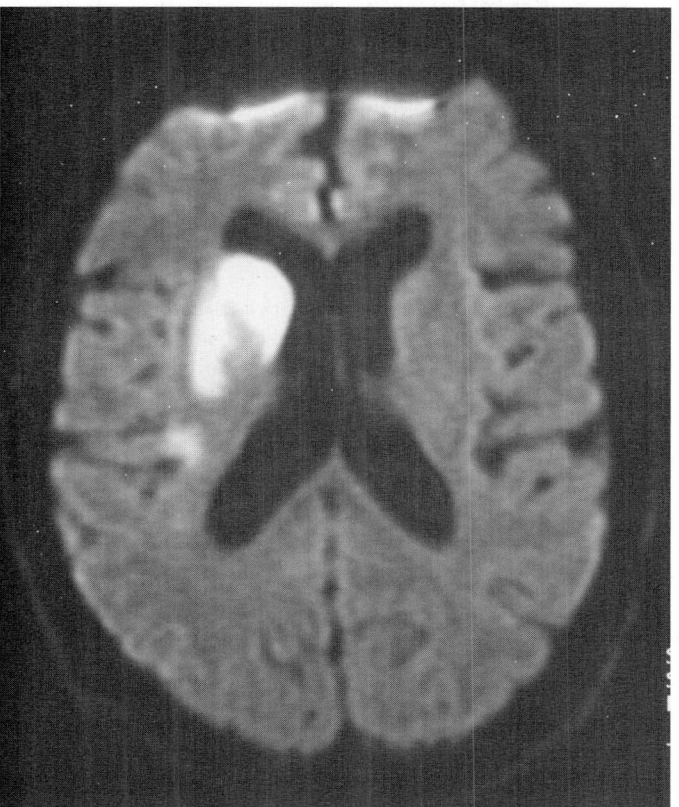

Fig. 35.38 Right striatocapsular infarct in a young woman with a right carotid occlusion secondary to dissection. T2-weighted MR brain scan.

et al. 2004). It may be difficult to distinguish between some partial anterior circulation syndrome and a 'lacunar' stroke.

Partial anterior circulation infarcts are caused by occlusion of a branch of the middle cerebral artery, or rarely the trunk of the anterior cerebral artery. They are usually a consequence of embolism from the heart or proximal atherothrombosis as in total anterior circulation infarcts (Fig. 35.20). Investigation should be prompt because of the high risk of recurrence. Anterior cerebral artery infarcts cause contralateral weakness predominantly of the lower limb, sometimes with cortical sensory loss, and aphasia if in the dominant hemisphere. Left, and rarely right, anterior cerebral infarcts can cause a curious dyspraxia of the left upper limb due to infarction of the corpus callosum disconnecting the right motor centres from the left language centres (Kazui *et al.* 1992). Bilateral leg and even additional bilateral arm weakness has been described when both anterior cerebral arteries are supplied from one stenosed internal carotid artery, or if both anterior cerebral arteries are occluded by embolism, so mimicking a brainstem or spinal cord syndrome (Borggreve *et al.* 1994).

Some anterior circulation syndromes, usually classified as partial anterior circulation syndromes, are caused by boundary zone infarcts. The rare anterior choroidal artery distribution infarcts, which can be defined only by the CT or MRI pattern, are probably due to microvascular disease as well as embolism, and can cause a partial anterior circulation syndromes or lacunar syndrome (Hupperts *et al.* 1994).

Lacunar syndrome

Lacunar syndromes are defined clinically. They are highly predictive of small, deep lesions affecting the motor and/or sensory pathways in the corona radiata, internal capsule, thalamus, cerebral peduncle, or pons. Although a few patients have partial anterior circulation infarcts (Bamford *et al.* 1987; Anzalone and Landi 1989; Arboix *et al.* 2007), the great majority have small infarcts which are sometimes visible on CT, more often on MRI. These are caused by presumed occlusion of a small perforating artery affected by intracranial small vessel disease (Fig. 35.20). There is no visual field defect, no new cortical defect, no impairment of consciousness, and nothing to suggest a brainstem syndrome such as diplopia, or crossed motor and sensory deficit.

The four main lacunar syndromes are:

- *Pure motor stroke* constitutes about 50 per cent of lacunar cases. It consists of a unilateral motor deficit involving two or three areas, the face, upper limb or lower limb, including the whole of each area which is affected. There are often sensory symptoms but no sensory signs. The lesion occurs at locations where the motor pathways are closely packed together and separate from other pathways: usually in the internal capsule or pons, sometimes the corona radiata or cerebral peduncle, and rarely in the medullary pyramid. There may be a flurry of immediately preceding transient ischaemic attacks, the so-called capsular warning syndrome (Donnan *et al.* 1996).

- *Pure sensory stroke* constitutes about 5 per cent of cases. It has the same distribution as pure motor stroke but the symptoms are of sensory loss, with or without sensory signs affecting all modalities equally, or sparing proprioception. The lesion is usually in the thalamus (Fig. 35.39) but can be in the brainstem.

- *Sensorimotor stroke* constitutes about 35 per cent of cases. It is the combination of a pure motor stroke with sensory signs in the affected body parts. The lesion is usually in the thalamus or internal capsule, but can be in the corona radiata or pons. A similar clinical picture can be caused by cortical infarcts leading to miss classification (Blecic *et al.* 1993).

- *Ataxic hemiparesis* constitutes about 10 per cent of cases. It is the combination of corticospinal and ipsilateral cerebellar-like dysfunction affecting the arm and/or leg. It includes the syndrome in which there is little more than dysarthria and one clumsy hand. The lesion is usually in the pons, internal capsule, or cerebral peduncle. Dysarthria, with or without upper motor neurone facial weakness, may also be a lacunar syndrome with similar lesion localization as ataxic hemiparesis, but there are other localizing possibilities as well.

Small deep infarcts in the subcortical white matter of the corona radiata may result from small vessel disease affecting the long medullary perforating arteries extending down from cortical branches of the middle cerebral artery, or embolism. Such centrum semiovale infarcts present as either a lacunar syndrome, or occasionally as a partial anterior cirulation syndrome with 'cortical' features (Read *et al.* 1998; Lammie and Wardlaw 1999). They are not, however, easy to classify or to distinguish from border zone infarcts deeper in the white matter lying between the arterial territories of the deep perforators from the first part of the middle cerebral artery and the superficial medullary perforators.

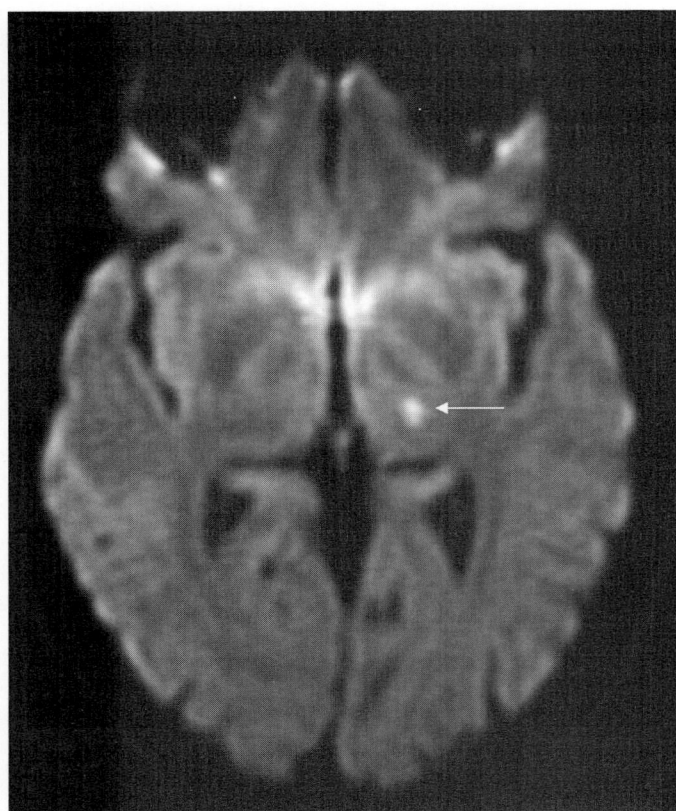

Fig. 35.39 Acute left thalamic lacunar infarct (arrow) in a patient with hemisensory symptoms. Diffusion-weighted MR brain scan.

Various other lacunar syndromes have been described with rather poor clinico-pathological-anatomical correlation; for example, chorea or hemiballismus usually appears to be due to a lesion in the contralateral subthalamic nucleus or elsewhere in the basal ganglia and tends to get better (Ghika and Bogousslavsky 2001).

Posterior circulation syndrome

Brainstem, cerebellar, thalamic, or occipital lobe signs normally indicate infarction in the distribution of the vertebrobasilar circulation or a localized haemorrhage. A combination of brainstem and occipital lobe signs is highly suggestive of infarction due to thromboembolism within the basilar and posterior cerebral artery territories. Occasionally, proximal posterior cerebral artery occlusion causes extensive temporal, thalamic, and perhaps midbrain infarction. This results in contralateral hemiparesis and sensory loss, and a marked cognitive deficit such as aphasia, as well as the expected homonymous hemianopia. This syndrome may be confused with occlusion of the middle cerebral artery or one of its branches (Argentino *et al.* 1996). This is the so-called 'walking total anterior circulation syndrome' because although it fulfils the definition of a walking total anterior circulation syndrome, the motor loss is mild. Cerebellar haematomas have fairly characteristic clinical features except in the case of massive haemorrhage, which is clinically indistinguishable from brainstem stroke, and very small haemorrhages, which may simulate a peripheral disorder of the vestibular system (Jensen and St Louis 2005).

The causes of infarction in the vertebrobasilar territory are heterogeneous. Often they are difficult to establish since vertebral angiography is seldom carried out and non-invasive arterial imaging is problematic. Some lacunar syndromes are due to small brainstem or thalamic infarcts as a consequence of small vessel occlusion: intracranial small vessel disease, or atheroma at the mouth of small perforating arteries. However, both small and large infarcts can be caused by embolism from the heart by atherothrombosis affecting the vertebral and basilar arteries, thrombotic occlusion complicating atheroma of the basilar artery or its major branches, or by low flow distal to vertebral and other arterial occlusions.

Although a large number of posterior circulation syndromes (Section 2.4.4) have been described, there is no clear association with a unique pattern of arterial occlusion, or prognosis. Syndromes include the 'top of the basilar' syndrome (Caplan 1980), various other midbrain syndromes (Bogousslavsky *et al.* 1994), the locked-in syndrome (Patterson and Grabois 1986), pontine syndromes (Bassetti *et al* 1996), lateral medullary syndromes (Kim *et al.* 1998), and medial medullary syndromes (Bassetti *et al.* 1997). Recognition of these is more an exercise in clinico-anatomical correlation rather than being very useful for clinical management. Because thalamic and cerebellar strokes can cause diagnostic confusion, and the latter may require surgical treatment, they are given separate consideration below.

Thalamic stroke

Small thalamic lesions may cause a pure sensory stroke or sensorimotor stroke, sometimes with ataxia in the same limbs (Schmahmann 2003). However, other deficits may occur in isolation, or in combination depending on which thalamic nuclei are involved. These include paralysis of upward gaze, small pupils, apathy, depressed consciousness, hypersomnolence, apathy, disorientation, visual hallucinations, aphasia and impairment of verbal memory attributable to the left thalamus, and visuospatial dysfunction attributable to the right thalamus. Occlusion of a single small branch of the proximal posterior cerebral artery can cause bilateral paramedian thalamic infarction with severe retrograde and anterograde amnesia.

Thalamic stroke should be considered when there is a sudden onset of behavioural disturbance. The diagnosis is often missed since patients are thought to have a primary psychiatric disorder, especially when neurologic dysfunction is lacking. Distinct behavioural patterns can be delineated on the basis of the four main arterial thalamic territories (Schmahmann 2003; Carrera and Bogousslavsky 2006):

- The anterior pattern consists mainly of perseverations, apathy, and amnesia. Paramedian infarction causes disinhibition and personality change, amnesia, and, in the case of extensive lesions, thalamic 'dementia'.

- After inferolateral lesions, executive dysfunction may develop but is often overlooked, although it may occasionally lead to severe long-term disability.

- Posterior lesions are known to cause cognitive dysfunction including neglect and aphasia but no specific behavioural syndrome has been reported.

Cerebellar strokes

Cerebellar strokes can be mild with sudden vertigo, nausea, imbalance, and horizontal nystagmus which soon recovers. They are frequently misdiagnosed as 'labyrinthitis'. More extensive infarction, or haemorrhage, causes additional ipsilateral limb and truncal ataxia, as well as dysarthria. Very severe strokes cause occipital headache, vomiting, and depressed consciousness, so making it impossible to detect limb or truncal ataxia. There are often additional brainstem signs such as ipsilateral facial weakness and sensory loss, a gaze palsy to the side of the lesion, ipsilateral deafness and tinnitus, and bilateral extensor plantar responses. These occur because of pressure from a large oedematous infarct or haematoma, or because an occluded artery supplying the cerebellum may also supply parts of the brainstem. Mass effect can obstruct CSF flow from the fourth ventricle to cause acute or subacute hydrocephalus. The consequent coma and meningism may be mistaken for subarachnoid haemorrhage. CT scan will reveal a haematoma but the signs of an infarct are more subtle, with disappearance or shift of the fourth ventricle due to mass effect before the low density of the lesion itself appears. MRI is more sensitive in infarction and provides detail of any additional brainstem involvement. It should be noted that initial mild symptoms may be misleading and patients may deteriorate rapidly. Accordingly urgent brain imaging is mandatory in suspected cerebellar stroke, irrespective of apparent clinical severity.

Patients who become acutely or subacutely comatose have a very poor prognosis. However, if there is little evidence of primary brainstem infarction, drainage of any hydrocephalus and/or decompression of the posterior fossa may sometimes be followed by relatively good-quality survival.

Boundary zone infarcts

These are infarcts in the border zones *between* arterial territories (Fig. 35.9):

◆ between the superficial territories of the middle cerebral artery and anterior cerebral artery in the frontoparasagittal region, the anterior boundary zone;

◆ between the superficial territories of the middle cerebral artery and posterior cerebral artery in the parieto-occipital region, the posterior boundary zone; and

◆ between the superficial medullary penetrators and deep lenticulostriate territories of the middle cerebral artery in the paraventricular white matter of the corona radiate, the subcortical boundary zone.

There is evidence that both low-flow and microembolism may be important in causing boundary zone infarction (Momjian-Mayor and Baron 2005). The evidence strongly favours a haemodynamic mechanism for internal boundary zone infarction, especially in the centrum semiovale. However, the relationship between cortical boundary zone infarction and haemodynamic compromise appears more complicated and artery-to-artery embolism may play an important role. Based on the high prevalence of microembolic signals documented by ultrasound in symptomatic carotid disease, embolism and hypoperfusion may play a synergistic role, with small emboli lodging in distal field arterioles being more likely to result in cortical micro-infarcts when chronic hypoperfusion prevails. Future studies combining imaging of brain perfusion, diffusion-weighted imaging, and ultrasound detection of microembolic signals should help resolve these issues.

Low flow may occur secondary to systemic hypotension, as during cardiac arrest. This results in bilateral infarcts, usually in the posterior boundary zones, causing cortical blindness, visual disorientation and agnosia, and amnesia. Alternatively, a relatively small drop in systemic blood pressure in the presence of internal carotid occlusion or stenosis may cause unilateral boundary zone infarction usually in the anterior and subcortical regions. This causes contralateral weakness of the leg more than the arm with sparing of the face, some impaired sensation in the same areas, and aphasia if the dominant hemisphere is affected. Unilateral posterior boundary zone infarcts are less common and cause contralateral hemianopia and cortical sensory loss, along with aphasia if the dominant hemisphere is affected.

Miscellaneous clinical features

Cranial nerves Unilateral supratentorial stroke lesions can cause contralateral weakness of the bulbar muscles with unilateral weakness of the palate and tongue, and forehead musculature resembling a lower motor-neurone rather than upper motor-neurone facial palsy. Because all these muscles have a strong bilateral upper motor-neurone innervation, this weakness tends to disappear quite quickly in most cases. The bulbar muscle weakness may be enough to cause significant dysphagia; thus dysphagia is not a symptom exclusive not confined to brainstem strokes. Dysarthria is common in supratentorial strokes, usually in proportion to any facial weakness, and is a defining feature of the clumsy hand-dysarthria syndrome, but it can be isolated with no localizing value (Ichikawa and Kageyama 1991). Any weakness of the sternomastoid muscle is ipsilateral to a supratentorial lesion so there is difficulty turning the head away from the side of the lesion. Lower cranial nerve lesions ipsilateral to a supratentorial infarct suggest dissection of the internal carotid artery. Third, fourth, and sixth cranial nerve lesions have rarely been described ipsilateral to internal carotid artery occlusion or dissection, presumably due to ischaemia of the nerve trunks.

Headache This is not uncommon around the time of stroke onset. It is more often severe in primary intracerebral haemorrhage, than ischaemic stroke, and more often with posterior than anterior circulation strokes. If the headache is localized at all, it tends to be over the site of the lesion. Headache is more common in cortical and posterior circulation than lacunar infarcts (Kumral *et al.* 1995). Severe unilateral neck, orbital, or scalp pain suggests internal carotid artery dissection, particularly if there is an ipsilateral Horner's syndrome. Severe occipital headache can occur with vertebral artery dissection. Headache is also a particular feature of venous infarcts. Unusual headache in the days before stroke would suggest giant cell arteritis or perhaps a mass lesion rather than a stroke.

Movement disorders Acute hemiparkinsonism contralateral to a basal ganglia stroke is rare. Contralateral chorea, hemiballismus, and sometimes tremor or dystonia are more common (D'Olhaberriague *et al.* 1995; Scott and Jankovic 1996; Giroud *et al.* 1997). Dystonia often develops gradually in a hemiplegic limb some weeks after the stroke, particularly in children and young adults. Rather nondescript 'limb-shaking' has been described in 'low flow' transient ischaemic attack patients and can occur in stroke, particularly in brainstem infarction.

35.7.3 Early deterioration after acute ischaemic stroke

Although stroke onset is usually abrupt, the neurological deficit often worsens over the following minutes, hours, and sometimes days. Deterioration may be caused by systemic (Table 35.15) or neurological factors (Table 35.16) (Karepov *et al.* 2006) but progressive non-stroke pathologies should also be reconsidered.

Haemorrhage into cerebral infarcts or haemorrhagic transformation

At post mortem, spontaneous petechial haemorrhages are very common in infarcts. During life they are seen on brain CT in around 15 per cent of patients in the absence of thrombolytic therapy (Fig. 35.40). However incidence rates must be considered in the context of the timing and mode of imaging, the definition of clinically significant haemorrhagic transformation, and the fact that there is interobserver variability in identifying haemorrhage (Khatri *et al.* 2007). Haemorrhagic transformation is said to be more common in cardioembolic infarcts (Alexandrov *et al.* 1997) perhaps because the infarcts are often large and patients are frequently anticoagulated. Haemorrhagic transformation is not often symptomatic unless there is confluent haematoma (Larrue *et al.* 1997; Berger *et al.* 2001).

Haemorrhagic transformation is of particular interest in acute stroke therapy since thrombolysis is associated with increased risk (Khatri *et al.* 2007). Other factors increasing risk of haemorrhagic transformation include heparin use (International Stroke Trial Collaborative Group 1997), oedema or mass effect, stroke severity, and age (Kent *et al.* 2001; Khatri *et al.* 2007) and possibly hyperglycaemia, timing of thrombolytic therapy, and successful recanalization. Thrombolytic therapy 3 h or more after stroke onset and late spontaneous recanalisation is associated with high rates of haemorrhage (Molina *et al.* 2001). The presence of cerebral microbleeds may also predicpose to haemorrhage following thrombolytic treatment, but data are limited.

Table 35.15 Systemic causes of neurological deterioration after stroke

Hypoxia	Pneumonia
	Pulmonary embolism
	Cardiac failure
	Chronic respiratory disease
Hypotension	Pulmonary embolism
	Cardiac failure
	Cardiac arrhythmia
	Pneumonia
	Dehydration
	Septicaemia
	Hypotensive drugs/vasodilators
	Bleeding peptic ulcer
Infection and fever	Pneumonia
	Urinary tract infection
	Septicaemia
Others	Water and electrolyte imbalance
	Hypo/hyperglycaemia
	Depression
	Sedatives/hypnotics
	Anticonvulsants

Table 35.16 Neurological causes of deterioration after stroke

	Ischaemic stroke	Primary intracerebral haemorrhage	Subarachnoid haemorrhage
Haemorrhagic transformation	+	−	−
Cerebral oedema	+	(+)	(+)
Brain shift (mass effect)	+	+	+
'Vasospasm'	−	−	+
Thrombus propagation	+	−	−
Recurrent embolism	+	−	−
Haemorrhage growth/recurrence	−	+	+
Hydrocephalus	(+)	+	+
Epileptic seizures	+	+	+

Microvascular damage leading to loss of vessel wall integrity is thought to be the cause of haemorrhage into an infarct. The mechanisms for this include plasmin-generated laminin degradation, matrix metalloproteinase activation, and transmigration of leukocytes through the vessel wall (Lapchak 2002; Wang *et al.* 2004). It has been suggested that tPA treatment may increase the risk of haemorrahic transformation not just by reperfusion but also through a direct effect on the molecular processes leading to vessel damage (Wang *et al.* 2004) (Section 35.10.3).

Peri-infarct oedema

Peri-infarct oedema reduces local cerebral blood flow and causes brain shift and herniation, the latter being the most common 'neurological' cause of death. This complication is a common explanation for worsening over the first few days and can often be detected by CT scan. Intravenous mannitol may reduce the deficit for a while but is unlikely to make a major impact on outcome. Recently, surgical decompression using hemicraniectomy has been shown to improve survival with satisfactory functional outcome in many patients.

Propagating thrombosis

Propagating thrombosis proximal or distal to a thrombotic, embolic, or any other type of occlusion, or within collateral vessels is often assumed to explain neurological deterioration if other causes have been excluded. However direct evidence is almost impossible to obtain, except perhaps with transcranial Doppler.

Recurrent embolisation

In theory recurrent embolization may cause deterioration. However, the distinction between propagating thrombosis and embolization is very difficult. There is only anecdotal evidence that full anticoagulation with intravenous heparin slows progression and improves outcome if no other cause of deterioration is evident (Slivka *et al.* 1989).

Epileptic seizures

Partial or generalized epileptic seizures occur for the first time in about 2 per cent of acute strokes at around the time of onset, rising to about 10 per cent at 5 years, more with large cortical infarcts or intracranial haemorrhage (Ferro and Pinto 2004). Seizures are more common in large strokes especially if haemorrhagic, and cortical as opposed to lacunar strokes. Cerebrovascular disease is

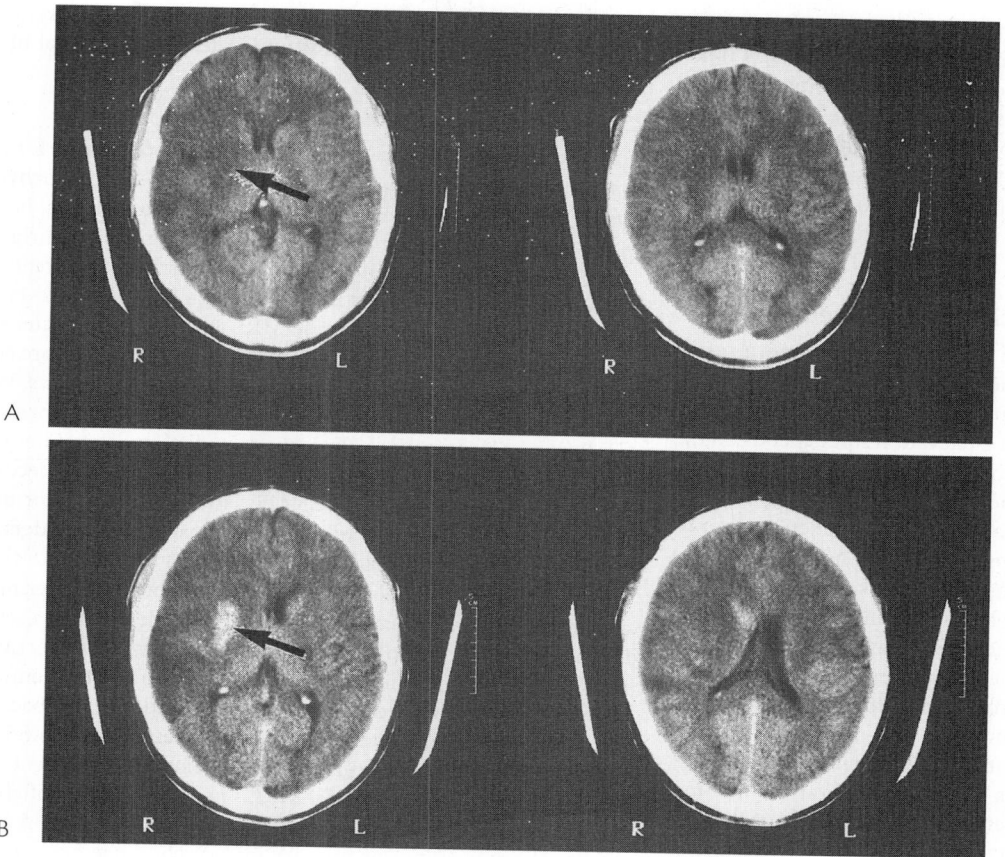

Fig. 35.40 An ischaemic stroke with a large middle cerebral artery infarct. (arrow) (A) Unenhanced-CT brain scan within 24 h. (B) The next day the scan shows haemorrhagic transformation (arrow) which would be very difficult to distinguish from primary intracerebral haemorrhage without knowledge of the initial scan.

the most common cause of epilepsy in the elderly and late onset epilepsy is a predictor of subsequent stroke (Cleary *et al*. 2004). Seizures may cause neurological deterioration or be mistaken for recurrent stroke. Intractable recurrent seizures are distinctly unusual.

35.8 Investigation of transient ischaemic attack and stroke

In patients with transient ischaemic attack or minor stroke there is no or little neurological deficit but, importantly, the risk of recurrent and possibly severe ischaemic events is high. Thus rapid investigation is required to establish the cause and to enable targeting of suitable secondary preventive treatment. In patients with more major stroke, investigations are required to exclude alternative diagnoses and to target treatment so as to reduce the neurological deficit and secondary complications of stroke.

Investigation of transient ischaemic attack is similar to investigation of stroke. Current consensus is that patients with transient ischaemic attack should have brain imaging although routine clinical practice varies considerably. Patients with stroke require brain imaging to distinguish between primary intracerebral haemorrhage and cerebral infarction since this distinction cannot be made reliably on clinical criteria alone (Hawkins *et al*. 1995). Recent developments in brain imaging, in particular new MRI sequences, and to a lesser extent CT techniques, have enabled visualization of the pathophysiological processes involved in brain infarction. They are being developed to select patients suitable for thrombolytic

treatment and may in the future enable targeting of treatments such as neuroprotection. Besides imaging investigation of stroke depends on whether the stroke is ischaemic or haemorrhagic and upon the stroke subtype as indicated by clinical features and the imaging findings.

35.8.1 Routine investigations

In general, all transient ischaemic attack and ischaemic stroke patients should have basic non-invasive first-line investigations within a few hours of presentation (Table 35.17). The yield of picking up a relevant abnormality may be very low for some tests,

Table 35.17 Baseline non-imaging tests for most transient ischaemic attack and ischaemic stroke patients CRP: C-reactive protein

Investigation	Disorders detected
Full blood count	Anaemia, polycythaemia, leukaemia, thrombocythaemia
ESR / CRP	Vasculitis, infective endocarditis, hyperviscosity, myxoma
Electrolytes	Hyponatraemia or hypokalaemia
Urea	Renal impairment
Plasma glucose	Diabetes, hypoglycaemia
Plasma lipids	Hyperlipidaemia
Urine analysis	Diabetes, renal disease, vasculitis
Electrocardiogram	Left ventricular hypertrophy, arrhythmia, conduction block, recent myocardial infarction

such as full blood count and erythrocyte sedimentation rate, but these are simple tests and the consequences of missing a treatable disorder, such as giant cell arteritis, are serious. Abnormalities are more likely on blood glucose testing, urine analysis, and electrocardiogram. Many patients are hypercholesterolaemic, although immediately after stroke there is a transient fall in plasma cholesterol, which will underestimate the usual level (Mendez *et al.* 1987; Woo *et al.* 1990). Brain imaging is discussed later in this section.

35.8.2 Second-line investigations for selected patients

Second-line investigations (Table 35.18) are usually more costly, invasive, and/or dangerous, so they must be targeted to patients most likely to gain from a change in management as a consequence of the test result. The likelihood of a relevant result depends on the selection of patients for the investigation, and a balance has to be struck between over- and under-investigation. This balance depends on the consequences of overlooking a particular diagnosis: missing severe carotid stenosis would be harmful because carotid endarterectomy reduces the risk of stroke, whereas missing the lupus anticoagulant whose relevance is unknown, and where the effect of any treatment is uncertain, may be of little consequence. Also, more investigation will generally be necessary in patients without evidence of atherothromboembolism, small vessel disease, or embolism from the heart.

Routine lumbar puncture is not indicated after stroke and may be dangerous in the presence of a large intracerebral haematoma, or oedematous infarct causing brain shift. Examination of the CSF may be necessary in cases of diagnostic uncertainty where there is a possibility of encephalitis or multiple sclerosis or if the stroke is thought to have been caused by infective endocarditis or by chronic meningitis due to syphilis, or tuberculosis. The CSF after stroke is usually acellular but there may be up to 100 cells per cubic millimetre. Levels above this suggest septic emboli to the brain (Powers 1986).

Routine EEG is not indicated in stroke but may be helpful where there is a possibility of encephalitis or generalized encephalopathy, or focal seizure activity.

35.8.3 Imaging the brain

In patients with transient ischaemic attack, brain imaging may be required to exclude other causes of transient neurological dysfunction such as tumour or subdural haematoma, particularly where symptoms are recurrent or there are additional clinical features such as headache or drowsiness. Previously, it was felt that CT brain imaging was not necessary in cases of uncomplicated transient ischaemic attack especially as such symptoms appear to be rarely caused by haemorrhage (Gunatilake 1998; Werring *et al.* 2005), but recent studies have demonstrated the usefulness of MRI in patients with suspected transient ischaemic attack, particularly diffusion-weighted imaging (Schulz *et al.* 2003; Schulz *et al.* 2004).

At present in most centres in most countries, CT remains the imaging modality used routinely to distinguish between haemorrhage and infarction in acute stroke: it is better tolerated, easier to perform in sick patients than MRI, and is still more widely available. Although conventional T1- and T2-weighted-MR sequences have a low sensitivity acutely for intracranial haemorrhage, newer MR sequences have greater sensitivity for haemorrhage than CT. Consequently, some centres advocate the use of multimodal MRI

as the imaging modality of choice in acute stroke (Chalela *et al.* 2007) although the use of perfusion CT to image cerebral blood flow is also becoming more widespread.

CT and conventional MRI: haemorrhage

CT shows haemorrhage as a hyperdense region often in the form of a space occupying mass. The sensitivity of CT for parenchymal haemorrhage is almost 100 per cent but small parenchymal haemorrhage or subarachnoid haemorrhage may be missed (Schriger *et al.* 1998). Only rarely has intracerebral haemorrhage been reported to cause focal symptoms lasting less than 24·h (Gunatilake 1998). The increased use of gradient echo MRI, with its much greater sensitivity for microbleeds than CT (Fig. 35.32) may reveal more cases where a small bleed has clearly caused transient symptoms. With increasing passage of time after the onset of the haemorrhage, CT becomes progressively less good at distinguishing between haemorrhage and infarction as the initial hyperdensity becomes iso- and then hypodense (Fig. 35.41). Thus MRI is superior more than a week after onset of haemorrhage, particularly in defining possible underlying pathology.

There is not usually a large amount of oedema early after intracerebral haemorrhage and such a finding should prompt a search for an underlying tumour or venous obstruction. Haemorrhages related to coagulopathies, anticoagulants, or to cerebral amyloid angiopathy are often inhomogeneous with fluid levels. The location and number of haemorrhages may provide clues as to the underlying cause and hence guide further investigation and treatment.

As stated above, conventional MRI is not the modality of choice for distinguishing between haemorrhage and infarction in acute stroke. The appearance of primary intracerebral haemorrhage on conventional MRI sequences at different time periods following stroke onset is complex since the T1 and T2 relaxation rate varies with the concentration of breakdown products of haemoglobin. As the haematoma ages, it passes from oxyhaemoglobin through deoxyhaemoglobin and methaemoglobin prior to red cell lysis and breakdown into ferritin and haemosiderin. Acute haematoma is characterized by central hypointensity on T2-and isointensity on T1-methaemoglobin formation leads to T1-shortening and to central hyperintensity on T1- with hyperintensity on T2-. Days and weeks after onset of bleeding, a hypointense border zone caused by the paramagnetic rim of methaemoglobin demarcates the border zone of the haematoma. The complexity and subtlety of these changes has resulted in a strong preference by clinicians and radiologists for CT over conventional MRI in the acute evaluation of intracerebral haemorrhage.

Conventional CT and MRI: ischaemic changes

Knowledge of the nature and time course of ischaemic changes on CT and conventional MRI is necessary as an aid to diagnosis, infarct localization, and selection of patients for acute stroke therapy. Whilst haemorrhage is visible almost immediately on CT, ischaemia takes longer to manifest although changes have been reported as early as 22 min in one study (von Kummer *et al.* 2001). Consistent with its end-artery vascular system, the striatocapsular area exhibits irreversible damage very early in patients with middle cerebral artery stem occlusion whereas the cortical areas usually fall within the penumbra. The earliest signs of ischaemia are of brain tissue swelling as shown by effacement of cortical sulci, asymmetry of the sylvian fissures, and ventricular distortion (Kucinski *et al.* 2002). Occasionally the segment of the artery occluded by the thrombus,

Table 35.18 Second-line investigations for selected ischaemic stroke and transient ischaemic attack patients

Investigation	Indications
Blood	
Liver function	Fever, malaise, raised ESR, suspected malignancy
Calcium	Recurrent focal neurological symptoms very rarely due to hypercalcaemia
Thyroid function tests	Atrial fibrillation
Activated partial thromboplastin time, dilute Russell's viper venom time anticardiolipin antibody*, antinuclear and other autoantibodies	Young (<50 years) and no other cause found, past history or family history of venous thrombosis, especially if unusual site (cerebral, mesenteric, hepatic veins) recurrent miscarriage, thrombocytopenia, cardiac valve vegetations, livedo reticularis, raised ESR, malaise, etc., positive syphilis serology
Serum proteins, serum protein electrophoresis, plasma viscosity	Raised ESR
Haemoglobin electrophoresis	Afro-Caribbean patients
Protein C and S, antithrombin III, activated protein C resistance, thrombin time†	Personal or family history of thrombosis (usually venous, particularly in unusual sites such as hepatic vein) at unusually young age
Blood cultures	Fever, cardiac murmur, haematuria, deranged liver function, raised ESR, malaise unexplained stroke in the young
HIV serology	Drug addict, homosexual, blood products transfusion, systemically unwell, lymphadenopathy, pneumonia, cytomegalovirus retinitis, etc.
Lipoprotein fractionation	Elevated cholesterol or strong family history Hyperlipoproteinaemia
Serum homocysteine	Marfanoid habitus, high myopia, dislocated lenses, osteoporosis, mental retardation, young patient
Leucocyte alpha-galactosidase A	Corneal opacities, cutaneous angiokeratomas, paraesthesias and pain, renal failure
Blood/CSF lactate	Young patient, basal ganglia calcification, epilepsy, MELAS/mitochondiral cytopathy, parieto-occipital ischaemia
Syphilis serology	Young patient, high risk of sexually transmitted diseases
Cardiac enzymes	History or electrocardiographic evidence of recent myocardial infarction
Drug screen	Young patient, no other obvious cause, cocaine/amphetamine, usage
Urine	
Amino acids	Marfanoid habitus, high myopia, dislocated lenses, osteoporosis, mental retardation
Drug screen	Young patient, no other obvious cause, cocaine/amphetamine, etc. usage
Imaging	
Chest X-ray	Hypertension, finger-clubbing, cardiac murmur, or abnormal electrocardiogram, ill patient
MRI	Suggestion of arterial dissection, uncertain diagnosis of stroke
Carotid ultrasound /MRA/CTA	Carotid transient ischaemic attack or mild ischaemic stroke
Cerebral angiography	Suspected arterial dissection (if unclear on MRI), arteritis, arteriovenous malformation or aneurysm
Arch aortography	Symptoms of subclavian steel and unequal brachial pulses and blood pressures
Cardiac	
Echocardiography (transthoracic or transoesophageal)	Possible cardiac source of embolism, or clinical, electrocardiogram or, CXR evidence of embologenic heart disease, aortic arch dissection
24-h electrocardiogram	Palpitations or blackout during a suspected transient ischaemic attack, suspicious resting electrocardiogram
Others	
EEG	Doubt about diagnosis of transient ischaemic attack or stroke: ?epilepsy
Body red cell mass	Raised haematocrit
Temporal artery biopsy	>60 years old, jaw claudication, headache, polymyalgia, malaise, anaemia, raised ESR

*Repeat to ensure persistently raised.

†Transient falls occur after stroke; so any low level must be repeated and family members investigated.

particularly in the case of the main trunk of middle cerebral artery, may appear hyperdense although interobserver agreement on such signs is only moderate (von Kummer *et al.* 1996; Grotta *et al.* 1999). Early brain swelling is followed by the development of parenchymal hypodensity corresponding to cytotoxic oedema and is seen on CT as loss of the normal grey/white matter differentiation in the cortex, insular ribbon, or basal ganglia.

The early ischaemic changes on CT are subtle and the CT may appear normal if performed in the first few hours. The sensitivity of CT within 5 h of ischaemic stroke was reported as 58 per cent in one early study (Horowitz *et al.* 1991) although a higher rate of 68 per cent has been reported within 2 h (von Kummer *et al.* 1994) and even higher of 75 per cent within 3 h with middle cerebral artery infarction (Barber *et al.* 2000). The interobserver reliability

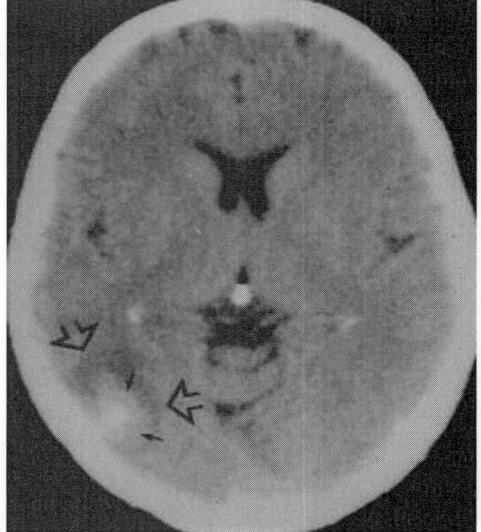 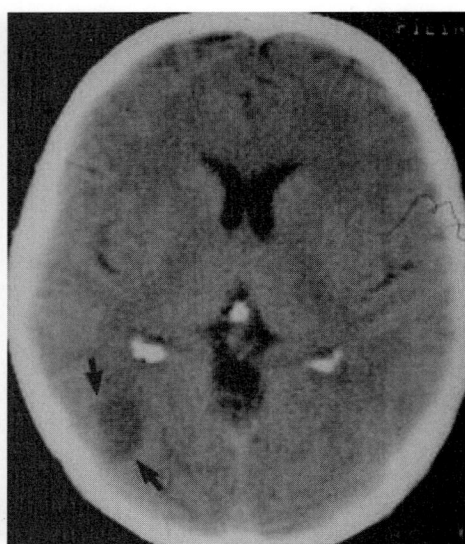

Fig. 35.41 Acute right cortical haemorrhage (arrowed) on the day of the stroke (left), which has become hypodense on repeat scanning seven days later (right). CT brain scan.

and reproducibility of CT in the estimation of the degree of ischaemic change is modest (von Kummer *et al.* 1996; Schriger *et al.* 1998; Grotta *et al.* 1999; von Kummer *et al.* 1997) although use of a systematic CT review system, the Alberta Stroke Programme Early CT Score, ASPECTS, by trained observers has better inter-rater reliability (Coutts *et al.* 2004). Sensitivity is less for small infarcts and infarcts in the posterior fossa and a significant minority of clinically definite strokes are not associated with an appropriate lesion on CT even after 2–3 days.

One to two days after stroke onset, the infarcted area appears as an ill-defined hypodense area as vasogenic oedema becomes predominant. Within 2 or 3 days the attenuation values become lower, the ischaemic area is better demarcated, and there may be evidence of mass effect. Later there may be ipsilateral ventricular dilatation due to loss of brain substance. Haemorrhagic transformation usually occurs a few days after stroke onset in large infarcts, but may develop within hours resulting in appearances very similar to primary intracerebral haemorrhage.

The site of any hypodensity relates to the underlying arterial distribution allowing for inter-individual differences in arterial anatomy. A small proportion of first ever stroke patients have focal hypodensities on CT in areas inconsistent with the presenting symptoms. Others have widespread diffuse periventricular hypodensity making any new infarcts difficult to delineate (Chodosh *et al.* 1988). Further, despite the temporal sequence of ischaemic changes on CT, it is often difficult to determine the age of an infarct from the CT appearance. Diffusion-weighted MR imaging overcomes these limitations.

In summary, the main role of CT in acute stroke is to exclude haemorrhage. Owing to the limitations in visualization of ischaemia, especially in the early stages, CT cannot be used to reliably stratify participants according to infarct location or size in trials of acute stroke therapy although it is currently used to exclude major completed infarction prior to early thrombolysis.

The earliest ischaemic change on conventional MRI, immediately detectable, is loss of the normal flow void in the affected artery, the MRI equivalent of the hyperdense artery sign on CT, and arterial enhancement if contrast has been used (Mohr *et al.* 1995). Subsequent changes are swelling on T_1-weighted images caused by cytotoxic oedema, which is present in up to half of patients within 6 h, hyperintensity on T2-weighted images from vasogenic oedema, present within 8 h, and T_1-weighted signal change, within 16 h (Yuh *et al.* 1991). Consequently, the sensitivity of conventional MRI is low in the first few hours following the onset of stroke symptoms with values similar to CT (Mohr *et al.* 1995; Mullins *et al.* 2002). In subacute ischaemic stroke, conventional magnetic resonance imaging has higher sensitivity than CT owing to its better spatial resolution and lack of posterior fossa artefact (Simmons *et al.* 1986; Bryan *et al.* 1991) but conventional MRI may still be normal in clinically definite stroke.

Conventional MRI is poor at distinguishing acute from chronic infarction. This is a particular problem in patients with multiple infarcts and in the elderly, in whom multiple T2-weighted abnormalities in the corona radiata, basal ganglia, and brainstem are common and in whom neurological symptoms may develop with intercurrent illness on a background of previous stroke. This, together with the poor sensitivity in the acute stroke period means that as for CT, conventional MRI is often unable to stratify patients according to infarct presence, ischaemic stroke subtype, size or location prior to therapy, or randomization in acute stroke trials. MRI and MR venography will help where there is a possibility of venous rather than arterial infarction.

New MRI techniques

There are limitations to the information on cerebrovascular pathophysiology *in vivo* that can be provided by CT and conventional MRI. Specifically, these include lack of sensitivity for acute ischaemic stroke, difficulty in determining infarct age, lack of demonstration of the ischaemic penumbra (Section 35.3.5), and low sensitivity and specificity for primary intracerebral haemorrhage in the case of MRI. New imaging techniques address some of the deficiencies and thereby impact on the management of patients with acute stroke.

Gradient echo MRI and primary intracerebral haemorrhage

Recent advances in new MRI sequences that allow further characterization of ischaemia have led to increased interest in using MRI to detect haemorrhage. This would avoid having to perform two imaging modalities, CT followed by MRI, in patients in whom further characterization of pathophysiology is necessary, for instance to determine selection for thrombolysis. Echoplanar gradient-echo T2*-weighted imaging, GRE MRI, which exploits the paramagnetic effects of deoxyhaemoglobin, has a high sensitivity for detecting primary intracerebral haemorrhage. Studies comparing CT and GRE MRI for detection of haemorrhage in acute stroke show that GRE MRI is as good or better than CT particularly for small or chronic bleeds (Schellinger et al. 1999; Fiebach et al. 2004; Kidwell et al. 2004; Chalela et al. 2007).

The sensitivity of GRE MRI for chronic haemorrhage, which appears as areas of signal loss, has enabled its use in the detection of cerebral microbleeds (Fig. 35.32). Cerebral microbleeds are associated with primary intracerebral haemorrhage and ischaemic stroke (Nighoghossian et al. 2002; Lee et al. 2004; Koennecke 2006; Cordonnier et al. 2007). There have been some small studies to investigate a possible relationship between the presence of cerebral microbleeds and the risk of haemorrhagic transformation after thrombolysis (Derex et al. 2005) but results are inconclusive and larger prospective studies are required (Koennecke 2006; Cordonnier et al. 2007).

Diffusion-weighted MRI

Diffusion-weighted imaging, or DWI, relies on changes in the Brownian motion of water molecules to generate contrast. During early ischaemia, there is decreased water proton movement caused by cytotoxic oedema as water moves from the less restricted extracellular environment into the more restricted intracellular environment. Reduced proton diffusion leads to a bright, high-signal diffusion-weighted imaging lesion. The degree of water proton restriction can be quantitatively measured using the apparent diffusion coefficient. In contrast to diffusion-weighted imaging, apparent diffusion coefficient maps depict reduced diffusion as a dark, low signal. The value of the apparent diffusion coefficient changes with time after stroke onset being reduced for the first few days, after which it rises, pseudonormalization, to become hyperintense in the chronic phase when there is vasogenic oedema and cellular necrosis. This allows diffusion-weighted imaging to distinguish between acute and chronic infarction unlike conventional MRI (Schaefer et al. 2005), although it should be noted that the diffusion-weighted imaging lesion persists for at least a week since it detects prolonged T2 signal 'T2 shine-through' so correct interpretation of diffusion-weighted imaging including identifying acute recurrence of ischaemia requires consideration of the apparent diffusion coefficient map.

Diffusion-weighted imaging has a high sensitivity for acute ischaemic stroke at around 90 per cent (Baird and Warach 1998) although other conditions such as seizure, encephalitis, and multiple sclerosis can all cause diffusion-weighted imaging lesions. Diffusion-weighted imaging is abnormal within minutes of stroke onset (Hjort et al. 2005a). Interobserver agreement is better for diffusion-weighted imaging than with conventional MRI with sensitivity and specificity of 95 per cent and nearly 100 per cent respectively (Lutsep et al. 1997; Lansberg et al. 2000). However there appears to be a lower sensitivity for diffusion-weighted imaging in posterior circulation acute stroke, with a 19–31 per cent false negative rate, particularly where lesions are small and within the first 24 h after stroke onset (Oppenheim et al. 2000).

The fact that diffusion-weighted imaging distinguishes between acute and chronic infarction and has high sensitivity in acute stroke makes it a valuable tool in the diagnosis and management of patients with acute stroke. Diffusion-weighted imaging has been reported to show localization in a different vascular territory from that initially suspected on the basis of clinical features and conventional MRI in 18 per cent of patients (Albers et al. 2000; Schulz et al. 2004). It can confirm that a new ischaemic cerebrovascular event has occurred in a confused elderly patient with previous strokes or in patients with non-specific symptoms such as confusion or dizziness. The presence of bilateral multiple acute infarcts as shown on diffusion-weighted imaging, may suggest cardioembolism prompting further cardiac investigation whereas one acute infarct with several old infarcts might be more suggestive of a thromboembolic event. Multiple recent infarcts in the anterior circulation of the same hemisphere (Fig. 35.42) suggest critical carotid stenosis (Gass et al. 2004) or proximal middle cerebral artery stenosis (Lee et al. 2005) and warrant urgent imaging of the anterior circulation vessels with referral for surgery where appropriate. Demonstration of posterior circulation infarction may prompt further assessment of the vertebrobasilar vessels. Symptomatic vertebrobasilar stenosis may warrant specific secondary preventive therapy.

Diffusion-weighted imaging is also valuable in the assessment of patients with transient ischaemic attack: of whom around 50 per cent have a focal abnormality if scanned within 24 h, and of whom 25 per cent do not have a lesion correlate on T2-weighted MRI (Kidwell et al. 1999; Ay et al. 2002). Presence of diffusion-weighted imaging abnormality is associated with symptom duration and cortical symptoms. Diffusion-weighted imaging also alters the attending physician's opinion regarding vascular localiaation, anatomical localization, and probable transient ischaemic attack mechanism in a significant number of patients (Albers et al. 2000; Gass et al. 2004; Schulz et al. 2004).

Not all patients with an initial diffusion weighted imaging lesion show a corresponding T2-weighted lesion on follow up MRI (Warach and Kidwell 2004). Taken together, these results suggest that around one-quarter of patients with transient ischaemic attack have cerebral infarction with transient signs in which diffusion-weighted imaging positivity corresponds to cytotoxic oedema which progresses to permanent parenchymal injury and increased tissue water content visible as a lesion on T2-weighted MRI. A distinct subset of patients, around one-fifth, have early diffusion-weighted imaging abnormality but no evidence of later T2-weighted abnormality. This suggests reversibility of the initial diffusion-weighted imaging abnormality if blood flow is restored early enough as seen in patients with stroke in whom the diffusion-weighted imaging lesion may regress with reperfusion. Reversibility of an initial diffusion-weighted imaging lesion is consistent with the development of cytotoxic oedema but without progression to permanent parenchymal injury. Patients with negative diffusion-weighted imaging may not have transient ischaemic attack or

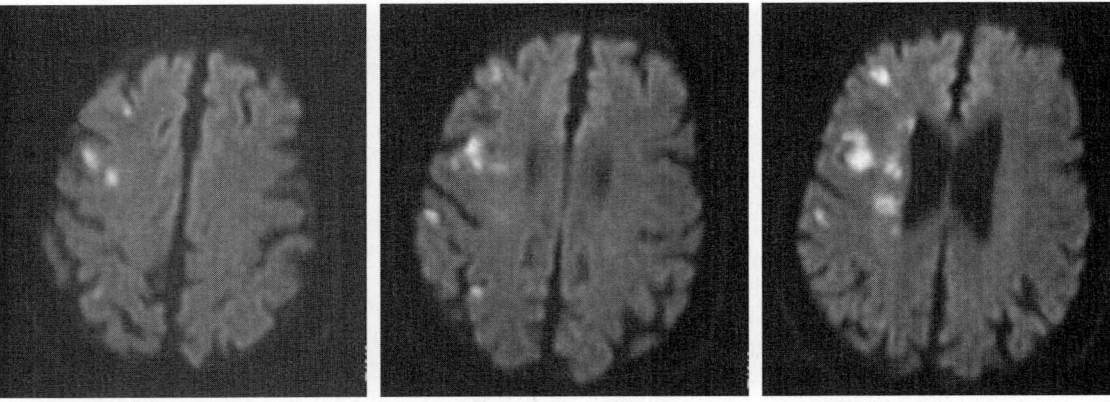

Fig. 35.42 Multiple areas of acute infarction in a patient with a severe right carotid stenosis. Diffusion-weighted MRI.

alternatively may have experienced a very brief period of ischaemia sufficient to disrupt neuronal activity but insufficient to cause cytotoxic oedema.

Many patients with transient ischaemic attack or minor stroke delay seeking medical attention. Often there is a further delay before they are seen by specialist stroke services. In these patients, a clear history may be more difficult to obtain, clinical signs may have resolved, and it may be difficult to make a definite diagnosis of a cerebral ischaemic event or to be certain of the vascular territory or territories involved. Recent studies suggest that diffusion-weighted imaging is also of use in diagnosis and management of patients presenting later with transient ischaemic attack or minor stroke symptoms (Schulz *et al.* 2003, 2004). Diffusion-weighted imaging detects clinically appropriate ischaemic lesions in a high proportion of minor stroke patients when scanned 2 weeks or more after their event (Schulz *et al.* 2004). Interobserver agreement for identifying recent ischaemic lesions in this patient group is much higher for diffusion-weighted imaging than for T2-weighted scans and diffusion-weighted imaging provides useful information over and above T2-imaging in about a third of patients, most commonly by increasing diagnostic certainty and by indicating the vascular territory involved. The presence of diffusion-weighted imaging lesions decreases with time since symptom onset and increases with NIH score and age and is positively associated with stroke rather than transient ischaemic attack, motor deficit, and dysarthria (Schulz *et al.* 2004).

Brain imaging in transient ischaemic attack also provides prognostic information. The presence of presumed recent infarction on CT in patients with transient ischaemic attack is associated with an increased likelihood of recurrent stroke (Douglas *et al.* 2003). Similarly, preliminary studies using diffusion-weighted imaging, suggest that the presence, absence, and pattern of diffusion-weighted imaging lesions in patients with transient ischaemic attack and minor stroke provide prognostic information (Purroy *et al.* 2004; Wen *et al.* 2004; Bang *et al.* 2005).

Diffusion and perfusion MRI and the ischaemic penumbra

There is a need to identify those patients with a small infarct but a large ischaemic penumbra in whom the risk benefit ratio of thrombolysis is likely to be favourable. Also it would be valuable to exclude from treatment patients at high risk of haemorrhagic transformation, or with small lacunar infarcts (Sections 35.10.3–5). Qualitative information on the ischaemic penumbra can be obtained using diffusion-weighted imaging in combination with perfusion-weighted imaging (Kidwell and Hsia 2006; Muir *et al.* 2006).

Perfusion-weighted imaging measures the relative blood flow rate through the brain and can be achieved using an injected contrast agent such as gadolinium or endogenous techniques. The latter have the advantage that they can be used for multiple repeat investigations but at present the level of contrast produced is less that that obtained with exogenous agents. Nearly all published studies of perfusion-weighted imaging in acute stroke have used exogenous contrast agents. In contrast to diffusion-weighted imaging lesion volumes, perfusion-weighted imaging lesion volumes are typically largest acutely and resolve over time. Consistent with earlier positron emission tomography studies, perfusion changes precede the development of diffusion-weighted imaging lesions and in the absence of reperfusion, diffusion-weighted imaging lesions progressively extend over 24 h into the area of reduced perfusion-weighted imaging.

Two patterns of diffusion-weighted imaging and perfusion-weighted imaging abnormalities have been observed in acute ischaemic stroke:

◆ perfusion-weighted imaging<diffusion-weighted imaging, where the volume of abnormal perfusion is less than that of the hyperintense diffusion-weighted imaging signal;

◆ perfusion-weighted imaging>diffusion-weighted imaging, where the volume of abnormal perfusion is greater than the volume of hyperintense diffusion- weighted imaging change.

When perfusion-weighted imaging>diffusion-weighted imaging is found, at least half of patients show an increase in diffusion-weighted imaging lesion volume over the 3–11 days after stroke onset (Sorensen *et al.* 1996; Baird *et al.* 1997; Barber *et al.* 1998a; van Everdingen *et al.* 1998; Beaulieu *et al.* 1999; Karonen *et al.* 1999). Lesion growth is amplified by hyperglycaemia, high haematocrit, old age, and hypoxia (Baird *et al.* 2003; Allport *et al.* 2005; Ay *et al.*

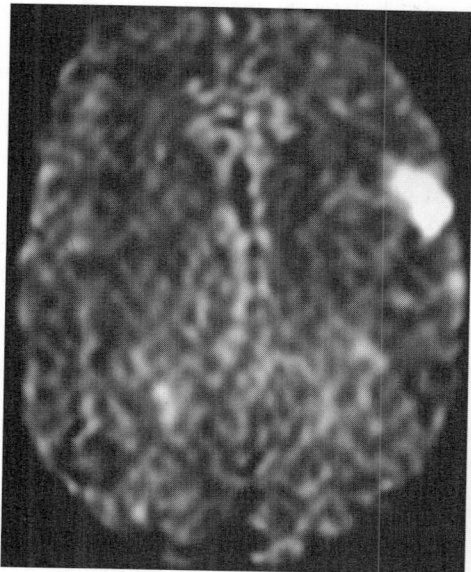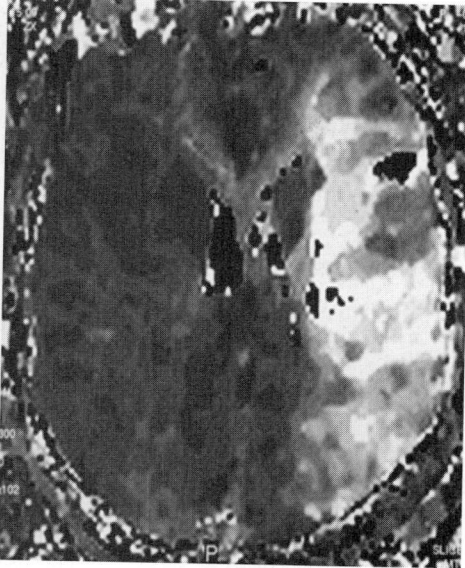

Fig. 35.43 Acute ischaemic stroke showing a perfusion/diffusion mismatch with a relatively small area of high signal on diffusion-weighted MRI (left) but a large area of reduced perfusion on perfusion weighted imaging (right).

2005; Singhal *et al.* 2005). When perfusion-weighted imaging< diffusion-weighted imaging is present initially, no significant lesion growth occurs (Barber *et al.* 1998a). Thus, it has been proposed that the tissue in which there is 'perfusion/diffusion mismatch' i.e. where there is reduced perfusion but not yet diffusion-weighted imaging signal change, represents the ischaemic penumbra and thus tissue that may be salvaged if perfusion can be restored quickly enough (Fig. 35.43).

However, it has recently become apparent that simple diffusion-weighted imaging/perfusion-weighted imaging mismatch does not delineate the ischaemic penumbra accurately. Diffusion-weighted imaging signal change does not correspond exactly to irreversibly infarcted tissue tending to overestimate it, thus part of the diffusion-weighted imaging lesion appears to lie within the ischaemic penumbra (Kidwell *et al.* 2003). Hence, patients with matched diffusion and perfusion deficits may actually have a significant penumbra. Diffusion weighted imaging deficits have been shown to be reversible both spontaneously (Lecouvet *et al.* 1999; Krueger *et al.* 2000) and following reperfusion with thrombolysis (Kidwell *et al.* 2000; Parsons *et al.* 2001). In some cases this initial resolution was temporary with development of recurrent lesions within a week (Kidwell *et al.* 2002). Regarding perfusion-weighted imaging, the size of the hypoperfused area will vary according to the method of measurement used: prolonged mean transit time, reduced cerebral blood volume. or reduced cerebral blood flow, of which only the latter reflects true hypoperfusion, this hampers comparisons between studies. Hypoperfused areas will include oligaemic areas as well as areas at risk of infarction and thus perfusion-weighted imaging will tend to overestimate the boundary of the penumbra. Given the fact that diffusion-weighted imaging change does not correspond exactly to irreversibly damaged tissue, severity of perfusion deficit has been proposed as a surrogate for subsequent infarction (Thijs *et al.* 2001; Fiehler *et al.* 2002; Shih *et al.* 2003) and as a potential tool for use in clinical selection of patients for thrombolysis. However, there is still no consensus on the criteria for patient selection (Hand *et al.* 2006b) (Section 35.10.4).

Perfusion CT

Recently, there has been increasing interest in the use of new CT methods (Wintermark and Bogousslavsky 2003) to examine cerebral blood flow. Cerebral blood flow measurement using CT can be done using existing CT-based technology, is easier to perform in sick patients than MRI, and is not contraindicated in those with pacemakers, ferromagnetic implants, mechanical heart valves, and those with claustrophobia. CT perfusion using exogenous contrast enables calculation of cerebral blood flow, cerebral blood volume, and mean transit time. However, quantification of cerebral blood flow is problematic, as it is in perfusion-weighted imaging, and most studies use ratios comparing values with homologous areas of the contralateral hemisphere which itself may show a reduced cerebral blood flow in the acute stroke period owing to diaschisis. In addition, current technology allows only limited brain coverage of 2–4 slices which means that large areas of ischaemia are imaged inadequately and small ones may be missed altogether.

Despite the limitations described above, there is evidence that perfusion CT values correlate with angiographic findings and diffusion-weighted imaging/perfusion-weighted imaging changes and can predict infarction and clinical outcome (Nabavi *et al.* 2002; Meuli 2004; Parsons *et al.* 2005; Wintermark *et al.* 2007), and may be able to identify patients likely to benefit from thrombolysis. It is likely that CT measurements of cerebral blood flow will become increasingly important in selecting patients for acute stroke therapy owing to the speed and relative ease of performing CT as compared to MRI.

35.8.4 Imaging the cerebral circulation

The various methods of imaging the cerebral circulation are described in detail in Section 3.2.

Imaging the cerebral circulation in ischaemic stroke

The main indications for imaging the cerebral circulation in transient ischaemic attack and ischaemic stroke include:

- selection of patients for carotid surgery;
- demonstration of arterial dissection;
- demonstration of vessel occlusion and thus consideration of thrombolysis;
- demonstration of intracranial venous thrombosis;
- demonstration of the site of large vessel stenosis in frequent vertebrobasilar ischaemia;
- subclavian steal syndrome;
- where there is suspicion of an aneurysm of the extra- or intracranial circulation large enough to contain embolizing thrombus;
- fibromuscular dysplasia and other vasculopathies; and
- cerebral vasculitis.

Cerebral circulation imaging in intracerebral haemorrhage

Cerebral angiography to reveal a vascular malformation or aneurysm is usually reserved for young patients with relatively mild strokes, given the predominance of bleeds due to either hypertension or amyloid angiopathy in older patients. Vascular malformations and aneurysms are most likely if the cerebral haemorrhage is lobar or intraventricular and there is no hypertension (Zhu *et al.* 1997). MRI can often be more sensitive than angiography in picking up vascular malformations, particularly cavernomas, and MR angiography is increasingly able to pick up aneurysms of a size likely to rebleed.

Imaging the carotid bifurcation

The aim of carotid bifurcation imaging is accurate measurement of carotid stenosis, and the plaque characteristics if possible, since this enables selection of patients for carotid endarterectomy. The 'gold-standard' is still intra-arterial selective catheter angiography, which has good inter-observer reliability (Rothwell *et al.* 1994a,b), but which is no longer used as the first-line investigation in most centres because of the cost and the procedural risk of stroke.

Angiography also provides useful information on atherothrombotic plaque surface morphology, particularly ulceration and occasionally intraluminal thrombus. However, there is only moderate inter-observer agreement in the angiographic diagnosis of 'ulceration' (Rothwell *et al.* 1998) and there is usually no data on the reproducibility of histological assessment in imaging–pathology comparisons (Lovett *et al.* 2005). Nevertheless, angiographic irregularity or ulceration of a stenotic plaque does predict an increased risk of stroke independent of the degree of carotid stenosis (Eliasziw *et al.* 1994; Rothwell *et al.* 2000a) and there is good correlation between angiographic findings and rigorously evaluated plaque histology (Lovett *et al.* 2004). What effect the angiographic demonstration of a 'floating thrombus' has on the risk of stroke is unknown (Martin *et al.* 1992). There are still problems in the standardization and reproducibility of non-invasive imaging of plaque morphology (Arnold *et al.* 1999; Gronholdt 1999; Fenster *et al.* 2006). Thus, there are currently no large prospective cohort studies showing that non-invasive imaging of plaque morphology can predict stroke.

The development of non-invasive methods of imaging the carotid bifurcation including duplex sonography, and CT or MR angiography have meant that catheter angiography is no longer routinely used for this purpose. Non-invasive measurements of carotid stenosis must be 'translated' to what would have been measured if angiography had been performed (Rothwell and Warlow 1996). Unfortunately, the literature remains poor comparing the accuracy of non-invasive methods versus angiography (Blakeley *et al.* 1995; U-King-Im *et al.* 2005; Chappell *et al.* 2006). Nonetheless, with stringent quality control and confirmation of stenosis by an independent observer, duplex sonography is now the most common way to diagnose severe carotid stenosis.

CT angiography continues to evolve as an imaging technique for the carotid arteries (Brink *et al.* 1997; Bartlett *et al.* 2006). It provides multiple viewing angles and three-dimensional reconstruction (Heiken *et al.* 1993; Leclerc *et al.* 1995; Nandalur *et al.* 2006) but requires the injection of contrast. Although non-invasive and safe, 'time of flight' MR angiography is not accurate enough to estimate carotid stenosis reliably at the present stage of development (Graves 1997; DeMarco *et al.* 2006). The severity of the stenosis tends to be overestimated, there may be a flow gap distal to a stenosis (Fig. 35.44), making precise stenosis measurement impossible, irregularity or ulceration are not well seen, and severe stenosis can be confused with occlusion (Siewert *et al.* 1995; Levi *et al.* 1996; Fox *et al.* 2005). However, image quality and reproducibility of measurement of stenosis are significantly improved with contrast-enhanced MRA (Mitra *et al.* 2006; DeMarco *et al.* 2006).

Cervical arterial dissection

Although arterial dissection tends to heal without any treatment, and optimal treatment is uncertain anyway, it is still important to make a definitive diagnosis. Until recently, the 'gold-standard' imaging technique was intra-arterial catheter angiography because ultrasound was neither specific nor sensitive enough. However, there is now a widespread consensus that cross-sectional MRI, to show thrombus within the widened arterial wall, combined with MRA, is the safest and best option (Fig. 35.26).

Imaging the posterior circulation

Contrast-enhanced MR or CT angiography are the most useful non-invasive methods of imaging the posterior circulation, although conventional arterial angiography is often still necessary in order to confirm or exclude significant stenosis.

Subclavian steal

Although asymptomatic subclavian steal is quite common, with reversed vertebral artery flow detected by ultrasound or vertebral angiography, symptomatic subclavian steal is rare. Innominate artery steal is even rarer, with retrograde vertebral artery flow distal to innominate rather than subclavian artery occlusion (Kempczinski and Hermann 1979; Grosveld *et al.* 1988).

Transcranial Doppler sonography

At present, transcranial Doppler sonography (Section 3.2.8) has a limited but increasing role in clinical management of transient ischaemic attack and ischaemic stroke. This includes monitoring during carotid endarterectomy, acceleration of clot lysis in combination with thrombolytic therapy in acute ischaemic stroke (Alexandrov *et al.* 2004; Kim *et al.* 2005), the diagnosis of patent foramen ovale, assessing stroke risk in patients with carotid stenosis (Babikian *et al.* 1997; Molloy and Markus 1999; Dittrich *et al.* 2006; Markus 2006), display of intracranial arterial occlusion and stenosis, and assessment of cerebrovascular reactivity.

Embolus detection may help in distinguishing cardiac and aortic arch from carotid emboli. In the former, emboli should be detected

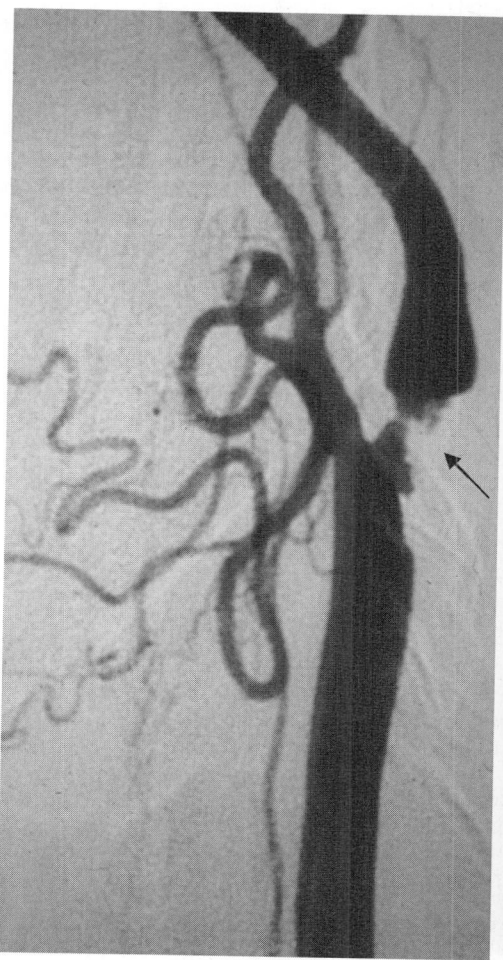
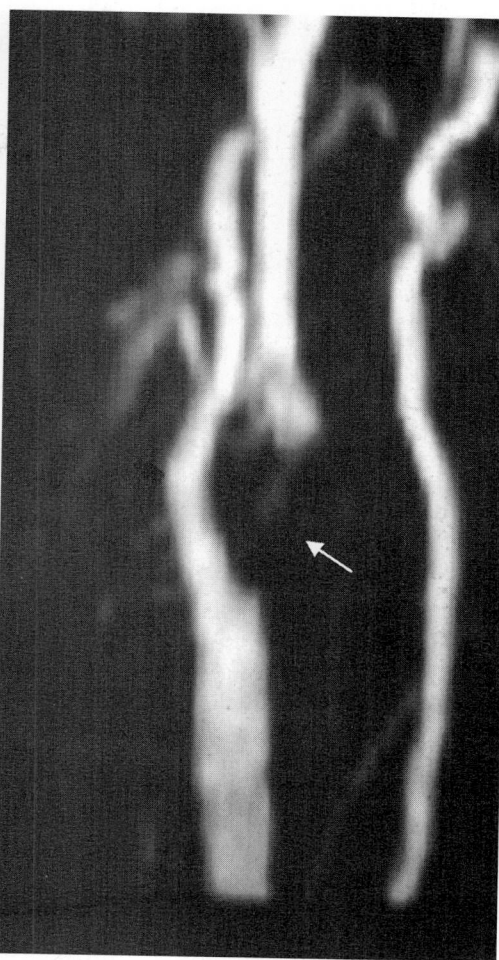

Fig. 35.44 A tight carotid stenosis shown clearly on catheter angiography (arrow) (left) but obscured by flow-void on time-of-flight MR angiography (arrow) (right).

in several arterial distributions, whereas in carotid disease emboli should be seen only in the one arterial distribution distal to the supposed embolic source (Markus *et al.* 1994; Sliwka *et al.* 1997). However, the frequency of emboli is usually so low and variable that their detection requires prolonged monitoring and automation (Markus *et al.* 1999; Markus 2006).

Transcranial Doppler sonography can also be used to assess cerebrovascular reactivity. Impaired reactivity may have some prognostic significance for identifying individual patients at particularly high risk of stroke from amongst those with carotid stenosis or occlusion, although the prognostic value is not yet clearly defined (Derdeyn *et al.* 1999, 2005).

35.9 General treatment of acute strokes

The general treatments described in this section are applicable to all patients with acute stroke regardless of aetiology. Treatment for specific subtypes is detailed later in this chapter. Therapy for acute stroke can be divided into:

- treatment of the acute event in which the aim is to minimize mortality, impairment and disability and reduce the complications of stroke; and

- prevention of recurrent stroke.

For patients who have suffered a major disabling stroke emphasis is placed, at least initially, on treatment whereas in patients with transient ischaemic attack or minor stroke, the emphasis is on prevention.

35.9.1 Non-neurological complications

Non-neurological complications after acute stroke are more frequent with increasing age, pre-stroke disability, stroke severity, and poor general nursing and other care (Table 35.19). To some extent the site of the lesion may also be relevant; for instance obstructive and central sleep apnoea might occur more often in brainstem stroke (Davenport *et al.* 1996a; van der Worp and Kappelle 1998). Early detection and prevention of complications depends on clinical monitoring (Table 35.20).

Pneumonia

Pneumonia is a common complication in elderly patients confined to bed. Chest infection is particularly common after stroke because of impairment in swallow and cough reflex, poor respiratory movement, and pulmonary embolism. The risks can be reduced by good nursing and chest physiotherapy. A pharyngeal airway may be required, particularly in drowsy patients or after a brainstem stroke and ventilation may be considered in certain patients.

Table 35.19 General medical complications of acute stroke

Pneumonia
Venous thromboembolism
Urinary incontinence and infection
Pressure sores
Cardiac arrhythmias, failure, myocardial infarction
Fluid imbalance, hyponatreamia
'Mechanical' problems
    spasticity
    contractures
    malalignment/subuxation/frozen shoulder
    falls and fractures
    osteoporosis
    ankle swelling
    peripheral nerve pressure palsies
Mood disorders
Gastric 'stress' ulceration and haemorrhage
Central post stroke 'thalamic' pain

Venous thromboembolism

About 50 per cent of hemiparetic patients in hospital develop a deep vein thrombosis in their paralysed leg, although this is not usually detectable clinically. However, a swollen and painful leg compromises rehabilitation. A resultant pulmonary embolism causes hypoxia, pneumonia, impacts on neurological recovery, and may cause death (Wijdicks and Scott 1997). Evidence from clinical trials of thromboembolism prophylaxis in a variety of postoperative conditions suggests that subcutaneous heparin, graduated compression stockings, and aspirin are effective in reducing the risk of deep venous thrombosis in bedridden patients (Andre *et al.* 2007) (Section 2.8.3). However, routine heparin prophylaxis is not recommended after stroke because of increased cerebral haemorrhage. Aspirin should be given after ischaemic stroke since this reduces the risk of recurrent stroke as well as of venous thromboembolism. There is no evidence that compression stockings are effective specifically after stroke but they are the method of choice for thromboprophylaxis after primary intracerebral haemorrhage. A large multicentre randomized controlled trial of low-dose heparin versus compression stockings in ischaemic stroke is currently ongoing (Dennis 2004).

Urinary incontinence

Incontinence is common after stroke, and may be permanent (Brittain *et al.* 1998). Catheterization is often required to maintain

Table 35.20 Routine monitoring in acute stroke

Vital signs	Respiratory rate and rhythm, and blood gases
	Heart rate and rhythm (often with electrocardiogram monitor)
	Blood pressure (normal arm)
	Temperature (normal axilla)
Neurological	Conscious level (Glasgow Coma Scale)
	Pupils
	Limb weakness
	Epileptic seizures
General	Fluid balance
	Electrolytes and urea
	Blood glucose
	Haematocrit

skin care, at least initially. Urinary infection is common owing immobility and the use of urinary catheters.

Pressure sores

Pressure sores may occur secondary to poor nursing, incontinence, or malnourishment. They may become infected and take months to heal, thus delaying rehabilitation. Pressure sores can be avoided by attention to pressure areas, use of appropriate mattresses and supports, and by regular turning of immobile patients (Effective Health Care 1995).

Cardiac complications

Electrocardiographic ST depression, T-wave flattening and inversion, U waves, and a prolonged Q-T interval are common but transient occurrences after acute ischaemic, and particularly after acute haemorrhagic stroke. They seldom cause clinical problems. Some abnormalities may have preceeded the stroke (Oppenheimer *et al.* 1990). It is unknown whether monitoring the electrocardiogram improves the prognosis, but it is often recommended, particularly after subarachnoid haemorrhage. Some patients have a rise in blood cardiac troponin after stroke and there is some evidence that this is more likely after involvement of the insula, particularly on the right, and may be linked to high circulating catecholamine levels (Cheshire and Saper 2006). However, it is not clear whether elevated cardiac troponin is an independent predictor of poor outcome after stroke.

Fluid imbalance

Patients unable to take sufficient fluids orally require fluid by nasogastric tube or intravenous hydration. Hyponatraemia, probably reflecting salt wasting and the stress response, is particularly common after subarachnoid haemorrhage and, in general, should be treated by plasma volume expansion and not fluid restriction. Urinary tract infection and dehydration may cause renal failure.

Mechanical problems

Spasticity, muscle contractures, painful shoulder and other joints of a paralysed limb, malalignment or subluxation of the shoulder, falls and fractures can all potentially be avoided by good nursing and physiotherapy. Osteoporosis in a paralysed limb presumably increases the risk of fractures, but may be unavoidable (Sato *et al.* 1998).

Acute gastric ulceration

Gastric ulceration with or without haemorrhage or perforation, is a well-recognized but rare complication in severe stroke and may be difficult to prevent (Davenport *et al.* 1996b).

Mood disorders

A systematic review of observational studies of post-stroke depression produced an estimated overall prevalence of 33 per cent among all stroke survivors (Hackett *et al.* 2005). Predictors of depression include severity of stroke and cognitive impairment (Hackett ML, Anderson CS (2005). It has been postulated that left-sided brain lesions are more likely to cause depression but this remains unproven (Bhogal *et al.* 2004). Mood disorder may impede rehabilitation and contribute to disability and handicap but usually improve with time. Treatment includes support and counselling and antidepressants.

Central post-stroke pain

Post-stroke, or 'thalamic', pain is a burning, severe, and paroxysmal pain exacerbated by touch and other stimuli (Section 17.2). Such post stroke pain is rare and usually occurs weeks or months after stroke (Nasreddine and Saver 1997; Frese *et al.* 2006). There are usually some sensory signs in the affected areas. The lesion is usually located in the contralateral thalamus but may lie elsewhere in the central sensory pathways. Treatment includes anticonvulsants, amitryptiline, and various sorts of counter stimulation but is often unsuccessful.

35.9.2 Generic treatments

Many of the interventions described below aim to prevent physiological changes such as hypotension, hyperglycaemia, and pyrexia. These might result in secondary ischaemic injury through excacerbating the flow/metabolism mismatch in the penumbral and oligaemic areas surrounding the infarcted core resulting in extension of infarction into these areas.

Stroke units

There is evidence that admission to an acute stroke unit reduces mortality and morbidity and dependency (Stroke Unit Trialists' Collaboration 2000). The benefits are seen for all ages, both sexes, and across the range of stroke severity. Compared to care on the general medical wards, care in a stroke unit reduces mortality, and increases the chances of independence (Stroke Unit Trialists' Collaboration 1997, 2000). Stroke unit care also reduces length of stay and is cost-effective in comparison to care on general wards with or without mobile specialist team input and domiciliary care (Patel *et al.* 2004; Moodie *et al.* 2006). The beneficial effects of stroke unit care are also maintained long-term (Fuentes *et al.* 2006).

Blood pressure

The optimal management of blood pressure in acute stroke is uncertain. Ischaemic and infarcted brain cannot autoregulate and so relatively modest increases in cerebral perfusion pressure can cause hyperaemia, increased cerebral blood flow, cerebral oedema, and haemorrhagic infarction whereas, a fall in cerebral perfusion pressure may exacerbate cerebral ischaemia. Blood pressure is often elevated on admission but tends to fall spontaneously during the first few days (Bath and Bath 1997). Arguments can be made for increasing or decreasing blood pressure after acute stroke, but reiable evidence from randomized trials of either policy is lacking (Blood pressure in Acute Stroke Collaboration 2001; Mistri *et al.* 2006). In practice, many clinicians continue existing antihypertensive therapy, although there is no evidence to support this, but only consider active treatment of blood pressure for sustained blood pressure of >220/120 mmHg or >185/105 mmHg in cerebral infarction and cerebral haemorrhage respectively as cerebral autoregulation is disturbed after stroke and excessive blood pressure lowering may compromise cerebral perfusion. Exceptions occur where there is coexistent hypertensive encephalopathy, aortic dissection, acute myocardial infarction, or severe left ventricular failure. However, the above guidelines are not based on good evidence and two large ongoing randomized controlled trials should provide more reliable guidance within the next few years (Willmot *et al.* 2006; Mistri *et al.* 2006).

Hypoxia

Damaged brain appears to have impaired responsiveness to PaCO2 and PaO2 as well as impaired autoregulation and perfusion reserve. This increases the likelihood of further 'secondary' insults such as systemic hypoxia, hypotension, and raised intracranial pressure (Cormio *et al.* 1997). There are good theoretical reasons why routine oxygen therary might be beneficial, but there are also potential deleterious effects (Singhal 2007) and so randomized trials are required. Oxygen should be given if hypoxia is detected. A randomized trial of routine oxygen supplementation is ongoing (Ali *et al.* 2006).

Hyperglycaemia

Hyperglycaemia after acute stroke is a common finding that has consistently been associated with an increased risk of death. The hypothesis that maintenance of euglycaemia post stroke would improve outcome was tested in the UK Glucose Insulin in Stroke Trial, GIST-UK (Gray *et al.* 2007). Infusion of glucose-potassium-insulin to maintain capillary glucose at 4–7 mmol/l showed no significant reduction in mortality or severe disability at 90 days. The authors concluded that treatment within the trial protocol was not associated with significant clinical benefit, although the study was underpowered and the difference in glycaemic control between the treatment groups was not large.

Fever

Fever may occur after stroke for a number of reasons (Table 35.21). Animal studies and observational data in man suggest that pyrexia increases infarct size and is associated with poor outcome. The converse is true for hypothermia but the risks and benefits of active cooling in acute stroke are yet to be determined (Reith *et al.* 1996; Hemmen and Lyden 2007; Sacco *et al.* 2007). Infection should be treated promptly but there is no evidence to support the routine use of antipyretics such as paracetamol for temperatures above 37.5°C although they are widely used in practice.

Dehydration and swallowing

Dehydration should be corrected with intravenous fluid replacement and nasogastric feeding should be instituted in those with unsatisfactory swallow. Impaired swallowing with risk of aspiration and pneumonia is common in drowsy patients with severe hemispheric strokes, and those with brainstem strokes. It almost always gets better in days or weeks (Hamdy *et al.* 1997). Swallowing difficulty is tested by asking the patient to sip some water, observing any tendency to choke in the next minute or so, and for added sensitivity using simple quantification (Hinds and Wiles 1998; Mari *et al.* 1997).

Nutrition

Feeding in the first few days may not be important but later the patient should be kept well nourished since poor nutrition may be

Table 35.21 Causes of fever after stroke

Infection	
	Urine
	Pneumonia
	Pressure sores
	Septicaemia
	Intravenous access site
Deep vein thrombosis and pulmonary embolism	
Infective endocarditis	
Drug reaction	

associated with worse outcome. There is no evidence to support early initiation of percutaneous endoscopic gastrostomy feeding in patients with unsafe swallow (Dennis *et al.* 2005).

Early mobilization

Mobilization may reduce complications including pneumonia, deep vein thrombosis, pulmonary embolism, and pressure ulcers. At present there are no reliable data to recommend the use of compression stockings or low dose heparin for deep vein thrombosis prohylaxis (Andre *et al.* 2007) but trials are ongoing (Dennis 2004).

35.10 Specific treatments for acute ischaemic stroke

Many of the trials of therapy in acute stroke did not distinguish between stroke subtypes other than by division into haemorragic and ischaemic stroke. Therefore, there is little evidence for different effectiveness for most acute ischaemic stroke treatments according to stroke subtype and location. On the other hand, stroke subtype determines patient selection for specific secondary preventive strategies. Thus better characterization of stroke will aid overall patient management (Section 35.14).

Therapies to reduce brain damage in ischaemic stroke may act by prevention of thrombus extension using antiplatelet agents, restoring blood flow and thus reducing infarction within the penumbra by thrombolysis, and increasing cerebral resistance to ischaemia by use of neuroprotective agents.

35.10.1 Antiplatelet therapy

Patients with acute stroke should be treated with aspirin as soon as practicable after brain imaging has excluded haemorrhage. One systematic review (Sandercock *et al.* 2003) including two very large randomized controlled trials (International Stroke Trial Collaborative Group 1997; CAST 1997) has clearly established that starting aspirin therapy within the first 48 h of acute ischaemic stroke avoids death or disability at 6 months for about 10 patients per 1000 patients treated. A further 10 patients per 1000 treated will make a complete recovery. Both intra-cranial and extra-cranial haemorrhage are reported with aspirin therapy, but the rates are low, and are off-set by the benefit of extra lives saved.

There is no clear consensus about whether aspirin should be given prior to brain imaging. This applies to situations where access to imaging is delayed, or where drugs could be administered by ambulance staff. Analysis of outcome in the subgroup of patients who were randomized and who received treatment prior to brain imaging, some of whom subsequently turned out to have primary intracerebral haemorrhage, did not show any obvious difference in risk and benenfit from that in the rest of the trial (International Stroke Trial Collaborative Group 1997).

There is no clear evidence that any particular dose of aspirin is more effective than others. However, the symptoms of aspirin toxicity, such as dyspepsia and constipation, are dose related, so the smallest effective dose should be used. A starting dose of 150–300 mg per day is advised for the acute phase of ischaemic stroke followed by long-term treatment with 75–150 mg per day. Patients intolerant of aspirin should be treated with clopidogrel if available, or if not, with dipyridamole. These newer agents are significantly more expensive than aspirin. The use of combination antiplatelet therapy is discussed further in the section on secondary prevention.

35.10.2 Anticoagulation

Immediate therapy with systemic anticoagulants including unfractionated heparin, low molecular weight heparin, heparinoids, or specific thrombin inhibitors in patients with acute ischaemic stroke is not associated with net short- or long-term benefit (International Stroke Trial Collaborative Group 1997; Berge 2007; Wong *et al.* 2007). These agents reduce the risk of deep venous thrombosis and pulmonary embolus, but are associated with a significant risk of intra-cranial haemorrhage, which is dose dependent. Patients in atrial fibrillation after presumed ischaemic stroke or transient ischaemic attack benefit from anticoagulation in the long-term to prevent further stroke. However, the best time to start therapy after an ischaemic stroke is unclear as the risk of haemorrhagic transformation is difficult to predict (International Stroke Trial Collaborative Group 1997; O'Donnell *et al.* 2006).

35.10.3 Intravenous thrombolysis

Thrombolysis aims to reduce the volume of infarcted brain by recanalizing the occluded vessel and restoring blood flow. Restoration of blood flow may not necessarily always be beneficial: studies in animals suggest that reperfusion of acutely ischaemic brain may actually be harmful through the release of free radicals and toxic products into the circulation. Secondly, thrombolysis will probably not be of benefit if infarction is completed or if the ischaemic penumbra is small. Finally, thrombolysis may cause haemorrhagic transformation of the infarct or extra-cranial bleeding.

There have been a number of trials of thrombolysis in acute stroke using various different thrombolytic agents, at different doses and all the trials had strict entry criteria including age cut offs. The relative effectiveness and the most effective doses of the various thrombolytic agents are unclear owing to the lack of comparative data. Systematic reviews and meta-analysis (Wardlaw *et al.* 2003) of intravenous thrombolytic therapy using streptokinase, urokinase, or rtPA, suggest some benefit from thrombolysis given within the first 6 h of stroke. There was a significant reduction in combined death or dependency at the end of follow up, with 64 fewer patients dead or dependent at the end of follow up per 1000 patients treated. This was despite a significant increase in the number of intra-cranial haemorrhages and deaths, an additional 73 haemorrhages per 1000 patients treated, in other words a 3.5-fold excess and in the numbers of deaths, amounting to an additional 46 deaths per 1000 patients treated. There was significant heterogeneity between trials making the overall estimate difficult to interpret but patients treated within 3 h had better outcome. Some studies included only patients with middle cerebral artery territory infarction such as ECASS, the European and Australian Acute Stroke Study (Hacke *et al.* 1998) whereas there was no such restriction in others, such as NINDS (National Institute of Neurological Disorders and Stroke rt-PA Study Group (1995). Numbers were too small to examine the effect of thrombolysis on stroke subtypes.

The NINDS study (National Institute of Neurological Disorders and Stroke rt-PA Study Group (1995) of r-tPA within 3 h of stroke onset in 624 patients, reported a larger treatment effect than that seen overall for all studies included in the systematic reviews. An additional 160 patients per 1000 patients treated were alive and independent at 3 months in the thrombolysis group representing a number needed to treat of 7 to prevent all death. This was despite an increase in the incidence of cerebral haemorrhage of an excess of 58 bleeds per 1000 patients treated. The improved outcome in the thrombolysis group was maintained at 1 year (Kwiatkowski *et al.* 1999).

The less striking benefit from thrombolysis indicated by the systematic reviews compared to that seen in the NINDS trial may result from methodological differences between trials. These included the fact that many of the earlier trials randomized patients up to 6 h from stroke onset, and the likely pathophysiological variability in the patients studied. Further, in the NINDS trial, only carefully selected patients were included and these strict inclusion and exclusion criteria have made these trial results difficult to generalize to clinical practice. One study (Jorgensen *et al.* 1999) showed that only 5 per cent of patients admitted to hospital in Denmark with ischaemic stroke would fulfil the entry criteria for this trial and that even when the time limit was ignored, only 45 per cent would be eligible for thrombolysis.

A pooled analysis of individual patient data from the NINDS, ECASS I and II and ATLANTIS trials (Hacke *et al.* 2004), which represents 99 per cent of patients randomized in trials of rt-PA in stroke, has confirmed that benefit from thrombolysis decreases with time since stroke onset, being most beneficial if given within 90 min, with an odds ratio of 2.8 for favourable outcome (Fig. 35.45), although benefit was still present at 4.5 h. However, it is clear that many patients will not benefit at this time point and conversely that other patients may benefit up to and beyond 6 h in view of the fact that the ischaemic penumbra may extend for much longer periods in some patients (Baron 2001). At present, only 1–6 per cent of patients receive thrombolysis in North America because of the perceived risks of treatment and the small fraction of patients who present within the 3-h time window.

Most of the data from the thrombolysis trials, together with most of the ongoing studies of diffusion-weighted imaging/ perfusion-weighted imaging and CT perfusion concern embolic middle cerebral artery territory strokes. Thus, there is uncertainty over whether lacunar or posterior circulation strokes should receive thrombolysis. The NINDS study found no difference in benefit for thrombolysis in lacunar as compared to other stroke but numbers were small and diagnosis of lacunar stroke was based on clinical syndrome and CT. Since lacunar stroke overall has a good prognosis and thrombolysis may be more likely to precipitate haemorrhage in the presence of small vessel disease, thrombolysis may not be beneficial but at present there are no data to guide management. Similarly there are few data on posterior circulation stroke although substantial diffusion-weighted imaging–perfusion-weighted imaging mismatch has been seen following basilar occlusion (Ostrem *et al.* 2004) and benefit of thrombolysis has been suggested as long as 7 h after stroke onset (Lindsberg *et al.* 2004; Montavont *et al.* 2004).

Several other uncertainties remain concerning thrombolysis after stroke including the length of the treatment window, the risk/ benefit ratio in older patients, the factors predicting intracranial haemorrhage, and the brain imaging appearances that predict response to treatment. This together with concern that there is a trend towards excess deaths with thrombolysis, has led to a lack of consensus regarding the use of thrombolysis and the criteria for selecting patients. Current guidelines vary but advise that thrombolysis should only be administered by a physician with expertise in stroke. It should be avoided in severe stroke, where there is early CT change indicating severe infarction or more than 3 h after stroke onset except as part of a randomized trial.

35.10.4 Radiological selection of patients for intravenous thrombolysis

Radiological investigation is essential in selecting patients for thrombolysis, principally by excluding haemorrhage. It also identifies those patients with large completed infarcts, a small ischaemic penumbra, or transient ischaemic attack in whom thrombolysis would not be beneficial. The majority of centres currently rely on CT rather than MRI as the first line investigation in stroke.

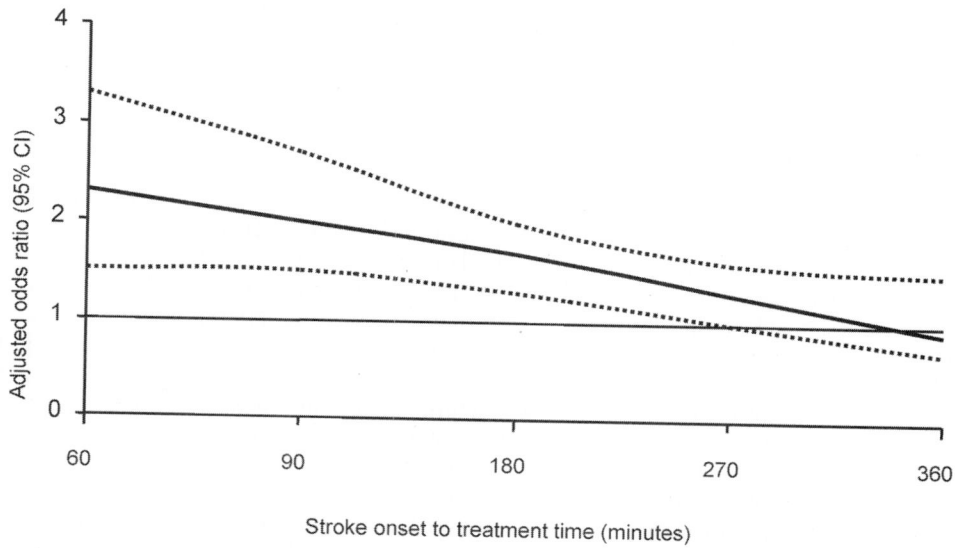

Fig. 35.45 Thrombolysis treatment for acute ischaemic stroke. Odds ratio (95 per cent CI) of a favourable outcome at 3 months follow-up in patients treated with thrombolysis compared with controls by time from stroke onset to treatment (minutes). This shows a decrease in benefit from treatment with time based on a pooled analysis of 2776 patients from randomized controlled trials (Hacke *et al.* 2004).

Retrospective analysis of CT scans has suggested that early signs of extensive infarction on CT, corresponding to a very poor ASPECTS score of <2 (Weir *et al.* 2006) are associated with 85 per cent mortality without thrombolysis and poor outcome, including haemorrhagic transformation, after thrombolysis (Von Kummer 1994, 1997; Hacke *et al.* 1998; Dzialowski *et al.* 2006). These observations have led to the introduction of the 'one-third rule', that is patients with signs of infarction of greater than one-third of the middle cerebral artery territory on CT should not be thrombolysed. However, overestimation of the degree of ischaemic change on CT may lead to thrombolytic treatment being withheld inappropriately (Barber *et al.* 1999).

The theory that ischaemic penumbra represents salvageable tissue has led to the proposal that thrombolysis is likely to be most effective in those patients with a small diffusion-weighted imaging lesion and a large perfusion-weighted imaging lesion; the so called DWI–PWI mismatch (Section 35.8.3). Such mismatch is present in around 70 per cent of patients with anterior circulation stroke scanned within 6 h of onset, and is strongly associated with middle cerebral artery occlusion. The area of DWI–PWI mismatch does not precisely delineate the penumbra in that the diffusion-weighted imaging area includes penumbra as well as infarct and the mismatch area contains oligaemia as well as penumbra. However, the mismatch area appears reasonably robust in representing tissue at risk of infarction: resolution of hypoperfusion from early revascularization prevents diffusion-weighted imaging expansion and use of r-tPA is associated with early resolution of perfusion-weighted imaging lesions in less than 36 h, reduced diffusion-weighted imaging lesion growth, smaller final stroke volumes, and better clinical outcome on follow up (Hjort *et al.* 2005b).

In patients with matched diffusion-weighted imaging–perfusion-weighted imaging lesions, it is unclear whether reperfusion will be of benefit since it is uncertain whether the diffusion-weighted imaging area still contains a significant penumbra or is already entirely irreversibly damaged. In these cases, MR angiographic findings and the clinical picture should be taken into account. A severe neurological deficit and middle cerebral artery occlusion predicts a malignant middle cerebral artery infarction. A diffusion-weighted imaging lesion with normal or increased perfusion indicates spontaneous recanalization and is inappropriate for thrombolysis.

MR and CT angiography have been used to assess patients in the acute phase after stroke and have been proposed as a possible means of selecting patients for thrombolysis. It has been observed that patients with a patent middle cerebral artery typically have smaller perfusion-weighted imaging volumes than patients with evidence of large vessel occlusion: absent middle cerebral artery flow on MR angiogram is significantly associated with a greater PWI/DWI mismatch, larger acute diffusion-weighted imaging lesions, secondary lesion growth, larger final infarct size, and worse final outcome (Barber *et al.* 1999). However, a significant minority of 30–35 per cent of patients with middle cerebral artery occlusion do not show progressive lesion enlargement and patients without middle cerebral artery occlusion may show significant PWI/DWI mismatch.

Although the recent advances in understanding of perfusion-weighted imaging and diffusion-weighted imaging changes in acute stroke suggest that these techniques may be used to select patients for thrombolytic therapy by demonstrating those that have residual salvageable tissue, at present there is no consensus on the exact criteria to use in patient selection (Hand *et al.* 2007). Many centres thrombolyse selected patients within 3 h of stroke onset according to local criteria including brain imaging findings, often CT based. There is ongoing interest in the use of diffusion-weighted imaging and perfusion-weighted imaging parameters to select patients in the 3–6 h time window or beyond particularly since thrombolysis has been shown to be potentially effective in improving clinical outcomes when given 3–9 h after stroke onset (Furlan *et al.* 2006). Given the good correlation of perfusion CT with diffusion-weighted imaging and perfusion-weighted imaging, there is similar interest in the use of CT to select patients for thrombolysis.

As seen earlier, patients with middle cerebral artery occlusion are likely to show evidence of a large ischaemic penumbra and thus to benefit from thrombolysis. This is not surprising given that the aim of thrombolysis is to restore cerebral perfusion. Occlusion of the middle cerebral artery can be demonstrated on MR or CT angiography at the time of diffusion-weighted imaging/perfusion-weighted imaging or perfusion CT imaging or with transcranial Doppler ultrasonography. Transcranial Doppler ultrasonography aimed at the occluded part of a vessel may help expose thrombi to the effect of thrombolytic agents and thus augment the effect of thrombolysis. The CLOTBUST II, ultrasound-enhanced systemic thrombolysis for acute ischemic stroke, trial tested this hypothesis (Alexandrov *et al.* 2004). Patients with acute ischaemic stroke and middle cerebral artery occlusion presenting within 3 h after stroke onset received intravenous rt-PA and were randomized to receive additional transcranial Doppler ultrasonography or placebo. The treatment group showed a significantly higher rate of recanalization and a nonsignificant trend towards more favourable outcome. This suggests that transcranial Doppler ultrasonography may indeed enhance the effectiveness of intravenous thrombolysis. There are two ongoing trials assessing this technique further: MUST, the Microbubbles and Ultrasound in Stroke Trial, and PULSE, the Pilot UltraSound Lysis Early treatment study.

35.10.5 Implementation of acute stroke imaging

Assuming equal access to diffusion-weighted imaging/perfusion-weighted imaging and CT perfusion, decisions regarding optimum imaging use can be dictated by the time elapsed since stroke onset and the symptom severity.

0–3 h after stroke onset

All existing trials used CT to select patients for thrombolysis. This remains the brain imaging method used in most centres for patients within the 0–3 h time window. Patients with haemorrhage or with signs of extensive infarction are in general excluded from thrombolysis. This latter exclusion remains a subject of debate since data from the NINDS trial (National Institute of Neurological Disorders and Stroke rt-PA Study Group (1995) do not support exclusion from thrombolysis on the basis of early ischaemic change alone.

In view of the uncertainties surrounding thrombolysis even within the 0–3 h time window, the ongoing Third International Stroke Trial (IST-3: www.ist3.com) is randomizing patients to alteplase, rt-PA, from 0 to 6 h. This trial has no upper age limit in contrast to previous thrombolysis trials and aims to establish the risk benefit ratio for thrombolysis in a broader selection of patients.

MRI takes longer than CT to perform and is less well tolerated in sick patients. However, it has been proposed that MRI may improve selection of patients for thrombolysis through greater sensitivity for excluding those likely to be harmed by treatment. At present, there are no randomized trial data although there are studies reporting that MRI is safe and feasible in acute stroke.

3–6 h after stroke onset

Ideally, patients presenting within the 3–6 h time window should be randomized to appropriate thrombolysis trials. However, some centres use diffusion-weighted imaging/perfusion-weighted imaging and the presence of mismatch to select patients for thrombolysis. There is some evidence that treatment with intravenous alteplase based on the presence of diffusion-weighted/perfusion-weighted imaging mismatch in the 3–6 h time window produces similar functional outcomes to alteplase treatment within 3 h based on CT criteria. However, it remains unclear how to identify individual patients at high risk of haemorrhagic transformation although there is evidence that low perfusion or low apparent diffusion coefficient values (Derex *et al.* 2005) are associated with a higher risk of haemorrhage.

There are several ongoing studies randomizing patients to thrombolysis following MRI, including DEFUSE, the DWI Evolution For Understanding Stroke Etiology using a 3–6 h treatment window using rt-PA, and EPITHET, the EchoPlanar Imaging THrombolysis Evaluation, which randomizes all patients to rt-PA from 3 to 6 h with the underlying hypothesis that those with mismatch will benefit most. The ECASS-III, European and Australian Cooperative Stroke Study III, of rt-PA 3–4 h after stroke onset is using CT screening rather than MRI.

35.10.6 Intra-arterial thrombolysis

Intra-arterial thrombolysis has been proposed as a treatment for acute ischaemic stroke since the 1980s and may potentially overcome many of the problems associated with patient selection for intravenous therapy. The first advantage is that it allows thrombolysis to be given only to those patients in whom vessel occlusion has been demonstrated. In 20 per cent of patients presenting within 6 h of stroke onset, no occlusion is identified. Second, recanalization rates appear to be higher for intra-arterial thrombolysis, at around 70 per cent, than for intravenous thrombolysis, around 34 per cent, although there has been no direct comparison of the two techniques. Third, since intra-arterial thrombolysis involves the use of small amounts of thrombolytic agent applied directly to the site of occlusion as compared to the relatively high doses used systemically in intravenous thrombolysis, intra-arterial thrombolysis may offer the potential to treat patients at increased risk of haemorrhagic complications more safely.

Two randomized controlled trials of intra-arterial thrombolysis were reported, the PROlyse in Acute Cerebral Thromboembolism trials, PROACT I (del Zoppo *et al.* 1998) and PROACT II (Furlan *et al.* 1999). PROACT I remains the only placebo-controlled double blind, multi-centre trial of intra-arterial thrombolysis in acute ischaemic stroke. Recanalization rate was 58 per cent in the active treatment group compared to 14 per cent in the placebo group. Two doses of subsequent heparin were used, the high-dose of 5000U bolus followed by 100 U/h group achieving a a recanalization rate of 80 per cent with a symptomatic intracranial haemorrhage rate of 27 per cent. The equivalent rates in the low-dose heparin group were 47 and 6 per cent respectively. In PROACT II, 180 patients, mean NIHSS score of 17, were randomly assigned to receive either 9 mg of intra-arterial thrombolysis plus low-dose intravenous heparin or low-dose intravenous heparin alone. The median time from onset of symptoms to the initiation of intra-arterial thrombolysis was 5.3 h. In the treated group, there was a 15 per cent absolute benefit in the number of patients who achieved a modified Rankin score of 2 or less at 90 days ($p=0.04$). On average, seven patients with middle cerebral artery occlusion would require intra-arterial thrombolysis for one to benefit. Symptomatic brain haemorrhage occurred in 10 per cent of the thrombolysis group and 2 per cent of the control group but there was no excess mortality.

Comparisons between the different intra-arterial thrombolysis trials and between intra-arterial thrombolysis and intravenous thrombolysis is hampered by differences in methodology and type of thrombolytic therapy. In addition, within the intra-arterial thrombolysis trials, thrombolytic delivery has varied between regional into a parent vessel of the thrombosed vessel, local into the affected artery, and into the thrombus itself or combinations of these methods. In addition, the infusion process has been variable, ranging from continuous to pulsed infusion. Some studies have allowed physical clot dispersion using the tip of the microcatheter whilst this was prohibited in others, for instance in the PROACT trials.

It may be feasible to combine intravenous thrombolysis with intra-arterial thrombolysis. In the pilot Emergency Management of Stroke, EMS, Bridging trial (Lewandowski *et al.* 1999) 70 per cent of patients still had a clot on angiography after intravenous t-PA. There was improved recanalization in patients who then received intra-arterial t-PA but also an increase in bleeding complications. Other trials are ongoing. Percutaneous angioplasty of the recanalized vessel following intra-arterial thrombolysis has also been attempted but at present there are no prospective comparisons of any of these combined techniques making their relative merits unclear. Intra-arterial urokinase has been used to treat ischaemic stroke occurring during insertion of coils for intravascular aneurysms but since the causes of vessel occlusion secondary to arterial cannulation include intimal dissection and arterial spasm as well as local thrombosis and emboli dislodged by the catheter, thrombolysis would be unlikely to be of benefit in many cases.

35.10.7 Surgical decompression for malignant middle cerebral artery infarction

Malignant middle cerebral artery territory infarction is defined as a large middle cerebral artery infarct with marked oedema and swelling leading to raised intracranial pressure and a high risk of coning (Fig. 35.46). This carries a mortality rate of around 80 per cent on medical treatment alone. Non-randomized studies had suggested a reduction in mortality with decompressive surgery, consisting of hemicraniectomy and duroplasty, without a major increase in the number of severely disabled survivors. Recently, data from three small randomized trials were pooled and showed that surgery within 48 h of stroke onset reduced case fatality from 71 to 22 per cent, and that 43 per cent of survivors had a modified Rankin score of <3 at 1-year follow-up (Fig. 35.47). The results were highly consistent across all three trials (Vahedi *et al.* 2007). Since the trials excluded patients older than 60 years, and existing non-randomized data suggest poor outcome in those over 50 years, the results cannot necessarily be generalized to older patients.

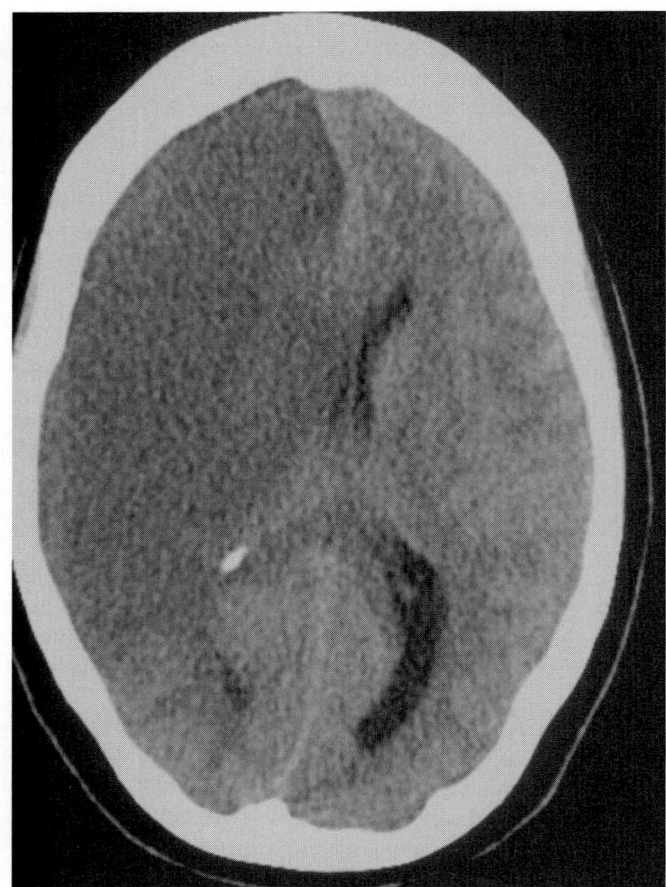

Fig. 35.46 The development of malignant middle cerebral artery infarction in a young woman who subsequently underwent hemicraniectomy. CT brain scan.

35.10.8 Neuroprotective interventions

Potential neuroprotctive agents include metalloprotease inhibitors that reduce vascular damage, anti-inflammatory drugs, and oxidative stress blockers. Several neuroprotective agents have shown promising results in animal studies but this has not in general translated into benefit in trials in humans (Savitz and Fisher 2007). The many possible reasons for the discrepancy in results between animals and humans include the delay before treatment, the heterogeneity of stroke in humans as compared to animal stroke models and small numbers and bias in animal studies (Muir and Teal 2005). Moreover, variation in the outcome measurements used in acute stroke trials make comparison between studies difficult and reanalysis of data using different methods may yield different results. In a recent large clinical trial of NXY-059, a free radical spin-trap agent, (Lees *et al.* 2006) which purported to show neuroprotective benefit in stroke was shown in reanalysis to be ineffective Koziol and Feng (2006). Further, although this agent had shown benefit in an initial trial, SAINT-1, this was not replicated in a subsequent trial, SAINT-2. The current consensus is that neuroprotective agents require more rigorous testing in clinically relevant animal models before testing in humans with appropriate time windows and outcome measures (Savitz and Fisher 2007).

Newer imaging techniques may allow better patient selection for neuroprotective agents. As discussed earlier, an initial diffusion-weighted MRI abnormality may resolve only to reappear in around 50 per cent of cases. Although the pathological processes underlying this phenomenon are not understood, this may be an area in which neuroprotection could play a role. Further, even when diffusion-weighted imaging lesions resolve without secondary reappearance, isolated necrotic neurons are seen on histological specimens in animal studies showing that partial ischaemic injury may be present even in regions with normal apparent diffusion coefficient and T2-weighted MRI values. Thus, neuroprotectants may also be of use in patients with fully reversible diffusion-weighted imaging lesions.

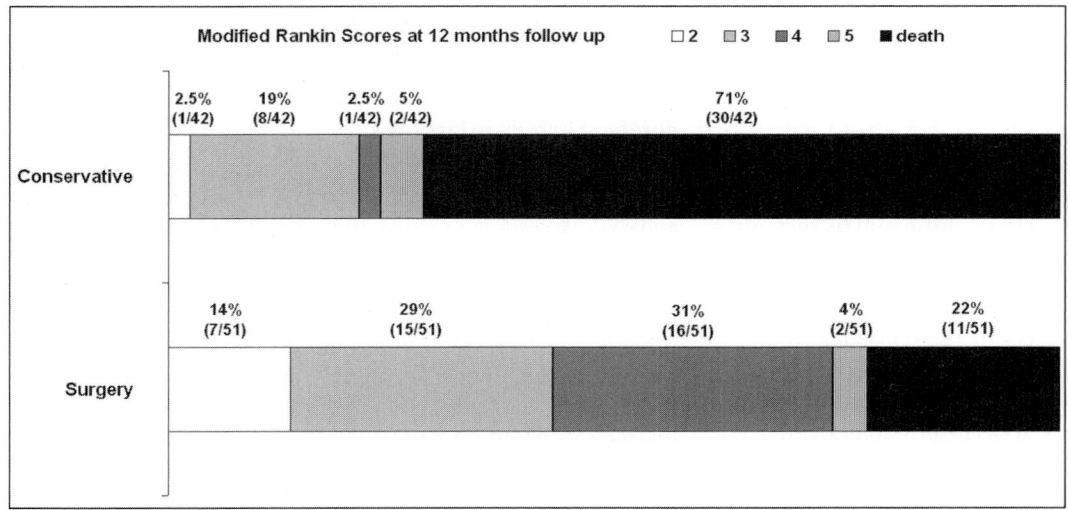

Fig. 35.47 Hemicraniectomy. A pooled analysis of data from three small randomized trials of hemicraniectomy versus medical treatment for malignant middle cerebral artery infarction (Vahedi *et al.* 2007). Surgery within 48 h of stroke onset reduced case fatality from 71 per cent to 22 per cent and left 43 per cent of survivors with only mild or moderate disability (modified Rankin score of <3) at 1-year follow-up.

Imaging of the ischaemic penumbra has been shown to be potentially useful in selecting patients for thrombolysis. It may allow extension of the thrombolysis window in selected patients. But it has also been proposed that tissue within the ischaemic penumbra represents a good target for neuroprotective agents which are unlikely to be as effective in tissue that is already severely damaged. Neuroprotective agents might allow protection of tissue within the penumbra while perfusion is restored and therefore extend the thrombolysis window. Alternatively, damage to the ischaemic penumbral tissue could be lessened in patients who did not reperfuse with thrombolyic therapy or in whom thrombolysis was contraindicated. Finally, although salvage of penumbral tissue is possible if reperfusion occurs, neurons within this area may still be at risk of delayed injury and this may represent a further area for neuroprotectant therapy.

Hypothermia causes a variety of responses in ischaemic brain that might confer neuroprotection, including significant alterations in metabolism, glutamate release and re-uptake, inflammation, and free radical generation (Krieger and Yenari 2004). This together with the observation that children and adults have survived prolonged immersion in cold water without neurological sequelae has led to the proposal that induced hypothermia may improve stroke outcome. Hypothermia may also lengthen the time window for thrombolysis. In animal models, hypothermia improves outcome in temporary cerebral vessel occlusion but effects after permanent occlusion are less consistent. In humans, two randomized trials have shown that mild hypothermia improves mortality and neurologic outcome in patients who suffer cardiac arrest (The Hypothermia after Cardiac Arrest Study Group 2002; Bernard *et al.* 2002). Pilot studies of hypothermia for stroke have been published but there are no large randomized trials and there is no consensus regarding timing, depth, and method of induction of hypothermia, including between surface and endovascular cooling (Hemmen and Lyden 2007). Anesthetized patients have tolerated hypothermia as long as 72 h, but few awake patients have been treated, partly because awake patients do not tolerate deep hypothermia.

35.11 Treatments for acute intracerebral haemorrhage

Brain imaging using CT has a high sensitivity for intracerebral haemorrhage. The location, characteristics, and number of haemorrhages can aid in the diagnosis of the underlying cause and thus may influence subsequent patient management. For instance, the presence of multiple lobar haemorrhages with surrounding oedema should prompt a search for underlying metastatic tumour whereas a basal ganglia haemorrhage in a hypertensive elderly person would not require further investigation. Repeat CT scanning is indicated if there is clinical deterioration, which may be caused by rebleeding, or by hydrocephalus; if there are no changes, the search for systemic disorders should be intensified.

It is vital to instigate the correct management of patients with cerebellar haemorrhage as soon as possible owing to the high risk of hydrocephalus and brainstem compression.

35.11.1 Non-surgical treatments
Recombinant factor VIIa
Recombinant factor VIIa is known to decrease the severity of haemorrhage in certain surgical settings. Since primary intracerebral haemorrhage has a tendency to enlarge over the first few hours it has been proposed that agents promoting haemostasis may be beneficial in primary intracerebral haemorrhage treatment. A phase 2 trial investigated the administration of rFVIIa within 3 h to patients with intracerebral haematoma without known structural or iatrogenic cause. This found that enlargement of the haematoma and poor outcome occurred significantly less often in patients treated with the active compound (Mayer *et al.* 2005), but the confidence limits were wide; the 95 per cent confidence interval of the number needed to treat to prevent a single death ranged between 5 and 166. Also the risk of thromboembolic complications was a potential concern; the rate of cerebral and myocardial infarction was 5 per cent in the active group and 0 per cent in the placebo group. A larger phase III trial, the FAST trial, Factor Seven for Acute Hemorrhagic Stroke Treatment, has recently failed to confirm ths earlier result, and no further trials are planned.

Cerebrospinal fluid drainage
Insertion of a cerebral ventricular catheter can be life-saving in patients with cerebellar haemorrhage and hydropcephalus. In patients with supratentorial haemorrhage and hydrocephalus, however, the benefits of CSF diversion are less certain. In patients with extensive intraventricular haemorrhage secondary to deep intraparenchymal or aneurysmal rupture, drainage of CSF is performed in some centres. Systematic reviews of observational studies suggest that this procedure may be helpful, especially when it is combined with instillation of fibrinolytic drugs. However only direct comparisons in a randomized trial can establish the value of these therapeutic measures (Nieuwkamp *et al.* 2000; Lapointe and Haines 2002; Lee *et al.* 2003; Fountas *et al.* 2005).

Hyperventilation
Hyperventilation decreases intracranial pressure because hypocapnia, usually down to values of the order of 4 kPa, causes vasoconstriction. However, this will not necessarily be beneficial in patients with intracerebral haemorrhage since brain ischaemia caused by compression is exchanged for ischaemia caused by vasoconstriction (Stocchetti *et al.* 2005). In head-injured patients, the single randomized controlled trial of prolonged hyperventilation not only failed to show any benefit but also raised concerns about potential harm (Muizelaar *et al.* 1991; Schierhout and Roberts 2000). There are no controlled trials of hyperventilation in patients with intracerebral haematomas.

Osmotic agents
Mannitol is widely used in patients with primary intracerebral haemorrhage and a depressed level of consciousness, to decrease intracranial pressure and alleviate the space-occupying effect of the haematoma in a deteriorating patient, although this practice is not backed up by randomized trials with clinical outcome rather than pressure as the outcome variable.

35.11.2 Surgical treatment
Supratentorial haemorrhage
There are four possible surgical procedures to treat intracerebral haematoma: simple aspiration, craniotomy with open surgery, endoscopic evacuation, and stereotactic aspiration. Open surgery remains the technique of choice at present.

In patients with large intracerebral haemorrhages and mass effect (Fig. 35.48) it is tempting to imagine surgical removal would be

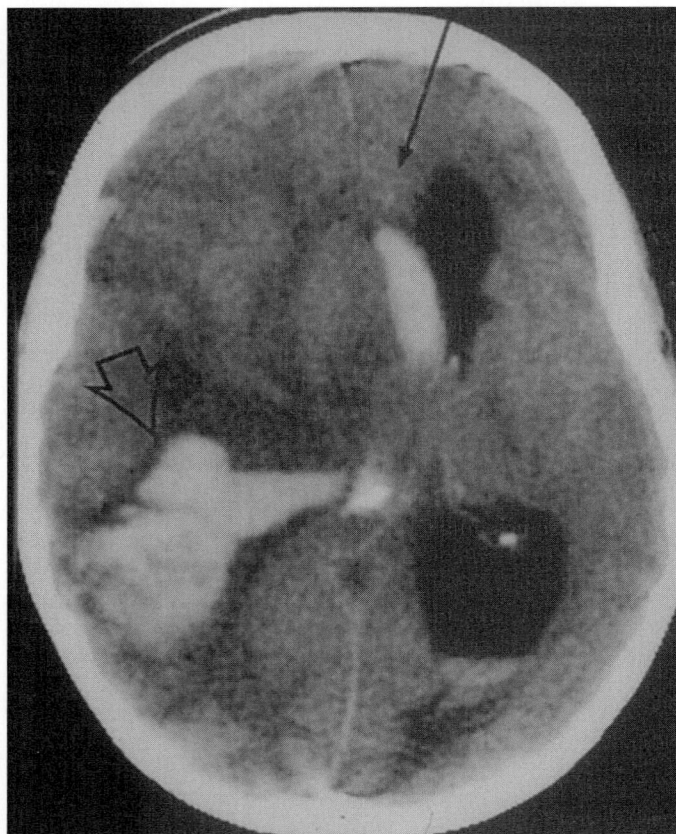

Fig. 35.48 Primary intracerebral haemorrhage (large arrow) with rupture into the ventricular system and considerable mass effect (arrow). CT brain scan.

very likely to improve outcome. Craniotomy with open surgery was studied in the large multicentre STICH trial (Mendelow *et al.* 2005). Nine preceding small randomized trials had produced conflicting results. The STICH trial randomized 1003 patients, all with supratentorial haemorrhages, four times as many patients were included as in all previous trials taken together. Patients with a haematoma of at least 2 cm in diameter and a Glasgow coma score of 5 or greater were randomized to initial conservative treatment versus early surgery if the admitting surgeon was uncertain of the benefit of surgery. Patients initially randomized to medical treatment could undergo surgery at a later time point if this was felt to be indicated. Surgical technique was left up to the discretion of the surgeon.

STICH reported no difference in outcome between those consigned to initial conservative or surgical management; most patients received open craniotomy. In a prespecified subgroup analysis, patients with superficial haematomas were more likely to have a favourable outcome than those with deep haematomas and there was a nonsignificant relative benefit for surgery in this group. The outcome for patients with Glasgow coma scores of 8 or below were uniformly poor and there was a suggestion that surgery increased the risk of poor outcome in these patients. It is possible that patients with deep seated haematomas might do better with less invasive and hence less traumatic methods of clot evacuation such as endoscopy. Further trials are needed to test these hypotheses.

The STICH trial did not address the question of whether operation improves outcome in patients who are deteriorating because of an expanding haematoma. A consecutive series of 26 such patients from the Mayo Clinic, of whom 24 had lobar haemorrhage, suggested that no patient regained independence, with or without operation, if corneal and oculocephalic reflexes had been lost. However approximately one-quarter of the remaining group, 6 out of 21, regained independence (Rabinstein *et al.* 2002).

Cerebellar haemorrhage

Over the past few decades, surgical evacuation has been felt to be life-saving in patients with cerebellar haematomas who have clinical evidence of progressive brainstem compression. This, together with the fact that the operative sequelae appear to be mild, makes a clinical trial of surgery versus best medical treatment now impossible. Certain patients can be managed conservatively, but there is uncertainty about the selection criteria. Indications for evacuation of cerebellar haematoma include haematoma greater than 3–4 cm and the combination of a depressed level of consciousness with signs of progressive brainstem compression, unless all brainstem reflexes have been lost for more than a few hours, in which case a fatal outcome is unavoidable (Wijdicks *et al.* 2000; Cohen *et al.* 2002; Jensen and St Louis 2005). Ventriculostomy alone may be adequate in some patients with clinical features of hydrocephalus without pontine compression such as gradual deterioration of consciousness with sustained downward gaze and small unreactive pupils).

Other surgical approaches

Aspiration not accompanied by any other intervention was attempted mainly in the 1950s but was subsequently abandoned because only small amounts of clot could be obtained, and because the procedure could precipitate 'blind' rebleeding. Stereotactic endoscopic evacuation is a promising technique, but its benefits and specific indications remain to be confirmed by clinical trials.

Since the 1990s, stereotactic aspiration of supratentorial haemorrhage without endoscopy, mostly combined with instillation of fibrinolytic agents, has been reported in several observational studies. Subsequently two controlled trials of this technique have been performed. A Japanese trial randomized 242 patients with putaminal haemorrhage and a moderately decreased level of consciousness involving eyes closed but opening to stimuli, between stereotactic haematoma evacuation and conservative treatment. This reported that outcome in terms of death and dependence was better in the surgical group (Hattori *et al.* 2004). A smaller trial of 70 patients in the Netherlands combined stereotactic aspiration with liquefaction by means of a plasminogen activating substance and found no conclusive differences in outcome (Teernstra *et al.* 2003).

35.11.3 Treatment of specific types of intracerebral haemorrhage

Pontine haemorrhages

Pontine haemorrhages are fatal in around 50 per cent of cases (Wijdicks and St Louis 1997). Those caused by cavernomas or arteriovenous malformations have a better outcome (Rabinstein *et al.* 2004). The management of patients with 'hypertensive' pontine haemorrhage is usually conservative, but some case reports have documented successful stereotactic aspiration. However, there is likely publication bias and the natural history of the condition is difficult to predict since patients with small haemorrhages do well with conservative management.

Lobar haemorrhage from presumed amyloid angiopathy

In patients with lobar haemorrhage from amyloid angiopathy and an impaired level of consciousness, the prognosis is poor after surgical intervention (McCarron et al. 1999). It remains unclear whether the outcome would have been better without surgery (Greene et al. 1990; Izumihara et al. 1999). Given these uncertainties and the danger that the operation provokes new haemorrhages from brittle vessels at distant sites (Brisman et al. 1996) conservative treatment seems the best option.

Cavernous angiomas

There is little doubt that haemorrhages, which may be recurrent, from these lesions are less destructive than brainstem haemorrhages from an arterial source (Rabinstein et al. 2004). Also the angioma may spontaneously regress (Yasui et al. 2005). Surgical treatment or stereotactic radiation is regularly performed but not supported by controlled studies and incomplete removal can prompt rebleeding (Kikuta et al. 2004). Stereotactic cobalt-generated radiation 'gamma knife' therapy has been applied to a large number of patients with cavernomas but it remains uncertain whether seizures or rebleeding are less likely to occur afterwards (Régis et al. 2000; Liu et al. 2005).

Dural arteriovenous fistulae

These malformations are heterogeneous. They may or may not be secondary to occlusion of a major sinus. Abnormal venous drainage from meningeal arteries may be channelled into a dural sinus or into superficial veins of the surface of the brain, cerebellum, or brainstem. Accordingly, different methods are required to occlude the fistula. Surgical techniques include selective ligation of leptomeningeal draining veins, resection of fistulous sinus tracts, and a cranial base approach with extradural bone removal. Endovascular techniques may consist of an approach from the arterial or venous side, recanalization and stenting of a venous sinus, or venous embolization via a craniotomy (Tomak et al. 2003). Finally stereotactic cobalt-generated radiation 'gamma knife' therapy is also used for obliterating arteriovenous fistulae (O'Leary et al. 2002; Pan et al. 2002), sometimes in conjunction with an endovascular approach (Friedman et al. 2001).

Moyamoya syndrome associated haemorrhage

Moyamoya syndrome (Section 35.4.5) causes gradual stenosis or occlusion of the terminal portions of the internal carotid arteries or middle cerebral arteries. This leads to formation of an abnormal collateral network of fragile collaterals, which occasionally rupture. It has been proposed that constructing a bypass to relieve the pressure on the collaterals would be beneficial, for example, between the superficial temporal artery and the middle cerebral artery (Kawaguchi et al. 2000), but whether this procedure prevents (re)bleeding remains uncertain. In fact, an aneurysm may form and rupture at the site of the arterial bypass (Nishimoto et al. 2005).

35.11.4 Iatrogenic intracerebral haemorrhage

Anticoagulation

Anticoagulants are associated with an increased rate of intracerebral haemorrhage. Observational data confirm that early normalisation of clotting status is associated with a relatively low rate of haematoma enlargement (Yasaka et al. 2003).

It remains uncertain when to reintroduce anticoagulants in patients with a strong indication for this treatment, such as those with artificial heart valves, since only anecdotal experience is available. In the Mayo clinic series, only one ischaemic stroke occurred in 52 patients with artificial heart valves in whom anticoagulants were discontinued for a median period of 10 days (Phan et al. 2000). By contrast, of seven similar patients from Heidelberg, three had large ischaemic strokes within a comparable period after stopping anticoagulation (Bertram et al. 2000). Reintroduction of anticoagulation between 1 and 2 weeks after the haemorrhage is probably reasonable (Estol and Kase 2003) and is supported by evidence of low rates of rebleeding (Wijdicks et al. 1998; Leker and Abramsky 1998).

Thrombolytic therapy

Thrombolytic therapy for myocardial infarction is rarely complicated by intracerebral haemorrhage, but the case fatality is high. Treatment includes control of hypertension and the infusion of coagulation factors. However the use of antifibrinolytic drugs is controversial. Surgical treatment is of unproven value and may be especially hazardous given that amyloid angiopathy may be a contributing factor. In patients with ischaemic stroke who develop major intracerebral haemorrhage after thrombolytic treatment, treatment remains unclear.

Aspirin

Aspirin treatment is associated with a small risk of intracerebral haemorrhage. It is reasonable to stop the drug once the diagnosis is made. The antiplatelet effect lasts until several days after discontinuing the drug. In trials of aspirin for the secondary prevention of cerebral ischaemia, some patients with small intracerebral haemorrhages were inadvertently included before CT scanning and did not come to obvious harm. However the number of cases was small and they were probably a rather atypical group (Keir et al. 2002).

35.12 Recovery and rehabilitation after stroke

The prognosis of hospitalized patients tends to be worse than that of patients in the population at large because mild strokes are more likely to be cared for at home. In the community, about 20 per cent of all first ever stroke patients are dead within a month. The prognosis is much better for ischaemic stroke than for intracranial haemorrhage, with about 10 and 50 per cent dying respectively (Bamford et al. 1990; Lovelock et al. 2007). Deaths in the first few days are almost all due to the brain lesion itself. Deaths after the first week are more likely to be due to the indirect consequences of the brain lesion, such as bronchopneumonia, pulmonary embolism, coincidental cardiac disease, or recurrence.

The various subtypes of ischaemic stroke have very different outcomes: patients with total anterior circulation infarction have just as poor an outcome as those with primary intracerebral haemorrhage (Fig. 35.49). The best single predictor of early death is impaired consciousness. Other predictors reflect the extent of the brain lesion and thus stroke severity, and the premorbid state of the patient: increasing age, pre-stroke handicap, size of any haematoma

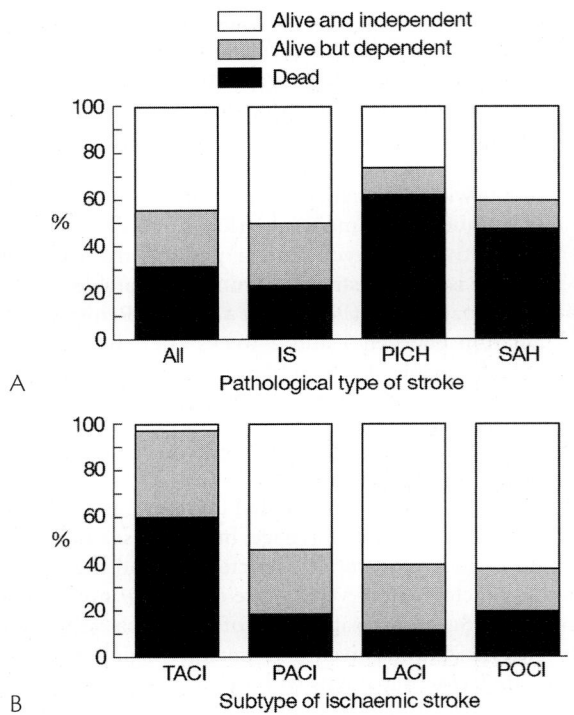

Fig. 35.49 Outcome of a first stroke. 1-year outcome in the 675 first-ever-in-a-lifetime of stroke patients in the Oxfordshire Community Stroke Project by (A) pathological type and (B) subtype of ischaemic stroke.
Key: IS: ischaemic stroke
LACI: lacunar infarction
PACI partial anterior circulation infarct
PICH: primary intracerebral haemorrhage
POCI: posterior circulation infarction
SAH: subarachnoid haemorrhage
TACI: total anterior circulation infarct (from Bamford et al. 1990, 1991).

and intraventricular extension of blood, midline shift, both high and low blood pressure, atrial fibrillation, fever, and high blood glucose. Visible infarction on CT adds some prognostic information (Wardlaw *et al.* 1998). Many of these variables are interrelated but prognostic models based on independent variables do not provide much more information than an experienced clinician's estimate (Counsell and Dennis 2001; Counsell *et al.* 2002).

After the first month, the risk of death becomes much less, about 7 per cent per annum, but it is still about twice that in the background population because stroke patients are particularly likely to die of a further stroke and even more likely to die of the consequences of ischaemic heart disease (van Wijk *et al.* 2005).

35.12.1 Recovery after stroke

Some degree of recovery occurs in the majority of patients after stroke. Although complete recovery is possible, the prognosis is difficult to predict in an individual patient.

Approximately two-thirds of stroke survivors become independent at 1 year, with little difference between ischaemic or haemorrhagic strokes (Fig. 35.49). However, within the ischaemic group, only about 5 per cent of patients with infarction of the whole middle cerebral artery territory are alive and independent at a year post-stroke, compared with 50 per cent of those with more restricted infarcts (Bamford *et al.* 1990). About 90 per cent of stroke survivors

return home, leaving only a small proportion in institutional care but, because stroke is so common, their absolute number is large (Legh-Smith *et al.* 1986; Chuang *et al.* 2005).

Early characteristics predicting death in patients with primary intracerebral haemorrhage are: level of consciousness assessed by the Glasgow Coma Scale, age, volume of haematoma, and intraventricular extension of haemorrhage. So far predictive models only apply to a small proportion of patients and are not sufficiently accurate to inform treatment decisions in routine clinical practice.

The mechanisms of recovery are incompletely understood. Acute resolution of oedema and recanalization of occluded vessels leading to resolution of penumbral dysfunction may contribute. In the subacute phase, changes in neuronal networks, neuronal plasticity, are important (Kreisel *et al.* 2007; Nudo 2007). The mechanisms share similarities with those involved in learning and memory. The rate of recovery of all impairments is maximal in the first few weeks, slows down after 2 or 3 months, and probably stops at about 6–12 months post-stroke (Pedersen *et al.* 1995; Kreisel *et al.* 2007). Later improvement in functional abilities, and particularly in social activities, is probably more to do with adaptation to disability and minimizing handicap rather than further recovery of physical impairments. Impaired quality of life is common even when patients appear to be little disabled.

Prediction of functional outcome for individuals immediately after stroke onset is difficult and clinical features predict outcome as well or better than radiological findings (Hand *et al.* 2006a). Prognostic scores have been developed but these are not particularly discriminating (German Stroke Study Collaboration 2004). About 2 weeks post-stroke, good prognostic signs include young age, an initially mild deficit, normal conscious level, good sitting balance, lack of cognitive impairment, urinary continence, and rapid improvement. Independent living is often also contingent on a high level of social support.

35.12.2 Strategies for rehabilitation

Stroke rehabilitation attempts to restore patients to their previous physical, mental, and social capability (Langton Hewer 1990; Brandstater 2005). Rehabilitation approaches include restoration of previous function, compensation by increasing function for a given impairment, environmental modification, prevention of complications such as recurrent stroke or shoulder pain, and maintenance or prevention of deterioration. Achieving optimal quality of life is the ultimate goal. Care in a designated ward area appears to be better than mobile specialist team care given on a general ward (Stroke Unit Trialists' Collaboration 2001). Although there is good evidence that increased time undergoing therapy is beneficial (Langhorne *et al.* 1996; Kwakkel *et al.* 2004; Kwakkel 2006), in general, patients spend very little of their time awake receiving therapy, only around 5 per cent in one study (Bernhardt *et al.* 2004). As little as 5 h a week extra therapy results in clinically important improvements in walking (Blennerhassett and Dite 2004). Rehabilitation should not necessarily be confined to patients in hospital and early discharge with multidisciplinary support achieves as good or better outcome (Langhorne and Holmqvist 2007).

The optimum time to start rehabilitation after stroke is not known but early rehabilitation within the first week is probably beneficial given the deleterious consequences of bed rest after stroke (Langhorne *et al.* 2000) and the benefits of early mobilization following other acute medical conditions including myocardial infarction. There is some evidence from animal studies that very early intense activity may increase lesion size. Early resumption of the upright posture may compromise perfusion in the penumbral zone. Thus it may be reasonable to delay mobilization for the first few days after stroke (Diserens *et al.* 2006). Although improvements in mobility and activities of daily living can occur months or years after stroke (Outpatient Service Trialists' 2003) benefits are greater when intervention occurs within the first 6 months.

There are insufficient data to determine which patient groups benefit most from rehabilitation. Although severely affected patients benefit from rehabilitaion and in fact receive the most inpatient and outpatient therapy, they have the worst functional outcomes (Alexander *et al.* 2001). Thus it is unclear at present how best to target the available rehabilitation resources most efficiently.

The evidence base to support particular rehabilitation strategies is limited owing to a lack of large randomized trials. Yet it is clear that multidisciplinary input facilitates recovery and reduces handicap. Key components of inpatient rehabilitation have been identified as (Stroke Unit Trialists' Collaboration 1997; Langhorne and Duncan 2001; Langhorne and Dennis 2004):

♦ coordinated, multidisciplinary care with regular team meetings;

♦ early institution of rehabilitation within 1–2 weeks;

♦ goal setting;

♦ early assessment of impairments and function;

♦ discharge planning with early assessment of discharge needs;

♦ staff with a specialist interest in stroke or rehabilitation;

♦ routine involvement of carers;

♦ close linking of nursing with other multidisciplinary care;

♦ regular programs of education and training for staff; and

♦ information provided about stroke, stroke recovery, and available services.

A systematic review of randomized control trials of inpatient multidisciplinary stroke rehabilitation has shown the benefits of rehabilitation, > 7 days after stroke, as distinct from the acute medical management aspects of acute stroke care in the first week after stroke (Langhorne and Duncan 2001) with a reduction in death (OR 0.66; 95 per cent CI 0.49–0.88) and death or dependency (OR 0.65; 95 per cent CI 0.50–0.85). For every 20 patients with stroke treated in a post-acute, > 7 days, multidisciplinary rehabilitation unit, 1 additional person returns home independent in activities of daily living.

There is also evidence that outpatient (Outpatient Service Trialists' 2003) and home- based rehabilitation (Early Supported Discharge Trialists 2005; Langhorne and Holmqvist 2007) are effective for patients who have returned to the community in preventing death, deterioration, and dependency, and may allow earlier hospital discharge.

Assessment tools used in rehabilitation

Assessment tools identify, measure, and record impairments, disabilities, handicaps, and quality of life. Assessment is the first step in rehabilitation and measurement of outcome is crucial for clinical trials, audit, and comparison of different institutions. The patient's present level of functioning must be compared with their premorbid level, taking account of the numerous co-morbidities often present in the elderly (Collen and Wade 1991).

Various assessment instruments are available that are designed to test different domains: (Lyden and Hantson 1998; Warlow *et al.* 1996c):

♦ The *motricity index* and *trunk control test* are useful measures of motor impairment (Collin and Wade 1990).

♦ The *walking speed test* is walking time measured over a standard distance, with or without a turn, using a stop-watch (Wade *et al.* 1987; Collen *et al.* 1990).

♦ The *Rivermead mobility index* is a measure of disability (Collen *et al.* 1991).

♦ The abbreviated *Hodkinson Mental Test Score* is recommended for overall cognitive function (Hodkinson 1972).

♦ The *Frenchay Aphasia Screening Test for aphasia* (Enderby *et al.* 1986).

♦ The *Star Cancellation Test* for neglect (Jehkonen *et al.* 1998).

♦ *Mood disorder* is difficult to measure, and many stroke patients are unable to complete the questionnaires (House *et al.* 1989). Nonetheless mood disorder is an important consequence of stroke.

♦ The *Barthel Activities of Daily Living index* is a good measure of all round disability although it has floor and ceiling-effects and takes no account of vision, hearing, and speaking.

♦ The *modified Rankin Scale* (Table 35.22), sometimes called the *Oxford Handicap Scale*, may be a better measure of overall disability and handicap than the Barthel Index (Bamford *et al.* 1989).

♦ For *clinical trial and audit purposes* in large samples, three simple questions can be assesssed face to face, by post, or on the telephone. This groups patients into those who are completely recovered, those who are still symptomatic but independent, those who are dependent, or dead (Dennis *et al.* 1997a, b).

♦ The *Frenchay Activity Index* can be used for social functioning, although this is influenced by gender and culture (Wade *et al.* 1985).

Table 35.22 Modified Rankin scale

Grade	
0	No symptoms
1	Minor symptoms which do not interfere with lifestyle
2	Minor handicap. Symptoms which lead to some restriction in lifestyle but do not interfere with the patients' ability to look after themselves
3	Moderate handicap. Symptoms which significantly restrict lifestyle and prevent totally independent existence
4	Moderately severe handicap. Symptoms which clearly prevent independent existence although not needing constant care and attention
5	Severe handicap. Totally dependent, requiring constant attention day and night
6	Dead

◆ *Short Form-36* is the most widely used generic instrument for assessment of 'quality of life' but the *EuroQol* is also used. Neither is reliable enough to monitor individuals over time but they are probably good enough to compare groups of patients (de Haan *et al.* 1993; Dorman *et al.* 1998).

Interventions

Stroke rehabilitaion teams usually consist of physiotherapists, occupational therapists, speech therapists, and social workers as well as nursing staff. The stroke rehabilitaion team provides a range of interventions and is also able to advise regarding return to work, driving, finance, benefits, and sexual activity. Information about local stroke clubs and other voluntary organizations should be given to patients and carers. The latter may also need care and support since they may experience high levels of burden (van Heugten *et al.* 2006).

Physiotherapy improves outcome after stroke, but it is unclear which type of approach is best (Pollock *et al.* 2007). Physiotherapy trains patients in reaching and manipulation, sitting, sit-to-stand, standing, and walking. Physiotherapists also assist with the use of foot-drop splints, sticks, and wheelchairs and instruct carers in transferring, lifting, walking, and exercises. They are also able to advise on the care of the hemiplegic arm, particularly the shoulder. There is good evidence that functional electric stimulation reduces shoulder pain, prevents shoulder subluxation, maintains range of movement, and improves upper limb activity (Foley *et al.* 2006). Routine corticosteroid injection or ultrasound for shoulder pain have not been shown to be beneficial and overhead arm pulleys and positional shoulder static stretches are harmful and should not be used (Gustafsson and McKenna 2006).

Occupational therapy prevents deterioration after stroke and facilitates patient independence but the exact nature of the occupational therapy intervention needed to achieve maximum benefit remains to be defined (Legg *et al.* 2006).

Speech therapy there is insufficient evidence to support or refute possible benefits of speech therapy after stroke for either aphasia (Greener *et al.* 2002) or speech apraxia (West *et al.* 2005) although patients and carers value such input and stroke care guidelines recommend speech therapy. Speech therapists also have a role in the management of dysarthria and swallowing.

Neglect is one of the most disabling impairments for patients after stroke. Although individual studies of rehabilitation specifically for neglect have shown benefit, a review of 15 such studies showed no impact on disability or discharge home (Bowen *et al.* 2002). Visuo-spatial-motor training in which the affected limb is moved to increase attention to that side has been shown to be effective in a single randomized controlled trial (Kalra *et al.* 1997).

Spasticity, if moderate to severe, may impede rehabilitation after stroke. Botulinum toxin has been shown to reduce tone and improve range of movement in small studies but robust evidence for improvement in function is lacking (van Kuijk *et al.* 2002).

Secondary consequences of neurological disability include painful shoulder, shoulder-hand syndrome, contractures, and falls. Physiotherapists and other members of the multidisciplinary team should have expertise in managing these problems.

35.12.3 Brain imaging and recovery after stroke

The relationship between radiological findings and functional outcome has been examined in a number of studies using different imaging modalities and outcome measures. The importance of these studies from the point of view of acute stroke management is that they may allow identification of those patients who have the potential for good functional recovery. Rehabilitative and therapeutic strategies, such as neuroprotection agents, could then be targeted to those patients. The majority of studies have correlated lesion size in the acute or subacute period, using conventional CT or MRI, with outcome measured using a variety of scales of impairment or disability or with combinations of these measures. Although a significant relationship between infarct size and outcome has been demonstrated in most studies this is not invariably the case. The major reasons for a lack of correlation relate to lack of sensitivity in detection of infarcts and more importantly to the fact that these studies ignore the importance of lesion location in determining outcome from stroke.

Lesion size on CT and MRI and functional outcome

Studies using CT to measure lesion size have shown conflicting results. The largest such study (Saver *et al.* 1999) found only a modest correlation ($r=0.5$), between subacute infarct volume and the National Institutes of Health Stroke Scale, NIHSS, at 3 months. In addition, although patients with large infarcts tended to have a poor outcome, the functional consequences of more moderately sized infarcts were more difficult to predict. T_2-weighted MRI is more sensitive than CT at detecting infarction and thus might be expected to be somewhat better prognostically. In a study relating T_2-weighted MRI and outcome, patients who were independent at 3 months had smaller strokes on MRI than patients who were dependent or dead (Saunders *et al.* 1995). However, there was considerable overlap between infarct sizes in the three groups and the outcome measures were rather coarse. Modest but significant correlations have also been shown between acute diffusion-weighted imaging lesion volume and neurologic outcome (Lovblad *et al.* 1997; Barber *et al.* 1998a; Tong *et al.* 1998) although the relationship may be stronger with perfusion-weighted imaging. Subacute perfusion-weighted imaging appears to be less closely related to functional outcome presumably because of the occurrence of spontaneous reperfusion over time.

One of the major reasons why outcome may not be correlated strongly with lesion volume is that volume measures do not take account of variations in lesion location or shape. Recently, evidence has been produced to suggest that the clinical consequences of an ischaemic lesion can be predicted if damage and outcome are measured within a specific functional system(Binkofski F, Seitz RJ, Arnold S *et al.* 1996; Pineiro *et al.* 2000).

MR spectroscopy and functional outcome

Cerebral damage following stroke has also been assessed by magnetic resonance spectroscopy. This technique uses the same methods as MRI but the signal obtained is converted into chemical as opposed to spatial information. Proton MR spectroscopy allows *in vivo* measurement of *N*-acetyl containing compounds, creatine, choline, and lactate. The majority of the *N*-acetyl signal comes from *N*-acetyl aspartate which is present in high concentrations in the brain. The function of *N*-acetyl aspartate is unclear but it is of particular interest in studies of the brain since it is located almost exclusively in neurons in the adult. Decreases in the *N*-acetyl aspartate resonance peak *in vivo* indicate neuronal or axonal injury or loss.

Early studies of MR spectroscopy in stroke showed increased lactate and decreased *N*-acetyl aspartate within the stroke lesion (Berkelbach van der Sprenkel *et al.* 1988; Bruhn *et al.* 1989). Subsequently, it was shown that the magnitude of neuronal damage as measured by *N*-acetyl aspartate loss from the infarcted region, correlated with disability and impairment in stroke patients (Ford *et al.* 1992; Federico *et al.* 1998). It remains unclear whether *N*-acetyl aspartate loss is a better prognostic indicator than other factors such as infarct volume as measured on imaging or indeed simple clinical tests. However, one study (Parsons *et al.* 2000) suggested that acute lactate/choline ratios correlate better with outcome than *N*-acetyl aspartate/choline ratios or infarct volume. In all the above studies, metabolite changes were measured from the centre of the infarcted region and thus were not representative of the total infarct damage. Also the chosen outcome measures were not necessarily relevant to the area of brain under study. These points were addressed in a study in which *N*-acetyl aspartate loss was measured in the descending motor pathways and correlated to a scale designed to measure motor impairment (Pendlebury *et al.* 1999). This study showed that *N*-acetyl aspartate loss in the descending motor pathways was significantly associated with motor deficit and with the maximum proportion of the descending motor pathway cross-sectional area occupied by stroke as described earlier.

Recovery after stroke and functional imaging

Studies of brain activation patterns after stroke have used electrical or magnetic brain stimulation of pathways or alternatively, imaging using positron emission tomography or functional MRI, fMRI. fMRI relies on the fact that deoxyhaemoglobin is paramagnetic whereas oxyhaemoglobin is not. During neuronal activation, the neuronal oxygen demand rises and local cerebral blood flow rises but to a level in excess of that required to supply the increased metabolic demand. Hence activation results in a reduced concentration of deoxyhaemoglobin and thus an increased signal on MRI. However, fMRI studies are technically demanding to perform, require awake and cooperative subjects and assume that the normal relationship between metabolism and blood flow is maintained in normal ageing and after a stroke. In fact, there are age-related decreases in cerebral blood flow and metabolic rate and blood flow/metabolism coupling is impaired within ischaemic tissue although blood flow increases have been reported to occur in the remaining tissue in response to brain activation (Chollet *et al.* 1991; Weiller *et al.* 1992). However, this does not necessarily indicate that the blood flow/metabolism coupling is normal in non-infarcted tissue. Certainly, there are widespread changes in the resting metabolic rate and perfusion of the brain after stroke, which persist into the chronic phase and may affect areas remote from the infarct.

Despite the limitations of fMRI outlined above, fMRI studies have shown similar findings to those of positron emission tomography studies in recovery after stroke (Yozbatiran and Cramer 2006; Rijntjes 2006). Increased ipsilateral sensorimotor cortical, bilateral supplementary motor area activation, and premotor cortical activation occurs after stroke with use of the affected hand, in comparison to the unaffected hand (Weiller *et al.* 1992; Cramer *et al.* 1997; Cao *et al.* 1998). Additionally there is posterior displacement of the ipsilesional focus of primary motor cortical activity (Pineiro *et al.* 2001). Contralateral sensorimotor cortical activation is enhanced by active rehabilitation (Cramer *et al.* 1997) and fluoxetine (Pariente *et al.* 2001), two interventions thought to improve motor recovery. It has been proposed that non-invasive functional imaging may in the future be used to select patients for specific rehabilitation therapy.

35.13 Primary prevention of stroke

Primary prevention of stroke focuses on control of the major risk factors for stroke, the most important being blood pressure. Most such primary prevention of cardiovascular disease is undertaken by primary care physicians and is not discussed here. The other specific interventions that are often considered in secondary care settings are reviewed below.

35.13.1 Antithrombotic treatment

Long-term antiplatelet treatment with aspirin for primary prevention of stroke and other vascular events is not generally recommended. Unless the risk of ischaemic vascular events is particularly high the small increase in risk of intra- and extracranial haemorrhage would offset any benefit (Antiplatelet Trialists' Collaboration 1994; Medical Research Council's General Practice Research Framework 1998; Bartolucci and Howard 2006).

Anticoagulation is sometimes used in the primary prevention of stroke in patients with non-valvular atrial fibrillation. The risk of stroke varies 20-fold among such patients, depending on age and presence of vascular risk factors. Low-risk individuals usually receive sufficient protection from daily aspirin whereas those at high risk of stroke benefit from warfarin where there is no contraindication. Several stroke risk stratification schemes have been published, the most widely used of which is the CHADS2 score (Hart *et al.* 2003). Patient preferences, availability of anticoagulation monitoring, and estimated bleeding risk are all key issues and several patient decision aids have also been developed (Man-Son-Hing *et al.* 2000; McAlister *et al.* 2005). Although elderly patients and those with comorbidity were often excluded from the initial warfarin trials, a recent pragmatic trial of warfarin in primary care showed that warfarin was better than aspirin 300 mg in patients aged 80–90 years with atrial fibrillation (Rash *et al.* 2007).

35.13.2 Asymptomatic carotid stenosis

Only about 20 per cent of ischaemic stroke patients have had a preceding transient ischaemic attack. Until the moment of stroke, any carotid stenosis is 'asymptomatic'. If these asymptomatic stenoses could be detected before the stroke, then unheralded stroke might be preventable by carotid endarterectomy. Asymptomatic carotid stenosis can be identified during screening programmes of apparently healthy people. A carotid bruit may be heard or an ultrasound examination may detect stenosis during the course of working up patients with angina, claudication, or non-focal neurological symptoms. Bilateral carotid imaging is undertaken in patients with unilateral carotid symptoms and may detect stenosis on the asymptomatic side. When patients are being worked up for major surgery below the neck a carotid bruit may be heard or ultrasound may reveal carotid stenosis.

Whether the benefits of carotid endarterectomy in patients with asymptomatic stenosis justify the risks and cost is still unclear

(Benavente *et al.* 1998; Chambers and Donnan 2005). This is particularly so in an era of improved medical treatments. There have been two large randomized trials of immediate endarterectomy plus medical treatment versus medical treatment alone, the asymptomatic carotid surgery trial, ACST, and the (Executive Committee for the Asymptomatic Carotid Atherosclerosis Study 1995; Halliday *et al.* 2004) and a few other smaller trials (Chambers and Donnan 2005). ACAS reported a 47 per cent relative reduction in the risk of ipsilateral stroke and perioperative death in the surgical arm despite a 5-year risk of ipsilateral stroke of only 11 per cent in the unoperated group. However, the absolute reduction in risk of stroke with endarterectomy was only about 1 per cent per year. The more pragmatic ACST randomized 3120 asymptomatic patients with >60 per cent asymptomatic carotid stenosis and reported a similar absolute reduction in 5-year risk of stroke with surgery to ACAS, of 5.3 per cent versus 5.1 per cent respectively. In addition, whereas ACAS had reported only a non-significant 2.7 per cent reduction (p=0.26) in the absolute risk of disabling or fatal stroke with surgery, ACST reported a significant 2.5 per cent (95 per cent CI: 0.8–4.3 per cent, *p*=0.004) absolute reduction. However the number needed to treat to prevent one disabling or fatal stroke after 5 years remains about 40.

The main differences between the ACAS and ACST were in the 30-day operative risks of death, at 0.14 per cent in ACAS versus 1.11 per cent in ACST (*p*=0.02), and in the combined operative risk of stroke and death, of 1.5 per cent in ACAS versus 3.0 per cent in ACST (*p*=0.04). ACAS only accepted surgeons with an excellent safety record, rejecting 40 per cent of initial applicants (Moore *et al.* 1996). A systematic review of all surgical case series published during 1990–2000 inclusive that reported the risks of stroke and death due to endarterectomy for asymptomatic stenosis found an overall risk of stroke and death of 3.0 per cent (2.5–3.5) in 28 studies, and 4.6 per cent (3.6–5.7) in 12 studies in which outcome was assessed by a neurologist (Bond *et al.* 2003a, b). Operative mortality was the same as in ACST at 1.1 per cent (0.9–1.4), some eight times higher than in ACAS. Similarly, the overall risk of stroke or death after endarterectomy performed for asymptomatic stenosis in 10 US states was 3.8 per cent, including 1 per cent mortality (Kresowik *et al.* 2004).

Thus, many clinicians do not routinely recommend endarterectomy for asymptomatic carotid stenosis (Warlow 1998). Furthermore its cost-effectiveness is questioned (Benade and Warlow 2002). Also insufficient data are currently available to allow reliable determination of the subgroups and individuals who are likely to benefit most. For example, neither ACST nor ACAS showed increasing benefit from surgery with increasing degree of stenosis within the 60–99 per cent range. This is in contrast to endarterectomy in patients with symptomatic carotid stenosis, in whom benefit increases with severity of stenosis. This failure might be because measurement of degree of stenosis was based on ultrasound rather than catheter angiography, or it may reflect a genuine difference in pathophysiology. However, ACAS and ACST did both report trends towards greater benefit from surgery in men, with no apparent benefit at 5-year follow-up in women (Executive Committee for the Asymptomatic Carotid Atherosclerosis Study 1995; Halliday *et al.* 2004). This sex difference is very pronounced when the results of both trials are combined (Rothwell 2004) (Fig. 35.50). It remains possible, however, that some benefit may accrue in women with longer follow-up.

Patients with contralateral occlusion are another group for whom the question of endarterectomy for asymptomatic stenosis is often raised. However, the ACAS trial found no long-term benefit from endarterectomy (Baker *et al.* 2000). Furthermore there was no additional benefit in patients with contralateral occlusion in ACST (Halliday *et al.* 2004).

Several investigations have been proposed to identify individual patients with asymptomatic carotid stenosis who are at particularly high risk of stroke without surgery. These include transcranial Doppler-detected emboli in the ipsilateral middle cerebral artery, impaired cerebral reactivity, the nature of the stenotic plaque on imaging, and the rate of plaque progression.

Some studies have reported prognostically useful information by measuring rates of micro-embolic signals detected on transcranial Doppler ultrasound scanning (Spence *et al.* 2005), whereas others

Subgroup	Events/Patients		OR	95% CI
	Surgical	Medical		
Males				
ACST	51/1021	97/1023	0.50	0.35–0.72
ACAS	18/544	38/547	0.46	0.26–0.81
TOTAL	69/1565	135/1570	0.49	0.36–0.66
Females				
ACST	31/539	34/537	0.90	0.55–1.49
ACAS	15/281	14/287	1.10	0.52–1.82
TOTAL	46/820	48/824	0.96	0.63–1.45

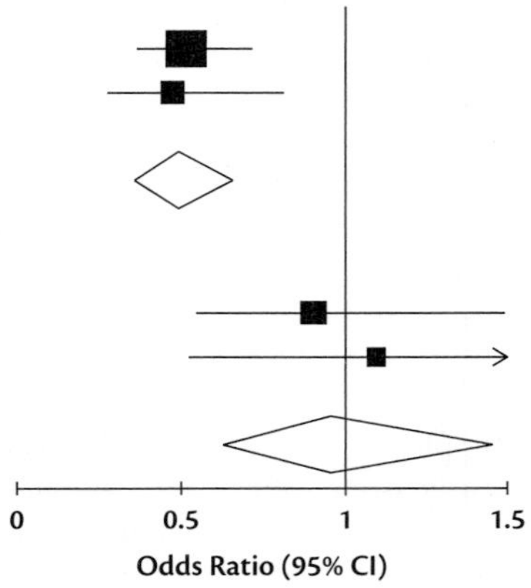

Fig. 35.50 Asymptomatic carotid stenosis . The effect of endarterectomy for severe on the risk of any stroke and operative death by sex in the Asymptomatic Carotid Surgery Trial (ACST) and the Asymptomatic Carotid Atherosclerosis study (ACAS).

Odds Ratio (95% CI)

have not (Abbott *et al.* 2005). A more definitive result should be available in 2008 from a large ongoing study (Markus and Cullinane 2000).

Several studies have suggested that increased plaque echolucency on ultrasound, a marker of plaque lipid and haemorrhage content, is associated with higher risks of stroke distal to a carotid stenosis. However, most of these studies were in patients with symptomatic stenosis and transient ischaemic attack was included in the primary end-point. A recent analysis of imaging data from a cohort study of 1115 patients with asymptomatic stenosis reported a stroke rate of 2 per cent per year in patients with plaques which were uniformly or partly echolucent and 0.14 per cent per year in the remaining patients (Nicolaides *et al.* 2005). However, plaque echolucency on ultrasound was not associated with increased benefit from endarterectomy in the ACST (Halliday *et al.* 2004).

Other methods of imaging might also be of prognostic value. These include multi-contrast-weighted MRI of plaque (Takaya *et al.* 2006), other MR-based methods of quantification of fibrous cap thickness (Cai *et al.* 2005), MR-visualization of plaque macrophages after their uptake of Ultra-Small Particles of Iron Oxide, USPIO (Tang *et al.* 2006), and various methods of assessment of cerebrovascular reserve (Markus and Cullinane 2001). However, large prospective studies are required to determine whether these imaging characteristics predict the risk of stroke.

35.14 Secondary prevention of stroke and other vascular disease

35.14.1 Early risk of recurrent ischaemic stroke after transient ischaemic attack or minor stroke

About 15–20 per cent of patients with stroke have had a preceding transient ischaemic attack (Rothwell and Warlow 2005). A similar proportion of major strokes have been preceded by minor strokes. Recent research has shown that the early risk of recurrent stroke after transient ischaemic attack or minor stroke is higher than previously thought (Coull *et al.* 2004) (Fig. 35.51). Recurrence rates

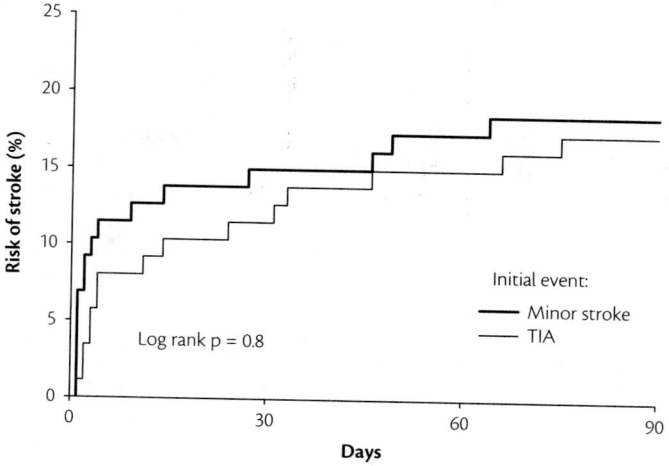

Fig. 35.51 Cumulative risk of stroke following a transient ischaemic attack (TIA) or minor stroke in the Oxford Vascular Study (Coull *et al.* 2004).

of up to 10 per cent at 7 days and 15–20 per cent at 1 month have been found in several studies (Johnston *et al.* 2000; Lovett *et al.* 2003; Coull *et al.* 2004; Hill *et al.* 2004). The timing of the preceding transient ischaemic attack in patients with stroke also indicates that the time-window for prevention is short (Fig. 35.52), with about 40 per cent of transient ischaemic attacks having occurred during the preceding 7 days (Rothwell and Warlow 2005).

Identification of high-risk patients

Patients with transient ischaemic attack and minor stroke are very heterogeneous in terms of symptoms, risk factors, underlying pathology, and early prognosis. Effective secondary prevention depends on identification of those at high risk so as to target preventative treatment. There is evidence that the clinical features of a transient ischaemic attack provide substantial prognostic information. Five risk factors are independently associated with high risk of recurrent stroke at 3 months in a large emergency cohort of patients with transient ischaemic attack. These include age over 60 years,

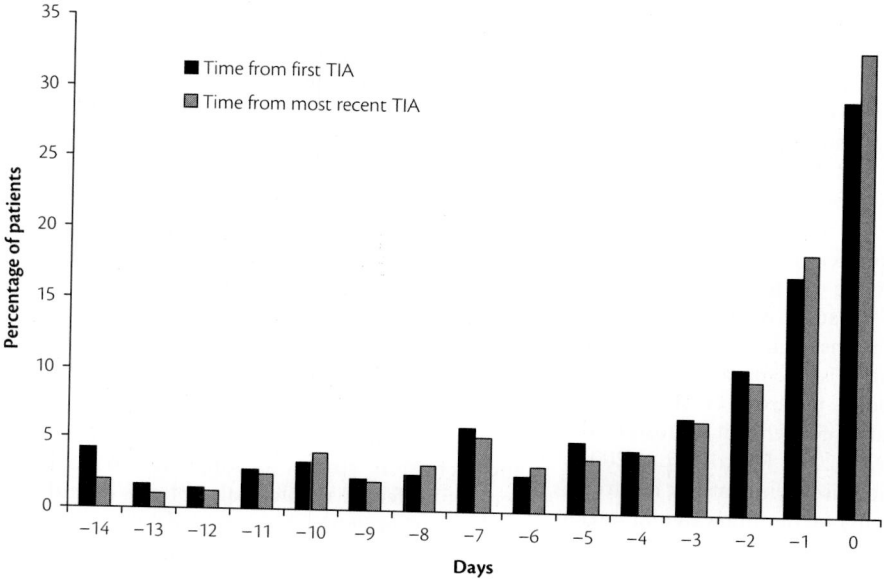

Fig. 35.52 Distribution of time from preceding transient ischaemic attack (TIA) to stroke for patients suffering a stroke who reported a transient ischaemic attack within previous 14 days (Rothwell and Warlow 2005).

symptom duration >10 min, motor weakness, speech impairment, and diabetes mellitus (Johnston *et al.* 2000). Recurrent stroke risk at 3 months varied from 0 per cent for those with none of these factors to 34 per cent for those with all five factors. Isolated sensory or visual symptoms were associated with low risk. There are similar predictors of stroke risk during the 7 days after transient ischaemic attack in two independent population-based studies (Rothwell *et al.* 2005c). A simple 6 point ABCD score, Age, Blood pressure, Clinical factors, Duration of symptoms, was highly predictive of the 7 day risk of stroke; 19/20 early recurrent strokes occurred in 27 per cent of patients with a score > 5 points (Rothwell *et al* 2005c). This ABCD scoring system was refined by the subsequent addition of Diabetes, the ABCD^2 score (Johnston *et al.* 2007):

◆ **A**ge >60 years (1 point);

◆ **B**lood pressure >140/90 (1 point);

◆ **C**linical features: unilateral weakness (2 points); speech impairment without weakness (1 point);

◆ **D**uration > 10 minutes (1 point); >60 minutes (2 points); and

◆ **D**iabetes (1 point).

Risk of recurrence by vascular territory

The early risk of stroke after transient ischaemic attack also depends on the vascular territory of the event. Monocular events are associated with a low risk of subsequent cerebral stroke (Hankey GJ 1991). Posterior circulation transient ischaemic attacks, which comprise about 25 per cent of all attacks, were thought for many years to be associated with a lower risk of stroke than carotid territory transient ischaemic attacks (Sivenius *et al.* 1991; Mohr *et al.* 1992; Caplan 1996). Correspondingly they were often managed less aggressively. However, recent work has shown that the risk of stroke in the first few weeks after a vertebrobasilar transient ischaemic attack is at least as high, and possibly higher, than that after carotid territory events (Flossmann and Rothwell 2003; Flossmann *et al.* 2006). This increases interest in imaging the posterior circulation and in angioplasty or stenting any atherothrombotic stenosis detected in the vertebral or proximal basilar arteries.

Risk by underlying pathology

Several population-based studies of stroke have shown that recurrent stroke risk is highest in those with large arterial territory stroke and lowest in those with lacunar stroke (Lovett *et al.* 2004). Although large artery pathology accounted for only 14 per cent of the initial strokes in a pooled analysis of data from four such studies, this group represented 37 per cent of the recurrences at 7 days (Lovett *et al.* 2004). Subtype differences in early recurrence risk are probably smaller in patients with transient ischaemic attack where some patients with small vessel disease can have a very high risk of early stroke, for instance the 'capsular warning syndrome'. Nevertheless, several other observations highlight the high early risk of stroke after large artery transient ischaemic attack, including the very high risk of stroke during delays to carotid endarterectomy in patients with recently symptomatic >50 per cent stenosis of the carotid artery (Fairhead and Rothwell 2005; Rantner *et al.* 2005). Patients with cardioembolic transient ischaemic attack or stroke, predominantly consisting of patients with non-valvular atrial fibrillation, are at intermediate early risk of recurrence (Lovett *et al.* 2004).

Imaging and prognosis

The presence of infarction on CT in patients with transient ischaemic attack and minor stroke is associated with increased risk of stroke in the medium- and long-term (van Swieten *et al.* 1992) and new infarction on CT performed within 48 h is highly predictive of recurrent stroke (Douglas *et al.* 2003). The presence and pattern of acute ischaemic lesions on diffusion-weighted MR brain imaging is of similar prognostic value (Purroy *et al.* 2004; Coutts *et al.* 2005). However, at present, it is unclear whether imaging provides additional prognostic information over and above the clinical characteristics in the risk scores. Focal motor weakness, speech disturbance, and symptoms lasting longer than 1 h are all associated with diffusion-weighted imaging lesions in patients with transient ischaemic attack (Redgrave *et al.* 2007a, b).

Detection of cerebral microemboli (Mackinnon *et al.* 2005) and detection of cerebral hypoperfusion (Markus and Cullinane 2001) are both of prognostic value in patients with large artery disease, but are not performed usually in routine clinical practice.

35.14.2 Cardiac risk after transient ischaemic attack or stroke

The risk of a serious cardiac event after stroke is substantial at about 3–5 per cent per annum, for instance fatal or non-fatal myocardial infarction or sudden presumed cardiac death. Heart disease is the most common cause of death after the first 30 days in patients who have suffered an ischaemic stroke. Therefore, the risk of occurrence of *all* serious vascular events after stroke should be considered together: the sum of stroke, myocardial infarction, and other vascular death. These are almost all potentially preventable by the control of vascular risk factors and antithrombotic drugs. In the long- term, cardiac death is more frequent than stroke death and as time goes by non-vascular deaths, such as cancer, become relatively more frequent (Dennis *et al.* 1990; Clarke *et al.* 2003; van Wijk *et al.* 2005).

35.14.3 Medical treatment

There is a substantial body of evidence relating to the effectivness of various treatments to reduce the medium and long-term risk of vascular events after transient ischaemic attack and stroke. Although these same treatments are also likely to be effective in reducing the early risk of stroke after transient ischaemic attack or minor stroke, there have been few randomized trials in the acute phase. A number of trials are ongoing and should report in 2008 (Rothwell *et al.* 2006). In the meantime, there is general agreement that immediate aspirin should be given, at 75–150 mg daily after a 300 mg loading dose, along with a statin and blood pressure lowering drugs as necessary. There are also reliable data to support the early use of carotid endarterectomy in patients with symptomatic carotid stenosis (Section 35.14.4).

Antiplatelet therapy

The available evidence suggests that antiplatelet therapy reduces the risk of recurrent vascular events after transient ischaemic attack and ischaemic stroke, although few trials have distinguished between different aetiological subtypes (Antithrombotic Trialists' Collaboration 2002). Most trial data concern aspirin but other antiplatelet agents such as clopidigrel (CAPRIE steering committee

1996) or extended-release dipyridamole (Sivenius et al. 1991) have also been shown to be effective.

The combination of aspirin and dipyridamole is more effective than aspirin alone (Diener et al. 1996; ESPRIT Study Group 2006). The combination results in a relative reduction in the risk of recurrent stroke of around 30 per cent compared to aspirin alone. In contrast, the combination of clopidogrel and aspirin was not superior to clopidogrel alone in secondary prevention after stroke or transient ischaemic attack in the MATCH trial (Diener et al. 2004). However, among patients randomized within the first week after the qualifying event there was a non-significant trend towards benefit from combination antiplatelet treatment. The risk of life-threatening haemorrhage was significantly higher in the combination antiplatelet group at 18 months but this difference did not become apparent until 3–4 months after randomization. Thus it is possible that a short course of clopidogrel, in addition to aspirin, might be effective in the acute phase after transient ischaemic attack and minor stroke.

Few trials of antiplatelet agents have distinguished between different vascular territories or mechanisms of stroke, but there are some data on antiplatelet agents in posterior circulation disease. The Canadian Cooperative Study (Barnett 1979) showed that aspirin reduced recurrent episodes of cerebral ischaemia and death in patients with vertebrobasilar events. The European Stroke Prevention Study, ESPS (Sivenius et al. 1991) of aspirin and immediate-release dipyridamole versus placebo appeared to show that patients with posterior circulation transient ischaemic attack benefited more than those with carotid disease but the numbers of events were too small to be certain.

Anticoagulation

Patients in atrial fibrillation who have a transient ischaemic attack or stroke without other clear aetiology should be anticoagulated if there are no contraindications (European Atrial fibrillation Trial Study Group 1993, 1995). Patients with presumed cardioembolic transient ischaemic attack or stroke secondary to other causes should certainly receive antithrombotic therapy. Also they may benefit from anticoagulation in certain circumstances, such as intracardiac mural thrombosis after myocardial infarction, although there have been no randomized trials in situations other than non-valvular atrial fibrillation.

Anticoagulation is not effective in secondary prevention of stroke for patients in sinus rhythm. Warfarin treatment to a target INR of 3–4.5 was associated with significant harm due to a large increase in major bleeding complications, especially intracerebral haemorrhage, in patients with previous transient ischaemic attack or ischaemic stroke in the SPIRIT trial (Algra et al. 1997). The subsequent WARSS Trial of aspirin versus warfarin for patients in sinus rhythm and without a cardioembolic source or >50 per cent carotid stenosis, showed no additional benefit for warfarin at a target INR of 1.4–2.8 (Redman and Allen 2002).

There has been uncertainty as to whether anticoagulation is preferable to antiplatelet treatment for the secondary prevention of ischaemia related to intracranial atherosclerosis. A retrospective analysis of 68 patients with a variety of symptomatic intracranial arterial stenoses appeared to show that warfarin was significantly better than aspirin in reducing the rate of stroke (Chimowitz et al. 1995). However, the subsequent randomized double blind WASID trial of warfarin, to a target INR of 2–3, versus aspirin to 1300 mg per day in patients with 50–99 per cent stenosis of a major intracranial artery, showed no significant benefit for warfarin over aspirin (Chimowitz et al. 2005). In fact, warfarin was associated with a significantly increased rate of adverse events including haemorrhage and, as a result of this, the study was stopped prematurely. However, patients receiving warfarin were in the therapeutic range for only about 63 per cent of the time. Therapeutic INR appeared to be associated with a much reduced incidence of ischaemic stroke and cardiac events suggesting that anticoagulation may provide increased benefit over aspirin if therapeutic INR can be maintained consistently.

Blood pressure and cholesterol lowering

There is good evidence from randomized trials to show that both blood pressure and cholesterol lowering are effective for secondary prevention of stroke. A trial of perindopril and indapamide (PROGRESS Collaborative Group (2001) showed that blood pressure reduction with an angiotensin converting enzyme inhibitor and diuretic starting several weeks or months after transient ischaemic attack or stroke reduces the risk of subsequent stroke by about a third. It is likely that this result can be generalized to most aetiological subtypes of transient ischaemic attack and ischaemic stroke, although many physicians are cautious about applying the PROGRESS results to patients with bilateral severe carotid stenosis or severe basilar or bilateral vertebral artery disease. Such patients may be at risk of borderzone infarction if their existing poor cerebral blood flow is further compromised by reduction in systemic blood pressure.

There is a positive association between cholesterol and risk of ischaemic stroke. Cholesterol lowering with statins reduces the risk of stroke in patients with previous stroke, coronary or peripheral vascular disease, or diabetes (MRC/BHF Heart Protection Study Collaborative Group (2002). However, this Heart Protection Study did not show a reduction in risk of recurrent stroke on statins (Collins et al. 2004;) possibly because patients were at low risk of stroke recurrence since the incident strokes occurred on average 4.6 years before the study onset. However, the subsequent Stroke Prevention by Aggressive Reduction in Cholesterol Levels, SPARCL, trial of atorvastatin in patients who had had a stroke or transient ischaemic attack within 1–6 months before study entry, showed a reduced overall stroke risk (Amarenco et al. 2006). However there was a significant concomitant increase in risk of intracerebral haemorrhage on statin treatment. Interestingly, the same increase in risk of haemorrhagic stroke had been found in the Heart Protection Study in the 3280 patients with previous stroke or transient ischaemic attack (Collins et al. 2004). Statins should not be used therefore in patients with previous intracerebral haemorrhage unless there is a strong indication related to the risk of ischaemic events.

35.14.4 Surgical treatment

Surgical treatments for secondary prevention of stroke in patients with large artery atherosclerosis can be divided into:

◆ endarterectomy where plaque is removed from a vessel; and

◆ by pass or anastomotic procedures designed to augment flow.

● the evidence for surgery is considerably more extensive in anterior circulation disease than in posterior circulation disease.

Carotid bifurcation stenosis

Significant atherosclerotic narrowing at, or around, the origin of the ipsilateral internal carotid artery is found in about 20–30 per cent of patients with a transient ischaemic attack or an ischaemic stroke (Fig. 35.53). That the carotid plaque is responsible for many of these strokes has been demonstrated by the observation that endarterectomy of severe symptomatic atherothrombotic stenosis markedly reduces the risk of subsequent ipsilateral carotid territory ischemic stroke. The risk of stroke is strongly related to the severity of ipsilateral carotid stenosis and to whether the stenosis is symptomatic or asymptomatic (Rothwell *et al.* 2000a; Fairhead *et al.* 2005; Cuffe and Rothwell 2006).

Although carotid endarterectomy is highly benefical in suitably selected patients, it has a variety of potential complications (Table 35.23). The most important of these are stroke and death. Death within a few days of surgery occurs in about 1–2 per cent of patients and is generally due to stroke, myocardial infarction, or some other complication of coronary heart disease or, rarely, to pulmonary embolism. Higher rates can be found in 'administrative data sets' which may be a more realistic reflection of routine practice than large randomized trials, but any comparisons are confounded by variation in casemix, particularly the proportion of patients with asymptomatic stenosis, who have a lower case fatality.

Risks of endarterectomy

The main complication of surgery is perioperative stroke. Most operative strokes are ischaemic and ipsilateral to the operated carotid artery. However intracranial haemorrhage accounts for about 5 per cent of per-operative strokes. Haemorrhage may be due to the increase in perfusion pressure and cerebral blood flow that occurs after removal of a severe internal carotid artery stenosis, particularly if cerebral autoregulation is defective as a consequence of a recent cerebral infarct. Antithrombotic drugs and uncontrolled hypertension may also play a part. In fact, transient cerebral hyperperfusion, ipsilateral but sometimes bilateral, lasting some days is quite common after carotid endarterectomy, and can occasionally cause ipsilateral transhemispheric cerebral oedema, focal epileptic seizures and headache, as well as intracerebral haemorrhage.

Myocardial infarction occurs in 1–2 per cent of patients during or soon after surgery, and more often if there is symptomatic coronary heart disease. Congestive cardiac failure, angina, and cardiac dysrhythmias are also occasional concerns.

Cranial or peripheral nerve injuries result from traction, pressure, or transaction. They occur in 5–20 per cent of cases, depending on how hard they are sought. However, these injuries seldom have any long-term consequence. If a simultaneous or staged bilateral carotid endarterectomy is done, then bilateral vocal cord paralysis or bilateral hypoglossal nerve damage are possible, both of which can cause airway obstruction. It is probably safer to do the operations a few weeks apart. Local infection may occur. Haematoma or, rarely, major haemorrhage, can result from leakage or rupture of the arteriotomy or patch, and can be life threatening if it causes tracheal compression. Aneurysm formation may occur weeks or years later. All these complications are rare.

Benefits of endarterectomy

The most reliable data on how the overall effect of endarterecomy relates to the degree of carotid stenosis come from a pooled analysis of data from the three largest randomized controlled trials of surgery for symptomatic carotid stenosis (Rothwell PM, Gutnokov SA, Eliasziw M *et al.* 2003a). The overall operative mortality was 1.1 per cent and the operative risk of stroke and death was 7.1 per cent and both operative risks were independent of the degree of carotid stenosis. Based on the method of measurement of angiographic degree of stenosis used in the NASCET Trial (North American Symptomatic Carotid Endarterectomy Trial Collaborators 1998) rather than the ECST Trial (European Carotid Surgery Trialists' Collaborative Group 1998) (Fig. 35.53), the pooled analysis showed that:

♦ surgery increased the 5-year risk of any stroke or operative death in patients with <30 per cent stenosis;

♦ surgery had no significant effect in patients with 30–49 per cent stenosis;

♦ surgery was of some benefit in patients with 50–69 per cent stenosis, with absolute reduction in 5 year risk of stroke of 7.8 per cent;

♦ surgery was highly beneficial in patients with ≥70 per cent stenosis without near-occlusion (Fig. 35.54); and

♦ there was no clear benefit in patients with the most severe disease of near-occlusions (Fig. 35.55), due to a low risk of stroke on medical treatment alone.

Qualitatively similar results were seen for disabling stroke (Fig. 35.54), but surgery had no consistent effect on survival.

Table 35.23 Complications of carotid endarterectomy

Ischaemic stroke (usually ipsilateral to the operated artery) due to:
  Embolism from the operation site during surgery
  Embolism from the operation site after surgery
  Carotid dissection
  Perioperative carotid occlusion
  Low cerebral blood flow during surgery
  Perioperative systemic hypotension

Haemorrhagic stroke (usually ipsilateral to the operated artery) due to:
  Perioperative hypertension
  Post-endarterectomy cerebral hyperperfusion

Death due to:
  Stroke
  Myocardial infarction
  Pulmonary embolism
  Rupture of arterial operation site

Myocardial infarction

Local complications
  Nerve injury (vagal, hypoglossal, marginal mandibular branch of facial, spinal accessory, greater auricular, transverse cervical nerves)
  Wound infection
  Neck haematoma
  Aneurysmal dilatation at operation site
  Patch disruption and haemorrhage

Others
  Deep venous thrombosis
  Transhemispheric cerebral oedema
  Headache, focal motor seizures
  Facial (parotid) pain
  Pain at vein donor site after vein patch angioplasty

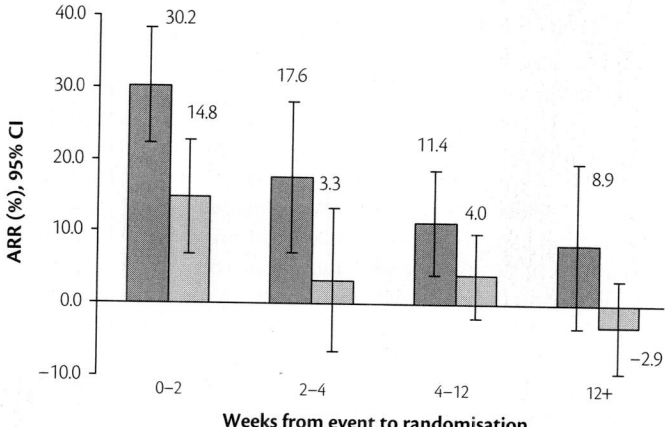

Fig. 35.57 Timing and the effectiveness of carotid surgery. Absolute reduction with surgery in the five year risk of ipsilateral carotid territory ischaemic stroke and any stroke or death within 30 days after trial surgery. Patients with 50–69 per cent stenosis (light bars) and ≥70 per cent stenosis (dark bars) without near-occlusion were stratified by the time from last symptomatic event to randomisation. This represents an analysis of pooled data from the two largest randomised trials of endarterectomy versus medical treatment for recently symptomatic carotids stenosis (Rothwell *et al.* 2004a). The numbers above the bars indicate the actual absolute risk reduction (ARR).

with surgery was considerable in those who were randomized within 2 weeks of their last event, but minimal in patients randomized later (Rothwell *et al.* 2004b).

The benefit from surgery is probably increased too in patients with stroke, intermediate in those with cerebral transient ischaemic attack, and lowest in those with retinal events (Rothwell *et al.* 2004b). Benefit is also greater in patients with irregular plaque than a smooth plaque (Rothwell *et al.* 2004b). These and the other subgroup observations described above are of some help in clinical practice, but individual patients frequently have several important risk factors, each of which interact in a way that cannot be described using univariate subgroup analysis alone. Clinicians need to weigh the often-conflicting effects of the important characteristics of an individual patient upon the likely benefit from treatment. One method is to base decisions on the predicted absolute risks of a poor outcome with each treatment option using prognostic models. A model for prediction of the risk of stroke on medical treatment in patients with recently symptomatic carotid stenosis was derived from the ECST (Rothwell and Warlow 1999, Rothwell *et al.* 2005b). The model was validated using data from NASCET and showed very good agreement between predicted and observed medical risk ($p<0.0001$) (Rothwell *et al.* 2005b). Importantly, the operative risk of stroke and death in patients who were randomized to surgery in NASCET was unrelated to the medical risk ($p=0.32$). Thus, when the operative risk and the small additional residual risk of stroke following successful endarterectomy were taken into account, benefit from endarterectomy at 5 years varied significantly across the quintiles ($p=0.001$) (Rothwell *et al.* 2005b). There was no benefit in patients in the lower three quintiles of predicted medical risk, moderate benefit in the fourth quintile, and substantial benefit in the highest quintile. A colour-coded risk table for the 5-year risk of ipsilateral ischaemic stroke in patients with recently symptomatic carotid stenosis receiving medical treatment has been derived

from the ECST model for use in routine clinical practice (Rothwell *et al.* 2005b).

Carotid occlusion and intracranial stenosis

About 10 per cent of patients with minor carotid ischaemic events have occlusion of the internal carotid artery, or stenosis of the well distal to the bifurcation, or middle cerebral artery occlusion or stenosis. The risk of recurrent stroke is high in patients with symptomatic middle cerebral artery stenosis or stenosis of other intracranial vessels (Bogousslavsky *et al.* 1986; Thijs and Alberts and 2000). Lesions of the distal carotid artery and the middle cerebral artery are surgically inaccessible although they may be treated by angioplasty or stenting. It is possible to increase cerebral perfusion in such situations, and also in patients with complete occlusion at the carotid bifurcation, by a bypass procedure in which a branch of the external carotid artery, usually the superficial temporal artery, is anastomosed via a skull burr hole to a cortical branch of the middle cerebral artery. This is known as extracranial–intracranial bypass.

This 'surgical collateral' has been shown to improve the blood supply in the distal middle cerebral artery bed. However there are several reasons why the procedure might not reduce the risk or severity of subsequent stroke. The artery feeding the anastomosis can take months to dilate into an effective collateral channel. Many patients have good collateral flow already from orbital collaterals or via the circle of Willis. Not all strokes distal to internal carotid or middle cerebral arterial occlusion or inaccessible stenosis are due to low flow. The risk of stroke in patients with internal carotid artery occlusion is not that high compared with severe and recently symptomatic internal carotid artery stenosis. Neither resting cerebral blood flow nor cerebral reactivity is necessarily depressed in these patients (Klijn *et al.* 1997; Powers *et al.* 2000).

The effectiveness of extracranial-intracranial bypass surgery was evaluated in one large randomized controlled trial, which failed to show any benefit from routine surgery (EC–IC Bypass Study Group 1985). However, it has been argued that patients with impaired cerebrovascular reactivity, or with maximal oxygen extraction, were not identified and it is perhaps these patients who might benefit from surgery (Derdeyn *et al.* 2005). A trial is now ongoing that will test the hypothesis that superficial temporal artery–middle cerebral artery anastomosis, when combined with the best medical therapy, can reduce ipsilateral ischaemic stroke by 40 per cent at 2 years in patients with symptomatic internal carotid artery occlusion and increased oxygen extraction fraction on positron emission tomography PET scanning (Grubb *et al.* 2003).

Posterior circulation disease

In contrast to symptomatic carotid disease, there have been no randomized controlled trials of interventional treatments for large artey disease in the posterior circulation (Coward *et al.* 2005b). Indeed the natural history of such disease is poorly documented. Numerous technically demanding surgical procedures have been tried, including endarterectomy, resection and anastomosis, resection and reimplantation, release of the vertebral artery from compressive fibrous bands or osteophytes, and various extra-to-intracranial bypass procedures (Hopkins *et al.* 1987; Spetzler *et al.* 1987; Malek *et al.* 1999). In the absence of randomized trials, published case series suggest that proximal vertebral reconstruction has a perioperative mortality of 0–4 per cent, with rates of stroke and death of 2.5–25 per cent (Eberhardt *et al.* 2006). For distal vertebral reconstruction a 2–8 per cent mortality rate has been reported.

Subclavian, and innominate, steal syndrome

Although subclavian, and innominate, steal is commonly detected ultrasonographically, it very rarely causes neurological symptoms and does not seem to lead on to ischaemic stroke. However, incapacitatingly frequent vertebrobasilar transient ischaemic attacks may occur in the presence of demonstrated unilateral or bilateral retrograde vertebral artery flow distal to severe subclavian or innominate stenosis. These may sometimes be relieved by endarterectomy or angioplasty of the subclavian artery; carotid-to-subclavian or femoral-to-subclavian bypass; transposition of the subclavian artery to the common carotid artery; transposition of the vertebral artery to the common carotid artery; and axillary-to-axillary artery bypass grafting. All these procedures carry a risk and it is not clear which is the most sensible. Irrespective of the neurological situation, some kind of vascular surgical procedure may be needed if the hand and arm become ischaemic distal to subclavian or innominate artery disease.

35.14.5 Endovascular treatment

Caroid bifurcation stenosis

Endovascular treatment was first used in the limbs in the 1960s and subsequently in the renal and coronary arteries. It was introduced more cautiously for treatment of carotid bifurcation stenosis in the early 1990s because of the procedural risk of stroke (Mathur *et al.* 1998) (Fig. 35.58). The endovascular approach is now widely used when carotid pathology makes endarterectomy difficult for instance with a high bifurcation or post-radiation stenosis. It is not always a feasible procedure because of contrast allergy, difficult vascular anatomy, or lumen thrombus.

Angioplasty and stenting is usually less unpleasant and less invasive than carotid endarterectomy, and generally more convenient and quicker. Being done under local anaesthetic, there may be less perioperative hypertension although cerebral haemorrhage and hyperperfusion have been reported (McCabe *et al.* 1999; Qureshi *et al.* 1999). It is less likely to cause nerve injuries, wound infection, venous thromboembolism, or myocardial infarction, and hospital stay may be shorter. Nonetheless there are also some potential disadvantages of stenting. The angioplasty balloon may dislodge atherothrombotic debris which then embolizes to the brain or eye,

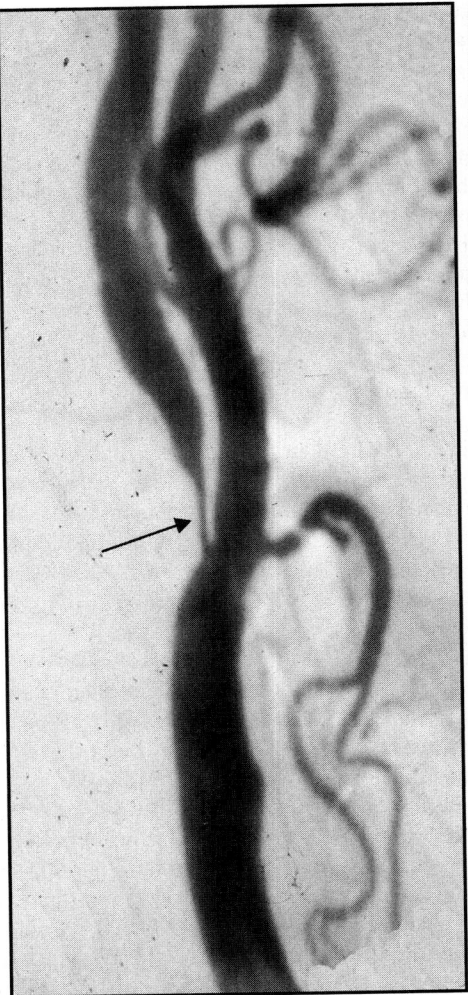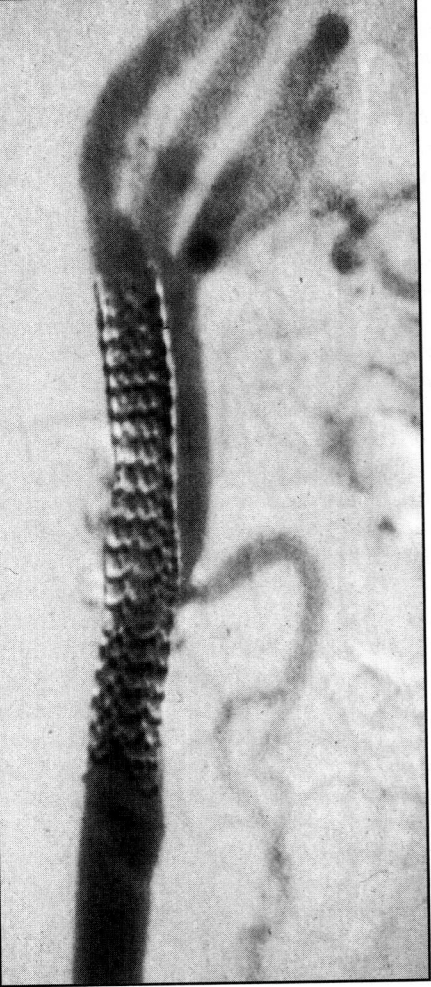

Fig. 35.58 Endovascular treatment of severe carotid stenosis (arrow). Selective arterial angiography before (left) and after (right) stenting.

although use of protection devices might help to reduce the risk of stroke due to peri-procedural embolization (Reimers *et al.* 2001). The procedure may cause arterial wall dissection at the time or afterwards. Late embolization might occur due to trombus formation on the damaged plaque. The angioplasty balloon may obstruct carotid blood flow for long enough to cause low-flow ischaemic stroke. Dilatation of the balloon may cause bradycardia or hypotension due to carotid sinus stimulation, or aneurysm formation and even arterial rupture due to over-distension of the arterial wall. Haematoma and aneurysm formation may also occur at the site of arterial cannulation in the groin. In the longer term, restenosis might be more problematic after stenting than after endarterectomy.

Prior to 2006 only five relatively small randomized comparisons of endarterectomy and angioplasty/stenting had been reported, totaling 1269 patients (Naylor *et al.* 1998; Alberts 2001; Brooks *et al.* 2001; CAVATAS Investigators 2001; Yadav *et al.* 2004). Taken together, the five trials suggested that angioplasty/stenting might have a higher procedural risk of stroke and death than endarterectomy (OR=1.33, 95 per cent CI=0.86–2.04) and a higher rate of restenosis (Coward *et al.* 2005a). However, since then improvements in endovascular techniques and cerebral protection might have reduced the procedural risks (Reimers *et al.* 2001), and several larger trials have been done, two of which reported initial results in 2006.

The SPACE Trial is the largest trial of carotid stenting versus endarterectomy to date, enrolling 1200 patients with recently symptomatic 50–99 per cent stenosis of the carotid. The procedural 30-day risk of stroke and death was not significantly higher in the angioplasty/stenting group than the endarterectomy group (OR = 1.1, 95 per cent CI = 0.7–1.7, p=0.09), with a similar trend for disabling ipsilateral stroke (4.01 per cent versus 2.91 per cent) (SPACE Collaborative Group 2006). The Endarterectomy Versus Angioplasty in Patients With Symptomatic Severe carotid Stenosis, EVA-3S, trial (Mas *et al.* 2006) in 527 patients with recently symptomatic carotid stenosis of 60–99 per cent was stopped early. A higher 30-day procedural risk of stroke and death was found after angioplasty/stenting at a planned interim analysis (9.6 per cent versus 3.9 per cent, p=0.01). There were also more local complications after angioplasty/stenting.

The results of further follow-up are awaited as are other large trials (Featherstone *et al.* 2004; Hobson *et al.* 2004; CaRESS Steering Committee 2005). Another important issue that is currently being addressed in a large randomized controlled trial (www.galatrial. com) is whether endarterectomy might have a lower operative risk with local versus general anaesthetic, for which there is some evidence (Rerkasem *et al.* 2004) Any advantage for local anaesthetic would have implications for comparisons with stenting since existing randomized controlled trials have mainly compared with endarterectomy under general anaesthetic.

Thus, the current position is that carotid stenting should be undertaken only in randomized controlled trials or in cases in which endarterectomy is technically difficult. Whichever intervention is used, early intervention and selection of patients based on predicted risk of stroke without intervention remain the keys to effective stroke prevention.

Posterior circulation disease

Angioplasty of stenoses in vertebral artery and proximal basilar artery is technically feasible but is not widely performed. Several case series have described angioplasty and stenting of symptomatic vertebral and basilar stenosis (Cloud *et al.* 2003). A review (Eberhardt *et al.* 2006) of more than 600 cases published up to 2005 provides useful information on perioperative complication rates, particularly the difference in complication rates in treatment of proximal versus distal vertebrobasilar artery lesions. In early studies, proximal lesions were treated primarily with angioplasty but this was associated with early restenosis in 15–30 per cent of cases. More recently stenting has been used for the proximal vertebral system, especially ostial lesions. Several series have reported low periprocedural or post-interventional stroke rates (Eberhardt *et al.* 2006). Pooling data from 20 reports in 313 patients, there was a perioperative stroke risk of 1.3 per cent and death rate of 0.3 per cent. However, the early restenosis rate was still about 25 per cent, albeit usually asymptomatic.

The complication rate for distal vertebrobasilar lesions treated with angioplasty and stenting is higher. Review (Eberhardt *et al.* 2006) and a prospective multicentre registry of stenting of symptomatic atherosclerotic lesions in the vertebral or intracranial arteries (SSYLVIA Study Investigators 2004) suggest peri-interventional complications rates of 7.1 per cent for stroke and 3.7 per cent for death for angioplasty for distal vertebrobasilar disease and risks of 3.2 per cent and 10.6 per cent respectively for angioplasty and stenting.

One randomized trial of stenting for vertebral artery disease was commenced (Coward *et al.* 2005b). The CAVATAS trial included both carotid and vertebral stenosis. However, only 16 patients were randomized between vertebral angioplasty or stenting and best medical treatment. Therefore, there is no robust data from randomized trials providing data on the safety and efficacy of vertebral artery stenting. Owing to the uncertainty over the risks and benefits' endovascular treatment in the posterior circulation therefore tends to be reserved for patients who continue to have symptoms despite antithrombotic therapy or in whom occlusion is likely to lead to severe stroke. That is likely where a dominant vertebral artery is involved or where there are no patent posterior communicating arteries.

Intracranial disease

Endovascular treatment for intracranial disease has not been subject to clinical trials, but there are reports that it may be of benefit to patients in whom medical therapy has failed (Marks *et al.* 1999; Connors and Wojak 1999).

35.14.6 Secondary prevention after primary intracerebral haemorrhage

Much less is known about the long-term risk of recurrence after primary intracerebral haemorrhage than after ischaemic stroke. In patients with primary intracerebral haemorrhage, about 25–50 per cent of recurrent strokes are further haemorrhages, depending on the underlying disease process. The absolute risk increased also depends on the various underlying causes such as arteriovenous malformation, cerebral amyloid angiopathy, poorly controlled hypertension, or coagulopathy. Although patients with primary intracerebral haemorrhage are at increased risk of ischaemic stroke as well as further haemorrhage most clincians do not recommend antiplatelet therapy unless there is a particularly high risk of coronary or other ischaemic vascular events. In contrast, most patients with primary intracerebral haemorrhage require blood pressure lowering medication, with the possible exception of those elderly patients with haemorrhages secondary to amyloid angiography in whom blood pressure is sometimes already rather low.

35.15 Cerebral venous thrombosis

Thrombosis in the dural sinuses or cerebral veins is much less common than cerebral arterial thromboembolism. It causes a variety of clinical syndromes, which often do not resemble stroke (Bousser and Ross Russell 1997). Whilst ischaemic arterial stroke and cerebral venous thrombosis share some causes (Southwick *et al.* 1986), others are specific to cerebral venous thrombosis (Table 35.24). A particularly high index of suspicion is required in women on the oral contraceptive pill (Saadatnia and Tajmirriahi 2007) and during the the puerperium. In the past, cerebral venous thrombosis was strongly associated with otitis media and mastoiditis, 'lateral sinus thrombosis' or 'otitic hydrocephalus' but the most common causes are now pregnancy and the puerperium, which constitute 5–20 per cent cerebral venous thrombosis in the developed world, the oral contraceptive pill, malignancy, dehydration, inflammatory disorders, and hereditary coagulation disorders. No cause is found in around 20 per cent of cases.

35.15.1 Clinical features

The superior sagittal sinus and the lateral sinuses are those most commonly affected cerebral venous thrombosis, These are followed by the straight sinuses (Fig. 35.10) and the cavernous sinuses (Stam 2005; Girot *et al.* 2007). Thrombosis of the Galenic system or isolated involvement of the cortical veins is infrequent. Cerebral venous thrombosis causes a rise in venous pressure leading to

Table 35.24 Causes of intracranial venous thrombosis

| **Local conditions affecting the cerebral veins and sinuses directly:** |
| Head injury (with or without fracture) |
| Intracranial surgery |
| Local sepsis (sinuses, ears, mastoids, scalp, nasopharynx) |
| Subdural empyema |
| Bacterial meningitis |
| Dural arteriovenous fistula |
| Tumour invasion of dural sinus (malignant meningitis, lymphoma, Skull base secondary, etc.) |
| Catheterization of jugular vein |
| Lumbar puncture |
| **Systemic disorders:** |
| Dehydration |
| Septicaemia |
| Pregnancy and the puerperium |
| Oral contraceptives / hormone replacement therapy |
| Haematological disorders |
| Prothrombotic states |
| Inflammatory vascular disorders |
| Homocysteinuria |
| Congestive cardiac failure |
| Inflammatory bowel disease |
| Androgen therapy |
| Antifibrinolytic drugs |
| Non-metastatic effect of extracranial malignancy |
| Nephrotic syndrome |

venous distension and oedema. This may be accompanied by raised intracranial pressure since the dural sinuses contain most of the arachnoid villi and granulations in which CSF absorption takes place. Occlusion of one of the larger venous sinuses is not likely to cause localized tissue damage unless there is involvement of cortical veins or the Galenic venous system since alternative drainage routes will suffice. Thrombosis in cerebral veins, with or without dural sinus thrombosis, causes multiple 'venous' infarcts which are congested, oedematous, and often haemorrhagic. Subarachnoid bleeding may occur. Transient neurological deficits may be caused by temporary ischaemia and oedema.

The incidence of cerebral venous thrombosis is uncertain since it has a wide range of clinical manifestations (Bousser 2000), whch may sometimes be obscured by the underlying disease process such as meningitis. It may be asymptomatic, and diagnosis depends on access to cerebral imaging. Cerebral venous thrombosis should be suspected when a patient develops signs of raised intracranial pressure with or without focal neurological deficits, papilloedema, and seizures, particularly when the CT brain scan is normal. Headache, which is often the presenting complaint, is present in 75 per cent of cases and has no specific characteristics: it may be acute or chronic, localized, or diffuse (Agostoni 2004). Papilloedema occurs in about 50 per cent. Focal deficits, seizures, and alterations in conscious level occur in about 30 per cent of cases. Isolated cranial nerve palsies have been described with transverse sinus thrombosis (Kuehnen *et al.* 1998). Cerebral venous thrombosis may be the underlying cause in patients with features suggestive of diffuse encephalopathy, stroke, and rarely subarachnoid haemorrhage (de Bruijn *et al.* 1996), psychosis, or migraine (Jacobs *et al.* 1996). It should be considered in all cases of apparent idiopathic intracranial hypertension particularly when the patient is male or a non obese female (Tehindrazanarivelo *et al.* 1992). (Section 26.5.5).

The progression of symptoms and signs in cerebral venous thrombosis is highly variable, ranging from less than 48 h to greater than 30 days. A gradual onset over days or weeks of headache, papilloedema and less frequently VI nerve palsy, tinnitus, and transient visual obscuration may occur. The prognosis of cerebral venous thrombosis is also variable and difficult to predict for an individual patient: a comatose patient may go on to make a complete recovery whereas a patient with few signs may gradually deteriorate and die. The current case fatality rate appears to be 10–20 per cent with a further 10–20 per cent surviving with persistent deficits (Bousser 2000; Girot *et al.* 2007). Independent predictors of death in one study were coma, mental disturbance, deep thrombosis, intracerebral haemorrhage, and posterior fossa lesions (Canhao *et al.* 2005). In the International Study on Cerebral Vein and Dural Sinus Thrombosis (Girot *et al.* 2007), independent predictors of a poor outcome, as defined by death or disability at 6 months, were older age, male gender, having a deep cerebral venous system thrombosis or a right lateral sinus thrombosis, and having a motor deficit. The prognosis for thrombosis of the deep cerebral veins is particularly poor.

Cavernous sinus thrombosis is a restricted form of cerebral venous thrombosis, usually associated with sepsis spreading from the veins in the face, nose, orbits, or sinuses (Ebright *et al.* 2001). In diabetics and immunocompromised hosts, fungal infection can be responsible, particularly mucormycosis (Section 42.12.8). The presentation is with unilateral orbital pain, periorbital oedema, chemosis, proptosis, reduced visual acuity and papilloedema.

The third, fourth, sixth, and upper two divisions of the fifth cranial nerves may be involved. Thrombus may propagate to the other cavernous sinus to cause bilateral signs. Septic meningitis and epidural empyema are occasional complications. The patients are generally severely toxic and ill. The differential diagnosis includes severe facial and orbital infection, and carotico-cavernous fistula (Section 13.4.5).

35.15.2 Diagnosis

Headache, papilloedema, and a normal CT scan should raise the possibility of cerebral venous thrombosis. Often cerebral venous thrombosis is not considered until other diagnoses have been excluded particularly when the presentation is atypical. However, it is not a diagnosis of exclusion and it must be confirmed on imaging.

Although brain CT with contrast is normal in up to 25 per cent of patients with proven cerebral venous thrombosis, it is a useful first-line investigation particularly in sick patients in whom MRI is difficult to undertake. Haemorrhagic and non-haemorrhagic infarcts outwith the usual arterial territories, oedema and intense contrast enhancement of the falx and tentorium may be seen. Sometimes there is subarachnoid blood, which is most unusual following either arterial infarcts or primary intracerebral haemorrhage (Bakac and Wardlaw 1997). Specific but less common changes include the 'empty delta sign', a triangular pattern of enhancement from dilated venous collateral channels surrounding a central relatively hypodense area of thrombosis, indicating superior sagittal sinus thrombosis (Fig. 35.59). Another is the 'cord sign' seen on a single slice only on non contrast enhanced CT scans in which fresh thrombus appears as increased density relative to grey matter in structures parallel to the scanning plane such as the straight sinus (Fig. 35.60).

MRI has greater sensitivity than CT for the changes of cerebral venous thrombosis (Bousser and Ross Russell 1997; Ferro *et al.* 2007). In the acute phase, at less than 3–5 days, the thrombus is isointense on both T1- and T2-weighted sequences. Subsequently the thrombus becomes hyperintense (Fig. 35.61). After 2–3 weeks, findings depend on whether or not the sinus remains occluded or whether it is partly or completely recanalized.

The imaging changes in patients with deep cerebral venous thrombosis are particularly striking, with bilateral deep haemorrhagic infarction (Fig. 35.62).

MR imaging and venography can now provide a definitive diagnosis in most patients although care must be taken to exclude artefacts.

Cerebral angiography with late venous views is the gold standard for the diagnosis of cerebral venous thrombosis. Nowadays this should only be performed in those cases where the diagnosis remains in doubt after MRI. There should be total or partial occlusion of at least one dural sinus on two projections. Often there is also occlusion of cerebral veins, late venous emptying, and evidence of venous collateral circulation. In subacute encephalopathies of uncertain cause, cerebral angiography or MR venography should always be done to rule out cerebral venous thrombosis before resorting to brain biopsy.

CSF is often abnormal in cerebral venous thrombosis: the pressure is usually raised and there may be elevated protein and pleocytosis especially in patients with focal signs. Lumbar puncture may be indicated in patients with isolated intracranial hypertension to lower CSF pressure when vision is threatened and to exclude meningeal infection.

EEG is abnormal in about 75 per cent of patients with cerebral venous thrombosis but changes are non-specific, with generalized slowing, often asymmetric, with superimposed epileptic activity.

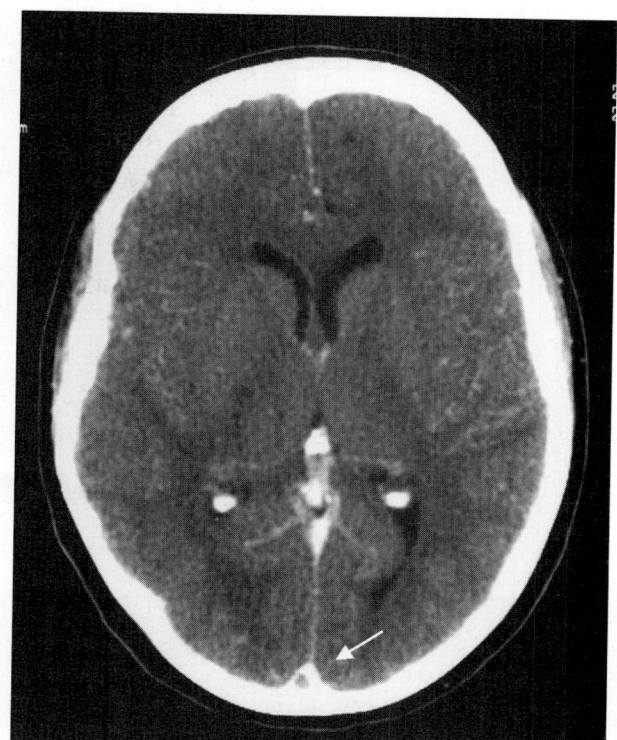

Fig. 35.59 Superior sagittal sinus thrombosis. Enhanced CT brain scan from a young woman with showing the 'empty delta sign'—a triangular pattern of enhancement surrounding a central relatively hypodense area of thrombosis (arrow).

35.15.3 Treatment

The general principles of stroke treatment apply. There is little evidence for specific treatments in cerebral venous thrombosis. Despite this most clinicians advocate anticoagulant therapy in line with guidelines for venous thromboembolism. This may prevent extension of the clot into neighbouring sinuses and veins, which is often accompanied by rapid deterioration (Einhaupl *et al.* 2006). Heparin has been reported to be safe and to improve prognosis in one small trial (Einhaupl *et al.* 1991) but this was not confirmed in a larger trial (de Bruijn *et al.* 1999). Anticoagulant therapy is associated with haemorrhage into venous infarcts but it is currently not possible to identify patients at high risk of this complication. Thrombolytic infusion has been used in patients with cerebral venous thrombosis but there are no randomized data (Kasner *et al.* 1998) and it should be reserved for patients who continue to deteriorate despite anticoagulation and supportive measures.

The European Federation of Neurological Societies guidelines (Einhaupl *et al.* 2006) advise that: 'patients with cerebral venous thrombosis without contraindications to anticoagulation should be treated either with body weight-adjusted subcutaneous

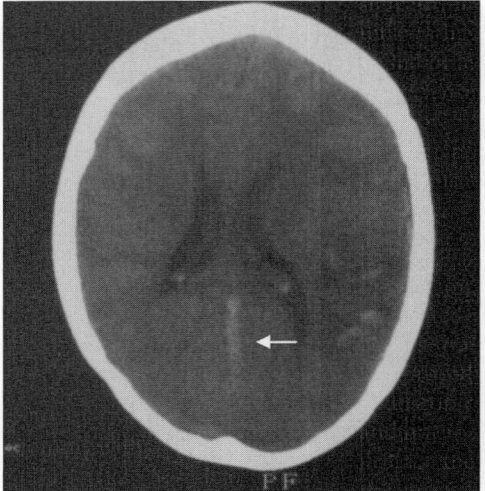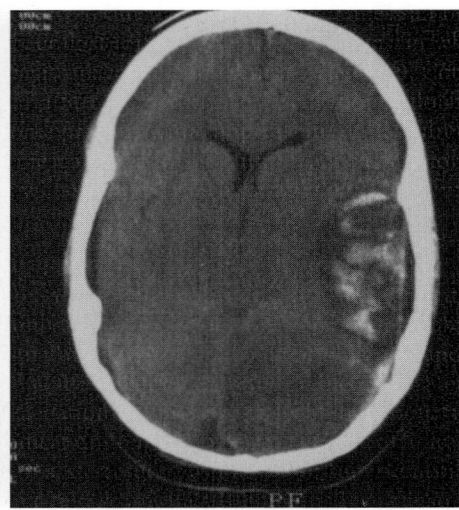

Fig. 35.60 Fresh thrombus in the straight sinus. Non-contrast CT brain scan showing the 'cord sign' (left) and a haemorrhagic left temporal lobe infarct (right).

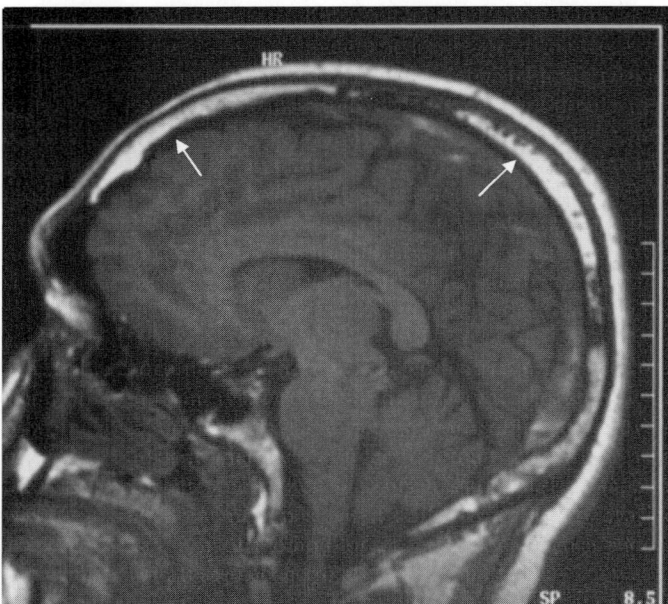

Fig. 35.61 Hyperintensity (arrowed) in the saggital sinus indicating subacute thrombosis. Saggital T1-weighted MRI.

low molecular weight or dose-adjusted intravenous heparins. Concomitant intracranial haemorrhage related to cerebral venous thrombosis is not a contraindication for heparin therapy. The optimal duration of oral anticoagulation after the acute phase is unclear. Oral anticoagulation may be given for 3 months if cerebral venous thrombosis was secondary to a transient risk factor, for 6–12 months in patients with idiopathic cerebral venous thrombosis and in those with 'mild' hereditary thrombophilia. An indefinite period of anticoagulation should be considered in patients with two or more episodes of cerebral venous thrombosis or in those with one episode of cerebral venous thrombosis and 'severe' hereditary thrombophilia. There is insufficient evidence to support the use of either systemic or local thrombolysis in patients with cerebral venous thrombosis. If patients deteriorate despite adequate

anticoagulation, and other causes of deterioration have been ruled out, thrombolysis may be a therapeutic option in selected cases, possibly in those without intracranial haemorrhage. There are no controlled data about the risks and benefits of certain therapeutic measures to reduce an elevated intracranial pressure with brain displacement in patients with severe cerebral venous thrombosis. Antioedema treatment, including hyperventilation, osmotic diuretics, and craniectomy, should be used as life saving interventions.'

Any underlying cause should be addressed: for example, patients with a definite thrombophilia should probably be anticoagulated for life, oral contraceptives should never be used again, but a further pregnancy may be safe (Preter *et al.* 1996).

35.16 Spontaneous subarachnoid haemorrhage

The incidence of subarachnoid haemorrhage increases with age and is about 5–10/100 000 population/annum. Nonetheless half of cases occur in those younger than 55 years. (Linn *et al.* 1996; Rothwell *et al.* 2004a). Risk factors for subarachnoid haemorrhage include hypertension, smoking, alcohol, oral contraceptives, and possibly coronary heart disease (Feigin *et al.* 2005a).The overall prognosis is poor: half of patients die and around one-third of survivors are left dependent (Hop *et al.* 1997; van Gijn *et al.* 2007). Coma on admission, old age, and a large amount of blood on the initial CT scan all are associated with a worse prognosis (Kassell *et al.* 1990a, b). Focal neurological deficits and, more commonly, cognitive deficits, behavioural disorders, seizures, anxiety, depression, and poor quality of life are frequent long-term sequelae (Hop *et al.* 1999). Chronic or repeated subarachnoid bleeding can produce the rare syndrome of superficial haemosiderosis of the central nervous system with sensorineural deafness, cerebellar ataxia, pyramidal signs, dementia, and bladder disturbance (Fearnley *et al.* 1995).

35.16.1 Causes

Around 85 per cent of cases of spontaneous subarachnoid haemorrhage are caused by ruptured aneurysm, 10 per cent are

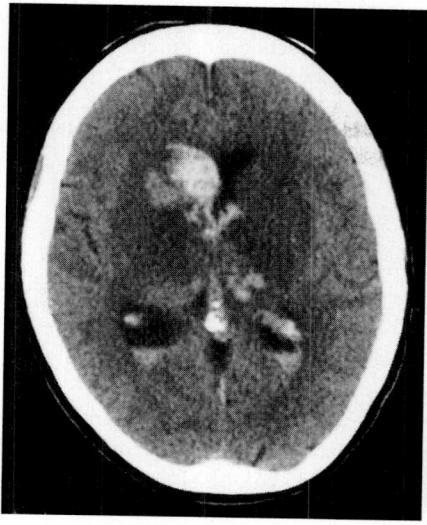
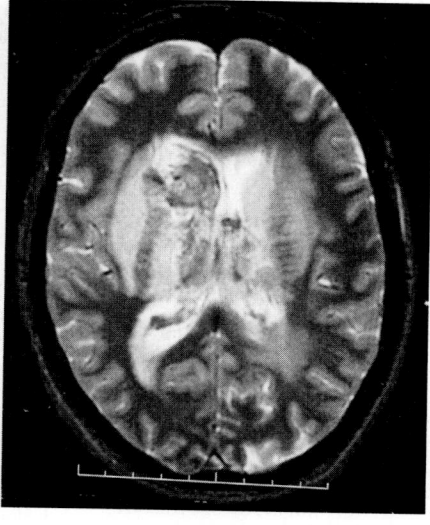

Fig. 35.62 Deep cerebral venous thrombosis with bilateral deep haemorrhagic infarction. CT (left) and MRI (right).

peri-mesencephalic subarachnoid haemorrhage, and the remainder are caused by rare disorders (Table 35.9) (Van Gijn and Rinkel 2001). The pattern of bleeding on CT is a clue to the underlying cause. Blood in the interhemispheric fissure suggests an anterior communicating artery aneurysm. In the sylvian fissure blood suggests internal carotid artery or middle cerebral artery aneurysm (Fig. 35.63).

Intracranial aneurysms are not congential but develop over the course of life. Around 10 per cent of subarachnoid haemorrhages are familial and candidate genes identified thus far include those coding for the extracellular matrix. Saccular aneurysms tend to occur at branching points on the circle of Willis and proximal cerebral arteries; about 40 per cent on the anterior communicating artery complex, 30 per cent on the posterior communicating artery or distal internal carotid artery, 20 per cent on the middle cerebral artery, and 10 per cent in the posterior circulation. About 25 per

cent occur at multiple sites. Aneurysms vary from a few millimetres to several centimetres in diameter, can enlarge with time and are an incidental finding in about 6 per cent of cerebral angiograms. This is almost certainly an over-estimate of the true rate which may be about 2 per cent (Rinkel *et al.* 1998).

Aneurysms may present with various clinical features:

- most commonly they present in middle life with subarachnoid haemorrhage;

- aneurysms may also cause primary intracerebral haemorrhage;

- adjacent structures may be compressed, such as the optic nerve, by an anterior communicating artery aneurysm; or compression of the third, fourth, and fifth cranial nerves from a distal internal carotid artery or posterior communicating artery aneurysm; or brainstem compression from a basilar artery aneurysm;

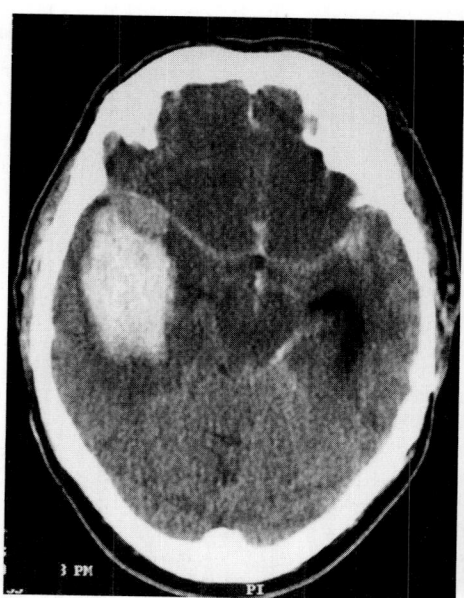
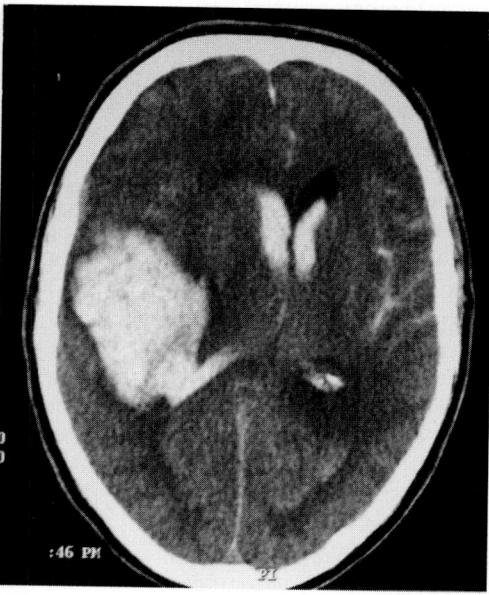

Fig. 35.63 Subarachnoid haemorrhage. CT brain scan from a patient with a ruptured right middle cerebral artery aneurysm showing widespread subarachnoid blood, with rupture into the right cerebral hemisphere and the ventricular system.

◆ with seizures;

◆ transient ischaemic attack or ischaemic stroke due to embolism of intra-aneurysmal thrombus; and

◆ with carotico-cavernous fistula from rupture of an intracavernous internal carotid artery aneurysm (Raps *et al.* 1993).

35.16.2 Clinical features

Subarachnoid haemorrhage may be provoked by exertion and rarely occurs during sleep (Ferro and Pinto 1994; Vermeer *et al.* 1997). The cardinal symptom is sudden severe headache, usually generalized. It is described as of instantaneous onset in around 50 per cent of cases but it may develop subacutely over 5 min or more (Linn *et al.* 1998), and may persist for weeks (Vermeulen *et al.* 1992; Warlow *et al.* 1996a; Schievink 1997). Headaches preceding subarachnoid haemorrhage, thought to be caused by 'warning leaks' or 'sentinel bleeds' are rare and overestimation of their importance is likely to have resulted from recall bias in hospital studies. Thus, the presence or absence of previous headache has no bearing on the diagnosis of subarachnoid haemorrhage and the terms 'warning leak' and 'sentinel bleed' should be abandoned. About a quarter of patients presenting with sudden severe headache will have subarachnoid haemorrhage, a further 40 per cent will have benign thunderclap headache (Section 18.5.6), about an eighth have some other serious neurological disorder, and the remainder have other headache syndromes (Linn *et al.* 1998).

Headache may be the only symptom in subarachnoid haemorrhage or there may be accompanying symptoms that may also be seen with other causes of sudden onset headache and so are not diagnostic. Patients are often irritable and photophobic. Loss of consciousness occurs in about half the patients but may only be brief. Nausea and vomiting are less common. Partial or generalized seizures occasionally occur at around the onset; since these do not occur in perimesencephalic haemorrhage or in thunderclap headache, their presence is a strong indicator of aneurismal rupture (Pinto *et al.* 1996). Early development of focal symptoms and signs suggest:

◆ an associated intracerebral haematoma; or

◆ local pressure from an aneurysm, such as posterior communicating artery aneurysm causing a third nerve palsy.

Later on focal symptoms are more likely to result from delayed cerebral ischaemia. Meningism develops over a few hours and pain may radiate down the legs mimicking sciatica but neck stiffness may be absent in unconscious patients. Preretinal and subhyaloid haemorrhages occur in a seventh of patients. There may be a mild fever and raised blood pressure and electrocardiographic changes that may be mistaken for myocardial infarction. Cardiac arrest occurs at onset of haemorrhage in about 3 per cent of patients, half of whom survive to independent existence with resuscitation (Toussaint *et al.* 2005). About 10 per cent of subarachnoid haemorrhage results in sudden death and about 15 per cent of patients die before receiving medical attention (Huang and van Gelder 2002). The patient's state can be graded using the World Federation of Neurological Surgeons Scale (Table 35.25).

35.16.3 Diagnosis

Since no clinical feature is specific to subarachnoid haemorrhage, the diagnosis must be excluded in anyone presenting with sudden

Table 35.25 World Federation of Neurological Surgeons' (WFNS) Scale for grading subarachnoid haemorrhage. For details of Glasgow Coma Scale (GCS), see Section 25.4.1

Grade	Glasgow coma scale	Motor or language deficit
I	15	Absent
II	14–13	Absent
III	14–13	Present
IV	12–7	Present or absent
V	6–3	Present or absent

onset severe headache lasting more than an hour and for which there is no alternative explanation. However, the differential diagnosis is wide (Table 35.26).

CT scan

Unenhanced CT scan is the quickest, most informative, and cost-effective confirmatory investigation to detect blood in the subarachnoid space (Warlow *et al.* 1996b). The sensitivity of CT for detecting subarachnoid blood depends on the amount of subarachnoid blood, the interval after symptom onset, the resolution of the scanner, and the skills of the radiologist. CT scanning misses around 2 per cent of subarachnoid haemorrhage within 12 h and this rises to 7 per cent at 24 h. Blood is almost completely reabsorbed within 10 days, and probably sooner with very mild subarachnoid haemorrhages (Brouwers *et al.* 1992). A false positive diagnosis of subarachnoid haemorrhage may be made in diffuse brain swelling when congested subarachnoid blood vessels cause a hyperdense appearance in the subarachnoid space.

CT provides a baseline for the diagnosis of later rebleeding, reveals any intracerebral, ventricular or subdural haematoma, or complicating hydrocephalus and may show calcification in the rim of an aneurysm. The pattern of bleeding may indicate the culprit aneurysm if multiple aneurysms are found on later angiography (Adams *et al.* 1983; Vermeulen and van Gijn 1990) or may

Table 35.26 Differential diagnosis of sudden unexpected headache.

With neck rigidity
Subarachnoid haemorrhage
Acute painful neck conditions
Meningitis/encephalitis
Cerebellar / brainstem stroke
Intraventricular haemorrhage
Recent head injury

Without neck rigidity
Migraine (Section 18.2)
Thunderclap headache (Section 18.5.6)
Sex headache (Section 18.5.4);
Benign exertional headache (Section 18.5.3)
Pituitary apoplexy (Section 27.7.1)
Phaeochromocytoma
Expanding intracranial aneurysm
Carotid or vertebral artery dissection (Section 35.4.4)
Intracranial venous thrombosis (Section 35.15)
Occipital neuralgia (Section 19.2.6)
Acute obstructive hydrocephalus (Section 26.4)

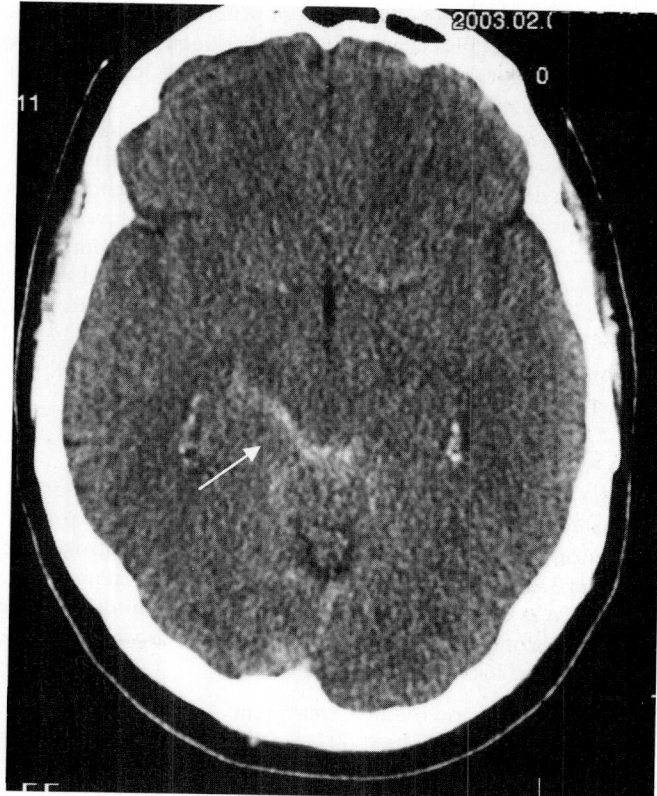

Fig. 35.64 Perimesencephalic subarachnoid haemorrhage. CT brain scan showing blood only in the basal cisterns (arrow).

show benign perimesencephalic haemorrhage (Rinkel *et al.* 1991) (Fig. 35.64). Evidence of primary or secondary head injury including brain contusion, soft tissue swelling of the scalp, and skull fracture may be present.

Lumbar puncture

All patients with suspected subarachnoid haemorrhage and a normal CT brain require lumbar puncture (van der Wee *et al.* 1995). The lumbar puncture should be delayed until at least 12 h after the onset of the headache, unless central nervous system infection is suspected, to allow haemoglobin to degrade into oxyhaemoglobin and bilirubin (van Gijn and Rinkel 2001). Bilirubin signifies subarachnoid haemorrhage since it is only synthesized *in vivo*, unlike oxyhaemoglobin which may result from a traumatic spinal tap. Acutely, the CSF glucose may be low, and the protein slightly raised with mild pleocytosis. The presence of 'xanthochromia', the yellow bilirubin pigment present in CSF following subarachnoid haemorrhage, may seen when the CSF is examined with the naked eye against a white background (Linn *et al.* 2005). However samples of CSF should also be sent for spectophotometry to detect bilirubin (UK National External Quality Assessment Scheme for Immunochemistry Working Group 2003). The least bloodstained sample of CSF should be taken to the laboratory and centrifuged immediately. The sample should be protected from light to prevent degradation of bilirubin. The estimation of red blood cell counts in serial samples does not reliably distinguish subarachnoid haemorrhage from traumatic tap (van Gijn and Rinkel 2001).

Subarachnoid haemorrhage presenting more than 2 weeks after onset

If CT and CSF examination are normal within 2 weeks of headache onset, then subarachnoid haemorrhage has been excluded. However, since xanthochromia is only detected in 70 per cent of cases after 3 weeks and only 40 per cent after 4 weeks, patients presenting beyond 2 weeks require investigation with CT or MR angiography or by catheter angiography.

Angiography

The chosen method of imaging of the cerebral circulation depends on the available technology. Multislice CT angiography is becoming more widespread because of its speed, tolerability, safety, and potential for 3-dimensional reconstructions. The sensitivity of CT angiography for aneurysms over 3-mm diameter is about 96 per cent but is less for smaller aneurysms. The sensitivity for detecting ruptured aneurysms with conventional angiography as the gold standard is currently around 95 per cent (Villablanca *et al.* 2002; Chappell *et al.* 2003; Wintermark *et al.* 2003). MR angiography has similar resolution to CT angiography but is less convenient and easy to use, especially in sick patients. Four-vessel catheter angiography may be required when non invasive imaging is negative or unavailable.

Perimesencephalic subarachnoid haemorrhage

Idiopathic perimesencephalic subarachnoid haemorrhage is restricted to the perimesencephalic cistern anterior to the midbrain (Fig. 35.64) (Schwartz and Solomon 1996). Patients present with acute headache that may be more gradual in onset than in aneurismal rupture (Linn FH, Rinkel GA, Algra A *et al.* 1998) but loss of consciousness, focal symptoms, and seizures are rare. The cause of perimesencephalic subarachnoid haemorrhage is usually unknown but it has an extremely good prognosis since rebleeding and vasospasm are unlikely (Rinkel *et al.* 1993). Aneurysms of the basilar and vertebral arteries may occasionally cause extravasation of blood into the midbrain cisterns in 2.5–5 per cent of cases of perimesencephalitic subarachnoid haemorrhage (Pinto *et al.* 1993). Such aneurysms may be excluded by performing high quality CT angiography (Ruigrok *et al.* 2000).

Angiogram negative subarachnoid haemorrhage

In up to 20 per cent of CT or CSF positive subarachnoid haemorrhage the cerebral angiogram shows no aneurysm, so called 'angiogram negative subarachnoid haemorrhage'. It should be noted that a traumatic lumbar puncture may be misdiagnosed as xanthochromic. Angiogram negative subarachnoid haemorrhage results when there is a false negative angiogram, 2–23 per cent, or when the subarachnoid haemorrhage was caused by something other than an intracerebral aneurysm (Rinkel *et al.* 1993). The pattern of subarachnoid bleeding is an important clue as to whether the bleed is likely to have been caused by an underlying aneurysm. In two-thirds of patients with angiogram negative subarachnoid haemorrhage, the CT shows perimesencephalic blood (Fig. 35.64). Patients with diffuse or anteriorly located blood on CT, which represents an aneurysmal pattern of haemorrhage, are at risk of rebleeding and repeat angiography should be performed. Repeat angiography is also required where the previous angiogram was technically inadequate, or if views of the cerebral vasculature were incomplete owing to vasospasm or haemorrhage.

Spinal subarachnoid haemorrhage

Spinal subarachnoid haemorrhage is very rare. It is caused by a vascular malformation, haemostatic failure, coarctation of the aorta, inflammatory vascular disease, mycotic aneurysm, or by vascular tumours such as ependymoma. Accumulating haematoma may compress the spinal cord. Suspicion is aroused if the cerebral angiogram is negative and the patient develops spinal cord signs.

35.16.4 Complications

Hydrocephalus

Hydrocephalus is due to blood obstructing CSF flow, occurs within days of onset in about 20 per cent of patients. It may cause clinical deterioration including a gradual reduction in conscious level. Patients with intraventricular blood or with extensive haemorrhage in the perimesencephalic cisterns are particularly predisposed to developing acute hydrocephalus. CT scanning will confirm the diagnosis. Temporary external ventricular drainage may lead to dramatic improvement but complications may occur including ventriculitis, and the risk of rebleeding of an untreated aneurysm may be slightly increased (Hellingman *et al.* 2007). Lumbar puncture may be performed in patients without a space occupying lesion or gross intrventricular haemorrhage but this requires certainty that the site of obstruction is in the subarachnoid space and not the ventricular system. In addition, it is not clear whether lumbar puncture increases the risk of rebleeding (Ruijs *et al.* 2005). Months or years after subarachnoid haemorrhage, organized thrombus and fibrosis in the CSF pathways can lead to the syndrome of normal pressure hydrocephalus (Section 26.5.5).

Delayed cerebral ischaemia

Delayed ischaemia secondary to vasospasm appears 4–14 days after onset in about 25 per cent of cases and has a bad prognosis. Loss of consciousness at onset, large quantities of subarachnoid or intraventricular blood on CT, hyponatraemia, and the use of antifibrinolytic drugs are all risk factors (Brouwers *et al.* 1992; Hop *et al.* 1999). Clinical onset is usually gradual with deteriorating conscious level and evolving focal neurological signs.

Hyponatraemia

Hyponatraemia occurs in about one-third of patients in the first week or two after subarachnoid haemorrhage and is related to the severity of the initial presentation. It is not usually caused by inappropriate antidiuretic hormone secretion but to 'salt wasting', in which there is excessive loss of salt and water by the kidneys with a decrease in plasma volume. Below a plasma sodium of about 125 mmol/l, correction is necessary by plasma volume expansion (Berendes *et al.* 1997).

Intracerebral haematoma

Intracerebral haematoma may cause a focal deficit and should be considered for removal if there is associated coma, clinical deterioration, and brain shift.

Long-term complications

Late rebleeding occurs in 2–3 per cent of patients in the first 10 years after clipping of an aneurysm, half of such bleeds being caused by newly developed aneurysms. After endovascular coiling, the long-term risks are unclear, being recorded at 0.7 per cent between 1 month and 1 year in the International Subarachnoid Aneurysm Trial, ISAT (Molyneux *et al.* 2005) and around 2–3 per cent in 1 month to 4 years in a Dutch cohort (Sluzewski *et al.* 2005)).

Epilepsy develops in 14–20 per cent of patients and putative risk factors include subdural haematoma, cerebral infarction, disability on discharge, ventricular drain insertion, and surgical treatment (Olafsson *et al.* 2000; Claassen *et al.* 2003).

Anosmia is a sequel in almost 30 per cent of patients, particularly after surgery and anterior communicating artery aneurysms.

Cognitive deficits and psychosocial dysfunction are common in the first year in patients who otherwise make a good recovery. They persist for years. In one study, a quarter of previously employed patients had stopped working, and another quarter worked shorter hours or in a position with reduced responsibility (Wermer *et al.* 2007). Changes in personality included increased irritability and emotionality. Overall, only 25 per cent of those living independently reported a complete absence of psychosocial problems.

35.16.5 Treatment

The aims of management are to identify the cause of the subarachnoid haemorrhage, to treat the source of the bleeding to prevent recurrence, to prevent the general complications of stroke, and to manage the complications of subarachnoid haemorrhage (Vermeulen *et al.* 1992; Wijdicks 1995). Close monitoring using the Glasgow coma score and pupillary responses as well as for the development of focal deficits is required.

General measures and medical treatment

The patient should be nursed in a quiet, darkened room. As with other stroke types, the management of raised blood pressure is controversial.

Secondary ischaemia is a frequent complication after subarachnoid haemorrhage, and is responsible for a substantial proportion of patients with poor outcome. The cause of secondary ischaemia is unknown, but hypovolaemia and fluid restriction are important risk factors. Hypovolaemia should be avoided and intravenous fluids given, at least 3 l per day, to reduce the likelihood of delayed ischaemia. Indeed, volume expansion therapy is frequently used in patients with subarachnoid haemorrhage to prevent or treat secondary ischaemia. However, the risks and benefits of volume expansion therapy have been studied properly in only two trials of patients with aneurysmal subarachnoid haemorrhage, with very small numbers (Rinkel *et al.* 2004). At present, there is no good evidence for the use of volume expansion therapy in patients with aneurysmal subarachnoid haemorrhage.

The risk of delayed cerebral ischaemia is also thought to be reduced and the overall outcome improved by prophylactic calcium blockers, specifically nimodipine, 60 mg 4-hourly, administered orally or by nasogastric tube, for 21 days (Rinkel *et al.* 2005). If this causes hypotension then the dose should be reduced. There is no good evidence to support intravenous nimodipine, which is particularly likely to cause hypotension. Evidence is inconclusive for other potentially neuroprotective drugs, such as nicardipine and magnesium (Rinkel *et al.* 2005). There is also no evidence of benefit from corticosteroids in patients with either subarachnoid haemorrhage or primary intracerebral haemorrhage (Feigin *et al.* 2005b).

Cardiac arrhythmias are common in the first few days, but seldom need treatment (Andreoli *et al.* 1987; Brouwers *et al.* 1989) although electrocardiographic monitoring is advisable. Neurogenic pulmonary oedema is rare but can occur very early causing diagnostic confusion. The mechanism is unclear but intensive cardiovascular monitoring and treatment are required (Parr *et al.* 1996).

The patient can be mobilized when the headache has resolved (Warlow *et al.* 1996b).

Surgical treatments for certain patients

Intracerebral extension of the haemorrhage occurs in at least a third of patients. Patients with a large haematoma and depressed consciousness might require immediate evacuation of the haematoma, preferably preceded by occlusion of the aneurysm (Niemann *et al.* 2003). Alternatively extensive craniectomy can be employed to allow expansion of the brain, as for malignant middle cerebral artery infarction (Smith *et al.* 2002) (Section 35.10.7). Subdural haematomas are rare but life threatening and should be removed.

Re-bleeding risk

About 10 per cent of untreated saccular aneurysms rebleed within hours and another 30 per cent within a few weeks (Brilstra *et al.* 2002). Subsequently, the rebleeding rate is about 2–3 per cent per annum. Deterioration is usually sudden, with reduced conscious level or fixed dilatation of the pupils in ventilated patients.

Ruptured arteriovenous malformations have a lower mortality than aneurysmal subarachnoid haemorrhage and are less likely to rebleed, certainly in the early period after the initial haemorrhage (Mast *et al.* 1997). It is unclear how to identify lesions at particularly high risk of bleeding or epilepsy (Duong *et al.* 1998).

Endovascular and surgical treatment

The purpose of occluding the source of subarachnoid haemorrhage is to prevent rebleeding. Occlusion may not be appropriate in severe cases or where there is significant co-morbidity. Neurosurgical 'clipping' was used routinely for ruptured saccular aneurysms but endovascular occlusion using detachable electrically released thrombogenic platinum coils 'coiling' is now the method of choice. For aneurysms suitable for either treatment, the International Subarachnoid Aneurysm Trial, ISAT, coiling confers an absolute risk reduction over clipping of about 7 per cent, with 25 per cent relative risk reduction, for dependency or death at 1 year (Molyneux *et al.* 2002). The risk of rebleeding from the ruptured aneurysm after 1 year was two per 1276 and zero per 1081 patient-years for patients allocated endovascular and neurosurgical treatment, respectively (Molyneux *et al.* 2002). The risk of epilepsy was also substantially lower in patients allocated to endovascular treatment (Molyneux *et al.* 2005).

The survival benefit in patients randomized to endovascular treatment in ISAT was still evident after 7 years follow-up. Although the risk of late rebleeding was low, it was more common after endovascular coiling than after neurosurgical clipping. Further follow-up is ongoing to determine the long-term risks of rebleeding, but the early benefits of endovascular treatment have convinced most centres to adopt this approach if feasible. Patients randomized in the trial were mostly young and had good WFNS score grades (Table 35.25) and small anterior circulation aneurysms, thus representing half to threee quarters of those with aneurysmal subarachnoid haemorrhage. The configuration of aneurysms of the middle cerebral artery is often less favourable for coiling. Aneurysm occlusion should be attempted as soon as practicable, preferably within 3–4 days, after onset of subarachnoid haemorrhage to prevent rebleeding.

Treatment of arteriovenous malformations and cavernomas is discussed in Section 35.11.2.

35.16.6 Unruptured aneurysms

Unruptured aneurysms in patients surviving a subarachnoid haemorrhage should be treated unless they are very small or difficult to reach. There is an assumed high risk of rupture in such aneurysms based on limited data (Wiebers *et al.* 2003). The adverse psychological impact of untreated aneurysms in patients who have survived a previous life-threatening aneurismal rupture is an equally important factor in determinng optimal treatment.

Unruptured aneurysms not associated with subarachnoid haemorrhage should normally be clipped or coiled if they are symptomatic, for instance, a IIIrd nerve palsy caused by a posterior communicating artery aneurysm (Fig. 35.65) (Raps *et al.* 1993). The optimal management of incidental unruptured asymptomatic aneurysms is unclear because the risk of rupture is low: up to 4 per cent per annum if over 10 mm in diameter, and less than 1 per cent per annum for smaller aneurysms (Rinkel *et al.* 1998;

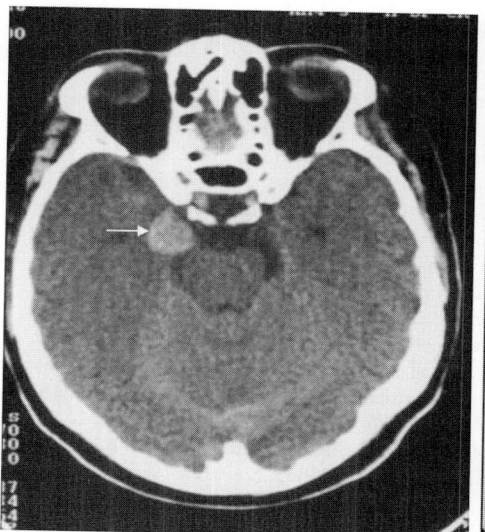
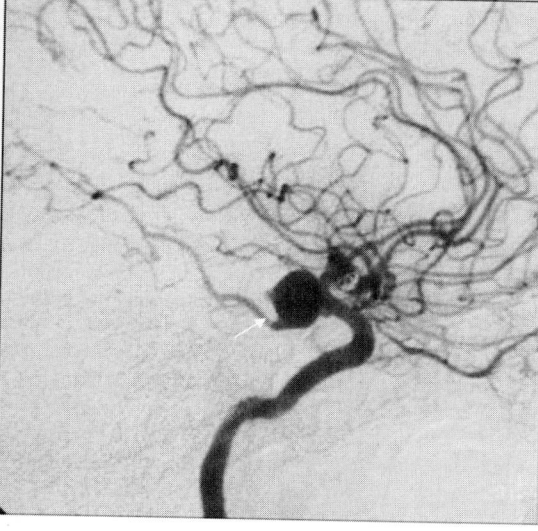

Fig. 35.65 Unruptured aneurysm (arrowed) of posterior communicating artery causing IIIrd cranial nerve palsy. CT brain imaging (left) and catheter angiography (right).

Table 35.27 Associations of intracranial saccular aneurysms

Polycystic kidney disease*
Fibromuscular dysplasia
Cervical artery dissection*
Coarctation of the aorta
Intracranial arteriovenous malformations*
Marfan's syndrome*
Ehlers–Danlos syndrome*
Pseudoxanthoma elasticum*
Neurofibromatosis type I*
Alpha I - antitrypsin deficiency*
Hereditary haemorrhagic telangiectasia*
Moyamoya syndrome
Klinefelter's syndrome
Progeria

* Can be familial.

International Study of Unruptured Intracranial Aneurysms Investigators 1998; Wermer *et al.* 2007). The risk of clipping or coiling is not insignificant (Raaymakers *et al.* 1998). The risks of rupture are higher with older age, female gender, increasing aneurysm size, posterior circulation aneurysm, and geographically with patients from Japan and Finland seemingly at higher risk (Wermer *et al.* 2007).

Individuals with an affected first-degree relative have a 5–12 times greater lifetime risk of subarachnoid haemorrhage than the general population, representing a lifetime risk of 2–5 per cent. However, the chances of finding an aneurysm by screening in an individual with a single affected relative is only 1.7 times higher than in the general population. This suggests that familial aneurysms have a higher rupture rate or grow faster than others. The indications for screening in those perceived to be at increased risk of subarachnoid haemorrhage are at present unclear (Teasdale *et al.* 2005). It could be argued that screening is not effective in those with only one affected relative but should be considered in those with two or more affected relatives, or those with polycystic kidney disease (Raaymakers 1999) or some of the other conditions associated with intracranial aneurysms (Table 35.27). Repeat screening should be discussed since the risk of finding an aneurysm after 5 years is around 7 per cent (Wermer *et al.* 2007). Patients should be referred to specialist clinics where an informed decision can be made on the basis of that individual's risks and benefits and their preferences. Screening for new aneurysms in those who have survived subarachnoid haemorrhage is not thought to be beneficial except in those with multiple aneurysms or who are very young at presentation.

References

Aaslid R, Lindegaard KF, Sorteberg W *et al.* (1989). Cerebral autoregulation dynamics in humans. *Stroke*, **20**, 45–52.

Abbott AL, Chambers BR, Stork JL *et al.* (2005). Embolic signals and prediction of ipsilateral stroke or transient ischemic attack in asymptomatic carotid stenosis: a multicenter prospective cohort study. *Stroke*, **36**, 1128–33.

Adachi T, Takagi M, Hoshino H *et al.* (1997). Effect of extracranial carotid artery stenosis and other risk factors for stroke on periventricular hyperintensity. *Stroke*, **28**, 2174–9.

Adams HP, Kassell NF, Torner JC *et al.* (1983). CT and clinical correlations in recent aneurysmal subarachnoid haemorrhage: a preliminary report of the Cooperative Aneurysm Study. *Neurology*, **33**, 981–8.

Adams HP Jr, Powers WJ, Grubb RL Jr (2001). Preview of a new trial of extracranial-to-intracranial arterial anastomosis: the carotid occlusion surgery study. *Neurosurg Clin Am*, **12**, 613.

Aerden L, Luijckx GJ, Ricci S *et al.* (2004). Validation of the Oxfordshire Community Stroke Project syndrome diagnosis derived from a standard symptom list in acute stroke. *J Neurol Sci*, **220**, 55–8.

Agostoni E (2004). Headache in cerebral venous thrombosis. *Neurol Sci*, **25**, S206–10.

Akins PT, Glen S, Nemeth PM *et al.* (1996). Carotid artery thrombus associated with severe iron-deficiency anaemia and thrombocytosis. *Stroke*, **27**, 1002–5.

Albers GW, Lansberg MG, Norbash AM *et al.* (2000). Yield of diffusion-weighted MRI for detection of potentially relevant findings in stroke patients. *Neurology*, **54**, 1562–7.

Alberts MJ (1991). Genetic aspects of cerebrovascular disease. *Stroke*, **22**, 276–80.

Alberts MJ (2001). Results of a multicantre prospective randomised trial of carotid artery stenting vs carotid endarterectomy. *Stroke*, **32**, 325.

Alexander H, Bugge C, Hagen S (2001). What is the association between the different components of stroke rehabilitation and health outcomes? *Clin Rehabil*, **15**, 207–15.

Alexandrov AV, Black SE, Ehrlich LE *et al.* (1997). Predictors of hemorrhagic transformation occurring spontaneously and on anticoagulants in patients with acute ischemic stroke. *Stroke*, **28**, 1198–202.

Alexandrov AV, Molina CA, Grotta JC *et al.* (2004). Ultrasound-enhanced systemic thrombolysis for acute ischemic stroke. *N Engl J Med*, **351**, 2170–8.

Algra A, Francke CL, Koehler PJJ (1997). A randomized trial of anticoagulants versus aspirin after cerebral ischaemia of presumed arterial origin. *Ann Neurol*, **42**, 857–65.

Al-Hakim M, Katirji MB, Osorio I *et al.* (1993). Cerebral venous thrombosis in paroxysmal nocturnal haemoglobinuria: report of two cases. *Neurology*, **43**, 742–6.

Ali K, Roffe C, Crome P (2006). What patients want: consumer involvement in the design of a randomized controlled trial of routine oxygen supplementation after acute stroke. *Stroke*, **37**, 865–71.

Allport LE, Parsons MW, Butcher KS *et al.* (2005). Elevated hematocrit is associated with reduced reperfusion and tissue survival in acute stroke. *Neurology*, **65**, 1382–7.

Amarenco P, Cohen A, Tzourio C *et al.* (1994). Atherosclerotic disease of the aortic arch and the risk of ischemic stroke. *N Engl J Med*, **331**, 1474–9.

Amarenco P (2005). Patent foramen ovale and the risk of stroke: smoking gun, guilty by association? *Heart*, **91**, 441–3.

Amarenco P, Bogousslavsky J, Callahan A 3rd *et al.* (2006). High-dose atorvastatin after stroke or transient ischemic attack. *N Engl J Med*, **355**, 549–9.

Anderson CS, Carter KN, Hackett ML *et al.* (2005). Study Group. Trends in stroke incidence in Auckland, New Zealand, during 1981 to 2003. *Stroke*, **36**, 2087–93.

Andre C, de Freitas GR, Fukujima MM (2007). Prevention of deep venous thrombosis and pulmonary embolism following stroke: a systematic review of published articles. *Eur J Neurol*, **14**, 21–32.

Andreoli A, di Pasquale G, Pinelli G *et al.* (1987). Subarachnoid haemorrhage: frequency and severity of cardiac arrhythmias. A survey of 70 cases studied in the acute phase. *Stroke*, **18**, 558–64.

Antiplatelet Trialists' Collaboration (1994). Collaborative overview of randomised trials of antiplatelet therapy I: Prevention of death, myocardial infarction and stroke by prolonged antiplatelet therapy in various categories of patients. *Br Med J*, **308**, 81–106.

Antithrombotic Trialists' Collaboration (2002). Collaborative meta-analysis of randomized trials of antiplatelet therapy for prevention of death, myocardial infarction and stroke in high risk patients. *BMJ*, **342**, 71–86.

Anzalone N, Landi G (1989). Non ischaemic causes of lacunar syndromes: prevalence and clinical findings. *J Neurol Neurosurg Psychiatry*, **52**, 1188–90.

Aoyagi M, Fukai N, Yamamoto M *et al.* (1996). Early development of intimal thickening in superficial temporal arteries in patients with Moyamoya disease. *Stroke*, **27**, 1750–4.

Arboix A, Besses C, Acin P *et al.* (1995). Ischaemic stroke as first manifestation of essential thrombocythemia. Report of six cases. *Stroke*, **26**, 1463–6.

Arboix A, Besses C (1997). Cerebrovascular disease as the initial clinical presentation of haematological disorders. *Eur Neurol*, **37**, 207–11.

Arboix A, Garcia-Eroles L, Massons J *et al.* (2007). Haemorrhagic pure motor stroke. *Eur J Neurol*, **14**, 219–3.

Argentino C, De Michele M, Fiorelli M *et al.* (1996). Posterior circulation infarcts simulating anterior circulation stroke: perspective of the acute phase. *Stroke*, **27**, 1306–9.

Arigoni M, Kneifel S, Fandino J *et al.* (2000). Simplified quantitative determination of cerebral perfusion reserve with H2(15)O PET and acetazolamide. *Eur J Nucl Med*, **27**, 1557–63.

Arnold JA, Modaresi KB, Thomas N *et al.* (1999). Carotid plaque characterization by duplex scanning: observer error may undermine current clinical trials. *Stroke*, **30**, 61–5.

Arnold M, Nedeltchev K, Remonda L *et al.* (2005). Recanalisation of middle cerebral artery occlusion after intra-arterial thrombolysis: different recanalisation grading systems and clinical functional outcome. *J Neurol Neurosurg Psychiatry*, **76**, 1373–6.

Arunodaya GR, Vani S, Shankar SK *et al.* (1997). Fibromuscular dysplasia with dissection of basilar artery presenting as 'locked-in-syndrome'. *Neurology*, **48**, 1605–8.

Astrup J, Siesjo BK, Symon L (1981). Thresholds in cerebral ischaemia. The ischaemic penumbra. *Stroke*, **12**, 723–5.

Atrial Fibrillation Investigators (1998). Echocardiographic predictors of stroke in patients with atrial fibrillation. *Arch Intern Med*, **158**, 1316–20.

Audebert HJ, Planck J, Eisenburg M *et al.* (2005). Cerebral ischemic infarction in paroxysmal nocturnal haemoglobinuria: report of 2 cases and updated review of 7 previously published patients. *J Neurol*, **252**, 1379–86.

Auer A, Felber S, Schmidauer C *et al.* (1998). Magnetic resonance angiographic and clinical features of extracranial vertebral artery dissection. *J Neurol Neurosurg Psychiatry*, **64**, 474–81.

Avendaño M, Kunst AE, Huisman M *et al.* (2004). Educational level and stroke mortality. A comparison of 10 European populations during the 1990s. *Stroke*, **35**, 432–7.

Ay H, Oliveira-Filho J, Buonanno FS *et al.* (2002). 'Footprints' of transient ischemic attacks: a diffusion-weighted MRI study. *Cerebrovascular Digest*, **14**, 177–86.

Ay H, Koroshetz WJ, Vangel M *et al.* (2005). Conversion of ischemic brain tissue into infarction increases with age. *Stroke*, **36**, 2632–6.

Ayas N, Wijdicks EFM (1995). Cardiac catheterization complicated by stroke: 14 patients. *Cerebrovasc Dis*, **5**, 304–7.

Babikian VL, Wijman CAC, Hyde C *et al.* (1997). Cerebral microembolism and early recurrent cerebral or retinal ischaemic events. *Stroke*, **28**, 1314–18.

Back T, Hemmen T, Schuler OG (2004). Lesion evolution in cerebral ischemia. *J Neurol*, **251**, 388–97.

Baigent C, Keech A, Kearney PM *et al.* (2005). Efficacy and safety of cholesterol-lowering treatment: prospective meta-analysis of data from 90056 participants in 14 randomised trials of statins. *Lancet*, **66**, 1267–78.

Baird AE, Benfield A, Schlaug G *et al.* (1997). Enlargement of human cerebral ischaemic lesion volumes measured by diffusion-weighted magnetic resonance imaging. *Ann Neurol*, **41**, 581–9.

Baird AE, Warach S (1998). Magnetic resonance imaging of acute stroke. *JCerebBlood Flow Metab*, **18**, 583–609.

Baird TA, Parsons MW, Phanh T *et al.* (2003). Persistent poststroke hyperglycemia is independently associated with infarct expansion and worse clinical outcome. *Stroke*, **34**, 2208–14.

Bakac G, Wardlaw JM (1997). Problems in the diagnosis of intracranial venous infarction. *Neuroradiology*, **39**, 566–70.

Baker WH, Howard VJ, Howard G *et al.* (2000). Effect of contralateral occlusion on long-term efficacy of endarterectomy in the asymptomatic carotid atherosclerosis study (ACAS). ACAS Investigators. *Stroke*, **31**, 2330–4.

Bamford J, Sandercock PAG, Jones L *et al.* (1987). The natural history of lacunar infarction: the Oxfordshire Community Stroke Project. *Stroke*, **18**, 545–51.

Bamford J, Sandercock PAG, Warlow CP *et al.* (1989). Interobserver agreement for the assessment of handicap in stroke patients. *Stroke*, **20**, 828.

Bamford J, Sandercock PAG, Dennis M *et al.* (1990). A prospective study of acute cerebrovascular disease in the community: the Oxfordshire Community Stroke Project 1981-86. 2. Incidence, case fatality rates and overall outcome at one year of cerebral infarction, primary intracerebral and subarachnoid haemorrhage. *J Neurol Neurosurg Psychiatry*, **53**, 16–22.

Bamford J, Sandercock P, Dennis M *et al.*(1991). Classification and natural history of clinically identifiable subtypes of cerebral infarction. *Lancet*, **337**, 1521–6.

Bang OY, Lee PH, Heo KG *et al.* (2005). Specific DWI lesion patterns predict prognosis after acute ischaemic stroke within the MCA territory. *J Neurol Neurosurg Psychiatry*, **76**, 1222–8.

Barber PA, Darby DG, Desmond PM *et al.* (1998a). Prediction of stroke outcome with echoplanar perfusion-and diffusion-weighted MRI. *Neurology*, **51**, 418–26.

Barber PA, Davis SM, Infeld B *et al.* (1998b). Spontaneous reperfusion after ischaemic stroke is associated with improved outcome. *Stroke*, **29**, 2522–8.

Barber PA, Davis SM, Darby DG *et al.* (1999). Absent middle cerebral artery flow predicts the presence and evolution of the ischemic penumbra. *Neurology*, **52**, 1125–32.

Barber PA, Demchuk AM, Zhang J *et al.* (2000). Validity and reliability of a quantitative computed tomography score in predicting outcome of hyperacute stroke before thrombolytic therapy. ASPECTS Study Group. Alberta Stroke Programme Early CT Score. *Lancet*, **355**, 1670–4.

Barker DJP (1995). Fetal origins of coronary heart disease. *Br Med J*, **311**, 171–4.

Barnett HJ (1979). The Canadian Cooperative Study of platelet suppressive drugs in transient cerebral ischemia. In: Price T, Nelson E, eds. *Cerebrovascular Disease: Proceedings of the Eleventh Princeton Conference.* pp. 221. Raven Press, New York, NY.

Baron JC (2001). Perfusion thresholds in human cerebral ischemia: historical perspective and therapeutic implications. *Cerebrovasc Dis*, **11**, 2–8.

Bartlett ES, Symons SP, Fox AJ (2006). Correlation of carotid stenosis diameter and cross-sectional areas with CT angiography. *Am J Neuroradiol*, **27**, 638–42.

Bartolucci AA, Howard G (2006). Meta-analysis of data from the six primary prevention trials of cardiovascular events using aspirin. *Am J Cardiol*, **98**, 746–50.

Bashir R, Jinkins J (1985). Cerebral infarction in a young female following snake bite. *Stroke*, **16**, 328–30.

Bassetti C, Bogousslavsky J, Barth A *et al.* (1996). Isolated infarcts of the pons. *Neurology*, **46**, 165–75.

Bassetti C, Bogousslavsky J, Mattle H *et al.* (1997). Medial medullary stroke: report of seven patients and review of the literature. *Neurology*, **48**, 882–90.

Bassetti C, Bogousslavsky J, Regli F (1993). Sensory syndromes in parietal stroke. *Neurology*, **43**, 1942–9.

Bath FJ, Bath PMW (1997). What is the correct management of blood pressure in acute stroke? The Blood Pressure in Acute Stroke Collaboration. *Cerebrovasc Dis*, **7**, 205–13.

Bathen J, Sparr S, Rokseth R (1978). Embolism in sinoatrial disease. *Acta Med Scand*, **203**, 7–11.

Beaulieu C, de Crespigny A, Tong DC *et al.* (1999). Longitudinal magnetic resonance imaging study of perfusion and diffusion in stroke: evolution of lesion volume and correlation with clinical outcome. *AnnNeurol*, **46**, 568–78.

Beck J, Garcia R, Heiss G *et al.* (1996). Periodontal disease and cardiovascular disease. *J Periodontol*, **67**, 1123–37.

Benade MM, Warlow CP (2002). Costs and benefits of carotid endarterectomy and associated preoperative arterial imaging: a systematic review of health economic literature. *Stroke*, **33**, 629–38.

Benavente O, Moher D, Pham BA (1998). Carotid endarterectomy for asymptomatic carotid stenosis: a meta-analysis. *Br Med J*, **317**, 1477–80.

Berendes E, Walter M, Cullen P *et al.* (1997). Secretion of brain natriuretic peptide in patients with aneurysmal subarachnoid haemorrhage. *Lancet*, **349**, 245–9.

Berge E (2007). Heparin for acute ischaemic stroke: a never-ending story? *Lancet Neurol*, **6**, 381–2.

Berger C, Fiorelli M, Steiner T *et al.* (2001). Hemorrhagic transformation of ischemic brain tissue: asymptomatic or symptomatic? *Stroke*, **32**, 1330–5.

Berkelbach van der Sprenkel JW, Luyten PR, van Rijen PC *et al.* (1988). Cerebral lactate detected by regional proton magnetic resonance spectroscopy in a patient with cerebral infarction. *Stroke*, **19**, 1556–60.

Berkovic SF, Bladin PF, Darby DG (1984). Metabolic disorders presenting as stroke. *Med J Aust*, **140**, 421–4.

Berlit P, Rakicky J, Tornow K (1991). Differential diagnosis of spontaneous and traumatic intracranial haemorrhage. *J Neurol Neurosurg Psychiatry*, **54**, 1118.

Bernard SAGT, Buist MD, Jones BM *et al.* (2002). Treatment of comatose survivors of out-of-hospital cardiac arrest with induced hypothermia. *New Engl J Med*, **346**, 557–63.

Bernhardt J, Dewey H, Thrift A *et al.* (2004). Inactive and alone: Physical activity within the first 14 days of acute stroke unit care. *Stroke*, **35**, 1005–9.

Bertram M, Bonsanto M, Hacke W *et al.* (2000). Managing the therapeutic dilemma: patients with spontaneous intracerebral hemorrhage and urgent need for anticoagulation. *J Neurol*, **247**, 209–14.

Bettermann K (2006). Transient global amnesia: the continuing quest for a source. *Arch Neurol*, **63**, 1336–8.

Bhatia M, Rothwell PM (2005). A systematic comparison of the published data available on risk factors for stroke, compared with risk factors for coronary heart disease. *Cerebrovasc Dis*, **20**, 180–6.

Bhogal SK, Teasell R, Foley N *et al.* (2004). Lesion location and poststroke depression: systematic review of the methodological limitations in the literature. *Stroke*, **35**, 794–802.

Bhopal R, Fischbacher C, Vartiainen E *et al.* (2005). Predicted and observed cardiovascular disease in South Asians: application of FINRISK, Framingham and SCORE models to Newcastle Heart Project data. *J Public Health (Oxf)*, **27**, 93–100.

Binkofski F, Seitz R J, Arnold S *et al.* (1996). Thalamic metbolism and corticospinal tract integrity determine motor recovery in stroke. *Ann Neurol*, **39**, 460–70.

Birns J, Markus H, Kalra L (2005). Blood pressure reduction for vascular risk: is there a price to be paid? *Stroke*, **36**, 1308–13.

Bitzer M, Topka H (1995). Progressive cerebral occlusive disease after radiation therapy. *Stroke*, **26**, 131–6.

Blackshear JL, Safford RE (2003). AFFIRM and RACE trials: implications for the management of atrial fibrillation. *Cardiol Electrophysiol Rev*, **7**, 366–9.

Blakeley DD, Oddone EZ, Hasselblad V *et al.* (1995). Non-invasive carotid artery testing: a meta-analytic review. *Ann Intern Med*, **122**, 360–7.

Blecic SA, Bogousslavsky J, Van Melle *et al.* (1993). Isolated sensorimotor stroke: a re-evaluation of clinical topographic and aetiological patterns. *Cerebrovasc Dis*, **3**, 357–63.

Blennerhassett J, Dite W (2004). Additional task-related practice improves mobility and upper limb function early after stroke: A randomised controlled trial. *Aust J Physiother*, **50**, 219–24.

Blesa R, Mohr E, Miletich RS *et al.* (1997). Changes in cerebral glucose metabolism with normal aging. *Eur J Neurol*, **4**, 8–14.

Blood pressure in Acute Stroke Collaboration (BASC) (2001). Interventions for deliberately altering blood pressure in acute stroke. *Cochrane Database Syst Rev*, **3**, CD000039.

Boesch SM, Plorer AL, Auer AJ *et al.* (2003). The natural course of Sneddon syndrome: clinical and magnetic resonance imaging findings in a prospective six year observation study. *J Neurol Neurosurg Psychiatry*, **74**, 542–4.

Bogousslavsky J, Hachinski VC, Boughner DR *et al.* (1986). Clinical predictors of cardiac and arterial lesions in carotid ischaemic attacks. *Arch Neurol*, **43**, 229–33.

Bogousslavsky J, Maeder P, Regli F *et al.* (1994). Pure midbrain infarction: clinical syndromes, MRI and aetiologic patterns. *Neurology*, **44**, 2032–40.

Boiten J, Rothwell PM, Slattery J *et al.* (1996). Ischaemic lacunar stroke in the European Carotid Surgery Trial. Risk factors, distribution of carotid stenosis, effect of surgery and type of recurrent stroke. *Cerebrovasc Dis*, **6**, 281–7.

Bonaa KH, Njlstad I, Ueland PM *et al* (2006). Homocysteine lowering and cardiovascular events after acute myocardial infarction. *N Engl J Med*, **354**, 1578–88.

Bond R, Rerkasem K, Rothwell PM (2003a). A systematic review of the risks of carotid endarterectomy in relation to the clinical indication and the timing of surgery. *Stroke*, **34**, 2290–301.

Bond R, Rerkasem K, Rothwell P M (2003b). High morbidity due to endarterectomy for asymptomatic carotid stenosis. *Cerebrovasc Dis*, **16**, 65.

Bond R, Rerkasem K, Cuffe R *et al.* (2005). A systematic review of the associations between age and sex and the operative risks of carotid endarterectomy. *Cerebrovasc Dis*, **20**, 69–77.

Bonita R, Stewart A, Beaglehole R (1990). International trends in stroke mortality 1970-1985. *Stroke*, **21**, 989–92.

Bonita R (1992). Epidemiology of stroke. *Lancet*, **339**, 342–4.

Bonita R, Broad JB, Beaglehole R (1993). Changes in stroke incidence and case-fatality in Auckland New Zealand 1981-1991. *Lancet*, **342**, 1470–3.

Bonita R, Broad JB, Beaglehole R (1997). Ethnic differences in stroke incidence and case fatality in Auckland New Zealand. *Stroke*, **28**, 758–61.

Boon A, Lodder J, Cheriex E *et al.* (1996). Risk of stroke in a cohort of 815 patients with calcification of the aortic valve with or without stenosis. *Stroke*, **27**, 847–51.

Borggreve F, De Deyn PP, Marien P *et al.* (1994). Bilateral infarction in the anterior cerebral artery vascular territory due to an unusual anomaly of the circle of Willis. *Stroke*, **25**, 1279–81.

Borhani Haghighi A, Pourmand R, Nikseresht AR (2005). Neuro-Behcet disease. A review. *Neurologist*, **11**, 80–9.

Bornebroek M, Westemdorp RGJ, Haan J *et al.* (1997). Mortality from hereditary cerebral haemorrhage with amyloidosis - Dutch type. The impact of sex parental transmission and year of birth. *Brain*, **120**, 2243–9.

Bots ML, van Swieten JC, Breteler MMB *et al.* (1993). Cerebral white matter lesions and atherosclerosis in the Rotterdam study. *Lancet*, **341**, 1232–7.

Bots ML, Looman SJ, Koudstaal PJ *et al.* (1996). Prevalence of stroke in the general population. The Rotterdam study. *Stroke*, **27**, 1499–501.

Bousser MG, Baron JC, Chiras J (1985). Ischaemic strokes and migraine. *Neuroradiology*, **27**, 583–7.

Bousser MG, Ross Russell R (1997). *Cerebral Venous Thrombosis*. Saunders, London.

Bousser MG, Kittner SJ (2000). Oral contraceptives and stroke. *Cephalagia*, **20**, 183–9.

Bousser MG (2000). Cerebral venous thrombosis: diagnosis and management. *J Neurol*, **247**, 252–8.

Bousser MG (2004). Estrogens, migraine and stroke. *Stroke*, **35**, 2652–6.

Bousser MG, Welch KM (2005). Relation between migraine and stroke. *Lancet Neurol*, **4**, 533–42.

Bowen A, Lincoln NB, Dewey ME (2002). Spatial neglect: is rehabilitation effective? *Stroke*, **33**, 2728–9.

Bowen J, Boudoulas H, Wooley CF (1987). Cardiovascular disease of connective tissue origin. *Am J Med*, **82**, 481–8.

Bowler JV, Wade JPH, Jones BE et al. (1995). Contribution of diaschisis to the clinical deficit in human cerebral infarction. *Stroke*, **26**, 1000–6.

Bragoni M, Di Piero V, Priori R et al. (1994). Sjogren's syndrome presenting as ischaemic stroke. *Stroke*, **25**, 2276–9.

Brandstater ME (2005). Stroke rehabilitation In: De Lisa JA, Gans BM, Walsh NE, eds. *Physical Medicine and Rehabilitation: Principles and Practice*. Lippincott Williams & Wilkins, Philadelphia.

Brass LM, Hartigan PM, Page WF et al. (1996). Importance of cerebrovascular disease in studies of myocardial infarction. *Stroke*, **27**, 1173–6.

Brey RL (2005). Antiphospholipid antibodies in young adults with stroke. *J Thromb Thrombolysis*, **20**, 105–12.

Brice JG, Dowsett DJ, Lowe RD (1964). Haemodynamic effects of carotid artery stenosis. *Br Med J*, **2**, 1363–6.

Brilstra EH, Algra A, Rinkel GJ et al. (2002). Effectiveness of neurosurgical clip application in patients with aneurismal subarachnoid hemorrhage. *J Neurosurg*, **97**, 1036–41.

Brink JA, McFarland EG, Heiken JP (1997). Helical/spiral computed body tomography. *Clin Radiol*, **52**, 489–503.

Brisman MH, Bederson JB, Sen CN et al. (1996). Intracerebral hemorrhage occurring remote from the craniotomy site. *Neurosurgery*, **39**, 1114–21.

British Heart Foundation Statistics Database (1998). *Coronary Heart Disease Statistics 1998*. British Heart Foundation, London.

Brittain KR, Peet SM, Castleden CM (1998). Stroke and incontinence. *Stroke*, **29**, 524–8.

Brooks WH, McClure RR, Jones MR et al. (2001). Carotid angioplasty and stenting versus carotid endarterectomy: randomized trial in a community hospital. *J Am Coll Cardiol*, **38**, 1589–95.

Brott T, Broderick J, Kothari R et al. (1997). Early haemorrhage growth in patients with intracerebral haemorrhage. *Stroke*, **28**, 1–5.

Brouwers JAM, Wijdicks EFM, Hasan D et al. (1989). Serial electrocardiographic recording in aneurysmal subarachnoid haemorrhage. *Stroke*, **20**, 1162–7.

Brouwers JAM, Wijdicks EFM, van Gijn J (1992). Infarction after aneurysm rupture does not depend on distribution or clearance rate of blood. *Stroke*, **23**, 374–9.

Brown MM, Wade JP, Marshall J (1985). Fundamental importance of arterial oxygen content in the regulation of cerebral blood flow in man. *Brain*, **108**, 81–93.

Brown RD, Wiebers DO, Torner JC et al. (1996). Incidence and prevalence of intracranial vascular malformations in Olmsted, County Minnesota 1965 to 1992. *Neurology*, **46**, 949–52.

Brugada R, Tapscott T, Czernuszewicz GZ et al. (1997). Identification of a Genetic Locus for Familial Atrial Fibrillation. *N Engl J Med*, **336**, 905–11.

Bruhn H, Frahm J, Gyngell ML et al. (1989). Cerebral metabolism in man after acute stroke: new observations using localized proton NMR spectroscopy. *Magn Reson Med*, **9**, 126–31.

Bruno A, Adams HP, Biller J et al. (1988). Cerebral infarction due to moyamoya disease in young adults. *Stroke*, **19**, 826–33.

Brunner E, Davey Smith G, Marmot M et al. (1996). Childhood social circumstances and psychosocial and behavioural factors as determinants of plasma fibrinogen. *Lancet*, **347**, 1008–13.

Bryan RN, Levy LM, Whitlow WD et al. (1991). Diagnosis of acute cerebral infarction: comparison of CT and MR imaging. *Am J Neuroradiol*, **12**, 611–20.

Butcher KS, Baird T, MacGregor L et al. (2004). Perihematomal edema in primary intracerebral hemorrhage is plasma derived. *Stroke*, **35**, 1879–85.

Byar DP, Corle DK (1988). Hormone therapy for prostate cancer: results of the Veterans Administration Cooperative Urological Research Group Studies. *NCI Monogr*, **7**, 165–70.

Byrne JV (2005). Cerebrovascular malformations. *Eur Radiol*, **15**, 448–52.

Cabanes L, Mas JL, Cohen A et al. (1993). Atrial septal aneurysm and patent foramen ovale as risk factors for cryptogenic stroke in patients less than 55 years of age. A study using transoesophageal echocardiography. *Stroke*, **24**, 1865–73.

Cai J, Hatsukami TS, Ferguson MS et al. (2005). In vivo quantitative measurement of intact fibrous cap and lipid-rich necrotic core size in atherosclerotic carotid plaque: comparison of high-resolution, contrast-enhanced magnetic resonance imaging and histology. *Circulation*, **112**, 3437–44.

Calguneri M, Ozturk MA, Ay H et al. (2004). Buerger's disease with multisystem involvement. A case report and a review of the literature. *Angiology*, **55**, 325–8.

Canhao P, Ferro JM, Lindgren AG et al. (2005). Causes and predictors of death in cerebral venous thrombosis. *Stroke*, **36**, 1720–5.

Cao Y, D'Olhaberriague L, Vikingstad EM et al. (1998). Pilot study of functional MRI to assess cerebral activation of motor function after poststroke hemiparesis. *Stroke*, **29**, 112–22.

Caplan LR, Sergay S (1976). Positional cerebral ischaemia. *J Neurol Neurosurg Psychiatry*, **39**, 385–91.

Caplan LR (1980). 'Top of the basilar' syndrome. *Neurology*, **30**, 72–9.

Caplan LR (1993). Brain embolism revisited. *Neurology*, **43**, 1281–7.

Caplan LR (1995). Binswanger's disease – revisited. *Neurology*, **45**, 626–33.

Caplan LR (1996). Posterior Circulation Disease. *Blackwell Science* p20–1.

Cappuccio FP (1997). Ethnicity and cardiovascular risk: variations in people of African ancestry and South Asian origin. *J Hum Hypertens*, **11**, 571–6

CAPRIE Steering Committee (1996). A randomised blinded trial of clopidogrel versus aspirin in patients at risk of ischaemic events (CAPRIE). *Lancet*, **348**, 1329–39.

Carney JA (1985). Differences between nonfamilial and familial cardiac myxoma. *Am J Surg Pathol*, **9**, 53–5.

CaRESS Steering Committee (2005). Carotid Revascularization Using Endarterectomy or Stenting Systems (CaRESS) phase I clinical trial: 1-year results. *J Vasc Surg*, **42**, 213–19.

Carrera E, Bogousslavsky J (2006). The thalamus and behaviour: effects of anatomically distinct strokes. *Neurology*, **66**, 1817–23

Casas JP, Hingorani AD, Bautista LE et al. (2004). Meta-analysis of genetic studies in ischemic stroke: thirty-two genes involving approximately 18,000 cases and 58,000 controls. *Arch Neurol*, **61**, 1652–61.

Casas JP, Cavalleri GL, Bautista LE et al. (2006). Endothelial nitric oxide synthase gene polymorphisms and cardiovascular disease: A HuGE review. *Am J Epidemiol*, **164**, 921–35.

Cascino GD, Westmoreland BF, Swanson TH et al. (1991). Seizure-associated speech arrest in elderly patients. *Mayo Clin Proc*, **66**, 254–8.

CAST (Chinese Acute Stroke Trial) Collaborative Group (1997). CAST: randomised placebo-controlled trial of early aspirin use in 20000 patients with acute ischaemic stroke. *Lancet*, **349**, 1641–9.

Castillo J, Dávalos A, Alvarez-Sabín J et al. (2002). Molecular signatures of brain injury after intracerebral hemorrhage. *Neurology*, **58**, 624–9.

Catala M, Rancurel G, Koskas F et al. (1993). Ischaemic stroke due to spontaneous extracranial vertebral giant aneurysm. *Cerebrovasc Dis*, **3**, 322–6.

CAVATAS Investigators (2001). Endovascular versus surgical treatment in patients with carotid stenosis in the carotid and vertebral artery. Transluminal angioplasty study: a randomised trial. *Lancet*, **357**, 1729–37.

Chabriat H, Levy C, Taillia H et al. (1998). Patterns of MRI lesions in CADASIL. *Neurology*, **51**, 452–7.

Chabriat H, Vahedi K, Iba-Zizen MT *et al.* (1995). Clinical spectrum of CADASIL: a study of 7 families. *Lancet*, **346**, 934–9.

Chalela JA, Kidwell CS, Nentwich LM *et al.* (2007). Magnetic resonance imaging and computed tomography in emergency assessment of patients with uspected acute stroke: a prospective comparison. *Lancet*, **369**, 293–8.

Chambers BR, Donnan GA (2005). Carotid endarterectomy for asymptomatic carotid stenosis. *Cochrane Database Syst Rev*: CD001923.

Chang CL, Donaghy M, Poulter N *et al.* (1999). Migraine and stroke in young women: case-control study. *Br Med J*, **318**, 13–8.

Chapin JE, Davis LE, Kornfeld M *et al.* (1995). Neurologic manifestations of intravascular lymphomatosis. *Acta Neurol Scand*, **91**, 494–9.

Chappell ET, Moure FC, Good MC (2003). Comparison of computed tomographic angiography with digital subtraction angiography in the diagnosis of cerebral aneurysms: a meta-analysis. *Neurosurgery*, **52**, 624–31.

Chappell F, Wardlaw J, Best JKK *et al.* (2006). Non-invasive imaging compared with intra-arterial angiography in the diagnosis of symptomatic carotid stenosis: a meta-analysis. *Lancet*, **376**, 1503–12.

Chaves C, Estol C, Esnaola MM *et al.* (2002). Spontaneous intracranial internal carotid artery dissection: report of 10 patients. *Arch Neurol*, **59**, 977–81.

Cheshire WP, Santos CC, Massey EW *et al.* (1996). Spinal cord infarction: aetiology and outcome. *Neurology*, **47**, 321–30.

Cheshire WP Jr, Saper CB (2006). The insular cortex and cardiac response to stroke. *Neurology*, **66**, 1296–7.

Chimowitz MI, Kokkinos J, Strong J *et al.* (1995). The Warfarin-Aspirin Symptomatic Intracranial Disease Study. *Neurology*, **45**, 1488–93.

Chimowitz MI, Lynn MJ, Howlett-Smith H *et al.* (2005). Comparison of warfarin and aspirin for symptomatic intracranial arterial stenosis. *N Engl J Med*, **352**, 1305–16.

Chiu D, Shedden P, Bratina P *et al.* (1998). Clinical features of moyamoya disease in the United States. *Stroke*, **29**, 1347–51.

Cho HJ, Song SK, Lee DW *et al.* (2007). Carotid-subclavian steal phenomenon. *Neurology*, **68**, 702.

Chodosh EH, Foulkes MA, Kase CS *et al.* (1988). Silent stroke in the NINCDS stroke data bank. *Neurology*, **38**, 1674–9.

Chuang KY, Wu SC, Yeh MC *et al.* (2005). Exploring the associations between long-term care and mortality rates among stroke patients. *J Nurs Res*, **13**, 66–74.

Claassen J, Peery S, Kreiter KT *et al.* (2003). Predictors and clinical impact of epilepsy after subarachnoid hemorrhage. *Neurology*, **60**, 208–14.

Clarke T, Murphy M, Rothwell PM (2003). Long-term risks of stroke, myocardial infarction, and vascular death in patients with a non-recent transient ischaemic attack. *JNNP*, **74**, 577–80.

Clarke R, Lewington S, Youngman L *et al.* (2002). Underestimation of the importance of blood pressure and cholesterol for coronary heart disease mortality in old age. *Eur Heart J*, **23**, 286–93.

Clatterbuck RE, Hsu FP, Spetzler RF (2005). Supratentorial arteriovenous malformations. *Neurosurgery*, **57**, 164–7.

Cleary P, Shorvon S, Tallis R (2004). Late-onset seizures as a predictor of subsequent stroke. *Lancet*, **363**, 1184–6.

Cloud GC, Crawley F, Clifton A *et al.* (2003). Vertebral artery origin angioplasty and primary stenting: safety and restenosis rates in a prospective series. *J Neurol Neurosurg Psychiatry*, **74**, 586–90.

CODICE Study Group (2006). Recurrent stroke is not associated with massive right-to-left shunt: preliminary results from the 3-year prospective Spanish Multicentre Centre (CODICE study). *Cerebrovasc Dis*, **21**, 1.

Cohen ZR, Ram Z, Knoller N *et al.* (2002). Management and outcome of non-traumatic cerebellar haemorrhage. *Cerebrovasc Dis*, **14**, 207–13.

Collen FM, Wade DT, Bradshaw CM (1990). Mobility after stroke: reliability of measures of impairment and disability. *Int Disabil Stud*, **12**, 6–9.

Collen FM, Wade DT (1991). Residual mobility problems after stroke. *Int Disabil Stud*, **13**, 12–15.

Collen FM, Wade DT, Robb GF *et al.* (1991). The Rivermead Mobility Index: a further development of the Rivermead Motor Assessment. *Int Disabil Stud*, **13**, 50–4.

Chollet F, DiPiero V, Wise RJ *et al.* (1991). The functional anatomy of motor recovery after stroke in humans: a study with positron emission tomography. *AnnNeurol*, **29**, 63–71.

Collin C, Wade D (1990). Assessing motor impairment after stroke: a pilot reliability study. *J Neurol Neurosurg Psychiatry*, **53**, 576–9.

Collins R, Armitage J, Parish S *et al.* (2004). Effects of cholesterol-lowering with simvastatin on stroke and other major vascular events in 20536 people with cerebrovascular disease or other high-risk conditions. *Lancet*, **363**, 757–67.

Comu S, Verstraeten T, Rinkoff JS *et al.* (1996). Neurological manifestations of acute posterior multifocal placoid pigment epitheliopathy. *Stroke*, **27**, 996–1001.

Connor MD, Walker R, Modi G *et al.* (2007). Burden of stroke in black populations in sub-Saharan Africa. *Lancet Neurol*, **6**, 269–78.

Connors JJ, Wojak JC (1999). Percutaneous transluminal angioplasty for intracranial atherosclerotic lesions: evolution of technique and short-term results. *J Neurosurg*, **91**, 415–23.

Cordonnier C, Al-Shahi Salman R, Wardlaw J (2007). Spontaneous brain microbleeds: systematic review subgroup analyses and standards for study design and reporting. *Brain* [Epub ahead of print].

Cormio M, Robertson CS, Narayan RK (1997). Secondary insults to the injured brain. *J Clin Neurosci*, **4**, 132–48.

Coull AJ, Lovett JK, Rothwell PM *et al.* (2004). Population based study of early risk of stroke after transient ischaemic attack or minor stroke: implications for public education and organisation of services. *BMJ*, **328**, 326.

Coull AJ, Silver LE, Bull LM *et al.* (2004). Study direct assessment of completeness of ascertainment in a stroke incidence study. *Stroke*, **35**, 2041–5.

Counsell and Dennis M (2001). Systematic review of prognostic models in patients with acute stroke. *Cerebrovasc Dis*, **12**, 159–70.

Counsell C, Dennis M, McDowall M *et al.* (2002). Predicting outcome after acute and subacute stroke: development and validation of new prognostic models. *Stroke*, **33**, 1041–7.

Coutts SB, Demchuk AM, Barber PA *et al.* (2004). Interobserver variation of ASPECTS in real time. *Stroke*, **35**, 103–5.

Coutts SB, Simon JE, Eliasziw M *et al.* (2005). Triaging transient ischemic attack and minor stroke patients using acute magnetic resonance imaging. *Ann Neurol*, **57**, 848–54.

Coward LJ, Featherstone RL, Brown MM (2005a). Safety and efficacy of endovascular treatment of carotid artery stenosis compared with carotid endarterectomy: a Cochrane systematic review of the randomized evidence. *Stroke*, **36**, 905–11.

Coward LJ, Featherstone R, Brown MM (2005b). Percutaneous transluminal angioplasty and stenting for vertebral artery stenosis. *Cochrane Database Syst Rev.*

Cramer SC, Robertson RL, Dooling EC *et al.* (1996). Moyamoya and Down syndrome. Clinical and radiological features. *Stroke*, **27**, 2131–5.

Cramer SC, Nelles G, Benson RR *et al.* (1997). A functional MRI study of subjects recovered from hemiparetic stroke. *Stroke*, **28**, 2518–7.

Cross SS (1991). How common is cholesterol embolism? *J Clin Pathol*, **44**, 859–61.

Cuffe RL, Rothwell PM (2006). Effect of nonoptimal imaging on the relationship between the measured degree of symptomatic carotid stenosis and risk of ischemic stroke. *Stroke*, **7**, 1785–91.

D'Cruz DP, Khamashta MA, Hughes GR (2007). Systemic lupus erythematosus. *Lancet*, **369**, 587–96.

D'Olhaberriague L, Arboix A, Marti-Vilalta JL *et al.* (1995). Movement disorders in ischaemic stroke: clinical study of 22 patients. *Eur J Neurol*, **2**, 553–7.

Dahlof B, Devereux RB, Kjeldsen SE et al. (2002). Cardiovascular morbidity and mortality in Losartan. Intervention For End Point Reduction in Hypertension (LIFE) study: a randomized trial against Atenolol. Lancet, 359, 995–1003.

Danesh J, Collins R, Peto R (1997). Chronic infections and coronary heart disease: is there a link? Lancet, 350, 430–6.

Danesh J, Youngman L, Clarke S et al. (1999). Helicobacter pylori infection and early onset myocardial infarction: case–control and sibling pairs study. Br Med J, 319, 1157–62.

Danesh J, Whincup P, Walker M et al. (2000). Chlamydia, pneumoniae IgG titres and coronary heart disease: prospective study and meta-analysis. Br Med J, 321, 208–13.

Danesh J, Whincup P, Walker M (2003). Chlamydia pneumoniae IgA titres and coronary heart disease: prospective study and meta-analysis. Eur Heart J, 24, 881.

Danesh J (2005). Antibiotics in the prevention of heart attacks. Lancet, 365, 365–7.

Danesh J, Lewington S, Thompson SG et al. (2005). Plasma fibrinogen level and the risk of major cardiovascular diseases and nonvascular mortality: an individual participant meta-analysis. J Am Med Assoc, 294, 1799–809.

Darby DG, Donnan GA, Saling MA et al. (1988). Primary intraventricular haemorrhage: clinical and neuropsychological findings in a prospective stroke series. Neurology, 38, 68–75.

Davenport RJ, Dennis MS, Wellwood I et al. (1996a). Complications after acute stroke. Stroke, 27, 415–20.

Davenport RJ, Dennis MS, Warlow CP (1996b). Gastrointestinal hemorrhage after acute stroke. Stroke, 27, 421–4.

Davey Smith G, Hart C, Blane D et al. (1997). Lifetime socioeconomic position and mortality: prospective observational study. Br Med J, 314, 547–52.

Davies-Jones GAB (1995). Neurological manifestations of haematological disorders in Neurology and General Medicine Aminoff M J, ed. 2nd edition, pp. 219–45. Churchill Livingstone, New York.

Davis JM, Zimmerman RA (1983). Injury of the carotid and vertebral arteries. Neuroradiology, 25, 55–69.

de Bray JM, Penisson-Besnier I, Dubas F et al. (1997). Extracranial and intracranial vertebrobasilar dissections: diagnosis and prognosis. J Neurol Neurosurg Psychiatry, 63, 46–51.

de Bruijn SFTM, Stam J, Kappelle LJ (1996). Thunderclap headache as first symptom of cerebral venous sinus thrombosis. Lancet, 348, 1623–5.

de Bruijn SFTM, Stam J (1999). Randomised placebo controlled trial of anticoagulant treatment with low molecular weight heparin for cerebral sinus thrombosis. Stroke, 30, 484–8.

De Haan R, Aaronson N, Limburg M et al. (1993). Measuring quality of life in stroke. Stroke, 24, 320–7.

DeMarco JK, Huston J 3rd, Nash AK (2006). Extracranial carotid MR imaging at 3T. Magn Reson Imaging Clin N Am, 14, 109–21.

De Vries JJ, den Dunnen WF, Timmerman EA et al. (2006). Acute posterior multifocal placoid pigment epitheliopathy with cerebral vasculitis: a multisystem granulomatous disease. Arch Ophthalmol, 124, 910–3.

del Zoppo GJ, Higashida RT, Furlan AJ et al. (1998). PROACT: a phase II randomized trial of recombinant pro-urokinase by direct arterial delivery in acute middle cerebral artery stroke PROACT Investigators Prolyse in Acute Cerebral Thromboembolism. Stroke, 29, 4–11.

Delalande S, de Seze J, Fauchais AL et al. (2004). Neurologic manifestations in primary Sjögren syndrome: a study of 82 patients. Medicine (Baltimore), 83, 280–91.

Dennis M, Bamford J, Sandercock P et al. (1990). The prognosis of transient ischaemic attacks in the Oxfordshire community stroke project. Stroke, 21, 848–53.

Dennis M, Warlow CP (1992). Migraine aura without headache: transient ischaemic attack or not? J Neurol Neurosurg Psychiatry, 55, 437–40.

Dennis M, Wellwood I, Warlow C (1997a). Are simple questions a valid measure of outcome after stroke? Cerebrovasc Dis, 7, 22–7.

Dennis M, Wellwood I, O'Rourke S et al. (1997b). How reliable are simple questions in assessing outcome after stroke? Cerebrovasc Dis, 7, 19–21.

Dennis MS (2004). Effective prophylaxis for deep vein thrombosis after stroke: low-dose anticoagulation rather than stockings alone: against. Stroke, 35, 2912–13.

Dennis MS, Lewis SC, Warlow C et al. (2005). Effect of timing and method of enteral tube feeding for dysphagic stroke patients (FOOD): a multicentre randomised controlled trial. Lancet, 365, 764–2.

Derdeyn CP, Grubb RL Jr, Powers WJ (1999). Cerebral hemodynamic impairment: methods of measurement and association with stroke risk. Neurology, 53, 251–9.

Derdeyn CP, Grubb RL Jr, Powers WJ (2005). Indications for cerebral revascularization for patients with atherosclerotic carotid occlusion. Skull Base, 15, 7–14.

Derex L, Hermier M, Adeleine P et al. (2005). Clinical and imaging predictors of intracerebral haemorrhage in stroke patients treated with intravenous tissue lasminogen activator. J Neurol Neurosurg Psychiatry, 76, 70–5.

Desai B, Toole JF (1975). Kinks, coils and carotids: A review. Stroke, 6, 649–53.

DeWeese JA, May AG, Lipchik EO et al. (1970). Anatomic and haemodynamic correlations in carotid artery stenosis. Stroke, 1, 149–57.

Dey AB, Stout NR, Kenny RA (1996). Cardiovascular syncope is the commonest cause of drop attacks in the older patient. European Journal of Cardiac Pacing and Electrophysiology, 6, 84–8.

di Pasquale G, Urbinati S, Pinelli G (1995). New echocardiographic markers of embolic risk in atrial fibrillation. Cerebrovasc Dis, 5, 315–22.

Diabetes Prevention Program Research Group (2002). Reduction in the incidence of type 2 diabetes with lifestyle intervention or Metformin. N Engl J Med, 346, 393–403.

Dichgans M, Mayer M, Müller-Myhsok B et al (1996). Identification of a Key Recombinant Narrows the CADASIL Gene Region to 8 cM and Argues against Allelism of CADASIL and Familial Hemiplegic Migraine. Genomics, 32, 151–4.

Dichgans M, Mayer M, Uttner I et al. (1998). The phenotypic spectrum of CADASIL: clinical findings in 102 cases. Ann Neurol, 44, 731–9.

Dichgans M, Markus HS (2005). Genetic association studies in stroke: methodological issues and proposed standard criteria. Stroke, 36, 2027–31.

Dichgans M (2007). Genetics of ischaemic stroke. Lancet Neurol, 6, 149–61.

Dickinson CJ (2001). Why are strokes related to hypertension? Classic studies and hypotheses revisited. J Hypertens, 19, 1515–21.

Diener HC, Cunha L, Forbes C et al.(1996). European Stroke Prevention Study. 2. Dipyridamole and acetylsalicylic acid in the secondary prevention of stroke. J Neurol Sci, 143, 1–13.

Diener HC, Bogousslavsky J, Brass LM et al. (2004). Aspirin and clopidogrel compared with clopidogrel alone after recent ischaemic stroke or transient ischaemic attack in high-risk patients (MATCH): randomised double-blind placebo-controlled trial. Lancet, 364, 331–7.

Diener HC, Weimar C, Katsarava Z (2005). Patent foramen ovale: paradoxical connection to migraine and stroke. Curr Opin Neurol, 18, 299–304.

Diserens K, Michel P, Bogousslavsky J (2006).Early mobilisation after stroke: Review of the literature. Cerebrovasc Dis, 22, 183–90.

Dittrich R, Ritter MA, Kaps M et al. (2006). The use of embolic signal detection in multicenter trials to evaluate antiplatelet efficacy: signal analysis and quality control mechanisms in the CARESS (Clopidogrel and Aspirin for Reduction of Emboli in Symptomatic carotid Stenosis) trial. Stroke, 37, 1065–9.

Dobyns WB, Garg BP (1991). Vascular abnormalities in epidermal nevus syndrome. Neurology, 41, 276–8.

Donaghy M, Chang CL, Poulter N et al. (2002). European Collaborators of the World Health Organisation Collaborative Study of Cardiovascular

Disease and Steroid Hormone Contraception. Duration, frequency, recency, and type of migraine and the risk of ischaemic stroke in women of childbearing age. *J Neurol Neurosurg Psychiatry*, **73**, 747–50.

Donnan GA, O'Malley HM, Quang L et al. (1996). The capsular warning syndrome: the high risk of early stroke. *Cerebrovasc Dis*, **6**, 202–7.

Dorman P, Slattery J, Farrell B et al. (1998). Qualitative comparison of the reliability of health status assessments with the EuroQol and SF-36 questionnaires after stroke. *Stroke*, **29**, 63–8.

Douglas AS, Allan TM, Rawles JM (1991). Composition of seasonality of disease. *Scott Med J*, **36**, 76–82.

Douglas VC, Johnston CM, Elkins J et al. (2003). Head computed tomography findings predict short-term stroke risk after transient ischemic attack. *Stroke*, **34**, 2894–8.

Dubroca C, Lacombe P, Domenga V et al. (2005). Impaired Vascular Mechanotransduction in a Transgenic Mouse Model of CADASIL Arteriopathy. *Stroke*, **36**, 113–7.

Duong DH, Young WL, Vang MC et al. (1998). Feeding artery pressure and venous drainage pattern are primary determinants of haemorrhage from cerebral arteriovenous malformations. *Stroke*, **29**, 1167–76.

Dusitanond P, Eikelboom JW, Hankey GJ et al. (2005). Homocysteine-lowering treatment with folic acid, cobalamin and pyridoxine does not reduce blood markers of inflammation, indothelial dysfunction or hypercoagulability in patients with previous transient ischemic attack or stroke: a randomized substudy of the VITATOPS trial. *Stroke*, **36**, 144–6.

Dutta M, Hanna E, Das P et al. (2006a). Incidence and prevention of ischemic stroke following myocardial infarction: review of current literature. *Cerebrovasc Dis*, **22**, 331–9.

Dutta M, Karas MG, Segal AZ et al. (2006b). Yield of transoesophageal echocardiography for nonbacterial thrombotic endocarditis and other cardiac sources of embolism in cancer patients with cerebral ischaemia. *Am J Cardiol*, **97**, 894–8.

Dzialowski I, Hill MD, Coutts SB et al. (2006). Extent of early ischemic changes on computed tomography (CT) before thrombolysis: prognostic value of the Alberta Stroke Program Early CT Score in ECASS II. *Stroke*, **37**, 973–8.

Eames PJ, Blake MJ, Dawson SL et al. (2002). Dynamic cerebral autoregulation and beat to beat blood pressure control are impaired in acute ischaemic stroke. *J Neurol Neurosurg Psychiatry*, **72**, 467–72.

Early Supported Discharge Trialists Services for reducing duration of hospital care for acute stroke patients (2005). Cochrane Database of Systematic Reviews:Art No: CD000443 DOI: 000410001002/14651858 CD14000443pub14651852.

Eastern Stroke and Coronary Heart Disease Collaborative Research Group (1998). Blood pressure, cholesterol and stroke in Eastern Asia. *Lancet*, **352**, 1801–7.

Eberhardt O, Naegele T, Raygrotzki S et al. (2006). Stenting of vertebrobasilar arteries in symptomatic atherosclerotic disease and acute occlusion: case series and review of the literature. *J Vasc Surg*, **43**, 1145–54.

Eberhardt RT, Dhadly M (2007). Giant cell arteritis: diagnosis management and cardiovascular implications. *Cariol Rev*, **15**, 55–61.

Ebke M, Dichgans M, Bergmann M et al. (1997). CADASIL: skin biopsy allows diagnosis in early stages. *Acta Neurol Scand*, **95**, 351–7.

Ebright JR, Pace MT, Niazi AF (2001). Septic thrombosis of the cavernous sinuses. *Arch Intern Med*, **161**, 2671–6.

EC-IC Bypass Study Group (1985). Failure of extracranial-intracranial arterial bypass to reduce the risk of ischaemic stroke: results of an international randomised trial. *N Engl J Med*, **313**, 1191–200.

Eckel RH, Grundy SM, Zimmet PZ (2005). The metabolic syndrome. *Lancet*, **365**, 1415–28.

Effective Health Care (1995). Vol 2 No 1 The prevention and treatment of pressure sores. Churchill Livingstone.

Einhaupl KM, Villringer A, Meister W et al. (1991). Heparin treatment in sinus venous thrombosis. *Lancet*, **338**, 597–600.

Einhaupl K, Bousser MG, de Bruijn SF et al. (2006). EFNS guideline on the treatment of cerebral venous and sinus thrombosis. *Eur J Neurol*, **13**, 553–9.

Ekinci EI, Donnan GA (2004). Neurological manifestations of cardiac myxoma: a review of the literature and report of cases. *Intern Med J*, **34**, 243–9.

Eliasziw M, Streifler JY, Fox AJ et al. (1994). Significance of plaque ulceration in symptomatic patients with high-grade carotid stenosis. *Stroke* **25**:304–8.

Elkind MS, Sciacca R, Boden-Albala B et al. (2006). Moderate alcohol consumption reduces risk of ischemic stroke: the Northern Manhattan Study. *Stroke*, **37**, 13–9.

Elliott WJ (1998). Circadian variation in the timing of stroke onset: a meta-analysis. *Stroke*, **29**, 992–6.

Enderby PM, Wood VA, Wade DT et al. (1986). The Frenchay Aphasia Screening Test: a short simple test for aphasia appropriate for non-specialists. *Int Rehabil Med*, **8**, 166–70.

ESPRIT Study Group, Halkes PH, van Gijn J et al. (2006). Aspirin plus dipyridamole versus aspirin alone after cerebral ischaemia of arterial origin (ESPRIT): randomised controlled trial. *Lancet*, **367**, 1665–73 Erratum in: *Lancet* 2007 7;369(9558), 274.

Estol CJ, Kase CS (2003). Need for Continued Use of Anticoagulants After Intracerebral Hemorrhage. *Curr Treat Options Cardiovasc Med*, **5**, 201–9.

European Atrial Fibrillation Trial Study Group (1993). Secondary prevention in non-rheumatic atrial fibrillation after transient ischaemic attack or minor stroke. *Lancet*, **342**, 1255–62.

European Atrial Fibrillation Trial Study Group (1995). Optimal oral anticoagulant therapy in patients with nonrheumatic atrial fibrillation and recent cerebral ischemia. *N Engl J Med*, **333**, 5–10.

European Carotid Surgery Trialists' Collaborative Group (1998). Randomised trial of endarterectomy for recently symptomatic carotid stenosis: final results of the MRC European Carotid Surgery Trial (ECST). *Lancet*, **351**, 1379–87.

Executive Committee for the Asymptomatic Carotid Atherosclerosis Study (1995). Endarterectomy for asymptomatic carotid artery stenosis. *JAMA*, **273**, 1421–8.

Fagerberg B, Gnarpe J, Gnarpe H et al. (1999). Chlamydia pneumoniae but not cytomegalovirus antibodies are associated with future risk of stroke and cardiovascular disease: a prospective study in middle-aged to elderly men with treated hypertension. *Stroke*, **30**, 299–305.

Fairhead JF, Mehta Z, Rothwell PM (2005). Population-based study of delays in carotid imaging and surgery and the risk of recurrent stroke. *Neurology*, **65**, 371–5.

Fairhead JF, Rothwell PM (2005). The need for urgency in identification and treatment of symptomatic carotid stenosis is already established. *Cerebrovasc Dis*, **19**, 355–8.

Farah S, Al-Shubaili A, Montaser A et al. (1998). Behcets syndrome: a report of 41 patients with emphasis on neurological manifestations. *J Neurol Neurosurg Psychiatry*, **64**, 382–4.

Farrell MA, Gilbert JJ, Kaufman JCE (1985). Fatal intracranial arterial dissection: clinical pathological correlation. *J Neurol Neurosurg Psychiatry*, **48**, 111–21.

Farquhar CM, Marjoribanks J, Lethaby A et al. (2005). Long term hormone therapy for perimenopausal and postmenopausal women. *Cochrane Database Syst Rev*, **3**, CD004143.

Fearnley JM, Stevens JM, Rudge P (1995). Superficial siderosis of the central nervous system. *Brain*, **118**, 1051–66.

Featherstone RL, Brown MM, Coward LJ (2004). International carotid stenting study: protocol for a randomised clinical trial comparing carotid stenting with endarterectomy in symptomatic carotid artery stenosis. *Cerebrovasc Dis*, **18**, 69–74.

Federico F, Simone IL, Lucivero V et al. (1998). Prognostic value of proton magnetic resonance spectroscopy in ischemic stroke. *Arch Neurol*, **55**, 489–94.

Feigin VL, Wiebers DO (1997). Environmental factors and stroke: a selective review. *J Stroke Cerebrovasc Dis*, **6**, 108–13.

Feigin VL, Wiebers DO, Nikitin YP *et al.* (1998). Risk factors for ischaemic stroke in a Russian community: a population-based case–control study. *Stroke*, **29**, 34–9.

Feigin VL, Lawes CM, Bennett DA *et al.* (2003). Stroke epidemiology: a review of population-based studies of incidence, prevalence, and case-fatality in the late 20th century. *Lancet Neurol*, **2**, 43–53.

Feigin VL, Rinkel GJ, Lawes CM *et al.* (2005a). Risk factors for subarachnoid hemorrhage: an updated systematic review of epidemiological studies. *Stroke*, **36**, 2773–80.

Feigin VL, Anderson N, Rinkel GJ (2005b). Corticosteroids for aneurysmal subarachnoid haemorrhage and primary intracerebral haemorrhage. *Cochrane Database Syst Rev*, **3**, CD004583.

Feigin V, Carter K, Hackett M *et al.* (2006). Ethnic disparities in incidence of stroke subtypes: Auckland Regional Community Stroke Study 2002-2003. *Lancet Neurol*, **5**, 130–9.

Feldmann E, Daneault N, Kwan E *et al.* (1990). Chinese-white differences in the distribution of occlusive cerebrovascular disease. *Neurology*, **40**, 1541–5.

Fenster A, Blake C, Gyacskov I *et al.* (2006). 3D ultrasound analysis of carotid plaque volume and surface morphology. *Ultrasonics*, **44**, e153–7.

Ferro JM, Pinto AN (1994). Sexual activity is a common precipitant of subarachnoid haemorrhage. *Cerebrovasc Dis*, **4**, 375.

Ferro JM (1998). Vasculitis of the central nervous system. *J Neurol*, **245**, 766–6.

Ferro JM, Pinto F (2004). Poststroke epilepsy: epidemiology pathophysiology and management. *Drugs Aging*, **21**, 639–53.

Ferro JM, Morgado C, Sousa R *et al.* (2007). Interobserver agreement in the magnetic resonance location of cerebral vein and dural sinus thrombosis. *Eur J Neurol*, **14**, 353–6.

Fiebach JB, Schellinger PD, Gass A *et al.* (2004). Stroke magnetic resonance imaging is accurate in hyperacute intracerebral hemorrhage: a multicenter study on the validity of stroke imaging. *Stroke*, **35**, 502–6.

Fiehler J, Foth M, Kucinski T *et al.* (2002). Severe ADC decreases do not predict irreversible tissue damage in humans. *Stroke*, **33**, 79–86.

Fieschi C, Argentino C, Lenzi GL *et al.* (1989). Clinical and instrumental evaluation of patients with ischaemic stroke within the first six hours. *J Neurol Sci*, **91**, 311–22.

Fine-Edelstein JS, Wolf PA, O'Leary *et al.* (1994). Precursors of extracranial carotid atherosclerosis in the Framingham Study. *Neurology*, **44**, 1046–50.

Fisher CM, Caplan LR (1971). Basilar artery branch occlusion: a cause of pontine infarction. *Neurology*, **21**, 900–5.

Fisher CM (1979). Capsular infarcts - the underlying vascular lesions. *Arch Neurol and Psychiatry*, **36**, 65–73.

Flemming KD, Brown RD Jr, Petty GW *et al.* (2004). Evaluation and management of transient ischaemic attack and minor cerebral infarction. *Mayo Clin Proc*, **79**, 1071–86.

Flis CM, Jager HR, Sidhu PS (2007). Carotid and vertebral artery dissections: clinical aspects imaging features and endovascular treatment. *Eur Radiol*, **17**, 20–834.

Flossmann E, Rothwell PM (2003). Prognosis of vertebrovasilar transient ischaemic attack and minor ischaemic stroke. *Brain*, **126**, 1940–54.

Flossmann E, Schulz UG, Rothwell PM (2004). Systematic review of methods and results of studies of the genetic epidemiology of ischemic stroke. *Stroke*, **35**, 212–7.

Flossmann E, Schulz UG, Rothwell PM (2005). Potential confounding by intermediate phenotypes in studies of the genetics of ischaemic stroke. *Cerebrovasc Dis*, **19**, 1–10.

Flossmann E, Touze E, Giles MF *et al.* (2006). The early risk of stroke after vertebrobasilar TIA is higher than after carotid TIA. *Cerebrivasc Dis*, **21**(suppl4), 6.

Foley N, Bhogal S, Foley N (2006). Painful hemiplegic shoulder. In: *Canadian Stroke Network Evidence-Based Reviews of Stroke Rehabilitation* [Online] 2006. Available at URL: http://ebrsrcom/index_homehtml. (Accessed 10 May 2007)

Ford CC, Griffey RH, Matwiyoff NA *et al.* (1992). Multivoxel 1H-MRS of stroke. *Neurology*, **42**, 1408–12.

Forouhi NG, Sattar N (2006). CVD risk factors and ethnicity-a homogeneous relationship? *Atheroscler Suppl*, **7**, 11–9.

Fountas KN, Kapsalaki EZ, Parish DC *et al.* (2005). Intraventricular administration of rt-PA in patients with intraventricular hemorrhage. *South Med J*, **98**, 767–73.

Fox AJ, Eliasziw M, Rothwell PM *et al.* (2005). Identification prognosis and management of patients with carotid artery near occlusion. *Am J Neuroradiol*, **26**, 2086–94.

Frackowiak RSJ, Jones T, Lenzi GL *et al.* (1980). Regional cerebral oxygen utilization and blood flow in normal man using oxygen-15 and positron emission tomography. *Acta Neurologica Scandinavia*, **62**, 336–44.

Frackowiak RSJ (1986). PET scanning: can it help resolve management issues in cerebral ischaemic disease? *Stroke*, **17**, 803–7.

Franz WM, Muller OJ, Katus HA (2001). Cardiomyopathies: from genetics to the prospect of treatment. *The Lancet*, **358**, 1627–37.

Frese A, Husstedt IW, Ringelstein EB *et al.* (2006). Pharmacologic treatment of central post-stroke pain. *Clin J Pain*, **22**, 252–60.

Friedman JA, Pollock BE, Nichols DA *et al.* (2001). Results of combined stereotactic, radiosurgery and transarterial embolization for dural arteriovenous fistulas of the transverse and sigmoid sinuses. *J Neurosurg*, **94**, 886–91.

Freyaldenhoven TE, Mrak RE, Rock L (2001.) Fibrocartilaginous embolisation. *Neurology*, **56**, 1354.

Fuentes B, Diez-Tejedor E, Ortega-Casarrubios MA *et al.* (2006). Consistency of the benefits of stroke units over years of operation: an 8-year effectiveness analysis. *Cerebrovasc Dis*, **21**, 173–9.

Furlan M, Marchal G, Viader F *et al.* (1996). Spontaneous neurological recovery after stroke and the fate of the ischaemic penumbra. *Ann Neurol*, **40**, 216–26.

Furlan A, Higashida R, Wechsler L *et al.* (1999). Intra-arterial prourokinase for acute ischemic stroke. The PROACT II study: a randomized controlled trial. Prolyse in Acute Cerebral Thromboembolism. *JAMA*, **282**, 2003–11.

Furlan AJ, Eyding D, Albers GW *et al.* (2006). Dose Escalation of Desmoteplase for Acute Ischemic Stroke (DEDAS): evidence of safety and efficacy 3 to 9 hours after stroke onset. *Stroke*, **37**, 1227–31.

Futrell N (1995). Inflammatory vascular disorders: diagnosis and treatment in ischaemic stroke. *Curr Opin Neurol*, **8**, 55–61.

Gabriel SR, Carmona L, Roque M *et al.* (2005). Hormone replacement therapy for preventing cardiovascular disease in post-menopausal women. *Cochrane Database Syst Rev*, **2**, CD002229.

Gan R, Sacco RL, Kargman DE *et al.* (1997). Testing the validity of the lacunar hypothesis: The Northern Manhattan Stroke Study experience. *Neurology*, **48**, 1204–11.

Gass A, Ay H, Szabo K *et al.* (2004). Diffusion-weighted MRI for the 'small stuff': the details of acute cerebral ischaemia. *Lancet Neurology*, **3**, 39–45.

Gates GC, Barnett HJM, Vinters HV *et al.* (1986). Primary intraventricular haemorrhage in adults. *Stroke*, **17**, 872–7.

Gautier JC, Durr A, Koussa S *et al.* (1991). Paradoxical cerebral embolism with a patent foramen ovale. A report of 29 patients. *Cerebrovasc Dis*, **1**, 193–202.

Gebel JM Jr, Jauch EC, Brott TG *et al.* (2002). Natural history of perihematomal edema in patients with hyperacute spontaneous intracerebral hemorrhage. *Stroke*, **33**, 2631–5.

Geddes JML, Fear J, Tennant A *et al.* (1996). Prevalence of self reported stroke in a population in northern England. *J Epidemiol Community Health*, **50**, 140–3.

Geisler BS, Brandhoff F, Fiehler J *et al.* (2006). Blood-oxygen-level-dependent MRI allows metabolic description of tissue at risk in acute stroke patients. *Stroke*, **37**, 1778–84.

Genta MS, Genta RM, Gabay C (2006). Systemic rheumatoid vasculitis: a review. *Semin Arthritis Rheum*, **36**, 88–98.

Georgiadis D, Schwarz S, Cencetti S *et al.* (2002). Noninvasive monitoring of hypertensive breakthrough of cerebral autoregulation in a patient with acute ischemic stroke. *Cerebrovasc Dis*, **14**, 129–32.

Georgiadis D, Oehler J, Schwarz S *et al.* (2004). Does acute occlusion of the carotid T invariably have a poor outcome? *Neurology*, **63**, 22–6.

German Stroke Study Collaboration (2004). Predicting outcome after acute ischemic stroke: an external validation of prognostic models. *Neurology*, **62**, 581–5.

Ghika J, Bogousslavsky J (2001). Abnormal movements. In: Bogousslavsky J, Caplan L, eds. *Stroke Syndromes*, pp. 162–81. Cambridge University Press.

Gillum RF (1999). Risk factors for stroke in blacks: a critical review. *Am J Epidemiol*, **150**, 1266–74.

Gil-Nunez AC, Vivancos-Mora J (2005). Blood pressure as a risk factor for stroke and the impact of antihypertensive treatment. *Cerebrovasc Dis*, **20**, 40–52.

Girot M, Ferro JM, Canhao P *et al.* (2007). Predictors of outcome in patients with cerebral venous thrombosis and intracerebral hemorrhage. *Stroke*, **38**, 337–42.

Giroud M, Lemesle M, Madinier G *et al.* (1997). Unilateral lenticular infarcts: radiological and clinical syndromes, aetiology and prognosis. *J Neurol Neurosurg Psychiatry*, **63**, 611–15.

GISSI-Prevenzione Investigators (1999). Dietary supplementation with n-3 polyunsaturated fatty acids and vitamin E after myocardial infarction: results of the GISSI-Prevenzione trial. *Lancet*, **354**, 447–55.

Glader CA, Stegmayr B, Boman J *et al.* (1999). Chlamydia pneumoniae antibodies and high lipoprotein (a) levels do not predict ischaemic cerebral infarctions: results from a nested case–control study in northern Sweden. *Stroke*, **30**, 2013–18.

Glass J (2006). Neurologic complications of lymphoma and leukaemia. *Semin Oncol*, **33**, 342–7.

Goldschmidt-Clermont PJ, Creager MA, Lorsordo DW *et al.* (2005). Atherosclerosis 2005: recent discoveries and novel hypotheses *Circulation*, **112**, 3348–53.

Goldstein LB, Hankey GJ (2006). Advances in primary stroke prevention. *Stroke*, **37**, 317–9.

Gomez CR (1990). Carotid plaque morphology and risk for stroke. *Stroke*, **21**, 148–151.

Gomez CR, Cruz-Flores S, Malkoff MD *et al.* (1996). Isolated vertigo as a manifestation of vertebrobasilar ischaemia. *Neurology*, **47**, 94–7.

Gorter JW, Stroke Prevention In Reversible Ischemia Trial (SPIRIT) European Atrial Fibrillation Trial (EAFT) Study Groups (1999). Major bleeding during anticoagulation after cerebral ischemia: patterns and risk factors. *Neurology*, **53**, 1319–27.

Gotto AM Jr (2005). Evolving concepts of dyslipidaemia, atherosclerosis and cardiovascular disease: the Louis F Bishop Lecture. *J Am College Cardiol*, **46**, 1219–24.

Graham GD, Kalvach P, Blamire AM *et al.* (1995). Clinical correlates of proton magnetic resonance spectroscopy findings after acute cerebral infarction. Stroke, **26**, 225–9.

Grau AJ, Buggle F, Becher H *et al.* (1998). Recent bacterial and viral infection is a risk factor for cerebrovascular ischaemia: clinical and biochemical studies. *Neurology*, **50**, 196–203.

Graves MJ (1997). Magnetic resonance angiography. *Br J Radiol*, **70**, 6–28.

Gray CS, Hildreth AJ, Sandercock PA *et al.* (2007). Glucose-potassium-insulin infusions in the management of post-stroke hyperglycaemia: the UK Glucose Insulin in Stroke Trial (GIST-UK). *Lancet Neurol*, **6**, 397–406.

Greene GM, Godersky JC Biller J *et al.* (1990). Surgical experience with cerebral amyloid angiopathy. *Stroke*, **21**, 1545–9.

Greener J, Enderby P, Whurr R (2002). Speech and language therapy for aphasia following stroke. *Cochrane Database Syst Rev*, **2**, CD000425.

Griewing B, Guo Y, Doherty C *et al.* (1995). Radiation-induced injury to the carotid artery: a longitudinal study. *Eur J Neurol*, **2**, 379–83.

Gronholdt MLM (1999). Ultrasound and lipoproteins as predictors of lipid-rich rupture-prone plaques in the carotid artery. Arterioscler Thromb *Vascular Biol*, **19**, 2–13.

Gross M, Banin E, Eliashar R *et al.* (2004). Susac Syndrome. *Otol Neurotol*, **25**, 470–3.

Grosveld WJ, Lawson JA, Eikelboom BC *et al.* (1988). Clinical and haemodynamic significance of innominate artery lesions evaluated by ultrasonography and digital angiography. *Stroke*, **19**, 958–62.

Grotta JC, Chiu D, Lu M *et al.* (1999). Agreement and variability in the interpretation of early CT changes in stroke patients qualifying for intravenous rtPA therapy. *Stroke*, **30**, 1528–33.

Grubb RL Jr and Powers WJ (2001). Risks of stroke and current indications for cerebral revascularization in patients with carotid occlusion. *Neurosurg Clin N Am*, **12**, 473–87.

Grubb RL Jr, Powers WJ, Derdeyn CP *et al.* (2003). The Carotid Occlusion Surgery Study. *Neurosurg Focus*, **14**, e9.

Grundy SM, Cleeman JI, Daniels SR *et al.* (2005). Diagnosis and management of the metabolic syndrome: an American Heart Association/National Heart Lung and Blood Institute Scientific Statement. *Circulation*, **112**, 2735–52.

Guadagno JV, Donnan GA, Markus R *et al* (2004). Imaging the ischaemic penumbra. *Curr Opin Neurol*, **17**, 61–7.

Gulcher JR, Gretarsdottir S, Helgadottir A *et al.* (2005). Genes contributing to risk for common forms of stroke. *Trends Mol Med*, **11**, 217–4.

Gulcher JR, Kong A, Gretarsdottir S *et al.* (2006). Reply to 'Many hypotheses but no replication for the association between PDE4D and stroke. *Nat Genet*, **38**, 1092–3.

Gullapalli D, Phillips LH 2nd (2004). Neurosarcoidosis. *Curr Neurol Neurosci Rep*, **4**, 441–7.

Gunatilake SB (1998). Rapid resolution of symptoms and signs of intracerebral haemorrhage: case reports. *Br Med J*, **316**, 1495–6.

Gurfinkel E, Bozovich G, Daroca A *et al.* (1997). Randomised trial of roxithromycin in non-Q-wave coronary syndromes: ROXIS Pilot Study. *Lancet*, **350**, 404–7.

Gustafsson L, McKenna K (2006). A programme of static positional stretches does not reduce hemiplegic shoulder pain or maintain shoulder range of motion - a randomized controlled trial. *Clin Rehabil*, **20**, 277.

Haapanen A, Koskenvuo M, Kaprio J *et al.* (1989). Carotid arteriosclerosis in identical twins discordant for cigarette smoking. *Circulation*, **80**, 10–16.

Hacke W, Kaste M, Fieschi C *et al.* (1998). Randomised double-blind placebo-controlled trial of thrombolytic therapy with intravenous alteplase in acute ischaemic stroke (ECASS II) Second European-Australasian Acute Stroke Study Investigators. *Lancet*, **352**, 1245–51.

Hacke W, Donnan G, Fieschi C *et al.* (2004). Association of outcome with early stroke treatment: pooled analysis of ATLANTIS ECASS and NINDS rt-PA stroke trials. *Lancet*, **363**, 768–4.

Hackett ML, Anderson CS (2005). Predictors of depression after stroke: a systematic review of observational studies. *Stroke*, **36**, 2296–301.

Hackett ML, Yapa C, Parag V *et al.* (2005). Frequency of depression after stroke: a systematic review of observational studies. *Stroke*, **36**, 1330–40.

Halliday A, Mansfield A, Marro J *et al.* (2004). Prevention of disabling and fatal strokes by successful carotid endarterectomy in patients without recent neurological symptoms: randomised controlled trial. *Lancet*, **363**, 1491–502.

Hamdy S, Aziz Q, Rothwell JC *et al.* (1997). Explaining oropharyngeal dysphagia after unilateral hemispheric stroke. *Lancet*, **350**, 686–92.

Hand PJ, Kwan J, Lindley RI *et al.* (2006a). Distinguishing between stroke and mimic at the bedside: the brain attack study. *Stroke*, **37**, 769–5.

Hand PJ, Wardlaw JM, Rivers CS et al. (2006b). MR diffusion-weighted imaging and outcome prediction after ischemic stroke. Neurology, 66, 1159–63.

Hankey GJ, Warlow C P (1990). Symptomatic carotid ischaemic events: safest and most cost effective way of selecting patients for angiography before carotid endarterectomy. Br Med J, 300, 1485–91.

Hankey GJ (1991). Isolated angiitis/angiopathy of the central nervous system. Cerebrovasc Dis, 1, 2–15.

Hankey GJ (1999). Smoking and risk of stroke. J Vasc Risk, 6, 207–11.

Hankey GJ (2002). Is homocysteine a causal and treatable risk factor for vascular diseases of the brain (cognitive impairment and stroke). Ann Neurol, 51, 279–81.

Hankey GJ, Eikelboom JW (2005). Homocysteine and stroke. Lancet, 365, 194–6.

Hankey GJ, Norman PE, Eikelboom JW (2006). Medical treatment of peripheral arterial disease. J Am Med Assoc, 295, 547–3.

Hardie K, Jamrozik K, Hankey GJ et al. (2005). Trends in five-year survival and risk of recurrent stroke after first-ever stroke in the Perth Community Stroke Study. Cerebrovasc Dis, 19, 179–85.

Harmsen P, Rosengren A, Tsipogiannia A et al. (1990). Risk factors for stroke in middle-aged men in Goteborg Sweden. Stroke, 21, 223–9.

Harrison CN, Linch DC, Machin SJ (1998). Desirability and problems of early diagnosis of essential thrombocythaemia. Lancet, 351, 846–7.

Hart CL, Hole DJ, Davey Smith G (2000a). Influence of socioeconomic circumstances in early and later life on stroke risk among men in a Scottish cohort study. Stroke, 31, 2093–97.

Hart CL, Hole DJ, Davey Smith G (2000b). The contribution of risk factors to stroke differentials by socioeconomic position in adulthood: the Renfrew/Paisley study. Am J Public Health, 90, 1788–91.

Hart RG, Foster JW, Lutner MF et al. (1990). Stroke in infective endocarditis. Stroke, 21, 695–700.

Hart RG, Benavente O, McBride R et al. (1999). Antithrombotic therapy to prevent stroke in patients with atrial fibrillation: a meta-analysis. Ann Intern Med, 131, 492–501.

Hart RG, Halperin JL, Pearce, LA et al. (2003). Lessons from the Stroke Prevention in Atrial Fibrillation trials. Ann Intern Med, 138, 831–8.

Harukuni I, Bhardwaj A (2006). Mechanisms of brain injury after global cerebral ischemia. Neurol Clin, 24, 1–21.

Hassan A, Markus HS (2000). Genetics and ischaemic stroke. Brain, 123, 1784–812.

Hassan A, Sham PC, Markus HS (2002). Planning genetic studies in human stroke: sample size estimates based on family history data. Neurology, 58, 1483–88.

Hatano S (1976). Experience from a multicentre stroke register: a preliminary report. Bulletin WHO, 54, 541–53.

Hatazawa J, Shimosegawa E, Osaki Y et al. (2004). Long-term angiotensin-converting enzyme inhibitor perindopril therapy improves cerebral perfusion reserve in patients with previous minor stroke. Stroke, 35, 2117–2.

Hattori N, Katayama Y, Maya Y et al. (2004). Impact of stereotactic hematoma evacuation on activities of daily living during the chronic period following spontaneous putaminal hemorrhage: a randomized study. J Neurosurg, 101, 417–20.

Hawkins GC, Bonita R, Broad JB et al. (1995). Inadequacy of clinical scoring systems to differentiate stroke subtypes in population-based studies. Stroke, 26, 1338–42.

He FJ, MacGregor GA (2004). The effect of longer-term modest salt reduction on blood pressure. The Cochrane Database Syst Rev Issue 1 Art No CD004937.

Heart Protection Study Collaborative Group (2002). MRC/BHF Heart Protection Study of cholesterol lowering with simvastatin in 20536 high-risk individuals: a randomised placebo-controlled trial. Lancet, 360, 7–22.

Heiken JP, Brink JA, Vannier MW (1993). Spiral (helical) CT. Radiology, 189, 647–56.

Heinzlef O, Cohen A, Amarenco P (1997). An update on aortic causes of ischemic stroke. Curr Opin Neurol, 10, 64–72.

Hellingman CA, van den Bergh, WM Beijer IS et al. (2007). Risk of rebleeding after treatment of acute hydrocephalus in patients with aneurysmal subarachnoid hemorrhage. Stroke, 38, 96–9.

Helms AK, Kittner SJ (2005). Pregnancy and stroke. CNS Spectr, 10, 580–7.

Hemmen TM, Lyden PD (2007). Induced hypothermia for acute stroke. Stroke, 38, 794–9.

Hender J, Harris DG, Bu H et al. (2006). Stroke in pregnancy. Br J Hosp Med (Lond), 67, 129–31.

Hennerici M, Aulich A, Sandmann W et al. (1981). Incidence of asymptomatic extracranial arterial disease. Stroke, 12, 750–8.

Heron E, Fornes P, Rance A et al. (1998). Brain involvement in scerloderma. Two autopsy cases. Stroke, 29, 719–21.

Hess DC, Nichols FT, Sethi KD et al. (1991). Transient cerebral ischaemia masquerading as paroxysmal dyskinesia. Cerebrovasc Dis, 1, 54–7.

Heye N, Hankey GJ (1996). Amphetamine-associated stroke. Cerebrovasc Dis, 6, 149–55.

Hietaharju A, Jantti V, Korpela M et al. (1993). Nervous system involvement in systemic lupus erythematous Sjögren syndrome and scleroderma. Acta Neurol Scand, 88, 299–308.

Hill MD, Yiannakoulias N, Jeerakathil T et al. (2004). The high risk of stroke immediately after transient ischemic attack: a population-based study. Neurology, 62, 2015–20.

Hillbom M, Numminen H, Juvela S (1999). Recent heavy drinking of alcohol and embolic stroke. Stroke, 30, 2307–12.

Hilton DA, Footitt D (2003). Neuropathological findings in Sneddon's syndrome. Neurology, 60, 1181–2.

Hinds NP, Wiles CM (1998). Assessment of swallowing and referral to speech and language therapists in acute stroke. Q J Med, 91, 829–35.

Hjort N, Christensen S, Solling C et al. (2005a). Ischemic injury detected by diffusion imaging 11 minutes after stroke. Ann Neurol, 58, 462–5.

Hjort N, Butcher K, Davis SM et al. (2005b). Magnetic resonance imaging criteria for thrombolysis in acute cerebral infarct. Stroke, 36, 388–97.

Hobson RW 2nd, Howard VJ, Roubin GS et al. (2004). Credentialing of surgeons as interventionalists for carotid artery stenting: experience from the lead-in phase of CREST. J Vasc Surg, 40, 952–7.

Hodges JR, Warlow CP (1990a). Syndromes of transient amnesia: towards a classification. A study of 153 cases. J Neurol Neurosurg Psychiatry, 53, 834–43.

Hodges JR, Warlow CP (1990b). The aetiology of transient global amnesia. A case-control study of 114 cases with prospective follow-up. Brain, 113, 639–57.

Hodkinson HM (1972). Evaluation of a mental test score for assessment of mental impairments in the elderly. Age Ageing, 1, 233–8.

Hoekstra-van Dalen RAH, Cillessen JPM, Kappelle LJ et al. (1996). Cerebral infarcts associated with migraine: clinical features, risk factors and follow up. J Neurol, 243, 511–15.

Hoffman M, Corr P, Robbs J (2000). Cerebrovascular findings in Takayasu disease. J Neuroimaging, 10, 84–90.

Hollander M, Hak AE, Koudstaal PJ et al. (2003). Comparison between measures of atherosclerosis and risk of stroke: the Rotterdam Study. Stroke, 34, 2367–72.

Holmes DR Jr (2004). Strokes and holes and headaches: are they a package deal? Lancet, 364, 1840–2.

Homer D, Ingall TJ, Baker HL et al. (1991). Serum lipids and lipoproteins are less powerful predictors of extracranial carotid artery atherosclerosis than are cigarette smoking and hypertension. Mayo Clinic Proc, 66, 259–67.

Homma S, Sacco RL (2005). Patent foramen ovale and stroke. Circulation, 112, 1063–72.

Hooper L, Capps N, Clements G et al. (2000a). Foods or supplements rich in omega-3 fatty acids for preventing cardiovascular disease in patients with ischaemic heart disease. (Protocol for a Cochrane Review) In: *The Cochrane Library* Oxford: Update Software.

Hooper L, Capps N, Clements G *et al.* (2000b). Anti-oxidant foods or supplements for preventing cardiovascular disease (Protocol for a Cochrane Review) In: *The Cochrane Library* Oxford: Update Software.

Hop JW, Rinkel GJE, Algra A *et al.* (1997). Case-fatality rates and functional outcome after subarachnoid haemorrhage: a systematic review. *Stroke*, **28**, 660–4.

Hop JW, Rinkel GJ, Algra A *et al.* (1999). Initial loss of consciousness and risk of delayed cerebral ischemia after aneurysmal subarachnoid hemorrhage. *Stroke*, **30**, 2268–71.

Hopkins LN, Martin NA, Hadley MN *et al.* (1987). Vertebrobasilar insufficiency. Part 2: microsurgical treatment of intracranial vertebrobasilar disease. *J Neurol*, **66**, 662–74.

Horowitz SH, Zito JL, Donnarumma R *et al.* (1991). Computed tomographic-angiographic findings within the first five hours of cerebral infarction. *Stroke*, **22**, 1245–53.

Hossman KA (1983). Experimental aspects of stroke in vascular disease of the central nervous system. In: Ross Russell RW, ed. pp. 73–100. Churchill Livingstone, Edinburgh.

House A, Dennis M, Hawton K *et al.* (1989). Methods of identifying mood disorders in stroke patients: experience in the Oxfordshire Community Stroke Project. *Age Ageing*, **18**, 371–9

Howard G, Wagenknecht LE, Burke GL *et al.* (1998). Cigarette smoking and progression of atherosclerosis: the Atherosclerosis Risk in Communities (ARIC) Study. *J Am Med Assoc*, **279**, 119–24.

Howard RS (1986). Brainstem haematoma due to presumed cryptic telangiectasia. *J Neurol Neurosurg Psychiatry*, **49**, 1241–5.

Howell SJL, Molyneux AJ (1988). Vertebrobasilar insufficiency in rheumatoid atlanto-axial subluxation: a case report with angiographic demonstration of left vertebral artery occlusion. *J Neurol*, **235**, 189–90.

Hsu KC, Wu YR, Lyu RK *et al* (2006). Aseptic meningitis and ischemic stroke in relapsing polychondritis. *Clin Rheumatol*, **25**, 265–7.

Huang CY, Chan FL, Yu YL *et al.* (1990). Cerebrovascular disease in Hong Kong Chinese. *Stroke*, **21**, 230–5.

Humphries SE, Morgan L (2004). Genetic risk factors for stroke and carotid atherosclerosis: insights into pathophysiology from candidate gene approaches. *Lancet Neurol*, **3**, 227–35.

Hupperts RMM, Lodder J, Heuts-van Raak EPM *et al.* (1994). Infarcts in the anterior choroidal artery territory. Anatomical distribution, clinical syndromes, presumed pathogenesis and early outcome. *Brain*, **117**, 825–34.

Hutchinson M, O'Riordan J, Javed M *et al.* (1995). Familial hemiplegic migraine and autosomal dominant ateriopathy with leukoencephalopathy (CADASIL). *Ann Neurol*, **38**, 817–24.

Hutter CM, Austin MA, Humphries SE (2004). Familial hypercholesterolemia, peripheral arterial disease and stroke: a HuGE minireview. *Am J Epidemiol*, **160**, 430–5.

Huang J, van Gelder JM (2002). The probability of sudden death from rupture of intracranial aneurysms: a meta-analysis. *Neurosurgery*, **51**, 1101–5.

Huxley R, Lewington S, Clarke R (2002). Cholesterol, coronary heart disease and stroke: a review of published evidence from observational studies and randomized controlled trials. *Semin Vasc Med*, **2**, 315–23.

Ichikawa K, Kageyama Y (1991). Clinical anatomic study of pure dysarthria. *Stroke*, **22**, 809–12.

Immink RV, van den Born BJ, van Montfrans GA *et al.* (2004). Impaired cerebral autoregulation in patients with malignant hypertension. *Circulation*, **110**, 2241–5.

International Stroke Trial Collaborative Group (1997). The International Stroke Trial (IST): a randomised trial of aspirin subcutaneous heparin both or neither among 19435 patients with acute ischaemic stroke. *Lancet*, **349**, 1569–81.

International Study of Unruptured Intracranial Aneurysms Investigators (1998). Unruptured intracranial aneurysms - risk of rupture and risks of surgical intervention. *N Engl J Med*, **339**, 1725–33.

Inzitari D (2003). Leukoariaosis: an independent risk factor for stroke? *Stroke*, **34**, 2067–71.

Inzitari D, Lamassa M, Amaducci L (1995). Stroke epidemiology in Europe. *Eur J Neurol*, **2**, 75–81.

Isaka Y, Nagano K, Narita M *et al.* (1997). High signal intensity of T2-weighted magnetic resonance imaging and cerebral hemodynamic reserve in carotid occlusive disease. *Stroke*, **28**, 354–7.

Iwama T, Hashimoto N, Murai BN *et al.* (1997). Intracranial rebleeding in moyamoya disease. *J Clin Neurosci*, **4**, 169–72.

Izumihara A, Ishihara T, Iwamoto N *et al.* (1999). Postoperative outcome of 37 patients with lobar intracerebral hemorrhage related to cerebral amyloid angiopathy. *Stroke*, **30**, 29–33.

Jackson C, Sudlow C (2005). Are lacunar strokes really different? A systematic review of differences in risk factor profiles between lacunar and nonlacunar infarcts. *Stroke*, **36**, 891–901.

Jacobs K, Moulin T, Bogousslavsky J *et al* (1996). The stroke syndrome of cortical vein thrombosis. *Neurology*, **47**, 376–82.

Jacobson DM, Terrence CF, Reinmuth OM (1986). The neurological manifestations of fat embolism. *Neurology*, **36**, 847–51.

Jakovljevic D, Salomaa V, Sivenius J *et al.* (1996). Seasonal variation in the occurrence of stroke in a Finnish adult population. The FINMONICA Stroke Register. *Stroke*, **27**, 1774–9.

Jamrozik K, Anderson CA, Stewart-Wynne EG (1994). The role of lifestyle factors in the etiology of stroke. A population-based case-control study in Perth Western Australia. *Stroke*, **25**, 51–9.

Jeanrenaud X, Kappenberger L (1991). Patent foramen ovale and stroke of unknown origin. *Cerebrovasc Dis*, **1**, 184–92.

Jehkonen M, Ahonen JP, Dastidar P *et al.* (1998). How to detect visual neglect in acute stroke. *Lancet*, **351**, 727–8.

Jennekens FG, Kater L (2002a). The central nervous system in systemic lupus erythematous. Part 1. Clinical syndromes: a literature investigation. *Rheumatology (Oxford)*, **41**, 605–18.

Jennekens FG, Kater L (2002b). The central nervous system in systemic lups erythematosus. Part 2. Pathogenic mechanisms of clinical syndromes: a literature investigation. *Rheumatology (Oxford)*, **41**, 619–30.

Jensen MB, St Louis EK (2005). Management of acute cerebellar stroke. *Arch Neurol*, **62**, 537–44.

Jensson O, Gudmundsson G, Arnason A *et al* (1987). Hereditary cystatin C (y-trace) amyloid angiopathy of the CNS causing cerebral haemorrhage. *Acta Neurologica Scandinavia*, **76**, 102–14.

Jespersen CM. Als-Nielsen B. Damgaard M *et al.* (2006). Randomised placebo controlled multicentre trial to assess short term clarithromycin for patients with stable coronary heart disease: CLARICOR trial. *BMJ*, **332**, 22–7.

Johansson B, Norrving B, Lindgren A (2000). Increased stroke incidence in Lund-Orup, Sweden between 1983 to 1985 and 1993 to 1995. *Stroke*, **31**, 481–6.

Johnston SC, Gress DR, Browner WS *et al.* (2000). Short-term prognosis after emergency department diagnosis of TIA. *JAMA*, **284**, 2901–6.

Johnston SC, Rothwell PM, Nguyen-Huynh MN *et al.* (2007). Validation and refinement of scores to predict very early stroke risk after transient ischaemic attack. *Lancet*, **369**, 283–92.

Jones EF, Kalman JM, Calafiore P *et al.* (1995). Proximal aortic atheroma. An independent risk factor for cerebral ischemia. *Stroke*, **26**, 218–4.

Jones HR, Siekert RG (1989). Neurological manifestations of infective endocarditis: review of clinical and therapeutic challenges. *Brain*, **112**, 1295–315.

Jorgensen HS, Nakayama H, Raaschou HO *et al.* (1994). Stroke in patients with diabetes. The Copenhagen Stroke Study. *Stroke*, **25**, 1977–84.

Jorgensen HS, Nakayama H, Kammersgaard LP *et al.* (1999). Predicted impact of intravenous thrombolysis on prognosis of general population of stroke patients: simulation model. *BMJ*, **319**, 288–9.

Joutel A, Corpechot C, Ducros A *et al.* (1996). Notch3 mutations in CADASIL, a hereditary adult-onset condition causing stroke and dementia. *Nature*, **383**, 707–10.

Joutel A, Vahedi K, Corpechot C *et al.* (1997). Strong clustering and stereotyped nature of Notch3 mutations in CADASIL patients. *The Lancet*, **350**, 1511–5.

Joutel A, Dodick DD, Parisi JE *et al* (2000). De novo mutation in the Notch3 gene causing CADASIL. *Ann Neurol*, **47**, 388–91.

Julius S, Kjeldsen SE, Weber M *et al.* (2004). Outcomes in hypertensive patients at high cardiovascular risk, treated with regimens based on Valsartan or Amlodipine: the VALUE randomized trail. *Lancet*, **363**, 2022–31.

Jung HH, Bassetti C, Tournier-Lasserve E *et al.* (1995). Cerebral autosomal dominant arteriopathy with subcortical infarcts and leukoencephalopathy: a clinicopathological and genetic study of a Swiss family. *J Neurol Neurosurg Psychiatry*, **59**, 138–43.

Kain K, Catto AJ, Young J *et al.* (2002). Increased fibrinogen von Willebrand factor and tissue plasminogen activator levels in insulin resistant South Asian patients with ischaemic stroke. *Atherosclerosis*, **163**, 371–6.

Kalashnikova LA, Nasonov EL, Stoyanovich LZ *et al.* (1994). Sneddon's syndrome and the primary antiphospholipid syndrome. *Cerebrovasc Dis*, **4**, 76–82.

Kalra L, Perez I, Gupta S *et al.* (1997). The influence of visual neglect on stroke rehabilitation. *Stroke*, **28**, 1386–91.

Kamata T, Yokata T, Furukawa T *et al.* (1994). Cerebral ischaemic attack caused by postprandial hypotension. *Stroke*, **25**, 511–13.

Kanter MC, Hart RG (1991). Neurologic complications of infective endocarditis. *Neurology*, **41**, 1015–20.

Kaplan GA, Keil JE (1993). Socioeconomic factors and cardiovascular disease: a review of the literature. *Circulation*, **88**, 1973–98.

Kaplan PW (1993). Focal seizures resembling transient ischaemic attacks due to subclinical ischaemia. *Cerebrovasc Dis*, **3**, 241–3.

Kaposzta Z, Young E, Bath PMW *et al.* (1999). Clinical application of asymptomatic embolic signal detection in acute stroke: a prospective study. *Stroke*, **30**, 1814–18.

Kaps M, Teschendorf U, Dorndorf W (1992). Haemodynamic studies in early stroke. *J Neurol*, **239**, 138–42.

Karepov VG, Gur AY, Bova I *et al.* (2006). Stroke-in-evolution: infarct-inherent mechanisms versus systemic causes. *Cerebrovasc Dis*, **21**, 42–6.

Karonen JO, Vanninen RL, Liu Y *et al.* (1999). Combined diffusion and perfusion MRI with correlation to single-photon emission CT in acute ischemic stroke. Ischemic penumbra predicts infarct growth. *Stroke*, **30**, 1583–90.

Kasner SE, Gurian JH, Grotta JC (1998). Urokinase treatment of sagittal sinus thrombosis with venous haemorrhagic infarction. *J Stroke Cerebrovasc Dis*, **7**, 421–5.

Kassell NF, Torner JC, Haley EC *et al.* (1990a). The International Cooperative Study on the Timing of Aneurysm Surgery Part 1: Overall management results. *J Neurol*, **73**, 18–36.

Kassell NF, Torner JC, Jane JA *et al.* (1990b). The international cooperative study on the timing of aneurysm surgery. Part 2. Surgical Results. *J Neurol*, **73**, 37–47.

Kattapong VJ, Hart BL, Davis LE (1995). Familial cerebral cavernous angiomas: clinical and radiologic studies. *Neurology*, **45**, 492–7.

Katzav A, Chapman J, Shoenfeld Y (2003). CNS dysfunction in the antiphospholipid syndrome. *Lupus*, **12**, 903–7.

Kawaguchi S, Okuno S, Sakaki T (2000). Effect of direct arterial bypass on the prevention of future stroke in patients with the hemorrhagic variety of moyamoya disease. *J Neurosurg*, **93**, 397–401.

Kazui S, Sawada T, Naritomi H *et al.* (1992). Left unilateral ideomotor apraxia in ischaemic stroke within the territory of the anterior cerebral artery. *Cerebrovasc Dis*, **2**, 35–9.

Keavney B, McKenzie C, Parish S *et al.* (2000). Large-scale test of hypothesised associations between the angiotensin-converting-enzyme, insertion/deletion polymorphism and myocardial infarction in about 5000 cases and 6000 controls. International Studies of Infarct Survival (ISIS) Collaborators. *Lancet*, **355**, 434–2.

Keir SL, Wardlaw JM, Sandercock PAG *et al.* (2002). Antithrombotic therapy in patients with any form of intracranial haemorrhage: A systematic review of the available controlled studies. *Cerebrovasc Dis*, **14**, 197–206.

Keli S, Bloemberg B, Kromhout D (1992). Predictive value of repeated systolic blood pressure measurements for stroke risk. The Zutphen Study. *Stroke*, **23**, 347–51.

Kelly-Hayes M, Wolf PA, Kase CS *et al.* (1995). Temporal patterns of stroke onset. The Framingham Study. *Stroke*, **26**, 1343–7.

Kempczinski R, Hermann G (1979). The innominate steal syndrome. *J Cardiovasc Surg*, **20**, 481–6.

Kent TA, Soukup VM, Fabian RH (2001). Heterogeneity affecting outcome from acute stroke therapy: making reperfusion worse. *Stroke*, **32**, 2318–7.

Keung YK, Owen J (2004). Iron deficiency and thrombosis: literature review. *Clin Appl Thromb Hemost*, **10**, 387–91.

Khatri P, Wechsler LR, Broderick JP (2007). Intracranial hemorrhage associated with revascularization therapies. *Stroke*, **38**, 431–0.

Khaw KT, Barrett-Connor E (1987). Dietary potassium and stroke-associated mortality. A 12-year prospective population study. *N Engl J Med*, **316**, 235–40.

Kidwell CS, Hsia AW (2006). Imaging of the brain and cerebral vasculature in patients with suspected stroke: advantages and disadvantages of CT and MRI. *Curr Neurol Neurosci Rep*, **6**, 9–16.

Kidwell CS, Alger JR, Di Salle F *et al.* (1999). Diffusion MRI in patients with transient ischemic attacks. *Stroke*, **30**, 1174–80.

Kidwell CS, Saver JL, Mattiello J *et al.* (2000). Thrombolytic reversal of acute human cerebral ischemic injury shown by diffusion/perfusion magnetic resonance imaging. *AnnNeurol*, **47**, 462–9.

Kidwell CS, Saver JL, Starkman S *et al.* (2002). Late secondary ischemic injury in patients receiving intraarterial thrombolysis. *AnnNeurol*, **52**, 698–703.

Kidwell CS, Alger JR, Saver JL (2003). Beyond mismatch: evolving paradigms in imaging the ischemic penumbra with multimodal magnetic resonance imaging. *Stroke*, **34**, 2729–35.

Kidwell CS, Chalela JA, Saver JL *et al.* (2004). Comparison of MRI and CT for detection of acute intracerebral hemorrhage. *JAMA*, **292**, 1823–30.

Kikuchi K, Murase K, Miki H *et al.* (2001). Measurement of cerebral hemodynamics with perfusion-weighted MR imaging: comparison with pre- and post-acetazolamide 133Xe-SPECT in occlusive carotid disease. *Am J Neuroradiol*, **22**, 248–54.

Kikuta K, Nozaki K, Takahashi JA *et al.* (2004). Postoperative evaluation of microsurgical resection for cavernous malformations of the brainstem. *J Neurosurg*, **101**, 607–12.

Klijn CJM, Kappelle LJ, Tulleken CAF *et al.* (1997). Symptomatic carotid artery occlusion. A reappraisal of haemodynamic factors. *Stroke*, **28**, 2084–93.

Kim JS, Lee JH, Choi CG (1998). Patterns of lateral medullary infarction. Vascular lesion - magnetic resonance imaging correlation of 34 cases. *Stroke*, **29**, 645–52.

Kim JS, Moon DH, Kim GE *et al.* (2000). Acetazolamide stress brain-perfusion SPECT predicts the need for carotid shunting during carotid endarterectomy. *J Nucl Med*, **41**, 1836–41.

Kim YS, Garami Z, Mikulik R *et al.* (2005). Early recanalization rates and clinical outcomes in patients with tandem internal carotid artery/middle cerebral artery occlusion and isolated middle cerebral artery occlusion. *Stroke*, **36**, 869–71.

Kitahara T, Ariga N, Yamaura A *et al.* (1979). Familial occurrence of moya-moya disease: report of three Japanese families. *J Neurol Neurosurg Psychiatry*, **42**, 208–14.

Kizer JR, Devereux RB (2005). Clinical practice. Patent foramen ovale in young adults with unexplained stroke. *N Engl J Med*, **353**, 2361–72.

Kluytmans M, van der Grond, KJ, van Everdingen KJ *et al.* (1999). Cerebral hemodynamics in relation to patterns of collateral flow. *Stroke*, **30**, 1432–9.

Kodama K (1993). Stroke trends in Japan. *Ann Epidemiol*, **3**, 524–8.

Koennecke H, Mast H, Trocio SH *et al.* (1998). Frequency and determinants of microembolic signals on transcranial Doppler in unselected patients with acute carotid territory ischaemia. A prospective study. *Cerebrovasc Dis*, **8**, 107–12.

Koennecke HC (2006). Cerebral microbleeds on MRI: prevalence associations and potential clinical implications. *Neurology*, **66**, 165–71.

Koren-Morag N, Tanne D, Graff E *et al.* (2002). Low and high density lipoprotein, cholesterol and ischaemic cerebrovascular disease. The bezafibrate Infarction Prevention Registry. *Arch Int Med*, **162**, 993–9.

Koziol JA, Feng AC (2006). On the analysis and interpretation of outcome measures in stroke clinical trials: lessons from the SAINT I study of NXY-059 for acute ischemic stroke. *Stroke*, **37**, 2644–7.

Krapf H, Skalej M, Voigt K (1999). Subarachnoid hemorrhage due to septic embolic infarction in infective endocarditis. *Cerebrovasc Dis*, **9**, 182–4.

Kreisel SH, Hennerici MG, Bazner H (2007). Pathophysiology of stroke rehabilitation: the natural course of clinical recovery, use-dependent plasticity and rehabilitative outcome. *Cerebrovasc Dis*, **23**, 243–55.

Kresowik TF, Bratzler DW, Kresowik RA *et al.* (2004). Multistate improvement in process and outcomes of carotid endarterectomy. *J Vasc Surg*, **39**, 372–80.

Krespi Y, Akman-Demir G, Poyraz M *et al.* (2001). Cerebral vasculitis and ischaemic stroke in Behcet's disease: report of one case and review of the literature. *Eur J Neurol*, **8**, 719–22.

Krieger DW, Yenari MA (2004). Therapeutic hypothermia for acute ischemic stroke: what do laboratory studies teach us? *Stroke*, **35**, 1482–9.

Krueger K, Kugel H, Grond M *et al.* (2000). Late resolution of diffusion-weighted MRI changes in a patient with prolonged reversible ischemic neurological deficit after thrombolytic therapy. *Stroke*, **31**, 2715–8.

Kucinski T, Vaterlein O, Glauche V *et al.* (2002). Correlation of apparent diffusion coefficient and computed tomography density in acute ischemic stroke. *Stroke*, **33**, 1786–91.

Kuehnen J, Schwartz A, Neff W *et al.* (1998). Cranial nerve syndrome in thrombosis of the transverse/sigmoid sinuses. *Brain*, **121**, 381–8.

Kumral E, Bogousslavsky J, Van Melle G *et al.* (1995). Headache at stroke onset: the Lausanne Stroke Registry. *J Neurol Neurosurg Psychiatry*, **58**, 490–2.

Kwa VIH, Franke CL, Verbeeten B *et al.* (1998). Silent intracerebral microhemorrhages in patients with ischaemic stroke for the Amsterdam Vascular Medicine Group. *Ann Neurol*, **44**, 372–7.

Kwakkel G, van Peppen R, Wagenaar RC *et al.* (2004). Effects of augmented exercise therapy time after stroke: A meta-analysis. *Stroke*, **35**, 2529–36.

Kwakkel G (2006). Impact of intensity of practice after stroke: Issues for consideration. *Disabil Rehabil*, **28**, 823–30.

Kwiatkowski TG, Libman RB, Frankel M (1999). Effects of tissue plasminogen activator for acute ischemic stroke at one year. National Institute of Neurological Disorders and Stroke Recombinant Tissue Plasminogen Activator Stroke Study Group. *N Engl J Med*, **340**, 1781–7.

Labauge P, Denier C, Bergametti F *et al.* (2007). Genetics of cavernous angiomas. *Lancet Neurol*, **6**, 237–44.

Labovitz DL, Boden-Albala B, Hauser WA *et al.* (2007). Lacunar infarct or deep intracerebral hemorrhage: who gets which? The Northern Manhattan Study. *Neurology*, **68**, 606–8.

Lamy C, Giannesini C, Zuber M *et al.* (2002). Clinical and imaging findings in cryptogenic stroke patients with and without patent foramen ovale: the PFO-ASA Study (Atrial Septal Aneurysm). *Stroke*, **33**, 706–11.

Lammie GA, Brannan F, Slattery J *et al.* (1997). Nonhypertensive cerebral small-vessel disease. An autopsy study. *Stroke*, **28**, 2222–9.

Lammie GA, Wardlaw JM (1999). Small centrum ovale infarcts - a pathological study. *Cerebrovasc Dis*, **9**, 82–90.

Langhorne P, Wagenaar R, Partridge C (1996). Physiotherapy after stroke: More is better? *Physiother Res Int*, **1**, 75–88.

Langhorne P, Stott D, Robertson L *et al.* (2000). Medical complications after stroke: A multicenter study. *Stroke*, **31**, 1223–9.

Langhorne P, Duncan P (2001). Does the organization of postacute stroke care really matter? *Stroke*, **32**, 268–74.

Langhorne P, Dennis MS (2004). Stroke units: The next 10 years. *Lancet*, **363**, 834–5.

Langhorne P, Holmqvist LW (2007). Early Supported Discharge Trialists. Early supported discharge after stroke. *J Rehabil Med*, **39**, 103–8

Langton Hewer R (1990). Rehabilitation after stroke. *Q J Med*, **76**, 659–74.

Lansberg MG, Norbash AM, Marks MP *et al.* (2000). Advantages of adding diffusion-weighted magnetic resonance imaging to conventional magnetic resonance imaging for evaluating acute stroke. *Arch Neurol*, **57**, 1311–6.

Lapchak PA (2002). Hemorrhagic transformation following ischemic stroke: significance causes and relationship to therapy and treatment. *Curr Neurol Neurosci Rep*, **2**, 38–43.

Lapointe M, Haines S (2002). Fibrinolytic therapy for intraventricular hemorrhage in adults. *Cochrane Database Syst Rev* CD003692.

Larrue V, von Kummer R, del Zoppo G *et al.* (1997). Haemorrhagic transformation in acute ischemic stroke. Potential contributing factors in the European Cooperative Acute Stroke Study. *Stroke*, **28**, 957–60.

Lassen NA, Fieschi C, Lenzi G L (1991). Ischaemic penumbra and neuronal death: Comments on the therapeutic window in acute stroke with particular reference to thrombolytic therapy. *Cerebrovasc Dis*, **1**, 32–5.

Lassen NA, Roland PE, Larsen B *et al.* (1977). Mapping of human cerebral functions: A study of the regional cerebral blood flow pattern during rest its reproducibility and the activations seen during basic sensory and motor functions. *Acta Neurol Scand*, **64**, 262–3.

Leclerc X, Godefroy O, Pruvo JP *et al.* (1995). Computed tomographic angiography for the evaluation of carotid artery stenosis. *Stroke*, **26**, 1577–81.

Lecouvet FE, Duprez TP, Raymackers JM *et al.* (1999). Resolution of early diffusion-weighted and FLAIR MRI abnormalities in a patient with TIA. *Neurology*, **52**, 1085–7.

Lee CD, Folsom AR, Blair SN (2003). Physical activity and stroke risk. A meta-analysis. *Stroke*, **34**, 2475–82.

Lee DK, Kim JS, Kwon SU *et al.* (2005). Lesion patterns and stroke mechanism in atherosclerotic middle cerebral artery disease: early diffusion-weighted imaging study. *Stroke*, **36**, 2583–8.

Lee MW, Pang KY, Ho WW *et al.* (2003). Outcome analysis of intraventricular thrombolytic therapy for intraventricular haemorrhage. *Hong Kong Med J*, **9**, 335–40.

Lee SH, Bae HJ, Kwon SJ *et al.* (2004). Cerebral microbleeds are regionally associated with intracerebral hemorrhage. *Neurology*, **62**, 72–6.

Leenders KL, Perani D, Lammertsma AA *et al.* (1990). Cerebral blood flow, blood volume and oxygen utilisation Normal values and effect of age. *Brain*, **113**, 27–47.

Lees KR, Zivin JA, Ashwood T *et al.* (2006). Stroke-Acute Ischemic NXY Treatment (SAINT I) Trial Investigators NXY-059 for acute ischemic stroke. *N Engl J Med*, **354**, 588–600.

Legg LA, Drummond AE, Langhorne P (2006). Occupational therapy for patients with problems in activities of daily living after stroke. *Cochrane Database Syst Rev*, **18**(4), CD0030585.

Legh-Smith J, Wade DT, Langton-Hewer R (1986). Services for stroke patients one year after stroke. *J Epidemiol Community Health*, **40**, 161–5.

Leira R, Davalos A, Silva Y *et al.* (2004). Early neurologic deterioration in intracerebral hemorrhage: predictors and associated factors. *Neurology*, **63**, 461–7.

Leker RR, Abramsky O (1998). Early anticoagulation in patients with prosthetic heart valves and intracerebral hematoma. *Neurology*, **50**, 1489–91.

Leppala JM, Virtamo J, Fogelholm R *et al.* (2000). Controlled trial of alpha-tocopherol and beta-carotene supplements on stroke incidence and mortality in male smokers. Arterioscler Thromb *Vascular Biol*, **20**, 230–5.

Leung SY, Ng THK, Yuen ST *et al.* (1993). Pattern of cerebral atherosclerosis in Hong Kong Chinese: Severity in intracranial and extracranial vessels. *Stroke*, **24**, 779–86.

Levi CR, Mitchell A, Fitt G *et al.* (1996). The accuracy of magnetic resonance angiography in the assessment of extracranial carotid artery occlusive disease. *Cerebrovasc Dis*, **6**, 231–6.

Lewandowski CA, Frankel M, Tomsick TA *et al.* (1999). Combined intravenous and intra-arterial r-TPA versus intra-arterial therapy of acute ischemic stroke: Emergency Management of Stroke (EMS) Bridging Trial. *Stroke*, **30**, 2598–605.

Lewington S, Clarke R, Qizilbash N *et al.* (2002). Age-specific relevance of usual blood pressure to vascular mortality: a meta-analysis of individual data for one million adults in 61 prospective studies. *Lancet*, **360**, 1903–13

Lewington S, Clarke R (2005). Combined effects of systolic blood pressure and total cholesterol on cardiovascular disease risk. *Circulation*, **112**, 3373–4.

Leys D, Pruvo JP, Scheltens P *et al.* (1992). Leukoaraiosis: relationship with the types of focal lesions occurring in acute cerebrovascular disorders. *Cerebrovasc Dis*, **2**, 169–76.

Leys D, Moulin Th, Stojkovic T *et al.* (1995). Follow-up of patients with history of cervical artery dissection. *Cerebrovasc Dis*, **5**, 43–9.

Leys D, Woimant F, Ferrieres J *et al.* (2006). Detection and management of associated atherothrombotic locations in patients with a recent atherothrombotic ischemic stroke: results of the DETECT survey. *Cerebrovasc Dis*, **21**, 60–6.

Lhermitte F, Gautier JC, Derouesne C (1970). Nature of occlusions of the middle cerebral artery. *Neurology*, **20**, 82–8.

Liang P, Hoffman GS (2005). Advances in the medical and surgical treatment of Takayasu arteritis. *Curr Opin Rheumatol*, **17**, 16–24.

Liebeskind DS (2003). Collateral circulation. *Stroke*, **34**, 2279–84.

Lim W, Crowther MA, Eikelboom JW (2006). Management of antiphospholipid antibody syndrome: a systematic review. *JAMA*, **295**, 1050–7.

Limburg M, Wijdicks EF, Li H (1998). Ischaemic stroke after surgical procedures: clinical features, neuroimaging and risk factors. *Neurology*, **50**, 895–901.

Lindgren A, Roijer A, Norrving B *et al.* (1994). Carotid artery and heart disease in subtypes of cerebral infarction. *Stroke*, **25**, 2356–62.

Lindsberg PJ, Soinne L, Tatlisumak T *et al.* (2004.) Long-term outcome after intravenous thrombolysis of basilar artery occlusion. *JAMA* 20, 292, 1862–6.

Linn FHH, Rinkel GJE, Algra A *et al.* (1996). Incidence of subarachnoid haemorrhage. Role of region, year and rate of computed tomography: a meta-analysis. *Stroke*, **27**, 625–9.

Linn FH, Rinkel GJ, Algra A *et al.* (1998). Headache characteristics in subarachnoid haemorrhage and benign thunderclap headache. *J Neurol Neurosurg Psychiatry*, **65**, 791–3.

Lip GY, Hee FL (2001). Paroxysmal atrial fibrillation. *Q J Med*, **94**, 665–78.

Lip GY (2005). Atrial fibrillation (recent onset). *Clinical Evidence*, **14**, 71–89.

Lip GY, Boos CJ (2006). Antithrombotic treatment in atrial fibrillation. *Heart*, **92**, 155–61.

Liu KD, Chung WY, Wu HM *et al.* (2005). Gamma knife surgery for cavernous hemangiomas: an analysis of 125 patients. *J Neurosurg*, **102**, 81–6.

Loh E, St John Sutton MS, Wun CC *et al.* (1997). Ventricular dysfunction and the risk of stroke after myocardial infarction. *N Engl J Med*, **336**, 251–7.

Longo DL, Witherspoon JM (1980). Focal neurologic symptoms in hypercalcaemia. *Neurology*, **30**, 200–1.

Lossos A, River Y, Eliakim A *et al.* (1995). Neurologic aspects of inflammatory bowel disease. *Neurology*, **45**, 416–21.

Lovblad KO, Baird AE, Schlaug G *et al.* (1997). Ischemic lesion volumes in acute stroke by diffusion-weighted magnetic resonance imaging correlate with clinical outcome. *Ann Neurol*, **42**, 164–70.

Lovelock CE, Molyneux AJ, Rothwell PM *et al.* (2007). Change in incidence and aetiology of intracerebral haemorrhage in Oxfordshire, UK, between 1981 and 2006: a population-based study. *Lancet Neurol*, **6**, 487–93.

Lovett JK, Dennis MS, Sandercock PA *et al.* (2003). Very early risk of stroke after a first transient ischemic attack. *Stroke*, **34**, e138–40.

Lovett JK, Coull A, Rothwell PM (2004). Early risk of recurrent stroke by aetiological subtype: implications for stroke prevention. *Neurology*, **62**, 569–74.

Lovett JK, Gallagher PJ, Hands LJ *et al.* (2004). Histological correlates of carotid plaque surface morphology on lumen contrast imaging. *Circulation*, **110**, 2190–7.

Lovett JK, Redgrave JN, Rothwell PM (2005). A critical appraisal of the performance reporting and interpretation of studies comparing carotid plaque imaging with histology. *Stroke*, **6**, 1091–7.

Lowe GDO, Lee AJ, Rumley A *et al.* (1997). Blood viscosity and risk of cardiovascular events: The Edinburgh Artery Study. *Br J Haematol*, **96**, 168–73.

Lucivero V, Mezzapesa DM, Petruzzellis M *et al.* (2004). Ischaemic stroke in progressive system sclerosis. *Neurol Sci*, **25**, 230–3.

Lutsep HL, Albers GW, DeCrespigny A *et al.* (1997). Clinical utility of diffusion-weighted magnetic resonance imaging in the assessment of ischemic stroke. *AnnNeurol*, **41**, 574–80.

Lyden PD, Hantson L (1998). Assessment scales for the evaluation of stroke patients. *J Stroke Cerebrovasc Dis*, **7**, 113–27.

Lynch GF, Gorelick PB (2000). Stroke in African Americans. *Neurol Clin*, **18**, 273–90.

MacDonald BK, Cockerell OC, Sander JWAS *et al.* (2000). The incidence and lifetime prevalence of neurological disorders in a prospective community-based study in the UK. *Brain*, **123**, 665–76.

Mackinnon AD, Aaslid R, Markus HS (2005). Ambulatory transcranial Doppler cerebral embolic signal detection in symptomatic and asymptomatic carotid stenosis. *Stroke*, **36**, 1726–30.

MacLaren K, Gillespie J, Shrestha S *et al.* (2005). Primary angitis of the central nervous system: emerging variants. *QJM*, **98**, 643–54.

Macleod MR, Amarenco P, Davis SM *et al.* (2004). Atheroma of the aortic arch: an important and poorly recognised factor in the aetiology of stroke. *Lancet Neurol*, **3**, 408–14.

MacMahon S, Peto R, Cutler J *et al.* (1990). Blood pressure, stroke and coronary heart disease. Part 1. Prolonged differences in blood pressure: prospective observational studies corrected for the regression dilution bias. *Lancet*, **335**, 765–74.

McAlister FA, Man-Son-Hing M, Straus SE *et al.* (2005). Impact of a patient decision aid on care among patients with nonvalvular atrial fibrillation: a cluster randomized trial. *Can Med Assoc J*, **17**, 496–501.

McCabe DJ, Brown MM, Clifton A (1999). Fatal cerebral reperfusion hemorrhage after carotid stenting. *Stroke*, **30**, 2483–6.

McCarron MO, Nicoll JA, Love S *et al.* (1999). Surgical intervention biopsy and APOE genotype in cerebral amyloid angiopathy-related haemorrhage. *Br J Neurosurg*, **13**, 462–7.

McDonald MJ, Brophy BP, Kneebone C (1998). Rendu-Osler-Weber syndrome: a current perspective on cerebral manifestations. *J Clin Neurosci*, **5**, 345–50.

McGowern PG, Burke GL, Sprafka JM *et al.* (1992). Trends in mortality, morbidity and risk factor levels for stroke from 1960-1990: the Minnesota Heart Survey. *JAMA*, **268**, 753–9.

Macko RF, Kittner SJ, Epstein A *et al.* (1999). Elevated tissue plasminogen activator, antigen and stroke risk. The stroke prevention in young women study. *Stroke*, **30**, 7–11.

Maia LF, MacKenzie IR, Feldmann HH (2007). Clinical phenotypes of cerebral amyloid angiopathy. *J Neurol Sci*, **257**, 23–30.

Malek AM, Higashida RT, Phatouros CC *et al.* (1999). Treatment of posterior circulation ischaemia with extracranial percutaneous balloon angioplasty and stent placement. *Stroke*, **30**, 2073–85.

Malouf R, Brust JCM (1985). Hypoglycaemia: Causes, neurological manifestations and outcome. *Ann Neurol*, **17**, 421–30.

Man-Son-Hing M, Laupacis A, O'Connor AM *et al.* (2000). Development of a decision aid for patients with atrial fibrillation who are considering antithrombotic therapy. *J Gen Intern Med*, **15**, 723–30.

Marchal G, Beaudouin V, Rioux P *et al.* (1996a). Prolonged persistence of substantial volumes of potentially viable brain tissue after stroke. A correlative PET-CT study with voxel-based data analysis. *Stroke*, **27**, 599–606.

Marchal G, Furlan M, Beaudouin V *et al* (1996b). Early spontaneous hyperperfusion after stroke. A marker of favourable tissue outcome? *Brain*, **119**, 409–19.

Mari F, Matei M, Ceravolo MG *et al.* (1997). Predictive value of clinical indices in detecting aspiration in patients with neurological disorders. *J Neurol Neurosurg Psychiatry*, **63**, 456–60.

Marinkovic SV, Milisavljevic MM, Kovacevic MS *et al.* (1985). Perforating branches of the middle cerebral artery. Micro-anatomy and clinical significance of their intracerebral segments. *Stroke*, **16**, 1022–9.

Marinkovic SV, Milisavljevic MM, Lolic-Draganic V *et al.* (1987). Distribution of the occipital branches of the posterior cerebral artery. Correlation with occipital lobe infarcts. *Stroke*, **18**, 728–32.

Marks MP, Marcellusm M, Norbashm AM *et al.* (1999). Outcome of angioplasty for atherosclerotic intracranial stenosis. *Stroke*, **30**, 1065–9.

Markus HS, Droste DW, Brown MM (1994). Detection of asymptomatic cerebral embolic signals with Doppler ultrasound. *Lancet*, **343**, 1011–12.

Markus HS, Hambley H (1998). Neurology and the blood: haematological abnormalities in ischaemic stroke. *J Neurol Neurosurg Psychiatry*, **64**, 150–9.

Markus HS, Mendall MA (1998). Helicobacter pylori infection: a risk factor for ischaemic cerebrovascular disease and carotid atheroma. *J Neurol Neurosurg Psychiatry*, **64**, 104–7.

Markus H, Cullinane M, Reid G (1999). Improved automated detection of embolic signals using a novel frequency filtering approach. *Stroke*, **30**, 1610–15.

Markus HS, Cullinane M (2000). Asymptomatic Carotid Emboli (ACES) Study. *Cerebrovasc Dis*, **10**, 3.

Markus H, Cullinane M (2001). Severely impaired cerebrovascular reactivity predicts stroke and TIA risk in patients with carotid artery stenosis and occlusion. *Brain*, **124**, 457–67.

Markus HS (2006). Can microemboli on transcranial Doppler identify patients at increased stroke risk? *Nature Clinical Practice in Cardiovascular Medicine*, **3**, 246–7.

Martin R, Bogousslavsky J, Miklossy J *et al.* (1992). Floating thrombus in the innominate artery as a cause of cerebral infarction in young adults. *Cerebrovasc Dis*, **2**, 177–81.

Martin R, Bogousslavsky J for the Lausanne Stroke Registry Group (1993). Mechanisms of late stroke after myocardial infarct: the Lausanne Stroke Registry. *J Neurol Neurosurg Psychiatry*, **56**, 760–4.

Martyn CN, Barker DJP, Osmond C (1996). Mothers' pelvic size, fetal growth and death from stroke and coronary heart disease in men in the UK. *Lancet*, **348**, 1264–8.

Mas JL, Arquizan C, Lamy C *et al.* (2001). Patent Foramen Ovale and Atrial Septal Aneurysm Study Group. Recurrent cerebrovascular events associated with patent foramen ovale atrial septal aneurysm or both. *N Engl J Med*, **345**, 1740–6.

Mas JL (2003). Specifics of patent foramen ovale. *Advanced Neurology*, **92**, 197–202.

Mas JL, Chatellier G, Beyssen B *et al.* (2006). Endarterectomy versus stenting in patients with symptomatic severe carotid stenosis. *NEJM*, **355**, 1660–71.

Mast H, Young WL, Koennecke HC *et al.* (1997). Risk of spontaneous haemorrhage after diagnosis of cerebral arteriovenous malformation. *Lancet*, **350**, 1065–8.

Masuda J, Yutani C, Waki R *et al.* (1992). Histopathological analysis of the mechanisms of intracranial haemorrhage complicating infective endocarditis. *Stroke*, **23**, 843–50.

Matarín M, *et al.* (2007). A genome-wide genotyping study in patients with ischaemic stroke: initial analysis and data release. *Lancet Neurol*, **6**, 414–20.

Matijevic N, Wu K (2006). Hypercoagulable states and strokes. *Curr Atheroscler Rep*, **8**, 324–9.

Mathur A, Roubin GS, Iyer SS *et al.* (1998). Predictors of stroke complicating carotid artery stenting. *Circulation*, **97**, 1239–45.

Mayer SA, Tatemichi TK, Spitz JL *et al.* (1994). Recurrent ischaemic events and diffuse white matter disease in patients with pseudoxanthoma elasticum. *Cerebrovasc Dis*, **4**, 294–7.

Mayer SA, Brun NC, Broderick J *et al.* (2005). Safety and feasibility of recombinant factor VIIa for acute intracerebral hemorrhage. *Stroke*, **36**, 74–9.

Mazzaglia G, Britton AR, Altmann DR *et al.* (2001). Exploring the relationship between alcohol consumption and non-fatal or fatal stroke: a systematic review. *Addiction*, **96**, 1743–56.

Mead GE, Lewis SC, Wardlaw JM *et al.* (1999). Should CT appearance of lacunar stroke influence patient management? *J Neurol Neurosurg Psychiatry*, **67**, 682–4.

Meade TW, Ruddock V, Stirling Y *et al.* (1993). Fibrinolytic activity, clotting factors and long-term incidence of ischaemic heart disease in the Northwick Park Heart Study. *Lancet*, **342**, 1076–9.

Medical Research Council's General Practice Research Framework (1998). Thrombosis prevention trial: randomised trial of low-intensity oral anticoagulation with warfarin and low-dose aspirin in the primary prevention of ischaemic heart disease in men at increased risk. *Lancet*, **351**, 233–41.

Medin J, Nordlund A, Ekberg K (2004). Increasing stroke incidence in Sweden between 1989 and 2000 among persons aged 30 to 65 years: evidence from the Swedish Hospital Discharge Register. *Stroke*, **35**, 1047–51.

Mendelow AD, Gregson BA, Fernandes HM *et al.* (2005). Early surgery versus initial conservative treatment in patients with spontaneous supratentorial intracerebral haematomas in the International Surgical Trial in Intracerebral Haemorrhage (STICH): a randomised trial. *Lancet*, **365**, 387–97.

Mendez I, Hachinski V, Wolfe B (1987). Serum lipids after stroke. *Neurology*, **37**, 507–11.

Menendez Gonzalez M, Rivera MM (2006). Transient global amnesia: increasing evidence of a venous etiology. *Arch Neurol*, **63**, 1334–6.

Menken M, Munsat TL, Toole JF (2000). The Global Burden of Disease Study Implications for neurology. *Arch Neurol*, **57**, 418–20.

Merrill JT (2007). Antiphospholipid syndrome: what's new in understanding antiphospholipid antibydo-related stroke? *Curr Rheumatol Rep*, **8**, 159–61.

Meschia JF (2003). Familial hypercholesterolemia: stroke and the broader perspective. *Stroke*, **34**, 22–5.

Meschia JF, Brott TG, Brown RD Jr (2005). Genetics of cerebrovascular disorders. *Mayo Clin Proc*, **80**, 122–32.

Messe SR, Cucchiara B, Luciano J *et al.* (2005). PFO management: neurologists vs cardiologists. *Neurology*, **65**, 172–3.

Metz H, Murray-Leslie RM, Bannister R G *et al.* (1961). Kinking of the internal carotid artery. *Lancet*, **1**, 424–6.

Meuli RA (2004). Imaging viable brain tissue with CT scan during acute stroke. *Cerebrovasc Dis*, **17**, 28–34.

Meyer JS, Obara K, Muramatsu K (1993). Diaschisis. *Neurol Res*, **15**, 362–6.

Milandre L, Pellissier JF, Boudouresques G *et al.* (1987). Non-hereditary multiple telangiectasias of the central nervous system. *J Neurol Sci*, **82**, 291–304.

Millikan C, Futrell N (1990). The fallacy of the lacune hypothesis. *Stroke*, **21**, 1251–7.

Mills JA (1994). Systemic lupus erythematosus. *N Engl J Med*, **330**, 1871–9.

Mistri AK, Robinson TG, Potter JF (2006). Pressor therapy in acute ischemic stroke: systematic review. *Stroke*, **37**, 1565–71.

Mitchell JRA, Schwartz CJ (1962). Relationship between arterial disease in different sites. A study of the aorta and coronary carotid and iliac arteries. *Br Med J*, **1**, 1293–301.

Mitra D, Connolly D, Jenkins S *et al.* (2006). Comparison of image quality, diagnostic confidence and interobserver variability in contrast enhanced MR angiography and 2D time of flight angiography in evaluation of carotid stenosis, *Br J Radiol*, **79**, 201–7.

Mitsias P, Levine SR (1994). Large cerebral vessel occlusive disease in systemic lupus erythematosus. *Neurology*, **44**, 385–93.

Mohr JP, Gautier JC, Pessin MS (1992). Internal carotid artery disease. In: Barnett HJM, Mohr JP, Stein BM, Yatsu FM, eds. *Stroke*, p. 311. Churchill Livingstone, New York.

Mohr JP, Biller J, Hilal SK *et al.* (1995). Magnetic resonance versus computed tomographic imaging in acute stroke. *Stroke*, **26**, 807–12.

Molina CA, Montaner J, Abilleira S *et al.* (2001). Timing of spontaneous recanalization and risk of hemorrhagic transformation in acute cardioembolic stroke. *Stroke*, **32**, 1079–84.

Molloy J, Markus HS (1999). Asymptomatic embolization predicts stroke and TIA risk in patients with carotid artery stenosis. *Stroke*, **30**, 1440–3.

Molyneux A, Kerr R, Stratton I *et al.* (2002). International Subarachnoid Aneurysm Trial (ISAT) of neurosurgical clipping versus endovascular coiling in 2143 patients with ruptured intracranial aneurysms: a randomised trial. *Lancet*, **360**, 1267–74.

Molyneux AJ, Kerr RS, Yu LM *et al.* (2005). International subarachnoid aneurysm trial (ISAT) of neurosurgical clipping versus endovascular coiling in 2143 patients with ruptured intracranial aneurysms: a randomised comparison of effects on survival, dependency, seizures, rebleeding subgroups and aneurysm occlusion. *Lancet*, **366**, 809–17.

Momjian-Mayor I, Baron JC (2005). The pathophysiology of watershed infarction in internal carotid artery disease: review of cerebral perfusion studies. *Stroke*, **36**, 567–77.

Montavont A, Nighoghossian N, Derex L *et al.* (2004). Intravenous r-TPA in vertebrobasilar acute infarcts. *Neurology*, **62**, 1854–6.

Moodie M, Cadilhac D, Pearce D *et al.* (2006). SCOPES Study Group Economic evaluation of Australian stroke services: a prospective multicenter study comparing dedicated stroke units with other care modalities. *Stroke*, **37**, 2790–5.

Moore PM (1997). Neuropsychiatric systemic lupus erythematous. Stress, stroke and seizures. *Ann N Y Acad Sci*, **823**, 1–17.

Moore WS, Young B, Baker WH *et al.* (1996). Surgical results: a justification of the surgeon selection process for the ACAS trial. The ACAS Investigators. *J Vasc Surg*, **23**, 323–8.

Moriwaki H, Uno H, Nagakane Y *et al* (2004). Losartan, an angiotensin II (AT1) receptor antagonist, preserves cerebral blood flow in hypertensive patients with a history of stroke. *J Hum Hypertens*; **18**, 693–9.

Mosso M, Georgiadis D, Baumgartner RW (2004). Progressive occlusive disease of large cerebral arteries and ischemic events in a patient with essential thrombocythemia. *Neurol Res*, **26**, 702–3.

MRC/BHF Heart Protection Study Collaborative Group (2002). Heart protection study of cholesterol lowering with simvastatin in 20,536 high-risk individuals: a randomised placebo-controlled trial. *Lancet*, **360**, 7–22.

Muir KW, Teal PA (2005). Why have neuro-protectants failed?: lessons learned from stroke trials. *J Neurol*, **252**, 1011–20.

Muir KW, Buchan A, von Kummer R *et al.* (2006). Imaging of acute stroke. *Lancet Neurol*, **5**, 755–68.

Muizelaar JP, Marmarou A, Ward JD *et al.* (1991). Adverse effects of prolonged hyperventilation in patients with severe head injury: a randomized clinical trial. *J Neurosurg*, **75**, 731–9.

Mullins ME, Schaefer PW, Sorensen AG *et al.* (2002). CT and conventional and diffusion-weighted MR imaging in acute stroke: study in 691 patients at presentation to the emergency department. *Radiology*, **224**, 353–60.

Mumford CJ, Fletcher NA. Ironside JW *et al.* (1996). Progressive ataxia, focal seizures and malabsorption syndrome in a 41 year old woman. *J Neurol Neurosurg Psychiatry*, **60**, 225–30.

Munoz DG (2006). Leukoaraiosis and ischemia: beyond the myth. *Stroke*, **37**, 1348–9.

Murray CJL, Lauer JA, Hutubessy RCW *et al.* (2003). Effectiveness and costs on interventions to lower systolic blood pressure: a global and regional analysis on reduction of cardiovascular risk. *Lancet*, **361**, 717–25.

Murray CJL, Lopez AD (1996). The Global Burden of Disease: a comprehensive assessment of mortality and disability from diseases, injuries and risk factors in 1990 and projected to 2020. Harvard University Press, Boston.

Nabavi DG, Kloska SP, Nam EM *et al.* (2002). MOSAIC: Multimodal Stroke Assessment Using Computed Tomography: novel diagnostic approach for the prediction of infarction size and clinical outcome. *Stroke*, **33**, 2819–6.

Nandalur KR, Baskurt E, Hagspiel KD *et al.* (2006). Carotid artery calcification on CT may independently predict stroke risk. *Am J Radiol*, **186**, 547–2.

Nasreddine ZS, Saver JL (1997). Pain after thalamic stroke: right diencephalic predominance and clinical features in 180 patients. *Neurology*, **48**, 1196–9.

National Institute of Neurological Disorders and Stroke rt-PA Stroke Study Group (1995). Tissue plasminogen activator for acute ischaemic stroke. *N Engl J Med*, **14**, 1581–7.

Nattel S, Opie L (2006). Controversies in atrial fibrillation. *Lancet*, **367**, 262–72.

Naylor AR, Bolia A, Abbott RJ *et al.* (1998). Randomized study of carotid angioplasty and stenting versus carotid endarterectomy: a stopped trial. *J Vasc Surg*, **28**, 326–34.

Naylor AR, Cuffe RL, Rothwell PM *et al.* (2003a). A systematic review of outcomes following staged and synchronous carotid endarterectomy and coronary artery bypass. *Eur J Vasc Endovasc Surg*, **25**, 380–9.

Naylor R, Cuffe RL, Rothwell PM *et al.* (2003b). A systematic review of outcome following synchronous carotid endarterectomy and coronary artery bypass: influence of surgical and patient variables. *Eur J Vasc Endovasc Surg*, **26**, 230–41.

Nazir FS, Overell JR, Bolster A *et al* (2004). The effect of losartan on global and focal cerebral perfusion and on renal function in hypertensives in mild early ischaemic stroke. *J Hypertens*; **22**, 989–5.

Neimann J, Haapaniemi HM, Hillbom M (2000). Neurological complications of drug abuse: pathophysiological mechanisms. *Eur J Neurol*, **7**, 595–606.

Nesher G (2000). Neurologic manifestations of giant cell arteritis. *Clini Exp Rheumatol*, **18**, S24–6.

Newman MF, Matthew JP, Grocott HP *et al.* (2006). Central nervous system injury associated with cardiac surgery. *Lancet*, **368**, 694–703.

Ngeh J, Gupta S, Goodbourn C *et al.* (2003). Chlamydia pneumoniae in elderly patients with stroke (C-PEPS): a case-control study on the seroprevalence of Chlamydia pneumoniae in elderly patients with acute cerebrovascular disease. *Cerebrovasc Dis*, **15**, 11–6.

Nicolai A, Lazzarino LG, Biasutti E (1996). Large striatocapsular infarcts: clinical features and risk factors. *J Neurol*, **243**, 44–50.

Nicolaides AN, Kakkos SK, Griffin M *et al.* (2005). Effect of image normalization on carotid plaque classification and the risk of ipsilateral hemispheric ischemic events: results from the asymptomatic carotid stenosis and risk of stroke study. *Vascular Surg*, **13**, 211–1.

Niemann DB, Wills AD, Maartens NF *et al.* (2003). Treatment of intracerebral hematomas caused by aneurysm rupture: coil placement followed by clot evacuation. *J Neurosurg*, **99**, 843–7.

Nieuwkamp DJ, de Gans K Rinkel GJE *et al.* (2000). Treatment and outcome of severe intraventricular extension in patients with subarachnoid or intracerebral hemorrhage: a systematic review of the literature. *J Neurol*, **247**, 117–21.

Nighoghossian N, Hermier M, Adeleine P *et al.* (2002). Old microbleeds are a potential risk factor for cerebral bleeding after ischemic stroke: a gradient-echo T2*-weighted brain MRI study. *Stroke*, **33**, 735–42.

Nishimoto T, Yuki K, Sasaki T *et al.* (2005). A ruptured middle cerebral artery aneurysm originating from the site of anastomosis 20 years after extracranial-intracranial bypass for moyamoya disease: case report. *Surg Neurol*, **64**, 261–5.

Nor AM, Davis J, Sen B *et al.* (2005). The recognition of Stroke in the Emergency Room (ROSIER) scale: development and validation of a stroke recognition instrument. *Lancet Neurol*, **4**, 727–34.

Norris JW, Hachinski VC (1982). Misdiagnosis of stroke. *Lancet*, **1**, 328–31.

North American Symptomatic Carotid Endarterectomy Trial Collaborators (1998). Benefit of carotid endarterectomy in patients with symptomatic moderate or severe stenosis. *N Engl J Med*, **339**, 1415–25.

North KN, Whiteman DAH, Pepin MG *et al.* (1995). Cerebrovascular complications in Ehlers-Danlos syndrome Type IV. *Ann Neurol*, **38**, 960–4.

Novak V, Chowdhary A, Farrar B *et al.* (2003). Altered cerebral vasoregulation in hypertension and stroke. *Neurology*, **60**, 1657–63.

Nudo RJ (2007). Post infarct cortical plasticity and behavioural recovery. *Stroke*, **38**, 840–5.

Numminen H, Kotila M, Waltimo O *et al.* (1996). Declining incidence and mortality rates of stroke in Finland from 1972 to 1991: results of 3 population-based stroke registers. *Stroke*, **27**, 1487–91.

O'Connell BK, Towfighi J, Brennan RW *et al.* (1985). Dissecting aneurysms of head and neck. *Neurology*, **35**, 993–7.

O'Connor AD, Rusyniak DE, Bruno A (2005). Cerebrovascular and cardiovascular complications of alcohol and sympathomimetic drug abuse. *Med Clin North Am*, **89**, 1343–58.

O'Connor MM, Mayberg MR (2000). Effects of radiation on cerebral vasculature: a review. *Neurosurgery*, **46**, 138–49.

O'Donnell MJ, Berge E, Sandset PM (2006). Are there patients with acute ischemic stroke and atrial fibrillation that benefit from low molecular weight heparin? *Stroke*, **37**, 452–5.

O'Laoire SA, Crockard A, Thomas DGT *et al.* (1982). Brain-stem haematoma. A report of six surgically treated cases. *J Neurol*, **56**, 222–7.

O'Leary S, Hodgson TJ, Coley SC *et al.* (2002). Intracranial dural arteriovenous malformations: results of stereotactic radiosurgery in 17 patients. *Clin Oncol R Coll Radiol*, **14**, 97–102.

Ogata J, Masuda J, Yutani C *et al.* (1994). Mechanisms of cerebral artery thrombosis: a histopathological analysis on eight necropsy cases. *J Neurol Neurosurg Psychiatry*, **57**, 17–21.

Ogata J, Yonemura K, Kimura K *et al.* (2005). Cerebral infarction associated with essential thrombocythemia: an autopsy case study. *Cerebrovasc Dis*, **19**, 201–5.

Olafsson E, Gudmundsson G, Hauser WA (2000). Risk of epilepsy in long-term survivors of surgery for aneurysmal subarachnoid hemorrhage: a population-based study in Iceland. *Epilepsia*, **41**, 1201–5.

Olesen J, Leonardi M (2003). The burden of brain diseases in Europe. *Eur J Neurol*, **10**, 471–7.

Olin JW, Shih A (2006). Thromboangiitis obliterans (Buerger's disease). *Curr Opin Rheumatol*, **18**, 18–24.

Oppenheim C, Stanescu R, Dormont D *et al.* (2000). False-negative diffusion-weighted MR findings in acute ischemic stroke. *Am J Neuroradiol*, **21**, 1434–40.

Oppenheimer SM, Cechetto DF, Hachinski VC (1990). Cerebrogenic cardiac arrhythmias. Cerebral electrocardiographic influences and their role in sudden death. *Arch Neurol*, **47**, 513–19.

Orencia AJ, Petty GW, Khandheria BK *et al.* (1995a). Risk of stroke with mitral valve prolapse in population-based cohort study. *Stroke*, **26**, 7–13.

Orencia AJ, Petty GW, Khandheria BK *et al.* (1995b). Mitral valve prolapse and the risk of stroke after initial cerebral ischaemia. *Neurology*, **45**, 1083–6.

Orencia AJ, Daviglus ML, Dyer AR *et al.* (1996). Fish consumption and stroke in men. 30-year findings of Chicago Western Electric Study. *Stroke*, **27**, 204–9.

Ostrem JL, Saver JL, Alger JR *et al.* (2004). Acute basilar artery occlusion: diffusion-perfusion MRI characterization of tissue salvage in patients receiving intra-arterial stroke therapies. *Stroke*, **35**, e30–4.

Outpatient Service Trialists' (2003). Therapy-based rehabilitation services for stroke patients at home Cochrane Database of Systematic Reviews 2003:Art No: CD002925 DOI: 002910/001002/14651858CD14002925.

Pan WH, Li LA, Tsai MJ (1995). Temperature extremes and mortality from coronary heart disease and cerebral infarction in elderly Chinese. *Lancet*, **345**, 353–5.

Pan DH, Chung WY, Guo WY *et al.* (2002). Stereotactic radiosurgery for the treatment of dural arteriovenous fistulas involving the transverse-sigmoid sinus. *J Neurosurg*, **96**, 823–9.

Pandey DK, Gorelick PB (2005). Epidemiology of stroke in African Americans and Hispanic Americans. *Med Clin North Am*, **89**, 739–52.

Pantoni L, Garcia JH (1997). Pathogenesis of leukoariaosis. A review. *Stroke*, **28**, 652–9.

Pariente J, Loubinoux I, Carel C *et al.* (2001). Fluoxetine modulates motor performance and cerebral activation of patients recovering from stroke. *AnnNeurol*, **50**, 718–29.

Parr MJA, Finfer SR, Morgan MK (1996). Reversible cardiogenic shock complicating subarachnoid haemorrhage. *Br Med J*, **313**, 681–3.

Parry SW, Steen N, Baptist M *et al.* (2006). Cerebral autoregulation is impaired in cardioinhibitory carotid sinus syndrome. *Heart*, **92**, 792–7.

Parsons MW, Li T, Barber PA *et al.* (2000). Combined (1)H MR spectroscopy and diffusion-weighted MRI improves the prediction of stroke outcome. *Neurology*, **55**, 498–505.

Parsons MW, Yang Q, Barber PA *et al.* (2001). Perfusion magnetic resonance imaging maps in hyperacute stroke: relative cerebral blood flow most accurately identifies tissue destined to infarct. *Stroke*, **32**, 1581–7.

Parsons MW, Pepper EM, Chan V *et al.* (2005). Perfusion computed tomography: prediction of final infarct extent and stroke outcome. *Ann Neurol*, **58**, 672–9.

Patel A, Knapp M, Evans A *et al.* (2004). Training care givers of stroke patients: economic evaluation. *BMJ*, **328**, 1102.

Patterson JR, Grabois M (1986). Locked-in syndrome: a review of 139 cases. *Stroke*, **17**, 758–64.

Paulson OB, Waldemar G (1990). ACE inhibitors and cerebral blood flow. *J Hum Hypertens*, **4**(4 Suppl), 69–72.

Pearson TC, Wetherley-Mein G (1978). Vascular occlusive episodes and venous haematocrit in primary proliferative polycythaemia. *Lancet*, **2**, 1219–22.

Pedersen PM, Jorgensen HS, Nakayama H *et al.* (1995). Aphasia in acute stroke: incidence determinants and recovery. *Ann Neurol*, **38**, 659–66.

Pendlebury ST, Blamire AM, Lee MA *et al* (1999). Axonal injury in the internal capsule correlates with motor impairment after stroke. *Stroke*; **30**, 956–62.

Pendlebury ST, Rothwell PM, Algra A *et al.* (2004). Underfunding of stroke research: a Europe-wide problem. *Stroke*, **35**, 2368–71.

Pendlebury ST (2007). Worldwide under-funding of stroke research. *Int J Stroke*, **2**, 80–4.

Petrovitch H, Curb D, Bloom-Marcus E (1995). Isolated systolic hypertension and risk of stroke in Japanese-American men. *Stroke*, **26**, 25–9.

Phan TG, Koh M, Wijdicks EFM (2000). Safety of discontinuation of anticoagulation in patients with intracranial hemorrhage at high thromboembolic risk. *Arch Neurol*, **57**, 1710–13.

Pineiro R, Pendlebury ST, Smith S *et al.* (2000). Relating MRI changes to motor deficit after ischemic stroke by segmentation of functional motor pathways. *Stroke*, **31**, 672–9.

Pineiro R, Pendlebury S, Johansen-Berg H *et al.* (2001). Functional MRI detects posterior shifts in primary sensorimotor cortex activation after stroke: evidence of local adaptive reorganization? *Stroke*, **32**, 1134–9.

Pinto AN, Ferro JM, Canhao P *et al.* (1993). How often is a perimesencephalic subarachnoid haemorrhage CT pattern caused by ruptured aneurysms? *Acta Neurochir (Wien)*, **124**, 79–81.

Pinto AN, Canhao P, Ferro JM (1996). Seizures at the onset of subarachnoid haemorrhage. *J Neurol*, **243**, 161–4.

Plant GT, Revesz T, Barnard RO *et al.* (1990). Familial cerebral amyloid angiopathy with non neuritic amyloid plaque formation. *Brain*, **113**, 721–47.

Pollock A, Baer G, Pomeroy V *et al.* (2007). Physiotherapy treatment approaches for the recovery of postural control and lower limb function following stroke. *Cochrane Database Syst Rev*, **24**, CD001920.

Powers WJ (1986). Should lumbar puncture be part of the routine evaluation of patients with cerebral ischaemia? *Stroke*, **17**, 332–3.

Powers WJ, Press GA, Grubb RL *et al.* (1987). The effect of haemodynamically significant carotid artery disease on the haemodynamic status of the cerebral circulation. *Ann Intern Med*, **106**, 27–34.

Powers WJ (1993). Acute hypertension after stroke: the scientific basis for treatment decisions. *Neurology*, **43**, 461–7.

Powers WJ, Derdeyn CP, Fritsch SM *et al.* (2000). Benign prognosis of never-symptomatic carotid occlusion. *Neurology*, **54**, 878–82.

Preter M, Tzourio C, Ameri A *et al.* (1996). Long-term prognosis in cerebral venous thrombosis: follow-up of 77 patients. *Stroke*, **27**, 243–6.

Prior AL, Wilson LA, Gosling RG *et al.* (1979). Retrograde cerebral embolism. *Lancet*, **2**, 1044–7.

PROGRESS Collaborative Group (2001). Randomised trial of a perindopril-based blood-pressure-lowering regimen among 6,105 individuals with previous stroke or transient ischaemic attacks. *Lancet*, **358**, 1033–41.

Prospective Studies Collaboration (1995). Cholesterol, diastolic blood pressure and stroke: 13000 strokes in 450000 people in 45 prospective cohorts. *Lancet*, **346**, 1647–53.

Purroy F, Montaner J, Rovira A *et al.* (2004). Higher risk of further vascular events among transient ischaemic attack patients with diffusion-weighted imaging acute lesions. *Stroke*, **35**, 2313–19.

Qizilbash N, Duffy S, Prentice CRM *et al.* (1997). Von Willebrand factor and risk of ischaemic stroke. *Neurology*, **49**, 1552–6.

Quinette P, Guillery-Girard B, DayanJ *et al.* (2006). What does transient global amnesia really mean? Review of the literature and thorough study of 142 cases. *Brain*, **129**, 1640–58.

Qureshi AI, Luft AR, Sharma M *et al.* (1999). Frequency and determinants of postprocedural hemodynamic instability after carotid angioplasty and stenting. *Stroke*, **30**, 2086–93.

Raaymakers TWM, Rinkel GJE, Limburg M *et al.* (1998). Mortality and morbidity of surgery for unruptured intracranial aneurysms: a meta-analysis. *Stroke*, **29**, 1531–8.

Raaymakers TW (1999). Aneurysms in relatives of patients with subarachnoid hemorrhage: frequency and risk factors. MARS Study Group. Magnetic Resonance Angiography in Relatives of patients with subarachnoid hemorrhage. Neurology, **53**, 982–8.

Rabinstein AA, Atkinson JL, Wijdicks EFM (2002). Emergency craniotomy in patients worsening due to expanded cerebral haematoma - To what purpose? *Neurology*, **58**, 1367–72.

Rabinstein AA, Tisch SH, McClelland RL *et al.* (2004). Cause is the main predictor of outcome in patients with pontine hemorrhage. *Cerebrovasc Dis*, **17**, 66–71.

Ramakrishna G, Malouf JF, Younge BR *et al.* (2005). Calcific retinal embolism as an indicator of severe unrecognised cardiovascular disease. *Heart*, **91**, 1154–7.

Rantner B, Pavelka M, Posch L (2005). Carotid endarterectomy after ischemic stroke-is there a justification for delayed surgery? *Eur J Vasc Endovasc Surg*, **30**, 36–40.

Raps EC, Rogers JD, Galetta SL *et al.* (1993). The clinical spectrum of unruptured intracranial aneurysms. *Arch Neurol*, **50**, 265–8.

Rash A, Downes T, Portner R *et al.* (2007). A randomised controlled trial of warfarin versus aspirin for stroke prevention in octogenarians with atrial fibrillation (WASPO). *Age Ageing*, **36**, 151–6.

Razvi SSM, Davidson R, Bone I *et al.* (2005). The prevalence of cerebral autosomal dominant arteriopathy with subcortical infarcts and leucoencephalopathy (CADASIL) in the west of Scotland. *J Neurol Neurosurg Psychiatry*, **76**, 739–41.

Razvi SS, Bone I (2006). Single gene disorders causing ischaemic stroke. *J Neurol*, **253**, 685–700.

Read SJ, Pettigrew L, Schimmel L *et al.* (1998). White matter medullary infarcts: acute subcortical infarction in the centrum ovale. *Cerebrovasc Dis*, **8**, 289–95.

Redgrave JN, Lovett JK, Gallagher PJ *et al.* (2006). Histological assessment of 526 symptomatic carotid plaques in relation to the nature and timing of ischemic symptoms: the Oxford plaque study. *Circulation*, **113**, 2320–8.

Redgrave JN, Schulz UG, Briley D *et al.* (2007a). Presence of Acute Ischaemic Lesions on Diffusion-Weighted Imaging Is Associated with Clinical Predictors of Early Risk of Stroke after Transient Ischaemic Attack. *Cerebrovasc Dis*, **24**, 86–90

Redgrave JN, Coutts SB, Schulz UG *et al.* (2007b). Systematic review of associations between the presence of acute ischemic lesions on diffusion-weighted imaging and clinical predictors of early stroke risk after transient ischemic attack. *Stroke*, **38**, 1482–8.

Redman AR, Allen LC (2002). Warfarin Versus Aspirin in the Secondary Prevention of Stroke: The WARSS Study. *Curr Atheroscler Rep*, **4**, 319–25.

Reed G, Devous M (1985). South-western Internal Medicine Conference: Cerebral blood flow autoregulation and hypertension. *Am J Med Sci*, **289**, 37–44.

Régis J, Bartolomei F, Kida Y *et al.* (2000). Radiosurgery for epilepsy associated with cavernous malformation: Retrospective study in 49 patients. *Neurosurgery*, **47**, 1091–7.

Reichart MD, Bogousslavsky J, Janzer RC (2000). Early lacunar strokes complicating polyarteritis nodosa: thrombotic microangiopathy. *Neurology*, **54**, 883–9.

Reimers B, Corvaja N, Moshiri S *et al.* (2001). Cerebral protection with filter devices during carotid artery stenting. *Circulation*, **104**, 12–15.

Reith J, Jorgensen HS, Pedersen PM *et al.* (1996). Body temperature in acute stroke: relation to stroke severity, infarct size, mortality and outcome. *Lancet*, **347**, 422–5.

Rerkasem K, Bond R, Rothwell PM (2004). Local versus general anaesthetic for carotid endarterectomy. *Cochrane Database Syst Rev*, **2**, CD000126.

Reynolds K, Lewis B, Nolen JD *et al.* (2003). Alcohol consumption and risk of stroke: a meta-analysis. *JAMA*, **289**, 579–88.

Rhodes (1996). Cholesterol crystal embolism: an important 'new' diagnosis for the general physician. *Lancet*, **347**, 1641.

Rijntjes M (2006). Mechanisms of recovery in stroke patients with hemiparesis or aphasia: new insights, old questions and the meaning of therapies. *Curr Opin Neurol*, **19**, 76–83.

Rinkel GJE, Wijdicks EFM, Hasan D *et al.* (1991). Outcome in patients with subarachnoid haemorrhage and negative angiography according to pattern of haemorrhage on computed tomography. *Lancet*, **338**, 964–8.

Rinkel GJE, van Gijn J, Wijdicks EFM (1993). Subarachnoid haemorrhage without detectable aneurysm. A review of the causes. *Stroke*, **24**, 1403–9.

Rinkel GJE, Djibuti M, Algra A *et al.* (1998). Prevalence and risk of rupture of intracranial aneurysms: a systematic review. *Stroke*, **29**, 251–6.

Rinkel GJ, Feigin VL, Algra A *et al.* (2004). Circulatory volume expansion therapy for aneurysmal subarachnoid haemorrhage. *Cochrane Database Syst Rev*, **4**, CD000483.

Rinkel GJ, Feigin VL, Algra A *et al.* (2005). Calcium antagonists for aneurysmal subarachnoid haemorrhage. *Cochrane Database Syst Rev*, **1**, CD000277.

Roach ES (2006). Transient global amnesia: look at mechanisms not causes. *Arch Neurol*, **63**, 1338–9.

Rogers LR (2003). Cerebrovascular complications in cancer patients. *Neurol Clin*, **21**, 167–92.

Roldan CA, Shively BK, Crawford MH (1996). An echocardiographic study of valvular heart disease associated with systematic lupus erythematosus. *N Engl J Med*, **335**, 1424–30.

Rolfs A, Bottcher T, Zschiesche M *et al.* (2005). Prevalence of Fabry disease in patients with cryptogenic stroke: a prospective study. *Lancet*, **366**, 1794–6.

Ronthal M, Gonzalez RG, Smith RN *et al.* (2003). Case records of the Massachusetts General Hospital. Weekly clinicopathological exercises. Case 21-2003. A 72 year old man with repetitive strokes in the posterior circulation. *N Engl J Med*, **349**, 170–80.

Rosand J, Eskey C, Chang Y (2002). Dynamic single-section CT demonstrates reduced cerebral blood flow in acute intracerebral hemorrhage. *Cerebrovasc Dis*, **14**, 214–20.

Rosand J, Bayley N, Rost N *et al.* (2006). Many hypotheses but no replication for the association between PDE4D and stroke. *Nat Genet*, **38**, 1091–2.

Ross R (1999). Atherosclerosis - an inflammatory disease. *N Engl J Med*, **340**, 115–26.

Ross Russell RW, Page NGR (1983). Critical perfusion of brain and retina. *Brain*, **106**, 419–34.

Rothwell PM (1994). Investigation of unilateral sensory or motor symptoms: frequency of neurological pathology depends on side of symptoms. *J Neurol Neurosurg Psychiatry*, **57**, 1401–2.

Rothwell PM, Gibson RJ, Slattery J *et al* (1994a). Equivalence of measurements of carotid stenosis: a comparison of three methods on 1001 angiograms. *Stroke*, **25**, 2435–9.

Rothwell PM, Gibson RJ, Slattery J *et al* (1994b). Prognostic value and reproducibility of measurements of carotid stenosis. A comparison of three methods on 1001 angiograms. European Carotid Surgery Trialists' Collaborative Group. *Stroke*, **25**, 2440–4.

Rothwell PM, Warlow CP (1996). Making sense of the measurement of carotid stenosis. *Cerebrovasc Dis*, **6**, 54–8.

Rothwell PM, Wroe SJ, Slattery J *et al* (1996). Is stroke incidence related to season or temperature? *Lancet*, **347**, 934–6.

Rothwell PM, Gibson RJ, Villagra R *et al* (1998). Measurement of carotid stenosis and assessment of plaque surface morphology: Do angiographic technique or image quality matter? *Clin Radiol*, **53**, 439–43.

Rothwell PM, Warlow CP on behalf of the European Carotid Surgery Trialists' Collaborative Group. (1999). Prediction of benefit from carotid endarterectomy in individual patients: a risk modeling study. *Lancet*, **353**, 2105–10.

Rothwell PM, Gibson R, Warlow CP (2000a). The interrelation between plaque surface morphology, degree of stenosis and the risk of ischaemic stroke in patients with symptomatic carotid stenosis. *Stroke*, **31**, 615–21.

Rothwell PM, Villagra R, Gibson R *et al* (2000b). Evidence of a chronic systemic cause of instability of atherosclerotic plaques. *Lancet*, **355**, 19–24.

Rothwell PM (2001a). The Interrelation between carotid, femoral and coronary artery disease. *Eur Heart J*, **22**, 11–4.

Rothwell PM (2001b). The high cost of not funding stroke research: a comparison with heart disease and cancer. *Lancet*, **19**, 1612–6.

Rothwell PM, Gutnikov SA, Eliasziw M *et al* (2003a). Pooled analysis of individual patient data from randomized controlled trials of endarterectomy for symptomatic carotid stenosis. *Lancet*, **361**, 107–16.

Rothwell PM, Howard SC, Spence D (2003c). Relationship between blood pressure and stroke risk in patients with symptomatic carotid occlusive disease. Stroke; **34**, 2583–90.

Rothwell PM, Gutnikov SA, Warlow CP for the ECST (2003b). Re-analysis of the final results of the European Carotid Surgery Trial. *Stroke*, **34**, 514–23.

Rothwell PM (2004). ACST: which subgroups will benefit most from carotid endarterectomy? *Lancet*, **364**, 1122–3.

Rothwell PM, Coull AJ, Giles MF *et al* (2004a). Change in stroke incidence, mortality, case-fatality, severity and risk factors in Oxfordshire, UK from 1981 to 2004 (Oxford Vascular Study). *Lancet*, **363**, 1925–33.

Rothwell PM, Eliasziw M, Gutnikov SA *et al* (2004b). Endarterectomy for symptomatic carotid stenosis in relation to clinical subgroups and timing of surgery. *Lancet*, **363**, 915–24.

Rothwell PM, Howard SC, Power DA *et al* (2004c). Fibrinogen and risk of ischaemic stroke and coronary events in 5183 patients with transient ischaemic attack and minor ischaemic stroke. *Stroke*, **35**, 2300–5.

Rothwell PM (2005). Prevention of stroke in patients with diabetes mellitus and the metabolic syndrome. *Cerebrovasc Dis*, **1**, 24–34.

Rothwell PM, Warlow CP (2005). Timing of TIAs preceding stroke: time window for prevention is very short. *Neurology*, **64**, 817–20.

Rothwell PM, Coull AJ, Silver LE *et al* (2005a). Population-based study of event-rate, incidence, case fatality and mortality for all acute vascular events in all arterial territories (Oxford Vascular Study). Lancet; **366**, 1773–83.

Rothwell PM, Z Mehta, SC Howard *et al* (2005b). From subgroups to individuals: general principles and the example of carotid endarterectomy. *Lancet*, **365**, 256–65.

Rothwell PM, Giles MF, Flossmann E *et al* (2005c). A simple score (ABCD) to identify individuals at high early risk of stroke after a transient ischaemic attack. Lancet; **366**, 29–36.

Rothwell PM, Buchan A, Johnston SC (2006). Recent advances in the management of transient ischaemic attacks and minor ischaemic strokes. Lancet Neurol; **5**, 323–1.

Rubba P, Mercuri M, Faccenda F *et al.* (1994). Premature carotid atherosclerosis: does it occur in both familial hypercholesterolemia and homocystinuria? Ultrasound assessment of arterial intima-media thickness and blood flow velocity. *Stroke*, **25**, 943–50.

Rubinstein SM, Peerdeman SM, van Tulder MW *et al.* (2005). A systematic review of the risk factors for cervical artery dissection. *Stroke*, **36**, 1575–80.

Ruby RJ, Burton JR (1977). Acute reversible hemiparesis and hyponatraemia. *Lancet*, **1**, 1212.

Ruff RL, Talman WT, Petito F (1981). Transient ischaemic attacks associated with hypotension in hypertensive patients with carotid artery stenosis. *Stroke*, **12**, 353–5.

Ruigrok YM, Rinkel GJ, Buskens E (2000). Perimesencephalic hemorrhage and CT angiography: A decision analysis. *Stroke*, **31**, 2976–83

Ruijs AC, Dirven CM, Algra A *et al.* (2005). The risk of rebleeding after external lumbar drainage in patients with untreated ruptured cerebral aneurysms. *Acta Neurochir (Wien)*, **147**, 1157–61.

Rutgers DR, Klijn CJ, Kappelle LJ *et al.* (2004). Recurrent stroke in patients with symptomatic carotid artery occlusion is associated with high-volume flow to the brain and increased collateral circulation. *Stroke*, **35**, 1345–9.

Saadatnia M, Tajmirriahi M (2007). Hormonal contraceptives as a risk factor for cerebral venous and sinus thrombosis. *Acta Neurol Scand*, **115**, 295–300.

Sabolek M, Bachus-Banaschak K, Bachus R et al. (2005). Multiple cerebral aneurysms as delayed complication of left cardiac myxoma: a case report and review. Acta Neurol Scand, 111, 345–50.

Sacco RL, Kargman DE, Gu Q et al (1995). Race-ethnicity and determinants of intracranial atherosclerotic cerebral infarction. The Northern Manhattan Stroke Study. Stroke, 26, 14–20.

Sacco RL (2001). Newer risk factors for stroke. Neurology, 57, S31–4.

Sacco RL, Boden-Albala B, Abel G et al. (2001). Race-ethnic disparities in the impact of stroke risk factors: the Northern Manhattan stroke study. Stroke, 32, 1725–31.

Sacco RL, Chong JY, Prabhakaran S et al. (2007). Experimental treatments for acute ischaemic stroke. Lancet, 369, 331–41.

Sagie A, Larson MG, Levy D (1993). The natural history of borderline isolated systolic hypertension. N Engl J Med, 329, 1912–17.

Salgado AV (1991). Central nervous system complications of infective endocarditis. Stroke, 22, 1461–3.

Samuelsson M, Lindell D, Norrving B (1996). Presumed pathogenetic mechanisms of recurrent stroke after lacunar infarction. Cerebrovasc Dis, 6, 128–36.

Sander K, Sander D (2005). New insights into transient global amnesia: recent imaging and clinical findings. Lancet Neurol, 4, 437–4.

Sandercock PAG, Molyneux A, Warlow C (1985). Value of computed tomography in patients with stroke: Oxfordshire Community Stroke Project. Br Med J, 290, 193–7.

Sandercock PAG, Bamford J, Dennis M et al. (1992). Atrial fibrillation and stroke: Frequency in different stroke types and influence on early and long term prognosis. The Oxfordshire community stroke project. Br Med J, 305, 1460–5.

Sandercock P, Gubitz G, Foley P et al. (2003). Antiplatelet therapy for acute ischaemic stroke. Cochrane Database Syst Rev, 2

Sanna G, Bertolaccini ML, Hughes GR (2005). Hughes syndrome the antiphospholipid syndrome: a new chapter in neurology. Ann N Y Acad Sci, 1051, 465–86.

Sarkari NBS, Holmes JM, Bickerstaff ER (1970). Neurological manifestations associated with internal carotid loops and kinks in children. J Neurol Neurosurg Psychiatry, 33, 194–200.

Sarti C, Stegmayr B, Tolonen H et al. (2003). Are changes in mortality from stroke caused by changes in stroke event rates or case fatality? Results from the WHO MONICA Project. Stroke, 34, 1833–40.

Sato Y, Kuno H, Kaji M et al. (1998). Increased bone resorption during the first year after stroke. Stroke, 29, 1373–7.

Saunders DE, Clifton AG, Brown MM (1995). Measurement of infarct size using MRI predicts prognosis in middle cerebral artery infarction. Stroke, 26, 2272–6.

Savage COS, Harper L, Adu D (1997). Primary systemic vasculitis. Lancet, 349, 553–8.

Saver JL, Johnston KC, Homer D et al. (1999). Infarct volume as a surrogate or auxiliary outcome measure in ischemic stroke clinical trials. The RANTTAS Investigators. Stroke, 30, 293–8.

Savitz SI, Fisher M (2007). Future of neuroprotection for acute stroke: In the aftermath of the SAINT trials. Ann Neurol, 61, 396–402.

Saxonhouse SJ, Curtis AB (2003). Risks and benefits of rate control versus maintenance of sinus rhythm. Am J Cardiol, 91, 27D–32D

Schaefer PW, Copen WA, Lev MH et al. (2005). Diffusion-weighted imaging in acute stroke. Neuroimaging Clin N Am, 15, 503–30.

Schellinger PD, Jansen O, Fiebach JB et al. (1999). A standardized MRI stroke protocol: comparison with CT in hyperacute intracerebral hemorrhage. Stroke, 30, 765–8.

Schierhout G, Roberts I (2000). Hyperventilation therapy for acute traumatic brain injury. Cochrane Database Syst Rev, CD000566.

Schievink WI, Michels VV, Piepgras DG (1994). Neurovascular manifestations of heritable connective tissue disorders. A review. Stroke, 25, 889–903.

Schievink WI (1997). Intracranial aneurysms. N Engl J Med, 336, 28–40.

Schievink WI (2001). Spontaneous dissection of the carotid and vertebral arteries. N Engl J Med, 344, 898–906.

Schimke RN, McKusick VA, Huang T et al. (1965). Homocystinuria Studies of 20 families with 38 affected members. J Am Med Assoc, 193, 711–19.

Schmahmann JD (2003). Vascular syndromes of the thalamus. Stroke, 34, 2264–78.

Schneider AT, Kissela B, Woo D et al. (2004). Ischemic stroke subtypes: a population-based study of incidence rates among blacks and whites. Stroke, 35, 1552–6.

Schriger DL, Kalafut M, Starkman S et al. (1998). Cranial computed tomography interpretation in acute stroke: physician accuracy in determining eligibility for thrombolytic therapy. JAMA, 279, 1293–7.

Schroeder T (1988). Haemodynamic significance of internal carotid artery disease. Acta Neurologica Scandinavia, 77, 353–72.

Schulz UGR, Rothwell PM (2001). Sex differences in carotid bifurcation anatomy and the distribution of atherosclerotic plaque. Stroke, 32, 1525–31.

Schulz UG, Rothwell PM (2002). Transient ischaemic attacks mimicking focal motor seizures. Postgrad Med J, 78, 246–7.

Schulz UGR, Rothwell PM (2003). Differences in vascular risk factors between aetiological subtypes of ischaemic stroke in population-based incidence studies. Stroke, 34, 2050–59.

Schulz UGR, Briley D, Meagher T et al. (2003). Abnormalities on diffusion weighted magnetic resonance imaging performed several weeks after a minor stroke or transient ischaemic attack. JNNP, 74, 734–8.

Schulz UG, Flossmann E, Rothwell PM (2004). Heritability of ischemic stroke in relation to age, vascular risk factors and subtypes of incident stroke in population-based studies. Stroke, 35, 819–24.

Schumann P, Touzani O, Young AR et al. (1998). Evaluation of the ratio of cerebral blood flow to cerebral blood volume as an index of local cerebral perfusion pressure. Brain, 121, 1369–79.

Schwartz TH, Solomon RA (1996). Perimesencephalic nonaneurysmal subarachnoid hemorrhage: review of the literature. Neurosurgery, 39, 433–0.

Scott BL, Jankovic J (1996). Delayed-onset progressive movement disorders after static brain lesions. Neurology, 46, 68–74.

Sedlaczek O. Hirsch JG. Grips E et al. (2004). Detection of delayed focal MR changes in the lateral hippocampus in transient global amnesia. Neurology, 62, 2165–70.

Segal JB, McNamara RL, Miller MR et al. (2001). Anticoagulants or antiplatelet therapy for non-rheumatic atrial fibrillation and flutter. Cochrane Database Syst Rev, 1, CD001938.

Seko Y (2007). Giant cell and Takayasu arteritis. Curr Opin Rheumatol, 19, 39–43.

Sempere AP, Duarte J, Cabezas C et al. (1998). Aetiopathogenesis of transient ischaemic attacks and minor ischaemic strokes: a community-based study in Segovia, Spain. Stroke, 29, 40–5.

Service FJ (1995). Hypoglycaemic disorders. N Engl J Med, 332, 1144–52.

Shanmugam V, Zimnowodzki S, Curtin J et al. (1997). Hypoglycaemic hemiplegia: insulinoma masquerading as stroke. J Stroke Cerebrovasc Dis, 6, 368–9.

Shaper AG, Phillips AN, Pocock SJ et al. (1991). Risk factors for stroke in middle-aged British men. Br Med J 302, 1111–15.

Sheldon JJ (1981). Blood Vessels of the Scalp and Brain. CIBA Pharmaceutical Company, New Jersey.

Sherman DG (2004). Reconsideration of TIA diagnostic criteria. Neurology, 62, S20–1.

Shih LC, Saver JL, Alger JR et al. (2003). Perfusion-weighted magnetic resonance imaging thresholds identifying core, irreversibly infarcted tissue. Stroke, 34, 1425–30.

Shiogai T, Uebo C, Makino M et al. (2002). Acetazolamide vasoreactivity in vascular dementia and persistent vegetative state evaluated by transcranial harmonic perfusion imaging and Doppler sonography. Ann N Y Acad Sci, 977, 445–53.

Shiogai T, Koshimura M, Murata Y *et al.* (2003). Acetazolamide vasoreactivity evaluated by transcranial harmonic perfusion imaging: relationship with transcranial Doppler sonography and dynamic CT. *Acta Neurochir Suppl*, **86**, 57–62.

Shiino A, Morita Y, Tsuji A *et al.* (2003). Estimation of cerebral perfusion reserve by blood oxygenation level-dependent imaging: comparison with single-photon emission computed tomography. *J Cereb Blood Flow Metab*, **23**, 121–35

Siddique MS, Fernandes HM, Wooldridge TD *et al.* (2002). Reversible ischemia around intracerebral hemorrhage: a single-photon emission computerized tomography study. *J Neurosurg*, **96**, 736–41.

Siewert B, Patel MR, Warach S (1995). Magnetic resonance angiography. *The Neurologist*, **1**, 167–84.

Silverstein A, Gilbert H, Wasserman LR (1962). Neurologic complications of polycythaemia. *Ann Intern Med*, **57**, 909–16.

Simmons Z, Biller J, Adams HP Jr *et al.* (1986). Cerebellar infarction: comparison of computed tomography and magnetic resonance imaging. *AnnNeurol*, **19**, 291–3.

Singhal S, Bevan S, Barrick T *et al.* (2004). The influence of genetic and cardiovascular risk factors on the CADASIL phenotype. *Brain*, **127**, 2031–8.

Singhal AB, Benner T, Roccatagliata L *et al.* (2005). A pilot study of normobaric oxygen therapy in acute ischemic stroke. *Stroke*, **36**, 797–802.

Singhal AB (2007). A review of oxygen therapy in ischemic stroke. *Neurol Res*, **29**, 173–83.

Siva A, Altintas A, Saip S (2004). Behcet's syndrome and the nervous system. *Curr Opin Neurol*, **17**, 347–57.

Sivenius J, Riekkinen PJ, Smets P *et al.* (1991). The European Stroke Prevention Study (ESPS): results by arterial distribution. *AnnNeurol*, **29**, 596–600.

Sliwka U, Lingnau A, Stohlmann WD *et al.* (1997). Prevalence and time course of microembolic signals in patients with acute stroke: a prospective study. *Stroke*, **28**, 358–63.

Slivka A, Levy D, Lapinski RH (1989). Risk associated with heparin withdrawal in ischaemic cerebrovascular disease. *J Neurol Neurosurg Psychiatry*, **52**, 1332–6.

Sloan MA, Haley EC (1990). The syndrome of bilateral hemispheric border zone ischaemia. *Stroke*, **21**, 1668–73.

Slovut DP, Olin JW (2004). Fibromuscular dysplasia. *N Engl J Med*, **350**, 1862–71.

Sluzewski M, van Rooij WJ, Beute GN *et al.* (2005). Late rebleeding of ruptured intracranial aneurysms treated with detachable coils. *Am J Neuroradiol*, **26**, 2542–9.

Smith EE, Eichler F (2006). Cerebral amyloid angiopathy and lobar intracerebral hemorrhage. *Arch Neurol*, **63**, 148–51.

Smith ER, Carter BS, Ogilvy CS (2002). Proposed use of prophylactic decompressive craniectomy in poor-grade aneurismal subarachnoid hemorrhage patients presenting with associated large sylvian haematomas. *Neurosurgery*, **51**, 117–24.

Snels IAK, Beckerman H, Twisk JWR *et al.* (2000). Effect of triamcinolone acetonide injections on hemiplegic shoulder pain: A randomized clinical trial. *Stroke*, **31**, 2396–401.

Sorensen AG, Buonanno FS, Gonzalez RG *et al.* (1996). Hyperacute stroke: evaluation with combined multisection diffusion-weighted and haemodynamically weighted echo-planar MR imaging. *Radiology*, **199**, 391–401.

Sotrel A, Lacson AG, Huff KR (1983). Childhood Kohlmeier-Degos disease with atypical skin lesions. *Neurology*, **33**, 1146–51.

Southwick FS, Richardson EP, Swartz M N (1986). Septic thrombosis of the dural venous sinuses. *Medicine (Baltimore)*, **65**, 82–06.

SPACE Collaborative Group (2006). 30 day results from the SPACE trial of stent-protected angioplasty versus carotid endarterectomy in symptomatic patients: a randomised non-inferiority trial. *Lancet*, **368**, 1239–47.

Spence JD, Tamayo A, Lownie SP *et al.* (2005). Absence of microemboli on transcranial Doppler identifies low-risk patients with asymptomatic carotid stenosis. *Stroke*, **36**, 2373–8.

Spencer TS, Campellone JV, Maldonado I *et al.* (2005). Clinical and magnetic resonance imaging manifestations of neurosarcoidosis. *Semin Arthritis Rheum*, **34**, 649–1.

Spetzler RF, Hadley MN, Martin NA *et al.* (1987). Vertebrobasilar insufficiency. Part 1: microsurgical treatment of extracranial vertebrobasilar disease. *J Neurol*, **66**, 648–61.

SSYLVIA Study Investigators (2004). Stenting of Symptomatic Atherosclerotic Lesions in the Vertebral or Intracranial Arteries (SSYLVIA): study results. *Stroke*, **35**, 1388–92

Stam J (2005). Thrombosis of the cerebral veins and sinuses. *N Engl J Med*, **352**, 1791–8.

Stegmayr B, Asplund K, Wester PO (1994). Trends in incidence, case fatality rate, and severity of stroke in Northern Sweden, 1985-1991. *Stroke*, **25**, 1738–45.

Steinberg D (1995). Clinical trials of antioxidants in atherosclerosis: are we doing the right thing? *Lancet*, **346**, 36–8.

Stemmermann GN, Hayashi T, Resch JA *et al.* (1984). Risk factors related to ischaemic and hemorrhagic cerebrovascular disease at autopsy: The Honolulu Heart Study. *Stroke*, **15**, 23–8.

Stephens N (1997). Anti-oxidant therapy for ischaemic heart disease: where do we stand? *Lancet*, **349**, 1710–11.

Stevens RJ, Coleman RL, Adler AI *et al.* (2004). Risk factors for myocardial infarction, case fatality and stroke case fatality in type 2 diabetes: UKPDS 66. *Diabetes Care*, **27**, 201–7.

Stewart SS, Ashizawa T, Dudley AW *et al.* (1988). Cerebral vasculitis in relapsing polychondritis. *Neurology*, **38**, 150–2.

Stocchetti N, Maas AI, Chieregato A *et al.* (2005). Hyperventilation in head injury: a review. *Chest*, **127**, 1812–27.

Stockhammer G, Felber SR, Zelger B *et al.* (1993). Sneddon's syndrome: diagnosis by skin biopsy and MRI in 17 patients. *Stroke*, **24**, 685–90.

Strachan DP, Carrington D, Mendall MA *et al.* (1999). Relation of Chlamydia pneumoniae serology to mortality and incidence of ischaemic heart disease over 13 years in the Caerphilly prospective heart disease study. *Br Med J* **318**, 1035–9.

Strandgaard S, Paulson OB (1992). Regulation of cerebral blood flow in health and disease. *J Cardiovasc Pharmacol*, **19**, S89–93.

Stroke Prevention in Atrial Fibrillation Investigators (1992). Predictors of thromboembolism in atrial fibrillation: II Echocardiographic features of patients at risk. *Ann Intern Med*, **116**, 6–12.

Stroke Prevention in Atrial Fibrillation Investigators (1995). Risk factors for thromboembolism during aspirin therapy in patients with atrial fibrillation: The Stroke Prevention in Atrial Fibrillation Study. *J Stroke Cerebrovasc Dis*, **5**, 147–57.

Stroke Unit Trialists' Collaboration (1997). Collaborative systematic review of the randomised trials of organised inpatient (stroke unit) care after stroke. *Br Med J*, **314**, 1151–9.

Stroke Unit Trialists' Collaboration (2000). Organised inpatient (stroke unit) care for stroke *Cochrane Database Syst Rev*, 2000; **2**, CD000197 Review Update in: *Cochrane Database Syst Rev*, 2002; **1**, CD000197.

Stroke Unit Trialists' Collaboration (2001). Organised inpatient (stroke unit) care for stroke Cochrane Database of Systematic Reviews 2001 Issue 3 Art No: CD000197 DOI: 101002/14651858CD000197.

Sturzenegger M, Mattle HP, Rivoir A *et al.* (1993). Ultrasound findings in spontaneous extracranial vertebral artery dissection. *Stroke*, **24**, 1910–21.

Sturzenegger M, Mattle HP, Rivoir A *et al.* (1995). Ultrasound findings in carotid artery dissection: analysis of 43 patients. *Neurology*, **45**, 691–8.

Subbiah P, Wijdicks E, Muenter M *et al.* (1996). Skin lesion with a fatal neurologic outcome (Degos' disease). *Neurology*, **46**, 636–40.

Sudlow CLM, Warlow CP (1996). Comparing stroke incidence worldwide. What makes studies comparable? *Stroke*, **27**, 550–8.

Sudlow CLM, Warlow CP (1997). Comparable studies of the incidence of stroke and its pathological types. Results from an International collaboration. *Stroke*, **28**, 491–9.

Sudlow C, Martinez Gonzalez NA, Kim J *et al.* (2006). Does apolipoprotein E genotype influence the risk of ischemic stroke, intracerebral hemorrhage, or subarachnoid hemorrhage? Systematic review and meta-analyses of 31 studies among 5961 cases and 17,965 controls. *Stroke*, **37**, 364–70.

Suk SH, Sacco RL, Boden-Albala B *et al.* (2003). Northern Manhattan Stroke Study. Abdominal obesity and risk of ischemic stroke: the Northern Manhattan Stroke Study. *Stroke*, **34**, 1586–92.

Sutherland GR, Auer RN (2006). Primary intracerebral hemorrhage. *J Clin Neurosci*, **13**, 511–7.

Switzer JA, Hess DC, Nichols FT *et al.* (2006). Pathophysiology and treatment of stroke in sickle-cell disease: present and future. *Lancet Neurol*, **5**, 501–12.

Symon L, Branston NM, Chikovani O (1979). Ischemic brain edema following middle cerebral artery occlusion in baboons: relationship between regional cerebral water content and blood flow at 1 to 2 hours. *Stroke*, **10**, 184–91.

Takaya N, Yuan C, Chu B *et al.* (2006). Association between carotid plaque characteristics and subsequent ischemic cerebrovascular events: a prospective assessment with MRI-initial results. *Stroke*, **37**, 818–23.

Takebayashi S, Kaneko M (1983). Electron microscopic studies of ruptured arteries in hypertensive intracerebral haemorrhage. *Stroke*, **14**, 28–36.

Tang T, Howarth SP, Miller SR *et al.* (2006). Assessment of inflammatory burden contralateral to the symptomatic carotid stenosis using high-resolution ultrasmall, superparamagnetic iron oxide-enhanced MRI. *Stroke*, **37**, 2266–70.

Tatlisumak T, Fisher M (1996). Hematologic disorders associated with ischemic stroke. *J Neurol Sci*, **140**, 1–11.

Teasdale GM, Wardlaw JM, White PM *et al.* (2005). The familial risk of subarachnoid haemorrhage. *Brain*, **128**, 1677–85.

Teernstra OP, Evers SM, Lodder J *et al.* (2003). Stereotactic treatment of intracerebral hematoma by means of a plasminogen activator: a multicenter randomized controlled trial (SICHPA). *Stroke*, **34**, 968–74.

Tegeler CH, Shi F, Morgan T (1991). Carotid stenosis in lacunar stroke. *Stroke*, **22**, 1124–8.

Tehindrazanarivelo A, Evrard S, Schaison M *et al* (1992). Prospective study of cerebral sinus venous thrombosis in patients presenting with benign intracranial hypertension. *Cerebrovasc Dis*, **2**, 22–7.

Thanvi B, Robinson T (2006). Sporadic cerebral amyloid angiopathy – an important cause of cerebral haemorrhage in older people. *Age Ageing*, **35**, 565–71.

The Diabetes Control and Complications Trial/Epidemiology of Diabetes Interventions and Complications Research Group (2003). Intensive diabetes therapy and carotid intima-media thickness in type I diabetes mellitus. *N Engl J Med*, **346**, 393–403.

The Hypothermia After Cardiac Arrest Study Group (2002). Mild therapeutic hypothermia to improve the neurologic outcome after cardiac arrest. *New Engl J Med*, **346**, 549–56.

Thie A, Goossens-Merkt H, Freitag J *et al.* (1993). Pulsatile tinnitus: clinical and angiological evaluation. *Cerebrovasc Dis*, **3**, 160–7.

Thijs VN, Albers GW (2000). Symptomatic intracranial atherosclerosis: outcome of patients who fail antithrombotic therapy. *Neurology*, **55**, 490–7.

Thijs VN, Adami A, Neumann-Haefelin T *et al.* (2001). Relationship between severity of MR perfusion deficit and DWI lesion evolution. *Neurology*, **57**, 1205–11.

Thom JT (1993). Stroke mortality trends: an international perspective. *Ann Epidemiol*, **3**, 509–18.

Tietjen GE (2005). The risk of stroke in patients with migraine and implications for migraine management. *CNS Drugs*, **19**, 683–92.

Tomak PR, Cloft HJ, Kaga A *et al.* (2003). Evolution of the management of tentorial dural arteriovenous malformations. *Neurosurgery*, **52**, 750–60.

Tong DC, Yenari MA, Albers GW *et al.* (1998). Correlation of perfusion- and diffusion-weighted MRI with NIHSS score in acute (<65 hour) ischemic stroke. *Neurology*, **50**, 864–70.

Toole JF, Malinow MR, Chambless LE *et al.* (2004). Lowering homocysteine in patients with ischemic stroke to prevent recurrent stroke, myocardial infarction and death: the Vitamin Intervention for Stroke Prevention (VISP) randomized controlled trial. *JAMA*, **291**, 565–75.

Tournier-Lasserve E, Joutel A, Melki J *et al.* (1993). Cerebral autosomal dominant arteriopathy with subcortical infarcts and leukoencephalopathy maps to chromosome 19q12. *Nat Genet*, **3**, 256–9.

Toussaint LG 3rd, Friedman JA, Wijdicks EF *et al.* (2005). Survival of cardiac arrest after aneurysmal subarachnoid hemorrhage. *Neurosurgery*, **57**, 25–31.

Touze E, Warlow CP, Rothwell PM (2006). Risk of coronary and other nonstroke vascular death in relation to the presence and extent of atherosclerotic disease at the carotid bifurcation. *Stroke*, **37**, 2904–9.

Traon AP, Costes-Salon MC, Galinier M *et al.* (2002). Dynamics of cerebral blood flow autoregulation in hypertensive patients. *J Neurol Sci*, **195**, 139–44.

Tunstall-Pedoe H, Kuulasmaa K, Amouyel P *et al.* (1994). Myocardial infarction and coronary deaths in the World Health Organization MONICA Project. Registration procedures, event rates and case-fatality rates in 38 populations from 21 countries in four continents. *Circulation*, **90**, 583–612.

Tuomilehto J, Rastenyte D (1999). Diabetes and glucose intolerance as risk factors for stroke. *J Cardiovasc Risk*, **6**, 241–9.

Tuomilehto J, Lindstrom J, Erikkson JG *et al.* (2001). Prevention of type 2 diabetes mellitus by changes in lifestyle among subjects with impaired glucose tolerance. *N Engl J Med*, **344**, 1343–50.

Turan TN, Stern BJ (2004). Stroke in pregnancy. *Neurol Clin*, **22**, 821–40.

U-King-Im JM, Hollingworth W, Trivedi RA *et al.* (2005). Cost-effectiveness of diagnostic strategies prior to carotid endarterectomy. *Ann Neurol*, **58**, 506–15.

UK National External Quality Assessment Scheme for Immunochemistry Working Group (2003). National guidelines for analysis of cerebrospinal fluid for bilirubin in suspected subarachnoid haemorrhage. *Ann Clin Biochem*, **40**, 481–8.

UK Prospective Diabetes Study (UKPDS) Group (1998). Intensive blood-glucose control with sulphonylureas or insulin, compared with conventional treatment and risk of complications in patients with type 2 diabetes (UKPDS 33). *Lancet*, **352**, 837–53.

Vahedi K, Hofmeijer J, Juettler E *et al.* (2007). Early decompressive surgery in malignant infarction of the middle cerebral artery: a pooled analysis of three randomised controlled trials. *Lancet Neurol*, **6**, 215–2.

van der Grond J, Eikelboom BC, Mali WPThM (1996). Flow-related anaerobic metabolic changes in patients with severe stenosis of the internal carotid artery. *Stroke*, **27**, 2026–32.

van der Meulen JH, Weststrate W, van Gijn J *et al.* (1992). Is cerebral angiography indicated in infective endocarditis? *Stroke*, **23**, 1662–7.

van der Schouw YT, van der Graaf Y, Steyerberg EW *et al.* (1996). Age at menopause as a risk factor for cardiovascular mortality. *Lancet*, **347**, 714–18.

van der Wee N, Rinkel GJE, Hasan D *et al.* (1995). Detection of subarachnoid haemorrhage on early CT: is lumbar puncture still needed after a negative scan? *J Neurol Neurosurg Psychiatry*, **58**, 357–9.

van der Worp HB, Kappelle LJ (1998). Complications of acute ischaemic stroke. *Cerebrovasc Dis*, **8**, 124–32.

van der Zwan A, Hillen B, Tulleken CAF *et al.* (1992). Variability of the territories of the major cerebral arteries. *J Neurol*, **77**, 927–40.

van Everdingen KJ, van der Grond J, Kappelle LJ *et al.* (1998). Diffusion-weighted magnetic resonance imaging in acute stroke. *Stroke*, **29**, 1783–90.

van Gijn J, Rinkel GJ (2001). Subarachnoid haemorrhage: diagnosis, causes and management. *Brain*, **124**, 249–78.

van Gijn J, Kerr RS, Rinkel GJ (2007). Subarachnoid haemorrhage. *Lancet*, **369**, 306–18.

van Heugten C, Visser-Meily A, Post M *et al.* (2006). Care for carers of stroke patients: evidence-based clinical practice guidelines. *J Rehabil Med*, **38**, 153–8.

van Kuijk AA, Geurts AC, Bevaart BJ *et al.* (2002). Treatment of upper extremity spasticity in stroke patients by focal neuronal or neuromuscular blockade: a systematic review of the literature. *J Rehabil Med*, **34**, 51–61.

van Oostenbrugge RJ, Freling G, Lodder J *et al.* (1996). Fatal stroke due to paradoxical fat embolism. *Cerebrovasc Dis*, **6**, 313–14.

van Swieten JC, Kappelle LJ, Algra A *et al.* (1992). Hypodensity of the cerebral white matter in patients with transient ischemic attack or minor stroke: influence on the rate of subsequent stroke. Dutch TIA Trial Study Group. *Ann Neurol*, **32**, 177–83.

van Wijk I, Kappelle LJ, van Gijn J *et al.* (2005). Long-term survival and vascular event risk after transient ischaemic attack or minor ischaemic stroke: a cohort study. *Lancet*, **365**, 2098–4.

Vermeer SE, Rinkel GJE, Algra A (1997). Circadian fluctuations in onset of subarachnoid haemorrhage. New data on aneurysmal and perimesencephalic haemorrhage and a systematic review. *Stroke*, **28**, 805–8.

Vermeulen M, Lindsay KW, van Gijn J (1992). *Subarachnoid Haemorrhage*. Saunders, London.

Vermeulen M, van Gijn J (1990). The diagnosis of subarachnoid haemorrhage. *J Neurol Neurosurg Psychiatry*, **53**, 365–72.

Vessey M, Painter R, Yeates D (2003). Mortality in relation to oral contraceptive use and smoking. *Lancet*, **362**, 185–91.

Vidailhet M, Piette JC, Wechsler B *et al.* (1990). Cerebral venous thrombosis in systemic lupus erythematosus. *Stroke*, **21**, 1226–31

Viles-Gonzalez JF, Fuster V, Badimon JJ (2004). Atherothrombosis: a widespread disease with unpredictable and life-threatening consequences. *Eur Heart J*, **25**, 1197–207.

Villablanca JP, Hooshi P, Martin N *et al.* (2002). Three-dimensional helical computerized tomography angiography in the diagnosis characterization and management of middle cerebral artery aneurysms: comparison with conventional angiography and intraoperative findings. *J Neurosurg*, **97**, 1322–32.

Viskin S, Long QT (1999). Syndromes and torsade de pointes. *The Lancet*, **354**, 1625–33.

Visy JM, Le Coz P, Chadefaux B *et al.* (1991). Homocystinuria due to 5 10-methylenetetrahydrofolate reductase deficiency revealed by stroke in adult siblings. *Neurology*, **41**, 1313–15.

Vollmer TL, Guarnaccia J, Harrington W *et al.* (1993). Idiopathic granulomatous angiitis of the central nervous system. Diagnostic challenges. *Arch Neurol*, **50**, 925–30.

von Kummer R, Meyding-Lamade U, Forsting M *et al.* (1994). Sensitivity and prognostic value of early CT in occlusion of the middle cerebral artery trunk. *Am J Neuroradiol*, **15**, 9–15.

von Kummer R, Holle R, Gizyska U *et al.* (1996). Interobserver agreement in assessing early CT signs of middle cerebral artery infarction. *Am J Neuroradiol*, **17**, 1743–8.

von Kummer R, Allen KL, Holle R *et al.* (1997). Acute stroke: usefulness of early CT findings before thrombolytic therapy. *Radiology*, **205**, 327–3.

von Kummer R, Bourquain H, Bastianello S *et al.* (2001). Early prediction of irreversible brain damage after ischemic stroke at CT. *Radiology*, **219**, 95–100.

Vongpatanasin W, Hillis LD, Lange RA (1996). Prosthetic heart valves. *N Engl J Med*, **335**, 407–16.

Wade DT, Legh-Smith J, Langton Hewer R (1985). Social activities after stroke: measurement and natural history using the Frenchay Activities Index. *Int Rehabil Med*, **7**, 176–81.

Wade DT, Wood VA, Heller A *et al.* (1987). Walking after stroke. Measurement and recovery over the first 3 months. *Scand J Rehabil Med*, **19**, 25–30.

Wahl A, Meier B, Haxel B *et al* (2001). Prognosis after percutaneous closure of patent foramen ovale for paradoxical embolism. *Neurology*, **57**, 1330–2

Wald DS, Law M, Morris JK (2002). Homocysteine and cardiovascular disease: evidence on causality from a meta-analysis. *BMJ*, **325**, 1202–6.

Wald NJ, Law MR (2003). A strategy to reduce cardiovascular disease by more than 80%. *BMJ*, **326**, 1419.

Wald DS, Law M, Morris JK (2004). The dose-response relationship between serum homocysteine and cardiovascular disease: implications for treatment and screening. *Eur J Cardiovasc Prevention Rehab*, **11**, 250–3.

Waldvogel D, Mattle HP, Sturzenegger M *et al.* (1998). Pulsatile tinnitus- a review of 84 patients. *J Neurol*, **245**, 137–42.

Wallis WE, Donaldson I, Scott RS *et al.* (1985). Hypoglycaemia masquerading as cerebrovascular disease (hypoglycaemic hemiplegia*). Ann Neurol*, **18**, 510–12.

Wang X, Tsuji K, Lee SR *et al.* (2004). Mechanisms of hemorrhagic transformation after tissue plasminogen activator reperfusion therapy for ischemic stroke. *Stroke*, **35**, 2726–30.

Wang X, Qin X, Demirtas H *et al.* (2007). Efficacy of folic acid supplementation in stroke prevention: a meta-analysis. *Lancet*, **369**, 1876–82.

Wannamethee SG, Perry IJ, Shaper AG (1999). Non-fasting serum glucose and insulin concentrations and the risk of stroke. *Stroke*, **30**, 1780–6.

Warach S, Kidwell CS (2004). The redefinition of TIA: the uses and limitations of DWI in acute ischemic cerebrovascular syndromes. *Neurology*, **62**, 359–60.

Wardlaw JM, Lewis SC, Dennis MS *et al.* (1998). Is visible infarction on computed tomography associated with an adverse prognosis in acute ischaemic stroke? *Stroke*, **29**, 315–1319.

Wardlaw JM, Dennis MS, Lindley RI *et al.* (1993). Does early reperfusion of a cerebral infarct influence cerebral infarct swelling in the acute stage or the final clinical outcome? *Cerebrovasc Dis*, **3**, 86–93.

Wardlaw JM, Merrick MV, Ferrington CM *et al.* (1996). Comparison of a simple isotope method of predicting likely middle cerebral artery occlusion with Transcranial Doppler ultrasound in acute ischaemic stroke. *Cerebrovasc Dis*, **6**, 32–9.

Wardlaw JM, Zoppo G, Yamaguchi T *et al.* (2003). Thrombolysis for acute ischaemic stroke. *Cochrane Database Syst Rev*, **3**, CD000213.

Warlow CP, Dennis MS, van Gijn J *et al.* (1996a). Stroke: a practical guide to management. *Chapter 5: What Pathological Type of Stroke is it?* pp. 146–89. Blackwell Scientific, Oxford.

Warlow CP, Dennis MS, van Gijn J *et al.* (1996b). Stroke: a practical guide to management. *Chapter 13: Specific Treatment of Aneurysmal Subarachnoid Haemorrhage*, pp. 438–68. Blackwell Scientific, Oxford.

Warlow CP, Dennis MS, van Gijn J *et al.* (1996c). Stroke: a practical guide to management. *Chapter 17: What are this Person's Problems? A Problem-based Approach to the General Management of Stroke*. pp. 477–544. Blackwell Scientific, Oxford.

Warlow CP (1998). Carotid endarterectomy for asymptomatic carotid stenosis: better data but the case is still not convincing. *Br Med J*, **317**, 1468.

Warner DS, Sheng H, Batinic-Haberle I (2004). Oxidants, antioxidants and the ischemic brain. J Exp Biol, **207**, 3221–1.

Weiller C, Chollet F, Friston KJ *et al.* (1992). Functional reorganization of the brain in recovery from striatocapsular infarction in man. *Ann Neurol*, **31**, 463–72.

Weir NU, Pexman JH, Hill MD *et al.* (2006). How well does ASPECTS predict the outcome of acute stroke treated with IV tPA? *Neurology*, **67**, 516–8.

Welin L, Svardsudd K, Wilhelmsen L *et al.* (1987). Analysis of risk factors for stroke in a cohort of men born in 1913. *N Engl J Med*, **317**, 521–6.

Wen HM, Lam WW, Rainer T *et al.* (2004). Multiple acute cerebral infarcts on diffusion-weighted imaging and risk of recurrent stroke. *Neurology*, **63**, 1317–9.

Wen PY, Feske SK, Teoh SK *et al.* (1997). Cerebral haemorrhage in a patient taking fenfluramine and phentermine for obesity. *Neurology*, **49**, 632–3.

Wermer MJ, Kool H, Albrecht KW *et al.* (2007). Aneurysm Screening after Treatment for Ruptured Aneurysms Study Group. Subarachnoid hemorrhage treated with clipping: long-term effects on employment, relationships, personality and mood. *Neurosurgery*, **60**, 91–7.

Werring DJ, Coward LJ, Losseff NA *et al.* (2005). Cerebral microbleeds are common in ischemic stroke but rare in TIA. *Neurology*, **65**, 1914–8.

West C, Hesketh A, Vail A *et al.* (2005). Interventions for apraxia of speech following stroke. *Cochrane Database Syst Rev*, **19**, CD004298.

Wheeler JG, Keavney BD, Watkins H *et al.* (2004). Four paraoxonase gene polymorphisms in 11212 cases of coronary heart disease and 12786 controls: meta-analysis of 43 studies. *Lancet*, **363**, 689–95.

Whelton PK, He J, Cutler JA *et al.* (1997). Effects of oral potassium on blood pressure: Meta-analysis of randomised controlled clinical trials. *J Am Med Assoc*, **277**, 1624–32.

White H, Boden-Albala B, Wang C *et al.* (2005). Ischemic stroke subtype incidence among whites, blacks and Hispanics: the Northern Manhattan Study. *Circulation*, **111**, 1327–31.

Wiebers DO, Whisnant JP, Huston J 3rd *et al.* (2003.) Unruptured intracranial aneurysms: natural history, clinical outcome and risks of surgical and endovascular treatment. *Lancet*, **362**, 103–10.

Wijdicks EFM (1995). Worse-case scenario: management in poor-grade aneurysmal subarachnoid haemorrhage. *Cerebrovasc Dis*, **5**, 163–9.

Wijdicks EFM, Scott JP (1997). Pulmonary embolism associated with acute stroke. *Mayo Clin Proc*, **72**, 297–300.

Wijdicks EF, St Louis E (1997). Clinical profiles predictive of outcome in pontine hemorrhage. *Neurology*, **49**, 1342–6.

Wijdicks EFM, Schievink WI, Brown RD *et al.* (1998). The dilemma of discontinuation of anticoagulation therapy for patients with intracranial hemorrhage and mechanical heart valves. *Neurosurgery*, **42**, 769–73.

Wijdicks EFM, Louis EKS, Atkinson JD *et al.* (2000). Clinician's biases toward surgery in cerebellar hematomas: an analysis of decision-making in 94 patients. *Cerebrovasc Dis*, **10**, 93–6.

Wijman CA, Gomes JA, Winter MR *et al.* (2004). Symptomatic and asymptomatic retinal embolism have different mechanisms. *Stroke*, **35**, 100–2.

Willmot M, Ghadami A, Whysall B *et al.* (2006). Transdermal glyceryl trinitrate lowers blood pressure and maintains cerebral blood flow in recent stroke. *Hypertension*, **47**, 1209–15.

Wintermark M, Uske A, Chalaron M *et al.* (2003). Multislice computerized tomography angiography in the evaluation of intracranial aneurysms: a comparison with intraarterial digital subtraction angiography. *J Neurosurg*, **98**, 828–36.

Wintermark M, Bogousslavsky J (2003). Imaging of acute ischemic brain injury: the return of computed tomography. *CurrOpinNeurol*, **16**, 59–63.

Wintermark M, Meuli R, Browaeys P *et al.* (2007). Comparison of CT perfusion and angiography and MRI in selecting stroke patients for acute treatment. *Neurology*, **68**, 694–7.

Wise RJS, Bernardi S, Frackowiak RSJ *et al.* (1983). Serial observations on the pathophysiology of acute stroke. The transition from ischaemia to infarction as reflected in regional oxygen extraction. *Brain*, **106**, 197–222.

Wityk RJ, Lehman D, Klag M *et al.* (1996). Race and sex differences in the distribution of cerebral atherosclerosis. *Stroke*, **27**, 1974–80.

Wolf PA, Abbott RD, Kannel WB (1991). Atrial fibrillation as an independent risk factor for stroke: the Framingham Study. *Stroke*, **22**, 983–8.

Wolf PA, Abbott RD, Kannel WB (1991a). Atrial fibrillation as an independent risk factor for stroke: the Framingham Study. *Stroke*, **22**, 983–8.

Wolf PA, D'Agostino RB, O'Neal MA *et al.* (1992). Secular trends in stroke incidence and mortality: the Framingham Study. *Stroke*, **23**, 1551–5.

Wolfe CDA (2000). The impact of stroke. *British Medical Bulletin*, **56**, 275–86.

Wolfe CD, Corbin DO, Smeeton NC *et al.* (2006a). Estimation of the risk of stroke in black populations in Barbados and South London. *Stroke*, **37**, 1986–90.

Wolfe CD, Corbin DO, Smeeton NC *et al.* (2006b). Poststroke survival for black-Caribbean populations in Barbados and South London. *Stroke*, **37**, 1991–6.

Wollner L, McCarthy ST, Soper NDW *et al.* (1979). Failure of cerebral autoregulation as a cause of brain dysfunction in the elderly. *Br Med J*, **1**, 1117–18.

Wong KS, Chen C, Ng PW *et al.* (2007). Low-molecular-weight heparin compared with aspirin for the treatment of acute ischaemic stroke in Asian patients with large artery occlusive disease: a randomised study. *Lancet Neurol*, **6**, 407–13.

Woo D, Gebel J, Miller R *et al.* (1999). Incidence rates of first-ever ischemic stroke subtypes among blacks: a population-based study. *Stroke*, **30**, 2517–2.

Woo J, Lam CWK, Kay R *et al.* (1990). Acute and long term changes in serum lipids after acute stroke. *Stroke*, **21**, 1407–11.

Wroe SJ, Sandercock P, Bamford J *et al.* (1992). Diurnal variation in the incidence of stroke: The Oxfordshire Community Stroke Project. *Br Med J*, **304**, 155–7.

Wyller TB, Bautz-Holter E, Holmen J (1994). Prevalence of stroke and stroke-related disability in North Trondelag County Norway. *Cerebrovasc Dis*, **4**, 421–7.

Yadav JS, Wholey MH, Kuntz RE *et al* (2004). Protected carotid-artery stenting versus endarterectomy in high-risk patients. *N Engl J Med*, **351**, 1493–501.

Yamauchi H, Fukuyama H, Nagahama Y *et al.* (1996). Evidence of misery perfusion and risk for recurrent stroke in major cerebral arterial occlusive diseases from PET. *J Neurol Neurosurg Psychiatry*, **61**, 18–25.

Yasaka M, Minematsu K, Naritomi H (2003). Predisposing factors for enlargement of intracerebral hemorrhage in patients treated with warfarin. *Thromb Haemost*, **89**, 278–83.

Yasui T, Komiyama M, Iwai Y *et al.* (2005). A brainstem cavernoma demonstrating a dramatic spontaneous decrease in size during follow-up: case report and review of the literature. *Surg Neurol*, **63**, 170–3.

Yonekawa Y, Kahn N (2003). Moyamoya disease. *Adv Neurol*, **92**, 113–8.

Yozbatiran N, Cramer SC (2006). Imaging motor recovery after stroke. *Neuro Rx*, **3**, 482–8.

Yuh WT, Crain MR, Loes DJ *et al.* (1991). MR imaging of cerebral ischemia: findings in the first 24 hours. *Am J Neuroradiol*, **12**, 621–9.

Yusuf S, Phil D, Sleight DM *et al.* (2000). Effects of an Angiotensin-coverting-enzyme inhibitor Ramipril on cardiovascular events in high-risk patients. *N Engl J Med*, **342**, 145–53.

Zajicek JP (2000). Neurosarcoidosis. *Curr Opin Neurol*, **13**, 323–5.

Zanette EM, Roberti C, Mancini G *et al.* (1995). Spontaneous middle cerebral artery reperfusion in ischaemic stroke. A follow-up study with transcranial Doppler. *Stroke*, **26**, 430–3.

Zeman AZJ, Boniface SJ, Hodges JR (1998). Transient epileptic amnesia: a description of the clinical and neuropsychological features in 10 cases and a review of the literature. *J Neurol Neurosurg Psychiatry*, **64**, 435–43.

Zemke D, Smith JL, Reeves MJ *et al.* (2004). Ischemia and ischemic tolerance in the brain: an overview. *Neurotoxicology*, **25**, 895–904.

Zhang F, Yin W, Chen J (2004). Apoptosis in cerebral ischemia: executional and regulatory signalling mechanisms. *Neurol Res*, **26**, 835–45.

Zhang X, Patel A, Horibe H *et al.* (2003). Cholesterol coronary heart disease and stroke in the Asia Pacific region. *Int J Epidemiol*, **32**, 563–72.

Zhang-Nunes SX, Maat-Schieman ML, van Duinen SG *et al.* (2006). The cerebral beta-amyloid angiopathies: hereditary and sporadic. *Brain Pathol*, **16**, 30–9.

Zheng Z, Yenari MA (2004). Post-ischemic inflammation: molecular mechanisms and therapeutic implications. *Neurol Res*, **26**, 884–92.

Zhu XL, Chan MSY, Poon WS (1997). Spontaneous intracranial haemorrhage: which patients need diagnostic cerebral angiography? A prospective study of 206 cases and review of the literature. *Stroke*, **28**, 1406–9.

Zuber M, Khoubesserian P, Meder JF *et al.* (1993). A 34-year delayed and focal postirradiation intracranial vasculopathy. *Cerebrovasc Dis*, **3**, 181–2.

Zukerman D, Selim R, Hochberg E (2006). Intravascular lymphoma: the oncologist's 'great imitator'. *Oncologist*, **11**, 496–502.

CHAPTER 36

Vasculitis and collagen vascular diseases

Neil Scolding

Contents

36.1 Introduction

That part of the clinical interface between neurology and general medicine occupied by inflammatory and immunological diseases is neither small nor medically trivial. Neurologists readily accept the challenges of 'primary' immune diseases of the nervous system (Table 36.1): these tend to be focussed on one particular target such as oligodendrocytes or the neuro-muscular junction present in predictable ways, and are amenable as a rule to rational, methodological diagnosis, and occasionally even treatment. This is proper neurology.

'Secondary' neurological involvement in diseases mainly considered systemic inflammatory conditions—for example, SLE, sarcoidosis, vasculitis, and Behçet's—is a rather different matter. It may be difficult enough to secure such a diagnosis even when systemic disease has previously been diagnosed and new neurological features need to be differentiated from iatrogenic disease, particularly

Table 36.1 Immunological and inflammatory diseases of the nervous system and muscle.

	Primary Disease	Secondary Disease
Brain (and spinal cord)	Multiple sclerosis	Vasculitides
Spinal cord	Inflammatory myelitides Stiff Man Syndrome	Lupus, rheumatoid disease; other connective tissue diseases, anti-cardiolipin syndromes
Peripheral nerve	Guillain–Barré syndrome and variants Chronic inflammatory demyelinating polyneuropathy Multifocal motor neuropathy	Behçet's Sarcoid Paraneoplasia
Neuromuscular junction	Myasthenia gravis	Organ-specific autoimmune disease (e.g. coeliac, Hashimoto's disease, etc.)
Muscle	Polymyositis Dermatomyositis	

drug side effects or the consequences of immune suppression. But all the diseases mentioned may present with and confine themselves wholly to the nervous system; they may mimic one another, and pursue erratic and unpredictable clinical courses. In central nervous system disease, diagnosis by tissue biopsy is potentially hazardous and unattractive. Few neurologists enjoy excesses of confidence or expertise when faced with such clinical problems: the cautious diagnostician is perplexed, and the evidence-based neuroprescriber confounded. Unsurprisingly, great variations in approaches to diagnosis and management are seen (Scolding *et al.* 2002b).

But rheumatologically inclined general, renal or respiratory physicians, comfortable when managing inflammation affecting their system or indeed other parts of the body designed to support the nervous system, are generally also ill at ease when faced with neurological features whose differential diagnosis may be large, particularly given the near universal diagnostic non-specificity of either imaging or CSF analysis.

Here then is the subject material for this chapter: the diagnosis and management of central nervous system involvement in inflammatory and immunological systemic diseases (Scolding 1999a). In not one of these neurological conditions has a single controlled therapeutic trial been reported, and much that is published on these conditions is misleading or inaccurate. And yet the frequency with which the diagnosis is only confirmed or even first emerges at autopsy bears stark witness to both the severity and evasiveness of these disorders.

36.2 Central nervous system vasculitis

36.2.1 Classification and histopathology

Vasculitis is not a diagnosis, still less a disease. It is best thought of as a histopathological description: intramural inflammation, often but not invariably with perivascular changes too, and necrosis of the blood vessel wall. It occurs in the context of a group of disorders, the vasculitides, whose classification is complex (Lie 1997b), with subdivisions into idiopathic primary vasculitic disorders, for example, giant cell arteritis and Wegener's granulomatosis, and those vasculitides secondary to collagen diseases, malignancy, viral infection and according to pathological features, largely by vessel size (Table 36.2).

Nervous system involvement can occur in any of the systemic vasculitides. Additionally, isolated vasculitis of the central or peripheral nervous system is recognized, where little or no inflammation is apparent outside the nervous system: termed central or peripheral nervous system angiitis (Calabrese *et al.* 1988).

36.2.2 Primary central nervous system vasculitis

In primary central nervous system vasculitis, vasculitis is confined to the brain and spinal cord, and there is no discernible systemic vasculitic or other disease. Although defined by this apparently exclusive distribution, autopsies have revealed subclinical extracranial involvement for instance of pulmonary arteries and abdominal viscera (Lie 1997b) presumably helping to explain the occasional features of fever, rigors, weight loss, or raised plasma viscosity.

The angiitic process is focal and segmental in distribution, and granulomatous, necrotizing, or lymphocytic in character, often with mixed morphologic types in individual patients, therefore rendering the common title of 'granulomatous angiitis' difficult to sustain.

Table 36.2 Classification of vasculitis according to the dominant vessels involved

	Primary	Secondary
Large arteries	Giant cell arteritis Takayasu's arteritis	Aortitis with rheumatoid disease; infection (e.g. syphilis)
Medium arteries	Classical polyarteritis nodosa Kawasaki disease	Infection (e.g. Hepatitis B)
Small vessels and medium arteries	Wegener's granulomatosis Churg–Strauss syndrome Microscopic polyangiitis	Vasculitis with rheumatoid disease, systemic lupus erythematosus, Sjögren's syndrome, drugs, infection (e.g. HIV)
Small vessels	Henoch–Schönlein purpura Essential cryoglobulinaemia Cutaneous leucocytoclastic vasculitis	Drugs (e.g. sulphonamides, etc.) Infection (e.g. Hepatitis C)

The clinical definition of cerebral vasculitis is not uniform (Scolding *et al.* 2002b), and this helps to explain significant differences in approach to diagnosis and therapy. Some define the disorder by its angiographic appearances, explicitly not requiring tissue confirmation. Indeed some have suggested a more favourable monophasic clinical course in the so-called 'benign angiopathy of the central nervous system'. This is a syndrome with normal, or only mildly abnormal CSF, and evidence of a vasculitic picture on angiography alone (Calabrese *et al.* 1993). The concept has been questioned in view of the recognized non-specificity of angiography, the fact that those cases not proceeding to biopsy are more likely to be the less severely affected, and because children satisfying 'benign angiopathy' criteria often do not have a temperate, monophasic course, and have required aggressive immunotherapy (Gallagher *et al.* 2001). Most, however, including the current author, believe a certain diagnosis of primary central nervous system angiitis must depend on positive biopsy.

Cogan's syndrome is an unusual disorder, mostly affecting young adults, characterized by recurrent episodes of interstitial keratitis and/or scleritis with vestibulo-auditary symptoms, which may be complicated by central or peripheral nervous involvement or systemic vasculitis (Vollertsen 1990).

Eale's disease is an isolated retinal vasculitis causing visual loss; again, neurological complications are rarely described (Dastur and Singhal 1976; Singhal and Dastur 1976).

36.2.3 Primary systemic vasculitis with neurological involvement

Virtually all of the systemic vasculitides may be complicated by neurological involvement; often they carry their own defining characteristics. Systemic disturbances—fever, night sweats, severe malaise, weight loss, rash, or arthropathy—are common.

Wegener's granulomatosis predominantly affects the upper and lower respiratory tracts, including the nose, causing the characteristic saddle nose deformity (Fig. 36.1). Renal disease occurs in 80 per cent of patients. The anti-neurophil cytoplasmic antibody, cANCA

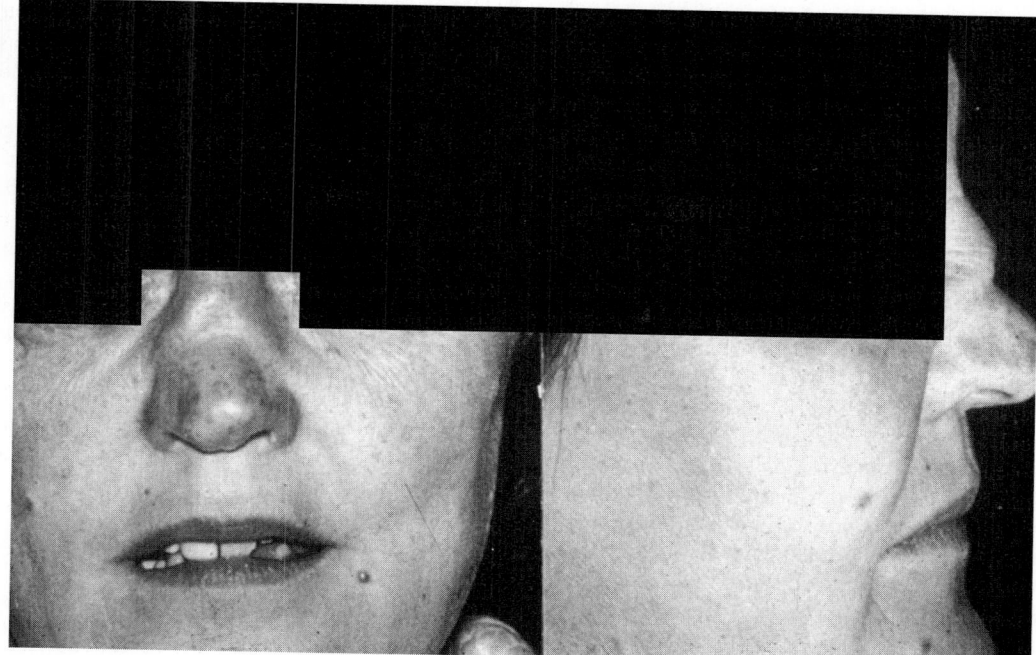

Fig. 36.1 Wegener's granulomatosis. Characteristic saddle nose deformity.

is positive, with proteinase-3 specificity, and biopsy shows a necrotizing, granulomatous vasculitis. Neurological involvement occurs in up to 35 per cent of patients (Nishino *et al.* 1993) but usually of the peripheral nervous system (Section 21.15.3), cerebral small-vessel vasculitis is rare. More likely is the unique contiguous extension of erosive granulomata from the sinuses or from remote metastatic granulomata to the central nervous system. Meningeal and middle ear disease may lead to significant cranial neuropathies especially of VII and VIII. Orbital pseudotumour may occur. Gadolinium-enhanced MR scanning may valuably reveal meningeal thickening and infiltration, offering a ready target for biopsy.

Microscopic polyangiitis is a pANCA-positive multisystem small vessel vasculitis. Renal disease is again common, indeed this vasculitis is occasionally confined to the kidney, but upper respiratory tract involvement is not, and granulomata are not seen. One study found mononeuritis multiplex in 55 per cent (Guillevin *et al.* 1999), but central nervous system disease was uncommon, in 11 per cent, and rarely contributes to mortality.

Classical polyarteritis nodosa is an unusual disorder with medium- and small-sized muscular artery involvement in multiple organs, excepting the lungs. Patients often present with renal failure and hypertension in 80 per cent. Gastrointestinal involvement occurs in up to 50 per cent, with abdominal pain due to visceral infarcts. Heart failure and myocardial infarction reflect cardiac involvement. Neurological abnormalities are prominent occurring in 50–60 per cent, but again mostly confined to the peripheral nervous system (Section 21.15.3). About 20–30 per cent have hepatitis B antigen or antibody in serum. Visceral angiography shows aneurysms or occlusions.

Churg–Strauss syndrome is characterized by hypereosinophilia with systemic vasculitis in individuals with recently developed atopy. Asthma and mononeuritis multiplex are frequent (Section 21.15.3). Rashes, with purpura, urticaria, and subcutaneous nodules,

are common. Glomerulonephritis may develop, and coronary, splanchnic and cerebral circulations are often involved. Central nervous system involvement is evident in only about 7 per cent of the cases (Sehgal *et al.* 1995). About 50 per cent of patients are positive for pANCA, 25 per cent positive for cANCA, and 25 per cent ANCA negative.

Henoch–Schönlein purpura is an immunologically mediated small vessel systemic vasculitis of children, affecting the skin, gastrointestinal tract, joints, and kidneys. Neurological involvement is well-described, with haemorrhage and/or encephalopathy, often apparently responsive to plasmapheresis, but although usually labelled as cerebral vasculitis, there is no reported tissue proof of the central nervous system process.

Kawasaki disease or the Mucocutaneous lymph node syndrome, usually affects children under 12 years, and is rare in the United Kingdom, <5/100 000/year, but far commoner in Japan. Coronary artery aneurysms occur in a fifth of untreated cases, and may result in myocardial infarction. Neurologically, there is commonly an aseptic meningitis, but hemiplegic strokes, encephalopathy, and facial palsy are also described. Pathologically, an acute systemic inflammatory vasculitis, with little or no fibrinoid necrosis, underlies the disease. Anti-endothelial cell antibodies may be involved.

36.2.4 Secondary central nervous system

Autoimmune and inflammatory disease. Neuropsychiatric disease is common in systemic lupus erythematosus (Section 36.3.1), occurring in 40 to 50 per cent (Scolding 1999c), but very rarely caused by a vasculitic process and the casually used 'lupus vasculitis' therefore is almost invariably an incorrectly applied epithet.

Sarcoidosis. This may be complicated by systemic vasculitis affecting small or large calibre vessels, with angiographic and indeed histological evidence of central nervous system vasculitis. Serum angiotensin converting enzyme and calcium levels are not always raised. CSF abnormalities are seen in 80 per cent of the cases, usually with an elevated protein and pleocytosis, and oligoclonal bands are positive in about 45 per cent of them. Cranial MRI shows non-specific white matter lesions or meningeal enhancement; whole body gallium scanning may reveal characteristic parotid gland and lung changes.

Seropositive rheumatoid disease is a well-recognized precipitant of cerebral vasculitis (Scolding 1999c), though skin involvement and mononeuritis multiplex are far more typical manifestations of rheumatoid vasculitis. There are also rare reports of central nervous system angiitis in *systemic sclerosis, Sjögren's syndrome*, and *mixed connective tissue disease*, though rarely without a preceding history of systemic symptoms, and even more rarely with tissue confirmation.

Behçet's disease is also considered in Section 36.5.1.

Infections. Infection-related vasculitis (Lie 1996) occurs through direct invasion of the vessel wall by Aspergillus, histoplasma or coccidioides, immune complex deposition involving hepatitis B, Epstein–Barr virus, cytomegalovirus, Lyme disease, syphilis, or malaria, and/or secondary cryoglobulinaemia. 'Hepatitis B and C, Epstein–Barr virus, cytomegalovirus, Lyme disease, syphilis, malaria, and coccidiomycosis all have been linked to mixed cryoglobulinaemia.

In HIV infection, cytomegalovirus and toxoplasma may precipitate vasculitis, and syphilitic cerebral vasculitis has re-emerged. Bacterial causes of meningoencephalitis, mycobacteria, pneumococci, and *H. influenzae*, may also trigger intracranial vasculitis.

In herpes zoster ophthalmicus, secondary, localized central nervous system vasculitis affecting the ipsilateral hemisphere occurs in approximately 0.5 per cent of the cases, probably by direct viral invasion of blood vessels (Hilt *et al*. 1983), producing single or multiple smooth-tapered segmental narrowing on angiography. The characteristic clinical picture is an acute monophasic hemiparesis contralateral to the, usually resolving, ocular disease. The latent period is usually 3–4 weeks. A CSF mononuclear pleocytosis and raised varicella-zoster antibody titre aid the diagnosis. A generalized necrotizing, granulomatous vasculitis can also occur.

Complications of shingles may affect children similarly, though there have been less frequent reports of chicken pox triggering cerebral or spinal vasculitis.

Coccidioides immitis spores which are endemic to the southwestern United States and Northern Mexico, can be inhaled with subsequent haematogenous spread, often to the meninges. Vasculitis involving the small penetrating branches of the major cerebral vessels, and consequent deep ischaemic infarction (de Carvalho *et al*. 1980), has been observed in up to 40 per cent of these cases, rarely with subarachnoid haemorrhage.

Malignancy, lymphomatoid granulomatosis, and malignant angioendothelioma. Leucocytoclastic vasculitis may occur in association with a variety of cancers as a paraneoplastic phenomenon. Central nervous system angiitis in the context of Hodgkin's disease is also reported. *Lymphomatoid granulomatosis* is a lymphomatous disorder centred on the vascular wall, with destructive change and secondary inflammatory infiltration lending the appearance of true vasculitis; the infiltrating neoplastic cell is derived from the T-lymphocyte. Cutaneous and pulmonary involvement are common, with nodular cavitating lung infiltrates, and neurological manifestations occur in 25–30 per cent of cases; they are the presenting feature in approximately 20 per cent of them. *Neoplastic or*

malignant angioendotheliosis is also a rare, nosologically separate disorder, wherein the neoplastic process is intravascular, i.e. within the lumen, and the lymphomatous cells are derived from the B-cell, and characteristically do not invade the vascular wall. The neurological features of each disorder are similar, largely representing those of cerebral vasculitic disease; in malignant angioendotheliomatosis lung involvement is not the rule; characteristic skin manifestations occur.

Drug- and toxin-induced cerebral vasculitis. The most compelling evidence relates to amphetamines (Section 5.3.4), with clinical and histological evidence of multisystem necrotizing vasculitis (Citron *et al*. 1970). Vasculitis may follow only a single dose, but repeated amphetamine exposure is the usual history. However, many reports include no tissue confirmation, with a diagnosis of 'vasculitis' based on angiography despite the fact that vasospasm causes identical angiographic changes. In cocaine abuse (Section 5.3.2), the significantly increased risk of ischaemic stroke results from vasospasm, probably from increased catecholamine release, and very seldom from vasculitis (Aggarwal *et al*. 1996). Co-injected contagion such as hepatitis C may cause vasculitis.

Rarely, an immune reaction against spontaneous amyloid deposits within the cerebral vasculature appear to precipitate a true central nervous system vasculitis, a recently described disorder which has been termed Aβ-related angiitis, ABRA (Scolding *et al*. 2005).

36.2.5 Clinical features, investigation, and diagnosis

Variably focal or multifocal infarction, or diffuse ischaemia, of any part of the central nervous system, explain the protean manifestations, wide variation in disease severity, and the absence of a pathognomic or even typical clinical picture. Consequently, most accounts of both primary and secondary intracranial vasculitis accurately but unhelpfully describe headaches, focal or generalized seizures, stroke-like episodes with hemispheric or brainstem deficits, acute or subacute encephalopathies, progressive cognitive changes, chorea, myoclonus and other movement disorders, and optic and other cranial neuropathies. The course can be acute or subacute, but chronic progressive presentations are well-described, as are spontaneous relapses and remissions. Systemic features such as fever, night sweats, livedo reticulares, or oligoarthropathy may also be present, often revealed only on direct questioning. Although included in many accounts, the clinical picture of conventional large vessel stroke, sufficiently like atheromatous thromboembolic stroke to cause diagnostic confusion, is profoundly uncommon.

Despite this clinical diversity of presentation, three broad categories, carrying neither pathological nor therapeutic implications, have been defined in the hope of improving recognition of the condition (Scolding *et al*. 1997):

- Acute or subacute encephalopathy, with confusion, drowsiness, or coma;
- Superficially resembling atypical multiple sclerosis, "multiple sclerosisplus", in clinical phenotype alone, with a relapsing-remitting course, with optic neuropathy and brainstem episodes, but other features less typical of multiple sclerosis such as seizures, severe headaches, or encephalopathic episodes;
- Intracranial mass lesions.

Clinical suspicion having been raised, the diagnosis of cerebral vasculitis involves exclusion of alternative possibilities, confirmation of intracranial vasculitis, and pursuit of the cause of the vasculitic process. No single simple investigation can confirm a diagnosis of cerebral vasculitis; some can exclude it. The diagnostic process can be difficult but practical approaches have been proposed. (Joseph *et al.* 2002)

Blood tests and serology. Anaemia is an infrequent finding, but leucocytosis is present in about 50 per cent of patients. The ESR and C-reactive protein levels are often raised, but lack specificity. Serological testing for ANA, ANCA or rheumatoid factor is vital, but of little value in confirming isolated cerebral vasculitis. 'False' ANCA positivity is sometimes seen in connective tissue disorders such as lupus and rarely in individuals without vasculitis.

Spinal fluid examination. CSF analysis is non-specific but again useful in suggesting an inflammatory central nervous system process, and excluding infection and malignant diseases. An elevated cell count, usually a lymphocytosis, and/or protein occurs in 50–80 per cent of the cases (Calabrese *et al.* 1988; Hankey 1991; Scolding 1999b). The CSF opening pressure is raised in almost half the cases of primary angiitis of the central nervous system. Oligoclonal immunoglobulin bands are found in up to 40–50 per cent of them (Scolding *et al.* 1997); band patterns vary discernibly over the course of disease, occasionally even disappearing altogether, helping to exclude multiple sclerosis when part of the differential diagnosis.

Imaging. MRI is a non-specific detector of vascular disease (Harris *et al.* 1994), disclosing of course the results of vascular inflammation, not vasculitis itself. Ischaemic areas, periventricular white matter lesions, haemorrhagic lesions, and parenchymal- or meningeal-enhancing areas can be seen. Sensitivity is by no means perfect: in one study, of 50 territories affected by vasculitis on contrast angiography, at least one- third were normal on MRI (Cloft *et al.* 1999). Unsurprisingly, therefore, histopathologically proven cerebral vasculitis with normal MR imaging is well-recognized.

MR angiography is valuable in imaging of large vessel vasculitides such as Takayasu's arteritis and classical polyarteritic nodosa, with potential to supplant contrast angiography (Atalay and Bluemke 2001), but has insufficient resolution to help in medium or small vessel cranial vasculitis.

Establishing the diagnostic value of contrast angiography is complicated by the many reports which use this as the 'gold standard' for diagnostic confirmation. However, correlative radiological-pathological studies show a false negative rate of 30–40 per cent (Hankey 1991; Hellmann *et al.* 1992; Alrawi *et al.* 1999) which is explained not least by the small size of the affected vessels.

When abnormalities are present, they include segmental multifocal narrowing, with areas of localized dilatation or beading (Fig. 36.2). In PACNS, single stenotic areas in multiple vessels are more frequent than multiple stenotic areas along a single vessel. Retrospective series suggest a specificity of around 33 per cent—an enormous number of other inflammatory, metabolic, malignant, and other vasculopathies accurately mimic angiitis.

Some have reported an angiographic risk of transient, in 10 per cent, or permanent neurological deficit, in 1 per cent (Hellmann *et al.* 1992). Although its diagnostic importance has been overemphasized, contrast angiography remains a valuable investigational tool.

Indium-labelled white cell nuclear scanning has a limited role, particularly in disclosing areas of sometimes unsuspected systemic inflammation (Scolding *et al.* 1997).

Ophthalmological examination. Careful ocular examination, including slit lamp study, is vital. Subclinical conjunctival, anterior or posterior inflammation, or retinal changes may, occasionally following conjunctival biopsy, confirm ocular, and thereby imply neurological, sarcoidosis, Behçet's disease, or other inflammatory disorders. Dynamic examination of the episcleral vasculature, and of red cell flow therein, using video slit lamp microscopic recording and low-dose fluorescein angiography, can be a useful additional investigation (Scolding *et al.* 1997).

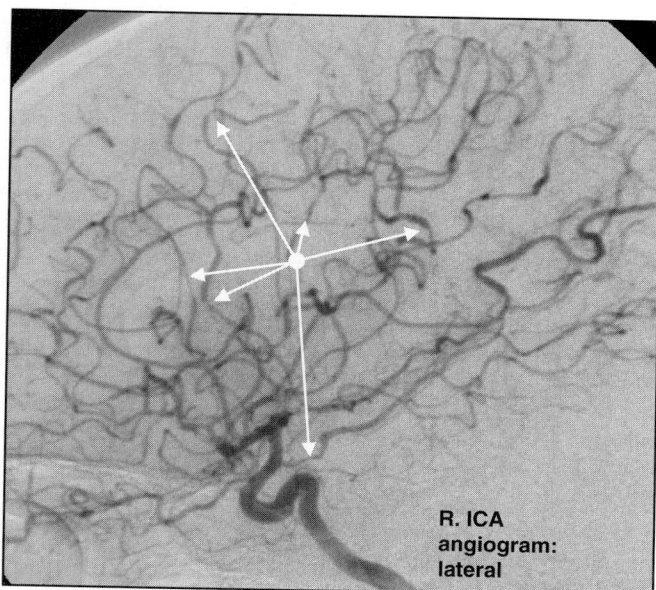

R. ICA angiogram: lateral

Fig. 36.2 Central nervous system vasculitis. Typical Contrast angiography changes in a 24 yr old patient. The arrows point to areas of segmental narrowing, irregularity or occlusion affecting multiple vessels. (Courtesy of Dr S Renowden.)

Table 36.3 Some neurological and systemic disorders which may mimic cerebral vasculitis

Other vasculopathies	Infections
Susac's syndrome	Lyme disease
Homocysteinuria	HIV
Ehlers–Danlos syndrome	Endocarditis
Radiation vasculopathy	Whipple's disease
Köhlmeyer–Degos disease	Viral encephalitis
Fibromuscular dysplasia	Legionella/mycoplasma pneumonia
Fabry's disease	
Moyamoya disease	**Tumours and malignancy**
Amyloid angiopathy	Atrial myxoma
Marfan's syndrome	Multifocal glioma
Pseudoxanthoma elasticum	Cerebral lymphoma
Viral or fungal vasculitis	Paraneoplastic disease

Other immune/inflammatory diseases	Other Multiple cholesterol emboli
Sarcoidosis	Thrombotic thrombocytopoenic
Lupus and anti-phospholipid disease	purpura
Behçet's syndrome	Cerebral sinus thrombosis
Multiple sclerosis/ADEM	Mitochondrial disease
Thyroid encephalopathy	

Key: CADASIL: Cerebral autosomal dominant arteriopathy with subcortical infarcts and leucoencephalopathy.
ADEM: Acute & isseminated encephalomyelitis.

Histopathology. Histopathological confirmation, biopsying an abnormal area of brain where possible, or 'blind' biopsy of meninges, and non-dominant temporal white and grey matter, is important. Currently, up to 75 per cent of reported cases are 'diagnosed' without histopathology (Alrawi *et al.* 1999). A recent series of 61 patients biopsied for suspected cerebral vasculitis reported that none suffered any significant procedure-related morbidity. In this study, 36 per cent were confirmed as having cerebral vasculitis, but importantly, 39 per cent biopsies showed alternative, unsuspected diagnoses: lymphoma in 6 cases, multiple sclerosis in 2, or infection in 7 cases, including toxoplasmosis, herpes, and also two cases of cerebral abscess. Biopsy failed to yield a clear diagnosis in 25 per cent of patients, though even here, biopsy arguably might not be described as 'non-contributory', at least helping to exclude the alternative diagnoses mentioned above. The decision not to biopsy must be balanced against the harmful effects of immunosuppressive drugs used, potentially, unnecessarily.

A vasculitic process having been confirmed, the specific defining characteristics of the primary and secondary vasculitides must be carefully sought.

36.2.6 Mechanisms of tissue damage in vasculitis

Microvascular ischaemia and infarction, including that within vasa vasora, results from three consequences of inflammation within the vascular wall: obstruction of the lumen, increased coagulation from changes on the endothelial surface, and alterations in vaso-motor tone. The development of a vasculitic process depends on interplay between cellular and humoral immune factors, most research interest having centred on the latter.

Antibody-dependent mechanisms:

- *Direct antibody attack.* In some systemic vasculitides, a pathogenic role for anti-endothelial cell antibodies in injuring or, paradoxically, activating endothelial cells is proposed (Salojin *et al.* 1996), though their lack of specificity and variable frequency of detection do raise questions about any truly causal role. Rarely, antibodies against amyloid-ß deposits may possibly precipitate cerebral vasculitis (Scolding *et al.* 2005)

- *Immune-complex-mediated vasculitis.* Immune complex deposition in the blood vessel wall triggers complement activation, thereby generating lytic, injurious membrane attack complexes, and triggering polymorph and macrophage recruitment. Hepatitis-B- and C-associated vasculitis are good examples

- *ANCA-related vasculitis.* ANCAs are antibodies directed against components of the neutrophil granules (Mohan and Kerr 2001). Cytoplasmic ANCA, c-ANCA, targets proteinase-3, and is associated with nearly 95 per cent specificity for Wegener's granulomatosis. Perinuclear ANCA, p-ANCA, directed at myeloperoxidase, is less specifically found in microscopic polyangiitis and Churg–Strauss syndrome. Such antibodies may play a role in generating and maintaining vascular inflammation (Xiao *et al.* 2002; Harper *et al.* 2004).

Cell-mediated damage. Evidence for cell-mediated involvement in tissue injury in vasculitis comes in part from studying microscopic polyarteritis nodosa and Wegener's granulomatosis. In both disorders, circulating T-cells responsive to PR-3 are found, and vascular lesions contain activated T-cells and antigen presenting major histocompatibility complex Class II positive dendritic cells (Lie 1997a). In primary central nervous system and peripheral nerve vasculitic lesions, the predominant infiltrate is one of CD4-positive and CD8-positive T-lymphocytes and monocytes (Griffith and Pusey 2001).

36.2.7 The treatment of cerebral vasculitis

Most would consider cerebral vasculitis a highly treatable condition, though controlled trials are there none, not least because of the rarity of the condition and the absence of unifying diagnostic criteria. Retrospective analyses, and lessons from systemic vasculitides (Adu *et al.* 1997; Jayne *et al.* 2003), lend significant support for the use of steroids with cyclophosphamide. In biopsy-proven disease, a reasonable induction regimen might comprise high-dose steroids with intravenous methyl prednisolone, 1g daily for 3 days, then oral prednisolone 60 mg/day, decreasing by 10 mg at weekly intervals to 10 mg/day if possible, coupled from the outset with cyclophosphamide 2.5 mg/kg/day reduced to 2 mg/kg in the elderly, or in cases of renal failure). This induction combination is suggested for 9–12 weeks. Pulsed weekly intravenous cyclophosphamide appears to differ insignificantly in efficacy from daily oral treatment, and may have fewer side effects. Careful monitoring of the blood count for evidence of bone marrow suppression should force a reduction in the cyclophosphamide dose if there is leucopoenia $< 4.0 \times 10^9$ or neutropoenia $< 2.0 \times 10^9$.

Cyclophosphamide is associated with haemorrhagic cystitis which is less likely with adequate hydration and MESNA cover, a 33-fold increase in bladder cancer, other malignancies, infertility, cardiotoxicity, and pulmonary fibrosis. In a report of 145 Wegener's disease patients treated with cyclophosphamide, and followed for 1333 patient-years, non-glomerular haematuria occurred in approximately 50 per cent of the patients, the majority of whom had macroscopic changes consistent with cyclophosphamide-induced bladder injury on cystoscopy (Talar-Williams *et al.* 1996). Seven of these, and none without haematuria, developed transitional cell bladder carcinoma; six had had a total cumulative dose over 100 g cyclophosphamide, and a duration of oral treatment exceeding 2.7 years.

Maintenance treatment, converting to alternate day prednisolone 10–20mg, and substituting cyclophosphamide with azathioprine 2 mg/kg/day, is commenced after induction, and continued for a further 10–12 months; it is then gradually withdrawn. Azathioprine is less toxic, but reversible bone marrow suppression can occur, hepatotoxicity is rare, and there is a small increased risk of malignancies. Deterioration, failure to respond initially, or intolerance of the above regimen may require the use of alternative agents. *Methotrexate* at 10–25 mg doses on a weekly basis may be used in conjunction with steroids, either during induction or maintenance. *Plasmapheresis* may be valuable in cryoglobulinaemia, and is also considered in severe systemic vasculitis causing pulmonary haemorrhage and severe glomerulonephritis, with 7–10 treatments over 14 days. Although there is little experience of its use in patients with intracranial disease, there is evidence of significant improvement when used in combination with steroids in cerebral disease associated with Henoch–Schönlein purpura.

A number of monoclonal antibodies directed against the CD52 (present on most lymphocytes), CD20 (B-cells), or against tumour necrosis factor are generating excitement as novel therapies in various inflammatory diseases including the vasculitides. Paradoxically the induction of vasculitis has also been reported with various of these agents (Mathieson *et al.* 1990; Unger *et al.* 2003; Booth *et al.* 2004; Mohan *et al.* 2004; Sneller 2005). Interferon-α can control not only hepatitis-C-associated hepatitis, but also associated cryoglobuli-naemia and vasculitis. Unfortunately, there is regular relapse within months of treatment withdrawal.

36.2.8 Giant cell arteritis

Giant cell arteritis includes two histologically similar but clinically distinct diseases: temporal arteritis (Section 18.7.1) and Takayasu's arteritis. Temporal arteritis is a chronic inflammatory disorder targeting large- and medium-sized arteries, which rarely affects individuals less than 55 years of age, women twice as commonly as men, with an overall prevalence of 100/105. It has an annual incidence of 17.4 per 100 000 in the over 50-year-old population, so that new onset unilateral or bilateral headache in this age group should alert the physician.

Classically it manifests as temporal headache with tender, pulseless, nodular temporal arteries, in addition, often elicited only on direct enquiry, there are symptoms of general malaise, jaw claudication, and features of polymyalgia rheumatica with stiffness and aching of the shoulder girdle, worse in the mornings, and malaise. Neuro-ophthalmological symptoms are widely feared, with blindness occurring in one-sixth of treated patients with the condition (Caselli et al. 1994). This occurs as a consequence of anterior ischaemic optic neuropathy following vasculitic involvement of the posterior ciliary arteries and/or the ophthalmic artery, from which they are derived. The typical picture comprises locally painless loss of acuity, commonly severe, often with an altitudinal field defect. The fundal appearances may be normal, although mild swelling can be seen. Conventionally, the inflammatory process is thought to involve only the extracranial vessels and rarely to extend beyond the point of penetration of the dura. A large study of 166 patients with biopsy-proven temporal arteritis demonstrated neurological involvement in 31 per cent, describing the comprehensive range of neurological manifestation more usually associated with other forms of intracranial vasculitis: neuropsychiatric syndromes, peripheral neuropathies, mononeuropathies, spinal cord lesions, neuro-otological syndromes, various pain syndromes, transient ischaemic attacks, and stroke. However most authorities would find almost all these pictures outside their common experience of giant cell arteritis. Infarction of the vertebrobasilar territory was relatively uncommon, but there have been isolated reports of temporal arteritis presenting as lateral medullary syndrome. The expected greater incidence of cerebrovascular disease in this senior subgroup may, however, be confounding.

The affected temporal artery (-ies) may be thickened and cord-like, often non-pulsatile, and tender. A raised ESR, often accompanied by a normochromic normocytic anaemia, *must* be followed by temporal artery biopsy. A specimen of several-centimetre length is recommended to help avoid false-negative results, which may occur because of the multifocal nature of the disorder. Histopathological examination of the vessel reveals an inflammatory infiltrate comprising mononuclear and giant cells; the latter phagocytose elastic laminae, causing characteristic fragmentation. Immunoglobulin and complement deposits are apparent in lesions, but activated T-cells predominate in the inflammatory infiltrate, suggesting cell-mediated immune damage. Vasculitic changes may still be apparent in biopsies taken 14 days or more after the commencement of steroids.

The ESR may be used to monitor treatment response. However, it has been pointed out that a low ESR in active disease is not excessively rare (Salvarani and Hunder 2001), and may perhaps be explained by an inability to mount an acute phase response, or by very localized arteritis. Measuring serum interleukin-6 levels is a promising alternative to the ESR. Recent work has also emphasized that an elevated platelet count should be considered a risk factor for permanent visual loss in temporal arteritis and should accentuate the need for urgent treatment (Lincoff et al. 2000).

Steroid-resistance is extremely rare. Fear of permanent blindness encourages most to prescribe an immediate starting dose of 60–80 mg of oral prednisolone daily although prospective studies have shown lower doses of 20 mg to be as effective. After 4–7 days on a high dose, gradual reduction by perhaps 5 mg weekly should be attempted to a maintenance dose of approximately 10 mg daily, using the clinical response and ESR or plasma viscosity as a guide. Most authorities recommend continuing steroids for a period of 12–24 months; some patients still require steroids 2–5 years later. The importance of preventing long-term consequences of cortico-steroids, in particular bone protection from osteoporosis, must be stressed. Azathioprine is often used as a steroid-sparing agent.

36.2.9 Takayasu's arteritis

Takayasu's arteritis was originally described in young Oriental women; it is now globally recognized. It is alternatively named 'pulse-less disease', since 98 per cent of affected individuals have at least one major arterial pulse absent, as a result of the characteristic involvement of the aorta and its large branches. The disease process is initially inflammatory, and later occlusive, during which phase most of the neurological abnormalities occur. Syncope is reported in at least 50 per cent of patients, but also seen are strokes, transient ischaemic attacks, and visual abnormalities, all exacerbated by hypertension. One should suspect this illness in a patient under the age of 40 years with symptoms of limb claudication, one or more absent pulses, systolic blood pressure difference between the arms of >10 mmHg, and the presence of arterial bruits. Early histological features of the disease include granulomatous changes in the media and adventitia of the aorta and its branches, later followed by intimal hyperplasia, medial degeneration, and sclerotic adventitial fibrosis.

36.3 Collagen vascular diseases and the central nervous system

36.3.1 Systemic lupus erythematosus

Systemic lupus erythematosus, or SLE, is one of a family of rheumatic disorders, which include rheumatoid arthritis, scleroderma, polymyositis, dermatomyositis, and Sjögren's syndrome. It is a chronic relapsing-remitting autoimmune disease of uncertain aetiology (Croker and Kimberly 2005). It may cause widespread organ involvement with consequences ranging from mild and transient to life-threatening. glomerulonephritis, pleurisy, and pneumonitis; pericarditis and so-called Libmann–Sachs endocarditis. Haematological disorders also occur: anaemia, thrombocytopoenia, leucocytopoenia, and the generation of circulating anticoagulants. Other laboratory abnormalities include the presence of a variety of auto-antibodies, including anti-nuclear antibodies and anti-native DNA antibodies.

The diagnosis, particularly for research and therapeutic trial purposes, is based on the widely accepted American College of Rheumatology revised diagnostic criteria (Table 36.4).

Table 36.4. American College of Rheumatology diagnostic criteria systemic lupus erythematosus (SLE). A person shall be said to have SLE if four or more of the 11 criteria are present, serially or simultaneously, during any interval of observation'

- malar flush
- discoid rash
- photosensitivity
- oral ulcers
- arthritis
- serositis (—pleurisy or pericarditis)
- renal disorder (proteinuria >0.5 g/24 h or cellular casts)
- neurological disorder (seizures, psychosis; other causes excluded)
- haematological disorder (haemolytic anaemia, leucopoenia or lymphopoenia on 2 or more occasions, or thrombocytopoenia)
- immunological disorder—LE cells, or anti-dsDNA or anti-Sm or persistent false positive syphilis serology
- anti-nuclear autoantibodies

Neurological complications. Neurological involvement in systemic lupus erythematosus is seen in perhaps 50 per cent of cases; neurological *presentation*, in perhaps 3 per cent. Central nervous system disease is much commoner than neuromuscular involvement, and is a poor prognostic sign, reducing the overall survival figures, and representing the third commonest cause of death, after renal and iatrogenic death.

An enormous variety of central nervous system complications can occur, reflecting two broad pathogenetic mechanisms:

- thromboembolic triggered either by changes in endothelial surfaces, or by coagulation disturbances, including lupus anticoagulant activity
- more direct autoimmune events affecting the target tissue— neurones or glia—in which soluble and cellular mediators are implicated (Scolding *et al.* 2002a; Meroni *et al.* 2003).

The classic histopathological studies of the nervous system in systemic lupus erythematosus are those of Johnson and Richardson (1968) and Ellis and Verity (1979). Studying between them almost a hundred cases, they reported strikingly concordant changes: small vessel infarction, particularly in the cerebral cortex and brainstem, correlating well with the clinical signs in most cases. These authors gave a clear and in essence unchallenged description (Scolding *et al.* 2002a) of small blood vessel changes, with evidence of necrosis of the vessel wall, extravasations of fibrin and red blood cells together with endothelial cell proliferation, hypertrophy, and the appearance of fibrin thrombi. These destructive and proliferative changes were responsible for the numerous areas of microinfarction and, less commonly, macroscopic haemorrhage. Importantly, vasculitis was found only rarely in approximately 5 per cent of cases.

Headache, including that associated with dural sinus thrombosis, acute or subacute encephalopathy, fits, myelitis which is often severe and dramatic, strokes and movement disorders, ataxia and brainstem abnormalities, and cranial and peripheral neuropathies are all seen in the context of systemic lupus erythematosus chorea, though often quoted as the classical neurological feature in systemic lupus erythematosus, occurs in no more than 4 per cent of patients. Aseptic meningitis and optic neuropathy are also both well-documented. Psychiatric and cognitive disturbances have also long been associated with systemic lupus erythematosus.

Stroke, the lupus anticoagulant, and the primary phospholipid syndrome. The thrombotic tendency in systemic lupus erythematosus with lupus anticoagulant manifests principally as stroke and recurrent spontaneous abortion. Intra-abdominal, deep venous, and peripheral arterial thrombosis are less common. Thrombocytopoenia is a key additional feature. Importantly, Hughes also showed that a similar clinical picture was associated with the presence of anticardiolipin antibodies and/or lupus anticoagulant in patients *without* serological or clinical evidence of systemic lupus erythematosus, and introduced the term 'antiphospholipid syndrome' or lupus anticoagulant.

Anticardialipin antibodies are an independent risk factor for stroke. Central nervous system thrombosis in patients with primary or secondary antiphospholipid syndrome takes the form of completed arterial stroke, repeated transient ischaemic attactes, multi-infarct dementia, and cerebral venous sinus thrombosis. Vascular visual problems, including amaurosis fugax, and ischaemic retinopathy, also occur. Chorea too is associated with anti-phospholipid antibodies; but the putative link with migraine may be factitious.

A very severe, indeed commonly fatal *acute ischaemic encephalopathy* is also described, with confusion, obtundation, spastic quadriparesis which is usually asymmetrical, with or without systemic disturbances usually dermatological and renal. CSF examination may show only a raised protein. The disorder may represent a focal variant of the recently described *catastrophic anti-phospholipid syndrome*, in which there is severe multi-organ failure and a mortality of the order of 60 per cent (Asherson 2005).

Diagnosis of central nervous system lupus. Serological tests lie at the heart of diagnosis. A number of studies have suggested a link between anti-ribosomal P antibodies and neurological or psychiatric systemic lupus erythematosus (Trysberg *et al.* 2004). CSF may contain a raised protein level and a neutrophil or lymphocyte pleocytosis. It is clearly vital in such cases to exclude infectious complications of immune suppressants or steroids, now a major cause of death in patients with systemic lupus erythematosus. Oligoclonal band analysis is positive in up to 50 per cent patients but, interestingly, can vanish with successful immunotherapy. MRI changes are common, though neither invariable nor specific. A skin biopsy can be extremely helpful in suspected lupus.

Management of neuropsychiatric lupus. Symptomatic therapies are vital in patients with encephalopathies, epilepsy, and/or psychiatric ailments. Disease-modifying therapeutic efforts fall into two categories depending on the presumed underlying mechanisms— stroke prevention in cerebral ischaemia, particularly that associated with anticardiolipin antibody, considered best achieved with moderate- to high-dose warfarin, and immunotherapy of 'other' central nervous system complications. Here, intravenous methyl prednisolone, followed by oral steroid treatment is the mainstay of treatment. Cyclophosphamide may be exhibited for severe or steroid-resistant disease, with azathioprine to maintain remission and spare steroids. Mycophenolate appears a more than reasonable substitute for both induction and maintenance therapy (Chan *et al.* 2000; Ong *et al.* 2005) Plasmapheresis synchronized with cyclophosphamide, and intravenous immunoglobulin, may prove useful, especially in more 'malignant' forms of the disease (Uthman *et al.* 2005). More recent attention has focussed on the promising role of the anti-B-cell monoclonal antibody Rituximab (Ginzler and Dvorkina 2005; Sfikakis *et al.* 2005; Thatayatikom *et al.* 2006)— perhaps not surprising in view of the pathognomic production of autoantibodies in systemic lupus erythematosus. The apparent increase in disease activity following anti-tumour necrosis factor-α therapy (De Bandt M. *et al.* 2005), whilst disappointing in the short term, may hold valuable clues as the cause of tissue damage.

There are both clinical and pathological similarities between microangiopathic complications of lupus and the syndrome of *thrombotic thrombocytopoenic purpura*. In this uncommon disorder, multi-organ involvement is also seen, with hepatic and renal disease, and fever, together with thrombocytopoenia and an associated purpuric rash, and other haemorrhagic complications. Neurologically, an encephalopathy occurs, often with fits, with or without focal deficits. Pathologically, there are widespread microangiopathic changes in the brain and systemically. Plasma exchange is commonly recommended.

36.3.2 Rheumatoid arthritis

An inflammatory peripheral neuropathy occurs in approximately 30 per cent seropositive rheumatoid cases, but central nervous system disease, either rheumatoid vasculitis, or deposition of rheumatoid nodules, is much less common. Pannus formation and cervical spine subluxation can result in cord compression. Recent excitement has emerged concerning the potential therapeutic role of anti-TNF and anti-B-cell therapies (Furst *et al.* 2005; Higashida *et al.* 2005).

36.3.3 Sjögren's syndrome

Sjögren's syndrome comprises a triad of (i) keratoconjunctivitis sicca, and (ii) xerostomia, occurring in (iii) the context of another connective tissue in approximately 50 per cent of cases, usually rheumatoid arthritis. Speckled antinuclear antibodies, anti-Ro, SS-A, or anti-La, SS-B, are present in most patients. Conventionally, the principal neurological manifestations are considered peripheral, with descriptions of both a mainly sensory neuropathy (Sections 21.13.1 and 21.18.10) including the classical trigeminal sensory neuropathy, and of myositis.

However, attention has also been drawn to apparent central nervous system complications, with seizures, focal stroke-like or brainstem neurological deficits, and an encephalopathy with or without an aseptic meningitis, often with raised CSF pressure, protein level, and white cell count, together with oligoclonal immunoglobulin bands. Psychiatric abnormalities may occur; spinal cord involvement may be seen, with acute transverse myelitis, chronic myelopathy, or intraspinal haemorrhage. Occasionally, the features resemble those of multiple sclerosis, optic neuropathy is particularly associated), though most such patients have additional features of peripheral neuropathy or myositis. Steroids and the usual more powerful immunosuppressants are recommended, again with no sound evidence base.

36.3.4 Systemic sclerosis

Systemic sclerosis results from the excessive deposition of collagen in the skin and other affected tissues, of unknown cause (Sakkas 2005). Scleroderma may exist in isolation, but in multisystem disease, it is accompanied by Raynaud's phenomenon, calcinosis and atrophy of subcutaneous tissues, telangiectasia, and oesophageal strictures. Neurological complications are not common. Peripheral nervous system disease predominates, particularly painful trigeminal neuropathy; myopathy, with an elevated creatine kinase also occurs. A myelopathy may be associated. No treatment is of proven benefit, but evidence implicating B-cells in the pathogenesis has stimulated

interest in the potential of monoclonal antibodies such as infliximab in therapy (Sakkas 2005).

36.3.5 Mixed connective tissue disease

In this disorder, features of scleroderma, polymyositis, and systemic lupus erythematosus coincide with high levels of antibodies directed against extractable nuclear antigens, ribonucleoproteins. Rheumatoid factor is also often present. Again, trigeminal neuralgia and/or sensory neuropathy are described.

36.3.6 Seronegative arthritides

Ankylosing spondylitis. Neurological involvement usually reflects advanced bony disease; a cauda equina syndrome is well-reported, unexplained, and difficult to treat.

Reiter's disease. Central nervous system disorders associated with this syndromic triad of seronegative arthropathy, non-specific urethritis, and conjunctivitis, usually following venereal infection or dysentery, include aseptic meningoencephalitis, seizures, and psychiatric disturbances, particularly paranoid psychosis, neurological features occurring in up to 25 per cent of patients. Cranial neuropathies, pyramidal signs, and myelopathy are also reported. Peripherally, radiculitis and polyneuritis occur. Cyclosporine may be of value in severe disease.

36.4 Sarcoidosis

36.4.1 Pathogenesis

Sarcoidosis is a multisystem granulomatous disease whose aetiology continues to defy explanation. Rational collaborative approaches to study have failed to elucidate causal factors (Semenzato 2005). Paradoxically, the wholly accidental finding of an exacerbation of sarcoid disease activity in patients treated with interferon-α (Leclerc *et al.* 2003) may ultimately prove informative, not least since the mononuclear phagocyte, the predominant cell type in the non-caseating epithelioid cell granuloma, is both highly responsive to, and produces interferon-α.

36.4.2 Clinical features of neurosarcoidosis

Pulmonary and/or systemic features in sarcoidosis are accompanied in approximately 5 per cent of patients, by nervous system involvement (Zajicek *et al.* 1999). Optic and other cranial neuropathies, especially the facial nerve, often due to meningeal infiltration, are the commoner manifestations. Aseptic meningitis may feature, and brainstem and spinal cord disease can occur. Cognitive and neuropsychiatric abnormalities are reported. The region of the pituitary gland appears peculiarly sensitive, and consequent endocrine, hypothalamic, or optic chiasmatic symptomatology are seen. Peripheral nerve and muscle involvement is also well-described (Section 21.18.6).

36.4.3 Investigation and diagnosis of neurosarcoidosis

The diagnosis can be difficult in isolated central nervous system disease, or patients with previously confirmed systemic sarcoidosis who have received chronic immune-modulating therapy and whose neurological presentation could therefore alternatively represent opportunistic infection or even lymphoma. Serum and CSF angiotensin converting enzyme levels may be elevated, but not reliably. The CSF may reveal wholly non-specific elevations of protein or cell count, and oligoclonal bands may be present. Whole body gallium scanning remains a useful indicator of systemic disease, and may disclose clinically cryptic areas of activity suitable and safe for biopsy. Cranial MRI may show multiple white matter lesions or meningeal enhancement and the latter also may be amenable to biopsy. Tissue diagnosis is invaluable where feasible.

36.4.4 Treatment and clinical course

The mainstay of medical treatment in neurosarcoidosis is corticosteroids, though response rates as low as 29 per cent have been reported. Methotrexate (Baughman *et al.* 2000), azathioprine, hydroxychloroquine, and cyclophosphamide (Doty *et al.* 2003) have been used in steroid-resistant cases though meta-analysis has confirmed the lack of an evidence base for even systemic disease (Paramothayan *et al.* 2003). Tumour necrosis factor inhibition appears promising (Pritchard *et al.* 2004; Baughman *et al.* 2005; Doty *et al.* 2005).

36.5 Other vasculopathies and inflammatory diseases

36.5.1 Behçet's disease

Behçet's disease is a chronic relapsing multisystem inflammatory disorder classically causing a triad of recurrent uveitis with oral and genital aphthous ulceration. Formal diagnostic criteria include recurrent oral ulceration occurring at least three times in one 12-month period as an absolute criterion, plus any two of (i) recurrent genital ulceration, (ii) anterior or posterior uveitis or retinal vasculitis, (iii) skin lesions, including erythema nodosum, or acneiform nodules, pseudofolliculitis or papulopustular lesions, or (iv) a positive pathergy test read at 24–48 h.

Neurological involvement is not uncommon and is well-documented (Akman-Demir *et al.* 1999). Cerebral venous sinus thrombosis is one of the more specific serious complications; others include sterile meningoencephalitis, encephalopathy, brainstem syndromes, cranial neuropathies, and cortical sensory and motor deficits. Psychiatric and progressive cognitive manifestations are reported. Investigation may reveal an active CSF, and oligoclonal IgA and IgM bands, but apparently not IgG, may be present. Evoked potentials may be diagnostically useful. MRI abnormalities are non-specific (Fig. 36.3).

Recent retrospective studies indicate an improved survival in patients with central nervous system Behçet's treated with steroids and immunosuppressants. The place of thalidomide in steroid-unresponsive Behçet's is currently under review; chlorambucil is often advocated. Monitoring treatment is difficult as neither the ESR nor C-reactive protein levels are useful; MRI might have such a role. Anti-tumour necrosis factor therapy is recently reported of value (Sarwar *et al.* 2005).

36.5.2 Vogt–Koyanagi–Harada syndrome

This is a rare inflammatory disorder, not truly vasculitic, whose principal features are ocular, with uveitis and retinal haemorrhages, and dermatological, with patches of depigmentation in the eyebrows, eyelashes, and scalp hair "poliosis", as well as vitiligo (Moorthy *et al.* 1995). The disorder is relatively benign, but senorineural deafness is common, and other cranial neuropathies can occur, particularly involving ocular motor nerves. However, aseptic meningitis is the main neurological manifestation. Neuropsychiatric changes may occur, as may signs of parenchymal involvement, including hemiplegia and transverse myelitis.

Spinal fluid analysis usually reveals a lymphocytic pleocytosis with a variably raised protein. CT and MRI may reveal characteristic choroidal changes, and in some patients MRI shows high signal density periventricular lesions (Ibanez and Pettigrew 1994). Both MR angiography and contrast angiography have shown vasculopathic changes suggestive of vasculitis (Ryan and Pettigrew 1995), but the neuropathological changes, whilst rarely reported, include inflammation, with perivascular cuffing, a brisk microglial reaction, and no true vasculitis (Alema *et al.* 1981).

A close Class II major histocompatibility complex antigen association has been reported to DR4 and DQ4 (Islam *et al.* 1994). The ocular aqueous fluid contains a predominance of activated T-lymphocytes, and peripheral cellular and humoral immune responses to uveal and retinal antigens have been described. Melanocytes represent the most obvious target and melanin-laden macrophages are found in the cerebrospinal fluid (Nakamura and Yano 1996). Considerable evidence of cytotoxic T-cell activity against these cells has been presented (Norose and Yano 1996; Sugita *et al.* 1996).

Conventional treatment comprises high-dose intravenous steroids (Ikeda *et al.* 1997); cyclosporine has been advocated in refractory cases. Intravenous immunoglobulin has also been suggested (Helveston and Gilmore 1996). It has been observed that 60 per cent of patients retain moderate to good vision (Rubsamen and Gass 1991).

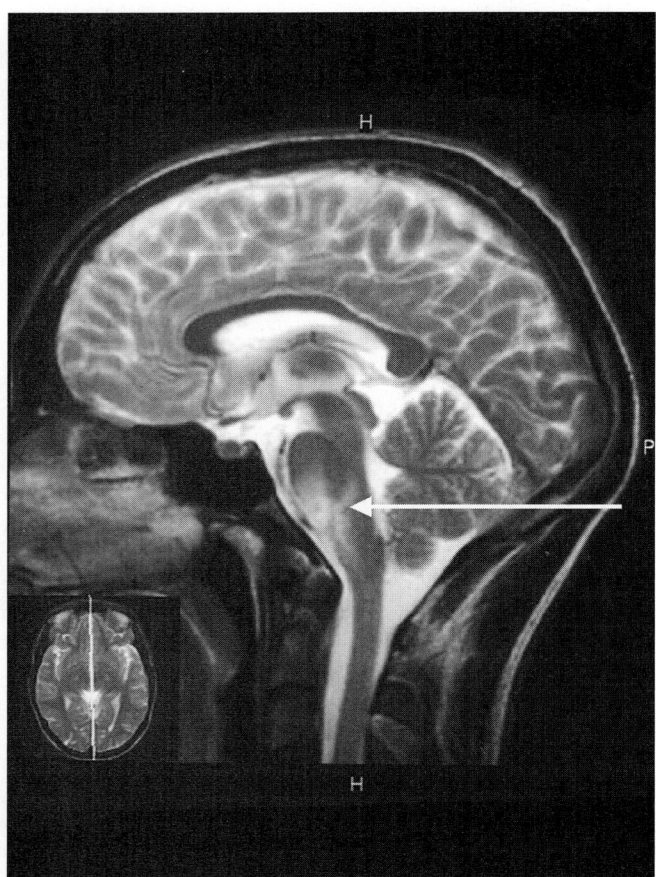

Fig. 36.3 Central nervous system Behçet's disease. MRI shining Ponto-medullary brain stem involvement (arrow).

36.5.3 Susac's syndrome

Susac's syndrome is an unusual, histologically non-inflammatory microvasculopathy, predominantly affecting females, and recognizable by a triad of deafness, retinal microinfarction, and encephalopathy (Susac *et al.* 1979; Susac 1994). Peripheral indices of systemic inflammation are not elevated, and spinal fluid examination shows no cellular reaction; oligoclonal immunoglobulin bands are not present. Most case reports initially described a monophasic, self-limiting course, but as clinical recognition of the syndrome increased, it became clear that recurrent attacks variably including one or more of the ocular, cerebral, or cochlear manifestations represented the more typical course. Chronic stable deficits in vision, hearing, and/or cognition commonly persist. High-dose steroids, and various immune suppressant or immune modulatory therapies have been exhibited and are empirically recommended. It is of unknown cause; one case is reported in which the Factor V Leiden mutation was present (Barker *et al.* 1999).

36.5.4 Köhlmeyer–Degos syndrome

Degos disease is a rare non-inflammatory occlusive vasculopathy particularly involving the skin, gastrointestinal tract, and the central nervous system, causing a combination of papular skin lesions, perforation of the gut and peritonitis, and haemorrhagic strokes (Dastur *et al.* 1981; McFarland *et al.* 1978; Subbiah *et al.* 1996).

References

Adu D, Pall A, Luqmani RA *et al.* (1997). Controlled trial of pulse versus continuous prednisolone and cyclophosphamide in the treatment of systemic vasculitis. *QJM*, **90**, 401–9.

Aggarwal SK, Williams V, Levine SR *et al.* (1996). Cocaine-associated intracranial hemorrhage: absence of vasculitis in 14 cases. *Neurology*, **46**, 1741–3.

Akman-Demir G, Serdaroglu P, Tasci B (1999). Clinical patterns of neurological involvement in Behcet's disease: evaluation of 200 patients. The Neuro-Behcet Study Group. *Brain*, **122**, 2171–82.

Alema G, Appicciutoli L, Corsi FM *et al.* (1981). Vogt-Koyanagi-Harada syndrome. Clinical and neuropathological report of a case. *Acta Neurol Napoli*, **3**, 680–6.

Alrawi A, Trobe J, Blaivas M *et al.* (1999). Brain biopsy in primary angiitis of the central nervous system. *Neurology*, **53**, 858–60.

Asherson RA (2005). Multiorgan failure and antiphospholipid antibodies: the catastrophic antiphospholipid (Asherson's) syndrome. *Immunobiology*, **210**, 727–33.

Atalay MK, Bluemke DA (2001). Magnetic resonance imaging of large vessel vasculitis. *Curr Opin Rheumatol*, **13**, 41–7.

Barker RA, Anderson JR, Meyer P *et al.* (1999). Microangiopathy of the brain and retina with hearing loss in a 50 year old woman: extending the spectrum of Susac's syndrome (In Process Citation). *J Neurol Neurosurg Psychiatry*, **66**, 641–3.

Baughman RP, Lower EE, Bradley DA *et al.* (2005). Etanercept for refractory ocular sarcoidosis: results of a double-blind randomized trial. *Chest*, **128**, 1062–47.

Baughman RP, Winget DB, Lower EE (2000). Methotrexate is steroid sparing in acute sarcoidosis: results of a double blind, randomized trial. *Sarcoidosis Vasc Diffuse Lung Dis*, **17**, 60–6.

Booth A, Harper L, Hammad T *et al.* (2004). Prospective study of TNFalpha blockade with infliximab in anti-neutrophil cytoplasmic antibody-associated systemic vasculitis. *J Am Soc Nephrol*, **15**, 717–21.

Calabrese LH, Gragg LA, Furlan AJ (1993). Benign angiopathy: a subset of angiographically defined primary angiitis of the central nervous system. *J Rheumatol*, **20**, 2046–50.

Calabrese LH, Mallek JA (1988). Primary angiitis of the central nervous system. Report of 8 new cases, review of the literature, and proposal for diagnostic criteria. *Medicine*, **67**, 20–39.

Caselli RJ, Hunder GG (1994). Neurologic complications of giant cell (temporal) arteritis. *Semin Neurol*, **14**, 349–53.

Chan TM, Li FK, Tang CSO *et al.* (2000). Efficacy of mycophenolate mofetil in reply to: patients with diffuse proliferative lupus nephritis. *N Engl J Med*, **343**, 1156–62.

Citron BP, Halpern M, McCarron M *et al.* (1970). Necrotizing angiitis associated with drug abuse. *N Engl J Med*, **283**, 1003–11.

Cloft HJ, Phillips CD, Dix JE *et al.* (1999). Correlation of angiography and MR imaging in cerebral vasculitis. *Acta Radiol*, **40**, 83–7.

Croker JA, Kimberly RP (2005). SLE: challenges and candidates in human disease. *Trends Immunol*, **26**, 580–6.

Dastur DK, Singhal BS (1976). Eales' disease with neurological involvement. Part 2. Pathology and pathogenesis. *J Neurol Sci*, **27**, 323–45.

Dastur DK, Singhal BS, Shroff HJ (1981). CNS involvement in malignant atrophic papulosis (Kohlmeier-Degos disease): vasculopathy and coagulopathy. *J Neurol Neurosurg Psychiatry*, **44**, 156–60.

De Bandt M., Sibilia J, Le L *et al.* (2005). Systemic lupus erythematosus induced by anti-tumour necrosis factor alpha therapy: a French national survey. *Arthritis Res Ther*, **7**, R545–51.

de Carvalho CA, Allen JN, Zafranis A *et al.* (1980). Coccidioidal meningitis complicated by cerebral arteritis and infarction. *Hum Pathol*, **11**, 293–6.

Doty JD, Mazur JE, Judson MA (2003). Treatment of corticosteroid-resistant neurosarcoidosis with a short-course cyclophosphamide regimen. *Chest*, **124**, 2023–6.

Doty JD, Mazur JE, Judson MA (2005). Treatment of sarcoidosis with infliximab. *Chest*, **127**, 1064–71.

Furst DE, Breedveld FC, Kalden JR *et al.* (2005). Updated consensus statement on biological agents, specifically tumour necrosis factor {alpha} (TNF {alpha}) blocking agents and interleukin-1 receptor antagonist (IL-1ra), for the treatment of rheumatic diseases, 2005. *Ann Rheum Dis*, **64** (Suppl 4), iv2–14, iv2–14.

Gallagher KT, Shaham B, Reiff A, *et al.* (2001). Primary angiitis of the central nervous system in children: 5 cases. *J Rheumatol*, **28**, 616–23.

Ginzler EM, Dvorkina O (2005). Newer therapeutic approaches for systemic lupus erythematosus. *Rheum Dis Clin North Am*, **31**, 315–28.

Griffith ME, Pusey CD (2001). Cellular aspects of vasculitis—T cell-mediated aspects. *Springer Semin Immunopathol*, **23**, 287–98.

Guillevin L, Durand-Gasselin B, Cevallos R *et al.* (1999). Microscopic polyangiitis: clinical and laboratory findings in eighty-five patients. *Arthritis Rheum*, **42**, 421–30.

Hankey G (1991). Isolated angiitis/angiopathy of the CNS. Prospective diagnostic and therapeutic experience. *Cerebrovasc Dis*, **1**, 2–15.

Harper L, Williams JM, Savage CO (2004). The importance of resolution of inflammation in the pathogenesis of ANCA-associated vasculitis. *Biochem Soc Trans*, **32**, 502–6.

Harris KG, Tran DD, Sickels WJ *et al.* (1994). Diagnosing intracranial vasculitis: The roles of MR and angiography. *Am J Neuroradiol*, **15**, 317–30.

Hellmann DB, Roubenoff R, Healy RA *et al.* (1992). Central nervous system angiography: Safety and predictors of a positive result in 125 consecutive patients evaluated for possible vasculitis. *J Rheumatol*, **19**, 568–72.

Helveston WR, Gilmore R (1996). Treatment of Vogt-Koyanagi-Harada syndrome with intravenous immunoglobulin. *Neurology*, **46**, 584–5.

Higashida J, Wun T, Schmidt S *et al.* (2005). Safety and efficacy of rituximab in patients with rheumatoid arthritis refractory to disease modifying antirheumatic drugs and anti-tumor necrosis factor-alpha treatment. *J Rheumatol*, **32**, 2109–15.

Hilt DC, Buchholz D, Krumholz A *et al.* (1983). Herpes zoster ophthalmicus and delayed contralateral hemiparesis caused by cerebral angiitis: diagnosis and management approaches. *Ann Neurol*, **14**, 543–53.

Ibanez HE, Grand MG, Meredith TA *et al.* (1994). Magnetic resonance imaging findings in Vogt-Koyanagi-Harada syndrome. *Retina*, **14**, 164–8.

Ikeda K, Suzuki S, Ichijo M *et al.* (1997). How high is high in steroid treatment of Vogt-Koyanagi-Harada syndrome? *Neurology*, **48**, 537.

Islam SM, Numaga J, Fujino Y *et al.* (1994). HLA class II genes in Vogt-Koyanagi-Harada disease. *Invest Ophthalmol Vis Sci*, **35**, 3890–6.

Jayne D, Rasmussen N, Andrassy K *et al.* (2003). A randomized trial of maintenance therapy for vasculitis associated with antineutrophil cytoplasmic autoantibodies. *N Engl J Med*, **349**, 36–44.

Joseph FG, Scolding NJ (2002). Cerebral vasculitis – a practical approach. *Practical Neurology*, **2**, 80–93.

Leclerc S, Myers RP, Moussalli J *et al.* (2003). Sarcoidosis and interferon therapy: report of five cases and review of the literature. *Eur J Intern Med*, **14**, 237–43.

Lie JT (1996). Vasculitis associated with infectious agents. *Curr Opin Rheumatol*, **8**, 26–9.

Lie JT (1997a). Biopsy diagnosis of systemic vasculitis. *Baillieres Clin Rheumatol*, **11**, 219–36.

Lie JT (1997b). Classification and histopathologic spectrum of central nervous system vasculitis. *Neurol Clin*, **15**, 805–19.

Lincoff NS, Erlich PD, Brass LS (2000). Thrombocytosis in temporal arteritis rising platelet counts: a red flag for giant cell arteritis. *J Neuroophthalmol*, **20**, 67–72.

Mathieson PW, Cobbold SP, Hale G *et al.* (1990). Monoclonal-antibody therapy in systemic vasculitis. *N Engl J Med*, **323**, 250–4.

McFarland HR, Wood WG, Drowns BV *et al.* (1978). Papulosis atrophicans maligna (Kohlmeier-Degos disease): a disseminated occlusive vasculopathy. *Ann Neurol*, **3**, 388–92.

Meroni PL, Tincani A, Sepp N *et al.* (2003). Endothelium and the brain in CNS lupus. *Lupus*, **12**, 919–28.

Mohan N, Edwards ET, Cupps TR *et al.* (2004). Leukocytoclastic vasculitis associated with tumor necrosis factor-alpha blocking agents. *J Rheumatol*, **31**, 1955–8.

Mohan N, Kerr GS (2001). Diagnosis of vasculitis. *Best Pract Res Clin Rheumatol*, **15**, 203–23.

Moorthy RS, Inomata H, Rao NA (1995). Vogt-Koyanagi-Harada syndrome. *Surv Ophthalmol*, **39**, 265–92.

Nakamura S, Nakazawa M, Yoshioka M *et al.* (1996). Melanin-laden macrophages in cerebrospinal fluid in Vogt-Koyanagi-Harada syndrome. *Arch Ophthalmol*, **114**, 1184–8.

Nishino H, Rubino FA, DeRemee RA *et al.* (1993). Neurological involvement in Wegener's granulomatosis: an analysis of 324 consecutive patients at the Mayo Clinic. *Ann Neurol*, **33**, 4–9.

Norose K, Yano A (1996). Melanoma specific Th1 cytotoxic T lymphocyte lines in Vogt-Koyanagi-Harada disease (see comments). *Br J Ophthalmol*, **80**, 1002–8.

Ong LM, Hooi LS, Lim TO *et al.* (2005). Randomized controlled trial of pulse intravenous cyclophosphamide versus mycophenolate mofetil in the induction therapy of proliferative lupus nephritis. *Nephrology (Carlton)*, **10**, 504–10.

Paramothayan S, Lasserson T, Walters EH (2003). Immunosuppressive and cytotoxic therapy for pulmonary sarcoidosis. *Cochrane Database Syst Rev*, CD003536.

Pritchard C, Nadarajah K (2004). Tumour necrosis factor alpha inhibitor treatment for sarcoidosis refractory to conventional treatments: a report of five patients. *Ann Rheum Dis*, **63**, 318–20.

Rubsamen PE, Gass JD (1991). Vogt-Koyanagi-Harada syndrome. Clinical course, therapy, and long-term visual outcome. *Arch Ophthalmol*, **109**, 682–7.

Ryan SJ, Pettigrew LC (1995). Cranial arteriopathy in familial Vogt-Koyanagi-Harada syndrome. *J Neuroimaging*, **5**, 244–5.

Sakkas LI (2005). New developments in the pathogenesis of systemic sclerosis. *Autoimmunity*, **38**, 113–6.

Salojin KV, Le TM, Nassovov EL *et al.* (1996). Anti-endothelial cell antibodies in patients with various forms of vasculitis. *Clin Exp Rheumatol*, **14**, 163–9.

Salvarani C, Hunder GG (2001). Giant cell arteritis with low erythrocyte sedimentation rate: frequency of occurence in a population-based study. *Arthritis Rheum*, **45**, 140–5.

Sarwar H, McGrath H Jr, Espinoza LR (2005). Successful Treatment of Long-standing Neuro-Behcet's Disease with Infliximab. *J Rheumatol*, **32**, 181–3.

Scolding NJ (1999b). Cerebral vasculitis. In Scolding NJ, ed. *Immunological and Inflammatory Diseases of the Central Nervous System*, pp. 210–58. Butterworth-Heinemann, Oxford.

Scolding NJ (1999c). Neurological complications of rheumatological and connective tissue disorders. In Scolding NJ, ed. *Immunological and Inflammatory Diseases of the Central Nervous System*, pp. 147–80. Butterworth-Heinemann, Oxford.

Scolding NJ (1999a). Immunological and Inflammatory Diseases of the Central Nervous System.Butterworth-Heinemann, Oxford.

Scolding NJ, Jayne DR, Zajicek JP *et al.* (1997). The syndrome of cerebral vasculitis: recognition, diagnosis and management. *QJM*, **90**, 61–73.

Scolding NJ, Joseph F, Kirby PA *et al.* (2005). A {beta}-related angiitis: primary angiitis of the central nervous system associated with cerebral amyloid angiopathy. *Brain*, **128**, 500–15.

Scolding NJ, Joseph FG (2002a). The neuropathology and pathogenesis of systemic lupus erythematosus. *Neuropathol Appl Neurobiol*, **28**, 173–89.

Scolding NJ, Wilson H, Hohlfeld R *et al.* (2002b). The recognition, diagnosis and management of cerebral vasculitis: a European survey. *Eur J Neurol*, **9**, 343–7.

Sehgal M, Swanson JW, DeRemee RA *et al.* (1995). Neurologic manifestations of Churg-Strauss syndrome. *Mayo Clin Proc*, **70**, 337–41.

Semenzato G (2005). ACCESS: A case control etiologic study of sarcoidosis. *Sarcoidosis Vasc Diffuse Lung Dis*, **22**, 83–6.

Sfikakis PP, Boletis JN, Lionaki S *et al.* (2005). Remission of proliferative lupus nephritis following B cell depletion therapy is preceded by down-regulation of the T cell costimulatory molecule CD40 ligand: an open-label trial. *Arthritis Rheum*, **52**, 501–13.

Singhal BS, Dastur DK (1976). Eales' disease with neurological involvement Part 1. Clinical features in 9 patients. *J Neurol Sci*, **27**, 313–21.

Sneller MC (2005). Rituximab and Wegener's granulomatosis: are B cells a target in vasculitis treatment? *Arthritis Rheum*, **52**, 1–5.

Subbiah P, Wijdicks E, Muenter M *et al.* (1996). Skin lesion with a fatal neurologic outcome (Degos' disease). *Neurology*, **46**, 636–40.

Sugita S, Sagawa K, Mochizuki M *et al.* (1996). Melanocyte lysis by cytotoxic T lymphocytes recognizing the MART-1 melanoma antigen in HLA-A2 patients with Vogt-Koyanagi-Harada disease. *Int Immunol*, **8**, 799–803.

Susac JO (1994). Susac's syndrome: the triad of microangiopathy of the brain and retina with hearing loss in young women. *Neurology*, **44**, 591–3.

Susac JO, Hardman JM, Selhorst JB (1979). Microangiopathy of the brain and retina. *Neurology*, **29**, 313–6.

Talar-Williams C, Hijazi YM, Walther MM *et al.* (1996). Cyclophosphamide-induced cystitis and bladder cancer in patients with Wegener granulomatosis (see comments). *Ann Intern Med*, **124**, 477–84.

Thatayatikom A, White AJ (2006). Rituximab: A promising therapy in systemic lupus erythematosus. *Autoimmun Rev*, **5**, 18–24.

Trysberg E, Tarkowski A (2004). Cerebral inflammation and degeneration in systemic lupus erythematosus. *Curr Opin Rheumatol*, **16**, 527–33.

Unger L, Kayser M, Nusslein HG (2003). Successful treatment of severe rheumatoid vasculitis by infliximab. *Ann Rheum Dis*, **62**, 587–8.

Uthman I, Shamseddine A, Taher A (2005). The role of therapeutic plasma exchange in the catastrophic antiphospholipid syndrome. *Transfus Apheresis Sci*, **33**, 11–7.

Vollertsen RS (1990). Vasculitis and Cogan's syndrome. *Rheum Dis Clin North Am*, **16**, 433–9.

Xiao H, Heeringa P, Hu P *et al.* (2002). Antineutrophil cytoplasmic autoantibodies specific for myeloperoxidase cause glomerulonephritis and vasculitis in mice. *J Clin Invest*, **110**, 955–63.

Zajicek JP, Scolding NJ, Foster O *et al.* (1999). Central nervous system sarcoidosis - diagnosis and management based on a large series. *QJM*, **92**, 103–17.

CHAPTER 37

Multiple sclerosis and other demyelinating diseases

Alastair Compston

Contents

37.1 Classification of demyelinating diseases

The oligodendrocyte–myelin unit subserves saltatory conduction of the nerve impulse in the healthy central nervous system. At one time, many disease processes were thought exclusively to target the structure and function of myelin. Therefore, they were designated 'demyelinating diseases'. But recent analyses, based mainly on pathological and imaging studies, (re)emphasize that axons are also directly involved in these disorders during both the acute and chronic phases. Another ambiguity is the extent to which these are inflammatory conditions. Here, distinctions should be made between inflammation, as a generic process, and autoimmunity in which rather a specific set of aetiological and mechanistic conditions pertain. And there are differences between disorders that are driven primarily by immune processes and those in which inflammation occurs in response to pre-existing tissue damage.

With these provisos, the pathological processes of demyelination and associated axonal dysfunction often account for episodic neurological symptoms and signs referable to white matter tracts of the brain, optic nerves, or spinal cord when these occur in young people. This is the clinical context in which the possibility of 'demyelinating disease' is usually considered by physicians and, increasingly, the informed patient. Neurologists will, with appropriate cautions, also be prepared to diagnose demyelinating disease in older patients presenting with progressive symptoms implicating these same pathways even when there is no suggestive past history. Both in its typical and atypical forms multiple sclerosis remains by far the commonest demyelinating disease. But acute disseminated encephalomyelitis, the leucodystrophies, and central pontine myelinolysis also need to be considered in particular circumstances; and multiple sclerosis itself has a differential diagnosis in which the relapsing-remitting course is mimicked by conditions not associated with direct injury to the axon–glial unit. Since our understanding of the cause, pathogenesis and features of demyelinating disease remains incomplete, classification combines aspects of the aetiology, clinical features, pathology, and laboratory components (Table 37.1). Whether the designation 'multiple sclerosis' encapsulates one or more conditions is now much debated. We anticipate that a major part of future studies in demyelinating disease will be further to resolve this question of disease heterogeneity leading to a new taxonomy based on mechanisms rather than clinical empiricism. But, for now, the variable ages of onset, unpredictable clinical course, protean clinical manifestations, and non-specific laboratory investigations continue to make demyelinating disease one of the more challenging diagnostic areas in clinical neurology.

Even though the primary involvement of tissue inflammation may still be debated, a simple pathological approach is to distinguish those demyelinating diseases attributable to perivascular

Table 37.1 Classification of demyelinating disease

Isolated demyelinating syndromes

Acute haemorrhagic leucoencephalomyelitis—Hurst's disease

Acute disseminated encephalomyelitis

Optic neuritis

Spinal Cord lesions

    acute necrotizing myelitis

    transverse myelitis

    chronic progressive myelopathy

    radiation myelopathy

    HTLV-1-associated myelopathy

    monophasic isolated demyelination—site unspecified

Multiple sclerosis

    relapsing-remitting

    secondary progressive

    primary progressive

    benign

    malignant or Marburg variant

    childhood

    silent multiple sclerosis

    Neuromyelitis optica or Devic's disease

    Balo's concentric sclerosis

    combined central and peripheral demyelination

Central pontine myelinolysis

Pontine

Extrapontine

Leucodystrophies

Adrenoleucodystrophy

    X-linked childhood adrenoleucodystrophy

    X-linked adult-onset adrenomyeloneuronopathy

    autosomal recessive neonatal adrenoleucodystrophy

    autosomal recessive Zellweger's syndrome

Metachromatic leucodystrophy

    late infantile

    juvenile

    adult onset

    multiple sulphatase deficiency

Krabbe's disease

    childhood onset

    late onset

Canavan's disease

Alexander's disease

Pelizaeus–Merbacher disease

    connatal form

    late onset

Vanishing white matter disease

Oculodentodigital syndrome

Reproduced from McAlpine's Multiple Sclerosis 4th edition 2005, with permission

inflammatory cell infiltration from the non-inflammatory forms of demyelination. Inflammation initially characterizes the focal and multifocal lesions associated with multiple sclerosis, the acute disseminated encephalomyelitides, and clinically isolated syndromes that usually affect the optic nerve, brainstem, or spinal cord. Other conditions, once separated, are now considered within the general context of multiple sclerosis. These include Balo's concentric sclerosis, neuromyelitis optica or Devic's disease, Marburg

type hyperacute demyelination, and certain mixed peripheral and central forms of demyelination.

In some young patients with progressive neurological disease, widespread demyelination results from genetically determined disorders of myelin formation. Mostly, these leucodystrophies are characterized by extensive non-inflammatory confluent areas of demyelination, although pathological examination may reveal focal inflammation. Because myelin is structurally abnormal in these conditions, they are often designated 'dysmyelinating' disorders. The most common leucodystrophies encountered in paediatric and adult neurological clinical practice are metachromatic leucodystrophy, adrenoleucodystrophy, and Krabbe's disease. Others are listed in Table 37.1. The availability of methods for molecular screening, improved understanding of the magnetic resonance imaging appearances, and recognition of phenocopies resulting from individual genetic mutations increasingly have led to precision but also expansion in classification of the leucodystrophies.

In the context of inflammatory demyelinating disease, magnetic resonance studies have established the important principle—not so easily gathered from clinical or retrospective neuropathological analyses—that patches of demyelination rarely occur in isolation. Many patients apparently experiencing clinically isolated episodes clearly have widespread lesions. Those in whom areas of demyelination are genuinely isolated show a lower conversion rate to clinically definite multiple sclerosis. The areas of damage in patients with leucodystrophy usually appear as diffuse abnormalities of the cerebral white matter, unlike the discrete lesions of multiple sclerosis and related syndromes associated with multifocal primary inflammation. Patients developing clinical evidence for extensive brainstem damage in the context of electrolyte imbalance, or following its correction, may develop a large area of pontine myelinolysis readily seen on magnetic resonance imaging. This is not a common condition but under the heading of central pontine myelinolysis should also be considered the even rarer cases of extrapontine myelinolysis and Marchiafava Bignami disease, consisting of demyelination in the corpus callosum and resulting from injudicious consumption of Italian red wine—a condition faithfully described in all textbooks but never knowingly encountered by their authors, this one included. Although the distribution and appearance of lesions in multiple sclerosis, acute disseminated encephalomyelitis, primary progressive inflammatory demyelination, the leucodystrophies, and central pontine myelinolysis can usually be distinguished, and some individual components of tissue injury depicted, conventional magnetic resonance and spectroscopic protocols do not reliably image each and every histological component of the pathological process.

37.2 The pathogenesis of demyelination

The evidence informing concepts on the pathogenesis of demyelinating disease derives from animal models, *in vitro* studies of glial cells and neurons, and especially the study of human tissue (Compston and Coles 2008). Most is known about multiple sclerosis, and the analyses of that disease provide a general understanding of the pathogenesis of human demyelinating disease (Lassmann and Wekerle 2005). The hallmark of demyelinating disease is formation of the sclerotic plaque. This represents the end stage of a process that, microscopically, involves inflammation, demyelination and remyelination, oligodendrocyte depletion and astrocytosis, and acute and chronic axon degeneration (Fig. 37.1). Although there is no shortage of opinion, the order and relationship of these separate components remain to be fully resolved.

Demyelination in multiple sclerosis occurs throughout the central nervous system white matter but certain zones are preferentially affected, and the grey matter is not spared. Plaques are clustered around the lateral ventricles, in the corpus callosum and floor of the aqueduct and fourth ventricle, and in the cortical or subcortical white matter. The optic nerve and brainstem are commonly involved. In the cervical portion of the spinal cord, lesions tend to be subpial and located where the denticulate ligament intermittently makes contact with the dorsal cord. These are to some extent areas containing extensive networks of veins, and where the blood brain barrier is relatively or absolutely defective. These associations are consistent with a primary role for vascular permeability in the pathogenesis of demyelinating disease. They have also been used, probably spuriously, to fuel the debate on whether trauma can precipitate demyelination.

Biopsy of acute lesions, usually prompted by radiological features that mimic tumour, and autopsy series of multiple sclerosis in which plaques of varying ages happen to be present, provide an opportunity to characterize the detailed cellular pathology of acute and chronic inflammatory demyelination. These are supplemented by autopsy series in which the tissue has necessarily reached a more advanced stage in evolution of the disease process. Both are studied using immunocytochemical reagents that discriminate many individual features of tissue injury. Very many recent publications have exploited these opportunities but those that provide contemporary analyses of the key features include Trapp *et al.* (1998) concerning

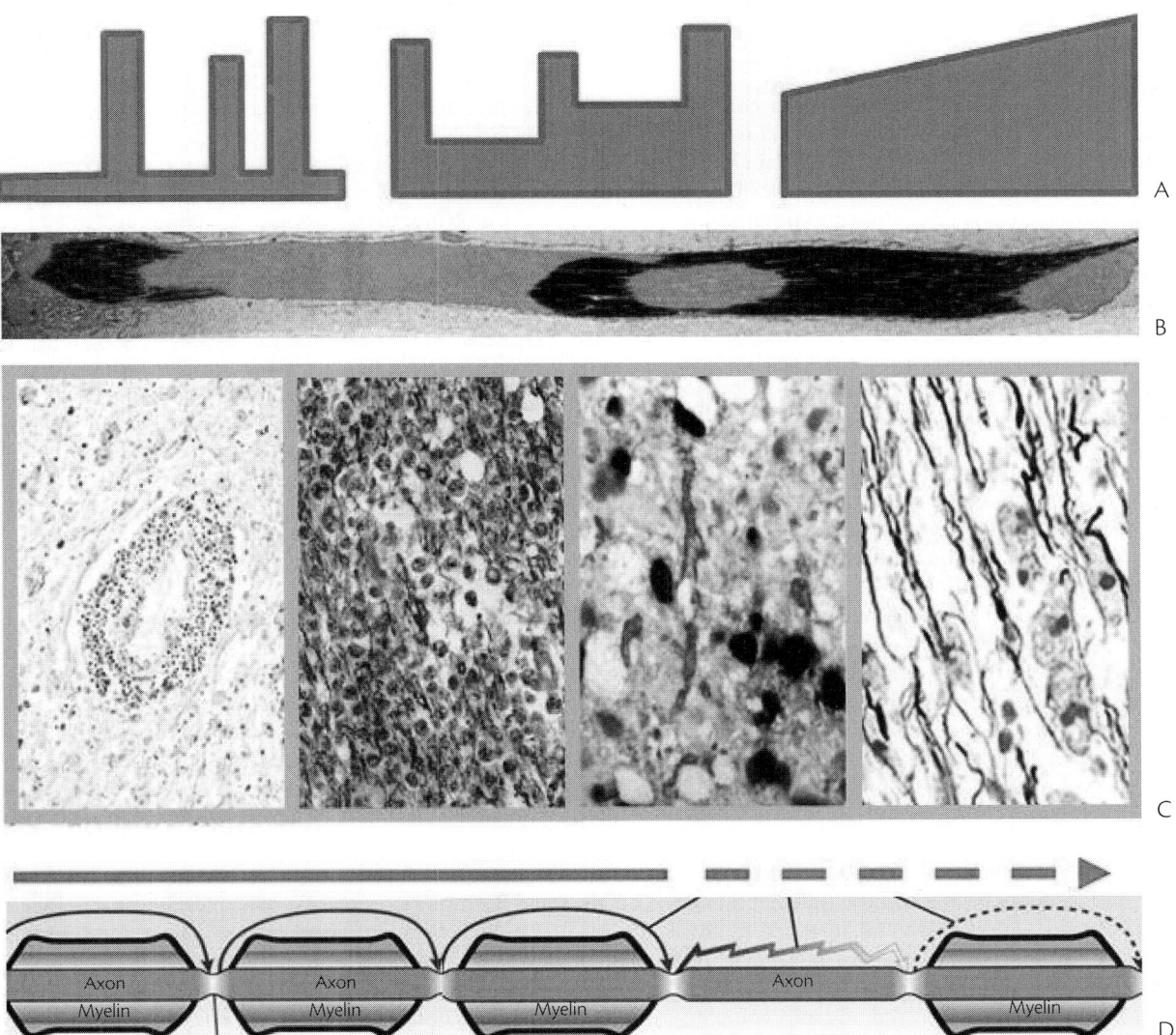

Fig. 37.1 The pathogenesis of multiple sclerosis. A. The clinical course is depicted. B. Macroscopic areas of demyelination in the optic nerve. C. The panels show perivascular infiltration by activated lymphocytes, microglial removal of myelin debris, remyelination, and axonal injury. D. Interruption in saltatory conduction. (See Plate 32.)

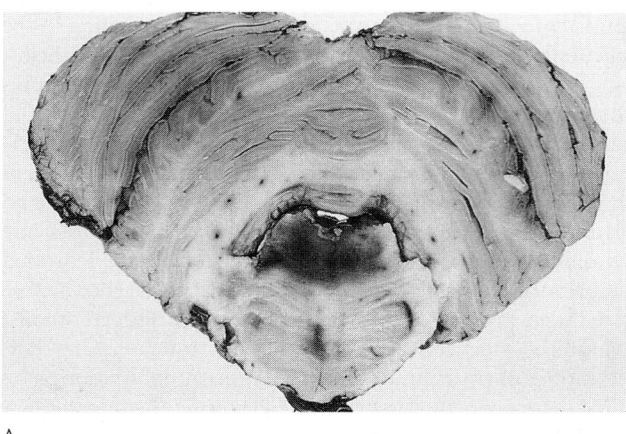

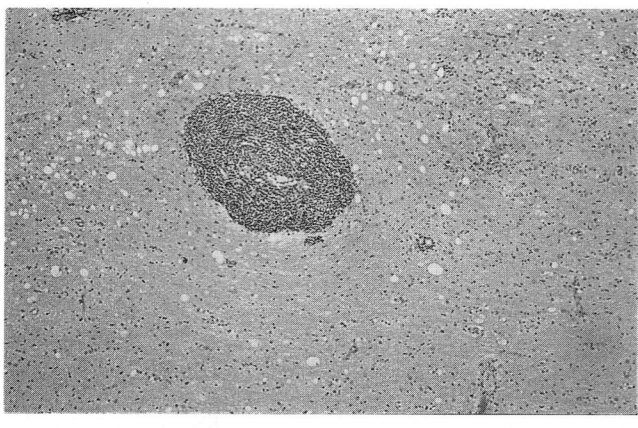

A

B

Fig. 37.2 Acute disseminated encephalomyelitis A. Macroscopic appearance of the brain in a patient dying from an illness lasting 10 days showing diffuse midbrain damage with a combination of demyelination and tentorial herniation due to cerebral oedema. B. Acute perivascular infiltration in the same patient.

axonal pathology, Kutzelnigg *et al.* (2005) concerning cortical pathology and progressive multiple sclerosis, and Patrikios *et al.* (2006) addressing remyelination. The acutely inflamed brain is visibly swollen and oedematous. In acute haemorrhagic encephalomyelitis, there is fibrinoid necrosis and inflammatory infiltration of vessels, oedema, perivenular petechial haemorrhage, and macrophage infiltration around vessels both in grey and white matter but without demyelination. In the less severe forms of acute disseminated encephalomyelitis, the histological appearances may be restricted to intense perivenular mononuclear cell infiltration without demyelination (Fig. 37.2), but more conspicuous are the inflammatory infiltrates of neutrophils, lymphocytes, plasma cells, microglia, and foamy myelin-debris containing macrophages indicating demyelination and phagocytosis. The pathological findings correlate with duration of the illness and show an evolution from diffuse vascular involvement to widespread demyelination.

The *Marburg type* of demyelination is characterized by destructive lesions showing large confluent zones of demyelination with extensive axonal loss and surrounding oedema, and intense inflammatory change that may affect both the central and peripheral nervous systems. In *Balo's concentric sclerosis*, symmetrical demyelination surrounded by normal or repaired myelin occurs in rings not unlike the onion bulb appearance of demyelinating peripheral neuropathies (Stadelamnn *et al.* 2005). In *neuromyelitis optica*, or Devic's disease, the brunt of the pathological process and its clinical expression are borne by the optic nerves and spinal cord which may show acute necrosis with limited evidence for remyelination. The course is now recognized often to be relapsing, and there is histological evidence for the lesions having arisen at different times. Demyelination may be more widespread than suggested by the clinical phenotype. The immunopathogenesis of neuromyelitis optica differs from some, but not all, cases of relapsing-remitting multiple sclerosis, showing conspicuous antibody and complement deposition in addition to lymphocyte infiltration (Lucchinetti *et al.* 2002).

When the brain is inflamed, lymphocytes cross the blood brain barrier and accumulate on the abluminal surface of venules. Lymphocytes that are not activated against brain antigen either return to the circulation or, in common with immune cells that have outlived their purpose elsewhere, die by apoptosis. Activated T-cells encountering antigen persist within the nervous system. The cells most probably contributing to local immune responses are initially confined to the perivascular or Virchow–Robin spaces and, with the exception of a few antibody producing plasma cells, do not contact the myelin sheaths. But as T- and B-lymphocytes, plasma cells, and macrophages accumulate around the vessels, pro-inflammatory cytokines, especially interferon-γ, drive an amplification of the immune response in which microglia are activated, leading to the release of yet more T-cell derived interferon-y, IFN-γ, and the recruitment of additional naive microglia. Contact is established between activated microglia and the oligodendrocyte–myelin unit if the latter is opsonized with ligands for receptors activated on the surface of microglia for Fc fragments and complement. Activated microglia deliver a lethal signal to the target oligodendrocyte producing a high concentration of cell surface bound tumour necrosis factor-alpha, TNFα (Zajicek *et al.* 1992). Together, these inflammatory processes lead to disruption of the myelin membrane with increased spacing, vesicular disruption, splitting, vacuolation, and fragmentation of the lamellae. All T-lymphocyte subsets are present in the lesions, although class I major histocompatibility complex-restricted CD8-positive cells predominate. The vast majority of these cells are cytotoxic T-cells with a Tc1 cytokine polarization. They are the major stimulus for microglial activation. The cellular pathology of the progressive stage is more uniform. The perivascular inflammatory infiltrates contain T and B lymphocytes, plasma cells, and some macrophages but in lower numbers than in active plaques of acute and relapsing multiple sclerosis. Diffuse and continuous damage in the progressive phase of multiple sclerosis is mediated by microglia driven by inflammatory mediators produced in the lymphocyte infiltrates. Thus, although the relative contributions vary, the core processes of T-cell and microglia mediated inflammatory tissue injury is present whether the illness is relapsing or has moved to the progressive phase.

Recently, much emphasis has been placed on the role of axonal injury as an additional pathological feature occurring at different stages in the evolution of multiple sclerosis. Three separate but inter-related processes are involved. Brief exposure of the (rat) spinal cord to nitric oxide donors produces reversible conduction block in normal, and especially hypomyelinated, axons (Redford *et al.* 1997). Hence, new symptoms may result from the direct effect of inflammatory mediators on conduction in normally myelinated axons. However, in the experimental setting, although axonal dysfunction is initially reversible, a separate and destructive sequence

of calcium dependent excitotoxic events follows more prolonged exposure to inflammatory mediators (Fig. 37.3). Electrically active axons are especially vulnerable and show irreversible conduction block, with demyelination and axonal degeneration, after prolonged exposure to nitric oxide (Smith *et al.* 2001). Immunohistochemical staining for phosphorylated amyloid precursor protein indicates that axonal injury occurs as part of the acute demyelinating lesion (Ferguson *et al.* 1997) with extensive axonal transection (Trapp *et al.* 1998). Soluble factors released by activated microglia impair mitochondrial cytochrome oxidase activity of neurons *in vitro*

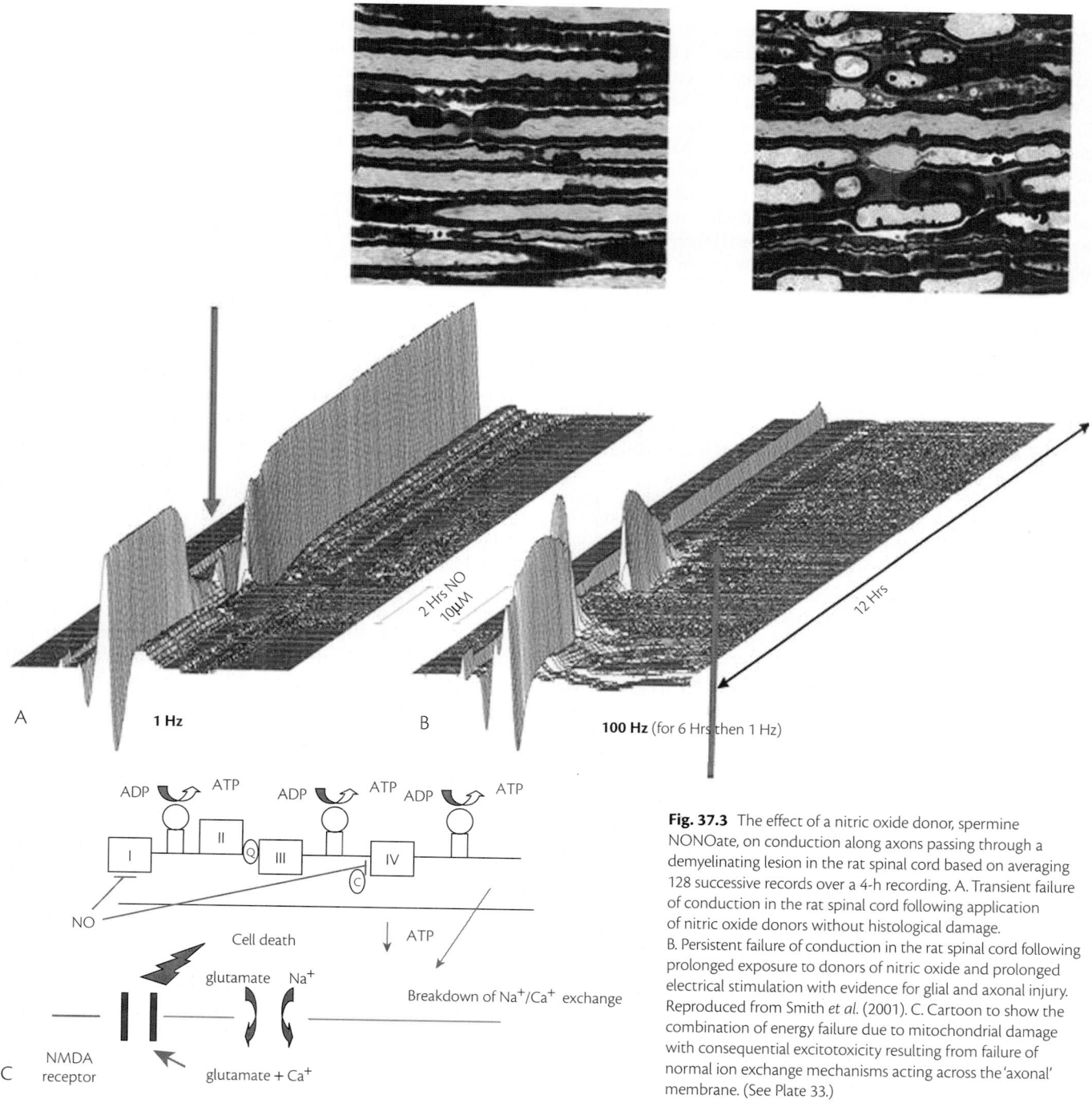

Fig. 37.3 The effect of a nitric oxide donor, spermine NONOate, on conduction along axons passing through a demyelinating lesion in the rat spinal cord based on averaging 128 successive records over a 4-h recording. A. Transient failure of conduction in the rat spinal cord following application of nitric oxide donors without histological damage. B. Persistent failure of conduction in the rat spinal cord following prolonged exposure to donors of nitric oxide and prolonged electrical stimulation with evidence for glial and axonal injury. Reproduced from Smith *et al.* (2001). C. Cartoon to show the combination of energy failure due to mitochondrial damage with consequential excitotoxicity resulting from failure of normal ion exchange mechanisms acting across the 'axonal' membrane. (See Plate 33.)

(Dutta *et al.* 2006). Although this neuronal dysfunction is initially reversible, a separate and lethal sequence of events follows more prolonged exposure to microglial soluble factors. In co-cultures of rat microglia and embryonic cortical neurons, inducible nitric oxide synthase or iNOS-derived nitric oxide alone is responsible for neuronal death from interferon-γ and lipopolysaccharide activated microglia. Neurons allowed to mature *in vitro* remain sensitive to nitric oxide but are rescued by blocking *N*-methyl-D-aspartate-receptor activation indicating that nitric oxide elicits an excitotoxic mechanism, most likely acting through neuronal glutamate release. Thus, similar concentrations of nitric oxide cause neuronal death by two distinct mechanisms: nitric oxide acts directly upon immature neurons but indirectly, via *N*-methyl-D-aspartate receptors, on more mature neurons (Golde *et al.* 2002).

Axonopathy continues throughout the course of the disease and contributes increasingly to disability during the chronic progressive phase of multiple sclerosis. The cortex, too, is severely affected in the progressive stage of the disease, showing large areas of demyelination associated with a variable amount of axonal and neuronal degeneration. In patients with primary and secondary progressive disease, pre-existing lesions grow slowly by radial expansion. In addition, there is diffuse injury of 'normal' white matter, reflected by a slowly progressive and widespread axonal injury in the absence of primary demyelination. In part this 'slow burn' is due to activated microglia. Thus, chronic progressive multiple sclerosis is not simply a focal white matter condition but one also characterized by widespread neurodegeneration. But it is important to emphasize that, although the intensity of the inflammatory response is less severe compared to that seen in fresh white matter plaques, and the focus is on activated microglia rather than T-lymphocyte infiltration, the diffuse central nervous system damage in progressive multiple sclerosis is invariably associated with brain inflammation (Kutzelnigg *et al.* 2005)

An additional contribution to axonal loss is loss of trophic support normally provided by oligodendrocytes and myelin. The *in vitro* evidence indicates that cells of the oligodendrocyte lineage support neuronal survival by both contact mediated and soluble mechanisms: insulin-like growth factor-1, IGF-1, contributes this effect through the PI3 kinase/Akt signalling pathway; conversely, differentiated oligodendrocytes increase neurofilament phosphorylation and axonal length due to an effect of glial cell derived nerve growth factor, GDNF, acting through the MAP kinase/Erk pathways (Wilkins *et al.* 2003). The *in vitro* data also suggest several mechanisms whereby the potential damage to viable but threatened neurons and myelinated axons in the vicinity of an inflammatory focus is limited through the activity of molecules that traditionally are considered to be neurotrophic or axon-guidance signals. IGF-2 and ciliary neurotrophic factor, CNTF, act *in vitro* through an autocrine loop to block oligodendrocyte toxicity caused by tumour necrosis factor alpha, TNFα, released from IFN-γ activated microglia (Nicholas *et al.* 2002). The activity of c-Jun kinase, JNK, stimulated by TNFα receptor ligation confirms that these factors inhibit TNFα signalling. Irrespective of activation, conditioned media do not induce JNK activity in oligodendrocytes unless CNTF or IGF-2 activity are also neutralized. The TNFα containing medium of IFNγ treated (activated) microglia is then toxic for oligodendrocytes. Conversely, separate *in vitro* experiments show that conditioned medium produced by stressed neurons is toxic for microglia. This effect is mediated by semaphorin 3A which binds the receptor complex made up by plexin and neuropilin and expressed by activated microglia (Majed *et al.* 2006). Although directed at the interplay between inflammatory cells and their mediators, and oligodendrocytes, this has implications for axonal survival secondary to demyelination. More directly, IGF-1 and GDNF modulate the direct effects of nitric oxide on survival of neurons and axonal injury mediated by exposure to nitric oxide *in vitro* (Wilkins and Compston 2005). This suggests that, at any one time, the amount of tissue injury reflects the interplay of active inflammation, the extent of existing neurodegeneration and the dynamic vulnerability of intact axons. It follows that although the absolute amount of inflammation may reduce with time, its impact is not altogether reduced given the increasing susceptibility of injured axons to residual inflammatory insult.

It is clear that the oligodendrocyte itself undergoes many acute changes in demyelinating disease; and that any depletion of oligodendrocytes must be replenished by the maturation of precursors if tissue repair is to occur. In a detailed analysis of acute lesions, Luchinetti *et al.* (1999) showed consistent variations, either an increase or decrease, in the number of oligodendrocytes within lesions from the same individual indicating a much more dynamic sequence of injury and recovery of the myelinating cell than had previously been assumed. The implication is of turnover with macrophage associated loss of healthy oligodendrocytes and recruitment of new progenitors that then undergo differentiation (see Chapter 7). Morphological criteria used as evidence for remyelination are myelin lamellae that are inappropriately thin for the corresponding axon, a short internode, and myelin embedded in a satellite cell with a membrane that is continuous from the surface of the cell around the axon, back to the surface again and compacted (Gledhill *et al.* 1973). Remyelination accounts for the appearance of 'shadow plaques' seen both in the acute and chronic stages of the disease. That said, remyelination is most active during the acute inflammatory process in which myelin debris is removed by macrophages, or microglia, and shadow plaques may co-exist with areas of active demyelination. Of course, successfully remyelinated plaques can be the target for new waves of demyelination, and these cycles may exhaust the capacity for tissue repair later in the course. The most comprehensive and encouraging survey of remyelination shows that the majority of plaques undergo successful remyelination in 20 per cent of cases, both in the acute and primary and secondary progressive phases of the disease; and the amount of remyelination appears to increase over time (Patrikios *et al.* 2006). Clearly, in many other instances remyelination is less successful and these inter-individual differences need to be better understood if myelin repair and the presumed associated axonal protection it offers, are to be manipulated as a treatment for multiple sclerosis.

Much attention has been devoted in recent years to the concept that different immunological and neurodegenerative processes are involved in the pathogenesis of tissue injury in groups of patients with multiple sclerosis (Lucchinetti *et al.* 2000; Lassmann and Wekerle 2005). In essence, the initial process of brain inflammation may subsequently be modified by additional immunological mechanisms, creating a state of disease heterogeneity. In some patients, the main factors associated with tissue injury are the T-cell infiltrates and macrophage associated tissue injury (Fig. 37.4, pattern 1). In others, antibody-mediated immune reactions against myelin, oligodendrocytes, and their progenitors and complement deposition amplify the demyelinating reaction and/or impair the recruitment of new oligodendrocytes and remyelination (Fig. 37.4, pattern 2). In a further pattern, hypoxia-like cell injury, resulting either from inflammation-induced vascular damage or macrophage toxins that impair mitochondrial function, is the substrate for tissue injury (Fig. 37.4, pattern 3). Finally, a genetic defect or polymorphism may change susceptibility of the target tissue to primary immune mediated injury of oligodendrocytes (Fig. 37.4, pattern 4). These distinct mechanisms may explain differences in the extent of demyelination, oligodendrocyte injury, remyelination, and axonal damage seen across the spectrum of multiple sclerosis and its atypical but related forms—neuromyelitis optica (Fig. 37.4, pattern 2) and Balo's concentric sclerosis (Fig. 37.4, pattern 3).

37.3 The pathophysiology of demyelinating disease

Against this complex background of structural injury to the myelinated axon, the pathophysiology of symptom production depends on alterations in saltatory conduction through pathways in the central nervous system that normally are myelinated. Although function may be preserved overall, even in the presence of structural damage, by redundancy in individual systems or tracts, strategically placed pathways soon lose their safety factor for conduction resulting in neurological symptoms and signs. In large measure, the manifestations of multiple sclerosis are not specific to the disease process or its physiological consequences but merely reflect the abnormalities to be expected from any process that disrupts physiological performance at that site. However, saltatory conduction may be compromised by partial demyelination in a variety of ways that are characteristic and account for some clinical manifestations of multiple sclerosis (Smith *et al.* 2005).

Partially demyelinated axons cannot transmit fast trains of impulse and this may explain those symptoms resulting from physiological fatigue. Depolarization may traverse the lesion but at reduced velocity. This abnormality does not of itself explain particular symptoms in multiple sclerosis but it does account for the characteristic delay in arrival of potentials evoked by sensory stimuli and recorded over appropriate cortical receptor zones. Partially demyelinated axons may discharge spontaneously, thus accounting for the many unpleasant distortions of sensation reported by a high proportion of patients. Increased mechanical sensitivity manifests as movement induced symptoms including flashes of light provoked by eye movement, and the electric sensation that spreads down the spine, limbs, or anterior chest wall after neck flexion—Lhermitte's symptom and sign—which has a less common motor counterpart in the spontaneous limb movements sometimes provoked by neck flexion in patients with multiple sclerosis. Spontaneous discharge in brainstem facial nerve neurons probably accounts for myokymia.

Increased temperature sensitivity, with a reduction in the safety factor for conduction in partially demyelinated axons, explains the temporary increase in severity of pre-existing symptoms,

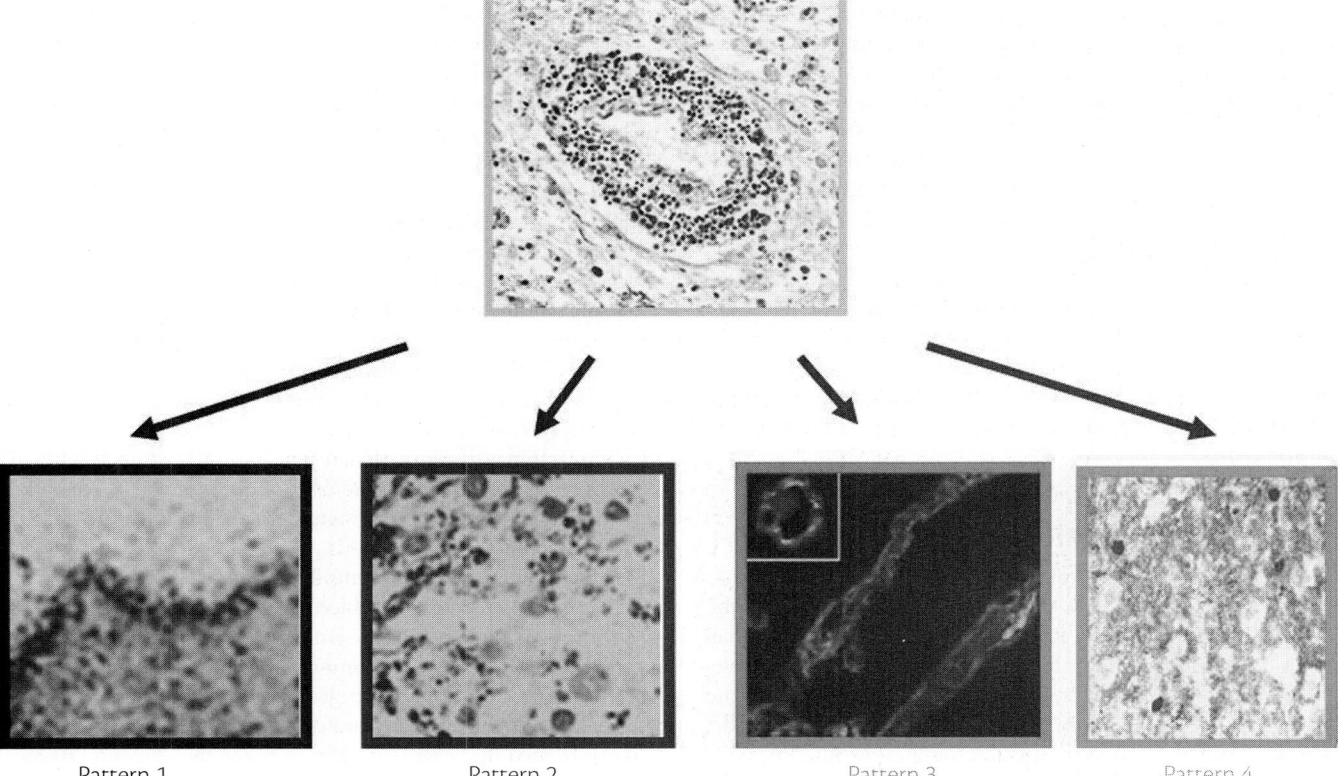

Fig. 37.4 Heterogeneity in multiple sclerosis plaques. Focal perivascular lymphocytic infiltration is associated with four patterns of further tissue injury. See text for details. (See Plate 34.)

experienced by many patients after exercise or immersion in hot water. Conversely, cold may improve performance—some patients even adopting complicated water cooled systems and others reporting that, for example, vision improves after eating ice-cream. Ephaptic transmission occurs between neighbouring and partially demyelinated axons giving rise to paroxysmal symptoms of demyelination usually manifesting as trigeminal neuralgia, ataxia and dysarthria, or tonic brainstem seizures. These are often triggered by touch or movement.

Despite these pathophysiological explanations for the symptoms of multiple sclerosis, there are many enigmatic aspects of the disease. Since symptoms can arise from areas that are not demyelinated, and demyelination may be clinically silent, discrepancies exist between the sites of the nervous system that are necessarily involved from an analysis of the symptoms and signs and the lesions that can be demonstrated by imaging or at autopsy.

There are several mechanisms of symptom recovery early in the course of multiple sclerosis. These include the resolution of conduction block in nerve fibres that were never demyelinated, re-establishment of conduction in persistently demyelinated axons, functional reorganization of surviving pathways, and remyelination. Youl *et al.* (1991) first showed that onset and recovery of conduction block and visual impairment in optic neuritis match the phase of acute inflammtion. Symptoms which recover may depend on the removal of cytokines, delivered by activated lymphocytes accumulating within the central nervous system that directly impair saltatory conduction. Function may be restored after demyelination by re-arrangement of sodium channels providing a variety of alternative patterns of ordered or partially disordered conduction.

The transitional phases between stable myelination, demyelination, and remyelination are associated with adaptations in ion exchange channels active at the nodes of Ranvier and across the naked axon. With demyelination, there is a redistribution of sodium channels that are swept ahead as the axon is remyelinated. Whilst this may eventually restore electrical activity, it has been suggested that the injurious properties of altered sodium and calcium exchange across the demyelinated axonal membrane, for which sodium channel blockade should be protective (Black *et al.* 2006), could prove especially hazardous during this period of remyelination, until such time as normal nodal arrangements are restored (Smith 2006). Experimentally, remyelinated axons restore both conduction of the nerve impulse and neurological function (Smith *et al.* 1981; Jeffrey and Blakemore 1997) and we can assume that there is a contribution to recovery from remyelination in multiple sclerosis.

37.4 Isolated demyelinating syndromes

Syndromes resulting from focal demyelination are often the first manifestation of an illness that subsequently recurs, leading to the diagnosis of multiple sclerosis. Rarely, the symptoms and signs may be truly monophasic and never recur. Clearly it is desirable to make this distinction ahead of events. Some guidance is available for the risk of multiple sclerosis following each type of isolated episode, considered according to the part affected, and basing outcome both on clinical features and laboratory investigations. This natural history varies with the different syndromes, ages of presentation, and duration of follow-up. Clinically definite multiple sclerosis can

reliably be diagnosed only when more than one demyelinating episode, separated by at least 1 month and with recovery between attacks, has occurred. The situation is more complex with respect to imaging studies. Whilst these have established that a high proportion of patients with clinically isolated demyelinating syndromes already have anatomically disseminated lesions at presentation, recent analyses suggest that patients with no lesion other than that attributable to the presenting episode have a lower conversion rate for multiple sclerosis. One cohort, comprising around 130 individuals, followed prospectively showed that 72 per cent of those with additional lesions had converted at 5 years compared with 6 per cent of patients without widespread lesions; comparable figures at 10 years were 83 per cent and 11 per cent: and 88 per cent and 19 per cent at 14 years; the most recent analysis benefiting from improved ascertainment at 20 years—a criticism of its immediate predecessor—shows conversion rates of 82 per cent and 21 per cent for those with and without additional imaging abnormalities at presentation, respectively (Fisniku *et al.* 2008).

37.4.1 Acute disseminated encephalomyelitis

Acute encephalopathy in children or young adults occurs as part of several specific metabolic and infectious illnesses which can be identified by laboratory investigations. These are distinct from acute or subacute encephalomyelitis, simultaneously affecting multiple parts of the nervous system, that follows an exanthematous or infectious illness or may emerge without apparent precipitation. The typical features of acute disseminated encephalomyelitis are headache, drowsiness and seizures, and one or more focal lesions producing hemisphere syndromes and disturbances of vision. In other cases, the clinical manifestations reflect damage either to the brainstem, optic nerves, or spinal cord but each may be involved together and the peripheral nervous system is sometimes affected. These cases were more common before the advent of public health programmes that reduced the frequency of predisposing exanthematous disorders; and fortunately they are seldom seen following vaccination itself, so that nowadays there is no distinct prodromal illness in most cases. Both in children and adults, a presumptive diagnosis of acute disseminated encephalomyelitis often therefore has to be made in the absence of an identifiable cause.

However, some patients, especially adults, recovering from the initial episode later relapse and follow a clinical course that is typical for multiple sclerosis. Thus, it is important to recognize that multiple sclerosis may have an encephalopathic presentation. In other situations, the illness remains monophasic but separate sites are involved sequentially in a step-wise fashion giving the appearance of an illness that is disseminated in time. For this reason, it is unwise to consider symptoms evolving over several weeks as necessarily signifying a relapsing illness sufficient for the diagnosis of multiple sclerosis.

The hyperacute form of acute disseminated encephalomyelitis (Hurst 1941) is usually preceded by a non-specific respiratory infection 3–10 days before the onset of neurological symptoms. Young adult males are most commonly affected, complaining initially of headache or dizziness and progressing over hours through stages of disorientation, confusion, and drowsiness to coma. The rate of progress is such that events usually overtake the detection of focal signs and this form of the disease is frequently fatal, although affected individuals may remain in a persistent vegetative state for several weeks and some survive with severe disability following treatments aimed at reducing intracranial pressure. The combination of pyrexia and a marked cerebrospinal fluid pleocytosis with a predominant neutrophil response mimics pyogenic infection of the central nervous system, but the course of acute disseminated encephalomyelitis is not influenced by anti-microbial treatment. In some cases the clinical and pathological features are entirely focal, suggesting rapidly growing tumour or herpes simplex encephalitis.

In the classical account of acute disseminated encephalomyelitis, Miller *et al.* (1956) reviewed several hundred cases gathered from the literature but made no attempt to separate diffuse from anatomically restricted forms of para-infectious neurological disease, including polyradiculitis. Some of their cases may have arisen from direct viral infection of the nervous system and related disorders that were not then recognized. The Newcastle experience indicated that about 1:1000 cases of exanthematous illness in childhood are complicated by acute disseminated encephalomyelitis, the risk being slightly higher with pertussis and scarlet fever than measles and rubella. No obvious differences are seen with acute disseminated encephalomyelitis complicating the various different childhood exanthematous illnesses. The post-rubella syndrome is less frequent and more severe but has a late progresssive variant, analagous to subacute sclerosing panencephalitis. Almost 50 per cent of post-varicella cases present with a pure cerebellar syndrome sometimes associated with involuntary movements. This carries a relatively good prognosis with a low rate of persistent disability. A wider range of causative organisms has been implicated and the individual case more usually develops in the context of a non-specific respiratory infection. New specific causes of acute disseminated encephalomyelitis have been described and these include the ECHO, Coxsackie and herpes viruses, mycoplasma, borreliosis, and cases associated with neoplasia and acquired immunodeficiency syndrome.

The symptomatology, course, and prognosis of acute disseminated encephalomyelitis in young adults are clinically similar. In the typical case, affected individuals develop headache, drowsiness, meningeal irritation, focal or generalized fits, and combinations of lesions indicating damage to the cerebrum, optic nerves, brainstem, or spinal cord about 10–20 days after the prodromal illness. They are often systemically ill with pyrexia and marked meningism both in encephalitic and myelitic forms of the disease. The symptoms and signs evolve over the course of a few days. The cerebrospinal fluid shows a mixed polymophonuclear and lymphocytic or predominantly mononuclear pleocytosis with raised protein and slight reduction in glucose; oligoclonal bands may be present, sometimes transiently. Whilst there is an appreciable mortality, the majority of patients survive and there is some evidence to suggest that outcome is influenced by early use of high dose corticosteroids but this has not been formally evaluated. Despite surviving the acute illness, patients may be left with persistent neurological deficits.

The magnetic resonance appearances of cerebral lesions in acute disseminated encephalomyelitis are multifocal asymmetric white matter abnormalities indistinguishable from the lesions of multiple sclerosis although many cases show extensive and rather symmetric changes in the cerebral or cerebellar white matter and in the basal ganglia. Serial or gadolinium-DTPA enhanced scans are more useful and suggest that whereas lesions persisting long after the clinical manifestations have resolved do not discriminate, the lesions of acute disseminated encephalomyelitis often resolve in contrast to the development of new lesions or the demonstration of areas with enhancement indicating disease activity typical of multiple sclerosis, and excluding acute monophasic demyelination (Dale *et al.* 2000). The proportion of cases presenting with an encephalpathic illness in childhood who eventually turn out to have multiple sclerosis, generally assumed to be low, may have been under-estimated: recent surveys suggest a figure of around 60 per cent (Mikaeloff *et al.* 2004) but these estimates are much affected by definitions and whether or not the entity of multiphasic, or relapsing, disseminated encephalomyelitis is accepted. For all these reasons, the interval that should be allowed before conclusions can safely be reached on the outcome after an episode consistent with acute disseminated encephalomyelitis is uncertain.

One reason for believing that acute disseminated encephalomyelitis arises from immune sensitization to brain antigens is that it has followed the use of vaccines containing central nervous system tissue. In its day, this form of the disease behaved much as other cases of acute disseminated encephalitis and the pathological features were also indistinguishable. Post vaccinial encephalomyelitis has become a rare disorder following alterations in the preparation of vaccines. The definitive series were collected several decades ago following episodes in which the need arose to vaccinate large numbers of individuals against smallpox as part of public health measures. The 62 cases studied clinico-pathologically and reported by de Vries (1960) had been collected over 34 years and were necessarily severe. In all, the neurological illness developed within 21 days of vaccination and, in fatal cases, death occurred at or soon after 13 days. In these hyperacute cases, the pathological findings mimic transitional forms of acute haemorrhagic or disseminated encephalomyelitis. The clinical illness starts with a vaccinial skin reaction and systemic symptoms that merge with the neurological manifestations, typically affecting the cerebrum but sometimes presenting as a myelitic disorder. Despite having a high mortality, post-vaccinial encephalomyelitis may recover spontaneously and completely.

37.4.2 Optic neuritis

Optic neuritis usually presents with pain on eye movement, followed or accompanied by blurred vision. Many patients first notice the visual loss on waking, or on accidentally closing one eye. Some report selective impairment of central vision with preservation of the peripheral field and awareness of movement. The symptoms usually evolve over hours or days and the degree of visual loss that accompanies the clinical nadir varies from slight blurring to blindness. A number of other visual symptoms are described. Some patients notice selective loss of colour intensity or perception, usually in the red range. Others describe disturbances of visual perception with persistence of images on re-fixation or flashes of light, known as phosphenes, provoked by eye movement. The usual pattern is that the pain disappears after a few days. Vision improves, rapidly at first and then more slowly. About 90 per cent of patients consider themselves to have made a full visual recovery but more formal assessments indicate that up to 50 per cent have persistent defects of vision, and colour perception frequently remains impaired. In one large series with prolonged follow-up, no difference was observed in the outcome for vision at 5 years between treatment groups and 6 per cent of all patients had poor visual recovery (Optic Neuritis Study Group 1997a).

The symptoms of bilateral simultaneous optic neuritis do not differ from unilateral disease. There may be a marked disparity in the extent to which each eye is affected; and the implications for recurrence of the neurological symptoms are not the same. The rate of onset varies and in some cases the loss of vision is slowly progressive. Particular care needs to be taken in that situation so as to exclude compression of the anterior visual pathway but it is also the case that structural lesions of the optic nerve can manifest as relapsing visual failure, closely mimicking optic neuritis. The diagnosis is one of exclusion. Ischaemic optic neuropathy needs to be considered in older patients. Relapsing visual loss may be due to sarcoidosis affecting the optic nerve or Eales' disease, in young women and men respectively. A careful family history should be taken since the onset of visual failure in Leber's hereditary optic neuropathy can be confused with bilateral sequential optic neuritis in men and mistakes may also occur in the context of so-called toxic amblyopia. The syndrome of a multiple sclerosis-like illness associated with mutations of mitochondrial DNA and manifesting as central nervous system demyelination with disproportionate involvement of the optic nerves has been recognized (Harding *et al.* 1992; Riordan-Eva *et al.* 1995). Chronic relapsing inflammatory optic neuropathy, known as CRION, describes a condition in which painful episodes suggestive of optic neuritis, responding to corticosteroids, occur in the relative or complete absence of other manifestations of demyelinating disease (Kidd *et al.* 2003). The variable density of the bony orbital walls means that sinus infection may spread to affect the optic nerve directly or as a consequence of local tissue oedema. Using present day protocols, the lesion responsible for optic neuritis can almost always be imaged, the nerve appearing swollen and showing a focal increase in magnetic resonance signal. Inflammation within the intracanalicular portion of the nerve and long lesions are associated with delayed or incomplete recovery of vision (Miller *et al.* 1988; Fig. 37.5).

Optic neuritis is one of the commonest presenting features of multiple sclerosis. Clinical involvement of the optic nerve occurs in >50 per cent of patients at some time and the visual pathway is invariably affected at autopsy. These statistics lead to a high level of anxiety in the informed patient that an episode of optic neuritis is likely to be the first manifestation of multiple sclerosis. But it has proved difficult to establish statistics for the rate of conversion that can reliably be applied to the individual since the reported series have varied in their selection criteria, and ascertainment biases have inevitably been introduced. Most likely, methodological factors and the nuances of classification account for geographical

differences in the estimated conversion rates. But the main factor determining the reported frequency of multiple sclerosis after optic neuritis is duration of follow-up. The risk is highest in the first 5 years but the proportion of cases with widespread demyelination increases steadily over time. Several retrospective and prospective series have been treated to life-table analysis in an attempt to compensate for the varying length of follow-up. This approach gives estimates of 38–78 per cent conversion depending on the location and actuarial time point. The most recent report brings up to date the follow-up of an 'ancient' series studied prospectively and showing a conversion rate to multiple sclerosis of 40 per cent at 30 years (Nilsson *et al.* 2005). Age at presentation influences the risk of multiple sclerosis developing after an attack of optic neuritis. In children, the disorder is commonly bilateral and the frequency of further symptoms of demyelination affecting the visual pathways or other parts of the nervous system much lower. Although less common than in children, bilateral simultaneous optic neuritis in adults also carries a lower risk of conversion to multiple sclerosis. In both situations the probable explanation is that these are anatomically restricted forms of acute disseminated encephalomyelitis.

Although most clinicians would not assign the same diagnostic significance to a second attack of optic neuritis as an episode of demyelination affecting another part of the central nervous system, recurrence carries an increased risk of subsequent conversion to multiple sclerosis. Other factors reported at one time or another to increase the risk of multiple sclerosis, include age at onset and female gender. Parkin *et al.* (1984) reviewed a series of cases with bilateral optic neuritis 25 years after presentation. Of the 6 adults with acute simultaneous optic neuritis, one had died from Devic's disease and another was thought to have had early probable multiple sclerosis but died from other causes at the age of 76 years. No other patient developed multiple sclerosis. Conversely, 20 patients had bilateral sequential optic neuritis within 3 months and, of these, 7 were known to have developed multiple sclerosis. Follow-up of 23 cases with acute or subacute simultaneous bilateral optic neuropathy revealed that, after a mean of 71 months, 4 were shown by genetic analysis to have a mutation in mitochondrial DNA typical of Leber's disease, and another 5 had developed multiple sclerosis; the rest remained undiagnosed (Morrissey *et al.* 1995).

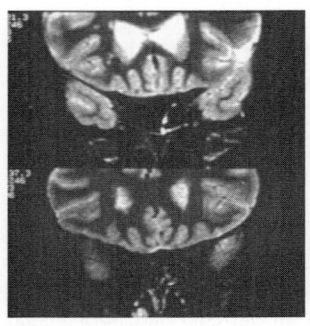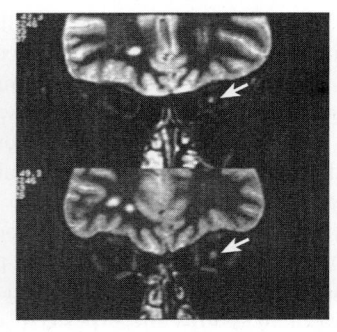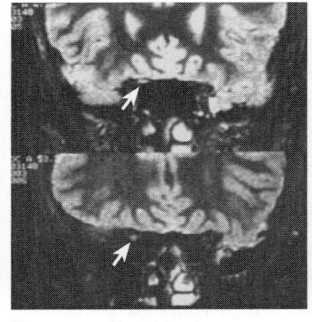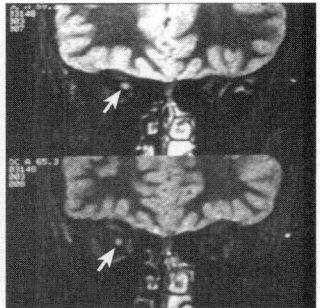

A   B

Fig. 37.5 Magnetic resonance STIR sequences showing lesions in the right optic nerve. A. Good prognosis—the lesion (arrow) does not extend posteriorly into the optic canal; white matter lesions are also seen in the frontal lobes. B. Poor prognosis—the optic nerve abnormality is seen in all four slices including the canal. Reproduced from Miller *et al.* (1988).

There is a confused literature on whether the treatment of acute optic neuritis either with intravenous methylprednisolone or oral prednisolone influences the risk of conversion to multiple sclerosis. Initially, optic neuritis recurred in 28 per cent of patients in the Optic Neuritis Study Group and there was an increased risk associated with exposure to oral prednisolone. But in a subsequent analysis, 30 per cent of the 308 patients enrolled between 1988 and 1991 converted to clinically definite multiple sclerosis and there were no differences depending on treatment (Optic Neuritis Study Group 1997b).

Given these uncertainties, attempts have been made to determine whether biomarkers characteristic of multiple sclerosis serve to identify patients with optic neuritis who are destined to develop widespread demyelination. Retinal vascular sheathing, or the presence of inflammatory cells in the ocular vitreous, occur in 30 per cent of patients with optic neuritis and are associated with a slightly increased risk of multiple sclerosis at follow-up (Lightman *et al.* 1987). Periventricular white matter abnormalities demonstrated by magnetic resonance imaging are found in 61 per cent of patients with optic neuritis, more of whom develop multiple sclerosis than those with normal brain scans (Fisniku *et al.* 2008). Human leukocyte antigen HLA-DR2 is present in a higher proportion of patients with optic neuritis who subsequently develop multiple sclerosis than isolated cases but the relative risk is low and HLA typing is not a useful prognostic marker. More patients with optic neuritis shown to have oligoclonal bands on cerebrospinal fluid electrophoresis at presentation subsequently develop multiple sclerosis than those with normal spinal fluid. In the Optic Neuritis Study Group (1997b), lesions detected by magnetic resonance imaging at presentation were associated with an increased rate of conversion—as expected. Conversely, all 185 patients with no prior symptoms or imaging abnormalities, lack of pain, relative preservation of acuity, and a swollen disc carried a relatively good prognosis for conversion to multiple sclerosis: in patients who underwent spinal fluid analysis at presentation, the presence of oligoclonal bands slightly increased the relative risk for developing multiple sclerosis within 2 years but imaging abnormalities were more discriminating predictors of widespread demyelination. Recent additions to this literature include a series of 146 patients assessed within 7 years of acute optic neuritis in whom MRI abnormalities, oligoclonal bands, and the presence of HLA-DR2 were risk factors for the development of multiple sclerosis (Söderström *et al.* 1998).

37.4.3 Isolated spinal cord syndromes

The spinal cord is especially vulnerable to demyelination. As with more generalized forms of inflammatory central nervous system disease, the relationship of isolated spinal demyelination to multiple sclerosis is complex. Acute or subacute cord involvement in patients with established multiple sclerosis is usually partial, sometimes conforming to the Brown Sequard syndrome but lesions may occur that exactly mimic isolated forms of transverse myelitis. This term is usually reserved for patients in whom the spinal lesion does not recur, but the distinction is only possible in retrospect. Even though a monophasic episode of cord inflammation may result in persistent disability, many patients recover fully and without sequalae. The diagnosis of transverse myelitis is often made by exclusion and the precipitating cause not identified, as with acute disseminated encephalomyelitis in adults. However, clinical and laboratory criteria can usefully distinguish the various conditions.

Acute necrotizing myelitis

The original description of necrotizing myelitis was made in men, rather older than most cases of transverse myelitis, with slowly progressive lumbar cord disease occurring in association with chronic respiratory disease (Foix and Alajounanine 1926). Now, the term acute necrotizing myelitis can reasonably be applied to patients developing severe inflammation of the thoracic or spinal cord, in whom flaccid areflexic paraplegia with anaesthesia and loss of sphincter control progress rapidly over hours. The inflammation is sufficient to cause severe pain with meningism and systemic symptoms, including pyrexia. The clinical presentation suggests compression, and contrast radiology or imaging often reveals a swollen cord with spinal block. Since the cerebrospinal fluid shows a marked polymorphonuclear pleocytosis, raised protein and lowered glucose concentrations, these patients are frequently thought to have pyogenic or tuberculous infection of the central nervous system and are therefore treated with appropriate antimicrobial therapy. For these reasons, there is often a reluctance to use corticosteroids but the course of acute necrotizing myelitis can be influenced by high dose intravenous methyl prednisolone. Acute necrotizing myelitis has an appreciable mortality but, in survivors, the systemic features resolve within weeks leaving significant handicap and disability. Several organisms have been implicated and acute necrotizing myelitis is also described after rabies vaccination, as a complication of acute lymphocytic leukaemias, lymphoma, hypernephroma, and other forms of carcinoma, and in acquired immunodeficiency syndrome.

Acute transverse myelitis

The majority of patients with transverse myelitis are not systemically ill and the neurological disorder usually evolves over a few days. Pain at the site of the lesion may be the initial symptom, followed by weakness in the legs and positive sensory symptoms with sphincter involvement. Over time, the weakness increases and may spread to involve one or both arms, usually in an asymmetric pattern and showing the flaccid areflexia characteristic of spinal shock. These features are infrequently seen when cord inflammation occurs in the context of multiple sclerosis and have been used to distinguish monophasic from relapsing disease. Sensory loss replaces the paraesthesia and there is often a band of unpleasant hyperaesthesia at the upper sensory level. As in other cases of incomplete focal spinal disease, this may not accurately reflect the site of spinal affection due to lamination of fibres in the spinothalamic pathways. Sphincter control is lost but, unlike patients with multiple sclerosis, the patient is usually unable to empty rather than fill the bladder. The need to exclude a structural cause for subacute cord injury occurring as a manifestation of transverse myelitis means that many patients necessarily undergo radiological investigation. This may demonstrate mild cord swelling, which is rarely sufficient to cause spinal block, but MRI usually shows a longitudinal rather than transverse area of increased signal (Fig. 37.6). The cerebrospinal fluid shows an increased mononuclear cell count, numerically intermediate between the marked pleocytosis of acute necrotizing myelitis and the abnormalities seen in patients with multiple sclerosis. The total protein is raised and oligoclonal bands may be present on electrophoresis but the glucose is usually normal.

Clinical guidelines can be used to distinguish transverse myelitis from multiple sclerosis. The most reliable indicator is the presence of spinal shock in acute monophasic lesions. Symmetry of the

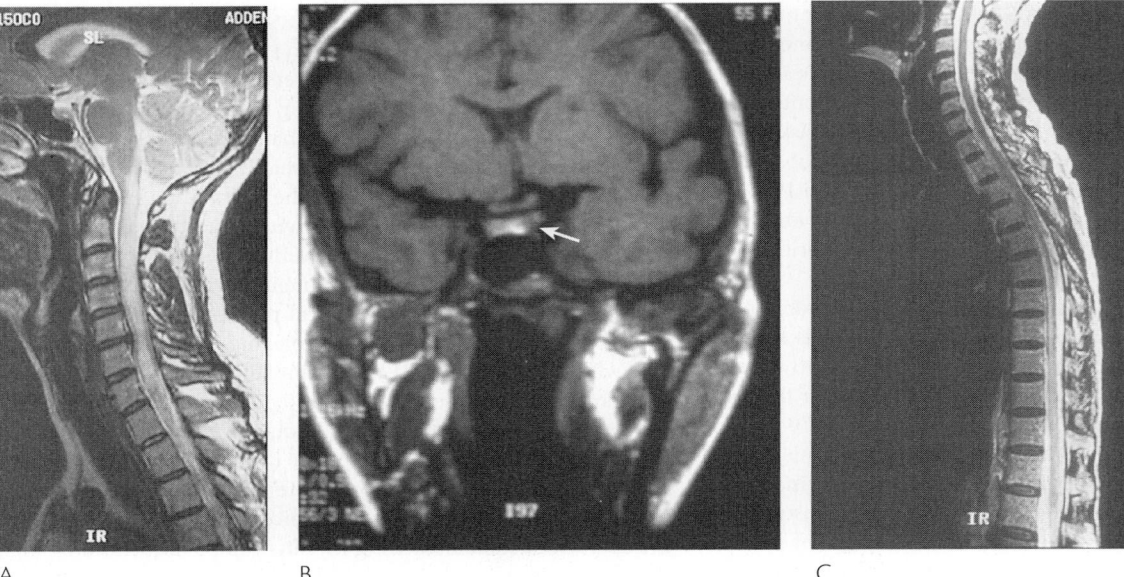

A   B   C

Fig. 37.6 Acute transverse myelitis. A. Swelling and high T_2-weighted MRI signal in the cervical cord and medulla B. T_2-weighted image showing high signal in a patient with bilateral optic neuritis due to neuromyelitis optica. C. Spinal cord imaging showing extensive abnormality over several segments in the same patient. The cerebral magnetic resonance scan showed a single periventricular abnormality.

motor and sensory manifestations of acute inflammatory spinal cord disease usefully identifies patients with transverse myelitis, whereas spinal involvement in multiple sclerosis is almost invariably asymmetric. Defined in this way, there is a low conversion rate from transverse myelitis to multiple sclerosis. Unlike acute disseminated encephalomyelitis, transverse myelitis is more common in adults than children, and in women than men. Transverse myelitis differs from acute disseminated encephalomyelitis in the peak age of onset and its more frequent occurrence as a manifestation of systemic vasculitis, immunodeficiency, and a range of specific infectious causes. It has a low rate of conversion to multiple sclerosis and, in this respect, resembles bilateral rather than unilateral optic neuritis in adults. Multiple sclerosis and transverse myelitis cannot reliably be distinguished on the basis of changes in CSF: as a generalization, the absence of oligoclonal bands and a cell count >100 lymphocytes/cmm favour transverse myelitis.

The series reported by Berman *et al.* (1981) is probably representative with respect to prognosis although a high proportion of this population-based sample of cases originally designated as transverse myelitis were later excluded from their study. Seventy two per cent of 62 patients in whom follow-up information was available made an adequate recovery over 3 months: 3 (5 per cent) died and 14 (23 per cent) were left with significant persistent disability. Incidence peaked in the second and third decades with a further bimodal increase in patients aged >70 years. Only one patient subsequently developed multiple sclerosis. Preceding infection was reported in one-third of patients, the majority having an upper respiratory infection, and other causes being herpes zoster or simplex virus infection, hepatitis, and smallpox vaccination. An identifiable infection was more common in young patients

suggesting that cases with spinal stroke, including examples due to collagen vascular disease, are sometimes erroneously diagnosed as having transverse myelitis. This is especially problematic in patients with sensory signs indicating an anterior cord lesion. Bakshi *et al.* (1998) compared MRI findings in 9 patients with acute transverse myelitis and 13 with myelitis due to multiple sclerosis. Cord swelling was non-discriminatory but long lesions involving multiple cord segments were characteristic of myelitis whereas the lesions involved on average only one or two cord segments in multiple sclerosis. Brain MRI was normal in 78 per cent with acute transverse myelitis but in only 15 per cent with multiple sclerosis.

Tyler *et al.* (1986) collated the following viral causes of transverse myelitis in man: picornaviruses, togaviruses, retroviruses, orthomyxoviruses, paramyxoviruses, bunyaviruses, arenaviruses, rhabdoviruses, hepatitis viruses, herpes viruses, and poxviruses. Subsequent isolated case reports emphasize the occurrence of transverse myelitis after infection with hepatitis A, hepatitis C, cytomegalovirus, herpes simplex type 2, toxoplasmosis, schistosomiasis, Borrelia, and Coxiella. Epstein–Barr virus infection has resulted in cases of myeloradiculitis and encephalomyeloradiculitis with the presence of viral DNA in cerebrospinal fluid suggesting a direct infectious mechanism.

37.4.4 Neuromyelitis optica

As originally described, neuromyelitis optica, or Devic's disease, is characterized by confluent demyelination of both optic nerves or the chiasm together with equally severe spinal cord damage. These sites may be affected simultaneously or sequentially and in either order, events usually being separated by several weeks or months.

It is now recognized that many cases of neuromyelitis optica follow a relapsing-remitting course.

Ideas on the nature of neuromyelitis optica, and its relationship to multiple sclerosis, have evolved. The largest series of 71 patients observed over 43 years at the Mayo Clinic adopted strict diagnostic criteria relating to the bilaterality of optic nerve involvement and the interval between defining events but usefully demonstrated a pattern of reasonable recovery when the optic nerves and spinal cord were affected in rapid sequence, and a less favourable prognosis with high mortality in relapsing patients with more than the two episodes. As now defined, neuromyelitis optica is characterized by bilateral optic neuritis and acute myelitis with no evidence for clinical disease elsewhere in the central nervous system: a spinal lesion extending over ≥ 3 segments and normal cerebral magnetic resonance imaging (Fig. 37.6); CSF pleocytosis of >50 wbc or 5 neutrophils but usually without oligoclonal bands; the course monophasic or relapsing and remitting and sometimes responding to plasma exchange (Wingerchuk et al. 2006). Pathologically, the extensive spinal lesions are associated with necrosis and cavitation, acute axonal injury, loss of oligodendrocytes, inflammatory infiltrates, and peri-vascular deposition of immunoglobulin, IgM, and complement. These are the features of a predominantly Th2 immune response with prominent humoral mechanisms (Lucchinetti et al. 2002). The whole story has been clarified by the demonstration that neuromyelitis optica is associated with an autoantibody directed against aquaporin-4 (Lennon et al. 2004, 2005; Misu et al. 2007; Roemer et al. 2007). The important therapeutic implication of these analyses is that affected individuals may respond usefully to plasma exchange (Weinshenker et al. 1999). Elsewhere, we have emphasized that neuromyelitis optica has been relatively more prevalent in Africans, Asians, Orientals, and Aboriginal Peoples with demyelinating disease than relapsing-remitting multiple sclerosis seen in western Europeans but with a change in phenotype following industrialization or, significantly, with exposure to the 'climate' of Europe during the years in which susceptibility factors leading to multiple sclerosis are believed to act (Cabre et al. 2006) an evolutionary hypothesis that links neuromyelitis optica in genetically more ancient populations to immunological and genetic maturation that results in the western-type disease of Europeans, now increasingly more prevalent in Japan, has been proposed (Compston 2007).

37.4.5 Other isolated syndromes

Clinically isolated brainstem episodes are often the harbingers of more widespread demyelination. The risks and predictors are summarized by O'Riordan et al. (1998). Of 16 such patients, 11 progressed to clinically definite multiple sclerosis; all had abnormal magnetic resonance imaging outside the affected site at presentation and only 1 patient with multiple lesions had not developed widespread demyelination at 10 years. Corresponding figures for conversion, with and without magnetic resonance imaging lesions, in patients with clinically isolated optic nerve and spinal cord lesions in this series were 25/28 and 1/14, and 10/15 and 2/8, at 10 years, respectively. The clinical symptoms and signs of isolated brainstem demyelination do not differ from those seen in multiple sclerosis typically consisting of dysequilibrium, disturbed eye movements, facial numbness, and dysarthria. But there may be severe headache which rightly prompts early investigation aimed at excluding a structural lesion.

37.5 Multiple sclerosis

37.5.1 Environmental factors

Epidemiological studies of multiple sclerosis have been used to generate aetiological hypotheses, to assess local needs for the provision of services and the allocation of resources, and to define the natural history of the disease. For northern Europeans, an approximate estimate of lifetime risk for multiple sclerosis is 1:400. Incidence measures the number of new diagnoses in a defined area over a given period. Prevalence defines the number of individuals in whom the diagnosis has been made, or those with a particular feature, in a population at risk and on a given occasion. Mortality describes the number of individuals with multiple sclerosis, as a proportion of the at-risk population dying within a defined area and over a given period. Self-evidently, morbidity statistics are dependent on classification and definition, and the success of case ascertainment. These vary inversely with the size and accessibility of the population at-risk. First surveys tend to underestimate prevalence and incidence, and higher figures are obtained with second and subsequent assessments, as awareness and vigilance improve amongst both the population at-risk and the investigators. Mortality estimates often fail to distinguish death due to multiple sclerosis or its direct complications, from other causes in people who nevertheless have the disease. In the most recent update of a prospective Danish series, Brønnum-Hansen et al. (2004) showed that life expectancy is reduced by around 10 years, 55 per cent patients with multiple sclerosis dying from the disease or related infections and with a somewhat higher than expected mortality from suicide and accidents. Assuming adequate ascertainment, incidence should equal mortality and prevalence will be the product of either statistic and disease duration unless there has been a change in aetiological factors. Given the gender differences and range of ages at onset, age- and sex-specific rates relating the number of affected individuals to a denominator confined to that proportion of the at-risk population born in the same decade should ideally be calculated. The usual way of dealing with variations in demography is to derive standardized prevalence ratios and to quote 95 per cent confidence intervals for each statistic.

Regional prevalence

Methodological factors have limited the value of many epidemiological studies. Most vulnerable have been the comparisons of prevalence between regions, and the serial studies of single places. Kurtzke (1975) first collated estimates for prevalence and suggested that the distribution of multiple sclerosis could broadly be classified into bands of low, medium and high prevalence. High risk (>25/10^5) was found throughout northern Europe, the northern United States, Canada, southern Australia and New Zealand. Medium risk (5–25/10^5) was found in southern Europe, the southern United States and northern Australia. Low risk (<5/10^5) areas included Asia, South America and many uncharted regions. Systematic updating of these figures shows that the absolute estimates have almost invariably risen and Kurtzke's original definitions of high, medium and low frequency bands no longer apply. Also, many apparent latitudinal gradients were evidently over-emphasized but the disease does show variations in distribution over quite small distances which may be informative with respect to aetiology (Compston and Confavreux 2005: Fig. 37.7).

Now, the epidemiological evidence indicates that, within Europe, multiple sclerosis is more common in southern Scandinavia than

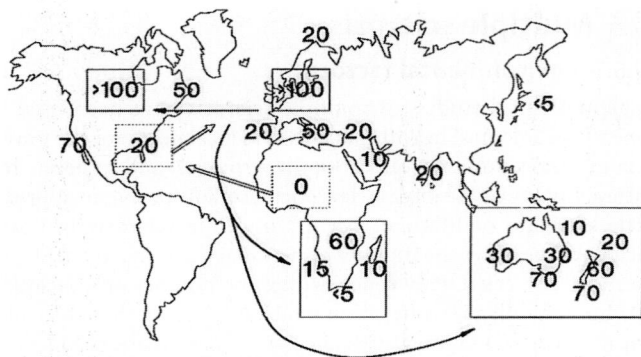

Fig. 37.7 A summary of epidemiological patterns in multiple sclerosis. Figures are estimates for prevalence. Solid lines with arrows represent migration vectors of northern Europeans. Open lines with arrows represent migration routes of Africans to the Caribbean and Missisipi delta and to the United Kingdom. In South Africa the numbers refer to English speaking whites migrating as adults (60), English speaking whites migrating as children (15), Afrikaaners (10), and Cape Coloureds (<5).

the north; in the Orkney and Shetland Islands than the Faroes or Iceland; in Italy, especially Sardinia, than Greece or Spain; in Sicily compared with neighbouring Malta; and in isolated parts of Finland. In North America, there is a diagonal gradient in frequency with rates highest in the midwest and lowest in the Mississippi Delta. A latitudinal gradient exists for the white Australian population, which shows higher rates in the south than the north. Multiple sclerosis has a frequency in the Far East one-twentieth of that seen in northern Europe. The rate is intermediate in India and Asia but even lower in Africa. The frequency of multiple sclerosis is considered to have fallen in some places, for instance Orkney, Denmark, southern Sweden, western Norway and Sardinia, whereas it is rising in other locations. The extent to which the latter is an artefact of ascertainment, or biologically driven, and the difficulty of making comparisons that are not confounded by the catching up of newly surveyed regions, with saturation in those that have been repeatedly scrutinized, cannot immediately be resolved. Meta-analysis of all surveys in multiple sclerosis where age and sex corrections can be made to the standardized European population, suggests that the perceived differences in distribution may in fact be artefacts of varying demographic structures in the populations at-risk (Zivadinov *et al.* 2003). With this manoeuvre, the north–south geographical gradient for age- and sex-adjusted prevalence lessens, and that for age-corrected incidence disappears altogether.

Changing incidence

Taking the big picture, Kurtzke (1993) concludes that whereas multiple sclerosis was geographically a regional disorder of males in the mid-20th century, subsequent experience shows more multiple sclerosis in places where the disease was already known to occur, diffusion into places where it was hitherto uncommon across even quite narrow confines within single land masses, and a steady increase amongst women and races other than Caucasians. Others have since emphasized the apparent rising prevalence in females (Orton *et al.* 2006). For Kurtzke, the regional increases within Asia have been in the former Soviet Union but not Japan, Korea or China. In Europe, the southern littoral of the Mediterranean and Israel have shown conspicuous increases in prevalence. The same trends

are apparent in Central and South America, notably Mexico, Argentina and Uruguay. But many of these focal trends are based on unpublished data, and hence vulnerable to all the factors that make for serial increase in morbidity statistics independent of an increase in incidence. The main factor determining patterns in the observed frequency of multiple sclerosis, excluding differences dependent on hospital- or population-based surveys, is when, as much as where, the study was carried out.

Migration studies

Populations are not stable, geographically or socially, and migrations involving relatively large numbers of people are known to have affected the distribution of multiple sclerosis. The interpretation of these studies has been to regard multiple sclerosis as an acquired disorder triggered by environmental conditions. Surveys in migrant northern Europeans support the environmental doctrine of causation in multiple sclerosis since, with the exception of Canada, parts of the world colonized from northern Europe show prevalence rates that are lower than the countries of origin. Dean (1967) first showed in South Africa that the age-corrected frequency is relatively high in immigrants from Europe, low in Afrikaaners, and intermediate in South African English-born both with respect to prevalence and incidence; no cases were observed in African blacks but a slightly higher rate was seen in the Cape-coloured population which has mixed African and European ancestry. In English speaking whites, those moving from northern Europe to southern Africa as adults retained the high frequency of the country of origin, whereas those migrating below the age of 15 years showed the lower rates characteristic of native-born inhabitants of southern Africa. The study of United Kingdom-born children of immigrants from the Indian subcontinent, Africa and the West Indies showed that the prevalence of multiple sclerosis in the United Kingdom-born children of West Indian, African and Asian immigrants approximates to that seen in similar age groups amongst Caucasians (Elian *et al.* 1990). Another important series of epidemiological studies showed no substantial difference in the frequency of multiple sclerosis amongst Orientals living in Japan, Hawaii or the west coast of North America despite more marked variations in frequency amongst Caucasians in some of these locations, indicating a protective effect for Orientals irrespective of environment (Detels *et al.* 1977). The study of multiple sclerosis in Israeli immigrants showed a difference in prevalence between northern European Ashkenazis and Asian and African Sephardis. The implication is that racially determined differences in risk for multiple sclerosis are modified by environment (Kahana *et al.* 1994). In United States military veterans, the mortality ratio was shown to depend not on whether individuals were born in southern, middle, or northern states, but on where they entered military service as young adults (Kurtzke 1993). The Australian epidemiological studies reporting on a large geographical area with a relatively homogeneous population mix usefully update these studies from the 1960s. The distribution of migrants from the United Kingdom and Ireland shows the same latitudinal gradient as for the entire population but with a bias introduced by the high prevalence for multiple sclerosis in the cohort from Hobart, Tasmania; that location apart, the rates are lower than for contemporary studies from the United Kingdom but with no differences dependent on age at migration, taking 15 years as the point of stratification (Hammond *et al.* 2000).

Point source epidemics. John Kurtzke has been the main proponent for the occurrence of point source epidemics of multiple sclerosis. Sceptics prefer the position that these merely follow cycles of increased ascertainment due to local enthusiasm rather than genuine changes in incidence arising from the introduction of transmissible aetiological factors into virgin populations. The fact that all the better documented examples have occurred in small island communities in the North Atlantic can be seen to support either view. In the Orkney and Shetland Islands, the incidence and prevalence of multiple sclerosis were at one time higher than anywhere else. Serial estimates between 1954 and 1974 showed that prevalence peaked at $309/10^5$ in Orkney and $184/10^5$ in Shetland (Poskanzer *et al.* 1980). Subsequently, these frequencies fell but prevalence remains higher than in places of comparable latitude. The original observations on multiple sclerosis in the Faroe Islands showed fewer cases than expected from comparisons with neighbouring Orkney and Shetland. The first survey of Iceland, performed retrospectively, showed increased quinquennial rates for incidence between 1945 and 1954 during which age-at-onset was also younger. This led to the conclusion that there has been a post-war epidemic of multiple sclerosis in Iceland but others prefer the interpretation that this is an artefact of improved case recognition (Benedikz *et al.* 1994). Based on a total of 41 cases with onset between 1943 and 1986, Kurtzke (1993) concludes that the critical factor determining the Faroes Islands' experience of multiple sclerosis was occupation by British troops between 1940 and 1945, the development of multiple sclerosis showing both a temporal and spatial relationship to villages where individuals lived who contributed to peaks of incidence.

Accordingly, one interpretation of the epidemiological patterns is that multiple sclerosis originated in Scandinavia, in central Norway or the south-central lake district of Sweden, in the early 18th century and diffused across the Baltic states and northern Europe including the British Isles over the next 100 years. From there, it was exported to North America and Australasia, to southern Africa and Italy. It is an attractive hypothesis but whereas for Kurtzke (1993) the factors being distributed are germs, for others they are genes.

Role of infections

Several studies have attempted to correlate exposure to viral illness in childhood with subsequent development of multiple sclerosis (see Granieri and Casetta 1997). The picture to emerge is that the risk is increased for individuals who develop a variety of exanthematous and other common viral disorders relatively late in childhood; this applies to measles, mumps, rubella and especially Epstein–Barr Virus infection. Rarely in the history of research in multiple sclerosis is there not a current microbial favourite for the cause of the disease based either on population serology, identification of particles in one tissue or another, recovery of defined organisms from body fluids, or relaxed speculation from the armchair. None has stood the test of time; and most have disappeared from interest often due to the subsequent recognition of technical artefacts as each tarnished microbe is displaced by new and more compelling germs eager to take their place on the rostrum. In the face of these difficulties in pinning the cause of multiple sclerosis on any one organism, some commentators have argued that the disease must have several triggers. But since the search for an environmental cause of multiple sclerosis remains stubbornly unproductive, sceptics point out that, in contrast to the interpretations offered by geneticists, who now have some toe-holds, proponents of the environmental doctrine still have little more than a smoking epidemiological gun in consolidating evidence for a microbial or physical agent as the dominant cause of multiple sclerosis. Frustrated by the low dividend from systematic searches for candidate infectious agents using sophisticated methods for virus detection, alternative theories have therefore been advanced involving the role of climate, diet, geomagnetism, sunlight, air pollutants, radioactive rocks, and toxins to account for the global pattern of the disease.

Age of exposure

Epidemiology suggests a third factor in the cascade of aetiology and pathogenesis. This is the age at which exposure to a critical trigger occurs in the susceptible individual. Patients with demyelinating disease report later age at infection by measles, mumps, rubella and, especially, Epstein–Barr virus infection compared with controls selected for the same frequency of the human leukocyte antigen, HLA-DR2 so as to match for at least one marker of genetic susceptibility (Martyn *et al.* 1993). Infectious mononucleosis after the age of 18 years carries a relative risk for multiple sclerosis of 7.9 (95 per cent CI 2–38). The hypothesis has some support from molecular immunology. Although confined to the study of a single T-lymphocyte clone harvested from the blood of an individual with multiple sclerosis, Lang *et al.* (2002) describe T-cell receptor specificity for a residue of myelin basic protein (amino acids 85–99) seen in the context of DRB1*1501 restriction, and an epitope of Epstein–Barr virus (residues 627–641) in association with DRB5*0101. Four T-cell receptor peptide contacts were identical for myelin basic protein and Epstein–Barr virus. Thus, there is molecular mimicry.

37.5.2 Genetic susceptibility

Multiple sclerosis has a familial recurrence rate of approximately 20 per cent and it is usually assumed that this is due to co-inheritance of susceptibility genes. The most comprehensive studies are from Canada, the United Kingdom, and Belgium (Sadovnick *et al.* 1988; Robertson *et al.* 1996; Carton *et al.* 1997). Meta-analysis of recurrence risk amongst 44 177 relatives of 2163 probands from these three population-based series shows that the age-adjusted risk is highest for siblings, 3 per cent, then parents, 2 per cent, and children, 2 per cent, with lower rates in second and third degree relatives. Overall, the reduction in risk changes from 3 per cent (relative risk 9.2) in first degree relatives to 1 per cent in second and third degree relatives (relative risks 3.4 and 2.9, respectively), compared to a background age-adjusted risk in northern European Caucasians of 0.3 per cent (Fig. 37.8). Four recent studies approximate to a population-based series of multiple sclerosis in twins. Two show consistency in demonstrating a higher clinical concordance rate in monozygotic (approximately 25 per cent) than dizygotic pairs (about 5 per cent: Mumford *et al.* 1994; Willer *et al.* 2003). These studies indicate that the relative risk for multiple sclerosis in the monozygotic twin partner of an affected proband is about 190. Studies from France and from Italy provide different results although, technically, all the surveys indicate rates for monozygotic and dizygotic twins that fall within the same confidence intervals (French Research Group on Multiple Sclerosis 1992; Ristori *et al.* 2006).

Considering individuals with multiple sclerosis who are adopted before the age of 1 year, and affected individuals who through adoption have non-biological siblings or children, the frequency of multiple sclerosis in non-biological parents, siblings, and

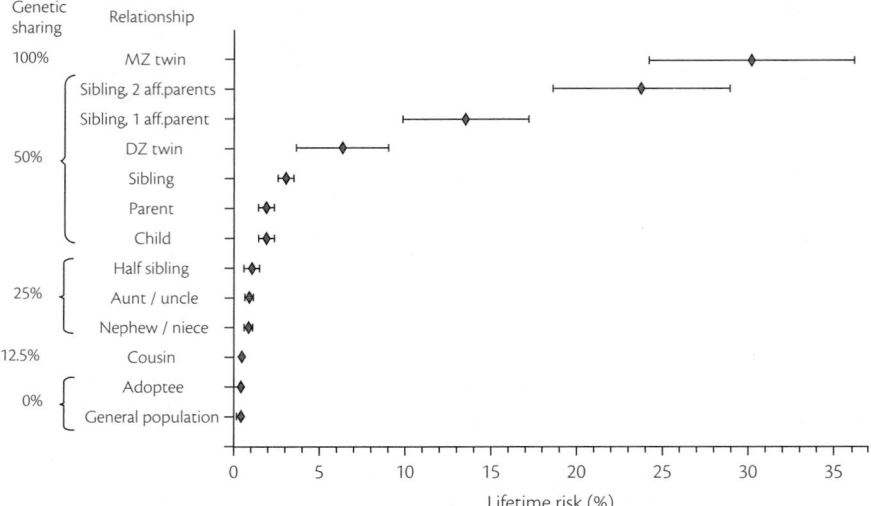

Fig. 37.8 Recurrence risks for multiple sclerosis in families. Age-adjusted recurrence risks for different relatives of probands with multiple sclerosis. Pooled data from population-based surveys. Estimated 95 per cent confidence intervals are shown. Figures on left show the degree of genetic sharing between relative and proband. (Adapted from Compston and Coles (2002).)

children is more or less identical to the population prevalence and lifetime risk for Europeans, and significantly lower than that expected from the study of recurrence risk in the biological relatives of index cases (Ebers *et al.* 1995). The age-adjusted risk for half-siblings is significantly lower than for full siblings and there is no difference in risk for half-siblings reared together or apart (Sadovnick *et al.* 1996). The recurrence risk is higher for the children of conjugal than single affected parents (Robertson *et al.* 1997; Ebers *et al.* 2000).

Population studies demonstrate an association between the class II MHC alleles DR15 and DQ6 and their corresponding genotypes DRB1*1501, DRB5*0101 and DQA1*0102, DQB2*0602 (Olerup and Hillert 1991). This relationship is most apparent in northern Europeans, and although the strength varies, the same association is seen in almost all other populations apart from Sardinians and some other Medditerranean populations in whom multiple sclerosis is associated with DR4 (DRB1*0405-DQA1*0301-DQB1*0302: Marrosu *et al.* 1992). For many years, extensive searches using association and linkage studies, yielded results implicating additional loci that could not be replicated but, as the scale of these enterprises has increased, genuine effects are at last being revealed. Three approaches have been used:

- whole genome screening aimed either at identifying linkage between a chromosomal region of interest, that may be wide and require narrowing in order to hone-in on the responsible gene (Sawcer *et al.* 2005), or association between a marker and the relevant susceptibility gene resulting from linkage disequilibrium (The Games Collaborative Group 2006; International multiple sclerosis genetics consortium 2007);

- linkage or association with putative candidate genes selected from *a priori* knowledge of the pathogenesis or some other selection process (Compston and Wekerle 2005);

- the study of highly informative populations such as genetic isolates or other groups in which the presence of multiple sclerosis offers an opportunity to characterize the effect in that population, and perhaps having implications for understanding

the disease more generally (Tienari *et al.* 1998; Reich *et al.* 2005).

Inevitably, the most attractive candidates have been genes encoding proteins involved in immunological pathways and structural molecules on which development and maintenance of the axon–glial unit depend. Additions to the list of susceptibility genes or loci include HLA-C (Yeo *et al.* 2007), IL-7Rα (Gregory *et al.* 2007) and IL-2Rα (International multiple sclerosis genetics consortium 2007). These genetic analyses are predicated on the assumption that multiple sclerosis is one disease. Mutations of mitochondrial DNA are responsible for a multiple sclerosis-like illness characterized by disproportionate involvement of the anterior visual pathway (Harding *et al.* 1992; Riordan-Eva *et al.* 1995) although mitochondrial genes do not contribute generally to susceptibility in multiple sclerosis.

Until recently, it was hard to take away from the literature a clear position on whether genetic factors influence the course of multiple sclerosis within families. Resolving this issue has a significance that goes beyond the important application of counseling affected individuals who look to the proximity of multiple sclerosis in an affected first-degree relative as a prognostic indicator since a whole industry of molecular genetic studies has been predicated on the fact that genes shaping the clinical course of multiple sclerosis exist and can be found by stratified case–control studies. Recently, Hensiek *et al.* (2007) evaluated 1083 families with ≥2 first-degree relatives having multiple sclerosis and showed concordance for age at onset in all families, and for year of onset and clinical course in affected siblings but not affected parents and their children. These influences on the natural history applied to all clinical subgroups of relapsing-remitting, and primary and secondary progressive MS, reflecting a familial effect on episodic and progressive phases of the disease but there was no concordance for disease severity. Therefore, familial factors do not significantly affect eventual disease severity but they increase the probability of a progressive clinical course, either from onset or after a phase of relapsing remitting disease. The familial effect is more likely to reflect genetic than environmental conditions in these families.

37.5.3 Clinical symptomatology

There are many authoritative clinical accounts of multiple sclerosis drawing on geographically defined population bases, hospital clinics, or extensive personal experience of the disease (McDonald and Compston 2005). The account that follows is organized by system and does not prioritize the frequency of individual manifestations of multiple sclerosis, or their importance for the individual.

Cognitive and affective symptoms

Periventricular plaques are almost invariably present in patients with multiple sclerosis from an early stage but it is usually held that these are not clinically significant. However, approximately 60 per cent of patients can be shown to have mental changes, and systematic testing has revealed that impairment of memory and learning ability may occur early in the course. These findings have been extended to patients with isolated demyelinating lesions, showing impaired cognitive ability due mainly to defects of visual and auditory attention. 'Cortical' presentation is well recognized (Zarei et al. 2003) and significant dementia may occasionally be an early component of the disease.

Amato et al. (2001) compared cognitive function in 50 patients early in the course of multiple sclerosis and after 4 years, with 70 controls. The patients performed worse at both assessments but, whereas defects of verbal memory and abstract reasoning remained stable, linguistic difficulties emerged in patients over time, and there was a poor correlation between cognitive and physical deficits. McIntosh-Michaelis et al. (1991) studied a population-based sample of more than 400 individuals, stratified for disability, and showed impaired intellect, working memory, or frontal lobe function in over 50 per cent. As expected, these deficits become more severe and prevalent in patients with chronic progressive multiple sclerosis, and as the disease advances cognitive and physical disabilities tend to converge. It has not proved easy to correlate cognitive abnormalities with regional brain lesions. Foong et al. (1997) were unable to disentangle the contribution from frontal lobe pathology to the general clouding of intellect, and found no correlation with regional magnetic resonance lesions. However, white matter lesions often affect the arcuate fasciculus in patients with prominent depressive symptoms (Pujol et al. 1997).

The cognitive defects usually occur in the absence of psychiatric symptoms although depression is more frequent in patients with multiple sclerosis than those with comparable neurological or medical disorders. Although it can be difficult to distinguish affective disease as a manifestation of multiple sclerosis from the reaction to physical deficits and their far-reaching implications, by comparing rates in first-degree relatives and unrelated individuals Sadovnick et al. (1996) showed that depression is reactive and not constitutional in patients with multiple sclerosis. The risk of suicide is increased about seven-fold in the first few years of diagnosis and in the descent from expectation to reality for the participants of clinical trials (Sadovnick et al. 1991). There is seldom much to be gained by confronting functional symptoms and exaggeration of physical deficits in multiple sclerosis but occasionally these are a major cause of handicap in the patient with otherwise minimally disabling disease. The usual methods of attempting to improve function by physical therapy and exhortation without confrontation may succeed.

Patients are occasionally seen with specific cognitive syndromes due to hypothalamic involvement including a Korsakoff state (Section 34.5) and features of the Klein–Levin syndrome, with bulimia, lack of social restraint, mental inertia, and mutism (Section 32.3.3). Disturbances of mood include hypomania, psychotic behaviour, the euphoria traditionally associated with multiple sclerosis, pathological laughter and crying arising from impaired central inhibition of facial and pharyngeal reflexes as part of a pseudo-bulbar palsy, and schizophrenia (Feinstein et al. 1993).

Special senses

Anosmia is reported in a high proportion of asymptomatic patients examined with more than usual clinical thoroughness (Pinching 1977). Clinical involvement of the visual pathway is almost invariable. The manifestations of optic neuritis are described above. As with so many features of multiple sclerosis, the episodic visual blurring described by patients early in the course may later evolve to fixed visual impairments, and slowly progressive visual deterioration. A few present with progressive loss of vision in one or both eyes—a clinical syndrome requiring special vigilance in the exclusion of other causes of visual failure (Ormerod and McDonald 1984). Rather few patients escape noticeable visual symptoms altogether, and many experience significant difficulties with reading late in the course. In addition to the sheathed appearance of retinal vessels on ophthalmoscopy (Lightman et al. 1987), symptoms due to uveitis are occasionally reported in association with multiple sclerosis (Biousse et al. 1999). Patchy involvement of the visual pathways and accumulated deficits from separate events, occasionally lead to unusual visual field defects. These include bitemporal hemianopia, hemianopic or junctional scotomata, visual slipping and post fixational blindness, homonymous hemianopia usually resulting from diffuse damage and associated with hemiplegia or aphasia, and visual perceptual defects (Plant et al. 1992).

Deafness occurs in multiple sclerosis, usually in established cases but occasionally at presentation. It is often associated with other clinical and electrophysiogical manifestations of brainstem disease although sudden unilateral or bilateral sequential loss of hearing with tinnitus have been described as the presenting symptom, as has cortical deafness. Feelings of unsteadiness are common, sometimes as part of an acute vestibular syndrome with severe positional vertigo, vomiting, ataxia, and the headache that often accompanies acute brainstem demyelination. Taste may be subjectively abnormal but genuine aguesia due to involvement of the tractus solitarius has not been systematically studied.

Eye movements

Abnormalities of eye movement are common in multiple sclerosis. They may occur in the absence of symptoms and so provide one method for detecting an additional site of demyelination in patients with clinically isolated lesions. The common symptomatic disorders are sixth nerve palsy, internuclear ophthalmoplegia, other horizontal and vertical gaze palsies, and the one-and-a-half syndrome in which there is horizontal gaze palsy to one side and impaired adduction to the other (Fig. 37.9). Isolated third and fourth nerve palsies are reported.

The commonest signs, often occurring without symptoms, are saccadic interruption of smooth pursuit, and first-degree symmetrical horizontal nystagmus. Vertical up-beating nystagmus is always associated with bilateral internuclear opthalmoplegia. Down-beating nystagmus has other important causes which can be confused with multiple sclerosis. Ocular flutter consists of bursts of horizontal saccadic oscillations without an intersaccadic interval. Opsoclonus, in which the movements occur in all directions, is equally disabling.

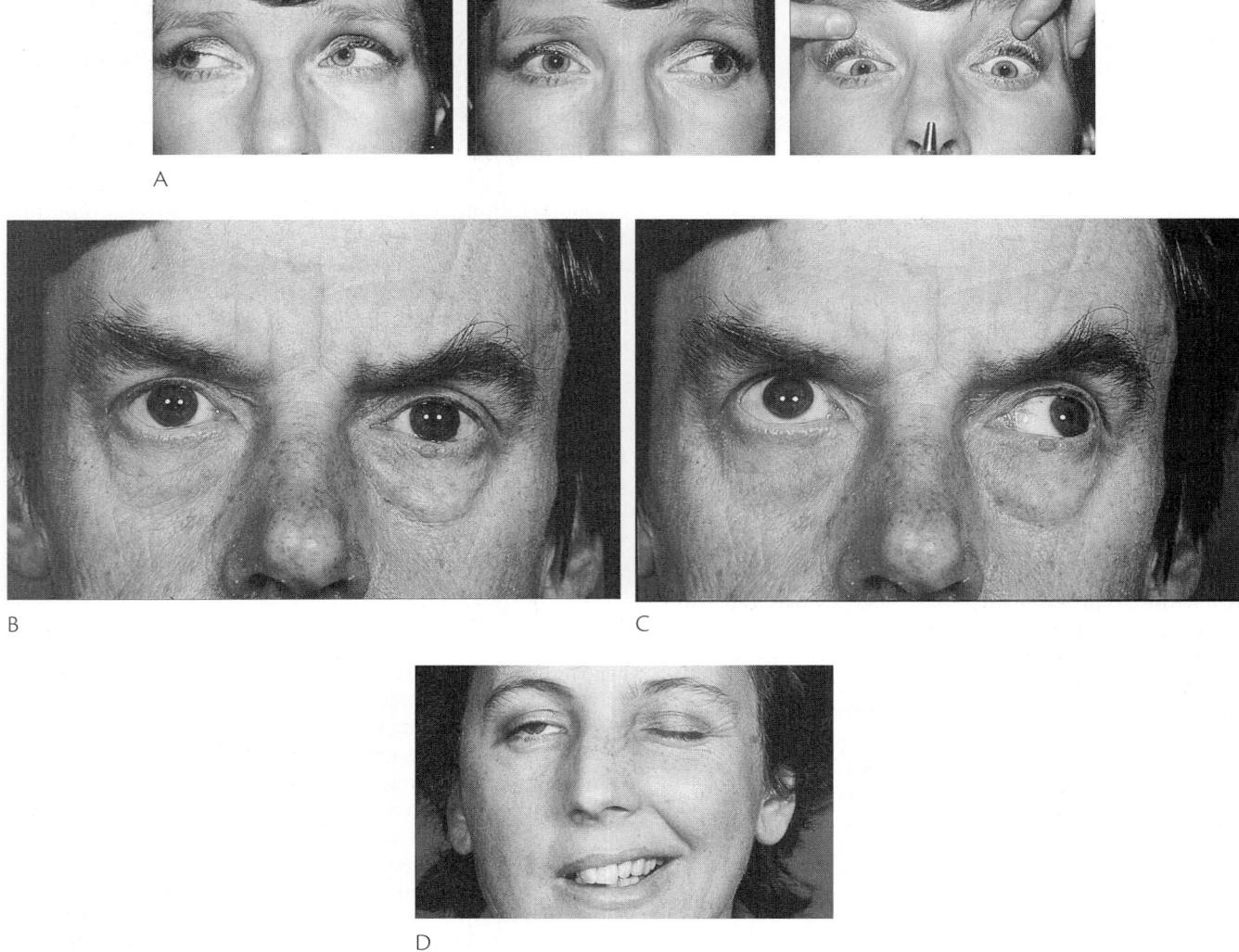

Fig. 37.9 A. Bilateral internuclear ophthalmoplegia; gaze to the right; gaze to the left; convergence (normal). B. The one-and-a-half syndrome; gaze to the right (absent). C. The one-and-a-half syndrome; gaze to the left (abduction only). D. Facial weakness during an acute episode of brainstem demyelination due to multiple sclerosis.

Ocular bobbing consists of an initial rapid downward eye movement followed by slow return to the neutral position and denotes cerebellar involvement. Abrupt displacement from the primary position during central fixation, known as square wave jerks, occurs with severe cerebellar deficits.

Involvement of other cranial nerves

Although isolated cranial nerve palsies are unusual manifestations of definite multiple sclerosis at any one time, Thomke *et al.* (1997) described these in around 2 per cent of patients: 1 third, 1 fourth, 12 sixth, 3 seventh, 6 vestibular, and 1 cochlear portion of the eighth nerve. Five per cent seen at presentation had an isolated cranial nerve lesion and this was the sole manifestation of a new relapse in a further 0.8 per cent of cases. Other than trigeminal neuralgia, isolated involvement of the fifth nerve is rare. In a large Japanese series, 20 per cent had facial palsy within the first 6 years of the illness and 5 per cent presented with facial weakness almost always with lesions

in the pontine tegmentum ipsilateral to the facial weakness (Fukazawa *et al.* 1997a) (Fig. 37.9). Hemifacial spasm and myokymia, a diffuse rippling of muscle fibres, are seen and, exceptionally, there may be unilateral involvement of the hypoglossal and recurrent laryngeal nerves.

Other brainstem manifestations

Extensive brainstem demyelination may produce disturbances of consciousness and respiratory failure distinct from the narcolepsy syndrome (Section 32.3.1) which is seen more frequently in patients with multiple sclerosis than expected by chance—an observation of immunogenetic interest in view of the strong HLA-DR2/DQ6 association with both conditions. Rare manifestations include the locked-in state (Section 33.5), persistent hiccough, cough syncope (Section 31.1.2), isolated unilateral abdominal paralysis, hypoglossal and laryngeal nerve palsies, optico-ciliary neuritis, and the lateral medullary syndrome.

Paroxysmal symptoms

These symptoms occurring in multiple sclerosis result from ephaptic transmission in partially demyelinated brainstem pathways. They are invariably brief but repetitive and last a few months before remitting. An individual patient may experience more than one type. Unlike the idiopathic variety, symptomatic trigeminal neuralgia may begin in the first division or bilaterally, at a younger age, and with associated signs of trigeminal involvement including motor weakness and sensory loss. It is usually associated with lesions of the dorsal root entry zone but may co-exist with compression of the fifth cranial nerve by ectatic vessels (Broggi *et al.* 2000). Paroxysmal dysarthria and ataxia with a clumsy arm, complex disturbances of sensation, and painful tetanic posturing of the limbs lasting 1 or 2 min are often triggered by movement and preceded by positive sensory symptoms on the side opposite to the muscular spasm. These are easily recognized and treated. Bursts of pain and paraesthesiae, sensory distortion, itching, cough and hiccough, painful extensor spasm, akinesia, kinesogenic choreoathetosis, and complex gaze palsies—any of which may respond to anticonvulsants, especially carbamazepine—also appear to be paroxysmal manifestations of multiple sclerosis (Fig. 37.10).

Cerebral hemisphere involvement

Multiple sclerosis may present as a cerebral mass lesion with hemiplegia, focal or generalized fits, and confusion. These symptoms are often due to large confluent space taking lesions. Understandably, they are confused with tumour and are often biopsied. Headache may occur, as with lesions in the posterior fossa. Other rare cerebral manifestations of multiple sclerosis are aphasia (Lacour *et al.* 2004), callosal disconnection syndromes, and cortical sensory loss. Epilepsy occurs in around 5 per cent patients with clinically definite multiple sclerosis and may manifest as tonic-clonic attacks with partial or focal seizures, or with epileptic status (Eriksson *et al.* 2002).

Motor symptoms and signs

Impaired mobility occurs in the majority of patients with multiple sclerosis at some stage but the pathophysiological basis and severity vary between cases. Weakness may develop gradually in one or more limbs, increasing with use. It is usually described as heaviness or clumsiness. Confusion may arise if the onset is sudden or the distribution hemiparetic thus mimicking a cerebrovascular episode. Although hemiplegia sometimes results from cord disease, spinal demyelination more usually causes progressive weakness in both legs especially when multiple sclerosis presents in older patients. Movements are slow, weakness differentially affecting extensors in the arms and flexors in the legs, and there are the expected reflex signs of upper motor neurone lesions. Motor disability in the limbs is often compounded by cerebellar ataxia and tremor, particularly in the arms, and by loss of postural sense and fatigue.

Spasticity forms only one component of the upper motor neurone syndrome but, depending on which descending corticospinal tracts are involved, it may dominate the clinical phenotype associated with damage to central motor pathways. In the progressive stage, increased muscle tone can manifest as painful spasms, both flexor and extensor, which disturb sleep and sometimes eject the patient from a wheelchair. As paraplegia in flexion and adduction develop, wheelchair and bladder management become difficult and the pain increases.

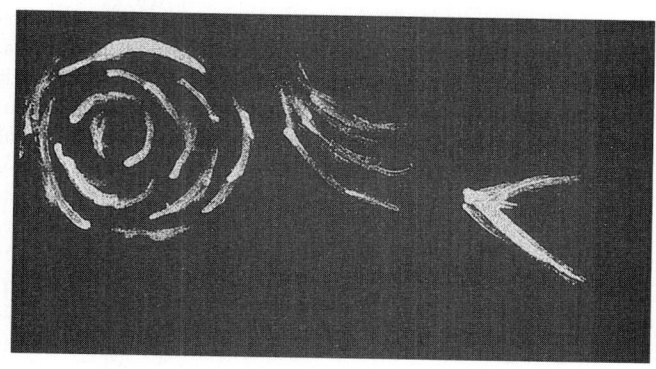

A

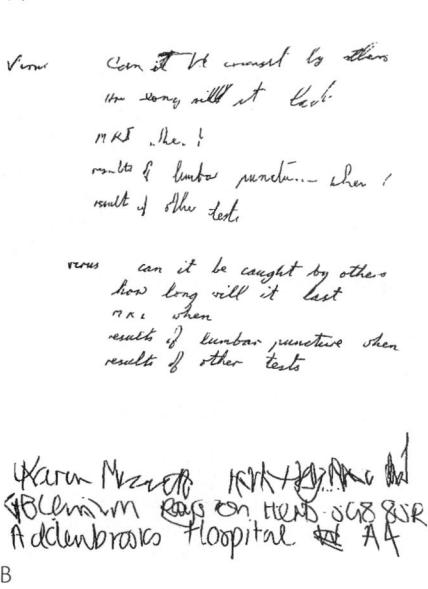

B

Fig. 37.10 A. Movement-induced phosphenes depicted by a patient with multiple sclerosis wishing to explain this *'unexpected'* symptom. B. Transient disturbance of handwriting in (two) patients with multiple sclerosis. *"virus–can it be caught–how long will it last–MRI when?–results of lumbar puncture when?–results of other tests'* [the usual questions of informed patients at the time of investigation]. *'Karen ********* Royston Herts ****** Addenbrooke's Hospital A4'.*

Signs of lower motor neurone involvement, with muscle wasting and loss of tendon reflexes, are seen in multiple sclerosis. This occurs either because of associated pressure palsies or extensive demyelination adjacent to the dorsal root entry zones, spinal cord grey matter, or the conus medullaris. The co-existence of peripheral neuropathy is well recognized (Thomas *et al.* 1987) with either the central or peripheral component dominating the clinical picture.

Equally disabling is fatigue—a poorly understood manifestation of multiple sclerosis described by a high proportion of patients to whom the question is put, and the dominant complaint of a few, often not otherwise disabled, young patients. Fatigue has been correlated with pyramidal tract involvement implicating a physical rather than psychological mechanism in multiple sclerosis. Alternative explanations based on physiological measurements, are that fatigue reflects impaired cortical drive despite sustained effort (Sheean *et al.* 1997).

Involvement of the cerebellum and its connections interferes with co-ordination of speech, bulbar function, eye movements, the individual limbs, or balance—usually in combination with corticospinal damage. Tremor is postural or kinetic, with several additional features that disrupt the rhythm or accuracy of goal directed movement, but rarely present at rest (Alusi *et al.* 1999). There are few clinical syndromes more disabling for the patient, more distressing for their carers, or therapeutically more frustrating for the neurologist than the combination of proximal upper limb tremor, titubation, and violent shaking of the trunk on attempted change in posture. This may result from lesions at several sites but especially from involvement of the superior cerebellar peduncle and its afferent connections with the thalamus.

Involuntary movements are an uncommon feature of multiple sclerosis but several stereotyped disorders are reported as symptoms rather than chance associations—a tricky distinction since several of these movement disorders are not known to have a firm anatomical substrate. The literature is usefully summarized by Tranchant *et al.* (1995), who also describe 14 additional cases in the context of 135 listed from other publications. The authors conclude that paroxysmal dystonias or tonic spasms, ballism or chorea, and palatal myoclonus may be manifestations of multiple sclerosis, whereas Parkinsonism, dystonia, and other types of myoclonus are likely to be co-incidental.

Sensory symptoms and signs

Altered sensation occurs at some stage in nearly every patient with multiple sclerosis. Certain patterns are characteristic but practically any part may be affected although these symptoms commonly arise from spinal involvement.

Demyelination of the dorsal or lumbar segments of the spinal cord produces paraesthesiae and numbness in the legs which usually start distally, ascending to a variable level on the trunk, indicating gradual extension of the evolving lesion through laminated fibres of the sensory pathways. Sacral sparing may occur but, alternatively, a characteristic sensory syndrome seen in patients with multiple sclerosis is numbness of the perineum and genitalia together with disturbed sphincter function.

Cervical spinal cord lesions tend differentially to affect the posterior columns producing perversions of normal feelings which are then perceived as tight, burning, twisting, tearing, or pulling sensations. These are more or less invariably unpleasant. Pain, frequently described as a manifestation of multiple sclerosis, often represents a complication of disability related to abnormal posture, spasm of the spinal musculature or legs, and osteoporosis arising from immobility and repeated courses of corticosteroids. Dysaesthetic limb pain is one of the most challenging problems encountered in pain relief clinics and is notoriously intractable. More disabling is the loss of proprioception and other forms of discriminative sensation which severely compromise function (Fig. 37.11). Pseudo-athetosis may be apparent in the outstretched hand, observed with the eyes closed, but a better test of discriminative sensation is the ability to manipulate buttons through a button-hole.

Lhermitte's symptom comprises an electric feeling in the trunk or limbs provoked by bending the neck, although other movements, such as coughing or laughing, can evoke the same effect. It indicates involvement of the posterior columns. Almost 50 per cent of patients experience the symptom and it may be a presenting

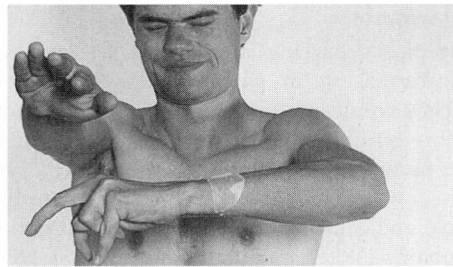

Fig. 37.11 On attempting to maintain posture with the eyes closed, the fingers develop 'pseudoathetosis' and the affected arm moves towards the opposite axilla.

feature, alone or in combination. Some patients describe a motor equivalent—involuntary movements of the legs provoked by flexion of the spine or neck.

Features of the Brown Sequard syndrome, resulting from involvement of the crossed spinothalamic tract, include the usual loss of thermal and pain sensation, but the diagnosis of multiple sclerosis should be questioned if there are painless burns. Non-specific tingling without accompanying signs is often encountered; and the commonest physical sign observed in the absence of symptoms is impaired vibration sense in the legs. Features of cortical sensory loss may occur with massive demyelination of the cerebral hemispheres.

Autonomic involvement

Bladder control depends on the ability of upper motor neurons, originating from the pontine tegmental micturition centre and set by the inferior frontal micturition centres to modulate spinal reflexes so as to promote either storage or emptying; curiously, each is right dominant in men and women. With uncoupling of reciprocal arrangements between the detrusor and sphincter, the bladder contracts against a closed sphincter leading to urgency and frequency with hesitancy or incomplete emptying and incontinence. In *conus medullaris* lesions, impaired emptying is more problematic than failure to fill and urinary hesitancy then occurs in association with retention. In other situations, the mechanisms of bladder filling and emptying become uncoupled leading to sphincter dyssynergia; attempts to promote filling may then further compromise emptying, and *vice versa*.

Bladder symptoms are described by 80 per cent patients, often in association with other manifestations of spinal cord disease but, in some patients, they are the predominant complaint and constitute the main factor interfering with aspects of daily living. Betts *et al.* (1993) report that, of 170 patients with multiple sclerosis referred because of urinary symptoms, 85 per cent had urgency, 82 per cent frequency, 63 per cent urge incontinence, 49 per cent hesitancy, and 14 per cent nocturnal enuresis. Only 2 patients had acute retention. An unstable bladder is more common in women; bladder symptoms are less prevalent in males in whom sexual impotence is the more frequent complaint.

Failure to control the rectal sphincter is much less of a problem than failure of emptying, and other disturbances of gastrointestinal function rarely occur. Up to 70 per cent of patients complain of constipation. Occasional faecal incontinence has probably been under-estimated in multiple sclerosis, occurring at least once in the previous 3 months in 51 per cent of a large clinic series (Hinds *et al.* 1990). However, altered bowel function features high on the list of subjective disabilities in people with multiple sclerosis. The pathophysiological mechanisms are not fully understood, but include involvement of the brain and spinal cord resulting in interruption of either or both the central afferent and efferent pathways on which normal bowel function depends. Emotional factors may also contribute.

The distinction is often made in clinical practice between those males who are impotent but retain reflex erections, in which case psychogenic factors are often infered, and those with erectile failure due to spinal cord demyelination. Assessment of libido may help in making this distinction which is important since both aspects can usefully be managed. The role of the autonomic nervous system in these situations is shown by the close association with other functional disturbances. Betts *et al.* (1994) found that a complaint of erectile failure in multiple sclerosis is always accompanied by urinary symptoms and by signs indicating disease affecting the spinal cord. Sexual function in women is more difficult to assess, but loss of libido, lack of lubrication, and failure of orgasm are common complaints and multiple sclerosis often has a profound effect on many aspects of female sexuality. In both sexes, alteration in genital sensation, fatigue, and spasticity in the lower limbs contribute to sexual difficulties and these difficulties increase with time.

Other autonomic features in multiple sclerosis include loss of thermoregulatory sweating, hyperthermia, hypothermia, fever, Horner's syndrome, abnormal cardiovascular responses, postural hypotension, atrial fibrillation, acute pulmonary oedema, fatal emaciation, and inappropriate secretion of anti-diuretic hormone (McDonald and Compston 2005).

Childhood multiple sclerosis

Even though criteria for the diagnosis of multiple sclerosis widely used in clinical practice limit the ages of onset to between 10–50 years, symptoms subsequently attributable to demyelination often first occur in childhood even though the significance may not become apparent until adult life. The majority first experience symptoms as teenagers but onset occurs at ≤ 16 years in around 5 per cent prevalent cases (Duquette *et al.* 1987; Bauer and Hanefeld 1993). Children with multiple sclerosis are usually girls and, if there is a special clinical flavour, it is the tendency to present with subacute encephalopathy. Fever and meningism, impaired conscious level due to cerebral oedema with swollen optic discs, and seizures all occur and the obvious distinction from acute disseminated encephalomyelitis can often only be made by the later occurrence of remission and relapse. Equally difficult is the separation of childhood multiple sclerosis from restricted forms of acute disseminated encephalomyelitis presenting as bilateral optic neuritis. In a retrospective series from the Mayo Clinic, from 1950 to 1988, from 13 per cent of 79 children, aged <16 years at presentation and reviewed a mean of 19 years later, had developed multiple sclerosis; actuarial analysis indicated conversion rates of 22 per cent and 26 per cent at 30 and 40 years, respectively (Lucchinetti *et al.* 1997) but there were no useful prognostic clinical features. Serial MRI is helpful in making the distinction from acute disseminated encephalomyelitis and predicting conversion to multiple sclerosis after an episode of isolated demyelination in childhood. Lesions perpendicular to the corpus callosum and well-defined discrete lesions are considered more specific but less sensitive than the application of

criteria for the radiological confirmation of multiple sclerosis in adults (Mikaeloff *et al.* 2004). There is no evidence that childhood onset multiple sclerosis evolves differently from adult onset relapsing-remitting disease (Simone *et al.* 2002). That said, children take longer from onset to reach the secondary progressive stage but do so at a younger age than adults (Renoux *et al.* 2007).

37.5.4 Clinical course and prognosis

The clinical course of multiple sclerosis is no less variable than the symptoms that may occur but some patterns can usefully be defined. Eighty per cent of patients present with episodic neurological symptoms that are multifocal or anatomically discrete and recover fully at first. Further episodes occur at a random frequency and for an unpredictable period but, in the majority of cases, seldom exceeding >1 year apart from brief bursts of more active disease. With time, recovery from each episode is incomplete and persistent symptoms begin to accumulate. Eventually, a proportion of patients cease to describe episodic manifestations and the course thereafter is slowly progressive. In 20 per cent of patients the illness is progressive from onset (Fig. 37.12). In a prevalent population, approximately one-third of patients are in a quiescent phase of the disease and not significantly disabled. A further third are slowly deteriorating and the remainder stable but variably disabled having had the disease for many years (Confavreux and Compston 2005). This clinical course usually evolves over several decades.

Relapsing and remitting disease

Many patients report new symptoms on waking. In others, the rate of onset of relapse varies from hours to days with symptoms usually accumulating in an anatomically coherent way. The commonest affected sites are the optic nerves, cervical portion of the spinal cord, and the brainstem. These will be affected in most patients at some stage in the illness. There is a tendency for old symptoms to recur in subsequent relapses, in part reflecting the reduced safety factor for transmission of the nerve impulse in hypomyelinated pathways. The rate of recovery is invariably slower than onset and several outcome patterns are recognized. Relapse may occur on an asymptomatic background or complicate existing disability. Recovery from an episode is often complete but, after the first few years, many patients experience step-wise accumulation in disability with each new relapse. Symptoms lasting less than 24 h are generally not categorized

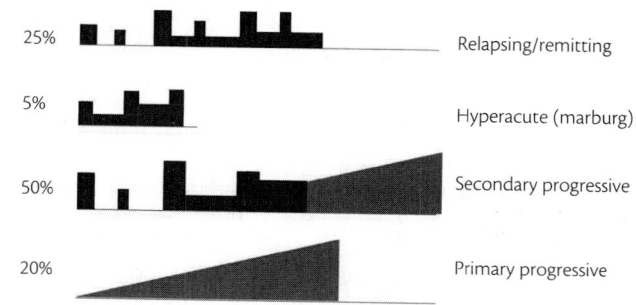

Fig. 37.12 The course of multiple sclerosis; relapsing/remitting with accumulating disability (50 per cent of patients), progressive from onset with or without superimposed relapses (20 per cent), hyperacute (5 per cent), and benign (25 per cent) forms of the disease are depicted.

as signifying a new relapse; and it is usual to regard those that evolve over a few weeks as part of the same episode.

Multiple sclerosis is not necessarily a disabling disease. In the occasional patient, a latent period of 2 or 3 decades occurs between the presenting episode and the next event. About 25 per cent patients have a form of the disease in which disability is minimal even after many years. Benign multiple sclerosis is epitomized by young females with predominantly sensory symptoms and complete recovery from individual episodes. Occasionally, however, patients with a mild clinical course suddenly deteriorate without useful recovery after a prolonged period of disease inactivity. Conversely, depending on social and personal circumstances, disproportionate handicap may arise from lesions that do not cause significant impairments. At the other extreme, patients may die early due to medullary involvement resulting in respiratory failure or massive confluent cerebral demyelination. In a few cases, the relapse rate is high and disability accumulates rapidly, producing immobility, lack of protective pharyngeal reflexes, bulbar failure, and bladder involvement all of which expose the patient to the hazards of immobility and life-threatening infections. This so-called 'Marburg variant' affects about 5 per cent of patients. But up to 15 per cent of all incident cases become severely disabled and dependent within a short time.

Progressive disease

Approximately 75 per cent of all patients with relapsing-remitting multiple sclerosis eventually enter the phase of secondary progression in which disabilities increase steadily despite the systematic reduction in relapse rate. In 20 per cent, the course is progressive from onset. This is the most common mode of presentation when multiple sclerosis develops beyond the 4th decade although relapsing disease is still occasionally seen in this age group.

Typically, primary progressive multiple sclerosis manifests as spinal disease but this diagnosis should only be made in patients who have been extensively investigated to exclude a structural cause for their spinal cord disease. Syndromes referable to progressive involvement of the optic nerves, cerebrum, and brainstem are also seen but less commonly. Secondary progressive multiple sclerosis may affect whichever system has borne the brunt of the disease earlier in the course. Again, this is typically the spinal cord and whereas sensory and motor sites may be involved at presentation in individuals who later develop secondary progressive multiple sclerosis, progression seems preferentially to target the corticospinal pathway. Confusion arises from the occasional overlap of the relapsing and progressive phases of multiple sclerosis. Progression may follow directly upon a severe relapse with partial recovery, be interrupted by further episodes, or occur without these temporary deviations. Recent epidemiological studies focus on the ages at which progression ensues, in order to try and understand the relationship between the relapsing and progressive phases of the clinical course and therefore between inflammation and axon degeneration. Kremenchutzky *et al.* (2006) conclude that the progressive phase of multiple sclerosis is an age-dependent degenerative process, at least in part; and that chronic axonal loss specific to the corticospinal tract is the pathologic substrate for progression, beginning early in the disease course and long before clinical symptoms manifest. Confavreux and Vukusic (2006) also conclude that the clinical phenotype and course of multiple sclerosis are age-dependent. Times to reach disability milestones, and the ages

at which these landmarks are reached, follow a predefined schedule not obviously influenced by relapses, whenever these may have occurred, or by the initial course of the disease, whatever its phenotype. Their analysis is that relapsing-remitting disease should be regarded as multiple sclerosis in which insufficient time has elapsed for conversion to secondary progression; secondary progressive multiple sclerosis as relapsing-remitting disease that has 'grown older'; and progressive from onset disease as multiple sclerosis 'amputated' from its usual preceding relapsing-remitting phase. When a detectable threshold of irreversible disability is reached, the disease enters a final common pathway where subsequent accumulation of disability becomes a self-perpetuating process, amnestic with respect to the prior history.

Prognosis

At presentation, every patient is anxious to know how the illness will progress. Since the outcome is variable and largely unpredictable, it is reasonable to err on the optimistic side when discussing prognosis early in the course. Nevertheless, certain clinical features can reliably be used in advising an individual patient on the medium- or long-term outlook. These opinions on prognosis are largely based on long-term natural history cohorts from London, Ontario in Canada, Lyon in France, and Göteborg in Sweden that differ somewhat in their authority but reach broadly similar conclusions. The extensive primary literature—together with other informative studies—is summarized by Confavreux and Compston (2005). The prognosis is relatively good when the features are exclusively or mainly relapsing disturbances of superficial sensation. These patients are often young women. Conversely, cases characterized by motor involvement, especially when co-ordination or balance are disturbed, have a less good prognosis. The outlook is also poor in older onset patients, usually males, compared with those developing multiple sclerosis in their youth. In some patients, a brisk start with several severe attacks settles down to a much more favourable subsequent course. With some ambiguity, the large population-based studies suggest that attack rate in the early years is of poor prognostic value. Thus, frequent and prolonged relapses with incomplete recovery, and a short interval between the initial episode and first relapse, carry a worse outlook. However, these effects are relatively weak and not seen in all studies. What is not in dispute is that the single most worrying feature in any patient with multiple sclerosis is entry into the progressive phase, whether from onset or after a number of relapses. The relatively poor prognosis for disability in primary progressive multiple sclerosis, whether or not there are associated relapses during the course as occurs in around 25 per cent patients, is further influenced by the rate of early deterioration and the number of systems contributing to the clinical phenotype. Although the estimated age at disease onset is later in those with primary progressive multiple sclerosis, movement across disability landmarks is even and uninfluenced by having previously had one, none or many previous episodes (Kremenchutzky *et al.* 2006).

Effect of pregnancy

Prospective surveys indicate that relapse rate is affected by pregnancy. A potential confound in the interpretation of these studies is the decision by women with severe disability not to embark on pregnancy and the corresponding preparedness of those with mild disease to start or extend their families. However, the evidence suggests approximately a three-fold higher risk in the 3–6 months after term than during pregnancy, and the attacks may be more severe. Runmarker and Andersen (1995) studied an inception cohort in Göteborg, Sweden and disposed off the hypothesis that the onset of multiple sclerosis is influenced by pregnancy; in fact, there was a conspicuous absence of onset bouts during pregnancy compared with non-pregnant epochs including the puerperal 8 months. The most recent large prospective study shows, for 222 completed pregnancies, a reduction in the pre-pregnancy relapse rate per quartile for each trimester with a subsequent increase in the puerperium. The clinical course was uninfluenced by breast feeding or epidural anaesthesia, and pregnant women showed the expected rate of clinical deterioration (Confavreux *et al.* 1998). Subsequent analysis of these data shows that, whereasthe management of pregnancy (breast feeding and epidural anaesthesia) and almost all characteristics of the disease preceding conception are not influential, women with greater disease activities in the year before pregnancy and those that do relapse whilst pregnant also have the highest risk of relapse in the puerperium (Vukusic *et al.* 2004: see Fig. 37.13). Multiple sclerosis does not of itself influence pregnancy, delivery, or the health of the infant.

Viral infections

By comparison with the unconfirmed studies on causation, there is unequivocal evidence from prospective studies that new episodes of demyelination are more likely to occur following presumed viral exposure (Sibley *et al.* 1985; Buljevac *et al.* 2002) especially upper respiratory adenovirus and gastrointestinal infections (Andersen *et al.* 1993). Nine per cent of presumed infections are followed by relapse; 27 per cent of new episodes are related to infection; and the relative risk for a new episode in the 4-week period after infection is 1.3. Conversely, recent evidence proposes that persistent parasitic infection protects from disease activity in multiple sclerosis by inducing cytokines and lymphocyte populations that promote T regulatory activity (Correale and Farez 2007).

Vaccinations

The consensus is that vaccinations are not a risk factor for the onset or relapse of multiple sclerosis (Confavreux *et al.* 2001), and there is no reason to advise people with multiple sclerosis to avoid vaccinations, including hepatitis B (Ascherio *et al.* 2001; DeStefano *et al.* 2003), although it may make sense to avoid vaccination

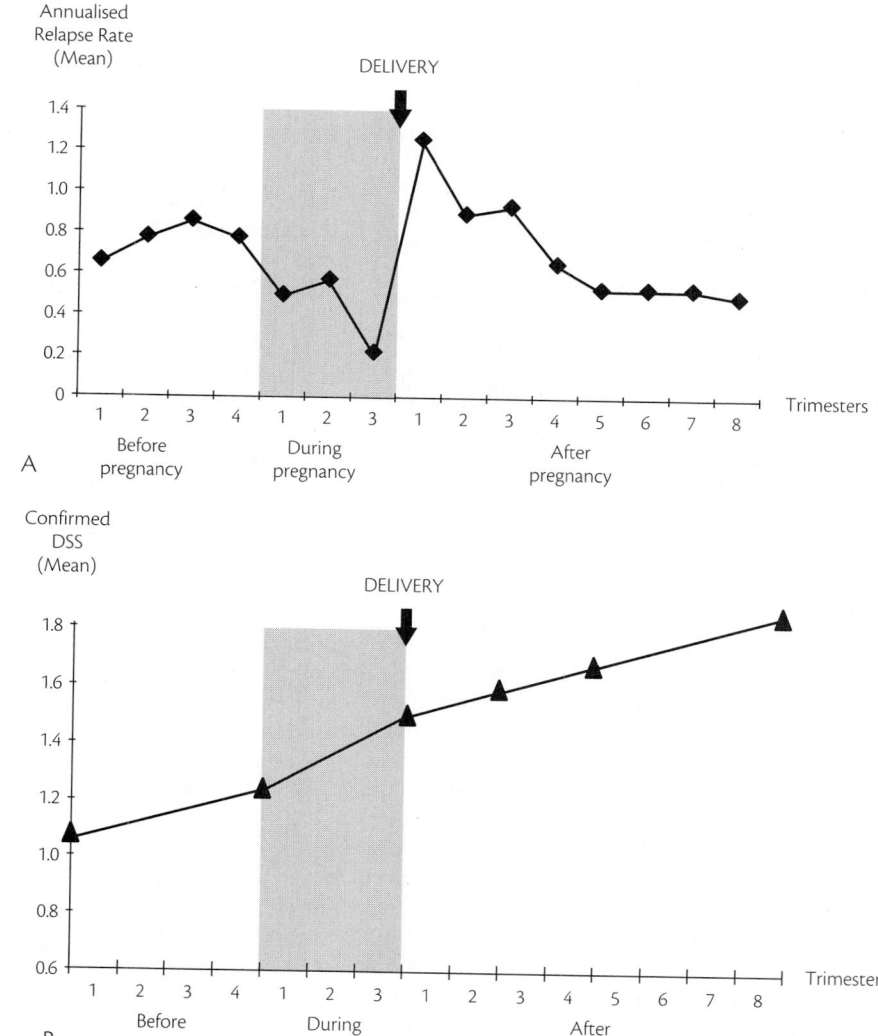

Fig. 37.13 A. Mean annualized relapse rate during the year before pregnancy, the pregnancy, and the 2 years after delivery, among 227 women with multiple sclerosis. B. Mean confirmed disability, according to DSS, during the year before pregnancy, the pregnancy, and the 2 years after delivery, among 227 women with multiple sclerosis. (Adapted from Vukusic *et al.* (2004).)

during known periods of disease activity and to take a somewhat higher threshold in patients receiving immunosuppressive treatments, especially when using vaccinations with live components. The suggestion that some vaccines, especially tetanus, are protective for the onset and activity of multiple sclerosis is less persuasive (Hernán *et al.* 2004).

Trauma

Several lines of evidence are considered relevant to the debate on whether trauma can trigger clinical manifestations of multiple sclerosis in someone who has the disease process, or alter the course in individuals who have already experienced symptoms. The key point is that the epidemiological evidence does not indicate an increased frequency of co-morbidity over and above the expected rate based on the frequency of head and neck trauma in the population at-risk and disease activity due to multiple sclerosis. Bamford *et al.* (1981) failed to show a relationship between trauma and onset of multiple sclerosis in a retrospective case series, and this was followed by a more systematic study of trauma and disease activity in 170 patients studied prospectively by monthly questionnaire and 3-monthly physical examination for 8 years (Sibley *et al.* 1991). Defining either the 3- or 6-month period following each event as at-risk, only electrical trauma showed an association with new episodes (defined as the occurrence of new manifestations lasting >48 h, in the absence of fever, or an exacerbation of old symptoms if there was a change in neurological examination) and all other forms of trauma were negatively correlated both with clinical exacerbations and disease progression. Later, Siva *et al.* (1993) identified trauma in the year preceding onset in only 3/223 incident patients from the Mayo Clinic series; specifically, they failed in prospective studies to show an increased risk after head trauma, limb fracture, and cervical or lumbar disc surgery. Sibley *et al.* (1991) found that patients with multiple sclerosis had more surgical procedures than controls, through complications of the disease, but relapse rate in the period at-risk and rate of progression were uninfluenced. Against this epidemiological background, further consideration of hypotheses for a causal link between trauma and multiple sclerosis might be considered redundant. Oppenheimer (1978) correlated demyelinating lesions in the cervical region with the regional distribution of spondylitic lesions but Kidd *et al.* (1993) used magnetic resonance imaging to show the maximum distribution of cervical cord lesions at different levels from the sites of compression by cervical spondylosis.

Diagnostic and disability scales

A number of diagnostic classifications and scales describing clinical disability and impairment, based on the clinical features of multiple sclerosis have been devised. Many clinicians intuitively use these discriminators when discussing the prognosis with individual patients. The Poser criteria (Poser *et al.* 1983) had as their core requirement for the diagnosis of multiple sclerosis two or more attacks affecting two or more necessarily separate sites within the central nervous system, including one or other optic nerves. Imaging, electrophysiology, and cerebrospinal fluid examination were used to supplement evidence for the diagnosis in situations where clinical criteria were not met, either through absence of the second episode or the second affected site. Subsequently, criteria were introduced and later modified that sought to bring forward the point of diagnosis by incorporating probabilistic information relating to MRI appearances at presentation, partly in the interests

of deciding on the introduction of disease modifying treatments, in situations where the clinical criteria for dissemination in time and space are not met (McDonald *et al.* 2001; Polman *et al.* 2005). These more minimal criteria require:

1. MRI dissemination in space based on three out of the following options

   ♦ ≥1 gadolimium enhancing lesions or ≥9 T2 lesions;

   ♦ ≥1 infratentorial lesion;

   ♦ ≥1 juxtacortical lesions; and

   ♦ ≥3 periventricular lesions.

2. Dissemination in time

   ♦ gadolinium enhancement ≥3 months after the onset of a clinical episode at another anatomical site;

   ♦ a new T2 lesion appearing after a reference scan and ≥30 days beyond a defining clinical episode;

   and for the diagnosis of primary progressive multiple sclerosis:

   1. ≥1 year of disease progression;

   2. Two of the following three options

   ♦ ≥9 T2 lesions, or ≥ T2 lesions with abnormal visual evoked potentials;

   ♦ ≥2 spinal cord lesions; and

   ♦ cerebrospinal fluid showing oligoclonal bands or an elevated IgG index.

The most extensively used scale for severity in multiple sclerosis is the Expanded Disability Status Scale or EDSS (Kurtzke 1983). It is excessively weighted to motor involvement and ordinal. The EDSS depends on subjective assessments of physical examination, does not assess handicap, and provides no information on how the present state has arisen, or what is to be expected. Despite coming in for a good deal of criticism, it has been extensively validated and is widely used in clinical trials. Subsequent introductions include the Scripps Neurological Rating Scale (Sipe *et al.* 1984); the Multiple Sclerosis Functional Composite comprising timed tests of walking, arm function, and cognitive ability (Cutter *et al.* 1999); the United Kingdom Neurological Disability Scale from Guy's which is a formidable questionnaire that comprehensively explores symptoms but with practice can be completed in a reasonably short time (Sharrack and Hughes 1999); the Multiple Sclerosis Impact scale (Hobart *et al.* 2001); and the Multiple Sclerosis Severity Scale that relates scores on the EDSS to the distribution of disabilities in patients with comparable disease durations (Roxburgh *et al.* 2005). The problem of how best to store this information remains unsolved but any system needs to be versatile without becoming excessively complex; the European Database in Multiple Sclerosis or EDMUS (Confavreux *et al.* 1992) has been adopted in Europe, whereas COSTAR (Paty *et al.* 1994) is preferred in North America.

37.5.5 **Investigation of multiple sclerosis**

Investigations are used in patients with multiple sclerosis to document the anatomical dissemination of lesions; to confirm the presence of intrathecal inflammation; to demonstrate the characteristic pathophysiological correlates of demyelination; and to exclude conditions that may mimic demyelinating disease. In many situations the clinical evidence is sufficient to establish the diagnosis and laboratory studies are superfluous but, as set out above, increasingly para-clinical features are embedded in diagnostic criteria and many neurologists chose to supplement the clinical evidence with these more objective abnormalities even when the diagnosis is not in doubt. Mistakes come to light in all large series of patients considered at one time to have multiple sclerosis and revisions are required in up to 10 per cent of cases placed on population-based registers.

Electrophysiology

The principles of using evoked potentials in clinical neurology are described in Section 3.4. The physiological consequences of demyelination serve to identify lesions in clinically unaffected pathways (Smith *et al.* 2005). The latency of the potential evoked by sensory or motor stimulation is characteristically delayed in patients with demyelinating disease whereas the amplitude is unaffected, except in the acute phase (Youl *et al.* 1991). Initially, electrical exploration was confined to the visual, auditory, and somatosensory systems but central motor conduction has subsequently been studied using the technique of magnetic brain stimulation, and event related potentials now also provide evidence for cognitive processing. The application of evoked potentials provided the first reliable and non-invasive means for demonstrating widespread involvement of the central nervous system in patients with isolated demyelination, and the techniques correlate with areas of histological damage but their role in revealing a second anatomical site involved sub-clinically has largely been replaced by imaging techniques. In that respect, evoked potentials provide little additional information in patients with clinical involvement of the pathway(s) under investigation; but they remain the best method for demonstrating that a lesion has resulted in delayed conduction of the nerve impulse and is therefore likely to be demyelinating in nature. In providing this clue to the pathology of the underlying lesion, evoked potentials remain an essential part of the diagnostic kit in multiple sclerosis.

Imaging

Magnetic resonance imaging, MRI, which depends on the relaxation time of protons exposed to a magnetic field, reliably demonstrates the lesions of multiple sclerosis (Fig. 37.14). Discrete, focal, or confluent areas of periventricular, callosal, pontine, medullary or cerebellar white matter abnormality are seen in 98 per cent of patients. Spinal cord abnormalities are also readily demonstrated, even in patients with normal cerebral scans. Unlike the cerebral lesions that may accumulate with age, these are rarely seen in normal individuals. Brain lesions are perhaps best demonstrated using fluid-attenuated inversion recovery, FLAIR, and this may also be more specific for the spinal cord. T_2-weighted lesions correspond to areas of histological damage but are not specific for any one pathological process. Nor can they be considered as diagnostic for multiple sclerosis. Increasingly, other imaging paradigms are described that characterize the separate components: inflammation using gadolinium-DTPA enhancement; axonal loss reflected by atrophy, T_1-hypointense lesions, and reduced *N*-acetyl aspartate on magnetic resonance spectroscopy; demyelination and remyelination signified by the magnetization transfer ratio; and astrocytosis as evidenced by T_2-weighted lesions. Despite lacking absolute specificity, the characteristic patterns and distributions of cerebral proton density and T_2-weighted lesions make the diagnosis of multiple sclerosis highly likely in the right context. These lesions are dynamic and the element that fades or disappears

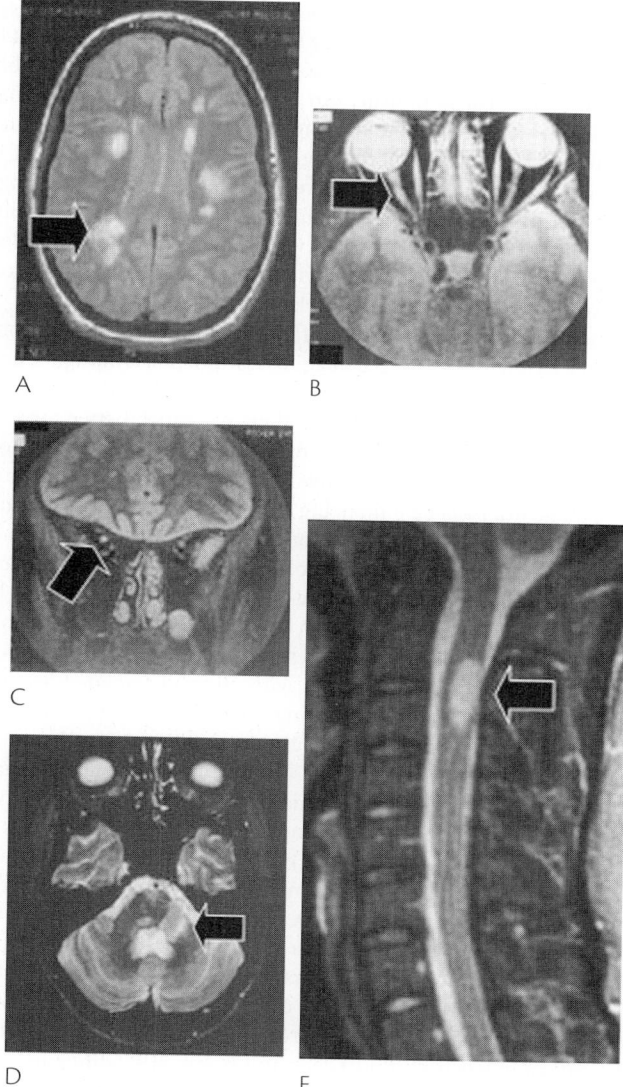

Fig. 37.14 T_2-weighted MRI showing high signal (arrows) affecting: A. Cerebrum. B. and C. Optic nerve D. Brainstem E. Spinal cord.

is mainly explained by alterations in water content due to resolution of oedema.

Serial studies of individual patients have established important principles concerning the dynamics of lesion formation. The earliest change is an increase in blood brain barrier permeability. New lesions are first recognizable as areas of gadolinium-DTPA enhancement lasting approximately 4 weeks, and preceding the onset of T_2-weighted magnetic resonance changes or symptoms by up to 2 weeks. As other features of the lesion develop, persistent enhancement is seen to occur as a ring around the edge of the lesion although a uniform increase in signal also occurs. This feature may recur in individual lesions, the cycles tending to complete within about 8 weeks. Enhancement is most likely to be seen during the relapsing-remitting phase of the disease. It may continue, but at a decreased frequency, during secondary progression but is less evident in patients with primary progressive multiple sclerosis. All serial studies have shown that magnetic resonance abnormalities occur many times more frequently than new clinical events, especially with the use of protocols for enhancing lesions. Some develop in areas that are strategically placed so as not readily to produce clinical symptoms. An alternative explanation is that many lesions never evolve to the stage at which function and structure are disrupted. As increasingly sophisticated methods for exploring integrity of the brain parenchyma are evolved, the range and extent of tissue damage becomes more clear; and, especially, alterations in 'normal appearing' white matter are now revealed (Filippi *et al.* 2003) that correlate with diffuse microglial activation and axonal loss (Lassmann and Wekerle 2005).

It has proved difficult to account for disability on the basis of conventional magnetic resonance imaging assessments although lesion load in and around the corticospinal tracts matches scores on the expanded disability status scale and there is a much better correlation between disability and imaging abnormalities dependent on axonal loss than white matter involvement both in the cerebrum and spinal cord. Isolated enhancing lesions can be detected in most patients with optic neuritis, and long intra-canalicular lesions are associated with poor visual outcome (Miller *et al.* 1988). Focal abnormalities, correlating with the clinical syndrome, can also be detected in patients with brainstem and spinal cord demyelination. Lesions that do not match the presenting clinical symptoms and signs are often present in patients with clinically isolated lesions. These are the harbingers of recurrent clinical activity. In one prospective series (Brex *et al.* 2002), the most recent estimate of conversion to multiple sclerosis at 20 years follow up in individuals presenting with clinically isolated syndromes affecting the optic nerve, brainstem or spinal cord and having additional imaging lesions at presentation is 82 per cent (Fisniku *et al.* 2008) (see above for details on the evolution of this cohort at intermediate time points). In the North American Optic Neuritis Treatment Trial cohort, follow-up after 10 years revealed conversion to clinically definite multiple sclerosis in 56 per cent of patients who had an abnormal brain MRI at presentation and in 22 per cent with a normal scan (Optic Neuritis Study Group 2003). Therefore, the majority of patients with clinically isolated syndromes and brain MRI abnormalities develop clinically definite multiple sclerosis (60–80 per cent), whereas a considerably smaller proportion (about 20 per cent) with a normal scan are seen to convert after prolonged follow-up.

Cerebrospinal fluid

CSF analysis provides qualitatively different but complementary information in patients suspected of having multiple sclerosis. There is an increase in cell count, usually due to a lymphocytic pleocytosis rarely exceeding 10 cells/cmm, especially during periods of clinical activity, in about 50 per cent of patients, and a modest rise in total protein, usually exceeding 1 g/l. More sensitive and specific are the increase in immunoglobulin concentration and the presence of oligoclonal bands on protein electrophoresis, after correction for leakage of serum proteins through the blood brain barrier. Both provide evidence for intrathecal immunoglobulin synthesis and are seen in about 90 per cent of patients—the precise level of detection being much influenced by technical aspects of the laboratory technique. Oligoclonal bands are seen in other diseases, many also associated with magnetic resonance abnormalities indistinguishable from the lesions of multiple sclerosis. Although several specificities have been demonstrated accounting for individual bands, the overall antigenicity of antibody appearing in cerebrospinal fluid has not been resolved.

Investigation of the individual patient

The patient with episodes disseminated in time, each of which can be attributed to demyelination, requires no investigation prior to establishing the diagnosis of clinically definite multiple sclerosis if presentation occurs between 20 and 50 years, separate anatomical sites have been affected, and the clinical phenotype is typical for multiple sclerosis. The diagnosis should always be considered more carefully in individuals not of Caucasian origin. Even in the typical case, most physicians would consolidate the clinical diagnosis with cerebral MRI and, perhaps, evoked potentials even if the threshold for examination of the cerebrospinal fluid is somewhat higher. Investigations become mandatory when these criteria are not met. The first investigation is cerebral magnetic resonance imaging. If this demonstrates typical radiological abnormalities, no further investigation is necessary but many neurologists would nevertheless advise spinal fluid examination in order to provide collateral evidence for intrathecal inflammation.

Imaging the cerebrum can be misleading in the acute phase of an isolated episode since, although the presence of additional cerebral lesions at presentation is associated with an increased risk of clinical recurrence, multiple lesions do not invariably correlate with further manifestations of demyelination. Conversely, recurrent demyelination undoubtedly does occur in patients with normal imaging at presentation. The erroneous diagnosis of multiple sclerosis, based on magnetic resonance imaging appearances, sometimes has to be reversed in patients with monophasic but anatomically diffuse episodes of demyelination and those with acute disseminated encephalomyelitis. A good case can therefore be made for waiting until further clinical events occur before consolidating the diagnosis of multiple sclerosis. If it proves necessary to resolve this uncertainty more rapidly, the diagnosis of multiple sclerosis can be made by comparing the results of serial magnetic resonance imaging with intervals of 3 months between scans. Abnormalities that persist are not in themselves sufficient for the diagnosis of multiple sclerosis but if new lesions are detected on a subsequent scan, or the spinal fluid becomes positive for oligoclonal bands, there is a high probability that further clinical episodes will occur, and recent criteria for the diagnosis of multiple sclerosis are met (Polman *et al.* 2005). The other approach is to delay investigation until a few months after the presenting episode and repeat the scan with gadolinium-DTPA enhancement to maximize the probability of showing active lesions. With time, if all laboratory investigations are normal and no further clinical episodes occur, the balance of evidence increasingly becomes in favour of the presenting episode being due to monophasic demyelination.

Outside the limits of 20–50 years, the probability of an alternative diagnosis increases and the threshold for investigation of disorders with which multiple sclerosis can be confused should fall. Spinal fluid examination is particularly valuable in older patients who present some years after first developing symptoms, and in those with a late-onset progressive syndrome that may be due to demyelination—situations in which cerebral white matter lesions otherwise suggestive of multiple sclerosis may be non-specific. The detection of oligoclonal bands is then highly informative and suggests inflammatory brain disease. Evoked potentials may also be delayed in those disorders which can be confused with multiple sclerosis and present in older patients with progressive disease referable to a single site.

Structural lesions can present with relapsing symptoms and for this reason imaging directed at the affected site is essential, even in the context of a relapsing history, when the previous events have not involved unambiguously separate sites in the central nervous system. Even then, errors can occasionally arise from the co-existence of more than one disease. For this reason, the past episodes need to be evaluated carefully before deciding that investigation is unnecessary, even in the patient with a relapsing history. The decision to investigate patients with an isolated episode of demyelination and a past history of other neurological symptoms depends on interpretation of the previous events. If these can reasonably be attributed with confidence to demyelination, management is the same as for the patient with disseminated lesions. If the previous episode is of doubtful significance, investigation follows the protocol for patients with a first episode of demyelination. Making the distinction between significant previous episodes and those that can safely be ignored requires judgement. Symptoms likely to be relevant in retrospectively establishing the diagnosis of multiple sclerosis include unilateral blurring of vision, numbness in an anatomical distribution or Lhermitte's symptom, double vision, and inability to use a limb. Symptoms from the past history mentioned by patients but usually turning out not to be important include events lasting less than 24 h, patchy numbness relieved by change in posture, giddiness, encephalitis or meningitis, bladder symptoms on straining, psychiatric symptoms, and epilepsy.

Thorough investigation is essential in all patients with slowly progressive symptoms referable to a single anatomical site. Some structural abnormalities, such as disc herniation with narrowing of the cervical canal and slight cord compression, are not necessarily significant—the presenting symptoms and signs of demyelination happening to arise at a site of minor but insignificant structural damage. Under these circumstances, and when imaging shows no structural lesion, it is necessary also to carry out cerebral MRI so as to demonstrate white matter abnormalities indicative of multifocal inflammatory brain disease. Imaging the head before the affected part may show white matter abnormalities, suggestive of multiple sclerosis, when there is a co-existing structural lesion. Spinal fluid analysis may establish the diagnosis of multiple sclerosis in patients having a progressive cord syndrome, and equivocal cord changes without cerebral white matter abnormalities on magnetic resonance imaging. In this situation, the presence of oligoclonal bands tilts the balance of probability in favour of demyelinating disease and may spare the patient unrewarding spinal surgery.

Where thorough investigation fails to provide evidence sufficient for the diagnosis of multiple sclerosis, the wise clinician will keep an open mind and repeat all the investigations after an interval. That said, the diagnosis can correctly be made even when imaging, spinal-fluid, and evoked potential studies are normal. However, the main priority in such cases is to exclude conditions that may mimic multiple sclerosis.

37.5.6 Differential diagnosis

Miller and Compston (2005) classify the differential diagnosis of multiple sclerosis into the following categories (Table 37.2):

- Diseases that may cause multiple lesions of the central nervous system and also often follow a relapsing-remitting course;

- Systematized diseases causing lesions in separate regions of the brain and spinal cord but usually with symmetrical manifestations and a progressive course;

- Isolated or monosymptomatic central nervous system syndromes often suggesting a single white matter lesion; and

- Pschologically determined symptoms which mimic the clinical manifestations and course of multiple sclerosis.

Isolated syndromes related to multiple sclerosis and acute disseminated encephalomyelitis are discussed in Section 37.4. Standing back from this list—many of which conditions will rarely be confused with multiple sclerosis—a real issue arises from the need to separate other inflammatory disorders absolutely or largely confined to the central nervous system. The cerebral or myelopathic form of systemic lupus erythematosus can occur in the relative absence of systemic manifestations and with only weakly positive serological abnormalities (Section 36.3.1). Primary Sjogren's syndrome (Section 36.3.3) can mimic multiple sclerosis and there is evidence that these two conditions co-exist more often than expected by chance. Sarcoidosis may present with widespread and relapsing involvement of the central nervous system showing typical magnetic resonance and cerebrospinal fluid abnormalities, and in the absence of characteristic pulmonary or cutaneous manifestations (Zajicek et al. 1999) (Section 36.4). The distinction from multiple sclerosis often cannot be made with confidence even in the presence of uveitis. A history of oro-genital ulceration in a patient with the clinical manifestations of multiple sclerosis should suggest the diagnosis of Behcet's disease (Section 36.5.1). Prominent headache and cognitive impairment, together with stroke-like episodes suggest a primary central nervous system vasculitis (Section 36.2.2), often supported by positive serology, but the phenotype is occasionally indistinguishable from multiple sclerosis (Scolding et al. 1997). Publicity relating to the anti-phospholipid syndrome has alerted patients to this alternative diagnosis, and although a syndrome of episodic focal white matter lesions resulting from pro-coagulant activity undoubtedly exists, headache, evidence for venous thromboses and a poor obstetric history are usually present. Not infrequently, the diagnosis of multiple sclerosis is preferred even when anti-phospholipid antibodies are detected. The same ambiguity arises when relapsing neurological symptoms and signs occur in the presence of anti-gliadin antibodies.

Direct infection of the nervous system may mimic the syndromes of acute isolated demyelination or multiple sclerosis; these include tuberculous and other potentially chronic meningitides, the protean neurological manifestations of acquired immunodeficiency syndrome, and Lyme disease. The characteristic painful polyradiculitis and facial palsy that epitomize borrelia infection (Section 42.5.2) in areas where this form of tick borne spirochaetal infection has hitherto been uncommon, should not cause confusion but the suggestion that borreliosis may produce a chronic or relapsing disorder of the central nervous system creates genuine diagnostic difficulty.

Care should be taken in making the diagnosis of multiple sclerosis when several members are affected in the same family. Pedigrees with hereditary spastic paraplegia can mimic familial multiple sclerosis (Section 23.4.2). Other familial disorders that can be confused with multiple sclerosis include the hereditary ataxias (Chapter 39) and adult onset leucodystrophies (Section 37.7.5).

Table 37.2 Differential diagnosis of multiple sclerosis

Diseases causing multiple lesions sometimes with a relapsing-remitting course	Phenylketonuria
	Leucoencephalopathy related to glue-sniffing
Acute disseminated encephalomyelitis	Multiple system atrophy
acute haemorrhagic encephalomyelitis	Paraneoplastic syndromes
post-vaccinial encephalomyelitis	Coeliac disease
Systemic lupus erythematosus	Myeloneuropathy from acquired copper deficiency
Anti-phospholipid antibody syndrome	Cerebellar ataxia with anti-GAD antibodies
Primary Sjögren's syndrome	Motor neuron disease and its variants
Behçet's disease	
Central nervous system vasculitis	**Isolated or monosymptomatic central nervous**
as part of a systemic vasculitis	**system syndromes**
isolated central nervous system vasculitis	
systemic sclerosis	***Spinal cord syndromes***
Susac syndrome	Chronic
Noninflammatory vascular disorders	compression
CADASIL	cervical spondylotic myelopathy
Sarcoidosis	Chiari malformation
Chronic infections	spinal dural arteriovenous malformation
Lyme disease	HTLV-1-associated myelopathy
meningovascular syphilis	noncompressive myelopathy
HIV encephalitis	primary lateral sclerosis
progressive multifocal leucoencephalopathy	amyotrophic lateral sclerosis
subacute sclerosing panencephalitis	Acute
Whipple's disease	compression
HTLV-1-associated myelopathy	transverse myelitis
Primary cerebral lymphoma	acute necrotizing myelitis
Mitochondrial disease	other myelitides
	spinal cord stroke
Systematized central nervous system diseases	
	Visual failure
Hereditary ataxias and paraplegias	Acute
adrenoleucodystrophy	anterior ischaemic optic neuropathy
metachromatic leucodystrophy	central serous retinopathy
globoid (Krabbe's) leucodystrophy	neuroretinitis
adult onset dominant leucodystrophy	Chronic
vanishing white matter disease	chronic relapsing inflammatory optic neuropathy
hereditary adult onset Alexander's disease	paraneoplastic optic neuritis
oculodentodigital syndrome	other disorders
Vitamin B12 deficiency	
nitrous oxide-related myelopathy	***Migratory sensory symptoms***
Cerebrotendinous xanthomatosis	***Central pontine myelinolysis***

Reproduced from McAlpine's Multiple sclerosis 4th edition (2006) with permission.

Although migraine, episodes of sudden onset indicating a vascular basis, early cognitive deficits, and a family history suggesting dominant inheritance do not immediately suggest multiple sclerosis, that is often the initial diagnosis in patients with Cerebral Autosomal Dominant Arteriopathy with Subcortical Infarcts and Leukoencephalopathy, or CADASIL, arising from mutation of the notch 3 gene encoded on chromosome 19p13. Pedigrees characterized by affected males showing maternal inheritance may be examples of X-linked adreno-leucodystrophy (Moser 1997). Patients with clinically definite multiple sclerosis occur in families which otherwise manifest the clinical and genetic features of Leber's hereditary optic atrophy (Harding *et al.* 1992; Riordan-Eva *et al.* 1995) (Section 12.5.2). Multiple sclerosis is unlikely to be confused with other mitochondrial cytopathies. Differences in the age and clinical manifestations of subacute combined degeneration of the spinal cord should prevent confusion with multiple sclerosis, although focal relapsing spinal syndromes, often accompanied by Lhermitte's symptom, may occur in B12 deficiency (Section 28.5.10). The least forgivable error in the context of demyelinating

disease is to accept the diagnosis of multiple sclerosis in patients with a progressive history in whom investigations have failed adequately to exclude a structural lesion—those at the foramen magnum being particularly well placed to confuse the unwary through appearing to produce independent spinal and brainstem symptoms (Section 28.3.2). Errors also occur when the progressive or relapsing symptoms of brainstem and spinal arteriovenous malformations are mistaken for multiple sclerosis (Section 28.5.11).

Increased public awareness of multiple sclerosis and its manifestations leads many individuals with sensory symptoms or dizziness to seek neurological advice. Frequent but brief symptoms not accompanied by physical signs can usually be dismissed but those who always ignore these complaints may occasionally be surprised by the findings of their more compliant colleagues. This is very different from the fabrication of spurious symptoms by individuals seeking the dignity of a neurological diagnosis in the setting of neurotic or psychiatric disease. Their management requires experience and firm handling. As in the whole of clinical neurology, a clear distinction has to be made between malingering and the tendency for any patient to exaggerate genuine manifestations of disease in order to get their symptomatic message across to the busy general practitioner or specialist.

37.5.7 Management

When initial discussion of the diagnosis takes place, it is usually possible to convey the impression of an illness that is not necessarily severe. Although the analysis that early intervention is needed to prevent mechanisms of tissue injury that may prove irreversible is reflected by the trend towards treatment at diagnosis or in the context of clinically isolated syndromes, many neurologists would still consider expectant management to be appropriate in the early stages. Whether early intervention does indeed alter the long-term outcome is not yet settled. Later, the case for treatment becomes unambiguous, especially since the impact of the licensed therapies and the additional dividend from those predicated on the early reduction of brain inflammation fades over time and may be nonexistent once secondary progression is well established. For the patient with established disability, as well as aiming to modify the long-term course of the disease, there may also be a need to improve quality of everyday life by masking individual symptoms, and to achieve temporary improvement at times of recent symptomatic deterioration.

Management of the acute episode

Since *corticosteroid* is effective in abbreviating acute episodes in multiple sclerosis (Milligan *et al.* 1987), the issue is not so much the choice of therapy but more the preferred protocol for its administration. Most variations have been compared and the trend has been towards short high dose oral regimens, not requiring hospital admission, and with fewer adverse effects. In the late 1990s, most United Kingdom based neurologists used corticosteroids at some time in their management of patients with multiple sclerosis, having a preference for intravenous methylprednisolone in the setting of an acute relapse. Meta-analysis of published trials using corticosteroids in acute relapses of multiple sclerosis showed no difference between high-dose and low-dose methylprednisolone regimens (Miller *et al.* 2000). There is evidence that *plasma exchange* given up to 1 month after onset can usefully reduce persistent severe deficits although this does not prevent subsequent disease activity (Weinshenker *et al.* 1999; Keegan *et al.* 2002) Subsequently, Keegan *et al.* (2005) reported that patients in whom brain biopsy showed antibody and complement deposition (pattern II) are most likely to improve from episodes of demyelination refractory to corticosteroids but this information is rarely available ahead of the need to decide on management.

The treatment of symptoms

Several manifestations of multiple sclerosis that cause persistent disability can usefully be improved by symptomatic treatment (see Noseworthy *et al.* 2005) (Table 37.3).

The bladder. Patients who have lost the ability to inhibit reflex bladder emptying and those with sphincter dysynergia often first train themselves to achieve reasonable bladder emptying by habit but soon become martyrs to the demographic anatomy of public lavatories, or elect not to venture away from domestic sanitary security. Thus, some people with multiple sclerosis are managed by their bladders, and not *vice versa*. Many tolerate occasional mild incontinence but soon require drugs that inhibit the detrusor response. The symptoms are usually dominated by failure to store

although, occasionally, difficulty with bladder emptying is the presenting problem. Abdominal pressure or perineal stimulation initially may trigger micturition.

Drug treatments aimed at improving bladder storage by inhibiting the detrusor include oxybutinin, tolterodine, and gabapentin; some patients are helped by drinking cranberry juice. A recent trial of delta-9-tetrahydrocannabinol (THC) was found not to benefit bladder function (Zajicek *et al.* 2003) although the continued enthusiasm of individual patients for this treatment is maintained. Ephedrine, phenylpranolamine, and imipramine stimulate the urethral sphincter and therefore also improve storage. Drugs that assist emptying by inhibition of the sphincter include phenoxybenzamine, diazepam, baclofen, and dantrium; and those that stimulate the detrusor are carbachol, bethanecol, and distigmine. A variety of intravesicular substances has been used to relieve failure to store of which the current favourite is botulinum toxin. Preliminary reports suggest that this is uncomplicated and stabilizes bladder function for ≥ 24 weeks (Smith *et al.* 2005) sometimes supplementing the use of intermittent catheterization (Schurch *et al.* 2005). Although the role remains to be determined in trials dedicated to patients with multiple sclerosis, we sense a growing enthusiasm for the use of intravesicular Botox amongst uroneurologists and their patients.

A simple means for reducing the volume of urine in the bladder is the use of intranasal desmopressin spray, especially at night. Meta-analysis suggests that this does have a useful role in reducing the number of voidings and residual urine volume in between emptying (Bosma *et al.* 2005).

But, before long, failure to fill and to empty co-exist: now, once a significant residual volume has been confirmed by ultrasound, the preferred treatment is clean self-intermittent catheterization. This is quickly taught and easily adopted by motivated patients retaining adequate vision and hand function. It can be performed discretely at work or at home using a short, low friction catheter which ensures complete bladder emptying with many social advantages and improved sleep. Permanent use of a catheter is preferable to constant dribbling incontinence, which leads to skin excoriation and aggravates other manifestations of spinal cord disease. Once the need for an indwelling catheter is established, more satisfactory management may be achieved by supra-pubic drainage, even if this also involves closure of the lower urinary tract, than an in-dwelling urethral catheter. Some patients require low-dose maintenance antibiotics for recurrent bladder infections; repeated infections promote the development of bladder stones and urinary tract obstruction with deleterious effects on renal function. Alternative measures for urinary management, now rarely used, include urinary diversion through an ileal conduit, insertion of an artificial mechanical sphincter, or electrical stimulation of the spinal nerve roots in an attempt to synchronize sphincter contraction and relaxation.

Constipation. The principle of managing constipation in multiple sclerosis is dietary alteration and the use of bulk laxatives, avoiding agents that act directly on the bowel wall. Loperamide may be useful where the predominant complaint is rectal urge incontinence.

Sexual function. The contribution of psychological factors should be considered in impotent males with multiple sclerosis; but in most cases, even when erections are still occurring, the problem is a direct consequence of spinal demyelination. The use of semi-rigid prostheses and vacuum pump-induced tumescence is likely to be used less

Table 37.3

Clinical feature	Treatment of symptom early in the course	Treatment of symptoms late in the course	Clinical feature	Treatment of symptom early in the course	Treatment of symptoms late in the course
Motor				Desmospressin	
Spasticity	Baclofen	Intrathecal baclofen	Bladder—improved emptying	Carcachol	Electrical stimulation
	Dantrium	Botulinum toxin			
	Tizanidine			Bethanacol	Bladder neck surgery
	Gabapentin	Transcranial magnetic stimulation		Imipramine	Artificial sphincters
				Phenlypranolamine	Urinary diversion
	Benzodiazepines			Ephedrine	Suprapubic catherization
		Phenol			
	iv methyl prednisolone	Tendon surgery			Urethral catherization
	Cannabinoids	Rhizotomy		Abdominal pressure	
		Nerve section		Perineal stimulation	
	Vigabatrin		Bladder—combined storage and emptying (bladder neck Dyssenergia)	Oxybutinin *and* clean self-intermittent catherization	Suprapubic catheterization
	Clonidine				Urethral catheterization
	Mexilitene				Urinary diversion
	Ivermectin		Bowel	Loperamide	Colostomy
Fatigue	Amantidine			Lactulose	Faecal containment
	Modafanil			Senna	
	Pemoline			Psyllium	
	Fluoxetine			Bulk laxatives	
Strength		4 aminopyridine		Mini-enemas	
		¾ diaminopyridine	Sexual function	Sildenafil	Mechanical devices
		Functional electrical stimulation		Alprostadil	Electrical ejacualtion
Tremor	Beta (ß) blockers	Lycra sleeve		Vardenagil	
	Primidone	Weighting		Papaverine	
	Glutethamide	Physical restraint		Yohimbine	
	Clonazepam			Phentolamine	
	Isoniazid	Deep brain stimulation		Propstaglandin E1	
	Ondanestron	Stereotactic thalamotomy	**Sensory**		
			Pain	Gabapentin	Nerve section
	Hyoscine			Pregabalin	Alcohol injection
	Carbamazepine			Amytriptilene	Sympathetic block
	Sodium valproate			Carbamazepine	
Sphincter control				Sodium valproate	Deep brain stimulation
Bladder—improved storage	Oxybutinin	Intravesical botox		Topiramate	
				Lamotrigine	
	Tolterodine	capsaicin		Tiagabine	
	Terazosin	lidocaine		Divalproex sodium	
	Propantheline	phenol		Cutaneous nerve stimulation (TENS)	
	Phenoxybenzamine		Paroxysmal	Carbamazepine	
	Prazosin			Phenytoin	
	Gabapentin			Gabapentin	
	Diazepam			Misoprostol	
	Cranberry juice		Dizziness and nystagmus	Cinnarizine	
	Flavoxate			Prochlorperazine	
	Dicyclamine			Memantine	
	Maprotilene				
	Empromium				
	Clean self-intermittent catherization				

with the availability of oral treatments for impotence. Sildenafil, or Viagra™, a phosphodiesterase inhibitor that acts by increasing local production of nitric oxide in response to sexual stmulation, offers a reasonably satisfactory method for restoring potency in males including those with spinal cord demyelination (Fowler *et al.* 2005). It has largely replaced other pharmaceuticals such as yomhibine, an α-2 adrenergic agonist, and self-administered cavernous injection of papaverine, and prostaglandin-E_1 or phentolamine which can be applied through the urethra. Electrical techniques can be used to achieve ejaculation and achieve fertility.

Spasticity. Baclofen is still the most widely used effective anti-spasticity agent. Tizanidine, which indirectly modulates the response to excitatory amino acids, reduces spasticity at an incremental dose of 20 mg daily (United Kingdom Tizanidine Trial Group 1994) but no clear preference emerges from comparative studies with baclofen. Dantrolene sodium has a direct effect on skeletal muscle and acts by uncoupling excitation–contraction mechanisms in individual fibres. Different sites of action can be exploited by combining drug use on the same or different occasions. Patients report that spasticity improves with the use of cannabis and many use this substance informally but the evidence remains ambiguous with inconclusive results from clinical trials (Zajicek *et al.* 2003). The best that can be said is that some outcome measures, especially those dependent on subjective assessments and selective components of individual scales may show positive effects, and medicinal use of cannabis is not legalized in many countries. Gabapentin, although usually prescribed for sensory discomfort, can contribute to a reduction in spasticity, as may benzodiazepines such as diazepam, clonazepam and tetrazepam by increasing pre-synaptic spinal inhibition (see Shakespeare *et al.* 2003). The dose of all these medications requires titration since each may cause muscle weakness and drowsiness.

Where spasticity proves resistant to drug treatment and continues to be a source of unpleasant symptoms, or prevents management of everyday activities by the patient or third parties, intrathecal baclofen carries the potential advantage of selectively reducing muscle tone in affected muscles whilst leaving others intact. It is mainly appropriate for patients with advanced disease and seems not to have any additional adverse effects, other than a slightly increased risk of seizures (Shuele *et al.* 2005), compared to systemic administration, apart from the hazards of implantation and the presence of an intrathecal catheter. Another approach is to use local injection of botulinum toxin; however, the number of muscles that must be treated, and the frequency of re-treatments, may reduce the practicalities of this approach (Hyman *et al.* 2000). Rarely, these days, does the need arise for surgical interruption of the reflex pathways or tenotomy and peripheral nerve block with phenol or alcohol. Magnetic stimulation may relieve spasticity but this is not used routinely even in specialist clinics (Nielsen *et al.* 1996).

Paroxysmal manifestations. Paroxysmal manifestations of multiple sclerosis are amongst the most satisfying to treat. Tonic brainstem attacks stop abruptly with the use of anticonvulsants that increase membrane stability (carbamazepine and phenytoin) and these drugs may also help patients with trigeminal neuralgia or the more refractory forms of pain arising from spinal demyelination. However, most neurologists would now prefer to use gabapentin or pregabalin to treat all forms of sensory discomfort whilst retaining carbamazepine for paroxysmal attacks; and, as mentioned, there may also be a dividend from reduced spasticity and improved bladder fucntion from the use of these gamma-aminobutyric acid agonists (Houtchens *et al.* 1997). Dysaesthetic sensations are coped with less well in the context of impaired mood but the benefit of antidepressants, especially amytriptilene, is usually attributed to neurochemical effects that are independent of affective state. This is also the preferred explanation for the subjective effect of cannabis acting on cannibinoid receptors and involving the rostral ventromedial medullary pathway that also subserves the analgesic effects of morphine. Many patients are helped by transcutaneous nerve stimulation, or TENS, alone or in combination with drug treatments. Alternative measures that may prove necessary for pain relief include destruction of nerve fibres using alcohol or phenol, differential nerve root section, cutaneous electrical stimulation of the dorsal spinal cord, or regional sympathetic blockade.

Tremor. The mainstay of pharmacological treatment for tremor is β blockers (propanalol, metoprolol, nadolol, and sotalol); alternatives include primidone, clonazepam, carbamazepine, isoniazid, ondansetron (a 5-hydroxytryptamine antagonist), and gabapentin. Peripheral physical restraint with weights is rarely successful but Lycra sleeves may have a role. More recently, there has been a resurgence of interest in stereotactic procedures involving stimulation of the ventrolateral thalamic nucleus producing results comparable to destructive procedures; but there are practical difficulties and, unlike the situation in extrapyramidal disease, the dividend is relatively small (Hooper and Whittle 1998; Alusi *et al.* 2001).

Unsteadiness. Unsteadiness arising from altered vestibular input may improve with the use of a vestibular sedative such as prochlorperazine or cinnarizine. Amongst the many that have been tried, the only potential drug treatments for symptomatic nystagmus are memantine (Starck *et al.* 1997) and gabapentin (Bandini *et al.* 2001). Local injection of botulinum toxin has been used. Fatigue is poorly understood and not explained by increased release of any biological marker of disease activity. Once treated with amphetamines, some patients respond to amantidine (Krupp *et al.* 1995). Now, the most prescribed medication is modafanil, introduced for the treatment of narcolepsy but with subjective efficacy in multiple sclerosis related fatigue (Stankoff *et al.* 2005).

4-aminopyridine. Many years of research experience on the pharmacological approach to increasing conduction properties of demyelinated axons leaves no doubt that use of 4-aminopyridine is limited by adverse effects including the risk of convulsions. Although these can improve vision and muscle strength (Schwid *et al.* 1997), they are not in routine clinical use.

Disability

For those who develop significant disabilities and impairments, comprehensive care includes access to physical and occupational therapists, social workers and other health-care staff with specific expertise in the management of chronic neurological illness (Section 6.4.3). Increasingly the available services and the management of disability are coordinated by dedicated nurse practitioners. The secondary complications of immobility with impaired cognitive, visual, bulbar, and bladder function may threaten communication, nutrition, skin care, and independence needed for all aspects of daily living. Many of these complications are best prevented by awareness and anticipation since they usually develop quickly yet take months to resolve. Minimizing handicap by attention to social, vocational, marital, sexual, and psychological aspects of the illness remains more important to most patients than drug treatment. In situations where the natural history has already led to loss of mobility, the early use of mechanical aids and home adaptations should be encouraged despite the associated stigma.

Measuring these interventions is not straightforward. Disability and handicap, but not impairment, improve for several months in patients with chronic progressive multiple sclerosis participating in a relatively brief programme of in-patient rehabilitation compared to matched controls merely placed on a waiting list or managed by exercises at home (Freeman *et al.* 1999; Wiles *et al.* 2001). Many of the effects can be attributed to the institution of appropriate treatment for managing symptoms, especially bladder dysfunction. Subjective improvements also derive from out-patient rehabilitation programmes (Pozzilli *et al.* 2002).

37.5.8 Disease modifying treatments

Azathioprine

The use of non-specific agents in multiple sclerosis established to most people's satisfaction that immunosuppression is a valid approach to treatment even if the magnitude of the effect failed unambiguously to establish a role for any one drug, alone or in combination. Based on meta-analysis of seven trials involving 793 participants, Yudkin *et al.* (1991) showed that the odds ratio for the probability of remaining free from relapse attributable to azathioprine at 2 years is 2.0 (95 per cent CI 1.4 to 2.9: $p<0.001$), whereas reduction in the expanded disability status score, EDSS, at −0.22 (95 per cent CI-0.43 to +0.003; $p<0.06$) is not significant. Discussion of this reduction in relapse rate with a more modest effect on disability was rightly cautious, emphasising the small effect, conceivably within the errors of clinical observation, of doubtful value to the individual patient, perhaps due to un-blinding, and potentially carrying serious long-term adverse effects due to the slightly increased risk of malignancy. As a result, azathioprine was not then routinely used to treat patients with multiple sclerosis. However, as perceptions of what is a 'useful' treatment for multiple sclerosis have altered, and yet not all patients wish or are eligible to receive regular injected treatments, the perceived efficacy of azathioprine seems more attractive. Some physicians have returned to its use. However, there have been no further attempts formally to evaluate efficacy or make comparisons with the newly introduced and much more expensive licensed therapies.

Other immunomodulatory drugs

In the 11th edition of this book, we listed several potential disease modifying drugs that had recently faded from interest because of an initial or eventual poor showing in clinical trials, or toxicity—expected and unexpected, real and perceived. Although pockets of prescribing no doubt persist, these casualties include cyclophosphamide, cyclosporine, linomide, anti-tumour necrosis factor α antibody, oral myelin, T-cell vaccination, methotrexate, acyclovir, sulfasalazine, and intravenous immunoglobulin.

Licensed medications for multiple sclerosis

Against this background, a number of medications are now licensed for use in multiple sclerosis. Others are in phase II/III trials. For several, the eventual status may take some time to be resolved. Standing back from a great deal of evidence based on clinical trials, some lack of restraint with respect to interpretation bumping up against hawkish critiques of the true efficacy and clinical usefulness of these medicines, together with realistic anxiety concerning adverse effects, all conditioned by aggressive marketing, the treatment of multiple sclerosis has become a fraught topic for discussion and decision-making in clinical prescribing. Our summary would be:

♦ The beta interferons or (IFN-β;) Rebif™, Avonex™ and Betaseron™, and glatiramer acetate; Copaxone™, reduce the frequency of new episodes in the relapsing-remitting phase of multiple sclerosis by up to 30 per cent;

♦ It follows that, in group studies, this effect will increase the interval between episodes;

♦ Therefore, patients with clinically isolated syndromes will show a delay in conversion to the diagnosis of multiple sclerosis, defined either by a second clinical episode or accumulation of new magnetic resonance imaging lesions;

♦ Since a proportion of episodes do not recover fully, this reduction will be associated with a reduced accumulation of fixed disabilities;

♦ In so far as there is carry-over of a significant inflammatory process, these licensed therapies may also have a detectable effect on the accumulation of disability into the secondary progressive phase of multiple sclerosis but this is small in magnitude and rarely noticeable to the individual patient;

♦ These considerations apart, there is no useful effect on disease progression;

♦ For this reason, and given the complex interplay between inflammatory and neurodegenerative mechanisms in multiple sclerosis, whether therapy limits brain atrophy or has a measurable and useful long-term disease modifying effect on clinical outcome in multiple sclerosis remain matters of opinion and debate;

♦ The therapies are generally well tolerated and, although idiosyncratic adverse effects are inevitably reported with increased frequency as the number of patient-years treatment experience increases, and some patients dislike the need for regular self-injection, safety is not the major issue in deciding whether to prescribe;

♦ Some patients may be high-responders who contribute disproportionately to group effects in clinical trials and, conversely, a significant proportion of patents are not helped by medication as a result of which there are many treatment failures—real and perceived;

♦ Whether there are differences in efficacy or immunogenicity between the three licensed therapies, useful dose response effects, or additive effects of other immunological agents are not unambiguously established; and

♦ The treatments are expensive and do not always meet standards for cost-effectiveness set by reimbursement agencies, as a result of which availability has been uneven.

As far as these medicines are concerned, those who were once dismissive of there being any discernable effects have to acknowledge that these therapies have made in-roads into the treatment of multiple sclerosis, and usefully encouraged yet more successful endeavours that are seen as clinically realistic and commercially attractive. As new therapies are provisionally identified that offer greater efficacy, so too the adverse effects profiles have changed. Therefore, a new dilemma arises: should a different approach to the clinical risk-benefit ratio be adopted leaving a few individuals badly compromised through having received a novel treatment that nevertheless proves more effective and for a greater number of recipients. There are no easy answers and opinions will differ on how best to proceed.

Glatiramer acetate. But to go back a step, based on the logic that immune suppression can be achieved by mimicking the antigenic challenge which initiates the disease process, the phase III placebo-controlled trial of glatiramer acetate copolymer-1: 20 mg by daily subcutaneous injection for 2 years, showed a treatment effect on reduction in relapse rate, more relapse free patients, and delay in time to relapse both in the initial analysis and a blinded extension but with a less clear-cut effect on disability (Johnson *et al.* 1995, 1998). This led to a license for glatiramer acetate in ambulant patients with relapsing–remitting multiple sclerosis aged ≥18 years and with ≥ 2 clinically significant relapses in the previous

2 years. Surprisingly, given the claims for clinical efficacy, the impact on MR lesions is rather modest although the proportion of lesions that go on to show markers of axonal loss may be reduced (Comi *et al.* 2001b); nor are dose response effects apparent. PROMISE, a trial of glatarimer acetate in primary progressive multiple sclerosis was terminated early when interim analysis indicated lack of efficacy (Wolinsky *et al.* 2007). A systematic review concluding that glatarimer acetate does not alter relapse rate or progression in multiple sclerosis (Munari *et al.* 2004) inevitably elicited vigorous counter-responses. Now, there is a suggestion that efficacy might be enhanced by prescribing a higher dose than that currently approved (Cohen *et al.* 2007).

Inteferons. The pivotal trials of interferon-ß1b, as Betaseron™ or Betaferon™ (IFNB Multiple Sclerosis Study Group 1993; IFNB Multiple Sclerosis Study Group and the University of British Columbia MS/MRI Analysis Group 1995) involved patients with active disease and pre-entry disability limited to scores on the Kurtzke expanded disability status scale of <5.5. There was a treatment effect on reduction in relapse rate over 4 years, although the main effect was achieved in the first year. The trial of inteferon-ß1a as Avonex™ given as 6 million units intra-muscularly on a weekly basis, involved patients with active relapsing-remitting multiple sclerosis and disabilities of Kurtzke expanded disability status scale <3.5 (Jacobs *et al.* 1996). Fewer treated patients had ≥3 exacerbations during the study than controls; annual exacerbation rates were reduced, as was the proportion of patients free from any relapse at 2 years. In the subsequent trial of inteferon-ß1a as Rebif™, a treatment effect for either of two doses was observed in patients with active relapsing remitting multiple sclerosis and an expanded disability status scale score of <5, on relapse rate, number of patients remaining free from relapse across the study, time to first relapse in those experiencing further episodes, and severity of those relapses that did occur (PRISMS; Prevention of Relapses and disability by Interferon-b1a Subcutaneously in Multiple Sclerosis 1998) (see Fig. 37.15).

The Avonex™ interferon-ß1a study differed from the trial of interferon-ß1b as Betaferon™ in showing a modest effect on disability and the probability of sustained progression. For Rebif™, an interferon-β1a, time to confirmed progression of >1 Kurtzke expanded disability status scale point increased in both treatment groups but, as for the trials of the interferon-ß1b, Betaferon™, and the interferon-ß1a, Avonex™, this was less marked than the effect on relapse rate.

The trial of inteferon-ß1b, Betaferon™, in secondary progressive multiple sclerosis was stopped early with placebo patients being switched to active treatment. The European Study Group on Interferon ß-1b in Secondary Progressive Multiple Sclerosis (1998) suggested that interferon-ß1b, Betaferon™, delayed progression by 1 expanded disability status scale point at around 9–12 months over a 2–3-year period, in patients with secondary progressive multiple sclerosis whether or not they continued to have superimposed relapses. This result led to an alteration in the product licence and extension of the prescribing indication for interferon-ß1b to include patients with secondary progressive multiple sclerosis. Imaging results from this study suggested that these clinical effects depended more on a reduction in new inflammatory lesions than other components of the disease process contributing to secondary progression (Miller *et al.* 1999). Three other trials have not reproduced this result (Goodkin 2000; SPECTRIMS Study group 2001; Cohen *et al.* 2002).

Apart from local injection site reactions, the main adverse effects of interferon-ß are flu-like symptoms and hyperthermia, perhaps due to cytokine release. Less predictable but potentially more serious adverse effects include autoimmune thyroid disease (Caraccio *et al.* 2005), capillary leak syndrome, anaphylactic shock, and thrombocytopaenic purpura (reviewed by Walther and Hohlfeld 1999).

In the initial study, 45 per cent of patients on high dose, 8 MIU, of the interferon-ß1b, Betaferon™, developed neutralizing activity, usually in the first year. The reported rate of 28 per cent was lower in the secondary progressive study but changes in assay and manufacturing process made these observations difficult to compare. In the trial of the interferon-ß1a, Avonex™, persistent neutralising anti-interferon activity was seen in 23 per cent at 2 years. Neutralizing antibodies developed in 24 per cent of patients treated

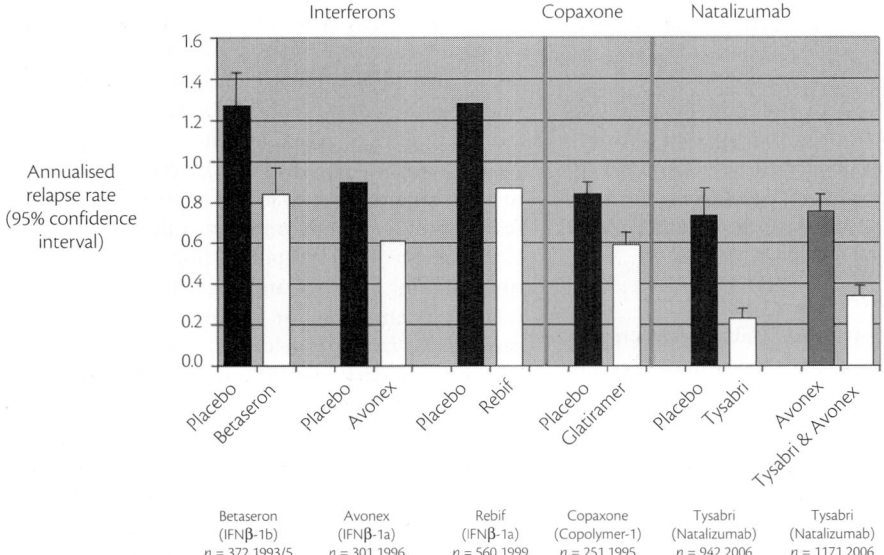

Fig. 37.15 Annualized relapse rates in the pivotal trials evaluating Betaseron™, Avonex™, Rebif™, and Nataluzimab™ in multiple sclerosis. See text for details.

with low and 13 per cent of those receiving high dose interferon-ß1a, Rebif™. Neutralizing antibodies to interferon-ß1a and interferon-ß1b are immunologically and biologically cross reactive suggesting that switching product may not escape the consequences for the immuno-reactive individual of exposure to the more immunogenic preparation. The most comprehensive account of the frequency and implications of neutralising antibodies indicates that one-third of treated patients develop antibodies, usually in the 1st year of treatment, that persist—the greatest incidence being seen with more frequent subcutaneous injection of interferon-ß1b and with an adverse effect on relapse activity but not progression (Sorensen *et al.* 2003).

The initial set of licences for the interferons were:

♦ Betaseron™—relapsing multiple sclerosis in the United States, and for relapsing and secondary progressive multiple sclerosis in the European Union;

♦ Avonex™—relapsing multiple sclerosis in the United States and the European Union; and

♦ Rebif™, high dose—relapsing multiple sclerosis in the United States and European Union.

Thereafter, a whole raft of further trials was launched to add precision to prescribing habits. These trials were designed to position the individual beta interferons and glatarimer acetate in an increasingly competitive market, with the emphasis on earlier and broader prescribing indications, dose responsiveness favouring higher exposure, and the use of MR surrogates to anticipate clinical activity. The major outcome was to add patients with clinically isolated syndromes to the licensed indications and to adjust market share but with much geographical variation. We refer readers to detailed critical analyses of these and other trials in Noseworthy *et al.* (2005). The key additional studies and claims are:

♦ OWIMS reporting added efficacy on MRI activity of high versus low once weekly dosing of Avonex™, an interferon-β1a (OWIMS 1999);

♦ CHAMPS reporting efficacy of Avonex™, an interferon-β1a, in delaying conversion to multiple sclerosis following a clinically isolated episode (Jacobs *et al.* 2000);

♦ ETOMS reporting efficacy of Rebif™, an interferon-β1a, in delaying conversion to multiple sclerosis following a clinically isolated episode (Comi *et al.* 2001b);

♦ EVIDENCE reporting added efficacy of alternate day high dose subcutaneous Rebif™, an interferon-β1a, compared with low does once weekly of Avonex™, an interferon-β1a (Panitch *et al.* 2002);

♦ INCOMIN reporting higher efficacy of alternate day Betaferon™, an interferon-β1b, compared with once weekly Avonex™, an interferon-β1a (Durelli *et al.* 2002); and

♦ BENEFIT reporting efficacy of Betaferon™, an interferon-β1b, in delaying conversion to multiple sclerosis following a clinically isolated episode (Kappos *et al.* 2006a).

Some dampening of enthusiasm arose with the systematic review of interferons in multiple sclerosis (Filippini *et al.* 2003) concluding that evidence only exists for a reduction in relapse frequency during the first year of treatment with no convincing benefit thereafter and no effect on disability. Furthermore, comparisons with

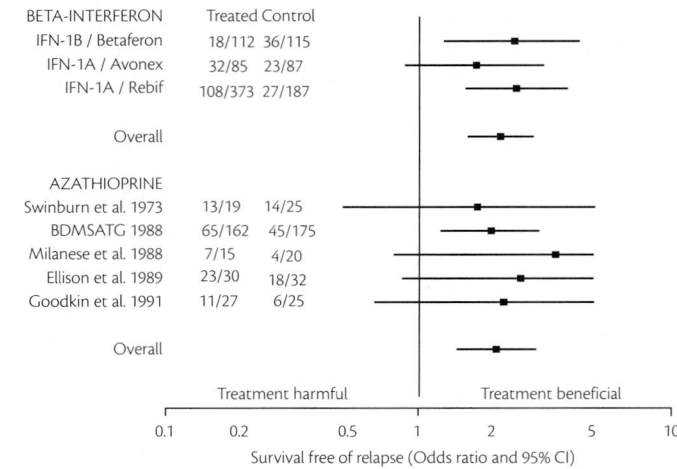

Fig. 37.16 The effect of treatment on the proportion of patients free from relapse at the end of 2 years (odds ratio and confidence intervals) in those randomized controlled trials of beta-interferon and azathioprine for which data are available. (Kindly provided by Professor Peter Rothwell.)

therapies discarded in an earlier period as not proving especially effective do not favour the interferons (Fig. 37.16)

The significant proportion of patients in whom therapy fails adequately to limit disease activity in relapsing multiple sclerosis, and the lack of efficacy in progressive multiple sclerosis have stimulated the assessment of drugs that may offer more benefit but at a greater risk of adverse effects. Two are now licensed.

Mitoxantrone. Mitoxantrone is an anthracenedione antineoplastic which intercalates with DNA and inhibits both DNA and RNA synthesis. It appears well tolerated in the short term, but carries the potential long-term risk of cumulative cardio-toxicity. It achieves a higher conversion to disease inactivity in patients with active disease than is seen with the interferons or copaxone and may carry this effect over into the progressive phase where the more active anti-inflammatory therapies could, in theory, still have a role. But for most observers, the evidence that progressive multiple sclerosis is a treatable condition using immunologically motivated therapies is overstated: rather, a paradigm shift is needed whereby drugs and interventions that target the neurodegenerative process are identified and prioritized. Edan *et al.* (1997) used monthly injections of methyl prednisolone before randomization to 6 months treatment with intravenous mitoxantrone, at 12 mg/m²/month, or no additional therapy. Millefiorini *et al.* (1997) reported a reduction in disability and relapse rate but not imaging abnormalites in patients receiving intravenous mitoxantrone for one year. Most recently, Hartung *et al.* (2002) reported improvement in a composite measure featuring relapses and disability in patients with worsening relapsing-remitting, or progressive relapsing, and secondary progressive multiple sclerosis. The magnitude of the effect was relatively modest; and although therapy was equally effective in patients with ongoing relapses with and without progression, those who did not relapse continued to show slow deterioration. Mitoxantrone is licensed in the United States for reducing neurologic disability and/or the frequency of clinical relapses in patients with secondary chronic progressive, progressive relapsing, or worsening relapsing-remitting multiple sclerosis but not those with primary progressive multiple sclerosis. Understandably, the Therapeutics and Technology Assessment Subcommittee of the American Academy of Neurology remains cautious pending confirmatory studies and more experience of safety (Goodin et al. 2003).

Anti-integrins. Lymphocyte migration involves the loose tethering of cells to the endothelial lining through binding of selectins and their ligands, which strengthens as integrins and immunoglobulin superfamily structures, VLA-4 and ICAM-1, are engaged. Migration follows the involvement of VLA-4 with CD31 and this is prevented by antibodies directed at the α4 chain of integrins with effects on inflammation and demyelination. In the initial double-blind, placebo-controlled study of a humanized anti-α4 integrin antibody in relapsing-remitting or progressive multiple sclerosis, there were fewer active magnetic resonance lesions in the 24 weeks after 2 pulses separated by 1 month but clinical relapses seemed to be unaffected. The phase 2 study involving patients with relapsing–remitting multiple sclerosis showed an early effect on MRI lesions and fewer clinical relapses but only during the period of monthly exposures to therapy with no carry-over effect (Miller et al. 2003). These lesions were less likely to mature into those with features of axonal loss. The phase III programme also seemed promising during interim analysis (Polman et al. 2006) and this led to a license for monthly natalizumab by intravenous infusion in relapsing forms of multiple sclerosis. Events were overtaken by the development of progressive multifocal leucoencephalopathy in two individuals participating in the trial combining the interferon-β1a, Avonex™, with natalizumab; this is now indicated only for active disease that is refractory to other licensed medication and not as first line therapy.

The future of treatment in multiple sclerosis

Much effort has gone into the design of treatments in multiple sclerosis that target specific components of the immune repertoire. Most target some aspect of T-cell function (Hohlfeld (1997). Screening for efficacy using MRI provides suggestive results for several such therapies (Bielekova et al. 2004; Kappos et al. 2006b), many of which are based on the use of chimaeric or fully humanized monoclonal antibodies. The advantage of using monoclonal antibodies that deplete or block T-cells lies in the possibility of prolonged stabilization of disease from pulsed therapy but there are concerns about the long-term consequences of immune modulation. These considerations apart, with the adoption of strategies for making antibodies more innocuous in terms of anti-idiotypic responses and the introduction of new specificities, greater use of monoclonal antibodies in multiple sclerosis can be anticipated. Initially, several specificities were tested and these included murine, chimaeric or humanized anti-CD6, -CD2, -CD3, and -CD4 antibodies. Since no one proved effective, this therapeutic stratagem might have been abandoned but the experience with anti-CD52, alemtuzumab, is more promising. Initially, treatment was used in patients with secondary progressive disease. Despite highly effective suppression of clinical and radiological disease activity, most patients became disabled from further progression of deficits acquired prior to treatment. Brain atrophy attributable to axon degeneration increased. Progression and atrophy correlated with the amount of brain inflammation in the pre-treatment phase but in the absence of on-going disease activity (Coles et al. 1999). This prompted the analysis that inflammation and neurodegeneration—relapses and

progression—dissociate as the pathogenesis matures indicating that effective immunosuppression must be used early in order to stabilize the cascade of events that culminates in irreversible disability. There followed open-label treatment of such cases (Coles et al. 2006) and a phase 2 single-blind comparison between alemtuzumab and Rebif™, an inteferon-β1a, each with very promising results (Coles and CAMMS223 Study Group 2008) (Fig. 37.17).

Bone marrow transplantation. Comparable perturbations of the immune system can be achieved with bone marrow transplantation, and with similar results. Initially, Fassas et al. (1997) reported on 15 patients with progressive disease who received haematologic stem cells mobilized with cytokines and drugs, after ablative immunotherapy. All survived the immediate procedure, despite significant toxicity. At 6 months follow-up, half of the group was deemed to have stabilized or improved. Protocols for minimising the hazards of harvesting and the procedure continue to be refined and most programmes use high-dose cyclophosphamide, 2–4 g/m², either alone or with granulocyte colony-stimulating factor to induce lymphopenia with infusion of autologuous stem cells from peripheral blood. Approximately 200 people with multiple sclerosis had been transplanted by mid-2003 (Burt et al. 2005). The evidence suggests that clinical and MRI evidence for stability or apparent improvement can be achieved but—as with alemtuzumab—progression is not arrested if the intervention is given later in the course. This clinical analysis is supported by pathological studies of cases studied at autopsy after bone marrow transplantation that show axonal loss in the absence of active inflammation (Metz et al. 2007).

Future treatment prospects. Taken together, the results of clinical trials support the hypothesis that inflammation is necessary for new lesion formation and conditions axon degeneration. The implication is that immunological therapies will best prevent progression of disability if given early in the course and before the cascade of events leading to axon degeneration is irretrievably established. This may explain the present limitations of immunotherapy in patients with secondary progressive multiple sclerosis. But it raises the dilemma of exposing individuals who may never develop disabilities from multiple sclerosis to the unpredictable hazards of prolonged immunosuppression as the price paid for stabilizing the disease process in those who are destined to do badly. The goals of future therapies in multiple sclerosis must confront the main unmet need of how to limit the neurodegenerative aspect of disease progression. Our analysis of the interplay between inflammation and neurodegeneration is that areas of focal damage, constituting plaques, are driven by inflammation and microglial activation. Regions of the central nervous system or the brain and spinal cord as a whole may be genetically susceptible to neurodegeneration but this vulnerability nevertheless must be exposed by inflammation.

This will require inhibiting the establishment of immunological chronicity manifesting as diffuse microglial activation; protecting intact axons from acute injury using anti-excitotoxic and membrane stabilizing agents; providing trophic support of persistently demyelinated axons using growth factors and strategies that might include pharmaceutical strategies for enhancing remyelination; and promoting plasticity and axon regeneration by manipulation of extracellular matrix molecules inhibitory environments. Most patients expect stem cell biology to deliver a dividend for remyelination and repair in multiple sclerosis. Validating the evidence

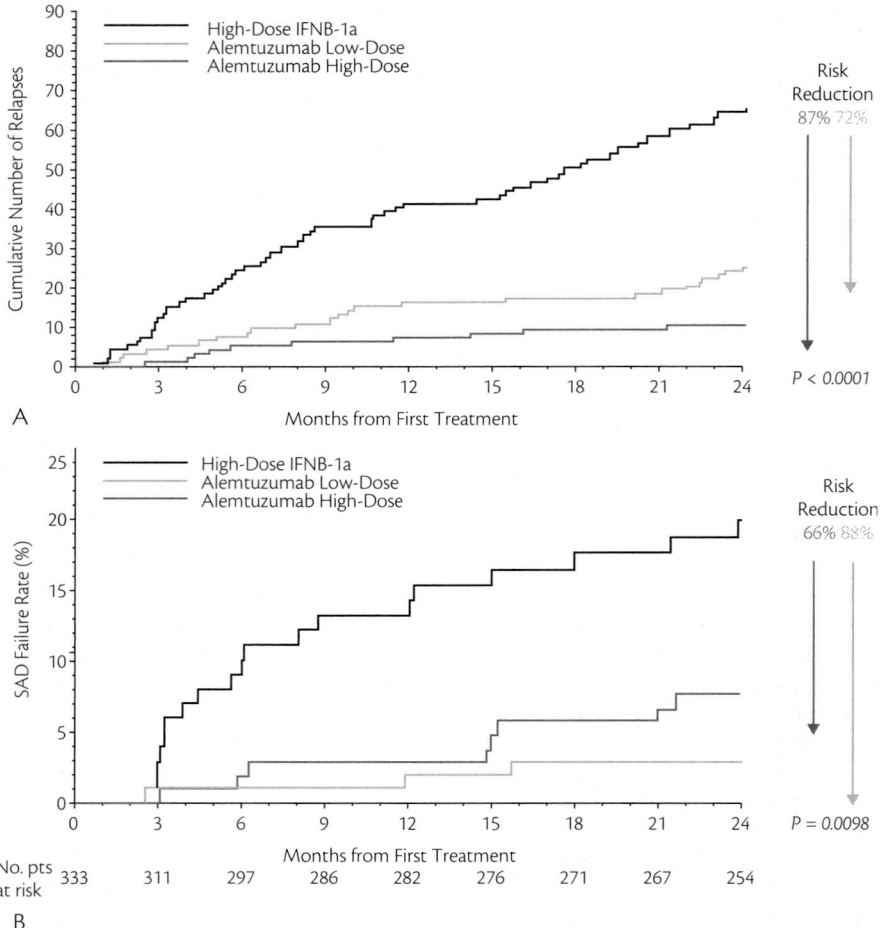

Fig. 37.17 Interim 2-year analyses in patients with early active relapsing remitting multiple sclerosis treated with alemtuzumab (Campath-1H) or Rebif™. A. Effects on of effects on relapse rate. B. Effects on sustained accumulation of disability.

already gathered in experimental studies will require several issues to be reconciled:

- Will cessation of the inflammatory process allow sufficient repair and reversal of deficits and does suppression of the inflammatory process inhibit remyelination?

- Is the potential for enhancing endogenous remyelination real enough to make the concept of exogenous rescue unnecessary?

- Is there a critical period when the naked axon can be rescued by reclothing it in myelin?

- How many axons must be remyelinated to achieve useful conduction through a critical pathway and can axon outgrowth be promoted in order to increase the 'theatre' of remyelinaton?

- Which intervention provides the best 'medicine' and how can this most effectively be delivered?

Answers must await a future edition of this book.

37.6 Central pontine myelinolysis

Central pontine myelinolysis was first described in four patients where autopsy studies revealed a single sharply outlined focus of myelin destruction in the rostral part of the pons indiscriminately involving all descending and ascending fibre pathways, with the exception of the ventrolateral tract, but sparing nerve fibres (Adams *et al.* 1959) (Fig. 37.18). With an improved understanding of the mechanism of pontine damage, and more judicious correction of the provocative electrolyte imbalances, the disorder has become somewhat rare. Although the original descriptive term—central pontine myelinolysis—remains valid, demyelination is often observed outside the pons (Wright *et al.* 1979). A high proportion of the cases first described had severe metabolic disturbances, often resulting from alcohol abuse (Section 5.2.6)—understandably causing confusion with Wernicke's encephalopathy—but the condition is now known to occur in a variety of general medical contexts. The theme that so often links these situations is a period of hyponatraemia. Thus disorders of thirst, vomiting, and those resulting in electrolyte imbalance should alert clinicians to the diagnosis. These observations led to the suggestion that the pons is unusually susceptible to changes in electrolyte balance but it is now believed that central pontine myelinolysis is most usually caused by overzealous correction of a low serum sodium (Sterns *et al.* 1986). A prospective study of electrolyte correction in patients with hyponatraemia confirmed that pontine damage correlates both with the degree of hyponatraemia and more especially with the speed at which it is corrected—starting levels of <110 mmol/l or

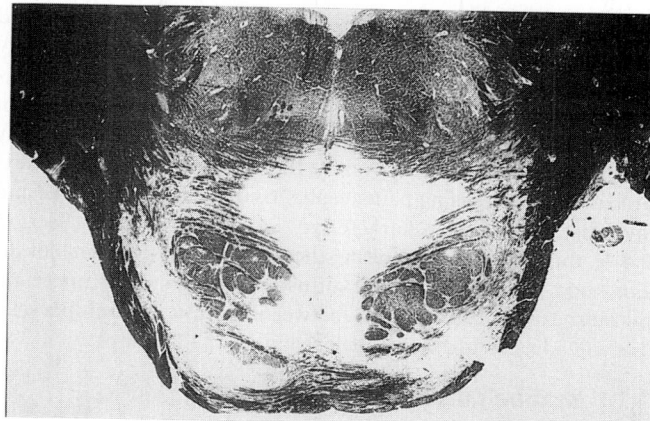

A

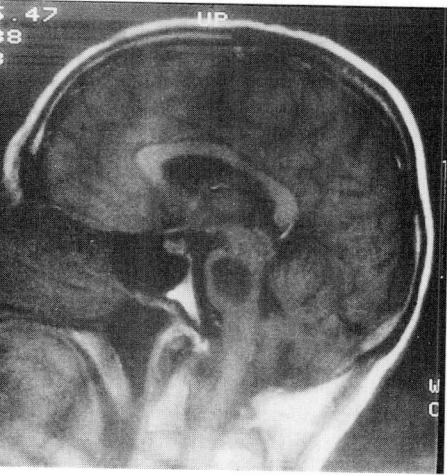

B

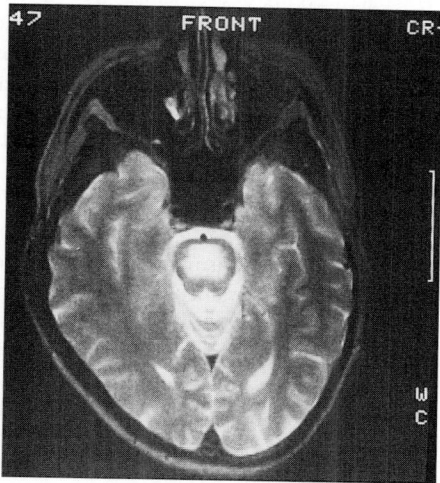

C

Fig. 37.18 Central pontine myelinolysis A. An area of confluent demyelination in the central pons occurring in a patient in whom the serum sodium rose from 112 to 145 mmol/l over 24 h following liver transplantation. B and C, saggital and collonal MRI of another patient with central pontine myelinolysis

rates of correction of >2 mmol/l/h increasing the risk of pontine damage (Brunner *et al.* 1990). Experimental studies have shown that rapid changes in sodium can better be tolerated when hyponatraemia has arisen acutely than in chronic electrolyte imbalance. Central pontine myelinolysis is occasionally seen as a result of sustained or rapidly corrected hypernatraemia.

Clinically, the process begins with damage to pathways placed centrally within the pons and it then spreads centrifugally. The fully evolved clinical picture is of flaccid paralysis with facial and bulbar weakness, disordered eye movements, profound imbalance, and alterations in consciousness. The recent literature highlights movement disorders and other extra-pyramidal manifestations of central pontine and extrapontine myelinolysis (Seiser *et al.* 1998). Some manifestations can be attributed to the poor medical condition of affected individuals. Since central pontine myelinolysis tends to occur following therapeutic intervention and correction of the serum sodium, other features of hyponatraemia, such as fits, tend not to be associated. In a series observed in hospitals in New York and Oxford, Ellis (1995) reported only one case of central pontine myelinolysis, having a striatal syndrome, amongst 184 presenting with hyponatraemia and managed in a variety of ways including rapid correction of serum sodium in some instances. Newell and Kleinschmidt-DeMasters (1996) carried out an autopsy based epidemiological survey of the prevalence and features of central pontine myelinolysis over a period in which the management of hyponatraemia changed but observed no alteration in incidence. Of 15 clinically undiagnosed cases, 5 of 6 with active lesions were associated with overzealous correction of the serum sodium.

The prognosis for clinical recovery is largely determined by the underlying metabolic disorder and the extent to which that can be managed; but with stabilization of the serum sodium, management of the bulbar failure and time, the prognosis for neurological recovery is good and the condition does not spontaneously relapse. Menger and Jorg (1999) retrospectively reviewed the outcome in 44 patients and concluded that death in 2, and poor recovery in 10 with residual deficits not affecting independence in a further 11, hinged on the general medical complications of the acute illness or its precipitating events rather than the severity of pontine myelinolysis. They did not identify any radiological or electrophysiological predictors of poor outcome.

The clinical features of brainstem disease, occurring in a patient known previously to have been hyponatraemic, are sufficiently distinctive not to cause diagnostic difficulties but uncertainty may arise if only a small reduction in serum sodium has occurred or the fall was not documented, in which case brain imaging may be informative. The acute changes can be recognized by computerized tomography or magnetic resonance imaging; persistent abnormal signals attributable to astrocytic proliferation and gliosis can be detected long after the clinical features have resolved.

Ultrastructural studies support the view that the condition is due to a metabolic process primarily directed at oligodendroglia. In a rat model of central pontine myelinolysis, the lesions are associated with massive accumulation of microglia expressing the pro-inflammatory cytokines tumour factor necrosis-α, interferon-γ, and nitric oxide synthase: inhibitors of microglial activation reduce the clinical signs and the amount of microglial infiltration within lesions (Takefuji *et al.* 2007).

37.7 Leucodystrophies

The leucodystrophies are a group of disorders, originally described on the basis of their non-inflammatory demyelinating neuropathology and including a heterogenous collection of conditions, many of which are now known to result from specific defects affecting genes that determine the synthesis, maintenance, and structure of myelin (Section 10.2) (Table 37.3). These are rare conditions but need to be considered in children and young adults with atypical syndromes combining physical and intellectual deficits, with or without peripheral nerve involvement, in whom imaging shows more confluent lesions confined to white matter (Fig. 37.19).

No attempt is made comprehensively to cover this miscellaneous group of predominantly paediatric developmental and neurodegenerative disorders but readers are referred to the full clinical, biochemical, and comprehensively illustrated imaging descriptions given by van der Knaap and Valk (2005) and other definitive accounts (Moser 1997). Here, the emphasis is mainly on those disorders which may be seen in adult neurological practice and their childhood forms.

It is apparent that several disorders once classified as leucodystrophies have specifically different aetiologies and are no longer included in the group of conditions. A good example is provided by *diffuse cerebral sclerosis* or *Schilder's disease* (Schilder 1912). As described in many earlier editions of this book, such children developed intellectual impairment and a gait disorder. Visual impairment was often an early symptom with homonymous hemianopia usually progressing to blindness in most cases as the pathological process spread. Progressive weakness of the extremities led eventually to spastic tetraplegia. Sensory loss of cortical type was common with loss of postural sensibility, appreciation of passive movement, and tactile discrimination but cutaneous sensation might also be involved, producing hemi-analgesia. Aphasia was noticeable in the early stages; later, this was masked by spastic dysarthria, and dysphagia due to pseudobulbar palsy eventually supervened. The mental changes were those of progressive dementia. Generalized or focal motor seizures were seen at any stage. The disease was progressive and usually fatal, although temporary remissions and, very rarely, arrest or recovery did occur. In describing this condition, we use the past tense because it has since become clear that diffuse cerebral sclerosis was used to identify an heterogeneous group of diseases affecting cerebral white matter. Not all have turned out to be genetically determined disorders of myelin formation and metabolism. Thus, of the diseases originally classified under this heading, familial sudanophilic diffuse sclerosis, Pelizaeus–Merzbacher disease (Section 10.2.10), Krabbe's disease or globoid cell leucodystrophy (Section 10.2.3), Canavan's diffuse sclerosis or spongy degeneration of the white matter (Section 10.2.9), Alexander's disease (Section 10.2.7), and metachromatic leucodystrophy (Section 10.2.2) are all dysmyelinating leucodystrophies. Some were probably early examples of cerebral autosomal dominant ateriopathy with subcortical infarcts and leukoencephalopathy, or CADASIL (Dichgans *et al.* 1998) or Balo's concentric sclerosis and many male cases probably suffered from adrenoleucodystrophy (Section 10.2.4). Some must have had subacute sclerosis panencephalitis (Section 42.3.7), and others, especially the relapsing syndromes, were almost certainly examples of Leigh's disease associated with mutations of mitochondrial DNA (Section 10.5.3). An extreme interpretation is that diffuse sclerosis was never anything more than an exceptionally severe and generalized variety of childhood multiple sclerosis. Relatively simple investigations subsequently proved useful in sorting out these childhood encephalomyelopathies. Electroencephalography helped to make the distinction from subacute sclerosing panencephalitis and the lipidoses, usually showing diffuse slow activity unlike the recurrent bizarre complexes of subacute sclerosing panencephalitis or the irregular spike and wave discharges that characterize Tay–Sachs disease (Section 10.4.2). But even after separating these specifically different conditions, the nosological status of diffuse sclerosis remains uncertain and most experts consider that, between them, acute multiple sclerosis and adrenoleucodystrophy account for all the cases.

37.7.1 Krabbe's disease

Globoid cell leucodystrophy is an autosomal recessive condition resulting from mutation of *GALC* at 14q24 with reduced activity of galatocerebrosidase (Section 10.2.3). It usually presents as an early-infantile disorder and late onset of globoid cell leucodystrophy is uncommon. Most cases present before the age of 5 years and so are rarely confused with childhood multiple sclerosis but onset and recognition may be delayed until adult life. The clinical picture is dominated by behavioural changes with light and noise startle; progressive intellectual and motor decline resulting in pyramidal, extrapyramidal, and cerebellar involvement; and epilepsy, visual failure, and peripheral neuropathy leading to severe disabilities with pyrexia and other autonomic features prior to the onset of a vegetative state. Visual evoked potentials are delayed and the spinal fluid has a raised protein but does not contain oligoclonal bands. Magnetic resonance imaging shows periventricular lesions and subsequently extensive white matter changes. Magnetic resonance spectroscopy is characterized by decreased *N*-acetyl aspartate and increased myoinositol and choline containing compounds (Brockmann *et al.* 2003).

Examination of peripheral blood leucocytes or skin fibroblasts shows a deficiency of β-galactocerebrosidase which leads to the accumulation of cerebroside in phagocytes resulting in the diagnostic feature of globoid cells; these represent the end stage of oligodendrocyte and myelin degradation triggered by psychosine accumulation. Although myelinating cells in the central and peripheral nervous systems are each affected, there seems to be differential susceptibility of oligodendrocytes and Schwann cells.

37.7.2 Adrenoleucodystrophy

The biochemical characterization of a peroxisomal defect separated adrenoleucodystrophy from the other dysmyelinating disorders although the X-linked pattern of inheritance had always been a distinctive feature (Section 10.2.4). There are several types each characterized by the accumulation of very long chain saturated fatty acids, VLCFAs, in all lipid-containing tissues and body fluids (Moser *et al.* 1981; Moser 1997). The various disorders arise from defective very long chain fatty acyl-coenzyme-A synthetase activity in peroxisomes (Wanders *et al.* 1988) due to mutations of *ABCD1* at Xq28, close to that for glucose-6-phosphate dehydrogenase deficiency and colour blindness (Aubourg *et al.* 1990). It encodes a peroxisomal membrane protein, now known as ALD protein, belonging to the ATP-binding cassette protein family and leading to the accumulation of membrane-like cytoplasmic inclusions in brain tissue. One hypothesis is that the protein product of the adrenoleucodystrophy gene normally anchors very long chain fatty

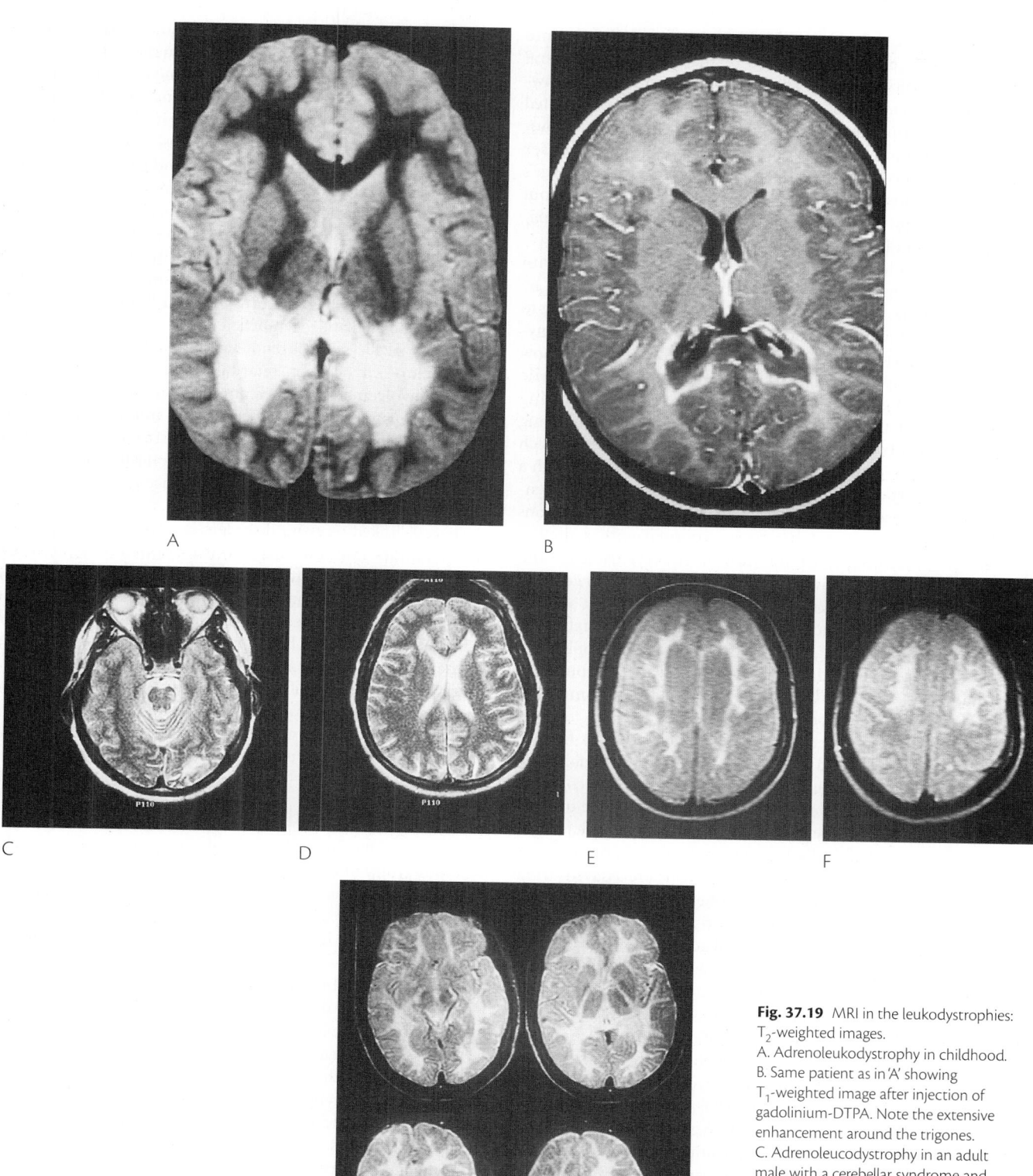

Fig. 37.19 MRI in the leukodystrophies: T_2-weighted images.
A. Adrenoleukodystrophy in childhood.
B. Same patient as in 'A' showing T_1-weighted image after injection of gadolinium-DTPA. Note the extensive enhancement around the trigones.
C. Adrenoleucodystrophy in an adult male with a cerebellar syndrome and Addison's disease; high signal abnormality in the cortico-spinal tract in the pons with atrophy of the cerebellum and midbrain. D. Bilateral high signal intensities adjacent to the anterior peri-ventricular region bilaterally. E and F. Metachromatic leukodystrophy; proton-density weighted image. G. Pelizaeus–Merzbacher disease, T_2-weighted images.

acids into the peroxisomal membrane or translocates these into peroxisomes. The diagnosis may also be made by ultrastructural examination of nerve biopsied from the skin or conjunctiva showing typical curved clefts and leaflets in Schwann cells. Four related syndromes share this biochemical abnormality: childhood adrenoleucodystrophy and adult-onset adrenomyeloneuropathy are sex linked, whereas neonatal adrenoleucodystrophy and Zellweger's syndrome are autosomal recessive disorders (Moser 1997). About 25 per cent of cases are of adult onset with a progressive myelopathy developing in the third or fourth decade.

X-linked adrenoleucodystrophy may present in childhood with behavioural disturbance, dementia, and epilepsy followed by involvement of the special senses and motor systems and leading to total disability within the first decade. Even though this is a dysmyelinating leucodystrophy, the aggressive childhood cases may show inflammatory demyelination with confluent symmetrical magnetic resonance imaging and spectroscopic abnormalities showing reduced NAA and increased choline. The proposal is that very long chain saturated fatty acids trigger the inflammatory response which is mediated by TNFα produced by activated microglia. Although a significant proportion of children later develop adrenal insufficiency, Addison's disease may precede the neurological manifestations by several years. Treatment has been proposed with a dietary supplement containing a 4:1 mixture of glyceryl trioleate and trierucate, popularly known as Lorenzo's oil. This lowers the plasma levels of very long chain saturated fatty acids but appears not to influence the phenotype in individuals with established neurological disease although there may yet be role in prophylaxis. Bone marrow transplantation is successful in early symptomatic cases and, in view of the inflammatory reaction, trials of immunosuppression have been proposed (see Moser 1997).

In adults, X-linked adrenomyeloneuropathy presents in men with spastic paraparesis and sensory loss in the legs; although clinicians may be alerted to unusual causes for this otherwise common neurological problem by the presence of peripheral nerve involvement, the diagnosis may be overlooked if adrenal insufficiency is not clinically or biochemically apparent at the time of presentation and unless very long chain saturated fatty acids are assayed. Some patients show a progressive cerebellar phenotype and intellectual function may be impaired with a subcortical pattern of dementia. The clinical phenotype of childhood and adult adrenoleucodystrophy, which are usually quite distinct, may both be seen in the same family. The availability of a biochemical assay that reliably identifies the peroxisomal defect in easily sampled body tissues or fluids, has inevitably led to the demonstration of cases with obscure clinical manifestations; these include focal cerebral lesions with imaging abnormalities, the Kluver Bucy syndrome, dementia, spinocerebellar degeneration, and olivopontocerebellar atrophy.

About 20 per cent of female heterozygotes develop relatively mild, and occasionally remitting, spastic limb weakness and sensory loss, but cerebellar involvement is rare, as is cerebral or adrenal involvement; the age of onset is usually in the fourth or fifth decade. A higher proportion of manifesting carriers have neurological signs without clinical disabilities. Although adrenal disease is almost never present in carriers, very long chain saturated fatty acids are usually elevated and brain imaging or measurement of the auditory-evoked brainstem responses may help in diagnosis of the carrier state (O'Neill *et al.* 1982). The spinal fluid is usually normal although

oligoclonal bands have been described, and white matter lesions are occasionally seen on magnetic resonance imaging (Ménage *et al.* 1994). Such cases may first be diagnosed as having multiple sclerosis, but the symmetry of the lesions and their diffuseness, even if predominantly placed posteriorly in the early stages, differ from the usual findings in multiple sclerosis. The correct diagnosis can be established on the family history and by finding an excess of very long chain saturated fatty acids in plasma or fibroblasts. Adrenal failure may precede neurological involvement or present as X-linked Addison's disease without nervous system involvement (Josien *et al.* 1993). Biochemical abnormalities include reduced urinary excretion of 17-hydroxycorticosteroids and 17-oxysteroids, with impaired response to corticotrophin but high basal levels; there may be associated androgen deficiency.

Autosomal recessive adrenoleucodystrophy presents in infancy with seizures, hypotonia, retardation, retinal degeneration, and hepatic involvement; females are more commonly affected than males. The pattern of organ involvement and mode of inheritance are similar in neonatal adrenoleucodystrophy and Zellweger's syndrome but these are thought to be separate disorders attributable to abnormal peroxisomal assembly or biogenesis (Raymond 2001) (Section 10.6).

These X-linked recessive disorders can produce a clinical picture which mimics multiple sclerosis; the distinction is most reliably made on the basis that affected patients have or subsequently develop the clinical and biochemical manifestations of Addison's disease.

The sensitivity and specificity of routine assay for very long chain saturated fatty acids has been reviewed (Moser *et al.* 1999): together, the level of hexacosanoic acid and its ratios to tetrasanoic and docosanoic acids are fully discriminating in homozygote males, irrespective of the clinical phenotype, from the day of birth if dietary supplements have not been given providing an opportunity for mass screening; there is a false negative rate of 15 per cent for heterozygotes.

37.7.3 Metachromatic leucodystrophy

The separation of metachromatic leucodystrophy from the heterogenous group of diffuse sclerosis was first made when metachromatic material was detected in urinary deposits (Austin 1957). It subsequently became clear that the diagnosis can be confirmed by demonstrating increased urinary sulphatide excretion with a deficiency of aryl-sulphatase A in urine, peripheral blood leucocytes, and skin fibroblasts, or by the demonstration of metachromatic material in peripheral nerve biopsies which show segmental demyelination and remyelination (Section 10.2.2). There is diffuse white matter involvement due to non-inflammatory demyelination with loss of oligodendrocytes, axon preservation, and reactive astrocytes which, together with macrophages, contain the metachromatic material especially in the most extensively demyelinated areas.

The gene for arylsulfatase A is encoded on chromosome 22 and the clinical phenotype varies with the amount of surviving enzyme depending on heterozygosity of the mutant allele; pseudodeficiency refers to those individuals with low but sufficient levels of arylsulfatase A who do not have a clinical phenotype. Some affected individuals have a genetic defect of the arylsulfatase A activator, encoded on chromosome 10, and this is associated with a more complex pattern of sphingomyelin storage, biochemically and in terms of the tissue distribution. The most common form of metachromatic leucodystrophy develops in late infancy with delayed walking and other motor milestones, due to the neuropathy which may be painful. There are also features of brainstem involvement and the emergence of diffuse upper motor neurone signs with reduced intellectual development, optic atrophy, and death within about 5 years from presentation. In older onset childhood cases, after several years of normal development, there are behavioural changes with poor school performance, ushering a cerebellar and upper motor neurone pattern of disability which then follows much the same course as in the younger cases although with less evidence for neuropathy. The early adult form of metachromatic leucodystrophy is rare, or perhaps seldom diagnosed, and tends to present with intellectual or emotional abnormalities; as with many other inherited disorders, onset after the age of 60 years has been described. Onset with dementia and behavioural disorders is usual with ataxia, paralysis, and optic atrophy only developing at late stages, although the presentation is occasionally with paraparesis or cerebellar ataxia (Hageman *et al.* 1995) and the condition can then more easily be mistaken for multiple sclerosis. Clinical evidence for peripheral neuropathy may be revealed by slowed nerve conduction. The full range of clinical manifestations has probably not yet been fully explored. Usually, the disorder is progressive with steady decline into dementia or persistent vegetative state but confusion may arise when the course is relapsing and remitting. Treatments have included dietary manipulation with reduced vitamin A and sulphur containing substances, and bone marrow transplantation but the successes are limited.

Multiple sulfatase deficiency combines the features of metachromatic leucodystrophy with mucopolysaccharidosis. It too has neonatal, early childhood, and juvenile forms. The pattern of combined motor and mental regression or lack of development reflecting widespread dysmyelination with peripheral neuropathy is associated with dysmorphic features and, organomegaly. The more severe phenotype also reflects extensive neuronal loss due to the combination of stored sulfatide, sulfated steroids, and mucopolysaccharides. The enzyme defects are complex involving many sulfatases other than arylsulfatatse A.

37.7.4 Pelizaeus–Merzbacher disease

The three phenotypes of X-linked Pelizaeus–Merzbacher disease usually present in childhood (Section 10.2.10). The clinical feature which may distinguish the otherwise ubiquitous motor and developmental delay with epilepsy is abnormal eye movements, dystonia and choreoathetosis, and laryngeal paralysis. Affected individuals often stabilize with severe disabilities and live into early adult life. In the earlier onset group, the connatal form, either sex may be affected with X-linked and autosomal pedigrees; the phenotype has many of the same features but the course is more accelerated. There are also transitional cases of intermediate prognosis. Some cases do not manifest until early adult life but here the blur with specifically different disorders becomes more apparent. Magnetic resonance imaging either fails to show myelin or depicts myelin which is immature with an atrophic brain.

The molecular defect is a mutation of the gene for proteolipid protein, *PLP*, encoded on Xq21.2 and, in severe cases, this leads to extensive loss of myelin in the central nervous system with patchy depletion of oligodendrocytes and axons. Proteolipid protein is normally involved in stabilizing the lamellar structure of central myelin. Biochemically, there is a reduction in sulfatide and cerebroside. Over 30 mutations have been described, and novel ones are still being described resulting in expression of truncated forms of proteolipid protein sufficient to cause extensive oligoldendrocyte loss and failure of myelination (Aoyagi *et al.* 1999), but no one can be detected in most patients. Osaka *et al.* (1999) found causative mutations in 6 members of 27 families, and correlated severity of the phenotype and closely related X-linked hereditary spastic paraplegia (Section 23.4.2) in which the closely related myelin protein DM20 may substitute for PLP, with these defects. Because more severe genetic defects of *PLP* and *DM20* are sometimes associated with a relatively mild phenotype, one mechanism for the dysmyelination is an increase in toxicity of mutant PLP through metabolic exertion of the compromised myelinating oligodendrocyte. Almost certainly, another genetic defect explains the non X-linked families.

37.7.5 Adult-onset dominant leucodystrophies

Forms of dominantly inherited leucodystrophy also occur exclusively in adults and may closely resemble chronic progressive multiple sclerosis (Eldridge *et al.* 1984; Schwankhaus *et al.* 1994). MRI shows diffuse, non-discrete, white matter disease and there are no oligoclonal bands in the spinal fluid. It remains uncertain whether all the adult-onset dominant leucodystrophies are one and the same disorder, and many are difficult to distinguish from the complicated hereditary spastic paraplegias. The various phenotypes are gradually being classified as their biochemical and genetic defects are characterized. Fukazawa *et al.* (1997b) described a family with spastic paraparesis, ataxia, and mild dementia, presenting in adulthood but with onset in childhood; diffuse white matter abnormalities were present on cerebral magnetic resonance, whereas pathognomic features of the other leucodystrophies were absent. Another addition to this group involves two siblings with behavioural abnormalities progressing to dementia with extensive white matter magnetic resonance abnormalities in whom brain biopsy showed glycolipid inclusions in macrophages unlike any other lysosomal storage disease (Simon *et al.* 1998). A new leucodystrophy recently described with autosomal recessive inheritance presents in childhood or adolescence with progressive pyramidal, cerebellar, and dorsal column dysfunction: imaging shows cerebral and cerebellar white matter abnormalities and marked involvement of brainstem and spinal cord tracts along with an elevated lactate in white matter (van der Knaap *et al.* 2003).

37.7.6 Vanishing white matter disease

This autosomal recessive leucodystrophy may present with relatively acute symptoms usually in childhood but occasionally in young adults (Section 10.2.5). The episodes manifest as altered consciousness, seizures, weakness, and ataxia. Examination shows signs of cognitive, pyramidal, and cerebellar dysfunction with MRI evidence for extensive symmetrical abnormality of cerebral white matter and tissue destruction such that large areas of cavitation are apparent, thus accounting for the nomenclature. Vanishing white matter disease is attributable to a mutation in any one of the five subunits making up the eukaryotic translation initiation factor gene mapping to 3q27 (van der Knaap *et al.* 2002). Abnormal protein synthesis leads to cell damage at times of increased metabolic demand, such as infection or minor head trauma.

References

Adams RD, Victor M, Mancall EL (1959). Central pontine myelinolysis. *Arch Neurol Psychiatry*, **81**, 154–72.

Alusi SH, Glickman S, Aziz TZ *et al.* (1999). Tremor in multiple sclerosis. *J Neurol Neurosurg Psychiatry*, **66**, 131–4.

Alusi SH, Aziz TZ, Glickman S *et al.* (2001). Stereotactic lesional surgery for the treatment of tremor in multiple sclerosis: a prospective case-controlled study. *Brain*, **124**, 1576–89.

Amato MP, Ponziani G, Siracusa G *et al.* (2001). Cognitive dysfunction in early-onset multiple sclerosis: a reappraisal after 10 years. *Arch Neurol*, **58**, 1602–6.

Andersen O, Lygner P-E, Bergstrom T *et al.* (1993). Viral infections trigger multiple sclerosis relapses: a prospective seroepidemiological study. *J Neurol*, **240**, 417–22.

Aoyagi Y, Kobayashi H, Tanaka K *et al.* (1999). A de novo splice donor site mutation causes in-frame deletion of 14 amino acids in the proteolipid protein in Pelizaeus-Merzbacher disease. *Ann Neurol*, **46**, 112–5.

Ascherio A, Munch M (2001). Epstein Barr virus and multiple sclerosis. *Epidemiology*, **11**, 220–4.

Aubourg P, Feil R, Guidoux S *et al.* (1990). The red-green visual pigment gene region in adrenoleucodystrophy. *Am J Hum Genet*, **46**, 459–69.

Austin (1957). Metachromatic form of diffuse cerebral sclerosis. 1. Diagnosis during life by urine sediment examination. *Neurology*, **7**, 415–26.

Bakshi R, Kinkel PR, Mechtler LL *et al.* (1998). Magnetic resonance imaging findings in 22 cases of myelitis: comparison between patients with and without multiple sclerosis. *Eur J Neurol*, **5**, 35–48.

Bamford CR, Sibley WA, Thies C *et al.* (1981). Trauma as an etiologic and aggravating factor in multiple sclerosis. *Neurology*, **31**, 1229–34.

Bandini F, Castello E, Mazzella L *et al.* (2001). Gabapentin but not vigabatrin is effective in the treatment of acquired nystagmus in multiple sclerosis: how valid is the GABAergic hypothesis? *J Neurol Neurosurg Psychiatry*, **71**, 107–10.

Bauer HJ, Hanefeld FA (1993). *Multiple Sclerosis, Its Impact from Childhood to Old Age*. Saunders, London.

Benedikz JG, Magnusson H, Gudmundsson G (1994). Multiple sclerosis in Iceland, with observations on the alleged epidemic in the Faroe Islands. *Ann Neurol*, **36** (Suppl 2), S175–9.

Berman M, Feldmann S, Alter M *et al.* (1981). Acute transverse myelitis: incidence and aetiologic considerations. *Neurology*, **31**, 966–71.

Betts CD, D'Mellow MT, Fowler CJ (1993). Urinary symptoms and the neurological features of bladder dysfunction in multiple sclerosis. *J Neurol Neurosurg Psychiatry*, **56**, 245–50.

Betts CD, Jones SJ, Fowler CG *et al.* (1994). Erectile dysfunction in multiple sclerosis: associated neurological and neurophysical deficits, and treatment of the condition. *Brain*, **117**, 1303–10.

Bielekova B, Richert N, Howard T *et al.* (2004). Humanized anti-CD25 (daclizumab) inhibits disease activity in multiple sclerosis patients failing to respond to interferon beta. *Proc Natl Acad Sci USA*, **101**, 8705–8.

Biousse V, Trichet C, Bloch-Michel E *et al.* (1999). Multiple sclerosis associated with uveitis in two large clinic-based series. *Neurology*, **52**, 179–81.

Black JA, Liu S, Hains BC *et al.* (2006). Long-term protection of central axons with phenytoin in monophasic and chronic-relapsing EAE. *Brain*, **129**, 3196–208.

Bosma R, Wynia K, Havlikova E *et al.* (2005). Efficacy of desmopressin in patients with multiple sclerosis suffering from bladder dysfunction: a meta-analysis. *Acta Neurol Scand*, **112**, 1–5.

Brex PA, Ciccarelli O, O'Riordan JI *et al.* (2002). A longitudinal study of abnormalities on MRI and disability from multiple sclerosis. *NEJM*, **346**, 158–64.

Brockmann K, Dechent P, Wilken B *et al.* (2003). Proton MRS profile of cerebral metabolic abnormalities in Krabbe disease. *Neurology*, **60**, 819–25.

Broggi G, Ferroli P, Franzini A *et al.* (2000). Microvascular decompression for trigeminal neuralgia: comments on a series of 250 cases, including 10 patients with multiple sclerosis. *J Neurol Neurosurg Psychiatry*, **68**, 59–64.

Brønnum-Hansen H, Koch-Henriksen N *et al.* (2004). Trends in survival and cause of death in Danish patients with multiple sclerosis. *Brain*, **127**, 844–50.

Brunner JE, Redmond JM, Haggar AM *et al.* (1990). Central pontine myelinolysis and pontine lesions after rapid correction of hyponatraemia: a prospective magnetic resonance imaging study. *Ann Neurol*, **27**, 61–6.

Buljevac D, Flach HZ, Hop WCJ *et al.* (2002). Prospective study on the relationship between infections and multiple sclerosis exacerbations. *Brain*, **125**, 952–60.

Burt RK, Cohen B, Rose J *et al.* (2005). Hematopoietic stem cell transplantation for multiple sclerosis. *Arch Neurol*, **62**, 860–4.

Cabre P, Signate A, Olindo S *et al.* (2005). Role of return migration in the emergence of multiple sclerosis in the French West Indies. *Brain*, **128**, 2899–910.

Caraccio N, Dardano A, Manfredonia F *et al.* (2005). Long-term follow-up of 106 multiple sclerosis patients undergoing interferon-beta 1a or 1b therapy: predictive factors of thyroid disease development and duration. *J Clin Endocrinol Metab*, **90**, 4133–7.

Carton H, Vlietinck R, Debruyne J *et al.* (1997). Recurrence risks of multiple sclerosis in relatives of patients in Flanders, Belgium. *J Neurol Neurosurg Psychiatry*, **62**, 329–33.

Cohen JA, Rovaris M, Goodman AD *et al.* (2007). Randomized, double-blind, dose-comparison study of glatiramer acetate in relapsing-remitting MS. *Neurology*, **68**, 939–44.

Cohen JA, Cutter GR, Fischer JS *et al.* (2002). Benefit of interferon beta-1a on MSFC progression in secondary progressive MS. *Neurology*, **59**, 679–87.

Coles AJ, Paolili A, Molyneux P *et al.* (1999). Monoclonal antibody treatment exposes three mechanisms underlying the clinical course in multiple sclerosis. *Ann Neurol*, **46**, 296–304.

Coles AJ, Cox A, Le Page E *et al.* (2006). The window of therapeutic opportunity in multiple sclerosis: evidence from monoclonal antibody therapy. *J Neurol*, **253**, 98–108.

Coles AJ, CAMMS223 Trial Investigators (2008). Alemtuzumab versus Interferon Beta-1a in early multiple sclerosis. *N Eng J Med*, **359**, 1786–1801.

Comi G, Filippi M, Wolinsky JS (2001a). European/Canadian multicenter, double-blind, randomized, placebo controlled study of the effects of glatiramer acetate on magnetic resonance imaging-measured disease activity and burden in patients with relapsing multiple sclerosis. *Ann Neurol*, **49**, 290–7.

Comi G, Filippi M, Barkhof F *et al.* (2001b). Effect of early interferon treatment on conversion to definite multiple sclerosis. *Lancet*, **357**, 1576–82.

Compston DAS (2007). Heterogeneity and complexity in demyelinating disease. *Brain*, **130**, 1178–80.

Compston DAS, Coles AJ (2008) Multiple sclerosis: seminar. *Lancet*, **372**, 1502–17.

Compston DAS, Confavreux C (2005). The distribution of multiple sclerosis. In Compston DAS, ed. *McAlpine's Multiple Sclerosis*, 4th edition, pp. 71–111. Elsevier, London.

Compston DAS, Wekerle H (2005). The genetics of multiple sclerosis. In Compston DAS ed. *McAlpine's Multiple Sclerosis*, 4th edition, pp. 113–81. Elsevier, London.

Confavreux C, Hutchinson M, Hours M *et al.* (1998). Rate of pregnancy-related relapse in multiple sclerosis. *N Engl J Med*, **339**, 285–91.

Confavreux C, Compston DAS, Hommes OR *et al.* (1992). EDMUS, an European database for multiple sclerosis. *J Neurol Neurosurg Psychiatry*, **55**, 671–6.

Confavreux C, Suissa S, Saddier P *et al.* (2001) Vaccinations and the risk of relapse in multiple sclerosis. Vaccines in Multiple Sclerosis Study Group. *NEJM*, **344**, 319–26.

Confavreux C, Compston DAS (2005). The natural history of multiple sclerosis. In Compston DAS, ed. *McAlpine's Multiple Sclerosis*, 4th edition, pp. 183–272. Elsevier, London.

Confavreux C, Vukusic S (2006). Age at disability milestones in multiple sclerosis. *Brain*, **129**, 595–605.

Correale J, Farez M (2007). Association between parasite infection and immune responses in multiple sclerosis. *Ann Neurol*, **61**, 97–108.

Cutter GR, Baier ML, Rudick RA *et al.* (1999). Development of a multiple sclerosis functional composite as a clinical trial outcome measure. *Brain*, **122**, 871–82.

Dale RC, de Sousa C, Chong WK et al. (2000). Acute disseminated encephalomyelitis, multiphasic disseminated encephalomyelitis and multiple sclerosis in children. *Brain*, **123**, 2407–22.

Dean G (1967). Annual incidence, prevalence and mortality of MS in white South African-born and in white immigrants to South Africa. *BMJ*, **2**, 724–30.

Detels R, Visscher B, Malmgrem RM et al. (1977). Evidence for lower susceptibility to multiple sclerosis in Japanese-Americans. *Am J Epidemiol*, **105**, 303–10.

DeStefano F, Verstraeten T, Jackson LA et al. (2003). Vaccinations and risk of central nervous system demyelinating diseases in adults. *Arch Neurol*, **60**, 504–9.

de Vries E (1960). *Postvaccinial Perivenous Encephalitis*. Elsevier, Amsterdam.

Dichgans M, Mayer M, Uttner I et al. (1998). The phenotypic spectrum of CADASIL: clinical findings in 102 cases. *Ann Neurol*, **44**, 731–9.

Duquette P, Murray TJ, Pleines J et al. (1987). Multiple sclerosis in childhood: clinical profile in 125 patients. *J Pediatr*, **111**, 359–63.

Durelli L, Verdun E, Barbero P et al. (2002) Every-other-day interferon beta-1b versus once-weekly interferon beta-1a for multiple sclerosis: results of a 2-year prospective randomised multicentre study (INCOMIN). *Lancet*, **359**, 1453–60.

Dutta R, McDonough J, Yin X et al. (2006). Mitochondrial dysfunction as a cause of axonal degeneration in multiple sclerosis patients. *Ann Neurol*, **59**, 478–89.

Ebers GC, Sadovnick AD, Risch NJ (1995). A genetic basis for familial aggregation in multiple sclerosis. *Nature*, **377**, 150–1.

Ebers GC, Yee IM, Sadovnick AD et al. (2000). Conjugal multiple sclerosis: population-based prevalence and recurrence risks in offspring. Canadian Collaborative Study Group. *Ann Neurol*, **48**, 927–31.

Edan G, Miller DH, Clanet M et al. (1997). Therapeutic effect of mitoxantrone combined with methylprednisolone in multiple sclerosis: a randomised multi-center study of active disease using MRI and clinical criteria. *J Neurol Neurosurg Psychiatry*, **62**, 112–8.

Elian M, Nightingale S, Dean G (1990) Multiple sclerosis among United Kingdom-born children of immigrants from the Indian subcontinent, Africa and the West Indies. *J Neurol Neurosurg Psychiatry*, **53**, 906–11.

Eldridge R, Anayiotos CP, Schlesinger S et al. (1984). Hereditary adult-onset leukodystrophy simulating chronic progressive multiple sclerosis. *NEJM*, **311**, 948–53.

Ellis SJ (1995). Severe hyponatraemia: complications and treatment. *QJM*, **88**, 905–9.

Eriksson M, Ben-Menachem E, Andersen O (2002). Epileptic seizures, cranial neuralgias and paroxysmal symptoms in remitting and progressive multiple sclerosis. *Mult Scler*, **8**, 495–9.

European Study Group in Interferon Beta-1b in Secondary Progressive MS (1998). Placebo-controlled multicentre randomised trial of interferon beta-1b in treatment of secondary progressive multiple sclerosis. *Lancet*, **352**, 1491–7.

Fassas A, Anagnostopolous A, Kasis A et al. (1997). Peripheral blood stem cell transplantation in the treatment of progressive multiple sclerosis: first results of a pilot study. *Bone Marrow Transplant*, **20**, 631–8.

Feinstein A, Ron M, Thompson A (1993). A serial study of psychometric and magnetic resonance imaging changes in multiple sclerosis. *Brain*, **116**, 569–602.

Ferguson B, Matyszak MK, Esiri MM et al. (1997). Axonal damage in acute multiple sclerosis lesions. *Brain*, **120**, 393–9.

Filippi M, Bozzali M, Rovaris M et al. (2003). Evidence for widespread axonal damage at the earliest clinical stage of multiple sclerosis. *Brain*, **126**, 433–7.

Filippini G, Munari L, Incorvaia B et al. (2003). Interferons in relapsing remitting multiple sclerosis: a systematic review. *Lancet*, **361**, 545–52.

Fisniku LK, Brex P, Altmann DR et al. (2008). Disability and T_2 MRI lesions: a 20-year follow-up of patients with relapse onset of multiple sclerosis. *Brain*, **131**, 808–17.

Foix C, Alajouanine T (1926). La myelité necrotique subaigue. *Rev Neurol*, **2**, 1–42.

Foong J, Rozewicz L, Quaghebeur G et al. (1997). Executive function in multiple sclerosis: the role of frontal lobe pathology. *Brain*, **120**, 15–26.

Fowler CJ, Miller JR, Sharief MK et al. (2005). A double blind, randomised study of sildenafil citrate for erectile dysfunction in men with multiple sclerosis. *J Neurol Neurosurg Psychiatry*, **76**, 700–5.

Freeman JA, Langdon DW, Hobart JC et al. (1999). Inpatient rehabilitation in multiple sclerosis: do the benefits carry over into the community? *Neurology*, **52**, 50–6.

French Research Group on Multiple Sclerosis (1992). Multiple sclerosis in 54 twinships: concordance rate is independent of zygosity. *Ann Neurol*, **32**, 724–7.

Fukazawa T, Moriwaka F, Hamada K et al. (1997a). Facial palsy in multiple sclerosis. *J Neurol*, **244**, 631–3.

Fukazawa T, Sasaki H, Kikuchi S et al. (1997b). Dominantly inherited leukodystrophy showing cerebellar deficits and spastic paraparesis: a new entity? *J Neurol*, **244**, 446–9.

Gledhill RF, Harrison BM, McDonald WI (1973). Pattern of remyelination in the CNS. *Nature*, **244**, 443–4.

Golde S, Chandran S, Brown GC et al. (2002). Different pathways for iNOS-mediated toxicity in vitro dependent on neuronal maturation and NMDA-receptor expression. *J Neurochem*, **82**, 269–82.

Goodin DS, Arnason B, Coyle P et al. (2003). The use of mitoxantrone (Novantrone) for the treatment of multiple sclerosis – Report of the Therapeutics and Technology Assessment Subcommittee of the American Academy of Neurology. *Neurology*, **61**, 1332–8.

Goodkin DE (2000). Interferon beta-1b in secondary progressive MS: clinical and MRI results of a 3-year randomized controlled trial. *Neurology*, **54**, 2352 (abstract).

Granieri E, Casetta I (1997). Common childhood and adolescent infections and multiple sclerosis. *Neurology*, **49** (Suppl 2), S42–54.

Gregory S, Schmidt S, Seth P et al. (2007). Allelic and functional association of the interleukin 7 receptor gene α chain (IL7R α) with multiple sclerosis. *Nat Genet*, **39**, 1083–91.

Hageman AT, Gabreils FJ, de Jong JG et al. (1995). Clinical symptoms of adult metachromatic leukodystrophy and arylsulfatase A pseudodeficiency. *Arch Neurol*, **52**, 408–13.

Hammond SR, English DR, McLeod JG (2000). The age-range of risk of developing multiple sclerosis: evidence from a migrant population in Australia. *Brain*, **123**, 968–74.

Harding AE, Sweeney MG, Miller DH et al. (1992). Occurrence of a multiple sclerosis-like illness in women who have a Leber's hereditary optic neuropathy mitochondrial DNA mutation. *Brain*, **115**, 979–89.

Hartung HP, Gonsette R, Konig N et al. (2002). Mitoxantrone in progressive multiple sclerosis: a placebo controlled, double-blind, randomised, multicentre trial. *Lancet*, **360**, 2018–25.

Hensiek AE, Seaman SR, Barcellos LF et al. (2007). Familial effects on the clinical course of multiple sclerosis. *Neurology*, **68**, 376–83.

Hernan MA, Jick SS, Olek MJ et al. (2004). Recombinant hepatitis B vaccine and the risk of multiple sclerosis: a prospective study. *Neurology*, **63**, 838–42.

Hinds J, Eidelman B, Wald A (1990). Prevalence of bowel dysfunction in multiple sclerosis. *Gastroenterology*, **98**, 1538–42.

Hobart JC, Lamping D, Fitzpatrick R et al. (2001). The Multiple Sclerosis Impact Scale (MSIS-29): a new patient based outcome measure. *Brain*, **124**, 962–73.

Hohlfeld R (1997). Biotechnological agents for the immunotherapy of multiple sclerosis: principles, problems and perspectives. *Brain*, **120**, 865–916.

Hooper J, Whittle IR (1998). Long term outcome after thalamotomy for movement disorders in multiple sclerosis. *Lancet*, **352**, 1984.

Houtchens MK, Richert JR, Sami A et al. (1997). Open label gabapentin treatment for pain in multiple sclerosis. *Mult Scler*, **3**, 250–3.

Hurst EW (1941). Acute haemorrhagic leuco-encephalitis, a previously undefined entity. *Med J Aust*, **2**, 1–6.

Hyman N, Barnes M, Bhakta B, Cozens A et al. (2000). Botulinum toxin (Dysport) treatment of hip adductor spasticity in multiple sclerosis: a prospective, randomised, double blind, placebo controlled, dose ranging study. *J Neurol Neurosurg Psychiatry*, **68**, 707–12.

IFNB Multiple Sclerosis Study Group (1993). Interferon beta-1b is effective in relapsing–remitting multiple sclerosis. 1.Clinical results of a multicenter, randomized, double-blind, placebo-controlled trial. *Neurology*, **43**, 655–61.

IFNB Multiple Sclerosis Study Group and the University of British Columbia MS/MRI Analysis Group (1995). Interferon beta-1b in the treatment of multiple sclerosis: final outcome of the randomised controlled trial. *Neurology*, **45**, 1277–85.

International multiple sclerosis genetics consortium. Risk alleles for multiple sclerosis identified by a genomewide association study. *New Eng J. Med* 2007: **30**, 851–62.

Jacobs LD, Cookfair DL, Rudick RA *et al*. (1996). Intramuscular interferon beta-1a for disease progression in relapsing multiple sclerosis. The Multiple Sclerosis Collaborative Research Group (MSCRG). *Ann Neurol*, **39**, 285–94.

Jacobs LD, Beck RW, Simon JH *et al*. and the CHAMPS Study Group (2000). Intramuscular interferon beta-1a therapy initiated during a first demyelinating event in multiple sclerosis. *NEJM*, **343**, 898–904.

Jeffery ND, Blakemore WF (1997). Locomotor deficits induced by experimental spinal cord demyelination are abolished by spontaneous remyelination. *Brain*, **120**, 27–37.

Johnson KP, Brooks BR, Cohen JA *et al*. (1995). Copolymer 1 reduces relapse rate and improves disability in relapsing–remitting multiple sclerosis: results of a phase III multicenter, double-blind placebo-cxontrolled trial. *Neurology*, **45**, 1268–76.

Johnson KP, Brooks BR, Cohen JA *et al*. (1998). Extended use of glatiramer acetate (Copaxone) is well tolerated and maintains its clinical effect on multiple sclerosis relapse rate and degree of disability. Copolymer 1 Multiple Sclerosis Study Group. *Neurology*, **50**, 701–8.

Josien E, Lefebvre V, Vermesch P *et al*. (1993). Adrénoleucomyéloneuropathy de l'adulte. *Rev Neurol*, **149**, 230–2.

Kahana E, Zilber N, Abramson JH *et al*. (1994). Multiple sclerosis: genetic versus environmental aetiology: epidemiology in Israel updated. *J Neurol*, **241**, 341–6.

Kappos L, polman CH, Freedman MS *et al*. (2006a) Treatment with interferon beta-1b delays conversion to clinically definite and McDonald MS in patients with clinically isolated syndromes. *Neurology*, **67**, 1242–9.

Kappos L, Antel J, Comi G *et al*. (2006b). Oral fingolimod (FTY720) for relapsing multiple sclerosis. *NEJM*, **355**, 1124–40.

Keegan M, Pineda AA, McClelland RL *et al*. (2002). Plasma exchange for severe attacks of CNS demyelination: predictors of response. *Neurology*, **58**, 143–6.

Keegan M, Konig F, McClelland R *et al*. (2005). Relation between humoral pathological changes in multiple sclerosis and response to therapeutic plasma exchange. *Lancet*, **366**, 579–82.

Kidd D, Thorpe JW, Thompson *et al*. (1993). Spinal cord MRI using multi-array coils and fast spin echo. II: Findings in multiple sclerosis. *Neurology*, **43**, 2632–7.

Kidd D, Burton B, Plant GT *et al*. (2003). Chronic relapsing inflammatory optic neuropathy (CRION). *Brain*, **126**, 278–84.

Kremenchutzky M, Rice GP, Baskerville J *et al*. (2006). The natural history of multiple sclerosis: a geographically based study 9: observations on the progressive phase of the disease. *Brain*, **129**, 584–94.

Krupp LB, Coyle PK, Doscher C *et al*. (1995). Fatigue therapy in multiple sclerosis: results of a double-blind, randomised, parallel trial of amantidine, pemoline and placebo. *Neurology*, **45**, 1956–61.

Kurtzke JF (1975). A reassessment of the distribution of multiple sclerosis. *Acta Neurol Scand*, **51**, 110–57.

Kurtzke JF (1983). Rating neurologic impairment in multiple sclerosis: an expanded disability status scale (EDSS). *Neurology*, **33**, 1444–52.

Kurtzke JF (1993). Epidemiologic evidence for multiple sclerosis as an infection. *Clin Microbiol Rev*, **6**, 382–427.

Kutzelnigg A, Lucchinetti CF, Stadelmann C *et al*. (2005). Cortical demyelination and diffuse white matter injury in multiple sclerosis. *Brain*, **128**, 2705–12.

Lacour A, De Seze J, Revenco E *et al*. (2004). Acute aphasia in multiple sclerosis: a multicenter study of 22 patients. *Neurology*, **62**, 974–7.

Lang HLE, Jacobsen H, Ikemizu S *et al*. (2002). A functional and structural basis for TCR cross-reactivity in multiple sclerosis. *Nature Immunol*, **3**, 940–3.

Lassmann H, Wekerle H (2005). The pathology of multiple sclerosis. In Compston DAS, ed. *McAlpine's Multiple Sclerosis*, 4th edition, pp. 557–99. Elsevier, London.

Lennon VA, Wingerchuk DM, Kryzer TJ *et al*. (2004). A serum autoantibody marker of neuromyelitis optica: distinction from multiple sclerosis. *Lancet*, **364**, 2106–12.

Lennon VA, Kryzer TJ, Pittock SJ *et al*. (2005). IgG marker of optic-spinal multiple sclerosis binds to the aquaporin-4 water channel. *J Exp Med*, **202**, 473–7.

Lightman S, McDonald WI, Bird AC *et al*. (1987). Retinal venous sheathing in optic neuritis: its significance for the pathogenesis of multiple sclerosis. *Brain*, **110**: 405–14.

Lucchinetti CF, Kiers L, O'Duffy A *et al*. (1997). Risk factors for developing multiple sclerosis after childhood optic neuritis. *Neurology*, **49**, 1413–8.

Lucchinetti C, Bruck W, Parisi J *et al*. (1999). A quantitative analysis of oligodendrocytes multiple sclerosis lesions: a study of 117 cases. *Brain*, **122**, 2279–95.

Lucchinetti CF, Brück W, Parisi J *et al*. (2000). Heterogeneity of multiple sclerosis lesions: implications for the pathogenesis of demyelination. *Ann Neurol*, **47**, 707–17.

Lucchinetti CF, Mandler R, McGavern D *et al*. (2002). A role for humoral mechanisms in the pathogenesis of Devic's neuromyelitis optica. *Brain*, **125**, 1450–61.

McDonald WI, Compston A, Edan G *et al*. (2001). Recommended diagnostic criteria for multiple sclerosis: guidelines from the International Panel on the diagnosis of multiple sclerosis. *Ann Neurol*, **50**, 121–7.

McDonald WI, Compston DAS (2005). The symptoms and signs of multiple sclerosis. In Compston DAS, ed. *McAlpine's Multiple Sclerosis*, 4th edition, pp. 287–346. Elsevier, London.

McIntosh-Michaelis SA, Roberts MH, Wilkinson SM *et al*. (1991). The prevalence of cognitive impairment in a community survey of multiple sclerosis. *Br J Clin Psychol*, **30**, 333–48.

Majed HH, Chandran S, Niclou SP *et al*. (2006). A novel role for Sema3A in neuroprotection from injury mediated by activated microglia. *J Neurosci*, **26**, 1730–8.

Marrosu MG, Muntoni F, Murru MR *et al*. (1992). HLA-DQB1 genotype in Sardinian multiple sclerosis: evidence for a key role of DQB1.0201 and DQB1.0302 alleles. *Neurology*, **42**, 883–6.

Martyn CN, Cruddas M, Compston DAS (1993). Symptomatic Epstein–Barr virus infection and multiple sclerosis. *J Neurol Neurosurg Psychiatry*, **56**, 167–8.

Ménage P, Carreau V, Tourbah A *et al*. (1994). Les adrénoleucodystrophies hétérozygotes symptomatiques de l'adulte: 10 cas. *Rev Neurol*, **49**, 445–54.

Menger H, Jorg J (1999). Outcome of central pontine myelinolysis. *J Neurol*, **246**, 700–5.

Metz I, Lucchinetti CF, Openshaw H *et al*. (2007). Autologous haematopoietic stem cell transplantation fails to stop demyelination and neurodegeneration in multiple sclerosis. *Brain*, **130**, 1254–62.

Mikaeloff Y, Adamsbaum C, Husson B *et al*. (2004). MRI prognostic factors for relapse after acute CNS inflammatory demyelination in childhood. *Brain*, **127**, 1942–7.

Millefiorini E, Gasperini C, Pozzilli C *et al*. (1997). Randomized placebo-controlled trial of mitoxanthrone in relapsing-remitting multiple sclerosis: 24 month clinical and MRI outcome. *J Neurol*, **244**, 153–9.

Miller HG, Stanton JB, Gibbons JL (1956). Parainfectious encephalomyelitis and related syndromes. *QJM*, **25**, 427–505.

Miller DH, Newton MR, van der Poel JC *et al*. (1988). Magnetic resonance imaging of the optic nerve in optic neuritis. *Neurology*, **38**, 175–9.

Miller DH, Molyneux PD, Barker GJ *et al*. (1999). Effect of interferon-beta1b on magnetic resonance imaging outcomes in secondary progressive multiple sclerosis: results of a European multicenter, randomized, double blind, placebo controlled trial. European Study Group on Interferon-beta1b in secondary progressive multiple sclerosis. *Ann Neurol*, **46**, 850–9.

Miller DH, Khan OA, Sheremata WA *et al*. (2003). A controlled trial of natalizumab for relapsing multiple sclerosis. *NEJM*, **348**, 15–23.

Miller DH, Compston DAS (2005). The differential diagnosis of multiple sclerosis. In Compston DAS, ed. *McAlpine's Multiple Sclerosis*, 4th edition, pp. 389–437. Elsevier, London.

Miller DM, Weinstock-Guttman B, Bethoux F *et al*. (2000). A meta-analysis of methylprednisolone in recovery from multiple sclerosis exacerbations. *Mult Scler*, **6**, 267–73.

Milligan NM, Newcombe R, Compston DAS (1987). A double blind controlled trial of high dose methylprednisolone in patients with multiple sclerosis. I. Clinical effects. *J Neurol Neurosurg Psychiatry*, **50**, 511–6.

Misu T, Fujihara K, Kakita A *et al.* (2007). Loss of aquaporin 4 in lesions of neuromyelitis optica: distinction from multiple sclerosis. *Brain*, **130**, 1224–34.

Morrissey SP, Borruat FX, Miller DH *et al.* (1995). Bilateral simultaneous optic neuritis in adults: clinical, imaging, serological and genetic studies. *J Neurol Neurosurg Psychiatry*, **58**, 70–4.

Moser HW (1997). Adrenoleucodystrophy: phenotype, genetics, pathogenesis, and therapy. *Brain*, **120**, 1485–508.

Moser HW, Moser AB, Frayer KK *et al.* (1981). Adrenoleukodystrophy: increased plasma content of saturated very long chain fatty acids. *Neurology*, **31**, 1241–9.

Moser AB, Kreiter N, Bezman L *et al.* (1999). Plasma very long chain fatty acids in 3000 peroxisome disease patients and 29000 controls. *Ann Neurol*, **45**, 100–10.

Mumford, CJ, Wood NW, Kellar-Wood HF *et al.* (1994). The British Isles survey of multiple sclerosis in twins. *Neurology*, **44**, 11–5.

Munari LM, Filippini G (2004). Lack of evidence for use of glatiramer acetate in multiple sclerosis. *Lancet Neurol*, **3**, 641.

Newell KL, Kleinschmidt-Demasters BK (1996). Central pontine myelinolysis at autopsy; a twelve year retrospective analysis. *J Neurol Sci*, **142**, 134–9.

Nicholas RS, Stevens S, Wing MG *et al.* (2002). Microglia-derived IGF-2 and CNTF prevent TNFa induced death of mature oligodendrocytes in vitro. *J Neuroimmunol*, **124**, 36–44.

Nielsen JF, Sinkjaer T, Jakobsen J (1996). Treatment of spasticity with repetitive magnetic stimulation: a double blind placebo-controlled study. *Mult Scler*, **2**, 227–32.

Nilsson P, Larsson EM, Maly-Sundgren P *et al.* (2005). Predicting the outcome of optic neuritis: evaluation of risk factors after 30 years of follow-up. *J Neurol*, **252**, 396–402.

Noseworthy J, Miller DH, Compston DAS (2005) Disease modifying treatments in multiple sclerosis. In Compston DAS, ed. *McAlpine's Multiple Sclerosis*, 4th edition, pp. 729–802.

Olerup O, Hillert J (1991). HLA class II-associated genetic susceptibility in multiple sclerosis: a critical evaluation. *Tissue Antigens*, **38**, 1–15.

Once Weekly Interferon for MS Study Group (OWIMS) (1999). Evidence of interferon beta-1a dose response in relapsing–remitting MS. *Neurology*, **53**, 679–86.

O'Neill BP, Moser HW, Saxena KM (1982). Familial x-linked Addison disease as an expression of adrenoleukodystrophy (ALD): elevated C26 fatty acids in cultured skin fibroblasts. *Neurology*, **32**, 543–7.

Oppenheimer D (1978). The cervical cord in multiple sclerosis. *Neuropathol Appl Neurobiol*, **4**, 151–62.

Optic Neuritis Study Group (1997a). Visual function 5 years after optic neuritis. *Arch Ophthalmol*, **115**, 1544–52.

Optic Neuritis Study Group (1997b). The 5-year risk of MS after optic neuritis: experience of the Optic Neuritis Treatment Trial. *Neurology*, **49**, 1404–13.

Optic Neuritis Study Group (2003). High- and low-risk profiles for the development of multiple sclerosis within 10 years after optic neuritis. *Arch Ophthalmol*, **121**, 944–9.

O'Riordan JI, Thompson AJ, Kingsley DPE *et al.* (1998). The prognostic value of brain MRI in clinically isolated syndromes of the CNS: a 10-year follow-up. *Brain*, **121**, 495–503.

Ormerod IEC, McDonald WI (1984). Multiple sclerosis presenting with progressive visual failure. *J Neurol Neurosurg Psychiatry*, **47**, 943–6.

Orton SM, Harrara B, Yee IM *et al.* Sex ratio of multiple sclerosis in Canada: a longitudinal study. *Cancer Neurology*, 2006: **5**, 932–6.

Osaka H, Kawanishi C, Inoue K *et al.* (1999). Pelizaeus-Merzbacher disease: three novel mutations and implication for locus heterogeneity. *Ann Neurol*, **45**, 59–64.

Panitch HS, Goodin DS, Francis G *et al.* (2002). Randomized, comparative study of interferon beta-1a treatment regimens in MS: the EVIDENCE Trial. *Neurology*, **59**, 1496–506.

Parkin PJ, Hierons R, McDonald WI (1984). Bilateral optic neuritis: a long-term follow up. *Brain*, **107**, 951–64.

Patrikios P, Stadelmann C, Kutzelnigg A *et al.* (2006). Remyelination is extensive in a subset of multiple sclerosis patients. *Brain*, **129**, 3165–72.

Paty DW, Studney D, Redekop K *et al.* (1994). MS COSTAR: a computerised patient record adapted for clinical research purposes. *Ann Neurol*, **36** (Suppl), S134–5.

Pinching A (1977). Clinical testing of olfaction reassessed. *Brain*, **100**, 377–88.

Plant GT, Kermode AG, Turano G *et al.* (1992). Symptomatic retrochiasmal lesions in multiple sclerosis: clinical features, visual evoked potentials, and magnetic resonance imaging. *Neurology*, **42**, 68–76.

Polman CH, Reingold SC, Edan G *et al.* (2005). Diagnostic criteria for multiple sclerosis: 2005 revisions to the "McDonald Criteria". *Ann Neurol*, **58**, 840–6.

Polman CH, O'Connor PW, Havrdova E *et al.* (2006). A randomized, placebo-controlled trial of natalizumab for relapsing multiple sclerosis. *NEJM*, **354**, 899–910.

Poser CM, Paty DW, Scheinberg L *et al.* (1983). New diagnostic criteria for multiple sclerosis: guidelines for research protocols. *Ann Neurol*, **13**, 227–31.

Poskanzer DC, Prenney LP, Sheridan JL *et al.* (1980). Multiple sclerosis in the Orkney and Shetland Islands. 1. Epidemiology, clinical factors and methodology. *J Epidemiol Community Health*, **34**, 229–39.

Pozzilli C, Brunetti M, Amicosante AM *et al.* (2002). Home based management in multiple sclerosis: results of a randomised controlled trial. *J Neurol Neurosurg Psychiatry*, **73**, 250–5.

PRISMS (Prevention of Relapses and Disability by Interferon-beta 1a Subsequently in Multiple Sclerosis) Study Group (1998). Randomised, double-blind, placebo-controlled study of interferon-beta 1a in relapsing–remitting multiple sclerosis: clinical results. *Lancet*, **352**, 1498–504.

Pujol J, Bello J, Deus J *et al.* (1997). Lesions in the left arcuate fasciculus region and depressive symptoms in multiple sclerosis. *Neurology*, **49**, 1105–10.

Raymond GV (2001). Peroxisomal disorders. *Curr Opin Neurol*, **14**, 783–7.

Redford EJ, Kapoor R, Smith KJ (1997). Nitric oxide donors reversibly block axonal conduction: demyelinated axons are especially susceptible. *Brain*, **120**, 2149–57.

Reich D, Patterson N, De Jager PL *et al.* (2005). A whole-genome admixture scan finds a candidate locus for multiple sclerosis susceptibility. *Nat Genet*, **37**, 1113–8.

Renoux C, Vukusic S, Mikaeloff H *et al.* Natural history of multiple sclerosis with childhood onset. *New Eng J. Med* 2007: **356**, 2603–13.

Riordan-Eva P, Sanders MD, Govan GG *et al.* (1995). The clinical features of Leber's hereditary optic neuropathy defined by the presence of a pathogenic mitochondrial DNA mutation. *Brain*, **118**, 319–37.

Ristori G, Cannoni S, Stazi MA *et al.* (2006). Multiple sclerosis in twins from continental Italy and Sardinia: a nationwide study. *Ann Neurol*, **59**, 27–34.

Robertson NP, Fraser M, Deans J *et al.* (1996). Age adjusted recurrence risks for relatives of patients with multiple sclerosis. *Brain*, **119**, 449–55.

Robertson NP, O'Riordan JI, Chataway J *et al.* (1997). Clinical characteristics and offspring recurrence rates of conjugal multiple sclerosis. *Lancet*, **349**, 1587–90.

Roemer SF, Parisi JE, Lennon VA *et al.* (2007). Pattern-specific loss of aquaporin-4 immunoreactivity distinguishes neuromyelitis optica from multiple sclerosis. *Brain*, **130**, 1194–205.

Roxburgh RH, Seaman SR, Masterman T *et al.* (2005). Multiple sclerosis severity score: ranking disability at similar duration to rate disease severity. *Neurology*, **64**, 1144–51.

Runmarker B, Andersen O (1995). Pregnancy is associated with a lower risk of onset and a better prognosis in multiple sclerosis. *Brain*, **118**, 253–61.

Sadovnick AD, Baird PA (1988). The familial nature of multiple sclerosis: age-corrected empiric recurrence risks for children and siblings of patients. *Neurology*, **38**, 990–1.

Sadovnick AD, Eisen K, Ebers GC *et al.* (1991). Cause of death in patients attending multiple sclerosis clinics. *Neurology*, **41**, 1193–6.

Sadovnick AD, Ebers GC, Dyment DA *et al.* (1996). Evidence for genetic basis of multiple sclerosis. *Lancet*, **347**, 1728–30.

Sawcer S, Ban M, Maranian M *et al.* (2005). A high-density screen for linkage in multiple sclerosis. *Am J Hum Genet*, **77**, 454–67.

Scolding NJ, Jayne DRW, Zajicek JP *et al.* (1997). Cerebral vasculitis – recognition, diagnosis and management. *QJM*, **90**, 61–73.

Schilder P (1912). Zur Kenntnis der sogenannten diffusen Sklerose (uber Encephalitis periaxalis diffusa). *Z Gesamte Neurol Psychiatr*, **10**, 1–60.

Schurch B, De Sèze M, Denys P *et al.* (2005). Botulinum toxin type A is a safe and effective treatment for neurogenic urinary incontinence: results of a single treatment, randomized, placebo-controlled 6-month study. *J Urol*, **174**, 196–200.

Schwankhaus JD, Katz DA, Eldridge R *et al.* (1994). Clinical and pathological features of an autosomal dominant, adult-onset leukodystrophy simulating chronic progressive multiple sclerosis. *Arch Neurol*, **51**, 757–66.

Schwid SR, Petrie MD, McDermott MP *et al.* (1997). Quantitative assessment of sustained-release 4-aminopyridine for symptomatic treatment of multiple sclerosis. *Neurology*, **48**, 817–21.

Secondary Progressive Efficacy Clinical Trial of Recombinant Interferon-beta-1a in MS (SPECTRIMS) Study Group (2001). Randomized controlled trial of interferon- beta-1a in secondary progressive MS: clinical results. *Neurology*, **56**, 1496–504.

Seiser A, Schwarz S, Aichinger-Steiner MM *et al.* (2002). Parkinsonism and dystonia in central pontine and extrapontine myelinolysis. *J Neurol Neurosurg Psychiatry*, **65**, 119–21.

Shakespeare DT, Boggild M, Young C (2003). Anti-spasticity agents for multiple sclerosis (Cochrane Review). *Cochrane Library*, **3**: update software.

Sharrack B, Hughes RAC (1999). The Guy's Neurological Disability Scale (GNDS): a new disability measure for multiple sclerosis. *Mult Scler*, **5**, 223–33.

Sheean GL, Murray NMF, Rothwell JC *et al.* (1997). An electrophysiological study of the mechansim of fatigue in multiple sclerosis. *Brain*, **120**, 299–315.

Schuele SU, Kellinghaus C, Shook SJ *et al.* (2005). Incidence of seizures in patients with multiple sclerosis treated with intrathecal baclofen. *Neurology*, **64**, 1086–7.

Sibley WA, Bamford CR, Clark K *et al.* (1991). A prospective study of physical trauma and multiple sclerosis. *J Neurol Neurosurg Psychiatry*, **54**, 584–9.

Sibley, WA, Bamford CR, Clark K (1985). Clinical viral infections and multiple sclerosis. *Lancet i*, 1313–5.

Simon DK, Rodriguez ML, Frosch MP *et al.* (1998). A unique familial leucodystrophy with adult onset dementia and abnormal glycolipid storage: a new lysosomal disease. *J Neurol Neurosurg Psychiatry*, **65**, 251–4.

Simone IL, Carrara D, Tortorella C *et al.* (2002). Course and prognosis in early-onset MS: comparison with adult-onset forms. *Neurology*, **59**, 1922–8.

Sipe JC, Knobler RL, Braheny SL *et al.* (1984). A neurologic rating scale (NRS) for use in multiple sclerosis. *Neurology*, **34**, 1368–72.

Siva A, Radhakrishnan K, Kurland LT *et al.* (1993). Trauma and multiple sclerosis: a population based cohort study from Olmsted County, Minnesota. *Neurology*, **43**, 1878–82.

Smith CP, Nishiguchi J, O'Leary M *et al.* (2005a). Single-institution experience in 110 patients with botulinum toxin An injection into bladder or urethra. *J Urol*, **65**, 37–41.

Smith KJ (2006). Axonal protection in multiple sclerosis—a particular need during remyelination? *Brain*, **129**, 3147–9.

Smith KJ, Blakemore WF, McDonald WI (1981). The restoration of conduction by central remyelination. *Brain*, **104**, 383–404.

Smith KJ, Kapoor R, Hall SM *et al.* (2001). Electrically active axons degenerate when exposed to nitric oxide. *Ann Neurol*, **49**, 470–6.

Smith KJ, McDonald WI, Miller DH *et al.* (2005b). The pathophysiology of multiple sclerosis. In Compston DAS, ed. *McAlpine's Multiple Sclerosis*, 4th edition, pp. 601–59. Elsevier, London.

Söderström M, Ya-Ping J, Hillert J *et al.* (1998). Optic neuritis: prognosis for multiple sclerosis from MRI, CSF and HLA findings. *Neurology*, **50**, 708–14.

Sorenson PS, Ross C, Clemmesen K *et al.* (2003). Clinical importance of neutralising antibodies against interferon beta in patients with relapsing–remitting multiple sclerosis. *Lancet*, **362**, 1184–91.

Stadelmann C, Ludwin S, Tabira T *et al.* (2005). Tissue preconditioning may explain concentric lesions in Balo's type of multiple sclerosis. *Brain*, **128**, 979–87.

Stankoff B, Waubant E, Confavreux C *et al.* (2005). Modafinil for fatigue in MS: a randomized placebo-controlled double-blind study. *Neurology*, **64**, 1139–43.

Starck M, Albrecht H, Pollmann W *et al.* (1997). Drug therapy for acquired pendular nystagmus in multiple sclerosis. *J Neurol*, **244**, 9–16.

Sterns RH, Riggs JE, Schochet SS (1986). Osmotic demyelination syndrome following correction of hyponatraemia. *NEJM*, **314**, 1535–42.

Takefuji S, Murase T, Sugimura Y *et al.* (2007). Role of microglia in the pathogenesis of osmotic-induced demyelination. *Exp Neurol*, **204**, 88–94.

The GAMES Collaborative Group (2006). Linkage disequilibrium screening for multiple sclerosis implicates *JAG1* and *POU2AF1* as susceptibility genes in Europeans. *J Neuroimmunol*, **179**, 108–16.

Thomas PK, Walker RWH, Rudge PR *et al.* (1987). Chronic demyelinating peripheral neuropathy associated with multifocal central nervous system demyelination. *Brain*, **110**, 53–76.

Thomke F, Lensch E, Ringel K *et al.* (1997). Isolated cranial nerve palsies in multiple sclerosis. *J Neurol Neurosurg Psychiatry*, **63**, 682–5.

Tienari P, Kuokkanen S, Pastinen T *et al.* (1998). Golli-MBP gene in multiple sclerosis. *J Neuroimmunol*, **81**, 158–67.

Tranchant C, Bhatia KP, Marsden CD (1995). Movement disorders in multiple sclerosis. *Mov Disord*, **10**, 418–23.

Trapp BD, Peterson J, Ransohof RM *et al.* (1998). Axonal transection in the lesions of multiple sclerosis. *N Engl J Med*, **338**, 278–85.

Tyler KL, Gross RA, Cascino GD (1986). Unusual viral causes of transverse myelitis: hepatitis A virus and cytomegalovirus. *Neurology*, **36**, 855–8.

United Kingdom Tizanidine Trial Group (1994). A double-blind, placebo-controlled trial of tizanidine in the treatment of spasticity caused by multiple sclerosis. *Neurology*, **44** (Suppl 9), S70–8.

van der Knaap MS, Leegwater PA, Konst AA *et al.* (2002). Mutations in each of the five subunits of translation initiation factor eIF2B can cause leukoencephalopathy with vanishing white matter. *Ann Neurol*, **51**, 264–70.

van der Knaap MS, van der Voorn P, Barkhof F *et al.* (2003). A new leukoencephalopathy with brainstem and spinal cord involvement and high lactate. *Ann Neurol*, **53**, 252–8.

van der Knaap MS, Valk J (2005). *Magnetic Resonance of Myelin, Myelination and Myelin Disorders*, 3rd edition, pp. 1084. Springer, Berlin.

Vukusic S, Hutchinson M, Hours M *et al.* (2004). Pregnancy and multiple sclerosis (the PRIMS study): clinical predictors of post-partum relapse. *Brain*, **127**, 1353–60.

Walther EU, Hohlfeld R (1999). Multiple sclerosis: side effects of interferon beta therapy and their management. *Neurology*, **53**, 1622–7.

Wanders RJA, van Roermund CWT, van Wijland MJA *et al.* (1988). X linked adrenoleukodystrophy: identification of the primary defect at the level of a deficient peroxisomal very long chain fatty acyl CoA synthetase using a newly developed method for the isolation of peroxisomes from skin fibroblasts. *J Inherit Metabol Dis*, **11** (Suppl 2), 173–7.

Weinshenker BG, O'Brien PC, Petterson TM *et al.* (1999). A randomised trial of plasma exchange in acute central nervous system inflammatory demyelinating disease. *Ann Neurol*, **46**, 878–86.

Wiles CM, Newcombe RG, Fuller KJ *et al.* (2001). Controlled randomised crossover trial of the effects of physiotherapy on mobility in chronic multiple sclerosis. *J Neurol Neurosurg Psychiatry*, **70**, 174–9.

Wilkins A, Majed H, Layfield R *et al.* (2003). Oligodendrocytes promote neuronal survival and axonal length by distinct intracellular mechanisms: a novel role for oligodendrocyte-derived glial cell line-derived neurotrophic factor. *J Neurosci*, **23**, 4967–74.

Wilkins A, Compston DAS (2005). Trophic factors attenuate nitric oxide mediated neuronal and axonal injury in vitro: roles and interactions of MAPkinase signaling pathways. *J Neurochem*, **92**, 1487–96.

Willer CJ, Dyment DA, Risch NJ *et al.* (2003). Twin concordance and sibling recurrence rates in multiple sclerosis. The Canadian collaborative study. *Proc Natl Acad Sci USA*, **100**, 12877–82.

Wingerchuk DM, Lennon VA, Pittock SJ *et al.* (2006). Revised diagnostic criteria for neuromyelitis optica. *Neurology*, **66**, 1485–9.

Wolinsky JS, Narayana PA, O'Connor P *et al.* (2007). Glatiramer acetate in primary progressive multiple sclerosis: results of a multinational, multicenter, double-blind, placebo-controlled trial. *Ann Neurol*, **61**(1), 14–24.

Wright DG, Laureno R, Victor M (1979). Pontine and extrapontine myelinolysis. *Brain*, **102**, 361–85.

Yeo TW, Walton A, Goris A, Fenoglio C *et al.* (2007). HLA-C: an independent susceptibility gene in multiple sclerosis. *Ann Neurol*, **61**, 228–36.

Youl BD, Turano G, Miller DH *et al.* (1991). The pathophysiology of acute optic neuritis: an association of gadolinium leakage with clinical and electrophysiological deficits. *Brain*, **114**, 2437–50.

Yudkin PL, Ellison GW, Ghezzi A *et al.* (1991). Overview of azathioprine treatment in multiple sclerosis. *Lancet*, **338**, 1051–5.

Zajicek JP, Wing M, Scolding NJ *et al.* (1992). Interactions between oligodendrocytes and microglia, a major role for complement and tumour necrosis factor in oligodendrocyte adherence and killing. *Brain*, **115**, 1611–31.

Zajicek JP, Scolding NJ, Foster O *et al.* (1999). Central nervous system sarcoidosis – diagnosis and management based on a large series. *QJM*, **92**, 103–17.

Zajicek JP, Fox P, Sanders H *et al.* (2003). Cannabinoids for treatment of spasticity and other symptoms related to multiple sclerosis (CAMS study): multicentre randomised placebo-controlled trial. *Lancet*, **362**, 1517–26.

Zarei M, Chandran S, Compston A *et al.* (2003). Cognitive presentation of multiple sclerosis: evidence for a cortical variant. *J Neurol Neurosurg Psychiatry*, **74**, 872–7.

Zivadinov R, Iona L, Monti-Bragadin L *et al.* (2003). The use of standardized incidence and prevalence rates in epidemiological studies on multiple sclerosis: a meta-analysis study. *Neuroepidemiology*, **22**, 65–74.

CHAPTER 38

Paraneoplastic disorders and neuroimmunology

Neil Scolding

Contents

38.1 Introduction

The extraordinary expansion in the field of neuroimmunology witnessed in the last decade is not just in the number of neurological disorders now considered to have an immune basis, nor the depth of understanding of disorders long known to be 'neuroimmune'. Nor is it in the number of antibodies discovered and now testable, nor in the range of new immune suppressant or modifying treatments now emerging or already available. It is of course all of these things, but it is also more than the sum of these parts. What we are currently privileged to witness is the coming together of immunological understanding, the neurobiology of disease, and rational immune therapy, or at least the beginning of this process.

To take one isolated example, neurogenetics and neurophysiology taught us about the clinical consequences of channel disruption; laboratory-based neuroimmunology showed antibodies to be capable of producing comparable acquired disease; and it seems likely that specific anti-B-cell humanized monoclonal antibodies offer the therapeutic potential to remove these channel-disrupting antibodies. Neither of these steps could be described in the last edition, and one can imagine similar dramatic changes will emerge before the next.

38.2 Neuroimmunology of the central nervous system

Immunology is at least in part a study of mechanisms of directed molecular and cellular injury, which has as one major and ultimate result damage to a specific target. To this end a variety of cytotoxic effectors may be recruited; the complex network of cells and soluble molecules that otherwise constitute the immune system is responsible for initiating, activating, coordinating, targeting, restraining, recording, and reproducing this response.

The conventional but perhaps oversimplified division into cellular and humoral immune responses remains a helpful way of classifying mechanisms of immune activity, and providing a structure within which to explore components and effectors, and of course this particularly applies to neurology, where we think of a number of the central nervous system (Section 38.3) and particularly peripheral nervous system disorders, with myasthenia as perhaps the prototypic disease, as being B-cell-mediated. In truth, however, while this division remains convenient, it is not sustainable to pigeon-hole either immune reactions or autoimmune diseases as 'T-cell-' or 'B-cell-' mediated—the two arms of the immune response are wholly and inextricably interdependent.

38.2.1 The components of the immune system

Humoral immunity. This depends on circulating antibodies which are immunoglobulins secreted by plasma cells following antigenic stimulation of *B-lymphocytes*. The combination of antibody with antigen may act to neutralize the target, recruit and activate macrophages, or bind and activate complement.

Macrophages. These are cells of the macrophage/monocyte lineage, and are present in most tissues, including the brain, where they are represented as *microglia*. They have a key role both in the initiation of immune responses (Section 38.2.2) and as effector cells. Macrophages are able to attack their target either by phagocytosis or by secreting a variety of toxic substances, including free oxygen radicals and tumour necrosis factor. They also have a role in immune regulation and secreting proteins and peptides, macrokines, which act locally and influence other immune cells (Block 2005).

Complement. It is a group of serum proteins which circulate in an inactive form; activation is initiated classically by antibody, but also by an alternative pathway, and results in the formation of vasoactive and chemotactic peptides and terminal membrane attack complexes, which may injure or lyse target cells.

Cellular immune responses are centred primarily around *T-lymphocytes* and these also respond very specifically to individual antigens; in fact, the surface T-cell antigen receptor which binds antigen is structurally related to immunoglobulin, and both derive from the same gene superfamily. All T-cell antigen receptors are dimeric complexes of either an α- and a β-chain, or a γ- and a δ-chain. Antigen specificity of T-cell antigen receptors is a function of these chains' tertiary structure; diversity is generated mostly by random recombination events affecting those groups of genes coding for peptides within the chains—closely reflecting mechanisms of immunoglobulin generation. Different T-cell antigen receptor can recognize the same epitope, and different epitopes may interact with the same T-cell antigen receptor.

The binding of antigen alone to T-lymphocytes is not, however, sufficient to activate the T-cell; an additional restriction is provided by the necessity for the antigen to be presented to the T-cell after antigen 'processing', which requires enzymatic breakdown of antigen to an appropriate form of 9–15 amino acids in length, and which can be effected only by *Antigen Presenting Cells* and also as a complex with a HLA molecule encoded by the major histocompatibility complex.

Class II major histocompatibility complex molecules, HLA-DR, are constitutively expressed only by antigen presenting macrophages and dendritic cells. Other cell types may be induced by lymphokines to express Class II antigens and so can acquire a role in T-cell stimulation; such cells are known as *non-professional* antigen presenting cells. Furthermore, even the formation of the trimolecular complex consisting of antigen, HLA-DR molecule, and T-cell receptor, is insufficient to allow T-cells to generate an antigen-specific response; binding is also required between a number of *accessory molecules* expressed by the antigen presenting cells and T-cell, including CD3 and either CD4 or CD8. Additionally, co-stimulatory signals, B7-1 and B7-2, provided exclusively by 'professional' antigen presenting cells, must also be present.

CD4-expressing T-cells, which recognize MHC Class II molecules, generally function as 'helper' cells: they help initiate immune responses including B-cell responses. CD4 cells fall into two main classes according to characteristic patterns of cytokine secretion. Activated 'Th1' cells release interferon-γ and interleukin-2, and lie at the heart of generating target-specific inflammatory reactions, a classical example being the tuberculin skin test. Th2 CD4 cells similarly secrete interleukin-2, but also interleukins 4, 5, 6, 10, and 13, and transforming growth factor β. They are vitally important in B-cell stimulation and maturation, and also act antagonistically to many Th1 cytokines and so help regulate the activity of Th1 cells. CD8 lymphocytes bind to major histocompatibility complex Class I molecules; they are often but not exclusively cytotoxic in activity.

The initial characterization of CD4 and CD8 as 'helper' and 'suppressor/cytotoxic' cells, whilst helpful to a degree, is simplistic to the point of inaccuracy. Precisely the same may be said of the Th1–Th2 division; there is more likely a spectrum of cytokine secretion, and these represent two extremes. It may here also be re-emphasized that antigen presentation by antigen presenting cells to T-cells is a prerequisite for the development of the great majority of B-cell antibody, as well as cellular immune reactions, stressing the fundamental importance of T-cells in the generation of immune responses. T-lymphocytes indeed play a key role in initiating and coordinating virtually every limb of the immune system, both through direct contact with other cells and by the secretion of numerous lymphokines, including interleukins and interferon gamma. Thus the fundamental distinction of 'T-cell' responses versus 'B-cell', or 'cellular' and 'humoral' immune reactions, is likewise a serious over-simplification.

38.2.2 Immune responses in the central nervous system

While the conventional dogma that the central nervous system is not routinely patrolled by lymphocytes is no longer tenable, any small volume of T-cell traffic would be unlikely to generate an immune response unless antigen-presenting cells were encountered, and also blood–brain barrier impairment seen, allowing other immune and inflammatory mediators to enter the central nervous system (Owens *et al.* 2002). The distribution of Class II major histocompatibility complex products gives some indication of individual cellular potential for antigen presentation and so may help identify the initial site of T-cell activation; and it is now clear that, contrary to initial reports, HLA antigens are expressed in the central nervous system.

In the normal central nervous system, microglia are the major cell population constitutively expressing Class II major histocompatibility complex molecules; this is not surprising since this population represents resident cells in the central nervous system of the macrophage-monocyte lineage, and is ultimately derived from the bone marrow. Interaction of T-cells with microglia, triggering secretion of cytokines, amplifies microglial Class II expression and acts on local endothelia, impairing blood–brain barrier function and recruiting further inflammatory and immunologically active cells.

In vitro, cytokines also induce the expression of Class I and II major histocompatibility complex products on astrocytes and cerebral vascular endothelial cells, both of which can present antigen to T-cells. Neither cell type appears normally to express Class II products *in vivo*, but cytokines such as γ-interferon generated by T-cell/microglia interactions could induce Class II expression. Astrocytes can express intercellular adhesion molecule, ICAM, and may also secrete interleukin-1. Astrocytes and endothelial cells could at least in theory therefore have a central role in *amplifying* T-cell reactions and, since both are involved in maintaining normal blood–brain barrier physiology, in augmenting blood–brain barrier damage. It must be emphasized, however, that most results supporting this role derive from *in vitro* studies, with rather sparse supportive *in vivo* evidence.

38.2.3 Tolerance and autoimmunity

In generating through stochastic recombination events to form T-cell receptors, the necessarily enormous diversity of epitope recognition required for the effective recognition of foreign antigens, chance dictates that many such receptors potentially recognize self antigens: against which an immune response would of course be detrimental, the underlying mechanism indeed of auto-immune diseases (Anderson *et al.* 2001). To help prevent such unhelpful behaviour, a number of mechanisms have evolved to allow 'tolerance', or active immune neglect, of self antigens. First, well over 90 per cent of T-cell antigen receptors generated in the thymus during T-cell genesis are clonally deleted, killed usually by apoptosis, and so never leave the thymus. Clonal deletion also occurs peripherally when, for example, very high levels of antigen are presented to the specific T-cell. The complex dance of antigen processing and presentation to the T-cell antigen receptor within a highly specific context needed for generating specific immune reactions also helps prevent inadvertent responsiveness to normally present self molecules.

Immune 'networks' also play a vital role in regulating the immune response. An immune response against a particular target triggers a secondary immune response directed against the very components of the primary response. This immune reactivity against specific antibodies and/or T-cell antigen receptors helps to suppress immunity in a highly target-selective manner. These are the anti-idiotypic responses and, complexly, in fact can in some circumstances stimulate the primary response too.

38.2.4 Principles of treatment in central nervous system autoimmune disease

Steroids are surely amongst the commonest drugs prescribed in neurological practice. Azathioprine and methotrexate are used with increasing enthusiasm in neurology, but almost always for unlicensed indications. Intravenous immunoglobulins in the past few years have come to consume up to 50 per cent of neurology unit drug budgets, though in all likelihood this is currently being overtaken by the no less expensive interferons and glatirimer being used in significantly increasing numbers of multiple sclerosis patients. On this varied background, brief comment on some of these agents is required.

Glucocorticoids are notorious for their risks and side effects: these include a predisposition to infections, hyperglycaemia, hypertension, obesity, glaucoma and cataract, delayed wound and fracture healing, osteoporosis, psychosis, peptic ulceration and erosions, intracranial hypertension, avascular necrosis and steroid myopathy, together with adrenal suppression. When used in the longer term, alternate day regimens are generally thought to help reduce side effects although the evidence for this is not strong.

Nonetheless, their beneficial effects of course help justify a widespread use. Corticosteroids reduce oedema acutely, and suppress inflammation subacutely. They act by entering cells, binding to cytoplasmic receptors, and forming complexes which bind to steroid-responsive sites in nuclear DNA, thus influencing the expression of various species of mRNA. Corticosteroids produce multiple effects on the immune system although it is believed that they spare B-lymphocytes *in vivo*.

Methotrexate is a dihydrofolate reductase inhibitor, administered as a rule orally, once weekly.

Azathioprine is also an antimetabolite, likewise used with the aim of interfering with normal immune cell function, particularly proliferation, and both are commonly used in the treatment of cancer. Azathioprine acts by inhibiting purine biosynthesis and hence decreases the rate of cell replication particularly among B- and T-lymphocytes, resulting in both T- and B-lymphocytopoenia. Azathioprine takes up to 6 months to exert its full therapeutic effect; it is very poorly tolerated in up to 50 per cent of patients, mostly because of upper gastrointestinal symptoms, but prospective enzymatic testing of thiopurine methyltransferase, TPMT, can help predict those who will tolerate azathioprine and is recommended by some authorities, albeit without Class IV evidence and some studies cast doubt on its value (Kader *et al.* 2000; Sayani *et al.* 2005)).

Azathioprine has been associated with a risk of cancer, particularly lymphoma, and of infections through leucopoenia, and with hepatitis and pneumonitis. In relation to perhaps the most serious of these, careful studies have in fact failed to show a significant increased risk of cancer with medium-term duration of use of azathioprine in multiple sclerosis for 3–5 years or less, though 10 years' adminstration or more may carry a more signficant risk (Confavreux, 1996; Taylor, 2004). Methotrexate carries a particular risk of hepatic fibrosis. Both drugs are also teratogenic, although careful analysis of published data suggest that evidence for this risk in patients, rather than experimental animals is poor, and the risk in fact very slight.

Mycophenolate is a more recently introduced anti-metabolite; it may be a more powerful immunosuppressant than azathioprine but carries a higher risk of malignancy.

Cyclophosphamide is yet more potent, acting as an alkylating agent, transferring alkyl groups to proteins, DNA, and RNA. It is thought particularly to affect B-cell proliferation, and so function. Bone marrow suppression, increased malignancy (especially urothelial), and haemorrhagic cystitis, teratogenicity, and sterility are amongst its more severe side effects (Section 36.2.6).

Agents such as *tacrolimus, or FK506, cyclosporine*, and *rapamycin* or *sirolimus* are relatively more recently introduced agents. Though structurally varied, they share the action of potently inhibiting calcineurin in the presence of their respective common ligands: the cytoplasmic immunophilins, cyclophilin, and FK506-binding protein. Immunophilins are in fact expressed in greater quantity in the central nervous system than in the immune system, and these drugs may well have useful neuroprotective effects in addition to their immune properties (Poulter *et al.* 2004) which are thought to lie more in suppressing T-cell function than B-cell. They are however significantly nephrotoxic, tacrolimus possibly more so than cyclosporin. Gastrointestinal disturbance, hepatic dysfunction, hypertension, arrhythmias, cardiomyopathy, psychosis, and encephalopathy also feature in their recognized side effects.

Mitoxantrone is a derivative of anthracyclin antibiotics, and acts as a cytotoxic agent. It exhibits dose-related cardiotoxicity and also carries a significant risk of malignancy, particularly drug-related acute myelogenous leukaemia of between 0.2 and 1 per cent. Nausea and alopecia also occur.

Immunomodulation A number of agents may be used to influence the activity of the immune system rather than simply suppress it. *Immunoglobulins given intravenously* have a variety of actions, interfering with anti-idiotypic networks, complement, and the activation and function of both T-cells, macrophages and microglia (Misra *et al.* 2005). Anaphylaxis or renal failure rarely occur as side effects; acute fulminant serum sickness is more likely in those with IgA deficiency.

Plasmapheresis superficially appears the obverse of intravenous imunglobulin therapy, its aim being to remove ciruclating immunoglobulins and also cytokines. The physical adventures of large vessel cannulation account for most potential hazards, but impaired electrolyte balance and haemostasis can also occur.

Interferon-β is a naturally occurring type 1 interferon made by many cell types, including dendritic cells and monocytes. It has a large number of effects on immune-active cells, on major histocompatibility complex Class I, with increases, and II, with decreases in expression, and on the blood–brain barrier. Precisely which effect(s) mediate its effect on relapses in multiple sclerosis is unknown (Billiau *et al.* 2004). Various side effects are reported, including commonly flu-like symptoms lasting from 3–4 to 24 h after delivery, injection site reactions, liver enzyme disturbances, and leucopoenias. Severe hepatitis is rarely reported, and depression may be precipitated or exacerbated.

Glatiramer acetate is also widely used in multiple sclerosis. It is a random copolymer of four amino acids in proportions resembling those found in myelin basic protein. Again, diverse actions are reported (Arnon *et al.* 2004), including the induction of glatiramer-specific T-cells and, interestingly, the stimulation of neurotrophin production by lymphocytes, but the mechanisms involved in the effects in multiple sclerosis are unknown. Urticarial and allergic reactions can occur but severe side effects are very uncommon.

38.3 Autoantibodies and central nervous system disease

38.3.1 Hashimoto's encephalopathy

This condition was first identified by Brain and co-authors in 1966. It presents variably either with episodic features: stroke-like events, relapsing encephalopathy, focal or generalized seizures, and/or psychotic episodes, or with a more insidious progressive pattern, with dementia, myoclonus in 50 per cent of cases, and also tremor. All occur in conjunction with high titres of anti-thyroid antibodies, usually anti-microsomal (Kothbauer *et al.* 1996).

The latter plainly are fundamental to the diagnosis, but their role in pathogenesis has yet to be established. It is possible, but thus far speculative, that anti-thyroid antibodies cross-react with brain antigens. Intrathecal synthesis of anti-thyroid antibodies, and the presence in CSF of immune complexes, has been demonstrated (Shaw *et al.* 1991; Ferracci *et al.* 2003). An alternative suggestion is that antibodies may, through immune complex formation, drive a putative vasculitic process; an association of Hashimoto's thyroiditis with vasculitis in other tissues has been reported, specifically, with giant cell arteritis and vasculitic peripheral neuropathy. However, post-mortem histopathological investigations have shown little change or only mild, diffuse perivascular lymphocytic infiltrates (Brain *et al.* 1966).

Hashimoto's encephalopathy exhibits female:male ratio of up to 9:1. Most cases are clinically and biochemically euthyroid at presentation. Imaging by CT or even MR is often normal, as is angiography, though isotope brain scanning may show patchy uptake. Very high titres of anti-thyroid antibodies are found, usually anti-microsomal. Spinal fluid examination may reveal a raised protein level and/or a raised cell count in 80 per cent of patients.

Most patients respond very well to steroid treatment; some have received further immunosuppressive therapy, such as cyclophosphamide, azathioprine, or plasmapheresis.

38.3.2 Dysthyroid eye disease

Dysthyroid eye disease is likely to be immunologically driven. Circulating Thyroid stimulation hormone receptor-stimulating antibodies cross-reactive with orbital fibroblasts are found. Thyroid-stimulating antibodies from patients with Graves' ophthalmopathy can also stimulate fibroblast collagen synthesis. The orbit and extraocular muscles are oedematous and infiltrated with inflammatory cells and glycosaminoglycans, resulting in proptosis and a restrictive ophthalmopathy (Section 13.4.2). Upgaze limitation is the commonest presenting sign, though other muscles are also involved and ocular irritation is frequent. Vision is occasionally threatened by a complicating infiltrative or compressive optic neuropathy. Steroid treatment and radiotherapy appear to be equally effective (Bartalena *et al.* 2005).

38.3.3 Coeliac disease

Coeliac disease, or non-tropical sprue, is an immunologically mediated disorder resulting from intolerance to dietary gluten; it causes weight loss with steatorrhoea and/or diarrhoea, and malabsorption. In common with other enteropathies, neurological sequelae of a predictable nature may complicate coeliac disease as a direct consequence of malabsorption. Central nervous system complications apparently unrelated to deficiency states may also occur in perhaps 10 per cent of patients. Rarely, vasculitis is responsible, but the cause of the most commonly described and distinctive central nervous system association, a progressive cerebellar or spinocerebellar degeneration, with eye movement disorders, myoclonus, and occasionally epilepsy, remains unresolved. This was first described in 1966 (Cooke *et al.* 1966), but a provocative report 30 years later proposed that a substantial proportion of patients with idiopathic ataxia had this disorder, often in the absence of bowel symptoms (Hadjivassiliou *et al.* 1998). Others, however, suggested the association is rare and may not be causal (Lock *et al.* 2005).

38.3.4 Stiff Man syndrome

Stiff Man, or stiff-person, syndrome (Sections 23.7.2; 40.10.3) is an uncommon disorder (Moersch *et al.* 1956; Dalakas *et al.* 2000) generally now agreed to be of autoimmune origin. It is associated with anti-GAD, glutamic acid decarboxylase antibodies which may, it is thought, cross-react with and affect a specific sub-population of spinal neurones. It is associated with diabetes mellitus, or with systemic autoimmune diseases, particularly lupus, and can be seen as a paraneoplastic disorder, though in the majority of cases it occurs in isolation.

It presents with adult onset slowness, aching discomfort and stiffness of muscles, mainly but not exclusively, axial, and with painful muscle cramps, progressing slowly over months and years (Dalakas *et al.* 2000). Spasms, often noise-, startle-, or action-induced, may be very severe; tendon and muscle rupture may occur. Walking may become clumsy ion reveals normal power and tendon reflexes, downgoing plantar responses, and no abnormalities either of sensation or, barring spasms, coordination. However, axial and abdominal wall rigidity is apparent, and there may be proximal limb muscle stiffness, agonists and antagonists acting simultaneously. A hysterical origin for the symptoms is often wrongly assumed. Asymmetrical contraction of the paraspinal muscles causes a characteristic lordotic and often scoliotic posture.

Brain and spinal cord imaging is normal. The spinal fluid is usually normal but for the common finding of oligoclonal immunoglobulin bands. Electrophysiological muscle examination reveals continuous muscle activity despite invitation to relax, with normal motor unit morphology; 'The patient was unable to relax during the examination' should raise suspicion. Importantly, voluntary contraction of antagonists fails to inhibit the activity in the muscle under examination. Abnormal activity, and likewise spasms, does not persist during sleep; its central origin is confirmed by its disappearance following pharmacological peripheral nerve block or spinal or general anaesthesia, in contrast to the abnormal activity demonstrable in neuromyotonic syndromes.

The syndrome is thought to result from an imbalance between excitatory catecholaminergic and descending inhibitory γ-amino butyric acid or GABA-ergic influences on spinal motor neurones. The finding of increased brainstem excitability is consistent with a widespread dysfunction of central inhibitory mechanisms (Molloy et al. 2002). Antibodies directed against glutamic acid decarboxylase, the enzyme responsible for producing GABA from glutamic acid, which therefore react with GABA-ergic neurones, and also with pancreatic islet β-cells, are present in 60 per cent of patients. A clonal B-cell response against glutamic acid decarboxylase is apparent within the CSF, partly accounting for the oligoclonal immunoglobulin bands (Dalakas et al. 2000).

Interestingly, anti-GAD antibodies are reported now also to occur in patients with Batten's disease (Section 10.3.2). Interpretations of this observation could not be more varied: either their occurrence in a hereditary neurodegenerative condition suggests they can arise as an epiphenomenon, or their presence suggests an additional and important immune component to the pathogenesis of Batten's disease (Dalakas 2005). The two possibilities are not, of course, mutually exclusive.

Benzodiazepines particularly, tizanidine, and also baclofen, and occasionally sodium valproate are used therapeutically. More experimental treatments have included intrathecal baclofen and paraspinal botulinum toxin. There is now Class 1b evidence for the value of intravenous immunoglobulin (Dalakas et al. 2001; Dalakas 2005).

In patients with cancer and Stiff Man syndrome, anti-neuronal antibodies of a different specificity, to a synaptic-vesicle-associated protein *amphiphysin*, may be found.

A more serious condition is *progressive encephalomyelitis with rigidity*, where stiffness is accompanied by cranial neuropathies, myoclonus, ataxia, diminished tendon jerks, and extensor plantar responses, MRI brainstem and spinal cord changes occur, and the CSF shows a pleomorphic leucocytosis. The course is substantially more aggressive, with death in 3–10 years. It is often paraneoplastic, and anti-GAD or anti-amphiphysin antibodies can be found.

38.3.5 Acquired channelopathies

The term 'channelopathy' first entered the neurological lexicography in relation to inherited disease: mutations in genes encoding voltage-gated calcium, chloride, sodium and potassium channels, and also of various ligand-gated channels, such as glycine, GABA, and nicotinic acetylcholine receptors all have been implicated in central nervous system and peripheral nervous system diseases ranging from epilepsy and ataxia to myotonia.

Subsequently, disorders arising as a consequence of antibodies directed against these ion channels came to be included in the new order as acquired channelopathies, although a detailed understanding of the pathophysiology of the archetypal disorder myasthenia gravis long preceded this terminology. Acquired neuromyotonia, Isaac's syndrome, related to antibodies to voltage-gated potassium channels, is also discussed elsewhere (Section 23.7.1).

A third neuromuscular disorder, the Lambert–Eaton myasthenic syndrome, is related to antibodies directed against voltage-gated P/Q-type calcium channel and is also more comprehensively described elsewhere (Section 24.10.2). Approximately two-thirds of patients exhibit Lambert–Eaton syndome as a paraneoplastic pheno-menon, the remaining third appearing to have a primary autoimmune antibody disorder. Up to 10 per cent of patients also have cerebellar ataxia. The same voltage-gated P/Q-type calcium channel is expressed by cerebellar Purkinje and granule cells, and such patients show Purkinje cell loss at post-mortem.

Another mix of a paraneoplastic central nervous system phenotype and neuromuscular channelopathy is *Morvan's syndrome*, an association of limbic encephalitis with neuromyotonia plus peripheral neuropathy and hyperhidrosis (Liguori et al. 2001). Here, antibodies to voltage-gated potassium channels, VGKCs, have been found in the sera and cerebrospinal fluid. Plasma exchange reduces the concentration of serum voltage-gated potassium channels antibodies and this is associated with clinical improvement. A significant proportion of such patients appears to have no tumour 'driving' antibody production. Antibodies to voltage-gated potassium channels have been increasingly associated with acquired acute or subacute epileptic and encephalopathic clinical presentations, particularly of an amnesic or limbic type (Vincent et al. 2004; McKnight et al. 2005), responsive to plasma exchange. These are in the main autoimmune, non-paraneoplastic, and potentially treatable.

38.4 Paraneoplastic syndromes

Paraneoplastic syndromes are a group of disorders caused by a malignancy, but not occurring as a direct structural consequence of the cancer or any metastases. The term is often used synonymously with the remote effects of cancer, and does not usually include more general non-metastatic manifestations of malignancy, such as fever, malaise, and lethargy, nutritional disorders, infection, and iatrogenic disorders.

Paraneoplastic syndromes may involve many organs apart from the nervous system:

- the skin: acanthosis nigricans, pruritis, ichthyosis
- the endocrine system: ectopic ACTH secretion, hypercalcaemia
- the blood: anaemia, coagulopathies system

Neurological paraneoplasia is not common, occurring perhaps in less than 1 per cent of patients with cancer, but may cause profound morbidity. Almost any part of the central or peripheral nervous system may be involved, and a number of classical and stereotyped syndromes have been described. Each may be associated with characteristic serum antibody (Table 38.1), and the combination of antibody plus clinical syndrome essentially secures the diagnosis. From the diagnostic perspective, however, it is important to note that often there is much overlap both in clinical syndrome and antibody association. Patients with small cell lung carcinoma, for example, often harbour several autoantibody types (Pittock *et al.* 2004).

Finally by way of introduction, these syndromes may precede symptoms more directly resulting from the tumour by months, or rarely years. Indeed often the cancer is identified only at post-mortem examination.

38.4.1 Subacute cerebellar degeneration

Paraneoplastic cerebellar degeneration may occur in various contexts of malignancy (Brain *et al.* 1965b). Many patients have cervical, uterine, ovarian, or breast cancer: they may harbour anti-Yo antibodies (Peterson *et al.* 1992). Hodgkin's disease is another precipitant, this is more common in men, and associated with anti-Tr antibodies (Trotter *et al.* 1976). Small cell lung cancer can also precipitate this disorder, often with no identifiable antibody, or occasionally with anti-Hu antibodies. In less than 50 per cent of cases, a prior diagnosis of cancer has been made.

The commonest presentation is with a subacutely progressive ataxic syndrome, usually in late middle age or above, reflecting the age range of the underlying cause. What has been termed a vermis phenotype is perhaps the most typical, with prominent gait and truncal ataxia and relative sparing of the limbs; disability is profound. It has been commonly observed that in many patients the syndrome may progress to a plateau and thereafter remain stable for considerable periods, irrespective of the progression or otherwise of the underlying tumour. Nystagmus, often downbeat, is usually present. Other neurological features may be present, including sensory symptoms, dysphagia, and extensor plantar responses. MRI can demonstrate cerebellar atrophy, but is more commonly normal at presentation.

Spinal fluid examination may reveal an elevated protein level; there may be a modestly raised lymphocyte count. The occasional presence of oligoclonal immunoglobulin bands completes a CSF pattern which is common to each of the paraneoplastic neurological syndromes.

38.4.2 Encephalomyelitis

A number of clinical phenotypes, often occurring in admixture with one another, and sharing a common underlying malignancy, which is small cell lung cancer, in 80 per cent cases, neuropathology, and frequent antibody association, anti-Hu, fall within the spectrum of encephalomyelitis:

◆ *Limbic encephalitis* While the commonest underlying malignancy causing this syndrome is small cell lung cancer (Brierly *et al.* 1960), many other tumours have at various times also proven culpable (Corsellis *et al.* 1968). The presentation is singular, with a relatively selective subacutely progressive amnesic syndrome. Magnetic imaging can reveal abnormal T2 high signal deep in the temporal lobes, with atrophy later apparent (Dirr *et al.* 1990). Epilepsy and psychiatric features may occasionally complicate the picture (Bakheit *et al.* 1990); accordingly, EEG changes of slow waves with or without spikes may occur.

◆ *Brainstem encephalitis* Here there is progressive vertigo, ophthalmoplegia, and nystagmus, and bulbar failure (Reddy *et al.* 1981). Pyramidal and extrapyramidal signs, ataxia, autonomic failure, and a sensory neuropathy are also commonly present. A cerebellar presentation is well-recognized, but this is usually distinguishable clinically from *subacute cerebellar degeneration*

(Section 38.4.1) by the additional presence of a neuropathy or other signs indicating more diffuse central nervous system involvement and antibody association (Table 38.1).

◆ *Myelitis* Occurring mostly in conjunction with signs of encephalitis, this may cause multifocal wasting and weakness (Section 28.5.12). The occasional finding of fasciculation may suggest amyotrophic lateral sclerosis in the absence of encephalitic features (Section 38.4.4), but sensory features are usually present and autonomic disturbances are also often apparent.

◆ *Necrotizing myelitis.* This usually occurs with haematological or lung cancers, with the more profound neurological picture of a subacute paraplegia and an active CSF being probably a distinct disorder of unrelated pathogenesis (Mancall *et al.* 1964). *Stiff Man syndrome* can also occur as a paraneoplastic disorder (Section 38.3.4).

38.4.3 Sensory neuronopathy

One of the commoner paraneoplastic manifestations (Denny-Brown 1948; Croft *et al.* 1965; Hughes *et al.* 1996), this syndrome is often grouped with the preceding encephalomyelitides, again sharing small cell lung carcinoma as the commonest precipitant and a common anti-Hu antibody association; indeed in up to 50 per cent of cases, it coexists with encephalitic features (Chalk *et al.* 1992). Autonomic failure may be prominent (Siemson *et al.* 1963). Approximately 1 per cent of patients with small cell cancers develop clinical or electrophysiological evidence of sensory neuronopathy (Elrington *et al.* 1991), but the disorder is also seen in breast and other cancers (Horwich *et al.* 1977; Hughes *et al.* 1996).

The neuronopathy is characteristically painful from the onset and accompanied by often distressing parasthaesia (Section 21.13.1). Symptoms commence in the extremities and progress, proximally sometimes very rapidly, and examination reveals the not unexpected findings of distal sensory loss to all modalities, including vibration and joint position sense, with areflexia and often sensory ataxia. Cranial sensory nerves may be involved. Electrophysiological testing shows absent sensory action potentials and no motor disturbance (Donofrio *et al.* 1989). The disorder is so-named, rather than as sensory neuropathy, on histopathological grounds, the dorsal root ganglia showing changes, not peripheral nerve. A clinically identical syndrome may occur in association with Sjögren's syndrome (Font *et al.* 1990).

38.4.4 Motor neurone disorders

The occurrence of a syndrome resembling amyotrophic lateral sclerosis phenotype, with mixed upper and lower motor neurone signs, in the context of cancer, has been intermittently reported since the first description four decades ago (Brain *et al.* 1965a). It has, however, remained controversial: the often quoted figure of up to 10 per cent of cases of amyotrophic lateral sclerosis associated with malignancy (Norris *et al.* 1965) certainly falls well outside most neurologists' experience.

A subacute motor neuronopathy without pyramidal signs, associated especially with lymphomatous malignancies is a more secure entity (Schold *et al.* 1979). The syndrome is progressive and painless, and bears some clinical resemblance to multifocal motor neuropathy with conduction block, though the latter most characteristically affects the upper limbs more than the lower; paraneoplastic motor disease mainly the reverse. The distinction is more reliable if made electrophysiologically, the self-evident features of the former contrasting with the normal motor and sensory conduction of paraneoplastic motor neuronopathy accompanied by electromyographic evidence of denervation. Neither upper motor neurones nor the bulbar lower motor neurones are usually affected. Nuchal and respiratory muscle weakness may occur; anti-Hu antibodies and subclinical loss of cerebellar Purkinje cells are reported (Verma *et al.* 1996).

Such a motor neuronopathy occurring with a limited myelitis might generate the clinical picture of amyotrophic lateral sclerosis. It has been suggested that the eleven cases of cancer-related motor neurone disease described by Brain in 1965 may have represented 'burnt out paraneoplastic encephalomyelitis' (Posner 1996). However, Younger *et al.* (1991) reported nine patients, all with lymphoma, eight of whom had clinical features suggesting amyotrophic lateral sclerosis (Younger *et al.* 1991). Others have reported series of patients with lymphoproliferative disorders and motor neurone disease with no motor neuropathy, most having 'definite or probable' upper motor neurone signs (Gordon *et al.* 1997). Of these only the small minority with purely lower motor neurone disease showed a neurological response to treatment for their lymphoproliferative disease. An association in very small numbers of patients of motor neurone disease with anti-Hu antibodies has also been reported, with a possible link with breast cancer (Forsyth *et al.* 1997).

Breast cancer may also be associated with two relatively discrete disorders, a sensorimotor neuropathy (Peterson *et al.* 1994), and the so-called 'numb chin syndrome', characterized by numbness in the distribution of the mental or alveolar branches of the mandibular nerve (Horton *et al.* 1973; Burt *et al.* 1992; Lossos *et al.* 1992). Other malignancies may precipitate the latter, which is thought to represent a direct effect of metastasis, rather than paraneoplasia.

38.4.5 Opsoclonus and myoclonus

This is a profoundly disabling, distressing syndrome, wherein opsoclonic eye movements cause profound vertigo, nausea and anorexia, and debility, usually confining the patient to bed (Section 13.2.5). The eye movement disorder consists of chaotic, involuntary partial or complete saccades which are arrhythmic, continuous, and randomly directed. The movements persist during sleep. Truncal and limb myoclonus may also be present, as may central ataxia (Anderson *et al.* 1988a). Some such patients have *anti-Ri* antibodies and breast or gynaecological cancers (Luque *et al.* 1991), but many harbour cancers and no obvious antibody (Bataller *et al.* 2001). In adults, small cell lung cancer is yet again the commonest associated malignancy, and a tendency to remit is described (Anderson *et al.* 1988a).

The syndrome was first described not in adults but in children with neuroblastoma (Kinsbourne 1962), the underlying cause in approximately 50 per cent of children with this neurological picture (Telander *et al.* 1989). It occurs too as a monophasic post-infectious process in adults, without cancer, usually following respiratory or gastrointestinal infection (Baringer *et al.* 1968). In one study of 58 patients, only 11 had identified cancers.

38.4.6 Retinal degeneration

This syndrome again is most commonly associated with small cell lung carcinoma, though other tumours may also be responsible: gynaecological malignancy and, perhaps particularly, melanoma (Section 12.3.2). Antibodies directed against the calcium-binding protein recoverin are found in some patients with paraneoplastic retinal degeneration (Polans *et al.* 1991; Thirkill *et al.* 1992). Paraneoplastic retinal degeneration precipitated by melanoma is clinically distinct (Kim *et al.* 1994). Serum antibodies are associated; they are directed not against recoverin but against bipolar retinal neurones (Weinstein *et al.* 1994). Antibodies directed against optic nerve cells are also described, and serum antibodies directed against enolase have been demonstrated in patients with retinopathy associated with a variety of malignancies (Adamus *et al.* 1996).

Photosensitivity, night blindness, visual scotomata, and retinal artery attenuation are typical features, though variations on this phenotype are well-described (Jacobson *et al.* 1990). Visual loss is painless, and may be monocular initially. Visual evoked responses usually reveal normal optic nerve function, electroretinography confirming the site of derangement. A paraneoplastic optic neuropathy, associated with anti-CV-2 antibodies, is also recognized.

38.4.7 Polymyositis

The association of inflammatory muscle disease with malignancy has for some years been considered overstated, but most now accept that dermatomyositis in particular is significantly linked with cancer (Sigurgeirsson 1992; Hill *et al.* 2001), though polymyositis only weakly. The clinical features and histopathology are not distinguishable from those of idiopathic polymyositis (Section 24.7.2).

38.4.8 Pathology and immunology

Neuropathology

The principal pathological picture associated with the described syndromes is strikingly similar, though with a few notable exceptions, across the spectrum of paraneoplastic conditions (Denny-Brown 1948). The main features may be divided into three groups:

- pronounced neuronal loss, with pyknotic changes which would currently be interpreted as indicating apoptotic cell death, and neuronophagia;

- inflammatory changes, including perivascular lymphocytic cuffing with parenchymal infiltration by lymphocytes and macrophages with the formation of microglial nodules;

- astrogliosis.

Thus in paraneoplastic cerebellar degeneration these changes occur in the cerebellar cortex (Brain *et al.* 1965b), where Purkinje cells are selectively lost. In encephalomyelitic syndromes, striking changes are found, according to the specific disorder, respectively in the limbic system (Corsellis *et al.* 1968), the brainstem (Henson *et al.* 1965), or the spinal cord (Mancall *et al.* 1964), while in subacute sensory neuronopathy, the pathology is centred upon the dorsal root ganglion (Denny-Brown 1948; Croft *et al.* 1967; Horwich *et al.* 1977). In later stages, the changes may extend proximally into the posterior columns, and distally into the peripheral nerve; demyelination is a marked feature in a minority of patients. In subacute motor neuronopathy, anterior horns in the spinal cord bear the brunt of the disease process; in contrast with amyotrophic lateral sclerosis, the lateral columns are not affected (Brain *et al.* 1965b; Henson *et al.* 1965). The findings in cancer-associated retinopathy are again typical, with loss of photoreceptors and retinal ganglion cells accompanying the inflammatory changes (Grunwald *et al.* 1987). It should also be noted that in most of these syndromes, less pronounced findings of a similar nature may be present diffusely in the cerebral hemispheres, brainstem, and spinal cord.

In paraneoplastic opsoclonus–myoclonus syndrome, the picture is a little less clear. Eye movement physiology might volunteer 'omnipause cells' as the target cell; these are inhibitory interneurones putatively located in the pontine reticular formation. In some patients, however, no clear abnormalities are identifiable here or indeed elsewhere (Ridley *et al.* 1987; Anderson *et al.* 1988a); in others, the findings are identical to those of paraneoplastic cerebellar degeneration (Ellenberger *et al.* 1968).

Humoral immunity

The most economical mechanistic interpretation of these changes would be that an immunological or inflammatory process is targeted upon certain neuronal sub-populations. The classical and obvious hypothesis has been that specific antibodies directed against tumour surface antigens arise as part of the anti-tumour immune response. Certain neuronal antigens may be expressed by tumour cells, and tumour-specific antibodies directed at these targets cross-react with and damage certain neuronal populations. Thus, small cell lung cancer cells may express P/Q type voltage-gated calcium channels, antibodies to which are considered responsible for causing Lambert–Eaton myasthenic syndrome; acquired channelopathy here meeting paraneoplasia.

Central nervous system paraneoplasia is less straightforward, however. Anti-Yo antibodies, for example, found in many patients with paraneoplastic cerebellar degeneration, and almost invariably in those with an underlying breast or gynaecological malignancy rather than those with small cell lung cancer, do indeed react with Purkinje cell antigens. They are also expressed in the tumour. However, the antigens are cytoplasmic, not surface-expressed; CD62 is a DNA-binding protein which directs gene transcription, of the leucine-zipper family (Sakai *et al.* 1990; Fathallah *et al.* 1991). How or indeed if these antibodies interact with these cryptic antigens *in vivo*, and whether this interaction is responsible for Purkinje cell loss, is unclear, but anti-Yo antibodies are not detected in non-paraneoplastic cerebellar degenerations (Anderson *et al.* 1988b;

Smith *et al.* 1988; Peterson *et al.* 1992), arguing against secondary generation triggered by other causes of Purkinje cell damage.

Purkinje cells exposed to anti-Yo/anti-Purkinje cell antibodies antibodies can specifically take up these immunoglobulin molecules (Graus *et al.* 1991; Greenlee *et al.* 1995) but, perplexingly, no cell injury is apparent in these cell culture models. There is little or no evidence from other immunological systems that internalized antibody can cause cell damage (Naparstek *et al.* 1993). Active immunization with recombinant Yo protein also fails to induce cerebellar disease (Tanaka *et al.* 1994; Sakai *et al.* 1995) while passive transfer of Yo/APCA-specific lymphocytes into immune-deficient mice similarly fails to replicate disease (Tanaka *et al.* 1995).

Likewise, anti-Hu-related antigens are also expressed in small cell lung cancer cells (King *et al.* 1996), and in neurons, but here ubiquitously by cells in the brain and spinal cord, and also in dorsal root ganglian sensory neurones and in autonomic ganglia (Dick *et al.* 1988; Altermatt *et al.* 1991). As with anti-Yo antibodies, intrathecal immunoglobulins of these specificities are found in patients with paraneoplasia. HuC, HuD, Hel-N1, and Hel-N2 share sequence homology with each other and appear to be RNA-binding proteins important in post-transcriptional gene processing (Dropcho *et al.* 1994). Again they are not expressed on the cell surface, but in the nucleus, and attempts to develop animal models of anti-Hu/ANNA-1-related disease by injecting antibody have repeatedly failed (Dick *et al.* 1988).

Anti-Ri antibodies react with breast tumour tissue from patients with opsoclonus, and are non-reactive with malignant breast tissue not associated with opsoclonus (Luque *et al.* 1991). They react with nuclear, not surface antigens, namely the RNA-binding protein Nova (Buckanovich *et al.* 1993).

However, in relation to the cancer-associated-retinopathy-associated antigen recoverin, although intracellularly located within photoreceptor cells in the retina, a different story is emerging. Recoverin expression in tumour tissue from a patient with cancer-associated retinopathy is found (Polans *et al.* 1995), and the presence of retinal deposits of immunoglobulin, with loss of

Table 38.1 The commoner central nervous system paraneoplastic syndromes, and their associated, tumours and antibodies.

Syndrome	Tumour	Antibody	Antigen
Subacute cerebellar degeneration	Breast and ovary	Anti-Yo antibodies	Purkinje cell cytoplasmic antigens (CDR34 and CDR 62)
	Hodgkin's lymphoma	Anti-Tr	Not known
Encephalomyelitis: Limbic encephalitis; subacute sensory neuronopathy; autonomic neuropathy	Small cell lung cancer	Anti-Hu antibodies	HuC, HuD, Hel-N1, and Hel-N2 in CNS, DRG, and autonomic ganglia neurons.
	Testicular tumours	Anti-Ma	Ma1-3
Opsoclonus/myoclonus syndrome	Breast	Anti-Ri antibodies	RNA-binding protein *NOVA*
Cancer-associated retinopathy	Small cell lung cancer	Recoverin antibodies	A photoreceptor calcium-binding protein
	Melanoma	Anti-bipolar retinal cell antibodies	
Optic neuropathy	Small cell lung cancer	CV-2 antibodies	Cytoplasm of neurons, oligodendrocytes, and retinal cells; found in optic and peripheral nerve axons
Paraneoplastic Stiff Man Syndrome	Breast, small cell lung cancer	Anti-amphiphysin	Amphiphysin (a synaptic vesicle-associated protein)

ganglion cells, in immunohistopathological studies of patients (Grunwald *et al.* 1987) provides further evidence for an autoimmune basis to cancer-associated retinopathy. In marked contrast to the various paraneoplastic antibodies and antigens described above, an animal model of cancer-associated retinopathy has been described: Lewis rats injected with a synthetic peptide fragment of recoverin develop photoreceptor cell degeneration (Polans *et al.* 1995). Furthermore, anti-recoverin antibodies cause apoptotic death of retinal cells *in vivo* (Adamus *et al.* 1997). Whether this offers mechanistic clues to other paraneoplastic disorders remains to be established. Alternatively, cell surface expression of anti-Hu-related antigens by small cell lung cancer cells has been reported (Tora *et al.* 1997), with obvious and important implications for pathogenesis.

Other possibilities are, however, considered. In paraneoplastic motor neurone disease, a possible opportunistic infectious aetiology has been suggested. The particular association with lymphoma offers immune paresis as a contribution to this suggestion; poliomyelitis represents an obvious example of viral disease of the anterior horn cell.

Cell-mediated immunity

Paraneoplastic anti-neuronal antibodies have assumed an important diagnostic role of great practical help in investigating patients. The difficulty in establishing a pathogenetic contribution has encouraged studies of T-cell involvement (Albert *et al.* 2000). In addition to B-cells, macrophages and T-lymphocytes are prominent in the lesions of paraneoplastic encephalomyelitis. CD4- and CD8-positive cells are present.

38.4.9 Diagnosis and treatment

The diagnosis is made on the basis of these characteristic clinical phenotypes arising in conjunction with pertinent antibodies found on serological testing. Some syndromes, for example, opsoclonus–myoclonus in an adult, are sufficiently suggestive of paraneoplasia to warrant extensive searching for malignancy even in the absence of antibodies. In most cases, whether with or without antibodies, the underlying malignancy will either already be known, or be disclosed by conventional detailed physical examination and 'first line' tests. Should this not be the case, FDG-positron emission tomography scanning now has a recognized role in revealing cryptic cancers (Rees *et al.* 2001).

The most obvious therapeutic approach is to treat the causative underlying malignancy (Batson *et al.* 1992): cure of the cancer should, surely, cure the neurology given time? Anecdotal reports attest to occasional striking neurological responses. The author has witnessed a severely amnesic patient with small cell lung cancer and limbic encephalitis improve in a clear, substantial, and carefully documented way following treatment of the lung tumour. Of course, when the tumour has generated neuromuscular disease, the greater capacity for repair of the affected neurological target allows tumour removal to be a successful means of treating the syndrome, for example, thymectomy in myasthenia, and lung tumour treatment in Lambert–Eaton myastheric syndrome.

For such to be the case consistently, however, one would have to postulate temporary mechanisms of neuronal dysfunction and not cell death. For syndromes persisting for many months this may be unlikely at least in central nervous system disease. The common clinical course of most paraneoplastic disorders of reaching a plateau after a period of subacute progression most likely suggests the acquisition of irreversible neuronal damage (Dropcho 1995). This may, help to explain 'stabilization' as an apparent partial response to well-timed interventions. Not surprisingly therefore, and despite these occasional reports of responding patients (Paone *et al.* 1980), when relatively large series of patients are studied, little objective support is seen for tumour removal as neurological therapy (Graus *et al.* 1995; Grisold *et al.* 1995).

Interestingly, the presence of anti-Hu antibodies in small cell lung cancer patients without paraneoplastic neurological disease appears to correlate with a better overall response to cancer treatment, consistent with an anti-tumoural protective role for the antibody (Graus *et al.* 1997).

This finding emphasizes the potential disadvantage of the alternative approach to treating neurological symptoms. This consists of immunotherapies directed not against the tumour, but towards the immune mediators putatively responsible for the neurological syndrome: that is, the antibodies themselves. Again considering for example the well-documented neuromuscular paraneoplastic disorders myasthenia gravis and Lambert–Eaton myasthenic syndrome, the clear response to plasmapheresis provides some support for this approach, though of course only as symptomatic therapy (Newsom Davis and Murray 1984). Of course, the incurability of many of the underlying malignancies in central nervous system paraneoplasia, combined with the extreme and unremitting distress, discomfort, and disability caused by many such disorders, and the absence of any other useful treatments, even palliative, must leave no possible therapeutic avenue unexplored.

Thus both conventional and novel immunotherapies have been exhibited. Steroids may help in paediatric opsoclonus–myoclonus (Boltshauser *et al.* 1979), and it has been suggested that adult opsoclonus–myoclonus may be a more benign disease, and more commonly steroid-responsive, than other paraneoplastic disorders (Dropcho *et al.* 1993). However, those with much experience have not found immunosuppression to be successful in other situations (Dalmau *et al.* 1996). Also the possible adverse effects of inadvertently suppressing anti-tumour immune surveillance and reactivity mitigate against non-specific immunosuppression.

Isolated case reports or small series attest to the possible merits of more immunoglobulin-specific approaches such as Intravenous immunoglobulin (Counsell *et al.* 1994) and plasma exchange (Weissman *et al.* 1989), but in larger series, no significant improvement emerged (Graus *et al.* 1995). In 18 patients suffering anti-Hu-related paraneoplastic disorders treatment with Intravenous immunoglobulin produced no benefit in seriously affected patients; (Vega *et al.* 1994), similarly, in a study of 22 patients with a variety of antibody-associated paraneoplastic disorders, no significant evidence of benefit was found (Uchuya *et al.* 1996); the same applies in respect of plasma exchange. Protein A immunoadsorptive columns have also been used in small numbers of patients (Cher *et al.* 1995; Nitschke *et al.* 1995).

The majority of patients with paraneoplastic diseases of the central nervous system do not recover neurologically; their disability may become static rather than continuing to progress, and their overall prognosis depends on that of the underlying malignancy. Non-immunological symptomatic treatment assumes great importance (Brady 1996).

References

Adamus G, Aptsiauri N, Guy J *et al.* (1996). The occurrence of serum autoantibodies against enolase in cancer-associated retinopathy. *Clin Immunol Immunopathol*, **78**, 120–9.

Adamus G, Machnicki M, Seigel GM (1997). Apoptotic retinal cell death induced by antirecoverin autoantibodies of cancer-associated retinopathy. *Invest Ophthalmol Vis Sci*, **38**, 283–91.

Albert ML, Austin LM, Darnell RB (2000). Detection and treatment of activated T cells in the cerebrospinal fluid of patients with paraneoplastic cerebellar degeneration. *Ann Neurol*, **47**, 9–17.

Altermatt HJ, Rodriguez M, Scheithauer BW *et al.* (1991). Paraneoplastic anti-Purkinje and type I anti-neuronal nuclear autoantibodies bind selectively to central, peripheral, and autonomic nervous system cells. *Lab Invest*, **65**, 412–20.

Anderson DE, Hafler DA (2001). Immune tolerance and the nervous system. *Adv Exp Med Biol*, **490**, 79–98.

Anderson NE, Budde SC, Rosenblum MK *et al.* (1988a). Opsoclonus, myoclonus, ataxia, and encephalopathy in adults with cancer: a distinct paraneoplastic syndrome. *Medicine Baltimore* b: 100–9.

Anderson NE, Rosenblum MK, Graus F *et al.* (1988b). Autoantibodies in paraneoplastic syndromes associated with small-cell lung cancer. *Neurology*, **38**, 1391–8.

Arnon R, Aharoni R (2004). Mechanism of action of glatiramer acetate in multiple sclerosis and its potential for the development of new applications. *Proc Natl Acad Sci USA*, **101** (Suppl 2), 14593–8.

Bakheit AM, Kennedy PG, Behan PO (1990). Paraneoplastic limbic encephalitis: clinico-pathological correlations. *J Neurol Neurosurg Psychiatry*, **53**, 1084–8.

Baringer JR, Sweeney VP, Winkler GF (1968). An acute syndrome of ocular oscillations and truncal myoclonus. *Brain*, **91**, 473–80.

Bartalena L, Marcocci C, Tanda ML *et al.* (2005). An update on medical management of Graves' ophthalmopathy. *J Endocrinol Invest*, **28**, 469–78.

Bataller L, Graus F, Saiz A *et al.* (2001). Clinical outcome in adult onset idiopathic or paraneoplastic opsoclonus-myoclonus. *Brain*, **124**, 437–43.

Batson OA, Fantle DM, Stewart JA (1992). Paraneoplastic encephalomyelitis. Dramatic response to chemotherapy alone. *Cancer*, **69**, 1291–3.

Billiau A, Kieseier BC, Hartung HP (2004). Biologic role of interferon beta in multiple sclerosis. *J Neurol*, **251** (Suppl 2), II10–4.: II10–II14.

Block ML, Hong JS (2005). Microglia and inflammation-mediated neurodegeneration: multiple triggers with a common mechanism. *Prog Neurobiol*, **76**, 77–98.

Boltshauser E, Deonna T, Hirt HR (1979). Myoclonic encephalopathy of infants or «dancing eyes syndrome». Report of 7 cases with long-term follow-up and review of the literature (cases with and without neuroblastoma). *Helv Paediatr Acta*, **34**, 119–33.

Brady AM (1996). Management of painful paraneoplastic syndromes. *Hematol Oncol Clin North Am*, **10**, 801–9.

Brain WR, Croft PB, Wilkinson M (1965a). Motor neurone disease as a manifestation of neoplasm (with a note on the course of classical motor neurone disease). *Brain*, **88**, 479–500.

Brain WR, Jellinek EH, Ball K (1966). Hashimoto's disease and encephalopathy. *Lancet*, **2**, 512–4.

Brain WR, Wilkinson M (1965b). Subacute cerebellar degeneration associated with neoplasms. *Brain*, **88**, 465–78.

Brierly JB, Corsellis JA, Hierons L *et al.* (1960). Subacute encephalitis of later adult life mainly affecting the limbic areas. *Brain*, **83**, 357–68.

Buckanovich RJ, Posner JB, Darnell RB (1993). Nova, the paraneoplastic Ri antigen, is homologous to an RNA-binding protein and is specifically expressed in the developing motor system. *Neuron*, **11**, 657–72.

Burt RK, Sharfman WH, Karp BI *et al.* (1992). Mental neuropathy (numb chin syndrome). A harbinger of tumor progression or relapse [see comments]. *Cancer*, **70**, 877–81.

Chalk CH, Windebank AJ, Kimmel DW *et al.* (1992). The distinctive clinical features of paraneoplastic sensory neuronopathy. *Can J Neurol Sci*, **19**, 346–51.

Cher LM, Hochberg FH, Teruya J *et al.* (1995). Therapy for paraneoplastic neurologic syndromes in six patients with protein A column immunoadsorption. *Cancer*, **75**, 1678–83.

Cooke WT, Smith WT (1966). Neurological disorders associated with adult coeliac disease. *Brain*, **89**, 683–722.

Corsellis JA, Goldberg GJ, Norton AR (1968). "Limbic encephalitis" and its association with carcinoma. *Brain*, **91**, 481–96.

Counsell CE, McLeod M, Grant R (1994). Reversal of subacute paraneoplastic cerebellar syndrome with intravenous immunoglobulin. *Neurology*, **44**, 1184–5.

Croft PB, Urich H, Wilkinson M (1967). Peripheral neuropathy of sensorimotor type associated with malignant disease. *Brain*, **90**, 31–66.

Croft PB, Wilkinson M (1965). The incidence of carcinomatous neuromyopathy in patients with various types of carcinoma. *Brain*, **88**, 427–34.

Dalakas MC (2005). The role of IVIg in the treatment of patients with stiff person syndrome and other neurological diseases associated with anti-GAD antibodies. *J Neurol*, **252** (Suppl 1), I19–I25.

Dalakas MC, Fujii M, Li M *et al.* (2001). High-dose intravenous immune globulin for stiff-person syndrome. *N Engl J Med*, **345**, 1870–6.

Dalakas MC, Fujii M, Li M *et al.* (2000). The clinical spectrum of anti-GAD antibody-positive patients with stiff-person syndrome. *Neurology*, **55**, 1531–5.

Dalmau J, Posner JB (1996). Neurological paraneoplastic syndromes. *Springer Semin Immunopathol*, **18**, 85–95.

Denny-Brown D (1948). Primary sensory neuropathy with muscular changes associated with carcinoma. *J Neurol Neurosurg Psychiatry*, **11**, 73–87.

Dick DJ, Harris JB, Falkous G *et al.* (1988). Neuronal anti-nuclear antibody in paraneoplastic sensory neuronopathy. *J Neurol Sci*, **85**, 1–8.

Dirr LY, Elster AD, Donofrio PD *et al.* (1990). Evolution of brain MRI abnormalities in limbic encephalitis. *Neurology*, **40**, 1304–6.

Donofrio PD, Alessi AG, Albers JW *et al.* (1989). Electrodiagnostic evolution of carcinomatous sensory neuronopathy. *Muscle Nerve*, **12**, 508–13.

Dropcho EJ (1995). Autoimmune central nervous system paraneoplastic disorders: mechanisms, diagnosis, and therapeutic options. *Ann Neurol*, **37** (Suppl 1), S102–13.

Dropcho EJ, King PH (1994). Autoantibodies against the Hel-N1 RNA-binding protein among patients with lung carcinoma: an association with type I anti-neuronal nuclear antibodies. *Ann Neurol*, **36**, 200–5.

Dropcho EJ, Kline LB, Riser J (1993). Antineuronal (anti-Ri) antibodies in a patient with steroid-responsive opsoclonus-myoclonus. *Neurology*, **43**, 207–11.

Ellenberger C, Campa JF, Netsky MG (1968). Opsoclonus and parenchymatous degeneration of the cerebellum. The cerebellar origin of an abnormal ocular movement. *Neurology*, **18**, 1041–6.

Elrington GM, Murray NM, Spiro SG *et al.* (1991). Neurological paraneoplastic syndromes in patients with small cell lung cancer. A prospective survey of 150 patients. *J Neurol Neurosurg Psychiatry*, **54**, 764–7.

Fathallah SH, Wolf S, Wong E *et al.* (1991) Cloning of a leucine-zipper protein recognized by the sera of patients with antibody-associated paraneoplastic cerebellar degeneration. *Proc Natl Acad Sci USA*, **88**, 3451–4.

Ferracci F, Moretto G, Candeago RM *et al.* (2003). Antithyroid antibodies in the CSF: their role in the pathogenesis of Hashimoto's encephalopathy. *Neurology*, **60**, 712–4.

Font J, Valls J, Cervera R, *et al.* (1990). Pure sensory neuropathy in patients with primary Sjogren's syndrome: clinical, immunological, and electromyographic findings. *Ann Rheum Dis*, **49**, 775–8.

Forsyth PA, Dalmau J, Graus F *et al.* (1997). Motor neuron syndromes in cancer patients [see comments]. *Ann Neurol*, **41**, 722–30.

Gordon PH, Rowland LP, Younger DS *et al.* (1997). Lymphoproliferative disorders and motor neuron disease: an update. *Neurology*, **48**, 1671–8.

Graus F, Dalmou J, Rene R *et al.* (1997). Anti-Hu antibodies in patients with small-cell lung cancer: association with complete response to therapy and improved survival. *J Clin Oncol*, **15**, 2866–72.

Graus F, Delattre JY (1995). Immune modulation of paraneoplastic neurologic disorders. *Clin Neurol Neurosurg*, **97**, 112–6.

Graus F, Illa I, Agusti M et al. (1991). Effect of intraventricular injection of an anti-Purkinje cell antibody (anti-Yo) in a guinea pig model. *J Neurol Sci*, **106**, 82–7.

Greenlee JE, Burns JB, Rose JW et al. (1995). Uptake of systemically administered human anticerebellar antibody by rat Purkinje cells following blood–brain barrier disruption. *Acta Neuropathol Berl*, **89**, 341–5.

Grisold W, Drlicek M, Liszka SU et al. (1995). Anti-tumour therapy in paraneoplastic neurological disease. *Clin Neurol Neurosurg*, **97**, 106–11.

Grunwald GB, Kornguth SE, Towfighi J et al. (1987). Autoimmune basis for visual paraneoplastic syndrome in patients with small cell lung carcinoma. Retinal immune deposits and ablation of retinal ganglion cells. *Cancer*, **60**, 780–6.

Hadjivassiliou M, Grunewald RA, Chattopadhyay AK et al. (1998). Clinical, radiological, neurophysiological, and neuropathological characteristics of gluten ataxia. *Lancet*, **352**, 1582–5.

Henson RA, Hoffman HL, Urich H (1965). Encephalomyelitis with carcinoma. *Brain*, **88**, 449–64.

Hill CL, Zhang Y, Sigurgeirsson B et al. (2001). Frequency of specific cancer types in dermatomyositis and polymyositis: a population-based study. *Lancet*, **357**, 96–100.

Horton J, Means ED, Cunningham TJ et al. (1973). The numb chin in breast cancer. *J Neurol Neurosurg Psychiatry*, **36**, 211–6.

Horwich MS, Cho L, Porro RS et al. (1977). Subacute sensory neuropathy: a remote effect of carcinoma. *Ann Neurol*, **2**, 7–19.

Hughes R, Sharrack B, Rubens R (1996). Carcinoma and the peripheral nervous system. *J Neurol*, **243**, 371–6.

Jacobson DM, Thirkill CE, Tipping SJ (1990). A clinical triad to diagnose paraneoplastic retinopathy. *Ann Neurol*, **28**, 162–7.

Kader HA, Wenner WJ Jr, Telega GW et al. (2000). Normal thiopurine methyltransferase levels do not eliminate 6-mercaptopurine or azathioprine toxicity in children with inflammatory bowel disease. *J Clin Gastroenterol*, **30**, 409–13.

Kim RY, Retsas S, Fitzke FW et al. (1994). Cutaneous melanoma-associated retinopathy. *Ophthalmology*, **101**, 1837–43.

King PH, Dropcho EJ (1996). Expression of Hel-N1 and Hel-N2 in small-cell lung carcinoma. *Ann Neurol*, **39**, 679–81.

Kinsbourne M (1962). Myoclonic encephalopathy of infants. *J Neurol Neurosurg Psychiatry*, **25**, 271–6.

Kothbauer MI, Sturzenegger M, Komor J et al. (1996). Encephalopathy associated with Hashimoto thyroiditis: diagnosis and treatment. *J Neurol*, **243**, 585–93.

Liguori R, Vincent A, Clover L et al. (2001). Morvan's syndrome: peripheral and central nervous system and cardiac involvement with antibodies to voltage-gated potassium channels. *Brain*, **124**, 2417–26.

Lock RJ, Pengiran Tengah DS, Unsworth DJ et al. (2005). Ataxia, peripheral neuropathy, and anti-gliadin antibody. Guilt by association? *J Neurol Neurosurg Psychiatry*, **76**, 1601–3.

Lossos A, Siegal T (1992). Numb chin syndrome in cancer patients: etiology, response to treatment, and prognostic significance [see comments]. *Neurology*, **42**, 1181–4.

Luque FA, Furneaux HM, Ferziger R et al. (1991). Anti-Ri: an antibody associated with paraneoplastic opsoclonus and breast cancer. *Ann Neurol*, **29**, 241–51.

Mancall EL, Rosales RK (1964). Necrotising myelopathy associated with visceral carcinoma. *Brain*, **87**, 639–64.

McKnight K, Jiang Y, Hart Y et al. (2005). Serum antibodies in epilepsy and seizure-associated disorders. *Neurology*, **65**, 1730–6.

Misra N, Bayry J, Ephrem A et al. (2005). Intravenous immunoglobulin in neurological disorders: a mechanistic perspective. *J Neurol*, **252** (Suppl 1), I1–I6.

Moersch FP, Woltman HW (1956). Progressive fluctuating muscular rigidity and spasm ("stiff man syndrome"): report of a case and observations in 13 other cases. *Mayo Clin Proc*, **31**, 421–7.

Molloy FM, Dalakas MC, Floeter MK (2002). Increased brainstem excitability in stiff-person syndrome. *Neurology*, **59**, 449–51.

Naparstek Y, Plotz PH (1993). The role of autoantibodies in autoimmune disease. *Annu Rev Immunol*, **11**, 79–104.

Newsom Davis J, Murray NM (1984). Plasma exchange and immunosuppressive drug treatment in the Lambert-Eaton myasthenic syndrome. *Neurology*, **34**, 480–5.

Nitschke M, Hochberg F, Dropcho E (1995). Improvement of paraneoplastic opsoclonus-myoclonus after protein A column therapy [letter]. *N Engl J Med*, **332**, 192.

Norris FH, Engel WK (1965). Carcinomatous amyotrophic lateral sclerosis. In Brain WR, Norris FH, eds. *The Remote Effects of Cancer on the Nervous System*, pp. 24–34. Grune and Stratton, New York.

Owens T, Babcock A (2002). Immune response induction in the central nervous system. *Front Biosci*, **7**, d427–38.

Paone JF, Jeyasingham K (1980). Remission of cerebellar dysfunction after pneumonectomy for bronchogenic carcinoma. *N Engl J Med*, **302**, 156.

Peterson K, Forsyth PA, Posner JB (1994). Paraneoplastic sensorimotor neuropathy associated with breast cancer. *J Neurooncol*, **21**, 159–70.

Peterson K, Rosenblum MK, Kotanides H et al. (1992). Paraneoplastic cerebellar degeneration. I. A clinical analysis of 55 anti-Yo antibody-positive patients. *Neurology*, **42**, 1931–7.

Pittock SJ, Kryzer TJ, Lennon VA (2004). Paraneoplastic antibodies coexist and predict cancer, not neurological syndrome. *Ann Neurol*, **56**, 715–9.

Polans AS, Buczylko J, Crabb J et al. (1991). A photoreceptor calcium binding protein is recognized by autoantibodies obtained from patients with cancer-associated retinopathy. *J Cell Biol*, **112**, 981–9.

Polans AS, Witkowska D, Haley TL et al. (1995). Recoverin, a photoreceptor-specific calcium-binding protein, is expressed by the tumor of a patient with cancer-associated retinopathy. *Proc Natl Acad Sci USA*, **92**, 9176–80.

Posner JB (1996). Paraneoplastic syndromes. In Marsden CD, Fenichel GM, Daroff, eds. pp. 1165–72. Butterworth Heinemann, London.

Poulter MO, Payne KB, Steiner JP (2004). Neuroimmunophilins: a novel drug therapy for the reversal of neurodegenerative disease? *Neuroscience*, **128**, 1–6.

Reddy RV, Vakili ST (1981). Midbrain encephalitis as a remote effect of a malignant neoplasm. *Arch Neurol*, **38**, 781–2.

Rees JH, Hain SF, Johnson MR et al. (2001). The role of [18F]fluoro-2-deoxyglucose-PET scanning in the diagnosis of paraneoplastic neurological disorders. *Brain*, **124**, 2223–31.

Ridley A, Kennard C, Scholtz CL et al. (1987). Omnipause neurons in two cases of opsoclonus associated with oat cell carcinoma of the lung. *Brain*, **110**, 1699–709.

Sakai K, Gofuku M, Kitagawa Y et al. (1995). Induction of anti-Purkinje cell antibodies in vivo by immunizing with a recombinant 52-kDa paraneoplastic cerebellar degeneration-associated protein. *J Neuroimmunol*, **60**, 135–41.

Sakai K, Mitchell DJ, Tsukamoto T et al. (1990). Isolation of a complementary DNA clone encoding an autoantigen recognized by an anti-neuronal cell antibody from a patient with paraneoplastic cerebellar degeneration [published erratum appears in *Ann Neurol* 1991 Nov; 30(5):738]. *Ann Neurol* **28**, 692–8.

Sayani FA, Prosser C, Bailey RJ et al. (2005). Thiopurine methyltransferase enzyme activity determination before treatment of inflammatory bowel disease with azathioprine: effect on cost and adverse events. *Can J Gastroenterol*, **19**, 147–51.

Schold SC, Cho ES, Somasundaram M et al. (1979). Subacute motor neuronopathy: a remote effect of lymphoma. *Ann Neurol*, **5**, 271–87.

Shaw PJ, Walls TJ, Newman PK et al. (1991). Hashimoto's encephalopathy: a steroid-responsive disorder associated with high anti-thyroid antibody titers—report of 5 cases. *Neurology*, **41**, 228–33.

Siemson JK, Meister L (1963). Bronchogenic carcinoma with severe orthostatic hypotension. *Ann Intern Med*, **8**, 669–72.

Sigurgeirsson B (1992). Skin disease and malignancy. An epidemiological study. *Acta Derm Venereol Suppl Stockh*, **178**, 1–110.

Smith JL, Finley JC, Lennon VA (1988). Autoantibodies in paraneoplastic cerebellar degeneration bind to cytoplasmic antigens of Purkinje cells in humans, rats and mice and are of multiple immunoglobulin classes. *J Neuroimmunol*, **18**, 37–48.

Tanaka K, Tanaka M, Igarashi S et al. (1995). Trial to establish an animal model of paraneoplastic cerebellar degeneration with anti-Yo antibody. 2. Passive transfer of murine mononuclear cells activated with recombinant Yo protein to paraneoplastic cerebellar degeneration lymphocytes in severe combined immunodeficiency mice. *Clin Neurol Neurosurg*, **97**, 101–5.

Tanaka K, Tanaka M, Onodera O *et al.* (1994). Passive transfer and active immunization with the recombinant leucine-zipper (Yo) protein as an attempt to establish an animal model of paraneoplastic cerebellar degeneration. *J Neurol Sci*, **127**, 153–8.

Telander RL, Smithson WA, Groover RV (1989). Clinical outcome in children with acute cerebellar encephalopathy and neuroblastoma. *J Pediatr Surg*, **24**, 11–4.

Thirkill CE, Tait RC, Tyler NK *et al.* (1992). The cancer-associated retinopathy antigen is a recoverin-like protein. *Invest Ophthalmol Vis Sci*, **33**, 2768–72.

Tinsley JA, Barth EM, Black JL *et al.* (1997). Psychiatric consultations in stiff-man syndrome. *J Clin Psychiatry*, **58**, 444–9.

Tora M, Graus F, de Bolos C *et al.* (1997). Cell surface expression of paraneoplastic encephalomyelitis/sensory neuronopathy-associated Hu antigens in small-cell lung cancers and neuroblastomas. *Neurology*, **48**, 735–41.

Trotter JL, Hendin BA, Osterland CK (1976). Cerebellar degeneration with Hodgkin disease. An immunological study. *Arch Neurol*, **33**, 660–1.

Uchuya M, Graus F, Vega F, *et al.* (1996). Intravenous immunoglobulin treatment in paraneoplastic neurological syndromes with antineuronal autoantibodies. *J Neurol Neurosurg Psychiatry*, **60**, 388–92.

Vega F, Graus F, Chen QM, *et al.* (1994). Intrathecal synthesis of the anti-Hu antibody in patients with paraneoplastic encephalomyelitis or sensory neuronopathy: clinical-immunologic correlation. *Neurology*, **44**, 2145–7.

Verma A, Berger JR, Snodgrass S *et al.* (1996). Motor neuron disease: a paraneoplastic process associated with anti-hu antibody and small-cell lung carcinoma. *Ann Neurol*, **40**, 112–6.

Vincent A, Buckley C, Schott JM *et al.* (2004). Potassium channel antibody-associated encephalopathy: a potentially immunotherapy-responsive form of limbic encephalitis. *Brain*, **127**, 701–12.

Weinstein JM, Kelman SE, Bresnick GH *et al.* (1994). Paraneoplastic retinopathy associated with antiretinal bipolar cell antibodies in cutaneous malignant melanoma. *Ophthalmology*, **101**, 1236–43.

Weissman DE, Gottschall JL (1989). Complete remission of paraneoplastic sensorimotor neuropathy: a case associated with small-cell lung cancer responsive to chemotherapy, plasma exchange, and radiotherapy. *J Clin Apheresis*, **5**, 3–6.

Younger DS, Rowland LP, Latov N *et al.* (1991). Lymphoma, motor neuron diseases, and amyotrophic lateral sclerosis [see comments]. *Ann Neurol*, **29**, 78–86.

CHAPTER 39

Tremor, ataxia, and cerebellar disorders

Nicholas Fletcher

Contents

39.1 **Tremor**

39.1.1 **Clinical diagnosis of tremor**

Tremors are characterized by rhythmic oscillations of one or more body parts. Although typically seen in the upper limbs, almost any area may be involved, including the trunk, head, facial muscles, and legs. Sometimes, tremor is not visible at all but may be heard or palpated, for example, in vocal or orthostatic tremor, respectively. In neurological practice, the diagnosis and treatment of tremor is an everyday problem. A common scenario is the distinction between essential tremor and Parkinson's disease; although this is normally straightforward, diagnostic error and consequent inappropriate medication are surprisingly frequent. The diagnosis of tremor depends crucially on the history, especially duration, exacerbating or relieving factors, associated symptoms, and family history. For example, a tremor that has occurred in a parent, has been present for 20 years, is alleviated by alcohol, and which is associated with little or no disability is much more likely to be essential tremor than Parkinson's disease. The response of the tremor to alcohol or sometimes to medication is helpful but can be difficult to gauge. It is important to note that anxiety will exacerbate any tremor, leading to the erroneous conclusion that it is psychogenic. Wilson's disease must always be kept in mind in the neurological clinic whenever a younger patient presents with tremor (or indeed almost any other movement disorder), particularly if associated with psychiatric changes, dystonia, Parkinsonism, or cerebellar features (see Section 40.8); there should be a low threshold for the use of the appropriate screening tests, especially a search for corneal Kayser-Fleischer rings and measurement of the serum caeruloplasmin. At one time, a tremor of the face, tongue, and hands was seen in association with neurosyphilis (general paresis); this possibility should still be considered when tremor is seen in conjunction with dementia or psychiatric changes. Patients are commonly referred to hospital with tremor while taking drugs known to cause this, frequently valproate, lithium, amiodarone, Adrenergic β-2 agonists or caffeine.

On examination, the appearance of the tremor and its relationship to the state of activation of the affected limb is important, along with any associated signs such as Parkinsonism, cerebellar dysfunction, or areflexia (see below). Frequency is often emphasized as a means to diagnosis; a slower 4–5 Hz tremor is said to be suggestive of Parkinsonism and a faster 8–12 Hz frequency is associated with essential tremor. In fact, there is considerable overlap of frequencies in different forms of tremor, and the frequency of any given tremor may vary with age and the body part affected (Bain 1993). Moreover, it is almost impossible to establish the frequency of a tremor clinically. Thus, tremor frequency is often unhelpful.

39.1.2 **Clinical and physiological classification of tremor**

Having taken an accurate and complete history, the tremor should first be classified by examination of its appearance and relationship to the activity in the affected limb (Table 39.1). This is not always simple or quick, and certain difficulties occur repeatedly. It may be difficult to demonstrate a rest tremor if the patient cannot relax fully; it may help to rest the affected arm on a pillow or the arm of a chair. Postural tremor may be obvious in one posture of the hands but not others; it is wise to test for this with the arms and hands in several positions, not just with the arms outstretched

Table 39.1 Classification of tremor by state of activity

	Features
Rest tremor	Tremor apparent with the limb supported against gravity and with no voluntary activation of muscles
Action tremor	Tremor during any voluntary movement (includes postural, kinetic, isometric, and task-specific tremors)
Postural tremor	Tremor during voluntary maintenance of a posture against gravity
Kinetic tremor	Tremor during any movement
Intention/ terminal tremor	Clinically obvious exacerbation of a kinetic tremor at the end of a goal directed movement as the target is approached, e.g. during finger nose testing
Task-specific tremor	Tremor appearing only or almost exclusively during a particular (usually skilled and precise) movement
Isometric tremor	Tremor appearing when a movement is opposed by a static force or object

and hands pronated. Task-specific tremors will only be seen during the relevant activity, such as writing or using an instrument or tool. The next step is to combine the data from the history and examination and arrive, if possible, at an aetiological classification (Table 39.2).

Neurophysiological measurements have not been particularly useful in the classification of tremor, but frequency analysis may occasionally help. Examples include primary orthostatic tremor, which has a very characteristic fast (14–18 Hz) frequency, and hysterical tremor, which may reveal marked frequency variation, a feature not seen with organic tremors (Cleeves *et al.* 1994).

39.1.3 **Physiological tremor**

This tremor is present in normal individuals but is usually asymptomatic. It may become more apparent during very precise delicate

Table 39.2 Classification of Tremor by Aetiology

Physiological tremor
Parkinsonian tremor
Essential tremor
Cerebellar tremor
Midbrain tremor (rubral tremor; Holmes' tremor)
Dystonic tremor (myorhythmia)
Orthostatic tremor
Cortical tremor*
Asterixis (flapping tremor, hepatic tremor)*
Neuropathic tremor
Toxic tremor
Posttraumatic tremor
Task specific tremor
Site specific tremor
Psychogenic tremor

* Not strictly a tremor, but due to positive or negative myoclonus.

movement or in association with anxiety or excitement, but is never severe. It can be very difficult in some individuals to distinguish exaggerated physiological tremor from mild essential tremor. There are two components to physiological tremor; a constant 8–12 Hz low-amplitude tremor is evident during limb postures and is so constant that it is likely to originate in the central nervous system. In addition, there is a peripheral *mechanical reflex component*, which is a passive mechanical oscillation produced by a complex interaction of the mechanical properties of the limbs, arterial pulsation, tetanic muscle contraction, stretch reflex feedback from muscle spindles, and motor neurone activation patterns. Physiological tremor is exacerbated by increased stimulation of beta adrenoceptors as a consequence of anxiety, thyrotoxicosis, hypoglycaemia, caffeine, and other beta agonists (e.g. salbutamol, amphetamine, or aminophylline); some drugs such as sodium valproate and lithium probably act via a similar mechanism. Enhanced physiological tremor can be treated if necessary, for example, in musicians, technicians, and other occupations requiring steadiness and dexterity, with propranolol (Hallett 1984).

39.1.4 Essential tremor

Essential tremor is one of the most common movement disorders with a population prevalence of up to 4 per cent in different population surveys (Louis *et al*. 1998a). The age of onset is bimodal with peaks at about 15 and 50 years, and the sexes are equally affected. Some essential tremor patients have a positive family history and, in these families, inheritance appears to be autosomal dominant with full penetrance by the age of 65 years (Bain *et al*. 1994). Twin studies indicate a significant genetic factor (Tanner *et al*. 2001; Lorenz *et al*. 2004), but essential tremor is genetically heterogeneous. At least two loci exist, *ETM2* at chromosome 2p22-25 (Higgins *et al*. 1997) and *ETM1* at 3q13 (Gulcher *et al*. 1997). A possible gene, *HS1-BP3*, has been identified at *ETM2* (Higgins *et al*. 2005), but this has been disputed (Deng *et al*. 2004). However, many cases are sporadic, and only about half of patients have affected relatives (Louis 2001). In these patients, there has been interest in environmental toxins such as lead or beta carboline alkaloids (Louis 2005).

The pathophysiology of essential tremor is poorly understood, but seems to involve bilateral cerebellar dysfunction (Jenkins *et al*. 1993; Pagan *et al*. 2003), the inferior olivary nucleus (Hallett *et al*. 1993), and abnormal rhythmic discharges of thalamic neurones (Hua *et al*. 1998). Neuropathological studies have been unrevealing (Louis 2005).

Clinically, sporadic and familial essential tremors are identical. Onset is typically in one or both arms and the tremor is mainly postural with a typical frequency of 4–12 Hz. In about 20 per cent of cases, there may be a mild kinetic or intention component. Although the onset may be unilateral, most cases eventually become bilateral (Bain *et al*. 1994). Although essential tremor may be asymmetrical (Louis *et al*. 1998b), it should be noted that persistently unilateral postural tremor may evolve into Parkinson's disease in later years (Chaudhuri *et al*. 2005).

The tremor slowly deteriorates over many years, but is usually confined to the upper limbs. In about a third of cases, the legs become affected. In severe essential tremor, there may be involvement of the head, of either yes-yes or no-no type, tongue, voice, jaw, and face in decreasing order of frequency. Isolated tremors of the legs, head, face, or voice do not occur (Bain *et al*. 1994). Essential tremor is characteristically relieved by 2–4 units of alcohol, but the effect lasts only a few hours, is often incomplete, and is evident in about two thirds of cases; accordingly, alcohol responsiveness is a useful diagnostic indicator but need not be present in every case.

Some patients have additional signs such as mild gait ataxia, consistent with the cerebellar dysfunction seen on functional imaging studies (Singer *et al*. 1994), and about 20 per cent have an element of rest tremor (Louis 2005). It is unlikely that such patients have Parkinson's disease because neuropathological studies have not confirmed this (Rajput *et al*. 1993). However, in some tense patients, it is very difficult to ensure the limbs are fully at rest.

Unfortunately, patients with essential tremor are still diagnosed incorrectly as having Parkinson's disease. Although some patients with Parkinson's disease can have a postural tremor, additional signs of bradykinesia and rigidity should be present; moreover, most patients with essential tremor present to the clinic after many years of shaking and are likely to have a family history, alcohol response, or both. It has to be said that in the majority of patients, Parkinson's disease and essential tremor should be easily differentiated. However, as noted already, one area of difficulty is the possibility of longstanding unilateral postural tremor evolving into Parkinson's disease after many years.

Essential tremor should be differentiated clinically and, if necessary, by appropriate laboratory tests from hyperthyroidism, drug- or caffeine-related tremor, dystonia, and Wilson's disease. Spinocerebellar ataxia, SCA12, can also cause a similar postural tremor (Holmes *et al*. 1999).

Many patients with essential tremor do not require treatment; in these cases, an explanation and reassurance are adequate. Many of these patients are referred to the outpatient department with a diagnosis of Parkinson's disease and often, antiparkinsonian medication has been administered unsuccessfully for years. These patients are usually greatly relieved by the correct diagnosis but require tactful handling and discontinuation of inappropriate drugs. However, essential tremor is not entirely benign; many patients have significant disability and handicap, especially in terms of employment. Propranolol at a dose of 80–320 mg/day will significantly relieve but not abolish tremor in about 70 per cent of cases (Koller *et al*. 1989), but is contraindicated in asthma, cardiac disease, or diabetes. The dose should be increased gradually. Long-acting propranolol is as effective as the standard preparation. Primidone is equally effective (Findley *et al*. 1985; Koller *et al*. 1989) but often poorly tolerated due to sedation; it is important to start at a very low dose, 62.5 mg once a day, and increase very slowly, by about 62.5 mg every 2 weeks, to a maximum of 750 mg/day. In practice, very few patients can tolerate high doses, and there is some evidence that a single, daily 250 mg dose is effective (Koller *et al*. 1986). Many other drugs have been tried in essential tremor; gabapentin has been effective in some patients (Gironell *et al*. 1999) but less so in others (Pahwa *et al*. 1998); topiramate was effective in a recent controlled trial (Ondo *et al*. 2006), but can have side effects including sedation, cognitive change, and psychosis (Zesiewicz *et al*. 2006). Benzodiazepines, calcium channel blockers, and neuroleptics have been used but with little evidence of benefit (Louis 2005). Some patients have been treated with local injections of botulinum toxin (Jankovic *et al*. 1996); although tremor is reduced, the benefit is temporary and the injections need to be repeated. This form of treatment requires further evaluation but is unlikely to become widespread.

In severe cases of essential tremor, stereotactic Vim nucleus deep brain stimulation, which is highly effective in suppressing tremor, can be carried out bilaterally (Limousin *et al.* 1999) and is as effective as thalamotomy, but safer (Schuurman *et al.* 2000). The duration of benefit is uncertain, with some patients experiencing a decline in tremor suppression over a few years (Kumar *et al.* 2003), but good results have persisted for over 5 years in other series (Rehncrona *et al.* 2003; Pahwa *et al.* 2006). Side effects include ataxia and dysarthria, in addition to hardware-related technical problems.

There have been several reports of possible relationships between essential tremor and other neurological disorders, especially Parkinson's disease (Koller *et al.* 1994; De Michele *et al.* 1995) and primary torsion dystonia (Fletcher *et al.* 1991). In Parkinson's disease, the difficulty arises because some patients have an additional postural component to the tremor and in some essential tremor patients, there may be tremor apparently 'at rest'. In addition, there are reports of an increased incidence of essential tremor among the relatives of Parkinson's disease patients. However, a detailed clinical analysis of essential tremor showed no convincing evidence of rest tremor or Parkinsonism (Bain *et al.* 1994). An earlier large study of patients with essential tremor, Parkinson's disease, and controls showed no evidence of any association (Cleeves *et al.* 1988). Moreover, there is no neuropathological evidence of Parkinson's disease in patients with essential tremor (Rajput *et al.* 1993), and positron emission tomography studies show no definite evidence of nigrostriatal dysfunction (Jenkins *et al.* 1993). While this issue is not entirely resolved, any association between essential tremor and Parkinson's disease seems tenuous (Louis 2005). In primary torsion dystonia, an upper limb postural tremor similar to essential tremor is often seen with torticollis or other forms of PTD; it may be the sole clinical abnormality in some primary torsion dystonia gene carriers. Despite this clinical overlap, the two disorders are genetically distinct; there is no genetic linkage between the primary torsion dystonia locus, DYT1, on chromosome 9 and essential tremor (Conway *et al.* 1993; Durr *et al.* 1993).

39.1.5 Parkinsonism and tremor

The pathophysiology and clinical features of Parkinsonian tremor are discussed in chapter 40. Many patients with Parkinson's disease have a postural tremor in addition to the typical Parkinsonian rest tremor; the former is very similar to essential tremor. In some cases, propranolol may be of benefit but the response is not as clear as that seen in essential tremor and primidone does not seem to be helpful (Cleeves *et al.* 1994). In multiple system atrophy and drug-induced Parkinsonism (see Sections 40.3.4 and 40.3.7), tremor is less common and usually postural, while tremor is unusual in other forms of Parkinsonism. Wilson's disease must again be emphasized as a vitally important diagnosis (see Section 40.8). Some patients with the fragile X syndrome but no history of cognitive impairment develop tremor with Parkinsonism and ataxia in adult life, the Fragile X Tremor Ataxia Syndrome, FXTAS (Hagerman *et al.* 2004b); (see Section 39.9.7). A family history of males with learning disability may be apparent but not in all cases, and some female cases of FXTAS have occurred (Hagerman *et al.* 2004c). MR scanning may show characteristic middle cerebellar peduncle signal changes (Hagerman *et al.* 2003).

Some patients present with a unilateral postural tremor for many years prior to developing Parkinson's disease (Chaudhuri *et al.* 2005); in some cases, an asymmetrical or unilateral essential

tremor-like disorder was present for 10–30 years before signs of Parkinson's disease appeared. In rare families with hereditary Parkinson's disease due to *PARK1* α-synuclein gene mutations, there can be individuals who develop essential tremor rather than Parkinsonism (Farrer *et al.* 1999). In some cases, patients with typical essential tremor develop Parkinson's disease after many years, probably coincidentally—the 'ET-PD' phenotype (Jankovic 2002). This confusing scenario can sometimes be distinguished from Parkinson's disease or essential tremor neurophysiologically with H-reflex studies (Sabbahi *et al.* 2002), but a clinical decision should be possible in many cases.

39.1.6 Rubral or midbrain tremor

This type of tremor was first described in 1904 by Holmes, who noted a low-frequency and large-amplitude upper limb tremor, present at rest but exacerbated by posture and movement, or kinetic/action, with an intention, or terminal component; the kinetic or action element of the tremor is worst followed by the postural component, and the rest tremor is milder. This combination of Parkinsonian and cerebellar features pointed to a lesion of the midbrain tegmentum in the region of the red nucleus and the cerebellothalamic pathway (see also Section 40.1.2). This localization has been confirmed by neuropathological studies. Midbrain tremor is usually severe and disabling, producing wild and uncontrollable arm tremor. Most cases are due to brainstem trauma, multiple sclerosis, vascular lesions, or cerebellar degenerations; a few are idiopathic and unilateral. Ipsilateral striatal dopamine deficiency due to nigrostrital tract damage has been demonstrated in some cases (Remy *et al.* 1995). Treatment is notoriously difficult; levodopa, propranolol, and anticholinergic agents may be tried, but the response to these drugs is unpredictable. In very severe cases, thalamic deep brain stimulation has been successful in Holmes' tremor caused by multiple sclerosis, vascular malformation, surgery, and trauma (Samadani *et al.* 2003; Nandi *et al.* 2004; Nikkhah *et al.* 2004; Foote *et al.* 2005).

39.1.7 Cerebellar tremor

Tremor is one of the cardinal features of cerebellar disease, along with dysmetria, dysdiadochokinesis, dysarthria, hypotonia, abnormal eye movement, ataxia, and decomposition of movement. Cerebellar disease produces a slow, high-amplitude *kinetic* tremor, which worsens during a movement leading to a terminal or intention component (see Table 39.1). This is seen clinically during finger-nose or heel-knee-shin testing and will usually be accompanied by dysmetria, past pointing, unless quite mild. In addition to the typical kinetic cerebellar tremor, there may be a large-amplitude postural tremor of the head, titubation, and sometimes the trunk and limbs. A hallmark of cerebellar tremor is that it mainly affects axial or proximal limb muscle groups. Pathology of the midline cerebellar vermis tends to produce ataxia of gait and titubation of the head and trunk, while lesions of the lateral cerebellar hemispheres or cerebellar outflow pathways are associated with the kinetic or intention limb tremor (Gilman *et al.* 1981). There is no specific treatment for cerebellar tremor.

39.1.8 Drug- and alcohol-induced tremor

An action tremor of the hands is commonly seen in alcoholics as a withdrawal symptom, the 'morning shakes'. This is usually relieved promptly by the first drink of the day. In full-blown withdrawal, the tremor can be severe and associated with prominent mental changes, fever, autonomic instability, and dehydration, 'delirium tremens' (Section 5.2.1). Beta adrenergic agonists such as salbutamol or aminophylline, used in asthma, are a common iatrogenic cause of an action tremor of the hands. In some individuals, a similar phenomenon is seen after caffeine ingestion. Neuroleptic drugs may cause tremor as a feature of drug-induced Parkinsonism; this is usually symmetrical and sometimes of action/postural type rather than a resting tremor (Section 40.3.4). In the 'rabbit syndrome', a particular form of drug-induced Parkinsonian tremor, there is a resting 4–6 Hz tremor of the lips. In some patients receiving long-term neuroleptic therapy, a persistent rest and postural tremor appears, which is increased by drug withdrawal and improved by dopamine depletion with terabenazine; these similarities to other tardive phenomena have led to the term tardive tremor for this phenomenon (Stacy *et al.* 1992). Tricyclic antidepressants are prone to cause an action tremor and sometimes Parkinsonism or rabbit syndrome (Vandel *et al.* 1997). Other causes of a postural or action tremor of the hands include lithium, sodium valproate, ciclosporin-A, and amiodarone.

39.1.9 Orthostatic tremor

This rare condition is seen mostly in older adults in whom there is a fast tremor of the legs and sometimes paraspinal muscles that develops only during standing and not when walking, sitting, or leaning on a support (Heilman 1984; Britton *et al.* 1995). The tremor tends to develop after standing for a minute or so, so that the patient feels increasingly unsteady and is in fear of falling and injury. Orthostatic tremor is fast and barely visible (Britton *et al.* 1992); it may be detected only by palpation of the legs while the patient is standing. There are no other features and in particular, an upper limb tremor is characteristically absent, although occasionally noted, and the gait is normal. The condition tends to worsen over several years, and the disorder does not appear to be familial. The tremor is much faster than essential tremor, typically 13–18 Hz (McManis *et al.* 1993), and this is one of the rare situations where tremor frequency measurement is diagnostically useful. The relationship between orthostatic tremor and essential tremor is uncertain, but the differences between the two suggest that they are separate disorders (Table 39.3)

Table 39.3 Differences between orthostatic tremor and essential tremor

	Orthostatic tremor	Essential tremor
Age of onset	Late	Bimodal
Main site affected	Legs	Arms
Main symptom	Unsteadiness	Shaking hands
Tremor visible?	No	Yes
Frequency	13–18 Hz	4–12 Hz
Familial	No	Often
Response to alcohol	No	Often
Response to beta blockers	No	Often
Response to primidone	Sometimes	Often
Response to clonazepam	Yes	Variable

Orthostatic tremor is not affected by peripheral stimuli and is likely to originate in the central nervous system, probably in the posterior fossa (Wu *et al.* 2001). Positron emission tomography has shown abnormally increased cerebellar activation, similar to that seen in essential tremor, suggesting that cerebellar overactivity may be a consequence of tremor caused by different mechanisms (Wills *et al.* 1996). Dopamine transporter SPECT scanning has shown reduced striatal signal consistent with a dopaminergic deficit (Katzenschlager *et al.* 2003).

Beta blockers are not effective, but the symptoms are improved with clonazepam (Heilman 1984). Levodopa may be effective in some patients (Wills *et al.* 1999), but this has not been consistently observed in other studies (Katzenschlager *et al.* 2003). Gabapentin may be effective, but formal trials are lacking (Evidente *et al.* 1998).

39.1.10 Site-specific tremors

In addition to patients who develop a focal onset of well-recognized tremors such as Parkinson's disease, in which tremor may be remarkably localized, even in only one digit, at onset, or the syndrome of painful feet and moving toes (Section 40.11.7), there is a miscellaneous group of disorders in which patients present with spontaneous tremor confined to one body part. It should be noted that focal tremor confined to the head, face, trunk, or legs is highly unusual in essential tremor, and in one recent study of familial essential tremor was not seen at all (Bain *et al.* 1994).

♦ Isolated head and trunk tremors are usually dystonic and not due to essential tremor (Rivest *et al.* 1990). Such tremors are not a feature of Parkinson's disease and, although cerebellar disease can lead to a slow nodding head tremor, titubation, additional cerebellar signs will be present in this situation. These focal tremors tend to be worse during standing or walking and absent when lying down, and there are often subtle dystonic movements of the trunk or neck in addition. A common outpatient situation is the occurrence of tremulous cervical dystonia in which the patient presents with a complex multidirectional head tremor but with slight additional torticollis or laterocollis as a clue to the true nature of the problem. It should be noted that dystonia may be jerky or tremulous and does not always produce fixed abnormal postures or spasms (see Section 40.4). Patients with isolated head and trunk tremors may be treated with anticholinergics or cervical botulinum toxin injections (Jankovic *et al.* 1991).

♦ Unilateral postural upper limb tremor is difficult to classify; some cases may be atypical essential tremor, others develop into Parkinson's disease, sometimes after many years (Chaudhuri *et al.* 2005), and others may be dystonic (Jain *et al.* 2006).

♦ Isolated leg resting tremor is usually due to Parkinson's disease, while focal lower limb tremor developing during standing is suggestive of primary orthostatic tremor (see Section 39.1.9).

♦ Focal tremor of the chin, geniospasm, may occur as a hereditary disorder (Danek 1993). Inheritance is autosomal dominant, and one gene locus on chromosome 9q13-21 has been identified (Jarman *et al.* 1997).

39.1.11 Task-specific tremors

In some patients, a focal upper limb tremor develops in association with a particular action. The best known example is primary writing tremor (Bain *et al.* 1995). Similar task-specific tremors have been reported with other activities such as playing golf, throwing darts, shooting, playing a musical instrument, drinking, or using tools (Soland *et al.* 1996). The cause of task-specific tremors is unclear. The striking task specificity is reminiscent of the focal dystonias such as writer's cramp or those of musicians (see Section 40.4.1), leading some to conclude that they are variants of focal dystonia, although studies of reciprocal inhibition are normal (Bain *et al.* 1995) unlike writer's cramp (Marsden *et al.* 1987). Electromyographic recordings of primary writing tremor show alternating agonist–antagonist bursts (Bain 1993). Others have regarded task-specific tremors as variants of essential tremor. At present, it is uncertain whether primary writing tremor is related to dystonia, essential tremor, or is a separate disorder.

In terms of response to drug therapy, some patients with focal task-specific tremors respond like essential tremor to propranolol or primidone, while others behave like a dystonia and improve with anticholinergic medication (Cleeves *et al.* 1994). Overall, the response to treatment is poor, and most patients are not improved by any medication. Botulinum toxin has been used in a few cases of primary writing tremor, and the use of thalamotomy has been described (Ohye *et al.* 1982). One patient was treated with thalamic deep brain stimulation (Racette *et al.* 2001). Although the occupational handicap in primary writing tremor may be considerable, it is difficult to recommend the use of stereotactic neurosurgery for this condition.

39.1.12 Tremor and dystonia

It has long been apparent that some patients with primary torsion dystonia, especially those with cervical dystonia, torticollis, can have a postural tremor of the hands that is indistinguishable from essential tremor (Jankovic *et al.* 1991). Other patients with focal dystonias have tremor in addition to dystonic spasms in the affected body part such as in writer's cramp (Sheehy *et al.* 1982) and axial dystonia (Rivest *et al.* 1990). Such tremors are also seen in patients with generalized primary torsion dystonia (Marsden *et al.* 1974) and may be the only clinical manifestation in otherwise asymptomatic primary torsion dystonia gene carriers (Fletcher *et al.* 1990). The tremor in these patients appears to be a manifestation of primary torsion dystonia and not to reflect coexisting essential tremor; the genetic loci for the two conditions are separate (see Section 39.1.4). Tremor and dystonia share a number of features, suggesting that they may arise by similar mechanisms. Not only do both sometimes appear in the same patients, but they share a tendency to task specificity (as seen in the task-specific tremors discussed above) and may both develop as a response to peripheral trauma. Other clues to a dystonic type of tremor include persistently focal tremor in one arm, jerky and irregular appearance, gestes antagonistes, or response to antidystonic drugs (Deuschl 2003). In addition to the coexistence of tremor and dystonia in some patients, it should be noted that dystonic movements themselves are sometimes deceptively tremulous, especially when the neck (see Section 39.1.10) or upper limbs are involved, a feature sometimes referred to as myorhythmia (Herz 1944).

39.1.13 Posttraumatic tremor

In addition to the severe midbrain tremor seen after traumatic brain injury (see Section 39.1.6), there are rare reports of tremor following mild head injuries (Biary *et al.* 1989) and also peripheral injury (Jankovic *et al.* 1988). The mechanism by which peripheral injuries lead to the onset of a tremor in the affected body part is unclear but a genetic predisposition, suggested by a mild pre-existing tremor or a family history of essential tremor, is possible, as suggested for posttraumatic dystonia (Section 40.4.16).

39.1.14 Neuropathic tremor

Tremor may be prominent in patients with peripheral neuropathy (Cleeves *et al.* 1994), and this possibility must be considered in all patients presenting with tremor. A postural tremor of the hands is the most common type but cases of rest tremor and intention tremor have also been reported. The majority of neuropathic tremors are seen with demyelinating neuropathies such as hereditary motor and sensory neuropathy type 1, chronic inflammatory demyelinating polyneuropathy, and dysproteinaemic neuropathy. Diabetes, alcoholism, and porphyria have occasionally been responsible (Said *et al.* 1982). The mechanism of neuropathic tremor is unclear. Treatment with propranolol may be helpful, along with management of the underlying neuropathy, if possible.

39.1.15 Palatal tremor

The term 'palatal myoclonus' is probably a misnomer as it clearly takes the form of a focal tremor (see Section 40.7.3). A rhythmic tremor of the soft palate is seen. Essential palatal tremor, myoclonus, is associated with ear clicking noises experienced by the patient, whereas the secondary palatal tremor is not. Treatment is difficult (Section 40.7.11).

39.2 Cerebellar disorders

39.2.1 Diagnosis

The anatomy and physiology of the cerebellum are described in Section 40.1, along with the rest of the motor system, and the principal regions of the cerebellum are shown in Fig. 39.1. Clinically, the recognition of cerebellar disease is usually straightforward.

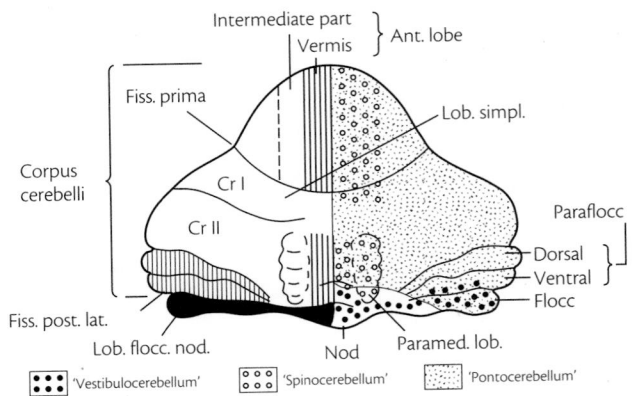

Fig. 39.1 A simplified anatomical representation of the cerebellum. On the left are the three main subdivisions, the archicerebellum (vestibulocerebellum) comprising the flocculonodular lobe (black); paleocerebellum (spinocerebellum) comprising the anterior and posterior vermis and the paraflocculus (vertical hatching); and the neocerebellum (pontocerebellum) comprising the central vermis and bulk of the hemispheres (white). On the right, the terminations of the vestibulocerebellar (heavy dots), spinocerebellar (open dots), and pontocerebellar (small dots) afferents are shown. (From Brodal A [1981] Neurological anatomy in relation to clinical medicine, Oxford University Press.)

A useful practical point is that lesions of the cerebellar vermis, orpaleocerebellum, tend to produce ataxia of gait, abnormal eye movements, and axial tremor, titubation only. More lateral lesions of the cerebellar hemispheres, the neocerebellum, are more likely to result in a full range of cerebellar deficits, namely gait ataxia, eye movement abnormalities, decomposition of movement, ataxia, tremor and dysmetria of the limbs, dysdiadochokinesis and hypotonia (Gilman *et al.* 1981). Dysarthria is seen with a wide range of cerebellar lesions and is of little localizing value.

It must be remembered that ataxia, clumsiness, disordered ocular motility, dysarthria, and even kinetic or intention tremor are not always caused by cerebellar disease. Diagnostic problems do arise in which a cerebellar disorder is suspected initially, but it eventually turns out that a different pathology is simulating cerebellar disease. Certain difficulties arise repeatedly in the clinic:

- Patients with peripheral neuropathy, especially Charcot–Marie–Tooth disease Type 1, chronic inflammatory demyelinating polyneuropathy, Guillain-Barré syndrome, and the Miller-Fisher syndrome may have prominent ataxia of gait. In early Guillain-Barré syndrome, particularly, the abnormal gait may be the presenting feature, before clear signs of peripheral nerve involvement are apparent. The ataxia is a consequence of impaired proprioception and may also lead to clumsiness and incoordination of the limbs; sometimes, a neuropathic tremor of the limbs may also resemble that seen in cerebellar disease. The absence of dysarthria, the normal eye movements, excepting the Miller Fisher syndrome, areflexia, and nerve conduction studies, will usually lead to the correct diagnosis. The distinction between early Friedreich's ataxia and hereditary motor and sensory neuropathy type I, in which the only early signs may be ataxia and areflexia, can be particularly difficult without nerve conduction studies or, increasingly, DNA analysis.

- Cervical spondylotic myelopathy often leads to impaired upper limb proprioception because of dorsal column involvement of the cervical cord; this can produce a combination of unsteadiness of gait and ataxic upper limbs. Clear pyramidal signs may not always be detectable, or if present may be attributed to one of the late-onset cerebellar degenerations.

- An unsteady, wide-based gait is seen with frontal lobe pathology, especially hydrocephalus, cerebrovascular disease causing Binswanger's encephalopathy, and mass lesions (see Section 40.11.10). Although this 'frontal gait disorder' can usually be identified clinically, the unsteadiness can be so severe (Nutt *et al.* 1993) that the distinction from a cerebellar syndrome is very difficult. The condition of primary progressive freezing of gait is similar (see Section 40.11.10).

- In progressive supranuclear palsy (see Section 40.3.9), falling and instability of gait may be prominent before other signs such as ophthalmoparesis have developed; a cerebellar disorder may initially be suspected. Even if Parkinsonism and abnormal eye movements are present, it can be difficult to exclude a sporadic late-onset cerebellar degeneration in the early stages.

- The distinction between cerebellar and pseudobulbar dysarthria is sometimes difficult. Accordingly, motor neurone disease, cerebrovascular disease, or multiple sclerosis may initially be confused with a cerebellar degeneration. Signs of pseudobulbar palsy such as emotional lability, slowing of tongue movement, and a brisk jaw jerk may be helpful.

- Wilson's disease may present with unsteadiness and clumsiness suggestive of a cerebellar disorder, the so called 'pseudosclerotic' presentation. Although very rare, this is an important clinical consideration because the condition is treatable (Section 40.8).

- Vitamin B12 deficiency may present with a progressive ataxia due to subacute combined degeneration of the spinal cord and loss of dorsal column proprioceptive function. This is, therefore, a spinal rather than a cerebellar ataxia, but B12 deficiency must always be considered in the ataxic patient if permanent and avoidable neurological damage is to be prevented.

39.2.2 Classification

The classification of the various causes of cerebellar ataxia has long been troublesome. Cerebellar diseases may be classified in terms of age of onset, aetiology, clinical features, neuropathology, inheritance and, increasingly, molecular genetic abnormalities. None of these approaches is entirely satisfactory for such a heterogeneous group of disorders in which there is a poor correlation between aetiology, clinical features, genotype, and neuropathology. A pragmatic clinical approach is to think of the ataxias initially in terms of mode and age of onset and with regard to any detectable underlying cause or inheritance. Aetiology may be suggested by the mode of onset and the age of the patient, as shown in Table 39.4. The principal causes of ataxia are shown, with the usual speed and age of onset indicated. This is a helpful clinical classification of ataxia, which is readily applied in the clinic or the acute ward setting. In practice, most cases of acute (or subacute) ataxia are due to toxins, principally drugs and alcohol, vascular lesions in older adults, paraneoplastic cerebellar degeneration, or demyelination in younger patients, while the most common cause of any chronic, slowly progressive ataxia is a degenerative cerebellar or spinal ataxia. It is important not to overlook a treatable cause of ataxia. The important possibilities are drugs and toxins, deficiency states of thiamine and vitamins B$_{12}$ or E, posterior fossa mass lesions, hypothyroidism, hydrocephalus, cholestanolosis, and Wilson's disease.

39.2.3 Investigation

Physical examination may reveal abnormalities that aid diagnosis. Features of the neurological examination are in Table 39.5, and general examination findings are considered in Table 39.6. The use of special investigations will be determined by the likely diagnostic possibilities. In patients with adult-onset cerebellar ataxia, MR scanning, thyroid function tests, a vitamin B$_{12}$ estimation, and a chest X-ray are usually adequate. If a paraneoplastic ataxia is suspected, investigations to reveal a malignancy are indicated. The EEG is helpful in some cases of prion disease. Nerve conduction studies may be needed to exclude a neuropathic ataxia along with a CSF examination if acute Guillain-Barré syndrome or chronic inflammatory demyelinating polyneuropathy are suspected. In younger patients, an electrocardiogram may be very helpful as it points very strongly to the diagnosis of Friedreich's ataxia in the clinic without the need for additional tests other than DNA analysis. Additional investigations such as lipids, white cell enzymes, immunoglobulins, alpha fetoprotein, caeruloplasmin, slit lamp examination for Kayser-Fleischer rings, muscle biopsy, lactate/pyruvate, very long chain fatty acids, and bone marrow

Table 39.4 Clinical classification of the ataxias by mode of onset, age, aetiology, and inheritance

	Inheritance	Childhood	Young adults	Older adults
Congenital				
Ataxic cerebral palsy		+		
Hereditary congenital ataxias	AD/AR/XL	+		
Acute/subacute-onset				
Infarction/haemorrhage				+
Demyelination (MS)			+	
Demyelination (ADEM)		+	+	
Postinfectious cerebellar ataxia		+	+	
Paraneoplastic		+	+	
Toxins		+	+	+
Thiamine deficiency (Wernicke)			+	+
Abscess/tumour		+	+	+
Basilar migraine		+	+	
Slower progressive				
Early-onset hereditary degenerative (<25 years)	AR (AD/XL rarely)	+	(+)	
Late-onset hereditary degenerative Ataxia (>25 years)	AD AR/XL rarely)		+	+
Sporadic idiopathic cerebellar degeneration				+
Tumour		+	+	+
Foramen magnum compression			+	+
Alcoholic cerebellar ataxia				+
Hydrocephalus		+	+	+
Hypothyroidism				+
Drugs, e.g. phenytoin				+
Prion disease	(occasionally AD)		+	+
Metabolic ataxias	AR/XL /mitochondrial	+	+	
Vitamin E deficiency	AR	+	+	
Intermittent				
Drugs/toxins			+	+
Multiple sclerosis			+	
Transient ischaemic attacks				+
Foramen magnum compression			+	+
Intermittent hydrocephalus			+	+
Metabolic ataxias	AR/XL /mitochondrial	+	+	
Periodic ataxias (hereditary)	AD	+	+	

Note: AD = autosomal dominant; ADEM = acute disseminated encephalomyelitis; AR = autosomal recessive; MS = multiple sclerosis; XL = X linked.

examination are usually indicated in younger patients if a neurometabolic disorder is suspected. The vitamin E level is important in younger patients, especially those with a phenotype resembling Friedreich's ataxia, even if cardiomyopathy is present. It is essential not to miss vitamin E deficiency, which is treatable; a sound clinical tip is to consider the diagnosis and request a vitamin E estimation in any patient with ataxia and areflexia. Cholestanolosis is also treatable. Although cholestanol estimations are not easily available, tendon xanthomas, cataracts, and a low serum cholesterol level provide suggestive clues.

Increasingly, DNA analysis is applied in the diagnosis of various hereditary cerebellar ataxias. This has led to a tendency to classify the hereditary ataxias, particularly the adult-onset dominant ataxias, by their underlying genetic mutations. Although this carries scientific appeal, such an approach is not very helpful in the clinic when the molecular diagnosis is as yet unknown and difficult to predict simply on the basis of the clinical phenotype. In this chapter, a modification of an earlier clinical classification of the hereditary ataxias will be used (Harding 1984). This is based on the age of onset, and is shown in Table 39.7.

39.3 Congenital early-onset ataxia

The congenital ataxias, or ataxic cerebral palsies, are a rare group of disorders characterized by a congenital neurological syndrome, of which the salient feature is cerebellar ataxia, a static non-progressive course that is an important distinguishing feature from other hereditary and idiopathic cerebellar degenerations, and varying associated clinical features. Some congenital cerebellar disorders such as Dandy-Walker syndrome and Chiari malformations do not cause congenital ataxia (Section 9.2.6). The first signs of congenital ataxia are usually motor delay and hypotonia, followed by cerebellar signs as the child starts to sit and walk. Among the congenital ataxias, there is considerable clinical and pathological heterogeneity. Some patients have a pure cerebellar syndrome of congenital onset, while in most cases there are additional features such as learning disability, spasticity, or other abnormalities (Steinlin 1998). Pathologically, there are several recognized forms of cerebellar hypoplasia affecting the hemispheres or confined to the vermis and, in some cases, the cerebellum is almost totally absent. Similar pathological appearances can be associated with widely different clinical features. For example, cerebellar aplasia may present as a severe neurological disorder during infancy or, surprisingly, as a mild congenital ataxia with clinical presentation in late adult life (Harding 1984). The aetiology of the congenital ataxias is mixed. Probably, about 50 per cent are hereditary, mostly autosomal recessive, but with occasional examples of X-linked and autosomal dominant inheritance (Bundey 1992). The remainder, which are difficult to distinguish clinically, are of unknown and presumably environmental origin. The high incidence of genetic disorders and usually normal perinatal histories among patients with congenital ataxia means that the term 'ataxic cerebral palsy' is preferably avoided.

39.3.1 Miscellaneous hereditary cerebellar hypoplasias

There are numerous clinically or pathologically defined hereditary congenital ataxia syndromes associated with hypoplasia of the cerebellum (Harding 1984). These conditions are rare, and some were described many years ago, before modern neuroimaging and

Table 39.5 Associated neurological/ocular features in ataxic disease

Sign	Association
Cognitive impairment:	
Learning disability	Congenital ataxias, ataxic cerebral palsy, early-onset hereditary degenerations, some metabolic ataxias
Dementia	Late-onset hereditary degenerations; idiopathic degenerations; prion disease; hydrocephalus; some metabolic ataxias; paraneoplastic syndromes; hypothyroidism
Ocular features:	
Retinopathy	Hereditary degenerations (SCA7); vitamin E deficiency; mitochondrial disease; abetalipoproteinaemia; hypobetalipoproteinaemia; hypobetalipoproteinaemia
Retinal angioma	Von Hippel Lindau disease
Optic atrophy	MS; hereditary degenerations; alcoholism; some metabolic ataxias; congenital ataxias; Leber's disease; Behr syndrome
Aniridia	Gillespie syndrome
Cataract	Marinesco–Sjogren syndrome; cholestanolosis
Internuclear ophthalmoplegia	Demyelination; Wernicke's encephalopathy; degenerations; posterior fossa tumour; stroke
Ophthalmoplegia	Degenerations; some metabolic ataxias; mitochondrial disease; hydrocephalus; Wernicke's encephalopathy
Vertical supranuclear gaze palsy	Niemann–Pick disease type C
Ocular apraxia	Ataxia telangiectasia; ataxia with oculomotor apraxia
Ptosis	Mitochondrial disease; degenerations; stroke
Downbeating nystagmus	Foramen magnum compression
Opsoclonus/ocular flutter	Paraneoplastic syndromes; drugs; postinfectious cerebellitis
Extrapyramidal features:	
Parkinsonism	Degenerations
Dystonia/chorea	Degenerations (multiple system atrophy); Friedreich's ataxia; ataxia telagiectasia; some metabolic ataxias; pontocerebellar hypoplasia type 2
Myoclonus	Ramsay Hunt syndrome; prion disease; drugs; dentaterubro-pallidoluysian atrophy
Tremor (essential tremor like)	Degeneration (SCA12)
Hypogonadism	Holmes' ataxia
Deafness	Degenerations; some metabolic ataxias; mitochondrial disease
Tendon xanthomas	Cholestanolosis
Cardiomyopathy	Friedreich's ataxia; vitamin E deficiency; mitochondrial disease
Respiratory irregularity	Joubert syndrome
Stupor/coma	Tumour; stroke; abscess; toxins and drugs; basilar migraine; Wernicke's encephalopathy; some metabolic ataxias
Headache	Tumour; stroke/transient ischaemic attack; abscess; basilar migraine
Pyramidal signs	Congenital ataxia; degenerations; Friedreich's ataxia; autosomal recessive ataxia of Charlevoix–Saguenay; X-linked ataxia; demyelination; foramen magnum compression; hydrocephalus; stroke/transient ischaemic attack; some metabolic ataxias
Areflexia	Degenerations (various hereditary forms and idiopathic); vitamin E deficiency; Wernicke's encephalopathy; some metabolic ataxias; neuropathic ataxia (see text); paraneoplastic; B_{12} deficiency
Muscle fasciculations/wasting	Degenerations (hereditary or idiopathic), especially pontocerebellar hypoplasia type 1; hexosaminidase deficiency; paraneoplastic

metabolic investigations were available. Any distinction between the following disorders and the other eponymous congenital ataxias is somewhat artificial, and many patients do not easily fit into any of these diagnostic entities (ten Donkelaar *et al.* 2003). Cerebellar dysgenesis can also occur in specific genetic disorders with a broader phenotype such as the Cornelia de Lange syndrome (Yamaguchi *et al.* 1999).

◆ Pontoneocerebellar hypoplasia causes cerebellar ataxia, learning disability, spasticity, microcephaly, and sometimes agensis of the corpus callosum. It is likely to be autosomal recessive. Characteristic neuroradiological findings have been reported (Goasdoue *et al.* 2001).

◆ Pontocerebellar hypoplasia is heterogeneous (Barth 2000). In type 1, there is additional anterior horn cell degeneration similar to a spinal muscular atrophy (Rudnik-Schoneborn *et al.* 2003). In type 2, there is additional dystonia (Grosso *et al.* 2002). Another form has been linked to a locus at chromosome 7q11-21 (Rajab *et al.* 2003).

◆ Granule cell layer hypoplasia is associated with congenital cerebellar ataxia of limbs and gait, nystagmus, strabismus, speech delay, short stature, and learning disability (Mathews *et al.* 1989; Pascual-Castroviejo *et al.* 1994). It is probably autosomal recessive.

◆ Various forms of mild congenital cerebellar ataxia with hypoplasia of the cerebellar vermis have been described. There are no

Table 39.6 Other associations with ataxic disease

Feature	Association
Dysmorphism	Some congenital and metabolic ataxias
Scoliosis	Various congenital and early-onset degenerations, especially Friedreich's ataxia
Skin changes:	
Pigmentation; scanty hair	Adrenoleucodystrophy
Telegiectasia	Ataxia telangiectasia
Photosensitivity	Xeroderma pigmentosum; ataxia telangiectasia; Cockayne syndrome; Harnup disease
Tendon xanthomas	Cholestanolosis
Malnutrition/cachexia	Wernicke's encephalopathy (alcoholism; malabsorption [acquired vitamin E deficiency]; sometimes paraneoplastic ataxia
Cardiac disease	Friedreich's ataxia; sometimes mitochondrial disease and vitamin E deficiency
Hypogonadism	Holmes syndrome; adrenoleucodystrophy; Marinesco–Sjogren syndrome

associated neurological features other than mild learning disability, which is not present in all cases. Inheritance can be autosomal dominant (Imamura *et al.* 1993), X-linked, or autosomal recessive.

◆ A more severe syndrome of cerebellar ataxia, learning disability, and spasticity is associated with vermis and cerebellar hemisphere hypoplasia; inheritance is autosomal recessive (al Shahwan *et al.* 1995).

◆ Cerebellar vermis hypoplasia, Oligophrenia, congenital Ataxia, Coloboma, Hepatic fibrocirrhosis, the 'COACH syndrome', presents with hepatic failure in infancy, due to hepatic fibrosis, and a congenital ataxia (Gentile *et al.* 1996). Inheritance is probably autosomal recessive but could be X-linked in some families. A similar disorder has been reported without coloboma (Coppola *et al.* 2002).

Table 39.7 Usual causes of ataxia at different ages

Congenital or early onset (<20 years)	Congenital ataxias	Hereditary Idiopathic
	Hereditary degenerative	Autosomal recessive X-lined
	Metabolic	Intermittent Progressive
	Mitochondrial DNA repair defects	
	Acquired	Infection Paraneoplastic Inflammatory
Adult onset (>20 years)	Hereditary degenerative	Autosomal dominant (SCA) X-linked (mitochondrial)
	Acquired	Sporadic/idiopathic Known cause

◆ An X-linked congenital cerebellar ataxia with ophthalmoplegia has been described, with a disease gene mapped to Xq23 (Illarioshkin *et al.* 1996).

◆ In another severe form of X-linked congenital ataxia with associated learning disability, the affected boys later develop myoclonus and retinal degeneration. A gene locus has been identified at Xp22 (des Portes *et al.* 1996).

39.3.2 **Paine syndrome**

The original description was of an X-linked congenital ataxia with cerebellar hypoplasia, microcephaly, developmental delay, spasticity, myoclonus, seizures, and optic atrophy (Paine 1960). There are other examples of this syndrome, but a clear distinction from various other X-linked cerebellar hypoplasias seems uncertain (Harding 1984).

39.3.3 **Gillespie syndrome**

Gillespie syndrome is characterized by partial aniridia, cerebellar ataxia, and mental retardation. The diagnosis is suggested by the discovery of fixed dilated pupils in a hypotonic infant. The ocular findings are specific to this disorder and are apparent from birth. Neurological involvement includes motor delay, hypotonia, disabling ataxia, and learning disability. There is cerebral and cerebellar atrophy, and white matter changes may be seen with MR scanning (Nelson *et al.* 1997). Gillespie syndrome seems to be genetically heterogeneous with both autosomal recessive and autosomal dominant forms.

39.3.4 **Marinesco–Sjogren syndrome**

This is a rare autosomal recessive disorder causing motor delay, cerebellar ataxia, cataract, and learning disability. In addition, there is often short stature, and tendon reflexes may be absent, normal, or brisk; plantar responses are usually extensor. In addition to the obvious central nervous system and ocular features, patients may have a peripheral neuropathy (Zimmer *et al.* 1992; Muller-Felber *et al.* 1998) and myopathy with fibre necrosis, atrophy, and rimmed vacuole formation (Sasaki *et al.* 1996). There have also been reports of hypogonadism (McLaughlin *et al.* 1996). MR imaging shows cerebellar atrophy with pituitary hypoplasia and white matter abnormalities (Georgy *et al.* 1998). Characteristic changes are seen in conjunctival biopsies (Zimmer *et al.* 1992). This is likely to be a lysosomal disorder, and inheritance is autosomal recessive (Lagier-Tourenne *et al.* 2003). Mutations of the SIL gene on chromosome 5q31 have been detected in Marinesco–Sjogren syndrome (Anttonen *et al.* 2005; Senderek *et al.* 2005).

39.3.5 **Joubert syndrome**

In this rare but striking autosomal recessive disorder, cerebellar vermis agenesis is associated with hypotonia and developmental delay, ataxia, severe learning disability, abnormal eye movements, and irregular respiration (Saraiva *et al.* 1992). The respiratory pattern is intermittent hyperpnoea alternating with apnoea; oculomotor abnormalities include nystagmus, slow saccades, and impaired pursuit movements (Maria *et al.* 1997). In a subgroup of families, the syndrome includes a combination of retinal dystrophy and renal cysts, suggesting possible genetic heterogeneity. Other associations include ocular colobomas, polydactyly, lingual tumours, and hepatic fibrosis (Lewis *et al.* 1994; Pellegrino *et al.* 1997). There are similarities with the 'COACH syndrome' (see Section 39.3.1). Cerebellar vermis agenesis and brainstem abnormalities are seen with MR imaging including the 'molar tooth sign' with vermian dysplasia (Romano *et al.* 2006). Affected children may die in infancy with severe respiratory abnormalities and developmental failure; others survive into adolescence with variable disability (Steinlin *et al.* 1997). Several autosomal recessive forms of Joubert's syndrome have been identified with mutations of the AHI1 (Parisi *et al.* 2006; Valente *et al.* 2006a), CEP290 (Sayer *et al.* 2006; Valente *et al.* 2006b), and *NPHP1* genes (Castori *et al.* 2005).

39.3.6 The disequilibrium syndrome

This syndrome has been described mainly in Scandinavia (Sanner 1973) but also in the Hutterite population of Canada (Glass *et al.* 2005). The affected children are grossly hypotonic with clumsiness and delayed motor milestones; a disabling congenital ataxia becomes apparent, and the child is unable to stand without falling. Walking is eventually achieved by about 10 years but is always abnormal. Speech delay and learning disability are common, and cataracts have been described. Neuroimaging studies may reveal cerebellar vermis hypoplasia. Inheritance is autosomal recessive.

39.4 Early-onset hereditary degenerative ataxias

The early-onset hereditary degenerative ataxias, of which Friedreich's ataxia is the prototype, are conveniently grouped together because they usually start in childhood, adolescence, or early adult life, normally before the age of 25 years. They are, therefore, distinguished from the congenital ataxias and the late-onset degenerative ataxias, which usually start after 25 years. It must be realized that sometimes these conditions present after 25 years and so the age of onset, which can be difficult to gauge, is only an approximate guide to diagnosis. However, even with the advent of molecular genetic classification of these disorders, a clinically based classification remains a useful approach and will be retained here. The term 'early-onset cerebellar ataxia' is widely used, but is potentially misleading in situations where the ataxia is largely spinal, as in Friedreich's ataxia.

39.4.1 Friedreich's ataxia

Friedreich's ataxia is an autosomal recessive disorder causing a degenerative ataxia that is principally of spinal origin. It is the most common of the degenerative ataxias, with a frequency in the population of approximately 1 in 50 000 (Harding 1984). Although rigorous clinical diagnostic criteria have been established, recent molecular genetic studies have shown that clinical variability is actually greater than previously suspected (Pandolfo 2003). The disorder is caused by an unstable trinucleotide GAA expansion, or occasionally a point mutation, of gene X25 located on chromosome 9q13. This gene encodes a 210 amino acid mitochondrial protein of unknown function, frataxin (Bradley et al 2000).

Pathology

The major findings are in the spinal cord where there is degeneration of the dorsal columns and spinocerebellar tracts and also of the pyramidal tracts (Harding 1984). The former is worst at the cervical level (Fig. 39.2), whereas the latter is more marked at the lumbar level. These findings suggest a distal axonopathy. In addition, there is loss of dorsal root ganglion cells with the depletion of large myelinated fibres in peripheral nerves. The dorsal roots are atrophic. Interestingly, the earliest changes seen in young children are in peripheral nerves where there is large fibre loss, suggesting damage to dorsal root ganglion sensory neurones. Neuropathological changes in the cerebellum, brainstem or cerebrum are absent or minimal.

Clinical features

In the classical form of Friedreich's ataxia, onset is usually between 8 and 16 years, but the range is wide, and one large series reported a mean of 10.5 years with 95 per cent confidence limits of 0–25 years (Harding 1981a). Almost all cases present with progressive gait ataxia but, occasionally, scoliosis or cardiac disease precede neurological involvement (Tsao *et al.* 1992). At presentation, there is also lower limb areflexia and electrophysiological evidence of a sensory axonal peripheral neuropathy. Additional signs including generalized areflexia, extensor plantar responses, dysarthria, pyramidal weakness of the legs, and loss of joint position and vibration sense appear later, usually within 5 years and almost always by 10 years from onset (Harding 1984). These core features are essential for the diagnosis of the classical form of Friedreich's ataxia (Table 39.8).

Scoliosis occurs in approximately 80 per cent of cases and can be severe in about 10 per cent of these patients. This tends to worsen at puberty, with increasing growth and can lead to pain and cardiorespiratory complications. Cardiomyopathy is common but not invariable. An abnormal electrocardiogram is seen in 65 per cent of cases (Harding *et al.* 1983), but isolated electrocardiogram recordings are normal in 25 per cent (Fig. 39.3). Echocardiography reveals

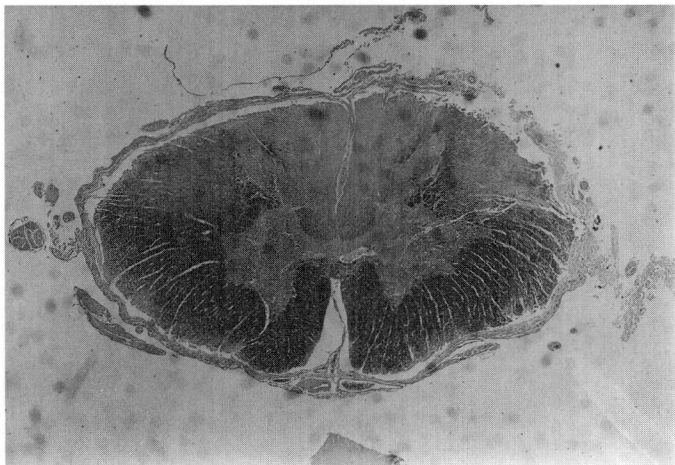

Fig. 39.2 The cervical spinal cord in Friedreich's ataxia (myelin stain) showing degeneration of the posterior columns. (Courtesy of Dr J Broome, Walton Centre, Liverpool.)

Table 39.8 Clinical features of Friedreich's ataxia

Classical Friedreich's ataxia		
Almost all cases	<5 years from onset	Progressive ataxia of gait
		Lower limb areflexia
		Extensor plantar responses
		Sensory axonal neuropathy detected on nerve conduction tests
	5–10 years from onset	Leg weakness
		Reduced vibration and joint position sense in lower limbs
		Dysarthria
		Generalized areflexia
Variable features	in >50%	Cardiomyopathy on ECG
		Scoliosis
	in <50%	Nystagmus
		Optic atrophy
		Deafness
		Diabetes
		Pes cavus
		Distal weakness and wasting
Variant Friedreich's ataxia		Late-onset disease (25–50 years)
		Mild, slowly progressive (Acadian) Friedreich's ataxia
		Lower limb spasticity and hyperreflexia in Friedreich's ataxia
		Movement disorders

hypertrophic changes or dilated cardiomyopathy in approximately 60 per cent of cases (Gunal *et al.* 1996). In patients who undergo continuous electrocardiogram monitoring, echocardiography, or nuclear ventriculography, the incidence of cardiac abnormalities is even higher. In some patients, the electrocardiogram becomes abnormal years after onset, even when earlier records have been normal. In contrast to the high frequency of electrocardiogram abnormalities, clinically apparent cardiomyopathy with signs of cardiac failure or evidence of arrhythmia is uncommon, probably developing in less than 15 per cent of cases (Harding *et al.* 1983).

There are several less common features of Friedreich's ataxia. Optic atrophy is seen in approximately 25 per cent of cases, but only in 5 per cent of patients is visual acuity significantly reduced. Nystagmus is seen in 20 per cent, but other forms of disordered eye movement such as jerky pursuit movements, dysmetric saccades, square wave jerks, and failure of vestibulo-ocular reflex suppression are more common (Moschner *et al.* 1994). Deafness develops in 10 per cent of cases. Distal wasting is seen in about half of the cases but is usually not associated with weakness. Diabetes develops in 10 per cent of patients and usually requires insulin therapy. In some patients, truncal ataxia makes sitting or standing still difficult, and there is a motor restlessness which may be mistaken for chorea (Harding 1984); this may occur very early in the disease. Another common feature of advanced Friedreich's ataxias the development of cold, cyanosed, and oedematous feet. Subtle cognitive and behavioural changes have been described, but these are usually subclinical (Mantovan *et al.* 2006).

The prognosis of classical Friedreich's ataxias is variable. On average, patients lose the ability to walk after 15 years from onset; over half of the cases are so affected by the age of 26 years and 95 per cent by 44 years of age. A few patients continue to walk independently into their fifties but this is unusual. Age at death is also highly variable, depending on associated features such as cardiac disease and diabetes (De Michele *et al.* 1996); some patients survive into the sixth and seventh decades.

The relationship between cases of atypical Friedreich's ataxia and the classical phenotype has long been controversial (Harding 1984). Variant forms have been identified:

♦ *Late-onset Friedreich's ataxia* develops after the age of 25 years, often in the fourth decade but sometimes as late as 51 years.

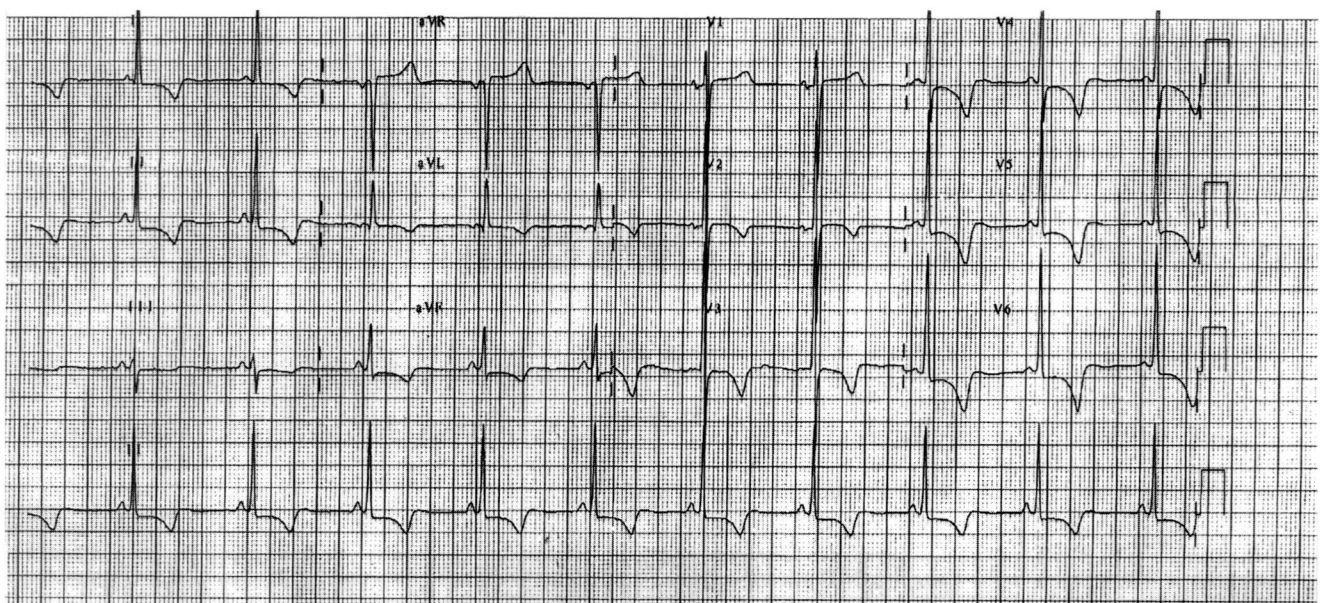

Fig. 39.3 The electrocardiogram in a case of Friedreich's ataxia. Note the widespread T wave abnormalities.

There is slower progression and a lower incidence of skeletal deformity (De Michele *et al.* 1994) and cardiomyopathy (Bhidayasiri *et al.* 2005). In these families, genetic linkage to the Friedreich's ataxia locus has been established (Klockgether *et al.* 1993; De Michele *et al.* 1994), and the GAA expansion mutation has now been detected in such patients who tend to have lower repeat numbers (Montermini *et al.* 1997b). There may be cerebellar atrophy on brain MR scans (Bhidayasiri *et al.* 2005). Very-late-onset Friedreich's ataxia after the age of 40 years may be mistaken for cerebellar multiple system atrophy; thus, Friedreich's ataxia should be considered within the differential diagnosis of 'idiopathic' ataxia, even at this age (Berciano *et al.* 2005).

◆ *Friedreich's ataxia with retained reflexes* differs from classical Friedreich's ataxia because of retained or increased reflexes, lower limb spasticity and extensor plantar responses. These features were previously regarded as incompatible with the diagnosis of Friedreich's ataxia and such patients were thought to have a separate condition: early-onset ataxia with retained reflexes (Harding 1984). However, some such patients have cardiomyopathy and, in some families, genetic linkage to the FA 9q13 locus was established (Palau *et al.* 1995; Klockgether *et al.* 1996). These patients also have the Friedreich's ataxia GAA mutation (Lamont *et al.* 1997) but repeat lengths seem similar to those seen in classical Friedreich's ataxia (Montermini *et al.* 1997b). Some patients have affected siblings with classical Friedreich's ataxia but have identical GAA repeat mutations indicating the role of other genes or environmental factors in Friedreich's ataxia expression (Armani *et al.* 2006).

◆ *Paraparesis* has been the presentation in some patients (Wilkinson *et al.* 2001) or tetraparesis without ataxia or neuropathy (Labauge 2002).

◆ *Acadian Friedreich's ataxias* is confined to a population of French descent living in the southern United States and Canada. The age of onset is similar, but progression is slower, with a lower incidence of cardiomyopathy. These patients sometimes have lower limb spasticity and extensor plantar responses. Linkage to the Friedreich's ataxia locus has been established (Richter *et al.* 1996), and the GAA repeat expansion has been confirmed in Acadian Friedreich's ataxia (Montermini *et al.* 1997b).

◆ *Other movement disorders*. Recently, the Friedreich's ataxia GAA expansion has been detected in patients presenting with ataxia and movement disorders such as generalized chorea (Hanna *et al.* 1998), myoclonus (Zhu *et al.* 2002), and dystonia (Hou *et al.* 2003). The frequency of this 'movement disorders' phenotype of Friedreich's ataxia is unclear. Motor restlessness has been noted in Friedreich's ataxia and attributed to imbalance rather than true chorea (Harding 1984).

Genetics

Friedreich's ataxia is inherited as an autosomal recessive trait. The carrier frequency is approximately 1 in 110, and parental consanguinity is present in 5–10 per cent of cases (Bundey 1992). Accordingly, the risk to siblings of affected children is 1 in 4. In 1988, the FA gene locus was established on chromosome 9q13 (Chamberlain *et al.* 1988). Subsequently, a gene, *X25*, was identified that contained a homozygous expansion of a GAA trinucleotide repeat sequence within the first intron of the gene in 96–97 per cent of patients (Campuzano *et al.* 1996). The few remaining patients were compound heterozygotes with one expanded allele

and a point mutation of the other (De Castro *et al.* 2000). In normal individuals, the gene contains 7–29 GAA repeats, increasing to between 66 and 1700 repeats in affected individuals (Durr *et al.* 1996; Schols *et al.* 1997). Larger GAA repeat numbers are associated with earlier age of onset (Mateo *et al.* 2003), more severe areflexia, and increased incidence of cardiomyopathy, optic atrophy, and deafness, while smaller expansions are likely to result in milder, later-onset disease (Durr *et al.* 1996; Gellera *et al.* 1997; Isnard *et al.* 1997; Montermini *et al.* 1997b). Although there is some correlation between larger repeat size and greater severity, this is not always reliable—some very large GAA repeat sizes (>800) can be associated with very-late-onset disease (Bidichandani *et al.* 2000) and there can be considerable phenotypic variability within families despite identical GAA repeat sizes (Armani *et al.* 2006). It has been shown that the smaller of the two frataxin gene GAA alleles is more predictive of some, but not all, clinical features such as age of onset (Mateo *et al.* 2004), while other aspects of Friedreich's ataxia are a feature of disease duration (Santoro *et al.* 2000). The mutation is unstable during meiosis and mitosis, leading to variability in repeat size both within the same family and in different tissues from the same individual. Somatic mosaicism has been noted within the nervous system but does not appear to explain the pattern of neuropathological changes (Montermini *et al.* 1997a). However, one patient with mild neurological features was found to have much smaller repeat sizes in peripheral nerve than in blood (Machkhas *et al.* 1998), indicating that somatic mosaicism might influence the Friedreich's ataxia phenotype.

The majority of parents of affected children are heterozygous for the GAA expansion but occasionally, one parent carries a premutation, a large allele of intermediate size (40–100 repeats), which has undergone further expansion during meiosis (Cossee *et al.* 1997; Delatycki *et al.* 1998). There are rare reports of patients inheriting Friedreich's ataxia from an affected parent (McGovern *et al.* 2000). Some of these are explained by misdiagnosis, but others arise when one parent has Friedreich's ataxia and the other is a heterozygote. The risk of such pseudodominant transmission to offspring is 1 in 220, based on a heterozygote frequency of 1 in 220 and a 1 in 2 risk of the healthy but heterozygous parent transmitting the expanded rather than the normal allele (Harding *et al.* 1981). Such pseudodominant transmission has been confirmed by direct mutation analysis (McGovern *et al.* 2000) but in other families a very mildly affected parent has been heterozygous for the GAA expansion (Lamont *et al.* 1997). These affected parents had a late onset, after the age of 40, of ataxia and dysarthria but preserved reflexes. Point mutations of the frataxin gene or other ataxia mutations were excluded and it is possible that in such situations, both parents are heterozygous but the 'affected' parent is a manifesting heterozygous carrier of the Friedreich's ataxia gene.

The gene product, frataxin, is a 210 amino acid protein. In Friedreich's ataxia the level of frataxin mRNA is reduced, leading to a cellular deficiency of the protein (Bidichandani *et al.* 1998). Frataxin is a mitochondrial protein (Koutnikova *et al.* 1997; Priller *et al.* 1997) which is involved in mitochondrial iron homeostasis (Babcock *et al.* 1997). Frataxin deficiency leads to intramitochondrial iron accumulation (Adamec *et al.* 2000) with abnormal mitochondrial respiration and regulation of mitochondrial DNA (Rotig *et al.* 1997; Wilson *et al.* 1997; Wilson *et al.* 1998). In cardiac muscle, there is reduced mitochondrial respiratory function with mitochondrial DNA depletion and increased oxidative stress as gauged by lowered aconitase levels (Bradley *et al.* 2000). Skeletal muscle shows similar but milder changes, but cerebellar and dorsal root ganglion tissue does not. Such abnormal muscle respiratory activity can be measured in vivo using magnetic resonance spectroscopy (Vorgerd *et al.* 2000) and has been used in clinical trials of anti-oxidative therapy. The clinical similarities between Friedreich's ataxia and some of the mitochondrial encephalomyopathies in terms of neurological involvement, cardiomyopathy, and deafness are consistent with the notion of disordered mitochondrial function (Calabrese *et al.* 2005).

Occasional patients with otherwise typical Friedreich's ataxia do not have GAA expansions within the frataxin gene (Schols *et al.* 1997). In these cases, some may have point mutations of both alleles but in others there is probably involvement of separate genes.

Investigations and differential diagnosis

Friedreich's ataxia should be strongly suspected in children, adolescents or young adults presenting with ataxia, areflexia, and extensor plantar responses. In the light of recent molecular genetic studies, the diagnosis must be considered in those with later age of onset than has hitherto been regarded as typical, as well as in patients with lower limb spasticity and retained reflexes. Features against the diagnosis include apparent autosomal dominant inheritance, dementia, Parkinsonism or dystonia, ophthalmoplegia, or a congenital onset. In the outpatient clinic, an electrocardiogram is both easily available and very useful; if abnormal, with evidence of T wave changes or ventricular hypertrophy, the diagnosis is very likely. It should be noted, however, that ataxia due to isolated vitamin E deficiency (Cavalier *et al.* 1998) or mitochondrial disease (Chinnery *et al.* 1997) can be associated with cardiomyopathy and that the electrocardiogram may be normal in Friedreich's ataxia.

Nerve conduction studies reveal normal conduction velocities but small or absent sensory nerve action potentials. This is an important means of distinguishing Friedreich's ataxia in which dysarthria and extensor plantar responses have not yet appeared, from early type 1 hereditary motor and sensory neuropathy in which there is also ataxia with areflexia and pes cavus but usually autosomal dominant inheritance and conduction velocities slowed to <40 m/s. In some patients, the nerve conduction studies can more closely resemble hereditary motor and sensory neuropathy, and DNA testing is then essential (Panas *et al.* 2002). It is essential to exclude vitamin B12 deficiency in which ataxia, areflexia, and extensor plantar responses are cardinal signs of subacute combined degeneration. Vitamin E deficiency can cause a neurological phenotype indistinguishable from Friedreich's ataxia but is treatable. It must be excluded by vitamin E estimation, especially in patients who appear to have Friedreich's ataxia, even with cardiomyopathy, but in whom DNA studies do not reveal a Friedreich's ataxia mutation (Hammans *et al.* 1998). Visual evoked potentials are frequently abnormal, even in those without clinically obvious optic atrophy and pyramidal tract dysfunction may be revealed by central motor conduction studies (Claus *et al.* 1988). CT and MR scans usually show normal cerebellar anatomy although slight atrophy of the vermis and medulla may be seen in late-onset or advanced Friedreich's ataxia (Ormerod *et al.* 1994). In contrast, the cervical cord is atrophic on MRI (Fig. 39.4). The presence of marked cerebellar atrophy, especially in the early stages of an ataxic illness, makes Friedreich's ataxia unlikely.

Increasingly, the diagnosis of Friedreich's ataxia is made by direct mutation screening of DNA samples to detect the GAA trinucleotide

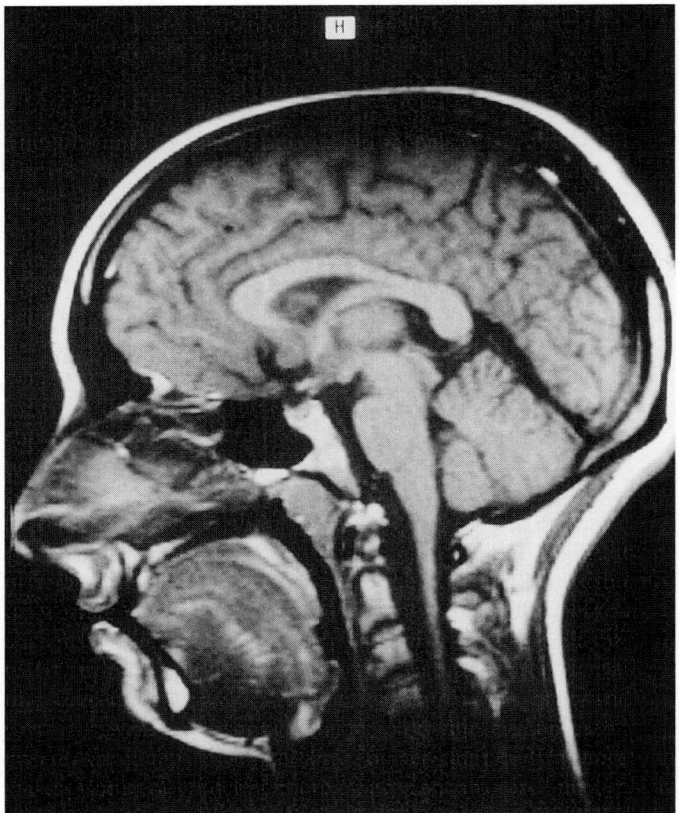

Fig. 39.4 Magnetic resonance scan of a patient with Friedreich's ataxia showing cervical spinal cord atrophy with preserved cerebellar anatomy.

expansion (Lamont *et al.* 1997). This prevents the need for neurophysiological and neuroradiological investigations, but a negative result should be followed up with additional investigations. The most important considerations are treatable conditions such as ataxia with vitamin E deficiency and vitamin B12 deficiency; laboratory tests for these disorders are mandatory in this situation. In addition, α-fetoprotein estimation and DNA testing for dominant spinocerebellar, 'SCA', mutations such as SCA3, which can sometimes present earlier than usual, resembling FA, should be considered (McCabe *et al.* 2000). Investigations for anti-gliadin antibodies, with or without coeliac disease, and testing for anti-GAD antibodies may resolve a few of these mutation-negative Friedreich's ataxia cases, but some are unexplained, indicating a degree of genetic heterogeneity within the phenotype.

If the patient is heterozygous for the GAA repeat with one expanded and one normal-length allele, there are several possibilities to consider. The patient may be one of the 3–4 per cent of Friedreich's ataxia cases who are compound heterozygotes with a GAA expansion and a point mutation of the other frataxin gene (De Castro *et al.* 2000). Alternatively, some cases with very late onset of mild ataxia may be manifesting carriers. However, unless compound heterozygosity can be verified, alternative diagnoses should be excluded by appropriate investigation.

Treatment

There is no effective treatment for Friedreich's ataxia. Various studies have reported negative results with a range of compounds.

The finding of increased intramitochondrial iron in mitochondria with defective frataxin function has led to the suggestion that iron chelation may be effective. However serum iron and ferritin are normal (Wilson *et al.* 1998), and the benefits of such an approach seem uncertain. It is important to discuss the prognosis openly with the patient, and genetic counselling is vital for both patients and their families. Physiotherapy and occupational therapy are important, along with suitable seating for those who have lost the ability to walk. Spasticity is treated with physiotherapy and baclofen, if needed; severe spasticity has been treated with intrathecal baclofen (Ben Smail *et al.* 2005). Orthopaedic advice is essential for those with scoliosis and foot deformity along with early diagnosis and treatment of cardiac complications and diabetes. Clinical observation for cardiac symptoms is probably more important and useful than regular cardiac screening investigations (English *et al.* 2006). Selected patients will require hearing aids and low visual aids. Dependent oedema and cyanosis of the feet can be very troublesome; elevation of the feet and prevention of secondary contractures are probably more helpful than diuretics. Orthopaedic surgery for equinovarus deformity can improve functional mobility in some cases (Delatycki *et al.* 2005). Friedreich's ataxia patients are often worse after a period of bed-rest during illness or after surgery; indeed, the diagnosis may first become apparent at such a time. It is important to encourage the earliest possible mobilization in these circumstances.

Experimental therapies have included L-carnitine, which may improve cellular energy metabolism, but no clinical endpoints have been assessed (Schols *et al.* 2005), and this intervention may be too late in the disease (Schols *et al.* 2004b). A trial of vitamin E with coenzyme Q10 improved cardiac and, to a lesser extent, skeletal muscle energy metabolism but without significant clinical effects (Cooper *et al.* 2003). Idebenone, another antioxidant agent, was ineffective (Buyse *et al.* 2003).

39.4.2 Early-onset ataxia with retained reflexes

This disorder is probably half as common as FA (Chio *et al.* 1993), from which it has previously been distinguished on clinical criteria (Harding 1984). The age of onset is similar, but progression is slower, with patients becoming wheelchair dependent after about 30 years from onset, in contrast to a mean of 15 years in Friedreich's ataxia. Upper limb and knee reflexes are preserved, and there is often lower limb spasticity. Skeletal deformity occurs but is usually mild. One series (Harding 1984) found no evidence of cardiomyopathy, optic atrophy, or diabetes, but more recent studies have detected these features in some cases (Klockgether *et al.* 1991; Montermini *et al.* 1997b). Neurophysiological studies reveal peripheral neuropathy in some but not all patients (Klockgether *et al.* 1991; Santoro *et al.* 1992). The frequency of skeletal deformity, sensory changes, distal wasting, and cardiomyopathy may be lower than in FA (Klockgether *et al.* 1991) but this is uncertain. Cerebellar atrophy is seen on MRI in some patients, but is absent in others (Klockgether *et al.* 1991; Ormerod *et al.* 1994). These findings indicate that early-onset cerebellar ataxia with retained reflexes is heterogeneous. Clinical genetic studies have suggested autosomal recessive inheritance, with increased parental consanguinity but slightly lower segregation ratios in siblings than the 0.25 expected (Filla *et al.* 1990; Chio *et al.* 1993).

It is now clear that some of these patients have Friedreich's ataxia (Espinos-Armero *et al.* 2005) but others do not (Geschwind *et al.* 1997b).

The Friedreich's ataxia with retained reflexes subgroup of Friedreich's ataxia (see Section 39.4.1) is more common with later onset (Bhidayasiri *et al.* 2005) but not exclusively. DNA testing for frataxin mutations is indicated in all patients with this presentation. The aetiology of the non-Friedreich's ataxia cases is unclear. However, the distinction from classical Friedreich's ataxia is important in terms of prognosis.

Some patients with early-onset ataxia with retained reflexes have the autosomal recessive spastic ataxia of Charlevoix–Saguenay (Section 39.4.3).

39.4.3 Autosomal recessive spastic ataxia of Charlevoix–Saguenay

This rare ataxia is seen mainly in Canada (Bouchard *et al.* 1998), but has also been described in Japan (Takiyama 2006). There is onset of ataxia and spasticity at the time the child starts to walk with subsequent progression. Additional signs include abnormal retinal myelination around blood vessels, dysarthria, peripheral neuropathy, and distal wasting. There is vermis atrophy on brain MR scans and slow nerve conduction velocities on nerve conduction tests (Bouchard *et al.* 1998). A gene encoding sacsin has been mapped to chromosome 13q11, in which several mutations have been detected (Engert *et al.* 2000; Grieco *et al.* 2004; Ouyang *et al.* 2006).

39.4.4 Early-onset ataxia with myoclonus

The combination of cerebellar ataxia with myoclonus is referred to as progressive myoclonic ataxia or the Ramsay Hunt syndrome. This confusing and highly heterogeneous syndrome is discussed fully in Section 40.7.8. Several neurometabolic disorders may present in this way, including mitochondrial disorders, neuronal ceroid lipofuscinosis, and sialidosis. Other cases are due to coeliac disease or Whipple's disease. An autosomal recessive disorder, Unverricht-Lundborg disease, also referred to as Baltic or Mediterranean myoclonus, is probably responsible for most otherwise unexplained cases in children, while autosomal dominant dentatorubropallidoluysian atrophy is discussed in Section 39.9.6.

39.4.5 Early-onset ataxia with hypogonadism, Holmes ataxia

This rare autosomal recessive disorder is heterogeneous. Onset is variable up to the age of 40 but usually before 30 years of age and so there can be overlap with the later-onset ataxias (Gironi *et al.* 2004). The ataxia is progressive, leading to severe disability after 15–20 years (Holmes 1907). Dysarthria is common, along with variable nystagmus, dementia, skeletal deformities, tremor, and peripheral neuropathy (De Michele *et al.* 1993); tendon reflexes may be absent or brisk and plantar responses are sometimes extensor. (Seminara *et al.* 2002). Pigmentary retinopathy has been described in a few cases. Neuroimaging studies reveal cerebellar atrophy and white matter lesions. The hypogonadism is clinically obvious from puberty onwards and is also heterogeneous, with hypogonadotrophic hypogonadism in some cases (Seminara *et al.* 2002) while others have gonadal failure (De Michele *et al.* 1993). Mitochondrial disorders may underlie some cases as there have been descriptions of this phenotype with cytochrome c oxidase (De Michele *et al.* 1993) and coenzyme Q10 deficiencies (Gironi *et al.* 2004). A disorder with similar clinical features has been described as the Boucher–Neuhauser syndrome (Limber *et al.* 1989). Cerebellar ataxia has occasionally been described in the Lawrence–Moon–Barted–Biedl syndrome, in which hypogonadism is a major feature, but this disorder is exceptionally rare, and the clinical features are distinctive (Green *et al.* 1989). Hypogonadism and ataxia also coexist in some cases of the Marinesco–Sjogren syndrome (McLaughlin *et al.* 1996) (see Section 39.3.4).

39.4.6 Early-onset ataxia with retinopathy

There have been several reports of cerebellar ataxia in association with pigmentary retinopathy. Harding (1984) reviewed these and grouped them together as a subtype of early-onset ataxia. The group is clearly heterogeneous, and the majority of cases are probably autosomal recessive. Additional clinical features have included dementia, deafness, peripheral neuropathy, and pyramidal signs. Many of these cases were described before modern biochemical investigations were available, and the diagnosis in some is questionable. Retinopathy is a feature of several metabolic ataxias, including abetalipoproteinaemia (Section 39.5.2), hypobetalipoproteinaemia (Matsuo *et al.* 1994), and other forms of acquired vitamin E deficiency (Harding 1987), as well as autosomal recessive ataxia with isolated vitamin E deficiency (Yokota *et al.* 1997) (Section 39.5.2). Retinopathy and ataxia can be prominent features of mitochondrial disease (Chinnery *et al.* 1997; Vedanarayanan 2003), especially the Kearns-Sayre (Holt *et al.* 1989) and NARP syndromes (Holt *et al.* 1990) (see Section 39.6). Finally, it should be noted that autosomal dominant cerebellar ataxia type II, due to the SCA7 mutation, sometimes presents in childhood, causing ataxia and retinopathy (see Section 39.9).

39.4.7 Early-onset ataxia with optic atrophy, Behr and Costeff syndromes

This subgroup of early-onset ataxia was suggested by Harding (Harding 1984), but is highly heterogeneous. Optic atrophy can be associated with classical and variant forms of Friedreich's ataxia (Section 39.4.1) and may also occur in combination with cerebellar signs in X-linked ataxia, leucodystrophies, and various forms of mitochondrial disease, including Leber's disease (Funakawa *et al.* 1995; Murakami *et al.* 1996), biotin deficiency (Rahman *et al.* 1997), neuronal ceroid lipofuscinosis, or demyelinating disease. Ataxia and optic atrophy are also features of the Wolfram syndrome (Scolding *et al.* 1996).

There remain, however, otherwise unexplained and apparently autosomal recessive examples of this syndrome (Harding 1984; Bundey 1992). The *Behr syndrome* comprises cerebellar ataxia, optic atrophy, spasticity, and mild learning disability; other features such as peripheral neuropathy and deafness have also been described. Inheritance is autosomal recessive and similar families have been reported by several authors (see (Harding 1984; Bundey 1992). Recently, patients with Behr syndrome have been discovered to have 3-methylglutaconic aciduria (Costeff *et al.* 1993; Elpeleg *et al.* 1994). Ataxia with optic atrophy and spasticity can be one form of a wide spectrum of phenotypes caused by methylglutaconic aciduria, type 3 or Costeff syndrome, and is caused by mutations of the *OPA3* gene on chromosome 19q13.2-q13.3 (Anikster *et al.* 2001).

39.4.8 Early-onset ataxia with deafness

A few families with ataxia and deafness have been reported (Harding 1984; Bundey 1992), and this is probably a heterogeneous

mitochondrial or autosomal recessive disorder (Pratap-Chand *et al.* 1995). Deafness, cardiomyopathy, retinopathy, and ataxia are common features of mitochondrial disorders, especially if there are other features such as short stature, ptosis, myopathy, or lactic acidosis (Chinnery *et al.* 1997). The May–White syndrome of ataxia, myoclonus, and deafness, a variant of the MERRF syndrome (see Section 40.7.8), has now been established as a mitochondrial disorder (Vaamonde *et al.* 1992).

39.4.9 Early-onset ataxia with extrapyramidal features

In contrast to the later-onset ataxias, the combination of ataxia with Parkinsonism, dystonia, or chorea is very uncommon. Some overlap in age of onset with the later-onset ataxias, such as type I autosomal dominant cerebellar ataxia or dentatorubropallidoluysian atrophy, accounts for some of these cases, but the olivopontocerebellar atrophy variant of multiple system atrophy (see Chapter 40.3.7) does not present in childhood or adolescence. Chorea has been described with cerebellar ataxia and other features in mitochondrial disease (Nelson *et al.* 1995). Friedreich's ataxia can cause extrapyramidal features in some patients.

39.4.10 Ataxia with oculomotor apraxia

Ataxia with oculomotor apraxia is an autosomal recessive disorder with ataxia, oculomotor apraxia with head thrusting, and sometimes chorea or dystonia (Nemeth *et al.* 2000; Barbot *et al.* 2001). There are similarities to ataxia telangiectasia but no skin changes, immunodeficiency, or DNA fragility. There are two forms:

◆ *Ataxia with oculomotor apraxia type 1* is characterized by cerebellar ataxia, axonal sensorimotor neuropathy, and cognitive decline (Le Ber *et al.* 2003b). There is considerable variability; many patients have oculomotor apraxia (86 per cent), chorea (79 per cent), hypoalbuminaemia (83 per cent), and hypercholesterolaemia (75 per cent). The chorea movements tend to be prominent early in the illness and reduce later but can dominate the condition with the A198V mutation. The neuropathy can be more prominent in adults. MR scanning shows cerebellar atrophy. Type 1 is caused by various mutations of the aprataxin, *APTX*, gene on chromosome 9p13 (Shimazaki *et al.* 2002). This condition has been referred to as 'early-onset ataxia with oculomotor apraxia and hypoalbuminemia' (Date *et al.* 2001).

◆ *Ataxia with oculomotor apraxia type 2* is linked to a gene on chromosome 9q34 (Nemeth *et al.* 2000). In this condition, there is a more consistent phenotype with ataxia, peripheral neuropathy (in 92 per cent), chorea, or dystonia (in 44 per cent) and less consistent oculomotor apraxia (in 56 per cent). Alpha fetoprotein is increased in all patients (Le Ber *et al.* 2004), and creatine kinase may also be increased. Type 2 is caused by various mutations of the senataxin, *SETX*, gene (Criscuolo *et al.* 2006). It accounts for about 8 per cent of non-Friedreich's early-onset cerebellar ataxia, indicating that it is the second most common cause of ataxia in children (Le Ber *et al.* 2004).

39.4.11 X-linked early-onset ataxia

X-linked cerebellar ataxia is very rare but has been reported by several authors (Harding 1984). Onset is usually in childhood or adolescence and the salient features are a spastic paraparesis, upper limb ataxia, dysarthria, and nystagmus. In some, ankle jerks were absent with neurophysiological evidence of a peripheralneuropathy.

There is also a more severe form of childhood-onset x-linked ataxia with spastic paraparesis (Apak *et al.* 1989). In other families, the prognosis is better with a slowly progressive pure cerebellar syndrome (Lutz *et al.* 1989). A rapidly fatal childhood-onset ataxia is associated with visual loss, deafness, and recurrent infections (Arts *et al.* 1993). Female carriers have ataxia and mild lower limb spasticity (Verhagen *et al.* 1996). Some of these patients have the rare autosomal recessive spastic ataxia of Charlevoix–Saguenay (Section 39.4.3), but the syndrome of X-linked early-onset spastic ataxia seems to be heterogeneous (Apak *et al.* 1989). Other families with additional cataracts (Guo *et al.* 2006) or sideroblastic anaemia (Maguire *et al.* 2001) have been reported. The latter is probably a mitochondrial disorder (Hellier *et al.* 2001) and is associated with missense mutation in the *ABC7* gene (Maguire *et al.* 2001). The Allan–Herndon–Dudley syndrome of spasticity, ataxia, and dysmorphism (Schwartz *et al.* 2005) is caused by mutations of the *MCT8* gene on chromosome Xq13.2.

It is important to exclude adrenoleucodystrophy in families presenting with an unexplained X-linked ataxic or paraplegic illness (Vianello *et al.* 2005); the same applies to sporadic male cases. Pelizaeus–Merzbacher disease, X-linked leucodystrophy, can also present with spasticity and ataxia in males (Golomb *et al.* 2004).

39.5 Metabolic ataxias

39.5.1 Intermittent metabolic ataxias

Hyperammonaemias

Hyperammonaemia, usually due to metabolic defects of the urea cycle, is the most common cause of intermittent metabolic ataxia. Various enzyme deficiencies may be involved. The most common is ornithine transcarbamylase deficiency, an X-linked disorder in which affected boys develop episodes of ataxia and encephalopathy with elevated ammonia levels (Schwarz *et al.* 1999). Female carriers may also be affected (Pridmore *et al.* 1995). The other urea cycle defects are autosomal recessive, including argininosuccinate synthetase deficiency, argininosuccinase deficiency, or argininosuccinic aciduria (Gerrits *et al.* 1993) and arginase deficiency. Some cases are due to the hyperornithinemia, hyperammonemia, and homocitrullinuria syndrome caused by mutations of the *ORNT1* and *ORNT2* mitochondrial ornithine transporter genes (Miyamoto *et al.* 2001; Salvi *et al.* 2001; Camacho *et al.* 2003). Hyperammonaemia may also occur in hyperornithinaemia (Tuchman *et al.* 1990). Ataxia may occur as a feature of the intermittent metabolic encephalopathy, often with cerebral oedema, seen in these patients. Such episodes may be precipitated by infection or intercurrent illness. Periodic hyperammonaemic encephalopathy with ataxia has also been reported after ureterosigmoidoscopy (Cascino *et al.* 1989).

Disorders of amino acid metabolism

Hartnup disease is caused by defective intestinal and renal transport of neutral amino acids and is an autosomal recessive inborn metabolic error. Affected children develop a pellagra like rash and an intermittent neurological syndrome with ataxia. The gene has been mapped to chromosome 5p15 and is due to mutations of the *SLC6A19* gene that encodes the Hartnup transporter protein (Kleta *et al.* 2004). Intermittent ataxia, similar to that seen in the hyperammonaemias, is also seen with nonketotic hyperglycinaemia (Nightingale *et al.* 1991), intermittent branched-chain ketoaciduria, and isovaleric aciduria (Harding 1984). Clinically, these intermittent metabolic ataxias are similar to the hyperammonaemias. Glutamic aciduria and hydroxyglutaric aciduria are associated with a progressive ataxia (Sawada *et al.* 1991; de Klerk *et al.* 1997).

Disorders of pyruvate and lactate metabolism

Intermittent ataxia also occurs in pyruvate dehydrogenase deficiency (Robinson *et al.* 1996). This condition is genetically and clinically heterogeneous. Most cases are X-linked, while a few are autosomal recessive. Clinically, there may be severe infantile lactic acidosis, intermittent acidosis with ataxia (Bindoff *et al.* 1989; Kinoshita *et al.* 1997), Leigh's syndrome or chronic neurological impairment with learning disability, ataxia, spasticity, seizures, and various cerebral malformations (Zeviani *et al.* 1994; Robinson *et al.* 1996). Intermittent ataxia may also occur in some forms of autosomal recessive pyruvate carboxylase deficiency. Biotinidase deficiency is a treatable disorder causing ataxia and lactic aciduria (Yang *et al.* 2003). Ataxia is a common feature of mitochondrial diseases (Vedanarayanan 2003) and is considered below.

39.5.2 Progressive metabolic ataxia

Various metabolic disorders are associated with a progressive ataxic syndrome (Parker *et al.* 2003). Often there are prominent additional neurological features that overshadow the ataxic component except for vitamin E deficiency, which resembles a degenerative ataxia (Section 21.22.5).

Cholestanolosis or cerebrotendinous xanthomatosis

This is a rare autosomal recessive disorder caused by mutations of the sterol 27-hydroxylase *CYP27A1* gene (Chen *et al.* 1998). Many different mutations have been described in *CYP27A1* (Gallus *et al.* 2006). Clinical features are bilateral Achilles tendon xanthomas, cataracts, low intelligence, spastic paraplegia, cerebellar signs, convulsions, peripheral neuropathy, foot deformity, premature cardiovascular disease, EEG abnormality and increased CSF protein (Berginer *et al.* 1989) (Section 21.8.9). MR scanning shows lesions in the dentate nuclei, brainstem, and basal ganglia (De Stefano *et al.* 2001). In some patients, there are xanthomas in other sites, or they may be absent and dementia is not invariable. Metabolically, serum cholesterol is reduced, and increased levels of cholestanol, the 5 alpha-dihydro derivative of cholesterol, accumulate in xanthomas, the nervous system, and in bile. Progression of cerebrotendinous xanthomatosis may be prevented with chenedeoxycholic acid and pravastatin (Kuriyama *et al.* 1994). In the absence of tendon xanthomas, cerebrotendinous xanthomatosis may resemble the Marinesco–Sjogren syndrome with cataract, ataxia, and spasticity (Siebner *et al.* 1996). In many cases, the spastic paraparesis is more prominent than the cerebellar features.

Leucodystrophies

Ataxia is a common feature of the complex neurological phenotypes seen in various leucodystrophies, including Krabbe's disease, metachromatic leucodystrophy, adrenoleucodystrophy, and Pelizaeus–Merzbacher disease (Section 10.2). In most cases, cerebellar signs are accompanied by and often overshadowed by other features such as dementia, spasticity, visual loss, and peripheral neuropathy (Sections 21.8; 37.7). Occasionally, a slowly progressive cerebellar phenotype occurs, not unlike progressive multiple sclerosis (Klemm *et al.* 1989). Diagnosis of leucodystrophy is suggested by distinctive clinical syndromes, especially in adrenoleucodystrophy and Pelizaeus–Merzbacher disease, and can be confirmed by white matter abnormalities on MR scanning, very long chain fatty acid levels in adrenoleucodystrophy, white cell enzyme estimations in Krabbe's and metachromatic leucodystrophy, and DNA analysis in Pelizaeus–Merzbacher disease. White matter lesions are also seen in dentatorubropallidoluysian atrophy and in 'vanishing white matter disease' due to mutations of the *eIF2B* gene (van der Knaap *et al.* 2006) and in the rare disorder of leucoencephalopathy, calcification, and cysts (Labrune *et al.* 1996).

Abetalipoproteinaemia and hypobetalipoproteinaemia

This disorder is an autosomal recessive trait, characterized by a failure in the synthesis of apolipoprotein-B-containing lipoproteins, the very low density lipoproteins, and chylomicrons (Hooper *et al.* 2005). There is a deficiency of microsomal triglyceride-transfer protein (Wetterau *et al.* 1992) caused by mutations of its MTP gene on chromosome 4q22-24 (Narcisi *et al.* 1995; Wang *et al.* 2000). The resulting fat malabsorption causes steatorrhoea and malabsorption of the fat-soluble vitamins A, D, E, and K. The patients are of small stature, with a pigmentary retinopathy and progressive cerebellar ataxia. Clinically, the neurological disorder is similar to Friedreich's ataxia with areflexia, doral column sensory loss, and peripheral neuropathy (Harding 1987) (Section 21.8.8). In some patients, the gastrointestinal symptoms are mild or subclinical. Laboratory investigations reveal peripheral blood acanthocytes, reduced serum cholesterol, and an abnormal lipoprotein electrophoretic pattern. Vitamin E levels are undetectable. Treatment with vitamin replacement, especially vitamin E, can prevent progression of the neurological damage. It is important to measure vitamin E levels regularly; oral doses may need to be large, and some patients require intramuscular therapy (Perlmutter *et al.* 1987).

Familial hypobetalipoproteinaemia is a distinct autosomal dominant condition causing less severe abnormalities of low-density and very-low-density lipoprotein assembly due to mutations of the apolipoprotein B-100 gene (Hooper *et al.* 2005). Serum cholesterol levels are reduced but neurological function is usually normal. In rare patients who have been homozygous for this disorder, a neurological syndrome similar to that seen in abetalipoproteinaemia has been reported (Harding 1987).

Ataxia with isolated vitamin E deficiency and other vitamin E deficiency states

A selective autosomal recessively inherited deficiency of vitamin E has been described without lipid abnormalities, acanthocytosis or other features of fat malabsorption (Harding 1987). In this disorder, mutations of the alpha-tocopherol transfer protein α-*TPP* gene on chromosome 8q13 cause a loss of α-TPP function with rapid loss of vitamin E from serum after absorption (Yokota *et al.* 1997). The resulting vitamin E deficiency is associated with a neurological disorder 'ataxia with isolated vitamin E deficiency', similar to Friedreich's ataxia with ataxia, dorsal column sensory loss, and peripheral neuropathy affecting large myelinated fibres (Section 21.22.5); some patients also have a cardiomyopathy (Hammans *et al.* 1998), and retinopathy has occasionally been described (Yokota *et al.* 1997; Usuki *et al.* 2000; Benomar *et al.* 2002). Dystonia has also been reported (Roubertie *et al.* 2003).

This condition must always be considered in the differential diagnosis of Friedreich's ataxia (see Section 39.4.1). An important difference from Friedreich's ataxia is that sensory nerve action potentials tend to be preserved or reduced, whereas they are absent in Friedreich's ataxia (Hammans *et al.* 1998). Cognitive function

may be affected by vitamin E deficiency, but this is uncertain (Koscik *et al.* 2005). Ataxia with isolated vitamin E deficiency may be treated with large doses of vitamin E as in abetalipoproteinaemia. Reversal of the ataxia is unlikely, but further deterioration may be prevented although spasticity and retinopathy have progressed in some cases (Mariotti *et al.* 2004).

Occasionally, vitamin E deficiency causes neurological damage similar to that seen in ataxia with isolated vitamin E deficiency or abetalipoproteinaemia as a result of other causes of intestinal malabsorption such as intestinal resection and cholestatic liver disease, usually in children (Muller *et al.* 1983; Harding 1987).

Partial hypoxanthine guanine phophoribosyltransferase deficiency

This deficiency is an X-linked disorder in which affected males develop hyperuricaemia with secondary complications. Severe hypoxanthine guanine phophoribosyltransferase deficiency leads to the Lesch–Nyhan syndrome of severe learning disability, dystonia, spasticity, and self mutilation (Section 10.3.3). Partial hypoxanthine guanine phophoribosyltransferase deficiency is associated with gouty arthritis, nephrolithiasis, and in about 20 per cent of cases, neurological features (Caskey *et al.* 1993). There is cerebellar ataxia with mild learning disability and spasticity, but no self mutilation. The variable phenotype deficiency is caused by different mutations of the hypoxanthine guanine phophoribosyltransferase gene on chromosome Xq26, allelic heterogeneity. Treatment with allopurinol is helpful for gouty arthritis and nephrolithiasis, but not the neurological features.

Neimann-Pick type C or juvenile dystonic lipidosis

This is an autosomal recessive disorder with a gene locus identified on chromosome 18q11-12 (Section 10.4.4). Intracellular cholesterol metabolism is impaired, and there is accumulation of cholesterol in lysosomes (Lossos *et al.* 1997; Sato *et al.* 1998). The clinical features are dementia, psychiatric changes, cataplexy, vertical supranuclear gaze palsy (Imrie *et al.* 2002), ataxia, extrapyramidal signs, spasticity, and sometimes splenomegaly (Fink *et al.* 1989). Some patients are severely affected in infancy, while in others there is a gradual later-onset presentation with dementia or psychosis (Campo *et al.* 1998). Neuroimaging is usually normal, but a multiple-sclerosis-like presentation with prominent dementia has been described (Grau *et al.* 1997). A helpful diagnostic testis filipin staining of cultured fibroblasts from a skin biopsy.

Hexosaminidase A deficiency

Hexosaminidase A deficiency is usually associated with Tay–Sachs disease, but atypical later-onset phenotypes exist (Section 10.4.2). One variant is characterized by progressive cerebellar ataxia, ophthalmoparesis, and anterior horn cell disease with neurogenic muscle weakness (Hund *et al.* 1997) (Section 23.3.5). Dementia may not be prominent in this form of the condition; inheritance is autosomal recessive.

Hereditary caeruloplasmin deficiency

This is a rare autosomal recessive condition characterized by dementia, dystonia, diabetes, chorea, and ataxia in various combinations (Logan 1996). Cerebellar ataxia has been a prominent feature in some cases (Miyajima *et al.* 2001). Pathologically, there is excessive iron deposition in the liver and also in basal ganglia, dentate, and thalamus; there are low signal changes on both T1- and T2-weighted MR scans affecting these regions (Kawanami *et al.* 1996) and also the cortex (Grisoli *et al.* 2005).

39.6 Ataxias caused by mitochondrial disorders

The mitochondrial diseases are a highly heterogeneous group of disorders caused by abnormal mitochondrial energy metabolism. The classification of mitochondrial disorders is complex (Schapira 2006). The clinical features of mitochondrial diseases are numerous, but for the neurologist, the two most important are central nervous system involvement and a characteristic metabolic myopathy. The latter is often but not invariably associated with characteristic 'ragged red fibres' or cytochrome oxidase negative fibres seen with different histochemical staining of muscle biopsy specimens (Section 24.6.3). Clinical clues to the existence of a mitochondrial disorder are in Table 39.9. Cerebellar ataxia is common in the various neurological syndromes associated with mitochondrial disease (Truong et al. 1990), especially those caused by the defects of mitochondrial (mt) DNA (Chinnery et al. 1997; Schapira 2006). There are several specific mitochondrial syndromes in which ataxia is prominent:

- *Chronic progressive external ophthalmoplegia* or the *Kearns–Sayre syndrome* characterized by chronic progressive external ophthalmoplegia, ptosis, metabolic myopathy, cerebellar ataxia, pigmentary retinopathy, and cardiac conduction defects. Some patients have dementia or deafness, and the CSF protein is typically raised. Usually, mtDNA analysis reveals large deletions, but some cases are due to the 3243 MELAS point mutation or other mtDNA point mutations (Zeviani et al. 1994). Some cases are caused by mutations of nuclear *ANT-1* gene, encoding a protein need for adenine nucleotide transport in mitochondria. Others are caused by mutations of the *C10orf2* gene encoding a mitochondrial regulatory protein *twinkle*. Chronic progressive external ophthalmoplegia can also be caused by mutations of nuclear mitochondrial DNA polymerase γ *POLG*, sometimes associated with Parkinsonism, ataxia, mental changes, deafness, cataracts, and myopathy.

- *MERRF syndrome* of Myoclonic Epilepsy with Ragged Red Fibres (Section 40.7.8) is often associated with cerebellar ataxia. It is usually caused by point mutations of mitochondrial DNA (8344 or

8356). A spinocerebellar degeneration may be caused by the 8344 MERRF mtDNA mutation (Howell et al. 1996).

- The 3243 MELAS mutation, *Mitochondrial Encephalopathy Lactic Acidosis and Stroke* like episodes may also occasionally cause a progressive cerebellar ataxia (Arai et al. 1997).

- *NARP syndrome* comprises Neurogenic weakness, due to peripheral neuropathy, cerebellar Ataxia, and Retinitis Pigmentosa; seizures and dementia may also occur. It is associated with the G8993 mtDNA point mutation (Holt et al. 1990).

- *Leber's hereditary optic neuropathy* is occasionally associated with central nervous system involvement, including progressive cerebellar ataxia (Murakami et al. 1996). It is caused by 11 778 or sometimes the 3460 and 14 484 mutations.

- *Maternally inherited Leigh's syndrome* has also been associated with point mutations of mitochondrial DNA (8344 or 8993) encoding proteins or RNAs and can cause ataxia.

- *Autosomal recessive Leigh's syndrome* can be caused by mutations of nuclear genes encoding proteins needed for assembly and regulation of cytochrome oxidase. Mutations of these genes (*SCO2*, *SURF1*, *COX10*, *COX15*, and *LRPPRC*) cause severe Leigh syndrome. A similar phenotype can arise due to mutations of the *BSC1L* gene encoding a protein needed for the assembly of mitochondrial complex III.

- Mutations of nuclear genes encoding mitochondrial respiratory chain oxidative phosphorylation, *OXPHOS, system complex I* or *II subunits* can cause Leigh's syndrome or ataxia.

- *SANDO* is a combination of Sensory Ataxia, Neuropathy, Dysarthria, and Ophthalmoplegia caused by *POLG* mutations.

- *Coenzyme Q10 deficiency* can cause ataxia with myopathy, seizures, pyramidal signs, and cognitive changes (Schapira 2006). Cerebellar atrophy is seen along with lactic acidosis, and improvement can be seen with coenzyme Q10 therapy (Musumeci et al. 2001; Lamperti et al. 2003). The muscle biopsy may not be diagnostic. A survey of 135 patients with genetically undiagnosed ataxia found 13 cases of early-onset ataxia with reduced coenzyme Q10 levels; clinical features were ataxia, seizures, pyramidal signs, and cognitive changes (Lamperti et al. 2003).

- *Friedreich's ataxia* due to deficiency of the mitochondrial protein frataxin has features of mitochondrial dysfunction (see Section 39.4.1).

- *Holmes' ataxia* also has features of a mitochondrial disorder (see Section 39.4.5).

Table 39.9 Clinical features suggestive of mitochondrial disease

Fatiguable proximal myopathy (especially in combination with CNS disease)
Ptosis, progressive ophthalmoplegia or pigmentary retinopathy, optic atrophy
Short stature
Deafness (sensorineural)
Myoclonus, seizures
Diabetes mellitus, hypoparathyroidism
Cardiomyopathy, cardiac conduction defects
Lipomas
Unexplained stroke before 40 years of age
Migraine
Lactic acidosis
Raised CSF protein/lactate
Muscle biopsy (ragged red fibres, COX negative fibres)
Adapted from Chinnery and Turnbull (1997)

39.7 Ataxias caused by DNA repair defects

39.7.1 Ataxia telangiectasia

Ataxia telangiectasia, the Louis-Barr syndrome, is an autosomal recessive condition caused by a variety of mutations of the *ATM*, or 'AT mutated', gene on chromosome 11q23 (Taylor et al. 2005) (Section 11.5). The population frequency is 1 in 100–300,000. ATM is a tumour suppressor gene encoding a large protein kinase with phosphatidyl-inositol 3 kinase activity, which is usually truncated in AT with consequent loss of function (Uhrhammer et al. 1998), defective DNA repair mechanisms, and increased chromosomal fragility. The normal function of ATM is unclear, but involves DNA repair and interaction with the cellular protein p53 (Khanna *et al.* 1998). Many different *ATM* gene mutations have been identified, and most patients are compound heterozygotes with two different mutations, one on each *ATM* allele (Wright *et al.* 1996).

Ataxia develops in early childhood, usually as walking starts, and is rapidly progressive with the appearance of cerebellar dysarthria and kinetic tremor. Most patients are in wheelchairs by the age of 15 years (Bott *et al.* 2006). There is a characteristic oculomotor apraxia with grossly reduced saccadic eye movements and the use head thrusts to achieve refixation (Stell *et al.* 1989). Many patients are small and dysmorphic, and mild learning disability is seen in a proportion of cases (Bundey 1992). Other neurological features in some patients include myoclonus (de Graaf *et al.* 1995), dystonia (Koepp *et al.* 1994), peripheral neuropathy (McFarlin *et al.* 1972), and anterior horn cell degeneration (Larnaout *et al.* 1998). The characteristic conjunctival telangiectasia (Fig. 39.5) usually appear during early childhood; but sometimes later, they may also appear on the face, ears, neck, flexor creases of the limbs, gums, and palate; occasionally, they are absent (Friedman *et al.* 1993). There is impaired immunity with thymic hypoplasia, low circulating levels of IgA, and frequently recurrent sinopulmonary and cutaneous infections. There is an increased risk of malignancy, especially lymphoma and leukaemia in patients and, to a lesser extent, heterozygous relatives (Bott *et al.* 2006).

Clinically helpful laboratory abnormalities include an elevated alpha-fetoprotein as well as low serum IgA levels. Neuroimaging typically shows cerebellar atrophy (Farina *et al.* 1994). In addition, chromosomal abnormalities are common, including abnormally increased chromosome fragility after gamma or X irradiation. DNA diagnosis is difficult due to the high degree of allelic heterogeneity in AT, with a large number of different *ATM* mutations in different families (Wright *et al.* 1996).

The prognosis is generally poor although life expectancy is variable, depending on the severity of the immunodeficiency and the development of malignancy. Generally, patients survive 20–25 years (Crawford *et al.* 2006). The recurrence risk to siblings is 1 in 4. Symptomatic management of infections and movement disorders is generally possible, while the management of malignancies is complicated by increased sensitivity to radiotherapy.

39.7.2 Xeroderma pigmentosum

This is a rare autosomal recessive disorder with a frequency of 1 in 250,000. It is characterized by severe photosensitive skin disease with cutaneous malignancies including basal and squamous carcinomas and melanomas (Section 11.6.7). Sunlight-exposed skin is atrophic and scaly with pigmentary changes, scaling, keratoses, and telangiectasia (Fig. 39.6). Conjunctivitis and keratitis also occur. Neurological features, which only develop in some Xeroderma pigmentosum patients, include peripheral neuropathy (Kanda *et al.* 1990) or more severe multi-system manifestations, the de Sanctis–Cacchione syndrome. The latter include ataxia, dementia, deafness, spasticity, and chorea (Mimaki *et al.* 1986). Xeroderma pigmentosum is genetically heterogeneous, with seven subtypes or complementation groups A to G. These are caused by mutations of several genes encoding DNA repair proteins including the *XPAC* gene on chromosome 9, Xeroderma pigmentosum type A; the *ERCC-3* gene on chromosome 2, Xeroderma pigmentosum type B; a DNA repair gene *XPCC*, Xeroderma pigmentosum type C; and the *ERCC-2* gene on chromosome 19, Xeroderma pigmentosum type D. Other Xeroderma pigmentosum genes have also been identified (Banfi *et al.* 1994). The complex genetics of Xeroderma pigmentosum are summarized elsewhere (Berneburg *et al.* 2001).

39.7.3 Cockayne syndrome

Cockayne syndrome is also a rare autosomal recessive condition due to inherited abnormality of DNA repair. Clinical features are dwarfism, microcephaly, ataxia, spasticity, retinopathy, deafness, and peripheral neuropathy (Section 11.6.8). Movement disorders can also occur (Hebb *et al.* 2006). There is a characteristic facial

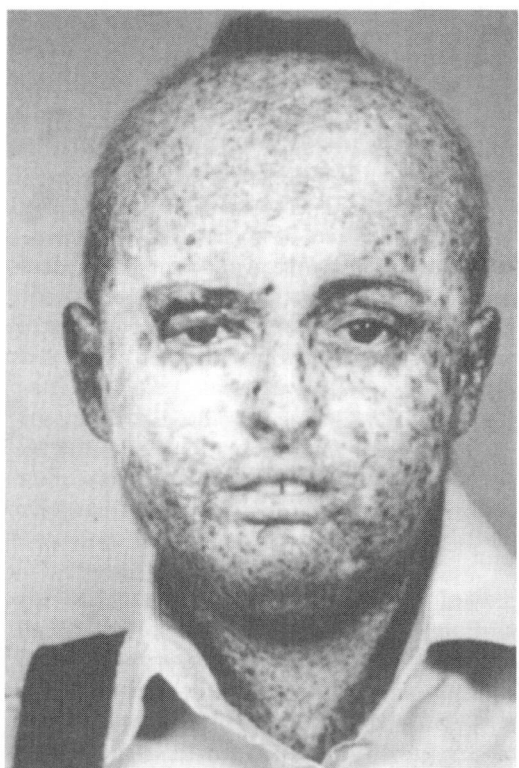

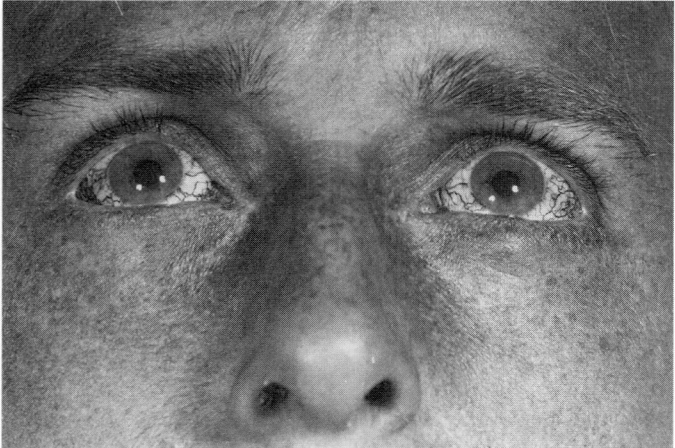

Fig. 39.5 Conjunctival telangiectasia in ataxia telangiectasia.

Fig. 39.6 Xeroderma pigmentosum. (From Harding 1984.)

appearance (Tan *et al.* 2005), the skin becomes atrophic, and cataracts appear (Nance *et al.* 1992). CT scans show intracranial calcification. Like Xeroderma pigmentosum, Cockayne syndrome is genetically heterogeneous and has been associated with several of the known Xeroderma pigmentosum mutations (Banfi *et al.* 1994). It appears that mutations of these genes may result in either Xeroderma pigmentosum, Cockayne syndrome, or a mixed picture.

39.8 Acquired ataxias of childhood

39.8.1 Infections and acute cerebellar ataxia

The most important postinfectious cerebellar syndrome is acute cerebellar ataxia of childhood (Gieron-Korthals *et al.* 1994). Most cases occur in children aged 1–6 years and can follow varicella, mumps, mycoplasma, Epstein-Barr virus, or nonspecific viral illnesses (Nussinovitch *et al.* 2003); usually the onset is approximately 6 days after the fever but after varicella the ataxia appears about 3 weeks after the rash. Other cases have followed plasmodium falciparum malaria (Senanayake *et al.* 1994) and streptococcal infection (Candler *et al.* 2006).

Ataxia develops within hours or days and mainly affects gait with cerebellar tremor, opsoclonus, or ocular flutter and myoclonus. The clinical features are similar to the paraneoplastic myoclonic encephalopathy of childhood associated with neuroblastoma (see Section 39.8.2). MR brain scans are normal, but the cerebrospinal fluid sometimes contains elevated protein and a lymphocytic pleocytosis. The prognosis is excellent; many patients recover in about 18 days (Martinez-Gonzalez *et al.* 2006), but recovery is sometimes more gradual over months.

Chronic progressive cerebellar ataxia is a feature of prion diseases (Section 42.8), subacute sclerosing panencephalitis (Section 42.3.7), and progressive rubella panencephalitis.

A similar disorder has been reported in older children and young adults. Cerebellar ataxia is also seen in approximately 10 per cent of cases of Legionnaire's disease, sometimes with confusion, cranial neuropathies, headache, and somnolence (Plaschke *et al.* 1997). Cerebellar ataxia after mycoplasma infection can be a result of an immunologically mediated acute disseminated encephalomyelitis (Komatsu *et al.* 1998) that may appear clinically similar.

39.8.2 Opsoclonus-myoclonus syndrome

This condition is typically seen in infants, but may occur in older children or adults and is heterogeneous. Most patients have a combination of opsoclonus, generalized myoclonus and cerebellar ataxia. Additional features in some cases include altered mental function, motor and language delay and behavioural changes (Koh *et al.* 1994). About 50 per cent of children have an underlying neuroblastoma, while the remainder are idiopathic. Only a small proportion of children with neuroblastoma develop opsoclonus-myoclonus-ataxia. Neuroblastoma-related cases are more likely to have benign tumour and have a better prognosis; there is also a greater chance of a thoracic neuroblastoma. Patients with neuroblastoma have IgG antibodies that bind to and are toxic to neuroblastoma cells (Korfei *et al.* 2005).

In idiopathic or neuroblastoma-related cases, the cerebrospinal fluid may show an inflammatory response but brain imaging is usually normal (Mitchell *et al.* 1990). Urinary catecholamine excretion may be increased in neuroblastoma cases, but this is not consistent, and the levels may be normal. In some cases, detailed

CT scanning of the abdomen and thorax is required to detect a tumour (Mitchell *et al.* 1990). Patients with both idiopathic and neuroblastoma-related opsoclonus-myoclonus have antibodies that bind to cerebellar neurones, indicating an autoimmune aetiology (Blaes *et al.* 2005). The neurological disorder often responds to steroids or ACTH and to treatment of an underlying neuroblastoma. Many children are left with chronic and persistent symptoms requiring long-term steroid therapy (Koh *et al.* 1994). Occasionally, there are frequent relapses over many years treated with intravenous immunoglobulin (Pranzatelli *et al.* 2002) or plasma exchange (Armstrong *et al.* 2005). In other children, the signs clear up spontaneously, and it is likely that these patients have had a postinfectious acute cerebellar ataxia.

A similar disorder occurs in adults. About half of these cases are idiopathic, and others are paraneoplastic, associated with anti-Ri antibodies (Caviness *et al.* 1995) (Section 38.4.5). As in children, some acute self-limiting cases are likely to be post-infectious acute cerebellar ataxia.

39.9 Late-onset hereditary cerebellar ataxias

39.9.1 Overview

The core feature of these disorders is a slowly progressive cerebellar syndrome with ataxia of gait and limbs, dysarthria, and nystagmus. There can be a 'pure' cerebellar syndrome, or there can be many combinations of this with additional non-cerebellar clinical features. In contrast to the autosomal recessive ataxias, which usually develop before the age of 25 years, patients with autosomal dominant cerebellar ataxia usually but not invariably develop symptoms after this age. Although there is some overlap in age of onset between the two groups, with some autosomal dominant cerebellar ataxia types sometimes starting in childhood and some autosomal recessive ataxias developing in adults, this broad generalization remains clinically useful when considering the differential diagnosis of hereditary degenerative ataxia. Unfortunately, the classification of the autosomal dominant ataxias must now be one of the most confusing issues in clinical neurology. Several classifications based on clinical features have been proposed, which have the advantage of being applicable in the clinic, often in advance of any investigation results (Harding 1984). With the discovery of several genes causing autosomal dominant cerebellar ataxia, confusingly referred to as spinocerebellar ataxia, or *SCA*, genes, a classification based on genotype has been proposed (Junck *et al.* 1996).The clinical classification proposed by Harding (Table 39.10) remains

Table 39.10 Clinical classification of the autosomal dominant cerebellar ataxias (from Harding 1984)

Type I: Autosomal dominant	Cerebellar ataxia with optic atrophy/ophthalmoplegia/dementia/pyramidal/extrapyramidal features/amyotrophy
Type II: Autosomal dominant	Cerebellar ataxia with pigmentary retinal degeneration ± ophthalmoplegia/pyramidal features
Type III: Autosomal dominant	'Pure' cerebellar ataxia with later onset ± pyramidal signs
Type IV: Autosomal dominant	Cerebellar ataxia with myoclonus and deafness

clinically useful but now requires modification to accommodate dentatorubropallidoluysian atrophy and episodic ataxia, while the patients in autosomal dominant cerebellar ataxia type IV probably had mitochondrial disease, and so this subgroup is obsolete. Unfortunately, one difficulty with any purely clinical classification is that the clinical signs may evolve over time, so that it may be difficult to classify individual patients confidently in the early years of the illness. An alternative genotypic classification based on the

spinocerebellar ataxias and other autosomal dominant ataxia genes is shown in Table 39.11. A problem with such a genotypic organization is that, while scientifically attractive, it cannot readily be applied in the clinic because the *SCA* gene mutations do not reliably define distinct clinical phenotypes (Subramony *et al.* 2001). A revised classification, incorporating Harding's clinical system of autosomal dominant cerebellar ataxia types I–III and recent molecular data is shown in Table 39.12 and will be used in this

Table 39.11 Genes causing autosomal dominant cerebellar ataxia with the more common types asterisked (*)

Gene	Locus	Mutation Type1	Gene product	Comments or notable features
SCA1*	6p23	CAG expansion	Ataxin 1	
SCA2*	12q24	CAG expansion	Ataxin 2	
SCA3*	14q24.3-q31	CAG expansion	Ataxin 3	
SCA4	16q22.1 16q22.1		Unknown PLEKHG4 (puratrophin1)	Rare families with sensory neuropathy Rare Japanese families with pure ataxia
SCA5	11cen		Ataxin 5	Three families (USA, France, and, Germany)
SCA6*	19p13	CAG expansion	α-1A calcium channel	Allelic to EA2
SCA7*	3p21.1-p12	CAG expansion	Ataxin 7	Retinopathy
SCA8*	13q21	CTG expansion	Ataxin 8	Unreliable DNA diagnosis
SCA10	22q13	ATTCT expansion	Ataxin 10	Rare; Mexican families; seizures
SCA11	15q14-21		Ataxin 11	One UK family
SCA12	5q31-q33	CAG expansion	PPP2R2B (protein kinase regulator)	Several families; prominent tremor
SCA13	19q13.3-4	Point mutations	KCNC3 (potassium channel)	Two families (France, Philippines)
SCA14*	19q13.4-qter	Point mutation	Protein kinase Cγ (PKCγ)	Several families (1.4% of ADCA in France)
SCA15	3p24.2-pter			Three families (Australia, Japan)
SCA16	8q22.1-q24.1			One family (Japan)
SCA17*	6q27	CAG/CAA expansion	TATA Binding Protein	Several families, can resemble HD (HDL4)
SCA18	7q22-q32			One family
SCA19	1p21-q21			Two families (Holland, China) probably alleleic with SCA22; myoclonus and tremor in some cases
SCA21	7p21-p15			One family (France)
SCA22	1p21-q23			One family (China) probably allelic with SCA19
SCA23	20p			One family (Holland)
SCA25	2p15-p21			One family (France);
SCA26	19p13.3			One family (Norwegian); locus adjacent to SCA6
SCA27	13q34	Point mutation	FGF14 (fibroblast growth factor 14)	One family (Holland)
SCA28	18p11.22-q11.2			One family (Italy)
DRPLA*	12p13.31	CAG expansion	Atrophin 1	
EA1*	12p13	Point mutations	KCNA1 potassium channel	
EA2*	19p13	Point mutations	α-1A calcium channel (CACNA1A)	Allelic with SCA6
EA3	1q42			One family
EA4	unknown			Two families
EA5	2q22-q23	Point mutation	β4 subunit calcium channel (CACNB4)	One family

Key: SCA = spinocerebellar ataxia; DRPLA = dentatorubropallidoluysian atrophy; EA = episodic ataxia.

Table 39.12 Clinical and molecular classification of the autosomal dominant cerebellar ataxias (ADCA)

	Associated genes identified
Autosomal dominant cerebellar ataxias type I	SCA1; SCA2; SCA3; SCA8; SCA12; SCA13; SCA14; SCA17; SCA18; SCA19; SCA21; SCA23;SCA25; SCA27; other unlinked loci
Autosomal dominant cerebellar ataxias type II	SCA7; other unlinked loci
Autosomal dominant cerebellar ataxias type III	SCA3; SCA5; SCA6; SCA10; SCA11; SCA15; SCA16; SCA22; SCA26; SCA28; other unlinked loci
Ataxia with sensory neuropathy (Biemond's ataxia)	SCA4
Dentatorubropallidoluysian atrophy	DRPLA gene (atrophin 1)
Episodic ataxia type 1 (with myokymia)	KCNA1
Episodic ataxia type 2	SCA6 / CACNL1A4
Episodic ataxia type 3 (vertigo and tinnitus)	
Episodic ataxia type 4 (vestibulocerebellar)	
Episodic ataxia type 5	CACNB4

Note: For ADCA definitions, see Table 39.10; abbreviations are as in Table 39.11.

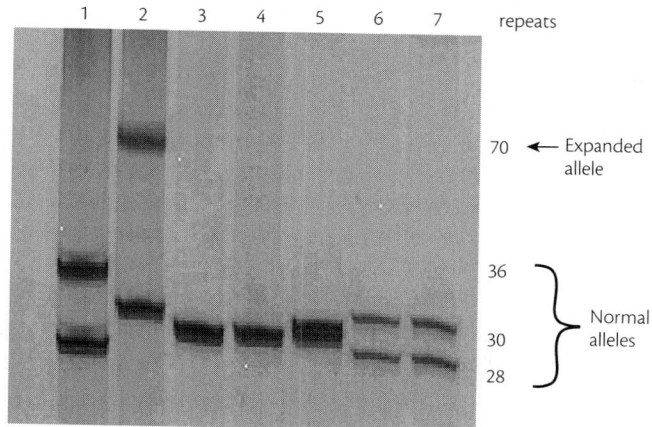

Fig. 39.7 Detection of CAG repeat expansion in the SCA1 locus by polymerase chain reaction amplification, polyacrylamide gel electrophoresis, and silver staining. DNA from different individuals has been electrophoresed in lanes 1–7; the mutant allele is shown in lane 2. (Courtesy of Dr R Mountford.)

chapter. Some clinical clues to particular *SCA* genes are shown in Table 39.13, but this is only an approximate guide, as there is considerable clinical overlap.

An interesting common feature of many *SCA* genes (Table 39.11) is pathological expansion of a CAG trinucleotide repeat sequence (Figs 39.7 and 39.8) (see Section 3.8). Increasing repeat length is correlated with earlier age of onset, and small repeat expansions may appear as sporadic cases without a family history (Schols *et al.* 2000). Repeat length can increase with each generation, causing progressively more severe disease and earlier onset—this effect is known as genetic anticipation. The mechanism by which trinucleotide or other sequence expansions occur is unclear but is similar to that seen in Huntington's disease (see Section 37.5.2).

Table 39.13 Additional clinical features in different autosomal dominant cerebellar ataxia type (ADCA) types

Various combinations of: supranuclear ophthalmoplegia, slow eye movements, optic atrophy, dementia, extrapyramidal features, pyramidal signs, amyotrophy, sensory motor peripheral neuropathy	SCA1; SCA2; SCA3; SCA8; SCA12; SCA13; SCA15; SCA16; SCA17; SCA18; SCA19; SCA21; SCA23; SCA25; SCA27; other unlinked loci
Retinopathy	SCA7
Sensory neuropathy	SCA4; SCA25
Seizures	SAC10; SCA17
Myoclonus	SCA14; SCA19
Orofacial dyskinesia	SCA27
Postural tremor (essential tremor like)	SCA12
Huntington's-like phenotype	SCA17
Psychiatric features	DRPLA; SCA12; SCA17; SCA27

The autosomal dominant cerebellar ataxias have an overall prevalence of about 3 per 100,000 (Schols *et al.* 2004a). Of the various subtypes, SCA3 is the most common, about 20 per cent, followed by SCA 1, 2, 6, 7 and 8; the others are very rare. About 30 per cent are unclassified, having no detectable mutation. The proportions of different genetic types and unclassified autosomal dominant cerebellar ataxia vary between different countries; SCA3 is more common in South America, while SCA2 is more common in Cuba (Schols *et al.* 2004a).

39.9.2 Autosomal dominant cerebellar ataxia type I

This is a heterogeneous category characterized by a progressive cerebellar syndrome with ataxia of gait and limbs, dysarthria, and nystagmus. Additional features including supranuclear ophthalmoplegia, slow eye movements, optic atrophy, dementia, extrapyramidal features, dysphagia, pyramidal signs, amyotrophy and peripheral neuropathy in various combinations (Harding 1984). Pathologically, most patients have olivopontocerebellar atrophy. In contrast to the olivopontocerebellar atrophy form of multiple system atrophy (see Section 40.3.7), glial and neuronal cytoplasmic inclusions are usually absent. Age of onset is typically between 30 and 49, with a median of 40 years (Klockgether *et al.* 1998) but a minority of cases develop in childhood or old age. The degree of dementia and any visual impairment due to optic atrophy are usually mild. Parkinsonism or dystonia are the usual extrapyramidal manifestations but chorea is occasionally seen. Tendon reflexes may be increased, normal, or absent, depending on the degree of pyramidal or peripheral nerve dysfunction. Anterior horn cell involvement is usually manifest as lingual, facial, or limb fasciculation. Bladder dysfunction is common. Progression is relentless, and treatment is entirely palliative. Most patients are wheelchair dependent after 17 years and survive up to 25 years after disease onset (Klockgether *et al.* 1998). Several subtypes within autosomal dominant cerebellar ataxia type I have been defined genotypically. Details of these mutations are shown in Table 39.11.

- *The SCA1 mutation* is present in up to 35 per cent of autosomal dominant cerebellar ataxia type I cases in some series

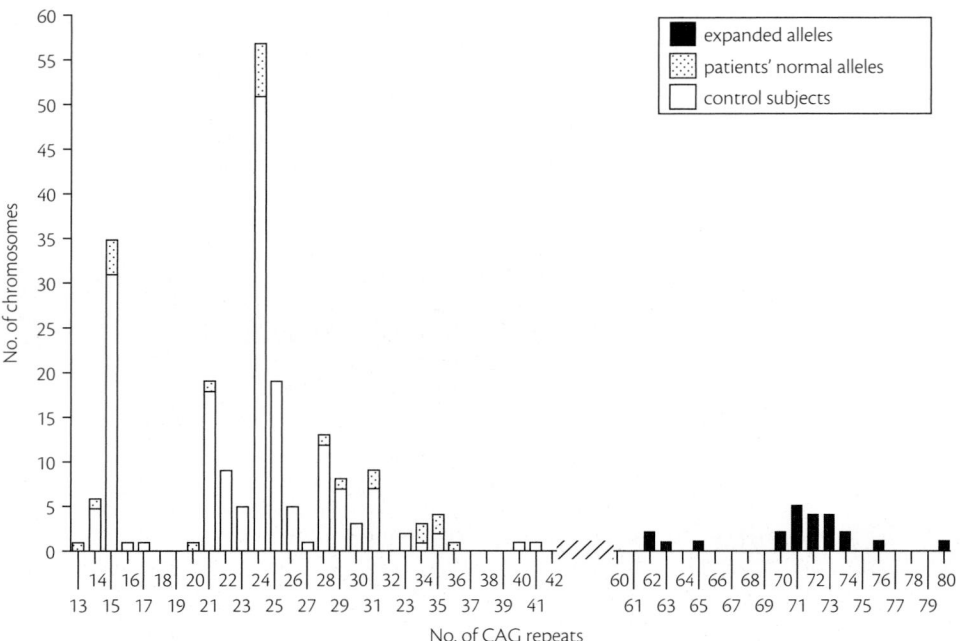

Fig. 39.8 Distributions of CAG repeat lengths in normal chromosomes of 91 control subjects, and normal and expanded alleles of 22 patients with the SCA3 mutation. (From Giunti *et al.* (1995).)

(Giunti *et al.* 1998; Klockgether *et al.* 1998), but the proportion varies considerably and is about 6 per cent overall (Schols *et al.* 2004a). As with other trinucleotide repeat expansions, increasing repeat length is associated with earlier age of onset, increased disease severity, and reduced survival (Banfi *et al.* 1994; Goldfarb *et al.* 1996), although one series was unable to demonstrate a relationship between repeat length and rate of disease progression (Klockgether *et al.* 1998). SCA1 families also demonstrate anticipation with increased severity and earlier onset in succeeding generations, especially with paternal transmission of the *SCA1* gene (Genis *et al.* 1995); the molecular basis for this is the tendency for the *SCA1* mutation, like other unstable trinucleotide repeats, to expand during meiosis, especially in spermatogenesis (Banfi *et al.* 1994). Clinically, SCA1 is highly variable; patients more commonly have optic atrophy, dysphagia, dysarthria, spasticity, and increased tendon reflexes compared with SCA2 or SCA3 cases (Burk *et al.* 1996). Hyporeflexia and ophthalmoplegia are less common, and fasciculations of the face and tongue may be seen. Magnetic resonance scanning reveals cerebellar and brainstem atrophy. The mechanism by which the *SCA1* mutation causes selective neurodegeneration is unknown. In normal alleles, the CAG repeat sequence is usually interrupted, while pathological expansions are continuous and incorporate a polyglutamine sequence into the translated ataxin1 protein similar to that seen with the Huntington's disease mutation and the Huntingtin protein. The function of ataxin1 is unclear, and the regional selectivity of the neuropathology in SCA1 patients difficult to explain; this does not appear to be due to somatic mosaicism of CAG repeat length within the brain (Lopes-Cendes *et al.* 1996).

♦ *The SCA2 mutation* has been reported in 21–40 per cent of autosomal dominant cerebellar ataxia type I cases in two large series (Giunti *et al.* 1998; Klockgether *et al.* 1998), but is more common in Cuba and accounts for about 15 per cent of SCA

mutations worldwide (Schols *et al.* 2004a). The effect of repeat length on age of onset (Fig. 39.9) and severity is similar to SCA1, and large expansions are associated with paternal transmission (Giunti *et al.* 1998). As with SCA1, pathological expansions are continuous, while normal alleles often have interruptions within the CAG repeat sequence. An anticipation effect is observed in SCA2 kindreds. Neurological deterioration in SCA2 patients is faster in females and with increasing CAG repeat lengths (Klockgether *et al.* 1998). Clinically, the SCA2 mutation is consistently more frequently associated with ophthalmoplegia, titubation, and areflexia (Durr *et al.* 1995; Filla *et al.* 1995; Burk *et al.* 1996; Giunti *et al.* 1998). Dementia and fasciculations have been

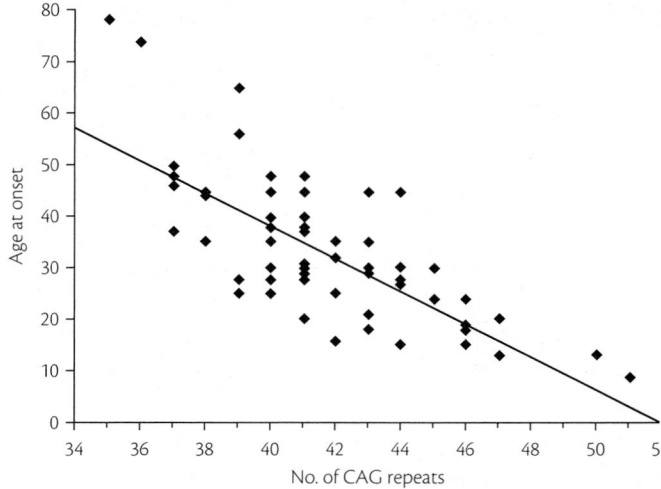

Fig. 39.9 Correlation between the number of SCA2 CAG repeats and the age of disease onset in years (from Giunti *et al.* (1998)).

reported more often than in SCA1 or SCA3 by some authors. The rate of progression is probably similar to other forms of autosomal dominant cerebellar ataxia type I. MRI reveals cerebellar and brainstem atrophy, the latter tending to be more severe than in SCA1 or 3 (Burk *et al.* 1996), as shown in Fig. 39.10. Occasionally, SCA2 can produce familial Parkinsonism (Shan *et al.* 2001).

- The frequency of the *SCA3 mutation* (Fig. 39.8) within autosomal dominant cerebellar ataxia type I is also uncertain, with estimates of 17–40 per cent in different series (Hammans 1996; Silveira *et al.* 1996; Giunti *et al.* 1998; Klockgether *et al.* 1998). Overall, *SCA3* accounts for 21 per cent of all *SCA* mutations (Schols *et al.* 2004a). It is very common (63 per cent) among

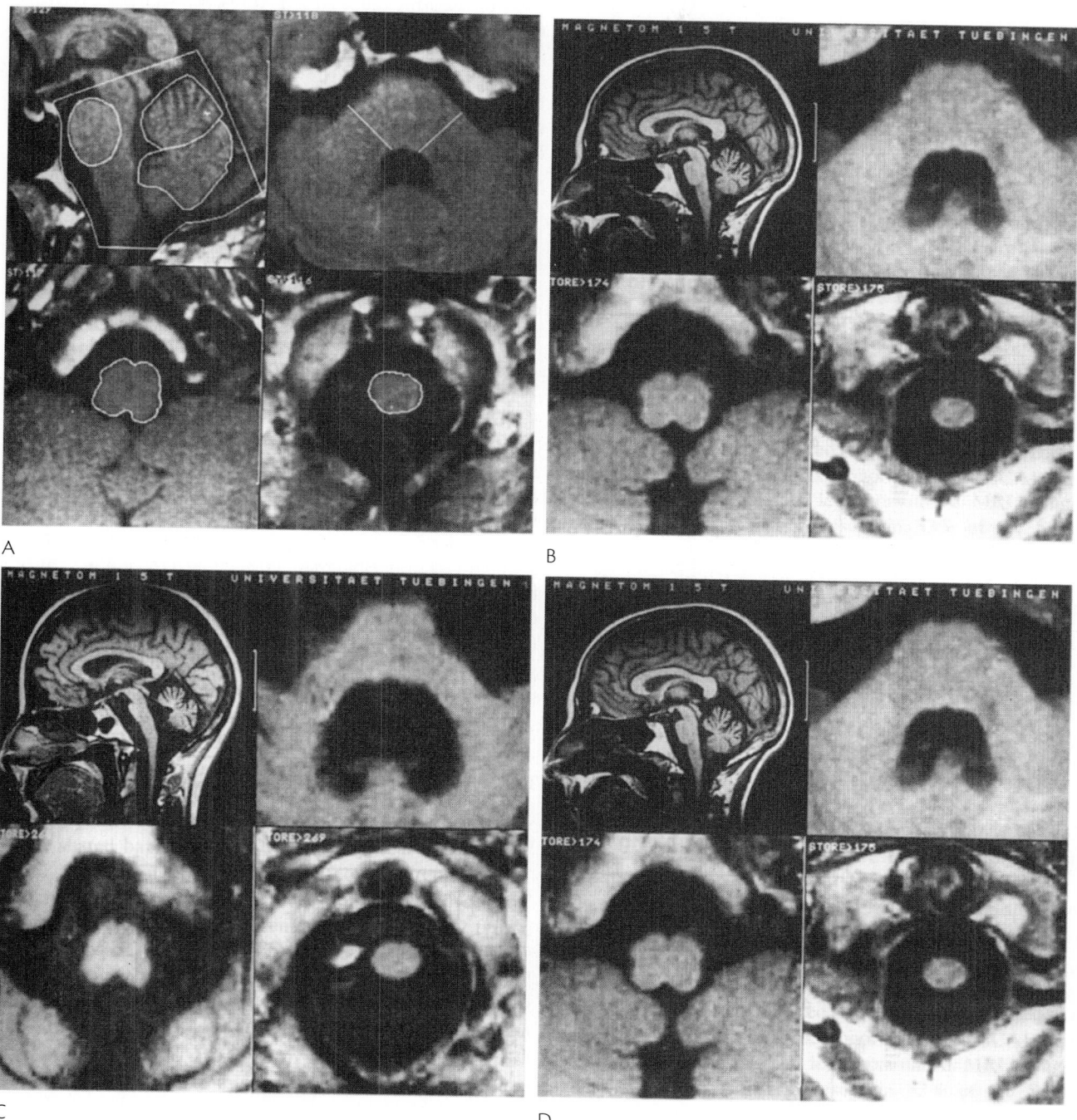

Fig. 39.10 Autosomal dominant cerebellar ataxias. T1-weighted MRI of infratentorial brain structures showing the posterior fossa in the midsagittal plane (upper left) and axial images at the level of the middle cerebellar peduncles (upper right), inferior olive complex (lower left) and cervicomedullary junction (lower right). A. MRI of 43-year-old healthy male. The areas for comparison in B, C, and D are outlined. B. MRI of a 44-year-old male SCA1 patient. C. MRI of a 38-year-old SCA2 patient. D. MRI of a 41-year-old male SCA3 patient. The degree of cerebellar, brainstem, and cervical cord atrophy is evident. (From Burk *et al.* [1996].)

ataxia patients in Brazil and Portugal (Silveira *et al.* 2002). CAG repeat size and age of onset are inversely related as with SCA 1 and 2, and anticipation is also observed in SCA3 families (Giunti *et al.* 1995; Burk *et al.* 1996; Higgins *et al.* 1996), but a clear effect of parental sex on repeat size has not been demonstrated so far. As with SCA2, females and those with larger repeat sizes tend to deteriorate more quickly (Klockgether *et al.* 1998). Clinically, autosomal dominant cerebellar ataxia type I patients with a SCA3 mutation do not differ significantly from those with the SCA1 gene, and magnetic resonance imaging appearances are also similar. It is now clear that the SCA3 mutation is also the genetic basis of Machado–Joseph disease (Matilla *et al.* 1995; Higgins *et al.* 1996). This condition, which is especially common among those of Portuguese or Azorean descent, was for many years regarded as a distinct clinical entity although Harding (1984) included Machado–Joseph disease within autosomal dominant cerebellar ataxia type I. Progressive cerebellar ataxia was the salient feature of Machado–Joseph disease, but a high frequency of Parkinsonism, dystonia, eyelid retraction, and bulbar fasciculation was considered characteristic (Rosenberg 1992), although all of these features are seen in SCA1, SCA2, and SCA3. Accordingly, Machado–Joseph disease does not, after all, appear to be clinically or genetically distinct from SCA3. Although the cerebellar syndrome is the most common presentation of the SCA3 mutation, there is considerable variation between and within families, as noted in earlier descriptions of Machado–Joseph disease. Some patients have pure cerebellar ataxia without prominent additional features (Ishikawa *et al.* 1996) or 'non cerebellar' presentations including levodopa responsive Parkinsonism with peripheral neuropathy (Tuite *et al.* 1995), dystonia, peripheral neuropathy, altered temperature sensation (Schols *et al.* 1996), restless legs syndrome (Schols *et al.* 1998), and spastic paraplegia (Sakai *et al.* 1996). It is therefore likely that if the whole SCA3 phenotype is considered, rather than just the usual cerebellar syndrome, the frequency of extrapyramidal features is greater with SCA3 than SCA1 or SCA2. The reasons for the phenotypic variability of SCA3 or the mechanism by which the polyglutamine sequence in ataxin 3 causes neurodegeneration are unclear, but as with SCA1, this cannot be accounted for simply by CAG repeat length variability (Lopes-Cendes *et al.* 1996). Generalized dystonia was the main clinical feature of a patient homozygous for the SCA3 mutation, suggesting that gene dosage also influences the phenotype (Lang *et al.* 1994). Ataxin 3 appears to be a ubiquitin binding protein that probably has abnormal binding properties with other proteins (Chai *et al.* 2004).

- *SCA8* is a CTG expansion mutation in the 3' untranslated region of a gene on chromosome 13q21. It is a rare cause of autosomal dominant cerebellar ataxia type 1; affected patients have had dysarthria, spasticity and sensory loss in addition to slowly progressive cerebellar ataxia (Koob *et al.* 1999). In other cases, there has been a pure cerebellar syndrome (Day *et al.* 2000). There is controversy about the CTG expansion associated with ataxia in SCA8; some authors regard it as a non-pathogenic polymorphism (Stevanin *et al.* 2000; Schols *et al.* 2003), while others disagree (Moseley *et al.* 2000). It has been detected in other ataxias, Alzheimer's disease (Sobrido *et al.* 2001), psychiatric disorders (Vincent *et al.* 2000), and vitamin E deficiency ataxia (Cellini *et al.* 2002). One patient with ataxia and a DNA test showing a

SCA8 expansion actually had pathologically confirmed multiple system atrophy (Factor *et al.* 2005). At present, the status of the SCA8 mutation is unclear, and DNA testing is therefore unreliable and difficult to interpret.

- *SCA12* is also a rare cause of the autosomal dominant cerebellar ataxia type 1 phenotype with onsets ranging from 8 to 55 years but typically in the fourth decade; An action and postural tremor affecting the arms and head similar to essential tremor is characteristic (O'Hearn *et al.* 2001), together with mild cerebellar ataxia, abnormal eye movements, Parkinsonism, and in some cases, dementia (Holmes *et al.* 1999). Psychiatric features may also occur. Neuroimaging shows cerebellar and cerebral atrophy. SCA12 is a CAG repeat expansion in the 5' region of the PPP2R2B protein kinase regulator gene (Holmes *et al.* 1999; O'Hearn *et al.* 2001).

- *SCA13* is very rare and has been described in only two families. A pure ataxia may occur, or there may be developmental delay, learning disability, or pyramidal signs (Herman-Bert *et al.* 2000). Imaging shows cerebellar and pontine atrophy. Point mutations of the KCNC3 potassium channel gene have been detected in this disorder (Waters *et al.* 2006). Onset is often in childhood, and SCA13 should probably be included in the congenital ataxias (see Section 39.3).

- *SCA14* is also very rare and can be associated with myoclonus, cognitive impairment, and early onset. It is caused by mutations within the protein kinase γ(PKC)γ gene (Klebe *et al.* 2005).

- *SCA17* is rare and highly variable. It can cause ataxia over a wide age range and is sometimes associated with psychiatric symptoms and chorea resembling Huntington's disease, Huntington-like disorder type 4 (see Section 40.5.3). Other features such as seizures, Parkinsonism, dystonia, pyramidal signs, and dementia may occur, and the condition is caused by CAG and CAA trinucleotide expansions of the TATA-binding protein TBP gene (Zuhlke *et al.* 2001).

- *SCA18* has been described in a single family, with ataxia combined with neuropathy and pyramidal signs.

- *SCA19* may rarely cause ataxia with or without myoclonus with a wide age of onset range (Schelhaas *et al.* 2001; Chung *et al.* 2003). It is probably allelic with SCA22 (Schelhaas *et al.* 2004).

- *SCA21, SCA23 and SCA25* have caused autosomal dominant cerebellar ataxia type I disorders in single families (Devos *et al.* 2001; Stevanin *et al.* 2004; Verbeek *et al.* 2004).

- *SCA27* is caused by point mutations of the fibroblast growth factor gene (Dalski *et al.* 2005). Clinical features include ataxia, psychiatric symptoms, cognitive impairment, tremor, and orofacial dyskinesia (van Swieten *et al.* 2003). Neuropathy has also occurred.

- Some autosomal dominant cerebellar ataxia type I families do not have any of the currently known SCA mutations. Although the exact proportion is unclear, it is probably small. Two large autosomal dominant cerebellar ataxia type I series demonstrated SCA1, 2, or 3 mutations in 74 and 90 per cent of families, respectively (Giunti *et al.* 1998; Klockgether *et al.* 1998).

39.9.3 Autosomal dominant cerebellar ataxia type II

This form of autosomal dominant cerebellar ataxia is characterized clinically by the presence of visual loss due to a pigmentary retinopathy with macular degeneration (Harding 1984). Pathologically, there is olivopontocerebellar atrophy with additional pregeniculate visual pathway involvement, but these findings are not unique (Giunti *et al.* 1999). Autosomal dominant cerebellar ataxia type II has been mapped to chromosome 3, and the pathogenic mutation *SCA7* identified as an unstable CAG trinucleotide repeat expansion (Benomar *et al.* 1995; Lindblad *et al.* 1996), as previously predicted clinically (Enevoldson *et al.* 1994). A few families do not have *SCA7* mutations (Giunti *et al.* 1999). The SCA7 CAG repeat is more unstable than any other trinucleotide repeat expansion, especially with paternal transmission; increased repeat size, as in other neurodegenerative CAG repeat disorders, is associated with earlier onset, greater severity, and the phenomenon of anticipation. The age of onset in patients with SCA7 is wider than in other forms of autosomal dominant cerebellar ataxia, and it is with this condition that the approximate guideline of autosomal recessive and autosomal dominant cerebellar ataxia usually developing either side of about 25 years of age is least dependable. Many patients are first affected in their thirties or later, but children may be affected with a more severe form of the disorder from infancy onwards. Occasional patients develop symptoms in their sixties, and there are some asymptomatic gene carriers indicating a gene penetrance of 95 per cent (Enevoldson *et al.* 1994). Patients with later onset show less rapid progression. The usual presenting features are ataxia or visual failure, but both are usually present within a few years. Visual loss tends to develop first in younger patients and can follow ataxia after many years in older patients (Schols *et al.* 2004a). Pyramidal signs, peripheral neuropathy (van de Warrenburg *et al.* 2004), and ophthalmoplegia are commonly associated while chorea, dementia, amyotrophy, and sensory loss are less common. With this combination of features, it is not surprising that misdiagnoses of multiple sclerosis have been reported. The visual failure is due to retinal degeneration with loss of central visual field and, eventually, severe visual loss in most cases. However, patients are often unaware of the insidious progression of visual loss until quite late in the disease and, although ophthalmoscopic examination reveals a pigmentary retinopathy in advanced cases (Fig. 39.11), the fundoscopic appearances may be normal for many years, leading to diagnostic difficulty. Careful ophthalmological examination, and especially electroretinogram recordings, may be needed to reveal the characteristic retinal involvement (Enevoldson *et al.* 1994). In contrast, visual-evoked potentials are less sensitive and may be misleadingly normal. Early-onset infantile cases are much more severe and rapidly fatal with wasting, weakness, fasciculations, dementia, and blindness; retinal pigmentary changes are usually prominent (Ansorge *et al.* 2004). Autosomal dominant cerebellar ataxia type II is probably heterogeneous; at least one family with this phenotype does not have the *SCA7* mutation (Giunti *et al.* 1999).

39.9.4 Autosomal dominant cerebellar ataxia type III

This form of autosomal dominant cerebellar ataxia typically has a later onset, after 50 years of age, and slow progression; there may or may not be inconspicuous additional pyramidal signs (Harding 1984). Such cases are not common, and autosomal dominant cerebellar ataxia type III is heterogeneous (see Table 39.12). Pathologically, there is cerebellar cortical atrophy that can lead to severe cerebellar atrophy but only very slow clinical deterioration (Frontali *et al.* 1992). One form of mild late-onset ataxia has been

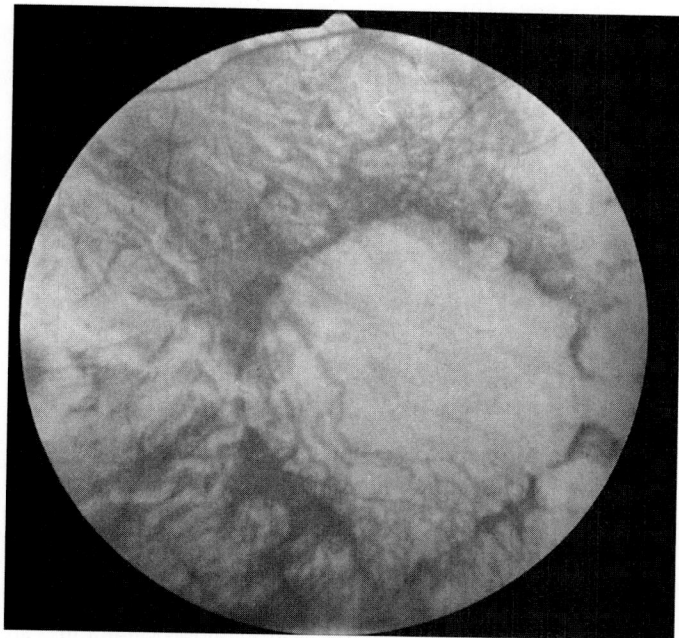

Fig. 39.11 Retinal appearance in autosomal dominant cerebellar ataxia type II (SCA7).

described in the descendants of President Lincoln, 'Lincoln ataxia', and a gene locus designated *SCA5* mapped to chromosome 11 (Ranum *et al.* 1994). *SCA5* has been detected in other countries and is very rare. A pure cerebellar autosomal dominant cerebellar ataxia type III phenotype may occur with several other *SCA* genes (see Table 39.12), particularly *SCA6* (Stevanin *et al.* 1997). The mild nature of some cases and later onset can lead to apparently sporadic cases of SCA6 with no clear family history of the disorder. Additional neurological signs are rare.

Families with an autosomal dominant cerebellar ataxia type III phenotype and epilepsy have a gene designated *SCA10* which has been mapped to chromosome 22 (Zu *et al.* 1999) but most of these patients are of Mexican descent (Matsuura *et al.* 2002). The mutation is a pentanucleotide ATTCT expansion in an intron of the *SCA10* gene (Matsuura *et al.* 2000).

An autosomal dominant cerebellar ataxia type III presentation has been described in a family with a pure ataxia linked to the *SCA 11* locus on chromosome 15 (Worth *et al.* 1999); the nature of this mutation is currently unknown.

Autosomal dominant cerebellar ataxia type III has also been reported in rare families with SCA15, SCA16, SCA22, SCA26, and SCA28 (Miyoshi *et al.* 2001; Storey *et al.* 2001; Chung *et al.* 2003; Hara *et al.* 2004; Yu *et al.* 2005; Cagnoli *et al.* 2006).

A pure form of cerebellar ataxia has been reported in Japan with linkage to the *SCA4* locus on chromosome 16 (Nagaoka *et al.* 2000). Sensory neuropathy does not occur in these patients who have a point mutation of the *PLEKHG4* puratrophin gene (Ohata *et al.* 2006). This seems distinct from the ataxia with the sensory neuropathy form of SCA4.

39.9.5 Autosomal dominant cerebellar ataxia with sensory neuropathy

This is a rare form of autosomal dominant cerebellar ataxia in which there is a prominent sensory peripheral neuropathy; it is likely to be the same disorder previously referred to as Biemond's ataxia (Harding 1984). Pathologically, there is degeneration of cerebellar Purkinje cells, dorsal columns, dorsal root ganglion cells, and peripheral nerves (Nachmanoff et al. 1997). The sensory symptoms affecting the face and limbs and rapid progression of the disorder are unlike autosomal dominant cerebellar ataxia types I, II, or III. In one family, linkage to a locus on chromosome 16, designated SCA4, has been reported (Flanigan et al. 1996) and the locus on chromosome 16 has been further characterized (Hellenbroich et al. 2003).

39.9.6 Dentatorubropallidoluysian atrophy

This rare autosomal dominant condition was until recently described mainly in Japan (IIzuka et al. 1984). In Europe, it is a very rare cause of autosomal dominant cerebellar ataxia or sporadic ataxia in adults, accounting for only 0.25 per cent of cases (Le Ber et al. 2003a). It has been reported elsewhere as the 'Haw River syndrome'. Dentatorubropallidoluysian atrophy is caused by a CAG repeat expansion mutation of the *atrophin 1* gene on chromosome 12 (Koide et al. 1994). Pathologically, there is degeneration in the dentate nucleus and external segment of the pallidum along with the projections of these areas to the red and subthalamic nuclei (Warner et al. 1994). However, the neuropathological findings are variable and may not suggest the diagnosis.

Most patients develop symptoms in adult life, although earlier onset in childhood may occur. The phenotype is highly variable both within and between families (Vinton et al. 2005); some patients have progressive cerebellar ataxia with prominent chorea and dystonia, while others have dementia with chorea, the pseudo-Huntingtonian type, or a progressive myoclonic epilepsy (IIzuka et al. 1984). Dentatorubropallidoluysian atrophy may present with adult-onset dystonia (Hatano et al. 2003). Other patients have various combinations of these subtypes, and a psychotic presentation has also been described. As expected for a trinucleotide expansion disorder, age of onset is earlier with increased repeat length (Koide et al. 1994) and dentatorubropallidoluysian atrophy families show anticipation and more severe manifestations following paternal transmission (Warner et al. 1995).

MRI shows cerebellar and brainstem atrophy as well as multiple white matter hyperintensities on T_2 sequences (Munoz et al. 2004), which may be related to severe epilepsy (Takamure et al. 2006). Dentatorubropallidoluysian atrophy needs to be considered in the differential diagnosis of Huntington's disease, especially if DNA testing is negative, autosomal dominant cerebellar ataxia type I cases with prominent dystonic or choreiform manifestations and progressive myoclonic epilepsy. However, a dentatorubropallidoluysian atrophy mutation is usually absent in patients without a family history (Warner et al. 1995). As with *SCA* mutations, the regional selectivity of the neuropathology in dentatorubropallidoluysian atrophy is not due to differences in CAG repeat length or gene expression in different brain regions (Nishiyama et al. 1997). There is possibly a defect of mitochondrial respiration in muscle based on results of in vivo magnetic resonance spectroscopy studies (Lodi et al. 2000). However, it is likely that accumulation of mutant atrophin in intraneuronal inclusions causes nuclear transcriptional abnormalities (Yamada et al. 2006).

39.9.7 Fragile X tremor ataxia syndrome

The Fragile X Tremor Ataxia Syndrome, FXTAS, affects about 1 in 4000 males and 1 in 8000 females and is the most common inherited form of mental retardation (Bardoni et al. 2002; Jin et al. 2003). Affected individuals have mental retardation, hyperkinetic or autistic behaviours, and a characteristic facial appearance with large ears and jaw; post-pubertal males have macro-orchidism. The condition is associated with a folate-sensitive fragile site on chromosome Xq27.3, where there is a CGG trinucleotide expansion in the 5' untranslated region of the *FMR-1* gene. Normal individuals have 7–55 repeats at this site; if there is an expansion of >200 repeats, there is hypermethylation of the site and inactivation of the gene with loss of FMR-P protein expression and the development of the syndrome.

In carriers of a premutation of 55–200 repeats, there may be milder and later features. About 40 per cent of males over the age of 50 (Loesch et al. 2005) develop the fragile X tremor ataxia syndrome, characterized by postural and action tremor, ataxia, and Parkinsonism (Hagerman et al. 2004a, b). Mild cognitive impairment may also occur. Neurologically, these older males present with what looks like a combination of mild ataxia, Parkinsonism, and essential tremor. There may be high-signal lesions within the middle cerebellar peduncles on brain MR scans, although this is not seen only in fragile X tremor ataxia syndrome (Storey et al. 2005). Severely affected males may have spasticity (Jacquemont et al. 2005), and some patients develop dementia (Bourgeois et al. 2006; Grigsby et al. 2006). There can be considerable variation in severity even within older men in the same family (Peters et al. 2006).

Fragile X tremor ataxia syndrome can present with Parkinsonism, tremor or ataxia in older males, with or without a family history of mental retardation in younger male relatives. It may be confused with Parkinson's disease, the cerebellar form of multiple system atrophy (see Section 40.3.7) or with late-onset idiopathic cerebellar ataxia and should now be considered in the differential diagnosis of these disorders. However, in a series of 414 males with Parkinson's disease, no cases of fragile X tremor ataxia syndrome were found (Toft et al. 2005) and in a series of 426 clinically diagnosed cases of cerebellar MSA only 4 per cent had a FXTAS premutation (Kamm et al. 2005).

Female premutation carriers may develop a mild form of fragile X tremor ataxia syndrome without cognitive impairment and also may develop premature ovarian failure (Hagerman et al. 2004a, c).

39.9.8 Episodic ataxias

In these disorders, there are attacks of ataxia, with or without persistent interictal cerebellar signs. Some of the episodic ataxias are due to defective ion channel function and should be distinguished from the intermittent metabolic ataxias (Section 39.5.1). An obvious clinical distinction from metabolic ataxia is a family history of autosomal dominant inheritance. There are two main types of episodic ataxias and a number of other rarer forms.

◆ *Episodic ataxia with myokymia,* or type 1, develops in early childhood with brief attacks lasting seconds or minutes. These may be triggered by startle or exercise and can be prevented with acetazolamide or phenytoin (Griggs et al. 1995). Persistent cerebellar signs or nystagmus do not always occur. The myokymia may be subtle and causes fine twitching or flickering in the face and hands. Some patients report episodic stiffening or jerking. It is caused by point mutations of the potassium channel gene on chromosome 12p (Browne et al. 1994; Comu et al. 1996). Acetazolamide may be effective in some cases (Gancher et al. 1986b; Brunt et al. 1990; Lubbers et al. 1995).

◆ *Episodic ataxia with nystagmus*, or type 2, generally starts later in childhood or adolescence, with more prolonged attacks of ataxia lasting hours or days. The ataxia may be accompanied by headache, nausea, vomiting, or vertigo (Subramony *et al.* 2003). Ataxic episodes are associated with fever, heat, stress, alcohol, or exertion but not startle, and there is no myokymia (Griggs *et al.* 1995). Some patients have atypical presentations such as late onset or dystonia (Imbrici *et al.* 2005; Spacey *et al.* 2005). Between attacks, there may be persistent nystagmus, usually downbeating (Gancher *et al.* 1986a), along with a slowly progressive persistent cerebellar syndrome. Some patients with progressive ataxia do not have acute attacks. Magnetic resonance imaging may show atrophy of the cerebellar vermis. Acetazolamide effectively prevents attacks. Episodic ataxia type 2 is caused by point mutations of the alpha 1A calcium channel gene, *CACNL1A4*, on chromosome 19p (Vahedi *et al.* 1995; Ophoff *et al.* 1996; Subramony *et al.* 2003). Interestingly, other point mutations of the same gene cause familial hemiplegic migraine, in which a mild persistent cerebellar ataxia may develop, while a small CAG repeat expansion within the *CACNL1A4* gene, the *SCA6* mutation, causes a slowly progressive autosomal dominant cerebellar ataxia type III phenotype. Some patients who clinically have episodic ataxia type 2 have a *SCA6* mutation rather than a point mutation in *CACNL1A4* (Geschwind *et al.* 1997a).

◆ *Episodic ataxia type 3* has been described in one family with attacks of ataxia, vertigo, and tinnitus responsive to acetazolamide (Steckley *et al.* 2001). Myokymia may occur. Although there has been confusion about the nomenclature, this condition is the same as the one linked to chromosome 1q42 by the same group (Cader *et al.* 2005).

◆ *Episodic ataxia type 4* has also been described as periodic vestibulocerebellar ataxia in two families. It is an autosomal dominant condition causing attacks of ataxia, diplopia, oscillopsia, and vertigo. A persistent cerebellar syndrome may eventually develop (Damji *et al.* 1996). Onset is variable and can be from the third to the sixth decade. This disorder is not linked to episodic ataxia types 1 or 2.

◆ *Episodic ataxia type 5* has been reported in one French Canadian family with autosomal dominant inheritance. There were attacks of ataxia and vertigo lasting several hours, and acetazolamide was effective. A point mutation of the α−4 subunit of the voltage-gated calcium channel gene caused the disorder (Escayg *et al.* 2000). In another family, the same mutation was associated with epilepsy.

39.10 Genetic counselling and DNA testing in adult-onset cerebellar ataxia

Patients with autosomal dominant cerebellar ataxia are faced with a genetic situation not unlike that confronting those with Huntington's disease. The prognosis is poor, with slow neurological deterioration and reduced life expectancy. The risk to offspring is 50 per cent, and siblings are also at significant risk, depending on their age. The situation is particularly complex in autosomal dominant cerebellar ataxia type II, SCA7, families in which mildly affected or even asymptomatic adults may have severely affected children who may die in infancy or early childhood (Enevoldson *et al.* 1994). Patients and family members should be offered genetic counselling

so that they are able to make fully informed reproductive decisions. In those patients with a family history of autosomal dominant cerebellar ataxia type I, DNA testing is often positive, with most patients carrying one of the *SCA* mutations (Moseley *et al.* 1998). In general, it is often impossible to predict the underlying genetic mutation from the clinical features in autosomal dominant cerebellar ataxia with any confidence. SCA6 is more likely to be found in those with a mild slowly progressive pure cerebellar ataxia, but the early stages of autosomal dominant cerebellar ataxia types I or II may be indistinguishable. It is reasonable to include the *SCA7* mutation in DNA analysis in autosomal dominant cerebellar ataxia even in the absence of visual failure or retinopathy, which may not be clinically detectable at onset in some cases, thereby making a clinical diagnosis of autosomal dominant cerebellar ataxia type II difficult initially. Many laboratories, therefore, offer screening for SCA1, 2, 3, 6, and 7, along with dentatorubropallidoluysian atrophy, which can also produce a mainly cerebellar clinical picture. SCA8 testing remains controversial due to difficulties with interpretation of the test (Schols *et al.* 2004a).

Although SCA and dentatorubropallidoluysian atrophy gene testing could be used for predictive diagnosis in healthy at-risk relatives and antenatal testing (Nance *et al.* 1994), this has not yet become common practice.

Genetic counselling in sporadic adult-onset ataxia is difficult. Care is needed when advising patients who have 'negative' but actually incomplete family histories; for example, where contact has been lost with relatives whose state of health is therefore uncertain or when a parent died prematurely. Certain clinical features such as optic atrophy, retinopathy, or ophthalmoplegia also increase the likelihood that the patient has autosomal dominant ataxia (see Section 39.9). Harding (1984) proposed that patients were advised that the risk of their ataxic illness being autosomal dominant was approximately 10 per cent, increasing to 20 per cent in the presence of an unreliable family history or the suspicious ophthalmological features already mentioned and 40 per cent with both factors present; risks to offspring are therefore 5, 10, and 20 per cent, respectively. The use of DNA diagnosis in sporadic cases has generally been unhelpful but not yet studied on a large scale. Theoretically, some patients below the age of 50 years may have a late presentation of autosomal recessive Friedreich's ataxia (see Section 39.4.1). Thus, testing for the frataxin gene GAA expansion in this situation may be helpful, although potentially misleading if a single Friedreich's ataxia allele is detected, given the population carrier frequency of about 1 per cent. Approximately 4–5 per cent of adult-onset ataxia, presenting after the age of 20 years, is due to Friedreich's ataxia (Moseley *et al.* 1998; Abele *et al.* 2002). *SCA* mutations have been detected in 5–10 per cent of cases (Moseley *et al.* 1998; Abele *et al.* 2002; Muzaimi *et al.* 2004). Older men with ataxia, especially if there is additional tremor or ataxia, may have the fragile X tremor ataxia syndrome, but the yield from DNA testing for this is low (Biancalana *et al.* 2005).

39.11 Other causes of cerebellar ataxia

39.11.1 Toxic and drug-induced ataxia

The most common drug-induced cerebellar syndrome is that seen with anticonvulsant therapy. Phenytoin is usually the cause, but carbamazepine and barbiturates may have similar effects and ataxia has been reported with low doses of gabapentin (Steinhoff *et al.* 1997). Patients develop ataxia, nystagmus, dysarthria, and drowsiness, often with serum anticonvulsant concentrations above the therapeutic range, but not always. A permanent cerebellar syndrome with cerebellar atrophy and loss of Purkinje cells may develop in patients with chronic phenytoin toxicity; this has been reported in a significant proportion of institutionalized patients with severe epilepsy (Young *et al.* 1994). Other drugs prone to cause cerebellar toxicity include benzodiazepines, piperazine, lithium, cyclosporin A, and cytotoxic drugs, especially cytosine arabinoside and 5-flurouracil. Lithium can cause tremor, cerebellar ataxia, or encephalopathy, even with normal serum concentrations, and a permanent cerebellar syndrome may also develop (Kores *et al.* 1997). In rare cases, lithium has produced an encephalopathy with ataxia and myoclonus similar to that seen with Creutzfeldt–Jacob disease (Fear 1992). Amiodarone can cause ataxia, tremor, and peripheral neuropathy.

A cerebellar syndrome may also occur following the use of illicit drugs, including marijuana, phencyclidine, and organic solvents. Usually, this is in the context of a toxic encephalopathy, but cerebellar atrophy has been detected in young drug abusers, suggesting permanent cerebellar damage; this is difficult to separate from the effects of alcohol excess, which is also common in this group (Aasly *et al.* 1993). Drug-induced ataxia is a common clinical problem and should always be suspected in any child or young adult presenting with acute cerebellar ataxia; in one series of 40 cases of acute childhood ataxia, 17 out of 35 drug screens were positive (Gieron-Korthals *et al.* 1994). Accidental exposure to organic solvents, acrylamide, and mercury may also produce cerebellar ataxia. Bismuth exposure may also cause a cerebellar syndrome (Gordon *et al.* 1995).

39.11.2 Alcoholic cerebellar ataxia

Ataxia is a cardinal feature of the clinical presentation of acute Wernicke's encephalopathy or the Wernicke–Korsakoff syndrome, along with ophthalmoplegia, nystagmus, and confusion (Section 5.2.3). This condition is due to thiamine deficiency in malnourished alcoholics and also occurs in other thiamine deficiency states. Pathologically, there is loss of Purkinje cells in the superior cerebellar vermis, and a permanent cerebellar ataxia commonly persists after thiamine replacement (Butterworth 1993). Alcoholics also develop a chronic cerebellar ataxia that usually evolves subacutely over weeks or months, but sometimes more insidiously (Section 5.2.4). Lower limb and gait ataxia are prominent, while upper limb involvement, dysarthria, and nystagmus are rare (Johnson-Greene *et al.* 1997). CT or MR scans typically show vermian atrophy, and pathologically, the appearances are similar to those seen in Wernicke–Korsakoff syndrome. Improvement with thiamine treatment is rarely complete. The role of a direct toxic effect of alcohol is more controversial, but cannot entirely be discounted. Although brain atrophy is seen in well-nourished alcoholics (Nicolas *et al.* 1997), most cases of alcoholic ataxia seem to be nutritional and should be treated urgently with parenteral thiamine.

39.11.3 Paraneoplastic cerebellar degeneration

This condition is rare and mainly occurs with small-cell tumours of the lung, breast, female genital tract, and with lymphoma (see

Section 38.4.1). Pathologically, there is severe depletion of cerebellar Purkinje cells with variable additional inflammatory changes. Typically, there is a subacute onset over weeks or sometimes months of rapidly progressive severe gait and limb ataxia with vertigo, nystagmus, oscillopsia, and dysarthria. Most patients are severely disabled within weeks or months (Peterson *et al.* 1992) and then stabilize; it is this severity and rapidity of the cerebellar ataxia, unlike most other forms of cerebellar degeneration, that is suggestive of paraneoplastic aetiology (Bolla *et al.* 1997). Most patients are severely affected within 12 weeks (Graus *et al.* 2004). Mental changes are common, and there may be evidence of other paraneoplastic syndromes (see chapter 35) such as opsoclonus, limbic encephalitis, or peripheral neuropathy (Anderson *et al.* 1988; Tsukamoto *et al.* 1993; Mason *et al.* 1997). About 40 per cent of patients with lung cancer and paraneoplastic cerebellar degeneration have associated Lambert–Eaton myaesthenic syndrome with anti-voltage-gated calcium channel antibodies (Graus *et al.* 2004). It has been suggested that up to 50 per cent of patients over the age of 50 who present with this type of rapid subacute cerebellar syndrome have a paraneoplastic disorder (Henson *et al.* 1982), although this is still a small proportion of adult-onset ataxic syndromes overall. In most cases, the ataxia appears months—but sometimes a few years—before the cancer is apparent.

Brain CT and MR scans are often normal, but cerebellar atrophy may develop later; a lymphocytic pleocytosis in the cerebrospinal fluid is commonly detected along with elevated protein and positive testing for oligoclonal bands. Many patients, especially women with breast, ovarian, or uterine cancer, have circulating antineuronal (anti-Hu) or more commonly anti-Purkinje-cell (anti-Yo or PCA-1) antibodies (Dropcho 1995), but the role of these is uncertain. A wide variety of additional auto-antibodies have been detected in these patients, including anti-Ri (Kikuchi *et al.* 2000), anti-GAD or VGCC (Trivedi *et al.* 2000), CV2 (CRMP5), Ma1, Ma2, ANNA-3 (Chan *et al.* 2001), and amphiphysin antibodies. (Pittock *et al.* 2005). In most cases, treatment of the underlying neoplasm does not influence the neurological symptoms, but there are occasional reports of some improvement with immunosuppression, intravenous immunoglobulin, or plasmapheresis (Moll *et al.* 1993; Dropcho 1995; Stark *et al.* 1995; David *et al.* 1996; Bataller *et al.* 2003; Phuphanich *et al.* 2006), but evidence for such an approach is weak (Vedeler *et al.* 2006).

Occasionally, a similar syndrome occurs without detecting any underlying malignancy. Careful follow-up and clinically directed re-investigation of these patients is wise in view of the delay before a malignancy can become apparent.

39.11.4 Acute disseminated encephalomyelitis and multiple sclerosis

Cerebellar ataxia is often prominent in these conditions (Sections 37.4.1, 37.5). In acute disseminated encephalomyelitis, there is usually an acute onset with depressed consciousness and signs of multifocal neurological involvement. Most cases follow viral illnesses, mycoplasma infection, or vaccination. Multiple sclerosis usually does not present with a progressive cerebellar syndrome, but cerebellar features are very common in advanced cases. In one series of adults with acquired cerebellar ataxia with a known cause, 35 per cent had multiple sclerosis (Abele *et al.* 2002).

39.11.5 Heatstroke

The acute neurological manifestations of heatstroke include confusion, drowsiness, seizures, and focal neurological signs (Section 5.10.1). The majority of patients recover fully, but some develop a persistent cerebellar syndrome with cerebellar atrophy, suggesting neuronal loss (Biary *et al.* 1995).

39.11.6 Prion disease

Cerebellar ataxia is a prominent feature of Creutzfeldt–Jacob disease, Gerstmann–Straussler–Scheinker syndrome, kuru, and new variant Creutzfeldt–Jacob disease (Zeidler *et al.* 1997) (Section 42.8). In about 10 per cent of cases of classical Creutzfeldt–Jacob disease, a cerebellar ataxia is the presenting feature, but is soon accompanied by myoclonus and dementia.

39.11.7 Hypothyroidism

Cerebellar ataxia has been described as a consequence of hypothyroidism (Jellinek *et al.* 1960). Although thyroid function tests are routinely requested as part of the laboratory evaluation of ataxic patients, only a few cases of hypothyroidism-related ataxia have ever been reported and usually in association with severe overt myxoedema. One of the patients originally described with ataxia attributed to myxoedema was subsequently found to have multiple system atrophy and a malignancy (Quinn *et al.* 1992).

39.11.8 Nutritional deficiency

In Europe, nutritional cerebellar ataxia is most commonly seen in chronic malnourished alcoholics. An identical syndrome may be seen in nutritional deficiency for other reasons, for example, in some haemodialysis patients and those with cachexia due to malignant disease. A seasonal ataxia is also endemic in some parts of the world, which has been shown to be caused by dietary thiamine deficiency (Adamolekun *et al.* 1994). A cerebellar ataxia is seen in pellagra due to deficiency of nicotinic acid, but is rare. Vitamin E deficiency has been discussed in Section 39.5.2.

39.11.9 Gluten ataxia

It is well recognized that patients with coeliac disease may develop a progressive spinocerebellar degeneration (Cooke *et al.* 1966). Other coeliac patients have developed progressive myoclonic ataxia, Ramsay Hunt syndrome, and opsoclonus myoclonus (Deconinck *et al.* 2006). In some of these cases, there has been clinically evident intestinal malabsorption but in others the diagnosis has been made only by small bowel biopsy or the presence of anti-gliadin antibodies (Lu *et al.* 1993; Chinnery *et al.* 1997).

Some reports have described patients with cerebellar ataxia associated with anti-gliadin antibodies but without evidence of coeliac disease, the so-called 'gluten ataxia'. Initially, 28 patients with progressive cerebellar ataxia associated with anti-gliadin antibodies were reported (Hadjivassiliou *et al.* 1998). Some had gastrointestinal symptoms or abnormal duodenal biopsies, but in 11 cases neither was present and the only clue to the diagnosis was the anti-gliadin titre. Clinically, the mean age of onset was 54 with a progressive gait and lower limb ataxia, usually without nystagmus or dysarthria; the cerebellar syndrome was similar to the Marie–Foix–Alajouanine type of idiopathic late-onset ataxia (see Section 39.12), although a later report stated that ocular signs, upper limb ataxia, and dysarthria were common

(Hadjivassiliou *et al.* 2003b). In a large series from the United Kingdom, anti-gliadin antibodies were detected in 41 per cent of 132 cases of sporadic cerebellar ataxia, in contrast to 10–15 per cent among controls, hereditary ataxia, or cerebellar type of multiple system atrophy (Hadjivassiliou *et al.* 2003b). Other reports have described anti-gliadin antibodies in 9 per cent (Luostarinen *et al.* 2001) and 12 per cent (Burk *et al.* 2001) of sporadic ataxia cases. It has been claimed that 'gluten ataxia' is the most common cause of idiopathic ataxia, and a gluten free diet has been proposed in these patients (Hadjivassiliou *et al.* 2003a).

However, another series did not confirm this, finding no anti-gliadin antibodies in 32 patients with idiopathic adult-onset ataxia (Combarros *et al.* 2000) and another series from the UK found anti-gliadin antibodies to be extremely rare in ataxia patients from a population of 2 million (Lock *et al.* 2005). Anti-gliadin antibodies are found in patients with various hereditary ataxias (Bushara *et al.* 2001; Abele *et al.* 2003) and in patients with ataxia due to multiple sclerosis and meningitis (Anheim *et al.* 2006). Other food antibodies, to egg or milk, are also found in neurological patients, suggesting that anti-gliadin antibodies are a non-specific finding and are not pathogenic (Lock *et al.* 2005). The concept of 'gluten ataxia' remains highly controversial and not universally accepted (Serratrice *et al.* 2004; Anheim *et al.* 2006).

39.11.10 Autoimmune ataxia

Antibodies to glutamic acid decarboxylase, GAD, are typically associated with the stiff person syndrome (see Sections 38.3.4; 40.11.3). However, there have been a number of reports of cerebellar ataxia in association with anti-GAD antibodies (Honnorat *et al.* 2001; Kono *et al.* 2001; Vianello *et al.* 2003). There can be evidence of CSF inflammation with oligoclonal bands and intrathecal anti-GAD antibodies (Honnorat *et al.* 2001; Vianello *et al.* 2003). This appears to be a very rare cause of sporadic adult-onset ataxia (Abele *et al.* 2002) but is important because it is sometimes treatable with steroids (Lauria *et al.* 2003) or intravenous immunoglobulin (Dalakas 2005). It is more common in women (Honnorat *et al.* 2001).

A cerebellar syndrome has also been associated with anti-thyroid antibodies (Manto 2002) as seen in Hashimoto's encephalopathy but appears to be very rare.

39.12 Idiopathic late-onset cerebellar ataxias

A common clinical problem is the occurrence of a progressive idiopathic cerebellar syndrome in adults without a family history. In the majority of these cases, there is no evidence of an underlying cause for the problem even after investigation (see Table 39.4). These patients appear to have an idiopathic neurodegenerative disorder affecting the cerebellum, frequently with additional widespread central and peripheral nervous system involvement (Harding 1984). There has been remarkably little attention paid to this category of adult-onset ataxic patients in contrast to the explosion of information about the autosomal dominant cerebellar ataxias. Idiopathic late-onset cerebellar ataxia is much more common than the hereditary forms. In the series of Harding (1984), there were equal numbers of autosomal dominant and sporadic cases of adult-onset ataxia. In two recent series, after exclusion of cases with a known cause, idiopathic adult-onset cerebellar ataxia was over four times more common than hereditary ataxia (Abele *et al.* 2002; Muzaimi *et al.* 2004).

Neuropathological appearances are variable, but most patients have olivopontocerebellar atrophy or cortical cerebellar atrophy. These findings are not specific to idiopathic late-onset cerebellar ataxia, as similar appearances are seen in various forms of autosomal dominant cerebellar ataxia and multiple system atrophy. Accordingly, the tendency to use the term 'olivopontocerebellar atrophy', a not entirely accurate neuropathological term introduced at the beginning of the twentieth century, as a diagnostic label for idiopathic late-onset ataxia is misleading (Harding 1987). Unfortunately, the term olivopontocerebellar atrophy does persist as a diagnostic term in patients with the cerebellar presentation of multiple system atrophy (see Section 40.3.7).

Age of onset tends to be later than in autosomal dominant cerebellar ataxia, with a mean of about 55 years (Abele *et al.* 2002; Muzaimi *et al.* 2004). Deterioration is relentless but variable; patients lose the ability to walk independently between 5 and 20 years from onset. An important clinical difference from autosomal dominant cerebellar ataxia is that optic atrophy and retinopathy are not observed and ophthalmoplegia is less common (Burk *et al.* 1997). Three main categories have been suggested (Harding 1981b) based on clinical features and similarities to categories of these cases appearing in the earlier literature.

- A 'Marie–Foix–Alajouanine' type presents with gait and lower limb ataxia but little or no upper limb involvement and infrequent dysarthria (Marie *et al.* 1922). Age of onset is usually in the fifties and pathologically there is marked vermian cerebellar atrophy. Extracerebellar degeneration is mild. This pure cerebellar form probably accounts for about a third of cases (Muzaimi *et al.* 2004).

- Other patients develop a progressive cerebellar syndrome with very severe postural and intention tremor, similar to the 'dyssynergia cerebellaris progressiva' described by Hunt (Hunt 1914). Onset is usually around 50 years of age.

- A 'Dejerine Thomas type' in which a progressive ataxia develops slightly earlier, between 35 and 55 years, and is associated with additional features including dysarthria, dysphagia, pyramidal signs, dementia, and sometimes ophthalmoplegia, Parkinsonism, or peripheral neuropathy with impaired vibration sense and absent ankle reflexes (Abele *et al.* 2002). Bladder dysfunction is not uncommon. These cases are similar to those described previously as 'sporadic olivopontocerebellar atrophy' (Dejerine *et al.* 1900), and there is clinical overlap with the cerebellar presentation of multiple system atrophy (Klockgether *et al.* 1990; Wenning *et al.* 1994).

- A rare type with bilateral vestibulopathy and sensory neuropathy has been described also (Migliaccio *et al.* 2004).

A major difficulty in the differential diagnosis of idiopathic late-onset cerebellar ataxia is in distinguishing the cerebellar presentation of multiple system atrophy, now the only condition correctly referred to clinically as olivopontocerebellar atrophy. This problem is discussed in Section 40.3.7.

Pronounced autonomic impairment, with postural hypotension and bladder dysfunction, is suggestive of multiple system atrophy rather than idiopathic late-onset cerebellar ataxia. A fast clinical progression is also more typical of multiple system atrophy with patients losing independent walking after about 6 years and surviving 9 years from onset (Klockgether *et al.* 1998). The occurrence of

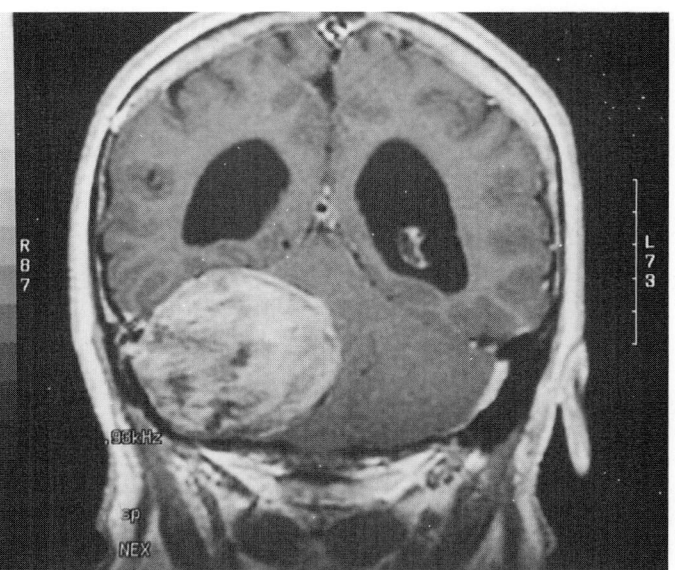

Fig. 39.12 Large posterior fossa meningioma on the right, gadolinium-enhanced magnetic resonance scan. The patient presented with a slowly evolving cerebellar ataxia without lateralizing signs or evidence of raised intracranial pressure.

cerebellar ataxia with two or more additional features such as autonomic impairment, Parkinsonism and pyramidal signs has been taken as evidence of multiple system atrophy in some studies (Abele *et al.* 2002). Ultimately, it may not be possible to resolve the matter without postmortem neuropathological examination, which may reveal characteristic glial and neuronal cytoplasmic inclusions in multiple system atrophy (Gilman *et al.* 1996).

A study of 112 adult-onset ataxic patients, of unknown cause with onset after the age of 20 years, found that 29 per cent had multiple system atrophy, 13 per cent had genetic mutations involving *Frataxin, SCA2, SCA3,* or *SCA6,* and 58 per cent were idiopathic; none had anti-GAD antibodies, and the incidence of anti-gliadin antibodies was similar in those with or without a discoverable cause (Abele *et al.* 2002). Another study in South East Wales described

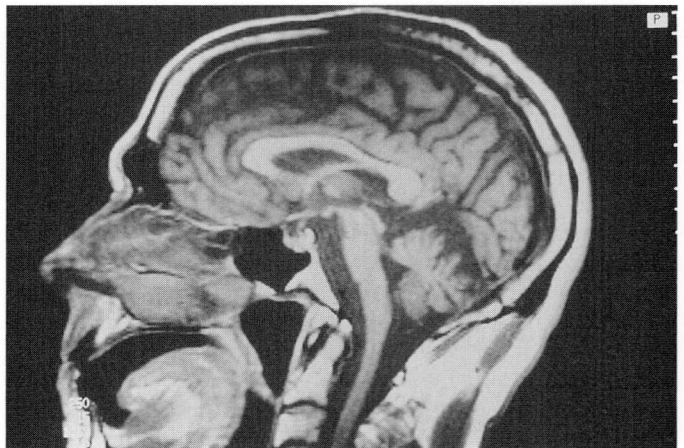

Fig. 39.13 Idiopathic late-onset cerebellar degeneration. Magnetic resonance scan showing severe cerebellar and brainstem atrophy. (Courtesy of Dr T P Enervoldson.)

76 patients with sporadic late-onset ataxia (Muzaimi *et al.* 2004). Only two additional patients had multiple system atrophy on clinical grounds. 83 per cent had idiopathic ataxia, while 17 per cent had late onset Friedreich's ataxia, episodic ataxia, or SCA6. Neither study analysed the cases for *FMR-1* mutations to detect the fragile X tremor ataxia syndrome. It was notable that in both studies the late-onset *SCA6* mutation could produce apparently sporadic late-onset ataxia.

Investigation of these patients is usually unrewarding but is important in order to exclude any treatable underlying condition. Neuroimaging is essential in order to exclude hydrocephalus or a posterior fossa mass lesion, which can mimic a cerebellar degeneration (Fig. 39.12). Most patients however, have cerebellar and brainstem atrophy (Wittkamper *et al.* 1993) (Fig. 39.13). It is important to exclude other potential causes of slowly progressive ataxia (see Table 39.4). An approach to investigations in this situation is discussed in Section 39.2. In most older adults, MR scanning and blood tests for thyroid function, vitamin B12 estimation, and basic investigations to exclude bronchial, ovarian, or breast malignancy is adequate unless other possibilities are suggested by the clinical presentation. Nerve conduction tests often show evidence of a peripheral neuropathy, while the EEG and CSF examination are rarely helpful. Anti-gliadin antibody estimation is probably of little value (Abele *et al.* 2002).

References

Aasly J, Storsaeter O, Nilsen G et al. (1993). Minor structural brain changes in young drug abusers. A magnetic resonance study. *Acta Neurol Scand*, **87**, 210–4.

Abele M, Burk K, Schols L et al. (2002). The aetiology of sporadic adult-onset ataxia. *Brain*, **125**, 961–8.

Abele M, Schols L, Schwartz S et al. (2003). Prevalence of antigliadin antibodies in ataxia patients. *Neurology*, **60**, 1674–5.

Adamec J, Rusnak F, Owen W et al. (2000). Iron-dependent self-assembly of recombinant yeast frataxin: implications for Friedreich ataxia. *Am J Hum Genet*, **67**, 549–62.

Adamolekun B, Adamolekun WE, Sonibare AD et al. (1994). A double-blind, placebo-controlled study of the efficacy of thiamine hydrochloride in a seasonal ataxia in Nigerians. *Neurology*, **44**, 549–51.

al Shahwan SA, Bruyn GW, al Deeb SM. (1995). Non-progressive familial congenital cerebellar hypoplasia. *J Neurol Sci*, **128**, 71–7.

Anderson NE, Budde-Steffen C, Rosenblum MK et al. (1988). Opsoclonus, myoclonus, ataxia, and encephalopathy in adults with cancer: a distinct paraneoplastic syndrome. *Medicine*, **67**, 100–9.

Anheim M, Degos B, Echaniz-Laguna A et al. (2006). Ataxia associated with gluten sensitivity, myth or reality?. *Rev Neurol (Paris)*, **162**, 214–21.

Anikster Y, Kleta R, Shaag A et al. (2001). Type III 3-methylglutaconic aciduria (optic atrophy plus syndrome, or Costeff optic atrophy syndrome): identification of the OPA3 gene and its founder mutation in Iraqi Jews. *Am J Hum Genet*, **69**, 1218–24.

Ansorge O, Giunti P, Michalik A et al. (2004). Ataxin-7 aggregation and ubiquitination in infantile SCA7 with 180 CAG repeats. *Ann Neurol*, **56**, 448–52.

Anttonen AK, Mahjneh I, Hamalainen RH et al. (2005). The gene disrupted in Marinesco-Sjogren syndrome encodes SIL1, an HSPA5 cochaperone. *Nat Genet*, **37**, 1309–11.

Apak S, Yuksel M, Ozmen M et al. (1989). Heterogeneity of X-linked recessive (spino)cerebellar ataxia with or without spastic diplegia. *Am J Med Genet*, **34**, 155–8.

Armani M, Zortea M, Pastorello E et al. (2006). Friedreich's ataxia: clinical heterogeneity in two sisters. *Neurol Sci*, **27**, 140–2.

Armstrong MB, Robertson PL, Castle VP (2005). Delayed, recurrent opsoclonus-myoclonus syndrome responding to plasmapheresis. *Pediatr Neurol*, **33**, 365–7.

Babcock M, de Silva D, Oaks R et al. (1997). Regulation of mitochondrial iron accumulation by Yfh1p, a putative homolog of frataxin. *Science*, **276**, 1709–12.

Bain P (1993). A combined clinical and neurophysiological approach to the study of patients with tremor. *J Neurol Neurosurg Psychiatry*, **56**, 839–44.

Bain PG, Findley LJ, Britton TC et al. (1995). Primary writing tremor. *Brain*, **118**, 1461–72.

Bain PG, Findley LJ, Thompson PD et al. (1994). A study of hereditary essential tremor. *Brain*, **117**, 805–24.

Barbot C, Coutinho P, Chorao R et al. (2001). Recessive ataxia with ocular apraxia: review of 22 Portuguese patients. *Arch Neurol*, **58**, 201–5.

Bardoni B, Mandel JL (2002). Advances in understanding of fragile X pathogenesis and FMRP function, and in identification of X linked mental retardation genes. *Curr Opin Genet Dev*, **12**, 284–93.

Barth PG (2000). Pontocerebellar hypoplasia--how many types? *Eur J Paediatr Neurol*, **4**, 161–2.

Bataller L, Dalmau J (2003). Paraneoplastic neurologic syndromes: approaches to diagnosis and treatment. *Semin Neurol*, **23**, 215–24.

Ben Smail D, Jacq C, Denys P et al. (2005). Intrathecal baclofen in the treatment of painful, disabling spasms in Friedreich's ataxia. *Mov Disord*, **20**, 758–9.

Benomar A, Krols L, Stevanin G et al. (1995). The gene for autosomal dominant cerebellar ataxia with pigmentary macular dystrophy maps to chromosome 3p12-p21.1. *Nat Genet*, **10**, 84–8.

Benomar A, Yahyaoui M, Meggouh F et al. (2002). Clinical comparison between AVED patients with 744 del A mutation and Friedreich ataxia with GAA expansion in 15 Moroccan families. *J Neurol Sci*, **198**, 25–9.

Berciano J, Infante J, Garcia A et al. (2005). Very late-onset Friedreich's ataxia with minimal GAA1 expansion mimicking multiple system atrophy of cerebellar type. *Mov Disord*, **20**, 1643–5.

Berneburg M, Lehmann AR (2001). Xeroderma pigmentosum and related disorders: defects in DNA repair and transcription. *Adv Genet*, **43**, 71–102.

Bhidayasiri R, Perlman SL, Pulst SM et al. (2005). Late-onset Friedreich ataxia: phenotypic analysis, magnetic resonance imaging findings, and review of the literature. *Arch Neurol*, **62**, 1865–9.

Biancalana V, Toft M, Le Ber I et al. (2005). FMR1 premutations associated with fragile X-associated tremor/ataxia syndrome in multiple system atrophy. *Arch Neurol*, **62**, 962–6.

Biary N, Cleeves L, Findley L et al. (1989). Post-traumatic tremor. *Neurology*, **39**, 103–6.

Biary N, Madkour MM, Sharif H (1995). Post-heatstroke parkinsonism and cerebellar dysfunction. *Clin Neurol Neurosurg*, **97**, 55–7.

Bidichandani S, Garcia C, Patel P et al. (2000). Very late-onset Friedreich ataxia despite large GAA triplet repeat expansions. *Arch Neurol*, **57**, 246–51.

Bidichandani SI, Ashizawa T, Patel PI (1998). The GAA triplet-repeat expansion in Friedreich ataxia interferes with transcription and may be associated with an unusual DNA structure. *Am J Hum Genet*, **62**, 111–21.

Blaes F, Fuhlhuber V, Korfei M et al. (2005). Surface-binding autoantibodies to cerebellar neurons in opsoclonus syndrome. *Ann Neurol*, **58**, 313–7.

Bolla L, Palmer RM (1997). Paraneoplastic cerebellar degeneration. Case report and literature review. *Arch Intern Med*, **157**, 1258–62.

Bott L, Thumerelle C, Cuvellier JC et al. (2006). Ataxia-telangiectasia: a review. *Arch Pediatr*, **13**, 293–8.

Bouchard JP, Richter A, Mathieu J et al. (1998). Autosomal recessive spastic ataxia of Charlevoix-Saguenay. *Neuromuscul Disord*, **8**, 474–9.

Bourgeois JA, Farzin F, Brunberg JA et al. (2006). Dementia with mood symptoms in a fragile X premutation carrier with the fragile X-associated tremor/ataxia syndrome: clinical intervention with donepezil and venlafaxine. *J Neuropsychiatry Clin Neurosci*, **18**, 171–7.

Bradley J, Blake J, Chamberlain S *et al.* (2000). Clinical, biochemical and molecular genetic correlations in Friedreich's ataxia. *Hum Mol Genet*, **9**, 275–82.

Britton TC, Thompson PD (1995). Primary orthostatic tremor [editorial]. *BMJ*, **310**, 143–4.

Britton TC, Thompson PD, van der Kamp W *et al.* (1992). Primary orthostatic tremor: further observations in six cases. *J Neurol*, **239**, 209–17.

Brunt ER, van Weerden TW (1990). Familial paroxysmal kinesigenic ataxia and continuous myokymia. *Brain*, **113**, 1361–82.

Bundey SE (1992). *Genetics and Neurology*. Churchill Livingstone, Edinburgh.

Burk K, Bosch S, Muller CA *et al.* (2001). Sporadic cerebellar ataxia associated with gluten sensitivity. *Brain*, **124**, 1013–9.

Burk K, Fetter M, Skalej M *et al.* (1997). Saccade velocity in idiopathic and autosomal dominant cerebellar ataxia. *J Neurol Neurosurg Psychiatry*, **62**, 662–4.

Bushara KO, Goebel SU, Shill H *et al.* (2001). Gluten sensitivity in sporadic and hereditary cerebellar ataxia. *Ann Neurol*, **49**, 540–3.

Butterworth RF (1993). Pathophysiology of cerebellar dysfunction in the Wernicke-Korsakoff syndrome. *Can J Neurol Sci*, **20**, S123–6.

Buyse G, Mertens L, Di Salvo G *et al.* (2003). Idebenone treatment in Friedreich's ataxia: neurological, cardiac, and biochemical monitoring. *Neurology*, **60**, 1679–81.

Cader MZ, Steckley JL, Dyment DA *et al.* (2005). A genome-wide screen and linkage mapping for a large pedigree with episodic ataxia. *Neurology*, **65**, 156–8.

Cagnoli C, Mariotti C, Taroni F *et al.* (2006). SCA28, a novel form of autosomal dominant cerebellar ataxia on chromosome 18p11.22-q11.2. *Brain*, **129**, 235–42.

Calabrese V, Lodi R, Tonon C *et al.* (2005). Oxidative stress, mitochondrial dysfunction and cellular stress response in Friedreich's ataxia. *J Neurol Sci*, **233**, 145–62.

Camacho JA, Rioseco-Camacho N, Andrade D *et al.* (2003). Cloning and characterization of human ORNT2: a second mitochondrial ornithine transporter that can rescue a defective ORNT1 in patients with the hyperornithinemia-hyperammonemia-homocitrullinuria syndrome, a urea cycle disorder. *Mol Genet Metab*, **79**, 257–71.

Campuzano V, Montermini L, Molto MD *et al.* (1996). Friedreich's ataxia: autosomal recessive disease caused by an intronic GAA triplet repeat expansion. *Science*, **271**, 1423–7.

Candler PM, Dale RC, Griffin S *et al.* (2006). Post-streptococcal opsoclonus-myoclonus syndrome associated with anti-neuroleukin antibodies. *J Neurol Neurosurg Psychiatry*, **77**, 507–12.

Castori M, Valente EM, Donati MA *et al.* (2005). NPHP1 gene deletion is a rare cause of Joubert syndrome related disorders. *J Med Genet*, **42**, e9.

Cavalier L, Ouahchi K, Kayden HJ *et al.* (1998). Ataxia with isolated vitamin E deficiency: heterogeneity of mutations and phenotypic variability in a large number of families. *Am J Hum Genet*, **62**, 301–10.

Cellini E, Piacentini S, Nacmias B *et al.* (2002). A family with spinocerebellar ataxia type 8 expansion and vitamin E deficiency ataxia. *Arch Neurol*, **59**, 1952–3.

Chai Y, Berke SS, Cohen RE *et al.* (2004). Poly-ubiquitin binding by the polyglutamine disease protein ataxin-3 links its normal function to protein surveillance pathways. *J Biol Chem*, **279**, 3605–11.

Chamberlain S, Shaw J, Rowland A *et al.* (1988). Mapping of mutation causing Friedreich's ataxia to human chromosome 9. *Nature*, **334**, 248–50.

Chan KH, Vernino S, Lennon VA (2001). ANNA-3 anti-neuronal nuclear antibody: marker of lung cancer-related autoimmunity. *Ann Neurol*, **50**, 301–11.

Chaudhuri KR, Buxton-Thomas M, Dhawan V *et al.* (2005). Long duration asymmetrical postural tremor is likely to predict development of Parkinson's disease and not essential tremor: clinical follow up study of 13 cases. *J Neurol Neurosurg Psychiatry*, **76**, 115–7.

Chinnery P, Turnbull D (1997). Clinical features, investigation and management of patients with defects of mitochondrial DNA. *J Neurol Neurosurg Psychiatry*, **63**, 559–63.

Chio A, Orsi L, Mortara P *et al.* (1993). Early onset cerebellar ataxia with retained tendon reflexes: prevalence and gene frequency in an Italian population. *Clin Genet*, **43**, 207–11.

Chung MY, Lu YC, Cheng NC *et al.* (2003). A novel autosomal dominant spinocerebellar ataxia (SCA22) linked to chromosome 1p21-q23. *Brain*, **126**, 1293–9.

Claus D, Harding A, Hess C *et al.* (1988). Central motor conduction in degenerative ataxic disorders: a magnetic stimulation study. *J Neurol Neurosurg Psychiatry*, **51**, 790–5.

Cleeves L, Findley LJ, Koller W (1988). Lack of association between essential tremor and Parkinson's disease. *Ann Neurol*, **24**, 23–6.

Cleeves L, Findley LJ, Marsden CD (1994). Odd tremors. In Marsden CD, Fahn S, eds. *Movement Disorders 3*, pp. 434–58. Butterworth Heinemann, Oxford.

Combarros O, Infante J, Lopez-Hoyos M *et al.* (2000). Celiac disease and idiopathic cerebellar ataxia. *Neurology*, **54**, 2346.

Conway D, Bain PG, Warner TT *et al.* (1993). Linkage analysis with chromosome 9 markers in hereditary essential tremor. *Mov Disord*, **8**, 374–6.

Cooke WT, Smith WT (1966). Neurological disorders associated with adult coeliac disease. *Brain*, **89**, 683–722.

Cooper JM, Schapira AH (2003). Friedreich's Ataxia: disease mechanisms, antioxidant and Coenzyme Q10 therapy. *Biofactors*, **18**, 163–71.

Coppola G, Vajro P, De Virgiliis S *et al.* (2002). Cerebellar vermis defect, oligophrenia, congenital ataxia, and hepatic fibrocirrhosis without coloboma and renal abnormalities: report of three cases. *Neuropediatrics*, **33**, 180–5.

Cossee M, Schmitt M, Campuzano V *et al.* (1997). Evolution of the Friedreich's ataxia trinucleotide repeat expansion: founder effect and premutations. *Proc Natl Acad Sci USA*, **94**, 7452–7.

Crawford TO, Skolasky RL, Fernandez R *et al.* (2006). Survival probability in ataxia telangiectasia. *Arch Dis Child*, **91**, 610–1.

Criscuolo C, Chessa L, Di Giandomenico S *et al.* (2006). Ataxia with oculomotor apraxia type 2: a clinical, pathologic, and genetic study. *Neurology*, **66**, 1207–10.

Dalakas MC (2005). The role of IVIg in the treatment of patients with stiff person syndrome and other neurological diseases associated with anti-GAD antibodies. *J Neurol*, **252** (Suppl 1), I19–25.

Dalski A, Atici J, Kreuz FR *et al.* (2005). Mutation analysis in the fibroblast growth factor 14 gene: frameshift mutation and polymorphisms in patients with inherited ataxias. *Eur J Hum Genet*, **13**, 118–20.

Damji KF, Allingham RR, Pollock SC *et al.* (1996). Periodic vestibulocerebellar ataxia, an autosomal dominant ataxia with defective smooth pursuit, is genetically distinct from other autosomal dominant ataxias. *Arch Neurol*, **53**, 338–44.

Danek A (1993). Geniospasm: hereditary chin trembling. *Mov Disord*, **8**, 335–8.

Date H, Onodera O, Tanaka H *et al.* (2001). Early-onset ataxia with ocular motor apraxia and hypoalbuminemia is caused by mutations in a new HIT superfamily gene. *Nat Genet*, **29**, 184–8.

David YB, Warner E, Levitan M *et al.* (1996). Autoimmune paraneoplastic cerebellar degeneration in ovarian carcinoma patients treated with plasmapheresis and immunoglobulin. A case report. *Cancer*, **78**, 2153–6.

Day JW, Schut LJ, Moseley ML *et al.* (2000). Spinocerebellar ataxia type 8: clinical features in a large family. *Neurology*, **55**, 649–57.

De Castro M, Garcia-Planells J, Monros E *et al.* (2000). Genotype and phenotype analysis of Friedreich's ataxia compound heterozygous patients. *Hum Genet*, **106**, 86–92.

De Michele G, Filla A, Cavalcanti F *et al.* (1994). Late onset Friedreich's disease: clinical features and mapping of mutation to the FRDA locus. *J Neurol Neurosurg Psychiatry*, **57**, 977–9.

De Michele G, Filla A, Marconi R et al. (1995). A genetic study of Parkinson's disease. *J Neural Transm Suppl*, **45**, 21–5.

De Michele G, Filla A, Striano S et al. (1993). Heterogeneous findings in four cases of cerebellar ataxia associated with hypogonadism (Holmes' type ataxia). *Clin Neurol Neurosurg*, **95**, 23–8.

De Michele G, Perrone F, Filla A et al. (1996). Age of onset, sex, and cardiomyopathy as predictors of disability and survival in Friedreich's disease: a retrospective study on 119 patients. *Neurology*, **47**, 1260–4.

De Stefano N, Dotti MT, Mortilla M et al. (2001). Magnetic resonance imaging and spectroscopic changes in brains of patients with cerebrotendinous xanthomatosis. *Brain*, **124**, 121–31.

Deconinck N, Scaillon M, Segers V et al. (2006). Opsoclonus-myoclonus associated with celiac disease. *Pediatr Neurol*, **34**, 312–4.

Dejerine J, Thomas A (1900). L'atrophie olivo-ponto-cerebelleuse. *Nouvelle iconographie de la Salpetriere*, **13**, 330–76.

Delatycki MB, Holian A, Corben L et al. (2005). Surgery for equinovarus deformity in Friedreich's ataxia improves mobility and independence. *Clin Orthop Relat Res*, **430**, 138–41.

Delatycki MB, Paris D, Gardner RJ et al. (1998). Sperm DNA analysis in a Friedreich ataxia premutation carrier suggests both meiotic and mitotic expansion in the FRDA gene. *J Med Genet*, **35**, 713–6.

Deng Y, Newman B, Dunne MP et al. (2004). Case-only study of interactions between genetic polymorphisms of GSTM1, P1, T1 and Z1 and smoking in Parkinson's disease. *Neurosci Lett*, **366**, 326–31.

des Portes V, Bachner L, Bruls T et al. (1996). X-linked neurodegenerative syndrome with congenital ataxia, late-onset progressive myoclonic encephalopathy and selective macular degeneration, linked to Xp22.33-pter. *Am J Med Genet*, **64**, 69–72.

Deuschl G (2003). Dystonic tremor. *Rev Neurol (Paris)*, **159**, 900–5.

Devos D, Schraen-Maschke S, Vuillaume I et al. (2001). Clinical features and genetic analysis of a new form of spinocerebellar ataxia. *Neurology*, **56**, 234–8.

Dropcho E (1995). Autoimmune central nervous system paraneoplastic disorders: mechanisms, diagnosis and therapeutic options. *Ann Neurol*, **37**, S102–13.

Durr A, Cossee M, Agid Y et al. (1996). Clinical and genetic abnormalities in patients with Friedreich's ataxia [see comments]. *N Engl J Med*, **335**, 1169–75.

Durr A, Stevanin G, Jedynak CP et al. (1993). Familial essential tremor and idiopathic torsion dystonia are different genetic entities. *Neurology*, **43**, 2212–4.

Enevoldson TP, Sanders MD, Harding AE (1994). Autosomal dominant cerebellar ataxia with pigmentary macular dystrophy. A clinical and genetic study of eight families. *Brain*, **117**, 445–60.

Engert JC, Berube P, Mercier J et al. (2000). ARSACS, a spastic ataxia common in northeastern Quebec, is caused by mutations in a new gene encoding an 11.5-kb ORF. *Nat Genet*, **24**, 120–5.

English KM, Gibbs JL (2006). Cardiac monitoring and treatment for children and adolescents with neuromuscular disorders. *Dev Med Child Neurol*, **48**, 231–5.

Escayg A, De Waard M, Lee DD et al. (2000). Coding and non coding variation of the human calcium channel beta (4)-subunit gene CACNB4 in patients with idiopathic generalised epilepsy and episodic ataxia. *Am J Hum Genet*, **66**, 1531–39.

Espinos-Armero C, Gonzalez-Cabo P, Palau-Martinez F (2005). Autosomal recessive cerebellar ataxias. Their classification, genetic features and pathophysiology. *Rev Neurol*, **41**, 409–22.

Evidente V, Adler C, Caviness J et al. (1998). Effective treatment of orthostatic tremor with gabapentin. *Mov Disord*, **13**, 829–31.

Factor SA, Qian J, Lava NS et al. (2005). False-positive SCA8 gene test in a patient with pathologically proven multiple system atrophy. *Ann Neurol*, **57**, 462–3.

Farrer M, Gwinn-Hardy K, Muenter M et al. (1999). A chromosome 4p haplotype segregating with Parkinson's disease and postural tremor. *Hum Mol Genet*, **8**, 81–5.

Fear C (1992). Drug induced Creutzfeldt Jacob like syndrome: a review. *Hum Psychopharmacol*, **7**, 89.

Filla A, De Michele G, Cavalcanti F et al. (1990). Clinical and genetic heterogeneity in early onset cerebellar ataxia with retained tendon reflexes. *J Neurol Neurosurg Psychiatry*, **53**, 667–70.

Findley LJ, Cleeves L, Calzetti S (1985). Primidone in essential tremor of the hands and head: a double blind controlled clinical study. *J Neurol Neurosurg Psychiatry*, **48**, 911–5.

Flanigan K, Gardner K, Alderson K et al. (1996). Autosomal dominant spinocerebellar ataxia with sensory axonal neuropathy (SCA4): clinical description and genetic localization to chromosome 16q22.1. *Am J Hum Genet*, **59**, 392–9.

Fletcher NA, Harding AE, Marsden CD (1990). A genetic study of idiopathic torsion dystonia in the United Kingdom. *Brain*, **113**, 379–95.

Fletcher NA, Harding AE, Marsden CD (1991). A case-control study of idiopathic torsion dystonia. *Mov Disord*, **6**, 304–9.

Foote KD, Okun MS (2005). Ventralis intermedius plus ventralis oralis anterior and posterior deep brain stimulation for posttraumatic Holmes tremor: two leads may be better than one: technical note. *Neurosurgery*, **56**(2 Suppl), E445; discussion E445.

Frontali M, Spadaro M, Giunti P et al. (1992). Autosomal dominant pure cerebellar ataxia. Neurological and genetic study. *Brain*, **115**, 1647–54.

Gallus GN, Dotti MT, Federico A (2006). Clinical and molecular diagnosis of cerebrotendinous xanthomatosis with a review of the mutations in the CYP27A1 gene. *Neurol Sci*, **27**, 143–9.

Gancher S, Nutt J. (1986a). Autosomal dominant episodic ataxia: a heterogeneous syndrome. *Mov Disord*, **1**, 239–53.

Gancher ST, Nutt JG (1986b). Autosomal dominant episodic ataxia: a heterogeneous syndrome. *Mov Disord*, **1**, 239–53.

Gellera C, Pareyson D, Castellotti B et al. (1997). Very late onset Friedreich's ataxia without cardiomyopathy is associated with limited GAA expansion in the X25 gene. *Neurology*, **49**, 1153–5.

Gentile M, Di Carlo A, Susca F et al. (1996). COACH syndrome: report of two brothers with congenital hepatic fibrosis, cerebellar vermis hypoplasia, oligophrenia, ataxia, and mental retardation. *Am J Med Genet*, **64**, 514–20.

Georgy BA, Snow RD, Brogdon BG et al. (1998). Neuroradiologic findings in Marinesco-Sjogren syndrome. *Am J Neuroradiol*, **19**, 281–3.

Geschwind DH, Perlman S, Figueroa KP et al. (1997a). Spinocerebellar ataxia type 6. Frequency of the mutation and genotype-phenotype correlations. *Neurology*, **49**, 1247–51.

Geschwind DH, Perlman S, Grody WW et al. (1997b). Friedreich's ataxia GAA repeat expansion in patients with recessive or sporadic ataxia. *Neurology*, **49**, 1004–9.

Gieron-Korthals MA, Westberry KR, Emmanuel PJ (1994). Acute childhood ataxia: 10-year experience. *J Child Neurol*, **9**, 381–4.

Gilman S, Bloedel JR, Lechtenberg R (1981). *Disorders of the Cerebellum*. FA Davis, Philadelphia.

Gilman S, Quinn NP (1996). The relationship of multiple system atrophy to sporadic olivopontocerebellar atrophy and other forms of idiopathic late-onset cerebellar atrophy. *Neurology*, **46**, 1197–9.

Gironell A, Kulisevsky J, Barbanoj M et al. (1999). A randomized placebo-controlled comparative trial of gabapentin and propranolol in essential tremor. *Arch Neurol*, **56**, 475–80.

Gironi M, Lamperti C, Nemni R et al. (2004). Late-onset cerebellar ataxia with hypogonadism and muscle coenzyme Q10 deficiency. *Neurology*, **62**, 818–20.

Giunti P, Stevanin G, Worth P et al. (1999). Molecular and clinical study of 18 families with ADCA type II: evidence for genetic heterogeneity and de novo mutation. *Am J Hum Genet*, **64**, 1594–603.

Glass HC, Boycott KM, Adams C et al. (2005). Autosomal recessive cerebellar hypoplasia in the Hutterite population. *Dev Med Child Neurol*, **47**, 691–5.

Goasdoue P, Rodriguez D, Moutard ML et al. (2001). Pontoneocerebellar hypoplasia: definition of MR features. *Pediatr Radiol*, **31**, 613–8.

Golomb MR, Walsh LE, Carvalho KS *et al.* (2004). Clinical findings in Pelizaeus-Merzbacher disease. *J Child Neurol*, **19**, 328–31.

Gordon MF, Abrams RI, Rubin DB *et al.* (1995). Bismuth subsalicylate toxicity as a cause of prolonged encephalopathy with myoclonus. *Mov Disord*, **10**, 220–2.

Graus F, Delattre JY, Antoine JC *et al.* (2004). Recommended diagnostic criteria for paraneoplastic neurological syndromes. *J Neurol Neurosurg Psychiatry*, **75**, 1135–40.

Green JS, Parfrey PS, Harnett JD *et al.* (1989). The cardinal manifestations of Bardet-Biedl syndrome, a form of Laurence-Moon-Biedl syndrome. *N Engl J Med*, **321**, 1002–9.

Grieco GS, Malandrini A, Comanducci G *et al.* (2004). Novel SACS mutations in autosomal recessive spastic ataxia of Charlevoix-Saguenay type. *Neurology*, **62**, 103–6.

Griggs RC, Nutt JG (1995). Episodic ataxias as channelopathies [editorial; comment]. *Ann Neurol*, **37**, 285–7.

Grigsby J, Leehey MA, Jacquemont S *et al.* (2006). Cognitive Impairment in a 65-year-old Male With the Fragile X-associated Tremor-Ataxia Syndrome (FXTAS). *Cogn Behav Neurol*, **19**, 165–71.

Grisoli M, Piperno A, Chiapparini L *et al.* (2005). MR imaging of cerebral cortical involvement in aceruloplasminemia. *Am J Neuroradiol*, **26**, 657–61.

Grosso S, Mostadini R, Cioni M *et al.* (2002). Pontocerebellar hypoplasia type 2: further clinical characterization and evidence of positive response of dyskinesia to levodopa. *J Neurol*, **249**, 596–600.

Gulcher JR, Jonsson P, Kong A *et al.* (1997). Mapping of a familial essential tremor gene, FET1, to chromosome 3q13. *Nat Genet*, **17**, 84–7.

Gunal N, Saraclar M, Ozkutlu S *et al.* (1996). Heart disease in Friedreich's ataxia: a clinical and echocardiographic study. *Acta Paediatr Jpn*, **38**, 308–11.

Guo X, Shen H, Xiao X *et al.* (2006). Cataracts, ataxia, short stature, and mental retardation in a Chinese family mapped to Xpter-q13.1. *J Hum Genet*, **51**, 695–700.

Hadjivassiliou M, Davies-Jones GA, Sanders DS *et al.* (2003a). Dietary treatment of gluten ataxia. *J Neurol Neurosurg Psychiatry*, **74**, 1221–4.

Hadjivassiliou M, Grunewald R, Sharrack B *et al.* (2003b). Gluten ataxia in perspective: epidemiology, genetic susceptibility and clinical characteristics. *Brain*, **126**, 685–91.

Hadjivassiliou M, Grunewald RA, Chattopadhyay AK *et al.* (1998). Clinical, radiological, neurophysiological, and neuropathological characteristics of gluten ataxia. *Lancet*, **352**, 1582–5.

Hagerman PJ, Greco CM, Hagerman RJ (2003). A cerebellar tremor/ataxia syndrome among fragile X premutation carriers. *Cytogenet Genome Res*, **100**, 206–12.

Hagerman PJ, Hagerman RJ (2004a). The fragile-X premutation: a maturing perspective. *Am J Hum Genet*, **74**, 805–16.

Hagerman PJ, Hagerman RJ (2004b). Fragile X-associated tremor/ataxia syndrome (FXTAS). *Ment Retard Dev Disabil Res Rev*, **10**, 25–30.

Hagerman RJ, Leavitt BR, Farzin F *et al.* (2004c). Fragile-X-associated tremor/ataxia syndrome (FXTAS) in females with the FMR1 premutation. *Am J Hum Genet*, **74**, 1051–6.

Hallett M (1984). Classification and treatment of tremor. *JAMA*, **266**, 1115–7.

Hallett M, Dubinsky RM (1993). Glucose metabolism in the brain of patients with essential tremor. *J Neurol Sci*, **114**, 45–8.

Hammans SR, Kennedy CR (1998). Ataxia with isolated vitamin E deficiency presenting as mutation negative Friedreich's ataxia. *J Neurol Neurosurg Psychiatry*, **64**, 368–70.

Hanna MG, Davis MB, Sweeney MG *et al.* (1998). Generalized chorea in two patients harboring the Friedreich's ataxia gene trinucleotide repeat expansion. *Mov Disord*, **13**, 339–40.

Hara K, Fukushima T, Suzuki T *et al.* (2004). Japanese SCA families with an unusual phenotype linked to a locus overlapping with SCA15 locus. *Neurology*, **62**, 648–51.

Harding A (1981a). Friedreich's ataxia: a clinical and genetic study of 90 families with an analysis of early diagnostic criteria and intrafamilial clustering of cases. *Brain*, **104**, 589–620.

Harding A (1981b). Idiopathic late onset cerebellar ataxia: a clinical and genetic study of 36 cases. *J Neurol Sci*, **51**, 259–71.

Harding A (1987). Olivopontocerebellar atrophy is not a useful concept. In Marsden C, Fahn S, eds. *Movement Disorders 2*, pp. 269–71. Butterworths, London.

Harding A, Hewer R (1983). The heart disease of Friedreich's ataxia: a clinical and electrocardiographic study of 115 patients, with an analysis of serial electrocardiographic changes in 30 cases. *QJM*, **208**, 489–502.

Harding A, Zilkha K (1981). Pseudodominant inheritance in Friedreich's ataxia. *J Med Genet*, **18**, 285–7.

Harding AE (1984). *The Hereditary Ataxias and Related Disorders*. Churchill Livingstone, Edinburgh.

Hatano T, Okuma Y, Iijima M *et al.* (2003). Cervical dystonia in dentatorubral-pallidoluysian atrophy. *Acta Neurol Scand*, **108**, 287–9.

Hebb MO, Gaudet P, Mendez I (2006). Deep brain stimulation to treat hyperkinetic symptoms of Cockayne syndrome. *Mov Disord*, **21**, 112–5.

Heilman K (1984). Orthostatic tremor. *Arch Neurol*, **41**, 880–1.

Hellenbroich Y, Bubel S, Pawlack H *et al.* (2003). Refinement of the spinocerebellar ataxia type 4 locus in a large German family and exclusion of CAG repeat expansions in this region. *J Neurol*, **250**, 668–71.

Hellier KD, Hatchwell E, Duncombe AS *et al.* (2001). X-linked sideroblastic anaemia with ataxia: another mitochondrial disease? *J Neurol Neurosurg Psychiatry*, **70**, 65–9.

Henson R, Urich H. (1982). *Cancer and the Nervous System*. Blackwell Scientific Publications, London.

Herman-Bert A, Stevanin G, Netter J *et al.* (2000). Mapping of spinocerebellar ataxia 13 to chromosome 19q13.3-q13.4 in a family with autosomal dominant cerebellar ataxia and mental retardation. *Am J Hum Genet*, **67**, 229–35.

Herz E (1944). Dystonia. I. Historical review; analysis of dystonic symptoms and physiologic mechanisms involved. *Arch Neurol Psychiatr*, **51**, 305–18.

Higgins JJ, Lombardi RQ, Pucilowska J *et al.* (2005). A variant in the HS1-BP3 gene is associated with familial essential tremor. *Neurology*, **64**, 417–21.

Higgins JJ, Pho LT, Nee LE (1997). A gene (ETM) for essential tremor maps to chromosome 2p22–p25. *Mov Disord*, **12**, 859–64.

Holmes SE, O'Hearn EE, McInnis MG *et al.* (1999). Expansion of a novel CAG trinucleotide repeat in the 5' region of PPP2R2B is associated with SCA12. *Nat Genet*, **23**, 391–2.

Honnorat J, Saiz A, Giometto B *et al.* (2001). Cerebellar ataxia with anti-glutamic acid decarboxylase antibodies: study of 14 patients. *Arch Neurol*, **58**, 225–30.

Hooper AJ, van Bockxmeer FM, Burnett JR (2005). Monogenic hypocholesterolaemic lipid disorders and apolipoprotein B metabolism. *Crit Rev Clin Lab Sci*, **42**, 515–45.

Hou JG, Jankovic J (2003). Movement disorders in Friedreich's ataxia. *J Neurol Sci*, **206**, 59–64.

Hua SE, Lenz FA, Zirh TA *et al.* (1998). Thalamic neuronal activity correlated with essential tremor. *J Neurol Neurosurg Psychiatry*, **64**, 273–6.

Hunt J (1914). Dyssynergia cerebellaris progressiva - a chronic progressive form of cerebellar tremor. *Brain*, **37**, 247–68.

IIzuka R, HirayamaK, Maehara K (1984). Dentato-rubro-pallidoluysian atrophy: a clinico-pathological study. *J Neurol Neurosurg Psychiatry*, **47**, 1288–98.

Illarioshkin SN, Tanaka H, Markova ED *et al.* (1996). X-linked nonprogressive congenital cerebellar hypoplasia: clinical description and mapping to chromosome Xq. *Ann Neurol*, **40**, 75–83.

Imamura S, Tachi N, Oya K. (1993). Dominantly inherited early-onset non-progressive cerebellar ataxia syndrome. *Brain Dev*, **15**, 372–6.

Imbrici P, Eunson LH, Graves TD *et al.* (2005). Late-onset episodic ataxia type 2 due to an in-frame insertion in CACNA1A. *Neurology*, **65**, 944–6.

Imrie J, Vijayaraghaven S, Whitehouse C *et al.* (2002). Niemann-Pick disease type C in adults. *J Inherit Metab Dis*, **25**, 491–500.

Isnard R, Kalotka H, Durr A *et al.* (1997). Correlation between left ventricular hypertrophy and GAA trinucleotide repeat length in Friedreich's ataxia. *Circulation*, **95**, 2247–9.

Jacquemont S, Orrico A, Galli L *et al.* (2005). Spastic paraparesis, cerebellar ataxia, and intention tremor: a severe variant of FXTAS? *J Med Genet*, **42**, e14.

Jain S, Lo SE, Louis ED (2006). Common misdiagnosis of a common neurological disorder: how are we misdiagnosing essential tremor? *Arch Neurol*, **63**, 1100–4.

Jankovic J (2002). Essential tremor: a heterogenous disorder. *Mov Disord*, **17**, 638–44.

Jankovic J, Leder S, Warner D *et al.* (1991). Cervical dystonia: clinical findings and associated movement disorders. *Neurology*, **41**, 1088–91.

Jankovic J, Schwartz K, Clemence W *et al.* (1996). A randomized, double-blind, placebo-controlled study to evaluate botulinum toxin type A in essential hand tremor. *Mov Disord*, **11**, 250–6.

Jankovic J, Van der Linden C (1988). Dystonia and tremor induced by peripheral trauma:predisposing factors. *J Neurol Neurosurg Psychiatry*, **51**, 1512–9.

Jarman PR, Wood NW, Davis MT *et al.* (1997). Hereditary geniospasm: linkage to chromosome 9q13-q21 and evidence for genetic heterogeneity. *Am J Hum Genet*, **61**, 928–33.

Jellinek E, Kelly R. (1960). Cerebellar syndrome in myxoedema. *Lancet*, ii, 225–7.

Jenkins IH, Bain PG, Colebatch JG *et al.* (1993). A positron emission tomography study of essential tremor: evidence for overactivity of cerebellar connections. *Ann Neurol*, **34**, 82–90.

Jin P, Warren ST (2003). New insights into fragile X syndrome: from molecules to neurobehaviors. *Trends Biochem Sci*, **28**, 152–8.

Johnson-Greene D, Adams KM, Gilman S *et al.* (1997). Impaired upper limb coordination in alcoholic cerebellar degeneration. *Arch Neurol*, **54**, 436–9.

Junck L, Fink J (1996). Macado-Joseph disease and SCA3: The genotype meets the phenotype. *Neurology*, **46**, 4–8.

Kamm C, Healy DG, Quinn NP *et al.* (2005). The fragile X tremor ataxia syndrome in the differential diagnosis of multiple system atrophy: data from the EMSA Study Group. *Brain*, **128**, 1855–60.

Katzenschlager R, Costa D, Gerschlager W *et al.* (2003). [123I]-FP-CIT-SPECT demonstrates dopaminergic deficit in orthostatic tremor. *Ann Neurol*, **53**, 489–96.

Kawanami T, Kato T, Daimon M *et al.* (1996). Hereditary caeruloplasmin deficiency: clinicopathological study of a patient. *J Neurol Neurosurg Psychiatry*, **61**, 506–9.

Kikuchi H, Yamada T, Okayama A *et al.* (2000). Anti-Ri-associated paraneoplastic cerebellar degeneration without opsoclonus in a patient with a neuroendocrine carcinoma of the stomach. *Fukuoka Igaku Zasshi*, **91**, 104–9.

Klebe S, Durr A, Rentschler A *et al.* (2005). New mutations in protein kinase Cgamma associated with spinocerebellar ataxia type 14. *Ann Neurol*, **58**, 720–9.

Kleta R, Romeo E, Ristic Z *et al.* (2004). Mutations in SLC6A19, encoding B0AT1, cause Hartnup disorder. *Nat Genet*, **36**, 999–1002.

Klockgether T, Chamberlain S, Wullner U *et al.* (1993). Late-onset Friedreich's ataxia. Molecular genetics, clinical neurophysiology, and magnetic resonance imaging. *Arch Neurol*, **50**, 803–6.

Klockgether T, Ludtke R, Kramer B *et al.* (1998). The natural history of degenerative ataxia: a retrospective study in 466 patients. *Brain*, **121**, 589–600.

Klockgether T, Petersen D, Grodd W *et al.* (1991). Early onset cerebellar ataxia with retained tendon reflexes. Clinical, electrophysiological and MRI observations in comparison with Friedreich's ataxia. *Brain*, **114**, 1559–73.

Klockgether T, Schroth G, Diener HC *et al.* (1990). Idiopathic cerebellar ataxia of late onset: natural history and MRI morphology. *J Neurol Neurosurg Psychiatry*, **53**, 297–305.

Klockgether T, Zuhlke C, Schulz JB *et al.* (1996). Friedreich's ataxia with retained tendon reflexes: molecular genetics, clinical neurophysiology, and magnetic resonance imaging. *Neurology*, **46**, 118–21.

Koide R, Ikeuchi T, Onodera O *et al.* (1994). Unstable expansion of CAG repeat in hereditary dentatorubral-pallidoluysian atrophy (DRPLA). *Nat Genet*, **6**, 14–8.

Koller WC, Busenbark K, Miner K. (1994). The relationship of essential tremor to other movement disorders: report on 678 patients. Essential Tremor Study Group. *Ann Neurol*, **35**, 717–23.

Koller WC, Royse VL (1986). Efficacy of primidone in essential tremor. *Neurology*, **36**, 121–4.

Koller WC, Vetere-Overfield B (1989). Acute and chronic effects of propranolol and primidone in essential tremor. *Neurology*, **39**, 1587–8.

Kono S, Miyajima H, Sugimoto M *et al.* (2001). Stiff-person syndrome associated with cerebellar ataxia and high glutamic acid decarboxylase antibody titer. *Intern Med*, **40**, 968–71.

Koob MD, Moseley ML, Schut LJ *et al.* (1999). An untranslated CTG expansion causes a novel form of spinocerebellar ataxia (SCA8). *Nat Genet*, **21**, 379–84.

Kores B, Lader MH (1997). Irreversible lithium neurotoxicity: an overview. *Clin Neuropharmacol*, **20**, 283–99.

Korfei M, Fuhlhuber V, Schmidt-Woll T *et al.* (2005). Functional characterisation of autoantibodies from patients with pediatric opsoclonus-myoclonus-syndrome. *J Neuroimmunol*, **170**, 150–7.

Koscik RL, Lai HJ, Laxova A *et al.* (2005). Preventing early, prolonged vitamin E deficiency: an opportunity for better cognitive outcomes via early diagnosis through neonatal screening. *J Pediatr*, **147**, S51–6.

Koutnikova H, Campuzano V, Foury F *et al.* (1997). Studies of human, mouse and yeast homologues indicate a mitochondrial function for frataxin. *Nat Genet*, **16**, 345–51.

Kumar R, Lozano AM, Sime E *et al.* (2003). Long-term follow-up of thalamic deep brain stimulation for essential and parkinsonian tremor. *Neurology*, **61**, 1601–4.

Labauge P. (2002). Very late onset Friedreich's presenting as spastic tetraparesis without ataxia or neuropathy. *Neurology*, **58**, 1136.

Labrune P, Lacroix C, Goutieres F *et al.* (1996). Extensive brain calcifications, leukodystrophy, and formation of parenchymal cysts: a new progressive disorder due to diffuse cerebral microangiopathy. *Neurology*, **46**, 1297–301.

Lagier-Tourenne C, Tranebaerg L, Chaigne D *et al.* (2003). Homozygosity mapping of Marinesco-Sjogren syndrome to 5q31. *Eur J Hum Genet*, **11**, 770–8.

Lamont PJ, Davis MB, Wood NW (1997). Identification and sizing of the GAA trinucleotide repeat expansion of Friedreich's ataxia in 56 patients. Clinical and genetic correlates. *Brain*, **120**, 673–80.

Lamperti C, Naini A, Hirano M *et al.* (2003). Cerebellar ataxia and coenzyme Q10 deficiency. *Neurology*, **60**, 1206–8.

Lauria G, Pareyson D, Pitzolu MG *et al.* (2003). Excellent response to steroid treatment in anti-GAD cerebellar ataxia. *Lancet Neurol*, **2**, 634–5.

Le Ber I, Bouslam N, Rivaud-Pechoux S *et al.* (2004). Frequency and phenotypic spectrum of ataxia with oculomotor apraxia 2: a clinical and genetic study in 18 patients. *Brain*, **127**, 759–67.

Le Ber I, Camuzat A, Castelnovo G *et al.* (2003a). Prevalence of dentatorubral-pallidoluysian atrophy in a large series of white patients with cerebellar ataxia. *Arch Neurol*, **60**, 1097–9.

Le Ber I, Moreira MC, Rivaud-Pechoux S *et al.* (2003b). Cerebellar ataxia with oculomotor apraxia type 1: clinical and genetic studies. *Brain*, **126**, 2761–72.

Lewis SM, Roberts EA, Marcon MA *et al.* (1994). Joubert syndrome with congenital hepatic fibrosis: an entity in the spectrum of oculo-encephalo-hepato-renal disorders. *Am J Med Genet*, **52**, 419–26.

Limber ER, Bresnick GH, Lebovitz RM *et al.* (1989). Spinocerebellar ataxia, hypogonadotropic hypogonadism, and choroidal dystrophy (Boucher-Neuhauser syndrome). *Am J Med Genet*, **33**, 409–14.

Limousin P, Speelman JD, Gielen F *et al.* (1999). Multicentre European study of thalamic stimulation in parkinsonian and essential tremor. *J Neurol Neurosurg Psychiatry*, **66**, 289–96.

Lindblad K, Savontaus ML, Stevanin G *et al.* (1996). An expanded CAG repeat sequence in spinocerebellar ataxia type 7. *Genome Res*, **6**, 965–71.

Lock RJ, Pengiran Tengah DS, Unsworth DJ *et al.* (2005). Ataxia, peripheral neuropathy, and anti-gliadin antibody. Guilt by association? *J Neurol Neurosurg Psychiatry*, **76**, 1601–3.

Lodi R, Schapira A, Manners D *et al.* (2000). Abnormal in vivo skeletal muscle energy metabolism in Huntington's disease and dentatorubropallidoluysian atrophy. *Ann Neurol*, **48**, 72–6.

Loesch DZ, Churchyard A, Brotchie P *et al.* (2005). Evidence for, and a spectrum of, neurological involvement in carriers of the fragile X pre-mutation: FXTAS and beyond. *Clin Genet*, **67**, 412–7.

Logan JI (1996). Hereitary deficiency of ferroxidase (aka caeruloplasmin) [editorial]. *J Neurol Neurosurg Psychiatry*, **61**, 431–2.

Lorenz D, Frederiksen H, Moises H *et al.* (2004). High concordance for essential tremor in monozygotic twins of old age. *Neurology*, **62**, 208–11.

Louis ED (2001). Etiology of essential tremor: should we be searching for environmental causes? *Mov Disord*, **16**, 822–9.

Louis ED (2005). Essential tremor. *Lancet Neurol*, **4**, 100–10.

Louis ED, Ottman R, Hauser WA (1998a). How common is the most common adult movement disorder? estimates of the prevalence of essential tremor throughout the world. *Mov Disord*, **13**, 5–10.

Louis ED, Wendt KJ, Pullman SL *et al.* (1998b). Is essential tremor symmetric? Observational data from a community-based study of essential tremor. *Arch Neurol*, **55**, 1553–9.

Lubbers WJ, Brunt ER, Scheffer H *et al.* (1995). Hereditary myokymia and paroxysmal ataxia linked to chromosome 12 is responsive to acetazolamide. *J Neurol Neurosurg Psychiatry*, **59**, 400–5.

Luostarinen LK, Collin PO, Peraaho MJ *et al.* (2001). Coeliac disease in patients with cerebellar ataxia of unknown origin. *Ann Med*, **33**, 445–9.

Machkhas H, Bidichandani SI, Patel PI *et al.* (1998). A mild case of Friedreich ataxia: lymphocyte and sural nerve analysis for GAA repeat length reveals somatic mosaicism. *Muscle Nerve*, **21**, 390–3.

Maguire A, Hellier K, Hammans S *et al.* (2001). X-linked cerebellar ataxia and sideroblastic anaemia associated with a missense mutation in the ABC7 gene predicting V411L. *Br J Haematol*, **115**, 910–7.

Manto MU (2002). Hashimoto's associated ataxia. *J Neurol Neurosurg Psychiatry*, **72**, 277–8.

Mantovan MC, Martinuzzi A, Squarzanti F *et al.* (2006). Exploring mental status in Friedreich's ataxia: a combined neuropsychological, behavioral and neuroimaging study. *Eur J Neurol*, **13**, 827–35.

Maria BL, Hoang KB, Tusa RJ *et al.* (1997). Joubert syndrome revisited: key ocular motor signs with magnetic resonance imaging correlation. *J Child Neurol*, **12**, 423–30.

Marie P, Foix C, Alajouanine T (1922). De l'atrophie cerebelleuse tardive a predominance corticale. *Rev Neurol*, **2**, 849–85.

Mariotti C, Gellera C, Rimoldi M *et al.* (2004). Ataxia with isolated vitamin E deficiency: neurological phenotype, clinical follow-up and novel mutations in TTPA gene in Italian families. *Neurol Sci*, **25**, 130–7.

Marsden CD, Harrison MGJ (1974). Idiopathic torsion dystonia (dystonia musculorum deformans). A review of forty-two patients. *Brain*, **97**, 793–810.

Marsden CD, Rothwell JC (1987). The physiology of idiopathic dystonia. *Can J Neurol Sci*, **14**, 521–7.

Martinez-Gonzalez MJ, Martinez-Gonzalez S, Garcia-Ribes A *et al.* (2006). Acute onset ataxia in infancy: its aetiology, treatment and follow-up. *Rev Neurol*, **42**, 321–4.

Mason WP, Graus F, Lang B *et al.* (1997). Small-cell lung cancer, paraneoplastic cerebellar degeneration and the Lambert-Eaton myasthenic syndrome. *Brain*, **120**, 1279–300.

Mateo I, Llorca J, Volpini V *et al.* (2003). GAA expansion size and age at onset of Friedreich's ataxia. *Neurology*, **61**, 274–5.

Mateo I, Llorca J, Volpini V *et al.* (2004). Expanded GAA repeats and clinical variation in Friedreich's ataxia. *Acta Neurol Scand*, **109**, 75–8.

Mathews KD, Afifi AK, Hanson JW (1989). Autosomal recessive cerebellar hypoplasia. *J Child Neurol*, **4**, 189–94.

Matsuura T, Ranum LP, Volpini V *et al.* (2002). Spinocerebellar ataxia type 10 is rare in populations other than Mexicans. *Neurology*, **58**, 983–4.

Matsuura T, Yamagata T, Burgess DL *et al.* (2000). Large expansion of the ATTCT pentanucleotide repeat in spinocerebellar ataxia type 10. *Nat Genet*, **26**, 191–4.

McCabe D, Ryan F, Moore D *et al.* (2000). Typical Friedreich's ataxia without GAA expansions and GAA expansion without typical Friedreich's ataxia. *J Neurol*, **247**, 346–55.

McGovern M, Stewart M, Morrison P *et al.* (2000). Early onset of Friedreich's ataxia in a compound heterozygote. *Arch Dis Child*, **83**, 74–5.

McLaughlin JF, Pagon RA, Weinberger E *et al.* (1996). Marinesco-Sjogren syndrome: clinical and magnetic resonance imaging features in three children [corrected and republished article originally printed in Dev Med Child Neurol 1996 Apr; 38(4):363–70]. *Dev Med Child Neurol*, **38**, 636–44.

McManis PG, Sharbrough FW (1993). Orthostatic tremor: clinical and electrophysiologic characteristics. *Muscle Nerve*, **16**, 1254–60.

Migliaccio AA, Halmagyi GM, McGarvie LA *et al.* (2004). Cerebellar ataxia with bilateral vestibulopathy: description of a syndrome and its characteristic clinical sign. *Brain*, **127**, 280–93.

Miyajima H, Kono S, Takahashi Y *et al.* (2001). Cerebellar ataxia associated with heteroallelic ceruloplasmin gene mutation. *Neurology*, **57**, 2205–10.

Miyamoto T, Kanazawa N, Kato S *et al.* (2001). Diagnosis of Japanese patients with HHH syndrome by molecular genetic analysis: a common mutation, R179X. *J Hum Genet*, **46**, 260–2.

Miyoshi Y, Yamada T, Tanimura M *et al.* (2001). A novel autosomal dominant spinocerebellar ataxia (SCA16) linked to chromosome 8q22.1-24.1. *Neurology*, **57**, 96–100.

Moll JW, Henzen-Logmans SC, Van der Meche FG *et al.* (1993). Early diagnosis and intravenous immune globulin therapy in paraneoplastic cerebellar degeneration [letter]. *J Neurol Neurosurg Psychiatry*, **56**, 112.

Montermini L, Kish SJ, Jiralerspong S *et al.* (1997a). Somatic mosaicism for Friedreich's ataxia GAA triplet repeat expansions in the central nervous system. *Neurology*, **49**, 606–10.

Montermini L, Richter A, Morgan K *et al.* (1997b). Phenotypic variability in Friedreich ataxia: role of the associated GAA triplet repeat expansion. *Ann Neurol*, **41**, 675–82.

Moschner C, Perlman, S. and Baloh, R. W. (1994). Comparison of oculomotor findings in the progressive ataxia syndromes. *Brain* **117**, 15–25.

Moseley ML, Benzow KA, Schut LJ *et al.* (1998). Incidence of dominant spinocerebellar and Friedreich triplet repeats among 361 ataxia families. *Neurology*, **51**, 1666–71.

Moseley ML, Schut LJ, Bird TD *et al.* (2000). Reply. *Nat Genet*, **24**, 215.

Muller-Felber W, Zafiriou D, Scheck R *et al.* (1998). Marinesco Sjogren syndrome with rhabdomyolysis. A new subtype of the disease. *Neuropediatrics*, **29**, 97–101.

Munoz E, Campdelacreu J, Ferrer I *et al.* (2004). Severe cerebral white matter involvement in a case of dentatorubropallidoluysian atrophy studied at autopsy. *Arch Neurol*, **61**, 946–9.

Musumeci O, Naini A, Slonim AE et al. (2001). Familial cerebellar ataxia with muscle coenzyme Q10 deficiency. Neurology, 56, 849–55.

Muzaimi MB, Thomas J, Palmer-Smith S et al. (2004). Population based study of late onset cerebellar ataxia in south east Wales. J Neurol Neurosurg Psychiatry, 75, 1129–34.

Nachmanoff DB, Segal RA, Dawson DM et al. (1997). Hereditary ataxia with sensory neuronopathy: Biemond's ataxia. Neurology, 48, 273–5.

Nagaoka U, Takashima M, Ishikawa K et al. (2000). A gene on SCA4 locus causes dominantly inherited pure cerebellar ataxia. Neurology, 54, 1971–5.

Nance MA, Sevenich EA, Schut LJ (1994). Knowledge of genetics and attitudes toward genetic testing in two hereditary ataxia (SCA 1) kindreds. Am J Med Genet, 54, 242–8.

Nandi D, Aziz TZ (2004). Deep brain stimulation in the management of neuropathic pain and multiple sclerosis tremor. J Clin Neurophysiol, 21, 31–9.

Narcisi TM, Shoulders CC, Chester SA et al. (1995). Mutations of the microsomal triglyceride-transfer-protein gene in abetalipoproteinemia. Am J Hum Genet, 57, 1298–310.

Nelson J, Flaherty M, Grattan-Smith P (1997). Gillespie syndrome: a report of two further cases. Am J Med Genet, 71, 134–8.

Nemeth AH, Bochukova E, Dunne E et al. (2000). Autosomal recessive cerebellar ataxia with oculomotor apraxia (ataxia-telangiectasia-like syndrome) is linked to chromosome 9q34. Am J Hum Genet, 67, 1320–6.

Nicolas JM, Estruch R, Salamero M et al. (1997). Brain impairment in well-nourished chronic alcoholics is related to ethanol intake. Ann Neurol, 41, 590–8.

Nikkhah G, Prokop T, Hellwig B et al. (2004). Deep brain stimulation of the nucleus ventralis intermedius for Holmes (rubral) tremor and associated dystonia caused by upper brainstem lesions. Report of two cases. J Neurosurg, 100, 1079–83.

Nishiyama K, Nakamura K, Murayama S et al. (1997). Regional and cellular expression of the dentatorubral-pallidoluysian atrophy gene in brains of normal and affected individuals. Ann Neurol, 41, 599–605.

Nussinovitch M, Prais D, Volovitz B et al. (2003). Post-infectious acute cerebellar ataxia in children. Clin Pediatr, 42, 581–4.

Nutt JG, Marsden CD, Thompson PD (1993). Human walking and higher level gait disorders, particularly in the elderly. Neurology, 43, 268–79.

O'Hearn E, Holmes SE, Calvert PC et al. (2001). SCA-12: Tremor with cerebellar and cortical atrophy is associated with a CAG repeat expansion. Neurology, 56, 299–303.

Ohata T, Yoshida K, Sakai H et al. (2006). A -16C>T substitution in the 5' UTR of the puratrophin-1 gene is prevalent in autosomal dominant cerebellar ataxia in Nagano. J Hum Genet, 51, 461–6.

Ohye C, Miyazaki M, Hirai T et al. (1982). Primary writing tremor treated by stereotactic surgery. J Neurol Neurosurg Psychiatry, 45, 988–97.

Ondo WG, Jankovic J, Connor GS et al. (2006). Topiramate in essential tremor: a double-blind, placebo-controlled trial. Neurology, 66, 672–7.

Ophoff RA, Terwindt GM, Vergouwe MN et al. (1996). Familial hemiplegic migraine and episodic ataxia type-2 are caused by mutations in the Ca2+ channel gene CACNL1A4. Cell, 87, 543–52.

Ormerod IE, Harding AE, Miller DH et al. (1994). Magnetic resonance imaging in degenerative ataxic disorders. J Neurol Neurosurg Psychiatry, 57, 51–7.

Ouyang Y, Takiyama Y, Sakoe K et al. (2006). Sacsin-related ataxia (ARSACS): expanding the genotype upstream from the gigantic exon. Neurology, 66, 1103–4.

Pagan FL, Butman JA, Dambrosia JM et al. (2003). Evaluation of essential tremor with multi-voxel magnetic resonance spectroscopy. Neurology, 60, 1344–7.

Pahwa R, Lyons K, Hubble J et al. (1998). Double-blind controlled trial of gabapentin in essential tremor. Mov Disord, 13, 465–7.

Pahwa R, Lyons KE, Wilkinson SB et al. (2006). Long-term evaluation of deep brain stimulation of the thalamus. J Neurosurg, 104, 506–12.

Paine RS (1960). Evaluation of familial biochemically determined mental retardation in children with special reference to aminoaciduria. N Engl J Med, 262, 658–65.

Palau F, De Michele G, Vilchez JJ et al. (1995). Early-onset ataxia with cardiomyopathy and retained tendon reflexes maps to the Friedreich's ataxia locus on chromosome 9q. Ann Neurol, 37, 359–62.

Panas M, Kalfakis N, Karadima G et al. (2002). Friedreich's ataxia mimicking hereditary motor and sensory neuropathy. J Neurol, 249, 1583–6.

Pandolfo M (2003). Friedreich ataxia. Semin Pediatr Neurol, 10, 163–72.

Parisi MA, Doherty D, Eckert ML et al. (2006). AHI1 mutations cause both retinal dystrophy and renal cystic disease in Joubert syndrome. J Med Genet, 43, 334–9.

Parker CC, Evans OB (2003). Metabolic disorders causing childhood ataxia. Semin Pediatr Neurol, 10, 193–9.

Pascual-Castroviejo I, Gutierrez M, Morales C et al. (1994). Primary degeneration of the granular layer of the cerebellum. A study of 14 patients and review of the literature. Neuropediatrics, 25, 183–90.

Pellegrino JE, Lensch MW, Muenke M et al. (1997). Clinical and molecular analysis in Joubert syndrome. Am J Med Genet, 72, 59–62.

Peters N, Kamm C, Asmus F et al. (2006). Intrafamilial variability in fragile X-associated tremor/ataxia syndrome. Mov Disord, 21, 98–102.

Peterson K, Rosenblum MK, Kotanides H et al. (1992). Paraneoplastic cerebellar degeneration. I. A clinical analysis of 55 anti-Yo antibody-positive patients. Neurology, 42, 1931–7.

Phuphanich S, Brock C. (2007). Neurologic improvement after high-dose intravenous immunoglobulin therapy in patients with paraneoplastic cerebellar degeneration associated with anti-purkinje cell antibody. J Neurooncol, 81, 67–9.

Pittock SJ, Lucchinetti CF, Parisi JE et al. (2005). Amphiphysin autoimmunity: paraneoplastic accompaniments. Ann Neurol, 58, 96–107.

Pranzatelli MR, Tate ED, Kinsbourne M et al. (2002). Forty-one year follow-up of childhood-onset opsoclonus-myoclonus-ataxia: cerebellar atrophy, multiphasic relapses, and response to IVIG. Mov Disord, 17, 1387–90.

Priller J, Scherzer CR, Faber PW et al. (1997). Frataxin gene of Friedreich's ataxia is targeted to mitochondria. Ann Neurol, 42, 265–9.

Quinn N, Barnard RO, Kelly RE (1992). Cerebellar syndrome in myxoedema revisited: a published case with carcinomatosis and multiple system atrophy at necropsy. J Neurol Neurosurg Psychiatry, 55, 616–8.

Racette BA, Dowling J, Randle J et al. (2001). Thalamic stimulation for primary writing tremor. J Neurol, 248, 380–2.

Rajab A, Mochida GH, Hill A et al. (2003). A novel form of pontocerebellar hypoplasia maps to chromosome 7q11–21. Neurology, 60, 1664–7.

Rajput AH, Rozdilsky B, Ang L et al. (1993). Significance of parkinsonian manifestations in essential tremor. Can J Neurol Sci, 20, 114–7.

Ranum LP, Schut LJ, Lundgren JK et al. (1994). Spinocerebellar ataxia type 5 in a family descended from the grandparents of President Lincoln maps to chromosome 11. Nat Genet, 8, 280–4.

Rehncrona S, Johnels B, Widner H et al. (2003). Long-term efficacy of thalamic deep brain stimulation for tremor: double-blind assessments. Mov Disord, 18, 163–70.

Remy P, de Recondo A, Defer G et al. (1995). Peduncular 'rubral' tremor and dopaminergic denervation: a PET study. Neurology, 45, 472–7.

Richter A, Poirier J, Mercier J et al. (1996). Friedreich ataxia in Acadian families from eastern Canada: clinical diversity with conserved haplotypes. Am J Med Genet, 64, 594–601.

Rivest J, Marsden CD (1990). Trunk and head tremor as isolated manifestations of dystonia. Mov Disord, 5, 60–5.

Romano S, Boddaert N, Desguerre I et al. (2006). Molar tooth sign and superior vermian dysplasia: a radiological, clinical, and genetic study. Neuropediatrics, 37, 42–5.

Rotig A, de Lonlay P, Chretien D et al. (1997). Aconitase and mitochondrial iron-sulphur protein deficiency in Friedreich ataxia. *Nat Genet*, **17**, 215–7.

Roubertie A, Biolsi B, Rivier F et al. (2003). Ataxia with vitamin E deficiency and severe dystonia: report of a case. *Brain Dev*, **25**, 442–5.

Rudnik-Schoneborn S, Sztriha L, Aithala GR et al. (2003). Extended phenotype of pontocerebellar hypoplasia with infantile spinal muscular atrophy. *Am J Med Genet A*, **117**, 10–7.

Sabbahi M, Etnyre B, Al-Jawayed I et al. (2002). H-reflex recovery curves differentiate essential tremor, Parkinson's disease, and the combination of essential tremor and Parkinson's disease. *J Clin Neurophysiol*, **19**, 245–51.

Said G, Bathien N, Cesaro P. (1982). Peripheral neuropathies and tremor. *Neurology*, **32**, 480–5.

Sakai T, Kawakami H (1996). Machado-Joseph disease: A proposal of spastic paraplegic subtype. *Neurology*, **46**, 846–7.

Salvi S, Dionisi-Vici C, Bertini E et al. (2001). Seven novel mutations in the ORNT1 gene (SLC25A15) in patients with hyperornithinemia, hyperammonemia, and homocitrullinuria syndrome. *Hum Mutat*, **18**, 460.

Samadani U, Umemura A, Jaggi JL et al. (2003). Thalamic deep brain stimulation for disabling tremor after excision of a midbrain cavernous angioma. Case report. *J Neurosurg*, **98**, 888–90.

Sanner G (1973). The dysequilibrium syndrome. *Neuropadiatrie*, **4**, 403–13.

Santoro L, Perretti A, Filla A et al. (1992). Is early onset cerebellar ataxia with retained tendon reflexes identifiable by electrophysiologic and histologic profile? A comparison with Friedreich's ataxia. *J Neurol Sci*, **113**, 43–9.

Santoro L, Perretti A, Lanzillo B et al. (2000). Influence of GAA expansion size and disease duration on central nervous system impairment in Friedreich's ataxia: contribution to the understanding of the pathophysiology of the disease. *Clin Neurophysiol*, **111**, 1023–30.

Saraiva JM, Baraitser M (1992). Joubert syndrome: a review. *Am J Med Genet*, **43**, 726–31.

Sasaki K, Suga K, Tsugawa S et al. (1996). Muscle pathology in Marinesco-Sjogren syndrome: a unique ultrastructural feature. *Brain Dev*, **18**, 64–7.

Sayer JA, Otto EA, O'Toole JF et al. (2006). The centrosomal protein nephrocystin-6 is mutated in Joubert syndrome and activates transcription factor ATF4. *Nat Genet*, **38**, 674–81.

Schapira AH (2006). Mitochondrial disease. *Lancet*, **368**, 70–82.

Schelhaas HJ, Ippel PF, Hageman G et al. (2001). Clinical and genetic analysis of a four-generation family with a distinct autosomal dominant cerebellar ataxia. *J Neurol*, **248**, 113–20.

Schelhaas HJ, Verbeek DS, Van de Warrenburg BP et al. (2004). SCA19 and SCA22: evidence for one locus with a worldwide distribution. *Brain*, **127**, E6; author reply E7.

Schols L, Amoiridis G, Epplen JT et al. (1996). Relations between genotype and phenotype in German patients with the Machado-Joseph disease mutation. *J Neurol Neurosurg Psychiatry*, **61**, 466–70.

Schols L, Amoiridis G, Przuntek H et al. (1997). Friedreich's ataxia. Revision of the phenotype according to molecular genetics. *Brain*, **120**, 2131–40.

Schols L, Bauer I, Zuhlke C et al. (2003). Do CTG expansions at the SCA8 locus cause ataxia? *Ann Neurol*, **54**, 110–5.

Schols L, Bauer P, Schmidt T et al. (2004a). Autosomal dominant cerebellar ataxias: clinical features, genetics, and pathogenesis. *Lancet Neurol*, **3**, 291–304.

Schols L, Haan J, Riess O et al. (1998). Sleep disturbance in spinocerebellar ataxias: is the SCA3 mutation a cause of restless legs syndrome? *Neurology*, **51**, 1603–7.

Schols L, Meyer C, Schmid G et al. (2004b). Therapeutic strategies in Friedreich's ataxia. *J Neural Transm Suppl*, **68**, 135–45.

Schols L, Szymanski S, Peters S et al. (2000). Genetic background of apparently idiopathic sporadic cerebellar ataxia. *Hum Genet*, **107**, 132–7.

Schols L, Zange J, Abele M et al. (2005). L-carnitine and creatine in Friedreich's ataxia. A randomized, placebo-controlled crossover trial. *J Neural Transm*, **112**, 789–96.

Schuurman PR, Bosch DA, Bossuyt PM et al. (2000). A comparison of continuous thalamic stimulation and thalamotomy for suppression of severe tremor. *N Engl J Med*, **342**, 461–8.

Schwartz CE, May MM, Carpenter NJ et al. (2005). Allan-Herndon-Dudley syndrome and the monocarboxylate transporter 8 (MCT8) gene. *Am J Hum Genet*, **77**, 41–53.

Schwarz S, Schwab S, Hoffmann GF (1999). Enzyme defects of the urea cycle in differential acute encephalopathy diagnosis in adulthood. Diagnosis and current therapy concepts. *Nervenarzt*, **70**, 111–8.

Seminara SB, Acierno JS Jr, Abdulwahid NA et al. (2002). Hypogonadotropic hypogonadism and cerebellar ataxia: detailed phenotypic characterization of a large, extended kindred. *J Clin Endocrinol Metab*, **87**, 1607–12.

Senanayake N, de Silva HJ (1994). Delayed cerebellar ataxia complicating falciparum malaria: a clinical study of 74 patients. *J Neurol*, **241**, 456–9.

Senderek J, Krieger M, Stendel C et al. (2005). Mutations in SIL1 cause Marinesco-Sjogren syndrome, a cerebellar ataxia with cataract and myopathy. *Nat Genet*, **37**, 1312–4.

Serratrice J, Attarian S, Disdier P et al. (2004). Neuromuscular diseases associated with antigliadin antibodies. A contentious concept. *Acta Myol*, **23**, 146–50.

Shan DE, Soong BW, Sun CM et al. (2001). Spinocerebellar ataxia type 2 presenting as familial levodopa-responsive parkinsonism. *Ann Neurol*, **50**, 812–5.

Sheehy MP, Marsden CD (1982). Writer's cramp - a focal dystonia. *Brain*, **105**, 462–80.

Shimazaki H, Takiyama Y, Sakoe K et al. (2002). Early-onset ataxia with ocular motor apraxia and hypoalbuminemia: the aprataxin gene mutations. *Neurology*, **59**, 590–5.

Silveira I, Miranda C, Guimaraes L et al. (2002). Trinucleotide repeats in 202 families with ataxia: a small expanded (CAG)n allele at the SCA17 locus. *Arch Neurol*, **59**, 623–9.

Singer C, Sanchez-Ramos J, Weiner WJ (1994). Gait abnormality in essential tremor. *Mov Disord*, **9**, 193–6.

Sobrido MJ, Cholfin JA, Perlman S et al. (2001). SCA8 repeat expansions in ataxia: a controversial association. *Neurology*, **57**, 1310–2.

Soland VL, Bhatia KP, Volonte MA et al. (1996). Focal task-specific tremors. *Mov Disord*, **11**, 665–70.

Spacey SD, Materek LA, Szczygielski BI et al. (2005). Two novel CACNA1A gene mutations associated with episodic ataxia type 2 and interictal dystonia. *Arch Neurol*, **62**, 314–6.

Stacy M, Jankovic J (1992). Tardive tremor. *Mov Disord*, **7**, 53–7.

Stark E, Wurster U, Patzold U et al. (1995). Immunological and clinical response to immunosuppressive treatment in paraneoplastic cerebellar degeneration. *Arch Neurol*, **52**, 814–8.

Steckley JL, Ebers GC, Cader MZ et al. (2001). An autosomal dominant disorder with episodic ataxia, vertigo, and tinnitus. *Neurology*, **57**, 1499–502.

Steinhoff BJ, Herrendorf G, Bittermann HJ et al. (1997). Isolated ataxia as an idiosyncratic side-effect under gabapentin. *Seizure*, **6**, 503–4.

Steinlin M (1998). Non-progressive congenital ataxias. *Brain Dev*, **20**, 199–208.

Steinlin M, Schmid M, Landau K et al. (1997). Follow-up in children with Joubert syndrome. *Neuropediatrics*, **28**, 204–11.

Stevanin G, Bouslam N, Thobois S et al. (2004). Spinocerebellar ataxia with sensory neuropathy (SCA25) maps to chromosome 2p. *Ann Neurol*, **55**, 97–104.

Stevanin G, Durr A, David G et al. (1997). Clinical and molecular features of spinocerebellar ataxia type 6. *Neurology*, **49**, 1243–6.

Stevanin G, Herman A, Durr A et al. (2000). Are (CTG)n expansions at the SCA8 locus rare polymorphisms? *Nat Genet*, **24**, 213; author reply 215.

Storey E, Billimoria P (2005). Increased T2 signal in the middle cerebellar peduncles on MRI is not specific for fragile X premutation syndrome. *J Clin Neurosci*, **12**, 42–3.

Storey E, Gardner RJ, Knight MA *et al.* (2001). A new autosomal dominant pure cerebellar ataxia. *Neurology*, **57**, 1913–5.

Subramony SH, Filla A (2001). Autosomal dominant spinocerebellar ataxias ad infinitum? *Neurology*, **56**, 287–89.

Subramony SH, Schott K, Raike RS *et al.* (2003). Novel CACNA1A mutation causes febrile episodic ataxia with interictal cerebellar deficits. *Ann Neurol*, **54**, 725–31.

Takamure M, Hirano M, Taoka T *et al.* (2006). White matter T2 hyperintensity development and clinical deterioration after status epilepticus in a patient with dentatorubral-pallidoluysian atrophy. *Clin Neurol Neurosurg*, **108**, 482–5.

Takiyama Y (2006). Autosomal recessive spastic ataxia of Charlevoix-Saguenay. *Neuropathology*, **26**, 368–75.

Tan WH, Baris H, Robson CD *et al.* (2005). Cockayne syndrome: the developing phenotype. *Am J Med Genet A*, **135**, 214–6.

Tanner CM, Goldman SM, Lyons KE *et al.* (2001). Essential tremor in twins: an assessment of genetic vs environmental determinants of etiology. *Neurology*, **57**, 1389–91.

Taylor AM, Byrd PJ (2005). Molecular pathology of ataxia telangiectasia. *J Clin Pathol*, **58**, 1009–15.

ten Donkelaar HJ, Lammens M, Wesseling P *et al.* (2003). Development and developmental disorders of the human cerebellum. *J Neurol*, **250**, 1025–36.

Toft M, Aasly J, Bisceglio G *et al.* (2005). Parkinsonism, FXTAS, and FMR1 premutations. *Mov Disord*, **20**, 230–3.

Trivedi R, Mundanthanam G, Amyes E *et al.* (2000). Autoantibody screening in subacute cerebellar ataxia. *Lancet*, **356**, 565–6.

Truong DD, Harding AE, Scaravilli F *et al.* (1990). Movement disorders in mitochondrial myopathies. A study of nine cases with two autopsy studies. *Mov Disord*, **5**, 109–17.

Tsao CY, Lo WD, Craenen J (1992). Congestive heart failure and cardiac thrombus as first presentations of Friedreich ataxia. *Pediatr Neurol*, **8**, 313–4.

Tsukamoto T, Mochizuki R, Mochizuki H *et al.* (1993). Paraneoplastic cerebellar degeneration and limbic encephalitis in a patient with adenocarcinoma of the colon. *J Neurol Neurosurg Psychiatry*, **56**, 713–6.

Tuite PJ, Rogaeva EA, St George-Hyslop PH *et al.* (1995). Dopa-responsive parkinsonism phenotype of Machado-Joseph disease: confirmation of 14q CAG expansion. *Ann Neurol*, **38**, 684–7.

Usuki F, Maruyama K (2000). Ataxia caused by mutations in the alpha-tocopherol transfer protein gene. *J Neurol Neurosurg Psychiatry*, **69**, 254–6.

Vahedi K, Joutel A, Van Bogaert P *et al.* (1995). A gene for hereditary paroxysmal cerebellar ataxia maps to chromosome 19p. *Ann Neurol*, **37**, 289–93.

Valente EM, Brancati F, Silhavy JL *et al.* (2006a). AHI1 gene mutations cause specific forms of Joubert syndrome-related disorders. *Ann Neurol*, **59**, 527–34.

Valente EM, Silhavy JL, Brancati F *et al.* (2006b). Mutations in CEP290, which encodes a centrosomal protein, cause pleiotropic forms of Joubert syndrome. *Nat Genet*, **38**, 623–5.

van de Warrenburg BP, Notermans NC, Schelhaas HJ *et al.* (2004). Peripheral nerve involvement in spinocerebellar ataxias. *Arch Neurol*, **61**, 257–61.

van der Knaap MS, Pronk JC, Scheper GC (2006). Vanishing white matter disease. *Lancet Neurol*, **5**, 413–23.

van Swieten JC, Brusse E, de Graaf BM *et al.* (2003). A mutation in the fibroblast growth factor 14 gene is associated with autosomal dominant cerebellar ataxia [corrected]. *Am J Hum Genet*, **72**, 191–9.

Vandel P, Bonin B, Leveque E *et al.* (1997). Tricyclic antidepressant-induced extrapyramidal side effects. *Eur Neuropsychopharmacol*, **7**, 207–12.

Vedanarayanan VV (2003). Mitochondrial disorders and ataxia. *Semin Pediatr Neurol*, **10**, 200–9.

Vedeler CA, Antoine JC, Giometto B *et al.* (2006). Management of paraneoplastic neurological syndromes: report of an EFNS Task Force. *Eur J Neurol*, **13**, 682–90.

Verbeek DS, van de Warrenburg BP, Wesseling P *et al.* (2004). Mapping of the SCA23 locus involved in autosomal dominant cerebellar ataxia to chromosome region 20p13-12.3. *Brain*, **127**, 2551–7.

Vianello M, Manara R, Betterle C *et al.* (2005). X-linked adrenoleukodystrophy with olivopontocerebellar atrophy. *Eur J Neurol*, **12**, 912–4.

Vianello M, Tavolato B, Armani M *et al.* (2003). Cerebellar ataxia associated with anti-glutamic acid decarboxylase autoantibodies. *Cerebellum*, **2**, 77–9.

Vincent J, Neves-Pereira M, Paterson A *et al.* (2000). An unstable trinucleotide repeat region on chromosome 13 implicated in spinocerebellar ataxia: a common expansion locus. *Am J Hum Genet*, **66**, 819–29.

Vinton A, Fahey MC, O'Brien TJ *et al.* (2005). Dentatorubral-pallidoluysian atrophy in three generations, with clinical courses from nearly asymptomatic elderly to severe juvenile, in an Australian family of Macedonian descent. *Am J Med Genet A*, **136**, 201–4.

Vorgerd M, Schols L, Hardt C *et al.* (2000). Mitochondrial impairment of human muscle in Friedreich ataxia in vivo. *Neuromuscul Disord*, **10**, 430–5.

Wang J, Hegele R (2000). Microsomal triglyceride transfer protein (MTP) gene mutations in Canadian subjects with abetalipoproteinemia. *Hum Mutat*, **15**, 294–5.

Warner T, Lennox G, Janota I *et al.* (1994). Autosomal dominant dentatorubropallidoluysian atrophy in the United Kingdom. *Mov Disord*, **9**, 289–96.

Warner TT, Williams LD, Walker RW *et al.* (1995). A clinical and molecular genetic study of dentatorubropallidoluysian atrophy in four European families. *Ann Neurol*, **37**, 452–9.

Waters MF, Minassian NA, Stevanin G *et al.* (2006). Mutations in voltage-gated potassium channel KCNC3 cause degenerative and developmental central nervous system phenotypes. *Nat Genet*, **38**, 447–51.

Wenning GK, Ben Shlomo Y, Magalhaes M *et al.* (1994). Clinical features and natural history of multiple system atrophy. An analysis of 100 cases. *Brain*, **117**, 835–45.

Wilkinson PA, Bradley JL, Warner TT (2001). Friedreich's ataxia presenting as an isolated spastic paraparesis. *J Neurol Neurosurg Psychiatry*, **71**, 709.

Wills A, Brusa L, Wang H *et al.* (1999). Levodopa may improve orthostatic tremor: case report and trial of treatment. *J Neurol Neurosurg Psychiatry*, **66**, 681–4.

Wills AJ, Thompson PD, Findley LJ *et al.* (1996). A positron emission tomography study of primary orthostatic tremor. *Neurology*, **46**, 747–52.

Wilson RB, Lynch DR, Fischbeck KH (1998). Normal serum iron and ferritin concentrations in patients with Friedreich's ataxia. *Ann Neurol*, **44**, 132–4.

Wilson RB, Roof DM (1997). Respiratory deficiency due to loss of mitochondrial DNA in yeast lacking the frataxin homologue. *Nat Genet*, **16**, 352–7.

Wittkamper A, Wessel K, Bruckmann H (1993). CT in autosomal dominant and idiopathic cerebellar ataxia. *Neuroradiology*, **35**, 520–4.

Worth PF, Giunti P, Gardner-Thorpe C *et al.* (1999). Autosomal dominant cerebellar ataxia type III: linkage in a large British family to a 7.6-cM region on chromosome 15q14-21.3. *Am J Hum Genet*, **65**, 420–6.

Wu YR, Ashby P, Lang AE (2001). Orthostatic tremor arises from an oscillator in the posterior fossa. *Mov Disord*, **16**, 272–9.

Yamada M, Shimohata M, Sato T *et al.* (2006). Polyglutamine disease: recent advances in the neuropathology of dentatorubral-pallidoluysian atrophy. *Neuropathology*, **26**, 346–51.

Yamaguchi K, Ishitobi F (1999). Brain dysgenesis in Cornelia de Lange syndrome. *Clin Neuropathol*, **18**, 99–105.

Yang YL, Yamaguchi S, Tagami Y *et al.* (2003). Diagnosis and treatment of biotinidase deficiency-clinical study of six patients. *Zhonghua Er Ke Za Zhi*, **41**, 249–51.

Yokota T, Shiojiri T, Gotoda T *et al.* (1997). Friedreich-like ataxia with retinitis pigmentosa caused by the His101Gln mutation of the alpha-tocopherol transfer protein gene. *Ann Neurol*, **41**, 826–32.

Young GB, Oppenheimer SR, Gordon BA *et al.* (1994). Ataxia in institutionalized patients with epilepsy. *Can J Neurol Sci*, **21**, 252–8.

Yu GY, Howell MJ, Roller MJ *et al.* (2005). Spinocerebellar ataxia type 26 maps to chromosome 19p13.3 adjacent to SCA6. *Ann Neurol*, **57**, 349–54.

Zeidler M, Stewart GE, Barraclough CR *et al.* (1997). New variant Creutzfeldt-Jakob disease: neurological features and diagnostic tests. *Lancet*, **350**, 903–7.

Zesiewicz TA, Tullidge A, Tidwell J *et al.* (2006). Topiramate-induced psychosis in patients with essential tremor: report of 2 cases. *Clin Neuropharmacol*, **29**, 168–9.

Zhu D, Burke C, Leslie A *et al.* (2002). Friedreich's ataxia with chorea and myoclonus caused by a compound heterozygosity for a novel deletion and the trinucleotide GAA expansion. *Mov Disord*, **17**, 585–9.

Zimmer C, Gosztonyi G, Cervos-Navarro J *et al.* (1992). Neuropathy with lysosomal changes in Marinesco-Sjogren syndrome: fine structural findings in skeletal muscle and conjunctiva. *Neuropediatrics*, **23**, 329–35.

Zu L, Figueroa KP, Grewal R *et al.* (1999). Mapping of a new autosomal dominant spinocerebellar ataxia to chromosome 22. *Am J Hum Genet*, **64**, 594–9.

Zuhlke C, Hellenbroich Y, Dalski A *et al.* (2001). Different types of repeat expansion in the TATA-binding protein gene are associated with a new form of inherited ataxia. *Eur J Hum Genet*, **9**, 160–4.

CHAPTER 40

Movement disorders

Nicholas Fletcher

Contents

40.1 Introduction to movement disorders and the motor system

40.1.1 The clinical approach to a patient with a movement disorder

Almost any neurological disorder can produce a disorder of movement but the 'movement disorders' include the akinetic rigid syndromes, hyperkinesias, and some tremors (Table 40.1). It can sometimes seem, especially with the use of videotape recordings, that diagnosis of movement disorders is mainly a matter of correct visual recognition. Such an approach is not reccommended and can lead to mistakes unless, as in other areas of medicine, the history is considered first and the physical signs second. Obvious examples include the family history in Huntington's disease, developmental history in dystonic cerebral palsy, and neuroleptic drug treatment in patients with tardive dyskinesia. In addition, a single disorder may give rise to several different types of involuntary movement. For example, Huntington's disease may give rise to an akinetic rigid state, chorea, myoclonus, tics, or dystonia. Patients with Parkinson's disease taking levodopa may show different types of movement disorder at different times of the day.

In akinetic rigid states the diagnostic issue will be whether the patient has idiopathic Parkinson's disease or one of the other Parkinsonian syndromes. With involuntary movements, the first step in diagnosis is to classify these as dystonia, tics, tremor, chorea, or myoclonus (Table 40.1). It must be remembered that involuntary movements are merely physical signs, not diagnostic entities, and that they do not always occur in a pure form; for example, patients with dystonia may have additional choreiform movements or tremor. If more than one form of abnormal movement seems to be present, the diagnosis should be based on the most obvious one. The next step is to decide on the cause of the movements and at

this stage the diagnosis must be based upon an accurate and complete history as noted above.

The movement disorders are often associated with abnormalities of the basal ganglia and, to some extent, vice versa. This is not entirely correct. Disturbances of basal ganglia function certainly have profound effects on movement with the development of bradykinesia, rigidity, tremor, or the various forms of dyskinesia. However, it is not correct when considering the pathophysiology of movement disorders to regard the basal ganglia as an isolated movement control centre. In fact, they are an important but poorly understood component of a much wider motor system. It is also important to remember that the basal ganglia are involved in the processing of limbic and other cognitive processes which may also be disturbed by basal ganglia dysfunction.

40.1.2 A brief overview of the motor system

The spinal anterior horn cells are influenced by descending pathways from the cerebral cortex and brainstem centres. These motor cortical and brainstem outflow areas are regulated by other cortical areas, the basal ganglia and cerebellum. The basal ganglia and the cerebellum are both characterized by parallel processing of inputs from cortical areas and subsequent projection back to cortex via thalamic relay nuclei (Alexander 1997).

Motor cortex

There are several motor cortical areas in the human; these are classified and numbered in a variety of ways but may be classified simply as the primary motor cortex, lateral premotor cortex, supplementary motor area, and motor regions of the cingulate cortex (Dum and Strick 1991).

◆ The *primary motor cortex*, Brodmann's area 4, pyramidal cells are the principal source of the pyramidal tract but also project to other motor cortical areas and to sensory cortex, thalamus, basal ganglia (corpus striatum), red nucleus, brainstem reticular nuclei, and to the cerebellum via the corticopontine fibres and

Table 40.1 Types of movement disorder

	Movement disorder	Aetiology
Hypokinesia	Akinetic rigid syndromes	Parkinson's disease Other Parkinsonian disorders
Hyperkinesia	Dystonia	Idiopathic torsion dystonia Other forms of hereditary dystonia Secondary dystonia
	Chorea	Huntington's disease Other causes of chorea
	Tics	Tourette's syndrome Other causes of tics
	Myoclonus	Cortical myoclonus Brainstem myoclonus Spinal myoclonus Other forms of myoclonus
	Tremor	Parkinsonian tremor Benign essential tremor Cerebellar tremor Other tremors

pontine nuclei. The primary motor cortex shows a somatotopic organization (Lemon 1988) but pyramidal cells may activate more than one muscle and a muscle may be affected by pyramidal cells spread over a wide cortical area. Primary motor cortex cells are activated prior to and during voluntary contralateral movement, as shown by functional imaging and neurophysiological event related recordings. Lesions of the primary motor cortex are associated principally with deficits of small accurate finger and hand movements. Such movements require motor cortex outputs not only to agonist muscle spinal motor neurones but also those needed to regulate the activity of antagonists and muscles needed to stabilize proximal joints.

- The *lateral premotor cortex*, the lateral part of Brodmann's area 6, is located on the lateral aspect of the frontal lobe, anterior to the primary motor cortex. There are two discreet motor areas (superior and inferior) within the lateral premotor cortex with separate motor cortical maps and corticospinal projections. These have a considerable output to the medial brainstem tegmentum which projects to the spinal cord in the ventromedial descending system. The premotor cortex is important for the control of axial and proximal limb movements and also of movements in response to external cues or instructions. The activation of the lateral premotor cortex begins prior to the execution of movement, suggesting an important role in preparation for movement, and may be bilateral (Remy *et al.* 1994). Lesions of the lateral premotor area produce proximal limb weakness and impairment of movements requiring use of the whole arm and hand. There is also impaired learning of motor responses to external cues.

- The *supplementary motor area*, the medial part of Brodmann's area 6, is on the inner aspect of the frontal lobe, anterior to area 4, and continuous with the lateral premotor cortex. It receives basal ganglia inputs and has outputs to primary motor cortex, corticospinal tract, basal ganglia, cerebellum, red nucleus, and reticular formation. Like the lateral premotor cortex, the supplementary motor area seems to be important for the preparation of movement, particularly internally generated, learned, sequential, or bimanual movements (Passingham 1987; Rothwell 1994). Blood flow experiments indicate that activation of supplementary motor area prior to movement may be bilateral (Remy *et al.* 1994). Supplementary motor area lesions lead to difficulty initiating speech and movement, impaired bimanual coordination, and the 'alien limb' phenomenon where the arm moves spontaneously in response to external stimuli, grasping at objects spontaneously. This may represent a dissociation of externally (lateral premotor cortex) and internally (supplementary motor area) cued movements.

- *Cingulate motor areas* have been identified in humans and non human primates (Rothwell 1994; Alexander 1997). They have outputs to spinal cord and the basal ganglia and receive inputs from frontal premotor cortex. Cingulate motor neurones are active in internally generated movements and cingulate activation is seen mainly with complex motor tasks. Stimulation of cingulate cortex produces complex motor activity.

Movement related cortical potentials

Surface scalp recordings in man have revealed bilateral activation of frontal motor cortex beginning about 1.5 s before movement onset. This early phase, the premovement or bereitschaftspotential arises from primary motor cortex and the supplementary motor area. Following the premovement potential, about 500 ms before the movement, a second potential, the negative slope, is detected. This is also bilateral and arises in the supplementary motor area and the contralateral primary motor cortex. In the 50–100 ms before movement, a movement potential occurs and arises from contralteral primary motor cortex (Neshige *et al.* 1988).

The descending pathways

The descending pathways are grouped into two divisions within the spinal cord. A dorsolateral group contains the lateral corticospinal tract along with the rubrospinal tract and the crossed reticulospinal tract from the lateral pontine tegmentum. These dorsolateral pathways are concerned with distal movements. The ventrolateral group is made up of the ventral corticospinal tract with the interstitiospinal, vestibulospinal, and reticulospinal (from medial pontine and medullary tegmentum) tracts and is involved in the control of proximal movements (Rothwell 1994; Alexander 1997).

The pyramidal tract comprises the corticobulbar and corticospinal tracts and in the human, the latter component contains about one million fibres on each side. About 60 per cent of the fibres in the corticospinal tract originate in the frontal motor areas. The remainder are from parietal and somatosensory cortex. The corticospinal tract influences spinal afferent pathways, inhibitory interneurones, and gamma efferent cells as well as the alpha motor neurones. There is considerable convergence and divergence within the corticospinal tract; cortical pyramidal cells project to several muscles and each muscle can be activated by cells from a large cortical motor field. Direct synapses with alpha motor neurones are particularly dense on cells concerned with finger movements and there is a close association between the corticospinal tract and the execution of fine finger movements. Lesions of the pyramidal tract tend to produce impaired manual dexterity with variable degrees of weakness.

Other descending pathways are poorly understood in man. The reticulospinal pathway appears to be important in the startle reflex (Brown *et al.* 1991a). Vestibulospinal pathways are concerned in part with postural control.

The regulation and coordination of movement; the cerebellum and basal ganglia

Neither the cerebellum nor the basal ganglia have direct efferent connections to the spinal motor apparatus and yet both are clearly involved in the correct execution and control of movement. This has long been evident from clinicopathological studies of patients with lesions of these structures. Such patients exhibit profound disturbances of movement as their primary symptoms. Anatomically, both structures are characterized by extensive inputs from the cerebral cortex, parallel internal circuitry, and projections back to the cortex via thalamic relay nuclei. Despite much information about the anatomy and cellular neurophysiology of the cerebellum and basal ganglia, their precise functions are not clear.

The cerebellum is made up of two hemispheres separated by a midline vermis. There are three lobes, the anterior, posterior, and the flocculonodular lobe. There are three deeply situated outflow nuclei in each hemisphere, the fastigial nucleus medially, the interposed made up of globose and emboliform, and the dentate nucleus laterally. The majority of cerebellar efferents leave via the superior peduncle. The inferior and middle peduncles are largely composed of afferent fibres from the spinal cord and brainstem nuclei. Within the cerebellum there are three phylogenetic regions:

The neocerebellum or pontocerebellum includes the bulk of the cerebellar hemispheres and the central portion of the vermis. It receives the bulk of the pontine inputs via the middle peduncle. The neocerebellum is principally concerned with the coordination of the ipsilateral limbs.

The paleocerebellum or spinocerebellum consists of the parafloccular lobes and the remainder of the vermis of the anterior and posterior lobes. It receives direct inputs from the spinal cord via the inferior peduncle and is probably concerned with the regulation of muscle tone.

The archicerebellum or vestibulocerebellum is composed of the flocculonodular lobe only and receives direct vestibular inputs via the inferior peduncle.

In fact, there is considerable overlap in the connections of these three regions and their separation on the basis of their inputs is only approximate. The cerebellum may also be subdivided on the basis of outflow projections. The lateral mass of each hemisphere projects to the dentate nucleus, the vermis projects mainly to the fastigial nucleus and an intermediate zone sends fibres to the interposed nucleus. The flocculonodular lobe, vermis, and certain hemisphere areas also project to the vestibular nuclei. Another intersting feature of cerebellar organization concerns sagittal strips of cortex which share inputs from the same regions of the olivary nuclei.

All *cerebellar afferents* terminate in both the cortex and the cerebellar nuclei. The cerebellar afferent systems are as follows (Brodal 1981; Ito 1984):

◆ The vast majority of cerebellar afferents are from the pontine nuclei, through the middle peduncle. These nuclei relay cortical inputs to the cerebellum from the ipsilateral frontal and parietal lobes. These are not collaterals of motor corticospinal fibres. The pontocerebellar projection is to most areas of the contralateral cerebellum and the interposed and dentate nuclei. Pontocerebellar afferents form mossy fibres within the cerebellum, synapsing in the granule cell layer of the cerebellar cortex and in the cerebellar nuclei.

◆ Direct spinocerebellar input is via the inferior peduncle along with afferents from the vestibular and brainstem reticular nuclei. These also enter the cerebellum as mossy fibres.

◆ The inferior olivary nucleus receives inputs from the spinal cord, cerebral motor cortex, and brainstem. It sends climbing fibres directly to cerebellar Purkinje cells and the cerebellar nuclei.

◆ Direct aminergic inputs arise from the locus coeruleus (noradrenergic), raphe nuclei (serotonergic), and midbrain ventral tegmentum (dopaminergic). These project directly to Purkinje cells (Ito 1984).

Cerebellar cortex

The cerebellar cortex contains a complex somatotopic representation which can only be detected in anaesthetized animals. There appear to be two maps, one in the anterior lobe and one in the posterior lobe. Midline structures are represented in the vermis and limbs laterally in the hemispheres.

The circuitry of the cerebellar cortex is located in three cortical layers; the superficial molecular layer, the Purkinje cell layer, and a deep granule cell layer. Afferents arrive as either mossy or climbing fibres, except for the small number of aminergic afferents, and all efferents are Purkinje cell axons which terminate in the

cerebellar or vestibular nuclei. There is some evidence that the firing characteristics of the Purkinje cells are altered by the pattern of climbing and parallel fibre activation and that this may form the basis of cerebellar learning (Ito 1984).

Efferents

The only efferent *cells* of the cerebellum are the Purkinje cells and they project to the cerebellar and vestibular nuclei; they are inhibitory. The fastigial nucleus, the main outflow of the medial cerebellar structures projects mainly to the vestibular nuclei, brainstem reticular formation, and the spinal cord. The interposed and dentate nuclei relay efferents from more lateral cerebellar structures to the contralateral red nucleus and thalamus via the superior peduncle. These axons also have descending branches to the lower brainstem reticular formation. In the thalamus, cerebellar efferents reach the ventrolateral and ventral posterolateral nuclei which then project to the primary motor cortex (area 4). Thalamic nucleus X also receives cerebellar efferents and projects to the lateral premotor cortex (area 6). Although cerebellar projections via the thalamus reach a wide area of cerebral cortex, the bulk of the output is to frontal motor cortex (Rothwell 1994).

Cerebellar function. The cerebellum receives inputs from the cerebral cortex and also from the spinal cord and vestibular system via the mossy and climbing systems respectively. This may enable a comparison of the motor command and the actual movement of the body, allowing correction and adjustment of the movement. Activation of lateral cerebellar cortical neurones occurs before movement, suggesting that cerebellar input to the frontal motor cortex is an important part of preparation for movement. There is evidence that the cerebellum is involved in timing and force of agonist and antagonist activation during movement; cerebellar dysfunction leads to delayed and prolonged agonist bursts and delayed antagonist activation, with dysmetria and dysdiadochokinesis. The cerebellum is also thought to be involved in the learning of motor tasks (Rothwell 1994).

There are three main theories of cerebellar function in relation to movement (Johnson and Montgomery 1997).

◆ A feedback control of the ongoing movement. This assumes continuous monitoring of afferent signals from the periphery by the cerebellum. This allows a continuous correction of the motor cortex activity by the cerebellar outflow, a feedback mechanism.

◆ A centre for the generation of precise motor plans concerning the timing and amplitude of movements which are then sent to the frontal motor cortex for execution, a feedforward mechanism.

◆ The efference copy theory integrates feedback and feedforward models and assumes that the cerebellum receives a copy of the movement command from the cerebral cortex and compares it to current and ongoing afferent signals from the periphery. The motor command is then adjusted and refined by the cerebellum to allow for the position and motion of the body prior to movement onset and during the execution of the movement itself.

40.2 **The basal ganglia**

The basal ganglia are thought to be important for motor control because of their anatomical connections, the neurophysiological properties of their neurones and the effects of anatomical lesions and alterations in basal ganglia neurotransmitter function on movement. The basal ganglia are made up of the putamen and caudate nucleus (neostriatum), the globus pallidus (paleostriatum), the substantia nigra, and the subthalamic nucleus. The inclusion of other nuclear structures such as the nucleus accumbens and parts of the olfactory tubercle (the ventral striatum) has been emphasized recently.

40.2.1 Anatomy and physiology

The putamen, globus pallidus, and caudate nucleus are large masses of grey matter lateral to the thalamus and separated from it by the internal capsule; anteriorly the internal capsule also separates the putamen from the head of the caudate nucleus (Fig. 40.1). The putamen lies lateral to the globus pallidus and these together form the lentiform nucleus with the head of the caudate nucleus lying anteriorly and the body curving over superiorly and then posteriorly in the floor of the lateral ventricle. The subthalamic nucleus and substantia nigra lie inferiorly and caudally in the diencephalon and rostral midbrain respectively.

The striatum takes its name from the histological appearance caused by efferent myelinated fibres, Wilson's pencils, traversing the putamen and caudate. The globus pallidus is subdivided into external and internal regions which have different connections. The substantia nigra is also divided into a dorsal pars compacta and a ventral pars reticulata; the latter is homologous to the internal globus pallidus.

Connections and functional loops

The striatum is the input region of the basal ganglia and receives excitatory corticostriatal fibres from all areas of the cerebral cortex but mainly the motor and somatosensory areas, prefrontal and limbic cortex; it also receives the dopaminergic nigrostratal projection. There are two pathways from the striatum to the internal globus pallidus, the source of the basal ganglia output; a direct pathway and an indirect circuit via external globus pallidus and the subthalamic nucleus (Fig. 40.2). The basal ganglia output from internal globus pallidus (and also ventral pars reticulata) is inhibitory and is mainly to the thalamus; there is a smaller output to the superior colliculus and the pedunculopontine nucleus. The thalamic output is excitatory and is thought to 'drive' the motor cortical outflow. The input–output connections of the basal ganglia therefore form a loop from cortex to basal ganglia and back to cortex. Several separate and parallel loops through the basal ganglia have been identified (motor, oculomotor, limbic, and prefrontal) depending on the cortical component of the circuit. Different areas of the striatum are involved with different loops (Fig. 40.3); for the motor loop the putamen is the striatal component. Within the striatum there are groups of cells which are more densely packed and contain more substance P, met-enkephalin, and opiate receptors but less acetylcholinesterase. These areas 'striosomes' appear to be involved in the processing of the limbic and prefrontal inputs and are seen mainly in the caudate rather than the putamen. The patches also receive most of the dopaminergic afferents from the dorsal pars compacta. Within the motor loop there are smaller but still separate loops related to different body regions. There is therefore considerable somatotopy within the putamen corresponding to the routes of the various parts of the motor loop. Neurophysiological studies indicate that the various loops remain separate during their course

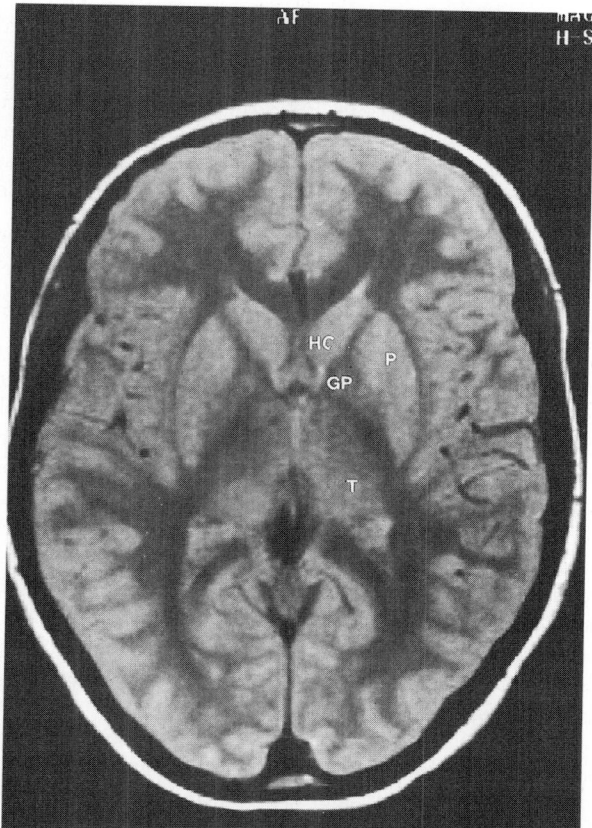

Fig. 40.1 MR brain scan showing the anatomy of the basal ganglia. The head of the caudate nucleus (HC), putamen (P), globus pallidus (GP), and thalamus (T) are indicated.

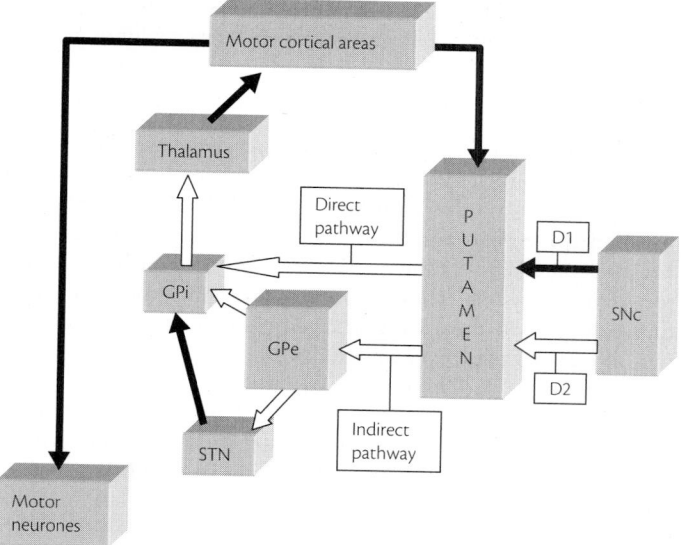

Fig. 40.2 The connections of the basal ganglia showing the direct and indirect pathways. Open arrows are inhibitory, filled, are excitatory.
Key: D1 and D2: D1-like and D2-like dopamine receptors
GPe: Globus pallidus pars externa
Gpi: Globus pallidus pars interna
SNc: Substantia nigra pars compacta
STN: Subthalamic nucleus

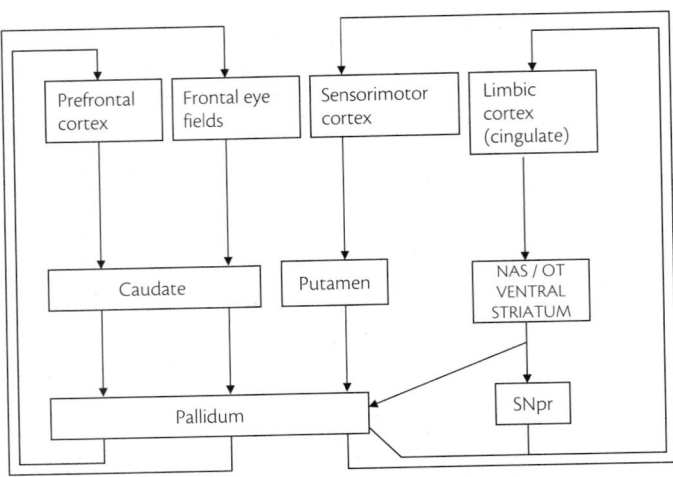

Fig. 40.3 Parallel loop circuits in the basal ganglia. NAS = nucleus accumbens septi; OT = olfactory tubercle; SNpr = substantia nigra pars reticulata.

through the basal ganglia although the anatomical studies suggest otherwise. The dendritic trees of pallidal neurones are very large, suggesting considerable potential for functional convergence. Moreover, the number of neurones within different basal ganglia regions also suggests convergence of 5000:1 from cortex to striatum; from striatum to pallidum the ratios are 300:1 and 100:1 to external globus pallidus and internal globus pallidus respectively (Alexander 1997). Nevertheless there appears to be very little if any functional convergence within the basal ganglia.

With regard to the functions of the direct and indirect pathways from putamen to internal globus pallidus within the motor loop, it can be seen from Fig. 40.2 that increased activity in the direct pathway will inhibit basal ganglia outflow and therefore increase motor cortex activation whereas indirect pathway activity increases subthalamic nucleus activity and therefore internal globus pallidus outflow with an opposite effect on motor cortex activity.

Transmitters

A complete overview of this topic is beyond the scope of this chapter. In simple terms, the corticostriatal projection is excitatory and glutamatergic onto the γ-amino butyric acid, GABA, containing medium spiny neurones, the major cell type in the striatum; the remainder are large cholinergic aspiny neurones which are inhibitory to the spiny cells. The projection from the putamen to the globus pallidus via the direct or indirect pathways is from the inhibitory GABAergic spiny neurones; the direct pathway medium spiny neurones also contain dynorphin and substance P while the indirect medium spiny neurones which project initially to the GPe utilize enkephalin as a cotransmitter. All other circuits within the basal ganglia (see Fig. 40.2) are inhibitory, utililizing GABA, with the exception of the excitatory glutamatergic subthalamic input to the internal globus pallidus.

Dopamine

The role of the dopaminergic nigrostriatal projection is complex. This pathway arises in the dorsal pars compacta and projects to striatal spiny neurones. Dopamine is inhibitory to the indirect pathway medium spiny neurones and excitatory to the direct pathway striatal neurones. The effect of dopaminergic input to the

striatum is therefore to reduce the inhibitory internal globus pallidus outflow to the thalamus, thereby releasing the excitatory thalamocortical projection to the motor cortical areas. This differential effect of dopamine on striatal medium spiny neurones is mediated via two distinct classes of dopamine receptors, D1-like (D1 and D5) and D2-like (D2, D3, D4). In the striatum, most receptors are either D1 or D2; D4 may be involved in limbic functions (Van Tol *et al.* 1991). D1 receptor stimulation leads to activation of adenyl cyclase and excites the direct pathway striatopallidal cells. D2 receptor binding inhibits adenyl cyclase and suppresses indirect pathway striatal cells. The effect of dopamine on the basal ganglia is therefore to drive the direct pathway activity and inhibit the indirect pathway, thereby facilitating movement (Fig. 40.2). There are some difficulties with this model, which is probably an oversimplification. In particular, D1 and D2 receptor stimulation is synergistic, each potentiating the effects of the other (Carlson *et al.* 1987). There are also D2 receptors on the presynaptic terminals of the nigrostriatal cells; stimulation of these receptors inhibits dopamine synthesis and release. This is clinically important in terms of the worsening of Parkinsonism seen with low brain concentrations of levodopa or dopamine agonists which preferentially activate presynaptic D2 autoreceptors (Merello and Lees 1992).

Acetylcholine

The role of acetylcholine in the basal ganglia is complex and uncertain. Both nicotinic and muscarinic receptors exist in the brain; the role of nicotinic, receptors is unclear but all five muscarinic receptor types, M1-M5, are found in the basal ganglia. It is likely that acetylcholine has a differential action on the direct and indirect pathways and interacts with dopaminergic activity. D1 and D2 receptor activation is followed by striatal acetylcholine release increasing or decreasing respectively. Anticholinergic drugs acting at muscarinic receptors are effective in several movement disorders especially dystonia and Parkinsonism. The mechanism of such clinical effects is unclear but must involve some degree of local basal ganglia cholinergic overactivity.

40.2.2 Pathophysiology of basal ganglia disorders

Parkinsonism

Loss of dopaminergic input to the striatum follows destruction of the dorsal pars compacta. The pathophysiological consequences have been studied in the MPTP treated primate model of Parkinsonism (DeLong 1990). As the action of dopamine is to activate the direct basal ganglia circuit and suppress the indirect pathway, the opposite effect follows striatal dopamine loss (Fig. 40.4). The same result may be produced by pharmacological depletion of brain dopamine or blockade of dopamine receptors. The effect of reduced direct and increased indirect pathway activity is to increase tonic activity in subthalamic nucleus and internal globus pallidus. This inhibits the thalamocortical neurones therefore suppressing motor cortex activity. This in turn leads to akinesia, bradykinesia, and rigidity due to abnormal premotor and motor cortex activation. Scalp recordings of movement related cortical potentials and cerebral blood flow studies in Parkinsonian patients tend to support this concept. In addition to altered tonic activity of subthalamic nucleus and internal globus pallidus neurones, the responses of these cells to peripheral movement are also altered, leading to abnormal feedback to the motor and premotor cortex. This leads to abnormal preparation and initiation of movement

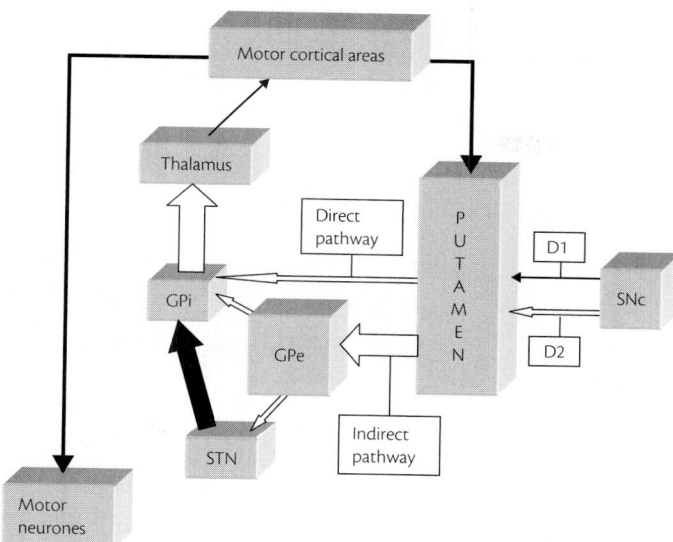

Fig. 40.4 The basal ganglia in Parkinson's disease. The width of the arrows indicates the activity of the relevant pathway. Open arrows are inhibitory, filled are excitatory.
Key: D1 and D2: D1-like and D2-like dopamine receptors
GPe: Globus pallidus pars externa
Gpi: Globus pallidus pars interna
SNc: Substantia nigra pars compacta
STN: Subthalamic nucleus

with small initial agonist bursts and prolonged movement times. The rigidity of Parkinson's disease is associated with abnormally increased long latency stretch reflexes, probably due to abnormal central processing of peripheral inputs. The mechanism by which the abnormalities of basal ganglia and motor cortex function are translated into bradykinesia and rigidity is poorly understood. An inhibitory effect on the pedunculopontine nucleus (Goldfarb *et al.* 1996) or underscaling of movement (Berardelli *et al.* 2001) may underlie akinesia.

Parkinsonian tremor is more difficult to explain. Intracerebral recordings in animal models and Parkinsonian patients reveal groups of neurones firing in bursts at the same frequency as the tremor in the thalamic ventralis intermedius nucleus and motor cortex and lesions of these areas abolish the tremor. This has been documented in Parkinsonian patients after stereotactic thalamic lesions or strokes involving the motor cortex or internal capsule. In animal models however, a rest tremor cannot be produced by a pure nigrostriatal lesion; a midbrain lesion involving nigrostriatal and cerebellothalamic pathways is required. Although a 4–6 Hz rest tremor is a feature of Parkinson's disease, it cannot be assumed that this is the result of damage to the nigrostriatal tract because other brain regions are affected by Parkinson's disease. Moreover, abnormal nigrostriatal function as measured by flurodopa positron emission tomography correlates reasonably with akinesia and rigidity but not with the severity of tremor (Otsuka *et al.* 1996) and suppression of Parkinsonian tremor is associated with reduced cerebellar activity (Deiber *et al.* 1993). Accordingly, it is not clear whether the thalamocortical oscillations underlying resting tremor are due to thalamic, cerebellar, or basal ganglia abnormalities. Probably all are involved. It is likely that the basal ganglia are involved because of the reduction in tremor seen after surgical

lesions of the basal ganglia outflow pathways although this is not as impressive as that seen after thalamic lesions (Obeso *et al.* 1997). In addition, basal ganglia pallidal and subthalamic nucleus neurones may show rythmic oscillatory activity in animals and Parkinsonian patients (Wichmann and De Long 2004).

Ballism/chorea

In Huntington's disease, degeneration of the striatal medium spiny neurones will lead to chorea if the source of the indirect GABA/enkephalin pathway is principally affected (Crossman *et al.* 1988). This leads to reduced internal globus pallidus output and thalamocortical hyperactivity as in hemiballismus. If the direct pathway GABA/substance P containing medium spiny neurones are involved later (or in some cases earlier), akinesia and rigidity are observed (Wichmann and DeLong 1997).

Hemiballismus is associated with lesions of the subthalamic nucleus, often vascular (Lee and Marsden 1994a; Vidakovic *et al.* 1994). This interupts the indirect pathway so that the internal globus pallidus is influenced only by the direct inhibitory striatopallidal pathway with consequent thalamic overactivity and increased thalamocortical motor drive. Reduced internal globus pallidus outflow releases the excitatory thalamocortical projection with increased motor cortical activity and involuntary movement. However, ballism may arise with lesions of several cortical regions and basal ganglia structures, not just the subthalamic nucleus (Postuma and Lang 2003) and may also arise with no discreet lesion as in severe hyperglycaemia. The subthalamic nucleus has extensive connections to regions other than the globus pallidus pars interna (Wichmann and DeLong 2003).

A difficult issue is the anti-dyskinetic effect of internal globus pallidus lesions in Parkinson's disease. If the 'rate model' scheme in Fig. 40.2 is correct this would not be expected. It is likely that abnormal firing patterns of bursting and excessive synchronization are involved with the generation of chorea, ballism, and other dyskinesias—hence the effect of pallidal lesions (Postuma and Lang 2003).

Dystonia

Although dystonia is commonly associated with basal ganglia lesions, it is difficult to explain on the basis of disordered basal ganglia circuitry. Dystonia is closely associated with putaminal or thalamic lesions (Bhatia *et al.* 1994; Kostic *et al.* 1996; Lehericy *et al.* 1996) but in some patients there are focal abnormalities in the brainstem (Gibb *et al.* 1988; Kulisevsky *et al.* 1988; Zweig *et al.* 1988; Esteban Munoz *et al.* 1996), cerebellum, spinal cord (Cammarota *et al.* 1995), or peripheral nerves. The mechanism by which dystonia occurs is poorly understood, with various levels of the central nervous system implicated. There is increased excitability of the motor cortex detectable with transcranial magnetic stimulation studies (Ridding *et al.* 1995; Ikoma *et al.* 1996) and positron emission tomography functional imaging (Eidelberg *et al.* 1995). Movement-related cortical potentials recorded over the premotor areas are reduced, suggesting abnormal motor programming, possibly secondary to abnormal basal ganglia output (Deuschl *et al.* 1995; Kaji *et al.* 1995b; Van der Kamp *et al.* 1995). Abnormal basal ganglia function has been difficult to detect but a position emission tomography study showed metabolic overactivity of the lentiform nucleus, probably due to overactivity of the direct putamenopallidal inhibitory pathway (Eidelberg *et al.* 1995). Possibly there is reduced 'surround inhibition' in the motor cortex due to imbalance of the direct and indirect basal ganglia pathway (Hallett 1998; Mink 2003). There is also evidence of increased interneuronal excitability in the brainstem of patients with cranial dystonia (Tolosa *et al.* 1988) and reduced reciprocal inhibition at spinal cord level, reflecting impaired spinal inhibitory circuitry (Marsden and Rothwell 1987). How these abnormalities could be secondary to a primary basal ganglia malfunction is unclear. There is even evidence of abnormal sensory processing in dystonia (Walsh *et al.* 2007) and vibration applied to an affected limb will activate dystonic postures and this effect can be abolished by intramuscular lignocaine (Kaji *et al.* 1995a). This must be due to abnormal central processing of Ia muscle spindle afferent inputs. It is possible that this explains the reduction of dystonia by so-called 'sensory tricks' (see Section 40.4.1).

As with chorea/dyskinesia, the rate model in Fig. 40.2 cannot explain the therapeutic effect of pallidotomy in dystonia; pallidal lesions would reduce pallidal output and so increase thalamic activity, exacerbating a dyskinetic disorder (Vitek 2002). Moreover, pallidal firing rates in dystonia can be increased (Hutchison *et al.* 2003). It is likely that dystonia is also a product of altered pallidal outflow patterns and altered activity in other brain regions outside the basal ganglia.

40.2.3 Neurosurgical interventions in basal ganglia disorders

It can be seen from Fig. 40.2 that there are several points at which the internal circuitry of the basal ganglia may be altered by surgical lesions. The most commonly used are the internal globus pallidus, thalamus ventralis intermedius, and subthalamic nucleus. The main indications for neurosurgical treatment of movement disorders are Parkinson's disease, dystonia, and tremor. Localized neuronal inactivation may be produced by a destructive electrolytic lesion or more commonly now an implanted deep brain (neuro) stimulation electrode to produce a reversible inhibitory effect.

Parkinson's disease

Parkinson's disease is associated with increased neuronal firing in the subthalamic nucleus and internal globus pallidus. Tremor related firing patterns are present in thalamic cells as well as the subthalamic nucleus and internal globus pallidus. As might be predicted from these observations, lesions or deep brain stimulation of the internal globus pallidus either by pallidotomy or pallidal deep brain stimulation, of the subthalamic nucleus by subthalamotomy or subthalamic deep brain stimulation or of the thalamus by thalamotomy or thalamic/Vim deep brain stimulation, all produce an antiparkinsonian effect (Obeso *et al.* 1997). The clinical effects of these targets (see Fig. 40.5) vary and deep brain stimulation is increasingly undertaken in preference to lesioning procedures. Thalamic deep brain stimulation is employed for treatment of medically intractable tremor, pallidal deep brain stimulation mainly for suppression of dyskinesia, and subthalamic nucleus-deep brain stimulation is of greater value in the relief of contralateral rigidity and akinesia.

A difficulty with the direct–indirect pathway model of basal ganglia function is that a thalamic lesion should exacerbate akinesia. This does not appear to be a problem after thalamotomy leading to the suggestion that the antiparkinsonian effect follows the removal of abnormal thalamocortical activity rather than merely a reduction of excessive pallidal outflow (Marsden *et al.* 1994). In fact,

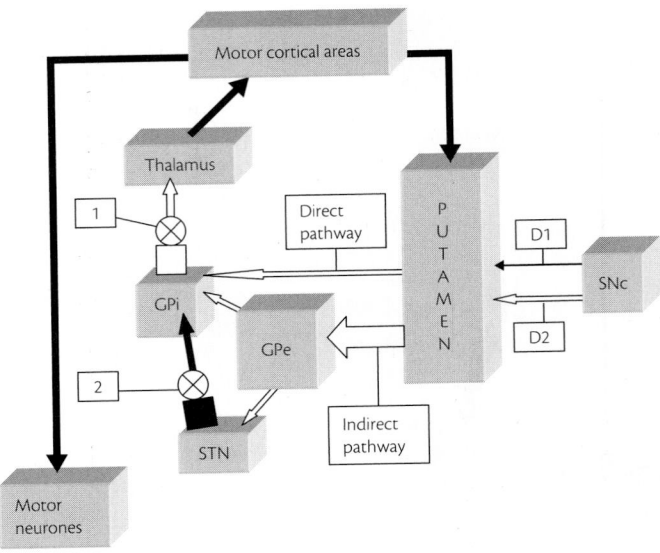

Fig. 40.5 The basal ganglia in Parkinson's disease, showing the effects of pallidal (1) and subthalamic (2) surgical intervention with correction of excessive inhibitory pallidothalamic or excitatory subthalamopallidal activity and restored motor cortical outflow.
Key: D1 and D2: D1-like and D2-like dopamine receptors
GPe: Globus pallidus pars externa
Gpi: Globus pallidus pars interna
SNc: Substantia nigra pars compacta
STN: Subthalamic nucleus

there is some evidence that thalamotomy does indeed reduce motor cortex activation with subtle impairment of finger dexterity contralaterally. This is overshadowed by tremor reduction (Boecker *et al.* 1997).

Dystonia

Although basal ganglia dysfunction in dystonia is less well understood, alteration of the pallidothalamic outflow to the motor cortex is likely. Lesions of the motor thalamus were effective in dystonia, especially of the contralateral distal musculature (Andrew *et al.* 1983) but thalamotomy is now rarely performed. Increasingly, pallidal deep brain stimulation is used to treat severe dystonia (Cif *et al.* 2003; Yianni *et al.* 2003) and probably acts by removal of abnormal basal ganglia output to the motor cortical areas.

40.3 The akinetic rigid disorders

In these disorders the clinical picture is one of Parkinsonism with slowness of movement and muscular rigidity. Usually akinetic rigid states are easily recognized but depression, catatonic schizophrenia, catalepsy, and frontal lobe dysfunction (abulia) may also cause striking immobility. Lesions of the descending corticospinal tracts leading to pseudobulbar palsy and spastic tetraparesis may, if mild, resemble an akinetic rigid state and cause initial diagnostic confusion; mild pyramidal lesions may cause only contralateral slowness of fine finger movements. Psychiatric disease, degenerative cerebral disorders, cerebrovascular disease and motor neurone disease may, on occasions, be mistaken for Parkinsonism.

Although the majority of akinetic rigid patients have idiopathic Parkinson's disease, other causes of Parkinsonism must be considered before this diagnosis is made (Table 40.2). Careful enquiry must be made regarding exposure to drugs and toxins as well as a family history. The onset and subsequent progression of the condition and, crucially, the response to any dopamine replacement therapy must be reviewed in order to distinguish Parkinson's disease from other Parkinsonian disorders.

40.3.1 Idiopathic Parkinson's disease

Definition

'Involuntary tremulous motion, with lessened muscular power, in parts not in action and even when supported; with a propensity to bend the trunk forward, and to pass from a walking to a running pace: the senses and intellects being uninjured'.

James Parkinson's original description of this disorder immediately suggests one of the most distinctive conditions in medicine. Parkinson's disease, or paralysis agitans, is one of the most frequently encountered neurological conditions and presents an increasing challenge to the health services of developed nations with ageing populations. Clinically, Parkinson's disease may be defined as:

- the presence of two out of the three cardinal features of bradykinesia, rigidity, and tremor; postural instability tends to occur later;

- a good clinical response to levodopa; and

- no 'atypical' features suggestive of another Parkinsonian syndrome.

Pathologically there is extensive loss of pigmented dopaminergic substantia nigra neurones and the presence of Lewy bodies.

Epidemiology

Numerous studies have yielded crude prevalence rates of 10–400 cases per 100 000 people (Zhang and Roman 1993). These estimates are difficult to compare due to different ascertainment methods. In Europe and North America the prevalence has been estimated to be between 100 and 250 per 100 000. In Europe, prevalence rates are similar in all countries and there is no major difference between men and women (De Rijk *et al.* 1997) although some studies have suggested that Parkinson's disease is slightly more common among men. There are many undetected cases and in one study, 24 per cent were diagnosed during the research survey (De Rijk *et al.* 1997).

The usual age of onset is after 50 years with the frequency rising steeply with age. Onset before 40 years *'early onset Parkinson's disease'* is unusual and before 20 is exceptional *'Juvenile Parkinson's disease'*. Advancing age remains the most powerful predictor of developing Parkinson's disease; the lifetime risk of developing Parkinson's disease is 2 per cent for men and 1.3 per cent for women (Elbaz *et al.* 2002).

Epidemiological studies have detected a variety of risk factors. These include rural residence, exposure to well water, pesticides and herbicides, farming, wood pulp mills, and iron and steel production (Tanner and Goldman 1996). The significance of these findings is unclear but a common thread is that Parkinson's disease might be due to exposure to an unidentified environmental toxin. Head trauma, inflammation, caucasian ancestry, Nocardia infection are possible risk factors in case control studies. Protective factors have also been identified: smoking reduces the risk of Parkinson's disease by 60 per cent and coffee drinking by 30 per cent (Hernan *et al.* 2002); alcohol and use of non-steroidal anti-inflammatory drugs may also be protective.

Nigral cell death

The cause of Parkinson's disease is the destruction of the pigmented dopaminergic neurones of the substantia nigra and some other brain regions. The mechanism appears to involve defective mitochondrial respiration and oxidative stress. This in turn may be due to toxic exposure, genetic factors, or an interaction of the two. However, known Parkinson's disease genes or environmental risk factors account only for a small minority of cases.

- Mitochondrial complex 1 is reduced in the substantia nigra of the brain in Parkinson's disease (Schapira *et al.* 1990; Schapira 1994) but not in other other brain regions. Although mitochondrial DNA deletions are seen, these appear similar to those associated with normal ageing. No pathogenic mutations have been detected and so the complex 1 deficiency is

Table 40.2 Akinetic rigid syndromes

Idiopathic Parkinson's disease		
Symptomatic Parkinsonism	*Common*	Drug induced Parkinsonism
	Rare	Cerebrovascular disease
		Cerebral trauma
		Toxic Parkinsonism
		Hydrocephalus
		Intracranial masses
		Creutzfeldt-Jacob Disease
		Post encephalitic parkinsonism
		Rett syndrome
		Post anoxic parkinsonism
Other akinetic rigid syndromes		Lewy body dementia
		Multiple system atrophy
		Progressive supranuclear palsy
		Corticobasal degeneration
		HemiParkinsonism-hemiatrophy
		Other non Lewy body parkinsonism
Metabolic disorders		Basal ganglia calcification
		Chronic hepatocerebral degeneration
		Gaucher's disease
Degenerative disorders		Alzheimer's disease
		Pick's disease
		Idiopathic late onset cerebellar degenerations
Hereditary disorders		Dopa responsive dystonia
		Huntington's disease
		Hallervorden–Spatz disease
		Hereditary cerebellar ataxias
		Wilson's disease
		X-linked dystonia-parkinsonism
		Pallidal degenerations
		Parkinsonism with hypoventilation
		Rapid onset dystonia Parkinsonism
		Dopa responsive dystonia
		Neuroacanthocytosis
		Fragile X tremor ataxia syndrome
		Frontotemporal dementia
		Kufor Rakeb syndrome

likely to be acquired. It is likely that reduction of complex 1 leads to impairment of neuronal energy metabolism, calcium homeostasis, oxidative stress, and ultimately premature cell death (Swerdlow *et al.* 1996). Rotenone inhibits mitochondrial complex 1 and produces a rat model of Parkinson's disease (Panov *et al.* 2005).

- Oxidative stress is increased in the substantia nigra. There is increased nigral superoxide dismutase activity due to increased superoxide production. This leads to more hydrogen peroxide and hydroxyl radical production, especially in the presence of iron which is concentrated in the substantia nigra. Increased lipid peroxidation has been detected in the substantia nigra in Parkinson's disease (Jenner and Olanow 1996) and there is a reduction in reduced glutathione, implying increased production of free radicals. There are increased levels of iron in the substantia nigra in Parkinson's disease and this also increases oxidative stress. The cause of iron accumulation in the brain is unclear but it is known that iron binds to neuromelanin. Dopamine metabolism by monoamine oxidase may increase oxidative stress because of the production of hydroxyl ions. This has raised concern about levodopa therapy but evidence of levodopa toxicity to neurones is conflicting and there is no evidence that the adverse effects of chronic levodopa treatment in Parkinson's disease are related to increased nigral cell death. Oxidative stress damages cell membranes, proteins, and both nuclear and mitochondrial DNA (Jenner and Olanow 1996). Oxidative stress may be associated with animal fat intake which is increased in Parkinson's disease (Logroscino *et al.* 1996) but there is no association with dietary antioxidants other than a possible weak protective effect with vitamin E intake (Etminan *et al.* 2005).

Environmental factors and the toxin hypothesis

An important role for environmental factors is suggested by the link with encephalitis, post-encephalitic Parkinsonism, forms of toxic Parkinsonism and low concordance in monozygotic twins. Associations with environmental factors suggest a role for some as yet unidentified environmental toxin. This has stimulated much interest in possible toxic compounds which might cause Parkinson's disease but the disorder is not new and so an industrial substance is unlikely to be responsible. Nor are these environmental associations consistent. In 1983 however, a number of cases of Parkinsonism caused by methyl-phenyl-tetrahydropyridine, MPTP, attracted considerable attention (Langston *et al.* 1983). Methyl-phenyl-tetrahydropyridine caused a rapidly progressive form of Parkinsonism which was clinically and pathologically very similar to idiopathic Parkinson's disease. Methyl-phenyl-tetrahydropyridine reproduces Parkinsonism reliably in experimental animals. Metabolism of the methyl-phenyl-tetrahydropyridine pro-toxin by monoamine oxidase B yields methylphenylpyridinium ion, MPP+, which is taken up into dopaminergic neurone terminals where it inhibits cellular mitochondrial complex 1 leading to cell death (Tanner 1994). Although the methyl-phenyl-tetrahydropyridine mechanism fits so well with other data concerning mitochondrial and oxidative factors, no similar toxin has been linked to sporadic Parkinson's disease. A very interesting finding has been an association with reduced activity of the ubiquitin proteasome system; this can be inhibited by epoxomicin—a product of actinomycetes found in soil and some root vegetables (McNaught *et al.* 2004b).

There is little evidence for heavy metal intoxication as a cause of Parkinson's disease; studies of aluminium, copper, iron, mercury, zinc, and manganese have been mostly negative. There has been some medical (and medicolegal) interest in manganese as a cause of Parkinson's disease in welders. Although a form of manganese induced toxic Parkinsonism exists (Section 40.3.5) it is clinically different to Parkinson's disease, tending to be symmetrical, subacute with mental changes, dystonia, myoclonus, and a poor response to levodopa (Josephs *et al.* 2005). Epidemiological surveys have not confirmed welding as a significant risk factor for Parkinson's disease (Goldman *et al.* 2005; Jankovic 2005; Kieburtz and Kurlan 2005).

Metabolic and genetic polymorphisms and susceptibility genes

It has been suggested that Parkinson's disease might be caused by slower detoxification of a pathogenic environmental agent in metabolically susceptible individuals. Heterozygous carriers of autosomal recessive forms of Parkinson's disease, *Parkin*, *DJ-1* and *Pink 1* genes, are not increased in case-control studies (Lincoln *et al.* 2003) but such gene carriers may show subtle signs (Pramstaller *et al.* 2005) and reduced dopaminergic activity on positron emission tomography scanning (Hilker *et al.* 2001). There is also an increased incidence of Parkinson's disease in heterozygous carriers of Gaucher's disease mutations of the glucocerebrosidase gene (Sato *et al.* 2005). Associations have been detected with various genetic polymorphisms of the *NAT2*, *MAOB*, ferritin light chain, *GSTT1*, and mitochondrial *tRNAGlu* genes (Tan *et al.* 2000) as well as tau H1 (Skipper *et al.* 2004). Polymorphisms of the alpha synuclein gene, *SNCA*, can increase risk (Mueller *et al.* 2005) while *UCHL1* polymorphism can be protective (Maraganore *et al.* 2004). Variants of the *semaphorin 5A*, *Park 10*, and *Park 11* genes are also associated with Parkinson's disease (Maraganore *et al.* 2005). Several genetic susceptibility-environmental interactions have been proposed including *CYP2D6* and pesticides (Elbaz *et al.* 2004), *P-gp* gene and blood brain barrier permeability to toxins (Kortekaas *et al.* 2005), glutathione–S–transferase and smoking (Deng *et al.* 2004) and complex 1 and environmental toxins (Panov *et al.* 2005).

Genetics of Parkinson's disease

Three main lines of evidence point to the role of hereditary factors in Parkinson's disease, in addition to the weak genetic associations already mentioned.

Family studies. The tendency for Parkinson's disease is to occur more commonly among the relatives of affected patients. This has been noted for many years but the exact proportion of patients with affected relatives has varied depending on the methods used and the definition of an affected relative. Some studies have included patients with postural tremor while others have not. Some studies have tended to use family history only and suggest that 10–40 per cent of patients have affected relatives. Among families with more than one affected individual, autosomal dominant inheritance with reduced penetrance seems likely with segregation ratios of 20–30 per cent among sibs and parents (Maraganore *et al.* 1991; Plante-Bordeneuve *et al.* 1995). A problem with such studies is the late age of onset of Parkinson's disease so that a normal relative may yet become affected with time. Positron emission tomography studies using flurodopa scans have indeed detected relatives with preclinical Parkinson's disease (Piccini *et al.* 1997).

Twin studies. Twin studies have shown very low concordance rates which is not consistent with a significant genetic factor. However, some co-twins only develop Parkinson's disease after many years and follow up studies have shown slightly higher concordance rates many years later. Such studies are also confounded by co-twins with pre-clinical Parkinson's disease which may be detected by positron emission tomography scanning (Burn *et al.* 1992; Piccini *et al.* 1999). The statistical inaccuracies of the twin method also make exclusion of a genetic contribution difficult.

Parkinson's disease genes. Families containing many affected individuals autosomal dominant inheritance (Golbe *et al.* 1990; Markopoulou *et al.* 1995; Wszolek *et al.* 1995) have been reported. In some of these families, a mutation of the gene for alpha synuclein, a presynaptic protein expressed in brain regions affected by Parkinson's disease, has been detected (Kruger *et al.* 1998; Polymeropoulos *et al.* 1997). Several other Parkinson's disease genes have now been identified, *PARK 1-11*, and a variety of other genetic disorders may also cause Parkinsonism (see Table 40.4).

Autosomal recessive Parkinson's disease tends to present earlier, before 40 years, is levodopa responsive, and is more likely to be associated with dystonia, brisk tendon reflexes, and overnight sleep benefit (Lucking *et al.* 2000).

- *Parkin, PARK2,* mutations (Kitada *et al.* 1998) cause an early onset, slowly progressive, symmetrical form of Parkinson's disease sometimes with dystonia; it is similar to idiopathic Parkinson's disease and accounts for 10–20 per cent of early onset Parkinson's disease. There are many different Parkin mutations which makes DNA diagnosis difficult and impractical. Parkin, a ubiquitin ligase, is widely expressed in neurones; mutations cause a loss of function affecting mitochondrial and oxidative functions.

- *Pink1,* or *PARK6* mutations (Bonifati *et al.* 2005) account for less than 10 per cent of early onset Parkinson's disease.

- *DJ1,* or *PARK7* mutations (Hedrich *et al.* 2004) account for only 1–2 per cent of early onset cases. This is a very rare cause of sporadic Parkinson's disease.

Autosomal dominant Parkinson's disease is more typical with usually later onset and is also levodopa responsive.

- *Alpha synuclein, SNCA,* or *PARK1* mutations are a very rare cause of Parkinson's disease (Berg *et al.* 2005) although a few autosomal dominant families have been reported. There is considerable genetic heterogeneity with several point mutations, duplications, and triplications described. It is likely that there is a gene dosage effect with *SNCA/PARK1* duplications causing typical Parkinson's disease and triplications causing more severe disease with dementia (Chartier-Harlin *et al.* 2004). The function of SNCA is unknown; it is a cytosolic and lipid bound protein possibly associated with vesicle function and transmitter release and is present in Lewy bodies.

- *PARK3* is a very rare cause of hereditary but otherwise typical Parkinson's disease (Gasser *et al.* 1998); it may be associated with the sepiapterin reductase gene (Sharma *et al.* 2006).

- *PARK5* has been described in only one family.

- *LRRK2/Dardarin, PARK8* mutations cause either sporadic or familial (autosomal dominant) Parkinson's disease (Gilks *et al.* 2005). Over 40 mutations have been described. *PARK8* causes <2

per cent of sporadic Parkinson's disease (Skipper *et al.* 2005) but is more common in Spain and Portugal where it accounts for 3–8 per cent of cases (Bras *et al.* 2005). Among familial dominant Parkinson's disease cases *LRRK2* is more common (3–13 per cent) (Khan *et al.* 2005) especially in North Africa where it accounts for 41 per cent (Lesage *et al.* 2005).

- *PARK9* causes the rare Kufor Rakeb syndrome—an atypical Parkinsonism with ophthalmoplegia and pyramidal signs (Williams *et al.* 2005b).

- *Nurr1, synphilin1, and Omi/HtrA2* mutations can cause autosomal dominant Parkinson's disease but these have no PARK designation and are extremely rare.

Pathology of Parkinson's disease

It is likely that Parkinson's disease symptoms are related to impaired nigral cell function and subsequent cellular loss as it is difficult to account for the natural history of the condition in terms of cell losses alone. The basic neuropathological feature is loss of pigmented dopaminergic neurones, astrocytic gliosis, dystrophic neurites, and the formation of Lewy bodies in the substantia nigra pars compacta. However, other brain regions are also affected—including other brainstem nuclei and the cerebral cortex. The pathological hallmark of Parkinson's disease is the *Lewy body* (Fig. 40.6), an eosinophilic intracellular inclusion. Typically there is a central core within a less intensely staining body surrounded by a pale halo. The significance of the Lewy body is unclear and the process by which it is formed is unknown. Lewy bodies have a complex composition but contain neurofilament proteins, presumably derived from the neuronal cytoskeleton as well as alpha synuclein and ubiquitin. Lewy bodies are not absolutely specific to Parkinson's disease as is sometimes suggested; they are found in a few other neurological disorders such as ataxia telangiectasia and Hallervorden–Spatz disease. Nor are they found only in the substantia nigra but in many areas of the nervous system including the locus coeruleus, dorsal motor nucleus of the vagus, and other brainstem regions as well as the nucleus basalis of Meynert, thalamus, cerebral cortex, intermediolateral columns of the spinal cord, and within the autonomic nervous system. The pale body is a more homogeneous granular inclusion which is specific to the substantia

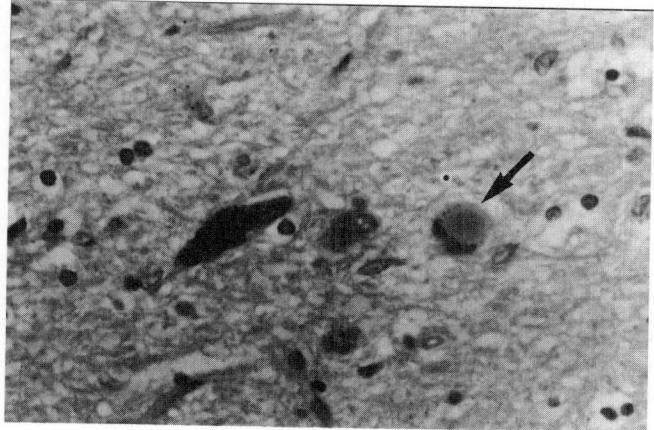

Fig. 40.6 A substantia nigra Lewy body (arrowed) in Parkinson's disease. (Courtesy of Dr J Broome.)

nigra and locus coeruleus. Lewy bodies are associated with devastating depletion of the melanin containing dopaminergic neurones of the pars compacta of the substantia nigra, SNc (Fig. 40.7). In symptomatic patients, at least 70 per cent or more of nigral cells are lost. Cell loss is also seen in other regions in association with Lewy bodies. The nigral cell loss is initially rapid and then slows down later in the disease; there is also preferential cell loss in the ventrolateral region of the SNc with relative preservation of the dorsal tier (Fearnley and Lees 1991). This differs from the pattern seen in normal ageing where the rate of cell loss is lower, the severity of active cell death appears much less, and the emphasis is on the dorsal tier of the SNc (Fearnley and Lees 1991).

The term 'incidental Lewy body disease' has been coined to describe the occurrence of Lewy bodies and other features of Parkinson's disease in the brains of people without clinical evidence of Parkinson's disease during life. The prevalence of incidental Lewy bodies approaches 10 per cent of those aged over 80 years. Such individuals presumably died during the presymptomatic phase of Parkinson's disease.

Braak has described six stages in the evolution of Parkinson's disease (Braak *et al.* 2004) with the pathology spreading up from the lower brainstem to the midbrain and into the cerebrum. During stages 1–2 patients are asymptomatic but inclusion body pathology appears initially in the medulla (dorsal motor nucleus of the vagus), brainstem reticular formation, the olfactory bulb, and anterior olfactory nucleus. In stages 3–4, the substantia nigra and other brainstem nuclei develop progressively severe pathological changes. At this point, most patients develop Parkinsonism. In stages 5–6, the neocortex is affected with advanced manifestations including cognitive decline.

Clinical features

The onset is insidious and the patient may have difficulty recalling when the first symptom appeared. Some patients mention malaise, tiredness and a general slowing up of physical activity. The symptoms are often attributed to depression or 'normal ageing'. The older patient, living alone, may simply start 'not coping' and may seem to be in a general decline. Writing becomes small and untidy, gradually fading away, leading to problems at work or when signing cheques. Musculoskeletal pain or stiffness especially in an upper limb may be prominent and cause diagnostic confusion. Some mention abnormal sensations such as pain and 'internal' tremor before it is clinically apparent (Shulman *et al.* 1996). Gradually the typical resting tremor, rigidity and slowness of movement appear.

Parkinson's disease is often regarded as an easy 'end of the bed' diagnosis but is highly variable in presentation and frequently misdiagnosed (Hughes *et al.* 1993b). Two principal forms are recognized. In the 'tremor dominant' form, the main feature is tremor, in the 'akinetic rigid' form, bradykinesia or gait disturbance is more prominent. The appearance of the patient is characteristic and becomes more so with time (Fig. 40.8). Bradykinesia causes a poverty of movement leading to an abnormal stillness and impassive appearance. The posture becomes one of flexion of the neck, spine elbows, wrists, hips and knees. The patient stands and walks on a narrow base. Facial expression is reduced with the mouth slightly open, the voice is quiet and monotonous and later there is a greasy skin. Some patients present with minimal and easily overlooked physical signs such as slightly reduced arm swing on one side while walking with barely noticeable wrist rigidity brought out by synkinesis of the contralateral arm; others present at a surprisingly late stage with gross Parkinsonism and considerable disability. The most characteristic presentation is with tremor, bradykinesia and rigidity of an upper limb. Unilateral onset is typical of Parkinson's disease and yet this often causes diagnostic confusion and the ordering of investigations to exclude a structural lesion of one cerebral hemisphere.

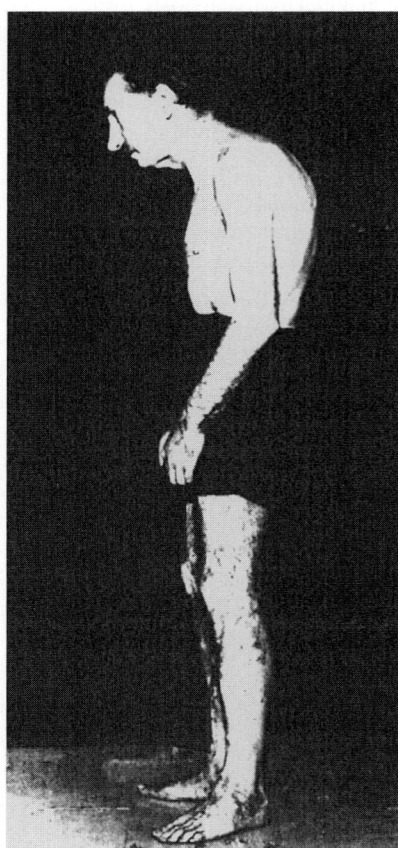

Fig. 40.8 A patient with Parkinson's disease. Note the typical posture with flexion of the elbows.

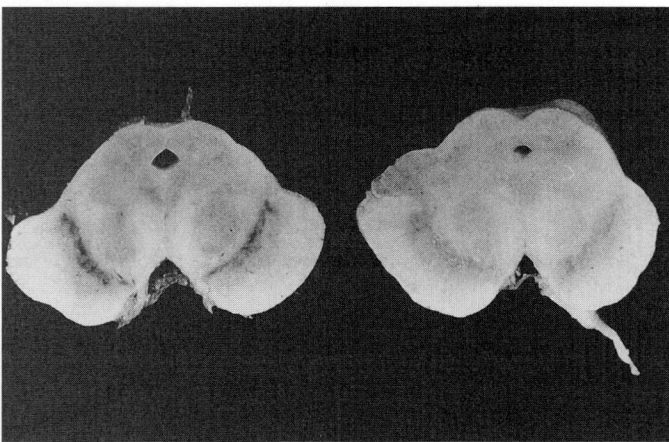

Fig. 40.7 A midbrain section from a normal brain (left) and in Parkinson's disease (right) showing depletion of the pigmented substantia nigra. (Courtesy of Dr J Broome.)

The most reliable diagnostic principle is the presence of two out of the three cardinal features of:

- tremor;
- bradykinesia; and
- rigidity.

To this could be added the presence of a good response to levodopa therapy which is observed in all but a handful of patients (Hughes *et al.* 1993b).

The Parkinsonian *tremor* is noticeable at rest and is less prominent or absent with action or posture. The frequency is slow, 4–7 Hz, and may involve only the thumb or one finger at first. In some patients the typical 'pillrolling' appearance is seen but this is not invariable. It is more evident with the patient distracted and can be brought out by asking the patient to carry out simple mental arithmetic or counting with the eyes shut and sitting quietly at rest. The tremor is increased by movement of the contralateral limbs or walking. Eventually the tremor spreads to the contralateral limbs but often remains more noticeable on the initially affected side. In some cases, the face and jaw are eventually involved. In addition to the classical rest tremor, there is often a faster postural or action tremor seen in the outstretched hands but this is not as prominent. Patients with a marked postural tremor, especially if this is the principal symptom are not likely to have Parkinson's disease.

Parkinsonian rigidity may be a smooth resistance to passive movement, 'lead pipe' or 'plastic' rigidity, or may have a ratchet like cogwheel effect due to additional postural tremor. Parkinsonian rigidity is more obvious in the extremities and only later becomes prominent in the axial musculature. It is detectable with slow passive limb movement unlike spasticity which is related to rapid movement. It can be brought out by contralateral limb movement, the Froment sign, and a useful point in the outpatient clinic is that the rigidity is often more noticeable at the wrist while the patient is standing or walking about.

Bradykinesia refers to the slowness of movement and is closely related to akinesia or absence of movement. Movement is slow and delayed, initially distally and later in the proximal muscles. There are many manifestations of bradykinesia including reduced facial expression, drooling of saliva, reduced blinking, reduced arm swing while walking, difficulty turning in bed or rising from a chair, slowness of gait and small shuffling steps. Repetitive movements show slowing and reduced amplitude; finger movements are particularly affected and become clumsy and laboured. Handwriting becomes small and spidery at an early stage, *micrographia*. This may be a presenting symptom.

Postural instability is often cited as a fourth core feature of Parkinson's disease but it is not often an early symptom. The patient has a tendency to topple forwards or backwards if pushed or pulled, *propulsion* or *retropulsion*.

When walking there is a liability to topple forward with faster steps and difficulty stopping, *festination*. Falls increase and the gait develops into the typical slow shuffling walk with small steps on a narrow base. There is a delay in the initiation of gait and the patient becomes intermittently stuck and unable to move, *freezing*, usually in confined spaces or at thresholds. Eventually it is impossible to walk unaided and without treatment a bedbound state develops. Postural instability and gait failure are the most disabling features of Parkinson's disease and the most resistant to dopamine replacement therapy although some improvement can be seen with levodopa even in the later stages of the condition. Parkinson's disease rarely presents initially with falling or gait disturbance and other disorders such as progressive supranuclear palsy, cerebrovascular disease, or multiple system atrophy should be considered if this is the case.

In addition to the four cardinal features of tremor, bradykinesia, rigidity, and postural instability, a number of other problems develop. Painful off period dystonia of one or both feet is a frequent symptom, especially in the early morning (Poewe *et al.* 1988). The foot locks into a plantar flexed and inverted posture which makes standing or walking difficult; there may also be stiffness of the calf. These spasms may last a few minutes or over an hour until antiParkinsonian medication is taken. In some patients the episodes appear later in the day as well, when the effects of medication have worn off before the next dose. Other patients develop dystonic movements of the face, trunk, or limbs as a complication of levodopa therapy and these spasms are seen after medication has been taken.

In early Parkinson's disease there are some mild neuropsychological abnormalities, which are consistent with impaired frontal lobe function but these are not clinically obvious. However, many Parkinson's disease patients eventually develop dementia, the proportion rising with advancing age and longer duration of disease (Aarsland *et al.* 2001). Once thought to be a feature of only a minority of cases, dementia is now seen increasingly in clinical practice. In a recent survey, 26 per cent of patients were demented at baseline, 51 per cent after 4 years follow up and 78 per cent at 8 years (Aarsland *et al.* 2003). Conventionally, a distinction is drawn between those who develop dementia and Parkinsonism within a year 'Dementia with Lewy bodies' (Section 40.3.2; 34.6.3) and those in whom the dementia appears after 12 months 'Parkinson's disease dementia'—however, there is no good pathological basis for this and the two conditions are probably the same (Aarsland *et al.* 2004). Most patients have cortical Lewy body pathology, some have Alzheimer pathology, and others have both (McKeith *et al.* 1996). However, cortical Lewy bodies are a key feature of both Parkinson's disease and dementia with Lewy bodies in most cases suggesting that cortical Lewy bodies are an integral part of Parkinson's disease (Colosimo *et al.* 2003) as suggested by Braak (Braak *et al.* 2004). In those who do develop dementia, cognitive slowing, depression, and prominent memory impairment are typical whereas aphasia and agnosias are rare; this pattern is said to indicate a 'subcortical' dementia. Other patients however, develop a pattern of cognitive impairment similar to Alzheimer's disease and the distinction between subcortical and cortical dementia is not clear cut. A characteristic clinical picture is the patient with advanced Parkinson's disease who develops fluctuating confusion, disorientation and visual hallucinations. This is often attributed by doctors, patients and relatives to drug side effects from dopamine agonists, amantadine, or anticholinergic drugs, whereas it is characteristic of Parkinson's disease with dementia and dementia with Lewy bodies (McKeith *et al.* 1996). Indeed, the appearance of visual hallucinations has been proposed as a diagnostic feature of Parkinson's disease and is probably largely a consequence of disease progression (Williams and Lees 2005). The development of visual hallucinations is not well correlated with medication and is more closely linked to age, duration of disease, and the presence of cognitive impairment; medication exposure is similar in hallucinating and non-hallucinating patients (Williams and Lees 2005). Visual hallucinations tend to worsen over time (Goetz *et al.* 2005b) and are associated with progression to more severe psychosis (Goetz *et al.* 2006) and dementia (Goetz *et al.* 1998). Visual hallucinations are a poor prognostic factor and a strong predictor of nursing home admission.

Depression is common in Parkinson's disease (Tandberg *et al.* 1996) and may exacerbate both physical and cognitive function. Patients may complain that their Parkinson's disease is worse in a non specific way or that their medication is not effective. Others complain of memory problems or insomnia. Demented patients are more likely to be depressed and depression can give a misleading impression of dementia. It is important to consider antidepressant treatment in patients who are deteriorating physically or mentally before concluding that this is due to disease progression or dementia.

Parkinsonian patients frequently complain of insomnia and this may be an early clue to the presence of depression. On the other hand, the condition itself also predisposes to sleep disturbance due to stiffness and immobility in bed. Patients complain that they cannot get into a comfortable position, adjust the bedclothes, or roll over. This leads to frustration and fatigue and can exacerbate depression. Other patients experience prominent limb pain or restlessness at night which prevents sleep.

In the later stages of the condition, speech becomes quiet and indistinct and significant dysarthria eventually occurs in about a half of patients. Dysphagia may be prominent in some patients with severe Parkinson's disease when it is usually accompanied by marked dysarthria. Some patients develop urinary frequency and urgency by day as well as nocturia. In older men this may be due to benign prostatic enlargement but Parkinson's disease can be associated with detrusor instability. Autonomic disturbances are seen in advanced disease but are not as severe or early as in multiple system atrophy (see Section 40.3.8). Postural hypotension is the most common manifestation but it may be difficult to assess the relative contributions of the disease and medications, especially dopamine agonists. The fall in blood pressure is rarely severe or disabling in idiopathic Parkinson's disease. Impotence is frequently reported and is probably secondary to autonomic involvement, depression, and physical immobility, especially at night. Constipation is very common in Parkinson's disease at most stages of the condition and is sometimes severe, requiring hospital admission. There are several factors underlying this; autonomic dysfunction, reduced exercise, dietary changes, and the effects of anticholinergic drugs all contribute to the problem. Patients often lose weight and this may be considerable in the later stages. The cause of this is unclear. Many patients notice increased sweating; in the later stages of the condition, especially in severe off periods in levodopa treated patients, there may be attacks of drenching sweating with immobility, fear, and often pain.

It is not widely appreciated that Parkinson's disease can be associated with severe pain in some patients (Quinn *et al.* 1986). The cause of the pain is often missed and unnecessary investigations and therapies are deployed to no effect other than the frustration of patient and doctor. The pain may precede the diagnosis for several months or even years or may appear later in relation to the timing of levodopa doses. There may or may not be associated spasms or dystonia with the pain which may occur at the beginning, middle, or end of the dose or in the intervening off periods.

Diagnosis and Investigation

Parkinson's disease remains a clinical diagnosis. Many cases are straightforward but diagnostic error is surprisingly common (Hughes *et al.* 1992). There are numerous causes of a Parkinsonian syndrome (Table 40.2) but many of these are obvious clinically. In clinical practice, the same diagnostic difficulties arise repeatedly:

- *Misleading early symptoms* In some cases the presenting symptoms may not immediately suggest the diagnosis or even seem neurological. Pain or weight loss may lead to inappropriate tests or treatment before the problem becomes more obvious. Surprisingly few physicians realize that an asymmetric onset is typical. The elderly patient with evolving Parkinson's disease may be placed in residential care due to inability to cope in the community and the diagnosis may be considered only at a very late stage. Patients regularly undergo psychiatric evaluation and treatment for depression or sometimes an anxiety neurosis before a careful physical examination reveals signs of bradykinesia or rigidity. In those who have received neuroleptic drugs, akinetic rigid features may be attributed to medication.

- *Multiple system atrophy* poses the greatest difficulty. The striatonigral degeneration form may remain indistiguishable from Parkinson's disease throughout the illness and may only be diagnosed postmortem (Hughes *et al.* 1993b). A poor response to levodopa is the single most reliable indicator of this condition and because of this a diagnosis of Parkinson's disease is always provisional until the response to levodopa has been evaluated at a follow up clinic visit. Even this is not absolutely reliable as some multiple system atrophy patients do respond well to levodopa and very occasional patients with a poor response are later found to have Parkinson's disease postmortem (Hughes *et al.* 1993b). In other patients, an atypical or absent tremor, pyramidal signs, cerebellar involvement, early orofacial dyskinesia, autonomic failure, or more rapid progression of disabilty will suggest multiple system atrophy. In general, a Parkinsonian illness with asymmetric onset, duration over 10 years, persisting levodopa responsiveness after 5 years, hallucinations, dementia, and severe dyskinesia all indicate Parkinson's disease rather than multiple system atrophy.

- *Benign essential tremor* is often misdiagnosed as Parkinson's disease. The postural tremor, length of the history (often many years by which time Parkinson's disease would be very severe), lack of additional neurological impairment, response to alchohol, and family history should all help to avoid this error.

- *Drug induced Parkinsonism* is common. A careful history will identify the offending agent. Although dopamine receptor blocking drugs are well known to cause Parkinsonism, others such as fluoxetine, calcium channel blockers, and amiodarone are less obvious culprits.

- Patients with a *frontal lobe gait apraxia* are often thought to have Parkinson's disease and receive unnecessary antiParkinsonian drugs. Usually these patients have cerebrovascular disease or normal pressure hydrocephalus. The gait disorder is different to that in Parkinson's disease with small shuffling steps on a wide base with falling and little additional evidence of Parkinsonism, especially in the upper body. In some older patients, there may be features of gait apraxia and otherwise typical Parkinson's disease. In this situation, there may be a combination of Parkinson's disease and cerebrovascular disease. This difficulty sometimes requires a therapeutic trial of levodopa to exclude a treatable Parkinsonian component to the patient's disability.

- Occasionally, *rare akinetic rigid disorders* such as progressive supranuclear palsy and corticobasal degeneration cause confusion. The early appearance of postural instability and falling, dysarthria, and eventually pyramidal signs and a supranuclear downgaze palsy (Litvan *et al.* 1997) are features of the former. The latter is characterized by striking asymmetry throughout the illness with cortical sensory loss, the alien limb phenomenon, supranuclear gaze palsy, and myoclonus. Pallidal degenerations (Jellinger 1968) may also cause an atypical Parkinsonian syndrome.

In order to reduce diagnostic error in Parkinson's disease, diagnostic criteria have been proposed (Hughes *et al.* 1993b) and widely accepted (Table 40.3).

Often, investigations are normal or misleading and must be interpreted in the light of the likely clinical diagnosis.

- CT brain scans are rarely helpful; in the elderly there may be cerebral atrophy but this is unlikely to be of significance. Although patients with multiple system atrophy may have cerebellar atrophy, ataxia is usually apparent clinically (Wenning *et al.* 1994b). CT may rarely be helpful in cases where hydrocephalus or a structural lesion (see Section 40.3.13) are suspected clinically or in patients with the hemiParkinsonism–hemiatrophy syndrome.

- MRI brain scans are particularly liable to cause confusion; many patients over the age of 60 years have subcortical vascular lesions which are unlikely to be relevant to the patient's akinetic rigid state (Section 40.3.11). However, in patients who are thought to have progressive supranuclear palsy the MR scan may show striking brainstem atrophy.

- There has been more interest in single photon emission computerized tomography 'SPECT' to visualize striatal D2 receptors using ^{123}I IBZM-SPECT (Schwarz *et al.* 1993) or more commonly presynaptic dopamine transporter (DAT) using beta-CIT or ^{123}I-IPT (Schwarz *et al.* 1993; Marek *et al.* 1996). While these methods may differentiate Parkinson's disease from normal individuals, essential tremor, psychogenic Parkinsonism or drug induced Parkinsonism, they will not distinguish Parkinson's disease from other degenerative conditions such as multiple system atrophy, progressive supranuclear palsy, and corticobasal degeneration. Moreover, they are affected by antiparkinsonian drugs, possibly by altered regulation of DAT and receptors by medication, so are unreliable in medicated patients.

- Although positron emission tomography 'PET' and magnetic resonance spectroscopy 'MRS' have been used in Parkinson's disease they are not widely used for diagnosis (Brooks *et al.* 1992; Burn *et al.* 1994; Federico *et al.* 1997).

- Apomorphine and levodopa 'challenge' tests are sometimes employed to assess dopaminergic responsiveness as a test for Parkinson's disease. These are unreliable and are not recommended (Clarke and Davies 2000).

- In younger patients with parkinsonism, it is particularly important to exclude Wilson's disease by appropriate biochemical and ophthalmological investigations.

Management

With the exception of a few, usually elderly, patients who present late, immediate drug treatment is often unneccesary. The most important thing at this point is a careful explanation of the diagnosis, treatment options and prognosis. Many clinics now employ a nurse specialist to provide additional support to the patient at the time of diagnosis and afterwards. Patients are often frightened by the mention of Parkinson's disease and worried that they should have consulted a doctor sooner. They should be reassured that this would not have influenced the situation. It is necessary to explain that no treatment will completely suppress all symptoms; this avoids premature drug therapy and pointless overmedication which is a common reaction to residual symptoms in treated patients.

Unfortunately, no currently available treatment prevents the progression of Parkinson's disease and the clinical deterioration of the patient. Treatment can be thought of in five ways:

1. neuroprotection in the patient not yet requiring suppression of symptoms;

2. early symptomatic therapy;

3. management of advanced disease with treatment related complications and drug resistant features;

Table 40.3 Parkinson's disease society brain bank criteria

1. **Criteria required to establish the presence of Parkinsonism**	Bradykinesia
	Plus one of the following:
	  Rigidity
	  Resting tremor
	  Postural instability

2. **Exclusion criteria for Parkinson's disease**	Repeated stroke or stepwise progression Repeated head injury Encephalitis Oculogyric crises Recent neuroleptic treatment Relevant toxic exposure >1 affected relative Sustained remission of symptoms Unilateral signs after 3 years Supranuclear gaze palsy Cerebellar signs Severe, early autonomic failure Severe, early dementia Pyramidal signs Mass lesion or hydrocephalus on CT scan No response to levodopa
3. **Positive criteria for Parkinson's disease (3 or more required)**	Unilateral onset Rest tremor Progressive disorder Persistent asymmetry Excellent (70–100%) response to levodopa Severe levodopa induced dyskinesia Response to levodopa lasting >5 years Clinical course over >10 years

Table 40.4 Parkinsonism genes

Gene	Inheritance	Locus	Gene product
PARK 1	Autosomal dominant	4q	Alpha synuclein (SNCA)
PARK 2	Autosomal recessive	6q	Parkin
PARK 3	Autosomal dominant	2p	?
PARK 4 (=PARK 1)*			
PARK 5	Autosomal dominant	4p	UCHL1
PARK 6	Autosomal recessive	1p	PINK-1
PARK 7	Autosomal recessive	1p	DJ-1
PARK 8	Autosomal dominant	12pq	LRRK2 / Dardarin
PARK 9 ** (Kufor Rakeb syndrome)	Autosomal dominant	1p	
PARK 10	Susceptibility	1p	?
PARK 11	Susceptibility	2q	?
-	Autosomal dominant	5q	Synphilin-1
-	Autosomal dominant	2q	Nurr1
-	Autosomal dominant	2p	Omi/htrA2
DYT3 (X linked dystonia parkinsonism)	X-linked	Xq	
DYT5 (dopa responsive dystonia)	Autosomal dominant	14q	GCH1
DYT12 (rapid onset dystonia parkinsonism)	Autosomal dominant	19q	Na/K ATPase alpha 3
SCA2 ataxia +/- parkinsonism	Autosomal dominant	12q	Ataxin 2
SCA3 ataxia +/- parkinsonism	Autosomal dominant	14q	Ataxin 3
SCA6 ataxia +/- parkinsonism	Autosomal dominant	19p	CACNA1A
Huntington's disease	Autosomal dominant	4p16.3	Huntingtin
PKAN / Hallervorden Spatz	Autosomal recessive	20p13	Pantothenate kinase
FXTAS ***	X-linked	Xq27	FMR protein
FTDP17 ****	Autosomal dominant	17q21	MAP Tau
Gaucher's disease	Susceptibility	1q21	Glucocerebrosidase

* PARK4 is the same gene as PARK1 and is no longer used.

** Kufor Rakeb syndrome is an atypical syndrome with ophthalmoplegia and pyramidal signs.

*** Fragile X tremor ataxia syndrome.

**** Frontotemporal dementia with Parkinsonism.

4. neurosurgical treatment; and

5. attempts at neuro-restoration.

Neuroprotection. In recent years there has been great controversy about whether any treatment influences the rate of nigral cell death and progression of the disease. To date, no drug has been shown to do this conclusively.

There are theoretical reasons to suppose that inhibition of type B monoamine oxidase might retard the progression of Parkinson's disease. Type B monoamine oxidase was shown to be involved in the conversion of MPTP to MPP+ (see above) and so might be similarly important in the activation of a theoretical pathogenic protoxin in Parkinson's disease and oxidative reactions caused by dopamine metabolism might also be slowed by type B monoamine oxidase inhibition. If patients with very early Parkinson's disease are treated with the type B monoamine oxidase inhibitor selegiline (deprenyl), the need for symptomatic treatment with levodopa is indeed delayed, by about 9 months (Parkinson Study Group 1993). This was interpreted as a neuroprotective effect and selegiline was, at one time, prescribed widely in early Parkinson's disease immediately after diagnosis. Subsequently, the role of selegiline became highly controversial. It has been argued that the delay in the need for levodopa was due to a weak antiparkinsonian effect rather than slowing of disease progression, i.e. a symptomatic rather than a neuroprotective action (Ward 1994). Others have not agreed with this interpretation and continue to propose a neuroprotective effect (Larsen *et al.* 1999). Whatever the mechanism of selegiline in Parkinson's disease, it has little if any influence on the clinical progression of the disease (Parkinson's Disease Research Group in the United Kingdom 1993; Parkinson Study Group 1996). There has been some concern about increased mortality in patients taking selegiline (Lees 1995) but this is also disputed and other (arguably less reliable) studies have not replicated this. Numerous reviews of this issue have concluded that there is no neuroprotective effect, no prevention of motor complications and no definite evidence of increased mortality (Miyasaki *et al.* 2002; Goetz *et al.* 2005a).

Rasagiline is a more recent type B monoamine oxidase inhibitor. One study, designed to allow for any symptomatic effect, reported a possible slowing of disease over 12 months in otherwise unmedicated patients but this is uncertain (Parkinson Study Group 2004). At present, a neuroprotective effect is unproven.

Vitamin E does not appear to have any effect on the progression of Parkinson's disease (Parkinson Study Group 1993).

Coenzyme Q_{10} given in a large dose of 1200 mg per day for 16 months was associated with a possible slight slowing of disease progression but lower doses were not effective (Shults *et al.* 2002). The results of this study were widely publicized but further study is needed before Q_{10} can be recommended in early Parkinson's disease.

Early symptomatic treatment. It follows that the use of medication for Parkinson's disease is for the control of symptoms rather than any effect on the disorder itself. Accordingly, there is no point in administering antiparkinsonian drugs until the patient's condition warrants this. This point will vary between patients and depends on social, psychological, and employment issues as well as the severity of the Parkinsonism. Many drugs are now available for the treatment of Parkinson's disease, with several new agents arriving within the last few years (Table 40.5).

Non-dopaminergic drugs. These may be considered in the mildly affected patient with early disease, usually in order to delay or avoid levodopa or a dopamine agonist (see below).

Anticholinergics have been used to treat Parkinson's disease since the time of Charcot. Modern agents such as orphenadrine or benzhexol (trihexphenidyl) cause a modest reduction in tremor and rigidity but are less active against akinesia. There is little evidence to support the commonly expressed view that anticholinergics are superior to levodopa for the suppression of tremor but they may be helpful in the early stages of Parkinson's

disease when tremor rather than akinesia is the main problem. Motor disability is slightly reduced but side effects are common, especially in the elderly in whom they are best avoided. It is worth trying an anticholinergic in patients with tremor dominant disease who are not helped by levodopa but the response is often disappointing.

Amantadine is also weakly antiParkinsonian (Obeso and Martinez-Lage 1992). Its mode of action is complex. It increases presynaptic dopamine reuptake and release and is a weak NMDA receptor antagonist, inhibiting the subthalamic glutamatergic output to the globus pallidus pars interna. It is mildly effective against Parkinsonian symptoms but not in all patients and the effect may wear off within months. Insomnia, agitation, confusion, ankle oedema, and livedo reticularis may occur. Amantadine may be tried in the early stages of Parkinson's disease in younger patients who are more likely to tolerate it.

Selegiline and rasagiline deliver mild but sometimes useful symptomatic benefit in early Parkinson's disease. The effects of selegiline have been discussed above; a similar symptomatic effect was seen with rasagiline in the TEMPO trial (Parkinson Study Group 2002).

Dopaminergic therapy. For patients with significant disability, dopaminergic treatment will be required. The dilemma (and controversy) is whether to advise levodopa or dopamine agonist monotherapy initially.

Levodopa is the most effective antiParkinsonian drug available. It is metabolized peripherally to 3-*O*-methyldopa and dopamine by catechol-*O*-methyltransferase and aromatic aminoacid decarboxylase respectively (see Fig. 40.9). Levodopa absorption is reduced by food because of reduced gastric emptying and competition from dietary protein derived large neutral amino acids. Residual levodopa crosses the blood brain barrier (where there is also competition from large neutral amino acids and 3-*O*-methyldopa) where it is converted to dopamine by aromatic aminoacid decarboxylase in surviving dopaminergic nigral neurones and also non-dopaminergic cells. Interestingly, the ability of the brain to convert levodopa to dopamine is not significantly reduced by progressive loss of residual nigrostriatal cells (Durso *et al.* 1997). Dopamine acts on striatal dopamine receptors and is then metabolized by striatal monoamine oxidase and catechol-*O*-methyltransferase (Fig. 40.9). In order to increase the bioavailability of orally administered levodopa it is given with a

Table 40.5 Drugs used in the treatment of Parkinson's disease

Anticholinergics	Trihexphenidyl (benzhexol) Procyclidine Orphenadrine Benztropine
Amantadine Levodopa	Levodopa + carbidopa* (co-careldopa) (Sinemet®) Levodopa + benzserazide* (co-beneldopa) (Madopar®) Both in standard and slow release formulations Madopar also in a rapid dispersible formulation
MAO inhibitors	Selegiline Rasagiline
COMT inhibitors	Entacapone Tolcapone
Combination tablets	Levodopa + carbidopa + entacapone (Stalevo®)
Dopamine agonists	Ergots : Pergolide Cabergoline Apomorphine Non ergots : Ropinirole Pramipexole Rotigotine (Lisuride, Bromocriptine are now used rarely)

* = peripheral decarboxylase inhibitor,
MAO = monoamine oxidase
COMT = catechol-O-methyl-transferase

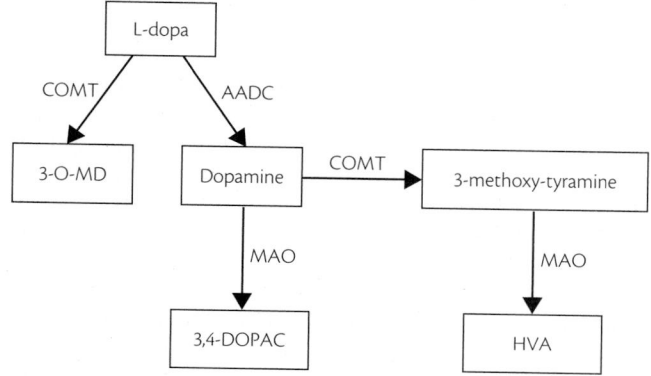

Fig. 40.9 The metabolism of levodopa. COMT = catechol-*O*-methyltransferase; MAO = monoamine oxidase; AAAD = aromatic amino acid decarboxylase; 3-*O*-MD = 3-*O*-methyldopa; DOPAC = dihydroxyphenylacetic acid; HVA = homovanillic acid.

peripherally acting decarboxylase inhibitor, either benzserazide, as Madopar, or carbidopa, as Sinemet. This reduces conversion to dopamine outside the brain and permits a lower oral dose with fewer side effects, especially nausea and vomiting.

Levodopa is highly effective against tremor, rigidity, and akinesia and a good response, with a 70 per cent or more improvement in symptoms, is a diagnostic feature of idiopathic Parkinson's disease. The onset of action is rapid, with most patients improving within days or weeks. The dose of levodopa required in early Parkinson's disease is variable, but most patients respond well to 300 mg per day, given with a peripheral decarboxylase inhibitor, and almost all to 600 mg. It is wise to start at 100 mg daily and increase slowly to 300 mg over 2 or 3 weeks. If there is no response to this dose, the diagnosis should be questioned. If there is no improvement on 600 mg daily, the patient is extremely unlikely to have Parkinson's disease. However, in levodopa responsive patients there is no definite maximum dose and many patients take and need more than 600 mg daily. Side effects are most unusual. Some patients develop nausea which can be reduced by slow initial dose increases and administration with food, which simply reduces bioavailability. If the problem persists, additional domperidone 20 mg tds is usually effective. Postural hypotension, confusion, cardiac arrythmias, and drowsiness are less common and a haemolytic anaemia is very rare.

It has long been recognized that levodopa has an immediate and short term response lasting hours, a 'short response', and a more long lasting effect lasting days, a 'long response' (Muenter and Tyce 1971). This must be borne in mind when levodopa is stopped—signs of Parkinson's disease may take days or weeks to re-emerge fully thus constituting a diagnostic trap for the unwary.

The initial dramatic response to levodopa is unfortunately not maintained. Within 5 years, about 40 per cent of patients experience problems due to instability of response and this occurs in almost all within 10 years, especially in younger patients (Ahlskog and Muenter 2001). There is progressive shortening of the response to each dose leading to the 'wearing off effect' or 'end of dose deterioration' with the reappearance of Parkinsonism before the next dose is due. These motor fluctuations become increasingly severe and rapid until eventually the patient frequently switches between a mobile 'ON' state and increasingly severe rigid 'OFF' periods. The daily levodopa requirement increases as does the dose frequency. In addition, the on periods are often associated with dyskinetic movements. Initially these are mild choreiform movements after each levodopa dose, 'peak dose dyskinesia', but later there may be severe dystonic and ballistic movements at the beginning and end of the dose, 'diphasic dyskinesia'. Such dyskinesias may lead to exhaustion and injury. This is more likely in younger patients. In patients under 60 years of age, 50 per cent have developed dyskinesia after 5 years in contrast to only 26 per cent among those aged over 60 (Kumar et al. 2005). Ultimately the patient, although still levodopa responsive, experiences a chaotic rollercoaster existence switching from severe dyskinesia to rigid immobility, sometimes within minutes.

The mechanisms underlying fluctuations and dyskinesias are incompletely understood (Nutt and Holford 1996). Disease progression leads to worsening of the underlying Parkinsonism thereby increasing the contrast between the medicated and unmedicated states. Increasing nigrostriatal tract destruction reduces presynaptic storage of dopamine reducing the long duration response. The short duration levodopa response, which probably reflects the action of dopamine synthesized and acting outside the nigrostiatal terminals, becomes increasingly evident, shorter, and more abrupt due to altered brain handling of levodopa and dopamine receptor activity. The 'negative response' also appears in which low striatal dopamine levels at the beginning and end of each levodopa dose worsen Parkinsonism by shutting off residual endogenous dopamine release via presynaptic dopamine autoreceptors. On periods are marred by dyskinesia, the threshold for which moves increasingly close to that for the on response, probably due to alterations in dopamine receptor activity and other mechanisms (Nutt and Holford 1996).

Although these problems only occur in levodopa treated patients, there has been controversy about whether they arise because of levodopa or disease progression. One view is that motor fluctuation and dyskinesia are related to the duration and dosage of levodopa treatment. This might be due to levodopa inducing alterations in its own pharmacological action or a direct neurotoxic action (Fahn 1996). An alternative view is that the changes in the clinical response to levodopa are due to disease progression (Caraceni 1994). There is however, no evidence that levodopa is toxic or that it accelerates Parkinson's disease (Agid et al. 1999). In a study comparing levodopa with placebo over 40 weeks, levodopa treated patients had improved even after a month without the medication indicating a long lasting benefit; there was no evidence of any clinical worsening or increased disease progression (Fahn et al. 2004). SPECT (DAT) scans indicated increased loss of dopamine transporter, a marker for nigral nerve terminals, but this was inconsistent with the clinical observations indicating a spurious finding due to the effect of levodopa on dopamine transporter regulation. Moreover, levodopa is associated with increased survival in Parkinson's disease, there is no animal or in vitro evidence of neurotoxicity and the disease does not appear to start in dopaminergic cells (Braak et al. 2004). There seems no justification to withhold levodopa in early Parkinson's disease.

Rarely, patients appear to develop a dysfunctional craving or overuse of levodopa—'dopamine dysregulation' (Evans et al. 2006).

Despite these findings, levodopa remains a therapeutic paradox. On the one hand it is the most effective treatment for Parkinson's disease. On the other it is associated with poorly understood long term problems and a probably unjustified reluctance to use it persists.

Dopamine agonist monotherapy has been used as an alternative to levodopa in order to delay the onset of long term levodopa related problems. It is vital to realize that agonist monotherapy is an *initial* alternative to levodopa which will almost certainly be needed after a few years. Early trials with bromocriptine (Rinne 1987; Hely et al. 1994; Montastruc et al. 1994) were followed by a plethora of agonist monotherapy trials with pergolide, cabergoline, and more recently pramipexole (Parkinson Study Group 2000; Holloway et al. 2004;), ropinirole (Adler et al. 1997; Rascol et al. 2000; Whone et al. 2003), and rotigotine (Watts et al. 2007). All of these studies show less motor fluctuation and dyskinesia than is seen with levodopa; however symptomatic benefit is less and there is a significant incidence of side effects. A new transdermal preparation—rotigotine—as monotherapy in early disease is similar in efficacy to ropinirole and pramipexole (Poewe and Luessi 2005). Most patients who start with dopamine agonist monotherapy will need additional levodopa within 5 years (Rascol et al. 2000).

Unfortunately, despite the lower incidence of motor complications, monotherapy with the currently available drugs fails in many patients because of inadequate antiParkinsonian efficacy, side effects, or both. Common adverse effects include:

◆ Nausea, which can usually be avoided by using additional domperidone for the first few weeks;

◆ Hypotension;

◆ Confusion; behavioural problems

◆ Hallucinations;

◆ Oedema; fibrotic reactions

◆ Somnolence or sudden onset sleep attacks which occur particularly with non-ergot agonists (Frucht et al. 1999).

Dopamine agonists must be used with caution in drivers, those with a history of cardiac disease, peptic ulceration, psychiatric illness and in the elderly and not at all in those with dementia. The ergot derivative agonists bromocriptine, cabergoline, and pergolide can occasionally be associated with erythromelalgia (an uncomfortable erythematous skin induration, usually around the ankles), retroperitoneal fibrosis, pleural effusions, and severe inflammatory lung disease. Patients must be monitored for such side effects but the best way of doing this is unclear. Pergolide and cabergoline have also been associated with restrictive cardiac valvulopathy in a significant number of patients (Van Camp et al. 2004; Pinero et al. 2005); caution is needed with dosage and regular screening with echocardiography is increasingly employed. Dopamine agonists can also be associated with behavioural disturbances including pathological gambling (Voon et al. 2006), hypersexuality, and the repetitive 'punding' behaviours occasionally seen with levodopa (Evans et al. 2004). Patients must be warned about these and monitored for their development.

The choice between levodopa and dopamine agonist monotherapy (see Table 40.6) in early Parkinson's disease remains difficult. It is important to emphasize that either is acceptable practice. A careful balanced discussion of the options with each patient is appropriate; the following points are noteworthy:

◆ There is no evidence that levodopa is harmful or toxic and there is no basis for the 'levodopa phobia' displayed by some patients and doctors. Consequent unrealistic avoidance or underdosing of levodopa is unhelpful to patients and can cause serious morbidity.

◆ It is not a choice between levodopa or agonists. All patients eventually need levodopa and the decision is which therapy to use initially. The aim is to delay levodopa therapy not to avoid it altogether. It is not appropriate to withhold levodopa unrealistically in patients doing badly on agonist monotherapy.

◆ Agonist monotherapy is associated with fewer motor complications in the short term, probably by delaying levodopa and/or reducing the dosage used.

◆ Levodopa can be associated with disabling dyskinesia, especially if high doses are used.

◆ Younger patients under 60 years are at significantly greater risk of dyskinesia and are more likely to tolerate agonists.

◆ Agonist monotherapy is ineffective in some patients and is unlikely to be adequate for more than 5 years.

◆ Dopamine agonists can cause serious side effects and should be used with caution in some patients including the elderly.

◆ It will not be possible to abolish all symptoms completely with either therapy and attempts to do so will be unsuccessful and may lead to side effects.

In general, agonist monotherapy is a realistic option for patients younger than 60 years who are at greater risk of dyskinesia. The patient must be willing to accept reduced relief of symptoms and a greater risk of adverse effects in return for a lower risk of fluctuation and dyskinesia. It is unclear how long this benefit lasts—at 10 years symptom control has been less effective in those who start with an agonist but there is little difference in the frequency of severe dyskinesia, suggesting that dyskinesia has been delayed rather than avoided (Lees et al. 2001). Much will depend on the patient's individual preference, medical history, and social situation, particularly the extent to which Parkinsonism threatens continued employment or independence. If agonist monotherapy is not tolerated, ineffective, contraindicated, or declined by the patient, levodopa should be used and it is important to use an effective dose—most patients will need at least 300 mg, with a peripheral decarboxylase inhibitor and some will need 600 mg.

In patients over 60 years, the risk of early fluctuation and dyskinesia is lower, agonist monotherapy is less likely to be tolerated and a rapid therapeutic response with levodopa is probably more likely to be needed in the face of disabling Parkinsonian symptoms to maintain activity and independence.

Medical management of advanced Parkinson's disease. Most of the difficulties in advanced Parkinson's disease are caused by alterations in levodopa responsiveness, with motor fluctuations and dyskinesias and the emergence of drug resistant features.

Motor fluctuations with end of dose deterioration/wearing off effects may be improved in several ways.

◆ Levodopa may be increased and given more frequently but this may be only a short term solution and there is a danger of increasingly frequent fluctuation with higher doses.

◆ Controlled release levodopa preparations prolong on periods and reduce the number of levodopa doses required. This is sometimes successful in mild, early fluctuation but can be associated with a slow onset of action, especially with the first dose of the day. The addition of a small dose, 62.5–125 mg, of conventional Sinemet or dispersible Madopar at this time may provide an initial 'kick in'.

Table 40.6 Characteristics of levodopa treatment and dopamine agonist monotherapy

	Levodopa	Agonist monotherapy
Initial antiparkinsonian efficacy	Excellent	Moderate or poor
Time to achieve full effect	Weeks	Months
Dose increase	Fast	Slow
Adverse effects	Rare	Common
Overall tolerance	Good	Poor
Fluctuation and dyskinesia	Common	Rare
Cost	Low	High

- Fluctuating patients are vulnerable to small changes in plasma levodopa concentrations and alterations in levodopa absorption. Taking levodopa without food and restriction of dietary protein can lead to more predictable and effective absorption and improved stability.

- Mild off periods respond to small supplemental levodopa doses in fast acting 'Madopar dispersible' tablets—these are dissolved in water and act within a few minutes but are short acting.

- The catechol-*O*-methyltransferase inhibitor entacapone 200 mg tds reduces the metabolism of levodopa to 3-*O*-methyldopa thereby prolonging the action of levodopa and reducing end of dose wearing off effects (Rinne *et al.* 1998). Levodopa requirements are reduced and on time increased in most patients. Side effects include increased dyskinesia and diarrhoea. Entacapone is available with levodopa and carbidopa as a combined tablet, Stalevo, for greater convenience (Findley *et al.* 2005).

- Tolcapone, 100 mg tds, is a more potent catechol-*O*-methyltransferase inhibitor and more effective than entacapone with reduced off time and fluctuation. It also causes diarrhoea in some patients and can lead to increased dyskinesia. However, there is also a small risk of severe hepatotoxicity (Borges 2005; Keating and Lyseng-Williamson 2005). Careful liver monitoring is needed and tolcapone is indicated only in patients with severe fluctuation who cannot use alternative therapies.

- Selegiline or rasagiline may be added to improve wearing off and increase on time in patients with motor fluctuations—the effect is similar to entacapone (Parkinson Study Group 2005; Rascol *et al.* 2005).

- A more effective option is to add a dopamine agonist to relieve off periods and fluctuations, preferably before the patient has started to take large doses of levodopa. A reasonable point at which to consider this is when the fluctuating patient is on 300–600 mg of levodopa. In patients taking larger doses, the addition of an agonist should allow a reduction in levodopa as well as improving control. In some studies, fluctuation and dyskinesia has been less in those taking low dose levodopa combined with a dopamine agonist than in those taking unrestricted doses of levodopa (Rinne 1985; Rinne 1987; Hely *et al.* 1994; Montastruc *et al.* 1994). It is therefore reasonable to consider the addition of a dopamine agonist in younger patients in whom 300 mg of levodopa is not adequate treatment. In older patients the risk of dyskinesia is less, side effects of agonists more common, and higher doses of levodopa, 600–800 mg daily, may be considered before an agonist is used. Several dopamine agonists are available (Table 40.5) with similar clinical effects. The long duration of action of cabergoline can be helpful in overnight Parkinsonism while pramipexole may be more effective for tremor (Kunig *et al.* 1999). Otherwise, there are no major differences in efficacy between these different agonists. The side effects of dopamine agonists are significant, as discussed above and shown in Table 40.7, and they should be used with caution, particularly the elderly, those with cardiac disease, and patients with a history of previous adverse reactions to other dopamine agonists, behavioural problems, or psychiatric illness; if there is a history of confusion or hallucination agonists should be avoided. Patients should be warned about driving should they develop somnolence or sudden onset sleep episodes and about gambling and

Table 40.7 Adverse effects of oral dopamine agonists

Type of agonist	Adverse effect
All	Nausea and vomiting Confusion or psychosis Visual hallucination Hypersexuality Hypotension Peptic ulceration
More likely with ergot derivatives	Erythromelalgia Pulmonary / retroperitoneal fibrosis Pleural effusion Cardiac valvulopathy
More likely with non ergot derivatives	Pathological gambling Somnolence / sudden onset of sleep Oedema

hypersexuality as these are not always disclosed spontaneously. Nausea and vomiting are very common in those starting agonist therapy and it is sensible to prescribe domperidone 20 mg tds in the first 4–6 weeks. Another common difficulty is deterioration of motor function in the first few weeks while the patient is gradually increasing the agonist but has not yet reached a therapeutic dosage (Kellett and Steiger 1999); patients should be warned of this and advised to carry on with the dose escalation until a therapeutic effect is reached. It is a common error to give supplemental agonists at too low a dosage.

Overnight Parkinsonism with troublesome nocturnal stiffness and immobility may respond to controlled release levodopa/carbidopa (Sinemet) taken before going to bed.

Severe off periods can be extremely unpleasant with severe rigidity and immobility, unpleasant limb restlessness, sweating, pain, autonomic abnormalities, and marked psychological distress (Riley and Lang 1993). Such attacks may be reduced if an orally active dopamine agonist is introduced or its dose increased but this may not be effective in very severe attacks. In this situation there are three alternatives: apomorphine, intraduodenal levodopa, or subthalamic nucleus deep brain stimulation.

- Apomorphine, a more powerful dopamine agonist, is usually effective (Frankel *et al.* 1990; Hughes *et al.* 1993a). Apomorphine must be given by subcutaneous injection, acts rapidly, and usually relieves such attacks within 10 min of injection and lasts 60–90 min. It is essential to administer domperidone 20 mg tds to prevent apomorphine induced vomiting. Many patients manage with a few injections per day but in those requiring more, a subcutaneous infusion may be preferable (Tyne *et al.* 2004). Patients often develop skin lesions consisting of nodules and small areas of skin necrosis at the injection sites, especially in those using subcutaneous infusions. Interestingly, psychosis appears to be less common with apomorphine than with other dopamine agonists and may be tolerated by patients in whom psychosis has prevented the use of other agonists.

- A recent effective alternative for severe fluctuations is intraduodenal infusion of a levodopa/carbidopa gel (Nyholm *et al.* 2005). Fluctuations are markedly reduced with no significant increase in dyskinesia. This therapy has yet to be used on a large scale and a percutaneous duodenostomy/jejunostomy tube is required.

♦ In some cases subthalamic deep brain stimulation should be considered.

Dyskinesias are difficult to treat. In patients taking large doses of levodopa, the introduction of a dopamine agonist and reduction in levodopa dosage is often effective. The aim should be to give as much dopaminergic treatment as possible in the form of the agonist and to reduce the levodopa to the lowest dose possible without an undue increase in off periods. It will not usually be possible to stop levodopa as this leads to unacceptable Parkinsonism. Moreover, some patients are unable to tolerate dopamine agonists because of side effects. In this situation there is little alternative to simply finding a dose of levodopa which provides the best balance between Parkinsonism and dyskinesia. Dyskinesias are often worse in the latter part of the day in which case omitting a midday dose of levodopa may be effective. It might be supposed that biphasic dyskinesias occurring at the beginning and end of levodopa doses might respond to an increase of levodopa but this is not the case; they usually respond in a similar fashion to peak dose dyskinesias to a reduction of levodopa intake. Amantadine 100–300 mg daily is often effective at reducing dyskinesia, possibly by inhibiting the glutamatergic pathway from the subthalamic nucleus to internal globus pallidus (Rajput *et al.* 1998). In some patients, despite amantadine, a dopamine agonist in high dose, discontinuation of any catechol-*O*-methyltransferase inhibitor and reduced levodopa intake, dyskinesia remains severe and any further levodopa reduction causes a severe off state. No further adjustment of medication is likely to help these patients for whom subthalamic or pallidal deep brain stimulation is now probably the treatment of choice.

Painful early morning foot dystonia is an off period symptom and responds to oral levodopa or apomorphine on waking, or taking a dopamine agonist or slow release levodopa at night. Baclofen or lithium may help in some cases.

Daytime somnolence is increasingly recognized in Parkinson's disease (Ondo *et al.* 2001). Initially reported with pramipexole and ropinirole (Frucht *et al.* 1999) it occurs with all dopamine agonists and also to some extent with levodopa. Excessive daytime sleepiness is probably multifactorial due to poor night time sleep, nocturnal symptoms of restless legs and periodic limb movements (Poewe and Hogl 2004; Adler 2005), cognitive impairment, and the effects of medication. It is important to ask patients about this and give appropriate warnings about driving and operating machinery.

Confusion and psychosis often occur in advanced disease with Parkinson's disease dementia, especially in older patients; they are an earlier feature in dementia with Lewy bodies. Hallucinations are usually visual and well formed with or without confusion, agitation, and delerium. Confusion is worse at night and may be worsened or precipitated by intercurrent illness, increased antiParkinsonian medication dosage especially with dopamine agonists and anticholinergics, or strange surroundings such as respite care or hospital admission. As mentioned earlier, care is needed in using anticholinergics or dopamine agonists in older Parkinsonian patients or those with any history of confusion or mental illness. Admission of such patients to hospital also requires caution.

The situation may be improved by stopping anticholinergics or dopamine agonists, followed if necessary by a reduction of levodopa. An increase in Parkinsonism is inevitable but it is often easier for the patient and carers to manage this than the psychosis. Most neuroleptic drugs worsen Parkinsonism due to dopamine receptor blockade but this is not seen with the atypical neuroleptic clozapine which often improves the mental state without a significant increase of Parkinsonism (Goetz *et al.* 2000; Molho and Factor 2001). Unfortunately clozapine is associated with a risk of agranulocytosis and regular blood counts are needed. Risperidone is of limited value as it worsens Parkinsonism as does olanzapine (Molho and Factor 1999). At present, the drug of choice is uncertain but quetiapine is a reasonable option (Juncos *et al.* 2004) as this has been associated with the least motor deterioration (Molho and Factor 2001) but this is based on limited data and further study of this area is badly needed. Rivastigmine has been used sucessfully to treat cognitive impairment and hallucinations (McKeith *et al.* 2000; Emre *et al.* 2004).

Depression may be difficult to detect as the patient often complains of non specific worsening of Parkinsonism, confusion, poor memory, or insomnia. Serotonin reuptake inhibitors are most commonly used and are effective with little evidence of significant worsening of Parkinsonism (Weintraub *et al.* 2005). Tricyclic antidepressants such as amitripyline or dothiepin are usually effective and are a useful treatment for insomnia in Parkinson's disease. The anticholinergic effect of tricyclics may also be useful overnight.

Postural instability with gait impairment, freezing, shuffling, and falling respond poorly if at all to medication. Marked gait failure carries a poor prognosis (Rajput *et al.* 1993). Physiotherapy may help, especially cueing techniques, (Nieuwboer *et al.* 2006) but such patients usually remain disabled and require adaptation of the living environment and mobility aids.

Dysarthria and dysphagia are resistant to antiParkinsonian drugs and can be managed only with speech therapy and in some cases a percutaneous gastrostomy tube may be needed. *Sialorrhoea* can be severe and distressing; anticholinergic drugs can help but their propensity to cause adverse effects has already been mentioned. In extreme cases, parotid radiotherapy may be considered. *Constipation* can be a major problem in all stages of Parkinson's disease and requires attention to diet, the use of laxatives especially macrogol 3350 (Eichhorn and Oertel 2001), and awareness of the exacerbating effects of some drugs, especially anticholinergics. Faecal impaction in the rectum with intermittent overflow diarrhoea often causes problems in immobile patients and requires regular manual rectal evacuation rather than inappropriate, and sometimes risky, use of laxatives and antidiarrhoeal drugs.

Autonomic dysfunction is not a feature of early Parkinson's disease but can develop in advanced cases. Bladder dysfunction due to detrusor instability is common in the later stages of Parkinson's disease. It can be difficult to distinguish from prostatism in older men and urodynamic testing may be needed. Anticholinergic drugs and adjustment of fluid intake are partially successful in most cases and patients with a significant post-micturition residual volume of urine of >100–150 ml require intermittent clean catheterization. Orthostatic hypotension is also seen, especially in those taking dopamine agonists. A head up tilt of the bed at night or oral fludrocortisone are helpful but it is wise to review the patient's medication beforehand. Management of autonomic impairment is discussed further in Section 40.3.8.

Neurosurgical treatment of Parkinson's disease. Before the advent of levodopa therapy in the late 1960s, there had been numerous attempts to improve Parkinsonism with neurosurgical lesions of the pyramidal and later extrapyramidal system. Initially these were associated with limited improvement and considerable morbidity. Thalamotomy was effective for tremor but had little impact on bradykinesia and rigidity. Pallidal lesions were not as effective for suppression of tremor and although Leksell had reported improved pallidotomy results in the 1960s, this issue was overshadowed by the discovery and introduction of levodopa. There are several reasons for a new enthusiasm for surgical treatment of Parkinson's disease. There is a better understanding of the functional anatomy of the basal ganglia and improved imaging and stereotactic methods allow greater surgical precision and the selection of physiologically logical targets. Most importantly however is the realization that chronic levodopa therapy often leads to intractable motor instability with increasing numbers of patients disabled by fluctuations and, more importantly, dyskinesia.

Thalamotomy of the nucleus ventralis intermedius and occasionally nucleus ventro-oral was used for many years in patients with severe tremor (Obeso *et al.* 1997). Patients with tremor dominant disease sometimes have a disappointing response to medical treatment—an exception to the general rule that idiopathic Parkinson's disease always responds to levodopa—and in these patients, early thalmotomy has been employed. Contralateral tremor is abolished after thalamotomy in 60–80 per cent of cases but there is little effect on akinesia or rigidity. Complications are uncommon with a unilateral lesion but can include cognitive change, dysarthria, hemisensory disturbances, ataxia, and limb incoordination; there is a high risk of dysarthria with bilateral lesions (Hauser *et al.* 1995). Thalamotomy has been largely abandoned in favour of thalamic deep brain stimulation of the ventralis intermedius nucleus (Schuurman *et al.* 2000).

Thalamic deep brain stimulation (Benabid *et al.* 1991) is as effective as thalamotomy in terms of tremor suppression but is slightly safer with fewer adverse effects (Schuurman *et al.* 2000). Several groups have reported favourable results with good tremor suppression in 85–100 per cent (Benabid *et al.* 1996; Koller *et al.* 1997; Limousin *et al.* 1999; Kumar *et al.* 2003; Pahwa *et al.* 2006). Tremor suppression is stable for over 5 years (Rehncrona *et al.* 2003; Pahwa *et al.* 2006) although patients can become disabled by motor fluctuations and dyskinesia which are not affected by thalamic deep brain stimulation (Kumar *et al.* 2003; Tarsy *et al.* 2005). Adverse effects include paresthesia in 45 per cent, pain in 41 per cent, and in patients receiving implants bilaterally dysarthria occurs in 75 per cent and balance difficulties in 56 per cent (Pahwa *et al.* 2006).

Pallidotomy is effective in those with severe fluctuation and dyskinesia including elderly patients (Uitti *et al.* 1997). Improved results followed the selection of the posteroventral lateral pallidum as a target, Leksell's pallidotomy, (Laitinen *et al.* 1992; Dogali *et al.* 1995; Iacono *et al.* 1995; Lozano *et al.* 1995; Obeso *et al.* 1996). Overall there is impressive suppression of contralateral in over 80 per cent and to a lesser degree ipsilateral dyskinesias and overall disability is reduced. Off period Parkinsonism is improved by about 20 per cent. Tremor is reduced by 60 per cent, rigidity by 40 per cent, and bradykinesia by 20 per cent and these effects appear to persist for 5 years or more (Fine *et al.* 2000); postural instability, gait failure, dysarthria are not significantly improved. Complications include haemorrhage, hemiparesis, and visual field loss but overall the risk of adverse effects has been low, less than 10 per cent. Bilateral pallidotomy has been used in some patients (Obeso

et al. 1997). There is concern about the risk of cognitive and behavioural changes after bilateral procedures. Increasingly pallidal or more commonly subthalamic deep brain stimulation is preferred to pallidotomy.

Pallidal deep brain stimulation has been undertaken in preference to lesioning procedures as it can more safely be performed bilaterally and because of the increased use of deep brain stimulation instead of lesioning in general (Siegfried and Lippitz 1994; Pahwa *et al.* 1997; Volkmann *et al.* 1998; Krack *et al.* 2000;). Motor Parkinsonism scores are reduced by over 50 per cent (Ghika *et al.* 1998; Volkmann *et al.* 1998) and dyskiesia is especially well suppressed in 80–90 per cent but the effect on off periods tends to decline over several years (Durif *et al.* 2002; Volkmann *et al.* 2004). There is no significant cognitive change after pallidal DBS (Ardouin *et al.* 1999).

Bilateral subthalamic deep brain stimulation (Kumar *et al.* 1998; Limousin *et al.* 1998) improves motor fluctuation and dyskinesia; drug requirements are reduced (Deuschl *et al.* 2006). Subthalamic stimulation has become the procedure of choice for patients with Parkinson's disease no longer responding to oral medication. Several trials have compared subthalamic and globus pallidus interna stimulation and although the motor effects are very similar (Volkmann *et al.* 2001; Minguez-Castellanos *et al.* 2005), with subthalamic stimulation there is a better response (Deep-Brain Stimulation for Parkinson's Disease Study Group 2001; Anderson *et al.* 2005) and reduction in medication (Volkmann *et al.* 2001). Improvements in motor scores of 60 per cent are seen in the 'off' state and on time is substantially increased with impressive suppression of dyskinesia. There is no significant improvement in the 'on' state except for suppression of dyskinesia. In contrast to pallidal or thalamic surgery, levodopa intake is reduced by up to 50 per cent (Volkmann *et al.* 2001; Anderson *et al.* 2005; Minguez-Castellanos *et al.* 2005) but in general, axial symptoms such as falls, postural instability, and freezing are not improved significantly. Dementia and dysarthria may be worsened. The benefit of subthalamic stimulation is maintained for up to 5 years (Krack *et al.* 2003; Rodriguez-Oroz *et al.* 2004). Complications include weight gain, dysarthria, diplopia, ataxia, paresthesiae, hemiparesis, eyelid apraxia, haemorrhage, and infection (Kumar *et al.* 1998; Saint-Cyr *et al.* 2000). There is also a small risk of failure due to electrode malpositioning. There is no significant effect on cognition except a slight reduction in verbal fluency (Ardouin *et al.* 1999). However, cognitive decline, confusion, and psychosis may occur in older patients (Saint-Cyr *et al.* 2000) and those with pre-existing cognitive impairment or hallucinations. Severe depression may also occur after subthalamic deep brain stimulation (Houeto *et al.* 2002). Selection of patients for subthalamic stimulation is crucial (Table 40.8). Patients with atypical Parkinsonism will not respond to deep brain stimulation; accordingly it is important to consider the degree of levodopa responsiveness, which may require review of old notes, and the presence of any atypical features. Patients with a poor response to drugs or rapid progression to severe disability within 5 years may have multiple system atrophy. There is no evidence that deep brain stimulation prevents subsequent progression of Parkinson's disease and there is no justification for surgery in mildly affected patients—they are unlikely to notice significant benefit. Age is controversial but several groups have reported inferior benefit and more complications (particularly cognitive) in older patients (Saint-Cyr *et al.* 2000; Charles *et al.* 2002;

Table 40.8 Subthalamic nucleus deep brain stimulation

Patients likely to benefit
Idiopathic Parkinson's disease
Good response to levodopa
No atypical features
Severe disease (motor fluctuations and dyskinesia)
Younger age (<70 years)
Normal cognition (no dementia or hallucinations)
No significant cognitive impairment on neuropsychology screening
No psychiatric features (depression, psychosis, behavioural abnormalities)
No prominent axial symptoms
Able to attend for postoperative care
Realistic expectations

Patients unlikely to benefit
Atypical Parkinsonism
Rapid progression and duration < 5years
Poor response to levodopa / unresponsive
Mild disease, absent / mild fluctuation, or dyskinesia
Dementia or hallucinations
Severe depression, psychosis, behavioural abnormality
Age over 70 years
Significant axial symptoms
Unable to attend regularly for programming
Unrealistic expectations

Welter *et al.* 2002; Russmann *et al.* 2004) although not all studies have confirmed this (Kleiner-Fisman *et al.* 2003). Patients with current severe depression, behavioural disturbances of gambling, hypersexuality, mania, or anxiety, dementia, or psychosis including isolated hallucinations should not undergo surgery as these problems may be exacerbated. In some cases, surgery may be considered later if the problem improves and has been absent for a reasonable period of time but this will require good judgement and a careful discussion with the patient and family about the balance of risks and benefit. Particular care should be taken with patients who have experienced confusion or hallucinations—these patients are likely to develop dementia and are unlikely to tolerate surgery unless the mental situation has returned to normal with no recurrence for several years—even then surgery is likely to be risky in terms of cognitive outcome. Many units advise full neuropsychological evaluation prior to deep brain stimulation and avoid surgery in patients with detectable cognitive impairment. The problem with this is that few patients with Parkinson's disease have 'normal' neuropsychology. Mild memory and dysexecutive deficits are common and those with no evidence of dementia may develop it later due to disease progression. The axial features of freezing, gait instability, and falling or non-motor symptoms such as dysarthria, autonomic failure, and dysphagia do not respond to deep brain stimulation; patients in whom these are prominent problems are unlikely to benefit from a stimulator. The ideal candidate is the patient under 70 years of age with levodopa responsive idiopathic Parkinson's disease, significant motor fluctuations, dyskinesia, or both, no significant axial or non-motor features, normal cognition and behaviour, and inadequate control with oral medication. Patients require frequent adjustment of electrical settings after deep brain stimulation and careful balancing of

medication over time. Many visits to hospital are needed, with patients needing to arrive in the off state and the patient must be able to undertake this—those who live far from the hospital may have problems. Finally, deep brain stimulation is not a cure for Parkinson's disease and despite the excellent results of subthalamic stimulation in most patients (Deuschl *et al.* 2006) they will still have Parkinsonism after surgery. It is vital that expectations are realistic.

Comparison of apomorphine therapy and subthalamic deep brain stimulation in advanced Parkinson's disease concluded that motor benefit and reduction of antiParkinsonian medication was superior after deep brain stimulation but behavioural outcomes such as apathy, depression and anxiety were worse (De Gaspari *et al.* 2006).

Neurorestoration. *Transplantation of foetal mesencephalic neurones* into the striatum of patients with advanced Parkinson's disease has been disappointing. Several groups have reported results in small series of patients (Lindvall *et al.* 1994; Peschanski *et al.* 1994; Freeman *et al.* 1995; Defer *et al.* 1996). Over 100 patients have now been treated worldwide but evaluation of the results is impaired by variable methodology and assessment. Foetal tissue has been unilaterally or bilaterally transplanted as cell suspensions or solid grafts into the putamen or caudate of the host striatum. Considerable improvement in all aspects of Parkinsonism have appeared 3–6 months after transplantation and have been maintained up to 3 years of follow up. Levodopa requirements have been reduced in some patients. Positron emission tomography has demonstrated striatal flurodopa accumulation and postmortem examination of the graft site in one patient has shown robust graft survival and reinnervation of the host striatum (Kordower *et al.* 1995). There has been uncertainty regarding the optimal method of grafting, the required number of foetal donors per graft, and the need for immunosuppression or adjuvant growth factors. At present, technical and ethical issues mean that foetal transplantation remains experimental. One study of 20 transplanted patients showed a modest 15 per cent improvement in motor scores but a high incidence of severe dyskinesia appearing over a year postoperatively (Freed *et al.* 2001). Modest motor benefit was seen in younger patients only. In another study of 34 patients, mild motor benefit was seen in only a few cases and dyskinesia occurred in 56 per cent (Olanow *et al.* 2003). Interest in foetal cell transplantation has declined in recent years. On the basis of these results, the prospects for stem cell therapy seem poor.

Glial cell line derived neurotrophic factor, GDNF, infusion into the striatum has also been tried in Parkinson's disease. Improvements similar to those seen with subthalamic deep brain stimulation were seen in small open trials of bilateral (Gill *et al.* 2003) and unilateral (Slevin *et al.* 2005) GDNF infusion but these results were not replicated in a larger double blind trial (Lang *et al.* 2006).

Other aspects of management. Not all othese problems are soluble with medical or neurosurgical therapy and advanced Parkinson's disease, even with modern treatment, remains a disabling and progressive condition. Physiotherapy, occupational therapy, and speech therapy are helpful in advanced cases with accumulating disability. Patients require intensive social support in the community and in some cases, residential care may be the only practical option. There is increasing emphasis on the development of specialist multidisciplinary clinics although proof of their value is lacking. Nurse practitioners, ideally working from specialist movement disorder units, provide invaluable additional support to patients, carers, and family doctors, especially those using apomorphine and with severe treatment related fluctuations. In the United Kingdom, patients and their families are assisted by the Parkinson's Disease Society who are able to provide information and support (http://www.parkinsons.org.uk/).

40.3.2 Dementia with Lewy Bodies

The relationship of this condition to Parkinson's disease is controversial as discussed above. The disorders share clinical and pathological characteristics and there is probably no meaningful difference between patients who develop dementia after a year or more of Parkinson's disease, termed Parkinson's disease dementia, and dementia with Lewy bodies in which the dementia and Parkinsonism appear together or within a year of each other (Aarsland *et al.* 2004). The frequency of dementia with Lewy bodies is unknown but 15 cases were identified in a single year at a United Kingdom centre, suggesting that this may be one of the more common forms of dementia (Byrne *et al.* 1989).

Originally, dementia with Lewy bodies was defined pathologically (Kosaka 1993) with Lewy bodies distributed throughout the cerebral cortex in addition to the substantia nigra and other brainstem nuclei. In fact, cortical Lewy bodies also occur in Parkinson's disease but not in the same numbers. Two forms of dementia with Lewy bodies are recognized, a common form with additional Alzheimer pathology, and a pure form in which this is absent. Those who regard it as a variant of Parkinson's disease have proposed the concept of 'Lewy body disease' incorporating three groups. Type A, the diffuse type, type B, the transitional or limbic type, and type C, the brainstem type, synonymous with Parkinson's disease (Kosaka 1993; McKeith *et al.* 1996). Indeed, cortical Lewy body formation is probably an integral part of Parkinson's disease (Colosimo *et al.* 2003; Braak *et al.* 2004).

Clinically, dementia with Lewy bodies is characterized by dementia and Parkinsonism. The Parkinsonian features are similar to idiopathic Parkinson's disease but there is a lower incidence of tremor and the response to levodopa is not as satisfactory. In 20 per cent of patients there may be additional myoclonus. Rare patients have had supranuclear gaze palsies similar to those seen in progressive supranuclear palsy (Fearnley *et al.* 1991). The dementia is striking and may precede or follow the development of Parkinsonism or sometimes occur alone, especially with the common form. Poor memory is usually the first sign but a key diagnostic clue is prominent fluctuation of the mental state. Often these patients are thought to have recurrent infections or strokes as the cause of an abrupt cognitive deterioration. Psychotic features such as hallucinations and paranoid delusions are common (Byrne *et al.* 1989).

The diagnosis of dementia with Lewy bodies may be obvious. It is likely in Parkinson's disease patients who develop dementia at an unusually early stage with prominent fluctuation in the severity of their confusion and hallucinations. Similarly it occurs in patients with this type of dementia who later develop Parkinsonism. The difficulty is with those who have the dementia alone, in whom it may be difficult to make the diagnosis in life, and in those with atypical features such as myoclonus or supranuclear gaze palsy in whom other diagnoses such as Creutzfeldt–Jacob disease and progressive supranuclear palsy may be suspected. A common problem is the patient with what appears to have been typical levodopa responsive Parkinson's disease for years who then develops confusion

and hallucinations. Whether these patients have Parkinson's disease with mental changes caused by additional dementia, so-called Parkinson's disease dementia, or dementia with Lewy bodies is usually impossible to determine clinically. Conventionally these are considered as separate disorders but the clinical similarities are such that this is widely doubted (Aarsland *et al.* 2004). Pathologically some of these patients have dementia with Lewy bodies and the two conditions are probably the same (Aarsland *et al.* 2004). Functional imaging with single photon emission tomography SPECT (Lobotesis *et al.* 2001) or positron emission tomography (Minoshima *et al.* 2001) reveals occipital hypoperfusion in dementia with Lewy bodies which may be useful if such scanning is available.

The management of dementia with Lewy bodies is difficult because dopaminergic treatment can exacerbate confusion and the effect of neuroleptics on Parkinsonism. Patients with dementia with Lewy bodies are often highly sensitive to dopamine receptor blocking drugs. Generally the management is similar to that of confusion and psychosis in Parkinson's disease (see Section 40.3.1). Quetiapine can be used for psychosis with no worsening of Parkinsonism (Fernandez *et al.* 2002). Rivastigmine may improve some patients (McKeith *et al.* 2000; Emre *et al.* 2004;), memantine has been tried with mixed results (Sabbagh *et al.* 2005), and donepezil can exacerbate Parkinsonism (Aarsland *et al.* 2002). Overall the prognosis is poor and the majority of these patients will require nursing home admission within a few years.

40.3.3 Early onset Parkinson's disease

Early onset Parkinson's disease develops before the age of 40 years. In Europe and North America this accounts for less than 10 per cent of cases but it is more common in Japan. Early onset forms are heterogeneous; some patients have idiopathic Parkinson's disease but others do not. There is an increased incidence of familial Parkinson's disease, especially autosomal recessive forms. Clinically, early onset patients have typical rest tremor, rigidity, and bradykinesia but there is an increased incidence of lower limb dystonia, prior to levodopa therapy, brisk tendon reflexes, and less dementia; autonomic features such as dizziness, abnormal sweating, and urinary frequency are slightly more common. A good response to levodopa is seen but fluctuation and dyskinesia are more common and earlier, sometimes within months of starting treatment (Quinn *et al.* 1987; Golbe 1991; Muthane *et al.* 1994). Despite these problems, progression is slower than in older patients. Two groups of early onset Parkinson's disease have been proposed. Those with onset before 20 years, termed juvenile Parkinsonism and between 20 and 40 years termed young onset Parkinson's disease (Quinn *et al.* 1987). A history of affected relatives, suggesting autosomal recessive or sometimes dominant inheritance, is more common in juvenile than young onset but there are few clinical differences (Muthane *et al.* 1994). Pathological studies are confusing; some early onset cases have had neuropathological changes of Parkinson's disease but others have showed atypical nigral degeneration without Lewy bodies, ballooned neurones, and neurofibrillary tangles. Neuropathological evidence of idiopathic Parkinson's disease has been more common in young onset rather than juvenile onset.

Early onset Parkinson's disease is genetically heterogeneous; Parkin, *PARK*2 mutations (Kitada *et al.* 1998) cause slowly progressive, symmetrical Parkinson's disease with dystonia accounting for 10–20 per cent of early onset Parkinson's disease. There are many different Parkin mutations which makes DNA diagnosis difficult. *Pink1, PARK6* mutations (Bonifati *et al.* 2005) account for less than 10 per cent of early onset Parkinsons disease and *DJ1, PARK7* mutations (Hedrich *et al.* 2004) account for only 1–2 per cent of cases.

In all young patients presenting with an akinetic rigid syndrome, it is essential to exclude Wilson's disease by biochemical and ophthalmological investigations. A liver biopsy is indicated if the results of these initial tests are in any way equivocal (Section 40.8). Dopa responsive dystonia must also be considered along with other rare causes of Parkinsonism including rigid Huntington's disease, Hallervorden–Spatz disease, neuroacanthocytosis, and the hereditary cerebellar degenerations (Table 40.2).

40.3.4 Drug induced Parkinsonism

The most common cause of drug induced Parkinsonism is exposure to neuroleptic medication. Overt Parkinsonism is seen in 10–15 per cent of patients taking neuropleptic drugs but this is almost certainly an underestimate (Arblaster *et al.* 1993). Drug induced Parkinsonism appears to be more common in older patients and females. It should be noted that some atypical limbic selective antipsychotics, clozapine and quetiapine, do not appear to cause Parkinsonism (Durif *et al.* 1997; Juncos *et al.* 2004). Clinically, drug induced Parkinsonism is indistinguishable from sporadic idiopathic Parkinson's disease and although a symmetrical and postural rather than resting tremor is more common this is not consistent. Some patients develop a rest tremor of the lips, the 'rabbit syndrome', without other signs of Parkinsonism. Obviously, the diagnosis depends on taking a careful history of recent medication. In the psychiatric patient on large doses of neuroleptic treatment, the diagnosis is all too clear but some cases are less obvious. Some patients take prochlorperazine for years for dizziness and mixtures of a tricyclic antidepressant and a neuroleptic are sometimes used in depression. In cases like these, chronic neuroleptic exposure may not immediately be evident. There is a poor correlation between duration and dosage of neuroleptic therapy and the development of Parkinsonism, suggesting that blockade of dopamine receptors is not the only mechanism. Some patients develop Parkinsonism rapidly within days of starting a neuropleptic while others tolerate large doses for years without problems.

Several factors suggest that some patients probably have subclinical Parkinson's disease which has been revealed by interference with dopaminergic neurotransmission. Not all patients obtain a remission after discontinuation of neuroleptics and some develop Parkinson's disease after an initial remission (Goetz 1983; Hardie and Lees 1988). Some have abnormal flurodopa positron emission tomography scans indicating presynaptic dopamine deficiency and persistence of Parkinsonism is more likely in these patients (Burn and Brooks 1993). Moreover, a few patients have had pathological evidence of Parkinson's disease postmortem (Rajput *et al.* 1982). However, the high frequency of drug induced Parkinsonism suggests that not all of these patients can have Parkinson's disease. Dopamine transporter DAT scanning with SPECT can distinguish between Parkinson's disease, with reduced DAT signal, and drug induced Parkinsonism, in which DAT is intact (Marshall and Grosset 2003).

A number of other drugs may sometimes cause drug induced Parkinsonism (Table 40.9). Dopamine depleting agents such as

Table 40.9 Causes of drug induced Parkinsonism

Dopamine receptor blocking drugs	Neuroleptics
	Metoclopramide
	Prochlorperazine
Dopamine depleting drugs	Tetrabenazine
	Reserpine
	Alpha methyldopa
Miscellaneous drugs	Cinnarizine
	Selective serotonin reuptake inhibitors
	Amiodarone
	Lithium

SSRI = selective serotonin reuptake inhibitior antidepressants.

reserpine, which is no longer available, and tetrabenazine cause Parkinsonism by reducing dopamine release at nigrostriatal terminals. Metoclopramide has a dopamine receptor blocking action and is, in this respect, similar to neuroleptic major tranquillizers. This drug is widely prescribed for nausea and may cause long lasting Parkinsonism especially in the presence of renal impairment (Sethi *et al.* 1989). Calcium channel blockers such as cinnarizine are important as they are widely prescribed and often for long periods. The mechanism by which they induce Parkinsonism is unclear. Cinnarizine is commonly prescribed for dizziness and is liable to cause Parkinsonism in the elderly. It is important to note the association of serotonin reuptake inhibitor antidepressants with Parkinsonism (Leo 1996). Depression commonly produces psychomotor slowing and so the Parkinsonian symptoms may be attributed erroneously to depression; moreover, depression is common in Parkinson's disease and selective serotonin reuptake inhibitors should be used with caution.

In most cases, drug induced Parkinsonism will remit after discontinuation of the causative agent but this may take weeks or months. It is not possible to decide whether Parkinsonism is drug induced or due to underlying idiopathic Parkinson's disease until a reasonable period has elapsed without medication. Some remissions have not occurred for years. This remains an area of difficulty but it is reasonable to expect most cases of drug induced Parkinsonism to show improvement within 6 months of drug withdrawal. Most of those in whom Parkinsonism persists are older and likely to have Parkinson's disease but occasionally younger patients fail to improve, suggesting permanent nigrostriatal damage (Melamed *et al.* 1991). In some patients, continuing neuroleptic treatment may be unavoidable because of severe psychiatric illness; in these patients, the use of an atypical neuroleptic such as clozapine or quetiapine may be preferable (Durif *et al.* 1997; Juncos *et al.* 2004). There is no evidence that the use of atypical limbic selective neuroleptics will allow reversal of already established drug induced Parkinsonism although this might be expected in view of their comparative freedom from extrapyramidal side effects.

In patients with drug induced Parkinsonism who await remission after withdrawal of the responsible drug or in whom this is not possible, antiParkinsonian drugs may be used. Levodopa or anticholinergics may be used but dopamine agonists must be avoided in psychotic patients in case the mental state is worsened. The 'rabbit syndrome' responds to anticholinergics.

40.3.5 **Toxic Parkinsonism**

The possible role of an exogenous toxin in Parkinson's disease and the effects of exposure to methyl-phenyl-tetrahydropyridine (MPTP) have already been discussed in Section 40.3.1. This form of toxin-induced Parkinson's disease is no longer seen, having only occurred in a few patients, but was clinically very similar to sporadic Parkinson's disease (Langston 1996). The response to levodopa was good but fluctuation and dyskinesia occurred very quickly. In some cases the disease slowly progressed as in Parkinson's disease.

Other cases of toxic Parkinsonism (Table 40.10) are very rare and do not closely resemble Parkinson's disease. Nevertheless the possibility of toxic Parkinsonism is often raised in the clinic by patients who have possibly been exposed to industrial toxins. Patients with clinically typical Parkinson's disease, with tremor and a good response to levodopa are extremely unlikely to have toxic Parkinsonism.

Manganese is one of the best known causes of a Parkinsonian syndrome and enters the differential diagnosis of the Parkinsonian patient with an industrial or mining background from time to time (Section 5.7.4). In manganese miners there is a brief psychotic reaction as a prodrome to the development of Parkinsonism but not in those with chronic lower level exposure. In both groups, Parkinsonism appears insidiously during or shortly after exposure (Jankovic 2005). Akinesia, rigidity, and gait disturbance are prominent and tremor is usually absent. Lower limb dystonia is common and cerebellar features may be seen. Pathologically, manganese causes degeneration of the globus pallidus and there is little improvement with levodopa (Pahwa 1997). Pallidal signal change may be seen on MR scanning (Josephs *et al.* 2005). Other toxic causes of Parkinsonism are very rare and share some of the salient features of manganism, namely a predominantly rigid and akinetic state with inconspicuous or absent tremor and no response to levodopa. Dystonia and spasticity are also noted frequently along with variable cognitive impairment (Pahwa 1997).

40.3.6 **Postencephalitic Parkinsonism**

This was most commonly seen after the large epidemic of encephalitis lethargica, Von Economo's encephalitis, which occurred in Europe and North America from 1917 until the 1930s. Many survivors from the epidemic had long term psychiatric and neurological impairments, most notably oculogyric crises, which occur in 20 per cent, and postencephalitic Parkinsonism, in about 50 per cent. Occasional cases of encephalitis lethargica still occur (Howard and Lees 1987) but are extremely rare. Parkinsonism sometimes appeared immediately but could take years to develop, after a latent interval. Bradykinesia and rigidity were prominent but tremor was variably present. Oculogyric crises, with forced vertical deviation of the eyes and accompanying anxiety or depression lasted for

Table 40.10 Causes of toxic Parkinsonism

MP TP
Manganese
Carbon monoxide
Carbon disulphide
Cyanide
Methanol

several minutes or hours and were characteristic of postencephalitic Parkinsonism. Additional features such as respiratory irregularity, behavioural disturbances, somnolence, ophthalmoparesis, and various hyperkinesias were also seen. Progression was very slow or negligible in many patients but late deterioration, many years later has been described. Pathologically there was severe neuronal loss and gliosis in the substantia nigra and widespread neurofibrillary tangle formation, especially in the brainstem. Anticholinergic treatment was often successful and well tolerated but the response to levodopa was less predicatable and side effects were common, mainly dyskinesias and agitation.

40.3.7 **Post traumatic Parkinsonism**

There are three aspects to the issue of head injury and Parkinson's disease which has been controversial for over a century. These are the roles of acute major brain injury, repeated concussion, and head injury without evidence of brain damage.

Occasionally a head injury can induce Parkinsonism secondary to direct basal ganglia or substantia nigra damage. Such cases are rare and a confident diagnosis requires evidence of significant brain damage and the early appearance of Parkinsonism. There should be radiological or later pathological evidence of basal ganglia or brainstem injury. It is likely that other neurological signs will be present.

Chronic repeated concussion as seen in boxers may result in a chronic traumatic encephalopathy, dementia pugilistica, or punch drunkenness. This produces akinetic rigid features and associated cognitive decline, ataxia, dysarthria, and neuropsychiatric abnormalities. CT scanning may reveal a cavum septum pellucidum and cerebral atrophy but other investigations are unhelpful. The response of the Parkinsonism to levodopa is variable. It will be noted that this akinetic rigid syndrome is unlikely to be confused with idiopathic Parkinson's disease.

Head trauma without significant brain injury is more controversially associated with Parkinson's disease. The symptoms of Parkinson's disease become worse in association with stress, intercurrent illness, or injury (Goetz and Stebbins 1991) but this is not due to any direct effect on the disease process and is almost always reversible.

40.3.8 **Multiple system atrophy**

Definition and prevalence

This highly variable disorder causes Parkinsonism, cerebellar dysfunction, pyramidal signs, and autonomic failure in various combinations. The aetiology is unknown and it does not appear to be hereditary. Unfortunately, there is widespread confusion regarding its definition. This is mainly because it was originally described under different diagnostic terms such as olivopontocerebellar atrophy, striatonigral degeneration, and the Shy–Drager syndrome whose clinical features were dominated by cerebellar ataxia, Parkinsonism, or autonomic failure respectively. However, it is important to note that each of these three syndromes actually featured all three of the key clinical manifestations of multiple system atrophy as shown in Table 40.11. The complex and confusing evolution of the multiple system atrophy concept has been reviewed in detail (Quinn 1994) but its constituent syndromes were assembled into the single condition of multiple system atrophy by Graham and Oppenheimer (1969) who reported a case of autonomic failure followed by cerebellar signs but

Table 40.11 Clinical variants of multiple system atrophy

	Major clinical feature	Additional features
Olivopontocerebellar atrophy	Ataxia	Pyramidal signs Parkinsonism Autonomic failure
Shy–Drager syndrome	Autonomic failure	Parkinsonism Muscle fasciculation and wasting*
Striatonigral degeneration	Parkinsonism	Pyramidal signs Autonomic failure Ataxia
Progressive autonomic failure	Autonomic failure	Parkinsonism Pyramidal signs Ataxia

* = observed in one of the two cases only

not Parkinsonism. Pathologically however, there were changes similar to those previously reported as olivopontocerebellar atrophy and striatonigral degeneration, despite the lack of clear akinetic rigid features in life, and cell loss in the intermediolateral columns of the spinal cord. This one case therefore combined the pathological features of three previously distinct entities. In retrospect, previous cases of the three constituent syndromes also had the pathological changes of multiple system atrophy. More recently, this concept of multiple system atrophy as a unitary disorder has been strengthened by the finding of a common neuropathological feature, glial and neuronal cytoplasmic inclusions detected with Beilschowsky silver or anti-ubiquitin stains, in patients from each of the different clinical subgroups (Quinn 1994).

Multiple system atrophy prevalence is 2–4 per 100 000 (Schrag et al. 1999; Vanacore 2005). Its cause is unknown but an association with farming suggests a possible toxic cause (Vanacore et al. 2005). In one study of patients selected on the basis of a clinical diagnosis of Parkinson's disease in life, 5 per cent actually had multiple system atrophy (Hughes et al. 1992). Of those with sporadic adult onset cerebellar ataxia, about 30 per cent have multiple system atrophy (Abele et al. 2002).

Pathology

Pathologically, there is neuronal loss and gliosis in the striatum, substantia nigra, locus coeruleus, pontine nuclei and/or middle cerebellar peduncles, cerebellar Purkinje cells, inferior olivary nuclei and intermediolateral columns, and Onuf's nucleus of the spinal cord. Not all of these are involved in every case but at least two areas are required for a pathological diagnosis along with the characteristic glial cytoplasmic inclusions common to all forms of multiple system atrophy (Papp et al. 1989). These inclusions, which contain alpha synuclein (Wenning and Jellinger 2005) and ubiquitin (Dickson et al. 1999), are also seen in neurones (Papp and Lantos 1994) and may be detected while neuronal loss is still very mild (Wenning et al. 1994c). It is possible that the inclusions are taken up by cells in the vicinity of other dying neurones (Wenning and Jellinger 2005). The significance of glial cytoplasmic inclusions is unclear but they are not seen in other neurological conditions, including Parkinson's disease and progressive supranuclear palsy with two exceptions: one case of dominantly inherited cere-

bellar ataxia, SCA type 1 (Gilman et al. 1996) and a previously reported case of adult onset hereditary cerebellar degeneration. Glial cytoplasmic inclusions were absent in all other cases of hereditary ataxia studied. Although the pathology of multiple system atrophy is variable, neuronal loss in the substantia nigra and striatum is usually seen in patients who had significant Parkinsonism in life.

Clinical features

The onset of multiple system atrophy is usually between 40 and 70 years and has not been described before 30. The diagnosis is primarily a clinical one (Quinn 2005). The first symptoms are usually due to Parkinsonism in 46 per cent and autonomic dysfunction which is almost as frequent occurring in 41 per cent; a few patients present with a cerebellar syndrome (5 per cent) or a mixed picture (7 per cent) (Wenning et al. 1994a). Eventually most patients develop Parkinsonism, cerebellar dysfunction, pyramidal and autonomic involvement in various combinations and with differing emphasis. Clinically, the majority of patients, about 80 per cent, have a mainly Parkinsonian picture, with or without additional cerebellar features, the striatonigral degeneration type. A predominantly cerebellar syndrome, the olivopontocerebellar atrophy type, is seen in aproximately 20 per cent of cases, usually with additional Parkinsonism. Autonomic impairment occurs in the majority of patients in both groups but is occasionally absent (Quinn 1994; Wenning et al. 1994a).

The Parkinsonism-dominant striatonigral degeneration form is commonly associated with atypical features such as an absent or jerky postural type of tremor and a poor response to levodopa often with unusual orofacial dyskinesia; dyskinesia in Parkinson's disease usually involves the limbs initially. Other features are symmetrical onset, early falling and unusually rapid progression, leading rapidly to the 'wheelchair sign'—an unusual feature in Parkinson's disease until the very advanced stages. Small myoclonic jerks of the fingers may be seen (Salazar et al. 2000). Pyramidal signs occur in 60 per cent of multiple system atrophy patients but are usually mild, with exaggerated tendon reflexes, extensor plantar responses, and sometimes spasticity. In most cases of striatonigral degeneration, the additional cerebellar signs, pyramidal or autonomic features will point to the correct diagnosis and not idiopathic Parkinson's disease. However, some multiple system atrophy patients have levodopa responsive Parkinsonism with asymmetrical onset, typical resting tremor, and no additional atypical features. Such cases are clinically indistinguishable from Parkinson's disease and may be detected only at postmortem (Hughes et al. 1992).

Autonomic failure develops in the geat majority of patients and is often a presenting feature. Urgency and frequency of micturition, impotence, and postural hypotension are the usual features; the latter rarely causes severe hypotension or syncope. Faecal incontinence is occasionally encountered and some patients develop urinary retention. It is rare for autonomic failure to be the only or dominant clinical problem in multiple system atrophy. It is almost always accompanied by Parkinsonism, cerebellar ataxia, or a combination of the two after a few years.

Cerebellar manifestations of multiple system atrophy include gait ataxia, dysarthria, limb ataxia, and nystagmus. About a half of patients overall have cerebellar involvement clinically but this is rarely the presenting feature. Of those patients, about 20 per cent, who have a predominantly cerebellar syndrome, the olivopontocerebellar atrophy type of multiple system atrophy, careful examination usually reveals evidence of autonomic involvement or Parkinsonism (Wenning *et al.* 1997). Although cerebellar involvement may not be evident during life, this is much more commonly detected pathologically (Wenning *et al.* 1995).

Other clinical signs may be seen in some patients. Some patients have an unusually severe and characteristic antecollis (Fig. 40.10), cold cyanotic extremities, or respiratory irregularities (Quinn 1994). Eye movement abnormalities include nystagmus and mild limitation of eye movement but never to the degree seen in progressive supranuclear palsy. Inspiratory stridor is a dramatic and well recognized manifestation in some patients (Wenning *et al.* 1994a). This is often severe enough to require tracheostomy and is occasionally a presenting symptom. Some patients have a high-pitched dysarthria which is difficult to classify and is prominent at a much earlier stage than in Parkinson's disease. A rare but curious feature in a few patients has been the emergence of multiple system atrophy, and sometimes of Parkinson's disease, in men with rapid eye movement sleep behaviour disorder (Schenck *et al.* 1996). In contrast to Parkinson's disease, treatment related psychosis does not occur in multiple system atrophy, and although mild frontal lobe cognitive deficits may be detectable (Dujardin *et al.* 2003), dementia is not seen. Peripheral nerve involvement is mild and almost always subclinical.

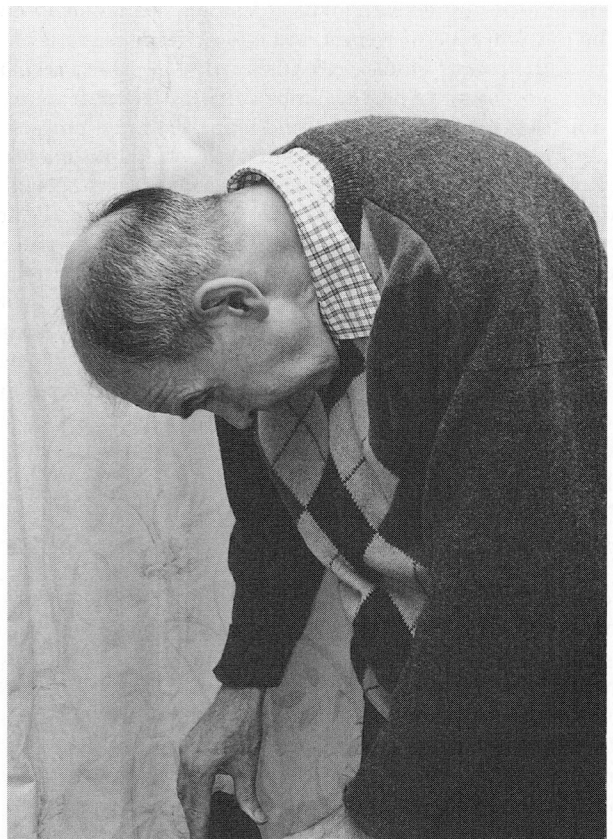

Fig. 40.10 Severe antecollis in a patient with multiple system atrophy of the striatonigral degeneration type.

Diagnosis

The diagnosis of multiple system atrophy may be difficult, especially in cases of levodopa responsive Parkinsonism and in cases with a cerebellar presentation. The latter group are in many respects similar to patients with sporadic late onset cerebellar degeneration in which cerebellar ataxia, Parkinsonism and pyramidal signs are common. It is the presence of autonomic impairment and lack of a family history that should suggest multiple system atrophy rather than other forms of idiopathic or hereditary late onset cerebellar degeneration (Gilman and Quinn 1996). The features of Parkinsonism which should suggest a diagnosis of multiple system atrophy have already been mentioned. A recurrent difficulty is the presence of mild autonomic problems such as urgency of micturition and postural hypotension in a patient with otherwise typical Parkinson's disease. It may be difficult to exclude multiple system atrophy in such cases but its autonomic features tend to be more severe, occur much earlier, and affect younger patients. Obviously the response to levodopa and the clinical features of such patients require careful scrutiny. The fragile X tremor ataxia syndrome (Section 39.9.7) may resemble multiple system atrophy with atypical Parkinsonism, ataxia, and a poor response to medication (Hagerman and Hagerman 2004).

Investigations are usually unhelpful. Although brain CT scanning often reveals cerebellar or brainstem atrophy this is usually apparent clinically (Wenning *et al.* 1994b). MRI will detect posterior fossa atrophy more reliably and may show altered putaminal signal, possibly reflecting altered striatal iron levels (Schwarz *et al.* 1996) although this is not yet adequately validated for clinical practice. MRI may show a characteristic pontine 'hot cross bun' appearance due to loss of pontocerebellar fibres (Watanabe *et al.* 2002). Positron emission tomography using 18Fdopa has been used to detect reduced nigrostriatal dopaminergic function in multiple system atrophy including cases with primarily Parkinsonian, cerebellar or autonomic presentations (Brooks *et al.* 1990; Rinne *et al.* 1995). However, positron emission tomography studies of nigrostriatal dopaminergic innervation and postsynaptic striatal D2 receptor density, using 18Fdopa and 11Craclopride ligands have not consistently discriminated multiple system atrophy from Parkinson's disease (Brooks *et al.* 1992; Burn *et al.* 1994). Analysis of postsynaptic striatal opioid receptor density, using 11C diprenorphine (Burn *et al.* 1995) or 18F deoxyglucose positron emission tomography to detect striatal and cerebellar hypometabolism have been studied in multiple system atrophy (Eidelberg *et al.* 1993a). The integrity of the striatal dopamine receptors, which should be intact in Parkinson's disease, can also be studied using 131I-iodobenzamide, single photon emission tomography (SPECT) scanning. Reduced striatal 131I-iodobenzamide binding is suggestive of multiple system atrophy rather than Parkinson's disease and has successfully predicted levodopa responsiveness (Oertel *et al.* 1993; Schwarz *et al.* 1993). Cerebral blood flow can be assessed with ^{123}I-IMP SPECT (Matsui *et al.* 2005a) with cerebellar, cortical, and occipital differences between multiple system atrophy and Parkinson's disease. Neuronal loss in the putamen has also been detected using proton magnetic resonance spectroscopy (Davie *et al.* 1995). Although positron emission tomography, SPECT, and magnetic resonance spectroscopy are promising diagnostic techniques, they are not easily available and not reliable enough for widespread application.

By contrast, external urethral sphincter electromyography is very useful as denervation changes are highly suggestive of multiple system atrophy (Pramstaller *et al.* 1995) and abnormal sphincter electromyography is detected in almost all cases of multiple system atrophy (Paviour *et al.* 2005). However, in a few multiple system atrophy patients, sphincter electromyography is normal; false positive results may occur after pelvic surgery, in multiparous women, and in progressive supranuclear palsy (Valldeoriola *et al.* 1995). In one series, sphincter electromyography did not distinguish multiple system atrophy from Parkinson's disease (Giladi *et al.* 2000). Autonomic dysfunction may be detected by autonomic function tests or urodynamic studies but in patients with significant autonomic failure, this should be evident clinically. The central autonomic dysfunction of multiple system atrophy can be distinguished from the peripheral autonomic impairment seen in Parkinson's disease by cardiac radionucleotide scanning using [123I]metaiodobenzylguanidine. In Parkinson's disease there is reduced cardiac but normal mediastinal uptake whereas in multiple system atrophy, both are reduced (Yoshita and Braune 2000; Matsui *et al.* 2005b). Probably the most reliable means of diagnosis in life is the progression of the illness over time.

Management and prognosis

The progression of multiple system atrophy is much more rapid than of Parkinson's disease and a rating scale has been developed (Wenning *et al.* 2004b). Deterioration of Parkinsonism in multiple system atrophy is seven times faster than in Parkinson's disease (Seppi *et al.* 2005) and walking aids and a wheelchair are required after a median of 3 and 5 years respectively (Watanabe *et al.* 2002). Median survival is about 10 years and most patients are dead after 15 years.

The management of multiple system atrophy is difficult. Parkinsonism may improve with levodopa in about a third of cases but rarely as well as in Parkinson's disease. Additional dopamine agonists or anticholinergics are usually unhelpful in patients who do not respond well to levodopa. Amantadine can help some patients (Wenning *et al.* 2004a). Postural hypotension often responds to head up tilt of the bed at night, fludrocortisone, or nocturnal desmopressin, elastic stockings are rarely helpful and are uncomfortable. Bladder dysfunction may respond to anticholinergic medication sometimes additionally with intermittent or permanent catheterization depending on residual manual dexterity. Stridor may require tracheostomy or respiratory support to prevent sudden death (Silber and Levine 2000; Ghorayeb *et al.* 2005). Sildefanil or papaverine may help male impotence. Like many neurologically disabled patients, those with multiple system atrophy need access to speech therapy, occupational therapy, social services, continence advisors, and physiotherapy when required. Useful information and support is available to patients from the Sarah Matheson Trust (http://www.msaweb.co.uk/home.htm).

Differentiation of autonomic failure in multiple system atrophy from primary autonomic failure

In multiple system atrophy the autonomic failure is caused by central nervous system involvement affecting the brainstem, intermediolateral columns, and Onuf's nucleus of the spinal cord. In these regions there is neuronal loss, gliosis, and glial cytoplasmic inclusions; Lewy bodies are absent. Parkinsonism, cerebellar ataxia, or both develop in all cases within a few years of onset. The progression of multiple system atrophy is often rapid and the prognosis is poor.

The distinction between multiple system atrophy and Parkinson's disease with autonomic involvement has already been mentioned above. Another point which may cause confusion is the distinction between multiple system atrophy and primary autonomic failure, a chronic autonomic neuropathy (Section 21.11.7) sometimes referred to as idiopathic orthostatic hypotension or the Bradbury Eccleston syndrome. This is a more slowly progressive, and sometimes static, disorder with slower onset and a better prognosis. There are no additional central nervous system features such as Parkinsonism, pyramidal involvement, or cerebellar signs and rapid eye movement sleep disorder does not occur (Plazzi *et al.* 1998). Postural hypotension, defined as a blood pressure reduction of 20 mmHg systolic or 10 mmHg diastolic within 3 min of standing (Consensus Statement 1996), occurs in both primary autonomic failure and multiple system atrophy but tends to be more severe in the former with a higher incidence of syncope and post prandial hypotension. Pathologically the main changes are in the autonomic ganglia of the peripheral nervous system, with degeneration of postganglionic sympathetic neurones and the presence of Lewy bodies in the ganglia or the post ganglionic neurones. There may be much less conspicuous and asymptomatic pathological involvement, including the presence of Lewy bodies, in brainstem nuclei, substantia nigra, and intermediolateral columns (Hague *et al.* 1997). The supine plasma noradrenaline level is reduced in primary autonomic failure but not in multiple system atrophy. Essentially, multiple system atrophy is characterized by central autonomic dyfunction and primary autonomic failure by peripheral autonomic impairment. This distinction may be detected by the use of growth hormone measurements after the administration of clonidine (Kimber *et al.* 1997); in normal individuals, Parkinson's disease, and primary autonomic failure, the level of growth hormone is increased by clonidine; this does not occur in multiple system atrophy. The cause of primary autonomic failure is unknown and its relationship to Parkinson's disease and dementia with Lewy bodies (Section 40.3.2) is unclear. In some patients, the distinction between primary autonomic failure and multiple system atrophy is very difficult; about 10 per cent of those who appear to have primary autonomic failure go on to develop other neurological features of multiple system atrophy. The treatment of the autonomic symptoms is similar in the two disorders. The noradrenaline precursor 3,4-DL-threo-dihydroxyphenylserine, the alpha adrenergic agonist midodrine, and erythropoietin can improve postural hypotension.

40.3.9 Progressive supranuclear palsy

Definition

This is a progressive degenerative disorder causing early gait instability and falling with axial rigidity, dementia, pseudobulbar palsy, pyramidal signs, and the characteristic eye movement disorder of a supranuclear downgaze palsy.

Frequency aetiology and pathology

Progressive supranuclear palsy is rare, with a prevalence estimated at 6 per 100 000 but is almost certainly underdiagnosed (Schrag *et al.* 1999). The aetiology is unknown. There is a high incidence of a similar illness in Guadeloupe associated with a toxin in tropical plants (Caparros-Lefebvre *et al.* 2002). Associations with both above and below average education and residence in small towns have been reported but are of doubtful significance. The main interest concerning the possible cause has focussed on accumulation of abnormal brain tau protein (Burn and Lees 2002).

The key neuropathological feature of progressive supranuclear palsy is tau containing neurofibrillary tangle formation, with variable neuronal loss and gliosis and in the globus pallidus, subthalamic nucleus, substantia nigra, and other brainstem nuclei, especially the superior colliculi and pretectal areas, which are involved in eye movement control, reticular formation, pedunculopontine nucleus, pontine nuclei, and the interstitial nucleus of Cajal (Hauw *et al.* 1994). Involvement of the pedunculopontine nucleus has been linked to the prominent gait instability of progressive supranuclear palsy and the marked axial and neck rigidity may be associated with involvement of the interstitial nucleus of Cajal. Disordered eye movement is probably secondary to the tectal and pretectal lesions, as well as the nucleus raphe interpositus (Revesz *et al.* 1996). Cortical involvement, especially of the frontal motor areas is common, in contrast to earlier reports (Verny *et al.* 1996). The neurofibrillary tangles contain abnormally phosphorylated tau protein, similar to that seen in Alzheimer's disease, but their distribution and morphology in progressive supranuclear palsy are different and there are also tau containing glial inclusions. There is no association between progressive supranuclear palsy and the *apo-e4* genotype (Tabaton *et al.* 1995) and no amyloid deposition. The main considerations regarding tau in progressive supranuclear palsy are:

- The prevalent tau isoforms are mainly 4-repeat isoforms in contrast to that seen in normal brain, in which there is a slight preponderance of 3-repeat tau, Pick's disease, only 3-repeat, and Alzheimer's disease, both 3- and 4-repeat.

- There is an association with one polymorphism, A0, of the *tau* gene on chromosome 17 (Conard *et al.* 1997). There is also an association with a more extended haplotype, H1, linked to A0 which is present in all progressive supranuclear palsy cases but 70 per cent of controls (Morris *et al.* 1999). The significance of this is uncertain.

- The majority of cases are sporadic but rarely autosomal dominant has been described (Rojo *et al.* 1999); in some other families, relatives of sporadic cases have had mild Parkinsonian features or abnormal dopaminergic function on positron emission tomography scanning (Piccini *et al.* 2001) indicating low penetrance autosomal dominant transmission of progressive supranuclear palsy. In a few unusual families, there have been mutations of the tau gene (Delisle *et al.* 1999; Stanford *et al.* 2000) but these are absent in sporadic cases and other families with hereditary progressive supranuclear palsy (Hoenicka *et al.* 1999).

Clinical features

The onset is always after the age of 40 years and the two most important features are early falling and a characteristic eye movement disorder. These are usually associated with frontal lobe impairment, pseudobulbar palsy, bradykinesia, rigidity and pyramidal signs (Burn and Lees 2002). Difficulty walking with falling, typically backwards, within a year of onset is the most common presenting feature. The second salient feature is a supranuclear downgaze palsy. Downgaze may initially be slowed or reduced and optokinetic responses to a downward moving stimulus are impaired (Garbutt *et al.* 2004). Saccades tend to be lost before pursuit movements. Other eye movement abnormalities such as impaired upward or lateral movements, internuclear ophthalmoplegia,

absent Bell's phenomenon, square wave jerks and impaired suppression of vestibulo-ocular reflexes are common but less specific and are seen in many other neurological disorders. Eventually there is a severe ophthalmoplegia in all directions but particularly downwards (Lees 1987). This leads to difficulty with eating, reading and safe descent of stairs. The face has a striking frozen staring expression with reduced blinking, frontalis overaction and a fixed gaze (Fig. 40.11). Eyelid abnormalities including blepharospasm, apraxia of eyelid opening due to blepharospasm of the pretarsal orbicularis oculi, apraxia of eyelid closure and ptosis are also common and contribute to the distinctive facial appearance. Speech is strained and dysarthric due to a pseudobulbar palsy. There may be palilalia, stuttering, explosive coughing and intermittent inspiratory sighs and emotional lability (Litvan *et al.* 1996a–d). Dysphagia frequently develops and is eventually a major determinant of survival. The patient develops a severe axial rigidity with extension of the neck and less prominent stiffness and slowness of the limbs. Tremor is usually absent but has been observed in a minority (Masucci and Kurtzke 1989). Pyramidal involvement with brisk limb reflexes and extensor plantar reactions, in addition to the pseudobulbar signs, eventually appear in most. Cognitive impairment commonly develops (Grafman *et al.* 1990; Litvan *et al.* 1996d) typically a frontal lobe syndrome with apathy, disinhibition, mood alterations, anxiety and psychomotor slowing, as well as perseveration and frontal release signs. Aggression and irritability are uncommon and psychiatric features are mild (Menza *et al.* 1995). Autonomic

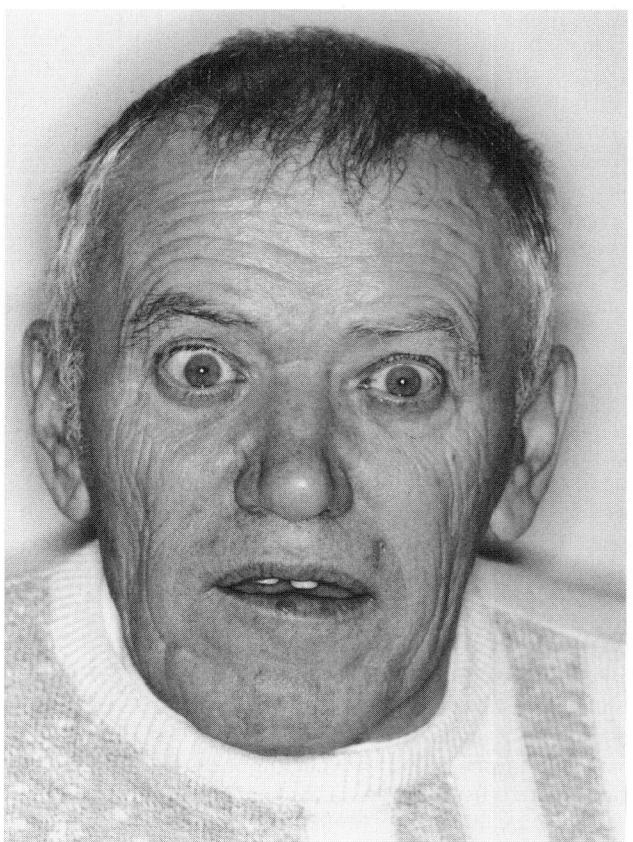

Fig. 40.11 The typical facial appearance of Steele–Richardson–Olszewski disease.

impairment is not a major feature but bladder dysfunction with urinary incontinence may be seen in more severe cases. Sleep disturbance is common.

Although the clinical picture of typical progressive supranuclear palsy is umistakable, atypical presentations occur commonly. While about half of patients have the typical presentation with falls, ophthalmoplegia and frontal lobe features, about a third present with an asymmetrical Parkinsonism with some initial levodopa responsiveness and tremor (Williams *et al.* 2005a). These two presentations have been termed 'Richardson's syndrome' and 'progressive supranuclear palsy-Parkinsonism'; the tau composition of brainstem neurofibrillary tangles differs in the two forms suggesting that progressive supranuclear palsy may well be heterogeneous.

Other atypical presentations occur: There may be prominent limb dystonia (Barclay and Lang 1997), resting tremor (Masucci and Kurtzke 1989; Collins *et al.* 1995), palatal myoclonus (Suyama *et al.* 1997) and apraxia (Burn and Lees 2002). In some cases there is an assymmetrical akinetic rigid picture which may resemble Parkinson's disease for years before the characteristic axial rigidity and ophthalmoplegia appear (Lees 1987; Collins *et al.* 1995). Some patients present with a syndrome of pure akinesia with striking slowness of movement affecting gait, speech, and writing but no other signs of the condition (Matsuo *et al.* 1991; Riley *et al.* 1994). Such patients do not always subsequently develop typical ocular and other progressive supranuclear palsy features. There are also pathologically confirmed cases in which the characteristic ophthalmoplegia was absent during life (Daniel *et al.* 1995).

Diagnosis

The diagnosis of progressive supranuclear palsy often proves difficult initially but is ultimately clinical. There may be vague complaints about vision, a change in speech, altered personality, and a general slowness and stiffness which may erroneously suggest Parkinson's disease. There may be no obvious physical cause for falling or difficulty with balance and walking. It may be several years before the suggestive physical signs develop to permit the correct diagnosis (Maher and Lees 1986).

In order to improve diagnostic accuracy, the NINDS-SPSP diagnostic criteria have been proposed which emphasize the early development of falling within a year of onset and a supranuclear downgaze palsy (Litvan *et al.* 1996a,b) as the most important diagnostic features and the diagnosis can scarcely be made without these features. Difficulty may arise in distinguishing progressive supranuclear palsy from Parkinson's disease, multiple system atrophy, dementia with Lewy bodies, corticobasal degeneration, and Pick's disease (Litvan *et al.* 1997). The axial rigidity, rarity of tremor, lack of response to levodopa, and early falling should avoid confusion with Parkinson's disease but this distinction may not be easy in the first few years of the disease. Hallucinations and mental fluctuations are not seen in contrast to dementia with Lewy bodies. Multiple system atrophy usually begins at an earlier age and is not associated with such a marked ophthalmoplegia. Autonomic impairment is more severe in multiple system atrophy but an abnormal external urethral sphincter electromyography may be noted in either disorder (Valldeoriola *et al.* 1995). Pick's disease is not associated with ophthalmoplegia or falling. There may be confusion with corticobasal degeneration in which an ophthalmoplegia, gait disorder, and rigidity may occur (Rinne *et al.* 1994b).

However in progressive supranuclear palsy, the ophthalmoplegia is much more severe, the falling more prominent, and striking asymmetry signs or the alien limb sign are not encountered. In atypical cases, especially those without ophthalmoplegia, the diagnosis may not be possible without prolonged follow up and in some cases may never become so.

On occasions, other conditions such as cerebrovascular disease, progressive subcortical gliosis (Will *et al.* 1988; Foster *et al.* 1992), dementia with Lewy bodies (Fearnley *et al.* 1991), Creutzfeldt–Jacob disease (Bertoni *et al.* 1992), cerebral tumour and basal ganglia calcification (Saver *et al.* 1994; Silbert *et al.* 1993) may closely resemble progressive supranuclear palsy. Hydrocephalus may cause gait impairment, dementia, and sometimes a vertical gaze palsy, but downgaze is usually spared and axial rigidity is not seen. It may be difficult, especially in advanced cases, to distinguish progressive supranuclear palsy from late onset cerebellar degeneration in which a severe ophthalmoplegia, inability to walk, and pyramidal signs may also occur; in this situation much will depend on the history and physical signs in the earlier years of the illness and the detection of cerebellar atrophy by neuroimaging.

Investigations in progressive supranuclear palsy are of limited value. MR scanning may reveal brainstem atrophy (Schrag *et al.* 2000a; Kato *et al.* 2003; Adachi *et al.* 2004) but the cerebellum appears normal and hydrocephalus, cerebral infarcts, mass lesions and other disorders associated with ophthalmoplegia such as Whipple's disease are easily excluded. Occasionally the 'eye of the tiger sign' seen in Hallervorden Spatz syndrome may be seen (Burn and Lees 2002). Neuropsychological evaluation can be helpful in confirming a clinical impression of frontal lobe dysfunction and frontal hypoperfusion may be detected by single photon emission tomography 'SPECT', blood flow scanning (Neary *et al.* 1987) although this is not diagnostically reliable (Burn and Lees 2002).

Specialized neurophysiological studies have shown loss of the normal auditory startle response, probably reflecting damaged brainstem reticular formation neurones (Rothwell *et al.* 1994). However, neurophysiological studies are unhelpful in the diagnosis (Burn and Lees 2002). In more specialized research studies, loss of striatal dopamine receptors detected by SPECT or positron emission tomography (Brooks *et al.* 1992; Schwarz *et al.* 1993) can help to exclude Parkinson's disease and striatal neuronal loss may be detected by magnetic resonance spectroscopy (Federico *et al.* 1997). However, such techniques are not yet reliable or easily available—presynaptic dopamine transporter (DAT) scanning using beta-CIT or ^{123}I-IPT DAT SPECT can detect presynaptic nigrostriatal degeneration but none of these methods reliably distinguishes PSP from Parkinson's disease or multiple system atrophy (Burn and Lees 2002).

Management and prognosis

The course of progressive supranuclear palsy is relentless with a median survival of 6 years (Maher and Lees 1986; Litvan *et al.* 1996c). A few atypical patients have a milder and more protracted illness which may partly resemble Parkinson's disease. Levodopa, other antiParkinsonian drugs, and antidepressants are of no real benefit (Kompoliti *et al.* 1998). Tricyclic antidepressants and amantadine may be helpful in some patients but the effect is modest and a variety of other medications including doepezil, zolpidem, physostigmine, and efaroxan an alpha 2 antagonist have all been ineffective (Burn and Lees 2002). There is no role for stereotactic surgery. Botulinum toxin may be used for blepharospasm and possibly for rigidity although experience in progressive supranuclear palsy is limited. Speech therapy may improve dysarthria and dysphagia but some patients eventually require gastrostomy feeding. Increasing physical disability may be improved by physiotherapy and occupational therapy. Community services are eventually required in most cases, including personal and respite care. In the United Kingdom, the Progressive Supranuclear Palsy Association (www.pspeur.org) can provide some information and support to patients and carers.

40.3.10 Corticobasal degeneration

This is a rare sporadic degenerative disorder of unknown aetiology. Very rare familial cases have been described (Brown *et al.* 1996a). It is characterized by a very asymmetrical akinetic rigid syndrome and localized cortical deficits such as cortical sensory loss, dyspraxia, myoclonus and the alien limb phenomenon (Mahapatra *et al.* 2004). There are clinical and pathological similarities to progressive supranuclear palsy and frototemporal dementia (Boeve *et al.* 2003).

Pathologically there is assymetric frontoparietal cortical atrophy which is particularly severe around the Rolandic fissure. Microscopically, there is cerebral cortical degeneration, with neuronal loss, gliosis, and ballooned or achromatic neurones; Pick bodies are not seen. There is also severe degeneration of the substantia nigra and variable involvement of the basal ganglia, brainstem, and cerebellar nuclei. Corticobasal degeneration shares some neuropathological features with progressive supranuclear palsy and Pick's disease including deposition of tau and an association with the Ao/H1 tau haplotype (Di Maria *et al.* 2000). The tau is mainly of the 4-repeat isoform and forms paired helical filaments (Mahapatra *et al.* 2004).

Typically the onset of corticobasal degeneration is after the age of 60 years. In most patients there is atypical levodopa resistant Parkinsonism with striking asymmetry (Riley *et al.* 1990). Rigidity is the main feature (Mahapatra *et al.* 2004) and with bradykinesia is usually confined to one limb, usually an arm, with additional dystonia and apraxia. The patient may hold the arm in an odd, flexed and abducted posture with little abnormality in the other limbs (Fig. 40.12). A painful flexion dystonia of the hand is typical (with a clenched fist appearance). In rarer cases, one leg is initially affected. Postural instability and falling may occur early and a rest tremor is seen in a third of cases (Watts *et al.* 1997; Wenning *et al.* 1998). The tremor is faster than that seen in Parkinson's disease and is typically jerky and irregular, sometimes with additional focal myoclonus. In a few cases, the initial presentation may resemble idiopathic Parkinson's disease. In addition to the akinetic rigid syndrome, various cortical features occur such as focal cortical sensory loss, stimulus sensitive focal myoclonus, typically in a rigid dyspraxic upper limb, apraxia, and aphasia. The supplementary motor area is often involved giving rise to the 'alien limb sign'. Usually this involves an arm which wanders about, uncontrollably moves and grasps independently and may interefere with the activity of the other hand, 'intermanual conflict'. Pyramidal signs, dystonia, myoclonus, and abnormal eye movements gradually appear along with increasing asymmetrical Parkinsonism. Gradually there is progression of the ipsilateral and later the contralateral limbs but marked persisting asymmetry is characteristic (Riley *et al.* 1990). A supranuclear eye movement disorder commonly affects mainly horizontal movements (Rinne *et al.* 1994b). Initiation of saccades

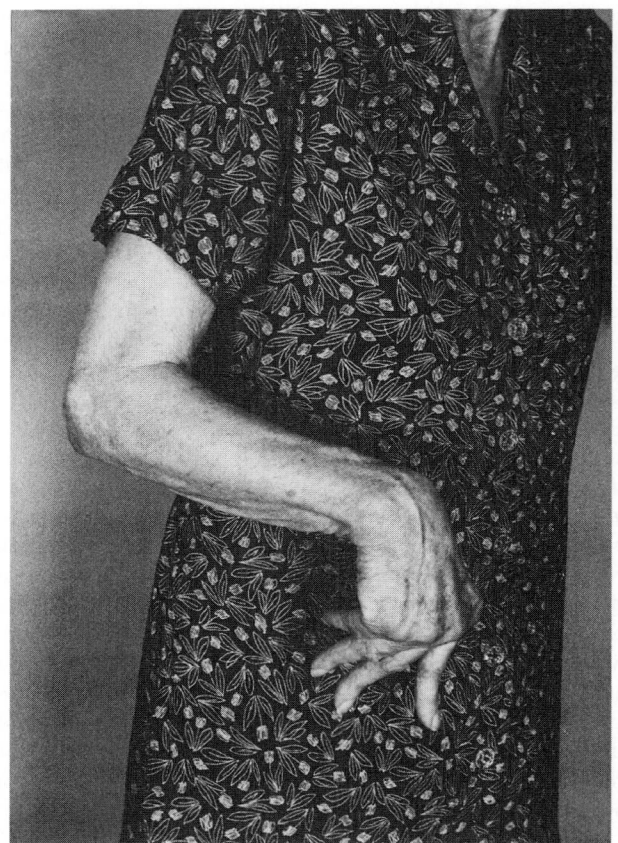

Fig. 40.12 Corticobasal degeneration. Note the appearance of the rigid, flexed arm which is characteristic of the condition.

is slow but their velocity and range are normal. Cognitive impairment is common (Mahapatra *et al.* 2004) and a frontoparietal dementia appears in many (Rinne *et al.* 1994b; Wenning *et al.* 1998) with frontal lobe deficits and apraxia (Pillon *et al.* 1995). In some cases, dementia or dysphasia are presenting features with little motor involvement (Bergeron *et al.* 1996). In these cases the presentation can resemble Pick's disease or frontotemporal dementia.

The diagnosis of corticobasal degeneration is often difficult and may not be possible until the disease progresses. Other conditions such as Alzheimer's disease (Doran *et al.* 2003) (Section 34.6.2), progressive supranuclear palsy (Section 40.3.9), Pick's disease (Section 34.6.4), and Creutzfeldt–Jacob disease (Sections 34.6.5; 42.9.6) have been diagnosed pathologically in patients clinically diagnosed with corticobasal degeneration (Boeve *et al.* 1999). It is distinguished from Parkinson's disease by the lack of response to levodopa and the various additional clinical features. The combination of an akinetic rigid syndrome with a vertical supranuclear gaze palsy will usually suggest progressive supranuclear palsy as the principal diagnosis. Early falling, axial and symmetrical rigidity, and ophthalmoplegia affecting downgaze more than horizontal movement are more suggestive of progressive supranuclear palsy but atypical cases of progressive supranuclear palsy are occasionally indistinguishable from corticobasal degeneration (Case records of the Massachusetts General Hospital 1993). Isolated focal cortical deficits causing apraxia, aphasia, cortical sensory loss, or an alien limb may be caused by structural lesions, Alzheimer's disease, or a focal cortical degeneration, although the additional akinetic rigid features or gait disturbance in corticobasal degeneration point to a more diffuse neurological disorder usually allowing the correct diagnosis. The myoclonus of Creutzfeldt–Jacob disease is usually more widespread and the rate of progression faster than in corticobasal degeneration. Although the typical presentations of corticobasal degeneration and Pick's disease are clearly distinguishable, atypical cases of either may cause difficulty. Pick's disease typically presents with a frontotemporal dementia with prominent conduct disorder and personality change. Some atypical cases may have Parkinsonism, gait disturbances, and focal cortical deficits. Similarly, some unusual corticobasal degeneration cases present with a frontal dementia and little motor dysfunction (Bergeron *et al.* 1996). Thus there is some clinical and pathological overlap between Pick's disease and corticobasal degeneration. The precise relationship between these two disorders is unclear.

Investigations are usually unhelpful and the diagnosis of corticobasal degeneration remains clinical. CT or MR scans may show asymmetrical frontoparietal cerebral atrophy. Positron emission tomography studies (Brooks 2000) show impaired nigrostriatal dopaminergic function and asymmetrical reductions in frontoparietal and thalamic blood flow and metabolic activity, corresponding to the areas of greatest neuronal degeneration (Sawle *et al.* 1991; Blin *et al.* 1992). Proton magnetic resonance spectroscopy may also detect cortical and subcortical involvement (Tedeschi *et al.* 1997).

The prognosis of corticobasal degeneration is poor, with most patients surviving between 5 and 10 years (Wenning *et al.* 1998). No effective treatment is available although clonazepam may help the myoclonus and tremor. Levodopa is usually ineffective, as are dopamine agonists although amantadine may help apraxia in some patients (Mahapatra *et al.* 2004).

40.3.11 The problem of 'cerebrovascular Parkinsonism'

Cerebrovascular disease as a cause of 'Parkinsonism' has been a controversial concept. Patients with multiple subcortical lacunar infarcts develop a disturbance of gait, often with additional pseudobulbar and limb pyramidal signs, emotional lability and dementia, also termed subacute ateriosclerotic encephalopathy or Binswanger's disease. The legs show a curious inability to relax when examined, giving rise to an impression of rigidity referred to either as paratonia or as 'gegenhalten', or counterholding. There is usually a history of strokes, hypertension, or other vascular risk factors. The progression is step-wise with sudden deteriorations as well as periods of stabilization (Caplan 1995). This form of cerebrovascular disease has been recognized for many years and has been referred to as 'arteriosclerotic Parkinsonism' (Critchley 1929). At first glance, the shuffling gait is superficially similar to that seen in Parkinson's disease with small steps, start hesitation, and freezing and has been described as 'lower half Parkinsonism' (Thompson and Marsden 1987). It differs from Parkinson's disease because the gait is wide based and there is no festination or lack of arm swing. Mild rigidity and bradykinesia may be seen but this is symmetrical and there is no tremor or response to levodopa. Although Binswanger's disease has some similarities to Parkinson's disease, a distinction can usually be made on the basis of these clinical differences. Another helpful diagnostic point is that the gait disturbance in cerebrovascular disease is more severe than would be predicted from the physical signs detected on the examination couch. This has given rise to the notion of an apraxia of gait. Impaired function of the periventricular white matter of the frontal lobes is a feature of other disorders associated with this sort of gait, such as hydrocephalus, as well as cerebrovascular disease, 'etat lacunaire'. CT scanning shows deep white matter periventricular low density changes and MRI reveals multiple lacunar subcortical infarcts. However, such neuroimaging abnormalities are common in the elderly, and must be interpreted with caution; the diagnosis of Binswanger's disease should be clinical and not radiological (Bennett *et al.* 1990). Although the description of Binswanger's disease as cerebrovascular Parkinsonism has become unfashionable, it is not unreasonable providing that the clinical features originally described (Critchley 1929) are borne in mind and not confused with those of idiopathic Parkinson's disease. A common dilemma is an elderly patient with cerebrovascular risk factors, a gait apraxia, and features of Parkinsonism. The possibility of both conditions occurring together cannot be excluded and a therapeutic trial of levodopa is the best policy.

Very occasionally, cerebrovascular disease produces an akinetic rigid syndrome similar to idiopathic Parkinson's disease (Hunter *et al.* 1978). Not only has levodopa responsive Parkinsonism occurred (Mark *et al.* 1995) but the correct diagnosis is sometimes made only at postmortem (Hughes *et al.* 1993b). It is difficult to be certain how often cerebrovascular disease produces true Parkinsonism but such cases are probably rare.

40.3.12 Pallidal degenerations

These rare conditions are pathologically and clinically heterogeneous and there is no agreed classification. Onset may be in children or adults. They are often sporadic but autosomal recessive or dominant forms are well described (Jellinger 1968).

A progressive akinetic-rigid syndrome with or without additional dystonia is a characteristic phenotype. In some patients the dystonia is the dominant feature. Pathologically there may be a pure pallidal degeneration (Hunt 1917; Aizawa *et al.* 1991) or degeneration of the pallidum and various combinations of other areas of the basal ganglia, especially the subthalamic nucleus, substantia nigra and striatum. Occasionally, Parkinsonism follows striatal necrosis (Miyoshi *et al.* 1969; Caparros-Lefebvre *et al.* 1997) inherited as a recessive or dominant trait and due to mitochondrial dysfunction (Thyagarajan *et al.* 1995). These patients usually have prominent dystonia but Parkinsonism can occur.

Pallidopyramidal degeneration causes Parkinsonism and pyramidal tract signs such as spasticity and extensor plantar responses (Davison 1954). There is sometimes a response to levodopa, even in pathologically confirmed cases. However, some other reports of levodopa responsive pallidopyramidal disease without pathological evidence of the diagnosis are difficult to evaluate as it is difficult to exclude dopa responsive dystonia (Section 40.4.4).

The *Kufor–Rakeb syndrome* is an autosomal recessive pallidal degeneration with nigro-striatal and pyramidal involvement. The onset is in young adults with Parkinsonism, spasticity, vertical gaze paresis, and dementia. The disorder is linked to a locus at chromosome 1p36 (Hampshire *et al.* 2001).

40.3.13 Other akinetic rigid states

Hereditary disorders

These are shown in Table 40.4.

- There are *rare familial forms of Parkinsonism* without typical Lewy body pathology (Section 40.3.1).

- *Parkinsonism with depression and hypoventilation* occurs in adult life with prominent apathy and mood change as well as respiratory irregularities which may be fatal (Tsuboi *et al.* 2002; Lechevalier *et al.* 2005). A good response to levodopa may be seen but the prognosis is poor. Inheritance is autosomal dominant. Mutation of the alpha synuclein gene has been described suggesting that this may be a form of PARK1 (Table 40.4) (Spira *et al.* 2001).

- A dominantly inherited Parkinsonism, FTDP17, with a frontal type dementia and amyotrophy has been linked to chromosome 17q21 (Wilhelmsen *et al.* 1994). Tau gene mutations are responsible for this and similar families.

- *Rapid onset dystonia Parkinsonism* is a striking disorder with a fast onset over hours or days of Parkinsonism and dystonia (Brashear *et al.* 1996,1997). It is considered under the dystonias (Section 40.4) there may be additional features such as pyramidal signs, dysarthria, ophthalmoparesis, psychiatric features, and dementia (Wszolek *et al.* 1992; Brashear *et al.* 1996). It is autosomal dominant, and the gene is located on chromosome 17q21 (Wijker *et al.* 1996; Kramer *et al.* 1999).

- *Neurodegeneration with brain iron accumulation* or Hallervorden–Spatz syndrome is now also referred to as pantothenate kinase associated neurodegeneration. It is an autosomal recessive disorder causing degeneration principally of the pallidum and substantia nigra pars reticulata. The usual clinical features are severe dystonia, dementia and spasticity in childhood, sometimes with retinal degeneration (Section 40.4.12) An adult onset form has been described too in which the presentation was familial adult onset Parkinsonism (Jankovic *et al.* 1985).

- A hereditary *X-linked dystonia-Parkinsonism*, Lubag, occurs in the Philippines and causes prominent dystonia with Parkinsonism (Section 40.4.4).

- Parkinsonism may also occur in *Huntington's disease*, especially the juvenile Westphal variant (Section 40.5.2).

- *Wilson's disease* causes a large number of movement disorders including Parkinsonism. It must always be considered in patients with early onset, probably under 50 years, or atypical Parkinsonism. Additional physical signs include corneal Kayser Fleischer rings, unusual tremor, psychiatric abnormality, and liver dysfunction (Sternlieb *et al.* 1987). Patients in whom this condition is suspected should be screened by appropriate ophthalmological and biochemical testing (Section 40.8).

- *Dopa responsive dystonia* may cause prominent Parkinsonism, and may present as mild, easily treated Parkinsonism in older patients. It may closely simulate young onset Parkinson's disease, with prominent lower limb dystonia (Section 40.4.4). However, there is an excellent and sustained response to levodopa without the fluctuation and dyskinesia so common later in Parkinson's disease. This excellent response to small doses of levodopa may be the only way to establish the diagnosis.

- Parkinsonism is commonly seen in the *adult onset cerebellar degenerations*, both sporadic and inherited . The akinetic rigid features are usually accompanied by other signs such as cerebellar ataxia, pyramidal signs, and peripheral neuropathy. Parkinsonism is especially typical of Machado–Joseph disease due to mutations of the *SCA3* gene, but may occur in other forms of degenerative cerebellar ataxia (see Section 39.2). Autosomal dominant cerebellar ataxia caused by mutations of the *SCA2* gene can cause familial Parkinsonism (Shan *et al.* 2001).

- The *fragile X tremor ataxia syndrome* can cause Parkinsonism along with ataxia, tremor, and cognitive decline (Hagerman *et al.* 2003; Hagerman *et al.* 2004) (Section 39.9.7).

- There is an increased incidence of Parkinsonism in carriers of *autosomal recessive Gaucher's disease* genes (Aharon-Peretz *et al.* 2004; Eblan *et al.* 2005) (Section 10.4.6). In addition, Parkinsonism can occur in affected homozygous patients.

- Parkinsonism may be seen with *mitochondrial disorders* including a form due to mitochondrial DNA polymerase γ *POLG* mutations (Schapira 2006) (Section 10.5.6). These patients have ophthalmoplegia and are rare; *POLG* mutations have not been found in sporadic Parkinson's disease (Taanman and Schapira 2005).

Degenerative disorders

The hemiparkinsonism–hemiatrophy syndrome is characterized by unilateral Parkinsonism with a variable degree of atrophy of the same side of the body, sometimes just the hand or alternatively the face and limbs (Klawans 1981). The Parkinsonism may become bilateral but is strikingly asymmetrical and there is sometimes contralateral cerebral atrophy detected on CT scans. Onset is typically in adult life and there may be slow progression. The response to levodopa is variable. Dopa responsive dystonia (Section 40.4.4) can produce a similar phenotype (Greene *et al.* 2000).

Alzheimer's disease (Section 34.6.2) can cause Parkinsonism in addition to dementia. This occurs in advanced cases reflecting a number of factors including neuroleptic medication and coexistent Lewy body Parkinsonism. Some such patients have a non-Lewy body degeneration of the substantia nigra (Morris *et al.* 1989). Some patients do not have any nigral abnormality when studied by positron emission tomography or at postmortem, indicating some other extranigral process.

Structural disease

Hydrocephalus (Section 26.4.3) typically presents with headache or with a combination of impaired gait, cognitive decline, and urinary incontinence. The gait disorder is similar to that seen in Binswanger's disease but some patients have had Parkinsonism. This seems to be due to the effects of the hydrocephalus as well as coexistent Parkinson's disease or other forms of degenerative Parkinsonism (Curran and Lang 1994). Some of these patients have responded to levodopa. Space occupying lesions such as tumours or subdural haematomas may very rarely cause Parkinsonism, possibly due to compression of a cerebral peduncle or distortion of the basal ganglia. These cases have resembled Parkinson's disease initially and diagnosis can be delayed by the understandable lack of brain scanning in otherwise typical Parkinsonism.

Miscellaneous rare causes of Parkinsonism

Cerebral anoxia (Section 5.4.2) may cause delayed Parkinsonism and basal ganglia necrosis (Straussberg *et al.* 1993). Such patients are more likely to develop dystonia. Creutzfeldt–Jacob disease usually causes rapidly progressive dementia with myoclonus and ataxia but Parkinsonism may occur occasionally. Neuroacanthocytosis (Section 40.5.10) and Rett syndrome (Section 9.6.2) are distinctive disorders which can have akinetic rigid features. Parkinsonism also occurs in basal ganglia calcification. This may be secondary to hypoparathyroidism or it may be a hereditary disorder with normal calcium metabolism (Martinelli *et al.* 1993). Chronic liver failure, with repeated episodes of hepatic encephalopathy may be followed by the development of a chronic hepatocerebral degeneration, in which Parkinsonism is a major feature. Parkinsonism is occasionally psychogenic (Factor *et al.* 1995).

40.4 Dystonias

40.4.1 Diagnosis and classification

Dystonia has been defined as '*a syndrome of sustained muscle contractions, frequently causing twisting and repetitive movements or abnormal postures*' (Fahn *et al.* 1987). It is important to remember that dystonia is a type of involuntary movement and is no more a diagnosis than spasticity, ataxia, or any other physical sign although it is often used as a shorthand for its most common cause, primary torsion dystonia. Dystonic movements are typically twisting and involve proximal more than distal muscles (Fig. 40.13). The eyes may close forcefully, blepharospasm; there may be jaw and tongue movements, oromandibular dystonia; or the neck may twist to one side, forward or backwards, torticollis, antecollis, and retrocollis. If there is involvement of the bulbar muscles, there may be dysarthria, dysphagia, and impaired facial and tongue movements but no increased jaw jerk, the so-called 'striatal' pseudobulbar palsy. In the limbs, an arm will often twist and pronate and there may be spasm of the hand, especially with writing, writer's cramp, or other fine movements. Dystonia in a leg typically causes inversion and plantar flexion of the foot, often with extension of the big toe. This makes interpretation of the plantar response unreliable: the striatal pseudobabinski response or striatal foot. The trunk may be involved with tremor, scoliosis, extension or flexion spasms: axial dystonia. Dystonic movements are often *action specific*, so that the hand may assume an abnormal posture only when writing and not during other actions such as shaving or typing. The gait may be abnormal with dystonic trunk and limb spasms and postures but the patient may be able to run or walk backwards normally. Action specificity may erroneously suggest a psychiatric disorder but is typical of dystonia. Sometimes there is *overflow dystonia* in which dystonic movements appear in other body parts which are not normally involved in the execution of a particular movement. Curiously, many patients display sensory tricks or *gestes antagonistiques* by which the movements may be suppressed, such as touching the face to correct torticollis, chewing to correct oromandibular dystonia, or holding a stick to correct a dystonic gait. There is considerable variation; the movements may be mobile and repetitive or fixed into abnormal postures. Sometimes, repetitive dystonic spasms lead to a tremulous appearance, myorrthymia, or jerking similar to myoclonus. The diagnosis of dystonia is clinical and may be difficult, requiring considerable experience.

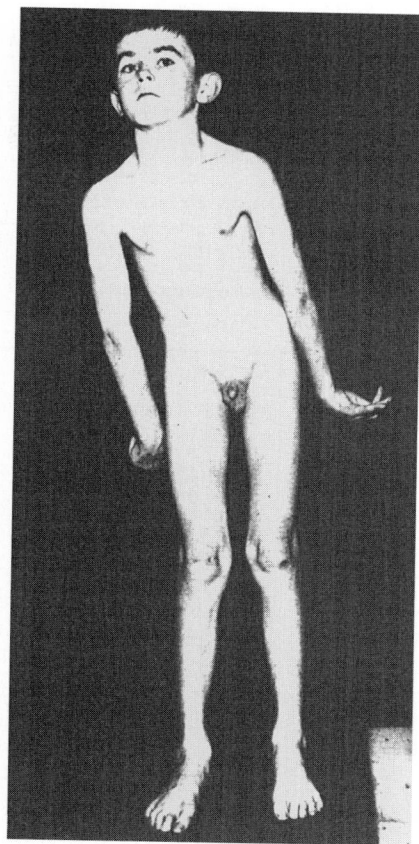

Fig. 40.13 A patient with generalized idiopathic torsion dystonia.

Dystonia is often described clinically in terms of its anatomical extent. *Generalized dystonia* affects the legs, trunk, and often other body parts; *segmental dystonia* involves adjacent areas of a body region such as the cranial musculature or an arm and the neck and *focal dystonia* affects only one site such as the neck, eyes, or a hand. *Hemidystonia* is usually due to a structural lesion of the contralateral basal ganglia. Dystonia can also be described in terms of *limb or non-limb* onset, usually focal or segmental axial, cranial or cervical onset.

Dystonia is often classified by age of onset; childhood, from 0 to 12 years, adolescent, between 13 and 20 years, and adults older than 21 years, types have been suggested but most cases fall into either *early onset*, before 26 years, or *older onset*, after age 26 years, types (Bressman 2004).

The most important classification of dystonia is by a combination of the types of dystonic disorders, dystonia genetics, and aetiology (Tables 40.12–40.14). In clinical practice, the main distinction is between primary torsion dystonia and any identifiable alternative cause. A secondary dystonia is more likely in those with childhood or adolescent onset and with generalized dystonia. Hemidystonia is due to a contralateral cerebral structural lesion in 80 per cent of cases (Marsden and Quinn 1990).

40.4.2 Primary torsion dystonia

Definition and prevalence

Also known as dystonia musculorum deformans, primary torsion dystonia causes a pure dystonic syndrome without any detectable

Table 40.12 Classification of dystonias

PRIMARY DYSTONIAS			
Autosomal dominant	*Mainly early / limb onset*	DYT1, others	
	Mixed types	DYT4, DYT6, DYT13, others	
	Mainly late / non-limb onset	DYT7, others	
Autosomal recessive	*Mainly early / limb onset*	DYT2	
Sporadic	*Mainly late onset focal*	Non-genetic	
SECONDARY DYSTONIAS			
Inherited	*'Dystonia plus'*	Dopa responsive dystonia	DYT5, DYT14, others
	Non-degenerative disorders	Myoclonus dystonia	DYT11, DYT15
		Rapid onset dystonia Parkinsonism	DYT12
	Degenerative	Autosomal dominant	Huntington's disease Spinocerebellar degenerations Dopa responsive dystonia Neuroferritinopathy Frontotemporal dementia e.g. FTDP17 Fahr syndrome Parkinson's disease (e.g. PARK1, 5, 8) Episodic choreoathetosis / spasticity CSE - DYT9 Paroxysmal dystonias DYT 8, DYT 10, others
		Autosomal recessive	Parkinson's disease (e.g. PARK2, 6, 7) Wilson's disease Neurodegeneration with brain iron accumulation Organic acidurias Gangliosidoses GM1 / GM21 Hexosaminidase A and B deficiency Lesch–Nyhan disease Neimann–Pick disease type C Metachromatic leucodystrophy Neuronal ceroid lipofuscinosis (Kuf's) Triosephosphate isomerase deficiency Hartnup disease Friedreich's ataxia Ataxia with vitamin E deficiency Ataxia telangiectasia Ataxia with ocular apraxia Others—see Table 40.14
		X-linked	X-linked dystonia-parkinsonism DYT3 Deafness-dystonia (Mohr–Tranebjaerg)
		Mitochondrial	Leber's disease Leigh's disease
Acquired		Parkinsonian disorders	Parkinson's disease Progressive supranuclear palsy Corticobasal degeneration Multiple system atrophy
		Other causes	Athetoid cerebral palsy Basal ganglia lesions e.g. stroke, tumour, AVM Cerebral anoxia Cerebral trauma Drugs Others— see Table 40.14

Table 40.13 Genetics of the dystonias

Dystonia gene designation	Locus	Inheritance	Clinical disorder	Gene
DYT1	9q34	Autosomal dominant	Primary torsion dystonia—early, limb onset	TorsinA
DYT2		Autosomal recessive	Primary torsion dystonia—1 family, early onset, rare	
DYT3	Xq13.1	X-linked recessive	X-linked dystonia Parkinsonism (Lubag)	Multiple transcript system
DYT4		Autosomal dominant	Primary torsion dystonia with whispering dysphonia	
DYT5	14q22.1	Autosomal dominant	Dopa responsive dystonia	GCH1
DYT6	8p	Autosomal dominant	Mixed form of primary torsion dystonia	
DYT7	18p	Autosomal dominant	Adult onset focal cervical primary torsion dystonia	
DYT8	2q	Autosomal dominant	Paroxysmal dystonic choreoathetosis	Myofibrillogenisis regulator
DYT9	1p21	Autosomal dominant	Episodic choreoathetosis / spasticity	
DYT10	16p-q	Autosomal dominant	Paroxysmal kinesiogenic choreoathetosis	
DYT11	7q21	Autosomal dominant	Myoclonus dystonia	Epsilon sarcoglycan gene (SGCE)
DYT12	19q13	Autosomal dominant	Rapid onset dystonia parkinsonism	Na/K ATPase α3
DYT13	1p36	Autosomal dominant	Mixed form of	
DYT14	14q13	Autosomal dominant	(non DYT5 linked)	
DYT15	18p11	Autosomal dominant	Myoclonus dystonia (non DYT11 linked)	

Key: GCHI, GTP cyclohydrolase 1

underlying cause clinically or after investigation and neuropathological abnormalities are absent. The concept of primary torsion dystonia was originally proposed to distinguish cases of dystonia without any obvious underlying cause from those due to Wilson's disease, lethargic encephalitis, or athetoid cerebral palsy (Herz 1944). It is the cause in most patients with dystonia.

The prevalence of primary torsion dystonia is approximately 330/million in the USA (Nutt *et al.* 1988). However, there is disagreement about prevalence due to methodological problems with completed surveys. The condition is more common in Ashkenazi Jews than would be expected by chance (Risch *et al.* 1995) almost certainly due to a genetic founder effect (Fletcher *et al.* 1990). Suggested prevalences are 111/million for early onset cases among Ashkenazi Jews in New York, 600/million for late onset primary torsion dystoniain northern England, and 3000/million for late onset dystonia in persons over 50 years of age in Italy (Defazio *et al.* 2004).

Diagnosis and clinical features

Although primary torsion dystonia may start in childhood, birth and early developmental milestones are normal and there should be no history of neonatal hypoxic-ischaemic encephalopathy or kernicterus. A detailed history is needed to distinguish primary from secondary dystonia caused by other neurological disorders or drugs (Table 40.14). The symptoms and level of disability are determined by the extent of the dystonia and the sites involved. Primary torsion dystonia is highly variable, ranging from severe disabling generalized dystonia in childhood, to mild focal dystonia appearing in later life. Although some patients can have a mild postural tremor, no other neurological abnormalities occur. Dystonia is worse with emotion or stress and improved by relaxation; it disappears in sleep.

There is a clear association between age, site of onset, and severity of dystonia. Generalized dystonia develops in the majority, 80 per cent, of childhood onset cases and about a third of adolescents. It is so rare in adult onset cases, occurring only in 1–2 per cent after age 20 years, that the diagnosis of primary torsion dystonia must be made only with great caution and after thorough investigation for a secondary cause. The great majority of patients with primary torsion dystonia starting in adult life develop segmental or focal dystonia (Marsden and Harrison 1974; Fahn *et al.* 1987).

The site of onset is also associated with age of onset and severity, so that early onset primary torsion dystonia usually starts in a leg, less commonly an arm. Later onset focal primary torsion dystonia usually starts in cranial or axial muscles or sometimes an arm. Consequently, most generalized cases start in childhood, focal cases appear mainly in adults in the 4th to 6th decades, and segmental cases seem to commence over a wide age range (Fletcher *et al.* 1990).

Initially there are intermittent action specific dystonic spasms but overflow dystonia (in other body areas during movement) and fixed dystonic postures may appear later. Affected children usually have a dystonic gait which may appear bizzare; a typical presentation at this age is inturning of a foot or a peculiar axial posture when walking or running. It is important to recognize the dystonic nature or action specificity of the movement to avoid a misdiagnosis of psychiatric disease. Subsequent progression to generalized dystonia is gradual over several years. Generalized primary torsion dystonia is severely incapacitating, leading to loss of walking and

Table 40.14 Secondary dystonias .The principal causes of secondary dystonia are shown in bolditalics. Other causes are rare

	Conditions
Rare metabolic disorders	***Wilson's disease*** [1]
	Caeruloplasmin deficiency
	Organic acidurias [1]
	GM1 gangliosidosis
	GM2 gangliosidosis
	Hexosaminidase A and B deficiency
	Lesch–Nyhan disease
	Neimann–Pick disease type C
	Metachromatic leucodystrophy
	Pelizaeus–Merzbacher disease
	Homocystinuria
	Neuronal ceroid lipofuscinosis (Kuf's)
	Triosephosphate isomerase deficiency
	Hartnup disease
Mitochondrial disorders	Leigh's disease [1]
	Leber's disease [1]
	Familial striatal necrosis [1]
Hereditary conditions	***Huntington's disease***
	Hereditary cerebellar degenerations
	Hallervorden Spatz disease [1]
	Pallidal degenerations
	Neuroacanthocytosis
	Fahr syndrome
	Ataxia telangiectasia
Degenerative disorders	***Parkinson's disease***
	Multiple system atrophy
	Steele–Richardson–Olszewski disease
	Corticobasal degeneration
	Rett syndrome
Drug induced dystonia	***Neuroleptics*** (tardive dystonia)
	Levodopa (in Parkinson's disease)
	Others
Acquired disorders	***Athetoid cerebral palsy*** [1]
	Basal ganglia lesions e.g. stroke, tumour, arteriovenous malformation
	Cerebral anoxia [1]
	Cerebral trauma
	Toxins [1] e.g. Manganese, CO, CS_2, methanol
	Osmotic myelinolysis
	Multiple sclerosis
	Hypoparathyroidism
	Encephalitis
	Haemolytic uraemic syndrome
	Wasp sting encephalopathy [1]
	Brainstem lesions
	Cervical cord lesions
	Peripheral trauma [2]

1 = associated with low density basal ganglia lesions on CT/MRI ; 2 = probably exacerbates or precipitates idiopathic torsion dystonia.

difficulty with upper limb function, speech, and swallowing. Such children may erroneously be thought to have athetoid cerebral palsy. Cognitive abilities, sphincter control, vision, hearing, sensory function and the peripheral nerves are not affected and seizures do not occur. Eventually, speech and swallowing are affected but this is rare in the early stages. Once generalized dystonia has developed

it is stable. Relentless and continuous deterioration is not seen in primary torsion dystonia and should suggest a secondary cause. In some cases there is deterioration due to additional pathology such as cervical spondylotic myelopathy in those with prominent cervical involvement.

In adults the usual presentations are with writer's cramp, torticollis, or one of the other cranial focal dystonias. In most cases the dystonia remains localized and focal but in some there may be gradual progression over 5–10 years to contiguous body areas so that a segmental pattern evolves. Clinically, the various forms of focal primary torsion dystonia are fragments of generalized dystonia:

◆ *Torticollis*, involves intermittent or fixed turning of the head to one side. Sometimes the head moves backwards, forward or laterally. In some patients this is a forceful slow movement, but in others it is tremulous. In such cases, the true nature of the tremor is indicated by an additional tilt or turn of the head due to the dystonia. Pain in the neck or shoulder is common and the torticollis may be more obvious when standing and walking than at rest.

◆ *Blepharospasm* causes forceful closure of the eyes with spasm of the orbicularis oculi muscles (Grandas *et al.* 1988). The eyes appear screwed up and the patient is unable to open them. This often occurs in bright sunlight or when watching television. In some patients the spasms are confined to the pretarsal section of the orbicularis oculi muscles so that the eyes close lightly, levator inhibition, or apraxia of eyelid opening (Elston 1992; Krack and Marion 1994).

◆ *Oromandibular dystonia* causes spasms of the jaw, tongue, and mouth. The jaw may open or forcefully close and the lip movements cause embarrassing facial movements. Chewing and eating are difficult along with speech. If part of a cranial segmental dystonia with cervical involvement or blepharospasm, this condition is referred to as Meige's disease or *Bruegel's syndrome*.

◆ *Laryngeal dystonia* causes altered speech but with normal appearance of the larynx and cords at rest. Such patients are often thought to have a psychogenic disorder after a series of normal otorhinolaryngological examinations. The vocal cords may go into adductor spasm with a strained strangulated speech or forceful abduction which causes a quiet whispery voice. The problem may be permanent or intermittent and some patients are still able to sing or whisper normally.

◆ *Writer's cramp* is the commonest upper limb dystonia. During writing there are involuntary movements of the fingers or wrist. Sometimes the whole arm adopts a dystonic posture. Writing may become impossible and some patients are forced to type or use the other hand which may itself become affected subsequently. This may affect writing alone, simple writer's cramp, or other actions such as shaving, dystonic writer's cramp. Similar action specific upper limb focal dystonia may be seen in typists, golfers, darts palyers, or musicians.

◆ *Axial dystonia* causes twisting, flexion, or extension of the spine, often worse when standing or walking but disappearing when lying supine. In some patients the trunk movement may be tremulous (Rivest and Marsden 1990). In all these forms of primary torsion dystonia remissions are uncommon and not usually sustained.

Genetics and aetiology

No consistent biochemical or pathological abnormality has been detected in primary torsion dystonia. Although altered dopaminergic function and basal ganglia lesions are frequently implicated in secondary dystonia, these have not been detected reliably in primary torsion dystonia. Some alterations of catecholamine levels have been reported (Hornykiewicz et al. 1988) but the significance of these is unclear and further studies are required. Positron emission tomography scanning has suggested mild alteration of nigrostriatal dopaminergic function but this is not marked (Playford et al. 1993). A basal ganglia mechanism underlying the dystonia is assumed on the basis of the frequent location of causative lesions in secondary dystonia but brainstem dysfunction may also be responsible (Zweig et al. 1988) and other areas of the central nervous system are sometimes implicated in secondary dystonia (Section 40.2.2).

Clinical observations and an increased incidence among Ashkenazi Jews have long indicated a genetic basis for the disorder. About 50 per cent of patients with generalized or segmental primary torsion dystonia have affected relatives (Fletcher et al. 1990). It is now established that in the majority of generalized and segmental cases or those affected as children or young adults, primary torsion dystonia is caused by an autosomal dominant gene with 30–40 per cent penetrance and highly variable expression. A recessive form of primary torsion dystonia, DYT 2, has been proposed (Table 40.13) but is unlikely. Similar inheritance has been demonstrated in Jewish and non-Jewish patients (Bressman et al. 1989; Fletcher et al. 1990). The higher incidence in the former is probably due to a genetic founder effect with a high gene frequency in Eastern Europe centuries ago (Fletcher et al. 1990). In families containing at least one case of early onset primary torsion dystonia, affected individuals may show only mild late onset focal dystonia or tremor and some obligate gene carriers are asymptomatic. There is also an increased family history of tremor and stuttering in such families suggesting that these may be minor manifestations of autosomal dominant primary torsion dystonia (Fletcher et al. 1991a). Sporadic cases are mainly due to new dominant mutations and inheritance from asymptomatic gene carrier parents; about 15 per cent may be non genetic phenocopies.

In the majority of families with early onset primary torsion dystonia the disorder is linked to a locus, DYT1 on chromosome 9q34 (Kramer et al. 1990; Warner et al. 1993). This gene has been cloned and sequenced revealing a 3 base pair CAG deletion in a gene encoding an ATP binding protein, TorsinA (Ozelius et al. 1997). TorsinA is a AAA+ protein; AAA+ proteins are a family of P-loop ATPases involved in the cellular processing of macromolecules (Ammelburg et al. 2006). Normally present on the endoplasmic reticulum, mutant TorsinA locates to the nuclear envelope (Goodchild and Dauer 2004). Neuropathological studies of primary torsion dystonia have shown perinuclear inclusions in midbrain reticular and periaqueductal grey matter (McNaught et al. 2004a).

Primary torsion dystonia can be considered as DYT1 related, early onset non-DYT1 and late onset non-DYT1:

- In *DYT1 related forms*, most patients have early onset, earlier than 26 years, dystonia; 65 per cent of such patients develop generalized or multifocal dystonia within 5 years (Edwards et al. 2003); 10 per cent are segmental, 25 per cent are focal, and cranial dystonia is unusual in less than 15 per cent of cases. DYT1 primary torsion dystonia accounts for 80 per cent of early onset primary torsion dystonia in Ashkenazi Jewish patients and less than 50 per cent among non-Jewish cases (Valente et al. 1998; Lebre et al. 1999; Bressman 2004;).

- In *early onset non-DYT1 forms*, there is a more variable range of dystonia with more cranial and cervical focal and segmental involvement; there are also more adult or late onset cases in these families. Families show a mixed pattern of early onset generalized and later onset focal or segmental cases (Bressman 2004). Some families have DYT6 or DYT13 genes (Almasy et al. 1997; Valente et al. 2001) but in most cases the genetic basis is unknown.

- In *late onset non-DYT1 forms*, adult onset segmental or focal primary torsion dystonia is more common and there is more cervical and cranial dystonia. The DYT1 gene is mostly not responsible for focal primary torsion dystonia in Jewish patients (Gasser et al. 1996). In Germany, autosomal dominant torticollis is associated with a gene, DYT7, on chromosome 18p31 (Leube et al. 1997) and there is evidence that a small number of patients with sporadic torticollis have inherited the same gene. Two studies have shown that approximately 25 per cent of focal cases have affected relatives compared to 50 per cent of those with generalized or segmental disease (Waddy et al. 1991; Stojanovic et al. 1995). Accordingly, focal primary torsion dystonia may be due to genes with lower penetrance or is more heterogeneous with a higher proportion of non genetic cases possibly caused by environmental factors. Late onset non-DYT1 primary torsion dystonia has been associated with a polymorphism of the D5 receptor gene but this has not been confirmed (Sibbing et al. 2003) although an association with a DYT1 haplotype is possible (Clarimon et al. 2005).

Overuse of a body part such as in musicians or manual workers has long been associated with the development of focal dystonia in the body part concerned. Such patients do not usually have the DYT1 mutation (Friedman et al. 2000). Moreover there is a long recognized association between dystonia and preceding local trauma (Fletcher et al. 1991b). This cannot be due to trauma alone because a history of injury is no more common among primary torsion dystonia patients than controls. An interaction between trauma and a pre-existing possibly genetic predisposition seems possible. Examples of local injury or pain followed by dystonia include limb dystonia after fractures and lacerations, oromandibular dystonia after dental surgery, blepharospasm after eye disease, and torticollis after neck injuries. It should be noted that existing primary torsion dystonia may be exacerbated by surgical procedures and orthopaedic correction of dystonic deformity is therefore unwise (Fletcher et al. 1991b).

Investigations and diagnosis

The diagnosis of primary torsion dystonia is clinical but investigations are required to exclude other causes of dystonia, particularly Wilson's disease in some patients (Fahn et al. 1987). Athetoid cerebral palsy is excluded by enquiring about the perinatal history and subsequent motor developmental delay. Many other causes of secondary dystonia are also readily apparent from the patient's history which should not suggest any other cause, especially exposure to neuroleptic drugs. In primary torsion dystonia there are no other neurological abnormalities, intellect is preserved and seizures do not occur. As has been mentioned, some patients do have a mild postural tremor.

Some features of a dystonia suggest a secondary cause rather than primary torsion dystonia:

- lower limb onset or generalization with onset after the age of 20 years;
- rapid progression or generalization;
- relentless deterioration without stabilization;
- marked fluctuation or intermittent symptoms;
- severe pain;
- early speech or swallowing involvement;
- hemidystonia;
- an onset with fixed dystonic postures; and
- any additional neurological features.

In patients with typical primary torsion dystonia, the degree of investigation will depend on age and severity. In younger patients, under 50, it is essential to exclude Wilson's disease by serum copper and caeruloplasmin estimations, urinary copper excretion and a slit lamp examination for Kayser Fleischer rings. If there is any doubt about the results a liver biopsy is indicated (Section 40.8). Most younger patients need MR brain scanning to exclude focal lesions, leucodystrophies, basal ganglia abnormalities (as seen in organic acidurias, mitochondrial disorders, neuroferritinopathy, pantothenate kinase deficiency) a blood count and film for acanthocytes, liver function tests, and calcium estimation. Older patients with typical late onset focal primary torsion dystonia probably do not reqire investigations. If there are atypical features to suggest a secondary dystonia, regardless of age, detailed investigation is required (Table 40.15) to look for a secondary cause.

Particular care is needed with torticollis. Occasionally this is not dystonic and is due to atlantoaxial rotatory subluxation (Subach *et al.* 1998). This can occur after trauma but this may be very slight and sometimes the condition is spontaneous (Grogaard *et al.* 1993). Cervical spine imaging is required in cases of acute or fixed torticollis to exclude this, especially if there is preceding trauma. Some cases of cervical dystonia can follow head or neck inflammation or infection of spinal structures or soft tissues, *Grisel's syndrome* (Berry and Moriarty 1999). Such cases can follow pharyngitis and occur in epidemics in some countries (Tsai *et al.* 1983). The neck dystonia can be accompanied by involvement of the tongue, mouth, and mandible.

Treatment

Oral drugs. It cannot be overemphasized that the diagnosis of primary torsion dystonia must not be made in younger patients without a therapeutic trial of levodopa to exclude dopa responsive dystonia (Section 40.4.4). An adequate trial consists of 600 mg of levodopa per day, combined with a decarboxylase inhibitor, for at least 3 months. If there is no response, treatment for primary torsion dystonia can be initiated. High dose anticholinergic therapy, usually with trihexphenidyl (benzhexol) is the treatment of choice with over 50 per cent of patients obtaining significant improvement (Burke *et al.* 1986). The dose of benzhexol is initially very small with gradual slow increases over several months. Doses of up

Table 40.15 Investigations in suspected secondary dystonia

MR brain scan
Nerve conduction studies
Evoked potentials (if clinically indicated)
Copper & caeruloplasmin levels 24-h urinary copper excretion Slit lamp examination for Kayser Fleischer rings
Biochemistry screen including calcium and liver function Blood count and film for acanthocytes CK level Alpha fetoprotein Immunoglobulins White cell enzymes Blood and urinary amino acids Urinary organic acids Blood gases, lactate, pyruvate CSF examination and lactate level Skin and rectal biopsies (storage diseases) Skin fibroblasts for filipin stains Muscle biopsy (mitochondrial diseases) DNA testing (mitochondrial diseases, Huntington's, *DRPLA*, *PANK2*, *SCA* mutations)

These investigations are intended as a guide only. The tests required in individual cases will be influenced by clinical probabilities and the age of the patient.

DRPLA = dentatorubropallidoluysian atrophy; PANK2 = pantothenate kinase 2; SCA = spinocerebellar ataxia.

to 120 mg per day may be required and a useful response is unlikely below 30 mg daily. The drug is tolerated better in children than adults due to dose related side effects including sedation, dry mouth and blurred vision. Occasionally patients develop chorea (Nomoto *et al.* 1987). A better response is seen in patients treated soon after the onset of the dystonia (Greene *et al.* 1988). In patients with severe generalized dystonia, 'triple therapy' with a combination of high dose benzhexol with tetrabenazine and a dopamine receptor blocker such as pimozide (Manji *et al.* 1998) may be effective but at the expense of side effects such as Parkinsonism and depression (Marsden *et al.* 1984a). Other drugs are of limited value. Baclofen may be helpful in some patients. It is more effective in children (Greene 1992) than adults. The effective dose is 40–180 mg per day and gradual titration is needed; abrupt withdrawal is inadvisable due to potential adverse effects. There are occasional reports of improvement with benzodiazepines, mainly clonazepam which can be effective in 15 per cent of cases, but these drugs have not been evaluated in great detail (Greene *et al.* 1988). Tetrabenazine alone may help in some cases (Jankovic and Orman 1988) but side effects are common, especially sedation and depression. In some very severely affected patients with generalized dystonia, intrathecal baclofen has been effective (Albright 1996; Paret *et al.* 1996) but in other patients the results were unimpressive (Ford *et al.* 1996). The effect seems variable but benefit may be seen in either primary or secondary dystonias (Walker *et al.* 2000; Hou *et al.* 2001).

Intramuscular botulinum toxin. This is the treatment of choice for focal primary torsion dystonia. The toxin binds to presynaptic cholinergic nerve terminals and cleaves proteins required for fusion of acetylcholine containing vesicles and the presynaptic membrane (Blasi *et al.* 1993; Tintner and Jankovic 2001). This leads to chemically induced denervation of the injected muscle. Weakness and atrophy appear after several days and the effect lasts for 3–6 months depending on the dose, the site treated, and the severity of the dystonia.

This treatment has been used for many years in a wide range of focal dystonias (Greene *et al.* 1994). In blepharospasm, over 90 per cent of patients improve for up to 4 months after periocular injections. Ptosis may occur in about 10 per cent of cases but is usually temporary; bruising and dryness of the eyes are occasionally troublesome. Some patients with pretarsal blepharospasm fail to respond unless the pretarsal orbicularis oculis muscle is specifically injected.

In torticollis, 70–80 per cent of patients are improved in terms of reduced dystonia and relief of local muscular pain. The pattern of the head movement will determine the sites of injection. The effect lasts 3–4 months after which the treatment is repeated. The results with botulinum toxin are superior to anticholinergic therapy (Brans *et al.* 1996). Side effects include dysphagia in 10–20 per cent of cases but this is usually mild and resolves within 2 weeks. Occasionally patients develop severe dysphagia requiring nasogastric feeding in hospital. All patients must be warned of this possibility and offered prompt review if swallowing is badly affected. Some patients develop antibodies to the toxin which usually causes loss of effect. This is more likely after larger doses and more frequent injections. Accordingly the minimum effective toxin dose should be used and patients should be treated as infrequently as their dystonia permits. In patients who become resistant to type A toxin, an immunologically different type B toxin may be effective in a third (Barnes *et al.* 2005).

Botulinum toxin injections may also improve writer's cramp (Cole *et al.* 1995; Wissel *et al.* 1996). Although most studies have used electromyographically guided injections, good results have been obtained without (Rivest *et al.* 1991). Although the dystonia is reduced, weakness of the arm is common. Electromyographically guided laryngeal injections have been effective in laryngeal dystonia (Greene *et al.* 1994) but adductor spasms respond better than whispering abductor dystonia. Other primary torsion dystonia patients with oromandibular dystonia, and dystonia of the lower limbs and axial muscles have been treated successfully with local botulinum toxin but these are not licensed indications.

Surgery. Some patients who are not improved by drugs or botulinum toxin may respond to stereotactic thalamotomy (Andrew *et al.* 1983; Cardoso *et al.* 1995) but the results are unpredictable and this is no longer employed. In patients with torticollis who do not respond to botulinum toxin, selective surgical denervation of the cervical muscles has been effective (Bertrand 1993). About half of patients can obtain useful benefit (Krauss *et al.* 1997; Ford *et al.* 1998).

There is increasing interest in bilateral pallidal deep brain stimulation (Kumar *et al.* 1999; Lozano and Abosch 2004) in both primary and some secondary dystonias. Several groups have shown excellent results in patients with primary torsion dystonia; results in secondary dystonias are less dramatic but nevertheless significant. Severity of dystonia is usually assessed by motor scores and functional disability e.g. the Burke–Fahn–Marsden Dystonia rating Scale (Burke *et al.* 1985). In a large European series (Cif *et al.* 2003) motor scores were reduced by 71 per cent in DYT1 primary torsion dystonia, 74 per cent in non-DYT1, and 31 per cent in secondary dystonia; functional scores were reduced in these groups by 63 per cent, 43 per cent, and 7 per cent respectively. Another smaller series (Yianni *et al.* 2003) reported motor reductions of 43 per cent in generalized primary torsion dystonia, 60 per cent in focal

primary torsion dystonia, and 37 per cent in secondary dystonia. Several groups have reported similar results (Bereznai *et al.* 2002; Vesper *et al.* 2002; Krauss *et al.* 2003; Kupsch *et al.* 2003). In a blinded trial of 22 patients, 14 obtained greater than 50 per cent improvement with 75 per cent improvement in a third; 2 children with secondary dystonias were worsened (Vidailhet *et al.* 2005). Cervical primary torsion dystonia also responds well to pallidal deep brain stimulation (Krauss *et al.* 2002; Yianni *et al.* 2003; Eltahawy *et al.* 2004b; Goto and Yamada 2004). Oromandibular primary torsion dystonia may also improve (Houser and Waltz 2005; Opherk *et al.* 2006).

40.4.3 X-linked dystonia (DYT3)

A form of X-linked dystonia, DYT3, has been reported in the Philippines. Affected men usually develop focal dystonia at 30–45 years of age and this usually generalizes after a few years. There is prominent dystonia of the legs and often the axial and cranial areas with Parkinsonism and tremor (Lee *et al.* 1976). Females are occasionally affected but not as severely. Pathologically there is gliosis and neuronal loss in the striatum and positron emission tomography studies reveal impaired striatal metabolism but normal nigrostriatal dopaminergic function (Eidelberg *et al.* 1993b). The gene has been located on chromosome Xq13 (Haberhausen *et al.* 1995) and the molecular basis of the disorder is probably mutation within a multiple transcript system (Nolte *et al.* 2003). Treatment is ineffective.

40.4.4 Dopa responsive dystonia (DYT5)

This is a rare disorder, probably accounting for 5–10 per cent of all childhood onset dystonia with a population frequency of about 0.5 per million (Nygaard 1995). It is nevertheless, one of the most important clinical diagnoses in clinical neurology as it is effectively cured by small doses of levodopa, even after many decades of severe neurological disability. Dopa responsive dystonia occurs in women three times more frequently and more severely than in men. Patients often have a family history suggesting autosomal dominant inheritance; penetrance of the Dopa responsive dystonia gene is approximately 45 per cent in women and 15 per cent in men. Typically the onset is in the first decade but some patients have a congenital onset with delayed motor milestones and others develop dopa responsive dystonia in adolescence or adult life. In children dystonia of the legs and feet causes stiffness of the lower limbs, inturning of the feet, and toe walking (Nygaard 1995). The gait may appear dystonic or spastic and pyramidal signs including spasticity, ankle clonus, and extensor plantar responses are often present along with mild Parkinsonian cogwheel rigidity and bradykinesia of the upper limbs. Most patients gradually progress to generalized dystonia but other neurological features such as cognitive impairment or seizures do not occur. In some childhood onset cases the dystonia is mild and remains confined to the feet and others have focal dystonia such as torticollis. In older patients there is often mild Parkinsonism (Nygaard *et al.* 1992), including rest tremor, showing an unusually good and sustained response to small doses of levodopa (Nygaard 1995). Diurnal fluctuation of symptoms with improvement in the morning or after sleep and increasing dystonia later in the day or after exercise is characteristic of dopa responsive dystonia but may be absent. Some atypical presentations include exercise related dystonia (Deonna *et al.* 1997), muscular weakness (Kong *et al.* 2001), scoliosis (Furukawa *et al.* 2000), oromandibular dystonia (Steinberger *et al.* 1999), and myoclonic dystonia (Leuzzi *et al.* 2002).

Investigations are unhelpful in this condition; cerebral imaging is normal and correct diagnosis depends entirely on a high index of suspicion and readiness to deploy a therapeutic trial of levodopa in cases where dopa responsive dystonia is even a remote clinical possiblility. There is a particular danger of an erroneous diagnosis of athetoid or spastic diplegic cerebral palsy (Nygaard *et al.* 1994) or hereditary spastic paraplegia from which the dystonia cannot always be reliably distinguished. This means that dopa responsive dystonia must be considered in any child or young adult with dystonia or unexplained paraparesis or gait disorder. The response to levodopa is dramatic; dystonia is rapidly abolished even after decades without treatment. Although many patients require only 50–100 mg of levodopa per day, with a decarboxylase inhibitor, this is variable; accordingly it is advisable to try at least 600 mg per day for 3 months before concluding that the patient does not have dopa responsive dystonia.

A further diagnostic difficulty is the distinction between dopa responsive dystonia and early onset Parkinson's disease in which prominent lower limb dystonia and Parkinsonism may also occur (Section 40.3.3). In dopa responsive dystonia, significant motor fluctuations and dyskinesias never develop whereas they appear at an early stage after starting levodopa in Parkinson's disease. Accordingly this distinction cannot always be made with confidence until the response to levodopa has been observed for a few years. It should be noted that a few patients with dopa responsive dystonia develop mild chorea on levodopa but this often disappears if the dose is reduced. An alternative but not widely available method is to examine the dopaminergic nigrostriatal system with positron emission tomography; this is normal in the dystonia but abnormal in Parkinson's disease (Nygaard *et al.* 1992; Naumann *et al.* 1997).

Biochemically dopa responsive dystonia is characterized by a striking reduction in central nervous system dopamine metabolism with reduced CSF dopamine metabolites and tetrahydrobiopterin, BH4 a cofactor for tyrosine hydroxylase-tetrahydrobiopterin (Le Witt *et al.* 1986). Pathologically there are normal numbers of nigral neurones but these are hypopigmented (Rajput *et al.* 1994); striatal dopamine is reduced with normal levels of tyrosine hydroxylase immunoreactivity but reduced enzyme acticvity. The condition is genetically heterogeneous and is caused by a large variety of point mutations of the GTP cyclohydrolase 1, *GCH1* gene (Ichinose *et al.* 2000) which is located on chromosome 14q (Nygaard *et al.* 1993). GCH1 is required for tetrahydrobiopterin, BH4, synthesis, hence the reduced tetrahydrobiopterin activity in the disorder. About 80–90 per cent of patients with dopa responsive dystonia have a detectable *GCH1* mutation (Hagenah *et al.* 2005). Some cases are caused by compound heterozygous mutations affecting the *GCH1* gene (Jarman *et al.* 1997). Other dopa responsive dystonia families do not have a detectable mutation in the *GCH1* gene (Bandmann *et al.* 1996) and the genetic basis of these cases is uncertain. One other family has been linked to a locus at 14q13(DYT14) (Grotzsch *et al.* 2002). There are very rare autosomal recessive forms of dopa responsive dystonia caused by mutations of the genes for tyrosial hydroxylase (Bartholome and Ludecke 1998; Furukawa 2003), 6-pyruvoyltetrahydrobiopterin synthetase (Hanihara *et al.* 1997), sepiapterin reductase (Steinberger *et al.* 2004), and aromatic acid decarboxylase (Swoboda *et al.* 2003). In these disorders the dystonia is more severe, with

onset in infancy or early childhood and with additional features such as seizures, hypotonia, developmental delay, and oculogyric crises.

Treatment is with small doses of levodopa in Sinemet or Madopar; sometimes larger doses are needed (Steinberger *et al.* 1999). Motor fluctuations and dyskinesia do not occur in dopa responsive dystonia—in contrast to early onset Parkinson's disease, especially if PARK2/Parkin related in which dystonia may be prominent but in which motor complications eventually develop.

40.4.5 Hereditary myoclonus-dystonia (DYT 11, DYT15)

This condition (hereditary myoclonus-dystonia) probably includes disorders previously referred to as familial or benign essential myoclonus (Fahn and Sjaastad 1991), myoclonic dystonia and hereditary dystonia with lightning jerks responsive to alcohol (Quinn 1996). Some patients with what appears to be primary torsion dystonia can have additional jerky myoclonic movements (Obeso *et al.* 1983) but some of these patients may have myoclonus dystonia. Onset is usually in childhood or early adult life with myoclonic jerks or jerky shock-like dystonic movements mainly of the neck, upper arms, and trunk (Asmus and Gasser 2004). There are additional dystonic features such as torticollis or writer's cramp in two-thirds. Lower limb involvement is rare. The diagnostic clue is the jerkiness of movements and their upper body emphasis. The movements often improve after alcohol. Psychiatric features are common including anxiety, depression, alcoholism, and obsessive-compulsive disorder (Saunders-Pullman *et al.* 2002). Inheritance is consistent with an autosomal dominant mechanism with reduced penetrance; paternal transmission leads to 90 per cent penetrance whereas maternal inheritance with only 5–10 per cent. The most common myoclonus dystonia gene, *DYT11*, has been mapped to 7q21 (Nygaard *et al.* 1999; Vidailhet *et al.* 2001); various mutations of the epsilon sarcoglycan, *SGCE*, gene at this locus cause 65 per cent of familial cases (Zimprich *et al.* 2001). Some families do not appear to have a 7q21 linked disorder and at least one other locus at 18p11, *DYT15* has been identified (Grimes *et al.* 2002). A similar phenotype has also been described with mutations of the *GCH1* gene (Section 40.4.4) (Leuzzi *et al.* 2002). Treatment is difficult; anticholinergics and clonazepam may help but drugs for myoclonus such as valproate, levetiracetam, and piracetam are ineffective (Asmus and Gasser 2004). Botulinum toxin may be considered for troublesome cervical dystonia. The condition may be improved with pallidal deep brain stimulation in severe cases (Cif *et al.* 2004; Magarinos-Ascone *et al.* 2005).

40.4.6 Rapid onset dystonia-Parkinsonism (DYT12)

Rapid onset dystonia-Parkinsonism is an autosomal dominant disorder causing dystonia and Parkinsonism, unusually acutely over hours or subacutely over days or weeks (Dobyns *et al.* 1993). Onset is from 5–55 years, can follow stress, intense exertion, or intercurrent illness and is often followed by long-lasting stability. The dystonia is usually of cranial, cervical, or upper limb type and may occur alone before later onset of Parkinsonism (Brashear *et al.* 1997). In some cases the onset may be more gradual (Brashear *et al.* 1996) and others have gradual onset before sudden deterioration later. Some patients have mainly upper body craniocervical dystonia without Parkinsonism (Kabakci *et al.* 2005). Psychiatric features or dysarthria may also occur (Pittock *et al.* 2000). The gene, *DYT12*, has been linked to chromosome 19q13 (Kramer *et al.* 1999; Zaremba *et al.* 2004) and rapid onset dystonia—Parkinsonism is now known to be associated with mutaions of the Na+/K+—ATPase alpha3 gene, *ATP1A3*, at this locus (de Carvalho Aguiar *et al.* 2004). Pathological examination of one case showed no clear neuropathological change (Pittock *et al.* 2000) suggesting that rapid onset dystonia-Parkinsonism is not a full neurodegeneration. AntiParkinsonisn therapy is usually ineffective.

40.4.7 Paroxysmal dystonias (DYT8, 9, 10, and others)

This is a heterogeneous group of disorders in which there are attacks of involuntary movements and complete recovery in between the episodes. The movements may be dystonic but are sometimes choreiform, ballistic, or mixed. For convenience they will be discussed with the dystonias. In some of the disorders the attacks are brief while others are more prolonged; some are triggered by movement while others are set off by other factors and paroxysmal dyskinesias may be familial or sporadic. In some cases, an underlying cause such as a focal brain lesion is apparent wheras others are idiopathic. Although other classifications have been proposed (Demirkiran and Jankovic 1995) the most clinically useful approach is to divide the paroxysmal dyskinesias as follows:

- brief attacks typically induced by movement: paroxysmal kinesigenic choreoathetosis;

- more prolonged, usually non-kinesigenic attacks: paroxysmal dystonic choreoathetosis;

- an intermediate form with longer attacks induced by prolonged exercise; and

- paroxysmal/episodic choreoathetosis with spasticity.

Paroxysmal nocturnal or hypnogenic dystonia is probably a form of nocturnal frontal lobe epilepsy (Fish 1994).

Paroxysmal kinesigenic choreoathetosis (DYT10)

This condition usually develops in childhood or adolescence and is more common in females. The attacks are brief, lasting seconds or a few minutes and may occur many times a day (Kertesz 1967). There is typically a prominent flurry of dystonic or choreiform movements which may be preceded by an altered sensation in the affected area such as stiffness or tingling. The movements may be unilateral or bilateral and although there may be falling, this is unusual. The majority of patients wth paroxysmal kinesigenic choreoathetosis report that the episodes are triggered by sudden movement or a startle such as suddenly running or jumping up quickly from a seat. Sometimes the movement trigger is absent or hyperventilation or anxiety may initiate attacks. There is no disturbance of consciousness or residual symptoms after the attack. There is a short refractory period during which another attack cannot be induced. A good response to carbamazepine or other anticonvulsants is characteristic and usually brings paroxysmal kinesigenic choreoathetosis easily under control (Fahn 1994a; Wein *et al.* 1996). The majority of cases are idiopathic and sporadic or familial with an autosomal dominant mode of inheritance. Investigations in such cases are all normal, including the EEG. Rarely paroxysmal kinesigenic choreoathetosis is secondary to some other underlying cerebral condition such as multiple sclerosis, head injury, stroke, or congenital cerebral malformations (Fahn 1994a). Most of the responsible lesions are in the thalamus,

internal capsule or basal ganglia but paroxysmal kinesigenic choreoathetosis may also follow brainstem or cervical cord lesions (Cosentino *et al.* 1996; Riley 1996). The attacks of paroxysmal kinesigenic choreoathetosis in multiple sclerosis often follow hyperventilation and are in many respects similar to the tonic spasms characteristic of that condition. There has been much debate as to whether paroxysmal kinesigenic choreoathetosis is a manifestation of epilepsy; this matter has never been entirely resolved. The preservation of consciousness and lack of post ictal features do not exclude epilepsy although the nature of the movements suggests a subcortical basal ganglia origin and paroxysmal kinesigenic choreoathetosis is not currently regarded as a manifestation of epilepsy. This has obvious implications for employment and driving. The gene for paroxysmal kinesigenic choreoathetosis has been mapped to chromosome 16p11.2-12.1 (Bennett *et al.* 2000). A gene for benign familial infantile convulsions has been mapped to the same region of chromosome 16. In some families paroxysmal kinesigenic choreoathetosis is combined with infantile convulsions, ICCA syndrome, which is also linked to the same locus on chromosome 16 (Lee *et al.* 1998) suggesting that these disorders are allelic (Swoboda *et al.* 2000).

Paroxysmal, non-kinesigenic, dystonic choreoathetosis (DYT8)

In paroxysmal, non-kinesigenic, dystonic choreoathetosis the attacks are also a mixture of dystonia and chorea but they are more prolonged, lasting minutes to hours (Mount and Reback 1940). The frequency is much lower, with only two or three attacks per day at most and often less than this. Onset is more variable than with paroxysmal kinesigenic choreoathetosis, ranging from childhood up to middle age. The episodes are often precipitated by alcohol, caffeine, stress, or fatigue but not movement. Falling may occur and a sensory aura is sometimes present. Most cases are idiopathic and may be autosomal dominantly inherited or sporadic. In one family some affected relatives only had exercise induced cramps while others had typical paroxysmal dystonic choreoathetosis attacks, but only after prolonged exertion similar to that reported the intermediate form (Schloesser *et al.* 1996). In famial or sporadic cases investigations are predictably normal and the diagnosis is clinical. Sporadic cases are often thought to be psychogenic; it may be difficult to exclude this possibility. Treatment is often unsuccessful but there may be a good response to clonazepam. Autosomal dominant paroxysmal dystonic choreoathetosis has been assigned to a locus on chromosome 2q36-37 (Fink *et al.* 1996). Mutations of the myofibrillogenesis regulator 1, *MR1* gene have been detected in paroxysmal dystonic choreoathetosis (Lee *et al.* 2004; Djarmati *et al.* 2005). Symptomatic paroxysmal dystonic choreoathetosis is less common than the hereditary form but has been reported with multiple sclerosis, a variety of other focal cerebral lesions, or hypoglycaemia (Shaw *et al.* 1996).

Intermediate paroxysmal dystonic choreoathetosis; paroxysmal exertional dyskinesia

The intermediate duration form is characterized by attacks which last 5–30 min. They are induced by movement in the form of prolonged exercise lasting several minutes (Lance 1977; Plant *et al.* 1984). There is no response to anticonvulsants or clonazepam. Most cases are familial with autosomal dominant inheritance, but a few sporadic cases have been reported. In one family there was a combination of rolandic epilepsy, writer's cramp, and paroxysmal dystonic choreoathetosis with linkage to chromosome 16 (Guerrini *et al.* 1999) suggesting a similarity to the ICCA / paroxysmal kinesigenic choreoathetosis locus. These conditions are possibly allelic disorders due to different mutations of the same gene.

Episodic choreoathetosis with spasticity, DYT9

A single family has been described with episodic dystonia with dysarthria, tingling, and diplopia followed by headache. Attacks lasted about 20 min and occurred twice a day or twice a year. Alcohol, stress, and fatigue could preceipitate the attacks. Some patients had lower limb pyramidal signs in between attacks. The disorder has been linked to chromosome 1p (Auburger *et al.* 1996) designated DYT9.

40.4.8 Dyskinetic athetoid cerebral palsy

Cerebral palsy is a clinically and pathologically heterogeneous group of congenital disorders caused by damage to the immature brain, characterized by abnormalities of motor control (Section 9.5). Various motor abnormalities and other neurological impairments are seen and cerebral palsy is divided into ataxic, spastic, and athetoid subtypes. In dyskinetic or athetoid cerebral palsy, the abnormal movements are principally dystonic and although the term athetosis was originally proposed to descibe the movement disorder in these patients, there is no logical reason to use it other than as a historical diagnostic label for dyskinetic cerebral palsy (Section 9.5.4). Some clinicians differentiate dystonic and dyskinetic cerebral palsy on the basis that the movements are slow or fixed in the former and more mobile in the latter but such variability has always been accommodated by the term dystonia. Nevertheless, the term dyskinetic cerebral palsy is in common usage and will be used here. It mainly occurs in infants born at term; the birth histories are variable.

In the past, severe neonatal jaundice, usually secondary to fetomaternal Rhesus or ABO blood group incompatibility, was a common cause of dyskinetic cerebral palsy, *kernicterus*. Affected children had severe dystonia, deafness, dental abnormalities, abnormal eye movements, and mental retardation. Pathological changes affected the basal ganglia and brainstem, status dysmyelinatus.

The other major group are those children with a history of significant perinatal asphyxia due to a clearly defined obstetric complication (Hagberg and Hagberg 1993). In these patients there will have been severe foetal distress and poor APGAR scores of less than 3 lasting 20 min with signs of significant neonatal hypoxic-ischaemic encephalopathy. These include hypotonia, impaired consciousness, irritability, and seizures. Pathologically there was damage attributed to ischaemia in the thalami and basal ganglia, status marmoratus (Friede 1989). In many patients however, these features are absent and the birth history is normal. The aetiology of the dyskinetic cerebral palsy in these patients is unknown. It seems unlikely that all cases of dyskinetic cerebral palsy are due to perinatal asphyxia or kernicterus. Not all patients have ischaemic lesions in the basal ganglia on MRI and neuropathological studies have been limited. Not all patients have ischaemic lesions at autopsy and those who do may not be representative of the majority of dyskinetic cerebral palsy patients who survive (Fletcher and Marsden 1996).

Clinically the patients may have signs of hypotonia in the neonatal period but this is variable. Some are normal initially but there is always abnormal motor development with delayed head control, sitting, and walking. Mild choreiform movements appear, often in the feet in the first few years and these graduallly develop during childhood into generalized dystonia. Some patients are able to walk but others are severely disabled (Kyllerman *et al.* 1982). Additional signs are variably present and include spasticity, seizures, eye movement abnormalities, dysarthria, and deafness; cognitive function is often normal. Some patients are stable for many years only to deteriorate markedly as adults. This may be due to additional cervical spondylotic myelopathy but others seem to have slowly progressive dystonia, spasticity and dysarthria (Fletcher and Marsden 1996). The reasons for this are unknown. Some patients with dyskinetic cerebral palsy have affected relatives and in single cases there is an increase in mean paternal age; this is indirect evidence that some of these cases may arise by fresh autosomal dominant genetic mutation (Fletcher and Foley 1993).

It is important not to diagnose dyskinetic cerebral palsy without considering alternative diagnoses. A therapeutic trial of levodopa is essential to exclude dopa responsive dystonia (Section 40.4.4) which may also cause motor delay and generalized dystonia. Some patients with dyskinetic cerebral palsy may improve partially with levodopa but not to the same extent (Fletcher *et al.* 1993). Other conditions which may cause marked dystonia in infancy are Lesch Nyhan disease, Pelizaeus Merzbacher disease, and the organic acidurias.

40.4.9 Delayed onset dystonia

It is well recognized that some brain lesions may be followed by the appearance of dystonia after an interval; the best known example of this is the delayed leucoencephalopathy and pallidal necrosis following anoxic brain damage caused by carbon monoxide poisoning (Lee *et al.* 1994). A similar delay in the appearance of dystonia may follow basal ganglia stroke, toxic brain damage, or head injury. In these situations, the dystonia appears after a variable delay ranging from a few days to several years and is often accompanied by spasticity and Parkinsonism. A more difficult concept has been the suggestion that dystonia may appear many years after perinatal brain damage with normal intervening development (Burke *et al.* 1982a; Saint Hilaire *et al.* 1991). Such patients often have dystonia similar to primary torsion dystonia but were diagnosed as delayed onset dystonia following cerebral anoxia at birth purely on the basis of an abnormal perinatal history. It is difficult to be certain whether this interpretation is correct because a history of mild obstetric complications is common in normal individuals and such perinatal factors are poor predictors of neurological abnormalities unless very severe. Accordingly, it is unwise to assign much diagnostic relevance to mild abnormalities of parturition or transient indicators of perinatal hypoxia especially in those with subsequently normal motor milestones.

40.4.10 Postencephalitic dystonia

Dystonia was often seen in patients who had survived an attack of encephalitis lethargica. Although Parkinsonism was the most common complication (Section 40.3.6) some patients developed cranial or limb dystonia. In the early part of the 20th century many cases of dystonia were attributed to encephalitis but the disorder is now very rare (Howard and Lees 1987).

40.4.11 Dystonia due to metabolic disorders

Many metabolic disorders may produce dystonia (Table 40.14). Most of these are very rare and produce complex clinical disorders in which the dystonia is often overshadowed by other features. The most important of these is *Wilson's disease* (Section 40.8). Other members of this group of dystonic disorders are rare and untreatable. There are several excellent reviews of these conditions (Barclay and Lang 1995; De Yebenes *et al.* 1996).

◆ *Hereditary caeruloplasmin deficiency* (Section 39.5.2) is very rare but is mentioned to distinguish it from Wilson's disease. Dementia, cranial dystonia, and diabetes are seen along with evidence of brain iron overload. There are abnormal MRI signals in the basal ganglia, brainstem, and dentate nuclei (Grisoli *et al.* 2005). Inheritance is autosomal recessive (Kawanami *et al.* 1996).

◆ *Organic acidurias.* Glutaric aciduria type 1 and methylmalonic aciduria usually present in infancy with a severe encephalopathy and developmental delay. Dystonia may be prominent and the diagnosis is established by urinary organic acid estimations. Basal ganglia necrosis may be seen with CT or MRI. In glutaric aciduria type 1 there may be sudden episodes of encephalopathy triggered by infection, anaesthesia, or other illness. There is no effective treatment but crises may be improved with hydration, glucose, and carnitine.

◆ *GM1 gangliosidosis* due to beta galactosidase deficiency may present in adult life with dysarthria, dystonia, and Parkinsonism (Section 10.4.3). Dysmorphism and macular lesions are not seen unlike the infantile form. There may be putaminal lesions on CT or MRI and the diagnosis can be made by measuring beta galactosidase in leucocytes.

◆ *GM2 gangliosidosis.* Tay–Sach's disease due to hexosaminidase deficiency usually presents in infancy with a severe encephalopathy with myoclonus and seizures (Section 10.4.2). Later onset forms also exist with dementia, psychiatric abnormalities, seizures, ataxia, and lower motor neurone findings. Dystonia may also occur but is not a prominent feature.

◆ *Lesch Nyhan disease* is an X-linked disorder caused by hypoxanthine–guanine phosphoribosyl transferase deficiency (Section 10.4.4). Affected infants develop mental retardation, delayed motor development, self mutilation, and dystonia (Jinnah *et al.* 2006). There is severe hyperuricaemia with renal stones and gouty arthritis. A partial deficiency of this enzyme also occurs with similar but less severe clinical features. Treatment with *S*-adenosylmethionine administration has been suggested (Glick 2006) and pallidal deep brain stimulation has been partially effective (Pralong *et al.* 2005).

◆ *Leucodystrophies* may be associated with dystonia but this is rare and only a minor feature of metachromatic leucodystrophy (Section 10.2). Dystonia is much more typical of Pelizaeus Merzbacher disease, X-linked sudanophilic leucodystrophy, which is characterized by nystagmus, ataxia, and dystonia in infancy. Leucocyte aryl-sulphatase A is reduced in metachromatic leucodystrophy, while Pelizaeus Merzbacher disease can now be diagnosed by mutation screening of the proteolipid protein, *PLP* gene on chromosome Xq22.

◆ *Neimann Pick disease type C* usually develops in infancy but a later onset form may occur with a vertical supranuclear gaze palsy, ataxia, and dystonia (Section 10.4.4). There are commonly psychiatric features and cataplexy (Imrie *et al.* 2002). There is an accumulation of sphingomyelin but sphingomyelinase levels are normal. Diagnosis is by bone marrow trephine to look for sea blue histiocytes. or more reliably by filipin staining of cultured fibroblasts from a skin biopsy. The *NPC1* gene is located on chromosome 18q11-12 (Fernandez-Valero *et al.* 2005; Tamura *et al.* 2006).

◆ *Homocystinuria* is occasionally associated with dystonia but the cause is unclear. There may be vascular damage to the basal ganglia or alterations in striatal neurotransmission.

◆ *Kuf's disease* is an adult onset form of neuronal ceroid lipofuscinosis (Section 10.3.2). There is no visual failure and affected adults develop either a progressive myoclonic epilepsy or a dementia with cranial dystonia. Demonstration of osmiophilic fingerprint profiles or granular osmiophilic deposits by electron microscopy in rectal, skin, or liver biopsy is required for the diagnosis. Urinary sediment dolichol levels are also elevated but this test is not routinely available. There is no treatment.

◆ *Mitochondrial disorders* are associated with a wide range of neurological features but dystonia is prominent in Leigh's disease, familial striatal necrosis, and in some cases of Leber's hereditary optic neuropathy (Hanna and Bhatia 1997) (Section 10.5). A diagnostic clue is the presence of striatal lesions on CT or MR scans. Leigh's disease may be caused by defects of oxidative phosphorylation, complex I, II, or IV deficiency, or pyruvate dehydrogenase deficiency and may be associated with mutations of nuclear or mitochondrial DNA. Dystonia may be prominent along with mental regression and brainstem signs especially respiratory irregularity, eye movement disorders, and ataxia. Familial striatal necrosis has also been linked to mitochondrial dysfunction (Solano *et al.* 2003) and usually presents with dystonia and Parkinsonism (Caparros-Lefebvre *et al.* 1997; Thyagarajan *et al.* 1995). Some patients with Lebers disease have also developed dystonia and striatal lesions (Marsden *et al.* 1986; Novotny *et al.* 1986). Clearly there is some overlap in the clinical features of these three forms of mitochondrial disease. The X-linked deafness–dystonia syndrome, Mohr–Tanebjaerg syndrome may also be related to mitochondrial dysfunction (Swerdlow *et al.* 2004).

40.4.12 Neurodegeneration with brain iron accumulation

The terminology of this disorder has become confusing. The old Hallervorden–Spatz syndrome eponym has become unpopular due to the unethical behaviour of Hallervorden and Spatz. Pantothenate kinase associated neurodegeneration 'PKAN' is not entirely satisfactory as only some cases actually have a defect of pantothenate kinase. Neurodegeneration with brain iron accumulation describes the main neuropathological feature of the disorders.

Neurodegeneration with brain iron accumulation is a rare autosomal recessive disorder characterized by progressive dystonia, spasticity, and dementia (Dooling *et al.* 1974). The onset is usually before 20 years of age but adult onset cases have occurred (Jankovic *et al.* 1985). Rigidity of the legs is an early feature with dystonia, chorea, spasticity, brisk reflexes, and extensor plantar responses (Swaiman 1991). Cognitive decline, retinitis pigmentosa, and sometimes seizures are additional features; relentless progression leads to generalized rigidity and dementia and survival is limited to a mean of 11 years. Some young onset cases show slower progression and later onset cases have presented with atypical Parkinsonism (Section 40.3.13).

A characteristic appearance on T2-weighted MRI scans of pallidal hypointensity with a high signal centre 'eye of the tiger sign' may be seen in many cases (Fig. 40.14). Peripheral blood acanthocytes, or sea blue histiocytes in the bone marrow, have been described. Pathologically there is increased iron deposition and axonal speroid formation in the globus pallidus and substantia nigra pars reticulata. There is no effective treatment.

Many cases of neurodegeneration with brain iron accumulation are associated with various mutations of the pantothenate kinase 2, *PANK2* gene (Zhou *et al.* 2001). It has been suggested that all classic early onset neurodegeneration with brain iron accumulation cases have *PANK2* mutations and the characteristic eye of the tiger sign on brain MR scans, while atypical later onset cases are less likely to have *PANK2* mutations and have normal brain MR scans (Hayflick *et al.* 2003). While it seems clear that earlier onset cases are more likely to have dystonia, dysarthria, cognitive decline, and gait failure, not all have *PANK2* mutations (Thomas *et al.* 2004). Probably only two-thirds of cases have *PANK2* mutations (Hartig *et al.* 2006). In addition, the suggestion that the eye of the tiger sign is pathognomonic of *PANK2* mutations (Hayflick *et al.* 2006) has been challenged. Some patients with the eye of the tiger sign do not have *PANK2* mutations (Valentino *et al.* 2006) and some *PANK2*

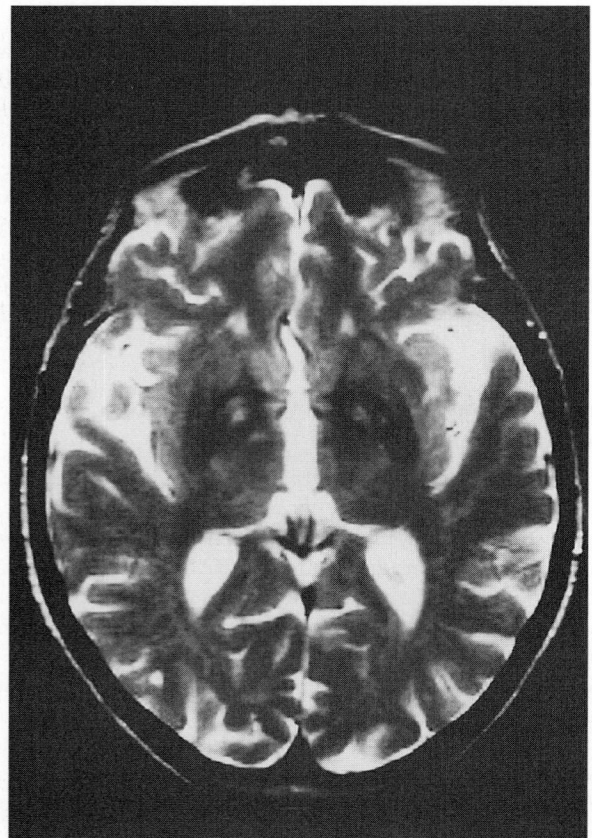

Fig. 40.14 Neurodegeneration with brain iron accumulation. MR brain scan of a patient with showing the characteristic pallidal signal changes, the 'eye of the tiger sign'.

mutation positive neurodegeneration with brain iron accumulation cases have no eye of the tiger sign (Zolkipli *et al.* 2006).

There are several reports of probable variants of neurodegeneration with brain iron accumulation . There can be atypical presentations with akinesia (Molinuevo *et al.* 2003) and even a motor neuron disease like picture (Vasconcelos *et al.* 2003). The HARP syndrome comprises Hypoprebetalipoproteinaemia, Acanthocytosis, Retinitis pigmentosa, and Pallidal degeneration (Orrell *et al.* 1995). The dystonia tends to affect cranial musculature and the characteristic brain MRI changes of neurodegeneration with brain iron accumulation have been detected. Another case with Parkinsonism, dementia, and suggestive MRI changes had pathological evidence of neurodegeneration with brain iron accumulation and Lewy bodies in the substantia nigra (Tuite *et al.* 1996). It is now clear that HARP syndrome and neurodegeneration with brain iron accumulation are the same disorder (Ching *et al.* 2002; Houlden *et al.* 2003).

Treatment of neurodegeneration with brain iron accumulation is difficult and largely symptomatic control of dystonia and seizures is all that is possible. Recently, pallidal deep brain stimulation has been used successfully to partially improve the symptoms of the condition (Castelnau *et al.* 2005). A trial of vitamin B5, pantothenic acid, has been proposed.

40.4.13 Basal ganglia calcification, Fahr's syndrome

This is a heterogeneous disorder in which there is idiopathic calcification of the basal ganglia and sometimes other brain areas together with neurological impairment (Manyam 2005). Calcification of the basal ganglia may be an incidental finding on CT scans and is usually mild and of no significance in this situation. Other specific neurological disorders such as cerebral anoxia, neurodegeneration with brain iron accumulation, cerebral infections, disorders of calcium homeostasis, the Cockayne syndrome (Sections 11.6.8; 39.7.3), and mitochondrial disorders may be associated with basal ganglia calcification and should be separated from the Fahr syndrome which refers to idiopathic cases which may be sporadic or autosomal dominant. Dystonia may be a prominent feature in these cases (Caraceni *et al.* 1974; Larsen *et al.* 1985) along with neuropsychological abnormalities consistent with basal ganglia dysfunction (Lopez-Villegas *et al.* 1996) and dementia (Modrego *et al.* 2005). The clinical phenotype may also include Parkinsonism, ataxia, and abnormal eye movements. In one multigenerational family, a gene locus was mapped to chromosome 14q, *IBGC1*, but the condition is genetically heterogeneous (Geschwind *et al.* 1999; Oliveira *et al.* 2004).

40.4.14 Deafness-dystonia, Mohr–Tranebjaerg syndrome

This is an X-linked recessive disorder causing deafness and dystonia (Swerdlow *et al.* 2004). Onset is in childhood or young adults and deafness occurs first. There may be additional cognitive decline, psychiatric features, spasticity, or optic atrophy. Linkage was established to Xq21-22 and mutations were described in a dystonia/deafness peptide gene (Jin *et al.* 1996) also known as the translocase of the inner mitochondrial membrane, the *TIMM8A* gene (Roesch *et al.* 2002; Binder *et al.* 2003). Dystonia in Mohr–Tranebjaerg syndrome is either generalized or upper body focal or segmental in type.

40.4.15 Dystonia in other hereditary and degenerative disorders

Other causes of dystonia in this group of conditions are shown in Tables 40.12 and 40.14 and are dealt with separately.

- A single family with autosomal dominant dystonia with *laryngeal dystonia* causing whispering dysphonia has been described (Ahmad *et al.* 1993). This form of hereditary dystonia is classified as DYT4 but no other cases have been described. Torticollis was the main feature other than dysphonia.

- Dystonia is common in *atypical Parkinsonian disorders*: multiple system atrophy, corticobasal degeneration and progressive supranuclear palsy (Section 40.3), and in some forms of autosomal recessive Parkinson's disease, especially if associated with *Parkin* gene mutations. Painful off-period foot dystonia is common in advanced idiopathic Parkinson's disease.

- It should be noted that dystonia is often prominent in advanced *Huntington's disease* especially in the rigid Westphal variant in younger onset, paternally inherited cases (Section 40.5.2).

- *Ataxia telangiectasia* may present with a mainly dystonic syndrome (Bodensteiner *et al.* 1980) and skin changes may be absent or inconspicuous (Sections 11.5; 39.7.1). Diagnosis is facilitated by alpha-fetoprotein and immunoglobulin estimations.

- Dystonia may be prominent in the *spinocerebellar degenerations*, especially ataxia with optic apraxia, ataxia with vitamin E deficiency (Roubertie *et al.* 2003) and SCA3; the late onset degenerative cerebellar degenerations are one of the more common causes of secondary dystonia in clinical practice. Friedreich's ataxia is occasionally associated with dystonia (Hou and Jankovic 2003).

- *Neuroacanthocytosis* is associated with both chorea and dystonia (Section 40.5.8).

- The *Rett syndrome* occurs mainly in girls and causes developmental regression, head growth deceleration, autism, stereotypic movements, especially of the hands, ataxia, bruxism, tremors, tics, respiratory abnormalities, scoliosis, and dystonia (Section 9.6.2). Later there may be signs of Parkinsonism (Fitzgerald *et al.* 1990). Occasionally severely affected males may be encountered (Shahbazian and Zoghbi 2001). Rett syndrome is caused by mutations of the *MECP2* gene on chromosome Xq28 (Amir *et al.* 1999). *MECP2* mutations are also seen in some males with non-specific X-linked mental retardation (Orrico *et al.* 2000) and some girls with autism (Carney *et al.* 2003) although this is not a common cause of autism (Lam *et al.* 2000).

40.4.16 Post-traumatic dystonia

There has been a frequently reported association between peripheral injuries or pain and the subsequent local development of dystonia. Although this has been noted for many years the mechanism is unclear and a causative link is speculative. Most cases have been in patients who develop various forms of focal adult onset post traumatic dystonia after local injuries (Section 40.4.2). Some of these patients may have a pre-existing genetic liability to dystonia, presumably post traumatic dystonia but this is unproven. Peripheral limb fractures, lacerations, and surgical procedures have figured prominently in these cases (Fletcher *et al.* 1991b) and patients with post traumatic dystonia should be warned of this possible risk

prior to surgical operations. These should, as a consequence, be avoided whenever possible, especially misguided attempts at surgical correction of dystonic postures. In a survey of 104 cases of post traumatic dystonia, 17 gave a history of initiation or exacerbation of dystonia by trauma; recall bias is a possible explanation but patients with post traumatic dystonia do not in fact give a history of injury more often than matched controls (Fletcher *et al.* 1991b). Focal dystonias may follow repetitive use of a limb which may act as a form of soft tissue trauma sometimes in association with ulnar neuropathy (Ross *et al.* 1995). Such patients do not appear to carry the *DYT1* gene (Gasser *et al.* 1996). In another group of patients focal dystonic spasms develop in association with pain and reflex sympathetic dystrophy after local injury (Marsden *et al.* 1984b).

Post-traumatic torticollis appears to differ from typical cervical dystonia in patients with a rapid onset of neck deformity after injury. There is a greater incidence of fixed laterocollis, pain, and poor response to botulinum toxin (Frei *et al.* 2004; O'Riordan and Hutchinson 2004). Psychogenic factors may be important (Sa *et al.* 2003) but this is uncertain.

The response to treatment is poor especially with fixed abnormal postures and contractures although sympathetic blockade may provide some benefit. It can be difficult in some of these patients to exclude a psychogenic element and there have been spontaneous remissions. It is therefore probably unwise to recommend destructive peripheral surgical procedures or deep brain stimulation in these patients. Physiotherapy, antidystonia medications, and pain management form a reasonable policy.

40.4.17 Drug induced dystonia

Acute transient and persistent tardive dystonia may follow the administration of neuroleptic drugs or other dopamine receptor blocking agents (Section 40.9). Dystonia is also a common complication of levodopa or dopamine agonist therapy in Parkinson's disease and multiple system atrophy (Section 40.3). Although the effects of drugs affecting dopaminergic function are the best known and most important examples of drug induced dystonia, many other agents have been implicated in small numbers of cases. These include tricyclic antidepressants, selective serotonin reuptake inhibitors, and buspirone. Some of these movement disorders have been persistent. Anticonvulsants have occasionally caused dystonia.

40.4.18 Dystonia caused by toxins

Various toxins may lead to dystonia, usually as part of a mixed extrapyramidal syndrome and with the characteristic features of a delay between the toxic exposure and the development of clinical signs and evidence from neuroimaging or neuropathological studies of damage to the globus pallidus. Toxic dystonia is reviewed in detail elsewhere (Chu *et al.* 1995). *Manganese* poisoning usually causes Parkinsonism with prominent dystonia especially involving gait (Section 40.3.5). Dystonia has been reported after *carbon monoxide* exposure but this is rare (Section 5.4.1). Most patients who survive the initial cerebral anoxia recover or are left with permanent neurological damage including mental changes, mutism, and rigidity due to spasticity and Parkinsonism. Some develop a delayed post anoxic encephalopathy in which Parkinsonism is usually much more prominent than dystonia (Lee and Marsden 1994b). Many of these patients also recover spontaneously but the neurological damage to the basal ganglia and cerebral white matter is progressive in some. Pallidal deep brain stimulation may be effective in rare cases (Ghika *et al.* 2002). *Carbon disulphide* may produce an encephalopathy with delerium and Parkinsonism as the main features but dystonia has occurred occasionally. Dystonia has also followed poisoning with *disulphiram*. *Cyanide* poisoning is usually fatal but Parkinsonism and sometimes dementia and dystonia have occurred in a few survivors. This is associated with striatal and pallidal necrosis which is visible on CT or MR scans.

40.4.19 Other secondary dystonias

Dystonia may develop after focal brain lesions of various types, including tumours, arteriovenous malformations, and vascular lesions (Section 40.2.2). These usually involve the thalamus or putamen. In some patients, hemidystonia may appear years after hemiparesis, with hemiatrophy and seizures, the 4 hemi syndrome (Thajeb 1996). Dystonia has occasionally occurred in multiple sclerosis (Coleman *et al.* 1988) and has to be considered in the differential diagnosis of patients with paroxysmal dystonia. Parkinsonism and dystonia have also followed osmotic brain damage (Maraganore *et al.* 1992). Occasionally dystonia may occur in surprising clinical settings including brainstem haemorrhage and spinal cord lesions (Cammarota *et al.* 1995; Esteban Munoz *et al.* 1996).

40.5 Chorea and ballism

40.5.1 Diagnosis and classification of chorea

Chorea may be recognized as irregular, low amplitude, rapid, movements of the extremities and the face. Mild chorea may be subtle and easily dismissed as simple fidgetiness. Patients often disguise the movements into apparently purposeful actions such as adjusting seating posture, clothing, or jewellery. More severe cases display obviously abnormal movements of the hands, feet, and face. Larger amplitude proximal choreiform movements may be violent and are referred to as ballism. Chorea may be partly suppressed voluntarily but incompletely and not for long. Facial grimacing, eyebrow movements, and respiratory noises may be evident. In tardive dyskinesia, the movements of the face and tongue are choreiform but in this situation they are stereotyped and repetitive. Chorea may coexist with other movements, particularly dystonia which is usually more proximal, sustained, and twisting (Section 40.4.1) but the distinction is sometimes difficult. Chorea is simulated by the irregular sinuous movements of the fingers in pseudoathetosis due to severe proprioceptive loss of the upper limbs, commonly caused by cervical myelopathy or large fibre peripheral neuropathy. The distinction is easily made by sensory testing during the neurological examination.

The causes of chorea are shown in Table 40.16. This list is by no means exhaustive but many other causes of chorea are rare, based on very small numbers of reported cases or single reports and so are of limited clinical value. The conditions in bold type are the important considerations, either because they are regular causes of chorea in the clinic or because of the importance of a diagnosis such as Wilson's disease. Some such as neuroferritinopathy, dentatorubral–pallidoluysian atrophy and benign hereditary chorea are rare but very important in the differential diagnosis of Huntington's disease and therefore potential causes of diagnostic error in the clinic.

40.5.2 Huntington's disease

Definition and prevalence

This is an autosomal dominant condition causing a movement disorder, usually chorea and mental changes. Huntington's disease

Table 40.16 Causes of chorea myokymia. Bolditalic type indicates the principal diagnostic considerations in cases of chorea.

Hereditary degenerative disorders	***Huntington's disease*** ***Benign hereditary chorea*** ***Dentatorubropallidoluysian atrophy*** ***Neuroferritinopathy*** Hereditary spinocerebellar ataxias Fahr's syndrome Ataxia telangiectasia Huntington's disease like phenotypes 1–4
Hereditary metabolic disorders	***Wilson's disease*** Lesch–Nyhan disease Mitochondrial disease
Endocrine disorders	***Thyrotoxicosis*** Pregnancy Addison's disease
Metabolic disorders	Hypo/hyperglycaemia Hypo/hypercalcaemia Non Wilsonian hepatolenticular degeneration
Haematological disorders	***Neuroacanthocytosis*** ***Antiphospholipid syndrome*** Polycythaemia rubra vera Sickle cell disease
Autoimmune conditions	***Systemic lupus erythematosus***
Infections	***Sydenham's chorea*** Subacute sclerosing panencephalitis Creutzfeldt–Jacob disease
Vascular disorders	***Lacunar infarction*** Post pump chorea
Drugs	Oral contraceptives ***Neuroleptics*** ***Levodopa*** Anticholinergics Other drugs
Other causes	Paroxysmal dyskinesias Focal lesions of the basal ganglia Senile chorea ***Athetoid cerebral palsy*** Toxin related encephalopathies

DRPLA = dentatorubropallidoluysian atrophy.

is progressive and invariably fatal; there is no effective treatment. The prevalence is 4–10 per 100 000 in the United Kingdom with slightly lower rates in some parts of Europe and in North America (Harper 1991). It is caused by an unstable trinucleotide, CAG, repeat expansion in the *IT15* gene on chromosome 4p16.3 (Huntington's disease collaborative research group 1993) which encodes a 348 kDa protein, Huntingtin, widely expressed in neural and other tissues.

Pathology

In advanced cases there is marked atrophy of the caudate and putamen with widening of the anterior horns of the lateral ventricles (Fig. 40.15). There is also atrophy of the cortex and subcortical white matter. There are varying degrees of atrophy and neuronal loss affecting the striatum and cerebral cortex and to a lesser extent

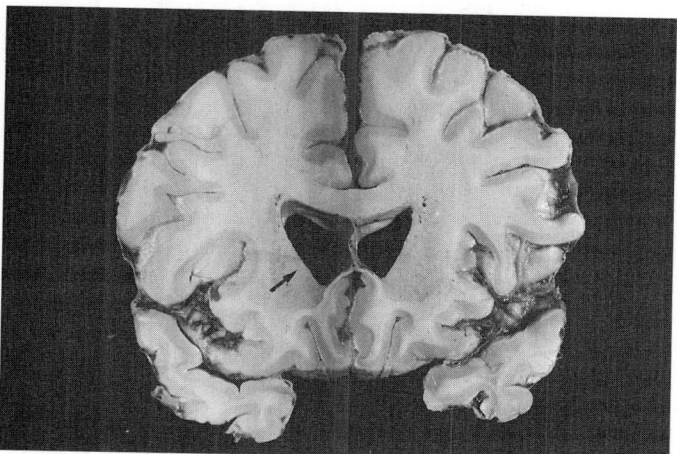

Fig. 40.15 Coronal brain section in Huntington's disease, showing almost total caudate atrophy (arrowed) and diltation of the adjacent lateral ventricles (Vonsattel grade 5). (Courtesy of Dr J Broome.)

in the thalami, other basal ganglia areas, brainstem, and sometimes cerebellum (Vonsattel *et al.* 1985). Although the degree of neuropathological change is related to age of onset, with older onset cases tending to have milder pathological changes, the relationship between clinical features and neuropathology is unreliable. Not only may clinically affected individuals, with the diagnosis confirmed by DNA analysis, have no detectable neuropathology but some asymptomatic individuals have had definite abnormalities. In the striatum, the main neuronal depletion affects medium spiny neurones with parallel reduction of striatal dopaminergic D1 and D2 receptors indicating simultaneous damage to the direct and indirect stiatopallidal pathways. At present, the mechanism by which mutant Huntingtin, which is widely expressed throughout the brain, induces regionally selective neuronal loss is unknown (Ross 2002). The increased polyglutamine repeat length in mutant Huntingtin increases protein aggregation and folding and the appearance of intracellular inclusions. However, these may, by sequestering the toxic Huntingtin, actually increase cellular survival (Arrasate *et al.* 2004).

Clinical features

Age of onset. Huntington's disease may start at almost any age from early childhood to advanced old age and the clinical features are considerably influenced by the age of onset. The disorder usually starts between 30 and 50 with a mean of about 40 years although the mean age of onset in the general population is approximately 5 years later than in hospital based series (Harper 1991). The juvenile Huntington's disease variant with onset before 20 years accounts for about 5 per cent of the total. Late onset disease, after age 60 years occurs in up to 25 per cent of cases. The onset is gradual and most patients are diagnosed after several years of progressive motor and mental changes. About two-thirds of patients develop motor neurological symptoms at the onset, usually chorea, while the remainder present with personality or cognitive changes (Di Maio *et al.* 1993a). Within this variable clinical picture, three broad phenotypes may be discerned: 'classical' Huntington's disease; the juvenile onset variant; and the elderly onset form.

The *motor disorder* of Huntington's disease is much more complex and variable than is often realized. In the classical forms,

chorea is prominent and, along with the mental changes, becomes gradually more evident as the disease progresses. Even at this stage there is additional bradykinesia which contributes to abnormal movement (Thompson *et al.* 1988). With time, there is increasing dystonia, rigidity, and bradykinesia with increasing immobility, postural instability, and eventual inability to walk. Spasticity with brisk tendon reflexes and extensor plantar responses may also appear. The gait often becomes wide based and staggering with falls, bradykinesia, and freezing. Eye movements are slowed at an early stage with difficulty initiating saccades, broken up pursuit movements, and impaired optokinetic responses. Dysarthria and dysphagia develop with a danger of aspiration pneumonia. Gradual weight loss is also common as Huntington's disease progresses and many patients in the terminal stages of the disorder are in a cachectic state. This is probably not due to hyperkinetic movements which become less prominent in the later stages and is not related to nutritional intake. Many patients develop insomnia at night and somnolence by day which is disruptive for carers and other family members. Incontinence occurs in 20 per cent of cases and epilepsy is slightly more common than would be expected by chance. It should be noted that even within the classical phenotype, the motor syndrome is variable. Not only is the severity of the chorea unpredictable but in some patients with the rigid form or Westphal variant of Huntington's disease the clinical picture is dominated by rigidity from the outset. These patients show a combination of dystonia and Parkinsonism, and chorea may be absent. Most patients with the Westphal variant have juvenile onset Huntington's disease but this is not always the case and the two are not synonymous. Other patients have myoclonus (Thompson *et al.* 1994) or tics resembling Tourette's syndrome (Jankovic and Ashizawa 1995).

The *mental changes* of classical Huntington's disease are also variable. Early on there are changes in personality with altered mood, irritability, apathy, and anxiety. There may be a decline in work performance, deteriorating marital or family relationships, or financial misjudgements. Depression is common, sometimes appearing before any other symptoms and often contributes significantly to impaired cognitive function (Morris 1991). Suicide and deliberate self harm are also common (Di Maio *et al.* 1993b) and relate to depression, the level of physical disability, and the extent of social support (Lipe *et al.* 1993). Dementia gradually becomes apparent in many patients and has been decribed as a subcortical dementia with slowing of thought processes, impaired long term memory, and frontal lobe deficits (Marshall and Shoulson 1997). This probably reflects impaired subcortical basal ganglia function, involving the caudate to frontal lobe cortical loop (Section 40.2.1) as well as cortical pathology. Although it is often assumed that Huntington's disease is invariably associated with dementia, the true frequency is difficult to establish due to vague and variable definitions of dementia and the confounding effects of depression. Some patients develop psychotic features, with hallucinations and delusions but this is uncommon. Others display obsessive compulsive features or altered sexual behaviour. It is common for there to be increasing apathy and social withdrawal (Marshall and Shoulson 1997) probably due to a combination of personality change, depression, and dementia. Survival in this form of Huntington's disease is approximately 15 years from the onset but it is important to note that even within 'classical' Huntington's disease, if such an entity can be defined, there is considerable variation. Some patients have a rapidly fatal illness lasting only a few years while others survive for up to 40 years (Harper 1991). Sometimes the illness is remarkably benign with mild chorea, absent mental changes, minimal progression and almost normal survival (MacMillan *et al.* 1993a). It is apparent that the progression of Huntington's disease depends on environmental and other genetic factors, not just the Huntingtin mutation (Wexler *et al.* 2004).

The *juvenile form* of Huntington's disease starts before the age of 20 years and sometimes in early childhood. It usually takes the form of the rigid or Westphal variant. Such patients have almost always inherited the disease from an affected father. In these cases the onset is typically with progressive behavioural change and cognitive decline which evolves into severe mental impairment. This is accompanied by striking Parkinsonism, dystonia, and rigidity. Eye movement abnormalities are prominent at an early stage along with brisk tendon reflexes, spasticity, and extensor plantar responses. Many of these patients develop seizures as well as dysarthria, dysphagia, and some have cerebellar signs or a postural tremor. Chorea can be present but is dominated by the severe rigidity and mental changes. Although juvenile Huntington's disease is thought to have a more rapid course than classical disease, survival is actually similar (Roos *et al.* 1993). Moreover, a few early onset cases display a mild phenotype with chorea and prolonged survival (MacMillan *et al.* 1993b). In patients with onset after 60 years the disease is entirely different. Chorea is the main feature and mental changes may be inconspicuous or absent. In these patients the disease runs a much more benign course but overall survival is similar due to other age related causes of death. Although there is variation within the three age related onset groups, the tendency for younger onset cases to have a more severe rigid form and older patients to have a milder illness with chorea and little rigidity is a useful clinical guide when considering the diagnosis of Huntington's disease.

Genetics

Huntington's disease is inherited as an autosomal dominant trait with full penetrance and 50 per cent risk to the offspring of affected individuals. This risk does not fall significantly until after the age of 40 years (Harper and Newcombe 1992). There is considerable variation in age of onset and severity even within families so this is difficult to predict for at risk relatives. Anticipation, with a tendency for earlier age of onset in each generation has long been observed. This is most noticeable with juvenile onset cases who usually have inherited the disorder from an affected father. Sporadic cases of Huntington's disease without a family history certainly occur but an unreliable family history, premature death of a parent, or suspected nonpaternity are often noteworthy and new mutations have been considered to be rare. With the discovery of the CAG repeat mutation in the *Huntingtin* gene, direct mutation testing on DNA samples is possible for both affected and at risk individuals. Normal chromosomes contain 11–30 CAG repeats in 99 per cent of instances while 99 per cent of HD disease gene-bearing chromosomes contain 36–121 repeats (Duayo *et al.* 1993; Kremer *et al.* 1994). Very occasional abnormal repeat numbers in control chromosomes have been attributed to Huntington's disease homozygosity, individuals unexpectedly found to have two mutations, incidentally discovered gene carriers and laboratory errors (MacMillan *et al.* 1993b; Kremer *et al.* 1994). About 1 per cent of those diagnosed with Huntington's disease have not had abnormal CAG repeat

numbers, some of whom are explicable by inaccurate diagnosis or laboratory errors. Only occasional patients appear to have clinically and pathologically typical Huntington's disease, presumably due to a different genetic mutation (Xuereb *et al.* 1996).

An autosomal dominant Huntington-like disorder may be caused by a gene on chromosome 20p (Xiang *et al.* 1998). An autosomal recessive disorder similar to Huntington's disease is caused by a gene on chromosome 4p15.3 (Kambouris *et al.* 2000). Trinucleotide expansions in other genes are unlikely (Vuillaume *et al.* 2000). Unlike alleles with a normal CAG repeat number, Huntington's disease CAG expansions are unstable and may lengthen or shorten during meiosis but with a preponderance of increased repeat numbers in affected offspring (Huntington's disease collaborative research group 1993). Most of these further increases are small but sometimes fathers transmit large increases suggesting that meiotic repeat length instability is greater in males. In sperm DNA there is therefore the combination of a stable normal repeat number and a spread of different abnormal allele sizes (MacDonald *et al.* 1993). Longer CAG repeat lengths are associated with earlier disease onset with juvenile onset cases tending to have the greatest trinucleotide expansions. Accordingly, juvenile onset cases are usually of paternal origin because of the greater tendency for CAG repeat enlargement in spermatogenesis. A given CAG repeat length however, is associated with a range of ages of onset and so DNA analysis cannot reliably predict the onset of the disease in relatives at risk of Huntington's disease (Craufurd and Dodge 1993).

It is important to note that a family history is not always apparent in Huntington's disease and this should not exclude the clinical diagnosis (Ramos-Arroyo *et al.* 2005). Among such patients, the Huntington's disease mutation is frequently detected, although some of these patients have unreliable family histories and may have had an affected parent (MacMillan *et al.* 1993a; Davis *et al.* 1994;). In other cases an allele of intermediate size, 30–35 repeats, in an unaffected father has spontaneously enlarged into the Huntington's disease range during spermatogenesis (Goldberg *et al.* 1993; Myers *et al.* 1993). Several examples of such new mutations arising in borderline paternal alleles have been described, usually in older fathers, indicating that there is a pool of premutations at the upper limit of the normal size range in the normal population from which new Huntington's disease mutations arise. In some reports the relevant paternal allele has been just within the Huntington's disease range suggesting that the father is in fact a gene carrier yet to develop elderly onset Huntington's disease.

Diagnosis

The main causes of chorea are shown in Table 40.16. Particular points are as follows:

◆ As with most movement disorders, exclusion of Wilson's disease is a priority in patients under the age of 50 years.

◆ A common difficulty is the combination of psychiatric illness and chorea in patients who have taken neuroleptic drugs. Some of these patients have tardive dyskinesia but others will have Huntington's disease and the distinction may be impossible without DNA testing.

◆ Neuroacanthocytosis, dentatorubral-pallidoluysian atrophy, benign hereditary chorea, late recurrences of rheumatic chorea, and new variant Creutzfeldt–Jacob disease (Zeidler *et al.* 1997) may resemble Huntington's disease closely.

♦ Confusion may arise in primary torsion dystonia and levodopa induced chorea in Parkinson's disease.

♦ Areflexia does not occur in Huntington's disease and suggests neuroacanthocytosis.

♦ Huntington's disease must be considered in women presenting with chorea in pregnancy or taking oral contaceptives.

♦ Thyrotoxicosis, systemic lupus erythematosis, and antiphosphol-ipid syndrome are excluded by blood tests.

♦ In patients with a wide based ataxic gait, extrapyramidal features, and abnormal eye movements it can be difficult to distinguish Huntington's disease from a degenerative cerebellar ataxia. In these patients, the correct diagnosis may only become clear with follow up observation or with DNA testing. SCA17 (Bauer *et al.* 2004) can resemble Huntington's disease closely.

♦ Neuroferritinopathy can cause a Huntington's disease like illness and is differentiated by neuroimaging and DNA testing (Section 40.5.4).

♦ Recently, a range of 'Huntington's disease like disorders' types 1, 2, 3 and 4 have been identified; as well as the familial dystonia with facial myokymia syndrome (see Section 40.5.3).

Investigations

Neuroimaging often shows caudate atrophy in clinically affected patients, although this can be absent and may occur in neuroacanthocytosis. Abnormal striatal function may be revealed by positron emission tomography scans showing reduced basal ganglia metabolism and dopamine receptor levels but these tests are neither specific nor routinely available. Consequently, most investigation is useful merely to exclude other diagnoses and should include a full blood count, thyroid function tests, a blood film for acanthocytes, creatine kinase, antinuclear factor, antiphospholipid antibodies, and calcium. MR or CT scans may be helpful in cases of cerebrovascular chorea, Fahr syndrome or focal lesions of the basal ganglia. Wison's disease must be excluded in younger patients with serum copper and caeruloplasmin estimations, urinary copper estimation and a slit lamp examination for Kayser–Fleischer rings. Many of the other causes of chorea in Table 40.16 are obvious from the history and so the need for additional tests varies. A definitive diagnosis of Huntington's disease can be made by DNA analysis and may be particularly useful in diagnostically difficult cases, saving the patient from repeated negative investigations. Care should be taken to provide adequate counselling of the patient and, whenever possible, the family before confirmatory genetic testing. Informed consent for testing should be obtained.

A genetic test is unwise without reasonable clinical grounds for the diagnosis, whatever the family history may suggest, as a positive test may not be diagnostically relevant in clinically dubious cases especially those presenting with behavioural or psychiatric problems. In those with an unexpectedly negative DNA test, the Huntington's disease-like conditions types 1–4, neuroacanthocytosis, neuroferritinopathy, and familial dystonia with facial myokymia syndrome (Fernandez *et al.* 2001) should be considered along with SCA17 (Bauer *et al.* 2004).

Asymptomatic relatives

Relatives of those with Huntington's disease often request medical advice because of concern over genetic risk or the significance of various symptoms. It is wise not to take a detailed history or examine such individuals unless specifically requested. They should realize that a clinical evaluation is a form of diagnostic test; such people may be already affected by mild chorea or psychiatric abnormalities. Genetic counselling is the main requirement in these individuals who are usually at 50 per cent risk of developing Huntington's disease. In sibs of apparently sporadic cases with definitely normal parents the risk is approximately 25 per cent as there may be a premutation in the father (Goldberg *et al.* 1993). In some presymptomatic gene carriers there are subtle physical signs or caudate atrophy on CT scans but it is usually impossible to predict an individual's genetic status. Some relatives request genetic testing which should not be carried out without detailed counselling within the framework of nationally agreed guidelines (Craufurd and Tyler 1992). Genetic testing of minors is considered unethical and should be tactfully refused if requested by parents. It is notable that many individuals who request presymptomatic testing are already showing early signs of Huntington's disease and the psychological impact of test results in truly asymptomatic carriers or non-carriers is limited (Horowitz *et al.* 2001).

Management

There is no effective treatment for Huntington's disease which usually progresses relentlessly. Psychological and social support together with genetic counselling are important. Chorea may be reduced by tetrabenazine which has been shown to be effective in a controlled trial (Huntington Study Group 2006). Neuroleptic drugs such as sulpiride or haloperidol are effective but not indicated unless the movements are severe; suppression of chorea does not always reduce disability and side effects may be troublesome (Shoulson 1981). Some neurologists use atypical neuroleptics for chorea suppression and mood stabilization (Bonelli and Hofmann 2004). Amantadine, 400 mg daily, may also be helpful (Verhagen Metman *et al.* 2002). Depression sometimes responds to antidepressant medication or electroconvulsive therapy. Mood stabilization can be attempted with valproate, serotonin reuptake inhibitors, carbamazepine, or beta blockers. In rigid patients, Parkinsonism may be improved modestly by levodopa. Psychotic features may be reduced with neuroleptic agents. Depending on the condition of the patient, referral for speech therapy, physiotherapy, and support from community social services may be indicated. Dysphagia may require a feeding gastrostomy, but the desirability of this always requires careful discussion with the patient and family. In some severely affected patients community care may not be realistic and admission to a nursing home is required. Management of Huntington's disease poses complex medical, social, and legal considerations (Morris and Tyler 1991).

Numerous treatment trials have been conducted in Huntington's disease. Minocycline, memantine, lamotrigine, riluzole, creatine, remacemide with coenzyme Q10, and the free radical scavenger OPC14117 have all been ineffective (Handley *et al.* 2006). Some trials have shown possible benefit: highly unsaturated fatty acids (HUFA) therapy can reduce chorea (Vaddadi *et al.* 2002) as can levetiracetam (de Tommaso *et al.* 2005). A trial of ethyl EPA was negative but with a possible trend to benefit which is being investigated further (Puri *et al.* 2005). Striatal foetal cell implants have been carried out in several small trials in Huntington's disease. The transplanted cells can survive in the host brain and do not appear to be affected by the disease in the short term (Freeman *et al.*

2000; Hauser *et al.* 2002; Rosser *et al.* 2002). Some resulting clinical and functional imaging improvements have been reported (Bachoud-Levi *et al.* 2000; Gaura *et al.* 2004) but not by others (Furtado *et al.* 2005).

40.5.3 Huntington's disease like phenotypes

Approximately 1–3 per cent of patients with a Huntington's disease-like illness do not have a detectable Huntingtin mutation on DNA testing. The differential diagnosis of chorea (Table 40.16 and Section 40.5.2) must be considered in this situation. Several recently described conditions are discussed here. Neuroferritinopathy and benign hereditary chorea are discussed in Sections 40.5.4 and 40.5.5. It should be noted that some patients with a condition resembling Huntington's disease are not currently classifiable (Richfield *et al.* 2002).

- *Huntington's disease-like phenotype 1* is a hereditary prion disease previously termed 'early onset prion disease with prominent psychiatric features' (Laplanche *et al.* 1999).

- *Huntington's disease-like phenotype 2* is an autosomal dominant Huntington's disease like illness caused by a trinucleotide repeat expansion in the junctophilin3 gene on chromosome 16q24 (Margolis *et al.* 2005). It occurs mostly in patients of African descent and causes chorea, dystonia or Parkinsonism coupled with dementia; acanthocytosis may occur (Walker *et al.* 2003).

- *Huntington's disease-like phenotype 3* is an autosomal recessive disorder with chorea, dystonia, myoclonus, ataxia or spasticity with dysarthria, seizures, and dementia. It is linked to a locus at chromosome 4 (Kambouris *et al.* 2000; Margolis *et al.* 2001).

- *Huntington's disease-like phenotype 4* is the SCA17 mutation on chromosome 6 (see chapter 39) which can produce a Huntington's disease like illness in some patients (Bauer *et al.* 2004).

- The *familial dystonia with facial myokymia syndrome* causes chorea with facial myokymia in children or adolescents. It is initially intermittent and becomes more persistent. Cognition is not affected (Fernandez *et al.* 2001).

- *Neuroferritinopathy* (Section 40.5.4), *dentatorubral-pallidoluysian atrophy* (Section 39.9.6), and *benign hereditary chorea* (Section 40.5.5) are also important differential diagnoses.

40.5.4 Neuroferritinopathy

Neuroferritinopathy is an autosomal dominant condition causing a variety of movement disorders including chorea resembling Huntington's disease, dystonia usually cranially, and Parkinsonism (Curtis *et al.* 2001). Some patients develop generalized or late onset dystonia (Chinnery *et al.* 2003; Mir *et al.* 2005). In others there can be spasticity, psychosis, or ataxia (Maciel *et al.* 2005). The most typical presentations are a Huntington's like illness or prominent bulbar dystonia. The condition is caused by a variety of pathogenic mutations of the ferritin light polypeptide gene on chromosome 19q13.3 (Curtis *et al.* 2001). There are low serum ferritin levels and there are iron and ferritin containing inclusions in brain (Mancuso *et al.* 2005) as well as in muscle and nerve (Schroder 2005). Brain MR scanning shows characteristic pallidal and nigral high signal lesions on T2-weighted images, sometimes with cyst formation (Fig. 40.16). Diagnosis is by DNA testing; there is no effective treatment.

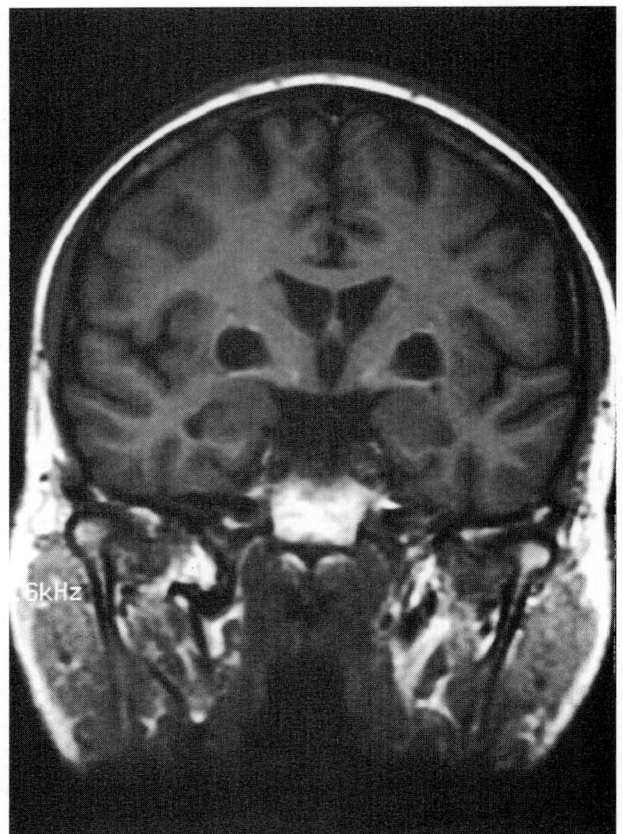

Fig. 40.16 MR scan appearance in pallidal necrosis. The patient had dystonia and Parkinsonism secondary to neuroferritinopathy. (Courtesy of Dr A Bowden.)

40.5.5 Benign hereditary chorea

This is an autosomal dominant disorder causing non progressive chorea without mental impairment (Pincus and Chutorian 1967). Penetrance of the gene is slightly reduced, especially in females. There is no progression, cognition is unaffected, and life expectancy is normal; the diagnosis should not be made if there are other neurological features or lack of a family history (Schrag *et al.* 2000b). The onset is in childhood whereas Huntington's disease beginning at this age is usually severe with a rigid presentation. Some reported cases have had ataxia, pyramidal signs, dysarthria, or tremor and in others there has been progression of the chorea and even cognitive decline. In patients with such additional signs the diagnosis should be regarded with suspicion as families thought to have benign hereditary chorea have turned out to have the Huntington's disease(MacMillan *et al.* 1993a) or other disorders such as ataxia telengiectasia or dystonia. The cause of benign hereditary chorea is mutations of the *TITF1* gene on chromosome 14q (Breedveld *et al.* 2002). Pathologically there is subtle loss of striatal *TITF1* associated cells (Kleiner-Fisman *et al.* 2005).

40.5.6 Drug induced chorea

Tardive dyskinesia caused by dopamine receptor blocking drugs is the most important example of drug induced chorea (Section 40.9.2). Chorea is also caused by levodopa in patients with Parkinson's disease and is the main type of involuntary movement in dyskinetic patients (Section 40.3.1). Other causes of drug induced chorea are rare and often based on one or a few reports only. The most commonly used drugs with this propensity are anticonvulsants, tricyclic antidepressants, selective serotonin reuptake inhibitor antidepressants, and calcium channel blockers. Anticholinergics occasionally cause chorea which may cause confusion in patients receiving these drugs for dystonia (Nomoto *et al.* 1987).

40.5.7 Sydenham's chorea

This condition was described in the 17th century by Dr Thomas Sydenham and the association with rheumatic fever noted by Richard Bright in 1831. Previously a common condition and referred to as rheumatic chorea or St Vitus' dance, it is now rare in developed countries but more common overseas especially in association with poverty (Walker *et al.* 2005). Sydenham's chorea is triggered by a preceding infection with a group A streptococcus although there may be no clear history of this. An immunological cross reaction between the streptococcal M protein and brain epitopes causes an autoimmune encephalitis with anti-basal ganglia antibodies (Martino and Giovannoni 2004). It is now clearthat Sydenham's chorea is part of a wider spectrum of neurological and neuropsychiatric sequelae of streptococcal infection (Section 40.10).

The condition occurs mainly in children and adolescents, more often in girls. There is a subacute onset of chorea, muscular weakness, hypotonia, and often prominent mental changes with emotional and behavioural abnormalities (Cardoso *et al.* 1997). The chorea is unilateral in 20 per cent and some patients develop frank hemiballismus (Vidakovic *et al.* 1994) (Fig. 40.17). There may be dysarthria, slow eye movements, and ataxia is common. Behavioural and psychological changes include personality changes, obsessive compulsive behaviours, irritability, emotional changes, aggression, anxiety, and mood disturbances (Maia *et al.* 2005). Occasionally there are tics and oculogyric crises (Mercadante *et al.* 1997). Carditis and arthritis often coexist with chorea (Cardoso *et al.* 1997).

The EEG shows slow wave changes while MR brain scans are usually normal but high signal lesions of the basal ganglia may be seen (Castillo *et al.* 1999). The streptococcal serological tests anti-Streptolysin O and DNase B levels are positive in about 80 per cent of cases; anti-basal ganglia antibodies are positive in all acute cases of Sydenham's chorea (Church *et al.* 2002) and many chronic cases (Harrison *et al.* 2004). Positron emission tomography scanning reveals increased striatal glucose metabolism (Weindl *et al.* 1993).

The illness is usually at its worst for several weeks followed by gradual resolution over 3–9 months. Some patients develop persistent chorea or mental changes and in others there is later recurrence of chorea (Korn-Lubetzki *et al.* 2004) usually within 2 years but recurrences after 10 years or even in old age are described; some patients have multiple relapses. This may cause diagnostic confusion with Huntington's disease. There may be weakly positive streptococcal serology and basal ganglia lesions on brain MR scans in recurrent cases (Moreau *et al.* 2005). Reactivation may be triggered by recurrent infection, pregnancy, or intercurrent illness. Relapses in older patients are possibly due to residual striatal damage and age related changes (Gibb and Lees 1989; Harrison *et al.* 2004).

In the acute stage the chorea may be reduced with neuroleptic medication such as haloperidol or sulpiride. Sodium valproate is sometimes effective. Immunotherapy with steroids, intravenous immunoglobulin, and plasma exchange have been used to shorten recovery time (Garvey *et al.* 2005). Penicillin prophylaxis is important

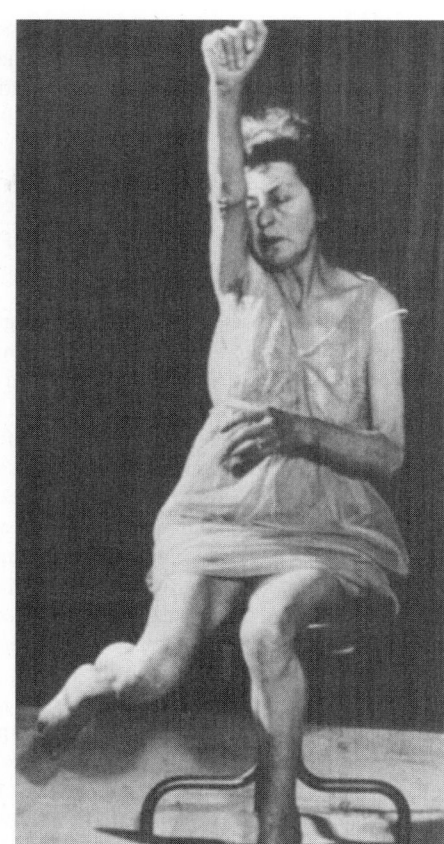

Fig. 40.17 A patient with right-sided hemiballismus. Note the large amplitude of the involuntary movements.

to prevent recurrences (Gebremariam 1999; Diaz-Grez *et al.* 2004) but duration of treatment is uncertain with recommendations for 5 years, up to age 18 years and lifelong—especially in those with a high risk of re-infection or severe cardiac or renal disease.

40.5.8 Chorea in systemic disease

Chorea may be seen as a complication of the *antiphospholipid syndrome* associated with false positive syphilis serology, anticardiolipin antibodies, and the lupus anticoagulant. Other features include thrombotic events such as stroke, deep venous thrombosis, migraine, recurrent abortion, and valvular heart disease. *Systemic lupus erythematosus* may also cause chorea, with or without the antiphospholipid syndrome. *Polycythemia rubra vera* is occasionally associated with chorea, probably also due to basal ganglia ischaemia. *Hyperthyroidism* and other metabolic derangements including hypoglycaemia and hyperglycaemia may occasionally cause choreiform movements. Bilateral or hemi-chorea has been described in non-ketotic hyperglycaemia with associated putaminal lesions on MR scans (Oh *et al.* 2002). There are rare reports of *paraneoplastic* chorea with inflammatory CSF changes and striatal lesions on neuroimaging (Vernino *et al.* 2002).

40.5.9 Chorea associated with pregnancy and oral contraceptives

Chorea occasionally occurs in pregnancy. This is much less common than it was, probably due to the decline in Sydenham's chorea.

Typically, affected women had a history of previous Sydenham's chorea several years previously but this was not invariable. Other possible causes to consider are Huntington's disease, antiphospholipid syndrome and systemic lupus erythematosus. Most cases resolve spontaneously after delivery. Chorea may occasionally develop in association with oestrogen containing oral contraceptives or topical creams (Nausieda *et al.* 1979) but the mechanism of this is unclear. The chorea resolves if the oestrogen is stopped.

40.5.10 Neuroacanthocytosis

Acanthocytosis may occur in association with neurological disease in abetaliporoteinaemia, hypobetaliporoteinaemia which usually causes cerebellar ataxia, neurodegeneration with brain iron accumulation (Section 40.4.12), and neuroacanthocytosis. In neuroacanthocytosis, or chorea-acanthocytosis, there are peripheral blood acanthocytes but normal serum lipoproteins in association with neurological involvement. The disorder is inherited as an autosomal recessive trait. Typically there is an onset of chorea in adult life, at around 35 years, with various additional features including orofacial dystonia, tics, Parkinsonism, cognitive impairment, psychiatric abnormalities, self mutilation, and seizures (Hardie *et al.* 1991; Kartsounis and Hardie 1996). Patients typically have areflexia due to an axonal peripheral neuropathy.

Pathologically there is atrophy of the stiatum with neuronal loss and gliosis but the cerebral cortex is spared. Involvement of the substantia nigra and pallidum is variable. The diagnosis is suggested by the combination of chorea, with or without the other movement disorders, and areflexia. A fresh wet blood film shows excessive numbers of acanthocytes (Fig. 40.18), the blood creatine kinase is elevated, and lipoproteins are normal. CT scans may show caudate atrophy similar to that seen in Huntington's disease and positron emission tomography scanning reveals reduced strital glucose metabolism and dopaminergic function. Neurochemically there is striatal dopamine depletion (De Yebenes *et al.* 1988).

Neuroacanthocytosis is caused by over 70 mutations of the *VPS13A/chorein* gene on chromosome 9q21 (Rampoldi *et al.* 2001; Danek and Walker 2005).

There is no effective treatment for the condition which is progressive over 3–22 years (Rinne *et al.* 1994a).

Some patients with the *McLeod syndrome* have similar neurological features (Danek *et al.* 2001). In this X-linked condition there

is acanthocytosis and abnormal expression of Kell blood group antigens with absent Kx antigen on erythrocytes. Several point mutations have been detected in the XK gene on the X chromosome. It is not known how many patients with neuroacanthocytosis have the McLeod syndrome but the latter has later onset, no orofacial dyskinesia, or self mutilation; however cognitive and psychiatric change is common as is a cardiomyopathy.

40.5.11 Other causes of chorea and ballism

Chorea occurs as a feature of several other neurological disorders including dentatorubral-pallidoluysian atrophy and other spinocerenellar degenerations, mitochondrial diseases (Nelson *et al.* 1995), various focal lesions of the basal ganglia (Lee *et al.* 1994) including tumours and paroxysmal dystonias (Section 40.4.5). Dentatorubro pallidoluysian atrophy is noteworthy because it may resemble Huntington's disease closely and is inherited as an autosomal dominant disorder (see Section 39.9.6). Chorea has been reported as a complication of cerebrovascular disease in older patients with multiple lacunar cerebral infarcts (Ristic *et al.* 2002), probably causing basal ganglia damage (Bhatia *et al.* 1994). Some patients develop chorea afer cardiopulmonary bypass usually for cardiac surgery in children. Recovery is variable and although the mechanism is unknown, basal ganglia ischaemia is possible. Senile chorea refers to the development of chorea in old age and is a heterogeneous condition. Many of these patients have Huntington's disease and some have cerebrovascular disease or late recurrence of Sydenham's chorea but others are unexplained.

Ballism is a clinically striking form of severe chorea which is commonly unilateral (Fig. 40.18) and associated with lesions of the subthalamic nucleus (Lee and Marsden 1994a). Vascular lesions account for most cases but imaging studies often fail to demonstrate subthalmic lesions and other causes including encephalitis, systemic lupus erythematosus, Sydenham's chorea, and metabolic disorders are sometimes responsible (Vidakovic *et al.* 1994).

40.6 Tic disorders

40.6.1 Diagnosis and classification

Tics are rapid brief jerks usually affecting the face, head, or upper body. They appear more semipurposeful and deliberate than dystonia, chorea, or myoclonus. They are repetitive and stereotyped and may be simple or complex. Simple motor tics are single brief movements such as blinking, head jerking, grimacing, or shrugging a shoulder. They may be brief or more sustained 'dystonic tics'. Complex motor tics involve more elaborate stereotyped actions such as picking at the body or objects, rubbing or manipulative movements, mimicry, or gestures. Vocal tics may be noises such as coughs, sniffs, or squeals. Complex vocal tics consist of words and phrases. Tics are involuntary although they can be suppressed more than other involuntary movements; this results in rising subjective tension followed by an exacerbation of tics as they are released. They are worse with anxiety or in certain social situations and relieved by distraction. Tics are preceded by an urge which is irresistable and relieved briefly by the movement. Some patients have more definite premonitory local sensations, 'sensory tics', and others describe an awareness of something in an external object 'phantom tics' which leads to a repetitive manipulation of the relevant object (Karp and Hallett 1996). The causes of tics are shown in Table 40.17.

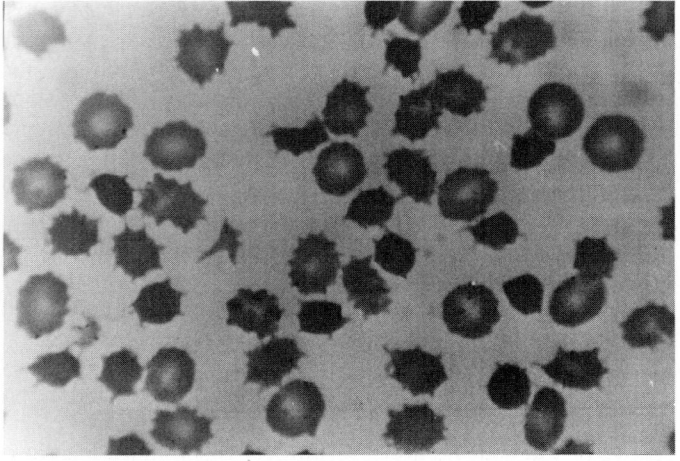

Fig. 40.18 Peripheral blood acanthocytes in neuroacanthocytosis.

Table 40.17 Causes of Tics

Primary tic disorders	Tourette's syndrome
	Transient tic disorder*
	Chronic single tic disorder*
	Chronic multiple tic disorder*
	Chronic multiple tic disorder*
Secondary tic disorders	Huntington's disease
	Neuroacanthocytosis
	Neurodegeneration with brain iron accumulation
	Sydenham's chorea
	Post-streptococcal psychiatric or movement disorders
	Drugs, mainly neuroleptics
	Encephalitis
	Focal basal ganglia lesions
	Rett syndrome

* may be variants of Tourette's syndrome.

Tourette's syndrome starts before the age of 21, typically in childhood, causing multiple motor and subsequently one or more vocal tics; there must be a waxing and waning course with increasing severity over time; duration more than one year and no evidence of any other underlying cause.

Transient tic disorder is characterized by duration less than one year with chronic motor or vocal tics. The relationship of these tic disorders to Tourette's syndrome is uncertain.

40.6.2 Pathophysiology of tics

The mechanism of tics is unknown but is suspected to involve abnormal basal ganglia function. This is because some diseases in which tics occur have definite basal ganglia pathology such as Huntington's disease and neuroacanthocytosis. Neuroimaging studies have shown slightly altered basal ganglia morphology in Tourette's syndrome (Singer *et al.* 1993; Hyde *et al.* 1995) and enlarged brainstem gray matter has been reported recently (Garraux *et al.* 2006). No clear pathogenetic conclusions are possible from these studies at present. Altered brain dopaminergic function may be involved as tics may be both suppressed by dopamine blockers and caused by dopamine receptor hypersensitivity after neuroleptic therapy (Bharucha and Sethi 1995). Positron emission tomography studies indicate abnormal nigrostriatal dopamine function (Singer *et al.* 2002). There is no premovement potential prior to tics, consistent with their involuntary nature (Obeso *et al.* 1981) and cortical excitability is increased, suggesting impaired inhibition (Ziemann *et al.* 1997). One pathological study has reported subtle alterations of parvalbumin positive cells in the basal ganglia of Tourette's syndrome patients (Kalanithi *et al.* 2005).

40.6.3 Gilles de la Tourette's syndrome

This condition was described in 1885 by Tourette in nine patients with motor and vocal tics. The frequency of is about 1–10 per 1000 children and adolescents (Singer 2005). The inheritance of Tourette's syndrome is controversial. Although an autosomal dominant model with reduced penetrance has been considered (Pauls and Leckman 1986), a polygenic model may be more pertinent (Comings *et al.* 2000) as a Mendelian model is unlikely (Seuchter *et al.* 2000). Very rare cases have been associated with mutations or

a chromosomal inversion affecting the *SLITRK1* gene on chromosome 13q31.1 (Abelson *et al.* 2005).

Tourette's syndrome has also been linked to group A beta haemolytic streptococcal infection, usually pharyngitis, because of similarities between Tourette's syndrome and paediatric autoimmune neuropsychiatric disorders associated with streptococcal infection (Section 40.10). Affected children develop neurological and behavioural symptoms with serological evidence of streptococcal infection (Swedo *et al.* 1998). It has been proposed that after such streptococcal infection, cross reacting anti-basal ganglia antibodies cause the neuropsychiatric disorder in children (Dale *et al.* 2005) although this has been disputed (Singer *et al.* 2004). In Tourette's syndrome, some patients have both serological evidence of the streptococcal infection and anti-basal ganglia antibodies (Church *et al.* 2003) but this has not been seen consistently (Singer *et al.* 2005). At present, routine screening of Tourette's syndrome patients for serological evidence of streptococcal infection or anti-basal ganglia antibodies is not recommended (Dale 2005). At present, there is inadequate evidence for an autoimmune cause of Tourette's syndrome (Singer 2005).

Onset is usually in childhood around 6–7 years and worsens until 9–11 years. The condition is worse in boys. The overall prognosis is variable. Tourette's syndrome tends to be worst during childhood and then improve during adolescence in over 80 per cent of cases (Bloch *et al.* 2006). Remissions may occur in later adult life but not in all cases; many adults have persistent tics but these are moderate or severe in only 24 per cent (Goetz *et al.* 1992). In one follow up study of adults with Tourette's syndrome, tics disappeared in 26 per cent, reduced in 46 per cent, were unchanged in 14 per cent, and worsened in 14 per cent (Erenberg *et al.* 1987).

Simple motor tics include jerking, blinking, grimacing, shrugging, or head movements. In complex tics there are more elaborate movements such as waving, touching, jumping, the mimicking of echopraxia, and obscene gestures of copropraxia. The movements may be mild and intermittent or continuously severe sufficient to cause physical injury such as cervical myelopathy. Vocal tics similarly vary from simple noises such as sniffing, coughing, grunting, whistling, to complex vocalizations such as echolalic repeating of others, palilalic repetition of the same syllable, word, or phrase, and coprolalicutterance of obscenities. Some patients report internal palilalic or coprolalic thoughts. Although coprophenomena are a well known feature of Tourette's syndrome, they are not required for the diagnosis and are seen only in a minority of patients. An important feature of the tics is that they are socially inappropriate and embarrasing to the patient. In most the repertoire of tics changes and the disorder shows exacerbations and periods of partial improvement either during the day or over months or years (Garcia-Ribes *et al.* 2003).

Tourette's syndrome patients often have psychological disturbances (Robertson 1994). Obsessive compulsive features occur in about 50 per cent of cases (Frankel *et al.* 1986; Jankovic 1987). Patients ruminate about recurring thoughts and have complex repetitive rituals which may impede normal daily routines. Compulsions and tics are closely interwoven in many of these patients. Compulsive touch or manipulation of objects or the body could be viewed either as compulsions or as complex motor or phantom tics. Many patients also display evidence of attention deficit hyperactivity disorder with impulsiveness, inattention, and distractability (Robertson 2006). This may cause serious educational, employment, and social difficulties. Self mutilation of various forms also occurs in Tourette's syndrome and is usually of a repetitive compulsive type (Robertson *et al.* 1989). Other psychological problems such as depression and anxiety are likely to be secondary to the handicaps experienced by these patients. Some patients have evidence of personality disorder with disruptive, aggressive rages and antisocial behaviour (Singer 2005). Intelligence is normal in Tourette's syndrome but educational difficulties are common, mainly due to the attentional disorder, erroneously suggesting a learning disability (Abwender *et al.* 1996).

Investigations are unhelpful and the diagnosis is clinical, depending largely on the history. One of the most useful diagnostic points is the early age of onset. A movement disorder appearing in adult life is unlikely to be Tourette's syndrome. This is a key distinction from the secondary tic disorders in Table 40.17 which start later and usually have other clinical differences. Tourette's syndrome is similarly differentiated from conditions such as hemifacial spasm or focal dystonias in which the onset will be much later, with different movements.

Management of Tourette's syndrome is difficult. Mildly affected patients do not necessarily need drug treatment. Explanation, reassurance, and information is often adequate especially in children. Only if tics are severe enough to cause significant disability should medication be considered but, with the exception of neuroleptic drugs, few controlled trials have been reported (Singer 2005). Clonidine, in doses up to 0.6 mg daily, is sometimes effective but this has been difficult to demonstrate in controlled studies especially at doses above 0.3 mg. It may be helpful if there is associated attention deficit hyperactivity disorder. Clonazepam can also be helpful and can be taken at night as a single dose. Baclofen is also effective in some patients (Singer 2005).

More severe tics usually require a neuroleptic such as sulpiride, haloperidol, or pimozide. Several controlled studies have shown neuroleptics to be the most effective therapy for tics but they are not completely suppressed and side effects may occur. Tardive effects are rare in Tourette's syndrome but can occur including tardive dyskinesia (Bharucha *et al.* 1988) and tardive dystonia. There is a risk of sudden cardiac death with neuroleptics, especially pimozide, due to cardiac QT interval prolongation and torsades de pointe arrhythmias (Sindo and Jorgensen 2002). Risperidone has been shown in several studies to be effective and well tolerated (Scahill *et al.* 2003) but can be associated with extrapyramidal effects similar to older neuroleptics. There has been some interest in the use of atypical neuroleptics such as olanzapine 2.5–20 mg daily (Budman *et al.* 2001; Van den Eynde *et al.* 2005) and quetiapine (50–100 mg daily) (Mukaddes and Abali 2003) but no informative trials have been completed and so the role of these drugs is currently uncertain.

Tetrabenazine is effective in Tourette's syndrome and is less likely to cause tardive dyskinesia; it may have side effects such as sedation or depression but is usually well tolerated (Kenney and Jankovic 2006). A reduction in tics has been reported with nicotine (Sanberg *et al.* 1997), levetiracetam (Awaad *et al.* 2005), and low dose dopamine agonists including pergolide 50–150 µg tds and ropinirole 250–500 µg bd (Gilbert *et al.* 2003; Anca *et al.* 2004). Some severe motor and vocal tics have been controlled with local injections of botulinum toxin (Awaad 1999; Singer 2005).

Obssessive-compulsive symptoms may be improved with selective serotonin reuptake inhibitors drugs such as fluoxetine 20–40 mg daily or citalopram 20–40 mg. Attention deficit hyperactivity disorder has been improved with tricyclics, methyphenidate (Kurlan and Trinidad 1995), or selegilene (Jankovic 1993). However stimulant drugs may worsen tics.

Although patients with paediatric autoimmune neuropsychiatric disorders associated with streptococcal infections improve with plasma exchange or intravenous immunoglobulin (Perlmutter *et al.* 1999), such improvement has not been seen in Tourette's syndrome using intravenous immunoglobulin (Hoekstra *et al.* 2004).

Surgical lesions including frontal leucotomy, anterior cingulotomy, limbic leucotomy, pallidotomy, and internal capsulotomy have been carried out with partial effect in patients with intolerable drug resistant symptoms. However it is difficult to recommend such procedures with little evidence of effectiveness. Deep brain stimulation of the pallidum or thalamus has been promisingly effective in some patients (Visser-Vandewalle *et al.* 2004; Diederich *et al.* 2005; Houeto *et al.* 2005; Ackermans *et al.* 2006).

40.6.4 Other tic disorders

Other causes of tics are shown in Table 40.17 and are discussed elsewhere. Transient isolated tics are common in normal children and last less than a year before disappearing. They are of no significance. Only if they persist for longer than a year and are combined with other motor and vocal tics can a diagnosis of Tourette's syndrome be made. The diagnosis in patients with chronic motor or vocal tics, but not both, is uncertain but such patients may have a variant form of Tourette's syndrome (Kurlan *et al.* 1988). Tics also occur in adolescents with developmental disorders such as Rett's syndrome (Sections 9.6.2; 40.4.15). Some of the behavioural features of Tourette's syndrome are seen in children with Asperger's syndrome some of whom have tics, indicating an overlap between these two disorders (Nass and Gutman 1997).

40.7 Myoclonus

40.7.1 Diagnosis and classification

Myoclonus refers to sudden shock-like jerks due to bursts of muscle activity lasting 50–300 ms. There may also be brief losses of muscle tone; these are referred to as negative myoclonus. In general, myoclonic jerks are involuntary, frequent, increased by movement, and disrupt voluntary movements. Negative myoclonus causes brief loss of muscle tone, asterixis, or postural lapses. The classifications of myoclonus are summarized in Table 40.18. These different classifications overlap; for example, cerebral anoxia or metabolic disturbances may cause cortical or reticular reflex myoclonus.

Although such a classification seems complex, most causes of myoclonus fit into one of the clinical settings shown in Table 40.19.

40.7.2 Cortical myoclonus

This is the most common type and produces myoclonus which may be focal or multifocal depending on the extent of the cortical pathological process. Due to the anatomy of cortical representation and the nature of cortical motor function, the face and hands are mainly affected with the myoclonus tending to be focal or multifocal (Caviness and Brown 2004). Occasionally, other lesions cause focal cortical myoclonus in a foot or leg. The jerks are often spontaneous but may be strikingly stimulus sensitive and exacerbated by movement. High frequency bursts are common, producing

Table 40.18 Approaches to the classification of myoclonus

Clinical features	Anatomical extent	Generalized, multifocal, segmental, focal
	Exacerbating factors	Spontaneous, action, reflex
	Effect on muscle activity	Positive, negative
	Appearance	Rhythmic, irregular
Pathophysiological mechanism and site of origin	Cortical myoclonus	
	Reticular reflex myoclonus	
	Other subcortical myoclonus	
	Palatal myoclonus	
	Spinal myoclonus	
	Peripheral myoclonus	
Aetiology	Physiological	Hypnic jerks, fragmentary myoclonus of sleep, hiccoughs
	Hereditary myoclonus-dystonia (Essential myoclonus)	
	Epilepsies	Progressive myoclonic epilepsies (Unverricht–Lundborg)
		Idiopathic generalized epilepsies
		Other myoclonic epilepsies
		Focal epilepsy e.g. epilepsia partialis continua
	Dementias	Alzheimer's disease, Creutzfeldt–Jacob disease, Frontotemporal dementia, Dementia with Lewy Bodies, Rett syndrome
	Metabolic disorders	Hepatic encephalopathy
		Uraemia, dialysis
		Hyponatraemia
		Hypo/hyperglycaemia
		Biotin deficiency*
		Mitochondrial disorders*
	Storage disorders	Lafora body disease*
		Neuronal storage diseases*
		Sialidosis
		Neuronal ceroid lipufuscinoses
		GM2 gangliosidosis
		Neimann–Pick type C
		Action myoclonus and renal failure*
	Infections of the central nervous system	Subacute, sclerosing panencephalitis, viral encephalitis, post streptococcal
	Parkinsonism and other basal ganglia disorders	Parkinson's disease, corticobasal degeneration, multiple system atrophy
		Huntington's disease, Huntington's disease-like type 3, dentatorubropallidoluiysian atrophy, dystonias
	Cerebellar syndromes	Spinocerebellar degenerations*, Friedreich's ataxia, ataxia telangiectasia' 'Ramsay Hunt syndrome'
	Posthypoxic myoclonus	
	Gastrointestinal	Whipple's disease*, Coeliac disease*
	Toxic encephalopathy	Drugs, toxins
	Inflammatory	Opsoclonus-myoclonus
	Focal lesions of the cortex, brainstem, spinal cord, or peripheral nervous system	Tumour, arteriovenous malformation, encephalitis, ischaemia, trauma, demyelination, degeneration

Conditions marked * account for most cases of 'progressive myoclonic ataxia' or the Ramsay Hunt syndrome (see Section 40.7.8).

localized rhythmic low amplitude jerks, cortical tremor. The electromyographic bursts during the jerks are short, 10–50 ms, and associated with a preceding, 10–40 ms, contralateral EEG cortical potential although usually this can be recorded only with specialized back-averaging techniques (Shibasaki and Hallett 2005). If there is stimulus sensitivity, abnormally large somatosensory evoked potentials can be recorded over the corresponding area of sensorimotor cortex (Hallett *et al.* 1979). In patients with generalized rather than multifocal cortical myoclonus there is more rapid spread of the cortical discharge due to impaired cortico-cortical and transcallosal inhibition (Brown *et al.* 1996b). Cranial nerve innervated musculature becomes activated in a strictly rostral-caudal sequence. Generalized or multifocal myoclonus may be spontaneous or triggered by tactile, visual, or auditory stimuli and may be seen with various forms of epilepsy or widespread cortical dysfunction due to anoxia, metabolic, toxic, or degenerative disorders. Focal cortical myoclonus is a form of epilepsy and if spontaneous and repetitive it is the same as epilepsia partialis continua. It may be spontaneous or stimulus sensitive. Focal myoclonus is a symptom of localized cortical pathology such as a structural lesion, encephalitis, or an area of focal degeneration such as in corticobasal degeneration.

Table 40.19 Clinical presentations of myoclonus

Normal individuals with jerks in sleep	Hypnic jerks Fragmentary myoclonus of sleep
Pure myoclonus +/- dystonic elements	Hereditary myoclonus-dystonia (= 'essential myoclonus')
Epilepsy + myoclonus	Idiopathic generalized epilepsy e.g. juvenile myoclonic epilepsy Childhood myoclonic epilepsy syndromes Focal motor seizures
Progressive myoclonic epilepsy Progressive myoclonic ataxia (Ramsay Hunt syndrome)	Unverricht–Lundborg disease Storage disorders Mitochondrial disorders Coeliac / Whipple's disease Dentatorubropallidoluysian atrophy
Dementia + myoclonus	Creutzfeldt–Jacob disease Alzheimer's disease Dementia with Lewy bodies Frontotemporal dementias Rett syndrome Subacute, progressive panencephalitis
Acute encephalopathy + myoclonus	Metabolic Toxic / drugs Acute central nervous system infection Post anoxic
Myoclonus within the context of another recognizable disorder	Parkinson's disease Corticobasal degeneration Multiple system atrophy Dystonias Huntington's / Huntington disease like disorders spinocerebellar degenerations Friedreich's ataxia Ataxia telengiectasia Opsoclonus myoclonus

40.7.3 Subcortical myoclonus

In this form of myoclonus the jerks are generated between the cortex and the spinal cord, usually in the brainstem. There are two forms; reticular reflex myoclonus and so-called palatal myoclonus; the latter is more properly regarded as a form of tremor.

In *reticular reflex myoclonus* the neuronal discharge arises in the brainstem reticular formation, spreads up the brainstem, and then induces a generalized cortical discharge. Cranial nerve innervated muscles are activated in a caudal-rostral sequence and the body jerks are generalized, mainly axial, very stimulus sensitive, and exacerbated by movement (Caviness and Brown 2004). This form of myoclonus is rare but can be seen post-anoxically, and in metabolic disorders, intoxications, brainstem encephalitis (Kullmann *et al.* 1996), and hyperekplexia (Brown *et al.* 1991b).

Palatal myoclonus is characterized by rhythmic movements of the palate at 0.5–3 Hz. There may be synchronous movements of the tongue, face, neck, or arms. In essential palatal myoclonus the movements are due to contraction of the tensor veli palatini and the patients report ear clicks; the condition disappears in sleep and there is no additional neurological impairment or evidence of an underlying cause. In symptomatic palatal myoclonus there are rythmic contractions of the levator veli palatini which are not associated with ear clicking and persist in sleep. This latter form of

palatal myoclonus is associated with lesions of the Guillain–Mollaret triangle, the dentate-red nucleus-inferior olivary circuit, and hypertrophy of the inferior olivary nucleus (Deuschl *et al.* 1996). Various brainstem lesions have been associated with symptomatic palatal myoclonus including strokes, demyelination, and degenerative disorders (Howard *et al.* 1993). A similar condition may affect the stapedius muscle causing 'middle ear myoclonus'; this causes a fluttering tinnitus and may occur alone or coexist with essential palatal myoclonus (Oliveira *et al.* 2003).

40.7.4 Spinal myoclonus

Localized spinal lesions may produce segmental or focal myoclonic jerking in body areas supplied by the affected spinal segments. The jerks are rhythmic and spontaneous and persist in sleep. Spinal myoclonus is not usually stimulus sensitive or aggravated by movement. A variety of spinal lesions may be responsible including myelitis, cervical sponylosis, tumours, demyelination, ischaemia, and arteriovenous malformations.

In propriospinal myoclonus there are flexion or extension jerks of the neck, trunk, hips, and knees which are both spontaneous and stimulus sensitive. The jerks are axial but unlike brainstem myoclonus, do not involve the face and are not stimulus sensitive (Caviness and Brown 2004). The myoclonus arises from a focal lesion and spreads up and down the cord to produce an extensive and long duration axial jerk (Brown *et al.* 1991c). Causes include trauma, myelitis, and tumour.

40.7.5 Myoclonus of peripheral origin

Spontaneous focal myoclonus in a limb or the trunk has occurred with nerve root, plexus or peripheral nerve lesions (Marsden 1994). Causes have included peripheral nerve tumours, radiotherapy, electrocution, surgery and spondylotic radiculopathy. Such lesions appear to induce repetitive discharges of groups of anterior horn cells by an unknown mechanism.

40.7.6 Essential myoclonus and hereditary myoclonus dystonia

In benign essential myoclonus there is the onset of sudden myoclonic jerks affecting the face, arms, and trunk . The myoclonus is sometimes sensitive to sound and may be synchronous and bilateral or asymmetrical and multifocal. Severity is variable ranging from minimal twitching to severe myoclonus and there is autosomal dominant inheritance with variable penetrance. The disorder is benign and non progressive and there are no additional neurological features. The EEG is normal and the myoclonus appears to be of subcortical origin (Quinn 1996). The jerks are characteristically sensitive to alcohol. Some of these patients also have signs of dystonia (Fahn and Sjaastad 1991) and families described as hereditary alcohol responsive myoclonic dystonia almost certainly have the same disorder (Quinn 1996). Many of these patients also have dystonic jerks, mainly of the upper body (Asmus and Gasser 2004) and the disorder is now known to be the same disorder as hereditary myoclonus-dystonia, DYT11 or DYT15 (Section 40.4.5).

40.7.7 Cerebral anoxic myoclonus

Myoclonus may be a prominent feature in patients who have survived severe cerebral anoxia, the *Lance Adams syndrome*, typically following cardiorespiratory arrest due to asthma (Polesin and

Stern 2006; Venkatesan and Frucht 2006). There may be additional cognitive impairment, epilepsy, spasticity, other movement disorders, and ataxia but this is variable. Most cases have stimulus sensitive, generalized or multifocal, cortical action myoclonus but reticular reflex myoclonus may occur occasionally (Brown *et al.* 1991d). Negative myoclonus with postural lapses also occurs and may be particularly disabling. Although some patients are severely affected due to action myoclonus, postural lapses, and other deficits, others may have a relatively pure myoclonic disorder and can improve gradually (Werhahn *et al.* 1997).

40.7.8 Spinocerebellar degenerations, progressive myoclonic epilepsies, and the Ramsay Hunt syndrome

This is a heterogeneous group of disorders which are grouped together because they share certain clinical features (Berkovic *et al.* 1986; Shahwan *et al.* 2005). Clinically there may be a combination of myoclonus and epilepsy with a variable degree of dementia or ataxia, a progressive myoclonic epilepsy. Alternatively myoclonus may be coupled with cerebellar ataxia and less prominent epilepsy and dementia, progressive myoclonic ataxia, or the Ramsay Hunt syndrome. It is important to note that these are variations of the same clinical presentation which differ in emphasis but share a common differential diagnosis (Table 40.20). Several of these causes are very rare and only the principal causes will be discussed. In some patients, especially those with the Ramsay Hunt syndrome, a definite clinical or pathological diagnosis is elusive (Marsden *et al.* 1990).

◆ *Unverricht–Lundborg disease* is an autosomal recessive disorder caused by mutations of the Cystatin B gene on chromosome 21q22.3 (Pennacchio *et al.* 1996; Lalioti *et al.* 1997;). The same disorder has been reported as Baltic, Finnish, and Mediterranean myoclonus. The main Cystatin B gene locus on chromosome 21 has been designated *EPM1A*. There is another gene, *EPM1B* on chromosome 12 (Berkovic *et al.* 2005; El-Shanti *et al.* 2006). The onset is at 8–13 years with severe epilepsy and myoclonus; cerebellar ataxia, dysarthria, and mild intellectual impairment develop later and are less prominent. Unverricht–Lundborg disease may be a common cause of myoclonus with no obvious underlying cause; in one series, 7 out of 21 cases of undiagnosed myoclonus in the Netherlands had *EPM1A* mutations (de Haan *et al.* 2004). The prognosis is poor but some patients survive into adult life. Levetiracetam may be effective (Kinrions *et al.* 2003).

◆ *Dentatorubro-pallidoluysian atrophy* may also cause a Ramsay Hunt syndrome with ataxia and myoclonus (Vinton *et al.* 2005) (see Section 39.9.6). Myoclonus may rarely occur in SCA2 and the olivopontocerebellar atrophy form of multiple system atrophy (Wenning *et al.* 1994a) but is usually inconspicuous.

◆ *Whipple's disease* may cause a variety of neurological syndromes including focal brain lesions, dementia, supranuclear gaze palsy, and myoclonus (Louis *et al.* 1996) (Section 42.5.8). These may occur independently of gastrointestinal malabsorption or other systemic features of the disorder. In some patients with neurological involvement, there is a rhythmic subcortical myoclonus affecting the extraocular and facial muscles, oculomasticatory myorhythmia, which may also affect the limbs. In most, but not all, the diagnosis is established by small bowel biopsy. It is possible to detect DNA from the causative organism *Tropheryma whippelii* in CSF, bowel, or brain tissue (Lynch *et al.* 1997; Anderson 2000). Treatment is with antibiotics but the optimal regimen has not been determined.

◆ *Coeliac disease* is also associated with a range of neurological complications (Hadjivassiliou *et al.* 2002) including peripheral neuropathy, encephalopathy, myelopathy, dementia, cerebellar ataxia, cerebral calcification, and seizures (Cooke and Smith 1966; Finelli *et al.* 1980). Some patients develop the Ramsay Hunt syndrome (Lu *et al.* 1993) in which there is stimulus sensitive cortical action myoclonus even without clear cortical pathology (Bhatia *et al.* 1995). Pathologically there may be degenerative changes or a central nervous system vasculitis (Rush *et al.* 1986; Mumford *et al.* 1996). The neurological features do not respond to a gluten free diet (Hadjivassiliou *et al.* 2002). Diagnosis is by small bowel biopsy and this should be considered in all cases of Ramsay Hunt syndrome even in the absence of gastrointestinal symptoms.

◆ *Mitochondrial disease* has many possible clinical presentations (Schapira 2006), including progressive myoclonic epilepsy, and is a cause of the Ramsay Hunt syndrome. When combined with the characteristic ragged red fibres on muscle biopsy this is referred to as 'Myoclonic Epilepsy with Ragged Red Fibres', MERRF. The age of onset varies from childhood to adult life with myoclonus, epilepsy, ataxia and sometimes deafness, muscle weakness, dementia, and short stature (Berkovic *et al.* 1989). The severity is variable within families and recurrence risks to relatives are low. It is associated with various mutations of mitochondrial mtDNA. The most common is an A to G transition at position 8344 of the mtDNA genome in the tRNA lysine gene; other cases have a T to C mutation at position 8356 or the A to G 3243 mutation seen more commonly with Mitochondrial Encephalomyopathy, Lactic Acidosis and Stroke like episodes, MELAS (Hanna and Bhatia 1997). *Ekbom's syndrome* and *May–White syndrome* are variants of MERRF, the former with additional subcutaneous lipomata and the latter with deafness. They are associated with the *MERRF 8344* mutation and the latter also with a C insertion at position 7472 (Hanna and Bhatia 1997).

Table 40.20 Causes of progressive myoclonic epilepsy and the Ramsay Hunt syndrome

Spinocerebellar degenerations	Unverricht–Lundborg disease Dentatorubropallidoluysian atrophy
Mitochondrial disorders	Myoclonic epilepsy with ragged red fibres May–White syndrome Ekbom syndrome
Intestinal disease	Coeliac disease Whipple's disease
Other causes	Lafora body disease Creutzfeldt–Jacob disease Sialidosis Neuronal ceroid lipofuscinosis Gaucher's disease GM2 gangliosidosis Biotin deficiency Neuroaxonal dystrophy Action myoclonus-renal failure syndrome

◆ *Lafora body disease* is an autosomal recessive disorder with onset in the second decade of myoclonus, seizures, frequently focal occipital attacks, and rapidly progressive dementia. A less rapidly progressive adult onset variant has been described (Footitt *et al.* 1997). The diagnosis is made by the presence of characteristic inclusions, Lafora bodies, in brain, liver, muscle, or skin biopsies. The biochemical defect is unclear and a gene locus on chromosome 6q24 has been identified by linkage studies (Sainz *et al.* 1997). Various mutations of the *EPM2A* gene have been identified but the function of the gene product, Laforin, is unknown (Minassian *et al.* 2000). It is now clear that Lafora disease is heterogeneous with at least three genes identified; *EPM2A*, *NHLRC1*, and another unknown locus (Ganesh *et al.* 2006). The prognosis is very poor with survival limited to a few years. Zonisamide may be helpful (Yoshimura *et al.* 2001).

◆ *Neuronal ceroid lipofuscinosis* takes various forms and may present in childhood with progressive myoclonic epilepsy, visual failure, and dementia (Berkovic *et al.* 1986) (Section 10.3.2). There is also an adult onset form, Kuf's disease, which presents with progressive myoclonic epilepsy (Sinha *et al.* 2004) or a dementia with extrapyramidal signs (see Section 40.4.9). There is no visual failure and the seizures may be photosensitive. Diagnosis is made by demonstrating characteristic fingerprint bodies or granular osmiophilic deposits in muscle or rectal biopsies (Pasquinelli *et al.* 2004) and can also be established by elevated urinary sediment dolichol levels (Berkovic *et al.* 1988).

◆ *Sialidosis* is caused by deficiency of alpha-N-neuraminidase, sialidase and has two main clinical variants. In late-onset, type I, the myoclonus is associated with bilateral macular cherry-red spots. In infantile-onset, type II, there is skeletal dysplasia, mental retardation, and hepatosplenomegaly but sometimes progressive myoclonic epilepsy. The sialidase, *NEU1* gene on chromosome 6p21.3 has different point mutations in sialidosis patients (Pshezhetsky *et al.* 1997; Seyrantepe *et al.* 2003). The diagnosis may be made by finding elevated urinary sialyloligosaccharides or deficiency of alpha-N-neuraminidase (Berkovic *et al.* 1986).

◆ The *action myoclonus renal failure syndrome* causes childhood or adult onset renal failure due to glomerulosclerosis with seizures, myoclonus, ataxia, and tremors (Badhwar *et al.* 2004). Inheritance is autosomal recessive and pathology is consistent with a storage disorder.

40.7.9 Prion diseases and other dementias

The combination of myoclonus with ataxia or dementia may be seen in several degenerative disorders. Cortical myoclonus is a prominent feature in Creutzfeldt–Jacob disease with ataxia and dementia (Section 39.11.6). A similar clinical picture is seen in subacute sclerosing panencephalitis. In some patients with Alzheimer's disease there may also be myoclonus, especially some familial cases with presenilin1 mutations (Janssen *et al.* 2000). Other dementias may also cause myoclonus including Huntington's disease (Thompson *et al.* 1994), dementia with Lewy bodies (Louis *et al.* 1997), and most strikingly in corticobasal degeneration (Section 40.3.10) (Mahapatra *et al.* 2004). In these disorders the myoclonus is produced by widespread cortical pathology.

40.7.10 Metabolic, toxic, and inflammatory encephalopathies

Myoclonus is prominent in metabolic encephalopathy, especially the 'negative' myoclonus which produces the characteristic 'flapping tremor', or asterixis, seen in hypercapnia and liver failure. The most likely metabolic causes of myoclonic encephalopathy are uraemia, hepatic failure, and hyponatraemia. Bismuth ingestion causes a severe myoclonic encephalopathy (Gordon *et al.* 1995). Drugs are occasionally responsible (Caviness and Brown 2004) such as tricyclic or selective serotonin reuptake inhibitor antidepressants, neuroleptics, quinolone antibiotics, and many anticonvulsants including gabapentin, pregabalin, and phenytoin. Lithium can cause prominent myoclonus (Fear 1992).

Myoclonic encephalopathies may occur as a paraneoplastic syndrome (Anderson *et al.* 1988) or as a post infectious phenomenon after beta haemolytic streptococcal infection (Smyth and Sinclair 2003; Candler *et al.* 2006) and viral disorders (Bhatia *et al.* 1992). Brainstem myoclonus may follow brainstem encephalitis.

The *opsoclonus-myoclonus syndrome* is a rare disorder in which there is a gradual or abrupt onset of opsoclonus, often with additional myoclonus (Section 38.4.5). Some patients have ataxia, tremor, dysarthria, or altered consciousness. In children, half of cases are paraneoplastic and usually improve after removal of the tumour, typically a neuroblastoma. The remainder are idiopathic and may improve with steroid therapy. In adults, half of cases are idiopathic and the remainder are postinfectious, paraneoplastic, toxic, or idiopathic (Caviness *et al.* 1995). Paraneoplastic cases may occur with breast carcinoma and anti-Ri antibodies or lung tumours and ANNA-2 antibodies among others.

40.7.11 Treatment of myoclonus

This is determined by the site of origin of the myoclonus. In most cases of action myoclonus, which is the most disabling form, the jerks have a cortical origin. Cortical myoclonus responds to clonazepam, valproate, primidone, and piracetam (Brown *et al.* 1993). Levetiracetam may also be effective (Caviness and Brown 2004) and zonisamide may be helpful in Lafora disease. Better results are seen when these drugs are used in combination and when the underlying cause is not progressive. Phenytoin may exacerbate myoclonus and vigabatrin is not usually effective. Severe focal cortical myoclonus may require surgical resection of a causative structural lesion while the myoclonus of corticobasal degeneration is resistant to treatment. Myoclonic epilepsies respond mainly to valproate, clonazepam, zonisamide, and lamotrigine (Wallace 1998).

Reticular reflex myoclonus responds less well to antimyoclonic drugs but may improve with clonazepam (Caviness and Brown 2004) or fluoxetine (Obeso 1995).

Palatal myoclonus has been treated with a wide range of drugs including anticholinergics, baclofen, valproate, carbamazepine, lamotrigine, piracetam, clonazepam, tetrabenazine, and local botulinum toxin (Penney *et al.* 2006). Surgery to the tensor veli palatine or Eustachian tube have also been tried.

In spinal myoclonus the underlying cause should be treated where possible; clonazepam is sometimes helpful. Other drugs have been tried occasionally such as levetiracetam, diazepam, tertabenazine, and carbamazepine but the efficacy of these agents is impossible to assess.

40.8 Wilson's disease

This is an autosomal recessive disorder in which an inability to excrete copper into bile leads to its accumulation in the liver and brain. Although rare it is one of the most important diagnoses in clinical neurology because early treatment may prevent or reverse otherwise permanent and fatal clinical manifestations. Unfortunately diagnosis is often delayed, with serious consequences.

40.8.1 Epidemiology, aetiology, and genetics

The prevalence of Wilson's disease is thought to be 30 per million, with about 1 per cent of the population being heterozygotes (Scheinberg and Sternlieb 1984). Many cases are thought to be unrecognized although this is controversial and the frequency may be lower (Reilly *et al.* 1993). The cause of the disorder is mutation of a gene on chromosome 13q14.3 which encodes the ATP7B protein, a copper transporting P-type ATPase (Bull *et al.* 1993; Tanzi *et al.* 1993). There are over 200 pathogenic mutations of the Wilson's disease gene (Thomas *et al.* 1995) and compound heterozygosity is common. DNA diagnosis is therefore impractical for clinical use. The effect of *ATP7B* gene mutations is a failure to excrete copper into bile due to defective cellular transport. The inevitable accumulation of copper in liver and subsequently brain causes tissue damage and the hepatic and neurological features of the disease.

40.8.2 Clinical features

There is no correlation between the presence of liver or brain pathology and the clinical presentation. Even though liver damage is present in all symptomatic cases of Wilson's disease, the initial presentation is neurological in 40 per cent. The remainder present with liver disease, 40 per cent, psychiatric manifestations in 20 per cent, or occasionally renal, haematological, or skeletal problems. The disease usually starts between 5 and 40 years although a few patients have become unwell in their fifties (Marsden 1987).

Neurological Wilson's disease may develop very gradually, sometimes with intermittent acute deteriorations, or sometimes explosively and with rapid progression. Whatever the rate of onset, the end result is deterioration to severe neurological disability. There are three main types (Marsden 1987; Sternlieb *et al.* 1987).

- A *dystonic form* presents with dysarthria, dysphagia, and drooling of saliva due to dystonia of the face and bulbar musculature. In contrast to a pseudobulbar palsy, the jaw jerk is not increased. Eventually there may be a characteristic fixed facial dystonia and sometimes stridor. Dystonia of the limbs leads to rigidity, abnormal postures, and a dystonic gait. Inexorably the patient becomes severely dysarthric, unable to swallow, and immobile.

- In other cases there is an *akinetic rigid* presentation with prominent resting or postural tremor and variable bradykinesia, rigidity, and facial immobility.

- The third group of neurological cases are those with a *cerebellar, pseudosclerotic,* presentation in which there is gait ataxia, dysarthria, loss of limb coordination, and titubation of the head. A tremor of the arms may be very severe, wing beating tremor, causing injury. There are usually additional extrapyramidal signs.

Chorea is sometimes seen in Wilson's disease but tics and myoclonus are not characteristic. Other less common manifestations include brisk tendon reflexes, extensor plantar responses, seizures, and jerky pursuit eye movements. Restriction of eye movements is rare and sensory loss, visual impairment, or paralysis are not seen. It should be noted that the initial neurological symptoms of Wilson's disease are often mild, sometimes for years, with minimal or no physical signs. They may be overlooked easily or misinterpreted as psychogenic, especially in the presence of psychiatric manifestations. Examples include a mild dysarthria, incoordination, subtle tremor, or dystonia and a slightly odd facial expression or gait. Late diagnosis is consequently common unless the condition is thought of in all patients under 50 with a movement disorder (Walshe and Yealland 1992).

Some patients present with liver disease, either acute hepatitis, fulminating liver failure, or cirrhosis. About 20 per cent initially develop insidious behavioural or psychiatric changes which may evolve into altered personality, depression, anxiety, or a psychosis with little to suggest the true diagnosis unless there is evidence of liver dysfunction or a neurological abnormality. Consequently, many patients are initially referred for psychiatric evaluation (Dening and Berrios 1989). Dementia may occur in Wilson's disease but is usually absent; it is difficult to assess the cognitive function of these severely disabled anarthric patients and a false impression of cognitive impairment is common. In many patients, severe psychiatric problems and behavioural outbursts may be secondary to the physical and social consequences of the disease but this distinction is difficult to make (Sternlieb *et al.* 1987). Occasionally patients with Wilson's disease come to medical attention because of a haemolytic anaemia, bleeding disorder, renal dysfunction, osteomalacia, rickets, or incidentally discovered laboratory tests.

In neurological Wilson's disease, patients almost always have corneal Kayser–Fleischer rings due to deposition of copper in Descemet's membrane (Fig. 40.19). The absence of Kayser–Fleischer rings in neurologically affected patients has occurred but is extremely rare (Demirkiran *et al.* 1996). The rings first appear in the superior cornea and may not be visible unless looked for specifically with a slit lamp by an experienced ophthalmologist who is familiar with their appearance. In patients with non-neurological presentations of Wilson's disease, Kayser–Fleischer rings are less reliable.

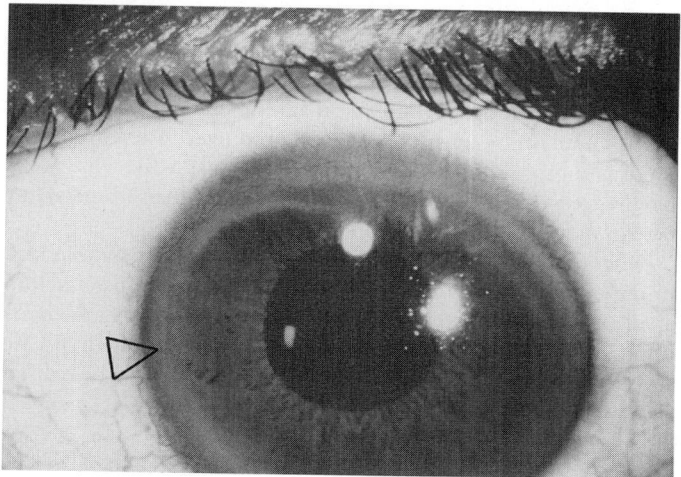

Fig. 40.19 A corneal Kayser Fleischer ring (arrowed) in Wilson's disease. The patient presented with writer's cramp, mild Parkinsonism, and a recent history of depression.

Not only may they be absent in hepatic Wilson's disease but some other forms of liver disease such as primary biliary cirrhosis may be associated with them.

40.8.3 **Laboratory diagnosis**

Liver function tests may be normal or only minimally altered even in the presence of significant liver disease. Specific laboratory diagnosis of Wilson's disease is required but requires specialized laboratories.

In 90 per cent of cases there is an associated deficiency of serum caeruloplasmin (Sternlieb *et al.* 1987) although the relevance of this to the pathophysiology of Wilson's disease is unknown. There is also an increase in urinary 24 hour copper excretion, usually exceeding 100 μg, but the total serum copper is misleading and of no value (Marsden 1987). It should be noted that these tests are occasionally difficult to interpret; caeruloplasmin is increased by oral contraceptives, pregnancy, and other forms of cholestatic liver disease and reduced in some Wilson's disease heterozygotes, protein losing states, and some forms of liver disease; in these situations a misleading result may occur. Caeruloplasmin may be absent in acaeruloplasminaemia, a disorder of iron storage (Miyajima *et al.* 2001). Accordingly, a normal caeruloplasmin level does not always exclude Wilson's disease. Twenty-four hour urinary copper output is always abnormal in symptomatic Wilson's disease, usually above 100 μg, if less than 50 μg it excludes the diagnosis (Brewer and Yuzbasiyan-Gurkan 1992). However, it is normal in many presymptomatic, clinically unaffected, cases and so is unreliable as a screening test. It may be elevated in some renal disorders and in cholestatic liver disease. In most patients however, measurement of caeruloplasmin, 24-h copper excretion, and examination of the eyes for Kayser–Fleischer rings will establish or exclude the diagnosis. If any test is abnormal or there is doubt about interpretation due to any of the above confounding factors, a liver biopsy is advisable.

In all cases of Wilson's disease there is increased dry liver weight copper concentration, >250 μg /g, and abnormal liver histology although histochemical staining for copper is notoriously unreliable. In advanced liver disease, the copper concentration may not be reliable. In suspected cases with a normal caeruloplasmin in whom a liver biopsy is suspect or cannot be done, a radiocopper loading test may demonstrate a failure of incorporation of copper into caeruloplasmin (Sternlieb *et al.* 1987).

A reasonable screening sequence in suspected cases or for siblings at risk is an ophthalmological assessment and measurement of caeruloplasmin, 24-h urine copper and liver function tests. If there is any doubt after these investigations, a liver biopsy is required, especially to distinguish affected individuals from healthy heterozygotes.

Other tests are of limited value. Brain MR scans may show lesions in the striatum of the basal ganglia, thalami, or brainstem (Roh *et al.* 1994) including the 'face of the giant panda sign' in the midbrain (Liebeskind *et al.* 2003). MR changes correlate with disease severity (Sinha *et al.* 2006). MR spectroscopy may also be helpful (Lucato *et al.* 2005).

40.8.4 **Treatment**

In established Wilson's disease or presymptomatic patients the only hope of cure is to reduce the total body copper content. Penicillamine remains the principal copper chelating drug, leading to increased urinary copper excretion and a negative copper balance. About two-thirds of patients with neurological Wilson's disease improve but the remainder are severely disabled or die despite treatment (Walshe and Yealland 1993). Improvement may be seen within weeks but usually within a few months of starting de-coppering therapy. A proportion of patients on penicillamine deteriorate neurologically in the early stages of treatment, possibly due to the mobilization of copper, even with gradual introduction of therapy. Such neurological deterioration can be serious (Svetel *et al.* 2001). In addition, side effects from penicillamine are common, including acute sensitivity reactions, agranulocytosis, skin lesions, nephrotic syndrome, Goodpastures syndrome, systemic lupus erythematosus, and myaesthenia (Marsden 1987). It is given with pyridoxine to reduce adverse effects and close clinical and laboratory monitoring of patients is required. Initially there is a marked cupruresis which eventually declines to about 500 μg / day. There is controversy about penicillamine therapy because of the potential for side effects and neurological worsening (LeWitt 1999).

Alternative methods of reducing copper levels are to be considered in patients unable to take penicillamine. These include trientene (Dahlman *et al.* 1995) and tetrathiomolybdate (Brewer *et al.* 2003). Trientene is safer with less risk of neurological deterioration (Brewer 1999). Tetrathiomolybdate is effective with a low risk of neurological deterioration but is not available except as an experimental agent. Oral zinc reduces copper absorbtion and has been used in presymptomatic cases, asymptomatic children (Marcellini *et al.* 2005), pregnancy, and for maintenance therapy in those who have completed chelation therapy (Brewer *et al.* 1989); its role in established Wilson's disease is less clear but zinc is not adequate for initial chelation in neurologically affected patients.

The treatment of Wilson's disease remains intensely controversial. Penicillamine is tried, tested, and highly effective but risky, as is the disease; trientene and zinc are safer and arguably as effective (Brewer 1999; Walshe 1999). Tetrathiomolybdate is effective and safe but unavailable, at least in the United Kingdom.

It is important that treatment is continued for life if disastrous clinical deterioration is to be avoided. The extrapyramidal symptoms are difficult to treat but anticholinergics may improve dystonia (Walshe and Yealland 1993). In patients with severe liver damage, transplantation has been successful (Schumacher *et al.* 2001).

40.9 **Drug induced movement disorders**

Numerous drugs may induce abnormal movements but the neuroleptic drugs and other dopamine receptor blocking agents are the most important. The effects of these drugs are summarized in Table 40.21. Drug induced Parkinsonism is discussed in Section 40.3.4.

40.9.1 **Acute neuroleptic induced movement disorders**
Acute dystonic reactions

These are most commonly seen in young adults, usually males, given neuroleptics or other dopamine receptor blocking drugs such as metoclopramide. There may be a family history of such reactions. The dystonia is often severe and dramatic with spasms affecting the eyes, face, mouth, and neck; oculogyric crises may occur. The attacks are often mistaken for a psychogenic reaction.

Table 40.21 Movement disorders caused by neuroleptic drugs

Acute reactions	Acute dystonia
	Acute akathisia
	Neuroleptic malignant syndrome
Tardive syndromes	Tardive dyskinesia
	Tardive akathisia
	Tardive dystonia
	Tardive tremor
	Tardive Tourettism
	Tardive myoclonus
Parkinsonism	
Withdrawal reactions	Withdrawal emergent syndrome
	Covert/withdrawal dyskinesia
	Withdrawal akathisia

The onset is within hours or a few days of starting treatment with the offending drug. The spasms persist for hours or days and are relieved by intravenous anticholinergics such as procyclidine, diazepam, or diphenhydramine (Miyasaki and Lang 1995). Such reactions occur in about 2.5 per cent of patients receiving dopamine receptor blockers (Rupniak *et al.* 1986) and other drugs are occasionally responsible.

Acute akathisia

Akathisia refers to an unpleasant subjective feeling of inner motor restlessness. Patients are typically very distressed and agitated. Some relief is obtained by moving around and affected patients may walk up and down, continuously move their legs, or repeatedly get up out of a chair. The onset is soon after exposure to neuroleptic drugs and is more common after higher doses. It is common, occuring in up to 50 per cent of patients receiving neuroleptics (Miyasaki and Lang 1995). The agitation caused by akathisia may be interpreted as worsening of the psychiatric disorder and may also threaten the patient's compliance with treatment. Some patients respond to a reduction in dosage or changing to a less potent neuroleptic. Others are relieved by anticholinergics and sometimes benzodiazepines but good trial data are lacking (Lima *et al.* 2002 2004). Beta blockers have been reported to be effective but one recent study did not confirm this (Sachdev and Loneragan 1993).

Neuroleptic malignant syndrome

The neuroleptic malignant syndrome is a potentially fatal idiosyncratic reaction to neuroleptic drugs (Section 5.5.2). It can occur after the initiation of treatment, following a change of drug or dose or at stable therapeutic doses (Reeves *et al.* 2001). It should be noted that it has been seen with risperidone and also with atypical neuroleptics (Ananth *et al.* 2004).

Neuroleptic malignant syndrome is rare, occurring in less than 1 per cent of patients on neuroleptic medication; occasionally it has been associated with other drugs (Haddad 1994) or withdrawal of dopaminergic antiparkinsonian treatment (Hashimoto *et al.* 2003). There is a rapid onset within 2–3 days of pyrexia, clouding of consciousness, rigidity, and autonomic disturbance (Buckley *et al.* 1995). There may be extrapyramidal signs, seizures, or delerium. Although the creatine kinase level is frequently elevated this is non

specific and there may be a leucocytosis or myoglobinuria. In some patients neuroleptic malignant syndrome develops more gradually and is less severe (Bristow and Kohen 1993). It should be considered in any patient with a fever while taking neuroleptic medication (Hall *et al.* 2005).

The differential diagnosis includes infection, an initially psychiatric presentation of encephalitis, malignant hyperthermia after anaesthesia, fulminating Wilson's disease, and lethal catatonia. The latter is clinically similar but rigidity occurs more gradually and it can occur without neuroleptic exposure. It has been suggested that neuroleptic malignant syndrome and lethal catatonia may be the same disorder (Buckley and Hutchinson 1995) and that catatonia may be a risk factor for neuroleptic malignant syndrome (Berardi *et al.* 2002). Although a reduction in central dopaminergic transmission seems the likely cause of neuroleptic malignant syndrome, similar clinical features are seen with the 'serotonin syndrome' in which there is increased central 5HT activity caused by an interaction between monoamine oxidase inhibitors and selective serotonin reuptake inhibitor antidepressant drugs (Sternbach 1991).

Management consists of stopping the causative drug and correction of dehydration, fever, acidosis, or autonomic disturbances. Dantrolene, bromocriptine, or a combination of the two are reported to be beneficial. Electroconvulsive therapy has been used successfully but there are no evidence based guidelines available (Susman 2001). Despite treatment, mortality remains high, up to 30 per cent. In those who recover, neuroleptics may be restarted after 2 weeks but with close supervision (Bristow and Kohen 1993; Miyasaki and Lang 1995).

40.9.2 Tardive neuroleptic induced movement disorders

Tardive dyskinesia

Tardive dyskinesia is a notorious complication of chronic treatment with neuroleptic drugs (Fernandez and Friedman 2003). All dopamine receptor blocking drugs, including metoclopramide have the potential to cause tardive dyskinesia but the newer atypical neuroleptics such as clozapine or olanzapine appear to be safer (Factor and Friedman 1997; Reus 1997). Tardive dyskinesia is often unmasked by a neuroleptic dose reduction or discontinuation of therapy but can develop on a stable dose. The likely cause is dopamine receptor supersensitivity as a result of chronic receptor blockade (Casey 2004) but other transmitters or a more complex imbalance of direct and indirect basal ganglia pathways may be involved (Feve *et al.* 1990). The reasons why only some patients are affected and for the delay in onset are unknown. Estimates of the frequency among patients treated with conventional neuroleptics vary; in one study 15–20 per cent developed tardive dyskinesia (Casey and Gerlach 1988) rising at about 5 per cent per year in the first 4 years of treatment (Gardos and Cole 1980). Other estimates are even higher including 17 per cent after 9 months of therapy (de Jesus Mari *et al.* 2004) and 32 per cent in a population of inpatients in Eastern Europe (Janno *et al.* 2004). The elderly, females, and those with affective disorders or organic brain disease are more likely to be affected; although anticholinergics make established tardive dyskinesia worse, they do not appear to confer increased risk. The significance of an association with the APOE2 allele and polymorphisms of the genes for the dopamine D2 receptor

(Liou *et al.* 2006), D3 receptor (Steen *et al.* 1997), D4 receptor, and the enzyme COMT (Srivastava *et al.* 2006) is uncertain. Atypical neuroleptic drugs such as clozapine or olanzapine are associated with a much lower risk of tardive dyskinesia (Nasrallah 2006) and are therefore increasingly preferred to older conventional agents. The reason for the greater safety of atypical agents is unclear; it may involve action at other non-dopamine receptor types, less potent or less prolonged dopamine receptor blockade or less induction of dopamine receptor supersensitivity (Casey 2004).

The movements are usually stereotypies or chorea involving the face and sometimes the trunk and limbs (Stacy *et al.* 1993). Typically there are repetitive tongue and lip smacking movements or grimacing. There may be additional trunk rocking or repetitive limb movement. In some patients tardive dyskinesia is mild and may be unnoticed but in others it is disabling. The movements differ from cranial dystonia as there is less other cranial or cervical involvement (Tan and Jankovic 2000). Rarely, similar but spontaneous dyskinesias have been reported in untreated psychotic patients. Whether these movements were indeed similar to tardive dyskinesia or were manifestations of an underlying condition such as Huntington's disease is unclear.

The treatment of tardive dyskinesia depends on whether the patient's psychiatric state requires continued neuroleptic medication. If this is essential, increased blockade of supersensitive dopamine receptors by a higher neuroleptic dose or switching to a higher potency drug may be effective but is a last resort (Tarsy 2000). In some patients, the use of an atypical neuroleptic such as clozapine may allow remission of the tardive dyskinesia while providing effective antipsychotic therapy (Factor and Friedman 1997). Dopamine depletion with tetrabenazine (Jankovic and Beach 1997) improves tardive dyskinesia in 90 per cent of cases but side effects are often troublesome, especially sedation and depression. In contrast to acute drug-induced Parkinsonism and tardive dystonia, anticholinergics often appear to make tardive dyskinesia worse and should be avoided or stopped (Tarsy 2000) although a systematic review was unable to confirm any effect of anticholinergic drugs either way (Soares and McGrath 2000).

If the causative drug can safely be stopped, tardive dyskinesia is said to resolve in about a third of cases but this may take months or even years. However, a recent systematic review was unable to confirm this (Soares-Weiser and Rathbone 2006). While a remission is awaited, the dyskinesias may be worse but can be suppressed with tetrabenazine. Vitamin E may be slightly beneficial (Soares and McGrath 2001b) and levetiracetam has been used successfully in a small open trial (Konitsiotis *et al.* 2006). Botulinum toxin can be used in severe cases with disfiguring movements (van Harten and Hovestadt 2006).

Low doses of levodopa or dopamine agonists have been used to hasten dopamine receptor downregulation but were unsuccessful (Miyasaki and Lang 1995). In resistant cases improvement may be seen with other drugs including GABA-ergic agents such as baclofen, sodium valproate, and progabide, benzodiazepines, and calcium channel blockers but there is no convincing evidence of their effectiveness (Soares and McGrath 2001a; Umbrich and Soares 2003; Soares *et al.* 2004).

Small numbers of patients have been treated successfully with pallidotomy (Weetman *et al.* 1997) or pallidal deep brain stimulation (Eltahawy *et al.* 2004a; Schrader *et al.* 2004).

Tardive dyskinesia is a preventable disorder; neuroleptic drugs should be used sparingly and only when absolutely necessary.

The use of limbic selective neuroleptics such as olanzapine or quetiapine offers the possibility of prevention (Reus 1997; Tran *et al.* 1997; Nasrallah 2006). At present, the frequency of tardive dyskinesia is probably declining (Tarsy and Baldessarini 2006).

Tardive dystonia

In some patients, dystonic movements develop, with or without additional tardive dyskinesia (Kang *et al.* 1988; Raja and Azzoni 1996). The dystonia is similar to that seen in primary torsion dystonia (see Section 40.4.2) but tends to involve the neck and trunk more severely and rarely starts in a leg (Fernandez and Friedman 2003; Skidmore and Reich 2005). Like primary torsion dystonia the extent of the dystonia is determined by age with generalized cases tending to be younger. Tardive dystonia can develop at any stage after starting neuroleptic medication, often within a year, and is often disabling. It occurs at all ages but is more common in younger males. Remissions after stopping neuroleptic medication are uncommon and may take years to appear. Like tardive dyskinesia, tardive dystonia responds to dopamine antagonists or tetrabenazine but by contrast it also improves with anticholinergics (Burke and Kang 1988; Miyasaki and Lang 1995). A switch to an atypical agent such as clozapine may allow a remission but this has not been studied formally (Friedman 1994; Trugman *et al.* 1994; Van Harten *et al.* 1996). Cervical dystonia or blepharospasm may be treated with botulinum toxin and there have been reports of improvement with bilateral pallidal deep brain stimulation in severe cases (Krause *et al.* 2004; Franzini *et al.* 2005; Trottenberg *et al.* 2005).

Tardive akathisia

Subjective motor restlessness similar to acute akathisia may also occur after chronic exposure to neuroleptics (Burke *et al.* 1982b). The patients are intensely restless and often unable to sit in a chair (Fernandez and Friedman 2003; Skidmore and Reich 2005). There may be concomitant tardive dyskinesia but more often any movements of the patient are clearly an attempt to obtain relief from the akathisia. Severe agitation may be mistaken for a worsening of the underlying psychiatric illness. Tardive akathisia, like other tardive phenomena, is often unmasked by a reduction of antipsychotic medication and is improved by restarting or increasing the neuroleptic while anticholinergics make the symptoms worse. This is the opposite pattern to that seen with acute akathisia. As with tardive dyskinesia, tetrabenazine is often helpful. Some patients improve with propranolol or opiates such as codeine or dextropropoxyphene (Burke *et al.* 1989). Occasionally, patients develop delayed acute akathisia after an increase in neuroleptic dosage, a switch to a higher potency drug or discontinuation of anticholinergic therapy; the precipitating changes in medication are helpful in making the distinction.

Other tardive phenomena

There have been rare reports of myoclonus, tics, and tremor as tardive effects (Stacy and Jankovic 1991; Bharucha and Sethi 1995). All have the hallmarks of tardive symptoms, namely appearance or exacerbation after neuroleptic reduction or discontinuation and improvement with tetrabenazine or increased dopamine receptor blockade. These reports are difficult to evaluate.

Withdrawal phenomena

In children, transient generalized chorea has been described after withdrawal of neuroleptic drugs. This 'withdrawal emergent syndrome' is analogous to tardive dyskinesia but lasts only a few weeks (Polizos and Engelhardt 1978). A similar situation in adults exists with the appearance of tardive dyskinesia for a few months after drug withdrawal, 'withdrawal dyskinesia', but this is simply tardive dyskinesia which has resolved after stopping the causative drug. Covert dyskinesia refers to persistent tardive dyskinesia which has been unmasked by stopping dopamine blockers. A similar phenomenon is seen with tardive akathisia which may appear after stopping a dopamine blocker, 'withdrawal akathisia' (Lang 1994).

40.9.3 Other drug induced movement disorders

Many different drugs cause dyskinesias or Parkinsonism and a detailed review is beyond the scope of this chapter. Detailed reviews exist elsewhere (Miyasaki and Lang 1995). Most cases involve antidepressants, stimulants, anticonvulsants, and calcium channel blockers. Tricyclic antidepressants have been associated with action tremor, chorea, myoclonus, and dystonia (Vandel et al. 1997). These effects are rare and are often associated with overt toxicity. Selective serotonin reuptake inhibitors such as fluoxetine may induce tardive dyskinesia, Parkinsonism, akathisia and dystonia (Leo 1996). Lithium frequently causes tremor and in cases of toxicity may cause myoclonus and encephalopathy resembling Creutzfeldt–Jacob disease (Fear 1992). Stimulants such as amphetamines, methylphenidate, and cocaine are associated with tremors, chorea, tics, and stereotypies (Klawans and Weiner 1974). Phenytoin may cause myoclonus, asterixis, chorea, or dystonia; this is usually in association with toxicity and is more common with underlying striatal pathology (Miyasaki and Lang 1995). Sodium valproate is a cause of postural tremor and dystonia may be caused by gabapentin (Reeves et al. 1996). The main risk with calcium antagonists is Parkinsonism (Section 40.3.4).

40.10 Post streptococcal movement disorders

The principal post streptococcal movement disorder is Sydenham's chorea (Section 40.5.7). However, the concept of neurological conditions associated with anti basal ganglia antibodies after group A beta haemolytic streptococcal infection has developed considerably and the clinical presentations are much wider than previously suspected. A theme of these disorders is the combination of a movement disorder with prominent neuropsychiatric and behavioural features. Tics, myoclonus, dystonia, Parkinsonism have been described in addition to 'atypical' movement disorders and forms of encephalitis (Dale 2003).

Some children develop neurological and behavioural symptoms, usually relapsing and remitting episodic tics and an obsessive compulsive, emotional, anxiety, or other behavioural disorder, with serological evidence, anti-streptolysin O or DNase B titres, of group A beta haemolytic streptococcal infection (Mell et al. 2005) at onset or in association with clinical exacerbations. This condition has been termed Paediatric Autoimmune Neuropsychiatric Disorders Associated with Streptococcal infection, PANDAS (Swedo et al. 1998). It has been proposed that the streptococcal infection, usually pharyngitis, initiates the formation of pathogenic cross-reacting anti-basal ganglia antibodies (Dale et al. 2005) although this has not been confirmed in other studies (Singer et al. 2004). Penicillin prophylaxis does not prevent relapses (Garvey et al. 1999) but this was probably due to poor compliance. Immunotherapy with plasma exchange or intravenous immunoglobulin does improve symptoms significantly (Perlmutter et al. 1999). At present, despite some highly suggestive findings, paediatric autoimmune neuropsychiatric disorders associated with streptococcal infection does not yet satisfy all criteria for an autoimmune disorder (Singer 2005). It is similar clinically to Tourette's syndrome and it is possible that a sub-group of Tourette's patients have a post-streptococcal disorder (Church et al. 2003). However, the link with Tourette's syndrome remains uncertain (Section 40.6.3).

A bewildering array of other movement disorders has been linked to group A beta-haemolytic streptococcus infection with anti-basal ganglia antibodies (Martino and Giovannoni 2004) including an encephalitis lethargica like illness (Dale et al. 2004; Vincent 2004), Parkinsonism (Ben-Pazi et al. 2003; McKee and Sussman 2005), various forms of dystonia (Edwards et al. 2004a), myoclonus (Smyth and Sinclair 2003), opsoclonus-myoclonus (Candler et al. 2006), striatal necrosis (Dale et al. 2002), post infectious encephalomyelitis (Dale et al. 2001), and even atypical movement disorders which were difficult to classify (Edwards et al. 2004b). Evidence of streptococcal infection and anti-basal ganglia antibodies has also been detected in children with obsessive compulsive disorder (Dale et al. 2005) although not in association with its exacerbations (Luo et al. 2004). At present the spectrum of neuropsychiatric and movement disorders associated with group A beta-hemolytic streptococcus is unresolved. Thus the therapeutic implications of finding a positive throat swab, serological evidence of streptococcal infection, or anti-basal ganglia antibodies in a patient with a movement disorder are unknown.

40.11 Miscellaneous movement disorders

A number of other disorders are conveniently considered along with the movement disorders. These often give rise to diagnostic problems unless their existence is remembered.

40.11.1 Restless legs syndrome and periodic limb movements of sleep

Restless legs syndrome causes unpleasant sensations in the lower limbs with restlessness and sleep disturbance. It is surprisingly common, occurring in about 10 per cent of the population (Hening et al. 2004a). Most cases are idiopathic but secondary restless legs syndrome can occur with uraemia, iron deficiency, lumbar radiculopathy, drugs including antidepressants, pregnancy, and some peripheral neuropathies. Onset is usually in older adults but this is not invariable and a family history of autosomal dominant inheritance is noted in 60 per cent of cases. Various loci have been identified by linkage studies but no gene has been characterized (Winkelmann and Ferini-Strambi 2006).

Diagnostic criteria have been suggested for the condition (Allen et al. 2003). There should be an urge to move the legs with or without unpleasant sensory alterations or dysaesthesiae in the legs, worsening of symptoms at rest, in the evening and at night, and partial relief with activity. There is often motor restlessness but restless legs syndrome may be entirely subjective and difficult to describe. The motor features include walking, stretching, or rubbing the legs and patients often get up at night and pace the room to obtain some relief. Sleep is disrupted with resulting daytime sleepiness. There is a tendency for restless leg syndrome to worsen with age.

Many patients develop frequent jerks of the legs and sometimes the arms while awaiting sleep at night. While asleep, there are characteristic slow, repetitive flexion movements of the legs, involving the ankle or the whole limb. These periodic limb movements of sleep occur in stage I or II sleep and last from 0.5 to 5 s. They also contribute to sleep fragmentation and daytime tiredness. About 80 per cent of patients with restless legs syndrome have periodic limb movements of sleep (Montplaisir *et al.* 1997). However, periodic limb movements of sleep are common in older patients and can occur as an isolated symptom.

The differential diagnosis of restless legs syndrome includes sensory disturbances due to peripheral neuropathy, akathisia in patients who have received neuroleptic drugs, and the painful feet and moving toes syndrome. Periodic limb movements of sleep should be distinguished from nocturnal frontal lobe seizures and rapid eye movement sleep parasomnias.

The mechanism of restless legs syndrome is unclear. Similar symptoms are seen in Parkinson's disease and drug induced akathisia. Dopaminergic drugs improve restless leg symptoms implicating abnormal central dopaminergic function. There is functional imaging evidence using positron emission tomography of reduced nigrostriatal dopamine and striatal D2 receptor levels (Turjanski *et al.* 1999). There is also some evidence of reduced brain iron storage (Connor *et al.* 2003) but the relevance of this to dopaminergic function is unclear.

Improvement may be obtained with levodopa, dopamine agonists, benzodiazepines, gabapentin, or opiates such as codeine; a useful algorithm is available (Silber *et al.* 2004). Dopaminergic medication is the most effective option (Hening *et al.* 2004b). Levodopa is effective in most patients but can lead to worsening of symptoms or 'augmentation' in some patients and a rebound effect the morning after an evening dose. Augmentation leads to earlier appearance of evening symptoms over time and increased severity. Regular use of levodopa, especially more than 200 mg is more likely to cause augmentation and so intermittent use is preferable. The mainstay of therapy is now the non-ergot dopamine agonists (Silber *et al.* 2004), mostly ropinirole or pramipexole. These drugs are usually well tolerated; caution is needed in the elderly and sedation may occur (Section 40.3.1). There are more trial data for ropinirole than pramipexole. The ergot agonists pergolide and cabergoline are less attractive due to the risk of fibrotic reactions. In patients who do not respond to or tolerate a dopamine agonist the options are to try another agonist, switch to gabapentin, or to continue the agonist and add gabapentin, a benzodiazepine such as temazepam or clonazepam, or opioids such as codeine, tramadol, oxycodone, or even methadone in very severe refractory cases (Silber *et al.* 2004).

40.11.2 **Other sleep related movements**

In normal individuals there may be generalized myoclonic jerks at the onset of sleep. Anxious individuals sometimes consult because of this but there is no pathological significance of these normal hypnic jerks. Some normal individuals also have brief multifocal myoclonic jerks of the limbs, referred to as physiological fragmentary myoclonus of sleep. This occurs in stages I, II, and rapid eye movement sleep (Fish 1994) and rarely causes symptoms. In some individuals there is an excessive amount of fragmentary myoclonus of sleep which also occurs in stages III and IV and is associated with a variety of sleep disturbances including sleep apnoea and narcolepsy (Broughton *et al.* 1985). Behavioural disturbances in non-rapid eye movement sleep such as night terrors

and sleepwalking and rapid eye movement sleep related behaviour disorder are discussed in Section 32.4.

Paroxysmal nocturnal dystonia is characterized by repetitive brief attacks of abnormal movements in sleep (Lugaresi *et al.* 1986). When first described it was thought to be a sleep related dyskinesia because the movements were thought to be dystonic and scalp EEG recordings were normal. However, it is now clear that this condition is actually a form of frontal lobe epilepsy in which attacks of bizzare motor activity from sleep are well recognized (Fish 1994).

40.11.3 **Stiff person syndrome**

In this condition there is a progressive rigidity of the trunk and proximal limbs with associated stimulus sensitive spasms (Moersch and Woltman 1956); women are more commonly affected (Section 23.7.2) (Murinson 2004). Many cases have been described and diagnostic criteria proposed (Thompson 1994). Onset is usually between 30 and 60 years with slowly progressive rigidity of the trunk and later the proximal limbs over many years (Dalakas *et al.* 2000). There is a hyperlordotic posture with striking rigidity of the lumbar paraspinal and abdominal muscles; the proximal limbs may also be affected. The lumbar spasm and hyperlordosis persist when supine, in contrast to axial dystonia which is reduced at rest. Painful muscle spasms are superimposed upon the fixed rigidity, sometimes triggered by stimuli or emotion leading to an erroneous impression of a psychogenic disorder. Some patients present with back pain and the spasms and stiffness are apparent on examination (Bastin *et al.* 2002). There are usually no other neurological signs although cerebellar ataxia has occurred (Kono *et al.* 2001).

Electromyography recordings of affected muscles reveal continuous motor unit activity with normal motor units and no sign of denervation or peripheral neuropathy. There is an association with diabetes in about a third of cases and 10 per cent of patients also have epilepsy. Other autoimmune disorders sometimes coexist and some cases are paraneoplastic (Levin 1997).

Pathophysiologically, the continuous motor unit activity persists during attempted relaxation but is abolished by sleep, spinal, or general anaesthesia and by peripheral nerve block; this indicates a central origin of the abnormal motor activity. Given the anatomical distribution of the rigidity in stiff person syndrome, this is probably within the spinal cord. A number of exteroceptive polysynaptic spinal reflexes are exaggerated in stiff person syndrome (Meinck *et al.* 1984) suggesting abnormal inhibitory interneuronal function within the cord. There may also be stimulus sensitive jerks of brainstem origin 'jerking stiff man syndrome' (Leigh *et al.* 1980) and abnormal brainstem excitability (Molloy *et al.* 2002) suggesting a more widespread defect of central nervous system inhibitory function.

Antibodies to glutamic acid decarboxylase are found in serum or CSF in approximately 80 per cent of cases of stiff person syndrome (Solimena *et al.* 1988; Rakocevic *et al.* 2004). The level of the antibodies does not correlate with symptom severity. Anti-glutamic acid decarboxylase antibodies cross react with pancreatic islet cells, gastric parietal cells, and thyroid antigens. Paraneoplastic cases are usually associated with anti-amphiphysin antibodies particularly in breast cancer (De Camilli *et al.* 1993) although other tumours such as lung, ovary, colon, and lymphoma have been associated and other antibodies can occur (McCabe *et al.* 2004). Not all anti-amphyphysin associated cases are paraneoplastic (Pittock *et al.* 2005) and anti-glutamic acid decarboxylase associated cases occasionally are (Thomas *et al.* 2005). In accord with the autoimmune inflammatory pathogenesis, oligoclonal bands and anti-glutamic acid decarboxylase antibodies may be present in CSF (Dalakas *et al.* 2001b) and white matter lesions are occasionally seen on MR brain scans. No significant neuropathological abnormalities have been described but in one case there were minor abnormalities of inhibitory neurones in the cerebellum and spinal cord (Warich-Kirches *et al.* 1997).

The spasms and stiffness are improved with high doses of diazepam, clonazepam, or baclofen and many patients remain ambulant (Barker *et al.* 1998); intravenous immunoglobulin is the main option in those who do not respond to medication (Dalakas 2005; Dalakas *et al.* 2001a). There have been reports of improvement with steroids and plasma exchange (Vicari *et al.* 1989) but other patients have not responded (Harding *et al.* 1989). Intrathecal baclofen has been successful in one patient (Seitz *et al.* 1995) but not in others (Silbert *et al.* 1995).

40.11.4 Progressive encephalomyelitis with rigidity, atypical stiff person syndrome, and the stiff limb syndrome

There are several disorders with stiffness, spasms, and continuous muscle unit activity which differ from typical stiff person syndrome despite some clinical and neurophysiological similarities. These are progressive encephalomyelitis with rigidity, 'jerking stiff man syndrome', and stiff limb syndrome (Brown and Marsden 1999). A clear separation of these disorders and stiff person syndrome is somewhat unclear.

Progressive encephalomyelitis with rigidity

This condition is also associated with severe muscular rigidity and spasms but is more rapidly progressive and there is clinical and pathological evidence of an active encephalomyelitis, mainly affecting the brainstem and spinal cord (Whitely *et al.* 1976). Some cases are paraneoplastic (Ishii *et al.* 2004) and one case was associated with HIV infection (Lannuzel *et al.* 2002). Typically there is a subacute onset over weeks or months of rigidity of one or more limbs, usually the legs. Unlike the stiff person syndrome, there may be additional pain, sensory loss, dysaesthesiae, jerking, and weakness due to segmental spinal pathology; there may also be loss of tendon reflexes and extensor plantar responses. The rigidity becomes progressively worse with severe stimulus sensitive spasms of the trunk and limbs. Brainstem involvement is eventually seen with bulbar and ocuolomotor abnormalities. Progression is much quicker than stiff person syndrome with death occurring between 3 weeks and 3 years from onset. The spasms are abolished by sleep, anaesthesia, or nerve blocks consistent with a central origin and electromyography may show additional denervation changes. Antiglutamic acid decarboxylase antibodies are usually absent but have been present in occasional atypical cases (Burn *et al.* 1991; Mitsumoto *et al.* 1991) and one one case was associated with antiamphiphysin antibodies (Ishii *et al.* 2004). The CSF contains a lymphocytic response and oligoclonal bands and areas of abnormal signal have been detected in the cerebral white matter and brainstem in some patients (Thompson 1994). Pathologically there are inflammatory changes in the brainstem and spinal cord, with long tract degeneration. The prognosis is usually poor with rapid fatal progression but diazepam and baclofen may help the distressing spasms. Immunoglobulin, steroids, and plasma exchange have occasionally been effective (Fogan 1996; Molina *et al.* 2000). The relationship between progressive encephalomyelitis with rigidity and stiff person syndrome is unclear. Some patients have a less rapid course more like stiff person syndrome (Armon *et al.* 1996) and stiff person syndrome may develop into progressive encephalomyelitis with rigidity (Gouider-Khouja *et al.* 2002) suggesting that the two disorders are within a spectrum of disorders associated with abnormal central nervous system hyperexcitability.

Jerking stiff man syndrome

In some cases of stiff person syndrome there has been additional stimulus sensitive generalized myoclonus indicating brainstem involvement (Leigh *et al.* 1980; Kullmann *et al.* 1996). Others have developed respiratory abnormalities with pathological evidence of brainstem inflammation (Mitsumoto *et al.* 1991; Armon *et al.* 1996; Kullmann *et al.* 1996) or oculomotor abnormalities (Thomas *et al.* 2005). Whether this is a form of stiff person syndrome in which there is additional clinical evidence of brainstem dysfunction such as myoclonus or respiratory irregularity or a separate disorder is unclear but as abnormal brainstem function is apparent in stiff person syndrome (Molloy *et al.* 2002) a firm distinction is difficult.

Stiff limb syndrome

These patients have rigidity of the distal lower limbs, continuous motor unit activity, stimulus sensitive spasms, without axial involvement or anti-glutamic acid decarboxylase antibodies but a slowly progressive course over many years (Brown *et al.* 1997). This 'stiff leg syndrome' is probably a chronic spinal interneuronitis, clinically overlapping stiff person syndrome and progressive encephalomyelitis with rigidity. Sunsequently other cases have been reported and some have anti-glutamic acid decarboxylase antibodies (Saiz *et al.* 1998; Gurol *et al.* 2001; Bartsch *et al.* 2003). The course tends to be long with gradual worsening over years; immunoglobulin can be effective (Souza-Lima *et al.* 2000).

Other forms of spinal rigidity

Occasionally, focal spinal cord lesions, usually cervical, such as neoplasms, syringomyelia, myelitis, trauma, and ischaemia may produce segmental rigidity with stimulus sensitive jerks (Brown *et al.* 1997).

40.11.5 Neuromyotonia and myokymia

Neuromyotonia refers to abnormal muscle stiffness caused by overactivity of peripheral nerve endings, peripheral nerve hyperexcitability (Section 23.7.1). Patients complain of twitching, stiffness and cramps, sweating, and sometimes sensory symptoms. On examination, there is clinical myokymia consisting of stiffness, rippling, or twitching of the stiff muscles pseudotetanic spasms, abnormal postures, and delayed relaxation but no percussion myotonia (Hart *et al.* 2002). Electromyography shows continuous motor unit activity which is abolished only by curare, indicating a peripheral nerve terminal origin, and there may be fasciculations, grouped motor unit discharges, also referred to as neurophysiological myokymia, and prolonged bursts of motor unit potentials. The presence of CSF oligoclonal bands and frequent improvement after plasma exchange indicates an autoimmune mechanism (Sinha *et al.* 1991; Newsom-Davis and Mills 1993) and antibodies against peripheral nerve voltage gated potassium channels occur in 40 per cent of cases (Hart *et al.* 2002). Antibody negative cases are clinically similar. Some cases are paraneoplastic or associated with a peripheral neuropathy (Layzer 1995). Treatment is with anticonvulsant drugs or plasma exchange (Hayat *et al.* 2000). Neuromyotonia also occurs in episodic ataxia type 2 caused by a mutation of the *KCNA1* potassium channel gene located at chromosome 12p13 (Comu *et al.* 1996).

40.11.6 Hemifacial spasm

This condition is included in this chapter as it is a common cause of involuntary facial movements and sometimes confused with tics and even dystonia of the face. There are irregular clonic and tonic contractions of one side of the face with simultaneous eye closure, forehead contraction, and elevation of the angle of the mouth (Section 20.2.4). There may be a mild facial weakness. The pattern of movement suggests an origin in the proximal part of the facial nerve in the posterior cranial fossa. Occasionally a tumour, arterio-venous malformation, or cyst is responsible but most cases are caused by microvascular compression of the facial nerve by a blood vessel which can be demonstrated by magnetic resonance tomo-graphic angiography (Bernardi *et al.* 1993). The relative roles of local ephaptic transmission at the site of compression, increased trigeminal reflex movement, and abnormal facial nucleus hyperex-citability are unclear (Misawa *et al.* 2006). Treatment is by microv-ascular surgical decompression (Moffat *et al.* 2005) or periocular botulinum toxin injections (Berardelli *et al.* 1993).

40.11.7 Painful feet and moving toes

In this rare condition there is pain in the lower limbs and continuous writhing movements of the toes (Dressler *et al.* 1994). Occasionally the upper limbs are affected and unilateral cases occur (Sanders *et al.* 1999). The pain is the most prominent and early symptom and is severe, continuous, and similar to causalgia. Most cases are due to lesions of the spinal roots, lumbosacral plexus, or peripheral nerves; the latter may be traumatic or due to a diffuse neuropathy (Dressler *et al.* 1994). The condition has also occurred after herpes zoster myelitis (Ikeda *et al.* 2004) but some cases are idiopathic. The move-ments are occasionally seen without pain (Walters *et al.* 1993) and are probably of central origin. Unless a surgically soluble lesion is identified, treatment is often ineffective although sympathetic blockade may be transiently helpful (Okuda *et al.* 1998).

40.11.8 Other movements originating in the peripheral nervous system

Various dyskinesias may appear after lesions of the peripheral nerves in addition to neuromyotonia, hemifacial spasm, and pain-ful legs–moving toes syndrome. Dystonia after peripheral injury or overuse has been discussed in Section 40.4.16 and myoclonus of peripheral origin in Section 40.7.5. A curious phenomenon is the involuntary movements of limb stumps after amputation, often with phantom limb pain. The jerks may be transient or persistent and the mechanism by which they appear is unclear (Marion *et al.* 1989). Abnormal writhing movements of the abdominal wall are also reported after local surgery or trauma (Marsden 1994). These may be a form of post traumatic dystonia.

40.11.9 Primary akinesia

Described mainly in Japan but also in the West, primary or pure akinesia refers to the development of striking isolated slowness of

gait, speech, and handwriting with no rigidity, tremor, dementia, or eye movement disorder (Matsuo *et al.* 1991; Riley *et al.* 1994). Some patients develop obsessive-compulsive symptoms (Miwa *et al.* 2001). A possible relationship with progressive supranuclear palsy has not been confirmed (Riley *et al.* 1994) and similar presenta-tions may occur with Lewy body disease (Quinn *et al.* 1989) and neurodegeneration with brain iron accumulation (Section 40.4.12) (Molinuevo *et al.* 2003). In most cases however, the cause of this disorder is unclear but there are clinical and pathological similari-ties to primary progressive freezing of gait (Section 40.11.10). Neuropathological studies suggest that pure akinesia is heterogene-ous; one patient had an unusual neuropathological syndrome of nigropallidal degeneration (Katayama *et al.* 1998) and another showed pallidonigroluysian atrophy (Konishi *et al.* 2005). There has been improvement in gait using L-threo-3,4dihydroxyphenyl-serine, 3-0-DOPS (Yamamoto *et al.* 1997) and a serotonin agonist tandospirone (Miyata *et al.* 2001). A specialized walking stick to act as a gait 'cue' has been effective (Asahi *et al.* 2001).

40.11.10 Gait apraxia and primary progressive freezing of gait

A disturbance of gait is a common effect of abnormal peripheral motor or sensory function, as well as with pyramidal, cerebellar, and extrapyramidal deficits. In some patients, the disorder of walk-ing involves higher level dysfunction, particularly of the frontal lobes. These disorders of gait often give rise to confusion, especially in the elderly in whom an erroneous diagnosis of Parkinson's dis-ease is often made (Section 2.6.4).

The best known of these is the wide based shuffling *frontal gait disorder* associated with Binswanger's disease (Thompson and Marsden 1987); a similar gait is seen with other frontal lobe lesions such as atrophy, tumours, and hydrocephalus. There is a dispropor-tionate degree of difficulty walking despite mild or absent neuro-logical signs when the patient is examined at rest. There is a shuffling marche a petit pas, freezing, and start hesitation as well as a variable degree of dysequilibrium. The legs are often held stiff when examined due to an inability to relax them; this becomes more evident as the examiner passively moves the limbs, paratonia, or gegenhalten. This gait is sometimes controversially referred to as *frontal lobe gait apraxia* due to the severe gait disturbance without corresponding motor or sensory deficit in the lower limbs (Nutt *et al.* 1993). In some patients with frontal pathology, the main deficit is a dysequilibrium of gait, the *frontal ataxia of Bruns or frontal dysequilibrium,* in which the patient may be unable to sit or stand independently (Nutt *et al.* 1993). Postural reflexes are lost or inappropriate.

Some older patients develop *primary progressive freezing of gait* in which there is a pure failure of gait initiation with start hesitation and freezing (Atchison *et al.* 1993). There is no evidence of any associated neurological disorder at onset and initially no falling. However, the condition is progressive with the development of falls, postural instability, and the need for a wheelchair within 5 years in many cases. Some patients develop signs of bradykinesia and rigidity but there is no response to dopaminergic drugs and functional imaging suggests frontal lobe dysfunction (Factor *et al.* 2002). Primary progressive freezing of gait is heterogeneous; neuropathological findings have included pallidonigroluysian degeneration as in pure akinesia (Section 40.11.9) and Lewy body changes while long term follow up has revealed clinical diagnoses of progressive supranuclear palsy and corticobasal degeneration in some cases (Factor *et al.* 2006). It is possible that the primary form of progressive supranuclear palsy is pallidonigroluysian degeneration and that along with pure akinesia, it is a form of primary pallidal degeneration.

40.11.11 Startle syndromes and hyperekplexia

The normal human auditory startle response is a polysynaptic reflex originating in the lower brainstem. The first activity is in the sternocleidomastoid, followed by the masseter and facial muscles and then by trunk and limb activation. There is conduction up the brainstem and caudally via slowly conducting reticulospinal pathways (Brown *et al.* 1991a). The auditory blink reflex precedes the startle response and is a separate phenomenon. Clinically there is eye closure, facial grimacing, flexion of the neck, trunk, and limbs, and autonomic changes with brief apnoea. Emotional and behavioural changes may follow.

Hereditary hyperekplexia

This is a disorder characterized by abnormal excitability of the normal startle reflex. Most cases are autosomal dominant, a few are autosomal recessive or X-linked, and others are sporadic. There are both major and minor forms of hyperekplexia. In the major form the onset is in infancy with generalized stiffness and excessive startle reactions. There may be repeated cardiorespiratory arrests especially in the sporadic major form (Bakker *et al.* 2006). In the older child, sudden noise or touch may cause two additional types of response; brief generalized tonic spasms with stiffening and falling and more frequent brief generalized startle responses (Brown *et al.* 1991b). The former may appear epileptic but there is no loss of consciousness. These attacks cause the child to adopt a cautious gait and muscle stiffness also contributes to a curious hesitant walk and posture. There may be brisk reflexes and clonus in the legs. There is an increased incidence of epilepsy and learning disability in hyperekplexia (Bakker *et al.* 2006), suggesting a widespread abnormality of cortical excitability. EEG and neuroimaging are usually normal. In the minor form, there is only an increased startle response and a later onset, after infancy (Matsumoto and Hallett 1994). Treatment is with clonazepam with little evidence for any alternative medications.

The major form of hereditary hyperekplexia is usually autosomal dominant and caused by mutations of the gene encoding the alpha 1 subunit of the glycine receptor, *GLRA1*, on chromosome 5q (Shiang *et al.* 1993). Numerous mutations of *GLRA1* have now been reported along with very rare mutations of the beta subunit, *GLRB*, *gephyrin*, *GPHN*, and *collybisin*, *ARHGEF9*, genes (Bakker *et al.* 2006). There are rare families with autosomal recessive hyperekplexia due to mutations of *GLRA1* (Coto *et al.* 2005) and compound heterozygous mutations of *GLRB*. The *ARHGEF9* mutation causes the very rare X-linked hyperekplexia. In sporadic cases and families with the minor form of hyperkplexia *GLRA1* muations are less commonly detected.

Symptomatic hyperekplexia

Symptomatic hyperekplexia can occur with a wide variety of cerebral and brainstem lesions as well as in multiple sclerosis, cerebrovascular disease, cerebral palsy, anoxic damage, trauma, and malformations (Bakker *et al.* 2006). Hyperekplexia must be distinguished from startle induced epilepsy which usually occurs in patients with congenital hemiparesis or severe generalized cerebral damage (Matsumoto and Hallett 1994).

Jumping, latah, and myriachit

In these conditions there is an excessive reaction to startle with a jump or start and then behavioural features such as automatic speech, echolalia, echopraxia, swearing, and aggression (Matsumoto and Hallett 1994). The cause of these conditions, which have similar core features, is unclear.

40.11.12 Mirror movements

Mirror movements are usually seen during voluntary finger movements which induce identical but involuntary movements on the contralateral side. They are usually seen in patients with a hemiparesis or a foramen magnum lesion and can occur in Klippel–Feil syndrome (Farmer 2005) (Section 9.2.2) and X-linked Kallmann's syndrome (Mayston *et al.* 1997) (Section 16.1.4). Such movements are normal during childhood and sometimes persist in healthy adults; they are more likely to be seen with increased effort, repetitive movement, and in the presence of pre-existing weakness. Some cases are familial. Mirror movements probably originate in the ipsilateral motor cortex, via the direct uncrossed corticospinal tract (Kanouchi *et al.* 1997).

40.11.13 Movement disorders in psychiatry

The majority of involuntary movements seen in psychiatric disease are tardive complications of neuroleptic treatment (Section 40.9). In addition, psychiatric features are prominent in some movement disorders and may be the presenting feature, such as Huntington's disease, Tourette's syndrome, dementia with Lewy bodies, and neuroacanthocytosis. Wilson's disease characteristically has a psychiatric presentation (Section 40.8). In some patients with conditions normally managed by psychiatrists, abnormal movements may also be a prominent feature and lead to diagnostic problems. Stereotypies are seen in autism and learning disability of various types. Rocking, rubbing, posturing, touching, bruxism, and self injury are typical (Jankovic 1994). Repetitive hand movements are seen in Rett's syndrome (Section 9.6.2) along with axial stereotypies and other dyskinesias such as dystonia and myoclonus (Fitzgerald *et al.* 1990). Some Rett's patients are Parkinsonian and others have ataxia, tremor, and respiratory irregularities. In schizophrenia there are spontaneous orofacial stereotypies similar to those seen in tardive dyskinesia (Fenton *et al.* 1994) as well as subtle abnormalities of fine motor control (Griffith *et al.* 1994). Mannerisms are normal actions performed in a bizarre or exaggerated way and are also a feature of schizophrenia.

Catatonia is a confusing term first proposed by Kahlbaum in 1863 to describe what he believed was a distinct psychiatric disease but what is now known to be a heterogeneous syndrome. It is best considered as a behavioural disorder associated with abnormal motor behaviour but is sometimes used to refer to a subtype of schizophrenia in which catatonic features occur (Joseph 1992). Clinically, catatonic patients may have reduced or increased motor activity; some show mutism, akinesia, and a curious 'waxy flexibility' of the limbs whereby the examiner can put them into fixed positions for long periods 'catalepsy'. These patients may appear uncooperative or even stuporose. Others have abnormal psychomotor hyperactivity which is difficult to distinguish from mania. Either form can occur with autonomic instability and in some cases may lead to coma and death, 'lethal catatonia'. In patients who have also received neuroleptic treatment, a diagnosis of neuroleptic malignant syndrome (Sections 5.5.2; 40.9.1) may be made

(Buckley and Hutchinson 1995) unless the onset of similar symptoms prior to medication is noted. Whether catatonia, lethal catatonia, and the neuroleptic malignant syndrome can truly be distinguished is unclear. Catatonia may be caused by schizophrenia, depression, mania, or neurological disorders such as encephalitis, toxic encephalopathies, epilepsy, cerebral tumours, or neuroleptic drugs (Joseph 1992). Treatment will depend on the underlying cause but in psychiatric disease, a good response is often seen with intravenous diazepam or electroconvulsive treatment.

40.11.14 Psychogenic movement disorders

The great majority of movement disorders are organic and a diagnosis of a psychogenic condition should be made with caution. Commonly, patients with dystonia are incorrectly thought to have a psychiatric illness especially if the clinician is unfamiliar with the curious action specificity of the spasms. In those with Tourette's syndrome and tardive dyskinesia or akathisia, the movements are often assumed to be a behavioural manifestation of associated mental illness. Nevertheless, true psychogenic movement disorders do occur either alone or in combination with an organic disorder (Hinson and Haren 2006) and have probably been underdiagnosed.

Psychogenic movement disorders can take many forms with tremors, dystonia, bradykinesia, myoclonus, and incoordination often combined with speech and gait abnormalities. Onset is typically abrupt, sometimes after injury, with rapid subsequent progression and disability. There may be unusual mixtures of movements, slowness, distractibility, and variability.

Psychogenic dystonia (Lang 1995) can be difficult to diagnose; some of these patients have even undergone stereotactic neurosurgery before the diagnosis was apparent (Batshaw et al. 1985; Fahn and Williams 1988). Psychogenic dystonia is more likely if there are features unlike primary torsion dystonia such as onset in a lower limb in an adult, a fixed posture rather than a mobile action specific spasm (Schrag et al. 2004), rapid progression or generalization and pain. Although pain is very common with focal cervical dystonia, or torticollis, organic dystonia in other parts of the body is rarely painful and severe pain suggests a psychogenic cause (Lang 1995; Vargas et al. 2000; Schrag et al. 2004). The distinction between post traumatic focal dystonia and a psychogenic disorder can be very difficult and at times impossible. Paroxysmal dyskinesias, resembling paroxysmal dystonic choreoathetosis are also sometimes psychogenic (Demirkiran and Jankovic 1995).

Psychogenic tremors are intermittent, variable in frequency, and may show entrainment (Hinson and Haren 2006) although it should be noted that Wilson's disease (Section 40.8) is notorious for presenting with bizarre tremors and psychiatric manifestations.

Psychogenic gait disorders may appear slow and laborious, odd postures, knee buckling, near falling, shaking, non organic weakness, or sensory signs and pain (Bhatia 2001; Thomas and Jankovic 2004). Other psychogenic movement disorder presentations include excessive and unusual startle reactions, facial spasms, and even Parkinsonism (Lang et al. 1995).

Whenever possible, a diagnosis of a psychogenic disorder should be supported by other evidence such as psychologically generated weakness or sensory loss, the disappearance of movements with distraction or volitional movements, excessive and apparently deliberate slowness, unusual variability, and unexplained exacerbations or remissions (Hinson and Haren 2006). Such patients may have a past history of multiple medical symptoms of an uncertain nature, excessive fatigue, secondary gain in the form of family dynamics or social security benefits and involvement in litigation of some sort. None of these features is diagnostic as any may be present with organic illness; it is the overall picture which is suggestive (Fahn 1994b).

Investigations are usually unhelpful but neurophysiology may halp with psychogenic tremors and jerks, demonstrating variability, unusual durations or frequencies, and entrainment. Dopamine transporter, DAT, scanning will reveal normal signal in psychogenic Parkinsonism (Gaig et al. 2006).

Treatment is difficult, especially in longstanding cases; psychiatric consultation is essential followed by treatment which may involve psychotherapy, cognitive behaviour therapy, stress management, relaxation, medication for additional depression or anxiety and rehabilitation (physiotherapy or occupational therapy). Outcome is variable (Hinson and Haren 2006).

References

Aarsland D, Andersen K, Larsen JP et al. (2001). Risk of dementia in Parkinson's disease: a community-based, prospective study. *Neurology*, **56**, 730–6.

Aarsland D, Laake K, Larsen JP et al. (2002). Donepezil for cognitive impairment in Parkinson's disease: a randomised controlled study. *J Neurol Neurosurg Psychiatry*, **72**, 708–12.

Aarsland D, Andersen K, Larsen JP et al. (2003). Prevalence and characteristics of dementia in Parkinson disease: an 8-year prospective study. *Arch Neurol*, **60**, 387–92.

Aarsland D, Ballard CG, Halliday G (2004). Are Parkinson's disease with dementia and dementia with Lewy bodies the same entity? *J Geriatr Psychiatry Neurol*, **17**, 137–45.

Abele M, Burk K, Schols L et al. (2002). The aetiology of sporadic adult-onset ataxia. *Brain*, **125**, 961–8.

Abelson JF, Kwan KY, O'Roak BJ et al. (2005). Sequence variants in SLITRK1 are associated with Tourette's syndrome. *Science*, **310**, 317–20.

Abwender DA, Como PG, Kurlan R et al. (1996). School problems in Tourette's syndrome. *Arch Neurol*, **53**, 509–11.

Ackermans L, Temel Y, Cath D et al. (2006). Deep brain stimulation in Tourette's syndrome: two targets? *Mov Disord*, **21**, 709–13.

Adachi M, Kawanami T, Ohshima H et al. (2004). Morning glory sign: a particular MR finding in progressive supranuclear palsy. *Magn Reson Med Sci*, **3**, 125–32.

Adler CH (2005). Nonmotor complications in Parkinson's disease. *Mov Disord* **20** (Suppl 11), S23–9.

Adler CH, Sethi KD, Hauser RA et al. (1997). Ropinirole for the treatment of early Parkinson's disease. The Ropinirole Study Group. *Neurology*, **49**, 393–9.

Agid Y, Ahlskog E, Albanese A et al. (1999). Levodopa in the treatment of Parkinson's disease: a consensus meeting. *Mov Disord*, **14**, 911–3.

Aharon-Peretz J, Rosenbaum H, Gershoni-Baruch R (2004). Mutations in the glucocerebrosidase gene and Parkinson's disease in Ashkenazi Jews. *N Engl J Med*, **351**, 1972–7.

Ahlskog JE, Muenter MD (2001). Frequency of levodopa-related dyskinesias and motor fluctuations as estimated from the cumulative literature. *Mov Disord*, **16**, 448–58.

Ahmad F, Davis MB, Waddy HM et al. (1993). Evidence for locus heterogeneity in autosomal dominant torsion dystonia. *Genomics*, **15**, 9–12.

Aizawa H, Kwak S, Shimizu T et al. (1991). A case of adult onset pure pallidal degeneration. I. Clinical manifestations and neuropathological observations. *J Neurol Sci*, **102**, 76–82.

Albright AL (1996). Intrathecal baclofen in cerebral palsy movement disorders. *J Child Neurol*, **11**, S29–35.

Alexander GE (1997). Anatomy of the basal ganglia and related motor structures. In: Watts RL, Koller WC, eds. *Movement Disorders: Neurologic Principles and Practice*, pp. 73–85. McGraw-Hill, New York.

Allen RP, Picchietti D, Hening WA *et al.* (2003). Restless legs syndrome: diagnostic criteria, special considerations, and epidemiology. A report from the restless legs syndrome diagnosis and epidemiology workshop at the National Institutes of Health. *Sleep Med*, **4**, 101–19.

Almasy L, Bressman SB, Raymond D *et al.* (1997). Idiopathic torsion dystonia linked to chromosome 8 in two Mennonite families. *Ann Neurol*, **42**, 670–3.

Ammelburg M, Frickely T, Lupas AN (2006). Classification of AAA+ proteins. *J Struct Biol*.

Amir RE, Van den Veyver IB, Wan M *et al.* (1999). Rett syndrome is caused by mutations in X-linked MECP2, encoding methyl-CpG-binding protein 2. *Nat Genet*, **23**, 185–8.

Ananth J, Parameswaran S, Gunatilake S *et al.* (2004). Neuroleptic malignant syndrome and atypical antipsychotic drugs. *J Clin Psychiatry*, **65**, 464–70.

Anca MH, Giladi N, Korczyn AD (2004). Ropinirole in Gilles de la Tourette syndrome. *Neurology*, **62**, 1626–7.

Anderson M (2000). Neurology of Whipple's disease. *J Neurol Neurosurg Psychiatry*, **68**, 2–5.

Anderson NE, Budde-Steffen C, Rosenblum MK *et al.* (1988). Opsoclonus, myoclonus, ataxia, and encephalopathy in adults with cancer: a distinct paraneoplastic syndrome. *Medicine (Baltimore)*, **67**, 100–9.

Anderson VC, Burchiel KJ, Hogarth P *et al.* (2005). Pallidal vs subthalamic nucleus deep brain stimulation in Parkinson disease. *Arch Neurol*, **62**, 554–60.

Andrew J, Fowler CJ, Harrison MJG (1983). Stereotaxic thalamotomy in 55 cases of dystonia. *Brain*, **106**, 981–1000.

Arblaster LA, Lakie M, Mutch WJ *et al.* (1993). A study of the early signs of drug induced parkinsonism. *J Neurol Neurosurg Psychiatry*, **56**, 301–3.

Ardouin C, Pillon B, Peiffer E *et al.* (1999). Bilateral subthalamic or pallidal stimulation for Parkinson's disease affects neither memory nor executive functions: a consecutive series of 62 patients. *Ann Neurol*, **46**, 217–23.

Armon C, Swanson JW, McLean JM *et al.* (1996). Subacute encephalomyelitis presenting as stiff-person syndrome: clinical, polygraphic, and pathologic correlations. *Mov Disord*, **11**, 701–9.

Arrasate M, Mitra S, Schweitzer ES *et al.* (2004). Inclusion body formation reduces levels of mutant huntingtin and the risk of neuronal death. *Nature*, **431**, 805–10.

Asahi T, Hirashima Y, Hamada H *et al.* (2001). A walking stick for a pure akinesia patient. *Neurorehabil Neural Repair*, **15**, 245–7.

Asmus F, Gasser T (2004). Inherited myoclonus-dystonia. *Adv Neurol*, **94**, 113–9.

Atchison PR, Thompson PD, Frackowiak RS *et al.* (1993). The syndrome of gait ignition failure: a report of six cases. *Mov Disord*, **8**, 285–92.

Auburger G, Ratzlaff T, Lunkes A *et al.* (1996). A gene for autosomal dominant paroxysmal choreoathetosis/spasticity (CSE) maps to the vicinity of a potassium channel gene cluster on chromosome 1p, probably within 2 cM between D1S443 and D1S197. *Genomics*, **31**, 90–4.

Awaad Y (1999). Tics in Tourette syndrome: new treatment options. *J Child Neurol*, **14**, 316–9.

Awaad Y, Michon AM, Minarik S (2005). Use of levetiracetam to treat tics in children and adolescents with Tourette syndrome. *Mov Disord*, **20**, 714–8.

Bachoud-Levi A, Bourdet C, Brugieres P *et al.* (2000). Safety and tolerability assessment of intrastriatal neural allografts in five patients with Huntington's disease. *Exp Neurol*, **161**, 194–202.

Badhwar A, Berkovic SF, Dowling JP *et al.* (2004). Action myoclonus-renal failure syndrome: characterization of a unique cerebro-renal disorder. *Brain*, **127**, 2173–82.

Bakker MJ, van Dijk JG, van den Maagdenberg AM *et al.* (2006). Startle syndromes. *Lancet Neurol*, **5**, 513–24.

Bandmann O, Nygaard TG, Surtees R *et al.* (1996). Dopa-responsive dystonia in British patients: new mutations of the GTP-cyclohydrolase I gene and evidence for genetic heterogeneity. *Hum Mol Genet*, **5**, 403–6.

Barclay CL, Lang AE (1995). Other secondary dystonias. In Tsui JKC, Calne DB, eds. *Handbook of Dystonia*, pp. 267–305. Marcel Dekker, New York.

Barclay CL, Lang AE (1997). Dystonia in progressive supranuclear palsy. *J Neurol Neurosurg Psychiatry*, **62**, 352–6.

Barker RA, Revesz T, Thom M *et al.* (1998). Review of 23 patients affected by the stiff man syndrome: clinical subdivision into stiff trunk (man) syndrome, stiff limb syndrome, and progressive encephalomyelitis with rigidity. *J Neurol Neurosurg Psychiatry*, **65**, 633–40.

Barnes MP, Best D, Kidd L *et al.* (2005). The use of botulinum toxin type-B in the treatment of patients who have become unresponsive to botulinum toxin type-A—initial experiences. *Eur J Neurol*, **12**, 947–55.

Bartholome K, Ludecke B (1998). Mutations in the tyrosine hydroxylase gene cause various forms of L-dopa-responsive dystonia. *Adv Pharmacol*, **42**, 48–9.

Bartsch T, Herzog J, Baron R *et al.* (2003). The stiff limb syndrome--a new case and a literature review. *J Neurol*, **250**, 488–90.

Bastin A, Gurmin V, Mediwake R *et al.* (2002). Stiff man syndrome presenting with low back pain. *Ann Rheum Dis*, **61**, 939–40.

Batshaw ML, Wachtel RC, Deckel AW *et al.* (1985). Munchausen's syndrome simulating torsion dystonia. *N Engl J Med*, **312**, 1437–9.

Bauer P, Laccone F, Rolfs A *et al.* (2004). Trinucleotide repeat expansion in SCA17/TBP in white patients with Huntington's disease-like phenotype. *J Med Genet*, **41**, 230–2.

Ben-Pazi H, Livne A, Shapira Y *et al.* (2003). Parkinsonian features after streptococcal pharyngitis. *J Pediatr*, **143**, 267–9.

Benabid AL, Pollak P, Gervason L *et al.* (1991). Long term suppression of tremor by chronic stimulation of the ventral intermediate thalamic nucleus. *Lancet*, **337**, 403–6.

Benabid AL, Pollak P, Gao D *et al.* (1996). Chronic electrical stimulation of the ventralis intermedius nucleus of the thalamus as a treatment of movement disorders. *J Neurosurg*, **84**, 203–14.

Bennett LB, Roach ES, Bowcock AM (2000). A locus for paroxysmal kinesigenic dyskinesia maps to human chromosome 16. *Neurology*, **54**, 125–30.

Bennett DA, Wilson RS, Gilley DW *et al.* (1990). Clinical diagnosis of Binswanger's disease. *J Neurol Neurosurg Psychiatry*, **53**, 961–5.

Berardelli A, Formica A, Mercuri B *et al.* (1993). Botulinum toxin treatment in patients with focal dystonia and hemifacial spasm. A multicenter study of the Italian Movement Disorder Group. *Ital J Neurol Sci*, **14**, 361–7.

Berardelli A, Rothwell JC, Thompson PD *et al.* (2001). Pathophysiology of bradykinesia in Parkinson's disease. *Brain*, **124**, 2131–46.

Berardi D, Dell'Atti M, Amore M *et al.* (2002). Clinical risk factors for neuroleptic malignant syndrome. *Hum Psychopharmacol*, **17**, 99–102.

Bereznai B, Steude U, Seelos K *et al.* (2002). Chronic high-frequency globus pallidus internus stimulation in different types of dystonia: a clinical, video, and MRI report of six patients presenting with segmental, cervical, and generalized dystonia. *Mov Disord*, **17**, 138–44.

Berg D, Niwar M, Maass S *et al.* (2005). Alpha-synuclein and Parkinson's disease: implications from the screening of more than 1,900 patients. *Mov Disord* 20, 1191–4.

Bergeron C, Pollanen MS, Weyer L *et al.* (1996). Unusual clinical presentations of cortical-basal ganglionic degeneration. *Ann Neurol*, **40**, 893–900.

Berkovic SF, Andermann F, Carpenter S *et al.* (1986). Progressive myoclonus epilepsies: specific causes and diagnosis. *N Engl J Med*, **315**, 296–305.

Berkovic SF, Carpenter S, Andermann F *et al.* (1988). Kufs' disease: a critical reappraisal. *Brain*, **111**, 27–62.

Berkovic SF, Carpenter S, Evans A *et al.* (1989). Myoclonus epilepsy and ragged-red fibres (MERRF). 1. A clinical, pathological, biochemical, magnetic resonance spectrographic and positron emission tomographic study. *Brain*, **112**, 1231–60.

Berkovic SF, Mazarib A, Walid S *et al.* (2005). A new clinical and molecular form of Unverricht-Lundborg disease localized by homozygosity mapping. *Brain*, **128**, 652–8.

Bernardi B, Zimmerman RA, Savino PJ *et al.* (1993). Magnetic resonance tomographic angiography in the investigation of hemifacial spasm. *Neuroradiology*, **35**, 606–11.

Berry DS, Moriarty RA (1999). Atlantoaxial subluxation related to pharyngitis: Grisel's syndrome. *Clin Pediatr (Phila)*, **38**, 673–5.

Bertoni JM, Brown P, Goldfarb LG *et al.* (1992). Familial Creutzfeldt-Jakob disease (codon 200 mutation) with supranuclear palsy. *JAMA*, **268**, 2413–5.

Bertrand CM (1993). Selective peripheral denervation for spasmodic torticollis: surgical technique, results, and observations in 260 cases. *Surg Neurol*, **40**, 96–103.

Bharucha KJ, Sethi KD (1995). Tardive tourettism after exposure to neuroleptic therapy. *Mov Disord*, **10**, 791–3.

Bharucha NE, Bharucha EP, Bharucha AE *et al.* (1988). Prevalence of Parkinson's disease in the Parsi community of Bombay, India. *Arch Neurol*, **45**, 1321–3.

Bhatia KP (2001). Psychogenic gait disorders. *Adv Neurol*, **87**, 251–4.

Bhatia K, Thompson PD, Marsden CD (1992). "Isolated" postinfectious myoclonus. *J Neurol Neurosurg Psychiatry*, **55**, 1089–91.

Bhatia KP, Brown P, Gregory R *et al.* (1995). Progressive myoclonic ataxia associated with coeliac disease. The myoclonus is of cortical origin, but the pathology is in the cerebellum. *Brain*, **118**, 1087–93.

Bhatia KP, Lera G, Luthert PJ *et al.* (1994). Vascular chorea: case report with pathology. *Mov Disord*, **9**, 447–50.

Binder J, Hofmann S, Kreisel S *et al.* (2003). Clinical and molecular findings in a patient with a novel mutation in the deafness-dystonia peptide (DDP1) gene. *Brain*, **126**, 1814–20.

Blasi J, Chapman ER, Link E *et al.* (1993). Botulinum neurotoxin A selectively cleaves the synaptic protein SNAP-25. *Nature*, **365**, 160–3.

Bloch MH, Peterson BS, Scahill L *et al.* (2006). Adulthood outcome of tic and obsessive-compulsive symptom severity in children with Tourette syndrome. *Arch Pediatr Adolesc Med*, **160**, 65–9.

Bodensteiner JB, Goldblum RM, Golman AS (1980). Progressive dystonia masking ataxia in ataxia telengiectasia. *Arch Neurol*, **37**, 464–5.

Boecker H, Wills AJ, Ceballos-Baumann A *et al.* (1997). Stereotactic thalamotomy in tremor-dominant Parkinson's disease: an H2(15)O PET motor activation study. *Ann Neurol*, **41**, 108–11.

Boeve BF, Lang AE, Litvan I (2003). Corticobasal degeneration and its relationship to progressive supranuclear palsy and frontotemporal dementia. *Ann Neurol*, 54 (Suppl 5), S15–9.

Boeve BF, Maraganore DM, Parisi JE *et al.* (1999). Pathologic heterogeneity in clinically diagnosed corticobasal degeneration. *Neurology*, **53**, 795–800.

Bonelli RM, Hofmann P (2004). A review of the treatment options for Huntington's disease. *Expert Opin Pharmacother*, **5**, 767–76.

Bonifati V, Rohe CF, Breedveld GJ *et al.* (2005). Early-onset parkinsonism associated with PINK1 mutations: frequency, genotypes, and phenotypes. *Neurology*, **65**, 87–95.

Borges N (2005). Tolcapone in Parkinson's disease: liver toxicity and clinical efficacy. *Expert Opin Drug Saf*, **4**, 69–73.

Braak H, Ghebremedhin E, Rub U *et al.* (2004). Stages in the development of Parkinson's disease-related pathology. *Cell Tissue Res*, **318**, 121–34.

Brans JW, Lindeboom R, Snoek JW *et al.* (1996). Botulinum toxin versus trihexyphenidyl in cervical dystonia: a prospective, randomized, double-blind controlled trial. *Neurology*, **46**, 1066–72.

Bras JM, Guerreiro RJ, Ribeiro MH *et al.* (2005). G2019S dardarin substitution is a common cause of Parkinson's disease in a Portuguese cohort. *Mov Disord*, **20**, 1653–5.

Brashear A, DeLeon D, Bressman SB *et al.* (1997). Rapid-onset dystonia-parkinsonism in a second family. *Neurology*, **48**, 1066–9.

Brashear A, Farlow MR, Butler IJ *et al.* (1996). Variable phenotype of rapid-onset dystonia-parkinsonism. *Mov Disord*, **11**, 151–6.

Breedveld GJ, van Dongen JW, Danesino C *et al.* (2002). Mutations in TITF-1 are associated with benign hereditary chorea. *Hum Mol Genet*, **11**, 971–9.

Bressman SB (2004). Dystonia genotypes, phenotypes, and classification. *Adv Neurol*, **94**, 101–7.

Bressman SB, de Leon D, Brin MF *et al.* (1989). Idiopathic dystonia among Ashkenazi Jews: evidence for autosomal dominant inheritance. *Ann Neurol*, **26**, 612–20.

Brewer GJ (1999). Penicillamine should not be used as initial therapy in Wilson's disease. *Mov Disord*, **14**, 551–4.

Brewer GJ, Hedera P, Kluin KJ *et al.* (2003). Treatment of Wilson disease with ammonium tetrathiomolybdate: III. Initial therapy in a total of 55 neurologically affected patients and follow-up with zinc therapy. *Arch Neurol*, **60**, 379–85.

Brewer GJ, Yuzbasiyan-Gurkan V (1992). Wilson disease. *Medicine (Baltimore)*, **71**, 139–64.

Brewer GJ, Yuzbasiyan-Gurkan V, Lee DY *et al.* (1989). Treatment of Wilson's disease with zinc. VI. Initial treatment studies. *J Lab Clin Med*, **114**, 633–8.

Bristow MF, Kohen D (1993). How "malignant" is the neuroleptic malignant syndrome? *BMJ*, **307**, 1223–4.

Brodal A (1981). *Neurological Anatomy in Relation to Clinical Medicine*. Oxford University Press, Oxford.

Brooks DJ (2000). Functional imaging studies in corticobasal degeneration. *Adv Neurol*, **82**, 209–15.

Brooks DJ, Ibanez V, Sawle GV *et al.* (1992). Striatal D2 receptor status in patients with Parkinson's disease, striatonigral degeneration, and progressive supranuclear palsy, measured with 11C-raclopride and positron emission tomography. *Ann Neurol*, **31**, 184–92.

Brooks DJ, Salmon EP, Mathias CJ *et al.* (1990). The relationship between locomotor disability, autonomic dysfunction, and the integrity of the striatal dopaminergic system in patients with multiple system atrophy, pure autonomic failure, and Parkinson's disease, studied with PET. *Brain*, **113**, 1539–52.

Broughton R, Tolentino MA, Krelina M (1985). Excessive fragmentary myoclonus in non REM sleep: a report of 38 cases. *Electroenceph Clin Neurophysiol*, **61**, 123–33.

Brown P, Marsden CD (1999). The stiff man and stiff man plus syndromes. *J Neurol*, **246**, 648–52.

Brown P, Rothwell JC, Thompson PD *et al.* (1991a). New observations on the normal auditory startle reflex in man. *Brain*, **114**, 1981–92.

Brown P, Rothwell JC, Thompson PD *et al.* (1991b). The hyperekplexias and their relationship to the normal startle reflex. *Brain*, **114**, 1903–28.

Brown P, Thompson PD, Rothwell JC *et al.* (1991c). Axial myoclonus of propriospinal origin. *Brain*, **114**, 197–214.

Brown P, Thompson PD, Rothwell JC *et al.* (1991d). A case of postanoxic encephalopathy with cortical action and brainstem reticular reflex myoclonus. *Mov Disord*, **6**, 139–44.

Brown J, Lantos PL, Roques P *et al.* (1996a). Familial dementia with swollen achromatic neurons and corticobasal inclusion bodies: a clinical and pathological study. *J Neurol Sci*, **135**, 21–30.

Brown P, Ridding MC, Werhahn KJ *et al.* (1996b). Abnormalities of the balance between inhibition and excitation in the motor cortex of patients with cortical myoclonus. *Brain*, **119**, 309–17.

Brown P, Rothwell JC, Marsden CD (1997). The stiff leg syndrome. *J Neurol Neurosurg Psychiatry*, **62**, 31–7.

Brown P, Steiger MJ, Thompson PD *et al.* (1993). Effectiveness of piracetam in cortical myoclonus. *Mov Disord*, **8**, 63–8.

Buckley PF and Hutchinson M (1995). Neuroleptic malignant syndrome [editorial]. *J Neurol Neurosurg Psychiatry*, **58**, 271–3.

Budman CL, Gayer A, Lesser M et al. (2001). An open-label study of the treatment efficacy of olanzapine for Tourette's disorder. *J Clin Psychiatry*, **62**, 290–4.

Bull PC, Thomas GR, Rommens JM et al. (1993). The Wilson disease gene is a putative copper transporting P-type ATPase similar to the Menkes gene [published erratum appears in *Nat Genet* 1994 Feb; 6(2), 214]. *Nat Genet*, **5**, 327–37.

Burke RE, Fahn S, Gold AP (1982a). Delayed onset dystonia in patients with static encephalopathy. *JNNP*, **43**, 787–97.

Burke RE, Fahn S, Jankovic J et al. (1982b). Tardive dystonia: Late onset and persistent dystonia caused by antipsychotic drugs. *Neurology*, **32**, 1335–46.

Burke RE, Fahn S, Marsden CD (1986). Torsion dystonia: A double blind prospective trial of high dosage trihexphenidyl. *Neurology*, **36**, 160–4.

Burke RE, Fahn S, Marsden CD et al. (1985). Validity and reliability of a rating scale for the primary torsion dystonias. *Neurology*, **35**, 73–7.

Burke RE, Kang UJ (1988). Tardive dystonia: clinical aspects and treatment. *Adv Neurol*, **49**, 199–210.

Burke RE, Kang UJ, Jankovic J et al. (1989). Tardive akathisia: An analysis of clinical features and response to open therapeutic trials. *Mov Disord*, **4**, 157–75.

Burn DJ, Brooks DJ (1993). Nigral dysfunction in drug-induced parkinsonism: an 18F-dopa PET study. *Neurology*, **43**, 552–6.

Burn DJ, Ball J, Lees AJ et al. (1991). A case of progressive encephalomyelitis with rigidity and positive antiglutamic acid decarboxylase antibodies. *J Neurol Neurosurg Psychiatry*, **54**, 449–51.

Burn DJ, Mark MH, Playford ED et al. (1992). Parkinson's disease in twins studied with 18F-dopa and positron emission tomography. *Neurology*, **42**, 1894–900.

Burn DJ, Rinne JO, Quinn NP et al. (1995). Striatal opioid receptor binding in Parkinson's disease, striatonigral degeneration and Steele-Richardson-Olszewski syndrome, A [11C]diprenorphine PET study. *Brain*, **118**, 951–8.

Burn DJ, Sawle GV, Brooks DJ (1994). Differential diagnosis of Parkinson's disease, multiple system atrophy, and Steele-Richardson-Olszewski syndrome: discriminant analysis of striatal 18F-dopa PET data. *J Neurol Neurosurg Psychiatry*, **57**, 278–84.

Byrne EJ, Lennox G, Lowe J et al. (1989). Diffuse Lewy body disease: clinical features in 15 cases. *J Neurol Neurosurg Psychiatry*, **52**, 709–17.

Cammarota A, Gershanik OS, Garcia S et al. (1995). Cervical dystonia due to spinal cord ependymoma: involvement of cervical cord segments in the pathogenesis of dystonia. *Mov Disord*, **10**, 500–3.

Candler PM, Dale RC, Griffin S et al. (2006). Post-streptococcal opsoclonus-myoclonus syndrome associated with anti-neuroleukin antibodies. *J Neurol Neurosurg Psychiatry*, **77**, 507–12.

Caparros-Lefebvre D, Destee A, Petit H (1997). Late onset familial dystonia: could mitochondrial deficits induce a diffuse lesioning process of the whole basal ganglia system? *J Neurol, Neurosurg Psychiatry*, **63** 196–203.

Caparros-Lefebvre D, Sergeant N, Lees A et al. (2002). Guadeloupean parkinsonism: a cluster of progressive supranuclear palsy-like tauopathy. *Brain*, **125**, 801–11.

Caplan LR (1995). Binswanger's disease--revisited. *Neurology*, 45:626–33.

Caraceni T (1994). A case for early levodopa treatment of Parkinson's disease. *Clin Neuropharmacol*, **17**, S38–42.

Caraceni B, Broggi G, Avanzini G (1974). Familial idiopathic basal ganglia calcification exhibiting dystonia musculorum deformans features. *Eur Neurol*, **12**, 351–9.

Cardoso F, Eduardo C, Silva AP et al. (1997). Chorea in fifty consecutive patients with rheumatic fever. *Mov Disord*, **12**, 701–3.

Cardoso F, Jankovic J, Grossman RG et al. (1995). Outcome after stereotactic thalamotomy for dystonia and hemiballismus. *Neurosurgery*, **36**, 501–7; discussion 507–8.

Carlson JH, Bergstrom DA, Walters JR (1987). Stimulation of both D1 and D2 receptors appears neccessary for full expression of postsynaptic effects of dopamine agonists: a neurophysiological study. *Brain Res*, **400**, 205–18.

Carney RM, Wolpert CM, Ravan SA et al. (2003). Identification of MeCP2 mutations in a series of females with autistic disorder. *Pediatr Neurol*, **28**, 205–11.

Case records of the Massachusetts General Hospital (1993). A 75 year old man with right sided rigidity, dysarthria and abnormal gait. Case 46-1993. *N Engl J Med*, **329**, 1560–7.

Casey DE (2004). Pathophysiology of antipsychotic drug-induced movement disorders. *J Clin Psychiatry*, 65 (Suppl 9), 25–8.

Casey DE, Gerlach J (1988). Tardive dyskinesia. *Acta Psychiatr Scand*, **77**, 369–78.

Castelnau P, Cif L, Valente EM et al. (2005). Pallidal stimulation improves pantothenate kinase-associated neurodegeneration. *Ann Neurol*, **57**, 738–41.

Castillo M, Kwock L, Arbelaez A (1999). Sydenham's chorea: MRI and proton spectroscopy. *Neuroradiology*, **41**, 943–5.

Caviness JN, Brown P (2004). Myoclonus: current concepts and recent advances. *Lancet Neurol*, **3**, 598–607.

Caviness JN, Forsyth PA, Layton DD et al. (1995). The movement disorder of adult opsoclonus. *Mov Disord*, **10**, 22–7.

Charles PD, Van Blercom N, Krack P et al. (2002). Predictors of effective bilateral subthalamic nucleus stimulation for PD. *Neurology*, **59**, 932–4.

Chartier-Harlin MC, Kachergus J, Roumier C et al. (2004). Alpha-synuclein locus duplication as a cause of familial Parkinson's disease. *Lancet*, **364**, 1167–9.

Ching KH, Westaway SK, Gitschier J et al. (2002). HARP syndrome is allelic with pantothenate kinase-associated neurodegeneration. *Neurology*, **58**, 1673–4.

Chinnery PF, Curtis AR, Fey C et al. (2003). Neuroferritinopathy in a French family with late onset dominant dystonia. *J Med Genet*, **40**, e69.

Chu N, Huang C, Lu C et al. (1995). Dystonia caused by toxins. In Tsui JKC, Calne DB, eds. *Handbook of Dystonia*, pp. 241–65. Marcel Dekker, New York.

Church AJ, Cardoso F, Dale RC et al. (2002). Anti-basal ganglia antibodies in acute and persistent Sydenham's chorea. *Neurology*, **59**, 227–31.

Church AJ, Dale RC, Lees AJ et al. (2003). Tourette's syndrome: a cross sectional study to examine the PANDAS hypothesis. *J Neurol Neurosurg Psychiatry*, **74**, 602–7.

Cif L, El Fertit H, Vayssiere N et al. (2003). Treatment of dystonic syndromes by chronic electrical stimulation of the internal globus pallidus. *J Neurosurg Sci*, **47**, 52–5.

Cif L, Valente EM, Hemm S et al. (2004). Deep brain stimulation in myoclonus-dystonia syndrome. *Mov Disord*, **19**, 724–7.

Clarimon J, Asgeirsson H, Singleton A et al. (2005). Torsin A haplotype predisposes to idiopathic dystonia. *Ann Neurol*, **57**, 765–7.

Clarke CE, Davies P (2000). Systematic review of acute levodopa and apomorphine challenge tests in the diagnosis of idiopathic Parkinson's disease. *J Neurol Neurosurg Psychiatry*, **69**, 590–4.

Cole R, Hallett M, Cohen LG (1995). Double-blind trial of botulinum toxin for treatment of focal hand dystonia. *Mov Disord*, **10**, 466–71.

Coleman RJ, Quinn NP, Marsden CD (1988). Multiple sclerosis presenting as adult onset dystonia. *Mov disord*, **3**, 329–32.

Collins SJ, Ahlskog JE, Parisi JE et al. (1995). Progressive supranuclear palsy: neuropathologically based diagnostic clinical criteria. *J Neurol Neurosurg Psychiatry*, **58**, 167–73.

Colosimo C, Hughes AJ, Kilford L et al. (2003). Lewy body cortical involvement may not always predict dementia in Parkinson's disease. *J Neurol Neurosurg Psychiatry*, **74**, 852–6.

Comings DE, Gade-Andavolu R, Gonzalez N et al. (2000). Comparison of the role of dopamine, serotonin, and noradrenaline genes in ADHD, ODD and conduct disorder: multivariate regression analysis of 20 genes. *Clin Genet*, **57**, 178–96.

Comu S, Giuliani M, Narayanan V (1996). Episodic ataxia and myokymia syndrome: a new mutation of potassium channel gene Kv1.1. *Ann Neurol*, **40**, 684–7.

Conard C, Andreadis A, Trojanowski JQ *et al.* (1997). Genetic evidence for the involvement of tau in progressive supranuclear palsy. *Ann Neurol*, **41**, 277–81.

Connor JR, Boyer PJ, Menzies SL *et al.* (2003). Neuropathological examination suggests impaired brain iron acquisition in restless legs syndrome. *Neurology*, **61**, 304–9.

Consensus Statement (1996). Consensus statement on the definition of orthostatic hypotension, pure autonomic failure, and multiple system atrophy. The Consensus Committee of the American Autonomic Society and the American Academy of Neurology. *Neurology*, **46**, 1470.

Cooke WT, Smith WT (1966). Neurological disorders associated with adult coeliac disease. *Brain*, **89**, 683–722.

Cosentino C, Torres L, Flores M *et al.* (1996). Paroxysmal kinesigenic dystonia and spinal cord lesion. *Mov Disord*, **11**, 453–5.

Coto E, Armenta D, Espinosa R *et al.* (2005). Recessive hyperekplexia due to a new mutation (R100H) in the GLRA1 gene. *Mov Disord* 20, 1626–9.

Craufurd D, Dodge A (1993). Mutation size and age at onset of Huntington's disease. *J Med Genet*, **30**, 1008–11.

Craufurd D, Tyler A (1992). Predictive testing for Huntington's disease: protocol of the UK Huntington's prediction consortium. *J Med Genet*, **29**, 915–8.

Critchley M (1929). Arteriosclerotic parkinsonism. *Brain*, **52**, 23–83.

Crossman AR, Mitchell IJ, Sambrook MA *et al.* (1988). Chorea and myoclonus in the monkey induced by gamma-aminobutyric acid antagonism in the lentiform complex. The site of drug action and a hypothesis for the neural mechanisms of chorea. *Brain*, **111**, 1211–33.

Curran T, Lang AE (1994). Parkinsonian syndromes associated with hydrocephalus: case reports, a review of the literature, and pathophysiological hypotheses. *Mov Disord*, **9**, 508–20.

Curtis AR, Fey C, Morris CM *et al.* (2001). Mutation in the gene encoding ferritin light polypeptide causes dominant adult-onset basal ganglia disease. *Nat Genet*, **28**, 350–4.

Dahlman T, Hartvig P, Lofholm M *et al.* (1995). Long-term treatment of Wilson's disease with triethylene tetramine dihydrochloride (trientine). *Q J Med*, **88**, 609–16.

Dalakas MC (2005). The role of IVIg in the treatment of patients with stiff person syndrome and other neurological diseases associated with anti-GAD antibodies. *J Neurol*, 252 (Suppl 1), I19–25.

Dalakas MC, Fujii M, Li M *et al.* (2000). The clinical spectrum of anti-GAD antibody-positive patients with stiff-person syndrome. *Neurology*, **55**, 1531–5.

Dalakas MC, Fujii M, Li M *et al.* (2001a). High-dose intravenous immune globulin for stiff-person syndrome. *N Engl J Med*, **345**, 1870–6.

Dalakas MC, Li M, Fujii M *et al.* (2001b). Stiff person syndrome: quantification, specificity, and intrathecal synthesis of GAD65 antibodies. *Neurology*, **57**, 780–4.

Dale RC (2003). Autoimmunity and the basal ganglia: new insights into old diseases. *Q J Med*, **96**, 183–91.

Dale RC (2005). Post-streptococcal autoimmune disorders of the central nervous system. *Dev Med Child Neurol*, **47**, 785–91.

Dale RC, Church AJ, Benton S *et al.* (2002). Post-streptococcal autoimmune dystonia with isolated bilateral striatal necrosis. *Dev Med Child Neurol*, **44**, 485–9.

Dale RC, Church AJ, Cardoso F *et al.* (2001). Poststreptococcal acute disseminated encephalomyelitis with basal ganglia involvement and auto-reactive antibasal ganglia antibodies. *Ann Neurol*, **50**, 588–95.

Dale RC, Church AJ, Surtees RA *et al.* (2004). Encephalitis lethargica syndrome: 20 new cases and evidence of basal ganglia autoimmunity. *Brain*, **127**, 21–33.

Dale RC, Heyman I, Giovannoni G *et al.* (2005). Incidence of anti-brain antibodies in children with obsessive-compulsive disorder. *Br J Psychiatry*, **187**, 314–9.

Danek A, Walker RH (2005). Neuroacanthocytosis. *Curr Opin Neurol*, **18**, 386–92.

Danek A, Rubio JP, Rampoldi L *et al.* (2001). McLeod neuroacanthocytosis: genotype and phenotype. *Ann Neurol*, **50**, 755–64.

Daniel SE, de Bruin VM, Lees AJ (1995). The clinical and pathological spectrum of Steele-Richardson-Olszewski syndrome (progressive supranuclear palsy): a reappraisal. *Brain*, **118**, 759–70.

Davie CA, Wenning GK, Barker GJ *et al.* (1995). Differentiation of multiple system atrophy from idiopathic Parkinson's disease using proton magnetic resonance spectroscopy. *Ann Neurol*, 37, 204–10.

Davis MB, Bateman D, Quinn NP *et al.* (1994). Mutation analysis in patients with possible but apparently sporadic Huntington's disease. *Lancet*, **344**, 714–7.

Davison C (1954). Pallidopyramidal disease. *J Neuropathol Exp Neurol*, **13**, 50–9.

De Camilli P, Thomas A, Cofiell R (1993). The synaptic vesicle associated protein amphiphysin is the 128 kD autoantigen of stiff man syndrome with breast cancer. *J Exp Med*, **178**, 2219–23.

de Carvalho Aguiar P, Sweadner KJ, Penniston JT *et al.* (2004). Mutations in the Na+/K+ -ATPase alpha3 gene ATP1A3 are associated with rapid-onset dystonia parkinsonism. *Neuron*, **43**, 169–75.

De Gaspari D, Siri C, Landi A *et al.* (2006). Clinical and neuropsychological follow up at 12 months in patients with complicated Parkinson's disease treated with subcutaneous apomorphine infusion or deep brain stimulation of the subthalamic nucleus. *J Neurol Neurosurg Psychiatry*, **77**, 450–3.

de Haan GJ, Halley DJ, Doelman JC *et al.* (2004). Univerricht-Lundborg disease: underdiagnosed in the Netherlands. *Epilepsia*, **45**, 1061–3.

de Jesus Mari J, Lima MS, Costa AN *et al.* (2004). The prevalence of tardive dyskinesia after a nine month naturalistic randomized trial comparing olanzapine with conventional treatment for schizophrenia and related disorders. *Eur Arch Psychiatry Clin Neurosci*, **254**, 356–61.

De Rijk MC, Tzourio C, Breteler MM *et al.* (1997). Prevalence of parkinsonism and Parkinson's disease in Europe: the EUROPARKINSON Collaborative Study. European Community Concerted Action on the Epidemiology of Parkinson's disease. *J Neurol Neurosurg Psychiatry*, **62**, 10–5.

de Tommaso M, Di Fruscolo O, Sciruicchio V *et al.* (2005). Efficacy of levetiracetam in Huntington disease. *Clin Neuropharmacol*, **28**, 280–4.

De Yebenes JG, Pernaute RS, Tabernero C (1996). Symptomatic dystonias. In Watts RL, Koller WC, eds. *Movement Disorders. Neurologic Principles and Practice*, pp. 455–75. McGraw-Hill, New York.

De Yebenes JG, Brin MF, Mena MA *et al.* (1988). Neurochemical findings in neuroacanthocytosis. *Mov Disord*, **3**, 300–12.

Deep-Brain Stimulation for Parkinson's Disease Study Group (2001). Deep-brain stimulation of the subthalamic nucleus or the pars interna of the globus pallidus in Parkinson's disease. *N Engl J Med*, **345**, 956–63.

Defazio G, Abbruzzese G, Livrea P *et al.* (2004). Epidemiology of primary dystonia. *Lancet Neurol*, **3**, 673–8.

Defer GL, Geny C, Ricolfi F *et al.* (1996). Long-term outcome of unilaterally transplanted parkinsonian patients. I. Clinical approach. *Brain*, **119**, 41–50.

Deiber MP, Pollak P, Passingham R *et al.* (1993). Thalamic stimulation and suppression of parkinsonian tremor. Evidence of a cerebellar deactivation using positron emission tomography. *Brain*, **116**, 267–79.

Delisle MB, Murrell JR, Richardson R *et al.* (1999). A mutation at codon 279 (N279K) in exon 10 of the Tau gene causes a tauopathy with dementia and supranuclear palsy. *Acta Neuropathol (Berl)*, **98**, 62–77.

DeLong MR (1990). Primate models of movement disorders of basal ganglia origin. *Trends Neurosci*, **13**, 281–5.

Demirkiran M, Jankovic J (1995). Paroxysmal dyskinesias: clinical features and classification. *Ann Neurol*, **38**, 571–9.

Demirkiran M, Jankovic J, Lewis RA *et al.* (1996). Neurologic presentation of Wilson disease without Kayser-Fleischer rings. *Neurology*, **46**, 1040–3.

Deng Y, Newman B, Dunne MP *et al.* (2004). Case-only study of interactions between genetic polymorphisms of GSTM1, P1, T1 and Z1 and smoking in Parkinson's disease. *Neurosci Lett*, **366**, 326–31.

Dening TR, Berrios GE (1989). Wilson's disease: psychiatric symptoms in 195 cases. *Arch Gen Psychiatry*, **46**, 1126–34.

Deonna T, Roulet E, Ghika J *et al.* (1997). Dopa-responsive childhood dystonia: a forme fruste with writer's cramp, triggered by exercise. *Dev Med Child Neurol*, **39**, 49–53.

Deuschl G, Schade-Brittinger C, Krack P *et al.* (2006). A randomized trial of deep-brain stimulation for Parkinson's disease. *N Engl J Med*, **355**, 896–908.

Deuschl G, Toro C, Matsumoto J *et al.* (1995). Movement-related cortical potentials in writer's cramp. *Ann Neurol*, **38**, 862–8.

Deuschl G, Toro C, Valls-Sole J *et al.* (1996). Symptomatic and essential palatal tremor. 3. Abnormal motor learning. *J Neurol Neurosurg Psychiatry*, **60**, 520–5.

Di Maio L, Squitieri F, Napolitano G *et al.* (1993a). Onset symptoms in 510 patients with Huntington's disease. *J Med Genet*, **30**, 289–92.

Di Maio L, Squitieri F, Napolitano G *et al.* (1993b). Suicide risk in Huntington's disease. *J Med Genet*, **30**, 293–5.

Di Maria E, Tabaton M, Vigo T *et al.* (2000). Corticobasal degeneration shares a common genetic background with progressive supranuclear palsy. *Ann Neurol*, **47**, 374–7.

Diaz-Grez F, Lay-Son L, del Barrio-Guerrero E *et al.* (2004). Sydenham's chorea. A clinical analysis of 55 patients with a prolonged follow-up. *Rev Neurol*, **39**, 810–5.

Dickson DW, Lin W, Liu WK *et al.* (1999). Multiple system atrophy: a sporadic synucleinopathy. *Brain Pathol*, **9**, 721–32.

Diederich NJ, Kalteis K, Stamenkovic M *et al.* (2005). Efficient internal pallidal stimulation in Gilles de la Tourette syndrome: a case report. *Mov Disord* **20**, 1496–9.

Djarmati A, Svetel M, Momcilovic D *et al.* (2005). Significance of recurrent mutations in the myofibrillogenesis regulator 1 gene. *Arch Neurol*, **62**, 1641.

Dobyns WB, Ozelius LJ, Kramer PL *et al.* (1993). Rapid-onset dystonia-parkinsonism. *Neurology*, **43**, 2596–602.

Dogali M, Fazzini E, Kolodny E *et al.* (1995). Stereotactic ventral pallidotomy for Parkinson's disease. *Neurology*, **45**, 753–61.

Dooling EC, Schoene WC, Richardson EP (1974). Hallervorden-Spatz syndrome. *Arch Neurol*, **30**, 70–83.

Doran M, du Plessis DG, Enevoldson TP *et al.* (2003). Pathological heterogeneity of clinically diagnosed corticobasal degeneration. *J Neurol Sci*, **216**, 127–34.

Dressler D, Thompson PD, Gledhill RF *et al.* (1994). The syndrome of painful legs and moving toes. *Mov Disord*, **9**, 13–21.

Duayo M, Ambrose C, Myers R *et al.* (1993). Trinucleotide repeat length instability and age of onset in Huntington's disease. *Nat Genet*, **4**, 387–92.

Dujardin K, Defebvre L, Krystkowiak P *et al.* (2003). Executive function differences in multiple system atrophy and Parkinson's disease. *Parkinsonism Relat Disord*, **9**, 205–11.

Dum RP, Strick PL (1991). the origin of corticospinal projections from premotor areas in the frontal lobe. *J Neurosci*, **11**, 667–89.

Durif F, Lemaire JJ, Debilly B *et al.* (2002). Long-term follow-up of globus pallidus chronic stimulation in advanced Parkinson's disease. *Mov Disord*, **17**, 803–7.

Durif F, Vidailhet M, Assal F *et al.* (1997). Low-dose clozapine improves dyskinesias in Parkinson's disease. *Neurology*, **48**, 658–62.

Durso R, Evans JE, Josephs E *et al.* (1997). Central levodopa metabolism in Parkinson's disease after administration of stable isotope labelled levodopa. *Ann Neurol*, **42**, 300–4.

Eblan MJ, Walker JM, Sidransky E (2005). The glucocerebrosidase gene and Parkinson's disease in Ashkenazi Jews. *N Engl J Med*, **352**, 728–31; author reply 728–31.

Edwards M, Wood N, Bhatia K (2003). Unusual phenotypes in DYT1 dystonia: a report of five cases and a review of the literature. *Mov Disord*, **18**, 706–11.

Edwards MJ, Dale RC, Church AJ *et al.* (2004a). A dystonic syndrome associated with anti-basal ganglia antibodies. *J Neurol Neurosurg Psychiatry*, **75**, 914–6.

Edwards MJ, Trikouli E, Martino D *et al.* (2004b). Anti-basal ganglia antibodies in patients with atypical dystonia and tics: a prospective study. *Neurology*, **63**, 156–8.

Eichhorn TE, Oertel WH (2001). Macrogol 3350/electrolyte improves constipation in Parkinson's disease and multiple system atrophy. *Mov Disord*, **16**, 1176–7.

Eidelberg D, Moeller JR, Ishikawa T *et al.* (1995). The metabolic topography of idiopathic torsion dystonia. *Brain*, **118**, 1473–84.

Eidelberg D, Takikawa S, Moeller JR *et al.* (1993a). Striatal hypometabolism distinguishes striatonigral degeneration from Parkinson's disease. *Ann Neurol*, **33**, 518–27.

Eidelberg D, Takikawa S, Wilhelmsen K *et al.* (1993b). Positron emission tomographic findings in Filipino X-linked dystonia-parkinsonism. *Ann Neurol*, **34**, 185–91.

El-Shanti H, Daoud A, Sadoon AA *et al.* (2006). A distinct autosomal recessive ataxia maps to chromosome 12 in an inbred family from Jordan. *Brain Dev*, **28**, 353–7.

Elbaz A, Bower JH, Maraganore DM *et al.* (2002). Risk tables for parkinsonism and Parkinson's disease. *J Clin Epidemiol*, **55**, 25–31.

Elbaz A, Levecque C, Clavel J *et al.* (2004). CYP2D6 polymorphism, pesticide exposure, and Parkinson's disease. *Ann Neurol*, **55**, 430–4.

Elston JS (1992). A new variant of blepharospasm. *J Neurol Neurosurg Psychiatry*, **55**, 369–71.

Eltahawy HA, Feinstein A, Khan F *et al.* (2004a). Bilateral globus pallidus internus deep brain stimulation in tardive dyskinesia: a case report. *Mov Disord* **19**, 969–72.

Eltahawy HA, Saint-Cyr J, Giladi N *et al.* (2004b). Primary dystonia is more responsive than secondary dystonia to pallidal interventions: outcome after pallidotomy or pallidal deep brain stimulation. *Neurosurgery*, **54**, 613–19; discussion 619–21.

Emre M, Aarsland D, Albanese A *et al.* (2004). Rivastigmine for dementia associated with Parkinson's disease. *N Engl J Med*, **351**, 2509–18.

Erenberg G, Cruse RP, Rothner AD (1987). The natural history of Tourette syndrome: a follow-up study. *Ann Neurol*, **22**, 383–5.

Esteban Munoz J, Tolosa E, Saiz A *et al.* (1996). Upper-limb dystonia secondary to a midbrain hemorrhage [letter]. *Mov Disord*, **11**, 96–9.

Etminan M, Gill SS, Samii A (2005). Intake of vitamin E, vitamin C, and carotenoids and the risk of Parkinson's disease: a meta-analysis. *Lancet Neurol*, **4**, 362–5.

Evans AH, Katzenschlager R, Paviour D *et al.* (2004). Punding in Parkinson's disease: its relation to the dopamine dysregulation syndrome. *Mov Disord* **19**, 397–405.

Evans AH, Pavese N, Lawrence AD *et al.* (2006). Compulsive drug use linked to sensitized ventral striatal dopamine transmission. *Ann Neurol*, **59**, 852–8.

Factor SA, Friedman JH (1997). The emerging role of clozapine in the treatment of movement disorders. *Mov Disord*, **12**, 483–96.

Factor SA, Higgins DS, Qian J (2006). Primary progressive freezing gait: a syndrome with many causes. *Neurology*, **66**, 411–4.

Factor SA, Jennings DL, Molho ES *et al.* (2002). The natural history of the syndrome of primary progressive freezing gait. *Arch Neurol*, **59**, 1778–83.

Factor SA, Podskalny GD, Molho ES (1995). Psychogenic movement disorders: frequency, clinical profile, and characteristics. *J Neurol Neurosurg Psychiatry*, **59**, 406–12.

Fahn S (1994a). The paroxysmal dyskinesias. In Marsden CD, Fahn S, eds. *Movement Disorders 3*, pp. 310–45. Butterworth-Heinemann, Oxford.

Fahn S (1994b). Psychogenic movement disorders. In Marsden CD, Fahn S, eds. *Movement Disorders 3*, pp. 359–72. Butterworth-Heinemann, Oxford.

Fahn S (1996). Is levodopa toxic? *Neurology*, **47**, S184–95.

Fahn S, Sjaastad O (1991). Hereditary essential myoclonus in a large Norwegian family. *Mov Disord*, **6**, 237–47.

Fahn S, Williams DT (1988). Psychogenic dystonia. *Adv Neurol*, **50**, 431–55.

Fahn S, Marsden CD, Calne DB (1987). Classification and investigation of dystonia. In Fahn S, Marsden CD, eds. *Movement Disorders, 2*, pp. 332–58. Butterworths, London.

Fahn S, Oakes D, Shoulson I *et al.* (2004). Levodopa and the progression of Parkinson's disease. *N Engl J Med*, **351**, 2498–508.

Farmer SF (2005). Mirror movements in neurology. *J Neurol Neurosurg Psychiatry*, **76**, 1330–.

Fear C (1992). Drug induced Creutzfeldt Jacob like syndrome: a review. *Hum Psychopharmacol*, **7**, 89.

Fearnley JM, Lees AJ (1991). Ageing and Parkinson's disease: substantia nigra regional selectivity. *Brain*, **114**, 2283–301.

Fearnley JM, Revesz T, Brooks DJ *et al.* (1991). Diffuse Lewy body disease presenting with a supranuclear gaze palsy. *J Neurol Neurosurg Psychiatry*, **54**, 159–61.

Federico F, Simone IL, Lucivero V *et al.* (1997). Proton magnetic resonance spectroscopy in Parkinson's disease and progressive supranuclear palsy. *J Neurol Neurosurg Psychiatry*, **62**, 239–42.

Fenton WS, Wyatt RJ, McGlashan TH (1994). Risk factors for spontaneous dyskinesia in schizophrenia. *Arch Gen Psychiatry*, **51**, 643–50.

Fernandez-Valero EM, Ballart A, Iturriaga C *et al.* (2005). Identification of 25 new mutations in 40 unrelated Spanish Niemann-Pick type C patients: genotype-phenotype correlations. *Clin Genet*, **68**, 245–54.

Fernandez HH, Friedman JH (2003). Classification and treatment of tardive syndromes. *Neurologist*, **9**, 16–27.

Fernandez M, Raskind W, Wolff J *et al.* (2001). Familial dyskinesia and facial myokymia (FDFM): a novel movement disorder. *Ann Neurol*, **49**, 486–92.

Fernandez HH, Trieschmann ME, Burke MA *et al.* (2002). Quetiapine for psychosis in Parkinson's disease versus dementia with Lewy bodies. *J Clin Psychiatry*, **63**, 513–5.

Feve A, Angelard B, Fenelon G *et al.* (1990). Neuroleptic induced tardive dyskinesia in the cebus monkey. *Mov Disord*, **7**, 32–37.

Findley LJ, Lees A, Apajasalo M *et al.* (2005). Cost-effectiveness of levodopa/carbidopa/entacapone (Stalevo) compared to standard care in UK Parkinson's disease patients with wearing-off. *Curr Med Res Opin*, **21**, 1005–14.

Fine J, Duff J, Chen R *et al.* (2000). Long-term follow-up of unilateral pallidotomy in advanced Parkinson's disease. *N Engl J Med*, **342**, 1708–14.

Finelli PF, McEntee WJ, Ambler M *et al.* (1980). Adult celiac disease presenting as cerebellar syndrome. *Neurology*, **30**, 245–9.

Fink JK, Rainer S, Wilkowski J *et al.* (1996). Paroxysmal dystonic choreoathetosis: tight linkage to chromosome 2q. *Am J Hum Genet*, **59**, 140–5.

Fish DR (1994). Epilepsy masquerading as a movement disorder. In Marsden CD, Fahn S, eds. *Movement Disorders 3*, pp. 346–58. Butterworth-Heinemann, Oxford.

Fitzgerald PM, Jankovic J, Glaze DG *et al.* (1990). Extrapyramidal involvement in Rett's syndrome. *Neurology*, **40**, 293–5.

Fletcher NA, Foley J (1993). Parental age, genetic mutation and cerebral palsy. *J Med Genet*, **30**, 44–6.

Fletcher NA, Marsden CD (1996). Dyskinetic cerebral palsy: A clinical and genetic study. *Dev Med Child Neurol*, **38**, 873–80.

Fletcher NA, Harding AE, Marsden CD (1990). A genetic study of idiopathic torsion dystonia in the United Kingdom. *Brain*, **113**, 379–95.

Fletcher NA, Harding AE, Marsden CD (1991a). A case-control study of idiopathic torsion dystonia. *Mov Disord*, **6**, 304–9.

Fletcher NA, Harding AE, Marsden CD (1991b). The relationship between trauma and idiopathic torsion dystonia. *J Neurol Neurosurg Psychiatry*, **54**, 713–7.

Fletcher NA, Thompson PD, Scadding JW *et al.* (1993). Successful treatment of childhood onset symptomatic dystonia with levodopa. *J Neurol Neurosurg Psychiatry*, **56**, 865–7.

Fogan L (1996). Progressive encephalomyelitis with rigidity responsive to plasmapheresis and immunosuppression. *Ann Neurol*, **40**, 451–3.

Footitt DR, Quinn N, Kocen RS *et al.* (1997). Familial Lafora body disease of late onset: report of four cases in one family and a review of the literature. *J Neurol*, **244**, 40–4.

Ford B, Greene P, Louis ED *et al.* (1996). Use of intrathecal baclofen in the treatment of patients with dystonia. *Arch Neurol*, **53**, 1241–6.

Ford B, Louis ED, Greene P *et al.* (1998). Outcome of selective ramisectomy for botulinum toxin resistant torticollis. *J Neurol Neurosurg Psychiatry*, **65**, 472–8.

Frankel M, Cummings JL, Robertson MM *et al.* (1986). Obsessions and compulsions in Gilles de la Tourette's syndrome. *Neurology*, **36**, 378–82.

Frankel JP, Lees AJ, Kempster PA *et al.* (1990). Subcutaneous apomorphine in the treatment of Parkinson's disease. *J Neurol, Neurosurg Psychiatry*, **53**, 96–101.

Franzini A, Marras C, Ferroli P *et al.* (2005). Long-term high-frequency bilateral pallidal stimulation for neuroleptic-induced tardive dystonia. Report of two cases. *J Neurosurg*, **102**, 721–5.

Freed CR, Greene PE, Breeze RE *et al.* (2001). Transplantation of embryonic dopamine neurons for severe Parkinson's disease. *N Engl J Med*, **344**, 710–9.

Freeman TB, Cicchetti F, Hauser RA *et al.* (2000). Transplanted fetal striatum in Huntington's disease: phenotypic development and lack of pathology. *Proc Natl Acad Sci USA*, **97**, 13877–82.

Freeman TB, Olanow CW, Hauser RA *et al.* (1995). Bilateral fetal nigral transplantation into the postcommissural putamen in Parkinson's disease. *Ann Neurol*, **38**, 379–88.

Frei KP, Pathak M, Jenkins S *et al.* (2004). Natural history of posttraumatic cervical dystonia. *Mov Disord* 19, 1492–8.

Friede RL (1989). *Developmental Neuropathology*, pp. 83–97. Springer, Berlin.

Friedman JH (1994). Clozapine treatment of psychosis in patients with tardive dystonia: report of three cases. *Mov Disord*, **9**, 321–4.

Friedman JR, Klein C, Leung J *et al.* (2000). The GAG deletion of the DYT1 gene is infrequent in musicians with focal dystonia. *Neurology*, **55**, 1417–8.

Frucht S, Rogers JD, Greene PE *et al.* (1999). Falling asleep at the wheel: motor vehicle mishaps in persons taking pramipexole and ropinirole. *Neurology*, **52**, 1908–10.

Furtado S, Sossi V, Hauser RA *et al.* (2005). Positron emission tomography after fetal transplantation in Huntington's disease. *Ann Neurol*, **58**, 331–7.

Furukawa Y (2003). Genetics and biochemistry of dopa-responsive dystonia: significance of striatal tyrosine hydroxylase protein loss. *Adv Neurol*, **91**, 401–10.

Furukawa Y, Kish SJ, Lang AE (2000). Scoliosis in a dopa-responsive dystonia family with a mutation of the GTP cyclohydrolase I gene. *Neurology*, **54**, 2187.

Gaig C, Marti MJ, Tolosa E *et al.* (2006). (123)I-Ioflupane SPECT in the diagnosis of suspected psychogenic Parkinsonism. *Mov Disord*.

Ganesh S, Puri R, Singh S *et al.* (2006). Recent advances in the molecular basis of Lafora's progressive myoclonus epilepsy. *J Hum Genet*, **51**, 1–8.

Garbutt S, Riley DE, Kumar AN *et al.* (2004). Abnormalities of optokinetic nystagmus in progressive supranuclear palsy. *J Neurol Neurosurg Psychiatry*, **75**, 1386–94.

Garcia-Ribes A, Marti-Carrera I, Martinez-Gonzalez MJ *et al.* (2003). Factors related to the short term remission of tics in children with Tourette syndrome. *Rev Neurol*, **37**, 901–3.

Gardos G, Cole JO (1980). Overview: Public health issues in tardive dyskinesia. *Am J Psychiatry*, **137**, 776–81.

Garraux G, Goldfine A, Bohlhalter S *et al.* (2006). Increased midbrain gray matter in Tourette's syndrome. *Ann Neurol*, **59**, 381–5.

Garvey MA, Perlmutter SJ, Allen AJ *et al.* (1999). A pilot study of penicillin prophylaxis for neuropsychiatric exacerbations triggered by streptococcal infections. *Biol Psychiatry*, **45**, 1564–71.

Garvey MA, Snider LA, Leitman SF et al. (2005). Treatment of Sydenham's chorea with intravenous immunoglobulin, plasma exchange, or prednisone. *J Child Neurol* 20, 424–9.

Gasser T, Bove CM, Ozelius LJ et al. (1996). Haplotype analysis at the DYT1 locus in Ashkenazi Jewish patients with occupational hand dystonia. *Mov Disord*, **11**, 163–6.

Gasser T, Muller-Myhsok B, Wszolek ZK et al. (1998). A susceptibility locus for Parkinson's disease maps to chromosome 2p13. *Nat Genet*, **18**, 262–5.

Gaura V, Bachoud-Levi AC, Ribeiro MJ et al. (2004). Striatal neural grafting improves cortical metabolism in Huntington's disease patients. *Brain*, **127**, 65–72.

Gebremariam A (1999). Sydenham's chorea: risk factors and the role of prophylactic benzathine penicillin G in preventing recurrence. *Ann Trop Paediatr* 19, 161–5.

Geschwind DH, Loginov M, Stern JM (1999). Identification of a locus on chromosome 14q for idiopathic basal ganglia calcification (Fahr disease). *Am J Hum Genet*, **65**, 764–72.

Ghika J, Villemure J, Fankhauser H et al. (1998). Efficiency and safety of bilateral contemporaneous pallidal stimulation (deep brain stimulation) in levodopa-responsive patients with Parkinson's disease with severe motor fluctuations: a 2-year follow-up review. *J Neurosurg*, **89**, 713–8.

Ghika J, Villemure JG, Miklossy J et al. (2002). Postanoxic generalized dystonia improved by bilateral Voa thalamic deep brain stimulation. *Neurology*, **58**, 311–3.

Ghorayeb I, Yekhlef F, Bioulac B et al. (2005). Continuous positive airway pressure for sleep-related breathing disorders in multiple system atrophy: long-term acceptance. *Sleep Med*, **6**, 359–62.

Gibb WR, Lees AJ (1989). Tendency to late recurrence following rheumatic chorea. *Neurology*, **39**, 999.

Gibb WR, Lees AJ, Marsden CD (1988). Pathological report of four patients presenting with cranial dystonias. *Mov Disord*, **3**, 211–21.

Giladi N, Simon ES, Korczyn AD et al. (2000). Anal sphincter EMG does not distinguish between multiple system atrophy and Parkinson's disease. *Muscle Nerve*, **23**, 731–4.

Gilbert DL, Dure L, Sethuraman G et al. (2003). Tic reduction with pergolide in a randomized controlled trial in children. *Neurology*, **60**, 606–11.

Gilks WP, Abou-Sleiman PM, Gandhi S et al. (2005). A common LRRK2 mutation in idiopathic Parkinson's disease. *Lancet*, **365**, 415–6.

Gill SS, Patel NK, Hotton GR et al. (2003). Direct brain infusion of glial cell line-derived neurotrophic factor in Parkinson disease. *Nat Med*, **9**, 589–95.

Gilman S, Quinn NP (1996). The relationship of multiple system atrophy to sporadic olivopontocerebellar atrophy and other forms of idiopathic late-onset cerebellar atrophy. *Neurology*, **46**, 1197–9.

Gilman S, Sima AA, Junck L et al. (1996). Spinocerebellar ataxia type 1 with multiple system degeneration and glial cytoplasmic inclusions. *Ann Neurol*, **39**, 241–55.

Glick N (2006). Dramatic reduction in self-injury in Lesch-Nyhan disease following S-adenosylmethionine administration. *J Inherit Metab Dis*.

Goetz CG (1983). Drug induced parkinsonism and idiopathic Parkinson's disease. *Arch Neurol*, **40**, 325–6.

Goetz CG, Stebbins GT (1991). Effects of head trauma from motor vehicle accidents on Parkinson's disease. *Ann Neurol*, **29**, 191–3.

Goetz CG, Blasucci LM, Leurgans S et al. (2000). Olanzapine and clozapine: Comparative effects on motor function in hallucinating PD patients. *Neurology*, **55**, 789–94.

Goetz CG, Fan W, Leurgans S et al. (2006). The malignant course of "benign hallucinations" in Parkinson disease. *Arch Neurol*, **63**, 713–6.

Goetz CG, Poewe W, Rascol O et al. (2005a). Evidence-based medical review update: pharmacological and surgical treatments of Parkinson's disease: 2001 to 2004. *Mov Disord* 20, 523–39.

Goetz CG, Tanner CM, Stebbins GT et al. (1992). Adult tics in Gilles de la Tourette's syndrome: description and risk factors. *Neurology*, **42**, 784–8.

Goetz CG, Vogel C, Tanner CM et al. (1998). Early dopaminergic drug-induced hallucinations in parkinsonian patients. *Neurology*, **51**, 811–4.

Goetz CG, Wuu J, Curgian LM et al. (2005b). Hallucinations and sleep disorders in PD: six-year prospective longitudinal study. *Neurology*, **64**, 81–6.

Golbe L (1991). Young onset Parkinson's disease: A clinical review. *Neurology*, **41**, 168–73.

Golbe LI, Di Iorio G, Bonavita V et al. (1990). A large kindred with autosomal dominant Parkinson's disease. *Ann Neurol*, **27**, 276–82.

Goldberg YP, Andrew SE, Theilmann J et al. (1993). Familial predisposition to recurrent mutations causing Huntington's disease: genetic risk to sibs of sporadic cases. *J Med Genet*, **30**, 987–90.

Goldfarb LG, Vasconcelos O, Platonov FA et al. (1996). Unstable triplet repeat and phenotypic variability of spinocerebellar ataxia type 1. *Ann Neurol*, **39**, 500–6.

Goldman SM, Tanner CM, Olanow CW et al. (2005). Occupation and parkinsonism in three movement disorders clinics. *Neurology*, **65**, 1430–5.

Goodchild RE, Dauer WT (2004). Mislocalization to the nuclear envelope: an effect of the dystonia-causing torsinA mutation. *Proc Natl Acad Sci USA*, **101**, 847–52.

Gordon MF, Abrams RI, Rubin DB et al. (1995). Bismuth subsalicylate toxicity as a cause of prolonged encephalopathy with myoclonus. *Mov Disord*, **10**, 220–2.

Goto S, Yamada K (2004). Long term continuous bilateral pallidal stimulation produces stimulation independent relief of cervical dystonia. *J Neurol Neurosurg Psychiatry*, **75**, 1506–7.

Gouider-Khouja N, Mekaouar A, Larnaout A et al. (2002). Progressive encephalomyelitis with rigidity presenting as a stiff-person syndrome. *Parkinsonism Relat Disord*, **8**, 285–8.

Grafman J, Litvan I, Gomez C et al. (1990). Frontal lobe function in progressive supranuclear palsy. *Arch Neurol*, **47**, 553–8.

Graham JG, Oppenheimer DR (1969). Orthostatic hypotension and nicotine sensitivity in a case of multiple system atrophy. *J Neurol, Neurosurg Psychiatry*, **32**, 28–34.

Grandas F, Elston J, Quinn N et al. (1988). Blepharospasm: a review of 264 patients. *J Neurol Neurosurg Psychiatry*, **51**, 767–72.

Greene P (1992). Baclofen in the treatment of dystonia. *Clin Neuropharmacol*, **15**, 276–88.

Greene PE, Shale H, Fahn S (1988). Analysis of open label trials in torsion dystonia using high dosages of anticholinergics and other drugs. *Mov Disord*, **3**, 46–60.

Greene PE, Bressman SB, Ford B et al. (2000). Parkinsonism, dystonia, and hemiatrophy. *Mov Disord*, **15**, 537–41.

Greene P, Fahn S, Brin MF et al (1994). Botulinum toxin therapy. In Marsden CD, Fahn S, eds. *Movement Disorders*, 3, pp. 477–502. Butterworth-Heinemann, Oxford.

Griffith JM, Adler LE, Freedman R (1994). Fine motor performance in schizophrenia. *Neuropsychobiology*, **29**, 179–84.

Grimes DA, Han F, Lang AE et al. (2002). A novel locus for inherited myoclonus-dystonia on 18p11. *Neurology*, **59**, 1183–6.

Grisoli M, Piperno A, Chiapparini L et al. (2005). MR imaging of cerebral cortical involvement in aceruloplasminemia. *AJNR Am J Neuroradiol*, **26**, 657–61.

Grogaard B, Dullerud R, Magnaes B (1993). Acute torticollis in children due to atlanto-axial rotary fixation. *Arch Orthop Trauma Surg*, **112**, 185–8.

Grotzsch H, Pizzolato GP, Ghika J et al. (2002). Neuropathology of a case of dopa-responsive dystonia associated with a new genetic locus, DYT14. *Neurology*, **58**, 1839–42.

Guerrini R, Bonanni P, Nardocci N *et al.* (1999). Autosomal recessive rolandic epilepsy with paroxysmal exercise-induced dystonia and writer's cramp: delineation of the syndrome and gene mapping to chromosome 16p12–11.2. *Ann Neurol*, **45**, 344–52.

Gurol ME, Ertas M, Hanagasi HA *et al.* (2001). Stiff leg syndrome: case report. *Mov Disord*, **16**, 1189–93.

Haberhausen G, Schmitt I, Kohler A *et al.* (1995). Assignment of the dystonia-parkinsonism syndrome locus, DYT3, to a small region within a 1.8-Mb YAC contig of Xq13.1. *Am J Hum Genet*, **57**, 644–50.

Haddad PM (1994). Neuroleptic malignant syndrome may be caused by other drugs. *BMJ*, **308**, 200.

Hadjivassiliou M, Grunewald RA, Davies-Jones GAB (2002). Gluten sensitivity as a neurological illness. *J Neurol Neurosurg Psychiatry*, **72**, 560–3.

Hagberg B, Hagberg G (1993). The origins of cerebral palsy. In David TJ, ed. *Recent Advances in Paediatrics*. Vol. XI, pp. 67–83. Churchill Livingstone, Edinburgh.

Hagenah J, Saunders-Pullman R, Hedrich K *et al.* (2005). High mutation rate in dopa-responsive dystonia: detection with comprehensive GCHI screening. *Neurology*, **64**, 908–11.

Hagerman PJ, Hagerman RJ (2004). Fragile X-associated tremor/ataxia syndrome (FXTAS). *Ment Retard Dev Disabil Res Rev*, **10**, 25–30.

Hagerman PJ, Greco CM, Hagerman RJ (2003). A cerebellar tremor/ataxia syndrome among fragile X premutation carriers. *Cytogenet Genome Res*, **100**, 206–12.

Hagerman RJ, Leavitt BR, Farzin F *et al.* (2004). Fragile-X-associated tremor/ataxia syndrome (FXTAS) in females with the FMR1 premutation. *Am J Hum Genet*, **74**, 1051–6.

Hague K, Lento P, Morgello S *et al.* (1997). The distribution of Lewy bodies in pure autonomic failure: autopsy findings and review of the literature. *Acta Neuropathol (Berl)*, **94**, 192–6.

Hall RC, Appleby B, Hall RC (2005). Atypical neuroleptic malignant syndrome presenting as fever of unknown origin in the elderly. *South Med J*, **98**, 114–7.

Hallett M (1998). The neurophysiology of dystonia. *Arch Neurol*, **55**, 601–3.

Hallett M, Chadwick D, Marsden CD (1979). Cortical reflex myoclonus. *Neurology*, **29**, 1107–25.

Hampshire DJ, Roberts E, Crow Y *et al.* (2001). Kufor-Rakeb syndrome, pallido-pyramidal degeneration with supranuclear upgaze paresis and dementia, maps to 1p36. *J Med Genet*, **38**, 680–2.

Handley OJ, Naji JJ, Dunnett SB *et al.* (2006). Pharmaceutical, cellular and genetic therapies for Huntington's disease. *Clin Sci (Lond)*, **110**, 73–88.

Hanihara T, Inoue K, Kawanishi C *et al.* (1997). 6-Pyruvoyl-tetrahydropterin synthase deficiency with generalized dystonia and diurnal fluctuation of symptoms: a clinical and molecular study. *Mov Disord*, **12**, 408–11.

Hanna MG, Bhatia KP (1997). Movement disorders and mitochondrial dysfunction. *Curr Opin Neurol*, **10**, 351–6.

Hardie RJ, Lees AJ (1988). Neuroleptic induced Parkinson's syndrome: Clinical features and results of treatment with levodopa. *J Neurol, Neurosurg Psychiatry*, **8**, 850–4.

Hardie RJ, Pullon HW, Harding AE *et al.* (1991). Neuroacanthocytosis. A clinical, haematological and pathological study of 19 cases. *Brain*, **114**, 13–49.

Harding AE, Thompson PD, Kocen RS *et al.* (1989). Plasma exchange and immunosuppression in the stiff man syndrome [letter]. *Lancet*, **2**, 915.

Harper PS (1991). The epidemiology of Huntington's disease. In Harper PS, ed. *Huntington's Disease*, pp. 251–80. W.B. Saunders, London.

Harper PS, Newcombe RG (1992). Age at onset and life table risks in genetic counselling for Huntington's disease. *J Med Genet*, **29**, 239–42.

Harrison NA, Church A, Nisbet A *et al.* (2004). Late recurrences of Sydenham's chorea are not associated with anti-basal ganglia antibodies. *J Neurol Neurosurg Psychiatry*, **75**, 1478–9.

Hart IK, Maddison P, Newsom-Davis J *et al.* (2002). Phenotypic variants of autoimmune peripheral nerve hyperexcitability. *Brain*, **125**, 1887–95.

Hartig MB, Hortnagel K, Garavaglia B *et al.* (2006). Genotypic and phenotypic spectrum of PANK2 mutations in patients with neurodegeneration with brain iron accumulation. *Ann Neurol*, **59**, 248–56.

Hashimoto T, Tokuda T, Hanyu N *et al.* (2003). Withdrawal of levodopa and other risk factors for malignant syndrome in Parkinson's disease. *Parkinsonism Relat Disord*, 9 (Suppl 1), S25–30.

Hauser RA, Freeman TB, Olanow CW (1995). Surgical therapies for Parkinson's disease. In Kurlan R, ed. *Treatment of Movement Disorders*, pp. 57–93. Lippincott, Philadelphia.

Hauser RA, Furtado S, Cimino CR *et al.* (2002). Bilateral human fetal striatal transplantation in Huntington's disease. *Neurology*, **58**, 687–95.

Hauw JJ, Daniel SE, Dickson D *et al.* (1994). Preliminary NINDS neuropathologic criteria for Steele-Richardson-Olszewski syndrome (progressive supranuclear palsy). *Neurology*, **44**, 2015–9.

Hayat GR, Kulkantrakorn K, Campbell WW *et al.* (2000). Neuromyotonia: autoimmune pathogenesis and response to immune modulating therapy. *J Neurol Sci*, **181**, 38–43.

Hayflick SJ, Hartman M, Coryell J *et al.* (2006). Brain MRI in neurodegeneration with brain iron accumulation with and without PANK2 mutations. *AJNR Am J Neuroradiol*, **27**, 1230–3.

Hayflick SJ, Westaway SK, Levinson B *et al.* (2003). Genetic, clinical, and radiographic delineation of Hallervorden-Spatz syndrome. *N Engl J Med*, **348**, 33–40.

Hedrich K, Djarmati A, Schafer N *et al.* (2004). DJ-1 (PARK7) mutations are less frequent than Parkin (PARK2) mutations in early-onset Parkinson disease. *Neurology*, **62**, 389–94.

Hely MA, Morris JG, Reid WG *et al.* (1994). The Sydney Multicentre Study of Parkinson's disease: a randomised, prospective five year study comparing low dose bromocriptine with low dose levodopa-carbidopa. *J Neurol Neurosurg Psychiatry*, **57**, 903–10.

Hening WA, Allen RP, Earley CJ *et al.* (2004b). An update on the dopaminergic treatment of restless legs syndrome and periodic limb movement disorder. *Sleep*, **27**, 560–83.

Hening W, Walters AS, Allen RP *et al.* (2004a). Impact, diagnosis and treatment of restless legs syndrome (RLS) in a primary care population: the REST (RLS epidemiology, symptoms, and treatment) primary care study. *Sleep Med*, **5**, 237–46.

Hernan MA, Takkouche B, Caamano-Isorna F *et al.* (2002). A meta-analysis of coffee drinking, cigarette smoking, and the risk of Parkinson's disease. *Ann Neurol*, **52**, 276–84.

Herz E (1944). Dystonia. I. Historical review; analysis of dystonic symptoms and physiologic mechanisms involved. *Arch Neurol Psychiat*, **51**, 305–18.

Hilker R, Klein C, Ghaemi M *et al.* (2001). Positron emission tomographic analysis of the nigrostriatal dopaminergic system in familial parkinsonism associated with mutations in the parkin gene. *Ann Neurol*, **49**, 367–76.

Hinson VK, Haren WB (2006). Psychogenic movement disorders. *Lancet Neurol*, **5**, 695–700.

Hoekstra PJ, Minderaa RB, Kallenberg CG (2004). Lack of effect of intravenous immunoglobulins on tics: a double-blind placebo-controlled study. *J Clin Psychiatry*, **65**, 537–42.

Hoenicka J, Perez M, Perez-Tur J *et al.* (1999). The tau gene A0 allele and progressive supranuclear palsy. *Neurology*, **53**, 1219–25.

Holloway RG, Shoulson I, Fahn S *et al.* (2004). Pramipexole vs levodopa as initial treatment for Parkinson disease: a 4-year randomized controlled trial. *Arch Neurol*, **61**, 1044–53.

Hornykiewicz O, Kish SJ, Becker LE *et al.* (1988). Biochemical evidence for brain neurotransmitter changes in idiopathic torsion dystonia (dystonia musculorum deformans). *Adv Neurol*, **50**, 157–65.

Horowitz MJ, Field NP, Zanko A *et al.* (2001). Psychological impact of news of genetic risk for Huntington disease. *Am J Med Genet*, **103**, 188–92.

Hou JG, Jankovic J (2003). Movement disorders in Friedreich's ataxia. *J Neurol Sci* 206, 59–64.

Hou JG, Ondo W, Jankovic J (2001). Intrathecal baclofen for dystonia. *Mov Disord*, 16, 1201–2.

Houeto JL, Karachi C, Mallet L *et al.* (2005). Tourette's syndrome and deep brain stimulation. *J Neurol Neurosurg Psychiatry*, 76, 992–5.

Houeto JL, Mesnage V, Mallet L *et al.* (2002). Behavioural disorders, Parkinson's disease and subthalamic stimulation. *J Neurol Neurosurg Psychiatry*, 72, 701–7.

Houlden H, Lincoln S, Farrer M *et al.* (2003). Compound heterozygous PANK2 mutations confirm HARP and Hallervorden-Spatz syndromes are allelic. *Neurology*, 61, 1423–6.

Houser M, Waltz T (2005). Meige syndrome and pallidal deep brain stimulation. *Mov Disord* 20, 1203–5.

Howard RS, Lees AJ (1987). Encephalitis lethargica. A report of four recent cases. *Brain*, 110, 19–33.

Howard RS, Greenwood R, Gawler J *et al.* (1993). A familial disorder associated with palatal myoclonus, other brainstem signs, tetraparesis, ataxia and Rosenthal fibre formation. *J Neurol Neurosurg Psychiatry*, 56, 977–81.

Hughes AJ, Bishop S, Kleedorfer B *et al.* (1993a). Subcutaneous apomorphine in Parkinson's disease: response to chronic administration for up to five years. *Mov Disord*, 8, 165–70.

Hughes AJ, Daniel SE, Blankson S *et al.* (1993b). A clinicopathologic study of 100 cases of Parkinson's disease. *Arch Neurol*, 50, 140–8.

Hughes AJ, Daniel SE, Kilford L *et al.* (1992). Accuracy of clinical diagnosis of idiopathic Parkinson's disease: a clinico-pathological study of 100 cases. *J Neurol Neurosurg Psychiatry*, 55, 181–4.

Hunt JR (1917). Progressive atrophy of the globus pallidus (primary atrophy of the pallidal system): A system disease of the paralysis agitans type, characterised by atrphy of the motor cells of the corpus striatum. A contribution to the functions of the corpus striatum. *Brain*, 40, 58–148.

Hunter R, Smith J, Thomson T *et al.* (1978). Hemiparkinsonism with infarction of the ipsilateral substantia nigra. *Neuropathol Appl Neurobiol*, 4, 297–301.

Huntington's disease collaborative research group (1993). A novel gene that is unstable and expanded on Huntington's disease chromosomes. *Cell*, 72, 971–83.

Huntington Study Group (2006). Tetrabenazine as antichorea therapy in Huntington disease: a randomized controlled trial. *Neurology*, 66, 366–72.

Hutchison WD, Lang AE, Dostrovsky JO *et al.* (2003). Pallidal neuronal activity: implications for models of dystonia. *Ann Neurol*, 53, 480–8.

Hyde TM, Stacey ME, Coppola R *et al.* (1995). Cerebral morphometric abnormalities in Tourette's syndrome: a quantitative MRI study of monozygotic twins. *Neurology*, 45, 1176–82.

Iacono RP, Shima F, Lonser RR *et al.* (1995). The results, indications, and physiology of posteroventral pallidotomy for patients with Parkinson's disease. *Neurosurgery*, 36, 1118–25; discussion 1125–7.

Ichinose H, Inagaki H, Suzuki T *et al.* (2000). Molecular mechanisms of hereditary progressive dystonia with marked diurnal fluctuation, Segawa's disease. *Brain Dev*, 22 (Suppl 1), S107–10.

Ikeda K, Deguchi K, Touge T *et al.* (2004). Painful legs and moving toes syndrome associated with herpes zoster myelitis. *J Neurol Sci*, 219, 147–50.

Ikoma K, Samii A, Mercuri B *et al.* (1996). Abnormal cortical motor excitability in dystonia. *Neurology*, 46, 1371–6.

Imrie J, Vijayaraghaven S, Whitehouse C *et al.* (2002). Niemann-Pick disease type C in adults. *J Inherit Metab Dis*, 25, 491–500.

Ishii A, Hayashi A, Ohkoshi N *et al.* (2004). Progressive encephalomyelitis with rigidity associated with anti-amphiphysin antibodies. *J Neurol Neurosurg Psychiatry*, 75, 661–2.

Ito M (1984). *The Cerebellum and Neural Control*. Raven Press, New York.

Jankovic J (1987). The neurology of tics. In Fahn S, Marsden CD, eds. *Movement Disorders 2*, pp. 383–405. Butterworths, London.

Jankovic J (1993). Deprenyl in attention deficit associated with Tourette's syndrome. *Arch Neurol*, 50, 286–8.

Jankovic J (1994). Stereotypies. In Fahn S, Marsden CD, eds. *Movement Disorders, 3*, pp. 503–17. Butterworth Heinemann, Oxford.

Jankovic J (2005). Searching for a relationship between manganese and welding and Parkinson's disease. *Neurology*, 64, 2021–8.

Jankovic J, Ashizawa T (1995). Tourettism associated with Huntington's disease. *Mov Disord*, 10, 103–5.

Jankovic J, Beach J (1997). Long-term effects of tetrabenazine in hyperkinetic movement disorders. *Neurology*, 48, 358–62.

Jankovic J, Kirkpatrick JB, Blomquist KA (1985). Late onset Hallervorden-Spatz disease presenting as familial parkinsonism. *Neurology*, 35, 227–34.

Jankovic J, Orman J (1988). Tetrabenazine therapy of dystonia, chorea, tics, and other dyskinesias. *Neurology*, 38, 391–4.

Janno S, Holi M, Tuisku K *et al.* (2004). Prevalence of neuroleptic-induced movement disorders in chronic schizophrenia inpatients. *Am J Psychiatry*, 161, 160–3.

Janssen JC, Hall M, Fox NC *et al.* (2000). Alzheimer's disease due to an intronic presenilin-1 (PSEN1 intron 4) mutation: A clinicopathological study. *Brain*, 123 (Pt 5), 894–907.

Jarman PR, Bandmann O, Marsden CD *et al.* (1997). GTP cyclohydrolase I mutations in patients with dystonia responsive to anticholinergic drugs. *J Neurol Neurosurg Psychiatry*, 63, 304–8.

Jellinger K (1968). Progressive pallidumatrophie. *J Neurol Sci*, 6, 19–44.

Jenner P, Olanow CW (1996). Oxidative stress and the pathogenesis of Parkinson's disease. *Neurology*, 47, S161–70.

Jin H, May M, Tranebjaerg L *et al.* (1996). A novel X-linked gene, DDP, shows mutations in families with deafness (DFN-1), dystonia, mental deficiency and blindness. *Nat Genet*, 14, 177–80.

Jinnah HA, Visser JE, Harris JC *et al.* (2006). Delineation of the motor disorder of Lesch-Nyhan disease. *Brain*, 129, 1201–17.

Johnson DS, Montgomery EB (1997). Pathophysiology of cerebellar disorders. In Watts RL, Koller WC, eds. *Movement Disorders: Neurologic Principles and Practice*, pp. 587–610. McGraw-Hill, New York.

Joseph AB (1992). Catatonia. In Joseph AB, Young RR, eds. *Movement Disorders in Neurology and Psychiatry*, pp. 335–42. Blackwell, Boston.

Josephs KA, Ahlskog JE, Klos KJ *et al.* (2005). Neurologic manifestations in welders with pallidal MRI T1 hyperintensity. *Neurology*, 64, 2033–9.

Juncos JL, Roberts VJ, Evatt ML *et al.* (2004). Quetiapine improves psychotic symptoms and cognition in Parkinson's disease. *Mov Disord* 19, 29–35.

Kabakci K, Isbruch K, Schilling K *et al.* (2005). Genetic heterogeneity in rapid onset dystonia-parkinsonism: description of a new family. *J Neurol Neurosurg Psychiatry*, 76, 860–2.

Kaji R, Rothwell JC, Katayama M *et al.* (1995a). Tonic vibration reflex and muscle afferent block in writer's cramp. *Ann Neurol*, 38, 155–62.

Kaji R, Shibasaki H, Kimura J (1995b). Writer's cramp: a disorder of motor subroutine? [editorial; comment]. *Ann Neurol*, 38, 837–8.

Kalanithi PS, Zheng W, Kataoka Y *et al.* (2005). Altered parvalbumin-positive neuron distribution in basal ganglia of individuals with Tourette syndrome. *Proc Natl Acad Sci USA*, 102, 13307–12.

Kambouris M, Bohlega S, Al-Tahan A *et al.* (2000). Localization of the gene for a novel autosomal recessive neurodegenerative Huntington-like disorder to 4p15.3. *Am J Hum Genet*, 66, 445–52.

Kang UJ, Burke RE, Fahn S (1988). Tardive dystonia. *Adv Neurol*, 50, 415–29.

Kanouchi T, Yokota T, Isa F *et al.* (1997). Role of the ipsilateral motor cortex in mirror movements. *J Neurol Neurosurg Psychiatry*, 62, 629–32.

Karp BI, Hallett M (1996). Extracorporeal phantom tics in Tourette's syndrome. *Neurology*, 46, 38–40.

Kartsounis LD, Hardie RJ (1996). The pattern of cognitive impairments in neuroacanthocytosis. A frontosubcortical dementia. *Arch Neurol*, 53, 77–80.

Katayama S, Watanabe C, Khoriyama T *et al.* (1998). Slowly progressive L-DOPA nonresponsive pure akinesia due to nigropallidal degeneration: a clinicopathological case study. *J Neurol Sci*, 161, 169–72.

Kato N, Arai K, Hattori T (2003). Study of the rostral midbrain atrophy in progressive supranuclear palsy. *J Neurol Sci*, **210**, 57–60.

Kawanami T, Kato T, Daimon M *et al.* (1996). Hereditary caeruloplasmin deficiency: clinicopathological study of a patient. *J Neurol Neurosurg Psychiatry*, **61**, 506–9.

Keating GM, Lyseng-Williamson KA (2005). Tolcapone: a review of its use in the management of Parkinson's disease. *CNS Drugs* 19, 165–84.

Kellett MW, Steiger MJ (1999). Deterioration in parkinsonism with low-dose pergolide. *J Neurology*, **246**, 309–11

Kenney C, Jankovic J (2006). Tetrabenazine in the treatment of hyperkinetic movement disorders. *Expert Rev Neurother*, **6**, 7–17.

Kertesz A (1967). Paroxysmal kinesigenic choreoathetosis. An entity with the paroxysmal choreoathetosis syndrome. Description of 10 cases including 1 autopsied. *Neurology*, **17**, 680–90.

Khan NL, Jain S, Lynch JM *et al.* (2005). Mutations in the gene LRRK2 encoding dardarin (PARK8) cause familial Parkinson's disease: clinical, pathological, olfactory and functional imaging and genetic data. *Brain*, **128**, 2786–96.

Kieburtz K, Kurlan R (2005). Welding and Parkinson disease: Is there a bond? *Neurology*, **64**, 2001–3.

Kimber JR, Watson L, Mathias CJ (1997). Distinction of idiopathic Parkinson's disease from multiple-system atrophy by stimulation of growth-hormone release with clonidine. *Lancet*, **349**, 1877–81.

Kinrions P, Ibrahim N, Murphy K *et al.* (2003). Efficacy of levetiracetam in a patient with Unverricht-Lundborg progressive myoclonic epilepsy. *Neurology*, **60**, 1394–5.

Kitada T, Asakawa S, Hattori N *et al.* (1998). Mutations in the parkin gene cause autosomal recessive juvenile parkinsonism. *Nature*, **392**, 605–8.

Klawans HL (1981). Hemiparkinsonism as a late complication of hemiatrophy: A new syndrome. *Neurology*, **31**, 625–8.

Klawans HL, Weiner WJ (1974). The effect of d amphetamine on choreiform movement disorders. *Neurology*, **24**, 314–8.

Kleiner-Fisman G, Calingasan NY, Putt M *et al.* (2005). Alterations of striatal neurons in benign hereditary chorea. *Mov Disord*, **20**, 1353–7.

Kleiner-Fisman G, Fisman DN, Sime E *et al.* (2003). Long-term follow up of bilateral deep brain stimulation of the subthalamic nucleus in patients with advanced Parkinson disease. *J Neurosurg*, **99**, 489–95.

Koller W, Pahwa R, Busenbark K *et al.* (1997). High-frequency unilateral thalamic stimulation in the treatment of essential and parkinsonian tremor. *Ann Neurol*, **42**, 292–9.

Kompoliti K, Goetz CG, Litvan I *et al.* (1998). Pharmacological therapy in progressive supranuclear palsy. *Arch Neurol*, **55**, 1099–102.

Kong CK, Ko CH, Tong SF *et al.* (2001). Atypical presentation of dopa-responsive dystonia: generalized hypotonia and proximal weakness. *Neurology*, **57**, 1121–4.

Konishi Y, Shirabe T, Katayama S *et al.* (2005). Autopsy case of pure akinesia showing pallidonigro-luysian atrophy. *Neuropathology*, **25**, 220–7.

Konitsiotis S, Pappa S, Mantas C *et al.* (2006). Levetiracetam in tardive dyskinesia: An open label study. *Mov Disord*.

Kono S, Miyajima H, Sugimoto M *et al.* (2001). Stiff-person syndrome associated with cerebellar ataxia and high glutamic acid decarboxylase antibody titer. *Intern Med*, **40**, 968–71.

Kordower JH, Freeman TB, Snow BJ (1995). Neuropathological evidence of graft survival and striatal reinnervation after the transplantation of fetal mesencephalic tissue in a patient with Parkinson's disease. *N Engl J Med*, **332**, 1118–24.

Korn-Lubetzki I, Brand A, Steiner I (2004). Recurrence of Sydenham chorea: implications for pathogenesis. *Arch Neurol*, **61**, 1261–4.

Kortekaas R, Leenders KL, van Oostrom JC *et al.* (2005). Blood-brain barrier dysfunction in parkinsonian midbrain in vivo. *Ann Neurol*, **57**, 176–9.

Kosaka K (1993). Dementia and neuropathology in Lewy body disease. *Adv Neurol*, **60**, 456–63.

Kostic VS, Stojanovic-Svetel M, Kacar A (1996). Symptomatic dystonias associated with structural brain lesions: report of 16 cases. *Can J Neurol Sci*, **23**, 53–6.

Krack P, Marion MH (1994). "Apraxia of lid opening," a focal eyelid dystonia: clinical study of 32 patients. *Mov Disord*, **9**, 610–5.

Krack P, Batir A, Van Blercom N *et al.* (2003). Five-year follow-up of bilateral stimulation of the subthalamic nucleus in advanced Parkinson's disease. *N Engl J Med*, **349**, 1925–34.

Krack P, Poepping M, Weinert D *et al.* (2000). Thalamic, pallidal, or subthalamic surgery for Parkinson's disease? *J Neurol*, 247 (Suppl 2), II122–34.

Kramer PL, de Leon D, Ozelius L *et al.* (1990). Dystonia gene in Ashkenazi Jewish population is located on chromosome 9q32-34. *Ann Neurol*, **27**, 114–20.

Kramer PL, Mineta M, Klein C *et al.* (1999). Rapid-onset dystonia-parkinsonism: linkage to chromosome 19q13. *Ann Neurol*, **46**, 176–82.

Krause M, Fogel W, Kloss M *et al.* (2004). Pallidal stimulation for dystonia. *Neurosurgery*, **55**, 1361–8; discussion 1368–70.

Krauss JK, Loher TJ, Pohle T *et al.* (2002). Pallidal deep brain stimulation in patients with cervical dystonia and severe cervical dyskinesias with cervical myelopathy. *J Neurol Neurosurg Psychiatry*, **72**, 249–56.

Krauss JK, Loher TJ, Weigel R *et al.* (2003). Chronic stimulation of the globus pallidus internus for treatment of non-dYT1 generalized dystonia and choreoathetosis: 2-year follow up. *J Neurosurg*, **98**, 785–92.

Krauss JK, Toups EG, Jankovic J *et al.* (1997). Symptomatic and functional outcome of surgical treatment of cervical dystonia. *J Neurol Neurosurg Psychiatry*, **63**, 642–8.

Kremer B, Goldberg P, Andrew SE *et al.* (1994). A worldwide study of the huntington's disease mutation. The sensitivity and specificity of measuring CAG repeats. *N Engl J Med*, **330**, 1401–6.

Kruger R, Kuhn W, Muller T *et al.* (1998). Ala30Pro mutation in the gene encoding alpha-synuclein in Parkinson's disease. *Nat Genet*, **18**, 106–8.

Kulisevsky J, Marti MJ, Ferrer I *et al.* (1988). Meige syndrome: neuropathology of a case. *Mov Disord*, **3**, 170–5.

Kullmann DM, Howard RS, Miller DH *et al.* (1996). Brainstem encephalopathy with stimulus-sensitive myoclonus leading to respiratory arrest, but with recovery: a description of two cases and review of the literature. *Mov Disord*, **11**, 715–8.

Kumar R, Dagher A, Hutchison WD *et al.* (1999). Globus pallidus deep brain stimulation for generalized dystonia: clinical and PET investigation. *Neurology*, **53**, 871–4.

Kumar R, Lozano A, Kim Y *et al.* (1998). Double-blind evaluation of subthalamic nucleus deep brain stimulation in advanced Parkinson's disease. *Neurology*, **51**, 850–5.

Kumar R, Lozano AM, Sime E *et al.* (2003). Long-term follow-up of thalamic deep brain stimulation for essential and parkinsonian tremor. *Neurology*, **61**, 1601–4.

Kumar N, Van Gerpen JA, Bower JH *et al.* (2005). Levodopa-dyskinesia incidence by age of Parkinson's disease onset. *Mov Disord* 20, 342–4.

Kunig G, Pogarell O, Moller JC *et al.* (1999). Pramipexole, a nonergot dopamine agonist, is effective against rest tremor in intermediate to advanced Parkinson's disease. *Clin Neuropharmacol*, **22**, 301–5.

Kupsch A, Klaffke S, Kuhn AA *et al.* (2003). The effects of frequency in pallidal deep brain stimulation for primary dystonia. *J Neurol*, **250**, 1201–5.

Kurlan R, Trinidad KS (1995). Treatment of tics. In Kurlan R, ed. *Treatment of Movement Disorders*, pp. 365–406. J.B. Lippincott, Philadelphia.

Kyllerman M, Bager B, Bensch J *et al.* (1982). Dyskinetic cerebral palsy I. Clinical categories, associated neurological abnormalities and incidences. *Acta Paediatr Scand*, **71**, 543–50.

Laitinen LV, Bergenheim AT, Hariz MI (1992). Leksell's posteroventral pallidotomy in the treatment of Parkinson's disease. *J Neurosurg*, **76**, 53–61.

Lalioti MD, Scott HS, Buresi C et al. (1997). Dodecamer repeat expansion in cystatin B gene in progressive myoclonus epilepsy. *Nature*, **386**, 847–51.

Lam CW, Yeung WL, Ko CH et al. (2000). Spectrum of mutations in the MECP2 gene in patients with infantile autism and Rett syndrome. *J Med Genet*, **37**, E41.

Lance JW (1977). Familial paroxysmal dystonic choreoathetosis and its differentiation from related syndromes. *Ann Neurol*, **2**, 285–93.

Lang AE (1994). Withdrawal akathisia: case reports and a proposed classification of chronic akathisia. *Mov Disord*, **9**, 188–92.

Lang AE (1995). Psychogenic dystonia: a review of 18 cases. *Can J Neurol Sci*, **22**, 136–43.

Lang AE, Koller WC, Fahn S (1995). Psychogenic parkinsonism. *Arch Neurol*, **52**, 802–10.

Lang AE, Gill S, Patel NK et al. (2006). Randomized controlled trial of intraputamenal glial cell line-derived neurotrophic factor infusion in Parkinson disease. *Ann Neurol*, **59**, 459–66.

Langston JW (1996). The etiology of Parkinson's disease with emphasis on the MPTP story. *Neurology*, **47**, S153–60.

Langston J, Ballard P, Tetrud J et al. (1983). Chronic parkinsonism in humans due to a product of meperidine-analog synthesis. *Science*, **219**, 979–80.

Lannuzel A, Hermann C, Yousry C et al. (2002). Encephalomyelitis with rigidity complicating human immunodeficiency virus infection. *Mov Disord*, **17**, 202–4.

Laplanche JL, Hachimi KH, Durieux I et al. (1999). Prominent psychiatric features and early onset in an inherited prion disease with a new insertional mutation in the prion protein gene. *Brain*, 122 (Pt 12), 2375–86.

Larsen JP, Boas J, Erdal JE (1999). Does selegiline modify the progression of early Parkinson's disease? Results from a five-year study. The Norwegian-Danish Study Group. *Eur J Neurol*, **6**, 539–47.

Larsen TA, Dunn HG, Jan JE et al. (1985). Dystonia and calcification of the basal ganglia. *Neurology*, **35**, 533–7.

Layzer RB (1995). Neuromyotonia: a new autoimmune disease [editorial; comment]. *Ann Neurol*, **38**, 701–2.

Le Witt PA and et al (1986). Terahydrobiopterin in dystonia: Identification of abnormal metabolism and therapeutic trials. *Neurology*, **36**, 760–4.

Lebre AS, Durr A, Jedynak P et al. (1999). DYT1 mutation in French families with idiopathic torsion dystonia. *Brain*, 122 (Pt 1), 41–5.

Lechevalier B, Chapon F, Defer G et al. (2005). Perry and Purdy's syndrome (familial and fatal parkinsonism with hypoventilation and athymhormia). *Bull Acad Natl Med*, **189**, 481–90; discussion 490–2.

Lees AJ (1987). The Steele-Richardson-Olszewski syndrome (Progressive supranuclear palsy). In Marsden CD, Fahn S, eds. *Mov Disord, 2*, pp. 272–87. Butterworths, London.

Lees AJ (1995). Comparison of therapeutic effects and mortality data of levodopa and levodopa combined with selegiline in patients with early, mild Parkinson's disease. Parkinson's Disease Research Group of the United Kingdom. *BMJ*, **311**, 1602–7.

Lee MS, Marsden CD (1994a). Movement disorders following lesions of the thalamus or subthalamic region. *Mov Disord*, **9**, 493–507.

Lee MS, Marsden CD (1994b). Neurological sequelae following carbon monoxide poisoning: Clinical course and outcome according to the clinical types and brain computed tomography scan findings. *Mov Disord*, **9**, 550–8.

Lees AJ, Katzenschlager R, Head J et al. (2001). Ten-year follow-up of three different initial treatments in de-novo PD: a randomized trial. *Neurology*, **57**, 1687–94.

Lee LV, Pascasio FM, Fuentes FD et al. (1976). Torsion dystonia in Panay, Philippines. *Adv Neurol*, **14**, 137–51.

Lee WL, Tay A, Ong HT et al. (1998). Association of infantile convulsions with paroxysmal dyskinesias (ICCA syndrome): confirmation of linkage to human chromosome 16p12-q12 in a Chinese family. *Hum Genet*, **103**, 608–12.

Lee HY, Xu Y, Huang Y et al. (2004). The gene for paroxysmal non-kinesigenic dyskinesia encodes an enzyme in a stress response pathway. *Hum Mol Genet*, **13**, 3161–70.

Lehericy S, Vidailhet M, Dormont D et al. (1996). Striatopallidal and thalamic dystonia. A magnetic resonance imaging anatomoclinical study. *Arch Neurol*, **53**, 241–50.

Leigh PN, Rothwell JC, Traub M et al. (1980). A patient with reflex myoclonus and muscle rigidity: "jerking stiff man syndrome". *JNNP*, **43**, 1125–31.

Lemon RN (1988). The output map of the primate motor cortex. *Trends Neurosci*, **11**, 501–6.

Leo RJ (1996). Movement disorders associated with the serotonin selective reuptake inhibitors. *J Clin Psychiatry*, **57**, 449–54.

Lesage S, Leutenegger AL, Ibanez P et al. (2005). LRRK2 haplotype analyses in European and North African families with Parkinson disease: a common founder for the G2019S mutation dating from the 13th century. *Am J Hum Genet*, **77**, 330–2.

Leube B, Hendgen T, Kessler KR et al. (1997). Evidence for DYT7 being a common cause of cervical dystonia (torticollis) in Central Europe. *Am J Med Genet*, **74**, 529–32.

Leuzzi V, Carducci C, Carducci C et al. (2002). Autosomal dominant GTP-CH deficiency presenting as a dopa-responsive myoclonus-dystonia syndrome. *Neurology*, **59**, 1241–3.

Levin KH (1997). Paraneoplastic neuromuscular syndromes. *Neurol Clin*, **15**, 597–614.

LeWitt PA (1999). Penicillamine as a controversial treatment for Wilson's disease. *Mov Disord*, **14**, 555–6.

Liebeskind DS, Wong S, Hamilton RH (2003). Faces of the giant panda and her cub: MRI correlates of Wilson's disease. *J Neurol Neurosurg Psychiatry*, **74**, 682–.

Lima AR, Soares-Weiser K, Bacaltchuk J et al. (2002). Benzodiazepines for neuroleptic-induced acute akathisia. *Cochrane Database Syst Rev*, CD001950.

Lima AR, Weiser KV, Bacaltchuk J et al. (2004). Anticholinergics for neuroleptic-induced acute akathisia. *Cochrane Database Syst Rev*, CD003727.

Limousin P, Krack P, Pollak P et al. (1998). Electrical stimulation of the subthalamic nucleus in advanced Parkinson's disease. *N Engl J Med*, **339**, 1105–11.

Limousin P, Speelman JD, Gielen F et al. (1999). Multicentre European study of thalamic stimulation in parkinsonian and essential tremor. *J Neurol Neurosurg Psychiatry*, **66**, 289–96.

Lincoln SJ, Maraganore DM, Lesnick TG et al. (2003). Parkin variants in North American Parkinson's disease: cases and controls. *Mov Disord*, **18**, 1306–11.

Lindvall O, Sawle G, Widner H et al. (1994). Evidence for long term survival and function of dopaminergic grafts in progressive Parkinson's disease. *Ann Neurol*, **35**, 172–80.

Liou YJ, Lai IC, Liao DL et al. (2006). The human dopamine receptor D2 (DRD2) gene is associated with tardive dyskinesia in patients with schizophrenia. *Schizophr Res*.

Lipe H, Schultze A, Bird TD (1993). Risk factors for suicide in Huntington's disease: a retrospective case controlled study. *Am J Med Genet*, **48**, 231–33.

Litvan I, Agid Y, Calne D et al. (1996a). Clinical research criteria for the diagnosis of progressive supranuclear palsy (Steele-Richardson-Olszewski syndrome): report of the NINDS-SPSP international workshop. *Neurology*, **47**, 1–9.

Litvan I, Agid Y, Jankovic J et al. (1996b). Accuracy of clinical criteria for the diagnosis of progressive supranuclear palsy (Steele-Richardson-Olszewski syndrome). *Neurology*, **46**, 922–30.

Litvan I, Campbell G, Mangone CA et al. (1997). Which clinical features differentiate progressive supranuclear palsy (Steele-Richardson-Olszewski syndrome) from related disorders? A clinicopathological study. *Brain*, **120**, 65–74.

Litvan I, Mangone CA, McKee A *et al.* (1996c). Natural history of progressive supranuclear palsy (Steele-Richardson-Olszewski syndrome) and clinical predictors of survival: a clinicopathological study. *J Neurol Neurosurg Psychiatry*, **60**, 615–20.

Litvan I, Mega MS, Cummings JL *et al.* (1996d). Neuropsychiatric aspects of progressive supranuclear palsy. *Neurology*, **47**, 1184–9.

Lobotesis K, Fenwick JD, Phipps A *et al.* (2001). Occipital hypoperfusion on SPECT in dementia with Lewy bodies but not AD. *Neurology*, **56**, 643–9.

Logroscino G, Marder K, Cote L *et al.* (1996). Dietary lipids and antioxidants in Parkinson's disease: a population-based, case-control study. *Ann Neurol*, **39**, 89–94.

Lopez-Villegas D, Kulisevsky J, Deus J *et al.* (1996). Neuropsychological alterations in patients with computed tomography-detected basal ganglia calcification. *Arch Neurol*, **53**, 251–6.

Louis ED, Klatka LA, Liu Y *et al.* (1997). Comparison of extrapyramidal features in 31 pathologically confirmed cases of diffuse Lewy body disease and 34 pathologically confirmed cases of Parkinson's disease. *Neurology*, **48**, 376–80.

Louis ED, Lynch T, Kaufmann P *et al.* (1996). Diagnostic guidelines in central nervous system Whipple's disease. *Ann Neurol*, **40**, 561–8.

Lozano AM, Abosch A (2004). Pallidal stimulation for dystonia. *Adv Neurol*, **94**, 301–8.

Lozano AM, Lang AE, Galvez-Jimenez N *et al.* (1995). Effect of GPi pallidotomy on motor function in Parkinson's disease. *Lancet*, **346**, 1383–7.

Lu CS, Thompson PD, Quinn NP *et al.* (1993). Ramsay Hunt syndrome and coeliac disease: a new association? *Mov Disord*, **1**, 209–19.

Lucato LT, Otaduy MC, Barbosa ER *et al.* (2005). Proton MR spectroscopy in Wilson disease: analysis of 36 cases. *AJNR Am J Neuroradiol*, **26**, 1066–71.

Lucking CB, Durr A, Bonifati V *et al.* (2000). Association between early-onset Parkinson's disease and mutations in the parkin gene. *N Engl J Med*, **342**, 1560–7.

Lugaresi E, Cirignotta F, Montagna P (1986). Nocturnal paroxysmal dystonia. *JNNP*, **49**, 375–80.

Luo F, Leckman JF, Katsovich L *et al.* (2004). Prospective longitudinal study of children with tic disorders and/or obsessive-compulsive disorder: relationship of symptom exacerbations to newly acquired streptococcal infections. *Pediatrics*, **113**, e578–85.

Lynch T, Odel J, Fredericks DN *et al.* (1997). Polymerase chain reaction-based detection of Tropheryma whippelii in central nervous system Whipple's disease. *Ann Neurol*, **42**, 120–4.

MacDonald ME, Barnes G, Srinidhi J *et al.* (1993). Gametic but not somatic instability of CAG repeat length in Huntington's disease. *J Med Genet*, **30**, 982–6.

Maciel P, Cruz VT, Constante M *et al.* (2005). Neuroferritinopathy: missense mutation in FTL causing early-onset bilateral pallidal involvement. *Neurology*, **65**, 603–5.

MacMillan JC, Morrison PJ, Nevin NC *et al.* (1993a). Identification of an expanded CAG repeat in the Huntington's disease gene (IT15) in a family reported to have benign hereditary chorea. *J Med Genet*, **30**, 1012–3.

MacMillan JC, Snell RG, Tyler A *et al.* (1993b). Molecular analysis and clinical correlations of the huntington's disease mutation. *Lancet*, **342**, 954–8.

Magarinos-Ascone CM, Regidor I, Martinez-Castrillo JC *et al.* (2005). Pallidal stimulation relieves myoclonus-dystonia syndrome. *J Neurol Neurosurg Psychiatry*, **76**, 989–91.

Mahapatra RK, Edwards MJ, Schott JM *et al.* (2004). Corticobasal degeneration. *Lancet Neurol*, **3**, 736–43.

Maher ER, Lees AJ (1986). The clinical features and natural history of the Steele-Richardson-Olszewski syndrome (progressive supranuclear palsy). *Neurology*, **36**, 1005–8.

Maia DP, Teixeira AL Jr, Quintao Cunningham MC *et al.* (2005). Obsessive compulsive behavior, hyperactivity, and attention deficit disorder in Sydenham chorea. *Neurology*, **64**, 1799–801.

Mancuso M, Davidzon G, Kurlan RM *et al.* (2005). Hereditary ferritinopathy: a novel mutation, its cellular pathology, and pathogenetic insights. *J Neuropathol Exp Neurol*, **64**, 280–94.

Manji H, Howard RS, Miller DH *et al.* (1998). Status dystonicus: the syndrome and its management. *Brain*, 121 (Pt 2), 243–52.

Manyam BV (2005). What is and what is not 'Fahr's disease'. *Parkinsonism Relat Disord*, **11**, 73–80.

Maraganore DM, Harding AE, Marsden CD (1991). A clinical and genetic study of familial Parkinson's disease. *Mov Disord*, **6**, 205–11.

Maraganore DM, de Andrade M, Lesnick TG *et al.* (2005). High-resolution whole-genome association study of Parkinson disease. *Am J Hum Genet*, **77**, 685–93.

Maraganore DM, Folger WN, Swanson JW *et al.* (1992). Movement disorders as sequelae of central pontine myelinolysis: Report of three cases. *Mov Disord*, **7**, 142–8.

Maraganore DM, Lesnick TG, Elbaz A *et al.* (2004). UCHL1 is a Parkinson's disease susceptibility gene. *Ann Neurol*, **55**, 512–21.

Marcellini M, Di Ciommo V, Callea F *et al.* (2005). Treatment of Wilson's disease with zinc from the time of diagnosis in pediatric patients: a single-hospital, 10-year follow-up study. *J Lab Clin Med*, **145**, 139–43.

Marek KL, Seibyl JP, Zoghbi SS *et al.* (1996). [123I] beta-CIT/SPECT imaging demonstrates bilateral loss of dopamine transporters in hemi-Parkinson's disease. *Neurology*, **46**, 231–7.

Margolis RL, Rudnicki DD, Holmes SE (2005). Huntington's disease like-2: review and update. *Acta Neurol Taiwan*, **14**, 1–8.

Margolis RL, O'Hearn E, Rosenblatt A *et al.* (2001). A disorder similar to Huntington's disease is associated with a novel CAG repeat expansion. *Ann Neurol*, **50**, 373–80.

Marion MH, Gledhill RF, Thompson PD (1989). Spasms of amputation stumps: a report of 2 cases. *Mov Disord*, **4**, 1354–8.

Mark MH, Sage JI, Walters AS *et al.* (1995). Binswanger's disease presenting as levodopa-responsive parkinsonism: clinicopathologic study of three cases. *Mov Disord*, **10**, 450–4.

Markopoulou K, Wszolek ZK, Pfeiffer RF (1995). A Greek-American kindred with autosomal dominant, levodopa-responsive parkinsonism and anticipation. *Ann Neurol*, **38**, 373–8.

Marsden CD (1987). Wilson's disease. *Q J Med*, **65**, 959–66.

Marsden CD (1994). Peripheral movement disorders. In Marsden CD, Fahn S, eds. *Movement Disorders*, 3, pp. 406–17. Butterworth-Heinemann, Oxford.

Marsden CD, Harrison MGJ (1974). Idiopathic torsion dystonia (dystonia musculorum deformans). A review of forty-two patients. *Brain*, **97**, 793–810.

Marsden CD, Quinn NP (1990). The dystonias. *BMJ*, **300**, 139–44.

Marsden CD, Rothwell JC (1987). The physiology of idiopathic dystonia. *Can J Neurol Sci*, **14**, 521–7.

Marsden CD, Marion M-H, Sheehy MP (1984a). The treatment of severe dystonia in children and adults. *JNNP*, **47**, 1166–73.

Marsden CD, Obeso JA, Traub MM (1984b). Muscle spasms associated with Sudeck's atrophy after injury. *BMJ*, **288**, 173–6.

Marsden CD, Harding AE, Obeso JA *et al.* (1990). Progressive myoclonic ataxia (the Ramsay Hunt syndrome). *Arch Neurol*, **47**, 1121–5.

Marsden CD, Lang AE, Quinn NP *et al.* (1986). Familial dystonia and visual failure with striatal CT lucencies. *JNNP*, **49**, 500–19.

Marshall V, Grosset D (2003). Role of dopamine transporter imaging in routine clinical practice. *Mov Disord*, **18**, 1415–23.

Marshall FJ, Shoulson I (1997). Clinical features and treatment of huntington's disease. In Watts RL, Koller WC, eds. *Movement Disorders: Neurologic Principles and Practice*, pp. 491–502. McGraw-Hill, New York.

Martinelli P, Giuliani S, Ippoliti M *et al.* (1993). Familial idiopathic strio-pallido-dentate calcifications with late onset extrapyramidal syndrome. *Mov Disord*, **8**, 220–2.

Martino D, Giovannoni G (2004). Antibasal ganglia antibodies and their relevance to movement disorders. *Curr Opin Neurol*, **17**, 425–32.

Masucci EF, Kurtzke JF (1989). Tremor in progressive supranuclear palsy. *Acta Neurol Scand*, **80**, 296–300.

Matsui H, Udaka F, Miyoshi T *et al.* (2005a). Brain perfusion differences between Parkinson's disease and multiple system atrophy with predominant parkinsonian features. *Parkinsonism Relat Disord*, **11**, 227–32.

Matsui H, Udaka F, Oda M *et al.* (2005b). Metaiodobenzylguanidine (MIBG) scintigraphy at various parts of the body in Parkinson's disease and multiple system atrophy. *Auton Neurosci*, **119**, 56–60.

Matsumoto J, Hallett M (1994). Startle syndromes. In Marsden CD, Fahn S, eds. *Mov Disord*, **3**, pp. 418–33. Butterworth-Heinemann, Oxford.

Matsuo H, Takashima H, Kishikawa M *et al.* (1991). Pure akinesia: an atypical manifestation of progressive supranuclear palsy. *J Neurol Neurosurg Psychiatry*, **54**, 397–400.

Mayston MJ, Harrison LM, Quinton R *et al.* (1997). Mirror movements in X-linked Kallmann's syndrome. I. A neurophysiological study. *Brain*, **120**, 1199–216.

McCabe DJ, Turner NC, Chao D *et al.* (2004). Paraneoplastic "stiff person syndrome" with metastatic adenocarcinoma and anti-Ri antibodies. *Neurology*, **62**, 1402–4.

McKee DH, Sussman JD (2005). Case report: severe acute Parkinsonism associated with streptococcal infection and antibasal ganglia antibodies. *Mov Disord* 20, 1661–3.

McKeith I, Del Ser T, Spano P *et al.* (2000). Efficacy of rivastigmine in dementia with Lewy bodies: a randomised, double-blind, placebo-controlled international study. *Lancet*, **356**, 2031–6.

McKeith LG, Galasko D, Kosaka K *et al.* (1996). Consensus guidelines for the clinical and pathologic diagnosis of dementia with Lewy bodies (DLB): report of the consortium on DLB international workshop. *Neurology*, **47**, 1113–24.

McNaught KS, Kapustin A, Jackson T *et al.* (2004a). Brainstem pathology in DYT1 primary torsion dystonia. *Ann Neurol*, **56**, 540–7.

McNaught KS, Perl DP, Brownell AL *et al.* (2004b). Systemic exposure to proteasome inhibitors causes a progressive model of Parkinson's disease. *Ann Neurol*, 56:149–62.

Meinck HM, Ricker K, Conrad B (1984). The stiff man syndrome: new pathophysiological aspects from abnormal exteroceptive reflexes and the response to clomipramine, clonidine and tizanidine. *JNNP*, **47**, 280–7.

Melamed E, Achiron A, Shapira A *et al.* (1991). Persistent and progressive parkinsonism after discontinuation of chronic neuroleptic therapy: an additional tardive syndrome? *Clin Neuropharmacol*, **14**, 273–8.

Mell LK, Davis RL, Owens D (2005). Association between streptococcal infection and obsessive-compulsive disorder, Tourette's syndrome, and tic disorder. *Pediatrics*, **116**, 56–60.

Menza MA, Cocchiola J, Golbe LI (1995). Psychiatric symptoms in progressive supranuclear palsy. *Psychosomatics*, **36**, 550–4.

Mercadante MT, Campos MC, Marques-Dias MJ *et al.* (1997). Vocal tics in Sydenham's chorea. *J Am Acad Child Adolesc Psychiatry*, **36**, 305–6.

Merello M, Lees AJ (1992). Beginning-of-dose motor deterioration following the acute administration of levodopa and apomorphine in Parkinson's disease. *J Neurol Neurosurg Psychiatry*, **55**, 1024–6.

Minassian BA, Ianzano L, Meloche M *et al.* (2000). Mutation spectrum and predicted function of laforin in Lafora's progressive myoclonus epilepsy. *Neurology*, **55**, 341–6.

Minguez-Castellanos A, Escamilla-Sevilla F, Katati MJ *et al.* (2005). Different patterns of medication change after subthalamic or pallidal stimulation for Parkinson's disease: target related effect or selection bias? *J Neurol Neurosurg Psychiatry*, **76**, 34–9.

Mink JW (2003). The Basal Ganglia and involuntary movements: impaired inhibition of competing motor patterns. *Arch Neurol*, **60**, 1365–8.

Minoshima S, Foster NL, Sima AA *et al.* (2001). Alzheimer's disease versus dementia with Lewy bodies: cerebral metabolic distinction with autopsy confirmation. *Ann Neurol*, **50**, 358–65.

Mir P, Edwards MJ, Curtis AR *et al.* (2005). Adult-onset generalized dystonia due to a mutation in the neuroferritinopathy gene. *Mov Disord*, **20**, 243–5.

Misawa S, Kuwabara S, Ogawara K *et al.* (2006). Abnormal muscle responses in hemifacial spasm: F waves or trigeminal reflexes? *J Neurol Neurosurg Psychiatry*, **77**, 216–8.

Mitsumoto H, Schwartzman MJ, Estes ML *et al.* (1991). Sudden death and paroxysmal autonomic dysfunction in stiff-man syndrome. *J Neurol*, **238**, 91–6.

Miwa H, Miwa T, Imai H *et al.* (2001). Obsessive-compulsive-like behavioral changes in pure akinesia. *Parkinsonism Relat Disord*, **7**, 315–7.

Miyajima H, Kono S, Takahashi Y *et al.* (2001). Cerebellar ataxia associated with heteroallelic ceruloplasmin gene mutation. *Neurology*, **57**, 2205–10.

Miyasaki JM, Lang AE (1995). Treatment of drug induced movement disorders. In Kurlan R, ed. *Treatment of Movement Disorders*, pp. 429–76. J.B. Lippincott, Philadelphia.

Miyasaki JM, Martin W, Suchowersky O *et al.* (2002). Practice parameter: initiation of treatment for Parkinson's disease: an evidence-based review: report of the Quality Standards Subcommittee of the American Academy of Neurology. *Neurology*, **58**, 11–7.

Miyata S, Hamamura T, Yoshinaga J *et al.* (2001). Amelioration of frozen gait by tandospirone, a serotonin 1A agonist, in a patient with pure akinesia developing resistance to L-threo-3,4-dihydroxyphenylserine. *Clin Neuropharmacol*, **24**, 232–4.

Miyoshi K, Matsuoka T, Mizushima S (1969). Familial holotopistic striatal necrosis. *Acta Neuropathol (Berlin)*, **13**, 240–9.

Modrego PJ, Mojonero J, Serrano M *et al.* (2005). Fahr's syndrome presenting with pure and progressive presenile dementia. *Neurol Sci*, **26**, 367–9.

Moersch FP, Woltman HW (1956). Progressive fluctuating muscular rigidity and spasm ("stiff man" syndrome): report of a case and some observations in 13 other cases. *Mayo Clinic Proceedings*, **31**, 421–7.

Moffat DA, Durvasula VS, Stevens King A *et al.* (2005). Outcome following retrosigmoid microvascular decompression of the facial nerve for hemifacial spasm. *J Laryngol Otol*, **119**, 779–83.

Molho ES, Factor SA (1999). Worsening of motor features of parkinsonism with olanzapine. *Mov Disord*, **14**, 1014–6.

Molho ES, Factor SA (2001). Parkinson's disease: the treatment of drug-induced hallucinations and psychosis. *Curr Neurol Neurosci Rep*, **1**, 320–8.

Molina JA, Porta J, Garcia-Morales I *et al.* (2000). Treatment with intravenous prednisone and immunoglobin in a case of progressive encephalomyelitis with rigidity. *J Neurol Neurosurg Psychiatry*, **68**, 395–6.

Molinuevo JL, Marti MJ, Blesa R *et al.* (2003). Pure akinesia: an unusual phenotype of Hallervorden-Spatz syndrome. *Mov Disord*, **18**, 1351–3.

Molloy FM, Dalakas MC, Floeter MK (2002). Increased brainstem excitability in stiff-person syndrome. *Neurology*, **59**, 449–51.

Montastruc JL, Rascol O, Senard JM *et al.* (1994). A randomised controlled study comparing bromocriptine to which levodopa was later added, with levodopa alone in previously untreated patients with Parkinson's disease: a five year follow up. *J Neurol Neurosurg Psychiatry*, **57**, 1034–8.

Montplaisir J, Boucher S, Poirier G *et al.* (1997). Clinical, polysomnographic, and genetic characteristics of restless legs syndrome: a study of 133 patients diagnosed with new standard criteria. *Mov Disord*, **12**, 61–5.

Moreau C, Devos D, Delmaire C *et al.* (2005). Progressive MRI abnormalities in late recurrence of Sydenham's chorea. *J Neurol*, **252**, 1341–4.

Morris M (1991). Psychiatric aspects of Huntington's disease. In Harper PS, ed. *Huntington's Disease*, pp. 81–126. W.B. Saunders, London.

Morris M, Tyler A (1991). Management and therapy. In Harper PS, ed. *Huntington's Disease*, pp. 205–50. W.B. Saunders, London.

Morris JC, Drazner M, Fulling K *et al.* (1989). Clinical and pathological aspects of parkinsonism in Alzheimer's disease. A role for extranigral factors? *Arch Neurol*, **46**, 651–7.

Morris HR, Janssen JC, Bandmann O *et al.* (1999). The tau gene A0 polymorphism in progressive supranuclear palsy and related neurodegenerative diseases. *J Neurol Neurosurg Psychiatry*, **66**, 665–7.

Mount LA, Reback S (1940). Familial paroxysmal choreoathetosis. *Arch Neurol Psychiatr*, **44**, 841–7.

Mueller JC, Fuchs J, Hofer A *et al.* (2005). Multiple regions of alpha-synuclein are associated with Parkinson's disease. *Ann Neurol*, **57**, 535–41.

Muenter MD, Tyce GM (1971). L-dopa therapy of Parkinson's disease: plasma L-dopa concentration, therapeutic response, and side effects. *Mayo Clin Proc*, **46**, 231–9.

Mukaddes NM, Abali O (2003). Quetiapine treatment of children and adolescents with Tourette's disorder. *J Child Adolesc Psychopharmacol*, **13**, 295–9.

Mumford CJ, Fletcher NA, Ironside JW *et al.* (1996). Progressive ataxia, focal seizures, and malabsorption syndrome in a 41 year old woman [clinical conference]. *J Neurol Neurosurg Psychiatry*, **60**, 225–30.

Murinson BB (2004). Stiff-person syndrome. *Neurologist*, **10**, 131–7.

Muthane UB, Swamy HS, Satishchandra P *et al.* (1994). Early onset Parkinson's disease: are juvenile- and young-onset different? *Mov Disord*, **9**, 539–44.

Myers RH, MacDonald ME, Koroshetz WJ *et al.* (1993). De novo expansion of a (CAG)n repeat in sporadic Huntington's disease. *Nat Genet*, **5**, 168–73.

Nasrallah HA (2006). Focus on lower risk of tardive dyskinesia with atypical antipsychotics. *Ann Clin Psychiatry*, **18**, 57–62.

Nass R, Gutman R (1997). Boys with Asperger's disorder, exceptional verbal intelligence, tics, and clumsiness. *Dev Med Child Neurol*, **39**, 691–5.

Naumann M, Pirker W, Reiners K *et al.* (1997). [123I]beta-CIT single-photon emission tomography in DOPA-responsive dystonia. *Mov Disord*, **12**, 448–51.

Nausieda PA, Koller WC, Weiner WJ *et al.* (1979). Chorea induced by oral contraceptives. *Neurology*, **29**, 1605–9.

Neary D, Snowdon JS, Shields RA *et al.* (1987). Single photon emission tomography using 99mTc-HM-PAO in the investigation of dementia. *J Neurol, Neurosurg Psychiatry*, **50**, 1101–9.

Nelson I, Hanna MG, Alsanjari N *et al.* (1995). A new mitochondrial DNA mutation associated with progressive dementia and chorea: a clinical, pathological, and molecular genetic study. *Ann Neurol*, **37**, 400–3.

Neshige R, Luders H, Shibasaki H (1988). Recording of movement related potentials from scalp and cortex in man. *Brain*, **111**, 719–36.

Newsom-Davis J, Mills KR (1993). Immunological associations of acquired neuromyotonia (Isaacs' syndrome). Report of five cases and literature review. *Brain*, **116**, 453–69.

Nieuwboer A, Kwakkel G, Rochester L *et al.* (2006). Cueing training in the home improves gait-related mobility in Parkinson's disease: The RESCUE-trial. *J Neurol Neurosurg Psychiatry*.

Nolte D, Niemann S, Muller U (2003). Specific sequence changes in multiple transcript system DYT3 are associated with X-linked dystonia parkinsonism. *Proc Natl Acad Sci USA*, **100**, 10347–52.

Nomoto M, Thompson PD, Sheehy MP *et al.* (1987). Anticholinergic induced chorea in the treatment of focal dystonia. *Mov Disord*, **2**, 53–6.

Novotny EJ, Singh G, Wallace DC *et al.* (1986). Lebers disease and dystonia: a mitochondrial disease. *Neurology*, **29**, 364–9.

Nutt JG, Holford NH (1996). The response to levodopa in Parkinson's disease: imposing pharmacological law and order. *Ann Neurol*, **39**, 561–73.

Nutt JG, Marsden CD, Thompson PD (1993). Human walking and higher level gait disorders, particularly in the elderly. *Neurology*, **43**, 268–79.

Nutt JG, Muenter MD, Aronson A *et al.* (1988). Epidemiology of focal and generalised dystonia in Rochester, Minnesota. *Mov Disord*, **3**, 188–94.

Nygaard TG (1995). Dopa-responsive dystonia. *Curr Opin Neurol*, **8**, 310–3.

Nygaard TG, Raymond D, Chen C *et al.* (1999). Localization of a gene for myoclonus-dystonia to chromosome 7q21-q31. *Ann Neurol*, **46**, 794–8.

Nygaard TG, Takahashi H, Heiman GA *et al.* (1992). Long-term treatment response and fluorodopa positron emission tomographic scanning of parkinsonism in a family with dopa-responsive dystonia. *Ann Neurol*, **32**, 603–8.

Nygaard TG, Waran SP, Levine RA *et al.* (1994). Dopa-responsive dystonia simulating cerebral palsy. *Pediatr Neurol*, **11**, 236–40.

Nygaard TG, Wilhelmsen KC, Risch NJ *et al.* (1993). Linkage mapping of dopa-responsive dystonia (DRD) to chromosome 14q. *Nat Genet*, **5**, 386–91.

Nyholm D, Nilsson Remahl AI, Dizdar N *et al.* (2005). Duodenal levodopa infusion monotherapy vs oral polypharmacy in advanced Parkinson disease. *Neurology*, **64**, 216–23.

O'Riordan S, Hutchinson M (2004). Cervical dystonia following peripheral trauma--a case-control study. *J Neurol*, **251**, 150–5.

Obeso JA (1995). Therapy of myoclonus. *Clin Neurosci*, **3**, 253–7.

Obeso JA, Martinez-Lage JM (1992). Anticholinergics and amantadine. In Koller WC, ed. *Handbook of Parkinson's Disease*, pp. 383–90. Marcel Dekker, New York.

Obeso JA, Guridi J, De Long M (1997). Surgery for Parkinson's disease [editorial]. *J Neurol Neurosurg Psychiatry*, **62**, 2–8.

Obeso JA, Rothwell JC, Marsden CD (1981). Simple tics in Gilles de la Tourette's syndrome are not prefaced by a normal premovement EEG potential. *J Neurol Neurosurg Psychiatry*, **44**, 735–8.

Obeso JA, Linazasoro G, Rothwell JC *et al.* (1996). Assessing the effects of pallidotomy in Parkinson's disease [letter; comment]. *Lancet*, **347**, 1490.

Obeso JA, Rothwell JC, Lang AE *et al.* (1983). Myoclonic dystonia. *Neurology*, **33**, 825–30.

Oertel WH, Schwarz J, Tatsch K *et al.* (1993). IBZM-SPECT as predictor for dopamimetic responsiveness of patients with de novo parkinsonian syndrome. *Adv Neurol*, **60**, 519–24.

Oh SH, Lee KY, Im JH *et al.* (2002). Chorea associated with non-ketotic hyperglycemia and hyperintensity basal ganglia lesion on T1-weighted brain MRI study: a meta-analysis of 53 cases including four present cases. *J Neurol Sci* **200**, 57–62.

Okuda Y, Suzuki K, Kitajima T *et al.* (1998). Lumbar epidural block for 'painful legs and moving toes' syndrome: a report of three cases. *Pain*, **78**, 145–7.

Olanow CW, Goetz CG, Kordower JH *et al.* (2003). A double-blind controlled trial of bilateral fetal nigral transplantation in Parkinson's disease. *Ann Neurol*, **54**, 403–14.

Oliveira CA, Negreiros Junior J, Cavalcante IC *et al.* (2003). Palatal and middle-ear myoclonus: a cause for objective tinnitus. *Int Tinnitus J*, **9**, 37–41.

Oliveira JR, Spiteri E, Sobrido MJ *et al.* (2004). Genetic heterogeneity in familial idiopathic basal ganglia calcification (Fahr disease). *Neurology*, **63**, 2165–7.

Ondo WG, Dat Vuong K, Khan H *et al.* (2001). Daytime sleepiness and other sleep disorders in Parkinson's disease. *Neurology*, **57**, 1392–6.

Opherk C, Gruber C, Steude U *et al.* (2006). Successful bilateral pallidal stimulation for Meige syndrome and spasmodic torticollis. *Neurology*, **66**, E14.

Orrell RW, Amrolia PJ, Heald A *et al.* (1995). Acanthocytosis, retinitis pigmentosa, and pallidal degeneration: a report of three patients, including the second reported case with hypoprebetalipoproteinemia (HARP syndrome). *Neurology*, **45**, 487–92.

Orrico A, Lam C, Galli L *et al.* (2000). MECP2 mutation in male patients with non-specific X-linked mental retardation. *FEBS Lett*, **481**, 285–8.

Otsuka M, Ichiya Y, Kuwabara Y *et al.* (1996). Differences in the reduced 18F-Dopa uptakes of the caudate and the putamen in Parkinson's disease: correlations with the three main symptoms. *J Neurol Sci*, **136**, 169–73.

Ozelius LJ, Hewett JW, Page CE et al. (1997). The early-onset torsion dystonia gene (DYT1) encodes an ATP-binding protein. Nat Genet, 17, 40–8.

Pahwa R (1997). Toxin induced parkinsonian syndromes. In Watts RL, Koller WC, eds. Movement Disorders: Neurologic Principles and Practice, pp. 315–23. McGraw Hill, New York.

Pahwa R, Lyons KE, Wilkinson SB et al. (2006). Long-term evaluation of deep brain stimulation of the thalamus. J Neurosurg, 104, 506–12.

Pahwa R, Wilkinson S, Smith D et al. (1997). High-frequency stimulation of the globus pallidus for the treatment of Parkinson's disease. Neurology, 49, 249–53.

Panov A, Dikalov S, Shalbuyeva N et al. (2005). Rotenone model of Parkinson disease: multiple brain mitochondria dysfunctions after short term systemic rotenone intoxication. J Biol Chem, 280, 42026–35.

Papp MI, Lantos PL (1994). The distribution of oligodendroglial inclusions in multiple system atrophy and its relevance to clinical symptomatology. Brain, 117, 235–43.

Papp MI, Kahn JE, Lantos PL (1989). Glial cytoplasmic inclusions in the CNS of patients with multiple system atrophy (striatonigral degeneration, olivopontocerebellar atrophy and Shy-Drager syndrome). J Neurol Sci, 94, 79–100.

Paret G, Tirosh R, Ben Zeev B et al. (1996). Intrathecal baclofen for severe torsion dystonia in a child. Acta Paediatr, 85, 635–7.

Parkinson's Disease Research Group in the United Kingdom (1993). Comparisons of therapeutic effects of levodopa, levodopa and selegiline, and bromocriptine in patients with early, mild Parkinson's disease: three year interim report. BMJ, 307, 469–72.

Parkinson Study Group (1993). Effects of tocopherol and deprenyl on the progression of disability in early Parkinson's disease. N Engl J Med, 328, 176–83.

Parkinson Study Group (1996). Impact of deprenyl and tocopherol treatment on Parkinson's disease in DATATOP patients requiring levodopa. Ann Neurol, 39, 37–45.

Parkinson Study Group (2000). Pramipexole vs levodopa as initial treatment for Parkinson disease: A randomized controlled trial. Parkinson Study Group. JAMA, 284, 1931–8.

Parkinson Study Group (2002). A controlled trial of rasagiline in early Parkinson disease: the TEMPO Study. Arch Neurol, 59, 1937–43.

Parkinson Study Group (2004). A controlled, randomized, delayed-start study of rasagiline in early Parkinson disease. Arch Neurol, 61, 561–6.

Parkinson Study Group (2005). A randomized placebo-controlled trial of rasagiline in levodopa-treated patients with Parkinson disease and motor fluctuations: the PRESTO study. Arch Neurol, 62, 241–8.

Pasquinelli G, Cenacchi G, Piane EL et al. (2004). The problematic issue of Kufs disease diagnosis as performed on rectal biopsies: a case report. Ultrastruct Pathol, 28, 43–8.

Passingham RE (1987). Two cortical systems for directing movement. Motor Areas of the Cerebral Cortex, pp. 151–64. John Wiley, Chichester.

Pauls DL, Leckman JF (1986). The inheritance of Gilles de la Tourette's syndrome and associated behaviours: Evidence for autosomal dominant transmission. N Engl J Med, 315, 993–7.

Paviour DC, Williams D, Fowler CJ et al. (2005). Is sphincter electromyography a helpful investigation in the diagnosis of multiple system atrophy? A retrospective study with pathological diagnosis. Mov Disord 20, 1425–30.

Pennacchio LA, Lehesjoki AE, Stone NE et al. (1996). Mutations in the gene encoding cystatin B in progressive myoclonus epilepsy (EPM1). Science, 271, 1731–4.

Penney SE, Bruce IA, Saeed SR (2006). Botulinum toxin is effective and safe for palatal tremor: A report of five cases and a review of the literature. J Neurol, 253, 857–60.

Perlmutter SJ, Leitman SF, Garvey MA et al. (1999). Therapeutic plasma exchange and intravenous immunoglobulin for obsessive-compulsive disorder and tic disorders in childhood. Lancet, 354, 1153–8.

Peschanski M, Defer G, JP NG et al. (1994). Bilateral motor improvement and alteration of L-dopa effect in two patients with Parkinson's disease following intrastriatal transplantation of foetal ventral mesencephalon. Brain, 117, 487–99.

Piccini P, Burn DJ, Ceravolo R et al. (1999). The role of inheritance in sporadic Parkinson's disease: evidence from a longitudinal study of dopaminergic function in twins. Ann Neurol, 45, 577–82.

Piccini P, de Yebenez J, Lees AJ et al. (2001). Familial progressive supranuclear palsy: detection of subclinical cases using 18F-dopa and 18fluorodeoxyglucose positron emission tomography. Arch Neurol, 58, 1846–51.

Piccini P, Morrish PK, Turjanski N et al. (1997). Dopaminergic function in familial Parkinson's disease: a clinical and 18F-dopa positron emission tomography study. Ann Neurol, 41, 222–9.

Pillon B, Blin J, Vidailhet M et al. (1995). The neuropsychological pattern of corticobasal degeneration: comparison with progressive supranuclear palsy and Alzheimer's disease. Neurology, 45, 1477–83.

Pincus JH, Chutorian A (1967). Familial benign chorea with intention tremor: a clinical entity. J Paediatr, 70, 724–9.

Pinero A, Marcos-Alberca P, Fortes J (2005). Cabergoline-related severe restrictive mitral regurgitation. N Engl J Med, 353, 1976–7.

Pittock SJ, Joyce C, O'Keane V et al. (2000). Rapid-onset dystonia-parkinsonism: a clinical and genetic analysis of a new kindred. Neurology, 55, 991–5.

Pittock SJ, Lucchinetti CF, Parisi JE et al. (2005). Amphiphysin autoimmunity: paraneoplastic accompaniments. Ann Neurol, 58, 96–107.

Plant GT, Williams AC, Earl CJ et al. (1984). Familial paroxysmal dystonia induced by excercise. J Neurol, Neurosurg Psychiatry, 47, 275–9.

Plante-Bordeneuve V, Taussig D, Thomas F et al. (1995). A clinical and genetic study of familial cases of Parkinson's disease. J Neurol Sci, 133, 164–72.

Playford ED, Fletcher NA, Sawle GV et al. (1993). Striatal [18F]dopa uptake in familial idiopathic dystonia. Brain, 116, 1191–9.

Plazzi G, Cortelli P, Montagna P et al. (1998). REM sleep behaviour disorder differentiates pure autonomic failure from multiple system atrophy with autonomic failure. J Neurol Neurosurg Psychiatry, 64, 683–5.

Poewe W, Hogl B (2004). Akathisia, restless legs and periodic limb movements in sleep in Parkinson's disease. Neurology, 63, S12–6.

Poewe W, Luessi F (2005). Clinical studies with transdermal rotigotine in early Parkinson's disease. Neurology, 65, S11–4.

Poewe WH, Lees AJ, Stern GM (1988). Dystonia in Parkinson's disease: clinical and pharmacological features. Ann Neurol, 23, 73–8.

Polesin A, Stern M (2006). Post-anoxic myoclonus: a case presentation and review of management in the rehabilitation setting. Brain Inj, 20, 213–7.

Polizos P, Engelhardt DM (1978). Dyskinetic phenomena in children treated with psychotropic medications. Psychpharmacol Bulletin, 14, 65–8.

Polymeropoulos MH, Lavedan C, Leroy E et al. (1997). Mutation in the alpha-synuclein gene identified in families with Parkinson's disease [see comments]. Science, 276, 2045–7.

Postuma RB, Lang AE (2003). Hemiballism: revisiting a classic disorder. Lancet Neurol, 2, 661–8.

Pralong E, Pollo C, Coubes P et al. (2005). Electrophysiological characteristics of limbic and motor globus pallidus internus (GPI) neurons in two cases of Lesch-Nyhan syndrome. Neurophysiol Clin, 35, 168–73.

Pramstaller PP, Schlossmacher MG, Jacques TS et al. (2005). Lewy body Parkinson's disease in a large pedigree with 77 Parkin mutation carriers. Ann Neurol, 58, 411–22.

Pramstaller PP, Wenning GK, Smith SJ et al. (1995). Nerve conduction studies, skeletal muscle EMG, and sphincter EMG in multiple system atrophy. J Neurol Neurosurg Psychiatry, 58, 618–21.

Pshezhetsky AV, Richard C, Michaud L et al. (1997). Cloning, expression and chromosomal mapping of human lysosomal sialidase and characterization of mutations in sialidosis. Nat Genet, 15, 316–20.

Puri BK, Leavitt BR, Hayden MR *et al.* (2005). Ethyl-EPA in Huntington disease: a double-blind, randomized, placebo-controlled trial. *Neurology*, **65**, 286–92.

Quinn N (1994). Multiple system atrophy. In Marsden CD, Fahn S, eds. *Movement Disorders 3*, pp. 262–81. Butterworth-Heinemann, Oxford.

Quinn NP (1996). Essential myoclonus and myoclonic dystonia. *Mov Disord*, **11**, 119–24.

Quinn NP (2005). How to diagnose multiple system atrophy. *Mov Disord* 20 (Suppl 12), S5–10.

Quinn N, Critchley P, Marsden CD (1987). Young onset Parkinson's disease. *Mov Disord*, **2**, 73–91.

Quinn NP, Lang AE, Koller WC *et al.* (1986). Painful Parkinson's disease. *Lancet*, **i**, 1366–69.

Quinn NP, Luthert P, Honavar M *et al.* (1989). Pure akinesia due to lewy body Parkinson's disease: a case with pathology. *Mov Disord*, **4**, 85–9.

Raja M, Azzoni A (1996). Tardive dystonia. Prevalence, risk factors and clinical features. *Ital J Neurol Sci*, **17**, 409–18.

Rajput AH, Gibb WR, Zhong XH *et al.* (1994). Dopa-responsive dystonia: pathological and biochemical observations in a case. *Ann Neurol*, **35**, 396–402.

Rajput AH, Pahwa R, Pahwa P *et al.* (1993). Prognostic significance of the onset mode in parkinsonism. *Neurology*, **43**, 829–30.

Rajput AH, Rajput A, Lang AE *et al.* (1998). New use for an old drug: amantadine benefits levodopa-induced dyskinesia. *Mov Disord*, **13**, 851.

Rajput AH, Rozdilsky B, Hornykiewicz O *et al.* (1982). Reversible drug induced parkinsonism: clinicopathologic study of two cases. *Arch Neurol*, **39**, 644–6.

Rakocevic G, Raju R, Dalakas MC (2004). Anti-glutamic acid decarboxylase antibodies in the serum and cerebrospinal fluid of patients with stiff-person syndrome: correlation with clinical severity. *Arch Neurol*, **61**, 902–4.

Ramos-Arroyo MA, Moreno S, Valiente A (2005). Incidence and mutation rates of Huntington's disease in Spain: experience of 9 years of direct genetic testing. *J Neurol Neurosurg Psychiatry*, **76**, 337–42.

Rampoldi L, Dobson-Stone C, Rubio JP *et al.* (2001). A conserved sorting-associated protein is mutant in chorea-acanthocytosis. *Nat Genet*, **28**, 119–20.

Rascol O, Brooks DJ, Korczyn AD *et al.* (2000). A five-year study of the incidence of dyskinesia in patients with early Parkinson's disease who were treated with ropinirole or levodopa. 056 Study Group. *N Engl J Med*, **342**, 1484–91.

Rascol O, Brooks DJ, Melamed E *et al.* (2005). Rasagiline as an adjunct to levodopa in patients with Parkinson's disease and motor fluctuations (LARGO, Lasting effect in Adjunct therapy with Rasagiline Given Once daily, study): a randomised, double-blind, parallel-group trial. *Lancet*, **365**, 947–54.

Reeves RR, Mack JE, Torres RA (2001). Neuroleptic malignant syndrome during a change from haloperidol to risperidone. *Ann Pharmacother*, **35**, 698–701.

Reeves AL, So EL, Sharbrough FW *et al.* (1996). Movement disorders associated with the use of gabapentin. *Epilepsia*, **37**, 988–90.

Rehncrona S, Johnels B, Widner H *et al.* (2003). Long-term efficacy of thalamic deep brain stimulation for tremor: double-blind assessments. *Mov Disord*, **18**, 163–70.

Reilly M, Daly L, Hutchinson M (1993). An epidemiological study of Wilson's disease in the Republic of Ireland. *J Neurol Neurosurg Psychiatry*, **56**, 298–300.

Remy P, Zilbovicius M, Leroy-Willig A *et al.* (1994). Movement- and task-related activations of motor cortical areas: a positron emission tomographic study. *Ann Neurol*, **36**, 19–26.

Reus VI (1997). Olanzapine: a novel atypical neuroleptic agent. *Lancet*, **349**, 1264–5.

Revesz T, Sangha H, Daniel SE (1996). The nucleus raphe interpositus in the Steele-Richardson-Olszewski syndrome (progressive supranuclear palsy). *Brain*, **119**, 1137–43.

Richfield EK, Vonsattel JP, MacDonald ME *et al.* (2002). Selective loss of striatal preprotachykinin neurons in a phenocopy of Huntington's disease. *Mov Disord*, **17**, 327–32.

Ridding MC, Sheean G, Rothwell JC *et al.* (1995). Changes in the balance between motor cortical excitation and inhibition in focal, task specific dystonia. *J Neurol Neurosurg Psychiatry*, **59**, 493–8.

Riley DE (1996). Paroxysmal kinesigenic dystonia associated with a medullary lesion. *Mov Disord*, **11**, 738–40.

Riley DE, Lang AE (1993). The spectrum of levodopa related fluctuations in parkinson's disease. *Neurology*, **43**, 1459–64.

Riley DE, Fogt N, Leigh RJ (1994). The syndrome of 'pure akinesia' and its relationship to progressive supranuclear palsy. *Neurology*, **44**, 1025–9.

Riley DE, Lang AE, Lewis A *et al.* (1990). Cortical-basal ganglionic degeneration. *Neurology*, **40**, 1203–12.

Rinne UK (1985). Combined bromocriptine-levodopa therapy early in Parkinson's disease. *Neurology*, **35**, 1196–8.

Rinne UK (1987). Early combination of bromocriptine and levodopa in the treatment of Parkinson's disease: A 5-year follow up. *Neurology*, **37**, 826–8.

Rinne JO, Burn DJ, Mathias CJ *et al.* (1995). Positron emission tomography studies on the dopaminergic system and striatal opioid binding in the olivopontocerebellar atrophy variant of multiple system atrophy. *Ann Neurol*, **37**, 568–73.

Rinne JO, Daniel SE, Scaravilli F *et al.* (1994a). The neuropathological features of neuroacanthocytosis. *Mov Disord*, **9**, 297–304.

Rinne UK, Larsen JP, Siden A *et al.* (1998). Entacapone enhances the response to levodopa in parkinsonian patients with motor fluctuations. Nomecomt Study Group. *Neurology*, **51**, 1309–14.

Rinne JO, Lee MS, Thompson PD *et al.* (1994b). Corticobasal degeneration. A clinical study of 36 cases. *Brain*, **117**, 1183–96.

Risch N, de Leon D, Ozelius L *et al.* (1995). Genetic analysis of idiopathic torsion dystonia in Ashkenazi Jews and their recent descent from a small founder population. *Nat Genet*, **9**, 152–9.

Ristic A, Marinkovic J, Dragasevic N *et al.* (2002). Long-term prognosis of vascular hemiballismus. *Stroke*, **33**, 2109–11.

Rivest J, Marsden CD (1990). Trunk and head tremor as isolated manifestations of dystonia. *Mov Disord*, **5**, 60–5.

Rivest J, Lees AJ, Marsden CD (1991). Writer's cramp: treatment with botulinum toxin injections. *Mov Disord*, **6**, 55–9.

Robertson MM (1994). Gilles de la Tourette syndrome: an update. *J Child Psychol Psychiatry*, **35**, 597–611.

Robertson MM (2006). Attention deficit hyperactivity disorder, tics and Tourette's syndrome: the relationship and treatment implications. A commentary. *Eur Child Adolesc Psychiatry*, **15**, 1–11.

Robertson MM, Trimble MR, Lees AJ (1989). Self injurious behaviour and the Gilles de la Tourette syndrome: a clinical study and review of the literature. *Psychol Med* 19, 611–25.

Rodriguez-Oroz MC, Zamarbide I, Guridi J *et al.* (2004). Efficacy of deep brain stimulation of the subthalamic nucleus in Parkinson's disease 4 years after surgery: double blind and open label evaluation. *J Neurol Neurosurg Psychiatry*, **75**, 1382–5.

Roesch K, Curran SP, Tranebjaerg L *et al.* (2002). Human deafness dystonia syndrome is caused by a defect in assembly of the DDP1/TIMM8a-TIMM13 complex. *Hum Mol Genet*, **11**, 477–86.

Roh JK, Lee TG, Wie BA *et al.* (1994). Initial and follow-up brain MRI findings and correlation with the clinical course in Wilson's disease. *Neurology*, **44**, 1064–8.

Rojo A, Pernaute RS, Fontan A *et al.* (1999). Clinical genetics of familial progressive supranuclear palsy. *Brain*, **122** (Pt 7), 1233–45.

Ross CA (2002). Polyglutamine pathogenesis: emergence of unifying mechanisms for Huntington's disease and related disorders. *Neuron*, **35**, 819–22.

Roos RA, Hermans J, Vegter-van der Vlis M *et al.* (1993). Duration of illness in Huntington's disease is not related to age at onset. *J Neurol Neurosurg Psychiatry*, **56**, 98–100.

Ross MH, Charness ME, Lee D *et al.* (1995). Does ulnar neuropathy predispose to focal dystonia? *Muscle Nerve*, **18**, 606–11.

Rosser AE, Barker RA, Harrower T *et al.* (2002). Unilateral transplantation of human primary fetal tissue in four patients with Huntington's disease: NEST-UK safety report ISRCTN no 36485475. *J Neurol Neurosurg Psychiatry*, **73**, 678–85.

Rothwell J (1994). *Control of Human Voluntary Movement.* Chapman & Hall, London.

Rothwell JC, Vidailhet M, Thompson PD *et al.* (1994). The auditory startle response in progressive supranuclear palsy. *J Neural Transm Suppl*, **42**, 43–50.

Roubertie A, Biolsi B, Rivier F *et al.* (2003). Ataxia with vitamin E deficiency and severe dystonia: report of a case. *Brain Dev*, **25**, 442–5.

Rupniak NM, Jenner P, Marsden CD (1986). Acute dystonia induced by neuroleptic drugs. *Psychopharmacology*, **88**, 403–19.

Rush PJ, Inman R, Bernstein M *et al.* (1986). Isolated vasculitis of the central nervous system in a patient with celiac disease. *Am J Med*, **81**, 1092–4.

Russmann H, Ghika J, Villemure JG *et al.* (2004). Subthalamic nucleus deep brain stimulation in Parkinson disease patients over age 70 years. *Neurology*, **63**, 1952–4.

Sa DS, Mailis-Gagnon A, Nicholson K *et al.* (2003). Posttraumatic painful torticollis. *Mov Disord*, **18**, 1482–91.

Sabbagh MN, Hake AM, Ahmed S *et al.* (2005). The use of memantine in dementia with Lewy bodies. *J Alzheimers Dis*, **7**, 285–9.

Sachdev P, Loneragan C (1993). Intravenous benztropine and propranolol challenges in acute neuroleptic-induced akathisia. *Clin Neuropharmacol*, **16**, 324–31.

Saint-Cyr JA, Trepanier LL, Kumar R *et al.* (2000). Neuropsychological consequences of chronic bilateral stimulation of the subthalamic nucleus in Parkinson's disease. *Brain*, **123**, 2091–108.

Saint Hilaire MH, Burke RE, Bressman SB *et al.* (1991). Delayed-onset dystonia due to perinatal or early childhood asphyxia. *Neurology*, **41**, 216–22.

Sainz J, Minassian BA, Serratosa JM *et al.* (1997). Lafora progressive myoclonus epilepsy: narrowing the chromosome 6q24 locus by recombinations and homozygosities [letter]. *Am J Hum Genet*, **61**, 1205–9.

Saiz A, Graus F, Valldeoriola F *et al.* (1998). Stiff-leg syndrome: a focal form of stiff-man syndrome. *Ann Neurol*, **43**, 400–3.

Salazar G, Valls-Sole J, Marti MJ *et al.* (2000). Postural and action myoclonus in patients with parkinsonian type multiple system atrophy. *Mov Disord*, **15**, 77–83.

Sanberg PR, Silver AA, Shytle RD *et al.* (1997). Nicotine for the treatment of Tourette's syndrome. *Pharmacol Ther*, **74**, 21–5.

Sanders P, Waddy HM, Thompson PD (1999). An 'annoying' foot: unilateral painful legs and moving toes syndrome. *Pain*, **82**, 103–4.

Sato C, Morgan A, Lang AE *et al.* (2005). Analysis of the glucocerebrosidase gene in Parkinson's disease. *Mov Disord* **20**, 367–70.

Saunders-Pullman R, Shriberg J, Heiman G *et al.* (2002). Myoclonus dystonia: possible association with obsessive-compulsive disorder and alcohol dependence. *Neurology*, **58**, 242–5.

Saver JL, Liu GT, Charness ME (1994). Idiopathic striopallidodentate calcification with prominent supranuclear abnormality of eye movement. *J Neuroophthalmol*, **14**, 29–33.

Scahill L, Leckman JF, Schultz RT *et al.* (2003). A placebo-controlled trial of risperidone in Tourette syndrome. *Neurology*, **60**, 1130–5.

Schapira AH (1994). Evidence for mitochondrial dysfunction in Parkinson's disease--a critical appraisal. *Mov Disord*, **9**, 125–38.

Schapira AH (2006). Mitochondrial disease. *Lancet*, **368**, 70–82.

Schapira AH, Mann VM, Cooper JM *et al.* (1990). Anatomic and disease specificity of NADH CoQ1 reductase (complex I) deficiency in Parkinson's disease. *J Neurochem*, **55**, 2142–5.

Scheinberg IH, Sternlieb I (1984). *Wilson's Disease.* W.B. Saunders, Philadelphia.

Schenck CH, Bundlie SR, Mahowald MW (1996). Delayed emergence of a parkinsonian disorder in 38% of 29 older men initially diagnosed with idiopathic rapid eye movement sleep behaviour disorder. *Neurology*, **46**, 388–93.

Schloesser DT, Ward TN, Williamson PD (1996). Familial paroxysmal dystonic choreoathetosis revisited. *Mov Disord*, **11**, 317–20.

Schrader C, Peschel T, Petermeyer M *et al.* (2004). Unilateral deep brain stimulation of the internal globus pallidus alleviates tardive dyskinesia. *Mov Disord* **19**, 583–5.

Schrag A, Ben-Shlomo Y, Quinn NP (1999). Prevalence of progressive supranuclear palsy and multiple system atrophy: a cross-sectional study. *Lancet*, **354**, 1771–5.

Schrag A, Good CD, Miszkiel K *et al.* (2000a). Differentiation of atypical parkinsonian syndromes with routine MRI. *Neurology*, **54**, 697–702.

Schrag A, Trimble M, Quinn N *et al.* (2004). The syndrome of fixed dystonia: an evaluation of 103 patients. *Brain*, **127**, 2360–72.

Schrag A, Quinn NP, Bhatia KP *et al.* (2000b). Benign hereditary chorea--entity or syndrome? *Mov Disord*, **15**, 280–8.

Schroder JM (2005). Ferritinopathy: diagnosis by muscle or nerve biopsy, with a note on other nuclear inclusion body diseases. *Acta Neuropathol (Berl)*, **109**, 109–14.

Schumacher G, Platz KP, Mueller AR *et al.* (2001). Liver transplantation in neurologic Wilson's disease. *Transplant Proc*, **33**, 1518–9.

Schuurman PR, Bosch DA, Bossuyt PM *et al.* (2000). A comparison of continuous thalamic stimulation and thalamotomy for suppression of severe tremor. *N Engl J Med*, **342**, 461–8.

Schwarz J, Tatsch K, Arnold G *et al.* (1993). 123I-iodobenzamide-SPECT in 83 patients with de novo parkinsonism. *Neurology*, **43**, S17–20.

Schwarz J, Weis S, Kraft E *et al.* (1996). Signal changes on MRI and increases in reactive microgliosis, astrogliosis, and iron in the putamen of two patients with multiple system atrophy. *J Neurol Neurosurg Psychiatry*, **60**, 98–101.

Seitz RJ, Blank B, Kiwit JC *et al.* (1995). Stiff-person syndrome with anti-glutamic acid decarboxylase autoantibodies: complete remission of symptoms after intrathecal baclofen administration. *J Neurol*, **242**, 618–22.

Seppi K, Yekhlef F, Diem A *et al.* (2005). Progression of parkinsonism in multiple system atrophy. *J Neurol*, **252**, 91–6.

Sethi KD, Patel B, Meador KJ (1989). Metoclopramide-induced parkinsonism. *South Med J*, **82**, 1581–2.

Seuchter SA, Hebebrand J, Klug B *et al.* (2000). Complex segregation analysis of families ascertained through Gilles de la Tourette syndrome. *Genet Epidemiol*, **18**, 33–47.

Seyrantepe V, Poupetova H, Froissart R *et al.* (2003). Molecular pathology of NEU1 gene in sialidosis. *Hum Mutat*, **22**, 343–52.

Shahbazian MD, Zoghbi HY (2001). Molecular genetics of Rett syndrome and clinical spectrum of MECP2 mutations. *Curr Opin Neurol*, **14**, 171–6.

Shahwan A, Farrell M, Delanty N (2005). Progressive myoclonic epilepsies: a review of genetic and therapeutic aspects. *Lancet Neurol*, **4**, 239–48.

Shan DE, Soong BW, Sun CM *et al.* (2001). Spinocerebellar ataxia type 2 presenting as familial levodopa-responsive parkinsonism. *Ann Neurol*, **50**, 812–5.

Sharma M, Mueller JC, Zimprich A *et al.* (2006). The sepiapterin reductase gene region reveals association in the PARK3 locus: analysis of familial and sporadic Parkinson's disease in European populations. *J Med Genet*, **43**, 557–62.

Shaw C, Haas L, Miller D *et al.* (1996). A case report of paroxysmal dystonic choreoathetosis due to hypoglycaemia induced by an insulinoma. *J Neurol Neurosurg Psychiatry*, **61**, 194–5.

Shiang R, Ryan SG, Zhu YZ *et al.* (1993). Mutations in the alpha 1 subunit of the inhibitory glycine receptor cause the dominant neurologic disorder, hyperekplexia. *Nat Genet*, **5**, 351–8.

Shibasaki H, Hallett M (2005). Electrophysiological studies of myoclonus. *Muscle Nerve*, **31**, 157–74.

Shoulson I (1981). Huntington's disease: Functional capacities in patients treated with neuroleptic and antidepressant drugs. *Neurology*, **31**, 1333–5.

Shulman LM, Singer C, Bean JA *et al.* (1996). Internal tremor in patients with Parkinson's disease. *Mov Disord*, **11**, 3–7.

Shults CW, Oakes D, Kieburtz K *et al.* (2002). Effects of coenzyme Q10 in early Parkinson disease: evidence of slowing of the functional decline. *Arch Neurol*, **59**, 1541–50.

Sibbing D, Asmus F, Konig IR *et al.* (2003). Candidate gene studies in focal dystonia. *Neurology*, **61**, 1097–101.

Siegfried J, Lippitz B (1994). Bilateral chronic electrostimulation of ventroposterolateral pallidum: a new therapeutic approach for alleviating all parkinsonian symptoms. *Neurosurgery*, **35**, 1126–9; discussion 1129–30.

Silber MH, Levine S (2000). Stridor and death in multiple system atrophy. *Mov Disord*, **15**, 699–704.

Silber MH, Ehrenberg BL, Allen RP *et al.* (2004). An algorithm for the management of restless legs syndrome. *Mayo Clin Proc*, **79**, 916–22.

Silbert PL, Gubbay SS, Khangure M (1993). Multifocal astrocytoma masquerading as possible progressive supranuclear palsy [letter]. *J Neurol Neurosurg Psychiatry*, **56**, 220–1.

Silbert PL, Matsumoto JY, McManis PG *et al.* (1995). Intrathecal baclofen therapy in stiff-man syndrome: a double-blind, placebo-controlled trial. *Neurology*, **45**, 1893–7.

Sindo I, Jorgensen JI (2002). [Treatment of tics in Tourette syndrome with atypical antipsychotic drugs]. *Ugeskr Laeger*, **164**, 3755–9.

Singer HS (2005). Tourette's syndrome: from behaviour to biology. *Lancet Neurol*, **4**, 149–59.

Singer HS, Hong JJ, Yoon DY *et al.* (2005). Serum autoantibodies do not differentiate PANDAS and Tourette syndrome from controls. *Neurology*, **65**, 1701–7.

Singer HS, Loiselle CR, Lee O *et al.* (2004). Anti-basal ganglia antibodies in PANDAS. *Mov Disord* 19, 406–15.

Singer HS, Reiss AL, Brown JE *et al.* (1993). Volumetric MRI changes in basal ganglia of children with Tourette's syndrome. *Neurology*, **43**, 950–6.

Singer HS, Szymanski S, Giuliano J *et al.* (2002). Elevated intrasynaptic dopamine release in Tourette's syndrome measured by PET. *Am J Psychiatry*, **159**, 1329–36.

Sinha S, Newsom-Davis J, Mills K *et al.* (1991). Autoimmune aetiology for acquired neuromyotonia (Isaacs' syndrome). *Lancet*, **338**, 75–7.

Sinha S, Satishchandra P, Santosh V *et al.* (2004). Neuronal ceroid lipofuscinosis: a clinicopathological study. *Seizure*, **13**, 235–40.

Sinha S, Taly AB, Ravishankar S *et al.* (2006). Wilson's disease: cranial MRI observations and clinical correlation. *Neuroradiology*, **48**, 613–21.

Skidmore F, Reich SG (2005). Tardive Dystonia. *Curr Treat Options Neurol*, **7**, 231–6.

Skipper L, Shen H, Chua E *et al.* (2005). Analysis of LRRK2 functional domains in nondominant Parkinson disease. *Neurology*, **65**, 1319–21.

Skipper L, Wilkes K, Toft M *et al.* (2004). Linkage disequilibrium and association of MAPT H1 in Parkinson disease. *Am J Hum Genet*, **75**, 669–77.

Slevin JT, Gerhardt GA, Smith CD *et al.* (2005). Improvement of bilateral motor functions in patients with Parkinson disease through the unilateral intraputaminal infusion of glial cell line-derived neurotrophic factor. *J Neurosurg*, **102**, 216–22.

Smyth P, Sinclair DB (2003). Multifocal myoclonus following group A streptococcal infection. *J Child Neurol*, **18**, 434–6.

Soares-Weiser K, Rathbone J (2006). Neuroleptic reduction and/or cessation and neuroleptics as specific treatments for tardive dyskinesia. *Cochrane Database Syst Rev*, CD000459.

Soares KV, McGrath JJ (2000). Anticholinergic medication for neuroleptic-induced tardive dyskinesia. *Cochrane Database Syst Rev*, CD000204.

Soares KV, McGrath JJ (2001a). Calcium channel blockers for neuroleptic-induced tardive dyskinesia. *Cochrane Database Syst Rev*, CD000206.

Soares KV, McGrath JJ (2001b). Vitamin E for neuroleptic-induced tardive dyskinesia. *Cochrane Database Syst Rev*, CD000209.

Soares K, Rathbone J, Deeks J (2004). Gamma-aminobutyric acid agonists for neuroleptic-induced tardive dyskinesia. *Cochrane Database Syst Rev*, CD000203.

Solano A, Roig M, Vives-Bauza C *et al.* (2003). Bilateral striatal necrosis associated with a novel mutation in the mitochondrial ND6 gene. *Ann Neurol*, **54**, 527–30.

Solimena M, Folli F, Denis-Donini S *et al.* (1988). Autoantibodies to glutamic acid decarboxylase in a patient with stiff man syndrome, epilepsy and type 1 diabetes mellitus. *N Engl J Med*, **318**, 1012–20.

Souza-Lima C, Ferraz H, Braz C *et al.* (2000). Marked improvement in a stiff-limb patient treated with intravenous immunoglobulin. *Mov Disord*, **15**, 358–9.

Spira PJ, Sharpe DM, Halliday G *et al.* (2001). Clinical and pathological features of a Parkinsonian syndrome in a family with an Ala53Thr alpha-synuclein mutation. *Ann Neurol*, **49**, 313–9.

Srivastava V, Varma PG, Prasad S *et al.* (2006). Genetic susceptibility to tardive dyskinesia among schizophrenia subjects: IV. Role of dopaminergic pathway gene polymorphisms. *Pharmacogenet Genomics*, **16**, 111–7.

Stacy M, Jankovic J (1991). Tardive dyskinesia. *Curr Opin Neurol Neurosurg*, **4**, 343–9.

Stacy M, Cardoso F, Jankovic J (1993). Tardive stereotypy and other movement disorders in tardive dyskinesias. *Neurology*, **43**, 937–41.

Stanford PM, Halliday GM, Brooks WS *et al.* (2000). Progressive supranuclear palsy pathology caused by a novel silent mutation in exon 10 of the tau gene: expansion of the disease phenotype caused by tau gene mutations. *Brain*, 123 (Pt 5), 880–93.

Steen VM, Lovlie R, MacEwan T *et al.* (1997). Dopamine D3-receptor gene variant and susceptibility to tardive dyskinesia in schizophrenic patients. *Mol Psychiatry*, **2**, 139–45.

Steinberger D, Blau N, Goriuonov D *et al.* (2004). Heterozygous mutation in 5'-untranslated region of sepiapterin reductase gene (SPR) in a patient with dopa-responsive dystonia. *Neurogenetics*, **5**, 187–90.

Steinberger D, Topka H, Fischer D *et al.* (1999). GCH1 mutation in a patient with adult-onset oromandibular dystonia. *Neurology*, **52**, 877–9.

Sternbach H (1991). The serotonin syndrome. *Am J Psychiatry*, **148**, 705–13.

Sternlieb I, Giblin DR, Scheinberg IH (1987). Wilson's disease. In Marsden CD, Fahn S, eds. *Movement Disorders 2*, pp. 288–302. Butterworths, London.

Stojanovic M, Cvetkovic D, Kostic VS (1995). A genetic study of idiopathic focal dystonias. *J Neurol*, **242**, 508–11.

Straussberg R, Shahar E, Gat R *et al.* (1993). Delayed parkinsonism associated with hypotension in a child undergoing open-heart surgery. *Dev Med Child Neurol*, **35**, 1011–4.

Subach BR, McLaughlin MR, Albright AL *et al.* (1998). Current management of pediatric atlantoaxial rotatory subluxation. *Spine*, **23**, 2174–9.

Susman VL (2001). Clinical management of neuroleptic malignant syndrome. *Psychiatr Q*, **72**, 325–36.

Suyama N, Kobayashi S, Isino H *et al.* (1997). Progressive supranuclear palsy with palatal myoclonus. *Acta Neuropathol (Berl)*, **94**, 290–3.

Svetel M, Sternic N, Pejovic S *et al.* (2001). Penicillamine-induced lethal status dystonicus in a patient with Wilson's disease. *Mov Disord*, **16**, 568–9.

Swaiman KF (1991). Hallervorden-Spatz syndrome and brain iron metabolism. *Arch Neurol*, **48**, 1285–93.

Swedo SE, Leonard HL, Garvey M *et al.* (1998). Pediatric autoimmune neuropsychiatric disorders associated with streptococcal infections: clinical description of the first 50 cases. *Am J Psychiatry*, **155**, 264–71.

Swerdlow RH, Juel VC, Wooten GF (2004). Dystonia with and without deafness is caused by TIMM8A mutation. *Adv Neurol*, **94**, 147–54.

Swerdlow RH, Parks JK, Miller SW *et al.* (1996). Origin and functional consequences of the complex I defect in Parkinson's disease. *Ann Neurol*, **40**, 663–71.

Swoboda KJ, Saul JP, McKenna CE et al. (2003). Aromatic L-amino acid decarboxylase deficiency: overview of clinical features and outcomes. Ann Neurol, 54 (Suppl 6), S49–55.

Swoboda KJ, Soong B, McKenna C et al. (2000). Paroxysmal kinesigenic dyskinesia and infantile convulsions: clinical and linkage studies. Neurology, 55, 224–30.

Taanman JW, Schapira AH (2005). Analysis of the trinucleotide CAG repeat from the DNA polymerase gamma gene (POLG) in patients with Parkinson's disease. Neurosci Lett, 376, 56–9.

Tabaton M, Rolleri M, Masturzo P et al. (1995). Apolipoprotein E epsilon 4 allele frequency is not increased in progressive supranuclear palsy. Neurology, 45, 1764–5.

Tamura H, Takahashi T, Ban N et al. (2006). Niemann-Pick type C disease: novel NPC1 mutations and characterization of the concomitant acid sphingomyelinase deficiency. Mol Genet Metab, 87, 113–21.

Tan EK, Jankovic J (2000). Tardive and idiopathic oromandibular dystonia: a clinical comparison. J Neurol Neurosurg Psychiatry, 68, 186–90.

Tan EK, Khajavi M, Thornby JI et al. (2000). Variability and validity of polymorphism association studies in Parkinson's disease. Neurology, 55, 533–8.

Tandberg E, Larsen JP, Aarsland D et al. (1996). The occurrence of depression in Parkinson's disease. A community-based study. Arch Neurol, 53, 175–9.

Tanner C (1994). Epidemiological clues to the cause of Parkinson's disease. In Marsden C, Fahn S, eds. Movement Disorders 3, pp. 124–46. Butterworth Heinemann, Oxford.

Tanner CM, Goldman SM (1996). Epidemiology of Parkinson's disease. Neurol Clin, 14, 317–35.

Tanzi RE, Petrukhin K, Chernov I et al. (1993). The Wilson disease gene is a copper transporting ATPase with homology to the Menkes disease gene. Nat Genet, 5, 344–50.

Tarsy D (2000). Tardive Dyskinesia. Curr Treat Options Neurol, 2, 205–14.

Tarsy D, Baldessarini RJ (2006). Epidemiology of tardive dyskinesia: is risk declining with modern antipsychotics? Mov Disord, 21, 589–98.

Tarsy D, Scollins L, Corapi K et al. (2005). Progression of Parkinson's disease following thalamic deep brain stimulation for tremor. Stereotact Funct Neurosurg, 83, 222–7.

Thajeb P (1996). The syndrome of delayed posthemiplegic hemidystonia, hemiatrophy, and partial seizure: clinical, neuroimaging, and motor-evoked potential studies. Clin Neurol Neurosurg, 98, 207–12.

Thomas M, Jankovic J (2004). Psychogenic movement disorders: diagnosis and management. CNS Drugs, 18, 437–52.

Thomas M, Hayflick SJ, Jankovic J (2004). Clinical heterogeneity of neurodegeneration with brain iron accumulation (Hallervorden-Spatz syndrome) and pantothenate kinase-associated neurodegeneration. Mov Disord 19, 36–42.

Thomas S, Critchley P, Lawden M et al. (2005). Stiff person syndrome with eye movement abnormality, myasthenia gravis, and thymoma. J Neurol Neurosurg Psychiatry, 76, 141–2.

Thomas GR, Forbes JR, Roberts EA et al. (1995). The Wilson disease gene: spectrum of mutations and their consequences [published erratum appears in Nat Genet 1995 Apr; 9(4), 451]. Nat Genet, 9, 210–7.

Thompson PD (1994). Stiff People. In Marsden CD, Fahn S, eds. Movement Disorders 3, pp. 373–405. Butterworth-Heinemann, Oxford.

Thompson PD, Marsden CD (1987). Gait disoder of subacute arteriosclerotic encephalopathy: Binswanger's disease. Mov Disord, 2, 1–8.

Thompson PD, Berardelli A, Rothwell JC et al. (1988). The coexistence of bradykinesia and chorea in Huntington's disease and its implications for theories of basal ganglia control of movement. Brain, 111, 223–44.

Thompson PD, Bhatia KP, Brown P et al. (1994). Cortical myoclonus in Huntington's disease. Mov Disord, 9, 633–41.

Thyagarajan D, Shanske S, Vazquez-Memije M et al. (1995). A novel mitochondrial ATPase 6 point mutation in familial bilateral striatal necrosis. Ann Neurol, 38, 468–72.

Tintner R, Jankovic J (2001). Focal dystonia: the role of botulinum toxin. Curr Neurol Neurosci Rep, 1, 337–45.

Tolosa E, Montserrat L, Bayes A (1988). Blink reflex studies in focal dystonias: enhanced excitability of brainstem interneurons in cranial dystonia and spasmodic torticollis. Mov Disord, 3, 61–9.

Tran PV, Dellva MA, Tollefson GD et al. (1997). Extrapyramidal symptoms and tolerability of olanzapine versus haloperidol in the acute treatment of schizophrenia. J Clin Psychiatry, 58, 205–11.

Trottenberg T, Volkmann J, Deuschl G et al. (2005). Treatment of severe tardive dystonia with pallidal deep brain stimulation. Neurology, 64, 344–6.

Trugman JM, Leadbetter R, Zalis ME et al. (1994). Treatment of severe axial tardive dystonia with clozapine: case report and hypothesis. Mov Disord, 9, 441–6.

Tsai N, Chen Y, Zhao XB et al. (1983). Acute infectious torticollis. Neurology, 33, 1344–6.

Tsuboi Y, Wszolek ZK, Kusuhara T et al. (2002). Japanese family with parkinsonism, depression, weight loss, and central hypoventilation. Neurology, 58, 1025–30.

Tuite PJ, Provias JP, Lang AE (1996). Atypical dopa responsive parkinsonism in a patient with megalencephaly, midbrain Lewy body disease, and some pathological features of Hallervorden-Spatz disease. J Neurol Neurosurg Psychiatry, 61, 523–7.

Turjanski N, Lees AJ, Brooks DJ (1999). Striatal dopaminergic function in restless legs syndrome: 18F-dopa and 11C-raclopride PET studies. Neurology, 52, 932–7.

Tyne HL, Parsons J, Sinnott A et al. (2004). A 10 year retrospective audit of long-term apomorphine use in Parkinson's disease. J Neurol, 251, 1370–4.

Uitti R, Wharen R, Turk M et al. (1997). Unilateral pallidotomy for Parkinson's disease: comparison of outcome in younger versus elderly patients. Neurology, 49, 1072–7.

Umbrich P, Soares KV (2003). Benzodiazepines for neuroleptic-induced tardive dyskinesia. Cochrane Database Syst Rev, CD000205.

Vaddadi KS, Soosai E, Chiu E et al. (2002). A randomised, placebo-controlled, double blind study of treatment of Huntington's disease with unsaturated fatty acids. Neuroreport, 13, 29–33.

Valente EM, Bentivoglio AR, Cassetta E et al. (2001). DYT13, a novel primary torsion dystonia locus, maps to chromosome 1p36.13–36.32 in an Italian family with cranial-cervical or upper limb onset. Ann Neurol, 49, 362–6.

Valente EM, Warner TT, Jarman PR et al. (1998). The role of DYT1 in primary torsion dystonia in Europe. Brain, 121, 2335–9.

Valentino P, Annesi G, Ciro Candiano IC et al. (2006). Genetic heterogeneity in patients with pantothenate kinase-associated neurodegeneration and classic magnetic resonance imaging eye-of-the-tiger pattern. Mov Disord, 21, 252–4.

Valldeoriola F, Valls-Sole J, Tolosa ES et al. (1995). Striated anal sphincter denervation in patients with progressive supranuclear palsy. Mov Disord, 10, 550–5.

Van Camp G, Flamez A, Cosyns B et al. (2004). Treatment of Parkinson's disease with pergolide and relation to restrictive valvular heart disease. Lancet, 363, 1179–83.

Van den Eynde F, Naudts KH, De Saedeleer S et al. (2005). Olanzapine in Gilles de la Tourette syndrome: beyond tics. Acta Neurol Belg, 105, 206–11.

Van der Kamp W, Rothwell JC, Thompson PD et al. (1995). The movement-related cortical potential is abnormal in patients with idiopathic torsion dystonia. Mov Disord, 10, 630–3.

Van Harten PN, Hovestadt A (2006). Botulinum toxin as a treatment for tardive dyskinesia. Mov Disord.

Van Harten PN, Kampuis DJ, Matroos GE (1996). Use of clozapine in tardive dystonia. Prog Neuropsychopharmacol Biol Psychiatry 20, 263–74.

Vanacore N (2005). Epidemiological evidence on multiple system atrophy. J Neural Transm, 112, 1605–12.

Vanacore N, Bonifati V, Fabbrini G *et al.* (2005). Case-control study of multiple system atrophy. *Mov Disord* 20, 158–63.

Vandel P, Bonin B, Leveque E *et al.* (1997). Tricyclic antidepressant-induced extrapyramidal side effects. *Eur Neuropsychopharmacol*, 7, 207–12.

Vargas AP, Carod-Artal FJ, Del Negro MC *et al.* (2000). [Psychogenic dystonia: report of 2 cases]. *Arq Neuropsiquiatr*, 58, 522–30.

Vasconcelos OM, Harter DH, Duffy C *et al.* (2003). Adult Hallervorden-Spatz syndrome simulating amyotrophic lateral sclerosis. *Muscle Nerve*, 28, 118–22.

Venkatesan A, Frucht S (2006). Movement disorders after resuscitation from cardiac arrest. *Neurol Clin*, 24, 123–32.

Verhagen Metman L, Morris MJ, Farmer C *et al.* (2002). Huntington's disease: a randomized, controlled trial using the NMDA-antagonist amantadine. *Neurology*, 59, 694–9.

Vernino S, Tuite P, Adler CH *et al.* (2002). Paraneoplastic chorea associated with CRMP-5 neuronal antibody and lung carcinoma. *Ann Neurol*, 51, 625–30.

Verny M, Duyckaerts C, Agid Y *et al.* (1996). The significance of cortical pathology in progressive supranuclear palsy. Clinico-pathological data in 10 cases. *Brain*, 119, 1123–36.

Vesper J, Klostermann F, Funk T *et al.* (2002). Deep brain stimulation of the globus pallidus internus (GPI) for torsion dystonia--a report of two cases. *Acta Neurochir Suppl*, 79, 83–8.

Vicari AM, Folli F, Pozza G *et al.* (1989). Plasmapheresis in the treatment of stiff-man syndrome [letter]. *N Engl J Med*, 320, 1499.

Vidailhet M, Tassin J, Durif F *et al.* (2001). A major locus for several phenotypes of myoclonus-dystonia on chromosome 7q. *Neurology*, 56, 1213–6.

Vidailhet M, Vercueil L, Houeto JL *et al.* (2005). Bilateral deep-brain stimulation of the globus pallidus in primary generalized dystonia. *N Engl J Med*, 352, 459–67.

Vidakovic A, Dragasevic N, Kostic VS (1994). Hemiballism: report of 25 cases. *J Neurol Neurosurg Psychiatry*, 57, 945–9.

Vincent A (2004). Encephalitis lethargica: part of a spectrum of post-streptococcal autoimmune diseases? *Brain*, 127, 2–3.

Vinton A, Fahey MC, O'Brien TJ *et al.* (2005). Dentatorubral-pallidoluysian atrophy in three generations, with clinical courses from nearly asymptomatic elderly to severe juvenile, in an Australian family of Macedonian descent. *Am J Med Genet A*, 136, 201–4.

Visser-Vandewalle V, Temel Y, van der Linden C *et al.* (2004). Deep brain stimulation in movement disorders. The applications reconsidered. *Acta Neurol Belg*, 104, 33–6.

Vitek JL (2002). Pathophysiology of dystonia: a neuronal model. *Mov Disord*, 17 (Suppl 3), S49–62.

Volkmann J, Allert N, Voges J *et al.* (2001). Safety and efficacy of pallidal or subthalamic nucleus stimulation in advanced PD. *Neurology*, 56, 548–51.

Volkmann J, Allert N, Voges J *et al.* (2004). Long-term results of bilateral pallidal stimulation in Parkinson's disease. *Ann Neurol*, 55, 871–5.

Volkmann J, Sturm V, Weiss P *et al.* (1998). Bilateral high-frequency stimulation of the internal globus pallidus in advanced Parkinson's disease. *Ann Neurol*, 44, 953–61.

Vonsattel J-P, Myers RH, Stevens TJ *et al.* (1985). Neuropathological classification of Huntington's disease. *J Neuropathol Exp Neurol*, 44, 559–77.

Voon V, Hassan K, Zurowski M *et al.* (2006). Prospective prevalence of pathologic gambling and medication association in Parkinson disease. *Neurology*, 66, 1750–2.

Vuillaume I, Meynieu P, Schraen-Maschke S *et al.* (2000). Absence of unidentified CAG repeat expansion in patients with Huntington's disease-like phenotype. *J Neurol Neurosurg Psychiatry*, 68, 672–5.

Waddy HM, Fletcher NA, Harding AE *et al.* (1991). A genetic study of idiopathic focal dystonias. *Ann Neurol*, 29, 320–4.

Walker KG, Lawrenson J, Wilmshurst JM (2005). Neuropsychiatric movement disorders following streptococcal infection. *Dev Med Child Neurol*, 47, 771–5.

Walker RH, Danisi FO, Swope DM *et al.* (2000). Intrathecal baclofen for dystonia: benefits and complications during six years of experience. *Mov Disord*, 15, 1242–7.

Walker RH, Jankovic J, O'Hearn E *et al.* (2003). Phenotypic features of Huntington's disease-like 2. *Mov Disord*, 18, 1527–30.

Wallace SJ (1998). Myoclonus and epilepsy in childhood: a review of treatment with valproate, ethosuximide, lamotrigine and zonisamide. *Epilepsy Res*, 29, 147–54.

Walsh R, O'Dwyer JP, Sheikh IH *et al.* (2007). Sporadic adult onset dystonia: sensory abnormalities as an endophenotype in unaffected relatives. *J Neurol Neurosurg Psychiatry*, 78, 980–3.

Walshe JM (1999). Penicillamine: the treatment of first choice for patients with Wilson's disease. *Mov Disord*, 14, 545–50.

Walshe JM, Yealland M (1992). Wilson's disease: the problem of delayed diagnosis. *J Neurol Neurosurg Psychiatry*, 55, 692–6.

Walshe JM, Yealland M (1993). Chelation treatment of neurological Wilson's disease. *Q J Med*, 86, 197–204.

Walters AS, Hening WA, Shah SK *et al.* (1993). Painless legs and moving toes: a syndrome related to painful legs and moving toes? *Mov Disord*, 8, 377–9.

Ward CD (1994). Does selegiline delay progression of Parkinson's disease? A critical re-evaluation of the DATATOP study. *J Neurol Neurosurg Psychiatry*, 57, 217–20.

Warich-Kirches M, Von Bossanyi P, Treuheit T *et al.* (1997). Stiff-man syndrome: possible autoimmune etiology targeted against GABA-ergic cells. *Clin Neuropathol*, 16, 214–9.

Warner TT, Fletcher NA, Davis MB *et al.* (1993). Linkage analysis in British and French families with idiopathic torsion dystonia. *Brain*, 116, 739–44.

Watanabe H, Saito Y, Terao S *et al.* (2002). Progression and prognosis in multiple system atrophy: an analysis of 230 Japanese patients. *Brain*, 125, 1070–83.

Watts RL, Mirra SS, Richardson EP (1997). Corticobasal ganglionic degeneration. In Marsden CD, Fahn S, eds. *Movement Disorders, 3*, pp. 282–299. Butterworth Heinemann, Oxford.

Watts RL, Jankovic J, Waters C *et al.* (2007). Randomized, blind, controlled trial of transdermal rotigotine in early Parkinson disease. *Neurology*, 68, 272–6.

Weetman J, Anderson IM, Gregory RP *et al.* (1997). Bilateral posteroventral pallidotomy for severe antipsychotic induced tardive dyskinesia and dystonia. *J Neurol Neurosurg Psychiatry*, 63, 554–6.

Wein T, Andermann F, Silver K *et al.* (1996). Exquisite sensitivity of paroxysmal kinesigenic choreoathetosis to carbamazepine. *Neurology*, 47, 1104–6.

Weindl A, Kuwert T, Leenders KL *et al.* (1993). Increased striatal glucose consumption in Sydenham's chorea. *Mov Disord*, 8, 437–44.

Weintraub D, Morales KH, Moberg PJ *et al.* (2005). Antidepressant studies in Parkinson's disease: a review and meta-analysis. *Mov Disord* 20, 1161–9.

Welter ML, Houeto JL, Tezenas du Montcel S *et al.* (2002). Clinical predictive factors of subthalamic stimulation in Parkinson's disease. *Brain*, 125, 575–83.

Wenning GK, Jellinger KA (2005). The role of alpha-synuclein in the pathogenesis of multiple system atrophy. *Acta Neuropathol (Berl)*, 109, 129–40.

Wenning GK, Ben-Shlomo Y, Magalhaes M *et al.* (1995). Clinicopathological study of 35 cases of multiple system atrophy. *J Neurol Neurosurg Psychiatry*, 58, 160–6.

Wenning GK, Ben Shlomo Y, Magalhaes M *et al.* (1994a). Clinical features and natural history of multiple system atrophy. An analysis of 100 cases. *Brain*, 117, 835–45.

Wenning GK, Colosimo C, Geser F *et al.* (2004a). Multiple system atrophy. *Lancet Neurol*, 3, 93–103.

Wenning GK, Jager R, Kendall B *et al.* (1994b). Is cranial computerized tomography useful in the diagnosis of multiple system atrophy? *Mov Disord*, 9, 333–6.

Wenning GK, Kraft E, Beck R et al. (1997). Cerebellar presentation of multiple system atrophy. Mov Disord, 12, 115–7.

Wenning GK, Litvan I, Jankovic J et al. (1998). Natural history and survival of 14 patients with corticobasal degeneration confirmed at postmortem examination. J Neurol, Neurosurg Psychiatry, 64, 184–9.

Wenning GK, Quinn NP, Magalhaes M et al. (1994c). Minimal change multiple system atrophy. Mov Disord, 9, 161–6.

Wenning GK, Tison F, Seppi K et al. (2004b). Development and validation of the Unified Multiple System Atrophy Rating Scale (UMSARS). Mov Disord 19, 1391–402.

Werhahn KJ, Brown P, Thompson PD et al. (1997). The clinical features and prognosis of chronic posthypoxic myoclonus. Mov Disord, 12, 216–20.

Wexler NS, Lorimer J, Porter J et al. (2004). Venezuelan kindreds reveal that genetic and environmental factors modulate Huntington's disease age of onset. Proc Natl Acad Sci USA, 101, 3498–503.

Whitely AM, Swash M, Urich H (1976). Progressive encephalomyelitis with rigidity: its relation to subacute myoclonic spinal interneuronitis and the stiff man syndrome. Brain, 99, 27–42.

Whone AL, Watts RL, Stoessl AJ et al. (2003). Slower progression of Parkinson's disease with ropinirole versus levodopa: The REAL-PET study. Ann Neurol, 54, 93–101.

Wichmann T, DeLong MR (1997). Physiology of the basal ganglia and pathophysiology of movement disorders of basal ganglia origin. In Watts RL, Koller WC, eds. Movement Disorders: Neurologic Principles and Practice, pp. 87–97. McGraw-Hill, New York.

Wichmann T, DeLong MR (2003). Functional neuroanatomy of the basal ganglia in Parkinson's disease. Adv Neurol, 91, 9–18.

Wichmann T, De Long MR (2004). Physiology of the basal ganglia and pathophysiology of movement disorders of basal ganglia origin. In Watts R, Koller W, eds. Movement Disorders, pp. 101–12. McGraw-Hill, New York.

Wijker M, Wszolek ZK, Wolters EC et al. (1996). Localization of the gene for rapidly progressive autosomal dominant parkinsonism and dementia with pallido-ponto-nigral degeneration to chromosome 17q21. Hum Mol Genet, 5, 151–4.

Wilhelmsen KC, Lynch T, Pavlou E et al. (1994). Localisation of disinhibition-dementia-parkinsonism-amyotrophy complex to 17q21-22. Am J Hum Genet, 55, 1159–65.

Williams DR, Lees AJ (2005). Visual hallucinations in the diagnosis of idiopathic Parkinson's disease: a retrospective autopsy study. Lancet Neurol, 4, 605–10.

Williams DR, de Silva R, Paviour DC et al. (2005a). Characteristics of two distinct clinical phenotypes in pathologically proven progressive supranuclear palsy: Richardson's syndrome and PSP-parkinsonism. Brain, 128, 1247–58.

Williams DR, Hadeed A, al-Din AS et al. (2005b). Kufor Rakeb disease: autosomal recessive, levodopa-responsive parkinsonism with pyramidal degeneration, supranuclear gaze palsy, and dementia. Mov Disord 20, 1264–71.

Winkelmann J, Ferini-Strambi L (2006). Genetics of restless legs syndrome. Sleep Med Rev, 10, 179–83.

Wissel J, Kabus C, Wenzel R et al. (1996). Botulinum toxin in writer's cramp: objective response evaluation in 31 patients. J Neurol Neurosurg Psychiatry, 61, 172–5.

Wszolek ZK, Pfeiffer RF, Bhatt MH et al. (1992). Rapidly progressive autosomal dominant parkinsonism and dementia with pallido-ponto-nigral degeneration. Ann Neurol, 32, 312–20.

Wszolek ZK, Pfeiffer B, Fulgham JR et al. (1995). Western Nebraska family (family D) with autosomal dominant parkinsonism. Neurology, 45, 502–5.

Xiang F, Almqvist EW, Huq M et al. (1998). A Huntington disease-like neurodegenerative disorder maps to chromosome 20p. Am J Hum Genet, 63, 1431–8.

Xuereb JH, MacMillan JC, Snell R et al. (1996). Neuropathological diagnosis and CAG repeat expansion in Huntington's disease. J Neurol Neurosurg Psychiatry, 60, 78–81.

Yamamoto M, Fujii S, Hatanaka Y (1997). Result of long-term administration of L-threo-3,4-dihydroxyphenylserine in patients with pure akinesia as an early symptom of progressive supranuclear palsy. Clin Neuropharmacol 20, 371–3.

Yianni J, Bain P, Giladi N et al. (2003). Globus pallidus internus deep brain stimulation for dystonic conditions: a prospective audit. Mov Disord, 18, 436–42.

Yoshimura I, Kaneko S, Yoshimura N et al. (2001). Long-term observations of two siblings with Lafora disease treated with zonisamide. Epilepsy Res, 46, 283–7.

Yoshita M, Braune S (2000). Cardiac uptake of [123I]MIBG separates PD from multiple system atrophy. Neurology, 54, 1877-a-1878.

Zaremba J, Mierzewska H, Lysiak Z et al. (2004). Rapid-onset dystonia-parkinsonism: a fourth family consistent with linkage to chromosome 19q13. Mov Disord 19, 1506–10.

Zeidler M, Stewart GE, Barraclough CR et al. (1997). New variant Creutzfeldt-Jakob disease: neurological features and diagnostic tests. Lancet, 350, 903–7.

Zhang ZX, Roman GC (1993). Worldwide occurrence of Parkinson's disease: an updated review. Neuroepidemiology, 12, 195–208.

Zhou B, Westaway SK, Levinson B et al. (2001). A novel pantothenate kinase gene (PANK2) is defective in Hallervorden-Spatz syndrome. Nat Genet, 28, 345–9.

Ziemann U, Paulus W, Rothenberger A (1997). Decreased motor inhibition in Tourette's disorder: evidence from transcranial magnetic stimulation. Am J Psychiatry, 154, 1277–84.

Zimprich A, Grabowski M, Asmus F et al. (2001). Mutations in the gene encoding epsilon-sarcoglycan cause myoclonus-dystonia syndrome. Nat Genet, 29, 66–9.

Zolkipli Z, Dahmoush H, Saunders DE et al. (2006). Pantothenate kinase 2 mutation with classic pantothenate-kinase-associated neurodegeneration without 'eye-of-the-tiger' sign on MRI in a pair of siblings. Pediatr Radiol, 36, 884–6.

Zweig RM, Hedreen JC, Jankel WR et al. (1988). Pathology in brainstem regions of individuals with primary dystonia. Neurology, 38, 702–6.

SECTION 9

Neurological infection

CHAPTER 41

Meningitis

Tom Solomon

Contents

41.1 Introduction

41.1.1 Introduction to neurological infectious diseases

Neurological infections pose different problems from much of neurological practice for several reasons. First we often know what causes the disease, second, we can often do something about it in terms of treatment. And third emergency treatment can be life saving, but conversely making mistakes can be disastrous. Neurological infectious diseases can be diagnostically challenging. Perhaps nothing illustrates this better than the fact that almost one-quarter of the case reports in the Lancet recently were patients with neurological infections (Solomon *et al.* 2005). Infections of the nervous system can be classified by anatomical location, infecting organism, or clinical syndrome. This and the following Chapters 42 and 43 on nervous system infections are introduced by presenting clinical syndromes, because this is the usual starting point for the clinician. Then the specific causative organisms are considered. Many organisms can cause more than one presenting neurological syndrome, and are thus mentioned in more than one section. However the main description is under the most important clinical presentation. Sufficient microbiological and epidemiological information is given to enable sensible discussions between neurologists and their colleagues, as well as with the patients and their families.

Microbiological pathogens cause:

◆ *meningitis*: inflammation of the meninges;

◆ *encephalitis*: inflammation of the brain parenchyma;

◆ *myelitis*: inflammation of the spinal cord;

◆ *radiculitis*: inflammation of the nerve roots;

◆ *space occupation* due to local supppuration or an abscess;

◆ or combinations of these.

As well as direct invasion of the nervous system by micro-organisms, some can cause indirect effects via toxins, such as tetanus, or by causing non-specific encephalopathy. Chapter 41 deals primarily with organisms causing meningitis; Chapter 42 looks at encephalitis and infectious encephalopathies including dementias; Chapter 43 describes infectious space occupying lesions, and ends with human immunodeficiency virus, HIV, which can present with the whole spectrum of neurological infectious syndromes.

41.1.2 Definition of meningitis

Meningitis is defined as inflammation of the brain meninges, characterized clinically by inflammatory cells in CSF. When there is concurrent parenchymal brain involvement the term *meningoencephalitis* is used, *meningoencephalomyelitis* implies that there is spinal cord involvement too.

Although increased cellularity in the CSF, or pleocytosis, is traditionally considered the hallmark of meningitis, some organisms, particularly fungi, can cause meningitis without a pleocytosis, especially in the immunocompromised. The advent of more sensitive methods of detecting viral nucleic acid in the CSF such as the polymerase chain reaction, have also shown that viral central nervous system infection can occur without an associated pleocytosis. When none of the common bacterial agents is easily identified the term *aseptic meningitis* is often used. The majority of such cases are caused by viruses (Table 41.1); non-viral causes of an aseptic meningitis picture include certain bacteria which are not readily cultured, and do not grow in standard culture media, such as *Borrelia burgdorferi*. The clinical presentations of meningitis can be broadly divided into the acute, recurrent, and chronic. The development of meningitis depends on the infecting organism, and also whether there is any particular host susceptibility.

41.1.3 Non-infectious causes of meningitis

Although this chapter focuses on infectious causes of meningitis, there is also a range of non-infectious conditions that can cause a meningitis-like presentation. Typically these present with a chronic aseptic meningitis picture, but they can also present acutely. Non-infectious causes include drugs, particularly non-steroidal anti-inflammatory drugs and antibiotics, and systemic illnesses such as systemic lupus erythematosis (Table 41.2). Occasionally patients will have a mild CSF pleocytosis after seizures, particularly prolonged. Migraines, and migraine-like syndromes can be associated with a CSF pleocytosis. These include the **HaNDL** syndrome, **H**eadache **a**ssociated with **N**eurological **D**eficit and CSF **L**ymphocytosis (Nakashima 2005).

41.1.4 Epidemiology of meningitis

Across the globe there are enormous variations in the epidemiology of meningitis, which depend on the prevalence of local organisms, and vaccination status of the population. In the developed world the epidemiology has changed in recent years with the introduction of vaccines, particularly those for measles, mumps, polio, *Neisseria meningitides*, and *Haemophilus influenzae* type B.

Whatever the location, bacteria and viruses still account for most cases of acute meningitis. In a study of a birth cohort of 12 000 children in Finland followed up to age 14, 174 central nervous system infections occurred in 167 children. The annual incidence of bacterial central nervous system infections was 36.3/100 000 and that of viral infections 688.0/100 000 (Rantakallio *et al.* 1986). In the United Kingdom, there have been few prospective studies assessing the epidemiology of meningitis as a whole. Data are available via the routine surveillance and reporting systems, though these are prone to under-reporting and bias. Nevertheless they do give some indication of the major pathogens and their relative importance (Davison *et al.* 2003). For example in one report of data from 2001, 48 per cent of 1216 clinically diagnosed meningitis cases in children in England were meningococcal meningitis, 12 per cent were other bacteria, and 19 per cent were viruses.

Causes and epidemiology of bacterial meningitis

The annual incidence of acute bacterial meningitis in adults in Western industrialized nations is estimated at 0.6–4 cases per 100 000 adults, but could be up to 10 times higher than this in less developed countries (Fitch *et al.* 2007). In adults, *Streptococcus pneumoniae*, the pneumococcus, followed by *Neisseria meningitides*, the meningococcus, are the most important causes being responsible for

Table 41.1 Causes of aseptic meningitis, by relative frequency and pathogen

	Viruses	Bacteria	Other
Common	Enteroviruses Herpes simplex virus type 2 Arboviruses* HIV (seroconversion illness)	Partially treated bacterial meningitis Parameningeal bacterial infections *Listeria monocytogenes*	Drugs (non-steroidal anti-inflammatory drugs, antibiotics, others)
Less common	Mumps Human herpes virus types 6 & 7 Lymphocytic choriomeningitis virus	Tuberculous meningitis *Mycoplasma pneumoniae*	Drugs (non-steroidal anti-inflammatory drugs, antibiotics, others) Fungi (e.g. *Cryptococcus* spp., *Candida* spp., *Aspergillus* spp.) Autoimmune disorders, vasculitis Sarcoid
Rare	Herpes simplex virus type 1 Varicella zoster virus Cytomegalovirus Epstein–Barr virus Influenza A and B viruses Parainfluenza virus Measles virus Rotavirus Coronavirus Encephalomyelitis virus Parvovirus B19 Lymphocytic choriomeningitis virus	*Borrelia burgdorferi* (Lyme disease) *Treponema pallidum* (syphillis) *Leptospira* spp *Brucella* spp *Rickettsia rickettsii* *Erlichia* spp. *Nocardia* spp.	*Toxoplasma gondii* Other fungi Malignancy Other parasites (e.g. *Angiostrongylus cantonensis*, *Naegleria fowleri*, *Acanthamoeba* spp.) Behcets disease

*Varies greatly depending on geographical location (Solomon *et al.* 2004)

Table 41.2 Non-infectious causes of meningitis. These typically present as a subacute or chronic aseptic meningitis picture

---

Drugs

Antibiotics

Trimethoprim, sulfamethoxazole, ciprofloxacin, penicillin, isoniazid, metronidazole, cephalosporins, pyrazinamide

Non-steroidal anti-inflammatory drugs

Ibuprofen, naproxen, diclofenac, ketoprofen, sulindac, tolmetin,

Immunomodulators

OKT3, azathioprine, immunoglobulin

Others

Ranitidine, Cytosine arabinoside, Phenazopyridine

Intracranial tumours

Craniopharyngioma

Dermoid / epidermoid cyst

Teratoma

Systemic illness

Systemic lupus erythematosis

Vogt–Koyanagi–Harada syndrome

Procedure-related

Post-neurosurgery

Intrathecal injections

Spinal anaethesia

Miscellaneous

Post-seizure

Migraine or migraine-like syndromes, including HaNDL (**H**eadache **a**ssociated with **N**eurological **D**eficit and CSF **L**ymphocytosis)

Mollaret's meningitis

---

80 per cent of all cases (Schuchat *et al.* 1997; van de Beek *et al.* 2004) (Table 41.3). Meningococcus occurs in young adults as well as children, but is rare in older adults. In children the major causes of

meningitis have historically been *Haemophilus influenzae* type B, or Hib, meningococcus, and pneumococcus. Aerobic gram negative bacteria such as *Escherichia coli*, *Klebsiella* species, *Pseudomonas* species, and *Salmonella* species occur in vulnerable patient groups, particularly the very young, the elderly, the immunocompromised, and those with head trauma or following neurosurgical procedures. These latter groups are also vulnerable to *Staphylococcus aureus* meningitis.

Penumococcal meningitis occurs with an incidence of 1.2 to 2.3 per 100 000 in European and American population-based studies (Weisfelt *et al.* 2006). It is more common in winter months, in males, and in the extremes of age. It is often associated with middle ear or sinus infection and is a particular problem in those who have reduced splenic function, other immune problems, and in alcoholics. Conjugate vaccines against *S. pneumoniae* may substantially reduce the burden of childhood pneumococcal meningitis, and even produce herd immunity in adults.

Haemophilis influenzae type B was the leading cause of bacterial meningitis in children until recently, with most cases occurring in those under 5 years. However, routine vaccination of children against *Haemophilis influenzae* type B, which began in 1992 in the United Kingdom, has virtually eradicated meningitis due to this bacterium in the developed world, though it remains an important cause in the developing world.

Meningococcal serogroups number 13 in total: A, B, and C count for more than 90 per cent of cases globally, but Y, W-135, and X are also important causes of outbreaks. Large outbreaks of meningitis A occur every few years across the meningitis belt, an area of sub-Saharan Africa which stretches from Senegal in the West to Ethiopia in the East. These occur in the dry months of April and May when the dry 'Harmattan' wind has ceased blowing. The largest outbreak was in 1996 with an estimated 250 000 cases and 25 000 deaths. Large outbreaks have also been seen in Brazil. In contrast the attack

Table 41.3 Acute bacterial meningitis. Pathogens and empirical treatment according to patient groups

Patient group	Bacterial pathogens	Recommended empirical antimicrobial therapy
Age		
< 1 month	Enterobacteriaceae (*Eshericia coli*, *Klebsiella pneumoniae*, *Enterobacter spp.*, *Proteus spp.*), *Streptococcus agalactiae* (and other group B streptococci), *Listeria monocytogenes*	3rd generation cephalosporin (e.g. Cefotaxime) plus ampicillin
1–23 months	*S. agalactiae*, *E. coli*, *Haemophillus influenzae*,* *Streptococcus pneumoniae*, *Neisseria meningitidis*	As above, plus vancomicin if penecillin-resistant pneumococcus is supected
2–50 years	*S. pneumoniae*, *N. meningitidis*	3rd generation cephalosporin (e.g.Cefotaxime) plus vancomicin if penicillin-resistant pneumococcus is suspected
>50 years	*S. pneumoniae*, *N. meningitidis*, *L. monocytogenes*, *Pseudomonas aeruginosa* (and other gram negative aerobic bacilli)	3rd generation cephalosporin (e.g.Cefotaxime) plus ampicillin, +/– vancomycin
Immunocompromised state	*S. pneumoniae*, *N. meningitidis*, *L. monocytogenes*, *P. aeruginosa* (and other gram negative aerobic bacilli)	Ceftazidime plus ampicillin +/– vancomycin
Basilar skull fracture	*S. pneumoniae*, *H. influenza*, Group A Beta-haemolytic streptococci	Ceftazidime plus vancomycin
Head trauma, post-neurosurgery	*Staphylococcus aureus*, *Staphylococcus epidermidis*, *P. aeruginas*, (other gram negative aerobic bacilli)	Ceftazidime plus vancomycin

*H. influenzae is now rarely seen in countries with *H. influenzae* type b vaccination

In some countries, because of increasing penicillin-resistance of *S. pneumoniae*, vancomycin (or rifampicin) is given to all patients with suspected pneumococcal meningitis, in addition to a 3rd generation cephalosporin, until the organism and its antibiotic sensitivities are known

Meropenem, is an alternative to cephalosporins

Gentamicin is often added to ampicillin if listeria is confirmed rates are much lower in the West, where serogroup A is not seen. For example in the United States the mean attack rate is between 1 and 2 per 100 000 per year; it is highest in children under 1 year at 17 per 100 000, dropping to 0.3 per 100 000 among adults (Jackson *et al.* 1993). Risk factors for disease in adults include myeloma, HIV, and congestive cardiac failure. Although in Europe and America, most cases are sporadic, outbreaks of meningococcal disease have been occurring with increasing frequency, particularly where young people are in close confinement such as colleges and universities, military camps, schools, and nurseries. There are sometimes also large outbreaks associated with the annual Haj pilgrimage to Mecca. Almost all secondary cases occur within 8 days of the index case. Nasopharyngeal carriage may persist for weeks to months. In England and Wales, cases of meningococcal meningitis and meningococcal septicaemia started to rise in 1995, because of an increase in serogroup C infections, particularly C2a strains (Davison *et al.* 2003). Although serogroup B accounts for the majority of cases in young children in England and Wales, serogroup C is associated with a higher case fatality rate, and is the predominant strain in older children. In November 1999, the United Kingdom became the first country to introduce the meningococcal serogroup C conjugate vaccine into the national immunization programme, as well as offering it to all children aged less than 18 years. The number of cases has fallen, whilst that due to serogroup B remains high. In temperate climates the peak incidence of meningococcal disease occurs in winter and spring. The approval of a conjugate vaccine against serogroups A, C, Y, and W135 of meningococcus may further reduce the incidence of meningococcal meningitis.

Listeria monocytogenes is an increasingly important cause of bacterial meningitis, being responsible for 8 per cent of cases in the USA, with 15–30 per cent mortality (Schuchat *et al.* 1997). Infants less than 1 month, adults above 60 years, alcoholics, and those immunosuppressed with corticosteroids or after transplants, are especially vulnerable, as are those with other chronic conditions such as diabetes.

Causes and epidemiology of viral meningitis

Before the introduction of the combined measles, mumps, and rubella, MMR vaccine in 1988, mumps was the commonest cause of viral meningitis in England and Wales. The incidence for all ages was an estimated 21 per 100 000 in 1987, but it had fallen to 1 per 100 000 by 1997. However, since 2003 there has been an increase in the number of notified and laboratory confirmed cases of mumps. This was mostly in people aged 15 and over, who never received MMR vaccine because they were told old, or only received a single dose. Like other respiratory viruses, mumps tends to peak in the winter and spring.

Enteroviruses are now thought to be the most common cause of viral meningitis (Maguire *et al.* 1999) (Table 41.1). In Western settings, group B coxsackieviruses and echoviruses account for more than 90 per cent of viral meningitis cases (Modlin, 1995a). There are more than 70 different types of enterovirus. They cause a wide range of clinical syndromes including enanthemas, exanthemas, respiratory, cardiac, and nervous system disease. The enteroviruses typically associated with aseptic meningitis include Coxsackie viruses, particularly type B, and the echoviruses which are named enteric cytopathic human orphan, ECHO, viruses because initially their relation to human disease was not known. Coxsackievirus B2 and B5, and echovirus serotypes 4, 6, 9, 11, 16, and 30 are the

most frequently implicated viruses. More recently characterized enterovirus serotypes are assigned type numbers, for instance enterovirus 68–71, because of limitations of the earlier classification system.

In the West aseptic meningitis caused by enteroviruses tends to peak in the summer months, for reasons that are not clear (Fig. 41.1). Large community wide outbreaks of aseptic meningitis caused by a single or multiple enterovirus serotypes can occur. Notable examples include an echovirus 30 outbreak in Japan affecting more than 4000 people, and an outbreak of mixed serotypes affecting nearly 5000 people in Romania (Yamashita *et al.* 1994; 2000). Of the newer enteroviruses, enterovirus 71 is particularly associated with neurological disease (McMinn 2002). Since the late 1990s, it has caused massive outbreaks of hand, foot, and mouth disease in children in the Asia Pacific region, with a significant number of cases of aseptic meningitis, encephalitis, and myelitis (Solomon *et al.* 2003). In one outbreak in Taiwan an estimated one and a half million people were affected (Ho *et al.* 1999). In the West enterovirus 71 has so far caused only small outbreaks or sporadic cases.

Enterovirus meningitis primarily affects children who are exposed to the virus for the first time. In one large study of central nervous system infections in a birth cohort of 12 000 Finnish children, the annual incidence of viral meningitis, most of which was attributed to enteroviruses, was 219 per 100 000 for children less than 1 year old, compared with 19 per 100 000 for children aged 1–4 (Rantakallio *et al.* 1986). By the time they reach adulthood, most individuals have already been exposed to the viruses, and thus are immune. However, occasional outbreaks of enterovirus central nervous system disease do occur in adults, caused by serotypes that have not been present in a community for several years.

Herpes viruses can also cause aseptic meningitis. Herpes simplex virus type 2 characteristically causes aseptic meningitis at the same time as primary genital herpes infection; in one study it occurred in 36 per cent of women and 14 per cent of men with a primary genital infection (Corey *et al.* 1983). Herpes simplex virus-2 meningitis may become recurrent (Section 41.3.1). Herpes simplex

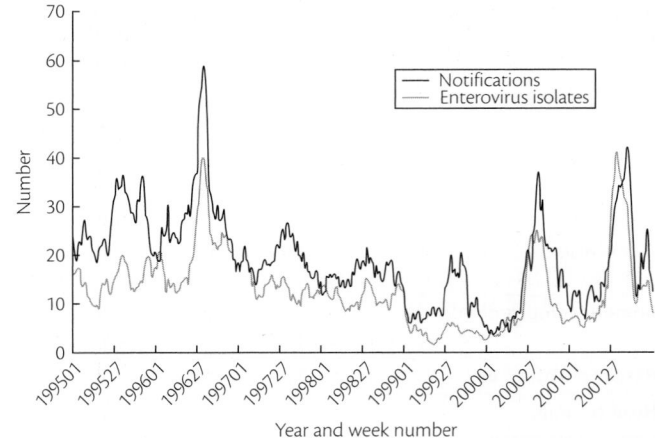

Fig. 41.1 Epidemiology of viral meningitis. Notifications of viral meningitis cases, and enterovirus isolates, shown as 5 week moving averages, to the Public Health Laboratory Services, England and Wales 1995–2001. Note the peak of meningitis notifications in the summer months, third quarter of the year, which is coincident with an increasing number of enterovirus isolates (from Davison and Ramsay 2003).

virus-1 can also cause aseptic meningitis, and overall herpes simplex virus accounts for 0.5–3 per cent of all aseptic meningitis cases (Corey *et al.* 1986). Cytomegalovirus and Epstein–Barr virus may also cause aseptic meningitis, especially in the immunocompromised. HIV itself is also an important cause of an aseptic meningitis syndrome, occurring in 5–10 per cent of patients during their primary HIV infection, as part of a 'seroconversion' syndrome (Hollander *et al.* 1987), but also occurring in those already infected (Section 43.3.2).

In one recent study of aseptic meningitis in adults, 95 (66 per cent) had a viral aetiological (Kupila *et al.* 2005), 38 (26 per cent) had enteroviruses, 24 (17 per cent) had herpes simplex virus-2, 12 (8 per cent) had varicella zoster virus, and 2 had herpes simplex virus-1.

41.1.5 Pathogenesis of meningitis

For both bacterial and viral meningitis the important steps in the pathogenesis are the means by which the host is initially infected, and in particular whether nasal carriage is an issue, the mode of entry into the central nervous system, the role of microbial versus host factors, and the development of subsequent complications.

Bacterial meningitis

For the common pathogens, meningoccocus, *H. influenzae*, and pneumococcus, the initial step in pathogenesis is a struggle between the microbes to colonize the nasopharyngeal mucosa, and the host's attempt to stop it. Some pathogens possess surface characteristics to assist colonization, such as the fimbriae of *N. meningitidis* and *H. influenzae*. The bacterial capsule of *H. influenzae* may also possess virulence factors, and this is certainly an important attribute of the polysaccharide capsule of *S. pneumoniae*. IgA antibodies are produced to limit mucosal adhesion, though in some circumstances they may facilitate it. Once attached to nasopharyngeal epithelial cells, meningococci have been shown to be transported across the cells in phagocytic vacuoles (Stephens *et al.* 1991). Bacteraemia follows, and for many of the major bacterial pathogens encapsulation is again thought to be the key to avoiding the host defence, by inhibiting neutrophil phagocytosis and the classic complement-mediated bactericidal activity. The importance of the alternative complement pathway for pneumococcus is demonstrated by the increased risk to pneumococcal disease in patients with impairment of the alternative pathway, for instance patients with sickle cell anaemia or a splenectomy.

The site of central nervous system invasion by meningeal bacterial pathogens is often unclear. Some have postulated that the dural venous sinus system may be important, whereas others implicate the cribriform plate or choroid plexus, both of which have high blood flow rates. For both meningococcus and *E. coli*, receptors are present on the cerebral capillaries and choroid plexus; critical binding proteins expressed on the bacterial surface have been shown for both these pathogens. In experimental models 'transcytosis' across microvascular cells has been shown for listeria, whereas for *Streptococcus suis*, a pathogen of pigs and humans, entry into the central nervous system may occur via migration of infected monocytes.

Within the subarachnoid space, bacterial survival is facilitated by the fact that CSF levels of complement components and specific antibody, which are essential for opsonization of encapsulated meningeal bacteria, are low or absent. The mechanism triggering the neutrophilic pleocytosis which develops following bacterial infection of the subarachnoid space is not completely understood. Complement components, and chemokines such as macrophage inflammatory proteins and interleukin-8, CXCL8, are likely to be important attractants. Molecules expressed on the surface of the vascular endothelial cells, including selectins and intracellular adhesion molecules, or ICAMs, and leukocyte receptors such as integrins, are thought to be critical to the rolling, activation, adhesion, and migration of leukocytes across the meninges, as they are in inflammation elsewhere. An interesting role for macrophages was demonstrated in one experimental model of pneumococcal meningitis by the injection of mannosylated clodronate liposomes, which depletes meningeal and perivascular macrophages, and resulted in increased illness (Polfliet *et al.* 2001). A range of mediators are triggered in response to bacterial cell wall components, and may contribute to inflammation in the subarachnoid space; these include interleukins, and chemokines, prostaglandins, cellular adhesion molecules, and reactive nitrogen intermediates; toll-like receptors, or TLRs, especially TLR-2, play a crucial role in the detection of microbial infection and activation of the inflammatory response. The increased permeability of the blood-brain barrier that follows is partly due to separation of the intercellular tight junctions. Matrix metalloproteinases are enzymes that degrade the extracellular matrix, and are thought to be involved in changes in blood-brain barrier permeability.

Pathogenesis of viral meningitis

Most viruses enter the body via the respiratory or gastrointestinal tract, notable exceptions being herpes simplex virus-2, and HIV, for which the genital mucosa is a common route of entry. There is typically viral replication at the site of entry and local lymph nodes, before the development of viraemia. Replication and then dissemination from the vascular tissues of liver, spleen and muscle, often produces a secondary, larger viraemia. Initial host defences against the infection may include IgA production at mucosal surfaces, and innate cytokine responses to viraemia. Many viruses replicate within human lymphocytes, which may thus protect them from some of these processes.

Viruses subsequently spread across the blood-brain barrier by replicating in the capillary endothelial cells of the cerebral vessels or choroid plexus. In some circumstances they may be carried across the barrier by infected leukocytes, the so-called 'Trojan horse' mechanism. Alternative postulated routes include spread via the olfactory nerves, or centripetal spread from peripheral nerves, a mechanism which may be important for poliovirus. Within the central nervous system viral replication results in an inflammatory cytokine response, which may include IL-6, IFNγ, and IL-1β (Lin *et al.* 2003; Winter *et al.* 2004). Immunoglobulins enter the central nervous system across the inflamed blood-brain barrier, but are also produced locally by B cells that differentiate into plasma cells. Persistent detection of antibody in the CSF, for example in mumps meningitis, suggests that viral persistence may be occurring. T- lymphocyte responses are important in viral clearance.

41.2 Clinical approach to acute meningitis

Acute central nervous system infections, including acute meningitis, are among the few acute neurological emergencies where the initial management can make a very great difference to the outcome. The key issues are to think of the diagnosis in the first place, to investigate appropriately, and to initiate treatment in a timely manner.

Unfortunately during the last two decades there has been a degree of controversy and confusion about what constitutes the best approach to investigation and treatment. However, data are becoming available which provide an evidence base to support the optimum approaches.

41.2.1 Clinical features

In immunocompetent adults and older children, acute meningitis classically presents with fever, headache, and signs of meningeal irritation, with or without altered consciousness. Whereas for bacterial infection altered consciousness is taken as a marker of the severity of the meningeal disease, for viral infection altered consciousness indicates that the brain parenchyma has been infected, and so the term meningoencephalitis is used. However, by no means all patients have the classic triad of fever, neck stiffness, and altered consciousness. One large prospective study of community acquired bacterial meningitis in adults, has shown that only 44 per cent of patients have these features (van de Beek *et al.* 2004). In one study the presence of all three features was more common in adults with pneumococcal meningitis than meningococcal meningitis, 58 per cent versus 27 per cent (van de Beek *et al.* 2006). The absence of all three of the classical features makes bacterial meningitis very unlikely in immunocompetent adults. However most patients have at least two of headache, fever, neck stiffness, and altered mental status (van de Beek *et al.* 2004).

Other signs of meningeal irritation, or meningism, include a positive Kernig and Brudzinski sign. The *Kernig sign* is positive if, when the patient is supine with the hip and knee flexed at 90°, forced extension of the knee causes pain in the lower back and/or posterior thigh. The *Brudzinski sign* is positive if forced flexion of the neck causes knee and hip flexion. These signs have low sensitivity but high specificity for meningitis (Attia *et al.* 1999). The sensitivity is particularly poor for patients with only a mild CSF pleocytosis, for example the severely immunocompromised, and those with fungal or viral meningitis (Thomas *et al.* 2002).

A range of features in the history may suggest a particular aetiology (Table 41.4). These include recent travel to the meningitis belt of Africa, or the Hajj in Saudi Arabia (2006), which are both risks for meningococcal disease, unusual occupational or recreational exposure, for example to leptospirosis from rat urine during freshwater activities, and details of ongoing disease epidemics, such as an outbreak of mumps or hand, foot, and mouth disease. Risk factors for HIV should be assessed, such as intravenous drug abuse, male homosexual sex, multiple heterosexual or homosexual partners, or sex with prostitutes; meningitis can be part of an HIV seroconversion illness. If meningitis is associated with genital ulcers, herpes simplex virus-2 may be the cause, especially if it is recurrent meningitis. It is also important to consider whether the patient is at risk of infection because of immunocompromise, recent head injury, or neurosurgery, as discussed below. The length of the history itself is a useful pointer towards the likely aetiology. Most of the common viruses and bacteria cause acute presentation; in contrast tuberculosis, fungi, parasites, and non-infectious causes such as carcinoma, are more likely to present chronically.

A thorough general medical examination is essential in patients with suspected meningitis. Look carefully for mucocutaneous stigmata that may point towards an aetiology. There may be a purpuric meningococcal rash, a measles rash, or lesions of hand, foot, and mouth disease (Fig. 41.2). Lymphadenopathy and tonsillitis are

Table 41.4 Questions to consider in approaching a patient with a suspected central nervous system infection

History

Travel, especially to Asia, Africa, South America

Risk factors for HIV—multiple sexual partners, male sex with males, sex workers; intravenous drug abuse; blood products;

Other immunocompromise—immunosuppressive drugs, chemotherapy, transplant recipients, malignancy

Exposure to insects—mosquitoes, (malaria, flaviviruses) ticks (flaviviruses, borrelia)

Exposure to sick animals—dogs, cats (rabies)

Ingestion of contaminated food—snails (Angiostrongylus), unpasteurized milk/dairy products (brucella, listeria), fresh produce/water (cysticercosis, typhoid);

Exposure to contaminated water (leptospirosis, schistosomiasis)

Illness in the community (measles, mumps)

Generalized illness (infectious mononucleosis)

Time of year

Examination

Lymphadenopathy, (HIV, tuberculosis) hepatosplenomegaly

Rash (meningococcal, viral)

Oral examination (hairy leukoplaqia, oral candida)

Intravenous drug injection sites

Other mucocutaneous disease (hand, foot, and mouth disease)

Physical compromise to the blood brain barrier (CSF leak, head injury, mastoiditis)

Ocular examination for papilloedema, retinitis (cytomegalovirus), uveitis, iritis

Sepsis (otitis, sinusitis, pneumonia)

consistent with infectious mononucleosis?. Parotid swelling and orchitis suggest mumps meningitis. Examine the ante-cubital fossas, and groins for possible intravenous drug injection sites, which may suggest an unusual organism. Look inside the mouth for oral candidiasis or hairy leukoplakia, found in immunocompromised HIV patients. Is there lymphadenopathy, seen in HIV or tuberculosis. If there is a history of head trauma look carefully for a CSF leak from the nose or ears suggestive of a base of skull fracture. Also examine the ears for infection, which is a risk factor for pneumococcal meningitis. Pneumonia is similarly a risk factor.

It is necessary to document the coma score, look for flexor or extensor responses to pain, and examine for papilloedema and focal neurological signs, particularly those suggestive of brain herniation (Fig. 41.3) (Kirkham 2001; Solomon 2002). These are important in determining the next step in the management, particularly what investigations to do next. Other focal signs may suggest a particular aetiology. Multiple cranial neuropathies are characteristic of some basal meningeal processes including tuberculous meningitis, carcinomatous meningitis, enterovirus meningitis, and sarcoid. Fundoscopic examination may reveal changes of tuberculoma, or cytomegalovirus retinitis.

41.2.2 Initial investigations and management

Examination of the CSF is the cardinal investigation for patients with any central nervous system infection, including meningitis. Introduced by Quincke in 1891 to diagnose and treat tuberculous meningitis, lumbar puncture, has since been used extensively in the investigation of many neurological conditions (Gorelick 2000).

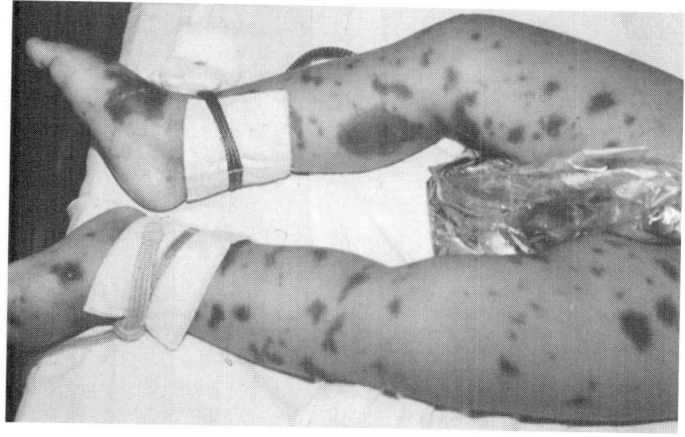

A

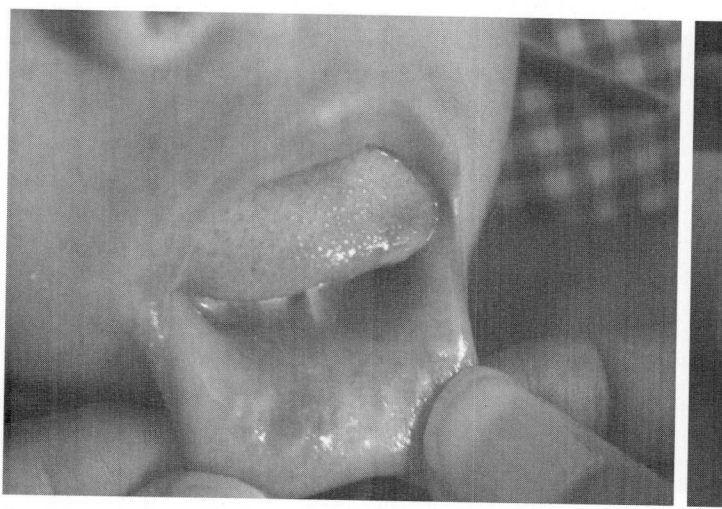

B

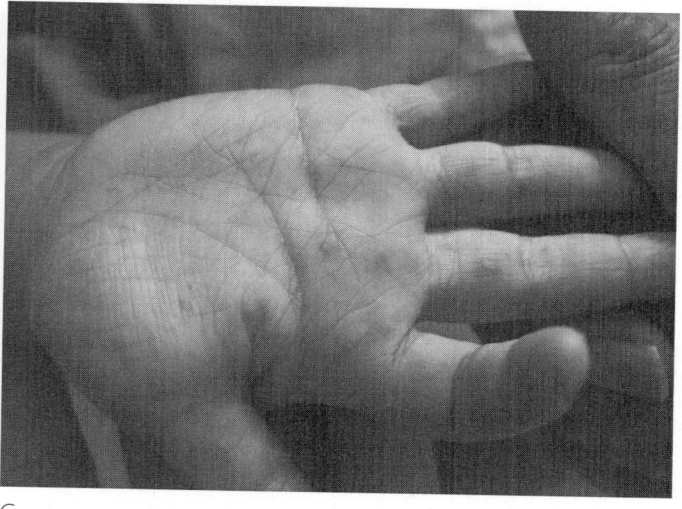

C

Fig. 41.2 Exanthema and enanthemas in meningitis. Meningococcal disease purpuric rash (a). Enterovirus 71, hand, foot, and mouth disease with ulcers on the tip of the tongue and lower lip (b) and palm vesicles (c). (Photographs by T Solomon.) (See Plate 35.)

CSF analysis and culture is the definitive method of diagnosing central nervous system infections, and until the late 1970s almost all patients admitted with suspected central nervous system infection in Western countries underwent a lumbar puncture (American College of Physicians 1986). In many developing countries this remains the practice. However, in the West the use of the lumbar puncture declined during the 1980s and 1990s, primarily because of increasing anxiety that the procedure may precipitate cerebral herniation, for example in the case of shift of the brain within the skull compartments, causing parenchymal damage.

Raised intracranial pressure and cerebral herniation are well-recognized complications of severe bacterial meningitis, whether or not patients receive a lumbar puncture (Dodge *et al.* 1965). High CSF opening pressure is not itself a cause of herniation, as demonstrated by the very high pressures in idiopathic intracranial hypertension (Section 26.5.5), which is treated by lumbar puncture. Herniation occurs when there are pressure differences between different brain compartments (Plum *et al.* 1982). The uncus of the temporal lobe may herniate through the tentorial opening, or the cerebellar tonsils may herniate through the foramen magnum, a process known as coning (Fig. 41.3). These processes damage the brainstem either directly, or indirectly to cause its blood supply resulting in ischaemia and haemorrhage. Although the evidence for a lumbar puncture causing herniation is only circumstantial (Horwitz *et al.* 1978; Slack 1982; Rennick *et al.* 1993), recommendations were produced in the 1980s as to which patients should not undergo an immediate lumbar puncture (Klein *et al.* 1986; Addy 1987; Mellor 1992). In patients with clinical signs that could indicate incipient brain shift or herniation it is recommended that lumbar puncture be delayed until a CT scan is performed. Meanwhile, in patients with obvious meningococcal disease, as indicated by a meningococcal rash, immediate treatment with antibiotics is recommended, because of the rapid deteriorations in patients with meningococcal shock (Wylie *et al.* 1997). However, rather than

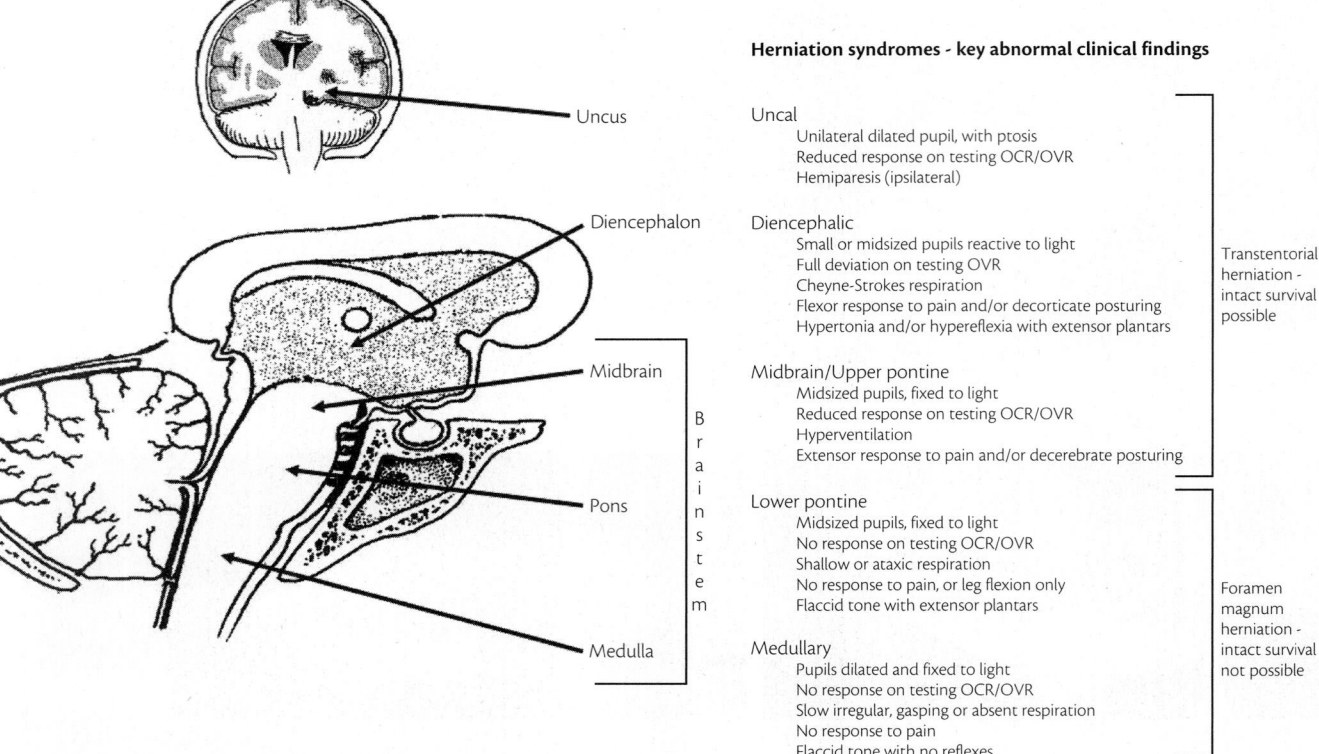

Herniation syndromes - key abnormal clinical findings

Uncal
  Unilateral dilated pupil, with ptosis
  Reduced response on testing OCR/OVR
  Hemiparesis (ipsilateral)

Diencephalic
  Small or midsized pupils reactive to light
  Full deviation on testing OVR
  Cheyne-Strokes respiration
  Flexor response to pain and/or decorticate posturing
  Hypertonia and/or hyperreflexia with extensor plantars

Midbrain/Upper pontine
  Midsized pupils, fixed to light
  Reduced response on testing OCR/OVR
  Hyperventilation
  Extensor response to pain and/or decerebrate posturing

Lower pontine
  Midsized pupils, fixed to light
  No response on testing OCR/OVR
  Shallow or ataxic respiration
  No response to pain, or leg flexion only
  Flaccid tone with extensor plantars

Medullary
  Pupils dilated and fixed to light
  No response on testing OCR/OVR
  Slow irregular, gasping or absent respiration
  No response to pain
  Flaccid tone with no reflexes

Transtentorial herniation - intact survival possible

Foramen magnum herniation - intact survival not possible

Fig. 41.3 Brain herniation syndromes Figure illustrating mid-sagittal section of the brain showing anatomy the and key abnormal clinical findings of midline herniation syndromes. Coronal section (top) showing herniation of the uncus of the temporal lobe; this compresses and causes a palsy of cranial nerve III and compression the contralateral peduncle to cause an ipsilateral hemiparesis. OCR = oculocephalic reflex; OVR = oculovestibular reflex (from Solomon and Keen 2004).

following these recommendations, in the 1990s practice seemed to drift towards managing all patients as if they had clinical signs of meningococcal septicaemia, and giving them blind antibiotics without investigations (Kneen *et al.* 2002). Most neurologists will be aware of the difficulty of subsequently trying to establish the aetiology in patients thus managed, who do not recover quickly.

Although there may be overlap in the clinical features, it is usually possible to distinguish patients with possible meningococcal disease, from those with other central nervous system infections (Fig. 41.4). Meningococcal disease occurs in children and young adults. Thus if someone in this age group appears to have a purpuric

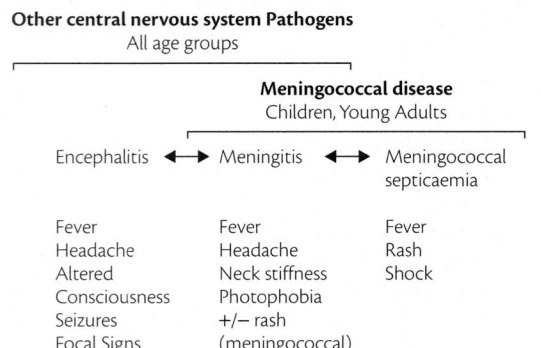

Fig. 41.4 The overlap of disease syndromes caused by meningococcus and other central nervous system pathogens.

meningococcal rash, with shock and features of septicaemia, they should be treated with intravenous cephalosporins. Management guidelines have been produced for both adults and children (www.britishinfectionsociety.org) (Begg *et al.* 1999). A venflon should be inserted to provide venous access, blood cultures can be drawn from this site, and then cefotaxine or ceftriaxone given. To maximize the chance of a microbiological diagnosis polymerase chain reaction of the skin lesions can be attempted, in addition to blood cultures. In such patients it is arguable that the examination of CSF adds little, or whether a lumbar puncture needs to be performed. However, even an apparent meningococcal rash is not always due to meningococcus, other gram negative organisms can cause a similar purpuric rash and the microbiological investigations are important. Because of the risk of rapid deterioration in patients with meningococcal septicaemia, general practitioners have been encouraged to give antibiotics immediately, whilst the patient awaits transfer to hospital. However it is unclear whether this is actually beneficial (Harnden *et al.* 2006).

Although in patients with suspected meningitis, a purpuric rash, and shock, immediate antibiotic treatment is indicated, this is not true for every patient with suspected meningitis. In most patients the emphasis should be on prompt investigation and diagnosis as a guide to appropriate treatment. Thus in patients with no contraindication a lumbar puncture should be performed without delay.

Imaging should precede lumbar puncture in adults who have moderate to severe impairment of consciousness, new onset seizures, or signs suggestive of space occupying lesions: papilloedema, focal

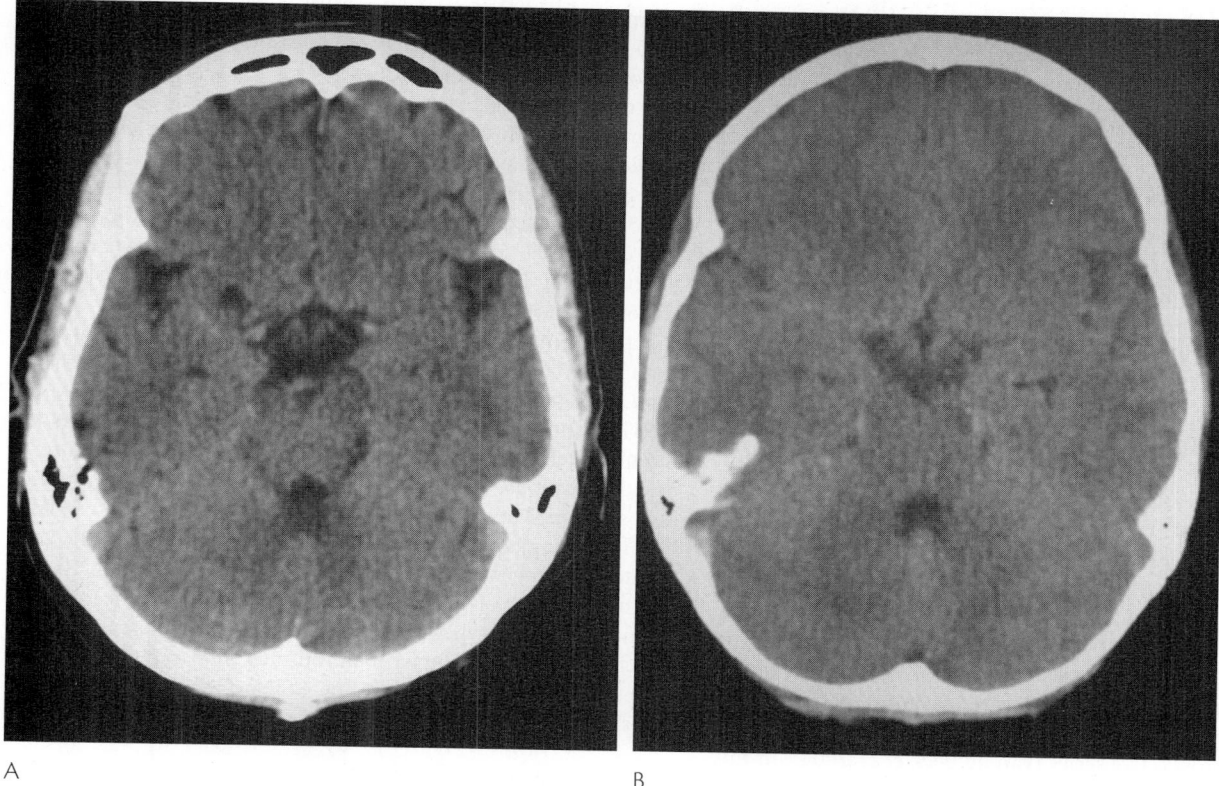

A   B

Fig. 41.5 Brain swelling on CT scan. (a) Normal Scan showing the sulcal pattern and space around the basal cisterns. (b) Swollen brain with loss of the sulci and reduced space around the basal cisterns.

neurological signs other than cranial neuropathies (Hasbun *et al.* 2001; Thomas *et al.* 2002; van Crevel *et al.* 2002; van de Beek *et al.* 2006). In addition to demonstrating a space occupying lesion, imaging may show diffuse brain swelling, with a lack of space or 'tightness' around the basal cisterns (Fig. 41.5). Other contraindications to a lumbar puncture include a bleeding disorder, warfarin treatment, or sepsis over the spine (Table 41.5).

Table 41.5 Contraindications to immediate lumbar puncture

Imaging is needed before lumbar puncture to exclude brain shift, swelling, or space occupying lesion when there are:
Focal neurological signs, other than cranial neuropathies
Papilloedema
Recent onset seizures
Moderate to severe impairment of consciousness
Hypertension with bradycardia
Immunocompromise (some patients)
Other contraindications
Bleeding disorder
Anticoagulant treatment
Sepsis over the spine

◆ There is no agreement on the depth of coma that necessitates imaging before lumbar puncture; some argue Glasgow coma score <12, others Glasgow coma score <10.
◆ Patients on warfarin should be treated with heparin instead, and the drug stopped shortly before lumbar puncture
◆ Imaging is preferable in patients with known severe immunocompromise (e.g. advanced AIDS)

Long delays in initiating antimicrobial therapy in acute bacterial meningitis, for example whilst waiting for imaging, are associated with a worse outcome. One retrospective study showed a delay of more than 6 hours between arrival in the emergency department and initiation of antimicrobial therapy was associated with death (Proulx *et al.* 2005). Thus in patients with suspected meningitis for whom imaging is indicated, antibiotic therapy should be started presumptively whilst the imaging and subsequent lumbar puncture are performed.

For patients that are having an immediate lumbar puncture, antibiotics should not be started until after the lumbar puncture. Recent evidence shows conclusively that prior treatment with antibiotics for just a few hours greatly reduces the chances of identifying the infecting organism (Kanegaye *et al.* 2001). Once the lumbar puncture has been performed, if the CSF is obviously cloudy, suggesting that bacterial meningitis is likely, or if the patient is deteriorating clinically, then antibiotics can be started (Section 41.2.6). However for stable patients with apparently clear CSF, decisions about starting therapy should await the initial CSF results, so that antibiotics are not given unnecessarily. In most hospitals the CSF cell count, biochemistry, and Gram stain result should be available within 1–2 h. The doctor performing the lumbar puncture must take responsibility for informing the laboratory that the sample is coming, finding out the result, and acting on it. Unfortunately the way health services are now organized, with shift systems and patients transferring between different teams, this sense of responsibility is not always apparent. In paediatrics the practice is slightly different, because unlike in adults, a normal CSF cell count cannot reliable rule out bacterial disease especially in young children; so there is a greater tendency to start antibiotics irrespective of the CSF, and only to stop them once CSF cultures have proven negative.

41.2.3 Cerebrospinal fluid findings

The CSF opening pressure and colour should be recorded, and samples taken for a cell count, protein, glucose, and microbiological investigations. In addition a plasma glucose should be taken immediately before the lumbar puncture, so that the CSF: plasma glucose ratio can be determined. Gram's staining of the CSF permits rapid identification of the causative organism in acute bacterial meningitis, with a sensitivity of 60–90 per cent, and specificity of >96 per cent. Bacterial antigen detection tests, although less sensitive, may be useful in patients for whom Gram's stain and culture are negative. Polymerase chain reaction has high sensitivity and specificity, and may be especially useful in patients with negative CSF cultures who have been pre-treated with antibiotics, though further refinements are needed before they can routinely be recommended (van de Beek *et al.* 2006). Investigations for viruses include viral culture, nucleic acid detection with polymerase chain reaction, and testing for antibodies against specific viruses.

The CSF opening pressure is elevated in most patients with bacterial meningitis, and was greater than 40 cm in 40 per cent of patients in one prospective series (van de Beek *et al.* 2004). In this series pressures >40 cm were associated with lower levels of consciousness, but not with adverse outcome. CSF pleocytosis of 100 to 10 000 white cells per mm³, elevated protein levels >0.5 g/l, and decreased CSF glucose ratios are usually present. Typically there is a predominance of polymorphonuclear cells in the CSF of 80–95 per cent, but lymphocyte predominance can occur, particularly if there has been pre-treatment with antibiotics (Table 41.6).

Normal or only marginally elevated CSF white cell counts occur in 5–10 per cent of patients, and are associated with an adverse outcome. (van de Beek *et al.* 2004).

In distinguishing bacterial from viral disease a range of parameters may be helpful One study identified the following as individual predictors of bacterial disease with >99 per cent certainty: a low CSF glucose of <1.9 mmol/l, a low CSF to plasma glucose ratio of <0.23, an elevated CSF protein 2.2 g/l, a CSF total leukocyte count of < 2000 × 10⁶/l, or a CSF polymorphonuclear leukocyte count of >1180 × 10⁶/L (Spanos *et al.* 1989). A model based on CSF–blood glucose ratio, total polymorphonuclear leukocyte count in CSF, age, and month of onset proved useful at distinguishing acute viral from bacterial disease in an independent test sample, and has since been validated in other retrospective studies (McKinney *et al.* 1994). A similar model based on blood and CSF parameters has recently been devised, and rigorously assessed and is found to be useful (Chavanet *et al.* 2007). Other markers that have been studied as possible indicators of bacterial as opposed to viral disease include CSF immunoglobulins, serum C-reactive protein, CSF lactic acid, lactate dehydrogenase, CSF interferon-gamma, interleukin 1β, and CSF tumour necrosis factor. Of all of these, CSF tumour necrosis factor may be the most specific for bacterial versus viral disease, but it is not clear that it is any better than routine CSF parameters. If more than 10 per cent of the cells in the CSF are eosinophils, then this raises the possibility of eosinophilic meningitis, caused by parasites.

41.2.4 Other investigations

A full blood count may reveal a neutrophilia which tends to be mild in viral disease, and more marked in bacterial disease. Lymphocytosis may occur with some viruses and parasites. Deranged renal and liver function could indicate leptospirosis. However, for the most part the initial blood investigations are not diagnostic. Blood cultures should be done on all patients before starting any treatment,

Table 41.6 Typical cerebrospinal fluid findings in central nervous system infections

	Viral meningo-encephalitis	Acute bacterial meningitis	Tuberculous meningitis	Fungal	Normal*
Opening pressure	Normal/High	High	High	High–very high	10–20 cm
Colour	'Gin' Clear	Cloudy	Cloudy/Yellow	Clear/Cloudy	Clear
Cells/mm³	Normal–High 0–1000	High–very high 1000–50 000	Slightly increased 25–500	Normal–high 0–1000	<5
Predominant white cell	Lymphocytes	Neutrophils	Lymphocytes	Lymphocytes	Lymphocytes
CSF/Plasma glucose ratio	Normal	Low	Low–very low (e.g.<30%)	Normal–low	66%
Protein (g/l)	Normal–high 0.5–1	High >1	High–very high 1–5	Normal–high 0.2–5.0	<0.5

*Normal values:

Normal CSF opening pressure is <20 cm for adults, <10 cm for children below age 8.

Although 66% is quoted as the normal glucose ratio, only values below 50% are taken as being significant in many settings

A bloody tap will falsely elevate the CSF white cell count, and protein. To correct for a bloody tap, subtract 1 white cell for every 700 red blood cells/mm³ in the CSF, and 0.1 g/dl of protein for every 1000 red blood cells

Some important exceptions:

In viral central nervous system infections, an early lumbar puncture may give predominantly neutrophils, or there may be no cells in early or late lumbar punctures.

In patients with acute bacterial meningitis that has been partially pre-treated with antibiotics (or patients <1 year old) the CSF cell count may not be very high and may be mostly lymphocytes.

Tuberculous meningitis may have predominant CSF polymorphs early on.

Listeria can give a similar CSF picture to TBM, but the history is shorter.

CSF findings in bacterial abscesses range from near normal to purulent, depending on location of the abscess, and whether there is associated meningitis, or rupture.

A cryptococcal antigen test (CRAG), and India ink stain should be performed on the CSF of all patients in whom cryptococcus is possible.

because they may be positive for bacteria even if the CSF is negative. Additional investigations include polymerase chain reaction of purpuric skin lesions for meningococci, and culture or polymerase chain reaction of skin vesicles, stool, urine, and throat swabs for enteroviruses.

41.2.5 Imaging

As indicated above, in most patients with suspected meningitis, CT or MR scanning is not needed. However in patients with a coma score less than 10, new onset of seizures, or focal neurological signs other than cranial neuropathies, imaging is needed to look for space occupying lesions such as abscesses, empyema, etc., or cerebral oedema, and midline shift. The demonstration of space around the basal cisterns on CT scan can also be helpful in deciding that it is safe to perform a lumbar puncture (Fig. 41.5). Meningeal enhancement may be demonstrated with gadolinium enhancement of T1-weighted MRI, particularly in chronic meningitis. Other findings, seen particularly in pneumococcal meningitis, include infarcts and venous sinus thrombosis.

41.2.6 Choice of antibiotics

In the United Kingdom parenteral treatment with a third generation cephalosporin, such as cefotaxime or ceftriaxone (2 g) is recommended as initial treatment for community acquired acute bacterial meningitis in adults. If a penicillin-resistant *Streptococcus pneumoniae* is suspected then vancomycin with or without rifampicin is also added. The use of steroids with antibiotics is discussed further below.

If the CSF cell count is normal, or consistent with an acute viral picture, affected by a lymphocytic CSF and a normal glucose ratio, antimicrobial treatment is not indicated in most patients. However, particular attention should be paid to CSF cell counts that in the right clinical context could indicate tuberculous meningitis, such as lymphocytic CSF with a low glucose ratio. Also, for patients at risk from listeria, such as those >55 years old and the chronically ill, ampicillin 2 g intravenously qds with or without gentamicin should be considered, especially if the CSF is lymphocytic with a normal/low glucose ratio, Antimicrobial treatment is further refined once the results of cultures and sensitivities are available. Treatment for tuberculous meningitis, and fungal disease is considered in Sections 41.5.3 and 41.5.9. Where meningococcal disease is suspected, based on a purpuric rash, or Gram's stain, patients should be subject to respiratory isolation for 24 h. In patients with severe penicillin allergy vancomicin and rifampicin, or chloramphenicol should be considered.

There are no established treatments for viral meningitis. Plecornaril was shown to reduce symptom duration in enterovirus meningitis (Section 41.5.6), but has not been used routinely. Aciclovir is sometimes given when herpes simplex virus, or varicella zoster virus are the aetiologic agents of meningitis, and there are anecdotal reports of good symptom resolution (Bergstrom *et al.* 1990), but whether this should become the standard treatment is not known (Tyler 2004).

41.2.7 The role of steroids

There is now relatively good evidence that glucocorticoids given shortly before the antibiotics improve the outcome for pneumococcal meningitis in adults, and *Haemophilus influenzae* meningitis in children (Jacob *et al.* 2006). In one trial they were given before or with the first dose of antibiotic therapy in patients with cloudy CSF, bacteria on the Gram's stain, or a CSF pleocytosis of more than 1000 cells/mm³ (de Gans *et al.* 2002). Unfavourable outcome consisting of death severe or moderate sequelae) was reduced from 25 to 15 per cent, with death dropping from 15 to 7 per cent. A systematic review of 632 patients in 5 trials confirmed these findings (van de Beek *et al.* 2004). The benefit was greatest in patients with pneumococcal meningitis. In patients with meningococcal meningitis there was a non-significant trend towards reduced mortality and sequelae (Jacob *et al.* 2006). In most studies, steroids were given to patients with cloudy CSF or initial CSF findings consistent with acute bacterial meningitis. There is no evidence supporting their 'blind' use in patients with suspected meningitis who have not had a lumbar puncture.

41.2.8 Complications

Patients with reduced consciousness should be managed in an intensive care setting.

Despite the initiation of antibiotic treatment, a proportion of patients deteriorate (Table 41.7). The pro-inflammatory cytokine cascade is thought to cause increased permeability of the blood-brain barrier, cerebral oedema, and raised intracranial pressure. A narrowing of ventricular size, and disappearance of the sylvian fissure may be seen on imaging. With more severe oedema the basal cisterns and sulci may be obliterated (Fig. 41.5). Intracranial monitoring and aggressive measures to reduce intracranial pressure (Grande *et al.* 2002), which may include osmotic diuretics are used, but there has been no systematic trial to address their efficacy.

Seizures should be treated symptomatically, but anticonvulsant prophylaxis for all patients is not recommended (van de Beek *et al.* 2006). Subtle motor status epilepticus is a rare, but important treatable cause of deterioration. Electroencephalograms should be performed in patients that have had seizures, and do not waken afterwards, especially if there is subtle twitching. Patients with pneumococcal meningitis are at risk of venous sinus thrombosis (Fig. 41.6). Septic shock, which may complicate bacterial meningitis from any cause, is sometimes complicated by adrenocorticoid insufficiency, for which low dose corticosteroids may be needed. Subdural effusions are common, and should be distinguished from subural empyemas, which often need surgical drainage.

41.2.9 Public health considerations

Meningitis is a notifiable disease whether or not the microbiological cause is determined. Therefore the public health authorities should be contacted. The local health protection unit should be notified,

Table 41.7 Causes of deterioration in patients with acute bacterial meningitis

Cerebral infarction
Venous sinus thrombosis and thrombophlebitis
Subdural empyema
Subdural effusion
Seizures, including subtle motor seizures
Cerebral oedema
Cerebral or cerebellar herniation (shift)
Septic shock
Aspiration pneumonia
Hyponatremia

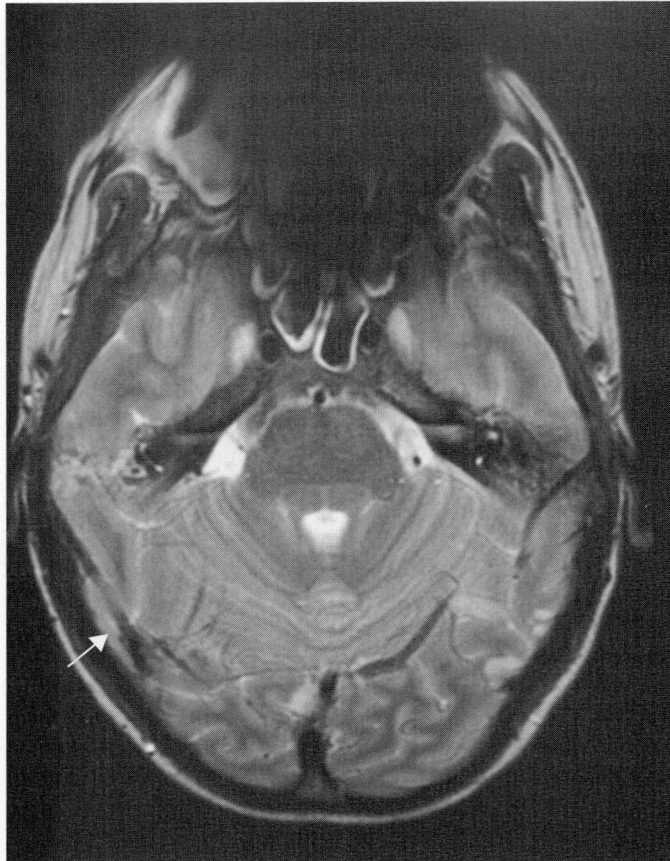

Fig. 41.6 Straight sinus thrombosis after pneumococcal meningitis. The right transverse sinus has thrombosis so that a flow void is not seen (arrow), compared with the left transverse sinus.

who will inform the consultant in communicable disease control (Table 41.8). Close contacts of patients with meningococcal disease either as meningitis or septicaemia, need prophylactic antibiotic treatment. The consultant in communicable disease control will pursue this. Close contacts are defined as those with intimate contact such as kissing, as well as those eating and sleeping in the same dwelling. In addition, health care workers involved in mouth to mouth resuscitation, intubation, etc. should be treated. Rifampicin is usually given, but other antibiotics can be used. Outbreaks of meningitis can attract considerable media attention. It is advisable to contact the hospital press officer as soon as possible, and decide on a communication plan.

41.3 **Recurrent and chronic meningitis**

In recurrent meningitis episodes of meningitis are separated by periods during which the CSF returns to normal. In contrast patients with chronic meningitis have persistently abnormal CSF, though they may present with recurrent symptoms separated by asymptomatic periods. Recurrent meningitis can be divided into recurrent pyogenic meningitis, with large numbers of CSF leukocytes, which are predominantly polymorphonuclear, or recurrent aseptic meningitis with fewer cells that are predominantly lymphocytes.

Table 41.8 Notifiable neurological infectious diseases

Acute encephalitis
Acute poliomyelitis
Anthrax
Diphtheria
Leptospirosis
Malaria
Measles
Meningitis
–Meningococcal
–Pneumococcal
–Haemophilus influenzae
–Viral
Meningococcal septicaemia (without meningitis)
Mumps
Paratyphoid fever
Rabies
Relapsing fever
Rubella
Tetanus
Tuberculosis
Typhoid fever

Diseases which pose a serious threat to the public health, such as meningitis, poliomyelitis, diphtheria, etc. should be notified to the local health protection unit by telephone, followed by written notification using the official form. The contact details of the local Health Protection Unit are available from most hospital switchboards, or from Health Protection Agency website, http://www.hpa.org.uk/lars_hpus.htm

41.3.1 **Recurrent aseptic meningitis**

In 1944 Mollaret described a syndrome of recurrent benign meningitis, that was associated with characteristic cells in the CSF. Mollaret's cells are large friable cells with faintly staining vacuolated cytoplasm, now thought to be activated macrophages. They are typically seen in the first 24 h of illness, but they are not always seen in benign recurrent meningitis, and indeed are not confined to this syndrome. Cells indistinguishable from Mollaret's have been seen in other forms of recurrent meningitis, such as that due to periodic discharge of the contents of an epidermoid tumour into the subarachnoid space; they have also been seen in other viral central nervous system infections, for example meningitis and encephalitis due to West Nile virus (Tyler 2004).

With the advent of molecular diagnostic techniques in the 1990s, it became clear that a large proportion of Mollaret's meningitis was actually due to or associated with reactivation of herpes simplex virus-2. The association was not made previously because culture of herpes simplex virus-2 from the CSF of patients with recurrent meningitis is normally negative, but polymerase chain reaction has greater sensitivity (Tyler 2004). Herpes simplex virus-2 also appears to be responsible for many cases of what was previously called 'benign lymphocytic meningitis'—a recurrent meningitis syndrome in which Mollaret's cells are not seen (Tedder *et al.* 1994). Approximately 20 per cent of patients with an initial episode of herpes simplex virus-2 meningitis will go on to have a recurrence, though in one study nearly half the patients had a proven or suspected episode within the first year (Aurelius *et al.* 2002).

Clinical features and diagnosis

Patients with recurrent aseptic meningitis present with repeated self-limiting episodes of fever, meningism, and severe headache, usually of 2–5 day duration, separated by symptom-free intervals. Women are affected more often than men. The interval between episodes may range from months to many years. Typically the episodes become less severe, as time passes, and the syndrome appears to eventually 'burn itself out' after several years. Approximately 30–40 per cent of patients with recurrent aseptic meningitis have a history of prior genital herpes, and in some patients the episodes of recurrent meningitis are associated with recurrent mucocutaneous lesions.

CSF culture is positive for virus in approximately 75 per cent of patients with herpes simplex virus-2 primary meningitis. After the initial episode, culture is almost never positive, but virus is detectable by polymerase chain reaction, which has become the investigation of choice. Serological means of diagnosis include the demonstration of seroconversion during primary infection, and the demonstration of intrathecal production of anti-herpes simplex virus-2 antibodies in recurrent disease.

Antiviral treatment

There have been several case reports of aciclovir treatment for primary or recurrent herpes simplex virus-2 meningitis associated with genital lesions. There was rapid improvement of symptoms, within 72 h, after initiating treatment with aciclovir 5–10 mg/kg three times daily, for up to 10 days (Bergstrom et al. 1990). Although it is hard to interpret such uncontrolled studies, treatment may be useful for those with frequent or severe recurrences. An important question, being addressed in a randomized multicentre trial in Sweden, is whether long-term treatment with the oral pro-drug valaciclovir will prevent recurrence following herpes simplex virus-2 meningitis (Tyler 2004). These developments, with the possible future treatment of recurrent aseptic meningitis, demonstrate the importance of trying to diagnose the cause of viral meningitis, wherever possible.

Non-viral causes

Drugs, connective tissue diseases, and parameningeal disease processes can all cause recurrent aseptic meningitis (Table 41.9). Non-steroidal anti-inflammatory drugs, and some antibiotics, particularly trimethoprim-sulfamethoxazole (cotrimoxazole), are well-recognized causes of aseptic meningitis. A range of

Table 41.9 Causes of recurrent meningitis

Pyogenic	Associated with immunocompromised states:
	HIV
	agammaglobulinaemia
	defects of complement system
	Associated with "mechanical compromise":
	congenital defects, e.g. skull base, dermal cysts
	head injury
	neurosurgery
Aseptic meningitis	
	Herpes simplex virus type 2
	Drug (see Table 42.2)
	Connective tissue diseases (e.g. systemic lupus erythematosus)
	Mollaret's
Parameningeal focus	
	Infection (sinusitis, mastoiditis, osteomyelitis, brain abscess)
	Tumour (epidermoid cyst, craniopharyngioma)

parameningeal infections and tumours may also inflame the meninges to cause an aseptic meningitis picture. Careful imaging is necessary to establish the cause.

41.3.2 Recurrent pyogenic meningitis

Clinical presentation

Recurrent pyogenic meningitis occurs because of a physical or immune susceptibility in an individual (Table 41.9). For example, immune deficient states, such as defects of the complement system, or agammaglobulinaemia are associated with recurrent meningococcal meningitis. HIV is associated with recurrent streptococcal and meningococcal infection (Molyneux et al. 2003). However a more common cause of recurrent pyogenic meningitis is an anatomical connection between the CSF space and the skin, or the nasopharynx, which may be congenital or acquired. Congenital risks for recurrent meningitis include defects in the skull base, middle ear, or a persistent dermal sinus anywhere along the vertebral column. These usually cause recurrent disease in childhood, but can present later. The most common acquired risks for recurrent pyogenic meningitis are head injury and neurosurgery. The classical injury is to the cribriform plate of the ethmoid bone (Ginsberg 2004).

It is necessary to take a detailed history about possible head trauma, which may have been mild, or history of other recurrent infections, which might suggest an immune deficient state. Careful examination of the skin over the vertebral column may reveal a dermoid sinus. If this reveals nothing, then the next step is to discuss with neuroradiology colleagues the best approach to imaging.

41.3.3 Chronic meningitis

Chronic meningitis is defined as a clinical syndrome of meningitis persisting for 4 weeks or more. Although 4 weeks is required for the strict definition, in practical terms one should begin considering the diagnosis in any patient who has persistent and undiagnosed meningitis (Swartz 1987). Some patients will have features of meningism only. In others there may be clinical features of an encephalopathy with clouding of consciousness, or cognitive impairment, suggesting a meningoencephalitis. The approach to these two patient groups is the same. Acute meningitis can sometimes have a prolonged recovery phase, however it is usually possible to distinguish this from chronic meningitis by the fact that in the former the patient and CSF are both improving. In chronic meningitis, although the patients' symptoms may wax and wane, the CSF abnormalities persist, reflecting ongoing disease activity. Typically the CSF in chronic meningitis is lymphocytic with a normal or low glucose ratio and elevated protein. Acute neurological deteriorations may occur secondary to cerebral oedema, hydrocephalus, cerebrovascular occlusions, or seizures.

There is a long list of infectious and non-infectious causes of chronic meningitis (Table 41.10). The most common infectious causes include tuberculosis and cryptococcus, whilst sarcoidosis and malignancy are among the most common non-infectious causes.

Clinical features

The clinical evaluation is broadly similar to that of a patient with acute meningitis. However, there are specific additional considerations. In the history it is important to establish possible exposures and risk factors. For example is there someone in the family with

Table 41.10 Chronic meningitis, and meningoencephalitis, causes, clues, and comments

Infectious cause	Clues and comments
Bacterial	
Tuberculosis	In both HIV positive and negative patients
Listeria	Rhomboencephalitis
Syphillis	Especially in HIV positive patients
Lyme neuroborreliosis	Neurological features several weeks after erythema migrans rash
Brucellosis	History of drinking unpasteurized milk?
Leptospirosis	Exposure to contaminated water?
Whipple's	Typically with neuropsychiatric and gastrointestinal features
Nocardia	Clinically presents like a fungal meningitis, but with neutrophilic CSF
Actinomyces	Typically has neutrophilic CSF
Fungal	
Cryptococcosis	Common in HIV positive patients
Candidiasis	In immunocompromised
Histoplasmosis	
Blastomycosis	
Coccidiomycosis	Especially in the chronically ill
Aspergillosis	Usually parenchymal granulomas with abscesses
Mucormycosis	Rhinocerebral disease
Viral	
HIV-1	
Cytomegalovirus	Typically has neutrophilic CSF in HIV infected patients
Measles (subacte sclerosing panencephalitis)	Cognitive decline and myoclonus
Lymphocytic choriomeningitis virus	
Mumps	
Enterovirus	In patients with agammaglobulinaemia

Non-infectious cause	Clues and comments
Neoplastic	
Neoplastic meningitis	Diagnosed by cytospin, neoplastic markers, evidence of primary
Non-Hodgkins lymphoma	
Primary brain tumours	
Acute leukaemia	
Systemic solid cancers	
Inflammatory / Connective Tissue Disease / Vasculitic	
Sarcoid	Associated with cranial neuropathies; a common cause
Connective tissue diseases (systemic lupus erythematosis)	Rashes, granulomas
Polyarteritis nodosum,	Autoantobidoes, especially anti-neutrophil cytoplasmic antibodies (ANCA)
Wegener's granulomatosis	
Sjogren's syndrome,	Associated with dry eyes, dry mouth
Behcet's	Oral and/or genital ulcers
Primary granulomatous central nervous system angiitis	Hard to diagnose without biopsy
Vogt–Koyanagi–Harada syndrome	Rare; young dark skinned people/Asians
Meningeal reaction	
to intrathecal drugs	
to systemic drugs	
to parameningeal infection—e.g. sinusitis, mastoiditis, osteomyelitis, brain abscess	
to parameningeal tumour—e.g. epidermoid cyst, craniopharyngioma	
to venous sinus thrombosis	

tuberculosis, or has there been recent travel to the Indian subcontinent or Africa? Has the patient consumed unpasteurized milk giving a risk of Brucella? Has the patient been generally unwell for months with fevers, weight loss, rashes, or joint problems suggestive of an underlying systemic illness, such as a neoplasm or connective tissue disease? Was there the rash of erythema chronica migrans, which may have disappeared by the time neurological features of lyme disease became present? Are there risk factors for HIV, either in the patient or their partner, such as intravenous drug abuse, male homosexuality, multiple sexual partners, or sexual partners in countries of high risk?

A thorough general medical examination is especially important in patients with chronic meningitis, because it may yield vital clues that suggest a systemic infection or an underlying medical condition. The skin must be examined carefully because several important causes of chronic meningitis may be associated with skin lesions, including sarcoid, vasculitis, secondary syphilis, acanthamoeba, and fungal infections such as cryptococcus, blastomycosis, and coccidioidomycosis. Subcutaneous nodules may be present, due to cysticercosis or metastatic cancer, such as from breast or melanoma). Lymphadenopathy is an important pointer to tuberculosis, or neoplasms. A detailed eye examination by an ophthalmologist using a split lamp may reveal choroidal tubercles, sarcoid granulomas, or uveitis suggestive of an underlying connective tissue disease.

Investigation

The laboratory investigations will be directed to some extent by the clinical situation. Particularly important are a full blood count with erythrocyte sedimentation rate, and a chest X-ray. A tuberculin test may also be useful.

Blood urine and sputum should be cultured even if there is no clinical evidence of systemic infection; they may be positive for tuberculosis, cryptococcus, brucella, histoplasmosis, or blastomycosis. Serological investigations need to include HIV and syphilis serology, in addition histoplasma, coccidioides, and borrelia serology should be performed if indicated. Rising serum antibodies may be helpful for diagnosing brucella and toxoplasma. A blood smear examination for trypanosomes may be diagnostic in some cases.

In addition to the routine CSF investigations, large CSF volumes of 5–10 ml, should be sent for culture for fastidious organisms including tuberculosis, Bartonella, and fungi. The clinical microbiologists should know that you are investigating chronic meningitis, since the culture techniques are different to those for acute meningitis, requiring growth for 4–6 weeks, and special culture conditions, such as anaeorboic culture. CSF should be tested for cryptococcal antigen using the CRAG test, which has now largely replaced the Indian ink stain. The laboratory should be asked to stain for eosinophils suggestive of angiostrongylus meningitis. CSF and serum angiotensin converting enzyme levels should be measured.

If there has been a relevant exposure history, CSF should be tested for antibodies to histoplasma, coccidioides, brucella, blastomyces, *Taenia solium*, and measles virus, as well as syphilis. Large CSF volumes should be centrifuged to look for malignant cells. B- and T-lymphocyte immunological markers may be helpful if lymphoma is suspected. CSF examination for infection and malignancy should be repeated regularly if no diagnosis has been made.

Cranial imaging should include an MRI with gadolinium which may show meningeal enhancement, parameningeal or parenchymal lesions. CT scan can also be helpful in looking for parameningeal

foci in the paranasal sinuses or mastoids, epidermoid cysts, and for bony anomalies or sinuses. Systemic scanning with CT of the chest and abdominal MRI may sometimes be helpful, revealing mass lesions.

Any skin lesions or enlarged lymph nodes should be biopsied. Muscle, liver, or bone marrow biopsies may also sometimes be helpful. Granulomas with caseation are found in tuberculosis, histoplasmosis, and coccidiomycosis. Focal necrosis is sometimes found in brucella. Biopsied material should always be sent for culture. Bone marrow biopsy and culture are especially useful for tuberculosis and histoplasmosis; liver biopsy may be useful to diagnose military tuberculosis, especially if liver enzymes are deranged. If a vasculitic cause is suspected angiography may be helpful.

Ultimately leptomeningeal and brain biopsy may be necessary if the diagnosis is not clear, especially if the patient is deteriorating. Studies have shown leptomeningeal brain biopsy gives a definitive diagnosis in 20–39 per cent of patients with chronic meningitis (Cheng *et al.* 1994; Anderson *et al.* 1995), the most common diagnoses at biopsy being sarcoid in 31 per cent and carcinomatous meningitis in 25 per cent. If the biopsy is directed towards an area of meningeal enhancement, as shown with a gadolinium enhanced MRI, the diagnostic yield is as high as 80 per cent (Cheng *et al.* 1994).

Management and treatment

Reassessment is important in such patients if they remain undiagnosed. This includes a thorough medical examination for new signs that may have become apparent, and repeat testing of large volumes of CSF, 5–10 ml. Therapeutic trials of empirical treatment are usually indicated if the cause remains elusive. Tuberculosis is the most common cause of chronic meningitis, and empirical treatment is advisable if there are risk factors, and a consistent CSF picture (Section 41.5.3), especially if there is evidence of current or previous systemic tuberculosis. Given the length of treatment required, and side effects, it is advisable to do everything possible to establish the diagnosis before empirical treatment is started. If cryptococcus, *Coccidioides* spp., or candida is suspected, empirical anti-fungal treatment, with amphoteracin followed by fluconazole or itraconazole may be appropriate (Section 41.5.9). Even after empirical treatment is started repeat CSF examination is necessary, both to look for possible alternative diagnoses, and to follow the response to treatment. Clinical improvement may be slow, and several months' treatment is usually necessary before deciding whether it has been effective, particularly regarding anti-fungal treatment. Traditionally empirical treatment with steroids has been avoided in patients with chronic meningitis because of the fear of causing deterioration if there is an infectious cause. Such treatment tended to be reserved for patients in whom tuberculosis was felt unlikely. In one series of patients from an area where tuberculosis is not endemic, none responded to anti-tuberculous treatment, but 50 per cent responded promptly to a course of corticosteroids (Smith *et al.* 1994). However recent studies have shown that adjuvant treatment with corticosteroids is indicated in tuberculosis (Section 41.5.3).

41.4 Meningitis in susceptible patient groups

Immunocompromise poses specific problems in regard both to meningitis, and other central nervous system infections. The initial diagnosis of meningitis can be difficult because the clinical features may be more subtle than in immunocompetent patients: There may be less inflammation so that signs of meningism, such as neck stiffness and Kernig sign may be absent. For the same reason CSF cell counts may be unexpectedly low, or normal. Second the range of organisms that infect the central nervous system is greater in the immunocompromised. To some extent this depends on the nature of the immune deficiency. In advanced HIV disease, the important causes of chronic meningitis to consider include tuberculous meningitis, cryptococcus, neurosyphilis, coccidioides, and candida. Listeria is a particular problem in pregnancy, as well as in HIV infected patients. Neutropenic patients, for example transplant or haematology patients, are at risk of *Peudomonas* spp., *Enterobacteriaeciae* spp., and aspergillus. Those with humoral immune defects, for example aggamaglobulinaemic patients, are at risk of recurrent enterovirus infections, meningococcus, pneumococcus, haemophilus, and enterobacteriaeciae.

Infection of CSF shunts is a relatively common problem among neurosurgical patients, with an incidence of 5–15 per cent in some series. Staphylococcal infections account for more than two-thirds of cases, particularly *Staphylococcus epidermidis*, and *Staphylococcus aureus*. Gram negative bacteria, including *E. coli*, *Klebsiella* spp., *Proteus* spp., and *Pseudomona* spp. account for up to 20 per cent of infections (Renier *et al.* 1984).

41.5 Specific causes of meningitis

41.5.1 Pneumococcal meningitis

Streptococcus pneumoniae, or pneumococci, are small 0.8 micron, non-spore forming, non-motile gram positive cocci that typically appear in clinical specimens as lancet-shaped pairs, 'diplococci,' with tapered ends in juxtaposition. On blood agar culture plates they are alpha-haemolytic meaning a greenish discolouration surrounds the colonies. Pneumococci are coated in a capsule of complex polysaccharides, and classified into more than 90 serotypes based on their capsular polysaccharide antigens. Capsular polysaccharide is essential for virulence, through its anti-phagocytic properties. Only a few of the 90 serotypes of pneumococcus are responsible for most of the invasive pneumococcal disease. In general these are the lower numbered serotypes, the higher numbers being common in carrier states. The pathogenic serotypes differ between adults and children, and between different geographical locations, but they also vary with time, which provides a major challenge to vaccine development. There are also other cell surface components including C polysaccharide antigens, cell wall antigens, and M- or R-protein antigens, whose role in the pathogenesis is undetermined.

As with the other major bacterial meningeal pathogens, pneumococcal meningitis often follows recent nasopharyngeal colonization by a virulent strain. Asymptomatic carriage of avirulent strains of pneumococcus is common, occurring for 5–75 per cent of normal adults; approximately 25 per cent acquire a new strain annually. The duration of carriage varies from weeks to months.

Clinical features

Most adults that develop pneumococcal meningitis have predisposing factors such as foci of pneumococcal infection elsewhere, immunocompromise, or a physical breach of the blood-brain barrier. Up to 25 per cent of patients with pneumococcal meningitis have coexistent pneumonia, and 30 per cent have acute otitis media. Acute sinusitis may also be an important antecedent event. HIV is an important risk factor for pneumococcal meningitis, which is much more common in patients with AIDS, than in the general population. Other immunocompromised states that increase the risk include asplenia due to sickle cell disease, Wiskott–Aldrich syndrome, childhood nephritic syndrome, multiple myeloma, chronic lymphocytic leukaemia, alcoholism, and cirrhosis.

Pneumococcal meningitis tends to be more severe than other forms of acute bacterial meningitis, as indicated by the high rate of 11–19 per cent of patients who are comatose on admission. About 29–42 per cent have focal neurological deficits, and 7–21 per cent have seizures (Weisfelt *et al.* 2006). Common focal deficits include aphasia, cranial neuropathies, especially hearing loss, and hemiparesis.

Diagnosis and treatment

CSF opening pressures are high in almost all patients with pneumococcal meningitis. The CSF usually shows the typical findings for bacterial meningitis (Table 41.6), Gram staining identifies the organism in 81–93 per cent of patients with pneumococcal meningitis, and CSF culture is positive for 76–88 per cent (Weisfelt *et al.* 2006) Bacterial antigen tests have limited sensitivity. Polymerase chain reaction for bacterial DNA is reported to show high sensitivity and specificity, though false positives still give rise to concern (van de Beek *et al.* 2006) Imaging of patients with pneumococcal meningitis shows hypodense lesions suspicious of infarction in 17–30 per cent, brain swelling in 20–39 per cent, and hydrocephalus in 5–16 per cent. Cortical vein or venous sinus thrombosis may also occur (Weisfelt *et al.* 2006). The antibiotic and adjunctive dexamethasone treatment of bacterial meningitis is discussed above.

Prognosis

Most adults with pneumococcal meningitis develop complications, which can be divided into intracranial and systemic. Intracranial complications are reported for up to 75 per cent of patients, and include seizures, brain infarction, brain swelling, hydrocephalus, and cranial nerve palsies (Durand *et al.* 1993; Hoen *et al.* 1993; Kastenbauer *et al.* 2003; Ostergaard *et al.* 2004, 2005; Weisfelt *et al.* 2006). Systemic complications, which occur in about 40 per cent of patients, include septic shock, cardiorespiratory failure, and disseminated intravascular coagulation. The mortality from recent series was 16–37 per cent (van de Beek *et al.* 2006).

Sequelae are common in pneumococcal meningitis, being reported for 29–72 per cent of patients. In a recent Dutch cohort which included 243 adult survivors of pneumococal meningitis, 22 per cent of patients had hearing loss, and 6 per cent had other cranial nerve palsies. Sequelae attributed to focal cerebral deficits, occurred in 11 per cent particularly aphasia, ataxia, hemiparesis, and monoparesis (Weisfelt *et al.* 2006). Death in younger patients tends to be associated with the neurological complications, whereas in older patients it tends to be associated with systemic complications (Weisfelt *et al.* 2006). Risk factors for unfavourable outcome in both morbidity and mortality were a low admission Glasgow coma score, cranial nerve palsies other than hearing loss, a high erythrocyte sedimentation rate, a CSF leukocyte count less than 1000×10^6/L, and a high CSF protein concentration. Other reported risk factors for a poor outcome include old age, seizures, a low thrombocyte count, low CSF glucose concentration, concurrent pneumonia, the absence of concurrent otitis, and the need for mechanical ventilation (de Gans *et al.* 2002).

Vaccines

There are currently two main vaccines used against pneumococcal disease, both targeted at the cell wall polysaccharide antigens. A 23-valent vaccine contains capsular polysaccharides of the 23 main serotypes, responsible for about 90 per cent of invasive pneumococcal infections. It is recommended for those over 65 years, and people in at-risk groups, such as those with asplenia, immunodeficiency, and renal failure. However, because the immune response to the vaccine is not good, especially in younger children with immunocompromise, a conjugate vaccine is recommended for this group. In conjugate vaccines the antigens of the pathogen are conjugated to carrier proteins that activate T-lymphocytes. A seven-valent conjugate vaccine, which contains 7 of the most common polysaccharides that affect children, is recommended for children in the 'at risk group', from 2 months to 5 years of age.

41.5.2 Meningococcal meningitis

Neisseria meningitidis is a non-spore forming, non-motile gram negative diplococcus that usually appears kidney shaped in clinical specimens. Based on the antigenic characteristics of the capsular polysaccharide, 13 serogroups are recognized, of which A, B, C, Y, and W-135 are the most important. Serogroup A is particularly notable for causing large outbreaks in the 'meningitis belt' of Africa, whereas B, C, and Y are more important in Europe and America. B isolates are responsible for most disease, in Europe, but in the UK C-2 which is an especially virulent strain, has been on the increase. In America B, C, and Y isolates each cause about 30 per cent of cases. Serogroup W-135 disease has been seen increasingly in young adults in America and parts of Africa.

The risk factors for invasive disease, as opposed to carriage, are not completely understood. In young children, the incidence is inversely proportional to the presence of serum bactericidal antibodies. Epidemiological risk factors include living in college or military dormitories, or household crowding, black race, low socio-economic status, active and passive cigarette smoking, and travel to countries with endemic or hyperendemic disease, especially during the right season (Fischer *et al.* 1997; Williams *et al.* 2004; Bilukha *et al.* 2005). In otherwise healthy individuals there is some evidence that an antecedent viral infection, especially influenza A, may predispose to invasive meningococcal disease. Other risk factors include impaired complement system, complement depletion due to nephrotic syndrome, hepatic failure, myeloma or systemic lupus erythematosis, or an asplenic state, HIV, and hepatitis C.

Clinical features

Meningococcus commonly causes two clinical syndromes, which may overlap: septicaemia and meningitis. Meningococcal septicaemia presents with fever, chills, malaise, prostration, and a rash; a non-blanching petechial purpuric rash is the classical finding, but urticarial and maculopapular rashes may also occur. Shock with disseminated intravascular coagulation, the Waterhouse–Friderichsen syndrome, can rapidly follow the development of meningococcaemia, leading to coma and death. Meningococcal meningitis is often clinically indistinguishable from other bacterial meningitides, though the classical triad of fever, neck stiffness, and altered consciousness is less common in meningococcal disease than pneumococcal meningitis. In infants with meningococcal meningitis the disease can present more slowly, without neck stiffness, but with a bulging fontanelle (Rosenstein *et al.* 2001).

A recent study of the clinical presentations of meningococcal disease in children and adolescents has highlighted that the symptoms in the first 4–6 h are non-specific, such as coryza and sore throat (Thompson *et al.* 2006). Early symptoms of sepsis, particularly leg pains, cold hands and feet, and abnormal skin colour, were present at a median time of 6 h, and were felt to be a helpful early indicator of severe disease. In contrast the classical features of meningococcal disease developed late from 12 to 22 h haemorrhagic rash, meningism, and impaired consciousness.

Diagnosis, treatment, and outcome

The diagnosis is confirmed by observation of the gram negative diplococcus in the CSF, or culture from the CSF, blood, or a skin lesion. Although most would support the use of pre-hospital parenteral antibiotic treatment by general practitioners as soon as meningococcal disease is suspected, it is still unclear whether this is beneficial (Harnden *et al.* 2006). Penicillin G and ampicillin have been the antimicrobial agents of choice for *N. meningitidis*, though reduced susceptibility to penicillin is being reported with increasing frequency. As a consequence 3rd generation cephalosporins are often used as first line treatment. Patients that are shocked require oxygen at high flow, volume resuscitation, and early transfer to an intensive care facility. Pre-emptive intubation is often now practised and inotropes and vasopressors are used for circulatory support as needed. The outcome of meningococcal meningitis is generally better than for streptococcal disease, with case fatality rates of 3–13 per cent, and 3–7 per cent morbidity rates. Patients with a septicaemic component fare worse than those with a purely meningitic presentation. Reduced consciousness, focal neurological deficits, seizures, thrombocytopenia, and moderate anaemia have been reported as poor prognostic indicators.

41.5.3 Tuberculous meningitis

Tuberculous meningitis is on the increase, both in the developing and developed world, because of the increased susceptibility of people with HIV to tuberculosis, increasing multi-drug resistance, and increasing travel. Globally more than 8 million people are estimated to develop tuberculosis annually, with approximately 2 million deaths. In the United Kingdom there are more than 6000 cases of tuberculosis annually. About 5–10 per cent of tuberculosis cases are thought to have central nervous system involvement.

Mycobacterium tuberculosis is an obligate aerobic bacterium whose only natural reservoir is humans. Humans become infected following inhalation of infectious particles. There is replication within macrophages in the alveoli for 2–4 weeks, during which there is virtually no immune response. During this period haematogenous spread throughout the body is believed to occur in all cases. The subsequent course depends on the host response. There may be complete elimination by macrophages, leaving no residua of infection except for a positive tuberculin skin test. Alternatively caseous foci surrounded by fibrous capsules, tubercles, may develop. Viable mycobacteria are controlled within tubercles, but not eliminated; thus they are ready to reactivate if the host's immunity wanes in the future. When the host immunity wanes, the tubercles grow with proliferation of the bacillus, and ultimately release the organisms and their antigenically potent products into the surrounding tissue. In the central nervous system tubercles are known as Rich foci, and are commonly subependymal, particularly in the Sylvian fissure, or subpial. Tuberculous meningitis occurs when the foci discharge their contents onto the meninges. Reactivation in the central nervous system may occur at the same time as reactivation elsewhere. Very few organisms need to be released for the meningitis to develop. Although most tuberculous meningitis results from reactivation, in immunocompromised hosts it may occur during primary infection, along with pulmonary or military tuberculosus.

Clinical features

Tuberculous meningitis typically causes a subacute meningitis, and so the features of acute meningitis—fever, headache, vomiting, meningismus—are often absent, especially in children (Thwaites et al. 2000). The duration of symptoms can vary from days to months; though in one series from Australia, half the patients presented within 2 weeks of the onset of symptoms. A recent contact with tuberculosis is suggestive of the diagnosis in children, as is evidence of extrameningeal tuberculous on clinical assessment.

The clinical features of tuberculosis are best understood in terms of the three major pathogenic processes that follow the release of the tubercle contents into the subarachnoid space: vasculitis, adhesions, and inflammation. An obliterative vasculitis of large and small vessels causes strokes in about 30 per cent of patients; this occurs particularly in the internal carotid, proximal middle cerebral and perforating vessels to the basal ganglia, leading to hemiparesis and movement disorders. Exudative adhesions may obstruct CSF flow, causing obstructive hydrocephalus, and may cause cranial neuropathies, especially of cranial nerves II, III, IV, VI, and VIII. The basal inflammatory process may extend into the parenchyma, causing an encephalitis with marked oedema. Seizures are common in children and the elderly, and may result from hydrocephalus, tuberculoma, oedema, or hyponatremia attributable to the syndrome of inappropriate antidiuretic hormone secretion. In children with progressive primary tuberculosis, encephalopathy may be due to demyelination (Udani et al. 1970). Tuberculous meningitis is especially common in HIV infected patients, the clinical features are similar, though extra-meningeal tuberculosis may be more common. Diagnosis in the elderly may be especially difficult because of the insidious presentation, and occasional absence of cells in the CSF.

The Modified United Kingdom Medical Research Council grading of tuberculous meningitis is:

- Grade I: Fully conscious with a Glasgow coma score of 15, with no focal neurological deficit;

- Grade II: Fully conscious with focal neurological deficit, or reduced consciousness reflected by a Glasgow coma score of 10–14, with or without focal neurological deficit;

- Grade III: Comatose with a Glasgow coma score of <10, with or without focal neurological deficit.

Given the lack of specific clinical manifestations, and the difficulty and delays in obtaining a microbiological diagnosis, studies have looked for clinical indicators that are predictive of tuberculous meningitis (Thwaites et al. 2005). Chest radiographs show active or previous tuberculosis in about 50 per cent of patients, but such findings lack specificity in settings with a high prevalence of pulmonary tuberculosis. Miliary tuberculosis on chest X-ray is more useful, because it indicates disseminated disease. In a study of Indian children five admission features were predictive of tuberculous meningitis: symptoms for more than 6 days; optic atrophy; focal neurological deficit; abnormal movements; and less than 50 per cent neutrophils in the CSF (Kumar et al. 1999). A similar

study of Vietnamese adults found five factors were predictive: illness duration of 6 days or more; peripheral white blood cell count $\leq 15\,000 \times 10^3$/ml; <90 per cent neutrophils in the CSF; total CSF white cell count of $< 750 \times 10^3$/ml; and age > 36 years (Thwaites et al. 2002). A diagnostic rule awarding points for these criteria proved useful in Vietnam, though its utility in other settings is unknown, particularly where the HIV incidence is high.

Diagnosis

The CSF in tuberculous meningitis typically shows a moderate lymphocytosis of 100–1000 cells/mm³ with a low to very low glucose ratio of <30 per cent, and a high or very high CSF protein, which may produce a yellow appearance. The more extreme glucose and protein levels tend to occur with longer disease duration. However within the first 10 days polymorphonuclear cells may predominate. In the elderly, or those who are HIV positive, an acellular CSF has been reported. Direct microscopy of CSF for acid- fast bacilli using the Ziehl–Nielson stain is still one of the most useful ways of diagnosing tuberculous meningitis. However meticulous microscopy of large CSF volumes of 5–10 ml is needed, with examination of repeat samples if necessary. Culture is also more likely to be positive with large volumes, though it can take several weeks to become positive. The role of the tuberculin test in diagnosing tuberculous meningitis is debated; it may be more useful in children than adults. Polymerase chain reaction of the CSF is used in some centres, but a meta-analysis showed the sensitivity was only about 50 per cent, though specificity was more than 95 per cent (Pai et al. 2003). Although polymerase chain reaction may be no better than conventional bacteriology for pre-treatment investigations, it may have a role once treatment has started. An immunological test of the ability of T-lymphocytes to produce interferon gamma in response to tuberculosis antigens, the ELIspot, has proved useful for diagnosis of peripheral tuberculosis (Ewer et al. 2003), but its role in tuberculous meningitis is unclear. A novel assay using an inverted light microscope to observe mycobacterial growth directly in culture may also offer hope for more rapid diagnosis in the future (Moore et al. 2006). CT and MRI may be helpful diagnostically: in one study basal enhancement, hydrocephalus, infarct, and tuberculoma were substantially more common in children with tuberculous meningitis, than those with pyogenic meningitis (Fig. 41.7).

Treatment

In patients with suspected tuberculous meningitis that are severely unwell, or deteriorating, treatment may need to be started before the diagnosis is confirmed. In those that appear stable, a short delay is justified, if this allows a repeat lumbar puncture with large CSF volumes to be taken. Tuberculous meningitis is usually treated with an initiation phase of four drugs, followed by 2 drugs for a prolonged continuation phase, which is typically 9–12 months or even longer depending on response to treatment.

The drugs recommended for the initial therapy in the UK are rifampicin, isoniazid, pyrazinamide, and ethambutol (Joint tuberculosis comittee 1998). Once the patient is established on oral therapy, the first three drugs are available in a single tablet, Rifater®. In the USA streptomycin is preferred as the fourth drug. For the continuation phase rifampicin and isoniazid are recommended, available combined as Rifinah® or Rimactazid®. There are daily treatment regimes for patients that can be relied upon to comply; for those that cannot, thrice weekly directly observed treatment regimes, 'DOTS', are used.

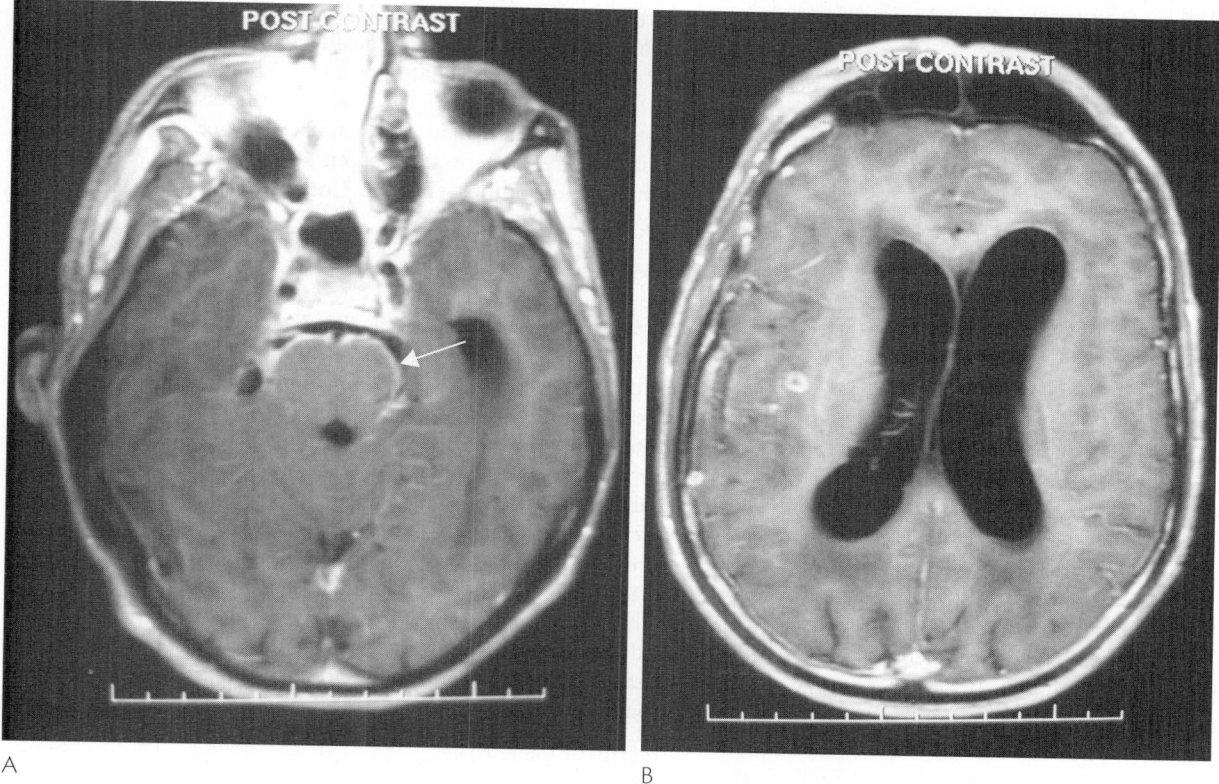

Fig. 41.7 Tuberculous meningitis. MRI with gadolinium enhancement showing (a) basal meningeal enhancement (arrow) (b) two small ring enhancing lesions in the right parietal cortex, diffuse meningeal enhancement, and hydrocephalus.

The initial phase drugs should be continued for 2 months. Where a positive culture for *M. tuberculosis* has been obtained, but susceptibility results are not available, treatment with pyrazinamide and ethambutol should be continued until full susceptibility is confirmed, even if this is for longer than 2 months.

Liver function tests should be checked at the start of treatment: Rifampicin often causes induction of liver enzymes with a mildly hepatotoxic picture, but sometimes can cause severe hepatotoxicity. Patients should be warned that this drug also turns urine, and contact lenses red. Renal function should also be checked at the start of treatment and ethambutol and streptomycin avoided if possible in patients with renal impairment. If liver and renal function is normal before treatment is started, then most physicians would not check them repeatedly, unless that patient became unwell. Pyridoxine treatment should be given to reduce the risk of the peripheral neuropathy caused by isoniazid. Patients should be warned that pyrazinamide occasionally causes a painful arthropathy. Ethambutol occasionally causes optic neuritis, so a patient's vision should be assessed, including colour vision, at the start of treatment. It is important to warn the patient to be alert for the development of visual disturbance and to report urgently should it occur. This warning should be clearly documented in the notes, for medicolegal reasons.

The use of adjunctive corticosteroids in tuberculous meningitis has been controversial, but recent evidence, along with analysis of earlier data indicates that steroids probably do reduce mortality in both adults and children at all stages of disease, though their effect

on sequelae is less certain (Jacob *et al.* 2006). A starting dose of oral prednisolone 60 mg/day for adults, or intravenous dexamethasone 0.4 mg/kg/day for adults or children is typical. Steroids should be continued for 4–8 weeks, and then tapered gently. It is unclear whether steroids should be used in patients with HIV.

Hydrocephalus is a common complication of tuberculous meningitis and requires treatment with diuretics, repeated lumbar puncture, or a shunt insertion if necessary. Some advocate that neurosurgical procedures be reserved for those with non-communicating hydrocephalus, or who have failed to respond to medical treatment, whilst others advocate early shunting in any patient with hydrocephalus. Tuberculous meningitis is a notifiable disease. Once the public health authorities should be informed, contact tracing will begin.

41.5.4 Syphilitic meningitis

Early in infection *Treponema pallidum* disseminates to the central nervous system (Section 42.5.1). It causes CSF abnormalities in 5–9 per cent of patients with primary syphilis, and occasionally can be isolated from the CSF. Syphilitic meningitis occurs in less than 3 per cent of untreated patients and is greatest in the first 2 years after infection. In contrast meningovascular syphilis is found in approximately 10 per cent of individuals with central nervous system involvement, occurring months to years after acquiring the infection (Simon 1985). The incidence of syphilis is increased in patients with HIV. The non-meningitic presentations of neurosyphilis, gummatous neurosyphilis, generalized paresis, and tabes dorsalis are discussed in Section 42.5.1.

Clinical features

Syphilitic meningitis presents as an acute aseptic meningitis with headache, nausea, and vomiting. Fever may be less prominent than in other acute meningitides, but cranial neuropathies occur in up to 45 per cent of patients, most commonly affecting cranial nerves VII and VIII, followed by II, III, V, and VI. In meningovascular syphilis there are focal neurological signs due to arteritis. Typically signs of a middle cerebral artery infarct follow weeks or months of 'prodromal' symptoms and signs such as headache, vertigo, and personality and behavioural changes. Some case series suggest patients coinfected with HIV may be more likely to progress to neurosyphilis, though this was not born out in a comparative study (Hutchinson *et al.* 1991).

Diagnosis of neurosyphilis

In patients with newly diagnosed syphilis a lumbar puncture is indicated in those where the duration of infection is not known or is likely to be greater than 24 months and in those with neurological signs. It is not necessary in early syphilis. Diagnosis of neurosyphilis is not straightforward. CSF cellular and protein abnormalities are found in 10–20 per cent of patients with primary syphilis and 30–70 per cent of those with secondary disease. In syphilitic meningitis there is a mild mononuclear pleocytosis, elevated protein and decreased glucose ratio. Frequently there are oligoclonal bands and anti-treponemal antibodies. Isolation of *T. pallidum* from the CSF is difficult and expensive.

The CSF Venereal Disease Research Laboratory, or VDRL, test, a 'non-treponemal' test, has high specificity, but a low sensitivity of only 30–70 per cent for neurosyphilis. Thus a positive test is sufficient to diagnose neurosyphilis, though a negative test does not exclude it. One potential problem with the VDRL test is contamination during a bloody tap, which can give false positives. The Rapid plasma reagin or RPR test is an alternative non-treponemal test. The fluorescent treponemal antibody absorption test, FTA-ABS, of the CSF is a treponemal test with high sensitivity but low specificity; so a negative test effectively rules out neurosyphilis, whereas a positive test is not so helpful. The microhaemagglutination assay, *T. pallidum*, MHA-TP, is an alternative 'treponemal' test. False-positives for VDRL tests occur in non-venereal treponemal infections, such as yaws and pinta, following smallpox immunization, in pregnancy, and other chronic conditions, including autoimmune disorders and malignancy. False positive RPR tests occur in tuberculous, pyogenic, and aseptic meningitis. Infection with *Borrelia burgdorferi*, another spirochaete, can cause false positive with the FTA-ABS test, but not the reagin test.

Treatment

In syphilis patients with CSF abnormalities but no neurological symptoms the aim of treatment is to prevent progression to clinical disease, and reverse the CSF abnormalities. In established disease the aim is to reverse the symptoms, or at least to arrest progress. The drug of choice is penicillin G, though there is controversy over the best dose, formulation, and duration (Section 42.5.1).

41.5.5 Lyme neuroborreliosis

Neurological disease caused by *Borrelia burgdorferi*, known as Lyme neuroborreliosis, often presents with meningitis. This follows the characteristic erythema migrans rash of lyme disease by 2–10 weeks, though it can occur without the rash (Pachner *et al.* 2007). It presents as a subacute meningitis, with headache and systemic features such as malaise, myalgias, fever, and weight loss; neck stiffness is less common. Typically there are recurrent attacks of several weeks, alternating with weeks of milder symptoms. About half the patients have cerebral symptoms such as somnolence, depression, impaired memory, and behavioural changes. Facial nerve palsies, occur in about 50 per cent of patients. Other cranial neuropathies and focal neurological signs also occur (Section 42.5.2).

Diagnosis

The diagnostic yields from culture and polymerase chain reaction are too low for routine practice. The diagnosis of neurological lyme borreliosis is therefore based on antibody detection in the appropriate clinical context. Lumbar puncture examination shows monocytic or lymphocytic CSF with a mild or moderately elevated protein. In Europe antibodies to *B. burgdorferi* are measured in the CSF and serum, and an antibody index greater than one is taken to be diagnostic (Pachner *et al.* 2007). In America the CSF antibody index approach has not been validated, and so a two-step approach to measuring antibody in the serum is taken. Initially a serum ELISA is performed, which has high sensitivity but low specificity. If this is positive, a more specific Western Blot is performed to confirm the diagnosis. Although antibody titres can be low early in disease, by the time patients develop neurological disease antibody should be present.

Treatment

Intravenous cephalosporins, such as ceftriaxone, are becoming the treatment of choice for lyme neuroborreliosis. Ceftriaxone has a long half life, and good blood-brain barrier penetration, and is given at 1–2 g per day for 14–28 days (Dattwyler *et al.* 1997); oral doxycycline at 200 mg for 14 days has also been shown to be effective (Pachner *et al.* 2007). In addition non-steroidal anti-inflammatory drugs may help relieve the arthralgias, myalgias, and headaches, but corticosteroids are not considered part of the routine therapy.

41.5.6 Enterovirus meningitis

Enteroviruses belong to the *enterovirus* genus in the *Picornaviridae* family. As the family name implies, these are among the smallest RNA viruses; pico = small (Spanish). The genus includes polioviruses, coxsackieviruses, echoviruses, and newer enteroviruses. These non-polio enteroviruses are associated with a much wider spectrum of clinical manifestations than polioviruses including exanthemas or rashes, enanthemas or mouth eruptions, conjunctivitis, respiratory infections, myocarditis, pericarditis, and pleurodynia which consists of attacks of intercostal muscle inflammation causing chest pain. Aseptic meningitis is the most common neurological manifestation of the non-polio enteroviruses, though they can also cause encephalitis and acute flaccid paralysis. Enteroviruses are spread via the faeco-oral route, particularly among children, though for some, respiratory spread has also been implicated. In any part of the world, each enterovirus 'season' is usually dominated by only a few serotypes, the predominant ones cycling with varying periodicity. In the United States where more detailed studies have been done, Coxsackieviruses A9, B2, and B4, and echoviruses 30, 11, 9, and 6 are often implicated. For the most part, the clinical features of viral meningitis in these patients do not vary. However some of the more recently described enteroviruses do display different clinical phenotypes. For example enterovirus 70, which was first recognized in 1969, has caused large outbreaks of acute haemorrhagic conjunctivitis, followed in some adults by acute flaccid paralysis, with radicular pain, paraesthesiae, and bulbar involvement in some cases (Kono *et al.* 1972, 1974; Wadia *et al.* 1983).

Enterovirus 71, the most recently identified enterovirus, is a cause of muco-cutaneous and central nervous system disease, including acute flaccid paralysis (Solomon *et al.* 2007). First isolated from young children with encephalitis and meningitis in California in 1969 (Schmidt *et al.* 1974), the virus has since caused epidemics in which some patients have hand, foot, and mouth disease, and others have aseptic meningitis (Blomberg *et al.* 1974). In the late 1970s an outbreak of enterovirus 71 in Bulgaria was associated with aseptic meningitis, and a high incidence of acute flaccid paralysis, indistinguishable from poliomyelitis, but with no associated hand, foot, and mouth disease (Shindarov *et al.* 1979). Half the patients with paralysis had encephalitis or cranial nerve involvement, and the mortality in these patients was high. In addition to young age and immunodeficiency, other possible risk factors for enterovirus meningitis include physical exercise during the infection, male sex, lower socioeconomic status, and overcrowding.

Clinical features

With the exception of enterovirus 70 and enterovirus 71, for most enteroviruses the clinical manifestations of meningitis depend more on the host's age and immune status, than on the infecting virus. Enterovirus infections in the neonatal period are usually severe, and systemic. Thus in addition to meningitis, echoviruses will often cause hepatic failure, whilst coxsackieviruses cause myocarditis; other complications include encephalitis and necrotising enterocolitis (Rotbart 1995). The mortality in neonates may be as high as 10 per cent. Maternal illness is often reported at the same time, but the question of whether the infection can be acquired transplacentally is unresolved.

In older children enterovirus meningitis is often biphasic with a non-specific febrile prodrome followed by a brief remission before fever returns with symptoms of meningeal involvement. Non-specific features of viral infection reported for children include vomiting, anorexia, rash, diarrhoea, cough, pharyngitis, and myalgia (Singer *et al.* 1980). Adults complain of nausea, vomiting, myalgia, and malaise (Rotbart *et al.* 1998). Neck stiffness and headache are present in adults, and children old enough to report it; photophobia is also common. Febrile seizures are common in children with enterovirus meningitis.

Hypogammaglobulinaemic patients

Individuals with antibody deficiencies, such as X-linked agammaglobulinaemia, or severe combined immunodeficiency, are at risk of developing chronic enterovirus meningitis or meningoencephalitis, because enteroviruses are primarily cleared by antibody-mediated mechanisms. Clinical features include chronic headache, weakness, lethargy, hearing loss, seizures, ataxia, paraesthesias, and coma (McKinney *et al.* 1987). In addition approximately 50 per cent of patients develop a dermatomyositis-like syndrome, thought to be due to direct viral invasion. Many patients with agammaglobulinaemia are treated with regular immunoglobulin, given via the intravenous, intrathecal or even intraventrical routes. This may result in clinical stabilization; recent studies show that even though enteroviruses can no longer be cultured from the CSF, they are often still detectable by polymerase chain reaction. During chronic gastrointestinal carriage of enteroviruses by agammaglobulinaemic patients, viral mutations may occur (Dunn *et al.* 2000), which in

the case of poliovirus vaccine strains could potentially have implications for the eradication strategy (MacLennan *et al.* 2004).

Diagnosis

In the majority of cases, CSF examination will reveal a lymphocytic pleocytosis of between 100 and 1000 cells (Singer *et al.* 1980), though polymorphonuclear cells may predominate in the first 1–2 days. Occasionally enterovirus infection may be demonstrated in patients with no pleocytosis, particularly young infants. Mild elevation of protein is common. The CSF glucose ratio is usually normal, but may be mildly reduced (Singer *et al.* 1980).

Traditionally, enterovirus meningitis has been diagnosed by isolating virus from the CSF, throat, or stool. The sensitivity of tissue culture of CSF for enteroviruses may be up to 65–75 per cent (Rotbart *et al.* 1995). Culture from the throat or stool is more likely to be positive, but because these are not sterile sites, a positive culture from here could represent recent coincidental infection from a virus that is still being shed. Shedding from the throat continues for about 1 week after infection, but from the rectum it may continue for several weeks. During recent enterovirus 71 outbreaks in Asia, where aseptic meningitis is often associated with hand, foot, and mouth disease, virus isolation from vesicles, in addition to throat swabs, proved useful (Ooi *et al.* 2007). The development of immunoassays to detect antibodies against enteroviruses has been hampered by the lack of a widely shared antigen.

In recent years detection of enterovirus RNA in the CSF using reverse transcriptase polymerase chain reaction has become the new standard for diagnosis of enteroviral meningitis (Rotbart *et al.* 1994). Polymerase chain reaction primers directed against highly conserved sections of the 5′ untranslated region of the viral genome will detect most enterovirus serotypes, with sensitivity and specificity greater than 95 per cent. Because it gives a diagnosis more quickly than viral culture, it can improve clinical management in terms of length of hospital stay, and reduced need for further investigations, and antibiotics (Rotbart *et al.* 1997). Such testing is now available through most routine virology laboratories (Lina *et al.* 1996). Although false positive results caused by the extreme sensitivity of the test, and contamination from earlier positive tests was a problem in the early days of polymerase chain reaction usage, newer amplification techniques using closed systems minimize this problem. Real-time polymerase chain reaction, which is an even more sensitive method for viral RNA detection is likely to be used increasingly in the future (Petitjean *et al.* 2006).

Treatment and outcome

Although antiviral drugs are becoming available for enterovirus meningitis, in most instances therapy is not routinely recommended. Pleconaril is a drug that binds within a hydrophobic pocket at the base of the receptor binding canyon in the viral capsid protein, thus inhibiting the virus from binding to its cellular receptor. The drug has broad activity against most enteroviruses at low concentrations of <0.1 μg per ml, and has good oral bioavailability. In phase III clinical trials pleconaril reduced symptoms of aseptic meningitis, particularly headache, by approximately 2 days, compared with placebo controls (Desmond *et al.* 2006). The drug has also been used in patients with chronic enterovirus infection due to agammaglobulinaemia, enterovirus myocarditis, poliovirus vaccine associated paralysis, and neonatal infection. Intravenous immunoglobulin is used in patients with chronic enterovirus meningitis

(McKinney *et al.* 1987), and may also be useful in patients with severe enterovirus 71 infection (Nolan *et al.* 2003).

The prognosis for patients with enterovirus meningitis is generally good. However one important exception is patients with hypogammaglobulinaemia, who are at risk of developing chronic meningitis. Most children with enterovirus meningitis make a good recovery in less than a week. Many feel better immediately after the lumbar puncture, presumably due to relief of raised intracranial pressure. Symptoms may persist longer in adults. Although the short-term recovery is good, several controlled long-term studies suggest that children who had neurological enterovirus disease in early life develop cognitive, developmental, and language abnormalities (Chang *et al.* 2007).

41.5.7 Mumps meningitis

Until the widespread use of live attenuated vaccines, mumps virus was the most common cause of viral meningitis in Western countries, including the United Kingdom. It is still an important cause of meningitis in the tropics, and in individuals in the West who are not vaccinated. For example during 2004–5 the United Kingdom experienced a nationwide epidemic of mumps, with 56 000 cases reported in England and Wales in 2005 (Bloom *et al.* 2005, 2006). The majority of cases resulted from a gap in mumps vaccination coverage during the 1990s, when shortages of the mumps-measles-rubella, MMR vaccine meant many children received a combined measles and rubella vaccine without the mumps vaccination. Mumps was added to routine vaccination, as part of MMR, in 1988. However, reduced uptake of the MMR in recent years, due to an unsubstantiated fear of a link to autism, means mumps meningitis is likely to continue.

Mumps virus is a single-stranded negative sense RNA virus, and is a member of the genus *Rubulavirus* in the family *Paramyxoviridae*. It exists as a single serotype

The virus is acquired via the respiratory route, and presumed to replicate in the upper respiratory mucosa before invading the parotid gland to give the parotitis for which the virus is best known. Viral spread to distant organs, including the pancreas, gonads, myocardium, breast, kidneys, and central nervous system, is thought to occur via a viraemia.

Clinical features

Symptomatic mumps is twice as frequent in males as females, and neurological involvement three times more common, for reasons that are not known. Studies have shown that more than 50 per cent of individuals with mumps parotitis have a CSF pleocytosis (Bang *et al.* 1995), but most of these people do not have clinical symptoms of meningitis. Clinical meningitis is reported for 1–10 per cent of people with mumps parotitis, 4–10 days into the illness. But in some patients the meningitis precedes the parotitis by up to a week, and others develop mumps virus meningitis with no parotid disease at all. Approximately 1 in 6000 mumps virus infections results in encephalitis.

In most cases, the clinical features of mumps meningitis are typical for those of other viral meningitides, with fever, headache, vomiting, and neck stiffness. However, occasionally orchitis or pancreatitis may provide the clue that a case of meningitis is due to mumps virus. Bradycardia, lethargy, and anaemia are also reported. A lumbar puncture typically reveals 10–500 leukocytes per mm^3, though

there may be several thousand. There is usually a CSF lymphocytosis, but up to 25 per cent may have polymorphonuclear predominance. The protein is often mildly elevated, and up to 30 per cent of patients may have a slightly reduced glucose ratio, which may be more common than for other viral meningitides. The symptoms usually resolve in 3–10 days, though the CSF abnormalities may persist for weeks. There is no specific antiviral treatment.

Diagnosis

The diagnosis of mumps virus infection is most often confirmed serologically, by detection of IgM antibody in a single serum sample using an enzyme-linked immunosorbant assay, or by showing a four-fold rise between acute and convalescent serum using complement fixation tests, haemagglutination inhibition assays, or neutralizations tests (Litman *et al.* 2004). IgM antibody can also be detected in the CSF in mumps meningitis. Virus can be isolated in the saliva from 2–3 days before the onset of parotitis to about 5 days after. It can also be isolated from CSF during the first 3 days of meningeal symptoms, but this has largely been replaced by polymerase chain reaction (Uchida *et al.* 2005). An elevated serum amylase may suggest mumps infection.

41.5.8 Herpes virus meningitis

The *Herpesviridae* family includes herpes simplex virus types 1 and 2, varicella zoster virus, Epstein–Barr virus, cytomegalovirus, and human herpesvirus types 6 and 7. All of these can cause viral central nervous system infections. For most of them viral meningitis is a less important presentation than encephalitis, in terms of number and severity of cases, and so they are covered in more detail in chapter 42. However herpes simplex virus-2 is an important cause of aseptic meningitis. It is usually acquired sexually, and is responsible for most cases of primary genital herpes. Symptoms of meningitis occur frequently during primary genital herpes infection, being reported for 35 per cent of women, and 11 per cent of men (Corey *et al.* 1983). Approximately 20 per cent of patients that develop an initial episode of herpes simplex virus-2 meningitis will go on to have a recurrence (Section 41.3.1).

Primary infection with herpes zoster virus results in varicella, chicken pox, whereas reactivation results in shingles (Section 21.14.5). Both these clinical syndromes can be associated with aseptic meningitis. In addition the virus can cause meningitis without cutaneous stigmata, particularly during reactivation; previously immunological assays were used to detect antibody, and more recently polymerase chain reaction has been used to confirm the presence of viral DNA

The neurological disease associated with human herpes viruses 6 and 7 is most commonly febrile seizures and encephalitis in children or the immunocompromised. However they can also occasionally cause meningitis.

41.5.9 Fungal meningitis

Fungal central nervous system infections are rare, but can be devastating. The incidence has increased due to the increased prevalence of immunocompromise, from HIV, transplants, corticosteroids, and chemotherapy. Fungal infections should be considered in all patients with a subacute or chronic meningitis, especially if they are immunocompromised. They can also present as space occupying lesions (Section 43.2). Although there are more than 100 000

Table 41.11 Fungal meningitis: patient groups, causes, and clinical features

Genus	Epidemiology/risk groups	Clinical features
Both Immunocompetent and immunocompromised hosts		
Cryptococcus	Common, especially in the immunocompromised,	Meningitis, raised pressure, hydrocephalus; rarely abscesses
Histoplasma	Rare; occurs in immunocompromised, elderly, very young	Meningitis and/or abscesses
Blastomyces	Occurs in non-immunocompromised	Meningitis; epidural or cranial abscesses common
Coccidioides	Southern and Central America; relatively common in the immunocompetent	Meningitis, granulomas, skin, bone, joint infection; hydrocephalus
Immunocompromised hosts only		
Candida	Common systemic infection, but rare in central nervous system; occurs in the immunocompromised, those with indwelling catheters, excessive antibiotic use, neonates, colonization of CSF shunts	Meningitis, microabscesses, vascular thrombosis, haemorrhage
Aspergillus	Severely immunocompromised; head injuries	Direct extension from nasal infection; microabscesses, not meningitis
Zygomycetes (mucormycosis)	Diabetics, neutropenics	Direct extension from nasal infection; black nasal discharge, cellulitis, cavernous sinus thrombosis, cranial neuropathies

fungi, only a small number cause central nervous system disease (Table 41.11), and of these *Cryptococcus neoformans* is by far the most common (Davis *et al.* 2006). *C. neoformans* central nervous system infections occur in 1 per 100 000 of the healthy population, but 20–50 per 1000 in HIV patients not receiving appropriate HIV therapy. *C. neoformans* is a ubiquitous fungus found in bird excrement and decaying plant matter. It has a predilection for the central nervous system, and so, unlike most other fungal infections, patients often do not have evidence for the fungus elsewhere.

Most pathogenic fungi are dimorphic, existing as a unicellular yeast phase, or a filamentous phase with hyphae that grow by extension and produce spores. Humans usually become infected through inhalation of spores, which subsequently cause fungaemia, and then invade the meninges or brain parenchyma. However Zygomycetes invasion occurs secondary to chronic sinusitis. Other less common routes include indwelling catheters, and skin wounds. In general fungi found in meninges are in the yeast phase, whilst those in the brain parenchyma are in the filamentous phase. Fungi can be divided into:

◆ Those which cause disease in both healthy and immunocompromised individuals: *Cryptococcus neoformans*, *Histoplasma capsulatum*, *Blastomyces dermaditidis*, *Coccidioides sp.*

◆ those which are only seen in susceptible individuals with immunocompromise or anatomical abnormalities: *Candida* species, *Aspergillus* species and *Zygomycetes* species, which is also known as mucormycosis.

Clinical features

The clinical presentation depends on whether the fungus invades the meninges, to cause a subacute or chronic meningitis, invades the brain parenchyma causing microabscesses or abscesses, or causes an arteritis which can thrombose or rarely rupture.

Fungal meningitis can be very difficult to distinguish from other subacute and chronics meningitides such as tuberculous meningitis. Patients typically prevent with a 1–3 week history of low grade fever, nausea, headache, neck stiffness, with progressive lethargy, malaise, and alteration in consciousness level. In patients with marked immunocompromise the only features may be confusion with or without low grade fever. As the disease progresses raised intracranial pressure may cause papilloedema. In addition basal meningeal exudates can lead to cranial neuropathies, and infarcts can cause hemiparesis and ataxia. Parenchymal invasion leads to abscesses, which may be surrounded by fibrous capsules and an inflammatory reaction thus forming a granuloma. The clinical presentation depends on the location of the lesions, but can include seizures, features of raised intracranial pressure, and focal signs. Abscesses in the basal ganglia and cerebellum are typical. *Blastomyces* and *Crytptococcus* species commonly cause abscesses as well as meningitis. Invasion of blood vessels causes angiitis, which can lead to cerebral infarcts.

Zygomycetes infection, also known as mucormycosis, often begins as a nasal or sinus infection, with a black nasal discharge, facial pain, and cellulitis. The fungus erodes into the cranium, causing a cavernous sinus thrombosis, with cranial neuropathies.

Diagnosis

The CSF findings in fungal meningitis are usually similar to those of other subacute meningitides such as tuberculous meningitis, with an elevated opening pressure, moderate lymphocytic CSF pleocytosis of 20 to 1000 cells/mm^3, low glucose, and elevated protein. A neutrophilic predominance may be found in *Blastomycosis dermatiditis*, Aspergillus or Zygomycetes, whereas eosinophilia may indicate coccidioidal or occasionally cryptococcal meningitis (Grosse *et al.* 2003).

Indian ink staining of CSF is positive in up to 80 per cent of cases of Cryptococcus. *Crytococcus neoformans* is isolated from the CSF in 75 per cent of cases of cryptococcal meningitis, and cryptococcal antigen testing is positive in most patients. Identification and isolation rates for other fungi in the CSF are much less. Removal of large volumes of CSF of about 10 ml on at least 3 occasions should be done to maximize the chance of identifying an organism.

Antibody testing of both serum and CSF is important for coccidiomycosis and histoplasmosis. For *Histoplasma capsulatum*, antigen can be detected in serum, urine, or CSF by enzyme-linked immunosorbant assay, in about 25 per cent of patients.

It is important therefore to search for fungi in the lungs, with X-ray and CT of the chest, as well as considering infection in the skin, bones, joints, and urine. Unusual skin or bone lesions should be biopsied. For most patients with subacute meningitis, the isolation of fungi from another site indicates the organism is causative; but one important exception is *Candida* sp., for which direct evidence of central nervous system infection must be sought. CSF antigen and antibody tests are becoming available for *Candida* species, but cerebral or meningeal biopsy may be required. For other fungi, biopsy of skin lesions, bone marrow aspiration, aspiration of any swollen joints may also be helpful. Given the difficulty of treatment it is essential to do all that is possible to obtain the right microbiological diagnosis.

Imaging may also be helpful. MRI with gadolinium enhancement often shows meningeal enhancement, especially in the basal cisterns; hydrocephalus, brain abscesses, and infarcts are also common.

Treatment

Fungal central nervous system infections are treated in three phases: an initial induction phase, a consolidation phase which depends on the host's immune status and response to treatment, and the maintenance phase which aims to prevent recurrences. Drugs commonly used include amphotericin B, which is a polyene class drug, flucytosine, which is a nucleoside analogue, and fluconazole and other azoles.

Amphotericin B is quite a toxic drug, that can cause fevers, chills, headache, and vomiting during infusions, and can lead cumulatively to renal failure and anaemia. Less toxic lipid formulations are available. The drug is usually given intravenously, but has also been given intrathecally. Flucytosine is given orally, and is especially useful in *Cryptococcus neoformans*, and *Candida* sp. However the drug is hepatotoxic and causes bone marrow suppression, and resistance often develops, so it cannot be used as a sole agent. The azoles which include fluconazole and ketoconazole, can be given orally or intravenously and are effective against many fungal infections. They cannot be given at the same time as amphoteracin because of antagonistic effects. Thus for the induction phase treatment of *Cryptococcal neoformans*, one of the commonest fungal central nervous system infections, amphoteracin B, or liposomal amphoteracin B is given intravenously for 6–10 weeks, with or without oral flucytosine for the first 2 weeks; this is followed by oral fluconazole for 8–10 weeks for the consolidation phase. Maintenance with fluconazole then continues for 1–2 years for otherwise healthy individuals, but is recommended to continue life long for the immunosuppressed at high risk (Davis *et al.* 2006). For coccidioides and candida infections, amphoteracin B, followed by fluconazole is used. For Histoplasma and Blastomyces amphoteracin followed by itraconazole is recommended.

Raised intracranial pressure is a common problem in AIDS patients with cryptococcal meningitis. It is associated with papilloedema, reduced visual acuity and hearing loss, confusion, and severe headaches. The pathogensis is unclear. Medical treatment with steroids, mannitol, and acetazolomide is not beneficial (Newton *et al.* 2002). Large volume CSF removal by lumbar puncture, to reduce the pressure by 50 per cent, should be done daily until the pressure normalizes. A ventricular drain may be needed if the pressure is very high, >40cm, or if there is obstructive hydrocephalus.

41.5.10 Listeria meningitis

The gram positive aerobic rod-shaped bacterium known as *Listeria monocytogenes* was first isolated from laboratory animals in the 1920s and recognized as a cause of human meningitis in the 1930s. It was named after the father of antisepsis, Lord Lister, in the 1940s. *L. monocytogenes* has many animal reservoirs, and is also found in the environment, especially in soil, and vegetation. It is common in dairy products, fresh meat, and vegetables, and in one study, 11 per cent of household refrigerators had listeria-contaminated products (Schuchat *et al.* 1992).

L. monocytogenes is probably commonly ingested, but inactivated by the gastrointenstinal tract of most people. However in susceptible individuals, systemic infection, and neurological disease can occur. Neonates, those over 60, pregnant women, and those immunosuppressed by malignancies, or chronic steroid therapy are particularly at risk, though approximately one-third of patients have no such risk factors. In western settings, listeria is probably the third most common cause of acute bacterial meningitis in adults after pneumococcus and meningococcus. Approximately 20–30 per cent of patients die (Mylonakis *et al.* 1998; Gerner-Smidt *et al.* 2005). So although it is a rare cause of food-borne illness, it accounts for 60 per cent of deaths attributed to food-borne disease.

Clinical features

Patients may present with fever, headache, vomiting, and meningism, which is clinically indistinguishable from other causes of acute bacterial meningitis. However, often there are features of a meningoencephalitis with drowsiness, coma, and seizures; in one series 77 per cent of patients had altered consciousness (Mylonakis *et al.* 1998). Tremors, myoclonus, and seizures are common, occurring in 15–25 per cent.

Rhomboencephalitis is described in up to 25 per cent of cases, and appears more commonly in the immunocompetent than the immunocompromised (Armstrong and Fung 1993). Clinically there are cranial neuropathies, especially affecting cranial nerves III, VI, VII, IX, and X. Occasionally patients present with cerebral abscesses or cerebritis.

Diagnosis, treatment, and outcome

As with most central nervous system infections, CSF examination is the key to diagnosis. The CSF typically has a moderate pleocytosis, of around 1000 cells, with lymphocyte predominance, a mildly reduced glucose ratio and high protein. However there can be a wide variation in these parameters, and in one study one-fifth of patients with listeria rhomboencephalitis had a normal CSF cell count (Armstrong and Fung 1993). Microbiological diagnosis can be difficult, but is easier if the laboratory is forewarned that listeria is suspected clinically. The initial gram stain is positive in only 10–40 per cent of patients, and can easily be misinterpreted as showing a pneumococcus. The organism grows slowly and is fastidious, but cultures are often positive eventually. However, if the lab is not pre-warned they can mistakenly think that they have grown a contaminating diphtheroid species. Blood cultures are positive in 40–80 per cent of cases. If listeria is suspected clinically based on risk factors or CSF findings, ampicillin with or without gentamicin should be given. In patients with penicillin allergy, vancomycin is used. However antibiotic treatment is often unsuccessful; the mortality is high at 20–30 per cent in those with risk factors, being lower in those who are otherwise healthy.

41.5.11 Eosinophilic angiostrongylus cantonensis meningitis

Infection with the larvae of the rat lung worm *Angiostrongylus cantonensis* is the most common cause of eosinophilic meningitis, defined as a pleocytosis with more than 10 per cent eosinophils. Humans become infected by eating infected intermediate hosts, such as snails, or contaminated fresh produce, such as lettuce. The larvae migrate to the central nervous system and mature into adult worms that migrate further. The parasite is common in Asia and the Pacific. Other nematode parasites that cause eosinophilic meningitis include *Gnathostoma spinigerum*, a gastrointestinal parasite in dogs and cats, which is common in Southeast Asia, China, and Japan but has been reported worldwide, and *Baylisascaris procyonis*, which is prevalent in raccoon populations in the United States.

Diagnosis and treatment

The diagnosis is often suspected based on a high peripheral eosinophilia and the history of ingestion of suspect food. There is usually a moderate CSF pleocytosis with up to 70 per cent eosinophils, increased protein, and normal glucose ratio. However, the CSF and peripheral eosinophilia may not be present initially (Slom *et al.* 2002). The larvae may be seen in the CSF, but culture and serological studies are not widely available. There is no consensus on whether or not eosinophilic meningitis caused by *A. cantonensis* should be treated with antihelminthics, though in Taiwan mebendazole or albendazole are used, often in combination with corticosteroids (Tsai *et al.* 2001).

References

Addy D (1987). When not to do a lumbar puncture. *Arch Dis Child*, **62**, 873–5.

American College of Physicians (1986). Health and Public Policy Committee. The diagnostic spinal tap. *Ann Intern Med*, **104**, 880–5.

Anderson NE, Willoughby EW, Synek BJ (1995). Leptomeningeal and brain biopsy in chronic meningitis. *Aust N Z J Med*, **25**, 703–6.

Armstrong RW, Fung PC (1993). Brainstem encephalitis (rhombencephalitis) due to Listeria monocytogenes: case report and review. *Clin Infect Dis*, **16**, 689–702.

Attia J, Hatala R, Cook DJ *et al.* (1999). The rational clinical examination. Does this adult patient have acute meningitis? *JAMA*, **282**, 175–81.

Aurelius E, Forsgren M, Gille E *et al.* (2002). Neurologic morbidity after herpes simplex virus type 2 meningitis: a retrospective study of 40 patients. *Scand J Infect Dis*, **34**, 278–83.

Bang E, Kjaer I, Christensen LR (1995). Etiologic aspects and orthodontic treatment of unilateral localized arrested tooth-development combined with hearing loss. *Am J Orthod Dentofacial Orthop*, **108**, 154–61.

Begg N, Cartwright KA, Cohen J *et al.* (1999). Consensus statement on diagnosis, investigation, treatment and prevention of acute bacterial meningitis in immunocompetent adults. British Infection Society Working Party. *J Infect*, **39**, 1–15.

Bergstrom T, *et al.* (1990). Treatment of primary and recurrent herpes simplex virus type 2 induced meningitis with acyclovir. *Scand J Infect Dis*, **22**, 239–40.

Bilukha OO, *et al.* (2005). Prevention and control of meningococcal disease. Recommendations of the Advisory Committee on Immunization Practices (ACIP). *MMWR Recomm Rep*, **54**, 1–21.

Bloom S, Wharton M (2005). Mumps outbreak among young adults in UK. *BMJ*, **331**, E363–4.

Blomberg, J, *et al.* (1974). "New enterovirus type associated with epidemic of aseptic meningitis and/or hand foot and mouth disease." *Lancet* **2**: 112.

Centres for Disease Control and Prevention (CDC) (2000). Outbreak of aseptic meningitis associated with multiple enterovirus serotypes—Romania, 1999. *MMWR Morb Mortal Wkly Rep*, **49**, 669–71.

Centres for Disease Control and Prevention (CDC) (2006). Mumps epidemic—United Kingdom, 2004-2005. *MMWR Morb Mortal Wkly Rep*, **55**, 173–5.

Chang LY, Huang LM, Gau SS *et al.* (2007). Neurodevelopment and cognition in children after enterovirus 71 infection. *N Engl J Med*, **356**, 1226–34.

Chavanet P, Schaller C, Levy C *et al.* (2007). Performance of a predictive rule to distinguish bacterial and viral meningitis. *J Infect*, **54**, 328–36.

Cheng TM, O'Neill BP, Scheithauer BW *et al.* (1994). Chronic meningitis: the role of meningeal or cortical biopsy. *Neurosurgery*, **34**, 590–5; discussion 6.

Corey L, Adams HG, Brown ZA *et al.* (1983). Genital herpes simplex virus infections: clinical manifestations, course, and complications. *Ann Intern Med*, **98**, 958–72.

Corey L, *et al.* (1986). Infections with herpes simplex viruses (2). *N Engl J Med*, **314**, 749–57.

Dattwyler RJ, Luft BJ, Kunkel MJ *et al.* (1997). Ceftriaxone compared with doxycycline for the treatment of acute disseminated Lyme disease. *N Engl J Med*, **337**, 289–94.

Davis LE, *et al.* (2006). Fungal infections. In Johnson RT, Griffin JW, McArthur JC, eds. *Current Therapy in Neurologic Diseases*, pp. 161–9. Philadelphia, Mosby Elesvier.

Davison KL, *et al.* (2003). The epidemiology of acute meningitis in children in England and Wales. *Arch Dis Child*, **88**, 662–4.

de Gans J, *et al.* (2002). Dexamethasone in adults with bacterial meningitis. *N Engl J Med*, **347**, 1549–56.

Desmond RA, Accortt NA, Talley L *et al.* (2006). Enteroviral meningitis: natural history and outcome of pleconaril therapy. *Antimicrob Agents Chemother*, **50**, 2409–14.

Dodge P, *et al.* (1965). Bacterial meningitis. A review of selected aspects. *N Eng J Med*, **272**, 898–902.

Dunn JJ, Romero JR, Wasserman R *et al.* (2000). Stable enterovirus 5' nontranslated region over a 7-year period in a patient with agammaglobulinemia and chronic infection. *J Infect Dis*, **182**, 298–301.

Durand ML, Calderwood SB, Weber DJ *et al.* (1993). Acute bacterial meningitis in adults. A review of 493 episodes. *N Engl J Med*, **328**, 21–8.

Ewer K, Deeks J, Alvarez L *et al.* (2003). Comparison of T-cell-based assay with tuberculin skin test for diagnosis of Mycobacterium tuberculosis infection in a school tuberculosis outbreak. *Lancet*, **361**, 1168–73.

Fischer M, Hedberg K, Cardosi P *et al.* (1997). Tobacco smoke as a risk factor for meningococcal disease. *Pediatr Infect Dis J*, **16**, 979–83.

Fitch MT, *et al.* (2007). Emergency diagnosis and treatment of adult meningitis. *Lancet Infect Dis*, **7**, 191–200.

Gerner-Smidt P, Ethelberg S, Schiellerup P *et al.* (2005). Invasive listeriosis in Denmark 1994-2003: a review of 299 cases with special emphasis on risk factors for mortality. *Clin Microbiol Infect*, **11**, 618–24.

Ginsberg L (2004). Difficult and recurrent meningitis. *J Neurol Neurosurg Psychiatry*, **75** (**Suppl 1**), i16–21.

Gorelick PB (2000). Neuroprotection in acute ischaemic stroke: a tale of for whom the bell tolls? *Lancet*, **355**, 1925–6.

Grande PO, Myhre EB, Nordstrom CH *et al.* (2002). Treatment of intracranial hypertension and aspects on lumbar dural puncture in severe bacterial meningitis. *Acta Anaesthesiol Scand*, **46**, 264–70.

Grosse P, Schulz J, Schmierer K (2003). Diagnostic pitfalls in eosinophilic cryptococcal meningoencephalitis. *Lancet Neurol*, **2**, 512.

Harnden A, Ninis N, Thompson M *et al.* (2006). Parenteral penicillin for children with meningococcal disease before hospital admission: case-control study. *BMJ*, **332**, 1295–8.

Hasbun R, Abrahams J, Jekel J *et al.* (2001). Computed tomography of the head before lumbar puncture in adults with suspected meningitis. *N Engl J Med*, **345**, 1727–33.

Ho M, Chen ER, Hsu KH *et al.* (1999). An epidemic of enterovirus 71 infection in Taiwan. Taiwan Enterovirus Epidemic Working Group. *N Engl J Med*, **341**, 929–35.

Hoen B, Viel JF, Gerard A *et al.* (1993). Mortality in pneumococcal meningitis: a multivariate analysis of prognostic factors. *Eur J Med*, **2**, 28–32.

Hollander H, *et al.* (1987). Human immunodeficiency virus-associated meningitis. Clinical course and correlations. *Am J Med*, **83**, 813–6.

Horwitz SJ, Boxerbaum B, O'Bell J (1978). Cerebral herniation in bacterial meningitis in childhood. *Ann Neurol*, **7**, 524–8.

Hutchinson CM, Rompalo AM, Reichart CA *et al.* (1991). Characteristics of patients with syphilis attending Baltimore STD clinics. Multiple high-risk subgroups and interactions with human immunodeficiency virus infection. *Arch Intern Med*, **151**, 511–6.

Jackson LA, *et al.* (1993). Laboratory-based surveillance for meningococcal disease in selected areas, United States, 1989-1991. *MMWR CDC Surveill Summ*, **42**, 21–30.

Jacob A, Solomon T, Garner P (2006). Corticosteroids in central nervous system. Evidence-based Neurology: Management of Neurological Disorders L. Cndelise, R. Hughes, A. Liberati B. Uitdehaas, C. Warlow Malden MA, Blackwell Publishing pp. 151–60.

Joint Tuberculosis Committee of the British Thoracic Society. (1998). Chemotherapy and management of tuberculosis in the United Kingdom: recommendations 1998. *Thorax*, **53**, 536–48.

Kanegaye JT, Soliemanzadeh P, Bradley JS (2001). Lumbar puncture in pediatric bacterial meningitis: defining the time interval for recovery of cerebrospinal fluid pathogens after parenteral antibiotic pretreatment. *Pediatrics*, **108**, 1169–74.

Kastenbauer S, *et al.* (2003). Pneumococcal meningitis in adults: spectrum of complications and prognostic factors in a series of 87 cases. *Brain*, **126**, 1015–25.

Kirkham FJ (2001). Non-traumatic coma in children. *Arch Dis Child*, **85**, 303–12.

Klein JO, Feigin RD, McCracken GH Jr (1986). Report of the task force on diagnosis and management of meningitis. *Pediatrics*, **78**, 959–82.

Kneen R, Solomon T, Appleton R (2002). The role of lumbar puncture in children with suspected central nervous system infection. *BMC Pediatrics*, **2**, 8.

Kumar R, Singh SN, Kohli N (1999). A diagnostic rule for tuberculous meningitis. *Arch Dis Child*, **81**, 221–4.

Kupila L, Vuorinen T, Vainionpaa R, *et al.* (2005). Diagnosis of enteroviral meningitis by use of polymerase chain reaction of cerebrospinal fluid, stool, and serum specimens. *Clin Infect Dis*, **40**, 982–7.

Lin TY, Hsia SH, Huang YC *et al.* (2003). Proinflammatory cytokine reactions in enterovirus 71 infections of the central nervous system. *Clin Infect Dis*, **36**, 269–74.

Lina B, Pozzetto B, Andreoletti L *et al.* (1996). Multicenter evaluation of a commercially available PCR assay for diagnosing enterovirus infection in a panel of cerebrospinal fluid specimens. *J Clin Microbiol*, **34**, 3002–6.

Litman N, *et al.* (2004). Mumps virus. In Mandell GL, Bennett JE, Dolin R. *Principles and Practice of Infectious Diseases*, pp. 2. Elsevier, Philadelphia.

MacLennan C, Dunn G, Huissoon AP *et al.* (2004). Failure to clear persistent vaccine-derived neurovirulent poliovirus infection in an immunodeficient man. *Lancet*, **363**, 1509–13.

Maguire HC, Atkinson P, Sharland M *et al.* (1999). Enterovirus infections in England and Wales: laboratory surveillance data: 1975 to 1994. *Commun Dis Public Health*, **2**, 122–5.

McKinney RE Jr, Katz SL, Wilfert CM (1987). Chronic enteroviral meningoencephalitis in agammaglobulinemic patients. *Rev Infect Dis*, **9**, 334–56.

McKinney WP, Heudebert GR, Harper SA *et al.* (1994). Validation of a clinical prediction rule for the differential diagnosis of acute meningitis. *J Gen Intern Med*, **9**, 8–12.

McMinn PC (2002). An overview of the evolution of enterovirus 71 and its clinical and public health significance. *FEMS Microbiol Rev*, **26**, 91–107.

Mellor D (1992). The place of computed tomography and lumbar puncture in suspected bacterial meningitis. *Arch Dis Child*, **67**, 1417–19.

Modlin, JF (1995). Coxsackieviruses echoviruses, and newer enteroviruses. *Principles and Practice of Infectious Diseases*. G. L. Mandell, J. E. Bennet and R. Dolin. New York, Churchill Livingstone Inc. **2**: 1621–36.

Molyneux EM, Tembo M, Kayira K *et al.* (2003). The effect of HIV infection on paediatric bacterial meningitis in Blantyre, Malawi. *Arch Dis Child*, **88**, 1112–8.

Moore DA, Evans CA, Gilman RH *et al.* (2006). Microscopic-observation drug-susceptibility assay for the diagnosis of TB. *N Engl J Med*, **355**, 1539–50.

Mylonakis E, Hohmann EL, Calderwood SB (1998). Central nervous system infection with Listeria monocytogenes. 33 years' experience at a general hospital and review of 776 episodes from the literature. *Medicine (Baltimore)*, **77**, 313–36.

Nakashima K (2005). Syndrome of transient headache and neurological deficits with cerebrospinal fluid lymphocytosis: HaNDL. *Intern Med*, **44**, 690–1.

Newton PN, Thai le H, Tip NQ *et al.* (2002). A randomized, double-blind, placebo-controlled trial of acetazolamide for the treatment of elevated intracranial pressure in cryptococcal meningitis. *Clin Infect Dis*, **35**, 769–72.

Nolan MA, Craig ME, Lahra MM *et al.* (2003). Survival after pulmonary edema due to enterovirus 71 encephalitis. *Neurology*, **60**, 1651–6.

Ooi MH, Solomon T, Podin Y *et al.* (2007). Evaluation of different clinical sample types in the diagnosis of human enterovirus 71 associated hand-foot-and-mouth disease. *J Clin Microbiol*, **45**, 1858–66.

Ostergaard C, Brandt C, Konradsen HB *et al.* (2004). Differences in survival, brain damage, and cerebrospinal fluid cytokine kinetics due to meningitis caused by 3 different Streptococcus pneumoniae serotypes: evaluation in humans and in 2 experimental models. *J Infect Dis*, **190**, 1212–20.

Ostergaard C, Konradsen HB, Samuelsson S (2005). Clinical presentation and prognostic factors of Streptococcus pneumoniae meningitis according to the focus of infection. *BMC Infect Dis*, **5**, 93.

Pachner AR, *et al.* (2007). Lyme neuroborreliosis: infection, immunity, and inflammation. *Lancet Neurol*, **6**, 544–52.

Pai M, Flores LL, Pai N *et al.* (2003). Diagnostic accuracy of nucleic acid amplification tests for tuberculous meningitis: a systematic review and meta-analysis. *Lancet Infect Dis*, **3**, 633–43.

Petitjean J, Vabret A, Dina J *et al.* (2006). Development and evaluation of a real-time RT-PCR assay on the LightCycler for the rapid detection of enterovirus in cerebrospinal fluid specimens. *J Clin Virol*, **35**, 278–84.

Plum F, *et al.* (1982). The pathological physiology of signs and symptoms in coma. In Plum F, Posner JB, eds. *The Diagnosis of Stupor and Coma*, pp. 1–73. FA Davis Co, Philadelphia, PA.

Polfliet MM, Zwijnenburg PJ, van Furth AM *et al.* (2001). Meningeal and perivascular macrophages of the central nervous system play a protective role during bacterial meningitis. *J Immunol*, **167**, 4644–50.

Proulx N, Frechette D, Toye B *et al.* (2005). Delays in the administration of antibiotics are associated with mortality from adult acute bacterial meningitis. *QJM*, **98**, 291–8.

Rantakallio P, Leskinen M, von Wendt L (1986). Incidence and prognosis of central nervous system infections in a birth cohort of 12,000 children. *Scand J Infect Dis*, **18**, 287–94.

Renier D, Lacombe J, Pierre-Kahn A *et al.* (1984). Factors causing acute shunt infection. Computer analysis of 1174 operations. *J Neurosurg*, **61**, 1072–8.

Rennick G, Shann F, de Campo J (1993). Cerebral herniation during bacterial meningitis in children. *BMJ*, **306**, 953–5.

Rosenstein NE, Perkins BA, Stephens DS et al. (2001). Meningococcal disease. N Engl J Med, 344, 1378–88.

Rotbart HA (1995). Enteroviral infections of the central nervous system. Clin Infect Dis, 20, 971–81.

Rotbart HA, et al. (1995). Laboratory diagnosis of enteroviral infection. In Rotbart HA, ed. Human Enterovirus Infection, pp. 401–18. American Society for Microbiology, Washington, DC.

Rotbart HA, Ahmed A, Hickey S et al. (1997). Diagnosis of enterovirus infection by polymerase chain reaction of multiple specimen types. Pediatr. Infect Dis J, 16, 409–11.

Rotbart HA, Brennan PJ, Fife KH et al. (1998). Enterovirus meningitis in adults. Clin Infect Dis, 27, 896–8.

Rotbart HA, Sawyer MH, Fast S et al. (1994). Diagnosis of enteroviral meningitis by using PCR with a colorimetric microwell detection assay. J Clin Microbiol, 32, 2590–2.

Schmidt, NJ, et al. (1974). "An apparently new enterovirus isolated from patients with disease of the central nervous system." Journal of Infectious Diseases 129: 304–9.

Schuchat A, Deaver KA, Wenger JD et al. (1992). Role of foods in sporadic listeriosis. I. Case-control study of dietary risk factors. The Listeria Study Group. JAMA, 267, 2041–5.

Schuchat A, Robinson K, Wenger JD et al. (1997). Bacterial meningitis in the United States in 1995. Active Surveillance Team. N Engl J Med, 337, 970–6.

Schwarzkopf A (1996). Listeria monocytogenes--aspects of pathogenicity. Pathol Biol, 44, 769–74.

Shindarov, LM, et al. (1979). "Epidemiological, clinical, and pathomorphological charicteristics of epidemic poliomyelitis-like disease caused by enterovirus 71." Journal of Hygiene, Epidemiology, Microbiology, and Immunology 23: 284–95.

Simon RP (1985). Neurosyphilis. Arch Neurol, 42, 606–13.

Singer JI, Maur PR, Riley JP et al. (1980). Management of central nervous system infections during an epidemic of enteroviral aseptic meningitis. J Pediatr, 96, 559–63.

Slack J (1982). Deaths from meningococcal infection in England and Wales in 1978. J R Coll Physicians Lond, 16, 40–4.

Slom TJ, Cortese MM, Gerber SI et al. (2002). An outbreak of eosinophilic meningitis caused by Angiostrongylus cantonensis in travelers returning from the Caribbean. N Engl J Med, 346, 668–75.

Smith JE, et al. (1994). Outcome of chronic idiopathic meningitis. Mayo Clin Proc, 69, 548–56.

Solomon T (2002). Neurological presentations. In Beeching N, Gill G, eds. Lecture Notes on Tropical Medicine. Blackwell Science, Oxford.

Solomon T, et al. (2005). Editorial: neurological infection in the Lancet Neurology. Lancet Neurology, 4, 139.

Solomon T, et al. (2004). Arthropod-borne viral encephalitides. InScheld M, Whitley RJ, Marra C, eds. Infections of the Central Nervous System. Lippincott Williams and Wilkins, Philiadelphia, PA.

Solomon T, et al. (2003). Infectious causes of acute flaccid paralysis. Curr Opin Infect Dis, 16, 375–81.

Solomon T, Ooi MH, Mallewa M (2007). Viral infections of lower motor neurons. In Eisen A, Shaw P, eds. Motor Neuron Disorders and Related Diseases A. Eisen and P. Shaw. Amsterdaw Elsevier 82: 179–206.

Spanos A, Harrell FE Jr, Durack DT (1989). Differential diagnosis of acute meningitis. An analysis of the predictive value of initial observations. JAMA, 262, 2700–7.

Stephens DS, et al. (1991). Pathogenic events during infection of the human nasopharynx with Neisseria meningitidis and Haemophilus influenzae. Rev Infect Dis, 13, 22–33.

Swartz MN (1987). Chronic meningitis--many causes to consider. N Engl J Med, 317, 957–9.

Tedder DG, Ashley R, Tyler KL et al. (1994). Herpes simplex virus infection as a cause of benign recurrent lymphocytic meningitis. Ann Intern Med, 121, 334–8.

Thomas KE, Hasbun R, Jekel J et al. (2002). The diagnostic accuracy of Kernig's sign, Brudzinski's sign, and nuchal rigidity in adults with suspected meningitis. Clin Infect Dis, 35, 46–52.

Thompson MJ, Ninis N, Perera R et al. (2006). Clinical recognition of meningococcal disease in children and adolescents. Lancet, 367, 397–403.

Thwaites GE, et al. (2005). Tuberculous meningitis: many questions, too few answers. Lancet Neurol, 4, 160–70.

Thwaites G, Chau TT, Mai NT et al. (2000). Tuberculous meningitis. J Neurol Neurosurg Psychiatry, 68, 289–99.

Thwaites GE, Chau TT, Stepniewska K et al. (2002). Diagnosis of adult tuberculous meningitis by use of clinical and laboratory features. Lancet, 360, 1287–92.

Tsai HC, Liu YC, Kunin CM et al. (2001). Eosinophilic meningitis caused by Angiostrongylus cantonensis: report of 17 cases. Am J Med, 111, 109–14.

Tyler KL (2004). Herpes simplex virus infections of the central nervous system: encephalitis and meningitis, including Mollaret's. Herpes, 11 (Suppl 2), 57A–64A.

Uchida K, Shinohara M, Shimada S et al. (2005). Rapid and sensitive detection of mumps virus RNA directly from clinical samples by real-time PCR. J Med Virol, 75, 470–4.

Udani PM, et al. (1970). Tuberculous encephalopathy with and without meningitis. Clinical features and pathological correlations. J Neurol Sci, 10, 541–61.

van Crevel H, Hijdra A, de Gans J (2002). Lumbar puncture and the risk of herniation: when should we first perform CT? J Neurol, 249, 129–37.

van de Beek D, de Gans J, McIntyre P et al. (2004a). Steroids in adults with acute bacterial meningitis: a systematic review. Lancet Infect Dis, 4, 139–43.

van de Beek D, de Gans J, Spanjaard L et al. (2004b). Clinical features and prognostic factors in adults with bacterial meningitis. N Engl J Med, 351, 1849–59.

van de Beek D, de Gans J, Tunkel AR et al. (2006). Community-acquired bacterial meningitis in adults. N Engl J Med, 354, 44–53.

Wadia, NH, et al. (1983). "Polio-like motor paralysis associated with acute haemorrhagic conjunctivitis in outbreak in 1981 in Bombay, India: Clinical and serological studies." Journal of Infectious Diseases 147: 660–8.

Weisfelt M, de Gans J, van der Poll T et al. (2006a). Pneumococcal meningitis in adults: new approaches to management and prevention. Lancet Neurol, 5, 332–42.

Weisfelt M, van de Beek D, Spanjaard L et al. (2006b). Clinical features, complications, and outcome in adults with pneumococcal meningitis: a prospective case series. Lancet Neurol, 5, 123–9.

Williams CJ, Willocks LJ, Lake IR et al. (2004). Geographic correlation between deprivation and risk of meningococcal disease: an ecological study. BMC Public Health, 4, 30.

Winter PM, Dung NM, Loan HT et al. (2004). Proinflammatory cytokines and chemokines in humans with Japanese encephalitis. J Infect Dis, 190, 1618–26.

World Health Organization (2006). Health conditions for travellers to Saudi Arabia for the pilgrimage to Mecca (Hajj). III. Poliomyelitis. Addendum. Wkly Epidemiol Rec, 81, 444.

Wylie P, Stevens D, Drake W et al. (1997). Epidemiology and clinical management of meningococcal disease in west Gloucestershire: retrospective, population based study. BMJ, 315, 774–9.

Yamashita K, Miyamura K, Yamadera S et al. (1994). Epidemics of aseptic meningitis due to echovirus 30 in Japan. A report of the National Epidemiological Surveillance of Infectious Agents in Japan. Jpn J Med Sci Biol, 47, 221–39.

CHAPTER 42

Encephalitis and infectious encephalopathies

Tom Solomon

Contents

42.1 **Introduction**

Micro-organisms can infect the central nervous system in a range of ways. Whereas Chapter 41 considered pathogens that primarily infect the meninges, in this and Chapter 43 organisms that affect the brain parenchyma itself are considered.

There are many different ways in which microbial organisms can do this, but after an initial assessment, which includes imaging and CSF examination, most patients will fall into one of the following syndromes:

◆ *Acute encephalitis*, in which there is reduced consciousness, an encephalopathy, with evidence of inflammation of the brain parenchyma, which is often caused by viruses; sometime this is associated with spinal cord inflammation, myelitis;

◆ *Encephalopathy*, without much evidence of brain parenchymal inflammation, which is often caused by intracellular bacteria or parasites;

◆ *Dementia*, in which there is a gradual cognitive decline, without a reduction in consciousness; and

◆ *Space occupying lesions*, which may be abscesses or other intracranial suppurations, granulomas, or cysts.

In this chapter the general approach to patients with viral encephalitis is discussed, before considering the specific viruses, followed by bacterial and parasitic causes of encephalitis or encephalopathy. For convenience, the section on viruses includes viral myelopathies such as HTLV-1; the section on bacteria includes neurological syndromes caused by bacterial toxins, such as tetanus. The chapter ends with infectious causes of dementia, and in particular prion diseases.

In Chapter 43 infectious causes of space occupying lesions are considered, along with human immunodeficiency virus, HIV, which can present with the whole range of neurological infectious syndromes, including meningitis, altered consciousness, space occupation, and many others.

42.2 **Encephalitis**

42.2.1 **Introduction**

The term 'encephalitis' implies inflammatory change affecting brain parenchyma, and comes from the Greek *enkephalon* meaning brain. Viruses are the most common infectious causes (Table 42.1), but other micro-organisms can cause a meningoencephalitis with inflammation, including protozoa such as toxoplasma, amoebae, and bacteria such as spirochaetes and rickettsiae (Table 42.2). In addition encephalitis can be associated with post infectious processes, as in acute disseminated encephalomyelitis (Section 37.4.1), paraneoplastic processes, such as paraneoplastic limbic encephalitis (Section 38.4.2), or a limbic encephalitis mediated by antibodies against voltage-gated potassium channels (Section 38.3.5). There are many metabolic conditions that can cause a reduction in consciousness, an encephalopathy, and thus enter the differential diagnosis of viral encephalitis (Table 42.2).

Encephalitis most often presents with a severe headache, and an encephalopathy, reflected by a reduction of consciousness, in the context of an acute febrile illness. Also seizures and focal neurological signs are common. The term encephalitis is not usually applied to processes whereby pathogens infect the brain parenchyma without an inflammatory process. For example most authorities do not use the term for HIV infection of the brain because it is associated with relatively little acute inflammation, but more of an insidious deterioration.

The pathogenesis of the inflammation varies with the organism; some viruses directly invade the brain parenchyma, especially the neurons, for example herpes simplex virus type-1. However the blood vessels are sometimes the primary site, for example in varicella zoster virus encephalitis. Often there is a para- or post-infectious immune-mediated component to the disease. Mumps virus can cause an acute viral encephalitis, or a delayed immune-mediated delayed encephalitis. Measles virus causes a post-infectious encephalitis, which can sometimes have a severe haemorrhagic component known as acute haemorrhagic leukoencephalitis. For other viruses a diffuse cerebral oedema is a major component in the pathogenesis, for instance influenza A virus. Acute disseminated encephalomyelitis is a white matter inflammation which is often associated with a systemic viral or other infection (Section 37.4.1).

Strictly speaking, encephalitis is a pathological diagnosis that should only be made if there is tissue confirmation, either on autopsy or brain biopsy. However, in practical terms most patients are diagnosed with encephalitis if they have the appropriate clinical presentation with surrogate markers of brain inflammation, such as inflammatory cells in the CSF, or inflammation shown on imaging; or if an organism is detected which is known to be a common cause of encephalitis.

The organisms that cause encephalitis often also cause meningitis, an associated meningeal reaction, or spinal cord inflammation known as myelitis, or nerve root involvement known as radiculitis; these terms are sometimes used in various combinations to reflect which part of the neuraxis is affected; for example meningoencephalitis, encephalomyelitis, meningoencephalomyelitis, myeloradiculitis, or meningoencephaloradiculitis.

The clinical features are not usually diagnostic, but they may reflect the site of central nervous system damage, and can sometimes be a clue as to the aetiology. For example altered behaviour, olfactory hallucinations, and partial seizures can indicate the fronto-temporal lobe involvement of herpes simplex virus-1. An acute febrile encephalopathy with lower cranial neuropathies and myoclonus suggests a basal meningoencephalitis, otherwise referred to as rhomboencephalitis, as seen with some enteroviruses, or listeria (Table 42.3). Tremors and other movement disorders may reflect basal ganglia involvement seen in the West Nile virus and other flaviviruses. Upper limb weakness and fasciculation suggests a cervical myelitis as seen in tick-borne encephalitis.

The management of patients with suspected encephalitis has been revolutionized in recent years with improved viral diagnostics and imaging studies, better antiviral and immunomodulatory therapies, and enhanced neurointensive care settings. This chapter begins with some general considerations on the epidemiology, pathogenesis, and clinical approach to encephalitis, before discussing some disease-specific considerations.

42.2.2 **Causes of encephalitis**

Almost all viruses that infect humans have been known to cause encephalitis. The important ones can be considered in a range of different ways. For example they can be classified taxonomically, according to whether they are primarily human or animal viruses, whether they occur sporadically or in epidemics, whether they affect the immunocompetent or immunocompromised, or whether

Table 42.1 Viral causes of acute encephalitis (after Solomon et al. 2004)

Groups	Viruses	Comments
DIRECTLY TRANSMITTED:		
Herpes viruses (family *Herpesviridae*)	Herpes simplex virus type 1	The most commonly diagnosed sporadic encephalitis
	Herpes simplex virus type 2	Causes meningitis (esp. recurrent); Meningoencephalitis occurs in the immunocompromised, or in neonates
	Varicella zoster virus type 1	Cerebellitis
	Epstein–Barr virus	Encephalitis in the immunocompromised
	Cytomegalovirus	Encephalitis in the immunocompromised; with retinitis, radiculitis; neutrophilic CSF with low glucose
	Human herpes virus 6 & 7	Febrile convulsions in children (after Roseola); Encephalitis in immunocompromised adults
Enteroviruses (family *Picornaviridae*)	Enterovirus 70	Epidemic haemorrhagic conjunctivitis, with central nervous system involvement
	Enterovirus 71	Epidemic hand foot and mouth disease, with aseptic meningitis, brainstem encephalitis, myelitis
	Poliovirus	Myelitis
	Coxsackieviruses, Echoviruses, Parechovirus	Mostly aseptic meningitis
Paramyxoviruses (family *Paramyxoviridae*)	Measles virus	Causes acute post-infectious encephalitis, subacute encephalitis, and subacute sclerosing panencephalitis
	Mumps virus	Parotitis, orchitis, pancreatitis
Others (rarer causes)	Influenza viruses, Adenovirus, Parvovirus B19, Lymphocytic choreomeningitis virus, Rubella virus,	
ZOONOTIC VIRUSES:	Rabies, other lyssaviruses	Transmitted by rabid dogs, cats, Daubenton's bats in United Kingdom
	Nipah virus	Transmitted in faeces of fruit bats
ARBOVIRUSES (most are also zoonotic):		
Flaviviruses (family *Flaviviridae*)	West Nile virus	North America, Southern Europe, associated with flaccid paralysis, Parkinsonian movement disorders
	Japanese encephalitis virus	Asia, associated with flaccid paralysis, Parkinsonian movement disorders
	Tick-borne encephalitis	Travel in Eastern Europe, Former Soviet Union; tick bite; upper limb flaccid paralysis
	Dengue	Causes fever, arthralgia, rash and haemorrhagic disease, occasional central nervous system disease
Alphaviruses (family *Togaviridae*)	Western, Eastern, and Venezuelan equine encephalitis viruses	Found in the Americas; encephalitis of horses and humans
Bunyaviruses	Chikungunya	Asia Pacific
Coltiviruses	La Crosse virus	Encephalitis in children in America
Vesiculoviruses	Colorado tick fever	
	Chandipura virus	Outbreaks in India

Note viral causes of chronic encephalitis such as JC viruses are not included here.

Table 42.2 Non-viral causes of encephalitis and encephalopathy (after Solomon *et al.* 2004)

DIRECT INFECTION OF THE BRAIN	REMOTE EFFECTS OF INFECTION
Bacteria	**Para/Post-Infectious Causes**
Small bacteria (mostly intracellular)	Acute disseminated encephalomyelitis
Mycoplasma	Viral illnesses with febrile convulsions
Chlamydia	Shigella
Rickettsia (including scrub typhus, rocky mountain spotted fever)	Viral infections associated with swollen fontanelle
Erlichiosis	Guillain–Barré syndrome
Coxiella (Q fever)	**NON INFECTIOUS DISEASES**
Bartonella (Cat Scratch fever)	**Vascular**
Tropherema whipplei (Whipple's disease)	Vasculitis
Brucellosis	Systemic lupus erythematosis
Typhoid fever	Behcet's disease
	Sub-arachnoid and sub-dural haemorrhage
	Ischaemic cerebrovascular accidents
Spirochetes	**Neoplastic**
Syphilis	Primary brain tumour
Lyme Neuroborreliosis	Metastases
Leptospirosis	Paraneoplastic limbic encephalitis
Borrelia recurrentis (Relapsing fever)	
Other bacteria	**Metabolic encephalopathy**
Subacute bacterial endocarditis	Hepatic encephalopathy
Listeria	Renal encephalopathy
Nocardia	Hypoglycaemia
Actinomycosis	Reye's syndrome
Parameningeal infection	Toxic encephalopathy (alcohol, drugs)
Abscess	Hashimoto's disease
Parasites	**Other**
Malaria	Drug reactions
Trypanosomiasis	Epilepsy
Amoebic encephalitis - *Naegleria fowleri*	Hysteria
Cysticercosis	Voltage- gated K$^+$ channel limbic encephalitis
Echinococcus	
Trichinosis	
Amoebiasis	
Fungi	
Cryptococcus	
Candidiasis	
Coccidiomycosis	
Histoplasmosis	
North American blastomycosis	

Almost every infectious and non-infectious condition can occasionally present with an encephalitis-like illness.

In this table some of the important conditions are listed that may need to be considered in the differential after the initial imaging and CSF investigations

Table 42.3 Brainstem encephalitis or rhomboencephalitis. Clues and cases (after Solomon *et al.* 2007)

Suggestive clinical features
Lower cranial nerve involvement
Myoclonus
Autonomic dysfunction
Locked-in syndrome
MRI changes in the brainstem, with gadolinium enhancement of basal meninges
Causes
Enteroviruses (especially EV71)
Flaviviruses, e.g. West Nile virus, Japanese encephalitis virus
Listeria
Tuberculosis
Brucella
Borrelia
Paraneoplastic syndromes

alpha, beta, and gamma herpes viruses, which following primary infection can become latent in a range of different cells, in some cases by incorporation into the human chromosome (Table 42.4). Thus they can cause encephalitis during the primary infection, or by reactivating following latency. This latency and ability to reactivate makes serological diagnosis especially difficult. There are also many important RNA viruses that cause encephalitis. They do not become latent, though some can cause chronic infection. Important RNA viruses that cause encephalitis include the enteroviruses, the paramyxoviruses of measles and mumps, and zoonotic viruses. Zoonotic viruses are naturally viruses of animals that also infect humans causing disease. Rabies is one of the most important zoonotic viruses; although geographically restricted it is sometimes seen among returning travellers. Many zoonotic viruses are arthropod-borne viruses, or arboviruses, being transmitted by insects or ticks. Important arboviral causes of central nervous system disease include Japanese encephalitis virus, West Nile virus, and tick-borne encephalitis virus.

42.2.3 **Epidemiology of encephalitis**

The epidemiology of encephalitis varies geographically, and is changing. Although some viral causes of encephalitis such as human simplex virus-1 appear to have the same incidence across the globe, others, such as the arthropod-borne viruses, or some zoonotic viruses such as rabies have marked geographical variations.

Table 42.4 Herpes viruses that infect the nervous system

	Characteristics
Alpha herpes viruses	Become latent in nerves
Herpes simplex virus type 1	Mostly transmitted via the oral mucosa
Herpes simplex virus type 2	Mostly transmitted via the genital route
Varicella zoster virus	Transmitted in droplet spread
Beta herpes viruses	Thought to become latent in T lymphocytes
Cytomegalovirus	Retinitis, encephalitis, radiculitis
Human herpes virus 6	
Human herpes virus 7	
Gamma herpes viruses	Become latent in B lymphocytes
Epstein–Barr virus	Drives central nervous system lymphomas in the immunosuppressed
Human herpes virus 8	Causes Kaposi's sarcoma in HIV patients

they are geographically restricted, or occur globally (Table 42.1). The herpes viruses herpes simplex virus-1 and 2, varicella zoster virus, Epstein–Barr virus, cytomegalovirus, and human herpes viruses-6 and 7 are all DNA viruses, and are some of the most important causes of sporadic encephalitis. They are divided into

The epidemiology of encephalitis has changed in recent decades. First, in the era of the acquired immune deficiency syndrome, or AIDS, viruses such as cytomegalovirus and Epstein–Barr virus which are associated with immunosuppression are of increasing importance. Second arthropod-borne viruses such as West Nile virus and Japanese encephalitis virus are spreading, to cause epidemic encephalitis in new areas. The contribution of climate change to these trends is debated (Gould *et al.* 2007). Third, disease control measures also have a marked impact on the epidemiology of encephalitis. For example vaccination against measles and mumps had meant that encephalitis caused by these viruses has been very rare in the United Kingdom, though in recent years a reduced uptake of vaccine has meant they are becoming more common again (English *et al.* 2006).

Meanwhile other viruses are causing unexpected outbreaks of neurological disease. For example enterovirus 71 has caused massive outbreaks of hand foot and mouth disease in Asia in recent years, which are often associated with aseptic meningitis, encephalitis, or myelitis (Huang *et al.* 1999; Solomon *et al.* 2003). Nipah virus is a morbillivirus in the same family as measles that was recognized for the first time in 1998 when it caused encephalitis in humans in Malaysia (Chua *et al.* 1999). The virus has also caused disease in Bangladesh and appears to be spreading.

The overall incidence of encephalitis is difficult to determine. Comparisons between studies are hampered by the fact that different definitions of encephalitis are used, most of them are retrospective, and diagnostic capabilities have changed over time. However, an analysis of some of the more important studies shows that the annual incidence of viral encephalitis is probably somewhere between 5 and 10 per 100 000, with a higher incidence among the young and the elderly. A study in Minnesota of 30 years of epidemiological data up to 1981 found an annual incidence of 7.4 per 100 000; encephalitis was more common in the summer months, and was most often cause by the arthropod-borne bunyavirus, California encephalitis virus, and by mumps (Beghi *et al.* 1984). In a study in the 1970s and 1980s of Finnish children the incidence of encephalitis was 8.8 per 100 000 (Rantala *et al.* 1989). The most commonly identified agents, based on virological and serological studies were varicella zoster virus, mumps, herpes simplex virus, and measles. The aetiology remained unknown in 39 per cent of children. No cases of encephalitis caused by mumps, measles, or rubella were found in the population after 1982, when vaccination against these viruses was introduced. A later prospective study in Finland found an overall incidence of 10.5/100 000 child-years with the highest figure in children < 1 year of age, 18.4/100 000 child-years (Koskiniemi *et al.* 1997). The microbiological diagnosis was considered as proven, with evidence of central nervous system infection, or as suggestive in which the clinical diagnosis was supported by serological or other evidence of infection in the periphery in more than 60 per cent of cases. Varicella zoster virus, respiratory, and enteroviruses comprised 61 per cent of these cases, and adenovirus, Epstein–Barr virus, herpes simplex virus, and rotavirus comprised 5 per cent each. In a study of 27 adults with encephalitis admitted to one centre in Sweden over 4 years, herpes simplex virus-1 was the most common cause occurring in 10 patients, followed by influenza A in 2 patients; there were one each with varicella zoster virus, and tick-borne encephalitis virus (Studahl *et al.* 1998). A study in Toronto, Canada found *Mycoplasma pneumonia* was common and accounted for 9 of 50 patients, 4 patients

had herpes simplex virus, and there were one each with human herpes virus-6, influenza A, and the tick-borne Powassan virus. Several patients had evidence of more than one infection, including two with evidence of both mycoplasma and an enterovirus, and one with human herpes virus-6 and influenza A (Kolski *et al.* 1998). In a retrospective study of hospitalization data for England over a 10-year period from 1989 to 1998 (Davison *et al.* 2003), 6414 encephalitis cases were identified, giving an incidence of approximately 1.5 per 100 000, which is lower than the other studies; however the paper recognized that it was probably underestimating the true incidence because of under reporting. In any series of encephalitis patients a large proportion remains undiagnosed. For example in the recent paediatric series from Toronto, Canada confirmed or probably diagnoses were made for 40 per cent of cases, and possible causes identified for a further 26 per cent of cases (Kolski *et al.* 1998).

42.2.4 Pathogenesis of viral encephalitis

The pathogenesis varies according to the virus concerned, but key issues in the pathogenesis are whether this is a primary viral infection or a reactivation of latent virus, how the virus crosses the blood–brain barrier and spreads, and which parts of the central nervous system are targeted. For some viruses most of the damage is caused by viral destruction of cells, whilst for others the host inflammatory and immunological response to infection makes an important contribution (Fig. 42.1). Sometimes these processes result in bystander damage of neighbouring cells that were not infected with virus. For most viruses, the brain parenchyma and neuronal cells are primarily infected, but for some, the blood vessels can be attacked giving a strong vasculitic component. Demyelination following infection can also contribute to the pathogenesis.

Entry into the central nervous system

In general viruses enter the body either:

- via mucosal surfaces of the respiratory, oral, gastrointestinal or genital tracts, or occasionally the cornea; or

- because they have been inoculated across the epidermis of the skin by a biting insect, by the bite or scratch of an animal, or via a needle.

Many of the viruses that are transmitted to humans via arthropods or animal bites are zoonotic viruses; human infections are coincidental, and not an important part of the virus' life cycle. Even for the few arboviruses where humans are the natural host, such as dengue, infection and disease of the central nervous system does not convey any evolutionary advantage to the virus, and is relatively rare. For the most part, those viruses transmitted across mucosal surfaces are those that have evolved to use humans as the natural hosts: the herpes viruses, enteroviruses, measles, and mumps.

There are thought to be three basic means by which virus enters the central nervous system:

- Viraemia with spread across the vascular endothelium. This applies to enteroviruses, such as poliovirus, and arboviruses such as West Nile virus. The virus initially replicates locally in the skin, or mucosal surface and possibly in local lymph tissue. This leads to a primary viraemia, during which more distant tissue of the reticulo-endothelial system is infected, including the liver and spleen. Replication in these tissues leads to a secondary viraemia,

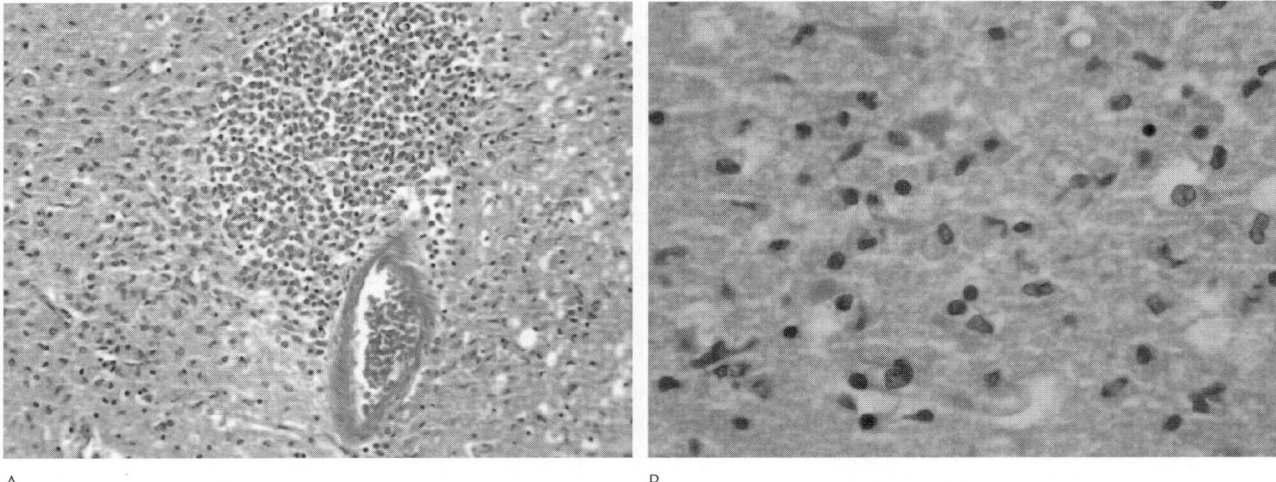

A B

Fig. 42.1 Histopathological picture of the temporal cortex of a man who died from herpes simplex virus encephalitis. (A) Intense perivascular inflammatory infiltrate consisting of activated microglia, macrophages and lymphocytes; haematoxylin, and eosin staining ×20. (B) High power view showing microglia and dead neurons with nuclear dissolution (karyolysis) and hypereosinophilia within the cytoplasm retaining the original pyramidal contour, haematoxylin and eosin staining ×40. (Pictures courtesy of Dr Daniel Crooks and Solomon *et al.* 2007.) (See Plate 36.)

and then spread into 'target' organs including the central nervous system. This occurs by replication in the vascular endothelium of the choroid plexus, of the meninges, and of the brain itself; alternatively virus may be passively transferred across the endothelium.

◆ Centripetal spread along nerves. This is thought to apply to herpes simplex virus following initial infection of the nasopharyngeal or genital mucosa, and to rabies virus following a bite or scratch from an infected dog or bat. The viruses are thought to hitch a ride along the neuron's axonal transport system. Herpes viruses become latent, ready for subsequent reactivation. Rabies virions continue to replicate, and spread centrifugally down axons, so that virions are secreted in saliva and other body fluids.

◆ Carriage within immune cells that traffic across the blood–brain barrier. This is demonstrated by HIV which enters the central nervous system 'Trojan horse-like' within leukocytes that are crossing the blood–brain barrier. HIV enters T lymphocytes using the cell surface protein CD4, along with the chemokine receptor CXCR4 as a co-receptor; 'Macrophage tropic' HIV viruses enter macrophages using CD4 and CCR5 as a coreceptor. Recent evidence suggests that macrophage tropic viruses, are associated with neurotropism, and HIV encephalopathy.

42.2.5 Clinical approach to the patient with acute encephalitis

Clinical features

Classically acute viral encephalitis presents with a febrile illness and reduction of consciousness, often associated with severe headache and seizures. The febrile illness is often a non-specific flu-like illness with malaise, myalgia, fatigue, and headaches; there may be nausea and vomiting or respiratory symptoms. However this can be very mild, and easily overlooked unless a careful history is taken.

The alteration in consciousness may range from subtle changes in behaviour, to confusion and disorientation, speech disturbances, lethargy, drowsiness, and deep coma. Thus occasionally patients with a mild encephalitis can be mistakenly thought to have a psychiatric illness (Table 42.5). Our ability to recognize milder forms of the disease corresponded to a shift in the diagnostic approach from brain biopsy, which was unlikely to be performed in people with mild behavioural problems, to CSF polymerase chain reaction, which is easily performed.

History

In most cases of acute encephalitis there are few or no clues as to the aetiology in the history. However, there may be a history of a recent exanthema such as measles or mumps; even if there is no such history in the patient, ask whether others in the family or in the community have been affected. In children, a recent rash may

Table 42.5 Reasons why encephalitis may not be diagnosed (after Solomon *et al.* 2007)

◆ Wrongly attributing a patient's fever and confusion to a urinary tract infection (based on a urine dipstick) or a chest infection (based on a few crepitations in the chest), without strong evidence

◆ Failure to realize that a patient has a febrile illness, simply because they are not febrile on admission

◆ Ignoring a relative's complaint that a patient is not quite right, is sleepy or lethargic, just because the Glasgow coma score is 15 (the coma score is a very crude tool)

◆ Wrongly attributing clouding of consciousness to drugs or alcohol, without good evidence to do so

◆ Failure to properly investigate a patient with a fever and seizure, following which they do not recover consciousness.

◆ Failure to do a lumbar puncture, even though there are no contraindications suggest chicken pox, parvovirus, or human herpes virus-6. Parotitis, or testicular pain and swelling, or abdomi\nal pain from pancreatitis may suggest mumps virus as the causative organism.

Ask specifically about risk factors for HIV, including male sex with males, multiple sexual partners, visiting prostitutes, intravenous drug abuse, blood transfusions, or having a partner who has any of these risk factors. However, it is advised to offer an HIV test to all patients with undiagnosed or unusual central nervous system infections, whether or not there are apparent risk factors. A travel history is essential, as are vaccination details. Ask if there has been any contact with animals, or whether there are sick animals in the neighbourhood. In the United States some outbreaks of West Nile virus were heralded by sick birds falling from the sky (Mostashari *et al.* 2003). The history of a dog bite in a rabies endemic area may be crucial, even if it appears trivial (Solomon *et al.* 2005). A bite or scratch from a bat may be important. In the United Kingdom and elsewhere in Europe rabies-like viruses, the European bat lyssaviruses, are carried by bats (Nathwani *et al.* 2003). Other risk factors relate to occupation and recreational activities: leptospirosis can be acquired from rat urine during swimming or fresh water sports. Tick-borne encephalitis virus is acquired whilst hiking in the forests of Austria, Germany, or Eastern Europe; Lyme disease is acquired in the same areas and from tick-bites in the New Forest in the United Kingdom and in the north-eastern USA.

Examination

On examination the priorities are to assess the level of consciousness, check that the airway is being protected, and treat any immediate complications such as generalized seizures. Look then for rashes, bites, and other stigmata that may suggest an aetiology, as outlined above. The general medical examination should also include an assessment of whether there might be alternative explanations of the coma; for example injection sites might indicate intravenous drug abuse; check the pockets for prescription medications, or other indications of underlying illness. Examine the chest and urine for infection. However be wary of ascribing reduced consciousness to a urinary or chest infection, especially in someone who is otherwise fit and well. Such patients need a lumbar puncture to rule out a central nervous system infection; especially if the only indication of infection elsewhere is 'one plus' of blood in the urine, or 'a few basal crepitations on chest examination, (Table 42.5).

Examine for evidence that this patient may have HIV or other immunocompromise, including oral hairy leukoplakia, candida in the mouth, cutaneous scars from previous zoster, or generalized

Table 42.6 Clues to HIV infection on examination and initial investigation

General	Weight loss, pyrexia of unknown origin, diarrhoea, generalized lymphadenopathy
Mouth	Oral hairy leukoplakia, candida, Kaposi sarcoma
Skin	Sebbhoraic dermatitis, Kaposi sarcoma, molluscam contageosum, zoster, herpes simplex
Fundoscopy	cotton wool spots, Cytomegalovirus toxoplasma; syphilis
Investigation	Low platelets; low white cell count; high ESR; Brain atrophy on MRI

lymphadenopathy (Table 42.6). Look for seizures including subtle motor seizures, and examine for tongue or cheek biting indicating that a seizure has been missed. Determine whether there is meningism, and look for focal neurological signs, including hemispheric signs suggestive of a mass lesion, or if there is flaccid paralysis suggestive of spinal cord involvement. Abnormal movements are seen in some forms of encephalitis, particularly those that involve the basal ganglia, such as flaviviruses or toxoplasma.

42.2.6 Investigations

A peripheral blood count in patients with encephalitis may show a leukocytosis or leukopenia; atypical lymphocytes are seen in Epstein–Barr virus infection. Hyponatremia is common, and due to the syndrome of inappropriate antidiuretic hormone. An elevated amylase may occur in mumps virus infection. Blood cultures should be part of the investigation of possible bacterial causes, and a chest X-ray performed to look for pulmonary infiltrates, for example in *Mycoplasma pneumonia*, legionella, or lymphocytic choriomeningitis virus. A low white cell count with low platelet count, and high ESR may be a clue that the patient has HIV.

A lumbar puncture is an essential investigation for any patient with suspected encephalitis, as it is for many other central nervous system infections. However a lumbar puncture may worsen the clinical condition by altering the pressure in the different compartments if patients have a space occupying lesion, marked brain swelling, or incipient brain shift involving herniation across the midline of the tentorium or through the foramen magnum. Raised intracranial pressure itself is not a problem; patients with idiopathic intracranial hypertension are treated by repeated lumbar puncture. In most patients with central nervous system infection the CSF opening pressure is elevated to some degree; the problem is the swelling or shift which is sometimes associated with the raised pressure. For this reason patients should be assessed for focal clinical signs which might indicate a space occupying lesion, brain swelling or shift; if the signs are present, a CT scan should be performed before deciding on lumbar puncture (Fig. 42.2). The indications for a CT include focal neurological signs, such as a hemiparesis which may indicate an abscess, deep coma, flexor or extensor posturing, papilloedema, or recent seizures. Some recommend imaging before lumbar puncture in all patients with known immunocompromise, especially advanced HIV disease, because they may have space occupying lesions without the usual clinical features.

If a lumbar puncture is to be delayed until after a CT, then it may be appropriate to start presumptive antimicrobial treatment whilst awaiting the results. There are no hard and fast rules about how long a delay is acceptable before treatment is initiated ((Solomon *et al.* 2007). For bacterial meningitis a delay of more than 6 h between arrival in hospital and initiation of antibiotic treatment is associated with a worse outcome (Proulx *et al.* 2005). If patients have a purpuric rash suggestive of meningococcal septicaemia, or are deteriorating rapidly, antibiotics should be started immediately (Heyderman *et al.* 2003). In herpes simplex virus-1 encephalitis, a bad outcome is associated with delay of 2 or more days between hospitalization and starting treatment (Raschilas *et al.* 2002). And given the relative rarity of herpes simplex virus encephalitis, it is probably reasonable, in a stable patient, to wait a few hours to see if the lumbar puncture result is consistent with viral encephalitis before deciding on antiviral treatment.

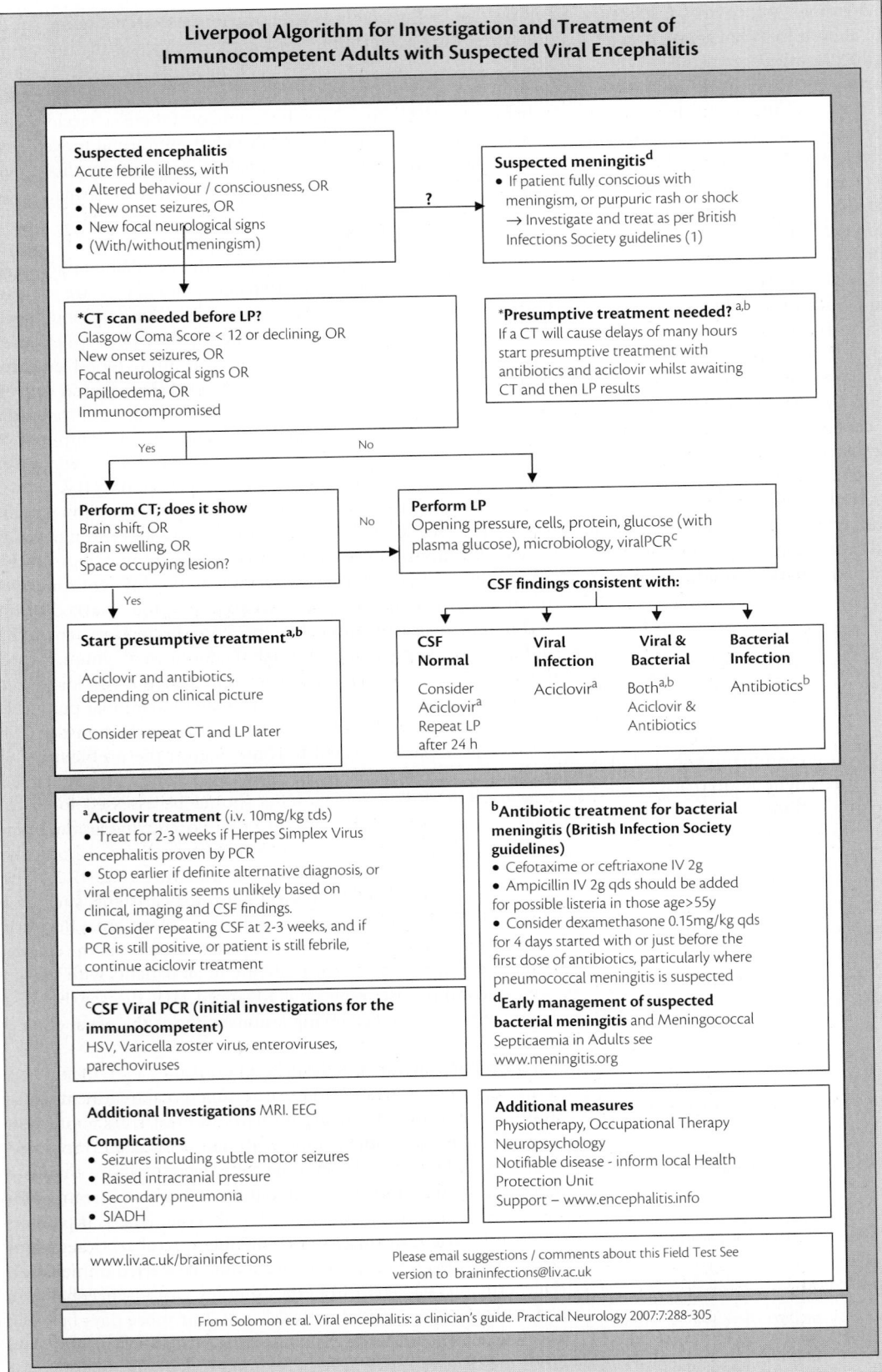

Liverpool Algorithm for Investigation and Treatment of Immunocompetent Adults with Suspected Viral Encephalitis

Suspected encephalitis
Acute febrile illness, with
- Altered behaviour / consciousness, OR
- New onset seizures, OR
- New focal neurological signs
- (With/without meningism)

?

Suspected meningitis[d]
- If patient fully conscious with meningism, or purpuric rash or shock → Investigate and treat as per British Infections Society guidelines (1)

*CT scan needed before LP?
Glasgow Coma Score < 12 or declining, OR
New onset seizures, OR
Focal neurological signs OR
Papilloedema, OR
Immunocompromised

*Presumptive treatment needed? [a,b]
If a CT will cause delays of many hours start presumptive treatment with antibiotics and aciclovir whilst awaiting CT and then LP results

Yes / No

Perform CT; does it show
Brain shift, OR
Brain swelling, OR
Space occupying lesion?

No →

Perform LP
Opening pressure, cells, protein, glucose (with plasma glucose), microbiology, viralPCR[c]

Yes

CSF findings consistent with:

Start presumptive treatment[a,b]

Aciclovir and antibiotics, depending on clinical picture

Consider repeat CT and LP later

CSF Normal	Viral Infection	Viral & Bacterial	Bacterial Infection
Consider Aciclovir[a] Repeat LP after 24 h	Aciclovir[a]	Both[a,b] Aciclovir & Antibiotics	Antibiotics[b]

[a]**Aciclovir treatment** (i.v. 10mg/kg tds)
- Treat for 2-3 weeks if Herpes Simplex Virus encephalitis proven by PCR
- Stop earlier if definite alternative diagnosis, or viral encephalitis seems unlikely based on clinical, imaging and CSF findings.
- Consider repeating CSF at 2-3 weeks, and if PCR is still positive, or patient is still febrile, continue aciclovir treatment

[b]**Antibiotic treatment for bacterial meningitis (British Infection Society guidelines)**
- Cefotaxime or ceftriaxone IV 2g
- Ampicillin IV 2g qds should be added for possible listeria in those age>55y
- Consider dexamethasone 0.15mg/kg qds for 4 days started with or just before the first dose of antibiotics, particularly where pneumococcal meningitis is suspected

[c]**CSF Viral PCR (initial investigations for the immunocompetent)**
HSV, Varicella zoster virus, enteroviruses, parechoviruses

[d]**Early management of suspected bacterial meningitis** and Meningococcal Septicaemia in Adults see www.meningitis.org

Additional Investigations MRI. EEG

Complications
- Seizures including subtle motor seizures
- Raised intracranial pressure
- Secondary pneumonia
- SIADH

Additional measures
Physiotherapy, Occupational Therapy
Neuropsychology
Notifiable disease - inform local Health Protection Unit
Support – www.encephalitis.info

www.liv.ac.uk/braininfections

Please email suggestions / comments about this Field Test See version to braininfections@liv.ac.uk

From Solomon et al. Viral encephalitis: a clinician's guide. Practical Neurology 2007:7:288-305

Fig. 42.2 Investigation and treatment of immunocompetent adults with suspected viral encephalitis; the Liverpool algorithm. (From Solomon *et al.* 2007.) Key: LP lumbar puncture; HSV herpes simplex virus; PCR polymerase chain reaction.

The CSF opening pressure is often mildly raised in encephalitis, and there is usually a mild to moderate CSF lymphocytosis of 5 to 1000 cells/mm³; however early in the infection the white cell count may be normal, or neutrophils may predominate. The CSF red cell count is usually normal, or mildly elevated, but it may be markedly raised in herpes simplex virus-1 encephalitis, which can be haemorrhagic, or in acute necrotizing haemorrhagic leukoencephalitis. The glucose ratio is usually normal in viral central nervous system infections, though it may be mildly reduced. The CSF protein is often mildly elevated to between 0.5 and 1.0g/l.

Diagnostic virology

The diagnosis of a viral central nervous system infection is based on demonstrating the virus within the central nervous system, for example from culture or polymerase chain reaction of CSF or brain tissue, or by demonstrating a specific antibody response in the CSF (Table 42.7). Less strong evidence comes from detecting virus elsewhere in the body for instance from throat, rectum or vesicles, or by showing an antibody response to the virus in the serum. In these instances the organism is presumed to be responsible for the clinical presentation, although there is always the possibility that it represents a coincidental infection.

The definitive diagnosis of viral encephalitis used to rely on brain biopsy (Whitley *et al.* 1989); however, many important viruses can now be detected using polymerase chain reaction (Dennett *et al.* 1991). This has been used extensively for the detection of herpes simplex virus, and other herpes viruses. Polymerase chain reaction tests for herpes simplex virus have overall sensitivity and specificity >95 per cent, but may be negative in the first few days of the illness, or after about 10 days (Cinque *et al.* 1996; Steiner *et al.* 2005). Initial investigations for patients with encephalitis should thus include polymerase chain reaction for herpes simplex virus, and varicella zoster virus, because these are potentially treatable with aciclovir. In immunocompromised patients Epstein–Barr virus and cytomegalovirus polymerase chain reaction should also be performed, and human herpes virus-6 and human herpes virus-7 considered. Mulitplex polymerase chain reactions are now available that will look for all of several viruses in a single reaction. Measles and mumps should be sought if there is a suggestive clinical indication, although they can occasionally cause encephalitis in patients with no other features. Other viruses to consider, especially in children, include adenoviruses, human herpes virus-6 and 7, respiratory syncitial virus, and influenza virus A and B; rotaviruses and parvovirus B19 also occasionally associated with central nervous system disease, especially in children (Barah *et al.* 2001; Goldwater *et al.* 2001). Polymerase chain reaction can also be used to detect *Chlamydia pneumoniae* in the CSF. CSF culture is now rarely performed, though it has the advantage over PCR of potentially being able to detect any viruses, whereas PCR is specific to the virus being looked for.

Newer more sensitive polymerase chain reaction methods, such as 'real-time' and quantitative polymerase chain reaction have improved the clinical utility of this test, and are becoming available for herpes viruses and enteroviruses (Rand *et al.* 2005). One problem is that the high sensitivity of some of the recent polymerase chain reaction assays for herpes viruses, such as Epstein–Barr virus, can make positive results difficult to interpret. Most of the adult population has been infected with this virus and carries it

Table 42.7 Staged approach to investigation of patients with suspected viral encephalitis

CSF polymerase chain reaction (PCR)

1. All patients
   Herpes simplex virus-1, Herpes simplex virus-2, Varicella zoster virus Enterovirus, parechovirus,

2. If indicated
   Epstein–Barr virus / cytomegalovirus (especially if immunocompromised)
   Human herpes virus 6, 7 (especially if immunocompromised, or children)
   Adenovirus, influenza A and B, rotavirus (children)
   Measles, mumps (if clinically indicated)
   Parvovirus B19
   Chlamydia (if clinically indicated)

3. Special circumstances
   Rabies, West Nile virus, tick-borne encephalitis virus (if appropriate exposure)

Antibody testing (where indicated)*

1. Viruses: IgM and IgG in CSF and serum (acute and convalescent), for antibodies against:
   Herpes simplex virus 1 and 2, varicella zoster virus, cytomegalovirus, human herpes virus 6 and 7, enteroviruses, respiratory syncytial virus, parvovirus B19, adenovirus, influenza A and B; *

2. If associated with atypical pneumonia, test serum for:
   Mycoplasma serology and cold agglutinins
   Chlamydia serology

Ancillary investigations (these establish systemic infection, but not necessarily the cause of the central nervous system disease)

- Throat swab, nasopharyngeal aspirate, rectal swab
   Polymerase chain reaction /culture of throat swab, rectal swab for enteroviruses
   Polymerase chain reaction of throat swab for mycoplasma, chlamydia
   Polymerase chain reaction / antigen detection of nasopharyngeal aspirate for respiratory viruses, adenovirus, influenza virus (especially children)

- Vesicle electron microscopy, polymerase chain reaction and culture in:
   Patients with herpetic lesions (for herpes simplex virus, varicella zoster virus)
   Children with hand foot and mouth disease (for enteroviruses)

Brain biopsy

   For culture, electron microscopy, polymerase chain reaction, and immunohistochemistry

* Antibody detection in the serum identifies infection (past or recent depending on the type of antibodies) but does not necessarily mean this virus has caused the central nervous system disease

in their lymphocytes. Therefore, there is debate about whether detection of the viruses by polymerase chain reaction of the CSF represents true pathogenic infection, rather than just the presence of infected lymphocytes (Tang *et al.* 1997). Where there is uncertainty about the significance of a result, the amount of virus in the CSF compared with the blood, as determined by quantitative-polymerase chain reaction, usually help resolve it. For example in a patient with HIV, a CSF cytomegalovirus polymerase chain reaction titre that is higher than that in the serum, or a rising serum titre is usually significant.

Antibody testing

Antibody testing continues to play an important role in the diagnosis of many viral central nervous system infections. Traditional techniques required the demonstration of a four-fold rise in antibody between acute and convalescent serum samples collected

2–4 weeks apart, and thus are not helpful in making an early diagnosis. In practice doctors often forget to take the convalescent samples.

Newer enzyme immunoassays can detect immunoglobulin IgM and IgG antibodies in the serum and CSF against many of the important RNA viruses, as well as *Mycoplasma pneumoniae*. Specific anti-viral IgM is often produced within a few days of a primary infection and can be measured by IgM enzyme immunoassays. The detection of virus specific IgM antibodies in the CSF in higher titres than in serum indicates local production of antibody in the central nervous system in response to infection. IgM does not normally cross the blood–brain barrier because of its size. However, if there is inflammation the barrier is leaky to IgM, and other immunoglobulins. In this circumstance, the ratio of CSF to serum for the specific IgM antibody can be compared to the ratio for immunoglobulin as a whole, to decide if this is local production rather than leak across an inflamed blood brain barrier. IgM detection is especially useful for flavivirus infections, but less so for herpes virus infections, which are often reactivations. In contrast to IgM, IgG is normally found in the CSF at a ratio of 1/200th of the serum concentration. Hence in a primary acute central nervous system infection, IgG rises later than IgM both in the CSF and serum. In reactivations and secondary infections, IgG tends to rise earlier and to a greater extent than IgM.

The detection of oligoclonal bands is sometimes a useful non-specific indicator that a patient has an inflammatory process in the central nervous system, rather than a non-inflammatory cause of encephalopathy. Immunoblotting of the bands against viral proteins has been used but mostly as a research tool to help determine the cause of the inflammation, for instance herpes simplex virus-1 or herpes simplex virus-2 (Cinque *et al.* 1996; Chataway *et al.* 2004). However, the detection of intrathecal anti-herpes simplex virus antibody has a sensitivity of 50 per cent by 10 days after clinical presentation, and is thus only considered useful for retrospective diagnosis.

Thus the approach for diagnosing herpes simplex virus infections, is to perform polymerase chain reaction on CSF acutely. If this is negative, but suspicion remains high, a lumbar puncture with polymerase chain reaction of the CSF it should be repeated after a few days. As indicated above polymerase chain reaction can sometimes be negative if the sample is taken early in the illness. However, if two CSFs are negative for herpes simplex virus by polymerase chain reaction then it is unlikely that the patient has this infection. If for logistic reasons CSFs were not taken at the appropriate time, or had not been sent for herpes simplex virus polymerase chain reaction, then it can useful to test a late CSF after more than 10 days of hospitalization, for the production of intrathecal antibodies against herpes simplex virus, by IgM, IgG, or immunoblotting of oligoclonal bands.

The list of all possible investigations for causes of viral central nervous system disease is clearly very long. Our practice is to do the initial CSF polymerase chain reaction for herpes simplex virus and varicella zoster virus as soon as the samples are received, because of their importance for early management. Polymerase chain reaction for enteroviruses and parechoviruses is also usually done at this stage. In addition, patients that are immunocompromised are investigated with cytomegalovirus and Epstein–Barr virus polymerase chain reaction. Further investigations are guided by the clinical picture, especially the patient's immune status

and the initial CSF findings. The opinion of a clinical virologist in assessing such patients is invaluable.

Investigation of other samples

When direct demonstration of central nervous system invasion, or local antibody production in the CSF has not been possible, other investigations can give important clause that a patient has been systemically infected with a virus or other pathogen that causes encephalitis. These investigations include investigating throat swabs, rectal swabs, nasopharyngeal aspirates, and urine with polymerase chain reaction, culture, antigen detection, or electron microscopy for respiratory viruses, enteroviruses, and other causes of exanthems (Table 42.7).

In general, virus detected in otherwise sterile sites such as vesicles, is more likely to be causal than virus from non-sterile sites. For example, enteroviruses detected in vesicle swabs indicate that they are temporally associated with the acute neurological disease, whereas their shedding from the rectum occurs for several weeks after an acute enterovirus infection which may itself have been coincidental and nothing to do with the acute neurological presentation (Ooi *et al.* 2007).

Brain biopsy

With the advent of CSF polymerase chain reaction, brain biopsy of patients with suspected encephalitis has become less common. Previously it was considered to be the gold standard for diagnosis of herpes simplex virus-1 encephalitis. It may still have a role in undiagnosed patients that are deteriorating. In one study approximately half of patients with suspected herpes simplex virus-1 encephalitis who had a brain biopsy had alternative diagnoses, of which 40 per cent were treatable (Whitley *et al.* 1989).

Imaging and EEG studies

A CT scan is often performed acutely to rule out any space occupying causes, such as cerebral abscess, of the presenting syndrome of fever, and reduced consciousness and to ensure there is no other contraindication to a lumbar puncture, such as marked swelling and brain shift. In herpes simplex virus encephalitis the scan may be normal initially, of there may be subtle swelling of the front-temporal region with loss of the normal gyral pattern; this is followed by hypodensity, and then later there may be high signal change indicating haemorrhagic transformation (Fig. 42.3).

MRI is generally more sensitive, showing high signal intensities in the affected brain areas (Fig. 42.3); though even MRIs can be normal if performed very early (Tien *et al.* 1993). Diffusion-weighted MRI may be especially useful at demonstrating early changes (McCabe *et al.* 2003).

An EEG usually shows non-specific diffuse high amplitude slow waves of encephalopathy, but it can be useful to look for subtle motor seizures. Periodic lateralized epileptiform discharges were once thought to be diagnostic of herpes simplex virus encephalitis (Fig. 42.4), but have since been seen in other conditions (Solomon *et al.* 2002).

42.2.7 Acute management

There are three elements to the acute management of a patient with encephalitis. The first is consideration of whether there is any antiviral or immunomodulatory treatment to halt or reverse the disease process; the second is to control the immediate complications of the encephalitis; and the third is to prevent some of the later complications.

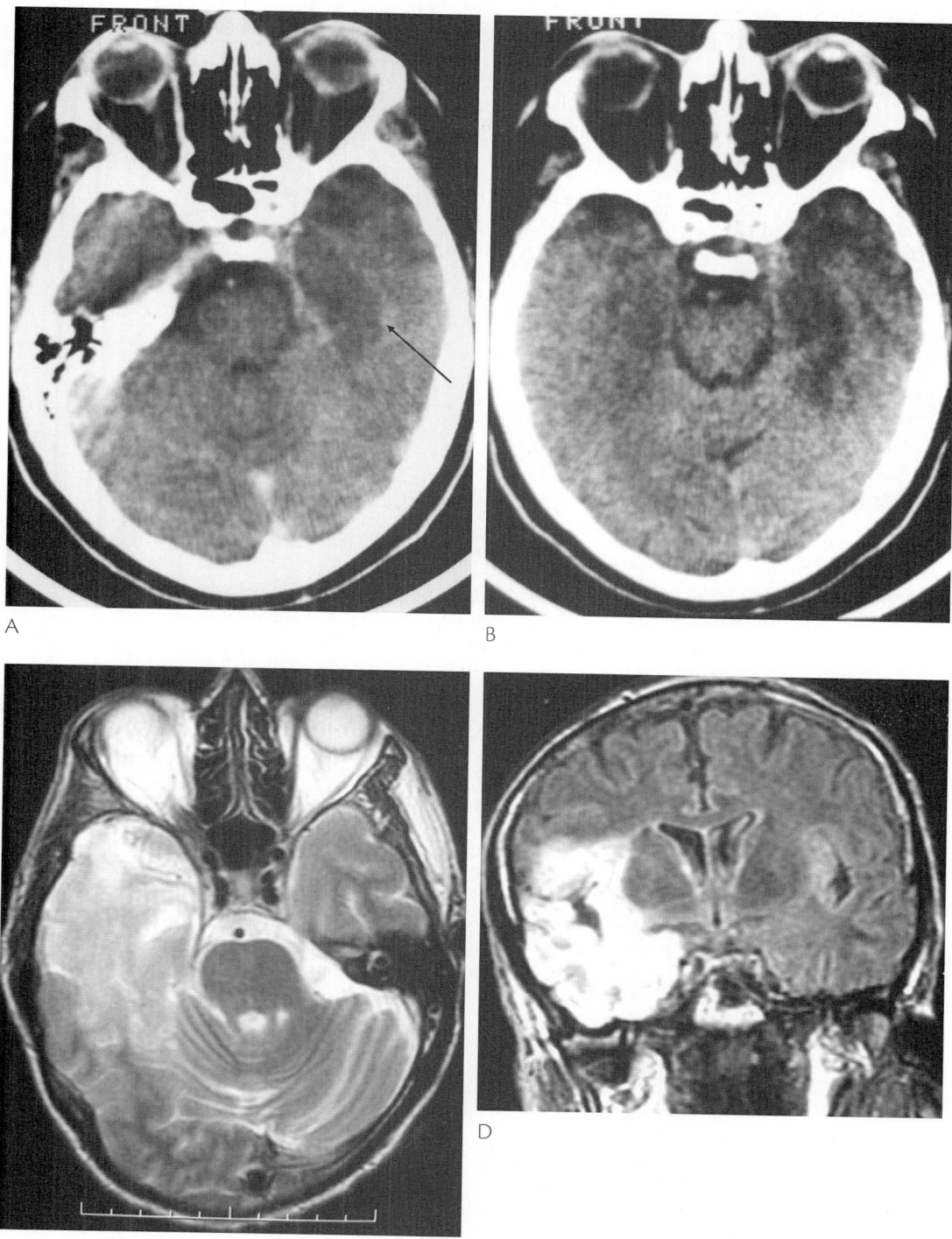

Fig. 42.3 Herpes simplex virus encephalitis in two patients.
(A) and (B). A middle-aged man with a 1-week history of flu-like illness, severe headache, and increasing confusion, had herpes simplex virus encephalitis confirmed by CSF polymerase chain reaction. (A) A low density area in the left temporal lobe, on CT scan with swelling and some contrast enhancement (arrow).
(B) Enhanced CT scan of the same patient 4 days later with more marked changes.
(C) and (D). A 52-year-old man with high fever, malaise, frontal headaches, and drowsiness, who had herpes simplex virus encephalitis diagnosed by polymerase chain reaction. T2 weighted MR brain scans showing high signal intensity in the right temporal lobe, (C) Axial FSE, (D) coronal Flair. (Pictures T Solomon.)

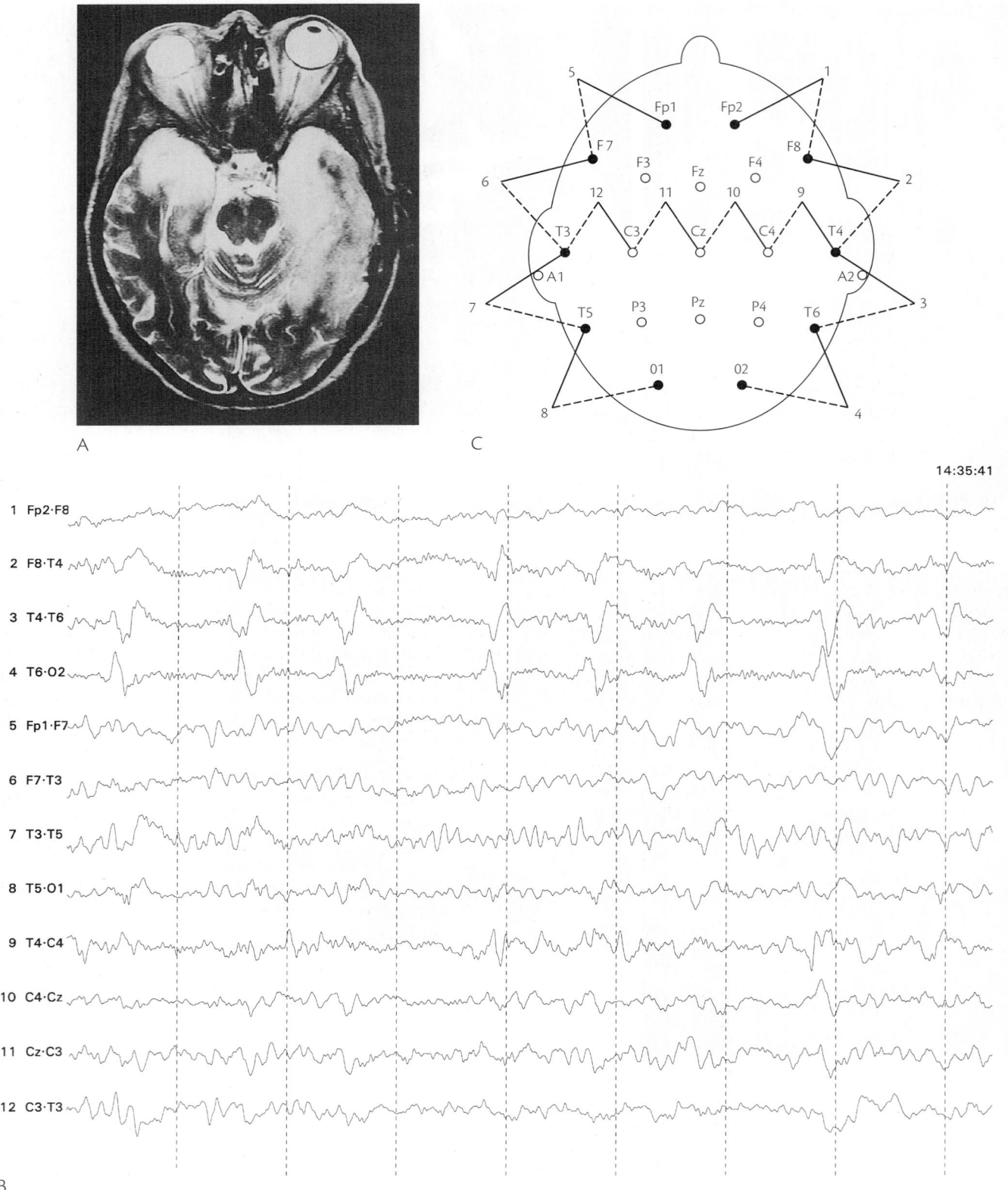

Fig. 42.4 Herpes simplex encephalitis. (A) MRI showing assymmetrical bilateral temporal lobe signal change. (B) 12 lead EEG recorded from the electrode pairs shown in (C) and demonstrating periodic lateralised epileptiform discharges, seen especially over leads 2, 3 and 4 recorded from the more affected temporal lobe.

Antiviral and immunomodulatory treatment

In most immunocompetent patients aciclovir should be given as soon as there is a strong suspicion of a viral encephalitis. This suspicion will be based on the clinical presentation and initial CSF and/or imaging findings.

Intravenous aciclovir at a dose of 10 mg/kg three times daily is effective treatment for herpes simplex virus encephalitis, and so should be started if the CSF picture is consistent with viral encephalitis. It reduces the risk of a fatal outcome from approximately 70 per cent to less than 20 per cent (Skoldenberg et al. 1984; Whitley et al. 1986). Ideally the decision to start aciclovir should await the CSF results. Thus a patient with fever and reduced consciousness needs prompt investigation with a lumbar puncture, if there is just a mild reduction in coma score, or a CT scan and then lumbar puncture if the coma score is markedly reduced or there are other indications. However, if there are likely to be long delays in arranging these investigations, then it is reasonable to start the aciclovir, and then proceed with the investigations. Antibiotics may also be needed if the patient has meningism or other features suggestive of meningitis.

However the indiscriminate use of aciclovir in all patients with fever and reduced consciousness without any investigations is not advised, because it puts patients at unnecessary risks from the rare side effects of the drug, such as renal failure, local inflammation, hepatitis, and bone marrow failure. In addition this approach risks missing other diagnoses that may require alternative treatments (Chataway et al. 2004). Although the mortality of herpes encephalitis has reduced with aciclovir treatment, the morbidity remains high: two-thirds of survivors have significant neuropsychiatric sequelae, including memory impairment in 69 per cent, personality and behavioural change in 45 per cent, dysphasia in 41 per cent, and epilepsy in up to 25 per cent (McGrath et al. 1997).

The original aciclovir trials were for a 10-day treatment, but most physicians would continue for 14–21 days, especially in patients with proven herpes encephalitis, because of the risk of relapse after 10 days (Yamada et al. 2003). The prognostic value of quantitative polymerase chain reaction detection of viral DNA at the end of a 3-week treatment, and prolonged high dose oral valaciclovir for 3 months is being evaluated by the American Collaborative Antiviral Study Group. In patients with brain swelling steroids and mannitol may be used to control raised intracranial pressure. A recent trial suggests steroids may be beneficial even in patients without marked swelling (Kamei et al. 2005). Their role in herpes simplex virus encephalitis merits further study (Openshaw et al. 2005). In patients with severe brain swelling, decompressive hemicraniectomy is sometimes performed.

Antibiotic treatment is sometimes also given, especially if the initial CSF findings could be consistent with bacterial disease. If listeria is suspected ampicillin and gentamicin should be given. The patients at risk of listeria are the elderly and the immunocompromised, who typically have a lymphocytic CSF with a normal or low glucose ratio. Aciclovir should not be stopped early, on the basis of a single negative polymerase chain reaction result, because false positives may occur. However if a definitive alternative diagnosis has become apparent then it is reasonable to stop aciclovir (Fig. 42.2). Use of oral valaciclovir may be reasonable after the first 10 days of intravenous treatment. In circumstances where ongoing intravenous treatment is proving difficult, as in a child who is now fully conscious (Chan et al. 2000). Valaciclovir is an ester of aciclovir, which is converted to aciclovir after absorption, and has good oral bioavailability. However oral aciclovir should not be used, because the levels achieved in the CSF are not high enough.

Other than for herpes simplex virus encephalitis, there are few large randomized controlled trials assessing the efficacy of antiviral treatments in viral central nervous system infections. However, for other conditions treatments are given based on our understanding of the pathogenesis, in vitro data, anecdotal reports, or small clinical series, though these sometimes provide conflicting data (Table 42.8). For example, aciclovir is also used in other herpes virus infections, particularly those caused by varicella zoster virus, where it is usually combined with steroid treatment, on account of the apparent immune basis of much of central nervous system pathology caused by varicella zoster virus. Severe cytomegalovirus and human herpes virus-6 infections are treated with ganciclovir, foscarnet, or cidofovir. Severe adenovirus infections have been treated with cidofovir or ribavirin and pleconaril has been used for severe enterovirus infections, particularly in the immunocompromised, although its overall role remains unclear (Webster 2005; Desmond et al. 2006). Interferon alpha has been used in West Nile virus and other flavivirus infections, but a randomized controlled trial in Japanese encephalitis showed it was not effective (Solomon et al. 2003).

If the clinical presentation and imaging findings suggest a post- or para-infectious encephalitis such as acute disseminated encephalomyelitis, acute haemorrhagic leukoencephalitis and diffuse encephalopathies associated with systemic viral infection, immunosuppressive drugs are given. High-dose steroid treatment is often the initial treatment, followed by intravenous immunoglobulin, plasmaphoresis, or further treatment with steroids, in patients that do not appear to be responding.

Treatment of immediate complications of infection

Standard intensive care measures for patients with encephalitis include oxygen given via a mask, paying attention to fluid and hydration, nasogastric or parenteral feeding, and treating the complications of infection such as pneumonia. Patients with an impaired gag reflex, or a moderate to severely reduced coma score should be intubated and ventilated.

Management of seizures and raised intracranial pressure

Seizures are common in encephalitis, particularly in children. In adults seizures can be useful in helping to distinguish acute viral encephalitis from para-infectious inflammatory encephalopathies. There may be obvious generalized tonic clonic seizures or subtle motor seizures, which may manifest as twitching of a digit, or around the mouth or eyes. An EEG should be performed if there is any uncertainty, and continuous monitoring may be useful in some settings.

Seizures can sometimes be the initial presenting feature of a patient with encephalitis in whom other features such as headache or a prodromal febrile phase was not apparent (Elbers et al. 2007). Although a mild increase in temperature can occur after a seizure, particularly a prolonged seizure, in most cases a patient admitted with a first seizure, who is subsequently found to have pyrexia should be investigated urgently for a suspected central nervous system infection.

Table 42.8 Treatment options to consider in encephalitis (modified from Boos *et al.* 2003)

Acute viral encephalitis	
Herpes simplex virus	Aciclovir
Varicella zoster virus (including cerebellitis)	Aciclovir plus corticosteroids
Human herpes virus-6	Ganciclovir, Foscarnet
Rabies	* Ketamine, Amantidine, Ribavirin
Subacute/chronic encephalitis	
In the immuncompromised	
Varicella zoster virus	Aciclovir
Cytomegalovirus	Ganciclovir, foscarnet,cidofovir
Measles inclusion body enephalitis	Ribavirin
Enterovirus	Pleconaril, specific immunoglobulin
Progressive multifocal leukoencephalopathy in HIV patients	Highly active antiretroviral treatment (HAART), cidofovir
In the immunocompetent	
Subacute sclerosing panencephalitis	Interferon alpha (intraventricular), ribavirin inosiplex
Progressive multifocal leukoencephalopathy	Consider cytosine arabinoside
Progressive rubella panencephalitis	Plasma exchange
Rasmussen's encephalitis	Intravenous immunoglobulin
Immune-mediated encephalitis	
Acute dissemintated encephalomyelitis	Steroids /intravenous immunoglobulin/plasmapharesis
Paraneoplastic	Treatment of underlying tumour, consider immunosuppression
Voltage-gated K channel limbic encephalitis	Intravenous immunoglobulin/plasma exchange plus corticosteroids

* considered experimental

Acute disseminated encephalomyelitis presents acutely; the others typically present with a subacute encephalitis.

Uncontrolled seizures lead to raised intracranial pressure, increased metabolic activity, acidosis, and vasodilatation, which in turn leads to further raised pressure; the resulting positive feedback cycle can ultimately precipitate brain shift and herniation. If seizures are not easily controlled with phenytoin and low doses of benzodiazepines, patients should be intubated and ventilated mechanically, so that higher doses of sedating anticonvulsive drugs, including benzodiazepines and phenobarbitol, can be used. Electroencephalographic monitoring should be used to look for ongoing ictal activity.

Standard measures to control raised intracranial pressure include nursing the patient at 30°, keeping the head straight to ensure there is no obstruction to venous return, and ventilating to keep a low pCO_2. Although there are not good data for viral encephalitis, data from other infectious encephalopathies suggest that osmotic diuretics give a short-term reduction in pressure (Newton *et al.* 1997). The role of anti-inflammatory drugs in viral encephalitis is uncertain (Openshaw *et al.* 2005; Jacob *et al.* 2006).

Management in the recovery period

To reduce the risk of deep venous thrombosis and pulmonary embolus, patients with reduced mobility should be fitted with anti-thrombosis stockings, and once it is clear that there is no major haemorrhagic component to the encephalitis, they should be given prophylactic heparin. Bed sores are a risk in immobile patients, and appropriate mattresses, and regular turning are needed. Patients with encephalitis are also at risk of secondary pneumonia, due to aspiration, and urinary tract infections. Passive and active limb movements will reduce the risk of limb contractures, which can occur in patients with limb weakness; splints and braces may be needed to facilitate mobilization.

At discharge, or soon after, a full neuropsychiatric assessment should be performed, including cognitive function, intelligence, memory, and speech assessment, because it will assist in determining the extent of any damage, and the help needed. Behavioural and psychiatric disturbances are common, and may include depression or disinhibition. Antidepressants and mild night sedatives may be necessary. Memory difficulties can be an especially important problem in herpes simplex virus encephalitis. A range of practical approaches can help to overcome these difficulties; these include simple measures such as the patient keeping a note-book and diary, labelling items around the house, and leaving messages as reminders. More sophisticated aids being developed include a neuropage system, which sends pager messages throughout the day, as reminders of what needs to be done.

Regular out-patient assessment following encephalitis is especially important in children. Other disabilities in the recovery period or afterwards include post-encephalitic Parkinsonism, (Section 40.3.6) which is seen following Von Economo's encephalitis, and encephalitis caused by flaviviruses, and seizures. The risk of seizures is greatest in those who had seizures during the acute period: in one study the cumulative risk of seizures at 5 years was 10 per cent for patients with no acute seizures, which increased to 20 per cent for those with acute seizures (Annegers *et al.* 1988).

42.3 Encephalitis: specific pathogens

42.3.1 Herpes simplex virus type 1 encephalitis

Incidence

Herpes simplex virus causes the most commonly diagnosed viral encephalitis in Western industrialized nations, with an annual incidence of 1 in 250 000 to 500 000 (Whitley 2006). Ninety per cent of these cases are due to herpes simplex virus-type 1, but about 10 per cent are herpes simplex virus-type 2. There is a bimodal distribution of cases, with one-third of cases occurring in those less than 20 years old and one half in those aged 50 years or more. In a recent study of 516 patients with clinical evidence of encephalitis, 7.4 per cent of cases were due to herpes simplex virus infection of which most were in those 40 years of age or older. (Koskiniemi *et al.* 1996). The peak incidence occurred in subjects aged between 60 and 64 years, for whom herpes simplex virus accounted for 37.5 per cent of all cases of encephalitis. The bimodal age distribution may reflect primary herpes simplex virus infection in the younger age group and reactivation of latent virus in older patients.

Pathogenesis

More than two-thirds of cases of encephalitis caused by herpes simplex virus are thought to be reactivation of latent virus: 70 per cent of patients with herpes simplex virus-1 encephalitis have antibody present at the start of the illness indicating that it is not a primary infection. However, encephalitis can also occur as a result of primary infection, or even secondary infection with a different virus strain (Whitley *et al.* 1982). Primary infection with herpes simplex virus-1 occurs via the oral mucosa where it may give ulcers or be asymptomatic. Serological studies show up to 90 per cent of adults have been so infected. Following primary infection the virus travels centripetally along the trigeminal nerve to cause latent infection of the trigeminal ganglion. Pathologically the damage principally affects the limbic system, and nearby structures. Thus the temporal lobes are most often affected, along with the inferior frontal lobes, and parietal occipital lobes; deeper structures include the amygdalae, hippocampus, insular cortex, and cingulate gyrus. The damage is often bilateral, but usually more marked on one side. Initially there is swelling, inflammation, and congestion, with petechial or larger haemorrhages, which may proceed to frank necrosis and liquefaction. Although it is clear that viral replication in the limbic system is critical, what is less clear is how and when the virus reaches that location. Postulated mechanisms include spread from the olfactory bulbs and tracts, reactivation of latent virus in the trigeminal ganglion and spread from there, or reactivation of virus that had already established latency within the brain itself (Esiri 1982).

Clinical features

The classical presentation of herpes simplex virus-1 encephalitis is generally as an acute flu-like prodrome, developing into an illness with high fever, severe headache, nausea, vomiting, and altered consciousness, and often associated with seizures and focal neurological signs. Ninety-one per cent of 93 adults with herpes simplex virus-1 encephalitis in one recent study were febrile on admission (Raschilas *et al.* 2002). Disorientation in 76 per cent, speech disturbances in 59 per cent, and behavioural changes in 41 per cent were the most common features, and one-third of patients had seizures(Raschilas *et al.* 2002). Alterations in higher mental function include lethargy, drowsiness, confusion, disorientation, and coma. With the advent of diagnosis by CSF polymerase chain reaction more subtle presentations of herpes simplex virus-1 encephalitis have been recognized, as described above (Fodor *et al.* 1998). Much of the management of patients with viral encephalitis, outlined above, is directed at diagnosing and treating herpes simplex virus-1 encephalitis because it is the most common sporadic cause of viral encephalitis, and it is treatable with acyclovir (Section 42.2.7).

42.3.2 Herpes simplex virus type 2 encephalitis
Incidence

Herpes simplex virus type 2 is usually transmitted via the genital mucosa, causing genital herpes in adults. In the United States approximately 20 per cent of individuals are sero-positive for herpes simplex virus-2. In England the figure is about 10 per cent, but rising. The neurological syndromes it causes include meningitis, encephalitis particularly in neonates, lumbosacral radiculitis, and meningitis, especially recurrent meningitis. Most cases of what was previously called Mollaret's meningitis is now thought to be due to herpes simplex virus-2 (Section 41.3.1).

Although herpes simplex virus-2 is best known as a cause of aseptic meningitis the virus also accounts for approximately 10 per cent of all encephalitis caused by herpes simplex viruses. Herpes simplex virus-2 encephalitis typically occurs in immunocompromised individuals, and in neonates, in whom it causes a disseminated infection. The encephalitis is thought to reflect primary infection, rather than reactivation (Aurelius *et al.* 1993). A number of cases of adult herpes simplex virus-2 meningoencephalitis have been reported in patients with AIDS (Schiff and Rosenblum 1998). Outside the neonatal period, herpes simplex virus-2 encephalitis is treated as for herpes simplex virus-1 encephalitis with 10 mg/kg acyclovir three times daily for 14–21 days on an empirical basis.

Neonatal encephalitis

Neonatal herpes simplex virus-2 infection can occur *in utero*, intrapartum, or postnatally. Acquisition during delivery, intrapartum infection, is the most common means of infection. This neonatal encephalitis usually occurs as a part of multi-organ disseminated infection, but can also occur without disseminated disease, though there is often a rash involving the skin, eyes, or mouth. Clinically there are seizures which may be generalized or focal, lethargy, irritability, tremors, poor feeding, temperature instability, and a bulging fontanelle. There is usually a CSF pleocytosis, and elevated protein, which may be very high. The diagnosis is confirmed by virus isolation or polymerase chain reaction detection in the CSF. The disease is treated with high dose acyclovir: 20 mg/kg every 8 h for 21 days.

42.3.3 Varicella zoster virus infections

Varicella zoster virus is notable for the great diversity of ways in which it can damage the nervous system including central syndromes such as a cerebellitis and encephalitis, vasculopathies affecting large or small vessels, or neuropathies such as Ramsay Hunt syndrome (Sections 19.2.4 and 20.2.2). The neurological complications can be divided into those associated with primary infection, and those associated with reactivation (Table 42.9).

Primary infection with varicella zoster virus causes chicken pox, or varicella, which in temperate climates occurs mostly in children under 10 years. Following this the virus becomes latent in neurons, from which it may emerge decades later to cause shingles, or zoster (Section 21.14.5). Most primary infection occurs via respiratory droplet spread from children with chicken pox, but virus shed at the time of shingles can also be infectious. Indeed this is how the two clinical syndromes of varicella and zoster were first linked: in 1909 von Bokay noted that chicken pox developed in children who had been exposed to adults with shingles (Von Bokay 1909). Nearly 50 years later Weller and colleagues confirmed that the two conditions were caused by the same virus (Weller 1953). Children who

Table 42.9 Neurological complications of varicella zoster virus (modified from Boos *et al.* 2003)

Complications of acute infection (varicella)
Cerebellitis
Acute encephalitis
Complications of viral reactivation (zoster)
Cranial neuropathies
Ramsay Hunt syndrome
Herpes zoster ophthalmicus
Trigeminal neuronitis
Optic and oculomotor neuropathies / retinitis
Mononeuritis of other cranial nerves,
Polyneuritis cranialis
Stroke syndromes
Herpes zoster ophthalmicus with delayed contralateral hemiparesis
Cervical zoster with posterior circulation infarcts
Granulomatous angiitis of the basilar artery
Encephalitis syndromes
Encephalitis
Diffuse small/medium vessel arteritis
Myelitis

are immunocompromised tend to have prolonged illness with chicken pox, and are more likely to develop complications, including neurological complications. Similarly immunocompromised adults are at increased risk of central nervous system complications.

Neurological complications of chicken pox

Neurological complications of primary infection usually follow the onset of the rash by a few days to a week, though they can occur before the rash, and very occasionally may not be associated with rash at all (Hausler *et al.* 2002).

Acute cerebellar ataxia. This is the most common neurological complication of chicken pox, occurring in approximately 1 per 4000 cases of chicken pox (Guess *et al.* 1986). The symptoms may develop from a few days before to several weeks after the onset of the rash, though they usually develop at the same time. Ataxia is typically accompanied by headache vomiting and lethargy; there may also be fever, nystagmus, slurred speech, and neck stiffness. The CSF may be normal, or show a moderate lymphocytic pleocytosis, with slightly elevated protein and normal glucose ratio. An EEG shows slowing in approximately 20 per cent of patients.

The disease is usually self-limiting, resolving within 1–3 weeks, though it may continue for several months. Varicella zoster virus has not been isolated from such cases, but can be detected by polymerase chain reaction. There is also intrathecal production of anti-varicella zoster virus antibodies. The relative contribution of viral cytopathology and immunologically mediated demyelination is not clear. The cerebellar syndrome is usually benign, and only requires supportive care.

Chicken pox encephalitis. Acute encephalitis or cerebritis is a rare complication of chicken pox, with an incidence of about 1–2 per 10 000 cases. The incidence is higher in adults over 20 or infants less than 1 year. It usually develops about 1 week after the rash onset, but may occur before the rash. Typically there is headache, fever, vomiting, reduced consciousness, and seizures. On CT there may be oedema, and MRI may show grey

or white matter abnormalities. Although there are no good studies, varicella zoster virus encephalitis is usually treated with aciclovir and steroids (Hausler *et al.* 2002). Approximately 5–10 per cent of cases are fatal. Pathological studies have shown diffuse oedema, perivascular infiltration of mononuclear cells, demyelination, and occasionally focal haemorrhage.

Other central nervous system complications associated with chicken pox include myelitis, aseptic meningitis, strokes, and rarely choreoathetosis. Facial nerve palsy, Guillain–Barré syndrome, and Reye's syndrome have also been described. However, since aspirin is no longer given to febrile children, the latter is seen much less often, and indeed may have been entirely attributable to the drug rather than the virus (Gilden 2004).

Neurological complications of varicella zoster virus reactivation

The central nervous system syndromes associated with varicella zoster virus emergence from latency appear to be primarily caused by immune-mediated reactions to the virus, rather than viral replication itself (Gilden *et al.* 2000). However viral replication may occur at a low level, particularly in immunocompromised patients, and may contribute to the pathogenesis. In addition a vasculopathy or vasculitis is an important pathogenic mechanism in some syndromes.

There is a wide range of neurological syndromes associated with varicella zoster virus reactivation (Table 42.9). Although they can occur at any age, they are more common in the elderly, those with advanced HIV disease, or with other immunocompromised states, particularly if cell-mediated immunity is impaired. Thus patients with leukaemia and lymphoproliferative disorders are vulnerable, and especially patients that have bone marrow suppression before transplant. Importantly these syndromes may occur without any cutaneous evidence of zoster, thus being equivalent to the zoster sine herpete, originally described for the pain of shingles without the rash.

Cranial neuropathies and brainstem encephalitis. Ramsay Hunt syndrome is the classical cranial neuropathy associated with varicella zoster virus reactivation (Section 19.2.4). There is a zoster rash over the auricle or in the ear canal, associated with facial pain and a facial palsy (Fig. 42.5). However, there may be extension of the rash to other areas supplied by the trigeminal nerve, or extension of the nerve dysfunction to ipsilateral cranial nerves; this often includes the VIIIth nerve, but may also involve cranial nerves V, VI, IX, or X. The term 'Ramsay Hunt plus' has been applied to such syndromes. Alternatively varicella zoster virus may cause single or multiple cranial neuropathies in the absence of a vesicular rash. Involvement of the VII nerve alone is thought to be due to compression of the nerve fibres as they pass through the inflamed geniculate ganglion, but this has not always been demonstrated pathologically, and more widespread involvement of other cranial nerves suggests extension of the inflammatory process to the brainstem.

The diagnosis of varizella zoster virus cranial neuropathy can be confirmed by isolating virus from vesicles, or by showing a rise in antibody titres in the serum; alternatively virus may be detected in the CSF by polymerase chain reaction, or there may be anti-varicella zoster virus antibodies. Although the evidence is limited, most would recommend treatment with acyclovir, either intravenously or orally, for up to 14 days, with or without corticosteroids

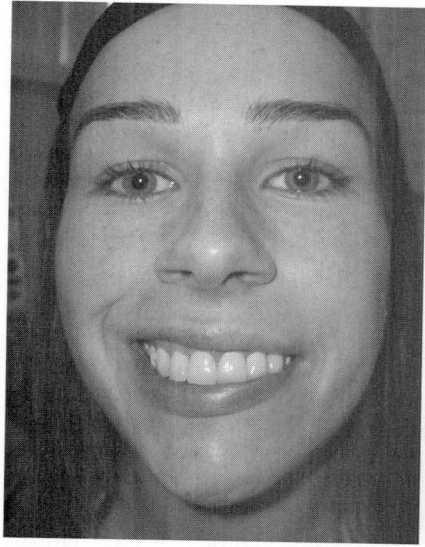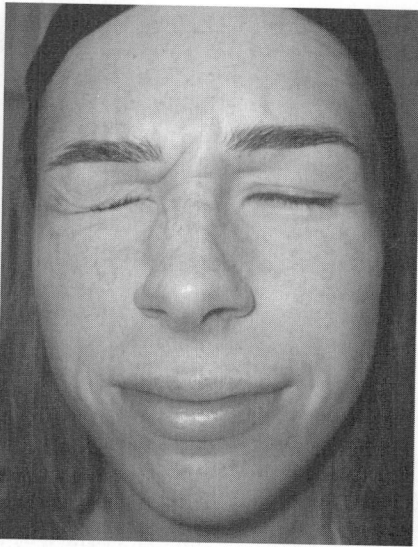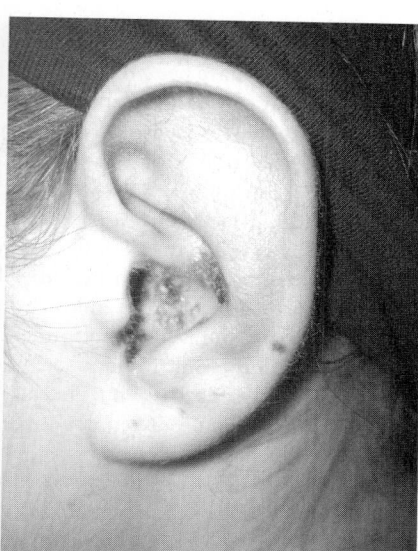

Fig. 42.5 Ramsay Hunt syndrome. This 25-year-old woman developed a left-sided facial paresis 1 week after the itchy painful vesicular rash of varicella zoster virus appeared over her left ear. (Pictures T Solomon.)

(Hato *et al.* 2003), especially if it can be started within a few days of symptom onset (Dworkin *et al.* 2007). In addition early use of amytriptyline significantly reduces post-herpetic neuralgia.

Stroke syndromes. A delayed contralateral hemiparesis following herpes zoster ophthalmica was the first recognized varicella zoster virus stroke-related syndrome, being described in immunocompetent elderly people (Verghese *et al.* 1986). A similar varicella zoster virus vasculopathy syndrome appears to be an important cause of stroke in children. In these patient groups major arteries are affected, whereas in the immunosuppressed a small vessel vasculopathy is important. Vascular syndromes in the posterior circulation, including pontine infarcts, have been reported in following zoster in a C2 root distribution. Pathologically a granulomatous inflammatory process in the wall of the thrombosed vessel is seen, and in some cases there is evidence of viral replication in the adventitia or smooth muscle of the vessel wall. Viral particles have also been seen in endothelial cells. The delay in the development of the stroke syndrome is postulated to represent the time it takes for the virus to spread from the trigeminal ganglion along the branches of the ophthalmic division of the trigeminal nerve to the cerebral arteries.

Typically, contralateral hemiplegia occurs in 8 weeks, ranging from 8 days to 6 months, after the rash. The CSF usually shows a pleocytosis with elevated protein, and digital subtraction angiography, or MR angiography can demonstrate segmental restrictions of flow. Virus cannot be isolated from the CSF. However CSF has on occasion been positive by polymerase chain reaction, and viral DNA has also been found in cerebral arteries. Hence a course of aciclovir is often given: 10 mg/kg three times daily for two weeks. Steroids are also often given, because of the inflammatory nature of the lesion; 60–80 mg for 3 to 5 days has been recommended (Gilden *et al.* 2000).

Encephalitis. Prior to the emergence of HIV, varicella zoster virus encephalitis was very rare. Typically it followed the rash by 1–3 weeks, and tended to occur in the elderly, or those with malignancies. However, it has been recognized as an important cause of encephalitis in patients with HIV. In both patient groups it may occur in the absence of a rash. It can be especially difficult to diagnose because there is not always fever or seizures, and there may not be a CSF pleocytosis (De La Blanchardiere *et al.* 2000). In addition to encephalitis there may be myelitis, radiculitis, and meningitis.

Small vessel vasculitis. A small, or sometimes medium, vessel vasculitis occurs in immunocompromised patients, particularly those with HIV, or organ transplant recipients. It has subacute presentation with headache, fever, hemiplagia, aphasia, visual field defects, altered consciousness, and seizures (Ryder *et al.* 1986). There is often no herpetic rash, though patients may have had the rash weeks or months previously. MRI shows multiple ischaemic and haemorrhagic infarcts of different sizes in the cortex and subcortical grey and white matter; there may also be demyelination. When these are lesions are predominantly in the white matter the MRI changes are reminiscent of progressive multifocal leukoencephalopathy, though they are smaller and more discrete. Varicella zoster virus vasculitis should be treated promptly with high dose aciclovir, with or without corticosteroids, though the evidence for their effect is limited.

Myelitis. This is a rare complication of shingles, occurring in less than one per thousand cases. It most often follows thoracic zoster, with neurological features occurring 1–2 weeks after the onset of the rash. However it can occur in the absence of a rash. There is unilateral motor and posterior column dysfunction, which may evolve to paraplegia with bladder and bowel involvement. Immunocompromised patients, particularly those with HIV, are at increased risk. MRI shows abnormal signal at the level of the lesion. The diagnosis is confirmed by polymerase chain reaction detection of virus, or anti-varicella zoster virus antibodies in the CSF. There is a variable response to aciclovir and corticosteroids (de Silva *et al.* 1996).

42.3.4 Cytomegalovirus infections

Cytomegalovirus infections have been recognized since the turn of the last century. The characteristic intranuclear inclusions found histologically in infected foetal tissue were misinterpreted as being the result of protozoan infection until the virus was isolated simultaneously in three laboratories in 1956. It was named cytomegalovirus after the cytopathic effect which it produced in cell culture. The virus is a beta herpes virus with a double stranded DNA core, and shares the property of latency with the other herpes viruses. Humans are the only hosts for the virus, but the age and incidence of infection depends on the socio-economic circumstances. Infection occurs earlier in those from poor socio-economic backgrounds, and in developing countries, with virtually 100 per cent of adults infected; in contrast in developed countries approximately 60 per cent of adults are infected.

The clinical features depend on the age at which individuals become first infected, or their immune status if they are already affected. Cytomegalovirus can cross the placenta, and if infants become infected *in utero* they can be extremely unwell, with high neonatal morbidity and mortality. However, when infants are infected during delivery, and when children are infected it does not cause overt disease. It is often spread in crèches and schools because it is shed in saliva and urine. Many of those not infected as children become infected during adulthood, through kissing and sexual intercourse. At this age it may cause an infectious mononucleosis-like syndrome with abnormal lymphocytes, splenomegaly, and impaired liver function, but no pharyngitis. A similar syndrome can occur if infection is acquired through blood transfusion. Immunosuppressed transplant recipients can develop cytomegalovirus disease either because they were already infected, the endogenous disease, or because the transplanted organ is infected, the exogenous disease. The latter forms are more severe.

Congenital infection

Cytomegalovirus is the most common pathogen to infect the foetus. This follows transplacental spread from maternal viraemia and affects up to 2.5 per cent of all live births. Fortunately, less than 5 per cent of those infected are symptomatic with evidence of cytomegalic inclusion disease although a proportion of those who are normal at birth develop problems later. Cytomegalic inclusion disease affects many organs and the infant may have low birth weight, hepatosplenomegaly, jaundice, thrombocytopaenia, choroidoretinitis and encephalitis with microcephaly, leading to mental retardation, seizures, spasticity, and deafness. Periventricular calcification is commonly seen on X-ray and on brain scanning with ventricular dilatation. If asymptomatic infected infants are followed up, neurological defects are found as they get older; up to 20 per cent have problems, including lower intelligence quotient, behavioural disorders, minor incoordination, defects in perceptual skills, and neural deafness.

Immunocompromised patients

Patients whose immune system is suppressed are particularly at risk of developing cytomegalovirus infection, both primary and reactivated, and this often involves the central nervous system. Almost all individuals with HIV are coinfected with cytomegalovirus. In a retrospective survey of encephalitis associated with cytomegalovirus, Arribas *et al.* (1996) found that only 3 per cent of cases were healthy before infection and 85 per cent had immune suppression from AIDS (Arribas *et al.* 1996). The rest were immune suppressed for different reasons. Primary infection in the immunocompromised is usually symptomatic with persistent pyrexia which may last for weeks, and is often fatal if bacterial or fungal infection supervenes. In some, disease may be limited to or accompanied by, severe choroidoretinitis.

Cytomegalovirus is a common opportunistic pathogen which causes a form of encephalitis and affects other organs in patients with advanced HIV (Drew 1988). Evidence of infection has been found in brain at necropsy in up to a third of AIDS patients coming to necropsy (Arribas *et al.* 1996). Clinically, patients with cytomegalovirus have a sub-acute or chronic encephalitis with confusions, disorientation, lethargy, and occasionally seizures (Holland *et al.* 1994). These features are often not distinguishable from HIV-associated dementia itself. However polyradiculopathy, or retinitis may be a clue that cytomegalovirus is contributing to an HIV patient's neurological status.

Unfortunately, it is impossible to determine in those patients who have dementia how much of their syndrome has been caused by HIV, or how much by cytomegalovirus or other pathogens. It is also difficult to correlate pathological findings with clinical course. There is often nothing specific about brain imaging (Clifford *et al.* 1996) and diagnosis requires the demonstration of cytomegalovirus DNA in CSF (Cohen 1996; Wildemann *et al.* 1998). A distinct syndrome of ventriculoencephalitis caused by cytomegalovirus in AIDS, characterized by simultaneous retinitis or other organ disease, mental change, large ventricles and periventricular enhancement on imaging, and CSF pleocytosis, is also characteristic (Kalayjian *et al.* 1994).

Often the discovery of a CSF polymorphonuclear pleocytosis, rather than a lymphocytic one is the clue that cytomegalovirus is involved. Diagnosis is by demonstration of virus in the CSF, usually by quantitative polymerase chain reaction or culture. A lumbosacral radiculopathy with pain and paraesthesia in a saddle distribution, which may spread to involve the lower limbs and sphincter is another common presentation (Section 21.14.2). Cytomegalovirus retinitis may precede or follow these syndromes. In other patients correlation with clinical symptoms may not be evident. Brainstem encephalitis, space occupying lesions, and myelitis due to cytomegalovirus have also been described.

The introduction of highly active antiretroviral therapy, HAART, has provided a means of reconstituting the immune system of those with HIV/AIDS in such a way as to allow cytomegalovirus infection to be controlled. In doing so, HAART has done much to reduce the mortality rate associated with cytomegalovirus disease in such patients. Despite this, the response to treatment in these patients remains suboptimal. As cytomegalovirus disease is always preceded by viraemia, treatment should be directed toward the prevention of cytomegalovirus disease.

However, if cytomegalovirus disease develops, intravenous ganciclovir is recommended as initial therapy and continued in a maintenance fashion, with its prodrug valganciclovir, which can be discontinued should the blood CD4 lymphocyte count remain above 100 cells/mm^3 for 6 months. Both ganciclovir and valganciclovir can cause bone marrow suppression. Use of foscarnet should be limited to ganciclovir-resistant cases due to the high level of toxicity associated with the drug and its intravenous mode of administration.

42.3.5 Human herpes viruses 6 and 7 encephalitis

Exanthem subitum also known as Roseola or sixth disease is a febrile illness in infants and children, in which rash occurs as the fever resolves. It is often associated with febrile seizures, which are sometimes associated with changes on MRI. It was subsequently realized that the cause was human herpes virus-6, which had first been isolated from patients with lymphoproliferative disorders. There are two antigenic variants, types 6A, and 6B, most frequently associated with encephalitis. Human herpes virus-6 and the closely related virus human herpes virus-7 are both beta herpes viruses related to cytomegalovirus. In children human herpes virus-6 is usually a primary infection, whereas in adults it is a reactivation. In adults human herpes virus-6 reactivation can be associated with encephalitis or encephalomyelitis, usually in the immunocompromised. The virus has been linked to multiple sclerosis.

42.3.6 Parvovirus B19 encephalitis

Erythema Infectiosum or 5th disease is a febrile exanthem caused by Parvovirus B19. This small DNA virus is spread between children via respiratory droplets. Infection typically causes facial 'slapped rash', with a fiery abrupt onset, few days later by a more generalized rash which includes the extremities. It occasionally causes encephalitis or encephalopathy, as well as meningitis or a stroke syndrome, particularly in children with sickle cell disease. Increasing numbers of central nervous system infections are being detected with polymerase chain reaction (Barah *et al.* 2001).

42.3.7 Measles virus encephalitides

Measles virus causes three distinct central nervous system syndromes:

◆ *A post-infectious acute disseminated encephalomyelitis* which typically follows acute measles by a few weeks;

◆ *a subacute 'inclusion body' encephalitis* which presents over months in the immunocompromised, several months after measles;

◆ *subacute sclerosing panencephalitis*, which presents in the immunocompetent, many years after the measles infection.

Measles is a highly infectious disease that affects children and young adults throughout the world. Before the development of an effective vaccine, 99 per cent of the population had been affected by the age of 20 and had antibodies to prove it; this remains the case in third world countries. Formerly a disease which affected the 5–9 years age group in the main, since the introduction of successful immunization, teenagers are now more frequently infected. In much of the developing world, where there is little immunization, and where there is much malnutrition, measles devastates the under twos with a mortality of up to 10 per cent. Infection confers lifelong immunity.

Measles virus is an RNA virus and a member of the genus *Morbillivirus*, in the family *Paramyxoviridae*. Other paramyxoviruses include Hendra virus, and Nipah virus, which was recently described in Malaysia (Chua *et al.* 2000), and can cause a late neurological presentation similar to subacute sclerosing panencephalitis (Wong *et al.* 2001).

Measles appears to be an antigenically stable virus and the development of neurological complications was thought to be due to variations in host susceptibility, age and immune status, rather than to viral properties. However, there is now evidence that the virus found in cases of subacute sclerosing panencephalitis differs from wild virus because it is has a deficient M protein, one of the virus-specific structural proteins. Antibody to M protein is low, while the antibody response to N and P proteins is particularly robust.

Measles is spread by aerosol droplets and respiratory inhalation and perhaps through the conjunctiva. Virus travels through the reticulo-endothelial system and the circulation. After an incubation period of 10 days, fever, malaise, conjunctival suffusion, and coryza appear. Koplik's spots, the pathognomonic sign of measles, appear as raised spots with white centres in the buccal and lower labial mucosa, fading as the rash develops at about 14 days on the face and behind the ears, and spreads cetrifugally to involve the trunk and then the extremities. After 3 or 4 days the rash begins to fade, antibody is formed, and most patients make a good and complete recovery. Patients have a high viraemia during the early stages. Viral shedding is maximal from a few days before the rash appears, until about 2 days after. Thus although isolation is used, most viral transmission has occurred before the diagnosis is made. Overall, the complications of measles are rare and include secondary infection of the rash, pneumonia, myocarditis, pericarditis, hepatitis, and neurological syndromes. Although overt neurological disease is rare, central nervous system involvement, even in uncomplicated disease, is common, as judged by the fact that 30 per cent of measles patients have a CSF pleocytosis, and 50 per cent have EEG abnormalities.

In addition to post-infectious encephalomyelitis, subacute encephalitis, and subacute sclerosing panencephalitis, other neurological complications of measles virus infection include polyneuritis, transverse myelitis, acute hemiplegia, toxic encephalopathy, and cerebellar ataxia (Ford 1928; Aarli 1974). Post-infectious encephalitis accounts for 95 per cent of cases of central nervous system disease. It occurs in patients with a normal immune system and who are over 2 years of age. The incidence approximates to 1:1000 of measles cases and rises with age. Subacute inclusion body encephalitis occurs in the immune compromised of any age and afflicts 1 in 10 of such patients who are affected by measles. Subacute sclerosing panencephalitis occurs once in a million cases of measles and affects immunologically competent individuals, with a predilection for those less than 2 years.

Post-infectious measles encephalitis

This affects males and females equally. Signs begin to develop within 8 days of the onset, as the exanthem begins to subside. Fever recurs and encephalitis develops rapidly with convulsions, abnormal movements, focal paresis, ataxia, myoclonus, paraparesis and rarely, a pure myelitis (Johnson *et al.* 1984). Approximately 15 per cent of patients die and more than 50 per cent have serious neurological sequelae. In contrast, uncomplicated measles carries a very good prognosis for complete recovery. There is no established treatment. Most would recommend high-dose corticosteroids to treat the inflammatory component to the disease. Ribavirin is effective against the virus *in vitro*, but has not been proven to be effective clinically, and most would not use it, given the disorder appears to be a post-infectious phenomenon.

The pathological findings of post-infectious measles encephalitis share similar features with other viral post-infectious encephalitides: perivenous demyelination, gliosis, perivascular cuffing, and in more severe cases, a frank haemorrhagic leuco-encephalitis. The pathogenesis is thought to be similar to that of experimental allergic encephalomyelitis—an autoimmune demyelinating disease triggered by measles. Measles virus has only rarely been found in brain and it has been postulated that the presence of virus triggers an abnormal immunoregulatory response to myelin basic protein.

Subacute measles encephalitis

This occurs only in those who have been immune suppressed. Most frequently these are children who have had lymphocytic leukaemia and received therapy for this. However it can also occur in lymphoma, renal transplantation, and people with HIV. The syndrome seems to develop in those who are already immunosuppressed when they encounter the virus for the first time. After an attack of measles, some 2–6 months later, symptoms insidiously develop. Occasionally it may occur in those with no known immunocompromise (Chadwick et al. 1982). Lethargy, mental confusion, epileptic seizures, myoclonus, and diffuse cerebral dysfunction, often affecting the parietal lobes, progress relentlessly to death in a few months (Hughes et al. 1993; Mustafa et al. 1993). It should thus be considered in any immunosuppressed child that develops seizures. Other causes of seizures in this patient group, including extension of the leukemia to the central nervous system, haemorrhagic complications of leukaemia, complications of radio or chemotherapy, and other opportunistic infections must be ruled out. EEG, CSF, and brain imaging show normal findings or non-specific changes only. The diagnosis can be hard to establish because antibody titres are low or absent, which reflects immune suppression. Brain biopsy may be necessary. Pathologically, inflammatory changes are seen throughout the brain and eosinophilic inclusions have been identified in the nuclei of neurons. There is no established treatment but ribavirin 20 mg/kg daily for 15 weeks was associated with recovery in one report (Mustafa et al. 1993). Virus antigen has been found in brain tissue. With increased frequency of measles immunization in a population, this complication has markedly receded.

Subacute sclerosing panencephalitis

Dawson was the first to describe this disease in 1933. It has rejoiced in a plethora of names in the past including inclusion body encephalitis, nodular panencephalitis, subacute sclerosing leukoencephalitis, and finally subacute sclerosing panencephalitis. A viral cause was postulated and in 1965 paramyxovirus structures were demonstrated in brain inclusion bodies (Bouteille et al. 1965). Two years later, the measles virus was cultivated from brain tissue and final confirmation of the relationship of measles to subacute sclerosing panencephalitis was obtained by the passage of the disease to animals (Katz et al. 1968). It is still not understood why subacute sclerosing panencephalitis, a progressive and fatal disease of the brain should develop in a very small number of all the patients affected by measles every year. It is not known whether central nervous system infection occurs at the time of primary measles infection and simmers on for years before producing clinical signs, or whether it lies dormant in an external site before spreading to the central nervous system not long before clinical signs become apparent. It seems likely that the former occurs, with virus entering the brain, probably by infection of cerebral capillary endothelium; measles virus-containing immune complexes have been found in blood vessel wall (Kirk et al. 1991). The virus then

and mutates to a less invasive and more persistent form, which has defective coding for M protein. It seems likely that the persisting viral infection triggers an immune-mediated response which is responsible for the widespread demyelination.

The incidence of subacute sclerosing panencephalitis varies from country to country. It occurs 3–4 times more frequently in boys than girls, in younger children in a family, in rural rather than urban communities, and in lower socio-economic groups. All racial groups are affected and it seems that there is a higher incidence around the eastern Mediterranean area and in Arabs. Where effective immunization programmes have been introduced, subacute sclerosing panencephalitis is disappearing. Lack of measles immunization is a significant risk factor (Miller et al. 1992). There does not seem to be a genetic predisposition.

Subacute sclerosing panencephalitis has the potential to reappear in developed countries when significant numbers of the population let immunization lapse, and measles epidemics occur. This was predicted following the 1988–90 measles epidemic in the United States (Gascon 1996), and a recently reported case of subacute sclerosing panencephalitis appears to be the first related to that epidemic (Dlugos et al. 2001). In the United Kingdom the incidence is less than 1 per million children per year (Anonymous 1979). Between 1970 and 1989 there were 290 cases. Here too its incidence in children has decreased with the widespread use of measles vaccine. However, in recent years it has been recognized more in adults (Miller et al. 1992), and it may become more common in the future because of the decreased uptake of measles vaccine following unsubstantiated fears over a link with autism (Elliman and Bedford 2001).

The most important risk factor for the development of subacute sclerosing panencephalitis is suffering from measles at an early age, particularly within the first year of life. Almost all children with subacute sclerosing panencephalitis have a history of measles and in those who do not, there is a history of household contact. The risk of contracting subacute sclerosing panencephalitis after measles is of the order of 5–9 cases per million and drops to one-tenth following measles immunization.

Subacute sclerosing penencephalitis is a disease of children who have been healthy. The average age of onset is 8–10 years and symptoms appear on average, 7 years after full recovery from measles. 90 per cent of cases occur before the age of 16. The onset is insidious and may be recognized only in retrospect after the occurrence of emotional lability, behavioural alteration, impairment of scholastic performance, clumsiness, cognitive defects, ataxia, and epileptic seizures. The onset of myoclonic jerks which may be focal and soon become generalized, heralds the second stage of the disease. They become frequent and repetitive and correlate with the giant complexes seen on EEG (Fig. 42.6) (Yaqub 1996). They may be provoked by sensory stimuli. Pyramidal and extrapyramidal signs with dystonia and dyskinesia appear. Akinetic seizures are common and there is inexorable progression to a vegetative state with dementia, decortication, and decerebration. Rarely, this may be interrupted by remission and exceptionally by stabilization of the disease. Progression is highly variable. Death may occur within weeks of onset, about half die within a year and most die within 2 years (Risk et al. 1979).

Once myoclonus is evident the clinical diagnosis seldom presents difficulty and investigations provide confirmatory results. The EEG is genuinely useful with high voltage stereotyped slow-wave

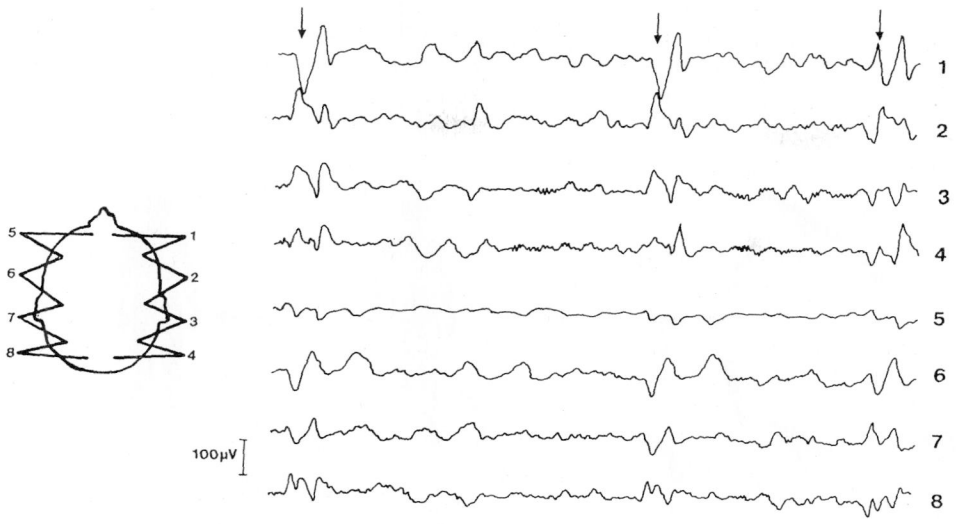

Fig. 42.6 Subacute sclerosing panencephalitis. EEG showing periodic activity (arrows) separated by periods of relatively featureless EEG activity.

100μV

complexes, often synchronous with clinical myoclonus, repeating every 8–10 sec. CT and MR brain scans are not diagnostic and may show white matter change and atrophy. CSF examination is most useful and demonstrates great elevation of the gamma globulin fraction of the protein content due to the presence of measles antibody which is found in high concentration in unconcentrated fluid. Polymerase chain reaction is reported to be useful (DeBiasi and Tyler 1999).

At necropsy, changes are evident in white and grey matter with neuronal degeneration, gliosis, proliferation of atrocytes, perivascular cuffing, lymphocytic and plasma cell infiltration and demyelination. Type A intranuclear inclusions are found in oligodendroglia and neurones. The changes are non specific.

A range of therapies have been tried for subacute sclerosing panencephalitis. Isoprinosine has both immune-modulating properties and antiviral properties that are dose dependent, and may be beneficial at a dose of 25–100 mg/kg per day in divided doses (Anlar *et al.* 1997). Interferon-alpha been administered systemically, intrathecally, and intraventricularly. Whilst interferon-alpha given systemically has been of no benefit, and intrathecal administration has produced variable results, the most promising results with interferon-alpha have occurred in patients treated by the intraventricular route. Doses of between 250 000 to 1 million IU biweekly have been used with apparent benefit (Panitch *et al.* 1986). Subsequently combinations of inosiplex and interferon alpha have also been used. Ribavirin, 1-ribofuranosy-1,2,4-thiazole-3-carboxamide, is a guanosine analogue that *in vitro* inhibits the replication of naturally occurring measles virus and measles virus derived from patients with subacute sclerosing panencephalitis (Murphy 1978). Although oral treatment did not appear effective, the drug has been given in combination with inosiplex and/or interferon alpha, with apparent benefit (Murphy 1978; Solomon *et al.* 2002).

42.3.8 Arboviral encephalitides

Altogether there are more than 500 arthropod-borne viruses, or 'arboviruses', that come from five different viral families: *Flaviviridae, Togaviridae, Bunyaviridae, Reoviridae, and Rhabdoviridae*

(Solomon *et al.* 2004). They are transmitted naturally between vertebrate hosts by insects, especially mosquitoes, or by ticks (Fig. 42.7). Many flaviviruses are zoonotic viruses, meaning their natural hosts are small animals or birds, rather than humans. Only a relatively small number of the arboviruses are responsible for causing diseases in humans. They cause three clinical syndromes (Table 42.10):

- Fever, arthralgia and rash;
- haemorrhagic fevers; and
- central nervous system infections.

The most important viruses are the flaviviruses, particularly Japanese encephalitis virus, West Nile virus, and tick-borne encephalitis virus. Dengue is a flavivirus best known for causing haemorrhagic fever, or a fever-rash syndrome. Its importance as a cause of neurological disease is being increasingly recognised, though in most cases this is due to a non-specific encephalopathic process rather than a true viral encephalitis. Other important arboviruses are found among: the alphaviruses, family *Togaviridae*, which cause encephalitis in horses and humans in America; the bunyaviruses, particularly Lacrosse virus; the coltivirus; the family *Reoviridae*, especially Colorado tick fever virus and the Vesciloviruses; the family Rhabdoviridae, principally Chandipura virus.

42.3.9 Flavivirus encephalitis: West Nile, Japanese encephalitis, and tick-borne encephalitis

Flaviviruses, of the genus, *Flavivirus*, and family *Flaviviridae* are named after the prototype yellow fever virus (Latin flavus = yellow), which was first isolated in 1927. Outbreaks of encephalitis were recognized in Japan from the 1870s, and Japanese encephalitis virus was isolated in 1935. Similar outbreaks of encephalitis were recognized in America in the 1930s, and in 1931 St Louis encephalitis virus was isolated from a fatal case in St Louis, Missouri. This virus was the most important flavivirus encephalitis in the United States until the appearance of West Nile virus in 1999. West Nile virus was originally isolated from a woman with a febrile illness in

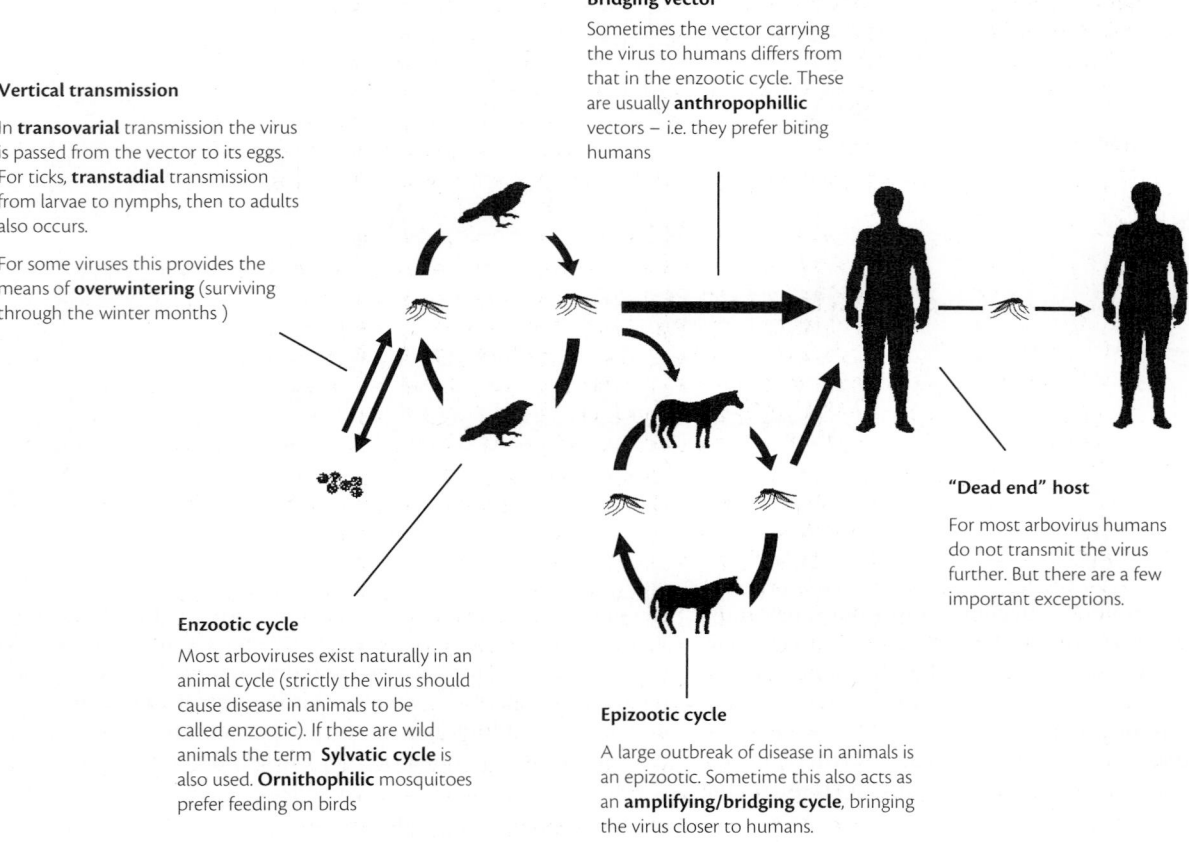

Vertical transmission

In **transovarial** transmission the virus is passed from the vector to its eggs. For ticks, **transtadial** transmission from larvae to nymphs, then to adults also occurs.

For some viruses this provides the means of **overwintering** (surviving through the winter months)

Bridging vector

Sometimes the vector carrying the virus to humans differs from that in the enzootic cycle. These are usually **anthropophillic** vectors – i.e. they prefer biting humans

"Dead end" host

For most arbovirus humans do not transmit the virus further. But there are a few important exceptions.

Enzootic cycle

Most arboviruses exist naturally in an animal cycle (strictly the virus should cause disease in animals to be called enzootic). If these are wild animals the term **Sylvatic cycle** is also used. **Ornithophilic** mosquitoes prefer feeding on birds

Epizootic cycle

A large outbreak of disease in animals is an epizootic. Sometime this also acts as an **amplifying/bridging cycle**, bringing the virus closer to humans.

Fig. 42.7 Overview of arboviral ecology. A hypothetical arboviral cycle, with an explanation of some commonly used terms. (Reproduced from Solomon 2004.)

Uganda in 1935, and until recent years was better known as a cause of a fever and rash syndrome that only occasionally caused central nervous system disease. In Australia, Murray Valley encephalitis virus was isolated in 1951, but is thought to have been responsible for earlier outbreaks of viral encephalitis called 'Australian X' disease (Breinl 1918; French 1952). These four viruses are all transmitted naturally between birds by Culex mosquitoes. Tick-borne encephalitis virus is a flavivirus transmitted naturally between small rodents by hard *Ixodes* ticks. Descriptions of a disease compatible with tick-borne encephalitis appeared from the 1930s, and the virus was first isolated by Russian scientists in the Far East in 1937 (Smorodintsev 1958). However across this vast geographical area, the disease was given a range of different names: Central European encephalitis, Russian spring-summer encephalitis, Far Eastern encephalitis, Biphasic milk fever, Taiga encephalitis, Kumlinge disease, and Fruhsommer-Meningoenzephalitis. Subsequently it was realized that they are essentially the same disease. Three closely related subtypes of tick-borne encephalitis virus exist, whose names reflect the geographical areas that they principally affect: European, Siberian, and Far Eastern. The tick-borne encephalitis group serocomplex, recently renamed the mammalian group of tick-borne flaviviruses, also includes Powassan virus, which is a rare cause of encephalitis in Canada, and louping ill virus, a very rare cause of central nervous system disease in the British Isles and Scandinavia.

Clinical and epidemiological features

The epidemiology of flavivirus encephalitis is governed by a complex interplay of climatic, entomological, human behavioural, viral, and host factors, that are not completely understood (Solomon *et al.* 2003). For the Japanese encephalitis serogroup viruses consisting of Japanese encephalitis virus, West Nile virus, St Louis encephalitis virus, and Murray Valley encephalitis virus, viruses are transmitted naturally in enzootic cycles involving birds and Culex mosquitoes. Pigs act as additional amplifying hosts for Japanese encephalitis virus itself. For tick-borne encephalitis virus and related viruses, the natural cycle involves small rodents and ticks. Humans become infected inadvertently when they encroach upon the flavivirus cycle, but they are considered 'dead end' hosts because they do not normally have sufficiently high or prolonged viraemias to transmit the virus further. In addition to arthropod-borne transmission, less common routes of transmission that have been documented include infected transplanted organs and blood products for West Nile virus (Iwamoto *et al.* 2003), transplacental transmission, for West Nile and Japanese encephalitis (Chaturvedi *et al.* 1980), and transmission by ingestion of unpasteurized goats' milk for tick-borne encephalitis virus.

Japanese encephalitis is mostly a disease of children, whereas in America, West Nile encephalitis and St Louis encephalitis are more likely to affect adults. This apparent paradox is probably largely

Table 42.10 Overview of the major arboviral encephalitides (modified from Solomon et al. 2004)

Family	Genus	Virus	Geographical distribution	Natural hosts	Main vectors (mosquitoes unless otherwise stated)	Human disease
Togaviridae	Alphavirus	Eastern equine encephalitis	North, Central, and South America	Freshwater swamp birds in N America, rodents and marsupials in S America	Culiseta melanura; Aedes species are bridging vectors	High mortality; CSF may resemble bacterial meningitis
		Western equine encephalitis	Western Canada, western and mid-west US	Passerine birds (sparrows, house finches); blacktail jackrabbit	Culex tarsalis; Ae. melanimon	No human cases since 1994, relatively low mortality
		Venezuelan equine encephalitis	Central and South America, occasionally extending further north	Rodents, aquatic birds	Culex (Melanoconion) species in enzootic cycle; Psorophora and Aedes species inoved in equine epizootics, and epidemics.	FAR syndrome common; CNS disease rare. Humans have high viraemias; virus also isolated from pharynx; Non-licensed live attenuated vaccine (TC-83) available
Flaviviridae	Flavivirus	West Nile	Africa, Middle East, Europe, Asia, America, Australia	Passeriform birds (jays, blackbirds, finches, warblers, sparrows, crows)	Culex pipiens and others	Nosocomial transmission also occurs; febrile syndrome common; disease is more severe in elderly and immunosuppressed
		Japanese encephalitis	Southeast Asia, Pacific Rim, Northern Australia, Asian Subcontinent	Herons, egrets, migrating birds; swine	Culex tritaeniorhynchus, C. gelidus and others	Most important arboviral encephalitis; seizures and Parkinsonian features common; inactivated vaccine (Biken) is licensed; live attenuated vaccine in China
		St Louis encephalitis	North, Central, and South America	Passeriform and columbiform birds	Culex tarsalis in the West; C. pipiens, C. quinquefasciatus, C. nigripalus in the East	Endemic disease in the western US; large outbreaks in the eastern USA
		Murray Valley encephalitis	New Guinea, Australia	Herons, egrets, aquatic birds	Culex annulirostris, and others	Mostly affects aboriginal children
		Tick-borne encephalitis	Europe, Siberia, Far East	Small forest rodents	Ixodes ricinis, I. persulcatus, and other ticks	Poliomyelitis-like paralysis of upper limbs is common. Inactivated vaccines are licensed
		Dengue	Almost every country between tropics of capricorn and cancer	Humans, and non-human primates	Aedes aegypti and Ae. Albopictus	FAR syndrome (dengue fever); syndrome of vascular leak and haemorrhage (dengue haemorrhagic fever); neurological disease
Bunyaviridae	Bunyavirus	La Crosse	Midwestern and Eastern USA	Chipmunks, squirrels	Aedes. triseriatus, Ae. albopictus	Major paediatric arboviral encephalitis in USA; fatalities are rare
		Jamestown Canyon	North America	White tailed deer	Aedes species, Culex. inornata	Encephalitis in the elderly
		California encephalitis	Western USA, Canada	Rabbits, other rodents	Ae. melanimon, Ae. dorsalis	Rare cause of human disease
		Tahyna	Europe, Russia	Domestic animals, rabbits	Ae. vexans, Culex. annulata	Newly recognized cause of febrile illness with CNS disease, especially in Russia
	Phlebovirus	Toscana	Mediterranean countries	Not known	Phlebotomus perniciosus sandflies	Mostly aseptic meningitis, encephalitis rare
		Rift Valley fever	Africa, Middle East	Livestock	Aedes and other mosquitoes; also direct transmission from livestock body fluids	Febrile illness with retinitis haemorrhagic fever, encephalitis (0–2%)
Reoviridae	Coltivirus	Colorado tick fever	Western USA and Canada	Squirrels, chipmunks, other small mammals	Dermacentor andersoni ticks	Meningitis in patient with leukopenia and thrombocytopenia
Rhabdoviridae	Vesiculovirus	Chandipura virus	India, West Africa		Phlebotomus sandflies	Encephalitis, with rapid onset and high mortality
	Vesiculovirus	Vesicular stomatitis virus	Noth and South America	Horses, cattle, pigs	Phelobomus sandflies and blackflies, possibly mosquitoes, but also direct transmission	Febrile illness, occasionally oral vesicles, CNS disease very rare

Ae — Aedes; Cq — Coquillidea; CNS — central nervous system; C — Culex; FAR — fever arthralgia rash; I —Ixodes; US — United States.

explained by the increased intensity of transmission of Japanese encephalitis in Asia. This means that most children in rural Asia meet the virus in childhood, and are immune by adulthood. In contrast when non-immune populations are exposed to flaviviruses, it is typically adults that are more likely to present with neurological disease. This applies to most St Louis encephalitis virus outbreaks in America, and West Nile virus outbreaks in America and southern Europe (Monath 1980), but also occurs when immunologically naïve adults meet Japanese encephalitis virus for the first time. This occurs when the virus spreads to new areas, (Solomon et al. 2000) or when immunologically naïve adults travel to endemic areas, for example holiday makers (Wittesjö et al. 1995), or military service personnel. The incidence of tick-borne encephalitis varies across northern Europe and Northern Asia, but typically affects adults with occupational or recreational exposure to ticks in forests.

Clinical features

Patients typically develop neurological disease after an incubation period of 5–15 days, and a short non-specific febrile prodrome. For most neurotropic flaviviruses, encephalitis is a more common presentation than aseptic meningitis or acute flaccid paralysis. Seizures are common in children with flavivirus encephalitis (Kumar et al. 1990; Burrow et al. 1998; Solomon et al. 2002), and may include subtle motor seizures, and status epilepticus (Wasay et al. 2000; Solomon et al. 2002). In addition movement disorders are common, and may include a parkinsonian syndrome, opisthotonus, generalized rigidity, choreoathetosis, orofacial dyskinesias, and myoclonic jerks (Misra et al. 1997; Solomon et al. 2002, 2003; Pepperell et al. 2003; Sejvar et al. 2003a). These disorders are thought to reflect involvement of the basal ganglia, particularly the thalamus and substantia nigra, as seen on magnetic resonance imaging and at autopsy (Zimmerman 1946; Bennet 1976; Brinker et al. 1980; Misra et al. 1997; Cerna et al. 1999; Weiss et al. 2001; Solomon et al. 2002, 2003; Bosanko et al. 2003; Kienzle et al. 2003). In some patients intention tremors and ataxia, may suggest cerebellar involvement (Pepperell et al. 2003).

Flaccid paralysis

Motor weakness is common in flavivirus encephalitis. As well as upper motor neuron weakness, which is reported for 30–50 per cent of patients, flaccid limb weakness, with reduced or absent reflexes, is also common. This is often associated with respiratory or bulbar paralysis, and is reported for approximately 20–60 per cent of patients (Misra et al. 1997; Burrow et al. 1998; Nash et al. 2001; Pepperell et al. 2003). In addition to causing flaccid weakness in comatose encephalitis patients, flaviviruses can also cause a poliomyelitis-like acute flaccid paralysis, in fully conscious patients. The earliest outbreaks of Murray Valley encephalitis were thought to be an 'aberrant form of poliomyelitis' (Breinl 1918), and poliomyelitis-like illness has been described for Japanese encephalitis virus and West Nile virus (Gadoth et al. 1979; Solomon et al. 1998; Leis et al. 2002; Li et al. 2003; Sejvar et al. 2003b). Flaccid paralysis is typically seen in a single lower limb, but there can be involvement of all four limbs, and respiratory muscle weakness too. In tick-borne encephalitis virus infection, flaccid paralysis usually affects the neck, and upper limbs to cause pain, sometimes with periodic muscle contractions and numbness, then upper limb weakness with winging of the scapula, wrist drop, or a 'hanged head' due to neck extensor weakness (Haglund et al. 2003). Muscle atrophy

begins after the second or third week, and persists (Kaiser 1999). Acute retention of urine, due to an atonic bladder, may be an early clue that paralysis is due to a flavivirus (Solomon et al. 1998; Solomon et al. 2002). In some patients the combination of upper and lower motor neuron damage can lead to bizarre mixtures of clinical signs, which may change hourly during the acute stages of infection.

Although anterior myelitis is the most important cause of paralysis for Japanese encephalitis serogroup viruses, in some patients flaccid paralysis reflects other pathologies such as Guillain–Barré syndrome, and other radiculopathies (Jeha et al. 2003; Park et al. 2003; Pepperell et al. 2003). A recent study of West Nile virus infection showed 74 per cent of 32 patients with paralysis had poliomyelitis-like paralysis and 13 per cent had Guillain–Barré syndrome (Sejvar et al. 2005). Similarly, some patients with tick-borne encephalitis virus infection develop a 'polyradiculoneurotic form' of disease: neuropathy occurs 1–2 weeks after the initial febrile phase, and is associated with a recurrence of fever, but there is usually complete recovery (Haglund et al. 2003).

In Russia a 'chronic form' of tick-borne encephalitis has been described, and is believed to be caused only by the Siberian viral subtype. Deterioration continues long after the acute disease, post mortem examination suggests chronic inflammation, and viral RNA may be detected by nucleic acid hybridization, or cultured (Gritsun et al. 2003). Because it was felt that some of these patients progressed to amyotrophic lateral sclerosis, there were extensive efforts to transmit this progressive neurological disease to non-human primates, but they were not successful (Haglund et al. 2003; Johnson and Cornblath 2003). Other patients with tick-borne encephalitis virus infection are asymptomatic following the initial tick bite, but present years later with a progressive form of disease, with virus being isolated at autopsy (Gritsun et al. 2003). Spontaneous regular contractions or myoclonic jerkings, of the limbs are seen in about a quarter of patients with neurological forms of tick-borne encephalitis virus disease, and may persist as epilepsia partialis continua, known as Kozhevnikov's epilepsy.

Investigation

Usually there is a moderate CSF pleocytosis of a few hundred lymphocytes/mm^3, protein is moderately elevated, and the glucose is normal. Approximately 50 per cent of Japanese encephalitis patients and 30 per cent of St Louis encephalitis patients have elevated CSF opening pressures. Attempts at isolating virus from the blood of patients with flavivirus encephalitis are usually negative because of the transient and low viraemias. Virus is occasionally isolated from the CSF of patients that do not yet have antibody, particularly those that subsequently die (Burke et al. 1985; Huang et al. 2002) and from post mortem brain tissue (Burke et al. 1985; George et al. 1987; Iwamoto et al. 2003). Viral RNA may occasionally be detected in the CSF by the reverse transcriptase polymerase chain reaction (Igarashi et al. 1994; Petersen et al. 2002). For West Nile virus real-time polymerase chain reaction has proved more useful (Lanciotti et al. 2000). IgM capture enzyme-linked immunosorbant assays that detect antibody are often positive on a single CSF or serum sample, and have therefore become the accepted standard for diagnosing flavivirus encephalitis (Solomon et al. 2000; Petersen et al. 2002). Not all patients have antibody on admission to hospital and the test should be repeated if initially negative. False positives can occur in areas where more than one flavivirus co-circulates, or for patients that have received a flavivirus vaccine, but this problem can be minimized by testing for antibody against various flaviviruses in parallel (Innis *et al.* 1989; Solomon *et al.* 1998; Martin *et al.* 2002). Investigations for neutralizing antibodies are more specific, but can only be performed in reference laboratories. Antibody may persist in the serum for many months after infection (Burke *et al.* 1985; Roehrig *et al.* 2003).

EEG in flavivirus encephalitis usually shows generalized or focal slowing. In addition EEG may reveal subtle motor statues epileptics, or periodic lateralized epileptiform discharges (Wasay *et al.* 2000; Solomon *et al.* 2002), which are often encountered in herpes simplex encephalitis. Nerve conduction studies in patients with myleitis caused by flaviviruses typically show reduced or absent compound muscle action potentials, with preserved sensory nerve action potentials and normal conduction velocities (Breinl 1918; McCordock *et al.* 1934; Zimmerman 1946; Newman *et al.* 1954; Solomon *et al.* 1998; Kelley *et al.* 2003; Li *et al.* 2003; Sejvar *et al.* 2003). Electromyography typically shows positive sharp waves and spontaneous fibrillations, consistent with denervation.

Brain CT of patients with flavivirus encephalitis shows low attenuation lesions, and MRI shows high signal on T2-weighted images in the basal ganglia, particularly the thalamus and substantia nigra (Fig. 42.8) (Zimmerman 1946; Bennet 1976; Brinker *et al.* 1980; Misra *et al.* 1997; Cerna *et al.* 1999; Weiss *et al.* 2001; Solomon *et al.* 2002, 2003; Bosanko *et al.* 2003; Kienzle *et al.* 2003). In patients with flaccid paralysis, magnetic resonance imaging of the spinal cord has shown high signal intensity on T2-weighted scans in the anterior horns of the spinal cord (Li *et al.* 2003).

Treatment

There is no established antiviral treatment for any flavivirus infection. For many years, the most promising compound was considered to be interferon alpha, which is produced naturally during flavivirus infection, and has efficacy in some animal models. It was reported to show promise in open clinical trials against Japanese encephalitis (Harinasuta *et al.* 1985), and St Louis encephalitis (Rahal *et al.* 2004), and, on this basis, has been given presumptively

to patients with West Nile encephalitis. However a randomized double-blind placebo-controlled trial of interferon alpha in Vietnamese children with Japanese encephalitis showed it had no effect on outcome (Solomon *et al.* 2003). Ribavirin and intravenous immunoglobulin have also been given presumptively to patients with West Nile encephalitis (Shimoni *et al.* 2001; Haley *et al.* 2003). There is supportive data from animal experiments for the use of immunoglobulin (Agrawal *et al.* 2003), and a clinical trial has been set up by the United States National Institute of Allergy and Infectious Diseases. Intravenous immunoglobulin has also been used for post-exposure prophylaxis against tick-borne encephalitis virus in those bitten by ticks in areas where the virus circulates (Chiba *et al.* 1999). Once symptoms have developed, treatment of flavivirus encephalitis consists of attending to the complications of infection, and avoiding bed sores and contractures with good nursing care and physiotherapy. Despite intensive care, severe neuropsychiatric sequelae are common in survivors of flavivirus encephalitis.

Prevention

Preventative measures include personal protection to avoid being bitten by infected mosquitoes or ticks, vector control, and other measures to reduce the amount of virus circulating naturally. However, vaccines offer the best protection. There are licensed formalin inactivated vaccines for tick-borne encephalitis virus (Kunz 2002) and Japanese encephalitis virus (Hoke *et al.* 1988), and a live attenuated vaccine for Japanese encephalitis virus, that has been used widely in China, and is being used increasingly across Asia (Barrett 2001; Bista *et al.* 2001; Monath 2002). There are as yet no licensed vaccines against West Nile virus for humans, though a formalin inactivated vaccine has been used to protect horses (Siger *et al.* 2004).

42.3.10 Neurological dengue disease
Epidemiology

Most patients affected by one of the four dengue viruses, 1–4, present with a fever arthralgia rash syndrome, or a viral hemorrhagic

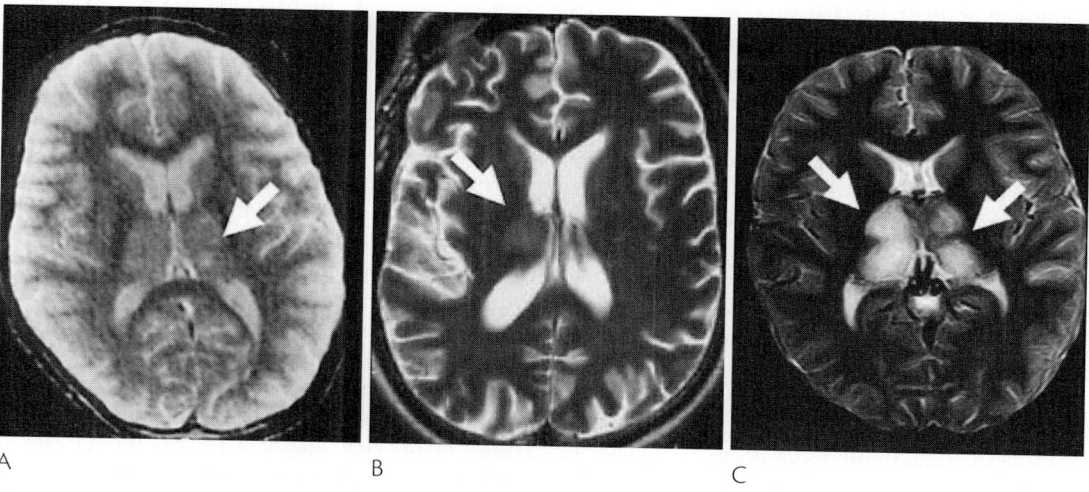

A   B   C

Fig. 42.8 Flavivirus encephalitis. T2-weighted MRI showing high signal intensity and swelling in the thalamus (arrows) of patients with Japanese encephalitis (A), West Nile encephalitis (B), and Murray Valley encephalitis (C). (From Solomon and Whitley 2004.)

fever (Solomon *et al.* 2001). The cause, and even existence, of neurological manifestations of dengue has been a longstanding controversy (Gubler *et al.* 1983; Nimmannitya 1987; Nimmannitya *et al.* 1987; Thisyakorn *et al.* 1999; Kankirawatana *et al.* 2000; Solomon *et al.* 2000). Most authorities now accept that dengue viruses can cause a non-specific encephalopathy, and occasionally encephalitis, even if the mechanism is not yet clear (Solomon *et al.* 2003). Dengue is unusual among arboviruses in that humans are the main natural host, though a sylvatic cycles exists in primates in southeast Asia and West Africa (Wang *et al.* 2000). The virus is transmitted between humans by *Aedes aegypti* and *Aedes albopictus* mosquitoes that breed in small collections of fresh water around the home, such as pools within rubbish, water containers, and tyres. Dengue is very widely distributed, being found in almost every country between the tropics (Halstead 1990), and neurological disease has been reported from virtually every country where dengue occurs (Gubler *et al.* 1983). It is now recognized that in parts of southeast Asia dengue is responsible for 1 in 20 patients admitted with a suspected central nervous system infection (Solomon *et al.* 2000), and as many as 1 in 5 of those with a clinical diagnosis of encephalitis (Kankirawatana *et al.* 2000).

Clinical features

Most patients with neurological dengue present with a reduced level of consciousness, alongside other signs of severe dengue infection, including the shock, vascular leakage, and haemorrhage which characterize dengue haemorrhagic fever. They may also have metabolic disturbances such as hyponatremia, and acidosis, and in many patients the encephalopathy is thought to be secondary to these complications. However in other patients with no rash, haemorrhage, or other signs of dengue fever, viral invasion across the blood–brain barrier causing encephalitis is thought to occur. In support of this is virus isolation, polymerase chain reaction detection, and detection of anti-dengue IgM antibody in the CSF of some patients (Lum *et al.* 1996; Hommel *et al.* 1998; Solomon *et al.* 2000; Chokephaibulkit *et al.* 2001). Clinically these patients present with a brief febrile illness followed by a reduction in consciousness level, which may range from lethargy, drowsiness, and irritability through to deep coma. Seizures are also common, particularly in young children (Familusi *et al.* 1972; Pancharoen *et al.* 2001). Pyramidal or long tract signs also occur, but the extra pyramidal tremors and tone abnormalities that characterize other arboviral encephalitides such as Japanese encephalitis and West Nile encephalitis (Solomon *et al.* 2002) are less common in dengue, though they have been described (George *et al.* 1988; Thakare *et al.* 1996). Presentations consistent with acute disseminated encephalomyelitis have also been described, some time after a dengue infection (Chimelli *et al.* 1990). Meningism occurs in up to 30 per cent of patients, usually as part of an encephalopathic illness. However, a simple viral meningitis due to dengue viruses is rare. Other neurological manifestations of dengue infection include mononeuropathies and polyradiculopathies (Solomon *et al.* 2003).

Investigation and diagnosis

Patients with dengue often have leucopenia and thrombocytopenia, as well as mildly elevated blood transaminases. In patients with dengue haemorrhagic fever an elevated hematocrit is found because of fluid loss from blood vessels due to increased vascular permeability (Nimmannitya 1987; Bethell *et al.* 2001). Hyponatremia is common particularly once fluid resuscitation has begun. CSF examination reveals a moderate lymphocytic pleocytosis in up to 30 per cent of patients with neurological disease. CT and MRI may show diffuse cerebral oedema, although focal abnormalities have also been reported (Lum *et al.* 1996; Thakare *et al.* 1996; Solomon *et al.* 2000; Cam *et al.* 2001). Dengue can be confirmed by isolating the virus from serum or CSF, polymerase chain reaction detection, or demonstrating IgM antibodies. Because dengue is so common in many parts of the world, and because IgM antibody may persist in the blood for up to 3 months, confirmation of antibody or virus in the CSF provides stronger evidence that dengue was the cause of the neurological symptoms (Solomon *et al.* 2003). In addition rigorous efforts must be made to rule out other cases.

Treatment and prevention

There are no effective drugs against dengue. Because patients with dengue haemorrhagic fever are especially likely to develop shock, fluid management is critical, and detailed guides are available (World Health Organization 1997). There are not yet any vaccines, but because there is no important enzootic cycle, and the disease is spread by peri-domestic mosquitoes, vector control measures can have an impact. These include educating people to remove *Aedes* breeding sites from around the house, by removing stagnant pools of water collected in tyres and other rubbish; by treating stored water with larvicide such as temephos, or by introducing the copepod *Mesocyclops* that feeds on *Aedes aegypti* larvae; by covering water storage containers to deny breeding mosquitoes access; and by ultra-low volume spraying of organophosphorus insecticides during epidemics (Rigau-Perez *et al.* 1998). In some settings legislation and fines for those that fail to remove *Aedes* breeding sites from the home have been effective. Personal protection with insect repellents is also recommended. The evidence for the efficacy of these measures is variable. The only undoubtedly effective vector control measure was the near eradication of *Aedes aegypti* from South America, using DDT, during the yellow fever campaign of the 1950s to 1970s. Since that campaign ended *Aedes* has reinfested South America (Monath 1997). Worldwide, *Aedes aegypti* continues to spread, and dengue is increasing as a global health problem (Gubler 2002).

42.3.11 Alphavirus encephalitis

The alphaviruses are mosquito-borne members of the genus *alphavirus*, in the family *Togaviridae*. The family is named after the cloak, or toga that envelopes the viruses (Griffin 2001). Alphaviruses cause two clinical syndromes. In Africa, 'old world' alphaviruses chikungunya virus, and o'nyong-nyong virus cause large outbreaks of fever arthralgia and rash; a similar syndrome, in which arthritis is predominant, is caused by Ross River virus in Australia. In recent years there have been very large outbreaks of Chikungunya across Asia and the Pacific (Bodenmann *et al.* 2006), and in some cases there has been neurological disease.

In the Americas, 'new world' alphaviruses, Eastern, Western, and Venezuelan equine encephalitis viruses, cause outbreaks of encephalitis in horses and humans. Eastern equine encephalitis is notable for having a higher case fatality rate than other viruses in the group, whereas Western equine encephalitis is remarkable for the fact that although it has caused large outbreaks in the past, it has now virtually disappeared as a disease of humans. Although Venezuelan equine encephalitis virus causes encephalitis in equines, in humans it causes large outbreaks of febrile disease, with only a small proportion developing central nervous system disease, which is usually mild. Phylogenetic analyses suggest that the alphaviruses arose in the new world, with subsequent introductions into the old world. The Western equine encephalitis lineage appears to have derived as a recombinant of an ancestral Eastern equine encephalitis-like and Sindbis-like viruses (Weaver *et al.* 1997).

Epidemiology

As implied by the names, Eastern equine encephalitis was originally found to affect the Eastern side of America, Western equine encephalitis the Western side, and Venezuelan equine encephalitis Venezuela and surrounding regions, though there has been considerable spread and overlap since then (Table 42.10) (Solomon *et al.* 2004). In 2007 the first case in a traveller returning to Europe was reported (Promed-mail 2007). The viruses are transmitted by a range of mosquito vectors, which for Venezuelan equine encephalitis differ according to whether there is endemic or epidemic disease (Beckwith *et al.* 2002). For Eastern equine encephalitis ornithophilic, that is bird-biting, mosquitoes are responsible for the natural cycle, but Aedes species, which bite birds and humans, act as bridging vectors, carrying the virus to humans (Fig. 42.7). Most Venezuelan equine encephalitis outbreaks occur when heavy rainfall and flooding expand the mosquito breeding habitats. One outbreak is thought to have originated from improperly activated equine vaccine strains as indicated by the fact that the epizootic strains are genetically almost identical to the vaccine strains (Weaver *et al.* 1999). During the 1995 outbreak of Venezuelan equine encephalitis in Venezuela and Colombia an estimated 85 000 human cases occurred, of which 3000 had neurological disease, and 300 were fatal (Weaver *et al.* 1996; Rivas *et al.* 1997). One-third of the human population seroconverted, and 8 per cent of the equines died. Unlike most of the other arboviral encephalitides, for Venezuelan equine encephalitis human viraemias are sufficiently high to infect mosquitoes, suggesting humans may not always be dead end hosts. In addition isolation of virus from the pharynx of up to 40 per cent of patients suggests direct spread between humans may be possible (Bowen *et al.* 1976). However no epidemiological evidence of such spread has even been demonstrated (Bowen *et al.* 1976; Rivas *et al.* 1997).

Clinical features

After an incubation period which is thought to range from 3 to 10 days, patients with Eastern equine encephalitis present with a few days of febrile prodrome, and then neurological disease. Early symptoms include headache, myalgia, photophobia, abnormal sensations, vomiting, dizziness and lethargy, followed by neck stiffness, a reduced level of consciousness, and seizures. On examination a high pyrexia >39°C is common and one-third of patients have signs of meningism. Clinical signs suggestive of brainstem involvement include gaze deviation, nystagmus, and papillary abnormalities. This may be because of inflammatory lesions in the brainstem, but uncal and subtentorial herniation have been seen at autopsy. A hemiparesis and limb spasticity suggest upper motor neuron involvement, whilst flaccid limbs suggest the involvement of lower motor neurons in the spinal cord. Seizures may be generalized or focal. In patients that remain conscious, aphasia and emotional lability may occur. Limb dysethesia and flaccid paralysis has also been reported in a fully conscious patient suggesting myelitis may occur without brain involvement (Clarke 1961). Approximately

one-third of patients die, but for those over 60 years the proportion is 50 per cent. One-third of survivors have moderate or severe sequelae (Deresiewicz *et al.* 1997).

The clinical features of Western equine encephalitis tend to be milder than those of Eastern equine encephalitis. Meningism is seen in half the patients, and although weakness and tremors are common, less than 10 per cent of patients develop coma (Bigler *et al.* 1976).The overall case fatality rate is approximately 4–10 per cent (MMWR 1988), being higher among the elderly. Western equine encephalitis is more severe in infants, with rapid progression from non-specific illness to convulsions and coma. Transplacental infection has been reported (Copps *et al.* 1959). The overall case fatality rate is 3 per cent, rising to 8 per cent for those over 50 years old (Longshore *et al.* 1956). Neurological sequelae are common in young children. Parkinsonian features including cogwheel rigidity and tremors have also been reported (Schultz *et al.* 1977).

Compared with the other alphavirus encephalitides, only a very small proportion of symptomatic Venezuelan equine encephalitis infections result in severe neurological disease. The incubation period is brief, from <1 to 5 days, and most patients then develop a febrile illness with severe headache, made worse by eye movements, photophobia, facial flushing, conjunctival injection, myalgia, arthralgia, nausea, vomiting, and dizziness. Somnolence and tremulousness occur frequently. More severe neurological features occur in 4–14 per cent of patients, particularly the young and the elderly. Seizures, particularly focal seizures, and raised intracranial pressure are common (Rivas *et al.* 1997; Molina *et al.* 1999). In 5–10 per cent of hospitalized patients cranial nerve palsies, motor weakness, paralysis, or cerebellar signs are reported. Respiratory infections, including interstitial pneumonia and tracheobronchitis are also common, and may result in secondary bacterial infection. Children are estimated to have 10 times the risk of neurological disease of adults, and younger children have a greater case fatality rate.

Investigation and diagnosis

There is often a peripheral leukocytosis in patients with Eastern equine encephalitis, with neutrophil predominance and left shift. In Eastern equine encephalitis, the CSF may resemble that of acute bacterial meningitis. There is a neutrophil pleocytosis, which may be as high as 5000 cells/mm³, the red cell count and protein are often elevated, and the glucose ratio is less than 50 per cent in half of the patients (Przelomski *et al.* 1988; Deresiewicz *et al.* 1997). In Western equine encephalitis the peripheral white cell count is usually normal or mildly elevated. Hyponatremia due to inappropriate antidiuretic hormone secretion has been reported. CSF opening pressures above 200 mm occur in two-thirds of patients. CSF white cell counts are usually less than 100 cells/mm³, but may range from less than 10 to 500 cells/mm³. There is usually lymphocyte predominance, a slightly elevated protein, and a normal glucose ratio. Rather like the arboviral fever arthralgia rash syndromes, Venezuelan equine encephalitis virus infection is typically associated with a leucopenia, and in one-third of patients, with elevated serum aspartate amino transferase levels (Bowen *et al.* 1976). When lumbar punctures are performed the CSF reveals a lymphocyte pleocytosis, which may reach several hundred cells/mm³, and is reported to be associated with an elevated CSF glucose.

CT scans show diffuse cerebral edema in three-quarters of patients with Eastern equine encephalitis, whereas they have been reported as normal in Western equine encephalitis. On MRI of Eastern equine encephalitis lesions are seen in the basal ganglia, thalamus, and brainstem, which do not enhance with gadolinium (Deresiewicz *et al.* 1997). EEGs show a mixture of background and focal slowing, and burst suppression. High voltage delta waves carry a poor prognosis (Przelomski *et al.* 1988). In Western equine encephalitis EEGs show diffuse slowing with focal delta activity in the temporal region, which may mimic herpes simplex virus-1 encephalitis (Bia *et al.* 1980).

Because viraemias in humans are usually low and transient in Eastern equine encephalitis and Western equine encephalitis, virus isolation from the blood is unusual, though it has been reported for Eastern equine encephalitis early in the illness (Clarke 1961). In fatal cases virus may be isolated from brain tissue, or detected in by polymerase chain reaction, or by immunohistochemical staining. In most cases, the diagnosis is made serologically by detecting antibody in the blood. Older serological tests such as haemagglutination inhibition, complement fixation, and neutralization test are being replaced by IgM capture ELISAs which detect antibody in a single blood or CSF sample, thus providing an earlier diagnosis. As expected, given the relatively high viraemias, Venezuelan equine encephalitis can be isolated from blood up to the 8th day of illness, and also detected by polymerase chain reaction (Bowen *et al.* 1976). Interestingly, virus can also be isolated from the pharynx. Antibody against Venezuelan equine encephalitis virus can be detected by IgM capture ELISAs or other serological tests (Watts *et al.* 1998).

Treatment and prevention

Since there are no antiviral drugs, treatment of alphavirus encephalitis is focused on the complications of infection particularly seizures and raised intracranial pressure, and pneumonia. There is no commercially available vaccine against Eastern equine encephalitis, though an inactivated vaccine is used by laboratory workers and others at high risk (Cole 1971). A similar vaccine is used to protect horses. A non-licensed inactivated vaccine is available against Western equine encephalitis for laboratory workers and others at high risk (Randall *et al.* 1947). Personal protective measures against mosquito bites should be employed, especially when entering marsh and woodland areas where the virus circulates. Equine vaccines for Venezuelan equine encephalitis have an important role in interrupting transmission during outbreaks, as well as in preventing epizootics. An inactivated vaccine has largely been replaced by a live vaccine, TC-83, attenuated by serial passage in guinea pig heart cells (Berge *et al.* 1961). Venezuelan equine encephalitis virus is considered a useful agent for biological warfare, because of its potential for droplet spread, and tendency to cause an incapacitating flu-like illness, rather than a fatal encephalitis, and for this reason the vaccine was produced by the US military.

42.3.12 Encephalitis caused by other arboviruses

Within the family *Bunyaviridae*, the *Bunyavirus* genus includes La Cross virus, and other California serogroup viruses that cause encephalitis, such as Jamestown Canyon virus and California encephalitis virus (Table 42.10). Lacrosse virus is a particularly important cause of central nervous system disease in children in America. The Bunyaviridae family also contains the *Phlebovirus* genus, which includes Toscana virus, an important emerging cause of central nervous system disease in southern Europe, and Rift Valley Fever, which causes febrile illness, and occasional encephalitis. The *Reoviridae* family includes Colorado tick fever virus, and similar members of the *Coltivirus* genus that causes febrile illness and central nervous system disease. The most important member of the Rhabdoviridae family is the lyssavirus rabies (Section 42.3.13). However the family also includes the genus vesiculovirus, whose members include vesicular stomatitis, and Chandipura virus (Table 42.10).

42.3.13 Rabies and other lyssaviruses

Introduction

For thousands of years rabies has been one of the most well-recognized and feared human diseases, being described by Aristotle in the 4th century BC, and probably also by the Chinese in the previous century (Warrell *et al.* 2004). The disease is caused by rabies virus, or related lyssaviruses. Worldwide approximately 55 000 cases of rabies occur every year, mostly in Asian and African countries (Fu *et al.* 2005). However occasional cases in travellers returning to Europe (Johnson *et al.* 2005; Solomon *et al.* 2005), and endemic cases in North America, serve as a reminder that the disease also affects the West. Rabies virus is transmitted naturally in enzootic cycles between mammals principally by bites and scratches. Humans most often become infected following dog bites, though other routes have been documented. There are effective pre- and post-exposure vaccines, but once symptoms develop, the disease is almost universally fatal. Approximately 80 per cent of patients present with encephalitic or 'furious rabies', which is characterized by hydrophobia and spasms; the remaining 20 per cent present with paralytic or 'dumb' rabies, and are of particular interest here.

Rabies virus belongs to genotype 1 in the genus *Lyssavirus*, family *Rhabdoviridae*. Other genotypes of lyssavirus include European bat lyssavirus types 1 and type 2, the latter of which has caused fatal human rabies in the United Kingdom (Fooks *et al.* 2003).

Rabies virus is principally transmitted to humans following animal bites, however there have been documented cases of transmission by the aerosol route, in laboratory workers (Winkler *et al.* 1973) and in an entomologist who frequently visited bat infested caves (Dutta *et al.* 1992). There has also been a single reported case of human transplacental transmission (Sipahioglu *et al.* 1985), and apparent transmission from a mother to her breastfeeding baby (Dutta *et al.* 1992). Other unusual routes of transmission include via corneal transplant (Javadi *et al.* 1996) and recently through transplantation of other organs (Srinivasan *et al.* 2005).

Any exposure to any mammal in an endemic country should be considered a potential source of rabies virus. The animal may not be obviously rabid. For example it is not uncommon for owners of a domestic dog to think it is gagging because of something stuck in the throat, rather than because it is rabid. Infected bats may simply appear 'friendly' or sleepy. All exposures should be taken seriously.

The true incidence of rabies is not known, because of under-reporting. However, in India 30 000 deaths were reported to the WHO in 1998 (Warrell 2003). Across Asia the reported incidence varies from 0.15 per 100 000 in Vietnam, to 3 per 100 000 in India. Only 204 cases were reported for the whole of Africa in 1998, probably due to gross under-reporting. In the United States, 32 deaths were reported between 1990 and 2000; 24 (75 per cent) of these were due to bat rabies viruses; 10 victims did recall contact with a bat, but only 2 reported a bite (Centers for Disease Control 2000).

Clinical features

Rabies presents after an incubation period which is typically between 20 and 90 days, but may range from a few days to more than 1 year (Jackson 2000). The early clinical features include localized pain and paraesthaesia which may cause pruritus or numbness at the site of the healed bite wound and are thought to reflect infection in the local peripheral sensory ganglia related to the site of viral entry (Jackson 2000). The pain can occasionally be excruciating and involve the whole leg (Solomon *et al.* 2005). This is followed by headache, often with fever.

In furious 'hydrophobic' rabies this progresses to periods of confusion, aggression and hallucinations, with intervening periods of lucidity. Sometimes patients are through to have psychiatric illness initially. However, once hydrophobia, fear of water, occurs, the diagnosis is usually obvious. Patients with hydrophobia might refuse to drink, despite being thirsty. If there is some uncertainty, then offering a patient a glass of water usually precipitates an inspiratory and extensor spasm, which helps clarify the diagnosis. Many patients also have aerophobia, fear of air, which can be confirmed by blowing on the cheek, or wafting air across the face; again this provokes a spasm.

The differential diagnosis includes hysterical pseudo-hydrophobia, usually in those who have been bitten by a dog and fear they have rabies, tetanus, and other brainstem encephalitides caused by enterovirus, borrelia, brucella, or mycoplasma. Muscle spasms caused by drugs should also be excluded, for example dystonia due to phenothiazines, or strychnine.

As the disease progresses there is autonomic stimulation, with excess salivation and frothing at the mouth, difficulty regulating the temperature, cardiac dysrythmias, and sometimes priapism in males. Cranial nerve involvement is common, particularly III, VII, and VIII. Complications include cardiac failure and hypotension, pneumothorax which may follow inspiratory spasms, seizures, diabetes insipidus, and the syndrome of inappropriate antidiuretic hormone secretion.

Paralytic rabies is harder to diagnose, and is often mistaken initially for other causes of acute flaccid paralysis, either because the history of exposure to a potentially rabid animal is not elicited, or not interpreted correctly. Other patients present with non-specific symptoms of headache, malaise, anorexia, and nausea, followed by ascending flaccid paralysis, with limb and then respiratory muscle involvement. Although, distinguishing clinically between paralytic rabies and Guillain–Barré syndrome can be difficult, there may be clues. For example, fever and headache at presentation, asymmetry of limb weakness, bladder involvement, and cells in the CSF suggest virus infection of the anterior horn cells in the spinal cord rather than immunologically mediated Guillain–Barré syndrome (Solomon *et al.* 2005).

Once the clinical features of rabies have developed, the disease is almost universally fatal. Without intensive care, patients with furious rabies die within 7 days of illness, whereas those with paralytic disease tend to have slower progression, and survive longer.

Diagnosis

Globally, most rabies is diagnosed clinically; this is not normally difficult for patients with furious rabies. Where possible, patients with suspected rabies are investigated by collecting saliva, CSF, serum, and a punch biopsy at the nape of the neck, which includes hair follicles containing peripheral nerve endings (Warrell 2003). The detection of rabies virus antigen in a skin biopsy, using a fluorescent antibody test, is one of the most reliable tests (Warrell 2003); corneal impression smears are less reliable, and so are no longer recommended. Detection of nucleic acid in saliva, and other samples using polymerase chain reaction is now used increasingly, and proving to be a rapid and reliable way of making the diagnosis (Smith *et al.* 2003; Solomon *et al.* 2005). Attempts should also be made to isolate the virus from the saliva and CSF. Because initial results may be negative, investigations should be repeated daily (Fooks *et al.* 2003). In some patients all ante-mortem testing will be negative, and the diagnosis is only made post-mortem by examining brain material. This can be obtained at autopsy, or by biopsy with a Vim-Silverman needle, of other long biopsy needle, such as that used for bone marrow aspiration (Warrell 2003; Solomon *et al.* 2005). In addition to virus detection, as described above, routine staining may reveal negri bodies. Virus infection can also be confirmed serologically by demonstrating the presence of neutralizing antibodies against rabies virus in patients not previously vaccinated. These antibodies are usually not present early in the illness until the second week of illness and sometimes will only become detectable just before death.

Management

Rapid diagnosis of rabies is important for the appropriate infection control and public health measures to be instituted, and it is a notifiable disease in the United Kingdom, and elsewhere (Department of Health 2004). Although there are no well-documented cases of human to human transmission except via organ transplantation (Srinivasan *et al.* 2005) barrier nursing is used, vaccination offered to relatives and exposed staff, and reassurance given to other staff members. Until recently, once clinical features developed, rabies was considered almost universally fatal. The only documented survivors (Jackson *et al.* 2003) had received some pre- or post-exposure vaccination, but had not received complete or prompt post-exposure vaccination, making them failures of the vaccination regime they had received (Hemachudha *et al.* 2002). However in 2004 a teenager in Wisconsin USA, who developed rabies following a bat bite, was successfully treated with a combination of ketamine, midazolam, ribavirin, and amantadine (Willoughby *et al.* 2005). Quite how the treatment worked is not clear. Ketamine has both anti-viral effects against rabies virus, but is also an NMDA receptor antagonist, and so is anti-exitotoxic. Her treatment was instituted on the 5th day of disease, when she was semi-obtunded, with cranial nerve signs, and ataxia.

Prevention

Preventative measures include eliminating infection in the animal vectors, pre- and post-exposure vaccination. Domestic-dog strains of rabies account for more than 90 per cent of the human disease worldwide. Rabies in dogs can be reduced by parenteral vaccination, fertility control, and clearing rubbish to reduce the food supply that maintains the population of stray dogs (Warrell and Warrell 2004).

Vaccination

Two types of the rabies vaccine are licensed for use in the United Kingdom and the USA: human diploid-cell vaccine Imovax Rabies™, and purified chick-embryo-cell vaccine Rabipur™. Pre-exposure vaccination, that is vaccination of travellers and others *before* they get bitten by a rabid animal, does not remove the need for revaccination after being bitten, but it simplifies the post-exposure regime, and makes it more likely to work (Warrell *et al.* 2004). An intramuscular regime has been the standard pre-exposure regime for many years, but recent studies have shown intradermal regimes are equally effective and require less vaccine (Warrell *et al.* 1988; World Health Organization 2005). There have not been any reported cases of rabies deaths in anyone who has had pre-exposure treatment followed by a booster dose after exposure (Warrell *et al.* 2004).

If someone has been bitten by a potentially rabid animal, they require immediate wound care, and then post-exposure vaccination. The wound care includes cleaning with soap and water, detergent, or plain water, followed by the application of ethanol, tincture or aqueous solution of iodine and removing any dead tissue (World Health Organization 2005). Wounds should be left open, rather than sutured, and a tetanus shot should be given. Modern post-exposure treatment is very effective. Failures of optimum treatment are uncommon. The standard anti-rabies vaccine regimen should be started as soon as possible. Recently intradermal regimes have been shown equally effective, and cheaper (World Health Organization 2005). Passive immunization with human rabies immunoglobulin, given at the same time as the active vaccination, has been shown to lower mortality after severe exposures, such as a savage bit to the head or neck; the vaccine affords protection for the first 7 days after a bite, while antibody is raised against the active vaccine. It may not be important for milder rabies exposure, for example a single bite on limbs, and indeed is not available in many countries. The full dose of rabies immunoglobulin, or as much as anatomically feasible, should be administered into and around the wound site. Any remainder should be injected intramuscularly at a site distant from the vaccine administrative site (World Health Organization 2005).

42.3.14 Progressive multifocal leukoencephalopathy

Progressive multifocal leucoencephalopathy is the only neurological disease in humans caused by a papovavirus. It was first described as a complication of lymphoproliferative conditions. Astrom *et al.* (1958) Cavanagh *et al.* (1959) suggested that progressive multifocal leukoencephalopathy may be caused by a virus and Zurhein *et al.* (1965), and separately, Silverman *et al.* (1965), showed with the electron microscope that the oligodendrocytic inclusions were packed with virions which were morphologically similar to papova viruses. 'Papova' derives from first syllable of the names of the major subvarieties of this type of DNA virus: the wart virus, *pa*pilloma; an oncogenic virus, *po*lyoma; and the *va*cuolating viruses. Padgett *et al.* (1971) grew and characterized a previously unknown polyoma virus from the brain of a patient with progressive multifocal leukoencephalopathy and named it 'JC' virus after the initials of the patient's name; it has nothing to do with Creutzfeldt–Jakob disease. Subsequent studies have confirmed that progressive multifocal leukoencephalopathy is caused by JC virus infection, or occasionally by a related virus, BK virus, which is also named after the initials of the first patient.

Progressive multifocal leukoencephalopathy has been associated with a range of immunocompromised conditions, such as sarcoidosis, carcinomatosis, organ transplantation, and immunosuppressive drug regimes. It remained a rare condition until the AIDS era.

HIV is now recognized to be a major risk factor for progressive multifocal leukoencephalopathy. In one recent series 3 per cent of all AIDS cases examined at autopsy had progressive multifocal leukoencephalopathy (Masliah *et al.* 2000). One immunosuppressive drug regime that has recently been associated with progressive multifocal leukoencephalopathy is the the humanized moncloncal antibody to alpha4 integrin, Natalizumab, which has been used in the treatment of multiple sclerosis and Crohns disease (Kleinschmidt-DeMasters *et al.* 2005).

Sero-epidemiological studies show that exposure to JC virus is ubiquitous among adults of all races in the world. Seroconversion takes place during childhood and does not result in disease unless the subject becomes immunocompromised. JC virus is unusual in that it causes no systemic disease, progressive multifocal leukoencephalopathy is its only clinical presentation. The kidney is the site of latent JC virus infection and it is thought that when the immune system is suppressed, JC virus enters the circulation and travels to the brain, lung, and lymphoreticular system (Houff *et al.* 1988). In the brain the oligodendrocytes become loaded with virus, with resultant cell destruction which leads to the breakdown of the myelin sheath and to the patchy demyelination which is so characteristic of the condition. It has been postulated that progressive multifocal leukoencephalopathy is due to reactivation of latent, previously, non-pathogenic JC virus infection in the brain. Mori *et al.* (1991) found evidence of JC virus DNA in oligodendroglia and astrocytes of elderly patients who had no evidence of progressive multifocal leukoencephalopathy (Mori *et al.* 1991). Others have found JC virus DNA in the brains of people, some of whom had AIDS and others did not, and who had no evidence of progressive multifocal leukoencephalopathy (Elsner *et al.* 1992; Quinlivan *et al.* 1992).

The pathology of progressive multifocal leukoencephalopathy is characteristic. JC virus produces a lysing infection of oligodendroglia which become large and spherical and have big inclusions. Astrocytes alter and show changes which are seen in malignancy. Inflammatory change is absent or minimal. There are multiple foci of white matter demyelination which coalesce as they enlarge. They are scattered throughout the cerebral hemispheres, cerebellum, and brainstem, and their location determines the clinical features. Initially they are most common at the junction of white and grey matter, which supports the notion that the virus reaches the brain by the bloodstream. Unlike multiple sclerosis the subpial and subependymal zones are relatively spared.

Progressive multifocal leukoencephalopathy is found throughout the world and rarely it may occur in childhood. Prior to AIDS it was a very rare disease of the middle aged and elderly. Sufferers had conditions which reduced cellular immunity such as carcinoma, lymphoreticular malignancy including Hodgkin's disease, lymphosarcoma and chronic lymphatic leukaemia, and granulomatous conditions such as tuberculosis and sarcoidosis. Others had been organ transplant recipients or patients treated with immunosuppresive drugs. In a few cases, no underlying condition was identified. With the evolution of AIDS, the picture changed. It became more common and affecting about 3–4 per cent of people with AIDS and frequently became initial AIDS-defining illness (Berger *et al.* 1987).

Interestingly, although many other opportunistic infections and tumours of the brain, have decreased significantly since the availability of Highly Active AntiRetroviral Therapy, HAART, for

HIV infection, the incidence of progressive multifocal leukoencephalopathy remains virtually unchanged (Sacktor 2002). However HAART has had an impact on the clinical presentation and diagnosis. For example, it is not uncommon for patients with HIV infection on HAART to have clinical and radiological features of progressive multifocal leukoencephalopathy, but without virus detected in the CSF; for this reason it has been proposed that such patients be considered 'possible progressive multifocal leukoencephalopathy' (Cinque et al. 2003).

In addition, progressive multifocal leukoencephalopathy has been shown to develop in HIV-infected patients shortly after the introduction of HAART, despite a recovery of the immune system. Therefore, progressive multifocal leukoencephalopathy could, in some cases, be another manifestation of the immune reconstitution inflammatory syndrome, discussed further below. Such inflammatory forms of progressive multifocal leukoencephalopathy now account for up to 18 per cent of cases, and usually have a favourable outcome (Du Pasquier et al. 2003).

Clinical features

The clinical features of progressive multifocal leukoencephalopathy are determined by the area of the brain which is involved. The onset is insidious and often difficult to recognize, even in retrospect, especially when it occurs in the evolution of an established predisposing disease. Much less commonly, progressive multifocal leukoencephalopathy may be the presenting manifestation of a previously undiagnosed illness which causes immune suppression, and rarely, the onset may be explosive and lead to the rapid demise of the patient. Such an explosive onset is seen more frequently in patients with AIDS. The most common symptoms and signs are mental disturbance, and impairment of awareness and of consciousness. Multiple enlarging, but not space occupying, lesions of white matter give rise to hemiparesis, parietal syndromes, visual pathway upset, pseudobulbar palsy, cortical blindness, dementia, and epileptic seizures; less commonly the white matter of the brainstem and cerebellum is affected with ataxia, nystagmus, and bulbar palsies. Raised intracranial pressure is seldom a problem and headache is uncommon.

Diagnosis

Brain imaging reveals widespread, multiple, and often confluent, non-enhancing white matter lesions that are not space occupying and have no mass effect or oedema (Berger et al. 1987) (Berger et al. 1987; Whiteman et al. 1993). CSF examination is usually unremarkable or it may show a slight pleocytosis and some rise of the protein content. It is now possible to detect and quantify JC virus in CSF using the polymerase chain reaction and this may correlate with survival rates (Yiannoutsos et al. 1999). High viral load of JC virus in the CSF is the norm. It may be necessary to resort to brain biopsy to diagnose progressive multifocal leukoencephalopathy, and also to exclude other pathologies such as lymphoma and other infection in AIDS patients. The characteristic pathological changes can usually be seen with light microscopy and virus particles can be seen with the electron microscope. A variety of immunocytological techniques and polymerase chain reaction can be used to demonstrate JC virus in tissue samples.

Treatment of progressive multifocal leukoencephalopathy remains difficult (Koralnik 2004). Cidofovir, which had looked promising in vitro and in animal models was shown not to be effective in HIV-positive patients in clinical trials. Other failed candidate drugs for progressive multifocal leukoencephalopathy in HIV-positive patients include interferon α2B and cytosine arabinoside, and steroids. Interestingly, however, cytosine arabinoside was clearly active in decreasing JC virus replication in vitro, and may have helped stabilize 7 out of 19 HIV-negative patients (36 per cent) with progressive multifocal leukoencephalopathy in a retrospective study (Aksamit 2001), despite significant bone marrow toxicity. Therefore, cytosine arabinoside should be considered in HIV-negative patients with progressive multifocal leukoencephalopathy.

In HIV-positive patients, HAART is now the only effective therapeutic option for progressive multifocal leukoencephalopathy. It probably acts indirectly by suppressing HIV replication, and thus allowing the immune system to keep JC virus replication under control. The mortality rate of progressive multifocal leukoencephalopathy is now between 30 and 50 per cent of patients during the first 3 months, although some patients stabilize and survive for many years. Survival of progressive multifocal leukoencephalopathy for more than 1 year has increased from 10 per cent before the HAART era to approximately 50 per cent of cases. Nevertheless, progressive multifocal leukoencephalopathy continues to occur in patients receiving HAART. It makes sense for HIV patients with progressive multifocal leukoencephalopathy to be on treatment regimes which include drugs with good blood–brain barrier penetration. Favourable prognostic factors include starting HAART at progressive multifocal leukoencephalopathy diagnosis in patients previously naive to antiretroviral agents, and a CD4 cell count greater than 100 cells/μl. However, as oligodendrocytes destroyed by JC virus are not replaced in the central nervous system, progressive multifocal leukoencephalopathy survivors are often left with devastating neurological sequelae.

42.3.15 Nipah virus encephalitis

Nipah virus is a recently identified paramyxovirus, of the family Paramyxoviridae, that was responsible for an outbreak of encephalitis in Malaysia and Singapore in 1998–1999. The virus has caused further outbreaks in Bangladesh since 2001 (Hsu et al. 2004). The virus is similar to Hendra virus, which caused disease in horses and their handlers in Australia in 1994. Nipah virus causes an illness in pigs, and appears to be directly transmitted to humans in close contact with pigs, such as farmers and abattoir workers. The virus is excreted in the urine of 'flying fox' fruit bats (Chua et al. 2002). During the original outbreak more than 250 people were affected with more than 100 deaths. Pathologically endothelial damage and vasculitis is seen, along with nuclear inclusions similar to those of other paramyxovirus infections, such as measles.

Human infection with Nipah virus causes pneumonitis and encephalitis characterized by a reduced level of consciousness, myoclonus, areflexia, and hypotonia (Goh et al. 2000). Magnetic resonance imaging shows increased signal intensity in the cortical white matter. Several reports have shown that patients who originally had mild or asymptomatic infections can present with encephalitis several months after the exposure (Wong et al. 2001; Tan et al. 2002). Clinically and immunologically these cases are reminiscent of subacute sclerosing panencephalitis caused by another measle's paramyxovirus (Wong et al. 2001). Some other individuals with an initial asymptomatic infection have characteristic small hyperintense lesions predominantly in the cerebral white matter on T2-weighted, suggesting subclinical damage (Lim et al. 2003).

Nipah virus encephalitis is diagnosed by isolating the virus from CSF, amplifying with polymerase chain reaction, or demonstrating IgM or IgG by enzyme linked immunosorbent assay. There is no treatment or vaccine, though ribavirin has been used in some cases.

42.4 Viral myelitis and myelopathy

The most important viral causes of spinal cord inflammation, or myelitis, are the enteroviruses, including poliovirus and enterovirus 71, the flaviviruses including Japanese encephalitis virus and tick-borne encephalitis virus, and the lyssaviruses, particularly rabies. Other viruses such as varicella zoster virus can occasionally cause myelitis, as outlined above. The retroviruses include HIV, which cause all manner of neurological infections, and is discussed in the next chapter, and human T-lymphotrophic virus 1, HTLV-1, for which the only important neurological syndrome is a myelopathy.

42.4.1 HTLV-1-associated myelopathy

Human T-lymphotrophic virus 1, HTLV-1-associated myelopathy, is also known as *tropical spastic paraparesis*. The latter term was coined by Mani *et al.* (1969) to describe a spastic paraplegic syndrome in India which was similar to the spastic form of Jamaican neuropathy which had been described by Cruickshank (1956) and Montgomery *et al.* (1964). The syndrome has a far wider geographical distribution than was first suspected: cases have been described from the USA, Caribbean islands, central and South America, Japan and the Far East, India and Central and South Africa. In Europe it is found in immigrants from the Caribbean in Britain, France, and Italy (Bucher *et al.* 1990). In 1985 IgG antibodies to HTLV-1 were found in the sera of 59 per cent of patients from Martinique with tropical spastic paraparesis and shortly after, antibodies to HTLV-1 were found in the CSF and sera of Jamaican patients with tropical spastic paraparesis. At much the same time a myelopathy associated with HTLV-1 infection was found in an endemic area in southern Japan, and was noted to be identical to tropical spastic paraparesis (Rodgers-Johnson *et al.* 1985). The clinical syndrome is now known as HTLV-1-associated myelopathy.

HTLV-1 is a type-C lymphotropic retrovirus which has a predilection for T4 lymphocytes. The virion is 110–140 nm in diameter, has a spherical core and a Mg^{2+}-dependant reverse transcriptase of molecular weight > 100 000. Lymphocytes which become infected are capable of indefinite growth and express specific surface viral antigens. The receptor has not yet been identified and transmission is from cell to cell. HTLV-1 was first isolated from a patient with a cutaneous T-cell lymphoma and it soon became recognized that HTLV-1 was associated with a unique form of adult T-cell leukaemia/lymphoma. Antibodies to HTLV-1 were found to be prevalent in areas where this T-cell malignancy was endemic, particularly in Japan, and in the West Indies where the association between HTLV-1 and tropical spastic paraparesis was made by Gessain *et al.* (1985). This association has been confirmed in many other areas of the world where HTLV-1-associated myelopathy is found, and its geographical distribution is roughly co-terminous with that of adult T-cell leukaemia/lymphoma. Both disorders have been described to co-exist in the same patient. Furthermore the same strain of HTLV-1 retrovirus has been isolated from separate patients with HTLV-1-associated myelopathy and adult T-cell leukaemia/lymphoma.

It is generally accepted now that HTLV-1 is the causal agent of HTLV-1-associated myelopathy because:

- there is intrathecal synthesis of antibody against HTLV-1 and oligoclonal immunoglobulin bands in CSF react with viral antigen;
- virus has been isolated from cultured peripheral blood and CSF mononuclear cells;
- HTLV-1 nucleic acid sequences have been identified in blood and CSF cells and viral antigen has been detected in CSF; and
- viral RNA has been detected in astrocytes in the central nervous system (Lehky *et al.* 1995).

The epidemiology of HTLV-1-associated myelopathy and HTLV-1 infection has been determined by using serological antibody detection. There is distinct geographical variation in the incidence of HTLV-1 positivity and HTLV-1-associated myelopathy. Southern Japan, the Caribbean, clusters in Panama, the Pacific coast of Colombia, and other areas of South America and equatorial regions of Africa, all have high rates. Studies of migrant populations demonstrate that infection is often acquired early in life and travels with the individual to non-endemic areas to cause the disease decades later.

Seropositivity rates vary from up to 10 per cent in regions of Japan, 5 per cent in Jamaica, Barbados, and Haiti, 6 per cent in Venezuela, and 1–2 per cent of blacks in the southern USA. Among whites, only intravenous drug abusers show high levels of antibodies (Bucher *et al.* 1990). An age-dependant rise in seropositivity is a consistent finding in all groups studied, with a peak at 65. There is a low rate of myelopathy in relation to the frequency of seropositivity and it has been estimated that HTLV-1 positive people have a less than 1 per cent chance of developing adult T-cell leukaemia/lymphoma or HTLV-1-associated myelopathy. It has been suggested that these observations may be explained by cumulative exposure to an agent of low infectivity over the individual's lifetime. Disease expression is probably modified by genetic factors, environmental factors or other acquired host factors.

The method of transmission is not evident in most cases. It can be passed on sexually from female to male, male to female most commonly, and male to male: from mother to child *in utero*, through breast milk, and via blood transfusion and intravenous drug abuse, this last being important for the transmission of HTLV-II. Infected cells are transmitted in the blood, concentrate in regional lymph nodes, and disseminate through the body. How they infect the central nervous system is not certain: possibilities include direct infection of endothelial cells by virus in the blood, or by invasion of infected marrow derived lymphocytes. There is no evidence to implicate an insect vector in transmission. The length of time between infection and the declaration of diseases is not known in the naturally occurring form of the disease, and may be as short as 6 months if it is transmitted by blood (Kaplan *et al.* 1991).

How the neurological syndrome is produced is not known. It could be that the virus induces a host response which triggers mechanisms causing tissue injury, or an immune response from on-going activation of immune cells, or there may be a direct toxic effect of the virus on oligodendroglia and astrocytes. The neuropathological changes seen in HTLV-1-associated myelopathy are essentially those of chronic progressive inflammation. The changes predominate in, but are not confined to, the spinal cord and the brunt falls on the thoracic cord. The meninges surrounding the cord are thickened (Iwasaki 1990). Chronic meningoencephalitis with perivascular cuffing and fibrosis, reactive astrocytic gliosis, and severe demyelination of pyramidal tracts and posterolateral columns leading to axonal loss, may all occur. Vasculitis and immune complex deposition in small vessels have also been described. Medulla, pons, and white matter of cerebrum and cerebellum may also have inflammatory changes.

Clinical manifestations

HTLV-1-associated myelopathy is of insidious onset and symptoms may be difficult to interpret in the early stages. It may rarely be seen in childhood but usually occurs after the age of 30 with a peak age of clinical onset after 40 years. Females predominate with a 2:1 bias, and HTLV-1-associated myelopathy is found in all socio-economic groups. There is difficulty with walking due to spasticity and weakness of the legs, usually symmetrical, with sensory symptoms of painful paraesthesiae in the legs, often with low back pain. Sphincter upset is common and is often an early feature. The course is slowly progressive over months and years and there are no remissions and no acute relapses. The physical signs are those of a spastic paraparesis with mild sensory loss, sphincter upset, and sometimes hyperreflexia of the upper limbs (Section 28.5.8). Less commonly, cerebellar signs and evidence of cerebral and cranial nerve have been noted. The course is insidious over months and years in more than 75 per cent, ranging from 6 months to 26 years with an average of 8 years (Bucher *et al.* 1990). In addition to the features of myelopathy, some cases have been described with lower motor neurone features when the disease may mimic amyotrophic lateral sclerosis.

Diagnosis

HTLV-1-associated myelopathy is suggested by the finding of a chronic myelopathy in one who lives in or who has migrated from an endemic area. MR imaging of the spine will exclude spinal compression and will demonstrate atrophy of the spinal cord and high intensity T2-weighted lesions in the cord. At least 50 per cent have lesions in brain white matter (Alcindor *et al.* 1992). Lesions have been found in the brain on MR scanning of people who have positive serology for HTLV-1 and have no neurological signs. The EEG may show non-specific slow wave abnormalities and visual, somatosensory, and brainstem auditory evoked response estimates may be prolonged. Most patients are seropositive to HTLV-1. Polymerase chain reaction may be necessary to distinguish positive results from HTLV-II in intravenous drug addicts. CSF examination may be normal but is more likely to demonstrate a raised mononuclear cell count and protein level. Over 70 per cent of cases have raised levels of intrathecal IgG antibody and HTLV-1 specific antibodies are demonstrated in virtually all. CSF neopterin levels are raised in 55 per cent of cases but only in 10 per cent of sera which suggests that there is a greater level of immune activation in the central nervous system than peripherally. A specific polymerase chain reaction test is available for CSF (DeBiasi *et al.* 1999).

The differential diagnosis of HTLV-1-associated myelopathy includes other myelopathies and causes of white matter lesions in the brain. Spinal compression is excluded by MR imaging, and if that is not available, by myelography. Weight is given to the suspected diagnosis if the patient lives in or has migrated from an endemic area, or if they have been exposed to blood transfusion, or if they abuse drugs.

Tropical ataxic neuropathies in Africa are usually associated with painful sensory loss and dermatitis, occur in relation to malnutrition, and are linked to ingestion of cassava and subsequent cyanide intoxication (Section 23.6.1). Acute tropical spastic paraparesis may be linked to lathyrism in the Indian sub-continent or malnutrition or malabsorption in Africa. The spinal from of multiple sclerosis may resemble HTLV-1-associated myelopathy; relapses and remissions and the occurrence of optic neuritis favour multiple sclerosis (Section 37.5). Brain and spinal cord imaging can be very similar in the two conditions. AIDS may also cause a myelopathy (Section 43.3.6) and AIDS shares common epidemiological features with HTLV-1 infection: sexual transmission, spread by blood transfusion and intravenous drug abuse, and mother to child infection. Dual infection with HTLV-1 and HIV has been described in AIDS patients with myelopathy in whom the myelopathy has been of HTLV-1-associated myelopathy type. It has been suggested that the presence of HIV increases the liklihood of the development of HTLV-1-associated myelopathy in a person with both infections (Berger 1991). The diagnosis of HTLV-1-associated myelopathy requires the demonstration of specific antibody or virus or virus genes in spinal fluid or CSF cells.

Treatment and prognosis

Attempts at treating HTLV-1-associated myelopathy have been disappointing. Corticosteroids have not produced consistent improvement. There are recent reports of improvement following treatment with high dose alpha-interferon, but the numbers are small (Yamasaki *et al.* 1997). Immunomodulation, azidothymidine, plasmaphoresis, and other treatments have not been consistently effective. The natural history of HTLV-1-associated myelopathy is slow and progressive over many years, with increasing spasticity and eventually, a wheelchair existence. In several patients, the disease arrests.

42.4.2 Poliomyelitis

The earliest record of a withered shortened leg with the characteristic appearance of poliomyelitis is an Egyptian stele of the 18th dynasty, 1580–1350 BC. The name is derived from the Greek 'polio' meaning grey and 'myelos' meaning marrow or spinal cord, and is descriptive of the pathological lesions that affect the grey matter, especially in the anterior horn of the spinal cord. Poliomyelitis is caused by poliovirus, which is a single-stranded RNA viruses belonging to the *Enterovirus* genus of the family Picornaviridae (Modlin 1995). There are three serotypes, distinguished on the basis of neutralization tests. Type 1 is the most important cause of paralytic disease. Polioviruses are carried in the human gastrointestinal tract, and transmitted via the faeco-oral route. Humans are the only natural hosts and reservoir.

Pathophysiology

Following ingestion the virus replicates in the gut wall, and adjacent deep lymph nodes before replicating further in the reticuloendothelial system resulting in viraemia. The route by which polioviruses then enter the central nervous system is uncertain; direct spread from the blood stream is postulated, but there is also evidence suggesting that the virus spreads from muscle up peripheral nerves to the central nervous system (Ren *et al.* 1992). Once the virus enters the nervous system there is direct neural spread, primarily between motor and autonomic neurones. Destruction of

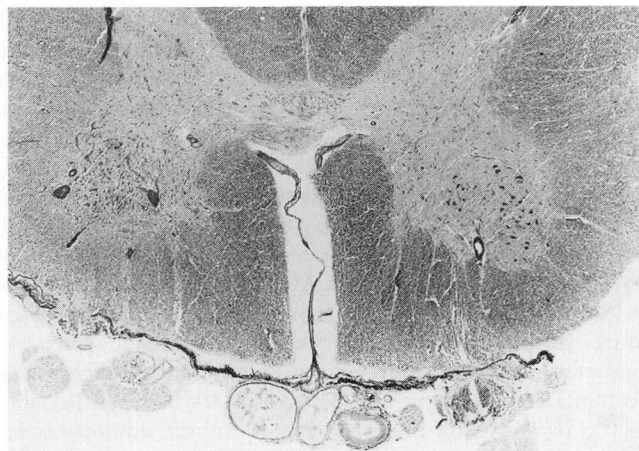

Fig. 42.9 Acute poliomyelitis. Transverse section of the cervical spinal cord, showing intense perivascular inflammatory cell cuffing in both anterior horns. There is parenchymal inflammation with loss of large motor neurons on the left, whereas the motor neuron population is preserved on the right. (Low power (×14) luxol fast blue/cresyl violet.) (Courtesy Professor M M Esiri and Blackwell Science Ltd.)

these neurons is accompanied by an inflammatory infiltrate of polymorphonuclear leukocytes, lymphocytes, and macrophages (Fig. 42.9). Primarily the grey matter of the anterior horns of the spinal cord, and the motor nuclei of the pons and medulla are affected (Bodian 1949). Involvement of the respiratory rhythm generating neurons in the medulla leads to respiratory failure (Baker *et al.* 1950). The distribution of lesions is similar in most cases, but the clinical features depend on their severity. The dorsal root ganglia are also commonly involved histologically, though this does not result in sensory deficits.

Epidemiology

Before the late 19th century, polio was predominantly a sporadic disease, which mostly affected children under 5 years. Subsequently large epidemics were recognized in Scandinavia, Western Europe, and the United States. This increased incidence was associated with a shift in the affected age groups: in the 1950s the peak incidence in the US was in 5–9-year olds with more than one-third of cases occurring in those over 15. In temperate regions infection tends to occur in late summer and autumn regions, whereas there is year round disease in tropical regions.

Epidemiological evidence supports the concept that before the 1900s the virus was ubiquitous resulting in mostly inapparent infection during early childhood (Bodian 1949). With improved hygiene infection was delayed until older ages, when the pool of susceptible children was large enough to support epidemics. In younger children passively transferred maternal antibody are postulated to play a role in producing inapparent as opposed to paralytic infection (Bodian 1949). Following the development of the inactivated and live attenuated polio vaccines, and their widespread use, the epidemiology has changed considerably.

Clinical features

The older literature describes five results of poliovirus infection:

- inapparent infection with no symptoms, which is the common outcome;

- 'abortive' poliomyelitis involving a mild febrile illness only;

- 'non-paralytic' poliomyelitis causing a viral meningitis syndrome;

- 'paralytic poliomyelitis' which may be spinal or bulbar, and is the most common reason for hospitalization; and

- 'encephalitis', which is rare.

Estimates of the ratio of apparent to inapparent infections vary from 1 in 60 to 1 in 1000 (Melnick *et al.* 1951; Nathanson *et al.* 1979).

Paralytic poliomyelitis occurs in 0.1 per cent of all poliovirus infections. In children there is typically a biphasic course (Horstmann 1949). Initially there is a non-specific febrile illness lasting 1–3 days that coincides with a viraemia, and is known as the 'minor illness'. Following this the patients may be asymptomatic for 2–5 days, before the 'major illness' starts. This begins with fever, headache, malaise, vomiting, and neck stiffness which are associated with a CSF pleocytosis. In older patients there is often spontaneous muscle pain, which may be relieved by walking (Weinstein *et al.* 1952). However exercise during the first 3 days of the major illness increases the incidence and severity of paralytic disease (Russell 1949). Tonsillectomy, pregnancy, and immunodeficiency are also thought to predispose to paralytic disease, or influence its severity. Sensory changes such as localized cutaneous hyperaesthesia and paraesthesias may occur at this stage. After 1–2 days there is frank paralysis and weakness, which may range from a single portion of one muscle to quadriplegia. The paralysis is flaccid, and although deep tendon reflexes may be brisk transiently, they soon become absent. The weakness is characterized by its asymmetrical distribution, which typically involves the legs more than the arms, and proximal muscles more than distal ones. Any combination of limbs may be paralysed, but the most frequent pattern is one leg, followed by one arm, or both legs and both arms. Paralysis tends to localize in a limb that has been the site of an intramuscular injection, or injury within 2–4 weeks before the onset of infection (Greenberg *et al.* 1952; Sutter *et al.* 1992). The weakness usually reaches a maximum over 2–3 days. Progression of the paralysis invariably halts when the patient becomes afebrile (Horstmann 1949). Muscle wasting commences within a week and fasciculations are prominent. Paralysis of the bladder, which is usually associated with leg paralysis, occurs in about one-quarter of adults, but is uncommon in children (Weinstein *et al.* 1952). Sensory loss is rare, and its occurrence may suggest an alternative diagnosis.

In bulbar poliomyelitis there is paralysis of the muscles innervated by the lower cranial nerves, particularly the IX and X, resulting in dysphagia, nasal speech, and occasionally dyspnoea (Modlin 1995). Its frequency has been reported as 5–35 per cent, and it is more common in adults (Weinstein *et al.* 1952). Rarely the medullary respiratory and vasomotor centres may be affected leading to irregular respiratory patterns, respiratory failure, cardiac dysrhythmias, and circulatory collapse (Baker *et al.* 1950).

Investigations, diagnosis, and treatment

The CSF typically shows a pleocytosis, often with elevated polymorphonuclear cells, a slightly elevated protein, and normal glucose ratio. Polioviruses can usually be isolated from throat secretions during the first week of illness, and from faeces subsequently. Virus is rarely isolated from the CSF in patients with paralysis. Although convalescent serum was at one time thought to be valuable therapy in poliomyelitis, subsequent trials showed there was no effect

(Horstmann 1950). Bedrest is advised for patients with paralysis, because of the risk that movement exacerbates the disease. Physiotherapy should be initiated once the progression of paralysis has ceased. Patients with respiratory failure receive positive pressure ventilation, which has replaced the 'iron lung' tank respirators which were used in the past for poliomyelitis.

Vaccination

In the 1950s, a formalin inactivated vaccine was developed by Salk, and a live attenuated vaccine by Sabin, and their usage in some countries was associated with a dramatic decline in cases of poliomyelitis. In 1988 the World Health Assembly resolved to eradicate polio by the year 2000 (Hinman *et al.* 1987; World Health Assembly 1988). The oral polio vaccine is a mixture of live attenuated poliovirus types 1, 2, and 3. Although the target of global eradication has not been achieved, the campaign has undoubtedly been a success. The number of countries where polio is endemic declined from 125 in 1988 to seven in 2002, and the estimated incidence of polio has decreased by more than 99 per cent (Anonymous 2003). In 2001 there were only 483 confirmed cases. However, there have been several setbacks in recent years; for example in 2004 there were 1255 confirmed cases from 16 countries. Reasons include difficulty immunizing in areas of ongoing conflict, poor compliance with immunization because of mistrust in some communities, natural disasters disrupting infrastructure, and programme financing. Despite this, there is still hope that ultimately the wild-type virus will be eradicated.

There is a very small risk of approximately 1 case per 2 million doses of the vaccine reverting to a virulent form to cause poliomyelitis in a recipient, or more commonly via faeco-oral spread to a household contact of a recipient. This is termed vaccine-associated paralytic poliomyelitis. Studies have shown that this reversion to virulence is associated with at least one per cent change of the amino acid sequence of the virus. In polio endemic areas this risk of the vaccine is greatly overshadowed by the risks of acquiring polio itself if you are not vaccinated. However, in areas where the disease has now been eradicated this is no longer the case, and so in the Americas, inactivated vaccine is once again being used more widely. Those with immunodeficiency should also be given inactivated vaccine because they too have an increased risk of reversion to virulence, and long-term shedding of virus.

Post-polio syndrome

Patients that have had poliomyelitis are at increased risk of a range of musculoskeletal symptoms, caused by mechanical complications of the weakness. However it has become clear that in addition to this there is a specific syndrome of late chronic progressive weakness and wasting known as the post-poliomyelitis syndrome, or post-poliomyelitis progressive muscular atrophy (Section 23.3.7). The cause is uncertain; however it is postulated that the giant motor units, which are the residue of poliomyelitis, age and die more quickly than normal motor units, thus resulting in late progressive atrophy.

42.5 Bacterial encephalopathy and encephalitis

Encephalopathy, a reduced level of consciousness, and encephalitis, which refers to inflammation of the brain parenchyma presenting with clinical features resembling viral encephalitis,

can be caused also by a range of bacteria, particularly atypical and intracellular bacteria. For some of these organisms, particularly the bacterial spirochaetes syphilis (Section 41.5.4) and borrelia (Section 41.5.5), the neurological manifestations are broad, and include meningitis, encephalopathy, and radiculopathy. Commonly, the encephalitis is accompanied by other clinical features which help to identify the aetiology such as the eschar which occurs in Mediterranean spotted fever, or the rashes of other rickettsial diseases. The disease may be peculiar to a particular geographical area, such as Rocky Mountain Spotted Fever, or it may occur in a specific season, or in certain climatic conditions, or in epidemic form such as typhus. It is therefore essential to take a travel history, especially in patients with an undiagnosed central nervous system infection; this includes paying attention to what activities were undertaken, and whether there were any insect bites. Few neurologists will have had experience of all these varieties of infectious encephalitis and phone or e-mail enquiries to infectious diseases and microbiology colleagues may prove useful. For many of these encephalitides, curative treatment is possible provided the diagnosis is made expeditiously.

The neurological diseases caused by spirochaetal gram negative bacteria include syphilis (Section 42.5.1), lyme disease (Section 45.5.2), leptospirosis (Section 45.5.3), and borrelia recurrentis (Section 45.5.4).

42.5.1 Syphilis

Syphilis derives from the name of a shepherd in a poem by Fracastorius in 1521, who was crippled (*Syphilos Gr.* maimed) by the disease, having offended the Gods. It is not known where the disease originated. Reports of infection of almost epidemic proportions began to appear in Europe at the end of the 15th and the beginning of the 16th centuries, and it was thought to originate from the New World. Dementia paralytica was described by the end of the 18th century and during the 19th century descriptions of general paresis of the insane, tabes dorsalis, and syphilitic arteritis appeared. In the early days of the 20th century, the causal organism *Treponema pallidum*, had been identified and linked to general paresis of the insane. The Wasserman complement fixation diagnostic test was devised in the first decade and treatment with Salvarsan and malaria was described. Penicillin was first used in 1943 and remains the mainstay of treatment.

The organism which causes syphilis is a noncultivable spirochaete gram negative bacterium, *Treponema pallidum*. It is a flexible, spiral organism with a corkscrew motion which can be observed by dark ground illumination through the microscope and is 5–15 μm in length. It is fastidious and requires heat, light, and moisture for survival, which are provided by contact with warm living tissue. It does not survive long without animal contact. Infection in humans occurs as a result of warm and moist contact, venereally in the vast majority. Congenital disease is caused by transplacental transfer. Treponemes spiral through mucous membranes which have been abraded, multiply locally and pass rapidly to regional lymph nodes, thence via lymph and blood throughout the body.

Central nervous system invasion is now thought to occur early in the course of the disease, as judged by detection of the organism in the CSF with polymerase chain reaction; this is followed by pleocytosis and then reactivity of the Veneral Diseases Reference Laboratory, or VDRL, test. Whereas the bacteria may be cleared from the CSF spontaneously in about 70 per cent of patients, and the pleocytosis resolves, VDRL reactivity of the CSF may persist. Invaded tissues react with lymphocytic and plasma cell infiltration causing endarteritis and giving rise eventually to fibrosis and scarification. This reaction may remain quiescent for years to be resurrected years later, producing new clinical manifestations.

By convention, congenital syphilis is separated from acquired syphilis, and of the latter, three phases of infection are recognized, primary, secondary, and tertiary. Neurological disease does not usually accompany primary infection although treponemes may be isolated from CSF at this stage. The central nervous system may be implicated in secondary syphilis and is a common site for tertiary syphilis. Uncertainties remain about the diagnostic criteria, optimum therapy, and response to treatment, especially in patients that are coinfected with HIV.

Syphilis affects every race throughout the world, both sexes and any age group. The true incidence is almost certainly underreported because of reluctance to admit to the disease and also because the clinical manifestations may be minimal at the onset or masked by another co-existing sexually acquired infection. Following the introduction of effective treatment with penicillin, the incidence diminished, and tertiary syphilis declined dramatically. However, this position has not been maintained, and syphilis continues to present a health hazard, particularly in HIV positive males. The natural history of syphilitic infection is greatly modified by antibiotic treatment which is curative in the majority, although it is impossible to predict who will develop late manifestations of the disease. This is particularly of concern in HIV-positive patients. The primary disease occurs within weeks, or at the most, 3 months of infection. Secondary syphilis follows after 1–6 months and tertiary syphilis can appear within 2–3 years or as long as 20 years or more, after the first infection.

- *Primary syphilis* usually takes the form of a chancre, or genital ulcer, 3–6 weeks after infection. This is associated with inguinal adenopathy. There may be other, co-existing sexually transmitted infection and this may obscure the clinical picture. Treated or untreated, the chancre and adenopathy resolve after some weeks.

- *Secondary syphilis* comes on 2–4 months later and is manifest by the development of skin lesions consisting of papular, macular, or vesicular rashes, condylomata lata around the genitals, and patchy alopecia. In this stage, rare central nervous system manifestations include acute meningitis. Following recovery from this phase, without treatment, patients may be asymptomatic for many years, as the disease lies latent, the only evidence of syphilitic infection being found in serological tests.

- *Tertiary syphilis develops*. In perhaps as many as 35 per cent of such cases. This affects skin, locomotor system, cardiovascular system, viscera, and the central nervous system. When the nervous system is involved, the term 'neurosyphilis' applies although clearly this may encompass the meningitis of secondary infection. Common usage infers that neurosyphilis means tertiary involvement of the central nervous system.

Neurosyphilis

Approximately 10 per cent of cases of tertiary syphilis affect the central nervous system. The clinical syndromes of neurosyphilis are several and more than one may occur in the same patient. Consequently, none of the clinical classifications is entirely satisfactory,

Table 42.11 Classification of neurosyphilis, modified from Wilson (1940)

Asymptomatic
Meningeal
Meningovascular
General paresis of the insane (GPI)
Tabes dorsalis
Optic atrophy
Spinal parenchymal
Gumma
Osteitis of skull and spine

but this is not of too great moment since the treatment for each is basically the same: penicillin with symptomatic management as necessary. The account of the clinical syndromes by Wilson (1940) remains unsurpassed and is the basis for the following classification (Table 42.11).

Asymptomatic neurosyphilis. This has no clinical manifestations and the diagnosis is based on the finding of abnormalities on CSF examination: sometimes a raised cell count and invariably positive immunological tests. This can be found at any time after infection, but especially in the first year. The significance of this form of the disease is that, without adequate treatment, and sometimes despite apparently adequate treatment, parenchymal neurosyphilis may develop later.

Meningeal syphilis. This may develop at any time following primary infection. Most commonly it takes the form of acute or subacute meningitis in the first year or two after the initial infection (Section 41.5.4). A more chronic form with meningeal granulomatous change may lead to multiple cranial nerve palsies, hydrocephalus, Argyll Robertson pupils, seizures, and focal signs (Merritt *et al.* 1935). CSF examination reveals meningitic change with raised cell count, protein, and reduced glucose content. Immunological tests for syphilis are positive.

Meningovascular syphilis. This follows initial infection by several months to several years, the average being about 7 years. It is brought about by endarteritis of large and medium cerebral vessels, giving rise to stroke syndromes from infarction. The meninges are also affected with granulomas and adhesions and spinal cord disease is not infrequent. Strokes in the young may be manifestations of meningovascular syphilis. Blood and CSF tests for syphilis are positive.

General paresis of the insane or dementia paralytica. This used to affect about 5 per cent of untreated syphilitics. Symptoms appear 10–15 years after primary infection and result from structural degeneration of the cerebrum with neuronal loss, cerebral atrophy, and glial proliferation. Onset is insidious with psychiatric symptoms predominating: memory loss, delusions, often of grandeur, disinhibition, emotional outbursts and erratic behaviour, deterioration of personal habits and hygiene, and disintegration of symbolic thought lead to professional disaster and social catastrophe. Further parenchymal damage results in clumsiness, inco-ordination, epilepsy, cranial nerve palsies, Argyll Robertson pupils, spastic paraparesis with extensor plantars, frank dementia, and sphincter incontinence. Blood and CSF tests are abnormal.

Tabes dorsalis or locomotor ataxia. This is more common in men, comes on later than general paresis of the insane, on average 10–20 years from first infection, and is often associated with other varieties of parenchymal disease. The initial symptoms are sensory. Lancinating, sudden, lightning pains affect the legs and less frequently, trunk and arms. They are repeated, exquisitely painful, and have a tendency to occur in clusters over several days or longer, at a time. Similar painful crises may disturb the abdominal viscera to cause severe abdominal pain, vomiting, paralytic ileus, and rarely may cause similar crises of bowel and bladder. Paraesthesiae and sensory diminution affect the feet and legs and sometimes the trunk in a cape or girdle distribution. Loss of posterior column function causes postural ataxia and unsteadiness of gait, more evident in the dark when visual clues are lost. Tendon reflexes are reduced and later lost, associated with hypotonia. Hypotonia leads to hypermobility of joints, which combined with sensory loss results in repeated and increasing trauma to joints. These become disorganized and grossly arthritic and swollen as first described by Charcot. Trophic ulceration occurs over areas of skin lacking sensation and which are subjected to pressure, usually in the feet. Argyll Robertson pupils are seen in more than half of cases of tabes and consist of eccentric and irregular pupils with depigmentation of the iris and failure of reaction to light but retention of the reaction of accommodation. Optic atrophy is common as are oculomotor palsies and result in a characteristic facial appearance with bilateral ptosis leading to compensatory frontalis over-reaction. Other forms of neurosyphilis commonly co-exist. In taboparesis there is a combination of signs of tabes dorsalis and upper motor neuron signs such as extensor plantars, and there may be other signs of general paresis of the insane. CSF and blood usually show positive changes of syphilis, but may be normal in advanced, 'burnt out' cases.

Syphilitic optic atrophy. This is the result of damage to the optic nerve fibres from chronic inflammatory changes and progresses insidiously to complete blindness on average 8 years after first infection and commonly accompanies tabes dorsalis.

Spinal parenchymal neurosyphilis. This is caused by meningo-myelitis, hypertrophic pachymengitis, and spinal cord infarction. Spinal cord syndromes incorporating radiculopathy appropriate to the level of disease are found. All spinal syndromes carry a high risk of vascular complication from physical intervention, and neurosurgery is almost always best avoided.

Gummas. These are rubbery nodules of inflammatory tissue and produce signs by virtue of their space occupying characteristics. They are exceptionally rare and are likely to be misdiagnosed as neoplasms or granulomas unless specific serological tests are undertaken (Punt 1983).

Syphilitic osteitis. Syphilitic osteitis of the skull and vertebrae is no more than a historical oddity nowadays. Signs are caused by secondary pressure on adjacent spinal cord or brain.

Atypical cases. Atypical cases of neurosyphilis are described also in which patients do not neatly fit into any of the traditional classifications, for example those with memory and behavioural changes and seizures, with temporal lobe changes on imaging reminiscent of herpes simplex encephalitis (Marano et al. 2004).

Associated HIV infection. It may be argued that neurosyphilis associated with HIV infection should not be categorized separately because the clinical manifestations incorporate each and several of the syndromes described above. However, it has been reported that HIV patients who have syphilis have a greater likelihood of developing secondary syphilis, are likely to have atypical manifestations of neurosyphilis, and atypical serological tests, including false positive tests for syphilis. There is disagreement about whether all HIV patients with secondary syphilis should have a lumbar puncture to look for evidence of neurological disease; in practice CSF examination is advisable HIV and syphilis are associated because they share common risk factors; up to 75 per cent of newly diagnosed syphilis patients in the developed world are HIV positive. Also, failure of early treatment for syphilis, thus leading to the early development of neurosyphilis, is more common in HIV-positive individuals, especially those with lower CD4 counts. It may be impossible to differentiate the neurological damage caused by HIV from that caused by syphilis and both syphilis and HIV may produce CSF abnormalities. It was thought that this response of such patients to treatment with penicillin was suboptimal. (Gordon et al. 1994). Recent studies indicate that this problem may not be as serious as once feared (Simon 1994). Nevertheless, the combination of the two pathologies in the same patient can be confusing, and produces a diagnostic scenario distinct from straightforward neurosyphilis (Johns et al. 1987; Katz et al. 1993).

Congenital syphilis. This is always contracted by transplacental passage of the treponeme. Abortion or stillbirth may ensue, or the infant may have syphilis at birth. Transplacental infection is more likely if the mother has primary or secondary disease. In most infants with early congenital syphilis, evident before 2 years of age, signs usually evolve over 2 weeks to 2 months or may appear later, manifesting as secondary syphilis in several organs. Nasal discharge or snuffles, is common. Hepatosplenomegaly, osteochondritis, choroidoretinitis, skin eruptions, and failure to thrive are amongst the features. In late congenital syphilis there may be no signs, infection being indicated by positive blood tests. Signs of infection include interstitial keratitis, choroidoretinitis, optic atrophy, perforation of the nasal septum and palate, anterior bowing of the tibia, painless arthropathy resulting in Clutton's joints, deafness, and peg-like deformities of the teeth known as Hutchinson's teeth.

Diagnosis

Darkfield microscopy of specimens from chancres and the identification of *Treponema pallidum* is the quickest and best test for the diagnosis of primary syphilis. Direct fluorescent antibody staining of the organism has about the same sensitivity. Polymerase chain reaction can also be used to detect treponemal genetic material in tissues and CSF but the tests lack sensitivity, especially in later stages of the disease. In later forms of syphilis the diagnosis is made by antibody tests of serum and CSF, which broadly fall into two groups:

◆ non-treponemal tests which rely on non-specific antibodies against cardiolipin which are elevated in the serum of many patients with syphillis;

◆ treponemal tests which utilize specific antibodies.

Serological diagnosis is not straightforward, especially for patients with neurosyphillis. The Venereal Disease Reference Laboratory test, or VDRL, and the rapid plasma reagin test, RPR, are both modifications of the original Wasserman test, and are examples of non-treponemal tests. They become positive within 4–6 weeks of infection and are positive in most patients with neurological syphilis, but may become negative in late neurosyphilis. False positives may occur with increasing age, pregnancy, malignancy, and other systemic and infectious diseases. The treponemal antibody tests such as the fluorescent treponemal antibody-absorbed, or FTA-ABS, *Treponema pallidum* haemagglutination assay, TPHA, the *Treponema pallidum* particle agglutination assay, TPPA, or microhaemagglutination test for *Treponema pallidum*, MHA-TP, are more specific, and have higher sensitivity. They are said to be positive in the serum of all patients with neurosyphillis, and so provide a good initial screen of patients with suspected neurosyphillis. In many centres the TPPA is increasingly used because of its ease, and slightly better sensitivity. The rapid plasma reagin test is the non-treponemal test often used. An enzyme linked immunosorption assay test is used as the first screen in many countries, and any positives have RPR and TPPA done.

Testing the blood with a treponemal antibody test is recommended as a first screen for patients with suspected neurosyphilis, because they are positive for all patients with neurosyphilis. If these tests are non-reactive neurosyphilis has effectively been excluded, except in patients with HIV. CSF examination is then the next step. Typically there is a lymphocytic CSF pleocytosis with moderately elevated protein. A CSF Venereal Disease Reference Laboratory test is positive in 30–70 per cent of patients with neurosyphilis, so reactivity here indicates the need for treatment. The CSF Venereal Disease Reference Laboratory test is highly specific, but false positives may be seen when the CSF is visibly tinged with blood, when it reflects serum antibody. A negative CSF Venereal Disease Reference Laboratory test does not exclude neurosyphilis. Most would consider that if there is a pleocytosis of >5 cells/μl, patients should still be treated for neurosyphilis. If there is no pleocytosis, but the protein is elevated to >45/dl, then neurosyphilis is still possible. In these patients a CSF treponemal antibody test, which is sensitive but not very specific, is helpful; if this test is positive patients should be treated; but if both CSF treponemal and non-treponemal antibody tests are negative, and there are no cells in the CSF, treatment is not indicated. In summary, if a treponemal test, such as TPPA or an enzyme linked immunosorption assay is negative, this effectively rules out neurosyphilis, but a positive test does not diagnose it. If a non-treponemal test, such as Venereal Disease Reference Laboratory test or rapid plasma reagin test is positive, this confirms neurosyphilis, but a negative does not rule it out.

In patients that are HIV positive the situation is a little different because mild CSF pleocytosis and elevated protein are common in HIV. For this reason it has been suggested that if the CSF Venereal Disease Reference Laboratory test is negative, then patients should not be treated unless the CSF cell count is greater than 20/μl, or there is a positive treponemal antibody test.

There are no diagnostic radiological features of neurosyphilis. Plain X-rays may reveal disorganization of joints in tabes dorsalis; similar changes may be seen in cases of syringomyelia and severe diabetic neuropathy. Brain imaging with MR or CT may reveal infarcts in meningovascular syphilis, general atrophic changes in general paresis of the insane and atrophic or pachymeningitic features on spinal imaging in cases of tabes. Conventional or MR angiography of the cerebral arteries in young stroke syndromes may show occlusion or variation in calibre of vessels, indicating arteritis.

Treatment

The treatment of all forms of neurosyphilis is penicillin. Many different dosage regimes have been advocated. Most agree that oral treatment is not appropriate and it is better to err on the side of too much. Twenty four million units of intravenous penicillin G daily, either as a continuous infusion or in divided doses, is recommended for 10–14 days. Many recommend following this with intramuscular penicillin G 2.4 million unit weekly for a further 3 weeks. In patients that are allergic to penicillin, ceftriaxone 2 g IV four times daily for 10–14 days is recommended if they have early neurosyphilis; erythromycin 500 mg every 6 h for 30 days is an alternative regime. However if allergic patients have late stage disease it is recommended that they undergo penicillin desensitization followed by a full treatment course if possible.

Penicillin is treponemicidal and lyses the organisms to release large quantities of antigenic material into the bloodstream. This may result in a Jarisch–Herxheimer type of reaction with confusion, delirium, convulsions, and pyrexia. Steroids reduce the severity of such reactions and some recommend they be administered routinely before the first dose of antibiotic is given and for 48 h thereafter.

The best response to treatment is in those that have a CSF pleocytosis before treatment. Treatment does not appear to reverse parenchymal damage in those with meningovascular disease, general paresis of the insane, or tabes dorsalis, however it should prevent disease progression in about 75 per cent of cases. To monitor the effect of treatment a CSF examination every 6 months is recommended. If the cell count has not started declining by 6 months, or returned to normal by 2 years, retreatment is recommended. In HIV positive patients, especially those with low CD4 counts, assessment every 3 months is advocated by some.

Symptomatic treatment is given for other manifestations: antiepileptic drugs for seizures, orthopaedic treatment for unstable joints. Lightning pains and crises in tabes may respond to treatment with carbamazepine, reinforced with amitryptiline at night, or gabapentin.

42.5.2 Lyme disease

Lyme disease is a form of borreliosis, characterized by cutaneous, arthritic, cardiological, and neurological manifestations, and takes its name from the Lyme district of Connecticut, USA. The term neuroborreliosis, or lyme neuroberreliosis is used for patients with neurological disease. In the early 1970s a cluster of juvenile cases were thought to have juvenile rheumatoid arthritis, and provided the stimulus for the description of the disease (Steere *et al.* 1977). It became evident that many of the cases had suffered a tick bite and developed an unusual skin lesion thereafter, erythema chronicum migrans, before going on to develop arthritis, and in some cases, cardiac and neurological complications. It became clear subsequently that it was an arthropod-borne infection, and that the skin lesions had been described previously in Sweden and associated with tick bites in the early 20th century. Furthermore, the connection between erythema chronicum migrans and arthritic and neurological disease had been documented from several places in Europe in the first half of the 20th century and it has since been demonstrated that the American and European diseases are caused by very similar spirochaetes. For clinicians one of the most difficult issues lies in establishing the diagnosis. It is not uncommon to be faced with a patient with neurological symptoms, and some vague history of an insect bite at some time in the past. In the United Kingdom nearly 800 cases were reported over a 12-year period, giving an incidence of less than 0.5/100 000 per year (Smith *et al.* 2000). Most cases came from Southern England, and 10 per cent were in travellers returning from overseas; 15 per cent of cases had neurological disease.

As more information accumulates, it is becoming evident that the manifestations of Lyme disease may occur worldwide. Endemic areas are found in USA, Europe, Central Asia, China and the Far East, and in Australia. It is anticipated that other continents will be found to harbour the disease. The causal organism, *Borrelia burgdorferi* in the USA, Europe and Asia whilst almost identical borrelia, *B garinii* and *B afzeli* are found only Europe and Asia. The species difference probably accounts for geographical variations of clinical manifestations; arthritis is more common in USA, cutaneous and neurological disease in Europe. The vectors for the disease are hard-bodied ticks of the genus *Ixodes* of which different species carry the infecting organism in different countries. What is common to them all is they are found in forest and scrub and are most active in late spring and early summer, which accounts for the seasonal variation of disease. Deer, mice, and many other small mammals and birds are the hosts. Humans become infected when bitten by an infected tick, whilst visiting woodlands for work or recreation. Clinical symptoms appear after an incubation period of 3–31 days and accord with tissue invasion; *Borreliae* have been isolated from skin, blood, CSF, and joint fluid. More chronic manifestations appear months to years later and are almost certainly due to autoimmune mechanisms.

Clinical features

Lyme disease is described nowadays as developing in three stages. Stage 1 is of early, localized infection. Stage 2 is of early, disseminated infection and stage 3 is of late or persisting infection. Not all patients pass through all of the stages and presentation varies. It is a multisystem disease which, if not recognized, can become chronic and disabling.

The early, localized stage is preceded by a tick bite, but this may be remembered by only about a third of all patients. After a week to a month, erythema chronicum migrans appears at the site of the bite, and may cause considerable itching. It is important to try and get a clear description of the rash, which is usually absent by the time patients see a neurologist. Erthyema chronicum migrans is usually a red, often raised leasion near or surrounding the site of the tick bite; it increases in diameter over several days, and represents an inflammatory reaction to the spirochete, as it spreads through the skin. It usually enlarges to over 5 cm, and is associated with malaise, lethargy, and fatigue, as the spirochete disseminates from the skin and causes a systemic reaction. There may be mild 'flu-like' reaction and regional lymphadenopathy and the skin lesion subsides. Erthyema chronicum migrans must be distinguished from a simple reaction to an insect bite, which may also be raised and itchy, but decreases in size over a week, and is only rarely associated with a systemic reaction.

Following haematogenous spread of the spirochaetes weeks to months after the bite, early, disseminated disease manifests with further cutaneous manifestations, adenopathy and fatigue, myalgia and mild meningo-encephalitis. With progression, the organisms localize to the nervous system in between 15–50 per cent, a higher proportion being found in Europe. Cardiac disturbance, usually with dysrhythmias, myocarditis, or pericarditis, occurs in a minority. In North America, recurrent arthropathies ensue in up to 40 per cent, much less in Europe. The knees and other large joints are most often affected; the hands only rarely so.

Late, persisting disease most commonly involves the musculoskeletal system with polyarthralgia followed by erosive bone disease. There is often a characteristic skin lesion, acrodermatitis chronica atrophicans, particularly in European cases. Neurological manifestations which occur in the early disseminated phase include polyneuritis and radiculitis, cranial neuropathies, and meningo-encephalitis. Radiculitis typically occurs unilaterally, and in the cervical or thoracic region; if it is associated with severe radicular pain it is known as Bannwarth's syndrome (Section 21.14.3). The seventh cranial nerve is most often affected; sometimes in association with other basal cranial nerves as part of a basal meningitis. Late manifestations include encephalomyelopathy, polyneuritis, optic neuritis, cerebral vasculitis, and mental and cognitive dysfunction. Polyneuritis is commonly asymmetrical, painful, and predominantly motor, and may resemble mononeuritis multiplex; Guillain–Barré syndrome like features may be seen (Garcia-Monco *et al.* 1995; Halperin 1998) (Section 21.14.3). Electrophysiology commonly shows signs of axonal disturbance, often in a radicular distribution and CSF examination may reveal a pleocytosis. Cranial neuropathies most frequently affect the seventh nerve and other nerves are involved less often, including the second. There is nothing specific about the meningo-encephalitis which is associated with CSF pleocytosis. The protein content may be raised and serological tests and polymerase chain reaction may be positive.

Diagnosis

If there is a history of tick bite and if it is followed by erythema chronicum migrans, the diagnosis of Lyme disease is not difficult. However, this sequence occurs in only a minority of cases. Therefore, any patient with meningoencephalitis, chronic lymphocytic meningitis, cranial neuritis including Bell's palsy, polyradiculitis, or polyneuritis may harbour the disease. If arthropathy co-exists, the chances of Lyme disease are higher. EEG, peripheral electrophysiology, and CT or MR brain scans may be abnormal, but the abnormalities are not specific. Diagnosis depends on the demonstration of raised antibody titres in blood and CSF. An enzyme linked immunosorbant assay, ELISA, is usually performed initially as a screening test. If this is negative in the serum of a patient with symptoms for 3 months or more, this virtually rules out the diagnosis (Pachner *et al.* 2007). Positives are reported as a ratio compared with 'normal sera', which can mean those just above to the cut-off value may be false positives. False positives may also occur with syphilis, infectious mononucleosis, rheumatoid, and autoimmune diseases. Hence any borderline results are tested with an immunoblot, which is specific for particular proteins of the organism. When there is neurological disease there is almost always a CSF pleocytosis, or antibody in the CSF, or both. The absence of either makes lyme neuroborreliosis very unlikely. In some patients, particularly in America, CSF antibody may be detectable at low levels. In most European patients there are higher levels, with a CSF to serum index greater than 1, indicating intrathecal production. Polymerase chain reaction detection of spirochete DNA has an uncertain role. It can be useful for examining joint fluid, and skin samples, but is often negative in the CSF.

Treatment

In order to reduce the chance of sequelae, treatment should be given as soon as possible after diagnosis. There have been few large-scale trials of antibiotics in Lyme disease. In one study of antibiotic therapy for Lyme neuroborreliosis, penicillin G, 20 million units/day in divided doses for 10 days was successful. Others have used ceftriaxone, which is thought to be better in view of its long half-life, resulting in maintenance of high concentrations in serum for a longer period of time, and its ability to penetrate the blood–brain barrier readily and maintain high CSF concentrations (Pachner *et al.* 2007). The dose of ceftriaxone is generally intravenous or intramuscular 1–2 g twice a day for 14–28 days. In a large European study, oral doxycycline was as effective as intravenous antibiotics in treating Lyme neuroborreliosis. The dose of doxycycline used was 200 mg orally daily for 14 days. Thus, oral therapy is a reasonable alternative to intravenous agents. Therapy is highly successful, and residual symptoms after therapy have been found predominantly in patients who had irreversible damage to the facial nerve or nerve roots before therapy. However, treatment failures have been seen in response to both doxycycline and ceftriaxone therapy and it is reasonable to try additional therapy with an alternative antibiotic under those circumstances.

A mild form of Jarisch–Herxheimer reaction may occur following initiation of antibiotic treatment, but this seldom requires steroid treatment. The outlook is good. In more chronic cases, a longer than 1-month course of antibiotics may be necessary. Patients with severe symptoms often improve quickly, whilst those with milder disease may take weeks to months to recover fully. The inflammatory response to the spirochete is an important part of the pathogenesis, and in some patients systemic symptoms of arthralgia, myalgia, and fatigue may persist. In some this may be a fibromyalgia-type syndrome; in others it may be distinguished from fibromyalgia by the absence of trigger spots, and absence of a sleep disorder. Systemic symptoms of inflammation often improve with simple over-the-counter non-steroidal anti-inflammatory drugs such as aspirin or brufen, or indometacin. Corticosteroids have been used by some clinicians to treat inflammatory syndromes in Lyme disease. However, corticosteroids may interfere with immune-mediated killing of spirochaetes, and thus should not be given unless a patient has undergone an adequate period of antibiotics to control the infection. Corticosteroids may be indicated in a small subset of patients who have inflammatory syndromes that persist after antibiotics and do not respond to non-steroidal anti-inflammatory drugs, but they are not considered standard therapy in patients with Lyme borreliosis. Antidepressants may help with sleep problems, and gentle exercise regimes may improve mobility.

42.5.3 Leptospirosis

Leptospirosis is a zoonotic disease which affects many species of mammals throughout the world. It is caused by several species of *Leptospira*, Gram negative, motile, helical, thin bacteria. These persist in the renal and genital tracts of carrier animals and are excreted in their fluids. They survive in water for a long time and in damp soils and animal tissues. Humans become infected by contact with infected tissues, urine, or water and transmission takes place through the skin, conjunctivae, or mucous membranes. Infection is more common in those who are exposed to rats, other rodents and livestock: sewer workers, veterinary attendants, and abattoir workers, or those who by recreation are infected by contaminated water: fresh water swimmers, water skiers, windsurfers. The disease is found throughout the world. In the USA and elsewhere dogs are the most important carriers in the urban setting. It appears to be spreading from its traditional rural base to cause epidmics in poor urban slum communities in developing countries (McBride *et al.* 2005).

The manifestations of leptospirosis vary according to the age and health of the patient, and with the character and type of the infecting organism. Subclinical infections are common and symptomatic disease is usually mild (Bharti *et al.* 2003). Only about 10 per cent develop full-blown Weil's disease involving the liver and kidneys. The incubation period ranges from 3 to 30 days and the clinical onset is abrupt with pyrexia, headache, myalgia and malaise, cough, chest pain, conjunctival suffusion, and photophobia. A maculopapular rash appears in 30 per cent of cases and there may be lymphadenopathy. This corresponds to the period of leptospiraemia. After a week, there is clinical improvement and the patient becomes afebrile for a day or two, following which the second phase of the illness occurs, corresponding to the development of specific antibodies. Lymphocytic meningitis with CSF cell counts in the hundreds occurs and resolves within a few days. It occurs in up to one-quarter of all cases (Bharti *et al.* 2003). CSF abnormalities may persist for several weeks. A minority go on to develop more severe manifestations of a generalized vasculitis which results in renal failure, jaundice, myocarditis, skin and mucosal haemorrhages, and culminates in hepato-renal failure and death, if there is no response to treatment (Edwards *et al.* 1960; Lecour *et al.* 1989). Mortality rates up to 40 per cent have been recorded in outbreaks associated with *L. icterohaemorrhagiae* and *copenhageni* infection.

The diagnosis of leptospirosis is usually made with serological tests. The microagglutination test, often referred to as the MAT, which detects agglutinating antibodies in the patients serum, was traditionally the method used. It is being replaced by IgM capture enzyme linked immunoadsorption assays which are usually become positive in the first week of illness. The organism may be cultured from blood and urine if samples are taken within the first fortnight. Polymerase chain reaction for leptospiral DNA in blood, urine and CSF is also used, and has the advantage over culture, that it remains positive, even if treatment has been started.

There is controversy over whether or not to treat mild disease, since most cases resolve spontaneously. A recent Cochrane review concluded that there was insufficient evidence to give clear guidance, though there was suggestive evidence to support the use of penicillin and doxycycline (Guidugli *et al.* 2000). Doxycycline is recommended for both prophylaxis and mild disease. Ampicillin and amoxicillin are also recommended in mild disease, whereas penicillin G and ampicillin are indicated for severe disease (Watt *et al.* 1988). In severe cases, dialysis and intensive care may be needed.

42.5.4 Borrelia recurrentis and other relapsing fevers

The relapsing fevers are caused by infection of humans by various species of *Borreliae*, which are Gram negative, helical bacteria.

Before 1980, the term Borreliosis was used synonymously with relapsing fever, but the recognition that Lyme disease, which is not a relapsing fever, was caused by *Borrelia burgdorferi* has led to the name 'relapsing fever' being resurrected to describe the syndromes caused by non-Lyme *Borreliae*.

Relapsing fever is a common bacterial infection in parts of West Africa (Vial *et al.* 2006). *Borrelia Crocidurae* is transmitted to man by ticks of the soft-shelled *Ornithodoros* species which harbour many species of *Borreliae*, and live in remote, forested areas at elevation, or by lice which transmit infection due to *B recurrentis*. The louse borne variety is now limited to the horn of Africa and the Sudan, while tick-borne disease is found in Africa, Asia, and the Americas. Ticks form an arthropod reservoir for the disease and become infectious when they feed on infected rodents. When they subsequently bite man they transmit *Borreliae* in their saliva. The clinical syndrome which is produced is the same, no matter which species of *Borrelia* is implicated.

The incubation period varies from 3 to 15 days with an average of a week. Onset is sudden with pyrexia, headache, rigors, myalgia, disturbance of sleep rhythm, confusion, delirium, and meningism. A petechial rash occurs in some and the conjunctivae are injected. In severe cases, thrombocytopaenia and consumptive coagulopathy may complicate the picture (Southern *et al.* 1969). Without treatment, the febrile episode resolves after a few days, only to recur a week later and this cycle may be repeated several times. There are clinical similarities to malaria and arboviral and rickettsial encephalitides. Diagnosis is by demonstration of the organism in blood smears, best taken during febrile attacks. Serum antibodies can be detected by enzyme linked immunoadsorption assays and immunoblotting techniques. Treatment is effective with antibiotics, and penicillin, erythromycin and chloramphenicol have all been given successfully. A Jarisch–Hexheimer reaction may occur.

42.5.5 Mycoplasma infection

Mycoplasmas are unique bacteria adapted to life in humans, animals, insects, and plants. They are named after the fungus-like growth pattern recognized for some species, but are in fact small bacteria of 200 nm which lack a cell wall, and thus are resistant to beta-lactam antibiotics. The two major mycoplasma organisms which are associated with neurological disease are *Mycoplasma pneumoniae* and *Mycoplasma hominis*; in addition Ureaplasma species, which are related organisms in the Mollicutes class, also can cause central nervous system disease. *Mycoplasma hominis* and Ureaplasma species are found in the genitourinary tract and may be a cause of neonatal meningitis; adult infections can also occur with *M. hominis*, but they are usually limited to the immunosuppressed; or follow trauma. The organism has been isolated from rare cases of brain abscess. *M. pneumoniae* has been associated with several neurological syndromes. The bacteria is transmitted by inhalation, and causes respiratory infection in most patients. In the majority who become symptomatic, an influenza-like illness creeps on after 2–3 weeks incubation with severe coughing, headache, and myalgia. However in up to 20 per cent of patients with central nervous system disease there are no symptoms or signs of a respiratory tract infection. Young adults, children, and adolescents are affected most frequently. Neurological complications have been reported in up to 7 per cent of hospitalized cases and occur from 3 to 30 days after the upper respiratory infection (Sterner *et al.* 1969). Aseptic meningitis, encephalitis, cerebellar syndrome,

Guillain–Barré polyneuropathy, transverse myelitis, and cranial nerve palsies, cerebellar syndromes, seizures, and status epilepticus have all been described. Non-neurological complications include myocarditis, pericarditis, renal syndromes, rashes, and autoimmune haemolytic anemia. The pathogenic mechanisms of neurological disease are not always clear; in some it seems that direct central nervous system invasion occurs, and in others there may be an immune-mediated form of post-infectious encephalitis perhaps with immune complex vasculopathy (Behan *et al.* 1986). Antibodies against the galactocerebroside component have been implicated (Talkington 2004).

Diagnosis can be made by the detection of cold haemagglutins in the serum of a patient with neurological complications after an upper respiratory infection. The traditional method is the complement fixation test, but this has limited sensitivity and specificity; newer methods are becoming available (Daxboeck 2006). Their absence does not exclude the diagnosis as their presence is fairly non-specific. Increase in serum antibody titres can be confirmatory. Polymerase chain reaction of CSF is likely to prove useful (DeBiasi *et al.* 1999). Culture often takes several weeks, and so is rarely attempted. Erythromycin, doxycycline, or tetracycline for 2 weeks is effective treatment for the acute respiratory infection and the response of neurological disease is variable and probably depends on the pathogenic mechanism. Steroids may be beneficial, especially in severe cases (Carpenter 2002). Plasmaphoresis or intravenous immunoglobulins may also be used. Perhaps one-third of patients have persisting neurological signs and severe meningoencephalitis carries a poor prognosis.

42.5.6 Chlamydial diseases

Three species of chlamydiae can cause human disease:

- *Chlamydia trachomatis* causes trachoma, chronic keraticoconjunctivits, in dry areas of the third world, and it is a major cause of sexually transmitted diseases throughout the world; gential chlamydia is the commonest sexually transmitted disease in the United Kingdom. Man is the only host and neurological complications are rare;

- *C. psittaci*, occurs in a wide range of birds and mammals and humans are infected accidentally via inhalation following exposure to infected birds;

- *C. pneumoniae* has been recognized as an organism separate from *C. psittaci* since the 1980s. Man is the only host. It produces clinical syndromes similar to that produced by *C. psittaci*.

Chlamydiae are coccoid, obligate, intracellular pathogens which contain DNA and RNA. *C. psittaci* causes infection in a large variety of birds, not just the psittacine family of parrots, budgerigars, and cockatoos, but also finches, pigeons, pheasants, seabirds, and poultry. The disease produced is more correctly termed 'ornithosis'. In birds, infection is commonly asymptomatic and the organism is shed in faeces, from the beak and contaminates the feathers. Human infection results from inhalation of the organism or by ingestion after handling contaminated plumage. Those occupationally exposed to birds such as pet shop owners and poultry workers are at particular risk, as are pigeon fanciers. However, even a short period of exposure may be sufficient to result in infection. Person to person transmission is rare. Laboratory personnel are at risk of potentially serious infection and should be warned of the suspected diagnosis when specimens are dispatched. The organism is pathogenic to the lung and infection can range from asymptomatic, to severe and potentially fatal, atypical pneumonia. Liver, spleen, meninges, brain, and heart may also be involved.

Pathologically, there are lymphocytic exudates in the pulmonary alveoli and interstitial spaces. The lungs become heavily congested and macrophages containing inclusion bodies are characteristic. When the brain is infected, there is oedema and congestion and with evidence of direct neuronal invasion, chromatolysis and intra-cytoplasmic inclusions in neurones and meningeal cells.

The incubation period varies from 4 to 15 days and in some, may be as long as a month. Clinical features may vary widely. Many merely have a mild upper respiratory tract infection which may not be recognized as anything more than a common cold. In others, there is abrupt onset of fever, rigors, and chills, with severe headache, myalgia, and a persistent dry cough. There is a relative bradycardia and few signs found on examination of the chest, which is at variance with the abnormalities often seen on chest X-ray. The white cell count is often normal and in as many as 25 per cent of cases there may be leucopaenia. In most, infection subsides within 2 weeks. In severe cases, endocarditis, hepatitis, and more severe respiratory involvement may supervene. A minority develop meningo-encephalitis which may progress to coma. Rarely, meningo-encephalitis may predominate from the outset (Crosse 1990). Other neurological syndromes have been described including transverse myelitis, cranial nerve palsies, and cerebellar disturbance. There are no specific features and diagnosis depends on the recognition of the atypical pneumonia, or that the patient has been exposed to potentially infected birds. Provided diagnosis and treatment are timely, mortality rates should not be above 5 per cent, although there is a greater risk in pregnant women.

Diagnosis is confirmed by isolation of the organism from respiratory secretions, which can be hazardous to laboratory staff, or by the demonstration of antibody rise in acute and convalescent sera. Polymerase chain reaction examination of CSF is now available, as are newer enzyme immunoassays.

Treatment of chlamydial infection is with tetracycline, erythromycin, or clarythromycin. Some of the newer quinolones are also effective. For neurological disease, parenteral administration is preferred, and a prolonged course is often necessary because of the risk of relapse. Steps should be taken to identify the source of infection with the co-operation of local public health authorities.

Aseptic meningitis and meningo-encephalitis have been described as rare accompaniments of the second or lymphatic stage of *lymphogranuloma venereum*, caused by certain strains of *C. trachomatis*. Recent reports have suggested that infection with *C. pneumoniae*, on the basis of serological testing, may play some part in the development of cerebral and coronary artery atherosclerosis. There has recently been interest in the possibility that it could similarly play a role in diverse neurologic diseases, including Alzheimer's disease, multiple sclerosis, and giant-cell arteritis (Yucesan *et al.* 2001).

42.5.7 Rickettsial and related infections

Rickettsiae are gram negative, cocco-bacillary, obligate intracellular bacteria that cause a range of diseases, including acute meningo-encephalitis. The classification of such pathogens is not straightforward; there is conflict between the desire to group them according to apparent clinical epidemiological and ecological similarity,

and recent developments in molecular taxonomic methods which have produced a reclassification of the Rickettsiales (Watt and Parola 2003). However, four groups of diseases are still commonly called rickettsioses:

- scrub typhus due to *Orientia tsutsugamushi*; diseases due to bacteria of the genus Rickettsia, including the spotted fever group and the typhus group;

- ehrlichioses due to bacteria within the family Anaplasmataceae;

- the ubiquitous Q fever due to Coxiella burnetii, which was recently removed from the Rickettsiales.

For many Rickettsiae, the definitive hosts are rodents and other mammals such as dogs, and the vectors which transmit them to man are ticks, lice, fleas, or mites. Rickettsiae are inoculated into the human body by the bite of the vector insect, or by scratching of contaminated insect material though skin or mucous membrane. They penetrate host cells where they multiply, then spread throughout the body via the bloodstream and lymphatics. Characteristically, they produce a widespread vasculitis which increases vascular permeability, produces oedema and endothelial cell injury, activates inflammatory and coagulation mechanisms, and may result in widespread organ damage due to thrombosis, haemorrhage, and shock. When this occurs in the brain, meningoencephalitis results. Q fever is probably acquired through inhalation of the organism from infected animal products, and possibly also by ingesting contaminated raw milk. The incubation period, progression, and clinical features of the different syndromes varies, but all generally share the features of high pyrexia, skin rash, headache, and myalgia, and if neurological features develop, meningoencephalitis is evident by the second week. Other similar diseases with a localized geographical distribution and caused by related rickettsial organisms are being continually described with slight variations on the same clinical theme.

Rocky Mountain spotted fever

This condition was first described from the Rocky Mountain region of the United States and it soon became evident that the same disease occurred throughout the Western hemisphere. The natural cycle of infection is maintained by small mammals and ixodid ticks which bite and infect humans. The wood tick and the dog tick, *Dermacentor andersoni* and *variabilis* respectively, are the principal vectors. There is a higher incidence of disease in children and males, probably reflecting proximity to dogs and occupational exposure. However, with increased foreign travel and exposure to insect vectors when following outdoor tourist activities, the rickettsial diseases may be encountered anywhere in the world and in any ethnic and social group. There is variation between the virulence of infection and the old, infirm, diabetic or alcoholic are said to suffer more severe disease. Blacks with glucose-6-phophate dehydrogenase deficiency develop particularly virulent disease. The incubation period is about 1 week and varies in proportion to the bite inoculum. Most patients can remember and report the tick bite. Rarely, a characteristic inoculation eschar, the 'tache noire', may be seen at the bite site.

Fever with rigors, chills, malaise, headache, myalgia, conjunctival injection and photophobia, and diffuse oedema come on suddenly. This is followed by a distinctive rash by the third to the fifth day of the illness: it is maculopapular, erythematous on the wrists and ankles, and spreads centrally to the trunk. It often becomes petechial and sometimes purpuric with haemorrhagic necrosis and gangrene in severe cases, which may mimic meningococcal septicaemia. Headache is invariable, and a variable proportion, perhaps as many as 25 per cent, of patients progress to develop encephalitis with drowsiness, confusion, convulsions, followed by stupor and coma if they are not treated appropriately. Most forms of focal, neurological disturbance have been described but these are uncommon (Helmick *et al.* 1984; Kirk *et al.* 1990). The overall mortality varies up to 10 per cent and is worsened by any delay in the institution of antibiotic treatment. Because of this, treatment should be instituted immediately on clinical suspicion; it may turn out to be fatal to await laboratory diagnosis.

Blood tests are seldom helpful in the acute stage. A rise in serum antibody titre is useful but takes time. CSF examination reveals non-specific changes with a rise in the cell count and protein content and may be normal. EEG will show non-specific slowing. Cerebral imaging with CT or MR is usually normal and may show non-specific white matter abnormality. Immunofluorescence of skin biopsy is said to be specific but not very sensitive and is not widely available. Polymerase chain reaction of CSF, blood, and skin biopsy specimens is also used in reference centres.

The treatments of choice are tetracycline or doxycycline for 7 days to a fortnight, for adults. Chloramphenicol should be considered for young children to avoid the teeth stains which may complicate treatment with tetracycline (Shaked 1991). Based on *in vivo* studies, and their effectiveness in Mediteranean spotted fever, fluoroquinolones may also be effective. Alternatively in children and pregnant women macrolides, including josamicin, but not erythromycin, can be used for 8 days (Watt *et al.* 2003). Suitable supportive treatment should be given and steroids have been used although the evidence is lacking that they are routinely efficacious.

Mediterranean spotted fever

This condition, also known as Boutonneuse fever, is caused by *R. conori* which is usually inoculated into humans by the brown dog tick *Rhipicephalus sanguineus*. Different vectors and hosts are found in central Europe and in Russia, the Ukraine, and around the Black and Caspian seas, and all cause different forms of spotted fevers. In the Mediterranean variety, at the site of inoculation the organisms reproduce and produce local inflammation with endothelial vasculitis which raises an erythematous papule. The centre of the papule becomes necrotic and dark coloured, the 'tache noire', which is characteristic. From there, spread takes place to local lymph nodes then to the general circulation. In 7 days or thereby, symptoms similar to those of Rocky Mountain spotted fever appear: fever, headache, myalgia, arthralgia, lymphadenopathy, and rash. Meningism and encephalopathic features occur in up to 20 per cent of cases. Laboratory investigations, EEG, and brain imaging produce similar appearances to those of Rocky Mountain spotted fever. Diagnosis is clinical, based on the findings of fever, rash, and eschar, and can be confirmed by antibody studies. Polymerase chain reaction diagnosis remains under evaluation. Treatment with tetracycline or chloramphenicol is given as described above. In addition fluoroquinolones have proved useful (Rolain *et al.* 2002) The outcome is usually excellent and mortality rates of about 1 per cent should be expected (Raoult *et al.* 1986). An excellent account of similar Rickettsial diseases and their epidemiology is to be found in Beati and Raoult (1998).

Epidemic and murine typhus

Epidemic and murine typhus are clinically similar. Epidemic typhus occurs in conditions of overcrowding and poor sanitation such as accompany war, natural disasters, and civil insurrection. It is caused by *R. prowazekii* and transmitted by the louse; humans become infected by aerosol inhalation or skin autoinoculation of infected louse faeces. Murine typhus is linked to the distribution of rats, is transmitted by the rat flea and is caused by *R. typhi*; in addition to inhalation or autoinoculation of flea faeces, humans may become infected by flea bites. The incubation period is about 2 weeks. The clinical syndrome of the two diseases resembles a severe form of Rocky Mountain spotted fever with the abrupt onset of fever, headache, myalgia and a spreading macular and petechial rash, delirium, and encephalopathy. White spots, the typhus nodules, may be evident in the retina on fundoscopy. Investigations and treatment are similar to those described above. The outcome from murine typhus is usually excellent. For epidemic typhus, mortality rates may be high, in proportion to the degree of debilitation of the population at the time of onset of the epidemic. Brill–Zinsser disease is a form of recrudescent typhus which may occur in people who have recovered from epidemic typhus years before. It may represent a diminution in host immunity. The clinical features are those of a mild form of epidemic typhus but the rash may be lacking, in which case diagnosis depends on eliciting a history of previous typhus infection. The response to doxycycline or tetracycline is usually good; macrolides may also be used.

Scrub typhus

Scrub typhus is found is found in the Far East, particularly in south-eastern Asia. It is caused by *Orienta*, formerly *Rickettsia*, *Tsutsugamuchi*. Although an obligate intracellular gram negative bacterium, it has a different cell wall structure and genetic make up to the rickettsiae. *O. tsutsugamushi* is transmitted to man by the bite of 'chiggers', mite larvae. Humans become infected in rural areas, for example when clearing land or logging, but it has also been acquired during ecotourism. The disease appears to be reemerging in Japan and China (Watt *et al.* 2003). The chigger bite is usually painless, and often in an area that is hard to examine, such as the genital region or axilla. The clinical features are of fever, general lymphadenopathy, and in half the cases an eschar at the site of the bite, which resembles a cigarette burn. Headache and myalgia are invariable and as many as 10 per cent develop features of encephalitis. Respiratory features due to pulmonary infiltrates ae also common. A rash on the trunk, spreading to the limbs, is common. Confirmation of the diagnosis is serological, though the commercially available Weil–Felix test lacks sensitivity and specificity; polymerase chain reaction has also been used. Treatment is with oral doxycycline for one week; tetracycline and chloramphenicol are alternatives. There is no vaccine, but prophylactic treatment with doxycycline has been used in those at risk.

Ehrlichiosis

Three species of *Ehrilchia* have been implicated in causing human disease. These are small intracellular coccobacilli like *Rickettsiae* and they are thought to transmit disease via tick vectors. Most commonly, *E. chaffeensis* has been the agent. In some regions, the vectors are the same ticks which carry the agents of Lyme disease and Babesiosis; co-infection may occur. Most cases have been described from the USA but cases have been reported from Europe, Africa and Asia. In the summer months, patients are bitten by a tick and develop a non-specific 'flu-like' illness 2 weeks later associated with a rash. In severe cases, disseminated intravascular coagulation and encephalopathy or meningitis may ensue. Diagnosis is by finding positive serology in the blood of these patients with an appropriate clinical syndrome and exposure to ticks. Polymerase chain reaction is both sensitive and specific (Dumler *et al.* 1995). Treatment is with doxycycline or rifampicin.

Q fever

Q fever is a worldwide zoonotic disease caused by *Coxiella burnetii*, a cocco-bacillus. It infects a large number of animal species, usually asymptomatically, where it localizes to the uterus and mammary glands. Direct or indirect exposure to infected animals may result in clinical disease in humans. Infection usually results from aerosol inhalation or ingestion of infected material for example from sheep or cows. Recent cases have occurred among military personel in the Middle East (Leung-Shea *et al.* 2006). This may take several forms which include an acute self-limiting febrile illness, pneumonia, endocarditis, hepatitis, aseptic meningitis, and encephalitis. In fact neurological complications are quite rare, although one group reported transient neurological features such as hallucinations, speech disturbance, and facial pain in 23 per cent of patients in one outbreak (Smith *et al.* 1993). Patients who are immune compromised may be more at risk of developing disease. The diagnosis can be made serologically with IgM and IgG enzyme linked immunosorption assays. Treatment is with doxycycline for up to 1 month.

42.5.8 Whipple's disease

In 1907, George Whipple published a case report of a man of 36 who developed weight loss, steatorrhoea, and arthropathy, and soon died. Necropsy revealed infiltration of the intestinal lymph glands by deposits of fats, fatty acids, mononuclear and polynuclear giant cells, and 'foamy' mononuclear cells in the intestinal mucosa. He named it an intestinal lypodystrophy and noted the presence of peculiar rod-shaped organisms in gland tissue, and he mused that they might be of aetiological significance (Whipple 1907). In succeeding years, electron microscopic observations have confirmed the presence of a rodshaped bacillus, weakly gram positive, 1–2 μm long with a thick outer wall which stains with periodic acid Schiff dyes. It had proven exceptionally difficult to culture and its presence in body tissues was inferred by the identification of a 16S rRNA gene sequence using polymerase chain reaction. Recently, however the bacteria has been cultured, and its complete genome sequenced (Raoult *et al.* 2000; Bentley *et al.* 2003). These are beginning to advance our understanding of the disease. *Tropheryma whipplei* has a very small genome for a bacteria, bears traits of a strictly host-adapted organism, and is adapted to evade the host immune response during chronic infections, with many variable cell surface proteins (Bentley *et al.* 2003).

The method of infection and pathogenesis remain poorly understood. Because of the gastrointestinal location of the disease, it is thought to be contracted by ingestion following which it disseminates through the body via lymph and blood. In addition to the gut, the central nervous system, eyes, heart, lungs, and skin and joints may be involved and the disease may present in any of these systems. As many as 15 per cent of cases do not have gastrointestinal symptoms and jejunal biopsy may be normal (Durand *et al.* 1997).

Clinical features

Whipple's disease is rare. It affects men more than women, and occurs at any age, most commonly with onset in the 40s. Usually there is diarrhoea with malabsorption, weight loss, wasting, pyrexia, sometimes of unknown origin, and lymphadenopathy. Onset is insidious and progress is often atypical. Between 5 and 40 per cent have neurological manifestations and in as many as 5 per cent the disease may be confined to the central nervous system (Keinath *et al.* 1985; Fleming *et al.* 1988). The most frequent neurological manifestations are dementia, ocular movement disturbance, movement disorders particularly myoclonus, hypothalamic upset, epilepsy, ataxia, meningitis, and focal cerebral signs. Headache is common. Ophthalmoplegia is of supranuclear type and affects vertical rather than horizontal movement. Internuclear ophthalmoplegia has been described. Movement disorders may affect the eyes; oculomasticatory myorhythmia and oculo-facial-skeletal myorhythmia are said to be diagnostic. The former consists of smooth convergent–divergent pendular oscillations of the eyes with synchronous rhythmic contractions of the jaw at about one per second. The latter is manifest by more diffuse rhythmic movements of the face and extremities. Certainly, the triad of dementia, ophthalmoplegia, and myoclonus is highly suggestive of Whipple's disease, although it probably only occurs in about 10 per cent of patients. Diagnostic guidelines for neurological Whipple's disease have been proposed (Louis *et al.* 1996).

Diagnosis

Whipple's is diagnosed by demonstration of a positive polymerase chain reaction against *T. whipplei* and/or periodic acid Schiff positive staining foamy macrophages in material from affected tissue. Appropriate tissue to investigate includes the duodenal mucosa, the CSF, and material from a brain biopsy. Gram positive bacteria may also be seen in the lamina propria of the intestine. There may also be periodic acid Schiff positive macrophages in lymph nodes, but this alone is not diagnostic, because it can also occur in tuberculosis, sarcoidosis, Gaucher disease, and berylliosis. A polyclonal antibody against *T. whipplei* can be used for immunohistochemical staining of tissues. CT and MR imaging of the brain may show atrophy, mass lesions and contrast enhancement, white matter high signal areas, ring enhancing lesions, and hydrocephalus. Differential diagnosis is wide and includes the dementias, encephalopathies, central nervous system vasculitides, demyelination, granulomatous disease, and chronic central nervous system infection. Creutzfeldt–Jakob disease merits exclusion if movement disorder is present with dementia.

Treatment

The recommended antibiotic regime is an induction period of 2–4 weeks, with a combination of parenteral penicillin and streptomycin, or a third generation cephalosporin such as ceftriaxone. Following this there should be long-term treatment for 1–2 years because of the high incidence of relapse. Long-term treatment should be with cotrimoxazole or a cephalosporin. Polymerase chain reaction is now recognized to be the best test to monitor progress and it is necessary to check CSF as well as bowel for negative results before discontinuing treatment (Ramzan *et al.* 1997). Untreated the disease is fatal; with treatment some patients may respond within a few months; other may take longer.

42.5.9 Infective endocarditis

Infective endocarditis is an infection of the heart in which the brunt falls on the endothelial lining. The full clinical spectrum was first comprehensively described by Sir William Osler in 1885. Heart valve leaflets are predominantly involved and in certain circumstances, chordae tendinae, mural endocardium or intracardiac foreign bodies such as prosthetic valves or pacing electrodes may be subject to infection. Because cardiac endothelium is resistant to infection, there is usually some pre-existing abnormality or anomaly to which sterile platelet and fibrin-containing thrombi can adhere and form, such as abnormal valves. If bacteraemia occurs, however transiently, organisms adhere to these previously sterile thrombi and multiply. The clinical manifestations of infective endocarditis are caused by local cardiac damage which these organisms induce and by the damage caused peripherally by embolization producing local infarction and often setting up one or multiple loci of infection. Sometimes, particularly virulent organisms such as *Staphylococcus aureus* will cause infection on normal cardiac endothelial structures. A large spectrum of organisms has been recognized to cause infective endocarditis and in practice most cases are caused by a small spectrum which share the ability to survive in the circulation, to adhere to endothelium, and to propagate in vegetations. The central nervous system is the site where many of these emboli lodge. Persisting bacteraemia is a constant feature of infective endocarditis and induces immunological changes that contribute to the development of vasculitis and other clinical manifestations.

It has been customary to recognize two forms of endocarditis, acute and subacute, the distinction between the two being based on the speed of evolution of the clinical picture which has usually been related to the virulence of the infecting organism:

◆ *Acute endocarditis* is associated with rapid progression over days, obvious toxicity and constitutional symptoms, rapid development of cardiac disease, and metastatic infection. There is often no evidence of pre-existing cardiac disease and the causal organisms are virulent: *Staphylococcus aureus*, Streptococci, gonococci, and Haemophilus.

◆ *Subacute endocarditis* is a more indolent condition which affects an already diseased heart, has less evidence of toxicity, and evolves over weeks or months. Peripheral manifestations are fewer and the organisms are less virulent. *Streptococcus viridans* is the most common but a wide range of organisms has been identified including enterococci and gram negative coccobacilli from the mouth and throat; transient bacteraemia is commonplace after teeth brushing and following iatrogenic instrumentation. Fungi and parasites rarely may cause endocarditis.

Occurrence

The incidence of infective endocarditis has remained surprisingly constant over the years at approximately 2–4 cases per 100 000 per annum in developed countries. However, the pattern of infection keeps changing, and the distinction between acute and sub-acute disease has blurred. Underlying conditions which predispose to infective endocarditis are rheumatic heart disease, particularly in third world countries; congenital heart disease with bicuspid aortic valve being important in older patients; degenerative disease with atheroma in the elderly; cardiac surgery with prosthetic valves; pacemakers; and systemic arterial shunts. Mitral valve prolapse has been described as a significant predisposing factor but this may have been exaggerated. To cause endocarditis, it is necessary for organisms to access the bloodstream and conditions which predispose to this include iatrogenic instrumentation, haemodialysis, dental manipulation, tissue trauma and burns, and intravenous drug addiction. Intravenous drug abuse carries a very high risk of septicaemia and is increasing as a cause of infective endocarditis. Immune suppression from AIDS and other causes permits cardiac colonization by a wide range of organisms including fungi. In perhaps as many as 5 per cent of patients, no organism can be detected, usually because of prior antibiotic administration. In such cases, diagnosis can be confirmed by serological tests or polymerase chain reaction.

General clinical features

The clinical manifestations of infective endocarditis derive from a combination of four main mechanisms:

◆ cardiac tissue destruction from valvar and other vegetations;

◆ distal embolization and infarction;

◆ metastatic infection from bacteraemia; and

◆ tissue damage due to immune complex deposition and local complement activation.

In addition, constitutional symptoms derive from bacteraemia. The signs and symptoms vary enormously. Fever is the most common sign and it may not be evident in the ill or elderly. Cardiac murmurs are next most common and are heard in 80 per cent of patients. They may change or be evanescent. Malaise, night sweats, weight loss, and arthralgia are common as is splenomegaly. Anaemia is invariable. Osler's nodes, which consist of tender nodules found in the pulp of the fingers, splinter haemorrhages and petechiae, finger clubbing and retinal haemorrhages were formerly commonplace but are seen less frequently now in the West because the diagnosis is usually made before they develop. Congestive cardiac failure has developed in as many as half of all patients. Peripheral embolism causes signs reflecting the organ involved. Renal dysfunction from immune-mediated glomerulonephritis may occur in as many as 15 per cent of patients. Neurological complications occur in 19 to 40 per cent with an average of 30 per cent; this has remained constant over the years. The presentation of endocarditis may be entirely neurological and any stroke in a young person should prompt a search for underlying endocarditis (Osler 1885; Lerner *et al.* 1966; Pruitt *et al.* 1978).

Neurological disorders

Neurological complications of endocarditis occur equally in the sexes and increase with age. They are usually associated with abnormalities of the valves on the left side of the heart:

◆ *Embolic stroke* is the most common. If the organism is virulent, stroke occurs earlier. Embolism takes place to the distribution of the middle cerebral artery in the vast majority of cases; consequently the signs are of hemiparesis, hemisensory deficit, and dysphasia. It is rare for the brainstem to be affected and rarer still for spinal cord. Emboli are often multiple so that signs may be complex. If the embolus is small and disintegrates, transient ischaemic attacks occur (Fig. 42.10).

◆ *Mycotic aneurysms* are caused by infected emboli lodging in the vasa vasorum causing local infection and arteritis which

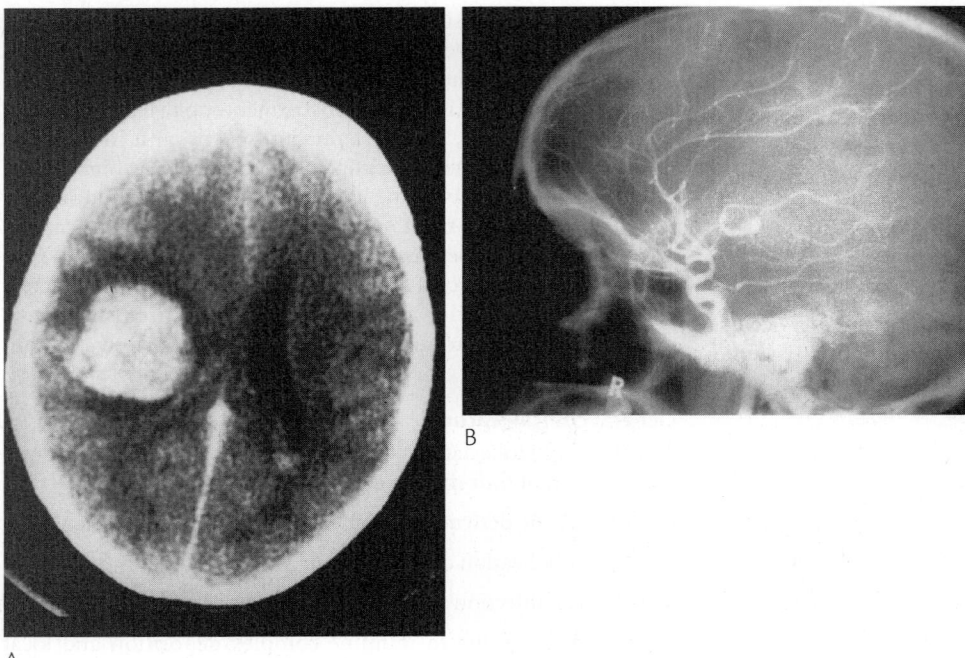

Fig. 42.10 Subacute bacterial endocarditis: cerebral imaging changes. (A) CT brain scan with contrast showing fronto-parietal haematoma. (B) Digital subtraction angiogram showing a right middle cerebral artery mycotic aneurysm which had ruptured to cause the intracerebral haematoma. (Pictures Milne Anderson.)

leads to weakening of the vessel wall and subsequent aneurysm formation. This is most frequent in middle cerebral artery territory, and they tend to be distal rather than proximal and grow rapidly. They may rupture into the brain or into the subarachnoid space. Thus intracranial haemorrhage occurs in about 5 per cent of endocarditis cases and may be intracerebral or subarachnoid. Although caused by rupture of mycotic aneurysms in most cases, in some it is due to transformation of infarction to haemorrhage by the use of anticoagulant drugs.

◆ *Brain abscesses* occur in up to 4 per cent of cases and are caused by emboli to the brain which set up a nidus of infection. They are often multiple. Meningitis, both septic and aseptic is recognized in as many as 10 per cent of cases who have neurological complications. Epileptic seizures occur in about 10 per cent and may be generalized or focal. Focal seizures imply underlying structural brain damage although this is usually caused by brain infarction.

◆ *Diffuse encephalopathy.* As many as 20 per cent of neurological cases are accompanied by a miscellany of symptoms and signs which have been loosely grouped under the term 'toxic encephalopathy' or 'septic encephalopathy'. These include confusion, confabulation, disorientation, personality change, and decrease of conscious level and coma. It is probable that many factors contribute to this, including anoxia, metabolic disturbance, and immune arteritis. Immune complex deposition has been postulated as a cause but remains unproven.

Diagnosis

The diagnosis of infective endocarditis is suggested by the clinical findings of changing cardiac murmurs and evidence of peripheral embolism in a toxic, ill patient, and is confirmed by the demonstration of infecting organisms in blood cultures and vegetations on cardiac valves by echocardiography. In more indolent infection, cardiac murmurs may not be so evident. Blood cultures are positive in more than 90 per cent of patients who have not received antibiotics. The more frequently cultures are taken, the higher is the positive yield. Search should be made for fastidious and exotic organisms and serological tests and polymerase chain reaction should be undertaken for organisms such as *Legionella*, *Brucella*, *Coxiella* and others. Normochromic, normocytic anaemia is common, there is peripheral leucocytosis and the ESR is invariably raised, often to values over 100 mm per hour. C-reactive protein levels are also elevated. Cardiac investigations including electrocardiogram and echocardiography are necessary. In addition to transthoracic echocardiography, transoesophageal echo is now often also used, because of its increased sensitivity. Urinalysis reveals proteinuria and haematuria in half of all patients. Brain imaging is indicated when there are neurological complications, and if the patient can tolerate it, MRI is to be preferred. Abscess formation, infarction, haemorrhage, cerebral oedema, and mycotic aneurysms may be identified by angiographic sequences. If this is not available, conventional arteriography will yield sufficient information to decide surgical strategy. CSF examination seldom provides sufficient information to alter clinical management and is contraindicated if there are focal lesions or signs of raised intracranial pressure. It may be necessary if there is neck stiffness and clinical suspicion of meningitis. Criteria for the diagnosis of infective endocarditis, the Duke criteria, have been defined (Durack *et al.* 1994).

Treatment

The treatment of infective endocarditis requires the eradication of the infecting organism and management of complications.

Antibiotics, or antifungal drugs if indicated, are administered, usually intravenously for long courses, as guided by infectious disease and microbiological advice. Cardiological expertise is required to monitor heart status, treat failure and advise on surgery for heart valve damage, which is often required.

The treatment of neurological complications is usually straightforward. Meningitis is treated with antibiotics. Multiple brain abscesses should be treated similarly unless they occupy space or fail to respond to antibiotics, in which case they should be drained surgically using stereotactic techniques. There is no general consensus regarding the treatment of cerebral emboli. Should a heart valve with a vegetation source of emboli be replaced? And if so, after how many episodes? Each case must be assessed individually. Anticoagulant treatment is not indicated in most patients because it increases the risk of haemorrhage. However for patients with mechanical prosthetic valves cautious anticoagulation may be necessary. The place of adjunctive treatment with aspirin has not yet been defined. Intracerebral haemorrhage does not require active surgical treatment unless the haematoma functions as an expanding space occupying lesion. Search should be made for an underlying mycotic aneurysm with angiography and conventional angiography retains the advantage over MR as it allows better delineation of the aneurysmal neck and surgical anatomy. These are uncommon lesions and complicate infective endocarditis in less than 5 per cent of cases (Pruitt *et al.* 1978; Salgado *et al.* 1989). The indications for surgical treatment for aneurysms remain controversial largely because of the observation that a proportion of unruptured aneurysms will resolve given antibiotic treatment over a prolonged period (Brust *et al.* 1990). Most agree that a mycotic aneurysm associated with intracerebral or subarachnoid haemorrhage should be treated surgically. For others it would seem reasonable to repeat angiography to determine if there is increase in size of the aneurysm, and if there is, proceed to surgical treatment provided the position of the aneurysm and the clinical state of the patient allow. In this circumstance, MR angiography has the advantage. The size of the aneurysm is not a good predictor of subsequent rupture.

Prognosis

The outcome of treatment for infective endocarditis depends upon the virulence of the infecting organism, the nature of any underlying cardiac lesion, any delay in diagnosis, and the presence of complications. Mycotic aneurysms carry a high mortality averaging 45 per cent whether they rupture or not. Meningitis and multiple cerebral abscesses have mortality rates approaching 80 per cent. Recurrent endocarditis occurs in up to 4 per cent of cases with a higher incidence in drug addicts. Chemoprophylaxis has been recommended for patient groups with cardiac lesions which are thought to predispose to the development of endocarditis, and surgical or dental procedures have been identified which may cause bacteraemia and lead to infection. However, the efficacy of chemoprophylaxis has not been proved.

42.5.10 Rheumatic fever

Rheumatic fever is the name given to the disease entity which follows upper respiratory infection by some types of Group A streptococci. Formerly a common infection of childhood world wide, its incidence diminished dramatically in Europe and North America in the 1950s and 1960s concurrently with improvement in social conditions and the increased availability of antibiotics, particularly penicillin. In these countries the disease is seldom seen, and consequently, is more difficult to diagnose. The annual incidence is believed to be in the region of 0.5 affected per 100 000 children of school age. The need to be constantly alert to its occurrence was demonstrated by outbreaks in the United States in the 1980s, which affected military institutions and reasonably affluent communities in five cities. In third world countries it is probable that the incidence is increasing and is of the order of 100 to 200 per 100 000 of school age children. Current understanding is that rheumatic fever is caused by a disturbed host autoimmune response to infection by particular serotypes of Group A β-haemolytic streptococci: M types 3, 5, 18, 19, 24, and others. A surface protein of the bacterial wall, protein M, carries specific epitopes and is instrumental in the disease process but the exact mechanisms are as yet, incompletely understood. Bacterial breakdown products are released from the pharynx and induce an immunological cross reaction with other tissues, particularly heart muscle and valves (Olivier 2000). It is believed that this reaction with other body tissues accounts for the various manifestations in other systems which are characteristic of rheumatic fever (Bisno 1991). It may be that a degree of genetic susceptibility predisposes to these reactions. The risk of developing rheumatic fever following tonsillitis or pharyngitis which has not been treated with antibiotics, is less than 1 per cent.

Any age group and either sex can be afflicted by rheumatic fever. The commonest age group at presentation is 5–15 years. It occurs more commonly in populations of lower socioeconomic status who are subject to overcrowding. Typically between 2 and 6 weeks following an attack of tonsillitis or pharyngitis symptoms of rheumatic fever appear. There is fever which may be irregular, malaise, and generalized toxicity.

General features

Carditis, polyarthritis, chorea, subcutaneous nodules, and erythema marginatum occur in frequencies which vary from population to population and with age. Young children tend to have carditis, which occurs in 40 per cent and includes myocarditis, pericarditis, and valvular disease. Older patients tend to have arthritis which is typically fleeting, polyarticular, and involves the large joints afflicting about 70 per cent of patients. About 10 per cent have subcutaneous nodules which appear on the extensor surfaces of the wrists, elbows, and knees, and 5 per cent have erythema marginatum. About 10 per cent develop chorea and this figure is higher in reports from some areas, particularly South America. Criteria for the diagnosis of rheumatic fever have been devised based on criteria proposed by Jones (1944) (Table 42.12).

Rheumatic chorea

Also known as Sydenham's chorea or St Vitus dance, rheumatic chorea is one of the common manifestations of rheumatic fever (Section 40.5.7). It affects children much more commonly than adults, and females more than males. It is not uncommon for there to be no obvious history of rheumatic fever and for the onset to be insidious. Consequently, Sydenham's chorea should be considered in the differential diagnosis of any movement disorder in a young person. At first, abnormal movements may be intermittent and tic like and they are often misinterpreted as being of hysterical origin. Onset can also be explosive with generalized and often violent, choreic movements which involve limbs and trunk and face. It has

Table 42.12 The Duckett Jones criteria for the diagnosis of rheumatic fever

Major criteria
Carditis
Polyarthritis
Chorea
Erythema marginatum
Subcutaneous nodules

Minor criteria
Arthralgia
Fever
Elevated ESR or CRP
Prolonged PR interval on electrocardiogram
Evidence of preceding group A streptococcal infection, e.g. positive results on throat cultures or rapid antigen test
Elevated or rising streptococcal antibody titre

If supported by evidence of preceding group A streptococcal infection, the presence of 2 major manifestations or 1 major and 2 minor manifestations indicates a high probability of acute rheumatic fever.

Failure to fulfill the Jones criteria should make the diagnosis doubtful, except in situations in which rheumatic fever is first discovered after a long latent period, e.g. Sydenham chorea or indolent carditis.

been said to subside during sleep but this varies. In 20 per cent only one side is affected. Commonly, the patient is more emotional than usual. The average duration of chorea is only 2–3 months but some persist for much longer, sometimes for years (Cardoso *et al.* 1999). It has been said that chorea is less likely to be followed by longterm cardiac problems but the incidence of accompanying carditis has been recorded at up to 30 per cent. There is a slight chance of recurrence and females who have had Sydenham's chorea may develop chorea later when they become pregnant or if they take the contraceptive pill.

Investigations to confirm a diagnosis of Sydenham's chorea include demonstration of previous streptococcal infection by serological means and application of the above diagnostic criteria for rheumatic fever. Cerebral imaging with CT or MR is usually normal. Sedation may be necessary for these procedures if movements are particularly violent. It has been assumed that some functional derangement of basal ganglia structures, perhaps on an immunological or vasculitic basis, underpins the mechanism of chorea in these cases. Antineuronal antibodies have been found in some and in others there has been an increase in CSF IgG but these findings are not consistent. CSF examination is not diagnostically useful. Few scan abnormalities have been documented. High signal on T2-weighted MR images have been reported in caudate and putamen (Cardoso *et al.* 1999) and vasculitic changes in major cerebral vessels (Ryan *et al.* 1999).

The diagnosis of Sydenham's chorea is not difficult if it occurs in the setting of rheumatic fever. If not, other conditions which bear exclusion are Huntington's disease in older patients (Section 40.5.2), Wilson's disease in teenagers and young adults (Section 40.8), and the syndrome of Gilles de la Tourette (Section 40.6.3). In younger patients white matter degenerations and metabolic encephalopathies may produce movement disorders, but intellect and mentation are usually disturbed with these conditions. Cerebral neoplasia must be excluded if the chorea is one sided. Pregnancy and exposure to oral contraceptives may cause chorea in young women (Section 40.5.9).

Treatment of Sydenham's chorea is initially that of rheumatic fever, followed by treatment of the movement disorder if it persists. A cardiac assessment should be carried out. β-haemolytic streptococci remain exquisitely sensitive to penicillin which should be administered. Other manifestations of rheumatic fever are treated with aspirin reinforced with steroids if necessary. There is no good evidence that the administration of steroids influences the chorea significantly. Haloperidol and tetrabenazine have been used to try to influence chorea.

42.5.11 Brucellosis

In humans, brucellosis is caused by infection with species that belong to the genus Brucella. It was first described as a clinical entity by Marston in 1861. Bruce, who was working for the Army Medical Service in Malta, cultured a micrococcus from the spleen of a fatal case and named it *Micrococcus melitensis* after the old name for Malta. The name was changed to *Brucella melitensis* in his honour in 1920. Zammit, a Maltese colleague of Bruce found antibodies against the organism in the serum of goats and Horrocks isolated the organism from the milk of an infected goat in 1905. Concurrently, Bang isolated *B. abortus* from cattle in Denmark and slightly later in 1914, Traum isolated *B. suis* from sows. Evans recognized the relationship between these organisms and their diseases in 1918 and she established that milk was the source of the infection.

Brucellosis is a disease of animals which spreads to man who ingests infected, unpasteurized milk, milk products such as cheese, or undercooked meat, or who comes in contact with the organism by occupational contact with animals. The bacterium enters through abrasions or by inhalation. There are six species of *Brucella*: *melitensis, abortus, suis, canis, neotomae,* and *ovis*. The last two do not cause human disease and *B. canis* very seldom does. *B. melitensis* infects sheep, goats, and camels, *B. abortus* infects cattle, and *B suis*, pigs and other species such as reindeer and hare. Brucellosis is found throughout the world and it is probable that the incidence is increasing. It is endemic in Asia, Africa, Central and South America, and in several Mediterranean countries. In Britain most cases are caused by *B. abortus* and *B. melintensis*. In the United States *B. suis* is the major pathogen, whilst in the Mediterranean it is *B. melitensis*.

Once the organism has entered the bloodstream it travels to regional lymph nodes, thence to the reticulo-endothelial system, particularly spleen, liver, bone marrow, and lymph nodes, where granulomas are formed with epithelioid and giant cells similar to sarcoidosis, and other chronic granulomatous conditions. These are full of bacteria and may progress to abscess formation. Endocarditis, osteomyelitis, and genitourinary tract granulomas occur rarely. Neurological involvement includes acute and chronic vascular inflammation, diffuse meningeal inflammation, and adhesions occasionally associated with non-caseating granulomas and encephalitic changes, brain oedema, and white matter degeneration. Perineural infiltration of nerve roots may take place (Larbrisseau *et al.* 1978). Bone of the vertebral column, discs, and joints may be invaded by granuloma.

Brucella organisms are non-motile, unencapsulated Gram negative coccobacilli which are facultatively intracellular. This last property makes it difficult to eradicate infection with antibiotics and prolonged courses are often necessary. Pasteurization kills the organism.

Clinical features

Brucellosis most commonly manifests as the acute form. Both sexes and any age are affected. The incubation period may be as short as 1–3 weeks, and may be as long as 6 months. Headache, weakness, malaise, back pain, and arthralgia are accompanied by fever, sweating, and shivering with a tendency to be more marked towards evenings. Progression may be insidious and evidence of disease spreading to other systems, particularly osteo-articular, ensues (Young 1983). Classically, intermittent and irregular 'undulant' fever with backache and sciatica is associated with lymphadenopathy and splenomegaly. The nervous system is affected in about 5 per cent of all cases (Shakir *et al.* 1987; Bahemuka *et al.* 1988; al Deeb *et al.* 1989). The clinical syndromes of neurobrucellosis are several and diverse and may affect any part of the central and peripheral nervous system. They include a meningoencephalitis, which may be acute or chronic, vasculitis with infarcts, granulomas, myelopathy, radiculopathy, and neuropathy.

The clinical features of meningitis and encephalitis are similar to those of other chronic forms such as tuberculosis or fungal meningitis, from which brucella can be distinguished only with further tests. The course is often chronic and relapsing. Focal granulomas can occur anywhere throughout the neuraxis and the signs reflect their location. Seizures are not uncommon and raised intracranial pressure may cause papilloedema. Vasculitis of the major cerebral vessels gives rise to focal signs as a result of infarction. Sub-arachnoid haemorrhage from rupture of mycotic aneurysms has been described (McLean *et al.* 1992) and embolization from endocarditis is a rare complication (McLean *et al.* 1992). Multiple mononeuropathy affecting the cranial nerves, particularly VIII, VII, V, and the ocular motor nerves, occurs due to perineural inflammation. Peripheral nerves may be similarly affected usually in the legs. Myelopathy and radiculopathy, often combined, have a tendency to occur in the lower thoracic and lumbosacral regions and have been caused by transverse myelitis, infarction, arachnoiditis, epidural abscess or spondylitis, and compression of cord and nerve roots. Central demyelination has been described, and papillitis and optic neuritis, and may represent an immunologically mediated reaction. Whether there is a specific psychiatric syndrome associated with brucellosis has been the subject of considerable argument. Undoubtedly many of the neurological syndromes are misdiagnosed or diagnosed late; more than one patient has been accused of malingering. Depression is common and organic brain damage may lead to psychiatric syndromes including psychoses.

The differential diagnosis of neurobrucellosis includes many syndromes. In the acute phase, bacterial and viral meningo-encephalitides, including herpes encephalitis, must be excluded. Chronic disease is similar to tuberculous or fungal meningitis and granulomas, tuberculous spondylitis and pyogenic infections, Lyme disease, syphilis, sarcoidosis, and Behcet's disease. The tendency of neurobrucellosis to exhibit a relapsing course can be very like multiple sclerosis and the scan appearances and CSF abnormalities may be similar.

Diagnosis

In an endemic area, diagnosis is usually not too difficult once it has been considered. Confirmation can be obtained by culture of the organism from blood, CSF, bone marrow, or lymph node. This is positive in about 25 per cent of cases. A presumptive diagnosis can be made by demonstrating rising titres of specific antibody in serum. Both IgM and IgG rise, the latter more slowly; discussion with the laboratory is advised to determine the most suitable assay. Raised titres of IgG are found in CSF in cases of meningitis, there is lymphocytic pleocytosis and raised protein content. Polymerase chain reaction of non-blood samples had a sensitivity and specificity of about 95 per cent in one study of brucellosis, which included five patients with neurological disease (Morata *et al.* 2001); it is also used on CSF (Matar *et al.* 1996). The radiological abnormalities of neurobrucellosis are not specific. Plain X-rays of the spine may indicate sacroileitis and erosion of vertebrae. CT scanning of brain may indicate granuloma formation or infarction, of the spine, bony erosion. MR imaging of brain will show white matter abnormalities to better advantage and of the spine, soft tissue granulomatous change and spinal block. Isotope bone scanning can highlight inflammatory changes in bone and joint before they become evident on CT or MR. None of these changes is specific and similar appearances can be seen with chronic pyogenic, tuberculous, and fungal infections.

Treatment

Treatment of brucellosis is with antibiotics. In the more chronic forms this can be difficult because of the ability of *Brucella* to remain protected within the monocyte–phagocyte environment. As a consequence, relapses are frequent. Commonly used regimes combine doxycycline with streptomycin or rifampicin for 6 weeks to 2 months with appropriate precautions for the use of streptomycin. For acute cases, a combination of tetracycline and rifampicin for 2 weeks would appear to be adequate. In more chronic cases, particularly those with polyradiculopathy and granulomas, antibiotic treatment may have to be given for 6 months or more. Repeat courses may be needed. It is prudent to cover the first few days of antibiotic treatment with steroids to diminish the risk of a Jarisch–Herxheimer reaction. Some reports suggest that concurrent prednisolone treatment is useful in those with arachnoiditis and cranial nerve palsies (al Deeb *et al.* 1989). With adequate and timely antibiotic treatment most cases make a complete recovery and the relapse rate should be less than 10 per cent. It is prudent to keep all cases under surveillance for a year after the cessation of treatment. Blood antibody levels fall but the clinical correlation between complete recovery and the levels is not absolute. Results with polymerase chain reaction monitoring are encouraging (Morata *et al.* 1999).

Prevention of brucellosis in humans depends upon eradication of infection in animals and upon the pasteurization of milk for human consumption. Immunization of people at risk, such as those involves in veterinary work and in animal husbandry, has been used, but has not been completely effective.

42.5.12 Bartonella

Cat scratch disease is caused by infection with *Bartonella henselae*, which can be cultured in cell free media, and is ubiquitous. Fleas carried by cats are the vectors. Most infections are self-limiting, but severe disease, including neurological disease can occur in the immunocompromised, and occasionally in the immunocompetent (Fouch *et al.* 2007). In immunocompromised patients, the onset is commonly insidious with malaise, myalgia, fatigue, and recurring headache and fever (Smith 1997). Progression to encephalopathy or meningo-encephalitis may follow; this may be impossible to distinguish from HIV-associated encephalopathy. In the immuno-competent by contrast, the onset may be quite acute. Bacillary angiomatosis may occur in both the immunocompetent and the immunosuppressed. A form of retinitis leading to sudden loss of visual acuity is a common feature. Diagnosis is serological, or by culture of CSF for fastidious organisms, and treatment is with doxycycline, or erythromycin (Adal *et al.* 1994; Smith 1997). Treatment for 3 months or more may be needed in those with HIV and relapse may require long-term treatment. *Bartonella bacilliformis* is the aetiological agent of Oroya fever, a bacteraemic infection limited to the Andes region and characterized by fever, headache, encephalopathy, lymphadenopathy, and anaemia. It carries a high mortality rate if untreated.

42.5.13 Septic encephalopathy

Septic encephalopathy is an ill-defined, yet well recognized state of disordered cerebral function which occurs in patients in the clinical setting of severe generalized infection. The incidence of sepsis and septic shock are not known because the entities are so ill defined but it is recognized that they are on the increase. Sepsis of various forms is the most common cause of death on intensive care units.

Septic encephalopathy is the commonest form of encephalopathy encountered in intensive care medicine and is present in 50–70 per cent of septic patients (Eidelman *et al.* 1996; Kramer *et al.* 2007). Factors which have contributed to this include the widespread use of antibiotics effective against gram negative and gram positive infections; an explosion in the population of patients who are immune suppressed from disease, immunotherapy, and chemotherapy; increased use of invasive procedures; and aggressive surgical interventions. In the order of 50 per cent of patients with septic shock have a bacteraemia, gram negative in the majority, gram positive and mixed in the remainder. Other patients have systemic infection with fungi and viruses. Most cases have an underlying cause which predisposes to the dissemination of infection such as tissue damage from trauma or burns, diabetes, drug abuse, AIDS, organ failure, and chemotherapy. In some there is no underlying cause and sepsis is the consequence of infection by a particularly virulent organism.

Pathophysiology

The pathophysiology of septic encephalopathy is poorly understood and is certainly multifactorial. Patients with underlying central nervous system disease are at greater risk. Mechanisms which have been postulated, and for which there is some evidence, are:

- brain microabscesses;
- diffuse microemboli,
- metabolic derangement;
- damage from bacteria and bacterial breakdown products;
- hypoxia and poor cerebral perfusion; and
- toxic effects of drugs.

Bacterial components are released into the circulation, endotoxin in Gram negative infection, exotoxins in the case of other organisms such as *Staphylococci* and *Streptococci*. These activate the complement system and the coagulation/fibrinolysis mechanisms

and trigger the breakdown of kallikreins to bradykinin which causes hypotension and increases vascular permeability. Cellular stimulation, phagocytic, neutrophil, and endothelial, releases cytokines which in turn stimulate or overstimulate, the inflammatory cascade. Inflammatory cytokines, such as interferon alpha, interleukin 1 beta or tumour necrosis factor-alpha, are elevated in patients with sepsis. These cytokines have an effect on the function of neurones, cerebral microvascular function, and the blood–brain barrier, possibly allowing increased access to the central nervous system of compounds which are usually restricted. This includes the aromatic amino acids, which may act as 'false neurotransmitters'; plasma concentrations of which have been correlated with the severity of encephalopathy (Sprung *et al.* 1991). The complex interaction of these and other mechanisms produces the clinical manifestations of sepsis which manifest in several organ systems, including the brain, to cause septic encephalopathy (Glauser *et al.* 1991).

Clinical features

The clinical signs of septic encephalopathy often appear early and are non-specific. Alteration of mental status is always present and is usually the first sign of brain disturbance. Drowsiness, confusion, disorientation, inattention, restlessness, and hallucinations, which progress to coma are usual. Spasticity or rigidity often accompanies coma. Epileptic seizures may occur but their presence and the declaration of focal signs should prompt a search for underlying structural brain pathology; septic encephalopathy is seldom accompanied by focal neurology (Young *et al.* 1990). Signs of generalized sepsis, hypothermia, hypotension, thrombocytopaenia, and involvement of other organ systems, hepatic, renal, respiratory or cardiac, are usual. Often, there is multiple organ failure. Search must be made for a focus of infection if it has not already been identified.

Diagnosis

Diagnosis of septic encephalopathy requires the exclusion of other causes of diffuse cerebral dysfunction. Anoxia and hypotension must be identified and corrected because their persistence will result in irreversible brain damage. Septic encephalopathy is usually reversible provided the underlying cause of sepsis can be treated, and treatment is given quickly. Mortality increases markedly with reduced Glasgow Coma Scale scores (Eidelman *et al.* 1996). Meningitis can be excluded by CSF examination; in most cases of septic encephalopathy the CSF findings are unremarkable. There is sometimes elevated CSF protein but a pleocytosis is very unusual, and should prompt investigations for other causes. Brain imaging is usually normal (Bolton *et al.* 1993). If it is clinically possible, brain MRI is to be preferred because of its ability to identify small areas of infarction, white matter abnormality, and microabscesses better than CT. The most sensitive indicator of brain dysfunction is the EEG. It can exclude significant encephalopathy and roughly quantify the degree of cerebral dysfunction; changes vary from excessive theta activity in mild encephalopathy, through delta then triphasic waves to suppression or burst suppression in the most severe encephalopathies (Young *et al.* 1992). The EEG is particularly useful in monitoring the progress of septic encephalopathy. Blood cultures should be done, together with a full blood count and coagulation screen, blood gas measurement, and appropriate tests for organ dysfunction.

Treatment

Treatment of septic encephalopathy is essentially that of the underlying condition. Antibiotics are given in high dose intravenously appropriate to the causal organism. Pus should be evacuated surgically. Cardiovascular and ventilatory support and the management of coagulopathy are imperative. It is important to review all medications to determine if any may be contributing to the syndrome. Therapeutic attempts to manipulate the inflammatory response have not produced much success to date. The prognosis for septic encephalopathy is very much the prognosis of the underlying cause and if this can be treated expeditiously and hypoxia and hypoperfusion of brain tissue can be avoided, prospects for complete recovery are good.

42.5.14 Tetanus

Tetanus, literally meaning 'muscular spasm', is the name given to a condition known since ancient times, caused by infection due to *Claustridium tetani*. Although not strictly an encephalopathy, since patients are usually fully conscious, this disease is discussed in this section for convenience. It was recognized by Hippocrates and Aretaeus of Cappodoccia gave a graphic account of the clinical manifestations. The organism was identified from the soil by Nicolaier in 1884 who recognized that it produced a toxin with similarities to strychnine. Two years later Rosenbach found the bacillus in man. Four years later, Behring and Kitasato were able to demonstrate that the disease could be prevented by immunization and in the same year Faber found a heat-labile toxin. Small amounts of toxin injected into animals would induce immunity which could be passively transferred to humans and the use of antitoxic serum prevented the development of tetanus in the wounded in the Great War. In countries where tetanus immunization has been introduced, the incidence of clinical tetanus has declined dramatically. There are probably in excess of one million cases of tetanus in the world each year. Neonatal tetanus accounts for about 50 per cent of all cases.

Clostridium tetani is an anaerobic gram positive bacillus which is slowly motile and produces a terminal spore as it matures, so causing an appearance like a tennis racket or distaff under the microscope, hence the name *Clostridium*. These spores are highly resistant to heat, disinfectants, and antibiotics, and may lie dormant in soil for many years. They are ubiquitous and are found in soils of all kinds, particularly if they have been enriched with moisture and manure, either human or animal. Human infection occurs as a consequence of contamination of cuts, burns, animal bites, or even insect bites. Any age group is at risk, but particularly neonates in undeveloped cultures who are born in unhygienic conditions when the umbilicus becomes contaminated by spores. Infection is more likely if the wound is particularly traumatic and associated with tissue damage. However, inoculation can take place via apparently innocuous pricks which penetrate the skin such as from rose thorns. Forgetting such incidents may account for the observation that no history of trauma is obtained in a small proportion of cases. Tetanus may follow childbirth or abortion, surgical operations, and body piercing. Otitis media and mastoiditis may be sources of infection. In the West, intravenous drug abusers are at risk, especially those that practise the 'skin popping' form of injection that involves raising a bleb under the skin. During one outbreak in recent years, 22 causes in heroin users were thought to be due to

contamination of the drug source itself (Beeching *et al.* 2005; Hahne *et al.* 2006).

The toxin

When the spores are introduced to damaged tissues they are activated to the vegetative form in anaerobic conditions and produce the tetanus toxin, tetanospasmin, which is neurotoxic, disseminates throughout the body, and causes clinical tetanus. Another toxin, tetanolysin, is also produced which has the ability to disrupt cell membranes and damage tissue adjacent to the wound. Its part in the pathogenesis of tetanus remains ill understood.

Tetanospasmin is produced by pathogenic strains of *C. tetani*. Strains which do not produce tetenospasmin do not cause tetanus. Tetanospasmin is a single polypeptide chain which is broken by endogenous proteases to produce a light and a heavy sub-chain which are linked co-valently by a disulphide bridge. The heavy sub-chain can be cleaved by papain treatment to produce a smaller fragment known as the C fragment, leaving the disulphide joined A–B fragment. The exact mechanisms by which these produce the manifestations of tetanus have not been fully elucidated. It is thought that fragment C binds to neural ganglioside receptors at presynaptic end plate junctions, and the light chain inhibits transmitter release. Fragment C may facilitate trans-synaptic transport of toxin. However it functions, once tetanospasmin has entered the presynaptic vesicle it can inhibit transmitter release for as long as several weeks. It can also be translocated by retrograde axonal transport and transynaptic migration to neurones in spinal cord and brain. Tetanospasmin also circulates in the blood and spreads to synaptic endplates throughout the body. Circulating toxin can be neutralized by antitoxin but once it has crossed the neural membrane it cannot be inactivated. Within the central nervous system it migrates to inhibitory cells where GABA release is prevented. This increases the activity of α-motoneurones resulting in muscular hypertonicity. Inhibitory reflexes are also disrupted which means that stimulation of the motor system produces intense contraction of all muscle groups, both agonist and antagonist, to cause tetanic muscle spasm. The sympathetic nervous system is affected, leading to signs of sympathetic over-activity.

Clinical features

Four clinical varieties of tetanus are recognized: local, cephalic, generalized, and neonatal. These reflect the host reaction to infection, the quantum of tetanospasmin production, and the site of action of the toxin The incubation period for the development of tetanus reflects the time from spore inoculation to toxin production and symptom onset, and this may be as short as 24 h or as long as several months but usually falls within 1–3 weeks. The shorter the incubation period, the more severe the infection and the poorer the prognosis. Distal wounds have a longer period than those proximate to the central nervous system and large wounds with a greater inoculum produce symptoms sooner.

Localized tetanus refers to rigidity and spasm of muscle in proximity to the site of inoculation. It may be associated with local weakness, and muscle pain, and it may persist for weeks. Such patients have some degree of immunity with sufficient circulating antibody to bind to toxin to prevent it reaching the central nervous system, but not enough to prevent local disturbance. If treated it has a very low mortality. The danger is that it may progress to generalized tetanus if not recognized and the toxin is not neutralized with antitoxin.

Cephalic tetanus is a localized form in which signs are limited to the lower cranial nerves. There is facial weakness, and trismus which consists of the grinding of teeth due to masseter spasm of the muscles (Greek 'to grate'). Dysphagia occurs and rarely the extraocular muscles are involved. Cephalic tetanus occurs in less than 1 per cent of cases and tends to be associated with head wounds and otitis media (Vakil *et al.* 1973; Dastur *et al.* 1977; deSouza *et al.* 1992).

Generalized tetanus is the most frequently recognized form of the disease to occur in adults. There is often ill-defined malaise for a day or so, followed by muscle spasms and trismus. The jaw muscles become tight on chewing and this progresses to inability to open the mouth fully because of muscle rigidity leading to 'lockjaw'. As the facial and jaw muscles become more rigid, facial expression becomes fixed in a grimace, the 'risus sardonicus'. Rigidity of abdominal and paraspinal muscles accompanies these changes and spasm spreads with neck retraction and opisthotonus. Recurrent spasms of muscles occur, either spontaneously or reflexly, in response to stimuli. Respiratory muscle involvement leads to periods of apnoea and cyanosis. Paraspinal muscle contraction may be of such severity that vertebral or other fractures are caused. Spasm of laryngeal and pharyngeal muscle may cause choking. Without treatment, the patient's sensorium and consciousness remain intact (Weinstein 1973). Autonomic dysfunction, with sympathetic hyperactivity may follow with tachycardia, hypertension, and cardiac dysrhythmia (Kerr *et al.* 1968) and sudden death (Trujillo *et al.* 1987). Provided supportive intensive care treatment is provided, improvement occurs in most patients within 4 weeks. If antitoxin is not given the disease persists for as long as tetanospasmin is produced. Patients who develop tetanus should be actively immunized because an adequate immune response is not always triggered by the disease as the amount of toxin produced may not have been sufficient.

Neonatal tetanus follows contamination of the umbilical stump in new-borns. It occurs in populations who have not been immunized because antibodies from the mother will transfer to the baby and confer protection in early life. After 5–7 days the baby becomes weak and irritable and fails to feed. Muscle rigidity, spasms, apnoea, and cyanosis follow and the infant rapidly becomes dehydrated and hypotensive. Autonomic disturbance ensues and is often fatal. Mortality rates often exceed 50 per cent.

Diagnosis and treatment

The diagnosis of tetanus must rely on clinical factors because there is no specific diagnostic test which will be positive in the early stages. Unfortunately, blood tests and electromyography do not help. Strychnine poisoning is the only condition which produces similar clinical features. Adverse reactions to psychotropic drugs may produce muscular rigidity and dystonia, sometimes to the severity of opisthotonus. Meningitis will feature neck stiffness, and the stiff man syndrome is decidedly more chronic.

Treatment is essentially supportive. Intensive care has greatly improved prognosis by preventing death from respiratory and cardiovascular failure (Trujillo *et al.* 1987). The wound which has provided the portal of entry should be thoroughly cleansed and damaged tissue removed. Muscular spasms should be controlled with antispastic agents. Benzodiazepines have been used most commonly and successful results have been reported with midazolam. Baclofen has been used, both systemically and intrathecally, and dantrolene may have a place. If these measures fail to control spasms, neuromuscular blockade becomes necessary. Concurrently, the immune status of the patient should be established by estimation of serum anti-tetanus antibodies. Human anti-tetanus immunoglobulin should be given to neutralize circulating tetanospasmin which has not yet entered the nervous system. Intrathecal use of immunoglobulin has been advocated but does not seem to be more useful than systemic treatment. Active immunization should be given. Autonomic overactivity is controlled with a combination of α- and β- adrenergic blockade. Magnesium has been used in recent years, with anecdotal evidence suggesting it may reduce spasms, and autonomic complications. A randomized placebo controlled trial showed the drug made no difference to mortality, need for ventilation, duration of stay in intensive care. However it did reduce the need for midazolam, pipecuronium, and verapamil.

Complications of tetanus include hypoxia, respiratory failure, and pneumonia. Isolated neuropathies of the lower cranial nerves and of peripheral nerves may occur, and may be part of a more generalized 'critical illness polyneuropathy' (Section 21.18.9). Fractures of vertebrae and long bones may follow violent muscle spasm as may rhabdomyolysis in severe cases.

Immunization

Tetanus can be prevented if active immunization of populations at risk is undertaken. In Britain tetanus immunization is combined with diphtheria and pertussis in the DTP vaccine; dosage schedules can be found in the British National Formulary. A series of three intramuscular injections at monthly intervals during the first months of life confers immunity for some years. Booster doses are given before 5 years and on leaving school. Any person suffering a wound which may have become contaminated with spores should have a booster dose if 10 years or more have elapsed since the last immunization.

42.6 Parasitic encephalitis and encephalopathy

Two parasites, *Plasmodium falciparum* and *Trypanosoma brucei* are major causes of central nervous system infection globally, particularly in Africa. The former causes cerebral malaria, a poorly understood encephalopathy, whilst the latter is responsible for sleeping sickness, a severe meningoencephalitis, with very high mortality rates. In addition, amoebi, such as *Naegleria fowleri* also occasionally case encephalitis, especially in the immunocompromised.

42.6.1 Cerebral malaria

Cerebral malaria is an acute encephalopathy caused by infection with the parasite *Plasmodium falciparum*. The three other malaria parasites, *P. vivax*, *P. malariae*, and *P. ovale* do not cause cerebral malaria, though they may cause febrile seizures in children. *P. falciparum* can cause a range of severe complications affecting many organs of the body. Thus there are many potential causes of reduced consciousness in someone infected with *P. falciparum*, including hypoglycaemia, acid–base disturbance, severe anaemia, and seizures. For research purposes the World Health Organization has defined cerebral malaria as unrousable coma, in other words a non-purposeful response to a painful stimulus, in patients in whom these metabolic deficits, and seizures have been corrected. However, for practical purposes any patient with *P. falciparum* infection and altered consciousness, including those that are drowsy or irritable, should be treated as if they have severe disease. Globally, malaria is one of the most important parasitic diseases of humans. In the region of two billion people live in malarial endemic zones, of whom 300 million or more are affected each year, the majority in Africa.

In 2002, there were 515 million cases of malaria in the world; 25 per cent in southeast Asia and 70 per cent in Africa, mostly sub-Saharan Africa (Idro *et al.* 2005), of whom an estimated three million die each year. The increase in global travel means that malaria can be seen in any country in the world, after having been contracted abroad in the tropics. Approximately 2000 cases are seen annually in the United Kingdom, and during 2005 there were 11 deaths.

The death rate from cerebral malaria varies from 10 to 50 per cent. Cerebral malaria can kill within 72 h if not recognized and treated. Delay in making the diagnosis contributes significantly to the high mortality rate. People who live in endemic areas may acquire a degree of immunity which diminishes with time. Those most at risk of developing severe malaria are the traveller who has no immunity and visits an endemic area to be exposed for the first time, the emigrant returning from a long stay abroad whose immunity has lapsed, pregnant women, children in malarious areas between the ages of 6 months and 3 years who have lost maternally acquired immunity, and those who have become immune suppressed. The presence of haemoglobin S offers some protection against falciparum infection. However, it is not necessary to have travelled to an endemic area to contract malaria. Occasional cases have been reported of cerebral malaria developing in people who have not been out of Britain and are thought to have acquired the disease by being bitten by infected mosquitoes that have been imported to the country by aircraft flying from endemic areas. Others may have been bitten while sitting in aircraft during stopovers in the tropics, so-called 'runway malaria' (Conlon *et al.* 1990).

Parasite life cycle

Malaria occurs in countries lying between the latitudes 60 degrees north and 40 degrees south. These include most of Africa, Central and South America, regions of the Middle East, Iran, the Indian sub-continent, South East Asia, the Philippines, Borneo, and Southern China. Infrequently, malaria may be contracted from infected blood transfused in endemic areas, or from needle sharing amongst drug addicts. The vast majority of cases of cerebral malaria result from humans being bitten by the female *Anopheles* mosquito which harbours the falciparum parasite. When these feed they inject sporozoites into the blood which circulate to be cleared rapidly by the liver where they enter the hepatocytes and reproduce asexually, forming schizonts. About 1–3 weeks later they rupture into the circulation as motile merozoites and quickly invade red blood cells allowing them to be seen on blood films under the microscope. They feed on haemoglobin and further asexual reproduction takes place. The progeny mature through the stages of merozoite and trophozoite, multiplying approximately ten-fold, following which they rupture out of the erythrocyte and parasitise other unaffected red cells. Cells of any age are affected so there is no limit on the potential degree of parasitaemia. If the parasite load is high severe haemolysis may ensue. This rupture and dispersal takes

place approximately every 48 h, and causes fever. Some of the merozoites develop into the sexual gametocyte stage and they are ingested by the female mosquito during her blood meal, before undergoing sexual replication within the mosquito to form infective sporozoites which are inoculated into another human when the mosquito next bites.

Pathophysiology

The manifestations of cerebral malaria result from a series of complex and ill-understood interactions within the cerebral vasculature which are mediated by vascular, humoral and haematological factors. The rheology of parasitised erythrocytes is disturbed in cerebral malaria and they adhere to the endothelium of brain capillaries and 'sludge' the circulation by mechanical occlusion. Changes take place in the erythrocyte wall which develops cone shaped knobs which adhere to endothelial cell membranes (MacPherson *et al.* 1985) and these cells are sequestrated. The molecular basis for these changes is not yet understood and several red cell surface molecules and receptor molecules are currently under study. Cytokines released by macrophages, tumour necrosis factor, interleukins and other vasoactive substances have been found to be raised in cases of cerebral malaria. Whether increased vascular permeability is important is the subject of dispute (Patnaik *et al.* 1994). The blood–brain barrier is disrupted (Brown *et al.* 1999). However the changes are brought about, they result in a fall in cerebral perfusion and hypoxia, shift to anaerobic metabolism, and increased production of CSF lactate. Hypoglycaemia from hepatic glycogen depletion, malabsoprtion of glucose from the gut, and increased metabolism of glucose by a large biomass of parasites, and metabolic acidosis may compound these changes, as may acute tubular necrosis and the adult respiratory distress syndrome. A proportion of patients with cerebral malaria have coexisting infections including pneumonia, urinary tract infections, septicaemia, and bacterial meningitis, each of which can exacerbate decline. Disseminated intravascular coagulation is another potential complication. Whether cerebral oedema occurs and intracranial pressure is significantly raised, has been the subject of debate. Cerebral oedema has been observed in necropsy specimens and it has been claimed that this has been an agonal manifestation. The CSF pressure has been found to be high at the time of lumbar puncture in a group of African children (Newton *et al.* 1991, 2000) yet Looareesuwan (1983) found in a CT study that there was no evidence of cerebral oedema and Warrell (1982) found that administration of dexamethazone did not improve prognosis. Recent studies have emphasized the contributon of convulsions, acidosis, and hypoglycaemia to P falciparum induced coma, particularly in children (Idro *et al.* 2005).

Clinical features

In non-endemic areas the diagnosis of cerebral malaria is not infrequently delayed or missed altogether, with fatal consequences, because people neglect to think of it. Any person from an endemic area or who has visited an endemic area, or even who may be exposed to aircraft from such a region, and who falls ill with a pyrexia and cerebral symptoms, should be considered to be a possible case of cerebral malaria and investigated and treated accordingly. It is also important to remember that falciparum malaria may affect organ systems other than the brain and the presentation may reflect this. In adults the symptoms and signs of cerebral involvement usually follow a few days of pyrexia and non-specific malaise which may be mistaken for influenza. The physician should be very wary of diagnosing influenza in someone who has recently returned from a malaria endemic area. Characteristically, there are paroxysms of chills, rigors, and fever, with headache and myalgia, which may recur. The temperature usually remains elevated. In some, the onset may be abrupt and patients who are very ill may be hypothermic rather than pyrexial. Cerebral malaria may occur at any time during falciparum infection, even after effective treatment has been started. Evidence of cerebral involvement manifests by a decrease in the conscious level gradually leading to coma. This may be punctuated by seizures, commonly non-focal, especially in children. Before coma supervenes, there may be signs of an organic brain syndrome with confusion, disorientation, agitation, and psychosis. Some develop focal weakness and movement disorders. Meningism and opisthotonus may occur and as cerebral involvement increase decerebrate and decorticate posturing may be seen. Retinal haemorrhages may be found, and bruxism is common. Anaemia and jaundice are frequent accompaniments and hepatosplenomegaly is common. Disseminated intravascular coagulation, pulmonary oedema, renal failure, and blackwater fever from haemolysis are serious complications. Prolonged coma and frequent convulsions should prompt a search for hypoglycaemia or other metabolic disturbance, and for pyogenic meningitis. Severely ill patients are liable to gram negative septicaemia.

In children, disease evolution is often much more rapid and they are even more likely to have epileptic seizures. Cough, diarrhoea, anorexia, and vomiting are common. Retinal haemorrhages are found with similar neurological signs to those in adults. Differentiation from bacterial meningitis can be difficult (Berkley *et al.* 1999). Recovery from coma with treatment is quicker in children who improve within 24 h whereas adults may take 3 days.

Diagnosis

The diagnosis of cerebral malaria should not be difficult, but there are some common pitfalls. One of the biggest mistakes is in not thinking of the diagnosis in the first place. An enquiry about travel should be made for *every* patient with an acute encephalopathy. Be wary of diagnosing flu in a returning traveller in which an urgent examination of blood films is required to exclude malaria. Make sure that the laboratory does thick films, which have the maximum chance of seeing parasites, and thin films, which allow the Plasmodium species to be identified. Another common mistake is to assume that because a traveller took anti-malarial chemoprophylaxis, malaria is not possible. Antimalarials do not provide complete protection, and are often not taken regularly; moreover unwary travellers may have allowed mosquitoes to bite them freely, in the mistaken belief that they were fully protected. Furthermore antimalarial chemoprophylactic drugs can make microscopic diagnosis harder; they lengthen the incubation period beyond the 1–2 weeks that is quoted in many textbooks, and they make it less likely that parasites will be seen on a blood film. In a similar way, do not be misled by the patient that 'cannot have malaria' because they were brought up in Africa, and have natural immunity. Such natural immunity is usually only partial, and wanes in those that have been living temporarily in non-endemic areas; the native African who has lived in the West for many years and then returns home for a holiday may well get malaria. However, because they have some residual immunity, once again the incubation period may be prolonged, and diagnosis by blood film examination may be difficult.

Sometimes, in partially treated patients, malaria pigment within neutrophils or monocytes may be easier to see than parasites. Mild thrombocytopenia of <100 000 platelets/μl is common in malaria, and may be a clue that a patient has malaria, even if the parasites cannot be seen. A normochromic normocytic anemia is also common, and a leucopenia is also sometimes seen. At least three blood films should be examined before declaring a patient to be negative for malaria. However, for most patients with severe malaria, the parasites will be easily identified on a thick film by an experienced laboratory. Newer rapid diagnostic methods include stick tests for malaria antigens, and the polymerase chain reaction, but they should not replace blood film examination. If it is not possible to access a prompt assessment of blood films by a suitable laboratory, treatment should be started straight away.

In patients in whom there is no contraindication, a lumbar puncture should be considered to exclude other pathologies. This is especially important in patients living in malaria endemic areas where many people have asymptomatic parasitemia, and coma may be due to a coexistent bacterial meningitis. In the West it may be preferable to perform CT scan before lumbar puncture, especially in patients with deep coma, or focal neurological signs. But the undertaking of a lumbar puncture should not delay treatment with antimalarial drugs. Although raised intracranial pressure, as determined by lumbar puncture opening pressure, is common in cerebral malaria, it is unclear whether this contributes to the pathophysiology.

A full blood count, urea, electrolytes, blood gases, and sugar, should be monitored. Hypoglycaemia is especially common in children, those on quinine, and pregnant women.

Treatment

General supportive measures include giving oxygen by mask, nursing the patient at an angle of 30°, and using antipyretics to control high fevers (Fig. 42.11). In an unconscious patient the airway should be secured, and a nasogastric tube passed to prevent aspiration pneumonia. If cerebral malaria seems likely, and there will be a delay in obtaining the malaria film result, anti-malarial treatment should be started.

Chloroquine used to be the drug of choice for the treatment of cerebral malaria, but the emergence of resistant parasites in many regions reduced its usefulness. For many quinine was the drug of choice, and it is still widely used, though the latest World Health Organization guidelines recommend artesunate as the first line treatment. Although oral quinine can be useful in non-severe malaria, intravenous quinine should be used in cerebral malaria (Fig. 42.11). Quinine is followed by a second drug, to prevent recrudescence of parasites. Until recently Fansidar® a combination of sulfadoxine and pyrimethamine, was usually given. However because of increasing resistance to Fansidar® many now give doxycycline, which can be started as soon the patient can swallow. Patients should be warned that quinine may cause an unpleasant buzzing in the ears, sometimes with nausea and vomiting; this 'cinchonism' will only be temporary. Doxycycline should not be used in children under 8 years, or pregnant women. Malarone, consisting of atovaquine-proguanil, is an alternative second line drug, but mefloquine should not be used because of an increased

Suspicion of Cerebral Malaria

- Always take a travel history
- Beware the patient that "cannot have malaria" because they "took prophylaxis" or are "immune"

Assessment

Clinical Assessment

- Coma score

- Seizures? – obvious or subtle

- Hyperventilation? - acidosis, pulmonary edema, or pneumonia

- Fluid balance - ?Deyhdration

Initial investigations

- Thick and thin blood film for parasites and pigment (repeated if necessary)

- Clues if blood film negative: ?thrombocytopenia; ?malaria pigment in neutrophils and monocytes

- Blood glucose

- Full blood count; urea and electrolytes; blood gasses or venous lactate for acidosis; blood culture

- Lumbar puncture if no contraindication

Treatment

General Measures
O$_2$, nurse at 30°, antipyretics secure airway, nasogastric tube

Quinine dihydrochloride salt*#	Adults	Children
Loading dose	20 mg/kg iv over 4 hours	20 mg/kg iv over 4 hours
Maintenance	10 mg/kg iv over 4 hours; starting 8 h after the start of loading dose, and repeated every 8 h; until patient can swallow, then	10 mg/kg iv over 4 hours; starting 12 h after the start of loading dose, and repeated every 12 h; until patient can swallow, then
Oral	10 mg/kg 8 hly to complete 7 day course	10 mg/kg 8 hly to complete 7 day course
2nd drug	**Doxycycline** 100mg 12 hly for 7 days	**Fansidar**® (Sulphadoxine 500mg - pyrimethamine 25mg) Aged up to 4 y 1/2 tablet; 5-6 y 1 tablet; 7-9 y 11/2 tablets; 10-14 y 2 tablets

Unexpected deterioration
Check for: hypoglycemia
             subtle motor seziures
             bacterial coinfection
? Resistance to quinine -
     consider using artemisinin derivatives

Other treatable complications
Anemia
Acidemia
Acute renal failure
Pulmonary edema
Hyperparasitemia

* If the patient originated in southeast Asia, consider using artesunate, which is now the WHO recommended treatment for severe malaria. Intravenous or intramuscular artesunate is given as a loading dose of 2.4mg/kg, then at at 12 and 24 hours, and then daily. Intramuscular artemether is given as a loading dose of 3.2 mg/kg followed by 1.6 mg/kg daily.

#Note that, unlike other antimalarial drugs, quinine doses are usually prescribed as doses of salt rather than base; mistakes here can lead to over-treatment

Fig. 42.11 Cerebral malaria. Algorithm showing the assessment and initial treatment of patients. (From Solomon and Blanchard 2006.)

risk of late neuropsychiatric sequelae. In pregnancy the adult intravenous and oral doses of quinine can be used, though there is an increased risk of hypoglycaemia. In parts of the tropics with limited resources quinine dihydrochloride can be given as a deep intramuscular injection.

In China, the herb qinghaosu, artemisinin, has been used to treat fevers for at least two thousand years. Its derivatives, including artesunate and the oil-soluble artemether, are often used to treat cerebral malaria in Asia. They are as effective as quinine, clearing parasites more quickly, are less toxic, and are easier to give. In addition they are effective against chloroquine resistant, and multi drug resistant *P. falciparum*. Artesunate 2.4 mg/kg given intravenously or intramuscularly on admission, then at 12 h and 24 h, then once a day is now the World Health Organization recommended choice for treatment of severe malaria in low transmission areas or outside malaria endemic areas. They are being used in the West increasingly, particularly for patients who became infected in mainland Southeast Asia, Thailand, Burma, Cambodia, Vietnam, where there is quinine resistance, or because of a failure to respond to treatment. These drugs are now also being recommended for severe malaria in travellers returning from Southeast Asia (Whitty *et al.* 2006). Intravenous artesunate is preferable to intramuscular artemether because the absorption of the latter can be erratic. Supportive treatment on an intensive care unit should be given paying particular attention to the control of epileptic seizures, monitoring anaemia, hypoglycaemia, acidosis renal function, and lung function. Antibiotics may be needed for gram negative septicaemia or meningitis. It may be necessary to treat cerebral oedema. If the parasite load is particularly high, exchange transfusion may be considered.

Prognosis

The mortality rate in adults and children is about 20 per cent, and most deaths happen within 24 h of admission, before antimalarial drugs may have had time to work. Neurological sequelae occur in 5–10 per cent of cases and take the form of epilepsy, mental retardation, spasticity, behavioural upset, focal signs if there has been brain infarction, and a cerebellar syndrome. Recent studies have reported that epilepsy is associated with cerebral malaria. In non-immune adults, the prevalence of <5 per cent and severity of subsequent neurological impairments is less than in children. In Vietnam, a self-limiting 'post-malaria neurological syndrome' consisting of acute confusional state, acute psychosis, generalized convulsions, or tremor occurred in up to 0.12 per cent of patients with *P. falciparum* malaria, and was associated with mefloquine treatment (Nguyen *et al.* 1996). Prophylaxis against malaria is possible by avoiding mosquito bites by the use of appropriate clothing and mosquito nets, and by taking prophylactic drugs before and during the time of visits to endemic areas. Patterns of drug resistance are varying in different parts of the world and advice should be sought from an expert in tropical diseases on the regime best suited to the geography.

42.6.2 African trypanosomiasis

Trypanosomes affect man and his livestock. There are two quite distinct forms of trypanosomiasis which occur on different continents, and cause diseases with quite characteristic features;

◆ *African trypanosomiasis* or *sleeping sickness* is caused by *Trypanosoma brucei* and affects the central nervous system in humans. It is a major cause of neurological disease in sub-Saharan Africa.

◆ *American trypanosomiasis* or *Chaga's disease* is caused by *Trypanosoma cruzi* and affects the heart and gut predominantly, with central nervous system features being relatively rare.

African trypanosomiasis is found in 36 countries throughout sub-Saharan Africa. There are an estimated 300–500 000 cases annually with at least 60 000 deaths and 60 million people living in areas at risk (Kennedy 2006). About 50 cases are reported annually outside Africa, including the occasional cases in travellers returning to the United Kingdom (Lejon *et al.* 2003).

There are two forms of African trypanosomiasis which correspond roughly to the epidemiological regions in which they occur between the latitudes of 15 degrees north and 15 degrees south in Africa. The Gambian or West African form is caused by Trypanosoma brucei gambiense and occurs in the western and central parts; the Rhodesian or East African from is caused by Trypanosoma brucei rhodesiense and occurs in the east; there is geographical overlap in mid-Africa where they meet. Trypanosomes are elongated motile protozoa, 10–30 μm long with a central nucleus and they are propelled by an undulating membrane and flagellum. Both forms are fatal if not treated. Gambian sleeping sickness has a slower and more protracted course than Rhodesian, which is thought to reflect greater adaption of the parasite to the host.

Trypanosomiasis life cycle

The parasite is transmitted by male and female tsetse flies of the species *Glossina*. The trypanosomes undergo a developmental cycle in the fly which remains infective for its life span of several months. Once injected into the human host, the trypanosomes multiply locally, cause inflammatory change which may result in a chancre, and spread via the lymphatics and blood stream throughout the body, including the central nervous system. They invade the intercellular and extracellular fluids of many organs, causing febrile attacks which relate to waves of parasitaemia, as they go. In the nervous system the meninges become congested and infiltrated by lymphocytes and large plasma cells which are full of immunoglobulin, known as Mott cells. There is a complex and incompletely understood relationship between the trypanosome and the body's immune system which allows the organism to alter its antigenic disposition when challenged by the host defences. This results in recurring production of immunoglobulin to each wave of parasitaemia, and may lead to a build up of large levels of gammaglobulin. The parasite can remain latent within the central nervous system, in the choroid plexus and elsewhere, and can re-invade the brain from time to time. There they cause a meningo-encephalitis.

Clinical features

The clinical features of Rhodesian sleeping sickness are more severe and acute than those of the Gambian variety, which tends to be more indolent. A chancre may form at the site of the bite, and lasts for 2–3 weeks. This is associated with the manifestations of haematological and lymphatic disease: weakness, myalgia, arthralgia, lymphadenopathy and splenomegaly, pruritis and skin rashes, recurrent fever and malaise. These features may persist for months or years. Neurological involvement is indicated by personality change, memory failure, persistent headache, and disturbance of the sleep rhythm with insomnia at night and somnolence during the day. Psychotic episodes and organic brain syndromes develop as the disease progresses, and pyramidal, extrapyramidal, cerebellar signs, together with convulsions, become evident. Mycardial and pericardial disease is often evident in Rhodesian sleeping sickness. Untreated, patients become increasingly obtunded and lapse into coma in months with the Rhodesian form, and after a longer period for the Gambian form.

Diagnosis and treatment

The diagnosis of can be difficult especially when malaria coexists. Demonstration of the parasite is necessary before commencing treatment, which itself can be quite toxic. Samples of the initial skin lesion, lymph nodes, blood, or centrifuged CSF should be concentrated and examined. Parasites are identified in the blood or lymph nodes, as is usually the case for *T. rhodesiense* disease. Diagnosis of *T. gambiense* infection relies mainly on serologic tests, such as the card agglutination trypanosomiasis test. The use of polymerase chain reaction to demonstrate trypanosome DNA in blood (Kabiri *et al.* 1999) and CSF (Truc *et al.* 1999) may prove useful. Detection of antigen in serum and CSF by enzyme linked immunosorption assays has shown high levels of accuracy (Nantulya *et al.* 1992). The CSF should be examined in all cases and any increase of cell count or elevation of protein level should be taken as evidence of central nervous system disease. Diagnosis of late stage neurological disease is particularly problematic as there is no widespread consensus as to its definition, but the WHO criteria are the demonstration of parasites within the CSF, or a CSF leucocyte count of >5/μl.

Treatment of cerebral trypanosomiasis is difficult and toxic. Three drugs are useful for the treatment of the blood and lymph node stage of the disease:

- *suramin*, which is effective for rhodesiense;
- *pentamidine* for gambiense forms; and
- *melarsoprol* which is an arsenical and is the only drug effective against central nervous system disease.

Melarsoprol is extremely toxic causing a severe post-treatment reactive encephalopathy in about 10 per cent of cases, and death in about 5 per cent of cases. However the disease is uniformly fatal if untreated. Steroid treatment may help if the reactive encephalopathy develops. Recent work suggests antagonists to substance P may be useful (Kennedy *et al.* 1997).

42.6.3 American trypanosomiasis

This from of trypanosomiasis, also known as Chaga's disease, occurs in South and Central America and as far north as southern Texas. It is cause by *Trypanosoma cruzi* which is transmitted by bugs of the *Triatoma* genus. Many different mammals function as hosts and there are some strain differences in the organism which lead to variations in virulence and disease syndromes. Humans are bitten by the bug and develop the acute stage of infection with local lymphatic invasion followed by secondary haematogenous dissemination allowing the organisms to lodge in other tissue sites. Haematogenous spread is recurrent and during this phase, which last for two or three months, there are symptoms and signs of generalized infection. Immune defences build up and the disease becomes quiescent for years, sometimes for the rest of the patient's life. Years later the chronic stage becomes evident, usually with cardiomyopathy or with gastrointestinal disease which leads to megaoesophagus or megacolon. Rarely, in the acute phase, there may be a sudden meningoencephalitis when the brain becomes invaded by trypanosomes. Sometimes this can take the form of localized necrotizing encephalitis. These acute meningoencephalitides carry a high mortality, sometimes approaching 50 per cent, and are seen more frequently in young children. In immune suppressed patients,

and in those with AIDS, fullblown encephalitis may occur at any time, in relation to variation of immune status. In chronic Chaga's disease, frank encephalitis is rare, but central nervous system involvement may be evident as minor cerebral dysfunction, epilepsy, behavioural change and cognitive disorders in children, polyneuropathy and stroke as a consequence of cardiomyopathy (Kirchhoff 1993; Pitella 1993). Diagnosis is by demonstration of the parasite in the bloodstream and detection of antibody or antigen in the blood. Drugs which are effective are niphotimox and benzonidazole, but treatment of central nervous system disease is often disappointing.

42.6.4 Amoebic encephalitis

Primary amoebic encephalitis results from infection by free living amoebae of the species *Naegleria fowleri*. This is a ubiquitous organism that lives in moist soils and occurs all over the world; most cases have been reported in children and young adults who have been swimming in infected water. Amoebae enter the nasal cavity, and ascend to the brain along the olfactory nerves and blood vessels to the meninges and spread thereafter to cause florid necrotizing inflammation. The clinical syndrome comes on suddenly with a severe meningoencephalitis which may be preceded by a change in smell or taste, but often has no distinguishing features. The clue to diagnosis is exposure to stagnant water (Carter 1968). Special examination of fresh, warm specimens of CSF are necessary to demonstrate motile trophozoites. Alternatively they may be seen in brain biopsy specimens. Prognosis is poor and most die rapidly. Survival has been described in patients treated with high doses of parenteral amphotercin. Rifampicin, miconazole, and tetracycline may enhance the effect of amphotercin.

Acanthamoeba and *Balamuthia* species can cause a subacute or chronic granulomatous meningoencephalitis, and tend to affect immunocompromised people but can cause disease in those whose immune response is intact. Amoebae are not found in the CSF and the diagnosis is not usually made during life unless there has been some reason to biopsy the brain or meninges. Prognosis is almost hopeless. There is no known effective treatment, but *in vitro B. mandrillaris* is sensitive to pentamadine and azithromycin, whilst *Acanthamoeba* sp. are sensitive to pentamadine, amphoteracin B, and other antifungals.

42.6.5 Other parasites

Isolated case reports have appeared of cases of meningoencephalitis from which various parasites have been identified. Most of these are medical curiosities, and usually, the offending pathogen has been discovered after the death of the patient. *Micronema, Lagochilascaris, Baylisascaris, Gnathostoma*, and other nematode parasites have all been implicated in such cases (Lowichik *et al.* 1995). *Angiostrongylus cantonensis*, which causes eosinophilic meningitis is considered in Section 41.5.11. Other parasites cause encephalitis rarely, but often enough to produce a recognizable clinical syndrome.

42.7 Post-infectious encephalitis and encephalopathy

42.7.1 Encephalitis lethargica

This condition was first described by von Economo in 1916 and his name has also been given to the diseases, von Economo's encephalitis (von Economo 1931). It is a central nervous system disorder presenting with pharyngitis followed by sleep disorder, basal ganglia signs particularly of parkinsonism (Section 40.3.6) and neuropsychiatric sequelae. It probably first appeared in 1915 and during the following decade and a half widespread epidemics occurred.

Since the 1916–1927 epidemic, only sporadic cases have been described (Rail *et al.* 1981; Blunt *et al.* 1997). Pathological studies revealed an encephalitis of the midbrain and basal ganglia, with lymphocyte infiltration, predominantly of plasma cells. The encephalitis lethargica epidemic occurred during the same time period as the 1918 influenza pandemic, and the two outbreaks have been linked in the medical literature. However, von Economo and other contemporary scientists thought that the 1918 influenza virus was not the cause of encephalitis lethargica. No virus has ever been isolated, and recent examination of archived encephalitis lethargica brain material has failed to demonstrate influenza RNA, adding to the evidence that encephalitis lethargica was not an invasive influenzal encephalitis (Dale *et al.* 2004).

Recently reported 20 patients were described with an encephalitis lethargica-like syndrome, often following pharyngeal infections, one of whom died (Dale *et al.* 2004). Group A Streptococcus, which was the commonest cause of pharyngitis, is a recognized cause of immune-mediated basal ganglia dysfunction, the classical phenotype being Sydenham's chorea (Husby *et al.* 1976; Church *et al.* 2002) (Section 40.5.7).

Motor tics in combination with a behavioural disorder have been described after streptococcal infections and have been termed Paediatric Autoimmune Neuropsychiatric Disorders Associated with Streptococcal Infections, referred to by the acronym PANDAS (Swedo *et al.* 1998) (Section 40.10). Antibodies induced after group A streptococcus infection are thought to cross-react with components of the basal ganglia, resulting in movement and psychiatric disorders. Autoantibodies reactive against basal ganglia and subthalamic neurons have been found in both Sydenham's chorea (Church *et al.* 2002), and PANDAS (Kiessling *et al.* 1993). Dale *et al.* were able to demonstrate similar autoantibodies reactive against discrete basal ganglia autoantigens in 95 per cent of these patients; reactivity was seen against several common basal ganglia antigens (Dale *et al.* 2004).

The pathological features of both epidemic encephalitis lethargica and contemporary encephalitis lethargica have shown perivascular lymphocytic cuffing, which predominantly involves the midbrain and basal ganglia (Kiley *et al.* 2001). A recent post-mortem case of a patient with the encephalitis lethargica phenotype demonstrated an unexpected excess of perivascular plasma cells which were distended by IgG. The authors concluded that a brisk humoral response was occurring (Kiley *et al.* 2001).

Clinical features

During the original encephalitis lethargica epidemic the sexes were equally affected and no age group was exempt. Most cases occurred in early adult life, and during spring. The onset was acute, sometimes fulminant, with headache, malaise, myalgia, delirium, and convulsions. Less acute cases would develop a characteristic sleep disturbance with severe lethargy by day from which they could be roused, and insomnia by night. In a substantial and increasing proportion of cases the evolution of the condition would become chronic, and extrapyramidal manifestations would supervene,

including frank parkinsonism with tremor, rigidity, and oculogyric crises. Some would have chorea or myoclonus. In the acute phase, pupillary abnormalities and disturbances of ocular movement were common. Epileptic seizures and signs of meningism were not common. CSF revealed no diagnostic features.

In the recent series the central features of the disease were sleep disorder, lethargy, extrapyramidal movements consisting of parkinsonism and dyskinesias, and a neuropsychiatric disturbance involving obsessive-compulsive disorder, catatonia, mutism, apathy, or conduct disorders.

The CSF was abnormal in about 50 per cent of epidemic cases, mild elevation of protein and mild lymphocytosis being characteristic (von Economo 1931). In 70–80 per cent of recent patients CSF pleocytosis and elevated CSF protein, with oligoclonal bands were found (Dale *et al.* 2004).

Brian MRI was abnormal in 40 per cent of patients; characteristic features were increased signal in the basal ganglia, substantia nigra, and tegmentum (Dale *et al.* 2004). The enhancement resolved after the acute stage in the few patients who had convalescent imaging.

Death rates varied, reaching 38 per cent in one large series, most patients died within the first month. Complete recovery occurred in only 25 per cent. Mental sequelae were common and included dementia in 25 per cent, depression, bradyphrenia and bradykinesis and inability to concentrate. Varying degrees of parkinsonism were common, often associated with oculogyric crises, and lethargy. Relapses of the acute attack were recorded up to 20 years later. No treatment regime has been shown to be consistently beneficial although steroids have benefited two recent cases (Blunt *et al.* 1997).

42.7.2 Vaccine-related encephalopathies

For many years, it has been recognized that certain individuals have developed neurological disorders, usually of an encephalitic or myelopathic nature, following apparent recovery from acute infections, mainly viral. Further studies demonstrated that there was an immunological basis for the syndrome and the term 'post infectious encephalomyelitis' was applied. Clinically and pathologically identical syndromes were seen following immunization or vaccination and the terms 'post-vaccinial' or 'post-immunization encephalomyelitis' given to them. The pathological substrate common to all is acute disseminated encephalomyelitis (Section 37.4.1) in some degree and there is overlap in particularly acute cases with acute haemorrhagic leukoencephalopathy. Excellent clinical accounts are to be found in Miller and Stanton (1954) and Spillane and Wells (1964). For the purposes of this discussion, the term 'vaccine-associated encephalopathy' is used.

The original descriptions related to cases which developed following vaccination against smallpox with vaccinia virus. With the demise of smallpox in 1977 and subsequent discontinuation of vaccinia immunization the incidence of vaccine-associated encephalopathy has fallen from reported figures of between 1:63 to 1:30,000 to nil. Other immunizations which have been associated with vaccine-associated encephalopathy and acute disseminated encephalomyelitis are measles, rabies, diphtheria/tetanus toxoid, pertussis, influenza, and Japanese B encephalitis. One ingredient common to all of these has been the presence of brain tissue containing myelin or myelin derived proteins. As vaccines have become modified and purified to be free of such immunogens, the incidence of vaccine-associated encephalopathy has diminished. It is felt that vaccine-associated encephalopathy resulting in acute dissemninated encephalomyelitis is a T-lymphocyte mediated autoimmune response against myelin basic protein (Hemachudha *et al.* 1987; Johnson *et al.* 1984). It is not clear why some subjects develop vaccine-associated encephalopathy while the majority exposed to the same agents do not. This may relate to genetic control of the immune response.

Pathologically, the changes in brain are of acute swelling and venous engorgement of white matter. There is perivascular oedema and mononuclear cell infiltration. Demyelination with relative axonal sparing occurs and lipid-laden macrophages are evident. Clinical features are variable, presumably in proportion to the vigour of the immune response. Days or more usually within 2–3 weeks following immunization, an encephalitic illness ensues, characterized by headache, fever, confusion, obtundation, epileptic seizures, sometimes focal neurological signs, hemiparesis, hemisensory disturbance, movement disorder, ataxia, optic neuritis, and progression to coma. There may be evidence of myelopathy. Mortality rates as high as 40 per cent have been reported with around 10 per cent being the norm. Significant neurological sequelae are common.

Routine blood investigations are seldom helpful. The EEG is abnormal showing diffuse slow wave activity, sometimes with localization and seizure activity, but this is quite non-specific. Brain imaging with CT, or preferably MR demonstrate white matter changes, which may enhance, ranging from circumscribed lumps invading grey matter and sometimes occupying space, to coalescing, infiltrative regions of white matter disturbance with oedema. These lesions are often widespread and multiple (Kepes 1993; Kesselring *et al.* 1990). CSF is usually abnormal with rise in protein, mononuclear pleocytosis, and increase in immunoglobulin content and myelin basic protein levels. No specific therapy is available. Supportive treatment with maintenance of vital functions, administration of anti-epileptic drugs as necessary, reduction of brain oedema with dexamethasone, osmotic agents and hyperventilation, and the management of intercurrent infection, should be given. Corticosteroids are usually given as anti-inflammatory agents but there is no good evidence that they are efficacious. It is advisable to persist with aggressive therapy in apparently hopeless circumstances because, despite the high mortality and morbidity, in some the outcome may be surprisingly good. With modern acellular vaccines and vaccines free of neural proteins the risk of development of encephalopathy following immunization is in most cases virtually nil, and in all cases less than the risk of neurological complications developing from the native infection itself.

42.8 Prion diseases

42.8.1 Infectious diseases causing dementia

For most of the bacterial and viral diseases that can present with dementia (Table 42.13) their other neurological syndromes are more important, and so they are covered under different sections in this and the following chapter. However it is for the prion diseases that dementia is the main, if not only, important neurological presentation (Section 34.6.5).

42.8.2 The range of prion disorders

The prion diseases, or transmissible spongiform encephalopathies, are progressive neurodegenerative diseases with transmissible

Table 42.13 Infectious diseases causing dementia

| **Bacteria** |
| Tuberculosis (Section 41.5.3) |
| Syphilis (Section 42.5.1) |
| Lyme neuroborreliosis (Section 42.5.2) |
| Whipple's disease (Section 42.5.8) |
| **Viruses** |
| HIV (Section 43.3.5) |
| Subacute sclerosing panencephalitis (Section 42.3.7) |
| **Fungi** |
| Cryptococcus (Section 41.5.9) |
| Other fungi |
| **Parasites** |
| Trypanosomiasis (Section 42.6.2) |
| **Prions** |
| Creutzfeldt–Jakob disease (Section 42.9.6) |
| Others |

properties. They are rare disorders affecting approximately one person per million per year. The term 'prion' was coined anagrammatically by Prusiner in 1982 and referred to a *proteinaceous infectious* particle that he believed to be the agent causing scrapie (Prusiner 1982). Scrapie is the name given in Scotland to a fatal disease of sheep which has been recognized for hundreds of years. Similar diseases have now been found in other animals: mink, deer, elk, kudu, puma, cheetah, domestic cats and others, and in cattle, in the form of bovine spongiform encephalopathy (Table 42.14). These diseases all share similar pathological features of brain histology: spongiform change in the cerebral hemispheres, and they have been found to be transmissible by inoculation into the same or other species. Since the 1920s spongiform change has been recognized in the brains of people dying from a form of rapidly progressive dementia described separately by Creutzfeldt and Jakob, now known as Creutzfeldt–Jakob disease, in a familial form of dementia with ataxia now known as the Gerstmann–Straussler–Scheinker syndrome, and in a curious form of progressive dementia associated with movement disorder, in a particular region of Papua New Guinea and known locally as 'Kuru'. In addition to their pathological similarities, these conditions have all shared the property of transmissibility to primates. Two more recently described prion disorders include fatal familial insomnia and new variant Creutzfeldt–Jakob disease. This whole group of diseases is caused, or triggered by, a group of infectious agents, called prions. The changes these induce in susceptible hosts to produce disease, are being unravelled. All of the spongiform encephalopathies have abnormal forms of prion protein, PrP, as constituents of the amyloid plaques which are found in the brain.

42.8.3 Scrapie

Scrapie is a disease of sheep that has been known to shepherds for centuries in Britain and on mainland Europe. It has many names which vary with locality, and the Scottish name, scrapie, is now that most widely used. Scrapie refers to the scraping or scratching which the affected animal does against fence posts or trees, in order to obtain relief from the irritation which it experiences. Progressive neurological deterioration and death follow within weeks or months.

Table 42.14 Prion diseases of humans and animals (modified from Knight *et al.* 2004)

Human diseases	Notes
Kuru	Confined to Papua New Guinea; related to cannibalistic mourning rituals
Creutzfeldt–Jakob disease	The most common human prion disease. First described in 1921
Sporadic	
- Brownell–Oppenheimer variant	Cerebellar syndrome
- Heidenheim form	Dyspraxia, agnosia, cortical blindness
Genetic	
Iatrogenic	
Variant (described 1996)	
Gerstmann–Sträussler–Scheinker syndrome	A rare autosomal dominant hereditary disease
Fatal Familial Insomnia	A rare autosomal dominant hereditary disease
Animal diseases	
Scrapie	Naturally occurring disease of sheep and goats
Transmissible mink encephalopathy	A disease of farmed mink
Chronic wasting disease	Affects deer confined to North America
Bovine spongiform encephalopathy	A disease of cattle first reported 1987
Bovine spongiform encephalopathy related diseases	Transmission to cats (Feline spongiform encephalopathy) and other animals

42.8.4 Kuru

In the 1950s a disease, new to western medicine, was being described from a remote area of Papua New Guinea (Gajdusek *et al.* 1957). Women and children were predominantly affected and developed a progressive disease of swift evolution over a year on average, characterized by ataxia of gait, truncal ataxia, dysarthria, tremor and titubation, usually with retention of intellect. Cerebellar deterioration became increasingly evident with disruption of eye movements and increasing disability, the appearance of pyramidal and extra-pyramidal signs, and various forms of movement disorder, but apparently not myoclonus. Emotional lability, generalized muscle wasting, and paralysis supervene and lead rapidly to death. Epidemiological studies linked the disease to the practice of cannibalism and it was felt that the higher incidence of the disease in women and children reflected their greater exposure to brain and offal of infected carcasses. Since cannibalism has waned so has the occurrence of kuru, although new cases are still occurring which may be a function of a particularly long incubation period (Collinge *et al.* 1992).

Hadlow (1957) recognized that the spongiform nature of the pathological change in kuru brains had similarities to the changes seen with scrapie, and suggested that the disease may be transmissible. Kuru was transmitted to monkeys after intracerebral inoculation, and 2 years later, Creutzfeldt–Jakob disease was transmitted to monkeys using similar methods (Gibbs *et al.* 1968).

42.8.5 Bovine spongiform encephalopathy

In Britain in 1985 and 1986, isolated cases of cattle suffering from a progressive central nervous system degeneration were being reported. By 1988 this had become an epidemic. Examination of the brains of these animals revealed widespread spongiform change and the disease was called bovine spongiform encephalopathy. Both pathological and clinical similarities to scrapie were noted. Bovine spongiform encephalopathy animals developed apprehension, hypersensitivity to stimuli, postural abnormalities, ataxia, and this progressed rapidly to death. Bovine spongiform encephalopathy has been shown to be transmissible to a range of animals including sheep, goats, mice, pigs, marmosets, monkeys, and mink. Following extensive epidemiological studies, the likely source of the disease in cattle was identified as cross-species transfer of scrapie infectious particles to cattle, in feed derived from sheep. In the 1970s a large number of surplus sheep had been slaughtered and their carcasses were processed to animal protein feed and other by-products. There was no longer a market for tallow and similar substances, so the process by which the carcasses were 'rendered' or dissolved, was carried out at lower temperatures than before, and certain solvents were omitted. As a result, it is thought that scrapie particles remained viable, or reproducible, and were concentrated in the protein feed, and so infected the cattle. Bovine spongiform encephalopathy has been found in many countries around the world, but in all of these countries, it seems likely that the source of infection has been meat or bone meal which originated from infected material in Britain.

Parallel to, and a little later than, the evolution of the bovine spongiform encephalopathy saga, rare cases were appearing in Britain of a new variant of Creutzfeldt–Jakob disease which had a different neuropathological profile, younger distribution, and atypical clinical features (Will *et al.* 1996) (Section 42.8.9). Cases were also found in other countries which had bovine spongiform encephalopathy. It has been suggested that the appearance of this variant Creutzfeldt–Jakob disease was causally linked to the bovine spongiform encephalopathy epidemic by human ingestion of bovine spongiform encephalopathy contaminated meat, representing a further illustration of cross-species infection.

42.8.6 Creutzfeldt–Jakob disease

In 1920, Creutzfeldt described the case of a 22-year-old woman with progressive dementia and in 1921 Jakob described five cases of dementia with their pathological findings. Since then the name Creutzfelt–Jakob disease has been ascribed to the syndrome. Creutzfeldt's case has been shown not to have suffered from the disease which now bears his name, while at least two of the cases of Jakob did have the proper syndrome. The transmissibility of the condition has been well established, and its forms are human examples of prion diseases (Table 42.14).

Creutzfeldt–Jakob disease is a rare, rapidly progressive dementia which occurs worldwide with a frequency in the order of 0.5 to 1 case per million population. It affects both sexes and most racial groups. With the exception of infrequent familial clusters, its distribution appears to be random. About 15 per cent of cases are familial. The onset is insidious and the incubation period is unknown but certainly can run into many years. The age of onset is between 60 and 70 years for sporadic cases, whereas it is less than 50 years in familial cases. In iatrogenic cases, when the time of inoculation is known, the incubation period has been as short as 2 years. The average age of cases at onset is 60 years but the range of onset is from 14 to 83 years.

Clinical features

The classical triad is dementia, ataxia, and myoclonus. However the initial symptoms are non-specific and variable, and no different from other forms of dementia; these include forgetfulness, fatigue, cognitive disturbance, depression and personality disorder, behavioural upset and derangement of sleep, weight loss, and malaise. Creutzfeldt–Jakob disease shares all of these symptoms with other dementias in the early stages, but it is much more quickly progressive. About 70 per cent of all sporadic cases are dead within 6 months. Familial cases may extend from one to several years. As other areas of the brain become affected, further signs appear and include sensory loss, visual disturbance usually cortical blindness, ataxia, weakness, muscle wasting, spasticity, dysarthia, and epileptic seizures. Extrapyramidal signs of rigidity and bradykinesis and abnormal movements may appear. Myoclonus appears at any stage of the disease, and does so in most cases of some stage. It may be assymetrical at onset, asynchronous and susceptible to startle, appearing in one limb before spreading. The appearance of myoclonus is not diagnostic but it is highly suggestive of Creutzfeldt–Jakob disease. Correspondingly the EEG shows characteristic repetitive sharp waves (Fig. 42.12). With progression, primitive reflexes appear together with autonomic disintegration, sometimes with central apnoea, wasting increases and the patient becomes vegetative and dies. Some rather atypical forms have been described. In the amyotrophic form, there is evidence of anterior horn cell degeneration. Doubt has been expressed that such cases may not be true Creutzfeldt–Jakob disease but amyotrophic lateral sclerosis with dementia (Section 34.6.4). Perhaps as many as 10 per cent of cases run a clinical course longer than 2 years, and some present with a cerebellar syndrome, the Brownell–Oppenheimer variant. When the brunt falls upon the parietal and occipital cortex, dyspraxia, agnosia, and cortical blindness indicate the Heidenheim form of the disease. There are no clinical features to distinguish the hereditary from the sporadic forms of Creutzfeldt–Jakob disease; they appear to be inherited on an autosomal dominant basis.

Iatrogenic Creutzfeldt–Jakob disease

Iatrogenic forms of Creutzfeldt–Jakob disease are no different clinically. Accidental inoculation of prions to humans has taken place by surgical insertion of infected graft material such as dura mater or cornea, injection of growth hormone or gonadotropins harvested from human cadavers, and by the use of inadequately sterilized neurosurgical instruments. With inoculation near to the brain, the incubation period is shorter and the disease is similar to classical Creutzfeldt–Jakob disease. Following peripheral injection such as

growth hormone the incubation period is longer and the clinical syndrome is more of a progressive ataxic syndrome. There is some evidence that there may be a genetic susceptibility to iatrogenic Creutzfeldt–Jakob disease. There does not seem to be an increased risk of development of prion diseases by occupational exposure as there has not been a higher incidence of Creutzfeldt–Jakob disease in veterinary workers, shepherds, and herdsmen.

42.8.7 Gerstmann–Straussler–Scheinker disease

This is a very rare form of autosomal dominant inherited progressive cerebellar ataxia with dementia. It begins between the ages of 20 and 50, with a mean age of onset of 42. Increasing clumsiness, ataxia, dysarthria, and dementia are followed by spasticity, weakness, rigidity, and tremor, sometimes myoclonic jerks and rapid progression to death within 1–5 years. Familial forms of Creutzfeldt–Jakob disease and Gerstmann–Straussler–Scheinker disease share many clinical features, and it may not be possible to distinguish them on clinical or even on pathological grounds.

42.8.8 Fatal familial insomnia

This disease is charicterized by weeks to months of intractable insomnia followed by dysfunction of the autonomic nervous system, with abnormal sweating, lacrimation, respiratory and/or blood pressure dysregulation. Subsequently there is ataxia, and eventually dementia, with most patients dying within 1–2 years. Pathologically there is neuronal loss and astrocytic gliosis with little or no spongiform change, localized primarily to the thalamus and inferior olivary nucleus.

42.8.9 Variant Creutzfeldt–Jakob disease

In 1996, Will *et al.* described 10 cases of Creutzfeldt–Jakob disease which differed from previous cases on clinical and on neuropathological grounds. The evidence indicates that its occurrence is linked to the ingestion of beef or beef products which have been contaminated by bovine spongiform encephalitis. These cases differed from normal Creutzfeldt–Jakob disease by having a much younger age of onset, typically teenagers and young adults, by presenting with psychiatric problems with behavioural changes, and by developing ataxia and cerebellar dysfunction early in the course of the disease (Section 39.11.6). Painful sensory upset with persistent limb and facial dysaesthesiae are prominent. Dementia, myoclonus, and movement disorder occur late. The duration of disease is significantly longer than with classical Creutzfeldt–Jakob disease. Up until February 2006, variant Creutzfeldt–Jakob disease has been confirmed in 160 patients resident in the United Kingdom and 28 elsewhere, some of whom have never visited the United Kingdom (Collee *et al.* 2006). Also, 16 cases have been reported in France, and 3 in Ireland.

42.8.10 Neuropathology

The changes in Creutzfeldt–Jakob disease are widespread throughout the neuraxis, with neuronal loss in all layers of the cortex, astrocytic proliferation, and marked spongiform changes particularly in the deeper cortical layers (Fig. 42.13a). There is little demyelination, and the spongiform change precedes neuronal loss and is due to the appearance of vacuoles within the cytoplasm of the neurophil. There is virtually no inflammatory change. Microglial proliferation with hyperplasia and hypertrophy, in relation to

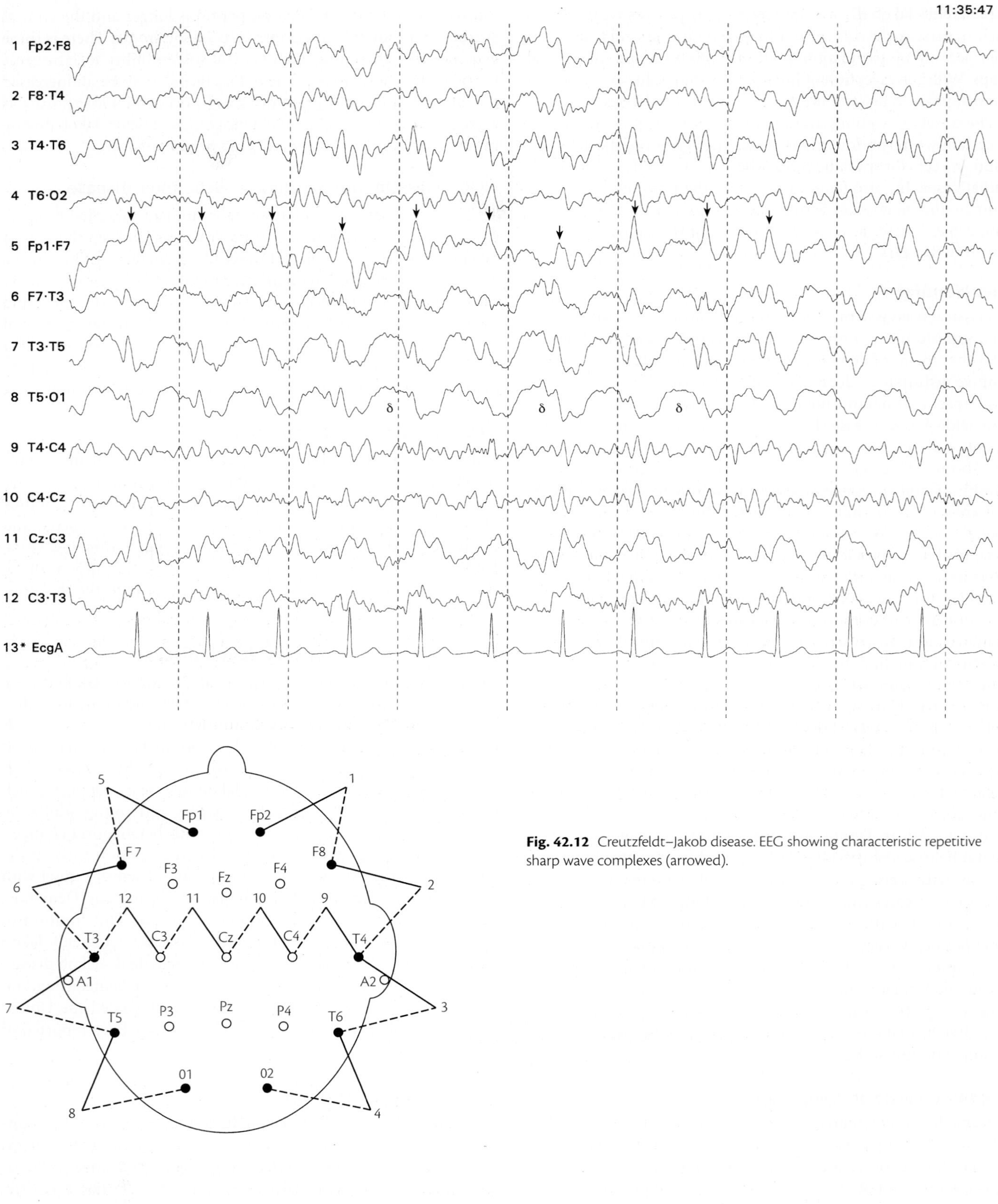

11:35:47

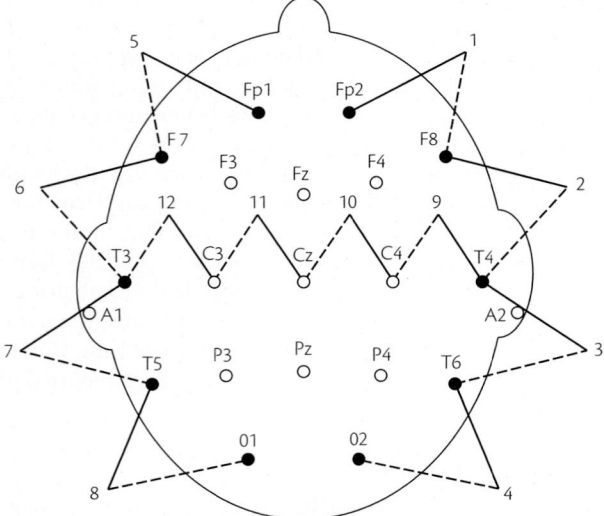

Fig. 42.12 Creutzfeldt–Jakob disease. EEG showing characteristic repetitive sharp wave complexes (arrowed).

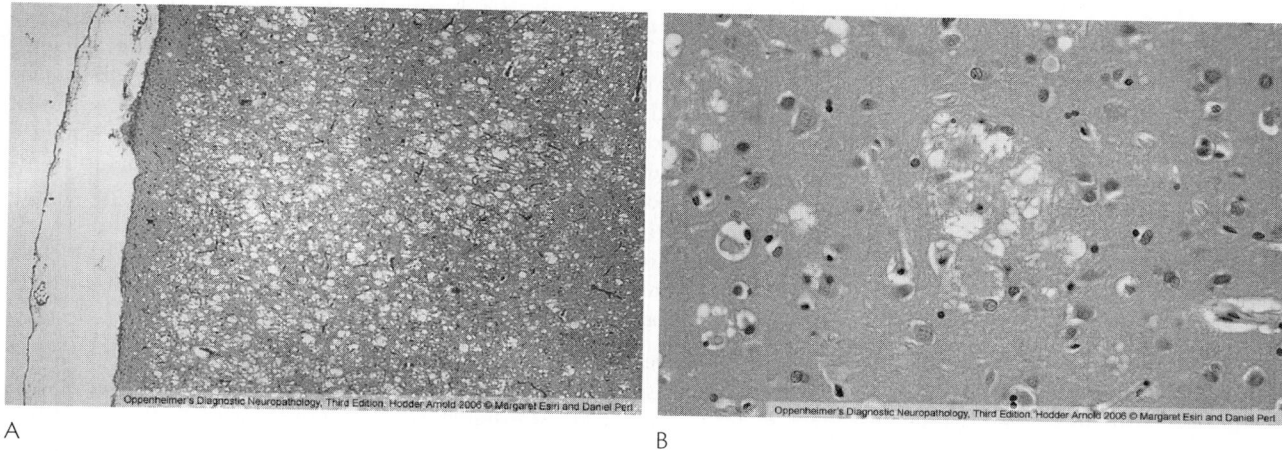

Fig. 42.13 Neuropathology of sporadic Creutzfeldt–Jakob disease. (A) Low power view showing cerebral cortical spongiform change. (B) Higher power view showing a florid amyloid plaque. (Reproduced from Esiri and Perl 2006.) (See Plate 37.)

amyloid plaques, is seen. Amyloid plaques are found within the extracellular space, particularly in the cerebellar cortex (Masters *et al.* 1978). There is a wide variation in the distribution and severity of the pathological lesions and it is recognized that spongiform change is not an exclusive feature of prion diseases (Bell *et al.* 1993). Amyloid plaques are seen in some cases of sporadic Creutzfeldt–Jakob disease, and with increasing concentration in kuru, inherited Creutzfeldt–Jakob disease, and Gerstmann–Straussler–Scheinker disease (Fig. 42.13b).

The pathological changes seen in variant Creutzfeldt–Jakob disease are rather different. There is spongiform change in the cerebral cortex and in the basal ganglia and thalamus but the striking change is of widespread plaque formation in cerebrum and cerebellum, the plaques closely resembling those seen in kuru. The amyloid plaques which are seen in the transmissible encephalopathies contain abnormal forms of prion protein.

42.8.11 The prion protein

Prion protein, PrP, is a normal protein found in most cell types and in the brain. The normal cellular form of the protein, designated PrP^C, is a glycoprotein with unknown function. Prion diseases result from neuronal accumulation of a misfolded isoform of the protein, designated PrP^{Sc} after scrapie disease of sheep. Once PrP^{Sc} is present in the brain, it replicates by an autocatalytic mechanism whereby PrP^{Sc} binds to the endogenous PrP^C and converts it to PrP^{Sc}. The coding gene for PrP is located on the short arm of chromosome 20, and the longer of its 2 exons contains the entire transcribed 253 codon region of the gene.

Various different codon mutations have been described for familial Creutzfeldt–Jakob disease, Gerstmann–Straussler–Scheinker disease, and fatal familial insomnia, and it seems likely that variations in the genotype give rise to abnormal forms of PrP. Most cases of sporadic Creutzfeldt–Jakob disease are homozygous for a common PrP protein polymorphism. Approximately 90 per cent of Creutzfeldt–Jakob disease patients are homozygous for either methionone or valine at codon 129, compared with 50 per cent of non-affected individuals. Interestingly nearly 100 per cent

of cases of variant Creutzfeldt–Jakob disease are homozygous for methionine at this position. Cases of variant Creutzfeldt–Jakob disease have shown homozygosity for methionine at codon 129. These abnormal forms of PrP then accumulate in the brain in amyloid plaques, and are in some way responsible for the different phenotypes of Creutzfeldt–Jakob disease, so that one insertion or mutation may give rise to the extrapyramidal from of the disease, and a different mutation will produce another clinical form. It is likely that these abnormal forms of PrP represent the major component of the infectious agent (Collinge and Palmer 1997).

42.8.12 Diagnosis and management

Differential diagnosis

Diagnosis of prion disease can be difficult because of its relative rarity, and the diverse clinical manifestations. When the presentation is gradual, which is especially true of the familial prion diseases, Creutzfeldt–Jakob disease must be distinguished from Alzheimer's disease (Section 34.6.2), Lewy body dementia (Sections 34.6.3; 40.3.2), frontotemporal dementia (Section 34.6.4), Huntington's disease (Section 40.5.2), spinocerebellar ataxia (Section 39.9), and progressive supranuclear palsy (Section 40.3.9), among other degenerative conditions. When it presents rapidly, sporadic Creutzfeldt–Jakob disease may be confused with subacute encephalopathies, caused by metabolic, toxic, infectious, or inflammatory conditions; these include viral, bacterial, and parasitic diseases, central nervous system vasculitis (Section 36.2), heavy metal poisoning especially bismuth ingestion, autoimmune diseases such as Hashimoto's encephalopathy (Section 38.3.1), and paraneoplastic syndromes (Section 38.4.2). On the other hand, the presentation can sometimes be very characteristic. A rapidly progressive dementia in which there is myoclonus is suggestive of sporadic Creutzfeldt–Jakob disease. The onset of insomnia in mid-life followed by autonomic disturbances and ataxia is suggestive of fatal insomnia. If ataxia and dysarthria are prominent, and dementia is a late feature, as in Gerstmann–Straussler–Scheinker syndrome and in 15–20 per cent of cases of sporadic Creutzfeldt–Jakob disease, then the diagnosis is likely to be achieved move slowly.

Diagnostic criteria

A definite diagnosis of Creutzfeldt–Jakob disease requires pathological confirmation. According to the World Health Organization criteria, a case is probable if there is progressive dementia with no alternative diagnosis, there are at least two of the following four clinical features: myoclonus, visual or cerebellar disturbance, pyramidal or extrapyramidal dysfunction, and akinetic mutism; and there is a typical EEG, or a positive CSF 14-3-3 assay, applicable to patients in whom the duration to death is less than 2 years (Knight *et al.* 2004).

Important features to establish in the diagnostic work up of a patient with suspected prion disease is whether there is any family history, and whether there could have been iatrogenic exposure, such as use of growth hormone gonadotrophin, dura mater grafts, corneal transplantation, or use of reusable intracranial electrodes. Recently probable transmission of variant Creutzfeldt–Jakob disease in blood products has been demonstrated (Hewitt *et al.* 2006).

Investigation

Investigations which are useful to diagnose Creutzfeldt–Jakob disease and other prion diseases are few. It is customary to carry out blood tests to exclude other forms of dementia; these should include vitamin B12, thyroid function tests, syphilis serology, and tests for anti-thyroid antibodies to exclude Hashimoto's encephalopathy. In addition paraneoplastic antibodies for limbic encephalitis, anti-Ri antibodies, and for a cerebellar syndrome anti-Yo antibodies should be considered. Investigations should be undertaken for possible vasculitis, including ESR, C-reactive protein and autoantibodies, especially in patients under 55 years. A heavy metal screen, copper and caeruloplasmin levels to exclude Wilson's disease in younger patients and investigations for Whipple's disease may be indicated.

MRI of the brain should be carried out to exclude other pathology, and to look for characteristic features of Creutzfeldt–Jakob disease. These include high signal intensities in the putamen and caudate nuclei on T2- and proton dense weighted images in many causes of sporadic and familial Creutzfeldt–Jakob disease. In variant Creutzfeldt–Jakob disease there may be characteristic hyperintensity of the pulvinar nucleus of the thalamus, the 'pulvinar' sign.

The EEG becomes abnormal in all cases later in the evolution of the disease by which time completely non-specific diffuse slow wave activity is seen. In 70 per cent of patients with sporadic Creutzfeldt–Jakob disease, rhythmic and periodic bursts of high-amplitude bi- and tri-phasic periodic sharp wave complexes may appear which may be synchronous with myoclonus (Fig. 42.12). If these are not seen initially, the EEG should be repeated weekly, if clinical suspicion is high. Such EEG changes are less common in familial Creutzfeldt–Jakob disease, and not seen in variant Creutzfeldt–Jakob disease, Gerstmann–Straussler–Scheinker disease, or fatal familial asomnia. These EEG changes are characteristic, but not pathognomonic (Chiofalo *et al.* 1980). They may also be seen in hepatic encephalopathy (Section 33.2.2), drug toxicity especially with Lithium (Section 5.5.3), and rarely in Alzheimer's disease (Section 34.6.2).

CSF examination is important to rule out other treatable causes. There may be a minor elevation of total protein. The 14-3-3 protein may be detectable, although this can also be positive in stroke, malignancy, and other encephalopathies; in the context of a patient with dementia it can be useful (Zerr *et al.* 1996). CSF neuron-specific

enolase, and the astroglial protein S100 are also considered useful by some centres. Histological examination of brain tissue obtained by surgical biopsy is sometimes helpful, though the benefit must be balanced against the dangers of transmission of prions. Most recently, demonstration of abnormal PrP in tonsil tissue obtained during life by biopsy with suitable precautions appears to be the most accurate method available (Hill *et al.* 1999). In practice, in Britain, if a case of prion disease is suspected clinically, contact should be made with the Creutzfeldt–Jakob Disease Surveillance Unit in Edinburgh, where advice is freely available on current diagnostic methodology.

Treatment

No treatment has been demonstrated to be effective to limit the progression of prion diseases. They are universally fatal. Currently quinacrine is undergoing clinical trials, and pentosan polysulfate has been given by intraventricular infusion to some patients.

Symptomatic treatment is given for seizures, spasticity, and myoclonus, and is generally ineffective. Because of the infectious nature of prion diseases, staff treating such patients must take strict precautions, and all secretions from the patients must be treated as potentially infectious. Special precautions are necessary because of the resistance of prions to standard disinfecting agents. Guidelines exist for the disposal of specimens, and surgical instruments, and for necropsy. People suspected of suffering from prion diseases should not donate blood and they are not suitable as donors for organ transplantation. If a hereditary form of prion disease is confirmed, genetic counselling should be offered.

References

Aarli JA (1974). Nervous complications of measles. Clinical manifestations and prognosis. *Eur Neurol*, **12**, 79–93.

Adal KA, Cockerell CJ, Petri WA Jr. (1994). Cat scratch disease, bacillary angiomatosis, and other infections due to Rochalimaea. *N Engl J Med*, **330**, 1509–15.

Agrawal AG, Petersen LR (2003). Human immunoglobulin as a treatment for west nile virus infection. *J Infect Dis*, **188**, 1–4.

Aksamit AJ (2001). Treatment of non-AIDS progressive multifocal leukoencephalopathy with cytosine arabinoside. *J Neurovirol*, **7**, 386–90.

al Deeb SM, Yaqub BA, Sharif HS et al. (1989). Neurobrucellosis: clinical characteristics, diagnosis, and outcome. *Neurology*, **39**, 498–501.

Alcindor F, Valderrama R, Canavaggio M et al. (1992). Imaging of human T-lymphotropic virus type I-associated chronic progressive myeloneuropathies. *Neuroradiology*, **35**, 69–74.

Anlar B, Yalaz K, Öktem F et al. (1997). Long-term follow-up of patients with subacute sclerosing panencephalitis treated with intraventricular alpha-interferon. *Neurology*, **48**, 526–8.

Annegers JF, Hauser WA, Beghi E et al. (1988). The risk of unprovoked seizures after encephalitis and meningitis. *Neurology*, **38**, 1407–10.

Anonymous (1979). Subacute sclerosing panencephalitis. *Br Med J*, **2**, 1096.

Anonymous (1988). Arboviral infections of the central nervous system–United States, 1987. *MMWR Morb Mortal Wkly Rep*, **37**, 506–8, 13–5.

Anonymous (2003). Progress toward global eradicatio of poliomyelitis, 2002. *MMWR Morb Mortal Wkly Rep*, **52**, 366–9.

Arribas JR, Storch GA, Clifford DB et al. (1996). Cytomegalovirus encephalitis. *Ann Intern Med*, **125**, 577–87.

Astrom KE, Mancall EL, Richardson EP Jr. (1958). Progressive multifocal leuko-encephalopathy; a hitherto unrecognized complication of chronic lymphatic leukaemia and Hodgkin's disease. *Brain*, **81**, 93–111.

Aurelius E, Johansson B, Skoldenberg B et al. (1993). Encephalitis in immunocompetent patients due to herpes simplex virus type 1 or 2 as determined by type-specific polymerase chain reaction and antibody assays of cerebrospinal fluid. *J Med Virol*, **39**, 179–86.

Bahemuka M, Shemena AR, Panayiotopoulos CP *et al.* (1988). Neurological syndromes of brucellosis. *J Neurol Neurosurg Psychiatry*, **51**, 1017–21.

Barah F, Vallely PJ, Chiswick ML, *et al.* (2001). Association of human parvovirus B19 infection with acute meningoencephalitis. *Lancet*, **358**, 729–30.

Baker AB, Matzke HA, Brown JR (1950). Poliomyelitis III. Bulbar poliomyelitis; a study of medullary function. *Archives of Neurology*, **1**, 257–81.

Barrett, AD (2001). Current status of flavivirus vaccines. *Ann N Y Acad Sci*, **951**, 262–71.

Beckwith WH, Sirpenski S, French RA *et al.* (2002). Isolation of eastern equine encephalitis virus and West Nile virus from crows during increased arbovirus surveillance in Connecticut, 2000. *Am J Trop Med Hyg*, **66**, 422–6.

Beeching NJ, Crowcroft NS (2005). Tetanus in injecting drug users. *BMJ*, **330**, 208–9.

Beghi E, Nicolosi A, Kurland LT *et al.* (1984). Encephalitis and aseptic meningitis, Olmsted County, Minnesota, 1950-1981: I. Epidemiology. *Ann Neurol*, **16**, 283–94.

Behan PO, Feldman RG, Segerra JM *et al.* (1986). Neurological aspects of mycoplasmal infection. *Acta Neurol Scand*, **74**, 314–22.

Bell JE, Ironside JW (1993). Neuropathology of spongiform encephalopathies in humans. *Br Med Bull*, **49**, 738–77.

Bennet NM (1976). Murray Valley encephalitis, 1974. Clinical features. *Med J Aust*, **2**, 446–54.

Bentley SD, Maiwald M, Murphy LD *et al.* (2003). Sequencing and analysis of the genome of the Whipple's disease bacterium Tropheryma whipplei. *Lancet*, **361**, 637–44.

Berge TO, Banks IS, Tigertt WD (1961). Attenuation of Venezuelan equine encephalomyelitis virus by in vitro cultivation in guinea-pig heart cells. *Am J Hygiene*, **73**, 209–18.

Berger JR (1991). Neurosyphilis in human immunodeficiency virus type 1-seropositive individuals. A prospective study. *Arch Neurol*, **48**, 700–2.

Berger JR, Kaszovitz B, Post MJ *et al.* (1987). Progressive multifocal leukoencephalopathy associated with human immunodeficiency virus infection. A review of the literature with a report of sixteen cases. *Ann Intern Med*, **107**, 78–87.

Berkley JA, Mwangi I, Mellington F *et al.* (1999). Cerebral malaria versus bacterial meningitis in children with impaired consciousness. *QJM*, **93**, 151–7.

Bethell DB, Gamble J, Pham PL *et al.* (2001). Noninvasive measurement of microvascular leakage in patients with dengue hemorrhagic fever. *Clin Infect Dis*, **32**, 243–53.

Bharti AR, Nally JE, Ricaldi JN *et al.* (2003). Leptospirosis: a zoonotic disease of global importance. *Lancet Infect Dis*, **3**, 757–71.

Bia FJ, Thornton GF, Main AJ *et al.* (1980). Western equine encephalitis mimicking herpes simplex encephalitis. *JAMA*, **244**, 367–9.

Bigler WJ, Lassing EB, Buff EE *et al.* (1976). Endemic eastern equine encephalomyelitis in Florida: a twenty-year analysis, 1955-1974. *Am J Trop Med Hyg*, **25**, 884–90.

Bisno AL (1991). Group A streptococcal infections and acute rheumatic fever. *N Engl J Med*, **325**, 783–93.

Bista MB, Banerjee MK, Shin SH *et al.* (2001). Efficacy of single-dose SA 14-14-2 vaccine against Japanese encephalitis: a case control study. *Lancet*, **358**, 791–5.

Blunt SB, Lane RJ, Turjanski N *et al.* (1997). Clinical features and management of two cases of encephalitis lethargica. *Mov Disord*, **12**, 354–9.

Bodenmann P, Genton B (2006). Chikungunya: an epidemic in real time. *Lancet*, **368**, 258.

Bodian D (1949). Histopathologic basis of clinical findings in poliomyelitis. *Am J Med*, **6**, 563–78.

Bolton CF, Young GB, Zochodne DW (1993). The neurological complications of sepsis. *Ann Neurol*, **33**, 94–100.

Boos J, Esiri MM (2003). *Viral Encephalitis in Humans*. ASM Press, Washington DC.

Bosanko CM, Gilroy J, Wang AM *et al.* (2003). West Nile virus encephalitis involving the substantia nigra: neuroimaging and pathologic findings with literature review. *Arch Neurol*, **60**, 1448–52.

Bouteille M, Fontaine C, Vedrenne C *et al.* (1965). Sur un cas d'encephalite subaigue a inclusions. Etude anotomo-clinique et ultrastructurale. *Rev Neurol*, **113**, 454–8.

Bowen GS, Calisher CH (1976). Virological and serological studies of Venezuelan equine encephalomyelitis in humans. *J Clin Microbiol*, **4**, 22–7.

Bowen GS, Fashinell TR, Dean PB *et al.* (1976). Clinical aspects of human Venezuelan equine encephalitis in Texas. *Bull Pan Am Health Organ*, **10**, 46–57.

Breinl A (1918). Clinical pathological and experimental observations on the mysterious disease, a clinically aberrant form of poliomyelitis. *Med J Aust*, **1**, 209–29.

Brinker KR, Monath TP (1980). The acute disease. In: Monath TP, ed. *St Louis Encephalitis*, pp. 503–34. American Public Health Associates, Washington DC.

Brown H, Turner G, Rogerson S *et al.* (1999). Cytokine expression in the brain in human cerebral malaria. *J Infect Dis*, **180**, 1742–6.

Brust JC, Dickinson PC, Hughes JE *et al.* (1990). The diagnosis and treatment of cerebral mycotic aneurysms. *Ann Neurol*, **27**, 238–46.

Bucher B, Poupard JA, Vernant JC *et al.* (1990). Tropical neuromyelopathies and retroviruses: a review. *Rev Infect Dis*, **12**, 890–9.

Burke DS, Lorsomrudee W, Leake CJ *et al.* (1985). Fatal outcome in Japanese encephalitis. *Am J Trop Med Hyg*, **34**, 1203–10.

Burke DS, Nisalak A, Ussery MA *et al.* (1985). Kinetics of IgM and IgG responses to Japanese encephalitis virus in human serum and cerebrospinal fluid. *J Infect Dis*, **151**, 1093–9.

Burrow JN, Whelan PI, Kilburn CJ *et al.* (1998). Australian encephalitis in the Northern Territory: clinical and epidemiological features, 1987-1996. *Aust N Z J Med*, **28**, 590–6.

Cam BV, Fonsmark L, Hue NB *et al.* (2001). Prospective case-control study of encephalopathy in children with dengue hemorrhagic fever. *Am J Trop Med Hyg*, **65**, 848–51.

Cardoso F, Vargas AP, Oliveira LD *et al.* (1999). Persistent Sydenham's chorea. *Mov Disord*, **14**, 805–7.

Carpenter TC (2002). Corticosteroids in the treatment of severe mycoplasma encephalitis in children. *Crit Care Med*, **30**, 925–7.

Carter RF (1968). Primary amoebic meningo-encephalitis: clinical, pathological and epidemiological features of six fatal cases. *J Pathol Bacteriol*, **96**, 1–25.

Cavanagh JB, Greenbaum D, Marshall AH *et al.* (1959). Cerebral demyelination associated with disorders of the reticuloendothelial system. *Lancet*, **2**, 524–9.

Centers for Disease Control (2000). Human rabies--California, Georgia, Minnesota, New York, and Wisconsin, 2000. *MMWR Morb Mortal Wkly Rep*, **49**, 1111–5.

Cerna F, Mehrad B, Luby JP *et al.* (1999). St. Louis encephalitis and the substantia nigra: MR imaging evaluation. *AJNR Am J Neuroradiol*, **20**, 1281–3.

Chadwick D, Martin S, Buxton PH *et al.* (1982). Measles virus and subacute neurological disorders: an unusual presentation of measles inclusion body encephalitis. *J Neurol Neurosurg Psychiatry*, 680–4.

Chan PK, Chow PC, Peiris JS *et al.* (2000). Use of oral valaciclovir in a 12-year-old boy with herpes simplex encephalitis. *Hong Kong Med J*, **6**, 119–21.

Chataway J, Davies NW, Farmer S *et al.* (2004). Herpes simplex encephalitis: an audit of the use of laboratory diagnostic tests. *QJM*, **97**, 325–30.

Chaturvedi UC, Mathur A, Chandra A *et al.* (1980). Transplacental infection with Japanese encephalitis virus. *J Infect Dis*, **141**, 712–15.

Chiba N, Osada M, Komoro K *et al.* (1999). Protection against tick-borne encephalitis virus isolated in Japan by active and passive immunization. *Vaccine*, **17**, 1532–9.

Chimelli L, Hahn MD, Netto MB *et al.* (1990). Dengue: neuropathological findings in 5 fatal cases from Brazil. *Clin Neuropathol*, **9**, 157–62.

Chiofalo N, Fuentes A, Galvez S (1980). Serial EEG findings in 27 cases of Creutzfeldt-Jakob disease. *Arch Neurol*, **37**, 143–5.

Chokephaibulkit K, Kankirawatana P, Apintanapong S *et al.* (2001). Viral etiologies of encephalitis in Thai children. *Pediatr Infect Dis J*, **20**, 216–8.

Chong HT, Kunjapan SR, Thayaparan T *et al.* (2002). Nipah encephalitis outbreak in Malaysia, clinical features in patients from Seremban. *Can J Neurol Sci*, **29**:83–7.

Chua KB, Bellini WJ, Rota PA *et al.* (2000). Nipah virus: A recently emergent deadly paramyxovirus. *Science*, **288**, 1432–5.

Chua KB, Goh KJ, Wong KT *et al.* (1999). Fatal encephalitis due to Nipah virus among pig-farmers in Malaysia. *Lancet*, **354**, 1257–9.

Chua KB, Koh CL, Hooi PS *et al.* (2002). Isolation of Nipah virus from Malaysian Island flying-foxes. *Microbes Infect*, **4**, 145–51.

Church AJ, Cardoso F, Dale RC *et al.* (2002). Anti-basal ganglia antibodies in acute and persistent Sydenham's chorea. *Neurology*, **59**, 227–31.

Cinque P, Cleator GM, Weber T *et al.* (1996). The role of laboratory investigation in the diagnosis and management of patients with suspected herpes simplex encephalitis: a consensus report. The EU concerted action on virus meningitis and encephalitis. *J Neurol Neurosurg Psychiatry*, **61**, 339–45.

Cinque P, Koralnik IJ, Clifford DB (2003). The evolving face of human immunodeficiency virus-related progressive multifocal leukoencephalopathy: defining a consensus terminology. *J Neurovirol*, **9 (Suppl 1)**, 88–92.

Clarke DH (1961). Two nonfatal human infections with the virus of eastern equine encephalitis. *Am J Trop Med Hyg*, **10**, 67–70.

Clifford DB, Arribas JR, Storch GA *et al.* (1996). Magnetic resonance brain imaging lacks sensitivity for AIDS associated cytomegalovirus encephalitis. *J Neurovirol*, **2**, 397–403.

Cohen BA (1996). Prognosis and response to therapy of cytomegalovirus encephalitis and meningomyelitis in AIDS. *Neurology*, **46**, 444–50.

Cole FE Jr. (1971). Inactivated eastern equine encephalomyelitis vaccine propagated in rolling-bottle cultures of chick embryo cells. *Appl Microbiol*, **22**, 842–5.

Collee JG, Bradley R, Liberski PP (2006). Variant CJD (vCJD) and bovine spongiform encephalopathy (BSE): 10 and 20 years on: part 2. *Folia Neuropathol*, **44**, 102–10.

Collinge J, Palmer MS (1992). Prion diseases. *Curr Opin Genet Dev*, **2**, 448–54.

Conlon CP, Berendt AR, Dawson K *et al.* (1990). Runway malaria. *Lancet*, **335**, 472–3.

Copps SC, Giddings LE (1959). Transplacental transmission of western equine encephalitis. *Pediatrics*, **24**, 31–3.

Crosse BA (1990). Psittacosis: a clinical review. *J Infect*, **21**, 251–9.

Cruickshank EK (1956). A neuropathic syndrome of uncertain origin; review of 100 cases. *West Indian Med J*, **5**, 147–58.

Dale RC, Church AJ, Surtees RA *et al.* (2004). Encephalitis lethargica syndrome: 20 new cases and evidence of basal ganglia autoimmunity. *Brain*, **127**, 21–33.

Dastur FD, Shahani MT, Dastoor DH *et al.* (1977). Cephalic tetanus: demonstration of a dual lesion. *J Neurol Neurosurg Psychiatry*, **40**, 782–6.

Davison KL, Crowcroft NS, Ramsay ME *et al.* (2003). Viral encephalitis in England, 1989-1998: what did we miss? *Emerg Infect Dis*, **9**, 234–40.

Daxboeck F (2006). Mycoplasma pneumoniae central nervous system infections. *Curr Opin Neurol*, **19**, 374–8.

De La Blanchardiere A, Rozenberg F, Caumes E *et al.* (2000). Neurological complications of varicella-zoster virus infection in adults with human immunodeficiency virus infection. *Scand J Infect Dis*, **32**, 263–9.

de Silva SM, Mark AS, Gilden DH *et al.* (1996). Zoster myelitis: improvement with antiviral therapy in two cases. *Neurology*, **47**, 929–31.

DeBiasi RL, Tyler KL (1999). Polymerase chain reaction in the diagnosis and management of central nervous system infections. *Arch Neurol*, **56**, 1215–9.

Dennett C, Klapper PE, Cleator GM *et al.* (1991). CSF pretreatment and the diagnosis of herpes encephalitis using the polymerase chain reaction. *J Virol Methods*, **34**, 101–4.

Department of Health (2004). Rabies - Draft Updated Chapter 27 of Immunisation Against Infectious Disease 1996 - The Green Book (accessed 15 August 2005 via http://www.dh.gov.uk/assetRoot/04/11/09/70/04110970.pdf), Department of Health.

Deresiewicz RL, Thaler SJ, Hsu L *et al.* (1997). Clinical and neuroradiographic manifestations of eastern equine encephalitis. *N Engl J Med*, **336**, 1867–74.

Desmond RA, Accortt NA, Talley L *et al.* (2006). Enteroviral meningitis: natural history and outcome of pleconaril therapy. *Antimicrob Agents Chemother*, **50**, 2409–14.

deSouza CE, Karnad DR, Tilve GH (1992). Clinical and bacteriological profile of the ear in otogenic tetanus: a case control study. *J Laryngol Otol*, **106**, 1051–4.

Dlugos DJ, Liu GT (2001). Subacute sclerosing panencephalitis in an American-born adult. *Clin Infect Dis*, **32**, 173–4.

Drew WL (1988). Cytomegalovirus infection in patients with AIDS. *J Infect Dis*, **158**, 449–56.

Du Pasquier RA, Koralnik IJ (2003). Inflammatory reaction in progressive multifocal leukoencephalopathy: harmful or beneficial? *J Neurovirol*, **9 (Suppl 1)**, 25–31.

Dumler JS, Bakken JS (1995). Ehrlichial diseases of humans: emerging tick-borne infections. *Clin Infect Dis*, **20**, 1102–10.

Durack DT, Lukes AS, Bright DK (1994). New criteria for diagnosis of infective endocarditis: utilization of specific echocardiographic findings. Duke Endocarditis Service. *Am J Med*, **96**, 200–9.

Durand DV, Lecomte C, Cathebras P *et al.* (1997). Whipple disease. Clinical review of 52 cases. The SNFMI Research Group on Whipple Disease. Societe Nationale Francaise de Medecine Interne. *Medicine (Baltimore)*, **76**, 170–84.

Dutta JK, Dutta TK, Das AK (1992). Human rabies: modes of transmission. *J Assoc Physicians India*, **40**, 322–4.

Dworkin RH, Johnson RW, Breuer J *et al.* (2007). Recommendations for the management of herpes zoster. *Clin Infect Dis*, **44 (Suppl 1)**, S1–26.

Edwards GA, Domm BM (1960). Human leptospirosis. *Medicine (Baltimore)*, **39**, 117–56.

Eidelman LA, Putterman D, Putterman C *et al.* (1996). The spectrum of septic encephalopathy. Definitions, etiologies, and mortalities. *JAMA*, **275**, 470–3.

Elbers JM, Bitnun A, Richardson SE *et al.* (2007). A 12-year prospective study of childhood herpes simplex encephalitis: is there a broader spectrum of disease? *Pediatrics*, **119**, e399–407.

Elliman D, Bedford H (2001). MMR vaccine: the continuing saga. *BMJ*, **322**, 183–4.

Elsner C, Dorries K (1992). Evidence of human polyomavirus BK and JC infection in normal brain tissue. *Virology*, **191**, 72–80.

English PM, Lang N, Raleigh A *et al.* (2006). Measles outbreak in Surrey. *BMJ*, **333**, 1021–2.

Esiri MM (1982). Herpes simplex encephalitis. An immunohistological study of the distribution of viral antigen within the brain. *J Neurol Sci*, **54**, 209–26.

Esiri MM, Perl D (2006). *Oppenheimer's Diagnostic Neuropathology*, 3rd edition. Arnold, London.

Familusi JB, Moore DL, Fomufod AK *et al.* (1972). Virus isolates from children with febrile convulsions in Nigeria. A correlation study of clinical and laboratory observations. *Clin Pediatr*, **11**, 272–6.

Fleming JL, Wiesner RH, Shorter RG (1988). Whipple's disease: clinical, biochemical, and histopathologic features and assessment of treatment in 29 patients. *Mayo Clin Proc*, **63**, 539–51.

Fodor PA, Levin MJ, Weinberg A *et al*. (1998). Atypical herpes simplex virus encephalitis diagnosed by PCR amplification of viral DNA from CSF. *Neurology*, **51**, 554–9.

Fooks AR, Johnson N, Brookes SM *et al*. (2003). Risk factors associated with travel to rabies endemic countries. *J Appl Microbiol*, **94 (Suppl)**, 31S–6S.

Fooks AR, McElhinney LM, Pounder DJ *et al*. (2003). Case report: isolation of a European bat lyssavirus type 2a from a fatal human case of rabies encephalitis. *J Med Virol*, **71**, 281–9.

Ford FR (1928). The nervous complications of measles. *Bull Johns Hopkins Hosp*, **43**, 140–55.

Fouch B, Coventry S (2007). A case of fatal disseminated Bartonella henselae infection (cat-scratch disease) with encephalitis. *Arch Pathol Lab Med*, **131**, 1591–4.

French EL (1952). Murray valley encephalitis: isolation and charicterisation of the aetiological agent. *Med J Aust*, **1**, 100–3.

Fu ZF, Jackson AC (2005). Neuronal dysfunction and death in rabies virus infection. *J Neurovirol*, **11**, 101–6.

Gadoth N, Weitzman S, Lehmann EE (1979). Acute anterior myelitis complicating West Nile fever. *Arch Neurol*, **36**, 172–3.

Gajdusek DC, Zigas V (1957). Degenerative disease of the central nervous system in New Guinea; the endemic occurrence of kuru in the native population. *N Engl J Med*, **257**, 974–8.

Garcia-Monco JC, Benach JL (1995). Lyme neuroborreliosis. *Ann Neurol*, **37**, 691–702.

Gascon GG (1996). Subacute sclerosing panencephalitis. *Semin Pediatr Neurol*, **3**, 260–9.

George A, Prasad SR, Rao JA *et al*. (1987). Isolation of Japanese encephalitis and West Nile viruses from fatal cases of encephalitis in Kolar district of Karnataka. *Indian J Med Res*, **86**, 131–4.

George R, Liam CK, Chua CT *et al*. (1988). Unusual clinical manifestations of dengue virus infection. *Southeast Asian J Trop Med Public Health*, **19**, 585–90.

Gessain A, Barin F, Vernant JC *et al*. (1985). Antibodies to human T-lymphotropic virus type-I in patients with tropical spastic paraparesis. *Lancet*, **2**, 407–10.

Gibbs CJ Jr, Gajdusek DC, Asher DM *et al*. (1968). Creutzfeldt-Jakob disease (spongiform encephalopathy): transmission to the chimpanzee. *Science*, **161**, 388–9.

Gilden D (2004). Varicella zoster virus and central nervous system syndromes. *Herpes*, **11 (Suppl 2)**, 89A–94A.

Gilden DH, Kleinschmidt-DeMasters BK *et al*. (2000). Neurologic complications of the reactivation of varicella-zoster virus. *N Engl J Med*, **342**, 635–45.

Glauser MP, Zanetti G, Baumgartner JD *et al*. (1991). Septic shock: pathogenesis. *Lancet*, **338**, 732–6.

Goh KJ, Tan CT, Chew NK *et al*. (2000). Clinical features of Nipah virus encephalitis among pig farmers in Malaysia. *N Engl J Med*, **342**, 1229–35.

Goldwater PN, Rowland K, Thesinger M *et al*. (2001). Rotavirus encephalopathy: pathogenesis reviewed. *J Paediatr Child Health*, **37**, 206–9.

Gordon MS, Eaton ME, George R *et al*. (1994). The response of symptomatic neurosyphilis to high-dose intravenous penicillin G in patients with human immunodeficiency virus infection. *New Engl J Med*, **331**, 1469–73.

Gould EA, Solomon T (2007). Pathogenic flaviviruses. *Lancet* (**In press**).

Greenberg M, Abramson H, Cooper HM *et al*. (1952). The relation between recent injections and paralytic poliomyelitis in children. *Am J Public Health Nations Health*, **42**, 142–52.

Griffin DE (2001). Alphaviruses. In: Knipe DM, Howley PM. *Fields Virology*, Vol. 1, pp. 917–62. Lippincott Williams & Wilkins, Philadelphia, USA.

Gritsun TS, Frolova TV, Zhankov AI *et al*. (2003). Characterization of a siberian virus isolated from a patient with progressive chronic tick-borne encephalitis. *J Virol*, **77**, 25–36.

Gubler DJ (2002). The global emergence/resurgence of arboviral diseases as public health problems. *Arch Med Res*, **33**, 330–42.

Gubler DJ, Kuno G, Wareman SH (1983). Neurological disorders associated with dengue fever infection. *Proceedings of the International Conference on Dengue/DHF*, University of Malaysia Press, Kualar Lumpar.

Guess HA, Broughton DD, Melton LJ III *et al*. (1986). Population-based studies of varicella complications. *Pediatrics*, **78**, 723–7.

Guidugli F, Castro AA, Atallah AN (2000). Systematic reviews on leptospirosis. *Rev Inst Med Trop Sao Paulo*, **42**, 47–9.

Hadlow WJ (1959). Scrapie and kuru. *Lancet*, **2**, 289–90.

Haglund M, Gunther G (2003). Tick-borne encephalitis-pathogenesis, clinical course and long-term follow-up. *Vaccine*, **21 (Suppl 1)**, S11–8.

Hahne SJ, White JM, Crowcroft NS *et al*. (2006). Tetanus in injecting drug users, United Kingdom. *Emerg Infect Dis*, **12**, 709–10.

Haley M, Retter AS, Fowler D *et al*. (2003). The role for intravenous immunoglobulin in the treatment of West Nile virus encephalitis. *Clin Infect Dis*, **37**, e88–90.

Halperin JJ (1998). Nervous system Lyme disease. *J Neurol Sci*, **153**, 182–91.

Halstead SB (1990). Global epidemiology of dengue hemorrhagic fever. *Southeast Asian J Trop Med Public Health*, **21**, 636–41.

Harinasuta C, Nimmanitya S, Titsyakorn U (1985). The effect of interferon alpha on two cases of Japanese encephalitis in Thailand. *Southeast Asian J Trop Med Public Health*, **16**, 332–6.

Hato N, Matsumoto S, Kisaki H *et al*. (2003). Efficacy of early treatment of Bell's palsy with oral acyclovir and prednisolone. *Otol Neurotol*, **24**, 948–51.

Hausler M, Schaade L, Kemeny S *et al*. (2002). Encephalitis related to primary varicella-zoster virus infection in immunocompetent children. *J Neurol Sci*, **195**, 111–6.

Helmick CG, Bernard KW, D'Angelo LJ (1984). Rocky Mountain spotted fever: clinical, laboratory, and epidemiological features of 262 cases. *J Infect Dis*, **150**, 480–8.

Hemachudha, T, *et al*. (1987). Myelin basic protein as an encephalitogen in encephalomyelitis and polyneuritis following rabies vaccination. *N Engl J Med*, **316**: 369–74.

Hemachudha T, Laothamatas J, Rupprecht CE (2002). Human rabies: a disease of complex neuropathogenetic mechanisms and diagnostic challenges. *Lancet Neurol*, **1**, 101–9.

Hewitt PE, Llewelyn CA, Mackenzie J, Will RG (2006). Creutzfeldt-Jakob disease and blood transfusion: results of the UK Transfusion Medicine Epidemiological Review study. *Vox Sang*, **91**, 221–30.

Heyderman RS, Lambert HP, O'Sullivan I *et al*. (2003). Early management of suspected bacterial meningitis and meningococcal septicaemia in adults. *J Infect*, **46**, 75–7.

Hill AF, Butterworth RJ, Joiner S *et al*. (1999). Investigation of variant Creutzfeldt-Jakob disease and other human prion diseases with tonsil biopsy samples. *Lancet*, **353**, 183–9.

Hinman AR, Foege WH, de Quadros CA *et al*. (1987). The case for global eradication of poliomyelitis. *Bull World Health Organ*, **65**, 835–40.

Hoke CH, Nisalak A, Sangawhipa N *et al*. (1988). Protection against Japanese encephalitis by inactivated vaccines. *N Engl J Med*, **319**, 608–14.

Holland NR, Power C, Mathews VP *et al*. (1994). Cytomegalovirus encephalitis in acquired immunodeficiency syndrome (AIDS). *Neurology*, **44**, 507–14.

Hommel D, Talarmin A, Deubel V *et al*. (1998). Dengue encephalitis in French Guiana. *Res Virol*, **149**, 235–8.

Horstmann DM (1949). Clinical aspects of acute poliomyelitis. *Am J Med*, **6**:592–605.

Horstmann DM (1950). Acute poliomyelitis relation of physical activity at the time of onset to the course of the disease. *J Am Med Assoc*, **142**, 236–41.

Houff SA, Major EO, Katz DA *et al.* (1988). Involvement of JC virus-infected mononuclear cells from the bone marrow and spleen in the pathogenesis of progressive multifocal leukoencephalopathy. *N Engl J Med*, **318**, 301–5.

Huang C, Slater B, Rudd R *et al.* (2002). First isolation of *West Nile virus* from a patient with encephalitis in the United States. *Emerg Infect Dis* **8**, 1367–71.

Huang CC, Liu CC, Chang YC *et al.* (1999). Neurologic complications in children with enterovirus 71 infection. *N Engl J Med*, **341**, 936–42.

Hughes I, Jenney ME, Newton RW *et al.* (1993). Measles encephalitis during immunosuppressive treatment for acute lymphoblastic leukaemia. *Arch Dis Child*, **68**, 775–8.

Husby G, van de Rijn I, Zabriskie JB *et al.* (1976). Antibodies reacting with cytoplasm of subthalamic and caudate nuclei neurons in chorea and acute rheumatic fever. *J Exp Med*, **144**, 1094–110.

Idro R, Jenkins NE, Newton CR (2005). Pathogenesis, clinical features, and neurological outcome of cerebral malaria. *Lancet Neurol*, **4**, 827–40.

Igarashi A, Tanaka M, Morita K *et al.* (1994). Detection of West Nile and Japanese encephalitis viral genome sequences in cerebrospinal fluid from acute encephalitis cases in Karachi, Pakistan. *Microbiol Immunol*, **38**, 827–30.

Innis BL, Nisalak A, Nimmannitya S *et al.* (1989). An enzyme-linked immunosorbent assay to characterize dengue infections where dengue and Japanese encephalitis co-circulate. *Am J Trop Med Hyg*, **40**, 418–27.

Iwamoto M, Jernigan DB, Guasch A *et al.* (2003). Transmission of West Nile virus from an organ donor to four transplant recipients. *N Engl J Med*, **348**, 2196–203.

Iwasaki Y (1990). Pathology of chronic myelopathy associated with HTLV-I infection (HAM/TSP). *J Neurol Sci*, **96**, 103–23.

Jackson AC (2000). Rabies. *Can J Neurol Sci*, **27**, 278–82.

Jackson AC, Warrell MJ, Rupprecht CE *et al.* (2003). Management of rabies in humans. *Clin Infect Dis*, **36**, 60–3.

Jacob A, Solomon T, Garner P (2006). Corticosteroids in central nervous system infections. In: Candelise L, Liberati A, Warlow C, Hughes R, Uitdehaag B, *Evidence-based Neurology: Management of Neurological Disorders*. Malden, MA, Blackwell publishing, 151–60.

Javadi MA, Fayaz A, Mirdehghan SA *et al.* (1996). Transmission of rabies by corneal graft. *Cornea*, **15**, 431–3.

Jeha LE, Sila CA, Lederman RJ, *et al.* (2003). West Nile virus infection: a new acute paralytic illness. *Neurology*, **61**, 55–9.

Johns DR, Tierney M, Felsenstein D (1987). Alteration in the natural history of neurosyphilis by concurrent infection with the human immunodeficiency virus. *N Engl J Med*, **316**, 1569–72.

Johnson N, Brookes SM, Fooks AR *et al.* (2005). Review of human rabies cases in the UK and in Germany. *Vet Rec*, **157**, 715.

Johnson RT, Cornblath DR (2003). Poliomyelitis and flaviviruses. *Ann Neurol*, **53**, 691–2.

Johnson RT, Griffin DE, Hirsch RL *et al.* (1984). Measles encephalomyelitis–clinical and immunologic studies. *N Engl J Med*, **310**, 137–41.

Jones TD (1944). Diagnosis of rheumatic fever. *JAMA*, **126**, 481–4.

Kabiri M, Franco JR, Simarro PP *et al.* (1999). Detection of Trypanosoma brucei gambiense in sleeping sickness suspects by PCR amplification of expression-site-associated genes 6 and 7. *Trop Med Int Health*, **4**, 658–61.

Kaiser R (1999). The clinical and epidemiological profile of tick-borne encephalitis in southern Germany 1994-98: a prospective study of 656 patients. *Brain*, **122**, 2067–78.

Kalayjian RC, Cohen ML, Bonomo RA *et al.* (1994). Cytomegalovirus ventriculoencephalitis in AIDS. A syndrome with distinct clinical and pathologic features. *Medicine*, **72**, 67–77.

Kamei S, Sekizawa T, Shiota H *et al.* (2005). Evaluation of combination therapy using aciclovir and corticosteroid in adult patients with herpes simplex virus encephalitis. *J Neurol Neurosurg Psychiatry*, **76**, 1544–9.

Kankirawatana P, Chokephaibulkit K, Puthavathana P *et al.* (2000). Dengue infection presenting with central nervous system manifestation. *J Child Psychol*, **15**, 544–7.

Kaplan JE, Litchfield B, Rouault C *et al.* (1991). HTLV-I-associated myelopathy associated with blood transfusion in the United States: epidemiologic and molecular evidence linking donor and recipient. *Neurology*, **41**, 192–7.

Katz DA, Berger JR, Duncan RC (1993). Neurosyphilis. A comparative study of the effects of infection with human immunodeficiency virus. *Arch Neurol*, **50**, 243–9.

Katz M, Rorke LB, Masland WS *et al.* (1968). Transmission of an encephalitogenic agent from brains of patients with subacute sclerosing panencephalitis to ferrets. Preliminary report. *N Engl J Med*, **279**, 793–8.

Keinath RD, Merrell DE, Vlietstra R *et al.* (1985). Antibiotic treatment and relapse in Whipple's disease. Long-term follow-up of 88 patients. *Gastroenterology*, **88**, 1867–73.

Kelley TW, Prayson RA, Isada CM (2003). Spinal cord disease in West Nile virus infection. *N Engl J Med*, **348**, 564–6.

Kennedy PG (2006). Human African trypanosomiasis-neurological aspects. *J Neurol*, **253**, 411–6.

Kennedy PG, Rodgers J, Jennings FW *et al.* (1997). A substance P antagonist, RP-67,580, ameliorates a mouse meningoencephalitic response to Trypanosoma brucei brucei. *Proc Natl Acad Sci USA*, **94**, 4167–70.

Kepes, JJ (1993). Large focal tumor-like demyelinating lesions of the brain: intermediate entity between multiple sclerosis and acute disseminated encephalomyelitis? A study of 31 patients. *Ann Neurol*, **33**: 18–27.

Kerr JH, Corbett JL, Prys-Roberts C *et al.* (1968). Involvement of the sympathetic nervous system in tetanus. Studies on 82 cases. *Lancet*, **2**, 236–41.

Kesselring, J, *et al.* (1990). Acute disseminated encephalomyelitis. MRI findings and the distinction from multiple sclerosis. *Brain*, **113 (Pt 2)**: 291–302.

Kienzle N, Boyes L (2003). Murray Valley encephalitis: Case report and review of neuroradiological features. *Australas Radiol*, **47**, 61–3.

Kiessling LS, Marcotte AC, Culpepper L (1993). Antineuronal antibodies in movement disorders. *Pediatrics*, **92**, 39–43.

Kiley M, Esiri MM (2001). A contemporary case of encephalitis lethargica. *Clin Neuropathol*, **20**, 2–7.

Kirchhoff LV (1993). American trypanosomiasis (Chagas' disease)--a tropical disease now in the United States. *N Engl J Med*, **329**, 639–44.

Kirk J, Zhou AL, McQuaid S, Cosby SL *et al.* (1991). Cerebral endothelial cell infection by measles virus in subacute sclerosing panencephalitis: ultrastructural and in situ hybridization evidence. *Neuropathol Appl Neurobiol*, **17**, 289–97.

Kirk LJ, Fine PD, Sexton JD *et al.* (1990). Rocky Mountain Spotted Fever: a clinical review based on 48 confirmed cases 1943-1986. *Medicine*, **69**, 35–45.

Kleinschmidt-DeMasters BK, Tyler KL (2005). Progressive multifocal leukoencephalopathy complicating treatment with natalizumab and interferon beta-1a for multiple sclerosis. *N Engl J Med*, **353**, 369–74.

Knight RSG, and Will RG (2004). Prion diseases. *J Neurol Neurosurg Psychiatry*, **75**.

Kolski H, Ford-Jones EL, Richardson S *et al.* (1998). Etiology of acute childhood encephalitis at The Hospital for Sick Children, Toronto, 1994-1995. *Clin Infect Dis*, **26**, 398–409.

Koralnik IJ (2004). New insights into progressive multifocal leukoencephalopathy. *Curr Opin Neurol*, **17**, 365–70.

Koskiniemi M, Korppi M, Mustonen K *et al.* (1997). Epidemiology of encephalitis in children. A prospective multicentre study. *Eur J Pediatr*, **156**, 541–5.

Koskiniemi M, Piiparinen H, Mannonen L *et al.* (1996). Herpes encephalitis is a disease of middle aged and elderly people: polymerase chain reaction for detection of herpes simplex virus in the CSF of 516 patients with encephalitis. The Study Group. *J Neurol Neurosurg Psychiatry*, **60**, 174–8.

Kramer AH, Bleck TP (2007). Neurocritical care of patients with central nervous system infections. *Curr Infect Dis Rep*, **9**, 308–14.

Kumar R, Mathur A, Kumar A *et al.* (1990). Clinical features and prognostic indicators of Japanese encephalitis in children in Lucknow (India). *Indian J Med Res*, **91**, 321–7.

Kunz C (2002). Vaccination against TBE in Austria: the success story continues. *Int J Med Microbiol*, **291 (Suppl 33)**, 56–7.

Lanciotti RS, Kerst AJ, Nasci RS *et al.* (2000). Rapid detection of West Nile virus from human clinical specimens, field-collected mosquitoes, and avian samples by a TaqMan reverse transcriptase-PCR assay. *J Clin Microbiol*, **38**, 4066–71.

Larbrisseau A, Maravi E, Aguilera F *et al.* (1978). The neurological complications of brucellosis. *Can J Neurol Sci*, **5**, 369–76.

Lecour H, Miranda M, Magro C *et al.* (1989). Human leptospirosis–a review of 50 cases. *Infection*, **17**, 8–12.

Lehky TJ, Fox CH, Koenig S *et al.* (1995). Detection of human T-lymphotropic virus type I (HTLV-I) tax RNA in the central nervous system of HTLV-I-associated myelopathy/tropical spastic paraparesis patients by in situ hybridization. *Ann Neurol*, **37**, 167–75.

Leis AA, Stokic DS, Polk JL *et al.* (2002). A poliomyelitis-like syndrome from West Nile virus infection. *N Engl J Med*, **347**, 1279–80.

Lejon V, Boelaert M, Jannin J *et al.* (2003). The challenge of Trypanosoma brucei gambiense sleeping sickness diagnosis outside Africa. *Lancet Infect Dis*, **3**, 804–8.

Lerner PI, Weinstein L (1966). Infective endocarditis in the antibiotic era. *N Engl J Med*, **274**, 199–206.

Leung-Shea C, Danaher PJ (2006). Q fever in members of the United States armed forces returning from Iraq. *Clin Infect Dis*, **43**, e77–82.

Li J, Loeb JA, Shy ME *et al.* (2003). Asymmetric flaccid paralysis: A neuromuscular presentation of West Nile virus infection. *Ann Neurol*, **53**, 703–10.

Lim CC, Lee WL, Leo YS *et al.* (2003). Late clinical and magnetic resonance imaging follow up of Nipah virus infection. *J Neurol Neurosurg Psychiatry*, **74**, 131–3.

LongshoreWA, Stevens IM, Hollister AC (1956). Epidemiologic observations on acute infectious encephalitis in California with special reference to the 1952 outbreak. *Am J Hyg*, **63**, 69–86.

Looareesuwan S, Warrell DA, White NJ *et al.* (1983). Do patients with cerebral malaria have cerebral oedema? A computed tomography study. *Lancet*, **1**, 434–7.

Louis ED, Lynch T, Kaufmann P *et al.* (1996). Diagnostic guidelines in central nervous system Whipple's disease. *Ann Neurol*, **40**, 561–8.

Lowichik A, Siegel JD (1995). Parasitic infections of the central nervous system in children. Part I: Congenital infections and meningoencephalitis. *J Child Neurol*, **10**, 4–17.

Lum LCS, Lam SK, Choy S *et al.* (1996). Dengue encephalitis: a true entity? *Am J Trop Med Hyg*, **54**, 256–9.

MacPherson GG, Warrell MJ, White NJ *et al.* (1985). Human cerebral malaria. A quantitative ultrastructural analysis of parasitized erythrocyte sequestration. *Am J Pathol*, **119**, 385–401.

Mani KS, Mani AJ, Montgomery RD (1969). A spastic paraplegic syndrome in South India. *J Neurol Sci*, **9**, 179–99.

Marano E, Briganti F, Tortora F *et al.* (2004). Neurosyphilis with complex partial status epilepticus and mesiotemporal MRI abnormalities mimicking herpes simplex encephalitis. *J Neurol Neurosurg Psychiatry*, **75**, 833.

Martin DA, Biggerstaff BJ, Allen B *et al.* (2002). Use of immunoglobulin m cross-reactions in differential diagnosis of human flaviviral encephalitis infections in the United States. *Clin Diagn Lab Immunol*, **9**, 544–9.

Masliah E, DeTeresa RM, Mallory ME *et al.* (2000). Changes in pathological findings at autopsy in AIDS cases for the last 15 years. *Aids*, **14**, 69–74.

Masters CL, Richardson EP Jr. (1978). Subacute spongiform encephalopathy (Creutzfeldt-Jakob disease). The nature and progression of spongiform change. *Brain*, **101**, 333–44.

Matar GM, Khneisser IA, Abdelnoor AM (1996). Rapid laboratory confirmation of human brucellosis by PCR analysis of a target sequence on the 31-kilodalton Brucella antigen DNA. *J Clin Microbiol*, **34**, 477–8.

McBride AJ, Athanazio DA, Reis MG *et al.* (2005). Leptospirosis. *Curr Opin Infect Dis*, **18**, 376–86.

McCabe K, Tyler K, Tanabe J (2003). Diffusion-weighted MRI abnormalities as a clue to the diagnosis of herpes simplex encephalitis. *Neurology*, **61**, 1015–6.

McCordock JA, Collier W, Gray SH (1934). The pathological changes of the St Louis type of acute encephalitis. *JAMA*, **103**, 822–5.

McGrath N, Anderson NE, Croxson MC *et al.* (1997). Herpes simplex encephalitis treated with acyclovir: diagnosis and long term outcome. *J Neurol Neurosurg Psychiatry*, **63**, 321–6.

McLean DR, Russell N, Khan MY (1992). Neurobrucellosis: clinical and therapeutic features. *Clin Infect Dis*, **15**, 582–90.

Melnick JL, Ledinko N (1951). Social serology; antibody levels in a normal young population during an epidemic of poliomyelitis. *Am J Hyg*, **54**, 354–82.

Merritt HH, Moore M (1935). Acute syphilitic meningitis. *Medicine*, **14**, 119–83.

Miller, HG *et al.* (1954). Neurological sequelae of prophylactic inoculation. *Q J Med*, **23**: 1–27.

Miller C, Farrington CP, Harbert K (1992). The epidemiology of subacute sclerosing panencephalitis in England and Wales 1970-89. *Int J Epidemiol*, **21**, 998–1006.

Misra UK, Kalita J (1997). Anterior horn cells are also involved in Japanese encephalitis. *Acta Neurologica Scandinavica*, **96**, 114–7.

Misra UK, Kalita J (1997). Movement disorders in Japanese encephalitis. *J Neurol*, **244**, 299–303.

Modlin J (1995). Poliovirus. In: Mandell GL, Bennet J, Dolin R, eds. *Principles and Practice of Infectious Diseases*. Churchill Livingstone, New York.

Molina OM, Morales MC, Soto ID *et al.* (1999). [Venezuelan equine encephalitis. 1995 outbreak: clinical profile of the case with neurologic involvement]. *Rev Neurol*, **29**, 296–8.

Monath TP (1980). Epidemiology. In: Monath TP, ed. *St Louis Encephalitis*, pp. 239–312. American Public Health Associates, Washington DC.

Monath TP (1997). Epidemiology of yellow fever: current status and speculations on future trends. In: Saluzzo JF, Dodet B, eds. *Factors in the Emergence of Arbovirus Diseases*, pp. 143–58. Elsevier, Paris.

Monath TP (2002). Japanese encephalitis vaccines: current vaccines and future prospects. *Curr Top Microbiol Immunol*, **267**, 105–38.

Montgomery RD, Cruickshank EK, Robertson WB *et al.* (1964). Clinical and pathological observations on Jamaican neuropathy: a report of 206 cases. *Brain*, **87**, 425–62.

Morata P, Queipo-Ortuno MI, Reguera JM *et al.* (1999). Posttreatment follow-up of brucellosis by PCR assay. *J Clin Microbiol*, **37**, 4163–6.

Morata P, Queipo-Ortuno MI, Reguera JM *et al.* (2001). Diagnostic yield of a PCR assay in focal complications of brucellosis. *J Clin Microbiol*, **39**, 3743–6.

Mori M, Kurata H, Tajima M *et al.* (1991). JC virus detection by in situ hybridization in brain tissue from elderly patients. *Ann Neurol*, **29**, 428–32.

Mostashari F, Kulldorff M, Hartman JJ *et al.* (2003). Dead bird clusters as an early warning system for West Nile virus activity. *Emerg Infect Dis*, **9**, 641–6.

Murphy MF (1978). In vitro inhibition of subacute sclerosing panencephalitis virus by the antiviral agent ribavirin. *J Infect Dis*, **138**, 249–51.

Mustafa MM, Weitman SD, Winick NJ *et al.* (1993). Subacute measles encephalitis in the young immunocompromised host: report of two cases diagnosed by polymerase chain reaction and treated with ribavirin and review of the literature. *Clin Infect Dis*, **16**, 654–60.

Nantulya VM, Doua F, Molisho S (1992). Diagnosis of Trypanosoma brucei gambiense sleeping sickness using an antigen detection enzyme-linked immunosorbent assay. *Trans R Soc Trop Med Hyg*, **86**, 42–5.

Nash D, Mostashari F, Fine A *et al.* (2001). The outbreak of West Nile virus infection in the New York City area in 1999. *N Engl J Med*, **344**, 1807–14.

Nathanson N, Martin JR (1979). The epidemiology of poliomyelitis: enigmas surrounding its appearance, epidemicity, and disappearance. *Am J Epidemiol*, **110**:672–92.

Nathwani D, McIntyre PG, White K *et al.* (2003). Fatal human rabies caused by European bat Lyssavirus type 2a infection in Scotland. *Clin Infect Dis*, **37**, 598–601.

Newman W, Southam CM (1954). Virus treatment in advanced cancer. A pathological study of fifty-seven cases. *Cancer*, **7**, 106–18.

Newton C, Hien TT, White NJ (2000). Cerebral Malaria. *J Neurol Neurosurg Psychiatry*, **69**, 433–41.

Newton, CR, Kirkham, FJ, Winstanley, PA, Pasvol, G, Peshu, N, Warrell, DA and Marsh, K (1991). Intracranial pressure in African children with cerebral malaria. *Lancet*, **337**: 573–6.

Newton CRJC, Crawley J, Sowumni A *et al.* (1997). Intracranial hypertension in Africans with cerebral malaria. *Arch Dis Child*, **76**, 219–26.

Nguyen TH, Day NP, Ly VC *et al.* (1996). Post-malaria neurological syndrome. *Lancet*, **348**, 917–21.

Nimmannitya S (1987). Clinical spectrum and management of dengue haemorrhagic fever. *Trans R Soc Trop Med Hyg*, **18**, 292–7.

Nimmannitya S, Thisyakorn U, Hemsrichart V (1987). Dengue haemorrhagic fever with unusual manifestations. *Southeast Asian J Trop Med Pub Health*, **18**, 398–406.

Olivier C (2000). Rheumatic fever--is it still a problem? *J Antimicrob Chemother*, **45 (Suppl)**, 13–21.

Ooi MH, Solomon T, Podin Y *et al.* (2007). Evaluation of different clinical sample types in the diagnosis of human enterovirus 71 associated hand-foot-and-mouth disease. *J Clin Microbiol*, **45**, 1858–66.

Openshaw H, Cantin EM (2005). Corticosteroids in herpes simplex virus encephalitis. *J Neurol Neurosurg Psychiatry*, **76**, 1469.

Osler W (1885). Gulstonian lectures on malignant endocarditis. *Lancet*, **1**, 415–18, 59–64, 505–8.

Pachner AR, Steiner I (2007). Lyme neuroborreliosis: infection, immunity, and inflammation. *Lancet Neurol*, **6**, 544–52.

Padgett, BL, *et al.* (1971). Cultivation of papova-like virus from human brain with progressive multifocal leucoencephalopathy. *Lancet*, **1**: 1257–60.

Pancharoen C, Thisyakorn U (2001). Neurological manifestations in dengue patients. *Southeast Asian J Trop Med Public Health*, **32**, 341–5.

Panitch HS, Gomez-Plascencia J *et al.* (1986). Subacute sclerosing panencephalitis: remission after treatment with intraventricular interferon. *Neurology*, **36**, 562–6.

Park M, Hui JS, Bartt RE (2003). Acute anterior radiculitis associated with West Nile virus infection. *J Neurol Neurosurg Psychiatry*, **74**, 823–5.

Patnaik JK, Das BS, Mishra SK *et al.* (1994). Vascular clogging, mononuclear cell margination, and enhanced vascular permeability in the pathogenesis of human cerebral malaria. *Am J Trop Med Hyg*, **51**, 642–7.

Pepperell C, Rau N, Krajden S *et al.* (2003). West Nile virus infection in 2002: morbidity and mortality among patients admitted to hospital in southcentral Ontario. *CMAJ*, **168**, 1399–405.

Petersen LR, Marfin AA (2002). West Nile virus: a primer for the clinician. *Ann Intern Med*, **137**, 173–9.

Petersen LR, Roehrig JT, Hughes JM (2002). West Nile Virus Encephalitis. *N Engl J Med*, **347**, 1225–6.

Pitella JE (1993). Central nervous system involvement in Chaga's disease: an updating. *Rev Inst Med Trop Sao Paulo*, **35**, 111–6.

Promed-mail. (2007). Eastern Equine Encephalitis - UK (Scotland) ex USA (20071008.3310). Retrieved 8th Oct 2007.

Proulx N, Frechette D, Toye B *et al.* (2005). Delays in the administration of antibiotics are associated with mortality from adult acute bacterial meningitis. *QJM*, **98**, 291–8.

Pruitt AA, Rubin RH, Karchmer AW *et al.* (1978). Neurologic complications of bacterial endocarditis. *Medicine*, **57**, 329–43.

Prusiner SB (1982). Novel proteinaceous infectious particles cause scrapie. *Science*, **216**, 136–44.

Przelomski MM, O'Rourke E, Grady GF *et al.* (1988). Eastern equine encephalitis in Massachusetts: a report of 16 cases, 1970-1984. *Neurology*, **38**, 736–9.

Punt J (1983). Multiple cerebral gummata. Case report. *J Neurosurg*, **58**, 959–61.

Quinlivan EB, Norris M, Bouldin TW *et al.* (1992). Subclinical central nervous system infection with JC virus in patients with AIDS. *J Infect Dis*, **166**, 80–5.

Rahal JJ, Anderson J, Rosenberg C *et al.* (2004). Effect of Interferon- alpha 2b therapy on St. Louis viral meningoencephalitis: Clinical and laboratory results of a pilot study. *J Infect Dis*, **190**, 1084–7.

Rail D, Scholtz C, Swash M (1981). Post-encephalitic Parkinsonism: current experience. *J Neurol Neurosurg Psychiatry*, **44**, 670–6.

Ramzan NN, Loftus E Jr, Burgart LJ *et al.* (1997). Diagnosis and monitoring of Whipple disease by polymerase chain reaction. *Ann Intern Med*, **126**, 520–7.

Rand K, Houck H, Lawrence R (2005). Real-time polymerase chain reaction detection of herpes simplex virus in cerebrospinal fluid and cost savings from earlier hospital discharge. *J Mol Diagn*, **7**, 511–6.

Randall R, Mills JW, Engel LL (1947). The preparation and properties of a purified equine encephalomyelitis vaccine. *J Immunol*, **55**, 41–52.

Rantala H, Uhari M (1989). Occurrence of childhood encephalitis: a population-based study. *Pediatr Infect Dis J*, **8**, 426–30.

Raoult D, Birg ML, La Scola B *et al.* (2000). Cultivation of the bacillus of Whipple's disease. *N Engl J Med*, **342**, 620–5.

Raoult D, Zuchelli P, Weiller PJ *et al.* (1986). Incidence, clinical observations and risk factors in the severe form of Mediterranean spotted fever among patients admitted to hospital in Marseilles 1983-1984. *J Infect*, **12**, 111–6.

Raschilas F, Wolff M, Delatour F *et al.* (2002). Outcome of and prognostic factors for herpes simplex encephalitis in adult patients: results of a multicenter study. *Clin Infect Dis*, **35**, 254–60.

Ren R, Racaniello VR (1992). Poliovirus spreads from muscle to the central nervous system by neural pathways. *J Infect Dis*, **166**, 747–52.

Rigau-Perez JG, Clark GG, Gubler DJ *et al.* (1998). Dengue and dengue haemorrhagic fever. *Lancet*, **352**, 971–7.

Risk WS, Haddad FS (1979). The variable natural history of subacute sclerosing panencephalitis: a study of 118 cases from the Middle East. *Arch Neurol*, **36**, 610–4.

Rivas F, Diaz LA, Cardenas VM *et al.* (1997). Epidemic Venezuelan equine encephalitis in La Guajira, Colombia, 1995. *J Infect Dis*, **175**, 828–32.

Rodgers-Johnson P, Gajdusek DC, Morgan OS *et al.* (1985). HTLV-I and HTLV-III antibodies and tropical spastic paraparesis. *Lancet*, **2**, 1247–8.

Rodriguez JJ, Parisien JP, Horvath CM (2002). Nipah virus V protein evades alpha and gamma interferons by preventing STAT1 and STAT2 activation and nuclear accumulation. *J Virol*, **76**, 11476–83.

Roehrig JT, Nash D, Maldin B *et al.* (2003). Persistence of virus-reactive serum immunoglobulin M antibody in confirmed West Nile Virus encephalitis cases. *Emerg Infect Dis*, **9**, 376–9.

Rolain JM, Stuhl L, Maurin M *et al.* (2002). Evaluation of antibiotic susceptibilities of three rickettsial species including Rickettsia felis by a quantitative PCR DNA assay. *Antimicrob Agents Chemother*, **46**, 2747–51.

Russell WR (1949). Paralytic poliomyelitis; the early symptoms and the effect of physical activity on the course of the disease. *Br Med J*, **1**, 465–71.

Ryan MM, Antony JH (1999). Cerebral vasculitis in a case of Sydenham's chorea. *J Child Neurol*, **14**, 815–8.

Ryder JW, Croen K, Kleinschmidt-DeMasters BK *et al.* (1986). Progressive encephalitis three months after resolution of cutaneous zoster in a patient with AIDS. *Ann Neurol*, **19**, 182–8.

Sacktor N (2002). The epidemiology of human immunodeficiency virus-associated neurological disease in the era of highly active antiretroviral therapy. *J Neurovirol*, **8 (Suppl 2)**, 115–21.

Salgado AV, Furlan AJ, Keys TF *et al.* (1989). Neurologic complications of endocarditis: a 12-year experience. *Neurology*, **39**, 173–8.

Schiff D, Rosenblum MK (1998). Herpes simplex encephalitis (HSE) and the immunocompromised: a clinical and autopsy study of HSE in the settings of cancer and human immunodeficiency virus-type 1 infection. *Hum Pathol*, **29**, 215–22.

Schultz DR, Barthal JS, Garrett G (1977). Western equine encephalitis with rapid onset of parkinsonism. *Neurology*, **27**, 1095–6.

Sejvar JJ, Bode AV, Marfin AA *et al.* (2005). West Nile virus-associated flaccid paralysis. *Emerg Infect Dis*, **11**, 1021–7.

Sejvar JJ, Haddad MB, Tierney BC *et al.* (2003a). Neurologic manifestations and outcome of West Nile virus infection. *JAMA*, **290**, 511–5.

Sejvar JJ, Leis AA, Stokic DS *et al.* (2003b). Acute flaccid paralysis and West Nile virus infection. *Emerg Infect Dis*, **9**, 788–93.

Shaked Y (1991). Rickettsial infection of the central nervous system: the role of prompt antimicrobial therapy. *QJM*, **79**, 301–6.

Shakir RA, Al-Din AS, Araj GF *et al.* (1987). Clinical categories of neurobrucellosis. A report on 19 cases. *Brain*, **110**, 213–23.

Shimoni Z, Niven MJ, Pitlick S *et al.* (2001). Treatment of West Nile virus encephalitis with intravenous immunoglobulin. *Emerg Infect Dis*, **7**, 759.

Siger L, Bowen RA, Karaca K *et al.* (2004). Assessment of the efficacy of a single dose of a recombinant vaccine against West Nile virus in response to natural challenge with West Nile virus-infected mosquitoes in horses. *Am J Vet Res*, **65**, 1459–62.

Silverman L, Rubinstein LJ (1965). Electron microscopic observations on a case of progressive multifocal leukoencephalopathy. *Acta Neuropathol*, **5**, 215–24.

Simon RP (1994). Neurosyphilis. *Neurology*, **44**, 2228–30.

Sipahioglu U, Alpaut S (1985). [Transplacental rabies in humans]. *Mikrobiyol Bul*, **19**, 95–9.

Skoldenberg B, Forsgren M, Alestig K *et al.* (1984). Acyclovir versus vidarabine in herpes simplex encephalitis. Randomised multicentre study in consecutive Swedish patients. *Lancet*, **2**, 707–11.

Smith DL (1997). Cat-scratch disease and related clinical syndromes. *Am Fam Physician*, **55**, 1783–9, 93–4.

Smith DL, Ayres JG, Blair I *et al.* (1993). A large Q fever outbreak in the West Midlands: clinical aspects. *Respir Med*, **87**, 509–16.

Smith J, McElhinney L, Parsons G *et al.* (2003). Case report: rapid ante-mortem diagnosis of a human case of rabies imported into the UK from the Philippines. *J Med Virol*, **69**, 150–5.

Smith R, O'Connell S, Palmer S (2000). Lyme disease surveillance in England and Wales, 1986 1998. *Emerg Infect Dis*, **6**, 404–7.

Smorodintsev AA (1958). Tick-borne spring-summer encephalitis. *Prog Med Virol*, **1**, 210–48.

Solomon T, Blanchard T (2006). Cerebral Malaria. In: Johnson RT, Griffin JW, McArthu JC, eds. *Current Therapy in Neurological Diseases*, 7th ed. Philadelphia, Elsevier, Mosby.

Solomon T, Mallewa MJ (2001). Dengue and other emerging flaviviruses. *J Infect*, **42**, 104–15.

Solomon T, Vaughn DW (2002). Clinical features and pathophysiology of Japanese encephalitis and West Nile virus infections. In: Mackenzie JS, Barrett AD, Deubel V, eds. *Current Topics in Microbiology and Immunology: Japanese Encephalitis and West Nile Virus Infections*, pp. 171–94. Springer-Verlag, Berlin.

Solomon T, Vaughn DW (2002). Pathogenesis and clinical features of Japanese encephalitis and West Nile virus infections. In: Mackenzie JS, Barrett AD, Deubel V. *Current Topics in Microbiology and Immunology:*

Japanese Encephalitis and West Nile Virus Infections, Vol. 267, pp. 171–94. Springer-Verlag, Berlin.

Solomon T, Barrett AD (2003). Dengue. In: Nath A, Berger J, eds. *Neurovirology*, pp. 469–516. Marcel Decker, New York, NY.

Solomon T, Willison H (2003). Infectious causes of acute flaccid paralysis. *Curr Opin Infect Dis*, **16**, 375–81.

Solomon T, Whitley RJ (2004). Arthropod-borne viral encephalitides. In: Scheld M, Whitley RJ, Marra C, eds. *Infections of the Central Nervous System*, Philidelphia, PA, Lippincott Williams and Wilkins.

Solomon T, Blanchard T. (2005) Cerebral Malaria. In: Johnson RT, Griffin JW, McArthu JC, eds. *Current Therapy in Neurological Diseases*, Elsevier, New York.

Solomon T, Kneen R, Dung NM *et al.* (1998). Poliomyelitis-like illness due to Japanese encephalitis virus. *Lancet*, **351**, 1094–7.

Solomon T, Thao LTT, Dung NM *et al.* (1998). Rapid diagnosis of Japanese encephalitis by using an IgM dot enzyme immunoassay. *J Clin Microbiol*, **36**, 2030–34.

Solomon T, Dung NM, Kneen R *et al.* (2000). Japanese encephalitis. *J Neurol Neurosurg Psychiatry*, **68**, 405–15.

Solomon T, Dung NM, Vaughn DW *et al.* (2000). Neurological manifestations of dengue infection. *Lancet*, **355**, 1053–59.

Solomon T, Dung NM, Kneen R *et al.* (2002). Seizures and raised intracranial pressure in Vietnamese patients with Japanese encephalitis. *Brain*, **125**, 1084–93.

Solomon T, Hart CA, Vinjamuri S *et al.* (2002). Treatment of subacute sclerosing panencephalitis with interferon-alpha, ribavirin, and inosiplex. *J Child Neurology*, **17**, 703–5.

Solomon T, Dung NM, Wills B *et al.* (2003). Interferon alfa-2a in Japanese encephalitis: a randomised double-blind placebo-controlled trial. *Lancet*, **361**, 821–6.

Solomon T, Fisher AF, Beasley DW *et al.* (2003). Natural and nosocomial infection in a patient with West Nile encephalitis and extrapyramidal movement disorders. *Clin Infect Dis*, **36**, E140–5.

Solomon T, Ooi MH, Beasley DW *et al.* (2003). West Nile encephalitis. *BMJ*, **326**, 865–9.

Solomon T, Marston D, Mallewa M *et al.* (2005). Paralytic rabies after a two week holiday in India. *BMJ*, **331**, 501–3.

Solomon T, Hart IJ, Beeching NJ (2007). Viral encephalitis: a clinician's guide. *Practical Neurology*, **7**, 288–305.

Southern PM, Sanford JP (1969). Relapsing fever: a clinical microbiological review. *Medicine*, **48**, 129–49.

Spillane, JD *et al.* (1964). The Neurology of Jennerian Vaccination. a Clinical Account of the Neurological Complications Which Occurred During the Smallpox Epidemic in South Wales in 1962. *Brain*, **87**: 1–44.

Sprung CL, Cerra FB, Freund HR *et al.* (1991). Amino acid alterations and encephalopathy in the sepsis syndrome. *Crit Care Med*, **19**, 753–7.

Srinivasan A, Burton EC, Kuehnert MJ *et al.* (2005). Transmission of rabies virus from an organ donor to four transplant recipients. *N Engl J Med*, **352**, 1103–11.

Steere, AC, *et al.* (1977). Lyme arthritis: an epidemic of oligoarticular arthritis in children and adults in three connecticut communities. *Arthritis Rheum*, **20**: 7–17.

Steiner I, Budka H, Chaudhuri A *et al.* (2005). Viral encephalitis: a review of diagnostic methods and guidelines for management. *Eur J Neurol*, **12**, 331–43.

Sterner G, Biberfeld G (1969). Central nervous system complications of Mycoplasma pneumoniae infection. *Scand J Infect Dis*, **1**, 203–8.

Studahl M, Bergstrom T, Hagberg L (1998). Acute viral encephalitis in adults--a prospective study. *Scand J Infect Dis*, **30**, 215–20.

Sutter RW, Patriarca PA, Suleiman AJ *et al.* (1992). Attributable risk of DTP (diphtheria and tetanus toxoids and pertussis vaccine) injection in provoking paralytic poliomyelitis during a large outbreak in Oman. *J Infect Dis*, **165**, 444–9.

Talkington DF (2004). Mycoplasmal and ureaplasmal infections. In: Scheld WM, Whitley RJ, Marra CM, eds. *Infections of the Central Nervous System*, pp. 605–11. Lippincott Williams and Wilkins.

Tan CT, Goh KJ, Wong KT *et al.* (2002). Relapsed and late-onset Nipah encephalitis. *Ann Neurol*, **51**:703–8.

Tang YW, Espy MJ, Persing DH *et al.* (1997). Molecular evidence and clinical significance of herpesvirus coinfection in the central nervous system. *J Clin Microbiol*, **35**, 2869–72.

Thakare J, Walhekar B, Banerjee K (1996). Haemorrhagic manifestations and encephalopathy in cases of dengue in India. *Southeast Asian J Trop Med Public Health*, **3**, 471–5.

Thisyakorn U, Thisyakorn C, Limpitikul W *et al.* (1999). Dengue infection with central nervous system manifestations. *Southeast Asian J Trop Med Public Health*, **30**, 504–6.

Tien RD, Felsberg GJ, Osumi AK (1993). Herpes virus infections of the CNS: MR findings. *AJR Am J Roentgenol*, **161**, 167–76.

Truc P, Jamonneau V, Cuny G *et al.* (1999). Use of polymerase chain reaction in human African trypanosomiasis stage determination and follow-up. *Bull World Health Organ*, **77**, 745–8.

Trujillo MH, Castillo A, Espana J *et al.* (1987). Impact of intensive care management on the prognosis of tetanus. Analysis of 641 cases. *Chest*, **92**, 63–5.

Vakil BJ, Singhal BS, Pandya SS *et al.* (1973). Cephalic tetanus. *Neurology*, **23**, 1091–6.

Verghese A, Sugar AM (1986). Herpes zoster ophthalmicus and granulomatous angiitis. An ill-appreciated cause of stroke. *J Am Geriatr Soc*, **34**, 309–12.

Vial L, Diatta G, Tall A *et al.* (2006). Incidence of tick-borne relapsing fever in west Africa: longitudinal study. *Lancet*, **368**, 37–43.

Von Bokay J (1909). Uber den atiologischen Zusammenhang der varizellen mit gewissen Fallen von Herpes Zoster. *Wien Klin Wochenschr*, **22**, 1323–42.

von Economo C (1931). *Encephalitis lethargica. Its sequelae and treatment. Translated by Newman KO.* Oxford University Press, London.

Wang E, Ni H, Xu R *et al.* (2000). Evolutionary relationships of endemic/epidemic and sylvatic dengue viruses. *J Virol*, **74**, 3227–34.

Warrell DA, Looareesuwan S, Warrell MJ *et al.* (1982). Dexamethasone proves deleterious in cerebral malaria. A double-blind trial in 100 comatose patients. *N Engl J Med*, **306**, 313–9.

Warrell DA, Warrell MJ (1988). Human rabies and its prevention: an overview. *Rev Infect Dis*, **10 (Suppl 4)**, S726–31.

Warrell MJ (2003). Rabies. In: Cook G, Zumlar A, eds. *Manson's Tropical Diseases*, pp. 807–21. Saunders, London.

Warrell MJ, Warrell DA (2004). Rabies and other lyssavirus diseases. *Lancet*, **363**, 959–69.

Wasay M, Diaz-Arrastia R, Suss RA *et al.* (2000). St Louis encephalitis: a review of 11 cases in a 1995 Dallas, Tex, epidemic. *Arch Neurol*, **57**, 114–8.

Watt G, Padre LP, Tuazon ML *et al.* (1988). Placebo-controlled trial of intravenous penicillin for severe and late leptospirosis. *Lancet*, **1**, 433–5.

Watt G, Parola P (2003). Scrub typhus and tropical rickettsioses. *Curr Opin Infect Dis*, **16**, 429–36.

Watts DM, Callahan J, Rossi C *et al.* (1998). Venezuelan equine encephalitis febrile cases among humans in the Peruvian Amazon River region. *Am J Trop Med Hyg*, **58**, 35–40.

Weaver SC, Kang W, Shirako Y *et al.* (1997). Recombinational history and molecular evolution of western equine encephalomyelitis complex alphaviruses. *J Virol*, **71**, 613–23.

Weaver SC, Pfeffer M, Marriott K *et al.* (1999). Genetic evidence for the origins of Venezuelan equine encephalitis virus subtype IAB outbreaks. *Am J Trop Med Hyg*, **60**, 441–8.

Weaver SC, Salas R, Rico-Hesse R *et al.* (1996). Re-emergence of epidemic Venezuelan equine encephalomyelitis in South America. VEE Study Group. *Lancet*, **348**, 436–40.

Webster AD (2005). Pleconaril--an advance in the treatment of enteroviral infection in immuno-compromised patients. *J Clin Virol*, **32**, 1–6.

Weinstein L (1973). Tetanus. *N Engl J Med*, **289**, 1293–6.

Weinstein L, Shelokov A, Seltser R *et al.* (1952). A comparison of the clinical features of poliomyelitis in adults and in children. *N Engl J Med*, **246**:297–302.

Weiss D, Carr D, Kellachan J *et al.* (2001). Clinical findings of West Nile virus infection in hospitalized patients, New York and New Jersey, 2000. *Emerg Infect Dis*, **7**, 654–8.

Weller TH (1953). Serial propagation in vitro of agents producing inclusion bodies derived from varicella and herpes zoster. *Proc Soc Exp Biol Med*, **83**, 340–6.

Whipple GH (1907). A hitherto undescribed disease characterized anatomically by deposits of fat and fatty acids in the intestinal mesenteric lymphatic tissues. *Johns Hopkins Hosp Bull*, **18**, 382–91.

Whitley R, Lakeman AD, Nahmias A *et al.* (1982). DNA restriction-enzyme analysis of herpes simplex virus isolates obtained from patients with encephalitis. *N Engl J Med*, **307**, 1060–2.

Whitley RJ (2006). Herpes simplex encephalitis: adolescents and adults. *Antiviral Res*, **71**, 141–8.

Whitley RJ, Alford CA, Hirsch MS *et al.* (1986). Vidarabine versus acyclovir therapy in herpes simplex encephalitis. *N Engl J Med*, **314**, 144–9.

Whitley RJ, Cobbs CG, Alford CA Jr *et al.* (1989). Diseases that mimic herpes simplex encephalitis. Diagnosis, presentation, and outcome. NIAD Collaborative Antiviral Study Group. *JAMA*, **262**, 234–9.

Whitty CJ, Lalloo D, Ustianowski A (2006). Malaria: an update on treatment of adults in non-endemic countries. *BMJ*, **333**, 241–5.

Wildemann B, Haas J, Lynen N, Stingele K *et al.* (1998). Diagnosis of cytomegalovirus encephalitis in patients with AIDS by quantitation of cytomegalovirus genomes in cells of cerebrospinal fluid. *Neurology*, **50**, 693–7.

Will RG, Ironside JW, Zeidler M *et al.* (1996). A new variant of Creutzfeldt–Jakob disease in the UK. *Lancet*, **347**, 921–5.

Willoughby RE Jr, Tieves KS, Hoffman GM *et al.* (2005). Survival after treatment of rabies with induction of coma. *N Engl J Med*, **352**, 2508–14.

Wilson SAK (1940). *Neurology*, Butterworth, London.

Winkler WG, Fashinell TR, Leffingwell L *et al.* (1973). Airborne rabies transmission in a laboratory worker. *JAMA*, **226**, 1219–21.

Wittesjö B, Eitrem R, Niklasson B *et al.* (1995). Japanese encephalitis after a 10-day holiday in Bali. *Lancet*, **345**, 856.

Wong SC, Ooi MH, Wong MN *et al.* (2001). Late presentation of Nipah virus encephalitis and kinetics of the humoral immune response. *J Neurol Neurosurg Psychiatry*, **71**, 552–4.

World Health Assembly (1988). *Global eradication of poliomyelitis by the year 2000.* World Health Organization, Geneva.

World Health Organization (1997). *Dengue Hemorrhagic Fever: diagnosis, treatment and control.* World Health Organization, Geneva.

World Health Organization (2005). *WHO Recommendations on Rabies Post-Exposure Treatment and the Correct Technique of Intradermal immunization against Rabies.* World Health Organization, Geneva.

Yamada S, Kameyama T, Nagaya S *et al.* (2003). Relapsing herpes simplex encephalitis: pathological confirmation of viral reactivation. *J Neurol Neurosurg Psychiatry*, **74**, 262–4.

Yamasaki K, Kira J, Koyanagi Y *et al.* (1997). Long-term, high dose interferon-alpha treatment in HTLV-I-associated myelopathy/tropical spastic paraparesis: a combined clinical, virological and immunological study. *J Neurol Sci*, **147**, 135–44.

Yaqub BA (1996). Subacute sclerosing panencephalitis (SSPE): early diagnosis, prognostic factors and natural history. *J Neurol Sci*, **139**, 227–34.

Yiannoutsos CT, Major EO, Curfman B *et al.* (1999). Relation of JC virus DNA in the cerebrospinal fluid to survival in acquired immunodeficiency syndrome patients with biopsy-proven progressive multifocal leukoencephalopathy. *Ann Neurol*, **45**, 816–21.

Young EJ (1983). Human brucellosis. *Rev Infect Dis*, **5**, 821–42.

Young GB, Bolton CF, Archibald YM *et al.* (1992). The electroencephalogram in sepsis-associated encephalopathy. *J Clin Neurophysiol*, **9**, 145–52.

Young GB, Bolton CF, Austin TW *et al.* (1990). The encephalopathy associated with septic illness. *Clin Invest Med*, **13**, 297–304.

Yucesan C, Sriram S (2001). Chlamydia pneumoniae infection of the central nervous system. *Curr Opin Neurol*, **14**, 355–9.

Zerr I, Bodemer M, Otto M *et al.* (1996). Diagnosis of Creutzfeldt-Jakob disease by two-dimensional gel electrophoresis of cerebrospinal fluid. *Lancet*, **348**, 846–9.

Zimmerman HM (1946). The pathology of Japanese B encephalitis. *Am J Pathol*, **22**, 965–91.

Zurhein G, Chou SM (1965). Particles resembling papova viruses in human eerebral demyelinating disease. *Science*, **148**, 1477–9.

CHAPTER 43

Intracranial space occupying infections and neurological HIV disease

Tom Solomon

Contents

Microbial pathogens can cause three types of space occupying lesion:

- *abscesses* are pus-filled, and usually caused by bacteria;
- *cysts* are fluid filled, and usually associated with parasitic diseases; and
- *granulomas* contain granulomatous tissue, and are caused by a range of bacteria, fungi, and parasites.

In addition there are many non-infectious causes of space occupying lesions. The number of lesions, their location, whether or not there is calcification, enhancement, and oedema all provide useful radiological clues as to their possible causation (Table 43.1).

This chapter begins with a discussion of space occupying lesions. It ends with a review of the neurological complications of HIV, which encompasses patients presenting with all central nervous system syndromes, including meningitis, reduced consciousness, myelopathy, and space occupying lesions.

43.1 Intracranial suppuration

43.1.1 General considerations

Brain abscess, subdural empyaema, and extradural abscess are all forms of intracranial suppuration. They share common clinical features: they occur relatively infrequently; they present as emergencies; and diagnosis is commonly delayed because of lack of familiarity with the clinical picture. Morbidity and mortality from each of these conditions remains high and the incidence of sequelae is considerable. This range of conditions also share common aetiological factors although there is evidence that patterns of infection are changing. It might be expected that any form of suppuration would be accompanied by obvious clinical signs of infection such as pyrexia, tachycardia, and malaise. While often this does happen, systemic manifestations of infection and inflammation are not always present when the site of suppuration is intracranial. This may obscure the true diagnosis. If there are overt foci of infection, this is a considerable aid to diagnosis, and specimens should be obtained from them. It is imperative that such foci be eradicated

Table 43.1 Intracranial space occupying lesions: radiological differential diagnosis

Single non-enhancing cystic lesion
Hydatid disease
Arachnoid cyst
Porencephaly
Cystic astrocytoma
Colloid cyst (3rd ventricle)

Several non-enhancing lesions
Multiple metastases
Hydatid disease (rare)

Enhancing lesions
Tuberculoma
Toxoplasma
Abscess
Mycosis
Early glioma
Metastases
Arteriovenous malformation
Primary central nervous system lymphoma

Lesions with calcifications
Tuberous sclerosis
Tuberculosis
Cytomegalovirus
Toxoplasma
Arteriovenous malformation

commensurate with treatment of the intracranial problem. The clinical profiles of brain abscess, subdural empyaema, and extradural abscess are sufficiently distinct to warrant separate descriptions.

43.1.2 Brain abscess

A brain abscess begins as a focal intracerebral infection, initially as a localized area of cerebritis, which develops into a collection of pus surrounded by a well-vascularized capsule. This section focuses on abscesses caused by bacterial infections. Those caused by fungi and protozoa are discussed under intracranial granulomas and cysts (Section 43.2).

The incidence of brain abscess is generally held to be about 1 in 10 000 of all general hospital admissions in the western world, approximating to 5 cases per million population per annum. All agree that this is an underestimate of the true incidence because a proportion are first diagnosed at necropsy. Also the increase of HIV infection and other causes of immune suppression has produced a large population vulnerable to opportunistic infection by an ever increasing range of pathogens. Probably the number of patients who develop brain abscess is rising. Additionally more cases are being diagnosed as imaging techniques have become more discriminating. Males are affected about twice as often as females although this sex difference is diminishing. The median age is 30–45 years with 25 per cent occurring under the age of 15. It is unusual for brain abscess to occur under the age of 2 years (McClelland *et al.* 1978; Seydoux *et al.* 1992). However, it may be that abscesses are misdiagnosed as meningitis in the very young age groups.

Pathogenesis

Bacteria reach the brain by three main routes; by the bloodstream; by extension from a contiguous focus of infection; or by being directly inoculated into the brain substance by trauma, or by neurosurgery which is not aseptic (Table 43.2). It seems unlikely that spread through the CSF plays a part. The route of infection cannot be identified with certainty in about 20 per cent of cases. Formerly the most common mechanism was spread from an adjacent infection such as otitis or paranasal sinusitis. However this now is being overtaken by blood-borne infection. Brain abscess is rare in bacterial endocarditis and experimental work has shown that if the blood–brain barrier remains intact, bacteraemia does not lead to brain abscess formation; bacteria cannot set up a nidus of infection in undamaged brain (Molinari *et al.* 1973). It is necessary for there to be a microscopic area of necrosis for an abscess to take root, either due to microinfarction from embolism or from hypoxaemia, or from adjacent infection causing thrombophlebitis.

Once started, an abscess develops through a series of stages which are well-documented and are reproducible under experimental conditions. Furthermore, the pathological changes of these stages of evolution correlate very well with clinical progression and CT brain scan appearances; it is necessary to bear in mind the timing of imaging in relation to the onset of symptoms when interpreting the scans. Other factors which influence the rate of development of an abscess are the quantum and duration of bacteraemia, the virulence of the infecting organism and the host resistance, and the magnitude of preceding embolism.

The first stage is of early cerebritis with perivascular inflammatory change and oedema surrounding the area of brain necrosis. Late cerebritis is characterized by necrosis of the centre; fibroblasts and neovascularity appear at the periphery and begin to form a capsule. Astrocytic reaction is induced and white matter oedema spreads. Early capsule formation continues with the development

Table 43.2 Origin of brain abscess—predisposition and bacteriology

	Site of abscess	Organism
Ear	Temporal lobe/cerebellum	Streptococci Enterobacteriaceae Bacillus fragilis
Paranasal sinuses	Frontal/deep temporal lobe Subdural empyaema	Streptococci Staphylococci Bacteroides Enterobacteriaceae
Dental sepsis/ instrumentation	Multiple	Mixed includes: Fusobacteria Bacteroides Streptococci
Chest Arterio-venous malformation Endocarditis	Multiple	Streptococci Staphylococci Bacteroides Fusobacterium
Immune suppression	Multiple	Fungi Toxoplasma Enterobacteriaceae Nocardia Mycobacteria Any organism

of further fibroblasts, peripheral vascularity, oedema, and reactive astrocytosis. Late capsule formation proceeds after about a fortnight as it thickens and puts down collagen (Britt *et al.* 1981, 1983).

The sources of primary infection which seed brain abscesses are many. There is a degree of correlation between the source of infection, the part of the brain in which the abscess grows, and the pathogenic organism (Table 43.2):

- *Otogenic infection*, which is diminishing in frequency in the western world, but not in the third world, spreads to the temporal lobe (Fig. 43.1) or cerebellum. Such abscesses are likely to be caused by a mixed flora which may include aerobic or anaerobic streptococci, enterobacteriaceae, and *Bacillus fragilis*.

- *Paranasal sinusitis* predisposes to frontal lobe and deep temporal lobe abscesses and to subdural empyaema. The offending organisms include streptococci, including *Strep millerii*, staphylococci, bacteroides, and enterobacteriaciae.

- *Blood-borne infection* may settle anywhere in the brain and may give rise to multiple and multi-loculated abscesses. The territory of the middle cerebral artery is favoured and the abscesses occur near the grey-white matter junction where the blood flow is not so rich. The source is often in the chest, and may be a lung abscess or bronchiectasis, congenital heart disease, particularly with right to left shunt, or pulmonary arteriovenous fistula. The organisms are varied and include streptococci, anaerobic, microaerophilic, and *viridans*, bacteroides, fusobacterium, and staphylococci.

- *Peridontal infection* is another cause of bacteraemia, either spontaneous or caused to flare up by dental manipulation. This gives rise to mixed infections with fusobacteria, bacteroides, and streptococci. Other procedures which may give rise to bacteraemia are instrumentation of the oesophagus, gastrointestinal and urinary tract, and to pelvic structures.

- *Trauma or neurosurgery* can cause implantation of organisms to form abscesses which occur in relation to the wound or tract of the injury; the organisms are staphylococci, streptococci, clostridia, or enterobacteriaceae.

- *Bacterial meningitis* seldom gives rise to a brain abscess. It may be that this happens more commonly in infants in which case the organism is likely to be a gram negative bacillus.

In patients who suffer immune suppression by virtue of disease or its treatment, the central nervous system is vulnerable to attack by a very wide range of organisms, most of which would not be virulent enough to cause infection in normal circumstances. Because of the lack of an immune reaction to infection, such patients may not exhibit the clinical signs that would normally be found, and, by the time of presentation, the infection may be overwhelming. In such circumstances, the lesions are often multiple. Fungi, toxoplasma, enterobacteriaciae, nocardia, and mycobacterial infection have all to be considered in addition to the usual pathogens (Table 43.2).

The ability to isolate an organism from abscess pus depends very much on close liaison with the microbiology laboratory and it is prudent to involve the microbiologists from the very beginning of the patient's work up. They will need to know of any obvious foci of infection, whether antibiotics have been administered, and if there are any underlying predisposing factors. Care should be taken that there is no delay in transport of the specimens to the laboratory and that proper culture media are set up for fastidious organisms and for anaerobes and fungi. The yield of positive cultures can be increased significantly when such measures are taken (de Louvois *et al.* 1977).

Clinical features

The presentation of brain abscesses is variable and depends on the virulence of the organism, the host defences, the age of the patient, the severity of the primary infection, and the site and number of brain abscesses. The onset may be fulminant, leading to death within days or even hours, particularly if an abscess ruptures into the cerebral ventricles. In others, the evolution may be over months or longer. In most, the duration of symptoms by the time of presentation is 2 weeks, and the prominent clinical manifestations are due to an expanding intracranial mass, rather than the infection itself. The classic triad of fever, headache, and focal neurological deficit occurs in less than 50 per cent of patients. Headache occurs in most patients and there are no specific features. It may be generalized or localized, continuous or intermittent. Pyrexia occurs in only about 50 per cent. Focal signs referrable to the site of the lesion often develop, in about half of all cases, and may include epileptic seizures which occur commonly and are more likely to be generalized. Brain abscesses occupy space and together with the surrounding oedema, lead to raised intracranial pressure. Some disturbance of the conscious level and sensorium is usual, ranging from lethargy to coma in severe cases. Abscesses of the posterior fossa often present with signs of raised intracranial pressure.

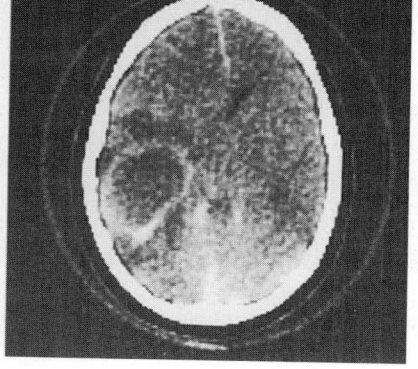

Fig. 43.1 Otogenic cerebral abscess in the temporal lobe shown by CT scan unenhanced (left) and contrast enhanced (right). (Courtesy of Dr Milne Anderson.)

Table 43.3 Infectious causes of granulomas and cysts

Bacteria
  Tuberculosis
  Nocardia
  Actinomycosis

Parasites
  Toxoplasmosis
  Cysticercosis
  Schistosomiasis
  Paragonimiasis
  Echinococcosis
  Amoebae

Fungi
  Aspergillus
  Blastomycosis
  Coccidiomycosis
  Mucormycosis

Prognosis is closely related to the level of consciousness at the time of admission. Neck stiffness may be evident and may suggest a degree of meningitis. If fever is present or if signs of infection elsewhere in the body are evident then the diagnosis is not usually too difficult to reach. Since these signs are absent as often as not, it is not uncommon to initially misdiagnose brain abscesses as intracranial tumours, infarction, haemorrhage with clot formation, necrotizing encephalitis, severe migraine, stroke, or meningitis.

CT brain scan typically shows a contrast enhancing ring surrounded by oedema (Fig. 43.1). Intracranial vasculitis and non-infective granulomas as well as malignant tumours and multiple sclerosis (Fig. 43.2) may all present clinically like an abscess, complete with raised temperature and scan appearances of ring enhancing lesions

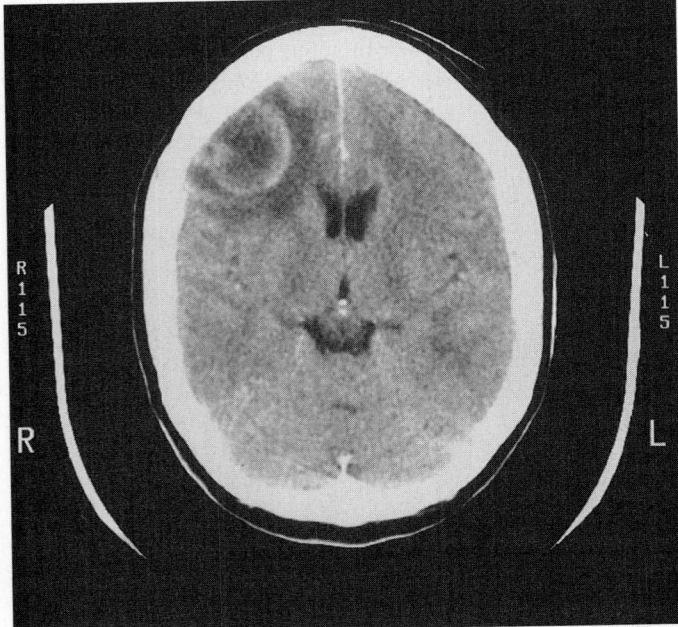

Fig. 43.2 CT brain scan of a patient with multiple sclerosis, showing a ring enhancing right frontal lesion, initially thought to be an abscess, which showed signs of acute multiple sclerosis on biopsy. (Courtesy of Dr Milne Anderson.)

(Nielsen *et al.* 1982; Chun *et al.* 1986). In young children and infants the clinical picture may be quite atypical. The head may enlarge from increasing intracranial pressure and may be associated with vomiting and epileptic seizures so mimicking a brain tumour or hydrocephalus from congenital abnormality.

43.1.3 Subdural empyaema

Pathogenesis

Subdural empyaema refers to the accumulation of pus in the potential space between the dura mater and arachnoid mater. It is more likely to collect over the convexities of the cerebral hemispheres, in the parafalcine region and above the tentorium cerebelli because brain is not as closely applied to the dura in these areas as it is in the region of the venous sinuses. Fewer than 10 per cent of subdural empyaemas occur in the posterior fossa. Subdural empyaema shares similar pathogenetic mechanisms with brain abscess and they not infrequently occur together. Subdural pus results from extension of infection from the ear or from paranasal sinusitis, or less frequently from osteomyelitis of the skull. Rarely, it may be caused by haematogenous spread or by direct implantation of organisms following cranial trauma. In young children the subdural effusions which accompany meningitis may become infected. The causal organisms reflect the underlying predisposition. In adults aerobic and anaerobic streptococci, staphylococci, and gram negative organisms predominate. In children and infants *Haemophilus influenzae*, pneumococci, and gram negative bacteria are the most frequent. It is important to remember that in circumstances favourable to the organism, any may become pathogenic, including fungi.

Clinical features

More than half of all patients are less than 20 years old and there is a male:female ratio of 2:1. The clinical presentation is acute, within 2 weeks, sometimes explosive, and usually includes overt signs of infection such as pyrexia, tachycardia, and rigors, perhaps with evidence of local infection; a painful tender face, pain over the mastoid process or tenderness on palpation over the paranasal sinuses. There may have been non-specific malaise for a few days. There are signs of raised intracranial pressure—drowsiness, decreased conscious level which may progress to coma, neck stiffness, papilloedema, and sixth or third cranial nerve palsies. Focal signs of hemiparesis, dysphasia, nystagmus, and ataxia may point to the region of the brain that is primarily affected, either by direct pressure and inflammation from the pus, or by vascular damage from superficial thrombophlebitis. Seizures are common, much more so than with intracerebral abscess, and are often focal. As many as two-thirds of all patients may be so afflicted. If subdural empyaema is suspected, anticonvulsants should be given straight away. A collection of pus in the parafalcine area of the brain may cause paraparesis and sphincter upset, mimicking a spinal cord lesion. The diagnosis should be considered in any patient with focal signs or symptoms accompanied by pyrexia and epileptic seizures (Bannister *et al.* 1981; Dill *et al.* 1995; Nathoo *et al.* 1999).

43.1.4 Extradural abscess

Also known as epidural brain abscess, this condition refers to the collection of pus in the extradural or epidural space. This represents the potential space which exists between the outside of the dura mater and the inside of the skull; normally there is no separation of the two. If sepsis occurs in adjacent structures such as the skull bones in mastoiditis or osteomyelitis, following trauma or paranasal infection, pus can spread to this space and strip the dura from the underlying bone, and in the process cause hyperaemia of the meninges and fibrin deposition and thrombophlebitis. Unfortunately, one of the increasingly frequent causes of introducing sepsis is neurosurgery (Hlavin *et al.* 1994). The organisms which cause extradural abscesses are those of the underlying cause of sepsis. Chief amongst them are streptococci, anaerobes, and staphylococci, but a variegation of organisms have been described including fungi.

Cranial extradural infection is rare and is less common than spinal extradural infection (Section 28.4.6). It may occur in any age group and in either sex. Unlike the other forms of intracranial suppuration, the signs of local infection tend to over-ride those of neurological dysfunction. Focal and neurological signs and raised intracranial pressure are uncommon. It is more likely that local pain, tenderness, and oedema will be found on examination of the skull. Localized headache is common as are signs of systemic infection. In fungal infections and in some of the more chronic bacterial infections, including tuberculosis, the evolution of the disease to hospital presentation may take months.

43.1.5 Investigation

A high index of suspicion is required in order to make an expeditious diagnosis of intracranial suppuration. In brain abscess signs of infection are often lacking, in subdural and extradural infection they are commonly present. All patients should have a full blood count and ESR examination carried out, together with electrolyte determination, C-reactive protein level, and blood cultures. Other blood tests should be carried out according to clinical indications. If there is an obvious site of systemic infection, specimens should be obtained for culture, if necessary by needle aspiration, and X-rays obtained according to clinical need. Lumbar puncture to examine CSF should not be undertaken if these pathologies are suspected, because each may be associated with a rise of intracranial pressure and associated shift of brain compartments.

The best way to diagnose suspected intracranial suppuration is to image with CT or MRI. MRI is in general more sensitive than CT but it does have limitations when the patient is acutely ill. The changes seen on scans must be interpreted with critical attention to the time at which they were undertaken in relation to the clinical evolution of the disease. It is always best to obtain the advice of a neuroradiologist if possible, because the changes may be subtle and similar to those produced by several other disease states. Repeat scans may be necessary to determine if there has been change in the radiological appearances for these can occur rapidly and when they happen, may serve to distinguish an abscess from an area of infarction or a tumour. It is also necessary to use contrast enhancement with CT, otherwise lesions may be missed (Fitzpatrick *et al.* 1999). On CT scan the earliest stages of the evolution of a cerebral abscess, cerebritis, appears as low density abnormality which may have some surrounding oedema and a little space occupation. Later, surrounding oedema and space occupation increase and the centre of the abscess develops a much lower density, and as the capsule of the abscess matures, ring enhancement following the injection of contrast media, takes place (Fig. 43.1). It is possible to identify the location and number of lesions, and the presence of hydrocephalus and mass effect.

Unfortunately, such imaging changes, although characteristic, are not specific for they may be seen with high grade gliomas and metastases and with some forms of granuloma. Haemorrhagic infarction and bleeding from arteriovenous anomalies has also caused confusion, as has relapsing multiple sclerosis. The condition of the paranasal sinuses, including the sphenoidal sinus, and mastoid air cells should be noted, and search made for skull fractures and cranial defects. MRI is superior to CT in the early detection of cerebritis and in determining the extent of white matter change. The application of newer techniques such as diffusion weighting MRI promises greater discrimination between abscesses and other pathology in the future (Schroth *et al.* 1987; Kim *et al.* 1998; Desprechins *et al.* 1999).

The appearances of subdural empyaema are of an extracerebral collection of low density with a rim of enhancement following contrast injection (Fig. 43.3). Extradural collections show as low density areas inside the skull with a thick densely enhancing ring at the periphery (Fig. 43.4). Changes of infection may be evident in the adjacent bone (Tsuchiya *et al.* 1992; Nathoo *et al.* 1999).

43.1.6 Management

The sooner that treatment with suitable antibiotics is given, the better is the prognosis for all forms of intracranial suppuration. If there is likely to be a delay before imaging or surgery or even transfer to a specialist unit, empirical antibiotic treatment should be given immediately. The regime can be altered later if necessary. Diligent search must be made for the source of infection, specimens taken, and the source eradicated. Supportive treatment aimed at the maintenance of circulation and the correction of hypoxaemia should be given. Electrolytic upset and derangement of serum osmolality are not uncommon and need to be monitored and corrected.

Raised intracranial pressure

Raised intracranial pressure is often a problem. There are no clear data on the effects of steroid treatment, beneficial or otherwise, when used to reduce intracranial pressure in these conditions. Most practitioners use steroids if the clinical situation is deteriorating, especially if the imaging studies show significant oedema and mass effect. More acutely, pressure monitoring is advisable if it is available, and hyperventilation and mannitol can be given in the short term. We do not know if such measures are consistently beneficial in the long term.

Seizures

Acute seizures should be treated with benzodiazepines, and prophylaxis should be started with phenytoin, carbamazepine, or valproate; treatment is also recommended for patients without seizures that have epileptic discharges on the EEG. However, whether all patients should receive prophylaxis is still debated, and there are few good data on which to base the decision. Studies of seizure risk following traumatic brain injury have shown prophylaxis is associated with a reduction of early, but not late seizures (The Brain Trauma Foundation 2000). For patients in whom prophylactic treatment has been started, it is recommended that treatment continues for 6–12 months, and that it is only withdrawn if the patient is seizure free, the EEG normal, and the CT shows only residual abscess but no active contrast enhancement, suggestive of ongoing inflammation (Mathisen *et al.* 1997).

The incidence of epileptic attacks is very high in subdural empyaema and some advocate that prophylactic anticonvulsants should be given as soon as that diagnosis is suspected. Whether they are needed in the longer term can be reviewed.

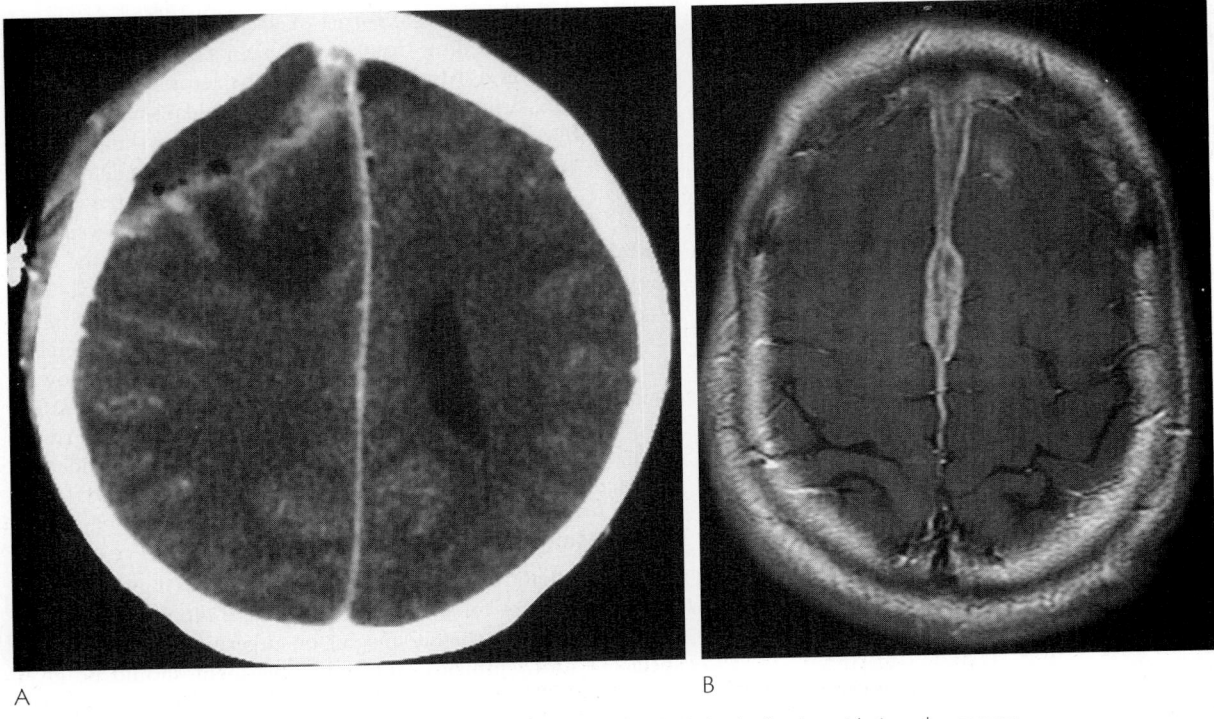

Fig. 43.3 Subdural empyaema. (A) CT scan with contrast showing right frontal subdural collection with rim enhancement and mass effect. (B) Axial T1-weighted MRI after gadolinium showing an enhancing bilateral parafalcine collection. (Courtesy Dr Ian Turnbull.)

Surgical aspiration

In all forms of intracranial suppuration, samples of pus and infected tissue should be obtained as soon as possible. There is no agreement on which form of surgery is best, or when it should be timed, in the treatment of brain abscess. If accessible and not in too risky a location, an abscess should be aspirated surgically.

Stereotactic techniques have improved precision and biopsy yield and have reduced surgical morbidity (Apuzzo *et al.* 1987; Barlas *et al.* 1999). The appearances should be monitored by repeat scanning and if the lesion or lesions do not show reduction in size, they should be re-aspirated, and the antimicrobial treatment regime re-assessed. In this respect, MR probably does have some advantage

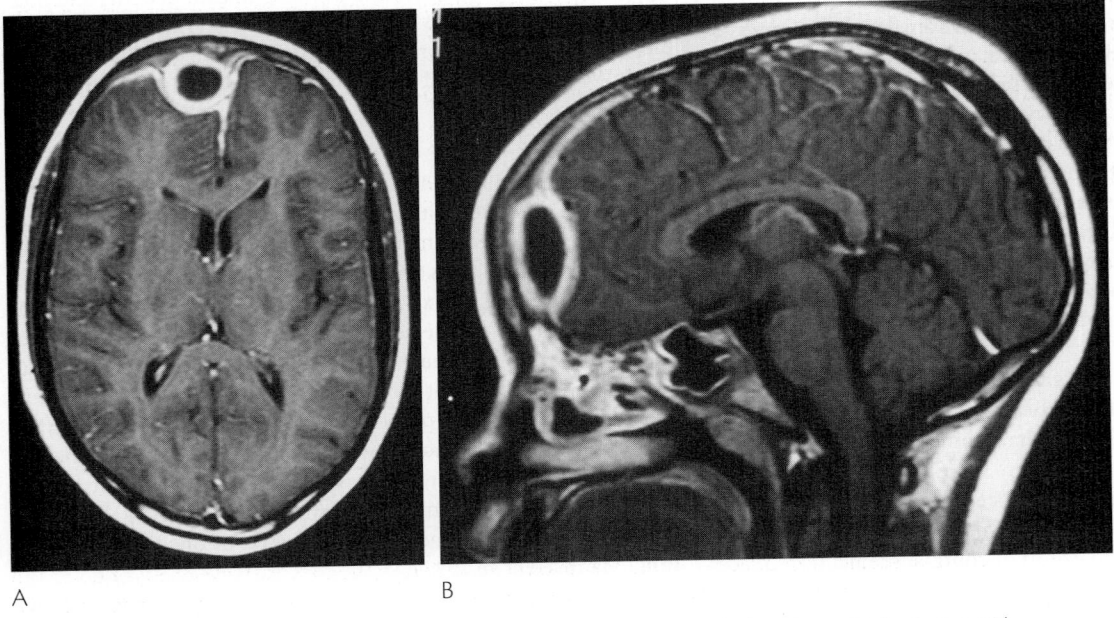

Fig. 43.4 Extradural abscess. (A) Axial T1-weighted MRI with gadolinium showing a right frontal extradural collection with uniform ring enhancement extending laterally over the dura. (B) Sagittal T1-weighted MRI with gadolinium showing the same collection. Note the enhancing material within the underlying frontal paranasal sinus, which is the most common source of this infection. (Courtesy Dr Ian Turnbull.)

over CT for monitoring progress; the scan detail is rather more precise and there is no known radiation hazard of repeat MR scanning. Multiple abscesses should be treated in this way too. There is some evidence that adequate reduction in abscess size and eventual resolution can be obtained by medical treatment alone without the need for aspiration (Boom *et al.* 1985). While this is possible, it is undoubtedly better to have a specimen of infected tissue to determine the nature and the antibiotic sensitivity of the infecting organism.

Immunosuppressed patients

If the patient is immune suppressed, the threshold for surgical aspiration is lower because the spectrum of organisms which may be causing the abscess is so much wider; in addition there is also a high incidence of non-infective space occupying lesions such as lymphoma. In patients with AIDS, toxoplasmosis is the likely cause of lesions which look like abscesses. However, the appearances are not specific. A reasonable course of action is to treat for toxoplasmosis with pyrimethamine and sulphadiazine, with folinic acid to counter pyrimethamine effects on haematogenesis, and to monitor progress closely. If the appearances are atypical, if blood serology for toxoplasmosis is negative, if there is clinical progression, or if there is no evidence of clinical and radiological improvement within a fortnight on treatment, aspiration should take place (Section 43.2).

Subdural and Extradural Collections

Surgical treatment is necessary for subdural empyaema. Burr-hole aspiration alone may be insufficient because it is possible to miss small loci of pus, and craniotomy with wide exploration of the exposed brain is recommended. (Bannister *et al.* 1981; Feuerman *et al.* 1989). Widespread exploration, excision of infected material and bone, and later cranioplasty, are used for extradural infection.

Choice of antibiotic

The choice of antibiotic will be determined by several factors which will include the likely causal organism, the ability of the drug to penetrate into CSF and into the pus in the abscess cavity, and its ability to be active in the presence of pus. Our knowledge of penetration and activity of the newer antibiotics is not good.

For community acquired abscesses in the immunocompetent, the recommended empirical treatment is now a third generation cephalosporin such as ceftriaxone or cefotaxime, which provide good gram positive and gram negative activity, plus metronidazole, which is effective against anaerobes. For infections acquired during cranial surgery or trauma there is a higher risk of methicillin-resistant *Staph. areus*, *Staph. epidermidis*, and multi-resistant enterobacteriaceae; thus meropenem plus vancomycin is recommended.

Immunosuppressed organ transplant recipients with brain abscesses should be treated with amphotericin B, and trimethoprim-sulfamethoxazole to cover the common pathogens in this risk group—*Aspergillus* species, *Candida* species, *Nocardia*, and *Toxoplasma gondii* (Table 43.2).

Further antibiotic treatment should be tailored according to the results of cultures and it is prudent to discuss the best treatment regime with one or all of bacteriologist, pharmacologist, and infectious diseases physician. The length of time for which drugs should be given and the frequency that blood levels should be checked, depend upon the circumstances of the individual patient and improvement of scan appearances. In this respect, it is important to recognize that abscesses which appear to be cured clinically may

show imaging changes, including contrast enhancement, for months afterwards. Parenteral treatment is usually given for 1–2 months, followed by a variable regime of oral treatment, if appropriate. Unfortunately, we do not have trial data on which to base decisions about antibiotic regimes.

Complications

The complications of an intracranial abscess include:

- seizures;
- obstructive hydrocephalus;
- intraventricular rupture of the abscess resulting in ventriculitis or meningitis; and
- brain oedema and herniation.

43.1.7 Outcome

The introduction of CT brought about a dramatic reduction in the mortality rate for brain abscess and for subdural empyaema, and to a lesser degree, for extradural haematoma. There had been a reduction with the introduction of antibiotics, but the ability to localize the lesions accurately, and to monitor their progress, has been the major factor in reducing death rates from 80 per cent in the pre-antibiotic era, to contemporary levels of less than 5 per cent. Doubtless many other factors have contributed to this improvement, including the introduction of stereotactic surgical techniques, the development of newer antibiotics, and improvement in intensive care. Nevertheless, it remains true that delay in diagnosis and the administration of antibiotics, has an adverse effect on the outcome. Increasing impairment of conscious level also carries a poor prognosis. Choice of the wrong antibiotic may be another factor (Yang 1981). The mortality rate for the treatment of subdural empyaema has been variably reported to be 20–25 per cent.

43.1.8 Sequelae

The major sequelae suffered by survivors of brain abscess are epilepsy, focal neurological deficit, and mental changes. Epilepsy has occurred in more than 50 per cent of survivors and is more frequent if the patient is young, if the abscess has been in the frontal, temporal, or parietal region, and if seizures have occurred in the acute phase. The longer the follow up, the greater is the number of survivors who will develop seizures. Generalized seizures are more common than focal and usually begin within 5 years. In the United Kingdom, because of the high risk of seizures in this patient group, the regulations require the patient to be banned from driving for a minimum period of 1 year from treatment of a cerebral abscess, or a subdural empyaema. For a professional driving licence, the prospective risk for the development of epilepsy is so high, that revocation of the licence is required, and restoration will not be contemplated until the applicant has been seizure free for 10 years. Focal deficits occur in between 25 and 50 per cent of survivors. Mental changes are almost certainly under-reported and affect at least 20 per cent.

After subdural empyaema, improvement from focal disturbance is much better. Epilepsy is a considerable problem. Cowie and Williams (1983) found that in their series, 63 per cent of cases had seizures during the acute phase of the illness (Cowie *et al.* 1983). Of these, only 29 per cent continued to have seizures. However, of those who had no seizures in the acute phase, 42 per cent went on to have fits. There is a case to recommend prophylactic treatment with anti-epileptic drugs to all patients with intracranial and subdural abscess, but there is no evidence that this protects against the development of epilepsy.

43.2 **Intracranial granulomas and cysts**

A range of bacteria, fungi, and parasites can cause intracranial granulomas and cysts (Table 43.3). In some cases, the same agents which cause meningitis and, sometimes encephalitis, can also induce foci of inflammation within brain parenchyma, which enlarge to cause solid granulomas or fluid-filled cysts which do not suppurate. The factors which lead the same infectious agent to give rise to meningitis in one patient, and a granuloma in another, or both in the same patient, are not fully understood. In such cases, it will be readily recognized that the clinical picture may be complicated. Granulomas and cysts often present with focal signs, features of space occupation and with seizures, and may be complicated by the clinical features of meningitis. The clue to granuloma formation is the appearance of focal signs in the context of meningo-encephalitis and the diagnosis can be confirmed by brain imaging.

The same clinical picture is caused by a wide variety of pathogens. It is usually not possible to differentiate these syndromes without cranial imaging, which has revolutionized management and greatly improved the prognosis for most of these conditions. Signs of infection elsewhere in the body, perhaps peculiar to a particular organism, the geographical location or the seasonal timing of an outbreak, or the association with an occupation or leisure pastime which may provoke exposure to a specific disease, can all provide clues to the cause. In a significant number of cases, even after these factors have been determined, there may still be a choice of diagnoses and it is necessary to revert to surgical biopsy of the lesion. As with other infections, delay in making a diagnosis leads to a poorer prognosis.

Other bacterial infections which may cause cerebral granulomas include tuberculosis (Section 43.2.1), nocardia (Section 43.2.2), actinomycosis (Section 43.2.3), syphilis (Section 42.5.1), and Brucella (Section 42.5.11).

43.2.1 **Tuberculoma**

The aetiology and pathogenesis of intracranial tuberculomas is similar to that of tuberculous meningitis (Section 41.5.3). Mycobacteria reach the brain via the bloodstream and from small tubercles; rather than rupturing into the subarachnoid space to cause meningitis. The tubercles continue to grow and become walled off by a fibrous capsule which surrounds a caseous core containing Langhans cells, epitheliod cells, lymphocytes, and sometimes organisms. The lesions may be single or multiple, and may vary in size from small nodules to a large, well-circumscribed tumour several centimetres in diameter which may be macroscopically indistinguishable from a meningioma—to the degree of inducing its own blood supply from the meninges. The consistency and degree of fibrous encapsulation can vary too, so that a tuberculoma may be mistaken for a glial tumour. Not all tuberculomas are discreet and nodular and it is possible to have an extensive multilocular tuberculoma, or a thin film of tuberculoma tissue over the surface of the brain adherent to the meninges, which is known as 'tuberculoma en plaque'. Rarely, but less so now than formerly, the centre of the tuberculoma may liquefy to become an abscess, which some authors suggest may cause a more florid and toxic clinical presentation.

Before the advent of antibiotic chemotherapy, tuberculomas were common, accounting for 34 per cent of space occupying lesions found in a necropsy series in Leeds (Garland et al. 1933). In the developing world where tuberculosis remains endemic, tuberculomas still account for a similar proportion of brain space occupying lesions, to the degree that it is often appropriate to treat for tuberculosis as soon as a space occupying lesion is found, and monitor the clinical status. Biopsy or surgical removal is carried out if there is no improvement or the patient deteriorates. Formerly, tuberculomas affected children in the main, but now the emphasis is shifting to adults and any age group may be affected. Approximately 10–15 per cent of cases of tuberculous meningitis are complicated by the development of tuberculomas, and this may happen even with adequate and appropriate chemotherapy (Fig. 43.5). Similarly, further tuberculomas may develop as the index one resolves (Fig. 43.6). In the majority of cases, tuberculomas develop as the sole neurological syndrome, and there is commonly no clinical evidence of tuberculosis elsewhere in the body. In such cases the diagnosis may be inferred by the racial origin of the patient or by travel from a known tuberculous region.

Clinical features

Tuberculomas and abscesses may occur in any area of the brain and the symptoms and signs correspond to the area of the brain affected. Multiple tuberculomas may produce a confusing clinical picture resembling multiple metastases. Presentation may be as a space occupying lesion with symptoms and signs of raised intracranial pressure, with focal neurological signs, or with epilepsy, either generalized or focal. Fever and signs of systemic infection are rare. The presentation of the syndrome is seldom acute, tending to evolve over weeks or months although exceptions do occur. There is usually no evidence of tuberculosis elsewhere in the body, although the diagnosis becomes easier if there is. Similarly, clinical and laboratory indices of infection or inflammation are lacking, more often than not. The patients often appear very well clinically. Indeed the discovery of a large space occupying lesion in a patient that is surprisingly well is felt by some to be almost pathognomonic of a tuberculoma. Brain imaging with CT or MRI has revolutionized the management of this condition. The appearances are not diagnostic and may mimic virtually any other form of intracranial space occupying lesion (Selvapandian et al. 1994). The usual appearance of a tuberculoma is of a low or equi-dense lesion surrounded by oedema, often with microscopic calcification, and a surrounding rim of enhancement following contrast. Lobulated rings and discs may also be seen.

Management

A search must be made for evidence of tuberculosis elsewhere in the body and must include chest X-ray. Mantoux testing is positive in about 2/3 of cases but this is not diagnostic. An immunological test of the ability of T-cells to produce interferon gamma in response to tuberculosis antigens, the ELIspot, has proved useful for diagnosis of peripheral tuberculosis (Ewer et al. 2003), but its role in tuberculoma is unclear. In those cases who do not show evidence of other tuberculous infection, a definitive diagnosis can be made only by examination of tissue obtained by biopsy. In patients who present with single or multiple lesions and in whom space occupation is not a problem, and who are of a recognized at-risk

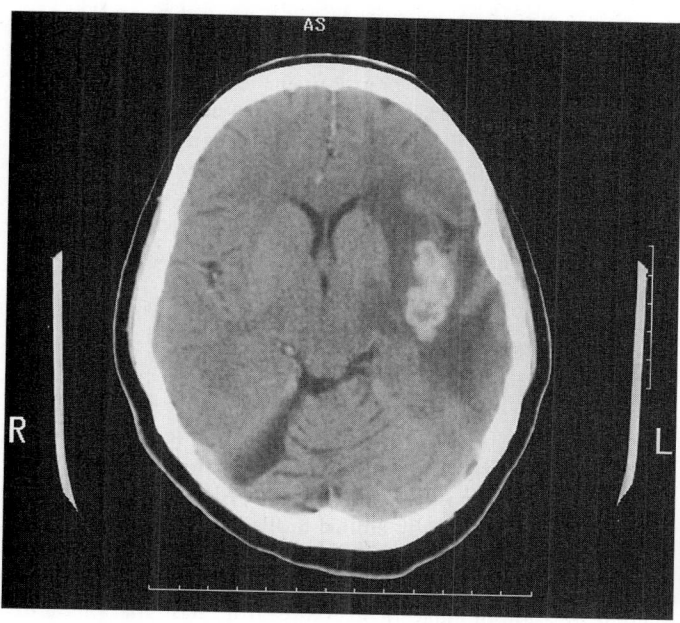

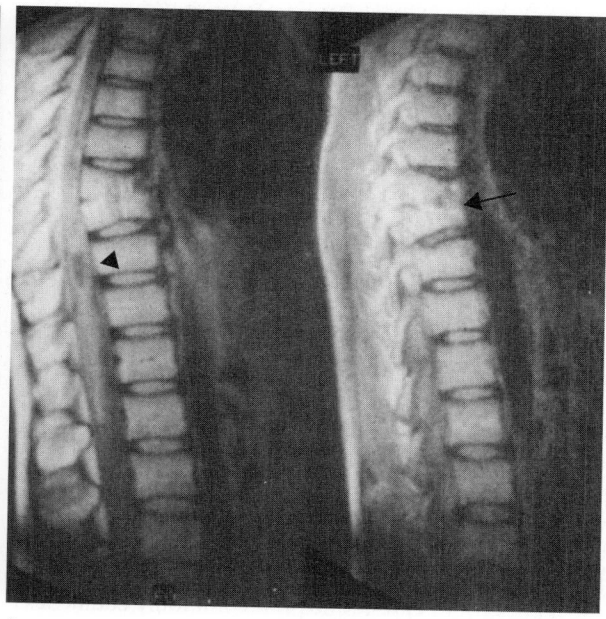

A

B

Fig. 43.5 Tuberculomas. (A) CT brain scan showing a developing tuberculoma in the left Sylvian fissure. The Indian lady had been treated with sensitivity-proven anti-tuberculous drugs throughout. (B) Sagittal section of an MRI of the thoracic spine showing a syrinx (arrowhead) which has developed 10 years after recovery from tuberculous meningitis. Bony involvement of adjacent vertebral bodies is evident (arrow). CSF dynamics were deranged at the foramen magnum due to meningeal adhesions. (Courtesy of Dr Milne Anderson.)

racial grouping, some would advocate treating presumptively with anti-tuberculous chemotherapy, and closely monitoring the appearance of the lesions with serial scans, and with frequent clinical assessments. If there is evidence of scan or clinical deterioration, then surgical biopsy to confirm the diagnosis is necessary. Others advocate the need for a tissue or microbiological diagnosis from the outset. This is becoming increasingly important with the spread of multidrug-resistant tuberculosis, and also markedly drug-resistant tuberculosis. This need not necessarily require

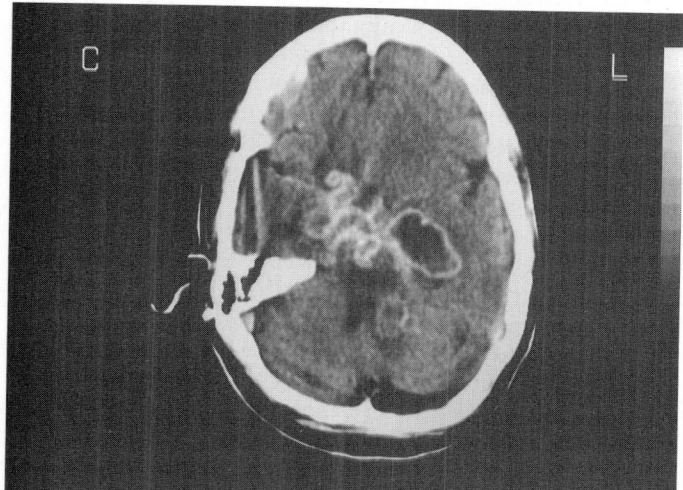

Fig. 43.6 Tuberculous abscess and tuberculomas which developed in an Indian lady who did not take prescribed anti-Tuberculous chemotherapy in her homeland. (Courtesy of Dr Milne Anderson.)

biopsy of the lesion; evidence of tuberculous in the cerebrospinal fluid or elsewhere in the body would suffice. Unfortunately, tuberculomas can enlarge or develop *de novo* despite anti-tuberculous chemotherapy, and this may be a source of confusion. The threshold for biopsy is much lower in patients with immune suppression or AIDS and who have lesions resembling abscesses. There does not seem to be a great risk of causing tuberculous meningitis or dissemination of tuberculosis to other parts of the body from these procedures, provided the patient is already on chemotherapy. Approximately 60 per cent of tissue specimens from tuberculomas have acid fast bacilli on smear, and 50 per cent will grow *M. tuberculosis*. However caseating granulomas will be seen in almost all cases.

Once the diagnosis of tuberculoma has been reached an anti-tuberculous drug regime is decided upon, appropriate to the circumstances of the patient, in the same way as treatment for tuberculous meningitis (Section 41.5.3). Discussion with an infectious diseases physician experienced in the treatment of tuberculosis who will have knowledge of local patterns of drug resistance, is always recommended. Notification to the public health services will result in tracing family and work place contacts. The length of time for which treatment should be given approximates to that for tuberculous meningitis and is influenced by improvement in scan appearances. Raised intracranial pressure can be treated with dexamethasone and other agents if needed, apparently without causing increased hazard of disease dissemination. Epilepsy is controlled with anticonvulsant drugs. Provided the diagnosis is made expeditiously, and a full course of anti-tuberculous treatment given, the outcome is usually excellent, and complete recovery is possible. The prognosis is not good for drug-resistant strains, or for patients with AIDS.

Occasionally the casesous core of a tuberculoma liquefies to form a tuberculous abscess, which is radiologically indistinguishable from other pyogenic abscesses. Clinically such patients are said to be more toxic than those with tuberculoma, and the response to treatment is poor.

43.2.2 Nocardiosis

Nocardia are aerobic gram positive bacteria which grow in tissue as a widely spreading mycelium; thus although they are bacteria, microbiologically they appear similar to fungi, and some of their clinical features are reminiscent of fungal infections. Nocardiosis is an opportunistic infection of man which occurs in patients who are immune-suppressed by reason of treatment or disease: lympho-reticular neoplasia, organ transplantation, connective tissue disease, malignancy, pulmonary alveolar proteinosis, chronic high dose glucocorticoid treatment, and HIV infection. Six species of *Nocardia* cause disease in man, *Nocardia asteroides* predominates and infection takes the form of invasive pulmonary disease acquired by inhalation, with multiple abscess formation which may disseminate via the bloodstream and seed to cause brain abscesses there in as many as 20 per cent of cases. Much less commonly cutaneous infection by the other species produces cellulitis and mycetoma.

The neurological manifestations of nocardial disease are those of brain abscess with signs of space occupation and raised intracranial pressure, focal deficit, and epilepsy. There is usually evidence of pulmonary infection ranging from localized pneumonitis to fulminating pneumonia and lung abscess. There is nothing particularly characteristic of nocardial brain abscess, and the clue to diagnosis comes from the underlying associated condition. Rarely, there may be meningitis or invasion of adjacent skull bone with osteomyelitis. Brain imaging with CT or MR shows characteristics of brain abscess with cystic space occupation, rim enhancement and surrounding oedema, and the lesions may be multiple. Diagnosis is by demonstration of the organism in pus obtained from a lesion by stereotactic biopsy. Some advocate surgical excision of the abscesses. Treatment is with sulphonamides and trimethoprim for at least a year. Minocycline or amikacin together with a third generation cephalosporin is an alternative combination (Mamelak *et al.* 1994). Mortality rates are high and vary from 30 to 60 per cent.

43.2.3 Actinomycosis

Actinomyces which cause actinomycosis are gram positive anaerobic bacteria and are normal commensals of the mouth and pharynx of humans. Virtually all human infection is caused by *Actinomyces israeli*. The organisms invade the tissue when trauma, usually minor, occurs to skin or mucous membrane. Risk factors include dental extraction and caries, trauma to the head, chronic middle ear disease, and osteomyelitis (Sundaram *et al.* 2004). Once in the tissues, they cause suppuration and inflammatory change and induce small granuloma formation which coalesces to form abscesses. These spread by invasion of adjacent tissue and pus is discharged through the sinuses which are formed. Neurological disease is rare, complicates no more than 3 per cent of all cases, and results from direct spread of craniofacial disease, or haematogenous dissemination from lung disease. The most common lesion is a brain abscess. Subdural and epidural abscesses are less common. Meningo-encephalitis and actinomycotic granulomas have also been described. Brain scan appearances are not diagnostic and mimic other kinds of abscess and skull osteomyelitis. Diagnosis is not difficult if there is evidence

of suppurating, discharging sinuses, and is confirmed by demonstration of the characteristic sulphur granules and demonstration of the organism in pus or biopsy tissue. Treatment is by surgical extirpation of diseased tissue as far as is possible, and drainage of pus. Penicillin is the drug of choice, and prolonged treatment for up to 1 year may be required. Tetracycline or erythromycin are alternatives. Secondary infection is common and requires treatment according to culture results. The overall mortality is about a quarter and a substantial proportion of survivors have neurological abnormalities (Smego 1987).

43.2.4 Aspergillosis

Aspergillus species are saprophytic fungal moulds which are found throughout the world in soil and water and in vegetable matter. Transmission to humans occurs through inhalation of the fungus, consequently the brunt of infection falls upon the lungs and the upper respiratory passages. In people with an intact immune system, infection is usually limited to the lungs and nasopharynx, where abscesses, granulomas, pneumonia, and chronic granulomatous disease develop. There is seldom spread elsewhere; sometimes there is invasion of adjacent structures from nasopharyngeal disease. In the immune compromised, invasive aspergillosis results and the disease spreads by the bloodstream to other organs including brain. Those conditions which predispose to invasive disease include neutropaenia, haematological malignancy, organ transplantation, chronic corticosteroid treatment, intravenous drug abuse, and other chronic debilitating states. Several varieties of *Aspergillus* can cause human disease, and the vast majority of cases are caused by *A. fumigatus* and *A. flavus*. Aspergillosis is found worldwide, in all races, both sexes, and in any age group from neonate to old age. Central nervous system disease usually results from haematogenous dissemination but it may also occur by direct extension from paranasal or orbital granulomas or via emissary veins. The pattern of neurological involvement varies: direct extension results in abscess formation in the frontal and temporal lobes. This may be single or multiple and can develop over weeks or months and is often associated with granuloma formation. By contrast haematogenous dissemination causes acute disease with multiple necrotizing abscesses. Because hyphae of the fungus invade blood vessels, such infection gives rise to thrombosis and infarction, and in some cases, true 'mycotic' aneurysms are formed. Stroke syndromes are therefore commonly found with *Aspergillus* infection.

The clinical syndromes of aspergillosis are many and reflect the area of brain affected and the underlying lesion. Abscesses and granulomas present as space occupying lesions with raised intracranial pressure and focal signs and epilepsy. Basal granulomas cause multiple cranial nerve palsies and ophthalmoplegia. Arterial invasion results in abrupt stroke syndromes. Rarely, meningitis may occur.

There are no specific features which point to a diagnosis of aspergillosis. It should be suspected in an immune suppressed patient who develops a brain abscess, granuloma, or stroke. Brain imaging will show abscess formation and infarction or haemorrhage, perhaps with paranasal granulomas (Ashdown *et al.* 1994; DeLone *et al.* 1999). Diagnosis is by demonstration of fungal elements in biopsy tissue. Serological tests have not been helpful. Recent reports suggest that polymerase chain reaction for *Aspergillus* DNA in CSF is useful (Kami *et al.* 1999) but lumbar puncture may not always be safe when there is space occupation.

Once the diagnosis has been established, treatment is with amphotercin and surgical extirpation of the lesion or lesions whenever possible (Denning *et al.* 1990) Concomitant therapy with rifampicin, ketoconazole, and itraconazole has been tried and individual reports have noted success. The outcome is often poor and the mortality rate exceeds 50 per cent for the immune compromised (Nadkarni *et al.* 2005).

43.2.5 Blastomycosis

Blastomyces dermatidis is a dimorphic yeast which was formerly thought to be peculiar to south and central USA and the Great Lakes region. Most cases have been recorded from these areas but reports from Africa and South America and rarer reports from other regions have confirmed that the organism is ubiquitous and probably resides in the soil. Infection is acquired by inhalation, occurs almost always in males who have worked outdoors or who have engaged in outdoor pursuits. It produces an acute chest infection which is usually self-limiting and has similarities to coccidiodomycosis and histoplasmosis. Increasingly cases are being seen in AIDS patients. In a few, dissemination to other organs takes place by the bloodstream. If the central nervous system is involved a chronic meningitic syndrome occurs and rarely, an abscess or granuloma may develop in the brain. Invariably, there is evidence of pulmonary or cutaneous disease. The clinical and imaging features are similar to other central nervous system granulomas and abscesses and diagnosis depends upon the demonstration of fungal material on biopsy specimens (Roos *et al.* 1987). Serodiagnostic and skin tests are suggestive but not conclusive. Treatment of central nervous system disease is with amphotercin, ketoconazole, and itraconazole.

43.2.6 Coccidioidomycosis

Coccidioidomycosis is caused by the dimorphic fungus *Coccidioides immitis* which exists in soil in southern USA, Mexico, and parts of South America. The spores are inhaled by people who work in or visit these areas and give rise to pulmonary infection which is asymptomatic in the majority. The others develop a self-limiting 'flu like illness with fever, cough, often accompanied by erythema nodosum, 1 week to 1 month after infection. In less than 1 per cent, spread beyond the respiratory system occurs, of whom one-third develop central nervous system involvement, almost always a form of chronic meningitis which comes on weeks or months later. Blacks, pregnant women, people on steroids, diabetics, and people with AIDS, are at greater risk of development of central nervous system complications (Bronnimann *et al.* 1987).

The clinical features of coccidioidomycotic meningitis are no different from any other form of meningitis. It produces a chronic mononuclear response with a high eosinophil count in the CSF. There is a granulomatous response in the meninges. Focal signs may be caused by granuloma formation, abscess, or from infarction consequent upon obliterative arteritis (Bouza *et al.* 1981) Brain imaging is necessary to demonstrate these features and to determine whether there is hydrocephalus prior to CSF examination. Diagnosis is by demonstration of raised complement fixing antibodies in CSF; the titres tend to reflect activity of disease and can be used to monitor the progress of therapy. It is possible to culture the fungus from CSF in up to one-third of cases, but this takes time.

Treatment is with fluconazole and amphotercin which may need to be instilled intrathecally. It is usually necessary to continue treatment for a prolonged time and relapse is common. The mortality rate is high.

43.2.7 Mucormycosis

Mucormycosis refers to infection with fungi of the Mucoraceae family. It is used interchangeably, although semantically inaccurately, with Phycomycosis and Zygomycosis; the class of fungi named Phycomycetes or Zygomycetes includes the family Mucoraceae. The major pathogens in this group are of the genera *Rhizopus*, *Absidia*, *Mucor*, and *Cunnunghamella*. They are found throughout the world and they grow in soil and decaying vegetable matter. Less than 5 per cent of cases occur in normal hosts. Seventy per cent occur in diabetics who become acidotic, and acidosis from other conditions can predispose to development of rhinocerebral mucormycosis. Other groups at risk are the severely malnourished, the immune suppressed, particularly those who have lymphoma and blood malignancy, and those who have suffered extensive trauma (Lehrer *et al.* 1980). Recently, there have been reports of association with treatment by deferoxamine (Vlasveld *et al.* 1991). The spores are inhaled or inoculated through damaged skin or mucosa. An acute suppurative pyogenic necrosis is produced with granuloma formation if the process is more chronic. These fungi have an affinity for arteries which they penetrate, inducing thrombosis and infarction and causing distal embolization and the formation of true mycotic aneurysms. Central nervous system disease occurs in about one-third of cases and the portal of entry is by the bloodstream or by contiguous spread from palate or paranasal sinuses and orbit.

The classical presentation is of painful proptosis asscociated with visual loss in a diabetic. The fungus then extends by the venous system into adjacent brain. Nasal ulcers and cutaneous necrosis are not uncommon. The patient is ill and there is usually pyrexia. As contiguous cranial nerves are picked off, multiple palsies occur and infarction of blood vessels leads to focal neurological signs. Abscesses distant from the site of local infection may be set up and cause confusing localizing signs. Progression of disease is rapid and secondary infection may occur.

Brain imaging with CT or MR will show invasion of paranasal and orbital spaces and brain, by granuloma, together with destructive bony erosion and areas of infarction (Fig. 43.7). Confirmation of the diagnosis is by demonstration and subsequent culture of the fungus from scrapings or biopsy specimens. CSF examination is not helpful, nor is blood serology. Treatment is by surgical extirpation of the fungal material insofar as is possible, and this may require the combined efforts of neurosurgeon, ear, nose and throat surgeon, maxillo-facial surgeon, and ophthalmic surgeon. At the same time antifungal agents should be administered, the best of which in this circumstance, is amphotercin B. Ketoconaxole has been reported to be effective. Equally important is prompt and aggressive treatment of the underlying predisposing condition such as diabetic ketoacidosis. A few survivors have been reported but rhinocerebral mucormycosis remains the most rapidly fatal of central nervous system fungal diseases.

43.2.8 Toxoplasmosis

Toxoplasmosis is caused by infection with the intracellular coccidian protozoan *Toxoplasma gondii*. The gondi is a North African rodent. Currently most clinically evident toxoplasmosis is associated with HIV infection, but transplant recipients, and patients

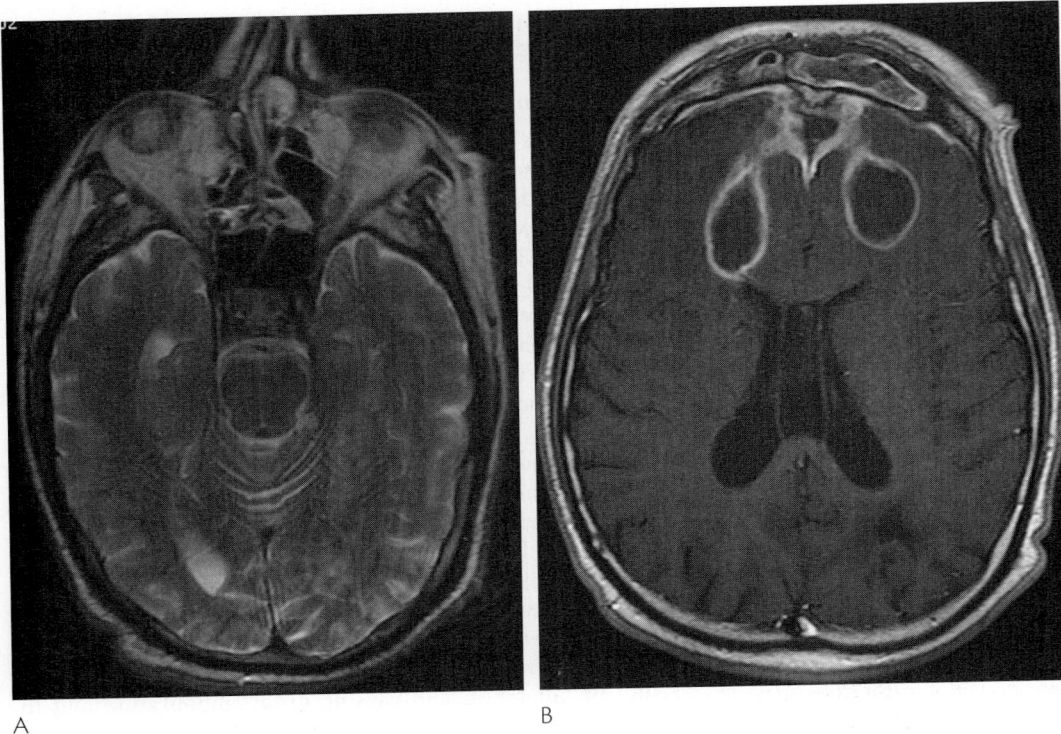

A   B

Fig. 43.7 Mucormycosis in an insulin-dependent diabetic patient. (A) Axial T2-weighted MRI showing a lesion of mixed signal in the left ethmoid paranasal sinus. The central low density represents the fungal ball, a valuable radiological clue to a fungal sinusitis. (B) Axial T1-weighted MRI with gadolinium showing diffuse enhancement of the frontal paranasal sinuses and adjacent leptomeninges and bilateral abscess cavities within the frontal lobes, with mass effect effacing the frontal horns of the lateral ventricles. (Courtesy Dr Ian Turnbull.)

with haematological malignancies such as Hodgkin's disease are also susceptible.

Life cycle of toxoplasma

Cats are the definitive hosts, and become infected by eating animals, usually rodents, that contain cysts in their tissues, or by oocysts passed in the faeces of other cats. The sexual phase of reproduction occurs in the cat gut, and results in the excretion of millions of oocysts. These may remain infectious in the environment for up to a year in the favourable conditions of a warm humid climate. The asexual phase of the parasite's life cycle occurs in intermediate hosts. After oocysts are ingested they penetrate the gut wall, replicate, and spread through the body to form cysts within the host cell cytoplasm. Here they lie indolent until, if the intermediate host is a rodent, they are ingested by a cat to complete the life cycle. In addition to rodents almost any warm blooded animal, including humans, can become infected by ingesting foodstuffs or water contaminated by cat droppings, or by eating infected meat which has not been fully cooked. In addition transplacental infection occurs. Most infections in normal immunocompetent hosts are of no clinical significance. Several factors influence human susceptibility including proximity to cats, dietary habits, climate, and sanitary provision. There is a wide variation in seropositivity throughout the world with highest rates being recorded in France, the United Kingdom, and USA.

Clinical features

The clinical manifestations of toxoplasma infection are influenced by the immunological competence of the host. In the majority of people who are immunologically competent, infection is asymptomatic. In a minority there is a benign, transient influenza like illness with malaise, lymphadenopathy, maculopapular rash, and myalgia. This may be complicated by hepatitis, meningo-encephalitis, or myocarditis. If the immune system is compromised as with AIDS, toxoplasmosis infection can be devastating, either by dissemination of primary infection, or more frequently, by re-activation of latent infection. Toxoplasma encephalitis almost always represents reactivation. Up to 40 per cent of AIDS patients are so infected and as many as 30 per cent die of toxoplasmosis in Europe. Cerebral toxoplasmosis usually occurs in patients whose CD4 lymphocyte counts are less than 200/μl and in as many as half, it is the AIDS defining event; that is the illness which changes their diagnosis from being someone who is HIV positive to someone who has AIDS. There may be racial and geographic susceptibility. As with other opportunistic infections, the incidence of toxoplasmosis has reduced since the advent of highly active antiretroviral therapy, HAART, for HIV infection. T. gondii can also cause choroidoretinitis, either as part of the primary infection or reactivation. Both choroidoretinitis and encephalitis can occur in neonates as a result of transplacental infection.

Toxoplasma in the brain can cause a localized, relatively indolent granuloma, multiple miliary granulomas, focal encephalitic change with necrosis and abscess formation with poorly defined capsules, or a diffuse necrotizing encephalitis, and combinations of all of these. Lesions are commonly multiple. Any part of the central nervous system can be affected and there is some predilection for deeper structures in the region of the basal ganglia and midline.

In AIDS patients there are multiple areas of focal necrotizing encephalitis which contain tissue cysts and extracellular tachyzoites (Luft *et al.* 1984).

The clinical syndromes approximate to a diffuse encephalitis or an enlarging intracranial mass lesion, or a combination of these. The onset is variable, from insidious to explosive with non-focal or focal symptoms and signs according to the location and nature of the process. Headache, confusion, fever, epileptic seizures, behavioural changes, and altered mental status are common. Focal neurological signs develop and with involvement of deep brain structures, extrapyramidal signs, and movement disorders have been described (Luft *et al.* 1993). When patients present with a more insidious global impairment, it may be difficult to differentiate from HIV-associated dementia (Section 43.3.5). Choroidoretinitis may occur at the same time, or after cerebral disease, but occasionally it may precede it.

Diagnosis

Patients who are suspected of having cerebral toxoplasmosis should have the diagnosis confirmed by a combination of brain imaging and serological testing. MRI is more sensitive than CT and will show lesions which CT has missed, thus it is the imaging modality of choice (Fig. 43.8). Single or more usually, multiple solid or cystic spherical lesions with ring enhancement, surrounding oedema, and space occupation, are common and occur in up to 90 per cent of patients. The appearances are not pathognomonic (Farkash *et al.*

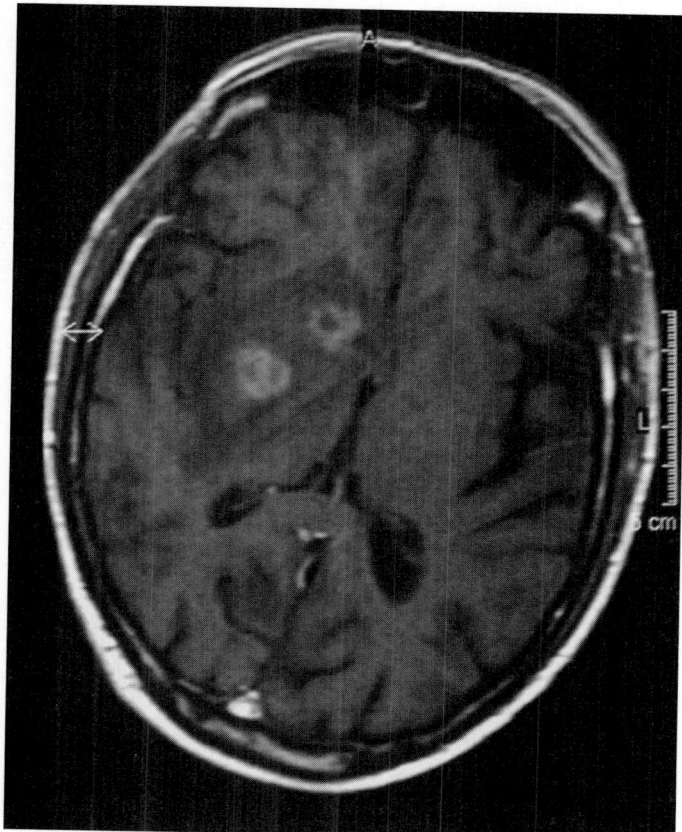

Fig. 43.8 Cerebral toxoplasmosis. Axial T1-weighted MRI following gadolinium showing two thick walled ring enhancing lesions within the right basal ganglia with mild local mass effect. This appearance and location are very typical of toxoplasmosis in HIV patients. (Courtesy Dr Ian Turnbull.)

1986). Multiple lesions favour toxoplasmosis, whereas a single lesion favours lymphomas, but there is overlap. In patients who are immunocompetent, single positive IgG blood titre is a sensitive indicator of prior infection. In AIDS and other immunocompromised patients, raised or rising titres do not necessarily indicate active infection. However, absence of an IgG response implies that the syndrome may not be due to toxoplasmosis and another cause should be sought, because <5 per cent of patients with toxoplasmosis have negative serology. Polymerase chain reaction testing of CSF has been disappointing: although the specificity is good at (96–100 per cent), the sensitivity is only about 50 per cent (Skiest 2002). Intrathecal antibody testing for *T. gondii* has also not proven very helpful. The CSF usually shows a mononuclear pleocytosis with normal glucose ratio. Definitive diagnosis rests with demonstration of the parasite in biopsy material from an affected area of brain. However, in practice most patients with multiple ring enhancing lesions who have a positive result on serological testing are treated empirically. If there has been no response to treatment within a couple of weeks then a biopsy should be considered. In patients with a single cerebral lesion, particularly if serology is negative, it is best to undertake biopsy before treatment, especially since lymphoma (Section 27.8.3) is high on the list of alternative diagnoses.

Treatment

The primary therapy for cerebral toxoplasmosis is pyrimethamine combined with sulphadiazine. This is given together with folinic acid, which prevents the haemopoietic toxicity of pyrimethamine, for 6 weeks. Following this, lifelong treatment with smaller doses for chronic disease suppression is necessary in AIDS patients. In patients who may react adversely to sulfadiazine, clindamycin may be used. There is evidence that in those patients who experience excellent virological and immunological responses to highly active antiretroviral therapy, HAART, lifelong treatment may not be necessary (Guex *et al.* 2000). Other drugs including azithromycin and clarithromycin are under evaluation. Space occupying lesions may require surgical decompression or aspiration.

Prevention of disease is difficult and avoidance of close contact with cats and ensuring that meat for human consumption is well cooked, may diminish the risk of infection in the vulnerable. Prophylaxis against toxoplasma is recommended in those with HIV, CD4 lymphocyte counts less than 200/mm³, and positive toxoplasma serology. But even this may not be necessary in those that respond well to HAART (Furrer *et al.* 2000).

43.2.9 Cysticercosis

Cysticercosis refers to infestation of humans by the larval form of the pork tapeworm, *Taenia solium*. Laennec invented the term 'cysticercus' from the Greek words 'kystis', meaning bladder, and 'kerkos', meaning tail. A bladder with a tail describes the appearance of the larva.

Life cycle of cysticercosis

Taenia solium is a cestode and a common tapeworm which infests the intestine of man. It occurs world-wide and particularly high rates of infestation have been reported from Central and South America, Africa, India, and the Far East. Pockets of infestation occur in areas of Europe such as Poland and there may be some relationship to the culinary habits of the populace who eat uncooked or undercooked pork. Improvements in veterinary hygiene practices have led to a marked reduction in cases in developed countries where cases may be found in immigrants and in people who have travelled to endemic areas. The adult tapeworm parasitizes the small intestine of man, the definitive host, and can grow to a length of some metres. Human intestinal infection with the adult tapeworm is called taeniasis. Segments of the worm containing ova are deposited in faeces which may contaminate the environment to be ingested by pigs, the intermediate host. In the pig the eggs hatch and the embryos migrate through intestinal wall, into the bloodstream, and are distributed to various tissues, particularly muscle, where the embryos develop into cysticerci. When infested pork is eaten by man, cysticerci, stimulated by gastric acid, penetrate the wall of the small bowel, attach themselves to the mucosa, and become adult *Taenia*, thus completing the life cycle. In this natural cycle then, humans do not develop cysticercosis. However, humans can also become the intermediate host by ingesting food contaminated by human faecal material if personal hygiene is poor; the eggs then hatch and form larvae which migrate through the tissue to form cysticerci. Most human infections are thought to occur within the home, infection occurring from a family member who is infected with an adult tapeworm.

The clinical manifestations of cysticercosis are determined by the number and location of cysts, and the immunological reaction to them. The incubation period is variable and may be as short as a few months or as long as several years. Most clinical cases occur between 20 and 50 years of age but any age or race and either sex may be affected. The clinical syndromes are classified according to the location, size, and associated features of the cysts (Table 43.4). The syndromes are not mutually exclusive and two or more forms of the disease may co-exist.

Clinical features

Cysticercosis encephalitis occurs in children when there is massive infiltration of cysts in the brain causing a generalized encephalitis. Intraventricular cysts are the most common extraparenchymal cyst. They usually cause obstruction to CSF flow and result in acute hydrocephalus with signs of raised intracranial pressure. If the lesion is pedunculated, acute attacks of hydrocephalus may result in the manner of colloid cysts. Sometimes they grow to be giant cysts of 1–2 cm in diameter. If they occur in the basal cisterns they may form a bunch of grape-like racemose cysts.

Spinal cysticercosis occurs in less than 5 per cent of cases and produces signs of spinal cord compression, often in the high cervical region. Meningeal cysticercosis results from a basal arachnoiditis and manifests as hydrocephalus complicated by cranial nerve palsies. In some, chronic meningitis is produced. It is only in recent years that it has been recognized that vascular occlusion may occur from obliteration of the vessels at the base of the brain or more peripherally, which have been affected by arachnoiditis so causing stroke syndromes in the distribution of these vessels. This is a frequent cause of stroke amongst the young in endemic areas (Del Brutto 1992). Occasionally infection of striated muscle can cause pseudohypertrophy and weakness.

The neurological syndromes of cysticercosis may mimic stroke, meningitis, tumour, abscess, raised intracranial pressure, and dementia, and diagnosis is greatly aided by a history of exposure through travel in endemic areas. However, there is adequate documentation of acquisition of the disease by societies who do not eat pork, by infection from immigrants from endemic areas (Schantz *et al.* 1992). Consequently, neurocysticercosis must be considered in the differential diagnosis of the above syndromes worldwide.

Diagnosis

Signs of cysticercosis elsewhere in the body aid diagnosis and subcutaneous intramuscular nodules (Fig. 43.9) should be sought; they may be found in between 5 and 50 per cent of cases dependent upon geographical location. Serological tests have not provided a

Table 43.4 Clinical syndromes caused by cysticercosis

Parenchymal cysts
  Seizures
  Focal signs
  Encephalitis-like presentation in children with multiple cysts

Extraparenchymal cysts
  Intraventricular cysts cause acute obstructive hydrocephalus
  Chronic hydrocephalus, arachnoiditis, ependymitis
  Giant cysts
  Racemose cysts in basal cisterns
  Meningitis with cranial neuropathies
  Arachnoiditis with strokes

Spinal cord disease

Striated muscle disease
  Weakness, pseudohypertrophy

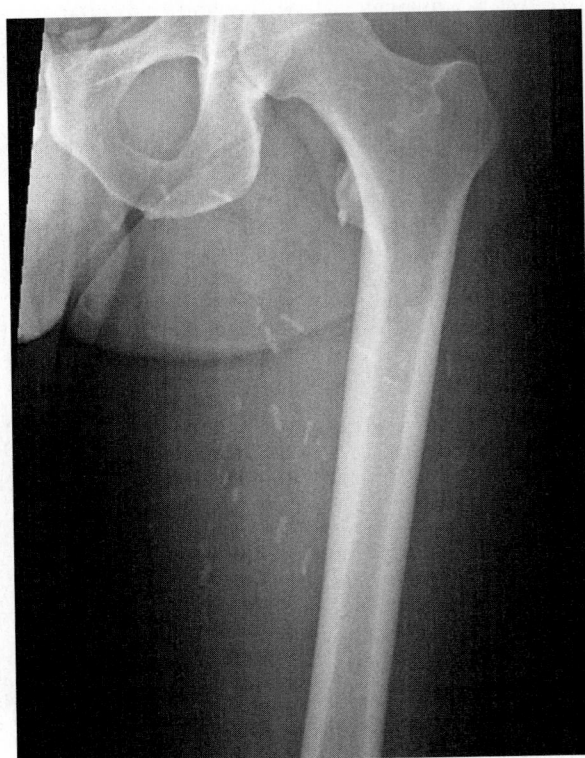

Fig. 43.9 Cysticercosis. X-ray showing calcified cysts of *Taenia solium* in the thigh. (Photo Tom Solomon.)

good index of activity of neurological disease and various tests to demonstrate antibody to or antigen from cysticerci have been developed. The most specific and sensitive test is the enzyme linked immuno-electrotransfer blotting antibody detection, EITB, estimation which can be carried out on serum and CSF. A CSF eosinophilia may be another clue. Brain imaging with CT or MR is essential for diagnosis of neurocysticercosis (Fig. 43.10). The appearances depend upon the degree of maturity of the cyst activity. On CT a viable cyst appears as a hypodense lesion with a hyperdense nodule, which is the live scolex. As this degenerates there is contrast enhancement at the edge of the cyst with oedema. A late degenerating cyst is isodense, but enhances with contrast and may be mistaken for tuberculosis. Eventually it forms a calcified scar. Extensive infestation with multiple cysts at different stages of development may be seen. Ventricular enlargement from obstruction either from cysts or basal arachnoiditis may be evident. Diagnosis can be confirmed by stereotactic biopsy and demonstration of larval components.

Treatment

Treatment is symptomatic with relief of intracranial hypertension by shunting, surgical removal of lesions which are space occupying, and the prescription of anti-epileptic drugs for seizures. Praziquantel and albendazole have been used as anti-parasitic drugs, but their use has been controversial. The arguments against using such drugs is that the natural history is for cysts to die anyway, and that treatment, by killing the parasites, may make inflammation worse. On the other hand earlier treatment may reduce the risk of scarring and subsequent epilepsy. A review of four trials found insufficient evidence either for or against treatment in terms of survival, cyst resolution, hydrocephalus, or subsequent seizures (Salinas et al. 2000). However, a more recent large randomized placebo controlled trial of albendazole with dexamethasone showed a non-significant reduction in total number of seizures, and in partial seizures, but a significant reduction in the number of seizures with secondary generalization (Garcia et al. 2004). There was also greater resolution of lesions on imaging, but more abdominal pain. On this basis most would now recommend albendazol treatment, typically at 15 mg/kg/day for 8 days, and with steroids such as dexamethasone. Albendazole has better CSF penetration than praziquantal, and is cheaper. Giant or basal cysts are treated for longer, for example 1 month. In cysticercosis encephalitis, steroids are given to control the inflammation, and anti-parasitic drugs are not given, for fear that killing the parasites will worsen inflammation.

The issue of when to stop anti-epileptic drugs also proves difficult. Some would advocate stopping the drugs once the oedema has resolved, even if there is calcification. However the relapse rate is high; up to 50 per cent of patients who have been seizure free for 2 years on anti-epileptic drugs will relapse off treatment (Carpio 2002).

43.2.10 Schistosomiasis

Schistosomiasis is endemic in the tropics and subtropics and affects more than 200 million people in more than 70 countries. Travellers to these countries are at risk of infection if they come into contact with the cercariae by swimming in, or drinking, contaminated water. Three species of schistosome infect man, *S. haematobium*, *S. mansoni*, and *S. japonicum*. Other species may cause human disease but only rarely. Schistosomes are digenetic trematodes which reproduce sexually in definitive hosts, man, and asexually in intermediate hosts, snails. The adult worms live in the portal or mesenteric veins of man, in the case of *S. mansoni* and *S. japonicum*, or the vesical plexus, *S. haematobium*. Eggs are produced which develop into miracidia which penetrate bowel or bladder and are excreted. In regions of poor sanitation they contaminate fresh water, infect susceptible snails, and develop in the tissues to become cercariae. These mature, leave the snail, and swim in the water.

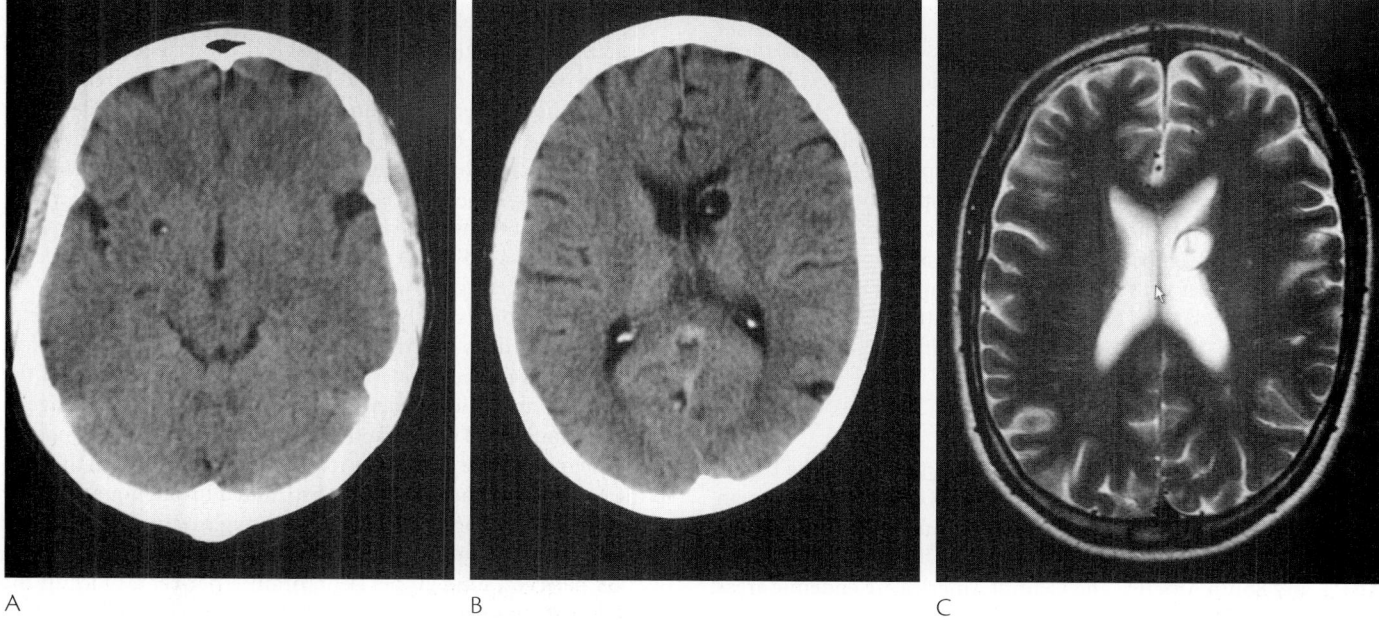

A   B   C

Fig. 43.10 Cysticercosis showing typical cystic lesions in the brain with central calcification on CT (A) and (B) and T2-weighted MRI (C). (Photo Tom Solomon.)

Humans are infected by swimming in, or drinking, contaminated water. Cercariae penetrate skin or gut mucosa, become shistosomula, and travel via the bloodstream and lymphatics to the lungs whence they are dispersed throughout the body in the circulation. They then settle in the mesenteric or vesical venous plexuses where they mature into adult worms. Chronic schistosomal infection results from the immunological reaction and granuloma formation incited by eggs trapped in the venous plexuses. This gives rise to hepatosplenic and urinary obstructive disease.

Central nervous system disease occurs in less than 5 per cent of cases and comes about when eggs travel through the pelvic venous plexuses to the venous plexuses surrounding the spinal cord to cause spinal schistosomiasis. Cerebral schistosomiasis probably results from transport of the eggs to the brain via the venous plexuses and not by arterial embolization. Each species of schistosome can affect the brain or the spinal cord. *S japonicum* favours brain almost exclusively, *S. haematobium* the spinal cord, and *S. mansoni* both (Ariizumi 1963; Pitella 1993). Schistosomal infestation is endemic in Africa, the Middle East, Egypt, Sudan, South America, South East Asia, China, the Philippines, and Japan.

Acute cerebral schistosomiasis presents as a fulminating encephalitis or encephalomyelitis with pyrexia and skin rashes, seizures, pyramidal signs, confusion, and neck stiffness. It may progress to coma. Examination of peripheral blood reveals an eosinophilic pleocytosis and CSF also shows a pleocytosis. It probably is a form of allergic encephalopathy and most patients recover.

Chronic cerebral schistosomiasis results from granuloma formation in the brain and may declare itself many years after the original infestation. Presentation is of a focal cerebral lesion which may occupy space. Epilepsy is common. Brain imaging with CT or MR reveals enhancing mass lesions often with surrounding oedema, an appearance also seen with other granulomas or tumours. Diagnosis depends upon the recognition of schistosomiasis elsewhere and the finding of eggs in stool or urine. Biopsy of the granuloma may show diagnostic changes and eggs. Serological tests using schistosomal antigen and enzyme linked immunosorption assays are useful.

Spinal schistosomiasis presents as a transverse myelitis, intrathecal granuloma, or conus syndrome with radiculopathy (Haribhai *et al.* 1991).

Treatment of schistosomiasis is effective. Praziquantel is the drug of choice and is effective against all three species. Metrifonate is effective against *S. haematobium* only. If space occupation is a problem, surgical removal of the lesions may be necessary. Otherwise if the diagnosis has been made with confidence, it may be sufficient to treat with praziquantel and monitor the resolution of granulomas with serial brain imaging (Watt *et al.* 1986). Steroids may be used to reduce oedema and modify any inflammatory reactions.

43.2.11 **Paragonimiasis**

Paragonimiasis is caused by infestation with the lung fluke trematode of the genus *Paragonimus*. Seven species cause disease in man. *P. westermani* is the predominant species causing disease in man. China, Japan, the Philippines, Korea and Thailand, regions of Africa, and South America and Central America are endemic areas. Man is infected accidentally by ingesting infested raw or undercooked crayfish or crabs. These have derived the infestation from freshwater snails that have picked up miracidia in the water, shed

by man from infected pulmonary cysts and the cycle is complete. After ingestion by man, the organisms penetrate the gut wall into the peritoneum and penetrate the diaphragm into lungs where they mature into flukes and lay eggs. These rupture into a bronchiole and eggs, blood, and debris are expectorated. The route by which the central nervous system becomes infected is by migration of the organism via the veins or vascular fascial sheaths through the jugular foramen. Dissemination may also take place by the bloodstream. The fluke or the eggs incite an inflammatory response which becomes granulomatous and fibrotic with time and may later calcify. The central nervous system is involved in less than 1 per cent of all cases and the clinical syndromes include meningo-encephalitis (Oh 1968), basal arachnoiditis, and multiple granuloma formation. Meningitis presents acutely. Granulomas are usually insidious in evolution producing focal signs, seizures, and space occupation. Optic atrophy caused by adhesive arachnoiditis is not uncommon. Brain imaging with CT or MR will demonstrate space occupying multiple granulomas with a tendency to coalesce. Ring-like enhancing lesions occurring in clusters are seen in the active stages (Cha *et al.* 1994). In more chronic lesions ring calcification may be seen. There may be an eosinophilic response in the blood. Chest X-ray is abnormal in most cases and eggs may be found in sputum and faeces. Complement fixation and immunoblot antibody tests are useful and may be done on blood and CSF. Treatment with praziquantel with steroids to cover any inflammatory reaction, is effective. In chronic cases residual signs and seizures are common.

43.2.12 **Echinococcosis**

Echinococcosis or hydatid disease is caused by the larval cysts or hydatids of the tapeworm *Echinococcus granulosus*. Hydatid is from the Greek for water droplet. Other species of *Echinococcus* have been known to cause disease in man, including *E. multilocularis*, but such infestation is rare and produces alveolar hydatid disease in which the cerebral cysts are multiple and multiloculated and filled with rather denser material. *E. granulosus* is a common parasite of the dog, which is the definitive host for the adult worm which lives in the small intestine. Eggs are discharged in faeces, and are ingested by farm animals, usually sheep. Inside the intermediate host, the organism penetrates gut wall and travels in the veins or lymphatics to other tissues where it forms an enlarging cyst. Dogs are infected by eating infected flesh or offal. Man is accidentally infected by coming into contact with the eggs in the faeces of dogs, commonly in the coats of dogs where the larvae lodge. The disease is widespread in those parts of the world where sheep are kept and where dogs are used to tend them—South America, Europe, Australia, New Zealand, Mediterranean countries, Britain, and Central Asia.

Children are particularly liable to infection, perhaps because of their affinity to dogs and their poor hygiene habit. Once ingested, the eggs penetrate the duodenal wall and travel in the veins of the portal system to the liver where most are trapped. Some manage to squeeze through and travel via the heart to the lungs where most of the surviving eggs are trapped. A few of these manage to get through to the general circulation and are distributed throughout the body with very few lodging in brain. This filtering mechanism accounts for the frequency with which different organs develop hydatids: liver 65 per cent, lung 20 per cent, brain 2 per cent. Once in the brain the larvae form a double layered cyst which slowly enlarges and may grow to a prodigious size (Fig. 43.11). Symptoms are of increasing intracranial pressure with focal signs appearing late if

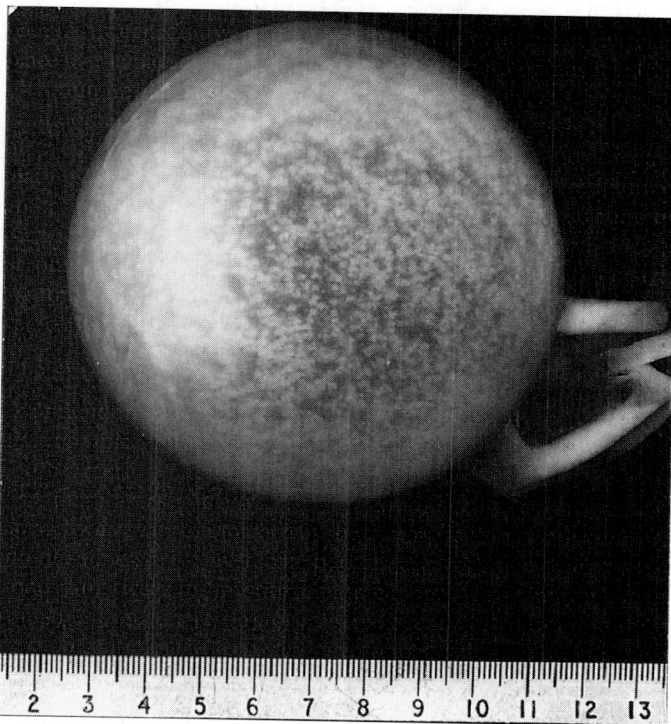

Fig. 43.11 Hydatid cyst, 9 cm in diameter, removed from the brain of an English boy of farming stock. (Courtesy Dr Milne Anderson.)

at all. Schroeder's (1941) diagnostic tetrad is useful: a child from a rural sheep rearing area, in good general health, with signs of raised intracranial pressure, and with ill-defined focal signs. In some cases the larvae lodge in bone of the skull or vertebrae and cysts form there to cause pressure on adjacent structures with resulting multiple cranial nerve palsies or paraparesis.

The diagnosis of cerebral hydatid disease is suggested by the finding of raised intracranial pressure in a person from a sheep rearing zone. The diagnosis is confirmed by brain imaging with CT or MR which will demonstrate one or sometimes multiple, large, spherical, space occupying cysts filled with fluid of the same radiological density as CSF (Ersahin *et al.* 1993). Immunological tests may be useful, but are not infrequently negative with brain cysts. CSF should not be examined because of raised intracranial pressure. Signs of hydatids elsewhere such as in liver or lung, lend support to the diagnosis. It is important that a hydatid cyst should be differentiated from other cystic brain lesions because the surgical technique for removal is specific. Great care should be taken not to puncture the cysts because to do so allows fluid containing daughter cysts to spread to form further hydatids. An ingenious method of removal was devised by Arana-Iniguez (1955) in which saline is injected between the outer layer of the hydatid and the brain, forcing the cyst to the surface where it can be delivered intact and removed. If cysts are removed intact the prognosis is excellent. Albendizole has been used for treatment of bony disease and for inoperable cysts but the results have not been encouraging.

43.2.13 Amoebal infestation

Cerebral amoebiasis occurs as a rare complication of amoebic dysentery and is to be distinguished from the meningoencephalitis caused by *Naegleria* (Section 42.6.4). Amoebic dysentery is caused by *Entamoeba histolytica* which colonizes the large bowel and is found in many countries throughout the world including Central and South America, Africa, South East Asia, and other areas. It may penetrate bowel and lead to hepatic abscesses in 10 per cent, or less commonly, lung abscesses. Rarely, it will lead to the formation of brain abscesses which are often multiple and invasive and destructive. Clinical features are of a focal expanding cerebral lesion in a patient who is ill and has hepatic involvement and a history of dysentery. Brain imaging shows changes of developing abscesses as described above. Diagnosis is usually made by demonstration of amoebic trophozoites in biopsy material and by demonstration of trophozoites or cysts in faecal samples. The prognosis is almost uniformly poor. Recommended treatment is by aspiration of abscess contents and systemic metronidazole.

43.2.14 Other parasitic infestations

In certain regions of the world and in exceptional circumstances, several other parasites may cause infection in the central nervous system leading to granulomas or cysts. The exceptional circumstances often include immune suppression of the host, or trauma which permits access to the neuraxis which would otherwise be denied. Sometimes travel to an inaccessible area invites exposure to exotic organisms. The clinical syndromes which result usually do not have pathognomonic features and the clue to the diagnosis in these cases depends upon recognition of the abnormal circumstances in which infection has been acquired. Commonly the correct diagnosis is not reached until after death.

Tapeworms. Infection by the larval from of the cestode tapeworm, *Spirometra*, causes sparganosis. Dogs and cats are parasitized and man is infected by drinking water contaminated by the larvae or by eating raw meat of infected snakes or frogs. Asia and South America are the main reservoirs. The larvae penetrate gut and migrate to other tissue, usually muscle, where they produce subcutaneous cysts. Only rarely is the central nervous system involved by a cerebral granuloma which functions as a space occupying mass. (Tsai *et al.* 1993). Diagnosis is by stereotactic biopsy. Treatment is by excision if possible. There is little information concerning the efficacy of treatment with albendazole or praziquantel. Spinal granulomas have also been described.

Coenurosis. This is caused by infestation with the larval form of some species of dog tapeworm, most commonly *Taenia multiceps* in Europe, Africa, and South America, and *T. serialis* in Europe and North America. Other species do not cause significant central nervous system infection. Sheep are intermediate hosts and man is infected in the same way as with cysticercosis. Cysts, usually solitary, develop in brain and cause a space occupying syndrome. Basal arachnoiditis may contribute to this. Imaging demonstrates a cystic mass in or around the ventricles (Pau *et al.* 1987). Treatment is by surgical excision.

43.3 Human Immunodeficiency Virus (HIV)

43.3.1 Historical overview

The recognition of five cases of pneumocystis pneumonia in previously healthy homosexual men, reported in 1981, launched the public recognition of acquired immunodeficiency syndrome, or AIDS, and a whole new era in infectious diseases, including neurological infections. It soon became evident that there was an epidemic of an immune-suppressive disease which was seen in homosexual males, intravenous drug abusers, haemophiliacs, and others. This rendered the sufferers liable to life-threatening infection with opportunistic organisms, and to Kaposi's sarcoma and other forms of neoplasm. An infectious agent was suspected. In 1983, a virus was isolated from a French patient with lymphadenopathy, found to be a retrovirus, and was designated lymphadenopathy associated virus (Barre-Sinoussi *et al.* 1983). The term 'Human T-cell Lymphotropic Virus' was coined to encompass a family of related viruses including leukaemia and AIDS retroviruses. The virus which they had isolated was confirmed as the cause of AIDS and was eventually designated Human Immunodeficiency Virus, now known colloquially as HIV. Most AIDS is due to HIV-1 but a second virus, HIV-2 is found mostly in western Africa, and is transmitted sexually or perinatally, and is less virulent.

As more cases of AIDS were identified, it became evident that many patients with HIV infection had milder syndromes of prodromal immunodeficiency, persistent generalized lymphadenopathy, and AIDS-related complex, which would progress after a variable period to full blown AIDS. The distinction between these syndromes has become blurred and their prognostic significance has diminished. Now it is recognized that HIV infection impairs the response of the immune system. This results in a progression of disease from the initial asymptomatic stages through to opportunistic infections of increasing severity, the development of malignancies, and the full blown clinical syndrome of AIDS. Previously particular illnesses were considered to be 'AIDS-defining'; now the CD4 lymphocyte count and HIV viral load are considered more helpful markers of disease progression.

In the near 3 decades since the disease's recognition there has been considerable progress in diagnostics and treatment:

◆ prevention of mother to child transmission;

◆ recognition of the importance of the plasma load and the CD4 count for prognosis; and

◆ in the development of treatment, particularly highly active antiretroviral therapy, or HAART.

Nevertheless the global burden of HIV remains extremely high, particularly in sub-Saharan Africa and parts of Asia.

43.3.2 The profile of HIV infection

Following initial HIV infection, there is a phase of viraemia with a high plasma virus load. Virus attaches to T lymphocytes and macrophages via the CD4 receptor. There is then a clinical lull that lasts for a variable number of years, even decades in some individuals, during which virus replicates at a slower rate and reduces the CD4 lymphocyte count. As these cells become depleted to dangerous levels, viral titres increase, immunity becomes further deranged, and opportunistic infections and malignancies occur.

Factors which influence the severity and the rate of development of AIDS are not fully understood. They include the strain and virulence of the virus, the presence of concurrent viral infection such as cytomegalovirus, Epstein–Barr virus, and herpes viruses; these coinfections may facilitate entry of HIV to cells. Genetic factors play a part and some individuals have not developed HIV infection despite high risk and sustained exposure. They have been found to have a homozygous deletion of the chemokine receptor CCR5 which makes their cells less pervious to HIV. Modern treatment regimes have led to a reduction of viral load and prolongation of survival.

Following infection, most patients generate an immune response with antibody production to a wide range of viral structural proteins. Most of the standard serological assay techniques can be employed to detect HIV antibody. HIV RNA can be detected in a quantitative fashion by the application of polymerase chain reaction and has been further refined by incorporation of branched DNA amplification methods. It is possible now to monitor the viral load accurately, and this correlates well with disease activity and with the response to anti-retroviral treatment (Mellors *et al.* 1996).

Most cases of HIV infection have been contracted by virus inoculation from infected body fluids, or tissues, usually blood or semen. This occurs through sexual intercourse, intravenous drug administration, transfusion of contaminated blood products, and by transplacental transmission from mother to child. The epidemiology of AIDS has changed. Formerly a disease of male homosexuals and drug addicts, then of people who had received contaminated blood products, it is now spreading by heterosexual activity throughout the world, with the highest rates of increase in Africa and Asia.

43.3.3 Neurological disease due to HIV

HIV is the most common viral infection of the nervous system. HIV is a neurotropic virus and invasion of the nervous system occurs at an early stage. Disease of the nervous system is common in HIV infected patients and occurs in as large a proportion as 70 per cent of all cases. Up to 90 per cent of patients who die with HIV have neuropathological changes at necropsy (Kure *et al.* 1991). Neurological disease is the first manifestation of severe disease in about 20 per cent of those who are HIV positive.

Neurological complications of HIV may stem from direct infection of any part of the nervous system by the virus causing a primary meningitis, encephalitis, myelitis, peripheral neuropathy, or myopathy (Table 43.5). Secondary involvement of the nervous system occurs as a consequence of immune suppression permitting the development of primary or secondary brain neoplasia, or the development of central nervous system infection from a wide host of organisms, chief amongst which are toxoplasmosis, cytomegalovirus infection, JC virus causing progressive multifocal leukoencephalopathy, fungal meningitis, and tuberculosis.

Suspected neurological disease: the clinical approach

There are several issues to consider in approaching an HIV positive patient with suspected neurological disease. It may be helpful to know how they initially presented and were diagnosed, and how they became infected: for example opportunistic infections in homosexual men are likely to be different to those in intravenous drug abusers. Knowing the country of origin may also be important.

Determining the current stage of the disease process is essential:

◆ What is the current CD4 count? In particular is it less than 200 cells/mm^3, at which opportunistic infections may occur?

◆ Is the viral load greater than 50 000 copies/mm^3?

◆ What medication has the patient been on?

Table 43.5 Neurological disease in HIV patients, by presenting clinical syndrome*

Meningitis
  Acute
    HIV seroconversion illness
    Pneumococcal (Section 41.5.1)
    Meningococcal (Section 41.5.2)
    E. coli
    Klebsiella
    Listeria (Section 41.5.10)
  Chronic
    Mycobacterium tuberculosis (Section 41.5.3)
    Mycobacterium avium complex
    Cryptococcus (Section 41.5.9)
    Syphilis (Section 41.5.4)
    Nocardia
    Candida (Section 41.5.9)

Encephalitis
  Cytomegalovirus (Section 42.3.4)
  JC virus (Progressive multifocal leukoencephalopathy) (Section 42.3.14)
  Herpes simplex virus (esp type 2) (Section 42.3.2)
  Varicella zoster virus (Section 42.3.3)
  Immune reconstitution inflammatory syndrome (IRIS)

Space occupying lesions
  Abscesses (Section 43.1.2)
    Tuberculous (Section 43.2.1)
    Listeria (Section 41.5.10)
    Nocardia (Section 43.2.2)
    E. coli
  Infective granulomas
    Toxoplasma (Section 43.2.8)
    Aspergillus (Section 43.2.4)
    Candida (Section 41.5.9)
    Cryptococcus (Section 41.5.9)
  Other
    JC virus (Progressive multifocal leukoencephalopathy) (Section 42.3.14)
    Varicella zoster virus (Section 42.3.3)
    Cytomegalovirus (Section 42.3.4)

Neoplastic
  Primary central nervous system lymphoma (Section 27.8.3)
  Glioma (Section 27.8.1)
  Metastatic Kaposi's sarcoma
  Metastatic lymphoma

Dementia
  HIV
  JC virus (progressive multifocal leukoencephalopathy (Section 42.3.14)
  Tuberculosis
  Syphilis (Section 42.5.1)
  Lymphoma

Stroke-like syndromes
  Varicella zoster virus vasculitis
  Venous thromboses, secondary to hypercoagulable state (Section 35.15)
  Secondary to basal meningeal tuberculosis (Section 41.5.3)
  Neurosyphilis (Section 41.5.4)

Retinitis
  Cytomegalovirus
  Syphilis

Myelopathy
  HIV
  Cytomegalovirus

Radiculopathy
  Cytomegalovirus (Sections 21.14.2 and 42.3.4)
  Herpes zoster (Sections 21.14.5 and 42.3.3)

Neuropathy
  HIV distal sensory neuropathy (Section 21.14.2)
  Herpes zoster (Section 21.14.5)
  Drugs, esp. nucleoside revere transcriptase inhibitors (Section 21.19.7)
  Cytomegalovirus vasculitic neuropathy

Myopathy
  HIV polymyositis (Section 24.7.4)
  Cytomegalovirus
  Drug induced, esp. nucleoside reverse transcriptase inhibitors
  Zidovudine (Section 24.9.5)

*The majority of presenting syndromes are shown here, but there is some overlap

◆ Ask about homeopathic and natural remedies, which are also popular among some HIV infected groups and may be toxic.

◆ Determine which other HIV-associated problems they may have: in particular hepatitis B and C, cytomegalovirus, toxoplasma, syphilis, and tuberculosis.

◆ Always do a detailed general medical examination, including examining the skin for Kaposi's sarcoma, and the oral cavity for candida or oral hairy leukoplakia (Table 42.6).

In the United Kingdom, the most common groups of patients seen with neurological presentations include:

◆ Newly infected HIV patients presenting with a neurological seroconversion illness.

◆ Patients unknowingly infected for some time, who now present with a secondary neurological complication, such as an infection or lymphoma.

◆ Patients with HIV diagnosed in the 1980s, who have been well controlled on HAART, but are now developing complications of drug therapy—particularly neuropathies.

◆ Patients with longstanding HIV infection, that has been controlled by HAART, but are now developing dementia, lymphoma, or other malignancies.

◆ HIV positive patients who happen to have coincidental neurological problems which are unrelated to their HIV (Manji & Miller 2004).

43.3.4 Investigation of associated central nervous system disease

All HIV patients with central nervous system syndromes need blood cultures, magnetic resonance imaging, and if there are no contraindications, a lumbar puncture (Table 43.6). Microscopy, culture, serological tests, and polymerase chain reaction for the diseases listed above should be carried out. Non-space occupying white matter abnormalities, diffuse or patchy, seen on scan may be caused by viral encephalitides including HIV encephalopathy and progressive multifocal leukoencephalopathy (Table 43.7). If a mass lesion or lesions are seen on the scan the most likely causes are toxoplasmic encephalitis, lymphoma, pyogenic abscess, parasitic or fungal

Table 43.6 Investigations to consider in HIV patients with neurological disease

---

Imaging

CT/MRI for space occupying lesions or atrophy

Chest X-ray (for tuberculosis)

Positron emission tomography scanning (if lymphoma suspected)

CSF

Microscopy, cells, protein, glucose ratio

PCR for: tuberculosis

Herper Simplex Virus 1 and 2

Varicella Zoster Virus

Cytomegalovirus (especially if radiculitis)

Epstein Barr Virus (if lymphoma suspected)

JC virus

Staining for acid fast bacillus and culture for tuberculosis

India ink for Cryptococcus

Cryptococcal antigen test (CRAG)

VDRL / TPHA

HIV viral load

Cytospin for malignancy

Blood tests

CD4 count

Blood cultures

Toxoplasma serology

HIV viral load

Cryptococcal antigen test (CRAG)

VDRL / TPHA

LDH (lymphoma)

PCR for CMV, EBV

---

granuloma, or tuberculoma. Infarction and haemorrhage are usually easy to distinguish from these. Metastatic neoplasia may cause confusion. If the mass or masses are causing significantly raised intracranial pressure, surgical decompression should be considered when possible, and representative tissue removed for histology and culture.

If raised intracranial pressure is not a problem the results of toxoplasma serology will influence management; if negative, stereotactic biopsy of a lesion should be undertaken. If serology is positive and the patient's condition is stable, it is reasonable to treat for toxoplasmosis, keep the patient under close observation, and re-scan within 2 weeks. If clinical progress and scan appearances continue to show improvement, it is reasonable to continue with pyrimethamine and sulphadiazine and have follow-up scans. If there is no clinical or radiological improvement after 2 weeks, or if there is obvious deterioration, stereotactic biopsy should be undertaken forthwith and the treatment regime adjusted according to results. If dexamethasone or other agents have been used to reduce brain oedema, suitable allowance must be made when assessing clinical and radiological improvement.

There have been some reports suggesting that ancillary radiological techniques such as single photon emission computed tomography, positron emission tomography, and manipulation of MR data can allow discrimination between white matter abnormalities caused by AIDS encephalopathy, progressive multifocal leukoencephalopathy, and lymphoma (Ernst *et al.* 1999).

Seroconversion illness

Acute HIV infection can be followed 2–6 weeks later by a flu-like seroconversion illness. This resembles infectious mononucleosis,

---

Table 43.7 Some common radiological abnormalities in patients with HIV (see Figs 43.12, 43.13, and 43.14)

	Toxoplasmosis	Lymphoma	Progressive multifocal leukoencephalopathy	Tuberculoma	HIV-associated dementia
Clinical features	Acute onset < 2 weeks with headache, fever, drowsiness, focal neurology, seizures in 30%	Subacute onset 2–8 weeks, with confusion, focal neurology, seizures in 15%	Gradual onset (weeks–months) cognitive decline, hemiparesis, hemianopia, dysphasia, ataxia; seizures rare, systemically well	Subacute presentation +/– meningitis, or dementia with or without focal signs, and seizures	Chronic onset of dementia, often with sphincter disturbance, may have associated movement disorders
Imaging					
Location	Space occupying lesions occur at cortical interface of grey and white matter, and basal ganglia	Space accupying lesions a anywhere, but especially periventricular	White matter lesions, extending up to edge, especially parieto-occipital, and frontal lobes. If contrast enhancement—good prognostic sign	Small multiple lesions, associated with meningeal enhancement (tuberculomas); or larger (tuberculous abscess)	No space occupying lesions, but may have diffuse high signal white matter change, with no enhancement. Usually just atrophy
Number	Multiple—usually	Single or few	Diffuse changes	Single or several	None
Oedema	Marked oedema	Moderate oedema	Usually not	Marked	None
Enhancement	Ring enhancing lesions	Homogeneous enhancement	Usually not	Ring enhancing	None
Ancillary investigations	Toxoplasma serology pos; CSF Toxoplasma PCR pos; CD4 < 200	CSF EBV PCR pos; Positron emission tomography scan pos	CSF JC PCR positive	CSF acid fast bacillus and culture; CXR; Tuberculin; Elispot	HIV viral load in CSF Diagnosis of exclusion of other causes
Treatment	Pyrimethamine Sulphadiazine Folinic acid	Radiotherapy/Palliation	HAART	Quadruple therapy plus steroids	HAART

CXR: Chest X-Ray, EBV: Epstein–Barr Virus, HAART: highly active artiretroviral therapy, PCR: polymerase chain reaction

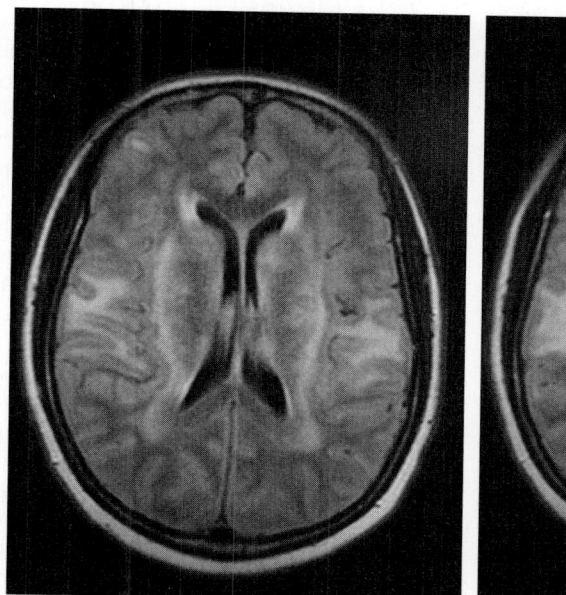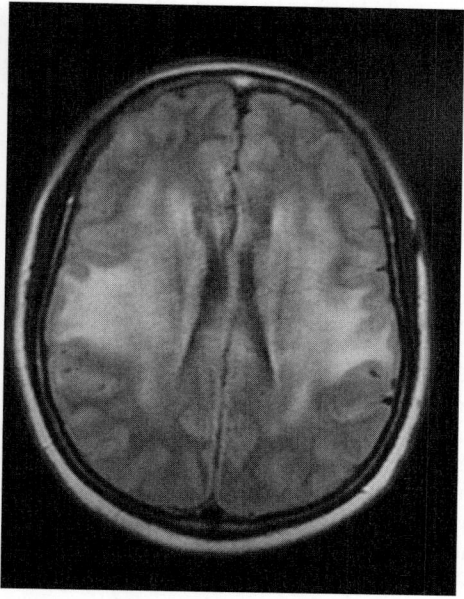

Fig. 43.12 HIV dementia. Diffuse sub-cortical white matter changes on T2-weighted MR FLAIR scan. (Photo Tom Solomon.)

occurs in a variable, but high, proportion of patients, and is often associated with malaise, lymphadenopathy, and a rash. In addition there may be pharyngitis, mucocutaneous lesions, oral candidiasis, and neurological complications. These include a self-limiting aseptic meningitis, which occurs in 10–15 per cent of those with a sero-conversion illness (Carne *et al.* 1985; Schacker *et al.* 1996), and occasionally encephalitis or even Guillain-Barre syndrome. The diagnosis is confirmed by finding viral RNA or p24 antigen in the blood, in the absence of antibodies. The virus may also be detected in the CSF. Although it is a self-limiting illness, some advocate early aggressive antiretroviral therapy with antiretroviral drugs which penetrate the central nervous system because this may substantially reduce the amount of virus in the CSF (Nath 2005). Occasionally white matter disorders resembling multiple sclerosis or acute disseminating encephalomyelitis may be seen at seroconversion.

The period of 'clinical latency'

Following the acute infection, a period described as 'clinical latency' occurs, during which there may be no or few clinical features. This is not a true viral latency, because virus continues to replicate and erode the host's immune system. The progress is monitored by CD4 lymphocyte counts and viral load. Above a CD4 count of 500 cells/mm^3 the only clinical manifestation may be persistent lymphadenopathy, but often there are no suggestive clinical features. As the CD4 count falls to between 200 and 500 cells/mm^3 patients develop non-life threatening bacterial infections similar to those of immunocompetent individuals, though perhaps with increasing frequency. Additional neurological complications during the period of latency include immune-mediated problems such as:

- an acute inflammatory demyelinating polyneuropathy, which is similar to Guillain–Barré syndrome (Section 21.10.1), but with a CSF pleocytosis up to 50 cells/µl;

- a chronic inflammatory demyelinating polyneuropathy (Section 21.11.2);

- mononeuritis multiplex (Section 21.15); or

- a multiple sclerosis-like leukoencephalitis (Section 37.4.1).

43.3.5 **HIV-associated dementia**

Not long after AIDS was first described, it was recognized that patients would develop neurological dysfunction which could not be attributed to infection or to neoplasia. A progressive dementing illness was recognized and has been called HIV-associated dementia, HIV encephalopathy, AIDS–dementia complex, and HIV-associated major cognitive/motor disorder (Navia *et al.* 1986; Janssen *et al.* 1991). It is difficult to know the exact frequency of occurrence of the syndrome, not least because it is likely that some cases may go undiagnosed because the signs may be subtle and similar syndromes may be caused by concurrent encephalitides. In the era prior to highly active antiretroviral therapy, HAART, it was estimated to affect at least 10–20 per cent of HIV patients; the more advanced the HIV disease, the higher the incidence of encephalopathy, and the more clinically severe. Rarely it can be the only presenting feature of HIV. It is one of the subcortical dementias and occurs most often in those whose CD4 cell counts have fallen below 200/mm^3 and who already have AIDS. By the terminal stages of infection with CD4 counts below 50/mm^3 almost all patients will have signs of HIV-associated dementia.

HIV infection in the central nervous system is largely confined to perivascular macrophages and microglia. The pathogenesis of dementia is thought to relate to toxic products of infected macrophages rather than the virus itself; these include quinolinic acid and cytokines such as tumour necrosis factor-a, eicosinoids, platelet-activating factor, or nitric oxide. However viral products such as gp120 and tat may contribute to the toxicity. Pathologically the development of dementia correlates with the degree of macrophage staining, rather than viral staining per se. The products of activated macrophages may be toxic to neurons, astrocytes, and oligodendrocytes.

With more subtle neuropsychological and neurophysiological assessment techniques it is possible to demonstrate mild cognitive defects in early HIV infection (Wilkie *et al.* 1990). Not all agree that such subtle abnormalities do herald decline into frank encephalopathy. In one study, a battery of investigations failed to demonstrate significant neurological deterioration in HIV patients until they developed AIDS and had a CD4 count below 350/mm^3 (Harrison *et al.* 1998). In children, an encephalopathic presentation often appears before opportunistic infections, and is associated with developmental delay, microcephaly, and progressive motor disturbance with pseudobulbar palsy and sometimes with seizures (Scott 1991). In the beginning, the syndrome affects cognition then motor function becomes disturbed. The patient often has no complaints, though family members and friends report apathy, lack of drive, mental slowing, difficulty with concentration, and poor memory. This is followed by more obvious cognitive defects, decline in work performance, depression, and frustration. Often, the responses are slow and there is difficulty in formulating verbal replies and motor responses. Disruption of sleep rhythm, lack of sexual drive, and sometimes seizures occur. Subtle motor signs may be found on examination, such as mild dysdiadochokinesis, general hyper-reflexia and the appearance of primitive snout, pout, and grasp reflexes. Progression to a full blown subcortical dementia ensues, paralleled by motor decline. Weakness may be spastic or extrapyramidal. Some exhibit movement disorders, including myoclonus, choreoathetosis, and marked ataxia of gait. Eye movement disorders are common. Deterioration is inexorable and leads to a vegetative state. Other manifestations of AIDS are often evident including wasting, lymphadenopathy, and alopecia. MRI shows diffuse subcortical white matter change (Fig. 43.12).

An important part of the assessment of patients with suspected HIV dementia, or its more subtle forms, includes the exclusion of other treatable causes of cognitive decline (Section 34.6.1). This includes an assessment for depressive pseudodementia, investigations for endocrine and nutritional causes, such as thyroid function tests, and Vitamin B$_{12}$ levels, and CSF and MRI examination to exclude opportunistic infections. MRI typically shows atrophy with diffuse white matter changes. A baseline neuropsychological assessment is useful.

Not uncommonly HIV dementia develops despite good control of HIV load in the plasma with highly active antiretroviral therpy, HAART. Once other causes of dementia have been excluded the CSF HIV viral load should be measured, and if it is high a change in the antiretroviral drug regimen considered. The mainstay of therapy for HIV dementia is aggressive antiretroviral therapy with agents that have good central nervous system penetration. However drug compliance can be poor in patients with cognitive impairment. The possible role of additional neuroprotective therapy is under investigation, including selegiline, valproic acid, and minocycline (Nath *et al.* 2006).

One of the consequences of highly active antiretroviral therapy, HAART, is the development of an immune reconstitution inflammatory syndrome. This develops a few weeks after starting HAART, and is associated with a dramatic rise in CD4 T-lymphocytes, and fall in viral load. Clinically there may be a worsening of underlying infection, or unmasking of a sub-clinical infection, with a high morbidity and mortality if untreated. The most common infections associated with immune reconstitution inflammatory syndrome are ophthalmic cytomegalovirus disease, disseminated infection with *Mycobacterium tuberculosis* or *Mycobacterium avium* complex, and central nervous system involvement with *Cryptococcus neoformans*.

43.3.6 HIV myelopathy

Myelopathy complicating AIDS is caused by several different pathologies (Section 28.5.8). Most common is a vacuolar myelopathy caused by the HIV virus, which is especially common in the terminal stages of the illness, when it may accompany HIV-associated dementia or neuropathy. Neuropathological signs of vacuolar myelopathy may be identified in as many of 30 per cent of patients dying of HIV, although not all have clinical evidence of the condition. The neuropathological changes are most marked in the thoracic spinal cord with vacuolation within the myelin of the dorsal and lateral columns, relative preservation of axons, microglial nodules, and multinucleate giant cells full of virus. Clinically, there is progressive spastic paraparesis over weeks to months which may be asymmetrical, and associated sensory and sphincter disturbance, and proprioceptive impairment (Petito *et al.* 1985). This is akin to the clinical features of sub-acute combined degeneration of the cord (Section 28.5.10). HIV myelopathy is a clinical diagnosis, the CSF being non-diagnostic, and the role of MRI being to exclude other causes of myelopathy. Vitamin B$_{12}$ is often given, though it has not been shown to be effective. Corticosteroids and intravenous immunoglobulin also appear ineffective.

43.3.7 HIV neuropathy

The most common neuropathy seen in AIDS is a distal, symmetrical, sensorimotor polyneuropathy in which painful dysaesthesiae are often predominant (Section 21.14.2). Neurophysiological studies suggest an axonal disturbance and this has been confirmed by biopsy. Autonomic neuropathy with postural hypotension, bowel disturbance, and cardiac dysrythmia often co-exists. Although direct neuronal infection by HIV virus has been suspected, it has not been proven and the polyneuropathy may be cytokine mediated. A similar disorder is caused by some antiretroviral drugs, especially the so-called 'D' drugs, ddi and stavudine (Section 21.19.7), or nutritional deficiencies (Section 21.22). Hence the need for neurotoxic antiretroviral drugs should be reviewed, and their dose reduced if possible. Symptomatic treatment includes simple analgesics, nonsteroidal anti-inflammatory drugs, antidepressants, and anti-convulsants.

Inflammatory, demyelinating polyneuropathies of both acute and chronic types (Sections 21.10.1 and 21.11.2), are described in association with HIV infection (Leger *et al.* 1989; Simpson *et al.* 1991). Mononeuritis multiplex may also occur (Section 21.15); in late disease this may be due to coinfection with cytomegalovirus, and so empirical treatment should be considered.

43.3.8 HIV myopathy

A primary HIV-associated myopathy occurs when the CD4 cell counts are below 500/µl. The clinical features are of limb girdle weakness, muscle tenderness, wasting and increase in creatine phosphokinase, exactly like idiopathic polymyositis (Simpson *et al.* 1988) (Section 24.7.4). Electromyographic and biopsy features are also similar. Zidovudine toxicity can cause a similar clinical syndrome (Section 24.9.5). A rapidly progressive neuromuscular syndrome has been associated with stavudine. Drug withdrawal may alleviate symptoms in some patients. In patients with a polymyositis syndrome steroids may be beneficial, though they should be used with caution because of their immunosuppressive effects. Intravenous immunoglobulin is an alternative.

43.3.9 Secondary complications of HIV

Other neurological manifestations of AIDS commonly co-exist in patients with AIDS encephalopathy. These include myelopathy; opportunistic infection with toxoplasma, cytomegalovirus, herpes, or progressive multifocal leukoencephalopathy which can cause a similar encephalitic picture; space occupying syndromes from abscess, granuloma, or tumour; metabolic encephalopathies; and cerebrovascular syndromes (Table 43.5). Investigations are therefore directed to exclude or confirm the presence of these other syndromes, as much as to identify AIDS encephalopathy. MR brain scanning is obligatory to exclude these conditions. The MRI findings in AIDS encephalopathy are not diagnostic and range from normal brain appearances through mild generalized atrophic changes with large ventricles and generous sulci, to gross cerebral atrophy and diffuse or focal white matter signal change (Levy *et al.* 1986). EEG is no more specific than brain imaging.

CSF examination is not diagnostic and is required to detect other co-existing infections. The CSF cell count and protein content may be raised, HIV specific antibody and oligoclonal bands are commonly present, and viral RNA can be demonstrated and quantified. Unfortunately, none of these findings confirms that the neurological syndrome is caused by HIV. Several surrogate markers in CSF for HIV encephalopathy which have been assessed include neopterin, β2 microglobulin, tumour necrosis factor, and matrix metalloproteinases; however as yet none has been shown to be useful in clinical practice.

Other secondary complications are brought about by immune suppression permitting infection or neoplasia to disrupt neurological function either directly or indirectly for instance through the side effects of drug treatment, metabolic derangements, or vascular impairment.

Secondary infections in HIV

Of the secondary complications of HIV, opportunistic infection is the most common. The risk of development of infection is in direct proportion to the degree of immune suppression as indicated by the CD4 cell count. The common pathogens are listed in Table 43.5.

Secondary viral infections

Cytomegalovirus infection is one of the most frequent viral opportunistic infections in patients with HIV and evidence of infection is found in 30 per cent of brains at necropsy. Cytomegalovirus encephalitis typically presents in patients with a CD4 count less than 50 cells/mm^3, with a progressive severe dementia, and ventriculitis on MRI (Section 42.3.4). A lumbosacral radiculopathy with pain and paraesthesia in a saddle distribution, which may spread to involve the lower limbs and sphincters is another common presentation (Section 21.14.2). Cytomegalovirus retinitis may precede or follow these syndromes. The presentation and evolution of encephalitis are measured in weeks (Holland *et al.* 1994). Rarely cytomegalovirus encephalitis may cause a mass cerebral lesion. CSF examination reveals polymorphonuclear leucocytosis and cytome-

galovirus can be cultured from CSF in about half of all cases. Polymerase chain reaction is often confirmatory. Treatment is with ganciclovir (Section 42.3.4).

Progressive multifocal leucoencephalopathy caused by the JC virus complicates HIV in up to 5 per cent of cases and is often the first severe infection (Section 42.3.14). *Herpes simplex virus* infections, especially type 2, and *varicella zoster virus* infections are also more common in patients with HIV (Sections 42.3.2 and 42.3.3).

Epstein–Barr virus is associated with the development of primary central nervous system lymphoma in patients with HIV (Section 27.8.3). Primary lymphoma develops in up to 4 per cent of AIDS patients and is their second most common cause of an intracranial space occupying lesion. It is a B cell lymphoma, with the cells almost always positive for Epstein–Barr virus, and it usually develops in those with CD4 cell counts less than 100/μl. The clinical features are of mental disturbance with confusion, hallucinations, memory disturbance, focal signs and epileptic seizures, obtundation, and features of raised intracranial pressure. There is often fever and meningeal involvement with neck stiffness. Lymphoma usually presents as a focal mass, but can be multifocal (Baumgartner *et al.* 1990). There may be prominent cerebral oedema seen on MRI (Fig. 43.13), which usually responds to treatment with dexamethasone. The prognosis is poor, but optimization of HAART therapy and whole-brain radiation can improve the outcome. If malignant cells are present in the CSF, intrathecal cytosine arabinoside is recommended.

Secondary bacterial infections

Tuberculous pulmonary infection is increasingly common in people with HIV, particularly in drug addicts and patients from Africa. HIV-positive intravenous drug users are also at greater risk of bacterial meningitis, than non-drug users. Extrapulmonary tuberculosis occurs in some 70 per cent and in a small minority this affects the central nervous system to cause meningitis which has the same features as in non-AIDS patients (Section 41.5.3). Tuberculomas and tuberculous abscesses also occur (Fig. 43.14). Atypical mycobacteria, including *M. avium-intracellulare* have a tendency to cause granulomas which may be multiple.

Syphilis is common in HIV, which seems to increase the risk of earlier development of neurosyphilis (Sections 41.5.4 and 42.5.1). The condition may be asymptomatic and detected on CSF examination (Berger 1991). If the definition of neurosyphilis is taken as a positive CSF VDRL test then the incidence of neurosyphilis, asymptomatic or symptomatic, approximates to 5 per cent in high-risk populations. The clinical features are the same as occur in non-HIV populations, although some unusual manifestations have been recorded. It remains undecided whether the co-existence of HIV infection alters the natural history of neurosyphilis; the response to medication may be impaired. (Marra *et al.* 1996) There is evidence that higher dose drug regimes are needed to obtain control of the infection and relapse is not infrequent following treatment with previously acceptable dosages (Malone *et al.* 1995).

Secondary fungal infections in HIV

Cryptococcal meningitis is one of the most common central nervous system infections in AIDS (Section 41.5.9). *Cryptococcus neoformans* is an encapsulated yeast which is found throughout the world in soil and animal and bird droppings. Most infections follow inhalation of the small yeast form of the fungus. This allows pulmonary infection which is usually without signs and infection

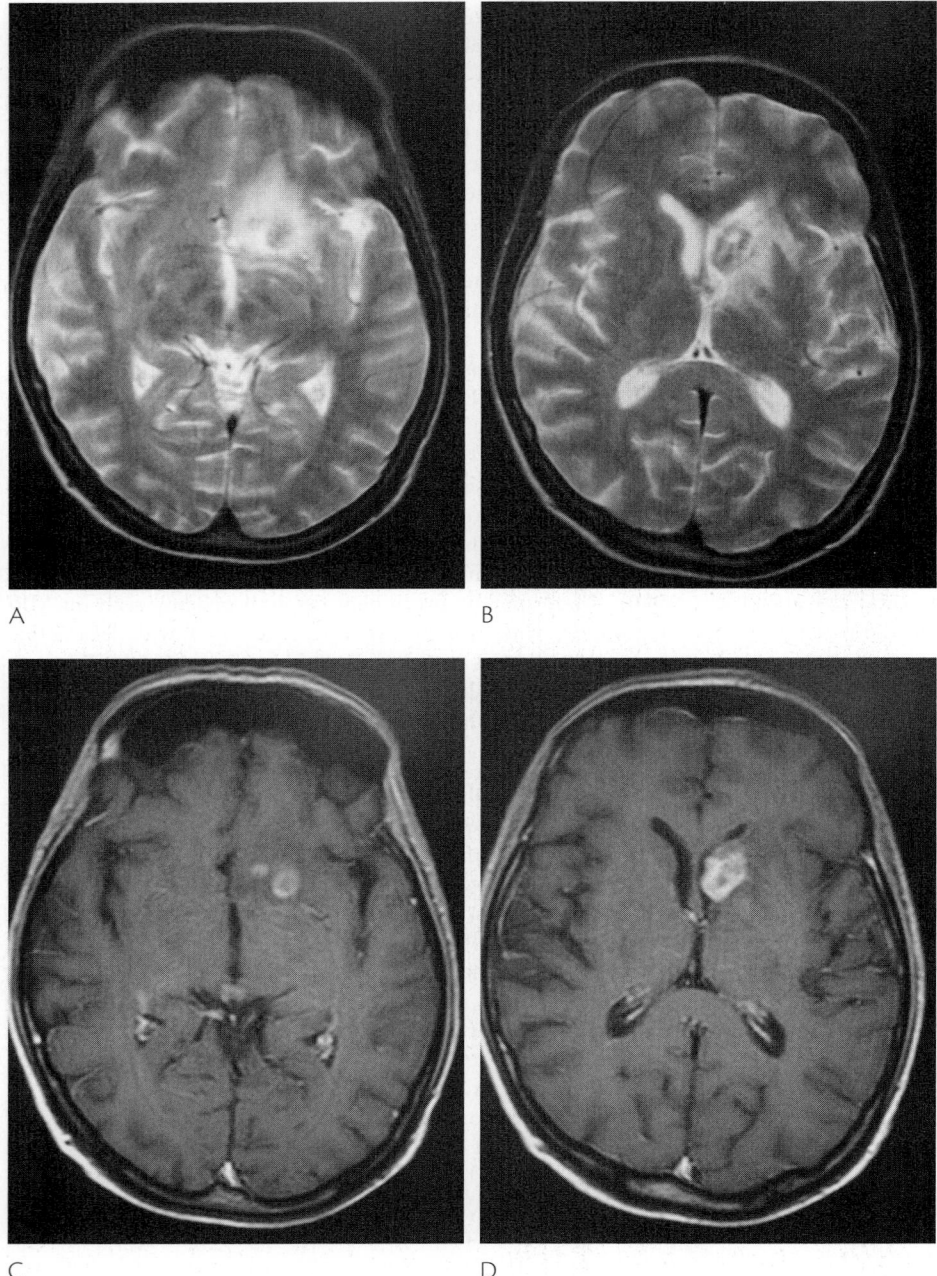

Fig. 43.13 Primary central nervous system lymphoma in a patient with HIV infection. MR scans showing a lobulated preventricular lesion with moderate oedema, and patchy gadolinium enhancement. (A) + (B): T2-weighted turbospin echo; (C) + (D): T1-weighted with gadolinium. (Photo: Tom Solomon.)

is limited to lung and lymph node. In the immunocompromised, an interstitial pneumonitis may occur. Reactivation of the primary pulmonary infection occurs when cellular immune defences fail and dissemination takes place via the bloodstream to other organs, most commonly the central nervous system. There a chronic form of meningitis occurs. A small number may develop cerebral granulomas.

Cryptococcal meningitis occurs in approximately 7 per cent of AIDS patients with advanced HIV. This incidence can be reduced if prophylactic antifungals are given. However anti-cryptococcal prophylaxis is now rarely advised in the highly active antiretroviral therapy, HAART, era. The clinical features are of non-specific debility and fatigue and the development over 3–4 weeks of neck stiffness, headache, fever, irritability, confusion, and neurological signs. Cranial nerve palsies may ensue. However, some patients have little more than a chronic headache, and so there should be a low threshold for performing a lumbar puncture. There may also be cryptococcal skin lesions, which are papules that must be distinguished from mollusculm contageosum and bacillary angiomatosis. Symptoms and signs may persist for months. In severe immune

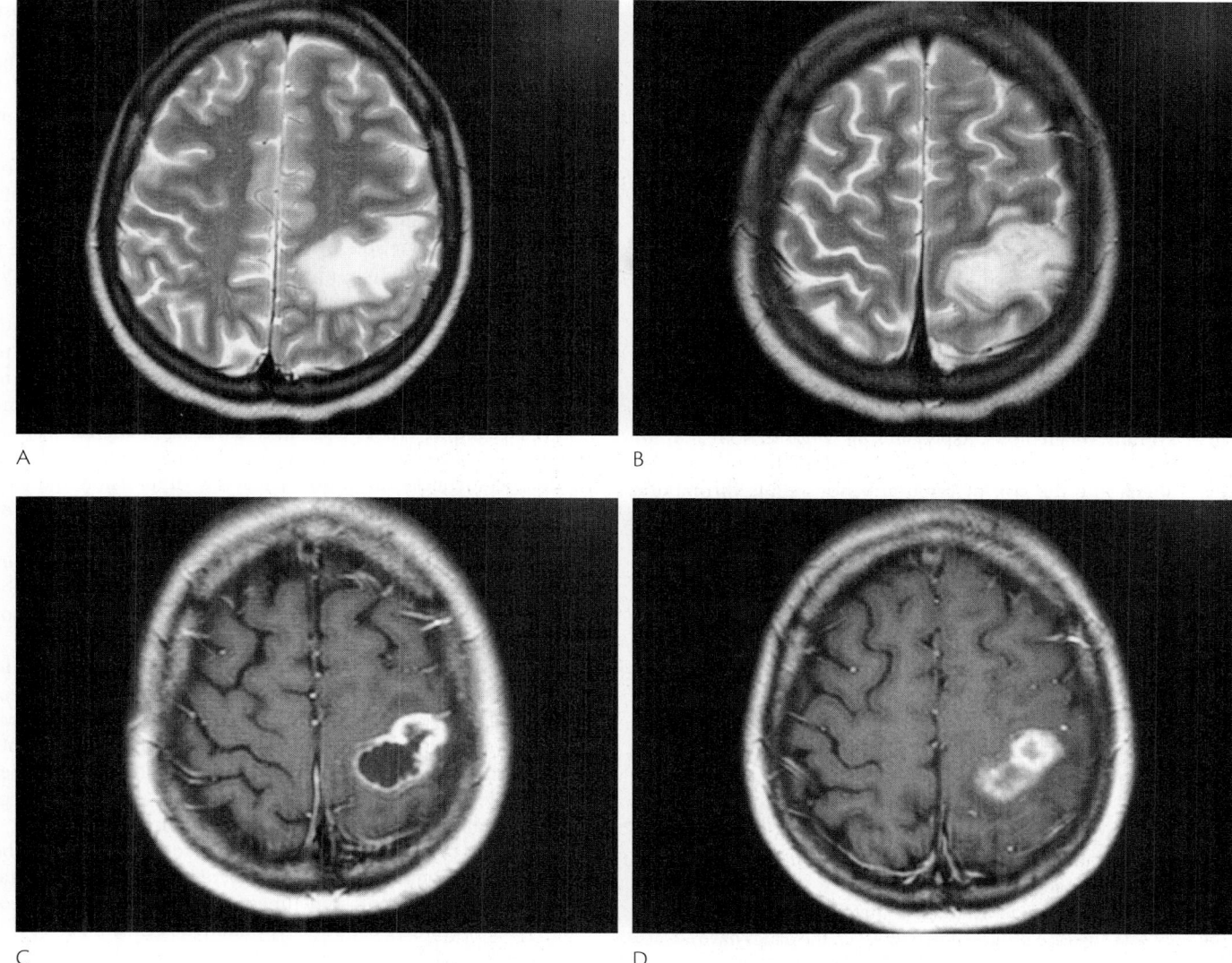

Fig. 43.14 Tuberculous abscess in a patient with advanced HIV infection. (A) + (B): T_2-turbo spin echo MRI. (C) + (D): T1-weighted MRI with gadolinium enhancement. (Photo: Tom Solomon.)

depression, onset may be acute. The differential diagnosis includes other causes of meningitis. A chest X-ray may show disease also. Diagnosis is by CSF examination which often reveals a very high lumbar puncture opening pressure, for example more than 600 mm, mononuclear pleocytosis, and raised protein content, but may be normal. The cryptococcal antigen, CRAG, test is positive in about 95 per cent of cases, whereas the presence of yeast forms under the microscope after staining with India ink is seen in about 75 per cent. In addition there may be a positive polymerase chain reaction. Cryptococci can often be cultured.

Treatment is with a combination of Amphotericin B combined with fluconazole for 2 weeks, followed by fluconazole for 3 months. A reduced maintenance dose may be continued for the rest of the patient's life, or withdrawn if the CD4 count rises above 200/mm^3 in response to highly active antiretroviral therapy, HAART. If maintenance therapy is not given as many as 80 per cent will relapse. Some patients have an idiopathic intracranial hypertension-like syndrome with raised pressure, headaches, and visual obscurations (Section 26.5.5). Treatment options include repeat lumbar puncture or insertion of a shunt.

Candida species, which are part of the normal human flora, can pass into the bloodstream of AIDS patients and cause a chronic form of meningitis. *Aspergillus* and *histoplasma* can also cause similar syndromes. The diagnosis of these fungal infections is by direct observation of the fungus in CSF, demonstration of fungal antigen, or a positive polymerase chain reaction in CSF. *Aspergillus* (Section 43.2.4) and *Mucorales* (Section 43.2.7) are more likely to produce a cerebral granuloma than meningitis in AIDS patients.

Secondary parasitic infection

Toxoplasmic encephalitis is estimated to develop in 20–30 per cent of HIV patients and this will be due to reactivation of latent infection in 95 per cent (Section 43.2.8). Toxoplasma encephalitis is the most common cause of focal intracranial masses in AIDS patients.

Definitive diagnosis can only be made by brain biopsy but a presumptive diagnosis may be made if there is a response to treatment with pyrimethamine and sulphadiazine.

Secondary malignancy

Metastatic malignant disease may affect the central nervous system in AIDS. Non-central nervous system lymphoma may spread to involve brain or spinal cord with malignant meningitis and multiple cranial nerve palsies as the most common manifestations. Exceptionally, Kaposi's sarcoma has spread to the central nervous system. Lymphomas are increasingly seen in those who have been on long-term highly active antiretroviral therapy, HAART.

Secondary cerebrovascular disease

Cerebrovascular disorders occur in AIDS due to several causes. Emboli may result from endocarditis. Infarction may follow thrombosis caused by cerebral vasculitis and infection with varicella zoster virus, syphilis, tuberculosis, Aspergillus, or Mucorales. In addition there may be changes in anticardiolipin and lupus anticoagulant levels increasing the risk of cerebral venous sinus thrombosis. Intracranial haemorrhage may be the end result of drug-induced thrombocytopaenia or disseminated intravascular coagulation.

43.3.10 The role of antiretroviral therapy

The decision concerning when to initiate therapy can be difficult. The virus itself cannot be eliminated, and so the aim is to control viral replication, and the development of AIDS. Conventionally treatment is started when patients have a CD4 count below 200–300/mm^3, or a rapidly falling count or the patient has an AIDS defining illness, such as an opportunistic infection (Manji and Miller 2004). The previous guide of starting treatment if the viral load is above 55 000 to 100 000 RNA copies per mm^3 is no longer followed. Treatment is indicated if the patient has an AIDS defining illness, such as an opportunistic infection (Manji and Miller 2004). However an approach of 'hit early and hard' is under consideration by some. The postulated advantage is that a reduction in viral burden may delay disease progression. However, the drugs have adverse effects, the regimens can be very inconvenient, and adherence may be poor. Furthermore drug resistance develops, and there is a need to hold some therapeutic options in reserve.

There is a bewildering and rapidly evolving array of drugs used to treat HIV infection. Understanding is further complicated by the fact that each drug is known by a generic name, a three letter abbreviation, and a trade name. Many drugs are used and sometimes formulated in combination. Broadly speaking there are three groups of drugs:

- nucleoside reverse transcriptase inhibitors, NRTIs, such as Zidovudine;
- non-nucleoside reverse transcriptase inhibitors, such as Nevirapine; and
- protease inhibitors, such as Idinavir.

The reverse transcriptase inhibitors inhibit the action of the viral reverse transcriptase enzyme, which converts the single stranded RNA genome into a complimentary DNA molecule. Protease inhibitors target the HIV protease which is required for post-translational cleavage of the Gag and Gag-Pol precursor polyproteins into the structural and functional proteins required for virion production. Newer drugs in development include CCR5 antagonists and integrase inhibitors. The former inhibit viral attachment to the cell membrane thus preventing viral entry into the cell. Integrase inhibitors, as their name suggests, inhibit the integrase enzyme. Enfuvirtide is a single injectable fusion inhibitor which tends to be used only in those with very limited treatment options.

Most antiretroviral regimes consist of at least three drugs; typically there are two NRTIs, plus either a protease inhibitor or a non-NRTI. The general aim is to spare at least one class of drugs for later in the disease process, when resistance has become a problem. Thus a common regimen is Kivexa™, which is a combination of two NRTIs—abacavir and lamivudine, plus efavirenz, Sustiva, a non-NRT; these can both be given once daily. Trizivir™ is a combination of three NRTIs—zidovudine, lamivudine, and abacavir.

Protease inhibitors' side effects include lipodystrophy with hyperlipidaemia, diarrhoea, and insulin resistance leading to hyperglycaemia. NTRIs can cause mitochondrial toxicity leading to lactic acidosis. Peripheral neuropathy is also a problem with some NRTIs (Section 21.19.7), particularly didanosine, and zalcitabine. Didanosine is associated also with peripheral neuropathy and pancreatitis. Zidovudine can produce marrow suppression and gastrointestinal upset. Drug interactions between different antiretroviral therapies is an increasing problem.

Of the many drugs available, those with better blood brain barrier penetrance include zidovudine or abacavir, which are nucleoside reverse transcriptase inhibitors and nevirapine, a non-nucleoside reverse transcriptase inhibitor.

Abacavir causes a hypersensitivity drug reaction especially in patients who are HLA-B*5701-positive, and so HLA testing is recommended before starting this drug. Because of increased viral resistance, resistance testing is now recommended for all patients before treatment is started. The latest British treatment guidelines are available from the British HIV Association (www.bhiva.org) and American guidelines are available (www.aidsmap.org).

Monitoring of CD4 lymphocyte counts and viral load can be helpful. Usually, within 1–2 months, there is a reduction in viral load of about 90 per cent, sometimes to below the level of detection. Typically the CD4 count rises by about 50 cells/mm^3.

Patients with HIV dementia are especially susceptible to psychoactive drugs, and so hypnotics and sedatives should be avoided. Clozapine is the preferred neuroleptic, along with small doses of haloperidol if needed in agitated or combative patients. Tricyclic antidepressants, at low doses, and selective serotonin reuptake inhibitors can be used if depression is a feature. Gabapentin and topiramate are the preferred anticonvulsants if seizures occur, because they do not interact with antiretroviral drugs. However valproate should be avoided because, *in vitro*, it can induce viral replication.

References

Apuzzo ML, Chandrasoma PT, Cohen D *et al.* (1987). Computed imaging stereotaxy: experience and perspective related to 500 procedures applied to brain masses. *Neurosurgery*, **20**, 930–7.

Arana-Iniguez R, San Julian J (1955). Hydatid cysts of the brain. *J Neurosurg*, **12**, 323–35.

Ariizumi M (1963). Cerebral schistosomiasis japonica: report of one operated case and fifty clinical cases. *Am J Trop Med Hyg*, **12**, 40–5.

Ashdown BC, Tien RD, Felsberg GJ (1994). Aspergillosis of the brain and paranasal sinuses in immunocompromised patients: CT and MR imaging findings. *AJR Am J Roentgenol*, **162**, 155–9.

Bannister G, Williams B, Smith S (1981). Treatment of subdural empyema. *J Neurosurg*, **55**, 82–8.

Barlas O, Sencer A, Erkan K *et al.* (1999). Stereotactic surgery in the management of brain abscess. *Surg Neurol*, **52**, 404–10; discussion 11.

Barre-Sinoussi F, Chermann JC, Rey F *et al.* (1983). Isolation of a T-lymphotropic retrovirus from a patient at risk for acquired immune deficiency syndrome (AIDS). *Science*, **220**, 868–71.

Baumgartner JE, Rachlin JR, Beckstead JH *et al.* (1990). Primary central nervous system lymphomas: natural history and response to radiation therapy in 55 patients with acquired immunodeficiency syndrome. *J Neurosurg*, **73**, 206–11.

Berger JR (1991). Neurosyphilis in human immunodeficiency virus type 1-seropositive individuals. A prospective study. *Arch Neurol*, **48**, 700–2.

Boom WH, Tuazon CU (1985). Successful treatment of multiple brain abscesses with antibiotics alone. *Rev Infect Dis*, **7**, 189–99.

Bouza E, Dreyer JS, Hewitt WL *et al.* (1981). Coccidioidal meningitis. An analysis of thirty-one cases and review of the literature. *Medicine*, **60**, 139–72.

Britt RH, Enzmann DR (1983). Clinical stages of human brain abscesses on serial CT scans after contrast infusion. Computerized tomographic, neuropathological, and clinical correlations. *J Neurosurg*, **59**, 972–89.

Britt RH, Enzmann DR, Yeager AS (1981). Neuropathological and computerized tomographic findings in experimental brain abscess. *J Neurosurg*, **55**, 590–603.

Bronnimann DA, Adam RD, Galgiani JN *et al.* (1987). Coccidioidomycosis in the acquired immunodeficiency syndrome. *Ann Intern Med*, **106**, 372–9.

Carne CA, Tedder RS, Smith A *et al.* (1985). Acute encephalopathy coincident with seroconversion for anti-HTLV-III. *Lancet*, **2**, 1206–8.

Carpio A (2002). Neurocysticercosis: an update. *Lancet Infect Dis*, **2**, 751–62.

Cha SH, Chang KH, Cho SY *et al.* (1994). Cerebral paragonimiasis in early active stage: CT and MR features. *AJR Am J Roentgenol*, **162**, 141–5.

Chun CH, Johnson JD, Hofstetter M *et al.* (1986). Brain abscess. A study of 45 consecutive cases. *Medicine*, **65**, 415–31.

Cowie R, Williams B (1983). Late seizures and morbidity after subdural empyema. *J Neurosurg*, **58**, 569–73.

de Louvois J, Gortvai P, Hurley R (1977). Antibiotic treatment of abscesses of the central nervous system. *Br Med J*, **2**, 985–7.

Del Brutto OH (1992). Cysticercosis and cerebrovascular disease: a review. *J Neurol Neurosurg Psychiatry*, **55**, 252–4.

DeLone DR, Goldstein RA, Petermann G *et al.* (1999). Disseminated aspergillosis involving the brain: distribution and imaging characteristics. *AJNR Am J Neuroradiol*, **20**, 1597–604.

Denning DW, Stevens DA (1990). Antifungal and surgical treatment of invasive aspergillosis: review of 2,121 published cases. *Rev Infect Dis*, **12**, 1147–201.

Desprechins B, Stadnik T, Koerts G *et al.* (1999). Use of diffusion-weighted MR imaging in differential diagnosis between intracerebral necrotic tumors and cerebral abscesses. *AJNR Am J Neuroradiol*, **20**, 1252–7.

Dill SR, Cobbs CG, McDonald CK (1995). Subdural empyema: analysis of 32 cases and review. *Clin Infect Dis*, **20**, 372–86.

Ernst T, Chang L, Witt M *et al.* (1999). Progressive multifocal leukoencephalopathy and human immunodeficiency virus-associated white matter lesions in AIDS: magnetization transfer MR imaging. *Radiology*, **210**, 539–43.

Ersahin Y, Mutluer S, Guzelbag E (1993). Intracranial hydatid cysts in children. *Neurosurgery*, **33**, 219–24; discussion 24–5.

Ewer K, Deeks J, Alvarez L *et al.* (2003). Comparison of T-cell-based assay with tuberculin skin test for diagnosis of Mycobacterium tuberculosis infection in a school tuberculosis outbreak. *Lancet*, **361**, 1168–73.

Farkash AE, Maccabee PJ, Sher JH *et al.* (1986). CNS toxoplasmosis in acquired immune deficiency syndrome: a clinical-pathological-radiological review of 12 cases. *J Neurol Neurosurg Psychiatry*, **49**, 744–8.

Feuerman T, Wackym PA, Gade GF *et al.* (1989). Craniotomy improves outcome in subdural empyema. *Surg Neurol*, **32**, 105–10.

Fitzpatrick MO, Gan P (1999). Contrast enhanced computed tomography in the early diagnosis of cerebral abscess. *BMJ*, **319**, 239–40.

Furrer H, Opravil M, Bernasconi E *et al.* (2000). Stopping primary prophylaxis in HIV-1-infected patients at high risk of toxoplasma encephalitis. Swiss HIV Cohort Study. *Lancet*, **355**, 2217–8.

Garcia HH, Pretell EJ, Gilman RH *et al.* (2004). A trial of antiparasitic treatment to reduce the rate of seizures due to cerebral cysticercosis. *N Engl J Med*, **350**, 249–58.

Garland HG, Armitage G (1933). Intracranial tuberculoma. *J Pathol Bacteriol*, **37**, 461–71.

Guex AC, Radziwill AJ, Bucher HC (2000). Discontinuation of secondary prophylaxis for toxoplasmic encephalitis in human immunodeficiency virus infection after immune restoration with highly active antiretroviral therapy. *Clin Infect Dis*, **30**, 602–3.

Haribhai HC, Bhigjee AI, Bill PL *et al.* (1991). Spinal cord schistosomiasis. A clinical, laboratory and radiological study, with a note on therapeutic aspects. *Brain*, **114**, 709–26.

Harrison MJ, Newman SP, Hall-Craggs MA *et al.* (1998). Evidence of CNS impairment in HIV infection: clinical, neuropsychological, EEG, and MRI/MRS study. *J Neurol Neurosurg Psychiatry*, **65**, 301–7.

Hlavin ML, Kaminski HJ, Fenstermaker RA *et al.* (1994). Intracranial suppuration: a modern decade of postoperative subdural empyema and epidural abscess. *Neurosurgery*, **34**, 974–80; discussion 80–1.

Holland NR, Power C, Mathews VP *et al.* (1994). Cytomegalovirus encephalitis in acquired immunodeficiency syndrome (AIDS). *Neurology*, **44**, 507–14.

Janssen RJ, Cornblath DR, Epstein LG *et al.* (1991). Nomenclature and research case definitions for neurologic manifestations of human immunodeficiency virus-type 1 (HIV-1) infection. *Neurology*, **41**, 778–85.

Kami M, Ogawa S, Kanda Y, Tanaka Y *et al.* (1999). Early diagnosis of central nervous system aspergillosis using polymerase chain reaction, latex agglutination test, and enzyme-linked immunosorbent assay. *Br J Haematol*, **106**, 536–7.

Kim YJ, Chang KH, Song IC *et al.* (1998). Brain abscess and necrotic or cystic brain tumor: discrimination with signal intensity on diffusion-weighted MR imaging. *AJR Am J Roentgenol*, **171**, 1487–90.

Kure K, Llena JF, Lyman WD *et al.* (1991). Human immunodeficiency virus-1 infection of the nervous system: an autopsy study of 268 adult, pediatric, and fetal brains. *Hum Pathol*, **22**, 700–10.

Leger JM, Bouche P, Bolgert F *et al.* (1989). The spectrum of polyneuropathies in patients infected with HIV. *J Neurol Neurosurg Psychiatry*, **52**, 1369–74.

Lehrer RI, Howard DH, Syperd PS *et al.* (1980). Mucormycosis. *Ann Intern Med*, **93**, 93–108.

Levy RM, Rosenbloom S, Perrett LV (1986). Neuroradiologic findings in AIDS: a review of 200 cases. *AJR Am J Roentgenol*, **147**, 977–83.

Luft BJ, Brooks RG, Conley FK *et al.* (1984). Toxoplasmic encephalitis in patients with acquired immune deficiency syndrome. *JAMA*, **252**, 913–7.

Luft BJ, Hafner R, Korzun AH *et al.* (1993). Toxoplasmic encephalitis in patients with the acquired immunodeficiency syndrome. Members of the ACTG 077p/ANRS 009 Study Team. *N Engl J Med*, **329**, 995–1000.

Malone JL, Wallace MR, Hendrick BB *et al.* (1995). Syphilis and neurosyphilis in a human immunodeficiency virus type-1 seropositive population: evidence for frequent serologic relapse after therapy. *Am J Med*, **99**, 55–63.

Mamelak AN, Obana WG, Flaherty JF *et al.* (1994). Nocardial brain abscess: treatment strategies and factors influencing outcome. *Neurosurgery*, **35**, 622–31.

Manji H, Miller R (2004). The neurology of HIV infection. *J Neurol Neurosurg Psychiatry*, **75** (**Suppl I**), i29–35.

Marra CM, Longstreth WT Jr, Maxwell CL *et al.* (1996). Resolution of serum and cerebrospinal fluid abnormalities after treatment of neurosyphilis. Influence of concomitant human immunodeficiency virus infection. *Sex Transm Dis*, **23**, 184–9.

Mathisen GE, Johnson JP (1997). Brain abscess. *Clin Infect Dis*, **25**, 763–79; quiz 80–1.

McClelland CJ, Craig BF, Crockard HA (1978). Brain abscesses in Northern Ireland: a 30 year community review. *J Neurol Neurosurg Psychiatry*, **41**, 1043–7.

Mellors JW, Rinaldo CR Jr, Gupta P *et al.* (1996). Prognosis in HIV-1 infection predicted by the quantity of virus in plasma. *Science*, **272**, 1167–70.

Molinari GF, Smith L, Goldstein MN *et al.* (1973). Brain abscess from septic cerebral embolism: an experimental model. *Neurology*, **23**, 1205–10.

Nadkarni T, Goel A (2005). Aspergilloma of the brain: an overview. *J Postgrad Med*, **51** (**Suppl 1**), S37–41.

Nath A (2005). Human immunodeficiency virus infections. In Johnson RT, Griffin JW, McArthur JC. *Current Therapy in Neurological Diseases*. Elsevier, Philadelphia.

Nath A, Sacktor N (2006). Influence of highly active antiretroviral therapy on persistence of HIV in the central nervous system. *Curr Opin Neurol*, **19**, 358–61.

Nathoo N, Nadvi SS, van Dellen JR *et al.* (1999). Intracranial subdural empyemas in the era of computed tomography: a review of 699 cases. *Neurosurgery*, **44**, 529–35; discussion 35–6.

Navia BA, Jordan BD, Price RW (1986). The AIDS dementia complex: I. Clinical features. *Ann Neurol*, **19**, 517–24.

Nielsen H, Gyldensted C, Harmsen A (1982). Cerebral abscess. Aetiology and pathogenesis, symptoms, diagnosis and treatment. A review of 200 cases from 1935-1976. *Acta Neurol Scand*, **65**, 609–22.

Oh SJ (1968). Paragonimus meningitis. *J Neurol Sci*, **6**, 419–33.

Pau A, Turtas S, Brambilla M *et al.* (1987). Computed tomography and magnetic resonance imaging of cerebral coenurosis. *Surg Neurol*, **27**, 548–52.

Petito CK, Navia BA, Cho ES *et al.* (1985). Vacuolar myelopathy pathologically resembling subacute combined degeneration in patients with the acquired immunodeficiency syndrome. *N Engl J Med*, **312**, 874–9.

Pitella JE (1993). Central nervous system involvement in Chaga's disease: an updating. *Rev Inst Med Trop Sao Paulo*, **35**, 111–6.

Roos KL, Bryan JP, Maggio WW *et al.* (1987). Intracranial blastomycoma. *Medicine*, **66**, 224–35.

Salinas R, Prasad K (2000). Drugs for treating neurocysticercosis (tapeworm infection of the brain). *Cochrane Database Syst Rev*: CD000215.

Schacker T, Collier AC, Hughes J *et al.* (1996). Clinical and epidemiologic features of primary HIV infection. *Ann Intern Med*, **125**, 257–64.

Schantz PM, Moore AC, Munoz JL *et al.* (1992). Neurocysticercosis in an Orthodox Jewish community in New York City. *N Engl J Med*, **327**, 692–5.

Schroeder AH (1941). Diagnostico del quiste hidatico cerebral y su tratamiento. *Anales Inst Neurologia (Montevideo)*, **3**, 11–38.

Schroth G, Kretzschmar K, Gawehn J *et al.* (1987). Advantage of magnetic resonance imaging in the diagnosis of cerebral infections. *Neuroradiology*, **29**, 120–6.

Scott GB (1991). HIV infection in children: clinical features and management. *J Acquir Immune Defic Syndr*, **4**, 109–15.

Selvapandian S, Rajshekhar V, Chandy MJ *et al.* (1994). Predictive value of computed tomography-based diagnosis of intracranial tuberculomas. *Neurosurgery*, **35**, 845–50; discussion 50.

Seydoux C, Francioli P (1992). Bacterial brain abscesses: factors influencing mortality and sequelae. *Clin Infect Dis*, **15**, 394–401.

Simpson DM, Bender AN (1988). Human immunodeficiency virus-associated myopathy: analysis of 11 patients. *Ann Neurol*, **24**, 79–84.

Simpson DM, Wolfe DE (1991). Neuromuscular complications of HIV infection and its treatment. *Aids*, **5**, 917–26.

Skiest DJ (2002). Focal neurological disease in patients with acquired immunodeficiency syndrome. *Clin Infect Dis*, **34**, 103–15.

Smego RA Jr. (1987). Actinomycosis of the central nervous system. *Rev Infect Dis*, **9**, 855–65.

Sundaram C, Purohit AK, Prasad VS *et al.* (2004). Cranial and intracranial actinomycosis. *Clin Neuropathol*, **23**, 173–7.

Tsai MD, Chang CN, Hoys, Wany AD. Cerebral sparganosis diagnosed and treated with stereolzeclic techniques. J Neurosurg 1983;**78**;129–32.

The Brain Trauma Foundation (2000). The Brain Trauma Foundation. The American Association of Neurological Surgeons. The Joint Section on Neurotrauma and Critical Care. Role of antiseizure prophylaxis following head injury. *J Neurotrauma*, **17**, 549–53.

Tsuchiya K, Makita K, Furui S *et al.* (1992). Contrast-enhanced magnetic resonance imaging of sub- and epidural empyemas. *Neuroradiology*, **34**, 494–6.

Vlasveld LT, van Asbeck BS (1991). Treatment with deferoxamine: a real risk factor for mucormycosis? *Nephron*, **57**, 487–8.

Watt G, Adapon B, Long GW *et al.* (1986). Praziquantel in treatment of cerebral schistosomiasis. *Lancet*, **2**, 529–32.

Wilkie FL, Eisdorfer C, Morgan R *et al.* (1990). Cognition in early human immunodeficiency virus infection. *Arch Neurol*, **47**, 433–40.

Yang SY (1981). Brain abscess: a review of 400 cases. *J Neurosurg*, **55**, 794–9.

Index

Numbers in italic refer to tables and/or illustrations separate from the text.